*(continued at back)*

# MACHINERY'S HANDBOOK

# MACHINERY'S HANDBOOK

*A Reference Book for the Mechanical Engineer, Draftsman, Toolmaker and Machinist*

*By* Erik Oberg, Franklin D. Jones *and*
Holbrook L. Horton

Paul B. Schubert, *Editor*

K. H. Moltrecht, *Assistant Editor*

Henry H. Ryffel, *Consulting Editor*

TWENTY-FIRST EDITION, *Second Printing, 1980*

INDUSTRIAL PRESS INC.
*200 Madison Ave., New York, N.Y. 10016*

# PREFACE

MACHINERY'S HANDBOOK, since the publication of its first edition in 1914, has continually increased in popularity throughout the world. It is now used extensively as a standard work of reference in all countries where machines or other mechanical products are designed and manufactured.

The aim of the publisher is to make each new edition of greater practical value than the preceding one. This has been accomplished by revising the HANDBOOK as frequently as is practicable for a voluminous work which must necessarily be printed in large editions to meet the constant demand for it both here and abroad.

The new material covers a large variety of subjects that are important to designers and builders of everything mechanical. Recent or revised engineering standards, both American and British, are included, together with a large amount of general information and mechanical data representing the latest designing and manufacturing practice.

In selecting material from the almost limitless supply of data pertaining to the mechanical field, the plan is to consider the requirements of the design and production departments of both large and small manufacturing plants as well as the needs of the jobbing shop, the trade school, and the technical school.

In this edition use of the trigonometric tables has been extended to give values for angles up to 360 degrees. Involute function values are now provided for angles from 0 through 90 degrees. An extensive table of hole coordinate dimension factors for jig boring also has been added. The plain bearings section has been thoroughly revised to fully consider current practice and requirements. The rolling element bearings section has been made more compact with greater orientation to the user, which should enhance its value. Other areas that have been reviewed to offer new and revised information are surface grinding, friction, cutting speeds and feeds, balancing rotating parts, and the three-wire measurement of acme and stub-acme screw threads. New metric information has been added from recently available ANSI standards for metric preferred limits and fits and metric retaining rings. Procedures for the measurement of relief angles on milling cutters and single-point threading tools are now explained in greater detail.

Many practical suggestions have been received from friends throughout the mechanical field, and their cooperation, which has proved to be invaluable, is highly appreciated. Directing the editor's attention to a possible defect or to the omission of some matter considered of general value often renders a service to the entire mechanical industry. For this reason and also because we desire to perfect the HANDBOOK as far as possible, all criticisms and suggestions about either revisions or the inclusion of new matter are welcome.

PAUL B. SCHUBERT, EDITOR

# ACKNOWLEDGMENTS

The editors wish to express their appreciation for the contributions in the 21st Edition, of Mr. Neil A. De Bruyne, to the "Friction," and "Plain and Rolling Contact Bearings" sections; of Dr. Francis T. Farago, to the "Surface Grinding" section and of all the other individuals such as Mr. Charles S. Davis, Mr. Harold Carlson, Mr. Earlwood T. Fortini, and Mr. Eugene Stamper who contributed generously in previous editions and whose material enhances the present edition.

Data and information from the following American National standards Institute (ANSI) Standards will be found in the HANDBOOK and have been extracted with permission of the publisher, the American Society of Mechanical Engineers, United Engineering Center, 345 East 47 Street, New York, New York, 10017:

B1.1 Unified Inch Screw Threads
B1.2 Gages and Gaging for Unified Screw Threads
B1.5 Acme Screw Threads
B1.7 Nomenclature, Definitions, and Letter Symbols for Screw Threads
B1.8 Stub Acme Screw Threads
B1.9 Buttress Inch Screw Threads
B1.10 Unified Miniature Screw Threads
B1.11 Microscope Objective Thread
B1.12 Class 5 Interference Fit Thread
B2.1 Pipe Threads (Except Dryseal)
B2.2 Dryseal Pipe Threads
B2.4 Hose Coupling Screw Threads
B4.1 Preferred Limits and Fits for Cylindrical Parts
B4.2 Preferred Metric Limits and Fits
B5.10 Machine Tapers
B5.18 Spindle Noses and Tool Shanks for Milling Machines
B5.20 Machine Pins
B6.1 Tooth Proportions for Coarse-Pitch Involute Spur Gears
B6.7 Tooth Proportions for Fine-Pitch Involute Spur and Helical Gears
B6.9 Design for Fine-Pitch Worm Gearing
B17.1 Keys and Keyseats
B17.2 Woodruff Keys and Keyseats
B18.1.1 Small Solid Rivets
B18.1.2 Large Rivets
B18.2.1 Square and Hex Bolts and Screws

On August 24, 1966, the American Standards Association was reconstituted as the United States of America Standards Institute and standards approved as American Standards were designated USA Standards. There was no change in their index identification or technical content. On October 6, 1969 the name was changed to American National Standards Institute. The present standards' designation is ANSI instead of ASA or USAS. Standards previously adopted by the American Standards Association are still referred to in the HANDBOOK by the designation ASA.

# GENERAL CONTENTS

# MATHEMATICAL TABLES

**Fractions of One Inch Converted to Decimal Inches and Millimetres**

| Fraction | Decimal | mm | Fraction | Decimal | mm |
|---|---|---|---|---|---|
| 1/64 | 0.015 625 | 0.396 9 | 33/64 | 0.515 625 | 13.096 9 |
| 1/32 | 0.031 25 | 0.793 8 | 17/32 | 0.531 25 | 13.493 8 |
| 3/64 | 0.046 875 | 1.190 6 | 35/64 | 0.546 875 | 13.890 6 |
| 1/16 | 0.062 5 | 1.587 5 | 9/16 | 0.562 5 | 14.287 5 |
| 5/64 | 0.078 125 | 1.984 4 | 37/64 | 0.578 125 | 14.684 4 |
| 3/32 | 0.093 75 | 2.381 3 | 19/32 | 0.593 75 | 15.081 3 |
| 7/64 | 0.109 375 | 2.778 1 | 39/64 | 0.609 375 | 15.478 1 |
| 1/8 | 0.125 | 3.175 | 5/8 | 0.625 | 15.875 |
| 9/64 | 0.140 625 | 3.571 9 | 41/64 | 0.640 625 | 16.271 9 |
| 5/32 | 0.156 25 | 3.968 8 | 21/32 | 0.656 25 | 16.668 8 |
| 11/64 | 0.171 875 | 4.365 6 | 43/64 | 0.671 875 | 17.065 6 |
| 3/16 | 0.187 5 | 4.762 5 | 11/16 | 0.687 5 | 17.462 5 |
| 13/64 | 0.203 125 | 5.159 4 | 45/64 | 0.703 125 | 17.859 4 |
| 7/32 | 0.218 75 | 5.556 3 | 23/32 | 0.718 75 | 18.256 3 |
| 15/64 | 0.234 375 | 5.953 1 | 47/64 | 0.734 375 | 18.653 1 |
| 1/4 | 0.25 | 6.35 | 3/4 | 0.75 | 19.05 |
| 17/64 | 0.265 625 | 6.746 9 | 49/64 | 0.765 625 | 19.446 9 |
| 9/32 | 0.281 25 | 7.143 8 | 25/32 | 0.781 25 | 19.843 8 |
| 19/64 | 0.296 875 | 7.540 6 | 51/64 | 0.796 875 | 20.240 6 |
| 5/16 | 0.312 5 | 7.937 5 | 13/16 | 0.812 5 | 20.637 5 |
| 21/64 | 0.328 125 | 8.334 4 | 53/64 | 0.828 125 | 21.034 4 |
| 11/32 | 0.343 75 | 8.731 3 | 27/32 | 0.843 75 | 21.431 3 |
| 23/64 | 0.359 375 | 9.128 1 | 55/64 | 0.859 375 | 21.828 1 |
| 3/8 | 0.375 | 9.525 | 7/8 | 0.875 | 22.225 |
| 25/64 | 0.390 625 | 9.921 9 | 57/64 | 0.890 625 | 22.621 9 |
| 13/32 | 0.406 25 | 10.318 8 | 29/32 | 0.906 25 | 23.018 8 |
| 27/64 | 0.421 875 | 10.715 6 | 59/64 | 0.921 875 | 23.415 6 |
| 7/16 | 0.437 5 | 11.112 5 | 15/16 | 0.937 5 | 23.812 5 |
| 29/64 | 0.453 125 | 11.509 4 | 61/64 | 0.953 125 | 24.209 4 |
| 15/32 | 0.468 75 | 11.906 3 | 31/32 | 0.968 75 | 24.666 3 |
| 31/64 | 0.484 375 | 12.303 1 | 63/64 | 0.984 375 | 25.003 1 |
| 1/2 | 0.5 | 12.7 | 1 | 1 | 25.4 |

## Powers, Roots and Reciprocals

| No. | Square | Cube | Sq. Root | Cube Root | Reciprocal | No. |
|---|---|---|---|---|---|---|
| 1 | 1 | 1 | 1.00000 | 1.00000 | 1.0000000 | 1 |
| 2 | 4 | 8 | 1.41421 | 1.25992 | 0.5000000 | 2 |
| 3 | 9 | 27 | 1.73205 | 1.44225 | 0.3333333 | 3 |
| 4 | 16 | 64 | 2.00000 | 1.58740 | 0.2500000 | 4 |
| 5 | 25 | 125 | 2.23607 | 1.70998 | 0.2000000 | 5 |
| 6 | 36 | 216 | 2.44949 | 1.81712 | 0.1666667 | 6 |
| 7 | 49 | 343 | 2.64575 | 1.91293 | 0.1428571 | 7 |
| 8 | 64 | 512 | 2.82843 | 2.00000 | 0.1250000 | 8 |
| 9 | 81 | 729 | 3.00000 | 2.08008 | 0.1111111 | 9 |
| 10 | 100 | 1,000 | 3.16228 | 2.15443 | 0.1000000 | 10 |
| 11 | 121 | 1,331 | 3.31662 | 2.22398 | 0.0909091 | 11 |
| 12 | 144 | 1,728 | 3.46410 | 2.28943 | 0.0833333 | 12 |
| 13 | 169 | 2,197 | 3.60555 | 2.35133 | 0.0769231 | 13 |
| 14 | 196 | 2,744 | 3.74166 | 2.41014 | 0.0714286 | 14 |
| 15 | 225 | 3,375 | 3.87298 | 2.46621 | 0.0666667 | 15 |
| 16 | 256 | 4,096 | 4.00000 | 2.51984 | 0.0625000 | 16 |
| 17 | 289 | 4,913 | 4.12311 | 2.57128 | 0.0588235 | 17 |
| 18 | 324 | 5,832 | 4.24264 | 2.62074 | 0.0555556 | 18 |
| 19 | 361 | 6,859 | 4.35890 | 2.66840 | 0.0526316 | 19 |
| 20 | 400 | 8,000 | 4.47214 | 2.71442 | 0.0500000 | 20 |
| 21 | 441 | 9,261 | 4.58258 | 2.75892 | 0.0476190 | 21 |
| 22 | 484 | 10,648 | 4.69042 | 2.80204 | 0.0454545 | 22 |
| 23 | 529 | 12,167 | 4.79583 | 2.84387 | 0.0434783 | 23 |
| 24 | 576 | 13,824 | 4.89898 | 2.88450 | 0.0416667 | 24 |
| 25 | 625 | 15,625 | 5.00000 | 2.92402 | 0.0400000 | 25 |
| 26 | 676 | 17,576 | 5.09902 | 2.96250 | 0.0384615 | 26 |
| 27 | 729 | 19,683 | 5.19615 | 3.00000 | 0.0370370 | 27 |
| 28 | 784 | 21,952 | 5.29150 | 3.03659 | 0.0357143 | 28 |
| 29 | 841 | 24,389 | 5.38516 | 3.07232 | 0.0344828 | 29 |
| 30 | 900 | 27,000 | 5.47723 | 3.10723 | 0.0333333 | 30 |
| 31 | 961 | 29,791 | 5.56776 | 3.14138 | 0.0322581 | 31 |
| 32 | 1,024 | 32,768 | 5.65685 | 3.17480 | 0.0312500 | 32 |
| 33 | 1,089 | 35,937 | 5.74456 | 3.20753 | 0.0303030 | 33 |
| 34 | 1,156 | 39,304 | 5.83095 | 3.23961 | 0.0294118 | 34 |
| 35 | 1,225 | 42,875 | 5.91608 | 3.27107 | 0.0285714 | 35 |
| 36 | 1,296 | 46,656 | 6.00000 | 3.30193 | 0.0277778 | 36 |
| 37 | 1,369 | 50,653 | 6.08276 | 3.33222 | 0.0270270 | 37 |
| 38 | 1,444 | 54,872 | 6.16441 | 3.36198 | 0.0263158 | 38 |
| 39 | 1,521 | 59,319 | 6.24500 | 3.39121 | 0.0256410 | 39 |
| 40 | 1,600 | 64,000 | 6.32456 | 3.41995 | 0.0250000 | 40 |
| 41 | 1,681 | 68,921 | 6.40312 | 3.44822 | 0.0243902 | 41 |
| 42 | 1,764 | 74,088 | 6.48074 | 3.47603 | 0.0238095 | 42 |
| 43 | 1,849 | 79,507 | 6.55744 | 3.50340 | 0.0232558 | 43 |
| 44 | 1,936 | 85,184 | 6.63325 | 3.53035 | 0.0227273 | 44 |
| 45 | 2,025 | 91,125 | 6.70820 | 3.55689 | 0.0222222 | 45 |
| 46 | 2,116 | 97,336 | 6.78233 | 3.58305 | 0.0217391 | 46 |
| 47 | 2,209 | 103,823 | 6.85565 | 3.60883 | 0.0212766 | 47 |
| 48 | 2,304 | 110,592 | 6.92830 | 3.63424 | 0.0208333 | 48 |
| 49 | 2,401 | 117,649 | 7.00000 | 3.65931 | 0.0204082 | 49 |
| 50 | 2,500 | 125,000 | 7.07107 | 3.68403 | 0.0200000 | 50 |

Powers, Roots and Reciprocals

| No. | Square | Cube | Sq. Root | Cube Root | Reciprocal | No. |
|---|---|---|---|---|---|---|
| 51 | 2,601 | 132,651 | 7.14143 | 3.70843 | 0.0196078 | 51 |
| 52 | 2,704 | 140,608 | 7.21110 | 3.73251 | 0.0192308 | 52 |
| 53 | 2,809 | 148,877 | 7.28011 | 3.75629 | 0.0188679 | 53 |
| 54 | 2,916 | 157,464 | 7.34847 | 3.77976 | 0.0185185 | 54 |
| 55 | 3,025 | 166,375 | 7.41620 | 3.80295 | 0.0181818 | 55 |
| 56 | 3,136 | 175,616 | 7.48331 | 3.82586 | 0.0178571 | 56 |
| 57 | 3,249 | 185,193 | 7.54983 | 3.84850 | 0.0175439 | 57 |
| 58 | 3,364 | 195,112 | 7.61577 | 3.87088 | 0.0172414 | 58 |
| 59 | 3,481 | 205,379 | 7.68115 | 3.89300 | 0.0169492 | 59 |
| 60 | 3,600 | 216,000 | 7.74597 | 3.91487 | 0.0166667 | 60 |
| 61 | 3,721 | 226,981 | 7.81025 | 3.93650 | 0.0163934 | 61 |
| 62 | 3,844 | 238,328 | 7.87401 | 3.95789 | 0.0161290 | 62 |
| 63 | 3,969 | 250,047 | 7.93725 | 3.97906 | 0.0158730 | 63 |
| 64 | 4,096 | 262,144 | 8.00000 | 4.00000 | 0.0156250 | 64 |
| 65 | 4,225 | 274,625 | 8.06226 | 4.02073 | 0.0153846 | 65 |
| 66 | 4,356 | 287,496 | 8.12404 | 4.04124 | 0.0151515 | 66 |
| 67 | 4,489 | 300,763 | 8.18535 | 4.06155 | 0.0149254 | 67 |
| 68 | 4,624 | 314,432 | 8.24621 | 4.08166 | 0.0147059 | 68 |
| 69 | 4,761 | 328,509 | 8.30662 | 4.10157 | 0.0144928 | 69 |
| 70 | 4,900 | 343,000 | 8.36660 | 4.12129 | 0.0142857 | 70 |
| 71 | 5,041 | 357,911 | 8.42615 | 4.14082 | 0.0140845 | 71 |
| 72 | 5,184 | 373,248 | 8.48528 | 4.16017 | 0.0138889 | 72 |
| 73 | 5,329 | 389,017 | 8.54400 | 4.17934 | 0.0136986 | 73 |
| 74 | 5,476 | 405,224 | 8.60233 | 4.19834 | 0.0135135 | 74 |
| 75 | 5,625 | 421,875 | 8.66025 | 4.21716 | 0.0133333 | 75 |
| 76 | 5,776 | 438,976 | 8.71780 | 4.23582 | 0.0131579 | 76 |
| 77 | 5,929 | 456,533 | 8.77496 | 4.25432 | 0.0129870 | 77 |
| 78 | 6,084 | 474,552 | 8.83176 | 4.27266 | 0.0128205 | 78 |
| 79 | 6,241 | 493,039 | 8.88819 | 4.29084 | 0.0126582 | 79 |
| 80 | 6,400 | 512,000 | 8.94427 | 4.30887 | 0.0125000 | 80 |
| 81 | 6,561 | 531,441 | 9.00000 | 4.32675 | 0.0123457 | 81 |
| 82 | 6,724 | 551,368 | 9.05539 | 4.34448 | 0.0121951 | 82 |
| 83 | 6,889 | 571,787 | 9.11043 | 4.36207 | 0.0120482 | 83 |
| 84 | 7,056 | 592,704 | 9.16515 | 4.37952 | 0.0119048 | 84 |
| 85 | 7,225 | 614,125 | 9.21954 | 4.39683 | 0.0117647 | 85 |
| 86 | 7,396 | 636,056 | 9.27362 | 4.41400 | 0.0116279 | 86 |
| 87 | 7,569 | 658,503 | 9.32738 | 4.43105 | 0.0114943 | 87 |
| 88 | 7,744 | 681,472 | 9.38083 | 4.44796 | 0.0113636 | 88 |
| 89 | 7,921 | 704,969 | 9.43398 | 4.46475 | 0.0112360 | 89 |
| 90 | 8,100 | 729,000 | 9.48683 | 4.48140 | 0.0111111 | 90 |
| 91 | 8,281 | 753,571 | 9.53939 | 4.49794 | 0.0109890 | 91 |
| 92 | 8,464 | 778,688 | 9.59166 | 4.51436 | 0.0108696 | 92 |
| 93 | 8,649 | 804,357 | 9.64365 | 4.53065 | 0.0107527 | 93 |
| 94 | 8,836 | 830,584 | 9.69536 | 4.54684 | 0.0106383 | 94 |
| 95 | 9,025 | 857,375 | 9.74679 | 4.56290 | 0.0105263 | 95 |
| 96 | 9,216 | 884,736 | 9.79796 | 4.57886 | 0.0104167 | 96 |
| 97 | 9,409 | 912,673 | 9.84886 | 4.59470 | 0.0103093 | 97 |
| 98 | 9,604 | 941,192 | 9.89949 | 4.61044 | 0.0102041 | 98 |
| 99 | 9,801 | 970,299 | 9.94987 | 4.62607 | 0.0101010 | 99 |
| 100 | 10,000 | 1,000,000 | 10.00000 | 4.64159 | 0.0100000 | 100 |

## Powers, Roots and Reciprocals

| No. | Square | Cube | Sq. Root | Cube Root | Reciprocal | No. |
|---|---|---|---|---|---|---|
| 101 | 10,201 | 1,030,301 | 10.0499 | 4.65701 | 0.0099010 | 101 |
| 102 | 10,404 | 1,061,208 | 10.0995 | 4.67233 | 0.0098039 | 102 |
| 103 | 10,609 | 1,092,727 | 10.1489 | 4.68755 | 0.0097087 | 103 |
| 104 | 10,816 | 1,124,864 | 10.1980 | 4.70267 | 0.0096154 | 104 |
| 105 | 11,025 | 1,157,625 | 10.2470 | 4.71769 | 0.0095238 | 105 |
| 106 | 11,236 | 1,191,016 | 10.2956 | 4.73262 | 0.0094340 | 106 |
| 107 | 11,449 | 1,225,043 | 10.3441 | 4.74746 | 0.0093458 | 107 |
| 108 | 11,664 | 1,259,712 | 10.3923 | 4.76220 | 0.0092593 | 108 |
| 109 | 11,881 | 1,295,029 | 10.4403 | 4.77686 | 0.0091743 | 109 |
| 110 | 12,100 | 1,331,000 | 10.4881 | 4.79142 | 0.0090909 | 110 |
| 111 | 12,321 | 1,367,631 | 10.5357 | 4.80590 | 0.0090090 | 111 |
| 112 | 12,544 | 1,404,928 | 10.5830 | 4.82028 | 0.0089286 | 112 |
| 113 | 12,769 | 1,442,897 | 10.6301 | 4.83459 | 0.0088496 | 113 |
| 114 | 12,996 | 1,481,544 | 10.6771 | 4.84881 | 0.0087719 | 114 |
| 115 | 13,225 | 1,520,875 | 10.7238 | 4.86294 | 0.0086957 | 115 |
| 116 | 13,456 | 1,560,896 | 10.7703 | 4.87700 | 0.0086207 | 116 |
| 117 | 13,689 | 1,601,613 | 10.8167 | 4.89097 | 0.0085470 | 117 |
| 118 | 13,924 | 1,643,032 | 10.8628 | 4.90487 | 0.0084746 | 118 |
| 119 | 14,161 | 1,685,159 | 10.9087 | 4.91868 | 0.0084034 | 119 |
| 120 | 14,400 | 1,728,000 | 10.9545 | 4.93242 | 0.0083333 | 120 |
| 121 | 14,641 | 1,771,561 | 11.0000 | 4.94609 | 0.0082645 | 121 |
| 122 | 14,884 | 1,815,848 | 11.0454 | 4.95968 | 0.0081967 | 122 |
| 123 | 15,129 | 1,860,867 | 11.0905 | 4.97319 | 0.0081301 | 123 |
| 124 | 15,376 | 1,906,624 | 11.1355 | 4.98663 | 0.0080645 | 124 |
| 125 | 15,625 | 1,953,125 | 11.1803 | 5.00000 | 0.0080000 | 125 |
| 126 | 15,876 | 2,000,376 | 11.2250 | 5.01330 | 0.0079365 | 126 |
| 127 | 16,129 | 2,048,383 | 11.2694 | 5.02653 | 0.0078740 | 127 |
| 128 | 16,384 | 2,097,152 | 11.3137 | 5.03968 | 0.0078125 | 128 |
| 129 | 16,641 | 2,146,689 | 11.3578 | 5.05277 | 0.0077519 | 129 |
| 130 | 16,900 | 2,197,000 | 11.4018 | 5.06580 | 0.0076923 | 130 |
| 131 | 17,161 | 2,248,091 | 11.4455 | 5.07875 | 0.0076336 | 131 |
| 132 | 17,424 | 2,299,968 | 11.4891 | 5.09164 | 0.0075758 | 132 |
| 133 | 17,689 | 2,352,637 | 11.5326 | 5.10447 | 0.0075188 | 133 |
| 134 | 17,956 | 2,406,104 | 11.5758 | 5.11723 | 0.0074627 | 134 |
| 135 | 18,225 | 2,460,375 | 11.6190 | 5.12993 | 0.0074074 | 135 |
| 136 | 18,496 | 2,515,456 | 11.6619 | 5.14256 | 0.0073529 | 136 |
| 137 | 18,769 | 2,571,353 | 11.7047 | 5.15514 | 0.0072993 | 137 |
| 138 | 19,044 | 2,628,072 | 11.7473 | 5.16765 | 0.0072464 | 138 |
| 139 | 19,321 | 2,685,619 | 11.7898 | 5.18010 | 0.0071942 | 139 |
| 140 | 19,600 | 2,744,000 | 11.8322 | 5.19249 | 0.0071429 | 140 |
| 141 | 19,881 | 2,803,221 | 11.8743 | 5.20483 | 0.0070922 | 141 |
| 142 | 20,164 | 2,863,288 | 11.9164 | 5.21710 | 0.0070423 | 142 |
| 143 | 20,449 | 2,924,207 | 11.9583 | 5.22932 | 0.0069930 | 143 |
| 144 | 20,736 | 2,985,984 | 12.0000 | 5.24148 | 0.0069444 | 144 |
| 145 | 21,025 | 3,048,625 | 12.0416 | 5.25359 | 0.0068966 | 145 |
| 146 | 21,316 | 3,112,136 | 12.0830 | 5.26564 | 0.0068493 | 146 |
| 147 | 21,609 | 3,176,523 | 12.1244 | 5.27763 | 0.0068027 | 147 |
| 148 | 21,904 | 3,241,792 | 12.1655 | 5.28957 | 0.0067568 | 148 |
| 149 | 22,201 | 3,307,949 | 12.2066 | 5.30146 | 0.0067114 | 149 |
| 150 | 22,500 | 3,375,000 | 12.2474 | 5.31329 | 0.0066667 | 150 |

## Powers, Roots and Reciprocals

| No. | Square | Cube | Sq. Root | Cube Root | Reciprocal | No. |
|-----|--------|------|----------|-----------|------------|-----|
| 151 | 22,801 | 3,442,951 | 12.2882 | 5.32507 | 0.0066225 | 151 |
| 152 | 23,104 | 3,511,808 | 12.3288 | 5.33680 | 0.0065789 | 152 |
| 153 | 23,409 | 3,581,577 | 12.3693 | 5.34848 | 0.0065359 | 153 |
| 154 | 23,716 | 3,652,264 | 12.4097 | 5.36011 | 0.0064935 | 154 |
| 155 | 24,025 | 3,723,875 | 12.4499 | 5.37169 | 0.0064516 | 155 |
| 156 | 24,336 | 3,796,416 | 12.4900 | 5.38321 | 0.0064103 | 156 |
| 157 | 24,649 | 3,869,893 | 12.5300 | 5.39469 | 0.0063694 | 157 |
| 158 | 24,964 | 3,944,312 | 12.5698 | 5.40612 | 0.0063291 | 158 |
| 159 | 25,281 | 4,019,679 | 12.6095 | 5.41750 | 0.0062893 | 159 |
| 160 | 25,600 | 4,096,000 | 12.6491 | 5.42884 | 0.0062500 | 160 |
| 161 | 25,921 | 4,173,281 | 12.6886 | 5.44012 | 0.0062112 | 161 |
| 162 | 26,244 | 4,251,528 | 12.7279 | 5.45136 | 0.0061728 | 162 |
| 163 | 26,569 | 4,330,747 | 12.7671 | 5.46256 | 0.0061350 | 163 |
| 164 | 26,896 | 4,410,944 | 12.8062 | 5.47370 | 0.0060976 | 164 |
| 165 | 27,225 | 4,492,125 | 12.8452 | 5.48481 | 0.0060606 | 165 |
| 166 | 27,556 | 4,574,296 | 12.8841 | 5.49586 | 0.0060241 | 166 |
| 167 | 27,889 | 4,657,463 | 12.9228 | 5.50688 | 0.0059880 | 167 |
| 168 | 28,224 | 4,741,632 | 12.9615 | 5.51785 | 0.0059524 | 168 |
| 169 | 28,561 | 4,826,809 | 13.0000 | 5.52877 | 0.0059172 | 169 |
| 170 | 28,900 | 4,913,000 | 13.0384 | 5.53966 | 0.0058824 | 170 |
| 171 | 29,241 | 5,000,211 | 13.0767 | 5.55050 | 0.0058480 | 171 |
| 172 | 29,584 | 5,088,448 | 13.1149 | 5.56130 | 0.0058140 | 172 |
| 173 | 29,929 | 5,177,717 | 13.1529 | 5.57205 | 0.0057803 | 173 |
| 174 | 30,276 | 5,268,024 | 13.1909 | 5.58277 | 0.0057471 | 174 |
| 175 | 30,625 | 5,359,375 | 13.2288 | 5.59344 | 0.0057143 | 175 |
| 176 | 30,976 | 5,451,776 | 13.2665 | 5.60408 | 0.0056818 | 176 |
| 177 | 31,329 | 5,545,233 | 13.3041 | 5.61467 | 0.0056497 | 177 |
| 178 | 31,684 | 5,639,752 | 13.3417 | 5.62523 | 0.0056180 | 178 |
| 179 | 32,041 | 5,735,339 | 13.3791 | 5.63574 | 0.0055866 | 179 |
| 180 | 32,400 | 5,832,000 | 13.4164 | 5.64622 | 0.0055556 | 180 |
| 181 | 32,761 | 5,929,741 | 13.4536 | 5.65665 | 0.0055249 | 181 |
| 182 | 33,124 | 6,028,568 | 13.4907 | 5.66705 | 0.0054945 | 182 |
| 183 | 33,489 | 6,128,487 | 13.5277 | 5.67741 | 0.0054645 | 183 |
| 184 | 33,856 | 6,229,504 | 13.5647 | 5.68773 | 0.0054348 | 184 |
| 185 | 34,225 | 6,331,625 | 13.6015 | 5.69802 | 0.0054054 | 185 |
| 186 | 34,596 | 6,434,856 | 13.6382 | 5.70827 | 0.0053763 | 186 |
| 187 | 34,969 | 6,539,203 | 13.6748 | 5.71848 | 0.0053476 | 187 |
| 188 | 35,344 | 6,644,672 | 13.7113 | 5.72865 | 0.0053191 | 188 |
| 189 | 35,721 | 6,751,269 | 13.7477 | 5.73879 | 0.0052910 | 189 |
| 190 | 36,100 | 6,859,000 | 13.7840 | 5.74890 | 0.0052632 | 190 |
| 191 | 36,481 | 6,967,871 | 13.8203 | 5.75897 | 0.0052356 | 191 |
| 192 | 36,864 | 7,077,888 | 13.8564 | 5.76900 | 0.0052083 | 192 |
| 193 | 37,249 | 7,189,057 | 13.8924 | 5.77900 | 0.0051813 | 193 |
| 194 | 37,636 | 7,301,384 | 13.9284 | 5.78896 | 0.0051546 | 194 |
| 195 | 38,025 | 7,414,875 | 13.9642 | 5.79889 | 0.0051282 | 195 |
| 196 | 38,416 | 7,529,536 | 14.0000 | 5.80879 | 0.0051020 | 196 |
| 197 | 38,809 | 7,645,373 | 14.0357 | 5.81865 | 0.0050761 | 197 |
| 198 | 39,204 | 7,762,392 | 14.0712 | 5.82848 | 0.0050505 | 198 |
| 199 | 39,601 | 7,880,599 | 14.1067 | 5.83827 | 0.0050251 | 199 |
| 200 | 40,000 | 8,000,000 | 14.1421 | 5.84804 | 0.0050000 | 200 |

## Powers, Roots and Reciprocals

| No. | Square | Cube | Sq. Root | Cube Root | Reciprocal | No. |
|---|---|---|---|---|---|---|
| 201 | 40,401 | 8,120,601 | 14.1774 | 5.85777 | 0.0049751 | 201 |
| 202 | 40,804 | 8,242,408 | 14.2127 | 5.86746 | 0.0049505 | 202 |
| 203 | 41,209 | 8,365,427 | 14.2478 | 5.87713 | 0.0049261 | 203 |
| 204 | 41,616 | 8,489,664 | 14.2829 | 5.88677 | 0.0049020 | 204 |
| 205 | 42,025 | 8,615,125 | 14.3178 | 5.89637 | 0.0048780 | 205 |
| 206 | 42,436 | 8,741,816 | 14.3527 | 5.90594 | 0.0048544 | 206 |
| 207 | 42,849 | 8,869,743 | 14.3875 | 5.91548 | 0.0048309 | 207 |
| 208 | 43,264 | 8,998,912 | 14.4222 | 5.92499 | 0.0048077 | 208 |
| 209 | 43,681 | 9,129,329 | 14.4568 | 5.93447 | 0.0047847 | 209 |
| 210 | 44,100 | 9,261,000 | 14.4914 | 5.94392 | 0.0047619 | 210 |
| 211 | 44,521 | 9,393,931 | 14.5258 | 5.95334 | 0.0047393 | 211 |
| 212 | 44,944 | 9,528,128 | 14.5602 | 5.96273 | 0.0047170 | 212 |
| 213 | 45,369 | 9,663,597 | 14.5945 | 5.97209 | 0.0046948 | 213 |
| 214 | 45,796 | 9,800,344 | 14.6287 | 5.98142 | 0.0046729 | 214 |
| 215 | 46,225 | 9,938,375 | 14.6629 | 5.99073 | 0.0046512 | 215 |
| 216 | 46,656 | 10,077,696 | 14.6969 | 6.00000 | 0.0046296 | 216 |
| 217 | 47,089 | 10,218,313 | 14.7309 | 6.00925 | 0.0046083 | 217 |
| 218 | 47,524 | 10,360,232 | 14.7648 | 6.01846 | 0.0045872 | 218 |
| 219 | 47,961 | 10,503,459 | 14.7986 | 6.02765 | 0.0045662 | 219 |
| 220 | 48,400 | 10,648,000 | 14.8324 | 6.03681 | 0.0045455 | 220 |
| 221 | 48,841 | 10,793,861 | 14.8661 | 6.04594 | 0.0045249 | 221 |
| 222 | 49,284 | 10,941,048 | 14.8997 | 6.05505 | 0.0045045 | 222 |
| 223 | 49,729 | 11,089,567 | 14.9332 | 6.06413 | 0.0044843 | 223 |
| 224 | 50,176 | 11,239,424 | 14.9666 | 6.07318 | 0.0044643 | 224 |
| 225 | 50,625 | 11,390,625 | 15.0000 | 6.08220 | 0.0044444 | 225 |
| 226 | 51,076 | 11,543,176 | 15.0333 | 6.09120 | 0.0044248 | 226 |
| 227 | 51,529 | 11,697,083 | 15.0665 | 6.10017 | 0.0044053 | 227 |
| 228 | 51,984 | 11,852,352 | 15.0997 | 6.10911 | 0.0043860 | 228 |
| 229 | 52,441 | 12,008,989 | 15.1327 | 6.11803 | 0.0043668 | 229 |
| 230 | 52,900 | 12,167,000 | 15.1658 | 6.12693 | 0.0043478 | 230 |
| 231 | 53,361 | 12,326,391 | 15.1987 | 6.13579 | 0.0043290 | 231 |
| 232 | 53,824 | 12,487,168 | 15.2315 | 6.14463 | 0.0043103 | 232 |
| 233 | 54,289 | 12,649,337 | 15.2643 | 6.15345 | 0.0042918 | 233 |
| 234 | 54,756 | 12,812,904 | 15.2971 | 6.16224 | 0.0042735 | 234 |
| 235 | 55,225 | 12,977,875 | 15.3297 | 6.17101 | 0.0042553 | 235 |
| 236 | 55,696 | 13,144,256 | 15.3623 | 6.17975 | 0.0042373 | 236 |
| 237 | 56,169 | 13,312,053 | 15.3948 | 6.18846 | 0.0042194 | 237 |
| 238 | 56,644 | 13,481,272 | 15.4272 | 6.19715 | 0.0042017 | 238 |
| 239 | 57,121 | 13,651,919 | 15.4596 | 6.20582 | 0.0041841 | 239 |
| 240 | 57,600 | 13,824,000 | 15.4919 | 6.21447 | 0.0041667 | 240 |
| 241 | 58,081 | 13,997,521 | 15.5242 | 6.22308 | 0.0041494 | 241 |
| 242 | 58,564 | 14,172,488 | 15.5563 | 6.23168 | 0.0041322 | 242 |
| 243 | 59,049 | 14,348,907 | 15.5885 | 6.24025 | 0.0041152 | 243 |
| 244 | 59,536 | 14,526,784 | 15.6205 | 6.24880 | 0.0040984 | 244 |
| 245 | 60,025 | 14,706,125 | 15.6525 | 6.25732 | 0.0040816 | 245 |
| 246 | 60,516 | 14,886,936 | 15.6844 | 6.26583 | 0.0040650 | 246 |
| 247 | 61,009 | 15,069,223 | 15.7162 | 6.27431 | 0.0040486 | 247 |
| 248 | 61,504 | 15,252,992 | 15.7480 | 6.28276 | 0.0040323 | 248 |
| 249 | 62,001 | 15,438,249 | 15.7797 | 6.29119 | 0.0040161 | 249 |
| 250 | 62,500 | 15,625,000 | 15.8114 | 6.29961 | 0.0040000 | 250 |

## Powers, Roots and Reciprocals

| No. | Square | Cube | Sq. Root | Cube Root | Reciprocal | No. |
|---|---|---|---|---|---|---|
| 251 | 63,001 | 15,813,251 | 15.8430 | 6.30799 | 0.0039841 | 251 |
| 252 | 63,504 | 16,003,008 | 15.8745 | 6.31636 | 0.0039683 | 252 |
| 253 | 64,009 | 16,194,277 | 15.9060 | 6.32470 | 0.0039526 | 253 |
| 254 | 64,516 | 16,387,064 | 15.9374 | 6.33303 | 0.0039370 | 254 |
| 255 | 65,025 | 16,581,375 | 15.9687 | 6.34133 | 0.0039216 | 255 |
| 256 | 65,536 | 16,777,216 | 16.0000 | 6.34960 | 0.0039063 | 256 |
| 257 | 66,049 | 16,974,593 | 16.0312 | 6.35786 | 0.0038911 | 257 |
| 258 | 66,564 | 17,173,512 | 16.0624 | 6.36610 | 0.0038760 | 258 |
| 259 | 67,081 | 17,373,979 | 16.0935 | 6.37431 | 0.0038610 | 259 |
| 260 | 67,600 | 17,576,000 | 16.1245 | 6.38250 | 0.0038462 | 260 |
| 261 | 68,121 | 17,779,581 | 16.1555 | 6.39068 | 0.0038314 | 261 |
| 262 | 68,644 | 17,984,728 | 16.1864 | 6.39883 | 0.0038168 | 262 |
| 263 | 69,169 | 18,191,447 | 16.2173 | 6.40696 | 0.0038023 | 263 |
| 264 | 69,696 | 18,399,744 | 16.2481 | 6.41507 | 0.0037879 | 264 |
| 265 | 70,225 | 18,609,625 | 16.2788 | 6.42316 | 0.0037736 | 265 |
| 266 | 70,756 | 18,821,096 | 16.3095 | 6.43123 | 0.0037594 | 266 |
| 267 | 71,289 | 19,034,163 | 16.3401 | 6.43928 | 0.0037453 | 267 |
| 268 | 71,824 | 19,248,832 | 16.3707 | 6.44731 | 0.0037313 | 268 |
| 269 | 72,361 | 19,465,109 | 16.4012 | 6.45531 | 0.0037175 | 269 |
| 270 | 72,900 | 19,683,000 | 16.4317 | 6.46330 | 0.0037037 | 270 |
| 271 | 73,441 | 19,902,511 | 16.4621 | 6.47127 | 0.0036900 | 271 |
| 272 | 73,984 | 20,123,648 | 16.4924 | 6.47922 | 0.0036765 | 272 |
| 273 | 74,529 | 20,346,417 | 16.5227 | 6.48715 | 0.0036630 | 273 |
| 274 | 75,076 | 20,570,824 | 16.5529 | 6.49507 | 0.0036496 | 274 |
| 275 | 75,625 | 20,796,875 | 16.5831 | 6.50296 | 0.0036364 | 275 |
| 276 | 76,176 | 21,024,576 | 16.6132 | 6.51083 | 0.0036232 | 276 |
| 277 | 76,729 | 21,253,933 | 16.6433 | 6.51868 | 0.0036101 | 277 |
| 278 | 77,284 | 21,484,952 | 16.6733 | 6.52652 | 0.0035971 | 278 |
| 279 | 77,841 | 21,717,639 | 16.7033 | 6.53434 | 0.0035842 | 279 |
| 280 | 78,400 | 21,952,000 | 16.7332 | 6.54213 | 0.0035714 | 280 |
| 281 | 78,961 | 22,188,041 | 16.7631 | 6.54991 | 0.0035587 | 281 |
| 282 | 79,524 | 22,425,768 | 16.7929 | 6.55767 | 0.0035461 | 282 |
| 283 | 80,089 | 22,665,187 | 16.8226 | 6.56541 | 0.0035336 | 283 |
| 284 | 80,656 | 22,906,304 | 16.8523 | 6.57314 | 0.0035211 | 284 |
| 285 | 81,225 | 23,149,125 | 16.8819 | 6.58084 | 0.0035088 | 285 |
| 286 | 81,796 | 23,393,656 | 16.9115 | 6.58853 | 0.0034965 | 286 |
| 287 | 82,369 | 23,639,903 | 16.9411 | 6.59620 | 0.0034843 | 287 |
| 288 | 82,944 | 23,887,872 | 16.9706 | 6.60385 | 0.0034722 | 288 |
| 289 | 83,521 | 24,137,569 | 17.0000 | 6.61149 | 0.0034602 | 289 |
| 290 | 84,100 | 24,389,000 | 17.0294 | 6.61911 | 0.0034483 | 290 |
| 291 | 84,681 | 24,642,171 | 17.0587 | 6.62671 | 0.0034364 | 291 |
| 292 | 85,264 | 24,897,088 | 17.0880 | 6.63429 | 0.0034247 | 292 |
| 293 | 85,849 | 25,153,757 | 17.1172 | 6.64185 | 0.0034130 | 293 |
| 294 | 86,436 | 25,412,184 | 17.1464 | 6.64940 | 0.0034014 | 294 |
| 295 | 87,025 | 25,672,375 | 17.1756 | 6.65693 | 0.0033898 | 295 |
| 296 | 87,616 | 25,934,336 | 17.2047 | 6.66444 | 0.0033784 | 296 |
| 297 | 88,209 | 26,198,073 | 17.2337 | 6.67194 | 0.0033670 | 297 |
| 298 | 88,804 | 26,463,592 | 17.2627 | 6.67942 | 0.0033557 | 298 |
| 299 | 89,401 | 26,730,899 | 17.2916 | 6.68688 | 0.0033445 | 299 |
| 300 | 90,000 | 27,000,000 | 17.3205 | 6.69433 | 0.0033333 | 300 |

## Powers, Roots and Reciprocals

| No. | Square | Cube | Sq. Root | Cube Root | Reciprocal | No. |
|---|---|---|---|---|---|---|
| 301 | 90,601 | 27,270,901 | 17.3494 | 6.70176 | 0.0033223 | 301 |
| 302 | 91,204 | 27,543,608 | 17.3781 | 6.70917 | 0.0033113 | 302 |
| 303 | 91,809 | 27,818,127 | 17.4069 | 6.71657 | 0.0033003 | 303 |
| 304 | 92,416 | 28,094,464 | 17.4356 | 6.72395 | 0.0032895 | 304 |
| 305 | 93,025 | 28,372,625 | 17.4642 | 6.73132 | 0.0032787 | 305 |
| 306 | 93,636 | 28,652,616 | 17.4929 | 6.73866 | 0.0032680 | 306 |
| 307 | 94,249 | 28,934,443 | 17.5214 | 6.74600 | 0.0032573 | 307 |
| 308 | 94,864 | 29,218,112 | 17.5499 | 6.75331 | 0.0032468 | 308 |
| 309 | 95,481 | 29,503,629 | 17.5784 | 6.76061 | 0.0032362 | 309 |
| 310 | 96,100 | 29,791,000 | 17.6068 | 6.76790 | 0.0032258 | 310 |
| 311 | 96,721 | 30,080,231 | 17.6352 | 6.77517 | 0.0032154 | 311 |
| 312 | 97,344 | 30,371,328 | 17.6635 | 6.78242 | 0.0032051 | 312 |
| 313 | 97,969 | 30,664,297 | 17.6918 | 6.78966 | 0.0031949 | 313 |
| 314 | 98,596 | 30,959,144 | 17.7200 | 6.79688 | 0.0031847 | 314 |
| 315 | 99,225 | 31,255,875 | 17.7482 | 6.80409 | 0.0031746 | 315 |
| 316 | 99,856 | 31,554,496 | 17.7764 | 6.81128 | 0.0031646 | 316 |
| 317 | 100,489 | 31,855,013 | 17.8045 | 6.81846 | 0.0031546 | 317 |
| 318 | 101,124 | 32,157,432 | 17.8326 | 6.82562 | 0.0031447 | 318 |
| 319 | 101,761 | 32,461,759 | 17.8606 | 6.83277 | 0.0031348 | 319 |
| 320 | 102,400 | 32,768,000 | 17.8885 | 6.83990 | 0.0031250 | 320 |
| 321 | 103,041 | 33,076,161 | 17.9165 | 6.84702 | 0.0031153 | 321 |
| 322 | 103,684 | 33,386,248 | 17.9444 | 6.85412 | 0.0031056 | 322 |
| 323 | 104,329 | 33,698,267 | 17.9722 | 6.86121 | 0.0030960 | 323 |
| 324 | 104,976 | 34,012,224 | 18.0000 | 6.86829 | 0.0030864 | 324 |
| 325 | 105,625 | 34,328,125 | 18.0278 | 6.87534 | 0.0030769 | 325 |
| 326 | 106,276 | 34,645,976 | 18.0555 | 6.88239 | 0.0030675 | 326 |
| 327 | 106,929 | 34,965,783 | 18.0831 | 6.88942 | 0.0030581 | 327 |
| 328 | 107,584 | 35,287,552 | 18.1108 | 6.89643 | 0.0030488 | 328 |
| 329 | 108,241 | 35,611,289 | 18.1384 | 6.90344 | 0.0030395 | 329 |
| 330 | 108,900 | 35,937,000 | 18.1659 | 6.91042 | 0.0030303 | 330 |
| 331 | 109,561 | 36,264,691 | 18.1934 | 6.91740 | 0.0030211 | 331 |
| 332 | 110,224 | 36,594,368 | 18.2209 | 6.92436 | 0.0030120 | 332 |
| 333 | 110,889 | 36,926,037 | 18.2483 | 6.93131 | 0.0030030 | 333 |
| 334 | 111,556 | 37,259,704 | 18.2757 | 6.93823 | 0.0029940 | 334 |
| 335 | 112,225 | 37,595,375 | 18.3030 | 6.94515 | 0.0029851 | 335 |
| 336 | 112,896 | 37,933,056 | 18.3303 | 6.95205 | 0.0029762 | 336 |
| 337 | 113,569 | 38,272,753 | 18.3576 | 6.95894 | 0.0029674 | 337 |
| 338 | 114,244 | 38,614,472 | 18.3848 | 6.96582 | 0.0029586 | 338 |
| 339 | 114,921 | 38,958,219 | 18.4120 | 6.97268 | 0.0029499 | 339 |
| 340 | 115,600 | 39,304,000 | 18.4391 | 6.97953 | 0.0029412 | 340 |
| 341 | 116,281 | 39,651,821 | 18.4662 | 6.98637 | 0.0029326 | 341 |
| 342 | 116,964 | 40,001,688 | 18.4932 | 6.99319 | 0.0029240 | 342 |
| 343 | 117,649 | 40,353,607 | 18.5203 | 7.00000 | 0.0029155 | 343 |
| 344 | 118,336 | 40,707,584 | 18.5472 | 7.00680 | 0.0029070 | 344 |
| 345 | 119,025 | 41,063,625 | 18.5742 | 7.01358 | 0.0028986 | 345 |
| 346 | 119,716 | 41,421,736 | 18.6011 | 7.02035 | 0.0028902 | 346 |
| 347 | 120,409 | 41,781,923 | 18.6279 | 7.02711 | 0.0028818 | 347 |
| 348 | 121,104 | 42,144,192 | 18.6548 | 7.03385 | 0.0028736 | 348 |
| 349 | 121,801 | 42,508,549 | 18.6815 | 7.04058 | 0.0028653 | 349 |
| 350 | 122,500 | 42,875,000 | 18.7083 | 7.04730 | 0.0028571 | 350 |

## Powers, Roots and Reciprocals

| No. | Square | Cube | Sq. Root | Cube Root | Reciprocal | No. |
|---|---|---|---|---|---|---|
| 351 | 123,201 | 43,243,551 | 18.7350 | 7.05400 | 0.0028490 | 351 |
| 352 | 123,904 | 43,614,208 | 18.7617 | 7.06070 | 0.0028409 | 352 |
| 353 | 124,609 | 43,986,977 | 18.7883 | 7.06738 | 0.0028329 | 353 |
| 354 | 125,316 | 44,361,864 | 18.8149 | 7.07404 | 0.0028249 | 354 |
| 355 | 126,025 | 44,738,875 | 18.8414 | 7.08070 | 0.0028169 | 355 |
| 356 | 126,736 | 45,118,016 | 18.8680 | 7.08734 | 0.0028090 | 356 |
| 357 | 127,449 | 45,499,293 | 18.8944 | 7.09397 | 0.0028011 | 357 |
| 358 | 128,164 | 45,882,712 | 18.9209 | 7.10059 | 0.0027933 | 358 |
| 359 | 128,881 | 46,268,279 | 18.9473 | 7.10719 | 0.0027855 | 359 |
| 360 | 129,600 | 46,656,000 | 18.9737 | 7.11379 | 0.0027778 | 360 |
| 361 | 130,321 | 47,045,881 | 19.0000 | 7.12037 | 0.0027701 | 361 |
| 362 | 131,044 | 47,437,928 | 19.0263 | 7.12694 | 0.0027624 | 362 |
| 363 | 131,769 | 47,832,147 | 19.0526 | 7.13349 | 0.0027548 | 363 |
| 364 | 132,496 | 48,228,544 | 19.0788 | 7.14004 | 0.0027473 | 364 |
| 365 | 133,225 | 48,627,125 | 19.1050 | 7.14657 | 0.0027397 | 365 |
| 366 | 133,956 | 49,027,896 | 19.1311 | 7.15309 | 0.0027322 | 366 |
| 367 | 134,689 | 49,430,863 | 19.1572 | 7.15960 | 0.0027248 | 367 |
| 368 | 135,424 | 49,836,032 | 19.1833 | 7.16610 | 0.0027174 | 368 |
| 369 | 136,161 | 50,243,409 | 19.2094 | 7.17258 | 0.0027100 | 369 |
| 370 | 136,900 | 50,653,000 | 19.2354 | 7.17905 | 0.0027027 | 370 |
| 371 | 137,641 | 51,064,811 | 19.2614 | 7.18552 | 0.0026954 | 371 |
| 372 | 138,384 | 51,478,848 | 19.2873 | 7.19197 | 0.0026882 | 372 |
| 373 | 139,129 | 51,895,117 | 19.3132 | 7.19840 | 0.0026810 | 373 |
| 374 | 139,876 | 52,313,624 | 19.3391 | 7.20483 | 0.0026738 | 374 |
| 375 | 140,625 | 52,734,375 | 19.3649 | 7.21125 | 0.0026667 | 375 |
| 376 | 141,376 | 53,157,376 | 19.3907 | 7.21765 | 0.0026596 | 376 |
| 377 | 142,129 | 53,582,633 | 19.4165 | 7.22405 | 0.0026525 | 377 |
| 378 | 142,884 | 54,010,152 | 19.4422 | 7.23043 | 0.0026455 | 378 |
| 379 | 143,641 | 54,439,939 | 19.4679 | 7.23680 | 0.0026385 | 379 |
| 380 | 144,400 | 54,872,000 | 19.4936 | 7.24316 | 0.0026316 | 380 |
| 381 | 145,161 | 55,306,341 | 19.5192 | 7.24950 | 0.0026247 | 381 |
| 382 | 145,924 | 55,742,968 | 19.5448 | 7.25584 | 0.0026178 | 382 |
| 383 | 146,689 | 56,181,887 | 19.5704 | 7.26217 | 0.0026110 | 383 |
| 384 | 147,456 | 56,623,104 | 19.5959 | 7.26848 | 0.0026042 | 384 |
| 385 | 148,225 | 57,066,625 | 19.6214 | 7.27479 | 0.0025974 | 385 |
| 386 | 148,996 | 57,512,456 | 19.6469 | 7.28108 | 0.0025907 | 386 |
| 387 | 149,769 | 57,960,603 | 19.6723 | 7.28736 | 0.0025840 | 387 |
| 388 | 150,544 | 58,411,072 | 19.6977 | 7.29363 | 0.0025773 | 388 |
| 389 | 151,321 | 58,863,869 | 19.7231 | 7.29989 | 0.0025707 | 389 |
| 390 | 152,100 | 59,319,000 | 19.7484 | 7.30614 | 0.0025641 | 390 |
| 391 | 152,881 | 59,776,471 | 19.7737 | 7.31238 | 0.0025575 | 391 |
| 392 | 153,664 | 60,236,288 | 19.7990 | 7.31861 | 0.0025510 | 392 |
| 393 | 154,449 | 60,698,457 | 19.8242 | 7.32483 | 0.0025445 | 393 |
| 394 | 155,236 | 61,162,984 | 19.8494 | 7.33104 | 0.0025381 | 394 |
| 395 | 156,025 | 61,629,875 | 19.8746 | 7.33723 | 0.0025316 | 395 |
| 396 | 156,816 | 62,099,136 | 19.8997 | 7.34342 | 0.0025253 | 396 |
| 397 | 157,609 | 62,570,773 | 19.9249 | 7.34960 | 0.0025189 | 397 |
| 398 | 158,404 | 63,044,792 | 19.9499 | 7.35576 | 0.0025126 | 398 |
| 399 | 159,201 | 63,521,199 | 19.9750 | 7.36192 | 0.0025063 | 399 |
| 400 | 160,000 | 64,000,000 | 20.0000 | 7.36806 | 0.0025000 | 400 |

MATHEMATICAL TABLES

Powers, Roots and Reciprocals

| No. | Square | Cube | Sq. Root | Cube Root | Reciprocal | No. |
|---|---|---|---|---|---|---|
| 401 | 160,801 | 64,481,201 | 20.0250 | 7.37420 | 0.0024938 | 401 |
| 402 | 161,604 | 64,964,808 | 20.0499 | 7.38032 | 0.0024876 | 402 |
| 403 | 162,409 | 65,450,827 | 20.0749 | 7.38644 | 0.0024814 | 403 |
| 404 | 163,216 | 65,939,264 | 20.0998 | 7.39254 | 0.0024752 | 404 |
| 405 | 164,025 | 66,430,125 | 20.1246 | 7.39864 | 0.0024691 | 405 |
| 406 | 164,836 | 66,923,416 | 20.1494 | 7.40472 | 0.0024631 | 406 |
| 407 | 165,649 | 67,419,143 | 20.1742 | 7.41080 | 0.0024570 | 407 |
| 408 | 166,464 | 67,917,312 | 20.1990 | 7.41686 | 0.0024510 | 408 |
| 409 | 167,281 | 68,417,929 | 20.2237 | 7.42291 | 0.0024450 | 409 |
| 410 | 168,100 | 68,921,000 | 20.2485 | 7.42896 | 0.0024390 | 410 |
| 411 | 168,921 | 69,426,531 | 20.2731 | 7.43499 | 0.0024331 | 411 |
| 412 | 169,744 | 69,934,528 | 20.2978 | 7.44102 | 0.0024272 | 412 |
| 413 | 170,569 | 70,444,997 | 20.3224 | 7.44703 | 0.0024213 | 413 |
| 414 | 171,396 | 70,957,944 | 20.3470 | 7.45304 | 0.0024155 | 414 |
| 415 | 172,225 | 71,473,375 | 20.3715 | 7.45904 | 0.0024096 | 415 |
| 416 | 173,056 | 71,991,296 | 20.3961 | 7.46502 | 0.0024038 | 416 |
| 417 | 173,889 | 72,511,713 | 20.4206 | 7.47100 | 0.0023981 | 417 |
| 418 | 174,724 | 73,034,632 | 20.4450 | 7.47697 | 0.0023923 | 418 |
| 419 | 175,561 | 73,560,059 | 20.4695 | 7.48292 | 0.0023866 | 419 |
| 420 | 176,400 | 74,088,000 | 20.4939 | 7.48887 | 0.0023810 | 420 |
| 421 | 177,241 | 74,618,461 | 20.5183 | 7.49481 | 0.0023753 | 421 |
| 422 | 178,084 | 75,151,448 | 20.5426 | 7.50074 | 0.0023697 | 422 |
| 423 | 178,929 | 75,686,967 | 20.5670 | 7.50666 | 0.0023641 | 423 |
| 424 | 179,776 | 76,225,024 | 20.5913 | 7.51257 | 0.0023585 | 424 |
| 425 | 180,625 | 76,765,625 | 20.6155 | 7.51847 | 0.0023529 | 425 |
| 426 | 181,476 | 77,308,776 | 20.6398 | 7.52437 | 0.0023474 | 426 |
| 427 | 182,329 | 77,854,483 | 20.6640 | 7.53025 | 0.0023419 | 427 |
| 428 | 183,184 | 78,402,752 | 20.6882 | 7.53612 | 0.0023364 | 428 |
| 429 | 184,041 | 78,953,589 | 20.7123 | 7.54199 | 0.0023310 | 429 |
| 430 | 184,900 | 79,507,000 | 20.7364 | 7.54784 | 0.0023256 | 430 |
| 431 | 185,761 | 80,062,991 | 20.7605 | 7.55369 | 0.0023202 | 431 |
| 432 | 186,624 | 80,621,568 | 20.7846 | 7.55953 | 0.0023148 | 432 |
| 433 | 187,489 | 81,182,737 | 20.8087 | 7.56535 | 0.0023095 | 433 |
| 434 | 188,356 | 81,746,504 | 20.8327 | 7.57117 | 0.0023041 | 434 |
| 435 | 189,225 | 82,312,875 | 20.8567 | 7.57698 | 0.0022989 | 435 |
| 436 | 190,096 | 82,881,856 | 20.8806 | 7.58279 | 0.0022936 | 436 |
| 437 | 190,969 | 83,453,453 | 20.9045 | 7.58858 | 0.0022883 | 437 |
| 438 | 191,844 | 84,027,672 | 20.9284 | 7.59436 | 0.0022831 | 438 |
| 439 | 192,721 | 84,604,519 | 20.9523 | 7.60014 | 0.0022779 | 439 |
| 440 | 193,600 | 85,184,000 | 20.9762 | 7.60590 | 0.0022727 | 440 |
| 441 | 194,481 | 85,766,121 | 21.0000 | 7.61166 | 0.0022676 | 441 |
| 442 | 195,364 | 86,350,888 | 21.0238 | 7.61741 | 0.0022624 | 442 |
| 443 | 196,249 | 86,938,307 | 21.0476 | 7.62315 | 0.0022573 | 443 |
| 444 | 197,136 | 87,528,384 | 21.0713 | 7.62888 | 0.0022523 | 444 |
| 445 | 198,025 | 88,121,125 | 21.0950 | 7.63461 | 0.0022472 | 445 |
| 446 | 198,916 | 88,716,536 | 21.1187 | 7.64032 | 0.0022422 | 446 |
| 447 | 199,809 | 89,314,623 | 21.1424 | 7.64603 | 0.0022371 | 447 |
| 448 | 200,704 | 89,915,392 | 21.1660 | 7.65172 | 0.0022321 | 448 |
| 449 | 201,601 | 90,518,849 | 21.1896 | 7.65741 | 0.0022272 | 449 |
| 450 | 202,500 | 91,125,000 | 21.2132 | 7.66309 | 0.0022222 | 450 |

### Powers, Roots and Reciprocals

| No. | Square | Cube | Sq. Root | Cube Root | Reciprocal | No. |
|---|---|---|---|---|---|---|
| 451 | 203,401 | 91,733,851 | 21.2368 | 7.66877 | 0.0022173 | 451 |
| 452 | 204,304 | 92,345,408 | 21.2603 | 7.67443 | 0.0022124 | 452 |
| 453 | 205,209 | 92,959,677 | 21.2838 | 7.68009 | 0.0022075 | 453 |
| 454 | 206,116 | 93,576,664 | 21.3073 | 7.68573 | 0.0022026 | 454 |
| 455 | 207,025 | 94,196,375 | 21.3307 | 7.69137 | 0.0021978 | 455 |
| 456 | 207,936 | 94,818,816 | 21.3542 | 7.69700 | 0.0021930 | 456 |
| 457 | 208,849 | 95,443,993 | 21.3776 | 7.70262 | 0.0021882 | 457 |
| 458 | 209,764 | 96,071,912 | 21.4009 | 7.70824 | 0.0021834 | 458 |
| 459 | 210,681 | 96,702,579 | 21.4243 | 7.71384 | 0.0021786 | 459 |
| 460 | 211,600 | 97,336,000 | 21.4476 | 7.71944 | 0.0021739 | 460 |
| 461 | 212,521 | 97,972,181 | 21.4709 | 7.72503 | 0.0021692 | 461 |
| 462 | 213,444 | 98,611,128 | 21.4942 | 7.73061 | 0.0021645 | 462 |
| 463 | 214,369 | 99,252,847 | 21.5174 | 7.73619 | 0.0021598 | 463 |
| 464 | 215,296 | 99,897,344 | 21.5407 | 7.74175 | 0.0021552 | 464 |
| 465 | 216,225 | 100,544,625 | 21.5639 | 7.74731 | 0.0021505 | 465 |
| 466 | 217,156 | 101,194,696 | 21.5870 | 7.75286 | 0.0021459 | 466 |
| 467 | 218,089 | 101,847,563 | 21.6102 | 7.75840 | 0.0021413 | 467 |
| 468 | 219,024 | 102,503,232 | 21.6333 | 7.76394 | 0.0021368 | 468 |
| 469 | 219,961 | 103,161,709 | 21.6564 | 7.76946 | 0.0021322 | 469 |
| 470 | 220,900 | 103,823,000 | 21.6795 | 7.77498 | 0.0021277 | 470 |
| 471 | 221,841 | 104,487,111 | 21.7025 | 7.78049 | 0.0021231 | 471 |
| 472 | 222,784 | 105,154,048 | 21.7256 | 7.78599 | 0.0021186 | 472 |
| 473 | 223,729 | 105,823,817 | 21.7486 | 7.79149 | 0.0021142 | 473 |
| 474 | 224,676 | 106,496,424 | 21.7715 | 7.79697 | 0.0021097 | 474 |
| 475 | 225,625 | 107,171,875 | 21.7945 | 7.80245 | 0.0021053 | 475 |
| 476 | 226,576 | 107,850,176 | 21.8174 | 7.80793 | 0.0021008 | 476 |
| 477 | 227,529 | 108,531,333 | 21.8403 | 7.81339 | 0.0020964 | 477 |
| 478 | 228,484 | 109,215,352 | 21.8632 | 7.81885 | 0.0020921 | 478 |
| 479 | 229,441 | 109,902,239 | 21.8861 | 7.82429 | 0.0020877 | 479 |
| 480 | 230,400 | 110,592,000 | 21.9089 | 7.82974 | 0.0020833 | 480 |
| 481 | 231,361 | 111,284,641 | 21.9317 | 7.83517 | 0.0020790 | 481 |
| 482 | 232,324 | 111,980,168 | 21.9545 | 7.84059 | 0.0020747 | 482 |
| 483 | 233,289 | 112,678,587 | 21.9773 | 7.84601 | 0.0020704 | 483 |
| 484 | 234,256 | 113,379,904 | 22.0000 | 7.85142 | 0.0020661 | 484 |
| 485 | 235,225 | 114,084,125 | 22.0227 | 7.85683 | 0.0020619 | 485 |
| 486 | 236,196 | 114,791,256 | 22.0454 | 7.86222 | 0.0020576 | 486 |
| 487 | 237,169 | 115,501,303 | 22.0681 | 7.86761 | 0.0020534 | 487 |
| 488 | 238,144 | 116,214,272 | 22.0907 | 7.87299 | 0.0020492 | 488 |
| 489 | 239,121 | 116,930,169 | 22.1133 | 7.87837 | 0.0020450 | 489 |
| 490 | 240,100 | 117,649,000 | 22.1359 | 7.88374 | 0.0020408 | 490 |
| 491 | 241,081 | 118,370,771 | 22.1585 | 7.88909 | 0.0020367 | 491 |
| 492 | 242,064 | 119,095,488 | 22.1811 | 7.89445 | 0.0020325 | 492 |
| 493 | 243,049 | 119,823,157 | 22.2036 | 7.89979 | 0.0020284 | 493 |
| 494 | 244,036 | 120,553,784 | 22.2261 | 7.90513 | 0.0020243 | 494 |
| 495 | 245,025 | 121,287,375 | 22.2486 | 7.91046 | 0.0020202 | 495 |
| 496 | 246,016 | 122,023,936 | 22.2711 | 7.91578 | 0.0020161 | 496 |
| 497 | 247,009 | 122,763,473 | 22.2935 | 7.92110 | 0.0020121 | 497 |
| 498 | 248,004 | 123,505,992 | 22.3159 | 7.92641 | 0.0020080 | 498 |
| 499 | 249,001 | 124,251,499 | 22.3383 | 7.93171 | 0.0020040 | 499 |
| 500 | 250,000 | 125,000,000 | 22.3607 | 7.93701 | 0.0020000 | 500 |

MATHEMATICAL TABLES

## Powers, Roots and Reciprocals

| No. | Square | Cube | Sq. Root | Cube Root | Reciprocal | No. |
|---|---|---|---|---|---|---|
| 501 | 251,001 | 125,751,501 | 22.3830 | 7.94229 | 0.0019960 | 501 |
| 502 | 252,004 | 126,506,008 | 22.4054 | 7.94757 | 0.0019920 | 502 |
| 503 | 253,009 | 127,263,527 | 22.4277 | 7.95285 | 0.0019881 | 503 |
| 504 | 254,016 | 128,024,064 | 22.4499 | 7.95811 | 0.0019841 | 504 |
| 505 | 255,025 | 128,787,625 | 22.4722 | 7.96337 | 0.0019802 | 505 |
| 506 | 256,036 | 129,554,216 | 22.4944 | 7.96863 | 0.0019763 | 506 |
| 507 | 257,049 | 130,323,843 | 22.5167 | 7.97387 | 0.0019724 | 507 |
| 508 | 258,064 | 131,096,512 | 22.5389 | 7.97911 | 0.0019685 | 508 |
| 509 | 259,081 | 131,872,229 | 22.5610 | 7.98434 | 0.0019646 | 509 |
| 510 | 260,100 | 132,651,000 | 22.5832 | 7.98957 | 0.0019608 | 510 |
| 511 | 261,121 | 133,432,831 | 22.6053 | 7.99479 | 0.0019569 | 511 |
| 512 | 262,144 | 134,217,728 | 22.6274 | 8.00000 | 0.0019531 | 512 |
| 513 | 263,169 | 135,005,697 | 22.6495 | 8.00520 | 0.0019493 | 513 |
| 514 | 264,196 | 135,796,744 | 22.6716 | 8.01040 | 0.0019455 | 514 |
| 515 | 265,225 | 136,590,875 | 22.6936 | 8.01559 | 0.0019417 | 515 |
| 516 | 266,256 | 137,388,096 | 22.7156 | 8.02078 | 0.0019380 | 516 |
| 517 | 267,289 | 138,188,413 | 22.7376 | 8.02596 | 0.0019342 | 517 |
| 518 | 268,324 | 138,991,832 | 22.7596 | 8.03113 | 0.0019305 | 518 |
| 519 | 269,361 | 139,798,359 | 22.7816 | 8.03629 | 0.0019268 | 519 |
| 520 | 270,400 | 140,608,000 | 22.8035 | 8.04145 | 0.0019231 | 520 |
| 521 | 271,441 | 141,420,761 | 22.8254 | 8.04660 | 0.0019194 | 521 |
| 522 | 272,484 | 142,236,648 | 22.8473 | 8.05175 | 0.0019157 | 522 |
| 523 | 273,529 | 143,055,667 | 22.8692 | 8.05689 | 0.0019120 | 523 |
| 524 | 274,576 | 143,877,824 | 22.8910 | 8.06202 | 0.0019084 | 524 |
| 525 | 275,625 | 144,703,125 | 22.9129 | 8.06714 | 0.0019048 | 525 |
| 526 | 276,676 | 145,531,576 | 22.9347 | 8.07226 | 0.0019011 | 526 |
| 527 | 277,729 | 146,363,183 | 22.9565 | 8.07737 | 0.0018975 | 527 |
| 528 | 278,784 | 147,197,952 | 22.9783 | 8.08248 | 0.0018939 | 528 |
| 529 | 279,841 | 148,035,889 | 23.0000 | 8.08758 | 0.0018904 | 529 |
| 530 | 280,900 | 148,877,000 | 23.0217 | 8.09267 | 0.0018868 | 530 |
| 531 | 281,961 | 149,721,291 | 23.0434 | 8.09776 | 0.0018832 | 531 |
| 532 | 283,024 | 150,568,768 | 23.0651 | 8.10284 | 0.0018797 | 532 |
| 533 | 284,089 | 151,419,437 | 23.0868 | 8.10791 | 0.0018762 | 533 |
| 534 | 285,156 | 152,273,304 | 23.1084 | 8.11298 | 0.0018727 | 534 |
| 535 | 286,225 | 153,130,375 | 23.1301 | 8.11804 | 0.0018692 | 535 |
| 536 | 287,296 | 153,990,656 | 23.1517 | 8.12310 | 0.0018657 | 536 |
| 537 | 288,369 | 154,854,153 | 23.1733 | 8.12814 | 0.0018622 | 537 |
| 538 | 289,444 | 155,720,872 | 23.1948 | 8.13319 | 0.0018587 | 538 |
| 539 | 290,521 | 156,590,819 | 23.2164 | 8.13822 | 0.0018553 | 539 |
| 540 | 291,600 | 157,464,000 | 23.2379 | 8.14325 | 0.0018519 | 540 |
| 541 | 292,681 | 158,340,421 | 23.2594 | 8.14828 | 0.0018484 | 541 |
| 542 | 293,764 | 159,220,088 | 23.2809 | 8.15329 | 0.0018450 | 542 |
| 543 | 294,849 | 160,103,007 | 23.3024 | 8.15831 | 0.0018416 | 543 |
| 544 | 295,936 | 160,989,184 | 23.3238 | 8.16331 | 0.0018382 | 544 |
| 545 | 297,025 | 161,878,625 | 23.3452 | 8.16831 | 0.0018349 | 545 |
| 546 | 298,116 | 162,771,336 | 23.3666 | 8.17330 | 0.0018315 | 546 |
| 547 | 299,209 | 163,667,323 | 23.3880 | 8.17829 | 0.0018282 | 547 |
| 548 | 300,304 | 164,566,592 | 23.4094 | 8.18327 | 0.0018248 | 548 |
| 549 | 301,401 | 165,469,149 | 23.4307 | 8.18824 | 0.0018215 | 549 |
| 550 | 302,500 | 166,375,000 | 23.4521 | 8.19321 | 0.0018182 | 550 |

Powers, Roots and Reciprocals

| No. | Square | Cube | Sq. Root | Cube Root | Reciprocal | No. |
|-----|--------|------|----------|-----------|------------|-----|
| 551 | 303,601 | 167,284,151 | 23.4734 | 8.19818 | 0.0018149 | 551 |
| 552 | 304,704 | 168,196,608 | 23.4947 | 8.20313 | 0.0018116 | 552 |
| 553 | 305,809 | 169,112,377 | 23.5160 | 8.20808 | 0.0018083 | 553 |
| 554 | 306,916 | 170,031,464 | 23.5372 | 8.21303 | 0.0018051 | 554 |
| 555 | 308,025 | 170,953,875 | 23.5584 | 8.21797 | 0.0018018 | 555 |
| 556 | 309,136 | 171,879,616 | 23.5797 | 8.22290 | 0.0017986 | 556 |
| 557 | 310,249 | 172,808,693 | 23.6008 | 8.22783 | 0.0017953 | 557 |
| 558 | 311,364 | 173,741,112 | 23.6220 | 8.23275 | 0.0017921 | 558 |
| 559 | 312,481 | 174,676,879 | 23.6432 | 8.23766 | 0.0017889 | 559 |
| 560 | 313,600 | 175,616,000 | 23.6643 | 8.24257 | 0.0017857 | 560 |
| 561 | 314,721 | 176,558,481 | 23.6854 | 8.24747 | 0.0017825 | 561 |
| 562 | 315,844 | 177,504,328 | 23.7065 | 8.25237 | 0.0017794 | 562 |
| 563 | 316,969 | 178,453,547 | 23.7276 | 8.25726 | 0.0017762 | 563 |
| 564 | 318,096 | 179,406,144 | 23.7487 | 8.26215 | 0.0017730 | 564 |
| 565 | 319,225 | 180,362,125 | 23.7697 | 8.26703 | 0.0017699 | 565 |
| 566 | 320,356 | 181,321,496 | 23.7908 | 8.27190 | 0.0017668 | 566 |
| 567 | 321,489 | 182,284,263 | 23.8118 | 8.27677 | 0.0017637 | 567 |
| 568 | 322,624 | 183,250,432 | 23.8328 | 8.28164 | 0.0017606 | 568 |
| 569 | 323,761 | 184,220,009 | 23.8537 | 8.28649 | 0.0017575 | 569 |
| 570 | 324,900 | 185,193,000 | 23.8747 | 8.29134 | 0.0017544 | 570 |
| 571 | 326,041 | 186,169,411 | 23.8956 | 8.29619 | 0.0017513 | 571 |
| 572 | 327,184 | 187,149,248 | 23.9165 | 8.30103 | 0.0017483 | 572 |
| 573 | 328,329 | 188,132,517 | 23.9374 | 8.30587 | 0.0017452 | 573 |
| 574 | 329,476 | 189,119,224 | 23.9583 | 8.31069 | 0.0017422 | 574 |
| 575 | 330,625 | 190,109,375 | 23.9792 | 8.31552 | 0.0017391 | 575 |
| 576 | 331,776 | 191,102,976 | 24.0000 | 8.32034 | 0.0017361 | 576 |
| 577 | 332,929 | 192,100,033 | 24.0208 | 8.32515 | 0.0017331 | 577 |
| 578 | 334,084 | 193,100,552 | 24.0416 | 8.32995 | 0.0017301 | 578 |
| 579 | 335,241 | 194,104,539 | 24.0624 | 8.33476 | 0.0017271 | 579 |
| 580 | 336,400 | 195,112,000 | 24.0832 | 8.33955 | 0.0017241 | 580 |
| 581 | 337,561 | 196,122,941 | 24.1039 | 8.34434 | 0.0017212 | 581 |
| 582 | 338,724 | 197,137,368 | 24.1247 | 8.34913 | 0.0017182 | 582 |
| 583 | 339,889 | 198,155,287 | 24.1454 | 8.35390 | 0.0017153 | 583 |
| 584 | 341,056 | 199,176,704 | 24.1661 | 8.35868 | 0.0017123 | 584 |
| 585 | 342,225 | 200,201,625 | 24.1868 | 8.36345 | 0.0017094 | 585 |
| 586 | 343,396 | 201,230,056 | 24.2074 | 8.36821 | 0.0017065 | 586 |
| 587 | 344,569 | 202,262,003 | 24.2281 | 8.37297 | 0.0017036 | 587 |
| 588 | 345,744 | 203,297,472 | 24.2487 | 8.37772 | 0.0017007 | 588 |
| 589 | 346,921 | 204,336,469 | 24.2693 | 8.38247 | 0.0016978 | 589 |
| 590 | 348,100 | 205,379,000 | 24.2899 | 8.38721 | 0.0016949 | 590 |
| 591 | 349,281 | 206,425,071 | 24.3105 | 8.39194 | 0.0016920 | 591 |
| 592 | 350,464 | 207,474,688 | 24.3311 | 8.39667 | 0.0016892 | 592 |
| 593 | 351,649 | 208,527,857 | 24.3516 | 8.40140 | 0.0016863 | 593 |
| 594 | 352,836 | 209,584,584 | 24.3721 | 8.40612 | 0.0016835 | 594 |
| 595 | 354,025 | 210,644,875 | 24.3926 | 8.41083 | 0.0016807 | 595 |
| 596 | 355,216 | 211,708,736 | 24.4131 | 8.41554 | 0.0016779 | 596 |
| 597 | 356,409 | 212,776,173 | 24.4336 | 8.42025 | 0.0016750 | 597 |
| 598 | 357,604 | 213,847,192 | 24.4540 | 8.42494 | 0.0016722 | 598 |
| 599 | 358,801 | 214,921,799 | 24.4745 | 8.42964 | 0.0016694 | 599 |
| 600 | 360,000 | 216,000,000 | 24.4949 | 8.43433 | 0.0016667 | 600 |

## Powers, Roots and Reciprocals

| No. | Square | Cube | Sq. Root | Cube Root | Reciprocal | No. |
|---|---|---|---|---|---|---|
| 601 | 361,201 | 217,081,801 | 24.5153 | 8.43901 | 0.0016639 | 601 |
| 602 | 362,404 | 218,167,208 | 24.5357 | 8.44369 | 0.0016611 | 602 |
| 603 | 363,609 | 219,256,227 | 24.5561 | 8.44836 | 0.0016584 | 603 |
| 604 | 364,816 | 220,348,864 | 24.5764 | 8.45303 | 0.0016556 | 604 |
| 605 | 366,025 | 221,445,125 | 24.5967 | 8.45769 | 0.0016529 | 605 |
| 606 | 367,236 | 222,545,016 | 24.6171 | 8.46235 | 0.0016502 | 606 |
| 607 | 368,449 | 223,648,543 | 24.6374 | 8.46700 | 0.0016474 | 607 |
| 608 | 369,664 | 224,755,712 | 24.6577 | 8.47165 | 0.0016447 | 608 |
| 609 | 370,881 | 225,866,529 | 24.6779 | 8.47629 | 0.0016420 | 609 |
| 610 | 372,100 | 226,981,000 | 24.6982 | 8.48093 | 0.0016393 | 610 |
| 611 | 373,321 | 228,099,131 | 24.7184 | 8.48556 | 0.0016367 | 611 |
| 612 | 374,544 | 229,220,928 | 24.7386 | 8.49018 | 0.0016340 | 612 |
| 613 | 375,769 | 230,346,397 | 24.7588 | 8.49481 | 0.0016313 | 613 |
| 614 | 376,996 | 231,475,544 | 24.7790 | 8.49942 | 0.0016287 | 614 |
| 615 | 378,225 | 232,608,375 | 24.7992 | 8.50403 | 0.0016260 | 615 |
| 616 | 379,456 | 233,744,896 | 24.8193 | 8.50864 | 0.0016234 | 616 |
| 617 | 380,689 | 234,885,113 | 24.8395 | 8.51324 | 0.0016207 | 617 |
| 618 | 381,924 | 236,029,032 | 24.8596 | 8.51784 | 0.0016181 | 618 |
| 619 | 383,161 | 237,176,659 | 24.8797 | 8.52243 | 0.0016155 | 619 |
| 620 | 384,400 | 238,328,000 | 24.8998 | 8.52702 | 0.0016129 | 620 |
| 621 | 385,641 | 239,483,061 | 24.9199 | 8.53160 | 0.0016103 | 621 |
| 622 | 386,884 | 240,641,848 | 24.9399 | 8.53618 | 0.0016077 | 622 |
| 623 | 388,129 | 241,804,367 | 24.9600 | 8.54075 | 0.0016051 | 623 |
| 624 | 389,376 | 242,970,624 | 24.9800 | 8.54532 | 0.0016026 | 624 |
| 625 | 390,625 | 244,140,625 | 25.0000 | 8.54988 | 0.0016000 | 625 |
| 626 | 391,876 | 245,314,376 | 25.0200 | 8.55444 | 0.0015974 | 626 |
| 627 | 393,129 | 246,491,883 | 25.0400 | 8.55899 | 0.0015949 | 627 |
| 628 | 394,384 | 247,673,152 | 25.0599 | 8.56354 | 0.0015924 | 628 |
| 629 | 395,641 | 248,858,189 | 25.0799 | 8.56808 | 0.0015898 | 629 |
| 630 | 396,900 | 250,047,000 | 25.0998 | 8.57262 | 0.0015873 | 630 |
| 631 | 398,161 | 251,239,591 | 25.1197 | 8.57715 | 0.0015848 | 631 |
| 632 | 399,424 | 252,435,968 | 25.1396 | 8.58168 | 0.0015823 | 632 |
| 633 | 400,689 | 253,636,137 | 25.1595 | 8.58620 | 0.0015798 | 633 |
| 634 | 401,956 | 254,840,104 | 25.1794 | 8.59072 | 0.0015773 | 634 |
| 635 | 403,225 | 256,047,875 | 25.1992 | 8.59524 | 0.0015748 | 635 |
| 636 | 404,496 | 257,259,456 | 25.2190 | 8.59975 | 0.0015723 | 636 |
| 637 | 405,769 | 258,474,853 | 25.2389 | 8.60425 | 0.0015699 | 637 |
| 638 | 407,044 | 259,694,072 | 25.2587 | 8.60875 | 0.0015674 | 638 |
| 639 | 408,321 | 260,917,119 | 25.2784 | 8.61325 | 0.0015649 | 639 |
| 640 | 409,600 | 262,144,000 | 25.2982 | 8.61774 | 0.0015625 | 640 |
| 641 | 410,881 | 263,374,721 | 25.3180 | 8.62222 | 0.0015601 | 641 |
| 642 | 412,164 | 264,609,288 | 25.3377 | 8.62671 | 0.0015576 | 642 |
| 643 | 413,449 | 265,847,707 | 25.3574 | 8.63118 | 0.0015552 | 643 |
| 644 | 414,736 | 267,089,984 | 25.3772 | 8.63566 | 0.0015528 | 644 |
| 645 | 416,025 | 268,336,125 | 25.3969 | 8.64012 | 0.0015504 | 645 |
| 646 | 417,316 | 269,586,136 | 25.4165 | 8.64459 | 0.0015480 | 646 |
| 647 | 418,609 | 270,840,023 | 25.4362 | 8.64904 | 0.0015456 | 647 |
| 648 | 419,904 | 272,097,792 | 25.4558 | 8.65350 | 0.0015432 | 648 |
| 649 | 421,201 | 273,359,449 | 25.4755 | 8.65795 | 0.0015408 | 649 |
| 650 | 422,500 | 274,625,000 | 25.4951 | 8.66239 | 0.0015385 | 650 |

Powers, Roots and Reciprocals

| No. | Square | Cube | Sq. Root | Cube Root | Reciprocal | No. |
|---|---|---|---|---|---|---|
| 651 | 423,801 | 275,894,451 | 25.5147 | 8.66683 | 0.0015361 | 651 |
| 652 | 425,104 | 277,167,808 | 25.5343 | 8.67127 | 0.0015337 | 652 |
| 653 | 426,409 | 278,445,077 | 25.5539 | 8.67570 | 0.0015314 | 653 |
| 654 | 427,716 | 279,726,264 | 25.5734 | 8.68012 | 0.0015291 | 654 |
| 655 | 429,025 | 281,011,375 | 25.5930 | 8.68455 | 0.0015267 | 655 |
| 656 | 430,336 | 282,300,416 | 25.6125 | 8.68896 | 0.0015244 | 656 |
| 657 | 431,649 | 283,593,393 | 25.6320 | 8.69338 | 0.0015221 | 657 |
| 658 | 432,964 | 284,890,312 | 25.6515 | 8.69778 | 0.0015198 | 658 |
| 659 | 434,281 | 286,191,179 | 25.6710 | 8.70219 | 0.0015175 | 659 |
| 660 | 435,600 | 287,496,000 | 25.6905 | 8.70659 | 0.0015152 | 660 |
| 661 | 436,921 | 288,804,781 | 25.7099 | 8.71098 | 0.0015129 | 661 |
| 662 | 438,244 | 290,117,528 | 25.7294 | 8.71537 | 0.0015106 | 662 |
| 663 | 439,569 | 291,434,247 | 25.7488 | 8.71976 | 0.0015083 | 663 |
| 664 | 440,896 | 292,754,944 | 25.7682 | 8.72414 | 0.0015060 | 664 |
| 665 | 442,225 | 294,079,625 | 25.7876 | 8.72852 | 0.0015038 | 665 |
| 666 | 443,556 | 295,408,296 | 25.8070 | 8.73289 | 0.0015015 | 666 |
| 667 | 444,889 | 296,740,963 | 25.8263 | 8.73726 | 0.0014993 | 667 |
| 668 | 446,224 | 298,077,632 | 25.8457 | 8.74162 | 0.0014970 | 668 |
| 669 | 447,561 | 299,418,309 | 25.8650 | 8.74598 | 0.0014948 | 669 |
| 670 | 448,900 | 300,763,000 | 25.8844 | 8.75034 | 0.0014925 | 670 |
| 671 | 450,241 | 302,111,711 | 25.9037 | 8.75469 | 0.0014903 | 671 |
| 672 | 451,584 | 303,464,448 | 25.9230 | 8.75904 | 0.0014881 | 672 |
| 673 | 452,929 | 304,821,217 | 25.9422 | 8.76338 | 0.0014859 | 673 |
| 674 | 454,276 | 306,182,024 | 25.9615 | 8.76772 | 0.0014837 | 674 |
| 675 | 455,625 | 307,546,875 | 25.9808 | 8.77205 | 0.0014815 | 675 |
| 676 | 456,976 | 308,915,776 | 26.0000 | 8.77638 | 0.0014793 | 676 |
| 677 | 458,329 | 310,288,733 | 26.0192 | 8.78071 | 0.0014771 | 677 |
| 678 | 459,684 | 311,665,752 | 26.0384 | 8.78503 | 0.0014749 | 678 |
| 679 | 461,041 | 313,046,839 | 26.0576 | 8.78935 | 0.0014728 | 679 |
| 680 | 462,400 | 314,432,000 | 26.0768 | 8.79366 | 0.0014706 | 680 |
| 681 | 463,761 | 315,821,241 | 26.0960 | 8.79797 | 0.0014684 | 681 |
| 682 | 465,124 | 317,214,568 | 26.1151 | 8.80227 | 0.0014663 | 682 |
| 683 | 466,489 | 318,611,987 | 26.1343 | 8.80657 | 0.0014641 | 683 |
| 684 | 467,856 | 320,013,504 | 26.1534 | 8.81087 | 0.0014620 | 684 |
| 685 | 469,225 | 321,419,125 | 26.1725 | 8.81516 | 0.0014599 | 685 |
| 686 | 470,596 | 322,828,856 | 26.1916 | 8.81945 | 0.0014577 | 686 |
| 687 | 471,969 | 324,242,703 | 26.2107 | 8.82373 | 0.0014556 | 687 |
| 688 | 473,344 | 325,660,672 | 26.2298 | 8.82801 | 0.0014535 | 688 |
| 689 | 474,721 | 327,082,769 | 26.2488 | 8.83228 | 0.0014514 | 689 |
| 690 | 476,100 | 328,509,000 | 26.2679 | 8.83656 | 0.0014493 | 690 |
| 691 | 477,481 | 329,939,371 | 26.2869 | 8.84082 | 0.0014472 | 691 |
| 692 | 478,864 | 331,373,888 | 26.3059 | 8.84509 | 0.0014451 | 692 |
| 693 | 480,249 | 332,812,557 | 26.3249 | 8.84934 | 0.0014430 | 693 |
| 694 | 481,636 | 334,255,384 | 26.3439 | 8.85360 | 0.0014409 | 694 |
| 695 | 483,025 | 335,702,375 | 26.3629 | 8.85785 | 0.0014388 | 695 |
| 696 | 484,416 | 337,153,536 | 26.3818 | 8.86210 | 0.0014368 | 696 |
| 697 | 485,809 | 338,608,873 | 26.4008 | 8.86634 | 0.0014347 | 697 |
| 698 | 487,204 | 340,068,392 | 26.4197 | 8.87058 | 0.0014327 | 698 |
| 699 | 488,601 | 341,532,099 | 26.4386 | 8.87481 | 0.0014306 | 699 |
| 700 | 490,000 | 343,000,000 | 26.4575 | 8.87904 | 0.0014286 | 700 |

## Powers, Roots and Reciprocals

| No. | Square | Cube | Sq. Root | Cube Root | Reciprocal | No. |
|-----|--------|------|----------|-----------|------------|-----|
| 701 | 491,401 | 344,472,101 | 26.4764 | 8.88327 | 0.0014265 | 701 |
| 702 | 492,804 | 345,948,408 | 26.4953 | 8.88749 | 0.0014245 | 702 |
| 703 | 494,209 | 347,428,927 | 26.5141 | 8.89171 | 0.0014225 | 703 |
| 704 | 495,616 | 348,913,664 | 26.5330 | 8.89592 | 0.0014205 | 704 |
| 705 | 497,025 | 350,402,625 | 26.5518 | 8.90013 | 0.0014184 | 705 |
| 706 | 498,436 | 351,895,816 | 26.5707 | 8.90434 | 0.0014164 | 706 |
| 707 | 499,849 | 353,393,243 | 26.5895 | 8.90854 | 0.0014144 | 707 |
| 708 | 501,264 | 354,894,912 | 26.6083 | 8.91274 | 0.0014124 | 708 |
| 709 | 502,681 | 356,400,829 | 26.6271 | 8.91693 | 0.0014104 | 709 |
| 710 | 504,100 | 357,911,000 | 26.6458 | 8.92112 | 0.0014085 | 710 |
| 711 | 505,521 | 359,425,431 | 26.6646 | 8.92531 | 0.0014065 | 711 |
| 712 | 506,944 | 360,944,128 | 26.6833 | 8.92949 | 0.0014045 | 712 |
| 713 | 508,369 | 362,467,097 | 26.7021 | 8.93367 | 0.0014025 | 713 |
| 714 | 509,796 | 363,994,344 | 26.7208 | 8.93784 | 0.0014006 | 714 |
| 715 | 511,225 | 365,525,875 | 26.7395 | 8.94201 | 0.0013986 | 715 |
| 716 | 512,656 | 367,061,696 | 26.7582 | 8.94618 | 0.0013966 | 716 |
| 717 | 514,089 | 368,601,813 | 26.7769 | 8.95034 | 0.0013947 | 717 |
| 718 | 515,524 | 370,146,232 | 26.7955 | 8.95450 | 0.0013928 | 718 |
| 719 | 516,961 | 371,694,959 | 26.8142 | 8.95866 | 0.0013908 | 719 |
| 720 | 518,400 | 373,248,000 | 26.8328 | 8.96281 | 0.0013889 | 720 |
| 721 | 519,841 | 374,805,361 | 26.8514 | 8.96696 | 0.0013870 | 721 |
| 722 | 521,284 | 376,367,048 | 26.8701 | 8.97110 | 0.0013850 | 722 |
| 723 | 522,729 | 377,933,067 | 26.8887 | 8.97524 | 0.0013831 | 723 |
| 724 | 524,176 | 379,503,424 | 26.9072 | 8.97938 | 0.0013812 | 724 |
| 725 | 525,625 | 381,078,125 | 26.9258 | 8.98351 | 0.0013793 | 725 |
| 726 | 527,076 | 382,657,176 | 26.9444 | 8.98764 | 0.0013774 | 726 |
| 727 | 528,529 | 384,240,583 | 26.9629 | 8.99176 | 0.0013755 | 727 |
| 728 | 529,984 | 385,828,352 | 26.9815 | 8.99589 | 0.0013736 | 728 |
| 729 | 531,441 | 387,420,489 | 27.0000 | 9.00000 | 0.0013717 | 729 |
| 730 | 532,900 | 389,017,000 | 27.0185 | 9.00411 | 0.0013699 | 730 |
| 731 | 534,361 | 390,617,891 | 27.0370 | 9.00822 | 0.0013680 | 731 |
| 732 | 535,824 | 392,223,168 | 27.0555 | 9.01233 | 0.0013661 | 732 |
| 733 | 537,289 | 393,832,837 | 27.0740 | 9.01643 | 0.0013643 | 733 |
| 734 | 538,756 | 395,446,904 | 27.0924 | 9.02053 | 0.0013624 | 734 |
| 735 | 540,225 | 397,065,375 | 27.1109 | 9.02462 | 0.0013605 | 735 |
| 736 | 541,696 | 398,688,256 | 27.1293 | 9.02871 | 0.0013587 | 736 |
| 737 | 543,169 | 400,315,553 | 27.1477 | 9.03280 | 0.0013569 | 737 |
| 738 | 544,644 | 401,947,272 | 27.1662 | 9.03689 | 0.0013550 | 738 |
| 739 | 546,121 | 403,583,419 | 27.1846 | 9.04097 | 0.0013532 | 739 |
| 740 | 547,600 | 405,224,000 | 27.2029 | 9.04504 | 0.0013514 | 740 |
| 741 | 549,081 | 406,869,021 | 27.2213 | 9.04911 | 0.0013495 | 741 |
| 742 | 550,564 | 408,518,488 | 27.2397 | 9.05318 | 0.0013477 | 742 |
| 743 | 552,049 | 410,172,407 | 27.2580 | 9.05725 | 0.0013459 | 743 |
| 744 | 553,536 | 411,830,784 | 27.2764 | 9.06131 | 0.0013441 | 744 |
| 745 | 555,025 | 413,493,625 | 27.2947 | 9.06537 | 0.0013423 | 745 |
| 746 | 556,516 | 415,160,936 | 27.3130 | 9.06942 | 0.0013405 | 746 |
| 747 | 558,009 | 416,832,723 | 27.3313 | 9.07347 | 0.0013387 | 747 |
| 748 | 559,504 | 418,508,992 | 27.3496 | 9.07752 | 0.0013369 | 748 |
| 749 | 561,001 | 420,189,749 | 27.3679 | 9.08156 | 0.0013351 | 749 |
| 750 | 562,500 | 421,875,000 | 27.3861 | 9.08560 | 0.0013333 | 750 |

## Powers, Roots and Reciprocals

| No. | Square | Cube | Sq. Root | Cube Root | Reciprocal | No. |
|---|---|---|---|---|---|---|
| 751 | 564,001 | 423,564,751 | 27.4044 | 9.08964 | 0.0013316 | 751 |
| 752 | 565,504 | 425,259,008 | 27.4226 | 9.09367 | 0.0013298 | 752 |
| 753 | 567,009 | 426,957,777 | 27.4408 | 9.09770 | 0.0013280 | 753 |
| 754 | 568,516 | 428,661,064 | 27.4591 | 9.10173 | 0.0013263 | 754 |
| 755 | 570,025 | 430,368,875 | 27.4773 | 9.10575 | 0.0013245 | 755 |
| 756 | 571,536 | 432,081,216 | 27.4955 | 9.10977 | 0.0013228 | 756 |
| 757 | 573,049 | 433,798,093 | 27.5136 | 9.11378 | 0.0013210 | 757 |
| 758 | 574,564 | 435,519,512 | 27.5318 | 9.11779 | 0.0013193 | 758 |
| 759 | 576,081 | 437,245,479 | 27.5500 | 9.12180 | 0.0013175 | 759 |
| 760 | 577,600 | 438,976,000 | 27.5681 | 9.12581 | 0.0013158 | 760 |
| 761 | 579,121 | 440,711,081 | 27.5862 | 9.12981 | 0.0013141 | 761 |
| 762 | 580,644 | 442,450,728 | 27.6043 | 9.13380 | 0.0013123 | 762 |
| 763 | 582,169 | 444,194,947 | 27.6225 | 9.13780 | 0.0013106 | 763 |
| 764 | 583,696 | 445,943,744 | 27.6405 | 9.14179 | 0.0013089 | 764 |
| 765 | 585,225 | 447,697,125 | 27.6586 | 9.14577 | 0.0013072 | 765 |
| 766 | 586,756 | 449,455,096 | 27.6767 | 9.14976 | 0.0013055 | 766 |
| 767 | 588,289 | 451,217,663 | 27.6948 | 9.15374 | 0.0013038 | 767 |
| 768 | 589,824 | 452,984,832 | 27.7128 | 9.15771 | 0.0013021 | 768 |
| 769 | 591,361 | 454,756,609 | 27.7308 | 9.16169 | 0.0013004 | 769 |
| 770 | 592,900 | 456,533,000 | 27.7489 | 9.16566 | 0.0012987 | 770 |
| 771 | 594,441 | 458,314,011 | 27.7669 | 9.16962 | 0.0012970 | 771 |
| 772 | 595,984 | 460,099,648 | 27.7849 | 9.17359 | 0.0012953 | 772 |
| 773 | 597,529 | 461,889,917 | 27.8029 | 9.17754 | 0.0012937 | 773 |
| 774 | 599,076 | 463,684,824 | 27.8209 | 9.18150 | 0.0012920 | 774 |
| 775 | 600,625 | 465,484,375 | 27.8388 | 9.18545 | 0.0012903 | 775 |
| 776 | 602,176 | 467,288,576 | 27.8568 | 9.18940 | 0.0012887 | 776 |
| 777 | 603,729 | 469,097,433 | 27.8747 | 9.19335 | 0.0012870 | 777 |
| 778 | 605,284 | 470,910,952 | 27.8927 | 9.19729 | 0.0012853 | 778 |
| 779 | 606,841 | 472,729,139 | 27.9106 | 9.20123 | 0.0012837 | 779 |
| 780 | 608,400 | 474,552,000 | 27.9285 | 9.20516 | 0.0012821 | 780 |
| 781 | 609,961 | 476,379,541 | 27.9464 | 9.20910 | 0.0012804 | 781 |
| 782 | 611,524 | 478,211,768 | 27.9643 | 9.21303 | 0.0012788 | 782 |
| 783 | 613,089 | 480,048,687 | 27.9821 | 9.21695 | 0.0012771 | 783 |
| 784 | 614,656 | 481,890,304 | 28.0000 | 9.22087 | 0.0012755 | 784 |
| 785 | 616,225 | 483,736,625 | 28.0179 | 9.22479 | 0.0012739 | 785 |
| 786 | 617,796 | 485,587,656 | 28.0357 | 9.22871 | 0.0012723 | 786 |
| 787 | 619,369 | 487,443,403 | 28.0535 | 9.23262 | 0.0012706 | 787 |
| 788 | 620,944 | 489,303,872 | 28.0713 | 9.23653 | 0.0012690 | 788 |
| 789 | 622,521 | 491,169,069 | 28.0891 | 9.24043 | 0.0012674 | 789 |
| 790 | 624,100 | 493,039,000 | 28.1069 | 9.24434 | 0.0012658 | 790 |
| 791 | 625,681 | 494,913,671 | 28.1247 | 9.24823 | 0.0012642 | 791 |
| 792 | 627,264 | 496,793,088 | 28.1425 | 9.25213 | 0.0012626 | 792 |
| 793 | 628,849 | 498,677,257 | 28.1603 | 9.25602 | 0.0012610 | 793 |
| 794 | 630,436 | 500,566,184 | 28.1780 | 9.25991 | 0.0012594 | 794 |
| 795 | 632,025 | 502,459,875 | 28.1957 | 9.26380 | 0.0012579 | 795 |
| 796 | 633,616 | 504,358,336 | 28.2135 | 9.26768 | 0.0012563 | 796 |
| 797 | 635,209 | 506,261,573 | 28.2312 | 9.27156 | 0.0012547 | 797 |
| 798 | 636,804 | 508,169,592 | 28.2489 | 9.27544 | 0.0012531 | 798 |
| 799 | 638,401 | 510,082,399 | 28.2666 | 9.27931 | 0.0012516 | 799 |
| 800 | 640,000 | 512,000,000 | 28.2843 | 9.28318 | 0.0012500 | 800 |

### Powers, Roots and Reciprocals

| No. | Square | Cube | Sq. Root | Cube Root | Reciprocal | No. |
|---|---|---|---|---|---|---|
| 801 | 641,601 | 513,922,401 | 28.3019 | 9.28704 | 0.0012484 | 801 |
| 802 | 643,204 | 515,849,608 | 28.3196 | 9.29091 | 0.0012469 | 802 |
| 803 | 644,809 | 517,781,627 | 28.3373 | 9.29477 | 0.0012453 | 803 |
| 804 | 646,416 | 519,718,464 | 28.3549 | 9.29862 | 0.0012438 | 804 |
| 805 | 648,025 | 521,660,125 | 28.3725 | 9.30248 | 0.0012422 | 805 |
| 806 | 649,636 | 523,606,616 | 28.3901 | 9.30633 | 0.0012407 | 806 |
| 807 | 651,249 | 525,557,943 | 28.4077 | 9.31018 | 0.0012392 | 807 |
| 808 | 652,864 | 527,514,112 | 28.4253 | 9.31402 | 0.0012376 | 808 |
| 809 | 654,481 | 529,475,129 | 28.4429 | 9.31786 | 0.0012361 | 809 |
| 810 | 656,100 | 531,441,000 | 28.4605 | 9.32170 | 0.0012346 | 810 |
| 811 | 657,721 | 533,411,731 | 28.4781 | 9.32553 | 0.0012330 | 811 |
| 812 | 659,344 | 535,387,328 | 28.4956 | 9.32936 | 0.0012315 | 812 |
| 813 | 660,969 | 537,367,797 | 28.5132 | 9.33319 | 0.0012300 | 813 |
| 814 | 662,596 | 539,353,144 | 28.5307 | 9.33702 | 0.0012285 | 814 |
| 815 | 664,225 | 541,343,375 | 28.5482 | 9.34084 | 0.0012270 | 815 |
| 816 | 665,856 | 543,338,496 | 28.5657 | 9.34466 | 0.0012255 | 816 |
| 817 | 667,489 | 545,338,513 | 28.5832 | 9.34847 | 0.0012240 | 817 |
| 818 | 669,124 | 547,343,432 | 28.6007 | 9.35229 | 0.0012225 | 818 |
| 819 | 670,761 | 549,353,259 | 28.6182 | 9.35610 | 0.0012210 | 819 |
| 820 | 672,400 | 551,368,000 | 28.6356 | 9.35990 | 0.0012195 | 820 |
| 821 | 674,041 | 553,387,661 | 28.6531 | 9.36370 | 0.0012180 | 821 |
| 822 | 675,684 | 555,412,248 | 28.6705 | 9.36751 | 0.0012165 | 822 |
| 823 | 677,329 | 557,441,767 | 28.6880 | 9.37130 | 0.0012151 | 823 |
| 824 | 678,976 | 559,476,224 | 28.7054 | 9.37510 | 0.0012136 | 824 |
| 825 | 680,625 | 561,515,625 | 28.7228 | 9.37889 | 0.0012121 | 825 |
| 826 | 682,276 | 563,559,976 | 28.7402 | 9.38268 | 0.0012107 | 826 |
| 827 | 683,929 | 565,609,283 | 28.7576 | 9.38646 | 0.0012092 | 827 |
| 828 | 685,584 | 567,663,552 | 28.7750 | 9.39024 | 0.0012077 | 828 |
| 829 | 687,241 | 569,722,789 | 28.7924 | 9.39402 | 0.0012063 | 829 |
| 830 | 688,900 | 571,787,000 | 28.8097 | 9.39780 | 0.0012048 | 830 |
| 831 | 690,561 | 573,856,191 | 28.8271 | 9.40157 | 0.0012034 | 831 |
| 832 | 692,224 | 575,930,368 | 28.8444 | 9.40534 | 0.0012019 | 832 |
| 833 | 693,889 | 578,009,537 | 28.8617 | 9.40911 | 0.0012005 | 833 |
| 834 | 695,556 | 580,093,704 | 28.8791 | 9.41287 | 0.0011990 | 834 |
| 835 | 697,225 | 582,182,875 | 28.8964 | 9.41663 | 0.0011976 | 835 |
| 836 | 698,896 | 584,277,056 | 28.9137 | 9.42039 | 0.0011962 | 836 |
| 837 | 700,569 | 586,376,253 | 28.9310 | 9.42414 | 0.0011947 | 837 |
| 838 | 702,244 | 588,480,472 | 28.9482 | 9.42789 | 0.0011933 | 838 |
| 839 | 703,921 | 590,589,719 | 28.9655 | 9.43164 | 0.0011919 | 839 |
| 840 | 705,600 | 592,704,000 | 28.9828 | 9.43539 | 0.0011905 | 840 |
| 841 | 707,281 | 594,823,321 | 29.0000 | 9.43913 | 0.0011891 | 841 |
| 842 | 708,964 | 596,947,688 | 29.0172 | 9.44287 | 0.0011876 | 842 |
| 843 | 710,649 | 599,077,107 | 29.0345 | 9.44661 | 0.0011862 | 843 |
| 844 | 712,336 | 601,211,584 | 29.0517 | 9.45034 | 0.0011848 | 844 |
| 845 | 714,025 | 603,351,125 | 29.0689 | 9.45407 | 0.0011834 | 845 |
| 846 | 715,716 | 605,495,736 | 29.0861 | 9.45780 | 0.0011820 | 846 |
| 847 | 717,409 | 607,645,423 | 29.1033 | 9.46152 | 0.0011806 | 847 |
| 848 | 719,104 | 609,800,192 | 29.1204 | 9.46525 | 0.0011792 | 848 |
| 849 | 720,801 | 611,960,049 | 29.1376 | 9.46897 | 0.0011779 | 849 |
| 850 | 722,500 | 614,125,000 | 29.1548 | 9.47268 | 0.0011765 | 850 |

MATHEMATICAL TABLES **19**

## Powers, Roots and Reciprocals

| No. | Square | Cube | Sq. Root | Cube Root | Reciprocal | No. |
|---|---|---|---|---|---|---|
| 851 | 724,201 | 616,295,051 | 29.1719 | 9.47640 | 0.0011751 | 851 |
| 852 | 725,904 | 618,470,208 | 29.1890 | 9.48011 | 0.0011737 | 852 |
| 853 | 727,609 | 620,650,477 | 29.2062 | 9.48381 | 0.0011723 | 853 |
| 854 | 729,316 | 622,835,864 | 29.2233 | 9.48752 | 0.0011710 | 854 |
| 855 | 731,025 | 625,026,375 | 29.2404 | 9.49122 | 0.0011696 | 855 |
| 856 | 732,736 | 627,222,016 | 29.2575 | 9.49492 | 0.0011682 | 856 |
| 857 | 734,449 | 629,422,793 | 29.2746 | 9.49861 | 0.0011669 | 857 |
| 858 | 736,164 | 631,628,712 | 29.2916 | 9.50231 | 0.0011655 | 858 |
| 859 | 737,881 | 633,839,779 | 29.3087 | 9.50600 | 0.0011641 | 859 |
| 860 | 739,600 | 636,056,000 | 29.3258 | 9.50969 | 0.0011628 | 860 |
| 861 | 741,321 | 638,277,381 | 29.3428 | 9.51337 | 0.0011614 | 861 |
| 862 | 743,044 | 640,503,928 | 29.3598 | 9.51705 | 0.0011601 | 862 |
| 863 | 744,769 | 642,735,647 | 29.3769 | 9.52073 | 0.0011587 | 863 |
| 864 | 746,496 | 644,972,544 | 29.3939 | 9.52441 | 0.0011574 | 864 |
| 865 | 748,225 | 647,214,625 | 29.4109 | 9.52808 | 0.0011561 | 865 |
| 866 | 749,956 | 649,461,896 | 29.4279 | 9.53175 | 0.0011547 | 866 |
| 867 | 751,689 | 651,714,363 | 29.4449 | 9.53542 | 0.0011534 | 867 |
| 868 | 753,424 | 653,972,032 | 29.4618 | 9.53908 | 0.0011521 | 868 |
| 869 | 755,161 | 656,234,909 | 29.4788 | 9.54274 | 0.0011507 | 869 |
| 870 | 756,900 | 658,503,000 | 29.4958 | 9.54640 | 0.0011494 | 870 |
| 871 | 758,641 | 660,776,311 | 29.5127 | 9.55006 | 0.0011481 | 871 |
| 872 | 760,384 | 663,054,848 | 29.5296 | 9.55371 | 0.0011468 | 872 |
| 873 | 762,129 | 665,338,617 | 29.5466 | 9.55736 | 0.0011455 | 873 |
| 874 | 763,876 | 667,627,624 | 29.5635 | 9.56101 | 0.0011442 | 874 |
| 875 | 765,625 | 669,921,875 | 29.5804 | 9.56466 | 0.0011429 | 875 |
| 876 | 767,376 | 672,221,376 | 29.5973 | 9.56830 | 0.0011416 | 876 |
| 877 | 769,129 | 674,526,133 | 29.6142 | 9.57194 | 0.0011403 | 877 |
| 878 | 770,884 | 676,836,152 | 29.6311 | 9.57557 | 0.0011390 | 878 |
| 879 | 772,641 | 679,151,439 | 29.6479 | 9.57921 | 0.0011377 | 879 |
| 880 | 774,400 | 681,472,000 | 29.6648 | 9.58284 | 0.0011364 | 880 |
| 881 | 776,161 | 683,797,841 | 29.6816 | 9.58647 | 0.0011351 | 881 |
| 882 | 777,924 | 686,128,968 | 29.6985 | 9.59009 | 0.0011338 | 882 |
| 883 | 779,689 | 688,465,387 | 29.7153 | 9.59372 | 0.0011325 | 883 |
| 884 | 781,456 | 690,807,104 | 29.7321 | 9.59734 | 0.0011312 | 884 |
| 885 | 783,225 | 693,154,125 | 29.7489 | 9.60095 | 0.0011299 | 885 |
| 886 | 784,996 | 695,506,456 | 29.7658 | 9.60457 | 0.0011287 | 886 |
| 887 | 786,769 | 697,864,103 | 29.7825 | 9.60818 | 0.0011274 | 887 |
| 888 | 788,544 | 700,227,072 | 29.7993 | 9.61179 | 0.0011261 | 888 |
| 889 | 790,321 | 702,595,369 | 29.8161 | 9.61540 | 0.0011249 | 889 |
| 890 | 792,100 | 704,969,000 | 29.8329 | 9.61900 | 0.0011236 | 890 |
| 891 | 793,881 | 707,347,971 | 29.8496 | 9.62260 | 0.0011223 | 891 |
| 892 | 795,664 | 709,732,288 | 29.8664 | 9.62620 | 0.0011211 | 892 |
| 893 | 797,449 | 712,121,957 | 29.8831 | 9.62980 | 0.0011198 | 893 |
| 894 | 799,236 | 714,516,984 | 29.8998 | 9.63339 | 0.0011186 | 894 |
| 895 | 801,025 | 716,917,375 | 29.9166 | 9.63698 | 0.0011173 | 895 |
| 896 | 802,816 | 719,323,136 | 29.9333 | 9.64057 | 0.0011161 | 896 |
| 897 | 804,609 | 721,734,273 | 29.9500 | 9.64415 | 0.0011148 | 897 |
| 898 | 806,404 | 724,150,792 | 29.9666 | 9.64774 | 0.0011136 | 898 |
| 899 | 808,201 | 726,572,699 | 29.9833 | 9.65132 | 0.0011123 | 899 |
| 900 | 810,000 | 729,000,000 | 30.0000 | 9.65489 | 0.0011111 | 900 |

## Powers, Roots and Reciprocals

| No. | Square | Cube | Sq. Root | Cube Root | Reciprocal | No. |
|---|---|---|---|---|---|---|
| 901 | 811,801 | 731,432,701 | 30.0167 | 9.65847 | 0.0011099 | 901 |
| 902 | 813,604 | 733,870,808 | 30.0333 | 9.66204 | 0.0011086 | 902 |
| 903 | 815,409 | 736,314,327 | 30.0500 | 9.66561 | 0.0011074 | 903 |
| 904 | 817,216 | 738,763,264 | 30.0666 | 9.66918 | 0.0011062 | 904 |
| 905 | 819,025 | 741,217,625 | 30.0832 | 9.67274 | 0.0011050 | 905 |
| 906 | 820,836 | 743,677,416 | 30.0998 | 9.67630 | 0.0011038 | 906 |
| 907 | 822,649 | 746,142,643 | 30.1164 | 9.67986 | 0.0011025 | 907 |
| 908 | 824,464 | 748,613,312 | 30.1330 | 9.68342 | 0.0011013 | 908 |
| 909 | 826,281 | 751,089,429 | 30.1496 | 9.68697 | 0.0011001 | 909 |
| 910 | 828,100 | 753,571,000 | 30.1662 | 9.69052 | 0.0010989 | 910 |
| 911 | 829,921 | 756,058,031 | 30.1828 | 9.69407 | 0.0010977 | 911 |
| 912 | 831,744 | 758,550,528 | 30.1993 | 9.69762 | 0.0010965 | 912 |
| 913 | 833,569 | 761,048,497 | 30.2159 | 9.70116 | 0.0010953 | 913 |
| 914 | 835,396 | 763,551,944 | 30.2324 | 9.70470 | 0.0010941 | 914 |
| 915 | 837,225 | 766,060,875 | 30.2490 | 9.70824 | 0.0010929 | 915 |
| 916 | 839,056 | 768,575,296 | 30.2655 | 9.71177 | 0.0010917 | 916 |
| 917 | 840,889 | 771,095,213 | 30.2820 | 9.71531 | 0.0010905 | 917 |
| 918 | 842,724 | 773,620,632 | 30.2985 | 9.71884 | 0.0010893 | 918 |
| 919 | 844,561 | 776,151,559 | 30.3150 | 9.72236 | 0.0010881 | 919 |
| 920 | 846,400 | 778,688,000 | 30.3315 | 9.72589 | 0.0010870 | 920 |
| 921 | 848,241 | 781,229,961 | 30.3480 | 9.72941 | 0.0010858 | 921 |
| 922 | 850,084 | 783,777,448 | 30.3645 | 9.73293 | 0.0010846 | 922 |
| 923 | 851,929 | 786,330,467 | 30.3809 | 9.73645 | 0.0010834 | 923 |
| 924 | 853,776 | 788,889,024 | 30.3974 | 9.73996 | 0.0010823 | 924 |
| 925 | 855,625 | 791,453,125 | 30.4138 | 9.74348 | 0.0010811 | 925 |
| 926 | 857,476 | 794,022,776 | 30.4302 | 9.74699 | 0.0010799 | 926 |
| 927 | 859,329 | 796,597,983 | 30.4467 | 9.75049 | 0.0010787 | 927 |
| 928 | 861,184 | 799,178,752 | 30.4631 | 9.75400 | 0.0010776 | 928 |
| 929 | 863,041 | 801,765,089 | 30.4795 | 9.75750 | 0.0010764 | 929 |
| 930 | 864,900 | 804,357,000 | 30.4959 | 9.76100 | 0.0010753 | 930 |
| 931 | 866,761 | 806,954,491 | 30.5123 | 9.76450 | 0.0010741 | 931 |
| 932 | 868,624 | 809,557,568 | 30.5287 | 9.76799 | 0.0010730 | 932 |
| 933 | 870,489 | 812,166,237 | 30.5450 | 9.77148 | 0.0010718 | 933 |
| 934 | 872,356 | 814,780,504 | 30.5614 | 9.77497 | 0.0010707 | 934 |
| 935 | 874,225 | 817,400,375 | 30.5778 | 9.77846 | 0.0010695 | 935 |
| 936 | 876,096 | 820,025,856 | 30.5941 | 9.78195 | 0.0010684 | 936 |
| 937 | 877,969 | 822,656,953 | 30.6105 | 9.78543 | 0.0010672 | 937 |
| 938 | 879,844 | 825,293,672 | 30.6268 | 9.78891 | 0.0010661 | 938 |
| 939 | 881,721 | 827,936,019 | 30.6431 | 9.79239 | 0.0010650 | 939 |
| 940 | 883,600 | 830,584,000 | 30.6594 | 9.79586 | 0.0010638 | 940 |
| 941 | 885,481 | 833,237,621 | 30.6757 | 9.79933 | 0.0010627 | 941 |
| 942 | 887,364 | 835,896,888 | 30.6920 | 9.80280 | 0.0010616 | 942 |
| 943 | 889,249 | 838,561,807 | 30.7083 | 9.80627 | 0.0010604 | 943 |
| 944 | 891,136 | 841,232,384 | 30.7246 | 9.80974 | 0.0010593 | 944 |
| 945 | 893,025 | 843,908,625 | 30.7409 | 9.81320 | 0.0010582 | 945 |
| 946 | 894,916 | 846,590,536 | 30.7571 | 9.81666 | 0.0010571 | 946 |
| 947 | 896,809 | 849,278,123 | 30.7734 | 9.82012 | 0.0010560 | 947 |
| 948 | 898,704 | 851,971,392 | 30.7896 | 9.82357 | 0.0010549 | 948 |
| 949 | 900,601 | 854,670,349 | 30.8058 | 9.82703 | 0.0010537 | 949 |
| 950 | 902,500 | 857,375,000 | 30.8221 | 9.83048 | 0.0010526 | 950 |

## Powers, Roots and Reciprocals

| No. | Square | Cube | Sq. Root | Cube Root | Reciprocal | No. |
|---|---|---|---|---|---|---|
| 951 | 904,401 | 860,085,351 | 30.8383 | 9.83392 | 0.0010515 | 951 |
| 952 | 906,304 | 862,801,408 | 30.8545 | 9.83737 | 0.0010504 | 952 |
| 953 | 908,209 | 865,523,177 | 30.8707 | 9.84081 | 0.0010493 | 953 |
| 954 | 910,116 | 868,250,664 | 30.8869 | 9.84425 | 0.0010482 | 954 |
| 955 | 912,025 | 870,983,875 | 30.9031 | 9.84769 | 0.0010471 | 955 |
| 956 | 913,936 | 873,722,816 | 30.9192 | 9.85113 | 0.0010460 | 956 |
| 957 | 915,849 | 876,467,493 | 30.9354 | 9.85456 | 0.0010449 | 957 |
| 958 | 917,764 | 879,217,912 | 30.9516 | 9.85799 | 0.0010438 | 958 |
| 959 | 919,681 | 881,974,079 | 30.9677 | 9.86142 | 0.0010428 | 959 |
| 960 | 921,600 | 884,736,000 | 30.9839 | 9.86485 | 0.0010417 | 960 |
| 961 | 923,521 | 887,503,681 | 31.0000 | 9.86827 | 0.0010406 | 961 |
| 962 | 925,444 | 890,277,128 | 31.0161 | 9.87169 | 0.0010395 | 962 |
| 963 | 927,369 | 893,056,347 | 31.0322 | 9.87511 | 0.0010384 | 963 |
| 964 | 929,296 | 895,841,344 | 31.0483 | 9.87853 | 0.0010373 | 964 |
| 965 | 931,225 | 898,632,125 | 31.0644 | 9.88195 | 0.0010363 | 965 |
| 966 | 933,156 | 901,428,696 | 31.0805 | 9.88536 | 0.0010352 | 966 |
| 967 | 935,089 | 904,231,063 | 31.0966 | 9.88877 | 0.0010341 | 967 |
| 968 | 937,024 | 907,039,232 | 31.1127 | 9.89217 | 0.0010331 | 968 |
| 969 | 938,961 | 909,853,209 | 31.1288 | 9.89558 | 0.0010320 | 969 |
| 970 | 940,900 | 912,673,000 | 31.1448 | 9.89898 | 0.0010309 | 970 |
| 971 | 942,841 | 915,498,611 | 31.1609 | 9.90238 | 0.0010299 | 971 |
| 972 | 944,784 | 918,330,048 | 31.1769 | 9.90578 | 0.0010288 | 972 |
| 973 | 946,729 | 921,167,317 | 31.1929 | 9.90918 | 0.0010277 | 973 |
| 974 | 948,676 | 924,010,424 | 31.2090 | 9.91257 | 0.0010267 | 974 |
| 975 | 950,625 | 926,859,375 | 31.2250 | 9.91596 | 0.0010256 | 975 |
| 976 | 952,576 | 929,714,176 | 31.2410 | 9.91935 | 0.0010246 | 976 |
| 977 | 954,529 | 932,574,833 | 31.2570 | 9.92274 | 0.0010235 | 977 |
| 978 | 956,484 | 935,441,352 | 31.2730 | 9.92612 | 0.0010225 | 978 |
| 979 | 958,441 | 938,313,739 | 31.2890 | 9.92950 | 0.0010215 | 979 |
| 980 | 960,400 | 941,192,000 | 31.3050 | 9.93288 | 0.0010204 | 980 |
| 981 | 962,361 | 944,076,141 | 31.3209 | 9.93626 | 0.0010194 | 981 |
| 982 | 964,324 | 946,966,168 | 31.3369 | 9.93964 | 0.0010183 | 982 |
| 983 | 966,289 | 949,862,087 | 31.3528 | 9.94301 | 0.0010173 | 983 |
| 984 | 968,256 | 952,763,904 | 31.3688 | 9.94638 | 0.0010163 | 984 |
| 985 | 970,225 | 955,671,625 | 31.3847 | 9.94975 | 0.0010152 | 985 |
| 986 | 972,196 | 958,585,256 | 31.4006 | 9.95311 | 0.0010142 | 986 |
| 987 | 974,169 | 961,504,803 | 31.4166 | 9.95648 | 0.0010132 | 987 |
| 988 | 976,144 | 964,430,272 | 31.4325 | 9.95984 | 0.0010121 | 988 |
| 989 | 978,121 | 967,361,669 | 31.4484 | 9.96320 | 0.0010111 | 989 |
| 990 | 980,100 | 970,299,000 | 31.4643 | 9.96655 | 0.0010101 | 990 |
| 991 | 982,081 | 973,242,271 | 31.4802 | 9.96991 | 0.0010091 | 991 |
| 992 | 984,064 | 976,191,488 | 31.4960 | 9.97326 | 0.0010081 | 992 |
| 993 | 986,049 | 979,146,657 | 31.5119 | 9.97661 | 0.0010070 | 993 |
| 994 | 988,036 | 982,107,784 | 31.5278 | 9.97996 | 0.0010060 | 994 |
| 995 | 990,025 | 985,074,875 | 31.5436 | 9.98331 | 0.0010050 | 995 |
| 996 | 992,016 | 988,047,936 | 31.5595 | 9.98665 | 0.0010040 | 996 |
| 997 | 994,009 | 991,026,973 | 31.5753 | 9.98999 | 0.0010030 | 997 |
| 998 | 996,004 | 994,011,992 | 31.5911 | 9.99333 | 0.0010020 | 998 |
| 999 | 998,001 | 997,002,999 | 31.6070 | 9.99667 | 0.0010010 | 999 |
| 1000 | 1,000,000 | 1,000,000,000 | 31.6228 | 10.00000 | 0.0010000 | 1000 |

### Powers, Roots and Reciprocals

| No. | Square | Cube | Sq. Root | Cube Root | Reciprocal | No. |
|---|---|---|---|---|---|---|
| 1001 | 1,002,001 | 1,003,003,001 | 31.6386 | 10.0033 | 0.0009990 | 1001 |
| 1002 | 1,004,004 | 1,006,012,008 | 31.6544 | 10.0067 | 0.0009980 | 1002 |
| 1003 | 1,006,009 | 1,009,027,027 | 31.6702 | 10.0100 | 0.0009970 | 1003 |
| 1004 | 1,008,016 | 1,012,048,064 | 31.6860 | 10.0133 | 0.0009960 | 1004 |
| 1005 | 1,010,025 | 1,015,075,125 | 31.7017 | 10.0166 | 0.0009950 | 1005 |
| 1006 | 1,012,036 | 1,018,108,216 | 31.7175 | 10.0200 | 0.0009940 | 1006 |
| 1007 | 1,014,049 | 1,021,147,343 | 31.7333 | 10.0233 | 0.0009930 | 1007 |
| 1008 | 1,016,064 | 1,024,192,512 | 31.7490 | 10.0266 | 0.0009921 | 1008 |
| 1009 | 1,018,081 | 1,027,243,729 | 31.7648 | 10.0299 | 0.0009911 | 1009 |
| 1010 | 1,020,100 | 1,030,301,000 | 31.7805 | 10.0332 | 0.0009901 | 1010 |
| 1011 | 1,022,121 | 1,033,364,331 | 31.7962 | 10.0365 | 0.0009891 | 1011 |
| 1012 | 1,024,144 | 1,036,433,728 | 31.8119 | 10.0398 | 0.0009881 | 1012 |
| 1013 | 1,026,169 | 1,039,509,197 | 31.8277 | 10.0431 | 0.0009872 | 1013 |
| 1014 | 1,028,196 | 1,042,590,744 | 31.8434 | 10.0465 | 0.0009862 | 1014 |
| 1015 | 1,030,225 | 1,045,678,375 | 31.8591 | 10.0498 | 0.0009852 | 1015 |
| 1016 | 1,032,256 | 1,048,772,096 | 31.8748 | 10.0531 | 0.0009843 | 1016 |
| 1017 | 1,034,289 | 1,051,871,913 | 31.8904 | 10.0563 | 0.0009833 | 1017 |
| 1018 | 1,036,324 | 1,054,977,832 | 31.9061 | 10.0596 | 0.0009823 | 1018 |
| 1019 | 1,038,361 | 1,058,089,859 | 31.9218 | 10.0629 | 0.0009814 | 1019 |
| 1020 | 1,040,400 | 1,061,208,000 | 31.9374 | 10.0662 | 0.0009804 | 1020 |
| 1021 | 1,042,441 | 1,064,332,261 | 31.9531 | 10.0695 | 0.0009794 | 1021 |
| 1022 | 1,044,484 | 1,067,462,648 | 31.9687 | 10.0728 | 0.0009785 | 1022 |
| 1023 | 1,046,529 | 1,070,599,167 | 31.9844 | 10.0761 | 0.0009775 | 1023 |
| 1024 | 1,048,576 | 1,073,741,824 | 32.0000 | 10.0794 | 0.0009766 | 1024 |
| 1025 | 1,050,625 | 1,076,890,625 | 32.0156 | 10.0826 | 0.0009756 | 1025 |
| 1026 | 1,052,676 | 1,080,045,576 | 32.0312 | 10.0859 | 0.0009747 | 1026 |
| 1027 | 1,054,729 | 1,083,206,683 | 32.0468 | 10.0892 | 0.0009737 | 1027 |
| 1028 | 1,056,784 | 1,086,373,952 | 32.0624 | 10.0925 | 0.0009728 | 1028 |
| 1029 | 1,058,841 | 1,089,547,389 | 32.0780 | 10.0957 | 0.0009718 | 1029 |
| 1030 | 1,060,900 | 1,092,727,000 | 32.0936 | 10.0990 | 0.0009709 | 1030 |
| 1031 | 1,062,961 | 1,095,912,791 | 32.1092 | 10.1023 | 0.0009699 | 1031 |
| 1032 | 1,065,024 | 1,099,104,768 | 32.1248 | 10.1055 | 0.0009690 | 1032 |
| 1033 | 1,067,089 | 1,102,302,937 | 32.1403 | 10.1088 | 0.0009681 | 1033 |
| 1034 | 1,069,156 | 1,105,507,304 | 32.1559 | 10.1121 | 0.0009671 | 1034 |
| 1035 | 1,071,225 | 1,108,717,875 | 32.1714 | 10.1153 | 0.0009662 | 1035 |
| 1036 | 1,073,296 | 1,111,934,656 | 32.1870 | 10.1186 | 0.0009653 | 1036 |
| 1037 | 1,075,369 | 1,115,157,653 | 32.2025 | 10.1218 | 0.0009643 | 1037 |
| 1038 | 1,077,444 | 1,118,386,872 | 32.2180 | 10.1251 | 0.0009634 | 1038 |
| 1039 | 1,079,521 | 1,121,622,319 | 32.2335 | 10.1283 | 0.0009625 | 1039 |
| 1040 | 1,081,600 | 1,124,864,000 | 32.2490 | 10.1316 | 0.0009615 | 1040 |
| 1041 | 1,083,681 | 1,128,111,921 | 32.2645 | 10.1348 | 0.0009606 | 1041 |
| 1042 | 1,085,764 | 1,131,366,088 | 32.2800 | 10.1381 | 0.0009597 | 1042 |
| 1043 | 1,087,849 | 1,134,626,507 | 32.2955 | 10.1413 | 0.0009588 | 1043 |
| 1044 | 1,089,936 | 1,137,893,184 | 32.3110 | 10.1446 | 0.0009579 | 1044 |
| 1045 | 1,092,025 | 1,141,166,125 | 32.3265 | 10.1478 | 0.0009569 | 1045 |
| 1046 | 1,094,116 | 1,144,445,336 | 32.3419 | 10.1510 | 0.0009560 | 1046 |
| 1047 | 1,096,209 | 1,147,730,823 | 32.3574 | 10.1543 | 0.0009551 | 1047 |
| 1048 | 1,098,304 | 1,151,022,592 | 32.3728 | 10.1575 | 0.0009542 | 1048 |
| 1049 | 1,100,401 | 1,154,320,649 | 32.3883 | 10.1607 | 0.0009533 | 1049 |
| 1050 | 1,102,500 | 1,157,625,000 | 32.4037 | 10.1640 | 0.0009524 | 1050 |

## Powers, Roots and Reciprocals

| No. | Square | Cube | Sq. Root | Cube Root | Reciprocal | No. |
|---|---|---|---|---|---|---|
| 1051 | 1,104,601 | 1,160,935,651 | 32.4191 | 10.1672 | 0.0009515 | 1051 |
| 1052 | 1,106,704 | 1,164,252,608 | 32.4345 | 10.1704 | 0.0009506 | 1052 |
| 1053 | 1,108,809 | 1,167,575,877 | 32.4500 | 10.1736 | 0.0009497 | 1053 |
| 1054 | 1,110,916 | 1,170,905,464 | 32.4654 | 10.1769 | 0.0009488 | 1054 |
| 1055 | 1,113,025 | 1,174,241,375 | 32.4808 | 10.1801 | 0.0009479 | 1055 |
| 1056 | 1,115,136 | 1,177,583,616 | 32.4962 | 10.1833 | 0.0009470 | 1056 |
| 1057 | 1,117,249 | 1,180,932,193 | 32.5115 | 10.1865 | 0.0009461 | 1057 |
| 1058 | 1,119,364 | 1,184,287,112 | 32.5269 | 10.1897 | 0.0009452 | 1058 |
| 1059 | 1,121,481 | 1,187,648,379 | 32.5423 | 10.1929 | 0.0009443 | 1059 |
| 1060 | 1,123,600 | 1,191,016,000 | 32.5576 | 10.1961 | 0.0009434 | 1060 |
| 1061 | 1,125,721 | 1,194,389,981 | 32.5730 | 10.1993 | 0.0009425 | 1061 |
| 1062 | 1,127,844 | 1,197,770,328 | 32.5883 | 10.2025 | 0.0009416 | 1062 |
| 1063 | 1,129,969 | 1,201,157,047 | 32.6037 | 10.2057 | 0.0009407 | 1063 |
| 1064 | 1,132,096 | 1,204,550,144 | 32.6190 | 10.2089 | 0.0009398 | 1064 |
| 1065 | 1,134,225 | 1,207,949,625 | 32.6343 | 10.2121 | 0.0009390 | 1065 |
| 1066 | 1,136,356 | 1,211,355,496 | 32.6497 | 10.2153 | 0.0009381 | 1066 |
| 1067 | 1,138,489 | 1,214,767,763 | 32.6650 | 10.2185 | 0.0009372 | 1067 |
| 1068 | 1,140,624 | 1,218,186,432 | 32.6803 | 10.2217 | 0.0009363 | 1068 |
| 1069 | 1,142,761 | 1,221,611,509 | 32.6956 | 10.2249 | 0.0009355 | 1069 |
| 1070 | 1,144,900 | 1,225,043,000 | 32.7109 | 10.2281 | 0.0009346 | 1070 |
| 1071 | 1,147,041 | 1,228,480,911 | 32.7261 | 10.2313 | 0.0009337 | 1071 |
| 1072 | 1,149,184 | 1,231,925,248 | 32.7414 | 10.2345 | 0.0009328 | 1072 |
| 1073 | 1,151,329 | 1,235,376,017 | 32.7567 | 10.2376 | 0.0009320 | 1073 |
| 1074 | 1,153,476 | 1,238,833,224 | 32.7719 | 10.2408 | 0.0009311 | 1074 |
| 1075 | 1,155,625 | 1,242,296,875 | 32.7872 | 10.2440 | 0.0009302 | 1075 |
| 1076 | 1,157,776 | 1,245,766,976 | 32.8024 | 10.2472 | 0.0009294 | 1076 |
| 1077 | 1,159,929 | 1,249,243,533 | 32.8177 | 10.2503 | 0.0009285 | 1077 |
| 1078 | 1,162,084 | 1,252,726,552 | 32.8329 | 10.2535 | 0.0009276 | 1078 |
| 1079 | 1,164,241 | 1,256,216,039 | 32.8481 | 10.2567 | 0.0009268 | 1079 |
| 1080 | 1,166,400 | 1,259,712,000 | 32.8634 | 10.2599 | 0.0009259 | 1080 |
| 1081 | 1,168,561 | 1,263,214,441 | 32.8786 | 10.2630 | 0.0009251 | 1081 |
| 1082 | 1,170,724 | 1,266,723,368 | 32.8938 | 10.2662 | 0.0009242 | 1082 |
| 1083 | 1,172,889 | 1,270,238,787 | 32.9090 | 10.2693 | 0.0009234 | 1083 |
| 1084 | 1,175,056 | 1,273,760,704 | 32.9242 | 10.2725 | 0.0009225 | 1084 |
| 1085 | 1,177,225 | 1,277,289,125 | 32.9393 | 10.2757 | 0.0009217 | 1085 |
| 1086 | 1,179,396 | 1,280,824,056 | 32.9545 | 10.2788 | 0.0009208 | 1086 |
| 1087 | 1,181,569 | 1,284,365,503 | 32.9697 | 10.2820 | 0.0009200 | 1087 |
| 1088 | 1,183,744 | 1,287,913,472 | 32.9848 | 10.2851 | 0.0009191 | 1088 |
| 1089 | 1,185,921 | 1,291,467,969 | 33.0000 | 10.2883 | 0.0009183 | 1089 |
| 1090 | 1,188,100 | 1,295,029,000 | 33.0151 | 10.2914 | 0.0009174 | 1090 |
| 1091 | 1,190,281 | 1,298,596,571 | 33.0303 | 10.2946 | 0.0009166 | 1091 |
| 1092 | 1,192,464 | 1,302,170,688 | 33.0454 | 10.2977 | 0.0009158 | 1092 |
| 1093 | 1,194,649 | 1,305,751,357 | 33.0606 | 10.3009 | 0.0009149 | 1093 |
| 1094 | 1,196,836 | 1,309,338,584 | 33.0757 | 10.3040 | 0.0009141 | 1094 |
| 1095 | 1,199,025 | 1,312,932,375 | 33.0908 | 10.3071 | 0.0009132 | 1095 |
| 1096 | 1,201,216 | 1,316,532,736 | 33.1059 | 10.3103 | 0.0009124 | 1096 |
| 1097 | 1,203,409 | 1,320,139,673 | 33.1210 | 10.3134 | 0.0009116 | 1097 |
| 1098 | 1,205,604 | 1,323,753,192 | 33.1361 | 10.3165 | 0.0009107 | 1098 |
| 1099 | 1,207,801 | 1,327,373,299 | 33.1512 | 10.3197 | 0.0009099 | 1099 |
| 1100 | 1,210,000 | 1,331,000,000 | 33.1662 | 10.3228 | 0.0009091 | 1100 |

### Powers, Roots and Reciprocals

| No. | Square | Cube | Sq. Root | Cube Root | Reciprocal | No. |
|---|---|---|---|---|---|---|
| 1101 | 1,212,201 | 1,334,633,301 | 33.1813 | 10.3259 | 0.0009083 | 1101 |
| 1102 | 1,214,404 | 1,338,273,208 | 33.1964 | 10.3291 | 0.0009074 | 1102 |
| 1103 | 1,216,609 | 1,341,919,727 | 33.2114 | 10.3322 | 0.0009066 | 1103 |
| 1104 | 1,218,816 | 1,345,572,864 | 33.2265 | 10.3353 | 0.0009058 | 1104 |
| 1105 | 1,221,025 | 1,349,232,625 | 33.2415 | 10.3384 | 0.0009050 | 1105 |
| 1106 | 1,223,236 | 1,352,899,016 | 33.2566 | 10.3415 | 0.0009042 | 1106 |
| 1107 | 1,225,449 | 1,356,572,043 | 33.2716 | 10.3447 | 0.0009033 | 1107 |
| 1108 | 1,227,664 | 1,360,251,712 | 33.2866 | 10.3478 | 0.0009025 | 1108 |
| 1109 | 1,229,881 | 1,363,938,029 | 33.3017 | 10.3509 | 0.0009017 | 1109 |
| 1110 | 1,232,100 | 1,367,631,000 | 33.3167 | 10.3540 | 0.0009009 | 1110 |
| 1111 | 1,234,321 | 1,371,330,631 | 33.3317 | 10.3571 | 0.0009001 | 1111 |
| 1112 | 1,236,544 | 1,375,036,928 | 33.3467 | 10.3602 | 0.0008993 | 1112 |
| 1113 | 1,238,769 | 1,378,749,897 | 33.3617 | 10.3633 | 0.0008985 | 1113 |
| 1114 | 1,240,996 | 1,382,469,544 | 33.3766 | 10.3664 | 0.0008977 | 1114 |
| 1115 | 1,243,225 | 1,386,195,875 | 33.3916 | 10.3695 | 0.0008969 | 1115 |
| 1116 | 1,245,456 | 1,389,928,896 | 33.4066 | 10.3726 | 0.0008961 | 1116 |
| 1117 | 1,247,689 | 1,393,668,613 | 33.4215 | 10.3757 | 0.0008953 | 1117 |
| 1118 | 1,249,924 | 1,397,415,032 | 33.4365 | 10.3788 | 0.0008945 | 1118 |
| 1119 | 1,252,161 | 1,401,168,159 | 33.4515 | 10.3819 | 0.0008937 | 1119 |
| 1120 | 1,254,400 | 1,404,928,000 | 33.4664 | 10.3850 | 0.0008929 | 1120 |
| 1121 | 1,256,641 | 1,408,694,561 | 33.4813 | 10.3881 | 0.0008921 | 1121 |
| 1122 | 1,258,884 | 1,412,467,848 | 33.4963 | 10.3912 | 0.0008913 | 1122 |
| 1123 | 1,261,129 | 1,416,247,867 | 33.5112 | 10.3943 | 0.0008905 | 1123 |
| 1124 | 1,263,376 | 1,420,034,624 | 33.5261 | 10.3973 | 0.0008897 | 1124 |
| 1125 | 1,265,625 | 1,423,828,125 | 33.5410 | 10.4004 | 0.0008889 | 1125 |
| 1126 | 1,267,876 | 1,427,628,376 | 33.5559 | 10.4035 | 0.0008881 | 1126 |
| 1127 | 1,270,129 | 1,431,435,383 | 33.5708 | 10.4066 | 0.0008873 | 1127 |
| 1128 | 1,272,384 | 1,435,249,152 | 33.5857 | 10.4097 | 0.0008865 | 1128 |
| 1129 | 1,274,641 | 1,439,069,689 | 33.6006 | 10.4127 | 0.0008857 | 1129 |
| 1130 | 1,276,900 | 1,442,897,000 | 33.6155 | 10.4158 | 0.0008850 | 1130 |
| 1131 | 1,279,161 | 1,446,731,091 | 33.6303 | 10.4189 | 0.0008842 | 1131 |
| 1132 | 1,281,424 | 1,450,571,968 | 33.6452 | 10.4219 | 0.0008834 | 1132 |
| 1133 | 1,283,689 | 1,454,419,637 | 33.6601 | 10.4250 | 0.0008826 | 1133 |
| 1134 | 1,285,956 | 1,458,274,104 | 33.6749 | 10.4281 | 0.0008818 | 1134 |
| 1135 | 1,288,225 | 1,462,135,375 | 33.6898 | 10.4311 | 0.0008811 | 1135 |
| 1136 | 1,290,496 | 1,466,003,456 | 33.7046 | 10.4342 | 0.0008803 | 1136 |
| 1137 | 1,292,769 | 1,469,878,353 | 33.7194 | 10.4373 | 0.0008795 | 1137 |
| 1138 | 1,295,044 | 1,473,760,072 | 33.7343 | 10.4403 | 0.0008787 | 1138 |
| 1139 | 1,297,321 | 1,477,648,619 | 33.7491 | 10.4434 | 0.0008780 | 1139 |
| 1140 | 1,299,600 | 1,481,544,000 | 33.7639 | 10.4464 | 0.0008772 | 1140 |
| 1141 | 1,301,881 | 1,485,446,221 | 33.7787 | 10.4495 | 0.0008764 | 1141 |
| 1142 | 1,304,164 | 1,489,355,288 | 33.7935 | 10.4525 | 0.0008757 | 1142 |
| 1143 | 1,306,449 | 1,493,271,207 | 33.8083 | 10.4556 | 0.0008749 | 1143 |
| 1144 | 1,308,736 | 1,497,193,984 | 33.8231 | 10.4586 | 0.0008741 | 1144 |
| 1145 | 1,311,025 | 1,501,123,625 | 33.8378 | 10.4617 | 0.0008734 | 1145 |
| 1146 | 1,313,316 | 1,505,060,136 | 33.8526 | 10.4647 | 0.0008726 | 1146 |
| 1147 | 1,315,609 | 1,509,003,523 | 33.8674 | 10.4678 | 0.0008718 | 1147 |
| 1148 | 1,317,904 | 1,512,953,792 | 33.8821 | 10.4708 | 0.0008711 | 1148 |
| 1149 | 1,320,201 | 1,516,910,949 | 33.8969 | 10.4739 | 0.0008703 | 1149 |
| 1150 | 1,322,500 | 1,520,875,000 | 33.9116 | 10.4769 | 0.0008696 | 1150 |

Powers, Roots and Reciprocals

| No. | Square | Cube | Sq. Root | Cube Root | Reciprocal | No. |
|---|---|---|---|---|---|---|
| 1151 | 1,324,801 | 1,524,845,951 | 33.9264 | 10.4799 | 0.0008688 | 1151 |
| 1152 | 1,327,104 | 1,528,823,808 | 33.9411 | 10.4830 | 0.0008681 | 1152 |
| 1153 | 1,329,409 | 1,532,808,577 | 33.9559 | 10.4860 | 0.0008673 | 1153 |
| 1154 | 1,331,716 | 1,536,800,264 | 33.9706 | 10.4890 | 0.0008666 | 1154 |
| 1155 | 1,334,025 | 1,540,798,875 | 33.9853 | 10.4921 | 0.0008658 | 1155 |
| 1156 | 1,336,336 | 1,544,804,416 | 34.0000 | 10.4951 | 0.0008651 | 1156 |
| 1157 | 1,338,649 | 1,548,816,893 | 34.0147 | 10.4981 | 0.0008643 | 1157 |
| 1158 | 1,340,964 | 1,552,836,312 | 34.0294 | 10.5011 | 0.0008636 | 1158 |
| 1159 | 1,343,281 | 1,556,862,679 | 34.0441 | 10.5042 | 0.0008628 | 1159 |
| 1160 | 1,345,600 | 1,560,896,000 | 34.0588 | 10.5072 | 0.0008621 | 1160 |
| 1161 | 1,347,921 | 1,564,936,281 | 34.0735 | 10.5102 | 0.0008613 | 1161 |
| 1162 | 1,350,244 | 1,568,983,528 | 34.0881 | 10.5132 | 0.0008606 | 1162 |
| 1163 | 1,352,569 | 1,573,037,747 | 34.1028 | 10.5162 | 0.0008598 | 1163 |
| 1164 | 1,354,896 | 1,577,098,944 | 34.1174 | 10.5192 | 0.0008591 | 1164 |
| 1165 | 1,357,225 | 1,581,167,125 | 34.1321 | 10.5223 | 0.0008584 | 1165 |
| 1166 | 1,359,556 | 1,585,242,296 | 34.1467 | 10.5253 | 0.0008576 | 1166 |
| 1167 | 1,361,889 | 1,589,324,463 | 34.1614 | 10.5283 | 0.0008569 | 1167 |
| 1168 | 1,364,224 | 1,593,413,632 | 34.1760 | 10.5313 | 0.0008562 | 1168 |
| 1169 | 1,366,561 | 1,597,509,809 | 34.1906 | 10.5343 | 0.0008554 | 1169 |
| 1170 | 1,368,900 | 1,601,613,000 | 34.2053 | 10.5373 | 0.0008547 | 1170 |
| 1171 | 1,371,241 | 1,605,723,211 | 34.2199 | 10.5403 | 0.0008540 | 1171 |
| 1172 | 1,373,584 | 1,609,840,448 | 34.2345 | 10.5433 | 0.0008532 | 1172 |
| 1173 | 1,375,929 | 1,613,964,717 | 34.2491 | 10.5463 | 0.0008525 | 1173 |
| 1174 | 1,378,276 | 1,618,096,024 | 34.2637 | 10.5493 | 0.0008518 | 1174 |
| 1175 | 1,380,625 | 1,622,234,375 | 34.2783 | 10.5523 | 0.0008511 | 1175 |
| 1176 | 1,382,976 | 1,626,379,776 | 34.2929 | 10.5553 | 0.0008503 | 1176 |
| 1177 | 1,385,329 | 1,630,532,233 | 34.3074 | 10.5583 | 0.0008496 | 1177 |
| 1178 | 1,387,684 | 1,634,691,752 | 34.3220 | 10.5612 | 0.0008489 | 1178 |
| 1179 | 1,390,041 | 1,638,858,339 | 34.3366 | 10.5642 | 0.0008482 | 1179 |
| 1180 | 1,392,400 | 1,643,032,000 | 34.3511 | 10.5672 | 0.0008475 | 1180 |
| 1181 | 1,394,761 | 1,647,212,741 | 34.3657 | 10.5702 | 0.0008467 | 1181 |
| 1182 | 1,397,124 | 1,651,400,568 | 34.3802 | 10.5732 | 0.0008460 | 1182 |
| 1183 | 1,399,489 | 1,655,595,487 | 34.3948 | 10.5762 | 0.0008453 | 1183 |
| 1184 | 1,401,856 | 1,659,797,504 | 34.4093 | 10.5791 | 0.0008446 | 1184 |
| 1185 | 1,404,225 | 1,664,006,625 | 34.4238 | 10.5821 | 0.0008439 | 1185 |
| 1186 | 1,406,596 | 1,668,222,856 | 34.4384 | 10.5851 | 0.0008432 | 1186 |
| 1187 | 1,408,969 | 1,672,446,203 | 34.4529 | 10.5881 | 0.0008425 | 1187 |
| 1188 | 1,411,344 | 1,676,676,672 | 34.4674 | 10.5910 | 0.0008418 | 1188 |
| 1189 | 1,413,721 | 1,680,914,269 | 34.4819 | 10.5940 | 0.0008410 | 1189 |
| 1190 | 1,416,100 | 1,685,159,000 | 34.4964 | 10.5970 | 0.0008403 | 1190 |
| 1191 | 1,418,481 | 1,689,410,871 | 34.5109 | 10.6000 | 0.0008396 | 1191 |
| 1192 | 1,420,864 | 1,693,669,888 | 34.5254 | 10.6029 | 0.0008389 | 1192 |
| 1193 | 1,423,249 | 1,697,936,057 | 34.5398 | 10.6059 | 0.0008382 | 1193 |
| 1194 | 1,425,636 | 1,702,209,384 | 34.5543 | 10.6088 | 0.0008375 | 1194 |
| 1195 | 1,428,025 | 1,706,489,875 | 34.5688 | 10.6118 | 0.0008368 | 1195 |
| 1196 | 1,430,416 | 1,710,777,536 | 34.5832 | 10.6148 | 0.0008361 | 1196 |
| 1197 | 1,432,809 | 1,715,072,373 | 34.5977 | 10.6177 | 0.0008354 | 1197 |
| 1198 | 1,435,204 | 1,719,374,392 | 34.6121 | 10.6207 | 0.0008347 | 1198 |
| 1199 | 1,437,601 | 1,723,683,599 | 34.6266 | 10.6236 | 0.0008340 | 1199 |
| 1200 | 1,440,000 | 1,728,000,000 | 34.6410 | 10.6266 | 0.0008333 | 1200 |

Powers, Roots and Reciprocals

| No. | Square | Cube | Sq. Root | Cube Root | Reciprocal | No. |
|-----|--------|------|----------|-----------|------------|-----|
| 1201 | 1,442,401 | 1,732,323,601 | 34.6554 | 10.6295 | 0.0008326 | 1201 |
| 1202 | 1,444,804 | 1,736,654,408 | 34.6699 | 10.6325 | 0.0008319 | 1202 |
| 1203 | 1,447,209 | 1,740,992,427 | 34.6843 | 10.6354 | 0.0008313 | 1203 |
| 1204 | 1,449,616 | 1,745,337,664 | 34.6987 | 10.6384 | 0.0008306 | 1204 |
| 1205 | 1,452,025 | 1,749,690,125 | 34.7131 | 10.6413 | 0.0008299 | 1205 |
| 1206 | 1,454,436 | 1,754,049,816 | 34.7275 | 10.6443 | 0.0008292 | 1206 |
| 1207 | 1,456,849 | 1,758,416,743 | 34.7419 | 10.6472 | 0.0008285 | 1207 |
| 1208 | 1,459,264 | 1,762,790,912 | 34.7563 | 10.6501 | 0.0008278 | 1208 |
| 1209 | 1,461,681 | 1,767,172,329 | 34.7707 | 10.6531 | 0.0008271 | 1209 |
| 1210 | 1,464,100 | 1,771,561,000 | 34.7851 | 10.6560 | 0.0008264 | 1210 |
| 1211 | 1,466,521 | 1,775,956,931 | 34.7994 | 10.6590 | 0.0008258 | 1211 |
| 1212 | 1,468,944 | 1,780,360,128 | 34.8138 | 10.6619 | 0.0008251 | 1212 |
| 1213 | 1,471,369 | 1,784,770,597 | 34.8281 | 10.6648 | 0.0008244 | 1213 |
| 1214 | 1,473,796 | 1,789,188,344 | 34.8425 | 10.6678 | 0.0008237 | 1214 |
| 1215 | 1,476,225 | 1,793,613,375 | 34.8569 | 10.6707 | 0.0008230 | 1215 |
| 1216 | 1,478,656 | 1,798,045,696 | 34.8712 | 10.6736 | 0.0008224 | 1216 |
| 1217 | 1,481,089 | 1,802,485,313 | 34.8855 | 10.6765 | 0.0008217 | 1217 |
| 1218 | 1,483,524 | 1,806,932,232 | 34.8999 | 10.6795 | 0.0008210 | 1218 |
| 1219 | 1,485,961 | 1,811,386,459 | 34.9142 | 10.6824 | 0.0008203 | 1219 |
| 1220 | 1,488,400 | 1,815,848,000 | 34.9285 | 10.6853 | 0.0008197 | 1220 |
| 1221 | 1,490,841 | 1,820,316,861 | 34.9428 | 10.6882 | 0.0008190 | 1221 |
| 1222 | 1,493,284 | 1,824,793,048 | 34.9571 | 10.6911 | 0.0008183 | 1222 |
| 1223 | 1,495,729 | 1,829,276,567 | 34.9714 | 10.6940 | 0.0008177 | 1223 |
| 1224 | 1,498,176 | 1,833,767,424 | 34.9857 | 10.6970 | 0.0008170 | 1224 |
| 1225 | 1,500,625 | 1,838,265,625 | 35.0000 | 10.6999 | 0.0008163 | 1225 |
| 1226 | 1,503,076 | 1,842,771,176 | 35.0143 | 10.7028 | 0.0008157 | 1226 |
| 1227 | 1,505,529 | 1,847,284,083 | 35.0286 | 10.7057 | 0.0008150 | 1227 |
| 1228 | 1,507,984 | 1,851,804,352 | 35.0428 | 10.7086 | 0.0008143 | 1228 |
| 1229 | 1,510,441 | 1,856,331,989 | 35.0571 | 10.7115 | 0.0008137 | 1229 |
| 1230 | 1,512,900 | 1,860,867,000 | 35.0714 | 10.7144 | 0.0008130 | 1230 |
| 1231 | 1,515,361 | 1,865,409,391 | 35.0856 | 10.7173 | 0.0008123 | 1231 |
| 1232 | 1,517,824 | 1,869,959,168 | 35.0999 | 10.7202 | 0.0008117 | 1232 |
| 1233 | 1,520,289 | 1,874,516,337 | 35.1141 | 10.7231 | 0.0008110 | 1233 |
| 1234 | 1,522,756 | 1,879,080,904 | 35.1283 | 10.7260 | 0.0008104 | 1234 |
| 1235 | 1,525,225 | 1,883,652,875 | 35.1426 | 10.7289 | 0.0008097 | 1235 |
| 1236 | 1,527,696 | 1,888,232,256 | 35.1568 | 10.7318 | 0.0008091 | 1236 |
| 1237 | 1,530,169 | 1,892,819,053 | 35.1710 | 10.7347 | 0.0008084 | 1237 |
| 1238 | 1,532,644 | 1,897,413,272 | 35.1852 | 10.7376 | 0.0008078 | 1238 |
| 1239 | 1,535,121 | 1,902,014,919 | 35.1994 | 10.7405 | 0.0008071 | 1239 |
| 1240 | 1,537,600 | 1,906,624,000 | 35.2136 | 10.7434 | 0.0008065 | 1240 |
| 1241 | 1,540,081 | 1,911,240,521 | 35.2278 | 10.7463 | 0.0008058 | 1241 |
| 1242 | 1,542,564 | 1,915,864,488 | 35.2420 | 10.7491 | 0.0008052 | 1242 |
| 1243 | 1,545,049 | 1,920,495,907 | 35.2562 | 10.7520 | 0.0008045 | 1243 |
| 1244 | 1,547,536 | 1,925,134,784 | 35.2704 | 10.7549 | 0.0008039 | 1244 |
| 1245 | 1,550,025 | 1,929,781,125 | 35.2846 | 10.7578 | 0.0008032 | 1245 |
| 1246 | 1,552,516 | 1,934,434,936 | 35.2987 | 10.7607 | 0.0008026 | 1246 |
| 1247 | 1,555,009 | 1,939,096,223 | 35.3129 | 10.7635 | 0.0008019 | 1247 |
| 1248 | 1,557,504 | 1,943,764,992 | 35.3270 | 10.7664 | 0.0008013 | 1248 |
| 1249 | 1,560,001 | 1,948,441,249 | 35.3412 | 10.7693 | 0.0008006 | 1249 |
| 1250 | 1,562,500 | 1,953,125,000 | 35.3553 | 10.7722 | 0.0008000 | 1250 |

## Powers, Roots and Reciprocals

| No. | Square | Cube | Sq. Root | Cube Root | Reciprocal | No. |
|---|---|---|---|---|---|---|
| 1251 | 1,565,001 | 1,957,816,251 | 35.3695 | 10.7750 | 0.0007994 | 1251 |
| 1252 | 1,567,504 | 1,962,515,008 | 35.3836 | 10.7779 | 0.0007987 | 1252 |
| 1253 | 1,570,009 | 1,967,221,277 | 35.3977 | 10.7808 | 0.0007981 | 1253 |
| 1254 | 1,572,516 | 1,971,935,064 | 35.4119 | 10.7837 | 0.0007974 | 1254 |
| 1255 | 1,575,025 | 1,976,656,375 | 35.4260 | 10.7865 | 0.0007968 | 1255 |
| 1256 | 1,577,536 | 1,981,385,216 | 35.4401 | 10.7894 | 0.0007962 | 1256 |
| 1257 | 1,580,049 | 1,986,121,593 | 35.4542 | 10.7922 | 0.0007955 | 1257 |
| 1258 | 1,582,564 | 1,990,865,512 | 35.4683 | 10.7951 | 0.0007949 | 1258 |
| 1259 | 1,585,081 | 1,995,616,979 | 35.4824 | 10.7980 | 0.0007943 | 1259 |
| 1260 | 1,587,600 | 2,000,376,000 | 35.4965 | 10.8008 | 0.0007937 | 1260 |
| 1261 | 1,590,121 | 2,005,142,581 | 35.5106 | 10.8037 | 0.0007930 | 1261 |
| 1262 | 1,592,644 | 2,009,916,728 | 35.5246 | 10.8065 | 0.0007924 | 1262 |
| 1263 | 1,595,169 | 2,014,698,447 | 35.5387 | 10.8094 | 0.0007918 | 1263 |
| 1264 | 1,597,696 | 2,019,487,744 | 35.5528 | 10.8122 | 0.0007911 | 1264 |
| 1265 | 1,600,225 | 2,024,284,625 | 35.5668 | 10.8151 | 0.0007905 | 1265 |
| 1266 | 1,602,756 | 2,029,089,096 | 35.5809 | 10.8179 | 0.0007899 | 1266 |
| 1267 | 1,605,289 | 2,033,901,163 | 35.5949 | 10.8208 | 0.0007893 | 1267 |
| 1268 | 1,607,824 | 2,038,720,832 | 35.6090 | 10.8236 | 0.0007886 | 1268 |
| 1269 | 1,610,361 | 2,043,548,109 | 35.6230 | 10.8265 | 0.0007880 | 1269 |
| 1270 | 1,612,900 | 2,048,383,000 | 35.6371 | 10.8293 | 0.0007874 | 1270 |
| 1271 | 1,615,441 | 2,053,225,511 | 35.6511 | 10.8322 | 0.0007868 | 1271 |
| 1272 | 1,617,984 | 2,058,075,648 | 35.6651 | 10.8350 | 0.0007862 | 1272 |
| 1273 | 1,620,529 | 2,062,933,417 | 35.6791 | 10.8378 | 0.0007855 | 1273 |
| 1274 | 1,623,076 | 2,067,798,824 | 35.6931 | 10.8407 | 0.0007849 | 1274 |
| 1275 | 1,625,625 | 2,072,671,875 | 35.7071 | 10.8435 | 0.0007843 | 1275 |
| 1276 | 1,628,176 | 2,077,552,576 | 35.7211 | 10.8463 | 0.0007837 | 1276 |
| 1277 | 1,630,729 | 2,082,440,933 | 35.7351 | 10.8492 | 0.0007831 | 1277 |
| 1278 | 1,633,284 | 2,087,336,952 | 35.7491 | 10.8520 | 0.0007825 | 1278 |
| 1279 | 1,635,841 | 2,092,240,639 | 35.7631 | 10.8548 | 0.0007819 | 1279 |
| 1280 | 1,638,400 | 2,097,152,000 | 35.7771 | 10.8577 | 0.0007813 | 1280 |
| 1281 | 1,640,961 | 2,102,071,041 | 35.7911 | 10.8605 | 0.0007806 | 1281 |
| 1282 | 1,643,524 | 2,106,997,768 | 35.8050 | 10.8633 | 0.0007800 | 1282 |
| 1283 | 1,646,089 | 2,111,932,187 | 35.8190 | 10.8661 | 0.0007794 | 1283 |
| 1284 | 1,648,656 | 2,116,874,304 | 35.8329 | 10.8690 | 0.0007788 | 1284 |
| 1285 | 1,651,225 | 2,121,824,125 | 35.8469 | 10.8718 | 0.0007782 | 1285 |
| 1286 | 1,653,796 | 2,126,781,656 | 35.8608 | 10.8746 | 0.0007776 | 1286 |
| 1287 | 1,656,369 | 2,131,746,903 | 35.8748 | 10.8774 | 0.0007770 | 1287 |
| 1288 | 1,658,944 | 2,136,719,872 | 35.8887 | 10.8802 | 0.0007764 | 1288 |
| 1289 | 1,661,521 | 2,141,700,569 | 35.9026 | 10.8831 | 0.0007758 | 1289 |
| 1290 | 1,664,100 | 2,146,689,000 | 35.9166 | 10.8859 | 0.0007752 | 1290 |
| 1291 | 1,666,681 | 2,151,685,171 | 35.9305 | 10.8887 | 0.0007746 | 1291 |
| 1292 | 1,669,264 | 2,156,689,088 | 35.9444 | 10.8915 | 0.0007740 | 1292 |
| 1293 | 1,671,849 | 2,161,700,757 | 35.9583 | 10.8943 | 0.0007734 | 1293 |
| 1294 | 1,674,436 | 2,166,720,184 | 35.9722 | 10.8971 | 0.0007728 | 1294 |
| 1295 | 1,677,025 | 2,171,747,375 | 35.9861 | 10.8999 | 0.0007722 | 1295 |
| 1296 | 1,679,616 | 2,176,782,336 | 36.0000 | 10.9027 | 0.0007716 | 1296 |
| 1297 | 1,682,209 | 2,181,825,073 | 36.0139 | 10.9055 | 0.0007710 | 1297 |
| 1298 | 1,684,804 | 2,186,875,592 | 36.0278 | 10.9083 | 0.0007704 | 1298 |
| 1299 | 1,687,401 | 2,191,933,899 | 36.0416 | 10.9111 | 0.0007698 | 1299 |
| 1300 | 1,690,000 | 2,197,000,000 | 36.0555 | 10.9139 | 0.0007692 | 1300 |

## Powers, Roots and Reciprocals

| No. | Square | Cube | Sq. Root | Cube Root | Reciprocal | No. |
|---|---|---|---|---|---|---|
| 1301 | 1,692,601 | 2,202,073,901 | 36.0694 | 10.9167 | 0.0007686 | 1301 |
| 1302 | 1,695,204 | 2,207,155,608 | 36.0832 | 10.9195 | 0.0007680 | 1302 |
| 1303 | 1,697,809 | 2,212,245,127 | 36.0971 | 10.9223 | 0.0007675 | 1303 |
| 1304 | 1,700,416 | 2,217,342,464 | 36.1109 | 10.9251 | 0.0007669 | 1304 |
| 1305 | 1,703,025 | 2,222,447,625 | 36.1248 | 10.9279 | 0.0007663 | 1305 |
| 1306 | 1,705,636 | 2,227,560,616 | 36.1386 | 10.9307 | 0.0007657 | 1306 |
| 1307 | 1,708,249 | 2,232,681,443 | 36.1525 | 10.9335 | 0.0007651 | 1307 |
| 1308 | 1,710,864 | 2,237,810,112 | 36.1663 | 10.9363 | 0.0007645 | 1308 |
| 1309 | 1,713,481 | 2,242,946,629 | 36.1801 | 10.9391 | 0.0007639 | 1309 |
| 1310 | 1,716,100 | 2,248,091,000 | 36.1939 | 10.9418 | 0.0007634 | 1310 |
| 1311 | 1,718,721 | 2,253,243,231 | 36.2077 | 10.9446 | 0.0007628 | 1311 |
| 1312 | 1,721,344 | 2,258,403,328 | 36.2215 | 10.9474 | 0.0007622 | 1312 |
| 1313 | 1,723,969 | 2,263,571,297 | 36.2353 | 10.9502 | 0.0007616 | 1313 |
| 1314 | 1,726,596 | 2,268,747,144 | 36.2491 | 10.9530 | 0.0007610 | 1314 |
| 1315 | 1,729,225 | 2,273,930,875 | 36.2629 | 10.9557 | 0.0007605 | 1315 |
| 1316 | 1,731,856 | 2,279,122,496 | 36.2767 | 10.9585 | 0.0007599 | 1316 |
| 1317 | 1,734,489 | 2,284,322,013 | 36.2905 | 10.9613 | 0.0007593 | 1317 |
| 1318 | 1,737,124 | 2,289,529,432 | 36.3043 | 10.9641 | 0.0007587 | 1318 |
| 1319 | 1,739,761 | 2,294,744,759 | 36.3180 | 10.9668 | 0.0007582 | 1319 |
| 1320 | 1,742,400 | 2,299,968,000 | 36.3318 | 10.9696 | 0.0007576 | 1320 |
| 1321 | 1,745,041 | 2,305,199,161 | 36.3456 | 10.9724 | 0.0007570 | 1321 |
| 1322 | 1,747,684 | 2,310,438,248 | 36.3593 | 10.9752 | 0.0007564 | 1322 |
| 1323 | 1,750,329 | 2,315,685,267 | 36.3731 | 10.9779 | 0.0007559 | 1323 |
| 1324 | 1,752,976 | 2,320,940,224 | 36.3868 | 10.9807 | 0.0007553 | 1324 |
| 1325 | 1,755,625 | 2,326,203,125 | 36.4005 | 10.9834 | 0.0007547 | 1325 |
| 1326 | 1,758,276 | 2,331,473,976 | 36.4143 | 10.9862 | 0.0007541 | 1326 |
| 1327 | 1,760,929 | 2,336,752,783 | 36.4280 | 10.9890 | 0.0007536 | 1327 |
| 1328 | 1,763,584 | 2,342,039,552 | 36.4417 | 10.9917 | 0.0007530 | 1328 |
| 1329 | 1,766,241 | 2,347,334,289 | 36.4555 | 10.9945 | 0.0007524 | 1329 |
| 1330 | 1,768,900 | 2,352,637,000 | 36.4692 | 10.9972 | 0.0007519 | 1330 |
| 1331 | 1,771,561 | 2,357,947,691 | 36.4829 | 11.0000 | 0.0007513 | 1331 |
| 1332 | 1,774,224 | 2,363,266,368 | 36.4966 | 11.0028 | 0.0007508 | 1332 |
| 1333 | 1,776,889 | 2,368,593,037 | 36.5103 | 11.0055 | 0.0007502 | 1333 |
| 1334 | 1,779,556 | 2,373,927,704 | 36.5240 | 11.0083 | 0.0007496 | 1334 |
| 1335 | 1,782,225 | 2,379,270,375 | 36.5377 | 11.0110 | 0.0007491 | 1335 |
| 1336 | 1,784,896 | 2,384,621,056 | 36.5513 | 11.0138 | 0.0007485 | 1336 |
| 1337 | 1,787,569 | 2,389,979,753 | 36.5650 | 11.0165 | 0.0007479 | 1337 |
| 1338 | 1,790,244 | 2,395,346,472 | 36.5787 | 11.0193 | 0.0007474 | 1338 |
| 1339 | 1,792,921 | 2,400,721,219 | 36.5923 | 11.0220 | 0.0007468 | 1339 |
| 1340 | 1,795,600 | 2,406,104,000 | 36.6060 | 11.0247 | 0.0007463 | 1340 |
| 1341 | 1,798,281 | 2,411,494,821 | 36.6197 | 11.0275 | 0.0007457 | 1341 |
| 1342 | 1,800,964 | 2,416,893,688 | 36.6333 | 11.0302 | 0.0007452 | 1342 |
| 1343 | 1,803,649 | 2,422,300,607 | 36.6470 | 11.0330 | 0.0007446 | 1343 |
| 1344 | 1,806,336 | 2,427,715,584 | 36.6606 | 11.0357 | 0.0007440 | 1344 |
| 1345 | 1,809,025 | 2,433,138,625 | 36.6742 | 11.0384 | 0.0007435 | 1345 |
| 1346 | 1,811,716 | 2,438,569,736 | 36.6879 | 11.0412 | 0.0007429 | 1346 |
| 1347 | 1,814,409 | 2,444,008,923 | 36.7015 | 11.0439 | 0.0007424 | 1347 |
| 1348 | 1,817,104 | 2,449,456,192 | 36.7151 | 11.0466 | 0.0007418 | 1348 |
| 1349 | 1,819,801 | 2,454,911,549 | 36.7287 | 11.0494 | 0.0007413 | 1349 |
| 1350 | 1,822,500 | 2,460,375,000 | 36.7423 | 11.0521 | 0.0007407 | 1350 |

## Powers, Roots and Reciprocals

| No. | Square | Cube | Sq. Root | Cube Root | Reciprocal | No. |
|---|---|---|---|---|---|---|
| 1351 | 1,825,201 | 2,465,846,551 | 36.7560 | 11.0548 | 0.0007402 | 1351 |
| 1352 | 1,827,904 | 2,471,326,208 | 36.7696 | 11.0575 | 0.0007396 | 1352 |
| 1353 | 1,830,609 | 2,476,813,977 | 36.7831 | 11.0603 | 0.0007391 | 1353 |
| 1354 | 1,833,316 | 2,482,309,864 | 36.7967 | 11.0630 | 0.0007386 | 1354 |
| 1355 | 1,836,025 | 2,487,813,875 | 36.8103 | 11.0657 | 0.0007380 | 1355 |
| 1356 | 1,838,736 | 2,493,326,016 | 36.8239 | 11.0684 | 0.0007375 | 1356 |
| 1357 | 1,841,449 | 2,498,846,293 | 36.8375 | 11.0712 | 0.0007369 | 1357 |
| 1358 | 1,844,164 | 2,504,374,712 | 36.8511 | 11.0739 | 0.0007364 | 1358 |
| 1359 | 1,846,881 | 2,509,911,279 | 36.8646 | 11.0766 | 0.0007358 | 1359 |
| 1360 | 1,849,600 | 2,515,456,000 | 36.8782 | 11.0793 | 0.0007353 | 1360 |
| 1361 | 1,852,321 | 2,521,008,881 | 36.8917 | 11.0820 | 0.0007348 | 1361 |
| 1362 | 1,855,044 | 2,526,569,928 | 36.9053 | 11.0847 | 0.0007342 | 1362 |
| 1363 | 1,857,769 | 2,532,139,147 | 36.9188 | 11.0875 | 0.0007337 | 1363 |
| 1364 | 1,860,496 | 2,537,716,544 | 36.9324 | 11.0902 | 0.0007331 | 1364 |
| 1365 | 1,863,225 | 2,543,302,125 | 36.9459 | 11.0929 | 0.0007326 | 1365 |
| 1366 | 1,865,956 | 2,548,895,896 | 36.9594 | 11.0956 | 0.0007321 | 1366 |
| 1367 | 1,868,689 | 2,554,497,863 | 36.9730 | 11.0983 | 0.0007315 | 1367 |
| 1368 | 1,871,424 | 2,560,108,032 | 36.9865 | 11.1010 | 0.0007310 | 1368 |
| 1369 | 1,874,161 | 2,565,726,409 | 37.0000 | 11.1037 | 0.0007305 | 1369 |
| 1370 | 1,876,900 | 2,571,353,000 | 37.0135 | 11.1064 | 0.0007299 | 1370 |
| 1371 | 1,879,641 | 2,576,987,811 | 37.0270 | 11.1091 | 0.0007294 | 1371 |
| 1372 | 1,882,384 | 2,582,630,848 | 37.0405 | 11.1118 | 0.0007289 | 1372 |
| 1373 | 1,885,129 | 2,588,282,117 | 37.0540 | 11.1145 | 0.0007283 | 1373 |
| 1374 | 1,887,876 | 2,593,941,624 | 37.0675 | 11.1172 | 0.0007278 | 1374 |
| 1375 | 1,890,625 | 2,599,609,375 | 37.0810 | 11.1199 | 0.0007273 | 1375 |
| 1376 | 1,893,376 | 2,605,285,376 | 37.0945 | 11.1226 | 0.0007267 | 1376 |
| 1377 | 1,896,129 | 2,610,969,633 | 37.1080 | 11.1253 | 0.0007262 | 1377 |
| 1378 | 1,898,884 | 2,616,662,152 | 37.1214 | 11.1280 | 0.0007257 | 1378 |
| 1379 | 1,901,641 | 2,622,362,939 | 37.1349 | 11.1307 | 0.0007252 | 1379 |
| 1380 | 1,904,400 | 2,628,072,000 | 37.1484 | 11.1334 | 0.0007246 | 1380 |
| 1381 | 1,907,161 | 2,633,789,341 | 37.1618 | 11.1361 | 0.0007241 | 1381 |
| 1382 | 1,909,924 | 2,639,514,968 | 37.1753 | 11.1387 | 0.0007236 | 1382 |
| 1383 | 1,912,689 | 2,645,248,887 | 37.1887 | 11.1414 | 0.0007231 | 1383 |
| 1384 | 1,915,456 | 2,650,991,104 | 37.2022 | 11.1441 | 0.0007225 | 1384 |
| 1385 | 1,918,225 | 2,656,741,625 | 37.2156 | 11.1468 | 0.0007220 | 1385 |
| 1386 | 1,920,996 | 2,662,500,456 | 37.2290 | 11.1495 | 0.0007215 | 1386 |
| 1387 | 1,923,769 | 2,668,267,603 | 37.2424 | 11.1522 | 0.0007210 | 1387 |
| 1388 | 1,926,544 | 2,674,043,072 | 37.2559 | 11.1548 | 0.0007205 | 1388 |
| 1389 | 1,929,321 | 2,679,826,869 | 37.2693 | 11.1575 | 0.0007199 | 1389 |
| 1390 | 1,932,100 | 2,685,619,000 | 37.2827 | 11.1602 | 0.0007194 | 1390 |
| 1391 | 1,934,881 | 2,691,419,471 | 37.2961 | 11.1629 | 0.0007189 | 1391 |
| 1392 | 1,937,664 | 2,697,228,288 | 37.3095 | 11.1655 | 0.0007184 | 1392 |
| 1393 | 1,940,449 | 2,703,045,457 | 37.3229 | 11.1682 | 0.0007179 | 1393 |
| 1394 | 1,943,236 | 2,708,870,984 | 37.3363 | 11.1709 | 0.0007174 | 1394 |
| 1395 | 1,946,025 | 2,714,704,875 | 37.3497 | 11.1736 | 0.0007168 | 1395 |
| 1396 | 1,948,816 | 2,720,547,136 | 37.3631 | 11.1762 | 0.0007163 | 1396 |
| 1397 | 1,951,609 | 2,726,397,773 | 37.3765 | 11.1789 | 0.0007158 | 1397 |
| 1398 | 1,954,404 | 2,732,256,792 | 37.3898 | 11.1816 | 0.0007153 | 1398 |
| 1399 | 1,957,201 | 2,738,124,199 | 37.4032 | 11 1842 | 0.0007148 | 1399 |
| 1400 | 1,960,000 | 2,744,000,000 | 37.4166 | 11.1869 | 0.0007143 | 1400 |

Powers, Roots and Reciprocals

| No. | Square | Cube | Sq. Root | Cube Root | Reciprocal | No. |
|---|---|---|---|---|---|---|
| 1401 | 1,962,801 | 2,749,884,201 | 37.4299 | 11.1896 | 0.0007138 | 1401 |
| 1402 | 1,965,604 | 2,755,776,808 | 37.4433 | 11.1922 | 0.0007133 | 1402 |
| 1403 | 1,968,409 | 2,761,677,827 | 37.4566 | 11.1949 | 0.0007128 | 1403 |
| 1404 | 1,971,216 | 2,767,587,264 | 37.4700 | 11.1975 | 0.0007123 | 1404 |
| 1405 | 1,974,025 | 2,773,505,125 | 37.4833 | 11.2002 | 0.0007117 | 1405 |
| 1406 | 1,976,836 | 2,779,431,416 | 37.4967 | 11.2028 | 0.0007112 | 1406 |
| 1407 | 1,979,649 | 2,785,366,143 | 37.5100 | 11.2055 | 0.0007107 | 1407 |
| 1408 | 1,982,464 | 2,791,309,312 | 37.5233 | 11.2082 | 0.0007102 | 1408 |
| 1409 | 1,985,281 | 2,797,260,929 | 37.5366 | 11.2108 | 0.0007097 | 1409 |
| 1410 | 1,988,100 | 2,803,221,000 | 37.5500 | 11.2135 | 0.0007092 | 1410 |
| 1411 | 1,990,921 | 2,809,189,531 | 37.5633 | 11.2161 | 0.0007087 | 1411 |
| 1412 | 1,993,744 | 2,815,166,528 | 37.5766 | 11.2188 | 0.0007082 | 1412 |
| 1413 | 1,996,569 | 2,821,151,997 | 37.5899 | 11.2214 | 0.0007077 | 1413 |
| 1414 | 1,999,396 | 2,827,145,944 | 37.6032 | 11.2241 | 0.0007072 | 1414 |
| 1415 | 2,002,225 | 2,833,148,375 | 37.6165 | 11.2267 | 0.0007067 | 1415 |
| 1416 | 2,005,056 | 2,839,159,296 | 37.6298 | 11.2293 | 0.0007062 | 1416 |
| 1417 | 2,007,889 | 2,845,178,713 | 37.6431 | 11.2320 | 0.0007057 | 1417 |
| 1418 | 2,010,724 | 2,851,206,632 | 37.6563 | 11.2346 | 0.0007052 | 1418 |
| 1419 | 2,013,561 | 2,857,243,059 | 37.6696 | 11.2373 | 0.0007047 | 1419 |
| 1420 | 2,016,400 | 2,863,288,000 | 37.6829 | 11.2399 | 0.0007042 | 1420 |
| 1421 | 2,019,241 | 2,869,341,461 | 37.6962 | 11.2425 | 0.0007037 | 1421 |
| 1422 | 2,022,084 | 2,875,403,448 | 37.7094 | 11.2452 | 0.0007032 | 1422 |
| 1423 | 2,024,929 | 2,881,473,967 | 37.7227 | 11.2478 | 0.0007027 | 1423 |
| 1424 | 2,027,776 | 2,887,553,024 | 37.7359 | 11.2505 | 0.0007022 | 1424 |
| 1425 | 2,030,625 | 2,893,640,625 | 37.7492 | 11.2531 | 0.0007018 | 1425 |
| 1426 | 2,033,476 | 2,899,736,776 | 37.7624 | 11.2557 | 0.0007013 | 1426 |
| 1427 | 2,036,329 | 2,905,841,483 | 37.7757 | 11.2583 | 0.0007008 | 1427 |
| 1428 | 2,039,184 | 2,911,954,752 | 37.7889 | 11.2610 | 0.0007003 | 1428 |
| 1429 | 2,042,041 | 2,918,076,589 | 37.8021 | 11.2636 | 0.0006998 | 1429 |
| 1430 | 2,044,900 | 2,924,207,000 | 37.8153 | 11.2662 | 0.0006993 | 1430 |
| 1431 | 2,047,761 | 2,930,345,991 | 37.8286 | 11.2689 | 0.0006988 | 1431 |
| 1432 | 2,050,624 | 2,936,493,568 | 37.8418 | 11.2715 | 0.0006983 | 1432 |
| 1433 | 2,053,489 | 2,942,649,737 | 37.8550 | 11.2741 | 0.0006978 | 1433 |
| 1434 | 2,056,356 | 2,948,814,504 | 37.8682 | 11.2767 | 0.0006974 | 1434 |
| 1435 | 2,059,225 | 2,954,987,875 | 37.8814 | 11.2793 | 0.0006969 | 1435 |
| 1436 | 2,062,096 | 2,961,169,856 | 37.8946 | 11.2820 | 0.0006964 | 1436 |
| 1437 | 2,064,969 | 2,967,360,453 | 37.9078 | 11.2846 | 0.0006959 | 1437 |
| 1438 | 2,067,844 | 2,973,559,672 | 37.9210 | 11.2872 | 0.0006954 | 1438 |
| 1439 | 2,070,721 | 2,979,767,519 | 37.9342 | 11.2898 | 0.0006949 | 1439 |
| 1440 | 2,073,600 | 2,985,984,000 | 37.9473 | 11.2924 | 0.0006944 | 1440 |
| 1441 | 2,076,481 | 2,992,209,121 | 37.9605 | 11.2950 | 0.0006940 | 1441 |
| 1442 | 2,079,364 | 2,998,442,888 | 37.9737 | 11.2977 | 0.0006935 | 1442 |
| 1443 | 2,082,249 | 3,004,685,307 | 37.9868 | 11.3003 | 0.0006930 | 1443 |
| 1444 | 2,085,136 | 3,010,936,384 | 38.0000 | 11.3029 | 0.0006925 | 1444 |
| 1445 | 2,088,025 | 3,017,196,125 | 38.0132 | 11.3055 | 0.0006920 | 1445 |
| 1446 | 2,090,916 | 3,023,464,536 | 38.0263 | 11.3081 | 0.0006916 | 1446 |
| 1447 | 2,093,809 | 3,029,741,623 | 38.0395 | 11.3107 | 0.0006911 | 1447 |
| 1448 | 2,096,704 | 3,036,027,392 | 38.0526 | 11.3133 | 0.0006906 | 1448 |
| 1449 | 2,099,601 | 3,042,321,849 | 38.0657 | 11.3159 | 0.0006901 | 1449 |
| 1450 | 2,102,500 | 3,048,625,000 | 38.0789 | 11.3185 | 0.0006897 | 1450 |

## Powers, Roots and Reciprocals

| No. | Square | Cube | Sq. Root | Cube Root | Reciprocal | No. |
|---|---|---|---|---|---|---|
| 1451 | 2,105,401 | 3,054,936,851 | 38.0920 | 11.3211 | 0.0006892 | 1451 |
| 1452 | 2,108,304 | 3,061,257,408 | 38.1051 | 11.3237 | 0.0006887 | 1452 |
| 1453 | 2,111,209 | 3,067,586,677 | 38.1182 | 11.3263 | 0.0006882 | 1453 |
| 1454 | 2,114,116 | 3,073,924,664 | 38.1314 | 11.3289 | 0.0006878 | 1454 |
| 1455 | 2,117,025 | 3,080,271,375 | 38.1445 | 11.3315 | 0.0006873 | 1455 |
| 1456 | 2,119,936 | 3,086,626,816 | 38.1576 | 11.3341 | 0.0006868 | 1456 |
| 1457 | 2,122,849 | 3,092,990,993 | 38.1707 | 11.3367 | 0.0006863 | 1457 |
| 1458 | 2,125,764 | 3,099,363,912 | 38.1838 | 11.3393 | 0.0006859 | 1458 |
| 1459 | 2,128,681 | 3,105,745,579 | 38.1969 | 11.3419 | 0.0006854 | 1459 |
| 1460 | 2,131,600 | 3,112,136,000 | 38.2099 | 11.3445 | 0.0006849 | 1460 |
| 1461 | 2,134,521 | 3,118,535,181 | 38.2230 | 11.3471 | 0.0006845 | 1461 |
| 1462 | 2,137,444 | 3,124,943,128 | 38.2361 | 11.3496 | 0.0006840 | 1462 |
| 1463 | 2,140,369 | 3,131,359,847 | 38.2492 | 11.3522 | 0.0006835 | 1463 |
| 1464 | 2,143,296 | 3,137,785,344 | 38.2623 | 11.3548 | 0.0006831 | 1464 |
| 1465 | 2,146,225 | 3,144,219,625 | 38.2753 | 11.3574 | 0.0006826 | 1465 |
| 1466 | 2,149,156 | 3,150,662,696 | 38.2884 | 11.3600 | 0.0006821 | 1466 |
| 1467 | 2,152,089 | 3,157,114,563 | 38.3014 | 11.3626 | 0.0006817 | 1467 |
| 1468 | 2,155,024 | 3,163,575,232 | 38.3145 | 11.3652 | 0.0006812 | 1468 |
| 1469 | 2,157,961 | 3,170,044,709 | 38.3275 | 11.3677 | 0.0006807 | 1469 |
| 1470 | 2,160,900 | 3,176,523,000 | 38.3406 | 11.3703 | 0.0006803 | 1470 |
| 1471 | 2,163,841 | 3,183,010,111 | 38.3536 | 11.3729 | 0.0006798 | 1471 |
| 1472 | 2,166,784 | 3,189,506,048 | 38.3667 | 11.3755 | 0.0006793 | 1472 |
| 1473 | 2,169,729 | 3,196,010,817 | 38.3797 | 11.3780 | 0.0006789 | 1473 |
| 1474 | 2,172,676 | 3,202,524,424 | 38.3927 | 11.3806 | 0.0006784 | 1474 |
| 1475 | 2,175,625 | 3,209,046,875 | 38.4057 | 11.3832 | 0.0006780 | 1475 |
| 1476 | 2,178,576 | 3,215,578,176 | 38.4187 | 11.3858 | 0.0006775 | 1476 |
| 1477 | 2,181,529 | 3,222,118,333 | 38.4318 | 11.3883 | 0.0006770 | 1477 |
| 1478 | 2,184,484 | 3,228,667,352 | 38.4448 | 11.3909 | 0.0006766 | 1478 |
| 1479 | 2,187,441 | 3,235,225,239 | 38.4578 | 11.3935 | 0.0006761 | 1479 |
| 1480 | 2,190,400 | 3,241,792,000 | 38.4708 | 11.3960 | 0.0006757 | 1480 |
| 1481 | 2,193,361 | 3,248,367,641 | 38.4838 | 11.3986 | 0.0006752 | 1481 |
| 1482 | 2,196,324 | 3,254,952,168 | 38.4968 | 11.4012 | 0.0006748 | 1482 |
| 1483 | 2,199,289 | 3,261,545,587 | 38.5097 | 11.4037 | 0.0006743 | 1483 |
| 1484 | 2,202,256 | 3,268,147,904 | 38.5227 | 11.4063 | 0.0006739 | 1484 |
| 1485 | 2,205,225 | 3,274,759,125 | 38.5357 | 11.4089 | 0.0006734 | 1485 |
| 1486 | 2,208,196 | 3,281,379,256 | 38.5487 | 11.4114 | 0.0006729 | 1486 |
| 1487 | 2,211,169 | 3,288,008,303 | 38.5616 | 11.4140 | 0.0006725 | 1487 |
| 1488 | 2,214,144 | 3,294,646,272 | 38.5746 | 11.4165 | 0.0006720 | 1488 |
| 1489 | 2,217,121 | 3,301,293,169 | 38.5876 | 11.4191 | 0.0006716 | 1489 |
| 1490 | 2,220,100 | 3,307,949,000 | 38.6005 | 11.4216 | 0.0006711 | 1490 |
| 1491 | 2,223,081 | 3,314,613,771 | 38.6135 | 11.4242 | 0.0006707 | 1491 |
| 1492 | 2,226,064 | 3,321,287,488 | 38.6264 | 11.4268 | 0.0006702 | 1492 |
| 1493 | 2,229,049 | 3,327,970,157 | 38.6394 | 11.4293 | 0.0006698 | 1493 |
| 1494 | 2,232,036 | 3,334,661,784 | 38.6523 | 11.4319 | 0.0006693 | 1494 |
| 1495 | 2,235,025 | 3,341,362,375 | 38.6652 | 11.4344 | 0.0006689 | 1495 |
| 1496 | 2,238,016 | 3,348,071,936 | 38.6782 | 11.4370 | 0.0006684 | 1496 |
| 1497 | 2,241,009 | 3,354,790,473 | 38.6911 | 11.4395 | 0.0006680 | 1497 |
| 1498 | 2,244,004 | 3,361,517,992 | 38.7040 | 11.4421 | 0.0006676 | 1498 |
| 1499 | 2,247,001 | 3,368,254,499 | 38.7169 | 11.4446 | 0.0006671 | 1499 |
| 1500 | 2,250,000 | 3,375,000,000 | 38.7298 | 11.4471 | 0.0006667 | 1500 |

Powers, Roots and Reciprocals

| No. | Square | Cube | Sq. Root | Cube Root | Reciprocal | No. |
|---|---|---|---|---|---|---|
| 1501 | 2,253,001 | 3,381,754,501 | 38.7427 | 11.4497 | 0.0006662 | 1501 |
| 1502 | 2,256,004 | 3,388,518,008 | 38.7556 | 11.4522 | 0.0006658 | 1502 |
| 1503 | 2,259,009 | 3,395,290,527 | 38.7685 | 11.4548 | 0.0006653 | 1503 |
| 1504 | 2,262,016 | 3,402,072,064 | 38.7814 | 11.4573 | 0.0006649 | 1504 |
| 1505 | 2,265,025 | 3,408,862,625 | 38.7943 | 11.4598 | 0.0006645 | 1505 |
| 1506 | 2,268,036 | 3,415,662,216 | 38.8072 | 11.4624 | 0.0006640 | 1506 |
| 1507 | 2,271,049 | 3,422,470,843 | 38.8201 | 11.4649 | 0.0006636 | 1507 |
| 1508 | 2,274,064 | 3,429,288,512 | 38.8330 | 11.4675 | 0.0006631 | 1508 |
| 1509 | 2,277,081 | 3,436,115,229 | 38.8458 | 11.4700 | 0.0006627 | 1509 |
| 1510 | 2,280,100 | 3,442,951,000 | 38.8587 | 11.4725 | 0.0006623 | 1510 |
| 1511 | 2,283,121 | 3,449,795,831 | 38.8716 | 11.4751 | 0.0006618 | 1511 |
| 1512 | 2,286,144 | 3,456,649,728 | 38.8844 | 11.4776 | 0.0006614 | 1512 |
| 1513 | 2,289,169 | 3,463,512,697 | 38.8973 | 11.4801 | 0.0006609 | 1513 |
| 1514 | 2,292,196 | 3,470,384,744 | 38.9102 | 11.4826 | 0.0006605 | 1514 |
| 1515 | 2,295,225 | 3,477,265,875 | 38.9230 | 11.4852 | 0.0006601 | 1515 |
| 1516 | 2,298,256 | 3,484,156,096 | 38.9358 | 11.4877 | 0.0006596 | 1516 |
| 1517 | 2,301,289 | 3,491,055,413 | 38.9487 | 11.4902 | 0.0006592 | 1517 |
| 1518 | 2,304,324 | 3,497,963,832 | 38.9615 | 11.4927 | 0.0006588 | 1518 |
| 1519 | 2,307,361 | 3,504,881,359 | 38.9744 | 11.4953 | 0.0006583 | 1519 |
| 1520 | 2,310,400 | 3,511,808,000 | 38.9872 | 11.4978 | 0.0006579 | 1520 |
| 1521 | 2,313,441 | 3,518,743,761 | 39.0000 | 11.5003 | 0.0006575 | 1521 |
| 1522 | 2,316,484 | 3,525,688,648 | 39.0128 | 11.5028 | 0.0006570 | 1522 |
| 1523 | 2,319,529 | 3,532,642,667 | 39.0256 | 11.5054 | 0.0006566 | 1523 |
| 1524 | 2,322,576 | 3,539,605,824 | 39.0384 | 11.5079 | 0.0006562 | 1524 |
| 1525 | 2,325,625 | 3,546,578,125 | 39.0512 | 11.5104 | 0.0006557 | 1525 |
| 1526 | 2,328,676 | 3,553,559,576 | 39.0640 | 11.5129 | 0.0006553 | 1526 |
| 1527 | 2,331,729 | 3,560,550,183 | 39.0768 | 11.5154 | 0.0006549 | 1527 |
| 1528 | 2,334,784 | 3,567,549,952 | 39.0896 | 11.5179 | 0.0006545 | 1528 |
| 1529 | 2,337,841 | 3,574,558,889 | 39.1024 | 11.5204 | 0.0006540 | 1529 |
| 1530 | 2,340,900 | 3,581,577,000 | 39.1152 | 11.5230 | 0.0006536 | 1530 |
| 1531 | 2,343,961 | 3,588,604,291 | 39.1280 | 11.5255 | 0.0006532 | 1531 |
| 1532 | 2,347,024 | 3,595,640,768 | 39.1408 | 11.5280 | 0.0006527 | 1532 |
| 1533 | 2,350,089 | 3,602,686,437 | 39.1535 | 11.5305 | 0.0006523 | 1533 |
| 1534 | 2,353,156 | 3,609,741,304 | 39.1663 | 11.5330 | 0.0006519 | 1534 |
| 1535 | 2,356,225 | 3,616,805,375 | 39.1791 | 11.5355 | 0.0006515 | 1535 |
| 1536 | 2,359,296 | 3,623,878,656 | 39.1918 | 11.5380 | 0.0006510 | 1536 |
| 1537 | 2,362,369 | 3,630,961,153 | 39.2046 | 11.5405 | 0.0006506 | 1537 |
| 1538 | 2,365,444 | 3,638,052,872 | 39.2173 | 11.5430 | 0.0006502 | 1538 |
| 1539 | 2,368,521 | 3,645,153,819 | 39.2301 | 11.5455 | 0.0006498 | 1539 |
| 1540 | 2,371,600 | 3,652,264,000 | 39.2428 | 11.5480 | 0.0006494 | 1540 |
| 1541 | 2,374,681 | 3,659,383,421 | 39.2556 | 11.5505 | 0.0006489 | 1541 |
| 1542 | 2,377,764 | 3,666,512,088 | 39.2683 | 11.5530 | 0.0006485 | 1542 |
| 1543 | 2,380,849 | 3,673,650,007 | 39.2810 | 11.5555 | 0.0006481 | 1543 |
| 1544 | 2,383,936 | 3,680,797,184 | 39.2938 | 11.5580 | 0.0006477 | 1544 |
| 1545 | 2,387,025 | 3,687,953,625 | 39.3065 | 11.5605 | 0.0006472 | 1545 |
| 1546 | 2,390,116 | 3,695,119,336 | 39.3192 | 11.5630 | 0.0006468 | 1546 |
| 1547 | 2,393,209 | 3,702,294,323 | 39.3319 | 11.5655 | 0.0006464 | 1547 |
| 1548 | 2,396,304 | 3,709,478,592 | 39.3446 | 11.5680 | 0.0006460 | 1548 |
| 1549 | 2,399,401 | 3,716,672,149 | 39.3573 | 11.5705 | 0.0006456 | 1549 |
| 1550 | 2,402,500 | 3,723,875,000 | 39.3700 | 11.5729 | 0.0006452 | 1550 |

## Powers, Roots and Reciprocals

| No. | Square | Cube | Sq. Root | Cube Root | Reciprocal | No. |
|---|---|---|---|---|---|---|
| 1551 | 2,405,601 | 3,731,087,151 | 39.3827 | 11.5754 | 0.0006447 | 1551 |
| 1552 | 2,408,704 | 3,738,308,608 | 39.3954 | 11.5779 | 0.0006443 | 1552 |
| 1553 | 2,411,809 | 3,745,539,377 | 39.4081 | 11.5804 | 0.0006439 | 1553 |
| 1554 | 2,414,916 | 3,752,779,464 | 39.4208 | 11.5829 | 0.0006435 | 1554 |
| 1555 | 2,418,025 | 3,760,028,875 | 39.4335 | 11.5854 | 0.0006431 | 1555 |
| 1556 | 2,421,136 | 3,767,287,616 | 39.4462 | 11.5879 | 0.0006427 | 1556 |
| 1557 | 2,424,249 | 3,774,555,693 | 39.4588 | 11.5903 | 0.0006423 | 1557 |
| 1558 | 2,427,364 | 3,781,833,112 | 39.4715 | 11.5928 | 0.0006418 | 1558 |
| 1559 | 2,430,481 | 3,789,119,879 | 39.4842 | 11.5953 | 0.0006414 | 1559 |
| 1560 | 2,433,600 | 3,796,416,000 | 39.4968 | 11.5978 | 0.0006410 | 1560 |
| 1561 | 2,436,721 | 3,803,721,481 | 39.5095 | 11.6003 | 0.0006406 | 1561 |
| 1562 | 2,439,844 | 3,811,036,328 | 39.5221 | 11.6027 | 0.0006402 | 1562 |
| 1563 | 2,442,969 | 3,818,360,547 | 39.5348 | 11.6052 | 0.0006398 | 1563 |
| 1564 | 2,446,096 | 3,825,694,144 | 39.5474 | 11.6077 | 0.0006394 | 1564 |
| 1565 | 2,449,225 | 3,833,037,125 | 39.5601 | 11.6102 | 0.0006390 | 1565 |
| 1566 | 2,452,356 | 3,840,389,496 | 39.5727 | 11.6126 | 0.0006386 | 1566 |
| 1567 | 2,455,489 | 3,847,751,263 | 39.5854 | 11.6151 | 0.0006382 | 1567 |
| 1568 | 2,458,624 | 3,855,122,432 | 39.5980 | 11.6176 | 0.0006378 | 1568 |
| 1569 | 2,461,761 | 3,862,503,009 | 39.6106 | 11.6200 | 0.0006373 | 1569 |
| 1570 | 2,464,900 | 3,869,893,000 | 39.6232 | 11.6225 | 0.0006369 | 1570 |
| 1571 | 2,468,041 | 3,877,292,411 | 39.6358 | 11.6250 | 0.0006365 | 1571 |
| 1572 | 2,471,184 | 3,884,701,248 | 39.6485 | 11.6274 | 0.0006361 | 1572 |
| 1573 | 2,474,329 | 3,892,119,517 | 39.6611 | 11.6299 | 0.0006357 | 1573 |
| 1574 | 2,477,476 | 3,899,547,224 | 39.6737 | 11.6324 | 0.0006353 | 1574 |
| 1575 | 2,480,625 | 3,906,984,375 | 39.6863 | 11.6348 | 0.0006349 | 1575 |
| 1576 | 2,483,776 | 3,914,430,976 | 39.6989 | 11.6373 | 0.0006345 | 1576 |
| 1577 | 2,486,929 | 3,921,887,033 | 39.7115 | 11.6398 | 0.0006341 | 1577 |
| 1578 | 2,490,084 | 3,929,352,552 | 39.7240 | 11.6422 | 0.0006337 | 1578 |
| 1579 | 2,493,241 | 3,936,827,539 | 39.7366 | 11.6447 | 0.0006333 | 1579 |
| 1580 | 2,496,400 | 3,944,312,000 | 39.7492 | 11.6471 | 0.0006329 | 1580 |
| 1581 | 2,499,561 | 3,951,805,941 | 39.7618 | 11.6496 | 0.0006325 | 1581 |
| 1582 | 2,502,724 | 3,959,309,368 | 39.7744 | 11.6520 | 0.0006321 | 1582 |
| 1583 | 2,505,889 | 3,966,822,287 | 39.7869 | 11.6545 | 0.0006317 | 1583 |
| 1584 | 2,509,056 | 3,974,344,704 | 39.7995 | 11.6570 | 0.0006313 | 1584 |
| 1585 | 2,512,225 | 3,981,876,625 | 39.8121 | 11.6594 | 0.0006309 | 1585 |
| 1586 | 2,515,396 | 3,989,418,056 | 39.8246 | 11.6619 | 0.0006305 | 1586 |
| 1587 | 2,518,569 | 3,996,969,003 | 39.8372 | 11.6643 | 0.0006301 | 1587 |
| 1588 | 2,521,744 | 4,004,529,472 | 39.8497 | 11.6668 | 0.0006297 | 1588 |
| 1589 | 2,524,921 | 4,012,099,469 | 39.8623 | 11.6692 | 0.0006293 | 1589 |
| 1590 | 2,528,100 | 4,019,679,000 | 39.8748 | 11.6717 | 0.0006289 | 1590 |
| 1591 | 2,531,281 | 4,027,268,071 | 39.8873 | 11.6741 | 0.0006285 | 1591 |
| 1592 | 2,534,464 | 4,034,866,688 | 39.8999 | 11.6765 | 0.0006281 | 1592 |
| 1593 | 2,537,649 | 4,042,474,857 | 39.9124 | 11.6790 | 0.0006277 | 1593 |
| 1594 | 2,540,836 | 4,050,092,584 | 39.9249 | 11.6814 | 0.0006274 | 1594 |
| 1595 | 2,544,025 | 4,057,719,875 | 39.9375 | 11.6839 | 0.0006270 | 1595 |
| 1596 | 2,547,216 | 4,065,356,736 | 39.9500 | 11.6863 | 0.0006266 | 1596 |
| 1597 | 2,550,409 | 4,073,003,173 | 39.9625 | 11.6888 | 0.0006262 | 1597 |
| 1598 | 2,553,604 | 4,080,659,192 | 39.9750 | 11.6912 | 0.0006258 | 1598 |
| 1599 | 2,556,801 | 4,088,324,799 | 39.9875 | 11.6936 | 0.0006254 | 1599 |
| 1600 | 2,560,000 | 4,096,000,000 | 40.0000 | 11.6961 | 0.0006250 | 1600 |

MATHEMATICAL TABLES

## Powers, Roots and Reciprocals

| No. | Square | Cube | Sq. Root | Cube Root | Reciprocal | No. |
|---|---|---|---|---|---|---|
| 1601 | 2,563,201 | 4,103,684,801 | 40.0125 | 11.6985 | 0.0006246 | 1601 |
| 1602 | 2,566,404 | 4,111,379,208 | 40.0250 | 11.7009 | 0.0006242 | 1602 |
| 1603 | 2,569,609 | 4,119,083,227 | 40.0375 | 11.7034 | 0.0006238 | 1603 |
| 1604 | 2,572,816 | 4,126,796,864 | 40.0500 | 11.7058 | 0.0006234 | 1604 |
| 1605 | 2,576,025 | 4,134,520,125 | 40.0625 | 11.7082 | 0.0006231 | 1605 |
| 1606 | 2,579,236 | 4,142,253,016 | 40.0749 | 11.7107 | 0.0006227 | 1606 |
| 1607 | 2,582,449 | 4,149,995,543 | 40.0874 | 11.7131 | 0.0006223 | 1607 |
| 1608 | 2,585,664 | 4,157,747,712 | 40.0999 | 11.7155 | 0.0006219 | 1608 |
| 1609 | 2,588,881 | 4,165,509,529 | 40.1123 | 11.7180 | 0.0006215 | 1609 |
| 1610 | 2,592,100 | 4,173,281,000 | 40.1248 | 11.7204 | 0.0006211 | 1610 |
| 1611 | 2,595,321 | 4,181,062,131 | 40.1373 | 11.7228 | 0.0006207 | 1611 |
| 1612 | 2,598,544 | 4,188,852,928 | 40.1497 | 11.7252 | 0.0006203 | 1612 |
| 1613 | 2,601,769 | 4,196,653,397 | 40.1622 | 11.7277 | 0.0006200 | 1613 |
| 1614 | 2,604,996 | 4,204,463,544 | 40.1746 | 11.7301 | 0.0006196 | 1614 |
| 1615 | 2,608,225 | 4,212,283,375 | 40.1871 | 11.7325 | 0.0006192 | 1615 |
| 1616 | 2,611,456 | 4,220,112,896 | 40.1995 | 11.7349 | 0.0006188 | 1616 |
| 1617 | 2,614,689 | 4,227,952,113 | 40.2119 | 11.7373 | 0.0006184 | 1617 |
| 1618 | 2,617,924 | 4,235,801,032 | 40.2244 | 11.7398 | 0.0006180 | 1618 |
| 1619 | 2,621,161 | 4,243,659,659 | 40.2368 | 11.7422 | 0.0006177 | 1619 |
| 1620 | 2,624,400 | 4,251,528,000 | 40.2492 | 11.7446 | 0.0006173 | 1620 |
| 1621 | 2,627,641 | 4,259,406,061 | 40.2616 | 11.7470 | 0.0006169 | 1621 |
| 1622 | 2,630,884 | 4,267,293,848 | 40.2741 | 11.7494 | 0.0006165 | 1622 |
| 1623 | 2,634,129 | 4,275,191,367 | 40.2865 | 11.7518 | 0.0006161 | 1623 |
| 1624 | 2,637,376 | 4,283,098,624 | 40.2989 | 11.7543 | 0.0006158 | 1624 |
| 1625 | 2,640,625 | 4,291,015,625 | 40.3113 | 11.7567 | 0.0006154 | 1625 |
| 1626 | 2,643,876 | 4,298,942,376 | 40.3237 | 11.7591 | 0.0006150 | 1626 |
| 1627 | 2,647,129 | 4,306,878,883 | 40.3361 | 11.7615 | 0.0006146 | 1627 |
| 1628 | 2,650,384 | 4,314,825,152 | 40.3485 | 11.7639 | 0.0006143 | 1628 |
| 1629 | 2,653,641 | 4,322,781,189 | 40.3609 | 11.7663 | 0.0006139 | 1629 |
| 1630 | 2,656,900 | 4,330,747,000 | 40.3733 | 11.7687 | 0.0006135 | 1630 |
| 1631 | 2,660,161 | 4,338,722,591 | 40.3856 | 11.7711 | 0.0006131 | 1631 |
| 1632 | 2,663,424 | 4,346,707,968 | 40.3980 | 11.7735 | 0.0006127 | 1632 |
| 1633 | 2,666,689 | 4,354,703,137 | 40.4104 | 11.7759 | 0.0006124 | 1633 |
| 1634 | 2,669,956 | 4,362,708,104 | 40.4228 | 11.7783 | 0.0006120 | 1634 |
| 1635 | 2,673,225 | 4,370,722,875 | 40.4351 | 11.7807 | 0.0006116 | 1635 |
| 1636 | 2,676,496 | 4,378,747,456 | 40.4475 | 11.7831 | 0.0006112 | 1636 |
| 1637 | 2,679,769 | 4,386,781,853 | 40.4599 | 11.7855 | 0.0006109 | 1637 |
| 1638 | 2,683,044 | 4,394,826,072 | 40.4722 | 11.7879 | 0.0006105 | 1638 |
| 1639 | 2,686,321 | 4,402,880,119 | 40.4846 | 11.7903 | 0.0006101 | 1639 |
| 1640 | 2,689,600 | 4,410,944,000 | 40.4969 | 11.7927 | 0.0006098 | 1640 |
| 1641 | 2,692,881 | 4,419,017,721 | 40.5093 | 11.7951 | 0.0006094 | 1641 |
| 1642 | 2,696,164 | 4,427,101,288 | 40.5216 | 11.7975 | 0.0006090 | 1642 |
| 1643 | 2,699,449 | 4,435,194,707 | 40.5339 | 11.7999 | 0.0006086 | 1643 |
| 1644 | 2,702,736 | 4,443,297,984 | 40.5463 | 11.8023 | 0.0006083 | 1644 |
| 1645 | 2,706,025 | 4,451,411,125 | 40.5586 | 11.8047 | 0.0006079 | 1645 |
| 1646 | 2,709,316 | 4,459,534,136 | 40.5709 | 11.8071 | 0.0006075 | 1646 |
| 1647 | 2,712,609 | 4,467,667,023 | 40.5832 | 11.8095 | 0.0006072 | 1647 |
| 1648 | 2,715,904 | 4,475,809,792 | 40.5956 | 11.8119 | 0.0006068 | 1648 |
| 1649 | 2,719,201 | 4,483,962,449 | 40.6079 | 11.8143 | 0.0006064 | 1649 |
| 1650 | 2,722,500 | 4,492,125,000 | 40.6202 | 11.8167 | 0.0006061 | 1650 |

## Powers, Roots and Reciprocals

| No. | Square | Cube | Sq. Root | Cube Root | Reciprocal | No. |
|-----|--------|------|----------|-----------|------------|-----|
| 1651 | 2,725,801 | 4,500,297,451 | 40.6325 | 11.8190 | 0.0006057 | 1651 |
| 1652 | 2,729,104 | 4,508,479,808 | 40.6448 | 11.8214 | 0.0006053 | 1652 |
| 1653 | 2,732,409 | 4,516,672,077 | 40.6571 | 11.8238 | 0.0006050 | 1653 |
| 1654 | 2,735,716 | 4,524,874,264 | 40.6694 | 11.8262 | 0.0006046 | 1654 |
| 1655 | 2,739,025 | 4,533,086,375 | 40.6817 | 11.8286 | 0.0006042 | 1655 |
| 1656 | 2,742,336 | 4,541,308,416 | 40.6940 | 11.8310 | 0.0006039 | 1656 |
| 1657 | 2,745,649 | 4,549,540,393 | 40.7063 | 11.8333 | 0.0006035 | 1657 |
| 1658 | 2,748,964 | 4,557,782,312 | 40.7185 | 11.8357 | 0.0006031 | 1658 |
| 1659 | 2,752,281 | 4,566,034,179 | 40.7308 | 11.8381 | 0.0006028 | 1659 |
| 1660 | 2,755,600 | 4,574,296,000 | 40.7431 | 11.8405 | 0.0006024 | 1660 |
| 1661 | 2,758,921 | 4,582,567,781 | 40.7554 | 11.8429 | 0.0006020 | 1661 |
| 1662 | 2,762,244 | 4,590,849,528 | 40.7676 | 11.8452 | 0.0006017 | 1662 |
| 1663 | 2,765,569 | 4,599,141,247 | 40.7799 | 11.8476 | 0.0006013 | 1663 |
| 1664 | 2,768,896 | 4,607,442,944 | 40.7922 | 11.8500 | 0.0006010 | 1664 |
| 1665 | 2,772,225 | 4,615,754,625 | 40.8044 | 11.8524 | 0.0006006 | 1665 |
| 1666 | 2,775,556 | 4,624,076,296 | 40.8167 | 11.8547 | 0.0006002 | 1666 |
| 1667 | 2,778,889 | 4,632,407,963 | 40.8289 | 11.8571 | 0.0005999 | 1667 |
| 1668 | 2,782,224 | 4,640,749,632 | 40.8412 | 11.8595 | 0.0005995 | 1668 |
| 1669 | 2,785,561 | 4,649,101,309 | 40.8534 | 11.8618 | 0.0005992 | 1669 |
| 1670 | 2,788,900 | 4,657,463,000 | 40.8656 | 11.8642 | 0.0005988 | 1670 |
| 1671 | 2,792,241 | 4,665,834,711 | 40.8779 | 11.8666 | 0.0005984 | 1671 |
| 1672 | 2,795,584 | 4,674,216,448 | 40.8901 | 11.8689 | 0.0005981 | 1672 |
| 1673 | 2,798,929 | 4,682,608,217 | 40.9023 | 11.8713 | 0.0005977 | 1673 |
| 1674 | 2,802,276 | 4,691,010,024 | 40.9145 | 11.8737 | 0.0005974 | 1674 |
| 1675 | 2,805,625 | 4,699,421,875 | 40.9268 | 11.8760 | 0.0005970 | 1675 |
| 1676 | 2,808,976 | 4,707,843,776 | 40.9390 | 11.8784 | 0.0005967 | 1676 |
| 1677 | 2,812,329 | 4,716,275,733 | 40.9512 | 11.8808 | 0.0005963 | 1677 |
| 1678 | 2,815,684 | 4,724,717,752 | 40.9634 | 11.8831 | 0.0005959 | 1678 |
| 1679 | 2,819,041 | 4,733,169,839 | 40.9756 | 11.8855 | 0.0005956 | 1679 |
| 1680 | 2,822,400 | 4,741,632,000 | 40.9878 | 11.8878 | 0.0005952 | 1680 |
| 1681 | 2,825,761 | 4,750,104,241 | 41.0000 | 11.8902 | 0.0005949 | 1681 |
| 1682 | 2,829,124 | 4,758,586,568 | 41.0122 | 11.8926 | 0.0005945 | 1682 |
| 1683 | 2,832,489 | 4,767,078,987 | 41.0244 | 11.8949 | 0.0005942 | 1683 |
| 1684 | 2,835,856 | 4,775,581,504 | 41.0366 | 11.8973 | 0.0005938 | 1684 |
| 1685 | 2,839,225 | 4,784,094,125 | 41.0488 | 11.8996 | 0.0005935 | 1685 |
| 1686 | 2,842,596 | 4,792,616,856 | 41.0609 | 11.9020 | 0.0005931 | 1686 |
| 1687 | 2,845,969 | 4,801,149,703 | 41.0731 | 11.9043 | 0.0005928 | 1687 |
| 1688 | 2,849,344 | 4,809,692,672 | 41.0853 | 11.9067 | 0.0005924 | 1688 |
| 1689 | 2,852,721 | 4,818,245,769 | 41.0974 | 11.9090 | 0.0005921 | 1689 |
| 1690 | 2,856,100 | 4,826,809,000 | 41.1096 | 11.9114 | 0.0005917 | 1690 |
| 1691 | 2,859,481 | 4,835,382,371 | 41.1218 | 11.9137 | 0.0005914 | 1691 |
| 1692 | 2,862,864 | 4,843,965,888 | 41.1339 | 11.9161 | 0.0005910 | 1692 |
| 1693 | 2,866,249 | 4,852,559,557 | 41.1461 | 11.9184 | 0.0005907 | 1693 |
| 1694 | 2,869,636 | 4,861,163,384 | 41.1582 | 11.9208 | 0.0005903 | 1694 |
| 1695 | 2,873,025 | 4,869,777,375 | 41.1704 | 11.9231 | 0.0005900 | 1695 |
| 1696 | 2,876,416 | 4,878,401,536 | 41.1825 | 11.9255 | 0.0005896 | 1696 |
| 1697 | 2,879,809 | 4,887,035,873 | 41.1947 | 11.9278 | 0.0005893 | 1697 |
| 1698 | 2,883,204 | 4,895,680,392 | 41.2068 | 11.9301 | 0.0005889 | 1698 |
| 1699 | 2,886,601 | 4,904,335,099 | 41.2189 | 11.9325 | 0.0005886 | 1699 |
| 1700 | 2,890,000 | 4,913,000,000 | 41.2311 | 11.9348 | 0.0005882 | 1700 |

## Powers, Roots and Reciprocals

| No. | Square | Cube | Sq. Root | Cube Root | Reciprocal | No. |
|---|---|---|---|---|---|---|
| 1701 | 2,893,401 | 4,921,675,101 | 41.2432 | 11.9372 | 0.0005879 | 1701 |
| 1702 | 2,896,804 | 4,930,360,408 | 41.2553 | 11.9395 | 0.0005875 | 1702 |
| 1703 | 2,900,209 | 4,939,055,927 | 41.2674 | 11.9418 | 0.0005872 | 1703 |
| 1704 | 2,903,616 | 4,947,761,664 | 41.2795 | 11.9442 | 0.0005869 | 1704 |
| 1705 | 2,907,025 | 4,956,477,625 | 41.2916 | 11.9465 | 0.0005865 | 1705 |
| 1706 | 2,910,436 | 4,965,203,816 | 41.3038 | 11.9489 | 0.0005862 | 1706 |
| 1707 | 2,913,849 | 4,973,940,243 | 41.3159 | 11.9512 | 0.0005858 | 1707 |
| 1708 | 2,917,264 | 4,982,686,912 | 41.3280 | 11.9535 | 0.0005855 | 1708 |
| 1709 | 2,920,681 | 4,991,443,829 | 41.3401 | 11.9559 | 0.0005851 | 1709 |
| 1710 | 2,924,100 | 5,000,211,000 | 41.3521 | 11.9582 | 0.0005848 | 1710 |
| 1711 | 2,927,521 | 5,008,988,431 | 41.3642 | 11.9605 | 0.0005845 | 1711 |
| 1712 | 2,930,944 | 5,017,776,128 | 41.3763 | 11.9628 | 0.0005841 | 1712 |
| 1713 | 2,934,369 | 5,026,574,097 | 41.3884 | 11.9652 | 0.0005838 | 1713 |
| 1714 | 2,937,796 | 5,035,382,344 | 41.4005 | 11.9675 | 0.0005834 | 1714 |
| 1715 | 2,941,225 | 5,044,200,875 | 41.4126 | 11.9698 | 0.0005831 | 1715 |
| 1716 | 2,944,656 | 5,053,029,696 | 41.4246 | 11.9722 | 0.0005828 | 1716 |
| 1717 | 2,948,089 | 5,061,868,813 | 41.4367 | 11.9745 | 0.0005824 | 1717 |
| 1718 | 2,951,524 | 5,070,718,232 | 41.4488 | 11.9768 | 0.0005821 | 1718 |
| 1719 | 2,954,961 | 5,079,577,959 | 41.4608 | 11.9791 | 0.0005817 | 1719 |
| 1720 | 2,958,400 | 5,088,448,000 | 41.4729 | 11.9815 | 0.0005814 | 1720 |
| 1721 | 2,961,841 | 5,097,328,361 | 41.4849 | 11.9838 | 0.0005811 | 1721 |
| 1722 | 2,965,284 | 5,106,219,048 | 41.4970 | 11.9861 | 0.0005807 | 1722 |
| 1723 | 2,968,729 | 5,115,120,067 | 41.5090 | 11.9884 | 0.0005804 | 1723 |
| 1724 | 2,972,176 | 5,124,031,424 | 41.5211 | 11.9907 | 0.0005800 | 1724 |
| 1725 | 2,975,625 | 5,132,953,125 | 41.5331 | 11.9931 | 0.0005797 | 1725 |
| 1726 | 2,979,076 | 5,141,885,176 | 41.5452 | 11.9954 | 0.0005794 | 1726 |
| 1727 | 2,982,529 | 5,150,827,583 | 41.5572 | 11.9977 | 0.0005790 | 1727 |
| 1728 | 2,985,984 | 5,159,780,352 | 41.5692 | 12.0000 | 0.0005787 | 1728 |
| 1729 | 2,989,441 | 5,168,743,489 | 41.5812 | 12.0023 | 0.0005784 | 1729 |
| 1730 | 2,992,900 | 5,177,717,000 | 41.5933 | 12.0046 | 0.0005780 | 1730 |
| 1731 | 2,996,361 | 5,186,700,891 | 41.6053 | 12.0069 | 0.0005777 | 1731 |
| 1732 | 2,999,824 | 5,195,695,168 | 41.6173 | 12.0093 | 0.0005774 | 1732 |
| 1733 | 3,003,289 | 5,204,699,837 | 41.6293 | 12.0116 | 0.0005770 | 1733 |
| 1734 | 3,006,756 | 5,213,714,904 | 41.6413 | 12.0139 | 0.0005767 | 1734 |
| 1735 | 3,010,225 | 5,222,740,375 | 41.6533 | 12.0162 | 0.0005764 | 1735 |
| 1736 | 3,013,696 | 5,231,776,256 | 41.6653 | 12.0185 | 0.0005760 | 1736 |
| 1737 | 3,017,169 | 5,240,822,553 | 41.6773 | 12.0208 | 0.0005757 | 1737 |
| 1738 | 3,020,644 | 5,249,879,272 | 41.6893 | 12.0231 | 0.0005754 | 1738 |
| 1739 | 3,024,121 | 5,258,946,419 | 41.7013 | 12.0254 | 0.0005750 | 1739 |
| 1740 | 3,027,600 | 5,268,024,000 | 41.7133 | 12.0277 | 0.0005747 | 1740 |
| 1741 | 3,031,081 | 5,277,112,021 | 41.7253 | 12.0300 | 0.0005744 | 1741 |
| 1742 | 3,034,564 | 5,286,210,488 | 41.7373 | 12.0323 | 0.0005741 | 1742 |
| 1743 | 3,038,049 | 5,295,319,407 | 41.7493 | 12.0346 | 0.0005737 | 1743 |
| 1744 | 3,041,536 | 5,304,438,784 | 41.7612 | 12.0369 | 0.0005734 | 1744 |
| 1745 | 3,045,025 | 5,313,568,625 | 41.7732 | 12.0392 | 0.0005731 | 1745 |
| 1746 | 3,048,516 | 5,322,708,936 | 41.7852 | 12.0415 | 0.0005727 | 1746 |
| 1747 | 3,052,009 | 5,331,859,723 | 41.7971 | 12.0438 | 0.0005724 | 1747 |
| 1748 | 3,055,504 | 5,341,020,992 | 41.8091 | 12.0461 | 0.0005721 | 1748 |
| 1749 | 3,059,001 | 5,350,192,749 | 41.8210 | 12.0484 | 0.0005718 | 1749 |
| 1750 | 3,062,500 | 5,359,375,000 | 41.8330 | 12.0507 | 0.0005714 | 1750 |

## Powers, Roots and Reciprocals

| No. | Square | Cube | Sq. Root | Cube Root | Reciprocal | No. |
|---|---|---|---|---|---|---|
| 1751 | 3,066,001 | 5,368,567,751 | 41.8450 | 12.0530 | 0.0005711 | 1751 |
| 1752 | 3,069,504 | 5,377,771,008 | 41.8569 | 12.0553 | 0.0005708 | 1752 |
| 1753 | 3,073,009 | 5,386,984,777 | 41.8688 | 12.0576 | 0.0005705 | 1753 |
| 1754 | 3,076,516 | 5,396,209,064 | 41.8808 | 12.0599 | 0.0005701 | 1754 |
| 1755 | 3,080,025 | 5,405,443,875 | 41.8927 | 12.0622 | 0.0005698 | 1755 |
| 1756 | 3,083,536 | 5,414,689,216 | 41.9047 | 12.0645 | 0.0005695 | 1756 |
| 1757 | 3,087,049 | 5,423,945,093 | 41.9166 | 12.0668 | 0.0005692 | 1757 |
| 1758 | 3,090,564 | 5,433,211,512 | 41.9285 | 12.0690 | 0.0005688 | 1758 |
| 1759 | 3,094,081 | 5,442,488,479 | 41.9404 | 12.0713 | 0.0005685 | 1759 |
| 1760 | 3,097,600 | 5,451,776,000 | 41.9524 | 12.0736 | 0.0005682 | 1760 |
| 1761 | 3,101,121 | 5,461,074,081 | 41.9643 | 12.0759 | 0.0005679 | 1761 |
| 1762 | 3,104,644 | 5,470,382,728 | 41.9762 | 12.0782 | 0.0005675 | 1762 |
| 1763 | 3,108,169 | 5,479,701,947 | 41.9881 | 12.0805 | 0.0005672 | 1763 |
| 1764 | 3,111,696 | 5,489,031,744 | 42.0000 | 12.0828 | 0.0005669 | 1764 |
| 1765 | 3,115,225 | 5,498,372,125 | 42.0119 | 12.0850 | 0.0005666 | 1765 |
| 1766 | 3,118,756 | 5,507,723,096 | 42.0238 | 12.0873 | 0.0005663 | 1766 |
| 1767 | 3,122,289 | 5,517,084,663 | 42.0357 | 12.0896 | 0.0005659 | 1767 |
| 1768 | 3,125,824 | 5,526,456,832 | 42.0476 | 12.0919 | 0.0005656 | 1768 |
| 1769 | 3,129,361 | 5,535,839,609 | 42.0595 | 12.0942 | 0.0005653 | 1769 |
| 1770 | 3,132,900 | 5,545,233,000 | 42.0714 | 12.0964 | 0.0005650 | 1770 |
| 1771 | 3,136,441 | 5,554,637,011 | 42.0833 | 12.0987 | 0.0005647 | 1771 |
| 1772 | 3,139,984 | 5,564,651,648 | 42.0951 | 12.1010 | 0.0005643 | 1772 |
| 1773 | 3,143,529 | 5,573,476,917 | 42.1070 | 12.1033 | 0.0005640 | 1773 |
| 1774 | 3,147,076 | 5,582,912,824 | 42.1189 | 12.1056 | 0.0005637 | 1774 |
| 1775 | 3,150,625 | 5,592,359,375 | 42.1307 | 12.1078 | 0.0005634 | 1775 |
| 1776 | 3,154,176 | 5,601,816,576 | 42.1426 | 12.1101 | 0.0005631 | 1776 |
| 1777 | 3,157,729 | 5,611,284,433 | 42.1545 | 12.1124 | 0.0005627 | 1777 |
| 1778 | 3,161,284 | 5,620,762,952 | 42.1663 | 12.1146 | 0.0005624 | 1778 |
| 1779 | 3,164,841 | 5,630,252,139 | 42.1782 | 12.1169 | 0.0005621 | 1779 |
| 1780 | 3,168,400 | 5,639,752,000 | 42.1900 | 12.1192 | 0.0005618 | 1780 |
| 1781 | 3,171,961 | 5,649,262,541 | 42.2019 | 12.1215 | 0.0005615 | 1781 |
| 1782 | 3,175,524 | 5,658,783,768 | 42.2137 | 12.1237 | 0.0005612 | 1782 |
| 1783 | 3,179,089 | 5,668,315,687 | 42.2256 | 12.1260 | 0.0005609 | 1783 |
| 1784 | 3,182,656 | 5,677,858,304 | 42.2374 | 12.1283 | 0.0005605 | 1784 |
| 1785 | 3,186,225 | 5,687,411,625 | 42.2493 | 12.1305 | 0.0005602 | 1785 |
| 1786 | 3,189,796 | 5,696,975,656 | 42.2611 | 12.1328 | 0.0005599 | 1786 |
| 1787 | 3,193,369 | 5,706,550,403 | 42.2729 | 12.1350 | 0.0005596 | 1787 |
| 1788 | 3,196,944 | 5,716,135,872 | 42.2847 | 12.1373 | 0.0005593 | 1788 |
| 1789 | 3,200,521 | 5,725,732,069 | 42.2966 | 12.1396 | 0.0005590 | 1789 |
| 1790 | 3,204,100 | 5,735,339,000 | 42.3084 | 12.1418 | 0.0005587 | 1790 |
| 1791 | 3,207,681 | 5,744,956,671 | 42.3202 | 12.1441 | 0.0005583 | 1791 |
| 1792 | 3,211,264 | 5,754,585,088 | 42.3320 | 12.1464 | 0.0005580 | 1792 |
| 1793 | 3,214,849 | 5,764,224,257 | 42.3438 | 12.1486 | 0.0005577 | 1793 |
| 1794 | 3,218,436 | 5,773,874,184 | 42.3556 | 12.1509 | 0.0005574 | 1794 |
| 1795 | 3,222,025 | 5,783,534,875 | 42.3674 | 12.1531 | 0.0005571 | 1795 |
| 1796 | 3,225,616 | 5,793,206,336 | 42.3792 | 12.1554 | 0.0005568 | 1796 |
| 1797 | 3,229,209 | 5,802,888,573 | 42.3910 | 12.1576 | 0.0005565 | 1797 |
| 1798 | 3,232,804 | 5,812,581,592 | 42.4028 | 12.1599 | 0.0005562 | 1798 |
| 1799 | 3,236,401 | 5,822,285,399 | 42.4146 | 12.1622 | 0.0005559 | 1799 |
| 1800 | 3,240,000 | 5,832,000,000 | 42.4264 | 12.1644 | 0.0005556 | 1800 |

## Powers, Roots and Reciprocals

| No. | Square | Cube | Sq. Root | Cube Root | Reciprocal | No. |
|---|---|---|---|---|---|---|
| 1801 | 3,243,601 | 5,841,725,401 | 42.4382 | 12.1667 | 0.0005552 | 1801 |
| 1802 | 3,247,204 | 5,851,461,608 | 42.4500 | 12.1689 | 0.0005549 | 1802 |
| 1803 | 3,250,809 | 5,861,208,627 | 42.4617 | 12.1712 | 0.0005546 | 1803 |
| 1804 | 3,254,416 | 5,870,966,464 | 42.4735 | 12.1734 | 0.0005543 | 1804 |
| 1805 | 3,258,025 | 5,880,735,125 | 42.4853 | 12.1757 | 0.0005540 | 1805 |
| 1806 | 3,261,636 | 5,890,514,616 | 42.4971 | 12.1779 | 0.0005537 | 1806 |
| 1807 | 3,265,249 | 5,900,304,943 | 42.5088 | 12.1802 | 0.0005534 | 1807 |
| 1808 | 3,268,864 | 5,910,106,112 | 42.5206 | 12.1824 | 0.0005531 | 1808 |
| 1809 | 3,272,481 | 5,919,918,129 | 42.5323 | 12.1846 | 0.0005528 | 1809 |
| 1810 | 3,276,100 | 5,929,741,000 | 42.5441 | 12.1869 | 0.0005525 | 1810 |
| 1811 | 3,279,721 | 5,939,574,731 | 42.5558 | 12.1891 | 0.0005522 | 1811 |
| 1812 | 3,283,344 | 5,949,419,328 | 42.5676 | 12.1914 | 0.0005519 | 1812 |
| 1813 | 3,286,969 | 5,959,274,797 | 42.5793 | 12.1936 | 0.0005516 | 1813 |
| 1814 | 3,290,596 | 5,969,141,144 | 42.5911 | 12.1959 | 0.0005513 | 1814 |
| 1815 | 3,294,225 | 5,979,018,375 | 42.6028 | 12.1981 | 0.0005510 | 1815 |
| 1816 | 3,297,856 | 5,988,906,496 | 42.6146 | 12.2003 | 0.0005507 | 1816 |
| 1817 | 3,301,489 | 5,998,805,513 | 42.6263 | 12.2026 | 0.0005504 | 1817 |
| 1818 | 3,305,124 | 6,008,715,432 | 42.6380 | 12.2048 | 0.0005501 | 1818 |
| 1819 | 3,308,761 | 6,018,636,259 | 42.6497 | 12.2071 | 0.0005498 | 1819 |
| 1820 | 3,312,400 | 6,028,568,000 | 42.6615 | 12.2093 | 0.0005495 | 1820 |
| 1821 | 3,316,041 | 6,038,510,661 | 42.6732 | 12.2115 | 0.0005491 | 1821 |
| 1822 | 3,319,684 | 6,048,464,248 | 42.6849 | 12.2138 | 0.0005488 | 1822 |
| 1823 | 3,323,329 | 6,058,428,767 | 42.6966 | 12.2160 | 0.0005485 | 1823 |
| 1824 | 3,326,976 | 6,068,404,224 | 42.7083 | 12.2182 | 0.0005482 | 1824 |
| 1825 | 3,330,625 | 6,078,390,625 | 42.7200 | 12.2205 | 0.0005479 | 1825 |
| 1826 | 3,334,276 | 6,088,387,976 | 42.7317 | 12.2227 | 0.0005476 | 1826 |
| 1827 | 3,337,929 | 6,098,396,283 | 42.7434 | 12.2249 | 0.0005473 | 1827 |
| 1828 | 3,341,584 | 6,108,415,552 | 42.7551 | 12.2272 | 0.0005470 | 1828 |
| 1829 | 3,345,241 | 6,118,445,789 | 42.7668 | 12.2294 | 0.0005467 | 1829 |
| 1830 | 3,348,900 | 6,128,487,000 | 42.7785 | 12.2316 | 0.0005464 | 1830 |
| 1831 | 3,352,561 | 6,138,539,191 | 42.7902 | 12.2338 | 0.0005461 | 1831 |
| 1832 | 3,356,224 | 6,148,602,368 | 42.8019 | 12.2361 | 0.0005459 | 1832 |
| 1833 | 3,359,889 | 6,158,676,537 | 42.8135 | 12.2383 | 0.0005456 | 1833 |
| 1834 | 3,363,556 | 6,168,761,704 | 42.8252 | 12.2405 | 0.0005453 | 1834 |
| 1835 | 3,367,225 | 6,178,857,875 | 42.8369 | 12.2427 | 0.0005450 | 1835 |
| 1836 | 3,370,896 | 6,188,965,056 | 42.8486 | 12.2450 | 0.0005447 | 1836 |
| 1837 | 3,374,569 | 6,199,083,253 | 42.8602 | 12.2472 | 0.0005444 | 1837 |
| 1838 | 3,378,244 | 6,209,212,472 | 42.8719 | 12.2494 | 0.0005441 | 1838 |
| 1839 | 3,381,921 | 6,219,352,719 | 42.8836 | 12.2516 | 0.0005438 | 1839 |
| 1840 | 3,385,600 | 6,229,504,000 | 42.8952 | 12.2539 | 0.0005435 | 1840 |
| 1841 | 3,389,281 | 6,239,666,321 | 42.9069 | 12.2561 | 0.0005432 | 1841 |
| 1842 | 3,392,964 | 6,249,839,688 | 42.9185 | 12.2583 | 0.0005429 | 1842 |
| 1843 | 3,396,649 | 6,260,024,107 | 42.9302 | 12.2605 | 0.0005426 | 1843 |
| 1844 | 3,400,336 | 6,270,219,584 | 42.9418 | 12.2627 | 0.0005423 | 1844 |
| 1845 | 3,404,025 | 6,280,426,125 | 42.9535 | 12.2649 | 0.0005420 | 1845 |
| 1846 | 3,407,716 | 6,290,643,736 | 42.9651 | 12.2672 | 0.0005417 | 1846 |
| 1847 | 3,411,409 | 6,300,872,423 | 42.9767 | 12.2694 | 0.0005414 | 1847 |
| 1848 | 3,415,104 | 6,311,112,192 | 42.9884 | 12.2716 | 0.0005411 | 1848 |
| 1849 | 3,418,801 | 6,321,363,049 | 43.0000 | 12.2738 | 0.0005408 | 1849 |
| 1850 | 3,422,500 | 6,331,625,000 | 43.0116 | 12.2760 | 0.0005405 | 1850 |

## Powers, Roots and Reciprocals

| No. | Square | Cube | Sq. Root | Cube Root | Reciprocal | No. |
|---|---|---|---|---|---|---|
| 1851 | 3,426,201 | 6,341,898,051 | 43.0232 | 12.2782 | 0.0005402 | 1851 |
| 1852 | 3,429,904 | 6,352,182,208 | 43.0349 | 12.2804 | 0.0005400 | 1852 |
| 1853 | 3,433,609 | 6,362,477,477 | 43.0465 | 12.2826 | 0.0005397 | 1853 |
| 1854 | 3,437,316 | 6,372,783,864 | 43.0581 | 12.2849 | 0.0005394 | 1854 |
| 1855 | 3,441,025 | 6,383,101,375 | 43.0697 | 12.2871 | 0.0005391 | 1855 |
| 1856 | 3,444,736 | 6,393,430,016 | 43.0813 | 12.2893 | 0.0005388 | 1856 |
| 1857 | 3,448,449 | 6,403,769,793 | 43.0929 | 12.2915 | 0.0005385 | 1857 |
| 1858 | 3,452,164 | 6,414,120,712 | 43.1045 | 12.2937 | 0.0005382 | 1858 |
| 1859 | 3,455,881 | 6,424,482,779 | 43.1161 | 12.2959 | 0.0005379 | 1859 |
| 1860 | 3,459,600 | 6,434,856,000 | 43.1277 | 12.2981 | 0.0005376 | 1860 |
| 1861 | 3,463,321 | 6,445,240,381 | 43.1393 | 12.3003 | 0.0005373 | 1861 |
| 1862 | 3,467,044 | 6,455,635,928 | 43.1509 | 12.3025 | 0.0005371 | 1862 |
| 1863 | 3,470,769 | 6,466,042,647 | 43.1625 | 12.3047 | 0.0005368 | 1863 |
| 1864 | 3,474,496 | 6,476,460,544 | 43.1741 | 12.3069 | 0.0005365 | 1864 |
| 1865 | 3,478,225 | 6,486,889,625 | 43.1856 | 12.3091 | 0.0005362 | 1865 |
| 1866 | 3,481,956 | 6,497,329,896 | 43.1972 | 12.3113 | 0.0005359 | 1866 |
| 1867 | 3,485,689 | 6,507,781,363 | 43.2088 | 12.3135 | 0.0005356 | 1867 |
| 1868 | 3,489,424 | 6,518,244,032 | 43.2204 | 12.3157 | 0.0005353 | 1868 |
| 1869 | 3,493,161 | 6,528,717,909 | 43.2319 | 12.3179 | 0.0005350 | 1869 |
| 1870 | 3,496,900 | 6,539,203,000 | 43.2435 | 12.3201 | 0.0005348 | 1870 |
| 1871 | 3,500,641 | 6,549,699,311 | 43.2551 | 12.3223 | 0.0005345 | 1871 |
| 1872 | 3,504,384 | 6,560,206,848 | 43.2666 | 12.3245 | 0.0005342 | 1872 |
| 1873 | 3,508,129 | 6,570,725,617 | 43.2782 | 12.3267 | 0.0005339 | 1873 |
| 1874 | 3,511,876 | 6,581,255,624 | 43.2897 | 12.3289 | 0.0005336 | 1874 |
| 1875 | 3,515,625 | 6,591,796,875 | 43.3013 | 12.3311 | 0.0005333 | 1875 |
| 1876 | 3,519,376 | 6,602,349,376 | 43.3128 | 12.3333 | 0.0005330 | 1876 |
| 1877 | 3,523,129 | 6,612,913,133 | 43.3244 | 12.3354 | 0.0005328 | 1877 |
| 1878 | 3,526,884 | 6,623,488,152 | 43.3359 | 12.3376 | 0.0005325 | 1878 |
| 1879 | 3,530,641 | 6,634,074,439 | 43.3474 | 12.3398 | 0.0005322 | 1879 |
| 1880 | 3,534,400 | 6,644,672,000 | 43.3590 | 12.3420 | 0.0005319 | 1880 |
| 1881 | 3,538,161 | 6,655,280,841 | 43.3705 | 12.3442 | 0.0005316 | 1881 |
| 1882 | 3,541,924 | 6,665,900,968 | 43.3820 | 12.3464 | 0.0005313 | 1882 |
| 1883 | 3,545,689 | 6,676,532,387 | 43.3935 | 12.3486 | 0.0005311 | 1883 |
| 1884 | 3,549,456 | 6,687,175,104 | 43.4051 | 12.3508 | 0.0005308 | 1884 |
| 1885 | 3,553,225 | 6,697,829,125 | 43.4166 | 12.3529 | 0.0005305 | 1885 |
| 1886 | 3,556,996 | 6,708,494,456 | 43.4281 | 12.3551 | 0.0005302 | 1886 |
| 1887 | 3,560,769 | 6,719,171,103 | 43.4396 | 12.3573 | 0.0005299 | 1887 |
| 1888 | 3,564,544 | 6,729,859,072 | 43.4511 | 12.3595 | 0.0005297 | 1888 |
| 1889 | 3,568,321 | 6,740,558,369 | 43.4626 | 12.3617 | 0.0005294 | 1889 |
| 1890 | 3,572,100 | 6,751,269,000 | 43.4741 | 12.3639 | 0.0005291 | 1890 |
| 1891 | 3,575,881 | 6,761,990,971 | 43.4856 | 12.3660 | 0.0005288 | 1891 |
| 1892 | 3,579,664 | 6,772,724,288 | 43.4971 | 12.3682 | 0.0005285 | 1892 |
| 1893 | 3,583,449 | 6,783,468,957 | 43.5086 | 12.3704 | 0.0005283 | 1893 |
| 1894 | 3,587,236 | 6,794,224,984 | 43.5201 | 12.3726 | 0.0005280 | 1894 |
| 1895 | 3,591,025 | 6,804,992,375 | 43.5316 | 12.3747 | 0.0005277 | 1895 |
| 1896 | 3,594,816 | 6,815,771,136 | 43.5431 | 12.3769 | 0.0005274 | 1896 |
| 1897 | 3,598,609 | 6,826,561,273 | 43.5546 | 12.3791 | 0.0005271 | 1897 |
| 1898 | 3,602,404 | 6,837,362,792 | 43.5660 | 12.3813 | 0.0005269 | 1898 |
| 1899 | 3,606,201 | 6,848,175,699 | 43.5775 | 12.3835 | 0.0005266 | 1899 |
| 1900 | 3,610,000 | 6,859,000,000 | 43.5890 | 12.3856 | 0.0005263 | 1900 |

## Powers, Roots and Reciprocals

| No. | Square | Cube | Sq. Root | Cube Root | Reciprocal | No. |
|---|---|---|---|---|---|---|
| 1901 | 3,613,801 | 6,869,835,701 | 43.6005 | 12.3878 | 0.0005260 | 1901 |
| 1902 | 3,617,604 | 6,880,682,808 | 43.6119 | 12.3900 | 0.0005258 | 1902 |
| 1903 | 3,621,409 | 6,891,541,327 | 43.6234 | 12.3921 | 0.0005255 | 1903 |
| 1904 | 3,625,216 | 6,902,411,264 | 43.6348 | 12.3943 | 0.0005252 | 1904 |
| 1905 | 3,629,025 | 6,913,292,625 | 43.6463 | 12.3965 | 0.0005249 | 1905 |
| 1906 | 3,632,836 | 6,924,185,416 | 43.6578 | 12.3986 | 0.0005247 | 1906 |
| 1907 | 3,636,649 | 6,935,089,643 | 43.6692 | 12.4008 | 0.0005244 | 1907 |
| 1908 | 3,640,464 | 6,946,005,312 | 43.6807 | 12.4030 | 0.0005241 | 1908 |
| 1909 | 3,644,281 | 6,956,932,429 | 43.6921 | 12.4051 | 0.0005238 | 1909 |
| 1910 | 3,648,100 | 6,967,871,000 | 43.7035 | 12.4073 | 0.0005236 | 1910 |
| 1911 | 3,651,921 | 6,978,821,031 | 43.7150 | 12.4095 | 0.0005233 | 1911 |
| 1912 | 3,655,744 | 6,989,782,528 | 43.7264 | 12.4116 | 0.0005230 | 1912 |
| 1913 | 3,659,569 | 7,000,755,497 | 43.7379 | 12.4138 | 0.0005227 | 1913 |
| 1914 | 3,663,396 | 7,011,739,944 | 43.7493 | 12.4160 | 0.0005225 | 1914 |
| 1915 | 3,667,225 | 7,022,735,875 | 43.7607 | 12.4181 | 0.0005222 | 1915 |
| 1916 | 3,671,056 | 7,033,743,296 | 43.7721 | 12.4203 | 0.0005219 | 1916 |
| 1917 | 3,674,889 | 7,044,762,213 | 43.7836 | 12.4225 | 0.0005216 | 1917 |
| 1918 | 3,678,724 | 7,055,792,632 | 43.7950 | 12.4246 | 0.0005214 | 1918 |
| 1919 | 3,682,561 | 7,066,834,559 | 43.8064 | 12.4268 | 0.0005211 | 1919 |
| 1920 | 3,686,400 | 7,077,888,000 | 43.8178 | 12.4289 | 0.0005208 | 1920 |
| 1921 | 3,690,241 | 7,088,952,961 | 43.8292 | 12.4311 | 0.0005206 | 1921 |
| 1922 | 3,694,084 | 7,100,029,448 | 43.8406 | 12.4332 | 0.0005203 | 1922 |
| 1923 | 3,697,929 | 7,111,117,467 | 43.8520 | 12.4354 | 0.0005200 | 1923 |
| 1924 | 3,701,776 | 7,122,217,024 | 43.8634 | 12.4376 | 0.0005198 | 1924 |
| 1925 | 3,705,625 | 7,133,328,125 | 43.8748 | 12.4397 | 0.0005195 | 1925 |
| 1926 | 3,709,476 | 7,144,450,776 | 43.8862 | 12.4419 | 0.0005192 | 1926 |
| 1927 | 3,713,329 | 7,155,584,983 | 43.8976 | 12.4440 | 0.0005189 | 1927 |
| 1928 | 3,717,184 | 7,166,730,752 | 43.9090 | 12.4462 | 0.0005187 | 1928 |
| 1929 | 3,721,041 | 7,177,888,089 | 43.9204 | 12.4483 | 0.0005184 | 1929 |
| 1930 | 3,724,900 | 7,189,057,000 | 43.9318 | 12.4505 | 0.0005181 | 1930 |
| 1931 | 3,728,761 | 7,200,237,491 | 43.9431 | 12.4526 | 0.0005179 | 1931 |
| 1932 | 3,732,624 | 7,211,429,568 | 43.9545 | 12.4548 | 0.0005176 | 1932 |
| 1933 | 3,736,489 | 7,222,633,237 | 43.9659 | 12.4569 | 0.0005173 | 1933 |
| 1934 | 3,740,356 | 7,233,848,504 | 43.9773 | 12.4591 | 0.0005171 | 1934 |
| 1935 | 3,744,225 | 7,245,075,375 | 43.9886 | 12.4612 | 0.0005168 | 1935 |
| 1936 | 3,748,096 | 7,256,313,856 | 44.0000 | 12.4634 | 0.0005165 | 1936 |
| 1937 | 3,751,969 | 7,267,563,953 | 44.0114 | 12.4655 | 0.0005163 | 1937 |
| 1938 | 3,755,844 | 7,278,825,672 | 44.0227 | 12.4676 | 0.0005160 | 1938 |
| 1939 | 3,759,721 | 7,290,099,019 | 44.0341 | 12.4698 | 0.0005157 | 1939 |
| 1940 | 3,763,600 | 7,301,384,000 | 44.0454 | 12.4719 | 0.0005155 | 1940 |
| 1941 | 3,767,481 | 7,312,680,621 | 44.0568 | 12.4741 | 0.0005152 | 1941 |
| 1942 | 3,771,364 | 7,323,988,888 | 44.0681 | 12.4762 | 0.0005149 | 1942 |
| 1943 | 3,775,249 | 7,335,308,807 | 44.0795 | 12.4784 | 0.0005147 | 1943 |
| 1944 | 3,779,136 | 7,346,640,384 | 44.0908 | 12.4805 | 0.0005144 | 1944 |
| 1945 | 3,783,025 | 7,357,983,625 | 44.1022 | 12.4826 | 0.0005141 | 1945 |
| 1946 | 3,786,916 | 7,369,338,536 | 44.1135 | 12.4848 | 0.0005139 | 1946 |
| 1947 | 3,790,809 | 7,380,705,123 | 44.1248 | 12.4869 | 0.0005136 | 1947 |
| 1948 | 3,794,704 | 7,392,083,392 | 44.1362 | 12.4891 | 0.0005133 | 1948 |
| 1949 | 3,798,601 | 7,403,473,349 | 44.1475 | 12.4912 | 0.0005131 | 1949 |
| 1950 | 3,802,500 | 7,414,875,000 | 44.1588 | 12.4933 | 0.0005128 | 1950 |

## Powers, Roots and Reciprocals

| No. | Square | Cube | Sq. Root | Cube Root | Reciprocal | No. |
|---|---|---|---|---|---|---|
| 1951 | 3,806,401 | 7,426,288,351 | 44.1701 | 12.4955 | 0.0005126 | 1951 |
| 1952 | 3,810,304 | 7,437,713,408 | 44.1814 | 12.4976 | 0.0005123 | 1952 |
| 1953 | 3,814,209 | 7,449,150,177 | 44.1928 | 12.4997 | 0.0005120 | 1953 |
| 1954 | 3,818,116 | 7,460,598,664 | 44.2041 | 12.5019 | 0.0005118 | 1954 |
| 1955 | 3,822,025 | 7,472,058,875 | 44.2154 | 12.5040 | 0.0005115 | 1955 |
| 1956 | 3,825,936 | 7,483,530,816 | 44.2267 | 12.5061 | 0.0005112 | 1956 |
| 1957 | 3,829,849 | 7,495,014,493 | 44.2380 | 12.5083 | 0.0005110 | 1957 |
| 1958 | 3,833,764 | 7,506,509,912 | 44.2493 | 12.5104 | 0.0005107 | 1958 |
| 1959 | 3,837,681 | 7,518,017,079 | 44.2606 | 12.5125 | 0.0005105 | 1959 |
| 1960 | 3,841,600 | 7,529,536,000 | 44.2719 | 12.5146 | 0.0005102 | 1960 |
| 1961 | 3,845,521 | 7,541,066,681 | 44.2832 | 12.5168 | 0.0005099 | 1961 |
| 1962 | 3,849,444 | 7,552,609,128 | 44.2945 | 12.5189 | 0.0005097 | 1962 |
| 1963 | 3,853,369 | 7,564,163,347 | 44.3058 | 12.5210 | 0.0005094 | 1963 |
| 1964 | 3,857,296 | 7,575,729,344 | 44.3170 | 12.5232 | 0.0005092 | 1964 |
| 1965 | 3,861,225 | 7,587,307,125 | 44.3283 | 12.5253 | 0.0005089 | 1965 |
| 1966 | 3,865,156 | 7,598,896,696 | 44.3396 | 12.5274 | 0.0005086 | 1966 |
| 1967 | 3,869,089 | 7,610,498,063 | 44.3509 | 12.5295 | 0.0005084 | 1967 |
| 1968 | 3,873,024 | 7,622,111,232 | 44.3621 | 12.5317 | 0.0005081 | 1968 |
| 1969 | 3,876,961 | 7,633,736,209 | 44.3734 | 12.5338 | 0.0005079 | 1969 |
| 1970 | 3,880,900 | 7,645,373,000 | 44.3847 | 12.5359 | 0.0005076 | 1970 |
| 1971 | 3,884,841 | 7,657,021,611 | 44.3959 | 12.5380 | 0.0005074 | 1971 |
| 1972 | 3,888,784 | 7,668,682,048 | 44.4072 | 12.5401 | 0.0005071 | 1972 |
| 1973 | 3,892,729 | 7,680,354,317 | 44.4185 | 12.5423 | 0.0005068 | 1973 |
| 1974 | 3,896,676 | 7,692,038,424 | 44.4297 | 12.5444 | 0.0005066 | 1974 |
| 1975 | 3,900,625 | 7,703,734,375 | 44.4410 | 12.5465 | 0.0005063 | 1975 |
| 1976 | 3,904,576 | 7,715,442,176 | 44.4522 | 12.5486 | 0.0005061 | 1976 |
| 1977 | 3,908,529 | 7,727,161,833 | 44.4635 | 12.5507 | 0.0005058 | 1977 |
| 1978 | 3,912,484 | 7,738,893,352 | 44.4747 | 12.5528 | 0.0005056 | 1978 |
| 1979 | 3,916,441 | 7,750,636,739 | 44.4860 | 12.5550 | 0.0005053 | 1979 |
| 1980 | 3,920,400 | 7,762,392,000 | 44.4972 | 12.5571 | 0.0005051 | 1980 |
| 1981 | 3,924,361 | 7,774,159,141 | 44.5084 | 12.5592 | 0.0005048 | 1981 |
| 1982 | 3,928,324 | 7,785,938,168 | 44.5197 | 12.5613 | 0.0005045 | 1982 |
| 1983 | 3,932,289 | 7,797,729,087 | 44.5309 | 12.5634 | 0.0005043 | 1983 |
| 1984 | 3,936,256 | 7,809,531,904 | 44.5421 | 12.5655 | 0.0005040 | 1984 |
| 1985 | 3,940,225 | 7,821,346,625 | 44.5533 | 12.5676 | 0.0005038 | 1985 |
| 1986 | 3,944,196 | 7,833,173,256 | 44.5646 | 12.5697 | 0.0005035 | 1986 |
| 1987 | 3,948,169 | 7,845,011,803 | 44.5758 | 12.5719 | 0.0005033 | 1987 |
| 1988 | 3,952,144 | 7,856,862,272 | 44.5870 | 12.5740 | 0.0005030 | 1988 |
| 1989 | 3,956,121 | 7,868,724,669 | 44.5982 | 12.5761 | 0.0005028 | 1989 |
| 1990 | 3,960,100 | 7,880,599,000 | 44.6094 | 12.5782 | 0.0005025 | 1990 |
| 1991 | 3,964,081 | 7,892,485,271 | 44.6206 | 12.5803 | 0.0005023 | 1991 |
| 1992 | 3,968,064 | 7,904,383,488 | 44.6318 | 12.5824 | 0.0005020 | 1992 |
| 1993 | 3,972,049 | 7,916,293,657 | 44.6430 | 12.5845 | 0.0005018 | 1993 |
| 1994 | 3,976,036 | 7,928,215,784 | 44.6542 | 12.5866 | 0.0005015 | 1994 |
| 1995 | 3,980,025 | 7,940,149,875 | 44.6654 | 12.5887 | 0.0005013 | 1995 |
| 1996 | 3,984,016 | 7,952,095,936 | 44.6766 | 12.5908 | 0.0005010 | 1996 |
| 1997 | 3,988,009 | 7,964,053,973 | 44.6878 | 12.5929 | 0.0005008 | 1997 |
| 1998 | 3,992,004 | 7,976,023,992 | 44.6990 | 12.5950 | 0.0005005 | 1998 |
| 1999 | 3,996,001 | 7,988,005,999 | 44.7102 | 12.5971 | 0.0005003 | 1999 |
| 2000 | 4,000,000 | 8,000,000,000 | 44.7214 | 12.5992 | 0.0005000 | 2000 |

## Squares of Mixed Numbers from 1/64 to 12, by 64ths

### I. Squares of Mixed Numbers from 1/64 to 6

|  | 0 | 1 | 2 | 3 | 4 | 5 |
|---|---|---|---|---|---|---|
| 1/64 | 0.00024 | 1.03149 | 4.06274 | 9.09399 | 16.12524 | 25.15649 |
| 1/32 | 0.00098 | 1.06348 | 4.12598 | 9.18848 | 16.25098 | 25.31348 |
| 3/64 | 0.00220 | 1.09595 | 4.18970 | 9.28345 | 16.37720 | 25.47095 |
| 1/16 | 0.00391 | 1.12891 | 4.25391 | 9.37891 | 16.50391 | 25.62891 |
| 5/64 | 0.00610 | 1.16235 | 4.31860 | 9.47485 | 16.63110 | 25.78735 |
| 3/32 | 0.00879 | 1.19629 | 4.38379 | 9.57129 | 16.75879 | 25.94629 |
| 7/64 | 0.01196 | 1.23071 | 4.44946 | 9.66821 | 16.88696 | 26.10571 |
| 1/8 | 0.01563 | 1.26563 | 4.51563 | 9.76563 | 17.01563 | 26.26563 |
| 9/64 | 0.01978 | 1.30103 | 4.58228 | 9.86353 | 17.14478 | 26.42603 |
| 5/32 | 0.02441 | 1.33691 | 4.64941 | 9.96191 | 17.27441 | 26.58691 |
| 11/64 | 0.02954 | 1.37329 | 4.71704 | 10.06079 | 17.40454 | 26.74829 |
| 3/16 | 0.03516 | 1.41016 | 4.78516 | 10.16016 | 17.53516 | 26.91016 |
| 13/64 | 0.04126 | 1.44751 | 4.85376 | 10.26001 | 17.66626 | 27.07251 |
| 7/32 | 0.04785 | 1.48535 | 4.92285 | 10.36035 | 17.79785 | 27.23535 |
| 15/64 | 0.05493 | 1.52368 | 4.99243 | 10.46118 | 17.92993 | 27.39868 |
| 1/4 | 0.06250 | 1.56250 | 5.06250 | 10.56250 | 18.06250 | 27.56250 |
| 17/64 | 0.07056 | 1.60181 | 5.13306 | 10.66431 | 18.19556 | 27.72681 |
| 9/32 | 0.07910 | 1.64160 | 5.20410 | 10.76660 | 18.32910 | 27.89160 |
| 19/64 | 0.08813 | 1.68188 | 5.27563 | 10.86938 | 18.46313 | 28.05688 |
| 5/16 | 0.09766 | 1.72266 | 5.34766 | 10.97266 | 18.59766 | 28.22266 |
| 21/64 | 0.10767 | 1.76392 | 5.42017 | 11.07642 | 18.73267 | 28.38892 |
| 11/32 | 0.11816 | 1.80566 | 5.49316 | 11.18066 | 18.86816 | 28.55566 |
| 23/64 | 0.12915 | 1.84790 | 5.56665 | 11.28540 | 19.00415 | 28.72290 |
| 3/8 | 0.14063 | 1.89063 | 5.64063 | 11.39063 | 19.14063 | 28.89063 |
| 25/64 | 0.15259 | 1.93384 | 5.71509 | 11.49634 | 19.27759 | 29.05884 |
| 13/32 | 0.16504 | 1.97754 | 5.79004 | 11.60254 | 19.41504 | 29.22754 |
| 27/64 | 0.17798 | 2.02173 | 5.86548 | 11.70923 | 19.55298 | 29.39673 |
| 7/16 | 0.19141 | 2.06641 | 5.94141 | 11.81641 | 19.69141 | 29.56641 |
| 29/64 | 0.20532 | 2.11157 | 6.01782 | 11.92407 | 19.83032 | 29.73657 |
| 15/32 | 0.21973 | 2.15723 | 6.09473 | 12.03223 | 19.96973 | 29.90723 |
| 31/64 | 0.23462 | 2.20337 | 6.17212 | 12.14087 | 20.10962 | 30.07837 |
| 1/2 | 0.25000 | 2.25000 | 6.25000 | 12.25000 | 20.25000 | 30.25000 |
| 33/64 | 0.26587 | 2.29712 | 6.32837 | 12.35962 | 20.39087 | 30.42212 |
| 17/32 | 0.28223 | 2.34473 | 6.40723 | 12.46973 | 20.53223 | 30.59473 |
| 35/64 | 0.29907 | 2.39282 | 6.48657 | 12.58032 | 20.67407 | 30.76782 |
| 9/16 | 0.31641 | 2.44141 | 6.56641 | 12.69141 | 20.81641 | 30.94141 |
| 37/64 | 0.33423 | 2.49048 | 6.64673 | 12.80298 | 20.95923 | 31.11548 |
| 19/32 | 0.35254 | 2.54004 | 6.72754 | 12.91504 | 21.10254 | 31.29004 |
| 39/64 | 0.37134 | 2.59009 | 6.80884 | 13.02759 | 21.24634 | 31.46509 |
| 5/8 | 0.39063 | 2.64063 | 6.89063 | 13.14063 | 21.39063 | 31.64063 |
| 41/64 | 0.41040 | 2.69165 | 6.97290 | 13.25415 | 21.53540 | 31.81665 |
| 21/32 | 0.43066 | 2.74316 | 7.05566 | 13.36816 | 21.68066 | 31.99316 |

The tables of squares of mixed numbers from 1/64 to 12 are arranged in as compact a manner as possible, and a few words may be necessary to explain their use. Assume, for example, that the square of 85/64 is required; 8 is located at the

## Squares of Mixed Numbers from 1/64 to 6 (Continued)

| | 0 | 1 | 2 | 3 | 4 | 5 |
|---|---|---|---|---|---|---|
| 43/64 | 0.45142 | 2.79517 | 7.13892 | 13.48267 | 21.82642 | 32.17017 |
| 11/16 | 0.47266 | 2.84766 | 7.22266 | 13.59766 | 21.97266 | 32.34766 |
| 45/64 | 0.49438 | 2.90063 | 7.30688 | 13.71313 | 22.11938 | 32.52563 |
| 23/32 | 0.51660 | 2.95410 | 7.39160 | 13.82910 | 22.26660 | 32.70410 |
| 47/64 | 0.53931 | 3.00806 | 7.47681 | 13.94556 | 22.41431 | 32.88306 |
| 3/4 | 0.56250 | 3.06250 | 7.56250 | 14.06250 | 22.56250 | 33.06250 |
| 49/64 | 0.58618 | 3.11743 | 7.64868 | 14.17993 | 22.71118 | 33.24243 |
| 25/32 | 0.61035 | 3.17285 | 7.73535 | 14.29785 | 22.86035 | 33.42285 |
| 51/64 | 0.63501 | 3.22876 | 7.82251 | 14.41626 | 23.01001 | 33.60376 |
| 13/16 | 0.66016 | 3.28516 | 7.91016 | 14.53516 | 23.16016 | 33.78516 |
| 53/64 | 0.68579 | 3.34204 | 7.99829 | 14.65454 | 23.31079 | 33.96704 |
| 27/32 | 0.71191 | 3.39941 | 8.08691 | 14.77441 | 23.46191 | 34.14941 |
| 55/64 | 0.73853 | 3.45728 | 8.17603 | 14.89478 | 23.61365 | 34.33228 |
| 7/8 | 0.76563 | 3.51563 | 8.26563 | 15.01563 | 23.76563 | 34.51563 |
| 57/64 | 0.79321 | 3.57446 | 8.35571 | 15.13696 | 23.91821 | 34.69946 |
| 29/32 | 0.82129 | 3.63379 | 8.44629 | 15.25879 | 24.07129 | 34.88379 |
| 59/64 | 0.84985 | 3.69360 | 8.53735 | 15.38110 | 24.22485 | 35.06860 |
| 15/16 | 0.87891 | 3.75391 | 8.62891 | 15.50391 | 24.37891 | 35.25391 |
| 61/64 | 0.90845 | 3.81470 | 8.72095 | 15.62720 | 24.53345 | 35.43970 |
| 31/32 | 0.93848 | 3.87598 | 8.81348 | 15.75098 | 24.68848 | 35.62598 |
| 63/64 | 0.96899 | 3.93774 | 8.90649 | 15.87524 | 24.84399 | 35.81274 |

## II. Squares of Mixed Numbers from 6 1/64 to 12

| | 6 | 7 | 8 | 9 | 10 | 11 |
|---|---|---|---|---|---|---|
| 1/64 | 36.18774 | 49.21899 | 64.25024 | 81.28149 | 100.31274 | 121.34399 |
| 1/32 | 36.37598 | 49.43848 | 64.50098 | 81.56348 | 100.62598 | 121.68848 |
| 3/64 | 36.56470 | 49.65845 | 64.75220 | 81.84595 | 100.93970 | 122.03345 |
| 1/16 | 36.75391 | 49.87891 | 65.00391 | 82.12891 | 101.25391 | 122.37891 |
| 5/64 | 36.94360 | 50.09985 | 65.25610 | 82.41235 | 101.56860 | 122.72485 |
| 3/32 | 37.13379 | 50.32129 | 65.50879 | 82.69629 | 101.88379 | 123.07129 |
| 7/64 | 37.32446 | 50.54321 | 65.76196 | 82.98071 | 102.19946 | 123.41821 |
| 1/8 | 37.51563 | 50.76563 | 66.01563 | 83.26563 | 102.51563 | 123.76563 |
| 9/64 | 37.70728 | 50.98853 | 66.26978 | 83.55103 | 102.83228 | 124.11353 |
| 5/32 | 37.89941 | 51.21191 | 66.52441 | 83.83691 | 103.14941 | 124.46191 |
| 11/64 | 38.09204 | 51.43579 | 66.77954 | 84.12329 | 103.46704 | 124.81079 |
| 3/16 | 38.28516 | 51.66016 | 67.03516 | 84.41016 | 103.78516 | 125.16016 |
| 13/64 | 38.47876 | 51.88501 | 67.29126 | 84.69751 | 104.10376 | 125.51001 |
| 7/32 | 38.67285 | 52.11035 | 67.54785 | 84.98535 | 104.42285 | 125.86035 |
| 15/64 | 38.86743 | 52.33618 | 67.80493 | 85.27368 | 104.74243 | 126.21118 |
| 1/4 | 39.06250 | 52.56250 | 68.06250 | 85.56250 | 105.06250 | 126.56250 |

top of its column, and 5/64 in the left-hand column. The square is then found to equal 65.25610. In the same way, the square of 3 9/16 is found to equal 10.16016.

Squares of Mixed Numbers from 6¹⁄₆₄ to 12 (Continued)

| | 6 | 7 | 8 | 9 | 10 | 11 |
|---|---|---|---|---|---|---|
| 17/64 | 39.25806 | 52.78931 | 68.32056 | 85.85181 | 105.38306 | 126.91431 |
| 9/32 | 39.45410 | 53.01660 | 68.57910 | 86.14160 | 105.70410 | 127.26660 |
| 19/64 | 39.65063 | 53.24438 | 68.83813 | 86.43188 | 106.02563 | 127.61938 |
| 5/16 | 39.84766 | 53.47266 | 69.09766 | 86.72266 | 106.34766 | 127.97266 |
| 21/64 | 40.04517 | 53.70142 | 69.35767 | 87.01392 | 106.67017 | 128.32642 |
| 11/32 | 40.24316 | 53.93066 | 69.61816 | 87.30566 | 106.99316 | 128.68066 |
| 23/64 | 40.44165 | 54.16040 | 69.87915 | 87.59790 | 107.31665 | 129.03540 |
| 3/8 | 40.64063 | 54.39063 | 70.14063 | 87.89063 | 107.64063 | 129.39063 |
| 25/64 | 40.84009 | 54.62134 | 70.40259 | 88.18384 | 107.96509 | 129.74634 |
| 13/32 | 41.04004 | 54.85254 | 70.66504 | 88.47754 | 108.29004 | 130.10254 |
| 27/64 | 41.24048 | 55.08423 | 70.92798 | 88.77173 | 108.61548 | 130.45923 |
| 7/16 | 41.44141 | 55.31641 | 71.19141 | 89.06641 | 108.94141 | 130.81641 |
| 29/64 | 41.64282 | 55.54907 | 71.45532 | 89.36157 | 109.26782 | 131.17407 |
| 15/32 | 41.84473 | 55.78223 | 71.71973 | 89.65723 | 109.59473 | 131.53223 |
| 31/64 | 42.04712 | 56.01587 | 71.98462 | 89.95337 | 109.92212 | 131.89087 |
| 1/2 | 42.25000 | 56.25000 | 72.25000 | 90.25000 | 110.25000 | 132.25000 |
| 33/64 | 42.45337 | 56.48462 | 72.51587 | 90.54712 | 110.57837 | 132.60962 |
| 17/32 | 42.65723 | 56.71973 | 72.78223 | 90.84473 | 110.90723 | 132.96973 |
| 35/64 | 42.86157 | 56.95532 | 73.04907 | 91.14282 | 111.23657 | 133.33032 |
| 9/16 | 43.06641 | 57.19141 | 73.31641 | 91.44141 | 111.56641 | 133.69141 |
| 37/64 | 43.27173 | 57.42798 | 73.58423 | 91.74048 | 111.89673 | 134.05298 |
| 19/32 | 43.47754 | 57.66504 | 73.85254 | 92.04004 | 112.22754 | 134.41504 |
| 39/64 | 43.68384 | 57.90259 | 74.12134 | 92.34009 | 112.55884 | 134.77759 |
| 5/8 | 43.89063 | 58.14063 | 74.39063 | 92.64063 | 112.89063 | 135.14063 |
| 41/64 | 44.09790 | 58.37915 | 74.66040 | 92.94165 | 113.22290 | 135.50415 |
| 21/32 | 44.30566 | 58.61816 | 74.93066 | 93.24316 | 113.55566 | 135.86316 |
| 43/64 | 44.51392 | 58.85767 | 75.20142 | 93.54517 | 113.88892 | 136.23267 |
| 11/16 | 44.72266 | 59.09766 | 75.47266 | 93.84766 | 114.22266 | 136.59766 |
| 45/64 | 44.93188 | 59.33813 | 75.74438 | 94.15063 | 114.55688 | 136.96313 |
| 23/32 | 45.14160 | 59.57910 | 76.01660 | 94.45410 | 114.89160 | 137.32910 |
| 47/64 | 45.35181 | 59.82056 | 76.28931 | 94.75806 | 115.22681 | 137.69556 |
| 3/4 | 45.56250 | 60.06250 | 76.56250 | 95.06250 | 115.56250 | 138.06250 |
| 49/64 | 45.77368 | 60.30493 | 76.83618 | 95.36743 | 115.89868 | 138.42993 |
| 25/32 | 45.98535 | 60.54785 | 77.11035 | 95.67285 | 116.23535 | 138.79785 |
| 51/64 | 46.19751 | 60.79126 | 77.38501 | 95.97876 | 116.57251 | 139.16626 |
| 13/16 | 46.41016 | 61.03516 | 77.66016 | 96.28516 | 116.91016 | 139.53516 |
| 53/64 | 46.62329 | 61.27954 | 77.93579 | 96.59204 | 117.24829 | 139.90454 |
| 27/32 | 46.83691 | 61.52441 | 78.21191 | 96.89941 | 117.58691 | 140.27441 |
| 55/64 | 47.05103 | 61.76978 | 78.48853 | 97.20728 | 117.92603 | 140.64478 |
| 7/8 | 47.26563 | 62.01563 | 78.76563 | 97.51563 | 118.26563 | 141.01563 |
| 57/64 | 47.48071 | 62.26196 | 79.04321 | 97.82446 | 118.60571 | 141.38696 |
| 29/32 | 47.69629 | 62.50879 | 79.32129 | 98.13379 | 118.94629 | 141.75879 |
| 59/64 | 47.91235 | 62.75610 | 79.59985 | 98.44360 | 119.28735 | 142.13110 |
| 15/16 | 48.12891 | 63.00391 | 79.87891 | 98.75391 | 119.62891 | 142.50391 |
| 61/64 | 48.34595 | 63.25220 | 80.15845 | 99.06470 | 119.97095 | 142.87720 |
| 31/32 | 48.56348 | 63.50098 | 80.43848 | 99.37598 | 120.31348 | 143.25098 |
| 63/64 | 48.78149 | 63.75024 | 80.71899 | 99.68774 | 120.65649 | 143.62524 |

## Squares and Cubes of Numbers from 1/32 to 100
### Advancing by 32nds to 2 ; from 2 to 10 by 16ths ; from 10 to 100 by 8ths

| No. | Square | Cube | No. | Square | Cube | No. | Square | Cube |
|---|---|---|---|---|---|---|---|---|
| 1/32 | 0.000977 | 0.000031 | 1 17/32 | 2.344727 | 3.590363 | 4 | 16.0000 | 64.0000 |
| 1/16 | 0.003906 | 0.000244 | 9/16 | 2.441406 | 3.814697 | 1/16 | 16.5039 | 67.0471 |
| 3/32 | 0.008789 | 0.000824 | 19/32 | 2.540039 | 4.048187 | 1/8 | 17.0156 | 70.1895 |
| 1/8 | 0.015625 | 0.001953 | 5/8 | 2.640625 | 4.291016 | 3/16 | 17.5352 | 73.4285 |
| 5/32 | 0.024414 | 0.003815 | 21/32 | 2.743164 | 4.543365 | 1/4 | 18.0625 | 76.7656 |
| 3/16 | 0.035156 | 0.006592 | 11/16 | 2.847656 | 4.805420 | 5/16 | 18.5977 | 80.2024 |
| 7/32 | 0.047852 | 0.010468 | 23/32 | 2.954102 | 5.077362 | 3/8 | 19.1406 | 83.7402 |
| 1/4 | 0.062500 | 0.015625 | 3/4 | 3.062500 | 5.359375 | 7/16 | 19.6914 | 87.3806 |
| 9/32 | 0.079102 | 0.022247 | 25/32 | 3.172852 | 5.651642 | 1/2 | 20.2500 | 91.1250 |
| 5/16 | 0.097656 | 0.030518 | 13/16 | 3.285156 | 5.954346 | 9/16 | 20.8164 | 94.9749 |
| 11/32 | 0.118164 | 0.040619 | 27/32 | 3.399414 | 6.267670 | 5/8 | 21.3906 | 98.9316 |
| 3/8 | 0.140625 | 0.052734 | 7/8 | 3.515625 | 6.591797 | 11/16 | 21.9727 | 102.9968 |
| 13/32 | 0.165039 | 0.067047 | 29/32 | 3.633789 | 6.926910 | 3/4 | 22.5625 | 107.1719 |
| 7/16 | 0.191406 | 0.083740 | 15/16 | 3.753906 | 7.273193 | 13/16 | 23.1602 | 111.4583 |
| 15/32 | 0.219727 | 0.102997 | 31/32 | 3.875977 | 7.630829 | 7/8 | 23.7656 | 115.8574 |
| 1/2 | 0.250000 | 0.125000 | 2 | 4.000000 | 8.00000 | 15/16 | 24.3789 | 120.3708 |
| 17/32 | 0.282227 | 0.149933 | 1/32 | 4.12598 | 8.38089 | 5 | 25.0000 | 125.0000 |
| 9/16 | 0.316406 | 0.177979 | 1/16 | 4.25391 | 8.77368 | 1/16 | 25.6289 | 129.7463 |
| 19/32 | 0.352539 | 0.209320 | 1/8 | 4.51563 | 9.59570 | 1/8 | 26.2656 | 134.6113 |
| 5/8 | 0.390625 | 0.244141 | 3/16 | 4.78516 | 10.46753 | 3/16 | 26.9102 | 139.5964 |
| 21/32 | 0.430664 | 0.282623 | 1/4 | 5.06250 | 11.39063 | 1/4 | 27.5625 | 144.7031 |
| 11/16 | 0.472656 | 0.324951 | 5/16 | 5.34766 | 12.36646 | 5/16 | 28.2227 | 149.9329 |
| 23/32 | 0.516602 | 0.371307 | 3/8 | 5.64063 | 13.39648 | 3/8 | 28.8906 | 155.2871 |
| 3/4 | 0.562500 | 0.421875 | 7/16 | 5.94141 | 14.48218 | 7/16 | 29.5664 | 160.7673 |
| 25/32 | 0.610352 | 0.476837 | 1/2 | 6.25000 | 15.62500 | 1/2 | 30.2500 | 166.3750 |
| 13/16 | 0.660156 | 0.535377 | 9/16 | 6.56641 | 16.82642 | 9/16 | 30.9414 | 172.1116 |
| 27/32 | 0.711914 | 0.600677 | 5/8 | 6.89063 | 18.08789 | 5/8 | 31.6406 | 177.9785 |
| 7/8 | 0.765625 | 0.669922 | 11/16 | 7.22266 | 19.41089 | 11/16 | 32.3477 | 183.9773 |
| 29/32 | 0.821289 | 0.744293 | 3/4 | 7.56250 | 20.79688 | 3/4 | 33.0625 | 190.1094 |
| 15/16 | 0.878906 | 0.823975 | 13/16 | 7.91016 | 22.24731 | 13/16 | 33.7852 | 196.3762 |
| 31/32 | 0.938477 | 0.909149 | 7/8 | 8.26563 | 23.76367 | 7/8 | 34.5156 | 202.7793 |
| 1 | 1.000000 | 1.000000 | 15/16 | 8.62891 | 25.34741 | 15/16 | 35.2539 | 209.3201 |
| 1 1/32 | 1.063477 | 1.096710 | 3 | 9.00000 | 27.00000 | 6 | 36.0000 | 216.0000 |
| 1 1/16 | 1.128906 | 1.199463 | 1/16 | 9.37891 | 28.72290 | 1/16 | 36.7539 | 222.8206 |
| 1 3/32 | 1.196289 | 1.308441 | 1/8 | 9.76563 | 30.51758 | 1/8 | 37.5156 | 229.7832 |
| 1 1/8 | 1.265625 | 1.423828 | 3/16 | 10.16016 | 32.38550 | 3/16 | 38.2852 | 236.8894 |
| 1 5/32 | 1.336914 | 1.545807 | 1/4 | 10.56250 | 34.32813 | 1/4 | 39.0625 | 244.1406 |
| 1 3/16 | 1.410156 | 1.674561 | 5/16 | 10.97266 | 36.34692 | 5/16 | 39.8477 | 251.5383 |
| 1 7/32 | 1.485352 | 1.810272 | 3/8 | 11.39063 | 38.44336 | 3/8 | 40.6406 | 259.0840 |
| 1 1/4 | 1.562500 | 1.953125 | 7/16 | 11.81641 | 40.61890 | 7/16 | 41.4414 | 266.7791 |
| 1 9/32 | 1.641602 | 2.103302 | 1/2 | 12.25000 | 42.87500 | 1/2 | 42.2500 | 274.6250 |
| 1 5/16 | 1.722656 | 2.260986 | 9/16 | 12.69141 | 45.21313 | 9/16 | 43.0664 | 282.6233 |
| 1 11/32 | 1.805664 | 2.426361 | 5/8 | 13.14063 | 47.63477 | 5/8 | 43.8906 | 290.7754 |
| 1 3/8 | 1.890625 | 2.599609 | 11/16 | 13.59766 | 50.14136 | 11/16 | 44.7227 | 299.0828 |
| 1 13/32 | 1.977539 | 2.780914 | 3/4 | 14.06250 | 52.73438 | 3/4 | 45.5625 | 307.5469 |
| 1 7/16 | 2.066406 | 2.970459 | 13/16 | 14.53516 | 55.41528 | 13/16 | 46.4102 | 316.1692 |
| 1 15/32 | 2.157227 | 3.168427 | 7/8 | 15.01563 | 58.18555 | 7/8 | 47.2656 | 324.9512 |
| 1 1/2 | 2.250000 | 3.375000 | 15/16 | 15.50391 | 61.04663 | 15/16 | 48.1289 | 333.8943 |

### Squares and Cubes of Numbers from 1/32 to 100 (Continued)

| No. | Square | Cube | No. | Square | Cube | No. | Square | Cube |
|---|---|---|---|---|---|---|---|---|
| 7 | 49.0000 | 343.0000 | 10 | 100.0000 | 1000.0000 | 16 | 256.0000 | 4096.000 |
| 1/16 | 49.8789 | 352.2698 | 1/8 | 102.5156 | 1037.9707 | 1/8 | 260.0156 | 4192.752 |
| 1/8 | 50.7656 | 361.7051 | 1/4 | 105.0625 | 1076.8906 | 1/4 | 264.0625 | 4291.016 |
| 3/16 | 51.6602 | 371.3074 | 3/8 | 107.6406 | 1116.7715 | 3/8 | 268.1406 | 4390.803 |
| 1/4 | 52.5625 | 381.0781 | 1/2 | 110.2500 | 1157.6250 | 1/2 | 272.2500 | 4492.125 |
| 5/16 | 53.4727 | 391.0188 | 5/8 | 112.8906 | 1199.4629 | 5/8 | 276.3906 | 4594.994 |
| 3/8 | 54.3906 | 401.1309 | 3/4 | 115.5625 | 1242.2969 | 3/4 | 280.5625 | 4699.422 |
| 7/16 | 55.3164 | 411.4158 | 7/8 | 118.2656 | 1286.1387 | 7/8 | 284.7656 | 4805.420 |
| 1/2 | 56.2500 | 421.8750 | 11 | 121.0000 | 1331.0000 | 17 | 289.0000 | 4913.000 |
| 9/16 | 57.1914 | 432.5100 | 1/8 | 123.7656 | 1376.8926 | 1/8 | 293.2656 | 5022.174 |
| 5/8 | 58.1406 | 443.3223 | 1/4 | 126.5625 | 1423.8281 | 1/4 | 297.5625 | 5132.953 |
| 11/16 | 59.0977 | 454.3132 | 3/8 | 129.3906 | 1471.8184 | 3/8 | 301.8906 | 5245.350 |
| 3/4 | 60.0625 | 465.4844 | 1/2 | 132.2500 | 1520.8750 | 1/2 | 306.2500 | 5359.375 |
| 13/16 | 61.0352 | 476.8372 | 5/8 | 135.1406 | 1571.0098 | 5/8 | 310.6406 | 5475.041 |
| 7/8 | 62.0156 | 488.3730 | 3/4 | 138.0625 | 1622.2344 | 3/4 | 315.0625 | 5592.359 |
| 15/16 | 63.0039 | 500.0935 | 7/8 | 141.0156 | 1674.5605 | 7/8 | 319.5156 | 5711.342 |
| 8 | 64.0000 | 512.0000 | 12 | 144.0000 | 1728.0000 | 18 | 324.0000 | 5832.000 |
| 1/16 | 65.0039 | 524.0940 | 1/8 | 147.0156 | 1782.5645 | 1/8 | 328.5156 | 5954.346 |
| 1/8 | 66.0156 | 536.3770 | 1/4 | 150.0625 | 1838.2656 | 1/4 | 333.0625 | 6078.391 |
| 3/16 | 67.0352 | 548.8503 | 3/8 | 153.1406 | 1895.1152 | 3/8 | 337.6406 | 6204.146 |
| 1/4 | 68.0625 | 561.5156 | 1/2 | 156.2500 | 1953.1250 | 1/2 | 342.2500 | 6331.625 |
| 5/16 | 69.0977 | 574.3743 | 5/8 | 159.3906 | 2012.3066 | 5/8 | 346.8906 | 6460.838 |
| 3/8 | 70.1406 | 587.4277 | 3/4 | 162.5625 | 2072.6719 | 3/4 | 351.5625 | 6591.797 |
| 7/16 | 71.1914 | 600.6775 | 7/8 | 165.7656 | 2134.2324 | 7/8 | 356.2656 | 6724.514 |
| 1/2 | 72.2500 | 614.1250 | 13 | 169.0000 | 2197.0000 | 19 | 361.0000 | 6859.000 |
| 9/16 | 73.3164 | 627.7717 | 1/8 | 172.2656 | 2260.9863 | 1/8 | 365.7656 | 6995.268 |
| 5/8 | 74.3906 | 641.6191 | 1/4 | 175.5625 | 2326.2031 | 1/4 | 370.5625 | 7133.328 |
| 11/16 | 75.4727 | 655.6687 | 3/8 | 178.8906 | 2392.6621 | 3/8 | 375.3906 | 7273.193 |
| 3/4 | 76.5625 | 669.9219 | 1/2 | 182.2500 | 2460.3750 | 1/2 | 380.2500 | 7414.875 |
| 13/16 | 77.6602 | 684.3801 | 5/8 | 185.6406 | 2529.3535 | 5/8 | 385.1406 | 7558.385 |
| 7/8 | 78.7656 | 699.0449 | 3/4 | 189.0625 | 2599.6094 | 3/4 | 390.0625 | 7703.734 |
| 15/16 | 79.8789 | 713.9177 | 7/8 | 192.5156 | 2671.1543 | 7/8 | 395.0156 | 7850.936 |
| 9 | 81.0000 | 729.0000 | 14 | 196.0000 | 2744.0000 | 20 | 400.0000 | 8000.000 |
| 1/16 | 82.1289 | 744.2932 | 1/8 | 199.5156 | 2818.1582 | 1/8 | 405.0156 | 8150.939 |
| 1/8 | 83.2656 | 759.7988 | 1/4 | 203.0625 | 2893.6406 | 1/4 | 410.0625 | 8303.766 |
| 3/16 | 84.4102 | 775.5183 | 3/8 | 206.6406 | 2970.4590 | 3/8 | 415.1406 | 8458.490 |
| 1/4 | 85.5625 | 791.4531 | 1/2 | 210.2500 | 3048.6250 | 1/2 | 420.2500 | 8615.125 |
| 5/16 | 86.7227 | 807.6047 | 5/8 | 213.8906 | 3128.1504 | 5/8 | 425.3906 | 8773.482 |
| 3/8 | 87.8906 | 823.9746 | 3/4 | 217.5625 | 3209.0469 | 3/4 | 430.5625 | 8934.172 |
| 7/16 | 89.0664 | 840.5642 | 7/8 | 221.2656 | 3291.3262 | 7/8 | 435.7656 | 9096.607 |
| 1/2 | 90.2500 | 857.3750 | 15 | 225.0000 | 3375.0000 | 21 | 441.0000 | 9261.000 |
| 9/16 | 91.4414 | 874.4084 | 1/8 | 228.7656 | 3460.0801 | 1/8 | 446.2656 | 9427.361 |
| 5/8 | 92.6406 | 891.6660 | 1/4 | 232.5625 | 3546.5781 | 1/4 | 451.5625 | 9595.703 |
| 11/16 | 93.8477 | 909.1492 | 3/8 | 236.3906 | 3634.5059 | 3/8 | 456.8906 | 9766.037 |
| 3/4 | 95.0625 | 926.8594 | 1/2 | 240.2500 | 3723.8750 | 1/2 | 462.2500 | 9,938.375 |
| 13/16 | 96.2852 | 944.7981 | 5/8 | 244.1406 | 3814.6973 | 5/8 | 467.6406 | 10,112.729 |
| 7/8 | 97.5156 | 962.9668 | 3/4 | 248.0625 | 3906.9844 | 3/4 | 473.0625 | 10,289.109 |
| 15/16 | 98.7539 | 981.3669 | 7/8 | 252.0156 | 4000.7480 | 7/8 | 478.5156 | 10,467.529 |

**Squares and Cubes of Numbers from 1/32 to 100 (Continued)**

| No. | Square | Cube | No. | Square | Cube | No. | Square | Cube |
|---|---|---|---|---|---|---|---|---|
| 22 | 484.0000 | 10,648.000 | 28 | 784.000 | 21,952.000 | 34 | 1156.000 | 39,304.000 |
| ⅛ | 489.5156 | 10,830.533 | ⅛ | 791.016 | 22,247.314 | ⅛ | 1164.516 | 39,739.096 |
| ¼ | 495.0625 | 11,015.140 | ¼ | 798.063 | 22,545.266 | ¼ | 1173.063 | 40,177.391 |
| ⅜ | 500.6406 | 11,201.834 | ⅜ | 805.141 | 22,845.865 | ⅜ | 1181.641 | 40,618.896 |
| ½ | 506.2500 | 11,390.625 | ½ | 812.250 | 23,149.125 | ½ | 1190.250 | 41,063.625 |
| ⅝ | 511.8906 | 11,581.525 | ⅝ | 819.391 | 23,455.057 | ⅝ | 1198.891 | 41,511.588 |
| ¾ | 517.5625 | 11,774.547 | ¾ | 826.563 | 23,763.672 | ¾ | 1207.563 | 41,962.797 |
| ⅞ | 523.2656 | 11,969.701 | ⅞ | 833.766 | 24,074.982 | ⅞ | 1216.266 | 42,417.264 |
| 23 | 529.0000 | 12,167.000 | 29 | 841.000 | 24,389.000 | 35 | 1225.000 | 42,875.000 |
| ⅛ | 534.7656 | 12,366.455 | ⅛ | 848.266 | 24,705.736 | ⅛ | 1233.766 | 43,336.018 |
| ¼ | 540.5625 | 12,568.078 | ¼ | 855.563 | 25,025.203 | ¼ | 1242.563 | 43,800.328 |
| ⅜ | 546.3906 | 12,771.881 | ⅜ | 862.891 | 25,347.412 | ⅜ | 1251.391 | 44,267.943 |
| ½ | 552.2500 | 12,977.875 | ½ | 870.250 | 25,672.375 | ½ | 1260.250 | 44,738.875 |
| ⅝ | 558.1406 | 13,186.072 | ⅝ | 877.641 | 26,000.104 | ⅝ | 1269.141 | 45,213.135 |
| ¾ | 564.0625 | 13,396.484 | ¾ | 885.063 | 26,330.609 | ¾ | 1278.063 | 45,690.734 |
| ⅞ | 570.0156 | 13,609.123 | ⅞ | 892.516 | 26,663.904 | ⅞ | 1287.016 | 46,171.686 |
| 24 | 576.0000 | 13,824.000 | 30 | 900.000 | 27,000.000 | 36 | 1296.000 | 46,656.000 |
| ⅛ | 582.0156 | 14,041.127 | ⅛ | 907.516 | 27,338.908 | ⅛ | 1305.016 | 47,143.689 |
| ¼ | 588.0625 | 14,260.516 | ¼ | 915.063 | 27,680.641 | ¼ | 1314.063 | 47,634.766 |
| ⅜ | 594.1406 | 14,482.178 | ⅜ | 922.641 | 28,025.209 | ⅜ | 1323.141 | 48,129.240 |
| ½ | 600.2500 | 14,706.125 | ½ | 930.250 | 28,372.625 | ½ | 1332.250 | 48,627.125 |
| ⅝ | 606.3906 | 14,932.369 | ⅝ | 937.891 | 28,722.900 | ⅝ | 1341.391 | 49,128.432 |
| ¾ | 612.5625 | 15,160.922 | ¾ | 945.563 | 29,076.047 | ¾ | 1350.563 | 49,633.172 |
| ⅞ | 618.7656 | 15,391.795 | ⅞ | 953.266 | 29,432.076 | ⅞ | 1359.766 | 50,141.357 |
| 25 | 625.0000 | 15,625.000 | 31 | 961.000 | 29,791.000 | 37 | 1369.000 | 50,653.000 |
| ⅛ | 631.2656 | 15,860.549 | ⅛ | 968.766 | 30,152.830 | ⅛ | 1378.266 | 51,168.111 |
| ¼ | 637.5625 | 16,098.453 | ¼ | 976.563 | 30,517.578 | ¼ | 1387.563 | 51,686.703 |
| ⅜ | 643.8906 | 16,338.725 | ⅜ | 984.391 | 30,885.256 | ⅜ | 1396.891 | 52,208.787 |
| ½ | 650.2500 | 16,581.375 | ½ | 992.250 | 31,255.875 | ½ | 1406.250 | 52,734.375 |
| ⅝ | 656.6406 | 16,826.416 | ⅝ | 1000.141 | 31,629.447 | ⅝ | 1415.641 | 53,263.479 |
| ¾ | 663.0625 | 17,073.859 | ¾ | 1008.063 | 32,005.984 | ¾ | 1425.063 | 53,796.109 |
| ⅞ | 669.5156 | 17,323.717 | ⅞ | 1016.016 | 32,385.498 | ⅞ | 1434.516 | 54,332.279 |
| 26 | 676.0000 | 17,576.000 | 32 | 1024.000 | 32,768.000 | 38 | 1444.000 | 54,872.000 |
| ⅛ | 682.5156 | 17,830.721 | ⅛ | 1032.016 | 33,153.502 | ⅛ | 1453.516 | 55,415.283 |
| ¼ | 689.0625 | 18,087.891 | ¼ | 1040.063 | 33,542.016 | ¼ | 1463.063 | 55,962.141 |
| ⅜ | 695.6406 | 18,347.521 | ⅜ | 1048.141 | 33,933.553 | ⅜ | 1472.641 | 56,512.584 |
| ½ | 702.2500 | 18,609.625 | ½ | 1056.250 | 34,328.125 | ½ | 1482.250 | 57,066.625 |
| ⅝ | 708.8906 | 18,874.213 | ⅝ | 1064.391 | 34,725.744 | ⅝ | 1491.891 | 57,624.275 |
| ¾ | 715.5625 | 19,141.297 | ¾ | 1072.563 | 35,126.422 | ¾ | 1501.563 | 58,185.547 |
| ⅞ | 722.2656 | 19,410.889 | ⅞ | 1080.766 | 35,530.170 | ⅞ | 1511.266 | 58,750.451 |
| 27 | 729.0000 | 19,683.000 | 33 | 1089.000 | 35,937.000 | 39 | 1521.000 | 59,319.000 |
| ⅛ | 735.7656 | 19,957.643 | ⅛ | 1097.266 | 36,346.924 | ⅛ | 1530.766 | 59,891.205 |
| ¼ | 742.5625 | 20,234.828 | ¼ | 1105.563 | 36,759.953 | ¼ | 1540.563 | 60,467.078 |
| ⅜ | 749.3906 | 20,514.568 | ⅜ | 1113.891 | 37,176.100 | ⅜ | 1550.391 | 61,046.631 |
| ½ | 756.2500 | 20,796.875 | ½ | 1122.250 | 37,595.375 | ½ | 1560.250 | 61,629.875 |
| ⅝ | 763.1406 | 21,081.760 | ⅝ | 1130.641 | 38,017.791 | ⅝ | 1570.141 | 62,216.822 |
| ¾ | 770.0625 | 21,369.234 | ¾ | 1139.063 | 38,443.359 | ¾ | 1580.063 | 62,807.484 |
| ⅞ | 777.0156 | 21,659.311 | ⅞ | 1147.516 | 38,872.092 | ⅞ | 1590.016 | 63,401.873 |

### Squares and Cubes of Numbers from 1/32 to 100 (Continued)

| No. | Square | Cube | No. | Square | Cube | No. | Square | Cube |
|---|---|---|---|---|---|---|---|---|
| 40 | 1600.000 | 64,000.000 | 46 | 2116.000 | 97,336.00 | 52 | 2704.000 | 140,608.00 |
| 1/8 | 1610.016 | 64,601.877 | 1/8 | 2127.516 | 98,131.66 | 1/8 | 2717.016 | 141,624.44 |
| 1/4 | 1620.063 | 65,207.516 | 1/4 | 2139.063 | 98,931.64 | 1/4 | 2730.063 | 142,645.77 |
| 3/8 | 1630.141 | 65,816.928 | 3/8 | 2150.641 | 99,735.96 | 3/8 | 2743.141 | 143,671.99 |
| 1/2 | 1640.250 | 66,430.125 | 1/2 | 2162.250 | 100,544.63 | 1/2 | 2756.250 | 144,703.13 |
| 5/8 | 1650.391 | 67,047.119 | 5/8 | 2173.891 | 101,357.65 | 5/8 | 2769.391 | 145,739.18 |
| 3/4 | 1660.563 | 67,667.922 | 3/4 | 2185.563 | 102,175.05 | 3/4 | 2782.563 | 146,780.17 |
| 7/8 | 1670.766 | 68,292.545 | 7/8 | 2197.266 | 102,996.83 | 7/8 | 2795.766 | 147,826.11 |
| 41 | 1681.000 | 68,921.000 | 47 | 2209.000 | 103,823.00 | 53 | 2809.000 | 148,877.00 |
| 1/8 | 1691.266 | 69,553.299 | 1/8 | 2220.766 | 104,653.58 | 1/8 | 2822.266 | 149,932.86 |
| 1/4 | 1701.563 | 70,189.453 | 1/4 | 2232.563 | 105,488.58 | 1/4 | 2835.563 | 150,993.70 |
| 3/8 | 1711.891 | 70,829.475 | 3/8 | 2244.391 | 106,328.01 | 3/8 | 2848.891 | 152,059.54 |
| 1/2 | 1722.250 | 71,473.375 | 1/2 | 2256.250 | 107,171.88 | 1/2 | 2862.250 | 153,130.38 |
| 5/8 | 1732.641 | 72,121.166 | 5/8 | 2268.141 | 108,020.20 | 5/8 | 2875.641 | 154,206.23 |
| 3/4 | 1743.063 | 72,772.859 | 3/4 | 2280.063 | 108,872.98 | 3/4 | 2889.063 | 155,287.11 |
| 7/8 | 1753.516 | 73,428.467 | 7/8 | 2292.016 | 109,730.25 | 7/8 | 2902.516 | 156,373.03 |
| 42 | 1764.000 | 74,088.000 | 48 | 2304.000 | 110,592.00 | 54 | 2916.000 | 157,464.00 |
| 1/8 | 1774.516 | 74,751.471 | 1/8 | 2316.016 | 111,458.25 | 1/8 | 2929.516 | 158,560.03 |
| 1/4 | 1785.063 | 75,418.891 | 1/4 | 2328.063 | 112,329.02 | 1/4 | 2943.063 | 159,661.14 |
| 3/8 | 1795.641 | 76,090.271 | 3/8 | 2340.141 | 113,204.30 | 3/8 | 2956.641 | 160,767.33 |
| 1/2 | 1806.250 | 76,765.625 | 1/2 | 2352.250 | 114,084.13 | 1/2 | 2970.250 | 161,878.63 |
| 5/8 | 1816.891 | 77,444.963 | 5/8 | 2364.391 | 114,968.49 | 5/8 | 2983.891 | 162,995.03 |
| 3/4 | 1827.563 | 78,128.297 | 3/4 | 2376.563 | 115,857.42 | 3/4 | 2997.563 | 164,116.55 |
| 7/8 | 1838.266 | 78,815.639 | 7/8 | 2388.766 | 116,750.92 | 7/8 | 3011.266 | 165,243.20 |
| 43 | 1849.000 | 79,507.000 | 49 | 2401.000 | 117,649.00 | 55 | 3025.000 | 166,375.00 |
| 1/8 | 1859.766 | 80,202.393 | 1/8 | 2413.266 | 118,551.67 | 1/8 | 3038.766 | 167,511.96 |
| 1/4 | 1870.563 | 80,901.828 | 1/4 | 2425.563 | 119,458.95 | 1/4 | 3052.563 | 168,654.08 |
| 3/8 | 1881.391 | 81,605.318 | 3/8 | 2437.891 | 120,370.85 | 3/8 | 3066.391 | 169,801.38 |
| 1/2 | 1892.250 | 82,312.875 | 1/2 | 2450.250 | 121,287.38 | 1/2 | 3080.250 | 170,953.88 |
| 5/8 | 1903.141 | 83,024.510 | 5/8 | 2462.641 | 122,208.54 | 5/8 | 3094.141 | 172,111.57 |
| 3/4 | 1914.063 | 83,740.234 | 3/4 | 2475.063 | 123,134.36 | 3/4 | 3108.063 | 173,274.48 |
| 7/8 | 1925.016 | 84,460.061 | 7/8 | 2487.516 | 124,064.84 | 7/8 | 3122.016 | 174,442.62 |
| 44 | 1936.000 | 85,184.000 | 50 | 2500.000 | 125,000.00 | 56 | 3136.000 | 175,616.00 |
| 1/8 | 1947.016 | 85,912.064 | 1/8 | 2512.516 | 125,939.85 | 1/8 | 3150.016 | 176,794.63 |
| 1/4 | 1958.063 | 86,644.266 | 1/4 | 2525.063 | 126,884.39 | 1/4 | 3164.063 | 177,978.52 |
| 3/8 | 1969.141 | 87,380.615 | 3/8 | 2537.641 | 127,833.65 | 3/8 | 3178.141 | 179,167.68 |
| 1/2 | 1980.250 | 88,121.125 | 1/2 | 2550.250 | 128,787.63 | 1/2 | 3192.250 | 180,362.13 |
| 5/8 | 1991.391 | 88,865.807 | 5/8 | 2562.891 | 129,746.34 | 5/8 | 3206.391 | 181,561.87 |
| 3/4 | 2002.563 | 89,614.672 | 3/4 | 2575.563 | 130,709.80 | 3/4 | 3220.563 | 182,766.92 |
| 7/8 | 2013.766 | 90,367.732 | 7/8 | 2588.266 | 131,678.01 | 7/8 | 3234.766 | 183,977.29 |
| 45 | 2025.000 | 91,125.000 | 51 | 2601.000 | 132,651.00 | 57 | 3249.000 | 185,193.00 |
| 1/8 | 2036.266 | 91,886.486 | 1/8 | 2613.766 | 133,628.77 | 1/8 | 3263.266 | 186,414.05 |
| 1/4 | 2047.563 | 92,652.203 | 1/4 | 2626.563 | 134,611.33 | 1/4 | 3277.563 | 187,640.45 |
| 3/8 | 2058.891 | 93,422.162 | 3/8 | 2639.391 | 135,598.69 | 3/8 | 3291.891 | 188,872.22 |
| 1/2 | 2070.250 | 94,196.375 | 1/2 | 2652.250 | 136,590.88 | 1/2 | 3306.250 | 190,109.38 |
| 5/8 | 2081.641 | 94,974.854 | 5/8 | 2665.141 | 137,587.88 | 5/8 | 3320.641 | 191,351.92 |
| 3/4 | 2093.063 | 95,757.609 | 3/4 | 2678.063 | 138,589.73 | 3/4 | 3335.063 | 192,599.86 |
| 7/8 | 2104.516 | 96,544.654 | 7/8 | 2691.016 | 139,596.44 | 7/8 | 3349.516 | 193,853.22 |

## Squares and Cubes of Numbers from 1/32 to 100 (Continued)

| No. | Square | Cube | No. | Square | Cube | No. | Square | Cube |
|---|---|---|---|---|---|---|---|---|
| 58 | 3364.000 | 195,112.00 | 64 | 4096.000 | 262,144.00 | 70 | 4900.000 | 343,000.00 |
| ⅛ | 3378.516 | 196,376.22 | ⅛ | 4112.016 | 263,683.00 | ⅛ | 4917.516 | 344,840.78 |
| ¼ | 3393.063 | 197,645.89 | ¼ | 4128.063 | 265,228.02 | ¼ | 4935.063 | 346,688.14 |
| ⅜ | 3407.641 | 198,921.02 | ⅜ | 4144.141 | 266,779.05 | ⅜ | 4952.641 | 348,542.08 |
| ½ | 3422.250 | 200,201.63 | ½ | 4160.250 | 268,336.13 | ½ | 4970.250 | 350,402.63 |
| ⅝ | 3436.891 | 201,487.71 | ⅝ | 4176.391 | 269,899.24 | ⅝ | 4987.891 | 352,269.78 |
| ¾ | 3451.563 | 202,779.30 | ¾ | 4192.563 | 271,468.42 | ¾ | 5005.563 | 354,143.55 |
| ⅞ | 3466.266 | 204,076.39 | ⅞ | 4208.766 | 273,043.67 | ⅞ | 5023.266 | 356,023.95 |
| 59 | 3481.000 | 205,379.00 | 65 | 4225.000 | 274,625.00 | 71 | 5041.000 | 357,911.00 |
| ⅛ | 3495.766 | 206,687.14 | ⅛ | 4241.266 | 276,212.42 | ⅛ | 5058.766 | 359,804.71 |
| ¼ | 3510.563 | 208,000.83 | ¼ | 4257.563 | 277,805.95 | ¼ | 5076.563 | 361,705.08 |
| ⅜ | 3525.391 | 209,320.07 | ⅜ | 4273.891 | 279,405.60 | ⅜ | 5094.391 | 363,612.13 |
| ½ | 3540.250 | 210,644.88 | ½ | 4290.250 | 281,011.38 | ½ | 5112.250 | 365,525.88 |
| ⅝ | 3555.141 | 211,975.26 | ⅝ | 4306.641 | 282,623.29 | ⅝ | 5130.141 | 367,446.32 |
| ¾ | 3570.063 | 213,311.23 | ¾ | 4323.063 | 284,241.36 | ¾ | 5148.063 | 369,373.48 |
| ⅞ | 3585.016 | 214,652.81 | ⅞ | 4339.516 | 285,865.59 | ⅞ | 5166.016 | 371,307.37 |
| 60 | 3600.000 | 216,000.00 | 66 | 4356.000 | 287,496.00 | 72 | 5184.000 | 373,248.00 |
| ⅛ | 3615.016 | 217,352.81 | ⅛ | 4372.516 | 289,132.60 | ⅛ | 5202.016 | 375,195.38 |
| ¼ | 3630.063 | 218,711.27 | ¼ | 4389.063 | 290,775.39 | ¼ | 5220.063 | 377,149.43 |
| ⅜ | 3645.141 | 220,075.37 | ⅜ | 4405.641 | 292,424.40 | ⅜ | 5238.141 | 379,110.43 |
| ½ | 3660.250 | 221,445.13 | ½ | 4422.250 | 294,079.63 | ½ | 5256.250 | 381,078.13 |
| ⅝ | 3675.391 | 222,820.56 | ⅝ | 4438.891 | 295,741.09 | ⅝ | 5274.391 | 383,052.62 |
| ¾ | 3690.563 | 224,201.67 | ¾ | 4455.563 | 297,408.80 | ¾ | 5292.563 | 385,033.92 |
| ⅞ | 3705.766 | 225,588.48 | ⅞ | 4472.266 | 299,082.76 | ⅞ | 5310.766 | 387,022.04 |
| 61 | 3721.000 | 226,981.00 | 67 | 4489.000 | 300,763.00 | 73 | 5329.000 | 389,017.00 |
| ⅛ | 3736.266 | 228,379.24 | ⅛ | 4505.766 | 302,449.52 | ⅛ | 5347.266 | 391,018.80 |
| ¼ | 3751.563 | 229,783.20 | ¼ | 4522.563 | 304,142.33 | ¼ | 5365.563 | 393,027.45 |
| ⅜ | 3766.891 | 231,192.91 | ⅜ | 4539.391 | 305,841.44 | ⅜ | 5383.891 | 395,042.97 |
| ½ | 3782.250 | 232,608.38 | ½ | 4556.250 | 307,546.88 | ½ | 5402.250 | 397,065.38 |
| ⅝ | 3797.641 | 234,029.60 | ⅝ | 4573.141 | 309,258.63 | ⅝ | 5420.641 | 399,094.67 |
| ¾ | 3813.063 | 235,456.61 | ¾ | 4590.063 | 310,976.73 | ¾ | 5439.063 | 401,130.86 |
| ⅞ | 3828.516 | 236,889.40 | ⅞ | 4607.016 | 312,701.19 | ⅞ | 5457.516 | 403,173.97 |
| 62 | 3844.000 | 238,328.00 | 68 | 4624.000 | 314,432.00 | 74 | 5476.000 | 405,224.00 |
| ⅛ | 3859.516 | 239,772.41 | ⅛ | 4641.016 | 316,169.19 | ⅛ | 5494.516 | 407,280.97 |
| ¼ | 3875.063 | 241,222.64 | ¼ | 4658.063 | 317,912.77 | ¼ | 5513.063 | 409,344.89 |
| ⅜ | 3890.641 | 242,678.71 | ⅜ | 4675.141 | 319,662.74 | ⅜ | 5531.641 | 411,415.77 |
| ½ | 3906.250 | 244,140.63 | ½ | 4692.250 | 321,419.13 | ½ | 5550.250 | 413,493.63 |
| ⅝ | 3921.891 | 245,608.40 | ⅝ | 4709.391 | 323,181.93 | ⅝ | 5568.891 | 415,578.46 |
| ¾ | 3937.563 | 247,082.05 | ¾ | 4726.563 | 324,951.17 | ¾ | 5587.563 | 417,670.30 |
| ⅞ | 3953.266 | 248,561.58 | ⅞ | 4743.766 | 326,726.86 | ⅞ | 5606.266 | 419,769.14 |
| 63 | 3969.000 | 250,047.00 | 69 | 4761.000 | 328,509.00 | 75 | 5625.000 | 421,875.00 |
| ⅛ | 3984.766 | 251,538.33 | ⅛ | 4778.266 | 330,297.61 | ⅛ | 5643.766 | 423,987.89 |
| ¼ | 4000.563 | 253,035.58 | ¼ | 4795.563 | 332,092.70 | ¼ | 5662.563 | 426,107.83 |
| ⅜ | 4016.391 | 254,538.76 | ⅜ | 4812.891 | 333,894.29 | ⅜ | 5681.391 | 428,234.82 |
| ½ | 4032.250 | 256,047.88 | ½ | 4830.250 | 335,702.38 | ½ | 5700.250 | 430,368.88 |
| ⅝ | 4048.141 | 257,562.95 | ⅝ | 4847.641 | 337,516.98 | ⅝ | 5719.141 | 432,510.01 |
| ¾ | 4064.063 | 259,083.98 | ¾ | 4865.063 | 339,338.11 | ¾ | 5738.063 | 434,658.23 |
| ⅞ | 4080.016 | 260,611.00 | ⅞ | 4882.516 | 341,165.78 | ⅞ | 5757.016 | 436,813.56 |

**Squares and Cubes of Numbers from 1/32 to 100 (Continued)**

| No. | Square | Cube | No. | Square | Cube | No. | Square | Cube |
|---|---|---|---|---|---|---|---|---|
| 76 | 5776.000 | 438,976.00 | 82 | 6724.000 | 551,368.00 | 88 | 7744.000 | 681,472.00 |
| 1/8 | 5795.016 | 441,145.56 | 1/8 | 6744.516 | 553,893.35 | 1/8 | 7766.016 | 684,380.13 |
| 1/4 | 5814.063 | 443,322.27 | 1/4 | 6765.063 | 556,426.39 | 1/4 | 7788.063 | 687,296.52 |
| 3/8 | 5833.141 | 445,506.12 | 3/8 | 6785.641 | 558,967.15 | 3/8 | 7810.141 | 690,221.18 |
| 1/2 | 5852.250 | 447,697.13 | 1/2 | 6806.250 | 561,515.63 | 1/2 | 7832.250 | 693,154.13 |
| 5/8 | 5871.390 | 449,895.30 | 5/8 | 6826.891 | 564,071.84 | 5/8 | 7854.391 | 696,095.37 |
| 3/4 | 5890.563 | 452,100.67 | 3/4 | 6847.563 | 566,635.80 | 3/4 | 7876.563 | 699,044.92 |
| 7/8 | 5909.766 | 454,313.23 | 7/8 | 6868.266 | 569,207.51 | 7/8 | 7898.766 | 702,002.79 |
| 77 | 5929.000 | 456,533.00 | 83 | 6889.000 | 571,787.00 | 89 | 7921.000 | 704,969.00 |
| 1/8 | 5948.266 | 458,759.99 | 1/8 | 6909.766 | 574,374.27 | 1/8 | 7943.266 | 707,943.55 |
| 1/4 | 5967.563 | 460,994.20 | 1/4 | 6930.563 | 576,969.33 | 1/4 | 7965.563 | 710,926.45 |
| 3/8 | 5986.891 | 463,235.66 | 3/8 | 6951.391 | 579,572.19 | 3/8 | 7987.891 | 713,917.72 |
| 1/2 | 6006.250 | 465,484.38 | 1/2 | 6972.250 | 582,182.88 | 1/2 | 8010.250 | 716,917.38 |
| 5/8 | 6025.641 | 467,740.35 | 5/8 | 6993.141 | 584,801.38 | 5/8 | 8032.641 | 719,925.42 |
| 3/4 | 6045.063 | 470,003.61 | 3/4 | 7014.063 | 587,427.73 | 3/4 | 8055.063 | 722,941.86 |
| 7/8 | 6064.516 | 472,274.15 | 7/8 | 7035.016 | 590,061.94 | 7/8 | 8077.516 | 725,966.72 |
| 78 | 6084.000 | 474,552.00 | 84 | 7056.000 | 592,704.00 | 90 | 8100.000 | 729,000.00 |
| 1/8 | 6103.516 | 476,837.16 | 1/8 | 7077.016 | 595,353.94 | 1/8 | 8122.516 | 732,041.72 |
| 1/4 | 6123.063 | 479,129.64 | 1/4 | 7098.063 | 598,011.77 | 1/4 | 8145.063 | 735,091.89 |
| 3/8 | 6142.641 | 481,429.46 | 3/8 | 7119.141 | 600,677.49 | 3/8 | 8167.641 | 738,150.52 |
| 1/2 | 6162.250 | 483,736.63 | 1/2 | 7140.250 | 603,351.13 | 1/2 | 8190.250 | 741,217.63 |
| 5/8 | 6181.891 | 486,051.15 | 5/8 | 7161.391 | 606,032.68 | 5/8 | 8212.891 | 744,293.21 |
| 3/4 | 6201.563 | 488,373.05 | 3/4 | 7182.563 | 608,722.17 | 3/4 | 8235.563 | 747,377.30 |
| 7/8 | 6221.266 | 490,702.33 | 7/8 | 7203.766 | 611,419.61 | 7/8 | 8258.266 | 750,469.89 |
| 79 | 6241.000 | 493,039.00 | 85 | 7225.000 | 614,125.00 | 91 | 8281.000 | 753,571.00 |
| 1/8 | 6260.766 | 495,383.08 | 1/8 | 7246.266 | 616,838.36 | 1/8 | 8303.766 | 756,680.64 |
| 1/4 | 6280.563 | 497,734.58 | 1/4 | 7267.563 | 619,559.70 | 1/4 | 8325.563 | 759,798.83 |
| 3/8 | 6300.391 | 500,093.51 | 3/8 | 7288.891 | 622,289.04 | 3/8 | 8349.391 | 762,925.57 |
| 1/2 | 6320.250 | 502,459.88 | 1/2 | 7310.250 | 625,026.38 | 1/2 | 8372.250 | 766,060.88 |
| 5/8 | 6340.141 | 504,833.70 | 5/8 | 7331.641 | 627,771.73 | 5/8 | 8395.141 | 769,204.76 |
| 3/4 | 6360.063 | 507,214.98 | 3/4 | 7353.063 | 630,525.11 | 3/4 | 8418.063 | 772,357.23 |
| 7/8 | 6380.016 | 509,603.75 | 7/8 | 7374.516 | 633,286.53 | 7/8 | 8441.016 | 775,518.31 |
| 80 | 6400.000 | 512,000.00 | 86 | 7396.000 | 636,056.00 | 92 | 8464.000 | 778,688.00 |
| 1/8 | 6420.016 | 514,403.75 | 1/8 | 7417.516 | 638,833.53 | 1/8 | 8487.016 | 781,866.31 |
| 1/4 | 6440.063 | 516,815.02 | 1/4 | 7439.063 | 641,619.14 | 1/4 | 8510.063 | 785,053.27 |
| 3/8 | 6460.141 | 519,233.80 | 3/8 | 7460.641 | 644,412.83 | 3/8 | 8533.141 | 788,248.87 |
| 1/2 | 6480.250 | 521,660.13 | 1/2 | 7482.250 | 647,214.63 | 1/2 | 8556.250 | 791,453.13 |
| 5/8 | 6500.391 | 524,093.99 | 5/8 | 7503.891 | 650,024.53 | 5/8 | 8579.391 | 794,666.06 |
| 3/4 | 6520.563 | 526,535.42 | 3/4 | 7525.563 | 652,842.55 | 3/4 | 8602.563 | 797,887.67 |
| 7/8 | 6540.766 | 528,984.42 | 7/8 | 7547.266 | 655,668.70 | 7/8 | 8625.766 | 801,117.98 |
| 81 | 6561.000 | 531,441.00 | 87 | 7569.000 | 658,503.00 | 93 | 8649.000 | 804,357.00 |
| 1/8 | 6581.266 | 533,905.17 | 1/8 | 7590.766 | 661,345.46 | 1/8 | 8672.266 | 807,604.74 |
| 1/4 | 6601.563 | 536,376.95 | 1/4 | 7612.563 | 664,196.08 | 1/4 | 8695.563 | 810,861.20 |
| 3/8 | 6621,891 | 538,856.35 | 3/8 | 7634.390 | 667,054.88 | 3/8 | 8718.891 | 814,126.41 |
| 1/2 | 6642.250 | 541,343.38 | 1/2 | 7656.250 | 669,921.88 | 1/2 | 8742.250 | 817,400.38 |
| 5/8 | 6662.641 | 543,838.04 | 5/8 | 7678.141 | 672,797.07 | 5/8 | 8765.641 | 820,683.10 |
| 3/4 | 6683.063 | 546,340.36 | 3/4 | 7700.063 | 675,680.48 | 3/4 | 8789.063 | 823,974.61 |
| 7/8 | 6703.516 | 548,850.34 | 7/8 | 7722.016 | 678,572.12 | 7/8 | 8812.516 | 827,274.90 |

| Fraction | Decimal | | | | | | | | | |
|---|---|---|---|---|---|---|---|---|---|---|
| 43/64 | 0.671875 | 0.4514 | 0.3033 | 1.48167 | 1.82729 | 1.65458 | | | | |
| 11/16 | 0.6875 | 0.4727 | 0.3250 | 1.51182 | 1.83727 | 1.67455 | 0.8292 | 0.8820 | 1.91864 | |
| 45/64 | 0.703125 | 0.4944 | 0.3476 | 1.54110 | 1.84703 | 1.69407 | 0.8385 | 0.8892 | 1.92352 | 1.94901 |
| 23/32 | 0.71875 | 0.5166 | 0.3713 | 1.56973 | 1.85058 | 1.71316 | 0.8478 | 0.8958 | 1.92829 | 1.95219 |
| 47/64 | 0.734375 | 0.5393 | 0.3961 | 1.59775 | 1.86592 | 1.73184 | 0.8570 | 0.9022 | 1.93296 | 1.95531 |
| 3/4 | 0.750 | 0.5625 | 0.4219 | 1.62518 | 1.87506 | 1.75012 | 0.8660 | 0.9086 | 1.93753 | 1.95835 |
| 49/64 | 0.765625 | 0.5862 | 0.4488 | 1.65205 | 1.88402 | 1.76803 | 0.8750 | 0.9148 | 1.94201 | 1.96134 |
| 25/32 | 0.78125 | 0.6104 | 0.4768 | 1.67837 | 1.89279 | 1.78558 | 0.8839 | 0.9210 | 1.94640 | 1.96426 |
| 51/64 | 0.796875 | 0.6350 | 0.5060 | 1.70417 | 1.90139 | 1.80278 | 0.8927 | 0.9271 | 1.95070 | 1.96713 |
| 13/16 | 0.8125 | 0.6602 | 0.5364 | 1.72947 | 1.90982 | 1.81965 | 0.9014 | 0.9331 | 1.95491 | 1.96994 |
| 53/64 | 0.828125 | 0.6858 | 0.5679 | 1.75429 | 1.91810 | 1.83619 | 0.9100 | 0.9391 | 1.95905 | 1.97270 |
| 27/32 | 0.84375 | 0.7119 | 0.6007 | 1.77864 | 1.92621 | 1.85243 | 0.9186 | 0.9449 | 1.96311 | 1.97540 |
| 55/64 | 0.859375 | 0.7385 | 0.6347 | 1.80255 | 1.93418 | 1.86837 | 0.9270 | 0.9507 | 1.96709 | 1.97806 |
| 7/8 | 0.875 | 0.7656 | 0.6699 | 1.82602 | 1.94201 | 1.88402 | 0.9354 | 0.9565 | 1.97100 | 1.98067 |
| 57/64 | 0.890625 | 0.7932 | 0.7065 | 1.84908 | 1.94969 | 1.89939 | 0.9437 | 0.9621 | 1.97485 | 1.98323 |
| 29/32 | 0.90625 | 0.8213 | 0.7443 | 1.87174 | 1.95725 | 1.91450 | 0.9520 | 0.9677 | 1.97862 | 1.98575 |
| 59/64 | 0.921875 | 0.8499 | 0.7835 | 1.89402 | 1.96467 | 1.92934 | 0.9601 | 0.9732 | 1.98234 | 1.98822 |
| 15/16 | 0.9375 | 0.8789 | 0.8240 | 1.91591 | 1.97197 | 1.94394 | 0.9682 | 0.9787 | 1.98599 | 1.99066 |
| 61/64 | 0.953125 | 0.9084 | 0.8659 | 1.93745 | 1.97915 | 1.95830 | 0.9763 | 0.9841 | 1.98957 | 1.99305 |
| 31/32 | 0.96875 | 0.9385 | 0.9091 | 1.95864 | 1.98621 | 1.97242 | 0.9843 | 0.9895 | 1.99311 | 1.99540 |
| 63/64 | 0.984375 | 0.9690 | 0.9539 | 1.97948 | 1.99316 | 1.98632 | 0.9922 | 0.9948 | 1.99658 | 1.99772 |

*To find the log of any mixed number convert the mixed number to a common fraction and subtract the log of the denominator from the log of the numerator. *Example:* Log 5⅞ = Log ⁴⁷⁄₈ = Log 47 − Log 8. (Log tables are on pages 125-142.)

### Squares and Cubes of Numbers from 1/32 to 100 (Continued)

| No. | Square | Cube | No. | Square | Cube | No. | Square | Cube |
|---|---|---|---|---|---|---|---|---|
| 94 | 8836.000 | 830,584.00 | 96 | 9216.000 | 884,736.00 | 98 | 9604.00 | 941,192.0 |
| 1/8 | 8859.516 | 833,901.91 | 1/8 | 9240.016 | 888,196.50 | 1/8 | 9628.52 | 944,798.1 |
| 1/4 | 8883.063 | 837,228.64 | 1/4 | 9264.063 | 891,666.02 | 1/4 | 9653.06 | 948,413.4 |
| 3/8 | 8906.641 | 840,564.21 | 3/8 | 9288.141 | 895,144.55 | 3/8 | 9677.64 | 952,037.9 |
| 1/2 | 8930.250 | 843,908.63 | 1/2 | 9312.250 | 898,632.13 | 1/2 | 9702.25 | 955,671.6 |
| 5/8 | 8953.891 | 847,261.90 | 5/8 | 9336.391 | 902,128.74 | 5/8 | 9726.89 | 959,314.6 |
| 3/4 | 8977.563 | 850,624.05 | 3/4 | 9360.563 | 905,634.42 | 3/4 | 9751.56 | 962,966.8 |
| 7/8 | 9001.266 | 853,995.08 | 7/8 | 9384.766 | 909,149.17 | 7/8 | 9776.27 | 966,628.3 |
| 95 | 9025.000 | 857,375.00 | 97 | 9409.000 | 912,673.00 | 99 | 9801.00 | 970,299.0 |
| 1/8 | 9048.766 | 860,763.83 | 1/8 | 9433.266 | 916,205.92 | 1/8 | 9825.77 | 973,979.0 |
| 1/4 | 9072.563 | 864,161.58 | 1/4 | 9457.563 | 919,747.95 | 1/4 | 9850.56 | 977,668.3 |
| 3/8 | 9096.391 | 867,568.26 | 3/8 | 9481.891 | 923,299.10 | 3/8 | 9875.39 | 981,366.9 |
| 1/2 | 9120.250 | 870,983.88 | 1/2 | 9506.250 | 926,859.38 | 1/2 | 9900.25 | 985,074.9 |
| 5/8 | 9144.141 | 874,408.45 | 5/8 | 9530.641 | 930,428.79 | 5/8 | 9925.14 | 988,792.1 |
| 3/4 | 9168.063 | 877,841.98 | 3/4 | 9555.063 | 934,007.36 | 3/4 | 9950.06 | 992,518.7 |
| 7/8 | 9192.016 | 881,284.50 | 7/8 | 9579.516 | 937,595.09 | 7/8 | 9975.02 | 996,254.7 |
| | | | | | | 100 | 10,000.00 | 1,000,000.0 |

### Table of Fractions of $\pi = 3.14159265$

| a | $\frac{\pi}{a}$ | a | $\frac{\pi}{a}$ | a | $\frac{\pi}{a}$ | a | $\frac{\pi}{a}$ | a | $\frac{\pi}{a}$ |
|---|---|---|---|---|---|---|---|---|---|
| 1 | 3.14159 | 21 | 0.14960 | 41 | 0.07662 | 61 | 0.05150 | 81 | 0.03879 |
| 2 | 1.57080 | 22 | 0.14280 | 42 | 0.07480 | 62 | 0.05067 | 82 | 0.03831 |
| 3 | 1.04720 | 23 | 0.13659 | 43 | 0.07306 | 63 | 0.04987 | 83 | 0.03785 |
| 4 | 0.78540 | 24 | 0.13090 | 44 | 0.07140 | 64 | 0.04909 | 84 | 0.03740 |
| 5 | 0.62832 | 25 | 0.12566 | 45 | 0.06981 | 65 | 0.04833 | 85 | 0.03696 |
| 6 | 0.52360 | 26 | 0.12083 | 46 | 0.06830 | 66 | 0.04760 | 86 | 0.03653 |
| 7 | 0.44880 | 27 | 0.11636 | 47 | 0.06684 | 67 | 0.04689 | 87 | 0.03611 |
| 8 | 0.39270 | 28 | 0.11220 | 48 | 0.06545 | 68 | 0.04620 | 88 | 0.03570 |
| 9 | 0.34907 | 29 | 0.10833 | 49 | 0.06411 | 69 | 0.04553 | 89 | 0.03530 |
| 10 | 0.31416 | 30 | 0.10472 | 50 | 0.06283 | 70 | 0.04488 | 90 | 0.03491 |
| 11 | 0.28560 | 31 | 0.10134 | 51 | 0.06160 | 71 | 0.04425 | 91 | 0.03452 |
| 12 | 0.26180 | 32 | 0.09817 | 52 | 0.06042 | 72 | 0.04363 | 92 | 0.03415 |
| 13 | 0.24166 | 33 | 0.09520 | 53 | 0.05928 | 73 | 0.04304 | 93 | 0.03378 |
| 14 | 0.22440 | 34 | 0.09240 | 54 | 0.05818 | 74 | 0.04245 | 94 | 0.03342 |
| 15 | 0.20944 | 35 | 0.08976 | 55 | 0.05712 | 75 | 0.04189 | 95 | 0.03307 |
| 16 | 0.19635 | 36 | 0.08727 | 56 | 0.05610 | 76 | 0.04134 | 96 | 0.03272 |
| 17 | 0.18480 | 37 | 0.08491 | 57 | 0.05512 | 77 | 0.04080 | 97 | 0.03239 |
| 18 | 0.17453 | 38 | 0.08267 | 58 | 0.05417 | 78 | 0.04028 | 98 | 0.03206 |
| 19 | 0.16535 | 39 | 0.08055 | 59 | 0.05325 | 79 | 0.03977 | 99 | 0.03173 |
| 20 | 0.15708 | 40 | 0.07854 | 60 | 0.05236 | 80 | 0.03927 | 100 | 0.03142 |

**Pi ($\pi$).** — The ratio of the circumference of a circle to its diameter, which is represented by the Greek letter pi ($\pi$), is an incommensurable quantity. The value 3.1416 is accurate enough for ordinary purposes and the value 22/7 is convenient for rough calculations. The fractions of $\pi$ given in the above table will be found convenient in certain calculations and also the values in the table of constants on page 88.

Table of Decimal Equivalents, Squares, Cubes, Square Roots, Cube Roots and Logarithms of Fractions from 1/64 to 1, by 64ths

| Fraction | Decimal Equivalent | Log. | Square | Log. | Cube | Log. | Sq. Root | Log. | Cube Root | Log. |
|---|---|---|---|---|---|---|---|---|---|---|
| 1/64 | 0.015625 | 2̄.19382 | 0.0002441 | 4̄.38764 | 0.000003815 | 6̄.58146 | 0.1250 | 1̄.09691 | 0.2500 | 1̄.39794 |
| 1/32 | 0.03125 | 2̄.49485 | 0.0009765 | 4̄.98970 | 0.00003052 | 5̄.48455 | 0.1768 | 1̄.24743 | 0.3150 | 1̄.49828 |
| 3/64 | 0.046875 | 2̄.67094 | 0.002197 | 3̄.34188 | 0.0001030 | 4̄.01282 | 0.2165 | 1̄.33547 | 0.3606 | 1̄.55698 |
| 1/16 | 0.0625 | 2̄.79588 | 0.003906 | 3̄.59176 | 0.0002441 | 4̄.38764 | 0.2500 | 1̄.39794 | 0.3969 | 1̄.59863 |
| 5/64 | 0.078125 | 2̄.89279 | 0.006104 | 3̄.78558 | 0.0004768 | 4̄.67837 | 0.2795 | 1̄.44640 | 0.4275 | 1̄.63093 |
| 3/32 | 0.09375 | 2̄.97197 | 0.008789 | 3̄.94394 | 0.0008240 | 4̄.91591 | 0.3062 | 1̄.48599 | 0.4543 | 1̄.65732 |
| 7/64 | 0.109375 | 1̄.03892 | 0.01196 | 2̄.07784 | 0.001308 | 3̄.11675 | 0.3307 | 1̄.51946 | 0.4782 | 1̄.67964 |
| 1/8 | 0.125 | 1̄.09691 | 0.015625 | 2̄.19382 | 0.001953 | 3̄.29073 | 0.3536 | 1̄.54846 | 0.5000 | 1̄.69897 |
| 9/64 | 0.140625 | 1̄.14806 | 0.01978 | 2̄.29613 | 0.002781 | 3̄.44419 | 0.3750 | 1̄.57403 | 0.5200 | 1̄.71602 |
| 5/32 | 0.15625 | 1̄.19382 | 0.02441 | 2̄.38764 | 0.003815 | 3̄.58146 | 0.3953 | 1̄.59691 | 0.5386 | 1̄.73127 |
| 11/64 | 0.171875 | 1̄.23521 | 0.02954 | 2̄.47043 | 0.005077 | 3̄.79564 | 0.4146 | 1̄.61761 | 0.5560 | 1̄.74507 |
| 3/16 | 0.1875 | 1̄.27300 | 0.03516 | 2̄.54600 | 0.006592 | 3̄.81900 | 0.4330 | 1̄.63650 | 0.5724 | 1̄.75767 |
| 13/64 | 0.203125 | 1̄.30776 | 0.04126 | 2̄.61553 | 0.008381 | 3̄.92329 | 0.4507 | 1̄.65388 | 0.5878 | 1̄.76925 |
| 7/32 | 0.21875 | 1̄.33995 | 0.04785 | 2̄.67990 | 0.01047 | 2̄.01984 | 0.4677 | 1̄.66997 | 0.6025 | 1̄.77998 |
| 15/64 | 0.234375 | 1̄.36991 | 0.05493 | 2̄.73982 | 0.01287 | 2̄.10973 | 0.4841 | 1̄.68496 | 0.6166 | 1̄.78997 |
| 1/4 | 0.250 | 1̄.39794 | 0.06250 | 2̄.79588 | 0.01563 | 2̄.19382 | 0.5000 | 1̄.69897 | 0.6300 | 1̄.79931 |
| 17/64 | 0.265625 | 1̄.42427 | 0.07056 | 2̄.84854 | 0.01874 | 2̄.27281 | 0.5154 | 1̄.71213 | 0.6428 | 1̄.80809 |
| 9/32 | 0.28125 | 1̄.44909 | 0.07910 | 2̄.89819 | 0.02225 | 2̄.34728 | 0.5303 | 1̄.72455 | 0.6552 | 1̄.81636 |
| 19/64 | 0.296875 | 1̄.47257 | 0.08813 | 2̄.94515 | 0.02617 | 2̄.41772 | 0.5449 | 1̄.73629 | 0.6671 | 1̄.82419 |
| 5/16 | 0.3125 | 1̄.49485 | 0.09766 | 2̄.98970 | 0.03052 | 2̄.48455 | 0.5590 | 1̄.74743 | 0.6786 | 1̄.83162 |

Table of Decimal Equivalents, Squares, Cubes, Etc., of Fractions

| Fraction | Decimal Equivalent | Log. | Square | Log. | Cube | Log. | Sq. Root | Log. | Cube Root | Log. |
|---|---|---|---|---|---|---|---|---|---|---|
| 21/64 | 0.328125 | 1̄.51604 | 0.1077 | 1̄.03308 | 0.03533 | 2̄.54812 | 0.5728 | 1̄.75802 | 0.6897 | 1̄.83868 |
| 11/32 | 0.34375 | 1̄.53624 | 0.1182 | 1̄.07249 | 0.04062 | 2̄.60873 | 0.5863 | 1̄.76812 | 0.7005 | 1̄.84541 |
| 23/64 | 0.359375 | 1̄.55555 | 0.1292 | 1̄.11110 | 0.04641 | 2̄.66664 | 0.5995 | 1̄.77777 | 0.7110 | 1̄.85185 |
| 3/8 | 0.375 | 1̄.57403 | 0.1406 | 1̄.14806 | 0.05273 | 2̄.72209 | 0.6124 | 1̄.78702 | 0.7211 | 1̄.85801 |
| 25/64 | 0.390625 | 1̄.59176 | 0.1526 | 1̄.18352 | 0.05960 | 2̄.77528 | 0.6250 | 1̄.79588 | 0.7310 | 1̄.86392 |
| 13/32 | 0.40625 | 1̄.60879 | 0.1650 | 1̄.21759 | 0.06705 | 2̄.82638 | 0.6374 | 1̄.80440 | 0.7406 | 1̄.86960 |
| 27/64 | 0.421875 | 1̄.62518 | 0.1780 | 1̄.25037 | 0.07508 | 2̄.87555 | 0.6495 | 1̄.81259 | 0.7500 | 1̄.87506 |
| 7/16 | 0.4375 | 1̄.64098 | 0.1914 | 1̄.28196 | 0.08374 | 2̄.92293 | 0.6614 | 1̄.82049 | 0.7591 | 1̄.88033 |
| 29/64 | 0.453125 | 1̄.65622 | 0.2053 | 1̄.31244 | 0.09304 | 2̄.96865 | 0.6731 | 1̄.82811 | 0.7681 | 1̄.88541 |
| 15/32 | 0.46875 | 1̄.67094 | 0.2197 | 1̄.34188 | 0.1030 | 1̄.01282 | 0.6847 | 1̄.83547 | 0.7768 | 1̄.89031 |
| 31/64 | 0.484375 | 1̄.68518 | 0.2346 | 1̄.37036 | 0.1136 | 1̄.05555 | 0.6960 | 1̄.84259 | 0.7853 | |
| 1/2 | 0.500 | 1̄.69897 | 0.2500 | 1̄.39794 | 0.1250 | 1̄.09691 | 0.7071 | | | |

Table of Decimal Equivalents, Squares, Cubes, Etc., of Fractions

| Fraction | Decimal Equivalent | Log.* | Square | Log. | Cube | Log. | Sq. Root | Log. | Cube Root | Log. |
|---|---|---|---|---|---|---|---|---|---|---|
| 33/64 | 0.515625 | 1̄.71233 | 0.2659 | 1̄.42467 | | | | | 0.8758 | 1̄.94243 |
| 17/32 | 0.53125 | 1̄.72530 | 0.28… | | | | | 0.91364 | | |
| 85/64 | 0.546875 | | | | | | | | | |

Circumferences and Areas of Circles*

| Diameter | Circumference | Area | Diameter | Circumference | Area | Diameter | Circumference | Area |
|---|---|---|---|---|---|---|---|---|
| 1/64 | 0.0491 | 0.0002 | 2 | 6.2832 | 3.1416 | 5 | 15.7080 | 19.635 |
| 1/32 | 0.0982 | 0.0008 | 1/16 | 6.4795 | 3.3410 | 1/16 | 15.9043 | 20.129 |
| 1/16 | 0.1963 | 0.0031 | 1/8 | 6.6759 | 3.5466 | 1/8 | 16.1007 | 20.629 |
| 3/32 | 0.2945 | 0.0069 | 3/16 | 6.8722 | 3.7583 | 3/16 | 16.2970 | 21.135 |
| 1/8 | 0.3927 | 0.0123 | 1/4 | 7.0686 | 3.9761 | 1/4 | 16.4934 | 21.648 |
| 5/32 | 0.4909 | 0.0192 | 5/16 | 7.2649 | 4.2000 | 5/16 | 16.6897 | 22.166 |
| 3/16 | 0.5890 | 0.0276 | 3/8 | 7.4613 | 4.4301 | 3/8 | 16.8861 | 22.691 |
| 7/32 | 0.6872 | 0.0376 | 7/16 | 7.6576 | 4.6664 | 7/16 | 17.0824 | 23.221 |
| 1/4 | 0.7854 | 0.0491 | 1/2 | 7.8540 | 4.9087 | 1/2 | 17.2788 | 23.758 |
| 9/32 | 0.8836 | 0.0621 | 9/16 | 8.0503 | 5.1572 | 9/16 | 17.4751 | 24.301 |
| 5/16 | 0.9817 | 0.0767 | 5/8 | 8.2467 | 5.4119 | 5/8 | 17.6715 | 24.850 |
| 11/32 | 1.0799 | 0.0928 | 11/16 | 8.4430 | 5.6727 | 11/16 | 17.8678 | 25.406 |
| 3/8 | 1.1781 | 0.1104 | 3/4 | 8.6394 | 5.9396 | 3/4 | 18.0642 | 25.967 |
| 13/32 | 1.2763 | 0.1296 | 13/16 | 8.8357 | 6.2126 | 13/16 | 18.2605 | 26.535 |
| 7/16 | 1.3744 | 0.1503 | 7/8 | 9.0321 | 6.4918 | 7/8 | 18.4569 | 27.109 |
| 15/32 | 1.4726 | 0.1726 | 15/16 | 9.2284 | 6.7771 | 15/16 | 18.6532 | 27.688 |
| 1/2 | 1.5708 | 0.1963 | 3 | 9.4248 | 7.0686 | 6 | 18.8496 | 28.274 |
| 17/32 | 1.6690 | 0.2217 | 1/16 | 9.6211 | 7.3662 | 1/8 | 19.2423 | 29.465 |
| 9/16 | 1.7671 | 0.2485 | 1/8 | 9.8175 | 7.6699 | 1/4 | 19.6350 | 30.680 |
| 19/32 | 1.8653 | 0.2769 | 3/16 | 10.0138 | 7.9798 | 3/8 | 20.0277 | 31.919 |
| 5/8 | 1.9635 | 0.3068 | 1/4 | 10.2102 | 8.2958 | 1/2 | 20.4204 | 33.183 |
| 21/32 | 2.0617 | 0.3382 | 5/16 | 10.4065 | 8.6179 | 5/8 | 20.8131 | 34.472 |
| 11/16 | 2.1598 | 0.3712 | 3/8 | 10.6029 | 8.9462 | 3/4 | 21.2058 | 35.785 |
| 23/32 | 2.2580 | 0.4057 | 7/16 | 10.7992 | 9.2806 | 7/8 | 21.5984 | 37.122 |
| 3/4 | 2.3562 | 0.4418 | 1/2 | 10.9956 | 9.6211 | 7 | 21.9911 | 38.485 |
| 25/32 | 2.4544 | 0.4794 | 9/16 | 11.1919 | 9.9678 | 1/8 | 22.3838 | 39.871 |
| 13/16 | 2.5525 | 0.5185 | 5/8 | 11.3883 | 10.321 | 1/4 | 22.7765 | 41.282 |
| 27/32 | 2.6507 | 0.5591 | 11/16 | 11.5846 | 10.680 | 3/8 | 23.1692 | 42.718 |
| 7/8 | 2.7489 | 0.6013 | 3/4 | 11.7810 | 11.045 | 1/2 | 23.5619 | 44.179 |
| 29/32 | 2.8471 | 0.6450 | 13/16 | 11.9773 | 11.416 | 5/8 | 23.9546 | 45.664 |
| 15/16 | 2.9452 | 0.6903 | 7/8 | 12.1737 | 11.793 | 3/4 | 24.3473 | 47.173 |
| 31/32 | 3.0434 | 0.7371 | 15/16 | 12.3700 | 12.177 | 7/8 | 24.7400 | 48.707 |
| 1 | 3.1416 | 0.7854 | 4 | 12.5664 | 12.566 | 8 | 25.1327 | 50.265 |
| 1/16 | 3.3379 | 0.8866 | 1/16 | 12.7627 | 12.962 | 1/8 | 25.5254 | 51.849 |
| 1/8 | 3.5343 | 0.9940 | 1/8 | 12.9591 | 13.364 | 1/4 | 25.9181 | 53.456 |
| 3/16 | 3.7306 | 1.1075 | 3/16 | 13.1554 | 13.772 | 3/8 | 26.3108 | 55.088 |
| 1/4 | 3.9270 | 1.2272 | 1/4 | 13.3518 | 14.186 | 1/2 | 26.7035 | 56.745 |
| 5/16 | 4.1233 | 1.3530 | 5/16 | 13.5481 | 14.607 | 5/8 | 27.0962 | 58.426 |
| 3/8 | 4.3197 | 1.4849 | 3/8 | 13.7445 | 15.033 | 3/4 | 27.4889 | 60.132 |
| 7/16 | 4.5160 | 1.6230 | 7/16 | 13.9408 | 15.466 | 7/8 | 27.8816 | 61.862 |
| 1/2 | 4.7124 | 1.7671 | 1/2 | 14.1372 | 15.904 | 9 | 28.2743 | 63.617 |
| 9/16 | 4.9087 | 1.9175 | 9/16 | 14.3335 | 16.349 | 1/8 | 28.6670 | 65.397 |
| 5/8 | 5.1051 | 2.0739 | 5/8 | 14.5299 | 16.800 | 1/4 | 29.0597 | 67.201 |
| 11/16 | 5.3014 | 2.2365 | 11/16 | 14.7262 | 17.257 | 3/8 | 29.4524 | 69.029 |
| 3/4 | 5.4978 | 2.4053 | 3/4 | 14.9226 | 17.721 | 1/2 | 29.8451 | 70.882 |
| 13/16 | 5.6941 | 2.5802 | 13/16 | 15.1189 | 18.190 | 5/8 | 30.2378 | 72.760 |
| 7/8 | 5.8905 | 2.7612 | 7/8 | 15.3153 | 18.665 | 3/4 | 30.6305 | 74.662 |
| 15/16 | 6.0868 | 2.9483 | 15/16 | 15.5116 | 19.147 | 7/8 | 31.0232 | 76.589 |

* All the figures given in the tables on pages 55 through 66 can be used for English units and those without common fractions can be used for metric units.

## Circumferences and Areas of Circles

| Diameter | Circumference | Area | Diameter | Circumference | Area | Diameter | Circumference | Area |
|---|---|---|---|---|---|---|---|---|
| 10 | 31.4159 | 78.540 | 16 | 50.2655 | 201.06 | 22 | 69.1150 | 380.13 |
| ⅛ | 31.8086 | 80.516 | ⅛ | 50.6582 | 204.22 | ⅛ | 69.5077 | 384.46 |
| ¼ | 32.2013 | 82.516 | ¼ | 51.0509 | 207.39 | ¼ | 69.9004 | 388.82 |
| ⅜ | 32.5940 | 84.541 | ⅜ | 51.4436 | 210.60 | ⅜ | 70.2931 | 393.20 |
| ½ | 32.9867 | 86.590 | ½ | 51.8363 | 213.82 | ½ | 70.6858 | 397.61 |
| ⅝ | 33.3794 | 88.664 | ⅝ | 52.2290 | 217.08 | ⅝ | 71.0785 | 402.04 |
| ¾ | 33.7721 | 90.763 | ¾ | 52.6217 | 220.35 | ¾ | 71.4712 | 406.49 |
| ⅞ | 34.1648 | 92.886 | ⅞ | 53.0144 | 223.65 | ⅞ | 71.8639 | 410.97 |
| 11 | 34.5575 | 95.033 | 17 | 53.4071 | 226.98 | 23 | 72.2566 | 415.48 |
| ⅛ | 34.9502 | 97.205 | ⅛ | 53.7998 | 230.33 | ⅛ | 72.6493 | 420.00 |
| ¼ | 35.3429 | 99.402 | ¼ | 54.1925 | 233.71 | ¼ | 73.0420 | 424.56 |
| ⅜ | 35.7356 | 101.62 | ⅜ | 54.5852 | 237.10 | ⅜ | 73.4347 | 429.13 |
| ½ | 36.1283 | 103.87 | ½ | 54.9779 | 240.53 | ½ | 73.8274 | 433.74 |
| ⅝ | 36.5210 | 106.14 | ⅝ | 55.3706 | 243.98 | ⅝ | 74.2201 | 438.36 |
| ¾ | 36.9137 | 108.43 | ¾ | 55.7633 | 247.45 | ¾ | 74.6128 | 443.01 |
| ⅞ | 37.3064 | 110.75 | ⅞ | 56.1560 | 250.95 | ⅞ | 75.0055 | 447.69 |
| 12 | 37.6991 | 113.10 | 18 | 56.5487 | 254.47 | 24 | 75.3982 | 452.39 |
| ⅛ | 38.0918 | 115.47 | ⅛ | 56.9414 | 258.02 | ⅛ | 75.7909 | 457.11 |
| ¼ | 38.4845 | 117.86 | ¼ | 57.3341 | 261.59 | ¼ | 76.1836 | 461.86 |
| ⅜ | 38.8772 | 120.28 | ⅜ | 57.7268 | 265.18 | ⅜ | 76.5763 | 466.64 |
| ½ | 39.2699 | 122.72 | ½ | 58.1195 | 268.80 | ½ | 76.9690 | 471.44 |
| ⅝ | 39.6626 | 125.19 | ⅝ | 58.5122 | 272.45 | ⅝ | 77.3617 | 476.26 |
| ¾ | 40.0553 | 127.68 | ¾ | 58.9049 | 276.12 | ¾ | 77.7544 | 481.11 |
| ⅞ | 40.4480 | 130.19 | ⅞ | 59.2976 | 279.81 | ⅞ | 78.1471 | 485.98 |
| 13 | 40.8407 | 132.73 | 19 | 59.6903 | 283.53 | 25 | 78.5398 | 490.87 |
| ⅛ | 41.2334 | 135.30 | ⅛ | 60.0830 | 287.27 | ⅛ | 78.9325 | 495.79 |
| ¼ | 41.6261 | 137.89 | ¼ | 60.4757 | 291.04 | ¼ | 79.3252 | 500.74 |
| ⅜ | 42.0188 | 140.50 | ⅜ | 60.8684 | 294.83 | ⅜ | 79.7179 | 505.71 |
| ½ | 42.4115 | 143.14 | ½ | 61.2611 | 298.65 | ½ | 80.1106 | 510.71 |
| ⅝ | 42.8042 | 145.80 | ⅝ | 61.6538 | 302.49 | ⅝ | 80.5033 | 515.72 |
| ¾ | 43.1969 | 148.49 | ¾ | 62.0465 | 306.35 | ¾ | 80.8960 | 520.77 |
| ⅞ | 43.5896 | 151.20 | ⅞ | 62.4392 | 310.24 | ⅞ | 81.2887 | 525.84 |
| 14 | 43.9823 | 153.94 | 20 | 62.8319 | 314.16 | 26 | 81.6814 | 530.93 |
| ⅛ | 44.3750 | 156.70 | ⅛ | 63.2246 | 318.10 | ⅛ | 82.0741 | 536.05 |
| ¼ | 44.7677 | 159.48 | ¼ | 63.6173 | 322.06 | ¼ | 82.4668 | 541.19 |
| ⅜ | 45.1604 | 162.30 | ⅜ | 64.0100 | 326.05 | ⅜ | 82.8595 | 546.35 |
| ½ | 45.5531 | 165.13 | ½ | 64.4026 | 330.06 | ½ | 83.2522 | 551.55 |
| ⅝ | 45.9458 | 167.99 | ⅝ | 64.7953 | 334.10 | ⅝ | 83.6449 | 556.76 |
| ¾ | 46.3385 | 170.87 | ¾ | 65.1880 | 338.16 | ¾ | 84.0376 | 562.00 |
| ⅞ | 46.7312 | 173.78 | ⅞ | 65.5807 | 342.25 | ⅞ | 84.4303 | 567.27 |
| 15 | 47.1239 | 176.71 | 21 | 65.9734 | 346.36 | 27 | 84.8230 | 572.56 |
| ⅛ | 47.5166 | 179.67 | ⅛ | 66.3661 | 350.50 | ⅛ | 85.2157 | 577.87 |
| ¼ | 47.9093 | 182.65 | ¼ | 66.7588 | 354.66 | ¼ | 85.6084 | 583.21 |
| ⅜ | 48.3020 | 185.66 | ⅜ | 67.1515 | 358.84 | ⅜ | 86.0011 | 588.57 |
| ½ | 48.6947 | 188.69 | ½ | 67.5442 | 363.05 | ½ | 86.3938 | 593.96 |
| ⅝ | 49.0874 | 191.75 | ⅝ | 67.9369 | 367.28 | ⅝ | 86.7865 | 599.37 |
| ¾ | 49.4801 | 194.83 | ¾ | 68.3296 | 371.54 | ¾ | 87.1792 | 604.81 |
| ⅞ | 49.8728 | 197.93 | ⅞ | 68.7223 | 375.83 | ⅞ | 87.5719 | 610.27 |

## Circumferences and Areas of Circles

| Diameter | Circumference | Area | Diameter | Circumference | Area | Diameter | Circumference | Area |
|---|---|---|---|---|---|---|---|---|
| 28 | 87.9646 | 615.75 | 34 | 106.814 | 907.92 | 40 | 125.664 | 1256.6 |
| ⅛ | 88.3573 | 621.26 | ⅛ | 107.207 | 914.61 | ⅛ | 126.056 | 1264.5 |
| ¼ | 88.7500 | 626.80 | ¼ | 107.600 | 921.32 | ¼ | 126.449 | 1272.4 |
| ⅜ | 89.1427 | 632.36 | ⅜ | 107.992 | 928.06 | ⅜ | 126.842 | 1280.3 |
| ½ | 89.5354 | 637.94 | ½ | 108.385 | 934.82 | ½ | 127.235 | 1288.2 |
| ⅝ | 89.9281 | 643.55 | ⅝ | 108.778 | 941.61 | ⅝ | 127.627 | 1296.2 |
| ¾ | 90.3208 | 649.18 | ¾ | 109.170 | 948.42 | ¾ | 128.020 | 1304.2 |
| ⅞ | 90.7135 | 654.84 | ⅞ | 109.563 | 955.25 | ⅞ | 128.413 | 1312.2 |
| 29 | 91.1062 | 660.52 | 35 | 109.956 | 962.11 | 41 | 128.805 | 1320.3 |
| ⅛ | 91.4989 | 666.23 | ⅛ | 110.348 | 969.00 | ⅛ | 129.198 | 1328.3 |
| ¼ | 91.8916 | 671.96 | ¼ | 110.741 | 975.91 | ¼ | 129.591 | 1336.4 |
| ⅜ | 92.2843 | 677.71 | ⅜ | 111.134 | 982.84 | ⅜ | 129.983 | 1344.5 |
| ½ | 92.6770 | 683.49 | ½ | 111.527 | 989.80 | ½ | 130.376 | 1352.7 |
| ⅝ | 93.0697 | 689.30 | ⅝ | 111.919 | 996.78 | ⅝ | 130.769 | 1360.8 |
| ¾ | 93.4624 | 695.13 | ¾ | 112.312 | 1003.8 | ¾ | 131.161 | 1369.0 |
| ⅞ | 93.8551 | 700.98 | ⅞ | 112.705 | 1010.8 | ⅞ | 131.554 | 1377.2 |
| 30 | 94.2478 | 706.86 | 36 | 113.097 | 1017.9 | 42 | 131.947 | 1385.4 |
| ⅛ | 94.6405 | 712.76 | ⅛ | 113.490 | 1025.0 | ⅛ | 132.340 | 1393.7 |
| ¼ | 95.0332 | 718.69 | ¼ | 113.883 | 1032.1 | ¼ | 132.732 | 1402.0 |
| ⅜ | 95.4259 | 724.64 | ⅜ | 114.275 | 1039.2 | ⅜ | 133.125 | 1410.3 |
| ½ | 95.8186 | 730.62 | ½ | 114.668 | 1046.3 | ½ | 133.518 | 1418.6 |
| ⅝ | 96.2113 | 736.62 | ⅝ | 115.061 | 1053.5 | ⅝ | 133.910 | 1427.0 |
| ¾ | 96.6040 | 742.64 | ¾ | 115.454 | 1060.7 | ¾ | 134.303 | 1435.4 |
| ⅞ | 96.9967 | 748.69 | ⅞ | 115.846 | 1068.0 | ⅞ | 134.696 | 1443.8 |
| 31 | 97.3894 | 754.77 | 37 | 116.239 | 1075.2 | 43 | 135.088 | 1452.2 |
| ⅛ | 97.7821 | 760.87 | ⅛ | 116.632 | 1082.5 | ⅛ | 135.481 | 1460.7 |
| ¼ | 98.1748 | 766.99 | ¼ | 117.024 | 1089.8 | ¼ | 135.874 | 1469.1 |
| ⅜ | 98.5675 | 773.14 | ⅜ | 117.417 | 1097.1 | ⅜ | 136.267 | 1477.6 |
| ½ | 98.9602 | 779.31 | ½ | 117.810 | 1104.5 | ½ | 136.659 | 1486.2 |
| ⅝ | 99.3529 | 785.51 | ⅝ | 118.202 | 1111.8 | ⅝ | 137.052 | 1494.7 |
| ¾ | 99.7456 | 791.73 | ¾ | 118.595 | 1119.2 | ¾ | 137.445 | 1503.3 |
| ⅞ | 100.138 | 797.98 | ⅞ | 118.988 | 1126.7 | ⅞ | 137.837 | 1511.9 |
| 32 | 100.531 | 804.25 | 38 | 119.381 | 1134.1 | 44 | 138.230 | 1520.5 |
| ⅛ | 100.924 | 810.54 | ⅛ | 119.773 | 1141.6 | ⅛ | 138.623 | 1529.2 |
| ¼ | 101.316 | 816.86 | ¼ | 120.166 | 1149.1 | ¼ | 139.015 | 1537.9 |
| ⅜ | 101.709 | 823.21 | ⅜ | 120.559 | 1156.6 | ⅜ | 139.408 | 1546.6 |
| ½ | 102.102 | 829.58 | ½ | 120.951 | 1164.2 | ½ | 139.801 | 1555.3 |
| ⅝ | 102.494 | 835.97 | ⅝ | 121.344 | 1171.7 | ⅝ | 140.194 | 1564.0 |
| ¾ | 102.887 | 842.39 | ¾ | 121.737 | 1179.3 | ¾ | 140.586 | 1572.8 |
| ⅞ | 103.280 | 848.83 | ⅞ | 122.129 | 1186.9 | ⅞ | 140.979 | 1581.6 |
| 33 | 103.673 | 855.30 | 39 | 122.522 | 1194.6 | 45 | 141.372 | 1590.4 |
| ⅛ | 104.065 | 861.79 | ⅛ | 122.915 | 1202.3 | ⅛ | 141.764 | 1599.3 |
| ¼ | 104.458 | 868.31 | ¼ | 123.308 | 1210.0 | ¼ | 142.157 | 1608.2 |
| ⅜ | 104.851 | 874.85 | ⅜ | 123.700 | 1217.7 | ⅜ | 142.550 | 1617.0 |
| ½ | 105.243 | 881.41 | ½ | 124.093 | 1225.4 | ½ | 142.942 | 1626.0 |
| ⅝ | 105.636 | 888.00 | ⅝ | 124.486 | 1233.2 | ⅝ | 143.335 | 1634.9 |
| ¾ | 106.029 | 894.62 | ¾ | 124.878 | 1241.0 | ¾ | 143.728 | 1643.9 |
| ⅞ | 106.421 | 901.26 | ⅞ | 125.271 | 1248.8 | ⅞ | 144.121 | 1652.9 |

# 58 MATHEMATICAL TABLES

## Circumferences and Areas of Circles

| Diameter | Circumference | Area | Diameter | Circumference | Area | Diameter | Circumference | Area |
|---|---|---|---|---|---|---|---|---|
| 46 | 144.513 | 1661.9 | 52 | 163.363 | 2123.7 | 58 | 182.212 | 2642.1 |
| 1/8 | 144.906 | 1670.9 | 1/8 | 163.756 | 2133.9 | 1/8 | 182.605 | 2653.5 |
| 1/4 | 145.299 | 1680.0 | 1/4 | 164.148 | 2144.2 | 1/4 | 182.998 | 2664.9 |
| 3/8 | 145.691 | 1689.1 | 3/8 | 164.541 | 2154.5 | 3/8 | 183.390 | 2676.4 |
| 1/2 | 146.084 | 1698.2 | 1/2 | 164.934 | 2164.8 | 1/2 | 183.783 | 2687.8 |
| 5/8 | 146.477 | 1707.4 | 5/8 | 165.326 | 2175.1 | 5/8 | 184.176 | 2699.3 |
| 3/4 | 146.869 | 1716.5 | 3/4 | 165.719 | 2185.4 | 3/4 | 184.569 | 2710.9 |
| 7/8 | 147.262 | 1725.7 | 7/8 | 166.112 | 2195.8 | 7/8 | 184.961 | 2722.4 |
| 47 | 147.655 | 1734.9 | 53 | 166.504 | 2206.2 | 59 | 185.354 | 2734.0 |
| 1/8 | 148.048 | 1744.2 | 1/8 | 166.897 | 2216.6 | 1/8 | 185.747 | 2745.6 |
| 1/4 | 148.440 | 1753.5 | 1/4 | 167.290 | 2227.0 | 1/4 | 186.139 | 2757.2 |
| 3/8 | 148.833 | 1762.7 | 3/8 | 167.683 | 2237.5 | 3/8 | 186.532 | 2768.8 |
| 1/2 | 149.226 | 1772.1 | 1/2 | 168.075 | 2248.0 | 1/2 | 186.925 | 2780.5 |
| 5/8 | 149.618 | 1781.4 | 5/8 | 168.468 | 2258.5 | 5/8 | 187.317 | 2792.2 |
| 3/4 | 150.011 | 1790.8 | 3/4 | 168.861 | 2269.1 | 3/4 | 187.710 | 2803.9 |
| 7/8 | 150.404 | 1800.1 | 7/8 | 169.253 | 2279.6 | 7/8 | 188.103 | 2815.7 |
| 48 | 150.796 | 1809.6 | 54 | 169.646 | 2290.2 | 60 | 188.496 | 2827.4 |
| 1/8 | 151.189 | 1819.0 | 1/8 | 170.039 | 2300.8 | 1/8 | 188.888 | 2839.2 |
| 1/4 | 151.582 | 1828.5 | 1/4 | 170.431 | 2311.5 | 1/4 | 189.281 | 2851.0 |
| 3/8 | 151.975 | 1837.9 | 3/8 | 170.824 | 2322.1 | 3/8 | 189.674 | 2862.9 |
| 1/2 | 152.367 | 1847.5 | 1/2 | 171.217 | 2332.8 | 1/2 | 190.066 | 2874.8 |
| 5/8 | 152.760 | 1857.0 | 5/8 | 171.609 | 2343.5 | 5/8 | 190.459 | 2886.6 |
| 3/4 | 153.153 | 1866.5 | 3/4 | 172.002 | 2354.3 | 3/4 | 190.852 | 2898.6 |
| 7/8 | 153.545 | 1876.1 | 7/8 | 172.395 | 2365.0 | 7/8 | 191.244 | 2910.5 |
| 49 | 153.938 | 1885.7 | 55 | 172.788 | 2375.8 | 61 | 191.637 | 2922.5 |
| 1/8 | 154.331 | 1895.4 | 1/8 | 173.180 | 2386.6 | 1/8 | 192.030 | 2934.5 |
| 1/4 | 154.723 | 1905.0 | 1/4 | 173.573 | 2397.5 | 1/4 | 192.423 | 2946.5 |
| 3/8 | 155.116 | 1914.7 | 3/8 | 173.966 | 2408.3 | 3/8 | 192.815 | 2958.5 |
| 1/2 | 155.509 | 1924.4 | 1/2 | 174.358 | 2419.2 | 1/2 | 193.208 | 2970.6 |
| 5/8 | 155.902 | 1934.2 | 5/8 | 174.751 | 2430.1 | 5/8 | 193.601 | 2982.7 |
| 3/4 | 156.294 | 1943.9 | 3/4 | 175.144 | 2441.1 | 3/4 | 193.993 | 2994.8 |
| 7/8 | 156.687 | 1953.7 | 7/8 | 175.536 | 2452.0 | 7/8 | 194.386 | 3006.9 |
| 50 | 157.080 | 1963.5 | 56 | 175.929 | 2463.0 | 62 | 194.779 | 3019.1 |
| 1/8 | 157.472 | 1973.3 | 1/8 | 176.322 | 2474.0 | 1/8 | 195.171 | 3031.3 |
| 1/4 | 157.865 | 1983.2 | 1/4 | 176.715 | 2485.0 | 1/4 | 195.564 | 3043.5 |
| 3/8 | 158.258 | 1993.1 | 3/8 | 177.107 | 2496.1 | 3/8 | 195.957 | 3055.7 |
| 1/2 | 158.650 | 2003.0 | 1/2 | 177.500 | 2507.2 | 1/2 | 196.350 | 3068.0 |
| 5/8 | 159.043 | 2012.9 | 5/8 | 177.893 | 2518.3 | 5/8 | 196.742 | 3080.3 |
| 3/4 | 159.436 | 2022.8 | 3/4 | 178.285 | 2529.4 | 3/4 | 197.135 | 3092.6 |
| 7/8 | 159.829 | 2032.8 | 7/8 | 178.678 | 2540.6 | 7/8 | 197.528 | 3104.9 |
| 51 | 160.221 | 2042.8 | 57 | 179.071 | 2551.8 | 63 | 197.920 | 3117.2 |
| 1/8 | 160.614 | 2052.8 | 1/8 | 179.463 | 2563.0 | 1/8 | 198.313 | 3129.6 |
| 1/4 | 161.007 | 2062.9 | 1/4 | 179.856 | 2574.2 | 1/4 | 198.706 | 3142.0 |
| 3/8 | 161.399 | 2073.0 | 3/8 | 180.249 | 2585.4 | 3/8 | 199.098 | 3154.5 |
| 1/2 | 161.792 | 2083.1 | 1/2 | 180.642 | 2596.7 | 1/2 | 199.491 | 3166.9 |
| 5/8 | 162.185 | 2093.2 | 5/8 | 181.034 | 2608.0 | 5/8 | 199.884 | 3179.4 |
| 3/4 | 162.577 | 2103.3 | 3/4 | 181.427 | 2619.4 | 3/4 | 200.277 | 3191.9 |
| 7/8 | 162.970 | 2113.5 | 7/8 | 181.820 | 2630.7 | 7/8 | 200.669 | 3204.4 |

Circumferences and Areas of Circles

| Diameter | Circumference | Area | Diameter | Circumference | Area | Diameter | Circumference | Area |
|---|---|---|---|---|---|---|---|---|
| 64 | 201.062 | 3217.0 | 70 | 219.911 | 3848.5 | 76 | 238.761 | 4536.5 |
| 1/8 | 201.455 | 3229.6 | 1/8 | 220.304 | 3862.2 | 1/8 | 239.154 | 4551.4 |
| 1/4 | 201.847 | 3242.2 | 1/4 | 220.697 | 3876.0 | 1/4 | 239.546 | 4566.4 |
| 3/8 | 202.240 | 3254.8 | 3/8 | 221.090 | 3889.8 | 3/8 | 239.939 | 4581.3 |
| 1/2 | 202.633 | 3267.5 | 1/2 | 221.482 | 3903.6 | 1/2 | 240.332 | 4596.3 |
| 5/8 | 203.025 | 3280.1 | 5/8 | 221.875 | 3917.5 | 5/8 | 240.725 | 4611.4 |
| 3/4 | 203.418 | 3292.8 | 3/4 | 222.268 | 3931.4 | 3/4 | 241.117 | 4626.4 |
| 7/8 | 203.811 | 3305.6 | 7/8 | 222.660 | 3945.3 | 7/8 | 241.510 | 4641.5 |
| 65 | 204.204 | 3318.3 | 71 | 223.053 | 3959.2 | 77 | 241.903 | 4656.6 |
| 1/8 | 204.596 | 3331.1 | 1/8 | 223.446 | 3973.1 | 1/8 | 242.295 | 4671.8 |
| 1/4 | 204.989 | 3343.9 | 1/4 | 223.838 | 3987.1 | 1/4 | 242.688 | 4686.9 |
| 3/8 | 205.382 | 3356.7 | 3/8 | 224.231 | 4001.1 | 3/8 | 243.081 | 4702.1 |
| 1/2 | 205.774 | 3369.6 | 1/2 | 224.624 | 4015.2 | 1/2 | 243.473 | 4717.3 |
| 5/8 | 206.167 | 3382.4 | 5/8 | 225.017 | 4029.2 | 5/8 | 243.866 | 4732.5 |
| 3/4 | 206.560 | 3395.3 | 3/4 | 225.409 | 4043.3 | 3/4 | 244.259 | 4747.8 |
| 7/8 | 206.952 | 3408.2 | 7/8 | 225.802 | 4057.4 | 7/8 | 244.652 | 4763.1 |
| 66 | 207.345 | 3421.2 | 72 | 226.195 | 4071.5 | 78 | 245.044 | 4778.4 |
| 1/8 | 207.738 | 3434.2 | 1/8 | 226.587 | 4085.7 | 1/8 | 245.437 | 4793.7 |
| 1/4 | 208.131 | 3447.2 | 1/4 | 226.980 | 4099.8 | 1/4 | 245.830 | 4809.0 |
| 3/8 | 208.523 | 3460.2 | 3/8 | 227.373 | 4114.0 | 3/8 | 246.222 | 4824.4 |
| 1/2 | 208.916 | 3473.2 | 1/2 | 227.765 | 4128.2 | 1/2 | 246.615 | 4839.8 |
| 5/8 | 209.309 | 3486.3 | 5/8 | 228.158 | 4142.5 | 5/8 | 247.008 | 4855.2 |
| 3/4 | 209.701 | 3499.4 | 3/4 | 228.551 | 4156.8 | 3/4 | 247.400 | 4870.7 |
| 7/8 | 210.094 | 3512.5 | 7/8 | 228.944 | 4171.1 | 7/8 | 247.793 | 4886.2 |
| 67 | 210.487 | 3525.7 | 73 | 229.336 | 4185.4 | 79 | 248.186 | 4901.7 |
| 1/8 | 210.879 | 3538.8 | 1/8 | 229.729 | 4199.7 | 1/8 | 248.579 | 4917.2 |
| 1/4 | 211.272 | 3552.0 | 1/4 | 230.122 | 4214.1 | 1/4 | 248.971 | 4932.7 |
| 3/8 | 211.665 | 3565.2 | 3/8 | 230.514 | 4228.5 | 3/8 | 249.364 | 4948.3 |
| 1/2 | 212.058 | 3578.5 | 1/2 | 230.907 | 4242.9 | 1/2 | 249.757 | 4963.9 |
| 5/8 | 212.450 | 3591.7 | 5/8 | 231.300 | 4257.4 | 5/8 | 250.149 | 4979.5 |
| 3/4 | 212.843 | 3605.0 | 3/4 | 231.692 | 4271.8 | 3/4 | 250.542 | 4995.2 |
| 7/8 | 213.236 | 3618.3 | 7/8 | 232.085 | 4286.3 | 7/8 | 250.935 | 5010.9 |
| 68 | 213.628 | 3631.7 | 74 | 232.478 | 4300.8 | 80 | 251.327 | 5026.5 |
| 1/8 | 214.021 | 3645.0 | 1/8 | 232.871 | 4315.4 | 1/8 | 251.720 | 5042.3 |
| 1/4 | 214.414 | 3658.4 | 1/4 | 233.263 | 4329.9 | 1/4 | 252.113 | 5058.0 |
| 3/8 | 214.806 | 3671.8 | 3/8 | 233.656 | 4344.5 | 3/8 | 252.506 | 5073.8 |
| 1/2 | 215.199 | 3685.3 | 1/2 | 234.049 | 4359.2 | 1/2 | 252.898 | 5089.6 |
| 5/8 | 215.592 | 3698.7 | 5/8 | 234.441 | 4373.8 | 5/8 | 253.291 | 5105.4 |
| 3/4 | 215.984 | 3712.2 | 3/4 | 234.834 | 4388.5 | 3/4 | 253.684 | 5121.2 |
| 7/8 | 216.377 | 3725.7 | 7/8 | 235.227 | 4403.1 | 7/8 | 254.076 | 5137.1 |
| 69 | 216.770 | 3739.3 | 75 | 235.619 | 4417.9 | 81 | 254.469 | 5153.0 |
| 1/8 | 217.163 | 3752.8 | 1/8 | 236.012 | 4432.6 | 1/8 | 254.862 | 5168.9 |
| 1/4 | 217.555 | 3766.4 | 1/4 | 236.405 | 4447.4 | 1/4 | 255.254 | 5184.9 |
| 3/8 | 217.948 | 3780.0 | 3/8 | 236.798 | 4462.2 | 3/8 | 255.647 | 5200.8 |
| 1/2 | 218.341 | 3793.7 | 1/2 | 237.190 | 4477.0 | 1/2 | 256.040 | 5216.8 |
| 5/8 | 218.733 | 3807.3 | 5/8 | 237.583 | 4491.8 | 5/8 | 256.433 | 5232.8 |
| 3/4 | 219.126 | 3821.0 | 3/4 | 237.976 | 4506.7 | 3/4 | 256.825 | 5248.9 |
| 7/8 | 219.519 | 3834.7 | 7/8 | 238.368 | 4521.5 | 7/8 | 257.218 | 5264.9 |

## Circumferences and Areas of Circles

| Diameter | Circumference | Area | Diameter | Circumference | Area | Diameter | Circumference | Area |
|---|---|---|---|---|---|---|---|---|
| 82 | 257.611 | 5281.0 | 88 | 276.460 | 6082.1 | 94 | 295.310 | 6939.8 |
| 1/8 | 258.003 | 5297.1 | 1/8 | 276.853 | 6099.4 | 1/8 | 295.702 | 6958.2 |
| 1/4 | 258.396 | 5313.3 | 1/4 | 277.246 | 6116.7 | 1/4 | 296.095 | 6976.7 |
| 3/8 | 258.789 | 5329.4 | 3/8 | 277.638 | 6134.1 | 3/8 | 296.488 | 6995.3 |
| 1/2 | 259.181 | 5345.6 | 1/2 | 278.031 | 6151.4 | 1/2 | 296.881 | 7013.8 |
| 5/8 | 259.574 | 5361.8 | 5/8 | 278.424 | 6168.8 | 5/8 | 297.273 | 7032.4 |
| 3/4 | 259.967 | 5378.1 | 3/4 | 278.816 | 6186.2 | 3/4 | 297.666 | 7051.0 |
| 7/8 | 260.359 | 5394.3 | 7/8 | 279.209 | 6203.7 | 7/8 | 298.059 | 7069.6 |
| 83 | 260.752 | 5410.6 | 89 | 279.602 | 6221.1 | 95 | 298.451 | 7088.2 |
| 1/8 | 261.145 | 5426.9 | 1/8 | 279.994 | 6238.6 | 1/8 | 298.844 | 7106.9 |
| 1/4 | 261.538 | 5443.3 | 1/4 | 280.387 | 6256.1 | 1/4 | 299.237 | 7125.6 |
| 3/8 | 261.930 | 5459.6 | 3/8 | 280.780 | 6273.7 | 3/8 | 299.629 | 7144.3 |
| 1/2 | 262.323 | 5476.0 | 1/2 | 281.173 | 6291.2 | 1/2 | 300.022 | 7163.0 |
| 5/8 | 262.716 | 5492.4 | 5/8 | 281.565 | 6308.8 | 5/8 | 300.415 | 7181.8 |
| 3/4 | 263.108 | 5508.8 | 3/4 | 281.958 | 6326.4 | 3/4 | 300.807 | 7200.6 |
| 7/8 | 263.501 | 5525.3 | 7/8 | 282.351 | 6344.1 | 7/8 | 301.200 | 7219.4 |
| 84 | 263.894 | 5541.8 | 90 | 282.743 | 6361.7 | 96 | 301.593 | 7238.2 |
| 1/8 | 264.286 | 5558.3 | 1/8 | 283.136 | 6379.4 | 1/8 | 301.986 | 7257.1 |
| 1/4 | 264.679 | 5574.8 | 1/4 | 283.529 | 6397.1 | 1/4 | 302.378 | 7276.0 |
| 3/8 | 265.072 | 5591.4 | 3/8 | 283.921 | 6414.8 | 3/8 | 302.771 | 7294.9 |
| 1/2 | 265.465 | 5607.9 | 1/2 | 284.314 | 6432.6 | 1/2 | 303.164 | 7313.8 |
| 5/8 | 265.857 | 5624.5 | 5/8 | 284.707 | 6450.4 | 5/8 | 303.556 | 7332.8 |
| 3/4 | 266.250 | 5641.2 | 3/4 | 285.100 | 6468.2 | 3/4 | 303.949 | 7351.8 |
| 7/8 | 266.643 | 5657.8 | 7/8 | 285.492 | 6486.0 | 7/8 | 304.342 | 7370.8 |
| 85 | 267.035 | 5674.5 | 91 | 285.885 | 6503.9 | 97 | 304.734 | 7389.8 |
| 1/8 | 267.428 | 5691.2 | 1/8 | 286.278 | 6521.8 | 1/8 | 305.127 | 7408.9 |
| 1/4 | 267.821 | 5707.9 | 1/4 | 286.670 | 6539.7 | 1/4 | 305.520 | 7428.0 |
| 3/8 | 268.213 | 5724.7 | 3/8 | 287.063 | 6557.6 | 3/8 | 305.913 | 7447.1 |
| 1/2 | 268.606 | 5741.5 | 1/2 | 287.456 | 6575.5 | 1/2 | 306.305 | 7466.2 |
| 5/8 | 268.999 | 5758.3 | 5/8 | 287.848 | 6593.5 | 5/8 | 306.698 | 7485.3 |
| 3/4 | 269.392 | 5775.1 | 3/4 | 288.241 | 6611.5 | 3/4 | 307.091 | 7504.5 |
| 7/8 | 269.784 | 5791.9 | 7/8 | 288.634 | 6629.6 | 7/8 | 307.483 | 7523.7 |
| 86 | 270.177 | 5808.8 | 92 | 289.027 | 6647.6 | 98 | 307.876 | 7543.0 |
| 1/8 | 270.570 | 5825.7 | 1/8 | 289.419 | 6665.7 | 1/8 | 308.269 | 7562.2 |
| 1/4 | 270.962 | 5842.6 | 1/4 | 289.812 | 6683.8 | 1/4 | 308.661 | 7581.5 |
| 3/8 | 271.355 | 5859.6 | 3/8 | 290.205 | 6701.9 | 3/8 | 309.054 | 7600.8 |
| 1/2 | 271.748 | 5876.5 | 1/2 | 290.597 | 6720.1 | 1/2 | 309.447 | 7620.1 |
| 5/8 | 272.140 | 5893.5 | 5/8 | 290.990 | 6738.2 | 5/8 | 309.840 | 7639.5 |
| 3/4 | 272.533 | 5910.6 | 3/4 | 291.383 | 6756.4 | 3/4 | 310.232 | 7658.9 |
| 7/8 | 272.926 | 5927.6 | 7/8 | 291.775 | 6774.7 | 7/8 | 310.625 | 7678.3 |
| 87 | 273.319 | 5944.7 | 93 | 292.168 | 6792.9 | 99 | 311.018 | 7697.7 |
| 1/8 | 273.711 | 5961.8 | 1/8 | 292.561 | 6811.2 | 1/8 | 311.410 | 7717.1 |
| 1/4 | 274.104 | 5978.9 | 1/4 | 292.954 | 6829.5 | 1/4 | 311.803 | 7736.6 |
| 3/8 | 274.497 | 5996.0 | 3/8 | 293.346 | 6847.8 | 3/8 | 312.196 | 7756.1 |
| 1/2 | 274.889 | 6013.2 | 1/2 | 293.739 | 6866.1 | 1/2 | 312.588 | 7775.6 |
| 5/8 | 275.282 | 6030.4 | 5/8 | 294.132 | 6884.5 | 5/8 | 312.981 | 7795.2 |
| 3/4 | 275.675 | 6047.6 | 3/4 | 294.524 | 6902.9 | 3/4 | 313.374 | 7814.8 |
| 7/8 | 276.067 | 6064.9 | 7/8 | 294.917 | 6921.3 | 7/8 | 313.767 | 7834.4 |

## Circumferences and Areas of Circles

| Diameter | Circumference | Area | Diameter | Circumference | Area | Diameter | Circumference | Area |
|---|---|---|---|---|---|---|---|---|
| 100 | 314.16 | 7,854.0 | 150 | 471.24 | 17,671.5 | 200 | 628.32 | 31,415.9 |
| 101 | 317.30 | 8,011.8 | 151 | 474.38 | 17,907.9 | 201 | 631.46 | 31,730.9 |
| 102 | 320.44 | 8,171.3 | 152 | 477.52 | 18,145.8 | 202 | 634.60 | 32,047.4 |
| 103 | 323.58 | 8,332.3 | 153 | 480.66 | 18,385.4 | 203 | 637.74 | 32,365.5 |
| 104 | 326.73 | 8,494.9 | 154 | 483.81 | 18,626.5 | 204 | 640.88 | 32,685.1 |
| 105 | 329.87 | 8,659.0 | 155 | 486.95 | 18,869.2 | 205 | 644.03 | 33,006.4 |
| 106 | 333.01 | 8,824.7 | 156 | 490.09 | 19,113.4 | 206 | 647.17 | 33,329.2 |
| 107 | 336.15 | 8,992.0 | 157 | 493.23 | 19,359.3 | 207 | 650.31 | 33,653.5 |
| 108 | 339.29 | 9,160.9 | 158 | 496.37 | 19,606.7 | 208 | 653.45 | 33,979.5 |
| 109 | 342.43 | 9,331.3 | 159 | 499.51 | 19,855.7 | 209 | 656.59 | 34,307.0 |
| 110 | 345.58 | 9,503.3 | 160 | 502.65 | 20,106.2 | 210 | 659.73 | 34,636.1 |
| 111 | 348.72 | 9,676.9 | 161 | 505.80 | 20,358.3 | 211 | 662.88 | 34,966.7 |
| 112 | 351.86 | 9,852.0 | 162 | 508.94 | 20,612.0 | 212 | 666.02 | 35,298.9 |
| 113 | 355.00 | 10,028.7 | 163 | 512.08 | 20,867.2 | 213 | 669.16 | 35,632.7 |
| 114 | 358.14 | 10,207.0 | 164 | 515.22 | 21,124.1 | 214 | 672.30 | 35,968.1 |
| 115 | 361.28 | 10,386.9 | 165 | 518.36 | 21,382.5 | 215 | 675.44 | 36,305.0 |
| 116 | 364.42 | 10,568.3 | 166 | 521.50 | 21,642.4 | 216 | 678.58 | 36,643.5 |
| 117 | 367.57 | 10,751.3 | 167 | 524.65 | 21,904.0 | 217 | 681.73 | 36,983.6 |
| 118 | 370.71 | 10,935.9 | 168 | 527.79 | 22,167.1 | 218 | 684.87 | 37,325.3 |
| 119 | 373.85 | 11,122.0 | 169 | 530.93 | 22,431.8 | 219 | 688.01 | 37,668.5 |
| 120 | 376.99 | 11,309.7 | 170 | 534.07 | 22,698.0 | 220 | 691.15 | 38,013.3 |
| 121 | 380.13 | 11,499.0 | 171 | 537.21 | 22,965.8 | 221 | 694.29 | 38,359.6 |
| 122 | 383.27 | 11,689.9 | 172 | 540.35 | 23,235.2 | 222 | 697.43 | 38,707.6 |
| 123 | 386.42 | 11,882.3 | 173 | 543.50 | 23,506.2 | 223 | 700.58 | 39,057.1 |
| 124 | 389.56 | 12,076.3 | 174 | 546.64 | 23,778.7 | 224 | 703.72 | 39,408.1 |
| 125 | 392.70 | 12,271.8 | 175 | 549.78 | 24,052.8 | 225 | 706.86 | 39,760.8 |
| 126 | 395.84 | 12,469.0 | 176 | 552.92 | 24,328.5 | 226 | 710.00 | 40,115.0 |
| 127 | 398.98 | 12,667.7 | 177 | 556.06 | 24,605.7 | 227 | 713.14 | 40,470.8 |
| 128 | 402.12 | 12,868.0 | 178 | 559.20 | 24,884.6 | 228 | 716.28 | 40,828.1 |
| 129 | 405.27 | 13,069.8 | 179 | 562.35 | 25,164.9 | 229 | 719.42 | 41,187.1 |
| 130 | 408.41 | 13,273.2 | 180 | 565.49 | 25,446.9 | 230 | 722.57 | 41,547.6 |
| 131 | 411.55 | 13,478.2 | 181 | 568.63 | 25,730.4 | 231 | 725.71 | 41,909.6 |
| 132 | 414.69 | 13,684.8 | 182 | 571.77 | 26,015.5 | 232 | 728.85 | 42,273.3 |
| 133 | 417.83 | 13,892.9 | 183 | 574.91 | 26,302.2 | 233 | 731.99 | 42,638.5 |
| 134 | 420.97 | 14,102.6 | 184 | 578.05 | 26,590.4 | 234 | 735.13 | 43,005.3 |
| 135 | 424.12 | 14,313.9 | 185 | 581.19 | 26,880.3 | 235 | 738.27 | 43,373.6 |
| 136 | 427.26 | 14,526.7 | 186 | 584.34 | 27,171.6 | 236 | 741.42 | 43,743.5 |
| 137 | 430.40 | 14,741.1 | 187 | 587.48 | 27,464.6 | 237 | 744.56 | 44,115.0 |
| 138 | 433.54 | 14,957.1 | 188 | 590.62 | 27,759.1 | 238 | 747.70 | 44,488.1 |
| 139 | 436.68 | 15,174.7 | 189 | 593.76 | 28,055.2 | 239 | 750.84 | 44,862.7 |
| 140 | 439.82 | 15,393.8 | 190 | 596.90 | 28,352.9 | 240 | 753.98 | 45,238.9 |
| 141 | 442.96 | 15,614.5 | 191 | 600.04 | 28,652.1 | 241 | 757.12 | 45,616.7 |
| 142 | 446.11 | 15,836.8 | 192 | 603.19 | 28,952.9 | 242 | 760.27 | 45,996.1 |
| 143 | 449.25 | 16,060.6 | 193 | 606.33 | 29,255.3 | 243 | 763.41 | 46,377.0 |
| 144 | 452.39 | 16,286.0 | 194 | 609.47 | 29,559.2 | 244 | 766.55 | 46,759.5 |
| 145 | 455.53 | 16,513.0 | 195 | 612.61 | 29,864.8 | 245 | 769.69 | 47,143.5 |
| 146 | 458.67 | 16,741.5 | 196 | 615.75 | 30,171.9 | 246 | 772.83 | 47,529.2 |
| 147 | 461.81 | 16,971.7 | 197 | 618.89 | 30,480.5 | 247 | 775.97 | 47,916.4 |
| 148 | 464.96 | 17,203.4 | 198 | 622.04 | 30,790.7 | 248 | 779.11 | 48,305.1 |
| 149 | 468.10 | 17,436.6 | 199 | 625.18 | 31,102.6 | 249 | 782.26 | 48,695.5 |

### Circumferences and Areas of Circles

| Diameter | Circumference | Area | Diameter | Circumference | Area | Diameter | Circumference | Area |
|---|---|---|---|---|---|---|---|---|
| 250 | 785.40 | 49,087.4 | 300 | 942.48 | 70,685.8 | 350 | 1099.56 | 96,211.3 |
| 251 | 788.54 | 49,480.9 | 301 | 945.62 | 71,157.9 | 351 | 1102.70 | 96,761.8 |
| 252 | 791.68 | 49,875.9 | 302 | 948.76 | 71,631.5 | 352 | 1105.84 | 97,314.0 |
| 253 | 794.82 | 50,272.6 | 303 | 951.90 | 72,106.6 | 353 | 1108.98 | 97,867.7 |
| 254 | 797.96 | 50,670.7 | 304 | 955.04 | 72,583.4 | 354 | 1112.12 | 98,423.0 |
| 255 | 801.11 | 51,070.5 | 305 | 958.19 | 73,061.7 | 355 | 1115.27 | 98,979.8 |
| 256 | 804.25 | 51,471.9 | 306 | 961.33 | 73,541.5 | 356 | 1118.41 | 99,538.2 |
| 257 | 807.39 | 51,874.8 | 307 | 964.47 | 74,023.0 | 357 | 1121.55 | 100,098 |
| 258 | 810.53 | 52,279.2 | 308 | 967.61 | 74,506.0 | 358 | 1124.69 | 100,660 |
| 259 | 813.67 | 52,685.3 | 309 | 970.75 | 74,990.6 | 359 | 1127.83 | 101,223 |
| 260 | 816.81 | 53,092.9 | 310 | 973.89 | 75,476.8 | 360 | 1130.97 | 101,788 |
| 261 | 819.96 | 53,502.1 | 311 | 977.04 | 75,964.5 | 361 | 1134.11 | 102,354 |
| 262 | 823.10 | 53,912.9 | 312 | 980.18 | 76,453.8 | 362 | 1137.26 | 102,922 |
| 263 | 826.24 | 54,325.2 | 313 | 983.32 | 76,944.7 | 363 | 1140.40 | 103,491 |
| 264 | 829.38 | 54,739.1 | 314 | 986.46 | 77,437.1 | 364 | 1143.54 | 104,062 |
| 265 | 832.52 | 55,154.6 | 315 | 989.60 | 77,931.1 | 365 | 1146.68 | 104,635 |
| 266 | 835.66 | 55,571.6 | 316 | 992.74 | 78,426.7 | 366 | 1149.82 | 105,209 |
| 267 | 838.81 | 55,990.2 | 317 | 995.88 | 78,923.9 | 367 | 1152.96 | 105,784 |
| 268 | 841.95 | 56,410.4 | 318 | 999.03 | 79,422.6 | 368 | 1156.11 | 106,362 |
| 269 | 845.09 | 56,832.2 | 319 | 1002.17 | 79,922.9 | 369 | 1159.25 | 106,941 |
| 270 | 848.23 | 57,255.5 | 320 | 1005.31 | 80,424.8 | 370 | 1162.39 | 107,521 |
| 271 | 851.37 | 57,680.4 | 321 | 1008.45 | 80,928.2 | 371 | 1165.53 | 108,103 |
| 272 | 854.51 | 58,106.9 | 322 | 1011.59 | 81,433.2 | 372 | 1168.67 | 108,687 |
| 273 | 857.65 | 58,534.9 | 323 | 1014.73 | 81,939.8 | 373 | 1171.81 | 109,272 |
| 274 | 860.80 | 58,964.6 | 324 | 1017.88 | 82,448.0 | 374 | 1174.96 | 109,858 |
| 275 | 863.94 | 59,395.7 | 325 | 1021.02 | 82,957.7 | 375 | 1178.10 | 110,447 |
| 276 | 867.08 | 59,828.5 | 326 | 1024.16 | 83,469.0 | 376 | 1181.24 | 111,036 |
| 277 | 870.22 | 60,262.8 | 327 | 1027.30 | 83,981.8 | 377 | 1184.38 | 111,628 |
| 278 | 873.36 | 60,698.7 | 328 | 1030.44 | 84,496.3 | 378 | 1187.52 | 112,221 |
| 279 | 876.50 | 61,136.2 | 329 | 1033.58 | 85,012.3 | 379 | 1190.66 | 112,815 |
| 280 | 879.65 | 61,575.2 | 330 | 1036.73 | 85,529.9 | 380 | 1193.81 | 113,411 |
| 281 | 882.79 | 62,015.8 | 331 | 1039.87 | 86,049.0 | 381 | 1196.95 | 114,009 |
| 282 | 885.93 | 62,458.0 | 332 | 1043.01 | 86,569.7 | 382 | 1200.09 | 114,608 |
| 283 | 889.07 | 62,901.8 | 333 | 1046.15 | 87,092.0 | 383 | 1203.23 | 115,209 |
| 284 | 892.21 | 63,347.1 | 334 | 1049.29 | 87,615.9 | 384 | 1206.37 | 115,812 |
| 285 | 895.35 | 63,794.0 | 335 | 1052.43 | 88,141.3 | 385 | 1209.51 | 116,416 |
| 286 | 898.50 | 64,242.4 | 336 | 1055.58 | 88,668.3 | 386 | 1212.65 | 117,021 |
| 287 | 901.64 | 64,692.5 | 337 | 1058.72 | 89,196.9 | 387 | 1215.80 | 117,628 |
| 288 | 904.78 | 65,144.1 | 338 | 1061.86 | 89,727.0 | 388 | 1218.94 | 118,237 |
| 289 | 907.92 | 65,597.2 | 339 | 1065.00 | 90,258.7 | 389 | 1222.08 | 118,847 |
| 290 | 911.06 | 66,052.0 | 340 | 1068.14 | 90,792.0 | 390 | 1225.22 | 119,459 |
| 291 | 914.20 | 66,508.3 | 341 | 1071.28 | 91,326.9 | 391 | 1228.36 | 120,072 |
| 292 | 917.35 | 66,966.2 | 342 | 1074.42 | 91,863.3 | 392 | 1231.50 | 120,687 |
| 293 | 920.49 | 67,425.6 | 343 | 1077.57 | 92,401.3 | 393 | 1234.65 | 121,304 |
| 294 | 923.63 | 67,886.7 | 344 | 1080.71 | 92,940.9 | 394 | 1237.79 | 121,922 |
| 295 | 926.77 | 68,349.3 | 345 | 1083.85 | 93,482.0 | 395 | 1240.93 | 122,542 |
| 296 | 929.91 | 68,813.4 | 346 | 1086.99 | 94,024.7 | 396 | 1244.07 | 123,163 |
| 297 | 933.05 | 69,279.2 | 347 | 1090.13 | 94,569.0 | 397 | 1247.21 | 123,786 |
| 298 | 936.19 | 69,746.5 | 348 | 1093.27 | 95,114.9 | 398 | 1250.35 | 124,410 |
| 299 | 939.34 | 70,215.4 | 349 | 1096.42 | 95,662.3 | 399 | 1253.50 | 125,036 |

## Circumferences and Areas of Circles

| Diameter | Circumference | Area | Diameter | Circumference | Area | Diameter | Circumference | Area |
|---|---|---|---|---|---|---|---|---|
| 400 | 1256.64 | 125,664 | 450 | 1413.72 | 159,043 | 500 | 1570.80 | 196,350 |
| 401 | 1259.78 | 126,293 | 451 | 1416.86 | 159,751 | 501 | 1573.94 | 197,136 |
| 402 | 1262.92 | 126,923 | 452 | 1420.00 | 160,460 | 502 | 1577.08 | 197,923 |
| 403 | 1266.06 | 127,556 | 453 | 1423.14 | 161,171 | 503 | 1580.22 | 198,713 |
| 404 | 1269.20 | 128,190 | 454 | 1426.28 | 161,883 | 504 | 1583.36 | 199,504 |
| 405 | 1272.35 | 128,825 | 455 | 1429.42 | 162,597 | 505 | 1586.50 | 200,296 |
| 406 | 1275.49 | 129,462 | 456 | 1432.57 | 163,313 | 506 | 1589.65 | 201,090 |
| 407 | 1278.63 | 130,100 | 457 | 1435.71 | 164,030 | 507 | 1592.79 | 201,886 |
| 408 | 1281.77 | 130,741 | 458 | 1438.85 | 164,748 | 508 | 1595.93 | 202,683 |
| 409 | 1284.91 | 131,382 | 459 | 1441.99 | 165,468 | 509 | 1599.07 | 203,482 |
| 410 | 1288.05 | 132,025 | 460 | 1445.13 | 166,190 | 510 | 1602.21 | 204,282 |
| 411 | 1291.19 | 132,670 | 461 | 1448.27 | 166,914 | 511 | 1605.35 | 205,084 |
| 412 | 1294.34 | 133,317 | 462 | 1451.42 | 167,639 | 512 | 1608.50 | 205,887 |
| 413 | 1297.48 | 133,965 | 463 | 1454.56 | 168,365 | 513 | 1611.64 | 206,692 |
| 414 | 1300.62 | 134,614 | 464 | 1457.70 | 169,093 | 514 | 1614.78 | 207,499 |
| 415 | 1303.76 | 135,265 | 465 | 1460.84 | 169,823 | 515 | 1617.92 | 208,307 |
| 416 | 1306.90 | 135,918 | 466 | 1463.98 | 170,554 | 516 | 1621.06 | 209,117 |
| 417 | 1310.04 | 136,572 | 467 | 1467.12 | 171,287 | 517 | 1624.20 | 209,928 |
| 418 | 1313.19 | 137,228 | 468 | 1470.27 | 172,021 | 518 | 1627.34 | 210,741 |
| 419 | 1316.33 | 137,885 | 469 | 1473.41 | 172,757 | 519 | 1630.49 | 211,556 |
| 420 | 1319.47 | 138,544 | 470 | 1476.55 | 173,494 | 520 | 1633.63 | 212,372 |
| 421 | 1322.61 | 139,205 | 471 | 1479.69 | 174,234 | 521 | 1636.77 | 213,189 |
| 422 | 1325.75 | 139,867 | 472 | 1482.83 | 174,974 | 522 | 1639.91 | 214,008 |
| 423 | 1328.89 | 140,531 | 473 | 1485.97 | 175,716 | 523 | 1643.05 | 214,829 |
| 424 | 1332.04 | 141,196 | 474 | 1489.11 | 176,460 | 524 | 1646.19 | 215,651 |
| 425 | 1335.18 | 141,863 | 475 | 1492.26 | 177,205 | 525 | 1649.34 | 216,475 |
| 426 | 1338.32 | 142,531 | 476 | 1495.40 | 177,952 | 526 | 1652.48 | 217,301 |
| 427 | 1341.46 | 143,201 | 477 | 1498.54 | 178,701 | 527 | 1655.62 | 218,128 |
| 428 | 1344.60 | 143,872 | 478 | 1501.68 | 179,451 | 528 | 1658.76 | 218,956 |
| 429 | 1347.74 | 144,545 | 479 | 1504.82 | 180,203 | 529 | 1661.90 | 219,787 |
| 430 | 1350.88 | 145,220 | 480 | 1507.96 | 180,956 | 530 | 1665.04 | 220,618 |
| 431 | 1354.03 | 145,896 | 481 | 1511.11 | 181,711 | 531 | 1668.19 | 221,452 |
| 432 | 1357.17 | 146,574 | 482 | 1514.25 | 182,467 | 532 | 1671.33 | 222,287 |
| 433 | 1360.31 | 147,254 | 483 | 1517.39 | 183,225 | 533 | 1674.47 | 223,123 |
| 434 | 1363.45 | 147,934 | 484 | 1520.53 | 183,984 | 534 | 1677.61 | 223,961 |
| 435 | 1366.59 | 148,617 | 485 | 1523.67 | 184,745 | 535 | 1680.75 | 224,801 |
| 436 | 1369.73 | 149,301 | 486 | 1526.81 | 185,508 | 536 | 1683.89 | 225,642 |
| 437 | 1372.88 | 149,987 | 487 | 1529.96 | 186,272 | 537 | 1687.04 | 226,484 |
| 438 | 1376.02 | 150,674 | 488 | 1533.10 | 187,038 | 538 | 1690.18 | 227,329 |
| 439 | 1379.16 | 151,363 | 489 | 1536.24 | 187,805 | 539 | 1693.32 | 228,175 |
| 440 | 1382.30 | 152,053 | 490 | 1539.38 | 188,574 | 540 | 1696.46 | 229,022 |
| 441 | 1385.44 | 152,745 | 491 | 1542.52 | 189,345 | 541 | 1699.60 | 229,871 |
| 442 | 1388.58 | 153,439 | 492 | 1545.66 | 190,117 | 542 | 1702.74 | 230,722 |
| 443 | 1391.73 | 154,134 | 493 | 1548.81 | 190,890 | 543 | 1705.88 | 231,574 |
| 444 | 1394.87 | 154,830 | 494 | 1551.95 | 191,665 | 544 | 1709.03 | 232,428 |
| 445 | 1398.01 | 155,528 | 495 | 1555.09 | 192,442 | 545 | 1712.17 | 233,283 |
| 446 | 1401.15 | 156,228 | 496 | 1558.23 | 193,221 | 546 | 1715.31 | 234,140 |
| 447 | 1404.29 | 156,930 | 497 | 1561.37 | 194,000 | 547 | 1718.45 | 234,998 |
| 448 | 1407.43 | 157,633 | 498 | 1564.51 | 194,782 | 548 | 1721.59 | 235,858 |
| 449 | 1410.58 | 158,337 | 499 | 1567.65 | 195,565 | 549 | 1724.73 | 236,720 |

## Circumferences and Areas of Circles

| Diameter | Circumference | Area | Diameter | Circumference | Area | Diameter | Circumference | Area |
|---|---|---|---|---|---|---|---|---|
| 550 | 1727.88 | 237,583 | 600 | 1884.96 | 282,743 | 650 | 2042.04 | 331,831 |
| 551 | 1731.02 | 238,448 | 601 | 1888.10 | 283,687 | 651 | 2045.18 | 332,853 |
| 552 | 1734.16 | 239,314 | 602 | 1891.24 | 284,631 | 652 | 2048.32 | 333,876 |
| 553 | 1737.30 | 240,182 | 603 | 1894.38 | 285,578 | 653 | 2051.46 | 334,901 |
| 554 | 1740.44 | 241,051 | 604 | 1897.52 | 286,526 | 654 | 2054.60 | 335,927 |
| 555 | 1743.58 | 241,922 | 605 | 1900.66 | 287,475 | 655 | 2057.74 | 336,955 |
| 556 | 1746.73 | 242,795 | 606 | 1903.81 | 288,426 | 656 | 2060.88 | 337,985 |
| 557 | 1749.87 | 243,669 | 607 | 1906.95 | 289,379 | 657 | 2064.03 | 339,016 |
| 558 | 1753.01 | 244,545 | 608 | 1910.09 | 290,333 | 658 | 2067.17 | 340,049 |
| 559 | 1756.15 | 245,422 | 609 | 1913.23 | 291,289 | 659 | 2070.31 | 341,083 |
| 560 | 1759.29 | 246,301 | 610 | 1916.37 | 292,247 | 660 | 2073.45 | 342,119 |
| 561 | 1762.43 | 247,181 | 611 | 1919.51 | 293,206 | 661 | 2076.59 | 343,157 |
| 562 | 1765.58 | 248,063 | 612 | 1922.65 | 294,166 | 662 | 2079.73 | 344,196 |
| 563 | 1768.72 | 248,947 | 613 | 1925.80 | 295,128 | 663 | 2082.88 | 345,237 |
| 564 | 1771.86 | 249,832 | 614 | 1928.94 | 296,092 | 664 | 2086.02 | 346,279 |
| 565 | 1775.00 | 250,719 | 615 | 1932.08 | 297,057 | 665 | 2089.16 | 347,323 |
| 566 | 1778.14 | 251,607 | 616 | 1935.22 | 298,024 | 666 | 2092.30 | 348,368 |
| 567 | 1781.28 | 252,497 | 617 | 1938.36 | 298,992 | 667 | 2095.44 | 349,415 |
| 568 | 1784.42 | 253,388 | 618 | 1941.50 | 299,962 | 668 | 2098.58 | 350,464 |
| 569 | 1787.57 | 254,281 | 619 | 1944.65 | 300,934 | 669 | 2101.73 | 351,514 |
| 570 | 1790.71 | 255,176 | 620 | 1947.79 | 301,907 | 670 | 2104.87 | 352,565 |
| 571 | 1793.85 | 256,072 | 621 | 1950.93 | 302,882 | 671 | 2108.01 | 353,618 |
| 572 | 1796.99 | 256,970 | 622 | 1954.07 | 303,858 | 672 | 2111.15 | 354,673 |
| 573 | 1800.13 | 257,869 | 623 | 1957.21 | 304,836 | 673 | 2114.29 | 355,730 |
| 574 | 1803.27 | 258,770 | 624 | 1960.35 | 305,815 | 674 | 2117.43 | 356,788 |
| 575 | 1806.42 | 259,672 | 625 | 1963.50 | 306,796 | 675 | 2120.58 | 357,847 |
| 576 | 1809.56 | 260,576 | 626 | 1966.64 | 307,779 | 676 | 2123.72 | 358,908 |
| 577 | 1812.70 | 261,482 | 627 | 1969.78 | 308,763 | 677 | 2126.86 | 359,971 |
| 578 | 1815.84 | 262,389 | 628 | 1972.92 | 309,748 | 678 | 2130.00 | 361,035 |
| 579 | 1818.98 | 263,298 | 629 | 1976.06 | 310,736 | 679 | 2133.14 | 362,101 |
| 580 | 1822.12 | 264,208 | 630 | 1979.20 | 311,725 | 680 | 2136.28 | 363,168 |
| 581 | 1825.27 | 265,120 | 631 | 1982.34 | 312,715 | 681 | 2139.42 | 364,237 |
| 582 | 1828.41 | 266,033 | 632 | 1985.49 | 313,707 | 682 | 2142.57 | 365,308 |
| 583 | 1831.55 | 266,948 | 633 | 1988.63 | 314,700 | 683 | 2145.71 | 366,380 |
| 584 | 1834.69 | 267,865 | 634 | 1991.77 | 315,696 | 684 | 2148.85 | 367,453 |
| 585 | 1837.83 | 268,783 | 635 | 1994.91 | 316,692 | 685 | 2151.99 | 368,528 |
| 586 | 1840.97 | 269,703 | 636 | 1998.05 | 317,690 | 686 | 2155.13 | 369,605 |
| 587 | 1844.11 | 270,624 | 637 | 2001.19 | 318,690 | 687 | 2158.27 | 370,684 |
| 588 | 1847.26 | 271,547 | 638 | 2004.34 | 319,692 | 688 | 2161.42 | 371,764 |
| 589 | 1850.40 | 272,471 | 639 | 2007.48 | 320,695 | 689 | 2164.56 | 372,845 |
| 590 | 1853.54 | 273,397 | 640 | 2010.62 | 321,699 | 690 | 2167.70 | 373,928 |
| 591 | 1856.68 | 274,325 | 641 | 2013.76 | 322,705 | 691 | 2170.84 | 375,013 |
| 592 | 1859.82 | 275,254 | 642 | 2016.90 | 323,713 | 692 | 2173.98 | 376,099 |
| 593 | 1862.96 | 276,184 | 643 | 2020.04 | 324,722 | 693 | 2177.12 | 377,187 |
| 594 | 1866.11 | 277,117 | 644 | 2023.19 | 325,733 | 694 | 2180.27 | 378,276 |
| 595 | 1869.25 | 278,051 | 645 | 2026.33 | 326,745 | 695 | 2183.41 | 379,367 |
| 596 | 1872.39 | 278,986 | 646 | 2029.47 | 327,759 | 696 | 2186.55 | 380,459 |
| 597 | 1875.53 | 279,923 | 647 | 2032.61 | 328,775 | 697 | 2189.69 | 381,553 |
| 598 | 1878.67 | 280,862 | 648 | 2035.75 | 329,792 | 698 | 2192.83 | 382,649 |
| 599 | 1881.81 | 281,802 | 649 | 2038.89 | 330,810 | 699 | 2195.97 | 383,746 |

## Circumferences and Areas of Circles

| Diameter | Circumference | Area | Diameter | Circumference | Area | Diameter | Circumference | Area |
|---|---|---|---|---|---|---|---|---|
| 700 | 2199.11 | 384,845 | 750 | 2356.19 | 441,786 | 800 | 2513.27 | 502,655 |
| 701 | 2202.26 | 385,945 | 751 | 2359.34 | 442,965 | 801 | 2516.42 | 503,912 |
| 702 | 2205.40 | 387,047 | 752 | 2362.48 | 444,146 | 802 | 2519.56 | 505,171 |
| 703 | 2208.54 | 388,151 | 753 | 2365.62 | 445,328 | 803 | 2522.70 | 506,432 |
| 704 | 2211.68 | 389,256 | 754 | 2368.76 | 446,511 | 804 | 2525.84 | 507,694 |
| 705 | 2214.82 | 390,363 | 755 | 2371.90 | 447,697 | 805 | 2528.98 | 508,958 |
| 706 | 2217.96 | 391,471 | 756 | 2375.04 | 448,883 | 806 | 2532.12 | 510,223 |
| 707 | 2221.11 | 392,580 | 757 | 2378.19 | 450,072 | 807 | 2535.27 | 511,490 |
| 708 | 2224.25 | 393,692 | 758 | 2381.33 | 451,262 | 808 | 2538.41 | 512,758 |
| 709 | 2227.39 | 394,805 | 759 | 2384.47 | 452,453 | 809 | 2541.55 | 514,028 |
| 710 | 2230.53 | 395,919 | 760 | 2387.61 | 453,646 | 810 | 2544.69 | 515,300 |
| 711 | 2233.67 | 397,035 | 761 | 2390.75 | 454,841 | 811 | 2547.83 | 516,573 |
| 712 | 2236.81 | 398,153 | 762 | 2393.89 | 456,037 | 812 | 2550.97 | 517,848 |
| 713 | 2239.96 | 399,272 | 763 | 2397.04 | 457,234 | 813 | 2554.11 | 519,124 |
| 714 | 2243.10 | 400,393 | 764 | 2400.18 | 458,434 | 814 | 2557.26 | 520,402 |
| 715 | 2246.24 | 401,515 | 765 | 2403.32 | 459,635 | 815 | 2560.40 | 521,681 |
| 716 | 2249.38 | 402,639 | 766 | 2406.46 | 460,837 | 816 | 2563.54 | 522,962 |
| 717 | 2252.52 | 403,765 | 767 | 2409.60 | 462,041 | 817 | 2566.68 | 524,245 |
| 718 | 2255.66 | 404,892 | 768 | 2412.74 | 463,247 | 818 | 2569.82 | 525,529 |
| 719 | 2258.81 | 406,020 | 769 | 2415.88 | 464,454 | 819 | 2572.96 | 526,814 |
| 720 | 2261.95 | 407,150 | 770 | 2419.03 | 465,663 | 820 | 2576.11 | 528,102 |
| 721 | 2265.09 | 408,282 | 771 | 2422.17 | 466,873 | 821 | 2579.25 | 529,391 |
| 722 | 2268.23 | 409,415 | 772 | 2425.31 | 468,085 | 822 | 2582.39 | 530,681 |
| 723 | 2271.37 | 410,550 | 773 | 2428.45 | 469,298 | 823 | 2585.53 | 531,973 |
| 724 | 2274.51 | 411,687 | 774 | 2431.59 | 470,513 | 824 | 2588.67 | 533,267 |
| 725 | 2277.65 | 412,825 | 775 | 2434.73 | 471,730 | 825 | 2591.81 | 534,562 |
| 726 | 2280.80 | 413,965 | 776 | 2437.88 | 472,948 | 826 | 2594.96 | 535,858 |
| 727 | 2283.94 | 415,106 | 777 | 2441.02 | 474,168 | 827 | 2598.10 | 537,157 |
| 728 | 2287.08 | 416,248 | 778 | 2444.16 | 475,389 | 828 | 2601.24 | 538,456 |
| 729 | 2290.22 | 417,393 | 779 | 2447.30 | 476,612 | 829 | 2604.38 | 539,758 |
| 730 | 2293.36 | 418,539 | 780 | 2450.44 | 477,836 | 830 | 2607.52 | 541,061 |
| 731 | 2296.50 | 419,686 | 781 | 2453.58 | 479,062 | 831 | 2610.66 | 542,365 |
| 732 | 2299.65 | 420,835 | 782 | 2456.73 | 480,290 | 832 | 2613.81 | 543,671 |
| 733 | 2302.79 | 421,986 | 783 | 2459.87 | 481,519 | 833 | 2616.95 | 544,979 |
| 734 | 2305.93 | 423,138 | 784 | 2463.01 | 482,750 | 834 | 2620.09 | 546,288 |
| 735 | 2309.07 | 424,292 | 785 | 2466.15 | 483,982 | 835 | 2623.23 | 547,599 |
| 736 | 2312.21 | 425,447 | 786 | 2469.29 | 485,216 | 836 | 2626.37 | 548,912 |
| 737 | 2315.35 | 426,604 | 787 | 2472.43 | 486,451 | 837 | 2629.51 | 550,226 |
| 738 | 2318.50 | 427,762 | 788 | 2475.58 | 487,688 | 838 | 2632.65 | 551,541 |
| 739 | 2321.64 | 428,922 | 789 | 2478.72 | 488,927 | 839 | 2635.80 | 552,858 |
| 740 | 2324.78 | 430,084 | 790 | 2481.86 | 490,167 | 840 | 2638.94 | 554,177 |
| 741 | 2327.92 | 431,247 | 791 | 2485.00 | 491,409 | 841 | 2642.08 | 555,497 |
| 742 | 2331.06 | 432,412 | 792 | 2488.14 | 492,652 | 842 | 2645.22 | 556,819 |
| 743 | 2334.20 | 433,578 | 793 | 2491.28 | 493,897 | 843 | 2648.36 | 558,142 |
| 744 | 2337.34 | 434,746 | 794 | 2494.42 | 495,143 | 844 | 2651.50 | 559,467 |
| 745 | 2340.49 | 435,916 | 795 | 2497.57 | 496,391 | 845 | 2654.65 | 560,794 |
| 746 | 2343.63 | 437,087 | 796 | 2500.71 | 497,641 | 846 | 2657.79 | 562,122 |
| 747 | 2346.77 | 438,259 | 797 | 2503.85 | 498,892 | 847 | 2660.93 | 563,452 |
| 748 | 2349.91 | 439,433 | 798 | 2506.99 | 500,145 | 848 | 2664.07 | 564,783 |
| 749 | 2353.05 | 440,609 | 799 | 2510.13 | 501,399 | 849 | 2667.21 | 566,116 |

## Circumferences and Areas of Circles

| Diameter | Circumference | Area | Diameter | Circumference | Area | Diameter | Circumference | Area |
|---|---|---|---|---|---|---|---|---|
| 850 | 2670.35 | 567,450 | 900 | 2827.43 | 636,173 | 950 | 2984.51 | 708,822 |
| 851 | 2673.50 | 568,786 | 901 | 2830.57 | 637,587 | 951 | 2987.65 | 710,315 |
| 852 | 2676.64 | 570,124 | 902 | 2833.72 | 639,003 | 952 | 2990.80 | 711,809 |
| 853 | 2679.78 | 571,463 | 903 | 2836.86 | 640,421 | 953 | 2993.94 | 713,306 |
| 854 | 2682.92 | 572,803 | 904 | 2840.00 | 641,840 | 954 | 2997.08 | 714,803 |
| 855 | 2686.06 | 574,146 | 905 | 2843.14 | 643,261 | 955 | 3000.22 | 716,303 |
| 856 | 2689.20 | 575,490 | 906 | 2846.28 | 644,683 | 956 | 3003.36 | 717,804 |
| 857 | 2692.34 | 576,835 | 907 | 2849.42 | 646,107 | 957 | 3006.50 | 719,306 |
| 858 | 2695.49 | 578,182 | 908 | 2852.57 | 647,533 | 958 | 3009.65 | 720,810 |
| 859 | 2698.63 | 579,530 | 909 | 2855.71 | 648,960 | 959 | 3012.79 | 722,316 |
| 860 | 2701.77 | 580,880 | 910 | 2858.85 | 650,388 | 960 | 3015.93 | 723,823 |
| 861 | 2704.91 | 582,232 | 911 | 2861.99 | 651,818 | 961 | 3019.07 | 725,332 |
| 862 | 2708.05 | 583,585 | 912 | 2865.13 | 653,250 | 962 | 3022.21 | 726,842 |
| 863 | 2711.19 | 584,940 | 913 | 2868.27 | 654,684 | 963 | 3025.35 | 728,354 |
| 864 | 2714.34 | 586,297 | 914 | 2871.42 | 656,118 | 964 | 3028.50 | 729,867 |
| 865 | 2717.48 | 587,655 | 915 | 2874.56 | 657,555 | 965 | 3031.64 | 731,382 |
| 866 | 2720.62 | 589,014 | 916 | 2877.70 | 658,993 | 966 | 3034.78 | 732,899 |
| 867 | 2723.76 | 590,375 | 917 | 2880.84 | 660,433 | 967 | 3037.92 | 734,417 |
| 868 | 2726.90 | 591,738 | 918 | 2883.98 | 661,874 | 968 | 3041.06 | 735,937 |
| 869 | 2730.04 | 593,102 | 919 | 2887.12 | 663,317 | 969 | 3044.20 | 737,458 |
| 870 | 2733.19 | 594,468 | 920 | 2890.27 | 664,761 | 970 | 3047.34 | 738,981 |
| 871 | 2736.33 | 595,835 | 921 | 2893.41 | 666,207 | 971 | 3050.49 | 740,506 |
| 872 | 2739.47 | 597,204 | 922 | 2896.55 | 667,654 | 972 | 3053.63 | 742,032 |
| 873 | 2742.61 | 598,575 | 923 | 2899.69 | 669,103 | 973 | 3056.77 | 743,559 |
| 874 | 2745.75 | 599,947 | 924 | 2902.83 | 670,554 | 974 | 3059.91 | 745,088 |
| 875 | 2748.89 | 601,320 | 925 | 2905.97 | 672,006 | 975 | 3063.05 | 746,619 |
| 876 | 2752.04 | 602,696 | 926 | 2909.11 | 673,460 | 976 | 3066.19 | 748,151 |
| 877 | 2755.18 | 604,073 | 927 | 2912.26 | 674,915 | 977 | 3069.34 | 749,685 |
| 878 | 2758.32 | 605,451 | 928 | 2915.40 | 676,372 | 978 | 3072.48 | 751,221 |
| 879 | 2761.46 | 606,831 | 929 | 2918.54 | 677,831 | 979 | 3075.62 | 752,758 |
| 880 | 2764.60 | 608,212 | 930 | 2921.68 | 679,291 | 980 | 3078.76 | 754,296 |
| 881 | 2767.74 | 609,595 | 931 | 2924.82 | 680,752 | 981 | 3081.90 | 755,837 |
| 882 | 2770.88 | 610,980 | 932 | 2927.96 | 682,216 | 982 | 3085.04 | 757,378 |
| 883 | 2774.03 | 612,366 | 933 | 2931.11 | 683,680 | 983 | 3088.19 | 758,922 |
| 884 | 2777.17 | 613,754 | 934 | 2934.25 | 685,147 | 984 | 3091.33 | 760,466 |
| 885 | 2780.31 | 615,143 | 935 | 2937.39 | 686,615 | 985 | 3094.47 | 762,013 |
| 886 | 2783.45 | 616,534 | 936 | 2940.53 | 688,084 | 986 | 3097.61 | 763,561 |
| 887 | 2786.59 | 617,927 | 937 | 2943.67 | 689,555 | 987 | 3100.75 | 765,111 |
| 888 | 2789.73 | 619,321 | 938 | 2946.81 | 691,028 | 988 | 3103.89 | 766,662 |
| 889 | 2792.88 | 620,717 | 939 | 2949.96 | 692,502 | 989 | 3107.04 | 768,214 |
| 890 | 2796.02 | 622,114 | 940 | 2953.10 | 693,978 | 990 | 3110.18 | 769,769 |
| 891 | 2799.16 | 623,513 | 941 | 2956.24 | 695,455 | 991 | 3113.32 | 771,325 |
| 892 | 2802.30 | 624,913 | 942 | 2959.38 | 696,934 | 992 | 3116.46 | 772,882 |
| 893 | 2805.44 | 626,315 | 943 | 2962.52 | 698,415 | 993 | 3119.60 | 774,441 |
| 894 | 2808.58 | 627,718 | 944 | 2965.66 | 699,897 | 994 | 3122.74 | 776,002 |
| 895 | 2811.73 | 629,124 | 945 | 2968.81 | 701,380 | 995 | 3125.88 | 777,564 |
| 896 | 2814.87 | 630,530 | 946 | 2971.95 | 702,865 | 996 | 3129.03 | 779,128 |
| 897 | 2818.01 | 631,938 | 947 | 2975.09 | 704,352 | 997 | 3132.17 | 780,693 |
| 898 | 2821.15 | 633,348 | 948 | 2978.23 | 705,840 | 998 | 3135.31 | 782,260 |
| 899 | 2824.29 | 634,760 | 949 | 2981.37 | 707,330 | 999 | 3138.45 | 783,828 |

**Diameters, Circumferences and Areas of Circles in Feet and Inches**

| Diam. Ft. | Diam. In. | Circum.* Ft. | Circum.* In. | Area Sq. In. | Area Sq. Ft. | Diam. Ft. | Diam. In. | Circum.* Ft. | Circum.* In. | Area Sq. In. | Area Sq. Ft. |
|---|---|---|---|---|---|---|---|---|---|---|---|
| 1 | 6 | 4 | 8½ | 254.469 | 1.7671 | 2 | 0 | 6 | 3⅜ | 452.389 | 3.1416 |
| | 6⅛ | 4 | 9 | 258.016 | 1.7918 | | 0¼ | 6 | 4⅛ | 461.863 | 3.2074 |
| | 6¼ | 4 | 9⅜ | 261.587 | 1.8166 | | 0½ | 6 | 5 | 471.435 | 3.2739 |
| | 6⅜ | 4 | 9¾ | 265.182 | 1.8415 | | 0¾ | 6 | 5¾ | 481.105 | 3.3410 |
| | 6½ | 4 | 10⅛ | 268.803 | 1.8667 | 2 | 1 | 6 | 6½ | 490.874 | 3.4088 |
| | 6⅝ | 4 | 10½ | 272.447 | 1.8920 | | 1¼ | 6 | 7⅜ | 500.740 | 3.4774 |
| | 6¾ | 4 | 10⅞ | 276.117 | 1.9175 | | 1½ | 6 | 8⅛ | 510.705 | 3.5466 |
| | 6⅞ | 4 | 11¼ | 279.810 | 1.9431 | | 1¾ | 6 | 8⅞ | 520.768 | 3.6164 |
| 1 | 7 | 4 | 11¾ | 283.529 | 1.9689 | 2 | 2 | 6 | 9⅝ | 530.929 | 3.6870 |
| | 7⅛ | 5 | 0⅛ | 287.272 | 1.9949 | | 2¼ | 6 | 10½ | 541.188 | 3.7583 |
| | 7¼ | 5 | 0½ | 291.039 | 2.0211 | | 2½ | 6 | 11¼ | 551.546 | 3.8302 |
| | 7⅜ | 5 | 0⅞ | 294.831 | 2.0474 | | 2¾ | 7 | 0 | 562.001 | 3.9028 |
| | 7½ | 5 | 1¼ | 298.648 | 2.0739 | 2 | 3 | 7 | 0⅞ | 572.555 | 3.9761 |
| | 7⅝ | 5 | 1⅝ | 302.489 | 2.1006 | | 3¼ | 7 | 1⅝ | 583.207 | 4.0501 |
| | 7¾ | 5 | 2 | 306.354 | 2.1275 | | 3½ | 7 | 2⅜ | 593.957 | 4.1247 |
| | 7⅞ | 5 | 2½ | 310.245 | 2.1545 | | 3¾ | 7 | 3⅛ | 604.806 | 4.2000 |
| 1 | 8 | 5 | 2⅞ | 314.159 | 2.1817 | 2 | 4 | 7 | 4 | 615.752 | 4.2761 |
| | 8⅛ | 5 | 3¼ | 318.099 | 2.2090 | | 4¼ | 7 | 4¾ | 626.797 | 4.3528 |
| | 8¼ | 5 | 3⅝ | 322.062 | 2.2365 | | 4½ | 7 | 5½ | 637.940 | 4.4301 |
| | 8⅜ | 5 | 4 | 326.051 | 2.2642 | | 4¾ | 7 | 6⅜ | 649.181 | 4.5082 |
| | 8½ | 5 | 4⅜ | 330.064 | 2.2921 | 2 | 5 | 7 | 7⅛ | 660.520 | 4.5869 |
| | 8⅝ | 5 | 4¾ | 334.101 | 2.3201 | | 5¼ | 7 | 7⅞ | 671.957 | 4.6665 |
| | 8¾ | 5 | 5¼ | 338.163 | 2.3480 | | 5½ | 7 | 8⅝ | 683.493 | 4.7465 |
| | 8⅞ | 5 | 5⅝ | 342.250 | 2.3767 | | 5¾ | 7 | 9½ | 695.126 | 4.8273 |
| 1 | 9 | 5 | 6 | 346.361 | 2.4053 | 2 | 6 | 7 | 10¼ | 706.858 | 4.9087 |
| | 9⅛ | 5 | 6⅜ | 350.496 | 2.4340 | | 6¼ | 7 | 11 | 718.688 | 4.9909 |
| | 9¼ | 5 | 6¾ | 354.656 | 2.4629 | | 6½ | 7 | 11⅞ | 730.617 | 5.0737 |
| | 9⅜ | 5 | 7⅛ | 358.841 | 2.4920 | | 6¾ | 8 | 0⅝ | 742.643 | 5.1572 |
| | 9½ | 5 | 7½ | 363.050 | 2.5212 | 2 | 7 | 8 | 1⅜ | 754.768 | 5.2414 |
| | 9⅝ | 5 | 7⅞ | 367.284 | 2.5506 | | 7¼ | 8 | 2⅛ | 766.990 | 5.3263 |
| | 9¾ | 5 | 8⅜ | 371.542 | 2.5802 | | 7½ | 8 | 3 | 779.311 | 5.4119 |
| | 9⅞ | 5 | 8¾ | 375.825 | 2.6099 | | 7¾ | 8 | 3¾ | 791.730 | 5.4981 |
| 1 | 10 | 5 | 9⅛ | 380.133 | 2.6398 | 2 | 8 | 8 | 4½ | 804.248 | 5.5851 |
| | 10⅛ | 5 | 9½ | 384.465 | 2.6699 | | 8¼ | 8 | 5⅜ | 816.863 | 5.6727 |
| | 10¼ | 5 | 9⅞ | 388.821 | 2.7001 | | 8½ | 8 | 6⅛ | 829.577 | 5.7610 |
| | 10⅜ | 5 | 10¼ | 393.202 | 2.7306 | | 8¾ | 8 | 6⅞ | 842.389 | 5.8499 |
| | 10½ | 5 | 10⅝ | 397.608 | 2.7612 | 2 | 9 | 8 | 7⅝ | 855.299 | 5.9396 |
| | 10⅝ | 5 | 11⅛ | 402.038 | 2.7919 | | 9¼ | 8 | 8½ | 868.307 | 6.0299 |
| | 10¾ | 5 | 11½ | 406.493 | 2.8229 | | 9½ | 8 | 9¼ | 881.413 | 6.1209 |
| | 10⅞ | 5 | 11⅞ | 410.972 | 2.8540 | | 9¾ | 8 | 10 | 894.618 | 6.2126 |
| 1 | 11 | 6 | 0¼ | 415.476 | 2.8852 | 2 | 10 | 8 | 10⅞ | 907.920 | 6.3050 |
| | 11⅛ | 6 | 0⅝ | 420.004 | 2.9167 | | 10¼ | 8 | 11⅝ | 921.321 | 6.3981 |
| | 11¼ | 6 | 1 | 424.557 | 2.9483 | | 10½ | 9 | 0⅜ | 934.820 | 6.4918 |
| | 11⅜ | 6 | 1⅜ | 429.134 | 2.9801 | | 10¾ | 9 | 1⅛ | 948.417 | 6.5862 |
| | 11½ | 6 | 1⅞ | 433.736 | 3.0121 | 2 | 11 | 9 | 2 | 962.113 | 6.6813 |
| | 11⅝ | 6 | 2¼ | 438.363 | 3.0442 | | 11¼ | 9 | 2¾ | 975.906 | 6.7771 |
| | 11¾ | 6 | 2⅝ | 443.014 | 3.0765 | | 11½ | 9 | 3½ | 989.798 | 6.8736 |
| | 11⅞ | 6 | 3 | 447.689 | 3.1090 | | 11¾ | 9 | 4¼ | 1003.788 | 6.9707 |

* Circumference to nearest ⅛ inch.

## MATHEMATICAL TABLES

### Diameters, Circumferences and Areas of Circles in Feet and Inches

| Diam. Ft. In. | Circum.* Ft. In. | Area Sq. In. | Area Sq. Ft. | Diam. Ft. In. | Circum.* Ft. In. | Area Sq. In. | Area Sq. Ft. |
|---|---|---|---|---|---|---|---|
| 3 0 | 9 5⅛ | 1017.88 | 7.069 | 4 0 | 12 6¾ | 1809.56 | 12.566 |
| 0¼ | 9 5⅞ | 1032.06 | 7.167 | 0¼ | 12 7⅝ | 1828.46 | 12.698 |
| 0½ | 9 6⅝ | 1046.35 | 7.266 | 0½ | 12 8⅜ | 1847.45 | 12.830 |
| 0¾ | 9 7½ | 1060.73 | 7.366 | 0¾ | 12 9⅛ | 1866.55 | 12.962 |
| 3 1 | 9 8¼ | 1075.21 | 7.467 | 4 1 | 12 10 | 1885.74 | 13.095 |
| 1¼ | 9 9 | 1089.79 | 7.568 | 1¼ | 12 10¾ | 1905.03 | 13.229 |
| 1½ | 9 9¾ | 1104.47 | 7.670 | 1½ | 12 11½ | 1924.42 | 13.364 |
| 1¾ | 9 10½ | 1119.24 | 7.773 | 1¾ | 13 0¼ | 1943.91 | 13.499 |
| 3 2 | 9 11⅜ | 1134.11 | 7.876 | 4 2 | 13 1⅛ | 1963.50 | 13.635 |
| 2¼ | 10 0⅛ | 1149.09 | 7.980 | 2¼ | 13 1⅞ | 1983.18 | 13.772 |
| 2½ | 10 1 | 1164.16 | 8.084 | 2½ | 13 2⅝ | 2002.96 | 13.909 |
| 2¾ | 10 1¾ | 1179.32 | 8.190 | 2¾ | 13 3⅜ | 2022.84 | 14.048 |
| 3 3 | 10 2½ | 1194.59 | 8.296 | 4 3 | 13 4¼ | 2042.82 | 14.186 |
| 3¼ | 10 3¼ | 1209.95 | 8.402 | 3¼ | 13 5 | 2062.90 | 14.326 |
| 3½ | 10 4⅛ | 1225.42 | 8.510 | 3½ | 13 5¾ | 2083.07 | 14.466 |
| 3¾ | 10 4⅞ | 1240.98 | 8.618 | 3¾ | 13 6⅝ | 2103.35 | 14.607 |
| 3 4 | 10 5⅝ | 1256.64 | 8.727 | 4 4 | 13 7⅜ | 2123.72 | 14.748 |
| 4¼ | 10 6½ | 1272.39 | 8.836 | 4¼ | 13 8⅛ | 2144.19 | 14.890 |
| 4½ | 10 7¼ | 1288.25 | 8.946 | 4½ | 13 8⅞ | 2164.75 | 15.033 |
| 4¾ | 10 8 | 1304.20 | 9.057 | 4¾ | 13 9¾ | 2185.42 | 15.177 |
| 3 5 | 10 8¾ | 1320.25 | 9.168 | 4 5 | 13 10½ | 2206.18 | 15.321 |
| 5¼ | 10 9⅝ | 1336.40 | 9.281 | 5¼ | 13 11¼ | 2227.05 | 15.466 |
| 5½ | 10 10⅜ | 1352.65 | 9.393 | 5½ | 14 0⅛ | 2248.01 | 15.611 |
| 5¾ | 10 11⅛ | 1369.00 | 9.507 | 5¾ | 14 0⅞ | 2269.06 | 15.757 |
| 3 6 | 11 0 | 1385.44 | 9.621 | 4 6 | 14 1⅝ | 2290.22 | 15.904 |
| 6¼ | 11 0¾ | 1401.98 | 9.736 | 6¼ | 14 2⅜ | 2311.48 | 16.052 |
| 6½ | 11 1½ | 1418.63 | 9.852 | 6½ | 14 3¼ | 2332.83 | 16.200 |
| 6¾ | 11 2¼ | 1435.36 | 9.968 | 6¾ | 14 4 | 2354.28 | 16.349 |
| 3 7 | 11 3⅛ | 1452.20 | 10.085 | 4 7 | 14 4¾ | 2375.83 | 16.499 |
| 7¼ | 11 3⅞ | 1469.14 | 10.202 | 7¼ | 14 5⅝ | 2397.48 | 16.649 |
| 7½ | 11 4⅝ | 1486.17 | 10.321 | 7½ | 14 6⅜ | 2419.22 | 16.800 |
| 7¾ | 11 5½ | 1503.30 | 10.440 | 7¾ | 14 7⅛ | 2441.07 | 16.952 |
| 3 8 | 11 6¼ | 1520.53 | 10.559 | 4 8 | 14 7⅞ | 2463.01 | 17.104 |
| 8¼ | 11 7 | 1537.86 | 10.680 | 8¼ | 14 8¾ | 2485.05 | 17.257 |
| 8½ | 11 7¾ | 1555.28 | 10.801 | 8½ | 14 9½ | 2507.19 | 17.411 |
| 8¾ | 11 8⅝ | 1572.81 | 10.922 | 8¾ | 14 10¼ | 2529.42 | 17.565 |
| 3 9 | 11 9⅜ | 1590.43 | 11.045 | 4 9 | 14 11⅛ | 2551.76 | 17.721 |
| 9¼ | 11 10⅛ | 1608.15 | 11.168 | 9¼ | 14 11⅞ | 2574.19 | 17.876 |
| 9½ | 11 11 | 1625.97 | 11.291 | 9½ | 15 0⅝ | 2596.72 | 18.033 |
| 9¾ | 11 11¾ | 1643.89 | 11.416 | 9¾ | 15 1⅜ | 2619.35 | 18.190 |
| 3 10 | 12 0½ | 1661.90 | 11.541 | 4 10 | 15 2¼ | 2642.08 | 18.348 |
| 10¼ | 12 1¼ | 1680.02 | 11.667 | 10¼ | 15 3 | 2664.91 | 18.506 |
| 10½ | 12 2⅛ | 1698.23 | 11.793 | 10½ | 15 3¾ | 2687.83 | 18.665 |
| 10¾ | 12 2⅞ | 1716.54 | 11.920 | 10¾ | 15 4⅝ | 2710.85 | 18.825 |
| 3 11 | 12 3⅝ | 1734.94 | 12.048 | 4 11 | 15 5⅜ | 2733.97 | 18.986 |
| 11¼ | 12 4½ | 1753.45 | 12.177 | 11¼ | 15 6⅛ | 2757.19 | 19.147 |
| 11½ | 12 5¼ | 1772.05 | 12.306 | 11½ | 15 6⅞ | 2780.51 | 19.309 |
| 11¾ | 12 6 | 1790.76 | 12.436 | 11¾ | 15 7¾ | 2803.92 | 19.472 |

* Circumference to nearest ⅛ inch.

**Diameters, Circumferences and Areas of Circles in Feet and Inches**

| Diam. | | Circum.* | | Area | Area | Diam. | | Circum.* | | Area | Area |
|---|---|---|---|---|---|---|---|---|---|---|---|
| Ft. | In. | Ft. | In. | Sq. In. | Sq. Ft. | Ft. | In. | Ft. | In. | Sq. In. | Sq. Ft. |
| 5 | 0 | 15 | 8½ | 2827.43 | 19.635 | 6 | 0 | 18 | 10¼ | 4071.50 | 28.274 |
| | 0¼ | 15 | 9¼ | 2851.04 | 19.799 | | 0¼ | 18 | 11 | 4099.83 | 28.471 |
| | 0½ | 15 | 10⅛ | 2874.75 | 19.964 | | 0½ | 18 | 11¾ | 4128.25 | 28.668 |
| | 0¾ | 15 | 10⅞ | 2898.56 | 20.129 | | 0¾ | 19 | 0½ | 4156.77 | 28.866 |
| 5 | 1 | 15 | 11⅝ | 2922.47 | 20.295 | 6 | 1 | 19 | 1⅜ | 4185.39 | 29.065 |
| | 1¼ | 16 | 0⅜ | 2946.47 | 20.462 | | 1¼ | 19 | 2⅛ | 4214.10 | 29.265 |
| | 1½ | 16 | 1¼ | 2970.57 | 20.629 | | 1½ | 19 | 2⅞ | 4242.92 | 29.465 |
| | 1¾ | 16 | 2 | 2994.77 | 20.797 | | 1¾ | 19 | 3¾ | 4271.83 | 29.665 |
| 5 | 2 | 16 | 2¾ | 3019.07 | 20.966 | 6 | 2 | 19 | 4½ | 4300.84 | 29.867 |
| | 2¼ | 16 | 3⅝ | 3043.47 | 21.135 | | 2¼ | 19 | 5¼ | 4329.95 | 30.069 |
| | 2½ | 16 | 4⅜ | 3067.96 | 21.305 | | 2½ | 19 | 6 | 4359.16 | 30.272 |
| | 2¾ | 16 | 5⅛ | 3092.56 | 21.476 | | 2¾ | 19 | 6⅞ | 4388.46 | 30.475 |
| 5 | 3 | 16 | 5⅞ | 3117.25 | 21.648 | 6 | 3 | 19 | 7⅝ | 4417.86 | 30.680 |
| | 3¼ | 16 | 6¾ | 3142.03 | 21.820 | | 3¼ | 19 | 8⅜ | 4447.37 | 30.884 |
| | 3½ | 16 | 7½ | 3166.92 | 21.993 | | 3½ | 19 | 9¼ | 4476.97 | 31.090 |
| | 3¾ | 16 | 8¼ | 3191.91 | 22.166 | | 3¾ | 19 | 10 | 4506.66 | 31.296 |
| 5 | 4 | 16 | 9 | 3216.99 | 22.340 | 6 | 4 | 19 | 10¾ | 4536.46 | 31.503 |
| | 4¼ | 16 | 9⅞ | 3242.17 | 22.515 | | 4¼ | 19 | 11½ | 4566.35 | 31.711 |
| | 4½ | 16 | 10⅝ | 3267.45 | 22.691 | | 4½ | 20 | 0⅜ | 4596.35 | 31.919 |
| | 4¾ | 16 | 11⅜ | 3292.83 | 22.867 | | 4¾ | 20 | 1⅛ | 4626.44 | 32.128 |
| 5 | 5 | 17 | 0¼ | 3318.31 | 23.044 | 6 | 5 | 20 | 1⅞ | 4656.63 | 32.338 |
| | 5¼ | 17 | 1 | 3343.88 | 23.221 | | 5¼ | 20 | 2¾ | 4686.91 | 32.548 |
| | 5½ | 17 | 1¾ | 3369.55 | 23.400 | | 5½ | 20 | 3½ | 4717.30 | 32.759 |
| | 5¾ | 17 | 2½ | 3395.33 | 23.579 | | 5¾ | 20 | 4¼ | 4747.78 | 32.971 |
| 5 | 6 | 17 | 3⅜ | 3421.19 | 23.758 | 6 | 6 | 20 | 5 | 4778.36 | 33.183 |
| | 6¼ | 17 | 4⅛ | 3447.16 | 23.939 | | 6¼ | 20 | 5⅞ | 4809.04 | 33.396 |
| | 6½ | 17 | 4⅞ | 3473.23 | 24.120 | | 6½ | 20 | 6⅝ | 4839.82 | 33.610 |
| | 6¾ | 17 | 5¾ | 3499.39 | 24.301 | | 6¾ | 20 | 7⅜ | 4870.70 | 33.824 |
| 5 | 7 | 17 | 6½ | 3525.65 | 24.484 | 6 | 7 | 20 | 8⅛ | 4901.67 | 34.039 |
| | 7¼ | 17 | 7¼ | 3552.01 | 24.667 | | 7¼ | 20 | 9 | 4932.74 | 34.255 |
| | 7½ | 17 | 8 | 3578.47 | 24.850 | | 7½ | 20 | 9¾ | 4963.91 | 34.472 |
| | 7¾ | 17 | 8⅞ | 3605.03 | 25.035 | | 7¾ | 20 | 10½ | 4995.18 | 34.689 |
| 5 | 8 | 17 | 9⅝ | 3631.68 | 25 220 | 6 | 8 | 20 | 11⅜ | 5026.55 | 34.907 |
| | 8¼ | 17 | 10⅜ | 3658.43 | 25.406 | | 8¼ | 21 | 0⅛ | 5058.01 | 35.125 |
| | 8½ | 17 | 11¼ | 3685.28 | 25.592 | | 8½ | 21 | 0⅞ | 5089.58 | 35.344 |
| | 8¾ | 18 | 0 | 3712.24 | 25.779 | | 8¾ | 21 | 1⅝ | 5121.24 | 35.564 |
| 5 | 9 | 18 | 0¾ | 3739.28 | 25.967 | 6 | 9 | 21 | 2½ | 5153.00 | 35.785 |
| | 9¼ | 18 | 1½ | 3766.43 | 26.156 | | 9¼ | 21 | 3¼ | 5184.86 | 36.006 |
| | 9½ | 18 | 2⅜ | 3793.67 | 26.345 | | 9½ | 21 | 4 | 5216.81 | 36.228 |
| | 9¾ | 18 | 3⅛ | 3821.01 | 26.535 | | 9¾ | 21 | 4⅞ | 5248.87 | 36.450 |
| 5 | 10 | 18 | 3⅞ | 3848.45 | 26.725 | 6 | 10 | 21 | 5⅝ | 5281.02 | 36.674 |
| | 10¼ | 18 | 4¾ | 3875.99 | 26.917 | | 10¼ | 21 | 6⅜ | 5313.27 | 36.898 |
| | 10½ | 18 | 5½ | 3903.63 | 27.109 | | 10½ | 21 | 7⅛ | 5345 62 | 37.122 |
| | 10¾ | 18 | 6¼ | 3931.36 | 27.301 | | 10¾ | 21 | 8 | 5378.06 | 37.348 |
| 5 | 11 | 18 | 7 | 3959.19 | 27.494 | 6 | 11 | 21 | 8¾ | 5410.61 | 37.574 |
| | 11¼ | 18 | 7⅞ | 3987.12 | 27.688 | | 11¼ | 21 | 9½ | 5443.25 | 37.800 |
| | 11½ | 18 | 8⅝ | 4015.15 | 27.883 | | 11½ | 21 | 10⅜ | 5475.99 | 38.028 |
| | 11¾ | 18 | 9⅜ | 4043.28 | 28.078 | | 11¾ | 21 | 11⅛ | 5508.83 | 38.256 |

* Circumference to nearest ⅛ inch.

**Diameters, Circumferences and Areas of Circles in Feet and Inches**

| Diam. Ft. | Diam. In. | Circum.* Ft. | Circum.* In. | Area Sq. Ft. | Diam. Ft. | Diam. In. | Circum.* Ft. | Circum.* In. | Area Sq. Ft. | Diam. Ft. | Diam. In. | Circum.* Ft. | Circum.* In. | Area Sq. Ft. |
|---|---|---|---|---|---|---|---|---|---|---|---|---|---|---|
| 7 | 0 | 21 | 11⅞ | 38.48 | 11 | 0 | 34 | 6¾ | 95.03 | 15 | 0 | 47 | 1½ | 176.71 |
| | 1 | 22 | 3 | 39.41 | | 1 | 34 | 9⅞ | 96.48 | | 1 | 47 | 4⅝ | 178.68 |
| | 2 | 22 | 6⅛ | 40.34 | | 2 | 35 | 1 | 97.93 | | 2 | 47 | 7¾ | 180.66 |
| | 3 | 22 | 9⅜ | 41.28 | | 3 | 35 | 4⅛ | 99.40 | | 3 | 47 | 10⅞ | 182 65 |
| | 4 | 23 | 0½ | 42.24 | | 4 | 35 | 7¼ | 100.88 | | 4 | 48 | 2 | 184.66 |
| | 5 | 23 | 3⅝ | 43.20 | | 5 | 35 | 10⅜ | 102.37 | | 5 | 48 | 5¼ | 186.67 |
| | 6 | 23 | 6¾ | 44.18 | | 6 | 36 | 1½ | 103.87 | | 6 | 48 | 8⅜ | 188.69 |
| | 7 | 23 | 9⅞ | 45.17 | | 7 | 36 | 4⅝ | 105.38 | | 7 | 48 | 11½ | 190.73 |
| | 8 | 24 | 1 | 46.16 | | 8 | 36 | 7⅞ | 106.90 | | 8 | 49 | 2⅝ | 192.77 |
| | 9 | 24 | 4⅛ | 47.17 | | 9 | 36 | 11 | 108.43 | | 9 | 49 | 5¾ | 194.83 |
| | 10 | 24 | 7¼ | 48.19 | | 10 | 37 | 2⅛ | 109.98 | | 10 | 49 | 8⅞ | 196.89 |
| | 11 | 24 | 10½ | 49.22 | | 11 | 37 | 5¼ | 111.53 | | 11 | 50 | 0 | 198.97 |
| 8 | 0 | 25 | 1⅝ | 50.27 | 12 | 0 | 37 | 8⅜ | 113.10 | 16 | 0 | 50 | 3⅛ | 201.06 |
| | 1 | 25 | 4¾ | 51.32 | | 1 | 37 | 11½ | 114.67 | | 1 | 50 | 6⅜ | 203.16 |
| | 2 | 25 | 7⅞ | 52.38 | | 2 | 38 | 2⅝ | 116.26 | | 2 | 50 | 9½ | 205.27 |
| | 3 | 25 | 11 | 53.46 | | 3 | 38 | 5⅞ | 117.86 | | 3 | 51 | 0⅝ | 207.39 |
| | 4 | 26 | 2⅛ | 54.54 | | 4 | 38 | 9 | 119.47 | | 4 | 51 | 3¾ | 209.53 |
| | 5 | 26 | 5¼ | 55.64 | | 5 | 39 | 0⅛ | 121.09 | | 5 | 51 | 6⅞ | 211.67 |
| | 6 | 26 | 8½ | 56.75 | | 6 | 39 | 3¼ | 122.72 | | 6 | 51 | 10 | 213.82 |
| | 7 | 26 | 11⅝ | 57.86 | | 7 | 39 | 6⅜ | 124.36 | | 7 | 52 | 1⅛ | 215.99 |
| | 8 | 27 | 2¾ | 58.99 | | 8 | 39 | 9½ | 126.01 | | 8 | 52 | 4⅜ | 218.17 |
| | 9 | 27 | 5⅞ | 60.13 | | 9 | 40 | 0⅝ | 127.68 | | 9 | 52 | 7½ | 220.35 |
| | 10 | 27 | 9 | 61.28 | | 10 | 40 | 3¾ | 129.35 | | 10 | 52 | 10⅝ | 222.55 |
| | 11 | 28 | 0⅛ | 62.44 | | 11 | 40 | 7 | 131.04 | | 11 | 53 | 1¾ | 224.76 |
| 9 | 0 | 28 | 3¼ | 63.62 | 13 | 0 | 40 | 10⅛ | 132.73 | 17 | 0 | 53 | 4⅞ | 226.98 |
| | 1 | 28 | 6⅜ | 64.80 | | 1 | 41 | 1¼ | 134.44 | | 1 | 53 | 8 | 229.21 |
| | 2 | 28 | 9⅝ | 66.00 | | 2 | 41 | 4⅜ | 136.16 | | 2 | 53 | 11⅛ | 231.45 |
| | 3 | 29 | 0¾ | 67.20 | | 3 | 41 | 7½ | 137.89 | | 3 | 54 | 2¼ | 233.71 |
| | 4 | 29 | 3⅞ | 68.42 | | 4 | 41 | 10⅝ | 139.63 | | 4 | 54 | 5½ | 235.97 |
| | 5 | 29 | 7 | 69.64 | | 5 | 42 | 1¾ | 141.38 | | 5 | 54 | 8⅝ | 238.24 |
| | 6 | 29 | 10⅛ | 70.88 | | 6 | 42 | 5 | 143.14 | | 6 | 54 | 11¾ | 240.53 |
| | 7 | 30 | 1¼ | 72.13 | | 7 | 42 | 8⅛ | 144.91 | | 7 | 55 | 2⅞ | 242.82 |
| | 8 | 30 | 4⅜ | 73.39 | | 8 | 42 | 11¼ | 146.69 | | 8 | 55 | 6 | 245.13 |
| | 9 | 30 | 7⅝ | 74.66 | | 9 | 43 | 2⅜ | 148.49 | | 9 | 55 | 9⅛ | 247.45 |
| | 10 | 30 | 10¾ | 75.94 | | 10 | 43 | 5½ | 150.29 | | 10 | 56 | 0¼ | 249.78 |
| | 11 | 31 | 1⅞ | 77.24 | | 11 | 43 | 8⅝ | 152.11 | | 11 | 56 | 3½ | 252.12 |
| 10 | 0 | 31 | 5 | 78.54 | 14 | 0 | 43 | 11¾ | 153.94 | 18 | 0 | 56 | 6⅝ | 254.47 |
| | 1 | 31 | 8⅛ | 79.85 | | 1 | 44 | 2⅞ | 155.78 | | 1 | 56 | 9¾ | 256.83 |
| | 2 | 31 | 11¼ | 81.18 | | 2 | 44 | 6⅛ | 157.63 | | 2 | 57 | 0⅞ | 259.20 |
| | 3 | 32 | 2⅜ | 82.52 | | 3 | 44 | 9¼ | 159.48 | | 3 | 57 | 4 | 261.59 |
| | 4 | 32 | 5½ | 83.86 | | 4 | 45 | 0⅜ | 161.36 | | 4 | 57 | 7⅛ | 263.98 |
| | 5 | 32 | 8¾ | 85.22 | | 5 | 45 | 3½ | 163.24 | | 5 | 57 | 10¼ | 266.39 |
| | 6 | 32 | 11⅞ | 86.59 | | 6 | 45 | 6⅝ | 165.13 | | 6 | 58 | 1⅜ | 268.80 |
| | 7 | 33 | 3 | 87.97 | | 7 | 45 | 9¾ | 167.03 | | 7 | 58 | 4⅝ | 271.23 |
| | 8 | 33 | 6⅛ | 89.36 | | 8 | 46 | 0⅞ | 168.95 | | 8 | 58 | 7¾ | 273.67 |
| | 9 | 33 | 9¼ | 90.76 | | 9 | 46 | 4 | 170.87 | | 9 | 58 | 10⅞ | 276.12 |
| | 10 | 34 | 0⅜ | 92.18 | | 10 | 46 | 7¼ | 172.81 | | 10 | 59 | 2 | 278.58 |
| | 11 | 34 | 3½ | 93.60 | | 11 | 46 | 10⅜ | 174.76 | | 11 | 59 | 5⅛ | 281.05 |

* Circumference to nearest ⅛ inch.

### Circumferences and Corresponding Diameters of Circles
(English or metric units)

| Circ. | Diam. | Circ. | Diam. | Circ. | Diam. | Circ. | Diam. | Circ. | Diam. |
|---|---|---|---|---|---|---|---|---|---|
| 1 | 0.3183 | 51 | 16.2338 | 101 | 32.149 | 151 | 48.065 | 201 | 63.980 |
| 2 | 0.6366 | 52 | 16.5521 | 102 | 32.468 | 152 | 48.383 | 202 | 64.299 |
| 3 | 0.9549 | 53 | 16.8704 | 103 | 32.786 | 153 | 48.701 | 203 | 64.617 |
| 4 | 1.2732 | 54 | 17.1887 | 104 | 33.104 | 154 | 49.020 | 204 | 64.935 |
| 5 | 1.5915 | 55 | 17.5070 | 105 | 33.423 | 155 | 49.338 | 205 | 65.254 |
| 6 | 1.9099 | 56 | 17.8254 | 106 | 33.741 | 156 | 49.656 | 206 | 65.572 |
| 7 | 2.2282 | 57 | 18.1437 | 107 | 34.059 | 157 | 49.975 | 207 | 65.890 |
| 8 | 2.5465 | 58 | 18.4620 | 108 | 34.377 | 158 | 50.293 | 208 | 66.208 |
| 9 | 2.8648 | 59 | 18.7803 | 109 | 34.696 | 159 | 50.611 | 209 | 66.527 |
| 10 | 3.1831 | 60 | 19.0986 | 110 | 35.014 | 160 | 50.930 | 210 | 66.845 |
| 11 | 3.5014 | 61 | 19.4169 | 111 | 35.332 | 161 | 51.248 | 211 | 67.163 |
| 12 | 3.8197 | 62 | 19.7352 | 112 | 35.651 | 162 | 51.566 | 212 | 67.482 |
| 13 | 4.1380 | 63 | 20.0535 | 113 | 35.969 | 163 | 51.885 | 213 | 67.800 |
| 14 | 4.4563 | 64 | 20.3718 | 114 | 36.287 | 164 | 52.203 | 214 | 68.118 |
| 15 | 4.7746 | 65 | 20.6901 | 115 | 36.606 | 165 | 52.521 | 215 | 68.437 |
| 16 | 5.0930 | 66 | 21.0085 | 116 | 36.924 | 166 | 52.839 | 216 | 68.755 |
| 17 | 5.4113 | 67 | 21.3268 | 117 | 37.242 | 167 | 53.158 | 217 | 69.073 |
| 18 | 5.7296 | 68 | 21.6451 | 118 | 37.561 | 168 | 53.476 | 218 | 69.392 |
| 19 | 6.0479 | 69 | 21.9634 | 119 | 37.879 | 169 | 53.794 | 219 | 69.710 |
| 20 | 6.3662 | 70 | 22.2817 | 120 | 38.197 | 170 | 54.113 | 220 | 70.028 |
| 21 | 6.6845 | 71 | 22.6000 | 121 | 38.515 | 171 | 54.431 | 221 | 70.346 |
| 22 | 7.0028 | 72 | 22.9183 | 122 | 38.834 | 172 | 54.749 | 222 | 70.665 |
| 23 | 7.3211 | 73 | 23.2366 | 123 | 39.152 | 173 | 55.068 | 223 | 70.983 |
| 24 | 7.6394 | 74 | 23.5549 | 124 | 39.470 | 174 | 55.386 | 224 | 71.301 |
| 25 | 7.9577 | 75 | 23.8732 | 125 | 39.789 | 175 | 55.704 | 225 | 71.620 |
| 26 | 8.2761 | 76 | 24.1916 | 126 | 40.107 | 176 | 56.023 | 226 | 71.938 |
| 27 | 8.5944 | 77 | 24.5099 | 127 | 40.425 | 177 | 56.341 | 227 | 72.256 |
| 28 | 8.9127 | 78 | 24.8282 | 128 | 40.744 | 178 | 56.659 | 228 | 72.575 |
| 29 | 9.2310 | 79 | 25.1465 | 129 | 41.062 | 179 | 56.977 | 229 | 72.893 |
| 30 | 9.5493 | 80 | 25.4648 | 130 | 41.380 | 180 | 57.296 | 230 | 73.211 |
| 31 | 9.8676 | 81 | 25.7831 | 131 | 41.699 | 181 | 57.614 | 231 | 73.530 |
| 32 | 10.1859 | 82 | 26.1014 | 132 | 42.017 | 182 | 57.932 | 232 | 73.848 |
| 33 | 10.5042 | 83 | 26.4197 | 133 | 42.335 | 183 | 58.251 | 233 | 74.166 |
| 34 | 10.8225 | 84 | 26.7380 | 134 | 42.654 | 184 | 58.569 | 234 | 74.485 |
| 35 | 11.1408 | 85 | 27.0563 | 135 | 42.972 | 185 | 58.887 | 235 | 74.803 |
| 36 | 11.4592 | 86 | 27.3747 | 136 | 43.290 | 186 | 59.206 | 236 | 75.121 |
| 37 | 11.7775 | 87 | 27.6930 | 137 | 43.608 | 187 | 59.524 | 237 | 75.439 |
| 38 | 12.0958 | 88 | 28.0113 | 138 | 43.927 | 188 | 59.842 | 238 | 75.758 |
| 39 | 12.4141 | 89 | 28.3296 | 139 | 44.245 | 189 | 60.161 | 239 | 76.076 |
| 40 | 12.7324 | 90 | 28.6479 | 140 | 44.563 | 190 | 60.479 | 240 | 76.394 |
| 41 | 13.0507 | 91 | 28.9662 | 141 | 44.882 | 191 | 60.797 | 241 | 76.713 |
| 42 | 13.3690 | 92 | 29.2845 | 142 | 45.200 | 192 | 61.115 | 242 | 77.031 |
| 43 | 13.6873 | 93 | 29.6028 | 143 | 45.518 | 193 | 61.434 | 243 | 77.349 |
| 44 | 14.0056 | 94 | 29.9211 | 144 | 45.837 | 194 | 61.752 | 244 | 77.668 |
| 45 | 14.3239 | 95 | 30.2394 | 145 | 46.155 | 195 | 62.070 | 245 | 77.986 |
| 46 | 14.6423 | 96 | 30.5577 | 146 | 46.473 | 196 | 62.389 | 246 | 78.304 |
| 47 | 14.9606 | 97 | 30.8761 | 147 | 46.792 | 197 | 62.707 | 247 | 78.623 |
| 48 | 15.2789 | 98 | 31.1944 | 148 | 47.110 | 198 | 63.025 | 248 | 78.941 |
| 49 | 15.5972 | 99 | 31.5127 | 149 | 47.428 | 199 | 63.344 | 249 | 79.259 |
| 50 | 15.9155 | 100 | 31.8310 | 150 | 47.746 | 200 | 63.662 | 250 | 79.577 |

### Segments of Circles for Radius = 1
(English or metric units)

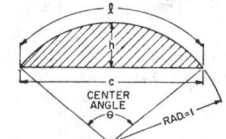

Length of arc, height of segment, length of chord, and area of segment for angles from 1 to 180 degrees and radius = 1. For other radii, multiply the values of $l$, $h$ and $c$ in the table by the given radius $r$, and the values for areas, by $r^2$, the square of the radius.

The values in the tables can be used for English or metric units.

| Center Angle $\theta$, Degrees | $l$ | $h$ | $c$ | Area of Segment $A$ | Center Angle $\theta$, Degrees | $l$ | $h$ | $c$ | Area of Segment $A$ |
|---|---|---|---|---|---|---|---|---|---|
| 1 | 0.01745 | 0.00004 | 0.01745 | 0.00000 | 46 | 0.803 | 0.0795 | 0.781 | 0.04176 |
| 2 | 0.03491 | 0.00015 | 0.03490 | 0.00000 | 47 | 0.820 | 0.0829 | 0.797 | 0.04448 |
| 3 | 0.05236 | 0.00034 | 0.05235 | 0.00001 | 48 | 0.838 | 0.0865 | 0.813 | 0.04731 |
| 4 | 0.06981 | 0.00061 | 0.06980 | 0.00003 | 49 | 0.855 | 0.0900 | 0.829 | 0.05025 |
| 5 | 0.08727 | 0.00095 | 0.08724 | 0.00006 | 50 | 0.873 | 0.0937 | 0.845 | 0.05331 |
| 6 | 0.10472 | 0.00137 | 0.10467 | 0.00010 | 51 | 0.890 | 0.0974 | 0.861 | 0.05649 |
| 7 | 0.12217 | 0.00187 | 0.12210 | 0.00015 | 52 | 0.908 | 0.1012 | 0.877 | 0.05978 |
| 8 | 0.13963 | 0.00244 | 0.13951 | 0.00023 | 53 | 0.925 | 0.1051 | 0.892 | 0.06319 |
| 9 | 0.15708 | 0.00308 | 0.15692 | 0.00032 | 54 | 0.942 | 0.1090 | 0.908 | 0.06673 |
| 10 | 0.17453 | 0.00381 | 0.17431 | 0.00044 | 55 | 0.960 | 0.1130 | 0.923 | 0.07039 |
| 11 | 0.19199 | 0.00460 | 0.19169 | 0.00059 | 56 | 0.977 | 0.1171 | 0.939 | 0.07417 |
| 12 | 0.20944 | 0.00548 | 0.20906 | 0.00076 | 57 | 0.995 | 0.1212 | 0.954 | 0.07808 |
| 13 | 0.22689 | 0.00643 | 0.22641 | 0.00097 | 58 | 1.012 | 0.1254 | 0.970 | 0.08212 |
| 14 | 0.24435 | 0.00745 | 0.24374 | 0.00121 | 59 | 1.030 | 0.1296 | 0.985 | 0.08629 |
| 15 | 0.26180 | 0.00856 | 0.26105 | 0.00149 | 60 | 1.047 | 0.1340 | 1.000 | 0.09059 |
| 16 | 0.27925 | 0.00973 | 0.27835 | 0.00181 | 61 | 1.065 | 0.1384 | 1.015 | 0.09502 |
| 17 | 0.29671 | 0.01098 | 0.29562 | 0.00217 | 62 | 1.082 | 0.1428 | 1.030 | 0.09958 |
| 18 | 0.31416 | 0.01231 | 0.31287 | 0.00257 | 63 | 1.100 | 0.1474 | 1.045 | 0.10428 |
| 19 | 0.33161 | 0.01371 | 0.33010 | 0.00302 | 64 | 1.117 | 0.1520 | 1.060 | 0.10911 |
| 20 | 0.34907 | 0.01519 | 0.34730 | 0.00352 | 65 | 1.134 | 0.1566 | 1.075 | 0.11408 |
| 21 | 0.36652 | 0.01675 | 0.36447 | 0.00408 | 66 | 1.152 | 0.1613 | 1.089 | 0.11919 |
| 22 | 0.38397 | 0.01837 | 0.38162 | 0.00468 | 67 | 1.169 | 0.1661 | 1.104 | 0.12443 |
| 23 | 0.40143 | 0.02008 | 0.39874 | 0.00535 | 68 | 1.187 | 0.1710 | 1.118 | 0.12982 |
| 24 | 0.41888 | 0.02185 | 0.41582 | 0.00607 | 69 | 1.204 | 0.1759 | 1.133 | 0.13535 |
| 25 | 0.43633 | 0.02370 | 0.43288 | 0.00686 | 70 | 1.222 | 0.1808 | 1.147 | 0.14102 |
| 26 | 0.45379 | 0.02563 | 0.44990 | 0.00771 | 71 | 1.239 | 0.1859 | 1.161 | 0.14683 |
| 27 | 0.47124 | 0.02763 | 0.46689 | 0.00862 | 72 | 1.257 | 0.1910 | 1.176 | 0.15279 |
| 28 | 0.48869 | 0.02970 | 0.48384 | 0.00961 | 73 | 1.274 | 0.1961 | 1.190 | 0.15889 |
| 29 | 0.50615 | 0.03185 | 0.50076 | 0.01067 | 74 | 1.292 | 0.2014 | 1.204 | 0.16514 |
| 30 | 0.52360 | 0.03407 | 0.51764 | 0.01180 | 75 | 1.309 | 0.2066 | 1.218 | 0.17154 |
| 31 | 0.54105 | 0.03637 | 0.53448 | 0.01301 | 76 | 1.326 | 0.2120 | 1.231 | 0.17808 |
| 32 | 0.55851 | 0.03874 | 0.55127 | 0.01429 | 77 | 1.344 | 0.2174 | 1.245 | 0.18477 |
| 33 | 0.57596 | 0.04118 | 0.56803 | 0.01566 | 78 | 1.361 | 0.2229 | 1.259 | 0.19160 |
| 34 | 0.59341 | 0.04370 | 0.58474 | 0.01711 | 79 | 1.379 | 0.2284 | 1.272 | 0.19859 |
| 35 | 0.61087 | 0.04628 | 0.60141 | 0.01864 | 80 | 1.396 | 0.2340 | 1.286 | 0.20573 |
| 36 | 0.62832 | 0.04894 | 0.61803 | 0.02027 | 81 | 1.414 | 0.2396 | 1.299 | 0.21301 |
| 37 | 0.64577 | 0.05168 | 0.63461 | 0.02198 | 82 | 1.431 | 0.2453 | 1.312 | 0.22045 |
| 38 | 0.66323 | 0.05448 | 0.65114 | 0.02378 | 83 | 1.449 | 0.2510 | 1.325 | 0.22804 |
| 39 | 0.68068 | 0.05736 | 0.66761 | 0.02568 | 84 | 1.466 | 0.2569 | 1.338 | 0.23578 |
| 40 | 0.69813 | 0.06031 | 0.68404 | 0.02767 | 85 | 1.484 | 0.2627 | 1.351 | 0.24367 |
| 41 | 0.71558 | 0.06333 | 0.70041 | 0.02976 | 86 | 1.501 | 0.2686 | 1.364 | 0.25171 |
| 42 | 0.73304 | 0.06642 | 0.71674 | 0.03195 | 87 | 1.518 | 0.2746 | 1.377 | 0.25990 |
| 43 | 0.75049 | 0.06958 | 0.73300 | 0.03425 | 88 | 1.536 | 0.2807 | 1.389 | 0.26825 |
| 44 | 0.76794 | 0.07282 | 0.74921 | 0.03664 | 89 | 1.553 | 0.2867 | 1.402 | 0.27675 |
| 45 | 0.78540 | 0.07612 | 0.76537 | 0.03915 | 90 | 1.571 | 0.2929 | 1.414 | 0.28540 |

## Segments of Circles for Radius = 1
(English or metric units)

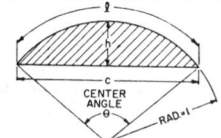

Length of arc, height of segment, length of chord, and area of segment for angles from 1 to 180 degrees and radius = 1. For other radii, multiply the values of $l$, $h$ and $c$ in the table by the given radius $r$, and the values for areas, by $r^2$, the square of the radius. The values in the table can be used for English or metric units.

| Center Angle θ, Degrees | $l$ | $h$ | $c$ | Area of Segment $A$ | Center Angle θ, Degrees | $l$ | $h$ | $c$ | Area of Segment $A$ |
|---|---|---|---|---|---|---|---|---|---|
| 91 | 1.588 | 0.2991 | 1.427 | 0.2942 | 136 | 2.374 | 0.6254 | 1.854 | 0.8395 |
| 92 | 1.606 | 0.3053 | 1.439 | 0.3032 | 137 | 2.391 | 0.6335 | 1.861 | 0.8546 |
| 93 | 1.623 | 0.3116 | 1.451 | 0.3123 | 138 | 2.409 | 0.6416 | 1.867 | 0.8697 |
| 94 | 1.641 | 0.3180 | 1.463 | 0.3215 | 139 | 2.426 | 0.6498 | 1.873 | 0.8850 |
| 95 | 1.658 | 0.3244 | 1.475 | 0.3309 | 140 | 2.443 | 0.6580 | 1.879 | 0.9003 |
| 96 | 1.676 | 0.3309 | 1.486 | 0.3405 | 141 | 2.461 | 0.6662 | 1.885 | 0.9158 |
| 97 | 1.693 | 0.3374 | 1.498 | 0.3502 | 142 | 2.478 | 0.6744 | 1.891 | 0.9314 |
| 98 | 1.710 | 0.3439 | 1.509 | 0.3601 | 143 | 2.496 | 0.6827 | 1.897 | 0.9470 |
| 99 | 1.728 | 0.3506 | 1.521 | 0.3701 | 144 | 2.513 | 0.6910 | 1.902 | 0.9627 |
| 100 | 1.745 | 0.3572 | 1.532 | 0.3803 | 145 | 2.531 | 0.6993 | 1.907 | 0.9786 |
| 101 | 1.763 | 0.3639 | 1.543 | 0.3906 | 146 | 2.548 | 0.7076 | 1.913 | 0.9945 |
| 102 | 1.780 | 0.3707 | 1.554 | 0.4010 | 147 | 2.566 | 0.7160 | 1.918 | 1.0105 |
| 103 | 1.798 | 0.3775 | 1.565 | 0.4117 | 148 | 2.583 | 0.7244 | 1.923 | 1.0266 |
| 104 | 1.815 | 0.3843 | 1.576 | 0.4224 | 149 | 2.601 | 0.7328 | 1.927 | 1.0428 |
| 105 | 1.833 | 0.3912 | 1.587 | 0.4333 | 150 | 2.618 | 0.7412 | 1.932 | 1.0590 |
| 106 | 1.850 | 0.3982 | 1.597 | 0.4444 | 151 | 2.635 | 0.7496 | 1.936 | 1.0753 |
| 107 | 1.868 | 0.4052 | 1.608 | 0.4556 | 152 | 2.653 | 0.7581 | 1.941 | 1.0917 |
| 108 | 1.885 | 0.4122 | 1.618 | 0.4669 | 153 | 2.670 | 0.7666 | 1.945 | 1.1082 |
| 109 | 1.902 | 0.4193 | 1.628 | 0.4784 | 154 | 2.688 | 0.7750 | 1.949 | 1.1247 |
| 110 | 1.920 | 0.4264 | 1.638 | 0.4901 | 155 | 2.705 | 0.7836 | 1.953 | 1.1413 |
| 111 | 1.937 | 0.4336 | 1.648 | 0.5019 | 156 | 2.723 | 0.7921 | 1.956 | 1.1580 |
| 112 | 1.955 | 0.4408 | 1.658 | 0.5138 | 157 | 2.740 | 0.8006 | 1.960 | 1.1747 |
| 113 | 1.972 | 0.4481 | 1.668 | 0.5259 | 158 | 2.758 | 0.8092 | 1.963 | 1.1915 |
| 114 | 1.990 | 0.4554 | 1.677 | 0.5381 | 159 | 2.775 | 0.8178 | 1.967 | 1.2084 |
| 115 | 2.007 | 0.4627 | 1.687 | 0.5504 | 160 | 2.793 | 0.8264 | 1.970 | 1.2253 |
| 116 | 2.025 | 0.4701 | 1.696 | 0.5629 | 161 | 2.810 | 0.8350 | 1.973 | 1.2422 |
| 117 | 2.042 | 0.4775 | 1.705 | 0.5755 | 162 | 2.827 | 0.8436 | 1.975 | 1.2592 |
| 118 | 2.059 | 0.4850 | 1.714 | 0.5883 | 163 | 2.845 | 0.8522 | 1.978 | 1.2763 |
| 119 | 2.077 | 0.4925 | 1.723 | 0.6012 | 164 | 2.862 | 0.8608 | 1.981 | 1.2934 |
| 120 | 2.094 | 0.5000 | 1.732 | 0.6142 | 165 | 2.880 | 0.8695 | 1.983 | 1.3105 |
| 121 | 2.112 | 0.5076 | 1.741 | 0.6273 | 166 | 2.897 | 0.8781 | 1.985 | 1.3277 |
| 122 | 2.129 | 0.5152 | 1.749 | 0.6406 | 167 | 2.915 | 0.8868 | 1.987 | 1.3449 |
| 123 | 2.147 | 0.5228 | 1.758 | 0.6540 | 168 | 2.932 | 0.8955 | 1.989 | 1.3621 |
| 124 | 2.164 | 0.5305 | 1.766 | 0.6676 | 169 | 2.950 | 0.9042 | 1.991 | 1.3794 |
| 125 | 2.182 | 0.5383 | 1.774 | 0.6813 | 170 | 2.967 | 0.9128 | 1.992 | 1.3967 |
| 126 | 2.199 | 0.5460 | 1.782 | 0.6950 | 171 | 2.985 | 0.9215 | 1.994 | 1.4140 |
| 127 | 2.217 | 0.5538 | 1.790 | 0.7090 | 172 | 3.002 | 0.9302 | 1.995 | 1.4314 |
| 128 | 2.234 | 0.5616 | 1.798 | 0.7230 | 173 | 3.019 | 0.9390 | 1.996 | 1.4488 |
| 129 | 2.251 | 0.5695 | 1.805 | 0.7372 | 174 | 3.037 | 0.9477 | 1.997 | 1.4662 |
| 130 | 2.269 | 0.5774 | 1.813 | 0.7514 | 175 | 3.054 | 0.9564 | 1.998 | 1.4836 |
| 131 | 2.286 | 0.5853 | 1.820 | 0.7658 | 176 | 3.072 | 0.9651 | 1.999 | 1.5010 |
| 132 | 2.304 | 0.5933 | 1.827 | 0.7803 | 177 | 3.089 | 0.9738 | 1.999 | 1.5184 |
| 133 | 2.321 | 0.6013 | 1.834 | 0.7950 | 178 | 3.107 | 0.9825 | 2.000 | 1.5359 |
| 134 | 2.339 | 0.6093 | 1.841 | 0.8097 | 179 | 3.124 | 0.9913 | 2.000 | 1.5533 |
| 135 | 2.356 | 0.6173 | 1.848 | 0.8245 | 180 | 3.142 | 1.0000 | 2.000 | 1.5708 |

**Exact and Approximate Formulas for Circular Segment Area.** — The areas of circular segments given in the table, pages 72 and 73, are based on the exact formula: $A = \frac{1}{2}[rl - c(r - h)]$. This and other formulas for segments are given on page 154.

In many cases, notably in calculating the area of an arch in construction work, only the length of the chord $c$ and the height of the segment $h$ may be known or can be measured directly. In such cases approximate formulas for obtaining $A$ directly in terms of $c$ and $h$ are useful since they eliminate the need to first calculate the radius $r$ or the angle $\theta$ in finding the area as would be the case if the exact formula is used.

An approximate formula which gives an error of about 0.1 per cent or less for circular segments ranging almost up to a semi-circle is:

$$A = \frac{4h^2}{3} \sqrt{\frac{c^2}{4h^2} + 0.392} \tag{1}$$

An approximate formula which gives an error of about 0.1 per cent when the ratio of $h$ to $c$ is ⅓ or less is:

$$A = \frac{2ch}{3} + \frac{h^3}{2c} \tag{2}$$

An approximate formula which is more accurate than Formula (2) for segments close to a semi-circle, i.e. with a ratio of $h$ to $c$ of from 0.454 to 0.500, is:

$$A = \frac{h^3}{2c} + 0.6604\, ch \tag{3}$$

## Lengths of Chords for Spacing off the Circumference of Circles

On the following pages are given tables of the lengths of chords for spacing off the circumference of circles. The object of these tables is to make possible the division of the periphery into a number of equal parts without trials with the dividers. The first table is calculated for circles having a diameter equal to 1. For circles of other diameters, the length of chord given in the table should be multiplied by the diameter of the circle. This first table may be used by tool-makers when setting "buttons" in circular formation. Assume that it is required to divide the periphery of a circle of 20 inches diameter into thirty-two equal parts. From the table the length of the chord is found to be 0.098017 inch, if the diameter of the circle were 1 inch. With a diameter of 20 inches the length of the chord for one division would be 20 × 0.098017 = 1.9603 inches. Another example in metric units: For a 100 millimetre diameter requiring 5 equal divisions, the length of the chord for one division would be 100 × 0.587785 = 58.7785 millimetres.

The two following pages give an additional table for the spacing off of circles, the table, in this case, being worked out for diameters from ⅟₁₆ inch to 14 inches. As an example, assume that it is required to divide a circle having a diameter of 6½ inches into seven equal parts. Find first, in the column headed "6" and in line with 7 divisions, the length of the chord for a 6-inch circle, which is 2.604 inches. Then find the length of the chord for a ½-inch diameter circle, 7 divisions, which is 0.217. The sum of these two values, 2.604 + 0.217 = 2.821 inches, is the length of the chord required for spacing off the circumference of a 6½-inch circle into seven equal divisions.

As another example, assume that it is required to divide a circle having a diameter of 9²³⁄₃₂ inches into 15 equal divisions. First find the length of the chord for a 9-inch circle, which is 1.871 inch. The length of the chord for a 2³⁄₃₂-inch circle can easily be estimated from the table by taking the value that is exactly between those given for 1¹⁄₁₆ and ¾ inch. The value for 1¹⁄₁₆ inch is 0.143, and for ¾ inch, 0.156. For 2³⁄₃₂, the value would be 0.150. Then, 1.871 + 0.150 = 2.021 inches.

## Lengths of Chords for Spacing Off the Circumference of Circles with a Diameter Equal to 1

For circles of other diameters multiply length given in table by
diameter of circle. (English or metric units)

| No. of Spaces | Length of Chord | No. of Spaces | Length of Chord | No. of Spaces | Length of Chord | No. of Spaces | Length of Chord |
|---|---|---|---|---|---|---|---|
| 3 | 0.866025 | 51 | 0.061561 | 99 | 0.031728 | 147 | 0.021370 |
| 4 | 0.707107 | 52 | 0.060378 | 100 | 0.031411 | 148 | 0.021225 |
| 5 | 0.587785 | 53 | 0.059241 | 101 | 0.031100 | 149 | 0.021083 |
| 6 | 0.500000 | 54 | 0.058145 | 102 | 0.030795 | 150 | 0.020942 |
| 7 | 0.433884 | 55 | 0.057089 | 103 | 0.030496 | 151 | 0.020804 |
| 8 | 0.382683 | 56 | 0.056070 | 104 | 0.030203 | 152 | 0.020667 |
| 9 | 0.342020 | 57 | 0.055088 | 105 | 0.029915 | 153 | 0.020532 |
| 10 | 0.309017 | 58 | 0.054139 | 106 | 0.029633 | 154 | 0.020399 |
| 11 | 0.281733 | 59 | 0.053222 | 107 | 0.029356 | 155 | 0.020267 |
| 12 | 0.258819 | 60 | 0.052336 | 108 | 0.029085 | 156 | 0.020137 |
| 13 | 0.239316 | 61 | 0.051479 | 109 | 0.028818 | 157 | 0.020009 |
| 14 | 0.222521 | 62 | 0.050649 | 110 | 0.028556 | 158 | 0.019882 |
| 15 | 0.207912 | 63 | 0.049846 | 111 | 0.028299 | 159 | 0.019757 |
| 16 | 0.195090 | 64 | 0.049068 | 112 | 0.028046 | 160 | 0.019634 |
| 17 | 0.183750 | 65 | 0.048313 | 113 | 0.027798 | 161 | 0.019512 |
| 18 | 0.173648 | 66 | 0.047582 | 114 | 0.027554 | 162 | 0.019391 |
| 19 | 0.164595 | 67 | 0.046872 | 115 | 0.027315 | 163 | 0.019272 |
| 20 | 0.156434 | 68 | 0.046183 | 116 | 0.027079 | 164 | 0.019155 |
| 21 | 0.149042 | 69 | 0.045515 | 117 | 0.026848 | 165 | 0.019039 |
| 22 | 0.142315 | 70 | 0.044865 | 118 | 0.026621 | 166 | 0.018924 |
| 23 | 0.136167 | 71 | 0.044233 | 119 | 0.026397 | 167 | 0.018811 |
| 24 | 0.130526 | 72 | 0.043619 | 120 | 0.026177 | 168 | 0.018699 |
| 25 | 0.125333 | 73 | 0.043022 | 121 | 0.025961 | 169 | 0.018588 |
| 26 | 0.120537 | 74 | 0.042441 | 122 | 0.025748 | 170 | 0.018479 |
| 27 | 0.116093 | 75 | 0.041876 | 123 | 0.025539 | 171 | 0.018371 |
| 28 | 0.111964 | 76 | 0.041325 | 124 | 0.025333 | 172 | 0.018264 |
| 29 | 0.108119 | 77 | 0.040789 | 125 | 0.025130 | 173 | 0.018158 |
| 30 | 0.104528 | 78 | 0.040266 | 126 | 0.024931 | 174 | 0.018054 |
| 31 | 0.101168 | 79 | 0.039757 | 127 | 0.024734 | 175 | 0.017951 |
| 32 | 0.098017 | 80 | 0.039260 | 128 | 0.024541 | 176 | 0.017849 |
| 33 | 0.095056 | 81 | 0.038775 | 129 | 0.024351 | 177 | 0.017748 |
| 34 | 0.092268 | 82 | 0.038303 | 130 | 0.024164 | 178 | 0.017648 |
| 35 | 0.089639 | 83 | 0.037841 | 131 | 0.023979 | 179 | 0.017550 |
| 36 | 0.087156 | 84 | 0.037391 | 132 | 0.023798 | 180 | 0.017452 |
| 37 | 0.084806 | 85 | 0.036951 | 133 | 0.023619 | 181 | 0.017356 |
| 38 | 0.082579 | 86 | 0.036522 | 134 | 0.023443 | 182 | 0.017261 |
| 39 | 0.080467 | 87 | 0.036102 | 135 | 0.023269 | 183 | 0.017166 |
| 40 | 0.078459 | 88 | 0.035692 | 136 | 0.023098 | 184 | 0.017073 |
| 41 | 0.076549 | 89 | 0.035291 | 137 | 0.022929 | 185 | 0.016981 |
| 42 | 0.074730 | 90 | 0.034899 | 138 | 0.022763 | 186 | 0.016889 |
| 43 | 0.072995 | 91 | 0.034516 | 139 | 0.022599 | 187 | 0.016799 |
| 44 | 0.071339 | 92 | 0.034141 | 140 | 0.022438 | 188 | 0.016710 |
| 45 | 0.069756 | 93 | 0.033774 | 141 | 0.022279 | 189 | 0.016621 |
| 46 | 0.068242 | 94 | 0.033415 | 142 | 0.022122 | 190 | 0.016534 |
| 47 | 0.066793 | 95 | 0.033063 | 143 | 0.021967 | 191 | 0.016447 |
| 48 | 0.065403 | 96 | 0.032719 | 144 | 0.021815 | 192 | 0.016362 |
| 49 | 0.064070 | 97 | 0.032382 | 145 | 0.021664 | 193 | 0.016277 |
| 50 | 0.062791 | 98 | 0.032052 | 146 | 0.021516 | 194 | 0.016193 |

## Table for Spacing Off the Circumference of Circles
(See page 74 for explanatory matter.)

| No. of Divisions | Degrees in Arc | 1/16 | 1/8 | 3/16 | 1/4 | 5/16 | 3/8 | 7/16 | 1/2 | 9/16 | 5/8 | 11/16 | 3/4 | 13/16 | 7/8 | 15/16 |
|---|---|---|---|---|---|---|---|---|---|---|---|---|---|---|---|---|
| | | Diameter of Circle to be Spaced Off — Length of Chord | | | | | | | | | | | | | | |
| 3 | 120 | 0.054 | 0.108 | 0.162 | 0.217 | 0.271 | 0.325 | 0.379 | 0.433 | 0.487 | 0.541 | 0.595 | 0.650 | 0.704 | 0.758 | 0.812 |
| 4 | 90 | 0.044 | 0.088 | 0.133 | 0.177 | 0.221 | 0.265 | 0.309 | 0.354 | 0.398 | 0.442 | 0.486 | 0.530 | 0.575 | 0.619 | 0.663 |
| 5 | 72 | 0.037 | 0.073 | 0.110 | 0.147 | 0.184 | 0.220 | 0.257 | 0.294 | 0.331 | 0.367 | 0.404 | 0.441 | 0.478 | 0.514 | 0.551 |
| 6 | 60 | 0.031 | 0.063 | 0.094 | 0.125 | 0.156 | 0.188 | 0.219 | 0.250 | 0.281 | 0.313 | 0.344 | 0.375 | 0.406 | 0.438 | 0.469 |
| 7 | 51 3/7 | 0.027 | 0.054 | 0.081 | 0.108 | 0.136 | 0.163 | 0.190 | 0.217 | 0.244 | 0.271 | 0.298 | 0.325 | 0.353 | 0.380 | 0.407 |
| 8 | 45 | 0.024 | 0.048 | 0.072 | 0.096 | 0.120 | 0.144 | 0.167 | 0.191 | 0.215 | 0.239 | 0.263 | 0.287 | 0.311 | 0.335 | 0.359 |
| 9 | 40 | 0.021 | 0.043 | 0.064 | 0.086 | 0.107 | 0.128 | 0.150 | 0.171 | 0.192 | 0.214 | 0.235 | 0.257 | 0.278 | 0.299 | 0.321 |
| 10 | 36 | 0.019 | 0.039 | 0.058 | 0.077 | 0.097 | 0.116 | 0.135 | 0.155 | 0.174 | 0.193 | 0.212 | 0.232 | 0.251 | 0.270 | 0.290 |
| 11 | 32 8/11 | 0.018 | 0.035 | 0.053 | 0.070 | 0.088 | 0.106 | 0.123 | 0.141 | 0.158 | 0.176 | 0.194 | 0.211 | 0.229 | 0.247 | 0.264 |
| 12 | 30 | 0.016 | 0.032 | 0.049 | 0.065 | 0.081 | 0.097 | 0.113 | 0.129 | 0.146 | 0.162 | 0.178 | 0.194 | 0.210 | 0.226 | 0.243 |
| 13 | 27 9/13 | 0.015 | 0.030 | 0.045 | 0.060 | 0.075 | 0.090 | 0.105 | 0.120 | 0.135 | 0.150 | 0.165 | 0.179 | 0.194 | 0.209 | 0.224 |
| 14 | 25 5/7 | 0.014 | 0.028 | 0.042 | 0.056 | 0.070 | 0.083 | 0.097 | 0.111 | 0.125 | 0.139 | 0.153 | 0.167 | 0.181 | 0.195 | 0.209 |
| 15 | 24 | 0.013 | 0.026 | 0.039 | 0.052 | 0.065 | 0.078 | 0.091 | 0.104 | 0.117 | 0.130 | 0.143 | 0.156 | 0.169 | 0.182 | 0.195 |
| 16 | 22 1/2 | 0.012 | 0.024 | 0.037 | 0.049 | 0.061 | 0.073 | 0.085 | 0.098 | 0.110 | 0.122 | 0.134 | 0.146 | 0.159 | 0.171 | 0.183 |
| 17 | 21 3/17 | 0.011 | 0.023 | 0.034 | 0.046 | 0.057 | 0.069 | 0.080 | 0.092 | 0.103 | 0.115 | 0.126 | 0.138 | 0.149 | 0.161 | 0.172 |
| 18 | 20 | 0.011 | 0.022 | 0.033 | 0.043 | 0.054 | 0.065 | 0.076 | 0.087 | 0.098 | 0.109 | 0.119 | 0.130 | 0.141 | 0.152 | 0.163 |
| 19 | 18 18/19 | 0.010 | 0.021 | 0.031 | 0.041 | 0.051 | 0.062 | 0.072 | 0.082 | 0.093 | 0.103 | 0.113 | 0.123 | 0.134 | 0.144 | 0.154 |
| 20 | 18 | 0.010 | 0.020 | 0.029 | 0.039 | 0.049 | 0.059 | 0.068 | 0.078 | 0.088 | 0.098 | 0.108 | 0.117 | 0.127 | 0.137 | 0.147 |
| 21 | 17 1/7 | 0.009 | 0.019 | 0.028 | 0.037 | 0.047 | 0.056 | 0.065 | 0.075 | 0.084 | 0.093 | 0.102 | 0.112 | 0.121 | 0.130 | 0.140 |
| 22 | 16 4/11 | 0.009 | 0.018 | 0.027 | 0.036 | 0.044 | 0.053 | 0.062 | 0.071 | 0.080 | 0.089 | 0.098 | 0.107 | 0.116 | 0.125 | 0.133 |
| 23 | 15 15/23 | 0.009 | 0.017 | 0.026 | 0.034 | 0.043 | 0.051 | 0.060 | 0.068 | 0.077 | 0.085 | 0.094 | 0.102 | 0.111 | 0.119 | 0.128 |
| 24 | 15 | 0.008 | 0.016 | 0.024 | 0.033 | 0.041 | 0.049 | 0.057 | 0.065 | 0.073 | 0.082 | 0.090 | 0.098 | 0.106 | 0.114 | 0.122 |
| 25 | 14 2/5 | 0.008 | 0.016 | 0.023 | 0.031 | 0.039 | 0.047 | 0.055 | 0.063 | 0.070 | 0.078 | 0.086 | 0.094 | 0.102 | 0.110 | 0.117 |
| 26 | 13 11/13 | 0.008 | 0.015 | 0.023 | 0.030 | 0.038 | 0.045 | 0.053 | 0.060 | 0.068 | 0.075 | 0.083 | 0.090 | 0.098 | 0.105 | 0.113 |
| 28 | 12 6/7 | 0.007 | 0.014 | 0.021 | 0.028 | 0.035 | 0.042 | 0.049 | 0.056 | 0.063 | 0.070 | 0.077 | 0.084 | 0.091 | 0.098 | 0.105 |
| 30 | 12 | 0.007 | 0.013 | 0.020 | 0.026 | 0.033 | 0.039 | 0.046 | 0.052 | 0.059 | 0.065 | 0.072 | 0.078 | 0.085 | 0.091 | 0.098 |
| 32 | 11 1/4 | 0.006 | 0.012 | 0.018 | 0.025 | 0.031 | 0.037 | 0.043 | 0.049 | 0.055 | 0.061 | 0.067 | 0.074 | 0.080 | 0.086 | 0.092 |

## Table for Spacing Off the Circumference of Circles

| No. of Divisions | Degrees in Arc | Diameter of Circle to be Spaced Off — Length of Chord | | | | | | | | | | | | | |
|---|---|---|---|---|---|---|---|---|---|---|---|---|---|---|
| | | 1 | 2 | 3 | 4 | 5 | 6 | 7 | 8 | 9 | 10 | 11 | 12 | 13 | 14 |
| 3 | 120 | 0.866 | 1.732 | 2.598 | 3.464 | 4.330 | 5.196 | 6.062 | 6.928 | 7.794 | 8.660 | 9.526 | 10.392 | 11.258 | 12.124 |
| 4 | 90 | 0.707 | 1.414 | 2.121 | 2.828 | 3.536 | 4.243 | 4.950 | 5.657 | 6.364 | 7.071 | 7.778 | 8.485 | 9.192 | 9.899 |
| 5 | 72 | 0.588 | 1.176 | 1.763 | 2.351 | 2.939 | 3.527 | 4.114 | 4.702 | 5.290 | 5.878 | 6.466 | 7.053 | 7.641 | 8.229 |
| 6 | 60 | 0.500 | 1.000 | 1.500 | 2.000 | 2.500 | 3.000 | 3.500 | 4.000 | 4.500 | 5.000 | 5.500 | 6.000 | 6.500 | 7.000 |
| 7 | 51 3/7 | 0.434 | 0.868 | 1.302 | 1.736 | 2.169 | 2.603 | 3.037 | 3.471 | 3.905 | 4.339 | 4.773 | 5.207 | 5.640 | 6.074 |
| 8 | 45 | 0.383 | 0.765 | 1.148 | 1.531 | 1.913 | 2.296 | 2.679 | 3.061 | 3.444 | 3.827 | 4.210 | 4.592 | 4.975 | 5.358 |
| 9 | 40 | 0.342 | 0.684 | 1.026 | 1.368 | 1.710 | 2.052 | 2.394 | 2.736 | 3.078 | 3.420 | 3.762 | 4.104 | 4.446 | 4.788 |
| 10 | 36 | 0.309 | 0.618 | 0.927 | 1.236 | 1.545 | 1.854 | 2.163 | 2.472 | 2.781 | 3.090 | 3.399 | 3.708 | 4.017 | 4.326 |
| 11 | 32 8/11 | 0.282 | 0.563 | 0.845 | 1.127 | 1.409 | 1.690 | 1.972 | 2.254 | 2.536 | 2.817 | 3.099 | 3.381 | 3.663 | 3.944 |
| 12 | 30 | 0.259 | 0.518 | 0.776 | 1.035 | 1.294 | 1.553 | 1.812 | 2.071 | 2.329 | 2.588 | 2.847 | 3.106 | 3.365 | 3.623 |
| 13 | 27 9/13 | 0.239 | 0.479 | 0.718 | 0.957 | 1.197 | 1.436 | 1.675 | 1.915 | 2.154 | 2.393 | 2.632 | 2.872 | 3.111 | 3.350 |
| 14 | 25 5/7 | 0.223 | 0.445 | 0.668 | 0.890 | 1.113 | 1.335 | 1.558 | 1.780 | 2.003 | 2.225 | 2.448 | 2.670 | 2.893 | 3.115 |
| 15 | 24 | 0.208 | 0.416 | 0.624 | 0.832 | 1.040 | 1.247 | 1.455 | 1.663 | 1.871 | 2.079 | 2.287 | 2.495 | 2.703 | 2.911 |
| 16 | 22 1/2 | 0.195 | 0.390 | 0.585 | 0.780 | 0.975 | 1.171 | 1.366 | 1.561 | 1.756 | 1.951 | 2.146 | 2.341 | 2.536 | 2.731 |
| 17 | 21 3/17 | 0.184 | 0.367 | 0.551 | 0.735 | 0.919 | 1.102 | 1.286 | 1.470 | 1.654 | 1.837 | 2.021 | 2.205 | 2.389 | 2.572 |
| 18 | 20 | 0.174 | 0.347 | 0.521 | 0.695 | 0.868 | 1.042 | 1.216 | 1.389 | 1.563 | 1.736 | 1.910 | 2.084 | 2.257 | 2.431 |
| 19 | 18 18/19 | 0.165 | 0.329 | 0.494 | 0.658 | 0.823 | 0.988 | 1.152 | 1.317 | 1.481 | 1.646 | 1.811 | 1.975 | 2.140 | 2.304 |
| 20 | 18 | 0.156 | 0.313 | 0.469 | 0.626 | 0.782 | 0.939 | 1.095 | 1.251 | 1.408 | 1.564 | 1.721 | 1.877 | 2.034 | 2.190 |
| 21 | 17 1/7 | 0.149 | 0.298 | 0.447 | 0.596 | 0.745 | 0.894 | 1.043 | 1.192 | 1.341 | 1.490 | 1.639 | 1.789 | 1.938 | 2.087 |
| 22 | 16 4/11 | 0.142 | 0.285 | 0.427 | 0.569 | 0.712 | 0.854 | 0.996 | 1.139 | 1.281 | 1.423 | 1.565 | 1.708 | 1.850 | 1.992 |
| 23 | 15 15/23 | 0.136 | 0.272 | 0.408 | 0.545 | 0.681 | 0.817 | 0.953 | 1.089 | 1.225 | 1.362 | 1.498 | 1.634 | 1.770 | 1.906 |
| 24 | 15 | 0.131 | 0.261 | 0.392 | 0.522 | 0.653 | 0.783 | 0.914 | 1.044 | 1.175 | 1.305 | 1.436 | 1.566 | 1.697 | 1.827 |
| 25 | 14 2/5 | 0.125 | 0.251 | 0.376 | 0.501 | 0.627 | 0.752 | 0.877 | 1.003 | 1.128 | 1.253 | 1.379 | 1.504 | 1.629 | 1.755 |
| 26 | 13 11/13 | 0.121 | 0.241 | 0.362 | 0.482 | 0.603 | 0.723 | 0.844 | 0.964 | 1.085 | 1.205 | 1.326 | 1.446 | 1.567 | 1.688 |
| 28 | 12 6/7 | 0.112 | 0.224 | 0.336 | 0.448 | 0.560 | 0.672 | 0.784 | 0.896 | 1.008 | 1.120 | 1.232 | 1.344 | 1.456 | 1.568 |
| 30 | 12 | 0.105 | 0.209 | 0.314 | 0.418 | 0.523 | 0.627 | 0.732 | 0.836 | 0.941 | 1.045 | 1.150 | 1.254 | 1.359 | 1.463 |
| 32 | 11 1/4 | 0.098 | 0.196 | 0.294 | 0.392 | 0.490 | 0.588 | 0.686 | 0.784 | 0.882 | 0.980 | 1.078 | 1.176 | 1.274 | 1.372 |

## Table 1. Hole Coordinate Dimension Factors for Jig Boring—Type "A" Hole Circles
(English or metric units)

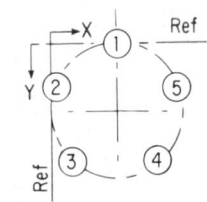

The diagram shows a type "A" circle for a 5-hole circle. Coordinates $x, y$ are given in the table for hole circles of from 3 to 28 holes. Dimensions are for holes numbered in a counterclockwise direction (as shown). Dimensions given are based upon a hole circle of unit diameter. For a hole circle of, say, 3-inch or 3-centimeter diameter, multiply table values by 3.

### 3 Holes

| Hole | $x$ | $y$ |
|---|---|---|
| 1 | 0.50000 | 0.00000 |
| 2 | 0.06699 | 0.75000 |
| 3 | 0.93301 | 0.75000 |

### 4 Holes

| Hole | $x$ | $y$ |
|---|---|---|
| 1 | 0.50000 | 0.00000 |
| 2 | 0.00000 | 0.50000 |
| 3 | 0.50000 | 1.00000 |
| 4 | 1.00000 | 0.50000 |

### 5 Holes

| Hole | $x$ | $y$ |
|---|---|---|
| 1 | 0.50000 | 0.00000 |
| 2 | 0.02447 | 0.34549 |
| 3 | 0.20611 | 0.90451 |
| 4 | 0.79389 | 0.90451 |
| 5 | 0.97553 | 0.34549 |

### 6 Holes

| Hole | $x$ | $y$ |
|---|---|---|
| 1 | 0.50000 | 0.00000 |
| 2 | 0.06699 | 0.25000 |
| 3 | 0.06699 | 0.75000 |
| 4 | 0.50000 | 1.00000 |
| 5 | 0.93301 | 0.75000 |
| 6 | 0.93301 | 0.25000 |

### 7 Holes

| Hole | $x$ | $y$ |
|---|---|---|
| 1 | 0.50000 | 0.00000 |
| 2 | 0.10908 | 0.18826 |
| 3 | 0.01254 | 0.61126 |
| 4 | 0.28306 | 0.95048 |
| 5 | 0.71694 | 0.95048 |
| 6 | 0.98746 | 0.61126 |
| 7 | 0.89092 | 0.18826 |

### 8 Holes

| Hole | $x$ | $y$ |
|---|---|---|
| 1 | 0.50000 | 0.00000 |
| 2 | 0.14645 | 0.14645 |
| 3 | 0.00000 | 0.50000 |
| 4 | 0.14645 | 0.85355 |
| 5 | 0.50000 | 1.00000 |
| 6 | 0.85355 | 0.85355 |
| 7 | 1.00000 | 0.50000 |
| 8 | 0.85355 | 0.14645 |

### 9 Holes

| Hole | $x$ | $y$ |
|---|---|---|
| 1 | 0.50000 | 0.00000 |
| 2 | 0.17861 | 0.11698 |
| 3 | 0.00760 | 0.41318 |
| 4 | 0.06699 | 0.75000 |
| 5 | 0.32899 | 0.96985 |
| 6 | 0.67101 | 0.96985 |
| 7 | 0.93301 | 0.75000 |
| 8 | 0.99240 | 0.41318 |
| 9 | 0.82139 | 0.11698 |

### 10 Holes

| Hole | $x$ | $y$ |
|---|---|---|
| 1 | 0.50000 | 0.00000 |
| 2 | 0.20611 | 0.09549 |
| 3 | 0.02447 | 0.34549 |
| 4 | 0.02447 | 0.65451 |
| 5 | 0.20611 | 0.90451 |
| 6 | 0.50000 | 1.00000 |
| 7 | 0.79389 | 0.90451 |
| 8 | 0.97553 | 0.65451 |
| 9 | 0.97553 | 0.34549 |
| 10 | 0.79389 | 0.09549 |

### 11 Holes

| Hole | $x$ | $y$ |
|---|---|---|
| 1 | 0.50000 | 0.00000 |
| 2 | 0.22968 | 0.07937 |
| 3 | 0.04518 | 0.29229 |
| 4 | 0.00509 | 0.57116 |
| 5 | 0.12213 | 0.82743 |
| 6 | 0.35913 | 0.97975 |
| 7 | 0.64087 | 0.97975 |
| 8 | 0.87787 | 0.82743 |
| 9 | 0.99491 | 0.57116 |
| 10 | 0.95482 | 0.29229 |
| 11 | 0.77032 | 0.07937 |

### 12 Holes

| Hole | $x$ | $y$ |
|---|---|---|
| 1 | 0.50000 | 0.00000 |
| 2 | 0.25000 | 0.06699 |
| 3 | 0.06699 | 0.25000 |
| 4 | 0.00000 | 0.50000 |
| 5 | 0.06699 | 0.75000 |
| 6 | 0.25000 | 0.93301 |
| 7 | 0.50000 | 1.00000 |
| 8 | 0.75000 | 0.93301 |
| 9 | 0.93301 | 0.75000 |
| 10 | 1.00000 | 0.50000 |
| 11 | 0.93301 | 0.25000 |
| 12 | 0.75000 | 0.06699 |

### 13 Holes

| Hole | $x$ | $y$ |
|---|---|---|
| 1 | 0.50000 | 0.00000 |
| 2 | 0.26764 | 0.05727 |
| 3 | 0.08851 | 0.21597 |
| 4 | 0.00365 | 0.43973 |
| 5 | 0.03249 | 0.67730 |
| 6 | 0.16844 | 0.87426 |
| 7 | 0.38034 | 0.98547 |
| 8 | 0.61966 | 0.98547 |
| 9 | 0.83156 | 0.87426 |
| 10 | 0.96751 | 0.67730 |
| 11 | 0.99635 | 0.43973 |
| 12 | 0.91149 | 0.21597 |
| 13 | 0.73236 | 0.05727 |

### 14 Holes

| Hole | $x$ | $y$ |
|---|---|---|
| 1 | 0.50000 | 0.00000 |
| 2 | 0.28306 | 0.04952 |
| 3 | 0.10908 | 0.18826 |
| 4 | 0.01254 | 0.38874 |
| 5 | 0.01254 | 0.61126 |
| 6 | 0.10908 | 0.81174 |
| 7 | 0.28306 | 0.95048 |
| 8 | 0.50000 | 1.00000 |
| 9 | 0.71694 | 0.95048 |
| 10 | 0.89092 | 0.81174 |
| 11 | 0.98746 | 0.61126 |
| 12 | 0.98746 | 0.38874 |
| 13 | 0.89092 | 0.18826 |
| 14 | 0.71694 | 0.04952 |

### 15 Holes

| Hole | $x$ | $y$ |
|---|---|---|
| 1 | 0.50000 | 0.00000 |
| 2 | 0.29663 | 0.04323 |
| 3 | 0.12843 | 0.16543 |
| 4 | 0.02447 | 0.34549 |
| 5 | 0.00274 | 0.55226 |
| 6 | 0.06699 | 0.75000 |
| 7 | 0.20611 | 0.90451 |
| 8 | 0.39604 | 0.98907 |
| 9 | 0.60396 | 0.98907 |
| 10 | 0.79389 | 0.90451 |
| 11 | 0.93301 | 0.75000 |
| 12 | 0.99726 | 0.55226 |
| 13 | 0.97553 | 0.34549 |
| 14 | 0.87157 | 0.16543 |
| 15 | 0.70337 | 0.04323 |

### 16 Holes

| Hole | $x$ | $y$ |
|---|---|---|
| 1 | 0.50000 | 0.00000 |
| 2 | 0.30866 | 0.03806 |
| 3 | 0.14645 | 0.14645 |
| 4 | 0.03806 | 0.30866 |
| 5 | 0.00000 | 0.50000 |
| 6 | 0.03806 | 0.69134 |
| 7 | 0.14645 | 0.85355 |
| 8 | 0.30866 | 0.96194 |
| 9 | 0.50000 | 1.00000 |
| 10 | 0.69134 | 0.96194 |
| 11 | 0.85355 | 0.85355 |
| 12 | 0.96194 | 0.69134 |
| 13 | 1.00000 | 0.50000 |
| 14 | 0.96194 | 0.30866 |
| 15 | 0.85355 | 0.14645 |
| 16 | 0.69134 | 0.03806 |

### 17 Holes

| Hole | $x$ | $y$ |
|---|---|---|
| 1 | 0.50000 | 0.00000 |
| 2 | 0.31938 | 0.03376 |
| 3 | 0.16315 | 0.13050 |
| 4 | 0.05242 | 0.27713 |
| 5 | 0.00213 | 0.45387 |
| 6 | 0.01909 | 0.63683 |
| 7 | 0.10099 | 0.80132 |
| 8 | 0.23678 | 0.92511 |
| 9 | 0.40813 | 0.99149 |
| 10 | 0.59187 | 0.99149 |
| 11 | 0.76322 | 0.92511 |
| 12 | 0.89901 | 0.80132 |
| 13 | 0.98091 | 0.63683 |
| 14 | 0.99787 | 0.45387 |
| 15 | 0.94758 | 0.27713 |
| 16 | 0.83685 | 0.13050 |
| 17 | 0.68062 | 0.03376 |

### 18 Holes

| Hole | $x$ | $y$ |
|---|---|---|
| 1 | 0.50000 | 0.00000 |
| 2 | 0.32899 | 0.03015 |
| 3 | 0.17861 | 0.11698 |
| 4 | 0.06699 | 0.25000 |
| 5 | 0.00760 | 0.41318 |
| 6 | 0.00760 | 0.58682 |
| 7 | 0.06699 | 0.75000 |
| 8 | 0.17861 | 0.88302 |
| 9 | 0.32899 | 0.96985 |
| 10 | 0.50000 | 1.00000 |
| 11 | 0.67101 | 0.96985 |
| 12 | 0.82139 | 0.88302 |
| 13 | 0.93301 | 0.75000 |
| 14 | 0.99240 | 0.58682 |
| 15 | 0.99240 | 0.41318 |
| 16 | 0.93301 | 0.25000 |
| 17 | 0.82139 | 0.11698 |
| 18 | 0.67101 | 0.03015 |

### 19 Holes

| Hole | $x$ | $y$ |
|---|---|---|
| 1 | 0.50000 | 0.00000 |
| 2 | 0.33765 | 0.02709 |
| 3 | 0.19289 | 0.10543 |
| 4 | 0.08142 | 0.22653 |
| 5 | 0.01530 | 0.37726 |
| 6 | 0.00171 | 0.54129 |
| 7 | 0.04211 | 0.70085 |
| … | … | … |

**Table 1** (*Concluded*). **Hole Coordinate Dimension Factors for Jig Boring—Type "A" Hole Circles** (English or metric units)

### 19 Holes (continued)

| x | y |
|---|---|
| x8 0.13214 | y8 0.83864 |
| x9 0.26203 | y9 0.93974 |
| x10 0.41770 | y10 0.99318 |
| x11 0.58230 | y11 0.99318 |
| x12 0.73797 | y12 0.93974 |
| x13 0.86786 | y13 0.83864 |
| x14 0.95789 | y14 0.70585 |
| x15 0.99829 | y15 0.54129 |
| x16 0.98470 | y16 0.37630 |
| x17 0.91858 | y17 0.22658 |
| x18 0.80711 | y18 0.10543 |
| x19 0.66235 | y19 0.02709 |

### 20 Holes

| x | y |
|---|---|
| x1 0.50000 | y1 0.00000 |
| x2 0.34549 | y2 0.02447 |
| x3 0.20611 | y3 0.09549 |
| x4 0.09549 | y4 0.20611 |
| x5 0.02447 | y5 0.34549 |
| x6 0.00000 | y6 0.50000 |
| x7 0.02447 | y7 0.65451 |
| x8 0.09549 | y8 0.79389 |
| x9 0.20611 | y9 0.90451 |
| x10 0.34549 | y10 0.97553 |
| x11 0.50000 | y11 1.00000 |
| x12 0.65451 | y12 0.97553 |
| x13 0.79389 | y13 0.90451 |
| x14 0.90451 | y14 0.79389 |
| x15 0.97553 | y15 0.65451 |
| x16 1.00000 | y16 0.50000 |
| x17 0.97553 | y17 0.34549 |
| x18 0.90451 | y18 0.20611 |
| x19 0.79389 | y19 0.09549 |
| x20 0.65451 | y20 0.02447 |

### 21 Holes

| x | y |
|---|---|
| x1 0.50000 | y1 0.00000 |
| x2 0.35262 | y2 0.02221 |
| x3 0.21834 | y3 0.08688 |
| x4 0.10908 | y4 0.18826 |
| x5 0.03456 | y5 0.31733 |
| x6 0.00140 | y6 0.46263 |
| x7 0.01254 | y7 0.61126 |
| x8 0.06699 | y8 0.75000 |
| x9 0.15991 | y9 0.86653 |
| x10 0.28306 | y10 0.95048 |
| x11 0.42548 | y11 0.99442 |
| x12 0.57452 | y12 0.99442 |
| x13 0.71694 | y13 0.95048 |
| x14 0.84009 | y14 0.86653 |
| x15 0.93301 | y15 0.75000 |
| x16 0.98746 | y16 0.61126 |
| x17 0.99860 | y17 0.46263 |
| x18 0.96544 | y18 0.31733 |
| x19 0.89092 | y19 0.18826 |
| x20 0.78166 | y20 0.08688 |
| x21 0.64738 | y21 0.02221 |

### 22 Holes

| x | y |
|---|---|
| x1 0.50000 | y1 0.00000 |
| x2 0.35913 | y2 0.02025 |
| x3 0.22968 | y3 0.07937 |
| x4 0.12213 | y4 0.17257 |
| x5 0.04518 | y5 0.29229 |
| x6 0.00509 | y6 0.42884 |
| x7 0.00509 | y7 0.57116 |
| x8 0.04518 | y8 0.70771 |
| x9 0.12213 | y9 0.82743 |
| x10 0.22968 | y10 0.92063 |
| x11 0.35913 | y11 0.97975 |
| x12 0.50000 | y12 1.00000 |
| x13 0.64087 | y13 0.97975 |
| x14 0.77032 | y14 0.92063 |
| x15 0.87787 | y15 0.82743 |
| x16 0.95482 | y16 0.70771 |
| x17 0.99491 | y17 0.57116 |
| x18 0.99491 | y18 0.42884 |
| x19 0.95482 | y19 0.29229 |
| x20 0.87787 | y20 0.17257 |
| x21 0.77032 | y21 0.07937 |
| x22 0.64087 | y22 0.02025 |

### 23 Holes

| x | y |
|---|---|
| x1 0.50000 | y1 0.00000 |
| x2 0.36510 | y2 0.01854 |
| x3 0.24021 | y3 0.07279 |
| x4 0.13458 | y4 0.15872 |
| x5 0.05606 | y5 0.26997 |
| x6 0.01046 | y6 0.39827 |
| x7 0.00117 | y7 0.53412 |
| x8 0.02887 | y8 0.66744 |
| x9 0.09152 | y9 0.78834 |
| x10 0.18446 | y10 0.88786 |
| x11 0.30080 | y11 0.95861 |
| x12 0.43192 | y12 0.99534 |
| x13 0.56808 | y13 0.99534 |
| x14 0.69920 | y14 0.95861 |
| x15 0.81554 | y15 0.88786 |
| x16 0.90848 | y16 0.78834 |
| x17 0.97113 | y17 0.66744 |
| x18 0.99883 | y18 0.53412 |
| x19 0.98954 | y19 0.39827 |
| x20 0.94394 | y20 0.26997 |
| x21 0.86542 | y21 0.15872 |
| x22 0.75979 | y22 0.07279 |
| x23 0.63490 | y23 0.01854 |

### 24 Holes

| x | y |
|---|---|
| x1 0.50000 | y1 0.00000 |
| x2 0.37059 | y2 0.01704 |
| x3 0.25000 | y3 0.06699 |
| x4 0.14645 | y4 0.14645 |
| x5 0.06699 | y5 0.25000 |
| x6 0.01704 | y6 0.37059 |
| x7 0.00000 | y7 0.50000 |
| x8 0.01704 | y8 0.62941 |
| x9 0.06699 | y9 0.75000 |
| x10 0.14645 | y10 0.85355 |
| x11 0.25000 | y11 0.93301 |
| x12 0.37059 | y12 0.98296 |
| x13 0.50000 | y13 1.00000 |
| x14 0.62941 | y14 0.98296 |
| x15 0.75000 | y15 0.93301 |
| x16 0.85355 | y16 0.85355 |
| x17 0.93301 | y17 0.75000 |
| x18 0.98296 | y18 0.62941 |
| x19 1.00000 | y19 0.50000 |
| x20 0.98296 | y20 0.37059 |
| x21 0.93301 | y21 0.25000 |
| x22 0.85355 | y22 0.14645 |
| x23 0.75000 | y23 0.06699 |
| x24 0.62941 | y24 0.01704 |

### 25 Holes

| x | y |
|---|---|
| x1 0.50000 | y1 0.00000 |
| x2 0.37566 | y2 0.01571 |
| x3 0.25912 | y3 0.06185 |
| x4 0.15773 | y4 0.13552 |
| x5 0.07784 | y5 0.23209 |
| x6 0.02447 | y6 0.34549 |
| x7 0.00089 | y7 0.46860 |
| x8 0.00886 | y8 0.59369 |
| x9 0.04759 | y9 0.71289 |
| x10 0.11474 | y10 0.81871 |
| x11 0.20611 | y11 0.90451 |
| x12 0.31594 | y12 0.96489 |
| x13 0.43733 | y13 0.99606 |
| x14 0.56267 | y14 0.99606 |
| x15 0.68406 | y15 0.96489 |
| x16 0.79389 | y16 0.90451 |
| x17 0.88526 | y17 0.81871 |
| x18 0.95241 | y18 0.71289 |
| x19 0.99114 | y19 0.59369 |
| x20 0.99911 | y20 0.46860 |
| x21 0.97553 | y21 0.34549 |
| x22 0.92216 | y22 0.23209 |
| x23 0.84227 | y23 0.13552 |
| x24 0.74088 | y24 0.06185 |
| x25 0.62434 | y25 0.01571 |

### 26 Holes

| x | y |
|---|---|
| x1 0.50000 | y1 0.00000 |
| x2 0.38034 | y2 0.01453 |
| x3 0.26764 | y3 0.05727 |
| x4 0.16844 | y4 0.12574 |
| x5 0.08851 | y5 0.21597 |
| x6 0.03249 | y6 0.32270 |
| x7 0.00365 | y7 0.43973 |
| x8 0.00085 | y8 0.56027 |
| x9 0.03249 | y9 0.64340 |
| x10 0.06699 | y10 0.75000 |
| x11 0.13631 | y11 0.84312 |
| x12 0.22525 | y12 0.91774 |
| x13 0.32899 | y13 0.96985 |
| x14 0.44195 | y14 0.99662 |
| x15 0.55805 | y15 0.99662 |
| x16 0.67101 | y16 0.96985 |
| x17 0.77475 | y17 0.91774 |
| x18 0.86369 | y18 0.84312 |
| x19 0.93301 | y19 0.75000 |
| x20 0.97899 | y20 0.64340 |
| x21 0.99915 | y21 0.52907 |
| x22 0.99240 | y22 0.41318 |
| x23 0.91149 | y23 0.21597 |
| x24 0.83156 | y24 0.12574 |
| x25 0.73236 | y25 0.05727 |
| x26 0.61966 | y26 0.01453 |

### 27 Holes

| x | y |
|---|---|
| x1 0.50000 | y1 0.00000 |
| x2 0.38469 | y2 0.01348 |
| x3 0.27560 | y3 0.05318 |
| x4 0.17861 | y4 0.11698 |
| x5 0.09894 | y5 0.20142 |
| x6 0.04089 | y6 0.30196 |
| x7 0.00760 | y7 0.41318 |
| x8 0.00085 | y8 0.52907 |
| x9 0.02101 | y9 0.64340 |
| x10 0.06699 | y10 0.75000 |
| x11 0.13631 | y11 0.84312 |
| x12 0.22525 | y12 0.91774 |
| x13 0.32899 | y13 0.95048 |
| x14 0.44195 | y14 0.98746 |
| x15 0.55805 | y15 1.00000 |
| x16 0.67101 | y16 0.96985 |
| x17 0.77475 | y17 0.95048 |
| x18 0.86369 | y18 0.89092 |
| x19 0.93301 | y19 0.89092 |
| x20 0.97899 | y20 0.81174 |
| x21 0.99915 | y21 0.61126 |
| x22 0.99240 | y22 0.50000 |
| x23 0.95911 | y23 0.38874 |
| x24 0.90106 | y24 0.28306 |
| x25 0.82139 | y25 0.18826 |
| x26 0.72440 | y26 0.10908 |
| x27 0.61531 | y27 0.04952 |

### 28 Holes

| x | y |
|---|---|
| x1 0.50000 | y1 0.00000 |
| x2 0.38874 | y2 0.01254 |
| x3 0.28306 | y3 0.04952 |
| x4 0.18826 | y4 0.10908 |
| x5 0.10908 | y5 0.18826 |
| x6 0.04952 | y6 0.28306 |
| x7 0.01254 | y7 0.38874 |
| x8 0.00000 | y8 0.50000 |
| x9 0.01254 | y9 0.61126 |
| x10 0.04952 | y10 0.71694 |
| x11 0.10908 | y11 0.81174 |
| x12 0.18826 | y12 0.89092 |
| x13 0.28306 | y13 0.95048 |
| x14 0.38874 | y14 0.98746 |
| x15 0.50000 | y15 1.00000 |
| x16 0.61126 | y16 0.98746 |
| x17 0.71694 | y17 0.95048 |
| x18 0.81174 | y18 0.89092 |
| x19 0.89092 | y19 0.81174 |
| x20 0.95048 | y20 0.71694 |
| x21 0.98746 | y21 0.61126 |
| x22 1.00000 | y22 0.50000 |
| x23 0.98746 | y23 0.38874 |
| x24 0.95048 | y24 0.28306 |
| x25 0.89092 | y25 0.18826 |
| x26 0.81174 | y26 0.10908 |
| x27 0.71694 | y27 0.04952 |
| x28 0.61126 | y28 0.01254 |

### Table 2. Hole Coordinate Dimension Factors for Jig Boring—Type "B" Hole Circles
(English or metric units)

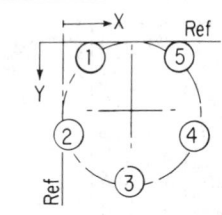

The diagram shows a type "B" circle for a 5-hole circle. Coordinates $x$, $y$ are given in the table for hole circles of from 3 to 28 holes. Dimensions are for holes numbered in a counterclockwise direction (as shown). Dimensions given are based upon a hole circle of unit diameter. For a hole circle of, say, 3-inch or 3-centimeter diameter, multiply table values by 3.

**3 Holes**

| | x | y |
|---|---|---|
| 1 | 0.06699 | 0.25000 |
| 2 | 0.50000 | 1.00000 |
| 3 | 0.93301 | 0.25000 |

**4 Holes**

| | x | y |
|---|---|---|
| 1 | 0.14645 | 0.14645 |
| 2 | 0.14645 | 0.85355 |
| 3 | 0.85355 | 0.85355 |
| 4 | 0.85355 | 0.14645 |

**5 Holes**

| | x | y |
|---|---|---|
| 1 | 0.20611 | 0.09549 |
| 2 | 0.02447 | 0.65451 |
| 3 | 0.50000 | 1.00000 |
| 4 | 0.97553 | 0.65451 |
| 5 | 0.79389 | 0.09549 |

**6 Holes**

| | x | y |
|---|---|---|
| 1 | 0.25000 | 0.06699 |
| 2 | 0.00000 | 0.50000 |
| 3 | 0.25000 | 0.93301 |
| 4 | 0.75000 | 0.93301 |
| 5 | 1.00000 | 0.50000 |
| 6 | 0.75000 | 0.06699 |

**7 Holes**

| | x | y |
|---|---|---|
| 1 | 0.28306 | 0.04952 |
| 2 | 0.01254 | 0.38874 |
| 3 | 0.10908 | 0.81174 |
| 4 | 0.50000 | 1.00000 |
| 5 | 0.89092 | 0.81174 |
| 6 | 0.98746 | 0.38874 |
| 7 | 0.71694 | 0.04952 |

**8 Holes**

| | x | y |
|---|---|---|
| 1 | 0.30866 | 0.03806 |
| 2 | 0.03806 | 0.30866 |
| 3 | 0.03806 | 0.69134 |
| 4 | 0.30866 | 0.69134 |
| 5 | 0.69134 | 0.69134 |
| 6 | 0.96194 | 0.69134 |
| 7 | 0.96194 | 0.30866 |
| 8 | 0.69134 | 0.03806 |

**9 Holes**

| | x | y |
|---|---|---|
| 1 | 0.32899 | 0.03015 |
| 2 | 0.06699 | 0.25000 |
| 3 | 0.00760 | 0.58682 |
| 4 | 0.17861 | 0.88302 |
| 5 | 0.50000 | 1.00000 |
| 6 | 0.82139 | 0.88302 |
| 7 | 0.99240 | 0.58682 |
| 8 | 0.93301 | 0.25000 |
| 9 | 0.67101 | 0.03015 |

**10 Holes**

| | x | y |
|---|---|---|
| 1 | 0.34549 | 0.02447 |
| 2 | 0.09549 | 0.20611 |
| 3 | 0.00000 | 0.50000 |
| 4 | 0.09549 | 0.79389 |
| 5 | 0.34549 | 0.97553 |
| 6 | 0.65451 | 0.97553 |
| 7 | 0.90451 | 0.79389 |
| 8 | 1.00000 | 0.50000 |
| 9 | 0.90451 | 0.20611 |
| 10 | 0.65451 | 0.02447 |

**11 Holes**

| | x | y |
|---|---|---|
| 1 | 0.35913 | 0.02025 |
| 2 | 0.12213 | 0.17257 |
| 3 | 0.00509 | 0.42884 |
| 4 | 0.04518 | 0.70771 |
| 5 | 0.22968 | 0.92063 |
| 6 | 0.50000 | 1.00000 |
| 7 | 0.77032 | 0.92063 |
| 8 | 0.95482 | 0.70771 |
| 9 | 0.99491 | 0.42884 |
| 10 | 0.87787 | 0.17257 |
| 11 | 0.64087 | 0.02025 |

**12 Holes**

| | x | y |
|---|---|---|
| 1 | 0.37059 | 0.01704 |
| 2 | 0.14645 | 0.14645 |
| 3 | 0.01704 | 0.37059 |
| 4 | 0.01704 | 0.62941 |
| 5 | 0.14645 | 0.85355 |
| 6 | 0.37059 | 0.98296 |
| 7 | 0.62941 | 0.98296 |
| 8 | 0.85355 | 0.85355 |
| 9 | 0.98296 | 0.62941 |
| 10 | 0.98296 | 0.37059 |
| 11 | 0.85355 | 0.14645 |
| 12 | 0.62941 | 0.01704 |

**13 Holes**

| | x | y |
|---|---|---|
| 1 | 0.38034 | 0.01453 |
| 2 | 0.16844 | 0.12574 |
| 3 | 0.03249 | 0.32270 |
| 4 | 0.00365 | 0.56027 |
| 5 | 0.08851 | 0.78403 |
| 6 | 0.26764 | 0.94273 |
| 7 | 0.50000 | 1.00000 |
| 8 | 0.73236 | 0.94273 |
| 9 | 0.91149 | 0.78403 |
| 10 | 0.99635 | 0.56027 |
| 11 | 0.96751 | 0.32270 |
| 12 | 0.83156 | 0.12574 |
| 13 | 0.61966 | 0.01453 |

**14 Holes**

| | x | y |
|---|---|---|
| 1 | 0.38874 | 0.01254 |
| 2 | 0.18826 | 0.10908 |
| 3 | 0.04952 | 0.28306 |
| 4 | 0.00000 | 0.50000 |
| 5 | 0.04952 | 0.71694 |
| 6 | 0.18826 | 0.89092 |
| 7 | 0.38874 | 0.98746 |
| 8 | 0.61126 | 0.98746 |
| 9 | 0.81174 | 0.89092 |
| 10 | 0.95048 | 0.71694 |
| 11 | 1.00000 | 0.50000 |
| 12 | 0.95048 | 0.28306 |
| 13 | 0.81174 | 0.10908 |
| 14 | 0.61126 | 0.01254 |

**15 Holes**

| | x | y |
|---|---|---|
| 1 | 0.39604 | 0.01093 |
| 2 | 0.20611 | 0.09549 |
| 3 | 0.06699 | 0.25000 |
| 4 | 0.00274 | 0.44774 |
| 5 | 0.02447 | 0.65451 |
| 6 | 0.12843 | 0.83457 |
| 7 | 0.29663 | 0.95677 |
| 8 | 0.50000 | 1.00000 |
| 9 | 0.70337 | 0.95677 |
| 10 | 0.87157 | 0.83457 |
| 11 | 0.97553 | 0.65451 |
| 12 | 0.99726 | 0.44774 |
| 13 | 0.93301 | 0.25000 |
| 14 | 0.79389 | 0.09549 |
| 15 | 0.60396 | 0.01093 |

**16 Holes**

| | x | y |
|---|---|---|
| 1 | 0.40245 | 0.00961 |
| 2 | 0.22221 | 0.08427 |
| 3 | 0.08427 | 0.22221 |
| 4 | 0.00961 | 0.40245 |
| 5 | 0.00961 | 0.59755 |
| 6 | 0.08427 | 0.77779 |
| 7 | 0.22221 | 0.91573 |
| 8 | 0.40245 | 0.99039 |
| 9 | 0.59755 | 0.99039 |
| 10 | 0.77779 | 0.91573 |
| 11 | 0.91573 | 0.77779 |
| 12 | 0.99039 | 0.59755 |
| 13 | 0.99039 | 0.40245 |
| 14 | 0.91573 | 0.22221 |
| 15 | 0.77779 | 0.08427 |
| 16 | 0.59755 | 0.00961 |

**17 Holes**

| | x | y |
|---|---|---|
| 1 | 0.40813 | 0.00851 |
| 2 | 0.23678 | 0.07489 |
| 3 | 0.10099 | 0.19868 |
| 4 | 0.01909 | 0.36317 |
| 5 | 0.00213 | 0.54613 |
| 6 | 0.05242 | 0.72287 |
| 7 | 0.16315 | 0.86950 |
| 8 | 0.31938 | 0.96643 |
| 9 | 0.50000 | 1.00000 |
| 10 | 0.68062 | 0.96643 |
| 11 | 0.83685 | 0.86950 |
| 12 | 0.94758 | 0.72287 |
| 13 | 0.99787 | 0.54613 |
| 14 | 0.98091 | 0.36317 |
| 15 | 0.89901 | 0.19868 |
| 16 | 0.76322 | 0.07489 |
| 17 | 0.59187 | 0.00851 |

**18 Holes**

| | x | y |
|---|---|---|
| 1 | 0.41318 | 0.00760 |
| 2 | 0.25000 | 0.06699 |
| 3 | 0.11698 | 0.17861 |
| 4 | 0.03015 | 0.32899 |
| 5 | 0.00000 | 0.50000 |
| 6 | 0.03015 | 0.67101 |
| 7 | 0.11698 | 0.82139 |
| 8 | 0.25000 | 0.93301 |
| 9 | 0.41318 | 0.99240 |
| 10 | 0.58682 | 0.99240 |
| 11 | 0.75000 | 0.93301 |
| 12 | 0.88302 | 0.82139 |
| 13 | 0.96985 | 0.67101 |
| 14 | 1.00000 | 0.50000 |
| 15 | 0.96985 | 0.32899 |
| 16 | 0.88302 | 0.17861 |
| 17 | 0.75000 | 0.06699 |
| 18 | 0.58682 | 0.00760 |

**19 Holes**

| | x | y |
|---|---|---|
| 1 | 0.41770 | 0.00682 |
| 2 | 0.26203 | 0.06026 |
| 3 | 0.13214 | 0.16136 |
| 4 | 0.04211 | 0.29915 |
| 5 | 0.00171 | 0.45871 |
| 6 | 0.01530 | 0.62274 |
| 7 | 0.08142 | 0.77347 |
| … | … | … |

**Table 2** (*Concluded*). **Hole Coordinate Dimension Factors for Jig Boring—Type "B" Hole Circles** (English or metric units)

| | |
|---|---|
| x8 | 0.19289 |
| y8 | 0.89457 |
| x9 | 0.33765 |
| y9 | 0.97291 |
| x10 | 0.50000 |
| y10 | 1.00000 |
| x11 | 0.66235 |
| y11 | 0.97291 |
| x12 | 0.80711 |
| y12 | 0.89457 |
| x13 | 0.91858 |
| y13 | 0.77347 |
| x14 | 0.98470 |
| y14 | 0.62274 |
| x15 | 0.99829 |
| y15 | 0.45871 |
| x16 | 0.95789 |
| y16 | 0.29915 |
| x17 | 0.86786 |
| y17 | 0.16136 |
| x18 | 0.73797 |
| y18 | 0.06026 |
| x19 | 0.58230 |
| y19 | 0.00682 |

**20 Holes**

| | |
|---|---|
| x1 | 0.42178 |
| y1 | 0.00616 |
| x2 | 0.27300 |
| y2 | 0.05450 |
| x3 | 0.14645 |
| y3 | 0.14645 |
| x4 | 0.05450 |
| y4 | 0.27300 |
| x5 | 0.00616 |
| y5 | 0.42178 |
| x6 | 0.00616 |
| y6 | 0.57822 |
| x7 | 0.05450 |
| y7 | 0.72700 |
| x8 | 0.14645 |
| y8 | 0.85355 |
| x9 | 0.27300 |
| y9 | 0.94550 |
| x10 | 0.42178 |
| y10 | 0.99384 |
| x11 | 0.57822 |
| y11 | 0.99384 |
| x12 | 0.72700 |
| y12 | 0.94550 |
| x13 | 0.85355 |
| y13 | 0.85355 |
| x14 | 0.94550 |
| y14 | 0.72700 |
| x15 | 0.99384 |
| y15 | 0.57822 |
| x16 | 0.99384 |
| y16 | 0.42178 |
| x17 | 0.94550 |
| y17 | 0.27300 |
| x18 | 0.85355 |
| y18 | 0.14645 |
| x19 | 0.72700 |
| y19 | 0.05450 |
| x20 | 0.57822 |
| y20 | 0.00616 |
| .. | .... |
| .. | .... |

**21 Holes**

| | |
|---|---|
| x1 | 0.42548 |
| y1 | 0.00558 |
| x2 | 0.28306 |
| y2 | 0.04952 |
| x3 | 0.15991 |
| y3 | 0.13347 |
| x4 | 0.06699 |
| y4 | 0.25000 |
| x5 | 0.01254 |
| y5 | 0.38874 |
| x6 | 0.00140 |
| y6 | 0.53737 |
| x7 | 0.03456 |
| y7 | 0.68267 |
| x8 | 0.10908 |
| y8 | 0.81174 |
| x9 | 0.21834 |
| y9 | 0.91312 |
| x10 | 0.35262 |
| y10 | 0.97779 |
| x11 | 0.50000 |
| y11 | 1.00000 |
| x12 | 0.64738 |
| y12 | 0.97779 |
| x13 | 0.78166 |
| y13 | 0.91312 |
| x14 | 0.89092 |
| y14 | 0.81174 |
| x15 | 0.96544 |
| y15 | 0.68267 |
| x16 | 0.99860 |
| y16 | 0.53737 |
| x17 | 0.98746 |
| y17 | 0.38874 |
| x18 | 0.93301 |
| y18 | 0.25000 |
| x19 | 0.84009 |
| y19 | 0.13347 |
| x20 | 0.71694 |
| y20 | 0.04952 |
| x21 | 0.57452 |
| y21 | 0.00558 |

**22 Holes**

| | |
|---|---|
| x1 | 0.42884 |
| y1 | 0.00509 |
| x2 | 0.29229 |
| y2 | 0.04518 |
| x3 | 0.17257 |
| y3 | 0.12213 |
| x4 | 0.07937 |
| y4 | 0.22968 |
| x5 | 0.02025 |
| y5 | 0.35913 |
| x6 | 0.00000 |
| y6 | 0.50000 |
| x7 | 0.02025 |
| y7 | 0.64087 |
| x8 | 0.07937 |
| y8 | 0.77032 |
| x9 | 0.17257 |
| y9 | 0.87787 |
| x10 | 0.29229 |
| y10 | 0.95482 |
| x11 | 0.42884 |
| y11 | 0.99491 |
| .. | .... |
| x12 | 0.57116 |
| y12 | 0.99491 |
| x13 | 0.70771 |
| y13 | 0.95482 |
| x14 | 0.82743 |
| y14 | 0.87787 |
| x15 | 0.92063 |
| y15 | 0.77032 |
| x16 | 0.97975 |
| y16 | 0.64087 |
| x17 | 1.00000 |
| y17 | 0.50000 |
| x18 | 0.97975 |
| y18 | 0.35913 |
| x19 | 0.92063 |
| y19 | 0.22968 |
| x20 | 0.82743 |
| y20 | 0.12213 |
| x21 | 0.70771 |
| y21 | 0.04518 |
| x22 | 0.57116 |
| y22 | 0.00509 |

**23 Holes**

| | |
|---|---|
| x1 | 0.43192 |
| y1 | 0.00466 |
| x2 | 0.30080 |
| y2 | 0.04139 |
| x3 | 0.18446 |
| y3 | 0.11214 |
| x4 | 0.09152 |
| y4 | 0.21166 |
| x5 | 0.02887 |
| y5 | 0.33256 |
| x6 | 0.00117 |
| y6 | 0.46588 |
| x7 | 0.01046 |
| y7 | 0.60173 |
| x8 | 0.05606 |
| y8 | 0.73003 |
| x9 | 0.13458 |
| y9 | 0.84128 |
| x10 | 0.24021 |
| y10 | 0.92721 |
| x11 | 0.36510 |
| y11 | 0.98146 |
| x12 | 0.50000 |
| y12 | 1.00000 |
| x13 | 0.63490 |
| y13 | 0.98146 |
| x14 | 0.75979 |
| y14 | 0.92721 |
| x15 | 0.86542 |
| y15 | 0.84128 |
| x16 | 0.94394 |
| y16 | 0.73003 |
| x17 | 0.98954 |
| y17 | 0.60173 |
| x18 | 0.99883 |
| y18 | 0.46588 |
| x19 | 0.97113 |
| y19 | 0.33256 |
| x20 | 0.90848 |
| y20 | 0.21166 |
| x21 | 0.81554 |
| y21 | 0.11214 |
| x22 | 0.69920 |
| y22 | 0.04139 |
| x23 | 0.56808 |
| y23 | 0.00466 |

**24 Holes**

| | |
|---|---|
| x1 | 0.43474 |
| y1 | 0.00428 |
| x2 | 0.30866 |
| y2 | 0.03806 |
| x3 | 0.19562 |
| y3 | 0.10332 |
| x4 | 0.10332 |
| y4 | 0.19562 |
| x5 | 0.03806 |
| y5 | 0.30866 |
| x6 | 0.00428 |
| y6 | 0.43474 |
| x7 | 0.00428 |
| y7 | 0.56526 |
| x8 | 0.03806 |
| y8 | 0.69134 |
| x9 | 0.10332 |
| y9 | 0.80438 |
| x10 | 0.19562 |
| y10 | 0.89668 |
| x11 | 0.30866 |
| y11 | 0.96194 |
| x12 | 0.43474 |
| y12 | 0.99572 |
| x13 | 0.56526 |
| y13 | 0.99572 |
| x14 | 0.69134 |
| y14 | 0.96194 |
| x15 | 0.80438 |
| y15 | 0.89668 |
| x16 | 0.89668 |
| y16 | 0.80438 |
| x17 | 0.96194 |
| y17 | 0.69134 |
| x18 | 0.99572 |
| y18 | 0.56526 |
| x19 | 0.99572 |
| y19 | 0.43474 |
| x20 | 0.96194 |
| y20 | 0.30866 |
| x21 | 0.89668 |
| y21 | 0.19562 |
| x22 | 0.80438 |
| y22 | 0.10332 |
| x23 | 0.69134 |
| y23 | 0.03806 |
| x24 | 0.56526 |
| y24 | 0.00428 |

**25 Holes**

| | |
|---|---|
| x1 | 0.43733 |
| y1 | 0.00394 |
| x2 | 0.31594 |
| y2 | 0.03511 |
| x3 | 0.20611 |
| y3 | 0.09549 |
| x4 | 0.11474 |
| y4 | 0.18129 |
| x5 | 0.04759 |
| y5 | 0.28711 |
| x6 | 0.00886 |
| y6 | 0.40631 |
| x7 | 0.00099 |
| y7 | 0.53140 |
| .. | .... |
| x8 | 0.02447 |
| y8 | 0.65451 |
| x9 | 0.07784 |
| y9 | 0.76791 |
| x10 | 0.15773 |
| y10 | 0.86448 |
| x11 | 0.25912 |
| y11 | 0.93815 |
| x12 | 0.37566 |
| y12 | 0.98429 |
| x13 | 0.50000 |
| y13 | 1.00000 |
| x14 | 0.62434 |
| y14 | 0.98429 |
| x15 | 0.74088 |
| y15 | 0.93815 |
| x16 | 0.84227 |
| y16 | 0.86448 |
| x17 | 0.92216 |
| y17 | 0.76791 |
| x18 | 0.97553 |
| y18 | 0.65451 |
| x19 | 0.99901 |
| y19 | 0.53140 |
| x20 | 0.99114 |
| y20 | 0.40631 |
| x21 | 0.95241 |
| y21 | 0.28711 |
| x22 | 0.88526 |
| y22 | 0.18129 |
| x23 | 0.79389 |
| y23 | 0.09549 |
| x24 | 0.68406 |
| y24 | 0.03511 |
| x25 | 0.56267 |
| y25 | 0.00394 |

**26 Holes**

| | |
|---|---|
| x1 | 0.43973 |
| y1 | 0.00365 |
| x2 | 0.32270 |
| y2 | 0.03249 |
| x3 | 0.21597 |
| y3 | 0.08851 |
| x4 | 0.12574 |
| y4 | 0.16844 |
| x5 | 0.05727 |
| y5 | 0.26764 |
| x6 | 0.01453 |
| y6 | 0.38034 |
| x7 | 0.00000 |
| y7 | 0.50000 |
| x8 | 0.01453 |
| y8 | 0.61966 |
| x9 | 0.05727 |
| y9 | 0.73236 |
| x10 | 0.12574 |
| y10 | 0.83156 |
| x11 | 0.21597 |
| y11 | 0.91149 |
| x12 | 0.32270 |
| y12 | 0.96751 |
| x13 | 0.43973 |
| y13 | 0.99635 |
| x14 | 0.56027 |
| y14 | 0.99635 |
| x15 | 0.67730 |
| y15 | 0.96751 |
| x16 | 0.78403 |
| y16 | 0.91149 |
| x17 | 0.87426 |
| y17 | 0.83156 |
| x18 | 0.94273 |
| y18 | 0.73236 |
| x19 | 0.98547 |
| y19 | 0.61966 |
| x20 | 1.00000 |
| y20 | 0.50000 |
| x21 | 0.98547 |
| y21 | 0.38034 |
| x22 | 0.94273 |
| y22 | 0.26764 |
| x23 | 0.87426 |
| y23 | 0.16844 |
| x24 | 0.78403 |
| y24 | 0.08851 |
| x25 | 0.67730 |
| y25 | 0.03249 |
| x26 | 0.56027 |
| y26 | 0.00365 |

**27 Holes**

| | |
|---|---|
| x1 | 0.44195 |
| y1 | 0.00338 |
| x2 | 0.32899 |
| y2 | 0.03015 |
| x3 | 0.22525 |
| y3 | 0.08226 |
| x4 | 0.13631 |
| y4 | 0.15688 |
| x5 | 0.06699 |
| y5 | 0.25000 |
| x6 | 0.02101 |
| y6 | 0.35660 |
| x7 | 0.00085 |
| y7 | 0.47093 |
| x8 | 0.00760 |
| y8 | 0.58682 |
| x9 | 0.04089 |
| y9 | 0.69804 |
| x10 | 0.09894 |
| y10 | 0.79858 |
| x11 | 0.17861 |
| y11 | 0.88302 |
| x12 | 0.27560 |
| y12 | 0.94682 |
| x13 | 0.38469 |
| y13 | 0.98652 |
| x14 | 0.50000 |
| y14 | 1.00000 |
| x15 | 0.61531 |
| y15 | 0.98652 |
| x16 | 0.72440 |
| y16 | 0.94682 |
| x17 | 0.82139 |
| y17 | 0.88302 |
| x18 | 0.90106 |
| y18 | 0.79858 |
| x19 | 0.95911 |
| y19 | 0.69804 |
| x20 | 0.99240 |
| y20 | 0.58682 |
| x21 | 0.99915 |
| y21 | 0.47093 |
| x22 | 0.97899 |
| y22 | 0.35660 |
| x23 | 0.93301 |
| y23 | 0.25000 |
| x24 | 0.86369 |
| y24 | 0.15688 |
| x25 | 0.77475 |
| y25 | 0.08226 |
| x26 | 0.67101 |
| y26 | 0.03015 |
| x27 | 0.55805 |
| y27 | 0.00338 |

**28 Holes**

| | |
|---|---|
| x1 | 0.44402 |
| y1 | 0.00314 |
| x2 | 0.33486 |
| y2 | 0.02806 |
| x3 | 0.23398 |
| y3 | 0.07664 |
| x4 | 0.14645 |
| y4 | 0.14645 |
| x5 | 0.07664 |
| y5 | 0.23398 |
| x6 | 0.02806 |
| y6 | 0.33486 |
| x7 | 0.00314 |
| y7 | 0.44402 |
| x8 | 0.00314 |
| y8 | 0.55598 |
| x9 | 0.02806 |
| y9 | 0.66514 |
| x10 | 0.07664 |
| y10 | 0.76602 |
| x11 | 0.14645 |
| y11 | 0.85355 |
| x12 | 0.23398 |
| y12 | 0.92336 |
| x13 | 0.33486 |
| y13 | 0.97194 |
| x14 | 0.44402 |
| y14 | 0.99686 |
| x15 | 0.55598 |
| y15 | 0.99686 |
| x16 | 0.66514 |
| y16 | 0.97194 |
| x17 | 0.76602 |
| y17 | 0.92336 |
| x18 | 0.85355 |
| y18 | 0.85355 |
| x19 | 0.92336 |
| y19 | 0.76602 |
| x20 | 0.97194 |
| y20 | 0.66514 |
| x21 | 0.99686 |
| y21 | 0.55598 |
| x22 | 0.99686 |
| y22 | 0.44402 |
| x23 | 0.97194 |
| y23 | 0.33486 |
| x24 | 0.92336 |
| y24 | 0.23398 |
| x25 | 0.85355 |
| y25 | 0.14645 |
| x26 | 0.76602 |
| y26 | 0.07664 |
| x27 | 0.66514 |
| y27 | 0.02806 |
| x28 | 0.55598 |
| y28 | 0.00314 |

## Table 3. Hole Coordinate Dimension Factors for Jig Boring—Type "A" Hole Circles, Central Coordinates (English or metric units)

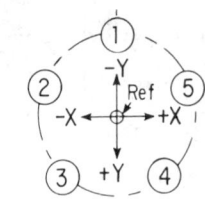

The diagram shows a type "A" circle for a 5-hole circle. Coordinates x, y are given in the table for hole circles of from 3 to 28 holes. Dimensions are for holes numbered in a counterclockwise direction (as shown). Dimensions given are based upon a hole circle of unit diameter. For a hole circle of, say, 3-inch or 3-centimeter diameter, multiply table values by 3.

### 3 Holes
| | x | y |
|---|---|---|
| x1, y1 | 0.00000 | −0.50000 |
| x2, y2 | −0.43301 | +0.25000 |
| x3, y3 | +0.43301 | +0.25000 |

### 4 Holes
| | x | y |
|---|---|---|
| x1, y1 | 0.00000 | −0.50000 |
| x2, y2 | −0.50000 | 0.00000 |
| x3, y3 | 0.00000 | +0.50000 |
| x4, y4 | +0.50000 | 0.00000 |

### 5 Holes
| | x | y |
|---|---|---|
| x1, y1 | 0.00000 | −0.50000 |
| x2, y2 | −0.47553 | −0.15451 |
| x3, y3 | −0.29389 | +0.40451 |
| x4, y4 | +0.29389 | +0.40451 |
| x5, y5 | +0.47553 | −0.15451 |

### 6 Holes
| | x | y |
|---|---|---|
| x1, y1 | 0.00000 | −0.50000 |
| x2, y2 | −0.43301 | −0.25000 |
| x3, y3 | −0.43301 | +0.25000 |
| x4, y4 | 0.00000 | +0.50000 |
| x5, y5 | +0.43301 | +0.25000 |
| x6, y6 | +0.43301 | −0.25000 |

### 7 Holes
| | x | y |
|---|---|---|
| x1, y1 | 0.00000 | −0.50000 |
| x2, y2 | −0.39092 | −0.31174 |
| x3, y3 | −0.48746 | +0.11126 |
| x4, y4 | −0.21694 | +0.45048 |
| x5, y5 | +0.21694 | +0.45048 |
| x6, y6 | +0.48746 | +0.11126 |
| x7, y7 | +0.39092 | −0.31174 |

### 8 Holes
| | x | y |
|---|---|---|
| x1, y1 | 0.00000 | −0.50000 |
| x2, y2 | −0.35355 | −0.35355 |
| x3, y3 | −0.50000 | 0.00000 |
| x4, y4 | −0.35355 | +0.35355 |
| x5, y5 | 0.00000 | +0.50000 |
| x6, y6 | +0.35355 | +0.35355 |
| x7, y7 | +0.50000 | 0.00000 |
| x8, y8 | +0.35355 | −0.35355 |

### 9 Holes
| | x | y |
|---|---|---|
| x1, y1 | 0.00000 | −0.50000 |
| x2, y2 | −0.32139 | −0.38302 |
| x3, y3 | −0.49240 | −0.08682 |
| x4, y4 | −0.43301 | +0.25000 |
| x5, y5 | −0.17101 | +0.46985 |
| x6, y6 | +0.17101 | +0.46985 |
| x7, y7 | +0.43301 | +0.25000 |
| x8, y8 | +0.49240 | −0.08682 |
| x9, y9 | +0.32139 | −0.38302 |

### 10 Holes
| | x | y |
|---|---|---|
| x1, y1 | 0.00000 | −0.50000 |
| x2, y2 | −0.29389 | −0.40451 |
| x3, y3 | −0.47553 | −0.15451 |
| x4, y4 | −0.47553 | +0.15451 |
| x5, y5 | −0.29389 | +0.40451 |
| x6, y6 | 0.00000 | +0.50000 |
| x7, y7 | +0.29389 | +0.40451 |
| x8, y8 | +0.47553 | +0.15451 |
| x9, y9 | +0.47553 | −0.15451 |
| x10, y10 | +0.29389 | −0.40451 |

### 11 Holes
| | x | y |
|---|---|---|
| x1, y1 | 0.00000 | −0.50000 |
| x2, y2 | −0.27032 | −0.42063 |
| x3, y3 | −0.45482 | −0.20771 |
| x4, y4 | −0.49491 | +0.07116 |
| x5, y5 | −0.37787 | +0.32743 |
| x6, y6 | −0.14087 | +0.47975 |
| x7, y7 | +0.14087 | +0.47975 |
| x8, y8 | +0.37787 | +0.32743 |
| x9, y9 | +0.49491 | +0.07116 |
| x10, y10 | +0.45482 | −0.20771 |
| x11, y11 | +0.27032 | −0.42063 |

### 12 Holes
| | x | y |
|---|---|---|
| x1, y1 | 0.00000 | −0.50000 |
| x2, y2 | −0.25000 | −0.43301 |
| x3, y3 | −0.43301 | −0.25000 |
| x4, y4 | −0.50000 | 0.00000 |
| x5, y5 | −0.43301 | +0.25000 |
| x6, y6 | −0.25000 | +0.43301 |
| x7, y7 | 0.00000 | +0.50000 |
| x8, y8 | +0.25000 | +0.43301 |
| x9, y9 | +0.43301 | +0.25000 |
| x10, y10 | +0.50000 | 0.00000 |
| x11, y11 | +0.43301 | −0.25000 |
| x12, y12 | +0.25000 | −0.43301 |

### 13 Holes
| | x | y |
|---|---|---|
| x1, y1 | 0.00000 | −0.50000 |
| x2, y2 | −0.23236 | −0.44273 |
| x3, y3 | −0.41149 | −0.28403 |
| x4, y4 | −0.49635 | −0.06027 |
| x5, y5 | −0.46751 | +0.17730 |
| x6, y6 | −0.33156 | +0.37426 |
| x7, y7 | −0.11966 | +0.48547 |
| x8, y8 | +0.11966 | +0.48547 |
| x9, y9 | +0.33156 | +0.37426 |
| x10, y10 | +0.46751 | +0.17730 |
| x11, y11 | +0.49635 | −0.06027 |
| x12, y12 | +0.41149 | −0.28403 |
| x13, y13 | +0.23236 | −0.44273 |

### 14 Holes
| | x | y |
|---|---|---|
| x1, y1 | 0.00000 | −0.50000 |
| x2, y2 | −0.21694 | −0.45048 |
| x3, y3 | −0.39092 | −0.31174 |
| x4, y4 | −0.48746 | −0.11126 |
| x5, y5 | −0.48746 | +0.11126 |
| x6, y6 | −0.39092 | +0.31174 |
| x7, y7 | −0.21694 | +0.45048 |
| x8, y8 | 0.00000 | +0.50000 |
| x9, y9 | +0.21694 | +0.45048 |
| x10, y10 | +0.39092 | +0.31174 |
| x11, y11 | +0.48746 | +0.11126 |
| x12, y12 | +0.48746 | −0.11126 |
| x13, y13 | +0.39092 | −0.31174 |
| x14, y14 | +0.21694 | −0.45048 |

### 15 Holes
| | x | y |
|---|---|---|
| x1, y1 | 0.00000 | −0.50000 |
| x2, y2 | −0.20337 | −0.45677 |
| x3, y3 | −0.37157 | −0.33457 |
| x4, y4 | −0.47553 | −0.15451 |
| x5, y5 | −0.49726 | +0.05226 |
| x6, y6 | −0.43301 | +0.25000 |
| x7, y7 | −0.29389 | +0.40451 |
| x8, y8 | −0.10396 | +0.48907 |
| x9, y9 | +0.10396 | +0.48907 |
| x10, y10 | +0.29389 | +0.40451 |
| x11, y11 | +0.43301 | +0.25000 |
| x12, y12 | +0.49726 | +0.05226 |
| x13, y13 | +0.47553 | −0.15451 |
| x14, y14 | +0.37157 | −0.33457 |
| x15, y15 | +0.20337 | −0.45677 |

### 16 Holes
| | x | y |
|---|---|---|
| x1, y1 | 0.00000 | −0.50000 |
| x2, y2 | −0.19134 | −0.46194 |
| x3, y3 | −0.35355 | −0.35355 |
| x4, y4 | −0.46194 | −0.19134 |
| x5, y5 | −0.50000 | 0.00000 |
| x6, y6 | −0.46194 | +0.19134 |
| x7, y7 | −0.35355 | +0.35355 |
| x8, y8 | −0.19134 | +0.46194 |
| x9, y9 | 0.00000 | +0.50000 |
| x10, y10 | +0.19134 | +0.46194 |
| x11, y11 | +0.35355 | +0.35355 |
| x12, y12 | +0.46194 | +0.19134 |
| x13, y13 | +0.50000 | 0.00000 |
| x14, y14 | +0.46194 | −0.19134 |
| x15, y15 | +0.35355 | −0.35355 |
| x16, y16 | +0.19134 | −0.46194 |

### 17 Holes
| | x | y |
|---|---|---|
| x1, y1 | 0.00000 | −0.50000 |
| x2, y2 | −0.18062 | −0.46624 |
| x3, y3 | −0.33685 | −0.36950 |
| x4, y4 | −0.44758 | −0.22287 |
| x5, y5 | −0.49787 | −0.04613 |
| x6, y6 | −0.48091 | +0.13683 |
| x7, y7 | −0.39901 | +0.30132 |
| x8, y8 | −0.26322 | +0.42511 |
| x9, y9 | −0.09187 | +0.49149 |
| x10, y10 | +0.09187 | +0.49149 |
| x11, y11 | +0.26322 | +0.42511 |
| x12, y12 | +0.39901 | +0.30132 |
| x13, y13 | +0.48091 | +0.13683 |
| x14, y14 | +0.49787 | −0.04613 |
| x15, y15 | +0.44758 | −0.22287 |
| x16, y16 | +0.33685 | −0.36950 |
| x17, y17 | +0.18062 | −0.46624 |

### 18 Holes
| | x | y |
|---|---|---|
| x1, y1 | 0.00000 | −0.50000 |
| x2, y2 | −0.17101 | −0.46985 |
| x3, y3 | −0.32139 | −0.38302 |
| x4, y4 | −0.43301 | −0.25000 |
| x5, y5 | −0.49240 | −0.08682 |
| x6, y6 | −0.49240 | +0.08682 |
| x7, y7 | −0.43301 | +0.25000 |
| x8, y8 | −0.32139 | +0.38302 |
| x9, y9 | −0.17101 | +0.46985 |
| x10, y10 | 0.00000 | +0.50000 |
| x11, y11 | +0.17101 | +0.46985 |
| x12, y12 | +0.32139 | +0.38302 |
| x13, y13 | +0.43301 | +0.25000 |
| x14, y14 | +0.49240 | +0.08682 |
| x15, y15 | +0.49240 | −0.08682 |
| x16, y16 | +0.43301 | −0.25000 |
| x17, y17 | +0.32139 | −0.38302 |
| x18, y18 | +0.17101 | −0.46985 |

### 19 Holes
| | x | y |
|---|---|---|
| x1, y1 | 0.00000 | −0.50000 |
| x2, y2 | −0.16235 | −0.47291 |
| x3, y3 | −0.30711 | −0.39457 |
| x4, y4 | −0.41858 | −0.27347 |
| x5, y5 | −0.48470 | −0.12274 |
| x6, y6 | −0.49829 | +0.04129 |
| x7, y7 | −0.45789 | +0.20085 |

Table 3 (*Concluded*). Hole Coordinate Dimension Factors for Jig Boring—Type "A" Hole Circles, Central Coordinates (English or metric units)

### 19 Holes (continued)

| Hole | x | y |
| --- | --- | --- |
| 8 | −0.36786 | +0.33864 |
| 9 | −0.23797 | +0.43974 |
| 10 | −0.08230 | +0.49318 |
| 11 | +0.08230 | +0.49318 |
| 12 | +0.23797 | +0.43974 |
| 13 | +0.36786 | +0.33864 |
| 14 | +0.45789 | +0.20085 |
| 15 | +0.49829 | +0.04129 |
| 16 | +0.48470 | −0.12274 |
| 17 | +0.41858 | −0.27347 |
| 18 | +0.30711 | −0.39457 |
| 19 | +0.16235 | −0.47291 |

### 20 Holes

| Hole | x | y |
| --- | --- | --- |
| 1 | 0.00000 | −0.50000 |
| 2 | −0.15451 | −0.47553 |
| 3 | −0.29389 | −0.40451 |
| 4 | −0.40451 | −0.29389 |
| 5 | −0.47553 | −0.15451 |
| 6 | −0.50000 | 0.00000 |
| 7 | −0.47553 | +0.15451 |
| 8 | −0.40451 | +0.29389 |
| 9 | −0.29389 | +0.40451 |
| 10 | −0.15451 | +0.47553 |
| 11 | 0.00000 | +0.50000 |
| 12 | +0.15451 | +0.47553 |
| 13 | +0.29389 | +0.40451 |
| 14 | +0.40451 | +0.29389 |
| 15 | +0.47553 | +0.15451 |
| 16 | +0.50000 | 0.00000 |
| 17 | +0.47553 | −0.15451 |
| 18 | +0.40451 | −0.29389 |
| 19 | +0.29389 | −0.40451 |
| 20 | +0.15451 | −0.47553 |

### 21 Holes

| Hole | x | y |
| --- | --- | --- |
| 1 | 0.00000 | −0.50000 |
| 2 | −0.14738 | −0.47779 |
| 3 | −0.28166 | −0.41312 |
| 4 | −0.39092 | −0.31174 |
| 5 | −0.46544 | −0.18267 |
| 6 | −0.49860 | −0.03737 |
| 7 | −0.48746 | +0.11116 |
| 8 | −0.43301 | +0.25000 |
| 9 | −0.34009 | +0.36653 |
| 10 | −0.21694 | +0.45048 |
| 11 | −0.07452 | +0.49442 |
| 12 | +0.07452 | +0.49442 |
| 13 | +0.21694 | +0.45048 |
| 14 | +0.34009 | +0.36653 |
| 15 | +0.43301 | +0.25000 |
| 16 | +0.48746 | +0.11126 |
| 17 | +0.49860 | −0.03737 |
| 18 | +0.46544 | −0.18267 |
| 19 | +0.39092 | −0.31174 |
| 20 | +0.28166 | −0.41312 |
| 21 | +0.14738 | −0.47779 |

### 22 Holes

| Hole | x | y |
| --- | --- | --- |
| 1 | 0.00000 | −0.50000 |
| 2 | −0.14087 | −0.47975 |
| 3 | −0.27032 | −0.42063 |
| 4 | −0.37787 | −0.32743 |
| 5 | −0.45482 | −0.20771 |
| 6 | −0.49491 | −0.07116 |
| 7 | −0.49491 | +0.07116 |
| 8 | −0.45482 | +0.20771 |
| 9 | −0.37787 | +0.32743 |
| 10 | −0.27032 | +0.42063 |
| 11 | −0.14087 | +0.47975 |
| 12 | 0.00000 | +0.50000 |
| 13 | +0.14087 | +0.47975 |
| 14 | +0.27032 | +0.42063 |
| 15 | +0.37787 | +0.32743 |
| 16 | +0.45482 | +0.20771 |
| 17 | +0.49491 | +0.07116 |
| 18 | +0.49491 | −0.07116 |
| 19 | +0.45482 | −0.20771 |
| 20 | +0.37787 | −0.32743 |
| 21 | +0.27032 | −0.42063 |
| 22 | +0.14087 | −0.47975 |

### 23 Holes

| Hole | x | y |
| --- | --- | --- |
| 1 | 0.00000 | −0.50000 |
| 2 | −0.13490 | −0.48146 |
| 3 | −0.25979 | −0.42721 |
| 4 | −0.36542 | −0.34128 |
| 5 | −0.44394 | −0.23003 |
| 6 | −0.48954 | −0.10173 |
| 7 | −0.49883 | +0.03412 |
| 8 | −0.47113 | +0.16744 |
| 9 | −0.40848 | +0.28834 |
| 10 | −0.31554 | +0.38786 |
| 11 | −0.19920 | +0.45861 |
| 12 | −0.06808 | +0.49534 |
| 13 | +0.06808 | +0.49534 |
| 14 | +0.19920 | +0.45861 |
| 15 | +0.31554 | +0.38786 |
| 16 | +0.40848 | +0.28834 |
| 17 | +0.47113 | +0.16744 |
| 18 | +0.49883 | +0.03412 |
| 19 | +0.48954 | −0.10173 |
| 20 | +0.44394 | −0.23003 |
| 21 | +0.36542 | −0.34128 |
| 22 | +0.25979 | −0.42721 |
| 23 | +0.13490 | −0.48146 |

### 24 Holes

| Hole | x | y |
| --- | --- | --- |
| 1 | 0.00000 | −0.50000 |
| 2 | −0.12941 | −0.48296 |
| 3 | −0.25000 | −0.43301 |
| 4 | −0.35355 | −0.35355 |
| 5 | −0.43301 | −0.25000 |
| 6 | −0.48296 | −0.12941 |
| 7 | −0.50000 | 0.00000 |
| 8 | −0.48296 | +0.12941 |
| 9 | −0.43301 | +0.25000 |
| 10 | −0.35355 | +0.35355 |
| 11 | −0.25000 | +0.43301 |
| 12 | −0.12941 | +0.48296 |
| 13 | 0.00000 | +0.50000 |
| 14 | +0.12941 | +0.48296 |
| 15 | +0.25000 | +0.43301 |
| 16 | +0.35355 | +0.35355 |
| 17 | +0.43301 | +0.25000 |
| 18 | +0.48296 | +0.12941 |
| 19 | +0.50000 | +0.00000 |
| 20 | +0.48296 | −0.12941 |
| 21 | +0.43301 | −0.25000 |
| 22 | +0.35355 | −0.35355 |
| 23 | +0.25000 | −0.43301 |
| 24 | +0.12941 | −0.48296 |

### 25 Holes

| Hole | x | y |
| --- | --- | --- |
| 1 | 0.00000 | −0.50000 |
| 2 | −0.12434 | −0.48429 |
| 3 | −0.24088 | −0.43815 |
| 4 | −0.34227 | −0.36648 |
| 5 | −0.42216 | −0.26791 |
| 6 | −0.47553 | −0.15451 |
| 7 | −0.49901 | −0.03140 |
| 8 | −0.49114 | +0.09369 |
| 9 | −0.45241 | +0.21289 |
| 10 | −0.38526 | +0.31871 |
| 11 | −0.29389 | +0.40451 |
| 12 | −0.18406 | +0.46489 |
| 13 | −0.06267 | +0.49606 |
| 14 | +0.06267 | +0.49606 |
| 15 | +0.18406 | +0.46489 |
| 16 | +0.29389 | +0.40451 |
| 17 | +0.38526 | +0.31871 |
| 18 | +0.45241 | +0.21289 |
| 19 | +0.49114 | +0.09369 |
| 20 | +0.49901 | −0.03140 |
| 21 | +0.47553 | −0.15451 |
| 22 | +0.42216 | −0.26791 |
| 23 | +0.34227 | −0.36648 |
| 24 | +0.24088 | −0.43815 |
| 25 | +0.12434 | −0.48429 |

### 26 Holes

| Hole | x | y |
| --- | --- | --- |
| 1 | 0.00000 | −0.50000 |
| 2 | −0.11966 | −0.48547 |
| 3 | −0.23236 | −0.44273 |
| 4 | −0.33156 | −0.37426 |
| 5 | −0.41149 | −0.28403 |
| 6 | −0.46751 | −0.17730 |
| 7 | −0.49635 | −0.06027 |
| 8 | −0.49635 | +0.06027 |
| 9 | −0.46751 | +0.17730 |
| 10 | −0.41149 | +0.28403 |
| 11 | −0.33156 | +0.37426 |
| 12 | −0.23236 | +0.44273 |
| 13 | −0.11966 | +0.48547 |
| 14 | 0.00000 | +0.50000 |
| 15 | +0.11966 | +0.48547 |
| 16 | +0.23236 | +0.44273 |
| 17 | +0.33156 | +0.37426 |
| 18 | +0.41149 | +0.28403 |
| 19 | +0.46751 | +0.17730 |
| 20 | +0.49635 | +0.06027 |
| 21 | +0.49635 | −0.06027 |
| 22 | +0.46751 | −0.17730 |
| 23 | +0.41149 | −0.28403 |
| 24 | +0.33156 | −0.37426 |
| 25 | +0.23236 | −0.44273 |
| 26 | +0.11966 | −0.48547 |

### 27 Holes

| Hole | x | y |
| --- | --- | --- |
| 1 | 0.00000 | −0.50000 |
| 2 | −0.11533 | −0.48652 |
| 3 | −0.22440 | −0.44682 |
| 4 | −0.32139 | −0.38302 |
| 5 | −0.40106 | −0.29858 |
| 6 | −0.45911 | −0.19804 |
| 7 | −0.49240 | −0.08682 |
| 8 | −0.49911 | +0.02907 |
| 9 | −0.47899 | +0.14340 |
| 10 | −0.43301 | +0.25000 |
| 11 | −0.36369 | +0.34312 |
| 12 | −0.27475 | +0.41774 |
| 13 | −0.17101 | +0.46985 |
| 14 | −0.05805 | +0.49662 |
| 15 | +0.05805 | +0.49662 |
| 16 | +0.17101 | +0.46985 |
| 17 | +0.27475 | +0.41774 |
| 18 | +0.36369 | +0.34312 |
| 19 | +0.43301 | +0.25000 |
| 20 | +0.47899 | +0.14340 |
| 21 | +0.49911 | +0.02907 |
| 22 | +0.49240 | −0.08682 |
| 23 | +0.45911 | −0.19804 |
| 24 | +0.40106 | −0.29858 |
| 25 | +0.32139 | −0.38302 |
| 26 | +0.22440 | −0.44682 |
| 27 | +0.11533 | −0.48652 |

### 28 Holes

| Hole | x | y |
| --- | --- | --- |
| 1 | 0.00000 | −0.50000 |
| 2 | −0.11126 | −0.48746 |
| 3 | −0.21694 | −0.45048 |
| 4 | −0.31174 | −0.39092 |
| 5 | −0.39092 | −0.31174 |
| 6 | −0.45048 | −0.21694 |
| 7 | −0.48746 | −0.11126 |
| 8 | −0.50000 | 0.00000 |
| 9 | −0.48746 | +0.11126 |
| 10 | −0.45048 | +0.21694 |
| 11 | −0.39092 | +0.31174 |
| 12 | −0.31174 | +0.39092 |
| 13 | −0.21694 | +0.45048 |
| 14 | −0.11126 | +0.48746 |
| 15 | 0.00000 | +0.50000 |
| 16 | +0.11126 | +0.48746 |
| 17 | +0.21694 | +0.45048 |
| 18 | +0.31174 | +0.39092 |
| 19 | +0.39092 | +0.31174 |
| 20 | +0.45048 | +0.21694 |
| 21 | +0.48746 | +0.11126 |
| 22 | +0.50000 | 0.00000 |
| 23 | +0.48746 | −0.11126 |
| 24 | +0.45048 | −0.21694 |
| 25 | +0.39092 | −0.31174 |
| 26 | +0.31174 | −0.39092 |
| 27 | +0.21694 | −0.45048 |
| 28 | +0.11126 | −0.48746 |

**Table 4.  Hole Coordinate Dimension Factors for Jig Boring—Type "B" Hole Circles**
**Central Coordinates** (English or metric units)

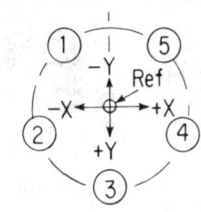

The diagram shows a type "B" circle for a 5-hole circle. Coordinates $x$, $y$ are given in the table for hole circles of from 3 to 28 holes. Dimensions are for holes numbered in a counterclockwise direction (as shown). Dimensions given are based upon a hole circle of unit diameter. For a hole circle of, say, 3-inch or 3-centimeter diameter, multiply table values by 3.

**3 Holes**

| Hole | $x$ | $y$ |
|---|---|---|
| 1 | −0.43301 | −0.25000 |
| 2 | 0.00000 | +0.50000 |
| 3 | +0.43301 | −0.25000 |

**4 Holes**

| Hole | $x$ | $y$ |
|---|---|---|
| 1 | −0.35355 | −0.35355 |
| 2 | −0.35355 | +0.35355 |
| 3 | +0.35355 | +0.35355 |
| 4 | +0.35355 | −0.35355 |

**5 Holes**

| Hole | $x$ | $y$ |
|---|---|---|
| 1 | −0.29389 | −0.40451 |
| 2 | −0.47553 | +0.15451 |
| 3 | 0.00000 | +0.50000 |
| 4 | +0.47553 | +0.15451 |
| 5 | +0.29389 | −0.40451 |

**6 Holes**

| Hole | $x$ | $y$ |
|---|---|---|
| 1 | −0.25000 | −0.43301 |
| 2 | −0.50000 | 0.00000 |
| 3 | −0.25000 | +0.43301 |
| 4 | +0.25000 | +0.43301 |
| 5 | +0.50000 | 0.00000 |
| 6 | +0.25000 | −0.43301 |

**7 Holes**

| Hole | $x$ | $y$ |
|---|---|---|
| 1 | −0.21694 | −0.45048 |
| 2 | −0.48746 | −0.11126 |
| 3 | −0.39092 | +0.31174 |
| 4 | 0.00000 | +0.50000 |
| 5 | +0.39092 | +0.31174 |
| 6 | +0.48746 | −0.11126 |
| 7 | +0.21694 | −0.45048 |

**8 Holes**

| Hole | $x$ | $y$ |
|---|---|---|
| 1 | −0.19134 | −0.46194 |
| 2 | −0.46194 | −0.19134 |
| 3 | −0.46194 | +0.19134 |
| 4 | −0.19134 | +0.46194 |
| 5 | +0.19134 | +0.46194 |
| 6 | +0.46194 | +0.19134 |
| 7 | +0.46194 | −0.19134 |
| 8 | +0.19134 | −0.46194 |

**9 Holes**

| Hole | $x$ | $y$ |
|---|---|---|
| 1 | −0.17101 | −0.46985 |
| 2 | −0.43301 | −0.25000 |
| 3 | −0.49240 | +0.08682 |
| 4 | −0.32139 | +0.38302 |
| 5 | 0.00000 | +0.50000 |
| 6 | +0.32139 | +0.38302 |
| 7 | +0.49240 | +0.08682 |
| 8 | +0.43301 | −0.25000 |
| 9 | +0.17101 | −0.46985 |

**10 Holes**

| Hole | $x$ | $y$ |
|---|---|---|
| 1 | −0.15451 | −0.47553 |
| 2 | −0.40451 | −0.29389 |
| 3 | −0.50000 | 0.00000 |
| 4 | −0.40451 | +0.29389 |
| 5 | −0.15451 | +0.47553 |
| 6 | +0.15451 | +0.47553 |
| 7 | +0.40451 | +0.29389 |
| 8 | +0.50000 | 0.00000 |
| 9 | +0.40451 | −0.29389 |
| 10 | +0.15451 | −0.47553 |

**11 Holes**

| Hole | $x$ | $y$ |
|---|---|---|
| 1 | −0.14087 | −0.47975 |
| 2 | −0.37787 | −0.32743 |
| 3 | −0.49491 | −0.07116 |
| 4 | −0.45482 | +0.20771 |
| 5 | −0.27032 | +0.42063 |
| 6 | 0.00000 | +0.50000 |
| 7 | +0.27032 | +0.42063 |
| 8 | +0.45482 | +0.20771 |
| 9 | +0.49491 | −0.07116 |
| 10 | +0.37787 | −0.32743 |
| 11 | +0.14087 | −0.47975 |

**12 Holes**

| Hole | $x$ | $y$ |
|---|---|---|
| 1 | −0.12941 | −0.48296 |
| 2 | −0.35355 | −0.35355 |
| 3 | −0.48296 | −0.12941 |
| 4 | −0.48296 | +0.12941 |
| 5 | −0.35355 | +0.35355 |
| 6 | −0.12941 | +0.48296 |
| 7 | +0.12941 | +0.48296 |
| 8 | +0.35355 | +0.35355 |
| 9 | +0.48296 | +0.12941 |
| 10 | +0.48296 | −0.12941 |
| 11 | +0.35355 | −0.35355 |
| 12 | +0.12941 | −0.48296 |

**13 Holes**

| Hole | $x$ | $y$ |
|---|---|---|
| 1 | −0.11966 | −0.48547 |
| 2 | −0.33156 | −0.37426 |
| 3 | −0.46751 | −0.17730 |
| 4 | −0.49635 | +0.06027 |
| 5 | −0.41149 | +0.28403 |
| 6 | −0.23236 | +0.44273 |
| 7 | 0.00000 | +0.50000 |
| 8 | +0.23236 | +0.44273 |
| 9 | +0.41149 | +0.28403 |
| 10 | +0.49635 | +0.06027 |
| 11 | +0.46751 | −0.17730 |
| 12 | +0.33156 | −0.37426 |
| 13 | +0.11966 | −0.48547 |

**14 Holes**

| Hole | $x$ | $y$ |
|---|---|---|
| 1 | −0.11126 | −0.48746 |
| 2 | −0.31174 | −0.39092 |
| 3 | −0.45048 | −0.21694 |
| 4 | −0.50000 | 0.00000 |
| 5 | −0.45048 | +0.21694 |
| 6 | −0.31174 | +0.39092 |
| 7 | −0.11126 | +0.48746 |
| 8 | +0.11126 | +0.48746 |
| 9 | +0.31174 | +0.39092 |
| 10 | +0.45048 | +0.21694 |
| 11 | +0.50000 | 0.00000 |
| 12 | +0.45048 | −0.21694 |
| 13 | +0.31174 | −0.39092 |
| 14 | +0.11126 | −0.48746 |

**15 Holes**

| Hole | $x$ | $y$ |
|---|---|---|
| 1 | −0.10396 | −0.48907 |
| 2 | −0.29389 | −0.40451 |
| 3 | −0.43301 | −0.25000 |
| 4 | −0.49726 | −0.05226 |
| 5 | −0.47553 | +0.15451 |
| 6 | −0.37157 | +0.33457 |
| 7 | −0.20337 | +0.45677 |
| 8 | 0.00000 | +0.50000 |
| 9 | +0.20337 | +0.45677 |
| 10 | +0.37157 | +0.33457 |
| 11 | +0.47553 | +0.15451 |
| 12 | +0.49726 | −0.05226 |
| 13 | +0.43301 | −0.25000 |
| 14 | +0.29389 | −0.40451 |
| 15 | +0.10396 | −0.48907 |

**16 Holes**

| Hole | $x$ | $y$ |
|---|---|---|
| 1 | −0.09755 | −0.49039 |
| 2 | −0.27779 | −0.41573 |
| 3 | −0.41573 | −0.27779 |
| 4 | −0.49039 | −0.09755 |
| 5 | −0.49039 | +0.09755 |
| 6 | −0.41573 | +0.27779 |
| 7 | −0.27779 | +0.41573 |
| 8 | −0.09755 | +0.49039 |
| 9 | +0.09755 | +0.49039 |
| 10 | +0.27779 | +0.41573 |
| 11 | +0.41573 | +0.27779 |
| 12 | +0.49039 | +0.09755 |
| 13 | +0.49039 | −0.09755 |
| 14 | +0.41573 | −0.27779 |
| 15 | +0.27779 | −0.41573 |
| 16 | +0.09755 | −0.49039 |

**17 Holes**

| Hole | $x$ | $y$ |
|---|---|---|
| 1 | −0.09187 | −0.49149 |
| 2 | −0.26322 | −0.42511 |
| 3 | −0.39901 | −0.30132 |
| 4 | −0.48091 | −0.13683 |
| 5 | −0.49787 | +0.04613 |
| 6 | −0.44758 | +0.22287 |
| 7 | −0.33685 | +0.36950 |
| 8 | −0.18062 | +0.46624 |
| 9 | 0.00000 | +0.50000 |
| 10 | +0.18062 | +0.46624 |
| 11 | +0.33685 | +0.36950 |
| 12 | +0.44758 | +0.22287 |
| 13 | +0.49787 | +0.04613 |
| 14 | +0.48091 | −0.13683 |
| 15 | +0.39901 | −0.30132 |
| 16 | +0.26322 | −0.42511 |
| 17 | +0.09187 | −0.49149 |

**18 Holes**

| Hole | $x$ | $y$ |
|---|---|---|
| 1 | −0.08682 | −0.49240 |
| 2 | −0.25000 | −0.43301 |
| 3 | −0.38302 | −0.32139 |
| 4 | −0.46985 | −0.17101 |
| 5 | −0.50000 | 0.00000 |
| 6 | −0.46985 | +0.17101 |
| 7 | −0.38302 | +0.32139 |
| 8 | −0.25000 | +0.43301 |
| 9 | −0.08682 | +0.49240 |
| 10 | +0.08682 | +0.49240 |
| 11 | +0.25000 | +0.43301 |
| 12 | +0.38302 | +0.32139 |
| 13 | +0.46985 | +0.17101 |
| 14 | +0.50000 | 0.00000 |
| 15 | +0.46985 | −0.17101 |
| 16 | +0.38302 | −0.32139 |
| 17 | +0.25000 | −0.43301 |
| 18 | +0.08682 | −0.49240 |

**19 Holes**

| Hole | $x$ | $y$ |
|---|---|---|
| 1 | −0.08230 | −0.49318 |
| 2 | −0.23797 | −0.43974 |
| 3 | −0.36786 | −0.33864 |
| 4 | −0.45789 | −0.20085 |
| 5 | −0.49829 | −0.04129 |
| 6 | −0.48470 | +0.12274 |
| 7 | −0.41858 | +0.27347 |
| 8 | −0.30709 | +0.39459 |
| 9 | −0.16235 | +0.47290 |
| 10 | 0.00000 | +0.50000 |
| 11 | +0.16235 | +0.47290 |
| 12 | +0.30709 | +0.39459 |
| 13 | +0.41858 | +0.27347 |
| 14 | +0.48470 | +0.12274 |
| 15 | +0.49829 | −0.04129 |
| 16 | +0.45789 | −0.20085 |
| 17 | +0.36786 | −0.33864 |
| 18 | +0.23797 | −0.43974 |
| 19 | +0.08230 | −0.49318 |

**Table 4** (*Concluded*). **Hole Coordinate Dimension Factors for Jig Boring—Type "B" Hole Circles, Central Coordinates** (English or metric units)

**19 Holes** (continued)

| Hole | x | y | Hole | x | y |
|---|---|---|---|---|---|
| x8 | −0.30711 | y8 +0.39457 | x14 | +0.48470 | y14 +0.12274 |
| x9 | −0.16235 | y9 +0.47291 | x15 | +0.49829 | y15 −0.04129 |
| x10 | 0.00000 | y10 +0.50000 | x16 | +0.45789 | y16 −0.20085 |
| x11 | +0.16235 | y11 +0.47291 | x17 | +0.36786 | y17 −0.33864 |
| x12 | +0.30711 | y12 +0.39457 | x18 | +0.23797 | y18 −0.43974 |
| x13 | +0.41858 | y13 +0.27347 | x19 | +0.08230 | y19 −0.49318 |

**20 Holes**

| Hole | x | y | Hole | x | y |
|---|---|---|---|---|---|
| x1 | −0.07822 | y1 −0.49384 | x11 | +0.07822 | y11 +0.49384 |
| x2 | −0.22700 | y2 −0.44550 | x12 | +0.22700 | y12 +0.44550 |
| x3 | −0.35355 | y3 −0.35355 | x13 | +0.35355 | y13 +0.35355 |
| x4 | −0.44550 | y4 −0.22700 | x14 | +0.44550 | y14 +0.22700 |
| x5 | −0.49384 | y5 −0.07822 | x15 | +0.49384 | y15 +0.07822 |
| x6 | −0.49384 | y6 +0.07822 | x16 | +0.49384 | y16 −0.07822 |
| x7 | −0.44550 | y7 +0.22700 | x17 | +0.44550 | y17 −0.22700 |
| x8 | −0.35355 | y8 +0.35355 | x18 | +0.35355 | y18 −0.35355 |
| x9 | −0.22700 | y9 +0.44550 | x19 | +0.22700 | y19 −0.44550 |
| x10 | −0.07822 | y10 +0.49384 | x20 | +0.07822 | y20 −0.49384 |

**21 Holes**

| Hole | x | y | Hole | x | y |
|---|---|---|---|---|---|
| x1 | −0.07452 | y1 −0.49442 | x12 | +0.14738 | y12 +0.47779 |
| x2 | −0.21694 | y2 −0.45048 | x13 | +0.28166 | y13 +0.41312 |
| x3 | −0.34009 | y3 −0.36653 | x14 | +0.39092 | y14 +0.31174 |
| x4 | −0.43301 | y4 −0.25000 | x15 | +0.46544 | y15 +0.18267 |
| x5 | −0.48746 | y5 −0.11126 | x16 | +0.49860 | y16 +0.03737 |
| x6 | −0.49860 | y6 +0.03737 | x17 | +0.48746 | y17 −0.11126 |
| x7 | −0.46544 | y7 +0.18267 | x18 | +0.43301 | y18 −0.25000 |
| x8 | −0.39092 | y8 +0.31174 | x19 | +0.34009 | y19 −0.36653 |
| x9 | −0.28166 | y9 +0.41312 | x20 | +0.21694 | y20 −0.45048 |
| x10 | −0.14738 | y10 +0.47779 | x21 | +0.07452 | y21 −0.49442 |
| x11 | 0.00000 | y11 +0.50000 | | | |

**22 Holes**

| Hole | x | y | Hole | x | y |
|---|---|---|---|---|---|
| x1 | −0.07116 | y1 −0.49491 | x12 | +0.07116 | y12 +0.49491 |
| x2 | −0.20771 | y2 −0.45482 | x13 | +0.20771 | y13 +0.45482 |
| x3 | −0.32743 | y3 −0.37787 | x14 | +0.32743 | y14 +0.37787 |
| x4 | −0.42063 | y4 −0.27032 | x15 | +0.42063 | y15 +0.27032 |
| x5 | −0.47975 | y5 −0.14087 | x16 | +0.47975 | y16 +0.14087 |
| x6 | −0.50000 | y6 0.00000 | x17 | +0.50000 | y17 0.00000 |
| x7 | −0.47975 | y7 +0.14087 | x18 | +0.47975 | y18 −0.14087 |
| x8 | −0.42063 | y8 +0.27032 | x19 | +0.42063 | y19 −0.27032 |
| x9 | −0.32743 | y9 +0.37787 | x20 | +0.32743 | y20 −0.37787 |
| x10 | −0.20771 | y10 +0.45482 | x21 | +0.20771 | y21 −0.45482 |
| x11 | −0.07116 | y11 +0.49491 | x22 | +0.07116 | y22 −0.49491 |

**23 Holes**

| Hole | x | y | Hole | x | y |
|---|---|---|---|---|---|
| x1 | −0.06808 | y1 −0.49534 | x13 | +0.13490 | y13 +0.48146 |
| x2 | −0.19920 | y2 −0.45861 | x14 | +0.25979 | y14 +0.42721 |
| x3 | −0.31554 | y3 −0.38786 | x15 | +0.36542 | y15 +0.34128 |
| x4 | −0.40848 | y4 −0.28834 | x16 | +0.44394 | y16 +0.23003 |
| x5 | −0.47113 | y5 −0.16744 | x17 | +0.48954 | y17 +0.10173 |
| x6 | −0.49883 | y6 −0.03412 | x18 | +0.49883 | y18 −0.03412 |
| x7 | −0.48954 | y7 +0.10173 | x19 | +0.47113 | y19 −0.16744 |
| x8 | −0.44394 | y8 +0.23003 | x20 | +0.40848 | y20 −0.28834 |
| x9 | −0.36542 | y9 +0.34128 | x21 | +0.31554 | y21 −0.38786 |
| x10 | −0.25979 | y10 +0.42721 | x22 | +0.19920 | y22 −0.45861 |
| x11 | −0.13490 | y11 +0.48146 | x23 | +0.06808 | y23 −0.49534 |
| x12 | 0.00000 | y12 +0.50000 | | | |

**24 Holes**

| Hole | x | y | Hole | x | y |
|---|---|---|---|---|---|
| x1 | −0.06526 | y1 −0.49572 | x13 | +0.06526 | y13 +0.49572 |
| x2 | −0.19134 | y2 −0.46194 | x14 | +0.19134 | y14 +0.46194 |
| x3 | −0.30438 | y3 −0.39668 | x15 | +0.30438 | y15 +0.39668 |
| x4 | −0.39668 | y4 −0.30438 | x16 | +0.39668 | y16 +0.30438 |
| x5 | −0.46194 | y5 −0.19134 | x17 | +0.46194 | y17 +0.19134 |
| x6 | −0.49572 | y6 −0.06526 | x18 | +0.49572 | y18 +0.06526 |
| x7 | −0.49572 | y7 +0.06526 | x19 | +0.49572 | y19 −0.06526 |
| x8 | −0.46194 | y8 +0.19134 | x20 | +0.46194 | y20 −0.19134 |
| x9 | −0.39668 | y9 +0.30438 | x21 | +0.39668 | y21 −0.30438 |
| x10 | −0.30438 | y10 +0.39668 | x22 | +0.30438 | y22 −0.39668 |
| x11 | −0.19134 | y11 +0.46194 | x23 | +0.19134 | y23 −0.46194 |
| x12 | −0.06526 | y12 +0.49572 | x24 | +0.06526 | y24 −0.49572 |

**25 Holes**

| Hole | x | y | Hole | x | y |
|---|---|---|---|---|---|
| x1 | −0.06267 | y1 −0.49606 | x14 | +0.12434 | y14 +0.48429 |
| x2 | −0.18406 | y2 −0.46489 | x15 | +0.24088 | y15 +0.43815 |
| x3 | −0.29389 | y3 −0.40451 | x16 | +0.34227 | y16 +0.36448 |
| x4 | −0.38526 | y4 −0.31871 | x17 | +0.42216 | y17 +0.26791 |
| x5 | −0.45241 | y5 −0.21289 | x18 | +0.47553 | y18 +0.15451 |
| x6 | −0.49114 | y6 −0.09369 | x19 | +0.49901 | y19 +0.03140 |
| x7 | −0.49901 | y7 +0.03140 | x20 | +0.49114 | y20 −0.09369 |
| x8 | −0.47553 | y8 +0.15451 | x21 | +0.45241 | y21 −0.21289 |
| x9 | −0.42216 | y9 +0.26791 | x22 | +0.38526 | y22 −0.31871 |
| x10 | −0.34227 | y10 +0.36448 | x23 | +0.29389 | y23 −0.40451 |
| x11 | −0.24088 | y11 +0.43815 | x24 | +0.18406 | y24 −0.46489 |
| x12 | −0.12434 | y12 +0.48429 | x25 | +0.06267 | y25 −0.49606 |
| x13 | 0.00000 | y13 +0.50000 | | | |

**26 Holes**

| Hole | x | y | Hole | x | y |
|---|---|---|---|---|---|
| x1 | −0.06027 | y1 −0.49635 | x14 | +0.06027 | y14 +0.49635 |
| x2 | −0.17730 | y2 −0.46751 | x15 | +0.17730 | y15 +0.46751 |
| x3 | −0.28403 | y3 −0.41149 | x16 | +0.28403 | y16 +0.41149 |
| x4 | −0.37426 | y4 −0.33156 | x17 | +0.37426 | y17 +0.33156 |
| x5 | −0.44273 | y5 −0.23236 | x18 | +0.44273 | y18 +0.23236 |
| x6 | −0.48547 | y6 −0.11966 | x19 | +0.48547 | y19 +0.11966 |
| x7 | −0.50000 | y7 0.00000 | x20 | +0.50000 | y20 0.00000 |
| x8 | −0.48547 | y8 +0.11966 | x21 | +0.48547 | y21 −0.11966 |
| x9 | −0.44273 | y9 +0.23236 | x22 | +0.44273 | y22 −0.23236 |
| x10 | −0.37426 | y10 +0.33156 | x23 | +0.37426 | y23 −0.33156 |
| x11 | −0.28403 | y11 +0.41149 | x24 | +0.28403 | y24 −0.41149 |
| x12 | −0.17730 | y12 +0.46751 | x25 | +0.17730 | y25 −0.46751 |
| x13 | −0.06027 | y13 +0.49635 | x26 | +0.06027 | y26 −0.49635 |

**27 Holes**

| Hole | x | y | Hole | x | y |
|---|---|---|---|---|---|
| x1 | −0.05805 | y1 −0.49662 | x15 | +0.11531 | y15 +0.48652 |
| x2 | −0.17101 | y2 −0.46985 | x16 | +0.22440 | y16 +0.44682 |
| x3 | −0.27475 | y3 −0.41774 | x17 | +0.32139 | y17 +0.38302 |
| x4 | −0.36369 | y4 −0.34312 | x18 | +0.40106 | y18 +0.29858 |
| x5 | −0.43301 | y5 −0.25000 | x19 | +0.45911 | y19 +0.19804 |
| x6 | −0.47899 | y6 −0.14340 | x20 | +0.49240 | y20 +0.08682 |
| x7 | −0.49915 | y7 −0.02907 | x21 | +0.49915 | y21 −0.02907 |
| x8 | −0.49240 | y8 +0.08682 | x22 | +0.47899 | y22 −0.14340 |
| x9 | −0.45911 | y9 +0.19804 | x23 | +0.43301 | y23 −0.25000 |
| x10 | −0.40106 | y10 +0.29858 | x24 | +0.36369 | y24 −0.34312 |
| x11 | −0.32139 | y11 +0.38302 | x25 | +0.27475 | y25 −0.41774 |
| x12 | −0.22440 | y12 +0.44682 | x26 | +0.17101 | y26 −0.46985 |
| x13 | −0.11531 | y13 +0.48652 | x27 | +0.05805 | y27 −0.49662 |
| x14 | 0.00000 | y14 +0.50000 | | | |

**28 Holes**

| Hole | x | y | Hole | x | y |
|---|---|---|---|---|---|
| x1 | −0.05598 | y1 −0.49686 | x15 | +0.05598 | y15 +0.49686 |
| x2 | −0.16514 | y2 −0.47194 | x16 | +0.16514 | y16 +0.47194 |
| x3 | −0.26602 | y3 −0.42336 | x17 | +0.26602 | y17 +0.42336 |
| x4 | −0.35355 | y4 −0.35355 | x18 | +0.35355 | y18 +0.35355 |
| x5 | −0.42336 | y5 −0.26602 | x19 | +0.42336 | y19 +0.26602 |
| x6 | −0.47194 | y6 −0.16514 | x20 | +0.47194 | y20 +0.16514 |
| x7 | −0.49686 | y7 −0.05598 | x21 | +0.49686 | y21 +0.05598 |
| x8 | −0.49686 | y8 +0.05598 | x22 | +0.49686 | y22 −0.05598 |
| x9 | −0.47194 | y9 +0.16514 | x23 | +0.47194 | y23 −0.16514 |
| x10 | −0.42336 | y10 +0.26602 | x24 | +0.42336 | y24 −0.26602 |
| x11 | −0.35355 | y11 +0.35355 | x25 | +0.35355 | y25 −0.35355 |
| x12 | −0.26602 | y12 +0.42336 | x26 | +0.26602 | y26 −0.42336 |
| x13 | −0.16514 | y13 +0.47194 | x27 | +0.16514 | y27 −0.47194 |
| x14 | −0.05598 | y14 +0.49686 | x28 | +0.05598 | y28 −0.49686 |

**Diameter of Circle Enclosing a Given Number of Smaller Circles.** — Four of many possible compact arrangements of circles within a circle are shown at A, B, C, and D in Fig. 1. To determine the diameter of the smallest enclosing circle for a particular number of enclosing circles all of the same size, three factors that influence the size of the enclosing circle should be considered. These are discussed in the paragraphs that follow which are based on the article, *How Many Wires Can Be Packed into a Circular Conduit*, by Jacques Dutka, Machinery, October 1956.

*1. Arrangement of Center or Core Circles:* The four most common arrangements of center or core circles are shown cross-sectioned in Fig. 1. It may seem, offhand, that the " A " pattern would require the smallest enclosing circle for a given number of enclosed circles but this is not always the case since the most compact arrangement will, in part, depend on the number of circles to be enclosed.

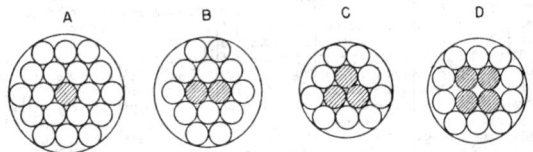

Fig. 1. Arrangements of Circles Within a Circle

*2. Diameter of Enclosing Circle When Outer Layer of Circles is Complete:* Successive, complete " layers " of circles may be placed around each of the central cores, Fig. 1, of 1, 2, 3, or 4 circles as the case may be. The number of circles contained in arrangements of complete " layers " around a central core of circles, as well as the diameter of the enclosing circle, may be obtained using the data in Table 1. Thus, for example, the " A " pattern in Fig. 1 shows, by actual count, a total of 19 circles arranged in two complete " layers " around a central core consisting of one circle; this agrees with the data shown in the left half of Table 1 for $n = 2$.

To determine the diameter of the enclosing circle, the data in the right half of Table 1 is used. Thus, for $n = 2$ and an " A " pattern, the diameter $D$ is 5 times the diameter $d$ of the enclosed circles.

*3. Diameter of Enclosing Circle When Outer Layer of Circles is not Complete:* In most cases it is possible to reduce the size of the enclosing circle from that required if the outer layer were complete. Thus, for example, the " B " pattern in Fig. 1 shows that the central core consisting of 2 circles is surrounded by 1 complete layer of 8 circles and 1 partial, outer layer of 4 circles so that the total number of circles enclosed is 14. If the outer layer was complete then (from Table 1) the total number of enclosed circles would be 24 and the diameter of the enclosing circle would be 6$d$; however, since the outer layer is composed of only 4 circles out of a possible 14 for a complete second layer, a smaller diameter of enclosing circle may be used. Table 2 shows that for a total of 14 enclosed circles arranged in a " B " pattern with the outer layer of circles incomplete, the diameter for the enclosing circle is 4.606$d$.

Table 2 can be used to determine the smallest enclosing circle for a given number of circles to be enclosed by direct comparison of the " A," " B," and " C " columns. For data outside the range of Table 2, use the formulas in Dr. Dutka's article.

*Approximate Formula When Number of Enclosed Circles is Large:* When a large number of circles are to be enclosed, the arrangement of the center circles has little effect on the diameter of enclosing circle. For numbers of circles greater than 10,000 the diameter of the enclosing circle may be calculated within 2 per cent from the formula: $D = d(1 + \sqrt{N \div 0.907})$. In this formula, $D$ = diameter of enclosing circle; $d$ = diameter of enclosed circles; and $N$ is the number of enclosed circles.

Table 1. Number of Circles Contained in Complete Layers of Circles and Diameter of Enclosing Circle (English or metric units)

| No. Complete Layers Over Core, n | Number of Circles in Center Pattern | | | | | | | |
|---|---|---|---|---|---|---|---|---|
| | 1 | 2 | 3 | 4 | 1 | 2 | 3 | 4 |
| | Arrangement of Circles in Center Pattern (see Fig. 1) | | | | | | | |
| | "A" | "B" | "C" | "D" | "A" | "B" | "C" | "D" |
| | Number of Circles, N, Enclosed | | | | Diameter, D, of Enclosing Circle* | | | |
| 0 | 1 | 2 | 3 | 4 | $d$ | $2d$ | $2.155d$ | $2.414d$ |
| 1 | 7 | 10 | 12 | 14 | $3d$ | $4d$ | $4.055d$ | $4.386d$ |
| 2 | 19 | 24 | 27 | 30 | $5d$ | $6d$ | $6.033d$ | $6.379d$ |
| 3 | 37 | 44 | 48 | 52 | $7d$ | $8d$ | $8.024d$ | $8.375d$ |
| 4 | 61 | 70 | 75 | 80 | $9d$ | $10d$ | $10.018d$ | $10.373d$ |
| 5 | 91 | 102 | 108 | 114 | $11d$ | $12d$ | $12.015d$ | $12.372d$ |
| n | ** | ** | ** | ** | ** | ** | ** | ** |

* Diameter $D$ is given in terms of $d$, the diameter of the enclosed circles.

** For $n$ complete layers over core, the number of enclosed circles $N$ for "A" center pattern is $3n^2 + 3n + 1$; for "B," $3n^2 + 5n + 2$; for "C," $3n^2 + 6n + 3$; for "D," $3n^2 + 7n + 4$; while the diameter $D$ of the enclosing circle for "A" center pattern is $(2n + 1)d$; for "B," $(2n + 2)d$; for "C," $(1 + 2\sqrt{n^2 + n + \frac{1}{3}})d$; and for "D," $(1 + \sqrt{4n^2 + 5.644n + 2})d$.

Table 2. Factors for Determining Diameter, D, of Smallest Enclosing Circle for Various Numbers, N, of Enclosed Circles* (English or metric units)

| No. N | Center Circle Pattern | | | No. N | Center Circle Pattern | | | No. N | Center Circle Pattern | | |
|---|---|---|---|---|---|---|---|---|---|---|---|
| | "A" | "B" | "C" | | "A" | "B" | "C" | | "A" | "B" | "C" |
| | Diameter Factor K | | | | Diameter Factor K | | | | Diameter Factor K | | |
| 2 | 3 | 2 | --- | 34 | 7 | 7.083 | 7.110 | 66 | 9.718 | 9.544 | 9.326 |
| 3 | 3 | 2.732 | 2.155 | 35 | 7 | 7.245 | 7.110 | 67 | 9.718 | 9.544 | 9.326 |
| 4 | 3 | 2.732 | 3.309 | 36 | 7 | 7.245 | 7.110 | 68 | 9.718 | 9.544 | 9.326 |
| 5 | 3 | 3.646 | 3.309 | 37 | 7 | 7.245 | 7.429 | 69 | 9.718 | 9.660 | 9.326 |
| 6 | 3 | 3.646 | 3.309 | 38 | 7.928 | 7.245 | 7.429 | 70 | 9.718 | 9.660 | 10.018 |
| 7 | 3 | 3.646 | 4.055 | 39 | 7.928 | 7.557 | 7.429 | 71 | 9.718 | 9.888 | 10.018 |
| 8 | 4.464 | 3.646 | 4.055 | 40 | 7.928 | 7.557 | 7.429 | 72 | 9.718 | 9.888 | 10.018 |
| 9 | 4.464 | 4 | 4.055 | 41 | 7.928 | 7.557 | 7.429 | 73 | 9.718 | 9.888 | 10.018 |
| 10 | 4.464 | 4 | 4.055 | 42 | 7.928 | 7.557 | 7.429 | 74 | 10.165 | 9.888 | 10.018 |
| 11 | 4.464 | 4.606 | 4.055 | 43 | 7.928 | 8 | 8.024 | 75 | 10.165 | 10 | 10.018 |
| 12 | 4.464 | 4.606 | 4.055 | 44 | 8.211 | 8 | 8.024 | 76 | 10.165 | 10 | 10.238 |
| 13 | 4.464 | 4.606 | 5.163 | 45 | 8.211 | 8 | 8.024 | 77 | 10.165 | 10.539 | 10.238 |
| 14 | 5 | 4.606 | 5.163 | 46 | 8.211 | 8 | 8.024 | 78 | 10.165 | 10.539 | 10.238 |
| 15 | 5 | 5.359 | 5.163 | 47 | 8.211 | 8 | 8.024 | 79 | 10.165 | 10.539 | 10.452 |
| 16 | 5 | 5.359 | 5.163 | 48 | 8.211 | 8 | 8.024 | 80 | 10.165 | 10.539 | 10.452 |
| 17 | 5 | 5.359 | 5.163 | 49 | 8.211 | 8.550 | 8 572 | 81 | 10.165 | 10.539 | 10.452 |
| 18 | 5 | 5.359 | 5.163 | 50 | 8.211 | 8.550 | 8.572 | 82 | 10.165 | 10.539 | 10.452 |
| 19 | 5 | 5.583 | 5.619 | 51 | 8.211 | 8.550 | 8.572 | 83 | 10 165 | 10.539 | 10.452 |
| 20 | 6.292 | 5.583 | 5.619 | 52 | 8.211 | 8.550 | 8.572 | 84 | 10.165 | 10 539 | 10.452 |
| 21 | 6.292 | 5.583 | 5.619 | 53 | 8.211 | 8.810 | 8.572 | 85 | 10.165 | 10.644 | 10.866 |
| 22 | 6.292 | 5.583 | 6.033 | 54 | 8.211 | 8.810 | 8.572 | 86 | 11 | 10.644 | 10.866 |
| 23 | 6.292 | 6 | 6.033 | 55 | 8.211 | 8.810 | 9.083 | 87 | 11 | 10.644 | 10.866 |
| 24 | 6.292 | 6 | 6.033 | 56 | 9 | 8.810 | 9 083 | 88 | 11 | 10.644 | 10.866 |
| 25 | 6.292 | 6.196 | 6.033 | 57 | 9 | 8.937 | 9.083 | 89 | 11 | 10.849 | 10.866 |
| 26 | 6.292 | 6.196 | 6.033 | 58 | 9 | 8.937 | 9.083 | 90 | 11 | 10.849 | 10.866 |
| 27 | 6.292 | 6.568 | 6.033 | 59 | 9 | 8.937 | 9.083 | 91 | 11 | 10.849 | 11.214 |
| 28 | 6.292 | 6.568 | 6.773 | 60 | 9 | 8.937 | 9.083 | 92 | 11.392 | 10.849 | 11.214 |
| 29 | 6.292 | 6.568 | 6.773 | 61 | 9 | 9.185 | 9.083 | 93 | 11.392 | 11 149 | 11.214 |
| 30 | 6.292 | 6.568 | 6.773 | 62 | 9 718 | 9.185 | 9.083 | 94 | 11.392 | 11 149 | 11.214 |
| 31 | 6.292 | 7 083 | 7.110 | 63 | 9.718 | 9.185 | 9.083 | 95 | 11.392 | 11.149 | 11.214 |
| 32 | 7 | 7 083 | 7 110 | 64 | 9.718 | 9.185 | 9.326 | 96 | 11.392 | 11.149 | 11.214 |
| 33 | 7 | 7.083 | 7.110 | 65 | 9.718 | 9.544 | 9.326 | 97 | 11.392 | 11.440 | 11.214 |

* The diameter $D$ of the enclosing circle is equal to the diameter factor, $K$, multiplied by $d$, the diameter of the enclosed circles or $D = K \times d$. For example, if the number of circles to be enclosed, $N$, is 12, and the center circle arrangement is "C," then for $d = 1\frac{1}{2}$ inches, $D = 4.055 \times 1\frac{1}{2} = 6.083$ inches. If $d = 50$ millimetres, then $D = 4.055 \times 50 = 202.750$ millimetres.

## Table of Commonly Used Constants

| Constant | Numerical Value | Logarithm | Constant | Numerical Value | Logarithm |
|---|---|---|---|---|---|
| $\pi$ | 3.141593 | 0.49715 | Weight in pounds of: | | |
| $2\pi$ | 6.283185 | 0.79818 | Water column, $1''\times1''\times1$ ft. | 0.4335 | $\bar{1}$.63699 |
| $\pi\div4$ | 0.785398 | $\bar{1}$.89509 | 1 U.S. gallon of water, 39.1° F. | 8.34 | 0.92117 |
| $\pi^2$ | 9.869604 | 0.99430 | 1 cu. ft. of water, 39.1° F... | 62.4245 | 1.79536 |
| $\pi^3$ | 31.006277 | 1.49145 | 1 cu. in. of water, 39.1° F... | 0.0361 | $\bar{2}$.55751 |
| $1\div\pi$ | 0.318310 | $\bar{1}$.50285 | 1 cu. ft. of air, 32° F., atmospheric pressure .......... | 0.08073 | $\bar{2}$.90703 |
| $1\div\pi^2$ | 0.101321 | $\bar{1}$.00570 | Volume in cu. ft. of: | | |
| $1\div\pi^3$ | 0.032252 | $\bar{2}$.50855 | 1 pound of water, 39.1° F... | 0.01602 | $\bar{2}$.20466 |
| $\sqrt{\pi}$ | 1.772454 | 0.24858 | 1 pound of air, 32° F., atmospheric pressure .......... | 12.387 | 1.09297 |
| $\sqrt[3]{\pi}$ | 1.464592 | 0.16572 | Volume in gallons of 1 pound | | |
| $g$ | 32.16 | 1.50732 | of water, 39.1° F......... | 0.1199 | $\bar{1}$.07882 |
| $g^2$ | 1034.266 | 3.01463 | Volume in cu. in. of 1 pound of | | |
| $2g$ | 64.32 | 1.80835 | water, 39.1° F. .......... | 27.70 | 1.44248 |
| $1\div2g$ | 0.01555 | $\bar{2}$.19165 | One cubic ft. in gallons ...... | 7.4805 | 0.87393 |
| $\sqrt{2g}$ | 8.01998 | 0.90417 | Atmospheric pressure in | | |
| $1\div\sqrt{g}$ | 0.17634 | $\bar{1}$.24635 | pounds per sq. in........ | 14.696 | 1.16720 |
| $\pi\div\sqrt{g}$ | 0.55398 | $\bar{1}$.74350 | | | |
| $e$ | 2.71828 | 0.43429 | | | |

## Useful Constants Multiplied and Divided by 1 to 10

| Constant | Multiplied by: | | | | | | | |
|---|---|---|---|---|---|---|---|---|
| | 2 | 3 | 4 | 5 | 6 | 7 | 8 | 9 |
| 0.7854 | 1.5708 | 2.3562 | 3.1416 | 3.9270 | 4.7124 | 5.4978 | 6.2832 | 7.0686 |
| 3.1416 | 6.2832 | 9.4248 | 12.566 | 15.708 | 18.850 | 21.991 | 25.133 | 28.274 |
| 14.7 | 29.4 | 44.1 | 58.8 | 73.5 | 88.2 | 102.9 | 117.6 | 132.3 |
| 32.16 | 64.32 | 96.48 | 128.64 | 160.80 | 192.96 | 225.12 | 257.28 | 289.44 |
| 64.32 | 128.64 | 192.96 | 257.28 | 321.60 | 385.92 | 450.24 | 514.56 | 578.88 |
| 144 | 288 | 432 | 576 | 720 | 864 | 1,008 | 1,152 | 1,296 |
| 778 | 1,556 | 2,334 | 3,112 | 3,890 | 4,668 | 5,446 | 6,224 | 7,002 |
| 1,728 | 3,456 | 5,184 | 6,912 | 8,640 | 10,368 | 12,096 | 13,824 | 15,552 |
| 33,000 | 66,000 | 99,000 | 132,000 | 165,000 | 198,000 | 231,000 | 264,000 | 297,000 |

| Constant | Divided by: | | | | | | | |
|---|---|---|---|---|---|---|---|---|
| | 2 | 3 | 4 | 5 | 6 | 7 | 8 | 9 |
| 0.7854 | 0.3927 | 0.2618 | 0.1964 | 0.1571 | 0.1309 | 0.1122 | 0.0982 | 0.0873 |
| 3.1416 | 1.5708 | 1.0472 | 0.7854 | 0.6283 | 0.5236 | 0.4488 | 0.3927 | 0.3491 |
| 14.7 | 7.350 | 4.900 | 3.675 | 2.940 | 2.450 | 2.100 | 1.838 | 1.633 |
| 32.16 | 16.080 | 10.720 | 8.040 | 6.432 | 5.360 | 4.594 | 4.020 | 3.573 |
| 64.32 | 32.160 | 21.440 | 16.080 | 12.864 | 10.720 | 9.189 | 8.040 | 7.147 |
| 144 | 72 | 48 | 36 | 28.800 | 24 | 20.571 | 18 | 16 |
| 778 | 389 | 259.33 | 194.50 | 155.60 | 129.67 | 111.14 | 97.25 | 86.44 |
| 1,728 | 864 | 576 | 432 | 345.60 | 288 | 246.86 | 216 | 192 |
| 33,000 | 16,500 | 11,000 | 8250 | 6600 | 5500 | 4714.3 | 4125 | 3666.7 |

## Surface and Volume of Spheres*

$d$ = diameter.  Surface = $\pi d^2$.  Volume = $\pi d^3 \div 6$.

| Diam. | Surface | Volume | Diam. | Surface | Volume | Diam. | Surface | Volume |
|---|---|---|---|---|---|---|---|---|
| 1/64 | 0.00077 | 0.000002 | 2 | 12.566 | 4.1888 | 6 1/2 | 132.73 | 143.79 |
| 1/32 | 0.00307 | 0.00002 | 2 1/16 | 13.364 | 4.5939 | 6 5/8 | 137.89 | 152.25 |
| 1/16 | 0.01227 | 0.00013 | 2 1/8 | 14.186 | 5.0243 | 6 3/4 | 143.14 | 161.03 |
| 3/32 | 0.02761 | 0.00043 | 2 3/16 | 15.033 | 5.4808 | 6 7/8 | 148.49 | 170.14 |
| 1/8 | 0.04909 | 0.00102 | 2 1/4 | 15.904 | 5.9641 | 7 | 153.94 | 179.59 |
| 5/32 | 0.07670 | 0.00200 | 2 5/16 | 16.800 | 6.4751 | 7 1/8 | 159.48 | 189.39 |
| 3/16 | 0.11045 | 0.00345 | 2 3/8 | 17.721 | 7.0144 | 7 1/4 | 165.13 | 199.53 |
| 7/32 | 0.15033 | 0.00548 | 2 7/16 | 18.665 | 7.5829 | 7 3/8 | 170.87 | 210.03 |
| 1/4 | 0.19635 | 0.00818 | 2 1/2 | 19.635 | 8.1812 | 7 1/2 | 176.71 | 220.89 |
| 9/32 | 0.24850 | 0.01165 | 2 9/16 | 20.629 | 8.8103 | 7 5/8 | 182.65 | 232.12 |
| 5/16 | 0.30680 | 0.01598 | 2 5/8 | 21.648 | 9.4708 | 7 3/4 | 188.69 | 243.73 |
| 11/32 | 0.37122 | 0.02127 | 2 11/16 | 22.691 | 10.164 | 7 7/8 | 194.83 | 255.71 |
| 3/8 | 0.44179 | 0.02761 | 2 3/4 | 23.758 | 10.889 | 8 | 201.06 | 268.08 |
| 13/32 | 0.51849 | 0.03511 | 2 13/16 | 24.850 | 11.649 | 8 1/8 | 207.39 | 280.85 |
| 7/16 | 0.60132 | 0.04385 | 2 7/8 | 25.967 | 12.443 | 8 1/4 | 213.82 | 294.01 |
| 15/32 | 0.69029 | 0.05393 | 2 15/16 | 27.109 | 13.272 | 8 3/8 | 220.35 | 307.58 |
| 1/2 | 0.78540 | 0.06545 | 3 | 28.274 | 14.137 | 8 1/2 | 226.98 | 321.56 |
| 17/32 | 0.88664 | 0.07850 | 3 1/16 | 29.465 | 15.039 | 8 5/8 | 233.71 | 335.95 |
| 9/16 | 0.99402 | 0.09319 | 3 1/8 | 30.680 | 15.979 | 8 3/4 | 240.53 | 350.77 |
| 19/32 | 1.1075 | 0.10960 | 3 3/16 | 31.919 | 16.957 | 8 7/8 | 247.45 | 366.02 |
| 5/8 | 1.2272 | 0.12783 | 3 1/4 | 33.183 | 17.974 | 9 | 254.47 | 381.70 |
| 21/32 | 1.3530 | 0.14798 | 3 5/16 | 34.472 | 19.031 | 9 1/8 | 261.59 | 397.83 |
| 11/16 | 1.4849 | 0.17014 | 3 3/8 | 35.785 | 20.129 | 9 1/4 | 268.80 | 414.40 |
| 23/32 | 1.6230 | 0.19442 | 3 7/16 | 37.122 | 21.268 | 9 3/8 | 276.12 | 431.43 |
| 3/4 | 1.7671 | 0.22089 | 3 1/2 | 38.485 | 22.449 | 9 1/2 | 283.53 | 448.92 |
| 25/32 | 1.9175 | 0.24967 | 3 5/8 | 41.282 | 24.942 | 9 5/8 | 291.04 | 466.88 |
| 13/16 | 2.0739 | 0.28085 | 3 3/4 | 44.179 | 27.612 | 9 3/4 | 298.65 | 485.30 |
| 27/32 | 2.2365 | 0.31451 | 3 7/8 | 47.173 | 30.466 | 9 7/8 | 306.35 | 504.21 |
| 7/8 | 2.4053 | 0.35077 | 4 | 50.265 | 33.510 | 10 | 314.16 | 523.60 |
| 29/32 | 2.5802 | 0.38971 | 4 1/8 | 53.456 | 36.751 | 10 1/8 | 330.06 | 563.86 |
| 15/16 | 2.7612 | 0.43143 | 4 1/4 | 56.745 | 40.194 | 10 1/2 | 346.36 | 606.13 |
| 31/32 | 2.9483 | 0.47603 | 4 3/8 | 60.132 | 43.846 | 10 3/4 | 363.05 | 650.47 |
| 1 | 3.1416 | 0.52360 | 4 1/2 | 63.617 | 47.713 | 11 | 380.13 | 696.91 |
| 1 1/16 | 3.5466 | 0.62804 | 4 5/8 | 67.201 | 51.801 | 11 1/4 | 397.61 | 745.51 |
| 1 1/8 | 3.9761 | 0.74551 | 4 3/4 | 70.882 | 56.115 | 11 1/2 | 415.48 | 796.33 |
| 1 3/16 | 4.4301 | 0.87680 | 4 7/8 | 74.662 | 60.663 | 11 3/4 | 433.74 | 849.40 |
| 1 1/4 | 4.9087 | 1.0227 | 5 | 78.540 | 65.450 | 12 | 452.39 | 904.78 |
| 1 5/16 | 5.4119 | 1.1838 | 5 1/8 | 82.516 | 70.482 | 12 1/4 | 471.44 | 962.51 |
| 1 3/8 | 5.9396 | 1.3612 | 5 1/4 | 86.590 | 75.766 | 12 1/2 | 490.87 | 1022.7 |
| 1 7/16 | 6.4918 | 1.5553 | 5 3/8 | 90.763 | 81.308 | 12 3/4 | 510.71 | 1085.2 |
| 1 1/2 | 7.0686 | 1.7671 | 5 1/2 | 95.033 | 87.114 | 13 | 530.93 | 1150.3 |
| 1 9/16 | 7.6699 | 1.9974 | 5 5/8 | 99.402 | 93.189 | 13 1/4 | 551.55 | 1218.0 |
| 1 5/8 | 8.2958 | 2.2468 | 5 3/4 | 103.87 | 99.541 | 13 1/2 | 572.56 | 1288.2 |
| 1 11/16 | 8.9462 | 2.5161 | 5 7/8 | 108.43 | 106.17 | 13 3/4 | 593.96 | 1361.2 |
| 1 3/4 | 9.6211 | 2.8062 | 6 | 113.10 | 113.10 | 14 | 615.75 | 1436.8 |
| 1 13/16 | 10.321 | 3.1177 | 6 1/8 | 117.86 | 120.31 | 14 1/4 | 637.94 | 1515.1 |
| 1 7/8 | 11.045 | 3.4515 | 6 1/4 | 122.72 | 127.83 | 14 1/2 | 660.52 | 1596.3 |
| 1 15/16 | 11.793 | 3.8082 | 6 3/8 | 127.68 | 135.66 | 14 3/4 | 683.49 | 1680.3 |

* All the figures given in the tables on pages 89 through 91 can be used for English units, and those without common fractions can be used for metric units.

### Surface and Volume of Spheres

| Diam. | Surface | Volume | Diam. | Surface | Volume | Diam. | Surface | Volume |
|---|---|---|---|---|---|---|---|---|
| 15 | 706.86 | 1,767.1 | 27½ | 2375.8 | 10,889 | 51 | 8,171.3 | 69,456 |
| 15¼ | 730.62 | 1,857.0 | 27¾ | 2419.2 | 11,189 | 51½ | 8,332.3 | 71,519 |
| 15½ | 754.77 | 1,949.8 | 28 | 2463.0 | 11,494 | 52 | 8,494.9 | 73,622 |
| 15¾ | 779.31 | 2,045.7 | 28¼ | 2507.2 | 11,805 | 52½ | 8,659.0 | 75,766 |
| 16 | 804.25 | 2,144.7 | 28½ | 2551.8 | 12,121 | 53 | 8,824.7 | 77,952 |
| 16¼ | 829.58 | 2,246.8 | 28¾ | 2596.7 | 12,443 | 53½ | 8,992.0 | 80,179 |
| 16½ | 855.30 | 2,352.1 | 29 | 2642.1 | 12,770 | 54 | 9,160.9 | 82,448 |
| 16¾ | 881.41 | 2,460.6 | 29½ | 2734.0 | 13,442 | 54½ | 9,331.3 | 84,759 |
| 17 | 907.92 | 2,572.4 | 30 | 2827.4 | 14,137 | 55 | 9,503.3 | 87,114 |
| 17¼ | 934.82 | 2,687.6 | 30½ | 2922.5 | 14,856 | 55½ | 9,676.9 | 89,511 |
| 17½ | 962.11 | 2,806.2 | 31 | 3019.1 | 15,599 | 56 | 9,852.0 | 91,952 |
| 17¾ | 989.80 | 2,928.2 | 31½ | 3117.2 | 16,366 | 56½ | 10,029 | 94,437 |
| 18 | 1017.9 | 3,053.6 | 32 | 3217.0 | 17,157 | 57 | 10,207 | 96,967 |
| 18¼ | 1046.3 | 3,182.6 | 32½ | 3318.3 | 17,974 | 57½ | 10,387 | 99,541 |
| 18½ | 1075.2 | 3,315.2 | 33 | 3421.2 | 18,817 | 58 | 10,568 | 102,160 |
| 18¾ | 1104.5 | 3,451.5 | 33½ | 3525.7 | 19,685 | 58½ | 10,751 | 104,825 |
| 19 | 1134.1 | 3,591.4 | 34 | 3631.7 | 20,580 | 59 | 10,936 | 107,536 |
| 19¼ | 1164.2 | 3,735.0 | 34½ | 3739.3 | 21,501 | 59½ | 11,122 | 110,293 |
| 19½ | 1194.6 | 3,882.4 | 35 | 3848.5 | 22,449 | 60 | 11,310 | 113,097 |
| 19¾ | 1225.4 | 4,033.7 | 35½ | 3959.2 | 23,425 | 60½ | 11,499 | 115,948 |
| 20 | 1256.6 | 4,188.8 | 36 | 4071.5 | 24,429 | 61 | 11,690 | 118,847 |
| 20¼ | 1288.2 | 4,347.8 | 36½ | 4185.4 | 25,461 | 61½ | 11,882 | 121,793 |
| 20½ | 1320.3 | 4,510.9 | 37 | 4300.8 | 26,522 | 62 | 12,076 | 124,788 |
| 20¾ | 1352.7 | 4,677.9 | 37½ | 4417.9 | 27,612 | 62½ | 12,272 | 127,832 |
| 21 | 1385.5 | 4,849.0 | 38 | 4536.5 | 28,731 | 63 | 12,469 | 130,924 |
| 21¼ | 1418.6 | 5,024.3 | 38½ | 4656.6 | 29,880 | 63½ | 12,668 | 134,066 |
| 21½ | 1452.2 | 5,203.7 | 39 | 4778.4 | 31,059 | 64 | 12,868 | 137,258 |
| 21¾ | 1486.2 | 5,387.4 | 39½ | 4901.7 | 32,269 | 64½ | 13,070 | 140,500 |
| 22 | 1520.5 | 5,575.3 | 40 | 5026.5 | 33,510 | 65 | 13,273 | 143,793 |
| 22¼ | 1555.3 | 5,767.5 | 40½ | 5153.0 | 34,783 | 65½ | 13,478 | 147,137 |
| 22½ | 1590.4 | 5,964.1 | 41 | 5281.0 | 36,087 | 66 | 13,685 | 150,533 |
| 22¾ | 1626.0 | 6,165.1 | 41½ | 5410.6 | 37,423 | 66½ | 13,893 | 153,980 |
| 23 | 1661.9 | 6,370.6 | 42 | 5541.8 | 38,792 | 67 | 14,103 | 157,479 |
| 23¼ | 1698.2 | 6,580.6 | 42½ | 5674.5 | 40,194 | 67½ | 14,314 | 161,031 |
| 23½ | 1734.9 | 6,795.2 | 43 | 5808.8 | 41,630 | 68 | 14,527 | 164,636 |
| 23¾ | 1772.1 | 7,014.4 | 43½ | 5944.7 | 43,099 | 68½ | 14,741 | 168,295 |
| 24 | 1809.6 | 7,238.2 | 44 | 6082.1 | 44,602 | 69 | 14,957 | 172,007 |
| 24¼ | 1847.5 | 7,466.8 | 44½ | 6221.1 | 46,140 | 69½ | 15,175 | 175,773 |
| 24½ | 1885.7 | 7,700.1 | 45 | 6361.7 | 47,713 | 70 | 15,394 | 179,594 |
| 24¾ | 1924.4 | 7,938.2 | 45½ | 6503.9 | 49,321 | 70½ | 15,615 | 183,470 |
| 25 | 1963.5 | 8,181.2 | 46 | 6647.6 | 50,965 | 71 | 15,837 | 187,402 |
| 25¼ | 2003.0 | 8,429.1 | 46½ | 6792.9 | 52,645 | 71½ | 16,061 | 191,389 |
| 25½ | 2042.8 | 8,682.0 | 47 | 6939.8 | 54,362 | 72 | 16,286 | 195,432 |
| 25¾ | 2083.1 | 8,939.9 | 47½ | 7088.2 | 56,115 | 72½ | 16,513 | 199,532 |
| 26 | 2123.7 | 9,202.8 | 48 | 7238.2 | 57,906 | 73 | 16,742 | 203,689 |
| 26¼ | 2164.8 | 9,470.8 | 48½ | 7389.8 | 59,734 | 73½ | 16,972 | 207,903 |
| 26½ | 2206.2 | 9,744.0 | 49 | 7543.0 | 61,601 | 74 | 17,203 | 212,175 |
| 26¾ | 2248.0 | 10,022 | 49½ | 7697.7 | 63,506 | 74½ | 17,437 | 216,505 |
| 27 | 2290.2 | 10,306 | 50 | 7854.0 | 65,450 | 75 | 17,671 | 220,893 |
| 27¼ | 2332.8 | 10,595 | 50½ | 8011.8 | 67,433 | 75½ | 17,908 | 225,341 |

## Surface and Volume of Spheres

| Diam. | Surface | Volume | Diam. | Surface | Volume | Diam. | Surface | Volume |
|---|---|---|---|---|---|---|---|---|
| 76 | 18,146 | 229,847 | 101 | 32,047 | 539,464 | 151 | 71,631 | 1,802,725 |
| 76½ | 18,385 | 234,414 | 102 | 32,685 | 555,647 | 152 | 72,583 | 1,838,778 |
| 77 | 18,627 | 239,040 | 103 | 33,329 | 572,151 | 153 | 73,542 | 1,875,309 |
| 77½ | 18,869 | 243,727 | 104 | 33,979 | 588,977 | 154 | 74,506 | 1,912,321 |
| 78 | 19,113 | 248,475 | 105 | 34,636 | 606,131 | 155 | 75,477 | 1,949,816 |
| 78½ | 19,359 | 253,284 | 106 | 35,299 | 623,615 | 156 | 76,454 | 1,987,799 |
| 79 | 19,607 | 258,155 | 107 | 35,968 | 641,431 | 157 | 77,437 | 2,026,271 |
| 79½ | 19,856 | 263,087 | 108 | 36,644 | 659,584 | 158 | 78,427 | 2,065,237 |
| 80 | 20,106 | 268,083 | 109 | 37,325 | 678,076 | 159 | 79,423 | 2,104,699 |
| 80½ | 20,358 | 273,141 | 110 | 38,013 | 696,910 | 160 | 80,425 | 2,144,661 |
| 81 | 20,612 | 278,262 | 111 | 38,708 | 716,090 | 161 | 81,433 | 2,185,125 |
| 81½ | 20,867 | 283,447 | 112 | 39,408 | 735,619 | 162 | 82,448 | 2,226,095 |
| 82 | 21,124 | 288,696 | 113 | 40,115 | 755,499 | 163 | 83,469 | 2,267,574 |
| 82½ | 21,382 | 294,009 | 114 | 40,828 | 775,735 | 164 | 84,496 | 2,309,565 |
| 83 | 21,642 | 299,387 | 115 | 41,548 | 796,328 | 165 | 85,530 | 2,352,071 |
| 83½ | 21,904 | 304,830 | 116 | 42,273 | 817,283 | 166 | 86,570 | 2,395,096 |
| 84 | 22,167 | 310,339 | 117 | 43,005 | 838,603 | 167 | 87,616 | 2,438,642 |
| 84½ | 22,432 | 315,914 | 118 | 43,744 | 860,290 | 168 | 88,668 | 2,482,713 |
| 85 | 22,698 | 321,555 | 119 | 44,488 | 882,347 | 169 | 89,727 | 2,527,311 |
| 85½ | 22,966 | 327,263 | 120 | 45,239 | 904,779 | 170 | 90,792 | 2,572,441 |
| 86 | 23,235 | 333,038 | 121 | 45,996 | 927,587 | 171 | 91,863 | 2,618,104 |
| 86½ | 23,506 | 338,881 | 122 | 46,759 | 950,776 | 172 | 92,941 | 2,664,305 |
| 87 | 23,779 | 344,791 | 123 | 47,529 | 974,348 | 173 | 94,025 | 2,711,046 |
| 87½ | 24,053 | 350,770 | 124 | 48,305 | 998,306 | 174 | 95,115 | 2,758,331 |
| 88 | 24,328 | 356,818 | 125 | 49,087 | 1,022,654 | 175 | 96,211 | 2,806,162 |
| 88½ | 24,606 | 362,935 | 126 | 49,876 | 1,047,394 | 176 | 97,314 | 2,854,543 |
| 89 | 24,885 | 369,121 | 127 | 50,671 | 1,072,531 | 177 | 98,423 | 2,903,477 |
| 89½ | 25,165 | 375,377 | 128 | 51,472 | 1,098,066 | 178 | 99,538 | 2,952,967 |
| 90 | 25,447 | 381,704 | 129 | 52,279 | 1,124,004 | 179 | 100,660 | 3,003,016 |
| 90½ | 25,730 | 388,101 | 130 | 53,093 | 1,150,347 | 180 | 101,788 | 3,053,628 |
| 91 | 26,016 | 394,569 | 131 | 53,913 | 1,177,098 | 181 | 102,922 | 3,104,805 |
| 91½ | 26,302 | 401,109 | 132 | 54,739 | 1,204,260 | 182 | 104,062 | 3,156,551 |
| 92 | 26,590 | 407,720 | 133 | 55,572 | 1,231,838 | 183 | 105,209 | 3,208,868 |
| 92½ | 26,880 | 414,404 | 134 | 56,410 | 1,259,833 | 184 | 106,362 | 3,261,761 |
| 93 | 27,172 | 421,160 | 135 | 57,256 | 1,288,249 | 185 | 107,521 | 3,315,231 |
| 93½ | 27,465 | 427,990 | 136 | 58,107 | 1,317,090 | 186 | 108,687 | 3,369,283 |
| 94 | 27,759 | 434,893 | 137 | 58,965 | 1,346,357 | 187 | 109,858 | 3,423,919 |
| 94½ | 28,055 | 441,870 | 138 | 59,828 | 1,376,055 | 188 | 111,036 | 3,479,142 |
| 95 | 28,353 | 448,921 | 139 | 60,699 | 1,406,187 | 189 | 112,221 | 3,534,956 |
| 95½ | 28,652 | 456,046 | 140 | 61,575 | 1,436,755 | 190 | 113,411 | 3,591,364 |
| 96 | 28,953 | 463,247 | 141 | 62,458 | 1,467,763 | 191 | 114,608 | 3,648,369 |
| 96½ | 29,255 | 470,524 | 142 | 63,347 | 1,499,214 | 192 | 115,812 | 3,705,973 |
| 97 | 29,559 | 477,874 | 143 | 64,242 | 1,531,111 | 193 | 117,021 | 3,764,181 |
| 97½ | 29,865 | 485,302 | 144 | 65,144 | 1,563,458 | 194 | 118,237 | 3,822,996 |
| 98 | 30,172 | 492,807 | 145 | 66,052 | 1,596,256 | 195 | 119,459 | 3,882,419 |
| 98½ | 30,481 | 500,388 | 146 | 66,966 | 1,629,511 | 196 | 120,687 | 3,942,456 |
| 99 | 30,791 | 508,047 | 147 | 67,887 | 1,663,224 | 197 | 121,922 | 4,003,108 |
| 99½ | 31,103 | 515,784 | 148 | 68,813 | 1,697,398 | 198 | 123,163 | 4,064,379 |
| 100 | 31,416 | 523,599 | 149 | 69,746 | 1,732,038 | 199 | 124,410 | 4,126,272 |
| 100½ | 31,731 | 531,492 | 150 | 70,686 | 1,767,146 | 200 | 125,664 | 4,188,790 |

Table for Finding Volume of Spherical Segments (English or metric units)

Multiply factor $C$ in table by the cube of the length of the chord of the segment; the product equals the volume.

| Center Angle of Segment, Deg. | $C$ | Center Angle of Segment, Deg. | $C$ | Center Angle of Segment, Deg. | $C$ | Center Angle of Segment, Deg. | $C$ | Center Angle of Segment, Deg. | $C$ |
|---|---|---|---|---|---|---|---|---|---|
| 3 | 0.0026 | 39 | 0.0341 | 75 | 0.0692 | 111 | 0.1128 | 147 | 0.1739 |
| 6 | 0.0051 | 42 | 0.0368 | 78 | 0.0724 | 114 | 0.1171 | 150 | 0.1802 |
| 9 | 0.0077 | 45 | 0.0396 | 81 | 0.0757 | 117 | 0.1215 | 153 | 0.1869 |
| 12 | 0.0103 | 48 | 0.0424 | 84 | 0.0791 | 120 | 0.1260 | 156 | 0.1937 |
| 15 | 0.0129 | 51 | 0.0452 | 87 | 0.0825 | 123 | 0.1306 | 159 | 0.2010 |
| 18 | 0.0155 | 54 | 0.0480 | 90 | 0.0860 | 126 | 0.1354 | 162 | 0.2085 |
| 21 | 0.0181 | 57 | 0.0509 | 93 | 0.0895 | 129 | 0.1403 | 165 | 0.2163 |
| 24 | 0.0207 | 60 | 0.0539 | 96 | 0.0932 | 132 | 0.1454 | 168 | 0.2246 |
| 27 | 0.0233 | 63 | 0.0568 | 99 | 0.0969 | 135 | 0.1507 | 171 | 0.2332 |
| 30 | 0.0260 | 66 | 0.0599 | 102 | 0.1008 | 138 | 0.1562 | 174 | 0.2423 |
| 33 | 0.0287 | 69 | 0.0629 | 105 | 0.1047 | 141 | 0.1619 | 177 | 0.2518 |
| 36 | 0.0314 | 72 | 0.0660 | 108 | 0.1087 | 144 | 0.1678 | 180 | 0.2618 |

*Example:* Find the volume of a spherical segment having a center angle of 30 degrees, if the length of the chord is (a) 10 inches, or (b) 10 centimetres.

$10^o \times 0.026 = 1000 \times 0.026 =$ (a) 26 cubic inches, or (b) 26 cubic centimetres.

## Prime Numbers—Factors

The *factors* of a given number are those numbers which when multiplied together give a product equal to that number; thus, 2 and 3 are factors of 6; and 5 and 7 are factors of 35.

A *prime number* is one which has no factors except itself and 1. Thus, 3, 5, 7, 11, etc., are prime numbers. A factor which is a prime number is called a *prime factor.*

The accompanying "Prime Number and Factor Table" gives the smallest prime factor of all odd numbers from 1 to 9600, and can be used for finding all the factors for numbers up to this limit. For example, find the factors of 931. In the column headed "900," and in the line indicated by "31" in the left-hand column, the smallest prime factor is found to be 7. As this leaves another factor 133 (since $931 \div 7 = 133$), find the smallest prime factor of this number. In the column headed "100" and in the line "33," this is found to be 7, leaving a factor 19. This latter is a prime number; hence, the factors of 931 are $7 \times 7 \times 19$. Where no factor is given for a number in the factor table, it indicates that the number is a prime number. Tables of prime numbers and factors are especially useful in calculating the gearing for unusual gear ratios and for spiral gear generating machines, etc.

For factoring, the following general rules will be found useful:

2 is a factor of any number the right-hand figure of which is an even number or 0. Thus, $28 = 2 \times 14$, and $210 = 2 \times 105$.

3 is a factor of any number the sum of the figures of which is evenly divisible by 3. Thus, 3 is a factor of 1869, because $1 + 8 + 6 + 9 = 24$, and $24 \div 3 = 8$.

4 is a factor of any number the two right-hand figures of which, considered as one number, are evenly divisible by 4. Thus, 1844 has a factor 4, because $44 \div 4 = 11$.

5 is a factor of any number the right-hand figure of which is 0 or 5. Thus, $85 = 5 \times 17$; $70 = 5 \times 14$.

### Prime Number and Factor Table

| From to | 0–100 | 100–200 | 200–300 | 300–400 | 400–500 | 500–600 | 600–700 | 700–800 | 800–900 | 900–1000 | 1000–1100 | 1100–1200 |
|---|---|---|---|---|---|---|---|---|---|---|---|---|
| 1 | P | P | 3 | 7 | P | 3 | P | P | 3 | 17 | 7 | 3 |
| 3 | P | P | 7 | 3 | 13 | P | 3 | 19 | 11 | 3 | 17 | P |
| 5 | P | 3 | 5 | 5 | 3 | 5 | 5 | 3 | 5 | 5 | 3 | 5 |
| 7 | P | P | 3 | P | 11 | 3 | P | 7 | 3 | P | 19 | 3 |
| 9 | 3 | P | 11 | 3 | P | P | 3 | P | P | 3 | P | P |
| 11 | P | 3 | P | P | 3 | 7 | 13 | 3 | P | P | 3 | 11 |
| 13 | P | P | 3 | P | 7 | 3 | P | 23 | 3 | 11 | P | 3 |
| 15 | 3 | 5 | 5 | 3 | 5 | 5 | 3 | 5 | 5 | 3 | 5 | 5 |
| 17 | P | 3 | 7 | P | 3 | 11 | P | 3 | 19 | 7 | 3 | P |
| 19 | P | 7 | 3 | 11 | P | 3 | P | P | 3 | P | P | 3 |
| 21 | 3 | 11 | 13 | 3 | P | P | 3 | 7 | P | 3 | P | 19 |
| 23 | P | 3 | P | 17 | 3 | P | 7 | 3 | P | 13 | 3 | P |
| 25 | 5 | 5 | 3 | 5 | 5 | 3 | 5 | 5 | 3 | 5 | 5 | 3 |
| 27 | 3 | P | P | 3 | 7 | 17 | 3 | P | P | 3 | 13 | 7 |
| 29 | P | 3 | P | 7 | 3 | 23 | 17 | 3 | P | P | 3 | P |
| 31 | P | P | 3 | P | P | 3 | P | 17 | 3 | 7 | P | 3 |
| 33 | 3 | 7 | P | 3 | P | 13 | 3 | P | 7 | 3 | P | 11 |
| 35 | 5 | 3 | 5 | 5 | 3 | 5 | 5 | 3 | 5 | 5 | 3 | 5 |
| 37 | P | P | 3 | P | 19 | 3 | 7 | 11 | 3 | P | 17 | 3 |
| 39 | 3 | P | P | 3 | P | 7 | 3 | P | P | 3 | P | 17 |
| 41 | P | 3 | P | 11 | 3 | P | P | 3 | 29 | P | 3 | 7 |
| 43 | P | 11 | 3 | 7 | P | 3 | P | P | 3 | 23 | 7 | 3 |
| 45 | 3 | 5 | 5 | 3 | 5 | 5 | 3 | 5 | 5 | 3 | 5 | 5 |
| 47 | P | 3 | 13 | P | 3 | P | P | 3 | 7 | P | 3 | 31 |
| 49 | 7 | P | 3 | P | P | 3 | 11 | 7 | 3 | 13 | P | 3 |
| 51 | 3 | P | P | 3 | 11 | 19 | 3 | P | 23 | 3 | P | P |
| 53 | P | 3 | 11 | P | 3 | 7 | P | 3 | P | P | 3 | P |
| 55 | 5 | 5 | 3 | 5 | 5 | 3 | 5 | 5 | 3 | 5 | 5 | 3 |
| 57 | 3 | P | P | 3 | P | P | 3 | P | P | 3 | 7 | 13 |
| 59 | P | 3 | 7 | P | 3 | 13 | P | 3 | P | 7 | 3 | 19 |
| 61 | P | 7 | 3 | 19 | P | 3 | P | P | 3 | 31 | P | 3 |
| 63 | 3 | P | P | 3 | P | P | 3 | 7 | P | 3 | P | P |
| 65 | 5 | 3 | 5 | 5 | 3 | 5 | 5 | 3 | 5 | 5 | 3 | 5 |
| 67 | P | P | 3 | P | P | 3 | 23 | 13 | 3 | P | 11 | 3 |
| 69 | 3 | 13 | P | 3 | 7 | P | 3 | P | 11 | 3 | P | 7 |
| 71 | P | 3 | P | 7 | 3 | P | 11 | 3 | 13 | P | 3 | P |
| 73 | P | P | 3 | P | 11 | 3 | P | P | 3 | 7 | 29 | 3 |
| 75 | 3 | 5 | 5 | 3 | 5 | 5 | 3 | 5 | 5 | 3 | 5 | 5 |
| 77 | 7 | 3 | P | 13 | 3 | P | P | 3 | P | P | 3 | 11 |
| 79 | P | P | 3 | P | P | 3 | 7 | 19 | 3 | 11 | 13 | 3 |
| 81 | 3 | P | P | 3 | 13 | 7 | 3 | 11 | P | 3 | 23 | P |
| 83 | P | 3 | P | P | 3 | 11 | P | 3 | P | P | 3 | 7 |
| 85 | 5 | 5 | 3 | 5 | 5 | 3 | 5 | 5 | 3 | 5 | 5 | 3 |
| 87 | 3 | 11 | 7 | 3 | P | P | 3 | P | P | 3 | P | P |
| 89 | P | 3 | 17 | P | 3 | 19 | 13 | 3 | 7 | 23 | 3 | 29 |
| 91 | 7 | P | 3 | 17 | P | 3 | P | 7 | 3 | P | P | 3 |
| 93 | 3 | P | P | 3 | 17 | P | 3 | 13 | 19 | 3 | P | P |
| 95 | 5 | 3 | 5 | 5 | 3 | 5 | 5 | 3 | 5 | 5 | 3 | 5 |
| 97 | P | P | 3 | P | 7 | 3 | 17 | P | 3 | P | P | 3 |
| 99 | 3 | P | 13 | 3 | P | P | 3 | 17 | 29 | 3 | 7 | 11 |

Prime Number and Factor Table

| From to | 1200 1300 | 1300 1400 | 1400 1500 | 1500 1600 | 1600 1700 | 1700 1800 | 1800 1900 | 1900 2000 | 2000 2100 | 2100 2200 | 2200 2300 | 2300 2400 |
|---|---|---|---|---|---|---|---|---|---|---|---|---|
| 1 | P | P | 3 | 19 | P | 3 | P | P | 3 | 11 | 31 | 3 |
| 3 | 3 | P | 23 | 3 | 7 | 13 | 3 | 11 | P | 3 | P | 7 |
| 5 | 5 | 3 | 5 | 5 | 3 | 5 | 5 | 3 | 5 | 5 | 3 | 5 |
| 7 | 17 | P | 3 | 11 | P | 3 | 13 | P | 3 | 7 | P | 3 |
| 9 | 3 | 7 | P | 3 | P | P | 3 | 23 | 7 | 3 | 47 | P |
| 11 | 7 | 3 | 17 | P | 3 | 29 | P | 3 | P | P | 3 | P |
| 13 | P | 13 | 3 | 17 | P | 3 | 7 | P | 3 | P | P | 3 |
| 15 | 3 | 5 | 5 | 3 | 5 | 5 | 3 | 5 | 5 | 5 | 3 | 5 |
| 17 | P | 3 | 13 | 37 | 3 | 17 | 23 | 3 | P | 29 | 3 | 7 |
| 19 | 23 | P | 3 | 7 | P | 3 | 17 | 19 | 3 | 13 | 7 | 3 |
| 21 | 3 | P | 7 | 3 | P | P | 3 | 17 | 43 | 3 | P | 11 |
| 23 | P | 3 | P | P | 3 | P | P | 3 | 7 | 11 | 3 | 23 |
| 25 | 5 | 5 | 3 | 5 | 3 | 5 | 5 | 3 | 5 | 5 | 5 | 3 |
| 27 | 3 | P | P | 3 | P | 11 | 3 | 41 | P | 3 | 17 | 13 |
| 29 | P | 3 | P | 11 | 3 | 7 | 31 | 3 | P | P | 3 | 17 |
| 31 | P | 11 | 3 | P | 7 | 3 | P | P | 3 | P | 23 | 3 |
| 33 | 3 | 31 | P | 3 | 23 | P | 3 | P | 19 | 3 | 7 | P |
| 35 | 5 | 3 | 5 | 5 | 3 | 5 | 5 | 3 | 5 | 5 | 3 | 5 |
| 37 | P | 7 | 3 | 29 | P | 3 | 11 | 13 | 3 | P | P | 3 |
| 39 | 3 | 13 | P | 3 | 11 | 37 | 3 | 7 | P | 3 | P | P |
| 41 | 17 | 3 | 11 | 23 | 3 | P | 7 | 3 | 13 | P | 3 | P |
| 43 | 11 | 17 | 3 | P | 31 | 3 | 19 | 29 | 3 | P | P | 3 |
| 45 | 3 | 5 | 5 | 3 | 5 | 5 | 3 | 5 | 5 | 3 | 5 | 5 |
| 47 | 29 | 3 | P | 7 | 3 | P | P | 3 | 23 | 19 | 3 | P |
| 49 | P | 19 | 3 | P | 17 | 3 | 43 | P | 3 | 7 | 13 | 3 |
| 51 | 3 | 7 | P | 3 | 13 | 17 | 3 | P | 7 | P | P | P |
| 53 | 7 | 3 | P | P | 3 | P | 17 | 3 | P | P | 3 | 13 |
| 55 | 5 | 5 | 3 | 5 | 5 | 3 | 5 | 5 | 3 | 5 | 5 | 3 |
| 57 | 3 | 23 | 31 | 3 | P | 7 | 3 | 19 | 11 | 3 | 37 | P |
| 59 | P | 3 | P | P | 3 | P | 11 | 3 | 29 | 17 | 3 | 7 |
| 61 | 13 | P | 3 | 7 | 11 | 3 | P | 37 | 3 | P | 7 | 3 |
| 63 | 3 | 29 | 7 | 3 | P | 41 | 3 | 13 | P | 3 | 31 | 17 |
| 65 | 5 | 3 | 5 | 5 | 3 | 5 | 5 | 3 | 5 | 5 | 3 | 5 |
| 67 | 7 | P | 3 | P | 3 | 3 | P | 7 | 3 | 11 | P | 3 |
| 69 | 3 | 37 | 13 | 3 | P | 29 | 3 | 11 | P | 3 | P | 23 |
| 71 | 31 | 3 | P | P | 3 | 7 | P | 3 | 19 | 13 | 3 | P |
| 73 | 19 | P | 3 | 11 | 7 | 3 | P | P | 3 | 41 | P | 3 |
| 75 | 3 | 5 | 5 | 5 | 3 | 5 | 5 | 3 | 5 | 5 | 3 | 5 |
| 77 | P | 3 | 7 | 19 | 3 | P | P | 3 | 31 | 7 | 3 | P |
| 79 | P | 7 | 3 | P | 23 | 3 | P | P | 3 | P | 43 | 3 |
| 81 | 3 | P | P | 3 | 41 | 13 | 3 | 7 | P | 3 | P | P |
| 83 | P | 3 | P | P | 3 | P | 7 | 3 | P | 37 | 3 | P |
| 85 | 5 | 5 | 3 | 5 | 5 | 3 | 5 | 5 | 3 | 5 | 5 | 3 |
| 87 | 3 | 19 | P | 3 | 7 | P | 3 | P | P | 3 | P | 7 |
| 89 | P | 3 | P | 7 | 3 | P | P | 3 | P | 11 | 3 | P |
| 91 | P | 13 | 3 | 37 | 19 | 3 | 31 | 11 | 3 | 7 | 29 | 3 |
| 93 | 3 | 7 | P | 3 | P | 11 | 3 | P | 7 | 3 | P | P |
| 95 | 5 | 3 | 5 | 5 | 3 | 5 | 5 | 3 | 5 | 5 | 3 | 5 |
| 97 | P | 11 | 3 | P | P | 3 | 7 | P | 3 | 13 | P | 3 |
| 99 | 3 | P | P | 3 | P | P | 7 | 3 | P | P | 3 | P |

## Prime Number and Factor Table

| From to | 2400 2500 | 2500 2600 | 2600 2700 | 2700 2800 | 2800 2900 | 2900 3000 | 3000 3100 | 3100 3200 | 3200 3300 | 3300 3400 | 3400 3500 | 3500 3600 |
|---|---|---|---|---|---|---|---|---|---|---|---|---|
| 1 | 7 | 41 | 3 | 37 | P | 3 | P | 7 | 3 | P | 19 | 3 |
| 3 | 3 | P | 19 | 3 | P | P | 3 | 29 | P | 3 | 41 | 31 |
| 5 | 5 | 3 | 5 | 5 | 3 | 5 | 5 | 3 | 5 | 5 | 3 | 5 |
| 7 | 29 | 23 | 3 | P | 7 | 3 | 31 | 13 | 3 | P | P | 3 |
| 9 | 3 | 13 | P | 3 | 53 | P | 3 | P | P | 3 | 7 | 11 |
| 11 | P | 3 | 7 | P | 3 | 41 | P | 3 | 13 | 7 | 3 | P |
| 13 | 19 | 7 | 3 | P | 29 | 3 | 23 | 11 | 3 | P | P | 3 |
| 15 | 3 | 5 | 5 | 3 | 5 | 5 | 3 | 5 | 5 | 3 | 5 | 5 |
| 17 | P | 3 | P | 11 | 3 | P | 7 | 3 | P | 31 | 3 | P |
| 19 | 41 | 11 | 3 | P | P | 3 | P | P | 3 | P | 13 | 3 |
| 21 | 3 | P | P | 3 | 7 | 23 | 3 | P | P | 3 | 11 | 7 |
| 23 | P | 3 | 43 | 7 | 3 | 37 | P | 3 | 11 | P | 3 | 13 |
| 25 | 5 | 5 | 3 | 5 | 5 | 3 | 5 | 5 | 3 | 5 | 5 | 3 |
| 27 | 3 | 7 | 37 | 3 | 11 | P | 3 | 53 | 7 | 3 | 23 | P |
| 29 | 7 | 3 | 11 | P | 3 | 29 | 13 | 3 | P | P | 3 | P |
| 31 | 11 | P | 3 | P | 19 | 3 | 7 | 31 | 3 | P | 47 | 3 |
| 33 | 3 | 17 | P | 3 | P | 7 | 3 | 13 | 53 | 3 | P | P |
| 35 | 5 | 3 | 5 | 5 | 3 | 5 | 5 | 3 | 5 | 5 | 3 | 5 |
| 37 | P | 43 | 3 | 7 | P | 3 | P | P | 3 | 47 | 7 | 3 |
| 39 | 3 | P | 7 | 3 | 17 | P | 3 | 43 | 41 | 3 | 19 | P |
| 41 | P | 3 | 19 | P | 3 | 17 | P | 3 | 7 | 13 | 3 | P |
| 43 | 7 | P | 3 | 13 | P | 3 | 17 | 7 | 3 | P | 11 | 3 |
| 45 | 3 | 5 | 5 | 3 | 5 | 5 | 3 | 5 | 5 | 3 | 5 | 5 |
| 47 | P | 3 | P | 41 | 3 | 7 | 11 | 3 | 17 | P | 3 | P |
| 49 | 31 | P | 3 | P | 7 | 3 | P | 47 | 3 | 17 | P | 3 |
| 51 | 3 | P | 11 | 3 | P | 13 | 3 | 23 | P | 3 | 7 | 53 |
| 53 | 11 | 3 | 7 | P | 3 | P | 43 | 3 | P | 7 | 3 | 11 |
| 55 | 5 | 5 | 3 | 5 | 5 | 3 | 5 | 5 | 3 | 5 | 5 | 3 |
| 57 | 3 | P | P | 3 | P | P | 3 | 7 | P | 3 | P | P |
| 59 | P | 3 | P | 31 | 3 | 11 | 7 | 3 | P | P | 3 | P |
| 61 | 23 | 13 | 3 | 11 | P | 3 | P | 29 | 3 | P | P | 3 |
| 63 | 3 | 11 | P | 3 | 7 | P | 3 | P | 13 | 3 | P | 7 |
| 65 | 5 | 3 | 5 | 5 | 3 | 5 | 5 | 3 | 5 | 5 | 3 | 5 |
| 67 | P | 17 | 3 | P | 47 | 3 | P | P | 3 | 7 | P | 3 |
| 69 | 3 | 7 | 17 | 3 | 19 | P | 3 | P | 7 | 3 | P | 43 |
| 71 | 7 | 3 | P | 17 | 3 | P | 37 | 3 | P | P | 3 | P |
| 73 | P | 31 | 3 | 47 | 13 | 3 | 7 | 19 | 3 | P | 23 | 3 |
| 75 | 3 | 5 | 5 | 3 | 5 | 5 | 3 | 5 | 5 | 3 | 5 | 5 |
| 77 | P | 3 | P | P | 3 | 13 | 17 | 3 | 29 | 11 | 3 | 7 |
| 79 | 37 | P | 3 | 7 | P | 3 | P | 11 | 3 | 31 | 7 | 3 |
| 81 | 3 | 29 | 7 | 3 | 43 | 11 | 3 | P | 17 | 3 | 59 | P |
| 83 | 13 | 3 | P | 11 | 3 | 19 | P | 3 | 7 | 17 | 3 | P |
| 85 | 5 | 5 | 3 | 5 | 5 | 3 | 5 | 5 | 3 | 5 | 5 | 3 |
| 87 | 3 | 13 | P | 3 | P | 29 | 3 | P | 19 | 3 | 11 | 17 |
| 89 | 19 | 3 | P | P | 3 | 7 | P | 3 | 11 | P | 3 | 37 |
| 91 | 47 | P | 3 | P | 7 | 3 | 11 | P | 3 | P | P | 3 |
| 93 | 3 | P | P | 3 | 11 | 41 | 3 | 31 | 37 | 3 | 7 | P |
| 95 | 5 | 3 | 5 | 5 | 3 | 5 | 5 | 3 | 5 | 5 | 3 | 5 |
| 97 | 11 | 7 | 3 | P | P | 3 | 19 | 23 | 3 | 43 | 13 | 3 |
| 99 | 3 | 23 | P | 3 | 13 | P | 3 | 7 | P | 3 | P | 59 |

## Prime Number and Factor Table

| From to | 3600 3700 | 3700 3800 | 3800 3900 | 3900 4000 | 4000 4100 | 4100 4200 | 4200 4300 | 4300 4400 | 4400 4500 | 4500 4600 | 4600 4700 | 4700 4800 |
|---|---|---|---|---|---|---|---|---|---|---|---|---|
| 1  | 13 | P  | 3  | 47 | P  | 3  | P  | 11 | 3  | 7  | 43 | 3  |
| 3  | 3  | 7  | P  | 3  | P  | 11 | 3  | 13 | 7  | 3  | P  | P  |
| 5  | 5  | 3  | 5  | 5  | 3  | 5  | 5  | 3  | 5  | 5  | 3  | 5  |
| 7  | P  | 11 | 3  | P  | P  | 3  | 7  | 59 | 3  | P  | 17 | 3  |
| 9  | 3  | P  | 13 | 3  | 19 | 7  | 3  | 31 | P  | 3  | 11 | 17 |
| 11 | 23 | 3  | 37 | P  | 3  | P  | P  | 3  | 11 | 13 | 3  | 7  |
| 13 | P  | 47 | 3  | 7  | P  | 3  | 11 | 19 | 3  | P  | 7  | 3  |
| 15 | 3  | 5  | 5  | 3  | 5  | 5  | 3  | 5  | 5  | 3  | 5  | 5  |
| 17 | P  | 3  | 11 | P  | 3  | 23 | P  | 3  | 7  | P  | 3  | 53 |
| 19 | 7  | P  | 3  | P  | P  | 3  | P  | 7  | 3  | P  | 31 | 3  |
| 21 | 3  | 61 | P  | 3  | P  | 13 | 3  | 29 | P  | 3  | P  | P  |
| 23 | P  | 3  | P  | P  | 3  | 7  | 41 | 3  | P  | P  | 3  | P  |
| 25 | 5  | 5  | 3  | 5  | 5  | 3  | 5  | 5  | 3  | 5  | 5  | 3  |
| 27 | 3  | P  | 43 | 3  | P  | P  | 3  | P  | 19 | 3  | 7  | 29 |
| 29 | 19 | 3  | 7  | P  | 3  | P  | P  | 3  | 43 | 7  | 3  | P  |
| 31 | P  | 7  | 3  | P  | 29 | 3  | P  | 61 | 3  | 23 | 11 | 3  |
| 33 | 3  | P  | P  | 3  | 37 | P  | 3  | 7  | 11 | 3  | 41 | P  |
| 35 | 5  | 3  | 5  | 5  | 3  | 5  | 5  | 3  | 5  | 5  | 3  | 5  |
| 37 | P  | 37 | 3  | 31 | 11 | 3  | 19 | P  | 3  | 13 | P  | 3  |
| 39 | 3  | P  | 11 | 3  | 7  | P  | 3  | P  | 23 | 3  | P  | 7  |
| 41 | 11 | 3  | 23 | 7  | 3  | 41 | P  | 3  | P  | 19 | 3  | 11 |
| 43 | P  | 19 | 3  | P  | 13 | 3  | P  | 43 | 3  | 7  | P  | 3  |
| 45 | 3  | 5  | 5  | 3  | 5  | 5  | 3  | 5  | 5  | 3  | 5  | 5  |
| 47 | 7  | 3  | P  | P  | 3  | 11 | 31 | 3  | P  | P  | 3  | 47 |
| 49 | 41 | 23 | 3  | 11 | P  | 3  | 7  | P  | 3  | P  | P  | 3  |
| 51 | 3  | 11 | P  | 3  | P  | 7  | 3  | 19 | P  | 3  | P  | P  |
| 53 | 13 | 3  | P  | 59 | 3  | P  | P  | 3  | 61 | 29 | 3  | 7  |
| 55 | 5  | 5  | 3  | 5  | 5  | 3  | 5  | 5  | 3  | 5  | 5  | 3  |
| 57 | 3  | 13 | 7  | 3  | P  | P  | 3  | P  | P  | 3  | P  | 67 |
| 59 | P  | 3  | 17 | 37 | 3  | P  | P  | 3  | 7  | 47 | 3  | P  |
| 61 | 7  | P  | 3  | 17 | 31 | 3  | P  | 7  | 3  | P  | 59 | 3  |
| 63 | 3  | 53 | P  | 3  | 17 | 23 | 3  | P  | P  | 3  | P  | 11 |
| 65 | 5  | 3  | 5  | 5  | 3  | 5  | 5  | 3  | 5  | 5  | 3  | 5  |
| 67 | 19 | P  | 3  | P  | 7  | 3  | 17 | 11 | 3  | P  | 13 | 3  |
| 69 | 3  | P  | 53 | 3  | 13 | 11 | 3  | 17 | 41 | 3  | 7  | 19 |
| 71 | P  | 3  | 7  | 11 | 3  | 43 | P  | 3  | 17 | 7  | 3  | 13 |
| 73 | P  | 7  | 3  | 29 | P  | 3  | P  | P  | 3  | 17 | P  | 3  |
| 75 | 3  | 5  | 5  | 3  | 5  | 5  | 3  | 5  | 5  | 3  | 5  | 5  |
| 77 | P  | 3  | P  | 41 | 3  | P  | 7  | 3  | 11 | 23 | 3  | 17 |
| 79 | 13 | P  | 3  | 23 | P  | 3  | 11 | 29 | 3  | 19 | P  | 3  |
| 81 | 3  | 19 | P  | 3  | 7  | 37 | 3  | 13 | P  | 3  | 31 | 7  |
| 83 | 29 | 3  | 11 | 7  | 3  | 47 | P  | 3  | P  | P  | 3  | P  |
| 85 | 5  | 5  | 3  | 5  | 5  | 3  | 5  | 5  | 3  | 5  | 5  | 3  |
| 87 | 3  | 7  | 13 | 3  | 61 | 53 | 3  | 41 | 7  | 3  | 43 | P  |
| 89 | 7  | 3  | P  | P  | 3  | 59 | P  | 3  | 67 | 13 | 3  | P  |
| 91 | P  | 17 | 3  | 13 | P  | 3  | P  | 7  | 3  | P  | P  | 3  |
| 93 | 3  | P  | 17 | 3  | P  | 7  | 3  | 23 | P  | 3  | 13 | P  |
| 95 | 5  | 3  | 5  | 5  | 3  | 5  | 5  | 3  | 5  | 5  | 3  | 5  |
| 97 | P  | P  | 3  | 7  | 17 | 3  | P  | P  | 3  | P  | 7  | 3  |
| 99 | 3  | 29 | 7  | 3  | P  | 13 | 3  | 53 | 11 | 3  | 37 | P  |

## Prime Number and Factor Table

| From to | 4800 4900 | 4900 5000 | 5000 5100 | 5100 5200 | 5200 5300 | 5300 5400 | 5400 5500 | 5500 5600 | 5600 5700 | 5700 5800 | 5800 5900 | 5900 6000 |
|---|---|---|---|---|---|---|---|---|---|---|---|---|
| 1 | P | 13 | 3 | P | 7 | 3 | 11 | P | 3 | P | P | 3 |
| 3 | 3 | P | P | 3 | 11 | P | 3 | P | 13 | 3 | 7 | P |
| 5 | 5 | 3 | 5 | 5 | 3 | 5 | 5 | 3 | 5 | 5 | 3 | 5 |
| 7 | 11 | 7 | 3 | P | 41 | 3 | P | P | 3 | 13 | P | 3 |
| 9 | 3 | P | P | 3 | P | P | 3 | 7 | 71 | 3 | 37 | 19 |
| 11 | 17 | 3 | P | 19 | 3 | 47 | 7 | 3 | 31 | P | 3 | 23 |
| 13 | P | 17 | 3 | P | 13 | 3 | P | 37 | 3 | 29 | P | 3 |
| 15 | 3 | 5 | 5 | 3 | 5 | 5 | 3 | 5 | 5 | 3 | 5 | 5 |
| 17 | P | 3 | 29 | 7 | 3 | 13 | P | 3 | 41 | P | 3 | 61 |
| 19 | 61 | P | 3 | P | 17 | 3 | P | P | 3 | 7 | 11 | 3 |
| 21 | 3 | 7 | P | 3 | 23 | 17 | 3 | P | 7 | 3 | P | 31 |
| 23 | 7 | 3 | P | 47 | 3 | P | 11 | 3 | P | 59 | 3 | P |
| 25 | 5 | 5 | 3 | 5 | 5 | 3 | 5 | 5 | 3 | 5 | 5 | 3 |
| 27 | 3 | 13 | 11 | 3 | P | 7 | 3 | P | 17 | 3 | P | P |
| 29 | 11 | 3 | 47 | 23 | 3 | 73 | 61 | 3 | 13 | 17 | 3 | 7 |
| 31 | P | P | 3 | 7 | P | 3 | P | P | 3 | 11 | 7 | 3 |
| 33 | 3 | P | 7 | 3 | P | P | 3 | 11 | 43 | 3 | 19 | 17 |
| 35 | 5 | 3 | 5 | 5 | 3 | 5 | 5 | 3 | 5 | 5 | 3 | 5 |
| 37 | 7 | P | 3 | 11 | P | 3 | P | 7 | 3 | P | 13 | 3 |
| 39 | 3 | 11 | P | 3 | 13 | 19 | 3 | 29 | P | 3 | P | P |
| 41 | 47 | 3 | 71 | 53 | 3 | 7 | P | 3 | P | P | 3 | 13 |
| 43 | 29 | P | 3 | 37 | 7 | 3 | P | 23 | 3 | P | P | 3 |
| 45 | 3 | 5 | 5 | 3 | 5 | 5 | 3 | 5 | 5 | 3 | 5 | 5 |
| 47 | 37 | 3 | 7 | P | 3 | P | 13 | 3 | P | 7 | 3 | 19 |
| 49 | 13 | 7 | 3 | 19 | 29 | 3 | P | 31 | 3 | P | P | 3 |
| 51 | 3 | P | P | 3 | 59 | P | 3 | 7 | P | 3 | P | 11 |
| 53 | 23 | 3 | 31 | P | 3 | 53 | 7 | 3 | P | 11 | 3 | P |
| 55 | 5 | 5 | 3 | 5 | 5 | 3 | 5 | 5 | 3 | 5 | 5 | 3 |
| 57 | 3 | P | 13 | 3 | 7 | 11 | 3 | P | P | 3 | P | 7 |
| 59 | 43 | 3 | P | 7 | 3 | 23 | 53 | 3 | P | 13 | 3 | 59 |
| 61 | P | 11 | 3 | 13 | P | 3 | 43 | 67 | 3 | 7 | P | 3 |
| 63 | 3 | 7 | 61 | 3 | 19 | 31 | 3 | P | 7 | 3 | 11 | 67 |
| 65 | 5 | 3 | 5 | 5 | 3 | 5 | 5 | 3 | 5 | 5 | 3 | 5 |
| 67 | 31 | P | 3 | P | 23 | 3 | 7 | 19 | 3 | 73 | P | 3 |
| 69 | 3 | P | 37 | 3 | 11 | 7 | 3 | P | P | 3 | P | 47 |
| 71 | P | 3 | 11 | P | 3 | 41 | P | 3 | 53 | 29 | 3 | 7 |
| 73 | 11 | P | 3 | 7 | P | 3 | 13 | P | 3 | 23 | 7 | 3 |
| 75 | 3 | 5 | 5 | 3 | 5 | 5 | 3 | 5 | 5 | 3 | 5 | 5 |
| 77 | P | 3 | P | 31 | 3 | 19 | P | 3 | 7 | 53 | 3 | 43 |
| 79 | 7 | 13 | 3 | P | P | 3 | P | 7 | 3 | P | P | 3 |
| 81 | 3 | 17 | P | 3 | P | P | 3 | P | 13 | 3 | P | P |
| 83 | 19 | 3 | 13 | 71 | 3 | 7 | P | 3 | P | P | 3 | 31 |
| 85 | 5 | 5 | 3 | 5 | 5 | 3 | 5 | 5 | 3 | 5 | 5 | 3 |
| 87 | 3 | P | P | 3 | 17 | P | 3 | 37 | 11 | 3 | 7 | P |
| 89 | P | 3 | 7 | P | 3 | 17 | 11 | 3 | P | 7 | 3 | 53 |
| 91 | 67 | 7 | 3 | 29 | 11 | 3 | 17 | P | 3 | P | 43 | 3 |
| 93 | 3 | P | 11 | 3 | 67 | P | 3 | 7 | P | 3 | 71 | 13 |
| 95 | 5 | 3 | 5 | 5 | 3 | 5 | 5 | 3 | 5 | 5 | 3 | 5 |
| 97 | 59 | 19 | 3 | P | P | 3 | 23 | 29 | 3 | 11 | P | 3 |
| 99 | 3 | P | P | 3 | 7 | P | 3 | 11 | 41 | 3 | 17 | 7 |

### Prime Number and Factor Table

| From to | 6000 6100 | 6100 6200 | 6200 6300 | 6300 6400 | 6400 6500 | 6500 6600 | 6600 6700 | 6700 6800 | 6800 6900 | 6900 7000 | 7000 7100 | 7100 7200 |
|---|---|---|---|---|---|---|---|---|---|---|---|---|
| 1 | 17 | P | 3 | P | 37 | 3 | 7 | P | 3 | 67 | P | 3 |
| 3 | 3 | 17 | P | 3 | 19 | 7 | 3 | P | P | 3 | 47 | P |
| 5 | 5 | 3 | 5 | 5 | 3 | 5 | 5 | 3 | 5 | 5 | 3 | 5 |
| 7 | P | 31 | 3 | 7 | 43 | 3 | P | 19 | 3 | P | 7 | 3 |
| 9 | 3 | 41 | 7 | 3 | 13 | 23 | 3 | P | 11 | 3 | 43 | P |
| 11 | P | 3 | P | P | 3 | 17 | 11 | 3 | 7 | P | 3 | 13 |
| 13 | 7 | P | 3 | 59 | 11 | 3 | 17 | 7 | 3 | 31 | P | 3 |
| 15 | 3 | 5 | 5 | 3 | 5 | 5 | 3 | 5 | 5 | 3 | 5 | 5 |
| 17 | 11 | 3 | P | P | 3 | 7 | 13 | 3 | 17 | P | 3 | 11 |
| 19 | 13 | 29 | 3 | 71 | 7 | 3 | P | P | 3 | 11 | P | 3 |
| 21 | 3 | P | P | 3 | P | P | 3 | 11 | 19 | 3 | 7 | P |
| 23 | 19 | 3 | 7 | P | 3 | 11 | 37 | 3 | P | 7 | 3 | 17 |
| 25 | 5 | 5 | 3 | 5 | 5 | 3 | 5 | 5 | 3 | 5 | 5 | 3 |
| 27 | 3 | 11 | 13 | 3 | P | 61 | 3 | 7 | P | 3 | P | P |
| 29 | P | 3 | P | P | 3 | P | 7 | 3 | P | 13 | 3 | P |
| 31 | 37 | P | 3 | 13 | 59 | 3 | 19 | 53 | 3 | 29 | 79 | 3 |
| 33 | 3 | P | 23 | 3 | 7 | 47 | 3 | P | P | 3 | 13 | 7 |
| 35 | 5 | 3 | 5 | 5 | 3 | 5 | 5 | 3 | 5 | 5 | 3 | 5 |
| 37 | P | 17 | 3 | P | 41 | 3 | P | P | 3 | 7 | 31 | 3 |
| 39 | 3 | 7 | 17 | 3 | 47 | 13 | 3 | 23 | 7 | 3 | P | 11 |
| 41 | 7 | 3 | 79 | 17 | 3 | 31 | 29 | 3 | P | 11 | 3 | 37 |
| 43 | P | P | 3 | P | 17 | 3 | 7 | 11 | 3 | 53 | P | 3 |
| 45 | 3 | 5 | 5 | 3 | 5 | 5 | 3 | 5 | 5 | 3 | 5 | 5 |
| 47 | P | 3 | P | 11 | 3 | P | 17 | 3 | 41 | P | 3 | 7 |
| 49 | 23 | 11 | 3 | 7 | P | 3 | 61 | 17 | 3 | P | 7 | 3 |
| 51 | 3 | P | 7 | 3 | P | P | 3 | 43 | 13 | 3 | 11 | P |
| 53 | P | 3 | 13 | P | 3 | P | P | 3 | 7 | 17 | 3 | 23 |
| 55 | 5 | 5 | 3 | 5 | 5 | 3 | 5 | 5 | 3 | 5 | 5 | 3 |
| 57 | 3 | 47 | P | 3 | 11 | 79 | 3 | 29 | P | 3 | P | 17 |
| 59 | 73 | 3 | 11 | P | 3 | 7 | P | 3 | 19 | P | 3 | P |
| 61 | 11 | 61 | 3 | P | 7 | 3 | P | P | 3 | P | 23 | 3 |
| 63 | 3 | P | P | 3 | 23 | P | 3 | P | P | 3 | 7 | 13 |
| 65 | 5 | 3 | 5 | 5 | 3 | 5 | 5 | 3 | 5 | 5 | 3 | 5 |
| 67 | P | 7 | 3 | P | 29 | 3 | 59 | 67 | 3 | P | 37 | 3 |
| 69 | 3 | 31 | P | 3 | P | P | 3 | 7 | P | 3 | P | 67 |
| 71 | 13 | 3 | P | 23 | 3 | P | 7 | 3 | P | P | 3 | 71 |
| 73 | P | P | 3 | P | P | 3 | P | 13 | 3 | 19 | 11 | 3 |
| 75 | 3 | 5 | 5 | 3 | 5 | 5 | 3 | 5 | 5 | 3 | 5 | 5 |
| 77 | 59 | 3 | P | 7 | 3 | P | 11 | 3 | 13 | P | 3 | P |
| 79 | P | 37 | 3 | P | 11 | 3 | P | P | 3 | 7 | P | 3 |
| 81 | 3 | 7 | 11 | 3 | P | P | 3 | P | 7 | 3 | 73 | 43 |
| 83 | 7 | 3 | 61 | 13 | 3 | 29 | 41 | 3 | P | P | 3 | 11 |
| 85 | 5 | 5 | 3 | 5 | 5 | 3 | 5 | 5 | 3 | 5 | 5 | 3 |
| 87 | 3 | 23 | P | 3 | 13 | 7 | 3 | 11 | 71 | 3 | 19 | P |
| 89 | P | 3 | 19 | P | 3 | 11 | P | 3 | 83 | 29 | 3 | 7 |
| 91 | P | 41 | 3 | 7 | P | 3 | P | P | 3 | P | 7 | 3 |
| 93 | 3 | 11 | 7 | 3 | 43 | 19 | 3 | P | 61 | 3 | 41 | P |
| 95 | 5 | 3 | 5 | 5 | 3 | 5 | 5 | 3 | 5 | 5 | 3 | 5 |
| 97 | 7 | P | 3 | P | 73 | 3 | 37 | 7 | 3 | P | 47 | 3 |
| 99 | 3 | P | P | 3 | 67 | P | 3 | 13 | P | 3 | 31 | 23 |

## Prime Number and Factor Table

| From to | 7200 7300 | 7300 7400 | 7400 7500 | 7500 7600 | 7600 7700 | 7700 7800 | 7800 7900 | 7900 8000 | 8000 8100 | 8100 8200 | 8200 8300 | 8300 8400 |
|---|---|---|---|---|---|---|---|---|---|---|---|---|
| 1 | 19 | 7 | 3 | 13 | 11 | 3 | 29 | P | 3 | P | 59 | 3 |
| 3 | 3 | 67 | 11 | 3 | P | P | 3 | 7 | 53 | 3 | 13 | 19 |
| 5 | 5 | 3 | 5 | 5 | 3 | 5 | 5 | 3 | 5 | 5 | 3 | 5 |
| 7 | P | P | 3 | P | P | 3 | 37 | P | 3 | 11 | 29 | 3 |
| 9 | 3 | P | 31 | 3 | 7 | 13 | 3 | 11 | P | 3 | P | 7 |
| 11 | P | 3 | P | 7 | 3 | 11 | 73 | 3 | P | P | 3 | P |
| 13 | P | 71 | 3 | 11 | 23 | 3 | 13 | 41 | 3 | 7 | 43 | 3 |
| 15 | 3 | 5 | 5 | 3 | 5 | 5 | 3 | 5 | 5 | 3 | 5 | 5 |
| 17 | 7 | 3 | P | P | 3 | P | P | 3 | P | P | 3 | P |
| 19 | P | 13 | 3 | 73 | 19 | 3 | 7 | P | 3 | 23 | P | 3 |
| 21 | 3 | P | 41 | 3 | P | 7 | 3 | 89 | 13 | 3 | P | 53 |
| 23 | 31 | 3 | 13 | P | 3 | P | P | 3 | 71 | P | 3 | 7 |
| 25 | 5 | 5 | 3 | 5 | 5 | 3 | 5 | 5 | 3 | 5 | 5 | 3 |
| 27 | 3 | 17 | 7 | 3 | 29 | P | 3 | P | 23 | 3 | 19 | 11 |
| 29 | P | 3 | 17 | P | 3 | 59 | P | 3 | 7 | 11 | 3 | P |
| 31 | 7 | P | 3 | 17 | 13 | 3 | 41 | 7 | 3 | 47 | P | 3 |
| 33 | 3 | P | P | 3 | 17 | 11 | 3 | P | 29 | 3 | P | 13 |
| 35 | 5 | 3 | 5 | 5 | 3 | 5 | 5 | 3 | 5 | 5 | 3 | 5 |
| 37 | P | 11 | 3 | P | 7 | 3 | 17 | P | 3 | 79 | P | 3 |
| 39 | 3 | 41 | 43 | 3 | P | 71 | 3 | 17 | P | 3 | 7 | 31 |
| 41 | 13 | 3 | 7 | P | 3 | P | P | 3 | 11 | 7 | 3 | 19 |
| 43 | P | 7 | 3 | 19 | P | 3 | 11 | 13 | 3 | 17 | P | 3 |
| 45 | 3 | 5 | 5 | 3 | 5 | 5 | 3 | 5 | 5 | 3 | 5 | 5 |
| 47 | P | 3 | 11 | P | 3 | 61 | 7 | 3 | 13 | P | 3 | 17 |
| 49 | 11 | P | 3 | P | P | 3 | 47 | P | 3 | 29 | 73 | 3 |
| 51 | 3 | P | P | 3 | 7 | 23 | 3 | P | 83 | 3 | 37 | 7 |
| 53 | P | 3 | 29 | 7 | 3 | P | P | 3 | P | 31 | 3 | P |
| 55 | 5 | 5 | 3 | 5 | 5 | 3 | 5 | 5 | 3 | 5 | 5 | 3 |
| 57 | 3 | 7 | P | 3 | 13 | P | 3 | 73 | 7 | 3 | 23 | 61 |
| 59 | 7 | 3 | P | P | 3 | P | 29 | 3 | P | 41 | 3 | 13 |
| 61 | 53 | 17 | 3 | P | 47 | 3 | 7 | 19 | 3 | P | 11 | 3 |
| 63 | 3 | 37 | 17 | 3 | 79 | 7 | 3 | P | 11 | 3 | P | P |
| 65 | 5 | 3 | 5 | 5 | 3 | 5 | 5 | 3 | 5 | 5 | 3 | 5 |
| 67 | 13 | 53 | 3 | 7 | 11 | 3 | P | 31 | 3 | P | 7 | 3 |
| 69 | 3 | P | 7 | 3 | P | 17 | 3 | 13 | P | 3 | P | P |
| 71 | 11 | 3 | 31 | 67 | 3 | 19 | 17 | 3 | 7 | P | 3 | 11 |
| 73 | 7 | 73 | 3 | P | P | 3 | P | 7 | 3 | 11 | P | 3 |
| 75 | 3 | 5 | 5 | 3 | 5 | 5 | 3 | 5 | 5 | 3 | 5 | 5 |
| 77 | 19 | 3 | P | P | 3 | 7 | P | 3 | 41 | 13 | 3 | P |
| 79 | 29 | 47 | 3 | 11 | 7 | 3 | P | 79 | 3 | P | 17 | 3 |
| 81 | 3 | 11 | P | 3 | P | 31 | 3 | 23 | P | 3 | 7 | 17 |
| 83 | P | 3 | 7 | P | 3 | 43 | P | 3 | 59 | 7 | 3 | 83 |
| 85 | 5 | 5 | 3 | 5 | 5 | 3 | 5 | 5 | 3 | 5 | 5 | 3 |
| 87 | 3 | 83 | P | 3 | P | 13 | 3 | 7 | P | 3 | P | P |
| 89 | 37 | 3 | P | P | 3 | P | 7 | 3 | P | 19 | 3 | P |
| 91 | 23 | 19 | 3 | P | P | 3 | 13 | 61 | 3 | P | P | 3 |
| 93 | 3 | P | 59 | 3 | 7 | P | 3 | P | P | 3 | P | 7 |
| 95 | 5 | 3 | 5 | 5 | 3 | 5 | 5 | 3 | 5 | 5 | 3 | 5 |
| 97 | P | 13 | 3 | 71 | 43 | 3 | 53 | 11 | 3 | 7 | P | 3 |
| 99 | 3 | 7 | P | 3 | P | 11 | 3 | 19 | 7 | 3 | 43 | 37 |

## Prime Number and Factor Table

| From / to | 8400–8500 | 8500–8600 | 8600–8700 | 8700–8800 | 8800–8900 | 8900–9000 | 9000–9100 | 9100–9200 | 9200–9300 | 9300–9400 | 9400–9500 | 9500–9600 |
|---|---|---|---|---|---|---|---|---|---|---|---|---|
| 1 | 31 | P | 3 | 7 | 13 | 3 | P | 19 | 3 | 71 | 7 | 3 |
| 3 | 3 | 11 | 7 | 3 | P | 29 | 3 | P | P | 3 | P | 13 |
| 5 | 5 | 3 | 5 | 5 | 3 | 5 | 5 | 3 | 5 | 5 | 3 | 5 |
| 7 | 7 | 47 | 3 | P | P | 3 | P | 7 | 3 | 41 | 23 | 3 |
| 9 | 3 | 67 | P | 3 | 23 | 59 | 3 | P | P | 3 | 97 | 37 |
| 11 | 13 | 3 | 79 | 31 | 3 | 7 | P | 3 | 61 | P | 3 | P |
| 13 | 47 | P | 3 | P | 7 | 3 | P | 13 | 3 | 67 | P | 3 |
| 15 | 3 | 5 | 5 | 3 | 5 | 5 | 3 | 5 | 5 | 3 | 5 | 5 |
| 17 | 19 | 3 | 7 | 23 | 3 | 37 | 71 | 3 | 13 | 7 | 3 | 31 |
| 19 | P | 7 | 3 | P | P | 3 | 29 | 11 | 3 | P | P | 3 |
| 21 | 3 | P | 37 | 3 | P | 11 | 3 | 7 | P | 3 | P | P |
| 23 | P | 3 | P | 11 | 3 | P | 7 | 3 | 23 | P | 3 | 89 |
| 25 | 5 | 5 | 3 | 5 | 5 | 3 | 5 | 5 | 3 | 5 | 5 | 3 |
| 27 | 3 | P | P | 3 | 7 | 79 | 3 | P | P | 3 | 11 | 7 |
| 29 | P | 3 | P | 7 | 3 | P | P | 3 | 11 | 19 | 3 | 13 |
| 31 | P | 19 | 3 | P | P | 3 | 11 | 23 | 3 | 7 | P | 3 |
| 33 | 3 | 7 | 89 | 3 | 11 | P | 3 | P | 7 | 3 | P | P |
| 35 | 5 | 3 | 5 | 5 | 3 | 5 | 5 | 3 | 5 | 5 | 3 | 5 |
| 37 | 11 | P | 3 | P | P | 3 | 7 | P | 3 | P | P | 3 |
| 39 | 3 | P | 53 | 3 | P | 7 | 3 | 13 | P | 3 | P | P |
| 41 | 23 | 3 | P | P | 3 | P | P | 3 | P | P | 3 | 7 |
| 43 | P | P | 3 | 7 | 37 | 3 | P | 41 | 3 | P | 7 | 3 |
| 45 | 3 | 5 | 5 | 3 | 5 | 5 | 3 | 5 | 5 | 3 | 5 | 5 |
| 47 | P | 3 | P | P | 3 | 23 | 83 | 3 | 7 | 13 | 3 | P |
| 49 | 7 | 83 | 3 | 13 | P | 3 | P | 7 | 3 | P | 11 | 3 |
| 51 | 3 | 17 | 41 | 3 | 53 | P | 3 | P | 11 | 3 | 13 | P |
| 53 | 79 | 3 | 17 | P | 3 | 7 | 11 | 3 | 19 | 47 | 3 | 41 |
| 55 | 5 | 5 | 3 | 5 | 5 | 3 | 5 | 5 | 3 | 5 | 5 | 3 |
| 57 | 3 | 43 | 11 | 3 | 17 | 13 | 3 | P | P | 3 | 7 | 19 |
| 59 | 11 | 3 | 7 | 19 | 3 | 17 | P | 3 | 47 | 7 | 3 | 11 |
| 61 | P | 7 | 3 | P | P | 3 | 13 | P | 3 | 11 | P | 3 |
| 63 | 3 | P | P | 3 | P | P | 3 | 7 | 59 | 3 | P | 73 |
| 65 | 5 | 3 | 5 | 5 | 3 | 5 | 5 | 3 | 5 | 5 | 3 | 5 |
| 67 | P | 13 | 3 | 11 | P | 3 | P | 89 | 3 | 17 | P | 3 |
| 69 | 3 | 11 | P | 3 | 7 | P | 3 | 53 | 13 | 3 | 17 | 7 |
| 71 | 43 | 3 | 13 | 7 | 3 | P | 47 | 3 | 73 | P | 3 | 17 |
| 73 | 37 | P | 3 | 31 | 19 | 3 | 43 | P | 3 | 7 | P | 3 |
| 75 | 3 | 5 | 5 | 3 | 5 | 5 | 3 | 5 | 5 | 3 | 5 | 5 |
| 77 | 7 | 3 | P | 67 | 3 | 47 | 29 | 3 | P | P | 3 | 61 |
| 79 | 61 | 23 | 3 | P | 13 | 3 | 7 | 67 | 3 | 83 | P | 3 |
| 81 | 3 | P | P | 3 | 83 | 7 | 3 | P | P | 3 | 19 | 11 |
| 83 | 17 | 3 | 19 | P | 3 | 13 | 31 | 3 | P | 11 | 3 | 7 |
| 85 | 5 | 5 | 3 | 5 | 5 | 3 | 5 | 5 | 3 | 5 | 5 | 3 |
| 87 | 3 | 31 | 7 | 3 | P | 11 | 3 | P | 37 | 3 | 53 | P |
| 89 | 13 | 3 | P | 11 | 3 | 89 | 61 | 3 | 7 | 41 | 3 | 43 |
| 91 | 7 | 11 | 3 | 59 | 17 | 3 | P | 7 | 3 | P | P | 3 |
| 93 | 3 | 13 | P | 3 | P | 17 | 3 | 29 | P | 3 | 11 | 53 |
| 95 | 5 | 3 | 5 | 5 | 3 | 5 | 5 | 3 | 5 | 5 | 3 | 5 |
| 97 | 29 | P | 3 | 19 | 7 | 3 | 11 | 17 | 3 | P | P | 3 |
| 99 | 3 | P | P | 3 | 11 | P | 3 | P | 17 | 3 | 7 | 29 |

## Rearrangement and Transposition of Terms in Formulas

A formula is a rule for a calculation expressed by using letters and signs instead of writing out the rule in words; by this means it is possible to condense, in a very small space, the essentials of long and cumbersome rules. The letters used in formulas simply stand in place of the figures which are to be substituted when solving a specific problem.

As an example, the formula for the horsepower transmitted by belting may be written:

$$P = \frac{SVW}{33,000}$$

in which
$P$ = horsepower transmitted;
$S$ = working stress of belt per inch of width, in pounds;
$V$ = velocity of belt in feet per minute;
$W$ = width of belt in inches.

If the working stress $S$, the velocity $V$, and the width $W$ are known, the horsepower can be found directly from this formula by inserting the given values. Assume $S = 33$; $V = 600$; and $W = 5$. Then:

$$P = \frac{33 \times 600 \times 5}{33,000} = 3.$$

Assume, however, that the horsepower $P$, the stress $S$, and the velocity $V$ are known, and that the width of belt, $W$, is to be found. The formula must then be rearranged so that the symbol $W$ will be on one side of the equals sign and all the known quantities on the other. The rearranged formula is as follows:

$$\frac{P \times 33,000}{SV} = W.$$

The quantities ($S$ and $V$) that were in the numerator on the right side of the equals sign are moved to the denominator on the left side, and "33,000" which was in the denominator on the right side of the equals sign is moved to the numerator on the other side. Symbols which are not part of a fraction, like "$P$" in the formula first given, are to be considered as being numerators (having the denominator 1).

Thus, any formula of the form $A = \dfrac{B}{C}$ can be rearranged as below:

$$A \times C = B, \quad \text{and} \quad C = \frac{B}{A}$$

Suppose a formula to be of the form:

$$A = \frac{B \times C}{D}$$

Then:
$$D = \frac{B \times C}{A}; \quad \frac{A \times D}{C} = B; \quad \frac{A \times D}{B} = C.$$

The method given is only directly applicable when all the quantities in the numerator or denominator are standing independently or are *factors of a product*. If connected by + or − signs, the entire numerator or denominator must be moved as a unit, thus,

Given:
$$\frac{B + C}{A} = \frac{D + E}{F}, \quad \text{to solve for } F$$

then
$$\frac{F}{A} = \frac{D + E}{B + C}$$

and
$$F = \frac{A(D + E)}{B + C}$$

A quantity preceded by a + or − sign can be transposed to the opposite side of the equals sign by changing its sign; if the sign is +, change it to − on the other side; if it is −, change it to +. This is called *transposition* of terms.

*Example:*
$$B + C = A - D; \text{ then } B + C + D = A;$$
$$B = A - D - C;$$
$$C = A - D - B;$$

## Order of Performing Arithmetic Operations

When several numbers or quantities in a formula are connected by signs indicating that additions, subtractions, multiplications, or divisions are to be made, the multiplications and divisions should be carried out first, in the order in which they appear, before the additions or subtractions are performed.

*Examples:*
$$10 + 26 \times 7 - 2 = 10 + 182 - 2 = 190.$$
$$18 \div 6 + 15 \times 3 = 3 + 45 = 48.$$
$$12 + 14 \div 2 - 4 = 12 + 7 - 4 = 15.$$

When it is required that certain additions and subtractions should precede multiplications and divisions, use is made of parentheses ( ) and brackets [ ]. These indicate that the calculation inside the parentheses or brackets should be carried out complete by itself before the remaining calculations are commenced. If one bracket is placed inside of another, the one inside is first calculated.

*Examples:*
$$(6 - 2) \times 5 + 8 = 4 \times 5 + 8 = 20 + 8 = 28.$$
$$6 \times (4 + 7) \div 22 = 6 \times 11 \div 22 = 66 \div 22 = 3.$$
$$2 + [10 \times 6(8 + 2) - 4] \times 2 = 2 + [10 \times 6 \times 10 - 4] \times 2$$
$$= 2 + [600 - 4] \times 2 = 2 + 596 \times 2 = 2 + 1192 = 1194.$$

The parentheses are considered as a sign of multiplication; for example, 6 (8 + 2) = 6 × (8 + 2).

The line or bar between the numerator and denominator in a fractional expression is to be considered as a division sign. For example,
$$\frac{12 + 16 + 22}{10} = (12 + 16 + 22) \div 10 = 50 \div 10 = 5.$$

In formulas the multiplication sign (×) is often left out between symbols or letters, the values of which are to be multiplied. Thus
$$AB = A \times B, \text{ and } \frac{ABC}{D} = (A \times B \times C) \div D$$

## Ratio and Proportion

The *ratio* between two quantities is the quotient obtained by dividing the first quantity by the second. For example, the ratio between 3 and 12 is $\frac{1}{4}$, and the ratio between 12 and 3 is 4. Ratio is generally indicated by the sign (:); thus 12 : 3 indicates the ratio of 12 to 3.

A *reciprocal* or *inverse* ratio is the reciprocal of the original ratio. Thus, the inverse ratio of 5 : 7 is 7 : 5.

In a *compound* ratio each term is the product of the corresponding terms in two or more simple ratios. Thus, when
$$8 : 2 = 4, \qquad 9 : 3 = 3, \qquad 10 : 5 = 2,$$
then the compound ratio is:
$$8 \times 9 \times 10 : 2 \times 3 \times 5 = 4 \times 3 \times 2,$$
$$720 : 30 = 24.$$

*Proportion* is the equality of ratios. Thus,

$$6 : 3 = 10 : 5, \quad \text{or} \quad 6 : 3 :: 10 : 5.$$

The first and last terms in a proportion are called the *extremes;* the second and third, the *means.* The product of the extremes is equal to the product of the means. Thus,

$$25 = 2 = 100 : 8 \quad \text{and} \quad 25 \times 8 = 2 \times 100.$$

If three terms in a proportion are known, the remaining term may be found by the following rules:

The first term is equal to the product of the second and third terms, divided by the fourth.

The second term is equal to the product of the first and fourth terms, divided by the third.

The third term is equal to the product of the first and fourth terms, divided by the second.

The fourth term is equal to the product of the second and third terms, divided by the first.

*Examples:* — Let $x$ be the term to be found, then,

$$x : 12 = 3.5 : 21 \qquad x = \frac{12 \times 3.5}{21} = \frac{42}{21} = 2.$$

$$\tfrac{1}{4} : x = 14 : 42 \qquad x = \frac{\tfrac{1}{4} \times 42}{14} = \frac{1}{4} \times 3 = \frac{3}{4}$$

$$5 : 9 = x : 63 \qquad x = \frac{5 \times 63}{9} = \frac{315}{9} = 35$$

$$\tfrac{1}{4} : \tfrac{7}{8} = 4 : x \qquad x = \frac{\tfrac{7}{8} \times 4}{\tfrac{1}{4}} = \frac{3\tfrac{1}{2}}{\tfrac{1}{4}} = 14.$$

If the second and third terms are the same, either is said to be the *mean proportional* between the other two. Thus, $8 : 4 = 4 : 2$, and 4 is the mean proportional between 8 and 2. The mean proportional between two numbers may be found by multiplying the numbers together, and extracting the square root of the product. Thus, the mean proportional between 3 and 12 is found as below:

$$3 \times 12 = 36, \quad \text{and} \quad \sqrt{36} = 6,$$

which is the mean proportional.

**Practical Examples Involving Simple Proportion.** — If it takes 18 days to assemble 4 lathes, how long would it require to assemble 14 lathes?

Let the number of days to be found be $x$. Then write out the proportion as below:

$$4 \quad : \quad 18 \quad = \quad 14 \quad : \quad x$$
$$\text{(lathes : days = lathes : days)}$$

Find now the fourth term by the rule given:

$$x = \frac{18 \times 14}{4} = 63 \text{ days.}$$

Thirty-four linear feet of bar stock are required for the blanks for 100 clamping bolts. How many feet of stock would be required for 912 bolts?

Let $x$ = total length of stock required for 912 bolts.

$$34 : 100 = x : 912$$
$$\text{(feet : bolts = feet : bolts)}$$

Then, the third term $x = \dfrac{34 \times 912}{100} = 310$ feet, approximately.

**Inverse Proportion.** — In an inverse proportion, as one of the items involved *increases*, the corresponding item in the proportion *decreases*, or vice versa. For example, a factory employing 270 men completes a given number of typewriters weekly, the number of working hours being 44 per week. How many men would be required for the same production if the working hours were reduced to 40 per week?

The time per week is in an inverse proportion to the number of men employed; the shorter the time, the more men. The inverse proportion is written:

$$270 : x = 40 : 44$$

(men, 44-hour basis: men, 40-hour basis = time, 40-hour basis: time, 44-hour basis)
Thus

$$\frac{270}{x} = \frac{40}{44} \quad \text{and} \quad x = \frac{270 \times 44}{40} = 297 \text{ men.}$$

**Problems Involving Both Simple and Inverse Proportions.** — If two groups of data are related both by direct (simple) and inverse proportions among the various quantities, then a simple mathematical relation that may be used in solving problems is as follows:

$$\frac{\text{Product of all directly proportional items in first group}}{\text{Product of all inversely proportional items in first group}}$$
$$= \frac{\text{Product of all directly proportional items in second group}}{\text{Product of all inversely proportional items in second group}}$$

*Example:* If a man capable of turning 65 studs in a day of 10 hours is paid $1.50 per hour, how much per hour ought a man be paid who turns 72 studs in a 9-hour day, if compensated in the same proportion?

The first group of data in this problem consists of the number of hours worked by the first man, his hourly wage, and the number of studs which he produces per day; the second group contains similar data for the second man except for his unknown hourly wage which may be indicated by $x$.

The labor cost per stud, as may be seen, is directly proportional to the number of hours worked and the hourly wage. These quantities, therefore, are used in the numerators of the fractions in the formula. The labor cost per stud is inversely proportional to the number of studs produced per day. (The greater the number of studs produced in a given time the less the cost per stud.) The numbers of studs per day, therefore, are placed in the denominators of the fractions in the formula. Thus,

$$\frac{10 \times 1.50}{65} = \frac{9 \times x}{72}$$

$$x = \frac{10 \times 1.50 \times 72}{65 \times 9} = \$1.85 \text{ per hour}$$

## Percentage

If out of 100 pieces made, 12 do not pass inspection, it is said that 12 per cent (12 on the hundred) are rejected. If a quantity of steel is bought for $100 and sold for $140, the profit is 40 per cent.

The per cent of gain or loss is found by dividing the amount of gain or loss by the *original* number of which the percentage is wanted, and multiplying the quotient by 100.

*Examples:* — Out of a total output of 280 castings a day, 30 castings are, on an average, rejected. What is the percentage of bad castings?

$$\frac{30}{280} \times 100 = 10.7 \text{ per cent.}$$

If by a new process 100 pieces can be made in the same time as 60 could formerly be made, what is the gain in output of the new process over the old, expressed in per cent?

Original number, 60; gain $100 - 60 = 40$. Hence,

$$\frac{40}{60} \times 100 = 66.7 \text{ per cent.}$$

Care should be taken always to use the original number, or the number of which the percentage is wanted, as the divisor in all percentage calculations. In the example just given, it is the percentage of gain over the old output 60 that is wanted, and not the percentage with relation to the new output 100. Mistakes are often made by overlooking this important point.

## Interest

Interest is the money paid for the use of money lent for a certain time. *Simple* interest is the interest paid on the principal (money lent) only. When simple interest that is due is not paid, and its amount is added to the interest-bearing principal, the interest calculated on this new principal is called *compound* interest. The compounding of the interest into the principal may take place yearly or oftener, according to circumstances.

**Simple Interest.** — The following formulas are applicable to the calculations involving simple interest. Let:

$P$ = principal or amount of money lent;
$p$ = per cent of interest;
$r$ = interest rate = the interest, expressed decimally, on \$1.00 for one year
= the per cent of interest divided by 100; thus, if the interest is 6 per cent, the rate $r = \frac{6}{100} = 0.06$;
$n$ = the number of years for which interest is calculated;
$I$ = the amount of interest for $n$ years at the given rate;
$P_n$ = principal with interest for $n$ years added, or the total amount after $n$ years.

Then:

Interest for $n$ years, $I = Prn$.
Total amount after $n$ years, $P_n = P + Prn = P(1 + rn)$.
Interest rate $r = I \div Pn$.
Number of years $n = I \div Pr$.
Principal, or amount lent $= I \div rn$.

*Example:* — Assume that \$250 has been loaned for three years at 6 per cent simple interest. Then: $P = 250$; $p = 6$; $r = p \div 100 = 0.06$; $n = 3$.

$$I = Prn = 250 \times 0.06 \times 3 = \$45.$$
$$P_n = P + I = 250 + 45 = \$295.$$

The accurate interest for one day is $\frac{1}{365}$ of the interest for one year. Banks, however, customarily take the year as composed of 12 months of 30 days, making a total of 360 days to a year.

**Compound Interest.** — The following formulas are applicable when compound interest is to be computed, using the same notation as for simple interest, and assuming that the interest is compounded annually.

The total amount after $n$ years, $P_n = P(1 + r)^n$.

The principal $P = \dfrac{P_n}{(1+r)^n}$ $\qquad$ The rate $r = \sqrt[n]{\dfrac{P_n}{P}} - 1$

The number of years during which the money is lent

$$n = \frac{\log P_n - \log P}{\log (1 + r)}$$

Logarithms are especially useful in calculating compound interest. To find the total amount $P_n$ of principal and interest after $n$ years, the formula just given can be transcribed as below:

$$\log P_n = \log P + n \log (1 + r).$$

If the interest is payable $q$ times a year, it will be computed $q$ times during each year, or $nq$ times during $n$ years. The rate for each compounding will be $r \div q$, if $r$ is the annual rate. Hence, at the end of $n$ years the amount due will be:

$$P_n = P \left(1 + \frac{r}{q}\right)^{nq}$$

Thus, if the term be five years, the interest be payable quarterly, and the annual rate be 6 per cent, then, $n = 5$; $q = 4$; $r = 0.06$; $r \div q = 0.06 \div 4 = 0.015$; and $nq = 5 \times 4 = 20$.

*Example:* — In what time will \$500 become \$1000 at 6 per cent interest compounded yearly?

$$P_n = 1000; \quad P = 500; \quad r = 0.06.$$

Substituting these values in the formula:

$$1000 = 500 (1 + 0.06)^n, \quad \text{or} \quad 2 = 1.06^n, \quad \text{and} \quad n \times \log 1.06 = \log 2.$$

Hence
$$n = \frac{0.30103}{0.02531} = 11.9 \text{ years.}$$

This is the number of years in which any principal will double itself at 6 per cent compound interest.

**Present Value and Discount.** — The present value $V$ of a given amount due in a given time, is the sum which placed at interest for the given time, will produce the given amount. Hence,

At simple interest, $V = \dfrac{P_n}{1 + nr}$

At compound interest, $V = \dfrac{P_n}{(1 + r)^n}$

in which $P_n$ is the amount due in $n$ years time, and $r$ is the rate of simple interest, or the per cent divided by 100.

The *true discount* $D$ is the difference between the amount due at the end of $n$ years and the present value, or,

At simple interest, $D = P_n - V = \dfrac{P_n n r}{1 + nr}$

At compound interest, $D = P_n - V = P_n \left[1 - \dfrac{1}{(1 + r)^n}\right]$

These formulas are for interest compounded annually. If the interest is payable and compounded semi-annually, or quarterly, modify the formulas as indicated in the formulas for compound interest.

*Example:* — Required the present value and discount of \$500 due in six months at 6 per cent simple interest. Here, $P_n = 500$; $n = {}^6/_{12}$ years $= \frac{1}{2}$; $r = 0.06$; then,

$$V = \frac{500}{1 + 0.5 \times 0.06} = \$485.44.$$

$$D = 500 - 485.44 = \$14.56.$$

*Example:* — Required the sum which placed at 5 per cent compound interest, will in three years produce \$5000. Here, $P_n = 5000$; $r = 0.05$; $n = 3$. Then,

$$V = \frac{5000}{(1 + 0.05)^3} = 4319.19.$$

Bank discount is calculated at simple interest on the total amount of a promissory note for the term of the note and on the basis of a year of 360 days.

**Annuities.** — An annuity is a fixed sum paid at regular intervals. In the formulas given below, yearly payments are assumed. It is customary to calculate annuities on the basis of compound interest.

If an annuity $A$ is to be paid out for $n$ consecutive years, the interest rate being then the present value $P$ of the annuity is:

$$P = A\, \frac{(1+r)^n - 1}{(1+r)^n r}$$

*Example:* — If an annuity of \$200 is to be paid for 10 years, what is the present amount of money that need be deposited if the interest is 5 per cent? Here,

$$A = 200; \quad r = 5 \div 100 = 0.05; \quad n = 10.$$

$$P = 200\, \frac{1.05^{10} - 1}{1.05^{10} \times 0.05} = 1544.36.$$

The annuity that a principal $P$, drawing interest at the rate $r$, will give for a period of $n$ years, is:

$$A = \frac{Pr\,(1+r)^n}{(1+r)^n - 1}$$

*Example:* — A sum of \$10,000 is placed at 4 per cent interest. What is the amount of the annuity which can be paid for 20 years out of this sum? Here,

$$P = 10,000; \quad r = 0.04; \quad n = 20.$$

$$A = \frac{10,000 \times 0.04 \times 1.04^{20}}{1.04^{20} - 1} = 735.82.$$

If at the beginning of each year a sum $A$ is set aside at an interest rate $r$, then the total value of the sum set aside, with interest, will be at the end of $n$ years:

$$P_n = A\, \frac{(1+r)\,[(1+r)^n - 1]}{r}$$

If at the end of each year a sum $A$ is set aside at an interest rate $r$, then the total value of the principal, with interest, at the end of $n$ years will be:

$$P_n = A\, \frac{(1+r)^n - 1}{r}$$

If a principal $P$ is increased or decreased by a sum $A$ at the end of each year, then the value of the principal after $n$ years will be:

$$P_n = P\,(1+r)^n \pm A\, \frac{(1+r)^n - 1}{r}$$

If the sum $A$ by which the principal $P$ is decreased each year is greater than the total yearly interest on the principal, then the principal, with the accumulated interest, will be entirely used up in $n$ years:

$$n = \frac{\log A - \log (A - Pr)}{\log (1 + r)}$$

**Sinking Funds.**— Amortization is "the extinction of a debt, usually by means of a sinking fund." The sinking fund is created by a fixed investment $S$ placed annually at compound interest for a term of years, and is hence an annuity of sufficient size to produce at the end of the term of years the amount necessary for the repayment of the principal of the debt, or to provide a definite sum for other purposes. Let:

$S$ = the annual investment;

$r$ = rate of interest (the per cent divided by 100);

$P$ = the amount of the sinking fund;

$n$ = the number of years for its creation.

Then:

$$P = S\frac{(1 + r)^n - 1}{r}, \quad \text{and} \quad S = \frac{Pr}{(1 + r)^n - 1}$$

which formulas correspond to those given above, where a sum $A$ was laid aside at the end of each year.

*Example:* — If $2000 is invested annually for 10 years, at 4 per cent compound interest, as a sinking fund, what would be the total amount of the fund at the expiration of the term? Here, $S = 2000$; $n = 10$; $r = 0.04$.

$$P = 2000 \frac{1.04^{10} - 1}{0.04} = 24,012.25$$

## Cost of Mixture

When an alloy is composed of several metals varying in price, the price per pound of the alloy can be found as in the following example: An alloy is composed of 50 pounds of copper at 14 cents a pound, 10 pounds of tin at 29 cents a pound, 20 pounds of zinc at 5 cents a pound, and 5 pounds of lead at 4 cents a pound. What is the cost of the alloy per pound, no account being taken of the cost of mixing it?

Multiply the number of pounds of each of the ingredients by its price per pound, add these products together, and divide the sum by the total weight of all the ingredients. The quotient is the price per pound of the alloy.

$50 \times 14 + 10 \times 29 + 20 \times 5 + 5 \times 4 = 700 + 290 + 100 + 20 = 1110$.

Total weight of metal in alloy = $50 + 10 + 20 + 5 = 85$.

Price per pound of alloy = $1110/85 = 13$ cents, approximately.

In general, let $a$, $b$, $c$ and $d$ be the weights of each of the ingredients, and $w$, $x$, $y$ and $z$ be their respective values per unit weight. Then the average price $P$ per unit weight of the alloy is found by the formula:

$$P = \frac{aw + bx + cy + dz}{a + b + c + d}$$

*Example:* — Find the average price per pound of an alloy containing 40 pounds of tin at 30 cents per pound, 48 pounds of lead at 4 cents per pound, 10 pounds of antimony at 8 cents per pound, and 2 pounds of copper at 15 cents per pound.

$$P = \frac{40 \times 30 + 48 \times 4 + 10 \times 8 + 2 \times 15}{40 + 48 + 10 + 2} = \frac{1502}{100} = 15.02 \text{ cents.}$$

**Formulas for Arithmetical Progression**

| To Find | Given | | | Use Equation |
|---|---|---|---|---|
| a | d | l | n | $a = l - (n-1)d$ |
| | d | n | S | $a = \dfrac{S}{n} - \dfrac{n-1}{2} \times d$ |
| | d | l | S | $a = \dfrac{d}{2} \pm \dfrac{1}{2}\sqrt{(2l+d)^2 - 8dS}$ |
| | l | n | S | $a = \dfrac{2S}{n} - l$ |
| d | a | l | n | $d = \dfrac{l-a}{n-1}$ |
| | a | n | S | $d = \dfrac{2S - 2an}{n(n-1)}$ |
| | a | l | S | $d = \dfrac{l^2 - a^2}{2S - l - a}$ |
| | l | n | S | $d = \dfrac{2nl - 2S}{n(n-1)}$ |
| l | a | d | n | $l = a + (n-1)d$ |
| | a | d | S | $l = -\dfrac{d}{2} \pm \dfrac{1}{2}\sqrt{8dS + (2a-d)^2}$ |
| | a | n | S | $l = \dfrac{2S}{n} - a$ |
| | d | n | S | $l = \dfrac{S}{n} + \dfrac{n-1}{2} \times d$ |
| n | a | d | l | $n = 1 + \dfrac{l-a}{d}$ |
| | a | d | S | $n = \dfrac{d-2a}{2d} \pm \dfrac{1}{2d}\sqrt{8dS + (2a-d)^2}$ |
| | a | l | S | $n = \dfrac{2S}{a+l}$ |
| | d | l | S | $n = \dfrac{2l+d}{2d} \pm \dfrac{1}{2d}\sqrt{(2l+d)^2 - 8dS}$ |
| S | a | d | n | $S = \dfrac{n}{2}[2a + (n-1)d]$ |
| | a | d | l | $S = \dfrac{a+l}{2} + \dfrac{l^2 - a^2}{2d} = \dfrac{a+l}{2}(l+d-a)$ |
| | a | l | n | $S = \dfrac{n}{2}(a+l)$ |
| | d | l | n | $S = \dfrac{n}{2}[2l - (n-1)d]$ |

## Arithmetical Progression

An arithmetical progression is a series of numbers in which each consecutive term differs from the preceding one by a fixed amount called the *common difference*, *d*. Thus, 1, 3, 5, 7, etc., is an arithmetical progression where the difference *d* is 2. The difference in this case is *added* to the preceding term, and the progression is called increasing. In the series 13, 10, 7, 4, etc., the difference is (− 3), and the progression is called decreasing. In any arithmetical progression (or part of progression) let

$a$ = the first term considered;
$l$ = the last term considered;
$n$ = the number of terms;
$d$ = the common difference;
$S$ = the sum of $n$ terms.

Then the general formulas are:

$$l = a + (n - 1)\, d \quad \text{and} \quad S = \frac{a + l}{2} \times n$$

In these formulas $d$ is positive in an increasing and negative in a decreasing progression. When any three of the five quantities above are given, the other two can be found by the formulas in the accompanying table of arithmetical progression.

*Example:* — In an arithmetical progression, the first term equals 5, and the last term 40. The difference is 7. Find the sum of the progression.

$$S = \frac{a + l}{2\,d}\,(l + d - a) = \frac{5 + 40}{2 \times 7}\,(40 + 7 - 5) = 135.$$

## Geometrical Progression

A geometrical progression or a geometrical series is a series in which each term is derived by multiplying the preceding term by a constant multiplier called the *ratio*. When the ratio is greater than 1, the progression is increasing; when smaller than 1, it is decreasing. Thus, 2, 6, 18, 54, etc., is an increasing geometrical progression with a ratio of 3, while 24, 12, 6, etc., is a decreasing progression with a ratio of ½.

In any geometrical progression (or part of progression) let

$a$ = the first term;
$l$ = the last (or nth) term;
$n$ = the number of terms;
$r$ = the ratio of the progression;
$S$ = the sum of $n$ terms.

Then the general formulas are:

$$l = ar^{n-1} \quad \text{and} \quad S = \frac{rl - a}{r - 1}$$

When any three of the five quantities above are given, the other two can be found by the formulas tabulated in the accompanying table. Geometrical progressions are used for finding the successive speeds in machine tool drives, in interest calculations, etc.

*Example:* — The lowest speed of a lathe is 20 R.P.M. The highest speed is 225 R.P.M. There are 18 speeds. Find the ratio between successive speeds.

$$\text{Ratio, } r = \sqrt[n-1]{\frac{l}{a}} = \sqrt[17]{\frac{225}{20}} = \sqrt[17]{11.25} = 1.153.$$

### Formulas for Geometrical Progression

| To Find | Given | | | Use Equation |
|---|---|---|---|---|
| a | $l$ | $n$ | $r$ | $a = \dfrac{l}{r^{n-1}}$ |
|   | $n$ | $r$ | $S$ | $a = \dfrac{(r-1)S}{r^n - 1}$ |
|   | $l$ | $r$ | $S$ | $a = lr - (r-1)S$ |
|   | $l$ | $n$ | $S$ | $a(S-a)^{n-1} = l(S-l)^{n-1}$ |
| l | $a$ | $n$ | $r$ | $l = ar^{n-1}$ |
|   | $a$ | $r$ | $S$ | $l = \dfrac{1}{r}[a + (r-1)S]$ |
|   | $a$ | $n$ | $S$ | $l(S-l)^{n-1} = a(S-a)^{n-1}$ |
|   | $n$ | $r$ | $S$ | $l = \dfrac{S(r-1)r^{n-1}}{r^n - 1}$ |
| n | $a$ | $l$ | $r$ | $n = \dfrac{\log l - \log a}{\log r} + 1$ |
|   | $a$ | $r$ | $S$ | $n = \dfrac{\log[a + (r-1)S] - \log a}{\log r}$ |
|   | $a$ | $l$ | $S$ | $n = \dfrac{\log l - \log a}{\log(S-a) - \log(S-l)} + 1$ |
|   | $l$ | $r$ | $S$ | $n = \dfrac{\log l - \log[lr - (r-1)S]}{\log r} + 1$ |
| r | $a$ | $l$ | $n$ | $r = \sqrt[n-1]{\dfrac{l}{a}}$ |
|   | $a$ | $n$ | $S$ | $r^n = \dfrac{Sr}{a} + \dfrac{a-S}{a}$ |
|   | $a$ | $l$ | $S$ | $r = \dfrac{S-a}{S-l}$ |
|   | $l$ | $n$ | $S$ | $r^n = \dfrac{Sr^{n-1}}{S-l} - \dfrac{l}{S-l}$ |
| S | $a$ | $n$ | $r$ | $S = \dfrac{a(r^n - 1)}{r - 1}$ |
|   | $a$ | $l$ | $r$ | $S = \dfrac{lr - a}{r - 1}$ |
|   | $a$ | $l$ | $n$ | $S = \dfrac{\sqrt[n-1]{l^n} - \sqrt[n-1]{a^n}}{\sqrt[n-1]{l} - \sqrt[n-1]{a}}$ |
|   | $l$ | $n$ | $r$ | $S = \dfrac{l(r^n - 1)}{(r-1)r^{n-1}}$ |

### Greek Letters

The Greek letters are frequently used in mathematical expressions and formulas. The Greek alphabet is given below.

| | | | | | | | | | | |
|---|---|---|---|---|---|---|---|---|---|---|
| A | $\alpha$ | Alpha | H | $\eta$ | Eta | N | $\nu$ | Nu | T | $\tau$ Tau |
| B | $\beta$ | Beta | $\Theta$ | $\vartheta\ \theta$ | Theta | $\Xi$ | $\xi$ | Xi | Y | $\upsilon$ Upsilon |
| $\Gamma$ | $\gamma$ | Gamma | I | $\iota$ | Iota | O | $o$ | Omicron | $\Phi$ | $\phi$ Phi |
| $\Delta$ | $\delta$ | Delta | K | $\kappa$ | Kappa | $\Pi$ | $\pi$ | Pi | X | $\chi$ Chi |
| E | $\epsilon$ | Epsilon | $\Lambda$ | $\lambda$ | Lambda | P | $\rho$ | Rho | $\Psi$ | $\psi$ Psi |
| Z | $\zeta$ | Zeta | M | $\mu$ | Mu | $\Sigma$ | $\sigma\ s$ | Sigma | $\Omega$ | $\omega$ Omega |

## Positive and Negative Numbers

The degrees on a thermometer scale extending upward from the zero point may be called *positive* and may be preceded by a plus sign; thus $+5$ degrees means 5 degrees above zero. The degrees below zero may be called *negative* and may be preceded by a minus sign; thus $-5$ degrees means 5 degrees below zero. In the same way, the ordinary numbers 1, 2, 3, etc., which are larger than 0, are called positive numbers; but numbers can be conceived of as extending in the other direction from 0, numbers that, in fact, are less than 0, and these are called negative. As these numbers must be expressed by the same figures as the positive numbers they are designated by a minus sign placed before them, thus: $(-3)$. A negative number should always be enclosed within parentheses whenever it is written in line with other numbers; for example: $17 + (-13) - 3 \times (-0.76)$.

Negative numbers are most commonly met with in the use of logarithms and natural trigonometric functions. The following rules govern calculations with negative numbers.

A negative number can be added to a positive number by subtracting its numerical value from the positive number.

*Example:*            $4 + (-3) = 4 - 3 = 1.$

A negative number can be subtracted from a positive number by adding its numerical value to the positive number.

*Example:*            $4 - (-3) = 4 + 3 = 7.$

A negative number can be added to a negative number by adding the numerical values and making the sum negative.

*Example:*            $(-4) + (-3) = -7.$

A negative number can be subtracted from a negative number by subtracting the numerical values and making the difference negative.

*Example:*            $(-4) - (-3) = -1.$

If in a subtraction the number to be subtracted is larger than the number from which it is to be subtracted, the calculation can be carried out by subtracting the smaller number from the larger, and indicating that the remainder is negative.

*Example:*            $3 - 5 = -(5 - 3) = -2.$

When a positive number is to be multiplied or divided by a negative number, multiply or divide the numerical values as usual; the product or quotient, respectively, is negative. The same rule is true if a negative number is multiplied or divided by a positive number.

*Examples:*        $4 \times (-3) = -12; \quad (-4) \times 3 = -12;$
               $15 \div (-3) = -5; \quad (-15) \div 3 = -5.$

When two negative numbers are to be multiplied by each other, the product is positive. When a negative number is divided by a negative number, the quotient is positive.

*Examples:*    $(-4) \times (-3) = 12;$    $(-4) \div (-3) = 1.333.$

The two last rules are often expressed for memorizing as follows: "Equal signs make plus, unequal signs make minus."

## Powers and Roots

The *square* of a number (or quantity) is the product of that number multiplied by itself. Thus, the square of 9 is $9 \times 9 = 81$. The square of a number is indicated by the *exponent* ($^2$), thus: $9^2 = 9 \times 9 = 81$.

The *cube* or *third power* of a number is the product obtained by using that number as a factor three times. Thus, the cube of 4 is $4 \times 4 \times 4 = 64$, and is written $4^3$.

If a number is used as a factor four or five times, the product is the fourth or fifth power. Thus $3^4 = 3 \times 3 \times 3 \times 3 = 81$, and $2^5 = 2 \times 2 \times 2 \times 2 \times 2 = 32$. A number can be raised to any power by using it as a factor the required number of times.

The *square root* of a given number is that number which, when multiplied by itself, will give a product equal to the given number. The square root of 16 (written $\sqrt{16}$) equals 4, because $4 \times 4 = 16$.

The *cube root* of a given number is that number which, when used as a factor three times, will give a product equal to the given number. Thus, the cube root of 64 (written $\sqrt[3]{64}$) equals 4, because $4 \times 4 \times 4 = 64$.

The fourth, fifth, etc., roots of a given number are those numbers which when used as factors four, five, etc., times, will give as a product the given number. Thus $\sqrt[4]{16} = 2$, because $2 \times 2 \times 2 \times 2 = 16$.

The multiplications required for raising numbers to powers and the extracting of roots are greatly facilitated by the use of logarithms. The extracting of the square root and cube root by the regular arithmetical methods is a slow and cumbersome operation, and any roots can be more rapidly found by using logarithms.

The tables of squares and cubes, and square roots and cube roots, found at the beginning of this book, give these values directly for all whole numbers up to 2000. For ordinary practical calculations the squares, cubes, etc., for fractional values between whole numbers can usually be estimated. These tables also give the *reciprocals* of numbers from 1 to 2000. The reciprocal of a number is the quotient obtained by dividing 1 by the number. Thus the reciprocal of 4 is $1 \div 4 = 0.25$. The reciprocal values given in the tables can be used to save labor in division, as the quotient can be obtained by multiplying the dividend by the reciprocal of the divisor. Thus, the reciprocal of 244 is 0.0040984. To divide 13 by 244, or to reduce $^{13}/_{244}$ to a decimal, multiply as follows: $13 \times 0.0040984 = 0.0532792$.

As the numbers in the second column of the tables, are the *squares* of the numbers in the first, it follows that the numbers in the first column are the *square roots* of the numbers in the second column. Similarly the numbers in the first column are the *cube roots* of the numbers or *cubes* in the third column. Hence the tables may be used for finding the roots of numbers beyond the direct range of the tables.

*Example:* — Find the square root of 9253 using the table. The table shows that 9216 is the square of 96; hence it is evident that the square root of the given number is a little over 96.

In the column of squares of numbers find a number the first four figures of which are nearest to the four figures in the given number. Thus, on page 21 we find in the column of squares the number 925444. The first four figures are within one of equalling the given number and this is the square of a number beginning with the figures 9 and 6; therefore the square root of 9253 is 96.2 nearly. The square root of 9253, accurate to three decimal places, is 96.192 so that the result obtained by the table is nearly correct. The indirect method of using the tables for determining *cube roots* is similar in principle to that just described for square roots.

## Powers of Ten Notation

Powers of ten notation is used to simplify calculations and insure accuracy, particularly with respect to the position of decimal points, and also simplifies the expression of numbers which are so large or so small as to be unwieldy. For example, the metric (SI) pressure unit pascal is equivalent to 0.00000986923 atmospheres or 0.0001450377 pound/inch². In powers of ten notation these figures are $9.86923 \times 10^{-6}$ atmospheres and $1.450377 \times 10^{-4}$ pounds/inch². The notation also facilitates adaptation of numbers for electronic data processing and computer readout.

**Expressing Numbers in Powers of Ten Notation.** — In this system of notation every number is expressed by two factors, one of which is some integer from 1 to 9 followed by a decimal and the other is some power of 10.

Thus, 10,000 is expressed as $1.0000 \times 10^4$ and 10,463 as $1.0463 \times 10^4$. The number 43 is expressed $4.3 \times 10$ and 568 is expressed $5.68 \times 10^2$.

In the case of decimals, the number 0.0001 which as a fraction is $\frac{1}{10,000}$ is expressed as $1 \times 10^{-4}$ and 0.0001463 is expressed as $1.463 \times 10^{-4}$. The decimal 0.498 is expressed as $4.98 \times 10^{-1}$ and 0.03146 is expressed as $3.146 \times 10^{-2}$.

**Rules for Converting any Number to Powers of Ten Notation.** — Any number can be converted to the powers of ten notation by means of one of two rules.

*Rule 1:* If the number is a whole number or a whole number and a decimal so that it has digits to the left of the decimal point, the decimal point is moved a sufficient number of places to the *left* to bring it to the immediate right of the first digit. With the decimal point shifted to this position, the number so written comprises the *first* factor when written in powers of ten notation.

The number of places that the decimal point is moved to the left, to bring it immediately to the right of the first digit is the *positive* index or power of 10 that comprises the *second* factor when written in powers of ten notation.

Thus, to write 4639 in this notation, the decimal point is moved three places to the left giving the two factors: $4.639 \times 10^3$. Similarly,

$$431.412 = 4.31412 \times 10^2$$
$$986388 = 9.86388 \times 10^5$$

*Rule 2:* If the number is a decimal, i.e., it has digits entirely to the right of the decimal point, then the decimal point is moved a sufficient number of places to the *right* to bring it immediately to the right of the first digit. With the decimal point shifted to this position, the number so written comprises the *first* factor when written in powers of ten notation.

The number of places that the decimal point is moved to the *right* to bring it immediately to the right of the first digit is the *negative* index or power of 10 that follows the number when written in powers of ten notation.

Thus, to bring the decimal point in 0.005721 to the immediate right of the first digit which is 5, it must be moved *three* places to the right, giving the two factors: $5.721 \times 10^{-3}$. Similarly,

$$0.469 = 4.69 \times 10^{-1}$$
$$0.0000516 = 5.16 \times 10^{-5}$$

## EQUATIONS

ions of the First Degree with Two Unknowns. — The form
ations is:

$$ax + by = c$$
$$a_1x + b_1y = c_1$$

$$x = \frac{cb_1 - c_1b}{ab_1 - a_1b} \qquad y = \frac{ac_1 - a_1c}{ab_1 - a_1b}$$

$$3x + 4y = 17$$
$$5x - 2y = 11$$

$$\frac{17 \times (-2) - 11 \times 4}{3 \times (-2) - 5 \times 4} = \frac{-34 - 44}{-6 - 20} = \frac{-78}{-26} = 3.$$

can now be most easily found by inserting the value of $x$ in one of

$$5 \times 3 - 2y = 11; \quad 2y = 15 - 11 = 4; \quad y = 2.$$

Quadratic Equations with One Unknown. — If the form of the
$+ bx + c = 0$, then

$$x = \frac{-b \pm \sqrt{b^2 - 4ac}}{2a}$$

Given the equation, $1x^2 + 6x + 5 = 0$, then $a = 1$; $b = 6$ and

$$\frac{\sqrt{6^2 - 4 \times 1 \times 5}}{2 \times 1} = \frac{(-6) + 4}{2} = -1; \quad \text{or} \quad \frac{(-6) - 4}{2} = -5$$

f the equation is $ax^2 + bx = c$, then

$$x = \frac{-b \pm \sqrt{b^2 + 4ac}}{2a}$$

A right-angle triangle has a hypotenuse 5 inches long and one si
inch longer than the other; find the lengths of the two sides.
ne side and $x + 1 = $ other side; then $x^2 + (x + 1)^2 = 5^2$ or $x^2 + x^2$
25; or $2x^2 + 2x = 24$; or $x^2 + x = 12$. Now referring to the ba
$^2 + bx = c$, we find, in this case, that $a = 1$; $b = 1$, and $c = 12$; hen

$$1 \pm \frac{\sqrt{1 + 4 \times 1 \times 12}}{2 \times 1} = \frac{(-1) + 7}{2} = 3 \quad \text{or} \quad x = \frac{(-1) - 7}{2} = -4$$

positive value (3) would apply in this case, the lengths of the two si
inches and $x + 1 = 4$ inches.

Equations. — If the given equation has the form: $x^3 + ax + b = 0$, th

$$x = \left( -\frac{b}{2} + \sqrt{\frac{a^3}{27} + \frac{b^2}{4}} \right)^{\frac{1}{3}} + \left( -\frac{b}{2} - \sqrt{\frac{a^3}{27} + \frac{b^2}{4}} \right)^{\frac{1}{3}}$$

quation $x^3 + px^2 + qx + r = 0$, may be reduced to the form $x_1^3 + ax_1 + b$
tituting $x_1 - \frac{p}{3}$ for $x$ in the given equation.

---

**Multiplying Numbers Written in Powers of Ten Notation.** — When multiplying two numbers written in the powers of ten notation together, the procedure is as follows:

1. Multiply the first factor of one number by the first factor of the other to obtain the first factor of the product.

2. Add the index of the second factor (which is some power of 10) of one number to the index of the second factor of the other number to obtain the index of the second factor (which is some power of 10) in the product. Thus

$$(4.31 \times 10^{-2}) \times (9.0125 \times 10) =$$
$$(4.31 \times 9.0125) \times 10^{-2+1} = 38.844 \times 10^{-1}$$
$$(5.986 \times 10^4) \times (4.375 \times 10^3) =$$
$$(5.986 \times 4.375) \times 10^{4+3} = 26.189 \times 10^7$$

in each case rounding the first factor off to three decimal places.

When multiplying several numbers written in this notation together, the procedure is the same. All of the first factors are multiplied together to get the first factor of the product and all of the indices of the respective powers of ten are added together, taking into account their respective signs, to get the index of the second factor of the product. Thus $(4.02 \times 10^{-3}) \times (3.987 \times 10) \times (4.863 \times 10^5) = (4.02 \times 3.987 \times 4.863) \times (10^{-3+1+5}) = 77.94 \times 10^3$ rounding off the first factor to two decimal places.

**Dividing Numbers Written in Powers of Ten Notation.** — When dividing one number by another when both are written in this notation, the procedure is as follows:

1. Divide the first factor of the dividend by the first factor of the divisor to get the first factor of the quotient.

2. Subtract the index of the second factor of the divisor from the index of the second factor of the dividend, taking into account their respective signs, to get the index of the second factor of the quotient. Thus

$$(4.31 \times 10^{-2}) \div (9.0125 \times 10) =$$
$$(4.31 \div 9.0125) \times (10^{-2-1}) = 0.4782 \times 10^{-3}$$

It can be seen, then, that where several numbers of different magnitudes are to be multiplied and divided this system of notation is helpful.

*Example:* Find the quotient of $\dfrac{250 \times 4698 \times 0.00039}{43678 \times 0.002 \times 0.0147}$

*Solution:* Changing all of these numbers to powers of ten notation and performing the operations indicated:

$$\frac{(2.5 \times 10^2) \times (4.698 \times 10^3) \times (3.9 \times 10^{-4})}{(4.3678 \times 10^4) \times (2 \times 10^{-3}) \times (1.47 \times 10^{-2})}$$
$$= \frac{(2.5 \times 4.698 \times 3.9)(10^{2+3-4})}{(4.3678 \times 2 \times 1.47)(10^{4-3-2})} = \frac{45.806 \times 10}{12.841 \times 10^{-1}}$$
$$= 3.5672 \times 10^{1-(-1)}$$
$$= 3.5672 \times 10^2$$
$$= 356.72$$

## Preferred Numbers

Preferred numbers are series of numbers selected to be used for standardization purposes in preference to any other numbers. Their use will lead to simplified practice and they should be employed whenever possible for individual standard sizes and ratings, or for a series, in applications similar to the following:

1. Important or characteristic linear dimensions, such as diameters and lengths, areas, volume, weights, capacities.

2. Ratings of machinery and apparatus in horsepower, kilowatts, kilovolt-amperes, voltages, currents, speeds, power-factors, pressures, heat units, temperatures, gas or liquid-flow units, weight-handling capacities, etc.

3. Characteristic ratios of figures for all kinds of units.

**American National Standard for Preferred Numbers.** — This ANSI Standard Z17.1-1973 covers basic series of preferred numbers which are independent of any measurement system and therefore can be used with metric or customary units.

The numbers are rounded values of the following five geometric series of numbers: $10^{N/5}$, $10^{N/10}$, $10^{N/20}$, $10^{N/40}$, and $10^{N/80}$, where $N$ is an integer in the series 0, 1, 2, 3, etc. The designations used for the five series are respectively R5, R10, R20, R40, and R80, where R stands for Renard (Charles Renard, originator of the first preferred number system) and the number indicates the root of 10 on which the particular series is based.

The R5 series gives 5 numbers approximately 60 per cent apart, the R10 series gives 10 numbers approximately 25 per cent apart, the R20 series gives 20 numbers approximately 12 per cent apart, the R40 series gives 40 numbers approximately 6 per cent apart, and the R80 series gives 80 numbers approximately 3 per cent apart. The number of sizes for a given purpose can be minimized by using first the R5 series and adding sizes from the R10 and R20 series as needed. The R40 and R80 series are used principally for expressing tolerances in sizes based on preferred numbers. Preferred numbers below 1 are formed by dividing the given numbers by 10, 100, etc., and numbers above 10 are obtained by multiplying the given numbers by 10, 100, etc. Sizes graded according to the system may not be exactly proportional to one another due to the fact that preferred numbers may differ from calculated values by $+1.26$ per cent to $-1.01$ per cent. Deviations from preferred numbers are used in some instances — for example, where whole numbers are needed, such as 32 instead of 31.5 for the number of teeth in a gear.

### Basic Series of Preferred Numbers (ANSI Z17.1-1973)

| Series Designation | | | | | | | | |
|---|---|---|---|---|---|---|---|---|
| R5 | R10 | R20 | R40 | R40 | R80 | R80 | R80 | R80 |
| Preferred Numbers | | | | | | | | |
| 1.00 | 1.00 | 1.00 | 1.00 | 3.15 | 1.00 | 1.80 | 3.15 | 5.60 |
| 1.60 | 1.25 | 1.12 | 1.06 | 3.35 | 1.03 | 1.85 | 3.25 | 5.80 |
| 2.50 | 1.60 | 1.25 | 1.12 | 3.55 | 1.06 | 1.90 | 3.35 | 6.00 |
| 4.00 | 2.00 | 1.40 | 1.18 | 3.75 | 1.09 | 1.95 | 3.45 | 6.15 |
| 6.30 | 2.50 | 1.60 | 1.25 | 4.00 | 1.12 | 2.00 | 3.55 | 6.30 |
| ... | 3.15 | 1.80 | 1.32 | 4.25 | 1.15 | 2.06 | 3.65 | 6.50 |
| ... | 4.00 | 2.00 | 1.40 | 4.50 | 1.18 | 2.12 | 3.75 | 6.70 |
| ... | 5.00 | 2.24 | 1.50 | 4.75 | 1.22 | 2.18 | 3.87 | 6.90 |
| ... | 6.30 | 2.50 | 1.60 | 5.00 | 1.25 | 2.24 | 4.00 | 7.10 |
| ... | 8.00 | 2.80 | 1.70 | 5.30 | 1.28 | 2.30 | 4.12 | 7.30 |
| ... | ... | 3.15 | 1.80 | 5.60 | 1.32 | 2.36 | 4.25 | 7.50 |
| ... | ... | 3.55 | 1.90 | 6.30 | 1.36 | 2.43 | 4.37 | 7.75 |
| ... | ... | 4.00 | 2.00 | 6.30 | 1.40 | 2.50 | 4.50 | 8.00 |
| ... | ... | 4.50 | 2.12 | 6.70 | 1.45 | 2.58 | 4.62 | 8.25 |
| ... | ... | 5.00 | 2.24 | 7.10 | 1.50 | 2.65 | 4.75 | 8.50 |
| ... | ... | 5.60 | 2.36 | 7.50 | 1.55 | 2.72 | 4.87 | 8.75 |
| ... | ... | 6.30 | 2.50 | 8.00 | 1.60 | 2.80 | 5.00 | 9.00 |
| ... | ... | 7.10 | 2.65 | 8.50 | 1.65 | 2.90 | 5.15 | 9.25 |
| ... | ... | 8.00 | 2.80 | 9.00 | 1.70 | 3.00 | 5.20 | 9.50 |
| ... | ... | 9.00 | 3.00 | 9.50 | 1.75 | 3.07 | 5.45 | 9.75 |

## Principal Algebr[aic]

$$a \times a = aa = a^2$$
$$a \times a \times a = aaa = a^3$$
$$a \times b = ab$$

$$a^2 b^2 = (ab)^2$$

$$a^2 a^3 = a^{2+3} = a^5$$
$$a^4 \div a^3 = a^{4-3} = a$$
$$a^0 = 1$$
$$a^2 - b^2 = (a+b)(a-b)$$
$$(a+b)^2 = a^2 + 2ab + b^2$$

$$\sqrt{a} \times \sqrt{a} = a$$
$$\sqrt[3]{a} \times \sqrt[3]{a} \times \sqrt[3]{a} = a$$
$$(\sqrt[3]{a})^3 = a$$
$$\sqrt[3]{a^2} = (\sqrt[3]{a})^2 = a^{\frac{2}{3}}$$
$$\sqrt[4]{\sqrt[3]{a}} = \sqrt[4 \times 3]{a} = \sqrt[3]{\sqrt[4]{a}}$$

**When**

| | | |
|---|---|---|
| $a \times b = x,$ | then | $\log a$ |
| $a \div b = x,$ | then | $\log a$ |
| $a^3 = x,$ | then | $3 \log a$ |
| $\sqrt[3]{a} = x,$ | then | $\dfrac{\log a}{3}$ |

## Equations

An equation is a statement of equality betwe[en] The unknown quantity in an equation is gener[ally] If there is more than one unknown quantity, the also selected at the end of the alphabet, as $y$, $z$, $u$, [...]

An equation of the first degree is one which cont[ains] in the first power, as $3x = 9$. A quadratic equat[ion] unknown quantity in the second, but no higher, power[...]

**Solving Equations of the First Degree with** [...] all the terms containing the unknown $x$ to one side [...] other terms to the other side. Combine and simpli[fy] possible, and divide both sides by the coefficient of the [...] given for transposition of formulas.)

*Example:*

$$22x - 11 = 15x + 10$$
$$22x - 15x = 10 + 11$$
$$7x = 21$$
$$x = 3$$

## 118

**Solution of Equa**[tion] **of the simplified eq**[uation]

**Then,**

*Example:*

$$x$$

The value of $y$ the equations:

**Solution of** [Quadratic] equation is $ax^2$ [...]

*Example:* [...] $c = 5$.

$$x = \frac{-6}{}$$

**If the form o**[f...]

*Example:* [...] which is one [...] Let $x = $ [...] $2x + 1 = $ [...] formula, $ax$ [...]

$$x = $$

Since the [...] are $x = 3$ [...]

**Cubic** [...]

The e[...] by sub[...]

# LOGARITHMS

The object of logarithms is to facilitate and shorten calculations involving multiplication, division, the extraction of roots and the obtaining of powers of numbers. A logarithm consists of two parts, a whole number and a decimal. The whole number, which may be either a positive or negative number, or zero, is called the *characteristic;* the decimal is called the *mantissa.* As a rule, the decimal or mantissa only is given in tables of logarithms. The characteristic is prefixed to the mantissa according to the following rules:

For 1 and for all numbers greater than 1, the characteristic is one less than the number of places to the left of the decimal point in the given number. For example, the characteristic of the logarithm of 237 is 2, and of 2536.5 is 3.

For numbers smaller than 1, that is for numbers wholly decimal, the characteristic is negative and its numerical value is one more than the number of ciphers between the decimal point and the first decimal which is not a cipher. For example, the characteristic of the logarithm of 0.036 is (− 2), and the characteristic of the logarithm of 0.0006 is (− 4). Instead of writing the minus sign (−) in front of the figure, as (− 2), it is frequently written over the figure, thus: ($\bar{2}$). This method

| N. | L. | 0 | 1 | 2 | 3 | 4 | | 5 | 6 | 7 | 8 | 9 | | P. P. |
|----|----|---|---|---|---|---|---|---|---|---|---|---|---|-------|
| 400 | 60 | 206 | 217 | 228 | 239 | 249 | | 260 | 271 | 282 | 293 | 304 | | |
| 401 | | 314 | 325 | 336 | 347 | 358 | | 369 | 379 | 390 | 401 | 412 | | |
| 402 | | 423 | 433 | 444 | 455 | 466 | | 477 | 487 | 498 | 509 | 520 | | |
| 403 | | 531 | 541 | 552 | 563 | 574 | | 584 | 595 | 606 | 617 | 627 | | |
| 404 | | 638 | 649 | 660 | 670 | 681 | | 692 | 703 | 713 | 724 | 735 | | |

Fig. 1

is used because the minus sign refers only to the characteristic and not to the mantissa, which is always positive.

The logarithmic tables in the following give, in the body of the tables, the mantissa of the logarithms of numbers from 1 to 10,000. When finding the mantissa, the decimal point in a number is disregarded. The mantissa of the logarithms of 2716, 271.6, 27.16, 2.716, or 0.02716, for example, is the same. The tables give directly the mantissa of logarithms of numbers with four figures or less; the logarithms for numbers with more than four figures can be approximated.

To find the logarithm of a number from the tables, locate the first three figures of the number in the left-hand column, and then find the fourth figure at the top of the columns of the page. Then follow the column down from this last figure until opposite the three first figures in the left-hand column. The figure thus found in the body of the table is the mantissa of the logarithm. If the number of which the logarithm is required does not contain four figures, annex ciphers to the right so as to obtain four figures. If the mantissa of the logarithm of 6 is required, for example, find the mantissa for 6000.

*Example:* — Find the logarithm of 4032. Locate 403 in the left-hand column of the logarithmic tables, then follow downward the column headed "2" at the top of the page, and find the required mantissa opposite 403. The mantissa is .60552 the "group" figures 60 being found in the column under " L " and prefixed to the figures 552 found directly in the column under "2." The characteristic of the logarithm being 3, log 4032 = 3.60552. (See Fig. 1.)

All the mantissas, or the numbers in the tables, are decimals, and the decimal point has, therefore, been omitted in the tables, since no confusion could arise from this; but it should always be put before the figures of the mantissa as soon as taken from the table.

In the tables it will be found that, in some cases, the figures are preceded by the sign (∗). The sign (∗) indicates that the two figures to be prefixed are those given in the next line below that in which the last three figures are read. For example the logarithm of 5018 is 3.70053, the two figures to be prefixed being 70 and not 69 as would ordinarily be the case. (See Fig. 2.)

**Finding a Number the Logarithm of which is Given.**—When a logarithm is given and it is required to find the corresponding number, find the first two figures of the mantissa in the column headed " L " in the tables; then find in the group of mantissas, all having the same first two figures, the remaining three figures. These may appear in any of the columns headed "0" to "9". The number heading the column in which the last three figures of the mantissa are found is the last figure in the number sought, and the number in the left-hand column, headed " N," in line with the last three figures of the mantissa, gives the three first figures in the number sought. When the actual figures in the number sought have been deter-

| N. | L. 0 | 1 | 2 | 3 | 4 | 5 | 6 | 7 | 8 | 9 | P. P. |
|---|---|---|---|---|---|---|---|---|---|---|---|
| 500 | 69 897 | 906 | 914 | 923 | 932 | 940 | 949 | 958 | 966 | 975 | |
| 501 | 984 | 992 | ∗001 | ∗010 | ∗018 | ∗027 | ∗036 | ∗044 | ∗053 | ∗062 | |
| 502 | 70 070 | 079 | 088 | 096 | 105 | 114 | 122 | 131 | 140 | 148 | |
| 503 | 157 | 165 | 174 | 183 | 191 | 200 | 209 | 217 | 226 | 234 | |
| 504 | 243 | 252 | 260 | 269 | 278 | 286 | 295 | 303 | 312 | 321 | |

Fig. 2

mined, locate the decimal point according to the rules given for the characteristic of logarithms. If the characteristic is greater than 3, add ciphers. For example, if the figures corresponding to a given mantissa are 3765 and the characteristic is 5, then the number sought has 6 figures to the left of the decimal point, and is 376,500. If the characteristic had been 3̄, then the number sought would, in this case, have been 0.003765. If the mantissa is not exactly obtainable in the tables, find the mantissa in the table which is the nearest to the one given and determine the number corresponding to this. In most cases, this gives results accurate enough. By interpolation, as will be explained later, more accurate results can be obtained.

If the three last figures of the mantissa, as found in the table, are preceded by a (∗), it indicates that these three figures belong to the group preceded by the two figures in the " L " column in the line next below.

*Example:* — Find the number the logarithm of which is 2.70053.

First find the two figures of the mantissa (70) in the column headed " L " in the tables. Then find the remaining three figures (053) in the mantissas which all have 70 for their first two figures. The (∗) in front of the figure ∗053 in the line next above that in which 70 is found indicates that these figures belong to the group preceded by 70. Therefore, the number corresponding to the logarithm 2.70053 is 501.8. (See Fig. 2.)

**Avoiding Use of Negative Characteristics.** — As previously explained, the logarithm of any number less than 1 has a negative characteristic and a positive mantissa. In many computations, the use of logarithms having negative characteristics is troublesome and frequently a source of error. A simple way to avoid this difficulty is to convert each logarithm having a negative characteristic into an equivalent logarithm having a positive characteristic. This is done according to the following method which is based on the principle that any number can be simultaneously added to and subtracted from the characteristic of a logarithm without changing its value, thus: Log 1 = 0.00000 = 10.00000 − 10; log 0.3 = $\overline{1}$.47712 = 9.47712 − 10; log 0.000478 = $\overline{4}$.67943 = 6.67943 − 10. Usually 10 or 20 are added to and subtracted from the logarithmic characteristic, but any convenient number may be so used.

**Multiplication by Logarithms.** — If two or more numbers are to be multiplied together, find the logarithms of the numbers to be multiplied, and add these logarithms. The sum is the logarithm of the product, and the number corresponding to this logarithm, as found from the logarithmic tables, is the required product.

*Example:* — Find the product of 2831 × 2.692 × 29.69 × 19.4.

This calculation is carried out by means of logarithms as follows:

$$
\begin{array}{rl}
\log 2831. & = 3.45194 \\
\log 2.692 & = 0.43008 \\
\log 29.69 & = 1.47261 \\
\log 19.4 & = \underline{1.28780} \\
& \phantom{=} 6.64243
\end{array}
$$

The product, as found from the tables, with ciphers added, then is 4,390,000.

In multiplication problems involving numbers less than 1, the method of avoiding the use of negative characteristics, previously outlined, simplifies the addition and tends to reduce the possibility of error. *Example:* Find the product of 0.002656 × 155.1 × 0.5853 × 7.968.

$$
\begin{array}{rll}
\log 0.002656 & = \overline{3}.42423 & = 7.42423 - 10 \\
\log 155.1 & = 2.19061 & = 2.19061 \\
\log 0.5853 & = \overline{1}.76738 & = 9.76738 - 10 \\
\log 7.968 & = 0.90135 & = \underline{0.90135} \\
& & 20.28357 - 20 = 0.28357
\end{array}
$$

Hence  0.002656 × 155.1 × 0.5853 × 7.968 = 1.9212

**Division by Logarithms.** — When dividing one number by another, subtract the logarithm of the divisor from the logarithm of the dividend; the remainder is the logarithm of the quotient. *Example:* — To find the quotient of 7658 ÷ 935.3.

$$
\begin{array}{rll}
\log 7658 & = & 3.88412 \\
-\log 935.3 & = & \underline{-2.97095} \\
\log (7658 \div 935.3) & = & 0.91317
\end{array}
$$

Hence  7658 ÷ 935.3 = 8.188

Instead of dividing 7658 by 935.3, the same answer would be obtained if 7658 were *multiplied* by the reciprocal of 935.3 or 1 ÷ 935.3. To do this by logarithms, the log of 7658 and the log of the reciprocal of 935.3 would be *added* together.

In order to use the method just outlined, it is necessary to know how to find the logarithm of the reciprocal of a number. This is done by simply subtracting the log of the number from the log of 1. To do this conveniently, some number, such as 10, is first added to and subtracted from the characteristic of the log of 1.

*Example:* — Find the log of the reciprocal of 935.3.

$$\begin{aligned}
\log \ \ 1 \ \ &= \ \ 0.00000 = 10.00000 - 10 \\
-\log 935.3 &= \ \ \underline{-2.97095} = \underline{-2.97095} \\
\log (1 \div 935.3) &= \ \ \ \ \ \ \ \ \ \ \ \ \ \ \ \ 7.02905 - 10 \\
\log (1 \div 935.3) &= \ \ \ \ \ \ \ \ \ \ \ \ \ \ \ \ \overline{3}.02905 \\
1 \div 935.3 &= \ \ \ \ \ \ \ \ \ \ \ \ \ \ \ \ 0.001069
\end{aligned}$$

Thus the quotient of $7658 \div 935.3$ can be found by adding the log of 7658 and the log of the reciprocal of 935.3:

$$\begin{aligned}
\log 7658 &= 3.88412 = \ \ 3.88412 \\
\log (1 \div 935.3) &= \overline{3}.02905 = \ \ \underline{7.02905 - 10} \\
& \ \ \ \ \ \ \ \ \ \ \ \ \ \ \ \ \ \ \ \ \ \ \ \ \ 10.91317 - 10 = 0.91317
\end{aligned}$$

Hence                               $7658 \div 935.3 = 8.188$

As is readily seen, this method is more cumbersome than the direct method, where there is only one factor each in the dividend and divisor, but it does greatly facilitate the solution of problems in division involving several factors in the dividend and the divisor. In such a problem the logarithm of each factor in the dividend is added to the logarithm of the reciprocal of each factor in the divisor.

*Example:* — Find the product of $\dfrac{0.0272 \times 27.1 \times 12.6}{2.371 \times 0.007}$

$$\begin{aligned}
\log \ 0.0272 \ \ &= \overline{2}.43457 = \ \ 8.43457 - 10 \\
\log 27.1 \ \ &= 1.43297 = \ \ 1.43297 \\
\log 12.6 \ \ &= 1.10037 = \ \ 1.10037 \\
\log (1 \div 2.371) \ \ &= \ \ \ \ \ \ \ \ \ \ \ \ \ \ \ 9.62507 - 10 \\
\log (1 \div 0.007) \ \ &= \ \ \ \ \ \ \ \ \ \ \ \ \ \ \ \underline{2.15490} \\
& \ \ \ \ \ \ \ \ \ \ \ \ \ \ \ \ \ \ \ \ \ \ 22.74788 - 20 = 2.74788
\end{aligned}$$

Hence the result is 559.6.

In problems in division where the divisor is larger than the dividend, the subtraction of logarithms is facilitated if some number is added to and subtracted from the log of the dividend. (This is the same method used to convert a logarithm with a negative characteristic to an equivalent logarithm with a positive characteristic except that in this case it serves to convert a logarithm with a positive characteristic to one with a larger positive characteristic but having the same value.)

*Example:* — To find the quotient of $43.2 \div 971.4$.

$$\begin{aligned}
\log \ \ 43.2 \ &= \ \ \ \ 1.63548 = 11.63548 - 10 \\
-\log 971.4 \ &= \ \ -2.98740 = \ \ \underline{-2.98740} \\
\log (43.2 \div 971.4) &= \ \ \ \ \ \ \ \ \ \ \ \ \ \ \ \ \ \ 8.64808 - 10 = \overline{2}.64808
\end{aligned}$$

Hence                         $43.2 \div 971.4 = 0.04447$

**Obtaining the Powers of Numbers.** — A number may be raised to any power by simply multiplying the logarithm of the number by the exponent of the number. The product gives the logarithm of the value of the power.

*Example 1.* — Find the value of $6.51^3$.

$$\begin{aligned}
\log 6.51 &= 0.81358 \\
3 \times 0.81358 &= 2.44074
\end{aligned}$$

The logarithm 2.44074 is then the logarithm of $6.51^3$. Hence $6.51^3$ equals the number corresponding to this logarithm, as found from the tables, or $6.51^3 = 275.9$.

*Example 2.* — Find the value of $12^{1.29}$.

$$\log 12 = 1.07918$$
$$1.29 \times 1.07918 = 1.39214$$

Hence, $12^{1.29} = 24.67$.

Raising a decimal to a decimal power presents a somewhat more difficult problem because of the negative characteristic of the logarithm and the fact that the logarithm must be multiplied by a decimal exponent. The method previously outlined for avoiding the use of negative characteristics is helpful in such cases.

*Example 3.* — Find the value of $0.0813^{0.46}$.

$$\log 0.0813 = \overline{2}.91009 = 8.91009 - 10$$
$$\log 0.0813^{0.46} = 0.46 \times (8.91009 - 10) = 4.09864 - 4.6$$

Subtracting and adding 0.6 to make the characteristic a whole number,

$$\begin{array}{r} 4.09864 - 4.6 \\ -0.6 \quad +0.6 \\ \hline \log 0.0813^{0.46} = 3.49864 - 4 = \overline{1}.49864 \end{array}$$

Hence
$$0.0813^{0.46} = 0.3152$$

**Extracting Roots by Logarithms.** — Roots of numbers, as for example $\sqrt[5]{37}$, can easily be extracted by means of logarithms. The small (⁵) in the radical ($\sqrt{\phantom{x}}$) of the root-sign is called the index of the root. Any root of a number may be found by dividing its logarithm by the index of the root; the quotient is the logarithm of the root.

*Example 1.* — Find $\sqrt[3]{276}$.

$$\log 276 = 2.44091$$
$$2.44091 \div 3 = 0.81364$$

Hence
$$\log \sqrt[3]{276} = 0.81364, \quad \text{and} \quad \sqrt[3]{276} = 6.511.$$

*Example 2.* — Find $\sqrt[3]{0.67}$.

$$\log 0.67 = \overline{1}.82607$$

In this case it is not possible to divide directly, because there is a negative characteristic and a positive mantissa. Here is another instance where the method of avoiding the use of negative characteristics, previously outlined, is helpful. The procedure in this case is to add and subtract some number to the characteristic which is evenly divisible by the index of the root. In this case the root index is 3. Thus 9 can be added to and subtracted from the characteristic, and the resulting logarithm divided by 3.

$$\log 0.67 = \overline{1}.82607 = 8.82607 - 9$$
$$\log \sqrt[3]{0.67} = \frac{8.82607 - 9}{3} = 2.94202 - 3$$
$$\log \sqrt[3]{0.67} = 2.94202 - 3 = \overline{1}.94202$$

Hence $\sqrt[3]{0.67} = 0.875$

*Example* 3. — Find $\sqrt[1.7]{0.2}$

$$\log 0.2 = \bar{1}.30103 = 16.30103 - 17$$

$$\log \sqrt[1.7]{0.2} = \frac{16.30103 - 17}{1.7} = 9.58884 - 10 = \bar{1}.58884$$

Hence
$$\sqrt[1.7]{0.2} = 0.388$$

**Interpolation.** — If the number for which the logarithm is required consists of five figures, it is possible by means of the small tables in the right-hand column of the logarithmic tables, headed "P.P." (proportional parts), to obtain the logarithm more accurately than by taking the nearest value for four figures. The logarithm of 1524.2, for example, is found as follows:

First find the difference between the nearest larger and the nearest smaller logarithms in the table. Log 1524 = 3.18298 and log 1525 = 3.18327. (See Fig. 3.) The difference is 0.00029. Then in the small table headed "29" in the right-hand

| N. | L. | 0 | 1 | 2 | 3 | ④ | 5 | 6 | 7 | 8 | 9 | P. P. | | |
|---|---|---|---|---|---|---|---|---|---|---|---|---|---|---|
| **150** | | 17 609 | 638 | 667 | 696 | 725 | 754 | 782 | 811 | 840 | 869 | | | |
| 151 | | 898 | 926 | 955 | 984 | *013 | *041 | *070 | *099 | *127 | *156 | | ㉙ | 28 |
| [152] | 18 | 184 | 213 | 241 | 270 | (298) | (327) | 355 | 384 | 412 | 441 | 1 | 2,9 | 2,8 |
| 153 | | 469 | 498 | 526 | 554 | 583 | 611 | 639 | 667 | 696 | 724 | 2 | 5,8 | 5,6 |
| 154 | | 752 | 780 | 808 | 837 | 865 | 893 | 921 | 949 | 977 | *005 | 3 | 8,7 | 8,4 |

Fig. 3

column, find the figure opposite 2 (2 being the last or fifth figure in the given number). This figure is 5.8. Add this to the mantissa of the smaller of the two logarithms already found, disregarding the decimal point in the mantissa, and considering it, for the while being, as a whole number. Then, 18298 + 5.8 = 18303.8, or approximately, 18304. This is the mantissa of the logarithm of 1524.2 and the complete logarithm is 3.18304.

To find a number more accurately than to four figures, when the mantissa cannot be found exactly in tne tables, find the mantissa which is nearest to, but less than, the given mantissa. Subtract this mantissa from the nearest larger mantissa in the tables and find in the right-hand column the small table headed by this difference. Then subtract the nearest smaller mantissa from the given logarithm and find the exact or approximate difference in the "proportional part" table. The corresponding figure in the left-hand column of the "proportional part" table is the fifth figure in the number sought, the other four figures being those corresponding to the logarithm next smaller than the given logarithm. In accordance with this rule, the number corresponding to the logarithm 4.46262 is found to be 29,015.

**Natural Logarithms.** — In certain formulas and in some branches of mathematical analysis, use is made of *natural* logarithms (formerly also called Napierian or *hyperbolic* logarithms). The base of this system, $e = 2.7182818284 +$, is the limit of a certain mathematical series. The logarithm of a number $A$ to the base $e$ is usually written $\log_e A$ or $\ln A$. Tables of natural logarithms for numbers ranging from 1 to 100 are given in this handbook after the tables of common logarithms. To obtain natural logs of numbers less than 1 or greater than 100, proceed as in the following examples: $\log_e 0.239 = \log_e 2.39 - \log_e 10$; $\log_e 0.0239 = \log_e 2.39 - 2 \log_e 10$; $\log_e 239 = \log_e 2.39 + 2 \log_e 10$; $\log_e 2390 = \log_e 2.39 + 3 \log_e 10$, etc.

To convert common logs to natural logs, and vice versa, use the following relations: natural log $= 2.3026 \times$ common log; common log $= 0.43429 \times$ natural log.

(.00000 to .17869)

| N. | L. 0 | 1 | 2 | 3 | 4 | 5 | 6 | 7 | 8 | 9 |
|---|---|---|---|---|---|---|---|---|---|---|
| 100 | 00 000 | 043 | 087 | 130 | 173 | 217 | 260 | 303 | 346 | 389 |
| 101 | 432 | 475 | 518 | 561 | 604 | 647 | 689 | 732 | 775 | 817 |
| 102 | 860 | 903 | 945 | 988 | *030 | *072 | *115 | *157 | *199 | *242 |
| 103 | 01 284 | 326 | 368 | 410 | 452 | 494 | 536 | 578 | 620 | 662 |
| 104 | 703 | 745 | 787 | 828 | 870 | 912 | 953 | 995 | *036 | *078 |
| 105 | 02 119 | 160 | 202 | 243 | 284 | 325 | 366 | 407 | 449 | 490 |
| 106 | 531 | 572 | 612 | 653 | 694 | 735 | 776 | 816 | 857 | 898 |
| 107 | 938 | 979 | *019 | *060 | *100 | *141 | *181 | *222 | *262 | *302 |
| 108 | 03 342 | 383 | 423 | 463 | 503 | 543 | 583 | 623 | 663 | 703 |
| 109 | 743 | 782 | 822 | 862 | 902 | 941 | 981 | *021 | *060 | *100 |
| 110 | 04 139 | 179 | 218 | 258 | 297 | 336 | 376 | 415 | 454 | 493 |
| 111 | 532 | 571 | 610 | 650 | 689 | 727 | 766 | 805 | 844 | 883 |
| 112 | 922 | 961 | 999 | *038 | *077 | *115 | *154 | *192 | *231 | *269 |
| 113 | 05 308 | 346 | 385 | 423 | 461 | 500 | 538 | 576 | 614 | 652 |
| 114 | 690 | 729 | 767 | 805 | 843 | 881 | 918 | 956 | 994 | *032 |
| 115 | 06 070 | 108 | 145 | 183 | 221 | 258 | 296 | 333 | 371 | 408 |
| 116 | 446 | 483 | 521 | 558 | 595 | 633 | 670 | 707 | 744 | 781 |
| 117 | 819 | 856 | 893 | 930 | 967 | *004 | *041 | *078 | *115 | *151 |
| 118 | 07 188 | 225 | 262 | 298 | 335 | 372 | 408 | 445 | 482 | 518 |
| 119 | 555 | 591 | 628 | 664 | 700 | 737 | 773 | 809 | 846 | 882 |
| 120 | 918 | 954 | 990 | *027 | *063 | *099 | *135 | *171 | *207 | *243 |
| 121 | 08 279 | 314 | 350 | 386 | 422 | 458 | 493 | 529 | 565 | 600 |
| 122 | 636 | 672 | 707 | 743 | 778 | 814 | 849 | 884 | 920 | 955 |
| 123 | 991 | *026 | *061 | *096 | *132 | *167 | *202 | *237 | *272 | *307 |
| 124 | 09 342 | 377 | 412 | 447 | 482 | 517 | 552 | 587 | 621 | 656 |
| 125 | 691 | 726 | 760 | 795 | 830 | 864 | 899 | 934 | 968 | *003 |
| 126 | 10 037 | 072 | 106 | 140 | 175 | 209 | 243 | 278 | 312 | 346 |
| 127 | 380 | 415 | 449 | 483 | 517 | 551 | 585 | 619 | 653 | 687 |
| 128 | 721 | 755 | 789 | 823 | 857 | 890 | 924 | 958 | 992 | *025 |
| 129 | 11 059 | 093 | 126 | 160 | 193 | 227 | 261 | 294 | 327 | 361 |
| 130 | 394 | 428 | 461 | 494 | 528 | 561 | 594 | 628 | 661 | 694 |
| 131 | 727 | 760 | 793 | 826 | 860 | 893 | 926 | 959 | 992 | *024 |
| 132 | 12 057 | 090 | 123 | 156 | 189 | 222 | 254 | 287 | 320 | 352 |
| 133 | 385 | 418 | 450 | 483 | 516 | 548 | 581 | 613 | 646 | 678 |
| 134 | 710 | 743 | 775 | 808 | 840 | 872 | 905 | 937 | 969 | *001 |
| 135 | 13 033 | 066 | 098 | 130 | 162 | 194 | 226 | 258 | 290 | 322 |
| 136 | 354 | 386 | 418 | 450 | 481 | 513 | 545 | 577 | 609 | 640 |
| 137 | 672 | 704 | 735 | 767 | 799 | 830 | 862 | 893 | 925 | 956 |
| 138 | 988 | *019 | *051 | *082 | *114 | *145 | *176 | *208 | *239 | *270 |
| 139 | 14 301 | 333 | 364 | 395 | 426 | 457 | 489 | 520 | 551 | 582 |
| 140 | 613 | 644 | 675 | 706 | 737 | 768 | 799 | 829 | 860 | 891 |
| 141 | 922 | 953 | 983 | *014 | *045 | *076 | *106 | *137 | *168 | *198 |
| 142 | 15 229 | 259 | 290 | 320 | 351 | 381 | 412 | 442 | 473 | 503 |
| 143 | 534 | 564 | 594 | 625 | 655 | 685 | 715 | 746 | 776 | 806 |
| 144 | 836 | 866 | 897 | 927 | 957 | 987 | *017 | *047 | *077 | *107 |
| 145 | 16 137 | 167 | 197 | 227 | 256 | 286 | 316 | 346 | 376 | 406 |
| 146 | 435 | 465 | 495 | 524 | 554 | 584 | 613 | 643 | 673 | 702 |
| 147 | 732 | 761 | 791 | 820 | 850 | 879 | 909 | 938 | 967 | 997 |
| 148 | 17 026 | 056 | 085 | 114 | 143 | 173 | 202 | 231 | 260 | 289 |
| 149 | 319 | 348 | 377 | 406 | 435 | 464 | 493 | 522 | 551 | 580 |
| 150 | 609 | 638 | 667 | 696 | 725 | 754 | 782 | 811 | 840 | 869 |

P. P.

| | 44 | 43 | 42 |
|---|---|---|---|
| 1 | 4.4 | 4.3 | 4.2 |
| 2 | 8.8 | 8.6 | 8.4 |
| 3 | 13.2 | 12.9 | 12.6 |
| 4 | 17.6 | 17.2 | 16.8 |
| 5 | 22.0 | 21.5 | 21.0 |
| 6 | 26.4 | 25.8 | 25.2 |
| 7 | 30.8 | 30.1 | 29.4 |
| 8 | 35.2 | 34.4 | 33.6 |
| 9 | 39.6 | 38.7 | 37.8 |

| | 41 | 40 | 39 |
|---|---|---|---|
| 1 | 4.1 | 4.0 | 3.9 |
| 2 | 8.2 | 8.0 | 7.8 |
| 3 | 12.3 | 12.0 | 11.7 |
| 4 | 16.4 | 16.0 | 15.6 |
| 5 | 20.5 | 20.0 | 19.5 |
| 6 | 24.6 | 24.0 | 23.4 |
| 7 | 28.7 | 28.0 | 27.3 |
| 8 | 32.8 | 32.0 | 31.2 |
| 9 | 36.9 | 36.0 | 35.1 |

| | 38 | 37 | 36 |
|---|---|---|---|
| 1 | 3.8 | 3.7 | 3.6 |
| 2 | 7.6 | 7.4 | 7.2 |
| 3 | 11.4 | 11.1 | 10.8 |
| 4 | 15.2 | 14.8 | 14.4 |
| 5 | 19.0 | 18.5 | 18.0 |
| 6 | 22.8 | 22.2 | 21.6 |
| 7 | 26.6 | 25.9 | 25.2 |
| 8 | 30.4 | 29.6 | 28.8 |
| 9 | 34.2 | 33.3 | 32.4 |

| | 35 | 34 | 33 |
|---|---|---|---|
| 1 | 3.5 | 3.4 | 3.3 |
| 2 | 7.0 | 6.8 | 6.6 |
| 3 | 10.5 | 10.2 | 9.9 |
| 4 | 14.0 | 13.6 | 13.2 |
| 5 | 17.5 | 17.0 | 16.5 |
| 6 | 21.0 | 20.4 | 19.8 |
| 7 | 24.5 | 23.8 | 23.1 |
| 8 | 28.0 | 27.2 | 26.4 |
| 9 | 31.5 | 30.6 | 29.7 |

| | 32 | 31 | 30 |
|---|---|---|---|
| 1 | 3.2 | 3.1 | 3.0 |
| 2 | 6.4 | 6.2 | 6.0 |
| 3 | 9.6 | 9.3 | 9.0 |
| 4 | 12.8 | 12.4 | 12.0 |
| 5 | 16.0 | 15.5 | 15.0 |
| 6 | 19.2 | 18.6 | 18.0 |
| 7 | 22.4 | 21.7 | 21.0 |
| 8 | 25.6 | 24.8 | 24.0 |
| 9 | 28.8 | 27.9 | 27.0 |

(.17609 to .30298)

| N. | L. | 0 | 1 | 2 | 3 | 4 | 5 | 6 | 7 | 8 | 9 |
|---|---|---|---|---|---|---|---|---|---|---|---|
| 150 | 17 | 609 | 638 | 667 | 696 | 725 | 754 | 782 | 811 | 840 | 869 |
| 151 | | 898 | 926 | 955 | 984 | *013 | *041 | *070 | *099 | *127 | *156 |
| 152 | 18 | 184 | 213 | 241 | 270 | 298 | 327 | 355 | 384 | 412 | 441 |
| 153 | | 469 | 498 | 526 | 554 | 583 | 611 | 639 | 667 | 696 | 724 |
| 154 | | 752 | 780 | 808 | 837 | 865 | 893 | 921 | 949 | 977 | *005 |
| 155 | 19 | 033 | 061 | 089 | 117 | 145 | 173 | 201 | 229 | 257 | 285 |
| 156 | | 312 | 340 | 368 | 396 | 424 | 451 | 479 | 507 | 535 | 562 |
| 157 | | 590 | 618 | 645 | 673 | 700 | 728 | 756 | 783 | 811 | 838 |
| 158 | | 866 | 893 | 921 | 948 | 976 | *003 | *030 | *058 | *085 | *112 |
| 159 | 20 | 140 | 167 | 194 | 222 | 249 | 276 | 303 | 330 | 358 | 385 |
| 160 | | 412 | 439 | 466 | 493 | 520 | 548 | 575 | 602 | 629 | 656 |
| 161 | | 683 | 710 | 737 | 763 | 790 | 817 | 844 | 871 | 898 | 925 |
| 162 | | 952 | 978 | *005 | *032 | *059 | *085 | *112 | *139 | *165 | *192 |
| 163 | 21 | 219 | 245 | 272 | 299 | 325 | 352 | 378 | 405 | 431 | 458 |
| 164 | | 484 | 511 | 537 | 564 | 590 | 617 | 643 | 669 | 696 | 722 |
| 165 | | 748 | 775 | 801 | 827 | 854 | 880 | 906 | 932 | 958 | 985 |
| 166 | 22 | 011 | 037 | 063 | 089 | 115 | 141 | 167 | 194 | 220 | 246 |
| 167 | | 272 | 298 | 324 | 350 | 376 | 401 | 427 | 453 | 479 | 505 |
| 168 | | 531 | 557 | 583 | 608 | 634 | 660 | 686 | 712 | 737 | 763 |
| 169 | | 789 | 814 | 840 | 866 | 891 | 917 | 943 | 968 | 994 | *019 |
| 170 | 23 | 045 | 070 | 096 | 121 | 147 | 172 | 198 | 223 | 249 | 274 |
| 171 | | 300 | 325 | 350 | 376 | 401 | 426 | 452 | 477 | 502 | 528 |
| 172 | | 553 | 578 | 603 | 629 | 654 | 679 | 704 | 729 | 754 | 779 |
| 173 | | 805 | 830 | 855 | 880 | 905 | 930 | 955 | 980 | *005 | *030 |
| 174 | 24 | 055 | 080 | 105 | 130 | 155 | 180 | 204 | 229 | 254 | 279 |
| 175 | | 304 | 329 | 353 | 378 | 403 | 428 | 452 | 477 | 502 | 527 |
| 176 | | 551 | 576 | 601 | 625 | 650 | 674 | 699 | 724 | 748 | 773 |
| 177 | | 797 | 822 | 846 | 871 | 895 | 920 | 944 | 969 | 993 | *018 |
| 178 | 25 | 042 | 066 | 091 | 115 | 139 | 164 | 188 | 212 | 237 | 261 |
| 179 | | 285 | 310 | 334 | 358 | 382 | 406 | 431 | 455 | 479 | 503 |
| 180 | | 527 | 551 | 575 | 600 | 624 | 648 | 672 | 696 | 720 | 744 |
| 181 | | 768 | 792 | 816 | 840 | 864 | 888 | 912 | 935 | 959 | 983 |
| 182 | 26 | 007 | 031 | 055 | 079 | 102 | 126 | 150 | 174 | 198 | 221 |
| 183 | | 245 | 269 | 293 | 316 | 340 | 364 | 387 | 411 | 435 | 458 |
| 184 | | 482 | 505 | 529 | 553 | 576 | 600 | 623 | 647 | 670 | 694 |
| 185 | | 717 | 741 | 764 | 788 | 811 | 834 | 858 | 881 | 905 | 928 |
| 186 | | 951 | 975 | 998 | *021 | *045 | *068 | *091 | *114 | *138 | *161 |
| 187 | 27 | 184 | 207 | 231 | 254 | 277 | 300 | 323 | 346 | 370 | 393 |
| 188 | | 416 | 439 | 462 | 485 | 508 | 531 | 554 | 577 | 600 | 623 |
| 189 | | 646 | 669 | 692 | 715 | 738 | 761 | 784 | 807 | 830 | 852 |
| 190 | | 875 | 898 | 921 | 944 | 967 | 989 | *012 | *035 | *058 | *081 |
| 191 | 28 | 103 | 126 | 149 | 171 | 194 | 217 | 240 | 262 | 285 | 307 |
| 192 | | 330 | 353 | 375 | 398 | 421 | 443 | 466 | 488 | 511 | 533 |
| 193 | | 556 | 578 | 601 | 623 | 646 | 668 | 691 | 713 | 735 | 758 |
| 194 | | 780 | 803 | 825 | 847 | 870 | 892 | 914 | 937 | 959 | 981 |
| 195 | 29 | 003 | 026 | 048 | 070 | 092 | 115 | 137 | 159 | 181 | 203 |
| 196 | | 226 | 248 | 270 | 292 | 314 | 336 | 358 | 380 | 403 | 425 |
| 197 | | 447 | 469 | 491 | 513 | 535 | 557 | 579 | 601 | 623 | 645 |
| 198 | | 667 | 688 | 710 | 732 | 754 | 776 | 798 | 820 | 842 | 863 |
| 199 | | 885 | 907 | 929 | 951 | 973 | 994 | *016 | *038 | *060 | *081 |
| 200 | 30 | 103 | 125 | 146 | 168 | 190 | 211 | 233 | 255 | 276 | 298 |

**P. P.**

| | 29 | 28 |
|---|---|---|
| 1 | 2.9 | 2.8 |
| 2 | 5.8 | 5.6 |
| 3 | 8.7 | 8.4 |
| 4 | 11.6 | 11.2 |
| 5 | 14.5 | 14.0 |
| 6 | 17.4 | 16.8 |
| 7 | 20.3 | 19.6 |
| 8 | 23.2 | 22.4 |
| 9 | 26.1 | 25.2 |

| | 27 | 26 |
|---|---|---|
| 1 | 2.7 | 2.6 |
| 2 | 5.4 | 5.2 |
| 3 | 8.1 | 7.8 |
| 4 | 10.8 | 10.4 |
| 5 | 13.5 | 13.0 |
| 6 | 16.2 | 15.6 |
| 7 | 18.9 | 18.2 |
| 8 | 21.6 | 20.8 |
| 9 | 24.3 | 23.4 |

| | 25 |
|---|---|
| 1 | 2.5 |
| 2 | 5.0 |
| 3 | 7.5 |
| 4 | 10.0 |
| 5 | 12.5 |
| 6 | 15.0 |
| 7 | 17.5 |
| 8 | 20.0 |
| 9 | 22.5 |

| | 24 | 23 |
|---|---|---|
| 1 | 2.4 | 2.3 |
| 2 | 4.8 | 4.6 |
| 3 | 7.2 | 6.9 |
| 4 | 9.6 | 9.2 |
| 5 | 12.0 | 11.5 |
| 6 | 14.4 | 13.8 |
| 7 | 16.8 | 16.1 |
| 8 | 19.2 | 18.4 |
| 9 | 21.6 | 20.7 |

| | 22 | 21 |
|---|---|---|
| 1 | 2.2 | 2.1 |
| 2 | 4.4 | 4.2 |
| 3 | 6.6 | 6.3 |
| 4 | 8.8 | 8.4 |
| 5 | 11.0 | 10.5 |
| 6 | 13.2 | 12.6 |
| 7 | 15.4 | 14.7 |
| 8 | 17.6 | 16.8 |
| 9 | 19.8 | 18.9 |

(.30103 to .39950)

| N. | L. 0 | 1 | 2 | 3 | 4 | 5 | 6 | 7 | 8 | 9 |
|---|---|---|---|---|---|---|---|---|---|---|
| 200 | 30 103 | 125 | 146 | 168 | 190 | 211 | 233 | 255 | 276 | 298 |
| 201 | 320 | 341 | 363 | 384 | 406 | 428 | 449 | 471 | 492 | 514 |
| 202 | 535 | 557 | 578 | 600 | 621 | 643 | 664 | 685 | 707 | 728 |
| 203 | 750 | 771 | 792 | 814 | 835 | 856 | 878 | 899 | 920 | 942 |
| 204 | 963 | 984 | *006 | *027 | *048 | *069 | *091 | *112 | *133 | *154 |
| 205 | 31 175 | 197 | 218 | 239 | 260 | 281 | 302 | 323 | 345 | 366 |
| 206 | 387 | 408 | 429 | 450 | 471 | 492 | 513 | 534 | 555 | 576 |
| 207 | 597 | 618 | 639 | 660 | 681 | 702 | 723 | 744 | 765 | 785 |
| 208 | 806 | 827 | 848 | 869 | 890 | 911 | 931 | 952 | 973 | 994 |
| 209 | 32 015 | 035 | 056 | 077 | 098 | 118 | 139 | 160 | 181 | 201 |
| 210 | 222 | 243 | 263 | 284 | 305 | 325 | 346 | 366 | 387 | 408 |
| 211 | 428 | 449 | 469 | 490 | 510 | 531 | 552 | 572 | 593 | 613 |
| 212 | 634 | 654 | 675 | 695 | 715 | 736 | 756 | 777 | 797 | 818 |
| 213 | 838 | 858 | 879 | 899 | 919 | 940 | 960 | 980 | *001 | *021 |
| 214 | 33 041 | 062 | 082 | 102 | 122 | 143 | 163 | 183 | 203 | 224 |
| 215 | 244 | 264 | 284 | 304 | 325 | 345 | 365 | 385 | 405 | 425 |
| 216 | 445 | 465 | 486 | 506 | 526 | 546 | 566 | 586 | 606 | 626 |
| 217 | 646 | 666 | 686 | 706 | 726 | 746 | 766 | 786 | 806 | 826 |
| 218 | 846 | 866 | 885 | 905 | 925 | 945 | 965 | 985 | *005 | *025 |
| 219 | 34 044 | 064 | 084 | 104 | 124 | 143 | 163 | 183 | 203 | 223 |
| 220 | 242 | 262 | 282 | 301 | 321 | 341 | 361 | 380 | 400 | 420 |
| 221 | 439 | 459 | 479 | 498 | 518 | 537 | 557 | 577 | 596 | 616 |
| 222 | 635 | 655 | 674 | 694 | 713 | 733 | 753 | 772 | 792 | 811 |
| 223 | 830 | 850 | 869 | 889 | 908 | 928 | 947 | 967 | 986 | *005 |
| 224 | 35 025 | 044 | 064 | 083 | 102 | 122 | 141 | 160 | 180 | 199 |
| 225 | 218 | 238 | 257 | 276 | 295 | 315 | 334 | 353 | 372 | 392 |
| 226 | 411 | 430 | 449 | 468 | 488 | 507 | 526 | 545 | 564 | 583 |
| 227 | 603 | 622 | 641 | 660 | 679 | 698 | 717 | 736 | 755 | 774 |
| 228 | 793 | 813 | 832 | 851 | 870 | 889 | 908 | 927 | 946 | 965 |
| 229 | 984 | *003 | *021 | *040 | *059 | *078 | *097 | *116 | *135 | *154 |
| 230 | 36 173 | 192 | 211 | 229 | 248 | 267 | 286 | 305 | 324 | 342 |
| 231 | 361 | 380 | 399 | 418 | 436 | 455 | 474 | 493 | 511 | 530 |
| 232 | 549 | 568 | 586 | 605 | 624 | 642 | 661 | 680 | 698 | 717 |
| 233 | 736 | 754 | 773 | 791 | 810 | 829 | 847 | 866 | 884 | 903 |
| 234 | 922 | 940 | 959 | 977 | 996 | *014 | *033 | *051 | *070 | *088 |
| 235 | 37 107 | 125 | 144 | 162 | 181 | 199 | 218 | 236 | 254 | 273 |
| 236 | 291 | 310 | 328 | 346 | 365 | 383 | 401 | 420 | 438 | 457 |
| 237 | 475 | 493 | 511 | 530 | 548 | 566 | 585 | 603 | 621 | 639 |
| 238 | 658 | 676 | 694 | 712 | 731 | 749 | 767 | 785 | 803 | 822 |
| 239 | 840 | 858 | 876 | 894 | 912 | 931 | 949 | 967 | 985 | *003 |
| 240 | 38 021 | 039 | 057 | 075 | 093 | 112 | 130 | 148 | 166 | 184 |
| 241 | 202 | 220 | 238 | 256 | 274 | 292 | 310 | 328 | 346 | 364 |
| 242 | 382 | 399 | 417 | 435 | 453 | 471 | 489 | 507 | 525 | 543 |
| 243 | 561 | 578 | 596 | 614 | 632 | 650 | 668 | 686 | 703 | 721 |
| 244 | 739 | 757 | 775 | 792 | 810 | 828 | 846 | 863 | 881 | 899 |
| 245 | 917 | 934 | 952 | 970 | 987 | *005 | *023 | *041 | *058 | *076 |
| 246 | 39 094 | 111 | 129 | 146 | 164 | 182 | 199 | 217 | 235 | 252 |
| 247 | 270 | 287 | 305 | 322 | 340 | 358 | 375 | 393 | 410 | 428 |
| 248 | 445 | 463 | 480 | 498 | 515 | 533 | 550 | 568 | 585 | 602 |
| 249 | 620 | 637 | 655 | 672 | 690 | 707 | 724 | 742 | 759 | 777 |
| 250 | 794 | 811 | 829 | 846 | 863 | 881 | 898 | 915 | 933 | 950 |

**P. P.**

| | 22 | 21 |
|---|---|---|
| 1 | 2.2 | 2.1 |
| 2 | 4.4 | 4.2 |
| 3 | 6.6 | 6.3 |
| 4 | 8.8 | 8.4 |
| 5 | 11.0 | 10.5 |
| 6 | 13.2 | 12.6 |
| 7 | 15.4 | 14.7 |
| 8 | 17.6 | 16.8 |
| 9 | 19.8 | 18.9 |

| | 20 |
|---|---|
| 1 | 2.0 |
| 2 | 4.0 |
| 3 | 6.0 |
| 4 | 8.0 |
| 5 | 10.0 |
| 6 | 12.0 |
| 7 | 14.0 |
| 8 | 16.0 |
| 9 | 18.0 |

| | 19 |
|---|---|
| 1 | 1.9 |
| 2 | 3.8 |
| 3 | 5.7 |
| 4 | 7.6 |
| 5 | 9.5 |
| 6 | 11.4 |
| 7 | 13.3 |
| 8 | 15.2 |
| 9 | 17.1 |

| | 18 |
|---|---|
| 1 | 1.8 |
| 2 | 3.6 |
| 3 | 5.4 |
| 4 | 7.2 |
| 5 | 9.0 |
| 6 | 10.8 |
| 7 | 12.6 |
| 8 | 14.4 |
| 9 | 16.2 |

| | 17 |
|---|---|
| 1 | 1.7 |
| 2 | 3.4 |
| 3 | 5.1 |
| 4 | 6.8 |
| 5 | 8.5 |
| 6 | 10.2 |
| 7 | 11.9 |
| 8 | 13.6 |
| 9 | 15.3 |

# COMMON LOGARITHMS

(.39794 to .47842)

| N. | L. 0 | 1 | 2 | 3 | 4 | 5 | 6 | 7 | 8 | 9 |
|---|---|---|---|---|---|---|---|---|---|---|
| 250 | 39 794 | 811 | 829 | 846 | 863 | 881 | 898 | 915 | 933 | 950 |
| 251 | 967 | 985 | *002 | *019 | *037 | *054 | *071 | *088 | *106 | *123 |
| 252 | 40 140 | 157 | 175 | 192 | 209 | 226 | 243 | 261 | 278 | 295 |
| 253 | 312 | 329 | 346 | 364 | 381 | 398 | 415 | 432 | 449 | 466 |
| 254 | 483 | 500 | 518 | 535 | 552 | 569 | 586 | 603 | 620 | 637 |
| 255 | 654 | 671 | 688 | 705 | 722 | 739 | 756 | 773 | 790 | 807 |
| 256 | 824 | 841 | 858 | 875 | 892 | 909 | 926 | 943 | 960 | 976 |
| 257 | 993 | *010 | *027 | *044 | *061 | *078 | *095 | *111 | *128 | *145 |
| 258 | 41 162 | 179 | 196 | 212 | 229 | 246 | 263 | 280 | 296 | 313 |
| 259 | 330 | 347 | 363 | 380 | 397 | 414 | 430 | 447 | 464 | 481 |
| 260 | 497 | 514 | 531 | 547 | 564 | 581 | 597 | 614 | 631 | 647 |
| 261 | 664 | 681 | 697 | 714 | 731 | 747 | 764 | 780 | 797 | 814 |
| 262 | 830 | 847 | 863 | 880 | 896 | 913 | 929 | 946 | 963 | 979 |
| 263 | 996 | *012 | *029 | *045 | *062 | *078 | *095 | *111 | *127 | *144 |
| 264 | 42 160 | 177 | 193 | 210 | 226 | 243 | 259 | 275 | 292 | 308 |
| 265 | 325 | 341 | 357 | 374 | 390 | 406 | 423 | 439 | 455 | 472 |
| 266 | 488 | 504 | 521 | 537 | 553 | 570 | 586 | 602 | 619 | 635 |
| 267 | 651 | 667 | 684 | 700 | 716 | 732 | 749 | 765 | 781 | 797 |
| 268 | 813 | 830 | 846 | 862 | 878 | 894 | 911 | 927 | 943 | 959 |
| 269 | 975 | 991 | *008 | *024 | *040 | *056 | *072 | *088 | *104 | *120 |
| 270 | 43 136 | 152 | 169 | 185 | 201 | 217 | 233 | 249 | 265 | 281 |
| 271 | 297 | 313 | 329 | 345 | 361 | 377 | 393 | 409 | 425 | 441 |
| 272 | 457 | 473 | 489 | 505 | 521 | 537 | 553 | 569 | 584 | 600 |
| 273 | 616 | 632 | 648 | 664 | 680 | 696 | 712 | 727 | 743 | 759 |
| 274 | 775 | 791 | 807 | 823 | 838 | 854 | 870 | 886 | 902 | 917 |
| 275 | 933 | 949 | 965 | 981 | 996 | *012 | *028 | *044 | *059 | *075 |
| 276 | 44 091 | 107 | 122 | 138 | 154 | 170 | 185 | 201 | 217 | 232 |
| 277 | 248 | 264 | 279 | 295 | 311 | 326 | 342 | 358 | 373 | 389 |
| 278 | 404 | 420 | 436 | 451 | 467 | 483 | 498 | 514 | 529 | 545 |
| 279 | 560 | 576 | 592 | 607 | 623 | 638 | 654 | 669 | 685 | 700 |
| 280 | 716 | 731 | 747 | 762 | 778 | 793 | 809 | 824 | 840 | 855 |
| 281 | 871 | 886 | 902 | 917 | 932 | 948 | 963 | 979 | 994 | *010 |
| 282 | 45 025 | 040 | 056 | 071 | 086 | 102 | 117 | 133 | 148 | 163 |
| 283 | 179 | 194 | 209 | 225 | 240 | 255 | 271 | 286 | 301 | 317 |
| 284 | 332 | 347 | 362 | 378 | 393 | 408 | 423 | 439 | 454 | 469 |
| 285 | 484 | 500 | 515 | 530 | 545 | 561 | 576 | 591 | 606 | 621 |
| 286 | 637 | 652 | 667 | 682 | 697 | 712 | 728 | 743 | 758 | 773 |
| 287 | 788 | 803 | 818 | 834 | 849 | 864 | 879 | 894 | 909 | 924 |
| 288 | 939 | 954 | 969 | 984 | *000 | *015 | *030 | *045 | *060 | *075 |
| 289 | 46 090 | 105 | 120 | 135 | 150 | 165 | 180 | 195 | 210 | 225 |
| 290 | 240 | 255 | 270 | 285 | 300 | 315 | 330 | 345 | 359 | 374 |
| 291 | 389 | 404 | 419 | 434 | 449 | 464 | 479 | 494 | 509 | 523 |
| 292 | 538 | 553 | 568 | 583 | 598 | 613 | 627 | 642 | 657 | 672 |
| 293 | 687 | 702 | 716 | 731 | 746 | 761 | 776 | 790 | 805 | 820 |
| 294 | 835 | 850 | 864 | 879 | 894 | 909 | 923 | 938 | 953 | 967 |
| 295 | 982 | 997 | *012 | *026 | *041 | *056 | *070 | *085 | *100 | *114 |
| 296 | 47 129 | 144 | 159 | 173 | 188 | 202 | 217 | 232 | 246 | 261 |
| 297 | 276 | 290 | 305 | 319 | 334 | 349 | 363 | 378 | 392 | 407 |
| 298 | 422 | 436 | 451 | 465 | 480 | 494 | 509 | 524 | 538 | 553 |
| 299 | 567 | 582 | 596 | 611 | 625 | 640 | 654 | 669 | 683 | 698 |
| 300 | 712 | 727 | 741 | 756 | 770 | 784 | 799 | 813 | 828 | 842 |

**P. P.**

18
| 1 | 1.8 |
|---|---|
| 2 | 3.6 |
| 3 | 5.4 |
| 4 | 7.2 |
| 5 | 9.0 |
| 6 | 10.8 |
| 7 | 12.6 |
| 8 | 14.4 |
| 9 | 16.2 |

17
| 1 | 1.7 |
|---|---|
| 2 | 3.4 |
| 3 | .5.1 |
| 4 | 6.8 |
| 5 | 8.5 |
| 6 | 10.2 |
| 7 | 11.9 |
| 8 | 13.6 |
| 9 | 15.3 |

16
| 1 | 1.6 |
|---|---|
| 2 | 3.2 |
| 3 | 4.8 |
| 4 | 6.4 |
| 5 | 8.0 |
| 6 | 9.6 |
| 7 | 11.2 |
| 8 | 12.8 |
| 9 | 14.4 |

15
| 1 | 1.5 |
|---|---|
| 2 | 3.0 |
| 3 | 4.5 |
| 4 | 6.0 |
| 5 | 7.5 |
| 6 | 9.0 |
| 7 | 10.5 |
| 8 | 12.0 |
| 9 | 13.5 |

14
| 1 | 1.4 |
|---|---|
| 2 | 2.8 |
| 3 | 4.2 |
| 4 | 5.6 |
| 5 | 7.0 |
| 6 | 8.4 |
| 7 | 9.8 |
| 8 | 11.2 |
| 9 | 12.6 |

(.47712 to .54518)

| N. | L. 0 | 1 | 2 | 3 | 4 | 5 | 6 | 7 | 8 | 9 |
|---|---|---|---|---|---|---|---|---|---|---|
| 300 | 47 712 | 727 | 741 | 756 | 770 | 784 | 799 | 813 | 828 | 842 |
| 301 | 857 | 871 | 885 | 900 | 914 | 929 | 943 | 958 | 972 | 986 |
| 302 | 48 001 | 015 | 029 | 044 | 058 | 073 | 087 | 101 | 116 | 130 |
| 303 | 144 | 159 | 173 | 187 | 202 | 216 | 230 | 244 | 259 | 273 |
| 304 | 287 | 302 | 316 | 330 | 344 | 359 | 373 | 387 | 401 | 416 |
| 305 | 430 | 444 | 458 | 473 | 487 | 501 | 515 | 530 | 544 | 558 |
| 306 | 572 | 586 | 601 | 615 | 629 | 643 | 657 | 671 | 686 | 700 |
| 307 | 714 | 728 | 742 | 756 | 770 | 785 | 799 | 813 | 827 | 841 |
| 308 | 855 | 869 | 883 | 897 | 911 | 926 | 940 | 954 | 968 | 982 |
| 309 | 996 | *010 | *024 | *038 | *052 | *066 | *080 | *094 | *108 | *122 |
| 310 | 49 136 | 150 | 164 | 178 | 192 | 206 | 220 | 234 | 248 | 262 |
| 311 | 276 | 290 | 304 | 318 | 332 | 346 | 360 | 374 | 388 | 402 |
| 312 | 415 | 429 | 443 | 457 | 471 | 485 | 499 | 513 | 527 | 541 |
| 313 | 554 | 568 | 582 | 596 | 610 | 624 | 638 | 651 | 665 | 679 |
| 314 | 693 | 707 | 721 | 734 | 748 | 762 | 776 | 790 | 803 | 817 |
| 315 | 831 | 845 | 859 | 872 | 886 | 900 | 914 | 927 | 941 | 955 |
| 316 | 969 | 982 | 996 | *010 | *024 | *037 | *051 | *065 | *079 | *092 |
| 317 | 50 106 | 120 | 133 | 147 | 161 | 174 | 188 | 202 | 215 | 229 |
| 318 | 243 | 256 | 270 | 284 | 297 | 311 | 325 | 338 | 352 | 365 |
| 319 | 379 | 393 | 406 | 420 | 433 | 447 | 461 | 474 | 488 | 501 |
| 320 | 515 | 529 | 542 | 556 | 569 | 583 | 596 | 610 | 623 | 637 |
| 321 | 651 | 664 | 678 | 691 | 705 | 718 | 732 | 745 | 759 | 772 |
| 322 | 786 | 799 | 813 | 826 | 840 | 853 | 866 | 880 | 893 | 907 |
| 323 | 920 | 934 | 947 | 961 | 974 | 987 | *001 | *014 | *028 | *041 |
| 324 | 51 055 | 068 | 081 | 095 | 108 | 121 | 135 | 148 | 162 | 175 |
| 325 | 188 | 202 | 215 | 228 | 242 | 255 | 268 | 282 | 295 | 308 |
| 326 | 322 | 335 | 348 | 362 | 375 | 388 | 402 | 415 | 428 | 441 |
| 327 | 455 | 468 | 481 | 495 | 508 | 521 | 534 | 548 | 561 | 574 |
| 328 | 587 | 601 | 614 | 627 | 640 | 654 | 667 | 680 | 693 | 706 |
| 329 | 720 | 733 | 746 | 759 | 772 | 786 | 799 | 812 | 825 | 838 |
| 330 | 851 | 865 | 878 | 891 | 904 | 917 | 930 | 943 | 957 | 970 |
| 331 | 983 | 996 | *009 | *022 | *035 | *048 | *061 | *075 | *088 | *101 |
| 332 | 52 114 | 127 | 140 | 153 | 166 | 179 | 192 | 205 | 218 | 231 |
| 333 | 244 | 257 | 270 | 284 | 297 | 310 | 323 | 336 | 349 | 362 |
| 334 | 375 | 388 | 401 | 414 | 427 | 440 | 453 | 466 | 479 | 492 |
| 335 | 504 | 517 | 530 | 543 | 556 | 569 | 582 | 595 | 608 | 621 |
| 336 | 634 | 647 | 660 | 673 | 686 | 699 | 711 | 724 | 737 | 750 |
| 337 | 763 | 776 | 789 | 802 | 815 | 827 | 840 | 853 | 866 | 879 |
| 338 | 892 | 905 | 917 | 930 | 943 | 956 | 969 | 982 | 994 | *007 |
| 339 | 53 020 | 033 | 046 | 058 | 071 | 084 | 097 | 110 | 122 | 135 |
| 340 | 148 | 161 | 173 | 186 | 199 | 212 | 224 | 237 | 250 | 263 |
| 341 | 275 | 288 | 301 | 314 | 326 | 339 | 352 | 364 | 377 | 390 |
| 342 | 403 | 415 | 428 | 441 | 453 | 466 | 479 | 491 | 504 | 517 |
| 343 | 529 | 542 | 555 | 567 | 580 | 593 | 605 | 618 | 631 | 643 |
| 344 | 656 | 668 | 681 | 694 | 706 | 719 | 732 | 744 | 757 | 769 |
| 345 | 782 | 794 | 807 | 820 | 832 | 845 | 857 | 870 | 882 | 895 |
| 346 | 908 | 920 | 933 | 945 | 958 | 970 | 983 | 995 | *008 | *020 |
| 347 | 54 033 | 045 | 058 | 070 | 083 | 095 | 108 | 120 | 133 | 145 |
| 348 | 158 | 170 | 183 | 195 | 208 | 220 | 233 | 245 | 258 | 270 |
| 349 | 283 | 295 | 307 | 320 | 332 | 345 | 357 | 370 | 382 | 394 |
| 350 | 407 | 419 | 432 | 444 | 456 | 469 | 481 | 494 | 506 | 518 |

P. P.

**15**

| 1 | 1.5 |
|---|---|
| 2 | 3.0 |
| 3 | 4.5 |
| 4 | 6.0 |
| 5 | 7.5 |
| 6 | 9.0 |
| 7 | 10.5 |
| 8 | 12.0 |
| 9 | 13.5 |

**14**

| 1 | 1.4 |
|---|---|
| 2 | 2.8 |
| 3 | 4.2 |
| 4 | 5.6 |
| 5 | 7.0 |
| 6 | 8.4 |
| 7 | 9.8 |
| 8 | 11.2 |
| 9 | 12.6 |

**13**

| 1 | 1.3 |
|---|---|
| 2 | 2.6 |
| 3 | 3.9 |
| 4 | 5.2 |
| 5 | 6.5 |
| 6 | 7.8 |
| 7 | 9.1 |
| 8 | 10.4 |
| 9 | 11.7 |

**12**

| 1 | 1.2 |
|---|---|
| 2 | 2.4 |
| 3 | 3.6 |
| 4 | 4.8 |
| 5 | 6.0 |
| 6 | 7.2 |
| 7 | 8.4 |
| 8 | 9.6 |
| 9 | 10.8 |

(.54407 to .60304)

| N. | L. 0 | 1 | 2 | 3 | 4 | 5 | 6 | 7 | 8 | 9 |
|---|---|---|---|---|---|---|---|---|---|---|
| 350 | 54 407 | 419 | 432 | 444 | 456 | 469 | 481 | 494 | 506 | 518 |
| 351 | 531 | 543 | 555 | 568 | 580 | 593 | 605 | 617 | 630 | 642 |
| 352 | 654 | 667 | 679 | 691 | 704 | 716 | 728 | 741 | 753 | 765 |
| 353 | 777 | 790 | 802 | 814 | 827 | 839 | 851 | 864 | 876 | 888 |
| 354 | 900 | 913 | 925 | 937 | 949 | 962 | 974 | 986 | 998 | *011 |
| 355 | 55 023 | 035 | 047 | 060 | 072 | 084 | 096 | 108 | 121 | 133 |
| 356 | 145 | 157 | 169 | 182 | 194 | 206 | 218 | 230 | 242 | 255 |
| 357 | 267 | 279 | 291 | 303 | 315 | 328 | 340 | 352 | 364 | 376 |
| 358 | 388 | 400 | 413 | 425 | 437 | 449 | 461 | 473 | 485 | 497 |
| 359 | 509 | 522 | 534 | 546 | 558 | 570 | 582 | 594 | 606 | 618 |
| 360 | 630 | 642 | 654 | 666 | 678 | 691 | 703 | 715 | 727 | 739 |
| 361 | 751 | 763 | 775 | 787 | 799 | 811 | 823 | 835 | 847 | 859 |
| 362 | 871 | 883 | 895 | 907 | 919 | 931 | 943 | 955 | 967 | 979 |
| 363 | 991 | *003 | *015 | *027 | *038 | *050 | *062 | *074 | *086 | *098 |
| 364 | 56 110 | 122 | 134 | 146 | 158 | 170 | 182 | 194 | 205 | 217 |
| 365 | 229 | 241 | 253 | 265 | 277 | 289 | 301 | 312 | 324 | 336 |
| 366 | 348 | 360 | 372 | 384 | 396 | 407 | 419 | 431 | 443 | 455 |
| 367 | 467 | 478 | 490 | 502 | 514 | 526 | 538 | 549 | 561 | 573 |
| 368 | 585 | 597 | 608 | 620 | 632 | 644 | 656 | 667 | 679 | 691 |
| 369 | 703 | 714 | 726 | 738 | 750 | 761 | 773 | 785 | 797 | 808 |
| 370 | 820 | 832 | 844 | 855 | 867 | 879 | 891 | 902 | 914 | 926 |
| 371 | 937 | 949 | 961 | 972 | 984 | 996 | *008 | *019 | *031 | *043 |
| 372 | 57 054 | 066 | 078 | 089 | 101 | 113 | 124 | 136 | 148 | 159 |
| 373 | 171 | 183 | 194 | 206 | 217 | 229 | 241 | 252 | 264 | 276 |
| 374 | 287 | 299 | 310 | 322 | 334 | 345 | 357 | 368 | 380 | 392 |
| 375 | 403 | 415 | 426 | 438 | 449 | 461 | 473 | 484 | 496 | 507 |
| 376 | 519 | 530 | 542 | 553 | 565 | 576 | 588 | 600 | 611 | 623 |
| 377 | 634 | 646 | 657 | 669 | 680 | 692 | 703 | 715 | 726 | 738 |
| 378 | 749 | 761 | 772 | 784 | 795 | 807 | 818 | 830 | 841 | 852 |
| 379 | 864 | 875 | 887 | 898 | 910 | 921 | 933 | 944 | 955 | 967 |
| 380 | 978 | 990 | *001 | *013 | *024 | *035 | *047 | *058 | *070 | *081 |
| 381 | 58 092 | 104 | 115 | 127 | 138 | 149 | 161 | 172 | 184 | 195 |
| 382 | 206 | 218 | 229 | 240 | 252 | 263 | 274 | 286 | 297 | 309 |
| 383 | 320 | 331 | 343 | 354 | 365 | 377 | 388 | 399 | 410 | 422 |
| 384 | 433 | 444 | 456 | 467 | 478 | 490 | 501 | 512 | 524 | 535 |
| 385 | 546 | 557 | 569 | 580 | 591 | 602 | 614 | 625 | 636 | 647 |
| 386 | 659 | 670 | 681 | 692 | 704 | 715 | 726 | 737 | 749 | 760 |
| 387 | 771 | 782 | 794 | 805 | 816 | 827 | 838 | 850 | 861 | 872 |
| 388 | 883 | 894 | 906 | 917 | 928 | 939 | 950 | 961 | 973 | 984 |
| 389 | 995 | *006 | *017 | *028 | *040 | *051 | *062 | *073 | *084 | *095 |
| 390 | 59 106 | 118 | 129 | 140 | 151 | 162 | 173 | 184 | 195 | 207 |
| 391 | 218 | 229 | 240 | 251 | 262 | 273 | 284 | 295 | 306 | 318 |
| 392 | 329 | 340 | 351 | 362 | 373 | 384 | 395 | 406 | 417 | 428 |
| 393 | 439 | 450 | 461 | 472 | 483 | 494 | 506 | 517 | 528 | 539 |
| 394 | 550 | 561 | 572 | 583 | 594 | 605 | 616 | 627 | 638 | 649 |
| 395 | 660 | 671 | 682 | 693 | 705 | 715 | 726 | 737 | 748 | 759 |
| 396 | 770 | 780 | 791 | 802 | 813 | 824 | 835 | 846 | 857 | 868 |
| 397 | 879 | 890 | 901 | 912 | 923 | 934 | 945 | 956 | 966 | 977 |
| 398 | 988 | 999 | *010 | *021 | *032 | *043 | *054 | *065 | *076 | *086 |
| 399 | 60 097 | 108 | 119 | 130 | 141 | 152 | 163 | 173 | 184 | 195 |
| 400 | 206 | 217 | 228 | 239 | 249 | 260 | 271 | 282 | 293 | 304 |

**P. P.**

**13**

| | |
|---|---|
| 1 | 1.3 |
| 2 | 2.6 |
| 3 | 3.9 |
| 4 | 5.2 |
| 5 | 6.5 |
| 6 | 7.8 |
| 7 | 9.1 |
| 8 | 10.4 |
| 9 | 11.7 |

**12**

| | |
|---|---|
| 1 | 1.2 |
| 2 | 2.4 |
| 3 | 3.6 |
| 4 | 4.8 |
| 5 | 6.0 |
| 6 | 7.2 |
| 7 | 8.4 |
| 8 | 9.6 |
| 9 | 10.8 |

**11**

| | |
|---|---|
| 1 | 1.1 |
| 2 | 2.2 |
| 3 | 3.3 |
| 4 | 4.4 |
| 5 | 5.5 |
| 6 | 6.6 |
| 7 | 7.7 |
| 8 | 8.8 |
| 9 | 9.9 |

**10**

| | |
|---|---|
| 1 | 1.0 |
| 2 | 2.0 |
| 3 | 3.0 |
| 4 | 4.0 |
| 5 | 5.0 |
| 6 | 6.0 |
| 7 | 7.0 |
| 8 | 8.0 |
| 9 | 9.0 |

(.60206 to .65408)

| N. | L. 0 | 1 | 2 | 3 | 4 | 5 | 6 | 7 | 8 | 9 | P. P. |
|---|---|---|---|---|---|---|---|---|---|---|---|
| 400 | 60 206 | 217 | 228 | 239 | 249 | 260 | 271 | 282 | 293 | 304 | |
| 401 | 314 | 325 | 336 | 347 | 358 | 369 | 379 | 390 | 401 | 412 | |
| 402 | 423 | 433 | 444 | 455 | 466 | 477 | 487 | 498 | 509 | 520 | |
| 403 | 531 | 541 | 552 | 563 | 574 | 584 | 595 | 606 | 617 | 627 | |
| 404 | 638 | 649 | 660 | 670 | 681 | 692 | 703 | 713 | 724 | 735 | |
| 405 | 746 | 756 | 767 | 778 | 788 | 799 | 810 | 821 | 831 | 842 | |
| 406 | 853 | 863 | 874 | 885 | 895 | 906 | 917 | 927 | 938 | 949 | |
| 407 | 959 | 970 | 981 | 991 | *002 | *013 | *023 | *034 | *045 | *055 | 11 |
| 408 | 61 066 | 077 | 087 | 098 | 109 | 119 | 130 | 140 | 151 | 162 | 1\|1.1 |
| 409 | 172 | 183 | 194 | 204 | 215 | 225 | 236 | 247 | 257 | 268 | 2\|2.2 |
| 410 | 278 | 289 | 300 | 310 | 321 | 331 | 342 | 352 | 363 | 374 | 3\|3.3 |
| 411 | 384 | 395 | 405 | 416 | 426 | 437 | 448 | 458 | 469 | 479 | 4\|4.4 |
| 412 | 490 | 500 | 511 | 521 | 532 | 542 | 553 | 563 | 574 | 584 | 5\|5.5 |
| 413 | 595 | 606 | 616 | 627 | 637 | 648 | 658 | 669 | 679 | 690 | 6\|6.6 |
| 414 | 700 | 711 | 721 | 731 | 742 | 752 | 763 | 773 | 784 | 794 | 7\|7.7 |
| 415 | 805 | 815 | 826 | 836 | 847 | 857 | 868 | 878 | 888 | 899 | 8\|8.8 |
| 416 | 909 | 920 | 930 | 941 | 951 | 962 | 972 | 982 | 993 | *003 | 9\|9.9 |
| 417 | 62 014 | 024 | 034 | 045 | 055 | 066 | 076 | 086 | 097 | 107 | |
| 418 | 118 | 128 | 138 | 149 | 159 | 170 | 180 | 190 | 201 | 211 | |
| 419 | 221 | 232 | 242 | 252 | 263 | 273 | 284 | 294 | 304 | 315 | |
| 420 | 325 | 335 | 346 | 356 | 366 | 377 | 387 | 397 | 408 | 418 | |
| 421 | 428 | 439 | 449 | 459 | 469 | 480 | 490 | 500 | 511 | 521 | 10 |
| 422 | 531 | 542 | 552 | 562 | 572 | 583 | 593 | 603 | 613 | 624 | 1\|1.0 |
| 423 | 634 | 644 | 655 | 665 | 675 | 685 | 696 | 706 | 716 | 726 | 2\|2.0 |
| 424 | 737 | 747 | 757 | 767 | 778 | 788 | 798 | 808 | 818 | 829 | 3\|3.0 |
| 425 | 839 | 849 | 859 | 870 | 880 | 890 | 900 | 910 | 921 | 931 | 4\|4.0 |
| 426 | 941 | 951 | 961 | 972 | 982 | 992 | *002 | *012 | *022 | *033 | 5\|5.0 |
| 427 | 63 043 | 053 | 063 | 073 | 083 | 094 | 104 | 114 | 124 | 134 | 6\|6.0 |
| 428 | 144 | 155 | 165 | 175 | 185 | 195 | 205 | 215 | 225 | 236 | 7\|7.0 |
| 429 | 246 | 256 | 266 | 276 | 286 | 296 | 306 | 317 | 327 | 337 | 8\|8.0 |
| 430 | 347 | 357 | 367 | 377 | 387 | 397 | 407 | 417 | 428 | 438 | 9\|9.0 |
| 431 | 448 | 458 | 468 | 478 | 488 | 498 | 508 | 518 | 528 | 538 | |
| 432 | 548 | 558 | 568 | 579 | 589 | 599 | 609 | 619 | 629 | 639 | |
| 433 | 649 | 659 | 669 | 679 | 689 | 699 | 709 | 719 | 729 | 739 | |
| 434 | 749 | 759 | 769 | 779 | 789 | 799 | 809 | 819 | 829 | 839 | |
| 435 | 849 | 859 | 869 | 879 | 889 | 899 | 909 | 919 | 929 | 939 | 9 |
| 436 | 949 | 959 | 969 | 979 | 988 | 998 | *008 | *018 | *028 | *038 | 1\|0.9 |
| 437 | 64 048 | 058 | 068 | 078 | 088 | 098 | 108 | 118 | 128 | 137 | 2\|1.8 |
| 438 | 147 | 157 | 167 | 177 | 187 | 197 | 207 | 217 | 227 | 237 | 3\|2.7 |
| 439 | 246 | 256 | 266 | 276 | 286 | 296 | 306 | 316 | 326 | 335 | 4\|3.6 |
| 440 | 345 | 355 | 365 | 375 | 385 | 395 | 404 | 414 | 424 | 434 | 5\|4.5 |
| 441 | 444 | 454 | 464 | 473 | 483 | 493 | 503 | 513 | 523 | 532 | 6\|5.4 |
| 442 | 542 | 552 | 562 | 572 | 582 | 591 | 601 | 611 | 621 | 631 | 7\|6.3 |
| 443 | 640 | 650 | 660 | 670 | 680 | 689 | 699 | 709 | 719 | 729 | 8\|7.2 |
| 444 | 738 | 748 | 758 | 768 | 777 | 787 | 797 | 807 | 816 | 826 | 9\|8.1 |
| 445 | 836 | 846 | 856 | 865 | 875 | 885 | 895 | 904 | 914 | 924 | |
| 446 | 933 | 943 | 953 | 963 | 972 | 982 | 992 | *002 | *011 | *021 | |
| 447 | 65 031 | 040 | 050 | 060 | 070 | 079 | 089 | 099 | 108 | 118 | |
| 448 | 128 | 137 | 147 | 157 | 167 | 176 | 186 | 196 | 205 | 215 | |
| 449 | 225 | 234 | 244 | 254 | 263 | 273 | 283 | 292 | 302 | 312 | |
| 450 | 321 | 331 | 341 | 350 | 360 | 369 | 379 | 389 | 398 | 408 | |

(.65321 to .69975)

| N. | L. 0 | 1 | 2 | 3 | 4 | 5 | 6 | 7 | 8 | 9 | P. P. |
|---|---|---|---|---|---|---|---|---|---|---|---|
| 450 | 65 321 | 331 | 341 | 350 | 360 | 369 | 379 | 389 | 398 | 408 | |
| 451 | 418 | 427 | 437 | 447 | 456 | 466 | 475 | 485 | 495 | 504 | |
| 452 | 514 | 523 | 533 | 543 | 552 | 562 | 571 | 581 | 591 | 600 | |
| 453 | 610 | 619 | 629 | 639 | 648 | 658 | 667 | 677 | 686 | 696 | |
| 454 | 706 | 715 | 725 | 734 | 744 | 753 | 763 | 772 | 782 | 792 | |
| 455 | 801 | 811 | 820 | 830 | 839 | 849 | 858 | 868 | 877 | 887 | |
| 456 | 896 | 906 | 916 | 925 | 935 | 944 | 954 | 963 | 973 | 982 | **10** |
| 457 | 992 | *001 | *011 | *020 | *030 | *039 | *049 | *058 | *068 | *077 | 1\|1.0 |
| 458 | 66 087 | 096 | 106 | 115 | 124 | 134 | 143 | 153 | 162 | 172 | 2\|2.0 |
| 459 | 181 | 191 | 200 | 210 | 219 | 229 | 238 | 247 | 257 | 266 | 3\|3.0 |
| 460 | 276 | 285 | 295 | 304 | 314 | 323 | 332 | 342 | 351 | 361 | 4\|4.0 |
| 461 | 370 | 380 | 389 | 398 | 408 | 417 | 427 | 436 | 445 | 455 | 5\|5.0 |
| 462 | 464 | 474 | 483 | 492 | 502 | 511 | 521 | 530 | 539 | 549 | 6\|6.0 |
| 463 | 558 | 567 | 577 | 586 | 596 | 605 | 614 | 624 | 633 | 642 | 7\|7.0 |
| 464 | 652 | 661 | 671 | 680 | 689 | 699 | 708 | 717 | 727 | 736 | 8\|8.0 |
| 465 | 745 | 755 | 764 | 773 | 783 | 792 | 801 | 811 | 820 | 829 | 9\|9.0 |
| 466 | 839 | 848 | 857 | 867 | 876 | 885 | 894 | 904 | 913 | 922 | |
| 467 | 932 | 941 | 950 | 960 | 969 | 978 | 987 | 997 | *006 | *015 | |
| 468 | 67 025 | 034 | 043 | 052 | 062 | 071 | 080 | 089 | 099 | 108 | |
| 469 | 117 | 127 | 136 | 145 | 154 | 164 | 173 | 182 | 191 | 201 | |
| 470 | 210 | 219 | 228 | 237 | 247 | 256 | 265 | 274 | 284 | 293 | **9** |
| 471 | 302 | 311 | 321 | 330 | 339 | 348 | 357 | 367 | 376 | 385 | 1\|0.9 |
| 472 | 394 | 403 | 413 | 422 | 431 | 440 | 449 | 459 | 468 | 477 | 2\|1.8 |
| 473 | 486 | 495 | 504 | 514 | 523 | 532 | 541 | 550 | 560 | 569 | 3\|2.7 |
| 474 | 578 | 587 | 596 | 605 | 614 | 624 | 633 | 642 | 651 | 660 | 4\|3.6 |
| 475 | 669 | 679 | 688 | 697 | 706 | 715 | 724 | 733 | 742 | 752 | 5\|4.5 |
| 476 | 761 | 770 | 779 | 788 | 797 | 806 | 815 | 825 | 834 | 843 | 6\|5.4 |
| 477 | 852 | 861 | 870 | 879 | 888 | 897 | 906 | 916 | 925 | 934 | 7\|6.3 |
| 478 | 943 | 952 | 961 | 970 | 979 | 988 | 997 | *006 | *015 | *024 | 8\|7.2 |
| 479 | 68 034 | 043 | 052 | 061 | 070 | 079 | 088 | 097 | 106 | 115 | 9\|8.1 |
| 480 | 124 | 133 | 142 | 151 | 160 | 169 | 178 | 187 | 196 | 205 | |
| 481 | 215 | 224 | 233 | 242 | 251 | 260 | 269 | 278 | 287 | 296 | |
| 482 | 305 | 314 | 323 | 332 | 341 | 350 | 359 | 368 | 377 | 386 | |
| 483 | 395 | 404 | 413 | 422 | 431 | 440 | 449 | 458 | 467 | 476 | |
| 484 | 485 | 494 | 502 | 511 | 520 | 529 | 538 | 547 | 556 | 565 | **8** |
| 485 | 574 | 583 | 592 | 601 | 610 | 619 | 628 | 637 | 646 | 655 | 1\|0.8 |
| 486 | 664 | 673 | 681 | 690 | 699 | 708 | 717 | 726 | 735 | 744 | 2\|1.6 |
| 487 | 753 | 762 | 771 | 780 | 789 | 797 | 806 | 815 | 824 | 833 | 3\|2.4 |
| 488 | 842 | 851 | 860 | 869 | 878 | 886 | 895 | 904 | 913 | 922 | 4\|3.2 |
| 489 | 931 | 940 | 949 | 958 | 966 | 975 | 984 | 993 | *002 | *011 | 5\|4.0 |
| 490 | 69 020 | 028 | 037 | 046 | 055 | 064 | 073 | 082 | 090 | 099 | 6\|4.8 |
| 491 | 108 | 117 | 126 | 135 | 144 | 152 | 161 | 170 | 179 | 188 | 7\|5.6 |
| 492 | 197 | 205 | 214 | 223 | 232 | 241 | 249 | 258 | 267 | 276 | 8\|6.4 |
| 493 | 285 | 294 | 302 | 311 | 320 | 329 | 338 | 346 | 355 | 364 | 9\|7.2 |
| 494 | 373 | 381 | 390 | 399 | 408 | 417 | 425 | 434 | 443 | 452 | |
| 495 | 461 | 469 | 478 | 487 | 496 | 504 | 513 | 522 | 531 | 539 | |
| 496 | 548 | 557 | 566 | 574 | 583 | 592 | 601 | 609 | 618 | 627 | |
| 497 | 636 | 644 | 653 | 662 | 671 | 679 | 688 | 697 | 705 | 714 | |
| 498 | 723 | 732 | 740 | 749 | 758 | 767 | 775 | 784 | 793 | 801 | |
| 499 | 810 | 819 | 827 | 836 | 845 | 854 | 862 | 871 | 880 | 888 | |
| 500 | 897 | 906 | 914 | 923 | 932 | 940 | 949 | 958 | 966 | 975 | |

(.69897 to .74107)

| N. | L. | 0 | 1 | 2 | 3 | 4 | 5 | 6 | 7 | 8 | 9 | P. P. |
|---|---|---|---|---|---|---|---|---|---|---|---|---|
| 500 | 69 897 | 906 | 914 | 923 | 932 | 940 | 949 | 958 | 966 | 975 | |
| 501 | 984 | 992 | *001 | *010 | *018 | *027 | *036 | *044 | *053 | *062 | |
| 502 | 70 070 | 079 | 088 | 096 | 105 | 114 | 122 | 131 | 140 | 148 | |
| 503 | 157 | 165 | 174 | 183 | 191 | 200 | 209 | 217 | 226 | 234 | |
| 504 | 243 | 252 | 260 | 269 | 278 | 286 | 295 | 303 | 312 | 321 | |
| 505 | 329 | 338 | 346 | 355 | 364 | 372 | 381 | 389 | 398 | 406 | |
| 506 | 415 | 424 | 432 | 441 | 449 | 458 | 467 | 475 | 484 | 492 | **9** |
| 507 | 501 | 509 | 518 | 526 | 535 | 544 | 552 | 561 | 569 | 578 | 1 \| 0.9 |
| 508 | 586 | 595 | 603 | 612 | 621 | 629 | 638 | 646 | 655 | 663 | 2 \| 1.8 |
| 509 | 672 | 680 | 689 | 697 | 706 | 714 | 723 | 731 | 740 | 749 | 3 \| 2.7 |
| 510 | 757 | 766 | 774 | 783 | 791 | 800 | 808 | 817 | 825 | 834 | 4 \| 3.6 |
| 511 | 842 | 851 | 859 | 868 | 876 | 885 | 893 | 902 | 910 | 919 | 5 \| 4.5 |
| 512 | 927 | 935 | 944 | 952 | 961 | 969 | 978 | 986 | 995 | *003 | 6 \| 5.4 |
| 513 | 71 012 | 020 | 029 | 037 | 046 | 054 | 063 | 071 | 079 | 088 | 7 \| 6.3 |
| 514 | 096 | 105 | 113 | 122 | 130 | 139 | 147 | 155 | 164 | 172 | 8 \| 7.2 |
| 515 | 181 | 189 | 198 | 206 | 214 | 223 | 231 | 240 | 248 | 257 | 9 \| 8.1 |
| 516 | 265 | 273 | 282 | 290 | 299 | 307 | 315 | 324 | 332 | 341 | |
| 517 | 349 | 357 | 366 | 374 | 383 | 391 | 399 | 408 | 416 | 425 | |
| 518 | 433 | 441 | 450 | 458 | 466 | 475 | 483 | 492 | 500 | 508 | |
| 519 | 517 | 525 | 533 | 542 | 550 | 559 | 567 | 575 | 584 | 592 | |
| 520 | 600 | 609 | 617 | 625 | 634 | 642 | 650 | 659 | 667 | 675 | |
| 521 | 684 | 692 | 700 | 709 | 717 | 725 | 734 | 742 | 750 | 759 | **8** |
| 522 | 767 | 775 | 784 | 792 | 800 | 809 | 817 | 825 | 834 | 842 | 1 \| 0.8 |
| 523 | 850 | 858 | 867 | 875 | 883 | 892 | 900 | 908 | 917 | 925 | 2 \| 1.6 |
| 524 | 933 | 941 | 950 | 958 | 966 | 975 | 983 | 991 | 999 | *008 | 3 \| 2.4 |
| 525 | 72 016 | 024 | 032 | 041 | 049 | 057 | 066 | 074 | 082 | 090 | 4 \| 3.2 |
| 526 | 099 | 107 | 115 | 123 | 132 | 140 | 148 | 156 | 165 | 173 | 5 \| 4.0 |
| 527 | 181 | 189 | 198 | 206 | 214 | 222 | 230 | 239 | 247 | 255 | 6 \| 4.8 |
| 528 | 263 | 272 | 280 | 288 | 296 | 304 | 313 | 321 | 329 | 337 | 7 \| 5.6 |
| 529 | 346 | 354 | 362 | 370 | 378 | 387 | 395 | 403 | 411 | 419 | 8 \| 6.4 |
| 530 | 428 | 436 | 444 | 452 | 460 | 469 | 477 | 485 | 493 | 501 | 9 \| 7.2 |
| 531 | 509 | 518 | 526 | 534 | 542 | 550 | 558 | 567 | 575 | 583 | |
| 532 | 591 | 599 | 607 | 616 | 624 | 632 | 640 | 648 | 656 | 665 | |
| 533 | 673 | 681 | 689 | 697 | 705 | 713 | 722 | 730 | 738 | 746 | |
| 534 | 754 | 762 | 770 | 779 | 787 | 795 | 803 | 811 | 819 | 827 | |
| 535 | 835 | 843 | 852 | 860 | 868 | 876 | 884 | 892 | 900 | 908 | **7** |
| 536 | 916 | 925 | 933 | 941 | 949 | 957 | 965 | 973 | 981 | 989 | 1 \| 0.7 |
| 537 | 997 | *006 | *014 | *022 | *030 | *038 | *046 | *054 | *062 | *070 | 2 \| 1.4 |
| 538 | 73 078 | 086 | 094 | 102 | 111 | 119 | 127 | 135 | 143 | 151 | 3 \| 2.1 |
| 539 | 159 | 167 | 175 | 183 | 191 | 199 | 207 | 215 | 223 | 231 | 4 \| 2.8 |
| 540 | 239 | 247 | 255 | 263 | 272 | 280 | 288 | 296 | 304 | 312 | 5 \| 3.5 |
| 541 | 320 | 328 | 336 | 344 | 352 | 360 | 368 | 376 | 384 | 392 | 6 \| 4.2 |
| 542 | 400 | 408 | 416 | 424 | 432 | 440 | 448 | 456 | 464 | 472 | 7 \| 4.9 |
| 543 | 480 | 488 | 496 | 504 | 512 | 520 | 528 | 536 | 544 | 552 | 8 \| 5.6 |
| 544 | 560 | 568 | 576 | 584 | 592 | 600 | 608 | 616 | 624 | 632 | 9 \| 6.3 |
| 545 | 640 | 648 | 656 | 664 | 672 | 679 | 687 | 695 | 703 | 711 | |
| 546 | 719 | 727 | 735 | 743 | 751 | 759 | 767 | 775 | 783 | 791 | |
| 547 | 799 | 807 | 815 | 823 | 830 | 838 | 846 | 854 | 862 | 870 | |
| 548 | 878 | 886 | 894 | 902 | 910 | 918 | 926 | 933 | 941 | 949 | |
| 549 | 957 | 965 | 973 | 981 | 989 | 997 | *005 | *013 | *020 | *028 | |
| 550 | 74 036 | 044 | 052 | 060 | 068 | 076 | 084 | 092 | 099 | 107 | |

(.74036 to .77880)

| N. | L. 0 | 1 | 2 | 3 | 4 | 5 | 6 | 7 | 8 | 9 |
|---|---|---|---|---|---|---|---|---|---|---|
| 550 | 74 036 | 044 | 052 | 060 | 068 | 076 | 084 | 092 | 099 | 107 |
| 551 | 115 | 123 | 131 | 139 | 147 | 155 | 162 | 170 | 178 | 186 |
| 552 | 194 | 202 | 210 | 218 | 225 | 233 | 241 | 249 | 257 | 265 |
| 553 | 273 | 280 | 288 | 296 | 304 | 312 | 320 | 327 | 335 | 343 |
| 554 | 351 | 359 | 367 | 374 | 382 | 390 | 398 | 406 | 414 | 421 |
| 555 | 429 | 437 | 445 | 453 | 461 | 468 | 476 | 484 | 492 | 500 |
| 556 | 507 | 515 | 523 | 531 | 539 | 547 | 554 | 562 | 570 | 578 |
| 557 | 586 | 593 | 601 | 609 | 617 | 624 | 632 | 640 | 648 | 656 |
| 558 | 663 | 671 | 679 | 687 | 695 | 702 | 710 | 718 | 726 | 733 |
| 559 | 741 | 749 | 757 | 764 | 772 | 780 | 788 | 796 | 803 | 811 |
| 560 | 819 | 827 | 834 | 842 | 850 | 858 | 865 | 873 | 881 | 889 |
| 561 | 896 | 904 | 912 | 920 | 927 | 935 | 943 | 950 | 958 | 966 |
| 562 | 974 | 981 | 989 | 997 | *005 | *012 | *020 | *028 | *035 | *043 |
| 563 | 75 051 | 059 | 066 | 074 | 082 | 089 | 097 | 105 | 113 | 120 |
| 564 | 128 | 136 | 143 | 151 | 159 | 166 | 174 | 182 | 189 | 197 |
| 565 | 205 | 213 | 220 | 228 | 236 | 243 | 251 | 259 | 266 | 274 |
| 566 | 282 | 289 | 297 | 305 | 312 | 320 | 328 | 335 | 343 | 351 |
| 567 | 358 | 366 | 374 | 381 | 389 | 397 | 404 | 412 | 420 | 427 |
| 568 | 435 | 442 | 450 | 458 | 465 | 473 | 481 | 488 | 496 | 504 |
| 569 | 511 | 519 | 526 | 534 | 542 | 549 | 557 | 565 | 572 | 580 |
| 570 | 587 | 595 | 603 | 610 | 618 | 626 | 633 | 641 | 648 | 656 |
| 571 | 664 | 671 | 679 | 686 | 694 | 702 | 709 | 717 | 724 | 732 |
| 572 | 740 | 747 | 755 | 762 | 770 | 778 | 785 | 793 | 800 | 808 |
| 573 | 815 | 823 | 831 | 838 | 846 | 853 | 861 | 868 | 876 | 884 |
| 574 | 891 | 899 | 906 | 914 | 921 | 929 | 937 | 944 | 952 | 959 |
| 575 | 967 | 974 | 982 | 989 | 997 | *005 | *012 | *020 | *027 | *035 |
| 576 | 76 042 | 050 | 057 | 065 | 072 | 080 | 087 | 095 | 103 | 110 |
| 577 | 118 | 125 | 133 | 140 | 148 | 155 | 163 | 170 | 178 | 185 |
| 578 | 193 | 200 | 208 | 215 | 223 | 230 | 238 | 245 | 253 | 260 |
| 579 | 268 | 275 | 283 | 290 | 298 | 305 | 313 | 320 | 328 | 335 |
| 580 | 343 | 350 | 358 | 365 | 373 | 380 | 388 | 395 | 403 | 410 |
| 581 | 418 | 425 | 433 | 440 | 448 | 455 | 462 | 470 | 477 | 485 |
| 582 | 492 | 500 | 507 | 515 | 522 | 530 | 537 | 545 | 552 | 559 |
| 583 | 567 | 574 | 582 | 589 | 597 | 604 | 612 | 619 | 626 | 634 |
| 584 | 641 | 649 | 656 | 664 | 671 | 678 | 686 | 693 | 701 | 708 |
| 585 | 716 | 723 | 730 | 738 | 745 | 753 | 760 | 768 | 775 | 782 |
| 586 | 790 | 797 | 805 | 812 | 819 | 827 | 834 | 842 | 849 | 856 |
| 587 | 864 | 871 | 879 | 886 | 893 | 901 | 908 | 916 | 923 | 930 |
| 588 | 938 | 945 | 953 | 960 | 967 | 975 | 982 | 989 | 997 | *004 |
| 589 | 77 012 | 019 | 026 | 034 | 041 | 048 | 056 | 063 | 070 | 078 |
| 590 | 085 | 093 | 100 | 107 | 115 | 122 | 129 | 137 | 144 | 151 |
| 591 | 159 | 166 | 173 | 181 | 188 | 195 | 203 | 210 | 217 | 225 |
| 592 | 232 | 240 | 247 | 254 | 262 | 269 | 276 | 283 | 291 | 298 |
| 593 | 305 | 313 | 320 | 327 | 335 | 342 | 349 | 357 | 364 | 371 |
| 594 | 379 | 386 | 393 | 401 | 408 | 415 | 422 | 430 | 437 | 444 |
| 595 | 452 | 459 | 466 | 474 | 481 | 488 | 495 | 503 | 510 | 517 |
| 596 | 525 | 532 | 539 | 546 | 554 | 561 | 568 | 576 | 583 | 590 |
| 597 | 597 | 605 | 612 | 619 | 627 | 634 | 641 | 648 | 656 | 663 |
| 598 | 670 | 677 | 685 | 692 | 699 | 706 | 714 | 721 | 728 | 735 |
| 599 | 743 | 750 | 757 | 764 | 772 | 779 | 786 | 793 | 801 | 808 |
| 600 | 815 | 822 | 830 | 837 | 844 | 851 | 859 | 866 | 873 | 880 |

P. P.

**8**

| 1 | 0.8 |
|---|---|
| 2 | 1.6 |
| 3 | 2.4 |
| 4 | 3.2 |
| 5 | 4.0 |
| 6 | 4.8 |
| 7 | 5.6 |
| 8 | 6.4 |
| 9 | 7.2 |

**7**

| 1 | 0.7 |
|---|---|
| 2 | 1.4 |
| 3 | 2.1 |
| 4 | 2.8 |
| 5 | 3.5 |
| 6 | 4.2 |
| 7 | 4.9 |
| 8 | 5.6 |
| 9 | 6.3 |

(.77815 to .81351)

| N. | L. 0 | 1 | 2 | 3 | 4 | 5 | 6 | 7 | 8 | 9 |
|---|---|---|---|---|---|---|---|---|---|---|
| 600 | 77 815 | 822 | 830 | 837 | 844 | 851 | 859 | 866 | 873 | 880 |
| 601 | 887 | 895 | 902 | 909 | 916 | 924 | 931 | 938 | 945 | 952 |
| 602 | 960 | 967 | 974 | 981 | 988 | 996 | *003 | *010 | *017 | *025 |
| 603 | 78 032 | 039 | 046 | 053 | 061 | 068 | 075 | 082 | 089 | 097 |
| 604 | 104 | 111 | 118 | 125 | 132 | 140 | 147 | 154 | 161 | 168 |
| 605 | 176 | 183 | 190 | 197 | 204 | 211 | 219 | 226 | 233 | 240 |
| 606 | 247 | 254 | 262 | 269 | 276 | 283 | 290 | 297 | 305 | 312 |
| 607 | 319 | 326 | 333 | 340 | 347 | 355 | 362 | 369 | 376 | 383 |
| 608 | 390 | 398 | 405 | 412 | 419 | 426 | 433 | 440 | 447 | 455 |
| 609 | 462 | 469 | 476 | 483 | 490 | 497 | 504 | 512 | 519 | 526 |
| 610 | 533 | 540 | 547 | 554 | 561 | 569 | 576 | 583 | 590 | 597 |
| 611 | 604 | 611 | 618 | 625 | 633 | 640 | 647 | 654 | 661 | 668 |
| 612 | 675 | 682 | 689 | 696 | 704 | 711 | 718 | 725 | 732 | 739 |
| 613 | 746 | 753 | 760 | 767 | 774 | 781 | 789 | 796 | 803 | 810 |
| 614 | 817 | 824 | 831 | 838 | 845 | 852 | 859 | 866 | 873 | 880 |
| 615 | 888 | 895 | 902 | 909 | 916 | 923 | 930 | 937 | 944 | 951 |
| 616 | 958 | 965 | 972 | 979 | 986 | 993 | *000 | *007 | *014 | *021 |
| 617 | 79 029 | 036 | 043 | 050 | 057 | 064 | 071 | 078 | 085 | 092 |
| 618 | 099 | 106 | 113 | 120 | 127 | 134 | 141 | 148 | 155 | 162 |
| 619 | 169 | 176 | 183 | 190 | 197 | 204 | 211 | 218 | 225 | 232 |
| 620 | 239 | 246 | 253 | 260 | 267 | 274 | 281 | 288 | 295 | 302 |
| 621 | 309 | 316 | 323 | 330 | 337 | 344 | 351 | 358 | 365 | 372 |
| 622 | 379 | 386 | 393 | 400 | 407 | 414 | 421 | 428 | 435 | 442 |
| 623 | 449 | 456 | 463 | 470 | 477 | 484 | 491 | 498 | 505 | 511 |
| 624 | 518 | 525 | 532 | 539 | 546 | 553 | 560 | 567 | 574 | 581 |
| 625 | 588 | 595 | 602 | 609 | 616 | 623 | 630 | 637 | 644 | 650 |
| 626 | 657 | 664 | 671 | 678 | 685 | 692 | 699 | 706 | 713 | 720 |
| 627 | 727 | 734 | 741 | 748 | 754 | 761 | 768 | 775 | 782 | 789 |
| 628 | 796 | 803 | 810 | 817 | 824 | 831 | 837 | 844 | 851 | 858 |
| 629 | 865 | 872 | 879 | 886 | 893 | 900 | 906 | 913 | 920 | 927 |
| 630 | 934 | 941 | 948 | 955 | 962 | 969 | 975 | 982 | 989 | 996 |
| 631 | 80 003 | 010 | 017 | 024 | 030 | 037 | 044 | 051 | 058 | 065 |
| 632 | 072 | 079 | 085 | 092 | 099 | 106 | 113 | 120 | 127 | 134 |
| 633 | 140 | 147 | 154 | 161 | 168 | 175 | 182 | 188 | 195 | 202 |
| 634 | 209 | 216 | 223 | 229 | 236 | 243 | 250 | 257 | 264 | 271 |
| 635 | 277 | 284 | 291 | 298 | 305 | 312 | 318 | 325 | 332 | 339 |
| 636 | 346 | 353 | 359 | 366 | 373 | 380 | 387 | 393 | 400 | 407 |
| 637 | 414 | 421 | 428 | 434 | 441 | 448 | 455 | 462 | 468 | 475 |
| 638 | 482 | 489 | 496 | 502 | 509 | 516 | 523 | 530 | 536 | 543 |
| 639 | 550 | 557 | 564 | 570 | 577 | 584 | 591 | 598 | 604 | 611 |
| 640 | 618 | 625 | 632 | 638 | 645 | 652 | 659 | 665 | 672 | 679 |
| 641 | 686 | 693 | 699 | 706 | 713 | 720 | 726 | 733 | 740 | 747 |
| 642 | 754 | 760 | 767 | 774 | 781 | 787 | 794 | 801 | 808 | 814 |
| 643 | 821 | 828 | 835 | 841 | 848 | 855 | 862 | 868 | 875 | 882 |
| 644 | 889 | 895 | 902 | 909 | 916 | 922 | 929 | 936 | 943 | 949 |
| 645 | 956 | 963 | 969 | 976 | 983 | 990 | 996 | *003 | *010 | *017 |
| 646 | 81 023 | 030 | 037 | 043 | 050 | 057 | 064 | 070 | 077 | 084 |
| 647 | 090 | 097 | 104 | 111 | 117 | 124 | 131 | 137 | 144 | 151 |
| 648 | 158 | 164 | 171 | 178 | 184 | 191 | 198 | 204 | 211 | 218 |
| 649 | 224 | 231 | 238 | 245 | 251 | 258 | 265 | 271 | 278 | 285 |
| 650 | 291 | 298 | 305 | 311 | 318 | 325 | 331 | 338 | 345 | 351 |

P. P.

**8**

| 1 | 0.8 |
|---|---|
| 2 | 1.6 |
| 3 | 2.4 |
| 4 | 3.2 |
| 5 | 4.0 |
| 6 | 4.8 |
| 7 | 5.6 |
| 8 | 6.4 |
| 9 | 7.2 |

**7**

| 1 | 0.7 |
|---|---|
| 2 | 1.4 |
| 3 | 2.1 |
| 4 | 2.8 |
| 5 | 3.5 |
| 6 | 4.2 |
| 7 | 4.9 |
| 8 | 5.6 |
| 9 | 6.3 |

**6**

| 1 | 0.6 |
|---|---|
| 2 | 1.2 |
| 3 | 1.8 |
| 4 | 2.4 |
| 5 | 3.0 |
| 6 | 3.6 |
| 7 | 4.2 |
| 8 | 4.8 |
| 9 | 5.4 |

COMMON LOGARITHMS

(.81291 to .84566)

| N. | L. 0 | 1 | 2 | 3 | 4 | 5 | 6 | 7 | 8 | 9 | P. P. |
|---|---|---|---|---|---|---|---|---|---|---|---|
| 650 | 81 291 | 298 | 305 | 311 | 318 | 325 | 331 | 338 | 345 | 351 | |
| 651 | 358 | 365 | 371 | 378 | 385 | 391 | 398 | 405 | 411 | 418 | |
| 652 | 425 | 431 | 438 | 445 | 451 | 458 | 465 | 471 | 478 | 485 | |
| 653 | 491 | 498 | 505 | 511 | 518 | 525 | 531 | 538 | 544 | 551 | |
| 654 | 558 | 564 | 571 | 578 | 584 | 591 | 598 | 604 | 611 | 617 | |
| 655 | 624 | 631 | 637 | 644 | 651 | 657 | 664 | 671 | 677 | 684 | |
| 656 | 690 | 697 | 704 | 710 | 717 | 723 | 730 | 737 | 743 | 750 | |
| 657 | 757 | 763 | 770 | 776 | 783 | 790 | 796 | 803 | 809 | 816 | |
| 658 | 823 | 829 | 836 | 842 | 849 | 856 | 862 | 869 | 875 | 882 | |
| 659 | 889 | 895 | 902 | 908 | 915 | 921 | 928 | 935 | 941 | 948 | |
| 660 | 954 | 961 | 968 | 974 | 981 | 987 | 994 | *000 | *007 | *014 | **7** |
| 661 | 82 020 | 027 | 033 | 040 | 046 | 053 | 060 | 066 | 073 | 079 | 1 \| 0.7 |
| 662 | 086 | 092 | 099 | 105 | 112 | 119 | 125 | 132 | 138 | 145 | 2 \| 1.4 |
| 663 | 151 | 158 | 164 | 171 | 178 | 184 | 191 | 197 | 204 | 210 | 3 \| 2.1 |
| 664 | 217 | 223 | 230 | 236 | 243 | 249 | 256 | 263 | 269 | 276 | 4 \| 2.8 |
| 665 | 282 | 289 | 295 | 302 | 308 | 315 | 321 | 328 | 334 | 341 | 5 \| 3.5 |
| 666 | 347 | 354 | 360 | 367 | 373 | 380 | 387 | 393 | 400 | 406 | 6 \| 4.2 |
| 667 | 413 | 419 | 426 | 432 | 439 | 445 | 452 | 458 | 465 | 471 | 7 \| 4.9 |
| 668 | 478 | 484 | 491 | 497 | 504 | 510 | 517 | 523 | 530 | 536 | 8 \| 5.6 |
| 669 | 543 | 549 | 556 | 562 | 569 | 575 | 582 | 588 | 595 | 601 | 9 \| 6.3 |
| 670 | 607 | 614 | 620 | 627 | 633 | 640 | 646 | 653 | 659 | 666 | |
| 671 | 672 | 679 | 685 | 692 | 698 | 705 | 711 | 718 | 724 | 730 | |
| 672 | 737 | 743 | 750 | 756 | 763 | 769 | 776 | 782 | 789 | 795 | |
| 673 | 802 | 808 | 814 | 821 | 827 | 834 | 840 | 847 | 853 | 860 | |
| 674 | 866 | 872 | 879 | 885 | 892 | 898 | 905 | 911 | 918 | 924 | |
| 675 | 930 | 937 | 943 | 950 | 956 | 963 | 969 | 975 | 982 | 988 | |
| 676 | 995 | *001 | *008 | *014 | *020 | *027 | *033 | *040 | *046 | *052 | |
| 677 | 83 059 | 065 | 072 | 078 | 085 | 091 | 097 | 104 | 110 | 117 | |
| 678 | 123 | 129 | 136 | 142 | 149 | 155 | 161 | 168 | 174 | 181 | |
| 679 | 187 | 193 | 200 | 206 | 213 | 219 | 225 | 232 | 238 | 245 | |
| 680 | 251 | 257 | 264 | 270 | 276 | 283 | 289 | 296 | 302 | 308 | **6** |
| 681 | 315 | 321 | 327 | 334 | 340 | 347 | 353 | 359 | 366 | 372 | 1 \| 0.6 |
| 682 | 378 | 385 | 391 | 398 | 404 | 410 | 417 | 423 | 429 | 436 | 2 \| 1.2 |
| 683 | 442 | 448 | 455 | 461 | 467 | 474 | 480 | 487 | 493 | 499 | 3 \| 1.8 |
| 684 | 506 | 512 | 518 | 525 | 531 | 537 | 544 | 550 | 556 | 563 | 4 \| 2.4 |
| 685 | 569 | 575 | 582 | 588 | 594 | 601 | 607 | 613 | 620 | 626 | 5 \| 3.0 |
| 686 | 632 | 639 | 645 | 651 | 658 | 664 | 670 | 677 | 683 | 689 | 6 \| 3.6 |
| 687 | 696 | 702 | 708 | 715 | 721 | 727 | 734 | 740 | 746 | 753 | 7 \| 4.2 |
| 688 | 759 | 765 | 771 | 778 | 784 | 790 | 797 | 803 | 809 | 816 | 8 \| 4.8 |
| 689 | 822 | 828 | 835 | 841 | 847 | 853 | 860 | 866 | 872 | 879 | 9 \| 5.4 |
| 690 | 885 | 891 | 897 | 904 | 910 | 916 | 923 | 929 | 935 | 942 | |
| 691 | 948 | 954 | 960 | 967 | 973 | 979 | 985 | 992 | 998 | *004 | |
| 692 | 84 011. | 017 | 023 | 029 | 036 | 042 | 048 | 055 | 061 | 067 | |
| 693 | 073 | 080 | 086 | 092 | 098 | 105 | 111 | 117 | 123 | 130 | |
| 694 | 136 | 142 | 148 | 155 | 161 | 167 | 173 | 180 | 186 | 192 | |
| 695 | 198 | 205 | 211 | 217 | 223 | 230 | 236 | 242 | 248 | 255 | |
| 696 | 261 | 267 | 273 | 280 | 286 | 292 | 298 | 305 | 311 | 317 | |
| 697 | 323 | 330 | 336 | 342 | 348 | 354 | 361 | 367 | 373 | 379 | |
| 698 | 386 | 392 | 398 | 404 | 410 | 417 | 423 | 429 | 435 | 442 | |
| 699 | 448 | 454 | 460 | 466 | 473 | 479 | 485 | 491 | 497 | 504 | |
| 700 | 510 | 516 | 522 | 528 | 535 | 541 | 547 | 553 | 559 | 566 | |

(.84510 to .87558)

| N. | L. 0 | 1 | 2 | 3 | 4 | 5 | 6 | 7 | 8 | 9 |
|---|---|---|---|---|---|---|---|---|---|---|
| 700 | 84 510 | 516 | 522 | 528 | 535 | 541 | 547 | 553 | 559 | 566 |
| 701 | 572 | 578 | 584 | 590 | 597 | 603 | 609 | 615 | 621 | 628 |
| 702 | 634 | 640 | 646 | 652 | 658 | 665 | 671 | 677 | 683 | 689 |
| 703 | 696 | 702 | 708 | 714 | 720 | 726 | 733 | 739 | 745 | 751 |
| 704 | 757 | 763 | 770 | 776 | 782 | 788 | 794 | 800 | 807 | 813 |
| 705 | 819 | 825 | 831 | 837 | 844 | 850 | 856 | 862 | 868 | 874 |
| 706 | 880 | 887 | 893 | 899 | 905 | 911 | 917 | 924 | 930 | 936 |
| 707 | 942 | 948 | 954 | 960 | 967 | 973 | 979 | 985 | 991 | 997 |
| 708 | 85 003 | 009 | 016 | 022 | 028 | 034 | 040 | 046 | 052 | 058 |
| 709 | 065 | 071 | 077 | 083 | 089 | 095 | 101 | 107 | 114 | 120 |
| 710 | 126 | 132 | 138 | 144 | 150 | 156 | 163 | 169 | 175 | 181 |
| 711 | 187 | 193 | 199 | 205 | 211 | 217 | 224 | 230 | 236 | 242 |
| 712 | 248 | 254 | 260 | 266 | 272 | 278 | 285 | 291 | 297 | 303 |
| 713 | 309 | 315 | 321 | 327 | 333 | 339 | 345 | 352 | 358 | 364 |
| 714 | 370 | 376 | 382 | 388 | 394 | 400 | 406 | 412 | 418 | 425 |
| 715 | 431 | 437 | 443 | 449 | 455 | 461 | 467 | 473 | 479 | 485 |
| 716 | 491 | 497 | 503 | 509 | 516 | 522 | 528 | 534 | 540 | 546 |
| 717 | 552 | 558 | 564 | 570 | 576 | 582 | 588 | 594 | 600 | 606 |
| 718 | 612 | 618 | 625 | 631 | 637 | 643 | 649 | 655 | 661 | 667 |
| 719 | 673 | 679 | 685 | 691 | 697 | 703 | 709 | 715 | 721 | 727 |
| 720 | 733 | 739 | 745 | 751 | 757 | 763 | 769 | 775 | 781 | 788 |
| 721 | 794 | 800 | 806 | 812 | 818 | 824 | 830 | 836 | 842 | 848 |
| 722 | 854 | 860 | 866 | 872 | 878 | 884 | 890 | 896 | 902 | 908 |
| 723 | 914 | 920 | 926 | 932 | 938 | 944 | 950 | 956 | 962 | 968 |
| 724 | 974 | 980 | 986 | 992 | 998 | *004 | *010 | *016 | *022 | *028 |
| 725 | 86 034 | 040 | 046 | 052 | 058 | 064 | 070 | 076 | 082 | 088 |
| 726 | 094 | 100 | 106 | 112 | 118 | 124 | 130 | 136 | 141 | 147 |
| 727 | 153 | 159 | 165 | 171 | 177 | 183 | 189 | 195 | 201 | 207 |
| 728 | 213 | 219 | 225 | 231 | 237 | 243 | 249 | 255 | 261 | 267 |
| 729 | 273 | 279 | 285 | 291 | 297 | 303 | 308 | 314 | 320 | 326 |
| 730 | 332 | 338 | 344 | 350 | 356 | 362 | 368 | 374 | 380 | 386 |
| 731 | 392 | 398 | 404 | 410 | 415 | 421 | 427 | 433 | 439 | 445 |
| 732 | 451 | 457 | 463 | 469 | 475 | 481 | 487 | 493 | 499 | 504 |
| 733 | 510 | 516 | 522 | 528 | 534 | 540 | 546 | 552 | 558 | 564 |
| 734 | 570 | 576 | 581 | 587 | 593 | 599 | 605 | 611 | 617 | 623 |
| 735 | 629 | 635 | 641 | 646 | 652 | 658 | 664 | 670 | 676 | 682 |
| 736 | 688 | 694 | 700 | 705 | 711 | 717 | 723 | 729 | 735 | 741 |
| 737 | 747 | 753 | 759 | 764 | 770 | 776 | 782 | 788 | 794 | 800 |
| 738 | 806 | 812 | 817 | 823 | 829 | 835 | 841 | 847 | 853 | 859 |
| 739 | 864 | 870 | 876 | 882 | 888 | 894 | 900 | 906 | 911 | 917 |
| 740 | 923 | 929 | 935 | 941 | 947 | 953 | 958 | 964 | 970 | 976 |
| 741 | 982 | 988 | 994 | 999 | *005 | *011 | *017 | *023 | *029 | *035 |
| 742 | 87 040 | 046 | 052 | 058 | 064 | 070 | 075 | 081 | 087 | 093 |
| 743 | 099 | 105 | 111 | 116 | 122 | 128 | 134 | 140 | 146 | 151 |
| 744 | 157 | 163 | 169 | 175 | 181 | 186 | 192 | 198 | 204 | 210 |
| 745 | 216 | 221 | 227 | 233 | 239 | 245 | 251 | 256 | 262 | 268 |
| 746 | 274 | 280 | 286 | 291 | 297 | 303 | 309 | 315 | 320 | 326 |
| 747 | 332 | 338 | 344 | 349 | 355 | 361 | 367 | 373 | 379 | 384 |
| 748 | 390 | 396 | 402 | 408 | 413 | 419 | 425 | 431 | 437 | 442 |
| 749 | 448 | 454 | 460 | 466 | 471 | 477 | 483 | 489 | 495 | 500 |
| 750 | 506 | 512 | 518 | 523 | 529 | 535 | 541 | 547 | 552 | 558 |

**P. P.**

**7**

| 1 | 0.7 |
|---|---|
| 2 | 1.4 |
| 3 | 2.1 |
| 4 | 2.8 |
| 5 | 3.5 |
| 6 | 4.2 |
| 7 | 4.9 |
| 8 | 5.6 |
| 9 | 6.3 |

**6**

| 1 | 0.6 |
|---|---|
| 2 | 1.2 |
| 3 | 1.8 |
| 4 | 2.4 |
| 5 | 3.0 |
| 6 | 3.6 |
| 7 | 4.2 |
| 8 | 4.8 |
| 9 | 5.4 |

**5**

| 1 | 0.5 |
|---|---|
| 2 | 1.0 |
| 3 | 1.5 |
| 4 | 2.0 |
| 5 | 2.5 |
| 6 | 3.0 |
| 7 | 3.5 |
| 8 | 4.0 |
| 9 | 4.5 |

COMMON LOGARITHMS

(.87506 to .90358)

| N. | L. 0 | 1 | 2 | 3 | 4 | 5 | 6 | 7 | 8 | 9 |
|---|---|---|---|---|---|---|---|---|---|---|
| 750 | 87 506 | 512 | 518 | 523 | 529 | 535 | 541 | 547 | 552 | 558 |
| 751 | 564 | 570 | 576 | 581 | 587 | 593 | 599 | 604 | 610 | 616 |
| 752 | 622 | 628 | 633 | 639 | 645 | 651 | 656 | 662 | 668 | 674 |
| 753 | 679 | 685 | 691 | 697 | 703 | 708 | 714 | 720 | 726 | 731 |
| 754 | 737 | 743 | 749 | 754 | 760 | 766 | 772 | 777 | 783 | 789 |
| 755 | 795 | 800 | 806 | 812 | 818 | 823 | 829 | 835 | 841 | 846 |
| 756 | 852 | 858 | 864 | 869 | 875 | 881 | 887 | 892 | 898 | 904 |
| 757 | 910 | 915 | 921 | 927 | 933 | 938 | 944 | 950 | 955 | 961 |
| 758 | 967 | 973 | 978 | 984 | 990 | 996 | *001 | *007 | *013 | *018 |
| 759 | 88 024 | 030 | 036 | 041 | 047 | 053 | 058 | 064 | 070 | 076 |
| 760 | 081 | 087 | 093 | 098 | 104 | 110 | 116 | 121 | 127 | 133 |
| 761 | 138 | 144 | 150 | 156 | 161 | 167 | 173 | 178 | 184 | 190 |
| 762 | 195 | 201 | 207 | 213 | 218 | 224 | 230 | 235 | 241 | 247 |
| 763 | 252 | 258 | 264 | 270 | 275 | 281 | 287 | 292 | 298 | 304 |
| 764 | 309 | 315 | 321 | 326 | 332 | 338 | 343 | 349 | 355 | 360 |
| 765 | 366 | 372 | 377 | 383 | 389 | 395 | 400 | 406 | 412 | 417 |
| 766 | 423 | 429 | 434 | 440 | 446 | 451 | 457 | 463 | 468 | 474 |
| 767 | 480 | 485 | 491 | 497 | 502 | 508 | 513 | 519 | 525 | 530 |
| 768 | 536 | 542 | 547 | 553 | 559 | 564 | 570 | 576 | 581 | 587 |
| 769 | 593 | 598 | 604 | 610 | 615 | 621 | 627 | 632 | 638 | 643 |
| 770 | 649 | 655 | 660 | 666 | 672 | 677 | 683 | 689 | 694 | 700 |
| 771 | 705 | 711 | 717 | 722 | 728 | 734 | 739 | 745 | 750 | 756 |
| 772 | 762 | 767 | 773 | 779 | 784 | 790 | 795 | 801 | 807 | 812 |
| 773 | 818 | 824 | 829 | 835 | 840 | 846 | 852 | 857 | 863 | 868 |
| 774 | 874 | 880 | 885 | 891 | 897 | 902 | 908 | 913 | 919 | 925 |
| 775 | 930 | 936 | 941 | 947 | 953 | 958 | 964 | 969 | 975 | 981 |
| 776 | 986 | 992 | 997 | *003 | *009 | *014 | *020 | *025 | *031 | *037 |
| 777 | 89 042 | 048 | 053 | 059 | 064 | 070 | 076 | 081 | 087 | 092 |
| 778 | 098 | 104 | 109 | 115 | 120 | 126 | 131 | 137 | 143 | 148 |
| 779 | 154 | 159 | 165 | 170 | 176 | 182 | 187 | 193 | 198 | 204 |
| 780 | 209 | 215 | 221 | 226 | 232 | 237 | 243 | 248 | 254 | 260 |
| 781 | 265 | 271 | 276 | 282 | 287 | 293 | 298 | 304 | 310 | 315 |
| 782 | 321 | 326 | 332 | 337 | 343 | 348 | 354 | 360 | 365 | 371 |
| 783 | 376 | 382 | 387 | 393 | 398 | 404 | 409 | 415 | 421 | 426 |
| 784 | 432 | 437 | 443 | 448 | 454 | 459 | 465 | 470 | 476 | 481 |
| 785 | 487 | 492 | 498 | 504 | 509 | 515 | 520 | 526 | 531 | 537 |
| 786 | 542 | 548 | 553 | 559 | 564 | 570 | 575 | 581 | 586 | 592 |
| 787 | 597 | 603 | 609 | 614 | 620 | 625 | 631 | 636 | 642 | 647 |
| 788 | 653 | 658 | 664 | 669 | 675 | 680 | 686 | 691 | 697 | 702 |
| 789 | 708 | 713 | 719 | 724 | 730 | 735 | 741 | 746 | 752 | 757 |
| 790 | 763 | 768 | 774 | 779 | 785 | 790 | 796 | 801 | 807 | 812 |
| 791 | 818 | 823 | 829 | 834 | 840 | 845 | 851 | 856 | 862 | 867 |
| 792 | 873 | 878 | 883 | 889 | 894 | 900 | 905 | 911 | 916 | 922 |
| 793 | 927 | 933 | 938 | 944 | 949 | 955 | 960 | 966 | 971 | 977 |
| 794 | 982 | 988 | 993 | 998 | *004 | *009 | *015 | *020 | *026 | *031 |
| 795 | 90 037 | 042 | 048 | 053 | 059 | 064 | 069 | 075 | 080 | 086 |
| 796 | 091 | 097 | 102 | 108 | 113 | 119 | 124 | 129 | 135 | 140 |
| 797 | 146 | 151 | 157 | 162 | 168 | 173 | 179 | 184 | 189 | 195 |
| 798 | 200 | 206 | 211 | 217 | 222 | 227 | 233 | 238 | 244 | 249 |
| 799 | 255 | 260 | 266 | 271 | 276 | 282 | 287 | 293 | 298 | 304 |
| 800 | 309 | 314 | 320 | 325 | 331 | 336 | 342 | 347 | 352 | 358 |

P. P.

**6**

| | |
|---|---|
| 1 | 0.6 |
| 2 | 1.2 |
| 3 | 1.8 |
| 4 | 2.4 |
| 5 | 3.0 |
| 6 | 3.6 |
| 7 | 4.2 |
| 8 | 4.8 |
| 9 | 5.4 |

**5**

| | |
|---|---|
| 1 | 0.5 |
| 2 | 1.0 |
| 3 | 1.5 |
| 4 | 2.0 |
| 5 | 2.5 |
| 6 | 3.0 |
| 7 | 3.5 |
| 8 | 4.0 |
| 9 | 4.5 |

(.90309 to .92988)

| N. | L. 0 | 1 | 2 | 3 | 4 | 5 | 6 | 7 | 8 | 9 |
|---|---|---|---|---|---|---|---|---|---|---|
| 800 | 90 309 | 314 | 320 | 325 | 331 | 336 | 342 | 347 | 352 | 358 |
| 801 | 363 | 369 | 374 | 380 | 385 | 390 | 396 | 401 | 407 | 412 |
| 802 | 417 | 423 | 428 | 434 | 439 | 445 | 450 | 455 | 461 | 466 |
| 803 | 472 | 477 | 482 | 488 | 493 | 499 | 504 | 509 | 515 | 520 |
| 804 | 526 | 531 | 536 | 542 | 547 | 553 | 558 | 563 | 569 | 574 |
| 805 | 580 | 585 | 590 | 596 | 601 | 607 | 612 | 617 | 623 | 628 |
| 806 | 634 | 639 | 644 | 650 | 655 | 660 | 666 | 671 | 677 | 682 |
| 807 | 687 | 693 | 698 | 703 | 709 | 714 | 720 | 725 | 730 | 736 |
| 808 | 741 | 747 | 752 | 757 | 763 | 768 | 773 | 779 | 784 | 789 |
| 809 | 795 | 800 | 806 | 811 | 816 | 822 | 827 | 832 | 838 | 843 |
| 810 | 849 | 854 | 859 | 865 | 870 | 875 | 881 | 886 | 891 | 897 |
| 811 | 902 | 907 | 913 | 918 | 924 | 929 | 934 | 940 | 945 | 950 |
| 812 | 956 | 961 | 966 | 972 | 977 | 982 | 988 | 993 | 998 | *004 |
| 813 | 91 009 | 014 | 020 | 025 | 030 | 036 | 041 | 046 | 052 | 057 |
| 814 | 062 | 068 | 073 | 078 | 084 | 089 | 094 | 100 | 105 | 110 |
| 815 | 116 | 121 | 126 | 132 | 137 | 142 | 148 | 153 | 158 | 164 |
| 816 | 169 | 174 | 180 | 185 | 190 | 196 | 201 | 206 | 212 | 217 |
| 817 | 222 | 228 | 233 | 238 | 243 | 249 | 254 | 259 | 265 | 270 |
| 818 | 275 | 281 | 286 | 291 | 297 | 302 | 307 | 312 | 318 | 323 |
| 819 | 328 | 334 | 339 | 344 | 350 | 355 | 360 | 365 | 371 | 376 |
| 820 | 381 | 387 | 392 | 397 | 403 | 408 | 413 | 418 | 424 | 429 |
| 821 | 434 | 440 | 445 | 450 | 455 | 461 | 466 | 471 | 477 | 482 |
| 822 | 487 | 492 | 498 | 503 | 508 | 514 | 519 | 524 | 529 | 535 |
| 823 | 540 | 545 | 551 | 556 | 561 | 566 | 572 | 577 | 582 | 587 |
| 824 | 593 | 598 | 603 | 609 | 614 | 619 | 624 | 630 | 635 | 640 |
| 825 | 645 | 651 | 656 | 661 | 666 | 672 | 677 | 682 | 687 | 693 |
| 826 | 698 | 703 | 709 | 714 | 719 | 724 | 730 | 735 | 740 | 745 |
| 827 | 751 | 756 | 761 | 766 | 772 | 777 | 782 | 787 | 793 | 798 |
| 828 | 803 | 808 | 814 | 819 | 824 | 829 | 834 | 840 | 845 | 850 |
| 829 | 855 | 861 | 866 | 871 | 876 | 882 | 887 | 892 | 897 | 903 |
| 830 | 908 | 913 | 918 | 924 | 929 | 934 | 939 | 944 | 950 | 955 |
| 831 | 960 | 965 | 971 | 976 | 981 | 986 | 991 | 997 | *002 | *007 |
| 832 | 92 012 | 018 | 023 | 028 | 033 | 038 | 044 | 049 | 054 | 059 |
| 833 | 065 | 070 | 075 | 080 | 085 | 091 | 096 | 101 | 106 | 111 |
| 834 | 117 | 122 | 127 | 132 | 137 | 143 | 148 | 153 | 158 | 163 |
| 835 | 169 | 174 | 179 | 184 | 189 | 195 | 200 | 205 | 210 | 215 |
| 836 | 221 | 226 | 231 | 236 | 241 | 247 | 252 | 257 | 262 | 267 |
| 837 | 273 | 278 | 283 | 288 | 293 | 298 | 304 | 309 | 314 | 319 |
| 838 | 324 | 330 | 335 | 340 | 345 | 350 | 355 | 361 | 366 | 371 |
| 839 | 376 | 381 | 387 | 392 | 397 | 402 | 407 | 412 | 418 | 423 |
| 840 | 428 | 433 | 438 | 443 | 449 | 454 | 459 | 464 | 469 | 474 |
| 841 | 480 | 485 | 490 | 495 | 500 | 505 | 511 | 516 | 521 | 526 |
| 842 | 531 | 536 | 542 | 547 | 552 | 557 | 562 | 567 | 572 | 578 |
| 843 | 583 | 588 | 593 | 598 | 603 | 609 | 614 | 619 | 624 | 629 |
| 844 | 634 | 639 | 645 | 650 | 655 | 660 | 665 | 670 | 675 | 681 |
| 845 | 686 | 691 | 696 | 701 | 706 | 711 | 716 | 722 | 727 | 732 |
| 846 | 737 | 742 | 747 | 752 | 758 | 763 | 768 | 773 | 778 | 783 |
| 847 | 788 | 793 | 799 | 804 | 809 | 814 | 819 | 824 | 829 | 834 |
| 848 | 840 | 845 | 850 | 855 | 860 | 865 | 870 | 875 | 881 | 886 |
| 849 | 891 | 896 | 901 | 906 | 911 | 916 | 921 | 927 | 932 | 937 |
| 850 | 942 | 947 | 952 | 957 | 962 | 967 | 973 | 978 | 983 | 988 |

P. P.

**6**

| 1 | 0.6 |
|---|---|
| 2 | 1.2 |
| 3 | 1.8 |
| 4 | 2.4 |
| 5 | 3.0 |
| 6 | 3.6 |
| 7 | 4.2 |
| 8 | 4.8 |
| 9 | 5.4 |

**5**

| 1 | 0.5 |
|---|---|
| 2 | 1.0 |
| 3 | 1.5 |
| 4 | 2.0 |
| 5 | 2.5 |
| 6 | 3.0 |
| 7 | 3.5 |
| 8 | 4.0 |
| 9 | 4.5 |

(.92942 to .95468)

| N. | L. 0 | 1 | 2 | 3 | 4 | 5 | 6 | 7 | 8 | 9 | P. P. |
|---|---|---|---|---|---|---|---|---|---|---|---|
| 850 | 92 942 | 947 | 952 | 957 | 962 | 967 | 973 | 978 | 983 | 988 | |
| 851 | 993 | 998 | *003 | *008 | *013 | *018 | *024 | *029 | *034 | *039 | |
| 852 | 93 044 | 049 | 054 | 059 | 064 | 069 | 075 | 080 | 085 | 090 | |
| 853 | 095 | 100 | 105 | 110 | 115 | 120 | 125 | 131 | 136 | 141 | |
| 854 | 146 | 151 | 156 | 161 | 166 | 171 | 176 | 181 | 186 | 192 | |
| 855 | 197 | 202 | 207 | 212 | 217 | 222 | 227 | 232 | 237 | 242 | **6** |
| 856 | 247 | 252 | 258 | 263 | 268 | 273 | 278 | 283 | 288 | 293 | 1\|0.6 |
| 857 | 298 | 303 | 308 | 313 | 318 | 323 | 328 | 334 | 339 | 344 | 2\|1.2 |
| 858 | 349 | 354 | 359 | 364 | 369 | 374 | 379 | 384 | 389 | 394 | 3\|1.8 |
| 859 | 399 | 404 | 409 | 414 | 420 | 425 | 430 | 435 | 440 | 445 | 4\|2.4 |
| 860 | 450 | 455 | 460 | 465 | 470 | 475 | 480 | 485 | 490 | 495 | 5\|3.0 |
| 861 | 500 | 505 | 510 | 515 | 520 | 526 | 531 | 536 | 541 | 546 | 6\|3.6 |
| 862 | 551 | 556 | 561 | 566 | 571 | 576 | 581 | 586 | 591 | 596 | 7\|4.2 |
| 863 | 601 | 606 | 611 | 616 | 621 | 626 | 631 | 636 | 641 | 646 | 8\|4.8 |
| 864 | 651 | 656 | 661 | 666 | 671 | 676 | 682 | 687 | 692 | 697 | 9\|5.4 |
| 865 | 702 | 707 | 712 | 717 | 722 | 727 | 732 | 737 | 742 | 747 | |
| 866 | 752 | 757 | 762 | 767 | 772 | 777 | 782 | 787 | 792 | 797 | |
| 867 | 802 | 807 | 812 | 817 | 822 | 827 | 832 | 837 | 842 | 847 | |
| 868 | 852 | 857 | 862 | 867 | 872 | 877 | 882 | 887 | 892 | 897 | |
| 869 | 902 | 907 | 912 | 917 | 922 | 927 | 932 | 937 | 942 | 947 | |
| 870 | 952 | 957 | 962 | 967 | 972 | 977 | 982 | 987 | 992 | 997 | |
| 871 | 94 002 | 007 | 012 | 017 | 022 | 027 | 032 | 037 | 042 | 047 | **5** |
| 872 | 052 | 057 | 062 | 067 | 072 | 077 | 082 | 086 | 091 | 096 | 1\|0.5 |
| 873 | 101 | 106 | 111 | 116 | 121 | 126 | 131 | 136 | 141 | 146 | 2\|1.0 |
| 874 | 151 | 156 | 161 | 166 | 171 | 176 | 181 | 186 | 191 | 196 | 3\|1.5 |
| 875 | 201 | 206 | 211 | 216 | 221 | 226 | 231 | 236 | 240 | 245 | 4\|2.0 |
| 876 | 250 | 255 | 260 | 265 | 270 | 275 | 280 | 285 | 290 | 295 | 5\|2.5 |
| 877 | 300 | 305 | 310 | 315 | 320 | 325 | 330 | 335 | 340 | 345 | 6\|3.0 |
| 878 | 349 | 354 | 359 | 364 | 369 | 374 | 379 | 384 | 389 | 394 | 7\|3.5 |
| 879 | 399 | 404 | 409 | 414 | 419 | 424 | 429 | 433 | 438 | 443 | 8\|4.0 |
| 880 | 448 | 453 | 458 | 463 | 468 | 473 | 478 | 483 | 488 | 493 | 9\|4.5 |
| 881 | 498 | 503 | 507 | 512 | 517 | 522 | 527 | 532 | 537 | 542 | |
| 882 | 547 | 552 | 557 | 562 | 567 | 571 | 576 | 581 | 586 | 591 | |
| 883 | 596 | 601 | 606 | 611 | 616 | 621 | 626 | 630 | 635 | 640 | |
| 884 | 645 | 650 | 655 | 660 | 665 | 670 | 675 | 680 | 685 | 689 | |
| 885 | 694 | 699 | 704 | 709 | 714 | 719 | 724 | 729 | 734 | 738 | |
| 886 | 743 | 748 | 753 | 758 | 763 | 768 | 773 | 778 | 783 | 787 | **4** |
| 887 | 792 | 797 | 802 | 807 | 812 | 817 | 822 | 827 | 832 | 836 | 1\|0.4 |
| 888 | 841 | 846 | 851 | 856 | 861 | 866 | 871 | 876 | 880 | 885 | 2\|0.8 |
| 889 | 890 | 895 | 900 | 905 | 910 | 915 | 919 | 924 | 929 | 934 | 3\|1.2 |
| 890 | 939 | 944 | 949 | 954 | 959 | 963 | 968 | 973 | 978 | 983 | 4\|1.6 |
| 891 | 988 | 993 | 998 | *002 | *007 | *012 | *017 | *022 | *027 | *032 | 5\|2.0 |
| 892 | 95 036 | 041 | 046 | 051 | 056 | 061 | 066 | 071 | 075 | 080 | 6\|2.4 |
| 893 | 085 | 090 | 095 | 100 | 105 | 109 | 114 | 119 | 124 | 129 | 7\|2.8 |
| 894 | 134 | 139 | 143 | 148 | 153 | 158 | 163 | 168 | 173 | 177 | 8\|3.2 |
| 895 | 182 | 187 | 192 | 197 | 202 | 207 | 211 | 216 | 221 | 226 | 9\|3.6 |
| 896 | 231 | 236 | 240 | 245 | 250 | 255 | 260 | 265 | 270 | 274 | |
| 897 | 279 | 284 | 289 | 294 | 299 | 303 | 308 | 313 | 318 | 323 | |
| 898 | 328 | 332 | 337 | 342 | 347 | 352 | 357 | 361 | 366 | 371 | |
| 899 | 376 | 381 | 386 | 390 | 395 | 400 | 405 | 410 | 415 | 419 | |
| 900 | 424 | 429 | 434 | 439 | 444 | 448 | 453 | 458 | 463 | 468 | |

(.95424 to .97813)

| N. | L. 0 | 1 | 2 | 3 | 4 | 5 | 6 | 7 | 8 | 9 |
|---|---|---|---|---|---|---|---|---|---|---|
| 900 | 95 424 | 429 | 434 | 439 | 444 | 448 | 453 | 458 | 463 | 468 |
| 901 | 472 | 477 | 482 | 487 | 492 | 497 | 501 | 506 | 511 | 516 |
| 902 | 521 | 525 | 530 | 535 | 540 | 545 | 550 | 554 | 559 | 564 |
| 903 | 569 | 574 | 578 | 583 | 588 | 593 | 598 | 602 | 607 | 612 |
| 904 | 617 | 622 | 626 | 631 | 636 | 641 | 646 | 650 | 655 | 660 |
| 905 | 665 | 670 | 674 | 679 | 684 | 689 | 694 | 698 | 703 | 708 |
| 906 | 713 | 718 | 722 | 727 | 732 | 737 | 742 | 746 | 751 | 756 |
| 907 | 761 | 766 | 770 | 775 | 780 | 785 | 789 | 794 | 799 | 804 |
| 908 | 809 | 813 | 818 | 823 | 828 | 832 | 837 | 842 | 847 | 852 |
| 909 | 856 | 861 | 866 | 871 | 875 | 880 | 885 | 890 | 895 | 899 |
| 910 | 904 | 909 | 914 | 918 | 923 | 928 | 933 | 938 | 942 | 947 |
| 911 | 952 | 957 | 961 | 966 | 971 | 976 | 980 | 985 | 990 | 995 |
| 912 | 999 | *004 | *009 | *014 | *019 | *023 | *028 | *033 | *038 | *042 |
| 913 | 96 047 | 052 | 057 | 061 | 066 | 071 | 076 | 080 | 085 | 090 |
| 914 | 095 | 099 | 104 | 109 | 114 | 118 | 123 | 128 | 133 | 137 |
| 915 | 142 | 147 | 152 | 156 | 161 | 166 | 171 | 175 | 180 | 185 |
| 916 | 190 | 194 | 199 | 204 | 209 | 213 | 218 | 223 | 227 | 232 |
| 917 | 237 | 242 | 246 | 251 | 256 | 261 | 265 | 270 | 275 | 280 |
| 918 | 284 | 289 | 294 | 298 | 303 | 308 | 313 | 317 | 322 | 327 |
| 919 | 332 | 336 | 341 | 346 | 350 | 355 | 360 | 365 | 369 | 374 |
| 920 | 379 | 384 | 388 | 393 | 398 | 402 | 407 | 412 | 417 | 421 |
| 921 | 426 | 431 | 435 | 440 | 445 | 450 | 454 | 459 | 464 | 468 |
| 922 | 473 | 478 | 483 | 487 | 492 | 497 | 501 | 506 | 511 | 515 |
| 923 | 520 | 525 | 530 | 534 | 539 | 544 | 548 | 553 | 558 | 562 |
| 924 | 567 | 572 | 577 | 581 | 586 | 591 | 595 | 600 | 605 | 609 |
| 925 | 614 | 619 | 624 | 628 | 633 | 638 | 642 | 647 | 652 | 656 |
| 926 | 661 | 666 | 670 | 675 | 680 | 685 | 689 | 694 | 699 | 703 |
| 927 | 708 | 713 | 717 | 722 | 727 | 731 | 736 | 741 | 745 | 750 |
| 928 | 755 | 759 | 764 | 769 | 774 | 778 | 783 | 788 | 792 | 797 |
| 929 | 802 | 806 | 811 | 816 | 820 | 825 | 830 | 834 | 839 | 844 |
| 930 | 848 | 853 | 858 | 862 | 867 | 872 | 876 | 881 | 886 | 890 |
| 931 | 895 | 900 | 904 | 909 | 914 | 918 | 923 | 928 | 932 | 937 |
| 932 | 942 | 946 | 951 | 956 | 960 | 965 | 970 | 974 | 979 | 984 |
| 933 | 988 | 993 | 997 | *002 | *007 | *011 | *016 | *021 | *025 | *030 |
| 934 | 97 035 | 039 | 044 | 049 | 053 | 058 | 063 | 067 | 072 | 077 |
| 935 | 081 | 086 | 090 | 095 | 100 | 104 | 109 | 114 | 118 | 123 |
| 936 | 128 | 132 | 137 | 142 | 146 | 151 | 155 | 160 | 165 | 169 |
| 937 | 174 | 179 | 183 | 188 | 192 | 197 | 202 | 206 | 211 | 216 |
| 938 | 220 | 225 | 230 | 234 | 239 | 243 | 248 | 253 | 257 | 262 |
| 939 | 267 | 271 | 276 | 280 | 285 | 290 | 294 | 299 | 304 | 308 |
| 940 | 313 | 317 | 322 | 327 | 331 | 336 | 340 | 345 | 350 | 354 |
| 941 | 359 | 364 | 368 | 373 | 377 | 382 | 387 | 391 | 396 | 400 |
| 942 | 405 | 410 | 414 | 419 | 424 | 428 | 433 | 437 | 442 | 447 |
| 943 | 451 | 456 | 460 | 465 | 470 | 474 | 479 | 483 | 488 | 493 |
| 944 | 497 | 502 | 506 | 511 | 516 | 520 | 525 | 529 | 534 | 539 |
| 945 | 543 | 548 | 552 | 557 | 562 | 566 | 571 | 575 | 580 | 585 |
| 946 | 589 | 594 | 598 | 603 | 607 | 612 | 617 | 621 | 626 | 630 |
| 947 | 635 | 640 | 644 | 649 | 653 | 658 | 663 | 667 | 672 | 676 |
| 948 | 681 | 685 | 690 | 695 | 699 | 704 | 708 | 713 | 717 | 722 |
| 949 | 727 | 731 | 736 | 740 | 745 | 749 | 754 | 759 | 763 | 768 |
| 950 | 772 | 777 | 782 | 786 | 791 | 795 | 800 | 804 | 809 | 813 |

P. P.

5
| 1 | 0.5 |
|---|---|
| 2 | 1.0 |
| 3 | 1.5 |
| 4 | 2.0 |
| 5 | 2.5 |
| 6 | 3.0 |
| 7 | 3.5 |
| 8 | 4.0 |
| 9 | 4.5 |

4
| 1 | 0.4 |
|---|---|
| 2 | 0.8 |
| 3 | 1.2 |
| 4 | 1.6 |
| 5 | 2.0 |
| 6 | 2.4 |
| 7 | 2.8 |
| 8 | 3.2 |
| 9 | 3.6 |

(.97772 to .00039)

| N. | L. 0 | 1 | 2 | 3 | 4 | 5 | 6 | 7 | 8 | 9 |
|---|---|---|---|---|---|---|---|---|---|---|
| 950 | 97 772 | 777 | 782 | 786 | 791 | 795 | 800 | 804 | 809 | 813 |
| 951 | 818 | 823 | 827 | 832 | 836 | 841 | 845 | 850 | 855 | 859 |
| 952 | 864 | 868 | 873 | 877 | 882 | 886 | 891 | 896 | 900 | 905 |
| 953 | 909 | 914 | 918 | 923 | 928 | 932 | 937 | 941 | 946 | 950 |
| 954 | 955 | 959 | 964 | 968 | 973 | 978 | 982 | 987 | 991 | 996 |
| 955 | 98 000 | 005 | 009 | 014 | 019 | 023 | 028 | 032 | 037 | 041 |
| 956 | 046 | 050 | 055 | 059 | 064 | 068 | 073 | 078 | 082 | 087 |
| 957 | 091 | 096 | 100 | 105 | 109 | 114 | 118 | 123 | 127 | 132 |
| 958 | 137 | 141 | 146 | 150 | 155 | 159 | 164 | 168 | 173 | 177 |
| 959 | 182 | 186 | 191 | 195 | 200 | 204 | 209 | 214 | 218 | 223 |
| 960 | 227 | 232 | 236 | 241 | 245 | 250 | 254 | 259 | 263 | 268 |
| 961 | 272 | 277 | 281 | 286 | 290 | 295 | 299 | 304 | 308 | 313 |
| 962 | 318 | 322 | 327 | 331 | 336 | 340 | 345 | 349 | 354 | 358 |
| 963 | 363 | 367 | 372 | 376 | 381 | 385 | 390 | 394 | 399 | 403 |
| 964 | 408 | 412 | 417 | 421 | 426 | 430 | 435 | 439 | 444 | 448 |
| 965 | 453 | 457 | 462 | 466 | 471 | 475 | 480 | 484 | 489 | 493 |
| 966 | 498 | 502 | 507 | 511 | 516 | 520 | 525 | 529 | 534 | 538 |
| 967 | 543 | 547 | 552 | 556 | 561 | 565 | 570 | 574 | 579 | 583 |
| 968 | 588 | 592 | 597 | 601 | 605 | 610 | 614 | 619 | 623 | 628 |
| 969 | 632 | 637 | 641 | 646 | 650 | 655 | 659 | 664 | 668 | 673 |
| 970 | 677 | 682 | 686 | 691 | 695 | 700 | 704 | 709 | 713 | 717 |
| 971 | 722 | 726 | 731 | 735 | 740 | 744 | 749 | 753 | 758 | 762 |
| 972 | 767 | 771 | 776 | 780 | 784 | 789 | 793 | 798 | 802 | 807 |
| 973 | 811 | 816 | 820 | 825 | 829 | 834 | 838 | 843 | 847 | 851 |
| 974 | 856 | 860 | 865 | 869 | 874 | 878 | 883 | 887 | 892 | 896 |
| 975 | 900 | 905 | 909 | 914 | 918 | 923 | 927 | 932 | 936 | 941 |
| 976 | 945 | 949 | 954 | 958 | 963 | 967 | 972 | 976 | 981 | 985 |
| 977 | 989 | 994 | 998 | *003 | *007 | *012 | *016 | *021 | *025 | *029 |
| 978 | 99 034 | 038 | 043 | 047 | 052 | 056 | 061 | 065 | 069 | 074 |
| 979 | 078 | 083 | 087 | 092 | 096 | 100 | 105 | 109 | 114 | 118 |
| 980 | 123 | 127 | 131 | 136 | 140 | 145 | 149 | 154 | 158 | 162 |
| 981 | 167 | 171 | 176 | 180 | 185 | 189 | 193 | 198 | 202 | 207 |
| 982 | 211 | 216 | 220 | 224 | 229 | 233 | 238 | 242 | 247 | 251 |
| 983 | 255 | 260 | 264 | 269 | 273 | 277 | 282 | 286 | 291 | 295 |
| 984 | 300 | 304 | 308 | 313 | 317 | 322 | 326 | 330 | 335 | 339 |
| 985 | 344 | 348 | 352 | 357 | 361 | 366 | 370 | 374 | 379 | 383 |
| 986 | 388 | 392 | 396 | 401 | 405 | 410 | 414 | 419 | 423 | 427 |
| 987 | 432 | 436 | 441 | 445 | 449 | 454 | 458 | 463 | 467 | 471 |
| 988 | 476 | 480 | 484 | 489 | 493 | 498 | 502 | 506 | 511 | 515 |
| 989 | 520 | 524 | 528 | 533 | 537 | 542 | 546 | 550 | 555 | 559 |
| 990 | 564 | 568 | 572 | 577 | 581 | 585 | 590 | 594 | 599 | 603 |
| 991 | 607 | 612 | 616 | 621 | 625 | 629 | 634 | 638 | 642 | 647 |
| 992 | 651 | 656 | 660 | 664 | 669 | 673 | 677 | 682 | 686 | 691 |
| 993 | 695 | 699 | 704 | 708 | 712 | 717 | 721 | 726 | 730 | 734 |
| 994 | 739 | 743 | 747 | 752 | 756 | 760 | 765 | 769 | 774 | 778 |
| 995 | 782 | 787 | 791 | 795 | 800 | 804 | 808 | 813 | 817 | 822 |
| 996 | 826 | 830 | 835 | 839 | 843 | 848 | 852 | 856 | 861 | 865 |
| 997 | 870 | 874 | 878 | 883 | 887 | 891 | 896 | 900 | 904 | 909 |
| 998 | 913 | 917 | 922 | 926 | 930 | 935 | 939 | 944 | 948 | 952 |
| 999 | 957 | 961 | 965 | 970 | 974 | 978 | 983 | 987 | 991 | 996 |
| 1000 | 00 000 | 004 | 009 | 013 | 017 | 022 | 026 | 030 | 035 | 039 |

P. P.

**5**

| 1 | 0.5 |
|---|---|
| 2 | 1.0 |
| 3 | 1.5 |
| 4 | 2.0 |
| 5 | 2.5 |
| 6 | 3.0 |
| 7 | 3.5 |
| 8 | 4.0 |
| 9 | 4.5 |

**4**

| 1 | 0.4 |
|---|---|
| 2 | 0.8 |
| 3 | 1.2 |
| 4 | 1.6 |
| 5 | 2.0 |
| 6 | 2.4 |
| 7 | 2.8 |
| 8 | 3.2 |
| 9 | 3.6 |

(See bottom of page 124 for numbers less than 1.)

| No. | Log$_e$ | No. | Log$_e$ | No. | Log$_e$ | No. | Log$_e$ | No. | Log$_e$ |
|-----|------|-----|------|-----|------|-----|------|-----|------|
| 1.01 | 0.0100 | 1.51 | 0.4121 | 2.01 | 0.6981 | 2.51 | 0.9203 | 3.01 | 1.1019 |
| 1.02 | 0.0198 | 1.52 | 0.4187 | 2.02 | 0.7031 | 2.52 | 0.9243 | 3.02 | 1.1053 |
| 1.03 | 0.0296 | 1.53 | 0.4253 | 2.03 | 0.7080 | 2.53 | 0.9282 | 3.03 | 1.1086 |
| 1.04 | 0.0392 | 1.54 | 0.4318 | 2.04 | 0.7129 | 2.54 | 0.9322 | 3.04 | 1.1119 |
| 1.05 | 0.0488 | 1.55 | 0.4383 | 2.05 | 0.7178 | 2.55 | 0.9361 | 3.05 | 1.1151 |
| 1.06 | 0.0583 | 1.56 | 0.4447 | 2.06 | 0.7227 | 2.56 | 0.9400 | 3.06 | 1.1184 |
| 1.07 | 0.0677 | 1.57 | 0.4511 | 2.07 | 0.7275 | 2.57 | 0.9439 | 3.07 | 1.1217 |
| 1.08 | 0.0770 | 1.58 | 0.4574 | 2.08 | 0.7324 | 2.58 | 0.9478 | 3.08 | 1.1249 |
| 1.09 | 0.0862 | 1.59 | 0.4637 | 2.09 | 0.7372 | 2.59 | 0.9517 | 3.09 | 1.1282 |
| 1.10 | 0.0953 | 1.60 | 0.4700 | 2.10 | 0.7419 | 2.60 | 0.9555 | 3.10 | 1.1314 |
| 1.11 | 0.1044 | 1.61 | 0.4762 | 2.11 | 0.7467 | 2.61 | 0.9594 | 3.11 | 1.1346 |
| 1.12 | 0.1133 | 1.62 | 0.4824 | 2.12 | 0.7514 | 2.62 | 0.9632 | 3.12 | 1.1378 |
| 1.13 | 0.1222 | 1.63 | 0.4886 | 2.13 | 0.7561 | 2.63 | 0.9670 | 3.13 | 1.1410 |
| 1.14 | 0.1310 | 1.64 | 0.4947 | 2.14 | 0.7608 | 2.64 | 0.9708 | 3.14 | 1.1442 |
| 1.15 | 0.1398 | 1.65 | 0.5008 | 2.15 | 0.7655 | 2.65 | 0.9746 | 3.15 | 1.1474 |
| 1.16 | 0.1484 | 1.66 | 0.5068 | 2.16 | 0.7701 | 2.66 | 0.9783 | 3.16 | 1.1506 |
| 1.17 | 0.1570 | 1.67 | 0.5128 | 2.17 | 0.7747 | 2.67 | 0.9821 | 3.17 | 1.1537 |
| 1.18 | 0.1655 | 1.68 | 0.5188 | 2.18 | 0.7793 | 2.68 | 0.9858 | 3.18 | 1.1569 |
| 1.19 | 0.1740 | 1.69 | 0.5247 | 2.19 | 0.7839 | 2.69 | 0.9895 | 3.19 | 1.1600 |
| 1.20 | 0.1823 | 1.70 | 0.5306 | 2.20 | 0.7885 | 2.70 | 0.9933 | 3.20 | 1.1632 |
| 1.21 | 0.1906 | 1.71 | 0.5365 | 2.21 | 0.7930 | 2.71 | 0.9969 | 3.21 | 1.1663 |
| 1.22 | 0.1989 | 1.72 | 0.5423 | 2.22 | 0.7975 | 2.72 | 1.0006 | 3.22 | 1.1694 |
| 1.23 | 0.2070 | 1.73 | 0.5481 | 2.23 | 0.8020 | 2.73 | 1.0043 | 3.23 | 1.1725 |
| 1.24 | 0.2151 | 1.74 | 0.5539 | 2.24 | 0.8065 | 2.74 | 1.0080 | 3.24 | 1.1756 |
| 1.25 | 0.2231 | 1.75 | 0.5596 | 2.25 | 0.8109 | 2.75 | 1.0116 | 3.25 | 1.1787 |
| 1.26 | 0.2311 | 1.76 | 0.5653 | 2.26 | 0.8154 | 2.76 | 1.0152 | 3.26 | 1.1817 |
| 1.27 | 0.2390 | 1.77 | 0.5710 | 2.27 | 0.8198 | 2.77 | 1.0188 | 3.27 | 1.1848 |
| 1.28 | 0.2469 | 1.78 | 0.5766 | 2.28 | 0.8242 | 2.78 | 1.0225 | 3.28 | 1.1878 |
| 1.29 | 0.2546 | 1.79 | 0.5822 | 2.29 | 0.8286 | 2.79 | 1.0260 | 3.29 | 1.1909 |
| 1.30 | 0.2624 | 1.80 | 0.5878 | 2.30 | 0.8329 | 2.80 | 1.0296 | 3.30 | 1.1939 |
| 1.31 | 0.2700 | 1.81 | 0.5933 | 2.31 | 0.8372 | 2.81 | 1.0332 | 3.31 | 1.1969 |
| 1.32 | 0.2776 | 1.82 | 0.5988 | 2.32 | 0.8416 | 2.82 | 1.0367 | 3.32 | 1.2000 |
| 1.33 | 0.2852 | 1.83 | 0.6043 | 2.33 | 0.8459 | 2.83 | 1.0403 | 3.33 | 1.2030 |
| 1.34 | 0.2927 | 1.84 | 0.6098 | 2.34 | 0.8502 | 2.84 | 1.0438 | 3.34 | 1.2060 |
| 1.35 | 0.3001 | 1.85 | 0.6152 | 2.35 | 0.8544 | 2.85 | 1.0473 | 3.35 | 1.2090 |
| 1.36 | 0.3075 | 1.86 | 0.6206 | 2.36 | 0.8587 | 2.86 | 1.0508 | 3.36 | 1.2119 |
| 1.37 | 0.3148 | 1.87 | 0.6259 | 2.37 | 0.8629 | 2.87 | 1.0543 | 3.37 | 1.2149 |
| 1.38 | 0.3221 | 1.88 | 0.6313 | 2.38 | 0.8671 | 2.88 | 1.0578 | 3.38 | 1.2179 |
| 1.39 | 0.3293 | 1.89 | 0.6366 | 2.39 | 0.8713 | 2.89 | 1.0613 | 3.39 | 1.2208 |
| 1.40 | 0.3365 | 1.90 | 0.6419 | 2.40 | 0.8755 | 2.90 | 1.0647 | 3.40 | 1.2238 |
| 1.41 | 0.3436 | 1.91 | 0.6471 | 2.41 | 0.8796 | 2.91 | 1.0682 | 3.41 | 1.2267 |
| 1.42 | 0.3507 | 1.92 | 0.6523 | 2.42 | 0.8838 | 2.92 | 1.0716 | 3.42 | 1.2296 |
| 1.43 | 0.3577 | 1.93 | 0.6575 | 2.43 | 0.8879 | 2.93 | 1.0750 | 3.43 | 1.2326 |
| 1.44 | 0.3646 | 1.94 | 0.6627 | 2.44 | 0.8920 | 2.94 | 1.0784 | 3.44 | 1.2355 |
| 1.45 | 0.3716 | 1.95 | 0.6678 | 2.45 | 0.8961 | 2.95 | 1.0818 | 3.45 | 1.2384 |
| 1.46 | 0.3784 | 1.96 | 0.6729 | 2.46 | 0.9002 | 2.96 | 1.0852 | 3.46 | 1.2413 |
| 1.47 | 0.3853 | 1.97 | 0.6780 | 2.47 | 0.9042 | 2.97 | 1.0886 | 3.47 | 1.2442 |
| 1.48 | 0.3920 | 1.98 | 0.6831 | 2.48 | 0.9083 | 2.98 | 1.0919 | 3.48 | 1.2470 |
| 1.49 | 0.3988 | 1.99 | 0.6881 | 2.49 | 0.9123 | 2.99 | 1.0953 | 3.49 | 1.2499 |
| 1.50 | 0.4055 | 2.00 | 0.6931 | 2.50 | 0.9163 | 3.00 | 1.0986 | 3.50 | 1.2528 |

| No. | Log$_e$ | No. | Log$_e$ | No. | Log$_e$ | No. | Log$_e$ | No. | Log$_e$ |
|---|---|---|---|---|---|---|---|---|---|
| 3.51 | 1.2556 | 4.01 | 1.3888 | 4.51 | 1.5063 | 5.01 | 1.6114 | 5.51 | 1.7066 |
| 3.52 | 1.2585 | 4.02 | 1.3913 | 4.52 | 1.5085 | 5.02 | 1.6134 | 5.52 | 1.7084 |
| 3.53 | 1.2613 | 4.03 | 1.3938 | 4.53 | 1.5107 | 5.03 | 1.6154 | 5.53 | 1.7102 |
| 3.54 | 1.2641 | 4.04 | 1.3962 | 4.54 | 1.5129 | 5.04 | 1.6174 | 5.54 | 1.7120 |
| 3.55 | 1.2669 | 4.05 | 1.3987 | 4.55 | 1.5151 | 5.05 | 1.6194 | 5.55 | 1.7138 |
| 3.56 | 1.2698 | 4.06 | 1.4012 | 4.56 | 1.5173 | 5.06 | 1.6214 | 5.56 | 1.7156 |
| 3.57 | 1.2726 | 4.07 | 1.4036 | 4.57 | 1.5195 | 5.07 | 1.6233 | 5.57 | 1.7174 |
| 3.58 | 1.2754 | 4.08 | 1.4061 | 4.58 | 1.5217 | 5.08 | 1.6253 | 5.58 | 1.7192 |
| 3.59 | 1.2782 | 4.09 | 1.4085 | 4.59 | 1.5239 | 5.09 | 1.6273 | 5.59 | 1.7210 |
| 3.60 | 1.2809 | 4.10 | 1.4110 | 4.60 | 1.5261 | 5.10 | 1.6292 | 5.60 | 1.7228 |
| 3.61 | 1.2837 | 4.11 | 1.4134 | 4.61 | 1.5282 | 5.11 | 1.6312 | 5.61 | 1.7246 |
| 3.62 | 1.2865 | 4.12 | 1.4159 | 4.62 | 1.5304 | 5.12 | 1.6332 | 5.62 | 1.7263 |
| 3.63 | 1.2892 | 4.13 | 1.4183 | 4.63 | 1.5326 | 5.13 | 1.6351 | 5.63 | 1.7281 |
| 3.64 | 1.2920 | 4.14 | 1.4207 | 4.64 | 1.5347 | 5.14 | 1.6371 | 5.64 | 1.7299 |
| 3.65 | 1.2947 | 4.15 | 1.4231 | 4.65 | 1.5369 | 5.15 | 1.6390 | 5.65 | 1.7317 |
| 3.66 | 1.2975 | 4.16 | 1.4255 | 4.66 | 1.5390 | 5.16 | 1.6409 | 5.66 | 1.7334 |
| 3.67 | 1.3002 | 4.17 | 1.4279 | 4.67 | 1.5412 | 5.17 | 1.6429 | 5.67 | 1.7352 |
| 3.68 | 1.3029 | 4.18 | 1.4303 | 4.68 | 1.5433 | 5.18 | 1.6448 | 5.68 | 1.7370 |
| 3.69 | 1.3056 | 4.19 | 1.4327 | 4.69 | 1.5454 | 5.19 | 1.6467 | 5.69 | 1.7387 |
| 3.70 | 1.3083 | 4.20 | 1.4351 | 4.70 | 1.5476 | 5.20 | 1.6487 | 5.70 | 1.7405 |
| 3.71 | 1.3110 | 4.21 | 1.4375 | 4.71 | 1.5497 | 5.21 | 1.6506 | 5.71 | 1.7422 |
| 3.72 | 1.3137 | 4.22 | 1.4398 | 4.72 | 1.5518 | 5.22 | 1.6525 | 5.72 | 1.7440 |
| 3.73 | 1.3164 | 4.23 | 1.4422 | 4.73 | 1.5539 | 5.23 | 1.6544 | 5.73 | 1.7457 |
| 3.74 | 1.3191 | 4.24 | 1.4446 | 4.74 | 1.5560 | 5.24 | 1.6563 | 5.74 | 1.7475 |
| 3.75 | 1.3218 | 4.25 | 1.4469 | 4.75 | 1.5581 | 5.25 | 1.6582 | 5.75 | 1.7492 |
| 3.76 | 1.3244 | 4.26 | 1.4493 | 4.76 | 1.5602 | 5.26 | 1.6601 | 5.76 | 1.7509 |
| 3.77 | 1.3271 | 4.27 | 1.4516 | 4.77 | 1.5623 | 5.27 | 1.6620 | 5.77 | 1.7527 |
| 3.78 | 1.3297 | 4.28 | 1.4540 | 4.78 | 1.5644 | 5.28 | 1.6639 | 5.78 | 1.7544 |
| 3.79 | 1.3324 | 4.29 | 1.4563 | 4.79 | 1.5665 | 5.29 | 1.6658 | 5.79 | 1.7561 |
| 3.80 | 1.3350 | 4.30 | 1.4586 | 4.80 | 1.5686 | 5.30 | 1.6677 | 5.80 | 1.7579 |
| 3.81 | 1.3376 | 4.31 | 1.4609 | 4.81 | 1.5707 | 5.31 | 1.6696 | 5.81 | 1.7596 |
| 3.82 | 1.3403 | 4.32 | 1.4633 | 4.82 | 1.5728 | 5.32 | 1.6715 | 5.82 | 1.7613 |
| 3.83 | 1.3429 | 4.33 | 1.4656 | 4.83 | 1.5748 | 5.33 | 1.6734 | 5.83 | 1.7630 |
| 3.84 | 1.3455 | 4.34 | 1.4679 | 4.84 | 1.5769 | 5.34 | 1.6752 | 5.84 | 1.7647 |
| 3.85 | 1.3481 | 4.35 | 1.4702 | 4.85 | 1.5790 | 5.35 | 1.6771 | 5.85 | 1.7664 |
| 3.86 | 1.3507 | 4.36 | 1.4725 | 4.86 | 1.5810 | 5.36 | 1.6790 | 5.86 | 1.7681 |
| 3.87 | 1.3533 | 4.37 | 1.4748 | 4.87 | 1.5831 | 5.37 | 1.6808 | 5.87 | 1.7699 |
| 3.88 | 1.3558 | 4.38 | 1.4770 | 4.88 | 1.5851 | 5.38 | 1.6827 | 5.88 | 1.7716 |
| 3.89 | 1.3584 | 4.39 | 1.4793 | 4.89 | 1.5872 | 5.39 | 1.6845 | 5.89 | 1.7733 |
| 3.90 | 1.3610 | 4.40 | 1.4816 | 4.90 | 1.5892 | 5.40 | 1.6864 | 5.90 | 1.7750 |
| 3.91 | 1.3635 | 4.41 | 1.4839 | 4.91 | 1.5913 | 5.41 | 1.6882 | 5.91 | 1.7766 |
| 3.92 | 1.3661 | 4.42 | 1.4861 | 4.92 | 1.5933 | 5.42 | 1.6901 | 5.92 | 1.7783 |
| 3.93 | 1.3686 | 4.43 | 1.4884 | 4.93 | 1.5953 | 5.43 | 1.6919 | 5.93 | 1.7800 |
| 3.94 | 1.3712 | 4.44 | 1.4907 | 4.94 | 1.5974 | 5.44 | 1.6938 | 5.94 | 1.7817 |
| 3.95 | 1.3737 | 4.45 | 1.4929 | 4.95 | 1.5994 | 5.45 | 1.6956 | 5.95 | 1.7834 |
| 3.96 | 1.3762 | 4.46 | 1.4951 | 4.96 | 1.6014 | 5.46 | 1.6974 | 5.96 | 1.7851 |
| 3.97 | 1.3788 | 4.47 | 1.4974 | 4.97 | 1.6034 | 5.47 | 1.6993 | 5.97 | 1.7867 |
| 3.98 | 1.3813 | 4.48 | 1.4996 | 4.98 | 1.6054 | 5.48 | 1.7011 | 5.98 | 1.7884 |
| 3.99 | 1.3838 | 4.49 | 1.5019 | 4.99 | 1.6074 | 5.49 | 1.7029 | 5.99 | 1.7901 |
| 4.00 | 1.3863 | 4.50 | 1.5041 | 5.00 | 1.6094 | 5.50 | 1.7047 | 6.00 | 1.7918 |

| No. | $\text{Log}_e$ | No. | $\text{Log}_e$ | No. | $\text{Log}_e$ | No. | $\text{Log}_e$ | No. | $\text{Log}_e$ |
|---|---|---|---|---|---|---|---|---|---|
| 6.01 | 1.7934 | 6.51 | 1.8733 | 7.01 | 1.9473 | 7.51 | 2.0162 | 8.01 | 2.0807 |
| 6.02 | 1.7951 | 6.52 | 1.8749 | 7.02 | 1.9488 | 7.52 | 2.0176 | 8.02 | 2.0819 |
| 6.03 | 1.7967 | 6.53 | 1.8764 | 7.03 | 1.9502 | 7.53 | 2.0189 | 8.03 | 2.0832 |
| 6.04 | 1.7984 | 6.54 | 1.8779 | 7.04 | 1.9516 | 7.54 | 2.0202 | 8.04 | 2.0844 |
| 6.05 | 1.8001 | 6.55 | 1.8795 | 7.05 | 1.9530 | 7.55 | 2.0215 | 8.05 | 2.0857 |
| 6.06 | 1.8017 | 6.56 | 1.8810 | 7.06 | 1.9544 | 7.56 | 2.0229 | 8.06 | 2.0869 |
| 6.07 | 1.8034 | 6.57 | 1.8825 | 7.07 | 1.9559 | 7.57 | 2.0242 | 8.07 | 2.0882 |
| 6.08 | 1.8050 | 6.58 | 1.8840 | 7.08 | 1.9573 | 7.58 | 2.0255 | 8.08 | 2.0894 |
| 6.09 | 1.8066 | 6.59 | 1.8856 | 7.09 | 1.9587 | 7.59 | 2.0268 | 8.09 | 2.0906 |
| 6.10 | 1.8083 | 6.60 | 1.8871 | 7.10 | 1.9601 | 7.60 | 2.0281 | 8.10 | 2.0919 |
| 6.11 | 1.8099 | 6.61 | 1.8886 | 7.11 | 1.9615 | 7.61 | 2.0295 | 8.11 | 2.0931 |
| 6.12 | 1.8116 | 6.62 | 1.8901 | 7.12 | 1.9629 | 7.62 | 2.0308 | 8.12 | 2.0943 |
| 6.13 | 1.8132 | 6.63 | 1.8916 | 7.13 | 1.9643 | 7.63 | 2.0321 | 8.13 | 2.0956 |
| 6.14 | 1.8148 | 6.64 | 1.8931 | 7.14 | 1.9657 | 7.64 | 2.0334 | 8.14 | 2.0968 |
| 6.15 | 1.8165 | 6.65 | 1.8946 | 7.15 | 1.9671 | 7.65 | 2.0347 | 8.15 | 2.0980 |
| 6.16 | 1.8181 | 6.66 | 1.8961 | 7.16 | 1.9685 | 7.66 | 2.0360 | 8.16 | 2.0992 |
| 6.17 | 1.8197 | 6.67 | 1.8976 | 7.17 | 1.9699 | 7.67 | 2.0373 | 8.17 | 2.1005 |
| 6.18 | 1.8213 | 6.68 | 1.8991 | 7.18 | 1.9713 | 7.68 | 2.0386 | 8.18 | 2.1017 |
| 6.19 | 1.8229 | 6.69 | 1.9006 | 7.19 | 1.9727 | 7.69 | 2.0399 | 8.19 | 2.1029 |
| 6.20 | 1.8245 | 6.70 | 1.9021 | 7.20 | 1.9741 | 7.70 | 2.0412 | 8.20 | 2.1041 |
| 6.21 | 1.8262 | 6.71 | 1.9036 | 7.21 | 1.9755 | 7.71 | 2.0425 | 8.21 | 2.1054 |
| 6.22 | 1.8278 | 6.72 | 1.9051 | 7.22 | 1.9769 | 7.72 | 2.0438 | 8.22 | 2.1066 |
| 6.23 | 1.8294 | 6.73 | 1.9066 | 7.23 | 1.9782 | 7.73 | 2.0451 | 8.23 | 2.1078 |
| 6.24 | 1.8310 | 6.74 | 1.9081 | 7.24 | 1.9796 | 7.74 | 2.0464 | 8.24 | 2.1090 |
| 6.25 | 1.8326 | 6.75 | 1.9095 | 7.25 | 1.9810 | 7.75 | 2.0477 | 8.25 | 2.1102 |
| 6.26 | 1.8342 | 6.76 | 1.9110 | 7.26 | 1.9824 | 7.76 | 2.0490 | 8.26 | 2.1114 |
| 6.27 | 1.8358 | 6.77 | 1.9125 | 7.27 | 1.9838 | 7.77 | 2.0503 | 8.27 | 2.1126 |
| 6.28 | 1.8374 | 6.78 | 1.9140 | 7.28 | 1.9851 | 7.78 | 2.0516 | 8.28 | 2.1138 |
| 6.29 | 1.8390 | 6.79 | 1.9155 | 7.29 | 1.9865 | 7.79 | 2.0528 | 8.29 | 2.1150 |
| 6.30 | 1.8405 | 6.80 | 1.9169 | 7.30 | 1.9879 | 7.80 | 2.0541 | 8.30 | 2.1163 |
| 6.31 | 1.8421 | 6.81 | 1.9184 | 7.31 | 1.9892 | 7.81 | 2.0554 | 8.31 | 2.1175 |
| 6.32 | 1.8437 | 6.82 | 1.9199 | 7.32 | 1.9906 | 7.82 | 2.0567 | 8.32 | 2.1187 |
| 6.33 | 1.8453 | 6.83 | 1.9213 | 7.33 | 1.9920 | 7.83 | 2.0580 | 8.33 | 2.1199 |
| 6.34 | 1.8469 | 6.84 | 1.9228 | 7.34 | 1.9933 | 7.84 | 2.0592 | 8.34 | 2.1211 |
| 6.35 | 1.8485 | 6.85 | 1.9242 | 7.35 | 1.9947 | 7.85 | 2.0605 | 8.35 | 2.1223 |
| 6.36 | 1.8500 | 6.86 | 1.9257 | 7.36 | 1.9961 | 7.86 | 2.0618 | 8.36 | 2.1235 |
| 6.37 | 1.8516 | 6.87 | 1.9272 | 7.37 | 1.9974 | 7.87 | 2.0631 | 8.37 | 2.1247 |
| 6.38 | 1.8532 | 6.88 | 1.9286 | 7.38 | 1.9988 | 7.88 | 2.0643 | 8.38 | 2.1258 |
| 6.39 | 1.8547 | 6.89 | 1.9301 | 7.39 | 2.0001 | 7.89 | 2.0656 | 8.39 | 2.1270 |
| 6.40 | 1.8563 | 6.90 | 1.9315 | 7.40 | 2.0015 | 7.90 | 2.0669 | 8.40 | 2.1282 |
| 6.41 | 1.8579 | 6.91 | 1.9330 | 7.41 | 2.0028 | 7.91 | 2.0681 | 8.41 | 2.1294 |
| 6.42 | 1.8594 | 6.92 | 1.9344 | 7.42 | 2.0042 | 7.92 | 2.0694 | 8.42 | 2.1306 |
| 6.43 | 1.8610 | 6.93 | 1.9359 | 7.43 | 2.0055 | 7.93 | 2.0707 | 8.43 | 2.1318 |
| 6.44 | 1.8625 | 6.94 | 1.9373 | 7.44 | 2.0069 | 7.94 | 2.0719 | 8.44 | 2.1330 |
| 6.45 | 1.8641 | 6.95 | 1.9387 | 7.45 | 2.0082 | 7.95 | 2.0732 | 8.45 | 2.1342 |
| 6.46 | 1.8656 | 6.96 | 1.9402 | 7.46 | 2.0096 | 7.96 | 2.0744 | 8.46 | 2.1353 |
| 6.47 | 1.8672 | 6.97 | 1.9416 | 7.47 | 2.0109 | 7.97 | 2.0757 | 8.47 | 2.1365 |
| 6.48 | 1.8687 | 6.98 | 1.9430 | 7.48 | 2.0122 | 7.98 | 2.0769 | 8.48 | 2.1377 |
| 6.49 | 1.8703 | 6.99 | 1.9445 | 7.49 | 2.0136 | 7.99 | 2.0782 | 8.49 | 2.1389 |
| 6.50 | 1.8718 | 7.00 | 1.9459 | 7.50 | 2.0149 | 8.00 | 2.0794 | 8.50 | 2.1401 |

(See bottom of page 124 for numbers greater than 100.)

| No. | Logₑ | No. | Logₑ | No. | Logₑ | No. | Logₑ | No. | Logₑ |
|---|---|---|---|---|---|---|---|---|---|
| 8.51 | 2.1412 | 9.01 | 2.1983 | 9.51 | 2.2523 | 10.25 | 2.3273 | 41 | 3.7136 |
| 8.52 | 2.1424 | 9.02 | 2.1994 | 9.52 | 2.2534 | 10.50 | 2.3514 | 42 | 3.7377 |
| 8.53 | 2.1436 | 9.03 | 2.2006 | 9.53 | 2.2544 | 10.75 | 2.3749 | 43 | 3.7612 |
| 8.54 | 2.1448 | 9.04 | 2.2017 | 9.54 | 2.2555 | 11.00 | 2.3979 | 44 | 3.7842 |
| 8.55 | 2.1459 | 9.05 | 2.2028 | 9.55 | 2.2565 | 11.25 | 2.4204 | 45 | 3.8067 |
| 8.56 | 2.1471 | 9.06 | 2.2039 | 9.56 | 2.2576 | 11.50 | 2.4423 | 46 | 3.8286 |
| 8.57 | 2.1483 | 9.07 | 2.2050 | 9.57 | 2.2586 | 11.75 | 2.4639 | 47 | 3.8501 |
| 8.58 | 2.1494 | 9.08 | 2.2061 | 9.58 | 2.2597 | 12.00 | 2.4849 | 48 | 3.8712 |
| 8.59 | 2.1506 | 9.09 | 2.2072 | 9.59 | 2.2607 | 12.25 | 2.5055 | 49 | 3.8918 |
| 8.60 | 2.1518 | 9.10 | 2.2083 | 9.60 | 2.2618 | 12.50 | 2.5257 | 50 | 3.9120 |
| 8.61 | 2.1529 | 9.11 | 2.2094 | 9.61 | 2.2628 | 12.75 | 2.5455 | 51 | 3.9318 |
| 8.62 | 2.1541 | 9.12 | 2.2105 | 9.62 | 2.2638 | 13.00 | 2.5649 | 52 | 3.9512 |
| 8.63 | 2.1552 | 9.13 | 2.2116 | 9.63 | 2.2649 | 13.25 | 2.5840 | 53 | 3.9703 |
| 8.64 | 2.1564 | 9.14 | 2.2127 | 9.64 | 2.2659 | 13.50 | 2.6027 | 54 | 3.9890 |
| 8.65 | 2.1576 | 9.15 | 2.2138 | 9.65 | 2.2670 | 13.75 | 2.6210 | 55 | 4.0073 |
| 8.66 | 2.1587 | 9.16 | 2.2148 | 9.66 | 2.2680 | 14.00 | 2.6391 | 56 | 4.0254 |
| 8.67 | 2.1599 | 9.17 | 2.2159 | 9.67 | 2.2690 | 14.25 | 2.6568 | 57 | 4.0431 |
| 8.68 | 2.1610 | 9.18 | 2.2170 | 9.68 | 2.2701 | 14.50 | 2.6741 | 58 | 4.0604 |
| 8.69 | 2.1622 | 9.19 | 2.2181 | 9.69 | 2.2711 | 14.75 | 2.6912 | 59 | 4.0775 |
| 8.70 | 2.1633 | 9.20 | 2.2192 | 9.70 | 2.2721 | 15.00 | 2.7081 | 60 | 4.0943 |
| 8.71 | 2.1645 | 9.21 | 2.2203 | 9.71 | 2.2732 | 15.50 | 2.7408 | 61 | 4.1109 |
| 8.72 | 2.1656 | 9.22 | 2.2214 | 9.72 | 2.2742 | 16.00 | 2.7726 | 62 | 4.1271 |
| 8.73 | 2.1668 | 9.23 | 2.2225 | 9.73 | 2.2752 | 16.50 | 2.8034 | 63 | 4.1431 |
| 8.74 | 2.1679 | 9.24 | 2.2235 | 9.74 | 2.2762 | 17.00 | 2.8332 | 64 | 4.1589 |
| 8.75 | 2.1691 | 9.25 | 2.2246 | 9.75 | 2.2773 | 17.50 | 2.8622 | 65 | 4.1744 |
| 8.76 | 2.1702 | 9.26 | 2.2257 | 9.76 | 2.2783 | 18.00 | 2.8904 | 66 | 4.1897 |
| 8.77 | 2.1713 | 9.27 | 2.2268 | 9.77 | 2.2793 | 18.50 | 2.9178 | 67 | 4.2047 |
| 8.78 | 2.1725 | 9.28 | 2.2279 | 9.78 | 2.2803 | 19.00 | 2.9444 | 68 | 4.2195 |
| 8.79 | 2.1736 | 9.29 | 2.2289 | 9.79 | 2.2814 | 19.50 | 2.9704 | 69 | 4.2341 |
| 8.80 | 2.1748 | 9.30 | 2.2300 | 9.80 | 2.2824 | 20.00 | 2.9957 | 70 | 4.2485 |
| 8.81 | 2.1759 | 9.31 | 2.2311 | 9.81 | 2.2834 | 21 | 3.0445 | 71 | 4.2627 |
| 8.82 | 2.1770 | 9.32 | 2.2322 | 9.82 | 2.2844 | 22 | 3.0910 | 72 | 4.2767 |
| 8.83 | 2.1782 | 9.33 | 2.2332 | 9.83 | 2.2854 | 23 | 3.1355 | 73 | 4.2905 |
| 8.84 | 2.1793 | 9.34 | 2.2343 | 9.84 | 2.2865 | 24 | 3.1781 | 74 | 4.3041 |
| 8.85 | 2.1804 | 9.35 | 2.2354 | 9.85 | 2.2875 | 25 | 3.2189 | 75 | 4.3175 |
| 8.86 | 2.1815 | 9.36 | 2.2364 | 9.86 | 2.2885 | 26 | 3.2581 | 76 | 4.3307 |
| 8.87 | 2.1827 | 9.37 | 2.2375 | 9.87 | 2.2895 | 27 | 3.2958 | 77 | 4.3438 |
| 8.88 | 2.1838 | 9.38 | 2.2386 | 9.88 | 2.2905 | 28 | 3.3322 | 78 | 4.3567 |
| 8.89 | 2.1849 | 9.39 | 2.2396 | 9.89 | 2.2915 | 29 | 3.3673 | 79 | 4.3694 |
| 8.90 | 2.1861 | 9.40 | 2.2407 | 9.90 | 2.2925 | 30 | 3.4012 | 80 | 4.3820 |
| 8.91 | 2.1872 | 9.41 | 2.2418 | 9.91 | 2.2935 | 31 | 3.4340 | 82 | 4.4067 |
| 8.92 | 2.1883 | 9.42 | 2.2428 | 9.92 | 2.2946 | 32 | 3.4657 | 84 | 4.4308 |
| 8.93 | 2.1894 | 9.43 | 2.2439 | 9.93 | 2.2956 | 33 | 3.4965 | 86 | 4.4543 |
| 8.94 | 2.1905 | 9.44 | 2.2450 | 9.94 | 2.2966 | 34 | 3.5264 | 88 | 4.4773 |
| 8.95 | 2.1917 | 9.45 | 2.2460 | 9.95 | 2.2976 | 35 | 3.5553 | 90 | 4.4998 |
| 8.96 | 2.1928 | 9.46 | 2.2471 | 9.96 | 2.2986 | 36 | 3.5835 | 92 | 4.5218 |
| 8.97 | 2.1939 | 9.47 | 2.2481 | 9.97 | 2.2996 | 37 | 3.6109 | 94 | 4.5433 |
| 8.98 | 2.1950 | 9.48 | 2.2492 | 9.98 | 2.3006 | 38 | 3.6376 | 96 | 4.5643 |
| 8.99 | 2.1961 | 9.49 | 2.2502 | 9.99 | 2.3016 | 39 | 3.6636 | 98 | 4.5850 |
| 9.00 | 2.1972 | 9.50 | 2.2513 | 10.00 | 2.3026 | 40 | 3.6889 | 100 | 4.6052 |

### Diameters of Circles and Sides of Squares of Equal Area

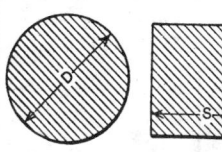

The table below will be found useful for determining the diameter of a circle of an area equal to that of a square, the side of which is known, or for determining the side of a square which has an area equal to that of a circle, the area or diameter of which is known. For example, if the diameter of a circle is 17½ inches, it is found from the table that the side of a square of the same area is 15.51 inches.

| Diam. of Circle, $D$ | Side of Square, $S$ | Area of Circle or Square | Diam. of Circle, $D$ | Side of Square, $S$ | Area of Circle or Square | Diam. of Circle, $D$ | Side of Square $S$ | Area of Circle or Square |
|---|---|---|---|---|---|---|---|---|
| ½ | 0.44 | 0.196 | 20½ | 18.17 | 330.06 | 40½ | 35.89 | 1288.25 |
| 1 | 0.89 | 0.785 | 21 | 18.61 | 346.36 | 41 | 36.34 | 1320.25 |
| 1½ | 1.33 | 1.767 | 21½ | 19.05 | 363.05 | 41½ | 36.78 | 1352.65 |
| 2 | 1.77 | 3.142 | 22 | 19.50 | 380.13 | 42 | 37.22 | 1385.44 |
| 2½ | 2.22 | 4.909 | 22½ | 19.94 | 397.61 | 42½ | 37.66 | 1418.63 |
| 3 | 2.66 | 7.069 | 23 | 20.38 | 415.48 | 43 | 38.11 | 1452.20 |
| 3½ | 3.10 | 9.621 | 23½ | 20.83 | 433.74 | 43½ | 38.55 | 1486.17 |
| 4 | 3.54 | 12.566 | 24 | 21.27 | 452.39 | 44 | 38.99 | 1520.53 |
| 4½ | 3.99 | 15.904 | 24½ | 21.71 | 471.44 | 44½ | 39.44 | 1555.28 |
| 5 | 4.43 | 19.635 | 25 | 22.16 | 490.87 | 45 | 39.88 | 1590.43 |
| 5½ | 4.87 | 23.758 | 25½ | 22.60 | 510.71 | 45½ | 40.32 | 1625.97 |
| 6 | 5.32 | 28.274 | 26 | 23.04 | 530.93 | 46 | 40.77 | 1661.90 |
| 6½ | 5.76 | 33.183 | 26½ | 23.49 | 551.55 | 46½ | 41.21 | 1698.23 |
| 7 | 6.20 | 38.485 | 27 | 23.93 | 572.56 | 47 | 41.65 | 1734.94 |
| 7½ | 6.65 | 44.179 | 27½ | 24.37 | 593.96 | 47½ | 42.10 | 1772.05 |
| 8 | 7.09 | 50.265 | 28 | 24.81 | 615.75 | 48 | 42.54 | 1809.56 |
| 8½ | 7.53 | 56.745 | 28½ | 25.26 | 637.94 | 48½ | 42.98 | 1847.45 |
| 9 | 7.98 | 63.617 | 29 | 25.70 | 660.52 | 49 | 43.43 | 1885.74 |
| 9½ | 8.42 | 70.882 | 29½ | 26.14 | 683.49 | 49½ | 43.87 | 1924.42 |
| 10 | 8.86 | 78.540 | 30 | 26.59 | 706.86 | 50 | 44.31 | 1963.50 |
| 10½ | 9.31 | 86.590 | 30½ | 27.03 | 730.62 | 50½ | 44.75 | 2002.96 |
| 11 | 9.75 | 95.033 | 31 | 27.47 | 754.77 | 51 | 45.20 | 2042.82 |
| 11½ | 10.19 | 103.87 | 31½ | 27.92 | 779.31 | 51½ | 45.64 | 2083.07 |
| 12 | 10.63 | 113.10 | 32 | 28.36 | 804.25 | 52 | 46.08 | 2123.72 |
| 12½ | 11.03 | 122.72 | 32½ | 28.80 | 829.58 | 52½ | 46.53 | 2164.75 |
| 13 | 11.52 | 132.73 | 33 | 29.25 | 855.30 | 53 | 46.97 | 2206.18 |
| 13½ | 11.96 | 143.14 | 33½ | 29.69 | 881.41 | 53½ | 47.41 | 2248.01 |
| 14 | 12.41 | 153.94 | 34 | 30.13 | 907.92 | 54 | 47.86 | 2290.22 |
| 14½ | 12.85 | 165.13 | 34½ | 30.57 | 934.82 | 54½ | 48.30 | 2332.83 |
| 15 | 13.29 | 176.71 | 35 | 31.02 | 962.11 | 55 | 48.74 | 2375.83 |
| 15½ | 13.74 | 188.69 | 35½ | 31.46 | 989.80 | 55½ | 49.19 | 2419.22 |
| 16 | 14.18 | 201.06 | 36 | 31.90 | 1017.88 | 56 | 49.63 | 2463.01 |
| 16½ | 14.62 | 213.82 | 36½ | 32.35 | 1046.35 | 56½ | 50.07 | 2507.19 |
| 17 | 15.07 | 226.98 | 37 | 32.79 | 1075.21 | 57 | 50.51 | 2551.76 |
| 17½ | 15.51 | 240.53 | 37½ | 33.23 | 1104.47 | 57½ | 50.96 | 2596.72 |
| 18 | 15.95 | 254.47 | 38 | 33.68 | 1134.11 | 58 | 51.40 | 2642.08 |
| 18½ | 16.40 | 268.80 | 38½ | 34.12 | 1164.16 | 58½ | 51.84 | 2687.83 |
| 19 | 16.84 | 283.53 | 39 | 34.56 | 1194.59 | 59 | 52.29 | 2733.97 |
| 19½ | 17.28 | 298.65 | 39½ | 35.01 | 1225.42 | 59½ | 52.73 | 2780.51 |
| 20 | 17.72 | 314.16 | 40 | 35.45 | 1256.64 | 60 | 53.17 | 2827.43 |

## Distance Across Corners of Squares and Hexagons

(English and metric units)

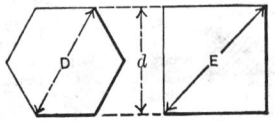

$$D = 1.154701\,d$$
$$E = 1.414214\,d$$

| d | D | E | d | D | E | d | D | E | d | D | E |
|---|---|---|---|---|---|---|---|---|---|---|---|
| 1/32 | 0.0361 | 0.0442 | 0.9 | 1.0392 | 1.2728 | 32 | 36.9504 | 45.2548 | 67 | 77.3649 | 94.7523 |
| 1/16 | 0.0722 | 0.0884 | 29/32 | 1.0464 | 1.2816 | 33 | 38.1051 | 46.6690 | 68 | 78.5196 | 96.1665 |
| 3/32 | 0.1083 | 0.1326 | 15/16 | 1.0825 | 1.3258 | 34 | 39.2598 | 48.0833 | 69 | 79.6743 | 97.5807 |
| 0.1 | 0.1155 | 0.1414 | 31/32 | 1.1186 | 1.3700 | 35 | 40.4145 | 49.4975 | 70 | 80.8290 | 98.9949 |
| 1/8 | 0.1443 | 0.1768 | 1.0 | 1.1547 | 1.4142 | 36 | 41.5692 | 50.9117 | 71 | 81.9837 | 100.409 |
| 5/32 | 0.1804 | 0.2210 | 2.0 | 2.3094 | 2.8284 | 37 | 42.7239 | 52.3259 | 72 | 83.1384 | 101.823 |
| 3/16 | 0.2165 | 0.2652 | 3.0 | 3.4641 | 4.2426 | 38 | 43.8786 | 53.7401 | 73 | 84.2931 | 103.238 |
| 0.2 | 0.2309 | 0.2828 | 4.0 | 4.6188 | 5.6569 | 39 | 45.0333 | 55.1543 | 74 | 85.4478 | 104.652 |
| 7/32 | 0.2526 | 0.3094 | 5.0 | 5.7735 | 7.0711 | 40 | 46.1880 | 56.5685 | 75 | 86.6025 | 106.066 |
| 1/4 | 0.2887 | 0.3536 | 6.0 | 6.9282 | 8.4853 | 41 | 47.3427 | 57.9828 | 76 | 87.7572 | 107.480 |
| 9/32 | 0.3248 | 0.3977 | 7.0 | 8.0829 | 9.8995 | 42 | 48.4974 | 59.3970 | 77 | 88.9119 | 108.894 |
| 0.3 | 0.3464 | 0.4243 | 8.0 | 9.2376 | 11.3137 | 43 | 49.6521 | 60.8112 | 78 | 90.0666 | 110.309 |
| 5/16 | 0.3608 | 0.4419 | 9.0 | 10.3923 | 12.7279 | 44 | 50.8068 | 62.2254 | 79 | 91.2213 | 111.723 |
| 11/32 | 0.3969 | 0.4861 | 10 | 11.5470 | 14.1421 | 45 | 51.9615 | 63.6396 | 80 | 92.3760 | 113.137 |
| 3/8 | 0.4330 | 0.5303 | 11 | 12.7017 | 15.5563 | 46 | 53.1162 | 65.0538 | 81 | 93.5307 | 114.551 |
| 0.4 | 0.4619 | 0.5657 | 12 | 13.8564 | 16.9706 | 47 | 54.2709 | 66.4680 | 82 | 94.6854 | 115.966 |
| 13/32 | 0.4691 | 0.5745 | 13 | 15.0111 | 18.3848 | 48 | 55.4256 | 67.8823 | 83 | 95.8401 | 117.380 |
| 7/16 | 0.5052 | 0.6187 | 14 | 16.1658 | 19.7990 | 49 | 56.5803 | 69.2965 | 84 | 96.9948 | 118.794 |
| 15/32 | 0.5413 | 0.6629 | 15 | 17.3205 | 21.2132 | 50 | 57.7350 | 70.7107 | 85 | 98.1495 | 120.208 |
| 0.5 | 0.5774 | 0.7071 | 16 | 18.4752 | 22.6274 | 51 | 58.8897 | 72.1249 | 86 | 99.3042 | 121.622 |
| 17/32 | 0.6134 | 0.7513 | 17 | 19.6299 | 24.0416 | 52 | 60.0444 | 73.5391 | 87 | 100.459 | 123.037 |
| 9/16 | 0.6495 | 0.7955 | 18 | 20.7846 | 25.4558 | 53 | 61.1991 | 74.9533 | 88 | 101.614 | 124.451 |
| 19/32 | 0.6856 | 0.8397 | 19 | 21.9393 | 26.8701 | 54 | 62.3538 | 76.3675 | 89 | 102.768 | 125.865 |
| 0.6 | 0.6928 | 0.8485 | 20 | 23.0940 | 28.2843 | 55 | 63.5085 | 77.7817 | 90 | 103.923 | 127.279 |
| 5/8 | 0.7217 | 0.8839 | 21 | 24.2487 | 29.6985 | 56 | 64.6632 | 79.1960 | 91 | 105.078 | 128.693 |
| 21/32 | 0.7578 | 0.9281 | 22 | 25.4034 | 31.1127 | 57 | 65.8179 | 80.6102 | 92 | 106.232 | 130.108 |
| 11/16 | 0.7939 | 0.9723 | 23 | 26.5581 | 32.5269 | 58 | 66.9726 | 82.0244 | 93 | 107.387 | 131.522 |
| 0.7 | 0.8083 | 0.9899 | 24 | 27.7128 | 33.9411 | 59 | 68.1273 | 83.4386 | 94 | 108.542 | 132.936 |
| 23/32 | 0.8299 | 1.0165 | 25 | 28.8675 | 35.3553 | 60 | 69.2820 | 84.8528 | 95 | 109.697 | 134.350 |
| 3/4 | 0.8660 | 1.0607 | 26 | 30.0222 | 36.7696 | 61 | 70.4367 | 86.2670 | 96 | 110.851 | 135.765 |
| 25/32 | 0.9021 | 1.1049 | 27 | 31.1769 | 38.1838 | 62 | 71.5914 | 87.6812 | 97 | 112.006 | 137.179 |
| 0.8 | 0.9238 | 1.1314 | 28 | 32.3316 | 39.5980 | 63 | 72.7461 | 89.0955 | 98 | 113.161 | 138.593 |
| 13/16 | 0.9382 | 1.1490 | 29 | 33.4863 | 41.0122 | 64 | 73.9008 | 90.5097 | 99 | 114.315 | 140.007 |
| 27/32 | 0.9743 | 1.1932 | 30 | 34.6410 | 42.4264 | 65 | 75.0555 | 91.9239 | 100 | 115.470 | 141.421 |
| 7/8 | 1.0104 | 1.2374 | 31 | 35.7957 | 43.8406 | 66 | 76.2102 | 93.3381 | ... | ... | ... |

A desired value not given directly in the table can be obtained by the simple addition of two or more values taken directly from the table. Further values can be obtained by shifting the decimal point.

*Example 1:* Find D when d = 2 5/16 inches. From the table, 2 = 2.3094, and 5/16 = 0.3608. Therefore, D = 2.3094 + 0.3608 = 2.6702 inches.

*Example 2:* Find E when d = 20.25 millimetres. From the table, 20 = 28.2843; 0.2 = 0.2828; and 0.05 = 0.0707 (obtained by shifting the decimal point one place to the left at d = 0.5). Thus, E = 28.2843 + 0.2828 + 0.0707 = 28.6378 millimetres.

## Formulas and Table for Regular Polygons

(English and metric units)

$N$ = number of sides.
$S$ = length of side.
$R$ = radius of circumscribed circle.
$r$ = radius of inscribed circle.
$A$ = area of polygon.
$α = 180° ÷ N$ = one-half center angle of one side.

Formulas:

$$A = (N × \cot α × S^2) ÷ 4 \qquad R = S ÷ (2 \sin α) \qquad S = 2R × \sin α$$
$$A = N × \sin α × \cos α × R^2 \qquad R = r ÷ \cos α \qquad * \qquad S = 2r × \tan α$$
$$A = N × \tan α × r^2 \qquad R = \sqrt{A ÷ (N \sin α \cos α)} \quad * \qquad S = 2\sqrt{(A × \tan α)} ÷ N$$
$$r = R × \cos α$$
$$r = (S × \cot α) ÷ 2 \quad *$$
$$r = \sqrt{(A × \cot α)} ÷ N$$

*These formulas may be used to calculate $R$, $S$, or $r$ needed to provide a required area $A$.

### Examples of Use of Table.

A regular hexagon is inscribed in a circle of 6 inches diameter. Find the area and the radius of an inscribed circle. — Here $R = 3$. From the table, area $(A) = 2.5981\,R^2 = 2.5981 × 9 = 23.3829$ square inches. Radius of inscribed circle, $r = 0.866\,R = 0.866 × 3 = 2.598$ inches.

An octagon is inscribed in a circle of 100 millimetres diameter. Thus $R = 50$. Find the area and radius of an inscribed circle. From the table, $A = 2.8284\,R^2 = 2.8284 × 2500 = 7071\ mm^2 = 70.7\ cm^2$. Radius of inscribed circle, $r = 0.9239\,R = 0.9239 × 50 = 46.195$ mm.

Thirty-two bolts are to be equally spaced on the periphery of a bolt-circle, 16 inches in diameter. Find the chordal distance between the bolts. — Chordal distance equals the side $(S)$ of a polygon with 32 sides. $R = 8$. Hence, $S = 0.196\,R = 0.196 × 8 = 1.568$ inch.

Sixteen bolts are to be equally spaced on the periphery of a bolt-circle, 250 millimetres diameter. Find the chordal distance between the bolts. — Chordal distance equals the side $(S)$ of a polygon with 16 sides. $R = 125$. Thus, $S = 0.3902\,R = 0.3902 × 125 = 48.775$ millimetres.

| No. of Sides | $A=$ | $A=$ | $A=$ | $R=$ | $R=$ | $S=$ | $S=$ | $r=$ | $r=$ | No. of Sides |
|---|---|---|---|---|---|---|---|---|---|---|
| 3 | $0.4330\,S^2$ | $1.2990\,R^2$ | $5.1962\,r^2$ | $0.5774\,S$ | $2.0000\,r$ | $1.7321\,R$ | $3.4641\,r$ | $0.5000\,R$ | $0.2887\,S$ | 3 |
| 4 | $1.0000\,S^2$ | $2.0000\,R^2$ | $4.0000\,r^2$ | $0.7071\,S$ | $1.4142\,r$ | $1.4142\,R$ | $2.0000\,r$ | $0.7071\,R$ | $0.5000\,S$ | 4 |
| 5 | $1.7205\,S^2$ | $2.3776\,R^2$ | $3.6327\,r^2$ | $0.8507\,S$ | $1.2361\,r$ | $1.1756\,R$ | $1.4531\,r$ | $0.8090\,R$ | $0.6882\,S$ | 5 |
| 6 | $2.5981\,S^2$ | $2.5981\,R^2$ | $3.4641\,r^2$ | $1.0000\,S$ | $1.1547\,r$ | $1.0000\,R$ | $1.1547\,r$ | $0.8660\,R$ | $0.8660\,S$ | 6 |
| 7 | $3.6339\,S^2$ | $2.7364\,R^2$ | $3.3710\,r^2$ | $1.1524\,S$ | $1.1099\,r$ | $0.8678\,R$ | $0.9631\,r$ | $0.9010\,R$ | $1.0383\,S$ | 7 |
| 8 | $4.8284\,S^2$ | $2.8284\,R^2$ | $3.3137\,r^2$ | $1.3066\,S$ | $1.0824\,r$ | $0.7654\,R$ | $0.8284\,r$ | $0.9239\,R$ | $1.2071\,S$ | 8 |
| 9 | $6.1818\,S^2$ | $2.8925\,R^2$ | $3.2757\,r^2$ | $1.4619\,S$ | $1.0642\,r$ | $0.6840\,R$ | $0.7279\,r$ | $0.9397\,R$ | $1.3737\,S$ | 9 |
| 10 | $7.6942\,S^2$ | $2.9389\,R^2$ | $3.2492\,r^2$ | $1.6180\,S$ | $1.0515\,r$ | $0.6180\,R$ | $0.6498\,r$ | $0.9511\,R$ | $1.5388\,S$ | 10 |
| 12 | $11.196\,S^2$ | $3.0000\,R^2$ | $3.2154\,r^2$ | $1.9319\,S$ | $1.0353\,r$ | $0.5176\,R$ | $0.5359\,r$ | $0.9659\,R$ | $1.8660\,S$ | 12 |
| 16 | $20.109\,S^2$ | $3.0615\,R^2$ | $3.1826\,r^2$ | $2.5629\,S$ | $1.0196\,r$ | $0.3902\,R$ | $0.3978\,r$ | $0.9868\,R$ | $2.5137\,S$ | 16 |
| 20 | $31.569\,S^2$ | $3.0902\,R^2$ | $3.1677\,r^2$ | $3.1962\,S$ | $1.0125\,r$ | $0.3129\,R$ | $0.3168\,r$ | $0.9877\,R$ | $3.1569\,S$ | 20 |
| 24 | $45.575\,S^2$ | $3.1058\,R^2$ | $3.1597\,r^2$ | $3.8306\,S$ | $1.0086\,r$ | $0.2611\,R$ | $0.2633\,r$ | $0.9914\,R$ | $3.7979\,S$ | 24 |
| 32 | $81.225\,S^2$ | $3.1214\,R^2$ | $3.1517\,r^2$ | $5.1011\,S$ | $1.0048\,r$ | $0.1960\,R$ | $0.1970\,r$ | $0.9952\,R$ | $5.0766\,S$ | 32 |
| 48 | $183.08\,S^2$ | $3.1326\,R^2$ | $3.1461\,r^2$ | $7.6449\,S$ | $1.0021\,r$ | $0.1308\,R$ | $0.1311\,r$ | $0.9979\,R$ | $7.6285\,S$ | 48 |
| 64 | $325.69\,S^2$ | $3.1365\,R^2$ | $3.1441\,r^2$ | $10.190\,S$ | $1.0012\,r$ | $0.0981\,R$ | $0.0983\,r$ | $0.9988\,R$ | $10.178\,S$ | 64 |

## Areas and Dimensions of Plane Figures

In the following tables are given the areas of plane figures, together with other formulas relating to their dimensions and properties; the surfaces of solids; and the volumes of solids. The notation used in the formulas is, as far as possible, given in the illustration accompanying them; where this has not been possible, it is given at the beginning of each set of formulas.

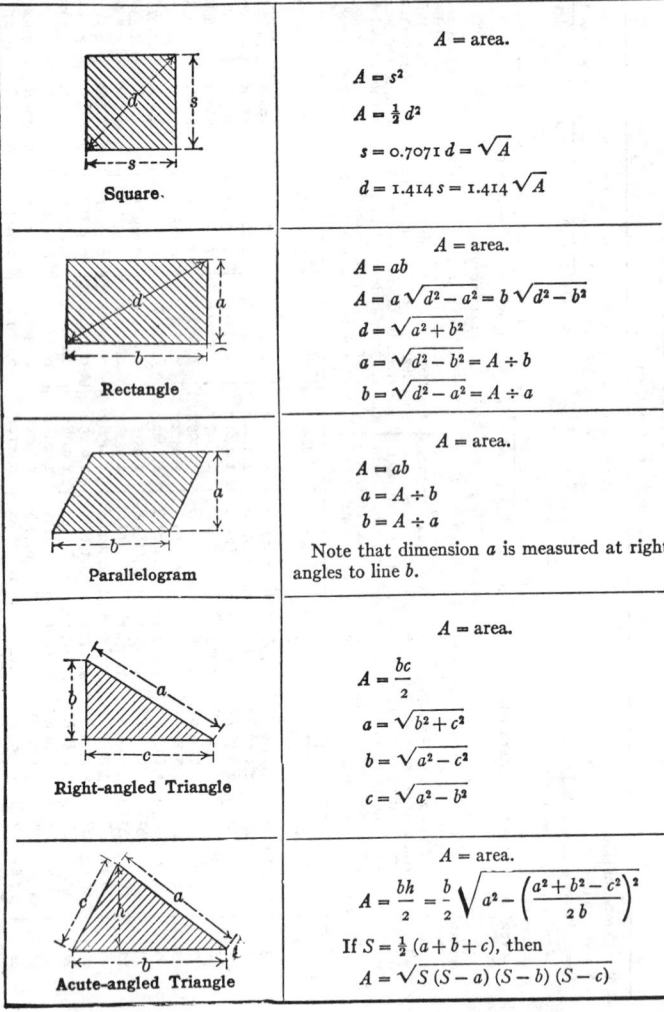

**Square**

$A$ = area.

$A = s^2$

$A = \tfrac{1}{2} d^2$

$s = 0.7071\, d = \sqrt{A}$

$d = 1.414\, s = 1.414 \sqrt{A}$

**Rectangle**

$A$ = area.

$A = ab$

$A = a \sqrt{d^2 - a^2} = b \sqrt{d^2 - b^2}$

$d = \sqrt{a^2 + b^2}$

$a = \sqrt{d^2 - b^2} = A \div b$

$b = \sqrt{d^2 - a^2} = A \div a$

**Parallelogram**

$A$ = area.

$A = ab$

$a = A \div b$

$b = A \div a$

Note that dimension $a$ is measured at right angles to line $b$.

**Right-angled Triangle**

$A$ = area.

$A = \dfrac{bc}{2}$

$a = \sqrt{b^2 + c^2}$

$b = \sqrt{a^2 - c^2}$

$c = \sqrt{a^2 - b^2}$

**Acute-angled Triangle**

$A$ = area.

$A = \dfrac{bh}{2} = \dfrac{b}{2} \sqrt{a^2 - \left(\dfrac{a^2 + b^2 - c^2}{2b}\right)^2}$

If $S = \tfrac{1}{2}(a + b + c)$, then

$A = \sqrt{S(S-a)(S-b)(S-c)}$

## Examples of the Use of the Formulas

(English and metric units)

Below are given examples, some in English and some in metric units, showing the use of the formulas on the opposite page. Each section corresponds to the opposite section on the previous page, and the illustration on that page should be referred to. The notation used in the illustrations is also used in the examples given.

**Square.** — Assume that the side $s$ of a square is 15 inches. Find the area and the length of the diagonal.

$$\text{Area} = A = s^2 = 15^2 = 225 \text{ square inches.}$$

$$\text{Diagonal} = d = 1.414\ s = 1.414 \times 15 = 21.21 \text{ inches.}$$

The area of a square is 625 square inches. Find the length of the side $s$ and the diagonal $d$.

$$s = \sqrt{A} = \sqrt{625} = 25 \text{ inches.}$$

$$d = 1.414 \sqrt{A} = 1.414 \times 25 = 35.35 \text{ inches.}$$

**Rectangle.** — The side $a$ of a rectangle is 12 centimetres, and the area 70.5 square centimetres. Find the length of the side $b$, and the diagonal $d$.

$$b = A \div a = 70.5 \div 12 = 5.875 \text{ centimetres.}$$

$$d = \sqrt{a^2 + b^2} = \sqrt{12^2 + 5.875^2} = \sqrt{178.516} = 13.361 \text{ centimetres.}$$

The sides of a rectangle are 30.5 and 11 centimetres long. Find the area.

$$\text{Area} = a \times b = 30.5 \times 11 = 335.5 \text{ square centimetres.}$$

**Parallelogram.** — The base $b$ of a parallelogram is 16 feet. The height $a$ is 5.5 feet. Find the area.

$$\text{Area} = A = a \times b = 5.5 \times 16 = 88 \text{ square feet.}$$

The area of a parallelogram is 12 square inches. The height is 1.5 inch. Find the length of the base $b$.

$$b = A \div a = 12 \div 1.5 = 8 \text{ inches.}$$

**Right-angled Triangle.** — The sides $b$ and $c$ in a right-angled triangle are 6 and 8 inches. Find side $a$ and the area.

$$a = \sqrt{b^2 + c^2} = \sqrt{6^2 + 8^2} = \sqrt{36 + 64} = \sqrt{100} = 10 \text{ inches.}$$

$$A = \frac{b \times c}{2} = \frac{6 \times 8}{2} = \frac{48}{2} = 24 \text{ square inches.}$$

If $a = 10$ and $b = 6$, had been known, but not $c$, the latter would have been found as follows:

$$c = \sqrt{a^2 - b^2} = \sqrt{10^2 - 6^2} = \sqrt{100 - 36} = \sqrt{64} = 8 \text{ inches.}$$

**Acute-angled Triangle.** — If $a = 10$, $b = 9$, and $c = 8$ centimetres, what is the area of the triangle?

$$A = \frac{b}{2} \sqrt{a^2 - \left(\frac{a^2 + b^2 - c^2}{2b}\right)^2} = \frac{9}{2} \sqrt{10^2 - \left(\frac{10^2 + 9^2 - 8^2}{2 \times 9}\right)^2} = 4.5 \sqrt{100 - \left(\frac{117}{18}\right)^2}$$

$$= 4.5 \sqrt{100 - 42.25} = 4.5\sqrt{57.75} = 4.5 \times 7.60 = 34.20 \text{ square centimetres.}$$

## Areas and Dimensions of Plane Figures

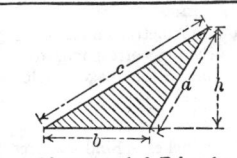

**Obtuse-angled Triangle**

$A$ = area.

$$A = \frac{bh}{2} = \frac{b}{2}\sqrt{a^2 - \left(\frac{c^2 - a^2 - b^2}{2b}\right)^2}$$

If $S = \frac{1}{2}(a + b + c)$, then

$$A = \sqrt{S(S - a)(S - b)(S - c)}$$

---

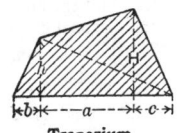

**Trapezoid**

$A$ = area.

$$A = \frac{(a + b)h}{2}$$

*Note:* In England, this figure is called a *trapezium* and the one below it is known as a *trapezoid*, the terms being reversed.

---

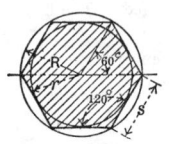

**Trapezium**

$A$ = area.

$$A = \frac{(H + h)a + bh + cH}{2}$$

A trapezium can also be divided into two triangles as indicated by the dotted line. The area of each of these triangles is computed, and the results added to find the area of the trapezium.

---

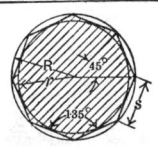

**Regular Hexagon**

$A$ = area;
$R$ = radius of circumscribed circle;
$r$ = radius of inscribed circle.
$A = 2.598\,s^2 = 2.598\,R^2 = 3.464\,r^2$
$R = s = 1.155\,r$
$r = 0.866\,s = 0.866\,R$
$s = R = 1.155\,r$

---

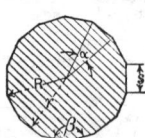

**Regular Octagon**

$A$ = area;
$R$ = radius of circumscribed circle;
$r$ = radius of inscribed circle.
$A = 4.828\,s^2 = 2.828\,R^2 = 3.314\,r^2$
$R = 1.307\,s = 1.082\,r$
$r = 1.207\,s = 0.924\,R$
$s = 0.765\,R = 0.828\,r$

---

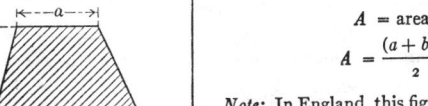

**Regular Polygon**

$A$ = area;　　　$n$ = number of sides.
$\alpha = 360° \div n$　　　$\beta = 180° - \alpha$

$$A = \frac{nsr}{2} = \frac{ns}{2}\sqrt{R^2 - \frac{s^2}{4}}$$

$$R = \sqrt{r^2 + \frac{s^2}{4}}; \quad r = \sqrt{R^2 - \frac{s^2}{4}}; \quad s = 2\sqrt{R^2 - r^2}$$

## Examples of the Use of the Formulas

(English and metric units)

---

**Obtuse-angled Triangle.** — The side $a = 5$, side $b = 4$, and side $c = 8$ inches. Find the area.

$$S = \tfrac{1}{2}(a + b + c) = \tfrac{1}{2}(5 + 4 + 8) = \tfrac{1}{2} \times 17 = 8.5$$

$$A = \sqrt{S(S-a)(S-b)(S-c)} = \sqrt{8.5(8.5-5)(8.5-4)(8.5-8)}$$

$$= \sqrt{8.5 \times 3.5 \times 4.5 \times 0.5} = \sqrt{66.937} = 8.18 \text{ square inches.}$$

---

**Trapezoid.** — Side $a = 23$ metres, side $b = 32$ metres, and height $h = 12$ metres. Find the area.

$$A = \frac{(a+b)h}{2} = \frac{(23+32)\,12}{2} = \frac{55 \times 12}{2} = \frac{660}{2} = 330 \text{ square metres.}$$

---

**Trapezium.** — Let $a = 10$, $b = 2$, $c = 3$, $h = 8$, and $H = 12$ inches. Find the area.

$$A = \frac{(H+h)a + bh + cH}{2} = \frac{(12+8)\,10 + 2 \times 8 + 3 \times 12}{2}$$

$$= \frac{20 \times 10 + 16 + 36}{2} = \frac{252}{2} = 126 \text{ square inches.}$$

---

**Regular Hexagon.** — The side $s$ of a regular hexagon is 40 millimetres. Find the area and the radius $r$ of the inscribed circle.

$A = 2.598\,s^2 = 2.598 \times 40^2 = 2.598 \times 1600 = 4156.8$ square millimetres.

$r = 0.866\,s = 0.866 \times 40 = 34.64$ millimetres.

What is the length of the side of a hexagon that is described about a circle of 50 millimetres radius? — Here $r = 50$. Hence,

$$s = 1.155\,r = 1.155 \times 50 = 57.75 \text{ millimetres.}$$

---

**Regular Octagon.** — Find the area and the length of the side of an octagon that is inscribed in a circle of 12 inches diameter.

Diameter of circumscribed circle = 12 inches; hence, $R = 6$ inches.

$$A = 2.828\,R^2 = 2.828 \times 6^2 = 2.828 \times 36 = 101.81 \text{ square inches.}$$

$$s = 0.765\,R = 0.765 \times 6 = 4.590 \text{ inches.}$$

---

**Regular Polygon.** — Find the area of a polygon having 12 sides, inscribed in a circle of 8 centimetres radius. The length of the side $s$ is 4.141 centimetres.

$$A = \frac{ns}{2}\sqrt{R^2 - \frac{s^2}{4}} = \frac{12 \times 4.141}{2}\sqrt{8^2 - \frac{4.141^2}{4}} = 24.846\sqrt{59.713}$$

$$= 24.846 \times 7.727 = 191.98 \text{ square centimetres.}$$

## Areas and Dimensions of Plane Figures

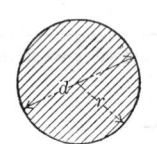

**Circle**

$A$ = area;   $C$ = circumference.

$A = \pi r^2 = 3.1416\, r^2 = 0.7854\, d^2$

$C = 2\pi r = 6.2832\, r = 3.1416\, d$

$r = C \div 6.2832 = \sqrt{A \div 3.1416} = 0.564\, \sqrt{A}$

$d = C \div 3.1416 = \sqrt{A \div 0.7854} = 1.128\, \sqrt{A}$

Length of arc for center-angle of $1° = 0.008727\, d$

Length of arc for center-angle of $n° = 0.008727\, nd$

---

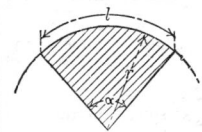

**Circular Sector**

$A$ = area;   $l$ = length of arc;   $\alpha$ = angle, in degrees.

$l = \dfrac{r \times \alpha \times 3.1416}{180} = 0.01745\, r\alpha = \dfrac{2A}{r}$

$A = \tfrac{1}{2}\, rl = 0.008727\, \alpha r^2$

$\alpha = \dfrac{57.296\, l}{r} \qquad r = \dfrac{2A}{l} = \dfrac{57.296\, l}{\alpha}$

---

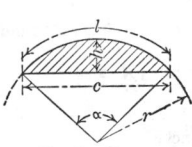

**Circular Segment**

$A$ = area;   $l$ = length of arc;   $\alpha$ = angle, in degrees.

$c = 2\sqrt{h(2r - h)} \qquad A = \tfrac{1}{2}\,[rl - c\,(r - h)]$

                              (see also p. 74)

$r = \dfrac{c^2 + 4h^2}{8h} \qquad l = 0.01745\, r\alpha$

$h = r - \tfrac{1}{2}\sqrt{4r^2 - c^2} \qquad \alpha = \dfrac{57.296\, l}{r}$

$h = r[1 - \cos(\alpha/2)]$

---

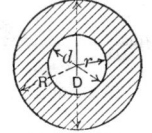

**Circular Ring**

$A$ = area.

$A = \pi\,(R^2 - r^2) = 3.1416\,(R^2 - r^2)$

$\quad = 3.1416\,(R + r)\,(R - r)$

$\quad = 0.7854\,(D^2 - d^2) = 0.7854\,(D + d)\,(D - d)$

---

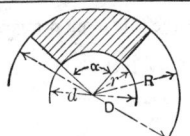

**Circular Ring Sector**

$A$ = area;    $\alpha$ = angle, in degrees.

$A = \dfrac{\alpha\pi}{360}\,(R^2 - r^2) = 0.00873\,\alpha\,(R^2 - r^2)$

$\quad = \dfrac{\alpha\pi}{4 \times 360}\,(D^2 - d^2) = 0.00218\,\alpha\,(D^2 - d^2)$

---

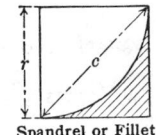

**Spandrel or Fillet**

$A$ = area.

$A = r^2 - \dfrac{\pi r^2}{4} = 0.215\, r^2$

$\quad = 0.1075\, c^2$

## Examples of the Use of the Formulas

(English and metric units)

---

**Circle.** — Find the area $A$ and circumference $C$ of a circle with a diameter of 2¾ inches.

$$A = 0.7854 \, d^2 = 0.7854 \times 2.75^2 = 0.7854 \times 2.75 \times 2.75 = 5.9396 \text{ square inches.}$$
$$C = 3.1416 \, d = 3.1416 \times 2.75 = 8.6394 \text{ inches.}$$

The area of a circle is 16.8 square inches. Find its diameter.

$$d = 1.128 \sqrt{A} = 1.128 \sqrt{16.8} = 1.128 \times 4.099 = 4.624 \text{ inches.}$$

---

**Circular Sector.** — The radius of a circle is 35 millimetres, and angle $\alpha$ of a sector of the circle is 60 degrees. Find the area of the sector and the length of arc $l$.

$$\begin{aligned} A &= 0.008727 \, \alpha r^2 = 0.008727 \times 60 \times 35^2 = 0.5236 \times 35 \times 35 \\ &= 641.41 \text{ square millimetres} \\ &= 6.41 \text{ square centimetres.} \end{aligned}$$

$$l = 0.01745 \, r\alpha = 0.01745 \times 35 \times 60 = 36.645 \text{ millimetres.}$$

---

**Circular Segment.** — The radius $r$ of a circular segment is 60 inches and the height $h$ is 8 inches. Find the length of the chord $c$.

$$c = 2 \sqrt{h(2r - h)} = 2 \sqrt{8 \times (2 \times 60 - 8)} = 2 \sqrt{896} = 2 \times 29.93 = 59.86 \text{ inches.}$$

If $c = 16$, and $h = 6$ inches, what is the radius of the circle of which the segment is a part?

$$r = \frac{c^2 + 4h^2}{8h} = \frac{16^2 + 4 \times 6^2}{8 \times 6} = \frac{256 + 144}{48} = \frac{400}{48} = 8\tfrac{1}{3} \text{ inches.}$$

---

**Circular Ring.** — Let the outside diameter $D = 12$ centimetres and the inside diameter $d = 8$ centimetres. Find area of ring.

$$\begin{aligned} A &= 0.7854 \, (D^2 - d^2) = 0.7854 \, (12^2 - 8^2) = 0.7854 \, (144 - 64) = 0.7854 \times 80 \\ &= 62.83 \text{ square centimetres.} \end{aligned}$$

By the alternative formula:

$$\begin{aligned} A &= 0.7854 \, (D + d)(D - d) = 0.7854 \, (12 + 8)(12 - 8) = 0.7854 \times 20 \times 4 \\ &= 62.83 \text{ square centimetres.} \end{aligned}$$

---

**Circular Ring Sector.** — Find the area, if the outside radius $R = 5$ inches, the inside radius $r = 2$ inches, and $\alpha = 72$ degrees.

$$\begin{aligned} A &= 0.00873 \, \alpha \, (R^2 - r^2) = 0.00873 \times 72 \, (5^2 - 2^2) \\ &= 0.6286 \, (25 - 4) = 0.6286 \times 21 = 13.2 \text{ square inches} \end{aligned}$$

---

**Spandrel or Fillet.** — Find the area of a spandrel, the radius of which is 0.7 inch.

$$A = 0.215 \, r^2 = 0.215 \times 0.7^2 = 0.215 \times 0.7 \times 0.7 = 0.105 \text{ square inch.}$$

If chord $c$ were given as 2.2 inches, what would be the area?

$$A = 0.1075 \, c^2 = 0.1075 \times 2.2^2 = 0.1075 \times 4.84 = 0.520 \text{ square inch.}$$

## Areas and Dimensions of Plane Figures

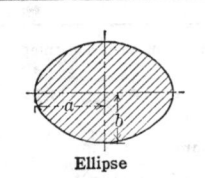

**Ellipse**

$A$ = area;   $P$ = perimeter or circumference.

$A = \pi ab = 3.1416\,ab.$

An approximate formula for the perimeter is:

$$P = 3.1416\sqrt{2(a^2 + b^2)}$$

A closer approximation is:

$$P = 3.1416\sqrt{2(a^2 + b^2) - \frac{(a-b)^2}{2.2}}$$

---

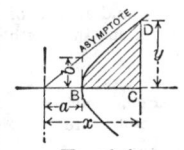

**Hyperbola**

$A$ = area $BCD$.

$$A = \frac{xy}{2} - \frac{ab}{2}\,\text{hyp. log}\left(\frac{x}{a} + \frac{y}{b}\right)$$

---

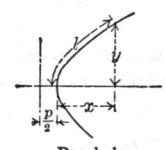

**Parabola**

$l$ = length of arc.

$$l = \frac{p}{2}\left[\sqrt{\frac{2x}{p}\left(1 + \frac{2x}{p}\right)} + \text{hyp. log}\left(\sqrt{\frac{2x}{p}} + \sqrt{1 + \frac{2x}{p}}\right)\right]$$

When $x$ is small in proportion to $y$, the following is a close approximation:

$$l = y\left[1 + \frac{2}{3}\left(\frac{x}{y}\right)^2 - \frac{2}{5}\left(\frac{x}{y}\right)^4\right], \quad \text{or} \quad l = \sqrt{y^2 + \frac{4}{3}x^2}$$

---

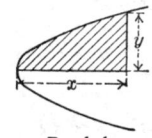

**Parabola**

$A$ = area.

$$A = \tfrac{2}{3}xy$$

(The area is equal to two-thirds of the rectangle which has $x$ for its base and $y$ for its height.)

---

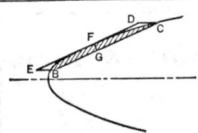

**Segment of Parabola**

$A$ = area.

Area $BFC = A = \tfrac{2}{3}$ area of parallelogram $BCDE$.

If $FG$ is the height of the segment, measured at right angles to $BC$, then:

Area of segment $BFC = \tfrac{2}{3}\,BC \times FG$

---

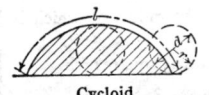

**Cycloid**

$A$ = area;   $l$ = length of cycloid.

$A = 3\pi r^2 = 9.4248\,r^2 = 2.3562\,d^2$

    $= 3 \times$ area of generating circle

$l = 8r = 4d$

## Examples of the Use of the Formulas
(English and metric units)

**Ellipse.** — The larger or major axis is 200 millimetres. The smaller or minor axis is 150 millimetres. Find the area and the approximate circumference. Here, then, $a = 100$, and $b = 75$.

$$A = 3.1416\, ab = 3.1416 \times 100 \times 75 = 23,562 \text{ square millimetres}$$
$$= 235.62 \text{ square centimetres.}$$

$$P = 3.1416 \sqrt{2\,(a^2 + b^2)} = 3.1416 \sqrt{2\,(100^2 + 75^2)} = 3.1416 \sqrt{2 \times 15,625}$$
$$= 3.1416 \sqrt{31,250} = 3.1416 \times 176.78 = 555.37 \text{ millimetres}$$
$$= 55.537 \text{ centimetres.}$$

**Hyperbola.** — The half-axes $a$ and $b$ are 3 and 2 inches, respectively. Find area shown shaded in illustration for $x = 8$ and $y = 5$.

Inserting the known values in the formula:

$$A = \frac{8 \times 5}{2} - \frac{3 \times 2}{2} \times \text{hyp. log} \left( \frac{8}{3} + \frac{5}{2} \right) = 20 - 3 \times \text{hyp. log } 5.167$$

$$= 20 - 3 \times 1.6423 = 20 - 4.927 = 15.073 \text{ square inches.}$$

**Parabola.** — If $x = 2$ and $y = 24$ feet, what is the approximate length $l$ of the parabolic curve?

$$l = y \left[ 1 + \frac{2}{3} \left( \frac{x}{y} \right)^2 - \frac{2}{5} \left( \frac{x}{y} \right)^4 \right] = 24 \left[ 1 + \frac{2}{3} \left( \frac{2}{24} \right)^2 - \frac{2}{5} \left( \frac{2}{24} \right)^4 \right]$$

$$= 24 \left[ 1 + \frac{2}{3} \times \frac{1}{144} - \frac{2}{5} \times \frac{1}{20,736} \right] = 24 \times 1.0046 = 24.11 \text{ feet.}$$

**Parabola.** — Let the dimension $x$ in the illustration be 15 centimetres, and $y$, 9 centimetres. Find the area of the shaded portion of the parabola.

$$A = \tfrac{2}{3} \times xy = \tfrac{2}{3} \times 15 \times 9 = 10 \times 9 = 90 \text{ square centimetres.}$$

**Segment of Parabola.** — The length of the chord $BC = 19.5$ inches. The distance between lines $BC$ and $DE$, measured at right angles to $BC$, is 2.25 inches. This is the height of the segment. Find the area.

$$\text{Area} = A = \tfrac{2}{3}\, BC \times FG = \tfrac{2}{3} \times 19.5 \times 2.25 = 29.25 \text{ square inches.}$$

**Cycloid.** — The diameter of the generating circle of a cycloid is 6 inches. Find the length $l$ of the cycloidal curve, and the area enclosed between the curve and the base line.

$$l = 4\,d = 4 \times 6 = 24 \text{ inches.}$$
$$A = 2.3562\, d^2 = 2.3562 \times 6^2 = 2.3562 \times 36 = 84.82 \text{ square inches.}$$

## Volumes of Solids

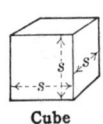

**Cube**

$V$ = volume.

$$V = s^3$$

$$s = \sqrt[3]{V}$$

---

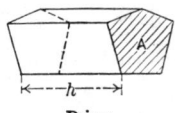

**Square Prism**

$V$ = volume.

$$V = abc$$

$$a = \frac{V}{bc} \qquad b = \frac{V}{ac} \qquad c = \frac{V}{ab}$$

---

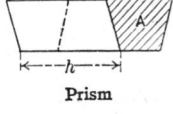

**Prism**

$V$ = volume; $A$ = area of end surface.

$$V = h \times A$$

The area $A$ of the end surface is found by the formulas for areas of plane figures on the preceding pages. Height $h$ must be measured perpendicular to end surface.

---

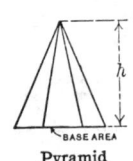

**Pyramid**

$V$ = volume.

$$V = \tfrac{1}{3} h \times \text{area of base.}$$

If the base is a regular polygon with $n$ sides, and $s$ = length of side, $r$ = radius of inscribed circle, and $R$ = radius of circumscribed circle, then:

$$V = \frac{nsrh}{6} = \frac{nsh}{6} \sqrt{R^2 - \frac{s^2}{4}}$$

---

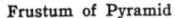

**Frustum of Pyramid**

$V$ = volume.

$$V = \frac{h}{3} \left( A_1 + A_2 + \sqrt{A_1 \times A_2} \right)$$

---

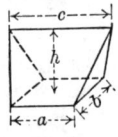

**Wedge**

$V$ = volume.

$$V = \frac{(2\,a + c)\,bh}{6}$$

## Examples of the Use of the Formulas

(English and metric units)

---

**Cube.** — The side of a cube equals 9.5 centimetres. Find its volume.

Volume = $V = s^3 = 9.5^3 = 9.5 \times 9.5 \times 9.5 = 857.375$ cubic centimetres.

The volume of a cube is 231 cubic centimetres. What is the length of the side?

$$s = \sqrt[3]{V} = \sqrt[3]{231} = 6.136 \text{ centimetres.}$$

---

**Square Prism.** — In a square prism, $a = 6$, $b = 5$, $c = 4$. Find the volume.

$$V = a \times b \times c = 6 \times 5 \times 4 = 120 \text{ cubic inches.}$$

How high should a box be made to contain 25 cubic feet, if it is 4 feet long and $2\frac{1}{2}$ feet wide? Here, $a = 4$, $c = 2.5$, and $V = 25$. Then,

$$b = \text{depth} = \frac{V}{ac} = \frac{25}{4 \times 2.5} = \frac{25}{10} = 2.5 \text{ feet.}$$

---

**Prism.** — A prism having for its base a regular hexagon with a side $s$ of 7.5 centimetres, is 25 centimetres high. Find the volume.

Area of hexagon = $A = 2.598\, s^2 = 2.598 \times 56.25 = 146.14$ square centimetres.

Volume of prism = $h \times A = 25 \times 146.14 = 3653.5$ cubic centimetres.

---

**Pyramid.** — A pyramid, having a height of 9 feet, has a base formed by a rectangle, the sides of which are 2 and 3 feet, respectively. Find the volume.

Area of base = $2 \times 3 = 6$ square feet; $h = 9$ feet.

Volume = $V = \frac{1}{3} h \times$ area of base $= \frac{1}{3} \times 9 \times 6 = 18$ cubic feet.

---

**Frustum of Pyramid.** — The pyramid in the previous example is cut off $4\frac{1}{2}$ feet from the base, the upper part being removed. The sides of the rectangle forming the top surface of the frustum are, then, 1 and $1\frac{1}{2}$ foot long, respectively. Find the volume of the frustum.

Area of top = $A_1 = 1 \times 1\frac{1}{2} = 1\frac{1}{2}$ sq. ft. Area of base = $A_2 = 2 \times 3 = 6$ sq. ft.

$$V = \frac{4.5}{3}\left(1.5 + 6 + \sqrt{1.5 \times 6}\right) = 1.5\left(7.5 + \sqrt{9}\right) = 1.5 \times 10.5 = 15.75 \text{ cubic feet.}$$

---

**Wedge.** — Let $a = 4$ inches, $b = 3$ inches, and $c = 5$ inches. The height $h = 4.5$ inches. Find the volume.

$$V = \frac{(2a+c)\,bh}{6} = \frac{(2 \times 4 + 5) \times 3 \times 4.5}{6} = \frac{(8+5) \times 13.5}{6} = \frac{13 \times 13.5}{6}$$

$$= \frac{175.5}{6} = 29.25 \text{ cubic inches.}$$

## Volumes of Solids

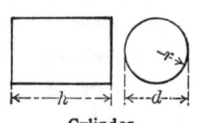

**Cylinder**

$V$ = volume;   $S$ = area of cylindrical surface.

$V = 3.1416\, r^2 h = 0.7854\, d^2 h$

$S = 6.2832\, rh = 3.1416\, dh$

Total area $A$ of cylindrical surface and end surfaces:

$A = 6.2832\, r\,(r + h) = 3.1416\, d\,(\tfrac{1}{2} d + h)$

---

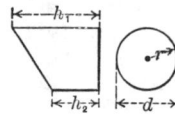

**Portion of Cylinder**

$V$ = volume;   $S$ = area of cylindrical surface.

$V = 1.5708\, r^2 (h_1 + h_2) = 0.3927\, d^2 (h_1 + h_2)$

$S = 3.1416\, r\,(h_1 + h_2) = 1.5708\, d\,(h_1 + h_2)$

---

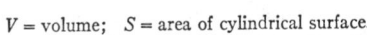

**Portion of Cylinder**

$V$ = volume;   $S$ = area of cylindrical surface.

$V = \left(\dfrac{2}{3} a^3 \pm b \times \text{area } ABC\right) \dfrac{h}{r \pm b}$

$S = (ad \pm b \times \text{length of arc } ABC)\, \dfrac{h}{r \pm b}$

Use + when base area is larger, and − when base area is less than one-half the base circle.

---

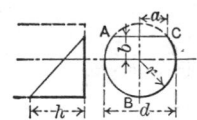

**Hollow Cylinder**

$V$ = volume.

$V = 3.1416\, h\,(R^2 - r^2) = 0.7854\, h\,(D^2 - d^2)$

$\quad = 3.1416\, ht\,(2\,R - t) = 3.1416\, ht\,(D - t)$

$\quad = 3.1416\, ht\,(2\,r + t) = 3.1416\, ht\,(d + t)$

$\quad = 3.1416\, ht\,(R + r) = 1.5708\, ht\,(D + d)$

---

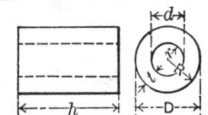

**Cone**

$V$ = volume;   $A$ = area of conical surface.

$V = \dfrac{3.1416\, r^2 h}{3} = 1.0472\, r^2 h = 0.2618\, d^2 h$

$A = 3.1416\, r \sqrt{r^2 + h^2} = 3.1416\, rs = 1.5708\, ds$

$s = \sqrt{r^2 + h^2} = \sqrt{\dfrac{d^2}{4} + h^2}$

---

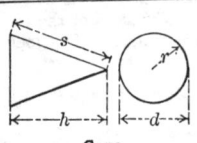

**Frustum of Cone**

$V$ = volume;   $A$ = area of conical surface.

$V = 1.0472\, h\,(R^2 + Rr + r^2) = 0.2618\, h\,(D^2 + Dd + d^2)$

$A = 3.1416\, s\,(R + r) = 1.5708\, s\,(D + d)$

$a = R - r \qquad s = \sqrt{a^2 + h^2} = \sqrt{(R - r)^2 + h^2}$

## Examples of the Use of the Formulas
(English and metric units)

---

**Cylinder.** — The diameter of a cylinder is $2\frac{1}{2}$ inches. The length or height is 20 inches. Find the volume, and the area of the cylindrical surface $S$.

$V = 0.7854\ d^2h = 0.7854 \times 2\frac{1}{2}^2 \times 20 = 0.7854 \times 6.25 \times 20 = 98.17$ cubic inches.
$S = 3.1416\ dh = 3.1416 \times 2\frac{1}{2} \times 20 = 157.08$ square inches.

---

**Portion of Cylinder.** — A cylinder 125 millimetres in diameter, is cut off at an angle, as shown in the illustration. Dimension $h_1 = 150$, and $h_2 = 100$ mm. Find the volume and the area $S$ of the cylindrical surface.

$V = 0.3927\ d^2\ (h_1 + h_2) = 0.3927 \times 125^2 \times (150 + 100)$
$\quad = 0.3927 \times 15,625 \times 250 = 1,533,984$ cubic millimetres $= 1534$ cm$^3$.
$S = 1.5708\ d\ (h_1 + h_2) = 1.5708 \times 125 \times 250$
$\quad = 49,087.5$ square millimetres $= 490.9$ square centimetres.

---

**Portion of Cylinder.** — Find the volume of a cylinder so cut off that line $AC$ passes through the center of the base circle — that is, the base area is a half-circle. The diameter of the cylinder $= 5$ inches, and height $h = 2$ inches.
In this case $a = 2.5$; $b = 0$; area $ABC = \frac{1}{2} \times 0.7854 \times 5^2 = 9.82$; $r = 2.5$.

$V = \left(\dfrac{2}{3} \times 2.5^3 + 0 \times 9.82\right) \dfrac{2}{2.5 + 0} = \dfrac{2}{3} \times 15.625 \times 0.8 = 8.33$ cubic inches.

---

**Hollow Cylinder.** — A cylindrical shell, 28 centimetres high, is 36 centimetres in outside diameter, and 4 centimetres thick. Find its volume.

$V = 3.1416\ ht\ (D - t) = 3.1416 \times 28 \times 4\ (36 - 4) = 3.1416 \times 28 \times 4 \times 32$
$\quad = 11,259.5$ cubic centimetres.

---

**Cone.** — Find the volume and area of conical surface of a cone, the base of which is a circle of 6 inches diameter, and the height of which is 4 inches.

$V = 0.2618\ d^2h = 0.2618 \times 6^2 \times 4 = 0.2618 \times 36 \times 4 = 37.7$ cubic inches.
$A = 3.1416\ r\ \sqrt{r^2 + h^2} = 3.1416 \times 3 \times \sqrt{3^2 + 4^2} = 9.4248 \times \sqrt{25}$
$\quad = 47.124$ square inches.

---

**Frustum of Cone.** — Find the volume of a frustum of a cone of the following dimensions: $D = 8$ centimetres; $d = 4$ centimetres; $h = 5$ centimetres.

$V = 0.2618 \times 5\ (8^2 + 8 \times 4 + 4^2) = 0.2618 \times 5\ (64 + 32 + 16)$
$\quad = 0.2618 \times 5 \times 112 = 146.61$ cubic centimetres.

## Volumes of Solids

| | |
|---|---|
| **Sphere** | $V$ = volume; $A$ = area of surface.<br><br>$V = \dfrac{4\pi r^3}{3} = \dfrac{\pi d^3}{6} = 4.1888\,r^3 = 0.5236\,d^3$<br><br>$A = 4\pi r^2 = \pi d^2 = 12.5664\,r^2 = 3.1416\,d^2$<br><br>$r = \sqrt[3]{\dfrac{3\,V}{4\pi}} = 0.6204\,\sqrt[3]{V}$ |
| **Spherical Sector** | $V$ = volume; $A$ = total area of conical and spherical surface.<br><br>$V = \dfrac{2\pi r^2 h}{3} = 2.0944\,r^2 h$<br><br>$A = 3.1416\,r\,(2\,h + \tfrac{1}{2}\,c)$<br><br>$c = 2\,\sqrt{h\,(2\,r - h)}$ |
| **Spherical Segment** | $V$ = volume; $A$ = area of spherical surface.<br><br>$V = 3.1416\,h^2\!\left(r - \dfrac{h}{3}\right) = 3.1416\,h\left(\dfrac{c^2}{8} + \dfrac{h^2}{6}\right)$<br><br>$A = 2\pi r h = 6.2832\,rh = 3.1416\left(\dfrac{c^2}{4} + h^2\right)$<br><br>$c = 2\,\sqrt{h\,(2\,r - h)}$;   $r = \dfrac{c^2 + 4\,h^2}{8\,h}$ |
| **Spherical Zone** | $V$ = volume; $A$ = area of spherical surface.<br><br>$V = 0.5236\,h\left(\dfrac{3\,c_1^{\,2}}{4} + \dfrac{3\,c_2^{\,2}}{4} + h^2\right)$<br><br>$A = 2\pi r h = 6.2832\,rh$<br><br>$r = \sqrt{\dfrac{c_2^{\,2}}{4} + \left(\dfrac{c_2^{\,2} - c_1^{\,2} - 4\,h^2}{8\,h}\right)^2}$ |
| **Spherical Wedge** | $V$ = volume; $A$ = area of spherical surface; $\alpha$ = center angle in degrees.<br><br>$V = \dfrac{\alpha}{360} \times \dfrac{4\pi r^3}{3} = 0.0116\,\alpha r^3$<br><br>$A = \dfrac{\alpha}{360} \times 4\pi r^2 = 0.0349\,\alpha r^2$ |
| **Hollow Sphere** | $V$ = volume.<br><br>$V = \dfrac{4\pi}{3}\,(R^3 - r^3) = 4.1888\,(R^3 - r^3)$<br><br>$= \dfrac{\pi}{6}\,(D^3 - d^3) = 0.5236\,(D^3 - d^3)$ |

## Examples of the Use of the Formulas

(English and metric units)

**Sphere.** — Find volume and surface of a sphere 6.5 centimetres diam.

$V = 0.5236\,d^3 = 0.5236 \times 6.5^3 = 0.5236 \times 6.5 \times 6.5 \times 6.5 = 143.79$ cm³.

$A = 3.1416\,d^2 = 3.1416 \times 6.5^2 = 3.1416 \times 6.5 \times 6.5 = 132.73$ cm².

The volume of a sphere is 64 cubic centimetres. Find its radius.

$r = 0.6204\,\sqrt[3]{64} = 0.6204 \times 4 = 2.4816$ centimetres.

---

**Spherical Sector.** — Find the volume of a sector of a sphere 6 inches in diameter, the height $h$ of the sector being 1.5 inch. Also find length of chord $c$. — Here $r = 3$, and $h = 1.5$.

$V = 2.0944\,r^2 h = 2.0944 \times 3^2 \times 1.5 = 2.0944 \times 9 \times 1.5 = 28.27$ cubic inches.

$c = 2\,\sqrt{h\,(2\,r - h)} = 2\,\sqrt{1.5\,(2 \times 3 - 1.5)} = 2\,\sqrt{6.75} = 2 \times 2.598$

$\quad\quad = 5.196$ inches.

---

**Spherical Segment.** — A segment of a sphere has the following dimensions: $h = 50$ millimetres; $c = 125$ millimetres. Find the volume $V$ and the radius of the sphere of which the segment is a part.

$V = 3.1416 \times 50 \times \left(\dfrac{125^2}{8} + \dfrac{50^2}{6}\right) = 157.08 \times \left(\dfrac{15,625}{8} + \dfrac{2500}{6}\right)$

$\quad\quad = 157.08 \times 2369.79 = 372,247$ cubic millimetres $= 372$ cm³.

$r = \dfrac{125^2 + 4 \times 50^2}{8 \times 50} = \dfrac{15,625 + 10,000}{400} = \dfrac{25,625}{400} = 64$ millimetres.

---

**Spherical Zone.** — In a spherical zone, let $c_1 = 3$; $c_2 = 4$; and $h = 1.5$ inch. Find the volume.

$V = 0.5236 \times 1.5 \times \left(\dfrac{3 \times 3^2}{4} + \dfrac{3 \times 4^2}{4} + 1.5^2\right) = 0.5236 \times 1.5 \times \left(\dfrac{27}{4} + \dfrac{48}{4} + 2.25\right)$

$\quad\quad = 0.5236 \times 1.5 \times 21 = 16.493$ cubic inches.

---

**Spherical Wedge.** — Find the area of the spherical surface and the volume of a wedge of a sphere. The diameter of the sphere is 100 millimeters, and the center angle $\alpha$ is 45 degrees.

$V = 0.0116 \times 45 \times 50^3 = 0.0116 \times 45 \times 125,000$

$\quad\quad = 65,250$ cubic millimetres $= 65.25$ cubic centimeters.

$A = 0.0349 \times 45 \times 50^2 = 3926.25$ square millimetres $= 39.26$ cm².

---

**Hollow Sphere.** — Find the volume of a hollow sphere, 8 inches in outside diameter, with a thickness of material of 1.5 inch.

Here $R = 4$; $r = 4 - 1.5 = 2.5$.

$V = 4.1888\,(4^3 - 2.5^3) = 4.1888\,(64 - 15.625) = 4.1888 \times 48.375$

$\quad\quad = 202.63$ cubic inches.

## Volumes of Solids

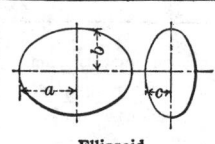

**Ellipsoid**

$V$ = volume.

$$V = \frac{4\pi}{3}\,abc = 4.1888\,abc$$

In an ellipsoid of revolution, or spheroid, where $c = b$:

$$V = 4.1888\,ab^2$$

---

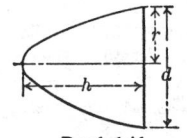

**Paraboloid**

$V$ = volume;   $V = \frac{1}{2}\pi r^2 h = 0.3927\,d^2 h$

$A$ = area;   $A = \frac{2\pi}{3\,p}\left[\sqrt{\left(\dfrac{d^2}{4} + p^2\right)^3} - p^3\right]$ in which

$$p = \frac{d^2}{8\,h}$$

---

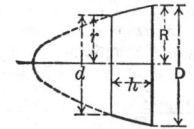

**Paraboloidal Segment**

$V$ = volume.

$$V = \frac{\pi}{2}\,h\,(R^2 + r^2) = 1.5708\,h\,(R^2 + r^2)$$

$$= \frac{\pi}{8}\,h\,(D^2 + d^2) = 0.3927\,h\,(D^2 + d^2)$$

---

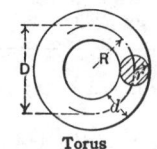

**Torus**

$V$ = volume;   $A$ = area of surface.

$$V = 2\,\pi^2\,Rr^2 = 19.739\,Rr^2$$

$$= \frac{\pi^2}{4}\,Dd^2 = 2.4674\,Dd^2$$

$$A = 4\,\pi^2\,Rr = 39.478\,Rr$$

$$= \pi^2 Dd \;\;= 9.8696\,Dd$$

---

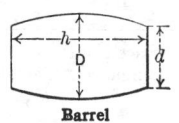

**Barrel**

$V$ = approximate volume.

If the sides are bent to the arc of a circle:

$$V = \tfrac{1}{12}\pi h\,(2\,D^2 + d^2) = 0.262\,h\,(2\,D^2 + d^2)$$

If the sides are bent to the arc of a parabola:

$$V = 0.209\,h\,(2\,D^2 + Dd + \tfrac{3}{4}\,d^2)$$

---

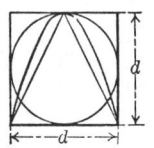

If $d$ = base diameter and height of a cone, a paraboloid and a cylinder, and the diameter of a sphere, then the volumes of these bodies are to each other as below:

Cone: paraboloid: sphere: cylinder = $\tfrac{1}{3} : \tfrac{1}{2} : \tfrac{2}{3} : 1$

## Examples of the Use of the Formulas
(English and metric units)

**Ellipsoid or Spheroid.** — Find the volume of a spheroid in which $a = 5$, and $b = c = 1.5$ inch.

$$V = 4.1888 \times 5 \times 1.5^2 = 4.1888 \times 5 \times 2.25 = 47.124 \text{ cubic inches.}$$

**Paraboloid.** — Find the volume of a paraboloid in which $h = 300$ millimetres and $d = 125$ millimetres.

$$V = 0.3927 \, d^2 h = 0.3927 \times 125^2 \times 300 = 0.3927 \times 15,625 \times 300$$
$$= 1,840,781 \text{ cubic millimetres} = 1,840.8 \text{ cubic centimetres.}$$

**Segment of Paraboloid.** — Find the volume of a segment of a paraboloid in which $D = 5$ inches, $d = 3$ inches, and $h = 6$ inches.

$$V = 0.3927 \, h \, (D^2 + d^2) = 0.3927 \times 6 \times (5^2 + 3^2) = 0.3927 \times 6 \times (25 + 9)$$
$$= 0.3927 \times 6 \times 34 = 80.11 \text{ cubic inches.}$$

**Torus.** — Find the volume and area of surface of a torus in which $d = 1.5$ and $D = 5$ inches.

$$V = 2.4674 \times 5 \times 1.5^2 = 2.4674 \times 5 \times 2.25 = 27.76 \text{ cubic inches.}$$
$$A = 9.8696 \times 5 \times 1.5 = 74.022 \text{ square inches}$$

**Barrel.** — Find the approximate contents of a barrel, the inside dimensions of which are $D = 60$ centimetres; $d = 50$ centimetres; $h = 120$ centimetres.

$$V = 0.262 \, h \, (2D^2 + d^2) = 0.262 \times 120 \times (2 \times 60^2 + 50^2)$$
$$= 0.262 \times 120 \times (7200 + 2500) = 0.262 \times 120 \times 9700$$
$$= 304,968 \text{ cubic centimetres} = 0.305 \text{ cubic metre.}$$

Assume, as an example, that the diameter of the base of a cone, paraboloid and cylinder is 2 inches, that the height is 2 inches, and that the diameter of a sphere is 2 inches. Then the volumes, written in formula-form, are as below:

| Cone | Paraboloid | Sphere | Cylinder | | | |
|---|---|---|---|---|---|---|
| $\dfrac{3.1416 \times 2^2 \times 2}{12}$ | $: \dfrac{3.1416 \times 2^2 \times 2}{8}$ | $: \dfrac{3.1416 \times 2^3}{6}$ | $: \dfrac{3.1416 \times 2^2 \times 2}{4}$ | $= \dfrac{1}{3}$ | $: \dfrac{1}{2} : \dfrac{2}{3}$ | $: 1$ |

**The Prismoidal Formula.** — The prismoidal formula is a general formula by which the volume of any prism, pyramid or frustum of a pyramid may be found.

$A_1$ = area at one end of the body;
$A_2$ = area at the other end;
$A_m$ = area of middle section between the two end surfaces;
$h$ = height of body.

Then, volume $V$ of the body is

$$V = \frac{h}{6}(A_1 + 4A_m + A_2)$$

**Pappus or Guldinus Rules.** — By means of these rules the area of any surface of revolution and the volume of any solid of revolution may be found. The area of the surface swept out by the revolution of a line $ABC$ (see illustration) about the axis $DE$ equals the length of the line multiplied by the length of the path

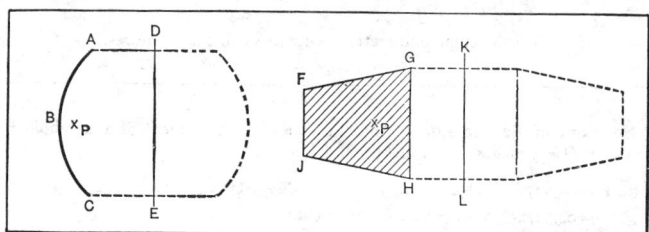

of its center of gravity, $P$. If the line is of such a shape that it is difficult to determine its center of gravity, then the line may be divided into a number of short sections, each of which may be considered as a straight line, and the areas swept out by these different sections, as computed by the rule given, may be added to find the total area. The line must lie wholly on one side of the axis of revolution and must be in the same plane.

The volume of a solid body formed by the revolution of a surface $FGHJ$ about axis $KL$ equals the area of the surface multiplied by the length of the path of its center of gravity. The surface must lie wholly on one side of the axis of revolution and in the same plane.

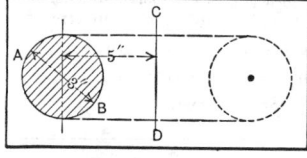

*Example:* — By means of these rules the area and volume of a cylindrical ring or torus may be found. The torus is formed by a circle $AB$ being rotated about axis $CD$. The center of gravity of the circle is at its center. Hence, with the dimensions given in the illustration, the length of the path of the center of gravity of the circle is 3.1416 × 10 = 31.416 inches. This multiplied by the length of the circumference of the circle, which is 3.1416 × 3 = 9.4248 inches, equals:

$$31.416 \times 9.4248 = 296.089 \text{ square inches}$$

which is the area of the torus.

The volume equals the area of the circle, which is 0.7854 × 9 = 7.0686 square inches, multiplied by the path of the center of gravity, which is 31.416, as before; hence,

$$\text{volume} = 7.0686 \times 31.416 = 222.067 \text{ cubic inches.}$$

**Example of Approximate Method for Finding the Area of a Surface of Revolution.** — The accompanying illustration is shown in order to give an example of the approximate method based on Guldinus' rule, that can be used for finding the area of a symmetrical body. In the illustration, the dimensions in common fractions are the known dimensions; those in decimals are found by actual measurements on a figure drawn to scale. The method for finding the area is as follows: First separate such areas as are cylindrical, conical or spherical, as these can be found by exact formulas. In the illustration $ABCD$ is a cylinder, the area of the surface of which can be easily found. The top area $EF$ is simply a circular area, and can thus be computed separately. The remainder of the surface generated by

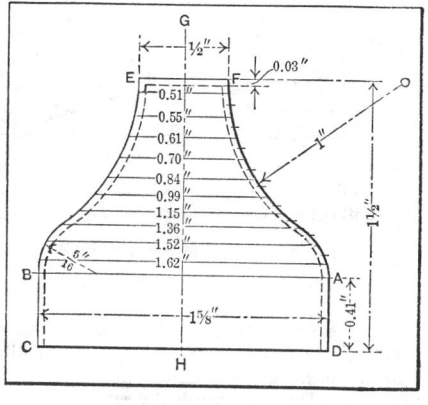

rotating line $AF$ about the axis $GH$ is found by the approximate method explained in the previous section. From point $A$, set off equal distances on line $AF$. In the present case each division indicated is $\frac{1}{8}$ inch long. From the central or middle point of each of these parts draw a line at right angles to the axis of rotation $GH$, measure the length of these lines or diameters (the length of each is given in decimals), add all these lengths together and multiply the sum by the length of one division set off on line $AF$ (in this case, $\frac{1}{8}$ inch), and multiply this product by $\pi$. This gives the approximate area of the surface of revolution.

In setting off divisions $\frac{1}{8}$ inch long along line $AF$, the last division does not reach exactly to point $F$, but only to a point 0.03 inch below it. The part 0.03 inch high at the top of the cup, can be considered as a cylinder of $\frac{1}{2}$ inch diameter and 0.03 inch height, the area of the cylindrical surface of which is easily computed. By adding the various surfaces together the total surface of the cup is found as below:

Cylinder, $1\frac{5}{8}$ inch diameter, 0.41 inch high........   2.093 square inches
Circle, $\frac{1}{2}$ inch diameter...........................   0.196 square inches
Cylinder, $\frac{1}{2}$ inch diameter, 0.03 inch high.........   0.047 square inches
Irregular surface................................   3.868 square inches

Total.....................................   6.204 square inches

**Area of Plane Surfaces of Irregular Outline.** — One of the most useful and accurate methods for determining the approximate area of a plane figure of irregular outline is known as *Simpson's Rule*. In applying Simpson's Rule to find an area the work is done in four steps:

1. Divide the area into an *even* number, $N$, of parallel strips of equal width $W$; for example, in the accompanying diagram the area has been divided into 8 strips of equal width;

2. Label the sides of the strips $V_0$, $V_1$, $V_2$, etc. up to $V_N$;

3. Measure the heights $V_0$, $V_1$, $V_2$, ... $V_N$ of the sides of the strips;

4. Substitute the heights $V_0$, $V_1$, etc. in the following formula to find the area $A$ of the figure:

$$A = \frac{W}{3}[(V_0 + V_N) + 4(V_1 + V_3 + \cdots V_{N-1}) + 2(V_2 + V_4 + \cdots V_{N-2})]$$

*Example:* The area of the accompanying figure was divided into 8 strips on a full-size drawing and the following data obtained. Calculate the area using Simpson's Rule.

$W = \frac{1}{2}''$
$V_0 = 0''$
$V_1 = \frac{3}{4}''$
$V_2 = 1\frac{1}{4}''$
$V_3 = 1\frac{1}{2}''$
$V_4 = 1\frac{5}{8}''$
$V_5 = 2\frac{1}{4}''$
$V_6 = 2\frac{1}{2}''$
$V_7 = 1\frac{3}{4}''$
$V_8 = \frac{1}{2}''$

Substituting the given data in the Simpson formula,

$$A = \frac{\frac{1}{2}}{3}[(0 + \frac{1}{2}) + 4(\frac{3}{4} + 1\frac{1}{2} + 2\frac{1}{4} + 1\frac{3}{4}) + 2(1\frac{1}{4} + 1\frac{5}{8} + 2\frac{1}{2})]$$
$$= \frac{1}{6}[(\frac{1}{2}) + 4(6\frac{1}{4}) + 2(5\frac{3}{8})] = \frac{1}{6}[36\frac{1}{4}]$$
$$= 6.04 \text{ square inches}$$

In applying Simpson's Rule it should be noted that the larger the number of strips into which the area is divided the more accurate the results obtained.

**Areas Enclosed by Cycloidal Curves.** — The area between a cycloid and the straight line upon which the generating circle rolls, equals three times the area of the generating circle (see diagram, page 156). The areas between epicycloidal and hypocycloidal curves and the "fixed circle" upon which the generating circle is rolled, may be determined by the following formulas, in which $a$ = radius of the fixed circle upon which the generating circle rolls; $b$ = radius of the generating circle; $A$ = the area for the epicycloidal curve; and $A_1$ = the area for the hypocycloidal curve.

$$A = \frac{3.1416\,b^2(3\,a + 2\,b)}{a}; \qquad A_1 = \frac{3.1416\,b^2(3\,a - 2\,b)}{a}$$

**Find the Contents of Cylindrical Tanks at Different Levels.** — In conjunction with the table "Segments of Circles for Radius = 1," presented on pages 72 and 73, the following relations can give a close approximation of the liquid contents, at any level, in a cylindrical tank.

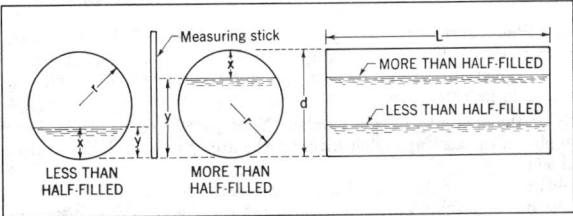

A long measuring rule calibrated in length units or simply a plain stick can be used for measuring contents at a particular level. In turn, the rule or stick can be graduated to serve as a volume gauge for the tank in question. The only requirements are: that the cross section of the tank is circular; the tank's dimensions are known; the gauge rod is inserted vertically through the top center of the tank so that it rests on the exact bottom of the tank; and that consistent English or metric units are used throughout the calculations.

(1) Tank Constant $= K = Cr^2L$ (remains the same for any given tank); (2) For a tank that is completely full: $V_T = \pi K$; (3) $V_s = KA$; (4) $V = V_s$, when tank is less than half full; (5) $V = V_T - V_s = V_T - KA$, when tank is more than half full.

Where

$C$ = liquid volume conversion factor, the exact value of which depends on the length and liquid volume units being used during measurement: 0.00433 U.S. gal/in.³; 7.48 U.S. gal/ft³; 0.00360 U.K. gal/in.³; 6.23 U.K. gal/ft³; 0.001 litres/cm³; or 1000 litres/m³;

$V_T$ = total volume of liquid tank can hold;

$V_s$ = volume formed by segment of circle having depth $= x$ in given tank (see diagram);

$V$ = volume of liquid at particular level in tank;

$d$ = diameter of tank; $L$ = length of tank; $r$ = radius of tank ($= \frac{1}{2}$ diameter);

$A$ = segment area of a corresponding unit circle taken from pages 72 or 73;

$y$ = actual depth of contents in tank as shown on a gauge rod or stick;

$x$ = depth of the segment of a circle to be considered in given tank. As can be seen in above diagram, $x$ is the actual depth of contents ($y$) when the tank is less than half full, and is the depth of the void ($d - y$) above the contents when tank is more than half full. From pages 72 and 73 it can also be seen that $h$, the height of a segment of a corresponding unit circle, is $x/r$.

*Example:* A tank is 20 feet long and 6 feet in diameter. Convert a long inch-stick into a gauge that is graduated at 1000 and 3000 U.S. gallons.

$$L = 20 \times 12 = 240 \text{ in.}; \quad r = 6/2 \times 12 = 36 \text{ in.}$$

From formula (1): $K = 0.00433(36)^2(240) = 1347$.

From formula (2): $V_T = 3.142 \times 1347 = 4232$ U.S. gal.

The 72-inch mark from the bottom on the inch-stick can be graduated for the rounded full volume "4230"; and the halfway point 36″ for 4230/2 or "2115." It can be seen that the 1000 gal mark would be below the halfway mark. From formulas (3) and (4):

$$A_{1000} = \frac{1000}{1347} = 0.7424; \text{ from page 73, } h \text{ can be interpolated as } 0.5724*; \text{ and}$$

$x = y = 36 \times 0.5724 = 20.61$.

Therefore, 1000 gal mark is graduated 20⅝″ from bottom of rod.

It can be seen that the 3000 mark would be above the halfway mark. Therefore, the circular segment considered is the cross section of the void space at the top of the tank. From formulas (3) and (5):

$$A_{3000} = \frac{4230 - 3000}{1347} = 0.9131; \quad h = 0.6648; \quad x = 36 \times 0.6648 = 23.93''.$$

Therefore, 3000 gal mark is 72.00 − 23.93 = 48.07, or at the 48¹⁄₁₆″ mark from the bottom.

---

* If the desired level of accuracy permits, interpolation can be omitted by choosing $h$ directly from the table for the value of $A$ nearest that calculated above.

## SOLUTION OF TRIANGLES

Any figure bounded by three straight lines is called a triangle. Any one of the three lines may be called the base, and the line drawn from the angle opposite the base at right angles to it is called the height or altitude of the triangle.

If all the three sides of a triangle are of equal length, the triangle is called *equilateral.* Each one of the three angles in an equilateral triangle equals 60 degrees. If two sides are of equal length, the triangle is an *isosceles* triangle. If one angle is a *right* or 90-degree angle, the triangle is a *right* or *right-angled* triangle. The side opposite the right angle is called the *hypotenuse.*

If all the angles are less than 90 degrees, the triangle is called an *acute* or *acute-angled* triangle. If one of the angles is larger than 90 degrees, the triangle is called an *obtuse-angled* triangle. Both acute and obtuse-angled triangles are known under the common name of *oblique-angled* triangles. The sum of the three angles in every triangle is 180 degrees.

The sides and angles of any triangle which are not known can be found when: 1. All the three sides; 2. Two sides and one angle; or, 3. One side and two angles, are given. In other words, if a triangle is considered as consisting of six parts, three angles and three sides, the unknown parts can be determined when any three parts are given, provided at least one of the given parts is a side.

**Functions of Angles.** — The functions of angles used in solving triangles are sine, cosine, tangent, cotangent, secant, and cosecant. These expressions are usually abbreviated as follows:

| | |
|---|---|
| sin = sine, | cot = cotangent, |
| cos = cosine, | sec = secant, |
| tan = tangent, | cosec = cosecant. |

If in a right-angled triangle (see the illustration in the table below), the lengths of the three sides are represented by $a$, $b$ and $c$, and the angles opposite each of these sides by $A$, $B$ and $C$, then the side $c$ opposite the right angle is the hypotenuse;

### Trigonometrical Functions of Angles

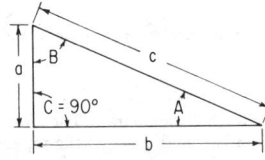

The *sine* of an angle equals the opposite side divided by the hypotenuse. Hence, $\sin B = b \div c$, and $\sin A = a \div c$.

The *cosine* of an angle equals the adjacent side divided by the hypotenuse. Hence, $\cos B = a \div c$, and $\cos A = b \div c$.

The *tangent* of an angle equals the opposite side divided by the adjacent side. Hence, $\tan B = b \div a$, and $\tan A = a \div b$.

The *cotangent* of an angle equals the adjacent side divided by the opposite side. Hence, $\cot B = a \div b$, and $\cot A = b \div a$.

The *secant* of an angle equals the hypotenuse divided by the adjacent side. Hence, $\sec B = c \div a$, and $\sec A = c \div b$.

The *cosecant* of an angle equals the hypotenuse divided by the opposite side. Hence, $\operatorname{cosec} B = c \div b$, and $\operatorname{cosec} A = c \div a$.

It should be noted that the functions of the angles can be found in this manner only when the triangle is right-angled.

side $b$ is called the *side adjacent* to angle $A$ and is also the *side opposite* to angle $B$; side $a$ is the side adjacent to angle $B$ and the side opposite to angle $A$  The meanings of the various functions of angles can be explained by the aid of a right-angled triangle.

The following relation exists between the angular functions of the two acute angles in a right-angled triangle:  The sine of angle $B$ equals the cosine of angle $A$; the tangent of angle $B$ equals the cotangent of angle $A$, and *vice versa*.  The sum of the two acute angles in a right-angled triangle always equals 90 degrees; hence, when one angle is known, the other can easily be found.  When any two angles together make 90 degrees, one is called the *complement* of the other, and in that case the sine of the one equals the cosine of the other, and the tangent of the one equals the cotangent of the other.

On page 177 a diagram, "Signs of Trigonometric Functions," is given. This diagram shows the proper sign (+ or −) for the trigonometric functions of angles in each of the four quadrants, o to 90, 90 to 180, 180 to 270, and 270 to 360 degrees. Thus, the cosine of an angle between 90 and 180 degrees is negative; the sine of the same angle is positive.

**The Law of Sines.** — In a triangle, any side is to any other side as the sine of the angle opposite the first side is to the sine of the angle opposite the other side; or, if $a$ and $b$ be the sides, and $A$ and $B$ the angles opposite them:

$$\frac{a}{b} = \frac{\sin A}{\sin B}$$

**The Law of Cosines.** — In a triangle, the square of any side is equal to the sum of the squares of the other two sides minus twice their product times the cosine of the included angle; or if $a$, $b$ and $c$ be the sides and the angle opposite side $a$ be denoted $A$, then:

$$a^2 = b^2 + c^2 - 2\,bc \cos A$$

These two laws, together with the proposition that the sum of the three angles equals 180 degrees, are the basis of all formulas relating to the solution of triangles.
Formulas for the solution of right-angled and oblique-angled triangles, arranged in tabular form, are given on the following pages.

**Use of Tables of Squares in Solving Right-angled Triangles.** — The tables of squares at the beginning of the book may be used to advantage in solving right-angled triangles.  Assume that the sides including the right angle are known, and that they are 1⅝ and 1⅞ inch, respectively.  Find the side opposite the right angle.

$$\text{Side to be found} = \sqrt{1.625^2 + 1.875^2}$$

From tables of squares:
$$1.625^2 = 2.640625$$
$$1.875^2 = 3.515625$$
$$6.156250$$

By looking up the figures 615 in the number column in the tables, and finding the square root, we get the figures 24.799. The square root of 616 is 24.819. Hence, by estimating, the square root of 615.625 = 24.812. As we want the square root of 6.15625, move the decimal point in the root one step to the left; then $\sqrt{6.15625} = 2.4812$. This is the length of the side opposite the right angle.

### Solution of Right-angled Triangles

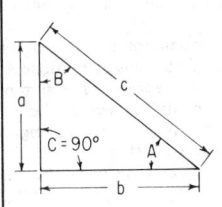

As shown in the illustration, the sides of the right-angled triangle are designated $a$ and $b$ and the hypotenuse, $c$. The angles opposite each of these sides are designated $A$ and $B$ respectively.

Angle $C$, opposite the hypotenuse $c$ is the right angle, and is therefore always one of the known quantities.

| Sides and Angles Known | Formulas for Sides and Angles to be Found | | |
|---|---|---|---|
| Side $a$; side $b$ .......... | $c = \sqrt{a^2 + b^2}$ | $\tan A = \dfrac{a}{b}$ | $B = 90° - A$ |
| Side $a$; hypotenuse $c$ ... | $b = \sqrt{c^2 - a^2}$ | $\sin A = \dfrac{a}{c}$ | $B = 90° - A$ |
| Side $b$; hypotenuse $c$ ... | $a = \sqrt{c^2 - b^2}$ | $\sin B = \dfrac{b}{c}$ | $A = 90° - B$ |
| Hypotenuse $c$; angle $B$ .. | $b = c \times \sin B$ | $a = c \times \cos B$ | $A = 90° - B$ |
| Hypotenuse $c$; angle $A$ .. | $b = c \times \cos A$ | $a = c \times \sin A$ | $B = 90° - A$ |
| Side $b$; angle $B$ ......... | $c = \dfrac{b}{\sin B}$ | $a = b \times \cot B$ | $A = 90° - B$ |
| Side $b$; angle $A$ ......... | $c = \dfrac{b}{\cos A}$ | $a = b \times \tan A$ | $B = 90° - A$ |
| Side $a$; angle $B$ ......... | $c = \dfrac{a}{\cos B}$ | $b = a \times \tan B$ | $A = 90° - B$ |
| Side $a$; angle $A$ ......... | $c = \dfrac{a}{\sin A}$ | $b = a \times \cot A$ | $B = 90° - A$ |

## Examples of the Solution of Right-angled Triangles

(English and metric units)

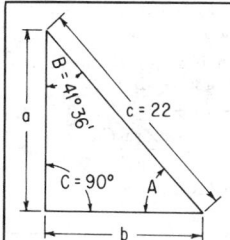

Hypotenuse and one angle known:

$$c = 22 \text{ inches}; \quad B = 41° 36'.$$

Then, by the formulas given on the preceding page:

$$a = c \times \cos B = 22 \times \cos 41° 36' = 22 \times 0.74780$$
$$= 16.4516 \text{ inches.}$$

$$b = c \times \sin B = 22 \times \sin 41° 36' = 22 \times 0.66393$$
$$= 14.6065 \text{ inches.}$$

$$A = 90° - B = 90° - 41° 36' = 48° 24'.$$

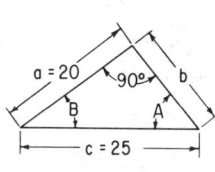

Hypotenuse and one side known:

$$c = 25 \text{ centimetres}; \quad a = 20 \text{ centimetres.}$$

From the formulas on the preceding page:

$$b = \sqrt{c^2 - a^2} = \sqrt{25^2 - 20^2} = \sqrt{625 - 400}$$
$$= \sqrt{225} = 15 \text{ centimetres.}$$

$$\sin A = \frac{a}{c} = \frac{20}{25} = 0.8$$

Hence, $A = 53° 8'.$
$$B = 90° - A = 90° - 53° 8' = 36° 52'.$$

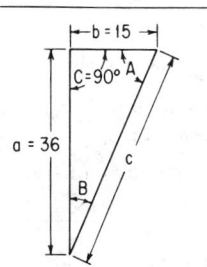

Two sides known:

$$a = 36 \text{ inches}; \quad b = 15 \text{ inches.}$$

Then, by the formulas given on the preceding page:

$$c = \sqrt{a^2 + b^2} = \sqrt{36^2 + 15^2} = \sqrt{1296 + 225}$$
$$= \sqrt{1521} = 39 \text{ inches.}$$

$$\tan A = \frac{a}{b} = \frac{36}{15} = 2.4$$

Hence, $A = 67° 23'.$
$$B = 90° - A = 90° - 67° 23' = 22° 37'.$$

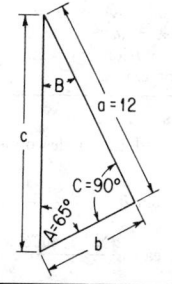

One side and one angle known:

$$a = 12 \text{ metres}; \quad A = 65°.$$

Then, by the formulas given on the preceding page:

$$c = \frac{a}{\sin A} = \frac{12}{\sin 65°} = \frac{12}{0.90631} = 13.2405 \text{ metres.}$$

$$b = a \times \cot A = 12 \times \cot 65° = 12 \times 0.46631$$
$$= 5.5957 \text{ metres.}$$

$$B = 90° - A = 90° - 65° = 25°.$$

### Solution of Oblique-angled Triangles

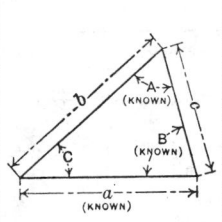

#### One side and two angles known.

Call the known side $a$, the angle opposite it $A$, and the other known angle $B$. Then:

$C = 180° - (A + B)$; or if angles $B$ and $C$ are given, but not $A$, then $A = 180° - (B + C)$.

$$C = 180° - (A + B)$$

$$b = \frac{a \times \sin B}{\sin A} \qquad c = \frac{a \times \sin C}{\sin A}$$

$$\text{Area} = \frac{a \times b \times \sin C}{2}$$

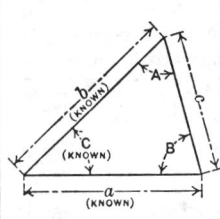

#### Two sides and the angle between them known.

Call the known sides $a$ and $b$, and the known angle between them $C$. Then:

$$\tan A = \frac{a \times \sin C}{b - (a \times \cos C)}$$

$$B = 180° - (A + C) \qquad c = \frac{a \times \sin C}{\sin A}$$

Side $c$ may also be found directly as below:

$$c = \sqrt{a^2 + b^2 - (2\,ab \times \cos C)}$$

$$\text{Area} = \frac{a \times b \times \sin C}{2}$$

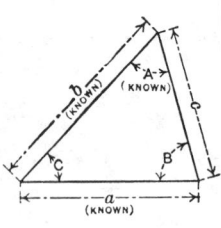

#### Two sides and the angle opposite one of the sides known.

Call the known angle $A$, the side opposite it $a$, and the other known side $b$. Then:

$$\sin B = \frac{b \times \sin A}{a} \qquad C = 180° - (A + B)$$

$$c = \frac{a \times \sin C}{\sin A} \qquad \text{Area} = \frac{a \times b \times \sin C}{2}$$

If in the above, angle $B >$ angle $A$ but $<90°$, then a second solution $B_2$, $C_2$, $c_2$ exists for which: $B_2 = 180° - B$; $C_2 = 180° - (A + B_2)$; $c_2 = (a \times \sin C_2) \div \sin A$; Area $= (a \times b \times \sin C_2) \div 2$. If $a \geqq b$, then the first solution only exists. If $a < b \times \sin A$, then no solution exists.

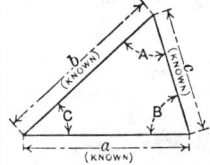

#### All three sides known.

Call the sides $a$, $b$ and $c$, and the angles opposite them, $A$, $B$ and $C$. Then:

$$\cos A = \frac{b^2 + c^2 - a^2}{2\,bc} \qquad \sin B = \frac{b \times \sin A}{a}$$

$$C = 180° - (A + B) \qquad \text{Area} = \frac{a \times b \times \sin C}{2}$$

**Examples of the Solution of Oblique-angled Triangles**

(English and metric units)

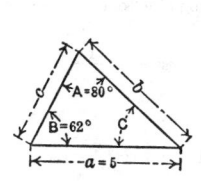

Sides and angles known:

$$a = 5 \text{ centimetres}; A = 80°; B = 62°$$

Then, by the formulas on the opposite page:

$$C = 180° - (80° + 62°) = 180° - 142° = 38°.$$

$$b = \frac{a \times \sin B}{\sin A} = \frac{5 \times \sin 62°}{\sin 80°} = \frac{5 \times 0.88295}{0.98481} = 4.483$$

centimetres.

$$c = \frac{a \times \sin C}{\sin A} = \frac{5 \times \sin 38°}{\sin 80°} = \frac{5 \times 0.61566}{0.98481} = 3.126$$

centimetres.

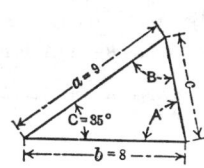

Sides and angles known:

$$a = 9 \text{ inches}; b = 8 \text{ inches}; C = 35°.$$

$$\tan A = \frac{a \times \sin C}{b - (a \times \cos C)} = \frac{9 \times \sin 35°}{8 - (9 \times \cos 35°)}$$

$$= \frac{9 \times 0.57358}{8 - (9 \times 0.81915)} = \frac{5.16222}{0.62765} = 8.22468.$$

Hence,          $A = 83° 4'.$

$$B = 180° - (A + C) = 180° - 118° 4' = 61° 56'.$$

$$c = \frac{a \times \sin C}{\sin A} = \frac{9 \times 0.57358}{0.99269} = 5.2 \text{ inches}.$$

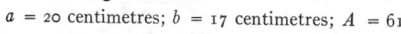

Sides and angles known:

$$a = 20 \text{ centimetres}; b = 17 \text{ centimetres}; A = 61°.$$

$$\sin B = \frac{b \times \sin A}{a} = \frac{17 \times \sin 61°}{20}$$

$$= \frac{17 \times 0.87462}{20} = 0.74343.$$

Hence,          $B = 48° 1'.$

$$C = 180° - (A + B) = 180° - 109° 1' = 70° 59'.$$

$$c = \frac{a \times \sin C}{\sin A} = \frac{20 \times \sin 70° 59'}{\sin 61°} = \frac{20 \times 0.94542}{0.87462}$$

$$= 21.62 \text{ centimetres}.$$

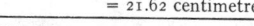

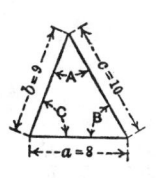

Sides known:

$$a = 8 \text{ inches}; b = 9 \text{ inches}; c = 10 \text{ inches}.$$

$$\cos A = \frac{b^2 + c^2 - a^2}{2 bc} = \frac{9^2 + 10^2 - 8^2}{2 \times 9 \times 10}$$

$$= \frac{81 + 100 - 64}{180} = \frac{117}{180} = 0.65000.$$

Hence,          $A = 49° 27'.$

$$\sin B = \frac{b \times \sin A}{a} = \frac{9 \times 0.75984}{8} = 0.85482.$$

Hence,          $B = 58° 44'.$

$$C = 180° - (A + B) = 180° - 108° 11' = 71° 49'.$$

**Trigonometric Identities.** — Trigonometric identities are formulas that show the relationship between different trigonometric functions. They may be used to change the form of some trigonometric expressions to simplify calculations. For example, if a formula has a term, $2 \sin A \cos A$, the equivalent but simpler term $\sin 2A$ may be substituted. The identities given below may themselves be combined or rearranged in various ways to form new identities.

1. *Basic:*    $\tan A = \dfrac{\sin A}{\cos A} = \dfrac{1}{\cot A}$    $\sec A = \dfrac{1}{\cos A}$    $\csc A = \dfrac{1}{\sin A}$

2. *Negative-Angle:*    $\sin -A = -\sin A$    $\cos -A = \cos A$    $\tan -A = -\tan A$

3. *Pythagorean:*    $\sin^2 A + \cos^2 A = 1$    $1 + \tan^2 A = \sec^2 A$    $1 + \cot^2 A = \csc^2 A$

4. *Sum and Difference of Angles:*

$$\tan (A + B) = \frac{\tan A + \tan B}{1 - \tan A \tan B} \qquad \cot (A + B) = \frac{\cot A \cot B - 1}{\cot B + \cot A}$$

$$\tan (A - B) = \frac{\tan A - \tan B}{1 + \tan A \tan B} \qquad \cot (A - B) = \frac{\cot A \cot B + 1}{\cot B - \cot A}$$

$$\sin (A + B) = \sin A \cos B + \cos A \sin B \qquad \cos (A + B) = \cos A \cos B - \sin A \sin B$$

$$\sin (A - B) = \sin A \cos B - \cos A \sin B \qquad \cos (A - B) = \cos A \cos B + \sin A \sin B$$

5. *Double-Angle:*    $\cos 2A = \cos^2 A - \sin^2 A = 2 \cos^2 A - 1 = 1 - 2 \sin^2 A$

$$\sin 2A = 2 \sin A \cos A \qquad \tan 2A = \frac{2 \tan A}{1 - \tan^2 A} = \frac{2}{\cot A - \tan A}$$

6. *Half-Angle:*    $\sin \frac{1}{2}A = \sqrt{\frac{1}{2}(1 - \cos A)}$    $\cos \frac{1}{2}A = \sqrt{\frac{1}{2}(1 + \cos A)}$

$$\tan \frac{1}{2}A = \sqrt{\frac{1 - \cos A}{1 + \cos A}} = \frac{1 - \cos A}{\sin A} = \frac{\sin A}{1 + \cos A}$$

7. *Product-to-Sum:*

$$\sin A \cos B = \frac{1}{2}[\sin (A + B) + \sin (A - B)]$$

$$\cos A \cos B = \frac{1}{2}[\cos (A + B) + \cos (A - B)]$$

$$\sin A \sin B = \frac{1}{2}[\cos (A - B) - \cos (A + B)]$$

$$\tan A \tan B = \frac{\tan A + \tan B}{\cot A + \cot B}$$

8. *Sum and Difference of Functions:*

$$\sin A + \sin B = 2[\sin \tfrac{1}{2}(A + B) \cos \tfrac{1}{2}(A - B)]$$

$$\sin A - \sin B = 2[\sin \tfrac{1}{2}(A - B) \cos \tfrac{1}{2}(A + B)]$$

$$\cos A + \cos B = 2[\cos \tfrac{1}{2}(A + B) \cos \tfrac{1}{2}(A - B)]$$

$$\cos A - \cos B = -2[\sin \tfrac{1}{2}(A + B) \sin \tfrac{1}{2}(A - B)]$$

$$\tan A + \tan B = \frac{\sin (A + B)}{\cos A \cos B} \qquad \cot A + \cot B = \frac{\sin (B + A)}{\sin A \sin B}$$

$$\tan A - \tan B = \frac{\sin (A - B)}{\cos A \cos B} \qquad \cot A - \cot B = \frac{\sin (B - A)}{\sin A \sin B}$$

**Tables of Trigonometric Functions.** — The numerical values for the natural or trigonometric functions for all angles from 0 to 360 degrees are given in the tables, pages 178 to 222. The chart below shows how to enter the table.

### How to Enter Table of Trigonometric Functions

| For Angles from | Enter Table for | | For Angles from | Enter Table for | |
|---|---|---|---|---|---|
| | Degrees and Function | Minutes | | Degrees and Function | Minutes |
| 0° to 45° | at top | at left | 180° to 225° | at top | at left |
| 45° to 90° | at bottom | at right | 225° to 270° | at bottom | at right |
| 90° to 135° | at bottom | at left | 270° to 315° | at bottom | at left |
| 135° to 180° | at top | at right | 315° to 360° | at top | at right |

*Examples:* The sine of 26° is 0.43837; of 126°, 0.80902; of 226°, −0.71934.

### Signs of Trigonometric Functions

This diagram shows the proper sign (+ or −) for the trigonometric functions of angles in each of the four quadrants of a complete circle.
*Examples:* The sine of 226° is −0.71934; of 326°, −0.55919.

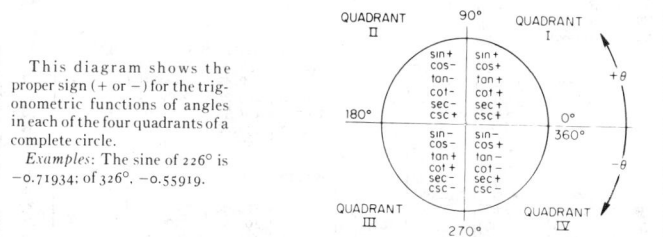

### Useful Relationships Among Angles

| Angle Function | $\theta$ | $-\theta$ | $90° \pm \theta$ | $180° \pm \theta$ | $270° \pm \theta$ | $360° \pm \theta$ |
|---|---|---|---|---|---|---|
| sin | $\sin \theta$ | $-\sin \theta$ | $+\cos \theta$ | $\mp \sin \theta$ | $-\cos \theta$ | $\pm \sin \theta$ |
| cos | $\cos \theta$ | $+\cos \theta$ | $\mp \sin \theta$ | $-\cos \theta$ | $\pm \sin \theta$ | $+\cos \theta$ |
| tan | $\tan \theta$ | $-\tan \theta$ | $\mp \cot \theta$ | $\pm \tan \theta$ | $\mp \cot \theta$ | $\pm \tan \theta$ |
| cot | $\cot \theta$ | $-\cot \theta$ | $\mp \tan \theta$ | $\pm \cot \theta$ | $\mp \tan \theta$ | $\pm \cot \theta$ |
| sec | $\sec \theta$ | $+\sec \theta$ | $\mp \csc \theta$ | $-\sec \theta$ | $\pm \csc \theta$ | $+\sec \theta$ |
| csc | $\csc \theta$ | $-\csc \theta$ | $+\sec \theta$ | $\mp \csc \theta$ | $-\sec \theta$ | $\pm \csc \theta$ |

*Examples:* $\cos 270° - \theta = -\sin \theta$; $\tan 90° + \theta = -\cot \theta$.

**Involute Functions.** — Involute functions are used in certain formulas relating to the design and measurement of gear teeth as well as measurement of threads over wires. Included in the trigonometric tables, pages 178 to 222, are values for the involute functions of angles of from 0 to 90 degrees.

**Sevolute Functions.** — Sevolute functions are used in certain spline calculations. They may be computed by subtracting the involute of an angle from the secant of the angle. Thus, sevolute 20° = sec 20° − inv 20° = 1.0642 − 0.014904 = 1.0493.

**Versed Sine and Versed Cosine.** — These functions are sometimes used in formulas for segments of a circle and may be obtained by using the trigonometric tables together with the relationships: versed sine $\theta$ = 1 − $\cos \theta$; and versed cosine $\theta$ = 1 − $\sin \theta$.

**0° or 180°**      Trigonometric Functions      **179° or 359°**

| M | Sine | Cosine | Tan. | Cotan. | Secant | Cosec. | Involute 0°–1° | READ UP | M |
|---|------|--------|------|--------|--------|--------|----------------|---------|---|
| 0 | 0.00000 | 1.0000 | 0.00000 | Infinite | 1.0000 | Infinite | .0000000 | Infinite | 60 |
| 1 | .00029 | .0000 | .00029 | 3437.7 | .0000 | 3437.7 | .0000000 | 3436.2 | 59 |
| 2 | .00058 | .0000 | .00058 | 1718.9 | .0000 | 1718.9 | .0000000 | 1717.3 | 58 |
| 3 | .00087 | .0000 | .00087 | 1145.9 | .0000 | 1145.9 | .0000000 | 1144.3 | 57 |
| 4 | .00116 | .0000 | .00116 | 859.44 | .0000 | 859.44 | .0000000 | 857.87 | 56 |
| 5 | 0.00145 | 1.0000 | 0.00145 | 687.55 | 1.0000 | 687.55 | .0000000 | 685.98 | 55 |
| 6 | .00175 | .0000 | .00175 | 572.96 | .0000 | 572.96 | .0000000 | 571.39 | 54 |
| 7 | .00204 | .0000 | .00204 | 491.11 | .0000 | 491.11 | .0000000 | 489.54 | 53 |
| 8 | .00233 | .0000 | .00233 | 429.72 | .0000 | 429.72 | .0000000 | 428.15 | 52 |
| 9 | .00262 | .0000 | .00262 | 381.97 | .0000 | 381.97 | .0000000 | 380.40 | 51 |
| 10 | 0.00291 | 1.00000 | 0.00291 | 343.77 | 1.0000 | 343.78 | .0000000 | 342.21 | 50 |
| 11 | .00320 | .99999 | .00320 | 312.52 | .0000 | 312.52 | .0000000 | 310.95 | 49 |
| 12 | .00349 | .99999 | .00349 | 286.48 | .0000 | 286.48 | .0000000 | 284.91 | 48 |
| 13 | .00378 | .99999 | .00378 | 264.44 | .0000 | 264.44 | .0900000 | 262.87 | 47 |
| 14 | .00407 | .99999 | .00407 | 245.55 | .0000 | 245.55 | .0000000 | 243.99 | 46 |
| 15 | 0.00436 | 0.99999 | 0.00436 | 229.18 | 1.0000 | 229.18 | .0000000 | 227.62 | 45 |
| 16 | .00465 | .99999 | .00465 | 214.86 | .0000 | 214.86 | .0000000 | 213.29 | 44 |
| 17 | .00495 | .99999 | .00495 | 202.22 | .0000 | 202.22 | .0000000 | 200.65 | 43 |
| 18 | .00524 | .99999 | .00524 | 190.98 | .0000 | 190.99 | .0000000 | 189.42 | 42 |
| 19 | .00553 | .99998 | .00553 | 180.93 | .0000 | 180.93 | .0000001 | 179.37 | 41 |
| 20 | 0.00582 | 0.99998 | 0.00582 | 171.89 | 1.0000 | 171.89 | .0000001 | 170.32 | 40 |
| 21 | .00611 | .99998 | .00611 | 163.70 | .0000 | 163.70 | .0000001 | 162.14 | 39 |
| 22 | .00640 | .99998 | .00640 | 156.26 | .0000 | 156.26 | .0000001 | 154.69 | 38 |
| 23 | .00669 | .99998 | .00669 | 149.47 | .0000 | 149.47 | .0000001 | 147.90 | 37 |
| 24 | .00698 | .99998 | .00698 | 143.24 | .0000 | 143.24 | .0000001 | 141.67 | 36 |
| 25 | 0.00727 | 0.99997 | 0.00727 | 137.51 | 1.0000 | 137.51 | .0000001 | 135.94 | 35 |
| 26 | .00756 | .99997 | .00756 | 132.22 | .0000 | 132.22 | .0000001 | 130.66 | 34 |
| 27 | .00785 | .99997 | .00785 | 127.32 | .0000 | 127.33 | .0000002 | 125.76 | 33 |
| 28 | .00814 | .99997 | .00815 | 122.77 | .0000 | 122.78 | .0000002 | 121.21 | 32 |
| 29 | .00844 | .99996 | .00844 | 118.54 | .0000 | 118.54 | .0000002 | 116.98 | 31 |
| 30 | 0.00873 | 0.99996 | 0.00873 | 114.59 | 1.0000 | 114.59 | .0000002 | 113.03 | 30 |
| 31 | .00902 | .99996 | .00902 | 110.89 | .0000 | 110.90 | .0000002 | 109.33 | 29 |
| 32 | .00931 | .99996 | .00931 | 107.43 | .0000 | 107.43 | .0000003 | 105.86 | 28 |
| 33 | .00960 | .99995 | .00960 | 104.17 | .0000 | 104.18 | .0000003 | 102.61 | 27 |
| 34 | .00989 | .99995 | .00989 | 101.11 | .0000 | 101.11 | .0000003 | 99.546 | 26 |
| 35 | 0.01018 | 0.99995 | 0.01018 | 98.218 | 1.0001 | 98.223 | .0000004 | 96.657 | 25 |
| 36 | .01047 | .99995 | .01047 | 95.489 | .0001 | 95.495 | .0000004 | 93.929 | 24 |
| 37 | .01076 | .99994 | .01076 | 92.908 | .0001 | 92.914 | .0000004 | 91.348 | 23 |
| 38 | .01105 | .99994 | .01105 | 90.463 | .0001 | 90.469 | .0000005 | 88.904 | 22 |
| 39 | .01134 | .99994 | .01135 | 88.144 | .0001 | 88.149 | .0000005 | 86.584 | 21 |
| 40 | 0.01164 | 0.99993 | 0.01164 | 85.940 | 1.0001 | 85.946 | .0000006 | 84.381 | 20 |
| 41 | .01193 | .99993 | .01193 | 83.844 | .0001 | 83.849 | .0000006 | 82.285 | 19 |
| 42 | .01222 | .99993 | .01222 | 81.847 | .0001 | 81.853 | .0000006 | 80.288 | 18 |
| 43 | .01251 | .99992 | .01251 | 79.943 | .0001 | 79.950 | .0000007 | 78.385 | 17 |
| 44 | .01280 | .99992 | .01280 | 78.126 | .0001 | 78.133 | .0000007 | 76.568 | 16 |
| 45 | 0.01309 | 0.99991 | 0.01309 | 76.390 | 1.0001 | 76.397 | .0000007 | 74.832 | 15 |
| 46 | .01338 | .99991 | .01338 | 74.729 | .0001 | 74.736 | .0000008 | 73.172 | 14 |
| 47 | .01367 | .99991 | .01367 | 73.139 | .0001 | 73.146 | .0000009 | 71.582 | 13 |
| 48 | .01396 | .99990 | .01396 | 71.615 | .0001 | 71.622 | .0000009 | 70.058 | 12 |
| 49 | .01425 | .99990 | .01425 | 70.153 | .0001 | 70.160 | .0000010 | 68.597 | 11 |
| 50 | 0.01454 | 0.99989 | 0.01455 | 68.750 | 1.0001 | 68.757 | .0000010 | 67.194 | 10 |
| 51 | .01483 | .99989 | .01484 | 67.402 | .0001 | 67.409 | .0000011 | 65.846 | 9 |
| 52 | .01513 | .99989 | .01513 | 66.105 | .0001 | 66.113 | .0000012 | 64.550 | 8 |
| 53 | .01542 | .99988 | .01542 | 64.858 | .0001 | 64.866 | .0000012 | 63.303 | 7 |
| 54 | .01571 | .99988 | .01571 | 63.657 | .0001 | 63.665 | .0000013 | 62.102 | 6 |
| 55 | 0.01600 | 0.99987 | 0.01600 | 62.499 | 1.0001 | 62.507 | .0000014 | 60.944 | 5 |
| 56 | .01629 | .99987 | .01629 | 61.383 | .0001 | 61.391 | .0000014 | 59.828 | 4 |
| 57 | .01658 | .99986 | .01658 | 60.306 | .0001 | 60.314 | .0000015 | 58.752 | 3 |
| 58 | .01687 | .99986 | .01687 | 59.266 | .0001 | 59.274 | .0000016 | 57.712 | 2 |
| 59 | .01716 | .99985 | .01716 | 58.261 | .0001 | 58.270 | .0000017 | 56.708 | 1 |
| 60 | 0.01745 | 0.99985 | 0.01746 | 57.290 | 1.0002 | 57.299 | .0000018 | 55.737 | 0 |
| M | Cosine | Sine | Cotan. | Tan. | Cosec. | Secant | READ DOWN | 89°–90° Involute | M |

**90° or 270°**          **89° or 269°**

**1° or 181°**        Trigonometric Functions        **178° or 358°**

| M | Sine | Cosine | Tan. | Cotan. | Secant | Cosec. | Involute 1°–2° | READ UP | M |
|---|------|--------|------|--------|--------|--------|---------|---------|---|
| 0 | 0.01745 | 0.99985 | 0.01746 | 57.290 | 1.0002 | 57.299 | .0000018 | 55.737 | 60 |
| 1 | .01774 | .99984 | .01775 | 56.351 | .0002 | 56.359 | .0000019 | 54.798 | 59 |
| 2 | .01803 | .99984 | .01804 | 55.442 | .0002 | 55.451 | .0000020 | 53.889 | 58 |
| 3 | .01832 | .99983 | .01833 | 54.561 | .0002 | 54.570 | .0000021 | 53.009 | 57 |
| 4 | .01862 | .99983 | .01862 | 53.709 | .0002 | 53.718 | .0000022 | 52.156 | 56 |
| 5 | 0.01891 | 0.99982 | 0.01891 | 52.882 | 1.0002 | 52.892 | .0000023 | 51.330 | 55 |
| 6 | .01920 | .99982 | .01920 | 52.081 | .0002 | 52.090 | .0000024 | 50.529 | 54 |
| 7 | .01949 | .99981 | .01949 | 51.303 | .0002 | 51.313 | .0000025 | 49.752 | 53 |
| 8 | .01978 | .99980 | .01978 | 50.549 | .0002 | 50.558 | .0000026 | 48.997 | 52 |
| 9 | .02007 | .99980 | .02007 | 49.816 | .0002 | 49.826 | .0000027 | 48.265 | 51 |
| 10 | 0.02036 | 0.99979 | 0.02036 | 49.104 | 1.0002 | 49.114 | .0000028 | 47.553 | 50 |
| 11 | .02065 | .99979 | .02066 | 48.412 | .0002 | 48.422 | .0000029 | 46.862 | 49 |
| 12 | .02094 | .99978 | .02095 | 47.740 | .0002 | 47.750 | .0000031 | 46.190 | 48 |
| 13 | .02123 | .99977 | .02124 | 47.085 | .0002 | 47.096 | .0000032 | 45.536 | 47 |
| 14 | .02152 | .99977 | .02153 | 46.449 | .0002 | 46.460 | .0000033 | 44.900 | 46 |
| 15 | 0.02181 | 0.99976 | 0.02182 | 45.829 | 1.0002 | 45.840 | .0000035 | 44.280 | 45 |
| 16 | .02211 | .99976 | .02211 | 45.226 | .0002 | 45.237 | .0000036 | 43.677 | 44 |
| 17 | .02240 | .99975 | .02240 | 44.639 | .0003 | 44.650 | .0000037 | 43.090 | 43 |
| 18 | .02269 | .99974 | .02269 | 44.066 | .0003 | 44.077 | .0000039 | 42.518 | 42 |
| 19 | .02298 | .99974 | .02298 | 43.508 | .0003 | 43.520 | .0000040 | 41.960 | 41 |
| 20 | 0.02327 | 0.99973 | 0.02328 | 42.964 | 1.0003 | 42.976 | .0000042 | 41.417 | 40 |
| 21 | .02356 | .99972 | .02357 | 42.433 | .0003 | 42.445 | .0000044 | 40.886 | 39 |
| 22 | .02385 | .99972 | .02386 | 41.916 | .0003 | 41.928 | .0000045 | 40.369 | 38 |
| 23 | .02414 | .99971 | .02415 | 41.411 | .0003 | 41.423 | .0000047 | 39.864 | 37 |
| 24 | .02443 | .99970 | .02444 | 40.917 | .0003 | 40.930 | .0000049 | 39.371 | 36 |
| 25 | 0.02472 | 0.99969 | 0.02473 | 40.436 | 1.0003 | 40.448 | .0000050 | 38.890 | 35 |
| 26 | .02501 | .99969 | .02502 | 39.965 | .0003 | 39.978 | .0000052 | 38.420 | 34 |
| 27 | .02530 | .99968 | .02531 | 39.506 | .0003 | 39.519 | .0000054 | 37.960 | 33 |
| 28 | .02560 | .99967 | .02560 | 39.057 | .0003 | 39.070 | .0000056 | 37.512 | 32 |
| 29 | .02589 | .99966 | .02589 | 38.618 | .0003 | 38.631 | .0000058 | 37.073 | 31 |
| 30 | 0.02618 | 0.99966 | 0.02619 | 38.188 | 1.0003 | 38.202 | .0000060 | 36.644 | 30 |
| 31 | .02647 | .99965 | .02648 | 37.769 | .0004 | 37.782 | .0000062 | 36.224 | 29 |
| 32 | .02676 | .99964 | .02677 | 37.358 | .0004 | 37.371 | .0000064 | 35.814 | 28 |
| 33 | .02705 | .99963 | .02706 | 36.956 | .0004 | 36.970 | .0000066 | 35.412 | 27 |
| 34 | .02734 | .99963 | .02735 | 36.563 | .0004 | 36.576 | .0000068 | 35.019 | 26 |
| 35 | 0.02763 | 0.99962 | 0.02764 | 36.178 | 1.0004 | 36.191 | .0000070 | 34.634 | 25 |
| 36 | .02792 | .99961 | .02793 | 35.801 | .0004 | 35.815 | .0000073 | 34.258 | 24 |
| 37 | .02821 | .99960 | .02822 | 35.431 | .0004 | 35.445 | .0000075 | 33.889 | 23 |
| 38 | .02850 | .99959 | .02851 | 35.070 | .0004 | 35.084 | .0000077 | 33.527 | 22 |
| 39 | .02879 | .99959 | .02881 | 34.715 | .0004 | 34.730 | .0000080 | 33.173 | 21 |
| 40 | 0.02908 | 0.99958 | 0.02910 | 34.368 | 1.0004 | 34.382 | .0000082 | 32.826 | 20 |
| 41 | .02938 | .99957 | .02939 | 34.027 | .0004 | 34.042 | .0000085 | 32.486 | 19 |
| 42 | .02967 | .99956 | .02968 | 33.694 | .0004 | 33.708 | .0000087 | 32.152 | 18 |
| 43 | .02996 | .99955 | .02997 | 33.366 | .0004 | 33.381 | .0000090 | 31.825 | 17 |
| 44 | .03025 | .99954 | .03026 | 33.045 | .0005 | 33.060 | .0000092 | 31.505 | 16 |
| 45 | 0.03054 | 0.99953 | 0.03055 | 32.730 | 1.0005 | 32.746 | .0000095 | 31.190 | 15 |
| 46 | .03083 | .99952 | .03084 | 32.421 | .0005 | 32.437 | .0000098 | 30.881 | 14 |
| 47 | .03112 | .99952 | .03114 | 32.118 | .0005 | 32.134 | .0000101 | 30.578 | 13 |
| 48 | .03141 | .99951 | .03143 | 31.821 | .0005 | 31.836 | .0000103 | 30.281 | 12 |
| 49 | .03170 | .99950 | .03172 | 31.528 | .0005 | 31.544 | .0000106 | 29.989 | 11 |
| 50 | 0.03199 | 0.99949 | 0.03201 | 31.242 | 1.0005 | 31.258 | .0000109 | 29.703 | 10 |
| 51 | .03228 | .99948 | .03230 | 30.960 | .0005 | 30.976 | .0000112 | 29.421 | 9 |
| 52 | .03257 | .99947 | .03259 | 30.683 | .0005 | 30.700 | .0000115 | 29.145 | 8 |
| 53 | .03286 | .99946 | .03288 | 30.412 | .0005 | 30.428 | .0000118 | 28.874 | 7 |
| 54 | .03316 | .99945 | .03317 | 30.145 | .0005 | 30.161 | .0000122 | 28.607 | 6 |
| 55 | 0.03345 | 0.99944 | 0.03346 | 29.882 | 1.0006 | 29.899 | .0000125 | 28.345 | 5 |
| 56 | .03374 | .99943 | .03376 | 29.624 | .0006 | 29.641 | .0000128 | 28.087 | 4 |
| 57 | .03403 | .99942 | .03405 | 29.371 | .0006 | 29.388 | .0000131 | 27.834 | 3 |
| 58 | .03432 | .99941 | .03434 | 29.122 | .0006 | 29.139 | .0000135 | 27.586 | 2 |
| 59 | .03461 | .99940 | .03463 | 28.877 | .0006 | 28.894 | .0000138 | 27.341 | 1 |
| 60 | 0.03490 | 0.99939 | 0.03492 | 28.636 | 1.0006 | 28.654 | .0000142 | 27.100 | 0 |
| M | Cosine | Sine | Cotan. | Tan. | Cosec. | Secant | READ DOWN | 88°–89° Involute | M |

**91° or 271°**        **88° or 268°**

Trigonometric Functions

| M | Sine | Cosine | Tan. | Cotan. | Secant | Cosec. | Involute 2°–3° | READ UP | M |
|---|------|--------|------|--------|--------|--------|----------------|---------|---|
| 0 | 0.03490 | 0.99939 | 0.03492 | 28.636 | 1.0006 | 28.654 | .0000142 | 27.100 | 60 |
| 1 | .03519 | .99938 | .03521 | 28.399 | .0006 | 28.417 | .0000145 | 26.864 | 59 |
| 2 | .03548 | .99937 | .03550 | 28.166 | .0006 | 28.184 | .0000149 | 26.631 | 58 |
| 3 | .03577 | .99936 | .03579 | 27.937 | .0006 | 27.955 | .0000153 | 26.402 | 57 |
| 4 | .03606 | .99935 | .03609 | 27.712 | .0007 | 27.730 | .0000157 | 26.177 | 56 |
| 5 | 0.03635 | 0.99934 | 0.03638 | 27.490 | 1.0007 | 27.508 | .0000160 | 25.955 | 55 |
| 6 | .03664 | .99933 | .03667 | 27.271 | .0007 | 27.290 | .0000164 | 25.737 | 54 |
| 7 | .03693 | .99932 | .03696 | 27.057 | .0007 | 27.075 | .0000168 | 25.523 | 53 |
| 8 | .03723 | .99931 | .03725 | 26.845 | .0007 | 26.864 | .0000172 | 25.311 | 52 |
| 9 | .03752 | .99930 | .03754 | 26.637 | .0007 | 26.655 | .0000176 | 25.103 | 51 |
| 10 | 0.03781 | 0.99929 | 0.03783 | 26.432 | 1.0007 | 26.451 | .0000180 | 24.899 | 50 |
| 11 | .03810 | .99927 | .03812 | 26.230 | .0007 | 26.249 | .0000185 | 24.697 | 49 |
| 12 | .03839 | .99926 | .03842 | 26.031 | .0007 | 26.050 | .0000189 | 24.498 | 48 |
| 13 | .03868 | .99925 | .03871 | 25.835 | .0007 | 25.854 | .0000193 | 24.303 | 47 |
| 14 | .03897 | .99924 | .03900 | 25.642 | .0008 | 25.661 | .0000198 | 24.110 | 46 |
| 15 | 0.03926 | 0.99923 | 0.03929 | 25.452 | 1.0008 | 25.471 | .0000202 | 23.920 | 45 |
| 16 | .03955 | .99922 | .03958 | 25.264 | .0008 | 25.284 | .0000207 | 23.733 | 44 |
| 17 | .03984 | .99921 | .03987 | 25.080 | .0008 | 25.100 | .0000211 | 23.549 | 43 |
| 18 | .04013 | .99919 | .04016 | 24.898 | .0008 | 24.918 | .0000216 | 23.367 | 42 |
| 19 | .04042 | .99918 | .04046 | 24.719 | .0008 | 24.739 | .0000220 | 23.188 | 41 |
| 20 | 0.04071 | 0.99917 | 0.04075 | 24.542 | 1.0008 | 24.562 | .0000225 | 23.012 | 40 |
| 21 | .04100 | .99916 | .04104 | 24.368 | .0008 | 24.388 | .0000230 | 22.838 | 39 |
| 22 | .04129 | .99915 | .04133 | 24.196 | .0009 | 24.216 | .0000235 | 22.666 | 38 |
| 23 | .04159 | .99913 | .04162 | 24.026 | .0009 | 24.047 | .0000240 | 22.497 | 37 |
| 24 | .04188 | .99912 | .04191 | 23.859 | .0009 | 23.880 | .0000245 | 22.330 | 36 |
| 25 | 0.04217 | 0.99911 | 0.04220 | 23.695 | 1.0009 | 23.716 | .0000250 | 22.166 | 35 |
| 26 | .04246 | .99910 | .04250 | 23.532 | .0009 | 23.553 | .0000256 | 22.004 | 34 |
| 27 | .04275 | .99909 | .04279 | 23.372 | .0009 | 23.393 | .0000261 | 21.844 | 33 |
| 28 | .04304 | .99907 | .04308 | 23.214 | .0009 | 23.235 | .0000266 | 21.686 | 32 |
| 29 | .04333 | .99906 | .04337 | 23.058 | .0009 | 23.079 | .0000272 | 21.530 | 31 |
| 30 | 0.04362 | 0.99905 | 0.04366 | 22.904 | 1.0010 | 22.926 | .0000277 | 21.377 | 30 |
| 31 | .04391 | .99904 | .04395 | 22.752 | .0010 | 22.774 | .0000283 | 21.225 | 29 |
| 32 | .04420 | .99902 | .04424 | 22.602 | .0010 | 22.624 | .0000288 | 21.075 | 28 |
| 33 | .04449 | .99901 | .04454 | 22.454 | .0010 | 22.476 | .0000294 | 20.928 | 27 |
| 34 | .04478 | .99900 | .04483 | 22.308 | .0010 | 22.330 | .0000300 | 20.782 | 26 |
| 35 | 0.04507 | 0.99898 | 0.04512 | 22.164 | 1.0010 | 22.187 | .0000306 | 20.638 | 25 |
| 36 | .04536 | .99897 | .04541 | 22.022 | .0010 | 22.044 | .0000312 | 20.496 | 24 |
| 37 | .04565 | .99896 | .04570 | 21.881 | .0010 | 21.904 | .0000318 | 20.356 | 23 |
| 38 | .04594 | .99894 | .04599 | 21.743 | .0011 | 21.766 | .0000324 | 20.218 | 22 |
| 39 | .04623 | .99893 | .04628 | 21.606 | .0011 | 21.629 | .0000330 | 20.081 | 21 |
| 40 | 0.04653 | 0.99892 | 0.04658 | 21.470 | 1.0011 | 21.494 | .0000336 | 19.946 | 20 |
| 41 | .04682 | .99890 | .04687 | 21.337 | .0011 | 21.360 | .0000343 | 19.813 | 19 |
| 42 | .04711 | .99889 | .04716 | 21.205 | .0011 | 21.229 | .0000349 | 19.681 | 18 |
| 43 | .04740 | .99888 | .04745 | 21.075 | .0011 | 21.098 | .0000356 | 19.551 | 17 |
| 44 | .04769 | .99886 | .04774 | 20.946 | .0011 | 20.970 | .0000362 | 19.423 | 16 |
| 45 | 0.04798 | 0.99885 | 0.04803 | 20.819 | 1.0012 | 20.843 | .0000369 | 19.296 | 15 |
| 46 | .04827 | .99883 | .04833 | 20.693 | .0012 | 20.717 | .0000376 | 19.171 | 14 |
| 47 | .04856 | .99882 | .04862 | 20.569 | .0012 | 20.593 | .0000382 | 19.047 | 13 |
| 48 | .04885 | .99881 | .04891 | 20.446 | .0012 | 20.471 | .0000389 | 18.925 | 12 |
| 49 | .04914 | .99879 | .04920 | 20.325 | .0012 | 20.350 | .0000396 | 18.804 | 11 |
| 50 | 0.04943 | 0.99878 | 0.04949 | 20.206 | 1.0012 | 20.230 | .0000403 | 18.684 | 10 |
| 51 | .04972 | .99876 | .04978 | 20.087 | .0012 | 20.112 | .0000411 | 18.566 | 9 |
| 52 | .05001 | .99875 | .05007 | 19.970 | .0013 | 19.995 | .0000418 | 18.449 | 8 |
| 53 | .05030 | .99873 | .05037 | 19.855 | .0013 | 19.880 | .0000425 | 18.334 | 7 |
| 54 | .05059 | .99872 | .05066 | 19.740 | .0013 | 19.766 | .0000433 | 18.220 | 6 |
| 55 | 0.05088 | 0.99870 | 0.05095 | 19.627 | 1.0013 | 19.653 | .0000440 | 18.107 | 5 |
| 56 | .05117 | .99869 | .05124 | 19.516 | .0013 | 19.541 | .0000448 | 17.996 | 4 |
| 57 | .05146 | .99867 | .05153 | 19.405 | .0013 | 19.431 | .0000455 | 17.886 | 3 |
| 58 | .05175 | .99866 | .05182 | 19.296 | .0013 | 19.322 | .0000463 | 17.777 | 2 |
| 59 | 05205 | .99864 | .05212 | 19.188 | .0014 | 19.214 | .0000471 | 17.669 | 1 |
| 60 | 0.05234 | 0.99863 | 0.05241 | 19.081 | 1.0014 | 19.107 | .0000479 | 17.563 | 0 |

| M | Cosine | Sine | Cotan. | Tan. | Cosec. | Secant | READ DOWN | 87°–88° Involute | M |

**3° or 183°**     Trigonometric Functions     **176° or 356°**

| M | Sine | Cosine | Tan. | Cotan. | Secant | Cosec. | Involute 3°-4° | READ UP | M |
|---|------|--------|------|--------|--------|--------|----------------|---------|---|
| 0 | 0.05234 | 0.99863 | 0.05241 | 19.081 | 1.0014 | 19.107 | .0000479 | 17.563 | 60 |
| 1 | .05263 | .99861 | .05270 | 18.976 | .0014 | 19.002 | .0000487 | 17.457 | 59 |
| 2 | .05292 | .99860 | .05299 | 18.871 | .0014 | 18.898 | .0000495 | 17.353 | 58 |
| 3 | .05321 | .99858 | .05328 | 18.768 | .0014 | 18.794 | .0000503 | 17.250 | 57 |
| 4 | .05350 | .99857 | .05357 | 18.666 | .0014 | 18.692 | .0000512 | 17.148 | 56 |
| 5 | 0.05379 | 0.99855 | 0.05387 | 18.564 | 1.0014 | 18.591 | .0000520 | 17.047 | 55 |
| 6 | .05408 | .99854 | .05416 | 18.464 | .0015 | 18.492 | .0000529 | 16.948 | 54 |
| 7 | .05437 | .99852 | .05445 | 18.366 | .0015 | 18.393 | .0000537 | 16.849 | 53 |
| 8 | .05466 | .99851 | .05474 | 18.268 | .0015 | 18.295 | .0000546 | 16.752 | 52 |
| 9 | .05495 | .99849 | .05503 | 18.171 | .0015 | 18.198 | .0000555 | 16.655 | 51 |
| 10 | 0.05524 | 0.99847 | 0.05533 | 18.075 | 1.0015 | 18.103 | .0000563 | 16.559 | 50 |
| 11 | .05553 | .99846 | .05562 | 17.980 | .0015 | 18.008 | .0000572 | 16.465 | 49 |
| 12 | .05582 | .99844 | .05591 | 17.886 | .0016 | 17.914 | .0000581 | 16.371 | 48 |
| 13 | .05611 | .99842 | .05620 | 17.793 | .0016 | 17.822 | .0000591 | 16.279 | 47 |
| 14 | .05640 | .99841 | .05649 | 17.702 | .0016 | 17.730 | .0000600 | 16.187 | 46 |
| 15 | .05669 | .99839 | .05678 | 17.611 | 1.0016 | 17.639 | .0000609 | 16.096 | 45 |
| 16 | .05698 | .99838 | .05708 | 17.521 | .0016 | 17.549 | .0000619 | 16.007 | 44 |
| 17 | .05727 | .99836 | .05737 | 17.431 | .0016 | 17.460 | .0000628 | 15.918 | 43 |
| 18 | .05756 | .99834 | .05766 | 17.343 | .0017 | 17.372 | .0000638 | 15.830 | 42 |
| 19 | .05785 | .99833 | .05795 | 17.256 | .0017 | 17.285 | .0000647 | 15.743 | 41 |
| 20 | 0.05814 | 0.99831 | 0.05824 | 17.169 | 1.0017 | 17.198 | .0000657 | 15.657 | 40 |
| 21 | .05844 | .99829 | .05854 | 17.084 | .0017 | 17.113 | .0000667 | 15.571 | 39 |
| 22 | .05873 | .99827 | .05883 | 16.999 | .0017 | 17.028 | .0000677 | 15.487 | 38 |
| 23 | .05902 | .99826 | .05912 | 16.915 | .0017 | 16.945 | .0000687 | 15.403 | 37 |
| 24 | .05931 | .99824 | .05941 | 16.832 | .0018 | 16.862 | .0000698 | 15.320 | 36 |
| 25 | 0.05960 | 0.99822 | 0.05970 | 16.750 | 1.0018 | 16.779 | .0000708 | 15.238 | 35 |
| 26 | .05989 | .99821 | .05999 | 16.668 | .0018 | 16.698 | .0000718 | 15.157 | 34 |
| 27 | .06018 | .99819 | .06029 | 16.587 | .0018 | 16.618 | .0000729 | 15.077 | 33 |
| 28 | .06047 | .99817 | .06058 | 16.507 | .0018 | 16.538 | .0000739 | 14.997 | 32 |
| 29 | .06076 | .99815 | .06087 | 16.428 | .0019 | 16.459 | .0000750 | 14.918 | 31 |
| 30 | .06105 | .99813 | .06116 | 16.350 | 1.0019 | 16.380 | .0000761 | 14.840 | 30 |
| 31 | .06134 | .99812 | .06145 | 16.272 | .0019 | 16.303 | .0000772 | 14.763 | 29 |
| 32 | .06163 | .99810 | .06175 | 16.195 | .0019 | 16.226 | .0000783 | 14.686 | 28 |
| 33 | .06192 | .99808 | .06204 | 16.119 | .0019 | 16.150 | .0000794 | 14.610 | 27 |
| 34 | .06221 | .99806 | .06233 | 16.043 | .0019 | 16.075 | .0000805 | 14.535 | 26 |
| 35 | .06250 | 0.99804 | 0.06262 | 15.969 | 1.0020 | 16.000 | .0000817 | 14.460 | 25 |
| 36 | .06279 | .99803 | .06291 | 15.895 | .0020 | 15.926 | .0000828 | 14.387 | 24 |
| 37 | .06308 | .99801 | .06321 | 15.821 | .0020 | 15.853 | .0000840 | 14.313 | 23 |
| 38 | .06337 | .99799 | .06350 | 15.748 | .0020 | 15.780 | .0000852 | 14.241 | 22 |
| 39 | .06366 | .99797 | .06379 | 15.676 | .0020 | 15.708 | .0000863 | 14.169 | 21 |
| 40 | 0.06395 | 0.99795 | 0.06408 | 15.605 | 1.0021 | 15.637 | .0000875 | 14.098 | 20 |
| 41 | .06424 | .99793 | .06438 | 15.534 | .0021 | 15.566 | .0000887 | 14.027 | 19 |
| 42 | .06453 | .99792 | .06467 | 15.464 | .0021 | 15.496 | .0000899 | 13.958 | 18 |
| 43 | .06482 | .99790 | .06496 | 15.394 | .0021 | 15.427 | .0000911 | 13.888 | 17 |
| 44 | .06511 | .99788 | .06525 | 15.325 | .0021 | 15.358 | .0000924 | 13.820 | 16 |
| 45 | .06540 | 0.99786 | 0.06554 | 15.257 | 1.0021 | 15.290 | .0000936 | 13.752 | 15 |
| 46 | .06569 | .99784 | .06584 | 15.189 | .0022 | 15.222 | .0000949 | 13.684 | 14 |
| 47 | .06598 | .99782 | .06613 | 15.122 | .0022 | 15.155 | .0000961 | 13.617 | 13 |
| 48 | .06627 | .99780 | .06642 | 15.056 | .0022 | 15.089 | .0000974 | 13.551 | 12 |
| 49 | .06656 | .99778 | .06671 | 14.990 | .0022 | 15.023 | .0000987 | 13.486 | 11 |
| 50 | 0.06685 | 0.99776 | 0.06700 | 14.924 | 1.0022 | 14.958 | .0001000 | 13.421 | 10 |
| 51 | .06714 | .99774 | .06730 | 14.860 | .0023 | 14.893 | .0001013 | 13.356 | 9 |
| 52 | .06743 | .99772 | .06759 | 14.795 | .0023 | 14.829 | .0001026 | 13.292 | 8 |
| 53 | .06773 | .99770 | .06788 | 14.732 | .0023 | 14.766 | .0001040 | 13.229 | 7 |
| 54 | .06802 | .99768 | .06817 | 14.669 | .0023 | 14.703 | .0001053 | 13.166 | 6 |
| 55 | 0.06831 | 0.99766 | 0.06847 | 14.606 | 1.0023 | 14.640 | .0001067 | 13.103 | 5 |
| 56 | .06860 | .99764 | .06876 | 14.544 | .0024 | 14.578 | .0001080 | 13.042 | 4 |
| 57 | .06889 | .99762 | .06905 | 14.482 | .0024 | 14.517 | .0001094 | 12.980 | 3 |
| 58 | .06918 | .99760 | .06934 | 14.421 | .0024 | 14.456 | .0001108 | 12.920 | 2 |
| 59 | .06947 | .99758 | .06963 | 14.361 | .0024 | 14.395 | .0001122 | 12.859 | 1 |
| 60 | 0.06976 | 0.99756 | 0.06993 | 14.301 | 1.0024 | 14.336 | .0001136 | 12.800 | 0 |
| M | Cosine | Sine | Cotan. | Tan. | Cosec. | Secant | READ DOWN | 86°-87° Involute | M |

| M | Sine | Cosine | Tan. | Cotan. | Secant | Cosec. | Involute 4°–5° | READ UP | M |
|---|---|---|---|---|---|---|---|---|---|
| 0 | 0.06976 | 0.99756 | 0.06993 | 14.301 | 1.0024 | 14.336 | .0001136 | 12.800 | 60 |
| 1 | .07005 | .99754 | .07022 | 14.241 | .0025 | 14.276 | .0001151 | 12.740 | 59 |
| 2 | .07034 | .99752 | .07051 | 14.182 | .0025 | 14.217 | .0001165 | 12.682 | 58 |
| 3 | .07063 | .99750 | .07080 | 14.124 | .0025 | 14.159 | .0001180 | 12.623 | 57 |
| 4 | .07092 | .99748 | .07110 | 14.065 | .0025 | 14.101 | .0001194 | 12.566 | 56 |
| 5 | 0.07121 | 0.99746 | 0.07139 | 14.008 | 1.0025 | 14.044 | .0001209 | 12.508 | 55 |
| 6 | .07150 | .99744 | .07168 | 13.951 | .0026 | 13.987 | .0001224 | 12.451 | 54 |
| 7 | .07179 | .99742 | .07197 | 13.894 | .0026 | 13.930 | .0001239 | 12.395 | 53 |
| 8 | .07208 | .99740 | .07227 | 13.838 | .0026 | 13.874 | .0001254 | 12.339 | 52 |
| 9 | .07237 | .99738 | .07256 | 13.782 | .0026 | 13.818 | .0001269 | 12.284 | 51 |
| 10 | 0.07266 | 0.99736 | 0.07285 | 13.727 | 1.0027 | 13.763 | .0001285 | 12.229 | 50 |
| 11 | .07295 | .99734 | .07314 | 13.672 | .0027 | 13.708 | .0001300 | 12.174 | 49 |
| 12 | .07324 | .99731 | .07344 | 13.617 | .0027 | 13.654 | .0001316 | 12.120 | 48 |
| 13 | .07353 | .99729 | .07373 | 13.563 | .0027 | 13.600 | .0001332 | 12.066 | 47 |
| 14 | .07382 | .99727 | .07402 | 13.510 | .0027 | 13.547 | .0001347 | 12.013 | 46 |
| 15 | 0.07411 | 0.99725 | 0.07431 | 13.457 | 1.0028 | 13.494 | .0001363 | 11.960 | 45 |
| 16 | .07440 | .99723 | .07461 | 13.404 | .0028 | 13.441 | .0001380 | 11.908 | 44 |
| 17 | .07469 | .99721 | .07490 | 13.352 | .0028 | 13.389 | .0001396 | 11.855 | 43 |
| 18 | .07498 | .99719 | .07519 | 13.300 | .0028 | 13.337 | .0001412 | 11.804 | 42 |
| 19 | .07527 | .99716 | .07548 | 13.248 | .0028 | 13.286 | .0001429 | 11.753 | 41 |
| 20 | 0.07556 | 0.99714 | 0.07578 | 13.197 | 1.0029 | 13.235 | .0001445 | 11.702 | 40 |
| 21 | .07585 | .99712 | .07607 | 13.146 | .0029 | 13.184 | .0001462 | 11.651 | 39 |
| 22 | .07614 | .99710 | .07636 | 13.096 | .0029 | 13.134 | .0001479 | 11.601 | 38 |
| 23 | .07643 | .99707 | .07665 | 13.046 | .0029 | 13.084 | .0001496 | 11.551 | 37 |
| 24 | .07672 | .99705 | .07695 | 12.996 | .0030 | 13.035 | .0001513 | 11.502 | 36 |
| 25 | 0.07701 | 0.99703 | 0.07724 | 12.947 | 1.0030 | 12.985 | .0001531 | 11.453 | 35 |
| 26 | .07730 | .99701 | .07753 | 12.898 | .0030 | 12.937 | .0001548 | 11.405 | 34 |
| 27 | .07759 | .99699 | .07782 | 12.850 | .0030 | 12.888 | .0001565 | 11.356 | 33 |
| 28 | .07788 | .99696 | .07812 | 12.801 | .0030 | 12.840 | .0001583 | 11.309 | 32 |
| 29 | .07817 | .99694 | .07841 | 12.754 | .0031 | 12.793 | .0001601 | 11.261 | 31 |
| 30 | 0.07846 | 0.99692 | 0.07870 | 12.706 | 1.0031 | 12.745 | .0001619 | 11.214 | 30 |
| 31 | .07875 | .99689 | .07899 | 12.659 | .0031 | 12.699 | .0001637 | 11.167 | 29 |
| 32 | .07904 | .99687 | .07929 | 12.612 | .0031 | 12.652 | .0001655 | 11.121 | 28 |
| 33 | .07933 | .99685 | .07958 | 12.566 | .0032 | 12.606 | .0001674 | 11.075 | 27 |
| 34 | .07962 | .99683 | .07987 | 12.520 | .0032 | 12.560 | .0001692 | 11.029 | 26 |
| 35 | 0.07991 | 0.99680 | 0.08017 | 12.474 | 1.0032 | 12.514 | .0001711 | 10.983 | 25 |
| 36 | .08020 | .99678 | .08046 | 12.429 | .0032 | 12.469 | .0001729 | 10.938 | 24 |
| 37 | .08049 | .99676 | .08075 | 12.384 | .0033 | 12.424 | .0001748 | 10.894 | 23 |
| 38 | .08078 | .99673 | .08104 | 12.339 | .0033 | 12.379 | .0001767 | 10.849 | 22 |
| 39 | .08107 | .99671 | .08134 | 12.295 | .0033 | 12.335 | .0001787 | 10.805 | 21 |
| 40 | 0.08136 | 0.99668 | 0.08163 | 12.251 | 1.0033 | 12.291 | .0001806 | 10.761 | 20 |
| 41 | .08165 | .99666 | .08192 | 12.207 | .0034 | 12.248 | .0001825 | 10.718 | 19 |
| 42 | .08194 | .99664 | .08221 | 12.163 | .0034 | 12.204 | .0001845 | 10.674 | 18 |
| 43 | .08223 | .99661 | .08251 | 12.120 | .0034 | 12.161 | .0001865 | 10.632 | 17 |
| 44 | .08252 | .99659 | .08280 | 12.077 | .0034 | 12.119 | .0001885 | 10.589 | 16 |
| 45 | 0.08281 | 0.99657 | 0.08309 | 12.035 | 1.0034 | 12.076 | .0001905 | 10.547 | 15 |
| 46 | .08310 | .99654 | .08339 | 11.992 | .0035 | 12.034 | .0001925 | 10.505 | 14 |
| 47 | .08339 | .99652 | .08368 | 11.950 | .0035 | 11.992 | .0001945 | 10.463 | 13 |
| 48 | .08368 | .99649 | .08397 | 11.909 | .0035 | 11.951 | .0001965 | 10.422 | 12 |
| 49 | .08397 | .99647 | .08427 | 11.867 | .0035 | 11.909 | .0001986 | 10.381 | 11 |
| 50 | 0.08426 | 0.99644 | 0.08456 | 11.826 | 1.0036 | 11.868 | .0002007 | 10.340 | 10 |
| 51 | .08455 | .99642 | .08485 | 11.785 | .0036 | 11.828 | .0002028 | 10.299 | 9 |
| 52 | .08484 | .99639 | .08514 | 11.745 | .0036 | 11.787 | .0002049 | 10.259 | 8 |
| 53 | .08513 | .99637 | .08544 | 11.705 | .0036 | 11.747 | .0002070 | 10.219 | 7 |
| 54 | .08542 | .99635 | .08573 | 11.664 | .0037 | 11.707 | .0002091 | 10.179 | 6 |
| 55 | 0.08571 | 0.99632 | 0.08602 | 11.625 | 1.0037 | 11.668 | .0002113 | 10.140 | 5 |
| 56 | .08600 | .99630 | .08632 | 11.585 | .0037 | 11.628 | .0002134 | 10.101 | 4 |
| 57 | .08629 | .99627 | .08661 | 11.546 | .0037 | 11.589 | .0002156 | 10.062 | 3 |
| 58 | .08658 | .99625 | .08690 | 11.507 | .0038 | 11.551 | .0002178 | 10.023 | 2 |
| 59 | .08687 | .99622 | .08720 | 11.468 | .0038 | 11.512 | .0002200 | 9.9847 | 1 |
| 60 | 0.08716 | 0.99619 | 0.08749 | 11.430 | 1.0038 | 11.474 | .0002222 | 9.9465 | 0 |

| M | Cosine | Sine | Cotan. | Tan. | Cosec. | Secant | READ DOWN | 85°–86° Involute | M |

| M | Sine | Cosine | Tan. | Cotan. | Secant | Cosec. | Involute 5°-6° | READ UP | M |
|---|---|---|---|---|---|---|---|---|---|
| 0 | 0.08716 | 0.99619 | 0.08749 | 11.430 | 1.0038 | 11.474 | .0002222 | 9.9465 | 60 |
| 1 | .08745 | .99617 | .08778 | 11.392 | .0038 | 11.436 | .0002244 | 9.9086 | 59 |
| 2 | .08774 | .99614 | .08807 | 11.354 | .0039 | 11.398 | .0002267 | 9.8710 | 58 |
| 3 | .08803 | .99612 | .08837 | 11.316 | .0039 | 11.360 | .0002289 | 9.8336 | 57 |
| 4 | .08831 | .99609 | .08866 | 11.279 | .0039 | 11.323 | .0002312 | 9.7965 | 56 |
| 5 | 0.08860 | 0.99607 | 0.08895 | 11.242 | 1.0039 | 11.286 | .0002335 | 9.7596 | 55 |
| 6 | .08889 | .99604 | .08925 | 11.205 | .0040 | 11.249 | .0002358 | 9.7230 | 54 |
| 7 | .08918 | .99602 | .08954 | 11.168 | .0040 | 11.213 | .0002382 | 9.6866 | 53 |
| 8 | .08947 | .99599 | .08983 | 11.132 | .0040 | 11.176 | .0002405 | 9.6504 | 52 |
| 9 | .08976 | .99596 | .09013 | 11.095 | .0041 | 11.140 | .0002429 | 9.6145 | 51 |
| 10 | 0.09005 | 0.99594 | 0.09042 | 11.059 | 1.0041 | 11.105 | .0002452 | 9.5788 | 50 |
| 11 | .09034 | .99591 | .09071 | 11.024 | .0041 | 11.069 | .0002476 | 9.5433 | 49 |
| 12 | .09063 | .99588 | .09101 | 10.988 | .0041 | 11.034 | .0002500 | 9.5081 | 48 |
| 13 | .09092 | .99586 | .09130 | 10.953 | .0042 | 10.998 | .0002524 | 9.4731 | 47 |
| 14 | .09121 | .99583 | .09159 | 10.918 | .0042 | 10.963 | .0002549 | 9.4383 | 46 |
| 15 | 0.09150 | 0.99580 | 0.09189 | 10.883 | 1.0042 | 10.929 | .0002573 | 9.4038 | 45 |
| 16 | .09179 | .99578 | .09218 | 10.848 | .0042 | 10.894 | .0002598 | 9.3694 | 44 |
| 17 | .09208 | .99575 | .09247 | 10.814 | .0043 | 10.860 | .0002622 | 9.3353 | 43 |
| 18 | .09237 | .99572 | .09277 | 10.780 | .0043 | 10.826 | .0002647 | 9.3014 | 42 |
| 19 | .09266 | .99570 | .09306 | 10.746 | .0043 | 10.792 | .0002673 | 9.2677 | 41 |
| 20 | 0.09295 | 0.99567 | 0.09335 | 10.712 | 1.0043 | 10.758 | .0002698 | 9.2342 | 40 |
| 21 | .09324 | .99564 | .09365 | 10.678 | .0044 | 10.725 | .0002723 | 9.2009 | 39 |
| 22 | .09353 | .99562 | .09394 | 10.645 | .0044 | 10.692 | .0002749 | 9.1679 | 38 |
| 23 | .09382 | .99559 | .09423 | 10.612 | .0044 | 10.659 | .0002775 | 9.1350 | 37 |
| 24 | .09411 | .99556 | .09453 | 10.579 | .0045 | 10.626 | .0002801 | 9.1023 | 36 |
| 25 | 0.09440 | 0.99553 | 0.09482 | 10.546 | 1.0045 | 10.593 | .0002827 | 9.0699 | 35 |
| 26 | .09469 | .99551 | .09511 | 10.514 | .0045 | 10.561 | .0002853 | 9.0376 | 34 |
| 27 | .09498 | .99548 | .09541 | 10.481 | .0045 | 10.529 | .0002879 | 9.0056 | 33 |
| 28 | .09527 | .99545 | .09570 | 10.449 | .0046 | 10.497 | .0002906 | 8.9737 | 32 |
| 29 | .09556 | .99542 | .09600 | 10.417 | .0046 | 10.465 | .0002932 | 8.9421 | 31 |
| 30 | 0.09585 | 0.99540 | 0.09629 | 10.385 | 1.0046 | 10.433 | .0002959 | 8.9106 | 30 |
| 31 | .09614 | .99537 | .09658 | 10.354 | .0047 | 10.402 | .0002986 | 8.8793 | 29 |
| 32 | .09642 | .99534 | .09688 | 10.322 | .0047 | 10.371 | .0003014 | 8.8482 | 28 |
| 33 | .09671 | .99531 | .09717 | 10.291 | .0047 | 10.340 | .0003041 | 8.8173 | 27 |
| 34 | .09700 | .99528 | .09746 | 10.260 | .0047 | 10.309 | .0003069 | 8.7866 | 26 |
| 35 | 0.09729 | 0.99526 | 0.09776 | 10.229 | 1.0048 | 10.278 | .0003096 | 8.7561 | 25 |
| 36 | .09758 | .99523 | .09805 | 10.199 | .0048 | 10.248 | .0003124 | 8.7257 | 24 |
| 37 | .09787 | .99520 | .09834 | 10.168 | .0048 | 10.217 | .0003152 | 8.6956 | 23 |
| 38 | .09816 | .99517 | .09864 | 10.138 | .0049 | 10.187 | .0003180 | 8.6656 | 22 |
| 39 | .09845 | .99514 | .09893 | 10.108 | .0049 | 10.157 | .0003209 | 8.6358 | 21 |
| 40 | 0.09874 | 0.99511 | 0.09923 | 10.078 | 1.0049 | 10.128 | .0003237 | 8.6061 | 20 |
| 41 | .09903 | .99508 | .09952 | 10.048 | .0049 | 10.098 | .0003266 | 8.5767 | 19 |
| 42 | .09932 | .99506 | .09981 | 10.019 | .0050 | 10.068 | .0003295 | 8.5474 | 18 |
| 43 | .09961 | .99503 | .10011 | 9.9893 | .0050 | 10.039 | .0003324 | 8.5183 | 17 |
| 44 | .09990 | .99500 | .10040 | 9.9601 | .0050 | 10.010 | .0003353 | 8.4893 | 16 |
| 45 | 0.10019 | 0.99497 | 0.10069 | 9.9310 | 1.0051 | 9.9812 | .0003383 | 8.4606 | 15 |
| 46 | .10048 | .99494 | .10099 | 9.9021 | .0051 | 9.9525 | .0003412 | 8.4320 | 14 |
| 47 | .10077 | .99491 | .10128 | 9.8734 | .0051 | 9.9239 | .0003442 | 8.4035 | 13 |
| 48 | .10106 | .99488 | .10158 | 9.8448 | .0051 | 9.8955 | .0003472 | 8.3752 | 12 |
| 49 | .10135 | .99485 | .10187 | 9.8164 | .0052 | 9.8672 | .0003502 | 8.3471 | 11 |
| 50 | 0.10164 | 0.99482 | 0.10216 | 9.7882 | 1.0052 | 9.8391 | .0003532 | 8.3192 | 10 |
| 51 | .10192 | .99479 | .10246 | 9.7601 | .0052 | 9.8112 | .0003563 | 8.2914 | 9 |
| 52 | .10221 | .99476 | .10275 | 9.7322 | .0053 | 9.7834 | .0003593 | 8.2638 | 8 |
| 53 | .10250 | .99473 | .10305 | 9.7044 | .0053 | 9.7558 | .0003624 | 8.2363 | 7 |
| 54 | .10279 | .99470 | .10334 | 9.6768 | .0053 | 9.7283 | .0003655 | 8.2090 | 6 |
| 55 | 0.10308 | 0.99467 | 0.10363 | 9.6493 | 1.0054 | 9.7010 | .0003686 | 8.1818 | 5 |
| 56 | .10337 | .99464 | .10393 | 9.6220 | .0054 | 9.6739 | .0003718 | 8.1548 | 4 |
| 57 | .10366 | .99461 | .10422 | 9.5949 | .0054 | 9.6469 | .0003749 | 8.1280 | 3 |
| 58 | .10395 | .99458 | .10452 | 9.5679 | .0054 | 9.6200 | .0003781 | 8.1012 | 2 |
| 59 | .10424 | .99455 | .10481 | 9.5411 | .0055 | 9.5933 | .0003813 | 8.0747 | 1 |
| 60 | 0.10453 | 0.99452 | 0.10510 | 9.5144 | 1.0055 | 9.5668 | .0003845 | 8.0483 | 0 |

| M | Cosine | Sine | Cotan. | Tan. | Cosec. | Secant | READ DOWN | 84°-85° Involute | M |

**6° or 186°**        Trigonometric Functions      **173° or 353°**

| M | Sine | Cosine | Tan. | Cotan. | Secant | Cosec. | Involute 6°–7° | READ UP | M |
|---|------|--------|------|--------|--------|--------|-----------------|---------|---|
| 0 | 0.10453 | 0.99452 | 0.10510 | 9.5144 | 1.0055 | 9.5668 | .0003845 | 8.0483 | 60 |
| 1 | .10482 | .99449 | .10540 | .4878 | .0055 | .5404 | .0003877 | 8.0220 | 59 |
| 2 | .10511 | .99446 | .10569 | .4614 | .0056 | .5141 | .0003909 | 7.9959 | 58 |
| 3 | .10540 | .99443 | .10599 | .4352 | .0056 | .4880 | .0003942 | 7.9699 | 57 |
| 4 | .10569 | .99440 | .10628 | .4090 | .0056 | .4620 | .0003975 | 7.9441 | 56 |
| 5 | 0.10597 | 0.99437 | 0.10657 | 9.3831 | 1.0057 | 9.4362 | .0004008 | 7.9184 | 55 |
| 6 | .10626 | .99434 | .10687 | .3572 | .0057 | .4105 | .0004041 | 7.8929 | 54 |
| 7 | .10655 | .99431 | .10716 | .3315 | .0057 | .3850 | .0004074 | 7.8675 | 53 |
| 8 | .10684 | .99428 | .10746 | .3060 | .0058 | .3596 | .0004108 | 7.8422 | 52 |
| 9 | .10713 | .99424 | .10775 | .2806 | .0058 | .3343 | .0004141 | 7.8171 | 51 |
| 10 | 0.10742 | 0.99421 | 0.10805 | 9.2553 | 1.0058 | 9.3092 | .0004175 | 7.7921 | 50 |
| 11 | .10771 | .99418 | .10834 | .2302 | .0059 | .2842 | .0004209 | 7.7673 | 49 |
| 12 | .10800 | .99415 | .10863 | .2052 | .0059 | .2593 | .0004244 | 7.7426 | 48 |
| 13 | .10829 | .99412 | .10893 | .1803 | .0059 | .2346 | .0004278 | 7.7180 | 47 |
| 14 | .10858 | .99409 | .10922 | .1555 | .0059 | .2100 | .0004313 | 7.6935 | 46 |
| 15 | 0.10887 | 0.99406 | 0.10952 | 9.1309 | 1.0060 | 9.1855 | .0004347 | 7.6692 | 45 |
| 16 | .10916 | .99402 | .10981 | .1065 | .0060 | .1612 | .0004382 | 7.6450 | 44 |
| 17 | .10945 | .99399 | .11011 | .0821 | .0060 | .1370 | .0004417 | 7.6210 | 43 |
| 18 | .10973 | .99396 | .11040 | .0579 | .0061 | .1129 | .0004453 | 7.5970 | 42 |
| 19 | .11002 | .99393 | .11070 | .0338 | .0061 | .0890 | .0004488 | 7.5732 | 41 |
| 20 | 0.11031 | 0.99390 | 0.11099 | 9.0098 | 1.0061 | 9.0652 | .0004524 | 7.5496 | 40 |
| 21 | .11060 | .99386 | .11128 | 8.9860 | .0062 | .0415 | .0004560 | 7.5260 | 39 |
| 22 | .11089 | .99383 | .11158 | .9623 | .0062 | .0179 | .0004596 | 7.5026 | 38 |
| 23 | .11118 | .99380 | .11187 | .9387 | .0062 | 8.9944 | .0004632 | 7.4793 | 37 |
| 24 | .11147 | .99377 | .11217 | .9152 | .0063 | .9711 | .0004669 | 7.4561 | 36 |
| 25 | 0.11176 | 0.99374 | 0.11246 | 8.8918 | 1.0063 | 8.9479 | .0004706 | 7.4330 | 35 |
| 26 | .11205 | .99370 | .11276 | .8686 | .0063 | .9248 | .0004743 | 7.4101 | 34 |
| 27 | .11234 | .99367 | .11305 | .8455 | .0064 | .9019 | .0004780 | 7.3873 | 33 |
| 28 | .11263 | .99364 | .11335 | .8225 | .0064 | .8790 | .0004817 | 7.3646 | 32 |
| 29 | .11291 | .99360 | .11364 | .7996 | .0064 | .8563 | .0004854 | 7.3420 | 31 |
| 30 | 0.11320 | 0.99357 | 0.11394 | 8.7769 | 1.0065 | 8.8337 | .0004892 | 7.3195 | 30 |
| 31 | .11349 | .99354 | .11423 | .7542 | .0065 | .8112 | .0004930 | 7.2972 | 29 |
| 32 | .11378 | .99351 | .11452 | .7317 | .0065 | .7888 | .0004968 | 7.2750 | 28 |
| 33 | .11407 | .99347 | .11482 | .7093 | .0066 | .7665 | .0005006 | 7.2528 | 27 |
| 34 | .11436 | .99344 | .11511 | .6870 | .0066 | .7444 | .0005045 | 7.2308 | 26 |
| 35 | 0.11465 | 0.99341 | 0.11541 | 8.6648 | 1.0066 | 8.7223 | .0005083 | 7.2089 | 25 |
| 36 | .11494 | .99337 | .11570 | .6427 | .0067 | .7004 | .0005122 | 7.1871 | 24 |
| 37 | .11523 | .99334 | .11600 | .6208 | .0067 | .6786 | .0005161 | 7.1655 | 23 |
| 38 | .11552 | .99330 | .11629 | .5989 | .0067 | .6569 | .0005200 | 7.1439 | 22 |
| 39 | .11580 | .99327 | .11659 | .5772 | .0068 | .6353 | .0005240 | 7.1225 | 21 |
| 40 | 0.11609 | 0.99324 | 0.11688 | 8.5555 | 1.0068 | 8.6138 | .0005280 | 7.1011 | 20 |
| 41 | .11638 | .99320 | .11718 | .5340 | .0068 | .5924 | .0005319 | 7.0799 | 19 |
| 42 | .11667 | .99317 | .11747 | .5126 | .0069 | .5711 | .0005359 | 7.0587 | 18 |
| 43 | .11696 | .99313 | .11777 | .4913 | .0069 | .5500 | .0005400 | 7.0377 | 17 |
| 44 | .11725 | .99310 | .11806 | .4701 | .0069 | .5289 | .0005440 | 7.0168 | 16 |
| 45 | 0.11754 | 0.99307 | 0.11836 | 8.4490 | 1.0070 | 8.5079 | .0005481 | 6.9960 | 15 |
| 46 | .11783 | .99303 | .11865 | .4280 | .0070 | .4871 | .0005522 | 6.9753 | 14 |
| 47 | .11812 | .99300 | .11895 | .4071 | .0070 | .4663 | .0005563 | 6.9546 | 13 |
| 48 | .11840 | .99296 | .11924 | .3863 | .0071 | .4457 | .0005604 | 6.9341 | 12 |
| 49 | .11869 | .99293 | .11954 | .3656 | .0071 | .4251 | .0005645 | 6.9137 | 11 |
| 50 | 0.11898 | 0.99290 | 0.11983 | 8.3450 | 1.0072 | 8.4047 | .0005687 | 6.8934 | 10 |
| 51 | .11927 | .99286 | .12013 | .3245 | .0072 | .3843 | .0005729 | 6.8732 | 9 |
| 52 | .11956 | .99283 | .12042 | .3041 | .0072 | .3641 | .0005771 | 6.8531 | 8 |
| 53 | .11985 | .99279 | .12072 | .2838 | .0073 | .3439 | .0005813 | 6.8331 | 7 |
| 54 | .12014 | .99276 | .12101 | .2636 | .0073 | .3238 | .0005856 | 6.8132 | 6 |
| 55 | 0.12043 | 0.99272 | 0.12131 | 8.2434 | 1.0073 | 8.3039 | .0005898 | 6.7934 | 5 |
| 56 | .12071 | .99269 | .12160 | .2234 | .0074 | .2840 | .0005941 | 6.7737 | 4 |
| 57 | .12100 | .99265 | .12190 | .2035 | .0074 | .2642 | .0005985 | 6.7540 | 3 |
| 58 | .12129 | .99262 | .12219 | .1837 | .0074 | .2446 | .0006028 | 6.7345 | 2 |
| 59 | .12158 | .99258 | .12249 | .1640 | .0075 | .2250 | .0006071 | 6.7151 | 1 |
| 60 | 0.12187 | 0.99255 | 0.12278 | 8.1443 | 1.0075 | 8.2055 | .0006115 | 6.6957 | 0 |
| M | Cosine | Sine | Cotan. | Tan. | Cosec. | Secant | READ DOWN | 83°–84° Involute | M |

**7° or 187°**  Trigonometric Functions  **172° or 352°**

| M | Sine | Cosine | Tan. | Cotan. | Secant | Cosec. | Involute 7°-8° | READ UP | M |
|---|------|--------|------|--------|--------|--------|----------------|---------|---|
| 0 | 0.12187 | .99255 | 0.12278 | 8.1443 | 1.0075 | 8.2055 | .0006115 | 6.6957 | 60 |
| 1 | .12216 | .99251 | .12308 | .1248 | .0075 | .1861 | .0006159 | 6.6765 | 59 |
| 2 | .12245 | .99248 | .12338 | .1054 | .0076 | .1668 | .0006203 | 6.6573 | 58 |
| 3 | .12274 | .99244 | .12367 | .0860 | .0076 | .1476 | .0006248 | 6.6383 | 57 |
| 4 | .12302 | .99240 | .12397 | .0667 | .0077 | .1285 | .0006292 | 6.6193 | 56 |
| 5 | 0.12331 | 0.99237 | 0.12426 | 8.0476 | 1.0077 | 8.1095 | .0006337 | 6.6004 | 55 |
| 6 | .12360 | .99233 | .12456 | .0285 | .0077 | .0905 | .0006382 | 6.5816 | 54 |
| 7 | .12389 | .99230 | .12485 | .0095 | .0078 | .0717 | .0006427 | 6.5629 | 53 |
| 8 | .12418 | .99226 | .12515 | 7.9906 | .0078 | .0529 | .0006473 | 6.5443 | 52 |
| 9 | .12447 | .99222 | .12544 | .9718 | .0078 | .0342 | .0006518 | 6.5258 | 51 |
| 10 | 0.12476 | 0.99219 | 0.12574 | 7.9530 | 1.0079 | 8.0156 | .0006564 | 6.5073 | 50 |
| 11 | .12504 | .99215 | .12603 | .9344 | .0079 | 7.9971 | .0006610 | 6.4890 | 49 |
| 12 | .12533 | .99211 | .12633 | .9158 | .0079 | .9787 | .0006657 | 6.4707 | 48 |
| 13 | .12562 | .99208 | .12662 | .8973 | .0080 | .9604 | .0006703 | 6.4525 | 47 |
| 14 | .12591 | .99204 | .12692 | .8789 | .0080 | .9422 | .0006750 | 6.4344 | 46 |
| 15 | 0.12620 | 0.99200 | 0.12722 | 7.8606 | 1.0081 | 7.9240 | .0006797 | 6.4164 | 45 |
| 16 | .12649 | .99197 | .12751 | .8424 | .0081 | .9059 | .0006844 | 6.3985 | 44 |
| 17 | .12678 | .99193 | .12781 | .8243 | .0081 | .8879 | .0006892 | 6.3806 | 43 |
| 18 | .12706 | .99189 | .12810 | .8062 | .0082 | .8700 | .0006939 | 6.3628 | 42 |
| 19 | .12735 | .99186 | .12840 | .7882 | .0082 | .8522 | .0006987 | 6.3451 | 41 |
| 20 | 0.12764 | 0.99182 | 0.12869 | 7.7704 | 1.0082 | 7.8344 | .0007035 | 6.3275 | 40 |
| 21 | .12793 | .99178 | .12899 | .7525 | .0083 | .8168 | .0007083 | 6.3100 | 39 |
| 22 | .12822 | .99175 | .12929 | .7348 | .0083 | .7992 | .0007132 | 6.2926 | 38 |
| 23 | .12851 | .99171 | .12958 | .7171 | .0084 | .7817 | .0007181 | 6.2752 | 37 |
| 24 | .12880 | .99167 | .12988 | .6996 | .0084 | .7642 | .0007230 | 6.2579 | 36 |
| 25 | 0.12908 | 0.99163 | 0.13017 | 7.6821 | 1.0084 | 7.7469 | .0007279 | 6.2407 | 35 |
| 26 | .12937 | .99160 | .13047 | .6647 | .0085 | .7296 | .0007328 | 6.2236 | 34 |
| 27 | .12966 | .99156 | .13076 | .6473 | .0085 | .7124 | .0007378 | 6.2065 | 33 |
| 28 | .12995 | .99152 | .13106 | .6301 | .0086 | .6953 | .0007428 | 6.1896 | 32 |
| 29 | .13024 | .99148 | .13136 | .6129 | .0086 | .6783 | .0007478 | 6.1727 | 31 |
| 30 | 0.13053 | 0.99144 | 0.13165 | 7.5958 | 1.0086 | 7.6613 | .0007528 | 6.1559 | 30 |
| 31 | .13081 | .99141 | .13195 | .5787 | .0087 | .6444 | .0007579 | 6.1391 | 29 |
| 32 | .13110 | .99137 | .13224 | .5618 | .0087 | .6276 | .0007629 | 6.1224 | 28 |
| 33 | .13139 | .99133 | .13254 | .5449 | .0087 | .6109 | .0007680 | 6.1058 | 27 |
| 34 | .13168 | .99129 | .13284 | .5281 | .0088 | .5942 | .0007732 | 6.0893 | 26 |
| 35 | 0.13197 | 0.99125 | 0.13313 | 7.5113 | 1.0088 | 7.5776 | .0007783 | 6.0729 | 25 |
| 36 | .13226 | .99122 | .13343 | .4947 | .0089 | .5611 | .0007835 | 6.0565 | 24 |
| 37 | .13254 | .99118 | .13372 | .4781 | .0089 | .5446 | .0007887 | 6.0402 | 23 |
| 38 | .13283 | .99114 | .13402 | .4615 | .0089 | .5282 | .0007939 | 6.0240 | 22 |
| 39 | .13312 | .99110 | .13432 | .4451 | .0090 | .5119 | .0007991 | 6.0078 | 21 |
| 40 | 0.13341 | 0.99106 | 0.13461 | 7.4287 | 1.0090 | 7.4957 | .0008044 | 5.9917 | 20 |
| 41 | .13370 | .99102 | .13491 | .4124 | .0091 | .4795 | .0008096 | 5.9757 | 19 |
| 42 | .13399 | .99098 | .13521 | .3962 | .0091 | .4635 | .0008150 | 5.9598 | 18 |
| 43 | .13427 | .99094 | .13550 | .3800 | .0091 | .4474 | .0008203 | 5.9439 | 17 |
| 44 | .13456 | .99091 | .13580 | .3639 | .0092 | .4315 | .0008256 | 5.9281 | 16 |
| 45 | 0.13485 | 0.99087 | 0.13609 | 7.3479 | 1.0092 | 7.4156 | .0008310 | 5.9123 | 15 |
| 46 | .13514 | .99083 | .13639 | .3319 | .0093 | .3998 | .0008364 | 5.8967 | 14 |
| 47 | .13543 | .99079 | .13669 | .3160 | .0093 | .3840 | .0008418 | 5.8811 | 13 |
| 48 | .13572 | .99075 | .13698 | .3002 | .0093 | .3684 | .0008473 | 5.8655 | 12 |
| 49 | .13600 | .99071 | .13728 | .2844 | .0094 | .3527 | .0008527 | 5.8500 | 11 |
| 50 | 0.13629 | 0.99067 | 0.13758 | 7.2687 | 1.0094 | 7.3372 | .0008582 | 5.8346 | 10 |
| 51 | .13658 | .99063 | .13787 | .2531 | .0095 | .3217 | .0008638 | 5.8193 | 9 |
| 52 | .13687 | .99059 | .13817 | .2375 | .0095 | .3063 | .0008693 | 5.8040 | 8 |
| 53 | .13716 | .99055 | .13846 | .2220 | .0095 | .2909 | .0008749 | 5.7888 | 7 |
| 54 | .13744 | .99051 | .13876 | .2066 | .0096 | .2757 | .0008805 | 5.7737 | 6 |
| 55 | 0.13773 | 0.99047 | 0.13906 | 7.1912 | 1.0096 | 7.2604 | .0008861 | 5.7586 | 5 |
| 56 | .13802 | .99043 | .13935 | .1759 | .0097 | .2453 | .0008917 | 5.7436 | 4 |
| 57 | .13831 | .99039 | .13965 | .1607 | .0097 | .2302 | .0008974 | 5.7287 | 3 |
| 58 | .13860 | .99035 | .13995 | .1455 | .0097 | .2152 | .0009031 | 5.7138 | 2 |
| 59 | .13889 | .99031 | .14024 | .1304 | .0098 | .2002 | .0009088 | 5.6990 | 1 |
| 60 | 0.13917 | 0.99027 | 0.14054 | 7.1154 | 1.0098 | 7.1853 | .0009145 | 5.6842 | 0 |
| M | Cosine | Sine | Cotan. | Tan. | Cosec. | Secant | READ DOWN | 82°-83° Involute | M |

**97° or 277°**  **82° or 262°**

| M | Sine | Cosine | Tan. | Cotan. | Secant | Cosec. | Involute 8°–9° | READ UP | M |
|---|------|--------|------|--------|--------|--------|----------------|---------|---|
| 0 | 0.13917 | 0.99027 | 0.14054 | 7.1154 | 1.0098 | 7.1853 | .0009145 | 5.6842 | 60 |
| 1 | .13946 | .99023 | .14084 | .1004 | .0099 | .1705 | .0009203 | 5.6695 | 59 |
| 2 | .13975 | .99019 | .14113 | .0855 | .0099 | .1557 | .0009260 | 5.6549 | 58 |
| 3 | .14004 | .99015 | .14143 | .0706 | .0100 | .1410 | .0009318 | 5.6403 | 57 |
| 4 | .14033 | .99011 | .14173 | .0558 | .0100 | .1263 | .0009377 | 5.6258 | 56 |
| 5 | 0.14061 | 0.99006 | 0.14202 | 7.0410 | 1.0100 | 7.1117 | .0009435 | 5.6113 | 55 |
| 6 | .14090 | .99002 | .14232 | .0264 | .0101 | .0972 | .0009494 | 5.5969 | 54 |
| 7 | .14119 | .98998 | .14262 | .0117 | .0101 | .0827 | .0009553 | 5.5826 | 53 |
| 8 | .14148 | .98994 | .14291 | 6.9972 | .0102 | .0683 | .0009612 | 5.5683 | 52 |
| 9 | .14177 | .98990 | .14321 | .9827 | .0102 | .0539 | .0009672 | 5.5541 | 51 |
| 10 | 0.14205 | 0.98986 | 0.14351 | 6.9682 | 1.0102 | 7.0396 | .0009732 | 5.5400 | 50 |
| 11 | .14234 | .98982 | .14381 | .9538 | .0103 | .0254 | .0009792 | 5.5259 | 49 |
| 12 | .14263 | .98978 | .14410 | .9395 | .0103 | .0112 | .0009852 | 5.5118 | 48 |
| 13 | .14292 | .98973 | .14440 | .9252 | .0104 | 6.9971 | .0009913 | 5.4979 | 47 |
| 14 | .14320 | .98969 | .14470 | .9110 | .0104 | .9830 | .0009973 | 5.4839 | 46 |
| 15 | 0.14349 | 0.98965 | 0.14499 | 6.8969 | 1.0105 | 6.9690 | .0010034 | 5.4701 | 45 |
| 16 | .14378 | .98961 | .14529 | .8828 | .0105 | .9550 | .0010096 | 5.4563 | 44 |
| 17 | .14407 | .98957 | .14559 | .8687 | .0105 | .9411 | .0010157 | 5.4425 | 43 |
| 18 | .14436 | .98953 | .14588 | .8547 | .0106 | .9273 | .0010219 | 5.4288 | 42 |
| 19 | .14464 | .98948 | .14618 | .8408 | .0106 | .9135 | .0010281 | 5.4152 | 41 |
| 20 | 0.14493 | 0.98944 | 0.14648 | 6.8269 | 1.0107 | 6.8998 | .0010343 | 5.4016 | 40 |
| 21 | .14522 | .98940 | .14678 | .8131 | .0107 | .8861 | .0010406 | 5.3881 | 39 |
| 22 | .14551 | .98936 | .14707 | .7994 | .0108 | .8725 | .0010469 | 5.3746 | 38 |
| 23 | .14580 | .98931 | .14737 | .7856 | .0108 | .8589 | .0010532 | 5.3612 | 37 |
| 24 | .14608 | .98927 | .14767 | .7720 | .0108 | .8454 | .0010595 | 5.3478 | 36 |
| 25 | 0.14637 | 0.98923 | 0.14796 | 6.7584 | 1.0109 | 6.8320 | .0010659 | 5.3345 | 35 |
| 26 | .14666 | .98919 | .14826 | .7448 | .0109 | .8186 | .0010722 | 5.3212 | 34 |
| 27 | .14695 | .98914 | .14856 | .7313 | .0110 | .8052 | .0010786 | 5.3080 | 33 |
| 28 | .14723 | .98910 | .14886 | .7179 | .0110 | .7919 | .0010851 | 5.2949 | 32 |
| 29 | .14752 | .98906 | .14915 | .7045 | .0111 | .7787 | .0010915 | 5.2818 | 31 |
| 30 | 0.14781 | 0.98902 | 0.14945 | 6.6912 | 1.0111 | 6.7655 | .0010980 | 5.2687 | 30 |
| 31 | .14810 | .98897 | .14975 | .6779 | .0112 | .7523 | .0011045 | 5.2557 | 29 |
| 32 | .14838 | .98893 | .15005 | .6646 | .0112 | .7392 | .0011111 | 5.2428 | 28 |
| 33 | .14867 | .98889 | .15034 | .6514 | .0112 | .7262 | .0011176 | 5.2299 | 27 |
| 34 | .14896 | .98884 | .15064 | .6383 | .0113 | .7132 | .0011242 | 5.2170 | 26 |
| 35 | 0.14925 | 0.98880 | 0.15094 | 6.6252 | 1.0113 | 6.7003 | .0011308 | 5.2042 | 25 |
| 36 | .14954 | .98876 | .15124 | .6122 | .0114 | .6874 | .0011375 | 5.1915 | 24 |
| 37 | .14982 | .98871 | .15153 | .5992 | .0114 | .6745 | .0011441 | 5.1788 | 23 |
| 38 | .15011 | .98867 | .15183 | .5863 | .0115 | .6618 | .0011508 | 5.1662 | 22 |
| 39 | .15040 | .98863 | .15213 | .5734 | .0115 | .6490 | .0011575 | 5.1536 | 21 |
| 40 | 0.15069 | 0.98858 | 0.15243 | 6.5606 | 1.0116 | 6.6363 | .0011643 | 5.1410 | 20 |
| 41 | .15097 | .98854 | .15272 | .5478 | .0116 | .6237 | .0011711 | 5.1285 | 19 |
| 42 | .15126 | .98849 | .15302 | .5350 | .0116 | .6111 | .0011779 | 5.1161 | 18 |
| 43 | .15155 | .98845 | .15332 | .5223 | .0117 | .5986 | .0011847 | 5.1037 | 17 |
| 44 | .15184 | .98841 | .15362 | .5097 | .0117 | .5861 | .0011915 | 5.0913 | 16 |
| 45 | 0.15212 | 0.98836 | 0.15391 | 6.4971 | 1.0118 | 6.5736 | .0011984 | 5.0790 | 15 |
| 46 | .15241 | .98832 | .15421 | .4846 | .0118 | .5612 | .0012053 | 5.0668 | 14 |
| 47 | .15270 | .98827 | .15451 | .4721 | .0119 | .5489 | .0012122 | 5.0546 | 13 |
| 48 | .15299 | .98823 | .15481 | .4596 | .0119 | .5366 | .0012192 | 5.0424 | 12 |
| 49 | .15327 | .98818 | .15511 | .4472 | .0120 | .5243 | .0012262 | 5.0303 | 11 |
| 50 | 0.15356 | 0.98814 | 0.15540 | 6.4348 | 1.0120 | 6.5121 | .0012332 | 5.0182 | 10 |
| 51 | .15385 | .98809 | .15570 | .4225 | .0120 | .4999 | .0012402 | 5.0062 | 9 |
| 52 | .15414 | .98805 | .15600 | .4103 | .0121 | .4878 | .0012473 | 4.9942 | 8 |
| 53 | .15442 | .98800 | .15630 | .3980 | .0121 | .4757 | .0012544 | 4.9823 | 7 |
| 54 | .15471 | .98796 | .15660 | .3859 | .0122 | .4637 | .0012615 | 4.9704 | 6 |
| 55 | 0.15500 | 0.98791 | 0.15689 | 6.3737 | 1.0122 | 6.4517 | .0012687 | 4.9586 | 5 |
| 56 | .15529 | .98787 | .15719 | .3617 | .0123 | .4398 | .0012758 | 4.9468 | 4 |
| 57 | .15557 | .98782 | .15749 | .3496 | .0123 | .4279 | .0012830 | 4.9350 | 3 |
| 58 | .15586 | .98778 | .15779 | .3376 | .0124 | .4160 | .0012903 | 4.9233 | 2 |
| 59 | .15615 | .98773 | .15809 | .3257 | .0124 | .4042 | .0012975 | 4.9117 | 1 |
| 60 | 0.15643 | 0.98769 | 0.15838 | 6.3138 | 1.0125 | 6.3925 | .0013048 | 4.9000 | 0 |
| M | Cosine | Sine | Cotan. | Tan. | Cosec. | Secant | READ DOWN | 81°–82° Involute | M |

**9° or 189°**  Trigonometric Functions  **170° or 350°**

| M | Sine | Cosine | Tan. | Cotan. | Secant | Cosec. | Involute 9°–10° | READ UP | M |
|---|---|---|---|---|---|---|---|---|---|
| 0 | 0.15643 | 0.98769 | 0.15838 | 6.3138 | 1.0125 | 6.3925 | .0013048 | 4.9000 | 60 |
| 1 | .15672 | .98764 | .15868 | .3019 | .0125 | .3807 | .0013121 | 4.8885 | 59 |
| 2 | .15701 | .98760 | .15898 | .2901 | .0126 | .3691 | .0013195 | 4.8769 | 58 |
| 3 | .15730 | .98755 | .15928 | .2783 | .0126 | .3574 | .0013268 | 4.8654 | 57 |
| 4 | .15758 | .98751 | .15958 | .2666 | .0127 | .3458 | .0013342 | 4.8540 | 56 |
| 5 | 0.15787 | 0.98746 | 0.15988 | 6.2549 | 1.0127 | 6.3343 | .0013416 | 4.8426 | 55 |
| 6 | .15816 | .98741 | .16017 | .2432 | .0127 | .3228 | .0013491 | 4.8312 | 54 |
| 7 | .15845 | .98737 | .16047 | .2316 | .0128 | .3113 | .0013566 | 4.8199 | 53 |
| 8 | .15873 | .98732 | .16077 | .2200 | .0128 | .2999 | .0013641 | 4.8086 | 52 |
| 9 | .15902 | .98728 | .16107 | .2085 | .0129 | .2885 | .0013716 | 4.7974 | 51 |
| 10 | 0.15931 | 0.98723 | 0.16137 | 6.1970 | 1.0129 | 6.2772 | .0013792 | 4.7862 | 50 |
| 11 | .15959 | .98718 | .16167 | .1856 | .0130 | .2659 | .0013868 | 4.7511 | 49 |
| 12 | .15988 | .98714 | .16196 | .1742 | .0130 | .2546 | .0013944 | 4.7640 | 48 |
| 13 | .16017 | .98709 | .16226 | .1628 | .0131 | .2434 | .0014020 | 4.7529 | 47 |
| 14 | .16046 | .98704 | .16256 | .1515 | .0131 | .2323 | .0014097 | 4.7419 | 46 |
| 15 | 0.16074 | 0.98700 | 0.16286 | 6.1402 | 1.0132 | 6.2211 | .0014174 | 4.7309 | 45 |
| 16 | .16103 | .98695 | .16316 | .1290 | .0132 | .2100 | .0014251 | 4.7199 | 44 |
| 17 | .16132 | .98690 | .16346 | .1178 | .0133 | .1990 | .0014329 | 4.7090 | 43 |
| 18 | .16160 | .98686 | .16376 | .1066 | .0133 | .1880 | .0014407 | 4.6982 | 42 |
| 19 | .16189 | .98681 | .16405 | .0955 | .0134 | .1770 | .0014485 | 4.6873 | 41 |
| 20 | 0.16218 | 0.98676 | 0.16435 | 6.0844 | 1.0134 | 6.1661 | .0014563 | 4.6765 | 40 |
| 21 | .16246 | .98671 | .16465 | .0734 | .0135 | .1552 | .0014642 | 4.6658 | 39 |
| 22 | .16275 | .98667 | .16495 | .0624 | .0135 | .1443 | .0014721 | 4.6551 | 38 |
| 23 | .16304 | .98662 | .16525 | .0514 | .0136 | .1335 | .0014800 | 4.6444 | 37 |
| 24 | .16333 | .98657 | .16555 | .0405 | .0136 | .1227 | .0014880 | 4.6338 | 36 |
| 25 | 0.16361 | 0.98652 | 0.16585 | 6.0296 | 1.0137 | 6.1120 | .0014960 | 4.6232 | 35 |
| 26 | .16390 | .98648 | .16615 | .0188 | .0137 | .1013 | .0015040 | 4.6126 | 34 |
| 27 | .16419 | .98643 | .16645 | .0080 | .0138 | .0906 | .0015120 | 4.6021 | 33 |
| 28 | .16447 | .98638 | .16674 | 5.9972 | .0138 | .0800 | .0015201 | 4.5916 | 32 |
| 29 | .16476 | .98633 | .16704 | .9865 | .0139 | .0694 | .0015282 | 4.5812 | 31 |
| 30 | 0.16505 | 0.98629 | 0.16734 | 5.9758 | 1.0139 | 6.0589 | .0015363 | 4.5708 | 30 |
| 31 | .16533 | .98624 | .16764 | .9651 | .0140 | .0483 | .0015445 | 4.5604 | 29 |
| 32 | .16562 | .98619 | .16794 | .9545 | .0140 | .0379 | .0015527 | 4.5501 | 28 |
| 33 | .16591 | .98614 | .16824 | .9439 | .0141 | .0274 | .0015609 | 4.5398 | 27 |
| 34 | .16620 | .98609 | .16854 | .9333 | .0141 | .0170 | .0015691 | 4.5295 | 26 |
| 35 | 0.16648 | 0.98604 | 0.16884 | 5.9228 | 1.0142 | 6.0067 | .0015774 | 4.5193 | 25 |
| 36 | .16677 | .98600 | .16914 | .9124 | .0142 | 5.9963 | .0015857 | 4.5091 | 24 |
| 37 | .16706 | .98595 | .16944 | .9019 | .0143 | .9860 | .0015941 | 4.4990 | 23 |
| 38 | .16734 | .98590 | .16974 | .8915 | .0143 | .9758 | .0016024 | 4.4888 | 22 |
| 39 | .16763 | .98585 | .17004 | .8811 | .0144 | .9656 | .0016108 | 4.4788 | 21 |
| 40 | 0.16792 | 0.98580 | 0.17033 | 5.8708 | 1.0144 | 5.9554 | .0016193 | 4.4687 | 20 |
| 41 | .16820 | .98575 | .17063 | .8605 | .0145 | .9452 | .0016277 | 4.4587 | 19 |
| 42 | .16849 | .98570 | .17093 | .8502 | .0145 | .9351 | .0016362 | 4.4487 | 18 |
| 43 | .16878 | .98565 | .17123 | .8400 | .0146 | .9250 | .0016447 | 4.4388 | 17 |
| 44 | .16906 | .98561 | .17153 | .8298 | .0146 | .9150 | .0016533 | 4.4289 | 16 |
| 45 | 0.16935 | 0.98556 | 0.17183 | 5.8197 | 1.0147 | 5.9049 | .0016618 | 4.4190 | 15 |
| 46 | .16964 | .98551 | .17213 | .8095 | .0147 | .8950 | .0016704 | 4.4092 | 14 |
| 47 | .16992 | .98546 | .17243 | .7994 | .0148 | .8850 | .0016791 | 4.3994 | 13 |
| 48 | .17021 | .98541 | .17273 | .7894 | .0148 | .8751 | .0016877 | 4.3896 | 12 |
| 49 | .17050 | .98536 | .17303 | .7794 | .0149 | .8652 | .0016964 | 4.3799 | 11 |
| 50 | 0.17078 | 0.98531 | 0.17333 | 5.7694 | 1.0149 | 5.8554 | .0017051 | 4.3702 | 10 |
| 51 | .17107 | .98526 | .17363 | .7594 | .0150 | .8456 | .0017139 | 4.3605 | 9 |
| 52 | .17136 | .98521 | .17393 | .7495 | .0150 | .8358 | .0017227 | 4.3509 | 8 |
| 53 | .17164 | .98516 | .17423 | .7396 | .0151 | .8261 | .0017315 | 4.3413 | 7 |
| 54 | .17193 | .98511 | .17453 | .7297 | .0151 | .8164 | .0017403 | 4.3317 | 6 |
| 55 | 0.17222 | 0.98506 | 0.17483 | 5.7199 | 1.0152 | 5.8067 | .0017492 | 4.3222 | 5 |
| 56 | .17250 | .98501 | .17513 | .7101 | .0152 | .7970 | .0017581 | 4.3127 | 4 |
| 57 | .17279 | .98496 | .17543 | .7004 | .0153 | .7874 | .0017671 | 4.3032 | 3 |
| 58 | .17308 | .98491 | .17573 | .6906 | .0153 | .7778 | .0017760 | 4.2938 | 2 |
| 59 | .17336 | .98486 | .17603 | .6809 | .0154 | .7683 | .0017850 | 4.2844 | 1 |
| 60 | 0.17365 | 0.98481 | 0.17633 | 5.6713 | 1.0154 | 5.7588 | .0017941 | 4.2750 | 0 |

| M | Cosine | Sine | Cotan. | Tan. | Cosec. | Secant | READ DOWN | 80°–81° Involute | M |
|---|---|---|---|---|---|---|---|---|---|

**99° or 279°**  **80° or 260°**

**10° or 190°**     Trigonometric Functions     **169° or 349°**

| M | Sine | Cosine | Tan. | Cotan. | Secant | Cosec. | Involute 10°–11° | READ UP | M |
|---|---|---|---|---|---|---|---|---|---|
| 0 | 0.17365 | 0.98481 | 0.17633 | 5.6713 | 1.0154 | 5.7588 | .0017941 | 4.2750 | 60 |
| 1 | .17393 | .98476 | .17663 | .6617 | .0155 | .7493 | .0018031 | 4.2657 | 59 |
| 2 | .17422 | .98471 | .17693 | .6521 | .0155 | .7398 | .0018122 | 4.2564 | 58 |
| 3 | .17451 | .98466 | .17723 | .6425 | .0156 | .7304 | .0018213 | 4.2471 | 57 |
| 4 | .17479 | .98461 | .17753 | .6329 | .0156 | .7210 | .0018305 | 4.2378 | 56 |
| 5 | 0.17508 | 0.98455 | 0.17783 | 5.6234 | 1.0157 | 5.7117 | .0018397 | 4.2286 | 55 |
| 6 | .17537 | .98450 | .17813 | .6140 | .0157 | .7023 | .0018489 | 4.2194 | 54 |
| 7 | .17565 | .98445 | .17843 | .6045 | .0158 | .6930 | .0018581 | 4.2103 | 53 |
| 8 | .17594 | .98440 | .17873 | .5951 | .0158 | .6838 | .0018674 | 4.2012 | 52 |
| 9 | .17623 | .98435 | .17903 | .5857 | .0159 | .6745 | .0018767 | 4.1921 | 51 |
| 10 | 0.17651 | 0.98430 | 0.17933 | 5.5764 | 1.0160 | 5.6653 | .0018860 | 4.1830 | 50 |
| 11 | .17680 | .98425 | .17963 | .5671 | .0160 | .6562 | .0018954 | 4.1740 | 49 |
| 12 | .17708 | .98420 | .17993 | .5578 | .0161 | .6470 | .0019048 | 4.1650 | 48 |
| 13 | .17737 | .98414 | .18023 | .5485 | .0161 | .6379 | .0019142 | 4.1560 | 47 |
| 14 | .17766 | .98409 | .18053 | .5393 | .0162 | .6288 | .0019237 | 4.1471 | 46 |
| 15 | 0.17794 | 0.98404 | 0.18083 | 5.5301 | 1.0162 | 5.6198 | .0019332 | 4.1382 | 45 |
| 16 | .17823 | .98399 | .18113 | .5209 | .0163 | .6107 | .0019427 | 4.1293 | 44 |
| 17 | .17852 | .98394 | .18143 | .5118 | .0163 | .6017 | .0019523 | 4.1204 | 43 |
| 18 | .17880 | .98389 | .18173 | .5026 | .0164 | .5928 | .0019619 | 4.1116 | 42 |
| 19 | .17909 | .98383 | .18203 | .4936 | .0164 | .5838 | .0019715 | 4.1028 | 41 |
| 20 | 0.17937 | 0.98378 | 0.18233 | 5.4845 | 1.0165 | 5.5749 | .0019812 | 4.0941 | 40 |
| 21 | .17966 | .98373 | .18263 | .4755 | .0165 | .5660 | .0019909 | 4.0853 | 39 |
| 22 | .17995 | .98368 | .18293 | .4665 | .0166 | .5572 | .0020006 | 4.0766 | 38 |
| 23 | .18023 | .98362 | .18323 | .4575 | .0166 | .5484 | .0020103 | 4.0679 | 37 |
| 24 | .18052 | .98357 | .18353 | .4486 | .0167 | .5396 | .0020201 | 4.0593 | 36 |
| 25 | 0.18081 | 0.98352 | 0.18384 | 5.4397 | 1.0168 | 5.5308 | .0020299 | 4.0507 | 35 |
| 26 | .18109 | .98347 | .18414 | .4308 | .0168 | .5221 | .0020398 | 4.0421 | 34 |
| 27 | .18138 | .98341 | .18444 | .4219 | .0169 | .5134 | .0020496 | 4.0335 | 33 |
| 28 | .18166 | .98336 | .18474 | .4131 | .0169 | .5047 | .0020596 | 4.0250 | 32 |
| 29 | .18195 | .98331 | .18504 | .4243 | .0170 | .4960 | .0020695 | 4.0165 | 31 |
| 30 | 0.18224 | 0.98325 | 0.18534 | 5.3955 | 1.0170 | 5.4874 | .0020795 | 4.0080 | 30 |
| 31 | .18252 | .98320 | .18564 | .3868 | .0171 | .4788 | .0020895 | 3.9995 | 29 |
| 32 | .18281 | .98315 | .18594 | .3781 | .0171 | .4702 | .0020995 | 3.9911 | 28 |
| 33 | .18309 | .98310 | .18624 | .3694 | .0172 | .4617 | .0021096 | 3.9827 | 27 |
| 34 | .18338 | .98304 | .18654 | .3607 | .0173 | .4532 | .0021197 | 3.9743 | 26 |
| 35 | 0.18367 | 0.98299 | 0.18684 | 5.3521 | 1.0173 | 5.4447 | .0021298 | 3.9660 | 25 |
| 36 | .18395 | .98294 | .18714 | .3435 | .0174 | .4362 | .0021400 | 3.9577 | 24 |
| 37 | .18424 | .98288 | .18745 | .3349 | .0174 | .4278 | .0021502 | 3.9494 | 23 |
| 38 | .18452 | .98283 | .18775 | .3263 | .0175 | .4194 | .0021605 | 3.9411 | 22 |
| 39 | .18481 | .98277 | .18805 | .3178 | .0175 | .4110 | .0021707 | 3.9329 | 21 |
| 40 | 0.18509 | 0.98272 | 0.18835 | 5.3093 | 1.0176 | 5.4026 | .0021810 | 3.9247 | 20 |
| 41 | .18538 | .98267 | .18865 | .3008 | .0176 | .3943 | .0021914 | 3.9165 | 19 |
| 42 | .18567 | .98261 | .18895 | .2924 | .0177 | .3860 | .0022017 | 3.9083 | 18 |
| 43 | .18595 | .98256 | .18925 | .2839 | .0178 | .3777 | .0022121 | 3.9002 | 17 |
| 44 | .18624 | .98250 | .18955 | .2755 | .0178 | .3695 | .0022226 | 3.8921 | 16 |
| 45 | 0.18652 | 0.98245 | 0.18986 | 5.2672 | 1.0179 | 5.3612 | .0022330 | 3.8840 | 15 |
| 46 | .18681 | .98240 | .19016 | .2588 | .0179 | .3530 | .0022435 | 3.8759 | 14 |
| 47 | .18710 | .98234 | .19046 | .2505 | .0180 | .3449 | .0022541 | 3.8679 | 13 |
| 48 | .18738 | .98229 | .19076 | .2422 | .0180 | .3367 | .0022646 | 3.8599 | 12 |
| 49 | .18767 | .98223 | .19106 | .2339 | .0181 | .3286 | .0022752 | 3.8519 | 11 |
| 50 | 0.18795 | 0.98218 | 0.19136 | 5.2257 | 1.0181 | 5.3205 | .0022859 | 3.8439 | 10 |
| 51 | .18824 | .98212 | .19166 | .2174 | .0182 | .3124 | .0022965 | 3.8360 | 9 |
| 52 | .18852 | .98207 | .19197 | .2092 | .0183 | .3044 | .0023073 | 3.8281 | 8 |
| 53 | .18881 | .98201 | .19227 | .2011 | .0183 | .2963 | .0023180 | 3.8202 | 7 |
| 54 | .18910 | .98196 | .19257 | .1929 | .0184 | .2883 | .0023288 | 3.8124 | 6 |
| 55 | 0.18938 | 0.98190 | 0.19287 | 5.1848 | 1.0184 | 5.2804 | .0023396 | 3.8045 | 5 |
| 56 | .18967 | .98185 | .19317 | .1767 | .0185 | .2724 | .0023504 | 3.7967 | 4 |
| 57 | .18995 | .98179 | .19347 | .1686 | .0185 | .2645 | .0023613 | 3.7889 | 3 |
| 58 | .19024 | .98173 | .19378 | .1606 | .0186 | .2566 | .0023722 | 3.7812 | 2 |
| 59 | .19052 | .98168 | .19408 | .1526 | .0187 | .2487 | .0023831 | 3.7735 | 1 |
| 60 | 0.19081 | 0.98163 | 0.19438 | 5.1446 | 1.0187 | 5.2408 | .0023941 | 3.7657 | 0 |
| M | Cosine | Sine | Cotan. | Tan. | Cosec. | Secant | READ DOWN | 79°–80° Involute | M |

**100° or 280°**                       **79° or 259°**

**11° or 191°**     Trigonometric Functions     **168° or 348°**

| M | Sine | Cosine | Tan. | Cotan. | Secant | Cosec. | Involute 11°–12° | READ UP | M |
|---|------|--------|------|--------|--------|--------|---------|---------|---|
| 0 | 0.19081 | 0.98163 | 0.19438 | 5.1446 | 1.0187 | 5.2408 | .0023941 | 3.7657 | 60 |
| 1 | .19109 | .98157 | .19468 | .1366 | .0188 | .2330 | .0024051 | 3.7581 | 59 |
| 2 | .19138 | .98152 | .19498 | .1286 | .0188 | .2252 | .0024161 | 3.7504 | 58 |
| 3 | .19167 | .98146 | .19529 | .1207 | .0189 | .2174 | .0024272 | 3.7428 | 57 |
| 4 | .19195 | .98140 | .19559 | .1128 | .0189 | .2097 | .0024383 | 3.7351 | 56 |
| 5 | 0.19224 | 0.98135 | 0.19589 | 5.1049 | 1.0190 | 5.2019 | .0024495 | 3.7275 | 55 |
| 6 | .19252 | .98129 | .19619 | .0970 | .0191 | .1942 | .0024607 | 3.7200 | 54 |
| 7 | .19281 | .98124 | .19649 | .0892 | .0191 | .1865 | .0024719 | 3.7124 | 53 |
| 8 | .19309 | .98118 | .19680 | .0814 | .0192 | .1789 | .0024831 | 3.7049 | 52 |
| 9 | .19338 | .98112 | .19710 | .0736 | .0192 | .1712 | .0024944 | 3.6974 | 51 |
| 10 | 0.19366 | 0.98107 | 0.19740 | 5.0658 | 1.0193 | 5.1636 | .0025057 | 3.6899 | 50 |
| 11 | .19395 | .98101 | .19770 | .0581 | .0194 | .1560 | .0025171 | 3.6825 | 49 |
| 12 | .19423 | .98096 | .19801 | .0504 | .0194 | .1484 | .0025285 | 3.6750 | 48 |
| 13 | .19452 | .98090 | .19831 | .0427 | .0195 | .1409 | .0025399 | 3.6676 | 47 |
| 14 | .19481 | .98084 | .19861 | .0350 | .0195 | .1333 | .0025513 | 3.6603 | 46 |
| 15 | 0.19509 | 0.98079 | 0.19891 | 5.0273 | 1.0196 | 5.1258 | .0025628 | 3.6529 | 45 |
| 16 | .19538 | .98073 | .19921 | .0197 | .0197 | .1183 | .0025744 | 3.6456 | 44 |
| 17 | .19566 | .98067 | .19952 | .0121 | .0197 | .1109 | .0025859 | 3.6382 | 43 |
| 18 | .19595 | .98061 | .19982 | .0045 | .0198 | .1034 | .0025975 | 3.6309 | 42 |
| 19 | .19623 | .98056 | .20012 | 4.9969 | .0198 | .0960 | .0026091 | 3.6237 | 41 |
| 20 | 0.19652 | 0.98050 | 0.20042 | 4.9894 | 1.0199 | 5.0886 | .0026208 | 3.6164 | 40 |
| 21 | .19680 | .98044 | .20073 | .9819 | .0199 | .0813 | .0026325 | 3.6092 | 39 |
| 22 | .19709 | .98039 | .20103 | .9744 | .0200 | .0739 | .0026443 | 3.6020 | 38 |
| 23 | .19737 | .98033 | .20133 | .9669 | .0201 | .0666 | .0026560 | 3.5948 | 37 |
| 24 | .19766 | .98027 | .20164 | .9594 | .0201 | .0593 | .0026678 | 3.5876 | 36 |
| 25 | 0.19794 | 0.98021 | 0.20194 | 4.9520 | 1.0202 | 5.0520 | .0026797 | 3.5805 | 35 |
| 26 | .19823 | .98016 | .20224 | .9446 | .0202 | .0447 | .0026916 | 3.5734 | 34 |
| 27 | .19851 | .98010 | .20254 | .9372 | .0203 | .0375 | .0027035 | 3.5663 | 33 |
| 28 | .19880 | .98004 | .20285 | .9298 | .0204 | .0302 | .0027154 | 3.5592 | 32 |
| 29 | .19908 | .97998 | .20315 | .9225 | .0204 | .0230 | .0027274 | 3.5521 | 31 |
| 30 | 0.19937 | 0.97992 | 0.20345 | 4.9152 | 1.0205 | 5.0159 | .0027394 | 3.5451 | 30 |
| 31 | .19965 | .97987 | .20376 | .9078 | .0205 | .0087 | .0027515 | 3.5381 | 29 |
| 32 | .19994 | .97981 | .20406 | .9006 | .0206 | .0016 | .0027636 | 3.5311 | 28 |
| 33 | .20022 | .97975 | .20436 | .8933 | .0207 | 4.9944 | .0027757 | 3.5241 | 27 |
| 34 | .20051 | .97969 | .20466 | .8860 | .0207 | .9873 | .0027879 | 3.5171 | 26 |
| 35 | 0.20079 | 0.97963 | 0.20497 | 4.8788 | 1.0208 | 4.9803 | .0028001 | 3.5102 | 25 |
| 36 | .20108 | .97958 | .20527 | .8716 | .0209 | .9732 | .0028123 | 3.5033 | 24 |
| 37 | .20136 | .97952 | .20557 | .8644 | .0209 | .9662 | .0028246 | 3.4964 | 23 |
| 38 | .20165 | .97946 | .20588 | .8573 | .0210 | .9591 | .0028369 | 3.4895 | 22 |
| 39 | .20193 | .97940 | .20618 | .8501 | .0210 | .9521 | .0028493 | 3.4827 | 21 |
| 40 | 0.20222 | 0.97934 | 0.20648 | 4.8430 | 1.0211 | 4.9452 | .0028616 | 3.4758 | 20 |
| 41 | .20250 | .97928 | .20679 | .8359 | .0212 | .9382 | .0028741 | 3.4690 | 19 |
| 42 | .20279 | .97922 | .20709 | .8288 | .0212 | .9313 | .0028865 | 3.4622 | 18 |
| 43 | .20307 | .97916 | .20739 | .8218 | .0213 | .9244 | .0028990 | 3.4555 | 17 |
| 44 | .20336 | .97910 | .20770 | .8147 | .0213 | .9175 | .0029115 | 3.4487 | 16 |
| 45 | 0.20364 | 0.97905 | 0.20800 | 4.8077 | 1.0214 | 4.9106 | .0029241 | 3.4420 | 15 |
| 46 | .20393 | .97899 | .20830 | .8007 | .0215 | .9037 | .0029367 | 3.4353 | 14 |
| 47 | .20421 | .97893 | .20861 | .7937 | .0215 | .8969 | .0029494 | 3.4286 | 13 |
| 48 | .20450 | .97887 | .20891 | .7867 | .0216 | .8901 | .0029620 | 3.4219 | 12 |
| 49 | .20478 | .97881 | .20921 | .7799 | .0217 | .8833 | .0029747 | 3.4152 | 11 |
| 50 | 0.20507 | 0.97875 | 0.20952 | 4.7729 | 1.0217 | 4.8765 | .0029875 | 3.4086 | 10 |
| 51 | .20535 | .97869 | .20982 | .7659 | .0218 | .8697 | .0030003 | 3.4020 | 9 |
| 52 | .20563 | .97863 | .21013 | .7591 | .0218 | .8630 | .0030131 | 3.3954 | 8 |
| 53 | .20592 | .97857 | .21043 | .7522 | .0219 | .8563 | .0030260 | 3.3888 | 7 |
| 54 | .20620 | .97851 | .21073 | .7453 | .0220 | .8496 | .0030389 | 3.3822 | 6 |
| 55 | 0.20649 | 0.97845 | 0.21104 | 4.7385 | 1.0220 | 4.8429 | .0030518 | 3.3757 | 5 |
| 56 | .20677 | .97839 | .21134 | .7317 | .0221 | .8362 | .0030648 | 3.3692 | 4 |
| 57 | .20706 | .97833 | .21164 | .7249 | .0222 | .8296 | .0030778 | 3.3627 | 3 |
| 58 | .20734 | .97827 | .21195 | .7181 | .0222 | .8229 | .0030908 | 3.3562 | 2 |
| 59 | .20763 | .97821 | .21225 | .7114 | .0223 | .8163 | .0031039 | 3.3497 | 1 |
| 60 | 0.20791 | 0.97815 | 0.21256 | 4.7046 | 1.0223 | 4.8097 | .0031171 | 3.3433 | 0 |

| M | Cosine | Sine | Cotan. | Tan. | Cosec. | Secant | READ DOWN | 78°–79° Involute | M |

**101° or 281°**     **78° or 258°**

| M | Sine | Cosine | Tan. | Cotan. | Secant | Cosec. | Involute 12°–13° | READ UP | M |
|---|---|---|---|---|---|---|---|---|---|
| 0 | 0.20791 | 0.97815 | 0.21256 | 4.7046 | 1.0223 | 4.8097 | .0031171 | 3.3433 | 60 |
| 1 | .20820 | .97809 | .21286 | .6979 | .0224 | .8032 | .0031302 | 3.3368 | 59 |
| 2 | .20848 | .97803 | .21316 | .6912 | .0225 | .7966 | .0031434 | 3.3304 | 58 |
| 3 | .20877 | .97797 | .21347 | .6845 | .0225 | .7901 | .0031566 | 3.3240 | 57 |
| 4 | .20905 | .97791 | .21377 | .6779 | .0226 | .7836 | .0031699 | 3.3177 | 56 |
| 5 | 0.20933 | .97784 | 0.21408 | 4.6712 | 1.0227 | 4.7771 | .0031832 | 3.3113 | 55 |
| 6 | .20962 | .97778 | .21438 | .6646 | .0227 | .7706 | .0031966 | 3.3050 | 54 |
| 7 | .20990 | .97772 | .21469 | .6580 | .0228 | .7641 | .0032100 | 3.2987 | 53 |
| 8 | .21019 | .97766 | .21499 | .6514 | .0228 | .7577 | .0032234 | 3.2923 | 52 |
| 9 | .21047 | .97760 | .21529 | .6448 | .0229 | .7512 | .0032369 | 3.2861 | 51 |
| 10 | 0.21076 | 0.97754 | 0.21560 | 4.6382 | 1.0230 | 4.7448 | .0032504 | 3.2798 | 50 |
| 11 | .21104 | .97748 | .21590 | .6317 | .0230 | .7384 | .0032639 | 3.2735 | 49 |
| 12 | .21132 | .97742 | .21621 | .6252 | .0231 | .7321 | .0032775 | 3.2673 | 48 |
| 13 | .21161 | .97735 | .21651 | .6187 | .0232 | .7257 | .0032911 | 3.2611 | 47 |
| 14 | .21189 | .97729 | .21682 | .6122 | .0232 | .7194 | .0033048 | 3.2549 | 46 |
| 15 | 0.21218 | 0.97723 | 0.21712 | 4.6057 | 1.0233 | 4.7130 | .0033185 | 3.2487 | 45 |
| 16 | .21246 | .97717 | .21743 | .5993 | .0234 | .7067 | .0033322 | 3.2426 | 44 |
| 17 | .21275 | .97711 | .21773 | .5928 | .0234 | .7004 | .0033460 | 3.2364 | 43 |
| 18 | .21303 | .97705 | .21804 | .5864 | .0235 | .6942 | .0033598 | 3.2303 | 42 |
| 19 | .21331 | .97698 | .21834 | .5800 | .0236 | .6879 | .0033736 | 3.2242 | 41 |
| 20 | 0.21360 | 0.97692 | 0.21864 | 4.5736 | 1.0236 | 4.6817 | .0033875 | 3.2181 | 40 |
| 21 | .21388 | .97686 | .21895 | .5673 | .0237 | .6755 | .0034014 | 3.2120 | 39 |
| 22 | .21417 | .97680 | .21925 | .5609 | .0238 | .6693 | .0034154 | 3.2060 | 38 |
| 23 | .21445 | .97673 | .21956 | .5546 | .0238 | .6631 | .0034294 | 3.1999 | 37 |
| 24 | .21474 | .97667 | .21986 | .5483 | .0239 | .6569 | .0034434 | 3.1939 | 36 |
| 25 | 0.21502 | 0.97661 | 0.22017 | 4.5420 | 1.0240 | 4.6507 | .0034575 | 3.1879 | 35 |
| 26 | .21530 | .97655 | .22047 | .5357 | .0240 | .6446 | .0034716 | 3.1819 | 34 |
| 27 | .21559 | .97648 | .22078 | .5294 | .0241 | .6385 | .0034858 | 3.1759 | 33 |
| 28 | .21587 | .97642 | .22108 | .5232 | .0241 | .6324 | .0035000 | 3.1699 | 32 |
| 29 | .21616 | .97636 | .22139 | .5169 | .0242 | .6263 | .0035142 | 3.1640 | 31 |
| 30 | 0.21644 | 0.97630 | 0.22169 | 4.5107 | 1.0243 | 4.6202 | .0035285 | 3.1581 | 30 |
| 31 | .21672 | .97623 | .22200 | .5045 | .0243 | .6142 | .0035428 | 3.1521 | 29 |
| 32 | .21701 | .97617 | .22231 | .4983 | .0244 | .6081 | .0035572 | 3.1463 | 28 |
| 33 | .21729 | .97611 | .22261 | .4922 | .0245 | .6021 | .0035716 | 3.1404 | 27 |
| 34 | .21758 | .97604 | .22292 | .4860 | .0245 | .5961 | .0035860 | 3.1345 | 26 |
| 35 | 0.21786 | 0.97598 | 0.22322 | 4.4799 | 1.0246 | 4.5901 | .0036005 | 3.1287 | 25 |
| 36 | .21814 | .97592 | .22353 | .4737 | .0247 | .5841 | .0036150 | 3.1229 | 24 |
| 37 | .21843 | .97585 | .22383 | .4676 | .0247 | .5782 | .0036296 | 3.1170 | 23 |
| 38 | .21871 | .97579 | .22414 | .4615 | .0248 | .5722 | .0036441 | 3.1112 | 22 |
| 39 | .21899 | .97573 | .22444 | .4555 | .0249 | .5663 | .0036588 | 3.1055 | 21 |
| 40 | 0.21928 | 0.97566 | 0.22475 | 4.4494 | 1.0249 | 4.5604 | .0036735 | 3.0997 | 20 |
| 41 | .21956 | .97560 | .22505 | .4434 | .0250 | .5545 | .0036882 | 3.0939 | 19 |
| 42 | .21985 | .97553 | .22536 | .4373 | .0251 | .5486 | .0037029 | 3.0882 | 18 |
| 43 | .22013 | .97547 | .22567 | .4313 | .0251 | .5428 | .0037177 | 3.0825 | 17 |
| 44 | .22041 | .97541 | .22597 | .4253 | .0252 | .5369 | .0037325 | 3.0768 | 16 |
| 45 | 0.22070 | 0.97534 | 0.22628 | 4.4194 | 1.0253 | 4.5311 | .0037474 | 3.0711 | 15 |
| 46 | .22098 | .97528 | .22658 | .4134 | .0253 | .5253 | .0037623 | 3.0654 | 14 |
| 47 | .22126 | .97521 | .22689 | .4075 | .0254 | .5195 | .0037773 | 3.0598 | 13 |
| 48 | .22155 | .97515 | .22719 | .4015 | .0255 | .5137 | .0037923 | 3.0541 | 12 |
| 49 | .22183 | .97508 | .22750 | .3956 | .0256 | .5079 | .0038073 | 3.0485 | 11 |
| 50 | 0.22212 | 0.97502 | 0.22781 | 4.3897 | 1.0256 | 4.5022 | .0038224 | 3.0429 | 10 |
| 51 | .22240 | .97496 | .22811 | .3838 | .0257 | .4964 | .0038375 | 3.0373 | 9 |
| 52 | .22268 | .97489 | .22842 | .3779 | .0258 | .4907 | .0038527 | 3.0317 | 8 |
| 53 | .22297 | .97483 | .22872 | .3721 | .0258 | .4850 | .0038679 | 3.0261 | 7 |
| 54 | .22325 | .97476 | .22903 | .3662 | .0259 | .4793 | .0038831 | 3.0206 | 6 |
| 55 | 0.22353 | 0.97470 | 0.22934 | 4.3604 | 1.0260 | 4.4736 | .0038984 | 3.0150 | 5 |
| 56 | .22382 | .97463 | .22964 | .3546 | .0260 | .4679 | .0039137 | 3.0095 | 4 |
| 57 | .22410 | .97457 | .22995 | .3488 | .0261 | .4623 | .0039291 | 3.0040 | 3 |
| 58 | .22438 | .97450 | .23026 | .3430 | .0262 | .4566 | .0039445 | 2.9985 | 2 |
| 59 | .22467 | .97444 | .23056 | .3372 | .0262 | .4510 | .0039599 | 2.9930 | 1 |
| 60 | 0.22495 | 0.97437 | 0.23087 | 4.3315 | 1.0263 | 4.4454 | .0039754 | 2.9876 | 0 |
| M | Cosine | Sine | Cotan. | Tan. | Cosec. | Secant | READ DOWN | 77°–78° Involute | M |

| M | Sine | Cosine | Tan. | Cotan. | Secant | Cosec. | Involute 13°–14° | READ UP | M |
|---|------|--------|------|--------|--------|--------|------------------|---------|---|
| 0 | 0.22495 | 0.97437 | 0.23087 | 4.3315 | 1.0263 | 4.4454 | .0039754 | 2.9876 | 60 |
| 1 | .22523 | .97430 | .23117 | .3257 | .0264 | .4398 | .0039909 | 2.9821 | 59 |
| 2 | .22552 | .97424 | .23148 | .3200 | .0264 | .4342 | .0040065 | 2.9767 | 58 |
| 3 | .22580 | .97417 | .23179 | .3143 | .0265 | .4287 | .0040221 | 2.9713 | 57 |
| 4 | .22608 | .97411 | .23209 | .3086 | .0266 | .4231 | .0040377 | 2.9659 | 56 |
| 5 | 0.22637 | 0.97404 | 0.23240 | 4.3029 | 1.0266 | 4.4176 | .0040534 | 2.9605 | 55 |
| 6 | .22665 | .97398 | .23271 | .2972 | .0267 | .4121 | .0040692 | 2.9551 | 54 |
| 7 | .22693 | .97391 | .23301 | .2916 | .0268 | .4066 | .0040849 | 2.9497 | 53 |
| 8 | .22722 | .97384 | .23332 | .2859 | .0269 | .4011 | .0041007 | 2.9444 | 52 |
| 9 | .22750 | .97378 | .23363 | .2803 | .0269 | .3956 | .0041166 | 2.9390 | 51 |
| 10 | 0.22778 | 0.97371 | 0.23393 | 4.2747 | 1.0270 | 4.3901 | .0041325 | 2.9337 | 50 |
| 11 | .22807 | .97365 | .23424 | .2691 | .0271 | .3847 | .0041484 | 2.9284 | 49 |
| 12 | .22835 | .97358 | .23455 | .2635 | .0271 | .3792 | .0041644 | 2.9231 | 48 |
| 13 | .22863 | .97351 | .23485 | .2580 | .0272 | .3738 | .0041804 | 2.9178 | 47 |
| 14 | .22892 | .97345 | .23516 | .2524 | .0273 | .3684 | .0041965 | 2.9126 | 46 |
| 15 | 0.22920 | 0.97338 | 0.23547 | 4.2468 | 1.0273 | 4.3630 | .0042126 | 2.9073 | 45 |
| 16 | .22948 | .97331 | .23578 | .2413 | .0274 | .3576 | .0042288 | 2.9021 | 44 |
| 17 | .22977 | .97325 | .23608 | .2358 | .0275 | .3522 | .0042450 | 2.8968 | 43 |
| 18 | .23005 | .97318 | .23639 | .2303 | .0276 | .3469 | .0042612 | 2.8916 | 42 |
| 19 | .23033 | .97311 | .23670 | .2248 | .0276 | .3415 | .0042775 | 2.8864 | 41 |
| 20 | 0.23062 | 0.97304 | 0.23700 | 4.2193 | 1.0277 | 4.3362 | .0042938 | 2.8812 | 40 |
| 21 | .23090 | .97298 | .23731 | .2139 | .0278 | .3309 | .0043101 | 2.8761 | 39 |
| 22 | .23118 | .97291 | .23762 | .2084 | .0278 | .3256 | .0043266 | 2.8709 | 38 |
| 23 | .23146 | .97284 | .23793 | .2030 | .0279 | .3203 | .0043430 | 2.8658 | 37 |
| 24 | .23175 | .97278 | .23823 | .1976 | .0280 | .3150 | .0043595 | 2.8606 | 36 |
| 25 | 0.23203 | 0.97271 | 0.23854 | 4.1922 | 1.0281 | 4.3098 | .0043760 | 2.8555 | 35 |
| 26 | .23231 | .97264 | .23885 | .1868 | .0281 | .3045 | .0043926 | 2.8504 | 34 |
| 27 | .23260 | .97257 | .23916 | .1814 | .0282 | .2993 | .0044092 | 2.8453 | 33 |
| 28 | .23288 | .97251 | .23946 | .1760 | .0283 | .2941 | .0044259 | 2.8402 | 32 |
| 29 | .23316 | .97244 | .23977 | .1706 | .0283 | .2889 | .0044426 | 2.8352 | 31 |
| 30 | 0.23345 | 0.97237 | 0.24008 | 4.1653 | 1.0284 | 4.2837 | .0044593 | 2.8301 | 30 |
| 31 | .23373 | .97230 | .24039 | .1600 | .0285 | .2785 | .0044761 | 2.8251 | 29 |
| 32 | .23401 | .97223 | .24069 | .1547 | .0286 | .2733 | .0044929 | 2.8201 | 28 |
| 33 | .23429 | .97217 | .24100 | .1493 | .0286 | .2681 | .0045098 | 2.8150 | 27 |
| 34 | .23458 | .97210 | .24131 | .1441 | .0287 | .2630 | .0045267 | 2.8100 | 26 |
| 35 | 0.23486 | 0.97203 | 0.24162 | 4.1388 | 1.0288 | 4.2579 | .0045437 | 2.8050 | 25 |
| 36 | .23514 | .97196 | .24193 | .1335 | .0288 | .2527 | .0045607 | 2.8001 | 24 |
| 37 | .23542 | .97189 | .24223 | .1282 | .0289 | .2476 | .0045777 | 2.7951 | 23 |
| 38 | .23571 | .97182 | .24254 | .1230 | .0290 | .2425 | .0045948 | 2.7902 | 22 |
| 39 | .23599 | .97176 | .24285 | .1178 | .0291 | .2375 | .0046120 | 2.7852 | 21 |
| 40 | 0.23627 | 0.97169 | 0.24316 | 4.1126 | 1.0291 | 4.2324 | .0046291 | 2.7803 | 20 |
| 41 | .23656 | .97162 | .24347 | .1074 | .0292 | .2273 | .0046464 | 2.7754 | 19 |
| 42 | .23684 | .97155 | .24377 | .1022 | .0293 | .2223 | .0046636 | 2.7705 | 18 |
| 43 | .23712 | .97148 | .24408 | .0970 | .0294 | .2173 | .0046809 | 2.7656 | 17 |
| 44 | .23740 | .97141 | .24439 | .0918 | .0294 | .2122 | .0046983 | 2.7607 | 16 |
| 45 | 0.23769 | 0.97134 | 0.24470 | 4.0867 | 1.0295 | 4.2072 | .0047157 | 2.7558 | 15 |
| 46 | .23797 | .97127 | .24501 | .0815 | .0296 | .2022 | .0047331 | 2.7510 | 14 |
| 47 | .23825 | .97120 | .24532 | .0764 | .0297 | .1973 | .0047506 | 2.7462 | 13 |
| 48 | .23853 | .97113 | .24562 | .0713 | .0297 | .1923 | .0047681 | 2.7413 | 12 |
| 49 | .23882 | .97106 | .24593 | .0662 | .0298 | .1873 | .0047857 | 2.7365 | 11 |
| 50 | 0.23910 | 0.97100 | 0.24624 | 4.0611 | 1.0299 | 4.1824 | .0048033 | 2.7317 | 10 |
| 51 | .23938 | .97093 | .24655 | .0560 | .0299 | .1774 | .0048210 | 2.7269 | 9 |
| 52 | .23966 | .97086 | .24686 | .0509 | .0300 | .1725 | .0048387 | 2.7221 | 8 |
| 53 | .23995 | .97079 | .24717 | .0459 | .0301 | .1676 | .0048564 | 2.7174 | 7 |
| 54 | .24023 | .97072 | .24747 | .0408 | .0302 | .1627 | .0048742 | 2.7126 | 6 |
| 55 | 0.24051 | 0.97065 | 0.24778 | 4.0358 | 1.0302 | 4.1578 | .0048921 | 2.7079 | 5 |
| 56 | .24079 | .97058 | .24809 | .0308 | .0303 | .1529 | .0049099 | 2.7031 | 4 |
| 57 | .24108 | .97051 | .24840 | .0257 | .0304 | .1481 | .0049279 | 2.6984 | 3 |
| 58 | .24136 | .97044 | .24871 | .0207 | .0305 | .1432 | .0049458 | 2.6937 | 2 |
| 59 | .24164 | .97037 | .24902 | .0158 | .0305 | .1384 | .0049638 | 2.6890 | 1 |
| 60 | 0.24192 | 0.97030 | 0.24933 | 4.0108 | 1.0306 | 4.1336 | .0049819 | 2.6843 | 0 |

| M | Cosine | Sine | Cotan. | Tan. | Cosec. | Secant | READ DOWN | 76°–77° Involute | M |
|---|--------|------|--------|------|--------|--------|-----------|------------------|---|

| M | Sine | Cosine | Tan. | Cotan. | Secant | Cosec. | Involute 14°-15° | READ UP | M |
|---|---|---|---|---|---|---|---|---|---|
| 0 | 0.24192 | 0.97030 | 0.24933 | 4.0108 | 1.0306 | 4.1336 | .0049819 | 2.6843 | 60 |
| 1 | .24220 | .97023 | .24964 | .0058 | .0307 | .1287 | .0050000 | 2.6797 | 59 |
| 2 | .24249 | .97015 | .24995 | .0009 | .0308 | .1239 | .0050182 | 2.6750 | 58 |
| 3 | .24277 | .97008 | .25026 | 3.9959 | .0308 | .1191 | .0050364 | 2.6703 | 57 |
| 4 | .24305 | .97001 | .25056 | .9910 | .0309 | .1144 | .0050546 | 2.6657 | 56 |
| 5 | 0.24333 | 0.96994 | 0.25087 | 3.9861 | 1.0310 | 4.1096 | .0050729 | 2.6611 | 55 |
| 6 | .24362 | .96987 | .25118 | .9812 | .0311 | .1048 | .0050912 | 2.6565 | 54 |
| 7 | .24390 | .96980 | .25149 | .9763 | .0311 | .1001 | .0051096 | 2.6519 | 53 |
| 8 | .24418 | .96973 | .25180 | .9714 | .0312 | .0954 | .0051280 | 2.6473 | 52 |
| 9 | .24446 | .96966 | .25211 | .9665 | .0313 | .0906 | .0051465 | 2.6427 | 51 |
| 10 | 0.24474 | 0.96959 | 0.25242 | 3.9617 | 1.0314 | 4.0859 | .0051650 | 2.6381 | 50 |
| 11 | .24503 | .96952 | .25273 | .9568 | .0314 | .0812 | .0051835 | 2.6336 | 49 |
| 12 | .24531 | .96945 | .25304 | .9520 | .0315 | .0765 | .0052021 | 2.6290 | 48 |
| 13 | .24559 | .96937 | .25335 | .9471 | .0316 | .0718 | .0052208 | 2.6245 | 47 |
| 14 | .24587 | .96930 | .25366 | .9423 | .0317 | .0672 | .0052395 | 2.6199 | 46 |
| 15 | 0.24615 | 0.96923 | 0.25397 | 3.9375 | 1.0317 | 4.0625 | .0052582 | 2.6154 | 45 |
| 16 | .24644 | .96916 | .25428 | .9327 | .0318 | .0579 | .0052770 | 2.6109 | 44 |
| 17 | .24672 | .96909 | .25459 | .9279 | .0319 | .0532 | .0052958 | 2.6064 | 43 |
| 18 | .24700 | .96902 | .25490 | .9232 | .0320 | .0486 | .0053147 | 2.6019 | 42 |
| 19 | .24728 | .96894 | .25521 | .9184 | .0321 | .0440 | .0053336 | 2.5975 | 41 |
| 20 | 0.24756 | 0.96887 | 0.25552 | 3.9136 | 1.0321 | 4.0394 | .0053526 | 2.5930 | 40 |
| 21 | .24784 | .96880 | .25583 | .9089 | .0322 | .0348 | .0053716 | 2.5886 | 39 |
| 22 | .24813 | .96873 | .25614 | .9042 | .0323 | .0302 | .0053907 | 2.5841 | 38 |
| 23 | .24841 | .96866 | .25645 | .8995 | .0324 | .0256 | .0054098 | 2.5797 | 37 |
| 24 | .24869 | .96858 | .25676 | .8947 | .0324 | .0211 | .0054289 | 2.5753 | 36 |
| 25 | 0.24897 | 0.96851 | 0.25707 | 3.8900 | 1.0325 | 4.0165 | .0054481 | 2.5709 | 35 |
| 26 | .24925 | .96844 | .25738 | .8854 | .0326 | .0120 | .0054674 | 2.5665 | 34 |
| 27 | .24954 | .96837 | .25769 | .8807 | .0327 | .0075 | .0054867 | 2.5621 | 33 |
| 28 | .24982 | .96829 | .25800 | .8760 | .0327 | .0029 | .0055060 | 2.5577 | 32 |
| 29 | .25010 | .96822 | .25831 | .8714 | .0328 | 3.9984 | .0055254 | 2.5533 | 31 |
| 30 | 0.25038 | 0.96815 | 0.25862 | 3.8667 | 1.0329 | 3.9939 | .0055448 | 2.5490 | 30 |
| 31 | .25066 | .96807 | .25893 | .8621 | .0330 | .9894 | .0055643 | 2.5446 | 29 |
| 32 | .25094 | .96800 | .25924 | .8575 | .0331 | .9850 | .0055838 | 2.5403 | 28 |
| 33 | .25122 | .96793 | .25955 | .8528 | .0331 | .9805 | .0056034 | 2.5360 | 27 |
| 34 | .25151 | .96786 | .25986 | .8482 | .0332 | .9760 | .0056230 | 2.5317 | 26 |
| 35 | 0.25179 | 0.96778 | 0.26017 | 3.8436 | 1.0333 | 3.9716 | .0056427 | 2.5274 | 25 |
| 36 | .25207 | .96771 | .26048 | .8391 | .0334 | .9672 | .0056624 | 2.5231 | 24 |
| 37 | .25235 | .96764 | .26079 | .8345 | .0334 | .9627 | .0056822 | 2.5188 | 23 |
| 38 | .25263 | .96756 | .26110 | .8299 | .0335 | .9583 | .0057020 | 2.5145 | 22 |
| 39 | .25291 | .96749 | .26141 | .8254 | .0336 | .9539 | .0057218 | 2.5103 | 21 |
| 40 | 0.25320 | 0.96742 | 0.26172 | 3.8208 | 1.0337 | 3.9495 | .0057417 | 2.5060 | 20 |
| 41 | .25348 | .96734 | .26203 | .8163 | .0338 | .9451 | .0057617 | 2.5018 | 19 |
| 42 | .25376 | .96727 | .26235 | .8118 | .0338 | .9408 | .0057817 | 2.4975 | 18 |
| 43 | .25404 | .96719 | .26266 | .8073 | .0339 | .9364 | .0058017 | 2.4933 | 17 |
| 44 | .25432 | .96712 | .26297 | .8028 | .0340 | .9320 | .0058218 | 2.4891 | 16 |
| 45 | 0.25460 | 0.96705 | 0.26328 | 3.7983 | 1.0341 | 3.9277 | .0058420 | 2.4849 | 15 |
| 46 | .25488 | .96697 | .26359 | .7938 | .0342 | .9234 | .0058622 | 2.4807 | 14 |
| 47 | .25516 | .96690 | .26390 | .7893 | .0342 | .9190 | .0058824 | 2.4765 | 13 |
| 48 | .25545 | .96682 | .26421 | .7848 | .0343 | .9147 | .0059027 | 2.4724 | 12 |
| 49 | .25573 | .96675 | .26452 | .7804 | .0344 | .9104 | .0059230 | 2.4682 | 11 |
| 50 | 0.25601 | 0.96667 | 0.26483 | 3.7760 | 1.0345 | 3.9061 | .0059434 | 2.4640 | 10 |
| 51 | .25629 | .96660 | .26515 | .7715 | .0346 | .9018 | .0059638 | 2.4599 | 9 |
| 52 | .25657 | .96653 | .26546 | .7671 | .0346 | .8976 | .0059843 | 2.4558 | 8 |
| 53 | .25685 | .96645 | .26577 | .7627 | .0347 | .8933 | .0060048 | 2.4516 | 7 |
| 54 | .25713 | .96638 | .26608 | .7583 | .0348 | .8890 | .0060254 | 2.4475 | 6 |
| 55 | 0.25741 | 0.96630 | 0.26639 | 3.7539 | 1.0349 | 3.8848 | .0060460 | 2.4434 | 5 |
| 56 | .25769 | .96623 | .26670 | .7495 | .0350 | .8806 | .0060667 | 2.4393 | 4 |
| 57 | .25798 | .96615 | .26701 | .7451 | .0350 | .8763 | .0060874 | 2.4353 | 3 |
| 58 | .25826 | .96608 | .26733 | .7408 | .0351 | .8721 | .0061081 | 2.4312 | 2 |
| 59 | .25854 | .96600 | .26764 | .7364 | .0352 | .8679 | .0061289 | 2.4271 | 1 |
| 60 | 0.25882 | 0 96593 | 0.26795 | 3.7321 | 1.0353 | 3.8637 | .0061498 | 2.4231 | 0 |
| M | Cosine | Sine | Cotan. | Tan. | Cosec. | Secant | READ DOWN | 75°-76° Involute | M |

**15° or 195°**  Trigonometric Functions  **164° or 344°**

| M | Sine | Cosine | Tan. | Cotan. | Secant | Cosec. | Involute 15°-16° | READ UP | M |
|---|---|---|---|---|---|---|---|---|---|
| 0 | 0.25882 | 0.96593 | 0.26795 | 3.7321 | 1.0353 | 3.8637 | .0061498 | 2.4231 | 60 |
| 1 | .25910 | .96585 | .26826 | .7277 | .0354 | .8595 | .0061707 | 2.4190 | 59 |
| 2 | .25938 | .96578 | .26857 | .7234 | .0354 | .8553 | .0061917 | 2.4150 | 58 |
| 3 | .25966 | .96570 | .26888 | .7191 | .0355 | .8512 | .0062127 | 2.4109 | 57 |
| 4 | .25994 | .96562 | .26920 | .7148 | .0356 | .8470 | .0062337 | 2.4069 | 56 |
| 5 | 0.26022 | 0.96555 | 0.26951 | 3.7105 | 1.0357 | 3.8428 | .0062548 | 2.4029 | 55 |
| 6 | .26050 | .96547 | .26982 | .7062 | .0358 | .8387 | .0062760 | 2.3989 | 54 |
| 7 | .26079 | .96540 | .27013 | .7019 | .0358 | .8346 | .0062972 | 2.3949 | 53 |
| 8 | .26107 | .96532 | .27044 | .6976 | .0359 | .8304 | .0063184 | 2.3909 | 52 |
| 9 | .26135 | .96524 | .27076 | .6933 | .0360 | .8263 | .0063397 | 2.3870 | 51 |
| 10 | 0.26163 | 0.96517 | 0.27107 | 3.6891 | 1.0361 | 3.8222 | .0063611 | 2.3830 | 50 |
| 11 | .26191 | .96509 | .27138 | .6848 | .0362 | .8181 | .0063825 | 2.3791 | 49 |
| 12 | .26219 | .96502 | .27169 | .6806 | .0363 | .8140 | .0064039 | 2.3751 | 48 |
| 13 | .26247 | .96494 | .27201 | .6764 | .0363 | .8100 | .0064254 | 2.3712 | 47 |
| 14 | .26275 | .96486 | .27232 | .6722 | .0364 | .8059 | .0064470 | 2.3672 | 46 |
| 15 | 0.26303 | 0.96479 | 0.27263 | 3.6680 | 1.0365 | 3.8018 | .0064686 | 2.3633 | 45 |
| 16 | .26331 | .96471 | .27294 | .6638 | .0366 | .7978 | .0064902 | 2.3594 | 44 |
| 17 | .26359 | .96463 | .27326 | .6596 | .0367 | .7937 | .0065119 | 2.3555 | 43 |
| 18 | .26387 | .96456 | .27357 | .6554 | .0367 | .7897 | .0065337 | 2.3516 | 42 |
| 19 | .26415 | .96449 | .27388 | .6512 | .0368 | .7857 | .0065555 | 2.3477 | 41 |
| 20 | 0.26443 | 0.96440 | 0.27419 | 3.6470 | 1.0369 | 3.7817 | .0065773 | 2.3439 | 40 |
| 21 | .26471 | .96433 | .27451 | .6429 | .0370 | .7777 | .0065992 | 2.3400 | 39 |
| 22 | .26500 | .96425 | .27482 | .6387 | .0371 | .7737 | .0066211 | 2.3361 | 38 |
| 23 | .26528 | .96417 | .27513 | .6346 | .0372 | .7697 | .0066431 | 2.3323 | 37 |
| 24 | .26556 | .96410 | .27545 | .6305 | .0372 | .7657 | .0066652 | 2.3285 | 36 |
| 25 | 0.26584 | 0.96402 | 0.27576 | 3.6264 | 1.0373 | 3.7617 | .0066873 | 2.3246 | 35 |
| 26 | .26612 | .96394 | .27607 | .6222 | .0374 | .7577 | .0067094 | 2.3208 | 34 |
| 27 | .26640 | .96386 | .27638 | .6181 | .0375 | .7538 | .0067316 | 2.3170 | 33 |
| 28 | .26668 | .96379 | .27670 | .6140 | .0376 | .7498 | .0067539 | 2.3132 | 32 |
| 29 | .26696 | .96371 | .27701 | .6100 | .0376 | .7459 | .0067762 | 2.3094 | 31 |
| 30 | 0.26724 | 0.96363 | 0.27732 | 3.6059 | 1.0377 | 3.7420 | .0067985 | 2.3056 | 30 |
| 31 | .26752 | .96355 | .27764 | .6018 | .0378 | .7381 | .0068209 | 2.3018 | 29 |
| 32 | .26780 | .96347 | .27795 | .5978 | .0379 | .7341 | .0068434 | 2.2981 | 28 |
| 33 | .26808 | .96340 | .27826 | .5937 | .0380 | .7302 | .0068659 | 2.2943 | 27 |
| 34 | .26836 | .96332 | .27858 | .5897 | .0381 | .7263 | .0068884 | 2.2906 | 26 |
| 35 | 0.26864 | 0.96324 | 0.27889 | 3.5856 | 1.0382 | 3.7225 | .0069110 | 2.2868 | 25 |
| 36 | .26892 | .96316 | .27921 | .5816 | .0382 | .7186 | .0069337 | 2.2831 | 24 |
| 37 | .26920 | .96308 | .27952 | .5776 | .0383 | .7147 | .0069564 | 2.2793 | 23 |
| 38 | .26948 | .96301 | .27983 | .5736 | .0384 | .7108 | .0069791 | 2.2756 | 22 |
| 39 | .26976 | .96293 | .28015 | .5696 | .0385 | .7070 | .0070019 | 2.2719 | 21 |
| 40 | 0.27004 | 0.96285 | 0.28046 | 3.5656 | 1.0386 | 3.7032 | .0070248 | 2.2682 | 20 |
| 41 | .27032 | .96277 | .28077 | .5616 | .0387 | .6993 | .0070477 | 2.2645 | 19 |
| 42 | .27060 | .96269 | .28109 | .5576 | .0388 | .6955 | .0070706 | 2.2608 | 18 |
| 43 | .27088 | .96261 | .28140 | .5536 | .0388 | .6917 | .0070936 | 2.2572 | 17 |
| 44 | .27116 | .96253 | .28172 | .5497 | .0389 | .6879 | .0071167 | 2.2535 | 16 |
| 45 | 0.27144 | 0.96246 | 0.28203 | 3.5457 | 1.0390 | 3.6840 | .0071398 | 2.2498 | 15 |
| 46 | .27172 | .96238 | .28234 | .5418 | .0391 | .6803 | .0071630 | 2.2462 | 14 |
| 47 | .27200 | .96230 | .28266 | .5379 | .0392 | .6765 | .0071862 | 2.2425 | 13 |
| 48 | .27228 | .96222 | .28297 | .5339 | .0393 | .6727 | .0072095 | 2.2389 | 12 |
| 49 | .27256 | .96214 | .28329 | .5300 | .0394 | .6689 | .0072328 | 2.2353 | 11 |
| 50 | 0.27284 | 0.96206 | 0.28360 | 3.5261 | 1.0394 | 3.6652 | .0072561 | 2.2316 | 10 |
| 51 | .27312 | .96198 | .28391 | .5222 | .0395 | .6614 | .0072796 | 2.2280 | 9 |
| 52 | .27340 | .96190 | .28423 | .5183 | .0396 | .6576 | .0073030 | 2.2244 | 8 |
| 53 | .27368 | .96182 | .28454 | .5144 | .0397 | .6539 | .0073266 | 2.2208 | 7 |
| 54 | .27396 | .96174 | .28486 | .5105 | .0398 | .6502 | .0073501 | 2.2172 | 6 |
| 55 | 0.27424 | 0.96166 | 0.28517 | 3.5067 | 1.0399 | 3.6465 | .0073738 | 2.2137 | 5 |
| 56 | .27452 | .96158 | .28549 | .5028 | .0400 | .6427 | .0073975 | 2.2101 | 4 |
| 57 | .27480 | .96150 | .28580 | .4989 | .0400 | .6390 | .0074212 | 2.2065 | 3 |
| 58 | .27508 | .96142 | .28612 | .4951 | .0401 | .6353 | .0074450 | 2.2030 | 2 |
| 59 | .27536 | .96134 | .28643 | .4912 | .0402 | .6316 | .0074688 | 2.1994 | 1 |
| 60 | 0.27564 | 0.96126 | 0.28675 | 3.4874 | 1.0403 | 3.6280 | .0074927 | 2.1959 | 0 |
| M | Cosine | Sine | Cotan. | Tan. | Cosec. | Secant | READ DOWN | 74°-75° Involute | M |

| M | Sine | Cosine | Tan. | Cotan. | Secant | Cosec. | Involute 16°–17° | READ UP | M |
|---|------|--------|------|--------|--------|--------|-------------------|---------|---|
| 0 | 0.27564 | 0.96126 | 0.28675 | 3.4874 | 1.0403 | 3.6280 | .0074927 | 2.1959 | 60 |
| 1 | .27592 | .96118 | .28706 | .4836 | .0404 | .6243 | .0075166 | 2.1923 | 59 |
| 2 | .27620 | .96110 | .28738 | .4798 | .0405 | .6206 | .0075406 | 2.1888 | 58 |
| 3 | .27648 | .96102 | .28769 | .4760 | .0406 | .6169 | .0075647 | 2.1853 | 57 |
| 4 | .27676 | .96094 | .28801 | .4722 | .0406 | .6133 | .0075888 | 2.1818 | 56 |
| 5 | 0.27704 | 0.96086 | 0.28832 | 3.4684 | 1.0407 | 3.6097 | .0076130 | 2.1783 | 55 |
| 6 | .27731 | .96078 | .28864 | .4646 | .0408 | .6060 | .0076372 | 2.1748 | 54 |
| 7 | .27759 | .96070 | .28895 | .4608 | .0409 | .6024 | .0076614 | 2.1713 | 53 |
| 8 | .27787 | .96062 | .28927 | .4570 | .0410 | .5988 | .0076857 | 2.1678 | 52 |
| 9 | .27815 | .96054 | .28958 | .4533 | .0411 | .5951 | .0077101 | 2.1643 | 51 |
| 10 | 0.27843 | 0.96046 | 0.28990 | 3.4495 | 1.0412 | 3.5915 | .0077345 | 2.1609 | 50 |
| 11 | .27871 | .96037 | .29021 | .4458 | .0413 | .5879 | .0077590 | 2.1574 | 49 |
| 12 | .27899 | .96029 | .29053 | .4420 | .0413 | .5843 | .0077835 | 2.1540 | 48 |
| 13 | .27927 | .96021 | .29084 | .4383 | .0414 | .5808 | .0078081 | 2.1505 | 47 |
| 14 | .27955 | .96013 | .29116 | .4346 | .0415 | .5772 | .0078327 | 2.1471 | 46 |
| 15 | 0.27983 | 0.96005 | 0.29147 | 3.4308 | 1.0416 | 3.5736 | .0078574 | 2.1437 | 45 |
| 16 | .28011 | .95997 | .29179 | .4271 | .0417 | .5700 | .0078822 | 2.1402 | 44 |
| 17 | .28039 | .95989 | .29210 | .4234 | .0418 | .5665 | .0079069 | 2.1368 | 43 |
| 18 | .28067 | .95981 | .29242 | .4197 | .0419 | .5629 | .0079318 | 2.1334 | 42 |
| 19 | .28095 | .95972 | .29274 | .4160 | .0420 | .5594 | .0079567 | 2.1300 | 41 |
| 20 | 0.28123 | 0.95964 | 0.29305 | 3.4124 | 1.0421 | 3.5559 | .0079817 | 2.1266 | 40 |
| 21 | .28150 | .95956 | .29337 | .4087 | .0421 | .5523 | .0080067 | 2.1233 | 39 |
| 22 | .28178 | .95948 | .29368 | .4050 | .0422 | .5488 | .0080317 | 2.1199 | 38 |
| 23 | .28206 | .95940 | .29400 | .4014 | .0423 | .5453 | .0080568 | 2.1165 | 37 |
| 24 | .28234 | .95931 | .29432 | .3977 | .0424 | .5418 | .0080820 | 2.1131 | 36 |
| 25 | 0.28262 | 0.95923 | 0.29463 | 3.3941 | 1.0425 | 3.5383 | .0081072 | 2.1098 | 35 |
| 26 | .28290 | .95915 | .29495 | .3904 | .0426 | .5348 | .0081325 | 2.1064 | 34 |
| 27 | .28318 | .95907 | .29526 | .3868 | .0427 | .5313 | .0081578 | 2.1031 | 33 |
| 28 | .28346 | .95898 | .29558 | .3832 | .0428 | .5279 | .0081832 | 2.0998 | 32 |
| 29 | .28374 | .95890 | .29590 | .3796 | .0429 | .5244 | .0082087 | 2.0964 | 31 |
| 30 | 0.28402 | 0.95882 | 0.29621 | 3.3759 | 1.0429 | 3.5209 | .0082342 | 2.0931 | 30 |
| 31 | .28429 | .95874 | .29653 | .3723 | .0430 | .5175 | .0082597 | 2.0898 | 29 |
| 32 | .28457 | .95865 | .29685 | .3687 | .0431 | .5140 | .0082853 | 2.0865 | 28 |
| 33 | .28485 | .95857 | .29716 | .3652 | .0432 | .5106 | .0083110 | 2.0832 | 27 |
| 34 | .28513 | .95849 | .29748 | .3616 | .0433 | .5072 | .0083367 | 2.0799 | 26 |
| 35 | 0.28541 | 0.95841 | 0.29780 | 3.3580 | 1.0434 | 3.5037 | .0083625 | 2.0766 | 25 |
| 36 | .28569 | .95832 | .29811 | .3544 | .0435 | .5003 | .0083883 | 2.0734 | 24 |
| 37 | .28597 | .95824 | .29843 | .3509 | .0436 | .4969 | .0084142 | 2.0701 | 23 |
| 38 | .28625 | .95816 | .29875 | .3473 | .0437 | .4935 | .0084401 | 2.0668 | 22 |
| 39 | .28652 | .95807 | .29906 | .3438 | .0438 | .4901 | .0084661 | 2.0636 | 21 |
| 40 | 0.28680 | 0.95799 | 0.29938 | 3.3402 | 1.0439 | 3.4867 | .0084921 | 2.0603 | 20 |
| 41 | .28708 | .95791 | .29970 | .3367 | .0439 | .4833 | .0085182 | 2.0571 | 19 |
| 42 | .28736 | .95782 | .30001 | .3332 | .0440 | .4799 | .0085444 | 2.0538 | 18 |
| 43 | .28764 | .95774 | .30033 | .3297 | .0441 | .4766 | .0085706 | 2.0506 | 17 |
| 44 | .28792 | .95766 | .30065 | .3261 | .0442 | .4732 | .0085969 | 2.0474 | 16 |
| 45 | 0.28820 | 0.95757 | 0.30097 | 3.3226 | 1.0443 | 3.4699 | .0086232 | 2.0442 | 15 |
| 46 | .28847 | .95749 | .30128 | .3191 | .0444 | .4665 | .0086496 | 2.0410 | 14 |
| 47 | .28875 | .95740 | .30160 | .3156 | .0445 | .4632 | .0086760 | 2.0378 | 13 |
| 48 | .28903 | .95732 | .30192 | .3122 | .0446 | .4598 | .0087025 | 2.0346 | 12 |
| 49 | .28931 | .95724 | .30224 | .3087 | .0447 | .4565 | .0087290 | 2.0314 | 11 |
| 50 | 0.28959 | 0.95715 | 0.30255 | 3.3052 | 1.0448 | 3.4532 | .0087556 | 2.0282 | 10 |
| 51 | .28987 | .95707 | .30287 | .3017 | .0449 | .4499 | .0087823 | 2.0250 | 9 |
| 52 | .29015 | .95698 | .30319 | .2983 | .0450 | .4465 | .0088090 | 2.0219 | 8 |
| 53 | .29042 | .95690 | .30351 | .2948 | .0450 | .4432 | .0088358 | 2.0187 | 7 |
| 54 | .29070 | .95681 | .30382 | .2914 | .0451 | .4399 | .0088626 | 2.0156 | 6 |
| 55 | 0.29098 | 0.95673 | 0.30414 | 3.2879 | 1.0452 | 3.4367 | .0088895 | 2.0124 | 5 |
| 56 | .29126 | .95664 | .30446 | .2845 | .0453 | .4334 | .0089164 | 2.0093 | 4 |
| 57 | .29154 | .95656 | .30478 | .2811 | .0454 | .4301 | .0089434 | 2.0061 | 3 |
| 58 | .29182 | .95647 | .30509 | .2777 | .0455 | .4268 | .0089704 | 2.0030 | 2 |
| 59 | .29209 | .95639 | .30541 | .2743 | .0456 | .4236 | .0089975 | 1.9999 | 1 |
| 60 | 0.29237 | 0.95630 | 0.30573 | 3.2709 | 1.0457 | 3.4203 | .0090247 | 1.9968 | 0 |
| M | Cosine | Sine | Cotan. | Tan. | Cosec. | Secant | READ DOWN | 73°–74° Involute | M |

| M | Sine | Cosine | Tan. | Cotan. | Secant | Cosec. | Involute 17°–18° | READ UP | M |
|---|------|--------|------|--------|--------|--------|------------------|---------|---|
| 0 | 0.29237 | 0.95630 | 0.30573 | 3.2709 | 1.0457 | 3.4203 | .0090247 | 1.9968 | 60 |
| 1 | .29265 | .95622 | .30605 | .2675 | .0458 | .4171 | .0090519 | 1.9937 | 59 |
| 2 | .29293 | .95613 | .30637 | .2641 | .0459 | .4138 | .0090792 | 1.9906 | 58 |
| 3 | .29321 | .95605 | .30669 | .2607 | .0460 | .4106 | .0091065 | 1.9875 | 57 |
| 4 | .29348 | .95596 | .30700 | .2573 | .0461 | .4073 | .0091339 | 1.9844 | 56 |
| 5 | 0.29376 | 0.95588 | 0.30732 | 3.2539 | 1.0462 | 3.4041 | .0091614 | 1.9813 | 55 |
| 6 | .29404 | .95579 | .30764 | .2506 | .0463 | .4009 | .0091889 | 1.9782 | 54 |
| 7 | .29432 | .95571 | .30796 | .2472 | .0463 | .3977 | .0092164 | 1.9751 | 53 |
| 8 | .29460 | .95562 | .30828 | .2438 | .0464 | .3945 | .0092440 | 1.9721 | 52 |
| 9 | .29487 | .95554 | .30860 | .2405 | .0465 | .3913 | .0092717 | 1.9690 | 51 |
| 10 | 0.29515 | 0.95545 | 0.30891 | 3.2371 | 1.0466 | 3.3881 | .0092994 | 1.9660 | 50 |
| 11 | .29543 | .95536 | .30923 | .2338 | .0467 | .3849 | .0093272 | 1.9629 | 49 |
| 12 | .29571 | .95528 | .30955 | .2305 | .0468 | .3817 | .0093551 | 1.9599 | 48 |
| 13 | .29599 | .95519 | .30987 | .2272 | .0469 | .3785 | .0093830 | 1.9568 | 47 |
| 14 | .29626 | .95511 | .31019 | .2238 | .0470 | .3754 | .0094109 | 1.9538 | 46 |
| 15 | 0.29654 | 0.95502 | 0.31051 | 3.2205 | 1.0471 | 3.3722 | .0094390 | 1.9508 | 45 |
| 16 | .29682 | .95493 | .31083 | .2172 | .0472 | .3691 | .0094670 | 1.9478 | 44 |
| 17 | .29710 | .95485 | .31115 | .2139 | .0473 | .3659 | .0094952 | 1.9448 | 43 |
| 18 | .29737 | .95476 | .31147 | .2106 | .0474 | .3628 | .0095234 | 1.9418 | 42 |
| 19 | .29765 | .95467 | .31178 | .2073 | .0475 | .3596 | .0095516 | 1.9388 | 41 |
| 20 | 0.29793 | 0.95459 | 0.31210 | 3.2041 | 1.0476 | 3.3565 | .0095799 | 1.9358 | 40 |
| 21 | .29821 | .95450 | .31242 | .2008 | .0477 | .3534 | .0096083 | 1.9328 | 39 |
| 22 | .29849 | .95441 | .31274 | .1975 | .0478 | .3502 | .0096367 | 1.9298 | 38 |
| 23 | .29876 | .95433 | .31306 | .1943 | .0479 | .3471 | .0096652 | 1.9269 | 37 |
| 24 | .29904 | .95424 | .31338 | .1910 | .0480 | .3440 | .0096937 | 1.9239 | 36 |
| 25 | 0.29932 | 0.95415 | 0.31370 | 3.1878 | 1.0480 | 3.3409 | .0097223 | 1.9209 | 35 |
| 26 | .29960 | .95407 | .31402 | .1845 | .0481 | .3378 | .0097510 | 1.9180 | 34 |
| 27 | .29987 | .95398 | .31434 | .1813 | .0482 | .3347 | .0097797 | 1.9150 | 33 |
| 28 | .30015 | .95389 | .31466 | .1780 | .0483 | .3317 | .0098085 | 1.9121 | 32 |
| 29 | .30043 | .95380 | .31498 | .1748 | .0484 | .3286 | .0098373 | 1.9092 | 31 |
| 30 | 0.30071 | 0.95372 | 0.31530 | 3.1716 | 1.0485 | 3.3255 | .0098662 | 1.9062 | 30 |
| 31 | .30098 | .95363 | .31562 | .1684 | .0486 | .3224 | .0098951 | 1.9033 | 29 |
| 32 | .30126 | .95354 | .31594 | .1652 | .0487 | .3194 | .0099241 | 1.9004 | 28 |
| 33 | .30154 | .95345 | .31626 | .1620 | .0488 | .3163 | .0099532 | 1.8975 | 27 |
| 34 | .30182 | .95337 | .31658 | .1588 | .0489 | .3133 | .0099823 | 1.8946 | 26 |
| 35 | 0.30209 | 0.95328 | 0.31690 | 3.1556 | 1.0490 | 3.3102 | .010012 | 1.8917 | 25 |
| 36 | .30237 | .95319 | .31722 | .1524 | .0491 | .3072 | .0100041 | 1.8888 | 24 |
| 37 | .30265 | .95310 | .31754 | .1492 | .0492 | .3042 | .010070 | 1.8859 | 23 |
| 38 | .30292 | .95301 | .31786 | .1460 | .0493 | .3012 | .010099 | 1.8830 | 22 |
| 39 | .30320 | .95293 | .31818 | .1429 | .0494 | .2981 | .010129 | 1.8801 | 21 |
| 40 | 0.30348 | 0.95284 | 0.31850 | 3.1397 | 1.0495 | 3.2951 | .010158 | 1.8773 | 20 |
| 41 | .30376 | .95275 | .31882 | .1366 | .0496 | .2921 | .010188 | 1.8744 | 19 |
| 42 | .30403 | .95266 | .31914 | .1334 | .0497 | .2891 | .010217 | 1.8715 | 18 |
| 43 | .30431 | .95257 | .31946 | .1303 | .0498 | .2861 | .010247 | 1.8687 | 17 |
| 44 | .30459 | .95248 | .31978 | .1271 | .0499 | .2831 | .010277 | 1.8658 | 16 |
| 45 | 0.30486 | 0.95240 | 0.32010 | 3.1240 | 1.0500 | 3.2801 | .010307 | 1.8630 | 15 |
| 46 | .30514 | .95231 | .32042 | .1209 | .0501 | .2772 | .010336 | 1.8602 | 14 |
| 47 | .30542 | .95222 | .32074 | .1178 | .0502 | .2742 | .010366 | 1.8573 | 13 |
| 48 | .30570 | .95213 | .32106 | .1146 | .0503 | .2712 | .010396 | 1.8545 | 12 |
| 49 | .30597 | .95204 | .32139 | .1115 | .0504 | .2683 | .010426 | 1.8517 | 11 |
| 50 | 0.30625 | 0.95195 | 0.32171 | 3.1084 | 1.0505 | 3.2653 | .010456 | 1.8489 | 10 |
| 51 | .30653 | .95186 | .32203 | .1053 | .0506 | .2624 | .010487 | 1.8461 | 9 |
| 52 | .30680 | .95177 | .32235 | .1022 | .0507 | .2594 | .010517 | 1.8433 | 8 |
| 53 | .30708 | .95168 | .32267 | .0991 | .0508 | .2565 | .010547 | 1.8405 | 7 |
| 54 | .30736 | .95159 | .32299 | .0961 | .0509 | .2535 | .010577 | 1.8377 | 6 |
| 55 | 0.30763 | 0.95150 | 0.32331 | 3.0930 | 1.0510 | 3.2506 | .010608 | 1.8349 | 5 |
| 56 | .30791 | .95142 | .32363 | .0899 | .0511 | .2477 | .010638 | 1.8321 | 4 |
| 57 | .30819 | .95133 | .32396 | .0868 | .0512 | .2448 | .010669 | 1.8293 | 3 |
| 58 | .30846 | .95124 | .32428 | .0838 | .0513 | .2419 | .010699 | 1.8266 | 2 |
| 59 | .30874 | .95115 | .32460 | .0807 | .0514 | .2390 | .010730 | 1.8238 | 1 |
| 60 | 0.30902 | 0.95106 | 0.32492 | 3.0777 | 1.0515 | 3.2361 | .010760 | 1.8210 | 0 |
| M | Cosine | Sine | Cotan. | Tan. | Cosec. | Secant | READ DOWN | 72°–73° Involute | M |

**18° or 198°**        Trigonometric Functions       **161° or 341°**

| M | Sine | Cosine | Tan. | Cotan. | Secant | Cosec. | Involute 18°–19° | READ UP | M |
|---|---|---|---|---|---|---|---|---|---|
| 0 | 0.30902 | 0.95106 | 0.32492 | 3.0777 | 1.0515 | 3.2361 | .010760 | 1.8210 | 60 |
| 1 | .30929 | .95097 | .32524 | .0746 | .0516 | .2332 | .010791 | 1.8183 | 59 |
| 2 | .30957 | .95088 | .32556 | .0716 | .0517 | .2303 | .010822 | 1.8155 | 58 |
| 3 | .30985 | .95079 | .32588 | .0686 | .0518 | .2274 | .010853 | 1.8128 | 57 |
| 4 | .31012 | .95070 | .32621 | .0655 | .0519 | .2245 | .010884 | 1.8101 | 56 |
| 5 | 0.31040 | 0.95061 | 0.32653 | 3.0625 | 1.0520 | 3.2217 | .010915 | 1.8073 | 55 |
| 6 | .31068 | .95052 | .32685 | .0595 | .0521 | .2188 | .010946 | 1.8046 | 54 |
| 7 | .31095 | .95043 | .32717 | .0565 | .0522 | .2159 | .010977 | 1.8019 | 53 |
| 8 | .31123 | .95033 | .32749 | .0535 | .0523 | .2131 | .011008 | 1.7992 | 52 |
| 9 | .31151 | .95024 | .32782 | .0505 | .0524 | .2102 | .011039 | 1.7965 | 51 |
| 10 | 0.31178 | 0.95015 | 0.32814 | 3.0475 | 1.0525 | 3.2074 | .011071 | 1.7938 | 50 |
| 11 | .31206 | .95006 | .32846 | .0445 | .0526 | .2045 | .011102 | 1.7911 | 49 |
| 12 | .31233 | .94997 | .32878 | .0415 | .0527 | .2017 | .011133 | 1.7884 | 48 |
| 13 | .31261 | .94988 | .32911 | .0385 | .0528 | .1989 | .011165 | 1.7857 | 47 |
| 14 | .31289 | .94979 | .32943 | .0356 | .0529 | .1960 | .011196 | 1.7830 | 46 |
| 15 | 0.31316 | 0.94970 | 0.32975 | 3.0326 | 1.0530 | 3.1932 | .011228 | 1.7803 | 45 |
| 16 | .31344 | .94961 | .33007 | .0296 | .0531 | .1904 | .011260 | 1.7776 | 44 |
| 17 | .31372 | .94952 | .33040 | .0267 | .0532 | .1876 | .011291 | 1.7750 | 43 |
| 18 | .31399 | .94943 | .33072 | .0237 | .0533 | .1848 | .011323 | 1.7723 | 42 |
| 19 | .31427 | .94933 | .33104 | .0208 | .0534 | .1820 | .011355 | 1.7697 | 41 |
| 20 | 0.31454 | 0.94924 | 0.33136 | 3.0178 | 1.0535 | 3.1792 | .011387 | 1.7670 | 40 |
| 21 | .31482 | .94915 | .33169 | .0149 | .0536 | .1764 | .011419 | 1.7644 | 39 |
| 22 | .31510 | .94906 | .33201 | .0120 | .0537 | .1736 | .011451 | 1.7617 | 38 |
| 23 | .31537 | .94897 | .33233 | .0090 | .0538 | .1708 | .011483 | 1.7591 | 37 |
| 24 | .31565 | .94888 | .33266 | .0061 | .0539 | .1681 | .011515 | 1.7565 | 36 |
| 25 | 0.31593 | 0.94878 | 0.33298 | 3.0032 | 1.0540 | 3.1653 | .011547 | 1.7538 | 35 |
| 26 | .31620 | .94869 | .33330 | .0003 | .0541 | .1625 | .011580 | 1.7512 | 34 |
| 27 | .31648 | .94860 | .33363 | 2.9974 | .0542 | .1598 | .011612 | 1.7486 | 33 |
| 28 | .31675 | .94851 | .33395 | .9945 | .0543 | .1570 | .011644 | 1.7460 | 32 |
| 29 | .31703 | .94842 | .33427 | .9916 | .0544 | .1543 | .011677 | 1.7434 | 31 |
| 30 | 0.31730 | 0.94832 | 0.33460 | 2.9887 | 1.0545 | 3.1515 | .011709 | 1.7408 | 30 |
| 31 | .31758 | .94823 | .33492 | .9858 | .0546 | .1488 | .011742 | 1.7382 | 29 |
| 32 | .31786 | .94814 | .33524 | .9829 | .0547 | .1461 | .011775 | 1.7356 | 28 |
| 33 | .31813 | .94805 | .33557 | .9800 | .0548 | .1433 | .011807 | 1.7330 | 27 |
| 34 | .31841 | .94795 | .33589 | .9772 | .0549 | .1406 | .011840 | 1.7304 | 26 |
| 35 | 0.31868 | 0.94786 | 0.33621 | 2.9743 | 1.0550 | 3.1379 | .011873 | 1.7278 | 25 |
| 36 | .31896 | .94777 | .33654 | .9714 | .0551 | .1352 | .011906 | 1.7253 | 24 |
| 37 | .31923 | .94768 | .33686 | .9686 | .0552 | .1325 | .011939 | 1.7227 | 23 |
| 38 | .31951 | .94758 | .33718 | .9657 | .0553 | .1298 | .011972 | 1.7201 | 22 |
| 39 | .31979 | .94749 | .33751 | .9629 | .0554 | .1271 | .012005 | 1.7176 | 21 |
| 40 | 0.32006 | 0.94740 | 0.33783 | 2.9600 | 1.0555 | 3.1244 | .012038 | 1.7150 | 20 |
| 41 | .32034 | .94730 | .33816 | .9572 | .0556 | .1217 | .012071 | 1.7125 | 19 |
| 42 | .32061 | .94721 | .33848 | .9544 | .0557 | .1190 | .012105 | 1.7100 | 18 |
| 43 | .32089 | .94712 | .33881 | .9515 | .0558 | .1163 | .012138 | 1.7074 | 17 |
| 44 | .32116 | .94702 | .33913 | .9487 | .0559 | .1137 | .012172 | 1.7049 | 16 |
| 45 | 0.32144 | 0.94693 | 0.33945 | 2.9459 | 1.0560 | 3.1110 | .012205 | 1.7024 | 15 |
| 46 | .32171 | .94684 | .33978 | .9431 | .0561 | .1083 | .012239 | 1.6998 | 14 |
| 47 | .32199 | .94674 | .34010 | .9403 | .0563 | .1057 | .012272 | 1.6973 | 13 |
| 48 | .32227 | .94665 | .34043 | .9375 | .0564 | .1030 | .012306 | 1.6948 | 12 |
| 49 | .32254 | .94656 | .34075 | .9347 | .0565 | .1004 | .012340 | 1.6923 | 11 |
| 50 | 0.32282 | 0.94646 | 0.34108 | 2.9319 | 1.0566 | 3.0977 | .012373 | 1.6898 | 10 |
| 51 | .32309 | .94637 | .34140 | .9291 | .0567 | .0951 | .012407 | 1.6873 | 9 |
| 52 | .32337 | .94627 | .34173 | .9263 | .0568 | .0925 | .012441 | 1.6848 | 8 |
| 53 | .32364 | .94618 | .34205 | .9235 | .0569 | .0898 | .012475 | 1.6823 | 7 |
| 54 | .32392 | .94609 | .34238 | .9208 | .0570 | .0872 | .012509 | 1.6798 | 6 |
| 55 | 0.32419 | 0.94599 | 0.34270 | 2.9180 | 1.0571 | 3.0846 | .012543 | 1.6774 | 5 |
| 56 | .32447 | .94590 | .34303 | .9152 | .0572 | .0820 | .012578 | 1.6749 | 4 |
| 57 | .32474 | .94580 | .34335 | .9125 | .0573 | .0794 | .012612 | 1.6724 | 3 |
| 58 | .32502 | .94571 | .34368 | .9097 | .0574 | .0768 | .012646 | 1.6699 | 2 |
| 59 | .32529 | .94561 | .34400 | .9070 | .0575 | .0742 | .012681 | 1.6675 | 1 |
| 60 | 0.32557 | 0.94552 | 0.34433 | 2.9042 | 1.0576 | 3.0716 | .012715 | 1.6650 | 0 |
| M | Cosine | Sine | Cotan. | Tan. | Cosec. | Secant | READ DOWN | 71°–72° Involute | M |

**Trigonometric Functions**

| M | Sine | Cosine | Tan. | Cotan. | Secant | Cosec. | Involute 19°–20° | READ UP | M |
|---|------|--------|------|--------|--------|--------|------------------|---------|---|
| 0 | 0.32557 | 0.94552 | 0.34433 | 2.9042 | 1.0576 | 3.0716 | .012715 | 1.6650 | 60 |
| 1 | .32584 | .94542 | .34465 | .9015 | .0577 | .0690 | .012750 | 1.6626 | 59 |
| 2 | .32612 | .94533 | .34498 | .8987 | .0578 | .0664 | .012784 | 1.6601 | 58 |
| 3 | .32639 | .94523 | .34530 | .8960 | .0579 | .0638 | .012819 | 1.6577 | 57 |
| 4 | .32667 | .94514 | .34563 | .8933 | .0580 | .0612 | .012854 | 1.6553 | 56 |
| 5 | 0.32694 | 0.94504 | 0.34596 | 2.8905 | 1.0582 | 3.0586 | .012888 | 1.6528 | 55 |
| 6 | .32722 | .94495 | .34628 | .8878 | .0583 | .0561 | .012923 | 1.6504 | 54 |
| 7 | .32749 | .94485 | .34661 | .8851 | .0584 | .0535 | .012958 | 1.6480 | 53 |
| 8 | .32777 | .94476 | .34693 | .8824 | .0585 | .0509 | .012993 | 1.6455 | 52 |
| 9 | .32804 | .94466 | .34726 | .8797 | .0586 | .0484 | .013028 | 1.6431 | 51 |
| 10 | 0.32832 | 0.94457 | 0.34758 | 2.8770 | 1.0587 | 3.0458 | .013063 | 1.6407 | 50 |
| 11 | .32859 | .94447 | .34791 | .8743 | .0588 | .0433 | .013098 | 1.6383 | 49 |
| 12 | .32887 | .94438 | .34824 | .8716 | .0589 | .0407 | .013134 | 1.6359 | 48 |
| 13 | .32914 | .94428 | .34856 | .8689 | .0590 | .0382 | .013169 | 1.6335 | 47 |
| 14 | .32942 | .94418 | .34889 | .8662 | .0591 | .0357 | .013204 | 1.6311 | 46 |
| 15 | 0.32969 | 0.94409 | 0.34922 | 2.8636 | 1.0592 | 3.0331 | .013240 | 1.6287 | 45 |
| 16 | .32997 | .94399 | .34954 | .8609 | .0593 | .0306 | .013275 | 1.6264 | 44 |
| 17 | .33024 | .94390 | .34987 | .8582 | .0594 | .0281 | .013311 | 1.6240 | 43 |
| 18 | .33051 | .94380 | .35020 | .8556 | .0595 | .0256 | .013346 | 1.6216 | 42 |
| 19 | .33079 | .94370 | .35052 | .8529 | .0597 | .0231 | .013382 | 1.6192 | 41 |
| 20 | 0.33106 | 0.94361 | 0.35085 | 2.8502 | 1.0598 | 3.0206 | .013418 | 1.6169 | 40 |
| 21 | .33134 | .94351 | .35118 | .8476 | .0599 | .0181 | .013454 | 1.6145 | 39 |
| 22 | .33161 | .94342 | .35150 | .8449 | .0600 | .0156 | .013490 | 1.6122 | 38 |
| 23 | .33189 | .94332 | .35183 | .8423 | .0601 | .0131 | .013526 | 1.6098 | 37 |
| 24 | .33216 | .94322 | .35216 | .8397 | .0602 | .0106 | .013562 | 1.6075 | 36 |
| 25 | 0.33244 | 0.94313 | 0.35248 | 2.8370 | 1.0603 | 3.0081 | .013598 | 1.6051 | 35 |
| 26 | .33271 | .94303 | .35281 | .8344 | .0604 | .0056 | .013634 | 1.6028 | 34 |
| 27 | .33298 | .94293 | .35314 | .8318 | .0605 | .0031 | .013670 | 1.6004 | 33 |
| 28 | .33326 | .94284 | .35346 | .8291 | .0606 | .0007 | .013707 | 1.5981 | 32 |
| 29 | .33353 | .94274 | .35379 | .8265 | .0607 | 2.9982 | .013743 | 1.5958 | 31 |
| 30 | 0.33381 | 0.94264 | 0.35412 | 2.8239 | 1.0608 | 2.9957 | .013779 | 1.5935 | 30 |
| 31 | .33408 | .94254 | .35445 | .8213 | .0610 | .9933 | .013816 | 1.5911 | 29 |
| 32 | .33436 | .94245 | .35477 | .8187 | .0611 | .9908 | .013852 | 1.5888 | 28 |
| 33 | .33463 | .94235 | .35510 | .8161 | .0612 | .9884 | .013889 | 1.5865 | 27 |
| 34 | .33490 | .94225 | .35543 | .8135 | .0613 | .9859 | .013926 | 1.5842 | 26 |
| 35 | 0.33518 | 0.94215 | 0.35576 | 2.8109 | 1.0614 | 2.9835 | .013963 | 1.5819 | 25 |
| 36 | .33545 | .94206 | .35608 | .8083 | .0615 | .9811 | .013999 | 1.5796 | 24 |
| 37 | .33573 | .94196 | .35641 | .8057 | .0616 | .9786 | .014036 | 1.5773 | 23 |
| 38 | .33600 | .94186 | .35674 | .8032 | .0617 | .9762 | .014073 | 1.5750 | 22 |
| 39 | .33627 | .94176 | .35707 | .8006 | .0618 | .9738 | .014110 | 1.5728 | 21 |
| 40 | 0.33655 | 0.94167 | 0.35740 | 2.7980 | 1.0619 | 2.9713 | .014148 | 1.5705 | 20 |
| 41 | .33682 | .94157 | .35772 | .7955 | .0621 | .9689 | .014185 | 1.5682 | 19 |
| 42 | .33710 | .94147 | .35805 | .7929 | .0622 | .9665 | .014222 | 1.5659 | 18 |
| 43 | .33737 | .94137 | .35838 | .7903 | .0623 | .9641 | .014259 | 1.5637 | 17 |
| 44 | .33764 | .94127 | .35871 | .7878 | .0624 | .9617 | .014297 | 1.5614 | 16 |
| 45 | 0.33792 | 0.94118 | 0.35904 | 2.7852 | 1.0625 | 2.9593 | .014334 | 1.5591 | 15 |
| 46 | .33819 | .94108 | .35937 | .7827 | .0626 | .9569 | .014372 | 1.5569 | 14 |
| 47 | .33846 | .94098 | .35969 | .7801 | .0627 | .9545 | .014409 | 1.5546 | 13 |
| 48 | .33874 | .94088 | .36002 | .7776 | .0628 | .9521 | .014447 | 1.5524 | 12 |
| 49 | .33901 | .94078 | .36035 | .7751 | .0629 | .9498 | .014485 | 1.5501 | 11 |
| 50 | 0.33929 | 0.94068 | 0.36068 | 2.7725 | 1.0631 | 2.9474 | .014523 | 1.5479 | 10 |
| 51 | .33956 | .94058 | .36101 | .7700 | .0632 | .9450 | .014560 | 1.5457 | 9 |
| 52 | .33983 | .94049 | .36134 | .7675 | .0633 | .9426 | .014598 | 1.5434 | 8 |
| 53 | .34011 | .94039 | .36167 | .7650 | .0634 | .9403 | .014636 | 1.5412 | 7 |
| 54 | .34038 | .94029 | .36199 | .7625 | .0635 | .9379 | .014674 | 1.5390 | 6 |
| 55 | 0.34065 | 0.94019 | 0.36232 | 2.7600 | 1.0636 | 2.9355 | .014713 | 1.5368 | 5 |
| 56 | .34093 | .94009 | .36265 | .7575 | .0637 | .9332 | .014751 | 1.5346 | 4 |
| 57 | .34120 | .93999 | .36298 | .7550 | .0638 | .9308 | .014789 | 1.5324 | 3 |
| 58 | .34147 | .93989 | .36331 | .7525 | .0640 | .9285 | .014827 | 1.5301 | 2 |
| 59 | .34175 | .93979 | .36364 | .7500 | .0641 | .9261 | .014866 | 1.5279 | 1 |
| 60 | 0.34202 | 0.93969 | 0.36397 | 2.7475 | 1.0642 | 2.9238 | .014904 | 1.5257 | 0 |
| M | Cosine | Sine | Cotan. | Tan. | Cosec. | Secant | READ DOWN | 70°–71° Involute | M |

**20° or 200°**     Trigonometric Functions     **159° or 339°**

| M | Sine | Cosine | Tan. | Cotan. | Secant | Cosec. | Involute 20°–21° | READ UP | M |
|---|------|--------|------|--------|--------|--------|------------------|---------|---|
| 0 | 0.34202 | 0.93969 | 0.36397 | 2.7475 | 1.0642 | 2.9238 | .014904 | 1.5257 | 60 |
| 1 | .34229 | .93959 | .36430 | .7450 | .0643 | .9215 | .014943 | 1.5236 | 59 |
| 2 | .34257 | .93949 | .36463 | .7425 | .0644 | .9191 | .014982 | 1.5214 | 58 |
| 3 | .34284 | .93939 | .36496 | .7400 | .0645 | .9168 | .015020 | 1.5192 | 57 |
| 4 | .34311 | .93929 | .36529 | .7376 | .0646 | .9145 | .015059 | 1.5170 | 56 |
| 5 | 0.34339 | 0.93919 | 0.36562 | 2.7351 | 1.0647 | 2.9122 | .015098 | 1.5148 | 55 |
| 6 | .34366 | .93909 | .36595 | .7326 | .0649 | .9099 | .015137 | 1.5126 | 54 |
| 7 | .34393 | .93899 | .36628 | .7302 | .0650 | .9075 | .015176 | 1.5105 | 53 |
| 8 | .34421 | .93889 | .36661 | .7277 | .0651 | .9052 | .015215 | 1.5083 | 52 |
| 9 | .34448 | .93879 | .36694 | .7253 | .0652 | .9029 | .015254 | 1.5061 | 51 |
| 10 | 0.34475 | 0.93869 | 0.36727 | 2.7228 | 1.0653 | 2.9006 | .015293 | 1.5040 | 50 |
| 11 | .34503 | .93859 | .36760 | .7204 | .0654 | .8983 | .015333 | 1.5018 | 49 |
| 12 | .34530 | .93849 | .36793 | .7179 | .0655 | .8960 | .015372 | 1.4997 | 48 |
| 13 | .34557 | .93839 | .36826 | .7155 | .0657 | .8938 | .015411 | 1.4975 | 47 |
| 14 | .34584 | .93829 | .36859 | .7130 | .0658 | .8915 | .015451 | 1.4954 | 46 |
| 15 | 0.34612 | 0.93819 | 0.36892 | 2.7106 | 1.0659 | 2.8892 | .015490 | 1.4933 | 45 |
| 16 | .34639 | .93809 | .36925 | .7082 | .0660 | .8869 | .015530 | 1.4911 | 44 |
| 17 | .34666 | .93799 | .36958 | .7058 | .0661 | .8846 | .015570 | 1.4890 | 43 |
| 18 | .34694 | .93789 | .36991 | .7034 | .0662 | .8824 | .015609 | 1.4869 | 42 |
| 19 | .34721 | .93779 | .37024 | .7009 | .0663 | .8801 | .015649 | 1.4847 | 41 |
| 20 | 0.34748 | 0.93769 | 0.37057 | 2.6985 | 1.0665 | 2.8779 | .015689 | 1.4826 | 40 |
| 21 | .34775 | .93759 | .37090 | .6961 | .0666 | .8756 | .015729 | 1.4805 | 39 |
| 22 | .34803 | .93748 | .37123 | .6937 | .0667 | .8733 | .015769 | 1.4784 | 38 |
| 23 | .34830 | .93738 | .37157 | .6913 | .0668 | .8711 | .015809 | 1.4763 | 37 |
| 24 | .34857 | .93728 | .37190 | .6889 | .0669 | .8688 | .015849 | 1.4742 | 36 |
| 25 | 0.34884 | 0.93718 | 0.37223 | 2.6865 | 1.0670 | 2.8666 | .015890 | 1.4721 | 35 |
| 26 | .34912 | .93708 | .37256 | .6841 | .0671 | .8644 | .015930 | 1.4700 | 34 |
| 27 | .34939 | .93698 | .37289 | .6818 | .0673 | .8621 | .015971 | 1.4679 | 33 |
| 28 | .34966 | .93688 | .37322 | .6794 | .0674 | .8599 | .016011 | 1.4658 | 32 |
| 29 | .34993 | .93677 | .37355 | .6770 | .0675 | .8577 | .016052 | 1.4637 | 31 |
| 30 | 0.35021 | 0.93667 | 0.37388 | 2.6746 | 1.0676 | 2.8555 | .016092 | 1.4616 | 30 |
| 31 | .35048 | .93657 | .37422 | .6723 | .0677 | .8532 | .016133 | 1.4595 | 29 |
| 32 | .35075 | .93647 | .37455 | .6699 | .0678 | .8510 | .016174 | 1.4575 | 28 |
| 33 | .35102 | .93637 | .37488 | .6675 | .0680 | .8488 | .016214 | 1.4554 | 27 |
| 34 | .35130 | .93626 | .37521 | .6652 | .0681 | .8466 | .016255 | 1.4533 | 26 |
| 35 | 0.35157 | 0.93616 | 0.37554 | 2.6628 | 1.0682 | 2.8444 | .016296 | 1.4513 | 25 |
| 36 | .35184 | .93606 | .37588 | .6605 | .0683 | .8422 | .016337 | 1.4492 | 24 |
| 37 | .35211 | .93596 | .37621 | .6581 | .0684 | .8400 | .016379 | 1.4471 | 23 |
| 38 | .35239 | .93585 | .37654 | .6558 | .0685 | .8378 | .016420 | 1.4451 | 22 |
| 39 | .35266 | .93575 | .37687 | .6534 | .0687 | .8356 | .016461 | 1.4430 | 21 |
| 40 | 0.35293 | 0.93565 | 0.37720 | 2.6511 | 1.0688 | 2.8334 | .016502 | 1.4410 | 20 |
| 41 | .35320 | .93555 | .37754 | .6488 | .0689 | .8312 | .016544 | 1.4389 | 19 |
| 42 | .35347 | .93544 | .37787 | .6464 | .0690 | .8291 | .016585 | 1.4369 | 18 |
| 43 | .35375 | .93534 | .37820 | .6441 | .0691 | .8269 | .016627 | 1.4349 | 17 |
| 44 | .35402 | .93524 | .37853 | .6418 | .0692 | .8247 | .016669 | 1.4328 | 16 |
| 45 | 0.35429 | 0.93514 | 0.37887 | 2.6395 | 1.0694 | 2.8225 | .016710 | 1.4308 | 15 |
| 46 | .35456 | .93503 | .37920 | .6371 | .0695 | .8204 | .016752 | 1.4288 | 14 |
| 47 | .35484 | .93493 | .37953 | .6348 | .0696 | .8182 | .016794 | 1.4268 | 13 |
| 48 | .35511 | .93483 | .37986 | .6325 | .0697 | .8161 | .016836 | 1.4248 | 12 |
| 49 | .35538 | .93472 | .38020 | .6302 | .0698 | .8139 | .016878 | 1.4227 | 11 |
| 50 | 0.35565 | 0.93462 | 0.38053 | 2.6279 | 1.0700 | 2.8117 | .016920 | 1.4207 | 10 |
| 51 | .35592 | .93452 | .38086 | .6256 | .0701 | .8096 | .017004 | 1.4187 | 9 |
| 52 | .35619 | .93441 | .38120 | .6233 | .0702 | .8075 | .017004 | 1.4167 | 8 |
| 53 | .35647 | .93431 | .38153 | .6210 | .0703 | .8053 | .017047 | 1.4147 | 7 |
| 54 | .35674 | .93420 | .38186 | .6187 | .0704 | .8032 | .017089 | 1.4127 | 6 |
| 55 | 0.35701 | 0.93410 | 0.38220 | 2.6165 | 1.0705 | 2.8010 | .017132 | 1.4107 | 5 |
| 56 | .35728 | .93400 | .38253 | .6142 | .0707 | .7989 | .017174 | 1.4087 | 4 |
| 57 | .35755 | .93389 | .38286 | .6119 | .0708 | .7968 | .017217 | 1.4067 | 3 |
| 58 | .35782 | .93379 | .38320 | .6096 | .0709 | .7947 | .017259 | 1.4048 | 2 |
| 59 | .35810 | .93368 | .38353 | .6074 | .0710 | .7925 | .017302 | 1.4028 | 1 |
| 60 | 0.35837 | 0.93358 | 0.38386 | 2.6051 | 1.0711 | 2.7904 | .017345 | 1.4008 | 0 |
| M | Cosine | Sine | Cotan. | Tan. | Cosec. | Secant | READ DOWN | 69°–70° Involute | M |

**21° or 201°** Trigonometric Functions **158° or 338°**

| M | Sine | Cosine | Tan. | Cotan. | Secant | Cosec. | Involute 21°–22° | READ UP | M |
|---|------|--------|------|--------|--------|--------|---------|---------|---|
| 0 | 0.35837 | 0.93358 | 0.38386 | 2.6051 | 1.0711 | 2.7904 | .017345 | 1.4008 | 60 |
| 1 | .35864 | .93348 | .38420 | .6028 | .0713 | .7883 | .017388 | 1.3988 | 59 |
| 2 | .35891 | .93337 | .38453 | .6006 | .0714 | .7862 | .017431 | 1.3969 | 58 |
| 3 | .35918 | .93327 | .38487 | .5983 | .0715 | .7841 | .017474 | 1.3949 | 57 |
| 4 | .35945 | .93316 | .38520 | .5961 | .0716 | .7820 | .017517 | 1.3929 | 56 |
| 5 | 0.35973 | 0.93306 | 0.38553 | 2.5938 | 1.0717 | 2.7799 | .017560 | 1.3910 | 55 |
| 6 | .36000 | .93295 | .38587 | .5916 | .0719 | .7778 | .017603 | 1.3890 | 54 |
| 7 | .36027 | .93285 | .38620 | .5893 | .0720 | .7757 | .017647 | 1.3871 | 53 |
| 8 | .36054 | .93274 | .38654 | .5871 | .0721 | .7736 | .017690 | 1.3851 | 52 |
| 9 | .36081 | .93264 | .38687 | .5848 | .0722 | .7715 | .017734 | 1.3832 | 51 |
| 10 | 0.36108 | 0.93253 | 0.38721 | 2.5826 | 1.0723 | 2.7695 | .017777 | 1.3812 | 50 |
| 11 | .36135 | .93243 | .38754 | .5804 | .0725 | .7674 | .017821 | 1.3793 | 49 |
| 12 | .36162 | .93232 | .38787 | .5782 | .0726 | .7653 | .017865 | 1.3774 | 48 |
| 13 | .36190 | .93222 | .38821 | .5759 | .0727 | .7632 | .017908 | 1.3754 | 47 |
| 14 | .36217 | .93211 | .38854 | .5737 | .0728 | .7612 | .017952 | 1.3735 | 46 |
| 15 | 0.36244 | 0.93201 | 0.38888 | 2.5715 | 1.0730 | 2.7591 | .017996 | 1.3716 | 45 |
| 16 | .36271 | .93190 | .38921 | .5693 | .0731 | .7570 | .018040 | 1.3697 | 44 |
| 17 | .36298 | .93180 | .38955 | .5671 | .0732 | .7550 | .018084 | 1.3677 | 43 |
| 18 | .36325 | .93169 | .38988 | .5649 | .0733 | .7529 | .018129 | 1.3658 | 42 |
| 19 | .36352 | .93159 | .39022 | .5627 | .0734 | .7509 | .018173 | 1.3639 | 41 |
| 20 | 0.36379 | 0.93148 | 0.39055 | 2.5605 | 1.0736 | 2.7488 | .018217 | 1.3620 | 40 |
| 21 | .36406 | .93137 | .39089 | .5583 | .0737 | .7468 | .018262 | 1.3601 | 39 |
| 22 | .36434 | .93127 | .39122 | .5561 | .0738 | .7447 | .018306 | 1.3582 | 38 |
| 23 | .36461 | .93116 | .39156 | .5539 | .0739 | .7427 | .018351 | 1.3563 | 37 |
| 24 | .36488 | .93106 | .39190 | .5517 | .0740 | .7407 | .018395 | 1.3544 | 36 |
| 25 | 0.36515 | 0.93095 | 0.39223 | 2.5495 | 1.0742 | 2.7386 | .018440 | 1.3525 | 35 |
| 26 | .36542 | .93084 | .39257 | .5473 | .0743 | .7366 | .018485 | 1.3506 | 34 |
| 27 | .36569 | .93074 | .39290 | .5452 | .0744 | .7346 | .018530 | 1.3487 | 33 |
| 28 | .36596 | .93063 | .39324 | .5430 | .0745 | .7325 | .018575 | 1.3469 | 32 |
| 29 | .36623 | .93052 | .39357 | .5408 | .0747 | .7305 | .018620 | 1.3450 | 31 |
| 30 | 0.36650 | 0.93042 | 0.39391 | 2.5386 | 1.0748 | 2.7285 | .018665 | 1.3431 | 30 |
| 31 | .36677 | .93031 | .39425 | .5365 | .0749 | .7265 | .018710 | 1.3412 | 29 |
| 32 | .36704 | .93020 | .39458 | .5343 | .0750 | .7245 | .018755 | 1.3394 | 28 |
| 33 | .36731 | .93010 | .39492 | .5322 | .0752 | .7225 | .018800 | 1.3375 | 27 |
| 34 | .36758 | .92999 | .39526 | .5300 | .0753 | .7205 | .018846 | 1.3356 | 26 |
| 35 | 0.36785 | 0.92988 | 0.39559 | 2.5279 | 1.0754 | 2.7185 | .018891 | 1.3338 | 25 |
| 36 | .36812 | .92978 | .39593 | .5257 | .0755 | .7165 | .018937 | 1.3319 | 24 |
| 37 | .36839 | .92967 | .39626 | .5236 | .0757 | .7145 | .018983 | 1.3301 | 23 |
| 38 | .36867 | .92956 | .39660 | .5214 | .0758 | .7125 | .019028 | 1.3282 | 22 |
| 39 | .36894 | .92945 | .39694 | .5193 | .0759 | .7105 | .019074 | 1.3264 | 21 |
| 40 | 0.36921 | 0.92935 | 0.39727 | 2.5172 | 1.0760 | 2.7085 | .019120 | 1.3245 | 20 |
| 41 | .36948 | .92924 | .39761 | .5150 | .0761 | .7065 | .019166 | 1.3227 | 19 |
| 42 | .36975 | .92913 | .39795 | .5129 | .0763 | .7046 | .019212 | 1.3208 | 18 |
| 43 | .37002 | .92902 | .39829 | .5108 | .0764 | .7026 | .019258 | 1.3190 | 17 |
| 44 | .37029 | .92892 | .39862 | .5086 | .0765 | .7006 | .019304 | 1.3172 | 16 |
| 45 | 0.37056 | 0.92881 | 0.39896 | 2.5065 | 1.0766 | 2.6986 | .019350 | 1.3153 | 15 |
| 46 | .37083 | .92870 | .39930 | .5044 | .0768 | .6967 | .019397 | 1.3135 | 14 |
| 47 | .37110 | .92859 | .39963 | .5023 | .0769 | .6947 | .019443 | 1.3117 | 13 |
| 48 | .37137 | .92849 | .39997 | .5002 | .0770 | .6927 | .019490 | 1.3099 | 12 |
| 49 | .37164 | .92838 | .40031 | .4981 | .0771 | .6908 | .019536 | 1.3080 | 11 |
| 50 | 0.37191 | 0.92827 | 0.40065 | 2.4960 | 1.0773 | 2.6888 | .019583 | 1.3062 | 10 |
| 51 | .37218 | .92816 | .40098 | .4939 | .0774 | .6869 | .019630 | 1.3044 | 9 |
| 52 | .37245 | .92805 | .40132 | .4918 | .0775 | .6849 | .019676 | 1.3026 | 8 |
| 53 | .37272 | .92794 | .40166 | .4897 | .0777 | .6830 | .019723 | 1.3008 | 7 |
| 54 | .37299 | .92784 | .40200 | .4876 | .0778 | .6811 | .019770 | 1.2990 | 6 |
| 55 | 0.37326 | 0.92773 | 0.40234 | 2.4855 | 1.0779 | 2.6791 | .019817 | 1.2972 | 5 |
| 56 | .37353 | .92762 | .40267 | .4834 | .0780 | .6772 | .019864 | 1.2954 | 4 |
| 57 | .37380 | .92751 | .40301 | .4813 | .0782 | .6752 | .019912 | 1.2936 | 3 |
| 58 | .37407 | .92740 | .40335 | .4792 | .0783 | .6733 | .019959 | 1.2918 | 2 |
| 59 | .37434 | .92729 | .40369 | .4772 | .0784 | .6714 | .020006 | 1.2900 | 1 |
| 60 | 0.37461 | 0.92718 | 0.40403 | 2.4751 | 1.0785 | 2.6695 | .020054 | 1.2883 | 0 |

| M | Cosine | Sine | Cotan. | Tan. | Cosec. | Secant | READ DOWN | 68°–69° Involute | M |
|---|--------|------|--------|------|--------|--------|-----------|---------|---|

**111° or 291°** **68° or 248°**

**22° or 202°**     Trigonometric Functions     **157° or 337°**

| M | Sine | Cosine | Tan. | Cotan. | Secant | Cosec. | Involute 22°–23° | READ UP | M |
|---|---|---|---|---|---|---|---|---|---|
| 0 | 0.37461 | 0.92718 | 0.40403 | 2.4751 | 1.0785 | 2.6695 | .020054 | 1.2883 | 60 |
| 1 | .37488 | .92707 | .40436 | .4730 | .0787 | .6675 | .020101 | 1.2865 | 59 |
| 2 | .37515 | .92697 | .40470 | .4709 | .0788 | .6656 | .020149 | 1.2847 | 58 |
| 3 | .37542 | .92686 | .40504 | .4689 | .0789 | .6637 | .020197 | 1.2829 | 57 |
| 4 | .37569 | .92675 | .40538 | .4668 | .0790 | .6618 | .020244 | 1.2812 | 56 |
| 5 | 0.37595 | 0.92664 | 0.40572 | 2.4648 | 1.0792 | 2.6599 | .020292 | 1.2794 | 55 |
| 6 | .37622 | .92653 | .40606 | .4627 | .0793 | .6580 | .020340 | 1.2776 | 54 |
| 7 | .37649 | .92642 | .40640 | .4606 | .0794 | .6561 | .020388 | 1.2759 | 53 |
| 8 | .37676 | .92631 | .40674 | .4586 | .0796 | .6542 | .020436 | 1.2741 | 52 |
| 9 | .37703 | .92620 | .40707 | .4566 | .0797 | .6523 | .020484 | 1.2723 | 51 |
| 10 | 0.37730 | 0.92609 | 0.40741 | 2.4545 | 1.0798 | 2.6504 | .020533 | 1.2706 | 50 |
| 11 | .37757 | .92598 | .40775 | .4525 | .0799 | .6485 | .020581 | 1.2688 | 49 |
| 12 | .37784 | .92587 | .40809 | .4504 | .0801 | .6466 | .020629 | 1.2671 | 48 |
| 13 | .37811 | .92576 | .40843 | .4484 | .0802 | .6447 | .020678 | 1.2653 | 47 |
| 14 | .37838 | .92565 | .40877 | .4464 | .0803 | .6429 | .020726 | 1.2636 | 46 |
| 15 | 0.37865 | 0.92554 | 0.40911 | 2.4443 | 1.0804 | 2.6410 | .020775 | 1.2619 | 45 |
| 16 | .37892 | .92543 | .40945 | .4423 | .0806 | .6391 | .020824 | 1.2601 | 44 |
| 17 | .37919 | .92532 | .40979 | .4403 | .0807 | .6372 | .020873 | 1.2584 | 43 |
| 18 | .37946 | .92521 | .41013 | .4383 | .0808 | .6354 | .020921 | 1.2567 | 42 |
| 19 | .37973 | .92510 | .41047 | .4362 | .0810 | .6335 | .020970 | 1.2549 | 41 |
| 20 | 0.37999 | 0.92499 | 0.41081 | 2.4342 | 1.0811 | 2.6316 | .021019 | 1.2532 | 40 |
| 21 | .38026 | .92488 | .41115 | .4322 | .0812 | .6298 | .021069 | 1.2515 | 39 |
| 22 | .38053 | .92477 | .41149 | .4302 | .0814 | .6279 | .021118 | 1.2498 | 38 |
| 23 | .38080 | .92466 | .41183 | .4282 | .0815 | .6260 | .021167 | 1.2481 | 37 |
| 24 | .38107 | .92455 | .41217 | .4262 | .0816 | .6242 | .021217 | 1.2463 | 36 |
| 25 | 0.38134 | 0.92443 | 0.41251 | 2.4242 | 1.0817 | 2.6223 | .021266 | 1.2446 | 35 |
| 26 | .38151 | .92432 | .41285 | .4222 | .0819 | .6205 | .021316 | 1.2429 | 34 |
| 27 | .38188 | .92421 | .41319 | .4202 | .0820 | .6186 | .021365 | 1.2412 | 33 |
| 28 | .38215 | .92410 | .41353 | .4182 | .0821 | .6168 | .021415 | 1.2395 | 32 |
| 29 | .38241 | .92399 | .41387 | .4162 | .0823 | .6150 | .021465 | 1.2378 | 31 |
| 30 | 0.38268 | 0.92388 | 0.41421 | 2.4142 | 1.0824 | 2.6131 | .021514 | 1.2361 | 30 |
| 31 | .38295 | .92377 | .41455 | .4122 | .0825 | .6113 | .021564 | 1.2344 | 29 |
| 32 | .38322 | .92366 | .41490 | .4102 | .0827 | .6095 | .021614 | 1.2327 | 28 |
| 33 | .38349 | .92355 | .41524 | .4083 | .0828 | .6076 | .021665 | 1.2310 | 27 |
| 34 | .38376 | .92343 | .41558 | .4063 | .0829 | .6058 | .021715 | 1.2294 | 26 |
| 35 | 0.38403 | 0.92332 | 0.41592 | 2.4043 | 1.0830 | 2.6040 | .021765 | 1.2277 | 25 |
| 36 | .38430 | .92321 | .41626 | .4023 | .0832 | .6022 | .021815 | 1.2260 | 24 |
| 37 | .38456 | .92310 | .41660 | .4004 | .0833 | .6003 | .021866 | 1.2243 | 23 |
| 38 | .38483 | .92299 | .41694 | .3984 | .0834 | .5985 | .021916 | 1.2226 | 22 |
| 39 | .38510 | .92287 | .41728 | .3964 | .0836 | .5967 | .021967 | 1.2210 | 21 |
| 40 | 0.38537 | 0.92276 | 0.41763 | 2.3945 | 1.0837 | 2.5949 | .022018 | 1.2193 | 20 |
| 41 | .38564 | .92265 | .41797 | .3925 | .0838 | .5931 | .022068 | 1.2176 | 19 |
| 42 | .38591 | .92254 | .41831 | .3906 | .0840 | .5913 | .022119 | 1.2160 | 18 |
| 43 | .38617 | .92243 | .41865 | .3886 | .0841 | .5895 | .022170 | 1.2143 | 17 |
| 44 | .38644 | .92231 | .41899 | .3867 | .0842 | .5877 | .022221 | 1.2127 | 16 |
| 45 | 0.38671 | 0.92220 | 0.41933 | 2.3847 | 1.0844 | 2.5859 | .022272 | 1.2110 | 15 |
| 46 | .38698 | .92209 | .41968 | .3828 | .0845 | .5841 | .022324 | 1.2093 | 14 |
| 47 | .38725 | .92198 | .42002 | .3808 | .0846 | .5823 | .022375 | 1.2077 | 13 |
| 48 | .38752 | .92186 | .42036 | .3789 | .0848 | .5805 | .022426 | 1.2060 | 12 |
| 49 | .38778 | .92175 | .42070 | .3770 | .0849 | .5788 | .022478 | 1.2044 | 11 |
| 50 | 0.38805 | 0.92164 | 0.42105 | 2.3750 | 1.0850 | 2.5770 | .022529 | 1.2028 | 10 |
| 51 | .38832 | .92152 | .42139 | .3731 | .0852 | .5752 | .022581 | 1.2011 | 9 |
| 52 | .38859 | .92141 | .42173 | .3712 | .0853 | .5734 | .022632 | 1.1995 | 8 |
| 53 | .38886 | .92130 | .42207 | .3692 | .0854 | .5716 | .022684 | 1.1978 | 7 |
| 54 | .38912 | .92119 | .42242 | .3673 | .0856 | .5699 | .022736 | 1.1962 | 6 |
| 55 | 0.38939 | 0.92107 | 0.42276 | 2.3654 | 1.0857 | 2.5681 | .022788 | 1.1946 | 5 |
| 56 | .38966 | .92096 | .42310 | .3635 | .0858 | .5663 | .022840 | 1.1930 | 4 |
| 57 | .38993 | .92085 | .42345 | .3616 | .0860 | .5646 | .022892 | 1.1913 | 3 |
| 58 | .39020 | .92073 | .42379 | .3597 | .0861 | .5628 | .022944 | 1.1897 | 2 |
| 59 | .39046 | .92062 | .42413 | .3578 | .0862 | .5611 | .022997 | 1.1881 | 1 |
| 60 | 0.39073 | 0.92050 | 0.42447 | 2.3559 | 1.0864 | 2.5593 | .023049 | 1.1865 | 0 |
| M | Cosine | Sine | Cotan. | Tan. | Cosec. | Secant | READ DOWN | 67°–68° Involute | M |

**Trigonometric Functions**

| M | Sine | Cosine | Tan. | Cotan. | Secant | Cosec. | Involute 23°–24° | READ UP | M |
|---|---|---|---|---|---|---|---|---|---|
| 0 | 0.39073 | 0.92050 | 0.42447 | 2.3559 | 1.0864 | 2.5593 | .023049 | 1.1865 | 60 |
| 1 | .39100 | .92043 | .42482 | .3539 | .0865 | .5576 | .023102 | 1.1849 | 59 |
| 2 | .39127 | .92028 | .42516 | .3520 | .0866 | .5559 | .023154 | 1.1833 | 58 |
| 3 | .39153 | .92016 | .42551 | .3501 | .0868 | .5541 | .023207 | 1.1817 | 57 |
| 4 | .39180 | .92005 | .42585 | .3483 | .0869 | .5523 | .023259 | 1.1800 | 56 |
| 5 | 0.39207 | 0.91994 | 0.42619 | 2.3464 | 1.0870 | 2.5506 | .023312 | 1.1784 | 55 |
| 6 | .39234 | .91982 | .42654 | .3445 | .0872 | .5488 | .023365 | 1.1768 | 54 |
| 7 | .39260 | .91971 | .42688 | .3426 | .0873 | .5471 | .023418 | 1.1752 | 53 |
| 8 | .39287 | .91959 | .42722 | .3407 | .0874 | .5454 | .023471 | 1.1736 | 52 |
| 9 | .39314 | .91948 | .42757 | .3388 | .0876 | .5436 | .023524 | 1.1721 | 51 |
| 10 | 0.39341 | 0.91936 | 0.42791 | 2.3369 | 1.0877 | 2.5419 | .023577 | 1.1705 | 50 |
| 11 | .39367 | .91925 | .42826 | .3351 | .0878 | .5402 | .023631 | 1.1689 | 49 |
| 12 | .39394 | .91914 | .42860 | .3332 | .0880 | .5384 | .023684 | 1.1673 | 48 |
| 13 | .39421 | .91902 | .42894 | .3313 | .0881 | .5367 | .023738 | 1.1657 | 47 |
| 14 | .39448 | .91891 | .42929 | .3294 | .0883 | .5350 | .023791 | 1.1641 | 46 |
| 15 | 0.39474 | 0.91879 | 0.42963 | 2.3276 | 1.0884 | 2.5333 | .023845 | 1.1626 | 45 |
| 16 | .39501 | .91868 | .42998 | .3257 | .0885 | .5316 | .023899 | 1.1610 | 44 |
| 17 | .39528 | .91856 | .43032 | .3238 | .0887 | .5299 | .023952 | 1.1594 | 43 |
| 18 | .39555 | .91845 | .43067 | .3220 | .0888 | .5282 | .024006 | 1.1578 | 42 |
| 19 | .39581 | .91833 | .43101 | .3201 | .0889 | .5264 | .024060 | 1.1563 | 41 |
| 20 | 0.39608 | 0.91822 | 0.43136 | 2.3183 | 1.0891 | 2.5247 | .024114 | 1.1547 | 40 |
| 21 | .39635 | .91810 | .43170 | .3164 | .0892 | .5230 | .024169 | 1.1531 | 39 |
| 22 | .39661 | .91799 | .43205 | .3146 | .0893 | .5213 | .024223 | 1.1516 | 38 |
| 23 | .39688 | .91787 | .43239 | .3127 | .0895 | .5196 | .024277 | 1.1500 | 37 |
| 24 | .39715 | .91775 | .43274 | .3109 | .0896 | .5180 | .024332 | 1.1485 | 36 |
| 25 | 0.39741 | 0.91764 | 0.43308 | 2.3090 | 1.0898 | 2.5163 | .024386 | 1.1469 | 35 |
| 26 | .39768 | .91752 | .43343 | .3072 | .0899 | .5146 | .024441 | 1.1454 | 34 |
| 27 | .39795 | .91741 | .43378 | .3053 | .0900 | .5129 | .024495 | 1.1438 | 33 |
| 28 | .39822 | .91729 | .43412 | .3035 | .0902 | .5112 | .024550 | 1.1423 | 32 |
| 29 | .39848 | .91718 | .43447 | .3017 | .0903 | .5095 | .024605 | 1.1407 | 31 |
| 30 | 0.39875 | 0.91706 | 0.43481 | 2.2998 | 1.0904 | 2.5078 | .024660 | 1.1392 | 30 |
| 31 | .39902 | .91694 | .43516 | .2980 | .0906 | .5062 | .024715 | 1.1377 | 29 |
| 32 | .39928 | .91683 | .43550 | .2962 | .0907 | .5045 | .024770 | 1.1361 | 28 |
| 33 | .39955 | .91671 | .43585 | .2944 | .0909 | .5028 | .024825 | 1.1346 | 27 |
| 34 | .39982 | .91660 | .43620 | .2925 | .0910 | .5012 | .024881 | 1.1331 | 26 |
| 35 | 0.40008 | 0.91648 | 0.43654 | 2.2907 | 1.0911 | 2.4995 | .024936 | 1.1315 | 25 |
| 36 | .40035 | .91636 | .43689 | .2889 | .0913 | .4978 | .024992 | 1.1300 | 24 |
| 37 | .40062 | .91625 | .43724 | .2871 | .0914 | .4962 | .025047 | 1.1285 | 23 |
| 38 | .40088 | .91613 | .43758 | .2853 | .0915 | .4945 | .025103 | 1.1270 | 22 |
| 39 | .40115 | .91601 | .43793 | .2835 | .0917 | .4928 | .025159 | 1.1254 | 21 |
| 40 | 0.40141 | 0.91590 | 0.43828 | 2.2817 | 1.0918 | 2.4912 | .025214 | 1.1239 | 20 |
| 41 | .40168 | .91578 | .43862 | .2799 | .0920 | .4895 | .025270 | 1.1224 | 19 |
| 42 | .40195 | .91566 | .43897 | .2781 | .0921 | .4879 | .025326 | 1.1209 | 18 |
| 43 | .40221 | .91555 | .43932 | .2763 | .0922 | .4862 | .025382 | 1.1194 | 17 |
| 44 | .40248 | .91543 | .43966 | .2745 | .0924 | .4846 | .025439 | 1.1179 | 16 |
| 45 | 0.40275 | 0.91531 | 0.44001 | 2.2727 | 1.0925 | 2.4830 | .025495 | 1.1164 | 15 |
| 46 | .40301 | .91519 | .44036 | .2709 | .0927 | .4813 | .025551 | 1.1149 | 14 |
| 47 | .40328 | .91508 | .44071 | .2691 | .0928 | .4797 | .025608 | 1.1134 | 13 |
| 48 | .40355 | .91496 | .44105 | .2673 | .0929 | .4780 | .025664 | 1.1119 | 12 |
| 49 | .40381 | .91484 | .44140 | .2655 | .0931 | .4764 | .025721 | 1.1104 | 11 |
| 50 | 0.40408 | 0.91472 | 0.44175 | 2.2637 | 1.0932 | 2.4748 | .025778 | 1.1089 | 10 |
| 51 | .40434 | .91461 | .44210 | .2620 | .0934 | .4731 | .025834 | 1.1074 | 9 |
| 52 | .40461 | .91449 | .44244 | .2602 | .0935 | .4715 | .025891 | 1.1059 | 8 |
| 53 | .40488 | .91437 | .44279 | .2584 | .0936 | .4699 | .025948 | 1.1044 | 7 |
| 54 | .40514 | .91425 | .44314 | .2566 | .0938 | .4683 | .026005 | 1.1030 | 6 |
| 55 | 0.40541 | 0.91414 | 0.44349 | 2.2549 | 1.0939 | 2.4667 | .026062 | 1.1015 | 5 |
| 56 | .40567 | .91402 | .44384 | .2531 | .0941 | .4650 | .026120 | 1.1000 | 4 |
| 57 | .40594 | .91390 | .44418 | .2513 | .0942 | .4634 | .026177 | 1.0985 | 3 |
| 58 | .40621 | .91378 | .44453 | .2496 | .0944 | .4618 | .026235 | 1.0971 | 2 |
| 59 | .40647 | .91366 | .44488 | .2478 | .0945 | .4602 | .026292 | 1.0956 | 1 |
| 60 | 0.40674 | 0.91355 | 0.44523 | 2.2460 | 1.0946 | 2.4586 | .026350 | 1.0941 | 0 |
| M | Cosine | Sine | Cotan. | Tan. | Cosec. | Secant | READ DOWN | 66°–67° Involute | M |

| M | Sine | Cosine | Tan. | Cotan. | Secant | Cosec. | Involute 24°–25° | READ UP | M |
|---|------|--------|------|--------|--------|--------|------------------|---------|---|
| 0 | 0.40674 | 0.91355 | 0.44523 | 2.2460 | 1.0946 | 2.4586 | .026350 | 1.0941 | 60 |
| 1 | .40700 | .91343 | .44558 | .2443 | .0948 | .4570 | .026407 | 1.0927 | 59 |
| 2 | .40727 | .91331 | .44593 | .2425 | .0949 | .4554 | .026465 | 1.0912 | 58 |
| 3 | .40753 | .91319 | .44627 | .2408 | .0951 | .4538 | .026523 | 1.0897 | 57 |
| 4 | .40780 | .91307 | .44662 | .2390 | .0952 | .4522 | .026581 | 1.0883 | 56 |
| 5 | 0.40806 | 0.91295 | 0.44697 | 2.2373 | 1.0953 | 2.4506 | .026639 | 1.0868 | 55 |
| 6 | .40833 | .91283 | .44732 | .2355 | .0955 | .4490 | .026697 | 1.0854 | 54 |
| 7 | .40860 | .91272 | .44767 | .2338 | .0956 | .4474 | .026756 | 1.0839 | 53 |
| 8 | .40886 | .91260 | .44802 | .2320 | .0958 | .4458 | .026814 | 1.0825 | 52 |
| 9 | .40913 | .91248 | .44837 | .2303 | .0959 | .4442 | .026872 | 1.0810 | 51 |
| 10 | 0.40939 | 0.91236 | 0.44872 | 2.2286 | 1.0961 | 2.4426 | .026931 | 1.0796 | 50 |
| 11 | .40966 | .91224 | .44907 | .2268 | .0962 | .4411 | .026989 | 1.0781 | 49 |
| 12 | .40992 | .91212 | .44942 | .2251 | .0963 | .4395 | .027048 | 1.0767 | 48 |
| 13 | .41019 | .91200 | .44977 | .2234 | .0965 | .4379 | .027107 | 1.0752 | 47 |
| 14 | .41045 | .91188 | .45012 | .2216 | .0966 | .4363 | .027166 | 1.0738 | 46 |
| 15 | 0.41072 | 0.91176 | 0.45047 | 2.2199 | 1.0968 | 2.4348 | .027225 | 1.0724 | 45 |
| 16 | .41098 | .91164 | .45082 | .2182 | .0969 | .4332 | .027284 | 1.0709 | 44 |
| 17 | .41125 | .91152 | .45117 | .2165 | .0971 | .4316 | .027343 | 1.0695 | 43 |
| 18 | .41151 | .91140 | .45152 | .2148 | .0972 | .4300 | .027402 | 1.0681 | 42 |
| 19 | .41178 | .91128 | .45187 | .2130 | .0974 | .4285 | .027462 | 1.0666 | 41 |
| 20 | 0.41204 | 0.91116 | 0.45222 | 2.2113 | 1.0975 | 2.4269 | .027521 | 1.0652 | 40 |
| 21 | .41231 | .91104 | .45257 | .2096 | .0976 | .4254 | .027581 | 1.0638 | 39 |
| 22 | .41257 | .91092 | .45292 | .2079 | .0978 | .4238 | .027640 | 1.0624 | 38 |
| 23 | .41284 | .91080 | .45327 | .2062 | .0979 | .4222 | .027700 | 1.0610 | 37 |
| 24 | .41310 | .91068 | .45362 | .2045 | .0981 | .4207 | .027760 | 1.0596 | 36 |
| 25 | 0.41337 | 0.91056 | 0.45397 | 2.2028 | 1.0982 | 2.4191 | .027820 | 1.0581 | 35 |
| 26 | .41363 | .91044 | .45432 | .2011 | .0984 | .4176 | .027880 | 1.0567 | 34 |
| 27 | .41390 | .91032 | .45467 | .1994 | .0985 | .4160 | .027940 | 1.0553 | 33 |
| 28 | .41416 | .91020 | .45502 | .1977 | .0987 | .4145 | .028000 | 1.0539 | 32 |
| 29 | .41443 | .91008 | .45538 | .1960 | .0988 | .4130 | .028060 | 1.0525 | 31 |
| 30 | 0.41469 | 0.90996 | 0.45573 | 2.1943 | 1.0989 | 2.4114 | .028121 | 1.0511 | 30 |
| 31 | .41496 | .90984 | .45608 | .1926 | .0991 | .4099 | .028181 | 1.0497 | 29 |
| 32 | .41522 | .90972 | .45643 | .1909 | .0992 | .4083 | .028242 | 1.0483 | 28 |
| 33 | .41549 | .90960 | .45678 | .1892 | .0994 | .4068 | .028302 | 1.0469 | 27 |
| 34 | .41575 | .90948 | .45713 | .1876 | .0995 | .4053 | .028363 | 1.0455 | 26 |
| 35 | 0.41602 | 0.90936 | 0.45748 | 2.1859 | 1.0997 | 2.4038 | .028424 | 1.0441 | 25 |
| 36 | .41628 | .90924 | .45784 | .1842 | .0998 | .4022 | .028485 | 1.0427 | 24 |
| 37 | .41655 | .90911 | .45819 | .1825 | .1000 | .4007 | .028546 | 1.0414 | 23 |
| 38 | .41681 | .90899 | .45854 | .1808 | .1001 | .3992 | .028607 | 1.0400 | 22 |
| 39 | .41707 | .90887 | .45889 | .1792 | .1003 | .3977 | .028668 | 1.0386 | 21 |
| 40 | 0.41734 | 0.90875 | 0.45924 | 2.1775 | 1.1004 | 2.3961 | .028729 | 1.0372 | 20 |
| 41 | .41760 | .90863 | .45960 | .1758 | .1006 | .3946 | .028791 | 1.0358 | 19 |
| 42 | .41787 | .90851 | .45995 | .1742 | .1007 | .3931 | .028852 | 1.0345 | 18 |
| 43 | .41813 | .90839 | .46030 | .1725 | .1009 | .3916 | .028914 | 1.0331 | 17 |
| 44 | .41840 | .90826 | .46065 | .1708 | .1010 | .3901 | .028976 | 1.0317 | 16 |
| 45 | 0.41866 | 0.90814 | 0.46101 | 2.1692 | 1.1011 | 2.3886 | .029037 | 1.0303 | 15 |
| 46 | .41892 | .90802 | .46136 | .1675 | .1013 | .3871 | .029099 | 1.0290 | 14 |
| 47 | .41919 | .90790 | .46171 | .1659 | .1014 | .3856 | .029161 | 1.0276 | 13 |
| 48 | .41945 | .90778 | .46206 | .1642 | .1016 | .3841 | .029223 | 1.0262 | 12 |
| 49 | .41972 | .90766 | .46242 | .1625 | .1017 | .3826 | .029285 | 1.0249 | 11 |
| 50 | 0.41998 | 0.90753 | 0.46277 | 2.1609 | 1.1019 | 2.3811 | .029348 | 1.0235 | 10 |
| 51 | .42024 | .90741 | .46312 | .1592 | .1020 | .3796 | .029410 | 1.0222 | 9 |
| 52 | .42051 | .90729 | .46348 | .1576 | .1022 | .3781 | .029472 | 1.0208 | 8 |
| 53 | .42077 | .90717 | .46383 | .1560 | .1023 | .3766 | .029535 | 1.0195 | 7 |
| 54 | .42104 | .90704 | .46418 | .1543 | .1025 | .3751 | .029598 | 1.0181 | 6 |
| 55 | 0.42130 | 0.90692 | 0.46454 | 2.1527 | 1.1026 | 2.3736 | .029660 | 1.0168 | 5 |
| 56 | .42156 | .90680 | .46489 | .1510 | .1028 | .3721 | .029723 | 1.0154 | 4 |
| 57 | .42183 | .90668 | .46525 | .1494 | .1029 | .3706 | .029786 | 1.0141 | 3 |
| 58 | .42209 | .90655 | .46560 | .1478 | .1031 | .3692 | .029849 | 1.0127 | 2 |
| 59 | .42235 | .90643 | .46595 | .1461 | .1032 | .3677 | .029912 | 1.0114 | 1 |
| 60 | 0.42262 | 0.90631 | 0.46631 | 2.1445 | 1.1034 | 2.3662 | .029975 | 1.0100 | 0 |
| M | Cosine | Sine | Cotan. | Tan. | Cosec. | Secant | READ DOWN | 65°–66° Involute | M |

**25° or 205°**      Trigonometric Functions      **154° or 334°**

| M | Sine | Cosine | Tan. | Cotan. | Secant | Cosec. | Involute 25°–26° | READ UP | M |
|---|---|---|---|---|---|---|---|---|---|
| 0 | 0.42262 | 0.90631 | 0.46631 | 2.1445 | 1.1034 | 2.3662 | .029975 | 1.0100 | 60 |
| 1 | .42288 | .90618 | .46666 | .1429 | .1035 | .3647 | .030039 | 1.0087 | 59 |
| 2 | .42315 | .90606 | .46702 | .1413 | .1037 | .3633 | .030102 | 1.0074 | 58 |
| 3 | .42341 | .90594 | .46737 | .1396 | .1038 | .3618 | .030166 | 1.0060 | 57 |
| 4 | .42367 | .90582 | .46772 | .1380 | .1040 | .3603 | .030229 | 1.0047 | 56 |
| 5 | 0.42394 | 0.90569 | 0.46808 | 2.1364 | 1.1041 | 2.3588 | .030293 | 1.0034 | 55 |
| 6 | .42420 | .90557 | .46843 | .1348 | .1043 | .3574 | .030357 | 1.0021 | 54 |
| 7 | .42446 | .90545 | .46879 | .1332 | .1044 | .3559 | .030420 | 1.0007 | 53 |
| 8 | .42473 | .90532 | .46914 | .1315 | .1046 | .3545 | .030484 | 0.9994 | 52 |
| 9 | .42499 | .90520 | .46950 | .1299 | .1047 | .3530 | .030549 | 0.9981 | 51 |
| 10 | 0.42525 | 0.90507 | 0.46985 | 2.1283 | 1.1049 | 2.3515 | .030613 | 0.9968 | 50 |
| 11 | .42552 | .90495 | .47021 | .1267 | .1050 | .3501 | .030677 | 0.9954 | 49 |
| 12 | .42578 | .90483 | .47056 | .1251 | .1052 | .3486 | .030741 | 0.9941 | 48 |
| 13 | .42604 | .90470 | .47092 | .1235 | .1053 | .3472 | .030806 | 0.9928 | 47 |
| 14 | .42631 | .90458 | .47128 | .1219 | .1055 | .3457 | .030870 | 0.9915 | 46 |
| 15 | 0.42657 | 0.90446 | 0.47163 | 2.1203 | 1.1056 | 2.3443 | .030935 | 0.9902 | 45 |
| 16 | .42683 | .90433 | .47199 | .1187 | .1058 | .3428 | .031000 | 0.9889 | 44 |
| 17 | .42709 | .90421 | .47234 | .1171 | .1059 | .3414 | .031065 | 0.9876 | 43 |
| 18 | .42736 | .90408 | .47270 | .1155 | .1061 | .3400 | .031130 | 0.9863 | 42 |
| 19 | .42762 | .90396 | .47305 | .1139 | .1062 | .3385 | .031195 | 0.9850 | 41 |
| 20 | 0.42788 | 0.90383 | 0.47341 | 2.1123 | 1.1064 | 2.3371 | .031260 | 0.9837 | 40 |
| 21 | .42815 | .90371 | .47377 | .1107 | .1066 | .3356 | .031325 | 0.9824 | 39 |
| 22 | .42841 | .90358 | .47412 | .1092 | .1067 | .3342 | .031390 | 0.9811 | 38 |
| 23 | .42867 | .90346 | .47448 | .1076 | .1069 | .3328 | .031456 | 0.9798 | 37 |
| 24 | .42894 | .90334 | .47483 | .1060 | .1070 | .3314 | .031521 | 0.9785 | 36 |
| 25 | 0.42920 | 0.90321 | 0.47519 | 2.1044 | 1.1072 | 2.3299 | .031587 | 0.9772 | 35 |
| 26 | .42946 | .90309 | .47555 | .1028 | .1073 | .3285 | .031653 | 0.9759 | 34 |
| 27 | .42972 | .90296 | .47590 | .1013 | .1075 | .3271 | .031718 | 0.9747 | 33 |
| 28 | .42999 | .90284 | .47626 | .0997 | .1076 | .3257 | .031784 | 0.9734 | 32 |
| 29 | .43025 | .90271 | .47662 | .0981 | .1078 | .3242 | .031850 | 0.9721 | 31 |
| 30 | 0.43051 | 0.90259 | 0.47698 | 2.0965 | 1.1079 | 2.3228 | .031917 | 0.9708 | 30 |
| 31 | .43077 | .90246 | .47733 | .0950 | .1081 | .3214 | .031983 | 0.9695 | 29 |
| 32 | .43104 | .90233 | .47769 | .0934 | .1082 | .3200 | .032049 | 0.9683 | 28 |
| 33 | .43130 | .90221 | .47805 | .0918 | .1084 | .3186 | .032116 | 0.9670 | 27 |
| 34 | .43156 | .90208 | .47840 | .0903 | .1085 | .3172 | .032182 | 0.9657 | 26 |
| 35 | 0.43182 | 0.90196 | 0.47876 | 2.0887 | 1.1087 | 2.3158 | .032249 | 0.9644 | 25 |
| 36 | .43209 | .90183 | .47912 | .0872 | .1089 | .3144 | .032315 | 0.9632 | 24 |
| 37 | .43235 | .90171 | .47948 | .0856 | .1090 | .3130 | .032382 | 0.9619 | 23 |
| 38 | .43261 | .90158 | .47984 | .0840 | .1092 | .3115 | .032449 | 0.9606 | 22 |
| 39 | .43287 | .90146 | .48019 | .0825 | .1093 | .3101 | .032516 | 0.9594 | 21 |
| 40 | 0.43313 | 0.90133 | 0.48055 | 2.0809 | 1.1095 | 2.3088 | .032583 | 0.9581 | 20 |
| 41 | .43340 | .90120 | .48091 | .0794 | .1096 | .3074 | .032651 | 0.9569 | 19 |
| 42 | .43366 | .90108 | .48127 | .0778 | .1098 | .3060 | .032718 | 0.9556 | 18 |
| 43 | .43392 | .90095 | .48163 | .0763 | .1099 | .3046 | .032785 | 0.9543 | 17 |
| 44 | .43418 | .90082 | .48198 | .0748 | .1101 | .3032 | .032853 | 0.9531 | 16 |
| 45 | 0.43445 | 0.90070 | 0.48234 | 2.0732 | 1.1102 | 2.3018 | .032920 | 0.9518 | 15 |
| 46 | .43471 | .90057 | .48270 | .0717 | .1104 | .3004 | .032988 | 0.9506 | 14 |
| 47 | .43497 | .90044 | .48306 | .0701 | .1106 | .2990 | .033056 | 0.9493 | 13 |
| 48 | .43523 | .90032 | .48342 | .0686 | .1107 | .2976 | .033124 | 0.9481 | 12 |
| 49 | .43549 | .90019 | .48378 | .0671 | .1109 | .2962 | .033192 | 0.9469 | 11 |
| 50 | 0.43575 | 0.90007 | 0.48414 | 2.0655 | 1.1110 | 2.2949 | .033260 | 0.9456 | 10 |
| 51 | .43602 | .89994 | .48450 | .0640 | .1112 | .2935 | .033328 | 0.9444 | 9 |
| 52 | .43628 | .89981 | .48486 | .0625 | .1113 | .2921 | .033397 | 0.9431 | 8 |
| 53 | .43654 | .89968 | .48521 | .0609 | .1115 | .2907 | .033465 | 0.9419 | 7 |
| 54 | .43680 | .89956 | .48557 | .0594 | .1117 | .2894 | .033534 | 0.9407 | 6 |
| 55 | 0.43706 | 0.89943 | 0.48593 | 2.0579 | 1.1118 | 2.2880 | .033602 | 0.9394 | 5 |
| 56 | .43733 | .89930 | .48629 | .0564 | .1120 | .2866 | .033671 | 0.9382 | 4 |
| 57 | .43759 | .89918 | .48665 | .0549 | .1121 | .2853 | .033740 | 0.9370 | 3 |
| 58 | .43785 | .89905 | .48701 | .0533 | .1123 | .2839 | .033809 | 0.9357 | 2 |
| 59 | .43811 | .89892 | .48737 | .0518 | .1124 | .2825 | .033878 | 0.9345 | 1 |
| 60 | 0.43837 | 0.89879 | 0.48773 | 2.0503 | 1.1126 | 2.2812 | .033947 | 0.9333 | 0 |

| M | Cosine | Sine | Cotan. | Tan. | Cosec. | Secant | READ DOWN | 64°–65° Involute | M |
|---|---|---|---|---|---|---|---|---|---|

**115° or 295°**      **64° or 244°**

| M | Sine | Cosine | Tan. | Cotan. | Secant | Cosec. | Involute 26°–27° | READ UP | M |
|---|---|---|---|---|---|---|---|---|---|
| 0 | 0.43837 | 0.89879 | 0.48773 | 2.0503 | 1.1126 | 2.2812 | .033947 | .93329 | 60 |
| 1 | .43863 | .89867 | .48809 | .0488 | .1128 | .2798 | .034016 | .93207 | 59 |
| 2 | .43889 | .89854 | .48845 | .0473 | .1129 | .2785 | .034086 | .93085 | 58 |
| 3 | .43916 | .89841 | .48881 | .0458 | .1131 | .2771 | .034155 | .92963 | 57 |
| 4 | .43942 | .89828 | .48917 | .0443 | .1132 | .2757 | .034225 | .92842 | 56 |
| 5 | 0.43968 | 0.89816 | 0.48953 | 2.0428 | 1.1134 | 2.2744 | .034294 | .92720 | 55 |
| 6 | .43994 | .89803 | .48989 | .0413 | .1136 | .2730 | .034364 | .92599 | 54 |
| 7 | .44020 | .89790 | .49026 | .0398 | .1137 | .2717 | .034434 | .92478 | 53 |
| 8 | .44046 | .89777 | .49062 | .0383 | .1139 | .2703 | .034504 | .92357 | 52 |
| 9 | .44072 | .89764 | .49098 | .0368 | .1140 | .2690 | .034574 | .92236 | 51 |
| 10 | 0.44098 | 0.89752 | 0.49134 | 2.0353 | 1.1142 | 2.2677 | .034644 | .92115 | 50 |
| 11 | .44124 | .89739 | .49170 | .0338 | .1143 | .2663 | .034714 | .91995 | 49 |
| 12 | .44151 | .89726 | .49206 | .0323 | .1145 | .2650 | .034785 | .91875 | 48 |
| 13 | .44177 | .89713 | .49242 | .0308 | .1147 | .2636 | .034855 | .91755 | 47 |
| 14 | .44203 | .89700 | .49278 | .0293 | .1148 | .2623 | .034926 | .91635 | 46 |
| 15 | 0.44229 | 0.89687 | 0.49315 | 2.0278 | 1.1150 | 2.2610 | .034996 | .91515 | 45 |
| 16 | .44255 | .89674 | .49351 | .0263 | .1151 | .2596 | .035067 | .91396 | 44 |
| 17 | .44281 | .89662 | .49387 | .0248 | .1153 | .2583 | .035138 | .91276 | 43 |
| 18 | .44307 | .89649 | .49423 | .0233 | .1155 | .2570 | .035209 | .91157 | 42 |
| 19 | .44333 | .89636 | .49459 | .0219 | .1156 | .2556 | .035280 | .91038 | 41 |
| 20 | 0.44359 | 0.89623 | 0.49495 | 2.0204 | 1.1158 | 2.2543 | .035352 | .90919 | 40 |
| 21 | .44385 | .89610 | .49532 | .0189 | .1159 | .2530 | .035423 | .90801 | 39 |
| 22 | .44411 | .89597 | .49568 | .0174 | .1161 | .2517 | .035494 | .90682 | 38 |
| 23 | .44437 | .89584 | .49604 | .0160 | .1163 | .2504 | .035566 | .90564 | 37 |
| 24 | .44464 | .89571 | .49640 | .0145 | .1164 | .2490 | .035637 | .90446 | 36 |
| 25 | 0.44490 | 0.89558 | 0.49677 | 2.0130 | 1.1166 | 2.2477 | .035709 | .90328 | 35 |
| 26 | .44516 | .89545 | .49713 | .0115 | .1168 | .2464 | .035781 | .90210 | 34 |
| 27 | .44542 | .89532 | .49749 | .0101 | .1169 | .2451 | .035853 | .90092 | 33 |
| 28 | .44568 | .89519 | .49786 | .0086 | .1171 | .2438 | .035925 | .89975 | 32 |
| 29 | .44594 | .89506 | .49822 | .0072 | .1172 | .2425 | .035997 | .89858 | 31 |
| 30 | 0.44620 | 0.89493 | 0.49858 | 2.0057 | 1.1174 | 2.2412 | .036069 | .89741 | 30 |
| 31 | .44646 | .89480 | .49894 | .0042 | .1176 | .2399 | .036142 | .89624 | 29 |
| 32 | .44672 | .89467 | .49931 | .0028 | .1177 | .2385 | .036214 | .89507 | 28 |
| 33 | .44698 | .89454 | .49967 | .0013 | .1179 | .2372 | .036287 | .89390 | 27 |
| 34 | .44724 | .89441 | .50004 | 1.9999 | .1180 | .2359 | .036359 | .89274 | 26 |
| 35 | 0.44750 | 0.89428 | 0.50040 | 1.9984 | 1.1182 | 2.2346 | .036432 | .89158 | 25 |
| 36 | .44776 | .89415 | .50076 | .9970 | .1184 | .2333 | .036505 | .89042 | 24 |
| 37 | .44802 | .89402 | .50113 | .9955 | .1185 | .2320 | .036578 | .88926 | 23 |
| 38 | .44828 | .89389 | .50149 | .9941 | .1187 | .2308 | .036651 | .88810 | 22 |
| 39 | .44854 | .89376 | .50185 | .9926 | .1189 | .2295 | .036724 | .88694 | 21 |
| 40 | 0.44880 | 0.89363 | 0.50222 | 1.9912 | 1.1190 | 2.2282 | .036798 | .88579 | 20 |
| 41 | .44906 | .89350 | .50258 | .9897 | .1192 | .2269 | .036871 | .88464 | 19 |
| 42 | .44932 | .89337 | .50295 | .9883 | .1194 | .2256 | .036945 | .88349 | 18 |
| 43 | .44958 | .89324 | .50331 | .9868 | .1195 | .2243 | .037018 | .88234 | 17 |
| 44 | .44984 | .89311 | .50368 | .9854 | .1197 | .2230 | .037092 | .88119 | 16 |
| 45 | 0.45010 | 0.89298 | 0.50404 | 1.9840 | 1.1198 | 2.2217 | .037166 | .88004 | 15 |
| 46 | .45036 | .89285 | .50441 | .9825 | .1200 | .2205 | .037240 | .87890 | 14 |
| 47 | .45062 | .89272 | .50477 | .9811 | .1202 | .2192 | .037314 | .87776 | 13 |
| 48 | .45088 | .89259 | .50514 | .9797 | .1203 | .2179 | .037388 | .87662 | 12 |
| 49 | .45114 | .89245 | .50550 | .9782 | .1205 | .2166 | .037462 | .87548 | 11 |
| 50 | 0.45140 | 0.89232 | 0.50587 | 1.9768 | 1.1207 | 2.2153 | .037537 | .87434 | 10 |
| 51 | .45166 | .89219 | .50623 | .9754 | .1208 | .2141 | .037611 | .87320 | 9 |
| 52 | .45192 | .89206 | .50660 | .9740 | .1210 | .2128 | .037686 | .87207 | 8 |
| 53 | .45218 | .89193 | .50696 | .9725 | .1212 | .2115 | .037761 | .87094 | 7 |
| 54 | .45243 | .89180 | .50733 | .9711 | .1213 | .2103 | .037835 | .86980 | 6 |
| 55 | 0.45269 | 0.89167 | 0.50769 | 1.9697 | 1.1215 | 2.2090 | .037910 | .86868 | 5 |
| 56 | .45295 | .89153 | .50806 | .9683 | .1217 | .2077 | .037985 | .86755 | 4 |
| 57 | .45321 | .89140 | .50843 | .9669 | .1218 | .2065 | .038060 | .86642 | 3 |
| 58 | .45347 | .89127 | .50879 | .9654 | .1220 | .2052 | .038136 | .86530 | 2 |
| 59 | .45373 | .89114 | .50916 | .9640 | .1222 | .2039 | .038211 | .86417 | 1 |
| 60 | 0.45399 | 0.89101 | 0.50953 | 1.9626 | 1.1223 | 2.2027 | .038287 | .86305 | 0 |

| M | Cosine | Sine | Cotan. | Tan. | Cosec. | Secant | READ DOWN | 63°–64° Involute | M |
|---|---|---|---|---|---|---|---|---|---|

| M | Sine | Cosine | Tan. | Cotan. | Secant | Cosec. | Involute 27°–28° | READ UP | M |
|---|------|--------|------|--------|--------|--------|------------------|---------|---|
| 0 | 0.45399 | 0.89101 | 0.50953 | 1.9626 | 1.1223 | 2.2027 | .038287 | .86305 | 60 |
| 1 | .45425 | .89087 | .50989 | .9612 | .1225 | .2014 | .038362 | .86193 | 59 |
| 2 | .45451 | .89074 | .51026 | .9598 | .1227 | .2002 | .038438 | .86082 | 58 |
| 3 | .45477 | .89061 | .51063 | .9584 | .1228 | .1989 | .038514 | .85970 | 57 |
| 4 | .45503 | .89048 | .51099 | .9570 | .1230 | .1977 | .038590 | .85858 | 56 |
| 5 | 0.45529 | 0.89035 | 0.51136 | 1.9556 | 1.1232 | 2.1964 | .038666 | .85747 | 55 |
| 6 | .45554 | .89021 | .51173 | .9542 | .1233 | .1952 | .038742 | .85636 | 54 |
| 7 | .45580 | .89008 | .51209 | .9528 | .1235 | .1939 | .038818 | .85525 | 53 |
| 8 | .45606 | .88995 | .51246 | .9514 | .1237 | .1927 | .038894 | .85414 | 52 |
| 9 | .45632 | .88981 | .51283 | .9500 | .1238 | .1914 | .038971 | .85303 | 51 |
| 10 | 0.45658 | 0.88968 | 0.51319 | 1.9486 | 1.1240 | 2.1902 | .039047 | .85193 | 50 |
| 11 | .45684 | .88955 | .51356 | .9472 | .1242 | .1890 | .039124 | .85082 | 49 |
| 12 | .45710 | .88942 | .51393 | .9458 | .1243 | .1877 | .039201 | .84972 | 48 |
| 13 | .45736 | .88928 | .51430 | .9444 | .1245 | .1865 | .039278 | .84862 | 47 |
| 14 | .45762 | .88915 | .51467 | .9430 | .1247 | .1852 | .039355 | .84752 | 46 |
| 15 | 0.45787 | 0.88902 | 0.51503 | 1.9416 | 1.1248 | 2.1840 | .039432 | .84643 | 45 |
| 16 | .45813 | .88888 | .51540 | .9402 | .1250 | .1828 | .039509 | .84533 | 44 |
| 17 | .45839 | .88875 | .51577 | .9388 | .1252 | .1815 | .039586 | .84424 | 43 |
| 18 | .45865 | .88862 | .51614 | .9375 | .1253 | .1803 | .039664 | .84314 | 42 |
| 19 | .45891 | .88848 | .51651 | .9361 | .1255 | .1791 | .039741 | .84205 | 41 |
| 20 | 0.45917 | 0.88835 | 0.51688 | 1.9347 | 1.1257 | 2.1779 | .039819 | .84096 | 40 |
| 21 | .45942 | .88822 | .51724 | .9333 | .1259 | .1766 | .039897 | .83987 | 39 |
| 22 | .45968 | .88808 | .51761 | .9319 | .1260 | .1754 | .039974 | .83879 | 38 |
| 23 | .45994 | .88795 | .51798 | .9306 | .1262 | .1742 | .040052 | .83770 | 37 |
| 24 | .46020 | .88782 | .51835 | .9292 | .1264 | .1730 | .040131 | .83662 | 36 |
| 25 | 0.46046 | 0.88768 | 0.51872 | 1.9278 | 1.1265 | 2.1718 | .040209 | .83554 | 35 |
| 26 | .46072 | .88755 | .51909 | .9265 | .1267 | .1705 | .040287 | .83446 | 34 |
| 27 | .46097 | .88741 | .51946 | .9251 | .1269 | .1693 | .040366 | .83338 | 33 |
| 28 | .46123 | .88728 | .51983 | .9237 | .1270 | .1681 | .040444 | .83230 | 32 |
| 29 | .46149 | .88715 | .52020 | .9223 | .1272 | .1669 | .040523 | .83123 | 31 |
| 30 | 0.46175 | 0.88701 | 0.52057 | 1.9210 | 1.1274 | 2.1657 | .040602 | .83015 | 30 |
| 31 | .46201 | .88688 | .52094 | .9196 | .1276 | .1645 | .040680 | .82908 | 29 |
| 32 | .46226 | .88674 | .52131 | .9183 | .1277 | .1633 | .040759 | .82801 | 28 |
| 33 | .46252 | .88661 | .52168 | .9169 | .1279 | .1621 | .040838 | .82694 | 27 |
| 34 | .46278 | .88647 | .52205 | .9155 | .1281 | .1609 | .040918 | .82587 | 26 |
| 35 | 0.46304 | 0.88634 | 0.52242 | 1.9142 | 1.1282 | 2.1596 | .040997 | .82480 | 25 |
| 36 | .46330 | .88620 | .52279 | .9128 | .1284 | .1584 | .041076 | .82374 | 24 |
| 37 | .46355 | .88607 | .52316 | .9115 | .1286 | .1572 | .041156 | .82267 | 23 |
| 38 | .46381 | .88593 | .52353 | .9101 | .1288 | .1560 | .041236 | .82161 | 22 |
| 39 | .46407 | .88580 | .52390 | .9088 | .1289 | .1549 | .041316 | .82055 | 21 |
| 40 | 0.46433 | 0.88566 | 0.52427 | 1.9074 | 1.1291 | 2.1537 | .041395 | .81949 | 20 |
| 41 | .46458 | .88553 | .52464 | .9061 | .1293 | .1525 | .041475 | .81844 | 19 |
| 42 | .46484 | .88539 | .52501 | .9047 | .1294 | .1513 | .041556 | .81738 | 18 |
| 43 | .46510 | .88526 | .52538 | .9034 | .1296 | .1501 | .041636 | .81632 | 17 |
| 44 | .46536 | .88512 | .52575 | .9020 | .1298 | .1489 | .041716 | .81527 | 16 |
| 45 | 0.46561 | 0.88499 | 0.52613 | 1.9007 | 1.1300 | 2.1477 | .041797 | .81422 | 15 |
| 46 | .46587 | .88485 | .52650 | .8993 | .1301 | .1465 | .041877 | .81317 | 14 |
| 47 | .46613 | .88472 | .52687 | .8980 | .1303 | .1453 | .041958 | .81212 | 13 |
| 48 | .46639 | .88458 | .52724 | .8967 | .1305 | .1441 | .042039 | .81107 | 12 |
| 49 | .46664 | .88445 | .52761 | .8953 | .1307 | .1430 | .042120 | .81003 | 11 |
| 50 | 0.46690 | 0.88431 | 0.52798 | 1.8940 | 1.1308 | 2.1418 | .042201 | .80898 | 10 |
| 51 | .46716 | .88417 | .52836 | .8927 | .1310 | .1406 | .042282 | .80794 | 9 |
| 52 | .46742 | .88404 | .52873 | .8913 | .1312 | .1394 | .042363 | .80690 | 8 |
| 53 | .46767 | .88390 | .52910 | .8900 | .1313 | .1382 | .042444 | .80586 | 7 |
| 54 | .46793 | .88377 | .52947 | .8887 | .1315 | .1371 | .042526 | .80482 | 6 |
| 55 | 0.46819 | 0.88363 | 0.52985 | 1.8873 | 1.1317 | 2.1359 | .042607 | .80378 | 5 |
| 56 | .46844 | .88349 | .53022 | .8860 | .1319 | .1347 | .042689 | .80275 | 4 |
| 57 | .46870 | .88336 | .53059 | .8847 | .1320 | .1336 | .042771 | .80172 | 3 |
| 58 | .46896 | .88322 | .53096 | .8834 | .1322 | .1324 | .042853 | .80068 | 2 |
| 59 | .46921 | .88308 | .53134 | .8820 | .1324 | .1312 | .042935 | .79965 | 1 |
| 60 | 0.46947 | 0.88295 | 0.53171 | 1.8807 | 1.1326 | 2.1301 | .043017 | .79862 | 0 |

| M | Cosine | Sine | Cotan. | Tan. | Cosec. | Secant | READ DOWN | 62°–63° Involute | M |
|---|--------|------|--------|------|--------|--------|-----------|-------------------|---|

**28° or 208°**      Trigonometric Functions      **151° or 331°**

| M | Sine | Cosine | Tan. | Cotan. | Secant | Cosec. | Involute 28°–29° | READ UP | M |
|---|---|---|---|---|---|---|---|---|---|
| 0 | 0.46947 | 0.88295 | 0.53171 | 1.8807 | 1.1326 | 2.1301 | .043017 | .79862 | 60 |
| 1 | .46973 | .88281 | .53208 | .8794 | .1327 | .1289 | .043100 | .79759 | 59 |
| 2 | .46999 | .88267 | .53246 | .8781 | .1329 | .1277 | .043182 | .79657 | 58 |
| 3 | .47024 | .88254 | .53283 | .8768 | .1331 | .1266 | .043264 | .79554 | 57 |
| 4 | .47050 | .88240 | .53320 | .8755 | .1333 | .1254 | .043347 | .79452 | 56 |
| 5 | 0.47076 | 0.88226 | 0.53358 | 1.8741 | 1.1334 | 2.1242 | .043430 | .79350 | 55 |
| 6 | .47101 | .88213 | .53395 | .8728 | .1336 | .1231 | .043513 | .79247 | 54 |
| 7 | .47127 | .88199 | .53432 | .8715 | .1338 | .1219 | .043596 | .79146 | 53 |
| 8 | .47153 | .88185 | .53470 | .8702 | .1340 | .1208 | .043679 | .79044 | 52 |
| 9 | .47178 | .88172 | .53507 | .8689 | .1342 | .1196 | .043762 | .78942 | 51 |
| 10 | 0.47204 | 0.88158 | 0.53545 | 1.8676 | 1.1343 | 2.1185 | .043845 | .78841 | 50 |
| 11 | .47229 | .88144 | .53582 | .8663 | .1345 | .1173 | .043929 | .78739 | 49 |
| 12 | .47255 | .88130 | .53620 | .8650 | .1347 | .1162 | .044012 | .78638 | 48 |
| 13 | .47281 | .88117 | .53657 | .8637 | .1349 | .1150 | .044096 | .78537 | 47 |
| 14 | .47306 | .88103 | .53694 | .8624 | .1350 | .1139 | .044180 | .78436 | 46 |
| 15 | 0.47332 | 0.88089 | 0.53732 | 1.8611 | 1.1352 | 2.1127 | .044264 | .78335 | 45 |
| 16 | .47358 | .88075 | .53769 | .8598 | .1354 | .1116 | .044348 | .78234 | 44 |
| 17 | .47383 | .88062 | .53807 | .8585 | .1356 | .1105 | .044432 | .78134 | 43 |
| 18 | .47409 | .88048 | .53844 | .8572 | .1357 | .1093 | .044516 | .78033 | 42 |
| 19 | .47434 | .88034 | .53882 | .8559 | .1359 | .1082 | .044601 | .77933 | 41 |
| 20 | 0.47460 | 0.88020 | 0.53920 | 1.8546 | 1.1361 | 2.1070 | .044685 | .77833 | 40 |
| 21 | .47486 | .88006 | .53957 | .8533 | .1363 | .1059 | .044770 | .77733 | 39 |
| 22 | .47511 | .87993 | .53995 | .8520 | .1365 | .1048 | .044855 | .77633 | 38 |
| 23 | .47537 | .87979 | .54032 | .8507 | .1366 | .1036 | .044939 | .77533 | 37 |
| 24 | .47562 | .87965 | .54070 | .8495 | .1368 | .1025 | .045024 | .77434 | 36 |
| 25 | 0.47588 | 0.87951 | 0.54107 | 1.8482 | 1.1370 | 2.1014 | .045110 | .77334 | 35 |
| 26 | .47614 | .87937 | .54145 | .8469 | .1372 | .1002 | .045195 | .77235 | 34 |
| 27 | .47639 | .87923 | .54183 | .8456 | .1374 | .0991 | .045280 | .77136 | 33 |
| 28 | .47665 | .87909 | .54220 | .8443 | .1375 | .0980 | .045366 | .77037 | 32 |
| 29 | .47690 | .87896 | .54258 | .8430 | .1377 | .0969 | .045451 | .76938 | 31 |
| 30 | 0.47716 | 0.87882 | 0.54296 | 1.8418 | 1.1379 | 2.0957 | .045537 | .76839 | 30 |
| 31 | .47741 | .87868 | .54333 | .8405 | .1381 | .0946 | .045623 | .76741 | 29 |
| 32 | .47767 | .87854 | .54371 | .8392 | .1383 | .0935 | .045709 | .76642 | 28 |
| 33 | .47793 | .87840 | .54409 | .8379 | .1384 | .0924 | .045795 | .76544 | 27 |
| 34 | .47818 | .87826 | .54446 | .8367 | .1386 | .0913 | .045881 | .76446 | 26 |
| 35 | 0.47844 | 0.87812 | 0.54484 | 1.8354 | 1.1388 | 2.0901 | .045967 | .76348 | 25 |
| 36 | .47869 | .87798 | .54522 | .8341 | .1390 | .0890 | .046054 | .76250 | 24 |
| 37 | .47895 | .87784 | .54560 | .8329 | .1392 | .0879 | .046140 | .76152 | 23 |
| 38 | .47920 | .87770 | .54597 | .8316 | .1393 | .0868 | .046227 | .76054 | 22 |
| 39 | .47946 | .87756 | .54635 | .8303 | .1395 | .0857 | .046313 | .75957 | 21 |
| 40 | 0.47971 | 0.87743 | 0.54673 | 1.8291 | 1.1397 | 2.0846 | .046400 | .75859 | 20 |
| 41 | .47997 | .87729 | .54711 | .8278 | .1399 | .0835 | .046487 | .75762 | 19 |
| 42 | .48022 | .87715 | .54748 | .8265 | .1401 | .0824 | .046575 | .75665 | 18 |
| 43 | .48048 | .87701 | .54786 | .8253 | .1402 | .0813 | .046662 | .75568 | 17 |
| 44 | .48073 | .87687 | .54824 | .8240 | .1404 | .0802 | .046749 | .75471 | 16 |
| 45 | 0.48099 | 0.87673 | 0.54862 | 1.8228 | 1.1406 | 2.0791 | .046837 | .75375 | 15 |
| 46 | .48124 | .87659 | .54900 | .8215 | .1408 | .0779 | .046924 | .75278 | 14 |
| 47 | .48150 | .87645 | .54938 | .8202 | .1410 | .0768 | .047012 | .75181 | 13 |
| 48 | .48175 | .87631 | .54975 | .8190 | .1412 | .0757 | .047100 | .75085 | 12 |
| 49 | .48201 | .87617 | .55013 | .8177 | .1413 | .0747 | .047188 | .74989 | 11 |
| 50 | 0.48226 | 0.87603 | 0.55051 | 1.8165 | 1.1415 | 2.0736 | .047276 | .74893 | 10 |
| 51 | .48252 | .87589 | .55089 | .8152 | .1417 | .0725 | .047364 | .74797 | 9 |
| 52 | .48277 | .87575 | .55127 | .8140 | .1419 | .0714 | .047452 | .74701 | 8 |
| 53 | .48303 | .87561 | .55165 | .8127 | .1421 | .0703 | .047541 | .74606 | 7 |
| 54 | .48328 | .87546 | .55203 | .8115 | .1423 | .0692 | .047630 | .74510 | 6 |
| 55 | 0.48354 | 0.87532 | 0.55241 | 1.8103 | 1.1424 | 2.0681 | .047718 | .74415 | 5 |
| 56 | .48379 | .87518 | .55279 | .8090 | .1426 | .0670 | .047807 | .74319 | 4 |
| 57 | .48405 | .87504 | .55317 | .8078 | .1428 | .0659 | .047896 | .74224 | 3 |
| 58 | .48430 | .87490 | .55355 | .8065 | .1430 | .0648 | .047985 | .74129 | 2 |
| 59 | .48456 | .87476 | .55393 | .8053 | .1432 | .0637 | .048074 | .74034 | 1 |
| 60 | 0.48481 | 0.87462 | 0.55431 | 1.8040 | 1.1434 | 2.0627 | .048164 | .73940 | 0 |
| M | Cosine | Sine | Cotan. | Tan. | Cosec. | Secant | READ DOWN | 61°–62° Involute | M |

**118° or 298°**                             **61° or 241°**

| M | Sine | Cosine | Tan. | Cotan. | Secant | Cosec. | Involute 29°–30° | READ UP | M |
|---|------|--------|------|--------|--------|--------|------------------|---------|---|
| 0 | 0.48481 | 0.87462 | 0.55431 | 1.8040 | 1.1434 | 2.0627 | .048164 | .73940 | 60 |
| 1 | .48506 | .87448 | .55469 | .8028 | .1435 | .0616 | .048253 | .73845 | 59 |
| 2 | .48532 | .87434 | .55507 | .8016 | .1437 | .0605 | .048343 | .73751 | 58 |
| 3 | .48557 | .87420 | .55545 | .8003 | .1439 | .0594 | .048432 | .73656 | 57 |
| 4 | .48583 | .87406 | .55583 | .7991 | .1441 | .0583 | .048522 | .73562 | 56 |
| 5 | 0.48608 | 0.87391 | 0.55621 | 1.7979 | 1.1443 | 2.0573 | .048612 | .73468 | 55 |
| 6 | .48634 | .87377 | .55659 | .7966 | .1445 | .0562 | .048702 | .73374 | 54 |
| 7 | .48659 | .87363 | .55697 | .7954 | .1446 | .0551 | .048792 | .73280 | 53 |
| 8 | .48684 | .87349 | .55736 | .7942 | .1448 | .0540 | .048883 | .73186 | 52 |
| 9 | .48710 | .87335 | .55774 | .7930 | .1450 | .0530 | .048973 | .73093 | 51 |
| 10 | 0.48735 | 0.87321 | 0.55812 | 1.7917 | 1.1452 | 2.0519 | .049063 | .72999 | 50 |
| 11 | .48761 | .87306 | .55850 | .7905 | .1454 | .0508 | .049154 | .72906 | 49 |
| 12 | .48786 | .87292 | .55888 | .7893 | .1456 | .0498 | .049245 | .72813 | 48 |
| 13 | .48811 | .87278 | .55926 | .7881 | .1458 | .0487 | .049336 | .72720 | 47 |
| 14 | .48837 | .87264 | .55964 | .7868 | .1460 | .0476 | .049427 | .72627 | 46 |
| 15 | 0.48862 | 0.87250 | 0.56003 | 1.7856 | 1.1461 | 2.0466 | .049518 | .72534 | 45 |
| 16 | .48888 | .87235 | .56041 | .7844 | .1463 | .0455 | .049609 | .72441 | 44 |
| 17 | .48913 | .87221 | .56079 | .7832 | .1465 | .0445 | .049701 | .72349 | 43 |
| 18 | .48938 | .87207 | .56117 | .7820 | .1467 | .0434 | .049792 | .72256 | 42 |
| 19 | .48964 | .87193 | .56156 | .7808 | .1469 | .0423 | .049884 | .72164 | 41 |
| 20 | 0.48989 | 0.87178 | 0.56194 | 1.7796 | 1.1471 | 2.0413 | .049976 | .72072 | 40 |
| 21 | .49014 | .87164 | .56232 | .7783 | .1473 | .0402 | .050068 | .71980 | 39 |
| 22 | .49040 | .87150 | .56270 | .7771 | .1474 | .0392 | .050160 | .71888 | 38 |
| 23 | .49065 | .87136 | .56309 | .7759 | .1476 | .0381 | .050252 | .71796 | 37 |
| 24 | .49090 | .87121 | .56347 | .7747 | .1478 | .0371 | .050344 | .71704 | 36 |
| 25 | 0.49116 | 0.87107 | 0.56385 | 1.7735 | 1.1480 | 2.0360 | .050437 | .71613 | 35 |
| 26 | .49141 | .87093 | .56424 | .7723 | .1482 | .0350 | .050529 | .71521 | 34 |
| 27 | .49166 | .87079 | .56462 | .7711 | .1484 | .0339 | .050622 | .71430 | 33 |
| 28 | .49192 | .87064 | .56501 | .7699 | .1486 | .0329 | .050715 | .71339 | 32 |
| 29 | .49217 | .87050 | .56539 | .7687 | .1488 | .0318 | .050808 | .71248 | 31 |
| 30 | 0.49242 | 0.87036 | 0.56577 | 1.7675 | 1.1490 | 2.0308 | .050901 | .71157 | 30 |
| 31 | .49268 | .87021 | .56616 | .7663 | .1491 | .0297 | .050994 | .71066 | 29 |
| 32 | .49293 | .87007 | .56654 | .7651 | .1493 | .0287 | .051087 | .70975 | 28 |
| 33 | .49318 | .86993 | .56693 | .7639 | .1495 | .0276 | .051181 | .70885 | 27 |
| 34 | .49344 | .86978 | .56731 | .7627 | .1497 | .0266 | .051274 | .70794 | 26 |
| 35 | 0.49369 | 0.86964 | 0.56769 | 1.7615 | 1.1499 | 2.0256 | .051368 | .70704 | 25 |
| 36 | .49394 | .86949 | .56808 | .7603 | .1501 | .0245 | .051462 | .70614 | 24 |
| 37 | .49419 | .86935 | .56846 | .7591 | .1503 | .0235 | .051556 | .70524 | 23 |
| 38 | .49445 | .86921 | .56885 | .7579 | .1505 | .0225 | .051650 | .70434 | 22 |
| 39 | .49470 | .86906 | .56923 | .7567 | .1507 | .0214 | .051744 | .70344 | 21 |
| 40 | 0.49495 | 0.86892 | 0.56962 | 1.7556 | 1.1509 | 2.0204 | .051838 | .70254 | 20 |
| 41 | .49521 | .86878 | .57000 | .7544 | .1510 | .0194 | .051933 | .70165 | 19 |
| 42 | .49546 | .86863 | .57039 | .7532 | .1512 | .0183 | .052027 | .70075 | 18 |
| 43 | .49571 | .86849 | .57078 | .7520 | .1514 | .0173 | .052122 | .69986 | 17 |
| 44 | .49596 | .86834 | .57116 | .7508 | .1516 | .0163 | .052217 | .69897 | 16 |
| 45 | 0.49622 | 0.86820 | 0.57155 | 1.7496 | 1.1518 | 2.0152 | .052312 | .69808 | 15 |
| 46 | .49647 | .86805 | .57193 | .7485 | .1520 | .0142 | .052407 | .69719 | 14 |
| 47 | .49672 | .86791 | .57232 | .7473 | .1522 | .0132 | .052502 | .69630 | 13 |
| 48 | .49697 | .86777 | .57271 | .7461 | .1524 | .0122 | .052597 | .69541 | 12 |
| 49 | .49723 | .86762 | .57309 | .7449 | .1526 | .0112 | .052693 | .69452 | 11 |
| 50 | 0.49748 | 0.86748 | 0.57348 | 1.7437 | 1.1528 | 2.0101 | .052788 | .69364 | 10 |
| 51 | .49773 | .86733 | .57386 | .7426 | .1530 | .0091 | .052884 | .69275 | 9 |
| 52 | .49798 | .86719 | .57425 | .7414 | .1532 | .0081 | .052980 | .69187 | 8 |
| 53 | .49824 | .86704 | .57464 | .7402 | .1533 | .0071 | .053076 | .69099 | 7 |
| 54 | .49849 | .86690 | .57503 | .7391 | .1535 | .0061 | .053172 | .69011 | 6 |
| 55 | 0.49874 | 0.86675 | 0.57541 | 1.7379 | 1.1537 | 2.0051 | .053268 | .68923 | 5 |
| 56 | .49899 | .86661 | .57580 | .7367 | .1539 | .0040 | .053365 | .68835 | 4 |
| 57 | .49924 | .86646 | .57619 | .7355 | .1541 | .0030 | .053461 | .68748 | 3 |
| 58 | .49950 | .86632 | .57657 | .7344 | .1543 | .0020 | .053558 | .68660 | 2 |
| 59 | .49975 | .86617 | .57696 | .7332 | .1545 | .0010 | .053655 | .68573 | 1 |
| 60 | 0.50000 | 0.86603 | 0.57735 | 1.7321 | 1.1547 | 2.0000 | .053751 | .68485 | 0 |

| M | Cosine | Sine | Cotan. | Tan. | Cosec. | Secant | READ DOWN | 60°–61° Involute | M |
|---|--------|------|--------|------|--------|--------|-----------|------------------|---|

**30° or 210°**    Trigonometric Functions    **149° or 329°**

| M | Sine | Cosine | Tan. | Cotan. | Secant | Cosec. | Involute 30°–31° | READ UP | M |
|---|---|---|---|---|---|---|---|---|---|
| 0 | 0.50000 | 0.86603 | 0.57735 | 1.7321 | 1.1547 | 2.0000 | .053751 | .68485 | 60 |
| 1 | .50025 | .86588 | .57774 | .7309 | .1549 | 1.9990 | .053849 | .68398 | 59 |
| 2 | .50050 | .86573 | .57813 | .7297 | .1551 | .9980 | .053946 | .68311 | 58 |
| 3 | .50076 | .86559 | .57851 | .7286 | .1553 | .9970 | .054043 | .68224 | 57 |
| 4 | .50101 | .86544 | .57890 | .7274 | .1555 | .9960 | .054140 | .68137 | 56 |
| 5 | 0.50126 | 0.86530 | 0.57929 | 1.7262 | 1.1557 | 1.9950 | .054238 | .68050 | 55 |
| 6 | .50151 | .86515 | .57968 | .7251 | .1559 | .9940 | .054336 | .67964 | 54 |
| 7 | .50176 | .86501 | .58007 | .7239 | .1561 | .9930 | .054433 | .67877 | 53 |
| 8 | .50201 | .86486 | .58046 | .7228 | .1563 | .9920 | .054531 | .67791 | 52 |
| 9 | .50227 | .86471 | .58085 | .7216 | .1565 | .9910 | .054629 | .67705 | 51 |
| 10 | 0.50252 | 0.86457 | 0.58124 | 1.7205 | 1.1566 | 1.9900 | .054728 | .67618 | 50 |
| 11 | .50277 | .86442 | .58162 | .7193 | .1568 | .9890 | .054826 | .67532 | 49 |
| 12 | .50302 | .86427 | .58201 | .7182 | .1570 | .9880 | .054924 | .67447 | 48 |
| 13 | .50327 | .86413 | .58240 | .7170 | .1572 | .9870 | .055023 | .67361 | 47 |
| 14 | .50352 | .86398 | .58279 | .7159 | .1574 | .9860 | .055122 | .67275 | 46 |
| 15 | 0.50377 | 0.86384 | 0.58318 | 1.7147 | 1.1576 | 1.9850 | .055221 | .67189 | 45 |
| 16 | .50403 | .86369 | .58357 | .7136 | .1578 | .9840 | .055320 | .67104 | 44 |
| 17 | .50428 | .86354 | .58396 | .7124 | .1580 | .9830 | .055419 | .67019 | 43 |
| 18 | .50453 | .86340 | .58435 | .7113 | .1582 | .9821 | .055518 | .66933 | 42 |
| 19 | .50478 | .86325 | .58474 | .7102 | .1584 | .9811 | .055617 | .66848 | 41 |
| 20 | 0.50503 | 0.86310 | 0.58513 | 1.7090 | 1.1586 | 1.9801 | .055717 | .66763 | 40 |
| 21 | .50528 | .86295 | .58552 | .7079 | .1588 | .9791 | .055817 | .66678 | 39 |
| 22 | .50553 | .86281 | .58591 | .7067 | .1590 | .9781 | .055916 | .66593 | 38 |
| 23 | .50578 | .86266 | .58631 | .7056 | .1592 | .9771 | .056016 | .66509 | 37 |
| 24 | .50603 | .86251 | .58670 | .7045 | .1594 | .9762 | .056116 | .66424 | 36 |
| 25 | 0.50628 | 0.86237 | 0.58709 | 1.7033 | 1.1596 | 1.9752 | .056217 | .66340 | 35 |
| 26 | .50654 | .86222 | .58748 | .7022 | .1598 | .9742 | .056317 | .66255 | 34 |
| 27 | .50679 | .86207 | .58787 | .7011 | .1600 | .9732 | .056417 | .66171 | 33 |
| 28 | .50704 | .86192 | .58826 | .6999 | .1602 | .9722 | .056518 | .66087 | 32 |
| 29 | .50729 | .86178 | .58865 | .6988 | .1604 | .9713 | .056619 | .66003 | 31 |
| 30 | 0.50754 | 0.86163 | 0.58905 | 1.6977 | 1.1606 | 1.9703 | .056720 | .65919 | 30 |
| 31 | .50779 | .86148 | .58944 | .6965 | .1608 | .9693 | .056821 | .65835 | 29 |
| 32 | .50804 | .86133 | .58983 | .6954 | .1610 | .9684 | .056922 | .65752 | 28 |
| 33 | .50829 | .86119 | .59022 | .6943 | .1612 | .9674 | .057023 | .65668 | 27 |
| 34 | .50854 | .86104 | .59061 | .6932 | .1614 | .9664 | .057124 | .65585 | 26 |
| 35 | 0.50879 | 0.86089 | 0.59101 | 1.6920 | 1.1616 | 1.9654 | .057226 | .65501 | 25 |
| 36 | .50904 | .86074 | .59140 | .6909 | .1618 | .9645 | .057328 | .65418 | 24 |
| 37 | .50929 | .86059 | .59179 | .6898 | .1620 | .9635 | .057429 | .65335 | 23 |
| 38 | .50954 | .86045 | .59218 | .6887 | .1622 | .9625 | .057531 | .65252 | 22 |
| 39 | .50979 | .86030 | .59258 | .6875 | .1624 | .9616 | .057633 | .65169 | 21 |
| 40 | 0.51004 | 0.86015 | 0.59297 | 1.6864 | 1.1626 | 1.9606 | .057736 | .65086 | 20 |
| 41 | .51029 | .86000 | .59336 | .6853 | .1628 | .9597 | .057838 | .65004 | 19 |
| 42 | .51054 | .85985 | .59376 | .6842 | .1630 | .9587 | .057940 | .64921 | 18 |
| 43 | .51079 | .85970 | .59415 | .6831 | .1632 | .9577 | .058043 | .64839 | 17 |
| 44 | .51104 | .85956 | .59454 | .6820 | .1634 | .9568 | .058146 | .64756 | 16 |
| 45 | 0.51129 | 0.85941 | 0.59494 | 1.6808 | 1.1636 | 1.9558 | .058249 | .64674 | 15 |
| 46 | .51154 | .85926 | .59533 | .6797 | .1638 | .9549 | .058352 | .64592 | 14 |
| 47 | .51179 | .85911 | .59573 | .6786 | .1640 | .9539 | .058455 | .64510 | 13 |
| 48 | .51204 | .85896 | .59612 | .6775 | .1642 | .9530 | .058558 | .64428 | 12 |
| 49 | .51229 | .85881 | .59651 | .6764 | .1644 | .9520 | .058662 | .64346 | 11 |
| 50 | 0.51254 | 0.85866 | 0.59691 | 1.6753 | 1.1646 | 1.9511 | .058765 | .64265 | 10 |
| 51 | .51279 | .85851 | .59730 | .6742 | .1648 | .9501 | .058869 | .64183 | 9 |
| 52 | .51304 | .85836 | .59770 | .6731 | .1650 | .9492 | .058973 | .64102 | 8 |
| 53 | .51329 | .85821 | .59809 | .6720 | .1652 | .9482 | .059077 | .64020 | 7 |
| 54 | .51354 | .85806 | .59849 | .6709 | .1654 | .9473 | .059181 | .63939 | 6 |
| 55 | 0.51379 | 0.85792 | 0.59888 | 1.6698 | 1.1656 | 1.9463 | .059285 | .63858 | 5 |
| 56 | .51404 | .85777 | .59928 | .6687 | .1658 | .9454 | .059390 | .63777 | 4 |
| 57 | .51429 | .85762 | .59967 | .6676 | .1660 | .9444 | .059494 | .63696 | 3 |
| 58 | .51454 | .85747 | .60007 | .6665 | .1662 | .9435 | .059599 | .63615 | 2 |
| 59 | .51479 | .85732 | .60046 | .6654 | .1664 | .9425 | .059704 | .63534 | 1 |
| 60 | 0.51504 | 0.85717 | 0.60086 | 1.6643 | 1.1666 | 1.9416 | .059809 | .63454 | 0 |
| M | Cosine | Sine | Cotan. | Tan. | Cosec. | Secant | READ DOWN | 59°–60° Involute | M |

**120° or 300°**    **59° or 239°**

**31° or 211°**  Trigonometric Functions  **148° or 328°**

| M | Sine | Cosine | Tan. | Cotan. | Secant | Cosec. | Involute 31°–32° | READ UP | M |
|---|---|---|---|---|---|---|---|---|---|
| 0 | 0.51504 | 0.85717 | 0.60086 | 1.6643 | 1.1666 | 1.9416 | .059809 | .63454 | 60 |
| 1 | .51529 | .85702 | .60126 | .6632 | .1668 | .9407 | .059914 | .63373 | 59 |
| 2 | .51554 | .85687 | .60165 | .6621 | .1670 | .9397 | .060019 | .63293 | 58 |
| 3 | .51579 | .85672 | .60205 | .6610 | .1672 | .9388 | .060124 | .63212 | 57 |
| 4 | .51604 | .85657 | .60245 | .6599 | .1675 | .9379 | .060230 | .63132 | 56 |
| 5 | 0.51628 | 0.85642 | 0.60284 | 1.6588 | 1.1677 | 1.9369 | .060335 | .63052 | 55 |
| 6 | .51653 | .85627 | .60324 | .6577 | .1679 | .9360 | .060441 | .62972 | 54 |
| 7 | .51678 | .85612 | .60364 | .6566 | .1681 | .9351 | .060547 | .62892 | 53 |
| 8 | .51703 | .85597 | .60403 | .6555 | .1683 | .9341 | .060653 | .62812 | 52 |
| 9 | .51728 | .85582 | .60443 | .6545 | .1685 | .9332 | .060759 | .62733 | 51 |
| 10 | 0.51753 | 0.85567 | 0.60483 | 1.6534 | 1.1687 | 1.9323 | .060866 | .62653 | 50 |
| 11 | .51778 | .85551 | .60522 | .6523 | .1689 | .9313 | .060972 | .62574 | 49 |
| 12 | .51803 | .85536 | .60562 | .6512 | .1691 | .9304 | .061079 | .62494 | 48 |
| 13 | .51828 | .85521 | .60602 | .6501 | .1693 | .9295 | .061186 | .62415 | 47 |
| 14 | .51852 | .85506 | .60642 | .6490 | .1695 | .9285 | .061292 | .62336 | 46 |
| 15 | 0.51877 | 0.85491 | 0.60681 | 1.6479 | 1.1697 | 1.9276 | .061400 | .62257 | 45 |
| 16 | .51902 | .85476 | .60721 | .6469 | .1699 | .9267 | .061507 | .62178 | 44 |
| 17 | .51927 | .85461 | .60761 | .6458 | .1701 | .9258 | .061614 | .62099 | 43 |
| 18 | .51952 | .85446 | .60801 | .6447 | .1703 | .9249 | .061721 | .62020 | 42 |
| 19 | .51977 | .85431 | .60841 | .6436 | .1705 | .9239 | .061829 | .61942 | 41 |
| 20 | 0.52002 | 0.85416 | 0.60881 | 1.6426 | 1.1707 | 1.9230 | .061937 | .61863 | 40 |
| 21 | .52026 | .85401 | .60921 | .6415 | .1710 | .9221 | .062045 | .61785 | 39 |
| 22 | .52051 | .85385 | .60960 | .6404 | .1712 | .9212 | .062153 | .61706 | 38 |
| 23 | .52076 | .85370 | .61000 | .6393 | .1714 | .9203 | .062261 | .61628 | 37 |
| 24 | .52101 | .85355 | .61040 | .6383 | .1716 | .9194 | .062369 | .61550 | 36 |
| 25 | 0.52126 | 0.85340 | 0.61080 | 1.6372 | 1.1718 | 1.9184 | .062478 | .61472 | 35 |
| 26 | .52151 | .85325 | .61120 | .6361 | .1720 | .9175 | .062586 | .61394 | 34 |
| 27 | .52175 | .85310 | .61160 | .6351 | .1722 | .9166 | .062695 | .61316 | 33 |
| 28 | .52200 | .85294 | .61200 | .6340 | .1724 | .9157 | .062804 | .61239 | 32 |
| 29 | .52225 | .85279 | .61240 | .6329 | .1726 | .9148 | .062913 | .61161 | 31 |
| 30 | 0.52250 | 0.85264 | 0.61280 | 1.6319 | 1.1728 | 1.9139 | .063022 | .61083 | 30 |
| 31 | .52275 | .85249 | .61320 | .6308 | .1730 | .9130 | .063131 | .61006 | 29 |
| 32 | .52299 | .85234 | .61360 | .6297 | .1732 | .9121 | .063241 | .60929 | 28 |
| 33 | .52324 | .85218 | .61400 | .6287 | .1735 | .9112 | .063350 | .60851 | 27 |
| 34 | .52349 | .85203 | .61440 | .6276 | .1737 | .9103 | .063460 | .60774 | 26 |
| 35 | 0.52374 | 0.85188 | 0.61480 | 1.6265 | 1.1739 | 1.9094 | .063570 | .60697 | 25 |
| 36 | .52399 | .85173 | .61520 | .6255 | .1741 | .9084 | .063680 | .60620 | 24 |
| 37 | .52423 | .85157 | .61561 | .6244 | .1743 | .9075 | .063790 | .60544 | 23 |
| 38 | .52448 | .85142 | .61601 | .6234 | .1745 | .9066 | .063901 | .60467 | 22 |
| 39 | .52473 | .85127 | .61641 | .6223 | .1747 | .9057 | .064011 | .60390 | 21 |
| 40 | 0.52498 | 0.85112 | 0.61681 | 1.6212 | 1.1749 | 1.9048 | .064122 | .60314 | 20 |
| 41 | .52522 | .85096 | .61721 | .6202 | .1751 | .9039 | .064232 | .60237 | 19 |
| 42 | .52547 | .85081 | .61761 | .6191 | .1753 | .9031 | .064343 | .60161 | 18 |
| 43 | .52572 | .85066 | .61801 | .6181 | .1756 | .9022 | .064454 | .60085 | 17 |
| 44 | .52597 | .85051 | .61842 | .6170 | .1758 | .9013 | .064565 | .60009 | 16 |
| 45 | 0.52621 | 0.85035 | 0.61882 | 1.6160 | 1.1760 | 1.9004 | .064677 | .59933 | 15 |
| 46 | .52646 | .85020 | .61922 | .6149 | .1762 | .8995 | .064788 | .59857 | 14 |
| 47 | .52671 | .85005 | .61962 | .6139 | .1764 | .8986 | .064900 | .59781 | 13 |
| 48 | .52696 | .84989 | .62003 | .6128 | .1766 | .8977 | .065012 | .59705 | 12 |
| 49 | .52720 | .84974 | .62043 | .6118 | .1768 | .8968 | .065123 | .59630 | 11 |
| 50 | 0.52745 | 0.84959 | 0.62083 | 1.6107 | 1.1770 | 1.8959 | .065236 | .59554 | 10 |
| 51 | .52770 | .84943 | .62124 | .6097 | .1773 | .8950 | .065348 | .59479 | 9 |
| 52 | .52794 | .84928 | .62164 | .6087 | .1775 | .8941 | .065460 | .59403 | 8 |
| 53 | .52819 | .84913 | .62204 | .6076 | .1777 | .8933 | .065573 | .59328 | 7 |
| 54 | .52844 | .84897 | .62245 | .6066 | .1779 | .8924 | .065685 | .59253 | 6 |
| 55 | 0.52869 | 0.84882 | 0.62285 | 1.6055 | 1.1781 | 1.8915 | .065798 | .59178 | 5 |
| 56 | .52893 | .84866 | .62325 | .6045 | .1783 | .8906 | .065911 | .59103 | 4 |
| 57 | .52918 | .84851 | .62366 | .6034 | .1785 | .8897 | .066024 | .59028 | 3 |
| 58 | .52942 | .84836 | .62406 | .6024 | .1788 | .8888 | .066137 | .58954 | 2 |
| 59 | .52967 | .84820 | .62446 | .6014 | .1790 | .8880 | .066250 | .58879 | 1 |
| 60 | 0.52992 | 0.84805 | 0.62487 | 1.6003 | 1.1792 | 1.8871 | .066364 | .58804 | 0 |
| M | Cosine | Sine | Cotan. | Tan. | Cosec. | Secant | 58°–59° Involute | READ DOWN | M |

**121° or 301°**  **58° or 238°**

## 32° or 212°　　Trigonometric Functions　　147° or 327°

| M | Sine | Cosine | Tan. | Cotan. | Secant | Cosec. | Involute 32°–33° | READ UP | M |
|---|------|--------|------|--------|--------|--------|------------------|---------|---|
| 0 | 0.52992 | 0.84805 | 0.62487 | 1.6003 | 1.1792 | 1.8871 | .066364 | .58804 | 60 |
| 1 | .53017 | .84789 | .62527 | .5993 | .1794 | .8862 | .066478 | .58730 | 59 |
| 2 | .53041 | .84774 | .62568 | .5983 | .1796 | .8853 | .066591 | .58656 | 58 |
| 3 | .53066 | .84758 | .62608 | .5972 | .1798 | .8844 | .066705 | .58581 | 57 |
| 4 | .53091 | .84743 | .62649 | .5962 | .1800 | .8836 | .066819 | .58507 | 56 |
| 5 | 0.53115 | 0.84728 | 0.62689 | 1.5952 | 1.1803 | 1.8827 | .066934 | .58433 | 55 |
| 6 | .53140 | .84712 | .62730 | .5941 | .1805 | .8818 | .067048 | .58359 | 54 |
| 7 | .53164 | .84697 | .62770 | .5931 | .1807 | .8810 | .067163 | .58285 | 53 |
| 8 | .53189 | .84681 | .62811 | .5921 | .1809 | .8801 | .067277 | .58211 | 52 |
| 9 | .53214 | .84666 | .62852 | .5911 | .1811 | .8792 | .067392 | .58138 | 51 |
| 10 | 0.53238 | 0.84650 | 0.62892 | 1.5900 | 1.1813 | 1.8783 | .067507 | .58064 | 50 |
| 11 | .53263 | .84635 | .62933 | .5890 | .1815 | .8775 | .067622 | .57991 | 49 |
| 12 | .53288 | .84619 | .62973 | .5880 | .1818 | .8766 | .067738 | .57917 | 48 |
| 13 | .53312 | .84604 | .63014 | .5869 | .1820 | .8757 | .067853 | .57844 | 47 |
| 14 | .53337 | .84588 | .63055 | .5859 | .1822 | .8749 | .067969 | .57771 | 46 |
| 15 | 0.53361 | 0.84573 | 0.63095 | 1.5849 | 1.1824 | 1.8740 | .068084 | .57698 | 45 |
| 16 | .53386 | .84557 | .63136 | .5839 | .1826 | .8731 | .068200 | .57625 | 44 |
| 17 | .53411 | .84542 | .63177 | .5829 | .1828 | .8723 | .068316 | .57552 | 43 |
| 18 | .53435 | .84526 | .63217 | .5818 | .1831 | .8714 | .068432 | .57479 | 42 |
| 19 | .53460 | .84511 | .63258 | .5808 | .1833 | .8706 | .068549 | .57406 | 41 |
| 20 | 0.53484 | 0.84495 | 0.63299 | 1.5798 | 1.1835 | 1.8697 | .068665 | .57333 | 40 |
| 21 | .53509 | .84480 | .63340 | .5788 | .1837 | .8688 | .068782 | .57261 | 39 |
| 22 | .53534 | .84464 | .63380 | .5778 | .1839 | .8680 | .068899 | .57188 | 38 |
| 23 | .53558 | .84448 | .63421 | .5768 | .1842 | .8671 | .069016 | .57116 | 37 |
| 24 | .53583 | .84433 | .63462 | .5757 | .1844 | .8663 | .069133 | .57044 | 36 |
| 25 | 0.53607 | 0.84417 | 0.63503 | 1.5747 | 1.1846 | 1.8654 | .069250 | .56972 | 35 |
| 26 | .53632 | .84402 | .63544 | .5737 | .1848 | .8646 | .069367 | .56900 | 34 |
| 27 | .53656 | .84386 | .63584 | .5727 | .1850 | .8637 | .069485 | .56828 | 33 |
| 28 | .53681 | .84370 | .63625 | .5717 | .1852 | .8629 | .069602 | .56756 | 32 |
| 29 | .53705 | .84355 | .63666 | .5707 | .1855 | .8620 | .069720 | .56684 | 31 |
| 30 | 0.53730 | 0.84339 | 0.63707 | 1.5697 | 1.1857 | 1.8612 | .069838 | .56612 | 30 |
| 31 | .53754 | .84324 | .63748 | .5687 | .1859 | .8603 | .069956 | .56540 | 29 |
| 32 | .53779 | .84308 | .63789 | .5677 | .1861 | .8595 | .070075 | .56469 | 28 |
| 33 | .53804 | .84292 | .63830 | .5667 | .1863 | .8586 | .070193 | .56398 | 27 |
| 34 | .53828 | .84277 | .63871 | .5657 | .1866 | .8578 | .070312 | .56326 | 26 |
| 35 | 0.53853 | 0.84261 | 0.63912 | 1.5647 | 1.1868 | 1.8569 | .070430 | .56255 | 25 |
| 36 | .53877 | .84245 | .63953 | .5637 | .1870 | .8561 | .070549 | .56184 | 24 |
| 37 | .53902 | .84230 | .63994 | .5627 | .1872 | .8552 | .070668 | .56113 | 23 |
| 38 | .53926 | .84214 | .64035 | .5617 | .1875 | .8544 | .070788 | .56042 | 22 |
| 39 | .53951 | .84193 | .64076 | .5607 | .1877 | .8535 | .070907 | .55971 | 21 |
| 40 | 0.53975 | 0.84182 | 0.64117 | 1.5597 | 1.1879 | 1.8527 | .071026 | .55900 | 20 |
| 41 | .54000 | .84167 | .64158 | .5587 | .1881 | .8519 | .071146 | .55829 | 19 |
| 42 | .54024 | .84151 | .64199 | .5577 | .1883 | .8510 | .071266 | .55759 | 18 |
| 43 | .54049 | .84135 | .64240 | .5567 | .1886 | .8502 | .071386 | .55688 | 17 |
| 44 | .54073 | .84120 | .64281 | .5557 | .1888 | .8494 | .071506 | .55618 | 16 |
| 45 | 0.54097 | 0.84104 | 0.64322 | 1.5547 | 1.1890 | 1.8485 | .071626 | .55547 | 15 |
| 46 | .54122 | .84088 | .64363 | .5537 | .1892 | .8477 | .071747 | .55477 | 14 |
| 47 | .54146 | .84072 | .64404 | .5527 | .1895 | .8468 | .071867 | .55407 | 13 |
| 48 | .54171 | .84057 | .64446 | .5517 | .1897 | .8460 | .071988 | .55337 | 12 |
| 49 | .54195 | .84041 | .64487 | .5507 | .1899 | .8452 | .072109 | .55267 | 11 |
| 50 | 0.54220 | 0.84025 | 0.64528 | 1.5497 | 1.1901 | 1.8443 | .072230 | .55197 | 10 |
| 51 | .54244 | .84009 | .64569 | .5487 | .1903 | .8435 | .072351 | .55127 | 9 |
| 52 | .54269 | .83994 | .64610 | .5477 | .1906 | .8427 | .072473 | .55057 | 8 |
| 53 | .54293 | .83978 | .64652 | .5468 | .1908 | .8419 | .072594 | .54988 | 7 |
| 54 | .54317 | .83962 | .64693 | .5458 | .1910 | .8410 | .072716 | .54918 | 6 |
| 55 | 0.54342 | 0.83946 | 0.64734 | 1.5448 | 1.1912 | 1.8402 | .072838 | .54849 | 5 |
| 56 | .54366 | .83930 | .64775 | .5438 | .1915 | .8394 | .072959 | .54779 | 4 |
| 57 | .54391 | .83915 | .64817 | .5428 | .1917 | .8385 | .073082 | .54710 | 3 |
| 58 | .54415 | .83899 | .64858 | .5418 | .1919 | .8377 | .073204 | .54641 | 2 |
| 59 | .54440 | .83883 | .64899 | .5408 | .1921 | .8369 | .073326 | .54572 | 1 |
| 60 | 0.54464 | 0.83867 | 0.64941 | 1.5399 | 1.1924 | 1.8361 | .073449 | .54503 | 0 |
| M | Cosine | Sine | Cotan. | Tan. | Cosec. | Secant | READ DOWN | 57°–58° Involute | M |

**33° or 213°**  Trigonometric Functions  **146° or 326°**

| M | Sine | Cosine | Tan. | Cotan. | Secant | Cosec. | Involute 33°–34° | READ UP | M |
|---|------|--------|------|--------|--------|--------|---------|---------|---|
| 0 | 0.54464 | 0.83867 | 0.64941 | 1.5399 | 1.1924 | 1.8361 | .073449 | .54503 | 60 |
| 1 | .54488 | .83851 | .64982 | .5389 | .1926 | .8353 | .073572 | .54434 | 59 |
| 2 | .54513 | .83835 | .65024 | .5379 | .1928 | .8344 | .073695 | .54365 | 58 |
| 3 | .54537 | .83819 | .65065 | .5369 | .1930 | .8336 | .073818 | .54296 | 57 |
| 4 | .54561 | .83804 | .65106 | .5359 | .1933 | .8328 | .073941 | .54228 | 56 |
| 5 | 0.54586 | 0.83788 | 0.65148 | 1.5350 | 1.1935 | 1.8320 | .074064 | .54159 | 55 |
| 6 | .54610 | .83772 | .65189 | .5340 | .1937 | .8312 | .074188 | .54090 | 54 |
| 7 | .54635 | .83756 | .65231 | .5330 | .1939 | .8303 | .074312 | .54022 | 53 |
| 8 | .54659 | .83740 | .65272 | .5320 | .1942 | .8295 | .074435 | .53954 | 52 |
| 9 | .54683 | .83724 | .65314 | .5311 | .1944 | .8287 | .074559 | .53885 | 51 |
| 10 | 0.54708 | 0.83708 | 0.65355 | 1.5301 | 1.1946 | 1.8279 | .074684 | .53817 | 50 |
| 11 | .54732 | .83692 | .65397 | .5291 | .1949 | .8271 | .074808 | .53749 | 49 |
| 12 | .54756 | .83676 | .65438 | .5282 | .1951 | .8263 | .074932 | .53681 | 48 |
| 13 | .54781 | .83660 | .65480 | .5272 | .1953 | .8255 | .075057 | .53613 | 47 |
| 14 | .54805 | .83645 | .65521 | .5262 | .1955 | .8247 | .075182 | .53546 | 46 |
| 15 | 0.54829 | 0.83629 | 0.65563 | 1.5253 | 1.1958 | 1.8238 | .075307 | .53478 | 45 |
| 16 | .54854 | .83613 | .65604 | .5243 | .1960 | .8230 | .075432 | .53410 | 44 |
| 17 | .54878 | .83597 | .65646 | .5233 | .1962 | .8222 | .075557 | .53343 | 43 |
| 18 | .54902 | .83581 | .65688 | .5224 | .1964 | .8214 | .075683 | .53275 | 42 |
| 19 | .54927 | .83565 | .65729 | .5214 | .1967 | .8206 | .075808 | .53208 | 41 |
| 20 | 0.54951 | 0.83549 | 0.65771 | 1.5204 | 1.1969 | 1.8198 | .075934 | .53141 | 40 |
| 21 | .54975 | .83533 | .65813 | .5195 | .1971 | .8190 | .076060 | .53073 | 39 |
| 22 | .54999 | .83517 | .65854 | .5185 | .1974 | .8182 | .076186 | .53006 | 38 |
| 23 | .55024 | .83501 | .65896 | .5175 | .1976 | .8174 | .076312 | .52939 | 37 |
| 24 | .55048 | .83485 | .65938 | .5166 | .1978 | .8166 | .076439 | .52872 | 36 |
| 25 | 0.55072 | 0.83469 | 0.65980 | 1.5156 | 1.1981 | 1.8158 | .076565 | .52805 | 35 |
| 26 | .55097 | .83453 | .66021 | .5147 | .1983 | .8150 | .076692 | .52739 | 34 |
| 27 | .55121 | .83437 | .66063 | .5137 | .1985 | .8142 | .076819 | .52672 | 33 |
| 28 | .55145 | .83421 | .66105 | .5127 | .1987 | .8134 | .076946 | .52605 | 32 |
| 29 | .55169 | .83405 | .66147 | .5118 | .1990 | .8126 | .077073 | .52539 | 31 |
| 30 | 0.55194 | 0.83389 | 0.66189 | 1.5108 | 1.1992 | 1.8118 | .077200 | .52472 | 30 |
| 31 | .55218 | .83373 | .66230 | .5099 | .1994 | .8110 | .077328 | .52406 | 29 |
| 32 | .55242 | .83356 | .66272 | .5089 | .1997 | .8102 | .077455 | .52340 | 28 |
| 33 | .55266 | .83340 | .66314 | .5080 | .1999 | .8094 | .077583 | .52274 | 27 |
| 34 | .55291 | .83324 | .66356 | .5070 | .2001 | .8086 | .077711 | .52207 | 26 |
| 35 | 0.55315 | 0.83308 | 0.66398 | 1.5061 | 1.2004 | 1.8078 | .077839 | .52141 | 25 |
| 36 | .55339 | .83292 | .66440 | .5051 | .2006 | .8070 | .077968 | .52076 | 24 |
| 37 | .55363 | .83276 | .66482 | .5042 | .2008 | .8062 | .078096 | .52010 | 23 |
| 38 | .55388 | .83260 | .66524 | .5032 | .2011 | .8055 | .078225 | .51944 | 22 |
| 39 | .55412 | .83244 | .66566 | .5023 | .2013 | .8047 | .078354 | .51878 | 21 |
| 40 | 0.55436 | 0.83228 | 0.66608 | 1.5013 | 1.2015 | 1.8039 | .078483 | .51813 | 20 |
| 41 | .55460 | .83212 | .66650 | .5004 | .2018 | .8031 | .078612 | .51747 | 19 |
| 42 | .55484 | .83195 | .66692 | .4994 | .2020 | .8023 | .078741 | .51682 | 18 |
| 43 | .55509 | .83179 | .66734 | .4985 | .2022 | .8015 | .078871 | .51616 | 17 |
| 44 | .55533 | .83163 | .66776 | .4975 | .2025 | .8007 | .079000 | .51551 | 16 |
| 45 | 0.55557 | 0.83147 | 0.66818 | 1.4966 | 1.2027 | 1.8000 | .079130 | .51486 | 15 |
| 46 | .55581 | .83131 | .66860 | .4957 | .2029 | .7992 | .079260 | .51421 | 14 |
| 47 | .55605 | .83115 | .66902 | .4947 | .2032 | .7984 | .079390 | .51356 | 13 |
| 48 | .55630 | .83098 | .66944 | .4938 | .2034 | .7976 | .079520 | .51291 | 12 |
| 49 | .55654 | .83082 | .66986 | .4928 | .2036 | .7968 | .079651 | .51226 | 11 |
| 50 | 0.55678 | 0.83066 | 0.67028 | 1.4919 | 1.2039 | 1.7960 | .079781 | .51161 | 10 |
| 51 | .55702 | .83050 | .67071 | .4910 | .2041 | .7953 | .079912 | .51096 | 9 |
| 52 | .55726 | .83034 | .67113 | .4900 | .2043 | .7945 | .080043 | .51032 | 8 |
| 53 | .55750 | .83017 | .67155 | .4891 | .2046 | .7937 | .080174 | .50967 | 7 |
| 54 | .55775 | .83001 | .67197 | .4882 | .2048 | .7929 | .080305 | .50903 | 6 |
| 55 | 0.55799 | 0.82985 | 0.67239 | 1.4872 | 1.2050 | 1.7922 | .080437 | .50838 | 5 |
| 56 | .55823 | .82969 | .67282 | .4863 | .2053 | .7914 | .080569 | .50774 | 4 |
| 57 | .55847 | .82953 | .67324 | .4854 | .2055 | .7906 | .080700 | .50710 | 3 |
| 58 | .55871 | .82936 | .67366 | .4844 | .2057 | .7898 | .080832 | .50646 | 2 |
| 59 | .55895 | .82920 | .67409 | .4835 | .2060 | .7891 | .080964 | .50582 | 1 |
| 60 | 0.55919 | 0.82904 | 0.67451 | 1.4826 | 1.2062 | 1.7883 | .081097 | .50518 | 0 |

| M | Cosine | Sine | Cotan. | Tan. | Cosec. | Secant | READ DOWN | 56°–57° Involute | M |
|---|--------|------|--------|------|--------|--------|-----------|------------------|---|

**123° or 303°**  **56° or 236°**

| M | Sine | Cosine | Tan. | Cotan. | Secant | Cosec. | Involute 34°–35° | READ UP | M |
|---|---|---|---|---|---|---|---|---|---|
| 0 | 0.55919 | 0.82904 | 0.67451 | 1.4826 | 1.2062 | 1.7883 | .081097 | .50518 | 60 |
| 1 | .55943 | .82887 | .67493 | .4816 | .2065 | .7875 | .081229 | .50454 | 59 |
| 2 | .55968 | .82871 | .67536 | .4807 | .2067 | .7868 | .081362 | .50390 | 58 |
| 3 | .55992 | .82855 | .67578 | .4798 | .2069 | .7860 | .081494 | .50326 | 57 |
| 4 | .56016 | .82839 | .67620 | .4788 | .2072 | .7852 | .081760 | .50199 | 56 |
| 5 | 0.56040 | 0.82822 | 0.67663 | 1.4779 | 1.2074 | 1.7844 | .081894 | .50135 | 55 |
| 6 | .56064 | .82806 | .67705 | .4770 | .2076 | .7837 | .082027 | .50072 | 54 |
| 7 | .56088 | .82790 | .67748 | .4761 | .2079 | .7829 | .082161 | .50009 | 53 |
| 8 | .56112 | .82773 | .67790 | .4751 | .2081 | .7821 | .082294 | .49945 | 52 |
| 9 | .56136 | .82757 | .67832 | .4742 | .2084 | .7814 | .082428 | .49882 | 51 |
| 10 | 0.56160 | 0.82741 | 0.67875 | 1.4733 | 1.2086 | 1.7806 | .082562 | .49819 | 50 |
| 11 | .56184 | .82724 | .67917 | .4724 | .2088 | .7799 | .082697 | .49756 | 49 |
| 12 | .56208 | .82708 | .67960 | .4715 | .2091 | .7791 | .082831 | .49693 | 48 |
| 13 | .56232 | .82692 | .68002 | .4705 | .2093 | .7783 | .082966 | .49630 | 47 |
| 14 | .56256 | .82675 | .68045 | .4696 | .2096 | .7776 | .083100 | .49568 | 46 |
| 15 | 0.56280 | 0.82659 | 0.68088 | 1.4687 | 1.2098 | 1.7768 | .083235 | .49505 | 45 |
| 16 | .56305 | .82643 | .68130 | .4678 | .2100 | .7761 | .083337 | .49442 | 44 |
| 17 | .56329 | .82626 | .68173 | .4669 | .2103 | .7753 | .083506 | .49380 | 43 |
| 18 | .56353 | .82610 | .68215 | .4659 | .2105 | .7745 | .083641 | .49317 | 42 |
| 19 | .56377 | .82593 | .68258 | .4650 | .2108 | .7738 | .083777 | .49255 | 41 |
| 20 | 0.56401 | 0.82577 | 0.68301 | 1.4641 | 1.2110 | 1.7730 | .083913 | .49192 | 40 |
| 21 | .56425 | .82561 | .68343 | .4632 | .2112 | .7723 | .084049 | .49130 | 39 |
| 22 | .56449 | .82544 | .68386 | .4623 | .2115 | .7715 | .084185 | .49068 | 38 |
| 23 | .56473 | .82528 | .68429 | .4614 | .2117 | .7708 | .084321 | .49006 | 37 |
| 24 | .56497 | .82511 | .68471 | .4605 | .2120 | .7700 | .084457 | .48944 | 36 |
| 25 | 0.56521 | 0.82495 | 0.68514 | 1.4596 | 1.2122 | 1.7693 | .084594 | .48882 | 35 |
| 26 | .56545 | .82478 | .68557 | .4586 | .2124 | .7685 | .084731 | .48820 | 34 |
| 27 | .56569 | .82462 | .68600 | .4577 | .2127 | .7678 | .084868 | .48758 | 33 |
| 28 | .56593 | .82446 | .68642 | .4568 | .2129 | .7670 | .085005 | .48697 | 32 |
| 29 | .56617 | .82429 | .68685 | .4559 | .2132 | .7663 | .085142 | .48635 | 31 |
| 30 | 0.56641 | 0.82413 | 0.68728 | 1.4550 | 1.2134 | 1.7655 | .085142 | .48635 | 30 |
| 31 | .56665 | .82396 | .68771 | .4541 | .2136 | .7648 | .085280 | .48574 | 29 |
| 32 | .56689 | .82380 | .68814 | .4532 | .2139 | .7640 | .085418 | .48512 | 28 |
| 33 | .56713 | .82363 | .68857 | .4523 | .2141 | .7633 | .085555 | .48451 | 27 |
| 34 | .56736 | .82347 | .68900 | .4514 | .2144 | .7625 | .085693 | .48389 | 26 |
| 35 | 0.56760 | 0.82330 | 0.68942 | 1.4505 | 1.2146 | 1.7618 | .085832 | .48328 | 25 |
| 36 | .56784 | .82314 | .68985 | .4496 | .2149 | .7610 | .085970 | .48267 | 24 |
| 37 | .56808 | .82297 | .69028 | .4487 | .2151 | .7603 | .086108 | .48206 | 23 |
| 38 | .56832 | .82281 | .69071 | .4478 | .2154 | .7596 | .086247 | .48145 | 22 |
| 39 | .56856 | .82264 | .69114 | .4469 | .2156 | .7588 | .086386 | .48084 | 21 |
| 40 | 0.56880 | 0.82248 | 0.69157 | 1.4460 | 1.2158 | 1.7581 | .086525 | .48023 | 20 |
| 41 | .56904 | .82231 | .69200 | .4451 | .2161 | .7573 | .086664 | .47962 | 19 |
| 42 | .56928 | .82214 | .69243 | .4442 | .2163 | .7566 | .086804 | .47902 | 18 |
| 43 | .56952 | .82198 | .69286 | .4433 | .2166 | .7559 | .086943 | .47841 | 17 |
| 44 | .56976 | .82181 | .69329 | .4424 | .2168 | .7551 | .087083 | .47780 | 16 |
| 45 | 0.57000 | 0.82165 | 0.69372 | 1.4415 | 1.2171 | 1.7544 | .087223 | .47720 | 15 |
| 46 | .57024 | .82148 | .69416 | .4406 | .2173 | .7537 | .087363 | .47660 | 14 |
| 47 | .57047 | .82132 | .69459 | .4397 | .2176 | .7529 | .087503 | .47599 | 13 |
| 48 | .57071 | .82115 | .69502 | .4388 | .2178 | .7522 | .087644 | .47539 | 12 |
| 49 | .57095 | .82098 | .69545 | .4379 | .2181 | .7515 | .087784 | .47479 | 11 |
| 50 | 0.57119 | 0.82082 | 0.69588 | 1.4370 | 1.2183 | 1.7507 | .087925 | .47419 | 10 |
| 51 | .57143 | .82065 | .69631 | .4361 | .2185 | .7500 | .088066 | .47359 | 9 |
| 52 | .57167 | .82048 | .69675 | .4352 | .2188 | .7493 | .088207 | .47299 | 8 |
| 53 | .57191 | .82032 | .69718 | .4344 | .2190 | .7485 | .088348 | .47239 | 7 |
| 54 | .57215 | .82015 | .69761 | .4335 | .2193 | .7478 | .088490 | .47179 | 6 |
| 55 | .57238 | 0.81999 | 0.69804 | 1.4326 | 1.2195 | 1.7471 | .088631 | .47119 | 5 |
| 56 | .57262 | .81982 | .69847 | .4317 | .2198 | .7463 | .088773 | .47060 | 4 |
| 57 | .57286 | .81965 | .69891 | .4308 | .2200 | .7456 | .088915 | .47000 | 3 |
| 58 | .57310 | .81949 | .69934 | .4299 | .2203 | .7449 | .089057 | .46940 | 2 |
| 59 | .57334 | .81932 | .69977 | .4290 | .2205 | .7442 | .089200 | .46881 | 1 |
| 60 | .57358 | 0.81915 | 0.70021 | 1.4281 | 1.2208 | 1.7434 | .089342 | .46822 | 0 |
| M | Cosine | Sine | Cotan. | Tan. | Cosec. | Secant | READ DOWN | 55°–56° Involute | M |

**35° or 215°**  Trigonometric Functions  **144° or 324°**

| M | Sine | Cosine | Tan. | Cotan. | Secant | Cosec. | Involute 35°–36° | READ UP | M |
|---|------|--------|------|--------|--------|--------|---------|---------|---|
| 0 | 0.57358 | 0.81915 | 0.70021 | 1.4281 | 1.2208 | 1.7434 | .089342 | .46822 | 60 |
| 1 | .57381 | .81899 | .70064 | .4273 | .2210 | .7427 | .089485 | .46762 | 59 |
| 2 | .57405 | .81882 | .70107 | .4264 | .2213 | .7420 | .089628 | .46703 | 58 |
| 3 | .57429 | .81865 | .70151 | .4255 | .2215 | .7413 | .089771 | .46644 | 57 |
| 4 | .57453 | .81848 | .70194 | .4246 | .2218 | .7406 | .089914 | .46585 | 56 |
| 5 | 0.57477 | 0.81832 | 0.70238 | 1.4237 | 1.2220 | 1.7398 | .090058 | .46526 | 55 |
| 6 | .57501 | .81815 | .70281 | .4229 | .2223 | .7391 | .090201 | .46467 | 54 |
| 7 | .57524 | .81798 | .70325 | .4220 | .2225 | .7384 | .090345 | .46408 | 53 |
| 8 | .57548 | .81782 | .70368 | .4211 | .2228 | .7377 | .090489 | .46349 | 52 |
| 9 | .57572 | .81765 | .70412 | .4202 | .2230 | .7370 | .090633 | .46291 | 51 |
| 10 | 0.57596 | 0.81748 | 0.70455 | 1.4193 | 1.2233 | 1.7362 | .090777 | .46232 | 50 |
| 11 | .57619 | .81731 | .70499 | .4185 | .2235 | .7355 | .090922 | .46173 | 49 |
| 12 | .57643 | .81714 | .70542 | .4176 | .2238 | .7348 | .091067 | .46115 | 48 |
| 13 | .57667 | .81698 | .70586 | .4167 | .2240 | .7341 | .091211 | .46057 | 47 |
| 14 | .57691 | .81681 | .70629 | .4158 | .2243 | .7334 | .091356 | .45998 | 46 |
| 15 | 0.57715 | 0.81664 | 0.70673 | 1.4150 | 1.2245 | 1.7327 | .091502 | .45940 | 45 |
| 16 | .57738 | .81647 | .70717 | .4141 | .2248 | .7320 | .091647 | .45882 | 44 |
| 17 | .57762 | .81631 | .70760 | .4132 | .2250 | .7312 | .091793 | .45824 | 43 |
| 18 | .57786 | .81614 | .70804 | .4124 | .2253 | .7305 | .091938 | .45766 | 42 |
| 19 | .57810 | .81597 | .70848 | .4115 | .2255 | .7298 | .092084 | .45708 | 41 |
| 20 | 0.57833 | 0.81580 | 0.70891 | 1.4106 | 1.2258 | 1.7291 | .092230 | .45650 | 40 |
| 21 | .57857 | .81563 | .70935 | .4097 | .2260 | .7284 | .092377 | .45592 | 39 |
| 22 | .57881 | .81546 | .70979 | .4089 | .2263 | .7277 | .092523 | .45534 | 38 |
| 23 | .57904 | .81530 | .71023 | .4080 | .2265 | .7270 | .092670 | .45476 | 37 |
| 24 | .57928 | .81513 | .71066 | .4071 | .2268 | .7263 | .092816 | .45419 | 36 |
| 25 | 0.57952 | 0.81496 | 0.71110 | 1.4063 | 1.2271 | 1.7256 | .092963 | .45361 | 35 |
| 26 | .57976 | .81479 | .71154 | .4054 | .2273 | .7249 | .093111 | .45304 | 34 |
| 27 | .57999 | .81462 | .71198 | .4045 | .2276 | .7242 | .093258 | .45246 | 33 |
| 28 | .58023 | .81445 | .71242 | .4037 | .2278 | .7235 | .093406 | .45189 | 32 |
| 29 | .58047 | .81428 | .71285 | .4028 | .2281 | .7228 | .093553 | .45132 | 31 |
| 30 | 0.58070 | 0.81412 | 0.71329 | 1.4019 | 1.2283 | 1.7221 | .093701 | .45074 | 30 |
| 31 | .58094 | .81395 | .71373 | .4011 | .2286 | .7213 | .093849 | .45017 | 29 |
| 32 | .58118 | .81378 | .71417 | .4002 | .2288 | .7206 | .093998 | .44960 | 28 |
| 33 | .58141 | .81361 | .71461 | .3994 | .2291 | .7199 | .094146 | .44903 | 27 |
| 34 | .58165 | .81344 | .71505 | .3985 | .2293 | .7192 | .094295 | .44846 | 26 |
| 35 | 0.58189 | 0.81327 | 0.71549 | 1.3976 | 1.2296 | 1.7185 | .094443 | .44789 | 25 |
| 36 | .58212 | .81310 | .71593 | .3968 | .2299 | .7179 | .094592 | .44733 | 24 |
| 37 | .58236 | .81293 | .71637 | .3959 | .2301 | .7172 | .094742 | .44676 | 23 |
| 38 | .58260 | .81276 | .71681 | .3951 | .2304 | .7165 | .094891 | .44619 | 22 |
| 39 | .58283 | .81259 | .71725 | .3942 | .2306 | .7158 | .095041 | .44563 | 21 |
| 40 | 0.58307 | 0.81242 | 0.71769 | 1.3934 | 1.2309 | 1.7151 | .095190 | .44506 | 20 |
| 41 | .58330 | .81225 | .71813 | .3925 | .2311 | .7144 | .095340 | .44450 | 19 |
| 42 | .58354 | .81208 | .71857 | .3916 | .2314 | .7137 | .095490 | .44393 | 18 |
| 43 | .58378 | .81191 | .71901 | .3908 | .2317 | .7130 | .095641 | .44337 | 17 |
| 44 | .58401 | .81174 | .71946 | .3899 | .2319 | .7123 | .095791 | .44281 | 16 |
| 45 | 0.58425 | 0.81157 | 0.71990 | 1.3891 | 1.2322 | 1.7116 | .095942 | .44225 | 15 |
| 46 | .58449 | .81140 | .72034 | .3882 | .2324 | .7109 | .096093 | .44169 | 14 |
| 47 | .58472 | .81123 | .72078 | .3874 | .2327 | .7102 | .096244 | .44113 | 13 |
| 48 | .58496 | .81106 | .72122 | .3865 | .2329 | .7095 | .096395 | .44057 | 12 |
| 49 | .58519 | .81089 | .72167 | .3857 | .2332 | .7088 | .096546 | .44001 | 11 |
| 50 | 0.58543 | 0.81072 | 0.72211 | 1.3848 | 1.2335 | 1.7081 | .096698 | .43945 | 10 |
| 51 | .58567 | .81055 | .72255 | .3840 | .2337 | .7075 | .096850 | .43889 | 9 |
| 52 | .58590 | .81038 | .72299 | .3831 | .2340 | .7068 | .097002 | .43833 | 8 |
| 53 | .58614 | .81021 | .72344 | .3823 | .2342 | .7061 | .097154 | .43778 | 7 |
| 54 | .58637 | .81004 | .72388 | .3814 | .2345 | .7054 | .097306 | .43722 | 6 |
| 55 | 0.58661 | 0.80987 | 0.72432 | 1.3806 | 1.2348 | 1.7047 | .097459 | .43667 | 5 |
| 56 | .58684 | .80970 | .72477 | .3798 | .2350 | .7040 | .097611 | .43611 | 4 |
| 57 | .58708 | .80953 | .72521 | .3789 | .2353 | .7033 | .097764 | .43556 | 3 |
| 58 | .58731 | .80936 | .72565 | .3781 | .2355 | .7027 | .097917 | .43501 | 2 |
| 59 | .58755 | .80919 | .72610 | .3772 | .2358 | .7020 | .098071 | .43446 | 1 |
| 60 | 0.58779 | 0.80902 | 0.72654 | 1.3764 | 1.2361 | 1.7013 | .098224 | .43390 | 0 |

| M | Cosine | Sine | Cotan. | Tan. | Cosec. | Secant | READ DOWN | 54°–55° Involute | M |
|---|--------|------|--------|------|--------|--------|-----------|---------|---|

**125° or 305°**  **54° or 234°**

| M | Sine | Cosine | Tan. | Cotan. | Secant | Cosec. | Involute 36°–37° | READ UP | M |
|---|------|--------|------|--------|--------|--------|-----------------|---------|---|
| 0 | 0.58779 | 0.80902 | 0.72654 | 1.3764 | 1.2361 | 1.7013 | .098224 | .43390 | 60 |
| 1 | .58802 | .80885 | .72699 | .3755 | .2363 | .7006 | .098378 | .43335 | 59 |
| 2 | .58826 | .80867 | .72743 | .3747 | .2366 | .6999 | .098532 | .43280 | 58 |
| 3 | .58849 | .80850 | .72788 | .3739 | .2369 | .6993 | .098686 | .43225 | 57 |
| 4 | .58873 | .80833 | .72832 | .3730 | .2371 | .6986 | .098840 | .43171 | 56 |
| 5 | 0.58896 | 0.80816 | 0.72877 | 1.3722 | 1.2374 | 1.6979 | .098994 | .43116 | 55 |
| 6 | .58920 | .80799 | .72921 | .3713 | .2376 | .6972 | .099149 | .43061 | 54 |
| 7 | .58943 | .80782 | .72966 | .3705 | .2379 | .6966 | .099304 | .43006 | 53 |
| 8 | .58967 | .80765 | .73010 | .3697 | .2382 | .6959 | .099459 | .42952 | 52 |
| 9 | .58990 | .80748 | .73055 | .3688 | .2384 | .6952 | .099614 | .42897 | 51 |
| 10 | 0.59014 | 0.80730 | 0.73100 | 1.3680 | 1.2387 | 1.6945 | .099769 | .42843 | 50 |
| 11 | .59037 | .80713 | .73144 | .3672 | .2390 | .6939 | .099925 | .42788 | 49 |
| 12 | .59061 | .80696 | .73189 | .3663 | .2392 | .6932 | .10008 | .42734 | 48 |
| 13 | .59084 | .80679 | .73234 | .3655 | .2395 | .6925 | .10024 | .42680 | 47 |
| 14 | .59108 | .80662 | .73278 | .3647 | .2397 | .6918 | .10039 | .42625 | 46 |
| 15 | 0.59131 | 0.80644 | 0.73323 | 1.3638 | 1.2400 | 1.6912 | .10055 | .42571 | 45 |
| 16 | .59154 | .80627 | .73368 | .3630 | .2403 | .6905 | .10070 | .42517 | 44 |
| 17 | .59178 | .80610 | .73413 | .3622 | .2405 | .6898 | .10086 | .42463 | 43 |
| 18 | .59201 | .80593 | .73457 | .3613 | .2408 | .6892 | .10102 | .42409 | 42 |
| 19 | .59225 | .80576 | .73502 | .3605 | .2411 | .6885 | .10118 | .42355 | 41 |
| 20 | 0.59248 | 0.80558 | 0.73547 | 1.3597 | 1.2413 | 1.6878 | .10133 | .42302 | 40 |
| 21 | .59272 | .80541 | .73592 | .3588 | .2416 | .6871 | .10149 | .42248 | 39 |
| 22 | .59295 | .80524 | .73637 | .3580 | .2419 | .6865 | .10165 | .42194 | 38 |
| 23 | .59318 | .80507 | .73681 | .3572 | .2421 | .6858 | .10181 | .42141 | 37 |
| 24 | .59342 | .80489 | .73726 | .3564 | .2424 | .6852 | .10196 | .42087 | 36 |
| 25 | 0.59365 | 0.80472 | 0.73771 | 1.3555 | 1.2427 | 1.6845 | .10212 | .42034 | 35 |
| 26 | .59389 | .80455 | .73816 | .3547 | .2429 | .6838 | .10228 | .41980 | 34 |
| 27 | .59412 | .80438 | .73861 | .3539 | .2432 | .6832 | .10244 | .41927 | 33 |
| 28 | .59436 | .80420 | .73906 | .3531 | .2435 | .6825 | .10260 | .41874 | 32 |
| 29 | .59459 | .80403 | .73951 | .3522 | .2437 | .6818 | .10276 | .41820 | 31 |
| 30 | 0.59482 | 0.80386 | 0.73996 | 1.3514 | 1.2440 | 1.6812 | .10292 | .41767 | 30 |
| 31 | .59506 | .80368 | .74041 | .3506 | .2443 | .6805 | .10308 | .41714 | 29 |
| 32 | .59529 | .80351 | .74086 | .3498 | .2445 | .6799 | .10323 | .41661 | 28 |
| 33 | .59552 | .80334 | .74131 | .3490 | .2448 | .6792 | .10339 | .41608 | 27 |
| 34 | .59576 | .80316 | .74176 | .3481 | .2451 | .6785 | .10355 | .41555 | 26 |
| 35 | 0.59599 | 0.80299 | 0.74221 | 1.3473 | 1.2453 | 1.6779 | .10371 | .41502 | 25 |
| 36 | .59622 | .80282 | .74267 | .3465 | .2456 | .6772 | .10388 | .41450 | 24 |
| 37 | .59646 | .80264 | .74312 | .3457 | .2459 | .6766 | .10404 | .41397 | 23 |
| 38 | .59669 | .80247 | .74357 | .3449 | .2462 | .6759 | .10420 | .41344 | 22 |
| 39 | .59693 | .80230 | .74402 | .3440 | .2464 | .6753 | .10436 | .41292 | 21 |
| 40 | 0.59716 | 0.80212 | 0.74447 | 1.3432 | 1.2467 | 1.6746 | .10452 | .41239 | 20 |
| 41 | .59739 | .80195 | .74492 | .3424 | .2470 | .6739 | .10468 | .41187 | 19 |
| 42 | .59763 | .80178 | .74538 | .3416 | .2472 | .6733 | .10484 | .41134 | 18 |
| 43 | .59786 | .80160 | .74583 | .3408 | .2475 | .6726 | .10500 | .41082 | 17 |
| 44 | .59809 | .80143 | .74628 | .3400 | .2478 | .6720 | .10516 | .41030 | 16 |
| 45 | 0.59832 | 0.80125 | 0.74674 | 1.3392 | 1.2480 | 1.6713 | .10533 | .40977 | 15 |
| 46 | .59856 | .80108 | .74719 | .3384 | .2483 | .6707 | .10549 | .40925 | 14 |
| 47 | .59879 | .80091 | .74764 | .3375 | .2486 | .6700 | .10565 | .40873 | 13 |
| 48 | .59902 | .80073 | .74810 | .3367 | .2489 | .6694 | .10581 | .40821 | 12 |
| 49 | .59926 | .80056 | .74855 | .3359 | .2491 | .6687 | .10598 | .40769 | 11 |
| 50 | 0.59949 | 0.80038 | 0.74900 | 1.3351 | 1.2494 | 1.6681 | .10614 | .40717 | 10 |
| 51 | .59972 | .80021 | .74946 | .3343 | .2497 | .6674 | .10630 | .40666 | 9 |
| 52 | .59995 | .80003 | .74991 | .3335 | .2499 | .6668 | .10647 | .40614 | 8 |
| 53 | .60019 | .79986 | .75037 | .3327 | .2502 | .6661 | .10663 | .40562 | 7 |
| 54 | .60042 | .79968 | .75082 | .3319 | .2505 | .6655 | .10679 | .40511 | 6 |
| 55 | 0.60065 | 0.79951 | 0.75128 | 1.3311 | 1.2508 | 1.6649 | .10696 | .40459 | 5 |
| 56 | .60089 | .79934 | .75173 | .3303 | .2510 | .6642 | .10712 | .40407 | 4 |
| 57 | .60112 | .79916 | .75219 | .3295 | .2513 | .6636 | .10729 | .40356 | 3 |
| 58 | .60135 | .79899 | .75264 | .3287 | .2516 | .6629 | .10745 | .40305 | 2 |
| 59 | .60158 | .79881 | .75310 | .3278 | .2519 | .6623 | .10762 | .40253 | 1 |
| 60 | 0.60182 | 0.79864 | 0.75355 | 1.3270 | 1.2521 | 1.6616 | .10778 | .40202 | 0 |
| M | Cosine | Sine | Cotan. | Tan. | Cosec. | Secant | READ DOWN | 53°–54° Involute | M |

**37° or 217°**    Trigonometric Functions    **142° or 322°**

| M | Sine | Cosine | Tan. | Cotan. | Secant | Cosec. | Involute 37°–38° | READ UP | M |
|---|------|--------|------|--------|--------|--------|------------------|---------|---|
| 0 | 0.60182 | 0.79864 | 0.75355 | 1.3270 | 1.2521 | 1.6616 | .10778 | .40202 | 60 |
| 1 | .60205 | .79846 | .75401 | .3262 | .2524 | .6610 | .10795 | .40151 | 59 |
| 2 | .60228 | .79829 | .75447 | .3254 | .2527 | .6604 | .10811 | .40100 | 58 |
| 3 | .60251 | .79811 | .75492 | .3246 | .2530 | .6597 | .10828 | .40049 | 57 |
| 4 | .60274 | .79793 | .75538 | .3238 | .2532 | .6591 | .10844 | .39998 | 56 |
| 5 | 0.60298 | 0.79776 | 0.75584 | 1.3230 | 1.2535 | 1.6584 | .10861 | .39947 | 55 |
| 6 | .60321 | .79758 | .75629 | .3222 | .2538 | .6578 | .10878 | .39896 | 54 |
| 7 | .60344 | .79741 | .75675 | .3214 | .2541 | .6572 | .10894 | .39845 | 53 |
| 8 | .60367 | .79723 | .75721 | .3206 | .2543 | .6565 | .10911 | .39794 | 52 |
| 9 | .60390 | .79706 | .75767 | .3198 | .2546 | .6559 | .10928 | .39743 | 51 |
| 10 | 0.60414 | 0.79688 | 0.75812 | 1.3190 | 1.2549 | 1.6553 | .10944 | .39693 | 50 |
| 11 | .60437 | .79671 | .75858 | .3182 | .2552 | .6546 | .10961 | .39642 | 49 |
| 12 | .60460 | .79653 | .75904 | .3175 | .2554 | .6540 | .10978 | .39592 | 48 |
| 13 | .60483 | .79635 | .75950 | .3167 | .2557 | .6534 | .10995 | .39541 | 47 |
| 14 | .60506 | .79618 | .75996 | .3159 | .2560 | .6527 | .11011 | .39491 | 46 |
| 15 | 0.60529 | 0.79600 | 0.76042 | 1.3151 | 1.2563 | 1.6521 | .11028 | .39441 | 45 |
| 16 | .60553 | .79583 | .76088 | .3143 | .2566 | .6515 | .11045 | .39390 | 44 |
| 17 | .60576 | .79565 | .76134 | .3135 | .2568 | .6508 | .11062 | .39340 | 43 |
| 18 | .60599 | .79547 | .76180 | .3127 | .2571 | .6502 | .11079 | .39290 | 42 |
| 19 | .60622 | .79530 | .76226 | .3119 | .2574 | .6496 | .11096 | .39240 | 41 |
| 20 | 0.60645 | 0.79512 | 0.76272 | 1.3111 | 1.2577 | 1.6489 | .11113 | .39190 | 40 |
| 21 | .60668 | .79494 | .76318 | .3103 | .2579 | .6483 | .11130 | .39140 | 39 |
| 22 | .60691 | .79477 | .76364 | .3095 | .2582 | .6477 | .11146 | .39090 | 38 |
| 23 | .60714 | .79459 | .76410 | .3087 | .2585 | .6471 | .11163 | .39040 | 37 |
| 24 | .60738 | .79441 | .76456 | .3079 | .2588 | .6464 | .11180 | .38991 | 36 |
| 25 | 0.60761 | 0.79424 | 0.76502 | 1.3072 | 1.2591 | 1.6458 | .11197 | .38941 | 35 |
| 26 | .60784 | .79406 | .76548 | .3064 | .2593 | .6452 | .11215 | .38891 | 34 |
| 27 | .60807 | .79388 | .76594 | .3056 | .2596 | .6446 | .11232 | .38841 | 33 |
| 28 | .60830 | .79371 | .76640 | .3048 | .2599 | .6439 | .11249 | .38792 | 32 |
| 29 | .60853 | .79353 | .76686 | .3040 | .2602 | .6433 | .11266 | .38742 | 31 |
| 30 | 0.60876 | 0.79335 | 0.76733 | 1.3032 | 1.2605 | 1.6427 | .11283 | .38693 | 30 |
| 31 | .60899 | .79318 | .76779 | .3024 | .2608 | .6421 | .11300 | .38643 | 29 |
| 32 | .60922 | .79300 | .76825 | .3017 | .2610 | .6414 | .11317 | .38594 | 28 |
| 33 | .60945 | .79282 | .76871 | .3009 | .2613 | .6408 | .11334 | .38545 | 27 |
| 34 | .60968 | .79264 | .76918 | .3001 | .2616 | .6402 | .11352 | .38496 | 26 |
| 35 | 0.60991 | 0.79247 | 0.76964 | 1.2993 | 1.2619 | 1.6396 | .11369 | .38446 | 25 |
| 36 | .61015 | .79229 | .77010 | .2985 | .2622 | .6390 | .11386 | .38397 | 24 |
| 37 | .61038 | .79211 | .77057 | .2977 | .2624 | .6383 | .11403 | .38348 | 23 |
| 38 | .61061 | .79193 | .77103 | .2970 | .2627 | .6377 | .11421 | .38299 | 22 |
| 39 | .61084 | .79176 | .77149 | .2962 | .2630 | .6371 | .11438 | .38251 | 21 |
| 40 | 0.61107 | 0.79158 | 0.77196 | 1.2954 | 1.2633 | 1.6365 | .11455 | .38202 | 20 |
| 41 | .61130 | .79140 | .77242 | .2946 | .2636 | .6359 | .11473 | .38153 | 19 |
| 42 | .61153 | .79122 | .77289 | .2938 | .2639 | .6353 | .11490 | .38104 | 18 |
| 43 | .61176 | .79105 | .77335 | .2931 | .2641 | .6346 | .11507 | .38055 | 17 |
| 44 | .61199 | .79087 | .77382 | .2923 | .2644 | .6340 | .11525 | .38007 | 16 |
| 45 | 0.61222 | 0.79069 | 0.77428 | 1.2915 | 1.2647 | 1.6334 | .11542 | .37958 | 15 |
| 46 | .61245 | .79051 | .77475 | .2907 | .2650 | .6328 | .11560 | .37910 | 14 |
| 47 | .61268 | .79033 | .77521 | .2900 | .2653 | .6322 | .11577 | .37861 | 13 |
| 48 | .61291 | .79016 | .77568 | .2892 | .2656 | .6316 | .11595 | .37813 | 12 |
| 49 | .61314 | .78998 | .77615 | .2884 | .2659 | .6310 | .11612 | .37765 | 11 |
| 50 | 0.61337 | 0.78980 | 0.77661 | 1.2876 | 1.2661 | 1.6303 | .11630 | .37716 | 10 |
| 51 | .61360 | .78962 | .77708 | .2869 | .2664 | .6297 | .11647 | .37668 | 9 |
| 52 | .61383 | .78944 | .77754 | .2861 | .2667 | .6291 | .11665 | .37620 | 8 |
| 53 | .61406 | .78926 | .77801 | .2853 | .2670 | .6285 | .11682 | .37572 | 7 |
| 54 | .61429 | .78908 | .77848 | .2846 | .2673 | .6279 | .11700 | .37524 | 6 |
| 55 | 0.61451 | 0.78891 | 0.77895 | 1.2838 | 1.2676 | 1.6273 | .11718 | .37476 | 5 |
| 56 | .61474 | .78873 | .77941 | .2830 | .2679 | .6267 | .11735 | .37428 | 4 |
| 57 | .61497 | .78855 | .77988 | .2822 | .2682 | .6261 | .11753 | .37380 | 3 |
| 58 | .61520 | .78837 | .78035 | .2815 | .2684 | .6255 | .11771 | .37332 | 2 |
| 59 | .61543 | .78819 | .78082 | .2807 | .2687 | .6249 | .11788 | .37285 | 1 |
| 60 | 0.61566 | 0.78801 | 0.78129 | 1.2799 | 1.2690 | 1.6243 | .11806 | .37237 | 0 |
| M | Cosine | Sine | Cotan. | Tan. | Cosec. | Secant | 52°–53° Involute | READ DOWN | M |

**127° or 307°**    **52° or 232°**

| M | Sine | Cosine | Tan. | Cotan. | Secant | Cosec. | Involute 38°–39° | READ UP | M |
|---|---|---|---|---|---|---|---|---|---|
| 0 | 0.61566 | 0.78801 | 0.78129 | 1.2799 | 1.2690 | 1.6243 | .11806 | .37237 | 60 |
| 1 | .61589 | .78783 | .78175 | .2792 | .2693 | .6237 | .11824 | .37189 | 59 |
| 2 | .61612 | .78765 | .78222 | .2784 | .2696 | .6231 | .11842 | .37142 | 58 |
| 3 | .61635 | .78747 | .78269 | .2776 | .2699 | .6225 | .11859 | .37094 | 57 |
| 4 | .61658 | .78729 | .78316 | .2769 | .2702 | .6219 | .11877 | .37047 | 56 |
| 5 | 0.61681 | 0.78711 | 0.78363 | 1.2761 | 1.2705 | 1.6213 | .11895 | .36999 | 55 |
| 6 | .61704 | .78694 | .78410 | .2753 | .2708 | .6207 | .11913 | .36952 | 54 |
| 7 | .61726 | .78676 | .78457 | .2746 | .2710 | .6201 | .11931 | .36905 | 53 |
| 8 | .61749 | .78658 | .78504 | .2738 | .2713 | .6195 | .11949 | .36858 | 52 |
| 9 | .61772 | .78640 | .78551 | .2731 | .2716 | .6189 | .11967 | .36810 | 51 |
| 10 | 0.61795 | 0.78622 | 0.78598 | 1.2723 | 1.2719 | 1.6183 | .11985 | .36763 | 50 |
| 11 | .61818 | .78604 | .78645 | .2715 | .2722 | .6177 | .12003 | .36716 | 49 |
| 12 | .61841 | .78586 | .78692 | .2708 | .2725 | .6171 | .12021 | .36669 | 48 |
| 13 | .61864 | .78563 | .78739 | .2700 | .2728 | .6165 | .12039 | .36622 | 47 |
| 14 | .61887 | .78550 | .78786 | .2693 | .2731 | .6159 | .12057 | .36575 | 46 |
| 15 | 0.61909 | 0.78532 | 0.78834 | 1.2685 | 1.2734 | 1.6153 | .12075 | .36529 | 45 |
| 16 | .61932 | .78514 | .78881 | .2677 | .2737 | .6147 | .12093 | .36482 | 44 |
| 17 | .61955 | .78496 | .78928 | .2670 | .2740 | .6141 | .12111 | .36435 | 43 |
| 18 | .61978 | .78478 | .78975 | .2662 | .2742 | .6135 | .12129 | .36388 | 42 |
| 19 | .62001 | .78460 | .79022 | .2655 | .2745 | .6129 | .12147 | .36342 | 41 |
| 20 | 0.62024 | 0.78442 | 0.79070 | 1.2647 | 1.2748 | 1.6123 | .12165 | .36295 | 40 |
| 21 | .62046 | .78424 | .79117 | .2640 | .2751 | .6117 | .12184 | .36249 | 39 |
| 22 | .62069 | .78405 | .79164 | .2632 | .2754 | .6111 | .12202 | .36202 | 38 |
| 23 | .62092 | .78387 | .79212 | .2624 | .2757 | .6105 | .12220 | .36156 | 37 |
| 24 | .62115 | .78369 | .79259 | .2617 | .2760 | .6099 | .12238 | .36110 | 36 |
| 25 | 0.62138 | 0.78351 | 0.79306 | 1.2609 | 1.2763 | 1.6093 | .12257 | .36063 | 35 |
| 26 | .62160 | .78333 | .79354 | .2602 | .2766 | .6087 | .12275 | .36017 | 34 |
| 27 | .62183 | .78315 | .79401 | .2594 | .2769 | .6082 | .12293 | .35971 | 33 |
| 28 | .62206 | .78297 | .79449 | .2587 | .2772 | .6076 | .12312 | .35925 | 32 |
| 29 | .62229 | .78279 | .79496 | .2579 | .2775 | .6070 | .12330 | .35879 | 31 |
| 30 | 0.62251 | 0.78261 | 0.79544 | 1.2572 | 1.2778 | 1.6064 | .12348 | .35833 | 30 |
| 31 | .62274 | .78243 | .79591 | .2564 | .2781 | .6058 | .12367 | .35787 | 29 |
| 32 | .62297 | .78225 | .79639 | .2557 | .2784 | .6052 | .12385 | .35741 | 28 |
| 33 | .62320 | .78206 | .79686 | .2549 | .2787 | .6046 | .12404 | .35695 | 27 |
| 34 | .62342 | .78188 | .79734 | .2542 | .2790 | .6040 | .12422 | .35649 | 26 |
| 35 | 0.62365 | 0.78170 | 0.79781 | 1.2534 | 1.2793 | 1.6035 | .12441 | .35604 | 25 |
| 36 | .62388 | .78152 | .79829 | .2527 | .2796 | .6029 | .12459 | .35558 | 24 |
| 37 | .62411 | .78134 | .79877 | .2519 | .2799 | .6023 | .12478 | .35512 | 23 |
| 38 | .62433 | .78116 | .79924 | .2512 | .2802 | .6017 | .12496 | .35467 | 22 |
| 39 | .62456 | .78098 | .79972 | .2504 | .2804 | .6011 | .12515 | .35421 | 21 |
| 40 | 0.62479 | 0.78079 | 0.80020 | 1.2497 | 1.2807 | 1.6005 | .12534 | .35376 | 20 |
| 41 | .62502 | .78061 | .80067 | .2489 | .2810 | .6000 | .12552 | .35330 | 19 |
| 42 | .62524 | .78043 | .80115 | .2482 | .2813 | .5994 | .12571 | .35285 | 18 |
| 43 | .62547 | .78025 | .80163 | .2475 | .2816 | .5988 | .12590 | .35240 | 17 |
| 44 | .62570 | .78007 | .80211 | .2467 | .2819 | .5982 | .12608 | .35194 | 16 |
| 45 | 0.62592 | 0.77988 | 0.80258 | 1.2460 | 1.2822 | 1.5976 | .12627 | .35149 | 15 |
| 46 | .62615 | .77970 | .80306 | .2452 | .2825 | .5971 | .12646 | .35104 | 14 |
| 47 | .62638 | .77952 | .80354 | .2445 | .2828 | .5965 | .12664 | .35059 | 13 |
| 48 | .62660 | .77934 | .80402 | .2437 | .2831 | .5959 | .12683 | .35014 | 12 |
| 49 | .62683 | .77916 | .80450 | .2430 | .2834 | .5953 | .12702 | .34969 | 11 |
| 50 | 0.62706 | 0.77897 | 0.80498 | 1.2423 | 1.2837 | 1.5948 | .12721 | .34924 | 10 |
| 51 | .62728 | .77879 | .80546 | .2415 | .2840 | .5942 | .12740 | .34879 | 9 |
| 52 | .62751 | .77861 | .80594 | .2408 | .2843 | .5936 | .12759 | .34834 | 8 |
| 53 | .62774 | .77843 | .80642 | .2401 | .2846 | .5930 | .12778 | .34790 | 7 |
| 54 | .62796 | .77824 | .80690 | .2393 | .2849 | .5925 | .12797 | .34745 | 6 |
| 55 | 0.62819 | 0.77806 | 0.80738 | 1.2386 | 1.2852 | 1.5919 | .12815 | .34700 | 5 |
| 56 | .62842 | .77788 | .80786 | .2378 | .2855 | .5913 | .12834 | .34656 | 4 |
| 57 | .62864 | .77769 | .80834 | .2371 | .2859 | .5907 | .12853 | .34611 | 3 |
| 58 | .62887 | .77751 | .80882 | .2364 | .2862 | .5902 | .12872 | .34567 | 2 |
| 59 | .62909 | .77733 | .80930 | .2356 | .2865 | .5896 | .12891 | .34522 | 1 |
| 60 | 0.62932 | 0.77715 | 0.80978 | 1.2349 | 1.2868 | 1.5890 | .12911 | .34478 | 0 |
| M | Cosine | Sine | Cotan. | Tan. | Cosec. | Secant | READ DOWN | 51°–52° Involute | M |

**39° or 219°**  Trigonometric Functions  **140° or 320°**

| M | Sine | Cosine | Tan. | Cotan. | Secant | Cosec. | Involute 39°–40° | READ UP | M |
|---|------|--------|------|--------|--------|--------|-------------------|---------|---|
| 0 | 0.62932 | 0.77715 | 0.80978 | 1.2349 | 1.2868 | 1.5890 | .12911 | .34478 | 60 |
| 1 | .62955 | .77696 | .81027 | .2342 | .2871 | .5884 | .12930 | .34434 | 59 |
| 2 | .62977 | .77678 | .81075 | .2334 | .2874 | .5879 | .12949 | .34389 | 58 |
| 3 | .63000 | .77660 | .81123 | .2327 | .2877 | .5873 | .12968 | .34345 | 57 |
| 4 | .63022 | .77641 | .81171 | .2320 | .2880 | .5867 | .12987 | .34301 | 56 |
| 5 | 0.63045 | 0.77623 | 0.81220 | 1.2312 | 1.2883 | 1.5862 | .13006 | .34257 | 55 |
| 6 | .63068 | .77605 | .81268 | .2305 | .2886 | .5856 | .13025 | .34213 | 54 |
| 7 | .63090 | .77586 | .81316 | .2298 | .2889 | .5850 | .13045 | .34169 | 53 |
| 8 | .63113 | .77568 | .81364 | .2290 | .2892 | .5845 | .13064 | .34125 | 52 |
| 9 | .63135 | .77550 | .81413 | .2283 | .2895 | .5839 | .13083 | .34081 | 51 |
| 10 | 0.63158 | 0.77531 | 0.81461 | 1.2276 | 1.2898 | 1.5833 | .13102 | .34037 | 50 |
| 11 | .63180 | .77513 | .81510 | .2268 | .2901 | .5828 | .13122 | .33993 | 49 |
| 12 | .63203 | .77494 | .81558 | .2261 | .2904 | .5822 | .13141 | .33949 | 48 |
| 13 | .63225 | .77476 | .81606 | .2254 | .2907 | .5816 | .13160 | .33906 | 47 |
| 14 | .63248 | .77458 | .81655 | .2247 | .2910 | .5811 | .13180 | .33862 | 46 |
| 15 | 0.63271 | 0.77439 | 0.81703 | 1.2239 | 1.2913 | 1.5805 | .13199 | .33818 | 45 |
| 16 | .63293 | .77421 | .81752 | .2232 | .2916 | .5799 | .13219 | .33775 | 44 |
| 17 | .63316 | .77402 | .81800 | .2225 | .2919 | .5794 | .13238 | .33731 | 43 |
| 18 | .63338 | .77384 | .81849 | .2218 | .2923 | .5788 | .13258 | .33688 | 42 |
| 19 | .63361 | .77366 | .81898 | .2210 | .2926 | .5783 | .13277 | .33645 | 41 |
| 20 | 0.63383 | 0.77347 | 0.81946 | 1.2203 | 1.2929 | 1.5777 | .13297 | .33601 | 40 |
| 21 | .63406 | .77329 | .81995 | .2196 | .2932 | .5771 | .13316 | .33558 | 39 |
| 22 | .63428 | .77310 | .82044 | .2189 | .2935 | .5766 | .13336 | .33515 | 38 |
| 23 | .63451 | .77292 | .82092 | .2181 | .2938 | .5760 | .13355 | .33471 | 37 |
| 24 | .63473 | .77273 | .82141 | .2174 | .2941 | .5755 | .13375 | .33428 | 36 |
| 25 | 0.63496 | 0.77255 | 0.82190 | 1.2167 | 1.2944 | 1.5749 | .13395 | .33385 | 35 |
| 26 | .63518 | .77236 | .82238 | .2160 | .2947 | .5744 | .13414 | .33342 | 34 |
| 27 | .63541 | .77218 | .82287 | .2153 | .2950 | .5738 | .13434 | .33299 | 33 |
| 28 | .63563 | .77199 | .82336 | .2145 | .2953 | .5732 | .13454 | .33256 | 32 |
| 29 | .63585 | .77181 | .82385 | .2138 | .2957 | .5727 | .13473 | .33213 | 31 |
| 30 | 0.63608 | 0.77162 | 0.82434 | 1.2131 | 1.2960 | 1.5721 | .13493 | .33171 | 30 |
| 31 | .63630 | .77144 | .82483 | .2124 | .2963 | .5716 | .13513 | .33128 | 29 |
| 32 | .63653 | .77125 | .82531 | .2117 | .2966 | .5710 | .13533 | .33085 | 28 |
| 33 | .63675 | .77107 | .82580 | .2109 | .2969 | .5705 | .13553 | .33042 | 27 |
| 34 | .63698 | .77088 | .82629 | .2102 | .2972 | .5699 | .13572 | .33000 | 26 |
| 35 | 0.63720 | 0.77070 | 0.82678 | 1.2095 | 1.2975 | 1.5694 | .13592 | .32957 | 25 |
| 36 | .63742 | .77051 | .82727 | .2088 | .2978 | .5688 | .13612 | .32915 | 24 |
| 37 | .63765 | .77033 | .82776 | .2081 | .2981 | .5683 | .13632 | .32872 | 23 |
| 38 | .63787 | .77014 | .82825 | .2074 | .2985 | .5677 | .13652 | .32830 | 22 |
| 39 | .63810 | .76996 | .82874 | .2066 | .2988 | .5672 | .13672 | .32787 | 21 |
| 40 | 0.63832 | 0.76977 | 0.82923 | 1.2059 | 1.2991 | 1.5666 | .13692 | .32745 | 20 |
| 41 | .63854 | .76959 | .82972 | .2052 | .2994 | .5661 | .13712 | .32703 | 19 |
| 42 | .63877 | .76940 | .83022 | .2045 | .2997 | .5655 | .13732 | .32661 | 18 |
| 43 | .63899 | .76921 | .83071 | .2038 | .3000 | .5650 | .13752 | .32618 | 17 |
| 44 | .63922 | .76903 | .83120 | .2031 | .3003 | .5644 | .13772 | .32576 | 16 |
| 45 | 0.63944 | 0.76884 | 0.83169 | 1.2024 | 1.3007 | 1.5639 | .13792 | .32534 | 15 |
| 46 | .63966 | .76866 | .83218 | .2017 | .3010 | .5633 | .13812 | .32492 | 14 |
| 47 | .63989 | .76847 | .83268 | .2009 | .3013 | .5628 | .13833 | .32450 | 13 |
| 48 | .64011 | .76828 | .83317 | .2002 | .3016 | .5622 | .13853 | .32408 | 12 |
| 49 | .64033 | .76810 | .83366 | .1995 | .3019 | .5617 | .13873 | .32366 | 11 |
| 50 | 0.64056 | 0.76791 | 0.83415 | 1.1988 | 1.3022 | 1.5611 | .13893 | .32324 | 10 |
| 51 | .64078 | .76772 | .83465 | .1981 | .3026 | .5606 | .13913 | .32283 | 9 |
| 52 | .64100 | .76754 | .83514 | .1974 | .3029 | .5601 | .13934 | .32241 | 8 |
| 53 | .64123 | .76735 | .83564 | .1967 | .3032 | .5595 | .13954 | .32199 | 7 |
| 54 | .64145 | .76717 | .83613 | .1960 | .3035 | .5590 | .13974 | .32158 | 6 |
| 55 | 0.64167 | 0.76698 | 0.83662 | 1.1953 | 1.3038 | 1.5584 | .13995 | .32116 | 5 |
| 56 | .64190 | .76679 | .83712 | .1946 | .3041 | .5579 | .14015 | .32075 | 4 |
| 57 | .64212 | .76661 | .83761 | .1939 | .3045 | .5573 | .14035 | .32033 | 3 |
| 58 | .64234 | .76642 | .83811 | .1932 | .3048 | .5568 | .14056 | .31992 | 2 |
| 59 | .64256 | .76623 | .83860 | .1925 | .3051 | .5563 | .14076 | .31950 | 1 |
| 60 | 0.64279 | 0.76604 | 0.83910 | 1.1918 | 1.3054 | 1.5557 | .14097 | .31909 | 0 |

| M | Cosine | Sine | Cotan. | Tan. | Cosec. | Secant | READ DOWN | 50°–51° Involute | M |
|---|--------|------|--------|------|--------|--------|-----------|------------------|---|

**129° or 309°**  **50° or 230°**

**40° or 220°**    Trigonometric Functions    **139° or 319°**

| M | Sine | Cosine | Tan. | Cotan. | Secant | Cosec. | Involute 40°–41° | READ UP | M |
|---|---|---|---|---|---|---|---|---|---|
| 0 | 0.64279 | 0.76604 | 0.83910 | 1.1918 | 1.3054 | 1.5557 | .14097 | .31909 | 60 |
| 1 | .64301 | .76586 | .83960 | .1910 | .3057 | .5552 | .14117 | .31868 | 59 |
| 2 | .64323 | .76567 | .84009 | .1903 | .3060 | .5546 | .14138 | .31826 | 58 |
| 3 | .64346 | .76548 | .84059 | .1896 | .3064 | .5541 | .14158 | .31785 | 57 |
| 4 | .64368 | .76530 | .84108 | .1889 | .3067 | .5536 | .14179 | .31744 | 56 |
| 5 | 0.64390 | 0.76511 | 0.84158 | 1.1882 | 1.3070 | 1.5530 | .14200 | .31703 | 55 |
| 6 | .64412 | .76492 | .84208 | .1875 | .3073 | .5525 | .14220 | .31662 | 54 |
| 7 | .64435 | .76473 | .84258 | .1868 | .3076 | .5520 | .14241 | .31621 | 53 |
| 8 | .64457 | .76455 | .84307 | .1861 | .3080 | .5514 | .14261 | .31580 | 52 |
| 9 | .64479 | .76436 | .84357 | .1854 | .3083 | .5509 | .14282 | .31539 | 51 |
| 10 | 0.64501 | 0.76417 | 0.84407 | 1.1847 | 1.3086 | 1.5504 | .14303 | .31498 | 50 |
| 11 | .64524 | .76398 | .84457 | .1840 | .3089 | .5498 | .14324 | .31457 | 49 |
| 12 | .64546 | .76380 | .84507 | .1833 | .3093 | .5493 | .14344 | .31417 | 48 |
| 13 | .64568 | .76361 | .84556 | .1826 | .3096 | .5488 | .14365 | .31376 | 47 |
| 14 | .64590 | .76342 | .84606 | .1819 | .3099 | .5482 | .14386 | .31335 | 46 |
| 15 | 0.64612 | 0.76323 | 0.84656 | 1.1812 | 1.3102 | 1.5477 | .14407 | .31295 | 45 |
| 16 | .64635 | .76304 | .84706 | .1806 | .3105 | .5472 | .14428 | .31254 | 44 |
| 17 | .64657 | .76286 | .84756 | .1799 | .3109 | .5466 | .14448 | .31214 | 43 |
| 18 | .64679 | .76267 | .84806 | .1792 | .3112 | .5461 | .14469 | .31173 | 42 |
| 19 | .64701 | .76248 | .84856 | .1785 | .3115 | .5456 | .14490 | .31133 | 41 |
| 20 | 0.64723 | 0.76229 | 0.84906 | 1.1778 | 1.3118 | 1.5450 | .14511 | .31092 | 40 |
| 21 | .64746 | .76210 | .84956 | .1771 | .3122 | .5445 | .14532 | .31052 | 39 |
| 22 | .64768 | .76192 | .85006 | .1764 | .3125 | .5440 | .14553 | .31012 | 38 |
| 23 | .64790 | .76173 | .85057 | .1757 | .3128 | .5435 | .14574 | .30971 | 37 |
| 24 | .64812 | .76154 | .85107 | .1750 | .3131 | .5429 | .14595 | .30931 | 36 |
| 25 | 0.64834 | 0.76135 | 0.85157 | 1.1743 | 1.3135 | 1.5424 | .14616 | .30891 | 35 |
| 26 | .64856 | .76116 | .85207 | .1736 | .3138 | .5419 | .14638 | .30851 | 34 |
| 27 | .64878 | .76097 | .85257 | .1729 | .3141 | .5413 | .14659 | .30811 | 33 |
| 28 | .64901 | .76078 | .85308 | .1722 | .3144 | .5408 | .14680 | .30771 | 32 |
| 29 | .64923 | .76059 | .85358 | .1715 | .3148 | .5403 | .14701 | .30731 | 31 |
| 30 | 0.64945 | 0.76041 | 0.85408 | 1.1708 | 1.3151 | 1.5398 | .14722 | .30691 | 30 |
| 31 | .64967 | .76022 | .85458 | .1702 | .3154 | .5392 | .14743 | .30651 | 29 |
| 32 | .64989 | .76003 | .85509 | .1695 | .3157 | .5387 | .14765 | .30611 | 28 |
| 33 | .65011 | .75984 | .85559 | .1688 | .3161 | .5382 | .14786 | .30572 | 27 |
| 34 | .65033 | .75965 | .85609 | .1681 | .3164 | .5377 | .14807 | .30532 | 26 |
| 35 | 0.65055 | 0.75946 | 0.85660 | 1.1674 | 1.3167 | 1.5372 | .14829 | .30492 | 25 |
| 36 | .65077 | .75927 | .85710 | .1667 | .3171 | .5366 | .14850 | .30453 | 24 |
| 37 | .65100 | .75908 | .85761 | .1660 | .3174 | .5361 | .14871 | .30413 | 23 |
| 38 | .65122 | .75889 | .85811 | .1653 | .3177 | .5356 | .14893 | .30374 | 22 |
| 39 | .65144 | .75870 | .85862 | .1647 | .3180 | .5351 | .14914 | .30334 | 21 |
| 40 | 0.65166 | 0.75851 | 0.85912 | 1.1640 | 1.3184 | 1.5345 | .14936 | .30295 | 20 |
| 41 | .65188 | .75832 | .85963 | .1633 | .3187 | .5340 | .14957 | .30255 | 19 |
| 42 | .65210 | .75813 | .86014 | .1626 | .3190 | .5335 | .14979 | .30216 | 18 |
| 43 | .65232 | .75794 | .86064 | .1619 | .3194 | .5330 | .15000 | .30177 | 17 |
| 44 | .65254 | .75775 | .86115 | .1612 | .3197 | .5325 | .15022 | .30137 | 16 |
| 45 | 0.65276 | 0.75756 | 0.86166 | 1.1606 | 1.3200 | 1.5320 | .15043 | .30098 | 15 |
| 46 | .65298 | .75738 | .86216 | .1599 | .3203 | .5314 | .15065 | .30059 | 14 |
| 47 | .65320 | .75719 | .86267 | .1592 | .3207 | .5309 | .15087 | .30020 | 13 |
| 48 | .65342 | .75700 | .86318 | .1585 | .3210 | .5304 | .15108 | .29981 | 12 |
| 49 | .65364 | .75680 | .86368 | .1578 | .3213 | .5299 | .15130 | .29942 | 11 |
| 50 | 0.65386 | 0.75661 | 0.86419 | 1.1571 | 1.3217 | 1.5294 | .15152 | .29903 | 10 |
| 51 | .65408 | .75642 | .86470 | .1565 | .3220 | .5289 | .15173 | .29864 | 9 |
| 52 | .65430 | .75623 | .86521 | .1558 | .3223 | .5283 | .15195 | .29825 | 8 |
| 53 | .65452 | .75604 | .86572 | .1551 | .3227 | .5278 | .15217 | .29786 | 7 |
| 54 | .65474 | .75585 | .86623 | .1544 | .3230 | .5273 | .15239 | .29747 | 6 |
| 55 | 0.65496 | 0.75566 | 0.86674 | 1.1538 | 1.3233 | 1.5268 | .15261 | .29709 | 5 |
| 56 | .65518 | .75547 | .86725 | .1531 | .3237 | .5263 | .15282 | .29670 | 4 |
| 57 | .65540 | .75528 | .86776 | .1524 | .3240 | .5258 | .15304 | .29631 | 3 |
| 58 | .65562 | .75509 | .86827 | .1517 | .3243 | .5253 | .15326 | .29593 | 2 |
| 59 | .65584 | .75490 | .86878 | .1510 | .3247 | .5248 | .15348 | .29554 | 1 |
| 60 | 0.65606 | 0.75471 | 0.86929 | 1.1504 | 1.3250 | 1.5243 | .15370 | .29516 | 0 |
| M | Cosine | Sine | Cotan. | Tan. | Cosec. | Secant | READ DOWN | 49°–50° Involute | M |

**130° or 310°**    **49° or 229°**

**41° or 221°**  Trigonometric Functions  **138° or 318°**

| M | Sine | Cosine | Tan. | Cotan. | Secant | Cosec. | Involute 41°–42° | READ UP | M |
|---|---|---|---|---|---|---|---|---|---|
| 0 | 0.65606 | 0.75471 | 0.86929 | 1.1504 | 1.3250 | 1.5243 | .15370 | .29516 | 60 |
| 1 | .65628 | .75452 | .86980 | .1497 | .3253 | .5237 | .15392 | .29477 | 59 |
| 2 | .65650 | .75433 | .87031 | .1490 | .3257 | .5232 | .15414 | .29439 | 58 |
| 3 | .65672 | .75414 | .87082 | .1483 | .3260 | .5227 | .15436 | .29400 | 57 |
| 4 | .65694 | .75395 | .87133 | .1477 | .3264 | .5222 | .15458 | .29362 | 56 |
| 5 | 0.65716 | 0.75375 | 0.87184 | 1.1470 | 1.3267 | 1.5217 | .15480 | .29324 | 55 |
| 6 | .65738 | .75356 | .87236 | .1463 | .3270 | .5212 | .15503 | .29286 | 54 |
| 7 | .65759 | .75337 | .87287 | .1456 | .3274 | .5207 | .15525 | .29247 | 53 |
| 8 | .65781 | .75318 | .87338 | .1450 | .3277 | .5202 | .15547 | .29209 | 52 |
| 9 | .65803 | .75299 | .87389 | .1443 | .3280 | .5197 | .15569 | .29171 | 51 |
| 10 | 0.65825 | 0.75280 | 0.87441 | 1.1436 | 1.3284 | 1.5192 | .15591 | .29133 | 50 |
| 11 | .65847 | .75261 | .87492 | .1430 | .3287 | .5187 | .15614 | .29095 | 49 |
| 12 | .65869 | .75241 | .87543 | .1423 | .3291 | .5182 | .15636 | .29057 | 48 |
| 13 | .65891 | .75222 | .87595 | .1416 | .3294 | .5177 | .15658 | .29019 | 47 |
| 14 | .65913 | .75203 | .87646 | .1410 | .3297 | .5172 | .15680 | .28981 | 46 |
| 15 | 0.65935 | 0.75184 | 0.87698 | 1.1403 | 1.3301 | 1.5167 | .15703 | .28943 | 45 |
| 16 | .65956 | .75165 | .87749 | .1396 | .3304 | .5162 | .15725 | .28906 | 44 |
| 17 | .65978 | .75146 | .87801 | .1389 | .3307 | .5156 | .15748 | .28868 | 43 |
| 18 | .66000 | .75126 | .87852 | .1383 | .3311 | .5151 | .15770 | .28830 | 42 |
| 19 | .66022 | .75107 | .87904 | .1376 | .3314 | .5146 | .15793 | .28792 | 41 |
| 20 | 0.66044 | 0.75088 | 0.87955 | 1.1369 | 1.3318 | 1.5141 | .15815 | .28755 | 40 |
| 21 | .66066 | .75069 | .88007 | .1363 | .3321 | .5136 | .15838 | .28717 | 39 |
| 22 | .66088 | .75050 | .88059 | .1356 | .3325 | .5131 | .15860 | .28680 | 38 |
| 23 | .66109 | .75030 | .88110 | .1349 | .3328 | .5126 | .15883 | .28642 | 37 |
| 24 | .66131 | .75011 | .88162 | .1343 | .3331 | .5121 | .15905 | .28605 | 36 |
| 25 | 0.66153 | 0.74992 | 0.88214 | 1.1336 | 1.3335 | 1.5116 | .15928 | .28567 | 35 |
| 26 | .66175 | .74973 | .88265 | .1329 | .3338 | .5111 | .15950 | .28530 | 34 |
| 27 | .66197 | .74953 | .88317 | .1323 | .3342 | .5107 | .15973 | .28493 | 33 |
| 28 | .66218 | .74934 | .88369 | .1316 | .3345 | .5102 | .15996 | .28455 | 32 |
| 29 | .66240 | .74915 | .88421 | .1310 | .3348 | .5097 | .16019 | .28418 | 31 |
| 30 | 0.66262 | 0.74896 | 0.88473 | 1.1303 | 1.3352 | 1.5092 | .16041 | .28381 | 30 |
| 31 | .66284 | .74876 | .88524 | .1296 | .3355 | .5087 | .16064 | .28344 | 29 |
| 32 | .66306 | .74857 | .88576 | .1290 | .3359 | .5082 | .16087 | .28307 | 28 |
| 33 | .66327 | .74838 | .88628 | .1283 | .3362 | .5077 | .16110 | .28270 | 27 |
| 34 | .66349 | .74818 | .88680 | .1276 | .3366 | .5072 | .16133 | .28233 | 26 |
| 35 | 0.66371 | 0.74799 | 0.88732 | 1.1270 | 1.3369 | 1.5067 | .16156 | .28196 | 25 |
| 36 | .66393 | .74780 | .88784 | .1263 | .3373 | .5062 | .16178 | .28159 | 24 |
| 37 | .66414 | .74760 | .88836 | .1257 | .3376 | .5057 | .16201 | .28122 | 23 |
| 38 | .66436 | .74741 | .88888 | .1250 | .3380 | .5052 | .16224 | .28085 | 22 |
| 39 | .66458 | .74722 | .88940 | .1243 | .3383 | .5047 | .16247 | .28048 | 21 |
| 40 | 0.66480 | 0.74703 | 0.88992 | 1.1237 | 1.3386 | 1.5042 | .16270 | .28012 | 20 |
| 41 | .66501 | .74683 | .89045 | .1230 | .3390 | .5037 | .16293 | .27975 | 19 |
| 42 | .66523 | .74664 | .89097 | .1224 | .3393 | .5032 | .16317 | .27938 | 18 |
| 43 | .66545 | .74644 | .89149 | .1217 | .3397 | .5027 | .16340 | .27902 | 17 |
| 44 | .66566 | .74625 | .89201 | .1211 | .3400 | .5023 | .16363 | .27865 | 16 |
| 45 | 0.66588 | 0.74606 | 0.89253 | 1.1204 | 1.3404 | 1.5018 | .16386 | .27828 | 15 |
| 46 | .66610 | .74586 | .89306 | .1197 | .3407 | .5013 | .16409 | .27792 | 14 |
| 47 | .66632 | .74567 | .89358 | .1191 | .3411 | .5008 | .16432 | .27755 | 13 |
| 48 | .66653 | .74548 | .89410 | .1184 | .3414 | .5003 | .16456 | .27719 | 12 |
| 49 | .66675 | .74528 | .89463 | .1178 | .3418 | .4998 | .16479 | .27683 | 11 |
| 50 | 0.66697 | 0.74509 | 0.89515 | 1.1171 | 1.3421 | 1.4993 | .16502 | .27646 | 10 |
| 51 | .66718 | .74489 | .89567 | .1165 | .3425 | .4988 | .16525 | .27610 | 9 |
| 52 | .66740 | .74470 | .89620 | .1158 | .3428 | .4984 | .16549 | .27574 | 8 |
| 53 | .66762 | .74451 | .89672 | .1152 | .3432 | .4979 | .16572 | .27538 | 7 |
| 54 | .66783 | .74431 | .89725 | .1145 | .3435 | .4974 | .16596 | .27501 | 6 |
| 55 | 0.66805 | 0.74412 | 0.89777 | 1.1139 | 1.3439 | 1.4969 | .16619 | .27465 | 5 |
| 56 | .66827 | .74392 | .89830 | .1132 | .3442 | .4964 | .16642 | .27429 | 4 |
| 57 | .66848 | .74373 | .89883 | .1126 | .3446 | .4959 | .16666 | .27393 | 3 |
| 58 | .66870 | .74353 | .89935 | .1119 | .3449 | .4954 | .16689 | .27357 | 2 |
| 59 | .66891 | .74334 | .89988 | .1113 | .3453 | .4950 | .16713 | .27321 | 1 |
| 60 | 0.66913 | 0.74314 | 0.90040 | 1.1106 | 1.3456 | 1.4945 | .16737 | .27285 | 0 |
| M | Cosine | Sine | Cotan. | Tan. | Cosec. | Secant | READ DOWN | 48°–49° Involute | M |

**131° or 311°**  **48° or 228°**

| M | Sine | Cosine | Tan. | Cotan. | Secant | Cosec. | Involute 42°–43° | READ UP | M |
|---|------|--------|------|--------|--------|--------|-------------------|---------|---|
| 0 | 0.66913 | 0.74314 | 0.90040 | 1.1106 | 1.3456 | 1.4945 | .16737 | .27285 | 60 |
| 1 | .66935 | .74295 | .90093 | .1100 | .3460 | .4940 | .16760 | .27250 | 59 |
| 2 | .66956 | .74276 | .90146 | .1093 | .3463 | .4935 | .16784 | .27214 | 58 |
| 3 | .66978 | .74256 | .90199 | .1087 | .3467 | .4930 | .16807 | .27178 | 57 |
| 4 | .66999 | .74237 | .90251 | .1080 | .3470 | .4925 | .16831 | .27142 | 56 |
| 5 | 0.67021 | 0.74217 | 0.90304 | 1.1074 | 1.3474 | 1.4921 | .16855 | .27107 | 55 |
| 6 | .67043 | .74198 | .90357 | .1067 | .3478 | .4916 | .16879 | .27071 | 54 |
| 7 | .67064 | .74178 | .90410 | .1061 | .3481 | .4911 | .16902 | .27035 | 53 |
| 8 | .67086 | .74159 | .90463 | .1054 | .3485 | .4906 | .16926 | .27000 | 52 |
| 9 | .67107 | .74139 | .90516 | .1048 | .3488 | .4901 | .16950 | .26964 | 51 |
| 10 | 0.67129 | 0.74120 | 0.90569 | 1.1041 | 1.3492 | 1.4897 | .16974 | .26929 | 50 |
| 11 | .67151 | .74100 | .90621 | .1035 | .3495 | .4892 | .16998 | .26893 | 49 |
| 12 | .67172 | .74080 | .90674 | .1028 | .3499 | .4887 | .17022 | .26858 | 48 |
| 13 | .67194 | .74061 | .90727 | .1022 | .3502 | .4882 | .17045 | .26823 | 47 |
| 14 | .67215 | .74041 | .90781 | .1016 | .3506 | .4878 | .17069 | .26787 | 46 |
| 15 | 0.67237 | 0.74022 | 0.90834 | 1.1009 | 1.3510 | 1.4873 | .17093 | .26752 | 45 |
| 16 | .67258 | .74002 | .90887 | .1003 | .3513 | .4868 | .17117 | .26717 | 44 |
| 17 | .67280 | .73983 | .90940 | .0996 | .3517 | .4863 | .17142 | .26682 | 43 |
| 18 | .67301 | .73963 | .90993 | .0990 | .3520 | .4859 | .17166 | .26646 | 42 |
| 19 | .67323 | .73944 | .91046 | .0983 | .3524 | .4854 | .17190 | .26611 | 41 |
| 20 | 0.67344 | 0.73924 | 0.91099 | 1.0977 | 1.3527 | 1.4849 | .17214 | .26576 | 40 |
| 21 | .67366 | .73904 | .91153 | .0971 | .3531 | .4844 | .17238 | .26541 | 39 |
| 22 | .67387 | .73885 | .91206 | .0964 | .3535 | .4840 | .17262 | .26506 | 38 |
| 23 | .67409 | .73865 | .91259 | .0958 | .3538 | .4835 | .17286 | .26471 | 37 |
| 24 | .67430 | .73846 | .91313 | .0951 | .3542 | .4830 | .17311 | .26436 | 36 |
| 25 | 0.67452 | 0.73826 | 0.91366 | 1.0945 | 1.3545 | 1.4825 | .17335 | .26401 | 35 |
| 26 | .67473 | .73806 | .91419 | .0939 | .3549 | .4821 | .17359 | .26367 | 34 |
| 27 | .67495 | .73787 | .91473 | .0932 | .3553 | .4816 | .17383 | .26332 | 33 |
| 28 | .67516 | .73767 | .91526 | .0926 | .3556 | .4811 | .17408 | .26297 | 32 |
| 29 | .67538 | .73747 | .91580 | .0919 | .3560 | .4807 | .17432 | .26262 | 31 |
| 30 | 0.67559 | 0.73728 | 0.91633 | 1.0913 | 1.3563 | 1.4802 | .17457 | .26228 | 30 |
| 31 | .67580 | .73708 | .91687 | .0907 | .3567 | .4797 | .17481 | .26193 | 29 |
| 32 | .67602 | .73688 | .91740 | .0900 | .3571 | .4792 | .17506 | .26159 | 28 |
| 33 | .67623 | .73669 | .91794 | .0894 | .3574 | .4788 | .17530 | .26124 | 27 |
| 34 | .67645 | .73649 | .91847 | .0888 | .3578 | .4783 | .17555 | .26089 | 26 |
| 35 | 0.67666 | 0.73629 | 0.91901 | 1.0881 | 1.3582 | 1.4778 | .17579 | .26055 | 25 |
| 36 | .67688 | .73610 | .91955 | .0875 | .3585 | .4774 | .17604 | .26021 | 24 |
| 37 | .67709 | .73590 | .92008 | .0869 | .3589 | .4769 | .17628 | .25986 | 23 |
| 38 | .67730 | .73570 | .92062 | .0862 | .3592 | .4764 | .17653 | .25952 | 22 |
| 39 | .67752 | .73551 | .92116 | .0856 | .3596 | .4760 | .17678 | .25918 | 21 |
| 40 | 0.67773 | 0.73531 | 0.92170 | 1.0850 | 1.3600 | 1.4755 | .17702 | .25883 | 20 |
| 41 | .67795 | .73511 | .92224 | .0843 | .3603 | .4750 | .17727 | .25849 | 19 |
| 42 | .67816 | .73491 | .92277 | .0837 | .3607 | .4746 | .17752 | .25815 | 18 |
| 43 | .67837 | .73472 | .92331 | .0831 | .3611 | .4741 | .17777 | .25781 | 17 |
| 44 | .67859 | .73452 | .92385 | .0824 | .3614 | .4737 | .17801 | .25747 | 16 |
| 45 | 0.67880 | 0.73432 | 0.92439 | 1.0818 | 1.3618 | 1.4732 | .17826 | .25713 | 15 |
| 46 | .67901 | .73413 | .92493 | .0812 | .3622 | .4727 | .17851 | .25679 | 14 |
| 47 | .67923 | .73393 | .92547 | .0805 | .3625 | .4723 | .17876 | .25645 | 13 |
| 48 | .67944 | .73373 | .92601 | .0799 | .3629 | .4718 | .17901 | .25611 | 12 |
| 49 | .67965 | .73353 | .92655 | .0793 | .3633 | .4713 | .17926 | .25577 | 11 |
| 50 | 0.67987 | 0.73333 | 0.92709 | 1.0786 | 1.3636 | 1.4709 | .17951 | .25543 | 10 |
| 51 | .68008 | .73314 | .92763 | .0780 | .3640 | .4704 | .17976 | .25509 | 9 |
| 52 | .68029 | .73294 | .92817 | .0774 | .3644 | .4700 | .18001 | .25475 | 8 |
| 53 | .68051 | .73274 | .92872 | .0768 | .3647 | .4695 | .18026 | .25442 | 7 |
| 54 | .68072 | .73254 | .92926 | .0761 | .3651 | .4600 | .18051 | .25408 | 6 |
| 55 | 0.68093 | 0.73234 | 0.92980 | 1.0755 | 1.3655 | 1.4686 | .18076 | .25374 | 5 |
| 56 | .68115 | .73215 | .93034 | .0749 | .3658 | .4681 | .18101 | .25341 | 4 |
| 57 | .68136 | .73195 | .93088 | .0742 | .3662 | .4677 | .18127 | .25307 | 3 |
| 58 | .68157 | .73175 | .93143 | .0736 | .3666 | .4672 | .18152 | .25273 | 2 |
| 59 | .68179 | .73155 | .93197 | .0730 | .3670 | .4667 | .18177 | .25240 | 1 |
| 60 | 0.68200 | 0.73135 | 0.93252 | 1.0724 | 1.3673 | 1.4663 | .18202 | .25206 | 0 |
| M | Cosine | Sine | Cotan. | Tan. | Cosec. | Secant | READ DOWN | 47°–48° Involute | M |

**43° or 223°**  Trigonometric Functions  **136° or 316°**

| M | Sine | Cosine | Tan. | Cotan. | Secant | Cosec. | Involute 43°-44° | READ UP | M |
|---|------|--------|------|--------|--------|--------|------------------|---------|---|
| 0 | 0.68200 | 0.73135 | 0.93252 | 1.0724 | 1.3673 | 1.4663 | .18202 | .25206 | 60 |
| 1 | .68221 | .73116 | .93306 | .0717 | .3677 | .4658 | .18228 | .25173 | 59 |
| 2 | .68242 | .73096 | .93360 | .0711 | .3681 | .4654 | .18253 | .25140 | 58 |
| 3 | .68264 | .73076 | .93415 | .0705 | .3684 | .4649 | .18278 | .25106 | 57 |
| 4 | .68285 | .73056 | .93469 | .0699 | .3688 | .4645 | .18304 | .25073 | 56 |
| 5 | 0.68306 | 0.73036 | 0.93524 | 1.0692 | 1.3692 | 1.4640 | .18329 | .25040 | 55 |
| 6 | .68327 | .73016 | .93578 | .0686 | .3696 | .4635 | .18355 | .25006 | 54 |
| 7 | .68349 | .72996 | .93633 | .0680 | .3699 | .4631 | .18380 | .24973 | 53 |
| 8 | .68370 | .72976 | .93688 | .0674 | .3703 | .4626 | .18406 | .24940 | 52 |
| 9 | .68391 | .72957 | .93742 | .0668 | .3707 | .4622 | .18431 | .24907 | 51 |
| 10 | 0.68412 | 0.72937 | 0.93797 | 1.0661 | 1.3711 | 1.4617 | .18457 | .24874 | 50 |
| 11 | .68434 | .72917 | .93852 | .0655 | .3714 | .4613 | .18482 | .24841 | 49 |
| 12 | .68455 | .72897 | .93906 | .0649 | .3718 | .4608 | .18508 | .24808 | 48 |
| 13 | .68476 | .72877 | .93961 | .0643 | .3722 | .4604 | .18534 | .24775 | 47 |
| 14 | .68497 | .72857 | .94016 | .0637 | .3726 | .4599 | .18559 | .24742 | 46 |
| 15 | 0.68518 | 0.72837 | 0.94071 | 1.0630 | 1.3729 | 1.4595 | .18585 | .24709 | 45 |
| 16 | .68539 | .72817 | .94125 | .0624 | .3733 | .4590 | .18611 | .24676 | 44 |
| 17 | .68561 | .72797 | .94180 | .0618 | .3737 | .4586 | .18637 | .24643 | 43 |
| 18 | .68582 | .72777 | .94235 | .0612 | .3741 | .4581 | .18662 | .24611 | 42 |
| 19 | .68603 | .72757 | .94290 | .0606 | .3744 | .4577 | .18688 | .24578 | 41 |
| 20 | 0.68624 | 0.72737 | 0.94345 | 1.0599 | 1.3748 | 1.4572 | .18714 | .24545 | 40 |
| 21 | .68645 | .72717 | .94400 | .0593 | .3752 | .4568 | .18740 | .24512 | 39 |
| 22 | .68666 | .72697 | .94455 | .0587 | .3756 | .4563 | .18766 | .24480 | 38 |
| 23 | .68688 | .72677 | .94510 | .0581 | .3759 | .4559 | .18792 | .24447 | 37 |
| 24 | .68709 | .72657 | .94565 | .0575 | .3763 | .4554 | .18818 | .24415 | 36 |
| 25 | 0.68730 | 0.72637 | 0.94620 | 1.0569 | 1.3767 | 1.4550 | .18844 | .24382 | 35 |
| 26 | .68751 | .72617 | .94676 | .0562 | .3771 | .4545 | .18870 | .24350 | 34 |
| 27 | .68772 | .72597 | .94731 | .0556 | .3775 | .4541 | .18896 | .24317 | 33 |
| 28 | .68793 | .72577 | .94786 | .0550 | .3778 | .4536 | .18922 | .24285 | 32 |
| 29 | .68814 | .72557 | .94841 | .0544 | .3782 | .4532 | .18948 | .24253 | 31 |
| 30 | 0.68835 | 0.72537 | 0.94896 | 1.0538 | 1.3786 | 1.4527 | .18975 | .24220 | 30 |
| 31 | .68857 | .72517 | .94952 | .0532 | .3790 | .4523 | .19001 | .24188 | 29 |
| 32 | .68878 | .72497 | .95007 | .0526 | .3794 | .4518 | .19027 | .24156 | 28 |
| 33 | .68899 | .72477 | .95062 | .0519 | .3797 | .4514 | .19053 | .24123 | 27 |
| 34 | .68920 | .72457 | .95118 | .0513 | .3801 | .4510 | .19080 | .24091 | 26 |
| 35 | 0.68941 | 0.72437 | 0.95173 | 1.0507 | 1.3805 | 1.4505 | .19106 | .24059 | 25 |
| 36 | .68962 | .72417 | .95229 | .0501 | .3809 | .4501 | .19132 | .24027 | 24 |
| 37 | .68983 | .72397 | .95284 | .0495 | .3813 | .4496 | .19159 | .23995 | 23 |
| 38 | .69004 | .72377 | .95340 | .0489 | .3817 | .4492 | .19185 | .23963 | 22 |
| 39 | .69025 | .72357 | .95395 | .0483 | .3820 | .4487 | .19212 | .23931 | 21 |
| 40 | 0.69046 | 0.72337 | 0.95451 | 1.0477 | 1.3824 | 1.4483 | .19238 | .23899 | 20 |
| 41 | .69067 | .72317 | .95506 | .0470 | .3828 | .4479 | .19265 | .23867 | 19 |
| 42 | .69088 | .72297 | .95562 | .0464 | .3832 | .4474 | .19291 | .23835 | 18 |
| 43 | .69109 | .72277 | .95618 | .0458 | .3836 | .4470 | .19318 | .23803 | 17 |
| 44 | .69130 | .72257 | .95673 | .0452 | .3840 | .4465 | .19344 | .23772 | 16 |
| 45 | 0.69151 | 0.72236 | 0.95729 | 1.0446 | 1.3843 | 1.4461 | .19371 | .23740 | 15 |
| 46 | .69172 | .72216 | .95785 | .0440 | .3847 | .4457 | .19398 | .23708 | 14 |
| 47 | .69193 | .72196 | .95841 | .0434 | .3851 | .4452 | .19424 | .23676 | 13 |
| 48 | .69214 | .72176 | .95897 | .0428 | .3855 | .4448 | .19451 | .23645 | 12 |
| 49 | .69235 | .72156 | .95952 | .0422 | .3859 | .4443 | .19478 | .23613 | 11 |
| 50 | 0.69256 | 0.72136 | 0.96008 | 1.0416 | 1.3863 | 1.4439 | .19505 | .23582 | 10 |
| 51 | .69277 | .72116 | .96064 | .0410 | .3867 | .4435 | .19532 | .23550 | 9 |
| 52 | .69298 | .72095 | .96120 | .0404 | .3871 | .4430 | .19558 | .23519 | 8 |
| 53 | .69319 | .72075 | .96176 | .0398 | .3874 | .4426 | .19585 | .23487 | 7 |
| 54 | .69340 | .72055 | .96232 | .0392 | .3878 | .4422 | .19612 | .23456 | 6 |
| 55 | 0.69361 | 0.72035 | 0.96288 | 1.0385 | 1.3882 | 1.4417 | .19639 | .23424 | 5 |
| 56 | .69382 | .72015 | .96344 | .0379 | .3886 | .4413 | .19666 | .23393 | 4 |
| 57 | .69403 | .71995 | .96400 | .0373 | .3890 | .4409 | .19693 | .23362 | 3 |
| 58 | .69424 | .71974 | .96457 | .0367 | .3894 | .4404 | .19720 | .23330 | 2 |
| 59 | .69445 | .71954 | .96513 | .0361 | .3898 | .4400 | .19747 | .23299 | 1 |
| 60 | 0.69466 | 0.71934 | 0.96569 | 1.0355 | 1.3902 | 1.4396 | .19774 | .23268 | 0 |
| M | Cosine | Sine | Cotan. | Tan. | Cosec. | Secant | READ DOWN | 46°-47° Involute | M |

**44° or 224°**     Trigonometric Functions     **135° or 315°**

| M | Sine | Cosine | Tan. | Cotan. | Secant | Cosec. | Involute 44°-45° | READ UP | M |
|---|------|--------|------|--------|--------|--------|--------|---------|---|
| 0 | 0.69466 | 0.71934 | 0.96569 | 1.0355 | 1.3902 | 1.4396 | .19774 | .23268 | 60 |
| 1 | .69487 | .71914 | .96625 | .0349 | .3906 | .4391 | .19802 | .23237 | 59 |
| 2 | .69508 | .71894 | .96681 | .0343 | .3909 | .4387 | .19829 | .23206 | 58 |
| 3 | .69529 | .71873 | .96738 | .0337 | .3913 | .4383 | .19856 | .23174 | 57 |
| 4 | .69549 | .71853 | .96794 | .0331 | .3917 | .4378 | .19883 | .23143 | 56 |
| 5 | 0.69570 | 0.71833 | 0.96850 | 1.0325 | 1.3921 | 1.4374 | .19910 | .23112 | 55 |
| 6 | .69591 | .71813 | .96907 | .0319 | .3925 | .4370 | .19938 | .23081 | 54 |
| 7 | .69612 | .71792 | .96963 | .0313 | .3929 | .4365 | .19965 | .23050 | 53 |
| 8 | .69633 | .71772 | .97020 | .0307 | .3933 | .4361 | .19992 | .23020 | 52 |
| 9 | .69654 | .71752 | .97076 | .0301 | .3937 | .4357 | .20020 | .22989 | 51 |
| 10 | 0.69675 | 0.71732 | 0.97133 | 1.0295 | 1.3941 | 1.4352 | .20047 | .22958 | 50 |
| 11 | .69696 | .71711 | .97189 | .0289 | .3945 | .4348 | .20075 | .22927 | 49 |
| 12 | .69717 | .71691 | .97246 | .0283 | .3949 | .4344 | .20102 | .22896 | 48 |
| 13 | .69737 | .71671 | .97302 | .0277 | .3953 | .4340 | .20130 | .22865 | 47 |
| 14 | .69758 | .71650 | .97359 | .0271 | .3957 | .4335 | .20157 | .22835 | 46 |
| 15 | 0.69779 | 0.71630 | 0.97416 | 1.0265 | 1.3961 | 1.4331 | .20185 | .22804 | 45 |
| 16 | .69800 | .71610 | .97472 | .0259 | .3965 | .4327 | .20212 | .22773 | 44 |
| 17 | .69821 | .71590 | .97529 | .0253 | .3969 | .4322 | .20240 | .22743 | 43 |
| 18 | .69842 | .71569 | .97586 | .0247 | .3972 | .4318 | .20268 | .22712 | 42 |
| 19 | .69862 | .71549 | .97643 | .0241 | .3976 | .4314 | .20296 | .22682 | 41 |
| 20 | 0.69883 | 0.71529 | 0.97700 | 1.0235 | 1.3980 | 1.4310 | .20323 | .22651 | 40 |
| 21 | .69904 | .71508 | .97756 | .0230 | .3984 | .4305 | .20351 | .22621 | 39 |
| 22 | .69925 | .71488 | .97813 | .0224 | .3988 | .4301 | .20379 | .22590 | 38 |
| 23 | .69946 | .71468 | .97870 | .0218 | .3992 | .4297 | .20407 | .22560 | 37 |
| 24 | .69966 | .71447 | .97927 | .0212 | .3996 | .4293 | .20435 | .22530 | 36 |
| 25 | 0.69987 | 0.71427 | 0.97984 | 1.0206 | 1.4000 | 1.4288 | .20463 | .22499 | 35 |
| 26 | .70008 | .71407 | .98041 | .0200 | .4004 | .4284 | .20490 | .22469 | 34 |
| 27 | .70029 | .71386 | .98098 | .0194 | .4008 | .4280 | .20518 | .22439 | 33 |
| 28 | .70049 | .71366 | .98155 | .0188 | .4012 | .4276 | .20546 | .22409 | 32 |
| 29 | .70070 | .71345 | .98213 | .0182 | .4016 | .4271 | .20575 | .22378 | 31 |
| 30 | 0.70091 | 0.71325 | 0.98270 | 1.0176 | 1.4020 | 1.4267 | .20603 | .22348 | 30 |
| 31 | .70112 | .71305 | .98327 | .0170 | .4024 | .4263 | .20631 | .22318 | 29 |
| 32 | .70132 | .71284 | .98384 | .0164 | .4028 | .4259 | .20659 | .22288 | 28 |
| 33 | .70153 | .71264 | .98441 | .0158 | .4032 | .4255 | .20687 | .22258 | 27 |
| 34 | .70174 | .71243 | .98499 | .0152 | .4036 | .4250 | .20715 | .22228 | 26 |
| 35 | 0.70195 | 0.71223 | 0.98556 | 1.0147 | 1.4040 | 1.4246 | .20743 | .22198 | 25 |
| 36 | .70215 | .71203 | .98613 | .0141 | .4044 | .4242 | .20772 | .22168 | 24 |
| 37 | .70236 | .71182 | .98671 | .0135 | .4048 | .4238 | .20800 | .22138 | 23 |
| 38 | .70257 | .71162 | .98728 | .0129 | .4052 | .4234 | .20828 | .22108 | 22 |
| 39 | .70277 | .71141 | .98786 | .0123 | .4057 | .4229 | .20857 | .22079 | 21 |
| 40 | 0.70298 | 0.71121 | 0.98843 | 1.0117 | 1.4061 | 1.4225 | .20885 | .22049 | 20 |
| 41 | .70319 | .71100 | .98901 | .0111 | .4065 | .4221 | .20914 | .22019 | 19 |
| 42 | .70339 | .71080 | .98958 | .0105 | .4069 | .4217 | .20942 | .21989 | 18 |
| 43 | .70360 | .71059 | .99016 | .0099 | .4073 | .4213 | .20971 | .21960 | 17 |
| 44 | .70381 | .71039 | .99073 | .0093 | .4077 | .4208 | .20999 | .21930 | 16 |
| 45 | 0.70401 | 0.71019 | 0.99131 | 1.0088 | 1.4081 | 1.4204 | .21028 | .21900 | 15 |
| 46 | .70422 | .70998 | .99189 | .0082 | .4085 | .4200 | .21056 | .21871 | 14 |
| 47 | .70443 | .70978 | .99247 | .0076 | .4089 | .4196 | .21085 | .21841 | 13 |
| 48 | .70463 | .70957 | .99304 | .0070 | .4093 | .4192 | .21114 | .21812 | 12 |
| 49 | .70484 | .70937 | .99362 | .0064 | .4097 | .4188 | .21142 | .21782 | 11 |
| 50 | 0.70505 | 0.70916 | 0.99420 | 1.0058 | 1.4101 | 1.4183 | .21171 | .21753 | 10 |
| 51 | .70525 | .70896 | .99478 | .0052 | .4105 | .4179 | .21200 | .21723 | 9 |
| 52 | .70546 | .70875 | .99536 | .0047 | .4109 | .4175 | .21229 | .21694 | 8 |
| 53 | .70567 | .70855 | .99594 | .0041 | .4113 | .4171 | .21257 | .21665 | 7 |
| 54 | .70587 | .70834 | .99652 | .0035 | .4118 | .4167 | .21286 | .21635 | 6 |
| 55 | 0.70608 | 0.70813 | 0.99710 | 1.0029 | 1.4122 | 1.4163 | .21315 | .21606 | 5 |
| 56 | .70628 | .70793 | .99768 | .0023 | .4126 | .4159 | .21344 | .21577 | 4 |
| 57 | .70649 | .70772 | .99826 | .0017 | .4130 | .4154 | .21373 | .21548 | 3 |
| 58 | .70670 | .70752 | .99884 | .0012 | .4134 | .4150 | .21402 | .21518 | 2 |
| 59 | .70690 | .70731 | .99942 | .0006 | .4138 | .4146 | .21431 | .21489 | 1 |
| 60 | 0.70711 | 0.70711 | 1.00000 | 1.0000 | 1.4142 | 1.4142 | .21460 | .21460 | 0 |

| M | Cosine | Sine | Cotan. | Tan. | Cosec. | Secant | READ DOWN | 45°-46° Involute | M |

## Use of Logarithms in Solving Triangles

The following tables "Logarithms of Trigonometrical Functions" may be used in the solution of triangles. The calculations are worked out in the same manner as with logarithms in general. In these tables, the characteristic is given in all cases, together with the mantissa. The complete logarithm of the functions, therefore, is found directly from the tables; however, as the values of the natural functions of sines and cosines, and of tangents for angles less than 45 degrees, are always less than 1, the characteristics would always be negative for these functions. In order to avoid these negative characteristics, the logarithm, as generally given, has had 10 added to its value; consequently, the actual value of the logarithm for "cosine 3 degrees," for example, is 9.99940 − 10.

When using these logarithms in calculations with other logarithms, the calcula-

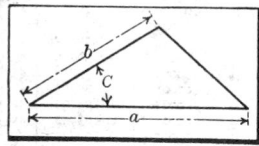

tions can be carried out exactly as explained under "LOGARITHMS." When writing down the logarithm taken from the tables of "Logarithms of Trigonometrical Functions," − 10 must be written after the value shown in the table if the computations are to be carried forward correctly. The examples which follow show the method of procedure.

*Example.* — Find the area of a triangle where the lengths of two sides are 53 and 82 inches, and the angle between them is 30 degrees.

The area is found by the formula (see accompanying illustration):

$$\text{Area} = \frac{a \times b \times \sin C}{2} = \frac{53 \times 82 \times \sin 30°}{2}$$

The logarithms of the numbers 53 and 82 are obtained from the regular tables in the usual manner. The logarithm for the sine of 30 degrees, as given in the table of "Logarithms of Trigonometrical Functions" is 9.69897 which, as previously explained, is written as 9.69897 − 10. The logarithms of the factors in the numerator are first added together and then the logarithm of the factor in the denominator is subtracted from their sum, as explained in the section on LOGARITHMS. (See "Division by Logarithms.") Proceed now to find the logarithm of the area.

First step:    log 53      = 1.72428          Second step:          13.33706 − 10
               log 82      = 1.91381                      −log 2 = −0.30103
               log sin 30° = 9.69897 − 10                         ─────────────
                            ─────────────                         13.03603 − 10
                            13.33706 − 10              or          3.03603

The logarithm of the area thus is 3.03603, and from a logarithmic table it is found, by interpolation, that the area equals 1086.5 square inches.

Angles $A$ and $C$ and side $a$ in a triangle are known. (See table "Solution of Oblique-angled Triangles.")  $A = 37° 42'$; $C = 68° 12'$; $a = 12$ inches.  Find side $c$.

$$c = \frac{a \times \sin C}{\sin A} = \frac{12 \times \sin 68° 12'}{\sin 37° 42'}$$

The solution is as follows:

          log 12              =      1.07918
          log sin 68° 12'     =      9.96778 − 10
                                    ─────────────
                                     11.04696 − 10
          −log sin 37° 42'    =  −( 9.78642 − 10)
              Thus, log c =        1.26054, and hence, c = 18.22 inches.

| M | Sine | Cosine | Tangent | Cotangent | Secant | Cosecant | M |
|---|------|--------|---------|-----------|--------|----------|---|
| 0 | Inf. Neg. | 10.00000 | Inf. Neg. | Infinite | 10.00000 | Infinite | 60 |
| 1 | 6.46373 | .00000 | 6.46373 | 13.53627 | .00000 | 13.53627 | 59 |
| 2 | .76476 | .00000 | .76476 | .23524 | .00000 | .23524 | 58 |
| 3 | .94085 | .00000 | .94085 | .05915 | .00000 | .05915 | 57 |
| 4 | 7.06579 | .00000 | 7.06579 | 12.93421 | .00000 | 12.93421 | 56 |
| 5 | 7.16270 | 10.00000 | 7.16270 | 12.83730 | 10.00000 | 12.83730 | 55 |
| 6 | .24188 | .00000 | .24188 | .75812 | .00000 | .75812 | 54 |
| 7 | .30882 | .00000 | .30882 | .69118 | .00000 | .69118 | 53 |
| 8 | .36682 | .00000 | .36682 | .63318 | .00000 | .63318 | 52 |
| 9 | .41797 | .00000 | .41797 | .58203 | .00000 | .58203 | 51 |
| 10 | 7.46373 | 10.00000 | 7.46373 | 12.53627 | 10.00000 | 12.53627 | 50 |
| 11 | .50512 | .00000 | .50512 | .49488 | .00000 | .49488 | 49 |
| 12 | .54291 | .00000 | .54291 | .45709 | .00000 | .45709 | 48 |
| 13 | .57767 | .00000 | .57767 | .42233 | .00000 | .42233 | 47 |
| 14 | .60985 | .00000 | .60986 | .39014 | .00000 | .39015 | 46 |
| 15 | 7.63982 | 10.00000 | 7.63982 | 12.36018 | 10.00000 | 12.36018 | 45 |
| 16 | .66784 | .00000 | .66785 | .33215 | .00000 | .33216 | 44 |
| 17 | .69417 | 9.99999 | .69418 | .30582 | .00001 | .30583 | 43 |
| 18 | .71900 | .99999 | .71900 | .28100 | .00001 | .28100 | 42 |
| 19 | .74248 | .99999 | .74248 | .25752 | .00001 | .25752 | 41 |
| 20 | 7.76475 | 9.99999 | 7.76476 | 12.23524 | 10.00001 | 12.23525 | 40 |
| 21 | .78594 | .99999 | .78595 | .21405 | .00001 | .21406 | 39 |
| 22 | .80615 | .99999 | .80615 | .19385 | .00001 | .19385 | 38 |
| 23 | .82545 | .99999 | .82546 | .17454 | .00001 | .17455 | 37 |
| 24 | .84393 | .99999 | .84394 | .15606 | .00001 | .15607 | 36 |
| 25 | 7.86166 | 9.99999 | 7.86167 | 12.13833 | 10.00001 | 12.13834 | 35 |
| 26 | .87870 | .99999 | .87871 | .12129 | .00001 | .12130 | 34 |
| 27 | .89509 | .99999 | .89510 | .10490 | .00001 | .10491 | 33 |
| 28 | .91088 | .99999 | .91089 | .08911 | .00001 | .08912 | 32 |
| 29 | .92612 | .99998 | .92613 | .07387 | .00002 | .07388 | 31 |
| 30 | 7.94084 | 9.99998 | 7.94086 | 12.05914 | 10.00002 | 12.05916 | 30 |
| 31 | .95508 | .99998 | .95510 | .04490 | .00002 | .04492 | 29 |
| 32 | .96887 | .99998 | .96889 | .03111 | .00002 | .03113 | 28 |
| 33 | .98223 | .99998 | .98225 | .01775 | .00002 | .01777 | 27 |
| 34 | .99520 | .99998 | .99522 | .00478 | .00002 | .00480 | 26 |
| 35 | 8.00779 | 9.99998 | 8.00781 | 11.99219 | 10.00002 | 11.99221 | 25 |
| 36 | .02002 | .99998 | .02004 | .97996 | .00002 | .97998 | 24 |
| 37 | .03192 | .99997 | .03194 | .96806 | .00003 | .96808 | 23 |
| 38 | .04350 | .99997 | .04353 | .95647 | .00003 | .95650 | 22 |
| 39 | .05478 | .99997 | .05481 | .94519 | .00003 | .94522 | 21 |
| 40 | 8.06578 | 9.99997 | 8.06581 | 11.93419 | 10.00003 | 11.93422 | 20 |
| 41 | .07650 | .99997 | .07653 | .92347 | .00003 | .92350 | 19 |
| 42 | .08696 | .99997 | .08700 | .91300 | .00003 | .91304 | 18 |
| 43 | .09718 | .99997 | .09722 | .90278 | .00003 | .90282 | 17 |
| 44 | .10717 | .99996 | .10720 | .89280 | .00004 | .89283 | 16 |
| 45 | 8.11693 | 9.99996 | 8.11696 | 11.88304 | 10.00004 | 11.88307 | 15 |
| 46 | .12647 | .99996 | .12651 | .87349 | .00004 | .87353 | 14 |
| 47 | .13581 | .99996 | .13585 | .86415 | .00004 | .86419 | 13 |
| 48 | .14495 | .99996 | .14500 | .85500 | .00004 | .85505 | 12 |
| 49 | .15391 | .99996 | .15395 | .84605 | .00004 | .84609 | 11 |
| 50 | 8.16268 | 9.99995 | 8.16273 | 11.83727 | 10.00005 | 11.83732 | 10 |
| 51 | .17128 | .99995 | .17133 | .82867 | .00005 | .82872 | 9 |
| 52 | .17971 | .99995 | .17976 | .82024 | .00005 | .82029 | 8 |
| 53 | .18798 | .99995 | .18804 | .81196 | .00005 | .81202 | 7 |
| 54 | .19610 | .99995 | .19616 | .80384 | .00005 | .80390 | 6 |
| 55 | 8.20407 | 9.99994 | 8.20413 | 11.79587 | 10.00006 | 11.79593 | 5 |
| 56 | .21189 | .99994 | .21195 | .78805 | .00006 | .78811 | 4 |
| 57 | .21958 | .99994 | .21964 | .78036 | .00006 | .78042 | 3 |
| 58 | .22713 | .99994 | .22720 | .77280 | .00006 | .77287 | 2 |
| 59 | .23456 | .99994 | .23462 | .76538 | .00006 | .76544 | 1 |
| 60 | 8.24186 | 9.99993 | 8.24192 | 11.75808 | 10.00007 | 11.75814 | 0 |
| M | Cosine | Sine | Cotangent | Tangent | Cosecant | Secant | M |

| M | Sine | Cosine | Tangent | Cotangent | Secant | Cosecant | M |
|---|------|--------|---------|-----------|--------|----------|---|
| 0 | 8.24186 | 9.99993 | 8.24192 | 11.75808 | 10.00007 | 11.75814 | 60 |
| 1 | .24903 | .99993 | .24910 | .75090 | .00007 | .75097 | 59 |
| 2 | .25609 | .99993 | .25616 | .74384 | .00007 | .74391 | 58 |
| 3 | .26304 | .99993 | .26312 | .73688 | .00007 | .73696 | 57 |
| 4 | .26988 | .99992 | .26996 | .73004 | .00008 | .73012 | 56 |
| 5 | 8.27661 | 9.99992 | 8.27669 | 11.72331 | 10.00008 | 11.72339 | 55 |
| 6 | .28324 | .99992 | .28332 | .71668 | .00008 | .71676 | 54 |
| 7 | .28977 | .99992 | .28986 | .71014 | .00008 | .71023 | 53 |
| 8 | .29621 | .99992 | .29629 | .70371 | .00008 | .70379 | 52 |
| 9 | .30255 | .99991 | .30263 | .69737 | .00009 | .69745 | 51 |
| 10 | 8.30879 | 9.99991 | 8.30888 | 11.69112 | 10.00009 | 11.69121 | 50 |
| 11 | .31495 | .99991 | .31505 | .68495 | .00009 | .68505 | 49 |
| 12 | .32103 | .99990 | .32112 | .67888 | .00010 | .67897 | 48 |
| 13 | .32702 | .99990 | .32711 | .67289 | .00010 | .67298 | 47 |
| 14 | .33292 | .99990 | .33302 | .66698 | .00010 | .66708 | 46 |
| 15 | 8.33875 | 9.99990 | 8.33886 | 11.66114 | 10.00010 | 11.66125 | 45 |
| 16 | .34450 | .99989 | .34461 | .65539 | .00011 | .65550 | 44 |
| 17 | .35018 | .99989 | .35029 | .64971 | .00011 | .64982 | 43 |
| 18 | .35578 | .99989 | .35590 | .64410 | .00011 | .64422 | 42 |
| 19 | .36131 | .99989 | .36143 | .63857 | .00011 | .63869 | 41 |
| 20 | 8.36678 | 9.99988 | 8.36689 | 11.63311 | 10.00012 | 11.63322 | 40 |
| 21 | .37217 | .99988 | .37229 | .62771 | .00012 | .62783 | 39 |
| 22 | .37750 | .99988 | .37762 | .62238 | .00012 | .62250 | 38 |
| 23 | .38276 | .99987 | .38289 | .61711 | .00013 | .61724 | 37 |
| 24 | .38796 | .99987 | .38809 | .61191 | .00013 | .61204 | 36 |
| 25 | 8.39310 | 9.99987 | 8.39323 | 11.60677 | 10.00013 | 11.60690 | 35 |
| 26 | .39818 | .99986 | .39832 | .60168 | .00014 | .60182 | 34 |
| 27 | .40320 | .99986 | .40334 | .59666 | .00014 | .59680 | 33 |
| 28 | .40816 | .99986 | .40830 | .59170 | .00014 | .59184 | 32 |
| 29 | .41307 | .99985 | .41321 | .58679 | .00015 | .58693 | 31 |
| 30 | 8.41792 | 9.99985 | 8.41807 | 11.58193 | 10.00015 | 11.58208 | 30 |
| 31 | .42272 | .99985 | .42287 | .57713 | .00015 | .57728 | 29 |
| 32 | .42746 | .99984 | .42762 | .57238 | .00016 | .57254 | 28 |
| 33 | .43216 | .99984 | .43232 | .56768 | .00016 | .56784 | 27 |
| 34 | .43680 | .99984 | .43696 | .56304 | .00016 | .56320 | 26 |
| 35 | 8.44139 | 9.99983 | 8.44156 | 11.55844 | 10.00017 | 11.55861 | 25 |
| 36 | .44594 | .99983 | .44611 | .55389 | .00017 | .55406 | 24 |
| 37 | .45044 | .99983 | .45061 | .54939 | .00017 | .54956 | 23 |
| 38 | .45489 | .99982 | .45507 | .54493 | .00018 | .54511 | 22 |
| 39 | .45930 | .99982 | .45948 | .54052 | .00018 | .54070 | 21 |
| 40 | 8.46366 | 9.99982 | 8.46385 | 11.53615 | 10.00018 | 11.53634 | 20 |
| 41 | .46799 | .99981 | .46817 | .53183 | .00019 | .53201 | 19 |
| 42 | .47226 | .99981 | .47245 | .52755 | .00019 | .52774 | 18 |
| 43 | .47650 | .99981 | .47669 | .52331 | .00019 | .52350 | 17 |
| 44 | .48069 | .99980 | .48089 | .51911 | .00020 | .51931 | 16 |
| 45 | 8.48485 | 9.99980 | 8.48505 | 11.51495 | 10.00020 | 11.51515 | 15 |
| 46 | .48896 | .99979 | .48917 | .51083 | .00021 | .51104 | 14 |
| 47 | .49304 | .99979 | .49325 | .50675 | .00021 | .50696 | 13 |
| 48 | .49708 | .99979 | .49729 | .50271 | .00021 | .50292 | 12 |
| 49 | .50108 | .99978 | .50130 | .49870 | .00022 | .49892 | 11 |
| 50 | 8.50504 | 9.99978 | 8.50527 | 11.49473 | 10.00022 | 11.49496 | 10 |
| 51 | .50897 | .99977 | .50920 | .49080 | .00023 | .49103 | 9 |
| 52 | .51287 | .99977 | .51310 | .48690 | .00023 | .48713 | 8 |
| 53 | .51673 | .99977 | .51696 | .48304 | .00023 | .48327 | 7 |
| 54 | .52055 | .99976 | .52079 | .47921 | .00024 | .47945 | 6 |
| 55 | 8.52434 | 9.99976 | 8.52459 | 11.47541 | 10.00024 | 11.47566 | 5 |
| 56 | .52810 | .99975 | .52835 | .47165 | .00025 | .47190 | 4 |
| 57 | .53183 | .99975 | .53208 | .46792 | .00025 | .46817 | 3 |
| 58 | .53552 | .99974 | .53578 | .46422 | .00026 | .46448 | 2 |
| 59 | .53919 | .99974 | .53945 | .46055 | .00026 | .46081 | 1 |
| 60 | 8.54282 | 9.99974 | 8.54308 | 11.45692 | 10.00026 | 11.45718 | 0 |

| M | Cosine | Sine | Cotangent | Tangent | Cosecant | Secant | M |

| M | Sine | Cosine | Tangent | Cotangent | Secant | Cosecant | M |
|---|------|--------|---------|-----------|--------|----------|---|
| 0 | 8.54282 | 9.99974 | 8.54308 | 11.45692 | 10.00026 | 11.45718 | 60 |
| 1 | .54642 | .99973 | .54669 | .45331 | .00027 | .45358 | 59 |
| 2 | .54999 | .99973 | .55027 | .44973 | .00027 | .45001 | 58 |
| 3 | .55354 | .99972 | .55382 | .44618 | .00028 | .44646 | 57 |
| 4 | .55705 | .99972 | .55734 | .44266 | .00028 | .44295 | 56 |
| 5 | 8.56054 | 9.99971 | 8.56083 | 11.43917 | 10.00029 | 11.43946 | 55 |
| 6 | .56400 | .99971 | .56429 | .43571 | .00029 | .43600 | 54 |
| 7 | .56743 | .99970 | .56773 | .43227 | .00030 | .43257 | 53 |
| 8 | .57084 | .99970 | .57114 | .42886 | .00030 | .42916 | 52 |
| 9 | .57421 | .99969 | .57452 | .42548 | .00031 | .42579 | 51 |
| 10 | 8.57757 | 9.99969 | 8.57788 | 11.42212 | 10.00031 | 11.42243 | 50 |
| 11 | .58089 | .99968 | .58121 | .41879 | .00032 | .41911 | 49 |
| 12 | .58419 | .99968 | .58451 | .41549 | .00032 | .41581 | 48 |
| 13 | .58747 | .99967 | .58779 | .41221 | .00033 | .41253 | 47 |
| 14 | .59072 | .99967 | .59105 | .40895 | .00033 | .40928 | 46 |
| 15 | 8.59395 | 9.99967 | 8.59428 | 11.40572 | 10.00033 | 11.40605 | 45 |
| 16 | .59715 | .99966 | .59749 | .40251 | .00034 | .40285 | 44 |
| 17 | .60033 | .99966 | .60068 | .39932 | .00034 | .39967 | 43 |
| 18 | .60349 | .99965 | .60384 | .39616 | .00035 | .39651 | 42 |
| 19 | .60662 | .99964 | .60698 | .39302 | .00036 | .39338 | 41 |
| 20 | 8.60973 | 9.99964 | 8.61009 | 11.38991 | 10.00036 | 11.39027 | 40 |
| 21 | .61282 | .99963 | .61319 | .38681 | .00037 | .38718 | 39 |
| 22 | .61589 | .99963 | .61626 | .38374 | .00037 | .38411 | 38 |
| 23 | .61894 | .99962 | .61931 | .38069 | .00038 | .38106 | 37 |
| 24 | .62196 | .99962 | .62234 | .37766 | .00038 | .37804 | 36 |
| 25 | 8.62497 | 9.99961 | 8.62535 | 11.37465 | 10.00039 | 11.37503 | 35 |
| 26 | .62795 | .99961 | .62834 | .37166 | .00039 | .37205 | 34 |
| 27 | .63091 | .99960 | .63131 | .36869 | .00040 | .36909 | 33 |
| 28 | .63385 | .99960 | .63426 | .36574 | .00040 | .36615 | 32 |
| 29 | .63678 | .99959 | .63718 | .36282 | .00041 | .36322 | 31 |
| 30 | 8.63968 | 9.99959 | 8.64009 | 11.35991 | 10.00041 | 11.36032 | 30 |
| 31 | .64256 | .99958 | .64298 | .35702 | .00042 | .35744 | 29 |
| 32 | .64543 | .99958 | .64585 | .35415 | .00042 | .35457 | 28 |
| 33 | .64827 | .99957 | .64870 | .35130 | .00043 | .35173 | 27 |
| 34 | .65110 | .99956 | .65154 | .34846 | .00044 | .34890 | 26 |
| 35 | 8.65391 | 9.99956 | 8.65435 | 11.34565 | 10.00044 | 11.34609 | 25 |
| 36 | .65670 | .99955 | .65715 | .34285 | .00045 | .34330 | 24 |
| 37 | .65947 | .99955 | .65993 | .34007 | .00045 | .34053 | 23 |
| 38 | .66223 | .99954 | .66269 | .33731 | .00046 | .33777 | 22 |
| 39 | .66497 | .99954 | .66543 | .33457 | .00046 | .33503 | 21 |
| 40 | 8.66769 | 9.99953 | 8.66816 | 11.33184 | 10.00047 | 11.33231 | 20 |
| 41 | .67039 | .99952 | .67087 | .32913 | .00048 | .32961 | 19 |
| 42 | .67308 | .99952 | .67356 | .32644 | .00048 | .32692 | 18 |
| 43 | .67575 | .99951 | .67624 | .32376 | .00049 | .32425 | 17 |
| 44 | .67841 | .99951 | .67890 | .32110 | .00049 | .32159 | 16 |
| 45 | 8.68104 | 9.99950 | 8.68154 | 11.31846 | 10.00050 | 11.31896 | 15 |
| 46 | .68367 | .99949 | .68417 | .31583 | .00051 | .31633 | 14 |
| 47 | .68627 | .99949 | .68677 | .31322 | .00051 | .31373 | 13 |
| 48 | .68886 | .99948 | .68938 | .31062 | .00052 | .31114 | 12 |
| 49 | .69144 | .99948 | .69196 | .30804 | .00052 | .30856 | 11 |
| 50 | 8.69400 | 9.99947 | 8.69453 | 11.30547 | 10.00053 | 11.30600 | 10 |
| 51 | .69654 | .99946 | .69708 | .30292 | .00054 | .30346 | 9 |
| 52 | .69907 | .99946 | .69962 | .30038 | .00054 | .30093 | 8 |
| 53 | .70159 | .99945 | .70214 | .29786 | .00055 | .29841 | 7 |
| 54 | .70409 | .99944 | .70465 | .29535 | .00056 | .29591 | 6 |
| 55 | 8.70658 | 9.99944 | 8.70714 | 11.29286 | 10.00056 | 11.29342 | 5 |
| 56 | .70905 | .99943 | .70962 | .29038 | .00057 | .29095 | 4 |
| 57 | .71151 | .99942 | .71208 | .28792 | .00058 | .28849 | 3 |
| 58 | .71395 | .99942 | .71453 | .28547 | .00058 | .28605 | 2 |
| 59 | .71638 | .99941 | .71697 | .28303 | .00059 | .28362 | 1 |
| 60 | 8.71880 | 9.99940 | 8.71940 | 11.28060 | 10.00060 | 11.28120 | 0 |
| M | Cosine | Sine | Cotangent | Tangent | Cosecant | Secant | M |

## 3° Logarithms of Trigonometrical Functions 176°

| M | Sine | Cosine | Tangent | Cotangent | Secant | Cosecant | M |
|---|------|--------|---------|-----------|--------|----------|---|
| 0 | 8.71880 | 9.99940 | 8.71940 | 11.28060 | 10.00060 | 11.28120 | 60 |
| 1 | .72120 | .99940 | .72181 | .27819 | .00060 | .27880 | 59 |
| 2 | .72359 | .99939 | .72420 | .27580 | .00061 | .27641 | 58 |
| 3 | .72597 | .99938 | .72659 | .27341 | .00062 | .27403 | 57 |
| 4 | .72834 | .99938 | .72896 | .27104 | .00062 | .27166 | 56 |
| 5 | 8.73069 | 9.99937 | 8.73132 | 11.26868 | 10.00063 | 11.26931 | 55 |
| 6 | .73303 | .99936 | .73366 | .26634 | .00064 | .26697 | 54 |
| 7 | .73535 | .99936 | .73600 | .26400 | .00064 | .26465 | 53 |
| 8 | .73767 | .99935 | .73832 | .26168 | .00065 | .26233 | 52 |
| 9 | .73997 | .99934 | .74063 | .25937 | .00066 | .26003 | 51 |
| 10 | 8.74226 | 9.99934 | 8.74292 | 11.25708 | 10.00066 | 11.25774 | 50 |
| 11 | .74454 | .99933 | .74521 | .25479 | .00067 | .25546 | 49 |
| 12 | .74680 | .99932 | .74748 | .25252 | .00068 | .25320 | 48 |
| 13 | .74906 | .99932 | .74974 | .25026 | .00068 | .25094 | 47 |
| 14 | .75130 | .99931 | .75199 | .24801 | .00069 | .24870 | 46 |
| 15 | 8.75353 | 9.99930 | 8.75423 | 11.24577 | 10.00070 | 11.24647 | 45 |
| 16 | .75575 | .99929 | .75645 | .24355 | .00071 | .24425 | 44 |
| 17 | .75795 | .99929 | .75867 | .24133 | .00071 | .24205 | 43 |
| 18 | .76015 | .99928 | .76087 | .23913 | .00072 | .23985 | 42 |
| 19 | .76234 | .99927 | .76306 | .23694 | .00073 | .23766 | 41 |
| 20 | 8.76451 | 9.99926 | 8.76525 | 11.23475 | 10.00074 | 11.23549 | 40 |
| 21 | .76667 | .99926 | .76742 | .23258 | .00074 | .23333 | 39 |
| 22 | .76883 | .99925 | .76958 | .23042 | .00075 | .23117 | 38 |
| 23 | .77097 | .99924 | .77173 | .22827 | .00076 | .22903 | 37 |
| 24 | .77310 | .99923 | .77387 | .22613 | .00077 | .22690 | 36 |
| 25 | 8.77522 | 9.99923 | 8.77600 | 11.22400 | 10.00077 | 11.22478 | 35 |
| 26 | .77733 | .99922 | .77811 | .22189 | .00078 | .22267 | 34 |
| 27 | .77943 | .99921 | .78022 | .21978 | .00079 | .22057 | 33 |
| 28 | .78152 | .99920 | .78232 | .21768 | .00080 | .21848 | 32 |
| 29 | .78360 | .99920 | .78441 | .21559 | .00080 | .21640 | 31 |
| 30 | 8.78568 | 9.99919 | 8.78649 | 11.21351 | 10.00081 | 11.21432 | 30 |
| 31 | .78774 | .99918 | .78855 | .21145 | .00082 | .21226 | 29 |
| 32 | .78979 | .99917 | .79061 | .20939 | .00083 | .21021 | 28 |
| 33 | .79183 | .99917 | .79266 | .20734 | .00083 | .20817 | 27 |
| 34 | .79386 | .99916 | .79470 | .20530 | .00084 | .20614 | 26 |
| 35 | 8.79588 | 9.99915 | 8.79673 | 11.20327 | 10.00085 | 11.20412 | 25 |
| 36 | .79789 | .99914 | .79875 | .20125 | .00086 | .20211 | 24 |
| 37 | .79990 | .99913 | .80076 | .19924 | .00087 | .20010 | 23 |
| 38 | .80189 | .99913 | .80277 | .19723 | .00087 | .19811 | 22 |
| 39 | .80388 | .99912 | .80476 | .19524 | .00088 | .19612 | 21 |
| 40 | 8.80585 | 9.99911 | 8.80674 | 11.19326 | 10.00089 | 11.19415 | 20 |
| 41 | .80782 | .99910 | .80872 | .19128 | .00090 | .19218 | 19 |
| 42 | .80978 | .99909 | .81068 | .18932 | .00091 | .19022 | 18 |
| 43 | .81173 | .99909 | .81264 | .18736 | .00091 | .18827 | 17 |
| 44 | .81367 | .99908 | .81459 | .18541 | .00092 | .18633 | 16 |
| 45 | 8.81560 | 9.99907 | 8.81653 | 11.18347 | 10.00093 | 11.18440 | 15 |
| 46 | .81752 | .99906 | .81846 | .18154 | .00094 | .18248 | 14 |
| 47 | .81944 | .99905 | .82038 | .17962 | .00095 | .18056 | 13 |
| 48 | .82134 | .99904 | .82230 | .17770 | .00096 | .17866 | 12 |
| 49 | .82324 | .99904 | .82420 | .17580 | .00096 | .17676 | 11 |
| 50 | 8.82513 | 9.99903 | 8.82610 | 11.17390 | 10.00097 | 11.17487 | 10 |
| 51 | .82701 | .99902 | .82799 | .17201 | .00098 | .17299 | 9 |
| 52 | .82888 | .99901 | .82987 | .17013 | .00099 | .17112 | 8 |
| 53 | .83075 | .99900 | .83175 | .16825 | .00100 | .16925 | 7 |
| 54 | .83261 | .99899 | .83361 | .16639 | .00101 | .16739 | 6 |
| 55 | 8.83446 | 9.99898 | 8.83547 | 11.16453 | 10.00102 | 11.16554 | 5 |
| 56 | .83630 | .99898 | .83732 | .16268 | .00102 | .16370 | 4 |
| 57 | .83813 | .99897 | .83916 | .16084 | .00103 | .16187 | 3 |
| 58 | .83996 | .99896 | .84100 | .15900 | .00104 | .16004 | 2 |
| 59 | .84177 | .99895 | .84282 | .15718 | .00105 | .15823 | 1 |
| 60 | 8.84358 | 9.99894 | 8.84464 | 11.15536 | 10.00106 | 11.15642 | 0 |

| M | Cosine | Sine | Cotangent | Tangent | Cosecant | Secant | M |

| M | Sine | Cosine | Tangent | Cotangent | Secant | Cosecant | M |
|---|------|--------|---------|-----------|--------|----------|---|
| 0 | 8.84358 | 9.99894 | 8.84464 | 11.15536 | 10.00106 | 11.15642 | 60 |
| 1 | .84539 | .99893 | .84646 | .15354 | .00107 | .15461 | 59 |
| 2 | .84718 | .99892 | .84826 | .15174 | .00108 | .15282 | 58 |
| 3 | .84897 | .99891 | .85006 | .14994 | .00109 | .15103 | 57 |
| 4 | .85075 | .99891 | .85185 | .14815 | .00109 | .14925 | 56 |
| 5 | 8.85252 | 9.99890 | 8.85363 | 11.14637 | 10.00110 | 11.14748 | 55 |
| 6 | .85429 | .99889 | .85540 | .14460 | .00111 | .14571 | 54 |
| 7 | .85605 | .99888 | .85717 | .14283 | .00112 | .14395 | 53 |
| 8 | .85780 | .99887 | .85893 | .14107 | .00113 | .14220 | 52 |
| 9 | .85955 | .99886 | .86069 | .13931 | .00114 | .14045 | 51 |
| 10 | 8.86128 | 9.99885 | 8.86243 | 11.13757 | 10.00115 | 11.13872 | 50 |
| 11 | .86301 | .99884 | .86417 | .13583 | .00116 | .13699 | 49 |
| 12 | .86474 | .99883 | .86591 | .13409 | .00117 | .13526 | 48 |
| 13 | .86645 | .99882 | .86763 | .13237 | .00118 | .13355 | 47 |
| 14 | .86816 | .99881 | .86935 | .13065 | .00119 | .13184 | 46 |
| 15 | 8.86987 | 9.99880 | 8.87106 | 11.12894 | 10.00120 | 11.13013 | 45 |
| 16 | .87156 | .99879 | .87277 | .12723 | .00121 | .12844 | 44 |
| 17 | .87325 | .99879 | .87447 | .12553 | .00121 | .12675 | 43 |
| 18 | .87494 | .99878 | .87616 | .12384 | .00122 | .12506 | 42 |
| 19 | .87661 | .99877 | .87785 | .12215 | .00123 | .12339 | 41 |
| 20 | 8.87829 | 9.99876 | 8.87953 | 11.12047 | 10.00124 | 11.12171 | 40 |
| 21 | .87995 | .99875 | .88120 | .11880 | .00125 | .12005 | 39 |
| 22 | .88161 | .99874 | .88287 | .11713 | .00126 | .11839 | 38 |
| 23 | .88326 | .99873 | .88453 | .11547 | .00127 | .11674 | 37 |
| 24 | .88490 | .99872 | .88618 | .11382 | .00128 | .11510 | 36 |
| 25 | 8.88654 | 9.99871 | 8.88783 | 11.11217 | 10.00129 | 11.11346 | 35 |
| 26 | .88817 | .99870 | .88948 | .11052 | .00130 | .11183 | 34 |
| 27 | .88980 | .99869 | .89111 | .10889 | .00131 | .11020 | 33 |
| 28 | .89142 | .99868 | .89274 | .10726 | .00132 | .10858 | 32 |
| 29 | .89304 | .99867 | .89437 | .10563 | .00133 | .10696 | 31 |
| 30 | 8.89464 | 9.99866 | 8.89598 | 11.10402 | 10.00134 | 11.10536 | 30 |
| 31 | .89625 | .99865 | .89760 | .10240 | .00135 | .10375 | 29 |
| 32 | .89784 | .99864 | .89920 | .10080 | .00136 | .10216 | 28 |
| 33 | .89943 | .99863 | .90080 | .09920 | .00137 | .10057 | 27 |
| 34 | .90102 | .99862 | .90240 | .09760 | .00138 | .09898 | 26 |
| 35 | 8.90260 | 9.99861 | 8.90399 | 11.09601 | 10.00139 | 11.09740 | 25 |
| 36 | .90417 | .99860 | .90557 | .09443 | .00140 | .09583 | 24 |
| 37 | .90574 | .99859 | .90715 | .09285 | .00141 | .09426 | 23 |
| 38 | .90730 | .99858 | .90872 | .09128 | .00142 | .09270 | 22 |
| 39 | .90885 | .99857 | .91029 | .08971 | .00143 | .09115 | 21 |
| 40 | 8.91040 | 9.99856 | 8.91185 | 11.08815 | 10.00144 | 11.08960 | 20 |
| 41 | .91195 | .99855 | .91340 | .08660 | .00145 | .08805 | 19 |
| 42 | .91349 | .99854 | .91495 | .08505 | .00146 | .08651 | 18 |
| 43 | .91502 | .99853 | .91650 | .08350 | .00147 | .08498 | 17 |
| 44 | .91655 | .99852 | .91803 | .08197 | .00148 | .08345 | 16 |
| 45 | 8.91807 | 9.99851 | 8.91957 | 11.08043 | 10.00149 | 11.08193 | 15 |
| 46 | .91959 | .99850 | .92110 | .07890 | .00150 | .08041 | 14 |
| 47 | .92110 | .99848 | .92262 | .07738 | .00152 | .07890 | 13 |
| 48 | .92261 | .99847 | .92414 | .07586 | .00153 | .07739 | 12 |
| 49 | .92411 | .99846 | .92565 | .07435 | .00154 | .07589 | 11 |
| 50 | 8.92561 | 9.99845 | 8.92716 | 11.07284 | 10.00155 | 11.07439 | 10 |
| 51 | .92710 | .99844 | .92866 | .07134 | .00156 | .07290 | 9 |
| 52 | .92859 | .99843 | .93016 | .06984 | .00157 | .07141 | 8 |
| 53 | .93007 | .99842 | .93165 | .06835 | .00158 | .06993 | 7 |
| 54 | .93154 | .99841 | .93313 | .06687 | .00159 | .06846 | 6 |
| 55 | 8.93301 | 9.99840 | 8.93462 | 11.06538 | 10.00160 | 11.06699 | 5 |
| 56 | .93448 | .99839 | .93609 | .06391 | .00161 | .06552 | 4 |
| 57 | .93594 | .99838 | .93756 | .06244 | .00162 | .06406 | 3 |
| 58 | .93740 | .99837 | .93903 | .06097 | .00163 | .06260 | 2 |
| 59 | .93885 | .99836 | .94049 | .05951 | .00164 | .06115 | 1 |
| 60 | 8.94030 | 9.99834 | 8.94195 | 11.05805 | 10.00166 | 11.05970 | 0 |

| M | Cosine | Sine | Cotangent | Tangent | Cosecant | Secant | M |
|---|--------|------|-----------|---------|----------|--------|---|

5°  **Logarithms of Trigonometrical Functions**  174°

| M | Sine | Cosine | Tangent | Cotangent | Secant | Cosecant | M |
|---|------|--------|---------|-----------|--------|----------|---|
| 0 | 8.94030 | 9.99834 | 8.94195 | 11.05805 | 10.00166 | 11.05970 | 60 |
| 1 | .94174 | .99833 | .94340 | .05660 | .00167 | .05826 | 59 |
| 2 | .94317 | .99832 | .94485 | .05515 | .00168 | .05683 | 58 |
| 3 | .94461 | .99831 | .94630 | .05370 | .00169 | .05539 | 57 |
| 4 | .94603 | .99830 | .94773 | .05227 | .00170 | .05397 | 56 |
| 5 | 8.94746 | 9.99829 | 8.94917 | 11.05083 | 10.00171 | 11.05254 | 55 |
| 6 | .94887 | .99828 | .95060 | .04940 | .00172 | .05113 | 54 |
| 7 | .95029 | .99827 | .95202 | .04798 | .00173 | .04971 | 53 |
| 8 | .95170 | .99825 | .95344 | .04656 | .00175 | .04830 | 52 |
| 9 | .95310 | .99824 | .95486 | .04514 | .00176 | .04690 | 51 |
| 10 | 8.95450 | 9.99823 | 8.95627 | 11.04373 | 10.00177 | 11.04550 | 50 |
| 11 | .95589 | .99822 | .95767 | .04233 | .00178 | .04411 | 49 |
| 12 | .95728 | .99821 | .95908 | .04092 | .00179 | .04272 | 48 |
| 13 | .95867 | .99820 | .96047 | .03953 | .00180 | .04133 | 47 |
| 14 | .96005 | .99819 | .96187 | .03813 | .00181 | .03995 | 46 |
| 15 | 8.96143 | 9.99817 | 8.96325 | 11.03675 | 10.00183 | 11.03857 | 45 |
| 16 | .96280 | .99816 | .96464 | .03536 | .00184 | .03720 | 44 |
| 17 | .96417 | .99815 | .96602 | .03398 | .00185 | .03583 | 43 |
| 18 | .96553 | .99814 | .96739 | .03261 | .00186 | .03447 | 42 |
| 19 | .96689 | .99813 | .96877 | .03123 | .00187 | .03311 | 41 |
| 20 | 8.96825 | 9.99812 | 8.97013 | 11.02987 | 10.00188 | 11.03175 | 40 |
| 21 | .96960 | .99810 | .97150 | .02850 | .00190 | .03040 | 39 |
| 22 | .97095 | .99809 | .97285 | .02715 | .00191 | .02905 | 38 |
| 23 | .97229 | .99808 | .97421 | .02579 | .00192 | .02771 | 37 |
| 24 | .97363 | .99807 | .97556 | .02444 | .00193 | .02637 | 36 |
| 25 | 8.97496 | 9.99806 | 8.97691 | 11.02309 | 10.00194 | 11.02504 | 35 |
| 26 | .97629 | .99804 | .97825 | .02175 | .00196 | .02371 | 34 |
| 27 | .97762 | .99803 | .97959 | .02041 | .00197 | .02238 | 33 |
| 28 | .97894 | .99802 | .98092 | .01908 | .00198 | .02106 | 32 |
| 29 | .98026 | .99801 | .98225 | .01775 | .00199 | .01974 | 31 |
| 30 | 8.98157 | 9.99800 | 8.98358 | 11.01642 | 10.00200 | 11.01843 | 30 |
| 31 | .98288 | .99798 | .98490 | .01510 | .00202 | .01712 | 29 |
| 32 | .98419 | .99797 | .98622 | .01378 | .00203 | .01581 | 28 |
| 33 | .98549 | .99796 | .98753 | .01247 | .00204 | .01451 | 27 |
| 34 | .98679 | .99795 | .98884 | .01116 | .00205 | .01321 | 26 |
| 35 | 8.98808 | 9.99793 | 8.99015 | 11.00985 | 10.00207 | 11.01192 | 25 |
| 36 | .98937 | .99792 | .99145 | .00855 | .00208 | .01063 | 24 |
| 37 | .99066 | .99791 | .99275 | .00725 | .00209 | .00934 | 23 |
| 38 | .99194 | .99790 | .99405 | .00595 | .00210 | .00806 | 22 |
| 39 | .99322 | .99788 | .99534 | .00466 | .00212 | .00678 | 21 |
| 40 | 8.99450 | 9.99787 | 8.99662 | 11.00338 | 10.00213 | 11.00550 | 20 |
| 41 | .99577 | .99786 | .99791 | .00209 | .00214 | .00423 | 19 |
| 42 | .99704 | .99785 | .99919 | .00081 | .00215 | .00296 | 18 |
| 43 | .99830 | .99783 | 9.00046 | 10.99954 | .00217 | .00170 | 17 |
| 44 | .99956 | .99782 | .00174 | .99826 | .00218 | .00044 | 16 |
| 45 | 9.00082 | 9.99781 | 9.00301 | 10.99699 | 10.00219 | 10.99918 | 15 |
| 46 | .00207 | .99780 | .00427 | .99573 | .00220 | .99793 | 14 |
| 47 | .00332 | .99778 | .00553 | .99447 | .00222 | .99668 | 13 |
| 48 | .00456 | .99777 | .00679 | .99321 | .00223 | .99544 | 12 |
| 49 | .00581 | .99776 | .00805 | .99195 | .00224 | .99419 | 11 |
| 50 | 9.00704 | 9.99775 | 9.00930 | 10.99070 | 10.00225 | 10.99296 | 10 |
| 51 | .00828 | .99773 | .01055 | .98945 | .00227 | .99172 | 9 |
| 52 | .00951 | .99772 | .01179 | .98821 | .00228 | .99049 | 8 |
| 53 | .01074 | .99771 | .01303 | .98697 | .00229 | .98926 | 7 |
| 54 | .01196 | .99769 | .01427 | .98573 | .00231 | .98804 | 6 |
| 55 | 9.01318 | 9.99768 | 9.01550 | 10.98450 | 10.00232 | 10.98682 | 5 |
| 56 | .01440 | .99767 | .01673 | .98327 | .00233 | .98560 | 4 |
| 57 | .01561 | .99765 | .01796 | .98204 | .00235 | .98439 | 3 |
| 58 | .01682 | .99764 | .01918 | .98082 | .00236 | .98318 | 2 |
| 59 | .01803 | .99763 | .02040 | .97960 | .00237 | .98197 | 1 |
| 60 | 9.01923 | 9.99761 | 9.02162 | 10.97838 | 10.00239 | 10.98077 | 0 |
| M | Cosine | Sine | Cotangent | Tangent | Cosecant | Secant | M |

95°                                                                                  84°

| M | Sine | Cosine | Tangent | Cotangent | Secant | Cosecant | M |
|---|------|--------|---------|-----------|--------|----------|---|
| 0 | 9.01923 | 9.99761 | 9.02162 | 10.97838 | 10.00239 | 10.98077 | 60 |
| 1 | .02043 | .99760 | .02283 | .97717 | .00240 | .97957 | 59 |
| 2 | .02163 | .99759 | .02404 | .97596 | .00241 | .97837 | 58 |
| 3 | .02283 | .99757 | .02525 | .97475 | .00243 | .97717 | 57 |
| 4 | .02402 | .99756 | .02645 | .97355 | .00244 | .97598 | 56 |
| 5 | 9.02520 | 9.99755 | 9.02766 | 10.97234 | 10.00245 | 10.97480 | 55 |
| 6 | .02639 | .99753 | .02885 | .97115 | .00247 | .97361 | 54 |
| 7 | .02757 | .99752 | .03005 | .96995 | .00248 | .97243 | 53 |
| 8 | .02874 | .99751 | .03124 | .96876 | .00249 | .97126 | 52 |
| 9 | .02992 | .99749 | .03242 | .96758 | .00251 | .97008 | 51 |
| 10 | 9.03109 | 9.99748 | 9.03361 | 10.96639 | 10.00252 | 10.96891 | 50 |
| 11 | .03226 | .99747 | .03479 | .96521 | .00253 | .96774 | 49 |
| 12 | .03342 | .99745 | .03597 | .96403 | .00255 | .96658 | 48 |
| 13 | .03458 | .99744 | .03714 | .96286 | .00256 | .96542 | 47 |
| 14 | .03574 | .99742 | .03832 | .96168 | .00258 | .96426 | 46 |
| 15 | 9.03690 | 9.99741 | 9.03948 | 10.96052 | 10.00259 | 10.96310 | 45 |
| 16 | .03805 | .99740 | .04065 | .95935 | .00260 | .96195 | 44 |
| 17 | .03920 | .99738 | .04181 | .95819 | .00262 | .96080 | 43 |
| 18 | .04034 | .99737 | .04297 | .95703 | .00263 | .95966 | 42 |
| 19 | .04149 | .99736 | .04413 | .95587 | .00264 | .95851 | 41 |
| 20 | 9.04262 | 9.99734 | 9.04528 | 10.95472 | 10.00266 | 10.95738 | 40 |
| 21 | .04376 | .99733 | .04643 | .95357 | .00267 | .95624 | 39 |
| 22 | .04490 | .99731 | .04758 | .95242 | .00269 | .95510 | 38 |
| 23 | .04603 | .99730 | .04873 | .95127 | .00270 | .95397 | 37 |
| 24 | .04715 | .99728 | .04987 | .95013 | .00272 | .95285 | 36 |
| 25 | 9.04828 | 9.99727 | 9.05101 | 10.94899 | 10.00273 | 10.95172 | 35 |
| 26 | .04940 | .99726 | .05214 | .94786 | .00274 | .95060 | 34 |
| 27 | .05052 | .99724 | .05328 | .94672 | .00276 | .94948 | 33 |
| 28 | .05164 | .99723 | .05441 | .94559 | .00277 | .94836 | 32 |
| 29 | .05275 | .99721 | .05553 | .94447 | .00279 | .94725 | 31 |
| 30 | 9.05386 | 9.99720 | 9.05666 | 10.94334 | 10.00280 | 10.94614 | 30 |
| 31 | .05497 | .99718 | .05778 | .94222 | .00282 | .94503 | 29 |
| 32 | .05607 | .99717 | .05890 | .94110 | .00283 | .94393 | 28 |
| 33 | .05717 | .99716 | .06002 | .93998 | .00284 | .94283 | 27 |
| 34 | .05827 | .99714 | .06113 | .93887 | .00286 | .94173 | 26 |
| 35 | 9.05937 | 9.99713 | 9.06224 | 10.93776 | 10.00287 | 10.94063 | 25 |
| 36 | .06046 | .99711 | .06335 | .93665 | .00289 | .93954 | 24 |
| 37 | .06155 | .99710 | .06445 | .93555 | .00290 | .93845 | 23 |
| 38 | .06264 | .99708 | .06556 | .93444 | .00292 | .93736 | 22 |
| 39 | .06372 | .99707 | .06666 | .93334 | .00293 | .93628 | 21 |
| 40 | 9.06481 | 9.99705 | 9.06775 | 10.93225 | 10.00295 | 10.93519 | 20 |
| 41 | .06589 | .99704 | .06885 | .93115 | .00296 | .93411 | 19 |
| 42 | .06696 | .99702 | .06994 | .93006 | .00298 | .93304 | 18 |
| 43 | .06804 | .99701 | .07103 | .92897 | .00299 | .93196 | 17 |
| 44 | .06911 | .99699 | .07211 | .92789 | .00301 | .93089 | 16 |
| 45 | 9.07018 | 9.99698 | 9.07320 | 10.92680 | 10.00302 | 10.92982 | 15 |
| 46 | .07124 | .99696 | .07428 | .92572 | .00304 | .92876 | 14 |
| 47 | .07231 | .99695 | .07536 | .92464 | .00305 | .92769 | 13 |
| 48 | .07337 | .99693 | .07643 | .92357 | .00307 | .92663 | 12 |
| 49 | .07442 | .99692 | .07751 | .92249 | .00308 | .92558 | 11 |
| 50 | 9.07548 | 9.99690 | 9.07858 | 10.92142 | 10.00310 | 10.92452 | 10 |
| 51 | .07653 | .99689 | .07964 | .92036 | .00311 | .92347 | 9 |
| 52 | .07758 | .99687 | .08071 | .91929 | .00313 | .92242 | 8 |
| 53 | .07863 | .99686 | .08177 | .91823 | .00314 | .92137 | 7 |
| 54 | .07968 | .99684 | .08283 | .91717 | .00316 | .92032 | 6 |
| 55 | 9.08072 | 9.99683 | 9.08389 | 10.91611 | 10.00317 | 10.91928 | 5 |
| 56 | .08176 | .99681 | .08495 | .91505 | .00319 | .91824 | 4 |
| 57 | .08280 | .99680 | .08600 | .91400 | .00320 | .91720 | 3 |
| 58 | .08383 | .99678 | .08705 | .91295 | .00322 | .91617 | 2 |
| 59 | .08486 | .99677 | .08810 | .91190 | .00323 | .91514 | 1 |
| 60 | 9.08589 | 9.99675 | 9.08914 | 10.91086 | 10.00325 | 10.91411 | 0 |
| M | Cosine | Sine | Cotangent | Tangent | Cosecant | Secant | M |

Logarithms of Trigonometrical Functions

| M | Sine | Cosine | Tangent | Cotangent | Secant | Cosecant | M |
|---|---|---|---|---|---|---|---|
| 0 | 9.08589 | 9.99675 | 9.08914 | 10.91086 | 10.00325 | 10.91411 | 60 |
| 1 | .08692 | .99674 | .09019 | .90981 | .00326 | .91308 | 59 |
| 2 | .08795 | .99672 | .09123 | .90877 | .00328 | .91205 | 58 |
| 3 | .08897 | .99670 | .09227 | .90773 | .00330 | .91103 | 57 |
| 4 | .08999 | .99669 | .09330 | .90670 | .00331 | .91001 | 56 |
| 5 | 9.09101 | 9.99667 | 9.09434 | 10.90566 | 10.00333 | 10.90899 | 55 |
| 6 | .09202 | .99666 | .09537 | .90463 | .00334 | .90798 | 54 |
| 7 | .09304 | .99664 | .09640 | .90360 | .00336 | .90696 | 53 |
| 8 | .09405 | .99663 | .09742 | .90258 | .00337 | .90595 | 52 |
| 9 | .09506 | .99661 | .09845 | .90155 | .00339 | .90494 | 51 |
| 10 | 9.09606 | 9.99659 | 9.09947 | 10.90053 | 10.00341 | 10.90394 | 50 |
| 11 | .09707 | .99658 | .10049 | .89951 | .00342 | .90293 | 49 |
| 12 | .09807 | .99656 | .10150 | .89850 | .00344 | .90193 | 48 |
| 13 | .09907 | .99655 | .10252 | .89748 | .00345 | .90093 | 47 |
| 14 | .10006 | .99653 | .10353 | .89647 | .00347 | .89994 | 46 |
| 15 | 9.10106 | 9.99651 | 9.10454 | 10.89546 | 10.00349 | 10.89894 | 45 |
| 16 | .10205 | .99650 | .10555 | .89445 | .00350 | .89795 | 44 |
| 17 | .10304 | .99648 | .10656 | .89344 | .00352 | .89696 | 43 |
| 18 | .10402 | .99647 | .10756 | .89244 | .00353 | .89598 | 42 |
| 19 | .10501 | .99645 | .10856 | .89144 | .00355 | .89499 | 41 |
| 20 | 9.10599 | 9.99643 | 9.10956 | 10.89044 | 10.00357 | 10.89401 | 40 |
| 21 | .10697 | .99642 | .11056 | .88944 | .00358 | .89303 | 39 |
| 22 | .10795 | .99640 | .11155 | .88845 | .00360 | .89205 | 38 |
| 23 | .10893 | .99638 | .11254 | .88746 | .00362 | .89107 | 37 |
| 24 | .10990 | .99637 | .11353 | .88647 | .00363 | .89010 | 36 |
| 25 | 9.11087 | 9.99635 | 9.11452 | 10.88548 | 10.00365 | 10.88913 | 35 |
| 26 | .11184 | .99633 | .11551 | .88449 | .00367 | .88816 | 34 |
| 27 | .11281 | .99632 | .11649 | .88351 | .00368 | .88719 | 33 |
| 28 | .11377 | .99630 | .11747 | .88253 | .00370 | .88623 | 32 |
| 29 | .11474 | .99629 | .11845 | .88155 | .00371 | .88526 | 31 |
| 30 | 9.11570 | 9.99627 | 9.11943 | 10.88057 | 10.00373 | 10.88430 | 30 |
| 31 | .11666 | .99625 | .12040 | .87960 | .00375 | .88334 | 29 |
| 32 | .11761 | .99624 | .12138 | .87862 | .00376 | .88239 | 28 |
| 33 | .11857 | .99622 | .12235 | .87765 | .00378 | .88143 | 27 |
| 34 | .11952 | .99620 | .12332 | .87668 | .00380 | .88048 | 26 |
| 35 | 9.12047 | 9.99618 | 9.12428 | 10.87572 | 10.00382 | 10.87953 | 25 |
| 36 | .12142 | .99617 | .12525 | .87475 | .00383 | .87858 | 24 |
| 37 | .12236 | .99615 | .12621 | .87379 | .00385 | .87764 | 23 |
| 38 | .12331 | .99613 | .12717 | .87283 | .00387 | .87669 | 22 |
| 39 | .12425 | .99612 | .12813 | .87187 | .00388 | .87575 | 21 |
| 40 | 9.12519 | 9.99610 | 9.12909 | 10.87091 | 10.00390 | 10.87481 | 20 |
| 41 | .12612 | .99608 | .13004 | .86996 | .00392 | .87388 | 19 |
| 42 | .12706 | .99607 | .13099 | .86901 | .00393 | .87294 | 18 |
| 43 | .12799 | .99605 | .13194 | .86806 | .00395 | .87201 | 17 |
| 44 | .12892 | .99603 | .13289 | .86711 | .00397 | .87108 | 16 |
| 45 | 9.12985 | 9.99601 | 9.13384 | 10.86616 | 10.00399 | 10.87015 | 15 |
| 46 | .13078 | .99600 | .13478 | .86522 | .00400 | .86922 | 14 |
| 47 | .13171 | .99598 | .13573 | .86427 | .00402 | .86829 | 13 |
| 48 | .13263 | .99596 | .13667 | .86333 | .00404 | .86737 | 12 |
| 49 | .13355 | .99595 | .13761 | .86239 | .00405 | .86645 | 11 |
| 50 | 9.13447 | 9.99593 | 9.13854 | 10.86146 | 10.00407 | 10.86553 | 10 |
| 51 | .13539 | .99591 | .13948 | .86052 | .00409 | .86461 | 9 |
| 52 | .13630 | .99589 | .14041 | .85959 | .00411 | .86370 | 8 |
| 53 | .13722 | .99588 | .14134 | .85866 | .00412 | .86278 | 7 |
| 54 | .13813 | .99586 | .14227 | .85773 | .00414 | .86187 | 6 |
| 55 | 9.13904 | 9.99584 | 9.14320 | 10.85680 | 10.00416 | 10.86096 | 5 |
| 56 | .13994 | .99582 | .14412 | .85588 | .00418 | .86006 | 4 |
| 57 | .14085 | .99581 | .14504 | .85496 | .00419 | .85915 | 3 |
| 58 | .14175 | .99579 | .14597 | .85403 | .00421 | .85825 | 2 |
| 59 | .14266 | .99577 | .14688 | .85312 | .00423 | .85734 | 1 |
| 60 | 9.14356 | 9.99575 | 9.14780 | 10.85220 | 10.00425 | 10.85644 | 0 |
| M | Cosine | Sine | Cotangent | Tangent | Cosecant | Secant | M |

| M | Sine | Cosine | Tangent | Cotangent | Secant | Cosecant | M |
|---|------|--------|---------|-----------|--------|----------|---|
| 0 | 9.14356 | 9.99575 | 9.14780 | 10.85220 | 10.00425 | 10.85644 | 60 |
| 1 | .14445 | .99574 | .14872 | .85128 | .00426 | .85555 | 59 |
| 2 | .14535 | .99572 | .14963 | .85037 | .00428 | .85465 | 58 |
| 3 | .14624 | .99570 | .15054 | .84946 | .00430 | .85376 | 57 |
| 4 | .14714 | .99568 | .15145 | .84855 | .00432 | .85286 | 56 |
| 5 | 9.14803 | 9.99566 | 9.15236 | 10.84764 | 10.00434 | 10.85197 | 55 |
| 6 | .14891 | .99565 | .15327 | .84673 | .00435 | .85109 | 54 |
| 7 | .14980 | .99563 | .15417 | .84583 | .00437 | .85020 | 53 |
| 8 | .15069 | .99561 | .15508 | .84492 | .00439 | .84931 | 52 |
| 9 | .15157 | .99559 | .15598 | .84402 | .00441 | .84843 | 51 |
| 10 | 9.15245 | 9.99557 | 9.15688 | 10.84312 | 10.00443 | 10.84755 | 50 |
| 11 | .15333 | .99556 | .15777 | .84223 | .00444 | .84667 | 49 |
| 12 | .15421 | .99554 | .15867 | .84133 | .00446 | .84579 | 48 |
| 13 | .15508 | .99552 | .15956 | .84044 | .00448 | .84492 | 47 |
| 14 | .15596 | .99550 | .16046 | .83954 | .00450 | .84404 | 46 |
| 15 | 9.15683 | 9.99548 | 9.16135 | 10.83865 | 10.00452 | 10.84317 | 45 |
| 16 | .15770 | .99546 | .16224 | .83776 | .00454 | .84230 | 44 |
| 17 | .15857 | .99545 | .16312 | .83688 | .00455 | .84143 | 43 |
| 18 | .15944 | .99543 | .16401 | .83599 | .00457 | .84056 | 42 |
| 19 | .16030 | .99541 | .16489 | .83511 | .00459 | .83970 | 41 |
| 20 | 9.16116 | 9.99539 | 9.16577 | 10.83423 | 10.00461 | 10.83884 | 40 |
| 21 | .16203 | .99537 | .16665 | .83335 | .00463 | .83797 | 39 |
| 22 | .16289 | .99535 | .16753 | .83247 | .00465 | .83711 | 38 |
| 23 | .16374 | .99533 | .16841 | .83159 | .00467 | .83626 | 37 |
| 24 | .16460 | .99532 | .16928 | .83072 | .00468 | .83540 | 36 |
| 25 | 9.16545 | 9.99530 | 9.17016 | 10.82984 | 10.00470 | 10.83455 | 35 |
| 26 | .16631 | .99528 | .17103 | .82897 | .00472 | .83369 | 34 |
| 27 | .16716 | .99526 | .17190 | .82810 | .00474 | .83284 | 33 |
| 28 | .16801 | .99524 | .17277 | .82723 | .00476 | .83199 | 32 |
| 29 | .16886 | .99522 | .17363 | .82637 | .00478 | .83114 | 31 |
| 30 | 9.16970 | 9.99520 | 9.17450 | 10.82550 | 10.00480 | 10.83030 | 30 |
| 31 | .17055 | .99518 | .17536 | .82464 | .00482 | .82945 | 29 |
| 32 | .17139 | .99517 | .17622 | .82378 | .00483 | .82861 | 28 |
| 33 | .17223 | .99515 | .17708 | .82292 | .00485 | .82777 | 27 |
| 34 | .17307 | .99513 | .17794 | .82206 | .00487 | .82693 | 26 |
| 35 | 9.17391 | 9.99511 | 9.17880 | 10.82120 | 10.00489 | 10.82609 | 25 |
| 36 | .17474 | .99509 | .17965 | .82035 | .00491 | .82526 | 24 |
| 37 | .17558 | .99507 | .18051 | .81949 | .00493 | .82442 | 23 |
| 38 | .17641 | .99505 | .18136 | .81864 | .00495 | .82359 | 22 |
| 39 | .17724 | .99503 | .18221 | .81779 | .00497 | .82276 | 21 |
| 40 | 9.17807 | 9.99501 | 9.18306 | 10.81694 | 10.00499 | 10.82193 | 20 |
| 41 | .17890 | .99499 | .18391 | .81609 | .00501 | .82110 | 19 |
| 42 | .17973 | .99497 | .18475 | .81525 | .00503 | .82027 | 18 |
| 43 | .18055 | .99495 | .18560 | .81440 | .00505 | .81945 | 17 |
| 44 | .18137 | .99494 | .18644 | .81356 | .00508 | .81863 | 16 |
| 45 | 9.18220 | 9.99492 | 9.18728 | 10.81272 | 10.00508 | 10.81780 | 15 |
| 46 | .18302 | .99490 | .18812 | .81188 | .00510 | .81698 | 14 |
| 47 | .18383 | .99488 | .18896 | .81104 | .00512 | .81617 | 13 |
| 48 | .18465 | .99486 | .18979 | .81021 | .00514 | .81535 | 12 |
| 49 | .18547 | .99484 | .19063 | .80937 | .00516 | .81453 | 11 |
| 50 | 9.18628 | 9.99482 | 9.19146 | 10.80854 | 10.00518 | 10.81372 | 10 |
| 51 | .18709 | .99480 | .19229 | .80771 | .00520 | .81291 | 9 |
| 52 | .18790 | .99478 | .19312 | .80688 | .00522 | .81210 | 8 |
| 53 | .18871 | .99476 | .19395 | .80605 | .00524 | .81129 | 7 |
| 54 | .18952 | .99474 | .19478 | .80522 | .00526 | .81048 | 6 |
| 55 | 9.19033. | 9.99472 | 9.19561 | 10.80439 | 10.00528 | 10.80967 | 5 |
| 56 | .19113 | .99470 | .19643 | .80357 | .00530 | .80887 | 4 |
| 57 | .19193 | .99468 | .19725 | .80275 | .00532 | .80807 | 3 |
| 58 | .19273 | .99466 | .19807 | .80193 | .00534 | .80727 | 2 |
| 59 | .19353 | .99464 | .19889 | .80111 | .00536 | .80647 | 1 |
| 60 | 9.19433 | 9.99462 | 9.19971 | 10.80029 | 10.00538 | 10.80567 | 0 |

| M | Cosine | Sine | Cotangent | Tangent | Cosecant | Secant | M |

## Logarithms of Trigonometrical Functions

| M | Sine | Cosine | Tangent | Cotangent | Secant | Cosecant | M |
|---|------|--------|---------|-----------|--------|----------|---|
| 0 | 9.19433 | 9.99462 | 9.19971 | 10.80029 | 10.00538 | 10.80567 | 60 |
| 1 | .19513 | .99460 | .20053 | .79947 | .00540 | .80487 | 59 |
| 2 | .19592 | .99458 | .20134 | .79866 | .00542 | .80408 | 58 |
| 3 | .19672 | .99456 | .20216 | .79784 | .00544 | .80328 | 57 |
| 4 | .19751 | .99454 | .20297 | .79703 | .00546 | .80249 | 56 |
| 5 | 9.19830 | 9.99452 | 9.20378 | 10.79622 | 10.00548 | 10.80170 | 55 |
| 6 | .19909 | .99450 | .20459 | .79541 | .00550 | .80091 | 54 |
| 7 | .19988 | .99448 | .20540 | .79460 | .00552 | .80012 | 53 |
| 8 | .20067 | .99446 | .20621 | .79379 | .00554 | .79933 | 52 |
| 9 | .20145 | .99444 | .20701 | .79299 | .00556 | .79855 | 51 |
| 10 | 9.20223 | 9.99442 | 9.20782 | 10.79218 | 10.00558 | 10.79777 | 50 |
| 11 | .20302 | .99440 | .20862 | .79138 | .00560 | .79698 | 49 |
| 12 | .20380 | .99438 | .20942 | .79058 | .00562 | .79620 | 48 |
| 13 | .20458 | .99436 | .21022 | .78978 | .00564 | .79542 | 47 |
| 14 | .20535 | .99434 | .21102 | .78898 | .00566 | .79465 | 46 |
| 15 | 9.20613 | 9.99432 | 9.21182 | 10.78818 | 10.00568 | 10.79387 | 45 |
| 16 | .20691 | .99429 | .21261 | .78739 | .00571 | .79309 | 44 |
| 17 | .20768 | .99427 | .21341 | .78659 | .00573 | .79232 | 43 |
| 18 | .20845 | .99425 | .21420 | .78580 | .00575 | .79155 | 42 |
| 19 | .20922 | .99423 | .21499 | .78501 | .00577 | .79078 | 41 |
| 20 | 9.20999 | 9.99421 | 9.21578 | 10.78422 | 10.00579 | 10.79001 | 40 |
| 21 | .21076 | .99419 | .21657 | .78343 | .00581 | .78924 | 39 |
| 22 | .21153 | .99417 | .21736 | .78264 | .00583 | .78847 | 38 |
| 23 | .21229 | .99415 | .21814 | .78186 | .00585 | .78771 | 37 |
| 24 | .21306 | .99413 | .21893 | .78107 | .00587 | .78694 | 36 |
| 25 | 9.21382 | 9.99411 | 9.21971 | 10.78029 | 10.00589 | 10.78618 | 35 |
| 26 | .21458 | .99409 | .22049 | .77951 | .00591 | .78542 | 34 |
| 27 | .21534 | .99407 | .22127 | .77873 | .00593 | .78466 | 33 |
| 28 | .21610 | .99404 | .22205 | .77795 | .00596 | .78390 | 32 |
| 29 | .21685 | .99402 | .22283 | .77717 | .00598 | .78315 | 31 |
| 30 | 9.21761 | 9.99400 | 9.22361 | 10.77639 | 10.00600 | 10.78239 | 30 |
| 31 | .21836 | .99398 | .22438 | .77562 | .00602 | .78164 | 29 |
| 32 | .21912 | .99396 | .22516 | .77484 | .00604 | .78088 | 28 |
| 33 | .21987 | .99394 | .22593 | .77407 | .00606 | .78013 | 27 |
| 34 | .22062 | .99392 | .22670 | .77330 | .00608 | .77938 | 26 |
| 35 | 9.22137 | 9.99390 | 9.22747 | 10.77253 | 10.00610 | 10.77863 | 25 |
| 36 | .22211 | .99388 | .22824 | .77176 | .00612 | .77789 | 24 |
| 37 | .22286 | .99385 | .22901 | .77099 | .00615 | .77714 | 23 |
| 38 | .22361 | .99383 | .22977 | .77023 | .00617 | .77639 | 22 |
| 39 | .22435 | .99381 | .23054 | .76946 | .00619 | .77565 | 21 |
| 40 | 9.22509 | 9.99379 | 9.23130 | 10.76870 | 10.00621 | 10.77491 | 20 |
| 41 | .22583 | .99377 | .23206 | .76794 | .00623 | .77417 | 19 |
| 42 | .22657 | .99375 | .23283 | .76717 | .00625 | .77343 | 18 |
| 43 | .22731 | .99372 | .23359 | .76641 | .00628 | .77269 | 17 |
| 44 | .22805 | .99370 | .23435 | .76565 | .00630 | .77195 | 16 |
| 45 | 9.22878 | 9.99368 | 9.23510 | 10.76490 | 10.00632 | 10.77122 | 15 |
| 46 | .22952 | .99366 | .23586 | .76414 | .00634 | .77048 | 14 |
| 47 | .23025 | .99364 | .23661 | .76339 | .00636 | .76975 | 13 |
| 48 | .23098 | .99362 | .23737 | .76263 | .00638 | .76902 | 12 |
| 49 | .23171 | .99359 | .23812 | .76188 | .00641 | .76829 | 11 |
| 50 | 9.23244 | 9.99357 | 9.23887 | 10.76113 | 10.00643 | 10.76756 | 10 |
| 51 | .23317 | .99355 | .23962 | .76038 | .00645 | .76683 | 9 |
| 52 | .23390 | .99353 | .24037 | .75963 | .00647 | .76610 | 8 |
| 53 | .23462 | .99351 | .24112 | .75888 | .00649 | .76538 | 7 |
| 54 | .23535 | .99348 | .24186 | .75814 | .00652 | .76465 | 6 |
| 55 | 9.23607 | 9.99346 | 9.24261 | 10.75739 | 10.00654 | 10.76393 | 5 |
| 56 | .23679 | .99344 | .24335 | .75665 | .00656 | .76321 | 4 |
| 57 | .23752 | .99342 | .24410 | .75590 | .00658 | .76248 | 3 |
| 58 | .23823 | .99340 | .24484 | .75516 | .00660 | .76177 | 2 |
| 59 | .23895 | .99337 | .24558 | .75442 | .00663 | .76105 | 1 |
| 60 | 9.23967 | 9.99335 | 9.24632 | 10.75368 | 10.00665 | 10.76033 | 0 |

| M | Cosine | Sine | Cotangent | Tangent | Cosecant | Secant | M |
|---|--------|------|-----------|---------|----------|--------|---|

| M | Sine | Cosine | Tangent | Cotangent | Secant | Cosecant | M |
|---|---|---|---|---|---|---|---|
| 0 | 9.23967 | 9.99335 | 9.24632 | 10.75368 | 10.00665 | 10.76033 | 60 |
| 1 | .24039 | .99333 | .24706 | .75294 | .00667 | .75961 | 59 |
| 2 | .24110 | .99331 | .24779 | .75221 | .00669 | .75890 | 58 |
| 3 | .24181 | .99328 | .24853 | .75147 | .00672 | .75819 | 57 |
| 4 | .24253 | .99326 | .24926 | .75074 | .00674 | .75747 | 56 |
| 5 | 9.24324 | 9.99324 | 9.25000 | 10.75000 | 10.00676 | 10.75676 | 55 |
| 6 | .24395 | .99322 | .25073 | .74927 | .00678 | .75605 | 54 |
| 7 | .24466 | .99319 | .25146 | .74854 | .00681 | .75534 | 53 |
| 8 | .24536 | .99317 | .25219 | .74781 | .00683 | .75464 | 52 |
| 9 | .24607 | .99315 | .25292 | .74708 | .00685 | .75393 | 51 |
| 10 | 9.24677 | 9.99313 | 9.25365 | 10.74635 | 10.00687 | 10.75323 | 50 |
| 11 | .24748 | .99310 | .25437 | .74563 | .00690 | .75252 | 49 |
| 12 | .24818 | .99308 | .25510 | .74490 | .00692 | .75182 | 48 |
| 13 | .24888 | .99306 | .25582 | .74418 | .00694 | .75112 | 47 |
| 14 | .24958 | .99304 | .25655 | .74345 | .00696 | .75042 | 46 |
| 15 | 9.25028 | 9.99301 | 9.25727 | 10.74273 | 10.00699 | 10.74972 | 45 |
| 16 | .25098 | .99299 | .25799 | .74201 | .00701 | .74902 | 44 |
| 17 | .25168 | .99297 | .25871 | .74129 | .00703 | .74832 | 43 |
| 18 | .25237 | .99294 | .25943 | .74057 | .00706 | .74763 | 42 |
| 19 | .25307 | .99292 | .26015 | .73985 | .00708 | .74693 | 41 |
| 20 | 9.25376 | 9.99290 | 9.26086 | 10.73914 | 10.00710 | 10.74624 | 40 |
| 21 | .25445 | .99288 | .26158 | .73842 | .00712 | .74555 | 39 |
| 22 | .25514 | .99285 | .26229 | .73771 | .00715 | .74486 | 38 |
| 23 | .25583 | .99283 | .26301 | .73699 | .00717 | .74417 | 37 |
| 24 | .25652 | .99281 | .26372 | .73628 | .00719 | .74348 | 36 |
| 25 | 9.25721 | 9.99278 | 9.26443 | 10.73557 | 10.00722 | 10.74279 | 35 |
| 26 | .25790 | .99276 | .26514 | .73486 | .00724 | .74210 | 34 |
| 27 | .25858 | .99274 | .26585 | .73415 | .00726 | .74142 | 33 |
| 28 | .25927 | .99271 | .26655 | .73345 | .00729 | .74073 | 32 |
| 29 | .25995 | .99269 | .26726 | .73274 | .00731 | .74005 | 31 |
| 30 | 9.26063 | 9.99267 | 9.26797 | 10.73203 | 10.00733 | 10.73937 | 30 |
| 31 | .26131 | .99264 | .26867 | .73133 | .00736 | .73869 | 29 |
| 32 | .26199 | .99262 | .26937 | .73063 | .00738 | .73801 | 28 |
| 33 | .26267 | .99260 | .27008 | .72992 | .00740 | .73733 | 27 |
| 34 | .26335 | .99257 | .27078 | .72922 | .00743 | .73665 | 26 |
| 35 | 9.26403 | 9.99255 | 9.27148 | 10.72852 | 10.00745 | 10.73597 | 25 |
| 36 | .26470 | .99252 | .27218 | .72782 | .00748 | .73530 | 24 |
| 37 | .26538 | .99250 | .27288 | .72712 | .00750 | .73462 | 23 |
| 38 | .26605 | .99248 | .27357 | .72643 | .00752 | .73395 | 22 |
| 39 | .26672 | .99245 | .27427 | .72573 | .00755 | .73328 | 21 |
| 40 | 9.26739 | 9.99243 | 9.27496 | 10.72504 | 10.00757 | 10.73261 | 20 |
| 41 | .26806 | .99241 | .27566 | .72434 | .00759 | .73194 | 19 |
| 42 | .26873 | .99238 | .27635 | .72365 | .00762 | .73127 | 18 |
| 43 | .26940 | .99236 | .27704 | .72296 | .00764 | .73060 | 17 |
| 44 | .27007 | .99233 | .27773 | .72227 | .00767 | .72993 | 16 |
| 45 | 9.27073 | 9.99231 | 9.27842 | 10.72158 | 10.00769 | 10.72927 | 15 |
| 46 | .27140 | .99229 | .27911 | .72089 | .00771 | .72860 | 14 |
| 47 | .27206 | .99226 | .27980 | .72020 | .00774 | .72794 | 13 |
| 48 | .27273 | .99224 | .28049 | .71951 | .00776 | .72727 | 12 |
| 49 | .27339 | .99221 | .28117 | .71883 | .00779 | .72661 | 11 |
| 50 | 9.27405 | 9.99219 | 9.28186 | 10.71814 | 10.00781 | 10.72595 | 10 |
| 51 | .27471 | .99217 | .28254 | .71746 | .00783 | .72529 | 9 |
| 52 | .27537 | .99214 | .28323 | .71677 | .00786 | .72463 | 8 |
| 53 | .27602 | .99212 | .28391 | .71609 | .00788 | .72398 | 7 |
| 54 | .27668 | .99209 | .28459 | .71541 | .00791 | .72332 | 6 |
| 55 | 9.27734 | 9.99207 | 9.28527 | 10.71473 | 10.00793 | 10.72266 | 5 |
| 56 | .27799 | .99204 | .28595 | .71405 | .00796 | .72201 | 4 |
| 57 | .27864 | .99202 | .28662 | .71338 | .00798 | .72136 | 3 |
| 58 | .27930 | .99200 | .28730 | .71270 | .00800 | .72070 | 2 |
| 59 | .27995 | .99197 | .28798 | .71202 | .00803 | .72005 | 1 |
| 60 | 9.28060 | 9.99195 | 9.28865 | 10.71135 | 10.00805 | 10.71940 | 0 |
| M | Cosine | Sine | Cotangent | Tangent | Cosecant | Secant | M |

**11°** Logarithms of Trigonometrical Functions **168°**

| M | Sine | Cosine | Tangent | Cotangent | Secant | Cosecant | M |
|---|------|--------|---------|-----------|--------|----------|---|
| 0 | 9.28060 | 9.99195 | 9.28865 | 10.71135 | 10.00805 | 10.71940 | 60 |
| 1 | .28125 | .99192 | .28933 | .71067 | .00808 | .71875 | 59 |
| 2 | .28190 | .99190 | .29000 | .71000 | .00810 | .71810 | 58 |
| 3 | .28254 | .99187 | .29067 | .70933 | .00813 | .71746 | 57 |
| 4 | .28319 | .99185 | .29134 | .70866 | .00815 | .71681 | 56 |
| 5 | 9.28384 | 9.99182 | 9.29201 | 10.70799 | 10.00818 | 10.71616 | 55 |
| 6 | .28448 | .99180 | .29268 | .70732 | .00820 | .71552 | 54 |
| 7 | .28512 | .99177 | .29335 | .70665 | .00823 | .71488 | 53 |
| 8 | .28577 | .99175 | .29402 | .70598 | .00825 | .71423 | 52 |
| 9 | .28641 | .99172 | .29468 | .70532 | .00828 | .71359 | 51 |
| 10 | 9.28705 | 9.99170 | 9.29535 | 10.70465 | 10.00830 | 10.71295 | 50 |
| 11 | .28769 | .99167 | .29601 | .70399 | .00833 | .71231 | 49 |
| 12 | .28833 | .99165 | .29668 | .70332 | .00835 | .71167 | 48 |
| 13 | .28896 | .99162 | .29734 | .70266 | .00838 | .71104 | 47 |
| 14 | .28960 | .99160 | .29800 | .70200 | .00840 | .71040 | 46 |
| 15 | 9.29024 | 9.99157 | 9.29866 | 10.70134 | 10.00843 | 10.70976 | 45 |
| 16 | .29087 | .99155 | .29932 | .70068 | .00845 | .70913 | 44 |
| 17 | .29150 | .99152 | .29998 | .70002 | .00848 | .70850 | 43 |
| 18 | .29214 | .99150 | .30064 | .69936 | .00850 | .70786 | 42 |
| 19 | .29277 | .99147 | .30130 | .69870 | .00853 | .70723 | 41 |
| 20 | 9.29340 | 9.99145 | 9.30195 | 10.69805 | 10.00855 | 10.70660 | 40 |
| 21 | .29403 | .99142 | .30261 | .69739 | .00858 | .70597 | 39 |
| 22 | .29466 | .99140 | .30326 | .69674 | .00860 | .70534 | 38 |
| 23 | .29529 | .99137 | .30391 | .69609 | .00863 | .70471 | 37 |
| 24 | .29591 | .99135 | .30457 | .69543 | .00865 | .70409 | 36 |
| 25 | 9.29654 | 9.99132 | 9.30522 | 10.69478 | 10.00868 | 10.70346 | 35 |
| 26 | .29716 | .99130 | .30587 | .69413 | .00870 | .70284 | 34 |
| 27 | .29779 | .99127 | .30652 | .69348 | .00873 | .70221 | 33 |
| 28 | .29841 | .99124 | .30717 | .69283 | .00876 | .70159 | 32 |
| 29 | .29903 | .99122 | .30782 | .69218 | .00878 | .70097 | 31 |
| 30 | 9.29966 | 9.99119 | 9.30846 | 10.69154 | 10.00881 | 10.70034 | 30 |
| 31 | .30028 | .99117 | .30911 | .69089 | .00883 | .69972 | 29 |
| 32 | .30090 | .99114 | .30975 | .69025 | .00886 | .69910 | 28 |
| 33 | .30151 | .99112 | .31040 | .68960 | .00888 | .69849 | 27 |
| 34 | .30213 | .99109 | .31104 | .68896 | .00891 | .69787 | 26 |
| 35 | 9.30275 | 9.99106 | 9.31168 | 10.68832 | 10.00894 | 10.69725 | 25 |
| 36 | .30336 | .99104 | .31233 | .68767 | .00896 | .69664 | 24 |
| 37 | .30398 | .99101 | .31297 | .68703 | .00899 | .69602 | 23 |
| 38 | .30459 | .99099 | .31361 | .68639 | .00901 | .69541 | 22 |
| 39 | .30521 | .99096 | .31425 | .68575 | .00904 | .69479 | 21 |
| 40 | 9.30582 | 9.99093 | 9.31489 | 10.68511 | 10.00907 | 10.69418 | 20 |
| 41 | .30643 | .99091 | .31552 | .68448 | .00909 | .69357 | 19 |
| 42 | .30704 | .99088 | .31616 | .68384 | .00912 | 69296 | 18 |
| 43 | .30765 | .99086 | .31679 | .68321 | .00914 | .69235 | 17 |
| 44 | .30826 | 99083 | .31743 | .68257 | .00917 | .69174 | 16 |
| 45 | 9.30887 | 9.99080 | 9.31806 | 10.68194 | 10.00920 | 10.69113 | 15 |
| 46 | .30947 | .99078 | .31870 | .68130 | .00922 | .69053 | 14 |
| 47 | .31008 | .99075 | .31933 | .68067 | .00925 | .68992 | 13 |
| 48 | .31068 | .99072 | .31996 | .68004 | .00928 | .68932 | 12 |
| 49 | .31129 | .99070 | .32059 | .67941 | .00930 | .68871 | 11 |
| 50 | 9.31189 | 9.99067 | 9.32122 | 10.67878 | 10.00933 | 10.68811 | 10 |
| 51 | .31250 | .99064 | .32185 | .67815 | .00936 | .68750 | 9 |
| 52 | .31310 | .99062 | .32248 | .67752 | .00938 | .68690 | 8 |
| 53 | .31370 | .99059 | .32311 | .67689 | .00941 | .68630 | 7 |
| 54 | .31430 | .99056 | .32373 | .67627 | .00944 | .68570 | 6 |
| 55 | 9.31490 | 9.99054 | 9.32436 | 10.67564 | 10.00946 | 10.68510 | 5 |
| 56 | .31549 | .99051 | .32498 | .67502 | .00949 | .68451 | 4 |
| 57 | .31609 | .99048 | .32561 | .67439 | .00952 | .68391 | 3 |
| 58 | .31669 | .99046 | .32623 | .67377 | .00954 | .68331 | 2 |
| 59 | .31728 | .99043 | .32685 | .67315 | .00957 | .68272 | 1 |
| 60 | 9.31788 | 9.99040 | 9.32747 | 10.67253 | 10.00960 | 10.68212 | 0 |

| M | Cosine | Sine | Cotangent | Tangent | Cosecant | Secant | M |
|---|--------|------|-----------|---------|----------|--------|---|

## 12°   Logarithms of Trigonometrical Functions   167°

| M | Sine | Cosine | Tangent | Cotangent | Secant | Cosecant | M |
|---|------|--------|---------|-----------|--------|----------|---|
| 0 | 9.31788 | 9.99040 | 9.32747 | 10.67253 | 10.00960 | 10.68212 | 60 |
| 1 | .31847 | .99038 | .32810 | .67190 | .00962 | .68153 | 59 |
| 2 | .31907 | .99035 | .32872 | .67128 | .00965 | .68093 | 58 |
| 3 | .31966 | .99032 | .32933 | .67067 | .00968 | .68034 | 57 |
| 4 | .32025 | .99030 | .32995 | .67005 | .00970 | .67975 | 56 |
| 5 | 9.32084 | 9.99027 | 9.33057 | 10.66943 | 10.00973 | 10.67916 | 55 |
| 6 | .32143 | .99024 | .33119 | .66881 | .00976 | .67857 | 54 |
| 7 | .32202 | .99022 | .33180 | .66820 | .00978 | .67798 | 53 |
| 8 | .32261 | .99019 | .33242 | .66758 | .00981 | .67739 | 52 |
| 9 | .32319 | .99016 | .33303 | .66697 | .00984 | .67681 | 51 |
| 10 | 9.32378 | 9.99013 | 9.33365 | 10.66635 | 10.00987 | 10.67622 | 50 |
| 11 | .32437 | .99011 | .33426 | .66574 | .00989 | .67563 | 49 |
| 12 | .32495 | .99008 | .33487 | .66513 | .00992 | .67505 | 48 |
| 13 | .32553 | .99005 | .33548 | .66452 | .00995 | .67447 | 47 |
| 14 | .32612 | .99002 | .33609 | .66391 | .00998 | .67388 | 46 |
| 15 | 9.32670 | 9.99000 | 9.33670 | 10.66330 | 10.01000 | 10.67330 | 45 |
| 16 | .32728 | .98997 | .33731 | .66269 | .01003 | .67272 | 44 |
| 17 | .32786 | .98994 | .33792 | .66208 | .01006 | .67214 | 43 |
| 18 | .32844 | .98991 | .33853 | .66147 | .01009 | .67156 | 42 |
| 19 | .32902 | .98989 | .33913 | .66087 | .01011 | .67098 | 41 |
| 20 | 9.32960 | 9.98986 | 9.33974 | 10.66026 | 10.01014 | 10.67040 | 40 |
| 21 | .33018 | .98983 | .34034 | .65966 | .01017 | .66982 | 39 |
| 22 | .33075 | .98980 | .34095 | .65905 | .01020 | .66925 | 38 |
| 23 | .33133 | .98978 | .34155 | .65845 | .01022 | .66867 | 37 |
| 24 | .33190 | .98975 | .34215 | .65785 | .01025 | .66810 | 36 |
| 25 | 9.33248 | 9.98972 | 9.34276 | 10.65724 | 10.01028 | 10.66752 | 35 |
| 26 | .33305 | .98969 | .34336 | .65664 | .01031 | .66695 | 34 |
| 27 | .33362 | .98967 | .34396 | .65604 | .01033 | .66638 | 33 |
| 28 | .33420 | .98964 | .34456 | .65544 | .01036 | .66580 | 32 |
| 29 | .33477 | .98961 | .34516 | .65484 | .01039 | .66523 | 31 |
| 30 | 9.33534 | 9.98958 | 9.34576 | 10.65424 | 10.01042 | 10.66466 | 30 |
| 31 | .33591 | .98955 | .34635 | .65365 | .01045 | .66409 | 29 |
| 32 | .33647 | .98953 | .34695 | .65305 | .01047 | .66353 | 28 |
| 33 | .33704 | .98950 | .34755 | .65245 | .01050 | .66296 | 27 |
| 34 | .33761 | .98947 | .34814 | .65186 | .01053 | .66239 | 26 |
| 35 | 9.33818 | 9.98944 | 9.34874 | 10.65126 | 10.01056 | 10.66182 | 25 |
| 36 | .33874 | .98941 | .34933 | .65067 | .01059 | .66126 | 24 |
| 37 | .33931 | .98938 | .34992 | .65008 | .01062 | .66069 | 23 |
| 38 | .33987 | .98936 | .35051 | .64949 | .01064 | .66013 | 22 |
| 39 | .34043 | .98933 | .35111 | .64889 | .01067 | .65957 | 21 |
| 40 | 9.34100 | 9.98930 | 9.35170 | 10.64830 | 10.01070 | 10.65900 | 20 |
| 41 | .34156 | .98927 | .35229 | .64771 | .01073 | .65844 | 19 |
| 42 | .34212 | .98924 | .35288 | .64712 | .01076 | .65788 | 18 |
| 43 | .34268 | .98921 | .35347 | .64653 | .01079 | .65732 | 17 |
| 44 | .34324 | .98919 | .35405 | .64595 | .01081 | .65676 | 16 |
| 45 | 9.34380 | 9.98916 | 9.35464 | 10.64536 | 10.01084 | 10.65620 | 15 |
| 46 | .34436 | .98913 | .35523 | .64477 | .01087 | .65564 | 14 |
| 47 | .34491 | .98910 | .35581 | .64419 | .01090 | .65509 | 13 |
| 48 | .34547 | .98907 | .35640 | .64360 | .01093 | .65453 | 12 |
| 49 | .34602 | .98904 | .35698 | .64302 | .01096 | .65398 | 11 |
| 50 | 9.34658 | 9.98901 | 9.35757 | 10.64243 | 10.01099 | 10.65342 | 10 |
| 51 | .34713 | .98898 | .35815 | .64185 | .01102 | .65287 | 9 |
| 52 | .34769 | .98896 | .35873 | .64127 | .01104 | .65231 | 8 |
| 53 | .34824 | .98893 | .35931 | .64069 | .01107 | .65176 | 7 |
| 54 | .34879 | .98890 | .35989 | .64011 | .01110 | .65121 | 6 |
| 55 | 9.34934 | 9.98887 | 9.36047 | 10.63953 | 10.01113 | 10.65066 | 5 |
| 56 | .34989 | .98884 | .36105 | .63895 | .01116 | .65011 | 4 |
| 57 | .35044 | .98881 | .36163 | .63837 | .01119 | .64956 | 3 |
| 58 | .35099 | .98878 | .36221 | .63779 | .01122 | .64901 | 2 |
| 59 | .35154 | .98875 | .36279 | .63721 | .01125 | .64846 | 1 |
| 60 | 9.35209 | 9.98872 | 9.36336 | 10.63664 | 10.01128 | 10.64791 | 0 |
| M | Cosine | Sine | Cotangent | Tangent | Cosecant | Secant | M |

| M | Sine | Cosine | Tangent | Cotangent | Secant | Cosecant | M |
|---|------|--------|---------|-----------|--------|----------|---|
| 0 | 9.35209 | 9.98872 | 9.36336 | 10.63664 | 10.01128 | 10.64791 | 60 |
| 1 | .35263 | .98869 | .36394 | .63606 | .01131 | .64737 | 59 |
| 2 | .35318 | .98867 | .36452 | .63548 | .01133 | .64682 | 58 |
| 3 | .35373 | .98864 | .36509 | .63491 | .01136 | .64627 | 57 |
| 4 | .35427 | .98861 | .36566 | .63434 | .01139 | .64573 | 56 |
| 5 | 9.35481 | 9.98858 | 9.36624 | 10.63376 | 10.01142 | 10.64519 | 55 |
| 6 | .35536 | .98855 | .36681 | .63319 | .01145 | .64464 | 54 |
| 7 | .35590 | .98852 | .36738 | .63262 | .01148 | .64410 | 53 |
| 8 | .35644 | .98849 | .36795 | .63205 | .01151 | .64356 | 52 |
| 9 | .35698 | .98846 | .36852 | .63148 | .01154 | .64302 | 51 |
| 10 | 9.35752 | 9.98843 | 9.36909 | 10.63091 | 10.01157 | 10.64248 | 50 |
| 11 | .35806 | .98840 | .36966 | .63034 | .01160 | .64194 | 49 |
| 12 | .35860 | .98837 | .37023 | .62977 | .01163 | .64140 | 48 |
| 13 | .35914 | .98834 | .37080 | .62920 | .01166 | .64086 | 47 |
| 14 | .35968 | .98831 | .37137 | .62863 | .01169 | .64032 | 46 |
| 15 | 9.36022 | 9.98828 | 9.37193 | 10.62807 | 10.01172 | 10.63978 | 45 |
| 16 | .36075 | .98825 | .37250 | .62750 | .01175 | .63925 | 44 |
| 17 | .36129 | .98822 | .37306 | .62694 | .01178 | .63871 | 43 |
| 18 | .36182 | .98819 | .37363 | .62637 | .01181 | .63818 | 42 |
| 19 | .36236 | .98816 | .37419 | .62581 | .01184 | .63764 | 41 |
| 20 | 9.36289 | 9.98813 | 9.37476 | 10.62524 | 10.01187 | 10.63711 | 40 |
| 21 | .36342 | .98810 | .37532 | .62468 | .01190 | .63658 | 39 |
| 22 | .36395 | .98807 | .37588 | .62412 | .01193 | .63605 | 38 |
| 23 | .36449 | .98804 | .37644 | .62356 | .01196 | .63551 | 37 |
| 24 | .36502 | .98801 | .37700 | .62300 | .01199 | .63498 | 36 |
| 25 | 9.36555 | 9.98798 | 9.37756 | 10.62244 | 10.01202 | 10.63445 | 35 |
| 26 | .36608 | .98795 | .37812 | .62188 | .01205 | .63392 | 34 |
| 27 | .36660 | .98792 | .37868 | .62132 | .01208 | .63340 | 33 |
| 28 | .36713 | .98789 | .37924 | .62076 | .01211 | .63287 | 32 |
| 29 | .36766 | .98786 | .37980 | .62020 | .01214 | .63234 | 31 |
| 30 | 9.36819 | 9.98783 | 9.38035 | 10.61965 | 10.01217 | 10.63181 | 30 |
| 31 | .36871 | .98780 | .38091 | .61909 | .01220 | .63129 | 29 |
| 32 | .36924 | .98777 | .38147 | .61853 | .01223 | .63076 | 28 |
| 33 | .36976 | .98774 | .38202 | .61798 | .01226 | .63024 | 27 |
| 34 | .37028 | .98771 | .38257 | .61743 | .01229 | .62972 | 26 |
| 35 | 9.37081 | 9.98768 | 9.38313 | 10.61687 | 10.01232 | 10.62919 | 25 |
| 36 | .37133 | .98765 | .38368 | .61632 | .01235 | .62867 | 24 |
| 37 | .37185 | .98762 | .38423 | .61577 | .01238 | .62815 | 23 |
| 38 | .37237 | .98759 | .38479 | .61521 | .01241 | .62763 | 22 |
| 39 | .37289 | .98756 | .38534 | .61466 | .01244 | .62711 | 21 |
| 40 | 9.37341 | 9.98753 | 9.38589 | 10.61411 | 10.01247 | 10.62659 | 20 |
| 41 | .37393 | .98750 | .38644 | .61356 | .01250 | .62607 | 19 |
| 42 | .37445 | .98746 | .38699 | .61301 | .01254 | .62555 | 18 |
| 43 | .37497 | .98743 | .38754 | .61246 | .01257 | .62503 | 17 |
| 44 | .37549 | .98740 | .38808 | .61192 | .01260 | .62451 | 16 |
| 45 | 9.37600 | 9.98737 | 9.38863 | 10.61137 | 10.01263 | 10.62400 | 15 |
| 46 | .37652 | .98734 | .38918 | .61082 | .01266 | .62348 | 14 |
| 47 | .37703 | .98731 | .38972 | .61028 | .01269 | .62297 | 13 |
| 48 | .37755 | .98728 | .39027 | .60973 | .01272 | .62245 | 12 |
| 49 | .37806 | .98725 | .39082 | .60918 | .01275 | .62194 | 11 |
| 50 | 9.37858 | 9.98722 | 9.39136 | 10.60864 | 10.01278 | 10.62142 | 10 |
| 51 | .37909 | .98719 | .39190 | .60810 | .01281 | .62091 | 9 |
| 52 | .37960 | .98715 | .39245 | .60755 | .01285 | .62040 | 8 |
| 53 | .38011 | .98712 | .39299 | .60701 | .01288 | .61989 | 7 |
| 54 | .38062 | .98709 | .39353 | .60647 | .01291 | .61938 | 6 |
| 55 | 9.38113 | 9.98706 | 9.39407 | 10.60593 | 10.01294 | 10.61887 | 5 |
| 56 | .38164 | .98703 | .39461 | .60539 | .01297 | .61836 | 4 |
| 57 | .38215 | .98700 | .39515 | .60485 | .01300 | .61785 | 3 |
| 58 | .38266 | .98697 | .39569 | .60431 | .01303 | .61734 | 2 |
| 59 | .38317 | .98694 | .39623 | .60377 | .01306 | .61683 | 1 |
| 60 | 9.38368 | 9.98690 | 9.39677 | 10.60323 | 10.01310 | 10.61632 | 0 |
| M | Cosine | Sine | Cotangent | Tangent | Cosecant | Secant | M |

## 14°     Logarithms of Trigonometrical Functions     165°

| M | Sine | Cosine | Tangent | Cotangent | Secant | Cosecant | M |
|---|------|--------|---------|-----------|--------|----------|---|
| 0 | 9.38368 | 9.98690 | 9.39677 | 10.60323 | 10.01310 | 10.61632 | 60 |
| 1 | .38418 | .98687 | .39731 | .60269 | .01313 | .61582 | 59 |
| 2 | .38469 | .98684 | .39785 | .60215 | .01316 | .61531 | 58 |
| 3 | .38519 | .98681 | .39838 | .60162 | .01319 | .61481 | 57 |
| 4 | .38570 | .98678 | .39892 | .60108 | .01322 | .61430 | 56 |
| 5 | 9.38620 | 9.98675 | 9.39945 | 10.60055 | 10.01325 | 10.61380 | 55 |
| 6 | .38670 | .98671 | .39999 | .60001 | .01329 | .61330 | 54 |
| 7 | .38721 | .98668 | .40052 | .59948 | .01332 | .61279 | 53 |
| 8 | .38771 | .98665 | .40106 | .59894 | .01335 | .61229 | 52 |
| 9 | .38821 | .98662 | .40159 | .59841 | .01338 | .61179 | 51 |
| 10 | 9.38871 | 9.98659 | 9.40212 | 10.59788 | 10.01341 | 10.61129 | 50 |
| 11 | .38921 | .98656 | .40266 | .59734 | .01344 | .61079 | 49 |
| 12 | .38971 | .98652 | .40319 | .59681 | .01348 | .61029 | 48 |
| 13 | .39021 | .98649 | .40372 | .59628 | .01351 | .60979 | 47 |
| 14 | .39071 | .98646 | .40425 | .59575 | .01354 | .60929 | 46 |
| 15 | 9.39121 | 9.98643 | 9.40478 | 10.59522 | 10.01357 | 10.60879 | 45 |
| 16 | .39170 | .98640 | .40531 | .59469 | .01360 | .60830 | 44 |
| 17 | .39220 | .98636 | .40584 | .59416 | .01364 | .60780 | 43 |
| 18 | .39270 | .98633 | .40636 | .59364 | .01367 | .60730 | 42 |
| 19 | .39319 | .98630 | .40689 | .59311 | .01370 | .60681 | 41 |
| 20 | 9.39369 | 9.98627 | 9.40742 | 10.59258 | 10.01373 | 10.60631 | 40 |
| 21 | .39418 | .98623 | .40795 | .59205 | .01377 | .60582 | 39 |
| 22 | .39467 | .98620 | .40847 | .59153 | .01380 | .60533 | 38 |
| 23 | .39517 | .98617 | .40900 | .59100 | .01383 | .60483 | 37 |
| 24 | .39566 | .98614 | .40952 | .59048 | .01386 | .60434 | 36 |
| 25 | 9.39615 | 9.98610 | 9.41005 | 10.58995 | 10.01390 | 10.60385 | 35 |
| 26 | .39664 | .98607 | .41057 | .58943 | .01393 | .60336 | 34 |
| 27 | .39713 | .98604 | .41109 | .58891 | .01396 | .60287 | 33 |
| 28 | .39762 | .98601 | .41161 | .58839 | .01399 | .60238 | 32 |
| 29 | .39811 | .98597 | .41214 | .58786 | .01403 | .60189 | 31 |
| 30 | 9.39860 | 9.98594 | 9.41266 | 10.58734 | 10.01406 | 10.60140 | 30 |
| 31 | .39909 | .98591 | .41318 | .58682 | .01409 | .60091 | 29 |
| 32 | .39958 | .98588 | .41370 | .58630 | .01412 | .60042 | 28 |
| 33 | .40006 | .98584 | .41422 | .58578 | .01416 | .59994 | 27 |
| 34 | .40055 | .98581 | .41474 | .58526 | .01419 | .59945 | 26 |
| 35 | 9.40103 | 9.98578 | 9.41526 | 10.58474 | 10.01422 | 10.59897 | 25 |
| 36 | .40152 | .98574 | .41578 | .58422 | .01426 | .59848 | 24 |
| 37 | .40200 | .98571 | .41629 | .58371 | .01429 | .59800 | 23 |
| 38 | .40249 | .98568 | .41681 | .58319 | .01432 | .59751 | 22 |
| 39 | .40297 | .98565 | .41733 | .58267 | .01435 | .59703 | 21 |
| 40 | 9.40346 | 9.98561 | 9.41784 | 10.58216 | 10.01439 | 10.59654 | 20 |
| 41 | .40394 | .98558 | .41836 | .58164 | .01442 | .59606 | 19 |
| 42 | .40442 | .98555 | .41887 | .58113 | .01445 | .59558 | 18 |
| 43 | .40490 | .98551 | .41939 | .58061 | .01449 | .59510 | 17 |
| 44 | .40538 | .98548 | .41990 | .58010 | .01452 | .59462 | 16 |
| 45 | 9.40585 | 9.98545 | 9.42041 | 10.57959 | 10.01455 | 10.59414 | 15 |
| 46 | .40634 | .98541 | .42093 | .57907 | .01459 | .59366 | 14 |
| 47 | .40682 | .98538 | .42144 | .57856 | .01462 | .59318 | 13 |
| 48 | .40730 | .98535 | .42195 | .57805 | .01465 | .59270 | 12 |
| 49 | .40778 | .98531 | .42246 | .57754 | .01469 | .59222 | 11 |
| 50 | 9.40825 | 9.98528 | 9.42297 | 10.57703 | 10.01472 | 10.59175 | 10 |
| 51 | .40873 | .98525 | .42348 | .57652 | .01475 | .59127 | 9 |
| 52 | .40921 | .98521 | .42399 | .57601 | .01479 | .59079 | 8 |
| 53 | .40968 | .98518 | .42450 | .57550 | .01482 | .59032 | 7 |
| 54 | .41016 | .98515 | .42501 | .57499 | .01485 | .58984 | 6 |
| 55 | 9.41063 | 9.98511 | 9.42552 | 10.57448 | 10.01489 | 10.58937 | 5 |
| 56 | .41111 | .98508 | .42603 | .57397 | .01492 | .58889 | 4 |
| 57 | .41158 | .98505 | .42653 | .57347 | .01495 | .58842 | 3 |
| 58 | .41205 | .98501 | .42704 | .57296 | .01499 | .58795 | 2 |
| 59 | .41252 | .98498 | .42755 | .57245 | .01502 | .58748 | 1 |
| 60 | 9.41300 | 9.98494 | 9.42805 | 10.57195 | 10.01506 | 10.58700 | 0 |

| M | Cosine | Sine | Cotangent | Tangent | Cosecant | Secant | M |
|---|--------|------|-----------|---------|----------|--------|---|

| M | Sine | Cosine | Tangent | Cotangent | Secant | Cosecant | M |
|---|------|--------|---------|-----------|--------|----------|---|
| 0 | 9.41300 | 9.98494 | 9.42805 | 10.57195 | 10.01506 | 10.58700 | 60 |
| 1 | .41347 | .98491 | .42856 | .57144 | .01509 | .58653 | 59 |
| 2 | .41394 | .98488 | .42906 | .57094 | .01512 | .58606 | 58 |
| 3 | .41441 | .98484 | .42957 | .57043 | .01516 | .58559 | 57 |
| 4 | .41488 | .98481 | .43007 | .56993 | .01519 | .58512 | 56 |
| 5 | 9.41535 | 9.98477 | 9.43057 | 10.56943 | 10.01523 | 10.58465 | 55 |
| 6 | .41582 | .98474 | .43108 | .56892 | .01526 | .58418 | 54 |
| 7 | .41628 | .98471 | .43158 | .56842 | .01529 | .58372 | 53 |
| 8 | .41675 | .98467 | .43208 | .56792 | .01533 | .58325 | 52 |
| 9 | .41722 | .98464 | .43258 | .56742 | .01536 | .58278 | 51 |
| 10 | 9.41768 | 9.98460 | 9.43308 | 10.56692 | 10.01540 | 10.58232 | 50 |
| 11 | .41815 | .98457 | .43358 | .56642 | .01543 | .58185 | 49 |
| 12 | .41861 | .98453 | .43408 | .56592 | .01547 | .58139 | 48 |
| 13 | .41908 | .98450 | .43458 | .56542 | .01550 | .58092 | 47 |
| 14 | .41954 | .98447 | .43508 | .56492 | .01553 | .58046 | 46 |
| 15 | 9.42001 | 9.98443 | 9.43558 | 10.56442 | 10.01557 | 10.57999 | 45 |
| 16 | .42047 | .98440 | .43607 | .56393 | .01560 | .57953 | 44 |
| 17 | .42093 | .98436 | .43657 | .56343 | .01564 | .57907 | 43 |
| 18 | .42140 | .98433 | .43707 | .56293 | .01567 | .57860 | 42 |
| 19 | .42186 | .98429 | .43756 | .56244 | .01571 | .57814 | 41 |
| 20 | 9.42232 | 9.98426 | 9.43806 | 10.56194 | 10.01574 | 10.57768 | 40 |
| 21 | .42278 | .98422 | .43855 | .56145 | .01578 | .57722 | 39 |
| 22 | .42324 | .98419 | .43905 | .56095 | .01581 | .57676 | 38 |
| 23 | .42370 | .98415 | .43954 | .56046 | .01585 | .57630 | 37 |
| 24 | .42416 | .98412 | .44004 | .55996 | .01588 | .57584 | 36 |
| 25 | 9.42461 | 9.98409 | 9.44053 | 10.55947 | 10.01591 | 10.57539 | 35 |
| 26 | .42507 | .98405 | .44102 | .55898 | .01595 | .57493 | 34 |
| 27 | .42553 | .98402 | .44151 | .55849 | .01598 | .57447 | 33 |
| 28 | .42599 | .98398 | .44201 | .55799 | .01602 | .57401 | 32 |
| 29 | .42644 | .98395 | .44250 | .55750 | .01605 | .57356 | 31 |
| 30 | 9.42690 | 9.98391 | 9.44299 | 10.55701 | 10.01609 | 10.57310 | 30 |
| 31 | .42735 | .98388 | .44348 | .55652 | .01612 | .57265 | 29 |
| 32 | .42781 | .98384 | .44397 | .55603 | .01616 | .57219 | 28 |
| 33 | .42826 | .98381 | .44446 | .55554 | .01619 | .57174 | 27 |
| 34 | .42872 | .98377 | .44495 | .55505 | .01623 | .57128 | 26 |
| 35 | 9.42917 | 9.98373 | 9.44544 | 10.55456 | 10.01627 | 10.57083 | 25 |
| 36 | .42962 | .98370 | .44592 | .55408 | .01630 | .57038 | 24 |
| 37 | .43008 | .98366 | .44641 | .55359 | .01634 | .56992 | 23 |
| 38 | .43053 | .98363 | .44690 | .55310 | .01637 | .56947 | 22 |
| 39 | .43098 | .98359 | .44738 | .55262 | .01641 | .56902 | 21 |
| 40 | 9.43143 | 9.98356 | 9.44787 | 10.55213 | 10.01644 | 10.56857 | 20 |
| 41 | .43188 | .98352 | .44836 | .55164 | .01648 | .56812 | 19 |
| 42 | .43233 | .98349 | .44884 | .55116 | .01651 | .56767 | 18 |
| 43 | .43278 | .98345 | .44933 | .55067 | .01655 | .56722 | 17 |
| 44 | .43323 | .98342 | .44981 | .55019 | .01658 | .56677 | 16 |
| 45 | 9.43367 | 9.98338 | 9.45029 | 10.54971 | 10.01662 | 10.56633 | 15 |
| 46 | .43412 | .98334 | .45078 | .54922 | .01666 | .56588 | 14 |
| 47 | .43457 | .98331 | .45126 | .54874 | .01669 | .56543 | 13 |
| 48 | .43502 | .98327 | .45174 | .54826 | .01673 | .56498 | 12 |
| 49 | .43546 | .98324 | .45222 | .54778 | .01676 | .56454 | 11 |
| 50 | 9.43591 | 9.98320 | 9.45271 | 10.54729 | 10.01680 | 10.56409 | 10 |
| 51 | .43635 | .98317 | .45319 | .54681 | .01683 | .56365 | 9 |
| 52 | .43680 | .98313 | .45367 | .54633 | .01687 | .56320 | 8 |
| 53 | .43724 | .98309 | .45415 | .54585 | .01691 | .56276 | 7 |
| 54 | .43769 | .98306 | .45463 | .54537 | .01694 | .56231 | 6 |
| 55 | 9.43813 | 9.98302 | 9.45511 | 10.54489 | 10.01698 | 10.56187 | 5 |
| 56 | .43857 | .98299 | .45559 | .54441 | .01701 | .56143 | 4 |
| 57 | .43901 | .98295 | .45606 | .54394 | .01705 | .56099 | 3 |
| 58 | .43946 | .98291 | .45654 | .54346 | .01709 | .56054 | 2 |
| 59 | .43990 | .98288 | .45702 | .54298 | .01712 | .56010 | 1 |
| 60 | 9.44034 | 9.98284 | 9.45750 | 10.54250 | 10.01716 | 10.55966 | 0 |
| M | Cosine | Sine | Cotangent | Tangent | Cosecant | Secant | M |

**Logarithms of Trigonometrical Functions**

| M | Sine | Cosine | Tangent | Cotangent | Secant | Cosecant | M |
|---|------|--------|---------|-----------|--------|----------|---|
| 0 | 9.44034 | 9.98284 | 9.45750 | 10.54250 | 10.01716 | 10.55966 | 60 |
| 1 | .44078 | .98281 | .45797 | .54203 | .01719 | .55922 | 59 |
| 2 | .44122 | .98277 | .45845 | .54155 | .01723 | .55878 | 58 |
| 3 | .44166 | .98273 | .45892 | .54108 | .01727 | .55834 | 57 |
| 4 | .44210 | .98270 | .45940 | .54060 | .01730 | .55790 | 56 |
| 5 | 9.44253 | 9.98266 | 9.45987 | 10.54013 | 10.01734 | 10.55747 | 55 |
| 6 | .44297 | .98262 | .46035 | .53965 | .01738 | .55703 | 54 |
| 7 | .44341 | .98259 | .46082 | .53918 | .01741 | .55659 | 53 |
| 8 | .44385 | .98255 | .46130 | .53870 | .01745 | .55615 | 52 |
| 9 | .44428 | .98251 | .46177 | .53823 | .01749 | .55572 | 51 |
| 10 | 9.44472 | 9.98248 | 9.46224 | 10.53776 | 10.01752 | 10.55528 | 50 |
| 11 | .44516 | .98244 | .46271 | .53729 | .01756 | .55484 | 49 |
| 12 | .44559 | .98240 | .46319 | .53681 | .01760 | .55441 | 48 |
| 13 | .44602 | .98237 | .46366 | .53634 | .01763 | .55398 | 47 |
| 14 | .44646 | .98233 | .46413 | .53587 | .01767 | .55354 | 46 |
| 15 | 9.44689 | 9.98229 | 9.46460 | 10.53540 | 10.01771 | 10.55311 | 45 |
| 16 | .44733 | .98226 | .46507 | .53493 | .01774 | .55267 | 44 |
| 17 | .44776 | .98222 | .46554 | .53446 | .01778 | .55224 | 43 |
| 18 | .44819 | .98218 | .46601 | .53399 | .01782 | .55181 | 42 |
| 19 | .44862 | .98215 | .46648 | .53352 | .01785 | .55138 | 41 |
| 20 | 9.44905 | 9.98211 | 9.46694 | 10.53306 | 10.01789 | 10.55095 | 40 |
| 21 | .44948 | .98207 | .46741 | .53259 | .01793 | .55052 | 39 |
| 22 | .44992 | .98204 | .46788 | .53212 | .01796 | .55008 | 38 |
| 23 | .45035 | .98200 | .46835 | .53165 | .01800 | .54965 | 37 |
| 24 | .45077 | .98196 | .46881 | .53119 | .01804 | .54923 | 36 |
| 25 | 9.45120 | 9.98192 | 9.46928 | 10.53072 | 10.01808 | 10.54880 | 35 |
| 26 | .45163 | .98189 | .46975 | .53025 | .01811 | .54837 | 34 |
| 27 | .45206 | .98185 | .47021 | .52979 | .01815 | .54794 | 33 |
| 28 | .45249 | .98181 | .47068 | .52932 | .01819 | .54751 | 32 |
| 29 | .45292 | .98177 | .47114 | .52886 | .01823 | .54708 | 31 |
| 30 | 9.45334 | 9.98174 | 9.47160 | 10.52840 | 10.01826 | 10.54666 | 30 |
| 31 | .45377 | .98170 | .47207 | .52793 | .01830 | .54623 | 29 |
| 32 | .45419 | .98166 | .47253 | .52747 | .01834 | .54581 | 28 |
| 33 | .45462 | .98162 | .47299 | .52701 | .01838 | .54538 | 27 |
| 34 | .45504 | .98159 | .47346 | .52654 | .01841 | .54496 | 26 |
| 35 | 9.45547 | 9.98155 | 9.47392 | 10.52608 | 10.01845 | 10.54453 | 25 |
| 36 | .45589 | .98151 | .47438 | .52562 | .01849 | .54411 | 24 |
| 37 | .45632 | .98147 | .47484 | .52516 | .01853 | .54368 | 23 |
| 38 | .45674 | .98144 | .47530 | .52470 | .01856 | .54326 | 22 |
| 39 | .45716 | .98140 | .47576 | .52424 | .01860 | .54284 | 21 |
| 40 | 9.45758 | 9.98136 | 9.47622 | 10.52378 | 10.01864 | 10.54242 | 20 |
| 41 | .45801 | .98132 | .47668 | .52332 | .01868 | .54199 | 19 |
| 42 | .45843 | .98129 | .47714 | .52286 | .01871 | .54157 | 18 |
| 43 | .45885 | .98125 | .47760 | .52240 | .01875 | .54115 | 17 |
| 44 | .45927 | .98121 | .47806 | .52194 | .01879 | .54073 | 16 |
| 45 | 9.45969 | 9.98117 | 9.47852 | 10.52148 | 10.01883 | 10.54031 | 15 |
| 46 | .46011 | .98113 | .47897 | .52103 | .01887 | .53989 | 14 |
| 47 | .46053 | .98110 | .47943 | .52057 | .01890 | .53947 | 13 |
| 48 | .46095 | .98106 | .47989 | .52011 | .01894 | .53905 | 12 |
| 49 | .46136 | .98102 | .48035 | .51965 | .01898 | .53864 | 11 |
| 50 | 9.46178 | 9.98098 | 9.48080 | 10.51920 | 10.01902 | 10.53822 | 10 |
| 51 | .46220 | .98094 | .48126 | .51874 | .01906 | .53780 | 9 |
| 52 | .46262 | .98090 | .48171 | .51829 | .01910 | .53738 | 8 |
| 53 | .46303 | .98087 | .48217 | .51783 | .01913 | .53697 | 7 |
| 54 | .46345 | .98083 | .48262 | .51738 | .01917 | .53655 | 6 |
| 55 | 9.46386 | 9.98079 | 9.48307 | 10.51693 | 10.01921 | 10.53614 | 5 |
| 56 | .46428 | .98075 | .48353 | .51647 | .01925 | .53572 | 4 |
| 57 | .46469 | .98071 | .48398 | .51602 | .01929 | .53531 | 3 |
| 58 | .46511 | .98067 | .48443 | .51557 | .01933 | .53489 | 2 |
| 59 | .46552 | .98063 | .48489 | .51511 | .01937 | .53448 | 1 |
| 60 | 9.46594 | 9.98060 | 9.48534 | 10.51466 | 10.01940 | 10.53406 | 0 |
| M | Cosine | Sine | Cotangent | Tangent | Cosecant | Secant | M |

| M | Sine | Cosine | Tangent | Cotangent | Secant | Cosecant | M |
|---|------|--------|---------|-----------|--------|----------|---|
| 0 | 9.46594 | 9.98060 | 9.48534 | 10.51466 | 10.01940 | 10.53406 | 60 |
| 1 | .46635 | .98056 | .48579 | .51421 | .01944 | .53365 | 59 |
| 2 | .46676 | .98052 | .48624 | .51376 | .01948 | .53324 | 58 |
| 3 | .46717 | .98048 | .48669 | .51331 | .01952 | .53283 | 57 |
| 4 | .46758 | .98044 | .48714 | .51286 | .01956 | .53242 | 56 |
| 5 | 9.46800 | 9.98040 | 9.48759 | 10.51241 | 10.01960 | 10.53200 | 55 |
| 6 | .46841 | .98036 | .48804 | .51196 | .01964 | .53159 | 54 |
| 7 | .46882 | .98032 | .48849 | .51151 | .01968 | .53118 | 53 |
| 8 | .46923 | .98029 | .48894 | .51106 | .01971 | .53077 | 52 |
| 9 | .46964 | .98025 | .48939 | .51061 | .01975 | .53036 | 51 |
| 10 | 9.47005 | 9.98021 | 9.48984 | 10.51016 | 10.01979 | 10.52995 | 50 |
| 11 | .47045 | .98017 | .49029 | .50971 | .01983 | .52955 | 49 |
| 12 | .47086 | .98013 | .49073 | .50927 | .01987 | .52914 | 48 |
| 13 | .47127 | .98009 | .49118 | .50882 | .01991 | .52873 | 47 |
| 14 | .47168 | .98005 | .49163 | .50837 | .01995 | .52832 | 46 |
| 15 | 9.47209 | 9.98001 | 9.49207 | 10.50793 | 10.01999 | 10.52791 | 45 |
| 16 | .47249 | .97997 | .49252 | .50748 | .02003 | .52751 | 44 |
| 17 | .47290 | .97993 | .49296 | .50704 | .02007 | .52710 | 43 |
| 18 | .47330 | .97989 | .49341 | .50659 | .02011 | .52670 | 42 |
| 19 | .47371 | .97986 | .49385 | .50615 | .02014 | .52629 | 41 |
| 20 | 9.47411 | 9.97982 | 9.49430 | 10.50570 | 10.02018 | 10.52589 | 40 |
| 21 | .47452 | .97978 | .49474 | .50526 | .02022 | .52548 | 39 |
| 22 | .47492 | .97974 | .49519 | .50481 | .02026 | .52508 | 38 |
| 23 | .47533 | .97970 | .49563 | .50437 | .02030 | .52467 | 37 |
| 24 | .47573 | .97966 | .49607 | .50393 | .02034 | .52427 | 36 |
| 25 | 9.47613 | 9.97962 | 9.49652 | 10.50348 | 10.02038 | 10.52387 | 35 |
| 26 | .47654 | .97958 | .49696 | .50304 | .02042 | .52346 | 34 |
| 27 | .47694 | .97954 | .49740 | .50260 | .02046 | .52306 | 33 |
| 28 | .47734 | .97950 | .49784 | .50216 | .02050 | .52266 | 32 |
| 29 | .47774 | .97946 | .49828 | .50172 | .02054 | .52226 | 31 |
| 30 | 9.47814 | 9.97942 | 9.49872 | 10.50128 | 10.02058 | 10.52186 | 30 |
| 31 | .47854 | .97938 | .49916 | .50084 | .02062 | .52146 | 29 |
| 32 | .47894 | .97934 | .49960 | .50040 | .02066 | .52106 | 28 |
| 33 | .47934 | .97930 | .50004 | .49996 | .02070 | .52066 | 27 |
| 34 | .47974 | .97926 | .50048 | .49952 | .02074 | .52026 | 26 |
| 35 | 9.48014 | 9.97922 | 9.50092 | 10.49908 | 10.02078 | 10.51986 | 25 |
| 36 | .48054 | .97918 | .50136 | .49864 | .02082 | .51946 | 24 |
| 37 | .48094 | .97914 | .50180 | .49820 | .02086 | .51906 | 23 |
| 38 | .48133 | .97910 | .50223 | .49777 | .02090 | .51867 | 22 |
| 39 | .48173 | .97906 | .50267 | .49733 | .02094 | .51827 | 21 |
| 40 | 9.48213 | 9.97902 | 9.50311 | 10.49689 | 10.02098 | 10.51787 | 20 |
| 41 | .48252 | .97898 | .50355 | .49645 | .02102 | .51748 | 19 |
| 42 | .48292 | .97894 | .50398 | .49602 | .02106 | .51708 | 18 |
| 43 | .48332 | .97890 | .50442 | .49558 | .02110 | .51668 | 17 |
| 44 | .48371 | .97886 | .50485 | .49515 | .02114 | .51629 | 16 |
| 45 | 9.48411 | 9.97882 | 9.50529 | 10.49471 | 10.02118 | 10.51589 | 15 |
| 46 | .48450 | .97878 | .50572 | .49428 | .02122 | .51550 | 14 |
| 47 | .48490 | .97874 | .50616 | .49384 | .02126 | .51510 | 13 |
| 48 | .48529 | .97870 | .50659 | .49341 | .02130 | .51471 | 12 |
| 49 | .48568 | .97866 | .50703 | .49297 | .02134 | .51432 | 11 |
| 50 | 9.48607 | 9.97861 | 9.50746 | 10.49254 | 10.02139 | 10.51393 | 10 |
| 51 | .48647 | .97857 | .50789 | .49211 | .02143 | .51353 | 9 |
| 52 | .48686 | .97853 | .50833 | .49167 | .02147 | .51314 | 8 |
| 53 | .48725 | .97849 | .50876 | .49124 | .02151 | .51275 | 7 |
| 54 | .48764 | .97845 | .50919 | .49081 | .02155 | .51236 | 6 |
| 55 | 9.48803 | 9.97841 | 9.50962 | 10.49038 | 10.02159 | 10.51197 | 5 |
| 56 | .48842 | .97837 | .51005 | .48995 | .02163 | .51158 | 4 |
| 57 | .48881 | .97833 | .51048 | .48952 | .02167 | .51119 | 3 |
| 58 | .48920 | .97829 | .51092 | .48908 | .02171 | .51080 | 2 |
| 59 | .48959 | .97825 | .51135 | .48865 | .02175 | .51041 | 1 |
| 60 | 9.48998 | 9.97821 | 9.51178 | 10.48822 | 10.02179 | 10.51002 | 0 |
| M | Cosine | Sine | Cotangent | Tangent | Cosecant | Secant | M |

| M | Sine | Cosine | Tangent | Cotangent | Secant | Cosecant | M |
|---|---|---|---|---|---|---|---|
| 0 | 9.48998 | 9.97821 | 9.51178 | 10.48822 | 10.02179 | 10.51002 | 60 |
| 1 | .49037 | .97817 | .51221 | .48779 | .02183 | .50963 | 59 |
| 2 | .49076 | .97812 | .51264 | .48736 | .02188 | .50924 | 58 |
| 3 | .49115 | .97808 | .51306 | .48694 | .02192 | .50885 | 57 |
| 4 | .49153 | .97804 | .51349 | .48651 | .02196 | .50847 | 56 |
| 5 | 9.49192 | 9.97800 | 9.51392 | 10.48608 | 10.02200 | 10.50808 | 55 |
| 6 | .49231 | .97796 | .51435 | .48565 | .02204 | .50769 | 54 |
| 7 | .49269 | .97792 | .51478 | .48522 | .02208 | .50731 | 53 |
| 8 | .49308 | .97788 | .51520 | .48480 | .02212 | .50692 | 52 |
| 9 | .49347 | .97784 | .51563 | .48437 | .02216 | .50653 | 51 |
| 10 | 9.49385 | 9.97779 | 9.51606 | 10.48394 | 10.02221 | 10.50615 | 50 |
| 11 | .49424 | .97775 | .51648 | .48352 | .02225 | .50576 | 49 |
| 12 | .49462 | .97771 | .51691 | .48309 | .02229 | .50538 | 48 |
| 13 | .49500 | .97767 | .51734 | .48266 | .02233 | .50500 | 47 |
| 14 | .49539 | .97763 | .51776 | .48224 | .02237 | .50461 | 46 |
| 15 | 9.49577 | 9.97759 | 9.51819 | 10.48181 | 10.02241 | 10.50423 | 45 |
| 16 | .49615 | .97754 | .51861 | .48139 | .02246 | .50385 | 44 |
| 17 | .49654 | .97750 | .51903 | .48097 | .02250 | .50346 | 43 |
| 18 | .49692 | .97746 | .51946 | .48054 | .02254 | .50308 | 42 |
| 19 | .49730 | .97742 | .51988 | .48012 | .02258 | .50270 | 41 |
| 20 | 9.49768 | 9.97738 | 9.52031 | 10.47969 | 10.02262 | 10.50232 | 40 |
| 21 | .49806 | .97734 | .52073 | .47927 | .02266 | .50194 | 39 |
| 22 | .49844 | .97729 | .52115 | .47885 | .02271 | .50156 | 38 |
| 23 | .49882 | .97725 | .52157 | .47843 | .02275 | .50118 | 37 |
| 24 | .49920 | .97721 | .52200 | .47800 | .02279 | .50080 | 36 |
| 25 | 9.49958 | 9.97717 | 9.52242 | 10.47758 | 10.02283 | 10.50042 | 35 |
| 26 | .49996 | .97713 | .52284 | .47716 | .02287 | .50004 | 34 |
| 27 | .50034 | .97708 | .52326 | .47674 | .02292 | .49966 | 33 |
| 28 | .50072 | .97704 | .52368 | .47632 | .02296 | .49928 | 32 |
| 29 | .50110 | .97700 | .52410 | .47590 | .02300 | .49890 | 31 |
| 30 | 9.50148 | 9.97696 | 9.52452 | 10.47548 | 10.02304 | 10.49852 | 30 |
| 31 | .50185 | .97691 | .52494 | .47506 | .02309 | .49815 | 29 |
| 32 | .50223 | .97687 | .52536 | .47464 | .02313 | .49777 | 28 |
| 33 | .50261 | .97683 | .52578 | .47422 | .02317 | .49739 | 27 |
| 34 | .50298 | .97679 | .52620 | .47380 | .02321 | .49702 | 26 |
| 35 | 9.50336 | 9.97674 | 9.52661 | 10.47339 | 10.02326 | 10.49664 | 25 |
| 36 | .50374 | .97670 | .52703 | .47297 | .02330 | .49626 | 24 |
| 37 | .50411 | .97666 | .52745 | .47255 | .02334 | .49589 | 23 |
| 38 | .50449 | .97662 | .52787 | .47213 | .02338 | .49551 | 22 |
| 39 | .50486 | .97657 | .52829 | .47171 | .02343 | .49514 | 21 |
| 40 | 9.50523 | 9.97653 | 9.52870 | 10.47130 | 10.02347 | 10.49477 | 20 |
| 41 | .50561 | .97649 | .52912 | .47088 | .02351 | .49439 | 19 |
| 42 | .50598 | .97645 | .52953 | .47047 | .02355 | .49402 | 18 |
| 43 | .50635 | .97640 | .52995 | .47005 | .02360 | .49365 | 17 |
| 44 | .50673 | .97636 | .53037 | .46963 | .02364 | .49327 | 16 |
| 45 | 9.50710 | 9.97632 | 9.53078 | 10.46922 | 10.02368 | 10.49290 | 15 |
| 46 | .50747 | .97628 | .53120 | .46880 | .02372 | .49253 | 14 |
| 47 | .50784 | .97623 | .53161 | .46839 | .02377 | .49216 | 13 |
| 48 | .50821 | .97619 | .53202 | .46798 | .02381 | .49179 | 12 |
| 49 | .50858 | .97615 | .53244 | .46756 | .02385 | .49142 | 11 |
| 50 | 9.50896 | 9.97610 | 9.53285 | 10.46715 | 10.02390 | 10.49104 | 10 |
| 51 | .50933 | .97606 | .53327 | .46673 | .02394 | .49067 | 9 |
| 52 | .50970 | .97602 | .53368 | .46632 | .02398 | .49030 | 8 |
| 53 | .51007 | .97597 | .53409 | .46591 | .02403 | .48993 | 7 |
| 54 | .51043 | .97593 | .53450 | .46550 | .02407 | .48957 | 6 |
| 55 | 9.51080 | 9.97589 | 9.53492 | 10.46508 | 10.02411 | 10.48920 | 5 |
| 56 | .51117 | .97584 | .53533 | .46467 | .02416 | .48883 | 4 |
| 57 | .51154 | .97580 | .53574 | .46426 | .02420 | .48846 | 3 |
| 58 | .51191 | .97576 | .53615 | .46385 | .02424 | .48809 | 2 |
| 59 | .51227 | .97571 | .53656 | .46344 | .02429 | .48773 | 1 |
| 60 | .51264 | 9.97567 | 9.53697 | 10.46303 | 10.02433 | 10.48736 | 0 |
| M | Cosine | Sine | Cotangent | Tangent | Cosecant | Secant | M |

**19°** Logarithms of Trigonometrical Functions **160°**

| M | Sine | Cosine | Tangent | Cotangent | Secant | Cosecant | M |
|---|------|--------|---------|-----------|--------|----------|---|
| 0 | 9.51264 | 9.97567 | 9.53697 | 10.46303 | 10.02433 | 10.48736 | 60 |
| 1 | .51301 | .97563 | .53738 | .46262 | .02437 | .48699 | 59 |
| 2 | .51338 | .97558 | .53779 | .46221 | .02442 | .48662 | 58 |
| 3 | .51374 | .97554 | .53820 | .46180 | .02446 | .48626 | 57 |
| 4 | .51411 | .97550 | .53861 | .46139 | .02450 | .48589 | 56 |
| 5 | 9.51447 | 9.97545 | 9.53902 | 10.46098 | 10.02455 | 10.48553 | 55 |
| 6 | .51484 | .97541 | .53943 | .46057 | .02459 | .48516 | 54 |
| 7 | .51520 | .97536 | .53984 | .46016 | .02464 | .48480 | 53 |
| 8 | .51557 | .97532 | .54025 | .45975 | .02468 | .48443 | 52 |
| 9 | .51593 | .97528 | .54065 | .45935 | .02472 | .48407 | 51 |
| 10 | 9.51629 | 9.97523 | 9.54106 | 10.45894 | 10.02477 | 10.48371 | 50 |
| 11 | .51666 | .97519 | .54147 | .45853 | .02481 | .48334 | 49 |
| 12 | .51702 | .97515 | .54187 | .45813 | .02485 | .48298 | 48 |
| 13 | .51738 | .97510 | .54228 | .45772 | .02490 | .48262 | 47 |
| 14 | .51774 | .97506 | .54269 | .45731 | .02494 | .48226 | 46 |
| 15 | 9.51811 | 9.97501 | 9.54309 | 10.45691 | 10.02499 | 10.48189 | 45 |
| 16 | .51847 | .97497 | .54350 | .45650 | .02503 | .48153 | 44 |
| 17 | .51883 | .97492 | .54390 | .45610 | .02508 | .48117 | 43 |
| 18 | .51919 | .97488 | .54431 | .45569 | .02512 | .48081 | 42 |
| 19 | .51955 | .97484 | .54471 | .45529 | .02516 | .48045 | 41 |
| 20 | 9.51991 | 9.97479 | 9.54512 | 10.45488 | 10.02521 | 10.48009 | 40 |
| 21 | .52027 | .97475 | .54552 | .45448 | .02525 | .47973 | 39 |
| 22 | .52063 | .97470 | .54593 | .45407 | .02530 | .47937 | 38 |
| 23 | .52099 | .97466 | .54633 | .45367 | .02534 | .47901 | 37 |
| 24 | .52135 | .97461 | .54673 | .45327 | .02539 | .47865 | 36 |
| 25 | 9.52171 | 9.97457 | 9.54714 | 10.45286 | 10.02543 | 10.47829 | 35 |
| 26 | .52207 | .97453 | .54754 | .45246 | .02547 | .47793 | 34 |
| 27 | .52242 | .97448 | .54794 | .45206 | .02552 | .47758 | 33 |
| 28 | .52278 | .97444 | .54835 | .45165 | .02556 | .47722 | 32 |
| 29 | .52314 | .97439 | .54875 | .45125 | .02561 | .47686 | 31 |
| 30 | 9.52350 | 9.97435 | 9.54915 | 10.45085 | 10.02565 | 10.47650 | 30 |
| 31 | .52385 | .97430 | .54955 | .45045 | .02570 | .47615 | 29 |
| 32 | .52421 | .97426 | .54995 | .45005 | .02574 | .47579 | 28 |
| 33 | .52456 | .97421 | .55035 | .44965 | .02579 | .47544 | 27 |
| 34 | .52492 | .97417 | .55075 | .44925 | .02583 | .47508 | 26 |
| 35 | 9.52527 | 9.97412 | 9.55115 | 10.44885 | 10.02588 | 10.47473 | 25 |
| 36 | .52563 | .97408 | .55155 | .44845 | .02592 | .47437 | 24 |
| 37 | .52598 | .97403 | .55195 | .44805 | .02597 | .47402 | 23 |
| 38 | .52634 | .97399 | .55235 | .44765 | .02601 | .47366 | 22 |
| 39 | .52669 | .97394 | .55275 | .44725 | .02606 | .47331 | 21 |
| 40 | 9.52705 | 9.97390 | 9.55315 | 10.44685 | 10.02610 | 10.47295 | 20 |
| 41 | .52740 | .97385 | .55355 | .44645 | .02615 | .47260 | 19 |
| 42 | .52775 | .97381 | .55395 | .44605 | .02619 | .47225 | 18 |
| 43 | .52811 | .97376 | .55434 | .44566 | .02624 | .47189 | 17 |
| 44 | .52846 | .97372 | .55474 | .44526 | .02628 | .47154 | 16 |
| 45 | 9.52881 | 9.97367 | 9.55514 | 10.44486 | 10.02633 | 10.47119 | 15 |
| 46 | .52916 | .97363 | .55554 | .44446 | .02637 | .47084 | 14 |
| 47 | .52951 | .97358 | .55593 | .44407 | .02642 | .47049 | 13 |
| 48 | .52986 | .97353 | .55633 | .44367 | .02647 | .47014 | 12 |
| 49 | .53021 | .97349 | .55673 | .44327 | .02651 | .46979 | 11 |
| 50 | 9.53056 | 9.97344 | 9.55712 | 10.44288 | 10.02656 | 10.46944 | 10 |
| 51 | .53092 | .97340 | .55752 | .44248 | .02660 | .46908 | 9 |
| 52 | .53126 | .97335 | .55791 | .44209 | .02665 | .46874 | 8 |
| 53 | .53161 | .97331 | .55831 | .44169 | .02669 | .46839 | 7 |
| 54 | .53196 | .97326 | .55870 | .44130 | .02674 | .46804 | 6 |
| 55 | 9.53231 | 9.97322 | 9.55910 | 10.44090 | 10.02678 | 10.46769 | 5 |
| 56 | .53266 | .97317 | .55949 | .44051 | .02683 | .46734 | 4 |
| 57 | .53301 | .97312 | .55989 | .44011 | .02688 | .46699 | 3 |
| 58 | .53336 | .97308 | .56028 | .43972 | .02692 | .46664 | 2 |
| 59 | .53370 | .97303 | .56067 | .43933 | .02697 | .46630 | 1 |
| 60 | 9.53405 | 9.97299 | 9.56107 | 10.43893 | 10.02701 | 10.46595 | 0 |
| M | Cosine | Sine | Cotangent | Tangent | Cosecant | Secant | M |

| M | Sine | Cosine | Tangent | Cotangent | Secant | Cosecant | M |
|---|---|---|---|---|---|---|---|
| 0 | 9.53405 | 9.97299 | 9.56107 | 10.43893 | 10.02701 | 10.46595 | 60 |
| 1 | .53440 | .97294 | .56146 | .43854 | .02706 | .46560 | 59 |
| 2 | .53475 | .97289 | .56185 | .43815 | .02711 | .46525 | 58 |
| 3 | .53509 | .97285 | .56224 | .43776 | .02715 | .46491 | 57 |
| 4 | .53544 | .97280 | .56264 | .43736 | .02720 | .46456 | 56 |
| 5 | 9.53578 | 9.97276 | 9.56303 | 10.43697 | 10.02724 | 10.46422 | 55 |
| 6 | .53613 | .97271 | .56342 | .43658 | .02729 | .46387 | 54 |
| 7 | .53647 | .97266 | .56381 | .43619 | .02734 | .46353 | 53 |
| 8 | .53682 | .97262 | .56420 | .43580 | .02738 | .46318 | 52 |
| 9 | .53716 | .97257 | .56459 | .43541 | .02743 | .46284 | 51 |
| 10 | 9.53751 | 9.97252 | 9.56498 | 10.43502 | 10.02748 | 10.46249 | 50 |
| 11 | .53785 | .97248 | .56537 | .43463 | .02752 | .46215 | 49 |
| 12 | .53819 | .97243 | .56576 | .43424 | .02757 | .46181 | 48 |
| 13 | .53854 | .97238 | .56615 | .43385 | .02762 | .46146 | 47 |
| 14 | .53888 | .97234 | .56654 | .43346 | .02766 | .46112 | 46 |
| 15 | 9.53922 | 9.97229 | 9.56693 | 10.43307 | 10.02771 | 10.46078 | 45 |
| 16 | .53957 | .97224 | .56732 | .43268 | .02776 | .46043 | 44 |
| 17 | .53991 | .97220 | .56771 | .43229 | .02780 | .46009 | 43 |
| 18 | .54025 | .97215 | .56810 | .43190 | .02785 | .45975 | 42 |
| 19 | .54059 | .97210 | .56849 | .43151 | .02790 | .45941 | 41 |
| 20 | 9.54093 | 9.97206 | 9.56887 | 10.43113 | 10.02794 | 10.45907 | 40 |
| 21 | .54127 | .97201 | .56926 | .43074 | .02799 | .45873 | 39 |
| 22 | .54161 | .97196 | .56965 | .43035 | .02804 | .45839 | 38 |
| 23 | .54195 | .97192 | .57004 | .42996 | .02808 | .45805 | 37 |
| 24 | .54229 | .97187 | .57042 | .42958 | .02813 | .45771 | 36 |
| 25 | 9.54263 | 9.97182 | 9.57081 | 10.42919 | 10.02818 | 10.45737 | 35 |
| 26 | .54297 | .97178 | .57120 | .42880 | .02822 | .45703 | 34 |
| 27 | .54331 | .97173 | .57158 | .42842 | .02827 | .45669 | 33 |
| 28 | .54365 | .97168 | .57197 | .42803 | .02832 | .45635 | 32 |
| 29 | .54399 | .97163 | .57235 | .42765 | .02837 | .45601 | 31 |
| 30 | 9.54433 | 9.97159 | 9.57274 | 10.42726 | 10.02841 | 10.45567 | 30 |
| 31 | .54466 | .97154 | .57312 | .42688 | .02846 | .45534 | 29 |
| 32 | .54500 | .97149 | .57351 | .42649 | .02851 | .45500 | 28 |
| 33 | .54534 | .97145 | .57389 | .42611 | .02855 | .45466 | 27 |
| 34 | .54567 | .97140 | .57428 | .42572 | .02860 | .45433 | 26 |
| 35 | 9.54601 | 9.97135 | 9.57466 | 10.42534 | 10.02865 | 10.45399 | 25 |
| 36 | .54635 | .97130 | .57504 | .42496 | .02870 | .45365 | 24 |
| 37 | .54668 | .97126 | .57543 | .42457 | .02874 | .45332 | 23 |
| 38 | .54702 | .97121 | .57581 | .42419 | .02879 | .45298 | 22 |
| 39 | .54735 | .97116 | .57619 | .42381 | .02884 | .45265 | 21 |
| 40 | 9.54769 | 9.97111 | 9.57658 | 10.42342 | 10.02889 | 10.45231 | 20 |
| 41 | .54802 | .97107 | .57696 | .42304 | .02893 | .45198 | 19 |
| 42 | .54836 | .97102 | .57734 | .42266 | .02898 | .45164 | 18 |
| 43 | .54869 | .97097 | .57772 | .42228 | .02903 | .45131 | 17 |
| 44 | .54903 | .97092 | .57810 | .42190 | .02908 | .45097 | 16 |
| 45 | 9.54936 | 9.97087 | 9.57849 | 10.42151 | 10.02913 | 10.45064 | 15 |
| 46 | .54969 | .97083 | .57887 | .42113 | .02917 | .45031 | 14 |
| 47 | .55003 | .97078 | .57925 | .42075 | .02922 | .44997 | 13 |
| 48 | .55036 | .97073 | .57963 | .42037 | .02927 | .44964 | 12 |
| 49 | .55069 | .97068 | .58001 | .41999 | .02932 | .44931 | 11 |
| 50 | 9.55102 | 9.97063 | 9.58039 | 10.41961 | 10.02937 | 10.44898 | 10 |
| 51 | .55136 | .97059 | .58077 | .41923 | .02941 | .44864 | 9 |
| 52 | .55169 | .97054 | .58115 | .41885 | .02946 | .44831 | 8 |
| 53 | .55202 | .97049 | .58153 | .41847 | .02951 | .44798 | 7 |
| 54 | .55235 | .97044 | .58191 | .41809 | .02956 | .44765 | 6 |
| 55 | 9.55268 | 9.97039 | 9.58229 | 10.41771 | 10.02961 | 10.44732 | 5 |
| 56 | .55301 | .97035 | .58267 | .41733 | .02965 | .44699 | 4 |
| 57 | .55334 | .97030 | .58304 | .41696 | .02970 | .44666 | 3 |
| 58 | .55367 | .97025 | .58342 | .41658 | .02975 | .44633 | 2 |
| 59 | .55400 | .97020 | .58380 | .41620 | .02980 | .44600 | 1 |
| 60 | 9.55433 | 9.97015 | 9.58418 | 10.41582 | 10.02985 | 10.44567 | 0 |
| M | Cosine | Sine | Cotangent | Tangent | Cosecant | Secant | M |

| M | Sine | Cosine | Tangent | Cotangent | Secant | Cosecant | M |
|---|------|--------|---------|-----------|--------|----------|---|
| 0 | 9.55433 | 9.97015 | 9.58418 | 10.41582 | 10.02985 | 10.44567 | 60 |
| 1 | .55466 | .97010 | .58455 | .41545 | .02990 | .44534 | 59 |
| 2 | .55499 | .97005 | .58493 | .41507 | .02995 | .44501 | 58 |
| 3 | .55532 | .97001 | .58531 | .41469 | .02999 | .44468 | 57 |
| 4 | .55564 | .96996 | .58569 | .41431 | .03004 | .44436 | 56 |
| 5 | 9.55597 | 9.96991 | 9.58606 | 10.41394 | 10.03009 | 10.44403 | 55 |
| 6 | .55630 | .96986 | .58644 | .41356 | .03014 | .44370 | 54 |
| 7 | .55663 | .96981 | .58681 | .41319 | .03019 | .44337 | 53 |
| 8 | .55695 | .96976 | .58719 | .41281 | .03024 | .44305 | 52 |
| 9 | .55728 | .96971 | .58757 | .41243 | .03029 | .44272 | 51 |
| 10 | 9.55761 | 9.96966 | 9.58794 | 10.41206 | 10.03034 | 10.44239 | 50 |
| 11 | .55793 | .96962 | .58832 | .41168 | .03038 | .44207 | 49 |
| 12 | .55826 | .96957 | .58869 | .41131 | .03043 | .44174 | 48 |
| 13 | .55858 | .96952 | .58907 | .41093 | .03048 | .44142 | 47 |
| 14 | .55891 | .96947 | .58944 | .41056 | .03053 | .44109 | 46 |
| 15 | 9.55923 | 9.96942 | 9.58981 | 10.41019 | 10.03058 | 10.44077 | 45 |
| 16 | .55956 | .96937 | .59019 | .40981 | .03063 | .44044 | 44 |
| 17 | .55988 | .96932 | .59056 | .40944 | .03068 | .44012 | 43 |
| 18 | .56021 | .96927 | .59094 | .40906 | .03073 | .43979 | 42 |
| 19 | .56053 | .96922 | .59131 | .40869 | .03078 | .43947 | 41 |
| 20 | 9.56085 | 9.96917 | 9.59168 | 10.40832 | 10.03083 | 10.43915 | 40 |
| 21 | .56118 | .96912 | .59205 | .40795 | .03088 | .43882 | 39 |
| 22 | .56150 | .96907 | .59243 | .40757 | .03093 | .43850 | 38 |
| 23 | .56182 | .96903 | .59280 | .40720 | .03097 | .43818 | 37 |
| 24 | .56215 | .96898 | .59317 | .40683 | .03102 | .43785 | 36 |
| 25 | 9.56247 | 9.96893 | 9.59354 | 10.40646 | 10.03107 | 10.43753 | 35 |
| 26 | .56279 | .96888 | .59391 | .40609 | .03112 | .43721 | 34 |
| 27 | .56311 | .96883 | .59429 | .40571 | .03117 | .43689 | 33 |
| 28 | .56343 | .96878 | .59466 | .40534 | .03122 | .43657 | 32 |
| 29 | .56375 | .96873 | .59503 | .40497 | .03127 | .43625 | 31 |
| 30 | 9.56408 | 9.96868 | 9.59540 | 10.40460 | 10.03132 | 10.43592 | 30 |
| 31 | .56440 | .96863 | .59577 | .40423 | .03137 | .43560 | 29 |
| 32 | .56472 | .96858 | .59614 | .40386 | .03142 | .43528 | 28 |
| 33 | .56504 | .96853 | .59651 | .40349 | .03147 | .43496 | 27 |
| 34 | .56536 | .96848 | .59688 | .40312 | .03152 | .43464 | 26 |
| 35 | 9.56568 | 9.96843 | 9.59725 | 10.40275 | 10.03157 | 10.43432 | 25 |
| 36 | .56599 | .96838 | .59762 | .40238 | .03162 | .43401 | 24 |
| 37 | .56631 | .96833 | .59799 | .40201 | .03167 | .43369 | 23 |
| 38 | .56663 | .96828 | .59835 | .40165 | .03172 | .43337 | 22 |
| 39 | .56695 | .96823 | .59872 | .40128 | .03177 | .43305 | 21 |
| 40 | 9.56727 | 9.96818 | 9.59909 | 10.40091 | 10.03182 | 10.43273 | 20 |
| 41 | .56759 | .96813 | .59946 | .40054 | .03187 | .43241 | 19 |
| 42 | .56790 | .96808 | .59983 | .40017 | .03192 | .43210 | 18 |
| 43 | .56822 | .96803 | .60019 | .39981 | .03197 | .43178 | 17 |
| 44 | .56854 | .96798 | .60056 | .39944 | .03202 | .43146 | 16 |
| 45 | 9.56886 | 9.96793 | 9.60093 | 10.39907 | 10.03207 | 10.43114 | 15 |
| 46 | .56917 | .96788 | .60130 | .39870 | .03212 | .43083 | 14 |
| 47 | .56949 | .96783 | .60166 | .39834 | .03217 | .43051 | 13 |
| 48 | .56980 | .96778 | .60203 | .39797 | .03222 | .43020 | 12 |
| 49 | .57012 | .96772 | .60240 | .39760 | .03228 | .42988 | 11 |
| 50 | 9.57044 | 9.96767 | 9.60276 | 10.39724 | 10.03233 | 10.42956 | 10 |
| 51 | .57075 | .96762 | .60313 | .39687 | .03238 | .42925 | 9 |
| 52 | .57107 | .96757 | .60349 | .39651 | .03243 | .42893 | 8 |
| 53 | .57138 | .96752 | .60386 | .39614 | .03248 | .42862 | 7 |
| 54 | .57169 | .96747 | .60422 | .39578 | .03253 | .42831 | 6 |
| 55 | 9.57201 | 9.96742 | 9.60459 | 10.39541 | 10.03258 | 10.42799 | 5 |
| 56 | .57232 | .96737 | .60495 | .39505 | .03263 | .42768 | 4 |
| 57 | .57264 | .96732 | .60532 | .39468 | .03268 | .42736 | 3 |
| 58 | .57295 | .96727 | .60568 | .39432 | .03273 | .42705 | 2 |
| 59 | .57326 | .96722 | .60605 | .39395 | .03278 | .42674 | 1 |
| 60 | 9.57358 | 9.96717 | 9.60641 | 10.39359 | 10.03283 | 10.42642 | 0 |

| M | Cosine | Sine | Cotangent | Tangent | Cosecant | Secant | M |
|---|--------|------|-----------|---------|----------|--------|---|

| M | Sine | Cosine | Tangent | Cotangent | Secant | Cosecant | M |
|---|---|---|---|---|---|---|---|
| 0 | 9.57358 | 9.96717 | 9.60641 | 10.39359 | 10.03283 | 10.42642 | 60 |
| 1 | .57389 | .96711 | .60677 | .39323 | .03289 | .42611 | 59 |
| 2 | .57420 | .96706 | .60714 | .39286 | .03294 | .42580 | 58 |
| 3 | .57451 | .96701 | .60750 | .39250 | .03299 | .42549 | 57 |
| 4 | .57482 | .96696 | .60786 | .39214 | .03304 | .42518 | 56 |
| 5 | 9.57514 | 9.96691 | 9.60823 | 10.39177 | 10.03309 | 10.42486 | 55 |
| 6 | .57545 | .96686 | .60859 | .39141 | .03314 | .42455 | 54 |
| 7 | .57576 | .96681 | .60895 | .39105 | .03319 | .42424 | 53 |
| 8 | .57607 | .96676 | .60931 | .39069 | .03324 | .42393 | 52 |
| 9 | .57638 | .96670 | .60967 | .39033 | .03330 | .42362 | 51 |
| 10 | 9.57669 | 9.96665 | 9.61004 | 10.38996 | 10.03335 | 10.42331 | 50 |
| 11 | .57700 | .96660 | .61040 | .38960 | .03340 | .42300 | 49 |
| 12 | .57731 | .96655 | .61076 | .38924 | .03345 | .42269 | 48 |
| 13 | .57762 | .96650 | .61112 | .38888 | .03350 | .42238 | 47 |
| 14 | .57793 | .96645 | .61148 | .38852 | .03355 | .42207 | 46 |
| 15 | 9.57824 | 9.96640 | 9.61184 | 10.38816 | 10.03360 | 10.42176 | 45 |
| 16 | .57855 | .96634 | .61220 | .38780 | .03366 | .42145 | 44 |
| 17 | .57885 | .96629 | .61256 | .38744 | .03371 | .42115 | 43 |
| 18 | .57916 | .96624 | .61292 | .38708 | .03376 | .42084 | 42 |
| 19 | .57947 | .96619 | .61328 | .38672 | .03381 | .42053 | 41 |
| 20 | 9.57978 | 9.96614 | 9.61364 | 10.38636 | 10.03386 | 10.42022 | 40 |
| 21 | .58008 | .96608 | .61400 | .38600 | .03392 | .41992 | 39 |
| 22 | .58039 | .96603 | .61436 | .38564 | .03397 | .41961 | 38 |
| 23 | .58070 | .96598 | .61472 | .38528 | .03402 | .41930 | 37 |
| 24 | .58101 | .96593 | .61508 | .38492 | .03407 | .41899 | 36 |
| 25 | 9.58131 | 9.96588 | 9.61544 | 10.38456 | 10.03412 | 10.41869 | 35 |
| 26 | .58162 | .96582 | .61579 | .38421 | .03418 | .41838 | 34 |
| 27 | .58192 | .96577 | .61615 | .38385 | .03423 | .41808 | 33 |
| 28 | .58223 | .96572 | .61651 | .38349 | .03428 | .41777 | 32 |
| 29 | .58253 | .96567 | .61687 | .38313 | .03433 | .41747 | 31 |
| 30 | 9.58284 | 9.96562 | 9.61722 | 10.38278 | 10.03438 | 10.41716 | 30 |
| 31 | .58314 | .96556 | .61758 | .38242 | .03444 | .41686 | 29 |
| 32 | .58345 | .96551 | .61794 | .38206 | .03449 | .41655 | 28 |
| 33 | .58375 | .96546 | .61830 | .38170 | .03454 | .41625 | 27 |
| 34 | .58406 | .96541 | .61865 | .38135 | .03459 | .41594 | 26 |
| 35 | 9.58436 | 9.96535 | 9.61901 | 10.38099 | 10.03465 | 10.41564 | 25 |
| 36 | .58467 | .96530 | .61936 | .38064 | .03470 | .41533 | 24 |
| 37 | .58497 | .96525 | .61972 | .38028 | .03475 | .41503 | 23 |
| 38 | .58527 | .96520 | .62008 | .37992 | .03480 | .41473 | 22 |
| 39 | .58557 | .96514 | .62043 | .37957 | .03486 | .41443 | 21 |
| 40 | 9.58588 | 9.96509 | 9.62079 | 10.37921 | 10.03491 | 10.41412 | 20 |
| 41 | .58618 | .96504 | .62114 | .37886 | .03496 | .41382 | 19 |
| 42 | .58648 | .96498 | .62150 | .37850 | .03502 | .41352 | 18 |
| 43 | .58678 | .96493 | .62185 | .37815 | .03507 | .41322 | 17 |
| 44 | .58709 | .96488 | .62221 | .37779 | .03512 | .41291 | 16 |
| 45 | 9.58739 | 9.96483 | 9.62256 | 10.37744 | 10.03517 | 10.41261 | 15 |
| 46 | .58769 | .96477 | .62292 | .37708 | .03523 | .41231 | 14 |
| 47 | .58799 | .96472 | .62327 | .37673 | .03528 | .41201 | 13 |
| 48 | .58829 | .96467 | .62362 | .37638 | .03533 | .41171 | 12 |
| 49 | .58859 | .96461 | .62398 | .37602 | .03539 | .41141 | 11 |
| 50 | 9.58889 | 9.96456 | 9.62433 | 10.37567 | 10.03544 | 10.41111 | 10 |
| 51 | .58919 | .96451 | .62468 | .37532 | .03549 | .41081 | 9 |
| 52 | .58949 | .96445 | .62504 | .37496 | .03555 | .41051 | 8 |
| 53 | .58979 | .96440 | .62539 | .37461 | .03560 | .41021 | 7 |
| 54 | .59009 | .96435 | .62574 | .37426 | .03565 | .40991 | 6 |
| 55 | 9.59039 | 9.96429 | 9.62609 | 10.37391 | 10.03571 | 10.40961 | 5 |
| 56 | .59069 | .96424 | .62645 | .37355 | .03576 | .40931 | 4 |
| 57 | .59098 | .96419 | .62680 | .37320 | .03581 | .40902 | 3 |
| 58 | .59128 | .96413 | .62715 | .37285 | .03587 | .40872 | 2 |
| 59 | .59158 | .96408 | .62750 | .37250 | .03592 | .40842 | 1 |
| 60 | 9.59188 | 9.96403 | 9.62785 | 10.37215 | 10.03597 | 10.40812 | 0 |
| M | Cosine | Sine | Cotangent | Tangent | Cosecant | Secant | M |

| M | Sine | Cosine | Tangent | Cotangent | Secant | Cosecant | M |
|---|------|--------|---------|-----------|--------|----------|---|
| 0 | 9.59188 | 9.96403 | 9.62785 | 10.37215 | 10.03597 | 10.40812 | 60 |
| 1 | .59218 | .96397 | .62820 | .37180 | .03603 | .40782 | 59 |
| 2 | .59247 | .96392 | .62855 | .37145 | .03608 | .40753 | 58 |
| 3 | .59277 | .96387 | .62890 | .37110 | .03613 | .40723 | 57 |
| 4 | .59307 | .96381 | .62926 | .37074 | .03619 | .40693 | 56 |
| 5 | 9.59336 | 9.96376 | 9.62961 | 10.37039 | 10.03624 | 10.40664 | 55 |
| 6 | .59366 | .96370 | .62996 | .37004 | .03630 | .40634 | 54 |
| 7 | .59396 | .96365 | .63031 | .36969 | .03635 | .40604 | 53 |
| 8 | .59425 | .96360 | .63066 | .36934 | .03640 | .40575 | 52 |
| 9 | .59455 | .96354 | .63101 | .36899 | .03646 | .40545 | 51 |
| 10 | 9.59484 | 9.96349 | 9.63135 | 10.36865 | 10.03651 | 10.40516 | 50 |
| 11 | .59514 | .96343 | .63170 | .36830 | .03657 | .40486 | 49 |
| 12 | .59543 | .96338 | .63205 | .36795 | .03662 | .40457 | 48 |
| 13 | .59573 | .96333 | .63240 | .36760 | .03667 | .40427 | 47 |
| 14 | .59602 | .96327 | .63275 | .36725 | .03673 | .40398 | 46 |
| 15 | 9.59632 | 9.96322 | 9.63310 | 10.36690 | 10.03678 | 10.40368 | 45 |
| 16 | .59661 | .96316 | .63345 | .36655 | .03684 | .40339 | 44 |
| 17 | .59690 | .96311 | .63379 | .36621 | .03689 | .40310 | 43 |
| 18 | .59720 | .96305 | .63414 | .36586 | .03695 | .40280 | 42 |
| 19 | .59749 | .96300 | .63449 | .36551 | .03700 | .40251 | 41 |
| 20 | 9.59778 | 9.96294 | 9.63484 | 10.36516 | 10.03706 | 10.40222 | 40 |
| 21 | .59808 | .96289 | .63519 | .36481 | .03711 | .40192 | 39 |
| 22 | .59837 | .96284 | .63553 | .36447 | .03716 | .40163 | 38 |
| 23 | .59866 | .96278 | .63588 | .36412 | .03722 | .40134 | 37 |
| 24 | .59895 | .96273 | .63623 | .36377 | .03727 | .40105 | 36 |
| 25 | 9.59924 | 9.96267 | 9.63657 | 10.36343 | 10.03733 | 10.40076 | 35 |
| 26 | .59954 | .96262 | .63692 | .36308 | .03738 | .40046 | 34 |
| 27 | .59983 | .96256 | .63726 | .36274 | .03744 | .40017 | 33 |
| 28 | .60012 | .96251 | .63761 | .36239 | .03749 | .39988 | 32 |
| 29 | .60041 | .96245 | .63796 | .36204 | .03755 | .39959 | 31 |
| 30 | 9.60070 | 9.96240 | 9.63830 | 10.36170 | 10.03760 | 10.39930 | 30 |
| 31 | .60099 | .96234 | .63865 | .36135 | .03766 | .39901 | 29 |
| 32 | .60128 | .96229 | .63899 | .36101 | .03771 | .39872 | 28 |
| 33 | .60157 | .96223 | .63934 | .36066 | .03777 | .39843 | 27 |
| 34 | .60186 | .96218 | .63968 | .36032 | .03782 | .39814 | 26 |
| 35 | 9.60215 | 9.96212 | 9.64003 | 10.35997 | 10.03788 | 10.39785 | 25 |
| 36 | .60244 | .96207 | .64037 | .35963 | .03793 | .39756 | 24 |
| 37 | .60273 | .96201 | .64072 | .35928 | .03799 | .39727 | 23 |
| 38 | .60302 | .96196 | .64106 | .35894 | .03804 | .39698 | 22 |
| 39 | .60331 | .96190 | .64140 | .35860 | .03810 | .39669 | 21 |
| 40 | 9.60359 | 9.96185 | 9.64175 | 10.35825 | 10.03815 | 10.39641 | 20 |
| 41 | .60388 | .96179 | .64209 | .35791 | .03821 | .39612 | 19 |
| 42 | .60417 | .96174 | .64243 | .35757 | .03826 | .39583 | 18 |
| 43 | .60446 | .96168 | .64278 | .35722 | .03832 | .39554 | 17 |
| 44 | .60474 | .96162 | .64312 | .35688 | .03838 | .39526 | 16 |
| 45 | 9.60503 | 9.96157 | 9.64346 | 10.35654 | 10.03843 | 10.39497 | 15 |
| 46 | .60532 | .96151 | .64381 | .35619 | .03849 | .39468 | 14 |
| 47 | .60561 | .96146 | .64415 | .35585 | .03854 | .39439 | 13 |
| 48 | .60589 | .96140 | .64449 | .35551 | .03860 | .39411 | 12 |
| 49 | .60618 | .96135 | .64483 | .35517 | .03865 | .39382 | 11 |
| 50 | 9.60646 | 9.96129 | 9.64517 | 10.35483 | 10.03871 | 10.39354 | 10 |
| 51 | .60675 | .96123 | .64552 | .35448 | .03877 | .39325 | 9 |
| 52 | .60704 | .96118 | .64586 | .35414 | .03882 | .39296 | 8 |
| 53 | .60732 | .96112 | .64620 | .35380 | .03888 | .39268 | 7 |
| 54 | .60761 | .96107 | .64654 | .35346 | .03893 | .39239 | 6 |
| 55 | 9.60789 | 9.96101 | 9.64688 | 10.35312 | 10.03899 | 10.39211 | 5 |
| 56 | .60818 | .96095 | .64722 | .35278 | .03905 | .39182 | 4 |
| 57 | .60846 | .96090 | .64756 | .35244 | .03910 | .39154 | 3 |
| 58 | .60875 | .96084 | .64790 | .35210 | .03916 | .39125 | 2 |
| 59 | .60903 | .96079 | .64824 | .35176 | .03921 | .39097 | 1 |
| 60 | 9.60931 | 9.96073 | 9.64858 | 10.35142 | 10.03927 | 10.39069 | 0 |

| M | Cosine | Sine | Cotangent | Tangent | Cosecant | Secant | M |

| M | Sine | Cosine | Tangent | Cotangent | Secant | Cosecant | M |
|---|------|--------|---------|-----------|--------|----------|---|
| 0 | 9.60931 | 9.96073 | 9.64858 | 10.35142 | 10.03927 | 10.39069 | 60 |
| 1 | .60960 | .96067 | .64892 | .35108 | .03933 | .39040 | 59 |
| 2 | .60988 | .96062 | .64926 | .35074 | .03938 | .39012 | 58 |
| 3 | .61016 | .96056 | .64960 | .35040 | .03944 | .38984 | 57 |
| 4 | .61045 | .96050 | .64994 | .35006 | .03950 | .38955 | 56 |
| 5 | 9.61073 | 9.96045 | 9.65028 | 10.34972 | 10.03955 | 10.38927 | 55 |
| 6 | .61101 | .96039 | .65062 | .34938 | .03961 | .38899 | 54 |
| 7 | .61129 | .96034 | .65096 | .34904 | .03966 | .38871 | 53 |
| 8 | .61158 | .96028 | .65130 | .34870 | .03972 | .38842 | 52 |
| 9 | .61186 | .96022 | .65164 | .34836 | .03978 | .38814 | 51 |
| 10 | 9.61214 | 9.96017 | 9.65197 | 10.34803 | 10.03983 | 10.38786 | 50 |
| 11 | .61242 | .96011 | .65231 | .34769 | .03989 | .38758 | 49 |
| 12 | .61270 | .96005 | .65265 | .34735 | .03995 | .38730 | 48 |
| 13 | .61298 | .96000 | .65299 | .34701 | .04000 | .38702 | 47 |
| 14 | .61326 | .95994 | .65333 | .34667 | .04006 | .38674 | 46 |
| 15 | 9.61354 | 9.95988 | 9.65366 | 10.34634 | 10.04012 | 10.38646 | 45 |
| 16 | .61382 | .95982 | .65400 | .34600 | .04018 | .38618 | 44 |
| 17 | .61411 | .95977 | .65434 | .34566 | .04023 | .38589 | 43 |
| 18 | .61438 | .95971 | .65467 | .34533 | .04029 | .38562 | 42 |
| 19 | .61466 | .95965 | .65501 | .34499 | .04035 | .38534 | 41 |
| 20 | 9.61494 | 9.95960 | 9.65535 | 10.34465 | 10.04040 | 10.38506 | 40 |
| 21 | .61522 | .95954 | .65568 | .34432 | .04046 | .38478 | 39 |
| 22 | .61550 | .95948 | .65602 | .34398 | .04052 | .38450 | 38 |
| 23 | .61578 | .95942 | .65636 | .34364 | .04058 | .38422 | 37 |
| 24 | .61606 | .95937 | .65669 | .34331 | .04063 | .38394 | 36 |
| 25 | 9.61634 | 9.95931 | 9.65703 | 10.34297 | 10.04069 | 10.38366 | 35 |
| 26 | .61662 | .95925 | .65736 | .34264 | .04075 | .38338 | 34 |
| 27 | .61689 | .95920 | .65770 | .34230 | .04080 | .38311 | 33 |
| 28 | .61717 | .95914 | .65803 | .34197 | .04086 | .38283 | 32 |
| 29 | .61745 | .95908 | .65837 | .34163 | .04092 | .38255 | 31 |
| 30 | 9.61773 | 9.95902 | 9.65870 | 10.34130 | 10.04098 | 10.38227 | 30 |
| 31 | .61800 | .95897 | .65904 | .34096 | .04103 | .38200 | 29 |
| 32 | .61828 | .95891 | .65937 | .34063 | .04109 | .38172 | 28 |
| 33 | .61856 | .95885 | .65971 | .34029 | .04115 | .38144 | 27 |
| 34 | .61883 | .95879 | .66004 | .33996 | .04121 | .38117 | 26 |
| 35 | 9.61911 | 9.95873 | 9.66038 | 10.33962 | 10.04127 | 10.38089 | 25 |
| 36 | .61939 | .95868 | .66071 | .33929 | .04132 | .38061 | 24 |
| 37 | .61966 | .95862 | .66104 | .33896 | .04138 | .38034 | 23 |
| 38 | .61994 | .95856 | .66138 | .33862 | .04144 | .38006 | 22 |
| 39 | .62021 | .95850 | .66171 | .33829 | .04150 | .37979 | 21 |
| 40 | 9.62049 | 9.95844 | 9.66204 | 10.33796 | 10.04156 | 10.37951 | 20 |
| 41 | .62076 | .95839 | .66238 | .33762 | .04161 | .37924 | 19 |
| 42 | .62104 | .95833 | .66271 | .33729 | .04167 | .37896 | 18 |
| 43 | .62131 | .95827 | .66304 | .33696 | .04173 | .37869 | 17 |
| 44 | .62159 | .95821 | .66337 | .33663 | .04179 | .37841 | 16 |
| 45 | 9.62186 | 9.95815 | 9.66371 | 10.33629 | 10.04185 | 10.37814 | 15 |
| 46 | .62214 | .95810 | .66404 | .33596 | .04190 | .37786 | 14 |
| 47 | .62241 | .95804 | .66437 | .33563 | .04196 | .37759 | 13 |
| 48 | .62268 | .95798 | .66470 | .33530 | .04202 | .37732 | 12 |
| 49 | .62296 | .95792 | .66503 | .33497 | .04208 | .37704 | 11 |
| 50 | 9.62323 | 9.95786 | 9.66537 | 10.33463 | 10.04214 | 10.37677 | 10 |
| 51 | .62350 | .95780 | .66570 | .33430 | .04220 | .37650 | 9 |
| 52 | .62377 | .95775 | .66603 | .33397 | .04225 | .37623 | 8 |
| 53 | .62405 | .95769 | .66636 | .33364 | .04231 | .37595 | 7 |
| 54 | .62432 | .95763 | .66669 | .33331 | .04237 | .37568 | 6 |
| 55 | 9.62459 | 9.95757 | 9.66702 | 10.33298 | 10.04243 | 10.37541 | 5 |
| 56 | .62486 | .95751 | .66735 | .33265 | .04249 | .37514 | 4 |
| 57 | .62513 | .95745 | .66768 | .33232 | .04255 | .37487 | 3 |
| 58 | .62541 | .95739 | .66801 | .33199 | .04261 | .37459 | 2 |
| 59 | .62568 | .95733 | .66834 | .33166 | .04267 | .37432 | 1 |
| 60 | 9.62595 | 9.95728 | 9.66867 | 10.33133 | 10.04272 | 10.37405 | 0 |

| M | Cosine | Sine | Cotangent | Tangent | Cosecant | Secant | M |

| M | Sine | Cosine | Tangent | Cotangent | Secant | Cosecant | M |
|---|------|--------|---------|-----------|--------|----------|---|
| 0 | 9.62595 | 9.95728 | 9.66867 | 10.33133 | 10.04272 | 10.37405 | 60 |
| 1 | .62622 | .95722 | .66900 | .33100 | .04278 | .37378 | 59 |
| 2 | .62649 | .95716 | .66933 | .33067 | .04284 | .37351 | 58 |
| 3 | .62676 | .95710 | .66966 | .33034 | .04290 | .37324 | 57 |
| 4 | .62703 | .95704 | .66999 | .33001 | .04296 | .37297 | 56 |
| 5 | 9.62730 | 9.95698 | 9.67032 | 10.32968 | 10.04302 | 10.37270 | 55 |
| 6 | .62757 | .95692 | .67065 | .32935 | .04308 | .37243 | 54 |
| 7 | .62784 | .95686 | .67098 | .32902 | .04314 | .37216 | 53 |
| 8 | .62811 | .95680 | .67131 | .32869 | .04320 | .37189 | 52 |
| 9 | .62838 | .95674 | .67163 | .32837 | .04326 | .37162 | 51 |
| 10 | 9.62865 | 9.95668 | 9.67196 | 10.32804 | 10.04332 | 10.37135 | 50 |
| 11 | .62892 | .95663 | .67229 | .32771 | .04337 | .37108 | 49 |
| 12 | .62918 | .95657 | .67262 | .32738 | .04343 | .37082 | 48 |
| 13 | .62945 | .95651 | .67295 | .32705 | .04349 | .37055 | 47 |
| 14 | .62972 | .95645 | .67327 | .32673 | .04355 | .37028 | 46 |
| 15 | 9.62999 | 9.95639 | 9.67360 | 10.32640 | 10.04361 | 10.37001 | 45 |
| 16 | .63026 | .95633 | .67393 | .32607 | .04367 | .36974 | 44 |
| 17 | .63052 | .95627 | .67426 | .32574 | .04373 | .36948 | 43 |
| 18 | .63079 | .95621 | .67458 | .32542 | .04379 | .36921 | 42 |
| 19 | .63106 | .95615 | .67491 | .32509 | .04385 | .36894 | 41 |
| 20 | 9.63133 | 9.95609 | 9.67524 | 10.32476 | 10.04391 | 10.36867 | 40 |
| 21 | .63159 | .95603 | .67556 | .32444 | .04397 | .36841 | 39 |
| 22 | .63186 | .95597 | .67589 | .32411 | .04403 | .36814 | 38 |
| 23 | .63213 | .95591 | .67622 | .32378 | .04409 | .36787 | 37 |
| 24 | .63239 | .95585 | .67654 | .32346 | .04415 | .36761 | 36 |
| 25 | 9.63266 | 9.95579 | 9.67687 | 10.32313 | 10.04421 | 10.36734 | 35 |
| 26 | .63292 | .95573 | .67719 | .32281 | .04427 | .36708 | 34 |
| 27 | .63319 | .95567 | .67752 | .32248 | .04433 | .36681 | 33 |
| 28 | .63345 | .95561 | .67785 | .32215 | .04439 | .36655 | 32 |
| 29 | .63372 | .95555 | .67817 | .32183 | .04445 | .36628 | 31 |
| 30 | 9.63398 | 9.95549 | 9.67850 | 10.32150 | 10.04451 | 10.36602 | 30 |
| 31 | .63425 | .95543 | .67882 | .32118 | .04457 | .36575 | 29 |
| 32 | .63451 | .95537 | .67915 | .32085 | .04463 | .36549 | 28 |
| 33 | .63478 | .95531 | .67947 | .32053 | .04469 | .36522 | 27 |
| 34 | .63504 | .95525 | .67980 | .32020 | .04475 | .36496 | 26 |
| 35 | 9.63531 | 9.95519 | 9.68012 | 10.31988 | 10.04481 | 10.36469 | 25 |
| 36 | .63557 | .95513 | .68044 | .31956 | .04487 | .36443 | 24 |
| 37 | .63583 | .95507 | .68077 | .31923 | .04493 | .36417 | 23 |
| 38 | .63610 | .95500 | .68109 | .31891 | .04500 | .36390 | 22 |
| 39 | .63636 | .95494 | .68142 | .31858 | .04506 | .36364 | 21 |
| 40 | 9.63662 | 9.95488 | 9.68174 | 10.31826 | 10.04512 | 10.36338 | 20 |
| 41 | .63689 | .95482 | .68206 | .31794 | .04518 | .36311 | 19 |
| 42 | .63715 | .95476 | .68239 | .31761 | .04524 | .36285 | 18 |
| 43 | .63741 | .95470 | .68271 | .31729 | .04530 | .36259 | 17 |
| 44 | .63767 | .95464 | .68303 | .31697 | .04536 | .36233 | 16 |
| 45 | 9.63794 | 9.95458 | 9.68336 | 10.31664 | 10.04542 | 10.36206 | 15 |
| 46 | .63820 | .95452 | .68368 | .31632 | .04548 | .36180 | 14 |
| 47 | .63846 | .95446 | .68400 | .31600 | .04554 | .36154 | 13 |
| 48 | .63872 | .95440 | .68432 | .31568 | .04560 | .36128 | 12 |
| 49 | .63898 | .95434 | .68465 | .31535 | .04566 | .36102 | 11 |
| 50 | 9.63924 | 9.95427 | 9.68497 | 10.31503 | 10.04573 | 10.36076 | 10 |
| 51 | .63950 | .95421 | .68529 | .31471 | .04579 | .36050 | 9 |
| 52 | .63976 | .95415 | .68561 | .31439 | .04585 | .36024 | 8 |
| 53 | .64002 | .95409 | .68593 | .31407 | .04591 | .35998 | 7 |
| 54 | .64028 | .95403 | .68626 | .31374 | .04597 | .35972 | 6 |
| 55 | 9.64054 | 9.95397 | 9.68658 | 10.31342 | 10.04603 | 10.35946 | 5 |
| 56 | .64080 | .95391 | .68690 | .31310 | .04609 | .35920 | 4 |
| 57 | .64106 | .95384 | .68722 | .31278 | .04616 | .35894 | 3 |
| 58 | .64132 | .95378 | .68754 | .31246 | .04622 | .35868 | 2 |
| 59 | .64158 | .95372 | .68786 | .31214 | .04628 | .35842 | 1 |
| 60 | 9.64184 | 9.95366 | 9.68818 | 10.31182 | 10.04634 | 10.35816 | 0 |
| M | Cosine | Sine | Cotangent | Tangent | Cosecant | Secant | M |

| M | Sine | Cosine | Tangent | Cotangent | Secant | Cosecant | M |
|---|------|--------|---------|-----------|--------|----------|---|
| 0 | 9.64184 | 9.95366 | 9.68818 | 10.31182 | 10.04634 | 10.35816 | 60 |
| 1 | .64210 | .95360 | .68850 | .31150 | .04640 | .35790 | 59 |
| 2 | .64236 | .95354 | .68882 | .31118 | .04646 | .35764 | 58 |
| 3 | .64262 | .95348 | .68914 | .31086 | .04652 | .35738 | 57 |
| 4 | .64288 | .95341 | .68946 | .31054 | .04659 | .35712 | 56 |
| 5 | 9.64313 | 9.95335 | 9.68978 | 10.31022 | 10.04665 | 10.35687 | 55 |
| 6 | .64339 | .95329 | .69010 | .30990 | .04671 | .35661 | 54 |
| 7 | .64365 | .95323 | .69042 | .30958 | .04677 | .35635 | 53 |
| 8 | .64391 | .95317 | .69074 | .30926 | .04683 | .35609 | 52 |
| 9 | .64417 | .95310 | .69106 | .30894 | .04690 | .35583 | 51 |
| 10 | 9.64442 | 9.95304 | 9.69138 | 10.30862 | 10.04696 | 10.35558 | 50 |
| 11 | .64468 | .95298 | .69170 | .30830 | .04702 | .35532 | 49 |
| 12 | .64494 | .95292 | .69202 | .30798 | .04708 | .35506 | 48 |
| 13 | .64519 | .95286 | .69234 | .30766 | .04714 | .35481 | 47 |
| 14 | .64545 | .95279 | .69266 | .30734 | .04721 | .35455 | 46 |
| 15 | 9.64571 | 9.95273 | 9.69298 | 10.30702 | 10.04727 | 10.35429 | 45 |
| 16 | .64596 | .95267 | .69329 | .30671 | .04733 | .35404 | 44 |
| 17 | .64622 | .95261 | .69361 | .30639 | .04739 | .35378 | 43 |
| 18 | .64647 | .95254 | .69393 | .30607 | .04746 | .35353 | 42 |
| 19 | .64673 | .95248 | .69425 | .30575 | .04752 | .35327 | 41 |
| 20 | 9.64698 | 9.95242 | 9.69457 | 10.30543 | 10.04758 | 10.35302 | 40 |
| 21 | .64724 | .95236 | .69488 | .30512 | .04764 | .35276 | 39 |
| 22 | .64749 | .95229 | .69520 | .30480 | .04771 | .35251 | 38 |
| 23 | .64775 | .95223 | .69552 | .30448 | .04777 | .35225 | 37 |
| 24 | .64800 | .95217 | .69584 | .30416 | .04783 | .35200 | 36 |
| 25 | 9.64826 | 9.95211 | 9.69615 | 10.30385 | 10.04789 | 1C.35174 | 35 |
| 26 | .64851 | .95204 | .69647 | .30353 | .04796 | .35149 | 34 |
| 27 | .64877 | .95198 | .69679 | .30321 | .04802 | .35123 | 33 |
| 28 | .64902 | .95192 | .69710 | .30290 | .04808 | .35098 | 32 |
| 29 | .64927 | .95185 | .69742 | .30258 | .04815 | .35073 | 31 |
| 30 | 9.64953 | 9.95179 | 9.69774 | 10.30226 | 10.04821 | 10.35047 | 30 |
| 31 | .64978 | .95173 | .69805 | .30195 | .04827 | .35022 | 29 |
| 32 | .65003 | .95167 | .69837 | .30163 | .04833 | .34997 | 28 |
| 33 | .65029 | .95160 | .69868 | .30132 | .04840 | .34971 | 27 |
| 34 | .65054 | .95154 | .69900 | .30100 | .04846 | .34946 | 26 |
| 35 | 9.65079 | 9.95148 | 9.69932 | 10.30068 | 10.04852 | 10.34921 | 25 |
| 36 | .65104 | .95141 | .69963 | .30037 | .04859 | .34896 | 24 |
| 37 | .65130 | .95135 | .69995 | .30005 | .04865 | .34870 | 23 |
| 38 | .65155 | .95129 | .70026 | .29974 | .04871 | .34845 | 22 |
| 39 | .65180 | .95122 | .70058 | .29942 | .04878 | .34820 | 21 |
| 40 | 9.65205 | 9.95116 | 9.70089 | 10.29911 | 10.04884 | 10.34795 | 20 |
| 41 | .65230 | .95110 | .70121 | .29879 | .04890 | .34770 | 19 |
| 42 | .65255 | .95103 | .70152 | .29848 | .04897 | .34745 | 18 |
| 43 | .65281 | .95097 | .70184 | .29816 | .04903 | .34719 | 17 |
| 44 | .65306 | .95090 | .70215 | .29785 | .04910 | .34694 | 16 |
| 45 | 9.65331 | 9.95084 | 9.70247 | 10.29753 | 10.04916 | 10.34669 | 15 |
| 46 | .65356 | .95078 | .70278 | .29722 | .04922 | .34644 | 14 |
| 47 | .65381 | .95071 | .70309 | .29691 | .04929 | .34619 | 13 |
| 48 | .65406 | .95065 | .70341 | .29659 | .04935 | .34594 | 12 |
| 49 | .65431 | .95059 | .70372 | .29628 | .04941 | .34569 | 11 |
| 50 | 9.65456 | 9.95052 | 9.70404 | 10.29596 | 10.04948 | 10.34544 | 10 |
| 51 | .65481 | .95046 | .70435 | .29565 | .04954 | .34519 | 9 |
| 52 | .65506 | .95039 | .70466 | .29534 | .04961 | .34494 | 8 |
| 53 | .65531 | .95033 | .70498 | .29502 | .04967 | .34469 | 7 |
| 54 | .65556 | .95027 | .70529 | .29471 | .04973 | .34444 | 6 |
| 55 | 9.65580 | 9.95020 | 9.70560 | 10.29440 | 10.04980 | 10.34420 | 5 |
| 56 | .65605 | .95014 | .70592 | .29408 | .04986 | .34395 | 4 |
| 57 | .65630 | .95007 | .70623 | .29377 | .04993 | .34370 | 3 |
| 58 | .65655 | .95001 | .70654 | .29346 | .04999 | .34345 | 2 |
| 59 | .65680 | .94995 | .70685 | .29315 | .05005 | .34320 | 1 |
| 60 | 9.65705 | 9.94988 | 9.70717 | 10.29283 | 10.05012 | 10.34295 | 0 |
| M | Cosine | Sine | Cotangent | Tangent | Cosecant | Secant | M |

| M | Sine | Cosine | Tangent | Cotangent | Secant | Cosecant | M |
|---|------|--------|---------|-----------|--------|----------|---|
| 0 | 9.65705 | 9.94988 | 9.70717 | 10.29283 | 10.05012 | 10.34295 | 60 |
| 1 | .65729 | .94982 | .70748 | .29252 | .05018 | .34271 | 59 |
| 2 | .65754 | .94975 | .70779 | .29221 | .05025 | .34246 | 58 |
| 3 | .65779 | .94969 | .70810 | .29190 | .05031 | .34221 | 57 |
| 4 | .65804 | .94962 | .70841 | .29159 | .05038 | .34196 | 56 |
| 5 | 9.65828 | 9.94956 | 9.70873 | 10.29127 | 10.05044 | 10.34172 | 55 |
| 6 | .65853 | .94949 | .70904 | .29096 | .05051 | .34147 | 54 |
| 7 | .65878 | .94943 | .70935 | .29065 | .05057 | .34122 | 53 |
| 8 | .65902 | .94936 | .70966 | .29034 | .05064 | .34098 | 52 |
| 9 | .65927 | .94930 | .70997 | .29003 | .05070 | .34073 | 51 |
| 10 | 9.65952 | 9.94923 | 9.71028 | 10.28972 | 10.05077 | 10.34048 | 50 |
| 11 | .65976 | .94917 | .71059 | .28941 | .05083 | .34024 | 49 |
| 12 | .66001 | .94911 | .71090 | .28910 | .05089 | .33999 | 48 |
| 13 | .66025 | .94904 | .71121 | .28879 | .05096 | .33975 | 47 |
| 14 | .66050 | .94898 | .71153 | .28847 | .05102 | .33950 | 46 |
| 15 | 9.66075 | 9.94891 | 9.71184 | 10.28816 | 10.05109 | 10.33925 | 45 |
| 16 | .66099 | .94885 | .71215 | .28785 | .05115 | .33901 | 44 |
| 17 | .66124 | .94878 | .71246 | .28754 | .05122 | .33876 | 43 |
| 18 | .66148 | .94871 | .71277 | .28723 | .05129 | .33852 | 42 |
| 19 | .66173 | .94865 | .71308 | .28692 | .05135 | .33827 | 41 |
| 20 | 9.66197 | 9.94858 | 9.71339 | 10.28661 | 10.05142 | 10.33803 | 40 |
| 21 | .66221 | .94852 | .71370 | .28630 | .05148 | .33779 | 39 |
| 22 | .66246 | .94845 | .71401 | .28599 | .05155 | .33754 | 38 |
| 23 | .66270 | .94839 | .71431 | .28569 | .05161 | .33730 | 37 |
| 24 | .66295 | .94832 | .71462 | .28538 | .05168 | .33705 | 36 |
| 25 | 9.66319 | 9.94826 | 9.71493 | 10.28507 | 10.05174 | 10.33681 | 35 |
| 26 | .66343 | .94819 | .71524 | .28476 | .05181 | .33657 | 34 |
| 27 | .66368 | .94813 | .71555 | .28445 | .05187 | .33632 | 33 |
| 28 | .66392 | .94806 | .71586 | .28414 | .05194 | .33608 | 32 |
| 29 | .66416 | .94799 | .71617 | .28383 | .05201 | .33584 | 31 |
| 30 | 9.66441 | 9.94793 | 9.71648 | 10.28352 | 10.05207 | 10.33559 | 30 |
| 31 | .66465 | .94786 | .71679 | .28321 | .05214 | .33535 | 29 |
| 32 | .66489 | .94780 | .71709 | .28291 | .05220 | .33511 | 28 |
| 33 | .66513 | .94773 | .71740 | .28260 | .05227 | .33487 | 27 |
| 34 | .66537 | .94767 | .71771 | .28229 | .05233 | .33463 | 26 |
| 35 | 9.66562 | 9.94760 | 9.71802 | 10.28198 | 10.05240 | 10.33438 | 25 |
| 36 | .66586 | .94753 | .71833 | .28167 | .05247 | .33414 | 24 |
| 37 | .66610 | .94747 | .71863 | .28137 | .05253 | .33390 | 23 |
| 38 | .66634 | .94740 | .71894 | .28106 | .05260 | .33366 | 22 |
| 39 | .66658 | .94734 | .71925 | .28075 | .05266 | .33342 | 21 |
| 40 | 9.66682 | 9.94727 | 9.71955 | 10.28045 | 10.05273 | 10.33318 | 20 |
| 41 | .66706 | .94720 | .71986 | .28014 | .05280 | .33294 | 19 |
| 42 | .66731 | .94714 | .72017 | .27983 | .05286 | .33269 | 18 |
| 43 | .66755 | .94707 | .72048 | .27952 | .05293 | .33245 | 17 |
| 44 | .66779 | .94700 | .72078 | .27922 | .05300 | .33221 | 16 |
| 45 | 9.66803 | 9.94694 | 9.72109 | 10.27891 | 10.05306 | 10.33197 | 15 |
| 46 | .66827 | .94687 | .72140 | .27860 | .05313 | .33173 | 14 |
| 47 | .66851 | .94680 | .72170 | .27830 | .05320 | .33149 | 13 |
| 48 | .66875 | .94674 | .72201 | .27799 | .05326 | .33125 | 12 |
| 49 | .66899 | .94667 | .72231 | .27769 | .05333 | .33101 | 11 |
| 50 | 9.66922 | 9.94660 | 9.72262 | 10.27738 | 10.05340 | 10.33078 | 10 |
| 51 | .66946 | .94654 | .72293 | .27707 | .05346 | .33054 | 9 |
| 52 | .66970 | .94647 | .72323 | .27677 | .05353 | .33030 | 8 |
| 53 | .66994 | .94640 | .72354 | .27646 | .05360 | .33006 | 7 |
| 54 | .67018 | .94634 | .72384 | .27616 | .05366 | .32982 | 6 |
| 55 | 9.67042 | 9.94627 | 9.72415 | 10.27585 | 10.05373 | 10.32958 | 5 |
| 56 | .67066 | .94620 | .72445 | .27555 | .05380 | .32934 | 4 |
| 57 | .67090 | .94614 | .72476 | .27524 | .05386 | .32910 | 3 |
| 58 | .67113 | .94607 | .72506 | .27494 | .05393 | .32887 | 2 |
| 59 | .67137 | .94600 | .72537 | .27463 | .05400 | .32863 | 1 |
| 60 | 9.67161 | 9.94593 | 9.72567 | 10.27433 | 10.05407 | 10.32839 | 0 |

| M | Cosine | Sine | Cotangent | Tangent | Cosecant | Secant | M |

| M | Sine | Cosine | Tangent | Cotangent | Secant | Cosecant | M |
|---|---|---|---|---|---|---|---|
| 0 | 9.67161 | 9.94593 | 9.72567 | 10.27433 | 10.05407 | 10.32839 | 60 |
| 1 | .67185 | .94587 | .72598 | .27402 | .05413 | .32815 | 59 |
| 2 | .67208 | .94580 | .72628 | .27372 | .05420 | .32792 | 58 |
| 3 | .67232 | .94573 | .72659 | .27341 | .05427 | .32768 | 57 |
| 4 | .67256 | .94567 | .72689 | .27311 | .05433 | .32744 | 56 |
| 5 | 9.67280 | 9.94560 | 9.72720 | 10.27280 | 10.05440 | 10.32720 | 55 |
| 6 | .67303 | .94553 | .72750 | .27250 | .05447 | .32697 | 54 |
| 7 | .67327 | .94546 | .72780 | .27220 | .05454 | .32673 | 53 |
| 8 | .67350 | .94540 | .72811 | .27189 | .05460 | .32650 | 52 |
| 9 | .67374 | .94533 | .72841 | .27159 | .05467 | .32626 | 51 |
| 10 | 9.67398 | 9.94526 | 9.72872 | 10.27128 | 10.05474 | 10.32602 | 50 |
| 11 | .67421 | .94519 | .72902 | .27098 | .05481 | .32579 | 49 |
| 12 | .67445 | .94513 | .72932 | .27068 | .05487 | .32555 | 48 |
| 13 | .67468 | .94506 | .72963 | .27037 | .05494 | .32532 | 47 |
| 14 | .67492 | .94499 | .72993 | .27007 | .05501 | .32508 | 46 |
| 15 | 9.67515 | 9.94492 | 9.73023 | 10.26977 | 10.05508 | 10.32485 | 45 |
| 16 | .67539 | .94485 | .73054 | .26946 | .05515 | .32461 | 44 |
| 17 | .67562 | .94479 | .73084 | .26916 | .05521 | .32438 | 43 |
| 18 | .67586 | .94472 | .73114 | .26886 | .05528 | .32414 | 42 |
| 19 | .67609 | .94465 | .73144 | .26856 | .05535 | .32391 | 41 |
| 20 | 9.67633 | 9.94458 | 9.73175 | 10.26825 | 10.05542 | 10.32367 | 40 |
| 21 | .67656 | .94451 | .73205 | .26795 | .05549 | .32344 | 39 |
| 22 | .67680 | .94445 | .73235 | .26765 | .05555 | .32320 | 38 |
| 23 | .67703 | .94438 | .73265 | .26735 | .05562 | .32297 | 37 |
| 24 | .67726 | .94431 | .73295 | .26705 | .05569 | .32274 | 36 |
| 25 | 9.67750 | 9.94424 | 9.73326 | 10.26674 | 10.05576 | 10.32250 | 35 |
| 26 | .67773 | .94417 | .73356 | .26644 | .05583 | .32227 | 34 |
| 27 | .67796 | .94410 | .73386 | .26614 | .05590 | .32204 | 33 |
| 28 | .67820 | .94404 | .73416 | .26584 | .05596 | .32180 | 32 |
| 29 | .67843 | .94397 | .73446 | .26554 | .05603 | .32157 | 31 |
| 30 | 9.67866 | 9.94390 | 9.73476 | 10.26524 | 10.05610 | 10.32134 | 30 |
| 31 | .67890 | .94383 | .73507 | .26493 | .05617 | .32110 | 29 |
| 32 | .67913 | .94376 | .73537 | .26463 | .05624 | .32087 | 28 |
| 33 | .67936 | .94369 | .73567 | .26433 | .05631 | .32064 | 27 |
| 34 | .67959 | .94362 | .73597 | .26403 | .05638 | .32041 | 26 |
| 35 | 9.67982 | 9.94355 | 9.73627 | 10.26373 | 10.05645 | 10.32018 | 25 |
| 36 | .68006 | .94349 | .73657 | .26343 | .05651 | .31994 | 24 |
| 37 | .68029 | .94342 | .73687 | .26313 | .05658 | .31971 | 23 |
| 38 | .68052 | .94335 | .73717 | .26283 | .05665 | .31948 | 22 |
| 39 | .68075 | .94328 | .73747 | .26253 | .05672 | .31925 | 21 |
| 40 | 9.68098 | 9.94321 | 9.73777 | 10.26223 | 10.05679 | 10.31902 | 20 |
| 41 | .68121 | .94314 | .73807 | .26193 | .05686 | .31879 | 19 |
| 42 | .68144 | .94307 | .73837 | .26163 | .05693 | .31856 | 18 |
| 43 | .68167 | .94300 | .73867 | .26133 | .05700 | .31833 | 17 |
| 44 | .68190 | .94293 | .73897 | .26103 | .05707 | .31810 | 16 |
| 45 | 9.68213 | 9.94286 | 9.73927 | 10.26073 | 10.05714 | 10.31787 | 15 |
| 46 | .68237 | .94279 | .73957 | .26043 | .05721 | .31763 | 14 |
| 47 | .68260 | .94273 | .73987 | .26013 | .05727 | .31740 | 13 |
| 48 | .68283 | .94266 | .74017 | .25983 | .05734 | .31717 | 12 |
| 49 | .68305 | .94259 | .74047 | .25953 | .05741 | .31695 | 11 |
| 50 | 9.68328 | 9.94252 | 9.74077 | 10.25923 | 10.05748 | 10.31672 | 10 |
| 51 | .68351 | .94245 | .74107 | .25893 | .05755 | .31649 | 9 |
| 52 | .68374 | .94238 | .74137 | .25863 | .05762 | .31626 | 8 |
| 53 | .68397 | .94231 | .74166 | .25834 | .05769 | .31603 | 7 |
| 54 | .68420 | .94224 | .74196 | .25804 | .05776 | .31580 | 6 |
| 55 | 9.68443 | 9.94217 | 9.74226 | 10.25774 | 10.05783 | 10.31557 | 5 |
| 56 | .68466 | .94210 | .74256 | .25744 | .05790 | .31534 | 4 |
| 57 | .68489 | .94203 | .74286 | .25714 | .05797 | .31511 | 3 |
| 58 | .68512 | .94196 | .74316 | .25684 | .05804 | .31488 | 2 |
| 59 | .68534 | .94189 | .74345 | .25655 | .05811 | .31466 | 1 |
| 60 | 9.68557 | 9.94182 | 9.74375 | 10.25625 | 10.05818 | 10.31443 | 0 |

| M | Cosine | Sine | Cotangent | Tangent | Cosecant | Secant | M |
|---|---|---|---|---|---|---|---|

**29°**  Logarithms of Trigonometrical Functions  **150°**

| M | Sine | Cosine | Tangent | Cotangent | Secant | Cosecant | M |
|---|---|---|---|---|---|---|---|
| 0 | 9.68557 | 9.94182 | 9.74375 | 10.25625 | 10.05818 | 10.31443 | 60 |
| 1 | .68580 | .94175 | .74405 | .25595 | .05825 | .31420 | 59 |
| 2 | .68603 | .94168 | .74435 | .25565 | .05832 | .31397 | 58 |
| 3 | .68625 | .94161 | .74465 | .25535 | .05839 | .31375 | 57 |
| 4 | .68648 | .94154 | .74494 | .25506 | .05846 | .31352 | 56 |
| 5 | 9.68671 | 9.94147 | 9.74524 | 10.25476 | 10.05853 | 10.31329 | 55 |
| 6 | .68694 | .94140 | .74554 | .25446 | .05860 | .31306 | 54 |
| 7 | .68716 | .94133 | .74583 | .25417 | .05867 | .31284 | 53 |
| 8 | .68739 | .94126 | .74613 | .25387 | .05874 | .31261 | 52 |
| 9 | .68762 | .94119 | .74643 | .25357 | .05881 | .31238 | 51 |
| 10 | 9.68784 | 9.94112 | 9.74673 | 10.25327 | 10.05888 | 10.31216 | 50 |
| 11 | .68807 | .94105 | .74702 | .25298 | .05895 | .31193 | 49 |
| 12 | .68829 | .94098 | .74732 | .25268 | .05902 | .31171 | 48 |
| 13 | .68852 | .94090 | .74762 | .25238 | .05910 | .31148 | 47 |
| 14 | .68875 | .94083 | .74791 | .25209 | .05917 | .31125 | 46 |
| 15 | 9.68897 | 9.94076 | 9.74821 | 10.25179 | 10.05924 | 10.31103 | 45 |
| 16 | .68920 | .94069 | .74851 | .25149 | .05931 | .31080 | 44 |
| 17 | .68942 | .94062 | .74880 | .25120 | .05938 | .31058 | 43 |
| 18 | .68965 | .94055 | .74910 | .25090 | .05945 | .31035 | 42 |
| 19 | .68987 | .94048 | .74939 | .25061 | .05952 | .31013 | 41 |
| 20 | 9.69010 | 9.94041 | 9.74969 | 10.25031 | 10.05959 | 10.30990 | 40 |
| 21 | .69032 | .94034 | .74998 | .25002 | .05966 | .30968 | 39 |
| 22 | .69055 | .94027 | .75028 | .24972 | .05973 | .30945 | 38 |
| 23 | .69077 | .94020 | .75058 | .24942 | .05980 | .30923 | 37 |
| 24 | .69100 | .94012 | .75087 | .24913 | .05988 | .30900 | 36 |
| 25 | 9.69122 | 9.94005 | 9.75117 | 10.24883 | 10.05995 | 10.30878 | 35 |
| 26 | .69144 | .93998 | .75146 | .24854 | .06002 | .30856 | 34 |
| 27 | .69167 | .93991 | .75176 | .24824 | .06009 | .30833 | 33 |
| 28 | .69189 | .93984 | .75205 | .24795 | .06016 | .30811 | 32 |
| 29 | .69212 | .93977 | .75235 | .24765 | .06023 | .30788 | 31 |
| 30 | 9.69234 | 9.93970 | 9.75264 | 10.24736 | 10.06030 | 10.30766 | 30 |
| 31 | .69256 | .93963 | .75294 | .24706 | .06037 | .30744 | 29 |
| 32 | .69279 | .93955 | .75323 | .24677 | .06045 | .30721 | 28 |
| 33 | .69301 | .93948 | .75353 | .24647 | .06052 | .30699 | 27 |
| 34 | .69323 | .93941 | .75382 | .24618 | .06059 | .30677 | 26 |
| 35 | 9.69345 | 9.93934 | 9.75411 | 10.24589 | 10.06066 | 10.30655 | 25 |
| 36 | .69368 | .93927 | .75441 | .24559 | .06073 | .30632 | 24 |
| 37 | .69390 | .93920 | .75470 | .24530 | .06080 | .30610 | 23 |
| 38 | .69412 | .93912 | .75500 | .24500 | .06088 | .30588 | 22 |
| 39 | .69434 | .93905 | .75529 | .24471 | .06095 | .30566 | 21 |
| 40 | 9.69456 | 9.93898 | 9.75558 | 10.24442 | 10.06102 | 10.30544 | 20 |
| 41 | .69479 | .93891 | .75588 | .24412 | .06109 | .30521 | 19 |
| 42 | .69501 | .93884 | .75617 | .24383 | .06116 | .30499 | 18 |
| 43 | .69523 | .93876 | .75647 | .24353 | .06124 | .30477 | 17 |
| 44 | .69545 | .93869 | .75676 | .24324 | .06131 | .30455 | 16 |
| 45 | 9.69567 | 9.93862 | 9.75705 | 10.24295 | 10.06138 | 10.30433 | 15 |
| 46 | .69589 | .93855 | .75735 | .24265 | .06145 | .30411 | 14 |
| 47 | .69611 | .93847 | .75764 | .24236 | .06153 | .30389 | 13 |
| 48 | .69633 | .93840 | .75793 | .24207 | .06160 | .30367 | 12 |
| 49 | .69655 | .93833 | .75822 | .24178 | .06167 | .30345 | 11 |
| 50 | 9.69677 | 9.93826 | 9.75852 | 10.24148 | 10.06174 | 10.30323 | 10 |
| 51 | .69699 | .93819 | .75881 | .24119 | .06181 | .30301 | 9 |
| 52 | .69721 | .93811 | .75910 | .24090 | .06189 | .30279 | 8 |
| 53 | .69743 | .93804 | .75939 | .24061 | .06196 | .30257 | 7 |
| 54 | .69765 | .93797 | .75969 | .24031 | .06203 | .30235 | 6 |
| 55 | 9.69787 | 9.93789 | 9.75998 | 10.24002 | 10.06211 | 10.30213 | 5 |
| 56 | .69809 | .93782 | .76027 | .23973 | .06218 | .30191 | 4 |
| 57 | .69831 | .93775 | .76056 | .23944 | .06225 | .30169 | 3 |
| 58 | .69853 | .93768 | .76086 | .23914 | .06232 | .30147 | 2 |
| 59 | .69875 | .93760 | .76115 | .23885 | .06240 | .30125 | 1 |
| 60 | 9.69897 | 9.93753 | 9.76144 | 10.23856 | 10.06247 | 10.30103 | 0 |
| M | Cosine | Sine | Cotangent | Tangent | Cosecant | Secant | M |

| M | Sine | Cosine | Tangent | Cotangent | Secant | Cosecant | M |
|---|------|--------|---------|-----------|--------|----------|---|
| 0 | 9.69897 | 9.93753 | 9.76144 | 10.23856 | 10.06247 | 10.30103 | 60 |
| 1 | .69919 | .93746 | .76173 | .23827 | .06254 | .30081 | 59 |
| 2 | .69941 | .93738 | .76202 | .23798 | .06262 | .30059 | 58 |
| 3 | .69963 | .93731 | .76231 | .23769 | .06269 | .30037 | 57 |
| 4 | .69984 | .93724 | .76261 | .23739 | .06276 | .30016 | 56 |
| 5 | 9.70006 | 9.93717 | 9.76290 | 10.23710 | 10.06283 | 10.29994 | 55 |
| 6 | .70028 | .93709 | .76319 | .23681 | .06291 | .29972 | 54 |
| 7 | .70050 | .93702 | .76348 | .23652 | .06298 | .29950 | 53 |
| 8 | .70072 | .93695 | .76377 | .23623 | .06305 | .29928 | 52 |
| 9 | .70093 | .93687 | .76406 | .23594 | .06313 | .29907 | 51 |
| 10 | 9.70115 | 9.93680 | 9.76435 | 10.23565 | 10.06320 | 10.29885 | 50 |
| 11 | .70137 | .93673 | .76464 | .23536 | .06327 | .29863 | 49 |
| 12 | .70159 | .93665 | .76493 | .23507 | .06335 | .29841 | 48 |
| 13 | .70180 | .93658 | .76522 | .23478 | .06342 | .29820 | 47 |
| 14 | .70202 | .93650 | .76551 | .23449 | .06350 | .29798 | 46 |
| 15 | 9.70224 | 9.93643 | 9.76580 | 10.23420 | 10.06357 | 10.29776 | 45 |
| 16 | .70245 | .93636 | .76609 | .23391 | .06364 | .29755 | 44 |
| 17 | .70267 | .93628 | .76639 | .23361 | .06372 | .29733 | 43 |
| 18 | .70288 | .93621 | .76668 | .23332 | .06379 | .29712 | 42 |
| 19 | .70310 | .93614 | .76697 | .23303 | .06386 | .29690 | 41 |
| 20 | 9.70332 | 9.93606 | 9.76725 | 10.23275 | 10.06394 | 10.29668 | 40 |
| 21 | .70353 | .93599 | .76754 | .23246 | .06401 | .29647 | 39 |
| 22 | .70375 | .93591 | .76783 | .23217 | .06409 | .29625 | 38 |
| 23 | .70396 | .93584 | .76812 | .23188 | .06416 | .29604 | 37 |
| 24 | .70418 | .93577 | .76841 | .23159 | .06423 | .29582 | 36 |
| 25 | 9.70439 | 9.93569 | 9.76870 | 10.23130 | 10.06431 | 10.29561 | 35 |
| 26 | .70461 | .93562 | .76899 | .23101 | .06438 | .29539 | 34 |
| 27 | .70482 | .93554 | .76928 | .23072 | .06446 | .29518 | 33 |
| 28 | .70504 | .93547 | .76957 | .23043 | .06453 | .29496 | 32 |
| 29 | .70525 | .93539 | .76986 | .23014 | .06461 | .29475 | 31 |
| 30 | 9.70547 | 9.93532 | 9.77015 | 10.22985 | 10.06468 | 10.29453 | 30 |
| 31 | .70568 | .93525 | .77044 | .22956 | .06475 | .29432 | 29 |
| 32 | .70590 | .93517 | .77073 | .22927 | .06483 | .29410 | 28 |
| 33 | .70611 | .93510 | .77101 | .22899 | .06490 | .29389 | 27 |
| 34 | .70633 | .93502 | .77130 | .22870 | .06498 | .29367 | 26 |
| 35 | 9.70654 | 9.93495 | 9.77159 | 10.22841 | 10.06505 | 10.29346 | 25 |
| 36 | .70675 | .93487 | .77188 | .22812 | .06513 | .29325 | 24 |
| 37 | .70697 | .93480 | .77217 | .22783 | .06520 | .29303 | 23 |
| 38 | .70718 | .93472 | .77246 | .22754 | .06528 | .29282 | 22 |
| 39 | .70739 | .93465 | .77274 | .22726 | .06535 | .29261 | 21 |
| 40 | 9.70761 | 9.93457 | 9.77303 | 10.22697 | 10.06543 | 10.29239 | 20 |
| 41 | .70782 | .93450 | .77332 | .22668 | .06550 | .29218 | 19 |
| 42 | .70803 | .93442 | .77361 | .22639 | .06558 | .29197 | 18 |
| 43 | .70824 | .93435 | .77390 | .22610 | .06565 | .29176 | 17 |
| 44 | .70846 | .93427 | .77418 | .22582 | .06573 | .29154 | 16 |
| 45 | 9.70867 | 9.93420 | 9.77447 | 10.22553 | 10.06580 | 10.29133 | 15 |
| 46 | .70888 | .93412 | .77476 | .22524 | .06588 | .29112 | 14 |
| 47 | .70909 | .93405 | .77505 | .22495 | .06595 | .29091 | 13 |
| 48 | .70931 | .93397 | .77533 | .22467 | .06603 | .29069 | 12 |
| 49 | .70952 | .93390 | .77562 | .22438 | .06610 | .29048 | 11 |
| 50 | 9.70973 | 9.93382 | 9.77591 | 10.22409 | 10.06618 | 10.29027 | 10 |
| 51 | .70994 | .93375 | .77619 | .22381 | .06625 | .29006 | 9 |
| 52 | .71015 | .93367 | .77648 | .22352 | .06633 | .28985 | 8 |
| 53 | .71036 | .93360 | .77677 | .22323 | .06640 | .28964 | 7 |
| 54 | .71058 | .93352 | .77706 | .22294 | .06648 | .28942 | 6 |
| 55 | 9.71079 | 9.93344 | 9.77734 | 10.22266 | 10.06656 | 10.28921 | 5 |
| 56 | .71100 | .93337 | .77763 | .22237 | .06663 | .28900 | 4 |
| 57 | .71121 | .93329 | .77791 | .22209 | .06671 | .28879 | 3 |
| 58 | .71142 | .93322 | .77820 | .22180 | .06678 | .28858 | 2 |
| 59 | .71163 | .93314 | .77849 | .22151 | .06686 | .28837 | 1 |
| 60 | 9.71184 | 9.93307 | 9.77877 | 10.22123 | 10.06693 | 10.28816 | 0 |
| M | Cosine | Sine | Cotangent | Tangent | Cosecant | Secant | M |

| M | Sine | Cosine | Tangent | Cotangent | Secant | Cosecant | M |
|---|------|--------|---------|-----------|--------|----------|---|
| 0 | 9.71184 | 9.93307 | 9.77877 | 10.22123 | 10.06693 | 10.28816 | 60 |
| 1 | .71205 | .93299 | .77906 | .22094 | .06701 | .28795 | 59 |
| 2 | .71226 | .93291 | .77935 | .22065 | .06709 | .28774 | 58 |
| 3 | .71247 | .93284 | .77963 | .22037 | .06716 | .28753 | 57 |
| 4 | .71268 | .93276 | .77992 | .22008 | .06724 | .28732 | 56 |
| 5 | 9.71289 | 9.93269 | 9.78020 | 10.21980 | 10.06731 | 10.28711 | 55 |
| 6 | .71310 | .93261 | .78049 | .21951 | .06739 | .28690 | 54 |
| 7 | .71331 | .93253 | .78077 | .21923 | .06747 | .28669 | 53 |
| 8 | .71352 | .93246 | .78106 | .21894 | .06754 | .28648 | 52 |
| 9 | .71373 | .93238 | .78135 | .21865 | .06762 | .28627 | 51 |
| 10 | 9.71393 | 9.93230 | 9.78163 | 10.21837 | 10.06770 | 10.28607 | 50 |
| 11 | .71414 | .93223 | .78192 | .21808 | .06777 | .28586 | 49 |
| 12 | .71435 | .93215 | .78220 | .21780 | .06785 | .28565 | 48 |
| 13 | .71456 | .93207 | .78249 | .21751 | .06793 | .28544 | 47 |
| 14 | .71477 | .93200 | .78277 | .21723 | .06800 | .28523 | 46 |
| 15 | 9.71498 | 9.93192 | 9.78306 | 10.21694 | 10.06808 | 10.28502 | 45 |
| 16 | .71519 | .93184 | .78334 | .21666 | .06816 | .28481 | 44 |
| 17 | .71539 | .93177 | .78363 | .21637 | .06823 | .28461 | 43 |
| 18 | .71560 | .93169 | .78391 | .21609 | .06831 | .28440 | 42 |
| 19 | .71581 | .93161 | .78419 | .21581 | .06839 | .28419 | 41 |
| 20 | 9.71602 | 9.93154 | 9.78448 | 10.21552 | 10.06846 | 10.28398 | 40 |
| 21 | .71622 | .93146 | .78476 | .21524 | .06854 | .28378 | 39 |
| 22 | .71643 | .93138 | .78505 | .21495 | .06862 | .28357 | 38 |
| 23 | .71664 | .93131 | .78533 | .21467 | .06869 | .28336 | 37 |
| 24 | .71685 | .93123 | .78562 | .21438 | .06877 | .28315 | 36 |
| 25 | 9.71705 | 9.93115 | 9.78590 | 10.21410 | 10.06885 | 10.28295 | 35 |
| 26 | .71726 | .93108 | .78618 | .21382 | .06892 | .28274 | 34 |
| 27 | .71747 | .93100 | .78647 | .21353 | .06900 | .28253 | 33 |
| 28 | .71767 | .93092 | .78675 | .21325 | .06908 | .28233 | 32 |
| 29 | .71788 | .93084 | .78704 | .21296 | .06916 | .28212 | 31 |
| 30 | 9.71809 | 9.93077 | 9.78732 | 10.21268 | 10.06923 | 10.28191 | 30 |
| 31 | .71829 | .93069 | .78760 | .21240 | .06931 | .28171 | 29 |
| 32 | .71850 | .93061 | .78789 | .21211 | .06939 | .28150 | 28 |
| 33 | .71870 | .93053 | .78817 | .21183 | .06947 | .28130 | 27 |
| 34 | .71891 | .93046 | .78845 | .21155 | .06954 | .28109 | 26 |
| 35 | 9.71911 | 9.93038 | 9.78874 | 10.21126 | 10.06962 | 10.28089 | 25 |
| 36 | .71932 | .93030 | .78902 | .21098 | .06970 | .28068 | 24 |
| 37 | .71952 | .93022 | .78930 | .21070 | .06978 | .28048 | 23 |
| 38 | .71973 | .93014 | .78959 | .21041 | .06986 | .28027 | 22 |
| 39 | .71994 | .93007 | .78987 | .21013 | .06993 | .28006 | 21 |
| 40 | 9.72014 | 9.92999 | 9.79015 | 10.20985 | 10.07001 | 10.27986 | 20 |
| 41 | .72034 | .92991 | .79043 | .20957 | .07009 | .27966 | 19 |
| 42 | .72055 | .92983 | .79072 | .20928 | .07017 | .27945 | 18 |
| 43 | .72075 | .92976 | .79100 | .20900 | .07024 | .27925 | 17 |
| 44 | .72096 | .92968 | .79128 | .20872 | .07032 | .27904 | 16 |
| 45 | 9.72116 | 9.92960 | 9.79156 | 10.20844 | 10.07040 | 10.27884 | 15 |
| 46 | .72137 | .92952 | .79185 | .20815 | .07048 | .27863 | 14 |
| 47 | .72157 | .92944 | .79213 | .20787 | .07056 | .27843 | 13 |
| 48 | .72177 | .92936 | .79241 | .20759 | .07064 | .27823 | 12 |
| 49 | .72198 | .92929 | .79269 | .20731 | .07071 | .27802 | 11 |
| 50 | 9.72218 | 9.92921 | 9.79297 | 10.20703 | 10.07079 | 10.27782 | 10 |
| 51 | .72238 | .92913 | .79326 | .20674 | .07087 | .27762 | 9 |
| 52 | .72259 | .92905 | .79354 | .20646 | .07095 | .27741 | 8 |
| 53 | .72279 | .92897 | .79382 | .20618 | .07103 | .27721 | 7 |
| 54 | .72299 | .92889 | .79410 | .20590 | .07111 | .27701 | 6 |
| 55 | 9.72320 | 9.92881 | 9.79438 | 10.20562 | 10.07119 | 10.27680 | 5 |
| 56 | .72340 | .92874 | .79466 | .20534 | .07126 | .27660 | 4 |
| 57 | .72360 | .92866 | .79495 | .20505 | .07134 | .27640 | 3 |
| 58 | .72381 | .92858 | .79523 | .20477 | .07142 | .27619 | 2 |
| 59 | .72401 | .92850 | .79551 | .20449 | .07150 | .27599 | 1 |
| 60 | 9.72421 | 9.92842 | 9.79579 | 10.20421 | 10.07158 | 10.27579 | 0 |

| M | Cosine | Sine | Cotangent | Tangent | Cosecant | Secant | M |
|---|--------|------|-----------|---------|----------|--------|---|

| M | Sine | Cosine | Tangent | Cotangent | Secant | Cosecant | M |
|---|------|--------|---------|-----------|--------|----------|---|
| 0 | 9.72421 | 9.92842 | 9.79579 | 10.20421 | 10.07158 | 10.27579 | 60 |
| 1 | .72441 | .92834 | .79607 | .20393 | .07166 | .27559 | 59 |
| 2 | .72461 | .92826 | .79635 | .20365 | .07174 | .27539 | 58 |
| 3 | .72482 | .92818 | .79663 | .20337 | .07182 | .27518 | 57 |
| 4 | .72502 | .92810 | .79691 | .20309 | .07190 | .27498 | 56 |
| 5 | 9.72522 | 9.92803 | 9.79719 | 10.20281 | 10.07197 | 10.27478 | 55 |
| 6 | .72542 | .92795 | .79747 | .20253 | .07205 | .27458 | 54 |
| 7 | .72562 | .92787 | .79776 | .20224 | .07213 | .27438 | 53 |
| 8 | .72582 | .92779 | .79804 | .20196 | .07221 | .27418 | 52 |
| 9 | .72602 | .92771 | .79832 | .20168 | .07229 | .27398 | 51 |
| 10 | 9.72622 | 9.92763 | 9.79860 | 10.20140 | 10.07237 | 10.27378 | 50 |
| 11 | .72643 | .92755 | .79888 | .20112 | .07245 | .27357 | 49 |
| 12 | .72663 | .92747 | .79916 | .20084 | .07253 | .27337 | 48 |
| 13 | .72683 | .92739 | .79944 | .20056 | .07261 | .27317 | 47 |
| 14 | .72703 | .92731 | .79972 | .20028 | .07269 | .27297 | 46 |
| 15 | 9.72723 | 9.92723 | 9.80000 | 10.20000 | 10.07277 | 10.27277 | 45 |
| 16 | .72743 | .92715 | .80028 | .19972 | .07285 | .27257 | 44 |
| 17 | .72763 | .92707 | .80056 | .19944 | .07293 | .27237 | 43 |
| 18 | .72783 | .92699 | .80084 | .19916 | .07301 | .27217 | 42 |
| 19 | 72803 | .92691 | .80112 | .19888 | .07309 | .27197 | 41 |
| 20 | 9.72823 | 9.92683 | 9.80140 | 10.19860 | 10.07317 | 10.27177 | 40 |
| 21 | .72843 | .92675 | .80168 | .19832 | .07325 | .27157 | 39 |
| 22 | .72863 | .92667 | .80195 | .19805 | .07333 | .27137 | 38 |
| 23 | .72883 | .92659 | .80223 | .19777 | .07341 | .27117 | 37 |
| 24 | .72902 | .92651 | .80251 | .19749 | .07349 | .27098 | 36 |
| 25 | 9.72922 | 9.92643 | 9.80279 | 10.19721 | 10.07357 | 10.27078 | 35 |
| 26 | .72942 | .92635 | .80307 | .19693 | .07365 | .27058 | 34 |
| 27 | .72962 | .92627 | .80335 | .19665 | .07373 | .27038 | 33 |
| 28 | .72982 | .92619 | .80363 | .19637 | .07381 | .27018 | 32 |
| 29 | .73002 | .92611 | .80391 | .19609 | .07389 | .26998 | 31 |
| 30 | 9.73022 | 9.92603 | 9.80419 | 10.19581 | 10.07397 | 10.26978 | 30 |
| 31 | .73041 | .92595 | .80447 | .19553 | .07405 | .26959 | 29 |
| 32 | .73061 | .92587 | .80474 | .19526 | .07413 | .26939 | 28 |
| 33 | .73081 | .92579 | .80502 | .19498 | .07421 | .26919 | 27 |
| 34 | .73101 | .92571 | .80530 | .19470 | .07429 | .26899 | 26 |
| 35 | 9.73121 | 9.92563 | 9.80558 | 10.19442 | 10.07437 | 10.26879 | 25 |
| 36 | .73140 | .92555 | .80586 | .19414 | .07445 | .26860 | 24 |
| 37 | .73160 | .92546 | .80614 | .19386 | .07454 | .26840 | 23 |
| 38 | .73180 | .92538 | .80642 | .19358 | .07462 | .26820 | 22 |
| 39 | .73200 | .92530 | .80669 | .19331 | .07470 | .26800 | 21 |
| 40 | 9.73219 | 9.92522 | 9.80697 | 10.19303 | 10.07478 | 10.26781 | 20 |
| 41 | .73239 | .92514 | .80725 | .19275 | .07486 | .26761 | 19 |
| 42 | .73259 | .92506 | .80753 | .19247 | .07494 | .26741 | 18 |
| 43 | .73278 | .92498 | .80781 | .19219 | .07502 | .26722 | 17 |
| 44 | .73298 | .92490 | .80808 | .19192 | .07510 | .26702 | 16 |
| 45 | 9.73318 | 9.92482 | 9.80836 | 10.19164 | 10.07518 | 10.26682 | 15 |
| 46 | .73337 | .92473 | .80864 | .19136 | .07527 | .26663 | 14 |
| 47 | .73357 | .92465 | .80892 | .19108 | .07535 | .26643 | 13 |
| 48 | .73377 | .92457 | .80919 | .19081 | .07543 | .26623 | 12 |
| 49 | .73396 | .92449 | .80947 | .19053 | .07551 | .26604 | 11 |
| 50 | 9.73416 | 9.92441 | 9.80975 | 10.19025 | 10.07559 | 10.26584 | 10 |
| 51 | .73435 | .92433 | .81003 | .18997 | .07567 | .26565 | 9 |
| 52 | .73455 | .92425 | .81030 | .18970 | .07575 | .26545 | 8 |
| 53 | .73474 | .92416 | .81058 | .18942 | .07584 | .26526 | 7 |
| 54 | .73494 | .92408 | .81086 | .18914 | .07592 | .26506 | 6 |
| 55 | 9.73513 | 9.92400 | 9.81113 | 10.18887 | 10.07600 | 10.26487 | 5 |
| 56 | .73533 | .92392 | .81141 | .18859 | .07608 | .26467 | 4 |
| 57 | .73552 | .92384 | .81169 | .18831 | .07616 | .26448 | 3 |
| 58 | .73572 | .92376 | .81196 | .18804 | .07624 | .26428 | 2 |
| 59 | .73591 | .92367 | .81224 | .18776 | .07633 | .26409 | 1 |
| 60 | 9.73611 | 9.92359 | 9.81252 | 10.18748 | 10.07641 | 10.26389 | 0 |

| M | Cosine | Sine | Cotangent | Tangent | Cosecant | Secant | M |
|---|--------|------|-----------|---------|----------|--------|---|

| M | Sine | Cosine | Tangent | Cotangent | Secant | Cosecant | M |
|---|------|--------|---------|-----------|--------|----------|---|
| 0 | 9.73611 | 9.92359 | 9.81252 | 10.18748 | 10.07641 | 10.26389 | 60 |
| 1 | .73630 | .92351 | .81279 | .18721 | .07649 | .26370 | 59 |
| 2 | .73650 | .92343 | .81307 | .18693 | .07657 | .26350 | 58 |
| 3 | .73669 | .92335 | .81335 | .18665 | .07665 | .26331 | 57 |
| 4 | .73689 | .92326 | .81362 | .18638 | .07674 | .26311 | 56 |
| 5 | 9.73708 | 9.92318 | 9.81390 | 10.18610 | 10.07682 | 10.26292 | 55 |
| 6 | .73727 | .92310 | .81418 | .18582 | .07690 | .26273 | 54 |
| 7 | .73747 | .92302 | .81445 | .18555 | .07698 | .26253 | 53 |
| 8 | .73766 | .92293 | .81473 | .18527 | .07707 | .26234 | 52 |
| 9 | .73785 | .92285 | .81500 | .18500 | .07715 | .26215 | 51 |
| 10 | 9.73805 | 9.92277 | 9.81528 | 10.18472 | 10.07723 | 10.26195 | 50 |
| 11 | .73824 | .92269 | .81556 | .18444 | .07731 | .26176 | 49 |
| 12 | .73843 | .92260 | .81583 | .18417 | .07740 | .26157 | 48 |
| 13 | .73863 | .92252 | .81611 | .18389 | .07748 | .26137 | 47 |
| 14 | .73882 | .92244 | .81638 | .18362 | .07756 | .26118 | 46 |
| 15 | 9.73901 | 9.92235 | 9.81666 | 10.18334 | 10.07765 | 10.26099 | 45 |
| 16 | .73921 | .92227 | .81693 | .18307 | .07773 | .26079 | 44 |
| 17 | .73940 | .92219 | .81721 | .18279 | .07781 | .26060 | 43 |
| 18 | .73959 | .92211 | .81748 | .18252 | .07789 | .26041 | 42 |
| 19 | .73978 | .92202 | .81776 | .18224 | .07798 | .26022 | 41 |
| 20 | 9.73997 | 9.92194 | 9.81803 | 10.18197 | 10.07806 | 10.26003 | 40 |
| 21 | .74017 | .92186 | .81831 | .18169 | .07814 | .25983 | 39 |
| 22 | .74036 | .92177 | .81858 | .18142 | .07823 | .25964 | 38 |
| 23 | .74055 | .92169 | .81886 | .18114 | .07831 | .25945 | 37 |
| 24 | .74074 | .92161 | .81913 | .18087 | .07839 | .25926 | 36 |
| 25 | 9.74093 | 9.92152 | 9.81941 | 10.18059 | 10.07848 | 10.25907 | 35 |
| 26 | .74113 | .92144 | .81968 | .18032 | .07856 | .25887 | 34 |
| 27 | .74132 | .92136 | .81996 | .18004 | .07864 | .25868 | 33 |
| 28 | .74151 | .92127 | .82023 | .17977 | .07873 | .25849 | 32 |
| 29 | .74170 | .92119 | .82051 | .17949 | .07881 | .25830 | 31 |
| 30 | 9.74189 | 9.92111 | 9.82078 | 10.17922 | 10.07889 | 10.25811 | 30 |
| 31 | .74208 | .92102 | .82106 | .17894 | .07898 | .25792 | 29 |
| 32 | .74227 | .92094 | .82133 | .17867 | .07906 | .25773 | 28 |
| 33 | .74246 | .92086 | .82161 | .17839 | .07914 | .25754 | 27 |
| 34 | .74265 | .92077 | .82188 | .17812 | .07923 | .25735 | 26 |
| 35 | 9.74284 | 9.92069 | 9.82215 | 10.17785 | 10.07931 | 10.25716 | 25 |
| 36 | .74303 | .92060 | .82243 | .17757 | .07940 | .25697 | 24 |
| 37 | .74322 | .92052 | .82270 | .17730 | .07948 | .25678 | 23 |
| 38 | .74341 | .92044 | .82298 | .17702 | .07956 | .25659 | 22 |
| 39 | .74360 | .92035 | .82325 | .17675 | .07965 | .25640 | 21 |
| 40 | 9.74379 | 9.92027 | 9.82352 | 10.17648 | 10.07973 | 10.25621 | 20 |
| 41 | .74398 | .92018 | .82380 | .17620 | .07982 | .25602 | 19 |
| 42 | .74417 | .92010 | .82407 | .17593 | .07990 | .25583 | 18 |
| 43 | .74436 | .92002 | .82435 | .17565 | .07998 | .25564 | 17 |
| 44 | .74455 | .91993 | .82462 | .17538 | .08007 | .25545 | 16 |
| 45 | 9.74474 | 9.91985 | 9.82489 | 10.17511 | 10.08015 | 10.25526 | 15 |
| 46 | .74493 | .91976 | .82517 | .17483 | .08024 | .25507 | 14 |
| 47 | .74512 | .91968 | .82544 | .17456 | .08032 | .25488 | 13 |
| 48 | .74531 | .91959 | .82571 | .17429 | .08041 | .25469 | 12 |
| 49 | .74549 | .91951 | .82599 | .17401 | .08049 | .25451 | 11 |
| 50 | 9.74568 | 9.91942 | 9.82626 | 10.17374 | 10.08058 | 10.25432 | 10 |
| 51 | .74587 | .91934 | .82653 | .17347 | .08066 | .25413 | 9 |
| 52 | .74606 | .91925 | .82681 | .17319 | .08075 | .25394 | 8 |
| 53 | .74625 | .91917 | .82708 | .17292 | .08083 | .25375 | 7 |
| 54 | .74644 | .91908 | .82735 | .17265 | .08092 | .25356 | 6 |
| 55 | 9.74662 | 9.91900 | 9.82762 | 10.17238 | 10.08100 | 10.25338 | 5 |
| 56 | .74681 | .91891 | .82790 | .17210 | .08109 | .25319 | 4 |
| 57 | .74700 | .91883 | .82817 | .17183 | .08117 | .25300 | 3 |
| 58 | .74719 | .91874 | .82844 | .17156 | .08126 | .25281 | 2 |
| 59 | .74737 | .91866 | .82871 | .17129 | .08134 | .25263 | 1 |
| 60 | 9.74756 | 9.91857 | 9.82899 | 10.17101 | 10.08143 | 10.25244 | 0 |
| M | Cosine | Sine | Cotangent | Tangent | Cosecant | Secant | M |

| M | Sine | Cosine | Tangent | Cotangent | Secant | Cosecant | M |
|---|------|--------|---------|-----------|--------|----------|---|
| 0 | 9.74756 | 9.91857 | 9.82899 | 10.17101 | 10.08143 | 10.25244 | 60 |
| 1 | .74775 | .91849 | .82926 | .17074 | .08151 | .25225 | 59 |
| 2 | .74794 | .91840 | .82953 | .17047 | .08160 | .25206 | 58 |
| 3 | .74812 | .91832 | .82980 | .17020 | .08168 | .25188 | 57 |
| 4 | .74831 | .91823 | .83008 | .16992 | .08177 | .25169 | 56 |
| 5 | 9.74850 | 9.91815 | 9.83035 | 10.16965 | 10.08185 | 10.25150 | 55 |
| 6 | .74868 | .91806 | .83062 | .16938 | .08194 | .25132 | 54 |
| 7 | .74887 | .91798 | .83089 | .16911 | .08202 | .25113 | 53 |
| 8 | .74906 | .91789 | .83117 | .16883 | .08211 | .25094 | 52 |
| 9 | .74924 | .91781 | .83144 | .16856 | .08219 | .25076 | 51 |
| 10 | 9.74943 | 9.91772 | 9.83171 | 10.16829 | 10.08228 | 10.25057 | 50 |
| 11 | .74961 | .91763 | .83198 | .16802 | .08237 | .25039 | 49 |
| 12 | .74980 | .91755 | .83225 | .16775 | .08245 | .25020 | 48 |
| 13 | .74999 | .91746 | .83252 | .16748 | .08254 | .25001 | 47 |
| 14 | .75017 | .91738 | .83280 | .16720 | .08262 | .24983 | 46 |
| 15 | 9.75036 | 9.91729 | 9.83307 | 10.16693 | 10.08271 | 10.24964 | 45 |
| 16 | .75054 | .91720 | .83334 | .16666 | .08280 | .24946 | 44 |
| 17 | .75073 | .91712 | .83361 | .16639 | .08288 | .24927 | 43 |
| 18 | .75091 | .91703 | .83388 | .16612 | .08297 | .24909 | 42 |
| 19 | .75110 | .91695 | .83415 | .16585 | .08305 | .24890 | 41 |
| 20 | 9.75128 | 9.91686 | 9.83442 | 10.16558 | 10.08314 | 10.24872 | 40 |
| 21 | .75147 | .91677 | .83470 | .16530 | .08323 | .24853 | 39 |
| 22 | .75165 | .91669 | .83497 | .16503 | .08331 | .24835 | 38 |
| 23 | .75184 | .91660 | .83524 | .16476 | .08340 | .24816 | 37 |
| 24 | .75202 | .91651 | .83551 | .16449 | .08349 | .24798 | 36 |
| 25 | 9.75221 | 9.91643 | 9.83578 | 10.16422 | 10.08357 | 10.24779 | 35 |
| 26 | .75239 | .91634 | .83605 | .16395 | .08366 | .24761 | 34 |
| 27 | .75258 | .91625 | .83632 | .16368 | .08375 | .24742 | 33 |
| 28 | .75276 | .91617 | .83659 | .16341 | .08383 | .24724 | 32 |
| 29 | .75294 | .91608 | .83686 | .16314 | .08392 | .24706 | 31 |
| 30 | 9.75313 | 9.91599 | 9.83713 | 10.16287 | 10.08401 | 10.24687 | 30 |
| 31 | .75331 | .91591 | .83740 | .16260 | .08409 | .24669 | 29 |
| 32 | .75350 | .91582 | .83768 | .16232 | .08418 | .24650 | 28 |
| 33 | .75368 | .91573 | .83795 | .16205 | .08427 | .24632 | 27 |
| 34 | .75386 | .91565 | .83822 | .16178 | .08435 | .24614 | 26 |
| 35 | 9.75405 | 9.91556 | 9.83849 | 10.16151 | 10.08444 | 10.24595 | 25 |
| 36 | .75423 | .91547 | .83876 | .16124 | .08453 | .24577 | 24 |
| 37 | .75441 | .91538 | .83903 | .16097 | .08462 | .24559 | 23 |
| 38 | .75459 | .91530 | .83930 | .16070 | .08470 | .24541 | 22 |
| 39 | .75478 | .91521 | .83957 | .16043 | .08479 | .24522 | 21 |
| 40 | 9.75496 | 9.91512 | 9.83984 | 10.16016 | 10.08488 | 10.24504 | 20 |
| 41 | .75514 | .91504 | .84011 | .15989 | .08496 | .24486 | 19 |
| 42 | .75533 | .91495 | .84038 | .15962 | .08505 | .24467 | 18 |
| 43 | .75551 | .91486 | .84065 | .15935 | .08514 | .24449 | 17 |
| 44 | .75569 | .91477 | .84092 | .15908 | .08523 | .24431 | 16 |
| 45 | 9.75587 | 9.91469 | 9.84119 | 10.15881 | 10.08531 | 10.24413 | 15 |
| 46 | .75605 | .91460 | .84146 | .15854 | .08540 | .24395 | 14 |
| 47 | .75624 | .91451 | .84173 | .15827 | .08549 | .24376 | 13 |
| 48 | .75642 | .91442 | .84200 | .15800 | .08558 | .24358 | 12 |
| 49 | .75660 | .91433 | .84227 | .15773 | .08567 | .24340 | 11 |
| 50 | 9.75678 | 9.91425 | 9.84254 | 10.15746 | 10.08575 | 10.24322 | 10 |
| 51 | .75696 | .91416 | .84280 | .15720 | .08584 | .24304 | 9 |
| 52 | .75714 | .91407 | .84307 | .15693 | .08593 | .24286 | 8 |
| 53 | .75733 | .91398 | .84334 | .15666 | .08602 | .24267 | 7 |
| 54 | .75751 | .91389 | .84361 | .15639 | .08611 | .24249 | 6 |
| 55 | 9.75769 | 9.91381 | 9.84388 | 10.15612 | 10.08619 | 10.24231 | 5 |
| 56 | .75787 | .91372 | .84415 | .15585 | .08628 | .24213 | 4 |
| 57 | .75805 | .91363 | .84442 | .15558 | .08637 | .24195 | 3 |
| 58 | .75823 | .91354 | .84469 | .15531 | .08646 | .24177 | 2 |
| 59 | .75841 | .91345 | .84496 | .15504 | .08655 | .24159 | 1 |
| 60 | 9.75859 | 9.91336 | 9.84523 | 10.15477 | 10.08664 | 10.24141 | 0 |

| M | Cosine | Sine | Cotangent | Tangent | Cosecant | Secant | M |

**35°** Logarithms of Trigonometrical Functions **144°**

| M | Sine | Cosine | Tangent | Cotangent | Secant | Cosecant | M |
|---|---|---|---|---|---|---|---|
| 0 | 9.75859 | 9.91336 | 9.84523 | 10.15477 | 10.08664 | 10.24141 | 60 |
| 1 | .75877 | .91328 | .84550 | .15450 | .08672 | .24123 | 59 |
| 2 | .75895 | .91319 | .84576 | .15424 | .08681 | .24105 | 58 |
| 3 | .75913 | .91310 | .84603 | .15397 | .08690 | .24087 | 57 |
| 4 | .75931 | .91301 | .84630 | .15370 | .08699 | .24069 | 56 |
| 5 | 9.75949 | 9.91292 | 9.84657 | 10.15343 | 10.08708 | 10.24051 | 55 |
| 6 | .75967 | .91283 | .84684 | .15316 | .08717 | .24033 | 54 |
| 7 | .75985 | .91274 | .84711 | .15289 | .08726 | .24015 | 53 |
| 8 | .76003 | .91266 | .84738 | .15262 | .08734 | .23997 | 52 |
| 9 | .76021 | .91257 | .84764 | .15236 | .08743 | .23979 | 51 |
| 10 | 9.76039 | 9.91248 | 9.84791 | 10.15209 | 10.08752 | 10.23961 | 50 |
| 11 | .76057 | .91239 | .84818 | .15182 | .08761 | .23943 | 49 |
| 12 | .76075 | .91230 | .84845 | .15155 | .08770 | .23925 | 48 |
| 13 | .76093 | .91221 | .84872 | .15128 | .08779 | .23907 | 47 |
| 14 | .76111 | .91212 | .84899 | .15101 | .08788 | .23889 | 46 |
| 15 | 9.76129 | 9.91203 | 9.84925 | 10.15075 | 10.08797 | 10.23871 | 45 |
| 16 | .76146 | .91194 | .84952 | .15048 | .08806 | .23854 | 44 |
| 17 | .76164 | .91185 | .84979 | .15021 | .08815 | .23836 | 43 |
| 18 | .76182 | .91176 | .85006 | .14994 | .08824 | .23818 | 42 |
| 19 | .76200 | .91167 | .85033 | .14967 | .08833 | .23800 | 41 |
| 20 | 9.76218 | 9.91158 | 9.85059 | 10.14941 | 10.08842 | 10.23782 | 40 |
| 21 | .76236 | .91149 | .85086 | .14914 | .08851 | .23764 | 39 |
| 22 | .76253 | .91141 | .85113 | .14887 | .08859 | .23747 | 38 |
| 23 | .76271 | .91132 | .85140 | .14860 | .08868 | .23729 | 37 |
| 24 | .76289 | .91123 | .85166 | .14834 | .08877 | .23711 | 36 |
| 25 | 9.76307 | 9.91114 | 9.85193 | 10.14807 | 10.08886 | 10.23693 | 35 |
| 26 | .76324 | .91105 | .85220 | .14780 | .08895 | .23676 | 34 |
| 27 | .76342 | .91096 | .85247 | .14753 | .08904 | .23658 | 33 |
| 28 | .76360 | .91087 | .85273 | .14727 | .08913 | .23640 | 32 |
| 29 | .76378 | .91078 | .85300 | .14700 | .08922 | .23622 | 31 |
| 30 | 9.76395 | 9.91069 | 9.85327 | 10.14673 | 10.08931 | 10.23605 | 30 |
| 31 | .76413 | .91060 | .85354 | .14646 | .08940 | .23587 | 29 |
| 32 | .76431 | .91051 | .85380 | .14620 | .08949 | .23569 | 28 |
| 33 | .76448 | .91042 | .85407 | .14593 | .08958 | .23552 | 27 |
| 34 | .76466 | .91033 | .85434 | .14566 | .08967 | .23534 | 26 |
| 35 | 9.76484 | 9.91023 | 9.85460 | 10.14540 | 10.08977 | 10.23516 | 25 |
| 36 | .76501 | .91014 | .85487 | .14513 | .08986 | .23499 | 24 |
| 37 | .76519 | .91005 | .85514 | .14486 | .08995 | .23481 | 23 |
| 38 | .76537 | .90996 | .85540 | .14460 | .09004 | .23463 | 22 |
| 39 | .76554 | .90987 | .85567 | .14433 | .09013 | .23446 | 21 |
| 40 | 9.76572 | 9.90978 | 9.85594 | 10.14406 | 10.09022 | 10.23428 | 20 |
| 41 | .76590 | .90969 | .85620 | .14380 | .09031 | .23410 | 19 |
| 42 | .76607 | .90960 | .85647 | .14353 | .09040 | .23393 | 18 |
| 43 | .76625 | .90951 | .85674 | .14326 | .09049 | .23375 | 17 |
| 44 | .76642 | .90942 | .85700 | .14300 | .09058 | .23358 | 16 |
| 45 | 9.76660 | 9.90933 | 9.85727 | 10.14273 | 10.09067 | 10.23340 | 15 |
| 46 | .76677 | .90924 | .85754 | .14246 | .09076 | .23323 | 14 |
| 47 | .76695 | .90915 | .85780 | .14220 | .09085 | .23305 | 13 |
| 48 | .76712 | .90906 | .85807 | .14193 | .09094 | .23288 | 12 |
| 49 | .76730 | .90896 | .85834 | .14166 | .09104 | .23270 | 11 |
| 50 | 9.76747 | 9.90887 | 9.85860 | 10.14140 | 10.09113 | 10.23253 | 10 |
| 51 | .76765 | .90878 | .85887 | .14113 | .09122 | .23235 | 9 |
| 52 | .76782 | .90869 | .85913 | .14087 | .09131 | .23218 | 8 |
| 53 | .76800 | .90860 | .85940 | .14060 | .09140 | .23200 | 7 |
| 54 | .76817 | .90851 | .85967 | .14033 | .09149 | .23183 | 6 |
| 55 | 9.76835 | 9.90842 | 9.85993 | 10.14007 | 10.09158 | 10.23165 | 5 |
| 56 | .76852 | .90832 | .86020 | .13980 | .09168 | .23148 | 4 |
| 57 | .76870 | .90823 | .86046 | .13954 | .09177 | .23130 | 3 |
| 58 | .76887 | .90814 | .86073 | .13927 | .09186 | .23113 | 2 |
| 59 | .76904 | .90805 | .86100 | .13900 | .09195 | .23096 | 1 |
| 60 | 9.76922 | 9.90796 | 9.86126 | 10.13874 | 10.09204 | 10.23078 | 0 |
| M | Cosine | Sine | Cotangent | Tangent | Cosecant | Secant | M |

**125°** **54°**

**36°**      **Logarithms of Trigonometrical Functions**      **143°**

| M | Sine | Cosine | Tangent | Cotangent | Secant | Cosecant | M |
|---|------|--------|---------|-----------|--------|----------|---|
| 0 | 9.76922 | 9.90796 | 9.86126 | 10.13874 | 10.09204 | 10.23078 | 60 |
| 1 | .76939 | .90787 | .86153 | .13847 | .09213 | .23061 | 59 |
| 2 | .76957 | .90777 | .86179 | .13821 | .09223 | .23043 | 58 |
| 3 | .76974 | .90768 | .86206 | .13794 | .09232 | .23026 | 57 |
| 4 | .76991 | .90759 | .86232 | .13768 | .09241 | .23009 | 56 |
| 5 | 9.77009 | 9.90750 | 9.86259 | 10.13741 | 10.09250 | 10.22991 | 55 |
| 6 | .77026 | .90741 | .86285 | .13715 | .09259 | .22974 | 54 |
| 7 | .77043 | .90731 | .86312 | .13688 | .09269 | .22957 | 53 |
| 8 | .77061 | .90722 | .86338 | .13662 | .09278 | .22939 | 52 |
| 9 | .77078 | .90713 | .86365 | .13635 | .09287 | .22922 | 51 |
| 10 | 9.77095 | 9.90704 | 9.86392 | 10.13608 | 10.09296 | 10.22905 | 50 |
| 11 | .77112 | .90694 | .86418 | .13582 | .09306 | .22888 | 49 |
| 12 | .77130 | .90685 | .86445 | .13555 | .09315 | .22870 | 48 |
| 13 | .77147 | .90676 | .86471 | .13529 | .09324 | .22853 | 47 |
| 14 | .77164 | .90667 | .86498 | .13502 | .09333 | .22836 | 46 |
| 15 | 9.77181 | 9.90657 | 9.86524 | 10.13476 | 10.09343 | 10.22819 | 45 |
| 16 | .77199 | .90648 | .86551 | .13449 | .09352 | .22801 | 44 |
| 17 | .77216 | .90639 | .86577 | .13423 | .09361 | .22784 | 43 |
| 18 | .77233 | .90630 | .86603 | .13397 | .09370 | .22767 | 42 |
| 19 | .77250 | .90620 | .86630 | .13370 | .09380 | .22750 | 41 |
| 20 | 9.77268 | 9.90611 | 9.86656 | 10.13344 | 10.09389 | 10.22732 | 40 |
| 21 | .77285 | .90602 | .86683 | .13317 | .09398 | .22715 | 39 |
| 22 | .77302 | .90592 | .86709 | .13291 | .09408 | .22698 | 38 |
| 23 | .77319 | .90583 | .86736 | .13264 | .09417 | .22681 | 37 |
| 24 | .77336 | .90574 | .86762 | .13238 | .09426 | .22664 | 36 |
| 25 | 9.77353 | 9.90565 | 9.86789 | 10.13211 | 10.09435 | 10.22647 | 35 |
| 26 | .77370 | .90555 | .86815 | .13185 | .09445 | .22630 | 34 |
| 27 | .77387 | .90546 | .86842 | .13158 | .09454 | .22613 | 33 |
| 28 | .77405 | .90537 | .86868 | .13132 | .09463 | .22595 | 32 |
| 29 | .77422 | .90527 | .86894 | .13106 | .09473 | .22578 | 31 |
| 30 | 9.77439 | 9.90518 | 9.86921 | 10.13079 | 10.09482 | 10.22561 | 30 |
| 31 | .77456 | .90509 | .86947 | .13053 | .09491 | .22544 | 29 |
| 32 | .77473 | .90499 | .86974 | .13026 | .09501 | .22527 | 28 |
| 33 | .77490 | .90490 | .87000 | .13000 | .09510 | .22510 | 27 |
| 34 | .77507 | .90480 | .87027 | .12973 | .09520 | .22493 | 26 |
| 35 | 9.77524 | 9.90471 | 9.87053 | 10.12947 | 10.09529 | 10.22476 | 25 |
| 36 | .77541 | .90462 | .87079 | .12921 | .09538 | .22459 | 24 |
| 37 | .77558 | .90452 | .87106 | .12894 | .09548 | .22442 | 23 |
| 38 | .77575 | .90443 | .87132 | .12868 | .09557 | .22425 | 22 |
| 39 | .77592 | .90434 | .87158 | .12842 | .09566 | .22408 | 21 |
| 40 | 9.77609 | 9.90424 | 9.87185 | 10.12815 | 10.09576 | 10.22391 | 20 |
| 41 | .77626 | .90415 | .87211 | .12789 | .09585 | .22374 | 19 |
| 42 | .77643 | .90405 | .87238 | .12762 | .09595 | .22357 | 18 |
| 43 | .77660 | .90396 | .87264 | .12736 | .09604 | .22340 | 17 |
| 44 | .77677 | .90386 | .87290 | .12710 | .09614 | .22323 | 16 |
| 45 | 9.77694 | 9.90377 | 9.87317 | 10.12683 | 10.09623 | 10.22306 | 15 |
| 46 | .77711 | .90368 | .87343 | .12657 | .09632 | .22289 | 14 |
| 47 | .77728 | .90358 | .87369 | .12631 | .09642 | .22272 | 13 |
| 48 | .77744 | .90349 | .87396 | .12604 | .09651 | .22256 | 12 |
| 49 | .77761 | .90339 | .87422 | .12578 | .09661 | .22239 | 11 |
| 50 | 9.77778 | 9.90330 | 9.87448 | 10.12552 | 10.09670 | 10.22222 | 10 |
| 51 | .77795 | .90320 | .87475 | .12525 | .09680 | .22205 | 9 |
| 52 | .77812 | .90311 | .87501 | .12499 | .09689 | .22188 | 8 |
| 53 | .77829 | .90301 | .87527 | .12473 | .09699 | .22171 | 7 |
| 54 | .77846 | .90292 | .87554 | .12446 | .09708 | .22154 | 6 |
| 55 | 9.77862 | 9.90282 | 9.87580 | 10.12420 | 10.09718 | 10.22138 | 5 |
| 56 | .77879 | .90273 | .87606 | .12394 | .09727 | .22121 | 4 |
| 57 | .77896 | .90263 | .87633 | .12367 | .09737 | .22104 | 3 |
| 58 | .77913 | .90254 | .87659 | .12341 | .09746 | .22087 | 2 |
| 59 | .77930 | .90244 | .87685 | .12315 | .09756 | .22070 | 1 |
| 60 | 9.77946 | 9.90235 | 9.87711 | 10.12289 | 10.09765 | 10.22054 | 0 |
| M | Cosine | Sine | Cotangent | Tangent | Cosecant | Secant | M |

| M | Sine | Cosine | Tangent | Cotangent | Secant | Cosecant | M |
|---|------|--------|---------|-----------|--------|----------|---|
| 0 | 9.77946 | 9.90235 | 9.87711 | 10.12289 | 10.09765 | 10.22054 | 60 |
| 1 | .77963 | .90225 | .87738 | .12262 | .09775 | .22037 | 59 |
| 2 | .77980 | .90216 | .87764 | .12236 | .09784 | .22020 | 58 |
| 3 | .77997 | .90206 | .87790 | .12210 | .09794 | .22003 | 57 |
| 4 | .78013 | .90197 | .87817 | .12183 | .09803 | .21987 | 56 |
| 5 | 9.78030 | 9.90187 | 9.87843 | 10.12157 | 10.09813 | 10.21970 | 55 |
| 6 | .78047 | .90178 | .87869 | .12131 | .09822 | .21953 | 54 |
| 7 | .78063 | .90168 | .87895 | .12105 | .09832 | .21937 | 53 |
| 8 | .78080 | .90159 | .87922 | .12078 | .09841 | .21920 | 52 |
| 9 | .78097 | .90149 | .87948 | .12052 | .09851 | .21903 | 51 |
| 10 | 9.78113 | 9.90139 | 9.87974 | 10.12026 | 10.09861 | 10.21887 | 50 |
| 11 | .78130 | .90130 | .88000 | .12000 | .09870 | .21870 | 49 |
| 12 | .78147 | .90120 | .88027 | .11973 | .09880 | .21853 | 48 |
| 13 | .78163 | .90111 | .88053 | .11947 | .09889 | .21837 | 47 |
| 14 | .78180 | .90101 | .88079 | .11921 | .09899 | .21820 | 46 |
| 15 | 9.78197 | 9.90091 | 9.88105 | 10.11895 | 10.09909 | 10.21803 | 45 |
| 16 | .78213 | .90082 | .88131 | .11869 | .09918 | .21787 | 44 |
| 17 | .78230 | .90072 | .88158 | .11842 | .09928 | .21770 | 43 |
| 18 | .78246 | .90063 | .88184 | .11816 | .09937 | .21754 | 42 |
| 19 | .78263 | .90053 | .88210 | .11790 | .09947 | .21737 | 41 |
| 20 | 9.78280 | 9.90043 | 9.88236 | 10.11764 | 10.09957 | 10.21720 | 40 |
| 21 | .78296 | .90034 | .88262 | .11738 | .09966 | .21704 | 39 |
| 22 | .78313 | .90024 | .88289 | .11711 | .09976 | .21687 | 38 |
| 23 | .78329 | .90014 | .88315 | .11685 | .09986 | .21671 | 37 |
| 24 | .78346 | .90005 | .88341 | .11659 | .09995 | .21654 | 36 |
| 25 | 9.78362 | 9.89995 | 9.88367 | 10.11633 | 10.10005 | 10.21638 | 35 |
| 26 | .78379 | .89985 | .88393 | .11607 | .10015 | .21621 | 34 |
| 27 | .78395 | .89976 | .88420 | .11580 | .10024 | .21605 | 33 |
| 28 | .78412 | .89966 | .88446 | .11554 | .10034 | .21588 | 32 |
| 29 | .78428 | .89956 | .88472 | .11528 | .10044 | .21572 | 31 |
| 30 | 9.78445 | 9.89947 | 9.88498 | 10.11502 | 10.10053 | 10.21555 | 30 |
| 31 | .78461 | .89937 | .88524 | .11476 | .10063 | .21539 | 29 |
| 32 | .78478 | .89927 | .88550 | .11450 | .10073 | .21522 | 28 |
| 33 | .78494 | .89918 | .88577 | .11423 | .10082 | .21506 | 27 |
| 34 | .78510 | .89908 | .88603 | .11397 | .10092 | .21490 | 26 |
| 35 | 9.78527 | 9.89898 | 9.88629 | 10.11371 | 10.10102 | 10.21473 | 25 |
| 36 | .78543 | .89888 | .88655 | .11345 | .10112 | .21457 | 24 |
| 37 | .78560 | .89879 | .88681 | .11319 | .10121 | .21440 | 23 |
| 38 | .78576 | .89869 | .88707 | .11293 | .10131 | .21424 | 22 |
| 39 | .78592 | .89859 | .88733 | .11267 | .10141 | .21408 | 21 |
| 40 | 9.78609 | 9.89849 | 9.88759 | 10.11241 | 10.10151 | 10.21391 | 20 |
| 41 | .78625 | .89840 | .88786 | .11214 | .10160 | .21375 | 19 |
| 42 | .78642 | .89830 | .88812 | .11188 | .10170 | .21358 | 18 |
| 43 | .78658 | .89820 | .88838 | .11162 | .10180 | .21342 | 17 |
| 44 | .78674 | .89810 | .88864 | .11136 | .10190 | .21326 | 16 |
| 45 | 9.78691 | 9.89801 | 9.88890 | 10.11110 | 10.10199 | 10.21309 | 15 |
| 46 | .78707 | .89791 | .88916 | .11084 | .10209 | .21293 | 14 |
| 47 | .78723 | .89781 | .88942 | .11058 | .10219 | .21277 | 13 |
| 48 | .78739 | .89771 | .88968 | .11032 | .10229 | .21261 | 12 |
| 49 | .78756 | .89761 | .88994 | .11006 | .10239 | .21244 | 11 |
| 50 | 9.78772 | 9.89752 | 9.89020 | 10.10980 | 10.10248 | 10.21228 | 10 |
| 51 | .78788 | .89742 | .89046 | .10954 | .10258 | .21212 | 9 |
| 52 | .78805 | .89732 | .89073 | .10927 | .10268 | .21195 | 8 |
| 53 | .78821 | .89722 | .89099 | .10901 | .10278 | .21179 | 7 |
| 54 | .78837 | .89712 | .89125 | .10875 | .10288 | .21163 | 6 |
| 55 | 9.78853 | 9.89702 | 9.89151 | 10.10849 | 10.10298 | 10.21147 | 5 |
| 56 | .78869 | .89693 | .89177 | .10823 | .10307 | .21131 | 4 |
| 57 | .78886 | .89683 | .89203 | .10797 | .10317 | .21114 | 3 |
| 58 | .78902 | .89673 | .89229 | .10771 | .10327 | .21098 | 2 |
| 59 | .78918 | .89663 | .89255 | .10745 | .10337 | .21082 | 1 |
| 60 | 9.78934 | 9.89653 | 9.89281 | 10.10719 | 10.10347 | 10.21066 | 0 |
| M | Cosine | Sine | Cotangent | Tangent | Cosecant | Secant | M |

| M | Sine | Cosine | Tangent | Cotangent | Secant | Cosecant | M |
|---|------|--------|---------|-----------|--------|----------|---|
| 0 | 9.78934 | 9.89653 | 9.89281 | 10.10719 | 10.10347 | 10.21066 | 60 |
| 1 | .78950 | .89643 | .89307 | .10693 | .10357 | .21050 | 59 |
| 2 | .78967 | .89633 | .89333 | .10667 | .10367 | .21033 | 58 |
| 3 | .78983 | .89624 | .89359 | .10641 | .10376 | .21017 | 57 |
| 4 | .78999 | .89614 | .89385 | .10615 | .10386 | .21001 | 56 |
| 5 | 9.79015 | 9.89604 | 9.89411 | 10.10589 | 10.10396 | 10.20985 | 55 |
| 6 | .79031 | .89594 | .89437 | .10563 | .10406 | .20969 | 54 |
| 7 | .79047 | .89584 | .89463 | .10537 | .10416 | .20953 | 53 |
| 8 | .79063 | .89574 | .89489 | .10511 | .10426 | .20937 | 52 |
| 9 | .79079 | .89564 | .89515 | .10485 | .10436 | .20921 | 51 |
| 10 | 9.79095 | 9.89554 | 9.89541 | 10.10459 | 10.10446 | 10.20905 | 50 |
| 11 | .79111 | .89544 | .89567 | .10433 | .10456 | .20889 | 49 |
| 12 | .79128 | .89534 | .89593 | .10407 | .10466 | .20872 | 48 |
| 13 | .79144 | .89524 | .89619 | .10381 | .10476 | .20856 | 47 |
| 14 | .79160 | .89514 | .89645 | .10355 | .10486 | .20840 | 46 |
| 15 | 9.79176 | 9.89504 | 9.89671 | 10.10329 | 10.10496 | 10.20824 | 45 |
| 16 | .79192 | .89495 | .89697 | .10303 | .10505 | .20808 | 44 |
| 17 | .79208 | .89485 | .89723 | .10277 | .10515 | .20792 | 43 |
| 18 | .79224 | .89475 | .89749 | .10251 | .10525 | .20776 | 42 |
| 19 | .79240 | .89465 | .89775 | .10225 | .10535 | .20760 | 41 |
| 20 | 9.79256 | 9.89455 | 9.89801 | 10.10199 | 10.10545 | 10.20744 | 40 |
| 21 | .79272 | .89445 | .89827 | .10173 | .10555 | .20728 | 39 |
| 22 | .79288 | .89435 | .89853 | .10147 | .10565 | .20712 | 38 |
| 23 | .79304 | .89425 | .89879 | .10121 | .10575 | .20696 | 37 |
| 24 | .79319 | .89415 | .89905 | .10095 | .10585 | .20681 | 36 |
| 25 | 9.79335 | 9.89405 | 9.89931 | 10.10069 | 10.10595 | 10.20665 | 35 |
| 26 | .79351 | .89395 | .89957 | .10043 | .10605 | .20649 | 34 |
| 27 | .79367 | .89385 | .89983 | .10017 | .10615 | .20633 | 33 |
| 28 | .79383 | .89375 | .90009 | .09991 | .10625 | .20617 | 32 |
| 29 | .79399 | .89364 | .90035 | .09965 | .10636 | .20601 | 31 |
| 30 | 9.79415 | 9.89354 | 9.90061 | 10.09939 | 10.10646 | 10.20585 | 30 |
| 31 | .79431 | .89344 | .90086 | .09914 | .10656 | .20569 | 29 |
| 32 | .79447 | .89334 | .90112 | .09888 | .10666 | .20553 | 28 |
| 33 | .79463 | .89324 | .90138 | .09862 | .10676 | .20537 | 27 |
| 34 | .79478 | .89314 | .90164 | .09836 | .10686 | .20522 | 26 |
| 35 | 9.79494 | 9.89304 | 9.90190 | 10.09810 | 10.10696 | 10.20506 | 25 |
| 36 | .79510 | .89294 | .90216 | .09784 | .10706 | .20490 | 24 |
| 37 | .79526 | .89284 | .90242 | .09758 | .10716 | .20474 | 23 |
| 38 | .79542 | .89274 | .90268 | .09732 | .10726 | .20458 | 22 |
| 39 | .79558 | .89264 | .90294 | .09706 | .10736 | .20442 | 21 |
| 40 | 9.79573 | 9.89254 | 9.90320 | 10.09680 | 10.10746 | 10.20427 | 20 |
| 41 | .79589 | .89244 | .90346 | .09654 | .10756 | .20411 | 19 |
| 42 | .79605 | .89233 | .90371 | .09629 | .10767 | .20395 | 18 |
| 43 | .79621 | .89223 | .90397 | .09603 | .10777 | .20379 | 17 |
| 44 | .79636 | .89213 | .90423 | .09577 | .10787 | .20364 | 16 |
| 45 | 9.79652 | 9.89203 | 9.90449 | 10.09551 | 10.10797 | 10.20348 | 15 |
| 46 | .79668 | .89193 | .90475 | .09525 | .10807 | .20332 | 14 |
| 47 | .79684 | .89183 | .90501 | .09499 | .10817 | .20316 | 13 |
| 48 | .79699 | .89173 | .90527 | .09473 | .10827 | .20301 | 12 |
| 49 | .79715 | .89162 | .90553 | .09447 | .10838 | .20285 | 11 |
| 50 | 9.79731 | 9.89152 | 9.90578 | 10.09422 | 10.10848 | 10.20269 | 10 |
| 51 | .79746 | .89142 | .90604 | .09396 | .10858 | .20254 | 9 |
| 52 | .79762 | .89132 | .90630 | .09370 | .10868 | .20238 | 8 |
| 53 | .79778 | .89122 | .90656 | .09344 | .10878 | .20222 | 7 |
| 54 | .79793 | .89112 | .90682 | .09318 | .10888 | .20207 | 6 |
| 55 | 9.79809 | 9.89101 | 9.90708 | 10.09292 | 10.10899 | 10.20191 | 5 |
| 56 | .79825 | .89091 | .90734 | .09266 | .10909 | .20175 | 4 |
| 57 | .79840 | .89081 | .90759 | .09241 | .10919 | .20160 | 3 |
| 58 | .79856 | .89071 | .90785 | .09215 | .10929 | .20144 | 2 |
| 59 | .79872 | .89060 | .90811 | .09189 | .10940 | .20128 | 1 |
| 60 | 9.79887 | 9.89050 | 9.90837 | 10.09163 | 10.10950 | 10.20113 | 0 |

| M | Cosine | Sine | Cotangent | Tangent | Cosecant | Secant | M |
|---|--------|------|-----------|---------|----------|--------|---|

39°     **Logarithms of Trigonometrical Functions**     140°

| M | Sine | Cosine | Tangent | Cotangent | Secant | Cosecant | M |
|---|------|--------|---------|-----------|--------|----------|---|
| 0  | 9.79887 | 9.89050 | 9.90837 | 10.09163 | 10.10950 | 10.20113 | 60 |
| 1  | .79903  | .89040  | .90863  | .09137   | .10960   | .20097   | 59 |
| 2  | .79918  | .89030  | .90889  | .09111   | .10970   | .20082   | 58 |
| 3  | .79934  | .89020  | .90914  | .09086   | .10980   | .20066   | 57 |
| 4  | .79950  | .89009  | .90940  | .09060   | .10991   | .20050   | 56 |
| 5  | 9.79965 | 9.88999 | 9.90966 | 10.09034 | 10.11001 | 10.20035 | 55 |
| 6  | .79981  | .88989  | .90992  | .09008   | .11011   | .20019   | 54 |
| 7  | .79996  | .88978  | .91018  | .08982   | .11022   | .20004   | 53 |
| 8  | .80012  | .88968  | .91043  | .08957   | .11032   | .19988   | 52 |
| 9  | .80027  | .88958  | .91069  | .08931   | .11042   | .19973   | 51 |
| 10 | 9.80043 | 9.88948 | 9.91095 | 10.08905 | 10.11052 | 10.19957 | 50 |
| 11 | .80058  | .88937  | .91121  | .08879   | .11063   | .19942   | 49 |
| 12 | .80074  | .88927  | .91147  | .08853   | .11073   | .19926   | 48 |
| 13 | .80089  | .88917  | .91172  | .08828   | .11083   | .19911   | 47 |
| 14 | .80105  | .88906  | .91198  | .08802   | .11094   | .19895   | 46 |
| 15 | 9.80120 | 9.88896 | 9.91224 | 10.08776 | 10.11104 | 10.19880 | 45 |
| 16 | .80136  | .88886  | .91250  | .08750   | .11114   | .19864   | 44 |
| 17 | .80151  | .88875  | .91276  | .08724   | .11125   | .19849   | 43 |
| 18 | .80166  | .88865  | .91301  | .08699   | .11135   | .19834   | 42 |
| 19 | .80182  | .88855  | .91327  | .08673   | .11145   | .19818   | 41 |
| 20 | 9.80197 | 9.88844 | 9.91353 | 10.08647 | 10.11156 | 10.19803 | 40 |
| 21 | .80213  | .88834  | .91379  | .08621   | .11166   | .19787   | 39 |
| 22 | .80228  | .88824  | .91404  | .08596   | .11176   | .19772   | 38 |
| 23 | .80244  | .88813  | .91430  | .08570   | .11187   | .19756   | 37 |
| 24 | .80259  | .88803  | .91456  | .08544   | .11197   | .19741   | 36 |
| 25 | 9.80274 | 9.88793 | 9.91482 | 10.08518 | 10.11207 | 10.19726 | 35 |
| 26 | .80290  | .88782  | .91507  | .08493   | .11218   | .19710   | 34 |
| 27 | .80305  | .88772  | .91533  | .08467   | .11228   | .19695   | 33 |
| 28 | .80320  | .88761  | .91559  | .08441   | .11239   | .19680   | 32 |
| 29 | .80336  | .88751  | .91585  | .08415   | .11249   | .19664   | 31 |
| 30 | 9.80351 | 9.88741 | 9.91610 | 10.08390 | 10.11259 | 10.19649 | 30 |
| 31 | .80366  | .88730  | .91636  | .08364   | .11270   | .19634   | 29 |
| 32 | .80382  | .88720  | .91662  | .08338   | .11280   | .19618   | 28 |
| 33 | .80397  | .88709  | .91688  | .08312   | .11291   | .19603   | 27 |
| 34 | .80412  | .88699  | .91713  | .08287   | .11301   | .19588   | 26 |
| 35 | 9.80428 | 9.88688 | 9.91739 | 10.08261 | 10.11312 | 10.19572 | 25 |
| 36 | .80443  | .88678  | .91765  | .08235   | .11322   | .19557   | 24 |
| 37 | .80458  | .88668  | .91791  | .08209   | .11332   | .19542   | 23 |
| 38 | .80473  | .88657  | .91816  | .08184   | .11343   | .19527   | 22 |
| 39 | .80489  | .88647  | .91842  | .08158   | .11353   | .19511   | 21 |
| 40 | 9.80504 | 9.88636 | 9.91868 | 10.08132 | 10.11364 | 10.19496 | 20 |
| 41 | .80519  | .88626  | .91893  | .08107   | .11374   | .19481   | 19 |
| 42 | .80534  | .88615  | .91919  | .08081   | .11385   | .19466   | 18 |
| 43 | .80550  | .88605  | .91945  | .08055   | .11395   | .19450   | 17 |
| 44 | .80565  | .88594  | .91971  | .08029   | .11406   | .19435   | 16 |
| 45 | 9.80580 | 9.88584 | 9.91996 | 10.08004 | 10.11416 | 10.19420 | 15 |
| 46 | .80595  | .88573  | .92022  | .07978   | .11427   | .19405   | 14 |
| 47 | .80610  | .88563  | .92048  | .07952   | .11437   | .19390   | 13 |
| 48 | .80625  | .88552  | .92073  | .07927   | .11448   | .19375   | 12 |
| 49 | .80641  | .88542  | .92099  | .07901   | .11458   | .19359   | 11 |
| 50 | 9.80656 | 9.88531 | 9.92125 | 10.07875 | 10.11469 | 10.19344 | 10 |
| 51 | .80671  | .88521  | .92150  | .07850   | .11479   | .19329   | 9 |
| 52 | .80686  | .88510  | .92176  | .07824   | .11490   | .19314   | 8 |
| 53 | .80701  | .88499  | .92202  | .07798   | .11501   | .19299   | 7 |
| 54 | .80716  | .88489  | .92227  | .07773   | .11511   | .19284   | 6 |
| 55 | 9.80731 | 9.88478 | 9.92253 | 10.07747 | 10.11522 | 10.19269 | 5 |
| 56 | .80746  | .88468  | .92279  | .07721   | .11532   | .19254   | 4 |
| 57 | .80762  | .88457  | .92304  | .07696   | .11543   | .19238   | 3 |
| 58 | .80777  | .88447  | .92330  | .07670   | .11553   | .19223   | 2 |
| 59 | .80792  | .88436  | .92356  | .07644   | .11564   | .19208   | 1 |
| 60 | 9.80807 | 9.88425 | 9.92381 | 10.07619 | 10.11575 | 10.19193 | 0 |
| M | Cosine | Sine | Cotangent | Tangent | Cosecant | Secant | M |

129°                          50°

**40°** **Logarithms of Trigonometrical Functions** **139°**

| M | Sine | Cosine | Tangent | Cotangent | Secant | Cosecant | M |
|---|---|---|---|---|---|---|---|
| 0 | 9.80807 | 9.88425 | 9.92381 | 10.07619 | 10.11575 | 10.19193 | 60 |
| 1 | .80822 | .88415 | .92407 | .07593 | .11585 | .19178 | 59 |
| 2 | .80837 | .88404 | .92433 | .07567 | .11596 | .19163 | 58 |
| 3 | .80852 | .88394 | .92458 | .07542 | .11606 | .19148 | 57 |
| 4 | .80867 | .88383 | .92484 | .07516 | .11617 | .19133 | 56 |
| 5 | 9.80882 | 9.88372 | 9.92510 | 10.07490 | 10.11628 | 10.19118 | 55 |
| 6 | .80897 | .88362 | .92535 | .07465 | .11638 | .19103 | 54 |
| 7 | .80912 | .88351 | .92561 | .07439 | .11649 | .19088 | 53 |
| 8 | .80927 | .88340 | .92587 | .07413 | .11660 | .19073 | 52 |
| 9 | .80942 | .88330 | .92612 | .07388 | .11670 | .19058 | 51 |
| 10 | 9.80957 | 9.88319 | 9.92638 | 10.07362 | 10.11681 | 10.19043 | 50 |
| 11 | .80972 | .88308 | .92663 | .07337 | .11692 | .19028 | 49 |
| 12 | .80987 | .88298 | .92689 | .07311 | .11702 | .19013 | 48 |
| 13 | .81002 | .88287 | .92715 | .07285 | .11713 | .18998 | 47 |
| 14 | .81017 | .88276 | .92740 | .07260 | .11724 | .18983 | 46 |
| 15 | 9.81032 | 9.88266 | 9.92766 | 10.07234 | 10.11734 | 10.18968 | 45 |
| 16 | .81047 | .88255 | .92792 | .07208 | .11745 | .18953 | 44 |
| 17 | .81061 | .88244 | .92817 | .07183 | .11756 | .18939 | 43 |
| 18 | .81076 | .88234 | .92843 | .07157 | .11766 | .18924 | 42 |
| 19 | .81091 | .88223 | .92868 | .07132 | .11777 | .18909 | 41 |
| 20 | 9.81106 | 9.88212 | 9.92894 | 10.07106 | 10.11788 | 10.18894 | 40 |
| 21 | .81121 | .88201 | .92920 | .07080 | .11799 | .18879 | 39 |
| 22 | .81136 | .88191 | .92945 | .07055 | .11809 | .18864 | 38 |
| 23 | .81151 | .88180 | .92971 | .07029 | .11820 | .18849 | 37 |
| 24 | .81166 | .88169 | .92996 | .07004 | .11831 | .18834 | 36 |
| 25 | 9.81180 | 9.88158 | 9.93022 | 10.06978 | 10.11842 | 10.18820 | 35 |
| 26 | .81195 | .88148 | .93048 | .06952 | .11852 | .18805 | 34 |
| 27 | .81210 | .88137 | .93073 | .06927 | .11863 | .18790 | 33 |
| 28 | .81225 | .88126 | .93099 | .06901 | .11874 | .18775 | 32 |
| 29 | .81240 | .88115 | .93124 | .06876 | .11885 | .18760 | 31 |
| 30 | 9.81254 | 9.88105 | 9.93150 | 10.06850 | 10.11895 | 10.18746 | 30 |
| 31 | .81269 | .88094 | .93175 | .06825 | .11906 | .18731 | 29 |
| 32 | .81284 | .88083 | .93201 | .06799 | .11917 | .18716 | 28 |
| 33 | .81299 | .88072 | .93227 | .06773 | .11928 | .18701 | 27 |
| 34 | .81314 | .88061 | .93252 | .06748 | .11939 | .18686 | 26 |
| 35 | 9.81328 | 9.88051 | 9.93278 | 10.06722 | 10.11949 | 10.18672 | 25 |
| 36 | .81343 | .88040 | .93303 | .06697 | .11960 | .18657 | 24 |
| 37 | .81358 | .88029 | .93329 | .06671 | .11971 | .18642 | 23 |
| 38 | .81372 | .88018 | .93354 | .06646 | .11982 | .18628 | 22 |
| 39 | .81387 | .88007 | .93380 | .06620 | .11993 | .18613 | 21 |
| 40 | 9.81402 | 9.87996 | 9.93406 | 10.06594 | 10.12004 | 10.18598 | 20 |
| 41 | .81417 | .87985 | .93431 | .06569 | .12015 | .18583 | 19 |
| 42 | .81431 | .87975 | .93457 | .06543 | .12025 | .18569 | 18 |
| 43 | .81446 | .87964 | .93482 | .06518 | .12036 | .18554 | 17 |
| 44 | .81461 | .87953 | .93508 | .06492 | .12047 | .18539 | 16 |
| 45 | 9.81475 | 9.87942 | 9.93533 | 10.06467 | 10.12058 | 10.18525 | 15 |
| 46 | .81490 | .87931 | .93559 | .06441 | .12069 | .18510 | 14 |
| 47 | .81505 | .87920 | .93584 | .06416 | .12080 | .18495 | 13 |
| 48 | .81519 | .87909 | .93610 | .06390 | .12091 | .18481 | 12 |
| 49 | .81534 | .87898 | .93636 | .06364 | .12102 | .18466 | 11 |
| 50 | 9.81549 | 9.87887 | 9.93661 | 10.06339 | 10.12113 | 10.18451 | 10 |
| 51 | .81563 | .87877 | .93687 | .06313 | .12123 | .18437 | 9 |
| 52 | .81578 | .87866 | .93712 | .06288 | .12134 | .18422 | 8 |
| 53 | .81592 | .87855 | .93738 | .06262 | .12145 | .18408 | 7 |
| 54 | .81607 | .87844 | .93763 | .06237 | .12156 | .18393 | 6 |
| 55 | 9.81622 | 9.87833 | 9.93789 | 10.06211 | 10.12167 | 10.18378 | 5 |
| 56 | .81636 | .87822 | .93814 | .06186 | .12178 | .18364 | 4 |
| 57 | .81651 | .87811 | .93840 | .06160 | .12189 | .18349 | 3 |
| 58 | .81665 | .87800 | .93865 | .06135 | .12200 | .18335 | 2 |
| 59 | .81680 | .87789 | .93891 | .06109 | .12211 | .18320 | 1 |
| 60 | 9.81694 | 9.87778 | 9.93916 | 10.06084 | 10.12222 | 10.18306 | 0 |

| M | Cosine | Sine | Cotangent | Tangent | Cosecant | Secant | M |
|---|---|---|---|---|---|---|---|

**130°** **49°**

| M | Sine | Cosine | Tangent | Cotangent | Secant | Cosecant | M |
|---|---|---|---|---|---|---|---|
| 0 | 9.81694 | 9.87778 | 9.93916 | 10.06084 | 10.12222 | 10.18306 | 60 |
| 1 | .81709 | .87767 | .93942 | .06058 | .12233 | .18291 | 59 |
| 2 | .81723 | .87756 | .93967 | .06033 | .12244 | .18277 | 58 |
| 3 | .81738 | .87745 | .93993 | .06007 | .12255 | .18262 | 57 |
| 4 | .81752 | .87734 | .94018 | .05982 | .12266 | .18248 | 56 |
| 5 | 9.81767 | 9.87723 | 9.94044 | 10.05956 | 10.12277 | 10.18233 | 55 |
| 6 | .81781 | .87712 | .94069 | .05931 | .12288 | .18219 | 54 |
| 7 | .81796 | .87701 | .94095 | .05905 | .12299 | .18204 | 53 |
| 8 | .81810 | .87690 | .94120 | .05880 | .12310 | .18190 | 52 |
| 9 | .81825 | .87679 | .94146 | .05854 | .12321 | .18175 | 51 |
| 10 | 9.81839 | 9.87668 | 9.94171 | 10.05829 | 10.12332 | 10.18161 | 50 |
| 11 | .81854 | .87657. | .94197 | .05803 | .12343 | .18146 | 49 |
| 12 | .81868 | .87646 | .94222 | .05778 | .12354 | .18132 | 48 |
| 13 | .81882 | .87635 | .94248 | .05752 | .12365 | .18118 | 47 |
| 14 | .81897 | .87624 | .94273 | .05727 | .12376 | .18103 | 46 |
| 15 | 9.81911 | 9.87613 | 9.94299 | 10.05701 | 10.12387 | 10.18089 | 45 |
| 16 | .81926 | .87601 | .94324 | .05676 | .12399 | .18074 | 44 |
| 17 | .81940 | .87590 | .94350 | .05650 | .12410 | .18060 | 43 |
| 18 | .81955 | .87579 | .94375 | .05625 | .12421 | .18045 | 42 |
| 19 | .81969 | .87568 | .94401 | .05599 | .12432 | .18031 | 41 |
| 20 | 9.81983 | 9.87557 | 9.94426 | 10.05574 | 10.12443 | 10.18017 | 40 |
| 21 | .81998 | .87546 | .94452 | .05548 | .12454 | .18002 | 39 |
| 22 | .82012 | .87535 | .94477 | .05523 | .12465 | .17988 | 38 |
| 23 | .82026 | .87524 | .94503 | .05497 | .12476 | .17974 | 37 |
| 24 | .82041 | .87513 | .94528 | .05472 | .12487 | .17959 | 36 |
| 25 | 9.82055 | 9.87501 | 9.94554 | 10.05446 | 10.12499 | 10.17945 | 35 |
| 26 | .82069 | .87490 | .94579 | .05421 | .12510 | .17931 | 34 |
| 27 | .82084 | .87479 | .94604 | .05396 | .12521 | .17916 | 33 |
| 28 | .82098 | .87468 | .94630 | .05370 | .12532 | .17902 | 32 |
| 29 | .82112 | .87457 | .94655 | .05345 | .12543 | .17888 | 31 |
| 30 | 9.82126 | 9.87446 | 9.94681 | 10.05319 | 10.12554 | 10.17874 | 30 |
| 31 | .82141 | .87434 | .94706 | .05294 | .12566 | .17859 | 29 |
| 32 | .82155 | .87423 | .94732 | .05268 | .12577 | .17845 | 28 |
| 33 | .82169 | .87412 | .94757 | .05243 | .12588 | .17831 | 27 |
| 34 | .82184 | .87401 | .94783 | .05217 | .12599 | .17816 | 26 |
| 35 | 9.82198 | 9.87390 | 9.94808 | 10.05192 | 10.12610 | 10.17802 | 25 |
| 36 | .82212 | .87378 | .94834 | .05166 | .12622 | .17788 | 24 |
| 37 | .82226 | .87367 | .94859 | .05141 | .12633 | .17774 | 23 |
| 38 | .82240 | .87356 | .94884 | .05116 | .12644 | .17760 | 22 |
| 39 | .82255 | .87345 | .94910 | .05090 | .12655 | .17745 | 21 |
| 40 | 9.82269 | 9.87334 | 9.94935 | 10.05065 | 10.12666 | 10.17731 | 20 |
| 41 | .82283 | .87322 | .94961 | .05039 | .12678 | .17717 | 19 |
| 42 | .82297 | .87311 | .94986 | .05014 | .12689 | .17703 | 18 |
| 43 | .82311 | .87300 | .95012 | .04988 | .12700 | .17689 | 17 |
| 44 | .82326 | .87288 | .95037 | .04963 | .12712 | .17674 | 16 |
| 45 | 9.82340 | 9.87277 | 9.95062 | 10.04938 | 10.12723 | 10.17660 | 15 |
| 46 | .82354 | .87266 | .95088 | .04912 | .12734 | .17646 | 14 |
| 47 | .82368 | .87255 | .95113 | .04887 | .12745 | .17632 | 13 |
| 48 | .82382 | .87243 | .95139 | .04861 | .12757 | .17618 | 12 |
| 49 | .82396 | .87232 | .95164 | .04836 | .12768 | .17604 | 11 |
| 50 | 9.82410 | 9.87221 | 9.95190 | 10.04810 | 10.12779 | 10.17590 | 10 |
| 51 | .82424 | .87209 | .95215 | .04785 | .12791 | .17576 | 9 |
| 52 | .82439 | .87198 | .95240 | .04760 | .12802 | .17561 | 8 |
| 53 | .82453 | .87187 | .95266 | .04734 | .12813 | .17547 | 7 |
| 54 | .82467 | .87175 | .95291 | .04709 | .12825 | .17533 | 6 |
| 55 | 9.82481 | 9.87164 | 9.95317 | 10.04683 | 10.12836 | 10.17519 | 5 |
| 56 | .82495 | .87153 | .95342 | .04658 | .12847 | .17505 | 4 |
| 57 | .82509 | .87141 | .95368 | .04632 | .12859 | .17491 | 3 |
| 58 | .82523 | .87130 | .95393 | .04607 | .12870 | .17477 | 2 |
| 59 | .82537 | .87119 | .95418 | .04582 | .12881 | .17463 | 1 |
| 60 | 9.82551 | 9.87107 | 9.95444 | 10.04556 | 10.12893 | 10.17449 | 0 |

| M | Cosine | Sine | Cotangent | Tangent | Cosecant | Secant | M |
|---|---|---|---|---|---|---|---|

**42°**    Logarithms of Trigonometrical Functions    **137°**

| M | Sine | Cosine | Tangent | Cotangent | Secant | Cosecant | M |
|---|------|--------|---------|-----------|--------|----------|---|
| 0 | 9.82551 | 9.87107 | 9.95444 | 10.04556 | 10.12893 | 10.17449 | 60 |
| 1 | .82565 | .87096 | .95469 | .04531 | .12904 | .17435 | 59 |
| 2 | .82579 | .87085 | .95495 | .04505 | .12915 | .17421 | 58 |
| 3 | .82593 | .87073 | .95520 | .04480 | .12927 | .17407 | 57 |
| 4 | .82607 | .87062 | .95545 | .04455 | .12938 | .17393 | 56 |
| 5 | 9.82621 | 9.87050 | 9.95571 | 10.04429 | 10.12950 | 10.17379 | 55 |
| 6 | .82635 | .87039 | .95596 | .04404 | .12961 | .17365 | 54 |
| 7 | .82649 | .87028 | .95622 | .04378 | .12972 | .17351 | 53 |
| 8 | .82663 | .87016 | .95647 | .04353 | .12984 | .17337 | 52 |
| 9 | .82677 | .87005 | .95672 | .04328 | .12995 | .17323 | 51 |
| 10 | 9.82691 | 9.86993 | 9.95698 | 10.04302 | 10.13007 | 10.17309 | 50 |
| 11 | .82705 | .86982 | .95723 | .04277 | .13018 | .17295 | 49 |
| 12 | .82719 | .86970 | .95748 | .04252 | .13030 | .17281 | 48 |
| 13 | .82733 | .86959 | .95774 | .04226 | .13041 | .17267 | 47 |
| 14 | .82747 | .86947 | .95799 | .04201 | .13053 | .17253 | 46 |
| 15 | 9.82761 | 9.86936 | 9.95825 | 10.04175 | 10.13064 | 10.17239 | 45 |
| 16 | .82775 | .86924 | .95850 | .04150 | .13076 | .17225 | 44 |
| 17 | .82788 | .86913 | .95875 | .04125 | .13087 | .17212 | 43 |
| 18 | .82802 | .86902 | .95901 | .04099 | .13098 | .17198 | 42 |
| 19 | .82816 | .86890 | .95926 | .04074 | .13110 | .17184 | 41 |
| 20 | 9.82830 | 9.86879 | 9.95952 | 10.04048 | 10.13121 | 10.17170 | 40 |
| 21 | .82844 | .86867 | .95977 | .04023 | .13133 | .17156 | 39 |
| 22 | .82858 | .86855 | .96002 | .03998 | .13145 | .17142 | 38 |
| 23 | .82872 | .86844 | .96028 | .03972 | .13156 | .17128 | 37 |
| 24 | .82885 | .86832 | .96053 | .03947 | .13168 | .17115 | 36 |
| 25 | 9.82899 | 9.86821 | 9.96078 | 10.03922 | 10.13179 | 10.17101 | 35 |
| 26 | .82913 | .86809 | .96104 | .03896 | .13191 | .17087 | 34 |
| 27 | .82927 | .86798 | .96129 | .03871 | .13202 | .17073 | 33 |
| 28 | .82941 | .86786 | .96155 | .03845 | .13214 | .17059 | 32 |
| 29 | .82955 | .86775 | .96180 | .03820 | .13225 | .17045 | 31 |
| 30 | 9.82968 | 9.86763 | 9.96205 | 10.03795 | 10.13237 | 10.17032 | 30 |
| 31 | .82982 | .86752 | .96231 | .03769 | .13248 | .17018 | 29 |
| 32 | .82996 | .86740 | .96256 | .03744 | .13260 | .17004 | 28 |
| 33 | .83010 | .86728 | .96281 | .03719 | .13272 | .16990 | 27 |
| 34 | .83023 | .86717 | .96307 | .03693 | .13283 | .16977 | 26 |
| 35 | 9.83037 | 9.86705 | 9.96332 | 10.03668 | 10.13295 | 10.16963 | 25 |
| 36 | .83051 | .86694 | .96357 | .03643 | .13306 | .16949 | 24 |
| 37 | .83065 | .86682 | .96383 | .03617 | .13318 | .16935 | 23 |
| 38 | .83078 | .86670 | .96408 | .03592 | .13330 | .16922 | 22 |
| 39 | .83092 | .86659 | .96433 | .03567 | .13341 | .16908 | 21 |
| 40 | 9.83106 | 9.86647 | 9.96459 | 10.03541 | 10.13353 | 10.16894 | 20 |
| 41 | .83120 | .86635 | .96484 | .03516 | .13365 | .16880 | 19 |
| 42 | .83133 | .86624 | .96510 | .03490 | .13376 | .16867 | 18 |
| 43 | .83147 | .86612 | .96535 | .03465 | .13388 | .16853 | 17 |
| 44 | .83161 | .86600 | .96560 | .03440 | .13400 | .16839 | 16 |
| 45 | 9.83174 | 9.86589 | 9.96586 | 10.03414 | 10.13411 | 10.16826 | 15 |
| 46 | .83188 | .86577 | .96611 | .03389 | .13423 | .16812 | 14 |
| 47 | .83202 | .86565 | .96636 | .03364 | .13435 | .16798 | 13 |
| 48 | .83215 | .86554 | .96662 | .03338 | .13446 | .16785 | 12 |
| 49 | .83229 | .86542 | .96687 | .03313 | .13458 | .16771 | 11 |
| 50 | 9.83242 | 9.86530 | 9.96712 | 10.03288 | 10.13470 | 10.16758 | 10 |
| 51 | .83256 | .86518 | .96738 | .03262 | .13482 | .16744 | 9 |
| 52 | .83270 | .86507 | .96763 | .03237 | .13493 | .16730 | 8 |
| 53 | .83283 | .86495 | .96788 | .03212 | .13505 | .16717 | 7 |
| 54 | .83297 | .86483 | .96814 | .03186 | .13517 | .16703 | 6 |
| 55 | 9.83310 | 9.86472 | 9.96839 | 10.03161 | 10.13528 | 10.16690 | 5 |
| 56 | .83324 | .86460 | .96864 | .03136 | .13540 | .16676 | 4 |
| 57 | .83338 | .86448 | .96890 | .03110 | .13552 | .16662 | 3 |
| 58 | .83351 | .86436 | .96915 | .03085 | .13564 | .16649 | 2 |
| 59 | .83365 | .86425 | .96940 | .03060 | .13575 | .16635 | 1 |
| 60 | 9.83378 | 9.86413 | 9.96966 | 10.03034 | 10.13587 | 10.16622 | 0 |
| M | Cosine | Sine | Cotangent | Tangent | Cosecant | Secant | M |

| M | Sine | Cosine | Tangent | Cotangent | Secant | Cosecant | M |
|---|------|--------|---------|-----------|--------|----------|---|
| 0 | 9.83378 | 9.86413 | 9.96966 | 10.03034 | 10.13587 | 10.16622 | 60 |
| 1 | .83392 | .86401 | .96991 | .03009 | .13599 | .16608 | 59 |
| 2 | .83405 | .86389 | .97016 | .02984 | .13611 | .16595 | 58 |
| 3 | .83419 | .86377 | .97042 | .02958 | .13623 | .16581 | 57 |
| 4 | .83432 | .86366 | .97067 | .02933 | .13634 | .16568 | 56 |
| 5 | 9.83446 | 9.86354 | 9.97092 | 10.02908 | 10.13646 | 10.16554 | 55 |
| 6 | .83459 | .86342 | .97118 | .02882 | .13658 | .16541 | 54 |
| 7 | .83473 | .86330 | .97143 | .02857 | .13670 | .16527 | 53 |
| 8 | .83486 | .86318 | .97168 | .02832 | .13682 | .16514 | 52 |
| 9 | .83500 | .86306 | .97193 | .02807 | .13694 | .16500 | 51 |
| 10 | 9.83513 | 9.86295 | 9.97219 | 10.02781 | 10.13705 | 10.16487 | 50 |
| 11 | .83527 | .86283 | .97244 | .02756 | .13717 | .16473 | 49 |
| 12 | .83540 | .86271 | .97269 | .02731 | .13729 | .16460 | 48 |
| 13 | .83554 | .86259 | .97295 | .02705 | .13741 | .16446 | 47 |
| 14 | .83567 | .86247 | .97320 | .02680 | .13753 | .16433 | 46 |
| 15 | 9.83581 | 9.86235 | 9.97345 | 10.02655 | 10.13765 | 10.16419 | 45 |
| 16 | .83594 | .86223 | .97371 | .02629 | .13777 | .16406 | 44 |
| 17 | .83608 | .86211 | .97396 | .02604 | .13789 | .16392 | 43 |
| 18 | .83621 | .86200 | .97421 | .02579 | .13800 | .16379 | 42 |
| 19 | .83634 | .86188 | .97447 | .02553 | .13812 | .16366 | 41 |
| 20 | 9.83648 | 9.86176 | 9.97472 | 10.02528 | 10.13824 | 10.16352 | 40 |
| 21 | .83661 | .86164 | .97497 | .02503 | .13836 | .16339 | 39 |
| 22 | .83674 | .86152 | .97523 | .02477 | .13848 | .16326 | 38 |
| 23 | .83688 | .86140 | .97548 | .02452 | .13860 | .16312 | 37 |
| 24 | .83701 | .86128 | .97573 | .02427 | .13872 | .16299 | 36 |
| 25 | 9.83715 | 9.86116 | 9.97598 | 10.02402 | 10.13884 | 10.16285 | 35 |
| 26 | .83728 | .86104 | .97624 | .02376 | .13896 | .16272 | 34 |
| 27 | .83741 | .86092 | .97649 | .02351 | .13908 | .16259 | 33 |
| 28 | .83755 | .86080 | .97674 | .02326 | .13920 | .16245 | 32 |
| 29 | .83768 | .86068 | .97700 | .02300 | .13932 | .16232 | 31 |
| 30 | 9.83781 | 9.86056 | 9.97725 | 10.02275 | 10.13944 | 10.16219 | 30 |
| 31 | .83795 | .86044 | .97750 | .02250 | .13956 | .16205 | 29 |
| 32 | .83808 | .86032 | .97776 | .02224 | .13968 | .16192 | 28 |
| 33 | .83821 | .86020 | .97801 | .02199 | .13980 | .16179 | 27 |
| 34 | .83834 | .86008 | .97826 | .02174 | .13992 | .16166 | 26 |
| 35 | 9.83848 | 9.85996 | 9.97851 | 10.02149 | 10.14004 | 10.16152 | 25 |
| 36 | .83861 | .85984 | .97877 | .02123 | .14016 | .16139 | 24 |
| 37 | .83874 | .85972 | .97902 | .02098 | .14028 | .16126 | 23 |
| 38 | .83887 | .85960 | .97927 | .02073 | .14040 | .16113 | 22 |
| 39 | .83901 | .85948 | .97953 | .02047 | .14052 | .16099 | 21 |
| 40 | 9.83914 | 9.85936 | 9.97978 | 10.02022 | 10.14064 | 10.16086 | 20 |
| 41 | .83927 | .85924 | .98003 | .01997 | .14076 | .16073 | 19 |
| 42 | .83940 | .85912 | .98029 | .01971 | .14088 | .16060 | 18 |
| 43 | .83954 | .85900 | .98054 | .01946 | .14100 | .16046 | 17 |
| 44 | .83967 | .85888 | .98079 | .01921 | .14112 | .16033 | 16 |
| 45 | 9.83980 | 9.85876 | 9.98104 | 10.01896 | 10.14124 | 10.16020 | 15 |
| 46 | .83993 | .85864 | .98130 | .01870 | .14136 | .16007 | 14 |
| 47 | .84006 | .85851 | .98155 | .01845 | .14149 | .15994 | 13 |
| 48 | .84020 | .85839 | .98180 | .01820 | .14161 | .15980 | 12 |
| 49 | .84033 | .85827 | .98206 | .01794 | .14173 | .15967 | 11 |
| 50 | 9.84046 | 9.85815 | 9.98231 | 10.01769 | 10.14185 | 10.15954 | 10 |
| 51 | .84059 | .85803 | .98256 | .01744 | .14197 | .15941 | 9 |
| 52 | .84072 | .85791 | .98281 | .01719 | .14209 | .15928 | 8 |
| 53 | .84085 | .85779 | .98307 | .01693 | .14221 | .15915 | 7 |
| 54 | .84098 | .85766 | .98332 | .01668 | .14234 | .15902 | 6 |
| 55 | 9.84112 | 9.85754 | 9.98357 | 10.01643 | 10.14246 | 10.15888 | 5 |
| 56 | .84125 | .85742 | .98383 | .01617 | .14258 | .15875 | 4 |
| 57 | .84138 | .85730 | .98408 | .01592 | .14270 | .15862 | 3 |
| 58 | .84151 | .85718 | .98433 | .01567 | .14282 | .15849 | 2 |
| 59 | .84164 | .85706 | .98458 | .01542 | .14294 | .15836 | 1 |
| 60 | 9.84177 | 9.85693 | 9.98484 | 10.01516 | 10.14307 | 10.15823 | 0 |
| M | Cosine | Sine | Cotangent | Tangent | Cosecant | Secant | M |

    **Logarithms of Trigonometrical Functions**    

| M | Sine | Cosine | Tangent | Cotangent | Secant | Cosecant | M |
|---|------|--------|---------|-----------|--------|----------|---|
| 0 | 9.84177 | 9.85693 | 9.98484 | 10.01516 | 10.14307 | 10.15823 | 60 |
| 1 | .84190 | .85681 | .98509 | .01491 | .14319 | .15810 | 59 |
| 2 | .84203 | .85669 | .98534 | .01466 | .14331 | .15797 | 58 |
| 3 | .84216 | .85657 | .98560 | .01440 | .14343 | .15784 | 57 |
| 4 | .84229 | .85645 | .98585 | .01415 | .14355 | .15771 | 56 |
| 5 | 9.84242 | 9.85632 | 9.98610 | 10.01390 | 10.14368 | 10.15758 | 55 |
| 6 | .84255 | .85620 | .98635 | .01365 | .14380 | .15745 | 54 |
| 7 | .84269 | .85608 | .98661 | .01339 | .14392 | .15731 | 53 |
| 8 | .84282 | .85596 | .98686 | .01314 | .14404 | .15718 | 52 |
| 9 | .84295 | .85583 | .98711 | .01289 | .14417 | .15705 | 51 |
| 10 | 9.84308 | 9.85571 | 9.98737 | 10.01263 | 10.14429 | 10.15692 | 50 |
| 11 | .84321 | .85559 | .98762 | .01238 | .14441 | .15679 | 49 |
| 12 | .84334 | .85547 | .98787 | .01213 | .14453 | .15666 | 48 |
| 13 | .84347 | .85534 | .98812 | .01188 | .14466 | .15653 | 47 |
| 14 | .84360 | .85522 | .98838 | .01162 | .14478 | .15640 | 46 |
| 15 | 9.84373 | 9.85510 | 9.98863 | 10.01137 | 10.14490 | 10.15627 | 45 |
| 16 | .84385 | .85497 | .98888 | .01112 | .14503 | .15615 | 44 |
| 17 | .84398 | .85485 | .98913 | .01087 | .14515 | .15602 | 43 |
| 18 | .84411 | .85473 | .98939 | .01061 | .14527 | .15589 | 42 |
| 19 | .84424 | .85460 | .98964 | .01036 | .14540 | .15576 | 41 |
| 20 | 9.84437 | 9.85448 | 9.98989 | 10.01011 | 10.14552 | 10.15563 | 40 |
| 21 | .84450 | .85436 | .99015 | .00985 | .14564 | .15550 | 39 |
| 22 | .84463 | .85423 | .99040 | .00960 | .14577 | .15537 | 38 |
| 23 | .84476 | .85411 | .99065 | .00935 | .14589 | .15524 | 37 |
| 24 | .84489 | .85399 | .99090 | .00910 | .14601 | .15511 | 36 |
| 25 | 9.84502 | 9.85386 | 9.99116 | 10.00884 | 10.14614 | 10.15498 | 35 |
| 26 | .84515 | .85374 | .99141 | .00859 | .14626 | .15485 | 34 |
| 27 | .84528 | .85361 | .99166 | .00834 | .14639 | .15472 | 33 |
| 28 | .84540 | .85349 | .99191 | .00809 | .14651 | .15460 | 32 |
| 29 | .84553 | .85337 | .99217 | .00783 | .14663 | .15447 | 31 |
| 30 | 9.84566 | 9.85324 | 9.99242 | 10.00758 | 10.14676 | 10.15434 | 30 |
| 31 | .84579 | .85312 | .99267 | .00733 | .14688 | .15421 | 29 |
| 32 | .84592 | .85299 | .99293 | .00707 | .14701 | .15408 | 28 |
| 33 | .84605 | .85287 | .99318 | .00682 | .14713 | .15395 | 27 |
| 34 | .84618 | .85274 | .99343 | .00657 | .14726 | .15382 | 26 |
| 35 | 9.84630 | 9.85262 | 9.99368 | 10.00632 | 10.14738 | 10.15370 | 25 |
| 36 | .84643 | .85250 | .99394 | .00606 | .14750 | .15357 | 24 |
| 37 | .84656 | .85237 | .99419 | .00581 | .14763 | .15344 | 23 |
| 38 | .84669 | .85225 | .99444 | .00556 | .14775 | .15331 | 22 |
| 39 | .84682 | .85212 | .99469 | .00531 | .14788 | .15318 | 21 |
| 40 | 9.84694 | 9.85200 | 9.99495 | 10.00505 | 10.14800 | 10.15306 | 20 |
| 41 | .84707 | .85187 | .99520 | .00480 | .14813 | .15293 | 19 |
| 42 | .84720 | .85175 | .99545 | .00455 | .14825 | .15280 | 18 |
| 43 | .84733 | .85162 | .99570 | .00430 | .14838 | .15267 | 17 |
| 44 | .84745 | .85150 | .99596 | .00404 | .14850 | .15255 | 16 |
| 45 | 9.84758 | 9.85137 | 9.99621 | 10.00379 | 10.14863 | 10.15242 | 15 |
| 46 | .84771 | .85125 | .99646 | .00354 | .14875 | .15229 | 14 |
| 47 | .84784 | .85112 | .99672 | .00328 | .14888 | .15216 | 13 |
| 48 | .84796 | .85100 | .99697 | .00303 | .14900 | .15204 | 12 |
| 49 | .84809 | .85087 | .99722 | .00278 | .14913 | .15191 | 11 |
| 50 | 9.84822 | 9.85074 | 9.99747 | 10.00253 | 10.14926 | 10.15178 | 10 |
| 51 | .84835 | .85062 | .99773 | .00227 | .14938 | .15165 | 9 |
| 52 | .84847 | .85049 | .99798 | .00202 | .14951 | .15153 | 8 |
| 53 | .84860 | .85037 | .99823 | .00177 | .14963 | .15140 | 7 |
| 54 | .84873 | .85024 | .99848 | .00152 | .14976 | .15127 | 6 |
| 55 | 9.84885 | 9.85012 | 9.99874 | 10.00126 | 10.14988 | 10.15115 | 5 |
| 56 | .84898 | .84999 | .99899 | .00101 | .15001 | .15102 | 4 |
| 57 | .84911 | .84986 | .99924 | .00076 | .15014 | .15089 | 3 |
| 58 | .84923 | .84974 | .99949 | .00051 | .15026 | .15077 | 2 |
| 59 | .84936 | .84961 | .99975 | .00025 | .15039 | .15064 | 1 |
| 60 | 9.84949 | 9.84949 | 10.00000 | 10.00000 | 10.15051 | 10.15051 | 0 |

| M | Cosine | Sine | Cotangent | Tangent | Cosecant | Secant | M |
|---|--------|------|-----------|---------|----------|--------|---|

                               

**Conversion Tables of Angular Measure.** — The accompanying tables of degrees, minutes, and seconds into radians; radians into degrees, minutes, and seconds; radians into degrees and decimals of a degree; and minutes and seconds into decimals of a degree and vice versa facilitate the conversion of measurements.

*Example:* The Degrees, Minutes, and Seconds into Radians Table is used to find the number of radians in 324 degrees, 25 minutes, 13 seconds as follows:

$$
\begin{array}{ll}
300 \text{ degrees} & = 5.235988 \text{ radians} \\
20 \text{ degrees} & = 0.349066 \text{ radian} \\
4 \text{ degrees} & = 0.069813 \text{ radian} \\
25 \text{ minutes} & = 0.007272 \text{ radian} \\
13 \text{ seconds} & = 0.000063 \text{ radian} \\
\hline
324°25'13'' & = 5.662202 \text{ radians}
\end{array}
$$

*Example:* The Radians into Degrees and Decimals of a Degree, and Radians into Degrees, Minutes and Seconds Tables are used to find the number of decimal degrees or degrees, minutes and seconds in 0.735 radian as follows:

$$
\begin{array}{ll}
0.7 \text{ radian} = 40.1070 \text{ degrees} \qquad & 0.7 \text{ radian} = 40° \ 6' \ 25'' \\
0.03 \text{ radian} = \ \ 1.7189 \text{ degrees} & 0.03 \text{ radian} = \ \ 1° 43' \ \ 8'' \\
0.005 \text{ radian} = \ \ 0.2865 \text{ degrees} & 0.005 \text{ radian} = \ \ 0° 17' 11'' \\
\hline
0.735 \text{ radian} = 43.1114 \text{ degrees} & 0.735 \text{ radian} = 41° 66' 44'' \text{ or } 42° 6' 44''
\end{array}
$$

### Degrees, Minutes, and Seconds into Radians
(Based on 180 degrees = $\pi$ radians)

| | | Degrees into Radians | | | | | | | |
|---|---|---|---|---|---|---|---|---|---|
| Deg. | Rad. | Deg. | Rad. | Deg. | Rad. | Deg. | Rad. | Deg. | Rad. | Deg. | Rad. |
| 1000 | 17.453293 | 100 | 1.745329 | 10 | 0.174533 | 1 | 0.017453 | 0.1 | 0.001745 | 0.01 | 0.000175 |
| 2000 | 34.906585 | 200 | 3.490659 | 20 | 0.349066 | 2 | 0.034907 | 0.2 | 0.003491 | 0.02 | 0.000349 |
| 3000 | 52.359878 | 300 | 5.235988 | 30 | 0.523599 | 3 | 0.052360 | 0.3 | 0.005236 | 0.03 | 0.000524 |
| 4000 | 69.813170 | 400 | 6.981317 | 40 | 0.698132 | 4 | 0.069813 | 0.4 | 0.006981 | 0.04 | 0.000698 |
| 5000 | 87.266463 | 500 | 8.726646 | 50 | 0.872665 | 5 | 0.087266 | 0.5 | 0.008727 | 0.05 | 0.000873 |
| 6000 | 104.719755 | 600 | 10.471976 | 60 | 1.047198 | 6 | 0.104720 | 0.6 | 0.010472 | 0.06 | 0.001047 |
| 7000 | 122.173048 | 700 | 12.217305 | 70 | 1.221730 | 7 | 0.122173 | 0.7 | 0.012217 | 0.07 | 0.001222 |
| 8000 | 139.626340 | 800 | 13.962634 | 80 | 1.396263 | 8 | 0.139626 | 0.8 | 0.013963 | 0.08 | 0.001396 |
| 9000 | 157.079633 | 900 | 15.707963 | 90 | 1.570796 | 9 | 0.157080 | 0.9 | 0.015708 | 0.09 | 0.001571 |
| 10000 | 174.532925 | 1000 | 17.453293 | 100 | 1.745329 | 10 | 0.174533 | 1.0 | 0.017453 | 0.10 | 0.001745 |

| | | Minutes into Radians | | | | | | | |
|---|---|---|---|---|---|---|---|---|---|
| Min. | Rad. | Min. | Rad. | Min. | Rad. | Min. | Rad. | Min. | Rad. | Min. | Rad. |
| 1 | 0.000291 | 11 | 0.003200 | 21 | 0.006109 | 31 | 0.009018 | 41 | 0.011926 | 51 | 0.014835 |
| 2 | 0.000582 | 12 | 0.003491 | 22 | 0.006400 | 32 | 0.009308 | 42 | 0.012217 | 52 | 0.015126 |
| 3 | 0.000873 | 13 | 0.003782 | 23 | 0.006690 | 33 | 0.009599 | 43 | 0.012508 | 53 | 0.015417 |
| 4 | 0.001164 | 14 | 0.004072 | 24 | 0.006981 | 34 | 0.009890 | 44 | 0.012799 | 54 | 0.015708 |
| 5 | 0.001454 | 15 | 0.004363 | 25 | 0.007272 | 35 | 0.010181 | 45 | 0.013090 | 55 | 0.015999 |
| 6 | 0.001745 | 16 | 0.004654 | 26 | 0.007563 | 36 | 0.010472 | 46 | 0.013381 | 56 | 0.016290 |
| 7 | 0.002036 | 17 | 0.004945 | 27 | 0.007854 | 37 | 0.010763 | 47 | 0.013672 | 57 | 0.016581 |
| 8 | 0.002327 | 18 | 0.005236 | 28 | 0.008145 | 38 | 0.011054 | 48 | 0.013963 | 58 | 0.016872 |
| 9 | 0.002618 | 19 | 0.005527 | 29 | 0.008436 | 39 | 0.011345 | 49 | 0.014254 | 59 | 0.017162 |
| 10 | 0.002909 | 20 | 0.005818 | 30 | 0.008727 | 40 | 0.011636 | 50 | 0.014544 | 60 | 0.017453 |

| | | Seconds into Radians | | | | | | | |
|---|---|---|---|---|---|---|---|---|---|
| Sec. | Rad. | Sec. | Rad. | Sec. | Rad. | Sec. | Rad. | Sec. | Rad. | Sec. | Rad. |
| 1 | 0.000005 | 11 | 0.000053 | 21 | 0.000102 | 31 | 0.000150 | 41 | 0.000199 | 51 | 0.000247 |
| 2 | 0.000010 | 12 | 0.000058 | 22 | 0.000107 | 32 | 0.000155 | 42 | 0.000204 | 52 | 0.000252 |
| 3 | 0.000015 | 13 | 0.000063 | 23 | 0.000112 | 33 | 0.000160 | 43 | 0.000208 | 53 | 0.000257 |
| 4 | 0.000019 | 14 | 0.000068 | 24 | 0.000116 | 34 | 0.000165 | 44 | 0.000213 | 54 | 0.000262 |
| 5 | 0.000024 | 15 | 0.000073 | 25 | 0.000121 | 35 | 0.000170 | 45 | 0.000218 | 55 | 0.000267 |
| 6 | 0.000029 | 16 | 0.000078 | 26 | 0.000126 | 36 | 0.000175 | 46 | 0.000223 | 56 | 0.000271 |
| 7 | 0.000034 | 17 | 0.000082 | 27 | 0.000131 | 37 | 0.000179 | 47 | 0.000228 | 57 | 0.000276 |
| 8 | 0.000039 | 18 | 0.000087 | 28 | 0.000136 | 38 | 0.000184 | 48 | 0.000233 | 58 | 0.000281 |
| 9 | 0.000044 | 19 | 0.000092 | 29 | 0.000141 | 39 | 0.000189 | 49 | 0.000238 | 59 | 0.000286 |
| 10 | 0.000048 | 20 | 0.000097 | 30 | 0.000145 | 40 | 0.000194 | 50 | 0.000242 | 60 | 0.000291 |

### Radians into Degrees and Decimals of a Degree
(Based on π radians = 180 degrees)

| Rad. | Deg. | Rad. | Deg. | Rad. | Deg. | Rad. | Deg. | Rad. | Deg. | Rad. | Deg. |
|---|---|---|---|---|---|---|---|---|---|---|---|
| 10 | 572.9578 | 1 | 57.2958 | 0.1 | 5.7296 | 0.01 | 0.5730 | 0.001 | 0.0573 | 0.0001 | 0.0057 |
| 20 | 1145.9156 | 2 | 114.5916 | 0.2 | 11.4592 | 0.02 | 1.1459 | 0.002 | 0.1146 | 0.0002 | 0.0115 |
| 30 | 1718.8734 | 3 | 171.8873 | 0.3 | 17.1887 | 0.03 | 1.7189 | 0.003 | 0.1719 | 0.0003 | 0.0172 |
| 40 | 2291.8312 | 4 | 229.1831 | 0.4 | 22.9183 | 0.04 | 2.2918 | 0.004 | 0.2292 | 0.0004 | 0.0229 |
| 50 | 2864.7890 | 5 | 286.4789 | 0.5 | 28.6479 | 0.05 | 2.8648 | 0.005 | 0.2865 | 0.0005 | 0.0286 |
| 60 | 3437.7468 | 6 | 343.7747 | 0.6 | 34.3775 | 0.06 | 3.4377 | 0.006 | 0.3438 | 0.0006 | 0.0344 |
| 70 | 4010.7046 | 7 | 401.0705 | 0.7 | 40.1070 | 0.07 | 4.0107 | 0.007 | 0.4011 | 0.0007 | 0.0401 |
| 80 | 4583.6624 | 8 | 458.3662 | 0.8 | 45.8366 | 0.08 | 4.5837 | 0.008 | 0.4584 | 0.0008 | 0.0458 |
| 90 | 5156.6202 | 9 | 515.6620 | 0.9 | 51.5662 | 0.09 | 5.1566 | 0.009 | 0.5157 | 0.0009 | 0.0516 |
| 100 | 5729.5780 | 10 | 572.9578 | 1.0 | 57.2958 | 0.10 | 5.7296 | 0.010 | 0.5730 | 0.0010 | 0.0573 |

### Radians into Degrees, Minutes and Seconds
(Based on π radians = 180 degrees)

| Rad. | Angle | Rad. | Angle | Rad. | Angle | Rad. | Angle | Rad. | Angle | Rad. | Angle |
|---|---|---|---|---|---|---|---|---|---|---|---|
| 10 | 572°57'28" | 1 | 57°17'45" | 0.1 | 5°43'46" | 0.01 | 0°34'23" | 0.001 | 0°3'26" | 0.0001 | 0°0'21" |
| 20 | 1145°54'56" | 2 | 114°35'30" | 0.2 | 11°27'33" | 0.02 | 1°8'45" | 0.002 | 0°6'53" | 0.0002 | 0°0'41" |
| 30 | 1718°52'24" | 3 | 171°53'14" | 0.3 | 17°11'19" | 0.03 | 1°43'8" | 0.003 | 0°10'19" | 0.0003 | 0°1'2" |
| 40 | 2291°49'52" | 4 | 229°10'59" | 0.4 | 22°55'6" | 0.04 | 2°17'31" | 0.004 | 0°13'45" | 0.0004 | 0°1'23" |
| 50 | 2864°47'20" | 5 | 286°28'44" | 0.5 | 28°38'52" | 0.05 | 2°51'53" | 0.005 | 0°17'11" | 0.0005 | 0°1'43" |
| 60 | 3437°44'48" | 6 | 343°46'29" | 0.6 | 34°22'39" | 0.06 | 3°26'16" | 0.006 | 0°20'38" | 0.0006 | 0°2'4" |
| 70 | 4010°42'16" | 7 | 401°4'14" | 0.7 | 40°6'25" | 0.07 | 4°0'39" | 0.007 | 0°24'4" | 0.0007 | 0°2'24" |
| 80 | 4583°39'44" | 8 | 458°21'58" | 0.8 | 45°50'12" | 0.08 | 4°35'1" | 0.008 | 0°27'30" | 0.0008 | 0°2'45" |
| 90 | 5156°37'13" | 9 | 515°39'43" | 0.9 | 51°33'58" | 0.09 | 5°9'24" | 0.009 | 0°30'56" | 0.0009 | 0°3'6" |
| 100 | 5729°34'41" | 10 | 572°57'28" | 1.0 | 57°17'45" | 0.10 | 5°43'46" | 0.010 | 0°34'23" | 0.0010 | 0°3'26" |

### Minutes and Seconds into Decimals of a Degree and Vice Versa
(Based on 1 second = 0.00027778 degrees)

| Minutes into Decimals of a Degree | | | | | | Seconds into Decimals of a Degree | | | | | |
|---|---|---|---|---|---|---|---|---|---|---|---|
| Min. | Deg. | Min. | Deg. | Min. | Deg. | Sec. | Deg. | Sec. | Deg. | Sec. | Deg. |
| 1 | 0.0167 | 21 | 0.3500 | 41 | 0.6833 | 1 | 0.0003 | 21 | 0.0058 | 41 | 0.0114 |
| 2 | 0.0333 | 22 | 0.3667 | 42 | 0.7000 | 2 | 0.0006 | 22 | 0.0061 | 42 | 0.0117 |
| 3 | 0.0500 | 23 | 0.3833 | 43 | 0.7167 | 3 | 0.0008 | 23 | 0.0064 | 43 | 0.0119 |
| 4 | 0.0667 | 24 | 0.4000 | 44 | 0.7333 | 4 | 0.0011 | 24 | 0.0067 | 44 | 0.0122 |
| 5 | 0.0833 | 25 | 0.4167 | 45 | 0.7500 | 5 | 0.0014 | 25 | 0.0069 | 45 | 0.0125 |
| 6 | 0.1000 | 26 | 0.4333 | 46 | 0.7667 | 6 | 0.0017 | 26 | 0.0072 | 46 | 0.0128 |
| 7 | 0.1167 | 27 | 0.4500 | 47 | 0.7833 | 7 | 0.0019 | 27 | 0.0075 | 47 | 0.0131 |
| 8 | 0.1333 | 28 | 0.4667 | 48 | 0.8000 | 8 | 0.0022 | 28 | 0.0078 | 48 | 0.0133 |
| 9 | 0.1500 | 29 | 0.4833 | 49 | 0.8167 | 9 | 0.0025 | 29 | 0.0081 | 49 | 0.0136 |
| 10 | 0.1667 | 30 | 0.5000 | 50 | 0.8333 | 10 | 0.0028 | 30 | 0.0083 | 50 | 0.0139 |
| 11 | 0.1833 | 31 | 0.5167 | 51 | 0.8500 | 11 | 0.0031 | 31 | 0.0086 | 51 | 0.0142 |
| 12 | 0.2000 | 32 | 0.5333 | 52 | 0.8667 | 12 | 0.0033 | 32 | 0.0089 | 52 | 0.0144 |
| 13 | 0.2167 | 33 | 0.5500 | 53 | 0.8833 | 13 | 0.0036 | 33 | 0.0092 | 53 | 0.0147 |
| 14 | 0.2333 | 34 | 0.5667 | 54 | 0.9000 | 14 | 0.0039 | 34 | 0.0094 | 54 | 0.0150 |
| 15 | 0.2500 | 35 | 0.5833 | 55 | 0.9167 | 15 | 0.0042 | 35 | 0.0097 | 55 | 0.0153 |
| 16 | 0.2667 | 36 | 0.6000 | 56 | 0.9333 | 16 | 0.0044 | 36 | 0.0100 | 56 | 0.0156 |
| 17 | 0.2833 | 37 | 0.6167 | 57 | 0.9500 | 17 | 0.0047 | 37 | 0.0103 | 57 | 0.0158 |
| 18 | 0.3000 | 38 | 0.6333 | 58 | 0.9667 | 18 | 0.0050 | 38 | 0.0106 | 58 | 0.0161 |
| 19 | 0.3167 | 39 | 0.6500 | 59 | 0.9833 | 19 | 0.0053 | 39 | 0.0108 | 59 | 0.0164 |
| 20 | 0.3333 | 40 | 0.6667 | 60 | 1 | 20 | 0.0056 | 40 | 0.0111 | 60 | 0.0167 |

*Example 1*: Convert 11'37" to decimals of a degree. From the left table, 11' = 0.1833 degree. From the right table, 37" = 0.0103 degree. Adding, 11'37" = 0.1833 + 0.0103 = 0.1936 degree.

*Example 2*: Convert 0.1234 degree to minutes and seconds. From the left table, 0.1167 degree = 7'. Subtracting 0.1167 from 0.1234 gives 0.0067. From the right table, 0.0067 = 24" so that 0.1234 = 7'24".

## Geometrical Propositions

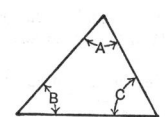

The sum of the three angles in a triangle always equals 180 degrees. Hence, if two angles are known, the third angle can always be found.

$$A + B + C = 180° \qquad A = 180° - (B + C)$$
$$B = 180° - (A + C) \qquad C = 180° - (A + B)$$

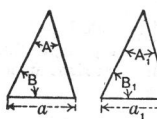

If one side and two angles in one triangle are equal to one side and similarly located angles in another triangle, then the remaining two sides and angle are also equal.

If $a = a_1$, $A = A_1$ and $B = B_1$, then the two other sides and the remaining angle are also equal.

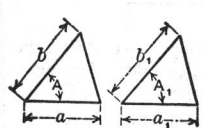

If two sides and the angle between them in one triangle are equal to two sides and a similarly located angle in another triangle, then the remaining side and angles are also equal.

If $a = a_1$, $b = b_1$ and $A = A_1$, then the remaining side and angles are also equal.

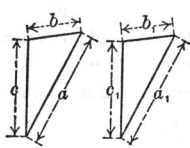

If the three sides in one triangle are equal to the three sides of another triangle, then the angles in the two triangles are also equal.

If $a = a_1$, $b = b_1$ and $c = c_1$, then the angles between the respective sides are also equal.

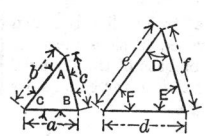

If the three sides of one triangle are proportional to corresponding sides in another triangle, then the triangles are called *similar*, and the angles in the one are equal to the angles in the other.

If $a : b : c = d : e : f$, then $A = D$, $B = E$ and $C = F$.

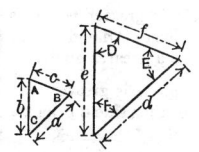

If the angles in one triangle are equal to the angles of another triangle, then the triangles are similar and their corresponding sides are proportional.

If $A = D$, $B = E$ and $C = F$, then $a : b : c = d : e : f$.

## Geometrical Propositions

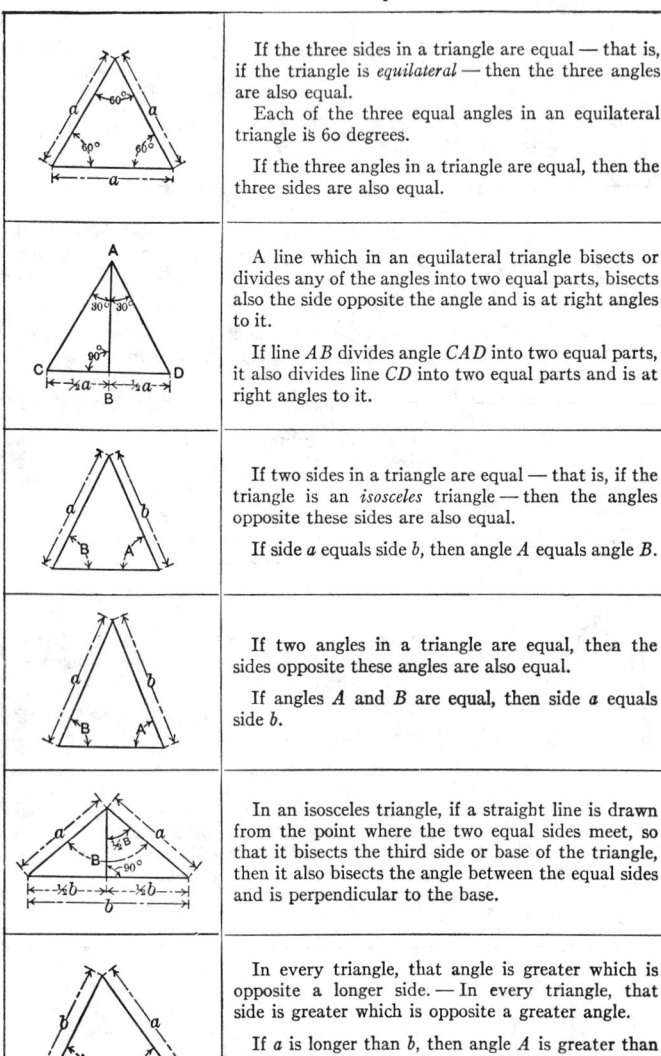

If the three sides in a triangle are equal — that is, if the triangle is *equilateral* — then the three angles are also equal.

Each of the three equal angles in an equilateral triangle is 60 degrees.

If the three angles in a triangle are equal, then the three sides are also equal.

A line which in an equilateral triangle bisects or divides any of the angles into two equal parts, bisects also the side opposite the angle and is at right angles to it.

If line $AB$ divides angle $CAD$ into two equal parts, it also divides line $CD$ into two equal parts and is at right angles to it.

If two sides in a triangle are equal — that is, if the triangle is an *isosceles* triangle — then the angles opposite these sides are also equal.

If side $a$ equals side $b$, then angle $A$ equals angle $B$.

If two angles in a triangle are equal, then the sides opposite these angles are also equal.

If angles $A$ and $B$ are equal, then side $a$ equals side $b$.

In an isosceles triangle, if a straight line is drawn from the point where the two equal sides meet, so that it bisects the third side or base of the triangle, then it also bisects the angle between the equal sides and is perpendicular to the base.

In every triangle, that angle is greater which is opposite a longer side. — In every triangle, that side is greater which is opposite a greater angle.

If $a$ is longer than $b$, then angle $A$ is greater than $B$. If angle $A$ is greater than $B$, then side $a$ is longer than $b$.

## Geometrical Propositions

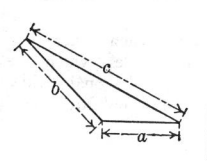

In every triangle, the sum of the lengths of two sides is always greater than the length of the third.

Side $a$ + side $b$ is always greater than side $c$.

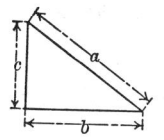

In a right-angled triangle, the square of the hypotenuse or the side opposite the right angle is equal to the sum of the squares on the two sides which form the right angle.

$$a^2 = b^2 + c^2.$$

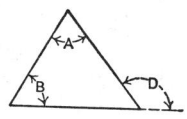

If one side of a triangle is produced, then the exterior angle is equal to the sum of the two interior opposite angles.

Angle $D$ = angle $A$ + angle $B$.

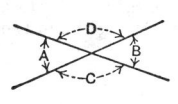

If two lines intersect, then the opposite angles formed by the intersecting lines are equal.

Angle $A$ = angle $B$.
Angle $C$ = angle $D$.

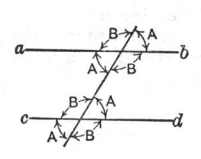

If a line intersects two parallel lines, then the corresponding angles formed by the intersecting line and the parallel lines are equal.

Lines $ab$ and $cd$ are parallel. Then all the angles designated $A$ are equal, and all those designated $B$ are equal.

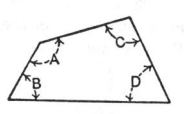

In any figure having four sides, the sum of the interior angles equals 360 degrees.

$A + B + C + D$ = 360 degrees.

## Geometrical Propositions

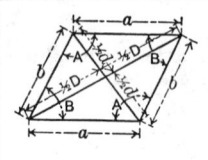

The sides which are opposite each other in a parallelogram are equal; the angles which are opposite each other are equal; the diagonal divides it into two equal parts. If two diagonals are drawn, they bisect each other.

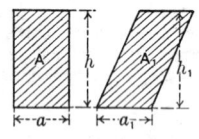

The areas of two parallelograms which have equal base and equal height, are equal.

If $a = a_1$ and $h = h_1$, then

$$\text{area } A = \text{area } A_1.$$

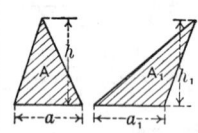

The areas of triangles having equal base and equal height are equal.

If $a = a_1$ and $h = h_1$, then

$$\text{area } A = \text{area } A_1.$$

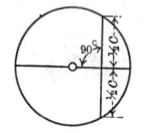

If a diameter of a circle is at right angles to a chord, then it bisects or divides the chord into two equal parts.

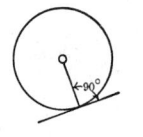

If a line is tangent to a circle, then it is also at right angles to a line drawn from the center of the circle to the point of tangency — that is, to a radial line through the point of tangency.

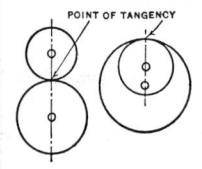

If two circles are tangent to each other, then the straight line which passes through the centers of the two circles must also pass through the point of tangency.

## Geometrical Propositions

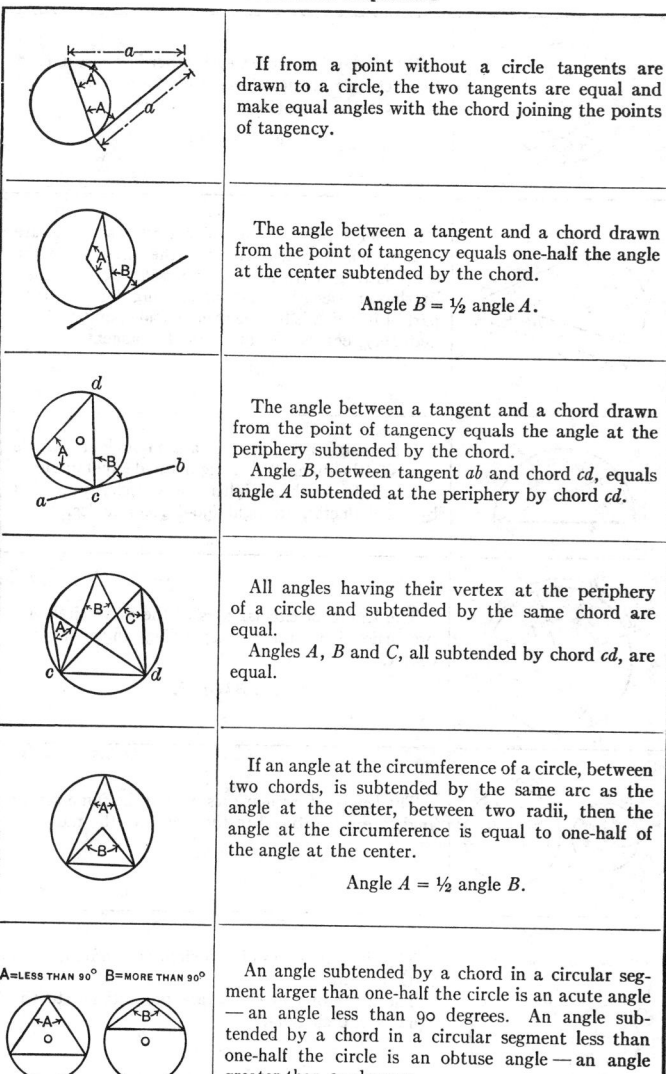

If from a point without a circle tangents are drawn to a circle, the two tangents are equal and make equal angles with the chord joining the points of tangency.

The angle between a tangent and a chord drawn from the point of tangency equals one-half the angle at the center subtended by the chord.

Angle $B = \frac{1}{2}$ angle $A$.

The angle between a tangent and a chord drawn from the point of tangency equals the angle at the periphery subtended by the chord.

Angle $B$, between tangent $ab$ and chord $cd$, equals angle $A$ subtended at the periphery by chord $cd$.

All angles having their vertex at the periphery of a circle and subtended by the same chord are equal.

Angles $A$, $B$ and $C$, all subtended by chord $cd$, are equal.

If an angle at the circumference of a circle, between two chords, is subtended by the same arc as the angle at the center, between two radii, then the angle at the circumference is equal to one-half of the angle at the center.

Angle $A = \frac{1}{2}$ angle $B$.

An angle subtended by a chord in a circular segment larger than one-half the circle is an acute angle — an angle less than 90 degrees. An angle subtended by a chord in a circular segment less than one-half the circle is an obtuse angle — an angle greater than 90 degrees.

**Geometrical Propositions**

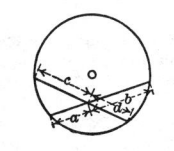

If two chords intersect each other in a circle, then the rectangle of the segments of the one equals the rectangle of the segments of the other.

$$a \times b = c \times d.$$

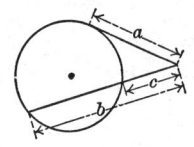

If from a point outside of a circle two lines are drawn, one of which intersects the circle while the other is tangent to it, then the rectangle contained by the total length of the intersecting line, and that part of it which is between the outside point and the periphery, equals the square of the tangent.

$$a^2 = b \times c.$$

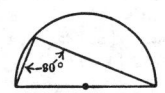

If a triangle is inscribed in a semi-circle, the angle opposite the diameter is a right (90-degree) angle.

All angles at the periphery of a circle, subtended by the diameter, are right (90-degree) angles.

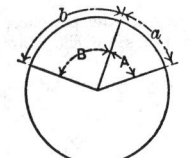

The length of circular arcs of the same circle are proportional to the corresponding angles at the center.

$$A : B = a : b.$$

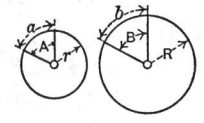

The length of circular arcs having the same center angle are proportional to the length of the radii.

If $A = B$, then $a : b = r : R.$

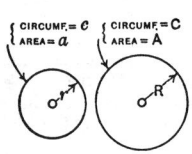

The circumferences of two circles are proportional to their radii.

The areas of two circles are proportional to the squares of their radii.

$$c : C = r : R.$$
$$a : A = r^2 : R^2.$$

## Geometrical Constructions

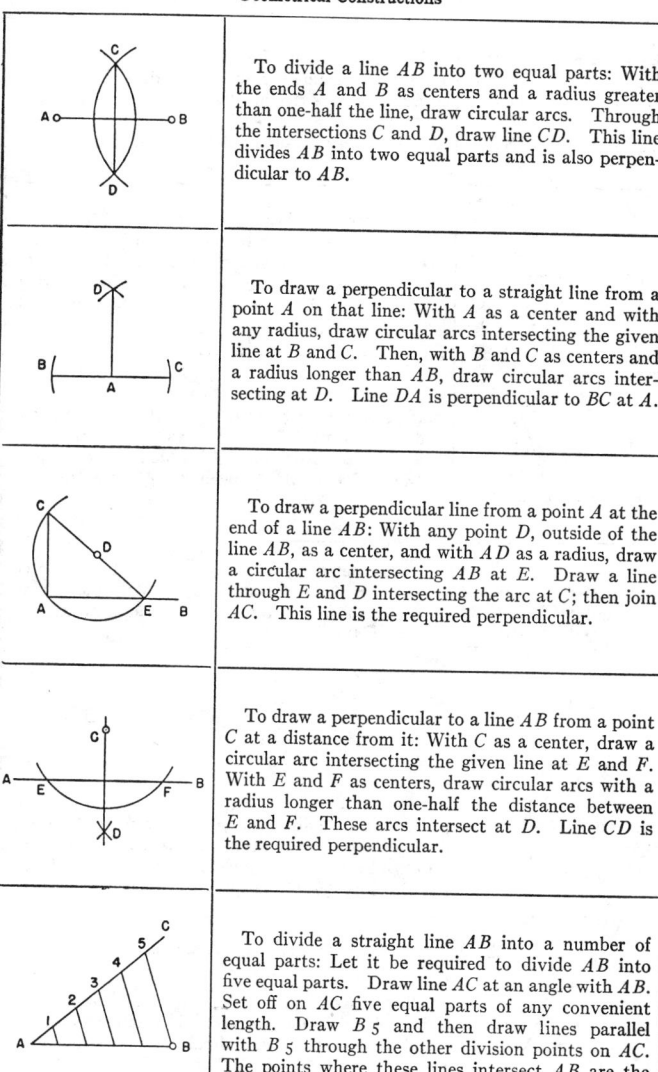

To divide a line *AB* into two equal parts: With the ends *A* and *B* as centers and a radius greater than one-half the line, draw circular arcs. Through the intersections *C* and *D*, draw line *CD*. This line divides *AB* into two equal parts and is also perpendicular to *AB*.

To draw a perpendicular to a straight line from a point *A* on that line: With *A* as a center and with any radius, draw circular arcs intersecting the given line at *B* and *C*. Then, with *B* and *C* as centers and a radius longer than *AB*, draw circular arcs intersecting at *D*. Line *DA* is perpendicular to *BC* at *A*.

To draw a perpendicular line from a point *A* at the end of a line *AB*: With any point *D*, outside of the line *AB*, as a center, and with *AD* as a radius, draw a circular arc intersecting *AB* at *E*. Draw a line through *E* and *D* intersecting the arc at *C*; then join *AC*. This line is the required perpendicular.

To draw a perpendicular to a line *AB* from a point *C* at a distance from it: With *C* as a center, draw a circular arc intersecting the given line at *E* and *F*. With *E* and *F* as centers, draw circular arcs with a radius longer than one-half the distance between *E* and *F*. These arcs intersect at *D*. Line *CD* is the required perpendicular.

To divide a straight line *AB* into a number of equal parts: Let it be required to divide *AB* into five equal parts. Draw line *AC* at an angle with *AB*. Set off on *AC* five equal parts of any convenient length. Draw *B* 5 and then draw lines parallel with *B* 5 through the other division points on *AC*. The points where these lines intersect *AB* are the required division points.

### Geometrical Constructions

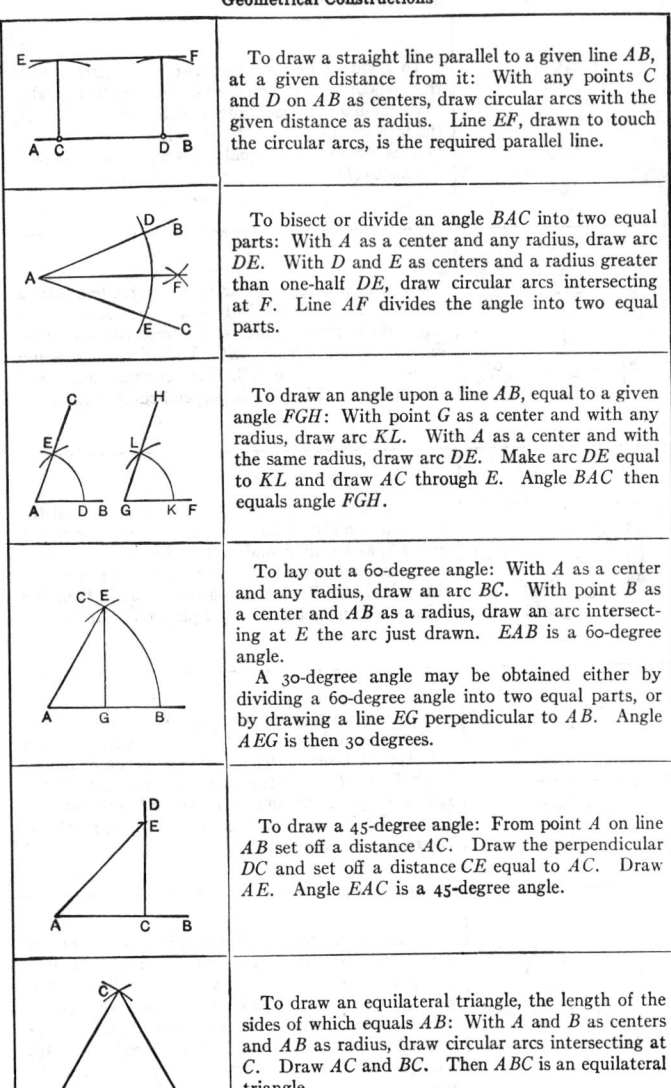

To draw a straight line parallel to a given line *AB*, at a given distance from it: With any points *C* and *D* on *AB* as centers, draw circular arcs with the given distance as radius. Line *EF*, drawn to touch the circular arcs, is the required parallel line.

To bisect or divide an angle *BAC* into two equal parts: With *A* as a center and any radius, draw arc *DE*. With *D* and *E* as centers and a radius greater than one-half *DE*, draw circular arcs intersecting at *F*. Line *AF* divides the angle into two equal parts.

To draw an angle upon a line *AB*, equal to a given angle *FGH*: With point *G* as a center and with any radius, draw arc *KL*. With *A* as a center and with the same radius, draw arc *DE*. Make arc *DE* equal to *KL* and draw *AC* through *E*. Angle *BAC* then equals angle *FGH*.

To lay out a 60-degree angle: With *A* as a center and any radius, draw an arc *BC*. With point *B* as a center and *AB* as a radius, draw an arc intersecting at *E* the arc just drawn. *EAB* is a 60-degree angle.

A 30-degree angle may be obtained either by dividing a 60-degree angle into two equal parts, or by drawing a line *EG* perpendicular to *AB*. Angle *AEG* is then 30 degrees.

To draw a 45-degree angle: From point *A* on line *AB* set off a distance *AC*. Draw the perpendicular *DC* and set off a distance *CE* equal to *AC*. Draw *AE*. Angle *EAC* is a 45-degree angle.

To draw an equilateral triangle, the length of the sides of which equals *AB*: With *A* and *B* as centers and *AB* as radius, draw circular arcs intersecting at *C*. Draw *AC* and *BC*. Then *ABC* is an equilateral triangle.

**Geometrical Constructions**

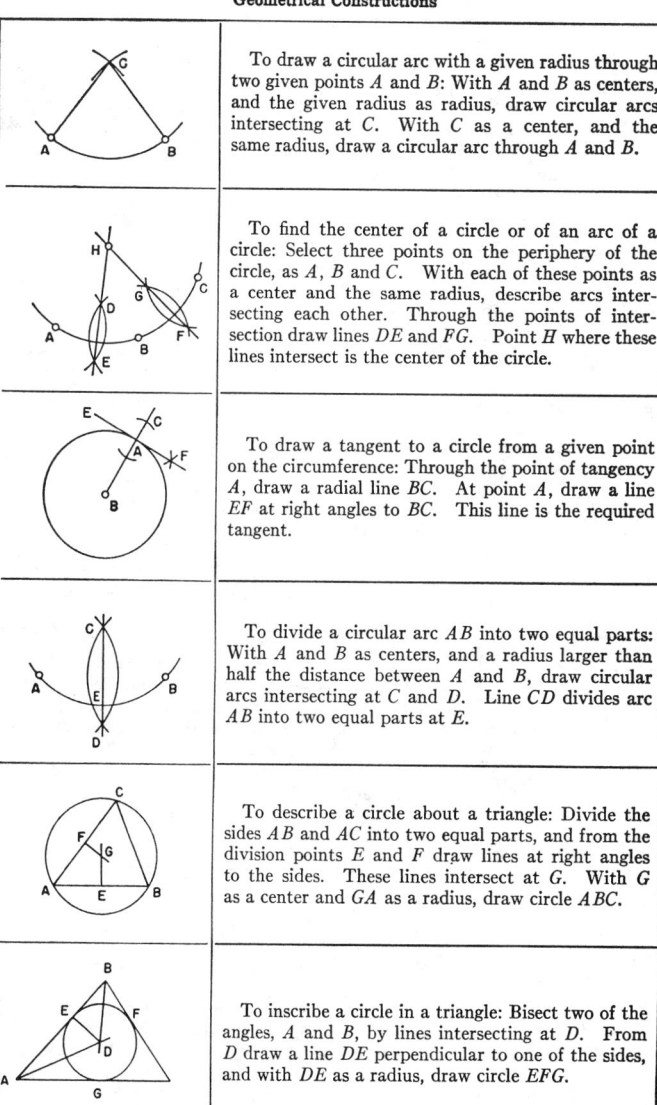

To draw a circular arc with a given radius through two given points A and B: With A and B as centers, and the given radius as radius, draw circular arcs intersecting at C. With C as a center, and the same radius, draw a circular arc through A and B.

To find the center of a circle or of an arc of a circle: Select three points on the periphery of the circle, as A, B and C. With each of these points as a center and the same radius, describe arcs intersecting each other. Through the points of intersection draw lines DE and FG. Point H where these lines intersect is the center of the circle.

To draw a tangent to a circle from a given point on the circumference: Through the point of tangency A, draw a radial line BC. At point A, draw a line EF at right angles to BC. This line is the required tangent.

To divide a circular arc AB into two equal parts: With A and B as centers, and a radius larger than half the distance between A and B, draw circular arcs intersecting at C and D. Line CD divides arc AB into two equal parts at E.

To describe a circle about a triangle: Divide the sides AB and AC into two equal parts, and from the division points E and F draw lines at right angles to the sides. These lines intersect at G. With G as a center and GA as a radius, draw circle ABC.

To inscribe a circle in a triangle: Bisect two of the angles, A and B, by lines intersecting at D. From D draw a line DE perpendicular to one of the sides, and with DE as a radius, draw circle EFG.

### Geometrical Problems

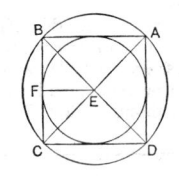

To describe a circle about a square and to inscribe a circle in a square: The center of both the circumscribed and inscribed circle is located at the point $E$, where the two diagonals of the square intersect. The radius of the circumscribed circle is $AE$, and of the inscribed circle, $EF$.

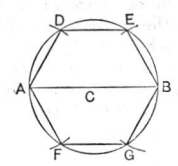

To inscribe a hexagon in a circle: Draw a diameter $AB$. With $A$ and $B$ as centers and with the radius of the circle as radius, describe circular arcs intersecting the given circle at $D, E, F$ and $G$. Draw lines $AD, DE$, etc., forming the required hexagon.

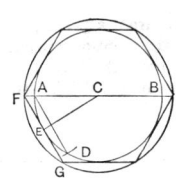

To describe a hexagon about a circle: Draw a diameter $AB$, and with $A$ as a center and the radius of the circle as radius, cut the circumference of the given circle at $D$. Join $AD$ and bisect it with radius $CE$. Through $E$, draw $FG$ parallel to $AD$ and intersecting line $AB$ at $F$. With $C$ as a center and $CF$ as radius, draw a circle. Within this circle inscribe the hexagon as in the preceding problem.

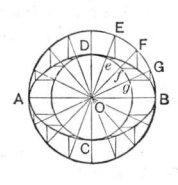

To describe an ellipse with the given axes $AB$ and $CD$: Describe circles with $O$ as a center and $AB$ and $CD$ as diameters. From a number of points, $E, F, G$, etc., on the outer circle draw radii intersecting the inner circle at $e, f, g$. From $E, F$ and $G$ draw lines perpendicular to $AB$, and from $e, f$ and $g$ draw lines parallel to $AB$. The intersections of these perpendicular and parallel lines are points on the curve of the ellipse.

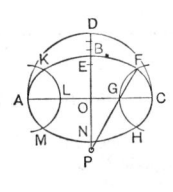

To construct an approximate ellipse by circular arcs: Let $AC$ be the major axis and $BN$ the minor. Draw half circle $ADC$ with $O$ as a center. Divide $BD$ into three equal parts and set off $BE$ equal to one of these parts. With $A$ and $C$ as centers and $OE$ as radius, describe circular arcs $KLM$ and $FGH$; with $G$ and $L$ as centers, and the same radius, describe arcs $FCH$ and $KAM$. Through $F$ and $G$ draw line $FP$, and with $P$ as a center draw the arc $FBK$. Arc $HNM$ is drawn in the same manner.

**Geometrical Constructions**

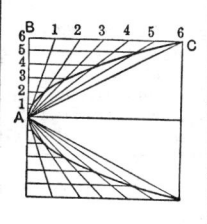

To construct a parabola: Divide line $AB$ into a number of equal parts and divide $BC$ into the same number of parts. From the division points on $AB$ draw horizontal lines. From the division points on $BC$ draw lines to point $A$. The points of intersection between lines drawn from points numbered alike are points on the parabola.

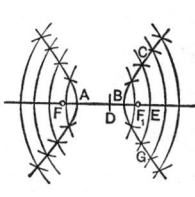

To construct a hyperbola: From focus $F$ lay off a distance $FD$ equal to the transverse axis, or the distance $AB$ between the two branches of the curve. With $F$ as a center and any distance $FE$ greater than $FB$ as a radius, describe a circular arc. Then with $F_1$ as a center and $DE$ as a radius, describe arcs intersecting at $C$ and $G$ the arc just described. $C$ and $G$ are points on the hyperbola. Any number of points can be found in a similar manner.

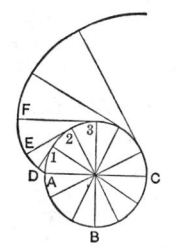

To construct an involute: Divide the circumference of the base circle $ABC$ into a number of equal parts. Through the division points 1, 2, 3, etc., draw tangents to the circle and make the lengths $D$ 1, $E$ 2, $F$ 3, etc., of these tangents equal to the actual length of the arcs $A$ 1, $A$ 2, $A$ 3, etc.

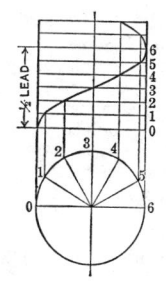

To construct a helix: Divide half the circumference of the cylinder on the surface of which the helix is to be described into a number of equal parts. Divide half the lead of the helix into the same number of equal parts. From the division points on the circle representing the cylinder draw vertical lines, and from the division points on the lead draw horizontal lines as shown. The intersections between lines numbered alike are points on the helix.

## Mathematical Signs and Commonly Used Abbreviations

| Sign | Meaning | Sign | Meaning |
|---|---|---|---|
| $+$ | Plus (sign of addition) | $\pi$ | Pi (3.1416) |
| $+$ | Positive | $\Sigma$ | Sigma (sign of summation) |
| $-$ | Minus (sign of subtraction) | $\omega$ | Omega (angles measured in radians) |
| $-$ | Negative | | |
| $\pm\ (\mp)$ | Plus or minus (minus or plus) | $g$ | Acceleration due to gravity (32.16 ft. per sec. per sec.) |
| $\times$ | Multiplied by (multiplication sign) | | |
| $\cdot$ | Multiplied by (multiplication sign) | $i$ (or $j$) | Imaginary quantity ($\sqrt{-1}$) |
| $\div$ | Divided by (division sign) | $\sin$ | Sine |
| $:$ | Divided by (division sign) | $\cos$ | Cosine |
| $:$ | Is to (in proportion) | $\tan$ | |
| $=$ | Equals | $(\mathrm{tg})$ | Tangent |
| $\neq$ | Is not equal to | $(\mathrm{tang})$ | |
| $\equiv$ | Is identical to | $\cot$ | Cotangent |
| $::$ | Equals (in proportion) | $(\mathrm{ctg})$ | |
| $\cong \approx \doteqdot$ | Approximately equals | $\sec$ | Secant |
| | | $\csc$ | Cosecant |
| $>$ | Greater than | $\operatorname{versin}$ | Versed sine |
| $<$ | Less than | $\operatorname{covers}$ | Coversed sine |
| $\geqq$ | Greater than or equal to | $\sin^{-1} a$ | Arc the sine of which is $a$ |
| $\leqq$ | Less than or equal to | $\arcsin a$ | |
| $\rightarrow$ | Approaches as a limit | $(\sin a)^{-1}$ | Reciprocal of $\sin a$ ($1 \div \sin a$) |
| $\propto$ | Varies directly as | | |
| $\therefore$ | Therefore | $\sinh x$ | Hyperbolic sine of $x$ |
| $\sqrt{\phantom{x}}$ | Square root | $\cosh x$ | Hyperbolic cosine of $x$ |
| $\sqrt[3]{\phantom{x}}$ | Cube root | $\Delta$ | Delta (increment of) |
| $\sqrt[4]{\phantom{x}}$ | 4th root | $\delta$ | Delta (variation of) |
| $\sqrt[n]{\phantom{x}}$ | $n$th root | $d$ | Differential (in calculus) |
| $a^2$ | $a$ squared (2d power of $a$) | $\int$ | Integral (in calculus) |
| $a^3$ | $a$ cubed (3d power of $a$) | | |
| $a^4$ | 4th power of $a$ | $\int_b^a$ | Integral between the limits $a$ and $b$ |
| $a^n$ | $n$th power of $a$ | | |
| $a^{-n}$ | $1 \div a^n$ | $!$ | $5! = 1 \times 2 \times 3 \times 4 \times 5$ |
| $\dfrac{1}{n}$ | Reciprocal value of $n$ | $\angle$ | Angle |
| $\log$ | Logarithm | $\llcorner$ | Right angle |
| | | $\perp$ | Perpendicular to |
| hyp. log | | $\triangle$ | Triangle |
| nat. log | Hyperbolic, natural or | $\odot$ | Circle |
| $\log_e$ | Napierian logarithm | $\Box$ | Parallelogram |
| $\ln$ | | $^\circ$ | Degree (circular arc or temperature) |
| $e$ | Base of hyp. logarithms (2.71828) | $'$ | Minutes or feet |
| | | $''$ | Seconds or inches |
| lim. | Limit value (of an expression) | $a'$ | $a$ prime |
| $\infty$ | Infinity | $a''$ | $a$ double prime |
| $\alpha$ | Alpha | $a_1$ | $a$ sub one |
| $\beta$ | Beta    commonly used | $a_2$ | $a$ sub two |
| $\gamma$ | Gamma   to denote angles | $a_n$ | $a$ sub $n$ |
| $\theta$ | Theta | $(\ )$ | Parentheses |
| $\phi$ | Phi | $[\ ]$ | Brackets |
| $\mu$ | Mu (coefficient of friction) | $\{\ \}$ | Braces |

# MECHANICS

Throughout the Mechanics section in this Handbook, both English and metric SI data and formulas are given to cover the requirements of working in either system of measurement. Except for the passage entitled "The Use of the Metric SI System in Mechanics Calculations", formulas and text relating exclusively to SI are given in bold face type.

**Definitions.** — The science of mechanics deals with the effects of forces in causing or preventing motion. *Statics* is that branch of mechanics which deals with bodies in equilibrium, i.e., the forces acting on them cause them to remain at rest or to move with uniform velocity. *Dynamics* is that branch of mechanics which deals with bodies not in equilibrium, i.e., the forces acting on them cause them to move with non-uniform velocity. *Kinetics* is that branch of dynamics which deals with both the forces acting on bodies and the motions which they cause. *Kinematics* is that branch of dynamics which deals only with the motions of bodies without reference to the forces that cause them.

Definitions of certain terms and quantities as used in mechanics follow:

A *force* may be defined simply as a push or a pull; the push or pull may result from the force of contact between bodies or from a force, such as magnetism or gravitation, in which no direct contact takes place.

*Matter* is any substance that occupies space; gases, liquids, solids, electrons, atoms, molecules, etc., all fit this definition.

*Inertia* is that property of matter which causes it to resist any change in its motion or state of rest.

*Mass* is a measure of the inertia of a body.

*Work*, in mechanics, is the product of force times distance and is expressed by a combination of units of force and distance, as foot-pounds, inch-pounds, meter-kilograms, etc. **The metric SI unit of work is the joule, which is the work done when the point of application of a force of one newton is displaced through a distance of one meter in the direction of the force.**

*Power*, in mechanics, is the product of force times distance divided by time; it measures the performance of a given amount of work in a given time. It is the rate of doing work and as such is expressed in foot-pounds per minute, foot-pounds per second, kilogram-meters per second, etc. **The metric SI unit is the watt, which is one joule per second.**

*Horsepower* is the unit of power that has been adopted for engineering work. One horsepower is equal to 33,000 foot-pounds per minute or 550 foot-pounds per second. The *kilowatt*, used in electrical work, equals 1.34 horsepower; or 1 horsepower equals 0.746 kilowatt. **However, in the metric SI, the term horsepower is not used, and the basic unit of power is the watt. This unit, and the derived units milliwatt and kilowatt, for example, are the same as those used in electrical work.**

*Torque* or *moment* of a force is a measure of the tendency of the force to rotate the body upon which it acts about an axis. The magnitude of the moment due to a force acting in a plane perpendicular to some axis is obtained by multiplying the force by the perpendicular distance from the axis to the line of action of the force. (If the axis of rotation is not perpendicular to the plane of the force, then the components of the force in a plane perpendicular to the axis of rotation are used to find the resultant moment of the force by finding the moment of each component and adding these component moments algebraically.) Moment or torque is commonly expressed in pound-feet, pound-inches, kilogram-meters, etc. **The metric SI unit is the newton-meter (N · m).**

*Velocity* is the time-rate of change of distance and is expressed as distance divided by time, that is, feet per second, miles per hour, centimeters per second, meters per second, etc.

*Acceleration* is defined as the time-rate of change of velocity and is expressed as velocity divided by time or as distance divided by time squared, that is, in feet per second, per second or feet per second **squared**; inches per second, per second or inches per second **squared**; centimeters per second, per second or centimeters per second **squared**; etc. **The metric SI unit is the meter per second squared.**

**Unit Systems.** — In mechanics calculations, both *absolute* and *gravitational* systems of units are employed. The fundamental units in absolute systems are *length, time,* and *mass,* and from these units, the dimension of force is derived. Two absolute systems which have been in use for many years are the cgs (centimeter-gram-second) and the MKS (meter-kilogram-second) systems. Another system, known as MKSA (meter-kilogram-second-ampere), links the MKS system of units of mechanics with electro magnetic units.

**The Conference General des Poids et Mesures (CGPM), which is the body responsible for all international matters concerning the metric system, adopted in 1954 a rationalized and coherent system of units based on the four MKSA units and including the kelvin as the unit of temperature, and the candela as the unit of luminous intensity.   In 1960, the CGPM formally named this system the 'Systeme International d'Unites', for which the abbreviation is SI in all languages.   In 1971, the 14th CGPM adopted a seventh base unit, the mole, which is the unit of quantity ("amount of substance").   Further details of the SI are given in the Weights and Measures section, and its application in mechanics calculations, contrasted with the use of the English system, is considered on page 285.**

The fundamental units in gravitational systems are *length, time,* and *force,* and from these units, the dimension of mass is derived. In the gravitational system most widely used in English measure countries, the units of length, time, and force are, respectively, the foot, the second, and the pound. The corresponding unit of mass, commonly called the *slug,* is equal to 1 pound second$^2$ per foot and is derived from the formula, $M = W \div g$ in which $M$ = mass in slugs, $W$ = weight in pounds, and $g$ = acceleration due to gravity, commonly taken as 32.16 feet per second$^2$. A body that weighs 32.16 lbs. on the surface of the earth has, therefore, a mass of one slug.

Many engineering calculations utilize a system of units consisting of the inch, the second, and the pound. The corresponding units of mass are pounds second$^2$ per inch and the value of $g$ is taken as 386 inches per second$^2$.

In a gravitational system that has been widely used in metric countries, the units of length, time, and force are, respectively, the meter, the second, and the kilogram. The corresponding units of mass are kilograms second$^2$ per meter and the value of $g$ is taken as 9.81 meters per second$^2$.

**Acceleration of Gravity $g$ Used in Mechanics Formulas.** — The acceleration of a freely falling body has been found to vary according to location on the earth's surface as well as with height, the value at the equator being 32.09 feet per second, per second while at the poles it is 32.26 ft/sec$^2$. In the United States it is customary to regard 32.16 as satisfactory for most practical purposes in engineering calculations.

*Standard Pound Force:* For use in defining the magnitude of a standard unit of force, known as the *pound force,* a fixed value of 32.1740 ft/sec$^2$, designated by the symbol $g_0$, has been adopted by international agreement. As a result of this agreement, whenever the term mass, $M$, appears in a mechanics formula and the substitution $M = W/g$ is made, use of the standard value $g_0 = 32.1740$ ft/sec$^2$ is implied although as stated previously, it is customary to use approximate values for $g$ except in those cases where extreme accuracy is required.

**The Use of the Metric SI System in Mechanics Calculations.** — The SI system is a development of the traditional metric system based on decimal arithmetic; fractions are avoided. For each physical quantity, units of different sizes are formed by multiplying or dividing a single base value by powers of 10. Thus, changes can be made very simply by adding zeros or shifting decimal points. For example, the meter is the basic unit of length; the kilometer is a multiple (1,000 meters); and the millimeter is a sub-multiple (one-thousandth of a meter).

In the older metric system, the simplicity of a series of units linked by powers of 10 is an advantage for plain quantities such as length, but this simplicity is lost as soon as more complex units are encountered. For example, in different branches of science and engineering, energy may appear as the erg, the calorie, the kilogram-meter, the liter-atmosphere, or the horsepower-hour. In contrast, the SI provides only one basic unit for each physical quantity, and universality is thus achieved.

There are seven base-units, and in mechanics calculations three are used, which are for the basic quantities of length, mass, and time, expressed as the meter (m), the kilogram (kg), and second (s). The other four base-units are the ampere (A) for electric current, the kelvin (K) for thermo-dynamic temperature, the candela (cd) for luminous intensity, and the mole (mol) for amount of substance.

The SI is a coherent system. A system of units is said to be coherent if the product or quotient of any two unit quantities in the system is the unit of the resultant quantity. For example, in a coherent system in which the foot is a unit of length, the square foot is the unit of area, whereas the acre is not. Further details of the SI, and definitions of the units, are given at the end of the book.

Other physical quantities are derived from the base-units. For example, the unit of velocity is the meter per second (m/s), which is a combination of the base-units of length and time. The unit of acceleration is the meter per second squared (m/s²). By applying Newton's second law of motion — force is proportional to mass multiplied by acceleration — the unit of force is obtained, which is the kg·m/s². This unit is known as the newton, or N. Work, or force times distance, is the kg·m²/s², which is the joule (1 joule = 1 newton-meter) and energy is also expressed in these terms. The abbreviation for joule is J. Power, or work per unit time, is the kg·m²/s³, which is the watt (1 watt = 1 joule per second = 1 newton-meter per second). The abbreviation for watt is W.

The coherence of SI units has two important advantages. The first, that of uniqueness and therefore universality, has been explained. The second is that it greatly simplifies technical calculations. Equations representing physical principles can be applied without introducing such numbers as 550 in power calculations, which, in the English system of measurement have to be used to convert units. Thus conversion factors largely disappear from calculations carried out in SI units, with a great saving in time and labor.

*Mass, weight, force, load.* SI is an absolute system (see page 284), and consequently it is necessary to make a clear distinction between mass and weight. The *mass* of a body is a measure of its inertia, whereas the weight of a body is the *force* exerted on it by gravity. In a fixed gravitational field, weight is directly proportional to mass, and the distinction between the two can be easily overlooked. However, if a body is moved to a different gravitational field, for example, that of the moon, its weight alters, but its mass remains unchanged. Since the gravitational field on earth varies from place to place by only a small amount, and weight is proportional to mass, it is practical to use the weight of unit mass as a unit of force, and this procedure is adopted in both the English and older metric systems of measurement. In common usage, they are given the same names, and we say that a mass of 1 pound has a weight of 1 pound. In the former case the pound is being used as a unit of mass, and in the latter case, as a unit of force. This procedure is convenient in some branches of engineering, but leads to confusion in others.

As mentioned earlier, Newton's second law of motion states that force is proportional to mass times acceleration. Because an unsupported body on the earth's surface falls with acceleration $g$ (32 ft/s² approximately), the pound (force) is that force which will impart an acceleration of $g$ ft/s² to a pound (mass). Similarly, the kilogram (force) is that force which will impart an acceleration of $g$ (9.8 meters per second² approximately), to a mass of one kilogram. In the SI, the *newton* is that force which will impart unit acceleration (1 m/s²) to a mass of one kilogram. It is therefore smaller than the kilogram (force) in the ratio 1:$g$ (about 1:9.8). This fact has important consequences in engineering calculations. The factor $g$ now disappears from a wide range of formulas in dynamics, but appears in many formulas in statics where it was formerly absent. It is however not quite the same $g$, for reasons which will now be explained.

In the article on page 321, the mass of a body is referred to as $M$, but it is immediately replaced in subsequent formulas by $W/g$, where $W$ is the weight in pounds (force), which leads to familiar expressions such as $WV^2/2g$ for kinetic energy. In this treatment, the $M$ which appears briefly is really expressed in terms of the slug (page 284), a unit normally used only in aeronautical engineering. In everyday engineers' language, weight and mass are regarded as synonymous and expressions such as $WV^2/2g$ are used without pondering the distinction. Nevertheless, on reflection it seems odd that $g$ should appear in a formula which has nothing to do with gravity at all. In fact the $g$ used here is not the true, local value of the acceleration due to gravity, but an arbitrary standard value which has been chosen as part of the definition of the pound (force) and is more properly designated $g_0$ (page 284). Its function is not to indicate the strength of the local gravitational field, but to convert from one unit to another.

In the SI the unit of mass is the *kilogram*, and the unit of force (and therefore weight) is the *newton*.

The following are typical statements in dynamics expressed in SI units:

A force of $R$ newtons acting on a mass of $M$ kilograms produces an acceleration of $R/M$ meters per second². The kinetic energy of a mass of $M$ kg moving with velocity $V$ m/s is ½ $MV^2$ kg (m/s)² or ½ $MV^2$ joules. The work done by a force of $R$ newtons moving a distance $L$ meters is $RL$ Nm, or $RL$ joules. If this work were converted entirely into kinetic energy we could write $RL$ = ½ $MV^2$ and it is instructive to consider the units. Remembering that the N is the same as the kg·m/s² we have (kg·m/s)² × m = kg (m/s)² which is obviously correct. It will be noted that $g$ does not appear anywhere in these statements.

In contrast, in many branches of engineering where the weight of a body is important, rather than its mass, using SI units $g$ does appear where formerly it was absent. Thus if a rope hangs vertically supporting a mass of $M$ kilograms the tension in the rope is $Mg$ N. Here $g$ is the acceleration due to gravity, and its units are m/s². The ordinary numerical value of 9.81 will be sufficiently accurate for most purposes on earth. The expression is still valid elsewhere, for example, on the moon, provided the proper value of $g$ is used. The maximum tension the rope can safely withstand (and other similar properties) will also be specified in terms of the newton, so that direct comparison may be made with the tension predicted.

Words like load and weight have to be used with greater care. In everyday language we might say "a lift carries a load of five people of average weight 70 kg", but in precise technical language we say that if the average mass is 70 kg, then the average weight is $70g$ N, and the total load (that is force) on the lift is $350g$ N.

If the lift starts to rise with acceleration $a$ m/s², the load becomes $350 (g + a)$ N; both $g$ and $a$ have units of m/s², the mass is in kg, so the load is in terms of kg·m/s², which is the same as the newton.

*Pressure and stress.* These quantities are expressed in terms of force per unit area. In the SI the unit is the pascal (Pa), which expressed in terms of SI derived and base units is the newton per meter squared $(N/m^2)$. The pascal is very small—it is only equivalent to $0.15 \times 10^{-3}$; lb/in²— hence the kilopascal (kPa = 1000 pascals), and the megapascal (MPa = $10^6$ pascals) may be more convenient multiples in practice. Thus, note: 1 newton per millimeter squared = 1 meganewton per meter squared = 1 megapascal.

In addition to the pascal, the bar, a non-SI unit, is in use in the field of pressure measurement in some countries, including England. Thus, in view of existing practice, the International Committee of Weights and Measures (CIPM) decided in 1969 to retain this unit for a limited time for use with those of SI. The bar = $10^5$ pascals and the hectobar = $10^7$ pascals.

**Scalar and Vector Quantities.** — The quantities dealt with in mechanics are of two kinds according to whether magnitude alone or direction as well as magnitude must be known in order to completely specify them. Quantities such as time, volume and density are completely specified when their magnitude is known. Such quantities are called *scalar* quantities. Quantities such as force, velocity, acceleration, moment and displacement which must, in order to be specified completely, have a specific direction as well as magnitude, are called *vector* quantities.

**Graphical Representation of Forces.** — A force has three characteristics which, when known, determine it. They are *direction, point of application*, and *magnitude*. The direction of a force is the direction in which it tends to move the body upon which it acts. The point of application is the place on the line of action where the force is applied. Forces may conveniently be represented by straight lines and arrow heads. The arrow head indicates the direction of the force, and the length of

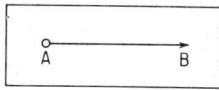

the line, its magnitude to any suitable scale. The point of application may be at any point on the line, but it is generally convenient to assume it to be at one end. In the accompanying illustration, a force is supposed to act along line $AB$ in a direction from left to right. The length of line $AB$ shows the magnitude of the force.

If point $A$ is the point of application, the force is exerted as a pull, but if point $B$ be assumed to be the point of application, it would indicate that the force is exerted as a push.

Velocities, moments, displacements, etc. may similarly be represented and manipulated graphically because they are all of the same class of quantities called vectors. (*See* Scalar and Vector Quantities.)

**Algebraic Composition and Resolution of Force Systems.** — The graphical methods shown on pages 289 and 290 are convenient for solving problems involving force systems in which all of the forces lie in the same plane and only a few forces are involved. If many forces are involved, however, or the forces do not lie in the same plane, it is better to use algebraic methods to avoid complicated space diagrams. Systematic procedures for solving force problems by algebraic methods are outlined beginning on page 291. In connection with the use of these procedures, it is necessary to define several terms applicable to force systems in general.

The single force which produces the same effect upon a body as two or more forces acting together is called their *resultant*. The separate forces which can be so combined are called the *components*. Finding the resultant of two or more forces is called the *composition of forces*, and finding two or more components of a given force, the *resolution of forces*. Forces are said to be *concurrent* when their lines of action can be extended to meet at a common point; forces that are *parallel* are, of course, *nonconcurrent*. Two forces having the same line of action are said to be *collinear*. Two forces equal in magnitude, parallel, and in opposite directions constitute a *couple.* Forces all in the same plane are said to be *coplanar;* if not in the same plane, they are called *noncoplanar* forces.

The *resultant* of a system of forces is the simplest equivalent system that can be determined. It may be a single force, a couple, or a noncoplanar force and a couple. This last type of resultant, a noncoplanar force and a couple, may be replaced, if desired, by two *skewed* forces (forces that are nonconcurrent, nonparallel, and noncoplanar). When the resultant of a system of forces is zero, the system is in equilibrium, that is, the body on which the force system acts remains at rest or continues to move with uniform velocity.

**Couples.** — If the forces $AB$ and $CD$ are equal and parallel but act in opposite directions, then the resultant equals o, or, in other words, the two forces have no resultant and are called a couple.

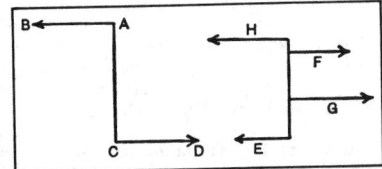

A couple tends to produce rotation. The measure of this tendency is called the moment of the couple and is the product of one of the forces multiplied by the distance between the two.

As a couple has no resultant, no single force can balance or counteract the tendency of the couple to produce rotation. To prevent the rotation of a body acted upon by a couple, two other forces are therefore required, forming a second couple. In the illustration, $E$ and $F$ form one couple and $G$ and $H$ are the balancing couple. The body on which they act is in equilibrium if the moments of the two couples are equal and tend to rotate the body in opposite directions. A couple may also be represented by a vector in the direction of the axis about which the couple acts. The length of the vector, to some scale, represents the magnitude of the couple, and the direction of the vector is that in which a right-hand screw would advance if it were to be rotated by the couple.

**Composition of a Single Force and Couple.** — A single force and a couple in the same plane or in parallel planes may be replaced by another single force equal and parallel to the first force, at a distance from it equal to the moment of the couple divided by the magnitude of the force. The new single force is located so that the moment of the resultant about the point of application of the original force is of the same sign as the moment of the couple.

In the figure, with the couple $N - N$ in the position shown, the resultant of $P, - N,$ and $N$ is $O$ (which equals $P$) acting on a line through point $L$ so that $(P - N) \times ac = N \times bc.$

Thus, it follows that,

$$ac = \frac{N(ac + bc)}{P} = \frac{\text{Moment of Couple}}{P}$$

### Graphical Composition and Resolution of Forces

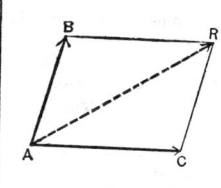

*Parallelogram of Forces.* — If two forces applied at a point are represented in magnitude and direction by the adjacent sides of a parallelogram ($AB$ and $AC$ in the accompanying illustration), their resultant will be represented in magnitude and direction by the diagonal $AR$ drawn from the intersection of the two component forces.

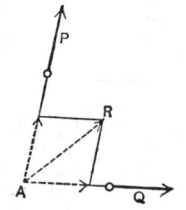

If two forces $P$ and $Q$ do not have the same point of application, but the lines indicating their directions intersect, the forces may be imagined as applied at the point of intersection between the lines (as at $A$), and the resultant of the two forces may be found by constructing the parallelogram of forces. Line $AR$ shows the direction and magnitude of the resultant, the point of application of which may be assumed to be any point on line $AR$ or its extension.

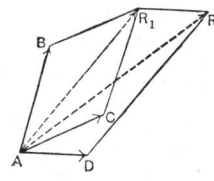

If the resultant of three or more forces having the same point of application is to be found, first find the resultant of any two of the forces ($AB$ and $AC$) and then find the resultant of the resultant just found ($AR_1$) and the third force ($AD$). If there be more than three forces, continue in this manner until the resultant of all the forces has been found.

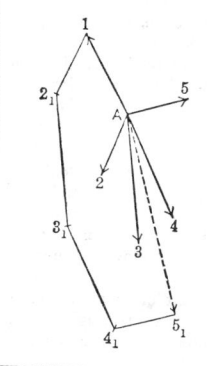

*Polygon of Forces.* — When several forces are applied at a point and act in a single plane, their resultant may be found more simply than by the method just described, as follows: From the extreme end of the line representing the first force, draw a line representing the second force, parallel to it and of the same length and in the direction of the second force. Then through the extreme end of this line draw a line parallel to, and of the same length and direction as the third force, and continue this until all the forces have been thus represented. Then draw a line from the point of application of the forces (as $A$) to the extreme point (as $5_1$) of the line last drawn. This line ($A\,5_1$) is the resultant of the forces.

### Graphical Composition and Resolution of Forces

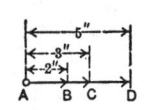

The resultant of two forces applied at the same point and acting in the same direction, is equal to the sum of the forces. For example, if the two forces $AB$ and $AC$, one equal to two and the other equal to three pounds, are applied at point $A$, then their resultant $AD$ equals the sum of these forces, or five pounds.

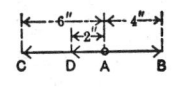

If two forces act in opposite directions, then their resultant is equal to their difference, and the direction of the resultant is the same as the direction of the greater of the two forces. For example: $AB$ and $AC$ are both applied at point $A$; then, if $AB$ equals four and $AC$ equals six pounds, the resultant $AD$ equals two pounds and acts in the direction of $AC$.

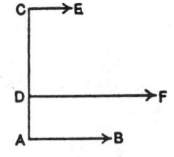

*Parallel Forces.* — If two forces are parallel and act in the same direction, then their resultant is parallel to both lines, is located between them, and is equal to the sum of the two components. The point of application of the resultant divides the line joining the points of application of the components inversely as the magnitude of the forces. Thus, $AB:CE = CD:AD$.

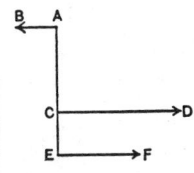

The resultant of two parallel forces acting in opposite directions is parallel to both lines, is located outside of them on the side of the greater of the components, has the same direction as the greater component, and is equal in magnitude to the difference between the two components. The point of application on the line $AC$ produced is found from the proportion:

$$AB : CD = CE : AE.$$

*Moment of a Force.* — The moment of a force with respect to a point is the product of the force multiplied by the perpendicular distance from the given point to the direction of the force. In the illustration, the moment of the force $P$ with relation to point $A$ is $P \times AB$. The perpendicular distance $AB$ is called the lever-arm of the force. The moment is the measure of the tendency of the force to produce rotation about the given point, which is termed the center of moments. If the force is measured in pounds and the distance in inches, the moment is expressed in inch-pounds. **In metric SI units, the moment is expressed in newton-meters (N·m), or newton-millimeters (N·mm).**

The moment of the resultant of any number of forces acting together in the same plane is equal to the algebraic sum of the moments of the separate forces.

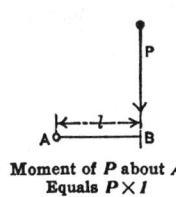

**Moment of $P$ about $A$ Equals $P \times l$**

**Table 1. Algebraic Solution of Force Systems — All Forces in the Same Plane**

Finding Two Concurrent Components of a Single Force

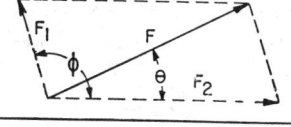

Case I: To find two components $F_1$ and $F_2$ at angles $\theta$ and $\phi$, $\phi$ not being 90°.

$$F_1 = \frac{F \sin \theta}{\sin \phi}$$

$$F_2 = \frac{F \sin (\phi - \theta)}{\sin \phi}$$

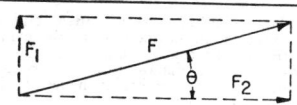

Case II: Components $F_1$ and $F_2$ form 90° angle.

$$F_1 = F \sin \theta$$

$$F_2 = F \cos \theta$$

Finding the Resultant of Two Concurrent Forces

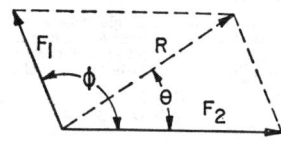

Case I: Forces $F_1$ and $F_2$ do not form 90° angle.

$$R = \frac{F_1 \sin \phi}{\sin \theta}, \text{ or, } R = \frac{F_2 \sin \phi}{\sin (\phi - \theta)}, \text{ or}$$

$$R = \sqrt{F_1{}^2 + F_2{}^2 + 2F_1F_2 \cos \phi}$$

$$\tan \theta = \frac{F_1 \sin \phi}{F_1 \cos \phi + F_2}$$

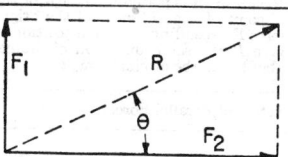

Case II: Forces $F_1$ and $F_2$ form 90° angle.

$$R = \frac{F_2}{\cos \theta}, \text{ or, } R = \frac{F_1}{\sin \theta}, \text{ or}$$

$$R = \sqrt{F_1{}^2 + F_2{}^2}$$

$$\tan \theta = \frac{F_1}{F_2}$$

Finding the Resultant of Three or More Concurrent Forces

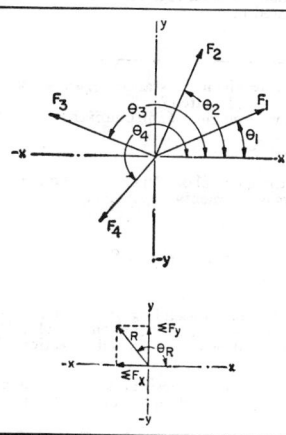

To determine resultant of forces $F_1$, $F_2$, $F_3$, etc. making angles, respectively, of $\theta_1$, $\theta_2$, $\theta_3$, etc. with the $x$ axis, find the $x$ and $y$ components $F_x$ and $F_y$ of each force and arrange in a table similar to that shown below for a system of three forces. Find the algebraic sum of the $F_x$ and $F_y$ components ($\Sigma F_x$ and $\Sigma F_y$) and use these to determine resultant $R$.

| Force | $F_x$ | $F_y$ |
|-------|-------|-------|
| $F_1$ | $F_1 \cos \theta_1$ | $F_1 \sin \theta_1$ |
| $F_2$ | $F_2 \cos \theta_2$ | $F_2 \sin \theta_2$ |
| $F_3$ | $F_3 \cos \theta_3$ | $F_3 \sin \theta_3$ |
|       | $\Sigma F_x$ | $\Sigma F_y$ |

$$R = \sqrt{(\Sigma F_x)^2 + (\Sigma F_y)^2}$$

$$\cos \theta_R = \frac{\Sigma F_x}{R}$$

$$\text{or, } \tan \theta_R = \frac{\Sigma F_y}{\Sigma F_x}$$

Table I (*Continued*).   Algebraic Solution of Force Systems — All Forces
in the Same Plane

| | |
|---|---|
| **Finding a Force and a Couple Which Together are Equivalent to a Single Force** | |
| 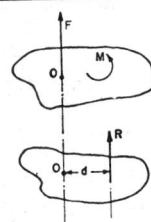 | To resolve a single force $F$ into a couple of moment $M$ and a force $P$ passing through any chosen point $O$ at a distance $d$ from the original force $F$, use the relations $$P = F$$ $$M = F \times d$$ The moment $M$ must, of course, tend to produce rotation about $O$ in the same direction as the original force. Thus, as seen in the diagram, $F$ tends to produce clockwise rotation; hence $M$ is shown clockwise. |

**Finding the Resultant of a Single Force and a Couple**

| | |
|---|---|
| 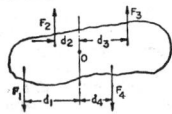 | The resultant of a single force $F$ and a couple $M$ is a single force $R$ equal in magnitude and direction to $F$ and parallel to it at a distance $d$ to the left or right of $F$. $$R = F$$ $$d = M \div R$$ Resultant $R$ is placed to the left or right of point of application $O$ of the original force $F$ depending on which position will give $R$ the same direction of moment about $O$ as the original couple $M$. |

**Finding the Resultant of a System of Parallel Forces**

| | |
|---|---|
| | To find the resultant of a system of coplanar parallel forces, proceed as indicated below. |

1. Select any convenient point $O$ from which perpendicular distances $d_1$, $d_2$, $d_3$, etc. to parallel forces $F_1$, $F_2$, $F_3$, etc. can be specified or calculated.

2. Find the algebraic sum of all the forces; this will give the magnitude of the resultant of the system.

$$R = \Sigma F = F_1 + F_2 + F_3 + \cdots$$

3. Find the algebraic sum of the moments of the forces about $O$; clockwise moments may be taken as negative and counterclockwise moments as positive:

$$\Sigma M_O = F_1 d_1 + F_2 d_2 + \cdots$$

4. Calculate the distance $d$ from $O$ to the line of action of resultant $R$:

$$d = \Sigma M_O \div R$$

This distance is measured to the left or right from $O$ depending on which position will give the moment of $R$ the same direction of rotation about $O$ as the couple $\Sigma M_O$, that is, if $\Sigma M_O$ is negative, then $d$ is left or right of $O$ depending on which direction will make $R \times d$ negative.

*Note Concerning Interpretation of Results:* If $R = 0$, then the resultant of the system is a couple $\Sigma M_O$; if $\Sigma M_O = 0$ then the resultant is a single force $R$; if both $R$ and $\Sigma M_O = 0$, then the system is in equilibrium.

Table I (*Continued*). Algebraic Solution of Force Systems — All Forces
in the Same Plane

Finding the Resultant of Forces Not Intersecting at a Common Point

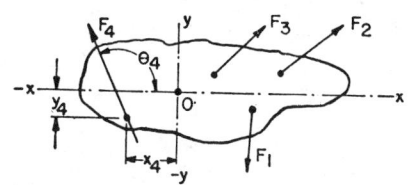

To determine the result-
ant of a coplanar, nonconcurrent, nonparallel force
system as shown in the
diagram, proceed as shown
below.

1. Draw a set of $x$ and $y$ coordinate axes through any convenient point $O$ in the plane
of the forces as shown in the diagram.

2. Determine the $x$ and $y$ coordinates of any convenient point on the line of action of
each force and the angle $\theta$, measured in a counterclockwise direction, that each line
of action makes with the positive $x$ axis. For example, in the diagram, coordinates
$x_4$, $y_4$, and $\theta_4$ are shown for $F_4$. Similar data should be known for each of the forces of
the system.

3. Calculate the $x$ and $y$ components ($F_x$, $F_y$) of each force and the moment of each
component about $O$. Counterclockwise moments are considered positive and clockwise moments as negative. Tabulate all results in a manner similar to that shown
below for a system of three forces and find $\Sigma F_x$, $\Sigma F_y$, $\Sigma M_O$ by algebraic addition.

| Force | Coordinates of $F$ | | | Components of $F$ | | Moment of $F$ about $O$ |
|---|---|---|---|---|---|---|
| $F$ | $x$ | $y$ | $\theta$ | $F_x$ | $F_y$ | $M_O = yF_x + xF_y$ |
| $F_1$ | $x_1$ | $y_1$ | $\theta_1$ | $F_1 \cos \theta_1$ | $F_1 \sin \theta_1$ | $y_1 F_1 \cos \theta_1 + x_1 F_1 \sin \theta_1$ |
| $F_2$ | $x_2$ | $y_2$ | $\theta_2$ | $F_2 \cos \theta_2$ | $F_2 \sin \theta_2$ | $y_2 F_2 \cos \theta_2 + x_2 F_2 \sin \theta_2$ |
| $F_3$ | $x_3$ | $y_3$ | $\theta_3$ | $F_3 \cos \theta_3$ | $F_3 \sin \theta_3$ | $y_3 F_3 \cos \theta_3 + x_3 F_3 \sin \theta_3$ |
| | | | | $\Sigma F_x$ | $\Sigma F_y$ | $\Sigma M_O$ |

4. Compute the resultant of the system and the angle $\theta_R$ it makes with the $x$ axis by
using the formulas:

$$R = \sqrt{(\Sigma F_x)^2 + (\Sigma F_y)^2}$$

$$\cos \theta_R = \Sigma F_x \div R, \quad \text{or,} \quad \tan \theta_R = \Sigma F_y \div \Sigma F_x$$

5. Calculate the distance $d$ from $O$ to the line of action of the resultant $R$:

$$d = \Sigma M_O \div R$$

Distance $d$ is in such direction from $O$ as will make the moment of $R$ about $O$ have
the same sign as $\Sigma M_O$.

*Note Concerning Interpretation of Results:* If $R = 0$, then the resultant is a couple
$\Sigma M_O$; if $\Sigma M_O = 0$, then $R$ passes through $O$; if both $R = 0$ and $\Sigma M_O = 0$, then the
system is in equilibrium.

Table I (*Continued*).   Algebraic Solution of Force Systems — All Forces
in the Same Plane

*Example:* Find the resultant of three coplanar nonconcurrent forces for which
the following data are given.

$$F_1 = 10 \text{ lbs}; x_1 = 5 \text{ in.}; y_1 = -1 \text{ in.}; \theta_1 = 270°.$$
$$F_2 = 20 \text{ lbs}; x_2 = 4 \text{ in.}; y_2 = 1.5 \text{ in.}; \theta_2 = 50°.$$
$$F_3 = 30 \text{ lbs}; x_3 = 2 \text{ in.}; y_3 = 2 \text{ in.}; \theta_3 = 60°.$$

$$F_{x_1} = 10 \cos 270° = 10 \times 0 = 0 \text{ lbs.}$$
$$F_{x_2} = 20 \cos 50° = 20 \times 0.64279 = 12.86 \text{ lbs.}$$
$$F_{x_3} = 30 \cos 60° = 30 \times 0.5000 = 15.00 \text{ lbs.}$$

$$F_{y_1} = 10 \times \sin 270° = 10 \times (-1) = -10.00 \text{ lbs.}$$
$$F_{y_2} = 20 \times \sin 50° = 20 \times 0.76604 = 15.32 \text{ lbs.}$$
$$F_{y_3} = 30 \times \sin 60° = 30 \times 0.86603 = 25.98 \text{ lbs.}$$

$$M_{O_1} = (-1) \times 0 + 5 \times (-10) = -50.00 \text{ in. lbs.}$$
$$M_{O_2} = 1.5 \times 12.86 + 4 \times 15.32 = 80.57 \text{ in. lbs.}$$
$$M_{O_3} = 2 \times 15 + 2 \times 25.98 = 81.96 \text{ in. lbs.}$$

*Note:* When working in metric SI units, pounds are replaced by newtons (N);
inches by meters or millimeters, and inch-pounds by newton-meters (N · m) or
newton-millimeters (N · mm).

| Force | Coordinates of $F$ | | | Components of $F$ | | Moment of $F$ about $O$ |
|---|---|---|---|---|---|---|
| $F$ | $x$ | $y$ | $\theta$ | $F_x$ | $F_y$ | |
| $F_1 = 10$ | 5 | $-1$ | 270° | 0 | $-10.00$ | $-50.00$ |
| $F_2 = 20$ | 4 | 1.5 | 50° | 12.86 | 15.32 | 80.57 |
| $F_3 = 30$ | 2 | 2 | 60° | 15.00 | 25.98 | 81.96 |
| | | | | 27.86 | 31.30 | 112.53 |

$$R = \sqrt{(27.86)^2 + (31.30)^2}$$
$$= 41.90 \text{ lbs.}$$

$$\tan \theta_R = \frac{31.30}{27.86} = 1.1235$$

$$\theta_R = 48° 20'$$

$$d = \frac{112.53}{41.90} = 2.69 \text{ inches}$$

measured as shown on the
diagram.

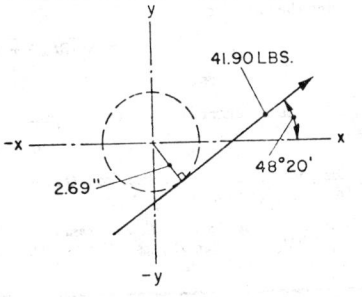

41.90 LBS.

48°20'

2.69"

### Table 2.  Algebraic Solution of Force Systems — Forces Not in Same Plane

---

Resolving a Single Force Into Its Three Rectangular Components

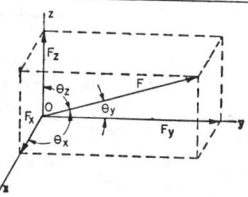

The diagram shows how a force $F$ may be resolved at any point $O$ on its line of action into three concurrent components each of which is perpendicular to the other two.

The $x$, $y$, $z$ components $F_x$, $F_y$, $F_z$ of force $F$ are determined from the accompanying relations in which $\theta_x$, $\theta_y$, $\theta_z$ are the angles which the force $F$ makes with the $x$, $y$, $z$ axes.

$$F_x = F \cos \theta_x$$
$$F_y = F \cos \theta_y$$
$$F_z = F \cos \theta_z$$
$$F = \sqrt{F_x{}^2 + F_y{}^2 + F_z{}^2}$$

---

Finding the Resultant of Any Number of Concurrent Forces

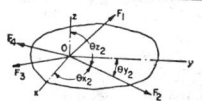

To find the resultant of any number of noncoplanar concurrent forces $F_1$, $F_2$, $F_3$, etc., use the procedure outlined below.

---

1. Draw a set of $x$, $y$, $z$ axes at $O$, the point of concurrency of the forces.  The angles each force makes measured counterclockwise from the positive $x$, $y$, and $z$ coordinate axes must be known in addition to the magnitudes of the forces.  For force $F_2$, for example, the angles are $\theta_{x2}$, $\theta_{y2}$, $\theta_{z2}$ as indicated on the diagram.

2. Apply the first three formulas given under the heading "Resolving a Single Force Into Its Three Rectangular Components" to each force to find its $x$, $y$, and $z$ components.  Tabulate these calculations as shown below for a system of three forces.  Algebraically add the calculated components to find $\Sigma F_x$, $\Sigma F_y$, and $\Sigma F_z$ which are the components of the resultant.

| Force | Angles | | | Components of Forces | | |
|---|---|---|---|---|---|---|
| $F$ | $\theta_x$ | $\theta_y$ | $\theta_z$ | $F_x$ | $F_y$ | $F_z$ |
| $F_1$ | $\theta_{x1}$ | $\theta_{y1}$ | $\theta_{z1}$ | $F_1 \cos \theta_{x1}$ | $F_1 \cos \theta_{y1}$ | $F_1 \cos \theta_{z1}$ |
| $F_2$ | $\theta_{x2}$ | $\theta_{y2}$ | $\theta_{z2}$ | $F_2 \cos \theta_{x2}$ | $F_2 \cos \theta_{y2}$ | $F_2 \cos \theta_{z2}$ |
| $F_3$ | $\theta_{x3}$ | $\theta_{y3}$ | $\theta_{z3}$ | $F_3 \cos \theta_{x3}$ | $F_3 \cos \theta_{y3}$ | $F_3 \cos \theta_{z3}$ |
| | | | | $\Sigma F_x$ | $\Sigma F_y$ | $\Sigma F_z$ |

3. Find the resultant of the system from the formula

$$R = \sqrt{(\Sigma F_x)^2 + (\Sigma F_y)^2 + (\Sigma F_z)^2}$$

4. Calculate the angles $\theta_{xR}$, $\theta_{yR}$, and $\theta_{zR}$ that the resultant $R$ makes with the respective coordinate axes:

$$\cos \theta_{xR} = \frac{\Sigma F_x}{R}$$

$$\cos \theta_{yR} = \frac{\Sigma F_y}{R}$$

$$\cos \theta_{zR} = \frac{\Sigma F_z}{R}$$

**Table 2** (*Continued*). **Algebraic Solution of Force Systems — Forces Not in Same Plane**

| Finding the Resultant of Parallel Forces Not in the Same Plane |
|---|

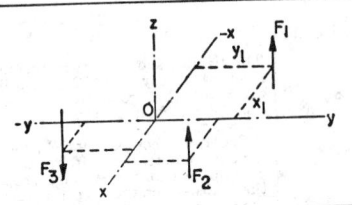

In the diagram, forces $F_1$, $F_2$, etc. represent a system of noncoplanar parallel forces. To find the resultant of such systems, use the procedure shown below.

**1.** Draw a set of $x$, $y$, and $z$ coordinate axes through any point $O$ in such a way that one of these axes, say the $z$ axis, is parallel to the lines of action of the forces. The $x$ and $y$ axes then will be perpendicular to the forces.

**2.** Set the distances of each force from the $x$ and $y$ axes in a table as shown below. For example, $x_1$ and $y_1$ are the $x$ and $y$ distances for $F_1$ shown in the diagram.

**3.** Calculate the moment of each force about the $x$ and $y$ axes and set the results in the table as shown for a system consisting of three forces. The algebraic sums of the moments $\Sigma M_x$ and $\Sigma M_y$ are then obtained. (In taking moments about the $x$ and $y$ axes, assign counterclockwise moments a plus (+) sign and clockwise moments a minus (−) sign. In deciding whether a moment is counterclockwise or clockwise, look from the positive side of the axis in question toward the negative side.)

| Force<br>$F$ | Coordinates of Force $F$ | | Moments $M_x$ and $M_y$ Due to $F$ | |
|---|---|---|---|---|
| | $x$ | $y$ | $M_x$ | $M_y$ |
| $F_1$ | $x_1$ | $y_1$ | $F_1 y_1$ | $F_1 x_1$ |
| $F_2$ | $x_2$ | $y_2$ | $F_2 y_2$ | $F_2 x_2$ |
| $F_3$ | $x_3$ | $y_3$ | $F_3 y_3$ | $F_3 x_3$ |
| $\Sigma F$ | | | $\Sigma M_x$ | $\Sigma M_y$ |

**4.** Find the algebraic sum $\Sigma F$ of all the forces; this will be the resultant $R$ of the system.

$$R = \Sigma F = F_1 + F_2 + \cdots$$

**5.** Calculate $x_R$ and $y_R$, the moment arms of the resultant:

$$x_R = \Sigma M_y \div R$$
$$y_R = \Sigma M_x \div R$$

These moment arms are measured in such direction along the $x$ and $y$ axes as will give the resultant a moment of the same direction of rotation as $\Sigma M_x$ and $\Sigma M_y$.

*Note Concerning Interpretation of Results:* If $\Sigma M_x$ and $\Sigma M_y$ are both 0, then the resultant is a single force $R$ along the $z$ axis; if $R$ also is 0, then the system is in equilibrium. If $R$ is 0 but $\Sigma M_x$ and $\Sigma M_y$ are not both 0, then the resultant is a couple

$$M_R = \sqrt{(\Sigma M_x)^2 + (\Sigma M_y)^2}$$

that lies in a plane parallel to the $z$ axis and making an angle $\theta_R$ measured in a counterclockwise direction from the positive $x$ axis and calculated from the following formula:

$$\sin \theta_R = \frac{\Sigma M_x}{M_R}$$

**Table 2** (*Continued*). **Algebraic Solution of Force Systems — Forces Not in Same Plane**

Finding the Resultant of Nonparallel Forces Not Meeting at a Common Point

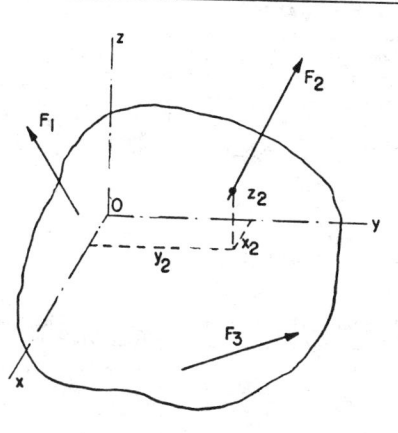

The diagram shows a system of noncoplanar, nonparallel, nonconcurrent forces $F_1$, $F_2$, etc. for which the resultant is to be determined. Generally speaking, the resultant will be a noncoplanar force and a couple which may be further combined, if desired, into two forces which are skewed.

Since this is the most general force system that can be devised, each of the other systems so far described represents a special, simpler case of this general force system. The method of solution described below for a system of three forces applies for any number of forces.

1. Select a set of coordinate $x$, $y$, and $z$ axes at any desired point $O$ in the body as shown in the diagram.

2. Determine the $x$, $y$, and $z$ coordinates of any convenient point on the line of action of each force as shown for $F_2$. Also determine the angles $\theta_x$, $\theta_y$, $\theta_z$ that each force makes with each coordinate axis. These angles are measured counterclockwise from the positive direction of the $x$, $y$, and $z$ axes. This data is tabulated, as shown in the table accompanying Step 3, for convenient use in subsequent calculations.

3. Calculate the $x$, $y$, and $z$ components of each force using the formulas given in the accompanying table. Add these components algebraically to get $\Sigma F_x$, $\Sigma F_y$ and $\Sigma F_z$ which are the components of the resultant, $R$, given by the formula,

$$R = \sqrt{(\Sigma F_x)^2 + (\Sigma F_y)^2 + (\Sigma F_z)^2}$$

| Force $F$ | Coordinates of Force $F$ | | | | | | Components of $F$ | | |
|---|---|---|---|---|---|---|---|---|---|
| | $x$ | $y$ | $z$ | $\theta_x$ | $\theta_y$ | $\theta_z$ | $F_x$ | $F_y$ | $F_z$ |
| $F_1$ | $x_1$ | $y_1$ | $z_1$ | $\theta_{x1}$ | $\theta_{y1}$ | $\theta_{z1}$ | $F_1 \cos \theta_{x1}$ | $F_1 \cos \theta_{y1}$ | $F_1 \cos \theta_{z1}$ |
| $F_2$ | $x_2$ | $y_2$ | $z_2$ | $\theta_{x2}$ | $\theta_{y2}$ | $\theta_{z2}$ | $F_2 \cos \theta_{x2}$ | $F_2 \cos \theta_{y2}$ | $F_2 \cos \theta_{z2}$ |
| $F_3$ | $x_3$ | $y_3$ | $z_3$ | $\theta_{x3}$ | $\theta_{y3}$ | $\theta_{z3}$ | $F_3 \cos \theta_{x3}$ | $F_3 \cos \theta_{y3}$ | $F_3 \cos \theta_{z3}$ |
| | | | | | | | $\Sigma F_x$ | $\Sigma F_y$ | $\Sigma F_z$ |

The resultant force $R$ makes angles of $\theta_{xR}, \theta_{yR}$, and $\theta_{zR}$ with the $x$, $y$, and $z$ axes, respectively, and passes through the selected point $O$. These angles are determined from the formulas,

$$\cos \theta_{xR} = \Sigma F_x \div R$$
$$\cos \theta_{yR} = \Sigma F_y \div R$$
$$\cos \theta_{zR} = \Sigma F_z \div R$$

**Table 2** *(Continued)*. Algebraic Solution of Force Systems — Forces Not in Same Plane

4. Calculate the moments $M_x$, $M_y$, $M_z$ about $x$, $y$, and $z$ axes, respectively due to the $F_x$, $F_y$, and $F_z$ components of each force and set them in tabular form. The formulas to use are given in the accompanying table.

| Force $F$ | Moments of Components of $F$ ($F_x$, $F_y$, $F_z$) about $x$, $y$, $z$ axes | | |
|---|---|---|---|
| | $M_x = yF_z + zF_y$ | $M_y = zF_x + xF_z$ | $M_z = xF_y + yF_x$ |
| $F_1$ | $M_{x1} = y_1F_{z1} + z_1F_{y1}$ | $M_{y1} = z_1F_{x1} + x_1F_{z1}$ | $M_{z1} = x_1F_{y1} + y_1F_{x1}$ |
| $F_2$ | $M_{x2} = y_2F_{z2} + z_2F_{y2}$ | $M_{y2} = z_2F_{x2} + x_2F_{z2}$ | $M_{z2} = x_2F_{y2} + y_2F_{x2}$ |
| $F_3$ | $M_{x3} = y_3F_{z3} + z_3F_{y3}$ | $M_{y3} = z_3F_{x3} + x_3F_{z3}$ | $M_{z3} = x_3F_{y3} + y_3F_{x3}$ |
| | $\Sigma M_x$ | $\Sigma M_y$ | $\Sigma M_z$ |

5. Add the component moments algebraically to get $\Sigma M_x$, $\Sigma M_y$ and $\Sigma M_z$ which are the components of the resultant couple, $M$, given by the formula,

$$M = \sqrt{(\Sigma M_x)^2 + (\Sigma M_y)^2 + (\Sigma M_z)^2}$$

The resultant couple $M$ will tend to produce rotation about an axis making angles of $\beta_x$, $\beta_y$, and $\beta_z$ with the $x$, $y$, $z$ axes, respectively. These angles are determined from the formulas,

$$\cos \beta_x = \frac{\Sigma M_x}{M}$$

$$\cos \beta_y = \frac{\Sigma M_y}{M}$$

$$\cos \beta_z = \frac{\Sigma M_z}{M}$$

General Method of Locating Resultant When Its Components are Known

To determine the position of the resultant force of a system of forces, proceed as follows:

From the origin, point $O$, of a set of coordinate axes $x$, $y$, $z$, lay off on the $x$ axis a length $A$ representing the algebraic sum $\Sigma F_x$ of the $x$ components of all the forces. From the end of line $A$ lay off a line $B$ representing $\Sigma F_y$, the algebraic sum of the $y$ components; this line $B$ is drawn in a direction parallel to the $y$ axis. From the end of line $B$ lay off a line $C$ representing $\Sigma F_z$. Finally, draw a line $R$ from $O$ to the end of $C$; $R$ will be the resultant of the system.

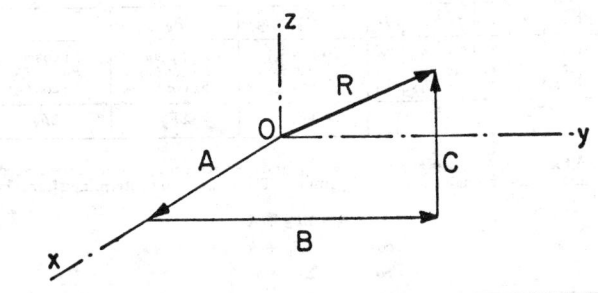

## Inclined Plane — Wedge

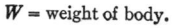

$W$ = weight of body.

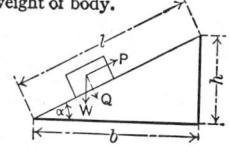

Neglecting friction:

$$P = W \times \frac{h}{l} = W \times \sin \alpha$$

$$W = P \times \frac{l}{h} = \frac{P}{\sin \alpha} = P \times \operatorname{cosec} \alpha$$

$$Q = W \times \frac{b}{l} = W \times \cos \alpha$$

If friction is taken into account, then force $P$ to pull body up is:

$$P = W \left( \mu \cos \alpha + \sin \alpha \right)$$

Force $P_1$ to pull body down is:

$$P_1 = W \left( \mu \cos \alpha - \sin \alpha \right)$$

Force $P_2$ to hold body stationary:

$$P_2 = W \left( \sin \alpha - \mu \cos \alpha \right)$$

in which $\mu$ is the coefficient of friction.

---

$W$ = weight of body.

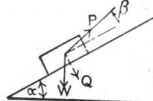

Neglecting friction:

$$P = W \times \frac{\sin \alpha}{\cos \beta}$$

$$W = P \times \frac{\cos \beta}{\sin \alpha}$$

$$Q = W \times \frac{\cos (\alpha + \beta)}{\cos \beta}$$

With friction:

Coefficient of friction = $\mu = \tan \phi$.

$$P = W \times \frac{\sin (\alpha + \phi)}{\cos (\beta - \phi)}$$

$W$ = weight of body.

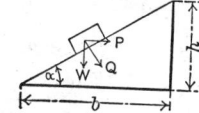

Neglecting friction:

$$P = W \times \frac{h}{b} = W \times \tan \alpha$$

$$W = P \times \frac{b}{h} = P \times \cot \alpha$$

$$Q = \frac{W}{\cos \alpha} = W \times \sec \alpha$$

With friction:

Coefficient of friction = $\mu = \tan \phi$.

$$P = W \tan (\alpha + \phi)$$

---

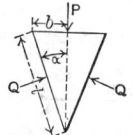

Neglecting friction:

$$P = 2 Q \times \frac{b}{l} = 2 Q \times \sin \alpha$$

$$Q = P \times \frac{l}{2 b} = \tfrac{1}{2} P \times \operatorname{cosec} \alpha$$

With friction:

Coefficient of friction = $\mu$.

$$P = 2 Q \left( \mu \cos \alpha + \sin \alpha \right)$$

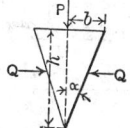

Neglecting friction:

$$P = 2 Q \times \frac{b}{h} = 2 Q \times \tan \alpha$$

$$Q = P \times \frac{h}{2 b} = \tfrac{1}{2} P \times \cot \alpha$$

With friction:

Coefficient of friction = $\mu = \tan \phi$.

$$P = 2 Q \tan (\alpha + \phi)$$

**Levers***

| Types of Levers | Examples |
|---|---|
| 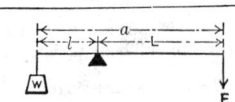 $$F : W = l : L \qquad F \times L = W \times l$$ $$F = \frac{W \times l}{L} \qquad W = \frac{F \times L}{l}$$ $$L = \frac{W \times a}{W + F} = \frac{W \times l}{F}; \quad l = \frac{F \times a}{W + F} = \frac{F \times L}{W}$$ | A pull of 80 pounds is exerted at the end of the lever, at $W$; $l = 12$ inches and $L = 32$ inches. Find the value of force $F$ required to balance the lever. $$F = \frac{80 \times 12}{32} = \frac{960}{32} = 30 \text{ pounds.}$$ If $F = 20$; $W = 180$; and $l = 3$; how long must $L$ be made to secure equilibrium? $$L = \frac{180 \times 3}{20} = 27.$$ |
| 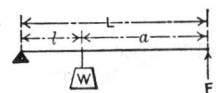 $$F : W = l : L \qquad F \times L = W \times l$$ $$F = \frac{W \times l}{L} \qquad W = \frac{F \times L}{l}$$ $$L = \frac{W \times a}{W - F} = \frac{W \times l}{F}; \quad l = \frac{F \times a}{W - F} = \frac{F \times L}{W}$$ | Total length $L$ of a lever is 25 inches. A weight of 90 pounds is supported at $W$; $l$ is 10 inches. Find the value of $F$. $$F = \frac{90 \times 10}{25} = 36 \text{ pounds.}$$ If $F = 100$ pounds, $W = 2200$ pounds, and $a = 5$ feet, what should $L$ equal to secure equilibrium? $$L = \frac{2200 \times 5}{2200 - 100} = 5.24 \text{ feet.}$$ |
| 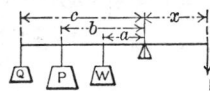 When three or more forces act on a lever: $$F \times x = W \times a + P \times b + Q \times c$$ $$x = \frac{W \times a + P \times b + Q \times c}{F}$$ $$F = \frac{W \times a + P \times b + Q \times c}{x}$$ | Let $W = 20$, $P = 30$, and $Q = 15$ pounds; $a = 4$, $b = 7$, and $c = 10$ inches. If $x = 6$ inches, find $F$. $$F = \frac{20 \times 4 + 30 \times 7 + 15 \times 10}{6} = 73\frac{1}{3} \text{ lbs.}$$ Assuming $F = 20$ in the example above, how long must lever arm $x$ be made? $$x = \frac{20 \times 4 + 30 \times 7 + 15 \times 10}{20} = 22 \text{ ins.}$$ |

* The above formulas are valid using metric SI units, with forces expressed in newtons, and lengths in meters. However, it should be noted that the weight of a mass $W$ kilograms is equal to a force of $Wg$ newtons, where $g$ is approximately $9.81$ m/s². Thus, supposing that in the first example $l = 0.4$ m, $L = 1.2$ m, and $W = 30$ kg, then the weight of $W$ is $30g$ N, so that the force $F$ required to balance the lever is

$$F = \frac{30g \times 0.4}{1.2} = 10g = 98.1 \text{ N.}$$

This force could be produced by suspending a mass of 10 kg at $F$.

**Toggle-joint.** — If arms $ED$ and $EH$ are of unequal length:

$$P = \frac{Fa}{b}$$

The relation between $P$ and $F$ changes constantly as $F$ moves downwards.

If arms $ED$ and $EH$ are equal:

$$P = \frac{Fa}{2h}$$

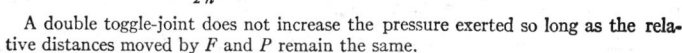

A double toggle-joint does not increase the pressure exerted so long as the relative distances moved by $F$ and $P$ remain the same.

### Toggle-joints with Equal Arms

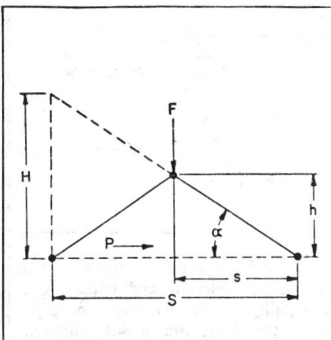

$F$ = force applied;
$P$ = resistance;
$\alpha$ = given angle.

$$2P\sin\alpha = F\cos\alpha;$$

$$\frac{P}{F} = \frac{\cos\alpha}{2\sin\alpha} = \text{coefficient};$$

or,     $P = F \times$ coefficient.

Equivalent expressions:

$$P = \frac{FS}{4h}; \qquad P = \frac{Fs}{H}, \text{ as per diagram.}$$

To use the table, measure angle $\alpha$, and find the coefficient in the table corresponding to the angle found. The coefficient is the ratio of the resistance to the force applied, and multiplying the force applied by the coefficient gives the resistance, neglecting friction.

| Angle | Coefficient | Angle | Coefficient | Angle | Coefficient | Angle | Coefficient |
|---|---|---|---|---|---|---|---|
| 0° 2′ | 862 | 0° 50′ | 34.4 | 2° 45′ | 10.4 | 8° 0′ | 3.58 |
| 0 4 | 456 | 0 55 | 31.2 | 2 50 | 10.1 | 8 30 | 3.35 |
| 0 6 | 285 | 1 0 | 28.6 | 3 0 | 9.54 | 9 0 | 3.15 |
| 0 8 | 216 | 1 10 | 24.6 | 3 15 | 8.81 | 9 30 | 2.99 |
| 0 10 | 171 | 1 15 | 22.9 | 3 30 | 8.17 | 10 0 | 2.84 |
| 0 12 | 143 | 1 20 | 21.5 | 3 45 | 7.63 | 11 0 | 2.57 |
| 0 14 | 122 | 1 30 | 19.1 | 4 0 | 7.25 | 12 0 | 2.35 |
| 0 15 | 115 | 1 40 | 17.2 | 4 15 | 6.73 | 13 0 | 2.17 |
| 0 16 | 107 | 1 45 | 16.4 | 4 30 | 6.35 | 14 0 | 2.00 |
| 0 18 | 95.4 | 1 50 | 15.6 | 4 45 | 6.02 | 15 0 | 1.87 |
| 0 20 | 85.8 | 2 0 | 14.3 | 5 0 | 5.71 | 16 0 | 1.74 |
| 0 25 | 68.6 | 2 10 | 13.2 | 5 30 | 5.19 | 17 0 | 1.64 |
| 0 30 | 57.3 | 2 15 | 12.7 | 6 0 | 4.76 | 18 0 | 1.54 |
| 0 35 | 49.1 | 2 20 | 12.5 | 6 30 | 4.39 | 19 0 | 1.45 |
| 0 40 | 42.8 | 2 30 | 11.5 | 7 0 | 4.07 | 20 0 | 1.37 |
| 0 45 | 38.2 | 2 40 | 10.7 | 7 30 | 3.79 | ..... | ....... |

## Wheels and Pulleys*

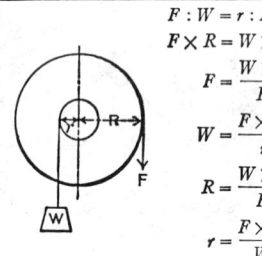

$$F : W = r : R$$

$$F \times R = W \times r$$

$$F = \frac{W \times r}{R}$$

$$W = \frac{F \times R}{r}$$

$$R = \frac{W \times r}{F}$$

$$r = \frac{F \times R}{W}$$

The radius of a drum on which is wound the lifting rope of a windlass is 2 inches. What force will be exerted at the periphery of a gear of 24 inches diameter, mounted on the same shaft as the drum and transmitting power to it, if one ton (2000 pounds) is to be lifted? Here $W = 2000$; $R = 12$; $r = 2$.

$$F = \frac{2000 \times 2}{12} = 333 \text{ pounds.}$$

---

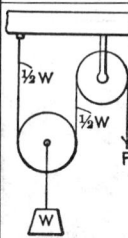

$$F = \tfrac{1}{2} W$$

The velocity with which weight $W$ will be raised equals one-half the velocity of the force applied at $F$.

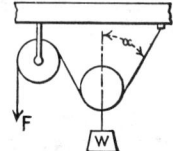

$$F : W = \sec \alpha : 2$$

$$F = \frac{W \times \sec \alpha}{2}$$

$$W = 2 F \times \cos \alpha$$

---

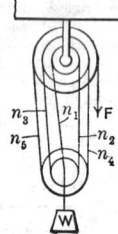

$n$ = number of strands or parts of rope ($n_1$, $n_2$, etc.).

$$F = \frac{1}{n} \times W$$

The velocity with which $W$ will be raised equals $\frac{1}{n}$ of the velocity of the force applied at $F$.

In the illustration is shown a combination of a double and triple block. The pulleys each turn freely on a pin as axis, and are drawn with different diameters, to show the parts of the rope more clearly. There are 5 parts of rope. Therefore, if 200 pounds is to be lifted, the force $F$ required at the end of the rope is:

$$F = \tfrac{1}{5} \times 200 = 40 \text{ pounds.}$$

---

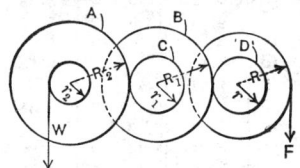

$A$, $B$, $C$ and $D$ are the pitch circles of gears.

$$F = \frac{W \times r \times n_1 \times r_2}{R \times R_1 \times R_2}$$

$$W = \frac{F \times R \times R_1 \times R_2}{r \times n_1 \times r_2}$$

Let the pitch diameters of gears $A$, $B$, $C$ and $D$ be 30, 28, 12 and 10 inches, respectively. Then $R_2 = 15$; $R_1 = 14$; $n_1 = 6$; and $r = 5$. Let $R = 12$, and $r_2 = 4$. Then the force $F$ required to lift a weight $W$ of 2000 pounds, friction being neglected, is:

$$F = \frac{2000 \times 5 \times 6 \times 4}{12 \times 14 \times 15} = 95 \text{ pounds.}$$

---

*Note: The above formulas are valid using metric SI units, with forces expressed in newtons, and lengths in meters or millimeters. (See note on page 300 concerning weight and mass).

## Differential Pulley — Screw

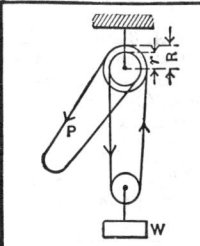

**Differential Pulley.** — In the differential pulley a chain must be used, engaging sprockets, so as to prevent the chain from slipping over the pulley faces.

$$P \times R = \tfrac{1}{2} W (R - r)$$

$$P = \frac{W (R - r)}{2 R}$$

$$W = \frac{2 PR}{R - r}$$

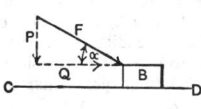

**Force Moving Body on Horizontal Plane.** — $F$ tends to move $B$ along line $CD$; $Q$ is the component which actually moves $B$; $P$ is the pressure, due to $F$, of the body on $CD$.

$$Q = F \times \cos \alpha; \qquad P = \sqrt{F^2 - Q^2}$$

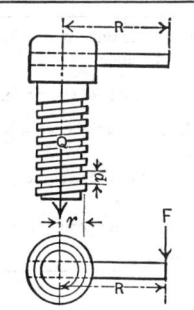

**Screw.** — $F$ = force at end of handle or wrench; $R$ = lever-arm of $F$; $r$ = pitch radius of screw; $p$ = lead of thread; $Q$ = load. Then, neglecting friction:

$$F = Q \times \frac{p}{6.2832\,R} \qquad Q = F \times \frac{6.2832\,R}{p}$$

If $\mu$ is the coefficient of friction, then:

For motion in direction of load $Q$ which *assists* it:

$$F = Q \times \frac{6.2832\,\mu r - p}{6.2832\,r + \mu p} \times \frac{r}{R}$$

For motion opposite load $Q$ which *resists* it:

$$F = Q \times \frac{p + 6.2832\,\mu r}{6.2832\,r - \mu p} \times \frac{r}{R}$$

**Center of Gravity.** — The center of gravity of a body, volume, area, or line is that point at which if the body, volume, area, or line were suspended it would be perfectly balanced in all positions. For symmetrical bodies of uniform material it is at the geometric center. The center of gravity of a uniform round rod, for example, is at the center of its diameter halfway along its length; the center of gravity of a sphere is at the center of the sphere. For solids, areas, and arcs that are not symmetrical, the determination of the center of gravity may be made experimentally or may be calculated by the use of formulas.

The tables that follow give such formulas for some of the more important shapes. For more complicated and unsymmetrical shapes the methods outlined on page 309 may be used.

*Example:* A piece of wire is bent into the form of a semi-circular arc of 10-inch radius. How far from the center of the arc is the center of gravity located?

Accompanying the third diagram on page 304 is a formula for the distance from the center of gravity of an arc to the center of the arc: $a = 2r \div \pi$. Therefore, in this case,

$$a = 2 \times 10 \div 3.1416 = 6.366 \text{ inches.}$$

Center of Gravity

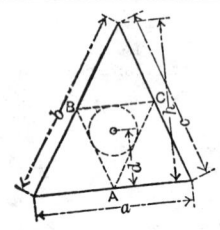

*Perimeter of a Triangle.* — If $A$, $B$ and $C$ are the middle points of the sides of the triangle, then the center of gravity is at the center of the circle that can be inscribed in triangle $ABC$. The distance $d$ of the center of gravity from side $a$ is:

$$d = \frac{h(b+c)}{2(a+b+c)}$$

where $h$ is the height perpendicular to $a$.

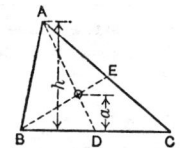

*Area of Triangle.* — The center of gravity is at the intersection of lines $AD$ and $BE$, which bisect the sides $BC$ and $AC$. The perpendicular distance from the center of gravity to any one of the sides is equal to one-third the height perpendicular to that side. Hence, $a = h \div 3$.

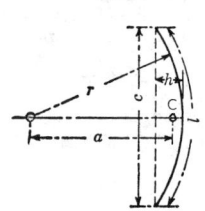

*Circular Arc.* — The center of gravity is on the line that bisects the arc, at a distance

$$a = \frac{r \times c}{l} = \frac{c(c^2 + 4h^2)}{8lh}$$ from the center of the circle.

For an arc equal to one-half the periphery:

$$a = 2r \div \pi = 0.6366\,r$$

For an arc equal to one-quarter of the periphery:

$$a = 2r\sqrt{2} \div \pi = 0.9003\,r$$

For an arc equal to one-sixth of the periphery:

$$a = 3r \div \pi = 0.9549\,r$$

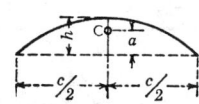

*Circular Arc* (approximate). —

$$a = \tfrac{2}{3}h$$

This formula is very nearly exact for all arcs less than one-quarter of the periphery. The error is only about one per cent for a quarter circle, and decreases for smaller arcs.

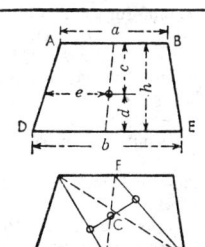

*Area of Trapezoid.* — The center of gravity is on the line joining the middle points of parallel lines $AB$ and $DE$.

$$c = \frac{h(a + 2b)}{3(a+b)} \qquad d = \frac{h(2a+b)}{3(a+b)}$$

$$e = \frac{a^2 + ab + b^2}{3(a+b)}$$

The trapezoid can also be divided into two triangles. The center of gravity is at the intersection of the line joining the centers of gravity of the triangles, and the middle line $FG$.

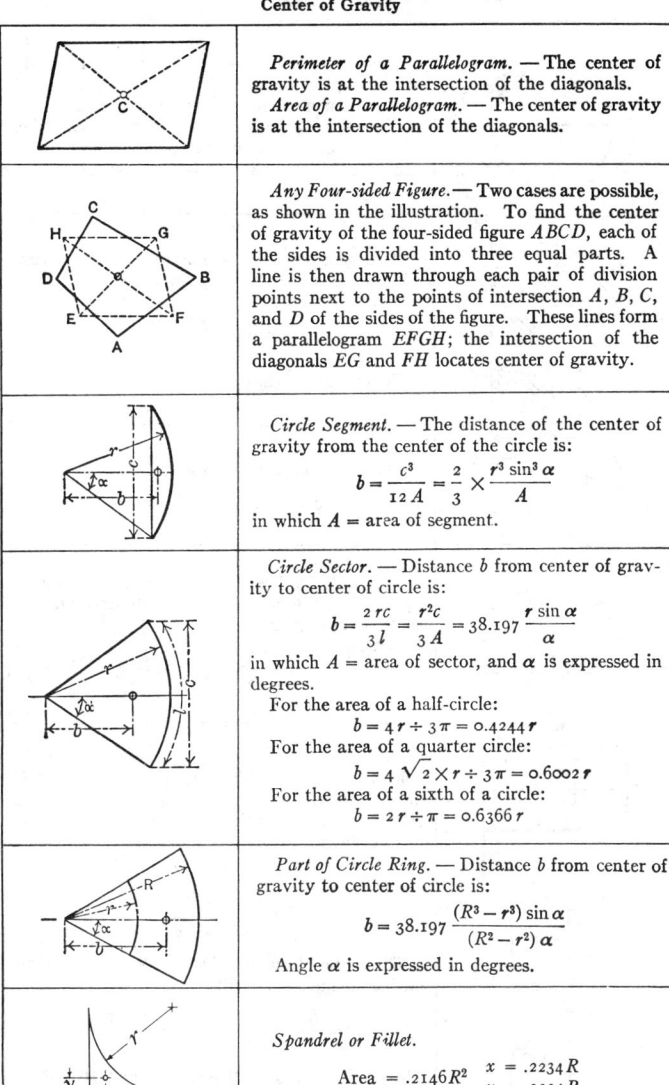

*Perimeter of a Parallelogram.* — The center of gravity is at the intersection of the diagonals.

*Area of a Parallelogram.* — The center of gravity is at the intersection of the diagonals.

*Any Four-sided Figure.* — Two cases are possible, as shown in the illustration. To find the center of gravity of the four-sided figure $ABCD$, each of the sides is divided into three equal parts. A line is then drawn through each pair of division points next to the points of intersection $A$, $B$, $C$, and $D$ of the sides of the figure. These lines form a parallelogram $EFGH$; the intersection of the diagonals $EG$ and $FH$ locates center of gravity.

*Circle Segment.* — The distance of the center of gravity from the center of the circle is:
$$b = \frac{c^3}{12A} = \frac{2}{3} \times \frac{r^3 \sin^3 \alpha}{A}$$
in which $A$ = area of segment.

*Circle Sector.* — Distance $b$ from center of gravity to center of circle is:
$$b = \frac{2rc}{3l} = \frac{r^2c}{3A} = 38.197 \frac{r \sin \alpha}{\alpha}$$
in which $A$ = area of sector, and $\alpha$ is expressed in degrees.

For the area of a half-circle:
$$b = 4r \div 3\pi = 0.4244r$$
For the area of a quarter circle:
$$b = 4\sqrt{2} \times r \div 3\pi = 0.6002r$$
For the area of a sixth of a circle:
$$b = 2r \div \pi = 0.6366r$$

*Part of Circle Ring.* — Distance $b$ from center of gravity to center of circle is:
$$b = 38.197 \frac{(R^3 - r^3) \sin \alpha}{(R^2 - r^2) \alpha}$$
Angle $\alpha$ is expressed in degrees.

*Spandrel or Fillet.*

Area = $.2146R^2$    $x = .2234R$
    $y = .2234R$

### Center of Gravity

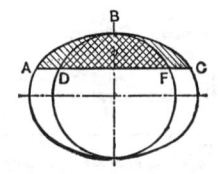

*Segment of an Ellipse.* — The center of gravity of an elliptic segment $ABC$, symmetrical about one of the axes, coincides with the center of gravity of the segment $DBF$ of a circle, the diameter of which is equal to that axis of the ellipse about which the elliptic segment is symmetrical.

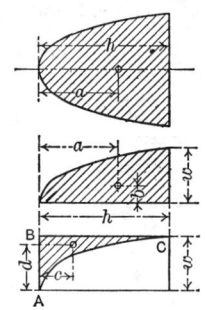

*Area of a Parabola.* — For the complete parabolic area, the center of gravity is on the center line or axis, and

$$a = \frac{3\,h}{5}$$

For one-half of the parabola:

$$a = \frac{3\,h}{5} \quad \text{and} \quad b = \frac{3\,w}{8}$$

For the complement area $ABC$:

$$c = 0.3\,h \quad \text{and} \quad d = 0.75\,w$$

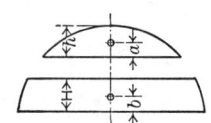

*Spherical Surface of Segments and Zones of Spheres.* — Distances $a$ and $b$ which determine the center of gravity, are:

$$a = \frac{h}{2} \qquad b = \frac{H}{2}$$

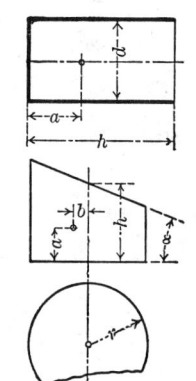

*Cylinder.* — The center of gravity of a solid cylinder (or prism) with parallel end surfaces, is located at the middle of the line that joins the centers of gravity of the end surfaces.

The center of gravity of a cylindrical surface or shell, with the base or end surface in one end, is found from:

$$a = \frac{2\,h^2}{4\,h + d}$$

The center of gravity of a cylinder cut off by an inclined plane is located by:

$$a = \frac{h}{2} + \frac{r^2 \tan^2 \alpha}{8\,h} \qquad b = \frac{r^2 \tan \alpha}{4\,h}$$

where $\alpha$ is the angle between the obliquely cut off surface and the base surface.

## Center of Gravity

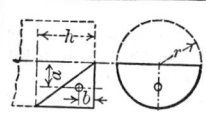

*Portion of Cylinder.* — For a solid portion of a cylinder, as shown, the center of gravity is determined by:

$$a = \tfrac{3}{16} \times 3.1416\, r \qquad b = \tfrac{3}{32} \times 3.1416\, h$$

For the cylindrical surface only:

$$a = \tfrac{1}{4} \times 3.1416\, r \qquad b = \tfrac{1}{8} \times 3.1416\, h$$

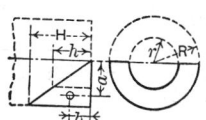

If the cylinder is hollow, the center of gravity of the solid shell is found by:

$$a = \tfrac{3}{16} \times 3.1416\, \frac{R^4 - r^4}{R^3 - r^3}; \quad b = \tfrac{3}{32} \times 3.1416\, \frac{H^4 - h^4}{H^3 - h^3}$$

---

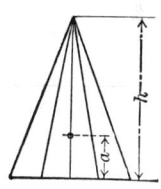

*Pyramid.* — In a solid pyramid the center of gravity is located on the line joining the apex with the center of gravity of the base surface, at a distance from the base equal to one-quarter of the height; or $a = \tfrac{1}{4}\, h$.

The center of gravity of the triangular surfaces forming the pyramid is located on the line joining the apex with the center of gravity of the base surface, at a distance from the base equal to one-third of the height; or $a = \tfrac{1}{3}\, h$.

---

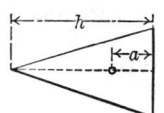

*Cone.* — The same rules apply as for the pyramid. For the solid cone:

$$a = \tfrac{1}{4}\, h$$

For the conical surface:

$$a = \tfrac{1}{3}\, h$$

---

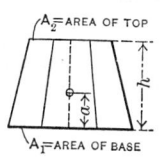

A₂ = AREA OF TOP

A₁ = AREA OF BASE

*Frustum of Pyramid.* — The center of gravity is located on the line that joins the centers of gravity of the end surfaces. If $A_1$ = area of base surface, and $A_2$ area of top surface,

$$a = \frac{h\left(A_1 + 2\sqrt{A_1 \times A_2} + 3\,A_2\right)}{4\left(A_1 + \sqrt{A_1 \times A_2} + A_2\right)}$$

---

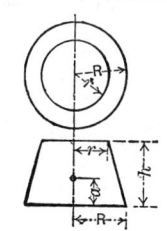

*Frustum of Cone.* — The same rules apply as for the frustum of a pyramid. For a solid frustum of a circular cone the formula below is also used:

$$a = \frac{h\left(R^2 + 2\,Rr + 3\,r^2\right)}{4\left(R^2 + Rr + r^2\right)}$$

The location of the center of gravity of the conical surface of a frustum of a cone is determined by:

$$a = \frac{h\left(R + 2\,r\right)}{3\left(R + r\right)}$$

### Center of Gravity

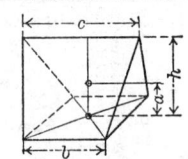

*Wedge.* — The center of gravity is on the line joining the center of gravity of the base with the middle point of the edge, and is located at:

$$a = \frac{h(b+c)}{2(2b+c)}$$

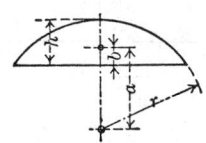

*Spherical Segment.* — The center of gravity of a solid segment is determined by:

$$a = \frac{3(2r-h)^2}{4(3r-h)}$$

$$b = \frac{h(4r-h)}{4(3r-h)}$$

For a half-sphere, $a = b = \frac{3}{8}r$

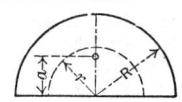

*Half of a Hollow Sphere.* — The center of gravity is located at:

$$a = \frac{3(R^4-r^4)}{8(R^3-r^3)}$$

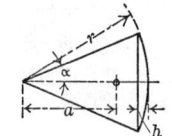

*Spherical Sector.* — The center of gravity of a solid sector is at:

$$a = \frac{3}{8}(1+\cos\alpha)\,r = \frac{3}{8}(2r-h)$$

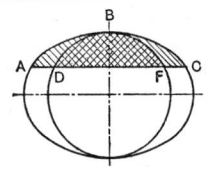

*Segment of Ellipsoid or Spheroid.* — The center of gravity of a solid segment $ABC$, symmetrical about the axis of rotation, coincides with the center of gravity of the segment $DBF$ of a sphere, the diameter of which is equal to the axis of rotation of the spheroid.

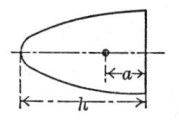

*Paraboloid.* — The center of gravity of a solid paraboloid of rotation is at:

$$a = \frac{1}{3}h$$

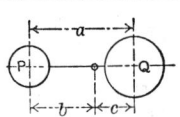

*Center of Gravity of Two Bodies.* — If the weights of the bodies are $P$ and $Q$, and the distance between their centers of gravity is $a$, then:

$$b = \frac{Qa}{P+Q} \qquad c = \frac{Pa}{P+Q}$$

**Center of Gravity of Figures of any Outline.** — If the figure is symmetrical about a center line, as in Fig. 1, the center of gravity will be located on that line. To find the exact location on that line, the simplest method is by taking moments with reference to any convenient axis at right angles to this center line. Divide the area into geometrical figures, the centers of gravity of which can be easily found. In this case, divide the figure into three rectangles $KLMN$, $EFGH$ and $OPRS$. Call the areas of these rectangles $A$, $B$ and $C$, respectively, and find the center of gravity of each. Then select any convenient axis, as $XX$, at right angles to the center line $YY$, and determine distances $a$, $b$ and $c$. The distance $y$ of the center of gravity of the complete figure from the axis $XX$ is then found from the equation:

$$y = \frac{Aa + Bb + Cc}{A + B + C}$$

As an example, assume that the area $A$ is 24 square inches, $B$, 14 square inches,

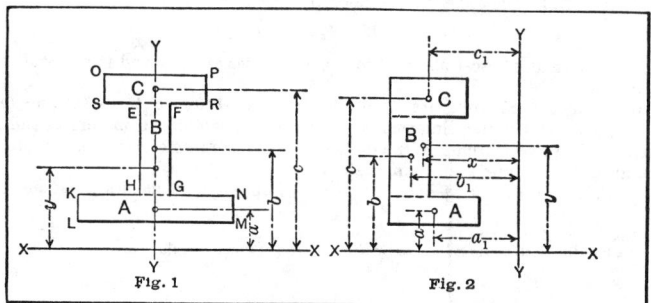

Fig. 1   Fig. 2

and $C$, 16 square inches, and that $a = 3$ inches, $b = 7.5$ inches, and $c = 12$ inches. Then:

$$y = \frac{24 \times 3 + 14 \times 7.5 + 16 \times 12}{24 + 14 + 16} = \frac{369}{54} = 6.83 \text{ inches.}$$

If the figure, the center of gravity of which is to be found, is not symmetrical about any axis, then moments must be taken with relation to two axes $XX$ and $YY$, as shown in Fig. 2. The figure is divided into convenient geometrical figures, the centers of gravity of which can be easily found, the same as before. The center of gravity is determined by the equations:

$$x = \frac{Aa_1 + Bb_1 + Cc_1}{A + B + C} \qquad y = \frac{Aa + Bb + Cc}{A + B + C}$$

As an example, let $A = 14$ square inches, $B = 18$ square inches, and $C = 20$ square inches. Let $a = 3$ inches, $b = 7$ inches, and $c = 11.5$ inches. Let $a_1 = 6.5$ inches, $b_1 = 8.5$ inches, and $c_1 = 7$ inches. Then:

$$x = \frac{14 \times 6.5 + 18 \times 8.5 + 20 \times 7}{14 + 18 + 20} = \frac{384}{52} = 7.38 \text{ inches.}$$

$$y = \frac{14 \times 3 + 18 \times 7 + 20 \times 11.5}{14 + 18 + 20} = \frac{398}{52} = 7.65 \text{ inches.}$$

In other words, the center of gravity is located at a distance of 7.65 inches from the axis $XX$ and 7.38 inches from the axis $YY$.

**Moments of Inertia.** — An important property of areas and solid bodies is the moment of inertia. Standard formulas are derived by multiplying elementary particles of area or mass by the squares of their distances from reference axes. Moments of inertia, therefore, depend on the location of reference axes. Values are minimum when these axes pass through the centers of gravity.

Three kinds of moments of inertia occur in engineering formulas:

(1) *Moments of inertia of plane areas*, $I$, in which the axis is in the plane of the area, are found in formulas for calculating deflections and stresses in beams. When dimensions are given in inches, the units of $I$ are inches[4]. Table of formulas for calculating the $I$ of common areas can be found in the Strength of Materials section.

(2) *Polar moments of inertia of plane areas*, $J$, in which the axis is at right angles to the plane of the area, occur in formulas for the torsional strength of shafting. When dimensions are given in inches, the units of $J$ are inches[4]. If moments of inertia, $I$, are known for a plane area with respect to both $x$ and $y$ axes, then the polar moment for the $z$ axis may be calculated using the equation,

$$J_z = I_x + I_y$$

Tables of formulas for calculating $J$ for common areas can be found in the Shafting section.

When metric SI units are used, the formulas referred to in (1) and (2) above, are valid if the dimensions are given consistently in meters or millimeters. If meters are used, the units of $I$ and $J$ are in meters[4]; if millimeters are used, these units are in millimeters[4].

(3) *Polar moments of inertia of masses*, $J_M$,[*] appear in dynamics equations in-

---

[*] In some books the symbol $I$ denotes the polar moment of inertia of masses; $J_M$ is used in this handbook to avoid confusion with moments of inertia of plane areas.

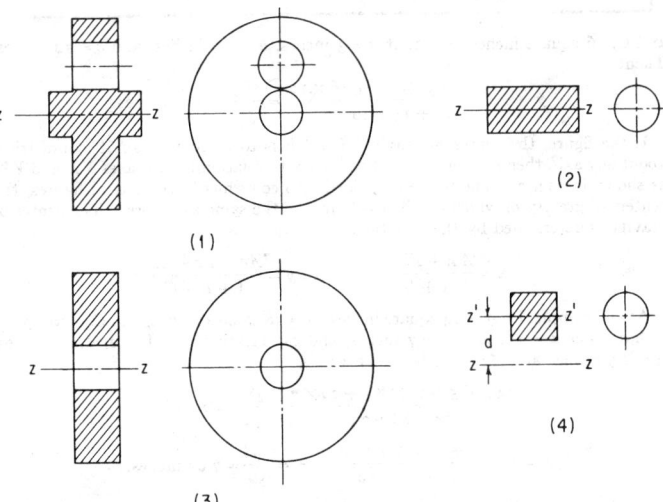

Moments of inertia of complex masses.

volving rotational motion. $J_M$ bears the same relationship to angular acceleration as mass does to linear acceleration. If units are in the foot-pound-second system, the units of $J_M$ are ft-lbs-sec² or slug-ft². (1 slug = 1 pound second² per foot.) If units are in the inch-pound-second system, the units of $J_M$ are inch-lbs-sec².

**If metric SI values are used, the units of $J_M$ are kilogram-meter squared.**

Formulas for calculating $J_M$ for various bodies are given beginning on page 312. If the polar moment of inertia $J$ is known for the area of a body of constant cross section, $J_M$ may be calculated using the equation,

$$J_M = \frac{\rho L}{g} J$$

where $\rho$ is the density of the material, $L$ the length of the part, and $g$ the gravitational constant. If dimensions are in the foot-pound-second system, $\rho$ is in lbs per ft³, $L$ is in ft, $g$ is 32.16 ft per sec², and $J$ is in ft⁴. If dimensions are in the inch-pound-second system, $\rho$ is in lbs per in³, $L$ in inches, $g$ is 386 inches per sec², and $J$ is in inches⁴.

**Using metric SI units, the above formula becomes $J_M = \rho L J$, where $\rho$ = the density in kilograms/meter³, $L$ = the length in meters, and $J$ = the polar moment of inertia in meters⁴. The units of $J_M$ are kg·m²**

*Moments of inertia of complex areas and masses* may be evaluated by the addition and subtraction of elementary areas and masses. For example, the accompanying figure shows a complex mass at (1); its mass polar moment of inertia can be determined by adding together the moments of inertia of the bodies shown at (2) and (3), and subtracting that at (4). Thus, $J_{M1} = J_{M2} + J_{M3} - J_{M4}$. All of these moments of inertia are with respect to the axis of rotation $z$-$z$. Formulas for $J_{M2}$ and $J_{M3}$ can be obtained from the tables beginning on page 312. The moment of inertia for the body at (4) can be evaluated by using the following transfer-axis equation: $J_{M4} = J_{M4}' + d^2 M$. The term $J_{M4}'$ is the moment of inertia with respect to axis $z'$-$z'$; it may be evaluated using the same equation that applies to $J_{M2}$. $d$ is the distance between the $z$-$z$ and the $z'$-$z'$ axes.

Similar calculations can be made when calculating $I$ and $J$ for complex areas. In these cases the appropriate transfer-axis equations are: $I = I' + d^2 A$ and $J = J' + d^2 A$. The primed term, $I'$ or $J'$, is with respect to the center of gravity of the corresponding area $A$; $d$ is the distance between the axis through the center of gravity and the axis to which $I$ or $J$ is referred.

**Radius of Gyration.** — The radius of gyration with reference to an axis is that distance from the axis at which the entire mass of a body may be considered as concentrated, the moment of inertia, meanwhile, remaining unchanged. If $W$ is the weight of a body; $J_M$, its moment of inertia with respect to some axis; and $k_o$, the radius of gyration with respect to the same axis, then:

$$k_o = \sqrt{\frac{J_M g}{W}} \quad \text{and} \quad J_M = \frac{W k_o^2}{g}$$

**When using metric SI units, the formulas are:**

$$k_o = \sqrt{\frac{J_M}{M}} \text{ and } J_M = M k_o^2,$$

**where $k_o$ = the radius of gyration in meters, $J_M$ = kilogram-meter squared, and $M$ = mass in kilograms.**

To find the radius of gyration of an area, as the cross-section of a beam, divide the moment of inertia of the area by the area and extract the square root.

## Moments Of Inertia, $J_M$

($J_M$ = polar moment of inertia of masses, see page 310.  $M$ = mass of body.)

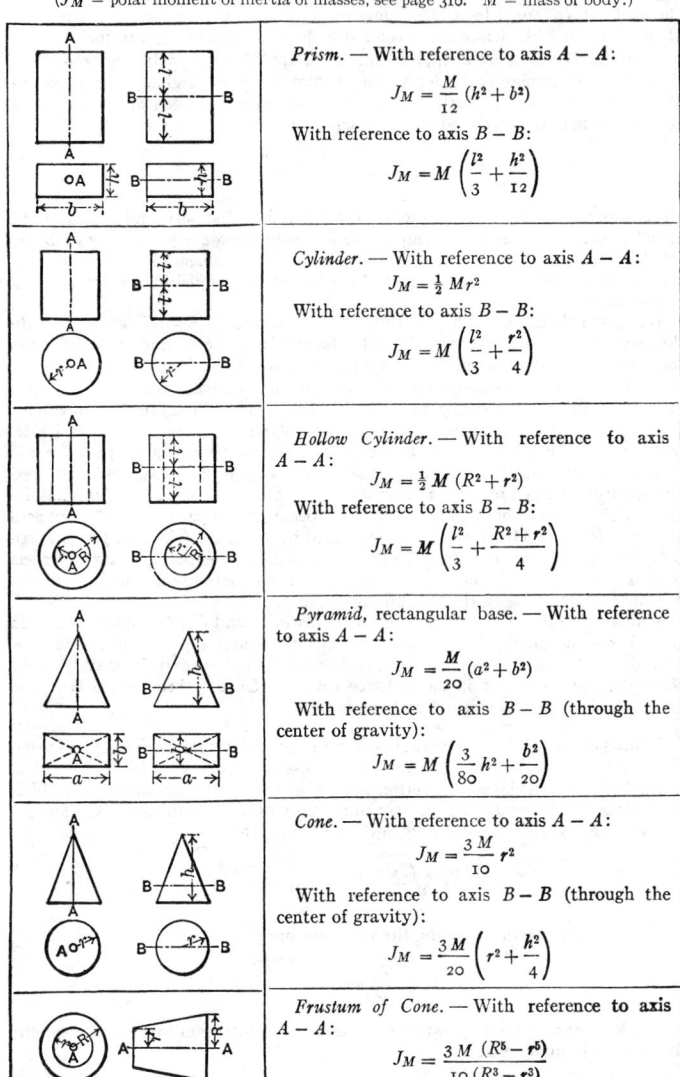

**Prism.** — With reference to axis $A - A$:

$$J_M = \frac{M}{12}\,(h^2 + b^2)$$

With reference to axis $B - B$:

$$J_M = M\left(\frac{l^2}{3} + \frac{h^2}{12}\right)$$

**Cylinder.** — With reference to axis $A - A$:

$$J_M = \tfrac{1}{2}\,Mr^2$$

With reference to axis $B - B$:

$$J_M = M\left(\frac{l^2}{3} + \frac{r^2}{4}\right)$$

**Hollow Cylinder.** — With reference to axis $A - A$:

$$J_M = \tfrac{1}{2}\,M\,(R^2 + r^2)$$

With reference to axis $B - B$:

$$J_M = M\left(\frac{l^2}{3} + \frac{R^2 + r^2}{4}\right)$$

**Pyramid**, rectangular base. — With reference to axis $A - A$:

$$J_M = \frac{M}{20}\,(a^2 + b^2)$$

With reference to axis $B - B$ (through the center of gravity):

$$J_M = M\left(\frac{3}{80}\,h^2 + \frac{b^2}{20}\right)$$

**Cone.** — With reference to axis $A - A$:

$$J_M = \frac{3\,M}{10}\,r^2$$

With reference to axis $B - B$ (through the center of gravity):

$$J_M = \frac{3\,M}{20}\left(r^2 + \frac{h^2}{4}\right)$$

**Frustum of Cone.** — With reference to axis $A - A$:

$$J_M = \frac{3\,M}{10}\frac{(R^5 - r^5)}{(R^3 - r^3)}$$

## Moments of Inertia, $J_M$

($J_M$ = polar moment of inertia of masses, see page 310.  $M$ = mass of body.)

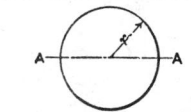

*Sphere.* — With reference to any axis through the center:

$$J_M = \tfrac{2}{5} M r^2$$

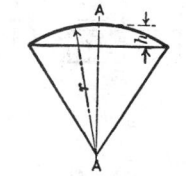

*Spherical Sector.* — With reference to axis $A - A$:

$$J_M = \frac{M}{5} (3 r h - h^2)$$

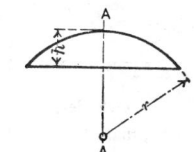

*Spherical Segment.* — With reference to axis $A - A$:

$$J_M = M \left( r^2 - \frac{3 r h}{4} + \frac{3 h^2}{20} \right) \frac{2 h}{3 r - h}$$

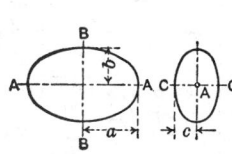

*Ellipsoid.* — With reference to axis $A - A$:

$$J_M = \frac{M}{5} (b^2 + c^2)$$

With reference to axis $B - B$:

$$J_M = \frac{M}{5} (a^2 + c^2)$$

With reference to axis $C - C$:

$$J_M = \frac{M}{5} (a^2 + b^2)$$

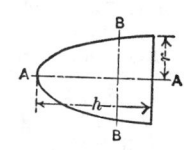

*Paraboloid.* — With reference to axis $A - A$:

$$J_M = \tfrac{1}{3} M r^2$$

With reference to axis $B - B$ (through the center of gravity):

$$J_M = M \left( \frac{r^2}{6} + \frac{h^2}{18} \right)$$

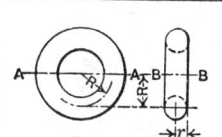

*Torus.* — With reference to axis $A - A$:

$$J_M = M \left( \frac{R^2}{2} + \frac{5 r^2}{8} \right)$$

With reference to axis $B - B$:

$$J_M = M \left( R^2 + \tfrac{3}{4} r^2 \right)$$

### Radius of Gyration

| | | |
|---|---|---|
| **Bar of Small Diameter.**<br>**Axis at end.**<br>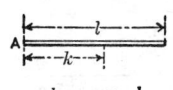<br>$k = 0.5773\,l$<br>$k^2 = \frac{1}{3}\,l^2$<br><br>**Axis at center.**<br>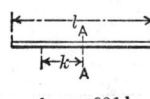<br>$k = 0.2886\,l$<br>$k^2 = \frac{1}{12}\,l^2$ | **Thin Circular Disk.**<br>**Axis through center.**<br>**Cylinder.**<br>**Axis through center.**<br>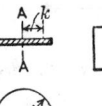<br><br>$k = 0.7071\,r$<br>$k^2 = \frac{1}{2}\,r^2$ | **Cylinder.**<br>**Axis, diameter at mid-length.**<br>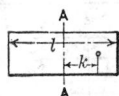 <br>$k = 0.289\sqrt{l^2 + 3r^2}$<br>$k^2 = \frac{l^2}{12} + \frac{r^2}{4}$ |
| **Bar of Small Diameter,<br>bent to Circular Shape.**<br>**Axis, a diameter of the ring.**<br>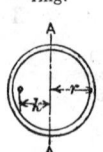<br>$k = 0.7071\,r$<br>$k^2 = \frac{1}{2}\,r^2$ | **Thin Circular Disk.**<br>**Axis its diameter.**<br>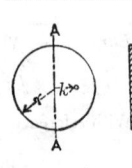<br>$k = \frac{1}{2}\,r$<br>$k^2 = \frac{1}{4}\,r^2$ | **Cylinder.**<br>**Axis, diameter at end.**<br>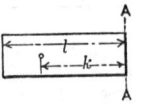<br>$k = 0.289\sqrt{4l^2 + 3r^2}$<br>$k^2 = \frac{l^2}{3} + \frac{r^2}{4}$ |
| **Bar of Small Diameter,<br>bent to Circular Shape.**<br>**Axis through center of ring.**<br>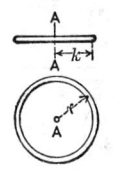<br>$k = r; \quad k^2 = r^2$ | **Parallelogram (Thin flat plate).**<br>**Axis at base.**<br>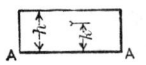<br>$k = 0.5773\,h; \quad k^2 = \frac{1}{3}\,h^2$<br><br>**Axis at mid-height.**<br>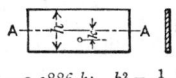<br>$k = 0.2886\,h; \quad k^2 = \frac{1}{12}\,h^2$ | **Thin, Flat, Circular Ring.**<br>**Axis its diameter.**<br>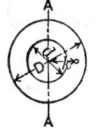 <br>$k = \frac{1}{4}\sqrt{D^2 + d^2}$<br>$k^2 = \frac{D^2 + d^2}{16}$ |

**Radius of Gyration**

Thin Hollow Cylinder.
Axis, diameter at mid-
length.

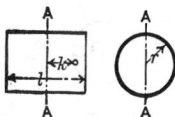

$$k = 0.289 \sqrt{l^2 + 6 r^2}$$

$$k^2 = \frac{l^2}{12} + \frac{r^2}{2}$$

Cylinder.
Axis at a distance.

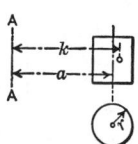

$$k = \sqrt{a^2 + \tfrac{1}{2} r^2}$$

$$k^2 = a^2 + \tfrac{1}{2} r^2$$

Parallelepiped.
Axis at distance from end.

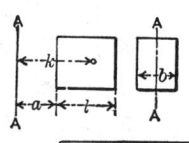

$$k = \sqrt{\frac{4 l^2 + b^2}{12} + a^2 + al}$$

Hollow Cylinder.
Longitudinal Axis.

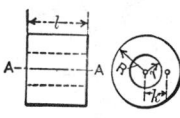

$$k = 0.7071 \sqrt{R^2 + r^2}$$

$$k^2 = \tfrac{1}{2} (R^2 + r^2)$$

Rectangular Prism.
Axis through center.

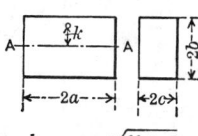

$$k = 0.577 \sqrt{b^2 + c^2}$$

$$k^2 = \tfrac{1}{3} (b^2 + c^2)$$

Cone.
Axis at base.

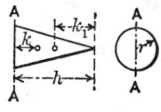

$$k = \sqrt{\frac{2 h^2 + 3 r^2}{20}}$$

Axis at apex.

$$k_1 = \sqrt{\frac{12 h^2 + 3 r^2}{20}}$$

Hollow Cylinder.
Axis, diameter at mid-
length.

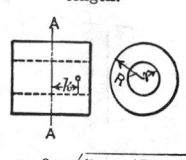

$$k = 0.289 \sqrt{l^2 + 3 (R^2 + r^2)}$$

$$k^2 = \frac{l^2}{12} + \frac{R^2 + r^2}{4}$$

Parallelepiped.
Axis at one end, central.

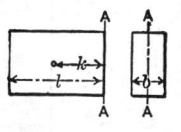

$$k = 0.289 \sqrt{4 l^2 + b^2}$$

$$k^2 = \frac{4 l^2 + b^2}{12}$$

Cone.
Axis through its center line.

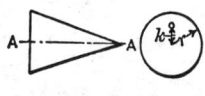

$$k = 0.5477 r$$

$$k^2 = 0.3 r^2$$

### Radius of Gyration

Frustum of Cone.
Axis at large end.

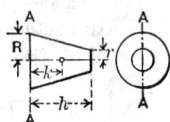

$$k = \sqrt{\frac{h^2}{10}\left(\frac{R^2 + 3\,Rr + 6\,r^2}{R^2 + Rr + r^2}\right) + \frac{3}{20}\left(\frac{R^5 - r^5}{R^3 - r^3}\right)}$$

Sphere.
Axis at a distance.

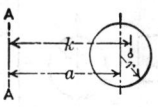

$$k = \sqrt{a^2 + \tfrac{2}{5}\,r^2}$$
$$k^2 = a^2 + \tfrac{2}{5}\,r^2$$

Sphere.
Axis its diameter.

$$k = 0.6325\,r; \quad k^2 = \tfrac{2}{5}\,r^2$$

Thin Spherical Shell.
$$k = 0.8165\,r; \quad k^2 = \tfrac{2}{3}\,r^2$$

Ellipsoid.
Axis through center.

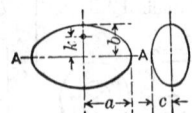

$$k = 0.447\,\sqrt{b^2 + c^2}$$
$$k^2 = \tfrac{1}{5}\,(b^2 + c^2)$$

Hollow Sphere.
Axis its diameter.

$$k = 0.6325\,\sqrt{\frac{R^5 - r^5}{R^3 - r^3}}$$

$$k^2 = \frac{2\,(R^5 - r^5)}{5\,(R^3 - r^3)}$$

Paraboloid.
Axis through center.

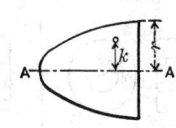

$$k = 0.5773\,r$$
$$k^2 = \tfrac{1}{3}\,r^2$$

When the axis, with reference to which the radius of gyration is taken, passes through the center of gravity, the radius of gyration is the least possible and is called the *principal* radius of gyration. If $k$ is the radius of gyration with respect to such an axis passing through the center of gravity of a body, then the radius of gyration, $k_o$, with respect to a parallel axis at a distance $d$ from the gravity axis is given by:

$$k_o = \sqrt{k^2 + d^2}$$

A table of radii of gyration for various bodies and axes is given beginning on page 314.

**Center and Radius of Oscillation.** — If a body oscillates about a horizontal axis which does not pass through its center of gravity, there will be a point on the line drawn from the center of gravity, perpendicular to the axis, the motion of which will be the same as if the whole mass were concentrated at that point. This point is called the *center of oscillation*. The *radius of oscillation* is the distance between the center of oscillation and the point of suspension. In a straight line, or in a bar of small diameter, suspended at one end and oscillating about it, the center of oscillation is at two-thirds the length of the rod from the end by which it is suspended.

When the vibrations are perpendicular to the plane of the figure, and the figure is suspended by the vertex of an angle or its uppermost point, the radius of oscillation of an isosceles triangle is equal to ¾ of the height of the triangle; of a circle, ⅚ of the diameter; of a parabola, ⁵⁄₇ of the height.

If the vibrations are in the plane of the figure, then the radius of oscillation of a circle equals ¾ of the diameter; of a rectangle, suspended at the vertex of one angle, ⅔ of the diagonal.

**Center of Percussion.** — For a body that moves without rotation, the resultant of all the forces acting on the body passes through the center of gravity. On the other hand, for a body that rotates about some *fixed axis*, the resultant of all the forces acting on it does not pass through the center of gravity of the body but through a point called the *center of percussion*. The center of percussion is useful in determining the position of the resultant in mechanics problems involving angular acceleration of bodies about a fixed axis.

*Finding the Center of Percussion when the Radius of Gyration and the Location of the Center of Gravity are Known:* The center of percussion lies on a line drawn through the center of rotation and the center of gravity. The distance from the axis of rotation to the center of percussion may be calculated from the following formula in which $q$ = distance from the axis of rotation to the center of percussion; $k_o$ = the radius of gyration of the body *with respect to the axis of rotation;* and $r$ = the distance from the axis of rotation to the center of gravity of the body.

$$q = k_o^2 \div r$$

## Velocity and Acceleration

Motion is a progressive change of position of a body. Velocity is the rate of motion, that is, the rate of change of position. When the velocity of a body is the same at every moment during which the motion takes place, the latter is called *uniform* motion. When the velocity is variable and constantly increasing, the rate at which it changes is called *acceleration;* that is, acceleration is the rate at which the velocity of a body changes in a unit of time, as the change in feet per second, in one second. When the motion is decreasing instead of increasing, it is called *retarded* motion, and the rate at which the motion is retarded is frequently called the *deceleration.* If the acceleration is uniform, the motion is called *uniformly accelerated* motion. An example of such motion is found in that of falling bodies.

**Motion with Constant Velocity.** — In the formulas that follow, $S$ = distance moved; $V$ = velocity; $t$ = time of motion, $\theta$ = angle of rotation, and $\omega$ = angular velocity; the usual units for these quantities are, respectively, feet, feet per second, seconds, radians, and radians per second. Any other consistent set of units may be employed.

*Constant Linear Velocity:*

$$S = V \times t; \quad V = S \div t; \quad t = S \div V$$

*Constant Angular Velocity:*

$$\theta = \omega t; \quad \omega = \theta \div t; \quad t = \theta \div \omega$$

*Relation between Angular Motion and Linear Motion:* The relation between the angular velocity of a rotating body and the linear velocity of a point at a distance feet from the center of rotation is:

$$V \text{ (ft per sec)} = r \text{ (ft)} \times \omega \text{ (radians per sec)}$$

Similarly, the distance moved by the point during rotation through angle $\theta$ is:

$$S \text{ (ft)} = r \text{ (ft)} \times \theta \text{ (radians)}$$

**Linear Motion with Constant Acceleration.** — The relations between distance, velocity, and time for linear motion with constant or uniform acceleration are given by the formulas in the accompanying table. In these formulas, the acceleration is assumed to be in the same direction as the initial velocity; hence, if the acceleration in a particular problem should happen to be in a direction opposite that of the initial velocity, then $a$ should be replaced by $-a$. Thus, for example, the formula $V_f = V_o + at$ becomes $V_f = V_o - at$ when $a$ and $V_o$ are opposite in direction.

*Example:* A car is moving at 60 mph when the brakes are suddenly locked and the

### Linear Motion with Constant Acceleration

| To Find | Known | Formula | To Find | Known | Formula |
|---|---|---|---|---|---|
| | | Motion Uniformly Accelerated From Rest | | | |
| $S$ | $a, t$ | $S = \frac{1}{2}at^2$ | $t$ | $S, V_f$ | $t = 2S \div V_f$ |
|  | $V_f, t$ | $S = \frac{1}{2}V_f t$ |  | $S, a$ | $t = \sqrt{2S \div a}$ |
|  | $V_f, a$ | $S = V_f{}^2 \div 2a$ |  | $a, V_f$ | $t = V_f \div a$ |
| $V_f$ | $a, t$ | $V_f = at$ | $a$ | $S, t$ | $a = 2S \div t^2$ |
|  | $S, t$ | $V_f = 2S \div t$ |  | $S, V_f$ | $a = V_f{}^2 \div 2S$ |
|  | $a, S$ | $V_f = \sqrt{2aS}$ |  | $V_f, t$ | $a = V_f \div t$ |
|  | | Motion Uniformly Accelerated From Initial Velocity $V_o$ | | | |
| $S$ | $a, t, V_o$ | $S = V_o t + \frac{1}{2}at^2$ | $t$ | $V_o, V_f, a$ | $t = (V_f - V_o) \div a$ |
|  | $V_o, V_f, t$ | $S = (V_f + V_o)t \div 2$ |  | $V_o, V_f, S$ | $t = 2S \div (V_f + V_o)$ |
|  | $V_o, V_f, a$ | $S = (V_f{}^2 - V_o{}^2) \div 2a$ |  | | |
|  | $V_f, a, t$ | $S = V_f t - \frac{1}{2}at^2$ | $a$ | $V_o, V_f, S$ | $a = (V_f{}^2 - V_o{}^2) \div 2S$ |
| $V_f$ | $V_o, a, t$ | $V_f = V_o + at$ |  | $V_o, V_f, t$ | $a = (V_f - V_o) \div t$ |
|  | $V_o, S, t$ | $V_f = (2S \div t) - V_o$ |  | $V_o, S, t$ | $a = 2(S - V_o t) \div t^2$ |
|  | $V_o, a, S$ | $V_f = \sqrt{V_o{}^2 + 2aS}$ |  | $V_f, S, t$ | $a = 2(V_f t - S) \div t^2$ |
|  | $S, a, t$ | $V_f = (S \div t) + \frac{1}{2}at$ | | | *Meaning of Symbols* |
| $V_o$ | $V_f, a, S$ | $V_o = \sqrt{V_f{}^2 - 2aS}$ | | | $S$ = distance moved in feet; |
|  | $V_f, S, t$ | $V_o = (2S \div t) - V_f$ | | | $V_f$ = final velocity, feet per second; |
|  | $V_f, a, t$ | $V_o = V_f - at$ | | | $V_o$ = initial velocity, feet per second; |
|  | $S, a, t$ | $V_o = (S \div t) - \frac{1}{2}at$ | | | $a$ = acceleration, feet per second per second; |
|  | | | | | $t$ = time of acceleration in seconds. |

car begins to skid.   If it takes 2 seconds to slow the car to 30 mph; at what rate is it being decelerated, how long is it before the car comes to a halt, and how far will it have traveled?

The initial velocity $V_o$ of the car is 60 mph or 88 ft/sec and the acceleration $a$ due to braking is opposite in direction to $V_o$ since the car is slowed to 30 mph or 44 ft/sec.

Since $V_o$, $V_f$ and $t$ are known, $a$ can be determined from the formula

$$a = (V_f - V_o) \div t = (44 - 88) \div 2$$
$$a = -22 \text{ ft/sec}^2$$

The time required to stop the car can be determined from the formula

$$t = (V_f - V_o) \div a = (0 - 88) \div -22$$
$$t = 4 \text{ seconds}$$

The distance traveled by the car is obtained from the formula

$$S = (V_f + V_o)t \div 2 = (0 + 88)4 \div 2$$
$$= 176 \text{ feet}$$

**Rotary Motion with Constant Acceleration.** — The relations between angle of rotation, angular velocity, and time for rotation with constant or uniform acceleration are given in the accompanying table.   In these formulas, the acceleration is assumed to be in the same direction as the initial angular velocity; hence, if the acceleration in a particular problem should happen to be in a direction opposite that of the initial angular velocity, then $\alpha$ should be replaced by $-\alpha$.   Thus, for example, the formula $\omega_f = \omega_o + \alpha t$ becomes $\omega_f = \omega_o - \alpha t$ when $\alpha$ and $\omega_o$ are opposite in direction.

**Rotary Motion with Constant Acceleration**

| To Find | Known | Formula | To Find | Known | Formula |
|---|---|---|---|---|---|
| Motion Uniformly Accelerated From Rest | | | | | |
| $\theta$ | $\alpha, t$ <br> $\omega_f, t$ <br> $\omega_f, \alpha$ | $\theta = \frac{1}{2}\alpha t^2$ <br> $\theta = \frac{1}{2}\omega_f t$ <br> $\theta = \omega_f^2 \div 2\alpha$ | $t$ | $\theta, \omega_f$ <br> $\theta, \alpha$ <br> $\alpha, \omega_f$ | $t = 2\theta \div \omega_f$ <br> $t = \sqrt{2\theta \div \alpha}$ <br> $t = \omega_f \div \alpha$ |
| $\omega_f$ | $\alpha, t$ <br> $\theta, t$ <br> $\alpha, \theta$ | $\omega_f = \alpha t$ <br> $\omega_f = 2\theta \div t$ <br> $\omega_f = \sqrt{2\alpha\theta}$ | $\alpha$ | $\theta, t$ <br> $\theta, \omega_f$ <br> $\omega_f, t$ | $\alpha = 2\theta \div t^2$ <br> $\alpha = \omega_f^2 \div 2\theta$ <br> $\alpha = \omega_f \div t$ |
| Motion Uniformly Accelerated From Initial Velocity $\omega_o$ | | | | | |
| $\theta$ | $\alpha, t, \omega_o$ <br> $\omega_o, \omega_f, t$ <br> $\omega_o, \omega_f, \alpha$ <br> $\omega_f, \alpha, t$ | $\theta = \omega_o t + \frac{1}{2}\alpha t^2$ <br> $\theta = (\omega_f + \omega_o)t \div 2$ <br> $\theta = (\omega_f^2 - \omega_o^2) \div 2\alpha$ <br> $\theta = \omega_f t - \frac{1}{2}\alpha t^2$ | $\alpha$ | $\omega_o, \omega_f, \theta$ <br> $\omega_o, \omega_f, t$ <br> $\omega_o, \theta, t$ <br> $\omega_f, \theta, t$ | $\alpha = (\omega_f^2 - \omega_o^2) \div 2\theta$ <br> $\alpha = (\omega_f - \omega_o) \div t$ <br> $\alpha = 2(\theta - \omega_o t) \div t^2$ <br> $\alpha = 2(\omega_f t - \theta) \div t^2$ |
| $\omega_f$ | $\omega_o, \alpha, t$ <br> $\omega_o, \theta, t$ <br> $\omega_o, \alpha, \theta$ <br> $\theta, \alpha, t$ | $\omega_f = \omega_o + \alpha t$ <br> $\omega_f = (2\theta \div t) - \omega_o$ <br> $\omega_f = \sqrt{\omega_o^2 + 2\alpha\theta}$ <br> $\omega_f = (\theta \div t) + \frac{1}{2}\alpha t$ | *Meaning of Symbols* | | |
| $\omega_o$ | $\omega_f, \alpha, \theta$ <br> $\omega_f, \theta, t$ <br> $\omega_f, \alpha, t$ <br> $\theta, \alpha, t$ | $\omega_o = \sqrt{\omega_f^2 - 2\alpha\theta}$ <br> $\omega_o = (2\theta \div t) - \omega_f$ <br> $\omega_o = \omega_f - \alpha t$ <br> $\omega_o = (\theta \div t) - \frac{1}{2}\alpha t$ | $\theta$ = angular distance of rotation, radians; <br> $\omega_f$ = final angular velocity, radians per second; <br> $\omega_o$ = initial angular velocity, radians per second; <br> $\alpha$ = angular acceleration, radians per second, per second; <br> $t$ = time in seconds. | | |
| $t$ | $\omega_o, \omega_f, \alpha$ <br> $\omega_o, \omega_f, \theta$ | $t = (\omega_f - \omega_o) \div \alpha$ <br> $t = 2\theta \div (\omega_f + \omega_o)$ | 1 degree = 0.01745 radians <br> (See conversion table on page 269) | | |

*Linear Acceleration of a Point on a Rotating Body:* A point on a body rotating about a fixed axis has a linear acceleration $a$ that is the resultant of two component accelerations. The first component is the centripetal or normal acceleration which is directed from the point $P$ toward the axis of rotation; its magnitude is $r\omega^2$ where $r$ is the radius from the axis to the point $P$ and $\omega$ is the angular velocity of the body at the time acceleration $a$ is to be determined. The second component of $a$ is the tangential acceleration which is equal to $r\alpha$ where $\alpha$ is the angular acceleration of the body.

The acceleration of point $P$ is the resultant of $r\omega^2$ and $r\alpha$ and is given by the formula

$$a = \sqrt{(r\omega^2)^2 + (r\alpha)^2}$$

When $\alpha = 0$, this formula reduces to: $a = r\omega^2$

*Example:* A flywheel on a press rotating at 120 rpm is slowed to 102 rpm during a punching operation that requires ¾ second for the punching portion of the cycle. What angular deceleration does the flywheel experience?

From the table on page 348, the angular velocities corresponding to 120 rpm and 102 rpm, respectively, are 12.57 and 10.68 radians per second. Therefore, using the formula

$$\alpha = (\omega_f - \omega_o) \div t$$
$$\alpha = (10.68 - 12.57) \div ¾ = -1.89 \div ¾$$
$$\alpha = -2.52 \text{ radians per second per second}$$

which is, from the table on page 348, 24 rpm per second, per second. The minus sign in the answer indicates that the acceleration $\alpha$ acts to slow the flywheel, that is, the flywheel is being decelerated.


## Force, Work, Energy, and Momentum

**Accelerations Resulting from Unbalanced Forces.** — In the section describing the resolution and composition of forces it was stated that when the resultant of a system of forces is zero, the system is in equilibrium, that is, the body on which the force system acts remains at rest or continues to move with uniform velocity. If, however, the resultant of a system of forces is not zero, the body on which the forces act will be accelerated in the direction of the unbalanced force. To determine the relation between the unbalanced force and the resulting acceleration, Newton's laws of motion must be applied. These laws may be stated as follows:

*First Law:* Every body continues in a state of rest or in uniform motion in a straight line, until it is compelled by a force to change its state of rest or motion.

*Second Law:* Change of motion is proportional to the force applied, and takes place along the straight line in which the force acts. The "force applied" represents the resultant of *all* the forces acting on the body. This law is sometimes worded: An unbalanced force acting on a body causes an acceleration of the body in the direction of the force and of magnitude proportional to the force and inversely proportional to the mass of the body. Stated as a formula, $R = Ma$ where $R$ is the resultant of *all* the forces acting on the body, $M$ is the mass of the body (mass = weight $W$ divided by acceleration due to gravity $g$), and $a$ is the acceleration of the body resulting from application of force $R$.

*Third Law:* To every action there is always an equal reaction, or, in other words, if a force acts to change the state of motion of a body, the body offers a resistance equal and directly opposite to the force.

Newton's second law may be used to calculate linear and angular accelerations of a body produced by unbalanced forces and torques acting on the body; however, it is necessary first to use the methods described under " Composition and Resolution of Forces " to determine the magnitude and direction of the resultant of *all* forces acting on the body. Then, for a body moving with pure translation,

$$R = Ma = \frac{W}{g}\, a$$

where $R$ is the resultant force in pounds acting on a body weighing $W$ pounds; $g$ is the gravitational constant, usually taken as 32.16 ft/sec², approximately; and $a$ is the resulting acceleration in ft/sec² of the body due to $R$ and in the same direction as $R$.

Using metric SI units, the formula is $R = Ma$, where $R$ = force in newtons (N), $M$ = mass in kilograms, and $a$ = acceleration in meters/second squared. It should be noted that the weight of a body of mass $M$ kg is $Mg$ N, where $g$ is approximately 9.81 m/s².

*Free Body Diagram:* In order to correctly determine the effect of forces on the motion of a body it is necessary to resort to what is known as a *free body diagram.* This diagram shows (1) the body removed or isolated from contact with all other bodies that exert force on the body and (2) *all* the forces acting on the body. Thus, for example, in diagram (a) the block being pulled up the plane is acted upon by certain forces; the free body diagram of this block is shown at (b). Note that all

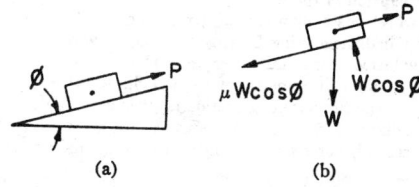

(a)          (b)

forces acting on the block are indicated. These forces include: (1) the force of gravity (weight); (2) the pull of the cable, $P$; (3) the normal component, $W \cos \phi$, of the force exerted on the block by the plane; and (4) the friction force, $\mu W \cos \phi$, of the plane on the block.

In preparing a free body diagram, it is important to realize that only those forces exerted *on* the body being considered are shown; forces exerted by the body on other bodies are disregarded. It is this feature that makes the free body diagram an invaluable aid in the solution of problems in mechanics.

*Example:* A 100-pound body is being hoisted by a winch, the tension in the hoisting cable being kept constant at 110 pounds. At what rate is the body accelerated?

Two forces are acting on the body, its weight, 100 pounds downward, and the pull of the cable, 110 pounds upward. The resultant force $R$, from a free body diagram, is therefore 110 − 100. Thus, applying Newton's second law,

$$110 - 100 = \frac{100}{32.16}\, a$$

$$a = \frac{32.16 \times 10}{100} = 3.216 \text{ ft/sec}^2 \text{ upward}$$

It should be noted that since in this problem the resultant force $R$ was positive (110 − 100 = +10), the acceleration $a$ is also positive, that is, $a$ is in the same direction as $R$, which is in accord with Newton's second law.

*Example using SI metric units:* A body of mass 50 kilograms is being hoisted by a winch, and the tension in the cable is 600 newtons. What is the acceleration? The weight of the 50 kg body is $50g$ N, where $g$ = approximately 9.81 m/s² (see note on page 300). Applying the formula $R = Ma$, the calculation is: $(600 - 50g) = 50a$. Thus,

$$a = \frac{600 - 50g}{50} = \frac{600 - 50 \times 9.81}{50} = 2.19 \text{ m/s}^2.$$

*Formulas Relating Torque and Angular Acceleration:* For a body rotating about a fixed axis the relation between the unbalanced torque acting to produce rotation and the resulting angular acceleration may be determined from any one of the following formulas, each based on Newton's second law:

$$T_o = J_M\alpha$$
$$T_o = Mk_o^2\alpha$$
$$T_o = \frac{Wk_o^2\alpha}{g} = \frac{Wk_o^2\alpha}{32.16}$$

where $T_o$ is the unbalanced torque in pounds-feet; $J_M$ in ft-lbs-sec² is the moment of inertia of the body about the axis of rotation; $k_o$ in feet is the radius of gyration of the body with respect to the axis of rotation, and $\alpha$ in radians per second, per second is the angular acceleration of the body.

*Example:* A flywheel has a diameter of 3 feet and weighs 1000 pounds. What torque must be applied, neglecting bearing friction, to accelerate the flywheel at the rate of 100 revolutions per minute, per second?

From page 312 the moment of inertia of a solid cylinder with respect to a gravity axis at right angles to the circular cross-section is given as $\frac{1}{2} Mr^2$. From page 348 100 rpm = 10.47 radians per second, hence an acceleration of 100 rpm per second = 10.47 radians per second, per second. Therefore, using the first of the preceding formulas,

$$T_o = J_M\alpha = \frac{1}{2} \frac{1000}{32.16} \left(\frac{3}{2}\right)^2 \times 10.47$$
$$= 366 \text{ ft-lbs}$$

Using metric SI units, the formulas are: $T_o = J_M\alpha = Mk_o^2\alpha$, where $T_o$ = torque in newton-meters; $J_M$ = the moment of inertia in kg · m², and $\alpha$ = the angular acceleration in radians per second squared.

*Example:* A flywheel has a diameter of 1.5 m, and a mass of 800 kg. What torque is needed to produce an angular acceleration of 100 revolutions per minute, per second? As in the preceding example, $\alpha$ = 10.47 rad/s². Thus:

$$J_M = \frac{1}{2}Mr^2 = \frac{1}{2} \times 800 \times 0.75^2 = 225 \text{ kg} \cdot \text{m}^2.$$

Therefore: $T_o = J_M\alpha = 225 \times 10.47 = 2356 \text{ N} \cdot \text{m}.$

**Energy.** — A body is said to possess energy when it is capable of doing work or overcoming resistance. The energy may be either mechanical or non-mechanical, the latter including chemical, electrical, thermal, and atomic energy.

Mechanical energy includes *kinetic energy* (energy possessed by a body because of its motion) and *potential energy* (energy possessed by a body because of its position in a field of force and/or its elastic deformation).

*Kinetic Energy:* The motion of a body may be one of pure translation, pure rotation, or a combination of rotation and translation. By translation is meant motion in which every line in the body remains parallel to its original position throughout the motion, that is, no rotation is associated with the motion of the body.

The kinetic energy of a translating body is given by the formula

$$\text{Kinetic Energy in ft lbs due to translation} = E_{KT} = \tfrac{1}{2}MV^2 = \frac{WV^2}{2g} \qquad \text{(a)}$$

where $M$ = mass of body ($= W \div g$); $V$ = velocity of the center of gravity of the body in feet per second; $W$ = weight of body in pounds; and $g$ = acceleration due to gravity = 32.16 feet per second, per second.

The kinetic energy of a body rotating about a fixed axis $O$ is expressed by the formula:

$$\text{Kinetic Energy in ft lbs due to rotation} = E_{KR} = \tfrac{1}{2}J_{MO}\omega^2 \qquad \text{(b)}$$

where $J_{MO}$ is the moment of inertia of the body about the fixed axis $O$ in pounds-feet-seconds$^2$, and $\omega$ = angular velocity in radians per second.

For a body that is moving with both translation and rotation, the total kinetic energy is given by the following formula as the sum of the kinetic energy due to translation of the center of gravity and the kinetic energy due to rotation about the center of gravity:

$$\begin{aligned}
\text{Total Kinetic Energy in ft lbs} = E_T &= \tfrac{1}{2}MV^2 + \tfrac{1}{2}J_{MG}\omega^2 \\
&= \frac{WV^2}{2g} + \frac{1}{2}J_{MG}\omega^2 \\
&= \frac{WV^2}{2g} + \frac{1}{2}\frac{Wk^2\omega^2}{g} \\
&= \frac{W}{2g}(V^2 + k^2\omega^2) \qquad \text{(c)}
\end{aligned}$$

where $J_{MG}$ is the moment of inertia of the body about its gravity axis in pounds-feet-seconds$^2$, $k$ is the radius of gyration in feet with respect to an axis through the center of gravity, and the other quantities are as previously defined.

In the metric SI system, energy is expressed as the joule (J). One joule = 1 newton-meter. The kinetic energy of a translating body is given by the formula $E_{KT} = \tfrac{1}{2}MV^2$, where $M$ = mass in kilograms, and $V$ = velocity in meters per second. Kinetic energy due to rotation is expressed by the formula $E_{KR} = \tfrac{1}{2}J_{MO}\omega^2$, where $J_{MO}$ = moment of inertia in kg·m$^2$, and $\omega$ = the angular velocity in radians per second. Total kinetic energy $E_T = \tfrac{1}{2}MV^2 + \tfrac{1}{2}J_{MO}\omega^2$ joules = $\tfrac{1}{2}M(V^2 + k^2\omega^2)$ joules, where $k$ = radius of gyration in meters.

*Potential Energy:* The most common example of a body having potential energy because of its position in a field of force is that of a body elevated to some height above the earth. Here the field of force is the gravitational field of the earth and the potential energy $E_{PF}$ of a body weighing $W$ pounds elevated to some height $S$ in feet above the surface of the earth is $WS$ foot-pounds. If the body is permitted to drop from this height its potential energy $E_{PF}$ will be converted to kinetic energy. Thus, after falling through height $S$ the kinetic energy of the body will be $WS$ ft-lbs.

In metric SI units, the potential energy $E_{PF}$ of a body of mass $M$ kg elevated to a height of $S$ meters, is $MgS$ joules. After it has fallen a distance $S$, the kinetic energy gained will thus be $MgS$ joules.

Another type of potential energy is elastic potential energy, such as possessed by a spring that has been compressed or extended. The amount of work in ft lbs done in compressing the spring $S$ feet is equal to $KS^2/2$, where $K$ is the spring constant in pounds per foot. Thus, when the spring is released to act against some resistance, it can perform $KS^2/2$ ft-lbs of work which is the amount of elastic potential energy $E_{PE}$ stored in the spring.

Using metric SI units, the amount of work done in compressing the spring a distance $S$ meters is $KS^2/2$ joules, where $K$ is the spring constant in newtons per meter.

**Work Performed by Forces and Couples.** — The work $U$ done by a force $F$ in moving an object along some path is the product of the distance $S$ the body is moved and the component $F \cos \alpha$ of the force $F$ in the direction of $S$.

$$U = FS \cos \alpha$$

where $U$ = work in ft-lbs; $S$ = distance moved in feet; $F$ = force in lbs; and $\alpha$ = angle between line of action of force and the path of $S$.

If the force is in the same direction as the motion, then $\cos \alpha = \cos 0 = 1$ and this formula reduces to:

$$U = FS$$

Similarly, the work done by a couple $M$ turning an object through an angle $\theta$ is:

$$U = M\theta$$

where $M$ = torque of couple in pounds-feet and $\theta$ = the angular rotation in radians.

**The above formulas can be used with metric SI units: $U$ is in joules; $S$ is in meters; $F$ is in newtons, and $M$ is in newton-meters.**

**Relation between Work and Energy.** — Theoretically, when work is performed on a body and there are no energy losses (such as due to friction, air resistance, etc.), the energy acquired by the body is equal to the work performed on the body; this energy may be either potential, kinetic, or a combination of both.

In actual situations, however, there may be energy losses that must be taken into account. Thus, the relation between work done on a body, energy losses, and the energy acquired by the body can be stated as:

$$\text{Work Performed} - \text{Losses} = \text{Energy Acquired}$$

$$U - \text{Losses} = E_T$$

*Example 1:* A 12-inch cube of steel weighing 490 pounds is being moved on a horizontal conveyer belt at a speed of 6 miles per hour (8.8 feet per second). What is the kinetic energy of the cube?

Since the block is not rotating, Formula (a) for the kinetic energy of a body moving with pure translation applies:

$$\text{Kinetic Energy} = \frac{WV^2}{2g}$$

$$= \frac{490 \times (8.8)^2}{2 \times 32.16} = 590 \text{ ft-lbs}$$

**A similar example using metric SI units is as follows: If a cube of mass 200 kg is being moved on a conveyor belt at a speed of 3 meters per second, what is the kinetic energy of the cube? It is:**

$$\text{Kinetic Energy} = \tfrac{1}{2}MV^2 = \tfrac{1}{2} \times 200 \times 3^2 = 900 \text{ joules.}$$

*Example 2:* If the conveyer in Example 1 is brought to an abrupt stop, how long would it take for the steel block to come to a stop and how far along the belt would it slide before stopping if the coefficient of friction $\mu$ between the block and the conveyer belt is 0.2 and the block slides without tipping over?

The only force acting to slow the motion of the block is the friction force between the block and the belt. This force $F$ is equal to the weight of the block, $W$, multiplied by the coefficient of friction; $F = \mu W = 0.2 \times 490 = 98$ lbs.

The time required to bring the block to a stop can be determined from the impulse-momentum formula (c) on page 326.

$$R \times t = \frac{W}{g}(V_f - V_o)$$

$$(-98)t = \frac{490}{32.16} \times (0 - 8.8)$$

$$t = \frac{490 \times 8.8}{98 \times 32.16} = 1.37 \text{ seconds}$$

The distance the block slides before stopping can be determined by equating the kinetic energy of the block and the work done by friction in stopping it:

Kinetic energy of block $(WV^2/2g)$ = Work done by friction $(F \times S)$

$$590 = 98 \times S$$

$$S = \frac{590}{98} = 6.0 \text{ feet}$$

If metric SI units are used, the calculation is as follows (for the cube of 200 kg mass): The friction force = $\mu$ multiplied by the weight $Mg$ where $g$ = approximately 9.81 m/s². Thus, $\mu Mg = 0.2 \times 200g = 392.4$ newtons. The time $t$ required to bring the block to a stop is $(-392.4)t = 200(0 - 3)$. Therefore,

$$t = \frac{200 \times 3}{392.4} = 1.53 \text{ s.}$$

The kinetic energy of the block is equal to the work done by friction, that is $392.4 \times S = 900$ joules. Thus, the distance $S$ which the block moves before stopping is

$$\frac{900}{392.4} = 2.29 \text{ m.}$$

**Force of a Blow.** — A body that weighs $W$ pounds and falls $S$ feet from an initial position of rest is capable of doing $WS$ foot-pounds of work. The work performed during its fall may be, for example, that necessary to drive a pile a distance $d$ into the ground. Neglecting losses in the form of dissipated heat and strain energy, the work done in driving the pile is equal to the product of the impact force acting on the pile and the distance $d$ which the pile is driven. Since the impact force is not accurately known, an average value, called the "average force of the blow," may be assumed. Equating the work done on the pile and the work done by the falling body, which in this case is a pile driver:

Average force of blow $\times d = WS$

or,                          Average force of blow $= \dfrac{WS}{d}$

where, $S$ = total height in feet through which the driver falls, including the distance $d$ that the pile is driven;

$W$ = weight of driver in pounds;

$d$ = distance in feet which pile is driven.

When using metric SI units, it should be noted that a body of mass $M$ kg has a weight of $Mg$ newtons, where $g$ = approximately 9.81 m/s². If the body falls a distance $S$ m, it can do work equal to $MgS$ J. The average force of the blow is $MgS/d$ N, where $d$ is the distance in meters that the pile is driven.

*Example:* — A pile driver weighing 200 pounds strikes the top of the pile after having fallen from a height of 20 feet. It forces the pile into the ground a distance of ½ foot. Before the ram is brought to rest, it will then have performed 200 × (20 + ½) = 4100 foot-pounds of work, and as this energy is expended in a distance of one-half foot, the average force of the blow equals 4100 ÷ ½ = 8200 pounds.

A similar example using metric SI units is as follows: A pile driver of mass 100 kg falls 10 meters and moves the pile a distance of 0.3 meters. The work done = $100g(10 + 0.3)$ J, and it is expended in 0.3 m. Thus, the average force is

$$\frac{100g \times 10.3}{0.3} = 33680 \text{ newtons, or } 33.68 \text{ kN.}$$

**Impulse and Momentum.** — The *linear momentum* of a body is defined as the product of the mass $M$ of the body and the velocity $V$ of the center of gravity of the body:

$$\text{Linear momentum} = MV, \quad \text{or,} \quad \text{since } M = W \div g$$

$$\text{Linear momentum} = \frac{WV}{g} \tag{a}$$

It should be noted that linear momentum is a vector quantity, the momentum being in the same direction as $V$.

*Linear impulse* is defined as the product of the resultant $R$ of *all* the forces acting on a body and the time $t$ that the resultant acts:

$$\text{Linear Impulse} = Rt \tag{b}$$

The change in the linear momentum of a body is numerically equal to the linear impulse that causes the change in momentum:

$$\text{Linear Impulse} = \text{change in Linear Momentum}$$

$$Rt = \frac{W}{g} V_f - \frac{W}{g} V_o = \frac{W}{g} (V_f - V_o) \tag{c}$$

where $V_f$, the final velocity of the body after time $t$ and $V_o$ the initial velocity of the body are both in the same direction as the applied force $R$. If $V_o$ and $V_f$ are in opposite directions, then the minus sign in the formula becomes a plus sign.

**In metric SI units, the formulas are: Linear Momentum** = $MV$ **kg·m/s,** where $M$ = mass in kg, and $V$ = velocity in meters per second; and **Linear Impulse** = $Rt$ **newton-seconds, where** $R$ = force in newtons, and $t$ = time in seconds. In formula (c) above, $W/g$ is replaced by $M$ when SI units are used.

*Example:* A 1000-pound block is pulled up a 2-degree incline by a cable exerting a constant force $F$ of 600 pounds. If the coefficient of friction $\mu$ between the block and the plane is 0.5, how fast will the block be moving up the plane 10 seconds after the pull is applied?

The resultant force $R$ causing the body to be accelerated up the plane is the difference between $F$, the force acting up the plane, and $P$, the force acting to resist motion up the plane. This latter force for a body on a plane is given by the formula at the top of page 299 as $P = W (\mu \cos \alpha + \sin \alpha)$ where $\alpha$ is the angle of the incline.

Thus,
$$R = F - P = F - W(\mu \cos \alpha + \sin \alpha)$$
$$= 600 - 1000(0.5 \cos 2° + \sin 2°)$$
$$= 600 - 1000(0.5 \times 0.99939 + 0.03490)$$
$$= 600 - 535$$
$$R = 65 \text{ pounds.}$$

Formula (c) can now be applied to determine the speed at which the body will be moving up the plane after 10 seconds.

$$Rt = \frac{W}{g} V_f - \frac{W}{g} V_o$$

$$65 \times 10 = \frac{1000}{32.2} V_f - \frac{1000}{32.2} \times 0$$

$$V_f = \frac{65 \times 10 \times 32.2}{1000} = 20.9 \text{ ft per sec}$$

$$= 14.3 \text{ miles per hour}$$

A similar example using metric SI units is as follows: A 500 kg block is pulled up a 2 degree incline by a constant force $F$ of 4 kN. The coefficient of friction $\mu$ between the block and the plane is 0.5. How fast will the block be moving 10 s after the pull is applied?

The resultant force $R$ is:
$$R = F - Mg(\mu \cos \alpha + \sin \alpha)$$
$$= 4000 - 500 \times 9.81(0.5 \times 0.99939 + 0.03490)$$
$$= 1378 \text{ N or } 1.378 \text{ kN.}$$

Formula (c) can now be applied to determine the speed at which the body will be moving up the plane after 10 seconds. Replacing $W/g$ by $M$ in the formula, the calculation is:

$$Rt = MV_f - MV_o$$
$$1378 \times 10 = 500(V_f - 0)$$
$$V_f = \frac{1378 \times 10}{500} = 27.6 \text{ m/s.}$$

*Angular Impulse and Momentum:* In a manner similar to that for linear impulse and moment, the formulas for angular impulse and momentum for a body rotating about a fixed axis are:

$$\text{Angular momentum} = J_M \omega \qquad (a)$$

$$\text{Angular impulse} = M_o t \qquad (b)$$

where $J_M$ is the moment of inertia of the body about the axis of rotation in pounds-feet-seconds², $\omega$ is the angular velocity in radians per second, $M_o$ is the torque in pounds-feet about the axis of rotation, and $t$ is the time in seconds that $M_o$ acts.

The change in angular momentum of a body is numerically equal to the angular impulse that causes the change in angular momentum:

$$\text{Angular Impulse} = \text{Change in Angular Momentum}$$

$$M_o t = J_M \omega_f - J_M \omega_o = J_M(\omega_f - \omega_o) \qquad (c)$$

where $\omega_f$ and $\omega_o$ are the final and initial angular velocities, respectively.

*Example:* A flywheel having a moment of inertia of 25 lbs-ft-sec² is revolving with an angular velocity of 10 radians per second when a constant torque of 20 lbs-ft is applied to reverse its direction of rotation. For what length of time must this constant torque act to stop the flywheel and bring it up to a reverse speed of 5 radians per second?

Applying formula (c),

$$M_o t = J_M(\omega_f - \omega_o)$$

$$20t = 25(10 - [-5]) = 250 + 125$$

$$t = 375 \div 20 = 18.8 \text{ seconds}$$

A similar example using metric SI units is as follows: A flywheel with a moment of inertia of 20 kilogram-meters² is revolving with an angular velocity of 10 radians per second when a constant torque of 30 newton-meters is applied to reverse its direction of rotation. For what length of time must this constant torque act to stop the flywheel and bring it up to a reverse speed of 5 radians per second? Applying formula (c), the calculation is:

$$M_o t = J_M(\omega_f - \omega_o),$$

$$30t = 20(10 - [-5]).$$

Thus, $t = \dfrac{20 \times 15}{30} = 10$ seconds.

**Formulas for Work and Power.** — The formulas in the accompanying table may be used to determine work and power in terms of the applied force and the velocity at the point of application of the force.

### Formulas for Work and Power

| To Find | Known | Formula | To Find | Known | Formula |
|---|---|---|---|---|---|
| $S$ | $P, t, F$ | $S = P \times t \div F$ | $P$ | $F, V$ | $P = F \times V$ |
| | $K, F$ | $S = K \div F$ | | $F, S, t$ | $P = F \times S \div t$ |
| | $t, F, hp$ | $S = 550 \times t \times hp \div F$ | | $K, t$ | $P = K \div t$ |
| | | | | $hp$ | $P = 550 \times hp$ |
| $V$ | $P, F$ | $V = P \div F$ | $K$ | $F, S$ | $K = F \times S$ |
| | $K, F, t$ | $V = K \div (F \times t)$ | | $P, t$ | $K = P \times t$ |
| | $F, hp$ | $V = 550 \times hp \div F$ | | $F, V, t$ | $K = F \times V \times t$ |
| | | | | $t, hp$ | $K = 550 \times t \times hp$ |
| $t$ | $F, S, P$ | $t = F \times S \div P$ | | | |
| | $K, F, V$ | $t = K \div (F \times V)$ | | $F, S, t$ | $hp = F \times S \div (550 \times t)$ |
| | $F, S, hp$ | $t = F \times S \div (550 \times hp)$ | $hp$ | $P$ | $hp = P \div 550$ |
| $F$ | $P, V$ | $F = P \div V$ | | $F, V$ | $hp = F \times V \div 550$ |
| | $K, S$ | $F = K \div S$ | | $K, t$ | $hp = K \div (550 \times t)$ |
| | $K, V, t$ | $F = K \div (V \times t)$ | | | |
| | $V, hp$ | $F = 550 \times hp \div V$ | | | |

*Meaning of Symbols:* $S$ = distance in feet; $V$ = constant or average velocity in feet per second; $t$ = time in seconds; $F$ = constant or average force in pounds; $P$ = power in foot-pounds per second; $K$ = work in foot-pounds; and $hp$ = horsepower.

*Note:* The metric SI unit of work is the joule (one joule = 1 newton-meter), and the unit of power is the watt (one watt = 1 joule per second = 1 N·m/s). The term horsepower is not used. Thus, those formulas above which involve horsepower and the factor 550 are not applicable when working in SI units. The remaining formulas can be used, and the units are: $S$ = distance in meters; $V$ = constant or average velocity in meters per second; $t$ = time in seconds; $F$ = force in newtons; $P$ = power in watts; $K$ = work in joules.

*Example:* — A casting weighing 300 pounds is to be lifted by means of an overhead crane. The casting is lifted 10 feet in 12 seconds. What is the horsepower developed? Here $F = 300$; $S = 10$; $t = 12$.

$$\text{hp} = \frac{F \times S}{550\,t} = \frac{300 \times 10}{550 \times 12} = 0.45.$$

A similar example using metric SI units, is as follows: A casting of mass 15 kg is lifted 4 meters in 15 seconds by means of a crane. What is the power? Here $F = 150g$ N, $S = 4$ m, and $t = 15$ s. Thus:

$$\text{Power} = \frac{FS}{t} = \frac{150g \times 4}{15}$$

$$= 392 \text{ watts or } 0.392 \text{ kW.}$$

## Centrifugal Force

**Centrifugal Force.** — When a body rotates about any axis other than one at its center of mass, it exerts an outward radial force called centrifugal force upon the axis or any arm or cord from the axis which restrains it from moving in a straight (tangential) line. In the following formulas:

$F$ = centrifugal force in pounds;

$W$ = weight of revolving body in pounds;

$v$ = velocity at radius $R$ on body in feet per second;

$n$ = number of revolutions per minute;

$g$ = acceleration due to gravity = 32.16 feet per second per second;

$R$ = perpendicular distance from axis of rotation to center of mass, or for practical use, to center of gravity of revolving body.

*Note:* If a body rotates about its own center of mass, $R$ equals zero and $v$ equals zero. This means that the *resultant* of the centrifugal forces of all the elements of the body is equal to zero or, in other words, no centrifugal force is exerted on the axis of rotation. The centrifugal force of any part or element of such a body is found by the equations given below where $R$ is the radius to the center of gravity of the part or element. In case of a flywheel rim, the mean radius of the rim meets practical requirements, as this is the radius to center of gravity of a thin radial section.

$$F = \frac{Wv^2}{gR} = \frac{Wv^2}{32.16\,R} = \frac{4\,WR\pi^2n^2}{60 \times 60\,g} = \frac{WRn^2}{2933} = 0.000341\,WRn^2$$

$$W = \frac{FRg}{v^2} = \frac{2933\,F}{Rn^2} \qquad\qquad v = \sqrt{\frac{FRg}{W}}$$

$$R = \frac{Wv^2}{Fg} = \frac{2933\,F}{Wn^2} \qquad\qquad n = \sqrt{\frac{2933\,F}{WR}}$$

(If $n$ is the number of revolutions per second instead of per minute, then $F = 1.227\,WRn^2$.)

If metric SI units are used in the foregoing formulas, $W/g$ is replaced by $M$, which is the mass in kilograms; $F$ = centrifugal force in newtons; $v$ =

velocity in meters per second; $n$ = number of revolutions per minute; and $R$ = the radius in meters. Thus:

$$F = Mv^2/R = \frac{Mn^2(2\pi R)^2}{60^2 R} = 0.01097\ MRn^2.$$

If the rate of rotation is expressed as $n_1$ = revolutions per second, then $F = 39.48\ MRn_1^2$; if it is expressed as $\omega$ radians per second, then $F = MR\omega^2$.

**Calculating Centrifugal Force.** — In the ordinary formula for centrifugal force, $F = 0.000341\ WRn^2$; the mean radius $R$ of the flywheel or pulley rim is given in feet. For small dimensions, it is more convenient to have the formula in the form:

$$F = 0.000028416\ Wrn^2$$

in which $F$ = centrifugal force, in pounds; $W$ = weight of rim, in pounds; $r$ = mean radius of rim, in inches; $n$ = number of revolutions per minute.

In this formula let $C = 0.000028416\ n^2$. This, then, is the centrifugal force of one pound, one inch from the axis. The formula can now be written in the form,

$$F = WrC$$

$C$ is calculated for various values of the revolutions per minute $n$, and the calculated values of $C$ are given in Table 1. To find the centrifugal force in any given case, simply find the value of $C$ in the table and multiply it by the product of $W$ and $r$, the four multiplications in the original formula given thus having been reduced to two.

*Example:* A cast-iron flywheel with a mean rim radius of 9 inches, is rotated at a speed of 800 revolutions per minute. If the weight of the rim is 20 pounds, what is the centrifugal force?

From Table 1, for $n$ = 800 revolutions per minute, the value of $C$ is 18.1862.

Thus,                     $F = WrC$
                          $= 20 \times 9 \times 18.1862$
                          $= 3273.52$ pounds

Using metric SI units, $0.01097n^2$ is the centrifugal force acting on a body of 1 kg mass rotating at $n$ revolutions per minute at a distance of 1 meter from the axis. If this value is designated $C_1$, then the centrifugal force of mass $M$ kg rotating at this speed at a distance from the axis of $R$ meters, is $C_1MR$ newtons. To simplify calculations, values for $C_1$ are given in Table 2. If it is required to work in terms of millimeters, the force is $0.001\ C_1MR_1$ newtons, where $R_1$ is the radius in millimeters.

Example: A steel pulley with a mean rim radius of 120 millimeters is rotated at a speed of 1100 revolutions per minute. If the mass of the rim is 5 kilograms, what is the centrifugal force?

From Table 2, for n = 2000 revolutions per minute, the value of $C_1$ is 13,269.1.

Thus             $F = 0.001\ C_1MR_1$
                 $= 0.001 \times 13,269.1 \times 5 \times 120$
                 $= 7961.50$ newtons

Table 1. Factors $C$ for Calculating Centrifugal Force (English units)

| $n$ | $C$ | $n$ | $C$ | $n$ | $C$ | $n$ | $C$ |
|---|---|---|---|---|---|---|---|
| 50 | 0.07104 | 100 | 0.28416 | 470 | 6.2770 | 5200 | 768.369 |
| 51 | 0.07391 | 101 | 0.28987 | 480 | 6.5470 | 5300 | 798.205 |
| 52 | 0.07684 | 102 | 0.29564 | 490 | 6.8227 | 5400 | 828.611 |
| 53 | 0.07982 | 103 | 0.30147 | 500 | 7.1040 | 5500 | 859.584 |
| 54 | 0.08286 | 104 | 0.30735 | 600 | 10.2298 | 5600 | 891.126 |
| 55 | 0.08596 | 105 | 0.31328 | 700 | 13.9238 | 5700 | 923.236 |
| 56 | 0.08911 | 106 | 0.31928 | 800 | 18.1862 | 5800 | 955.914 |
| 57 | 0.09232 | 107 | 0.32533 | 900 | 23.0170 | 5900 | 989.161 |
| 58 | 0.09559 | 108 | 0.33144 | 1000 | 28.4160 | 6000 | 1022.980 |
| 59 | 0.09892 | 109 | 0.33761 | 1100 | 34.3834 | 6100 | 1057.360 |
| 60 | 0.10230 | 110 | 0.34383 | 1200 | 40.9190 | 6200 | 1092.310 |
| 61 | 0.10573 | 115 | 0.37580 | 1300 | 48.0230 | 6300 | 1127.830 |
| 62 | 0.10923 | 120 | 0.40921 | 1400 | 55.6954 | 6400 | 1163.920 |
| 63 | 0.11278 | 125 | 0.44400 | 1500 | 63.9360 | 6500 | 1200.580 |
| 64 | 0.11639 | 130 | 0.48023 | 1600 | 72.7450 | 6600 | 1237.800 |
| 65 | 0.12006 | 135 | 0.51788 | 1700 | 82.1222 | 6700 | 1275.590 |
| 66 | 0.12378 | 140 | 0.55695 | 1800 | 92.0678 | 6800 | 1313.960 |
| 67 | 0.12756 | 145 | 0.59744 | 1900 | 102.5820 | 6900 | 1352.890 |
| 68 | 0.13140 | 150 | 0.63936 | 2000 | 113.6640 | 7000 | 1392.380 |
| 69 | 0.13529 | 160 | 0.72745 | 2100 | 125.3150 | 7100 | 1432.450 |
| 70 | 0.13924 | 170 | 0.82122 | 2200 | 137.5330 | 7200 | 1473.090 |
| 71 | 0.14325 | 180 | 0.92067 | 2300 | 150.3210 | 7300 | 1514.290 |
| 72 | 0.14731 | 190 | 1.02590 | 2400 | 163.6760 | 7400 | 1556.060 |
| 73 | 0.15143 | 200 | 1.1367 | 2500 | 177.6000 | 7500 | 1598.400 |
| 74 | 0.15561 | 210 | 1.2531 | 2600 | 192.0920 | 7600 | 1641.310 |
| 75 | 0.15984 | 220 | 1.3753 | 2700 | 207.1530 | 7700 | 1684.780 |
| 76 | 0.16413 | 230 | 1.5032 | 2800 | 222.7810 | 7800 | 1728.830 |
| 77 | 0.16848 | 240 | 1.6358 | 2900 | 238.9790 | 7900 | 1773.440 |
| 78 | 0.17288 | 250 | 1.7760 | 3000 | 255.7400 | 8000 | 1818.620 |
| 79 | 0.17734 | 260 | 1.9209 | 3100 | 273.7080 | 8100 | 1864.370 |
| 80 | 0.18186 | 270 | 2.0715 | 3200 | 290.9800 | 8200 | 1910.690 |
| 81 | 0.18644 | 280 | 2.2278 | 3300 | 309.4500 | 8300 | 1957.580 |
| 82 | 0.19107 | 290 | 2.3898 | 3400 | 328.4890 | 8400 | 2005.030 |
| 83 | 0.19576 | 300 | 2.5574 | 3500 | 348.0960 | 8500 | 2053.060 |
| 84 | 0.20050 | 310 | 2.7308 | 3600 | 368.2710 | 8600 | 2101.650 |
| 85 | 0.20530 | 320 | 2.9098 | 3700 | 389.0150 | 8700 | 2150.810 |
| 86 | 0.21016 | 330 | 3.0945 | 3800 | 410.3270 | 8800 | 2200.540 |
| 87 | 0.21508 | 340 | 3.2849 | 3900 | 432.2070 | 8900 | 2250.830 |
| 88 | 0.22005 | 350 | 3.4809 | 4000 | 454.6560 | 9000 | 2301.700 |
| 89 | 0.22508 | 360 | 3.6823 | 4100 | 477.6730 | 9100 | 2353.130 |
| 90 | 0.23017 | 370 | 3.8901 | 4200 | 501.2580 | 9200 | 2405.130 |
| 91 | 0.23531 | 380 | 4.1032 | 4300 | 525.4120 | 9300 | 2457.700 |
| 92 | 0.24051 | 390 | 4.3220 | 4400 | 550.1340 | 9400 | 2510.840 |
| 93 | 0.24577 | 400 | 4.5466 | 4500 | 575.4240 | 9500 | 2564.540 |
| 94 | 0.25108 | 410 | 4.7767 | 4600 | 601.2830 | 9600 | 2618.820 |
| 95 | 0.25645 | 420 | 5.0126 | 4700 | 627.7090 | 9700 | 2673.660 |
| 96 | 0.26188 | 430 | 5.2541 | 4800 | 654.7050 | 9800 | 2729.070 |
| 97 | 0.26737 | 440 | 5.5013 | 4900 | 682.2680 | 9900 | 2785.050 |
| 98 | 0.27291 | 450 | 5.7542 | 5000 | 710.4000 | 10000 | 2841.600 |
| 99 | 0.27851 | 460 | 6.0128 | 5100 | 739.1000 | | |

MECHANICS

Table 2. Factors $C_1$ for Calculating Centrifugal Force (Metric SI units)

| $n$ | $C_1$ | $n$ | $C_1$ | $n$ | $C_1$ | $n$ | $C_1$ |
|---|---|---|---|---|---|---|---|
| 50 | 27.4156 | 100 | 109.662 | 470 | 2,422.44 | 5200 | 296,527 |
| 51 | 28.5232 | 101 | 111.867 | 480 | 2,526.62 | 5300 | 308,041 |
| 52 | 29.6527 | 102 | 114.093 | 490 | 2,632.99 | 5400 | 319,775 |
| 53 | 30.8041 | 103 | 116.341 | 500 | 2,741.56 | 5500 | 331,728 |
| 54 | 31.9775 | 104 | 118.611 | 600 | 3,947.84 | 5600 | 343,901 |
| 55 | 33.1728 | 105 | 120.903 | 700 | 5,373.45 | 5700 | 356,293 |
| 56 | 34.3901 | 106 | 123.217 | 800 | 7,018.39 | 5800 | 368,904 |
| 57 | 35.6293 | 107 | 125.552 | 900 | 8,882.64 | 5900 | 381,734 |
| 58 | 36.8904 | 108 | 127.910 | 1000 | 10,966.2 | 6000 | 394,784 |
| 59 | 38.1734 | 109 | 130.290 | 1100 | 13,269.1 | 6100 | 408,053 |
| 60 | 39.4784 | 110 | 132.691 | 1200 | 15,791.4 | 6200 | 421,542 |
| 61 | 40.8053 | 115 | 145.028 | 1300 | 18,532.9 | 6300 | 435,250 |
| 62 | 42.1542 | 120 | 157.914 | 1400 | 21,493.8 | 6400 | 449,177 |
| 63 | 43.5250 | 125 | 171.347 | 1500 | 24,674.0 | 6500 | 463,323 |
| 64 | 44.9177 | 130 | 185.329 | 1600 | 28,073.5 | 6600 | 477,689 |
| 65 | 46.3323 | 135 | 199.860 | 1700 | 31,692.4 | 6700 | 492,274 |
| 66 | 47.7689 | 140 | 214.938 | 1800 | 35,530.6 | 6800 | 507,078 |
| 67 | 49.2274 | 145 | 230.565 | 1900 | 39,588.1 | 6900 | 522,102 |
| 68 | 50.7078 | 150 | 246.740 | 2000 | 43,864.9 | 7000 | 537,345 |
| 69 | 52.2102 | 160 | 280.735 | 2100 | 48,361.1 | 7100 | 552,808 |
| 70 | 53.7345 | 170 | 316.924 | 2200 | 53,076.5 | 7200 | 568,489 |
| 71 | 55.2808 | 180 | 355.306 | 2300 | 58,011.3 | 7300 | 584,390 |
| 72 | 56.8489 | 190 | 395.881 | 2400 | 63,165.5 | 7400 | 600,511 |
| 73 | 58.4390 | 200 | 438.649 | 2500 | 68,538.9 | 7500 | 616,850 |
| 74 | 60.0511 | 210 | 483.611 | 2600 | 74,131.7 | 7600 | 633,409 |
| 75 | 61.6850 | 220 | 530.765 | 2700 | 79,943.8 | 7700 | 650,188 |
| 76 | 63.3409 | 230 | 580.113 | 2800 | 85,975.2 | 7800 | 667,185 |
| 77 | 65.0188 | 240 | 631.655 | 2900 | 92,226.0 | 7900 | 684,402 |
| 78 | 66.7185 | 250 | 685.389 | 3000 | 98,696.0 | 8000 | 701,839 |
| 79 | 68.4402 | 260 | 741.317 | 3100 | 105,385 | 8100 | 719,494 |
| 80 | 70.1839 | 270 | 799.438 | 3200 | 112,294 | 8200 | 737,369 |
| 81 | 71.9494 | 280 | 859.752 | 3300 | 119,422 | 8300 | 755,463 |
| 82 | 73.7369 | 290 | 922.260 | 3400 | 126,770 | 8400 | 773,777 |
| 83 | 75.5463 | 300 | 986.960 | 3500 | 134,336 | 8500 | 792,310 |
| 84 | 77.3777 | 310 | 1,053.85 | 3600 | 142,122 | 8600 | 811,062 |
| 85 | 79.2310 | 320 | 1,122.94 | 3700 | 150,128 | 8700 | 830,034 |
| 86 | 81.1062 | 330 | 1,194.22 | 3800 | 158,352 | 8800 | 849,225 |
| 87 | 83.0034 | 340 | 1,267.70 | 3900 | 166,796 | 8900 | 868,635 |
| 88 | 84.9225 | 350 | 1,343.36 | 4000 | 175,460 | 9000 | 888,264 |
| 89 | 86.8635 | 360 | 1,421.22 | 4100 | 184,342 | 9100 | 908,113 |
| 90 | 88.8264 | 370 | 1,501.28 | 4200 | 193,444 | 9200 | 928,182 |
| 91 | 90.8113 | 380 | 1,583.52 | 4300 | 202,766 | 9300 | 948,469 |
| 92 | 92.8182 | 390 | 1,667.96 | 4400 | 212,306 | 9400 | 968,976 |
| 93 | 94.8469 | 400 | 1,754.60 | 4500 | 222,066 | 9500 | 989,702 |
| 94 | 96.8976 | 410 | 1,843.42 | 4600 | 232,045 | 9600 | 1,010,650 |
| 95 | 98.9702 | 420 | 1,934.44 | 4700 | 242,244 | 9700 | 1,031,810 |
| 96 | 101.065 | 430 | 2,027.66 | 4800 | 252,662 | 9800 | 1,053,200 |
| 97 | 103.181 | 440 | 2,123.06 | 4900 | 263,299 | 9900 | 1,074,800 |
| 98 | 105.320 | 450 | 2,220.66 | 5000 | 274,156 | 10000 | 1,096,620 |
| 99 | 107.480 | 460 | 2,320.45 | 5100 | 285,232 | ... | ... |

# FLYWHEELS

Flywheels may be classified either as *balance wheels* or as *flywheel pulleys*. The object of all flywheels is to equalize the energy exerted and the work done and thereby prevent excessive or sudden changes of speed. The permissible speed variation is an important factor in all flywheel designs. The allowable speed change varies considerably for different classes of machinery; for instance, it is about 1 or 2 per cent in modern steam engines, while in punching and shearing machinery a speed variation of 20 per cent may be allowed.

As the function of a balance wheel is to absorb and equalize energy in case the resistance to motion, or driving power, varies throughout the cycle, the rim section is generally quite heavy and is designed with reference to the energy that must be stored in it to prevent excessive speed variations and also with reference to the strength necessary to withstand safely the stresses resulting from the required speed. The rims of most balance wheels are either square or nearly square in section, but flywheel pulleys are commonly made wide to accommodate a belt and relatively thin in a radial direction, although this is not an invariable rule.

Flywheels, in general, may either be formed of a solid or one-piece section, or they may be of sectional construction. Flywheels in diameters up to about eight feet are usually cast solid, the hubs being divided in some cases to relieve cooling stresses. Flywheels ranging from, say, eight feet to fifteen feet in diameter, are commonly cast in half sections, and the larger sizes in several sections, the number of which may equal the number of arms in the wheel. The sectional flywheels may be divided into two general classes. One class includes cast wheels which are formed of sections principally because a solid casting would be too large to transport readily. The second class includes wheels of sectional construction which, by reason of the materials used and the special arrangement of the sections, enables much higher peripheral speeds to be obtained safely than would be possible with ordinary sectional wheels of the type not designed especially for high speeds. Various designs have been built to withstand the extreme stresses encountered in some classes of service. The rims in some cases are laminated, being partly or entirely formed of numerous segment-shaped steel plates. Another type of flywheel, which is superior to an ordinary sectional wheel, has a solid cast-iron rim connected to the hub by disk-shaped steel plates instead of cast spokes. Steel wheels may be divided into three distinct types, including (1) those having the center and rim built up entirely of steel plates, (2) those having a cast-iron center and steel rim, and (3) those having a cast-steel center and rim formed of steel plates. Wheels having wire-wound rims have been used to a limited extent when extremely high speeds have been necessary.

When the rim is formed of sections held together by joints it is very important to design these joints properly. The ordinary bolted and flanged rim joints located between the arms average about 20 per cent of the strength of a solid rim and about 25 per cent is the maximum strength obtainable for a joint of this kind. However, by placing the joints at the ends of the arms instead of between them, an efficiency of 50 per cent of the strength of the rim may be obtained. This is due to the fact that the joint is not subjected to the outward bending stresses between the arms but is directly supported by the arm, the end of which is secured to the rim just beneath the joint. When the rim sections of heavy balance wheels are held together by steel links shrunk into place, an efficiency of 60 per cent may be obtained; and by using a rim of box or I-section, a link type of joint connection may have an efficiency of 100 per cent.

**Energy Due to Changes of Velocity.** — When a flywheel absorbs energy from a variable driving force, as in the case of a steam engine, the velocity increases; and when this stored energy is given out, the velocity diminishes. When the driven member of a machine encounters a variable resistance in performing its work, as when the punch of a punching machine is passing through a steel plate, the flywheel gives up energy while the punch is at work, and, consequently, the speed of the flywheel is reduced. The total energy that a flywheel would give out if brought to a standstill is given by the formula:

$$E = \frac{Wv^2}{2g} = \frac{Wv^2}{64.32}$$

in which  $E$ = total energy of flywheel, in foot-pounds;

$W$ = weight of flywheel rim, in pounds;

$v$ = velocity at mean radius of flywheel rim, in feet per second;

$g$ = acceleration due to gravity = 32.16.

If the velocity of a flywheel changes, the energy it will absorb or give up is proportional to the difference between the squares of its initial and final speeds, and is equal to the difference between the energy which it would give out if brought to a full stop and that which is still stored in it at the reduced velocity. Hence:

$$E_1 = \frac{Wv_1^2}{2g} - \frac{Wv_2^2}{2g} = \frac{W(v_1^2 - v_2^2)}{64.32}$$

in which  $E_1$ = energy in foot-pounds which a flywheel will give out while the speed is reduced from $v_1$ to $v_2$;

$W$ = weight of flywheel rim, in pounds;

$v_1$ = velocity at mean radius of flywheel rim before any energy has been given out, in feet per second;

$v_2$ = velocity of flywheel rim at end of period during which the energy has been given out, in feet per second.

Ordinarily, the effect of the arms and hub does not enter into flywheel calculations, and only the weight of the rim is considered. In computing the velocity, the mean radius of the rim is commonly used.

Using metric SI units, the formulas are  $E = \frac{1}{2}MV^2$ , and  $E_1 = \frac{1}{2}M(v_1^2 - v_2^2)$ , where  $E$  and  $E_1$  are in joules;  $M$  = the mass of the rim in kilograms; and  $v$ ,  $v_1$ , and  $v_2$  = velocities in meters per second. Note: In the SI, the unit of mass is the kilogram. If the weight of the flywheel rim is given in kilograms, the value referred to is the mass. Should the weight be given in newtons, then

$$M = \frac{W}{g},$$

where  $g$  is approximately 9.81 meters per second squared.

**General Procedure in Flywheel Design.** — The general method of designing a flywheel is to determine first the value of  $E_1$  or the energy the flywheel must either supply or absorb for a given change in velocity, which, in turn, varies for different classes of service. The mean diameter of the flywheel may be assumed, or it may be fixed within certain limits by the general design of the machine. Ordinarily the speed of the flywheel shaft is known, at least approximately; the values of  $v_1$  and  $v_2$  can then be determined, the latter depending upon the allowable percentage of speed variation. When these values are known, the weight of the rim and the cross-sectional area required to obtain this weight may be computed. The general procedure will be illustrated more in detail by considering the design of flywheels for punching and shearing machinery.

**Flywheels for Presses, Punches, Shears, Etc.** — In these classes of machinery, the work that the machine performs is of an intermittent nature and is done during a small part of the time required for the driving shaft of the machine to make a complete revolution. In order to distribute the work of the machine over the entire period of revolution of the driving shaft, a heavy-rimmed flywheel is placed on the shaft, giving the belt an opportunity to perform an almost uniform amount of work during the whole revolution. During the greater part of the revolution of the driving shaft, the belt power is used to accelerate the speed of the flywheel. During the part of the revolution when the work is done, the energy thus stored up in the flywheel is given out at the expense of its velocity. The problem is to determine the weight and cross-sectional area of the rim when the conditions affecting the design of the flywheel are known.

*Example:* — A flywheel is required for a punching machine capable of punching ¾-inch holes through structural steel plates ¾ inch thick. This machine (see accompanying diagram) is of the general type having a belt-driven shaft at the rear which carries a flywheel and a pinion that meshes with a large gear on the main shaft at the top of the machine. It is assumed that the relative speeds of the pinion and large gear are 7 to 1, respectively, and that the slide is to make 30 working strokes per minute. The preliminary lay-out shows that the flywheel should have a mean diameter (see enlarged detail) of about 30 inches. Find the weight of the flywheel and the size of the rim.

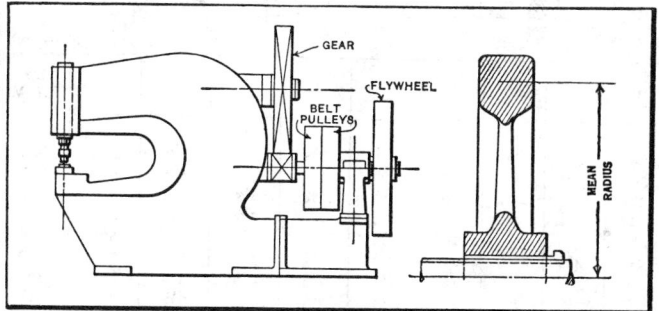

*Energy Supplied by Flywheel.* — The energy which the flywheel must give up for a given change in velocity, and the weight of rim necessary to supply that energy, must be determined. The maximum pressure for shearing a ¾-inch hole through ¾-inch structural steel equals approximately the circumference of the hole multiplied by the thickness of the stock multiplied by the tensile strength, which is nearly the same as the shearing resistance of the steel. Thus, in this case, 3.1416 × ¾ × ¾ × 60,000 = 106,000 pounds. The average pressure will be much less than the maximum. Some designers assume that the average pressure is about one-half the maximum, although experiments show that the material is practically sheared off when the punch has entered the sheet a distance equal to about one-third the sheet thickness. On this latter basis, the average energy $E_a$ in foot-pounds is 2200 in this case. Thus:

$$E_a = \frac{106,000 \times \frac{1}{3} \times \frac{3}{4}}{12} = \frac{106,000}{4 \times 12} = 2200 \text{ foot-pounds.}$$

If the efficiency of the machine is taken at 85 per cent, the energy required will equal 2200 ÷ 0.85 = 2600 foot-pounds nearly. Assume that the energy supplied by the belt while the punch is at work is determined by calculation to equal 175 foot-pounds. Then the flywheel must supply 2600 − 175 = 2425 foot-pounds = $E_1$.

### Dimensions of Flywheels for Punches and Shears

(Maximum number of revolutions per minute given in table should never be exceeded for cast-iron flywheels.)

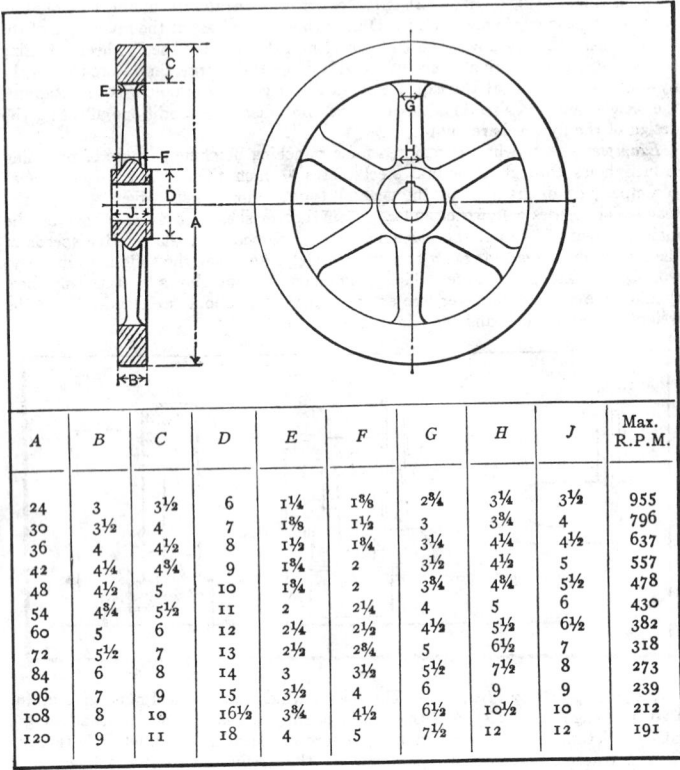

| A | B | C | D | E | F | G | H | J | Max. R.P.M. |
|---|---|---|---|---|---|---|---|---|---|
| 24 | 3 | 3½ | 6 | 1¼ | 1⅜ | 2¾ | 3¼ | 3½ | 955 |
| 30 | 3½ | 4 | 7 | 1⅜ | 1½ | 3 | 3¾ | 4 | 796 |
| 36 | 4 | 4½ | 8 | 1½ | 1¾ | 3¼ | 4¼ | 4½ | 637 |
| 42 | 4¼ | 4¾ | 9 | 1¾ | 2 | 3½ | 4½ | 5 | 557 |
| 48 | 4½ | 5 | 10 | 1¾ | 2 | 3¾ | 4¾ | 5½ | 478 |
| 54 | 4¾ | 5½ | 11 | 2 | 2¼ | 4 | 5 | 6 | 430 |
| 60 | 5 | 6 | 12 | 2¼ | 2½ | 4½ | 5½ | 6½ | 382 |
| 72 | 5½ | 7 | 13 | 2½ | 2¾ | 5 | 6½ | 7 | 318 |
| 84 | 6 | 8 | 14 | 3 | 3½ | 5½ | 7½ | 8 | 273 |
| 96 | 7 | 9 | 15 | 3½ | 4 | 6 | 9 | 9 | 239 |
| 108 | 8 | 10 | 16½ | 3¾ | 4½ | 6½ | 10½ | 10 | 212 |
| 120 | 9 | 11 | 18 | 4 | 5 | 7½ | 12 | 12 | 191 |

*Rim Velocity at Mean Radius.* — When the mean radius of the flywheel is known, the velocity of the rim at the mean radius, in feet per second, is:

$$v = \frac{2 \times 3.1416 \times R \times n}{60}$$

in which    $v$ = velocity at mean radius of flywheel, in feet per second;
$R$ = mean radius of flywheel rim, in feet;
$n$ = number of revolutions per minute.

According to the preliminary lay-out the mean diameter in this case should be about 30 inches and the driving shaft is to make 210 R.P.M.; hence,

$$v = \frac{2 \times 3.1416 \times 1.25 \times 210}{60} = 27.5 \text{ feet per second.}$$

*Weight of Flywheel Rim.* — Assuming that the allowable variation in velocity when punching is about 15 per cent, and values of $v_1$ and $v_2$ are respectively 27.5 and 23.4 feet per second (27.5 × 0.85 = 23.4), the weight of a flywheel rim necessary to supply a given amount of energy in foot-pounds while the speed is reduced from $v_1$ to $v_2$ would be:

$$W = \frac{E_1 \times 64.32}{v_1^2 - v_2^2} = \frac{2425 \times 64.32}{27.5^2 - 23.4^2} = 750 \text{ pounds.}$$

*Size of Rim for Given Weight.* — Since 1 cubic inch of cast iron weighs 0.26 pound, a flywheel rim weighing 750 pounds contains $\frac{750}{0.26} = 2884$ cubic inches. The cross-sectional area of the rim in square inches equals the total number of cubic inches divided by the mean circumference, or $\frac{2884}{94.25} = 31$ square inches nearly, which is approximately the area of a rim 5⅛ inches wide and 6 inches deep.

**Simplified Flywheel Calculations.** — Calculations for designing the flywheels of punches and shears are simplified by the following formulas and the accompanying table of constants applying to different percentages of speed reduction. In these formulas let:

H.P. = horsepower required;
$N$ = number of strokes per minute;
$E$ = total energy required per stroke, in foot-pounds;
$E_1$ = energy given up by flywheel, in foot-pounds;
$T$ = time in seconds per stroke;
$T_1$ = time in seconds of actual cut;
$W$ = weight of flywheel rim, in pounds;
$D$ = mean diameter of flywheel rim, in feet;
$R$ = maximum allowable speed of flywheel in revolutions per minute;
$C$ and $C_1$ = values as given in table;
$a$ = width of flywheel rim;
$b$ = depth of flywheel rim;
$y$ = ratio of depth to width of rim.

$$\text{H.P.} = \frac{EN}{33,000} = \frac{E}{T \times 550} \qquad E_1 = E\left(1 - \frac{T_1}{T}\right)$$

$$W = \frac{E_1}{CD^2R^2} \qquad a = \sqrt{\frac{1.22\,W}{12\,Dy}} \qquad b = ay$$

For cast-iron flywheels, with a maximum stress of 1000 pounds per square inch:

$$W = C_1E_1 \qquad R = 1940 \div D$$

**Values of C and C₁ in the Previous Formulas**

| Per Cent Reduction | $C$ | $C_1$ | Per Cent Reduction | $C$ | $C_1$ |
|---|---|---|---|---|---|
| 2½ | 0.00000213 | 0.1250 | 10 | 0.00000810 | 0.0328 |
| 5 | 0.00000426 | 0.0625 | 15 | 0.00001180 | 0.0225 |
| 7½ | 0.00000617 | 0.0432 | 20 | 0.00001535 | 0.0173 |

*Example* 1: — A hot slab shear is required to cut a slab $4 \times 15$ inches which, at a shearing stress of 6000 pounds per square inch, gives a pressure between the knives of 360,000 pounds. The total energy required for the cut will then be $360,000 \times \frac{4}{12} = 120,000$ foot-pounds. The shear is to make 20 strokes per minute; the actual cutting time is 0.75 second, and the balance of the stroke is 2.25 seconds.

The flywheel is to have a mean diameter of 6 feet 6 inches and is to run at a speed of 200 R.P.M.; the reduction in speed to be 10 per cent per stroke when cutting.

$$\text{H.P.} = \frac{120,000 \times 20}{33,000} = 72.7 \text{ horsepower;}$$

$$E_1 = 120,000 \times \left(1 - \frac{0.75}{3}\right) = 90,000 \text{ foot-pounds;}$$

$$W = \frac{90,000}{0.0000081 \times 6.5^2 \times 200^2} = 6570 \text{ pounds.}$$

**Assuming a ratio of 1.22 between depth and width of rim,**

$$a = \sqrt{\frac{6570}{12 \times 6.5}} = 9.18 \text{ inches;}$$

$$b = 1.22 \times 9.18 = 11.2 \text{ inches;}$$

or size of rim, say, $9 \times 11\frac{1}{2}$ inches.

*Example* 2: — Suppose that the flywheel in Example 1 is to be made with a stress of 1000 pounds, due to centrifugal force, per square inch of rim section.

$$C_1 \text{ for 10 per cent} = 0.0328;$$

$$W = 0.0328 \times 90,000 = 2950 \text{ pounds.}$$

$$R = \frac{1940}{D}. \quad \text{If } D = 6 \text{ feet, } R = \frac{1940}{6} = 323 \text{ R.P.M.}$$

**Assuming a ratio of 1.22 between depth and width of rim, as before:**

$$a = \sqrt{\frac{2950}{12 \times 6}} = 6.4 \text{ inches;}$$

$$b = 1.22 \times 6.4 = 7.8 \text{ inches;}$$

or size of rim, say, $6\frac{1}{4} \times 8$ inches.

**Centrifugal Stresses in Flywheel Rims.** — In general, high speed is desirable for flywheels in order to avoid using wheels which are unnecessarily large and heavy. The centrifugal tension or hoop tension stress which tends to rupture a flywheel rim of given area, depends solely upon the rim velocity, and is independent of the rim radius. The bursting velocity of a flywheel, based on hoop stress alone (not considering bending stresses), is related to the tensile stress in the flywheel rim by the following formula which is based on the centrifugal force formula from mechanics.

$$V = \sqrt{10 \times s} \quad \text{or,} \quad s = V^2 \div 10$$

where $V$ = velocity of outside circumference of rim in feet per second, and $s$ is the tensile strength of the rim material in pounds per square inch.

For cast iron having a tensile strength of 19,000 pounds per square inch the bursting speed would be:

$$V = \sqrt{10 \times 19,000} = 436 \text{ feet per second}$$

*Built-up Flywheels:* Flywheels built up of solid disks of rolled steel plate stacked and bolted together on a through shaft have greater speed capacity than other types. The maximum hoop stress is at the bore and is given by the formula,

$$s = 0.0194 \ V^2[4.333 + (d/D)^2]$$

In this formula, $s$ and $V$ are the stress and velocity as previously defined and $d$ and $D$ are the bore and outside diameters, respectively.

Assuming the plates to be of steel having a tensile strength of 60,000 pounds per square inch and a safe working stress of 24,000 pounds per square inch (using a factor of safety of 2.5 on stress or $\sqrt{2.5}$ on speed) and taking the worst condition (when $d$ approaches $D$), the safe rim speed for this type of flywheel is 500 feet per second or 30,000 feet per minute.

**Combined Stresses in Flywheels.** — The bending stresses in the rim of a flywheel may exceed the centrifugal (hoop tension) stress predicted by the simple formula $s = V^2 \div 10$ by a considerable amount. By taking into account certain characteristics of flywheels, relatively simple formulas have been developed to determine the stress due to the combined effect of hoop tension and bending stress. Some of the factors that influence the magnitude of the maximum combined stress acting at the rim of a flywheel are:

*1. The number of spokes.* Increasing the number of spokes decreases the rim span between spokes and hence decreases the bending moment. Thus an eight-spoke wheel can be driven to a considerably higher speed before bursting than a six-spoke wheel having the same rim.

*2. The relative thickness of the spokes.* If the spokes were extremely thin, like wires, they could offer little constraint to the rim in expanding to its natural diameter under centrifugal force, and hence would cause little bending stress. Conversely, if the spokes were extremely heavy in proportion to the rim, they would restrain the rim thereby setting up heavy bending stresses at the junctions of the rim and spokes.

*3. The relative thickness of the rim to the diameter.* If the rim is quite thick (i.e. has a large section modulus in proportion to span), its resistance to bending will be great and bending stress small. Conversely, thin rims with a section modulus small in comparison with diameter or span have little resistance to bending, thus are subject to high bending stresses.

*4. Residual stresses.* These include shrinkage stresses, impact stresses, and stresses caused by operating torques and imperfections in the material. Residual stresses are taken into account by the use of a suitable factor of safety. (See " Factors of Safety for Flywheels.")

The formulas that follow give the maximum combined stress at the rim of flywheels having 6, 8, and 10 spokes. These formulas are for flywheels with *rectangular rim sections* and take into account the first three of the four factors listed as influencing the magnitude of the combined stress in flywheels.

For 6 spokes:

$$s = \frac{V^2}{10}\left[1 + \left(\frac{0.56\ B - 1.81}{3\ Q + 3.14}\right)Q\right]$$

For 8 spokes:

$$s = \frac{V^2}{10}\left[1 + \left(\frac{0.42\ B - 2.53}{4\ Q + 3.14}\right)Q\right]$$

For 10 spokes:

$$s = \frac{V^2}{10}\left[1 + \left(\frac{0.33\ B - 3.22}{5\ Q + 3.14}\right)Q\right]$$

In these formulas, $s$ = maximum combined stress in pounds per square inch; $Q$ = ratio of mean spoke cross-section area to rim cross-section area; $B$ = ratio of outside diameter of rim to rim thickness; and $V$ = velocity of flywheel rim in feet per second.

**Thickness of Cast Iron Flywheel Rims.** — The mathematical analysis of the stresses in flywheel rims is not conclusive owing to the uncertainty of shrinkage stresses in castings or the strength of the joint in the case of sectional wheels. When a flywheel of ordinary design is revolving at high speed, the tendency of the rim is to bend or bow outward between the arms, and the bending stresses may be serious, especially if the rim is wide and thin and the spokes are rather widely spaced. When the rims are thick, this tendency does not need to be considered, but in the case of a thin rim, running at a high rate of speed, the stress in the middle might become sufficiently great to cause the wheel to fail. The proper thickness of a cast-iron rim to resist this tendency is given for solid rims by Formula 1 and for a jointed rim by Formula 2.

$$t = \frac{0.475 \, d}{n^2 \left( \dfrac{6000}{v^2} - \dfrac{1}{10} \right)} \quad (1); \qquad t = \frac{0.95 \, d}{n^2 \left( \dfrac{6000}{v^2} - \dfrac{1}{10} \right)} \quad (2)$$

In these formulas, $t$ = thickness of rim, in inches; $d$ = diameter of flywheel, in inches; $n$ = number of arms; $v$ = peripheral speed, in feet per second.

**Tables of Safe Speeds for Flywheels.** — The accompanying Table 1, prepared by T. C. Rathbone of The Fidelity and Casualty Company of New York, gives general recommendations for safe rim speeds for flywheels of various constructions. Table 2 shows the number of revolutions per minute corresponding to the rim speeds in Table 1.

**Table 1. Safe Rim Speeds for Flywheels\***

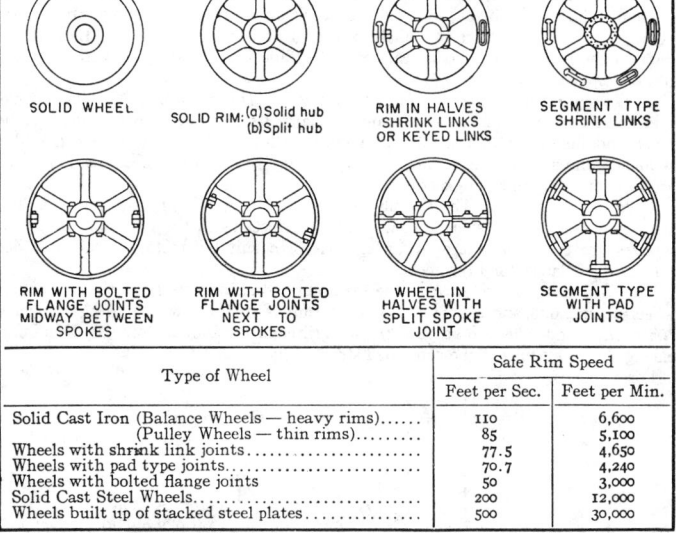

| Type of Wheel | Safe Rim Speed | |
|---|---|---|
| | Feet per Sec. | Feet per Min. |
| Solid Cast Iron (Balance Wheels — heavy rims)...... | 110 | 6,600 |
| (Pulley Wheels — thin rims)........ | 85 | 5,100 |
| Wheels with shrink link joints...................... | 77.5 | 4,650 |
| Wheels with pad type joints......................... | 70.7 | 4,240 |
| Wheels with bolted flange joints | 50 | 3,000 |
| Solid Cast Steel Wheels............................ | 200 | 12,000 |
| Wheels built up of stacked steel plates............... | 500 | 30,000 |

\* To find the safe speed in revolutions per minute, divide the safe rim speed in feet per minute by 3.14 times the outside diameter of the flywheel rim in feet. For flywheels up to 15 feet in diameter, see Table 2.

Table 2. Safe Speeds of Rotation for Flywheels*

| Outside Diameter of Rim, in Feet | Safe Rim Speed in Feet per Minute (from Table 1) | | | | | | |
|---|---|---|---|---|---|---|---|
| | 6,600 | 5,100 | 4,650 | 4,240 | 3,000 | 12,000 | 30,000 |
| | Safe Speed of Rotation in Revolutions per Minute | | | | | | |
| 1 | 2100 | 1623 | 1480 | 1350 | 955 | 3820 | 9549 |
| 2 | 1050 | 812 | 740 | 676 | 478 | 1910 | 4775 |
| 3 | 700 | 541 | 493 | 450 | 318 | 1273 | 3183 |
| 4 | 525 | 406 | 370 | 338 | 239 | 955 | 2387 |
| 5 | 420 | 325 | 296 | 270 | 191 | 764 | 1910 |
| 6 | 350 | 271 | 247 | 225 | 159 | 637 | 1592 |
| 7 | 300 | 232 | 211 | 193 | 136 | 546 | 1364 |
| 8 | 263 | 203 | 185 | 169 | 119 | 478 | 1194 |
| 9 | 233 | 180 | 164 | 150 | 106 | 424 | 1061 |
| 10 | 210 | 162 | 148 | 135 | 96 | 382 | 955 |
| 11 | 191 | 148 | 135 | 123 | 87 | 347 | 868 |
| 12 | 175 | 135 | 123 | 113 | 80 | 318 | 796 |
| 13 | 162 | 125 | 114 | 104 | 73 | 294 | 735 |
| 14 | 150 | 116 | 106 | 97 | 68 | 273 | 682 |
| 15 | 140 | 108 | 99 | 90 | 64 | 255 | 637 |

* Safe speeds of rotation are based on safe rim speeds shown in Table 1.

**Factors of Safety for Flywheels.** — Cast-iron flywheels are commonly designed with a factor of safety of 10 to 13. A factor of safety of 10 applied to the tensile strength of a flywheel material is equivalent to a factor of safety of $\sqrt{10}$ or 3.16 on the speed of the flywheel due to the fact that the stress on the rim of a flywheel increases as the square of the speed. Thus, a flywheel operating at a speed twice that for which it was designed would undergo rim stresses four times as great as at the design speed.

**Safe Speed Formulas for Flywheels and Pulleys.** — No simple formula can possibly accommodate all of the various types and proportions of flywheels and pulleys and at the same time provide a uniform factor of safety for each. Because of considerations of safety, such a formula would penalize the better constructions to accommodate the weaker designs.

One formula that has been used to check the maximum rated operating speed of flywheels and pulleys and which takes into account material properties, construction, rim thickness, and joint efficiences is the following:

$$N = \frac{CAMEK}{D}$$

In this formula,

$N$ = maximum rated operating speed in revolutions per minute

$C$ = 1.0 for wheels driven by a constant speed electric motor (i.e. a–c squirrel-cage induction motor or a–c synchronous motor, etc.)

= 0.90 for wheels driven by variable speed motors, engines or turbines where overspeed is not over 110 per cent of rated operating speed.

$A$ = 0.90 for 4 arms or spokes

1.00 for 6 arms or spokes

1.08 for 8 arms or spokes

1.50 for disc type

$M$ = 1.00 for cast iron of 20,000 psi tensile strength, or unknown

1.12 for cast iron of 25,000 psi tensile strength

1.22 for cast iron of 30,000 psi tensile strength

1.32 for cast iron of 35,000 psi tensile strength

2.20 for nodular iron of 60,000 psi tensile strength

2.45 for cast steel of 60,000 psi tensile strength

2.75 for plate or forged steel of 60,000 psi tensile strength

$E$ = joint efficiency
    1.0 for solid rim
    0.85 for link or prison joints
    0.75 for split rim — bolted joint at arms
    0.70 for split rim — bolted joint between arms

$K$ = 1355 for rim thickness equal to 1 per cent of outside diameter
    1650 for rim thickness equal to 2 per cent of outside diameter
    1840 for rim thickness equal to 3 per cent of outside diameter
    1960 for rim thickness equal to 4 per cent of outside diameter
    2040 for rim thickness equal to 5 per cent of outside diameter
    2140 for rim thickness equal to 7 per cent of outside diameter
    2225 for rim thickness equal to 10 per cent of outside diameter
    2310 for rim thickness equal to 15 per cent of outside diameter
    2340 for rim thickness equal to 20 per cent of outside diameter

$D$ = outside diameter of rim in feet

*Example:* A six-spoke solid cast iron balance wheel 8 feet in diameter has a rectangular rim 10 inches thick. What is the safe speed, in revolutions per minute, if driven by a constant speed motor?

In this case, $C = 1$; $A = 1$; $M = 1$, since tensile strength is unknown; $E = 1$; $K = 2225$ since the rim thickness is approximately 10 per cent of the wheel diameter; $D = 8$ feet. Thus,

$$N = \frac{1 \times 1 \times 1 \times 2225}{8} = 278 \text{ rpm}$$

(*Note*: This safe speed is slightly greater than the value of 263 rpm obtainable directly from Tables 1 and 2.)

**Tests to Determine Flywheel Bursting Speeds.** — Tests made by Prof. C. H. Benjamin, to determine the bursting speeds of flywheels, showed the following results:

*Cast-iron Wheels with Solid Rims.* — Cast-iron wheels having solid rims burst at a rim speed of 395 feet per second, corresponding to a centrifugal tension of about 15,600 pounds per square inch.

*Wheels with Jointed Rims.* — Four wheels were tested with joints and bolts inside the rim, after the familiar design ordinarily employed for band wheels, but with the joints located at points one-fourth of the distance from one arm to the next, these being the points of least bending moment, and, consequently, the points at which the deflection due to centrifugal force would be expected to have the least effect. The tests, however, did not bear out this conclusion. The wheels burst at a rim speed of 194 feet per second, corresponding to a centrifugal tension of about 3750 pounds per square inch. These wheels, therefore, were only about one-quarter as strong as the wheels with solid rims, and burst at practically the same speed as wheels in a previous series of tests in which the rim joints were midway between the arms.

*Bursting Speed for Link Joints.* — Another type of wheel with deep rim, fastened together at the joints midway between the arms by links shrunk into recesses, after the manner of flywheels for massive engines, gave much superior results. This wheel burst at a speed of 256 feet per second, indicating a centrifugal tension of about 6600 pounds per square inch.

*Wheel having Tie-rods.* — Tests were made on a band wheel having joints inside the rim, midway between the arms, and in all respects like others of this design previously tested, except that tie-rods were used to connect the joints with the hub. It burst at a speed of 225 feet per second, showing an increase of strength of from 30 to 40 per cent over similar wheels without the tie-rods.

*Wheel Rim of I-section.* — Several wheels of special design, not in common use, were also tested, the one giving the greatest strength being an English wheel, with solid rim of I-section, made of high-grade cast iron and with the rim tied to the hub by steel wire spokes. These spokes were adjusted to have a uniform tension. The wheel gave way at a rim speed of 424 feet per second, which is slightly higher than the speed of rupture of the solid rim wheels with ordinary style of spokes.

*Tests on Flywheel of Special Construction.* — A test was made on a flywheel 49 inches in diameter and weighing about 900 pounds. The rim was 6¾ inches wide and 1⅛ inches thick, and was built of ten segments, the material being steel casting. Each joint was secured by three " prisoners " of an I-section on the outside face, by link prisoners on each edge, and by a dovetailed bronze clamp on the inside, fitting over lugs on the rim. The arms were of phosphor-bronze, twenty in number, ten on each side, and were a cross in section. These arms came midway between the rim joints and were bolted to plane faces on the polygonal hub. The rim was further reinforced by a system of diagonal bracing, each section of the rim being supported at five points on each side, in such a way as to relieve it almost entirely from bending. The braces, like the arms, were of phosphor-bronze, and all bolts and connecting links of steel. This wheel was designed as a model of a proposed 30-foot flywheel. On account of the excessive air resistance the wheel was enclosed at the sides between sheet-metal disks. This wheel burst at 1775 revolutions per minute or at a linear speed of 372 feet per second. The hub and main spokes of the wheel remained nearly in place, but parts of the rim were found two hundred feet away. This sudden failure of the rim casing was unexpected, as it was thought the flange bolts would be the parts to give way first. The tensile strength of the casing at the point of fracture was about four times the strength of the wheel rim at a solid section.

**Stresses in Rotating Disks.** — When a disk of uniform width is rotated, the max. stress $S_t$ is tangential and at the bore of the hub, and the tangential stress is always greater than the radial stress at the same point on the disk. If $S_t$ = maximum tangential stress in pounds per sq. in.; $w$ = weight of material, lb. per cu. in.; $N$ = rev. per min.; $m$ = Poisson's ratio = 0.3 for steel; $R$ = outer radius of disk, inches; $r$ = inner radius of disk or radius of bore, inches.

$$S_t = 0.0000071 w N^2 [(3 + m)R^2 + (1 - m)r^2]$$

**Steam Engine Flywheels.** — The variable amount of energy during each stroke and the allowable percentage of speed variation are of especial importance in designing steam engine flywheels. The earlier the point of cut-off, the greater the variation in energy and the larger the flywheel that will be required. The weight of the reciprocating parts and the length of the connecting-rod also affect the variation. The following formula is used for computing the weight of the flywheel rim:

Let    $W$ = weight of rim in pounds;

        $D$ = mean diameter of rim in feet;

        $N$ = number of revolutions per minute;

        $\dfrac{1}{n}$ = allowable variation in speed (from 1/60 to 1/100);

        $E$ = excess and deficiency of energy in foot-pounds;

        $c$ = factor of energy excess, from the accompanying table;

    H.P. = indicated horsepower.

**Then,** if the indicated horsepower is given:

$$W = \frac{387,587,500 \times cn \times \text{H.P.}}{D^2 N^3} \tag{1}$$

If the work in foot-pounds is given, then:

$$W = \frac{11{,}745\,nE}{D^2N^2} \qquad (2)$$

In the second formula, $E$ equals the average work in foot-pounds done by the engine in one revolution, multiplied by the decimal given in the accompanying

### Factors for Engine Flywheel Calculations

| Condensing Engines | | | | | | |
|---|---|---|---|---|---|---|
| Fraction of stroke at which steam is cut off........ | ⅓ | ¼ | ⅕ | ⅙ | ⅐ | ⅛ |
| Factor of energy excess... | 0.163 | 0.173 | 0.178 | 0.184 | 0.189 | 0.191 |

| Non-condensing Engines | | | | |
|---|---|---|---|---|
| Steam cut off at...................... | ½ | ⅓ | ¼ | ⅕ |
| Factor of energy excess................. | 0.160 | 0.186 | 0.209 | 0.232 |

table, "Factors for Engine Flywheel Calculations," which covers both the condensing and non-condensing engines:

*Example 1.* — A non-condensing engine of 150 indicated horsepower is to make 200 revolutions per minute, with a speed variation of 2 per cent. The average cut-off is to be at one-quarter stroke, and the flywheel is to have a mean diameter of 6 feet. Required, the necessary weight of rim in pounds.

From the table $c = 0.209$, and from the data given H.P. = 150; $N = 200$; $\frac{1}{n} =$ ⅟₅₀ or $n = 50$; $D = 6$. Substituting these values in equation (1):

$$W = \frac{387{,}587{,}500 \times 0.209 \times 50 \times 150}{6^2 \times 200^3} = 2110 \text{ pounds, nearly.}$$

*Example 2.* — A condensing engine, 24 × 42 inches, cuts off at one-third stroke and has a mean effective pressure of 50 pounds per square inch. The flywheel is to be 18 feet in mean diameter and make 75 revolutions per minute with a variation of 1 per cent. Required, weight of rim.

The work done on the piston in one revolution is equal to the pressure on the piston multiplied by the distance traveled or twice the stroke in feet. The area of the piston in this case is 452.4 square inches, and twice the stroke is 7 feet. The work done on the piston in one revolution is, therefore, 452.4 × 50 × 7 = 158,340 foot-pounds. From the table $c = 0.163$, and therefore:

$$E = 158{,}340 \times 0.163 = 25{,}810 \text{ foot-pounds.}$$

From the data given: $n = 100$; $D = 18$; $N = 75$. Substituting these values in equation (2):

$$W = \frac{11{,}745 \times 100 \times 25{,}810}{18^2 \times 75^2} = 16{,}650 \text{ pounds, nearly.}$$

**Spokes or Arms of Flywheels.** — Flywheel arms are usually of elliptical cross-section. The major axis of the ellipse is in the plane of rotation to give the arms greater resistance to bending stresses and reduce the air resistance which may be considerable at high velocity. The stresses in the arms may be severe, due to the inertia of a heavy rim when sudden load changes occur. The strength of the arms should equal three-fourths the strength of the shaft in torsion.

If $W$ equals the width of the arm at the hub (length of major axis) and $D$ equals the shaft diameter, then $W$ equals $1.3\,D$ for a wheel having 6 arms; and for an 8-arm wheel $W$ equals $1.2\,D$. The thickness of the arm at the hub (length of minor axis) equals one-half the width. The arms usually taper toward the rim. The cross-sectional area at the rim should not be less than two-thirds the area at the hub.

**Flywheels for Motor-driven Planers.** — The primary function of a flywheel for a motor-driven planer is not so much for maintaining a constant speed as relieving the motor from excessive shocks at the points of reversal. Tests made at the Worcester Polytechnic Institute with a 36- by 36- by 10-foot planer driven by a 10-horsepower induction motor showed a current consumption of 1.85 kilowatt-hour when no flywheel was used and the length of the stroke was five feet. With a ten-foot stroke, the consumption was 1.63 kilowatt-hour without a flywheel. When a flywheel was used, the consumption was 1.3 and 1.24 kilowatt-hour, respectively, for the two lengths of stroke mentioned. Thus, with the flywheel, 29.5 per cent less power was required with the short stroke, and 24 per cent less power with the long stroke. The flywheel was also an advantage in increasing the average rate of production and preventing "slow-downs" and tardy reversals.

**Critical Speed of Rotating Body.** — If a body or disk mounted upon a shaft rotates about it, the center of gravity of the body or disk must be at the center of the shaft, if a perfect running balance is to be obtained. In most cases, however, the center of gravity of the disk will be slightly removed from the center of the shaft, owing to the difficulty of perfect balancing. Now, if the shaft and disk be rotated, the centrifugal force generated by the heavier side will be greater than that generated by the lighter side geometrically opposite to it, and the shaft will deflect toward the heavier side, causing the center of the disk to rotate in a small circle. These conditions hold true up to a comparatively high speed; but a point is eventually reached (at several thousand revolutions per minute) when momentarily there will be excessive vibration, and then the parts will run quietly again. The speed at which this occurs is called the *critical speed* of the wheel, and the phenomenon itself is called the *settling* of the wheel. The explanation of the settling is that at this speed the axis of rotation changes, and the wheel and shaft, instead of rotating about their geometrical center, begin to rotate about an axis through their center of gravity. The shaft itself is then deflected so that for every revolution its geometrical center traces a circle around the center of gravity of the rotating mass.

Critical speeds depend upon the magnitude or location of the load or loads carried by the shaft, the length of the shaft, its diameter and the kind of supporting bearings. The normal operating speed of a machine may or may not be higher than the critical speed. For instance, some steam turbines exceed the critical speed, although they do not run long enough at the critical speed for the vibrations to build up to an excessive amplitude. The practice of the General Electric Co. at Schenectady is to keep below the critical speeds. It is assumed that the maximum speed of a machine may be within 20 per cent of the critical speed without vibration troubles. Thus, in a design of steam turbine sets, critical speed is a factor that determines the size of the shafts, both for the generators and turbines. While a machine may run very close to the critical speed, the alignment and play of the bearings, the balance and construction generally, will require extra care, resulting in a more expensive machine; moreover, while such a machine may run smoothly for a considerable time, any looseness or play that may develop later, causing a slight unbalance, will immediately set up excessive vibrations.

**Formulas for Critical Speeds.** — The critical speed formulas given in the accompanying table (from the paper on Critical Speed Calculation presented

## Critical Speed Formulas

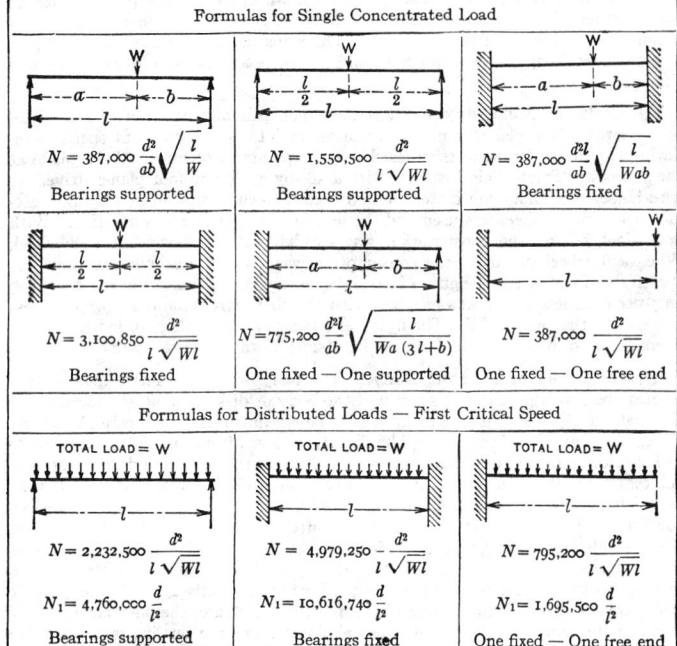

Formulas for Single Concentrated Load

$N = 387,000 \dfrac{d^2}{ab} \sqrt{\dfrac{l}{W}}$

Bearings supported

$N = 1,550,500 \dfrac{d^2}{l \sqrt{Wl}}$

Bearings supported

$N = 387,000 \dfrac{d^2 l}{ab} \sqrt{\dfrac{l}{Wab}}$

Bearings fixed

$N = 3,100,850 \dfrac{d^2}{l \sqrt{Wl}}$

Bearings fixed

$N = 775,200 \dfrac{d^2 l}{ab} \sqrt{\dfrac{l}{Wa(3l+b)}}$

One fixed — One supported

$N = 387,000 \dfrac{d^2}{l \sqrt{Wl}}$

One fixed — One free end

Formulas for Distributed Loads — First Critical Speed

TOTAL LOAD = W

$N = 2,232,500 \dfrac{d^2}{l \sqrt{Wl}}$

$N_1 = 4,760,000 \dfrac{d}{l^2}$

Bearings supported

TOTAL LOAD = W

$N = 4,979,250 \dfrac{d^2}{l \sqrt{Wl}}$

$N_1 = 10,616,740 \dfrac{d}{l^2}$

Bearings fixed

TOTAL LOAD = W

$N = 795,200 \dfrac{d^2}{l \sqrt{Wl}}$

$N_1 = 1,695,500 \dfrac{d}{l^2}$

One fixed — One free end

$N$ = critical speed, R.P.M.; $N_1$ = critical speed of shaft alone; $d$ = diameter of shaft, in inches; $W$ = load applied to shaft, in pounds; $l$ = distance between centers of bearings, in inches; $a$ and $b$ = distances from bearings to load.

before the A.S.M.E. by S. H. Weaver) apply to (1) shafts with single concentrated loads and (2) shafts carrying uniformly distributed loads. These formulas also cover different conditions as regards bearings. If the bearings are self-aligning or very short, the shaft is considered supported at the ends; whereas, if the bearings are long and rigid, the shaft is considered fixed. These formulas, for both concentrated and distributed loads, apply to vertical shafts as well as horizontal shafts, the critical speeds having the same value in both cases. The data required for the solution of critical speed problems are the same as for shaft deflection. As the shaft is usually of variable diameter and its stiffness is increased by a long hub, an ideal shaft of uniform diameter and equal stiffness must be assumed.

In calculating critical speeds, the weight of the shaft is either neglected or, say, one-half to two-thirds of the weight is added to the concentrated load. The formulas apply to steel shafts having a modulus of elasticity $E = 29,000,000$. While a shaft carrying a number of loads or a distributed load may have an infinite number of critical speeds, ordinarily it is the first critical speed that is of importance in engineering work, which is the speed obtained by the formulas given in the table for distributed loads.

**Angular Velocity of Rotating Bodies.** — The angular velocity of a rotating body equals the angle through which the body turns in a unit of time. Angular velocity is commonly expressed in terms of revolutions per minute, but in certain engineering applications it is necessary to express it as radians per second. By definition there are $2\pi$ radians in 360 degrees, or one revolution, so that one radian = $360 \div 2\pi = 57.3$ degrees. To convert angular velocity in revolutions per minute, $n$, to angular velocity in radians per second, $\omega$, multiply by $\pi$ and divide by 30:

$$\omega = \frac{\pi n}{30} \tag{1}$$

The table on page 348 may be used to obtain angular velocity in radians per second for all numbers of revolutions per minute from 1 to 499.

*Example:* To find the angular velocity in radians per second of a flywheel making 97 revolutions per minute, locate 90 in the left-hand column and 7 at the top of the columns; at the intersection of the two lines, the angular velocity is read off as equal to 10.16 radians per second.

**Linear Velocity of Points on a Rotating Body.** — The linear velocity, $v$, of any point on a rotating body expressed in feet per second may be found by multiplying the angular velocity of the body in radians per second, $\omega$, by the radius, $r$, in feet from the center of rotation to the point:

$$v = \omega r \tag{2}$$

The metric SI units are $v$ = meters per second; $\omega$ = radians per second, $r$ = meters.

## Pendulums

**Types of Pendulums.** — A *compound* or *physical* pendulum consists of any rigid body suspended from a fixed horizontal axis about which the body may oscillate in a vertical plane due to the action of gravity.

A *simple* or *mathematical* pendulum is similar to a compound pendulum except that the mass of the body is concentrated at a single point which is suspended from a fixed horizontal axis by a weightless cord. Actually, a simple pendulum cannot be constructed since it is impossible to have either a weightless cord or a body whose mass is entirely concentrated at one point. A good approximation, however, consists of a small, heavy bob suspended by a light, fine wire. If these conditions are not met by the pendulum, it should be considered as a compound pendulum.

A *conical* pendulum is similar to a simple pendulum except that the weight suspended by the cord moves at a uniform speed around the circumference of a circle in a horizontal plane instead of oscillating back and forth in a vertical plane. The principle of the conical pendulum is employed in the Watt fly-ball governor.

A *torsional* pendulum in its simplest form consists of a disk fixed to a slender shaft, the other end of which is fastened to a fixed frame. When the disc is twisted through some angle and released, it will then oscillate back and forth about the axis of the rod because of the torque exerted by the rod.

**Pendulum Formulas.** — From the formulas that follow, the period of vibration or time required for one complete cycle back and forth may be determined for the types of pendulums shown in the accompanying diagram:

For a *simple* pendulum,

$$T = 2\pi \sqrt{\frac{l}{g}} \tag{1}$$

## Angular Velocity in Revolutions per Minute Converted to Radians per Second

| R.P.M. | Angular Velocity in Radians per Second | | | | | | | | | |
|---|---|---|---|---|---|---|---|---|---|---|
| | 0 | 1 | 2 | 3 | 4 | 5 | 6 | 7 | 8 | 9 |
| 0 | 0.00 | 0.10 | 0.21 | 0.31 | 0.42 | 0.52 | 0.63 | 0.73 | 0.84 | 0.94 |
| 10 | 1.05 | 1.15 | 1.26 | 1.36 | 1.47 | 1.57 | 1.67 | 1.78 | 1.88 | 1.99 |
| 20 | 2.09 | 2.20 | 2.30 | 2.41 | 2.51 | 2.62 | 2.72 | 2.83 | 2.93 | 3.04 |
| 30 | 3.14 | 3.25 | 3.35 | 3.46 | 3.56 | 3.66 | 3.77 | 3.87 | 3.98 | 4.08 |
| 40 | 4.19 | 4.29 | 4.40 | 4.50 | 4.61 | 4.71 | 4.82 | 4.92 | 5.03 | 5.13 |
| 50 | 5.24 | 5.34 | 5.44 | 5.55 | 5.65 | 5.76 | 5.86 | 5.97 | 6.07 | 6.18 |
| 60 | 6.28 | 6.39 | 6.49 | 6.60 | 6.70 | 6.81 | 6.91 | 7.02 | 7.12 | 7.23 |
| 70 | 7.33 | 7.43 | 7.54 | 7.64 | 7.75 | 7.85 | 7.96 | 8.06 | 8.17 | 8.27 |
| 80 | 8.38 | 8.48 | 8.59 | 8.69 | 8.80 | 8.90 | 9.01 | 9.11 | 9.21 | 9.32 |
| 90 | 9.42 | 9.53 | 9.63 | 9.74 | 9.84 | 9.95 | 10.05 | 10.16 | 10.26 | 10.37 |
| 100 | 10.47 | 10.58 | 10.68 | 10.79 | 10.89 | 11.00 | 11.10 | 11.20 | 11.31 | 11.41 |
| 110 | 11.52 | 11.62 | 11.73 | 11.83 | 11.94 | 12.04 | 12.15 | 12.25 | 12.36 | 12.46 |
| 120 | 12.57 | 12.67 | 12.78 | 12.88 | 12.98 | 13.09 | 13.19 | 13.30 | 13.40 | 13.51 |
| 130 | 13.61 | 13.72 | 13.82 | 13.93 | 14.03 | 14.14 | 14.24 | 14.35 | 14.45 | 14.56 |
| 140 | 14.66 | 14.76 | 14.87 | 14.97 | 15.08 | 15.18 | 15.29 | 15.39 | 15.50 | 15.60 |
| 150 | 15.71 | 15.81 | 15.92 | 16.02 | 16.13 | 16.23 | 16.34 | 16.44 | 16.55 | 16.65 |
| 160 | 16.75 | 16.86 | 16.96 | 17.07 | 17.17 | 17.28 | 17.38 | 17.49 | 17.59 | 17.70 |
| 170 | 17.80 | 17.91 | 18.01 | 18.12 | 18.22 | 18.33 | 18.43 | 18.53 | 18.64 | 18.74 |
| 180 | 18.85 | 18.95 | 19.06 | 19.16 | 19.27 | 19.37 | 19.48 | 19.58 | 19.69 | 19.79 |
| 190 | 19.90 | 20.00 | 20.11 | 20.21 | 20.32 | 20.42 | 20.52 | 20.63 | 20.73 | 20.84 |
| 200 | 20.94 | 21.05 | 21.15 | 21.26 | 21.36 | 21.47 | 21.57 | 21.68 | 21.78 | 21.89 |
| 210 | 21.99 | 22.10 | 22.20 | 22.30 | 22.41 | 22.51 | 22.62 | 22.72 | 22.83 | 22.93 |
| 220 | 23.04 | 23.14 | 23.25 | 23.35 | 23.46 | 23.56 | 23.67 | 23.77 | 23.88 | 23.98 |
| 230 | 24.09 | 24.19 | 24.29 | 24.40 | 24.50 | 24.61 | 24.71 | 24.82 | 24.92 | 25.03 |
| 240 | 25.13 | 25.24 | 25.34 | 25.45 | 25.55 | 25.66 | 25.76 | 25.87 | 25.97 | 26.07 |
| 250 | 26.18 | 26.28 | 26.39 | 26.49 | 26.60 | 26.70 | 26.81 | 26.91 | 27.02 | 27.12 |
| 260 | 27.23 | 27.33 | 27.44 | 27.54 | 27.65 | 27.75 | 27.85 | 27.96 | 28.06 | 28.17 |
| 270 | 28.27 | 28.38 | 28.48 | 28.59 | 28.69 | 28.80 | 28.90 | 29.01 | 29.11 | 29.22 |
| 280 | 29.32 | 29.43 | 29.53 | 29.64 | 29.74 | 29.84 | 29.95 | 30.05 | 30.16 | 30.26 |
| 290 | 30.37 | 30.47 | 30.58 | 30.68 | 30.79 | 30.89 | 31.00 | 31.10 | 31.21 | 31.31 |
| 300 | 31.42 | 31.52 | 31.62 | 31.73 | 31.83 | 31.94 | 32.04 | 32.15 | 32.25 | 32.36 |
| 310 | 32.46 | 32.57 | 32.67 | 32.78 | 32.88 | 32.99 | 33.09 | 33.20 | 33.30 | 33.41 |
| 320 | 33.51 | 33.61 | 33.72 | 33.82 | 33.93 | 34.03 | 34.14 | 34.24 | 34.35 | 34.45 |
| 330 | 34.56 | 34.66 | 34.77 | 34.87 | 34.98 | 35.08 | 35.19 | 35.29 | 35.39 | 35.50 |
| 340 | 35.60 | 35.71 | 35.81 | 35.92 | 36.02 | 36.13 | 36.23 | 36.34 | 36.44 | 36.55 |
| 350 | 36.65 | 36.76 | 36.86 | 36.97 | 37.07 | 37.18 | 37.28 | 37.38 | 37.49 | 37.59 |
| 360 | 37.70 | 37.80 | 37.91 | 38.01 | 38.12 | 38.22 | 38.33 | 38.43 | 38.54 | 38.64 |
| 370 | 38.75 | 38.85 | 38.96 | 39.06 | 39.16 | 39.27 | 39.37 | 39.48 | 39.58 | 39.69 |
| 380 | 39.79 | 39.90 | 40.00 | 40.11 | 40.21 | 40.32 | 40.42 | 40.53 | 40.63 | 40.74 |
| 390 | 40.84 | 40.94 | 41.05 | 41.15 | 41.26 | 41.36 | 41.47 | 41.57 | 41.68 | 41.78 |
| 400 | 41.89 | 41.99 | 42.10 | 42.20 | 42.31 | 42.41 | 42.52 | 42.62 | 42.73 | 42.83 |
| 410 | 42.93 | 43.04 | 43.14 | 43.25 | 43.35 | 43.46 | 43.56 | 43.67 | 43.77 | 43.88 |
| 420 | 43.98 | 44.09 | 44.19 | 44.30 | 44.40 | 44.51 | 44.61 | 44.71 | 44.82 | 44.92 |
| 430 | 45.03 | 45.13 | 45.24 | 45.34 | 45.45 | 45.45 | 45.66 | 45.76 | 45.87 | 45.97 |
| 440 | 46.08 | 46.18 | 46.29 | 46.39 | 46.50 | 46.60 | 46.70 | 46.81 | 46.91 | 47.02 |
| 450 | 47.12 | 47.23 | 47.33 | 47.44 | 47.54 | 47.65 | 47.75 | 47.86 | 47.96 | 48.07 |
| 460 | 48.17 | 48.28 | 48.38 | 48.48 | 48.59 | 48.69 | 48.80 | 48.90 | 49.01 | 49.11 |
| 470 | 49.22 | 49.32 | 49.43 | 49.53 | 49.64 | 49.74 | 49.85 | 49.95 | 50.06 | 50.16 |
| 480 | 50.26 | 50.37 | 50.47 | 50.58 | 50.68 | 50.79 | 50.89 | 51.00 | 51.10 | 51.21 |
| 490 | 51.31 | 51.42 | 51.52 | 51.63 | 51.73 | 51.84 | 51.94 | 52.05 | 52.15 | 52.26 |

### Four Types of Pendulum

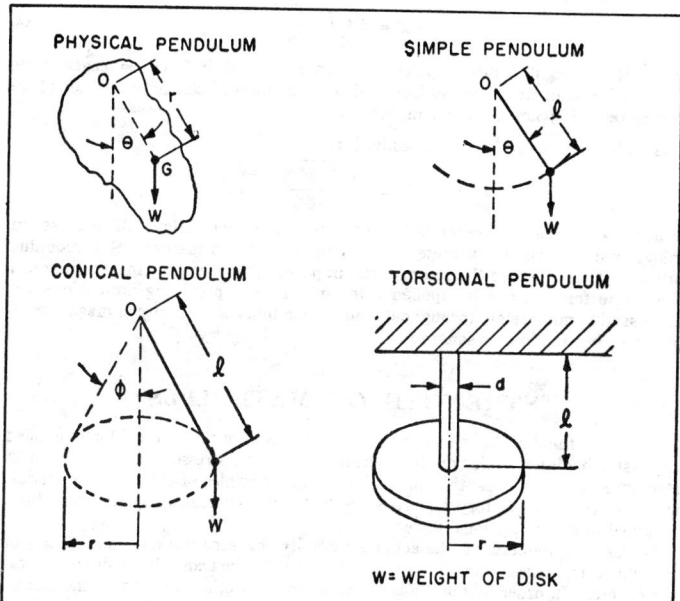

where $T$ = period in seconds for one complete cycle; $g$ = acceleration due to gravity = 32.17 feet per second per second (approximately); and $l$ is the length of the pendulum in feet as shown on the accompanying diagram.

For a *physical* or *compound* pendulum,

$$T = 2\pi \sqrt{\frac{k_0^2}{gr}} \qquad (2)$$

where $k_0$ = radius of gyration of the pendulum about the axis of rotation, in feet, and $r$ is the distance from the axis of rotation to the center of gravity, in feet.

The metric SI units that can be used in the two above formulas are $T$ = seconds; $g$ = approximately 9.81 meters per second squared, which is the value for acceleration due to gravity; $l$ = the length of the pendulum in meters; $k_0$ = the radius of gyration in meters, and $r$ = the distance from the axis of rotation to the center of gravity, in meters.

Formulas (1) and (2) are accurate when the angle of oscillation $\theta$ shown in the diagram is very small. For $\theta$ equal to 22 degrees, these formulas give results that are too small by 1 per cent; for $\theta$ equal to 32 degrees, by 2 per cent.

For a *conical* pendulum, the time in seconds for one revolution is:

$$T = 2\pi \sqrt{\frac{l \cos \phi}{g}}; \quad \text{or,} \quad T = 2\pi \sqrt{\frac{r \cot \phi}{g}} \qquad \text{(3a) and (3b)}$$

For a *torsional* pendulum consisting of a thin rod and a disk as shown in the figure

$$T = \tfrac{2}{3} \sqrt{\frac{\pi W r^2 l}{g d^4 G}} \tag{4}$$

where $W$ = weight of disk in pounds; $r$ = radius of disk in feet; $l$ = length of rod in feet; $d$ = diameter of rod in feet; and $G$ = modulus of elasticity in shear of the rod material in pounds per square inch.

The formula using metric SI units is:

$$T = 8 \sqrt{\frac{\pi M r^2 l}{d^4 G}}$$

**Where $T$ = time in seconds for one complete oscillation; $M$ = mass in kilograms; $r$ = radius in meters; $l$ = length of rod in meters; $G$ = modulus of elasticity in shear of the rod material in pascals (newtons per meter squared). The same formula can be applied using millimeters, providing dimensions are expressed in millimeters throughout, and the modulus of elasticity in megapascals (newtons per millimeter squared).**

# STRENGTH OF MATERIALS

Strength of materials deals with the relations between the external forces applied to elastic bodies and the resulting deformations and stresses. In the design of structures and machines, the application of the principles of strength of materials is necessary if satisfactory materials are to be utilized and adequate proportions obtained to resist functional forces.

Forces are produced by the action of gravity, by accelerations and impacts of moving parts, by gasses and fluids under pressure, by the transmission of mechanical power, etc. In order to analyze the stresses and deflections of a body, the magnitudes, directions and points of application of forces acting on the body must be known. Information given in the Mechanics section provides the basis for evaluating force systems.

The time element in the application of a force on a body is an important consideration. Thus a force may be static or change so slowly that its maximum value can be treated as if it were static; it may be suddenly applied, as in the case of impact; or it may have a repetitive or cyclic behavior.

Also important is the environment in which forces act on a machine or part. Such factors as high and low temperatures; the presence of corrosive gasses, vapors and liquids; radiation, etc. may have a marked effect on how well parts are able to resist stresses.

**Throughout the Strength of Materials section in this Handbook, both English and metric SI data and formulas are given to cover the requirements of working in either system of measurement. Formulas and text relating exclusively to SI units are given in bold-face type.**

**Mechanical Properties of Materials.** — Many mechanical properties of materials are determined from tests, some of which give relationships between stresses and strains as shown by the curves in the accompanying figures.

*Stress* is force per unit area and is usually expressed in pounds per square inch. If the stress tends to stretch or lengthen the material, it is called *tensile* stress; if to compress or shorten the material, a *compressive* stress; and if to shear the material, a *shearing* stress. Tensile and compressive stresses always act at right-angles to (normal to) the area being considered; shearing stresses are always in the plane of the area (at right-angles to compressive or tensile stresses).

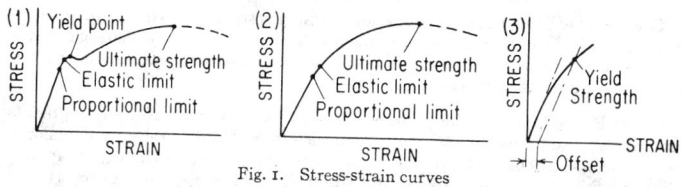

Fig. 1.  Stress-strain curves

In the SI, the unit of stress is the pascal (Pa), the newton per meter squared (N/m²). The megapascal (newton per millimeter squared) is often an appropriate sub-multiple for use in practice.

*Unit strain* is the amount by which a dimension of a body changes when the body is subjected to a load, divided by the original value of the dimension. The simpler term *strain* is often used instead of unit strain.

*Proportional limit* is the point on a stress-strain curve at which it begins to deviate from the straight-line relationship between stress and strain.

*Elastic limit* is the maximum stress to which a test specimen may be subjected and still return to its original length upon release of the load. A material is said to be stressed within the *elastic region* when the working stress does not exceed the elastic limit, and to be stressed within the *plastic region* when the working stress does exceed the elastic limit. The elastic limit for steel is for all purposes the same as its proportional limit.

*Yield point* is a point on the stress-strain curve at which there is a sudden increase in strain without a corresponding increase in stress. Not all materials have a yield point.

*Yield strength*, $S_y$, is the maximum stress that can be applied without permanent deformation of the test specimen. This is the value of the stress at the elastic limit for materials for which there is an elastic limit. Because of the difficulty in determining the elastic limit, and since many materials do not have an elastic region, yield strength is often determined by the offset method as illustrated by the accompanying figure at (3). Yield strength in such a case is the stress value on the stress-strain curve corresponding to a definite amount of permanent set or strain, usually 0.1 or 0.2 per cent of the original dimension.

*Ultimate strength*, $S_u$, (also called *tensile strength*) is the maximum stress value obtained on a stress-strain curve.

*Modulus of elasticity*, E, (also called *Young's modulus*) is the ratio of unit stress to unit strain within the proportional limit of a material in tension or compression.

*Modulus of elasticity in shear*, G, is the ratio of unit stress to unit strain within the proportional limit of a material in shear.

*Poisson's ratio*, $\mu$, is the ratio of lateral strain to longitudinal strain for a given material subjected to uniform longitudinal stresses within the proportional limit. The term is found in certain equations associated with strength of materials. Values of Poisson's ratio for common materials are as follows:

| | | | |
|---|---|---|---|
| Aluminum | 0.334 | Nickel silver | 0.322 |
| Beryllium copper | 0.285 | Phosphor bronze | 0.349 |
| Brass | 0.340 | Rubber | 0.500 |
| Cast iron, gray | 0.211 | Steel, cast | 0.265 |
| Copper | 0.340 | high carbon | 0.295 |
| Inconel | 0.290 | mild | 0.303 |
| Lead | 0.431 | nickel | 0.291 |
| Magnesium | 0.350 | Wrought iron | 0.278 |
| Monel metal | 0.320 | Zinc | 0.331 |

**Compressive Properties.** — From compression tests, *compressive yield strength*, $S_{cy}$, and *compressive ultimate strength*, $S_{cu}$, are determined. Ductile materials under compression loading merely swell or buckle without fracture, hence do not have a compressive ultimate strength.

**Shear Properties.** — The properties of *shear yield strength*, $S_{sy}$, *shear ultimate strength*, $S_{su}$, and the *modulus of rigidity*, $G$ are determined by direct shear and torsional tests. The modulus of rigidity is also known as the modulus of elasticity in shear. It is the ratio of the shear stress, $\tau$, to the shear strain, $\gamma$, in radians, within the proportional limit: $G = \tau/\gamma$.

**Fatigue Properties.** — When a material is subjected to many cycles of stress reversal or fluctuation (variation in magnitude without reversal), failure may occur, even though the maximum stress at any cycle is considerably less than the value at which failure would occur if the stress were constant. Fatigue properties are determined by subjecting test specimens to stress cycles and counting the number of cycles to failure. From a series of such tests in which maximum stress values are progressively reduced, S–N diagrams can be plotted as illustrated by the accompanying figures. The S–N diagram at (1) shows the behavior of a material for which there is an *endurance limit*, $S_{en}$. Endurance limit is the stress value at which the number of cycles to failure is infinite. Steels have endurance limits that vary according to hardness, composition, and quality; but many non-ferrous metals do not. The S–N diagram at (2) does not have an endurance limit. For a metal that does not have an endurance limit, it is standard practice to specify fatigue strength as the stress value corresponding to a specific number of stress reversals, usually 100,000,000 or 500,000,000.

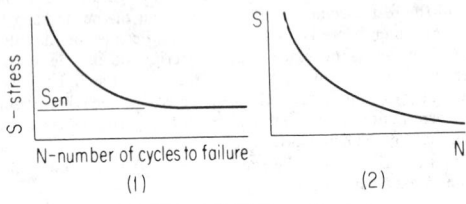

Fig. 2. S–N diagrams

**Factors of Safety.** — There is always a risk that the working stress to which a member is subjected will exceed the strength of its material. The purpose of a factor of safety is to minimize this risk.

Factors of safety can be incorporated into design calculations in many ways. For most calculations the following equation is used:

$$s_w = \frac{S_m}{f_s} \tag{1}$$

$f_s$ is the factor of safety, $S_m$ is the strength of the material in pounds per square inch, and $s_w$ is the allowable working stress, also in pounds per square inch. Since the factor of safety is greater than 1, the allowable working stress will be less than the strength of the material.

In general, $S_m$ is based on yield strength for ductile materials, ultimate strength for brittle materials, and fatigue strength for parts subjected to cyclic stressing.

Most strength values are obtained by testing standard specimens at 68°F. in normal atmospheres. If, however, the character of the stress or environment differs significantly from that used in obtaining standard strength data, then special data must be obtained. If special data are not available, standard data must be suitably modified.

General recommendations for values of factors of safety are given in the following table.

| $f_s$ | Application |
|---|---|
| 1.3–1.5 | For use with highly reliable materials where loading and environmental conditions are not severe, and where weight is an important consideration. |
| 1.5–2 | For applications using reliable materials where loading and environmental conditions are not severe. |
| 2 –2.5 | For use with ordinary materials where loading and environmental conditions are not severe. |
| 2.5–3 | For less tried and for brittle materials where loading and environmental conditions are not severe. |
| 3 –4 | For applications in which material properties are not reliable and where loading and environmental conditions are not severe, or where reliable materials are to be used under difficult loading and environmental conditions. |

**Working Stress.** — Calculated working stresses are the products of calculated nominal stress values and stress concentration factors. Calculated nominal stress values are based on the assumption of idealized stress distributions. Such nominal stresses may be simple stresses, combined stresses, or cyclic stresses. Depending on the nature of the nominal stress, one of the following equations apply:

$$(2) \quad s_w = K\sigma \qquad (4) \quad s_w = K\sigma' \qquad (6) \quad s_w = K\sigma_{cy}$$
$$(3) \quad s_w = K\tau \qquad (5) \quad s_w = K\tau' \qquad (7) \quad s_w = K\tau_{cy}$$

where: $K$ is a stress concentration factor; $\sigma$ and $\tau$ are, respectively, simple normal (tensile or compressive) and shear stresses; $\sigma'$ and $\tau'$ are combined normal and shear stresses; $\sigma_{cy}$ and $\tau_{cy}$ are cyclic normal and shear stresses.

Where there is uneven stress distribution, as illustrated in the accompanying table of simple stresses for Cases 3, 4 and 6, the maximum stress is the one to which the stress concentration factor is applied in computing working stresses. The location of the maximum stress in each case is discussed under the section "Simple Stresses" and the formulas for these maximum stresses are given in the table of simple stresses on page 359.

**Stress Concentration Factors.** — Stress concentration is related to type of material, the nature of the stress, environmental conditions, and the geometry of parts. When stress concentration factors are not available that specifically match all of the foregoing conditions, then the following equation may be used:

$$K = 1 + q(K_t - 1) \qquad (8)$$

$K_t$ is a theoretical stress concentration factor that is a function only of the geometry of a part and the nature of the stress; $q$ is the *index of sensitivity* of the material. If the geometry is such as to provide no theoretical stress concentration, $K_t = 1$.

Curves for evaluating $K_t$ are on pages 355 to 358. For constant stresses in cast iron and in ductile materials, $q = 0$ (hence $K = 1$). For constant stresses in brittle materials such as hardened steel, $q$ may be taken as 0.15; for very brittle materials such as steels that have been quenched but not drawn, $q$ may be taken as 0.25.

When stresses are suddenly applied (impact stresses) $q$ ranges from 0.4 to 0.6 for ductile materials; for cast iron it is taken as 0.5, and 1 for brittle materials.

**Simple Stresses.** — Simple stresses are produced by constant conditions of loading on elements that can be represented as beams, rods, or bars. The table on page 359 summarizes information pertaining to the calculation of simple stresses. Following is an explanation of the symbols used in simple stress formulae:

$\sigma$ = simple normal (tensile or compressive) stress in pounds per square inch
$\tau$ = simple shear stress in pounds per square inch
$F$ = external force in pounds
$V$ = shearing force in pounds
$M$ = bending moment in inch-pounds
$T$ = torsional moment in inch-pounds
$A$ = cross-sectional area in square inches
$Z$ = section modulus in inches$^3$
$Z_p$ = polar section modulus in inches$^3$
$I$ = moment of inertia in inches$^4$
$J$ = polar moment of inertia in inches$^4$
$a$ = area of the web of wide flange and I beams in square inches
$y$ = perpendicular distance from axis through center of gravity of cross-sectional area to stressed fiber in inches
$c$ = radial distance from center of gravity to stressed fiber in inches

**SI metric units can be applied in the calculations in place of the English units of measurement without changes to the formulas. The SI units are the newton (N), which is the unit of force; the meter; the meter squared; the pascal (Pa) which is the newton per meter squared ($N/m^2$); and the newton-meter ($N \cdot m$) for moment of force. Often in design work using the metric system, the millimeter is employed rather than the meter. In such instances, the dimensions can be converted to meters before the stress calculations are begun. Alternatively, the same formulas can be applied using millimeters in place of the meter, providing the treatment is consistent throughout. In such instances, stress and strength properties must be expressed in megapascals (MPa), which is the same as newtons per millimeter squared ($N/mm^2$), and moments in newton-millimeters ($N \cdot mm$). Note: $1 \ N/mm^2 = 1 \ N/10^{-6}m^2 = 10^6 \ N/m^2 = 1$ meganewton/$m^2 = 1$ megapascal.**

For direct tension and direct compression loading, Cases 1 and 2 in the Table of Simple Stresses on page 359, the force $F$ must act along a line through the center of gravity of the section at which the stress is calculated. The equation for direct compression loading applies only to members for which the ratio of length to least radius of gyration is relatively small, approximately 20, otherwise the member must be treated as a column.

The tables on pages 404 to 419 give equations for calculating stresses due to bending for common types of beams and conditions of loading. Where these tables are not applicable, stress may be calculated using Equation (11) in the table on page 359. In using this equation it is necessary to determine the value of the bending moment at the point where the stress is to be calculated. For beams of constant cross-section, stress is ordinarily calculated at the point coinciding with the maximum value of bending moment. Bending loading results in the characteristic stress distribution shown in the table for Case 3. It will be noted that the maximum stress values are at the surfaces farthest from the neutral plane. One of the surfaces is stressed in tension and the other in compression. It is for this reason that the $\pm$ sign is used in Equation (11). Numerous tables for evaluating section modulii are given in the following pages.

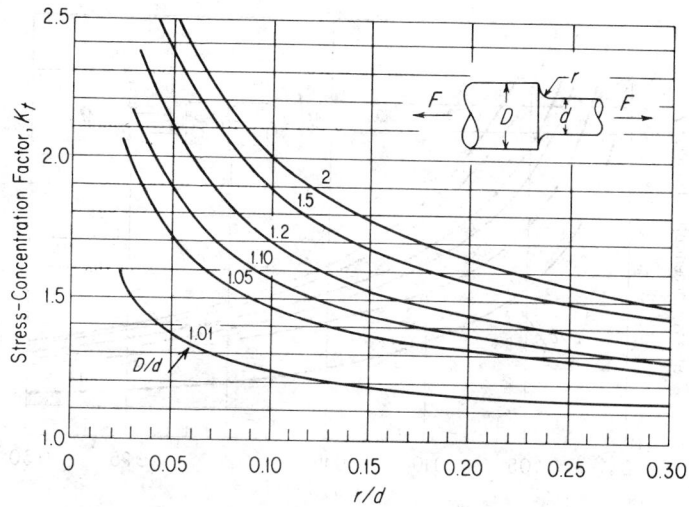

Fig. 3. Stress-concentration factor, $K_t$, for a filleted shaft in tension*

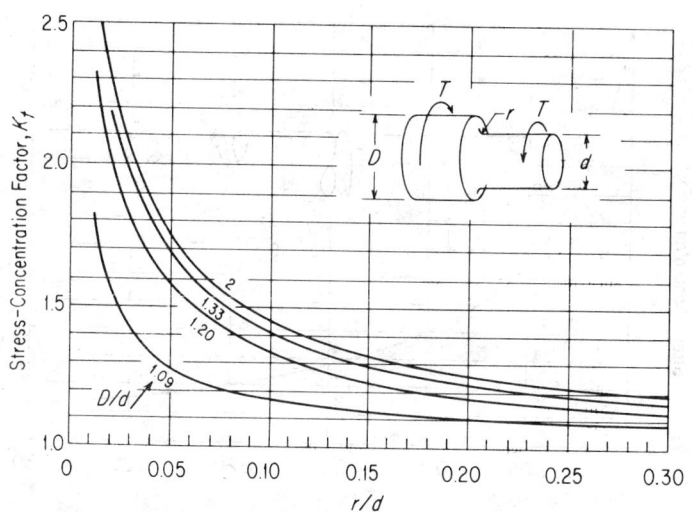

Fig. 4. Stress-concentration factor, $K_t$, for a filleted shaft in torsion*

* See footnote to Fig. 9.

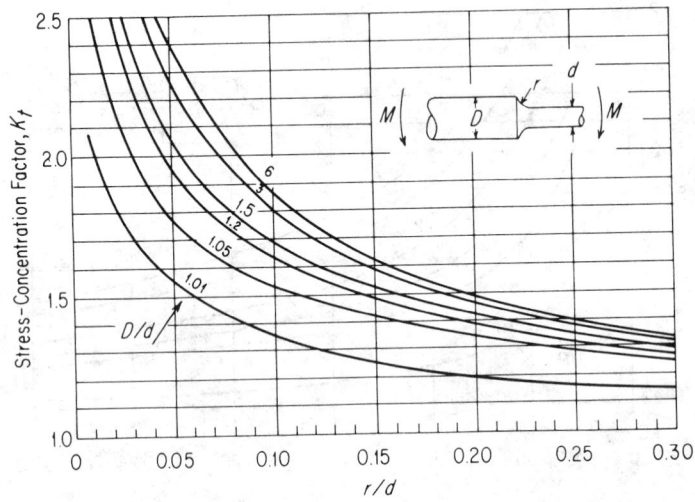

Fig. 5. Stress-concentration factor, $K_t$, for a shaft with shoulder fillet in bending*

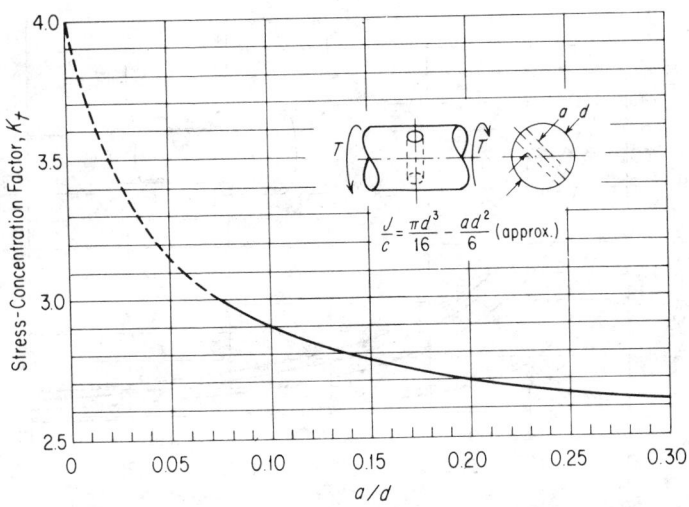

Fig. 6. Stress-concentration factor, $K_t$, for a shaft, with a transverse hole, in torsion*

* See footnote to Fig. 9.

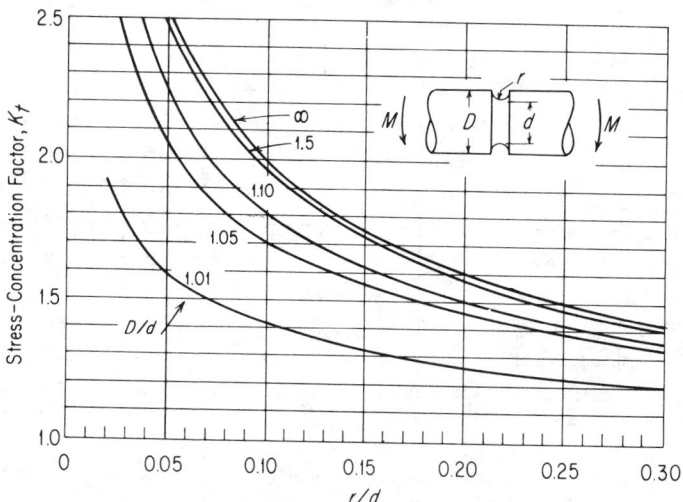

Fig. 7.   Stress-concentration factor, $K_t$, for a grooved shaft in bending*

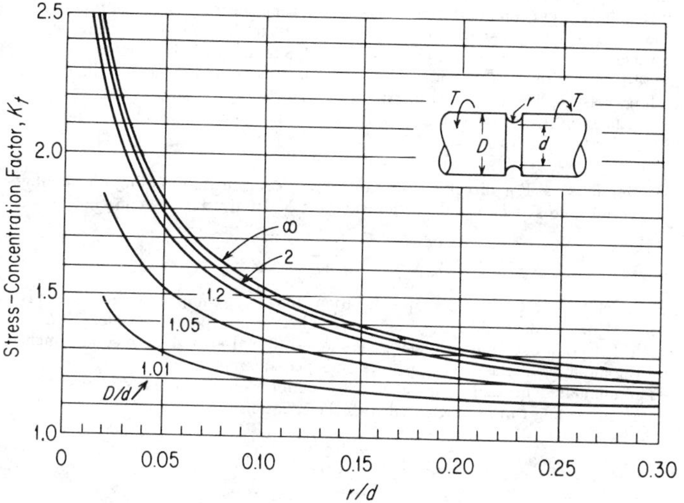

Fig. 8.   Stress-concentration factor, $K_t$, for a grooved shaft in torsion*

\* See footnote to Fig. 9.

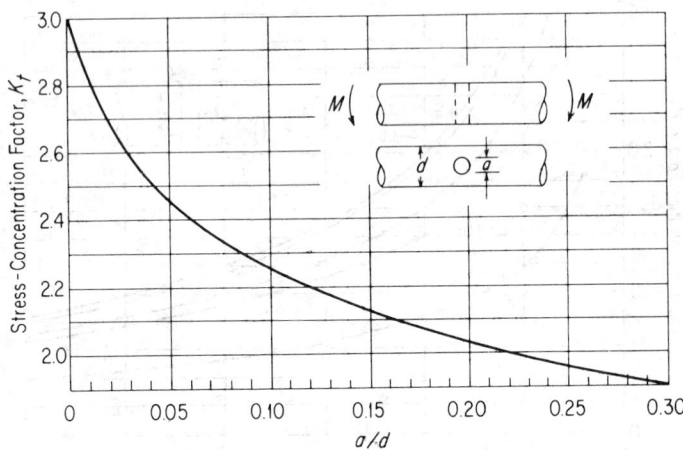

Fig. 9. Stress-concentration factor, $K_t$, for a shaft, with a transverse hole, in bending[*]

Shear stresses caused by bending have maximum values at neutral planes and zero values at the surfaces farthest from the neutral axis, as indicated by the stress distribution diagram shown for Case 4 in the table on page 359. Values for $V$ in Equations (12), (13) and (14) can be determined from shearing force diagrams. The shearing force diagram shown in Case 4 corresponds to the bending moment diagram for Case 3. As shown in this diagram, the value taken for $V$ is represented by the greatest vertical distance from the $x$ axis. The shear stress caused by direct shear loading, Case 5, has a uniform distribution. However, the shear stress caused by torsion loading, Case 6, has a zero value at the axis and a maximum value at the surface farthest from the axis.

**Deflections.** — For direct tension and direct compression loading on members with uniform cross sections, deflection can be calculated using the equation,

$$e = \frac{FL}{AE} \qquad (17)$$

For direct tension loading, $e$ is an elongation; for direct compression loading, $e$ is a contraction. Deflection is in inches when the load $F$ is in pounds, the length $L$ over which deflection occurs is in inches, the cross-sectional area $A$ is in square inches, and the modulus of elasticity $E$ is in pounds per square inch.

The angular deflection of members with uniform circular cross sections subject to torsion loading can be calculated with the equation,

$$\theta = \frac{TL}{GJ} \qquad (18)$$

[*] Source: R. E. Peterson, Design Factors for Stress Concentration, *Machine Design*, vol. 23, 1951. For other stress concentration charts, see Lipson and Juvinall, *The Handbook of Stress and Strength*, The Macmillan Co., 1963.

## Table of Simple Stresses*

| Case | Type of Loading | Illustration | Stress Distribution | Stress Equations |
|------|-----------------|--------------|---------------------|------------------|
| 1 | Direct tension | F ←——→ F | Uniform | $\sigma = \dfrac{F}{A}$   (9) |
| 2 | Direct compression | F —→←— F | Uniform | $\sigma = -\dfrac{F}{A}$   (10) |
| 3 | Bending | Bending moment diagram | $-\sigma$ / $+\sigma$   Neutral plane | $\sigma = \pm\dfrac{M}{Z} = \pm\dfrac{My}{I}$   (11) |
| 4 | Bending | Shearing force diagram | Neutral plane | For beams of rectangular cross-section: $\tau = \dfrac{3V}{2A}$  (12) <br> For beams of solid circular cross-section: $\tau = \dfrac{4V}{3A}$  (13) <br> For wide flange and I beams (approximately): $\tau = \dfrac{V}{a}$  (14) |
| 5 | Direct shear | F↓ / ↑F | Uniform | $\tau = \dfrac{F}{A}$   (15) |
| 6 | Torsion | T | $\tau$ | $\tau = \dfrac{T}{Z_p} = \dfrac{Tc}{J}$   (16) |

*See page 354 for explanation of the stress equation symbols, in terms of English and metric SI units of measurement.

The angular deflection $\theta$ is in radians when the torsional moment $T$ is in inch-pounds, the length $L$ over which the member is twisted is in inches, the modulus of rigidity $G$ is in pounds per square inch, and the polar moment of inertia $J$ is in inches.[4]

Metric SI units can be used in Equations (17) and (18), where $F$ = force in newtons (N); $L$ = length over which deflection or twisting occurs in meters; $A$ = cross-sectional area in meters squared; $E$ = the modulus of elasticity in (newtons per meter squared); $\theta$ = radians; $T$ = the torsional moment in newton-meters (N·m); $G$ = modulus of rigidity, in pascals; and $J$ = the polar moment of inertia in meters[4]. If the load ($F$) is applied as a weight, it should be noted that the weight of a mass $M$ kilograms is $Mg$ newtons, where $g$ = 9.81 m/s². Millimeters can be used in the calculations in place of meters, providing the treatment is consistent throughout.

**Combined Stresses.**— A member may be loaded in such a way that a combination of simple stresses act at a point. Three general cases occur, examples of which are shown in the accompanying illustration.

*Superposition of stresses:* The figure at (1) illustrates a common situation that results in simple stresses combining by superposition at points $a$ and $b$. The equal and opposite forces $F_1$ will cause a compressive stress $\sigma_1 = -F_1/A$. Force $F_2$ will cause a bending moment $M$ to exist in the plane of points $a$ and $b$. The resulting stress $\sigma_2 = \pm M/Z$. The combined stress at point $a$,

$$\sigma_a' = -\frac{F_1}{A} - \frac{M}{Z}; \quad \text{and at } b, \quad \sigma_b' = -\frac{F_1}{A} + \frac{M}{Z} \quad \text{(19 and 20)}$$

where the minus sign indicates a compressive stress and the plus sign a tensile stress. Thus, the stress at $a$ will be compressive and at $b$ either tensile or compressive depending on which term in the equation for $\sigma_b'$ has the greatest value.

*Normal stresses at right angles:* This is shown in the figure at (2). This combination of stresses occurs, for example, in tanks subjected to internal or external pressure. The principal normal stress is either $\sigma_x' = F_1/A_1$ or $\sigma_y' = F_2/A_2$, depending on whichever is greatest. Because of the combination of $\sigma_x$ and $\sigma_y$ there will be a maximum shear stress,

$$\tau' = \frac{\sigma_x - \sigma_y}{2} \quad (21)$$

*Normal and shear stresses:* The example in the figure at (3) shows a member subjected to a torsional shear stress, $\tau = T/Z_p$, and a direct compressive stress, $\sigma = -F/A$. At some point $a$ on the member the principal normal stresses are calculated using the equation,

$$\sigma' = \frac{\sigma}{2} \pm \sqrt{\left(\frac{\sigma}{2}\right)^2 + \tau^2} \quad (22)$$

The maximum shear stress is calculated by using the equation,

$$\tau' = \sqrt{\left(\frac{\sigma}{2}\right)^2 + \tau^2} \quad (23)$$

The point $a$ should ordinarily be selected where stress is a maximum value. For the example shown in the figure at (3), the point $a$ can be anywhere on the cylindrical surface since the combined stress has the same value anywhere on that surface.

The calculations involved in the use of equations (22) and (23) may be simplified by using the accompanying table. This table gives factors by means of which the principal normal stress $\sigma'$ (tensile or compressive) and the maximum shear stress $\tau'$ may be determined if the simple normal stress $\sigma$, and the simple shear stress $\tau$

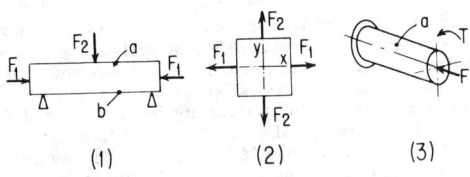

Fig. 10. Types of combined loading

## Shear Stresses Combined with Tension or Compression Stresses

| Ratio of Shear to Tensile Stress, $\tau/\sigma$ | Tension Factor $x$ | Shear Factor $y$ | Ratio of Shear to Tensile Stress, $\tau/\sigma$ | Tension Factor $x$ | Shear Factor $y$ |
|---|---|---|---|---|---|
| 0.05 | 1.002 | 10.05 | 0.80 | 1.44 | 1.18 |
| 0.10 | 1.010 | 5.10  | 0.85 | 1.49 | 1.16 |
| 0.15 | 1.022 | 3.48  | 0.90 | 1.53 | 1.14 |
| 0.20 | 1.038 | 2.69  | 0.95 | 1.57 | 1.13 |
| 0.25 | 1.059 | 2.24  | 1.00 | 1.62 | 1.12 |
| 0.30 | 1.083 | 1.94  | 1.05 | 1.66 | 1.11 |
| 0.35 | 1.110 | 1.74  | 1.10 | 1.71 | 1.10 |
| 0.40 | 1.140 | 1.60  | 1.15 | 1.75 | 1.09 |
| 0.45 | 1.173 | 1.49  | 1.20 | 1.80 | 1.08 |
| 0.50 | 1.207 | 1.41  | 1.25 | 1.85 | 1.08 |
| 0.55 | 1.243 | 1.35  | 1.30 | 1.89 | 1.07 |
| 0.60 | 1.281 | 1.30  | 1.35 | 1.94 | 1.07 |
| 0.65 | 1.320 | 1.26  | 1.40 | 1.99 | 1.06 |
| 0.70 | 1.360 | 1.23  | 1.45 | 2.03 | 1.06 |
| 0.75 | 1.401 | 1.20  | 1.50 | 2.08 | 1.05 |

| Ratio of Tensile to Shear Stress, $\sigma/\tau$ | Shear Factor $y$ | Tension Factor $x$ | Ratio of Tensile to Shear Stress, $\sigma/\tau$ | Shear Factor $y$ | Tension Factor $x$ |
|---|---|---|---|---|---|
| 0.05 | 1.0003 | 20.50 | 0.80 | 1.077 | 1.85 |
| 0.10 | 1.0012 | 10.51 | 0.85 | 1.086 | 1.78 |
| 0.15 | 1.0028 | 7.18  | 0.90 | 1.096 | 1.72 |
| 0.20 | 1.0050 | 5.52  | 0.95 | 1.107 | 1.67 |
| 0.25 | 1.0078 | 4.53  | 1.00 | 1.118 | 1.62 |
| 0.30 | 1.0112 | 3.87  | 1.05 | 1.129 | 1.57 |
| 0.35 | 1.0152 | 3.40  | 1.10 | 1.141 | 1.54 |
| 0.40 | 1.0198 | 3.04  | 1.15 | 1.153 | 1.50 |
| 0.45 | 1.0250 | 2.77  | 1.20 | 1.166 | 1.47 |
| 0.50 | 1.0308 | 2.56  | 1.25 | 1.179 | 1.44 |
| 0.55 | 1.0371 | 2.38  | 1.30 | 1.193 | 1.42 |
| 0.60 | 1.0440 | 2.24  | 1.35 | 1.206 | 1.39 |
| 0.65 | 1.0515 | 2.11  | 1.40 | 1.221 | 1.37 |
| 0.70 | 1.0595 | 2.01  | 1.45 | 1.235 | 1.35 |
| 0.75 | 1.0680 | 1.92  | 1.50 | 1.250 | 1.33 |

are known. For example, assume that a shaft is loaded in direct compression and torsion such that the normal stress due to the direct compression loading is 12,000 pounds per square inch and the shear stress due to torsion loading is 9000 pounds per square inch. The ratio $\tau/\sigma = 0.75$. From the table the tension (or compression) factor $x = 1.401$ and the shear factor $y = 1.20$. Hence, the principal normal stress $\sigma' = x\sigma = 1.401 \times 12,000 = 16,810$ psi, and the maximum shear stress $\tau' = y\tau = 1.20 \times 9000 = 10,800$ psi.

Beginning below, another table, the Table of Combined Stresses, lists equations for maximum nominal tensile or compressive (normal) stresses, and maximum nominal shear stresses for common machine elements. These equations were derived using general equations (19), (20), (22), and (23). The equations apply to the critical points indicated on the figures. Cases 1 through 4 are cantilever beams. These may be loaded with a combination of a vertical and horizontal force, or by a single oblique force. If the single oblique force $F$ and the angle $\theta$ are given, then horizontal and vertical forces can be calculated using the equations $F_x = F \cos \theta$ and $F_y = F \sin \theta$. In cases 9 and 10 of the table, the equations for $\sigma_a'$ can give a tensile and a compressive stress because of the $\pm$ sign in front of the radical. Equations involving direct compression are valid only if machine elements have relatively short lengths with respect to their sections, otherwise column equations apply.

*Calculation of worst stress condition:* Stress failure can occur at any critical point if either the tensile, compressive, or shear stress properties of the material are exceeded by the corresponding working stress. It is necessary to evaluate the factor of safety for each possible failure condition.

The following rules apply to calculations using equations in the Table of Simple Stresses, and to calculations based on equations (19) and (20). *Rule 1:* For every calculated normal stress there is a corresponding induced shear stress; the value of the shear stress is equal to half that of the normal stress. *Rule 2:* For every calculated shear stress there is a corresponding induced normal stress; the value of the normal stress is equal to that of the shear stress. The Table of Combined Stresses includes equations for calculating both maximum nominal tensile or compressive stresses, and maximum nominal shear stresses.

### Table of Combined Stresses — 1

| Case | Type of Beam and Loading* | Maximum Nominal Tens. or Comp. Stress | Maximum Nominal Shear Stress |
|------|---------------------------|---------------------------------------|------------------------------|
| 1 | | $\sigma_a' = \dfrac{1.273}{d^2}\left(\dfrac{8LF_y}{d} - F_x\right)$ <br><br> $\sigma_b' = -\dfrac{1.273}{d^2}\left(\dfrac{8LF_y}{d} + F_x\right)$ | $\tau_a' = 0.5\sigma_a'$ <br><br> $\tau_b' = 0.5\sigma_b'$ |
| 2 | | $\sigma_a' = \dfrac{1.273}{d^2}\left(F_x + \dfrac{8LF_y}{d}\right)$ <br><br> $\sigma_b' = \dfrac{1.273}{d^2}\left(F_x - \dfrac{8LF_y}{d}\right)$ | $\tau_a' = 0.5\sigma_a'$ <br><br> $\tau_b' = 0.5\sigma_b'$ |
| 3 | | $\sigma_a' = \dfrac{1}{bh}\left(\dfrac{6LF_y}{h} - F_x\right)$ <br><br> $\sigma_b' = -\dfrac{1}{bh}\left(\dfrac{6LF_y}{h} + F_x\right)$ | $\tau_a' = 0.5\sigma_a'$ <br><br> $\tau_b' = 0.5\sigma_b'$ |

* *Case 1:* Circular cantilever beam in direct compression and bending. *Case 2:* Circular cantilever beam in direct tension and bending. *Case 3:* Rectangular cantilever beam in direct compression and bending. See pages 353 and 360 for combined stress symbols.

## Table of Combined Stresses — 2

| Case | Type of Beam and Loading* | Maximum Nominal Tens. or Comp. Stress | Maximum Nominal Shear Stress |
|---|---|---|---|
| 4 | | $\sigma_a' = \dfrac{1}{bh}\left(F_x + \dfrac{6LF_y}{h}\right)$ <br><br> $\sigma_b' = \dfrac{1}{bh}\left(F_x - \dfrac{6LF_y}{h}\right)$ | $\tau_a' = 0.5\sigma_a'$ <br><br> $\tau_b' = 0.5\sigma_b'$ |
| 5 | | $\sigma_a' = -\dfrac{1.273}{d^2}\left(\dfrac{2LF_y}{d} + F_x\right)$ <br><br> $\sigma_b' = \dfrac{1.273}{d^2}\left(\dfrac{2LF_y}{d} - F_x\right)$ | $\tau_a' = 0.5\sigma_a'$ <br><br> $\tau_b' = 0.5\sigma_b'$ |
| 6 | | $\sigma_a' = \dfrac{1.273}{d^2}\left(F_x - \dfrac{2LF_y}{d}\right)$ <br><br> $\sigma_b' = \dfrac{1.273}{d^2}\left(F_x + \dfrac{2LF_y}{d}\right)$ | $\tau_a' = 0.5\sigma_a'$ <br><br> $\tau_b' = 0.5\sigma_b'$ |
| 7 | | $\sigma_a' = -\dfrac{1}{bh}\left(\dfrac{3LF_y}{2h} + F_x\right)$ <br><br> $\sigma_b' = \dfrac{1}{bh}\left(\dfrac{3LF_y}{2h} - F_x\right)$ | $\tau_a' = 0.5\sigma_a'$ <br><br> $\tau_b' = 0.5\sigma_b'$ |
| 8 | | $\sigma_a' = \dfrac{1}{bh}\left(F_x - \dfrac{3LF_y}{2h}\right)$ <br><br> $\sigma_b' = \dfrac{1}{bh}\left(F_x + \dfrac{3LF_y}{2h}\right)$ | $\tau_a' = 0.5\sigma_a'$ <br><br> $\tau_b' = 0.5\sigma_b'$ |
| 9 | a anywhere on surface | $\sigma_a' = -\dfrac{0.637}{d^2}\left[F \pm \sqrt{F^2 + \left(\dfrac{8T}{d}\right)^2}\right]$ | $\tau_a' = -\dfrac{0.637}{d^2}\sqrt{F^2 + \left(\dfrac{8T}{d}\right)^2}$ |
| 10 | a anywhere on surface | $\sigma_a' = \dfrac{0.637}{d^2}\left[F \pm \sqrt{F^2 + \left(\dfrac{8T}{d}\right)^2}\right]$ | $\tau_a' = \dfrac{0.637}{d^2}\sqrt{F^2 + \left(\dfrac{8T}{d}\right)^2}$ |
| 11 | | $\sigma_a' = \dfrac{1.273F}{d^2}\left(1 - \dfrac{8e}{d}\right)$ <br><br> $\sigma_b' = \dfrac{1.273F}{d^2}\left(1 + \dfrac{8e}{d}\right)$ | $\tau_a' = 0.5\sigma_a'$ <br><br> $\tau_b' = 0.5\sigma_b'$ |

* *Case 4:* Rectangular cantilever beam in direct tension and bending. *Case 5:* Circular beam or shaft in direct compression and bending. *Case 6:* Circular beam or shaft in direct tension and bending. *Case 7:* Rectangular beam or shaft in direct compression and bending. *Case 8:* Rectangular beam or shaft in direct tension and bending. *Case 9:* Circular shaft in direct compression and torsion. *Case 10:* Circular shaft in direction tension and torsion. *Case 11:* Offset link, circular cross section, in direct tension. See pages 353 and 360 for combined stress symbols.

## Table of Combined Stresses — 3

| Case | Type of Beam and Loading* | Maximum Nominal Tens. or Comp. Stress | Maximum Nominal Shear Stress |
|---|---|---|---|
| 12 | | $\sigma_a' = \dfrac{1.273F}{d^2}\left(\dfrac{8e}{d} - 1\right)$ <br><br> $\sigma_b' = -\dfrac{1.273F}{d^2}\left(\dfrac{8e}{d} + 1\right)$ | $\tau_a' = 0.5\sigma_a'$ <br><br> $\tau_b' = 0.5\sigma_b'$ |
| 13 | | $\sigma_a' = \dfrac{F}{bh}\left(1 - \dfrac{6e}{h}\right)$ <br><br> $\sigma_b' = \dfrac{F}{bh}\left(1 + \dfrac{6e}{h}\right)$ | $\tau_a' = 0.5\sigma_a'$ <br><br> $\tau_b' = 0.5\sigma_b'$ |
| 14 | | $\sigma_a' = \dfrac{F}{bh}\left(\dfrac{6e}{h} - 1\right)$ <br><br> $\sigma_b' = -\dfrac{F}{bh}\left(\dfrac{6e}{h} + 1\right)$ | $\tau_a' = 0.5\sigma_a'$ <br><br> $\tau_b' = 0.5\sigma_b'$ |

* *Case 12:* Offset link, circular cross section, in direct compression. *Case 13:* Offset link, rectangular section, in direct tension. *Case 14:* Offset link, rectangular section, in direct compression. See pages 353 and 360 for combined stress symbols.

Formulas from the simple and combined stress tables, as well as tension and shear factors, can be applied without change in calculations using metric SI units. Stresses are given in newtons per meter squared ($N/m^2$) or in $N/mm^2$.

Sample Calculations. — The following examples illustrate some typical strength of materials calculations, using both English and metric SI units of measurement.

*Example 1(a):* A round bar made from SAE 1025 low carbon steel is to support a direct tension load of 50,000 pounds. Using a factor of safety of 4, and assuming that the stress concentration factor $K = 1$, a suitable standard diameter is to be determined. Calculations are to be based on a yield strength of 40,000 psi.

Because the factor of safety and strength of the material are known, the allowable working stress $s_w$ may be calculated using Equation (1): $40,000/4 = 10,000$ psi. The relationship between working stress $s_w$ and nominal stress $\sigma$ is given by equation (2). Since $K = 1$, $\sigma = 10,000$ psi. Applying Equation (9) in the Table of Simple Stresses, the area of the bar can be solved for: $A = 50,000/10,000 = 5$ square inches. The next largest standard diameter corresponding to this area is $2\frac{9}{16}$ inches.

*Example 1(b):* A similar example to that given in 1(a), using metric SI units is as follows. A round steel bar of 300 meganewtons/meter² yield strength, is to withstand a direct tension of 200 kilonewtons. Using a safety factor of 4, and assuming that the stress concentration factor $K = 1$, a suitable diameter is to be determined.

Because the factor of safety and the strength of the material are known, the allowable working stress $s_w$ may be calculated using Equation (1): $300/4 = 75$ meganewtons/meter$^2$. The relationship between working stress and nominal stress $\sigma$ is given by Equation (2). Since $K = 1$, $\sigma = 75$ MN/m$^2$. Applying Equation (9) in the Table of Simple Stresses, the area of the bar can be determined from:

$$A = \frac{200 \text{ kN}}{75 \text{ MN/m}^2} = \frac{200,000 \text{ N}}{75,000,000 \text{ N/m}^2} = 0.00267 \text{ m}^2.$$

The diameter corresponding to this area is 0.058 meters, which, rounded up would be 0.06 m.

Millimeters can be employed in the calculations in place of meters, providing the treatment is consistent throughout. In this instance the diameter would be 60 mm.

*Note:* If the tension in the bar is produced by hanging a mass of $M$ kg from the end of it, the value is $Mg$ newtons, where $g =$ approximately 9.81 meters per second$^2$.

*Example 2(a):* What would the total elongation of the bar in Example 1(a) be if its length were 60 inches? Applying Equation (17),

$$e = \frac{50,000 \times 60}{5.157 \times 30,000,000} = 0.019 \text{ inch}$$

*Example 2(b):* What would be the total elongation of the bar in Example 1(b) if its length were 1.5 meters? The problem is solved by applying Equation (17) in which $F = 200$ kilonewtons; $L = 1.5$ meters; $A = \pi 0.06^2/4 = 0.00283$ m$^2$. Assuming a modulus of elasticity $E$ of 200 giganewtons/meter$^2$, then the calculation is:

$$e = \frac{200,000 \times 1.5}{0.00283 \times 200,000,000,000} = 0.000530 \text{ m}.$$

The calculation is less unwieldly if carried out using millimeters in place of meters; then $F = 200$ kN; $L = 1500$ mm; $A = 2830$ mm$^2$, and $E = 200,000$ N/mm$^2$. Thus:

$$e = \frac{200,000 \times 1500}{2830 \times 200,000} = 0.530 \text{ mm}.$$

*Example 3(a):* Determine the size for the section of a square bar which is to be held firmly at one end and is to support a load of 3000 pounds at the outer end. The bar is to be 30 inches long and is to be made from SAE 1045 medium carbon steel with a yield point of 60,000 psi. A factor of safety of 3 and a stress concentration factor of 1.3 are to be used.

From Equation (1) the allowable working stress $s_w = 60,000/3 = 20,000$ psi. The applicable equation relating working stress and nominal stress is equation (2); hence, $\sigma = 20,000/1.3 = 15,400$ psi. The member must be treated as a cantilever beam subject to a bending moment of $30 \times 3000$ or 90,000 inch-pounds. Solving Equation (11) in the Table of Simple Stresses for section modulus: $Z = 90,000/15,400 = 5.85$ inch.[3] Since the section modulus for a square section with neutral axis equidistant from either side is $a^3/6$, where $a$ is the dimension of the square, $a = \sqrt[3]{35.1} = 3.27$ inches. The size of the bar can therefore be $3\frac{5}{16}$ inches.

*Example 3(b):* A similar example to that given in 3(a), using metric SI units is as follows. Determine the size for the section of a square bar which is to be held firmly at one end and is to support a load of 1600 kilograms at the outer end. The bar is to be 1 meter long, and is to be made from steel with a yield strength of 500 newtons/mm². A factor of safety of 3, and a stress concentration factor of 1.3 are to be used. The calculation can be performed using millimeters throughout.

From Equation (1) the allowable working stress $s_w$ = 500 N/mm²/3 = 167 N/mm². The formula relating working stress and nominal stress is Equation (2); hence $\sigma$ = 167/1.3 = 128 N/mm². Since a mass of 1600 kg equals a weight of 1600 $g$ newtons, where $g$ = 9.81 meters/second², the force acting on the bar is 15,700 newtons. The bending moment on the bar, which must be treated as a cantilever beam is thus 1000 mm × 15,700 N = 15,700-000 Nmm. Solving Equation (11) in the Table of Simple Stresses for section modulus: $Z$ = $M/\sigma$ = 15,700,000/128 = 123,000 mm³. Since the section modulus for a square section with neutral axis equidistant from either side is $a^3/6$, where $a$ is the dimension of the square,

$$a = \sqrt[3]{6 \times 123,000} = 90.4 \text{ mm}.$$

*Example 4(a):* Find the working stress in a 2-inch diameter shaft through which a transverse hole ¼ inch in diameter has been drilled. The shaft is subject to a torsional moment of 80,000 inch-pounds and is made from hardened steel so that the index of sensitivity $q$ = 0.2.

The polar section modulus is calculated using the equation shown in the stress concentration curve for a Round Shaft in Torsion with Transverse Hole, page 356,

$$\frac{J}{c} = Z_p = \frac{\pi 2^3}{16} - \frac{2^2}{4 \times 6} = 1.4 \text{ inches.}^3$$

The nominal shear stress due to the torsion loading is computed using Equation (16) in the Table of Simple Stresses:

$$\tau = 80,000/1.4 = 57,200 \text{ psi.}$$

Referring to the previously mentioned stress concentration curve on page 356, $K_t$ is 2.82 since $d/D$ is 0.125. The stress concentration factor may now be calculated by means of Equation (8): $K$ = 1 + 0.2(2.82 − 1) = 1.36. Working stress calculated with Equation (3) is $s_w$ = 1.36 × 57,200 = 77,800 psi.

*Example 4(b):* A similar example to that given in 4(a), using metric SI units is as follows. Find the working stress in a 50 mm diameter shaft through which a transverse hole 6 mm in diameter has been drilled. The shaft is subject to a torsional moment of 8000 newton-meters, and has an index of sensitivity of $g$ = 0.2. If the calculation is made in millimeters, the torsional moment is 8,000,000 N·mm.

The polar section modulus is calculated using the equation shown in the stress concentration curve for a Round Shaft with Transverse Hole, page 356:

$$\frac{J}{c} = Z_p = \frac{\pi 50^3}{16} - \frac{6 \times 50^2}{6}$$

$$= 24,544 - 2500 = 22,044 \text{ mm}^3.$$

The nominal shear stress due to torsion loading is computed using Equation (16) in the Table of Simple Stresses:

$$\tau = 8,000,000/22,044 = 363 \text{ N/mm}^2 = 363 \text{ megapascals.}$$

Referring to the previously mentioned stress concentration curve on page 356, $K_t$ is 2.85, since $a/d$ = 6/50 = 0.12. The stress concentration factor may now be calculated by means of Equation (8): $K$ = 1 + 0.2(2.85 − 1) = 1.37. From Equation (3), working stress $s_w$ = 1.37 × 363 = 497 N/mm² = 497 megapascals.

*Example 5 (a):* For Case 3 in the Table of Combined Stresses, calculate the least factor of safety if: a 5052-H32 aluminum beam is 10 inches long; one inch wide, and 2 inches high. Yield strengths are 23,000 psi tension; 21,000 psi compression; 13,000 psi shear. The stress concentration factor is 1.5; $F_y$ is 600 lbs; $F_x$ 500 lbs. From Table of Combined Stresses, Case 3:

$$\sigma_b' = -\frac{1}{1 \times 2}\left(\frac{6 \times 10 \times 600}{2} + 500\right) = -9250 \text{ psi (in compression)}$$

The other formulas for Case 3 give $\sigma_a' = 8750$ psi (in tension); $\tau_a' = 4375$ psi, and $\tau_b' = 4625$ psi. Using equation (4) for the nominal compressive stress of 9250 psi: $S_w = 1.5 \times 9250 = 13,900$ psi. From equation (1) $f_s = 21,000/13,900 = 1.51$. Applying equations (1), (4) and (5) in appropriate fashion to the other calculated nominal stress values for tension and shear will show that the factor of safety of 1.51, governed by the compressive stress at $b$ on the beam, is minimum.

*Example 5 (b):* What maximum $F$ can be applied in Case 3 if the aluminum beam is 200 mm long; 20 mm wide; 40 mm high; $\theta = 30°$; $f_s = 2$, governing for compression, $K = 1.5$, and $S_m = 144 \text{N/mm}^2$ for compression.

From equation (1) $S_w = -144/2 = -72 \text{N/mm}^2$. Therefore, from equation (4), $\sigma_b' = -72/1.5 = -48 \text{N/mm}^2$. Since $F_x = F \cos 30° = 0.866 F$, and $F_y = F \sin 30° = 0.5 F$:

$$-48 = -\frac{1}{20 \times 40}\left(0.866F + \frac{6 \times 200 \times 0.5F}{40}\right)$$
$$F = 2420 \text{ N}$$

**Moment of Inertia of Built-up Sections.** — The usual method of calculating the moment of inertia of a built-up section involves the calculations of the moment of inertia for each element of the section about its own neutral axis, and the transferring of this moment of inertia to the previously found neutral axis of the whole built-up section. A much simpler method that can be used in the case of any section which can be divided into rectangular elements bounded by lines parallel

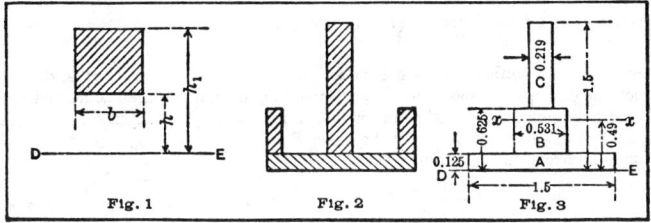

Fig. 1          Fig. 2          Fig. 3

and perpendicular to the neutral axis is the so-called tabular method based upon the formula: $I = \dfrac{b\,(h_1{}^3 - h^3)}{3}$ in which $I$ = the moment of inertia about axis $DE$, Fig. 1, and $b$, $h$ and $h_1$ are dimensions as given in the same illustration.

The method may be illustrated by applying it to the section shown in Fig. 2, and for simplicity of calculation shown " massed " in Fig. 3. The calculation may then be tabulated as shown in the accompanying table. The distance from the axis $DE$ to the neutral axis $xx$ (which will be designated as $d$) is found by dividing the sum of the geometrical moments by the area. The moment of inertia about the neutral axis is then found in the usual way by subtracting the area multiplied by $d^2$ from the moment of inertia about the axis $DE$.

**Tabulated Calculation of Moment of Inertia**

| Section | Breadth $b$ | Height $h_1$ | Area $b(h_1 - h)$ | $h_1^2$ | Moment $\dfrac{b(h_1^2 - h^2)}{2}$ | $h_1^3$ | $I$ about axis $DE$ $\dfrac{b(h_1^3 - h^3)}{3}$ |
|---------|---------|---------|---------|---------|---------|---------|---------|
| A | 1.500 | 0.125 | 0.187 | 0.016 | 0.012 | 0.002 | 0.001 |
| B | 0.531 | 0.625 | 0.266 | 0.391 | 0.100 | 0.244 | 0.043 |
| C | 0.219 | 1.500 | 0.191 | 2.250 | 0.203 | 3.375 | 0.228 |
| | $A = 0.644$ | | | $M = 0.315$ | | $I_{DE} = 0.272$ | |

The distance $d$ from $DE$, the axis through the base of the configuration, to the neutral axis $xx$ is:

$$d = \frac{M}{A} = \frac{0.315}{0.644} = 0.49$$

The moment of inertia of the entire section with reference to the neutral axis $xx$ is:

$$I_N = I_{DE} - Ad^2$$
$$= 0.272 - 0.644 \times 0.49^2$$
$$= 0.117$$

**Tables of Moments of Inertia, Section Moduli, etc.** — On the following pages are given tables of the moments of inertia and other properties of forty-two different cross-sections. The tables give the area of the section and the distance $y$ from the neutral axis to the extreme fiber, in each case. In some cases, where the formulas for the section modulus and radius of gyration are very lengthy, the formula for the section modulus, for example, has been simply given as $\dfrac{I}{y}$. The radius of gyration is sometimes given as $\sqrt{\dfrac{I}{A}}$ to save space.

**Stresses and Deflections in a Loaded Ring.** — For *thin* rings, that is, rings in which the dimension $d$ shown in the accompanying diagram is small compared with $D$, the maximum stress in the ring is due primarily to bending moments produced by the forces $P$. The maximum stress due to bending is:

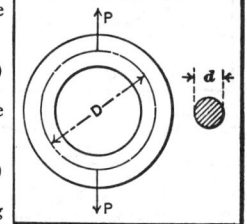

$$S = \frac{PDd}{4\pi I} \qquad (1)$$

For a ring of circular cross section where $d$ is the diameter of the bar from which the ring is made,

$$S = \frac{1.621\,PD}{d^3}, \quad \text{or,} \quad P = \frac{0.617 S\,d^3}{D} \qquad (2)$$

The increase in the vertical diameter of the ring due to the load $P$ is:

Increase in vertical diameter $= \dfrac{0.0186\,PD^3}{EI}$ inches $\qquad (3)$

The *decrease* in the horizontal diameter will be about 92% of the increase in the vertical diameter given by Formula (3). In the above formulas, $P$ = load on ring in pounds; $D$ = mean diameter of ring in inches; $S$ = tensile stress in pounds per square inch, $I$ = moment of inertia of section in inches[4]; and $E$ = modulus of elasticity of material in pounds per square inch.

**Strength of Taper Pins.** — The mean diameter of taper pin required to safely transmit a known turning moment, may be found from the formulas:

$$d = 1.13 \sqrt{\frac{PR}{DS}} \qquad (1), \quad \text{and} \quad d = 283 \sqrt{\frac{\text{H.P.}}{NDS}} \qquad (2)$$

in which formulas $PR$ = turning moment in inch-pounds; $S$ = safe unit stress in pounds per square inch; H.P. = horsepower transmitted; $N$ = number of revolutions per minute; and $d$ and $D$ denote dimensions shown in the figure.

Formula (1) can be used with metric SI units where $d$ and $D$ denote dimensions shown in the figure in millimeters; $PR$ = turning moment in newton-millimeters (N·mm); and $S$ = safe unit stress in newtons per millimeter$^2$ (N/mm$^2$). Formula (2) is replaced by:

$$d = 110.3 \sqrt{\frac{\text{Power}}{NDS}},$$

where $d$ and $D$ denote dimensions shown in the figure in millimeters; $S$ = safe unit stress in N/mm$^2$; $N$ = number of revolutions per minute, and Power = power transmitted in watts.

*Examples:* — A lever secured to a 2-inch round shaft by a steel tapered pin (dimension $d = \frac{3}{8}$ inch) has a pull of 50 pounds at a 30-inch radius from shaft center. Find $S$, the unit working stress on the pin. By rearranging Formula (1):

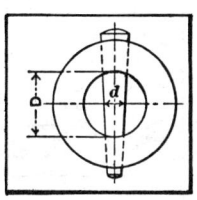

$$S = \frac{1.27\,PR}{Dd^2} = \frac{1.27 \times 50 \times 30}{2 \times (\frac{3}{8})^2} = 6770$$

pounds per square inch (nearly), which is a safe unit working stress for machine steel in shear.

Let $P$ = 50 pounds, $R$ = 30 inches, $D$ = 2 inches, and $S$ = 6000 pounds unit working stress. Using Formula (1) to find $d$:

$$d = 1.13 \sqrt{\frac{PR}{DS}} = 1.13 \sqrt{\frac{50 \times 30}{2 \times 6000}} = 1.13 \sqrt{\frac{1}{8}} = 0.4 \text{ inch.}$$

A similar example using SI units is as follows: A lever secured to a 50 mm round shaft by a steel tapered pin ($d$ = 10 mm) has a pull of 200 newtons at a radius of 800 mm. Find $S$, the working stress on the pin. By rearranging Formula (1):

$$S = \frac{1.27PR}{Dd^2} = \frac{1.27 \times 200 \times 800}{50 \times 10^2} = 40.6 \text{ N/mm}^2 = 40.6 \text{ megapascals.}$$

If a shaft of 50 mm diameter is to transmit power of 12 kilowatts at a speed of 500 rpm, find the mean diameter of the pin for a material having a safe unit stress of 40 N/mm$^2$. Using the formula:

$$d = 110.3 \sqrt{\frac{\text{Power}}{NDS}}, \qquad \text{then } d = 110.3 \sqrt{\frac{12,000}{500 \times 50 \times 40}}$$

$$= 110.3 \times 0.1096 = 12.09 \text{ mm.}$$

## Moments of Inertia, Section Moduli, etc., of Sections

| Section<br>$A$ = area<br>$y$ = distance from axis to extreme fiber | Moment of Inertia<br>$I$ | Section Modulus<br>$Z = \dfrac{I}{y}$ | Radius of Gyration<br>$k = \sqrt{\dfrac{I}{A}}$ |
|---|---|---|---|
| <br>$A = a^2$;　$y = \tfrac{1}{2}a$ | $\dfrac{a^4}{12}$ | $\dfrac{a^3}{6}$ | $\dfrac{a}{\sqrt{12}} = 0.289\,a$ |
| <br>$A = a^2$;　$y = a$ | $\dfrac{a^4}{3}$ | $\dfrac{a^3}{3}$ | $\dfrac{a}{\sqrt{3}} = 0.577\,a$ |
| <br>$A = a^2$<br>$y = \dfrac{a}{\sqrt{2}} = 0.707\,a$ | $\dfrac{a^4}{12}$ | $\dfrac{a^3}{6\sqrt{2}} = 0.118\,a^3$ | $\dfrac{a}{\sqrt{12}} = 0.289\,a$ |
| <br>$A = a^2 - b^2$;　$y = \tfrac{1}{2}a$ | $\dfrac{a^4 - b^4}{12}$ | $\dfrac{a^4 - b^4}{6a}$ | $\sqrt{\dfrac{a^2 + b^2}{12}}$<br>$= 0.289\sqrt{a^2 + b^2}$ |
| <br>$A = a^2 - b^2$<br>$y = \dfrac{a}{\sqrt{2}} = 0.707\,a$ | $\dfrac{a^4 - b^4}{12}$ | $\dfrac{\sqrt{2}\,(a^4 - b^4)}{12a}$<br>$= 0.118\,\dfrac{a^4 - b^4}{a}$ | $\sqrt{\dfrac{a^2 + b^2}{12}}$<br>$= 0.289\sqrt{a^2 + b^2}$ |
| <br>$A = bd$;　$y = \tfrac{1}{2}d$ | $\dfrac{bd^3}{12}$ | $\dfrac{bd^2}{6}$ | $\dfrac{d}{\sqrt{12}} = 0.289\,d$ |

## Moments of Inertia, Section Moduli, etc., of Sections

| $A$ = area<br>$y$ = distance from axis to extreme fiber | Moment of Inertia<br>$I$ | Section Modulus<br>$Z = \dfrac{I}{y}$ | Radius of Gyration<br>$k = \sqrt{\dfrac{I}{A}}$ |
|---|---|---|---|
| $A = bd; \quad y = d$ | $\dfrac{bd^3}{3}$ | $\dfrac{bd^2}{3}$ | $\dfrac{d}{\sqrt{3}} = 0.577\,d$ |
| $A = bd - hk$<br>$y = \frac{1}{2}d$ | $\dfrac{bd^3 - hk^3}{12}$ | $\dfrac{bd^3 - hk^3}{6\,d}$ | $\sqrt{\dfrac{bd^3 - hk^3}{12\,(bd - hk)}}$<br>$= 0.289\sqrt{\dfrac{bd^3 - hk^3}{bd - hk}}$ |
| $A = bd$<br>$y = \dfrac{bd}{\sqrt{b^2 + d^2}}$ | $\dfrac{b^3 d^3}{6\,(b^2 + d^2)}$ | $\dfrac{b^2 d^2}{6\sqrt{b^2 + d^2}}$ | $\dfrac{bd}{\sqrt{6\,(b^2 + d^2)}}$<br>$= 0.408\,\dfrac{bd}{\sqrt{b^2 + d^2}}$ |
| $A = bd$<br>$y = \frac{1}{2}(d\cos\alpha + b\sin\alpha)$ | $\dfrac{bd}{12}(d^2\cos^2\alpha + b^2\sin^2\alpha)$ | $\dfrac{bd}{6}\left(\dfrac{d^2\cos^2\alpha}{d\cos\alpha} + \dfrac{b^2\sin^2\alpha}{b\sin\alpha}\right)$ | $\sqrt{\dfrac{d^2\cos^2\alpha + b^2\sin^2\alpha}{12}}$<br>$= 0.289 \times$<br>$\sqrt{d^2\cos^2\alpha + b^2\sin^2\alpha}$ |
| $A = \frac{1}{2}bd; \quad y = \frac{2}{3}d$ | $\dfrac{bd^3}{36}$ | $\dfrac{bd^2}{24}$ | $\dfrac{d}{\sqrt{18}} = 0.236\,d$ |
| $A = \frac{1}{2}bd; \quad y = d$ | $\dfrac{bd^3}{12}$ | $\dfrac{bd^2}{12}$ | $\dfrac{d}{\sqrt{6}} = 0.408\,d$ |

Moments of Inertia, Section Moduli, etc.

| Section | Area of Section, $A$ | Distance from Neutral Axis to Extreme Fiber, $y$ |
|---|---|---|
| | $\dfrac{d\,(a+b)}{2}$ | $\dfrac{d\,(a+2\,b)}{3\,(a+b)}$ |
| | $\dfrac{3\,d^2 \tan 30°}{2} = 0.866\,d^2$ | $\dfrac{d}{2}$ |
| | $\dfrac{3\,d^2 \tan 30°}{2} = 0.866\,d^2$ | $\dfrac{d}{2 \cos 30°} = 0.577\,d$ |
| | $2\,d^2 \tan 22\frac{1}{2}° = 0.828\,d^2$ | $\dfrac{d}{2}$ |
| | $\dfrac{\pi d^2}{4} = 0.7854\,d^2$ | $\dfrac{d}{2}$ |
| | $\dfrac{\pi\,(D^2 - d^2)}{4}$ $= 0.7854\,(D^2 - d^2)$ | $\dfrac{D}{2}$ |
| | $\dfrac{\pi d^2}{8} = 0.393\,d^2$ | $\dfrac{(3\pi - 4)\,d}{6\pi}$ $= 0.288\,d$ |
| | $\dfrac{\pi\,(R^2 - r^2)}{2}$ $= 1.5708\,(R^2 - r^2)$ | $\dfrac{4\,(R^3 - r^3)}{3\pi\,(R^2 - r^2)}$ $= 0.424\,\dfrac{R^3 - r^3}{R^2 - r^2}$ |

## Moments of Inertia, Section Moduli, etc.

| Moment of Inertia, $I$ | Section Modulus. $Z = \dfrac{I}{y}$ | Radius of Gyration, $k = \sqrt{\dfrac{I}{A}}$ |
|---|---|---|
| $\dfrac{d^3\,(a^2 + 4\,ab + b^2)}{36\,(a+b)}$ | $\dfrac{d^2\,(a^2 + 4\,ab + b^2)}{12\,(a + 2\,b)}$ | $\sqrt{\dfrac{d^2\,(a^2 + 4\,ab + b^2)}{18\,(a+b)^2}}$ |
| $\dfrac{A}{12}\left[\dfrac{d^2\,(1 + 2\cos^2 30°)}{4\cos^2 30°}\right]$ $= 0.06\,d^4$ | $\dfrac{A}{6}\left[\dfrac{d\,(1 + 2\cos^2 30°)}{4\cos^2 30°}\right]$ $= 0.12\,d^3$ | $\sqrt{\dfrac{d^2\,(1 + 2\cos^2 30°)}{48\cos^2 30°}}$ $= 0.264\,d$ |
| $\dfrac{A}{12}\left[\dfrac{d^2\,(1 + 2\cos^2 30°)}{4\cos^2 30°}\right]$ $= 0.06\,d^4$ | $\dfrac{A}{6.9}\left[\dfrac{d\,(1 + 2\cos^2 30°)}{4\cos^2 30°}\right]$ $= 0.104\,d^3$ | $\sqrt{\dfrac{d^2\,(1 + 2\cos^2 30°)}{48\cos^2 30°}}$ $= 0.264\,d$ |
| $\dfrac{A}{12}\left[\dfrac{d^2\,(1 + 2\cos^2 22\frac{1}{2}°)}{4\cos^2 22\frac{1}{2}°}\right]$ $= 0.055\,d^4$ | $\dfrac{A}{6}\left[\dfrac{d\,(1 + 2\cos^2 22\frac{1}{2}°)}{4\cos^2 22\frac{1}{2}°}\right]$ $= 0.109\,d^3$ | $\sqrt{\dfrac{d^2\,(1 + 2\cos^2 22\frac{1}{2}°)}{48\cos^2 22\frac{1}{2}°}}$ $= 0.257\,d$ |
| $\dfrac{\pi d^4}{64} = 0.049\,d^4$ | $\dfrac{\pi d^3}{32} = 0.098\,d^3$ | $\dfrac{d}{4}$ |
| $\dfrac{\pi\,(D^4 - d^4)}{64}$ $= 0.049\,(D^4 - d^4)$ | $\dfrac{\pi\,(D^4 - d^4)}{32\,D}$ $= 0.098\,\dfrac{D^4 - d^4}{D}$ | $\dfrac{\sqrt{D^2 + d^2}}{4}$ |
| $\dfrac{(9\pi^2 - 64)\,d^4}{1152\,\pi}$ $= 0.007\,d^4$ | $\dfrac{(9\pi^2 - 64)\,d^3}{192\,(3\pi - 4)}$ $= 0.024\,d^3$ | $\dfrac{\sqrt{(9\pi^2 - 64)\,d^2}}{12\,\pi}$ $= 0.132\,d$ |
| $0.1098\,(R^4 - r^4)$ $-\,\dfrac{0.283\,R^2 r^2\,(R - r)}{R + r}$ | $\dfrac{I}{y}$ | $\sqrt{\dfrac{I}{A}}$ |

## Moments of Inertia, Section Moduli, etc.

| Section | Area of Section, $A$ | Distance from Neutral Axis to Extreme Fiber, $y$ |
|---|---|---|
| | $\pi ab = 3.1416\, ab$ | $a$ |
| | $\pi (ab - cd)$ $= 3.1416\,(ab - cd)$ | $a$ |
| | $dt + 2\,a\,(s + n)$ | $\dfrac{d}{2}$ |
| | $dt + 2\,a\,(s + n)$ | $\dfrac{b}{2}$ |
| | $dt + a\,(s + n)$ | $\dfrac{d}{2}$ |
| | $dt + a\,(s + n)$ | $b - [b^2 s + \dfrac{hl^2}{2} + \dfrac{g}{3}(b - t)^2$ $\times (b + 2\,t)] \div A$ in which $g =$ slope of flange $= \dfrac{h - l}{2\,(b - t)}$ |
| | $\dfrac{l\,(T + t)}{2} + Tn + a\,(s + n)$ | $d - [3\,s^2\,(b - T)$ $+ 2\,am\,(m + 3\,s) + 3\,Td^2$ $- l\,(T - t)(3\,d - l)] \div 6\,A$ |

## Moments of Inertia, Section Moduli, etc.

| Moment of Inertia, $I$ | Section Modulus, $Z = \dfrac{I}{y}$ | Radius of Gyration, $k = \sqrt{\dfrac{I}{A}}$ |
|---|---|---|
| $\dfrac{\pi a^3 b}{4} = 0.7854\, a^3 b$ | $\dfrac{\pi a^2 b}{4} = 0.7854\, a^2 b$ | $\dfrac{a}{2}$ |
| $\dfrac{\pi}{4}(a^3 b - c^3 d)$ $= 0.7854\,(a^3 b - c^3 d)$ | $\dfrac{\pi(a^3 b - c^3 d)}{4a}$ $= 0.7854\,\dfrac{a^3 b - c^3 d}{a}$ | $\dfrac{1}{2}\sqrt{\dfrac{a^3 b - c^3 d}{ab - cd}}$ |
| $\dfrac{1}{12}\left[bd^3 - \dfrac{1}{4g}(h^4 - l^4)\right]$ in which $g =$ slope of flange $= \dfrac{h-l}{b-t} = \dfrac{1}{6}$ for standard I-beams. | $\dfrac{1}{6d}\left[bd^3 - \dfrac{1}{4g}(h^4 - l^4)\right]$ | $\sqrt{\dfrac{\dfrac{1}{12}\left[bd^3 - \dfrac{1}{4g}(h^4 - l^4)\right]}{dt + 2a(s+n)}}$ |
| $\dfrac{1}{12}\left[b^3(d-h) + lt^3 + \dfrac{g}{4}(b^4 - t^4)\right]$ in which $g =$ slope of flange (see above). | $\dfrac{1}{6b}\left[b^3(d-h) + lt^3 + \dfrac{g}{4}(b^4 - t^4)\right]$ | $\sqrt{\dfrac{I}{A}}$ |
| $\dfrac{1}{12}\left[bd^3 - \dfrac{1}{8g}(h^4 - l^4)\right]$ in which $g =$ slope of flange $= \dfrac{h-l}{2(b-t)} = \dfrac{1}{6}$ for standard channels | $\dfrac{1}{6d}\left[bd^3 - \dfrac{1}{8g}(h^4 - l^4)\right]$ | $\sqrt{\dfrac{\dfrac{1}{12}\left[bd^3 - \dfrac{1}{8g}(h^4 - l^4)\right]}{dt + a(s+n)}}$ |
| $\dfrac{1}{3}\left[2sb^3 + lt^3 + \dfrac{g}{2}(b^4 - t^4)\right]$ $- A(b - y)^2$ in which $g =$ slope of flange (see above). | $\dfrac{I}{y}$ | $\sqrt{\dfrac{I}{A}}$ |
| $\dfrac{1}{12}[l^2(T + 3t) + 4bn^3$ $- 2am^3] - A(d - y - n)^2$ | $\dfrac{I}{y}$ | $\sqrt{\dfrac{I}{A}}$ |

**Moments of Inertia, Section Moduli, etc.**

| Section | Area of Section, $A$ | Distance from Neutral Axis to Extreme Fiber, $y$ |
|---|---|---|
| | $\dfrac{l\,(T+t)}{2} + Tn$ <br> $+ a\,(s+n)$ | $\dfrac{b}{2}$ |
| | $t\,(2\,a - t)$ | $a - \dfrac{a^2 + at - t^2}{2\,(2\,a - t)}$ |
| | $t\,(2\,a - t)$ | $\dfrac{a^2 + at - t^2}{2\,(2\,a - t)\cos 45°}$ |
| | $bd - h\,(b - t)$ | $\dfrac{d}{2}$ |
| | $bd - h\,(b - t)$ | $\dfrac{b}{2}$ |
| | $bd - h\,(b - t)$ | $\dfrac{d}{2}$ |
| | $bd - h\,(b - t)$ | $b - \dfrac{2\,b^2 s + ht^2}{2\,bd - 2\,h\,(b - t)}$ |
| | $dt + s\,(b - t)$ | $\dfrac{d}{2}$ |

Moments of Inertia, Section Moduli, etc.

| Moment of inertia, $I$ | Section Modulus, $Z = \dfrac{I}{y}$ | Radius of Gyration, $\kappa = \sqrt{\dfrac{I}{A}}$ |
|---|---|---|
| $\dfrac{sb^3 + mT^3 + lt^3}{12}$ $+ \dfrac{am[2a^2 + (2a + 3T)^2]}{36}$ $+ \dfrac{l(T-t)[(T-t)^2 + 2(T+2t)^2]}{144}$ | $\dfrac{I}{y}$ | $\sqrt{\dfrac{I}{A}}$ |
| $\tfrac{1}{3}[ty^3 + a(a-y)^3$ $- (a-t)(a-y-t)^3]$ | $\dfrac{I}{y}$ | $\sqrt{\dfrac{I}{A}}$ |
| $\tfrac{1}{3}[2x^4 - 2(x-t)^4$ $+ t[a - (2x - \tfrac{1}{2}t)]^3]$ in which $x = \dfrac{a^2 + at - t^2}{2(2a-t)}$ | $\dfrac{I}{y}$ | $\sqrt{\dfrac{I}{A}}$ |
| $\dfrac{bd^3 - h^3(b-t)}{12}$ | $\dfrac{bd^3 - h^3(b-t)}{6d}$ | $\sqrt{\dfrac{bd^3 - h^3(b-t)}{12[bd - h(b-t)]}}$ |
| $\dfrac{2sb^3 + ht^3}{12}$ | $\dfrac{2sb^3 + ht^3}{6b}$ | $\sqrt{\dfrac{2sb^3 + ht^3}{12[bd - h(b-t)]}}$ |
| $\dfrac{bd^3 - h^3(b-t)}{12}$ | $\dfrac{bd^3 - h^3(b-t)}{6d}$ | $\sqrt{\dfrac{bd^3 - h^3(b-t)}{12[bd - h(b-t)]}}$ |
| $\dfrac{2sb^3 + ht^3}{3} - A(b-y)^2$ | $\dfrac{I}{y}$ | $\sqrt{\dfrac{I}{A}}$ |
| $\dfrac{td^3 + s^3(b-t)}{12}$ | $\dfrac{td^3 + s^3(b-t)}{6d}$ | $\sqrt{\dfrac{td^3 + s^3(b-t)}{12[td + s(b-t)]}}$ |

Moments of Inertia, Section Moduli, etc.

| Section | Area of Section, $A$ | Distance from Neutral Axis to Extreme Fiber, $y$ |
|---|---|---|
| | $bs + ht + as$ | $d - [td^2 + s^2(b-t) + s(a-t)(2d-s)] \div 2A$ |
| | $bs + ht$ | $d - \dfrac{d^2 t + s^2(b-t)}{2(bs+ht)}$ |
| | $bs + \dfrac{h(T+t)}{2}$ | $d - [3bs^2 + 3ht(d+s) + h(T-t)(h+3s)] \div 6A$ |
| | $t(a+b-t)$ | $b - \dfrac{t(2d+a)+d^2}{2(d+a)}$ |
| | $t(a+b-t)$ | $a - \dfrac{t(2c+b)+c^2}{2(c+b)}$ |
| | $t[b+2(a-t)]$ | $\dfrac{b}{2}$ |
| | $t[b+2(a-t)]$ | $\dfrac{2a-t}{2}$ |

## Moments of Inertia, Section Moduli, etc.

| Moment of Inertia, $I$ | Section Modulus, $Z = \dfrac{I}{y}$ | Radius of Gyration, $k = \sqrt{\dfrac{I}{A}}$ |
|---|---|---|
| $\frac{1}{3}\,[b\,(d-y)^3 + ay^3$ $\quad - (b-t)\,(d-y-s)^3$ $\quad - (a-t)\,(y-s)^3]$ | $\dfrac{I}{y}$ | $\sqrt{\dfrac{I}{A}}$ |
| $\frac{1}{3}\,[ty^3 + b\,(d-y)^3$ $\quad - (b-t)\,(d-y-s)^3]$ | $\dfrac{I}{y}$ | $\sqrt{\dfrac{1}{3\,(bs+ht)}\,[ty^3 + b\,(d-y)^3}$ $\overline{\quad - (b-t)\,(d-y-s)^3]}$ |
| $\frac{1}{12}\,[4\,bs^3 + h^3\,(3\,t+T)]$ $\quad - A\,(d-y-s)^2$ | $\dfrac{I}{y}$ | $\sqrt{\dfrac{I}{A}}$ |
| $\frac{1}{3}\,[ty^3 + a\,(b-y)^3$ $\quad - (a-t)\,(b-y-t)^3]$ | $\dfrac{I}{y}$ | $\sqrt{\dfrac{1}{3\,t\,(a+b-t)}\,[ty^3 + a(b-y)^3}$ $\overline{\quad -(a-t)\,(b-y-t)^3]}$ |
| $\frac{1}{3}\,[ty^3 + b\,(a-y)^3$ $\quad - (b-t)\,(a-y-t)^3]$ | $\dfrac{I}{y}$ | $\sqrt{\dfrac{1}{3\,t\,(a+b-t)}\,[ty^3 + b(a-y)^3}$ $\overline{\quad -(b-t)\,(a-y-t)^3]}$ |
| $\dfrac{ab^3 - c\,(b-2\,t)^3}{12}$ | $\dfrac{ab^3 - c\,(b-2\,t)^3}{6\,b}$ | $\sqrt{\dfrac{ab^3 - c\,(b-2\,t)^3}{12\,t\,[b+2\,(a-t)]}}$ |
| $\dfrac{b\,(a+c)^3 - 2\,c^3d - 6\,a^2cd}{12}$ | $\dfrac{b\,(a+c)^3 - 2\,c^3d - 6\,a^2cd}{6\,(2\,a-t)}$ | $\sqrt{\dfrac{b\,(a+c)^3 - 2\,c^3d - 6\,a^2cd}{12\,t\,[b+2\,(a-t)]}}$ |

**Section Modulus, Area, etc., of Sections for Punch and Shear Frames.** — Machine frames cannot be standardized so as to permit tables of section modulus, area, etc., of the sections to be made up in the same way as for standard structural steel sections, but it is possible to arrange a table for punch and shear frames so as to simplify the work of selecting proper sections. A table of these quantities is, therefore, given. To illustrate the use of the table, take as an example the punch frame shown diagrammatically in Fig. 1. The distance from the center line of the punch to the back of the gap is 24 inches. Assume that the maximum pressure $P$, tending to force the jaws apart, is that due to punching a 1-inch circular hole in soft steel plate 1 inch thick, or say about 157,000 pounds. Consider the section at $TX$. The action of $P$ is such as to produce a tensile stress on the section to the left of the neutral axis $N$, with a compressive stress to the right of $N$, both due to

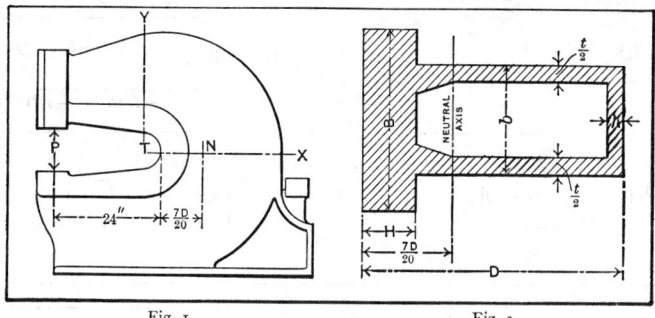

Fig. 1                                    Fig. 2

flexure; and, besides, there is a tensile stress distributed uniformly over the section. It is usually sufficient to determine the maximum tensile stress to the left of $N$.

Maximum tensile stress = flexure tensile stress + uniformly distributed tensile stress.

$$\text{Flexure tensile stress} = \frac{\text{Moment of } P \text{ about } N}{\text{Tensile section modulus of section } TX}$$

$$\text{Uniformly distributed tensile stress} = \frac{P}{\text{Area of section } TX}$$

Assume that $D$, in Fig. 2, is about 30 inches. Then $\frac{7\,D}{20} = 10\frac{1}{2}$ inches. If 3000 pounds per square inch is the allowable fiber stress, and the stress due to flexure only is considered, the required section modulus for tension would be:

$$\frac{157,000 \times (24 + 10\frac{1}{2})}{3000} = 1800, \text{ about.}$$

As no allowance has been made for the additional tensile stress uniformly distributed over the section, a section must be selected the section modulus of which is somewhat greater than 1800, say about 2500.

The table of "Properties of Sections for Punch and Shear Frames" gives a great variety of shapes and proportions, the only dimensions common to all being the depth of section, $D$, and the distance of the neutral axis from the extreme tension fiber. This location of the neutral axis represents average practice and insures an economical distribution of metal. Generous fillets and rounded corners should, of course, be used on the actual section.

Suppose that a deep narrow section is desired, similar to that in Fig. 2, in which the dimensions are as follows: $D = 10$ inches; $B = 7$ inches; $b = 4$ inches; $H = 1.98$ inch; $\frac{1}{2} t = h = \frac{5}{8}$ inch; $F$ (area) $= 26.3$ square inches; and $Z_t$ (section modulus for tension) $= 78.6$. This section, taken from the table, is not, of course, large enough, but a similar one which is large enough may be easily found as follows:

If two sections $A$ and $B$ are similar in all respects, then

$$\frac{\text{Area of } A}{\text{Area of } B} = \frac{(\text{Any dimension of } A)^2}{(\text{Corresponding dimension of } B)^2},$$

and

$$\frac{\text{Section modulus of } A}{\text{Section modulus of } B} = \frac{(\text{Any dimension of } A)^3}{(\text{Corresponding dimension of } B)^3};$$

$$\frac{\text{Required section modulus}}{\text{Modulus of section from the table}} = \frac{(\text{Required } D)^3}{(D \text{ of section from the table})^3}$$

Hence, $\dfrac{2500}{78.6} = \dfrac{(\text{Required } D)^3}{(10)^3}$. This last equation solved for $D$ gives $D = 31.7$ inches, nearly. As the large section will be exactly similar to the small one, the area of the large section is found from the equation:

$$\frac{\text{Area of required section}}{26.3} = \frac{(31.7)^2}{(10)^2}$$

and area $= 264$ square inches, about.

The neutral axis will be $\frac{7}{20} \times 31.7$, or $11.1$ inches from the extreme tension fiber.

This trial section may now be tested for the maximum stress on it.

Flexure tensile stress $= \dfrac{157,000 \times (24 + 11.1)}{2500} = 2200$ pounds per square inch, about.

Uniformly distributed tensile stress $= \dfrac{157,000}{264} = 600$ pounds per square inch, about.

The total tensile stress is, therefore, 2800 pounds per square inch.

If this result had not been near enough to the 3000 pounds per square inch assumed, another section could have been selected and worked out in the same way to get a closer result. The depth $D$ of the required section being 31.7 inches as compared with 10 inches of the similar section given in the table, each of the dimensions of the required section will be 3.17 times the corresponding one given in the table. Hence $D = 31.7$, say 32 inches; $B = 22.2$, say 22 inches; $b = 12.7$, say 13 inches; $H = 6.3$, say $6\frac{1}{2}$ inches, $\frac{1}{2} t = h = 1.98$, say 2 inches. The webs thicken gradually from the neutral axis to the tension flange so as to avoid too sudden a change in the section. For selecting the section at $TY$, Fig. 1, the procedure would be the same except that there would be no uniformly distributed stress to be added as in the case of section $TX$.

The method here used to determine the maximum stress on the curved section $TX$ is actually correct only in the case of *straight* sections; for curved sections, various investigators have showed that the maximum stress is greater than indicated by straight beam formulas by a factor of up to 3 or more.

Since determination of the dimensions for a curved beam based on the use of theoretically correct formulae is a laborious trial-and-error process, it is common practice to use the method outlined and to apply an ample factor of safety.

A simplified method for the determination of stresses in a curved beam of known proportions is shown on page 424.

Properties of Sections for Punch and Shear Frames

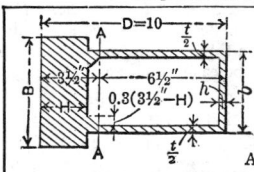

$Z_c$ = Section Modulus for Compression;
$Z_t$ = Section Modulus for Tension;
$F$ = Area of Section;
$I$ = Moment of Inertia about Gravity Axis $A - A$.

All dimensions in inches

| B | b | $h = \frac{1}{2}t$ | H | F | I | $Z_c$ | $Z_t$ |
|---|---|---|---|---|---|---|---|
| 10 | 10 | ¼ | 0.57 | 15.36 | 228.51 | 35.20 | 65.40 |
|  |  | ⅜ | 1.10 | 23.43 | 311.78 | 47.95 | 89.10 |
|  |  | ½ | 1.80 | 31.82 | 397.83 | 61.20 | 113.70 |
|  | 9 | ¼ | 0.51 | 14.66 | 200.77 | 30.89 | 57.36 |
|  |  | ⅜ | 0.99 | 21.64 | 290.32 | 44.66 | 82.95 |
|  |  | ½ | 1.61 | 29.56 | 371.95 | 57.22 | 106.27 |
|  |  | ⅝ | 2.41 | 38.69 | 438.24 | 67.44 | 125.21 |
|  | 8 | ¼ | 0.44 | 13.87 | 180.98 | 27.80 | 51.50 |
|  |  | ⅜ | 0.88 | 20.41 | 272.07 | 41.90 | 77.70 |
|  |  | ½ | 1.38 | 27.27 | 345.47 | 53.20 | 98.50 |
|  |  | ⅝ | 2.04 | 35.20 | 410.37 | 63.10 | 117.00 |
|  |  | ¾ | 3.50 | 49.63 | 462.98 | 71.20 | 132.00 |
|  | 7 | ¼ | 0.38 | 13.15 | 172.76 | 26.50 | 49.30 |
|  |  | ⅜ | 0.77 | 19.21 | 261.46 | 40.30 | 74.60 |
|  |  | ½ | 1.24 | 25.69 | 320.52 | 49.30 | 91.50 |
|  |  | ⅝ | 1.74 | 32.24 | 378.80 | 58.30 | 108.00 |
|  |  | ¾ | 2.34 | 39.42 | 428.57 | 65.90 | 122.60 |
| 9 | 9 | ¼ | 0.59 | 14.66 | 204.06 | 31.40 | 58.30 |
|  |  | ⅜ | 1.20 | 21.57 | 291.47 | 44.80 | 83.40 |
|  |  | ½ | 2.00 | 30.68 | 363.50 | 55.80 | 103.60 |
|  | 8 | ¼ | 0.50 | 13.83 | 185.91 | 28.60 | 53.65 |
|  |  | ⅜ | 1.00 | 20.34 | 268.15 | 41.28 | 76.65 |
|  |  | ½ | 1.70 | 28.07 | 338.66 | 52.15 | 96.81 |
|  |  | ⅝ | 2.60 | 37.12 | 403.27 | 62.10 | 115.12 |
|  | 7 | ¼ | 0.42 | 13.00 | 173.33 | 26.60 | 49.80 |
|  |  | ⅜ | 0.89 | 19.25 | 250.41 | 38.40 | 71.60 |
|  |  | ½ | 1.42 | 25.65 | 317.24 | 48.80 | 90.50 |
|  |  | ⅝ | 2.11 | 33.03 | 375.00 | 57.70 | 107.10 |
|  |  | ¾ | 3.06 | 42.13 | 420.36 | 64.60 | 120.10 |
|  | 6 | ¼ | 0.36 | 12.40 | 161.70 | 24.90 | 46.20 |
|  |  | ⅜ | 0.75 | 17.92 | 226.20 | 34.80 | 64.70 |
|  |  | ½ | 1.23 | 23.89 | 290.50 | 44.70 | 83.00 |
|  |  | ⅝ | 1.79 | 30.22 | 345.70 | 53.10 | 100.60 |
|  |  | ¾ | 2.62 | 38.26 | 392.00 | 60.30 | 112.00 |
| 8 | 8 | ¼ | 0.62 | 14.09 | 187.46 | 28.90 | 53.55 |
|  |  | ⅜ | 1.20 | 20.50 | 268.38 | 41.30 | 76.70 |
|  |  | ½ | 2.10 | 28.79 | 336.75 | 51.80 | 96.20 |
|  | 7 | ¼ | 0.55 | 13.69 | 172.25 | 26.50 | 49.20 |
|  |  | ⅜ | 1.10 | 19.50 | 248.56 | 38.30 | 71.10 |
|  |  | ½ | 1.90 | 27.07 | 315.01 | 48.50 | 90.00 |

### Properties of Sections for Punch and Shear Frames

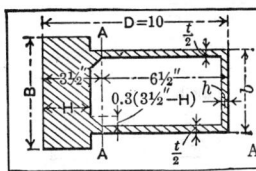

$Z_c$ = Section Modulus for Compression;
$Z_t$ = Section Modulus for Tension;
$F$ = Area of Section;
$I$ = Moment of Inertia about Gravity Axis
$A - A$.

All dimensions in inches

| B | b | $h = ½ t$ | H | F | I | $Z_c$ | $Z_t$ |
|---|---|---|---|---|---|---|---|
| 8 | 7 | ⅝ | 3.00 | 36.41 | 377.05 | 58.0 | 108.0 |
|   | 6 | ¼ | 0.43 | 12.43 | 155.80 | 23.9 | 44.5 |
|   |   | ⅜ | 0.91 | 18.07 | 221.14 | 34.0 | 63.1 |
|   |   | ½ | 1.50 | 24.20 | 283.77 | 43.5 | 80.8 |
|   |   | ⅝ | 2.36 | 31.79 | 339.75 | 52.3 | 96.9 |
|   | 5 | ¼ | 0.33 | 11.61 | 139.76 | 21.42 | 39.9 |
|   |   | ⅜ | 0.75 | 16.81 | 204.37 | 31.41 | 58.4 |
|   |   | ½ | 1.25 | 22.27 | 261.65 | 40.25 | 75.2 |
|   |   | ⅝ | 2.00 | 29.00 | 310.74 | 49.7 | 88.8 |
|   |   | ¾ | 2.75 | 35.66 | 350.65 | 54.0 | 100.0 |
| 7 | 7 | ¼ | 0.70 | 13.52 | 169.40 | 26.04 | 48.4 |
|   |   | ⅜ | 1.40 | 19.99 | 243.80 | 37.5 | 69.6 |
|   |   | ½ | 2.44 | 27.28 | 310.80 | 47.8 | 88.8 |
|   | 6 | ¼ | 0.55 | 12.55 | 147.20 | 22.6 | 42.1 |
|   |   | ⅜ | 1.14 | 18.27 | 220.03 | 33.8 | 62.8 |
|   |   | ½ | 2.00 | 25.17 | 282.60 | 43.4 | 80.8 |
|   | 5 | ¼ | 0.41 | 12.28 | 136.92 | 20.0 | 39.1 |
|   |   | ⅜ | 0.95 | 16.98 | 203.40 | 31.0 | 58.0 |
|   |   | ½ | 1.65 | 23.03 | 258.64 | 40.0 | 73.8 |
|   |   | ⅝ | 2.55 | 30.36 | 302.60 | 46.5 | 86.5 |
|   | 4 | ¼ | 0.31 | 10.96 | 124.20 | 19.12 | 35.5 |
|   |   | ⅜ | 0.76 | 15.71 | 183.00 | 28.15 | 52.3 |
|   |   | ½ | 1.31 | 20.80 | 232.20 | 35.7 | 66.4 |
|   |   | ⅝ | 1.98 | 26.30 | 275.00 | 42.3 | 78.6 |
| 6 | 6 | ¼ | 0.68 | 12.56 | 156.25 | 24.0 | 44.6 |
|   |   | ⅜ | 1.56 | 18.79 | 222.10 | 32.6 | 60.5 |
|   |   | ½ | 3.50 | 30.00 | 275.00 | 42.3 | 78.6 |
|   | 5 | ¼ | 0.53 | 11.68 | 180.83 | 29.0 | 51.6 |
|   |   | ⅜ | 1.27 | 17.25 | 214.79 | 33.0 | 61.4 |
|   |   | ½ | 2.35 | 24.15 | 253.85 | 39.0 | 73.4 |
|   | 4 | ¼ | 0.38 | 10.88 | 125.59 | 19.3 | 35.8 |
|   |   | ⅜ | 1.00 | 15.84 | 181.36 | 27.9 | 51.8 |
|   |   | ½ | 1.80 | 20.87 | 228.96 | 35.2 | 65.4 |
| 5 | 5 | ¼ | 0.73 | 11.76 | 139.73 | 21.5 | 40.0 |
|   |   | ⅜ | 1.70 | 17.28 | 196.98 | 30.2 | 56.2 |
|   | 4 | ¼ | 0.55 | 10.97 | 125.40 | 19.3 | 35.9 |
|   |   | ⅜ | 1.45 | 16.13 | 181.52 | 28.0 | 52.0 |
|   | 3 | ¼ | 0.84 | 10.16 | 110.14 | 17.25 | 32.0 |
|   |   | ⅜ | 1.12 | 14.90 | 167.70 | 25.7 | 47.6 |
|   |   | ½ | 2.10 | 19.99 | 198.57 | 30.5 | 56.8 |

## Moments of Inertia and Section Moduli for Rectangles (Metric units)

Moment of inertia and section modulus values shown are for rectangles 1 millimeter wide. To obtain moment of inertia or section modulus for rectangle of given length of side, multiply appropriate value in table by given width. .(See bottom of page 370 for basic formulas.)

| Lgth. of Side (mm) | Moment of Inertia | Section Modulus | Lgth. of Side (mm) | Moment of Inertia | Section Modulus | Lgth. of Side (mm) | Moment of Inertia | Section Modulus |
|---|---|---|---|---|---|---|---|---|
| 5 | 10.4167 | 4.16667 | 56 | 14634.7 | 522.667 | 107 | 102087 | 1908.17 |
| 6 | 18.0000 | 6.00000 | 57 | 15432.8 | 541.500 | 108 | 104976 | 1944.00 |
| 7 | 28.5833 | 8.16667 | 58 | 16259.3 | 560.667 | 109 | 107919 | 1980.17 |
| 8 | 42.6667 | 10.6667 | 59 | 17114.9 | 580.167 | 110 | 110917 | 2016.67 |
| 9 | 60.7500 | 13.5000 | 60 | 18000.0 | 600.000 | 111 | 113969 | 2053.50 |
| 10 | 83.3333 | 16.6667 | 61 | 18915.1 | 620.167 | 112 | 117077 | 2090.67 |
| 11 | 110.917 | 20.1667 | 62 | 19860.7 | 640.667 | 113 | 120241 | 2128.17 |
| 12 | 144.000 | 24.0000 | 63 | 20837.3 | 661.500 | 114 | 123462 | 2166.00 |
| 13 | 183.083 | 28.1667 | 64 | 21845.3 | 682.667 | 115 | 126740 | 2204.17 |
| 14 | 228.667 | 32.6667 | 65 | 22885.4 | 704.167 | 116 | 130075 | 2242.67 |
| 15 | 281.250 | 37.5000 | 66 | 23958.0 | 726.000 | 117 | 133468 | 2281.50 |
| 16 | 341.333 | 42.6667 | 67 | 25063.6 | 748.167 | 118 | 136919 | 2320.67 |
| 17 | 409.417 | 48.1667 | 68 | 26202.7 | 770.667 | 119 | 140430 | 2360.17 |
| 18 | 486.000 | 54.0000 | 69 | 27375.8 | 793.500 | 120 | 144000 | 2400.00 |
| 19 | 571.583 | 60.1667 | 70 | 28583.3 | 816.667 | 121 | 147630 | 2440.17 |
| 20 | 666.667 | 66.6667 | 71 | 29825.9 | 840.167 | 122 | 151321 | 2480.67 |
| 21 | 771.750 | 73.5000 | 72 | 31104.0 | 864.000 | 123 | 155072 | 2521.50 |
| 22 | 887.333 | 80.6667 | 73 | 32418.1 | 888.167 | 124 | 158885 | 2562.67 |
| 23 | 1013.92 | 88.1667 | 74 | 33768.7 | 912.667 | 125 | 162760 | 2604.17 |
| 24 | 1152.00 | 96.0000 | 75 | 35156.3 | 937.500 | 126 | 166698 | 2646.00 |
| 25 | 1302.08 | 104.167 | 76 | 36581.3 | 962.667 | 127 | 170699 | 2688.17 |
| 26 | 1464.67 | 112.667 | 77 | 38044.4 | 988.167 | 128 | 174763 | 2730.67 |
| 27 | 1640.25 | 121.500 | 78 | 39546.0 | 1014.00 | 130 | 183083 | 2816.67 |
| 28 | 1829.33 | 130.667 | 79 | 41086.6 | 1040.17 | 132 | 191664 | 2904.00 |
| 29 | 2032.42 | 140.167 | 80 | 42666.7 | 1066.67 | 135 | 205031 | 3037.50 |
| 30 | 2250.00 | 150.000 | 81 | 44286.8 | 1093.50 | 138 | 219006 | 3174.00 |
| 31 | 2482.58 | 160.167 | 82 | 45947.3 | 1120.67 | 140 | 228667 | 3266.67 |
| 32 | 2730.67 | 170.667 | 83 | 47648.9 | 1148.17 | 143 | 243684 | 3408.17 |
| 33 | 2994.75 | 181.500 | 84 | 49392.0 | 1176.00 | 147 | 264710 | 3601.50 |
| 34 | 3275.33 | 192.667 | 85 | 51177.1 | 1204.17 | 150 | 281250 | 3750.00 |
| 35 | 3572.92 | 204.167 | 86 | 53004.7 | 1232.67 | 155 | 310323 | 4004.17 |
| 36 | 3888.00 | 216.000 | 87 | 54875.3 | 1261.50 | 160 | 341333 | 4266.67 |
| 37 | 4221.08 | 228.167 | 88 | 56789.3 | 1290.67 | 165 | 374344 | 4537.50 |
| 38 | 4572.67 | 240.667 | 89 | 58747.4 | 1320.17 | 170 | 409417 | 4816.67 |
| 39 | 4943.25 | 253.500 | 90 | 60750.0 | 1350.00 | 175 | 446615 | 5104.17 |
| 40 | 5333.33 | 266.667 | 91 | 62797.6 | 1380.17 | 180 | 486000 | 5400.00 |
| 41 | 5743.42 | 280.167 | 92 | 64890.7 | 1410.67 | 185 | 527635 | 5704.17 |
| 42 | 6174.00 | 294.000 | 93 | 67029.8 | 1441.50 | 190 | 571583 | 6016.67 |
| 43 | 6625.58 | 308.167 | 94 | 69215.3 | 1472.67 | 195 | 617906 | 6337.50 |
| 44 | 7098.67 | 322.667 | 95 | 71447.9 | 1504.17 | 200 | 666667 | 6666.67 |
| 45 | 7593.75 | 337.500 | 96 | 73728.0 | 1536.00 | 210 | 771750 | 7350.00 |
| 46 | 8111.33 | 352.667 | 97 | 76056.1 | 1568.17 | 220 | 887333 | 8066.67 |
| 47 | 8651.92 | 368.167 | 98 | 78432.7 | 1600.67 | 230 | 1013920 | 8816.67 |
| 48 | 9216.00 | 384.000 | 99 | 80858.3 | 1633.50 | 240 | 1152000 | 9600.00 |
| 49 | 9804.08 | 400.167 | 100 | 83333.3 | 1666.67 | 250 | 1302080 | 10416.7 |
| 50 | 10416.7 | 416.667 | 101 | 85858.4 | 1700.17 | 260 | 1464670 | 11266.7 |
| 51 | 11054.3 | 433.5 | 102 | 88434.0 | 1734.00 | 270 | 1640250 | 12150.0 |
| 52 | 11717.3 | 450.667 | 103 | 91060.6 | 1768.17 | 280 | 1829330 | 13066.7 |
| 53 | 12406.4 | 468.167 | 104 | 93738.7 | 1802.67 | 290 | 2032410 | 14016.7 |
| 54 | 13122.0 | 486.000 | 105 | 96468.8 | 1837.5 | 300 | 2250000 | 15000.0 |
| 55 | 13864.6 | 504.167 | 106 | 99251.3 | 1872.67 | ... | ... | ... |

## Section Moduli for Rectangles

Section modulus values shown are for rectangles 1 inch wide. To obtain section modulus for rectangle of given length of side, multiply value in table by given width.

| Length of Side | Section Modulus | Length of Side | Section Modulus | Length of Side | Section Modulus | Length of Side | Section Modulus |
|---|---|---|---|---|---|---|---|
| 1/8 | 0.0026 | 2 3/4 | 1.26 | 12 | 24.00 | 25 | 104.2 |
| 3/16 | 0.0059 | 3 | 1.50 | 12 1/2 | 26.04 | 26 | 112.7 |
| 1/4 | 0.0104 | 3 1/4 | 1.76 | 13 | 28.17 | 27 | 121.5 |
| 5/16 | 0.0163 | 3 1/2 | 2.04 | 13 1/2 | 30.38 | 28 | 130.7 |
| 3/8 | 0.0234 | 3 3/4 | 2.34 | 14 | 32.67 | 29 | 140.2 |
| 7/16 | 0.032 | 4 | 2.67 | 14 1/2 | 35.04 | 30 | 150 |
| 1/2 | 0.042 | 4 1/2 | 3.38 | 15 | 37.5 | 32 | 171 |
| 5/8 | 0.065 | 5 | 4.17 | 15 1/2 | 40.0 | 34 | 193 |
| 3/4 | 0.094 | 5 1/2 | 5.04 | 16 | 42.7 | 36 | 216 |
| 7/8 | 0.128 | 6 | 6.00 | 16 1/2 | 45.4 | 38 | 241 |
| 1 | 0.167 | 6 1/2 | 7.04 | 17 | 48.2 | 40 | 267 |
| 1 1/8 | 0.211 | 7 | 8.17 | 17 1/2 | 51.0 | 42 | 294 |
| 1 1/4 | 0.260 | 7 1/2 | 9.38 | 18 | 54.0 | 44 | 323 |
| 1 3/8 | 0.315 | 8 | 10.67 | 18 1/2 | 57.0 | 46 | 353 |
| 1 1/2 | 0.375 | 8 1/2 | 12.04 | 19 | 60.2 | 48 | 384 |
| 1 5/8 | 0.440 | 9 | 13.50 | 19 1/2 | 63.4 | 50 | 417 |
| 1 3/4 | 0.510 | 9 1/2 | 15.04 | 20 | 66.7 | 52 | 451 |
| 1 7/8 | 0.586 | 10 | 16.67 | 21 | 73.5 | 54 | 486 |
| 2 | 0.67 | 10 1/2 | 18.38 | 22 | 80.7 | 56 | 523 |
| 2 1/4 | 0.84 | 11 | 20.17 | 23 | 88.2 | 58 | 561 |
| 2 1/2 | 1.04 | 11 1/2 | 22.04 | 24 | 96.0 | 60 | 600 |

## Section Moduli and Moments of Inertia for Round Shafts*

| Diam. | Section Modulus | Moment of Inertia | Diam. | Section Modulus | Moment of Inertia | Diam. | Section Modulus | Moment of Inertia |
|---|---|---|---|---|---|---|---|---|
| 1/8 | 0.00019 | 0.00001 | 27/64 | 0.0074 | 0.00155 | 23/32 | 0.0364 | 0.01308 |
| 9/64 | 0.00027 | 0.00002 | 7/16 | 0.0082 | 0.00180 | 47/64 | 0.0388 | 0.01425 |
| 5/32 | 0.00037 | 0.00003 | 29/64 | 0.0091 | 0.00207 | 3/4 | 0.0413 | 0.01550 |
| 11/64 | 0.00050 | 0.00004 | 15/32 | 0.0101 | 0.00237 | 49/64 | 0.0440 | 0.01684 |
| 3/16 | 0.00065 | 0.00006 | 31/64 | 0.0111 | 0.00270 | 25/32 | 0.0467 | 0.01825 |
| 13/64 | 0.00082 | 0.00008 | 1/2 | 0.0123 | 0.00306 | 51/64 | 0.0496 | 0.01976 |
| 7/32 | 0.00102 | 0.00011 | 33/64 | 0.0134 | 0.00346 | 13/16 | 0.0526 | 0.02135 |
| 15/64 | 0.00126 | 0.00015 | 17/32 | 0.0147 | 0.00390 | 53/64 | 0.0557 | 0.02305 |
| 1/4 | 0.00153 | 0.00019 | 35/64 | 0.0160 | 0.00438 | 27/32 | 0.0588 | 0.02483 |
| 17/64 | 0.00183 | 0.00024 | 9/16 | 0.0174 | 0.00491 | 55/64 | 0.0622 | 0.02673 |
| 9/32 | 0.00218 | 0.00031 | 37/64 | 0.0189 | 0.00547 | 7/8 | 0.0656 | 0.02872 |
| 19/64 | 0.00256 | 0.00038 | 19/32 | 0.0205 | 0.00609 | 57/64 | 0.0692 | 0.03083 |
| 5/16 | 0.00299 | 0.00047 | 39/64 | 0.0222 | 0.00676 | 29/32 | 0.0728 | 0.03305 |
| 21/64 | 0.00344 | 0.00057 | 5/8 | 0.0239 | 0.00748 | 59/64 | 0.0767 | 0.03539 |
| 11/32 | 0.00398 | 0.00068 | 41/64 | 0.0258 | 0.00825 | 15/16 | 0.0807 | 0.03785 |
| 23/64 | 0.0045 | 0.00082 | 21/32 | 0.0277 | 0.00909 | 61/64 | 0.0849 | 0.04044 |
| 3/8 | 0.0052 | 0.00097 | 43/64 | 0.0297 | 0.00999 | 31/32 | 0.0891 | 0.04316 |
| 25/64 | 0.0058 | 0.00114 | 11/16 | 0.0318 | 0.01095 | 63/64 | 0.0934 | 0.04601 |
| 13/32 | 0.0066 | 0.00133 | 45/64 | 0.0341 | 0.01198 | ... | ... | ... |

* The diameters and values given on pages 386 to 390 are valid as English or metric units.

In this and succeeding tables, the *Polar Section Modulus* for a shaft of given diameter can be obtained by multiplying its Section Modulus by 2. Similarly its *Polar Moment of Inertia* can be obtained by multiplying its Moment of Inertia by 2.

### Section Moduli and Moments of Inertia for Round Shafts

| Diam. | Section Modulus | Moment of Inertia | Diam. | Section Modulus | Moment of Inertia | Diam. | Section Modulus | Moment of Inertia |
|---|---|---|---|---|---|---|---|---|
| 1.00 | 0.0981 | 0.0490 | 1.50 | 0.3313 | 0.2485 | 2.00 | 0.7854 | 0.7854 |
| 1.01 | 0.1011 | 0.0510 | 1.51 | 0.3380 | 0.2552 | 2.01 | 0.7972 | 0.8012 |
| 1.02 | 0.1041 | 0.0531 | 1.52 | 0.3447 | 0.2620 | 2.02 | 0.8092 | 0.8172 |
| 1.03 | 0.1072 | 0.0552 | 1.53 | 0.3516 | 0.2689 | 2.03 | 0.8212 | 0.8335 |
| 1.04 | 0.1104 | 0.0574 | 1.54 | 0.3585 | 0.2761 | 2.04 | 0.8334 | 0.8501 |
| 1.05 | 0.1136 | 0.0596 | 1.55 | 0.3655 | 0.2833 | 2.05 | 0.8457 | 0.8669 |
| 1.06 | 0.1169 | 0.0619 | 1.56 | 0.3727 | 0.2907 | 2.06 | 0.8582 | 0.8839 |
| 1.07 | 0.1202 | 0.0643 | 1.57 | 0.3799 | 0.2982 | 2.07 | 0.8707 | 0.9012 |
| 1.08 | 0.1236 | 0.0667 | 1.58 | 0.3872 | 0.3059 | 2.08 | 0.8834 | 0.9188 |
| 1.09 | 0.1271 | 0.0692 | 1.59 | 0.3946 | 0.3137 | 2.09 | 0.8962 | 0.9366 |
| 1.10 | 0.1307 | 0.0718 | 1.60 | 0.4021 | 0.3217 | 2.10 | 0.9092 | 0.9547 |
| 1.11 | 0.1342 | 0.0745 | 1.61 | 0.4097 | 0.3298 | 2.11 | 0.9222 | 0.9729 |
| 1.12 | 0.1379 | 0.0772 | 1.62 | 0.4173 | 0.3380 | 2.12 | 0.9354 | 0.9915 |
| 1.13 | 0.1416 | 0.0800 | 1.63 | 0.4251 | 0.3465 | 2.13 | 0.9487 | 1.0103 |
| 1.14 | 0.1454 | 0.0829 | 1.64 | 0.4330 | 0.3550 | 2.14 | 0.9621 | 1.0295 |
| 1.15 | 0.1493 | 0.0859 | 1.65 | 0.4410 | 0.3638 | 2.15 | 0.9757 | 1.0488 |
| 1.16 | 0.1532 | 0.0888 | 1.66 | 0.4490 | 0.3727 | 2.16 | 0.9894 | 1.0685 |
| 1.17 | 0.1572 | 0.0919 | 1.67 | 0.4572 | 0.3818 | 2.17 | 1.0031 | 1.0884 |
| 1.18 | 0.1613 | 0.0951 | 1.68 | 0.4655 | 0.3910 | 2.18 | 1.0171 | 1.1086 |
| 1.19 | 0.1654 | 0.0984 | 1.69 | 0.4738 | 0.4004 | 2.19 | 1.0311 | 1.1291 |
| 1.20 | 0.1696 | 0.1018 | 1.70 | 0.4823 | 0.4100 | 2.20 | 1.0454 | 1.1499 |
| 1.21 | 0.1739 | 0.1052 | 1.71 | 0.4908 | 0.4197 | 2.21 | 1.0596 | 1.1709 |
| 1.22 | 0.1782 | 0.1087 | 1.72 | 0.4995 | 0.4296 | 2.22 | 1.0741 | 1.1923 |
| 1.23 | 0.1826 | 0.1123 | 1.73 | 0.5083 | 0.4397 | 2.23 | 1.0887 | 1.2139 |
| 1.24 | 0.1871 | 0.1160 | 1.74 | 0.5171 | 0.4499 | 2.24 | 1.1034 | 1.2358 |
| 1.25 | 0.1917 | 0.1198 | 1.75 | 0.5261 | 0.4603 | 2.25 | 1.1183 | 1.2580 |
| 1.26 | 0.1963 | 0.1237 | 1.76 | 0.5352 | 0.4710 | 2.26 | 1.1332 | 1.2806 |
| 1.27 | 0.2011 | 0.1277 | 1.77 | 0.5444 | 0.4818 | 2.27 | 1.1483 | 1.3034 |
| 1.28 | 0.2058 | 0.1317 | 1.78 | 0.5536 | 0.4927 | 2.28 | 1.1636 | 1.3265 |
| 1.29 | 0.2107 | 0.1359 | 1.79 | 0.5630 | 0.5039 | 2.29 | 1.1790 | 1.3499 |
| 1.30 | 0.2157 | 0.1402 | 1.80 | 0.5726 | 0.5153 | 2.30 | 1.1945 | 1.3737 |
| 1.31 | 0.2207 | 0.1445 | 1.81 | 0.5821 | 0.5268 | 2.31 | 1.2101 | 1.3977 |
| 1.32 | 0.2258 | 0.1490 | 1.82 | 0.5918 | 0.5385 | 2.32 | 1.2259 | 1.4234 |
| 1.33 | 0.2309 | 0.1535 | 1.83 | 0.6016 | 0.5505 | 2.33 | 1.2418 | 1.4468 |
| 1.34 | 0.2362 | 0.1582 | 1.84 | 0.6115 | 0.5626 | 2.34 | 1.2579 | 1.4718 |
| 1.35 | 0.2415 | 0.1630 | 1.85 | 0.6216 | 0.5749 | 2.35 | 1.2741 | 1.4971 |
| 1.36 | 0.2469 | 0.1679 | 1.86 | 0.6317 | 0.5875 | 2.36 | 1.2904 | 1.5227 |
| 1.37 | 0.2524 | 0.1729 | 1.87 | 0.6419 | 0.6002 | 2.37 | 1.3069 | 1.5487 |
| 1.38 | 0.2580 | 0.1780 | 1.88 | 0.6524 | 0.6132 | 2.38 | 1.3235 | 1.5750 |
| 1.39 | 0.2636 | 0.1832 | 1.89 | 0.6628 | 0.6263 | 2.39 | 1.3403 | 1.6016 |
| 1.40 | 0.2694 | 0.1886 | 1.90 | 0.6734 | 0.6397 | 2.40 | 1.3572 | 1.6286 |
| 1.41 | 0.2752 | 0.1940 | 1.91 | 0.6840 | 0.6532 | 2.41 | 1.3742 | 1.6559 |
| 1.42 | 0.2811 | 0.1995 | 1.92 | 0.6948 | 0.6670 | 2.42 | 1.3914 | 1.6836 |
| 1.43 | 0.2870 | 0.2052 | 1.93 | 0.7057 | 0.6810 | 2.43 | 1.4087 | 1.7116 |
| 1.44 | 0.2931 | 0.2110 | 1.94 | 0.7168 | 0.6953 | 2.44 | 1.4262 | 1.7399 |
| 1.45 | 0.2993 | 0.2170 | 1.95 | 0.7279 | 0.7097 | 2.45 | 1.4438 | 1.7686 |
| 1.46 | 0.3055 | 0.2230 | 1.96 | 0.7392 | 0.7244 | 2.46 | 1.4615 | 1.7977 |
| 1.47 | 0.3118 | 0.2292 | 1.97 | 0.7505 | 0.7393 | 2.47 | 1.4794 | 1.8271 |
| 1.48 | 0.3182 | 0.2355 | 1.98 | 0.7620 | 0.7544 | 2.48 | 1.4975 | 1.8526 |
| 1.49 | 0.3247 | 0.2419 | 1.99 | 0.7736 | 0.7698 | 2.49 | 1.5156 | 1.8870 |

## Section Moduli and Moments of Inertia for Round Shafts

| Diam. | Section Modulus | Moment of Inertia | Diam. | Section Modulus | Moment of Inertia | Diam. | Section Modulus | Moment of Inertia |
|---|---|---|---|---|---|---|---|---|
| 2.50 | 1.5340 | 1.9175 | 3.00 | 2.6510 | 3.9761 | 3.50 | 4.2090 | 7.3662 |
| 2.51 | 1.5525 | 1.9483 | 3.01 | 2.6773 | 4.0293 | 3.51 | 4.2455 | 7.4507 |
| 2.52 | 1.5711 | 1.9796 | 3.02 | 2.7041 | 4.0831 | 3.52 | 4.2818 | 7.5360 |
| 2.53 | 1.5899 | 2.0112 | 3.03 | 2.7310 | 4.1375 | 3.53 | 4.3184 | 7.6220 |
| 2.54 | 1.6088 | 2.0431 | 3.04 | 2.7581 | 4.1924 | 3.54 | 4.3552 | 7.7087 |
| 2.55 | 1.6279 | 2.0755 | 3.05 | 2.7855 | 4.2478 | 3.55 | 4.3922 | 7.7962 |
| 2.56 | 1.6471 | 2.1083 | 3.06 | 2.8130 | 4.3038 | 3.56 | 4.4294 | 7.8845 |
| 2.57 | 1.6665 | 2.1414 | 3.07 | 2.8406 | 4.3604 | 3.57 | 4.4669 | 7.9734 |
| 2.58 | 1.6860 | 2.1749 | 3.08 | 2.8685 | 4.4175 | 3.58 | 4.5045 | 8.0631 |
| 2.59 | 1.7057 | 2.2088 | 3.09 | 2.8965 | 4.4751 | 3.59 | 4.5424 | 8.1536 |
| 2.60 | 1.7260 | 2.2432 | 3.10 | 2.9250 | 4.5333 | 3.60 | 4.5804 | 8.2448 |
| 2.61 | 1.7455 | 2.2779 | 3.11 | 2.9531 | 4.5921 | 3.61 | 4.6187 | 8.3367 |
| 2.62 | 1.7656 | 2.3130 | 3.12 | 2.9817 | 4.6514 | 3.62 | 4.6572 | 8.4296 |
| 2.63 | 1.7859 | 2.3485 | 3.13 | 3.0104 | 4.7113 | 3.63 | 4.6959 | 8.5231 |
| 2.64 | 1.8064 | 2.3844 | 3.14 | 3.0394 | 4.7718 | 3.64 | 4.7347 | 8.6174 |
| 2.65 | 1.8270 | 2.4208 | 3.15 | 3.0685 | 4.8330 | 3.65 | 4.7740 | 8.7125 |
| 2.66 | 1.8478 | 2.4575 | 3.16 | 3.0978 | 4.8946 | 3.66 | 4.8133 | 8.8084 |
| 2.67 | 1.8686 | 2.4947 | 3.17 | 3.1274 | 4.9568 | 3.67 | 4.8529 | 8.9050 |
| 2.68 | 1.8897 | 2.5322 | 3.18 | 3.1570 | 5.0197 | 3.68 | 4.8926 | 9.0025 |
| 2.69 | 1.9110 | 2.5702 | 3.19 | 3.1869 | 5.0832 | 3.69 | 4.9325 | 9.1007 |
| 2.70 | 1.9320 | 2.6087 | 3.20 | 3.2170 | 5.1472 | 3.70 | 4.9730 | 9.1998 |
| 2.71 | 1.9539 | 2.6476 | 3.21 | 3.2472 | 5.2119 | 3.71 | 5.0133 | 9.2996 |
| 2.72 | 1.9756 | 2.6868 | 3.22 | 3.2777 | 5.2771 | 3.72 | 5.0540 | 9.4003 |
| 2.73 | 1.9975 | 2.7266 | 3.23 | 3.3083 | 5.3430 | 3.73 | 5.0948 | 9.5018 |
| 2.74 | 2.0195 | 2.7668 | 3.24 | 3.3391 | 5.4094 | 3.74 | 5.1359 | 9.6041 |
| 2.75 | 2.0417 | 2.8074 | 3.25 | 3.3701 | 5.4765 | 3.75 | 5.1771 | 9.7072 |
| 2.76 | 2.0641 | 2.8484 | 3.26 | 3.4014 | 5.5442 | 3.76 | 5.2187 | 9.8112 |
| 2.77 | 2.0866 | 2.8899 | 3.27 | 3.4328 | 5.6126 | 3.77 | 5.2605 | 9.9160 |
| 2.78 | 2.1093 | 2.9319 | 3.28 | 3.4644 | 5.6815 | 3.78 | 5.3024 | 10.0216 |
| 2.79 | 2.1321 | 2.9743 | 3.29 | 3.4961 | 5.7511 | 3.79 | 5.3444 | 10.1286 |
| 2.80 | 2.1550 | 3.0172 | 3.30 | 3.5280 | 5.8214 | 3.80 | 5.3870 | 10.2350 |
| 2.81 | 2.1783 | 3.0605 | 3.31 | 3.5603 | 5.8923 | 3.81 | 5.4297 | 10.3436 |
| 2.82 | 2.2016 | 3.1043 | 3.32 | 3.5926 | 5.9638 | 3.82 | 5.4726 | 10.4526 |
| 2.83 | 2.2251 | 3.1486 | 3.33 | 3.6252 | 6.0363 | 3.83 | 5.5156 | 10.5624 |
| 2.84 | 2.2488 | 3.1933 | 3.34 | 3.6580 | 6.1088 | 3.84 | 5.5590 | 10.6732 |
| 2.85 | 2.2727 | 3.2385 | 3.35 | 3.6909 | 6.1823 | 3.85 | 5.6025 | 10.7848 |
| 2.86 | 2.2966 | 3.2842 | 3.36 | 3.7241 | 6.2564 | 3.86 | 5.6462 | 10.8970 |
| 2.87 | 2.3208 | 3.3304 | 3.37 | 3.7575 | 6.3312 | 3.87 | 5.6903 | 11.0110 |
| 2.88 | 2.3452 | 3.3771 | 3.38 | 3.7909 | 6.4067 | 3.88 | 5.7345 | 11.1250 |
| 2.89 | 2.3697 | 3.4242 | 3.39 | 3.8246 | 6.4829 | 3.89 | 5.7789 | 11.2400 |
| 2.90 | 2.3940 | 3.4719 | 3.40 | 3.8590 | 6.5597 | 3.90 | 5.8240 | 11.3560 |
| 2.91 | 2.4192 | 3.5200 | 3.41 | 3.8928 | 6.6372 | 3.91 | 5.8685 | 11.4730 |
| 2.92 | 2.4442 | 3.5686 | 3.42 | 3.9272 | 6.7154 | 3.92 | 5.9137 | 11.5910 |
| 2.93 | 2.4695 | 3.6178 | 3.43 | 3.9617 | 6.7943 | 3.93 | 5.9590 | 11.7100 |
| 2.94 | 2.4949 | 3.6674 | 3.44 | 3.9965 | 6.8739 | 3.94 | 6.0046 | 11.8290 |
| 2.95 | 2.5204 | 3.7175 | 3.45 | 4.0314 | 6.9542 | 3.95 | 6.0505 | 11.9500 |
| 2.96 | 2.5461 | 3.7682 | 3.46 | 4.0666 | 7.0352 | 3.96 | 6.0966 | 12.0690 |
| 2.97 | 2.5720 | 3.8196 | 3.47 | 4.1019 | 7.1168 | 3.97 | 6.1429 | 12.1930 |
| 2.98 | 2.5981 | 3.8711 | 3.48 | 4.1375 | 7.1976 | 3.98 | 6.1894 | 12.3170 |
| 2.99 | 2.6243 | 3.9233 | 3.49 | 4.1732 | 7.2824 | 3.99 | 6.2361 | 12.4410 |

## Section Moduli and Moments of Inertia for Round Shafts

| Diam. | Section Modulus | Moment of Inertia | Diam. | Section Modulus | Moment of Inertia | Diam. | Section Modulus | Moment of Inertia |
|---|---|---|---|---|---|---|---|---|
| 4.00 | 6.2830 | 12.566 | 4.50 | 8.946 | 20.129 | 5.00 | 12.272 | 30.680 |
| 4.01 | 6.3304 | 12.692 | 4.51 | 9.006 | 20.308 | 5.01 | 12.345 | 30.926 |
| 4.02 | 6.3779 | 12.820 | 4.52 | 9.066 | 20.489 | 5.02 | 12.420 | 31.173 |
| 4.03 | 6.4256 | 12.948 | 4.53 | 9.126 | 20.671 | 5.03 | 12.493 | 31.423 |
| 4.04 | 6.4736 | 13.077 | 4.54 | 9.186 | 20.854 | 5.04 | 12.568 | 31.673 |
| 4.05 | 6.5217 | 13.207 | 4.55 | 9.247 | 21.039 | 5.05 | 12.644 | 31.925 |
| 4.06 | 6.5701 | 13.337 | 4.56 | 9.308 | 21.224 | 5.06 | 12.718 | 32.179 |
| 4.07 | 6.6188 | 13.469 | 4.57 | 9.370 | 21.411 | 5.07 | 12.794 | 32.434 |
| 4.08 | 6.6677 | 13.602 | 4.58 | 9.431 | 21.599 | 5.08 | 12.870 | 32.691 |
| 4.09 | 6.7169 | 13.736 | 4.59 | 9.493 | 21.788 | 5.09 | 12.946 | 32.949 |
| 4.10 | 6.7660 | 13.871 | 4.60 | 9.556 | 21.979 | 5.10 | 13.023 | 33.209 |
| 4.11 | 6.8159 | 14.007 | 4.61 | 9.618 | 22.170 | 5.11 | 13.099 | 33.470 |
| 4.12 | 6.8657 | 14.143 | 4.62 | 9.681 | 22.363 | 5.12 | 13.177 | 33.733 |
| 4.13 | 6.9164 | 14.281 | 4.63 | 9.744 | 22.557 | 5.13 | 13.254 | 33.997 |
| 4.14 | 6.9663 | 14.420 | 4.64 | 9.807 | 22.753 | 5.14 | 13.332 | 34.263 |
| 4.15 | 7.0169 | 14.560 | 4.65 | 9.870 | 22.950 | 5.15 | 13.410 | 34.530 |
| 4.16 | 7.0677 | 14.701 | 4.66 | 9.934 | 23.148 | 5.16 | 13.488 | 34.799 |
| 4.17 | 7.1188 | 14.843 | 4.67 | 9.998 | 23.347 | 5.17 | 13.567 | 35.070 |
| 4.18 | 7.1702 | 14.985 | 4.68 | 10.063 | 23.548 | 5.18 | 13.645 | 35.342 |
| 4.19 | 7.2217 | 15.129 | 4.69 | 10.127 | 23.750 | 5.19 | 13.725 | 35.615 |
| 4.20 | 7.2740 | 15.274 | 4.70 | 10.193 | 23.953 | 5.20 | 13.804 | 35.891 |
| 4.21 | 7.3256 | 15.420 | 4.71 | 10.258 | 24.157 | 5.21 | 13.884 | 36.168 |
| 4.22 | 7.3779 | 15.568 | 4.72 | 10.323 | 24.363 | 5.22 | 13.964 | 36.446 |
| 4.23 | 7.4305 | 15.715 | 4.73 | 10.389 | 24.570 | 5.23 | 14.045 | 36.726 |
| 4.24 | 7.4833 | 15.865 | 4.74 | 10.455 | 24.779 | 5.24 | 14.125 | 37.008 |
| 4.25 | 7.5364 | 16.015 | 4.75 | 10.522 | 24.989 | 5.25 | 14.206 | 37.291 |
| 4.26 | 7.5898 | 16.166 | 4.76 | 10.588 | 25.200 | 5.26 | 14.287 | 37.576 |
| 4.27 | 7.6433 | 16.319 | 4.77 | 10.655 | 25.412 | 5.27 | 14.369 | 37.863 |
| 4.28 | 7.6972 | 16.472 | 4.78 | 10.722 | 25.626 | 5.28 | 14.451 | 38.151 |
| 4.29 | 7.7513 | 16.626 | 4.79 | 10.790 | 25.841 | 5.29 | 14.534 | 38.440 |
| 4.30 | 7.8060 | 16.782 | 4.80 | 10.857 | 26.058 | 5.30 | 14.616 | 38.732 |
| 4.31 | 7.8602 | 16.938 | 4.81 | 10.925 | 26.275 | 5.31 | 14.699 | 39.025 |
| 4.32 | 7.9149 | 17.096 | 4.82 | 10.994 | 26.495 | 5.32 | 14.782 | 39.320 |
| 4.33 | 7.9701 | 17.255 | 4.83 | 11.062 | 26.715 | 5.33 | 14.866 | 39.617 |
| 4.34 | 8.0254 | 17.415 | 4.84 | 11.131 | 26.937 | 5.34 | 14.949 | 39.915 |
| 4.35 | 8.0810 | 17.576 | 4.85 | 11.200 | 27.160 | 5.35 | 15.034 | 40.215 |
| 4.36 | 8.1369 | 17.738 | 4.86 | 11.269 | 27.385 | 5.36 | 15.118 | 40.516 |
| 4.37 | 8.1930 | 17.902 | 4.87 | 11.339 | 27.611 | 5.37 | 15.202 | 40.819 |
| 4.38 | 8.2494 | 18.066 | 4.88 | 11.409 | 27.839 | 5.38 | 15.288 | 41.124 |
| 4.39 | 8.3060 | 18.231 | 4.89 | 11.479 | 28.067 | 5.39 | 15.373 | 41.431 |
| 4.40 | 8.3630 | 18.398 | 4.90 | 11.550 | 28.298 | 5.40 | 15.459 | 41.739 |
| 4.41 | 8.4200 | 18.566 | 4.91 | 11.621 | 28.530 | 5.41 | 15.545 | 42.049 |
| 4.42 | 8.4775 | 18.735 | 4.92 | 11.692 | 28.763 | 5.42 | 15.631 | 42.361 |
| 4.43 | 8.5351 | 18.905 | 4.93 | 11.763 | 28.997 | 5.43 | 15.718 | 42.674 |
| 4.44 | 8.5930 | 19.077 | 4.94 | 11.835 | 29.233 | 5.44 | 15.805 | 42.990 |
| 4.45 | 8.6513 | 19.249 | 4.95 | 11.907 | 29.471 | 5.45 | 15.893 | 43.307 |
| 4.46 | 8.7097 | 19.423 | 4.96 | 11.979 | 29.710 | 5.46 | 15.980 | 43.626 |
| 4.47 | 8.7685 | 19.598 | 4.97 | 12.052 | 29.950 | 5.47 | 16.068 | 43.946 |
| 4.48 | 8.8274 | 19.773 | 4.98 | 12.124 | 30.192 | 5.48 | 16.157 | 44.268 |
| 4.49 | 8.8867 | 19.950 | 4.99 | 12.198 | 30.435 | 5.49 | 16.245 | 44.592 |

### Section Moduli and Moments of Inertia for Round Shafts

| Diam. | Section Modulus | Moment of Inertia | Diam. | Section Modulus | Moment of Inertia | Diam. | Section Modulus | Moment of Inertia |
|---|---|---|---|---|---|---|---|---|
| 5.5 | 16.3338 | 44.9180 | 30 | 2650.72 | 39760.8 | 54.5 | 15892.4 | 433068 |
| 6 | 21.2058 | 63.6173 | 30.5 | 2785.48 | 42478.5 | 55 | 16333.8 | 449180 |
| 6.5 | 26.9613 | 87.6241 | 31 | 2924.73 | 45333.2 | 55.5 | 16783.4 | 465738 |
| 7 | 33.6740 | 117.859 | 31.5 | 3068.54 | 48329.5 | 56 | 17241.1 | 482750 |
| 7.5 | 41.4175 | 155.316 | 32 | 3216.99 | 51471.9 | 56.5 | 17707.0 | 500223 |
| 8 | 50.2655 | 201.062 | 32.5 | 3370.16 | 54765.0 | 57 | 18181.3 | 518167 |
| 8.5 | 60.2916 | 256.239 | 33 | 3528.11 | 58213.8 | 57.5 | 18663.9 | 536588 |
| 9 | 71.5694 | 322.062 | 33.5 | 3690.92 | 61822.9 | 58 | 19155.1 | 555497 |
| 9.5 | 84.1726 | 399.820 | 34 | 3858.66 | 65597.2 | 58.5 | 19654.8 | 574901 |
| 10 | 98.1748 | 490.874 | 34.5 | 4031.41 | 69541.9 | 59 | 20163.0 | 594810 |
| 10.5 | 113.650 | 596.660 | 35 | 4209.24 | 73661.8 | 59.5 | 20680.0 | 615230 |
| 11 | 130.671 | 718.688 | 35.5 | 4392.23 | 77962.1 | 60 | 21205.8 | 636173 |
| 11.5 | 149.312 | 858.541 | 36 | 4580.44 | 82448.0 | 60.5 | 21740.3 | 657645 |
| 12 | 169.646 | 1017.88 | 36.5 | 4773.96 | 87124.7 | 61 | 22283.8 | 679656 |
| 12.5 | 191.748 | 1198.42 | 37 | 4972.85 | 91997.7 | 61.5 | 22836.3 | 702215 |
| 13 | 215.690 | 1401.99 | 37.5 | 5177.19 | 97072.2 | 62 | 23397.8 | 725332 |
| 13.5 | 241.547 | 1630.44 | 38 | 5387.05 | 102354 | 62.5 | 23968.5 | 749014 |
| 14 | 269.392 | 1885.74 | 38.5 | 5602.50 | 107848 | 63 | 24548.3 | 773272 |
| 14.5 | 299.298 | 2169.91 | 39 | 5823.63 | 113561 | 63.5 | 25137.4 | 798114 |
| 15 | 331.340 | 2485.05 | 39.5 | 6050.50 | 119497 | 64 | 25735.9 | 823550 |
| 15.5 | 365.591 | 2833.33 | 40 | 6283.19 | 125664 | 64.5 | 26343.8 | 849589 |
| 16 | 402.124 | 3216.99 | 40.5 | 6521.76 | 132066 | 65 | 26961.3 | 876241 |
| 16.5 | 441.013 | 3638.36 | 41 | 6766.30 | 138709 | 65.5 | 27588.2 | 903514 |
| 17 | 482.333 | 4099.83 | 41.5 | 7016.88 | 145600 | 66 | 28224.9 | 931420 |
| 17.5 | 526.155 | 4603.86 | 42 | 7273.57 | 152745 | 66.5 | 28871.2 | 959967 |
| 18 | 572.555 | 5153.00 | 42.5 | 7536.45 | 160150 | 67 | 29527.3 | 989166 |
| 18.5 | 621.606 | 5749.85 | 43 | 7805.58 | 167820 | 67.5 | 30193.3 | 1019030 |
| 19 | 673.381 | 6397.12 | 43.5 | 8081.05 | 175763 | 68 | 30869.3 | 1049560 |
| 19.5 | 727.954 | 7097.55 | 44 | 8362.92 | 183984 | 68.5 | 31555.3 | 1080770 |
| 20 | 785.398 | 7853.98 | 44.5 | 8651.27 | 192491 | 69 | 32251.3 | 1112670 |
| 20.5 | 845.788 | 8669.33 | 45 | 8946.18 | 201289 | 69.5 | 32957.5 | 1145270 |
| 21 | 909.197 | 9546.56 | 45.5 | 9247.71 | 210385 | 70 | 33674.0 | 1178590 |
| 21.5 | 975.698 | 10488.8 | 46 | 9555.94 | 219787 | 70.5 | 34400.7 | 1212630 |
| 22 | 1045.37 | 11499.0 | 46.5 | 9870.95 | 229500 | 71 | 35137.8 | 1247390 |
| 22.5 | 1118.27 | 12580.6 | 47 | 10192.8 | 239531 | 71.5 | 35885.4 | 1282900 |
| 23 | 1194.49 | 13736.7 | 47.5 | 10521.6 | 249887 | 72 | 36643.5 | 1319170 |
| 23.5 | 1274.10 | 14970.7 | 48 | 10857.3 | 260576 | 72.5 | 37412.3 | 1356190 |
| 24 | 1357.17 | 16286.0 | 48.5 | 11200.2 | 271604 | 73 | 38191.7 | 1393990 |
| 24.5 | 1443.77 | 17686.2 | 49 | 11550.2 | 282979 | 73.5 | 38981.8 | 1432580 |
| 25 | 1533.98 | 19174.8 | 49.5 | 11907.4 | 294707 | 74 | 39782.8 | 1471960 |
| 25.5 | 1627.87 | 20755.4 | 50 | 12271.9 | 306796 | 74.5 | 40594.6 | 1512150 |
| 26 | 1725.52 | 22431.8 | 50.5 | 12643.7 | 319253 | 75 | 41417.5 | 1553160 |
| 26.5 | 1827.00 | 24207.7 | 51 | 13023.0 | 332086 | 75.5 | 42251.4 | 1594990 |
| 27 | 1932.37 | 26087.1 | 51.5 | 13409.8 | 345302 | 76 | 43096.4 | 1637660 |
| 27.5 | 2041.73 | 28073.8 | 52 | 13804.2 | 358908 | 76.5 | 43952.6 | 1681190 |
| 28 | 2155.13 | 30171.9 | 52.5 | 14206.2 | 372913 | 77 | 44820.0 | 1725570 |
| 28.5 | 2272.66 | 32385.4 | 53 | 14616.0 | 387323 | 77.5 | 45698.8 | 1770830 |
| 29 | 2394.38 | 34718.6 | 53.5 | 15033.5 | 402147 | 78 | 46589.0 | 1816970 |
| 29.5 | 2520.38 | 37175.6 | 54 | 15459.0 | 417393 | 78.5 | 47490.7 | 1864010 |

### Section Moduli and Moments of Inertia for Round Shafts

| Diam. | Section Modulus | Moment of Inertia | Diam. | Section Modulus | Moment of Inertia | Diam. | Section Modulus | Moment of Inertia |
|---|---|---|---|---|---|---|---|---|
| 79 | 48404.0 | 1911960 | 103.5 | 108848 | 5632890 | 128 | 205887 | 13176790 |
| 79.5 | 49328.9 | 1960820 | 104 | 110433 | 5742530 | 128.5 | 208310 | 13383890 |
| 80 | 50265.5 | 2010620 | 104.5 | 112034 | 5853760 | 129 | 210751 | 13593420 |
| 80.5 | 51213.9 | 2061360 | 105 | 113650 | 5966600 | 129.5 | 213211 | 13805400 |
| 81 | 52174.1 | 2113050 | 105.5 | 115281 | 6081070 | 130 | 215690 | 14019850 |
| 81.5 | 53146.3 | 2165710 | 106 | 116928 | 6197170 | 130.5 | 218188 | 14236790 |
| 82 | 54130.4 | 2219350 | 106.5 | 118590 | 6314930 | 131 | 220706 | 14456230 |
| 82.5 | 55126.7 | 2273980 | 107 | 120268 | 6434350 | 131.5 | 223243 | 14678200 |
| 83 | 56135.1 | 2329610 | 107.5 | 121962 | 6555470 | 132 | 225799 | 14902720 |
| 83.5 | 57155.7 | 2386250 | 108 | 123672 | 6678280 | 132.5 | 228375 | 15129810 |
| 84 | 58188.6 | 2443920 | 108.5 | 125398 | 6802820 | 133 | 230970 | 15359480 |
| 84.5 | 59233.9 | 2502630 | 109 | 127139 | 6929080 | 133.5 | 233584 | 15591750 |
| 85 | 60291.6 | 2562390 | 109.5 | 128897 | 7057100 | 134 | 236219 | 15826650 |
| 85.5 | 61361.8 | 2623220 | 110 | 130671 | 7186880 | 134.5 | 238873 | 16064200 |
| 86 | 62444.7 | 2685120 | 110.5 | 132461 | 7318450 | 135 | 241547 | 16304410 |
| 86.5 | 63540.2 | 2748110 | 111 | 134267 | 7451810 | 135.5 | 244241 | 16547300 |
| 87 | 64648.4 | 2812210 | 111.5 | 136090 | 7586990 | 136 | 246954 | 16792890 |
| 87.5 | 65769.4 | 2877410 | 112 | 137929 | 7723990 | 136.5 | 249688 | 17041210 |
| 88 | 66903.4 | 2943750 | 112.5 | 139784 | 7862850 | 137 | 252442 | 17292280 |
| 88.5 | 68050.3 | 3011220 | 113 | 141656 | 8003570 | 137.5 | 255216 | 17546100 |
| 89 | 69210.2 | 3079850 | 113.5 | 143545 | 8146170 | 138 | 258010 | 17802720 |
| 89.5 | 70383.2 | 3149650 | 114 | 145450 | 8290660 | 138.5 | 260825 | 18062130 |
| 90 | 71569.4 | 3220620 | 114.5 | 147373 | 8437070 | 139 | 263660 | 18324370 |
| 90.5 | 72768.9 | 3292790 | 115 | 149312 | 8585410 | 139.5 | 266516 | 18589460 |
| 91 | 73981.7 | 3366170 | 115.5 | 151268 | 8735700 | 140 | 269392 | 18857410 |
| 91.5 | 75207.9 | 3440760 | 116 | 153241 | 8887950 | 140.5 | 272288 | 19128250 |
| 92 | 76447.5 | 3516590 | 116.5 | 155231 | 9042190 | 141 | 275206 | 19401990 |
| 92.5 | 77700.7 | 3593660 | 117 | 157238 | 9198420 | 141.5 | 278144 | 19678670 |
| 93 | 78967.6 | 3671990 | 117.5 | 159263 | 9356670 | 142 | 281103 | 19958290 |
| 93.5 | 80248.1 | 3751600 | 118 | 161304 | 9516950 | 142.5 | 284083 | 20240880 |
| 94 | 81542.4 | 3832490 | 118.5 | 163364 | 9679280 | 143 | 287083 | 20526460 |
| 94.5 | 82850.5 | 3914690 | 119 | 165440 | 9843690 | 143.5 | 290105 | 20815050 |
| 95 | 84172.6 | 3998200 | 119.5 | 167534 | 10010170 | 144 | 293148 | 21106680 |
| 95.5 | 85508.6 | 4083040 | 120 | 169646 | 10178760 | 144.5 | 296213 | 21401360 |
| 96 | 86858.8 | 4169220 | 120.5 | 171775 | 10349470 | 145 | 299298 | 21699110 |
| 96.5 | 88223.0 | 4256760 | 121 | 173923 | 10522320 | 145.5 | 302405 | 21999960 |
| 97 | 89601.5 | 4345670 | 121.5 | 176088 | 10697320 | 146 | 305533 | 22303930 |
| 97.5 | 90994.2 | 4435970 | 122 | 178271 | 10874500 | 146.5 | 308683 | 22611030 |
| 98 | 92401.3 | 4527660 | 122.5 | 180471 | 11053870 | 147 | 311854 | 22921300 |
| 98.5 | 93822.8 | 4620780 | 123 | 182690 | 11235450 | 147.5 | 315047 | 23234750 |
| 99 | 95258.9 | 4715320 | 123.5 | 184927 | 11419250 | 148 | 318262 | 23551400 |
| 99.5 | 96709.5 | 4811300 | 124 | 187182 | 11605310 | 148.5 | 321499 | 23871280 |
| 100 | 98174.8 | 4908740 | 124.5 | 189456 | 11793620 | 149 | 324757 | 24194410 |
| 100.5 | 99654.8 | 5007650 | 125 | 191748 | 11984220 | 149.5 | 328038 | 24520800 |
| 101 | 101150 | 5108050 | 125.5 | 194058 | 12177130 | 150 | 331340 | 24850490 |
| 101.5 | 102659 | 5209960 | 126 | 196387 | 12372350 | ..... | ....... | ......... |
| 102 | 104184 | 5313380 | 126.5 | 198734 | 12569910 | ..... | ....... | ......... |
| 102.5 | 105724 | 5418330 | 127 | 201100 | 12769820 | ..... | ....... | ......... |
| 103 | 107278 | 5524830 | 127.5 | 203484 | 12972110 | ..... | ....... | ......... |

**Length of Angles Bent to Circular Shape.** — To calculate the length of an angle-iron used either inside or outside of a tank or smokestack, the following table of constants may be used: Assume, for example, that a stand-pipe, 20 feet inside diameter, is provided with a 3 by 3 by ⅜ inch angle-iron on the inside near the top. The circumference of a circle 20 feet in diameter is 754 inches. From the table of constants, find the constant for a 3 by 3 by ⅜ inch angle-iron, which is 4.319. The length of the angle then is 754 − 4.319 = 749.681 inches. Should the angle be on the outside, add the constant instead of subtracting it; thus, 754 + 4.319 = 758.319 inches.

| Size of Angle | Const. | Size of Angle | Const. | Size of Angle | Const. |
|---|---|---|---|---|---|
| ¼ × 2 × 2 | 2.879 | ⁵⁄₁₆ × 3 × 3 | 4.123 | ½ × 5 × 5 | 6.804 |
| ⁵⁄₁₆ × 2 × 2 | 3.076 | ⅜ × 3 × 3 | 4.319 | ⅜ × 6 × 6 | 7.461 |
| ⅜ × 2 × 2 | 3.272 | ½ × 3 × 3 | 4.711 | ½ × 6 × 6 | 7.854 |
| ¼ × 2½ × 2½ | 3.403 | ⅜ × 3½ × 3½ | 4.843 | ¾ × 6 × 6 | 8.639 |
| ⁵⁄₁₆ × 2½ × 2½ | 3.600 | ½ × 3½ × 3½ | 5.235 | ½ × 8 × 8 | 9.949 |
| ⅜ × 2½ × 2½ | 3.796 | ⅜ × 4 × 4 | 5.366 | ¾ × 8 × 8 | 10.734 |
| ½ × 2½ × 2½ | 4.188 | ½ × 4 × 4 | 5.758 | 1 × 8 × 8 | 11.520 |
| ¼ × 3 × 3 | 3.926 | ⅜ × 5 × 5 | 6.414 | .......... | ...... |

**Standard Designations of Rolled Steel Shapes.** — Through a joint effort, the American Iron and Steel Institute (AISI) and the American Institute of Steel Construction (AISC) have changed most of the designations for their hot-rolled structural steel shapes. The present designations, standard for steel producing and fabricating industries, should be used when designing, detailing, and ordering steel. The accompanying table compares the present designations with the previous descriptions.

**Hot-Rolled Structural Steel Shape Designations** (AISI and AISC)*

| Present Designation | Type of Shape | Previous Designation |
|---|---|---|
| W 24 × 76 | W shape | 24 WF 76 |
| W 14 × 26 | W shape | 14 B 26 |
| S 24 × 100 | S shape | 24 I 100 |
| M 8 × 18.5 | M shape | 8 M 18.5 |
| M 10 × 9 | M shape | 10 JR 9.0 |
| M 8 × 34.3 | M shape | 8 × 8 M 34.3 |
| C 12 × 20.7 | American Standard Channel | 12 [ 20.7 |
| MC 12 × 45 | Miscellaneous Channel | 12 × 4 [ 45.0 |
| MC 12 × 10.6 | Miscellaneous Channel | 12 JR [ 10.6 |
| HP 14 × 73 | HP shape | 14 BP 73 |
| L 6 × 6 × ¾ | Equal Leg Angle | ∠ 6 × 6 × ¾ |
| L 6 × 4 × ⅝ | Unequal Leg Angle | ∠ 6 × 4 × ⅝ |
| WT 12 × 38 | Structural Tee cut from W shape | ST 12 WF 38 |
| WT 7 × 13 | Structural Tee cut from W shape | ST 7 B 13 |
| ST 12 × 50 | Structural Tee cut from S shape | ST 12 I 50 |
| MT 4 × 9.25 | Structural Tee cut from M shape | ST 4 M 9.25 |
| MT 5 × 4.5 | Structural Tee cut from M shape | ST 5 JR 4.5 |
| MT 4 × 17.15 | Structural Tee cut from M shape | ST 4 M 17.15 |
| PL ½ × 18 | Plate | PL 18 × ½ |
| Bar 1 ⊡ | Square Bar | Bar 1 ⊡ |
| Bar 1¼ ⏀ | Round Bar | Bar 1¼ ⏀ |
| Bar 2½ × ½ | Flat Bar | Bar 2½ × ½ |
| Pipe 4 Std. | Pipe | Pipe 4 Std. |
| Pipe 4 X - Strong | Pipe | Pipe 4 X - Strong |
| Pipe 4 XX - Strong | Pipe | Pipe 4 XX - Strong |
| TS 4 × 4 × .375 | Structural Tubing: Square | Tube 4 × 4 × .375 |
| TS 5 × 3 × .375 | Structural Tubing: Rectangular | Tube 5 × 3 × .375 |
| TS 3 OD × .250 | Structural Tubing: Circular | Tube 3 OD × .250 |

* Data taken from the "Manual of Steel Construction," 7th Edition, 1970, with permission of the American Institute of Steel Construction.

## Steel Wide-Flange Sections — 1*

Wide-flange sections are designated, in order, by a section letter, nominal depth of the member in inches, and the nominal weight in pounds per foot; thus:

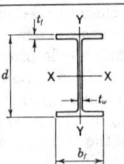

### W 18 x 64

indicates a wide-flange section having a nominal depth of 18 inches, and a nominal weight per foot of 64 pounds. Actual geometry for each section can be obtained from the values below.

| Designation | Area, $A$ | Depth, $d$ | Flange | | Web Thickness, $t_w$ | Axis X-X | | | Axis Y-Y | | |
|---|---|---|---|---|---|---|---|---|---|---|---|
| | | | Width, $b_f$ | Thickness, $t_f$ | | $I$ | $S$ | $r$ | $I$ | $S$ | $r$ |
| | in.² | in. | in. | in. | in. | in.⁴ | in.³ | in. | in.⁴ | in.³ | in. |
| †W 27×177 | 52.2 | 27.31 | 14.090 | 1.190 | 0.725 | 6740 | 494 | 11.4 | 556 | 78.9 | 3.26 |
| ×160 | 47.1 | 27.08 | 14.023 | 1.075 | 0.658 | 6030 | 446 | 11.3 | 495 | 70.6 | 3.24 |
| ×145 | 42.7 | 26.88 | 13.965 | 0.975 | 0.600 | 5430 | 404 | 11.3 | 443 | 63.5 | 3.22 |
| ×114 | 33.6 | 27.28 | 10.070 | 0.932 | 0.570 | 4090 | 300 | 11.0 | 159 | 31.6 | 2.18 |
| ×102 | 30.0 | 27.07 | 10.018 | 0.827 | 0.518 | 3610 | 267 | 11.0 | 139 | 27.7 | 2.15 |
| × 94 | 27.7 | 26.91 | 9.990 | 0.747 | 0.490 | 3270 | 243 | 10.9 | 124 | 24.9 | 2.12 |
| × 84 | 24.8 | 26.69 | 9.963 | 0.636 | 0.463 | 2830 | 212 | 10.7 | 105 | 21.1 | 2.06 |
| W 24×160 | 47.1 | 24.72 | 14.091 | 1.135 | 0.656 | 5120 | 414 | 10.4 | 530 | 75.2 | 3.35 |
| ×145 | 42.7 | 24.49 | 14.043 | 1.020 | 0.608 | 4570 | 373 | 10.3 | 67.1 | | 3.32 |
| ×130 | 38.3 | 24.25 | 14.000 | 0.900 | 0.565 | 4020 | 332 | 10.2 | 412 | 58.9 | 3.28 |
| ×120 | 35.4 | 24.31 | 12.088 | 0.930 | 0.556 | 3650 | 300 | 10.2 | 274 | 45.4 | 2.78 |
| ×110 | 32.5 | 24.16 | 12.042 | 0.855 | 0.510 | 3330 | 276 | 10.1 | 249 | 41.4 | 2.77 |
| ×100 | 29.5 | 24.00 | 12.000 | 0.775 | 0.468 | 3000 | 250 | 10.1 | 223 | 37.2 | 2.75 |
| × 94 | 27.7 | 24.29 | 9.061 | 0.872 | 0.516 | 2690 | 221 | 9.86 | 108 | 23.9 | 1.98 |
| × 84 | 24.7 | 24.09 | 9.015 | 0.772 | 0.470 | 2370 | 197 | 9.79 | 94.5 | 21.0 | 1.95 |
| × 76 | 22.4 | 23.91 | 8.985 | 0.682 | 0.440 | 2100 | 176 | 9.69 | 82.6 | 18.4 | 1.92 |
| × 68 | 20.0 | 23.71 | 8.961 | 0.582 | 0.416 | 1820 | 153 | 9.53 | 70.0 | 15.6 | 1.87 |
| × 61 | 18.0 | 23.72 | 7.023 | 0.591 | 0.419 | 1540 | 130 | 9.25 | 34.3 | 9.76 | 1.38 |
| × 55 | 16.2 | 23.55 | 7.000 | 0.503 | 0.396 | 1340 | 114 | 9.10 | 28.9 | 8.25 | 1.34 |
| W 21×142 | 41.8 | 21.46 | 13.132 | 1.095 | 0.659 | 3410 | 317 | 9.03 | 414 | 63.0 | 3.15 |
| ×127 | 37.4 | 21.24 | 13.061 | 0.985 | 0.588 | 3020 | 284 | 8.99 | 366 | 56.1 | 3.13 |
| ×112 | 33.0 | 21.00 | 13.000 | 0.865 | 0.527 | 2620 | 250 | 8.92 | 317 | 48.8 | 3.19 |
| × 96 | 28.3 | 21.14 | 9.038 | 0.935 | 0.575 | 2100 | 198 | 8.61 | 115 | 25.5 | 2.02 |
| × 82 | 24.2 | 20.86 | 8.962 | 0.795 | 0.499 | 1760 | 169 | 8.53 | 95.6 | 21.3 | 1.99 |
| × 73 | 21.5 | 21.24 | 8.295 | 0.740 | 0.455 | 1600 | 151 | 8.64 | 70.6 | 17.0 | 1.81 |
| × 68 | 20.0 | 21.13 | 8.270 | 0.685 | 0.430 | 1480 | 140 | 8.60 | 64.7 | 15.7 | 1.80 |
| × 62 | 18.3 | 20.99 | 8.240 | 0.615 | 0.400 | 1330 | 127 | 8.54 | 57.5 | 13.9 | 1.77 |
| × 55 | 16.2 | 20.80 | 8.215 | 0.522 | 0.375 | 1140 | 110 | 8.40 | 48.3 | 11.8 | 1.73 |
| × 49 | 14.4 | 20.82 | 6.520 | 0.532 | 0.368 | 971 | 93.3 | 8.21 | 24.7 | 7.57 | 1.31 |
| × 44 | 13.0 | 20.66 | 6.500 | 0.451 | 0.348 | 843 | 81.6 | 8.07 | 20.7 | 6.38 | 1.27 |
| W 18×114 | 33.5 | 18.48 | 11.833 | 0.991 | 0.595 | 2040 | 220 | 7.79 | 274 | 46.3 | 2.86 |
| ×105 | 30.9 | 18.32 | 11.792 | 0.911 | 0.554 | 1850 | 202 | 7.75 | 249 | 42.3 | 2.84 |
| × 96 | 28.2 | 18.16 | 11.750 | 0.831 | 0.512 | 1680 | 185 | 7.70 | 225 | 38.3 | 2.82 |
| × 85 | 25.0 | 18.32 | 8.838 | 0.911 | 0.526 | 1440 | 157 | 7.57 | 105 | 23.8 | 2.05 |
| × 77 | 22.7 | 18.16 | 8.787 | 0.831 | 0.475 | 1290 | 142 | 7.54 | 94.1 | 21.4 | 2.04 |
| × 70 | 20.6 | 18.00 | 8.750 | 0.751 | 0.438 | 1160 | 129 | 7.50 | 84.0 | 19.2 | 2.02 |
| × 64 | 18.9 | 17.87 | 8.715 | 0.686 | 0.403 | 1050 | 118 | 7.46 | 75.8 | 17.4 | 2.00 |
| × 60 | 17.7 | 18.25 | 7.558 | 0.695 | 0.416 | 986 | 108 | 7.47 | 50.1 | 13.3 | 1.68 |
| × 55 | 16.2 | 18.12 | 7.532 | 0.630 | 0.390 | 891 | 98.4 | 7.42 | 45.0 | 11.9 | 1.67 |
| × 50 | 14.7 | 18.00 | 7.500 | 0.570 | 0.358 | 802 | 89.1 | 7.38 | 40.2 | 10.7 | 1.65 |
| × 45 | 13.2 | 17.86 | 7.477 | 0.499 | 0.335 | 706 | 79.0 | 7.30 | 34.8 | 9.32 | 1.62 |
| × 40 | 11.8 | 17.90 | 6.018 | 0.524 | 0.316 | 612 | 68.4 | 7.21 | 19.1 | 6.34 | 1.27 |
| × 35 | 10.3 | 17.71 | 6.000 | 0.429 | 0.298 | 513 | 57.9 | 7.05 | 15.5 | 5.16 | 1.23 |

\* Data taken from the "Manual of Steel Construction," 7th Edition, 1970, with permission of the American Institute of Steel Construction.
† Consult the AISC Manual, noted above, for W steel shapes having nominal depths greater than 27 in.
*Symbols:* $I$ = moment of inertia; $S$ = section modulus; $r$ = radius of gyration.

## Steel Wide-Flange Sections — 2*

Wide-flange sections are designated, in order, by a section letter, nominal depth of the member in inches, and the nominal weight in pounds per foot; thus:

**W 16 x 78**

indicates a wide-flange section having a nominal depth of 16 inches, and a nominal weight per foot of 78 pounds. Actual geometry for each section can be obtained from the values below.

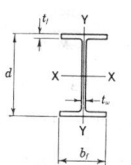

| Designation | Area, A | Depth, d | Flange Width, $b_f$ | Flange Thickness, $t_f$ | Web Thickness, $t_w$ | Axis X-X I | Axis X-X S | Axis X-X r | Axis Y-Y I | Axis Y-Y S | Axis Y-Y r |
|---|---|---|---|---|---|---|---|---|---|---|---|
| | in.² | in. | in. | in. | in. | in.⁴ | in.³ | in. | in.⁴ | in.³ | in. |
| W 16× 96 | 28.2 | 16.32 | 11.533 | 0.875 | 0.535 | 1360 | 166 | 6.93 | 224 | 38.8 | 2.82 |
| × 88 | 25.9 | 16.16 | 11.502 | 0.795 | 0.504 | 1220 | 151 | 6.87 | 202 | 35.1 | 2.79 |
| × 78 | 23.0 | 16.32 | 8.586 | 0.875 | 0.529 | 1050 | 128 | 6.75 | 92.5 | 21.6 | 2.01 |
| × 71 | 20.9 | 16.16 | 8.543 | 0.795 | 0.486 | 941 | 116 | 6.71 | 82.8 | 19.4 | 1.99 |
| × 64 | 18.8 | 16.00 | 8.500 | 0.715 | 0.443 | 836 | 104 | 6.66 | 73.3 | 17.3 | 1.97 |
| × 58 | 17.1 | 15.86 | 8.464 | 0.645 | 0.407 | 748 | 94.4 | 6.62 | 65.3 | 15.4 | 1.96 |
| × 50 | 14.7 | 16.25 | 7.073 | 0.628 | 0.380 | 657 | 80.8 | 6.68 | 37.1 | 10.5 | 1.59 |
| × 45 | 13.3 | 16.12 | 7.039 | 0.563 | 0.346 | 584 | 72.5 | 6.64 | 32.8 | 9.32 | 1.57 |
| × 40 | 11.8 | 16.00 | 7.000 | 0.503 | 0.307 | 517 | 64.6 | 6.62 | 28.8 | 8.23 | 1.56 |
| × 36 | 10.6 | 15.85 | 6.992 | 0.428 | 0.299 | 447 | 56.5 | 6.50 | 24.4 | 6.99 | 1.52 |
| × 31 | 9.13 | 15.84 | 5.525 | 0.442 | 0.275 | 374 | 47.2 | 6.40 | 12.5 | 4.51 | 1.17 |
| × 26 | 7.67 | 15.65 | 5.500 | 0.345 | 0.250 | 300 | 38.3 | 6.25 | 9.59 | 3.49 | 1.12 |
| W 14×730 | 215 | 22.44 | 17.889 | 4.910 | 3.069 | 14400 | 1280 | 8.18 | 4720 | 527 | 4.69 |
| ×665 | 196 | 21.67 | 17.646 | 4.522 | 2.826 | 12500 | 1150 | 7.99 | 4170 | 496 | 4.62 |
| ×605 | 178 | 20.94 | 17.418 | 4.157 | 2.598 | 10900 | 1040 | 7.81 | 3680 | 423 | 4.55 |
| ×550 | 162 | 20.26 | 17.206 | 3.818 | 2.386 | 9450 | 933 | 7.64 | 3260 | 378 | 4.49 |
| ×500 | 147 | 19.63 | 17.008 | 3.501 | 2.188 | 8250 | 840 | 7.49 | 2880 | 339 | 4.43 |
| ×455 | 134 | 19.05 | 16.828 | 3.213 | 2.008 | 7220 | 758 | 7.35 | 2560 | 304 | 4.37 |
| ×426 | 125 | 18.69 | 16.695 | 3.033 | 1.875 | 6610 | 707 | 7.26 | 2360 | 283 | 4.34 |
| ×398 | 117 | 18.31 | 16.590 | 2.843 | 1.770 | 6010 | 657 | 7.17 | 2170 | 262 | 4.31 |
| ×370 | 109 | 17.94 | 16.475 | 2.658 | 1.655 | 5450 | 608 | 7.08 | 1990 | 241 | 4.27 |
| ×342 | 101 | 17.56 | 16.365 | 2.468 | 1.545 | 4910 | 559 | 6.99 | 1810 | 221 | 4.24 |
| ×320 | 94.1 | 16.81 | 16.710 | 2.093 | 1.890 | 4140 | 493 | 6.63 | 1640 | 196 | 4.17 |
| ×314 | 92.3 | 17.19 | 16.235 | 2.283 | 1.415 | 4400 | 512 | 6.90 | 1630 | 201 | 4.20 |
| ×287 | 84.4 | 16.81 | 16.130 | 2.093 | 1.310 | 3910 | 465 | 6.81 | 1470 | 182 | 4.17 |
| ×264 | 77.6 | 16.50 | 16.025 | 1.938 | 1.205 | 3530 | 427 | 6.74 | 1330 | 166 | 4.14 |
| ×246 | 72.3 | 16.25 | 15.945 | 1.813 | 1.125 | 3230 | 397 | 6.68 | 1230 | 154 | 4.12 |
| ×237 | 69.7 | 16.12 | 15.910 | 1.748 | 1.090 | 3080 | 382 | 6.65 | 1170 | 148 | 4.11 |
| ×228 | 67.1 | 16.00 | 15.865 | 1.688 | 1.045 | 2940 | 368 | 6.62 | 1120 | 142 | 4.10 |
| ×219 | 64.4 | 15.87 | 15.825 | 1.623 | 1.005 | 2800 | 353 | 6.59 | 1070 | 136 | 4.08 |
| ×211 | 62.1 | 15.75 | 15.800 | 1.563 | 0.980 | 2670 | 339 | 6.56 | 1030 | 130 | 4.07 |
| ×202 | 59.4 | 15.63 | 15.750 | 1.503 | 0.930 | 2540 | 325 | 6.54 | 980 | 124 | 4.06 |
| ×193 | 56.7 | 15.50 | 15.710 | 1.438 | 0.890 | 2400 | 310 | 6.51 | 930 | 118 | 4.05 |
| ×184 | 54.1 | 15.38 | 15.660 | 1.378 | 0.840 | 2270 | 296 | 6.49 | 883 | 113 | 4.04 |
| ×176 | 51.7 | 15.25 | 15.640 | 1.313 | 0.820 | 2150 | 282 | 6.45 | 838 | 107 | 4.02 |
| ×167 | 49.1 | 15.12 | 15.600 | 1.248 | 0.780 | 2020 | 267 | 6.42 | 790 | 101 | 4.01 |
| ×158 | 46.5 | 15.00 | 15.550 | 1.188 | 0.730 | 1900 | 253 | 6.40 | 745 | 95.8 | 4.00 |
| ×150 | 44.1 | 14.88 | 15.515 | 1.128 | 0.695 | 1790 | 240 | 6.37 | 703 | 90.6 | 3.99 |
| ×142 | 41.8 | 14.75 | 15.500 | 1.063 | 0.680 | 1670 | 227 | 6.32 | 660 | 85.2 | 3.97 |

* Data taken from the "Manual of Steel Construction," 7th Edition, 1970, with permission of the American Institute of Steel Construction.

*Symbols:* $I$ = moment of inertia; $S$ = section modulus; $r$ = radius of gyration.

## Steel Wide-Flange Sections — 3*

Wide-flange sections are designated, in order, by a section letter, nominal depth of the member in inches, and the nominal weight in pounds per foot; thus:

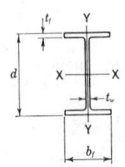

### W 14 x 38

indicates a wide-flange section having a nominal depth of 14 inches, and a nominal weight per foot of 38 pounds. Actual geometry for each section can be obtained from the values below.

| Designation | Area, $A$ | Depth, $d$ | Flange Width, $b_f$ | Flange Thickness, $t_f$ | Web Thickness, $t_w$ | Axis X-X $I$ | Axis X-X $S$ | Axis X-X $r$ | Axis Y-Y $I$ | Axis Y-Y $S$ | Axis Y-Y $r$ |
|---|---|---|---|---|---|---|---|---|---|---|---|
| | in.² | in. | in. | in. | in. | in.⁴ | in.³ | in. | in.⁴ | in.³ | in. |
| W 14×136 | 40.0 | 14.75 | 14.740 | 1.063 | 0.660 | 1590 | 216 | 6.31 | 568 | 77.0 | 3.77 |
| ×127 | 37.3 | 14.62 | 14.690 | 0.998 | 0.610 | 1480 | 202 | 6.29 | 528 | 71.8 | 3.76 |
| ×119 | 35.0 | 14.50 | 14.650 | 0.938 | 0.570 | 1370 | 189 | 6.26 | 492 | 67.1 | 3.75 |
| ×111 | 32.7 | 14.37 | 14.620 | 0.873 | 0.540 | 1270 | 176 | 6.23 | 455 | 62.2 | 3.73 |
| ×103 | 30.3 | 14.25 | 14.575 | 0.813 | 0.495 | 1170 | 164 | 6.21 | 420 | 57.6 | 3.72 |
| × 95 | 27.9 | 14.12 | 14.545 | 0.748 | 0.465 | 1060 | 151 | 6.17 | 384 | 52.8 | 3.71 |
| × 87 | 25.6 | 14.00 | 14.500 | 0.688 | 0.420 | 967 | 138 | 6.15 | 350 | 48.2 | 3.70 |
| × 84 | 24.7 | 14.18 | 12.023 | 0.778 | 0.451 | 928 | 131 | 6.13 | 225 | 37.5 | 3.02 |
| × 78 | 22.9 | 14.06 | 12.000 | 0.718 | 0.428 | 851 | 121 | 6.09 | 207 | 34.5 | 3.00 |
| × 74 | 21.8 | 14.19 | 10.072 | 0.783 | 0.450 | 797 | 112 | 6.05 | 133 | 26.5 | 2.48 |
| × 68 | 20.0 | 14.06 | 10.040 | 0.718 | 0.418 | 724 | 103 | 6.02 | 121 | 24.1 | 2.46 |
| × 61 | 17.9 | 13.91 | 10.000 | 0.643 | 0.378 | 641 | 92.2 | 5.98 | 107 | 21.5 | 2.45 |
| × 53 | 15.6 | 13.94 | 8.062 | 0.658 | 0.370 | 542 | 77.8 | 5.90 | 57.5 | 14.3 | 1.92 |
| × 48 | 14.1 | 13.81 | 8.031 | 0.593 | 0.339 | 485 | 70.2 | 5.86 | 51.3 | 12.8 | 1.91 |
| × 43 | 12.6 | 13.68 | 8.000 | 0.528 | 0.308 | 429 | 62.7 | 5.82 | 45.1 | 11.3 | 1.89 |
| × 38 | 11.2 | 14.12 | 6.776 | 0.513 | 0.313 | 386 | 54.7 | 5.88 | 26.6 | 7.86 | 1.54 |
| × 34 | 10.0 | 14.00 | 6.750 | 0.453 | 0.287 | 340 | 48.6 | 5.83 | 23.3 | 6.89 | 1.52 |
| × 30 | 8.83 | 13.86 | 6.733 | 0.383 | 0.270 | 290 | 41.9 | 5.74 | 19.5 | 5.80 | 1.49 |
| × 26 | 7.67 | 13.89 | 5.025 | 0.418 | 0.255 | 244 | 35.1 | 5.64 | 8.86 | 3.53 | 1.08 |
| × 22 | 6.49 | 13.72 | 5.000 | 0.335 | 0.230 | 198 | 28.9 | 5.53 | 7.00 | 2.80 | 1.04 |
| W 12×190 | 55.9 | 14.38 | 12.670 | 1.736 | 1.060 | 1890 | 263 | 5.82 | 590 | 93.1 | 3.25 |
| ×161 | 47.4 | 13.88 | 12.515 | 1.486 | 0.905 | 1540 | 222 | 5.70 | 486 | 77.7 | 3.20 |
| ×133 | 39.1 | 13.38 | 12.365 | 1.236 | 0.755 | 1220 | 183 | 5.59 | 390 | 63.1 | 3.16 |
| ×120 | 35.3 | 13.12 | 12.320 | 1.106 | 0.710 | 1070 | 163 | 5.51 | 345 | 56.0 | 3.13 |
| ×106 | 31.2 | 12.88 | 12.230 | 0.986 | 0.620 | 931 | 145 | 5.46 | 301 | 49.2 | 3.11 |
| × 99 | 29.1 | 12.75 | 12.192 | 0.921 | 0.582 | 859 | 135 | 5.43 | 278 | 45.7 | 3.09 |
| × 92 | 27.1 | 12.62 | 12.155 | 0.856 | 0.545 | 789 | 125 | 5.40 | 256 | 42.2 | 3.08 |
| × 85 | 25.0 | 12.50 | 12.105 | 0.796 | 0.495 | 723 | 116 | 5.38 | 235 | 38.9 | 3.07 |
| × 79 | 23.2 | 12.38 | 12.080 | 0.736 | 0.470 | 663 | 107 | 5.34 | 216 | 35.8 | 3.05 |
| × 72 | 21.2 | 12.25 | 12.040 | 0.671 | 0.430 | 597 | 97.5 | 5.31 | 195 | 32.4 | 3.04 |
| × 65 | 19.1 | 12.12 | 12.000 | 0.606 | 0.390 | 533 | 88.0 | 5.28 | 175 | 29.1 | 3.02 |
| × 58 | 17.1 | 12.19 | 10.014 | 0.641 | 0.359 | 476 | 78.1 | 5.28 | 107 | 21.4 | 2.51 |
| × 53 | 15.6 | 12.06 | 10.000 | 0.576 | 0.345 | 426 | 70.7 | 5.23 | 96.1 | 19.2 | 2.48 |
| × 50 | 14.7 | 12.19 | 8.077 | 0.641 | 0.371 | 395 | 64.7 | 5.18 | 56.4 | 14.0 | 1.96 |
| × 45 | 13.2 | 12.06 | 8.042 | 0.576 | 0.336 | 351 | 58.2 | 5.15 | 50.0 | 12.4 | 1.94 |
| × 40 | 11.8 | 11.94 | 8.000 | 0.516 | 0.294 | 310 | 51.9 | 5.13 | 44.1 | 11.0 | 1.94 |
| × 36 | 10.6 | 12.24 | 6.565 | 0.540 | 0.305 | 281 | 46.0 | 5.15 | 25.5 | 7.77 | 1.55 |
| × 31 | 9.13 | 12.09 | 6.525 | 0.465 | 0.265 | 239 | 39.5 | 5.12 | 21.6 | 6.61 | 1.54 |
| × 27 | 7.95 | 11.96 | 6.497 | 0.400 | 0.237 | 204 | 34.2 | 5.07 | 18.3 | 5.63 | 1.52 |
| × 22 | 6.47 | 12.31 | 4.030 | 0.424 | 0.260 | 156 | 25.3 | 4.91 | 4.64 | 2.31 | 0.847 |
| × 19 | 5.59 | 12.16 | 4.007 | 0.349 | 0.237 | 130 | 21.3 | 4.82 | 3.76 | 1.88 | 0.820 |
| × 16.5 | 4.87 | 12.00 | 4.000 | 0.269 | 0.230 | 105 | 17.6 | 4.65 | 2.88 | 1.44 | 0.770 |
| × 14 | 4.12 | 11.91 | 3.968 | 0.224 | 0.198 | 88.0 | 14.8 | 4.62 | 2.34 | 1.18 | 0.754 |

* Data taken from the "Manual of Steel Construction," 7th Edition, 1970, with permission of the American Institute of Steel Construction.

### Steel Wide-Flange Sections — 4*

Wide-flange sections are designated, in order, by a section letter, nominal depth of the member in inches, and the nominal weight in pounds per foot; thus:

$$W\ 8\ x\ 67$$

indicates a wide-flange section having a nominal depth of 8 inches, and a nominal weight per foot of 67 pounds. Actual geometry for each section can be obtained from the values below.

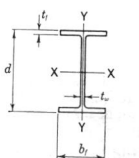

| Designation | Area, A | Depth, d | Flange | | Web Thickness, $t_w$ | Axis X-X | | | Axis Y-Y | | |
| | | | Width, $b_f$ | Thickness, $t_f$ | | I | S | r | I | S | r |
| --- | --- | --- | --- | --- | --- | --- | --- | --- | --- | --- | --- |
| | in.² | in. | in. | in. | in. | in.⁴ | in.³ | in. | in.⁴ | in.³ | in. |
| W 10×112 | 32.9 | 11.38 | 10.415 | 1.248 | 0.755 | 719 | 126 | 4.67 | 235 | 45.2 | 2.67 |
| ×100 | 29.4 | 11.12 | 10.345 | 1.118 | 0.685 | 625 | 112 | 4.61 | 207 | 39.9 | 2.65 |
| × 89 | 26.2 | 10.88 | 10.275 | 0.998 | 0.615 | 542 | 99.7 | 4.55 | 181 | 35.2 | 2.63 |
| × 77 | 22.7 | 10.62 | 10.195 | 0.868 | 0.535 | 457 | 86.1 | 4.49 | 153 | 30.1 | 2.60 |
| × 72 | 21.2 | 10.50 | 10.170 | 0.808 | 0.510 | 421 | 80.1 | 4.46 | 142 | 27.9 | 2.59 |
| × 66 | 19.4 | 10.38 | 10.117 | 0.748 | 0.457 | 382 | 73.7 | 4.44 | 129 | 25.5 | 2.58 |
| × 60 | 17.7 | 10.25 | 10.075 | 0.683 | 0.415 | 344 | 67.1 | 4.41 | 116 | 23.1 | 2.57 |
| × 54 | 15.9 | 10.12 | 10.028 | 0.618 | 0.368 | 306 | 60.4 | 4.39 | 104 | 20.7 | 2.56 |
| × 49 | 14.4 | 10.00 | 10.000 | 0.558 | 0.340 | 273 | 54.6 | 4.35 | 93.0 | 18.6 | 2.54 |
| × 45 | 13.2 | 10.12 | 8.022 | 0.618 | 0.350 | 249 | 49.1 | 4.33 | 53.2 | 13.3 | 2.00 |
| × 39 | 11.5 | 9.94 | 7.990 | 0.528 | 0.318 | 210 | 42.2 | 4.27 | 44.9 | 11.2 | 1.98 |
| × 33 | 9.71 | 9.75 | 7.964 | 0.433 | 0.292 | 171 | 35.0 | 4.20 | 36.5 | 9.16 | 1.94 |
| × 29 | 8.54 | 10.22 | 5.799 | 0.500 | 0.289 | 158 | 30.8 | 4.30 | 16.3 | 5.61 | 1.38 |
| × 25 | 7.36 | 10.08 | 5.762 | 0.430 | 0.252 | 133 | 26.5 | 4.26 | 13.7 | 4.76 | 1.37 |
| × 21 | 6.20 | 9.90 | 5.750 | 0.340 | 0.240 | 107 | 21.5 | 4.15 | 10.8 | 3.75 | 1.32 |
| × 19 | 5.61 | 10.25 | 4.020 | 0.394 | 0.250 | 96.3 | 18.8 | 4.14 | 4.28 | 2.13 | 0.874 |
| × 17 | 4.99 | 10.12 | 4.010 | 0.329 | 0.240 | 81.9 | 16.2 | 4.05 | 3.55 | 1.77 | 0.844 |
| × 15 | 4.41 | 10.00 | 4.000 | 0.269 | 0.230 | 68.9 | 13.8 | 3.95 | 2.88 | 1.44 | 0.809 |
| × 11.5 | 3.39 | 9.87 | 3.950 | 0.204 | 0.180 | 52.0 | 10.5 | 3.92 | 2.10 | 1.06 | 0.787 |
| W 8×67 | 19.7 | 9.00 | 8.287 | 0.933 | 0.575 | 272 | 60.4 | 3.71 | 88.6 | 21.4 | 2.12 |
| ×58 | 17.1 | 8.75 | 8.222 | 0.808 | 0.510 | 227 | 52.0 | 3.65 | 74.9 | 18.2 | 2.10 |
| ×48 | 14.1 | 8.50 | 8.117 | 0.683 | 0.405 | 184 | 43.2 | 3.61 | 60.9 | 15.0 | 2.08 |
| ×40 | 11.8 | 8.25 | 8.077 | 0.558 | 0.365 | 146 | 35.5 | 3.53 | 49.0 | 12.1 | 2.04 |
| ×35 | 10.3 | 8.12 | 8.027 | 0.493 | 0.315 | 126 | 31.1 | 3.50 | 42.5 | 10.6 | 2.03 |
| ×31 | 9.12 | 8.00 | 8.000 | 0.433 | 0.288 | 110 | 27.4 | 3.47 | 37.0 | 9.24 | 2.01 |
| ×28 | 8.23 | 8.06 | 6.540 | 0.463 | 0.285 | 97.8 | 24.3 | 3.45 | 21.6 | 6.61 | 1.62 |
| ×24 | 7.06 | 7.93 | 6.500 | 0.398 | 0.245 | 82.5 | 20.8 | 3.42 | 18.2 | 5.61 | 1.61 |
| ×20 | 5.89 | 8.14 | 5.268 | 0.378 | 0.248 | 69.4 | 17.0 | 3.43 | 9.22 | 3.50 | 1.25 |
| ×17 | 5.01 | 8.00 | 5.250 | 0.308 | 0.230 | 56.6 | 14.1 | 3.36 | 7.44 | 2.83 | 1.22 |
| ×15 | 4.43 | 8.12 | 4.015 | 0.314 | 0.245 | 48.1 | 11.8 | 3.29 | 3.40 | 1.69 | 0.876 |
| ×13 | 3.83 | 8.00 | 4.000 | 0.254 | 0.230 | 39.6 | 9.90 | 3.21 | 2.72 | 1.36 | 0.842 |
| ×10 | 2.96 | 7.90 | 3.940 | 0.204 | 0.170 | 30.8 | 7.80 | 3.23 | 2.08 | 1.06 | 0.839 |
| W 6×25 | 7.35 | 6.37 | 6.080 | 0.456 | 0.320 | 53.3 | 16.7 | 2.69 | 17.1 | 5.62 | 1.53 |
| ×20 | 5.88 | 6.20 | 6.018 | 0.367 | 0.258 | 41.5 | 13.4 | 2.66 | 13.3 | 4.43 | 1.51 |
| ×16 | 4.72 | 6.25 | 4.030 | 0.404 | 0.260 | 31.7 | 10.2 | 2.59 | 4.42 | 2.19 | 0.967 |
| ×15.5 | 4.56 | 6.00 | 5.995 | 0.269 | 0.235 | 30.1 | 10.0 | 2.57 | 9.67 | 3.23 | 1.46 |
| ×12 | 3.54 | 6.00 | 4.000 | 0.279 | 0.230 | 21.7 | 7.25 | 2.48 | 2.98 | 1.49 | 0.918 |
| × 8.5 | 2.51 | 5.83 | 3.940 | 0.194 | 0.170 | 14.8 | 5.08 | 2.43 | 1.98 | 1.01 | 0.889 |
| W 5×18.5 | 5.43 | 5.12 | 5.025 | 0.420 | 0.265 | 25.4 | 9.94 | 2.16 | 8.89 | 3.54 | 1.28 |
| ×16 | 4.70 | 5.00 | 5.000 | 0.360 | 0.240 | 21.3 | 8.53 | 2.13 | 7.51 | 3.00 | 1.26 |
| W 4×13 | 3.82 | 4.16 | 4.060 | 0.345 | 0.280 | 11.3 | 5.45 | 1.72 | 3.76 | 1.85 | 0.991 |

* Data taken from the "Manual of Steel Construction," 7th Edition, 1970, with permission of the American Institute of Steel Construction.

*Symbols:* I = moment of inertia; S = section modulus; r = radius of gyration.

### Steel S Sections*

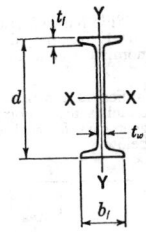

"S" is the section symbol for "I" Beams. S shapes are designated, in order, by their section letter, actual depth in inches, and nominal weight in pounds per foot. Thus:

**S 5 x 14.75**

indicates an S shape (or I beam) having a depth of 5 inches and a nominal weight of 14.75 pounds per foot.

| Designation | Area $A$ | Depth $d$ | Flange Width $b_f$ | Flange Thickness $t_f$ | Web Thickness $t_w$ | Axis X-X $I$ | $S$ | $r$ | Axis Y-Y $I$ | $S$ | $r$ |
|---|---|---|---|---|---|---|---|---|---|---|---|
| | in.² | in. | in. | in. | in. | in.⁴ | in.³ | in. | in.⁴ | in.³ | in. |
| S 24×120 | 35.3 | 24.00 | 8.048 | 1.102 | 0.798 | 3030 | 252 | 9.26 | 84.2 | 20.9 | 1.54 |
| ×105.9 | 31.1 | 24.00 | 7.875 | 1.102 | 0.625 | 2830 | 236 | 9.53 | 78.2 | 19.8 | 1.58 |
| ×100 | 29.4 | 24.00 | 7.247 | 0.871 | 0.747 | 2390 | 199 | 9.01 | 47.8 | 13.2 | 1.27 |
| ×90 | 26.5 | 24.00 | 7.124 | 0.871 | 0.624 | 2250 | 187 | 9.22 | 44.9 | 12.6 | 1.30 |
| ×79.9 | 23.5 | 24.00 | 7.001 | 0.871 | 0.501 | 2110 | 175 | 9.47 | 42.3 | 12.1 | 1.34 |
| S 20×95 | 27.9 | 20.00 | 7.200 | 0.916 | 0.800 | 1610 | 161 | 7.60 | 49.7 | 13.8 | 1.33 |
| ×85 | 25.0 | 20.00 | 7.053 | 0.916 | 0.653 | 1520 | 152 | 7.79 | 46.2 | 13.1 | 1.36 |
| ×75 | 22.1 | 20.00 | 6.391 | 0.789 | 0.641 | 1280 | 128 | 7.60 | 29.6 | 9.28 | 1.16 |
| ×65.4 | 19.2 | 20.00 | 6.250 | 0.789 | 0.500 | 1180 | 118 | 7.84 | 27.4 | 8.77 | 1.19 |
| S 18×70 | 20.6 | 18.00 | 6.251 | 0.691 | 0.711 | 926 | 103 | 6.71 | 24.1 | 7.72 | 1.08 |
| ×54.7 | 16.1 | 18.00 | 6.001 | 0.691 | 0.461 | 804 | 89.4 | 7.07 | 20.8 | 6.94 | 1.14 |
| S 15×50 | 14.7 | 15.00 | 5.640 | 0.622 | 0.550 | 486 | 64.8 | 5.75 | 15.7 | 5.57 | 1.03 |
| ×42.9 | 12.6 | 15.00 | 5.501 | 0.622 | 0.411 | 447 | 59.6 | 5.95 | 14.4 | 5.23 | 1.07 |
| S 12×50 | 14.7 | 12.00 | 5.477 | 0.659 | 0.687 | 305 | 50.8 | 4.55 | 15.7 | 5.74 | 1.03 |
| ×40.8 | 12.0 | 12.00 | 5.252 | 0.659 | 0.462 | 272 | 45.4 | 4.77 | 13.6 | 5.16 | 1.06 |
| ×35 | 10.3 | 12.00 | 5.078 | 0.544 | 0.428 | 229 | 38.2 | 4.72 | 9.87 | 3.89 | 0.980 |
| ×31.8 | 9.35 | 12.00 | 5.000 | 0.544 | 0.350 | 218 | 36.4 | 4.83 | 9.36 | 3.74 | 1.00 |
| S 10×35 | 10.3 | 10.00 | 4.944 | 0.491 | 0.594 | 147 | 29.4 | 3.78 | 8.36 | 3.38 | 0.901 |
| ×25.4 | 7.46 | 10.00 | 4.661 | 0.491 | 0.311 | 124 | 24.7 | 4.07 | 6.79 | 2.91 | 0.954 |
| S 8×23 | 6.77 | 8.00 | 4.171 | 0.425 | 0.441 | 64.9 | 16.2 | 3.10 | 4.31 | 2.07 | 0.798 |
| ×18.4 | 5.41 | 8.00 | 4.001 | 0.425 | 0.271 | 57.6 | 14.4 | 3.26 | 3.73 | 1.86 | 0.831 |
| S 7×20 | 5.88 | 7.00 | 3.860 | 0.392 | 0.450 | 42.4 | 12.1 | 2.69 | 3.17 | 1.64 | 0.734 |
| ×15.3 | 4.50 | 7.00 | 3.662 | 0.392 | 0.252 | 36.7 | 10.5 | 2.86 | 2.64 | 1.44 | 0.766 |
| S 6×17.25 | 5.07 | 6.00 | 3.565 | 0.359 | 0.465 | 26.3 | 8.77 | 2.28 | 2.31 | 1.30 | 0.675 |
| ×12.5 | 3.67 | 6.00 | 3.332 | 0.359 | 0.232 | 22.1 | 7.37 | 2.45 | 1.82 | 1.09 | 0.705 |
| S 5×14.75 | 4.34 | 5.00 | 3.284 | 0.326 | 0.494 | 15.2 | 6.09 | 1.87 | 1.67 | 1.01 | 0.620 |
| ×10 | 2.94 | 5.00 | 3.004 | 0.326 | 0.214 | 12.3 | 4.92 | 2.05 | 1.22 | 0.809 | 0.643 |
| S 4×9.5 | 2.79 | 4.00 | 2.796 | 0.293 | 0.326 | 6.79 | 3.39 | 1.56 | 0.903 | 0.646 | 0.569 |
| ×7.7 | 2.26 | 4.00 | 2.663 | 0.293 | 0.193 | 6.08 | 3.04 | 1.64 | 0.764 | 0.574 | 0.581 |
| S 3×7.5 | 2.21 | 3.00 | 2.509 | 0.260 | 0.349 | 2.93 | 1.95 | 1.15 | 0.586 | 0.468 | 0.516 |
| ×5.7 | 1.67 | 3.00 | 2.330 | 0.260 | 0.170 | 2.52 | 1.68 | 1.23 | 0.455 | 0.390 | 0.522 |

* Data taken from the "Manual of Steel Construction," 7th Edition, 1970, with permission of the American Institute of Steel Construction.

## American Standard Steel Channels*

American Standard Channels are designated, in order, by a section letter, actual depth in inches, and by nominal weight per foot in pounds. Thus:

$$C\ 7 \times 14.75$$

indicates an American Standard Channel with a depth of 7 inches and a nominal weight of 14.75 pounds per foot.

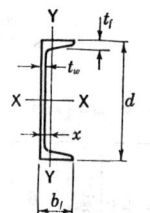

| Designation | Area $A$ | Depth $d$ | Flange Width $b_f$ | Flange Aver. Thickness $t_f$ | Web Thickness $t_w$ | Axis X-X $I$ | Axis X-X $S$ | Axis X-X $r$ | Axis Y-Y $I$ | Axis Y-Y $S$ | Axis Y-Y $r$ | $x$ |
|---|---|---|---|---|---|---|---|---|---|---|---|---|
| | in.² | in. | in. | in. | in. | in.⁴ | in.³ | in. | in.⁴ | in.³ | in. | in. |
| C 15×50 | 14.7 | 15.00 | 3.716 | 0.650 | 0.716 | 404 | 53.8 | 5.24 | 11.0 | 3.78 | 0.867 | 0.799 |
| ×40 | 11.8 | 15.00 | 3.520 | 0.650 | 0.520 | 349 | 46.5 | 5.44 | 9.23 | 3.36 | 0.886 | 0.778 |
| ×33.9 | 9.96 | 15.00 | 3.400 | 0.650 | 0.400 | 315 | 42.0 | 5.62 | 8.13 | 3.11 | 0.904 | 0.787 |
| C 12×30 | 8.82 | 12.00 | 3.170 | 0.501 | 0.510 | 162 | 27.0 | 4.29 | 5.14 | 2.06 | 0.763 | 0.674 |
| ×25 | 7.35 | 12.00 | 3.047 | 0.501 | 0.387 | 144 | 24.1 | 4.43 | 4.47 | 1.88 | 0.780 | 0.674 |
| ×20.7 | 6.09 | 12.00 | 2.942 | 0.501 | 0.282 | 129 | 21.5 | 4.61 | 3.88 | 1.73 | 0.799 | 0.698 |
| C 10×30 | 8.82 | 10.00 | 3.033 | 0.436 | 0.673 | 103 | 20.7 | 3.42 | 3.94 | 1.65 | 0.669 | 0.649 |
| ×25 | 7.35 | 10.00 | 2.886 | 0.436 | 0.526 | 91.2 | 18.2 | 3.52 | 3.36 | 1.48 | 0.676 | 0.617 |
| ×20 | 5.88 | 10.00 | 2.739 | 0.436 | 0.379 | 78.9 | 15.8 | 3.66 | 2.81 | 1.32 | 0.691 | 0.606 |
| ×15.3 | 4.49 | 10.00 | 2.600 | 0.436 | 0.240 | 67.4 | 13.5 | 3.87 | 2.28 | 1.16 | 0.713 | 0.634 |
| C 9×20 | 5.88 | 9.00 | 2.648 | 0.413 | 0.448 | 60.9 | 13.5 | 3.22 | 2.42 | 1.17 | 0.642 | 0.583 |
| ×15 | 4.41 | 9.00 | 2.485 | 0.413 | 0.285 | 51.0 | 11.3 | 3.40 | 1.93 | 1.01 | 0.661 | 0.586 |
| ×13.4 | 3.94 | 9.00 | 2.433 | 0.413 | 0.233 | 47.9 | 10.6 | 3.48 | 1.76 | 0.962 | 0.668 | 0.601 |
| C 8×18.75 | 5.51 | 8.00 | 2.527 | 0.390 | 0.487 | 44.0 | 11.0 | 2.82 | 1.98 | 1.01 | 0.599 | 0.565 |
| ×13.75 | 4.04 | 8.00 | 2.343 | 0.390 | 0.303 | 36.1 | 9.03 | 2.99 | 1.53 | 0.853 | 0.615 | 0.553 |
| ×11.5 | 3.38 | 8.00 | 2.260 | 0.390 | 0.220 | 32.6 | 8.14 | 3.11 | 1.32 | 0.781 | 0.625 | 0.571 |
| C 7×14.75 | 4.33 | 7.00 | 2.299 | 0.366 | 0.419 | 27.2 | 7.78 | 2.51 | 1.38 | 0.779 | 0.564 | 0.532 |
| ×12.25 | 3.60 | 7.00 | 2.194 | 0.366 | 0.314 | 24.2 | 6.93 | 2.60 | 1.17 | 0.702 | 0.571 | 0.525 |
| ×9.8 | 2.87 | 7.00 | 2.090 | 0.366 | 0.210 | 21.3 | 6.08 | 2.72 | 0.968 | 0.625 | 0.581 | 0.541 |
| C 6×13 | 3.83 | 6.00 | 2.157 | 0.343 | 0.437 | 17.4 | 5.80 | 2.13 | 1.05 | 0.642 | 0.525 | 0.514 |
| ×10.5 | 3.09 | 6.00 | 2.034 | 0.343 | 0.314 | 15.2 | 5.06 | 2.22 | 0.865 | 0.564 | 0.529 | 0.500 |
| ×8.2 | 2.40 | 6.00 | 1.920 | 0.343 | 0.200 | 13.1 | 4.38 | 2.34 | 0.692 | 0.492 | 0.537 | 0.512 |
| C 5×9 | 2.64 | 5.00 | 1.885 | 0.320 | 0.325 | 8.90 | 3.56 | 1.83 | 0.632 | 0.449 | 0.489 | 0.478 |
| ×6.7 | 1.97 | 5.00 | 1.750 | 0.320 | 0.190 | 7.49 | 3.00 | 1.95 | 0.478 | 0.378 | 0.493 | 0.484 |
| C 4×7.25 | 2.13 | 4.00 | 1.721 | 0.296 | 0.321 | 4.59 | 2.29 | 1.47 | 0.432 | 0.343 | 0.450 | 0.459 |
| ×5.4 | 1.59 | 4.00 | 1.584 | 0.296 | 0.184 | 3.85 | 1.93 | 1.56 | 0.319 | 0.283 | 0.449 | 0.458 |
| C 3×6 | 1.76 | 3.00 | 1.596 | 0.273 | 0.356 | 2.07 | 1.38 | 1.08 | 0.305 | 0.268 | 0.416 | 0.455 |
| ×5 | 1.47 | 3.00 | 1.498 | 0.273 | 0.258 | 1.85 | 1.24 | 1.12 | 0.247 | 0.233 | 0.410 | 0.438 |
| ×4.1 | 1.21 | 3.00 | 1.410 | 0.273 | 0.170 | 1.66 | 1.10 | 1.17 | 0.197 | 0.202 | 0.404 | 0.437 |

* Data taken from the "Manual of Steel Construction," 7th Edition, 1970, with permission of the American Institute of Steel Construction.

*Symbols:* $I$ = moment of inertia; $S$ = section modulus; $r$ = radius of gyration; $X$ = distance from center of gravity of section to outer face of structural shape.

### Steel Angles with Equal Legs†

These angles are commonly designated by section symbol, width of each leg, and thickness, thus:

## L 3×3× ¼

indicates a 3 x 3-inch angle of ¼-inch thickness.

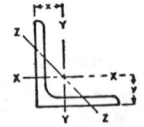

| Size | Thickness | Weight per Foot | Area | Axis X-X & Y-Y I | r | x or y | Z-Z r |
|---|---|---|---|---|---|---|---|
| in. | in. | lb. | in.² | in.⁴ | in. | in. | in. |
| **8 x 8** | 1⅛ | 56.9 | 16.7 | 98.0 | 2.42 | 2.41 | 1.56 |
| | 1 | 51.0 | 15.0 | 89.0 | 2.44 | 2.37 | 1.56 |
| | ⅞ | 45.0 | 13.2 | 79.6 | 2.45 | 2.32 | 1.57 |
| | ¾ | 38.9 | 11.4 | 69.7 | 2.47 | 2.28 | 1.58 |
| | ⅝ | 32.7 | 9.61 | 59.4 | 2.49 | 2.23 | 1.58 |
| | 9⁄16 | 29.6 | 8.68 | 54.1 | 2.50 | 2.21 | 1.59 |
| | ½ | 26.4 | 7.75 | 48.6 | 2.50 | 2.19 | 1.59 |
| **6 x 6** | 1 | 37.4 | 11.00 | 35.5 | 1.80 | 1.86 | 1.17 |
| | ⅞ | 33.1 | 9.73 | 31.9 | 1.81 | 1.82 | 1.17 |
| | ¾ | 28.7 | 8.44 | 28.2 | 1.83 | 1.78 | 1.17 |
| | ⅝ | 24.2 | 7.11 | 24.2 | 1.84 | 1.73 | 1.18 |
| | 9⁄16 | 21.9 | 6.43 | 22.1 | 1.85 | 1.71 | 1.18 |
| | ½ | 19.6 | 5.75 | 19.9 | 1.86 | 1.68 | 1.18 |
| | 7⁄16 | 17.2 | 5.06 | 17.7 | 1.87 | 1.66 | 1.19 |
| | ⅜ | 14.9 | 4.36 | 15.4 | 1.88 | 1.64 | 1.19 |
| | 5⁄16 | 12.4 | 3.65 | 13.0 | 1.89 | 1.62 | 1.20 |
| **5 x 5** | ⅞ | 27.2 | 7.98 | 17.8 | 1.49 | 1.57 | .97 |
| | ¾ | 23.6 | 6.94 | 15.7 | 1.51 | 1.52 | .98 |
| | ⅝ | 20.0 | 5.86 | 13.6 | 1.52 | 1.48 | .98 |
| | ½ | 16.2 | 4.75 | 11.3 | 1.54 | 1.43 | .98 |
| | 7⁄16 | 14.3 | 4.18 | 10.0 | 1.55 | 1.41 | .99 |
| | ⅜ | 12.3 | 3.61 | 8.7 | 1.56 | 1.39 | .99 |
| | 5⁄16 | 10.3 | 3.03 | 7.4 | 1.57 | 1.37 | .99 |
| **4 x 4** | ¾ | 18.5 | 5.44 | 7.7 | 1.19 | 1.27 | .78 |
| | ⅝ | 15.7 | 4.61 | 6.7 | 1.20 | 1.23 | .78 |
| | ½ | 12.8 | 3.75 | 5.8 | 1.22 | 1.18 | .78 |
| | 7⁄16 | 11.3 | 3.31 | 5.0 | 1.23 | 1.16 | .78 |
| | ⅜ | 9.8 | 2.86 | 4.4 | 1.23 | 1.14 | .79 |
| | 5⁄16 | 8.2 | 2.40 | 3.7 | 1.24 | 1.12 | .79 |
| | ¼ | 6.6 | 1.94 | 3.0 | 1.25 | 1.09 | .80 |
| **3½ x 3½** | ½ | 11.1 | 3.25 | 3.6 | 1.06 | 1.06 | .68 |
| | 7⁄16 | 9.8 | 2.87 | 3.3 | 1.07 | 1.04 | .68 |
| | ⅜ | 8.5 | 2.48 | 2.9 | 1.07 | 1.01 | .69 |
| | 5⁄16 | 7.2 | 2.09 | 2.5 | 1.08 | .99 | .69 |
| | ¼ | 5.8 | 1.69 | 2.1 | 1.09 | .97 | .69 |

| Size | Thickness | Weight per Foot | Area | Axis X-X & Y-Y I | r | x or y | Z-Z r |
|---|---|---|---|---|---|---|---|
| in. | in. | lb. | in.² | in.⁴ | in. | in. | in. |
| **3 x 3** | ½ | 9.4 | 2.75 | 2.2 | .90 | .93 | .58 |
| | 7⁄16 | 8.3 | 2.43 | 2.0 | .91 | .91 | .58 |
| | ⅜ | 7.2 | 2.11 | 1.8 | .91 | .89 | .59 |
| | 5⁄16 | 6.1 | 1.78 | 1.5 | .92 | .87 | .59 |
| | ¼ | 4.9 | 1.44 | 1.2 | .93 | .84 | .59 |
| | 3⁄16 | 3.71 | 1.09 | .96 | .94 | .82 | .60 |
| **2½ x 2½** | ½ | 7.7 | 2.25 | 1.2 | .74 | .81 | .49 |
| | ⅜ | 5.9 | 1.73 | .98 | .75 | .76 | .49 |
| | 5⁄16 | 5.0 | 1.46 | .85 | .76 | .74 | .49 |
| | ¼ | 4.1 | 1.19 | .70 | .77 | .72 | .49 |
| | 3⁄16 | 3.07 | .90 | .55 | .78 | .69 | .50 |
| **2 x 2** | ⅜ | 4.7 | 1.36 | .48 | .59 | .64 | .39 |
| | 5⁄16 | 3.92 | 1.15 | .42 | .60 | .61 | .39 |
| | ¼ | 3.19 | .94 | .35 | .61 | .59 | .39 |
| | 3⁄16 | 2.44 | .72 | .27 | .62 | .57 | .39 |
| | ⅛ | 1.65 | .48 | .19 | .63 | .55 | .40 |
| **1¾ x 1¾** | ¼ | 2.77 | .81 | .23 | .53 | .53 | .34 |
| | 3⁄16 | 2.12 | .62 | .18 | .54 | .51 | .34 |
| | ⅛ | 1.44 | .42 | .13 | .55 | .48 | .35 |
| **1½ x 1½** | ¼ | 2.34 | .69 | .14 | .45 | .47 | .29 |
| | 3⁄16 | 1.80 | .53 | .11 | .46 | .44 | .29 |
| | 5⁄32 | 1.52 | .44 | .094 | .46 | .43 | .30 |
| **1¼ x 1¼** | ¼ | 1.92 | .56 | .077 | .37 | .40 | .24 |
| | 3⁄16 | 1.48 | .43 | .061 | .38 | .38 | .24 |
| | ⅛ | 1.01 | .30 | .044 | .38 | .36 | .25 |
| **1 x 1** | ¼ | 1.49 | .44 | .037 | .29 | .34 | .20 |
| | 3⁄16 | 1.16 | .34 | .030 | .30 | .32 | .20 |
| | ⅛ | .80 | .23 | .022 | .30 | .30 | .20 |
| | .. | ... | .... | ... | ... | ... | ... |
| | .. | ... | .... | ... | ... | ... | ... |

† Data taken from the "Manual of Steel Construction," 7th Edition, 1970, with permission of the American Institute of Steel Construction.

Meaning of symbols: I = moment of inertia; r = radius of gyration.

## Steel Angles with Unequal Legs†

These angles are commonly designated by section symbol, width of each leg, and thickness, thus:

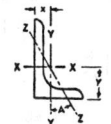

# L 7 x 4 x ½

indicates a 7 x 4-inch angle of ½-inch thickness.

| Size | Thickness | Weight per Foot | Area | Axis X-X | | | | Axis Y-Y | | | | Axis Z-Z | |
|---|---|---|---|---|---|---|---|---|---|---|---|---|---|
| | | | | $I$ | $S$ | $r$ | $y$ | $I$ | $S$ | $r$ | $x$ | $r$ | Tan $A$ |
| in. | in. | lb. | in.² | in.⁴ | in.³ | in. | in. | in.⁴ | in.³ | in. | in. | in. | |
| 9 x 4 | 1 | 40.8 | 12.00 | 97.0 | 17.6 | 2.84 | 3.50 | 12.0 | 4.0 | 1.00 | 1.00 | .83 | .203 |
| | ⅞ | 36.1 | 10.61 | 86.8 | 15.7 | 2.86 | 3.45 | 10.8 | 3.6 | 1.01 | .95 | .84 | .208 |
| | ¾ | 31.3 | 9.19 | 76.1 | 13.6 | 2.88 | 3.41 | 9.6 | 3.1 | 1.02 | .91 | .84 | .212 |
| | ⅝ | 26.3 | 7.73 | 64.9 | 11.5 | 2.90 | 3.36 | 8.3 | 2.6 | 1.04 | .86 | .85 | .216 |
| | ⁹⁄₁₆ | 23.8 | 7.00 | 59.1 | 10.4 | 2.91 | 3.33 | 7.6 | 2.4 | 1.04 | .83 | .85 | .218 |
| | ½ | 21.3 | 6.25 | 53.2 | 9.3 | 2.92 | 3.31 | 6.9 | 2.2 | 1.05 | .81 | .85 | .220 |
| 8 x 6 | 1 | 44.2 | 13.00 | 80.8 | 15.1 | 2.49 | 2.65 | 38.8 | 8.9 | 1.73 | 1.65 | 1.28 | .543 |
| | ⅞ | 39.1 | 11.48 | 72.3 | 13.4 | 2.51 | 2.61 | 34.9 | 7.9 | 1.74 | 1.61 | 1.28 | .547 |
| | ¾ | 33.8 | 9.94 | 63.4 | 11.7 | 2.53 | 2.56 | 30.7 | 6.9 | 1.76 | 1.56 | 1.29 | .551 |
| | ⅝ | 28.5 | 8.36 | 54.1 | 9.9 | 2.54 | 2.52 | 26.3 | 5.9 | 1.77 | 1.52 | 1.29 | .554 |
| | ⁹⁄₁₆ | 25.7 | 7.56 | 49.3 | 9.0 | 2.55 | 2.50 | 24.0 | 5.3 | 1.78 | 1.50 | 1.30 | .556 |
| | ½ | 23.0 | 6.75 | 44.3 | 8.0 | 2.56 | 2.47 | 21.7 | 4.8 | 1.79 | 1.47 | 1.30 | .558 |
| | ⁷⁄₁₆ | 20.2 | 5.93 | 39.2 | 7.1 | 2.57 | 2.45 | 19.3 | 4.2 | 1.80 | 1.45 | 1.31 | .560 |
| 8 x 4 | 1 | 37.4 | 11.00 | 69.6 | 14.1 | 2.52 | 3.05 | 11.6 | 3.9 | 1.03 | 1.05 | .85 | .247 |
| | ⅞ | 33.1 | 9.73 | 62.5 | 12.5 | 2.53 | 3.00 | 10.5 | 3.5 | 1.04 | 1.00 | .85 | .253 |
| | ¾ | 28.7 | 8.44 | 54.9 | 10.9 | 2.55 | 2.95 | 9.4 | 3.1 | 1.05 | .95 | .85 | .258 |
| | ⅝ | 24.2 | 7.11 | 46.9 | 9.2 | 2.57 | 2.91 | 8.1 | 2.6 | 1.07 | .91 | .86 | .262 |
| | ⁹⁄₁₆ | 21.9 | 6.43 | 42.8 | 8.4 | 2.58 | 2.88 | 7.4 | 2.4 | 1.07 | .88 | .86 | .265 |
| | ½ | 19.6 | 5.75 | 38.5 | 7.5 | 2.59 | 2.86 | 6.7 | 2.2 | 1.08 | .86 | .86 | .267 |
| | ⁷⁄₁₆ | 17.2 | 5.06 | 34.1 | 6.6 | 2.60 | 2.83 | 6.0 | 1.9 | 1.09 | .84 | .87 | .269 |
| 7 x 4 | ⅞ | 30.2 | 8.86 | 42.9 | 9.7 | 2.20 | 2.55 | 10.2 | 3.5 | 1.07 | 1.05 | .86 | .318 |
| | ¾ | 26.2 | 7.69 | 37.8 | 8.4 | 2.22 | 2.51 | 9.1 | 3.0 | 1.09 | 1.01 | .86 | .324 |
| | ⅝ | 22.1 | 6.48 | 32.4 | 7.1 | 2.24 | 2.46 | 7.8 | 2.6 | 1.10 | .96 | .86 | .329 |
| | ⁹⁄₁₆ | 20.0 | 5.87 | 29.6 | 6.5 | 2.24 | 2.44 | 7.2 | 2.4 | 1.11 | .94 | .87 | .332 |
| | ½ | 17.9 | 5.25 | 26.7 | 5.8 | 2.25 | 2.42 | 6.5 | 2.1 | 1.11 | .92 | .87 | .335 |
| | ⁷⁄₁₆ | 15.8 | 4.62 | 23.7 | 5.1 | 2.26 | 2.39 | 5.8 | 1.9 | 1.12 | .89 | .88 | .337 |
| | ⅜ | 13.6 | 3.98 | 20.6 | 4.4 | 2.27 | 2.37 | 5.1 | 1.6 | 1.13 | .87 | .88 | .340 |
| 6 x 4 | ⅞ | 27.2 | 7.98 | 27.7 | 7.2 | 1.86 | 2.12 | 9.8 | 3.4 | 1.11 | 1.12 | .86 | .421 |
| | ¾ | 23.6 | 6.94 | 24.5 | 6.3 | 1.88 | 2.08 | 8.7 | 3.0 | 1.12 | 1.08 | .86 | .428 |
| | ⅝ | 20.0 | 5.86 | 21.1 | 5.3 | 1.90 | 2.03 | 7.5 | 2.5 | 1.13 | 1.03 | .86 | .435 |
| | ⁹⁄₁₆ | 18.1 | 5.31 | 19.3 | 4.8 | 1.90 | 2.01 | 6.9 | 2.3 | 1.14 | 1.01 | .87 | .438 |
| | ½ | 16.2 | 4.75 | 17.4 | 4.3 | 1.91 | 1.99 | 6.3 | 2.1 | 1.15 | .99 | .87 | .440 |
| | ⁷⁄₁₆ | 14.3 | 4.18 | 15.5 | 3.8 | 1.92 | 1.96 | 5.6 | 1.9 | 1.16 | .96 | .87 | .443 |
| | ⅜ | 12.3 | 3.61 | 13.5 | 3.3 | 1.93 | 1.94 | 4.9 | 1.6 | 1.17 | .94 | .88 | .446 |
| | ⁵⁄₁₆ | 10.3 | 3.03 | 11.4 | 2.8 | 1.94 | 1.92 | 4.2 | 1.4 | 1.17 | .92 | .88 | .448 |
| | ¼ | 8.3 | 2.44 | 9.3 | 2.3 | 1.95 | 1.89 | 3.4 | 1.1 | 1.18 | .89 | .89 | .451 |
| 6 x 3½ | ½ | 15.3 | 4.50 | 16.6 | 4.2 | 1.92 | 2.08 | 4.3 | 1.6 | .97 | .83 | .76 | .344 |
| | ⅜ | 11.7 | 3.42 | 12.9 | 3.2 | 1.94 | 2.04 | 3.3 | 1.2 | .99 | .79 | .77 | .350 |
| | ⁵⁄₁₆ | 9.8 | 2.87 | 10.9 | 2.7 | 1.95 | 2.01 | 2.9 | 1.0 | 1.00 | .76 | .77 | .352 |
| | ¼ | 7.9 | 2.31 | 8.9 | 2.2 | 1.96 | 1.99 | 2.3 | .85 | 1.01 | .74 | .78 | .355 |
| 5 x 3½ | ¾ | 19.8 | 5.81 | 13.9 | 4.3 | 1.55 | 1.75 | 5.6 | 2.2 | .98 | 1.00 | .75 | .464 |
| | ⅝ | 16.8 | 4.92 | 12.0 | 3.7 | 1.56 | 1.70 | 4.8 | 1.9 | .99 | .95 | .75 | .472 |
| | ½ | 13.6 | 4.00 | 10.0 | 3.0 | 1.58 | 1.66 | 4.1 | 1.6 | 1.01 | .91 | .76 | .479 |
| | ⁷⁄₁₆ | 12.0 | 3.53 | 8.9 | 2.6 | 1.59 | 1.63 | 3.6 | 1.4 | 1.01 | .88 | .76 | .482 |
| | ⅜ | 10.4 | 3.05 | 7.8 | 2.3 | 1.60 | 1.61 | 3.2 | 1.2 | 1.02 | .86 | .76 | .486 |
| | ⁵⁄₁₆ | 8.7 | 2.56 | 6.6 | 1.9 | 1.61 | 1.59 | 2.7 | 1.0 | 1.03 | .84 | .77 | .489 |
| | ¼ | 7.0 | 2.06 | 5.4 | 1.6 | 1.62 | 1.56 | 2.3 | .83 | 1.04 | .81 | .77 | .492 |

† See footnotes at end of table on page 397

### Steel Angles with Unequal Legs (continued)

| Size | Thickness | Weight per Foot | Area | Axis X-X | | | | Axis Y-Y | | | | Axis Z-Z | |
|---|---|---|---|---|---|---|---|---|---|---|---|---|---|
| | | | | I | S | r | y | I | S | r | z | r | Tan A |
| in. | in. | lb. | in.² | in.⁴ | in.³ | in. | in. | in.⁴ | in.³ | in. | in. | in. | |
| 5 x 3 | ½ | 12.8 | 3.75 | 9.5 | 2.9 | 1.59 | 1.75 | 2.6 | 1.1 | .83 | .75 | .65 | .357 |
| | 7/16 | 11.3 | 3.31 | 8.4 | 2.6 | 1.60 | 1.73 | 2.3 | 1.0 | .84 | .73 | .65 | .361 |
| | 3/8 | 9.8 | 2.86 | 7.4 | 2.2 | 1.61 | 1.70 | 2.0 | .89 | .84 | .70 | .65 | .364 |
| | 5/16 | 8.2 | 2.40 | 6.3 | 1.9 | 1.61 | 1.68 | 1.8 | .75 | .85 | .68 | .66 | .368 |
| | ¼ | 6.6 | 1.94 | 5.1 | 1.5 | 1.62 | 1.66 | 1.4 | .61 | .86 | .66 | .66 | .371 |
| 4 x 3½ | 5/8 | 14.7 | 4.30 | 6.4 | 2.4 | 1.22 | 1.29 | 4.5 | 1.8 | 1.03 | 1.04 | .72 | .745 |
| | ½ | 11.9 | 3.50 | 5.3 | 1.9 | 1.23 | 1.25 | 3.8 | 1.5 | 1.04 | 1.00 | .72 | .750 |
| | 7/16 | 10.6 | 3.09 | 4.8 | 1.7 | 1.24 | 1.23 | 3.4 | 1.4 | 1.05 | .98 | .72 | .753 |
| | 3/8 | 9.1 | 2.67 | 4.2 | 1.5 | 1.25 | 1.21 | 3.0 | 1.2 | 1.06 | .96 | .73 | .755 |
| | 5/16 | 7.7 | 2.25 | 3.6 | 1.3 | 1.26 | 1.18 | 2.6 | 1.0 | 1.07 | .93 | .73 | .757 |
| | ¼ | 6.2 | 1.81 | 2.9 | 1.0 | 1.27 | 1.16 | 2.1 | .81 | 1.07 | .91 | .73 | .759 |
| 4 x 3 | 5/8 | 13.6 | 3.98 | 6.0 | 2.3 | 1.23 | 1.37 | 2.9 | 1.4 | .85 | .87 | .64 | .534 |
| | ½ | 11.1 | 3.25 | 5.1 | 1.9 | 1.25 | 1.33 | 2.4 | 1.1 | .86 | .83 | .64 | .543 |
| | 7/16 | 9.8 | 2.87 | 4.5 | 1.7 | 1.25 | 1.30 | 2.2 | 1.0 | .87 | .80 | .64 | .547 |
| | 3/8 | 8.5 | 2.48 | 4.0 | 1.5 | 1.26 | 1.28 | 1.9 | .87 | .88 | .78 | .64 | .551 |
| | 5/16 | 7.2 | 2.09 | 3.4 | 1.2 | 1.27 | 1.26 | 1.7 | .73 | .89 | .76 | .65 | .554 |
| | ¼ | 5.8 | 1.69 | 2.8 | 1.0 | 1.28 | 1.24 | 1.4 | .60 | .90 | .74 | .65 | .558 |
| 3½ x 3 | ½ | 10.2 | 3.00 | 3.5 | 1.5 | 1.07 | 1.13 | 2.3 | 1.1 | .88 | .88 | .62 | .714 |
| | 7/16 | 9.1 | 2.65 | 3.1 | 1.3 | 1.08 | 1.10 | 2.1 | .98 | .89 | .85 | .62 | .718 |
| | 3/8 | 7.9 | 2.30 | 2.7 | 1.1 | 1.09 | 1.08 | 1.9 | .85 | .90 | .83 | .62 | .721 |
| | 5/16 | 6.6 | 1.93 | 2.3 | .95 | 1.10 | 1.06 | 1.6 | .72 | .90 | .81 | .63 | .724 |
| | ¼ | 5.4 | 1.56 | 1.9 | .78 | 1.11 | 1.04 | 1.3 | .59 | .91 | .79 | .63 | .727 |
| 3½ x 2½ | ½ | 9.4 | 2.75 | 3.2 | 1.4 | 1.09 | 1.20 | 1.4 | .76 | .70 | .70 | .53 | .486 |
| | 7/16 | 8.3 | 2.43 | 2.9 | 1.3 | 1.09 | 1.18 | 1.2 | .68 | .71 | .68 | .54 | .491 |
| | 3/8 | 7.2 | 2.11 | 2.6 | 1.1 | 1.10 | 1.16 | 1.1 | .59 | .72 | .66 | .54 | .496 |
| | 5/16 | 6.1 | 1.78 | 2.2 | .93 | 1.11 | 1.14 | .94 | .50 | .73 | .64 | .54 | .501 |
| | ¼ | 4.9 | 1.44 | 1.8 | .75 | 1.12 | 1.11 | .78 | .41 | .74 | .61 | .54 | .506 |
| 3 x 2½ | ½ | 8.5 | 2.50 | 2.1 | 1.0 | .91 | 1.00 | 1.3 | .74 | .72 | .75 | .52 | .667 |
| | 7/16 | 7.6 | 2.21 | 1.9 | .93 | .92 | .98 | 1.2 | .66 | .73 | .73 | .52 | .672 |
| | 3/8 | 6.6 | 1.92 | 1.7 | .81 | .93 | .96 | 1.0 | .58 | .74 | .71 | .52 | .676 |
| | 5/16 | 5.6 | 1.62 | 1.4 | .69 | .94 | .93 | .90 | .49 | .74 | .68 | .53 | .680 |
| | ¼ | 4.5 | 1.31 | 1.2 | .56 | .95 | .91 | .74 | .40 | .75 | .66 | .53 | .684 |
| | 3/16 | 3.4 | 1.00 | 0.9 | .43 | .95 | .89 | .58 | .31 | .76 | .64 | .53 | .688 |
| 3 x 2 | ½ | 7.7 | 2.25 | 1.9 | 1.0 | .92 | 1.08 | .67 | .47 | .55 | .58 | .43 | .414 |
| | 7/16 | 6.8 | 2.00 | 1.7 | .89 | .93 | 1.06 | .61 | .42 | .55 | .56 | .43 | .421 |
| | 3/8 | 5.9 | 1.73 | 1.5 | .78 | .94 | 1.04 | .54 | .37 | .56 | .54 | .43 | .428 |
| | 5/16 | 5.0 | 1.46 | 1.3 | .66 | .95 | 1.02 | .47 | .32 | .57 | .52 | .43 | .435 |
| | ¼ | 4.1 | 1.19 | 1.1 | .54 | .96 | .99 | .39 | .26 | .57 | .49 | .44 | .440 |
| | 3/16 | 3.07 | .90 | .84 | .42 | .97 | .97 | .31 | .20 | .58 | .47 | .44 | .446 |
| 2½ x 2 | 3/8 | 5.3 | 1.55 | .91 | .55 | .77 | .83 | .51 | .36 | .58 | .58 | .42 | .614 |
| | 5/16 | 4.5 | 1.31 | .79 | .47 | .78 | .81 | .45 | .31 | .58 | .56 | .42 | .620 |
| | ¼ | 3.62 | 1.06 | .65 | .38 | .78 | .79 | .37 | .25 | .59 | .54 | .42 | .626 |
| | 3/16 | 2.75 | .81 | .51 | .29 | .79 | .76 | .29 | .20 | .60 | .51 | .43 | .631 |
| 2½ x 1½ | 5/16 | 3.92 | 1.15 | .71 | .44 | .79 | .90 | .19 | .17 | .41 | .40 | .32 | .349 |
| | ¼ | 3.19 | .94 | .59 | .36 | .79 | .88 | .16 | .14 | .42 | .38 | .32 | .357 |
| | 3/16 | 2.44 | .72 | .46 | .28 | .80 | .85 | .13 | .11 | .42 | .35 | .33 | .364 |
| 2 x 1½ | ¼ | 2.77 | .81 | .32 | .24 | .62 | .66 | .15 | .14 | .43 | .41 | .32 | .543 |
| | 3/16 | 2.12 | .62 | .25 | .18 | .63 | .64 | .12 | .11 | .44 | .39 | .32 | .551 |
| | 1/8 | 1.44 | .42 | .17 | .13 | .64 | .62 | .085 | .075 | .45 | .37 | .33 | .558 |
| 2 x 1¼ | ¼ | 2.55 | .75 | .30 | .23 | .63 | .71 | .089 | .097 | .34 | .33 | .27 | .378 |
| | 3/16 | 1.96 | .57 | .23 | .18 | .64 | .69 | .071 | .075 | .35 | .31 | .27 | .387 |
| | 1/8 | 1.33 | .39 | .16 | .12 | .65 | .66 | .050 | .052 | .36 | .29 | .27 | .396 |
| 1¾ x 1¼ | ¼ | 2.34 | .69 | .20 | .18 | .54 | .60 | .085 | .095 | .35 | .35 | .27 | .486 |
| | 3/16 | 1.80 | .53 | .16 | .14 | .55 | .58 | .068 | .051 | .36 | .33 | .27 | .496 |
| | 1/8 | 1.23 | .36 | .11 | .094 | .56 | .56 | .049 | .051 | .37 | .31 | .27 | .506 |

**Aluminum Association Standard Structural Shapes***

I-BEAMS / CHANNELS

| Depth | Width | Weight per Foot | Area | Flange Thickness | Web Thickness | Fillet Radius | Axis X-X | | | Axis Y-Y | | | |
|---|---|---|---|---|---|---|---|---|---|---|---|---|---|
| | | | | | | | $I$ | $S$ | $r$ | $I$ | $S$ | $r$ | $x$ |
| in. | in. | lb. | in.² | in. | in. | in. | in.⁴ | in.³ | in. | in.⁴ | in.³ | in. | in. |
| | | | | | | I-BEAMS | | | | | | | |
| 3.00 | 2.50 | 1.637 | 1.392 | 0.20 | 0.13 | 0.25 | 2.24 | 1.49 | 1.27 | 0.52 | 0.42 | .61 | .... |
| 3.00 | 2.50 | 2.030 | 1.726 | 0.26 | 0.15 | 0.25 | 2.71 | 1.81 | 1.25 | 0.68 | 0.54 | .63 | .... |
| 4.00 | 3.00 | 2.311 | 1.965 | 0.23 | 0.15 | 0.25 | 5.62 | 2.81 | 1.69 | 1.04 | 0.69 | .73 | .... |
| 4.00 | 3.00 | 2.793 | 2.375 | 0.29 | 0.17 | 0.25 | 6.71 | 3.36 | 1.68 | 1.31 | 0.87 | .74 | .... |
| 5.00 | 3.50 | 3.700 | 3.146 | 0.32 | 0.19 | 0.30 | 13.94 | 5.58 | 2.11 | 2.29 | 1.31 | .85 | .... |
| 6.00 | 4.00 | 4.030 | 3.427 | 0.29 | 0.19 | 0.30 | 21.99 | 7.33 | 2.53 | 3.10 | 1.55 | .95 | .... |
| 6.00 | 4.00 | 4.692 | 3.990 | 0.35 | 0.21 | 0.30 | 25.50 | 8.50 | 2.53 | 3.74 | 1.87 | .97 | .... |
| 7.00 | 4.50 | 5.800 | 4.932 | 0.38 | 0.23 | 0.30 | 42.89 | 12.25 | 2.95 | 5.78 | 2.57 | 1.08 | .... |
| 8.00 | 5.00 | 6.181 | 5.256 | 0.35 | 0.23 | 0.30 | 59.69 | 14.92 | 3.37 | 7.30 | 2.92 | 1.18 | .... |
| 8.00 | 5.00 | 7.023 | 5.972 | 0.41 | 0.25 | 0.30 | 67.78 | 16.94 | 3.37 | 8.55 | 3.42 | 1.20 | .... |
| 9.00 | 5.50 | 8.361 | 7.110 | 0.44 | 0.27 | 0.30 | 102.02 | 22.67 | 3.79 | 12.22 | 4.44 | 1.31 | .... |
| 10.00 | 6.00 | 8.646 | 7.352 | 0.41 | 0.25 | 0.40 | 132.09 | 26.42 | 4.24 | 14.78 | 4.93 | 1.42 | .... |
| 10.00 | 6.00 | 10.286 | 8.747 | 0.50 | 0.29 | 0.40 | 155.79 | 31.16 | 4.22 | 18.03 | 6.01 | 1.44 | .... |
| 12.00 | 7.00 | 11.672 | 9.925 | 0.47 | 0.29 | 0.40 | 255.57 | 42.60 | 5.07 | 26.90 | 7.69 | 1.65 | .... |
| 12.00 | 7.00 | 14.292 | 12.153 | 0.62 | 0.31 | 0.40 | 317.33 | 52.89 | 5.11 | 35.48 | 10.14 | 1.71 | .... |
| | | | | | | CHANNELS | | | | | | | |
| 2.00 | 1.00 | 0.577 | 0.491 | 0.13 | 0.13 | 0.10 | 0.288 | 0.288 | 0.766 | 0.045 | 0.064 | 0.303 | 0.298 |
| 2.00 | 1.25 | 1.071 | 0.911 | 0.26 | 0.17 | 0.15 | 0.546 | 0.546 | 0.774 | 0.139 | 0.178 | 0.391 | 0.471 |
| 3.00 | 1.50 | 1.135 | 0.965 | 0.20 | 0.13 | 0.25 | 1.41 | 0.94 | 1.21 | 0.22 | 0.22 | 0.47 | 0.49 |
| 3.00 | 1.75 | 1.597 | 1.358 | 0.26 | 0.17 | 0.25 | 1.97 | 1.31 | 1.20 | 0.42 | 0.37 | 0.55 | 0.62 |
| 4.00 | 2.00 | 1.738 | 1.478 | 0.23 | 0.15 | 0.25 | 3.91 | 1.95 | 1.63 | 0.60 | 0.45 | 0.64 | 0.65 |
| 4.00 | 2.25 | 2.331 | 1.982 | 0.29 | 0.19 | 0.25 | 5.21 | 2.60 | 1.62 | 1.02 | 0.69 | 0.72 | 0.78 |
| 5.00 | 2.25 | 2.212 | 1.881 | 0.26 | 0.15 | 0.30 | 7.88 | 3.15 | 2.05 | 0.98 | 0.64 | 0.72 | 0.73 |
| 5.00 | 2.75 | 3.089 | 2.627 | 0.32 | 0.19 | 0.30 | 11.14 | 4.45 | 2.06 | 2.05 | 1.14 | 0.88 | 0.95 |
| 6.00 | 2.50 | 2.834 | 2.410 | 0.29 | 0.17 | 0.30 | 14.35 | 4.78 | 2.44 | 1.53 | 0.90 | 0.80 | 0.79 |
| 6.00 | 3.25 | 4.030 | 3.427 | 0.35 | 0.21 | 0.30 | 21.04 | 7.01 | 2.48 | 3.76 | 1.76 | 1.05 | 1.12 |
| 7.00 | 2.75 | 3.205 | 2.725 | 0.29 | 0.17 | 0.30 | 22.09 | 6.31 | 2.85 | 2.10 | 1.10 | 0.88 | 0.84 |
| 7.00 | 3.50 | 4.715 | 4.009 | 0.38 | 0.21 | 0.30 | 33.79 | 9.65 | 2.90 | 5.13 | 2.23 | 1.13 | 1.20 |
| 8.00 | 3.00 | 4.147 | 3.526 | 0.35 | 0.19 | 0.30 | 37.40 | 9.35 | 3.26 | 3.25 | 1.57 | 0.96 | 0.93 |
| 8.00 | 3.75 | 5.789 | 4.923 | 0.41 | 0.25 | 0.35 | 52.69 | 13.17 | 3.27 | 7.13 | 2.82 | 1.20 | 1.22 |
| 9.00 | 3.25 | 4.983 | 4.237 | 0.35 | 0.23 | 0.35 | 54.41 | 12.09 | 3.58 | 4.40 | 1.89 | 1.02 | 0.93 |
| 9.00 | 4.00 | 6.970 | 5.927 | 0.44 | 0.29 | 0.35 | 78.31 | 17.40 | 3.63 | 9.61 | 3.49 | 1.27 | 1.25 |
| 10.00 | 3.50 | 6.136 | 5.218 | 0.41 | 0.25 | 0.35 | 83.22 | 16.64 | 3.99 | 6.33 | 2.56 | 1.10 | 1.02 |
| 10.00 | 4.25 | 8.360 | 7.109 | 0.50 | 0.31 | 0.40 | 116.15 | 23.23 | 4.04 | 13.02 | 4.47 | 1.35 | 1.34 |
| 12.00 | 4.00 | 8.274 | 7.036 | 0.47 | 0.29 | 0.40 | 159.76 | 26.63 | 4.77 | 11.03 | 3.86 | 1.25 | 1.14 |
| 12.00 | 5.00 | 11.822 | 10.053 | 0.62 | 0.35 | 0.45 | 239.69 | 39.95 | 4.88 | 25.74 | 7.60 | 1.60 | 1.61 |

* Structural sections are available in 6061-T6 aluminum alloy. Data supplied by The Aluminum Association.

**Size of Rail Necessary to Carry a Given Load.** — The following formulas may be employed for determining the size of rail and wheel suitable for carrying a given load. Let, $A$ = the width of the head of the rail in inches; $B$ = width of the tread of the rail in inches; $C$ = the wheel-load in pounds; $D$ = the diameter of the wheel in inches.

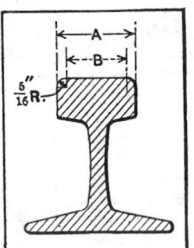

Then the width of the tread of the rail in inches is found from the formula:

$$B = \frac{C}{1250\,D} \qquad (1)$$

The width $A$ of the head equals $B + \frac{5}{8}$ inch. The diameter $D$ of the smallest track wheel that will safely carry the load is found from the formula:

$$D = \frac{C}{A \times K} \qquad (2)$$

in which $K$ = 600 to 800 for steel castings; $K$ = 300 to 400 for cast iron.

As an example, assume that the wheel-load in a given case is 10,000 pounds; the diameter of the wheel is 20 inches; and the material steel casting. Determine the size of rail necessary to carry this load. From Formula (1):

$$B = \frac{10,000}{1250 \times 20} = 0.4 \text{ inch.}$$

Hence the width of the rail required equals $0.4 + \frac{5}{8}$ inch = 1.025 inch. Determine also whether a wheel 20 inches in diameter is large enough to safely carry the load. From Formula (2):

$$D = \frac{10,000}{1.025 \times 600} = 16\frac{1}{4} \text{ inches.}$$

This is the smallest diameter of track wheel that will safely carry the load; hence a 20-inch wheel is ample.

## Beams

**Reaction at the Supports.** — When a beam is loaded by vertical loads or forces, the sum of the reactions at the supports equals the sum of the loads. In a simple beam, when the loads are symmetrically placed with reference to the supports, or when the load is uniformly distributed, the reaction at each end will equal one-half of the sum of the loads. When the loads are not symmetrically placed, the reaction at each support may be ascertained from the fact that the algebraic sum of the moments must equal zero. In the accompanying illustration, if moments are taken about the support to the left, then: $R_2 \times 40 - 8000 \times 10 - 10,000 \times 16 - 20,000 \times 20 = 0$; $R_2 = 16,000$ pounds.

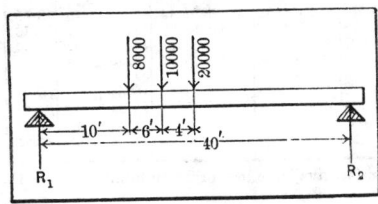

Moments taken about the support at the right will, in the same way, give

$$R_1 = 22,000 \text{ pounds.}$$

The sum of the reactions equals 38,000 pounds, which is also the sum of the loads. If part of the load is uniformly distributed over the beam, this part is first equally divided between the two supports, or the uniform load may be considered as concentrated at its center of gravity.

If metric SI units are used for the calculations, distances may be expressed in meters or millimeters, providing the treatment is consistent, and loads in newtons. *Note:* If the load is given in kilograms, the value referred to is the mass. A mass of $M$ kilograms has a weight (applies a force) of $Mg$ newtons, where $g$ = approximately 9.81 meters per second$^2$.

**Stresses and Deflections in Beams.** — On the following pages is given an extensive table of formulas for stresses and deflections in beams, shafts, etc. It is assumed that all the dimensions are in inches, all loads in pounds, and all stresses in pounds per square inch. The formulas are also valid using metric SI units, with all dimensions in millimeters, all loads in newtons, and stresses and moduli in newtons per millimeter$^2$ ($N/mm^2$). *Note:* A load due to the weight of a mass of $M$ kilograms is $Mg$ newtons, where $g$ = approximately 9.81 meters per second$^2$. In the tables:

$E$ = modulus of elasticity of the material;

$I$ = moment of inertia of the cross-section of the beam;

$Z$ = section modulus of the cross-section of the beam = $I \div$ distance from neutral axis to extreme fiber;

$W$ = load on beam;

$s$ = stress in extreme fiber, or maximum stress in the cross-section considered, due to load $W$. A positive value of $s$ denotes tension in the upper fibers and compression in the lower ones (as in a cantilever). A negative value of $s$ denotes the reverse (as in a beam supported at the ends). The greatest safe load is that value of $W$ which causes a maximum stress equal to, but not exceeding, the greatest safe value of $s$;

$y$ = deflection measured from the position occupied if the load causing the deflection were removed. A positive value of $y$ denotes deflection below this position; a negative value, deflection upward;

$u, v, w, x$ = variable distances along the beam from a given support to any point.

If there are several kinds of loads, as, for instance, a uniform load and a load at any point, or separate loads at different points, the total stress and the total deflection at any point is found by adding together the various stresses or deflections at the point considered due to each load acting by itself. If the stress or deflection due to any one of the loads is negative, it must be subtracted instead of added.

**Remarks Relative to the Use of the Tables.** — In the diagrammatical illustrations of the beams and their loading, the values indicated near, but below, the supports are the "reactions" or upward forces at the supports. For Cases 1 to 12, inclusive, the reactions, as well as the formulas for the stresses, are the same whether the beam is of constant or variable cross-section. For the other cases, the reactions and the stresses given are for constant cross-section beams only.

The bending moment at any point in inch-pounds is $s \times Z$ and can be found by omitting the divisor $Z$ in the formula for the stress given in the tables. A positive value of the bending moment denotes tension in the upper fibers and compression in the lower ones. A negative value denotes the reverse. The value of $W$ corresponding to a given stress is found by transposition of the formula. For example, in Case 1, the stress at the critical point is $s = -Wl \div 8Z$. From this we find $W = -8Zs \div l$. Of course, the negative sign of $W$ may be ignored.

Stresses and Deflections in Beams

| Type of Beam | Stresses | |
|---|---|---|
| | General Formula for Stress at any Point | Stresses at Critical Points |
| **Case 1. — Supported at Both Ends, Uniform Load** TOTAL LOAD W | $s = -\dfrac{W}{2\,Zl}\,x\,(l-x)$ | Stress at center, $-\dfrac{Wl}{8\,Z}$ If cross-section is constant, this is the maximum stress. |
| **Case 2. — Supported at Both Ends, Load at Center** | Between each support and load, $s = -\dfrac{Wx}{2\,Z}$ | Stress at center, $-\dfrac{Wl}{4\,Z}$ If cross-section is constant, this is the maximum stress. |
| **Case 3. — Supported at Both Ends, Load at any Point** $a+b=l$ | For segment of length $a$, $s = -\dfrac{Wbx}{Zl}$ For segment of length $b$, $s = -\dfrac{Wav}{Zl}$ | Stress at load, $-\dfrac{Wab}{Zl}$ If cross-section is constant, this is the maximum stress. |
| **Case 4. — Supported at Both Ends, Two Symmetrical Loads** | Between each support and adjacent load, $s = -\dfrac{Wx}{Z}$ Between loads, $s = -\dfrac{Wa}{Z}$ | Stress at each load, and at all points between, $-\dfrac{Wa}{Z}$ |
| **Case 5. — Both Ends Overhanging Supports Symmetrically, Uniform Load** TOTAL LOAD W $L = l + 2c$ | Between each support and adjacent end, $s = \dfrac{W}{2\,ZL}\,(c-u)^2$ Between supports, $s = \dfrac{W}{2\,ZL}\,[c^2 - x\,(l-x)]$ | Stress at each support, $\dfrac{Wc^2}{2\,ZL}$ Stress at center, $\dfrac{W}{2\,ZL}\,(c^2 - \tfrac{1}{4}\,l^2)$ If cross-section is constant, the greater of these is the maximum stress. If $l$ is greater than $2\,c$, the stress is zero at points $\sqrt{\tfrac{1}{4}\,l^2 - c^2}$ on both sides of the center. If cross-section is constant and if $l = 2.828\,c$, the stresses at supports and center are equal and opposite, and are $\pm\dfrac{WL}{46.62\,Z}$ |

## Stresses and Deflections in Beams

| Deflections (*See footnote*) | |
|---|---|
| General Formula for Deflection at any Point | Deflections at Critical Points |
| $$y = \frac{Wx\,(l-x)}{24\,EIl}\,[l^2 + x\,(l-x)]$$ | Maximum deflection, at center, $$\frac{5}{384}\,\frac{Wl^3}{EI}$$ |
| Between each support and load, $$y = \frac{Wx}{48\,EI}\,(3\,l^2 - 4\,x^2)$$ | Maximum deflection, at load, $\dfrac{Wl^3}{48\,EI}$ |
| For segment of length $a$, $$y = \frac{Wbx}{6\,EIl}\,(l^2 - x^2 - b^2)$$ For segment of length $b$, $$y = \frac{Wav}{6\,EIl}\,(l^2 - v^2 - a^2)$$ | Deflection at load, $\dfrac{Wa^2b^2}{3\,EIl}$ Let $a$ be the length of the shorter segment and $b$ of the longer one. The maximum deflection is in the longer segment, at $$v = b\sqrt{\frac{1}{3} + \frac{2\,a}{3\,b}} = v_1, \text{ and is } \frac{Wav_1^3}{3\,EIl}$$ |
| Between each support and adjacent load, $$y = \frac{Wx}{6\,EI}\,[3\,a\,(l-a) - x^2]$$ Between loads, $$y = \frac{Wa}{6\,EI}\,[3\,v\,(l-v) - a^2]$$ | Maximum deflection at center, $$\frac{Wa}{24\,EI}\,(3\,l^2 - 4\,a^2)$$ Deflection at loads $\dfrac{Wa^2}{6\,EI}\,(3\,l - 4\,a)$ |
| Between each support and adjacent end, $$y = \frac{Wu}{24\,EIL}\,[6\,c^2\,(l+u) - u^2\,(4\,c - u) - l^3]$$ Between supports, $$y = \frac{Wx\,(l-x)}{24\,EIL}\,[x\,(l-x) + l^3 - 6\,c^2]$$ | Deflection at ends, $$\frac{Wc}{24\,EIL}[3\,c^2\,(c + 2\,l) - l^3]$$ Deflection at center, $$\frac{Wl^2}{384\,EIL}\,(5\,l^2 - 24\,c^2)$$ If $l$ is between $2\,c$ and $2.449\,c$, there are maximum upward deflections at points $\sqrt{3\,(\tfrac{1}{4}\,l^2 - c^2)}$ on both sides of the center, which are, $-\dfrac{W}{96\,EIL}\,(6\,c^2 - l^2)^2$ |

The deflections apply only to cases where the cross-section of the beam is constant for its entire length

### Stresses and Deflections in Beams

| Type of Beam | Stresses | |
|---|---|---|
| | General Formula for Stress at any Point | Stresses at Critical Points |
| **Case 6. — Both Ends Overhanging Supports Unsymmetrically, Uniform Load**<br><br>TOTAL LOAD W<br><br>$\frac{W}{2l}(l-d+c)$    $\frac{W}{2l}(l+d-c)$ | For overhanging end of length $c$,<br>$$s = \frac{W}{2\,ZL}(c-u)^2$$<br>Between supports,<br>$$s = \frac{W}{2\,ZL}\left\{ c^2\left(\frac{l-x}{l}\right) + d^2\frac{x}{l} - x\,(l-x) \right\}$$<br>For overhanging end of length $d$,<br>$$s = \frac{W}{2\,ZL}(d-w)^2$$ | Stress at support next end of length $c$,<br>$$\frac{Wc^2}{2\,ZL}$$<br>Critical stress between supports is at<br>$$x = \frac{l^2+c^2-d^2}{2\,l} = x_1$$<br>and is<br>$$\frac{W}{2\,ZL}(c^2 - x_1^2)$$<br>Stress at support next end of length $d$,<br>$$\frac{Wd^2}{2\,ZL}$$<br>If cross-section is constant, the greatest of these three is the maximum stress.<br>If $x_1 > c$, the stress is zero at points $\sqrt{x_1^2 - c^2}$ on both sides of $x = x_1$. |
| **Case 7. — Both Ends Overhanging Supports, Load at any Point Between**<br><br>$\frac{Wb}{l}$    $\frac{Wa}{l}$<br>$(a+b=l)$ | Between supports:<br>For segment of length $a$, $s = -\dfrac{Wbx}{Zl}$<br>For segment of length $b$, $s = -\dfrac{Wav}{Zl}$<br>Beyond supports $s = 0$. | Stress at load,<br>$$-\frac{Wab}{Zl}$$<br>If cross-section is constant, this is the maximum stress. |
| **Case 8. — Both Ends Overhanging Supports, Single Overhanging Load**<br><br>$\frac{W(c+l)}{l}$    $-\frac{Wc}{l}$ | Between load and adjacent support,<br>$$s = \frac{W}{Z}(c-u)$$<br>Between supports,<br>$$s = \frac{Wc}{Zl}(l-x)$$<br>Between unloaded end and adjacent support, $s = 0$. | Stress at support adjacent to load, $\dfrac{Wc}{Z}$<br>If cross-section is constant, this is the maximum stress.<br>Stress is zero at other support. |
| **Case 9. — Both Ends Overhanging Supports, Symmetrical Overhanging Loads**<br> | Between each load and adjacent support,<br>$$s = \frac{W}{Z}(c-u)$$<br>Between supports,<br>$$s = \frac{Wc}{Z}$$ | Stress at supports and at all points between, $\dfrac{Wc}{Z}$<br>If cross-section is constant, this is the maximum stress. |

## Stresses and Deflections in Beams

| Deflections (*See footnote at beginning of Table*) | |
| --- | --- |
| General Formula for Deflections at any Point | Deflections at Critical Points |

For overhanging end of length $c$,

$$y = \frac{Wu}{24\,EIL}\,[2\,l\,(d^2 + 2\,c^2) + 6\,c^2 u - u^2\,(4\,c - u) - l^3]$$

Between supports,

$$y = \frac{Wx\,(l-x)}{24\,EIL}\left\{ x\,(l-x) + l^2 - 2\,(d^2 + c^2) - \frac{2}{l}\,[d^2 x + c^2\,(l-x)] \right\}$$

For overhanging end of length $d$,

$$y = \frac{Ww}{24\,EIL}\,[2\,l\,(c^2 + 2\,d^2) + 6\,d^2 w - w^2\,(4\,d - w) - l^3]$$

Deflection at end $c$,

$$\frac{Wc}{24\,EIL}\,[2\,l\,(d^2 + 2\,c^2) + 3\,c^3 - l^3]$$

Deflection at end $d$,

$$\frac{Wd}{24\,EIL}\,[2\,l\,(c^2 + 2\,d^2) + 3\,d^3 - l^3]$$

This case is so complicated that convenient general expressions for the critical deflections between supports cannot be obtained.

---

Between supports, same as Case 3.
For overhanging end of length $c$,

$$y = -\frac{Wabu}{6\,EIl}\,(l+b)$$

For overhanging end of length $d$,

$$y = -\frac{Wabw}{6\,EIl}\,(l+a)$$

Between supports, same as Case 3.

Deflection at end $c$, $-\dfrac{Wabc}{6\,EIl}\,(l+b)$

Deflection at end $d$, $-\dfrac{Wabd}{6\,EIl}\,(l+a)$

---

Between load and adjacent support,

$$y = \frac{Wu}{6\,EI}\,(3\,cu - u^2 + 2\,cl)$$

Between supports,

$$y = -\frac{Wcx}{6\,EIl}\,(l-x)(2\,l-x)$$

Between unloaded end and adjacent support, $y = \dfrac{Wclw}{6\,EI}$

Deflection at load, $\dfrac{Wc^2}{3\,EI}\,(c+l)$

Maximum upward deflection is at

$$x = 0.42265\,l,\text{ and is } -\frac{Wcl^2}{15.55\,EI}$$

Deflection at unloaded end, $\dfrac{Wcld}{6\,EI}$

---

Between each load and adjacent support, $y = \dfrac{Wu}{6\,EI}\,[3\,c\,(l+u) - u^2]$

Between supports, $y = -\dfrac{Wcx}{2\,EI}\,(l-x)$

Deflections at loads, $\dfrac{Wc^2}{6\,EI}\,(2\,c+3\,l)$

Deflection at center, $-\dfrac{Wcl^2}{8\,EI}$

---

The above expressions involve the usual approximations of the theory of flexure, and hold only for small deflections. Exact expressions for deflections of any magnitude are as follows:

Between supports the curve is a circle of radius $r = \dfrac{EI}{Wc}$; $y = \sqrt{r^2 - \frac{1}{4}\,l^2} - \sqrt{r^2 - (\frac{1}{2}\,l - x)^2}$

Deflection at center, $\sqrt{r^2 - \frac{1}{4}\,l^2} - r$

## Stresses and Deflections in Beams

| Type of Beam | Stresses | |
| --- | --- | --- |
| | General Formula for Stress at any Point | Stresses at Critical Points |
| **Case 10. — Fixed at One End, Uniform Load**<br> | $s = \dfrac{W}{2\,Zl}\,(l-x)^2$ | Stress at support,<br>$\dfrac{Wl}{2\,Z}$<br>If cross-section is constant, this is the maximum stress. |
| **Case 11. — Fixed at One End, Load at Other**<br> | $s = \dfrac{W}{Z}\,(l-x)$ | Stress at support, $\dfrac{Wl}{Z}$<br>If cross-section is constant, this is the maximum stress. |
| **Case 12. — Fixed at One End, Intermediate Load**<br> | Between support and load,<br>$s = \dfrac{W}{Z}\,(l-x)$<br>Beyond load, $s = 0$. | Stress at support, $\dfrac{Wl}{Z}$<br>If cross-section is constant, this is the maximum stress. |
| **Case 13. — Fixed at One End, Supported at the Other, Uniform Load**<br> | $s = \dfrac{W(l-x)}{2\,Zl}(\tfrac{1}{4}\,l-x)$ | Maximum stress at point of fixture, $\dfrac{Wl}{8\,Z}$<br>Stress is zero at<br>$x = \tfrac{1}{4}\,l$<br>Greatest negative stress is at $x = \tfrac{5}{8}\,l$ and is $-\dfrac{9}{128}\dfrac{Wl}{Z}$ |
| **Case 14. — Fixed at One End, Supported at the Other, Load at Center**<br> | Between point of fixture and load,<br>$s = \dfrac{W}{16\,Z}(3\,l-11\,x)$<br>Between support and load,<br>$s = -\dfrac{5}{16}\dfrac{Wv}{Z}$ | Maximum stress at point of fixture, $\dfrac{3}{16}\dfrac{Wl}{Z}$<br>Stress is zero at<br>$x = \dfrac{3}{11}\,l$<br>Greatest negative stress at center, $-\dfrac{5}{32}\dfrac{Wl}{Z}$ |

## Stresses and Deflections in Beams

| Deflections (*See footnote at beginning of Table*) | |
| --- | --- |
| General Formula for Deflection at any Point | Deflections at Critical Points |
| $y = \dfrac{Wx^2}{24\,EIl}\,[2\,l^2 + (2\,l - x)^2]$ | Maximum deflection, at end, $\dfrac{Wl^3}{8\,EI}$ |
| $y = \dfrac{Wx^2}{6\,EI}\,(3\,l - x)$ | Maximum deflection, at end, $\dfrac{Wl^3}{3\,EI}$ |
| Between support and load, $$y = \frac{Wx^2}{6\,EI}\,(3\,l - x)$$ Beyond load, $$y = \frac{Wl^2}{6\,EI}\,(3\,v - l)$$ | Deflection at load, $\dfrac{Wl^3}{3\,EI}$ <br><br> Maximum deflection, at end, $$\frac{Wl^2}{6\,EI}\,(2\,l + 3\,b)$$ |
| $y = \dfrac{Wx^2\,(l - x)}{48\,EIl}\,(3\,l - 2\,x)$ | Maximum deflection is at $x = 0.5785\,l$, and is $\dfrac{Wl^3}{185\,EI}$ <br><br> Deflection at center, $\dfrac{Wl^3}{192\,EI}$ <br><br> Deflection at point of greatest negative stress, at $x = \dfrac{5}{8}\,l$ is $\dfrac{Wl^3}{187\,EI}$ |
| Between point of fixture and load, $$y = \frac{Wx^2}{96\,EI}\,(9\,l - 11\,x)$$ Between support and load, $$y = \frac{Wv}{96\,EI}\,(3\,l^2 - 5\,v^2)$$ | Maximum deflection is at $v = 0.4472\,l$, and is $\dfrac{Wl^3}{107.33\,EI}$ <br><br> Deflection at load, $\dfrac{7}{768}\,\dfrac{Wl^3}{EI}$ |

Stresses and Deflections in Beams

| Type of Beam | Stresses | |
|---|---|---|
| | General Formula for Stress at any Point | Stresses at Critical Points |
| **Case 15.** — Fixed at One End, Supported at the Other, Load at any Point<br><br>$m = (l + a)(l + b) + al$<br>$n = al(l + b)$<br><br><br>$\dfrac{Wab(l+b)}{2l^3}$<br>$W\left[1 - \dfrac{a^2}{2l^3}(3l - a)\right]$ $\dfrac{Wa^2(3l-a)}{2l^3}$ | Between point of fixture and load,<br>$$s = \frac{Wb}{2Zl^3}(n - mx)$$<br><br>Between support and load,<br>$$s = -\frac{Wa^2y}{2Zl^3}(3l - a)$$ | Greatest positive stress, at point of fixture, $\dfrac{Wab}{2Zl^2}(l + b)$<br>Greatest negative stress, at load, $-\dfrac{Wa^2b}{2Zl^3}(3l - a)$<br>If $a < 0.5858\,l$, the first is the maximum stress. If $a = 0.5858\,l$, the two are equal and are $\pm\dfrac{Wl}{5.83\,Z}$. If $a > 0.5858\,l$, the second is the maximum stress.<br>Stress is zero at $x = \dfrac{n}{m}$ |
| **Case 16.** — Fixed at One End, Free but Guided at the Other, Uniform Load<br><br>TOTAL LOAD W<br><br>$\dfrac{Wl}{3}$ $\dfrac{Wl}{6}$ | $$s = \frac{Wl}{Z}\left\{\frac{1}{3} - \frac{x}{l} + \frac{1}{2}\left(\frac{x}{l}\right)^2\right\}$$ | Maximum stress, at support, $\dfrac{Wl}{3Z}$<br>Stress is zero for $x = 0.4227\,l$<br>Greatest negative stress, at free end, $-\dfrac{Wl}{6Z}$ |
| **Case 17.** — Fixed at One End, Free but Guided at the Other, with Load<br><br><br>$\dfrac{Wl}{2}$ $\dfrac{Wl}{2}$ | $$s = \frac{W}{Z}\left(\tfrac{1}{2}l - x\right)$$ | Stress at support, $\dfrac{Wl}{2Z}$<br>Stress at free end, $-\dfrac{Wl}{2Z}$<br>These are the maximum stresses and are equal and opposite. Stress is zero at center. |
| **Case 18.** — Fixed at Both Ends, Uniform Load<br><br>TOTAL LOAD W<br><br>$\dfrac{Wl}{12}$ $\dfrac{Wl}{12}$<br>$\dfrac{W}{2}$ $\dfrac{W}{2}$ | $$s = \frac{Wl}{2Z}\left\{\frac{1}{6} - \frac{x}{l} + \left(\frac{x}{l}\right)^2\right\}$$ | Maximum stress, at ends, $\dfrac{Wl}{12Z}$<br>Stress is zero at $x = 0.7887\,l$ and at $x = 0.2113\,l$<br>Greatest negative stress, at center, $-\dfrac{Wl}{24Z}$ |

## Stresses and Deflections in Beams

| Deflections (*See footnote at beginning of Table*) | |
|---|---|
| General Formula for Deflections at any Point | Deflections at Critical Points |

| | |
|---|---|
| Between point of fixture and load, $$y = \frac{Wx^2b}{12\,EIl^3}\,(3\,n - mx)$$ Between support and load, $$y = \frac{Wa^2v}{12\,EIl^3}\,[3\,l^2b - v^2\,(3\,l - a)]$$ | Deflection at load, $\dfrac{Wa^3b^2}{12\,EIl^3}\,(3\,l + b)$ If $a < 0.5858\,l$, maximum deflection is between load and support, at $$v = l\sqrt{\frac{b}{2\,l+b}} \text{ and is } \frac{Wa^2b}{6\,EI}\sqrt{\frac{b}{2\,l+b}}$$ If $a = 0.5858\,l$, maximum deflection is at load and is $\dfrac{Wl^3}{101.9\,EI}$ If $a > 0.5858\,l$, maximum deflection is between load and point of fixture, at $$z = \frac{2\,n}{m}, \text{ and is } \frac{Wbn^3}{3\,EIm^2l^3}$$ |
| $$y = \frac{Wx^2}{24\,EIl}\,(2\,l - x)^2$$ | Maximum deflection, at free end, $$\frac{Wl^3}{24\,EI}$$ |
| $$y = \frac{Wx^2}{12\,EI}\,(3\,l - 2\,x)$$ | Maximum deflection, at free end, $$\frac{Wl^3}{12\,EI}$$ |
| $$y = \frac{Wx^2}{24\,EIl}\,(l - x)^2$$ | Maximum deflection, at center, $$\frac{Wl^3}{384\,EI}$$ |

**Stresses and Deflections in Beams**

| Type of Beam | Stresses | |
|---|---|---|
| | General Formula for Stress at any Point | Stresses at Critical Points |
| **Case 19. — Fixed at Both Ends, Load at Center** | Between each end and load, $$s = \frac{W}{2Z}\left(\tfrac{1}{4}l - x\right)$$ | Stress at ends $\frac{Wl}{8Z}$; at load $-\frac{Wl}{8Z}$. These are the maximum stresses and are equal and opposite. Stress is zero at $x = \tfrac{1}{4}l$ |
| **Case 20. — Fixed at Both Ends, Load at any Point** | For segment of length $a$, $$s = \frac{Wb^2}{Zl^2}[al - x(l + 2a)]$$ For segment of length $b$, $$s = \frac{Wa^2}{Zl^2}[bl - v(l + 2b)]$$ | Stress at end next segment of length $a$, $\frac{Wab^2}{Zl^2}$ Stress at end next segment of length $b$, $\frac{Wa^2b}{Zl^2}$ Maximum stress is at end next shorter segment. Stress is zero for $x = \frac{al}{l+2a}$ and $v = \frac{bl}{l+2b}$ Greatest negative stress, at load, $-\frac{2\,Wa^2b^2}{Zl^3}$ |
| **Case 21. — Continuous Beam, with Two Equal Spans, Uniform Load** TOTAL LOAD ON EACH SPAN, W | $$s = \frac{W(l-x)}{2Zl}\left(\tfrac{1}{4}l - x\right)$$ | Maximum stress at point $A$, $\frac{Wl}{8Z}$ Stress is zero at $x = \tfrac{1}{4}l$. Greatest negative stress is at $x = \tfrac{5}{8}$ and is, $-\frac{9}{128}\frac{Wl}{Z}$ |
| **Case 22. — Continuous Beam, with Two Unequal Spans, Unequal, Uniform Loads** TOTAL LOAD W₁    TOTAL LOAD W₂ $\frac{l_1 W_1(3l_1+4l_2)-W_2 l_2^2}{8l_1(l_1+l_2)}$    $\frac{l_2 W_2(3l_2+4l_1)-W_1 l_1^2}{8l_2(l_1+l_2)}$ $\left(\frac{W_1+W_2}{2}\right)+\frac{1}{8}\left(\frac{W_1 l_1}{l_2}+\frac{W_2 l_2}{l_1}\right)$ | Between $R_1$ and $R$, $$s = \frac{l_1-x}{Z}\left\{\frac{(l_1-x)W_1}{2l_1} - R_1\right\}$$ Between $R_2$ and $R$, $$s = \frac{l_2-u}{Z}\left\{\frac{(l_2-u)W_2}{2l_2} - R_2\right\}$$ | Stress at support $R$, $\frac{W_1 l_1^2 + W_2 l_2^2}{8Z(l_1+l_2)}$ Greatest stress in the first span is at $x = \frac{l_1}{W_1}(W_1 - R_1)$, and is, $-\frac{R_1^2 l_1}{2ZW_1}$ Greatest stress in the second span is at $u = \frac{l_1}{W_2}(W_2 - R_2)$, and is, $-\frac{R_2^2 l_1}{2ZW_2}$ |

**Stresses and Deflections in Beams**

| Deflections (*See footnote at beginning of Table*) | |
|---|---|
| General Formula for Deflections at any Point | Deflections at Critical Points |
| $$y = \frac{Wx^2}{48\,EI}\,(3l - 4x)$$ | Maximum deflection, at load, $$\frac{Wl^3}{192\,EI}$$ |
| For segment of length $a$, $$y = \frac{Wx^2b^2}{6\,EIl^3}\,[2a(l-x) + l(a-x)]$$ For segment of length $b$, $$y = \frac{Wv^2a^2}{6\,EIl^3}\,[2b(l-v) + l(b-v)]$$ | Deflection at load, $\dfrac{Wa^3b^3}{3\,EIl^3}$ Let $b$ be the length of the longer segment and $a$ of the shorter one. The maximum deflection is in the longer segment, at $v = \dfrac{2bl}{l + 2b}$, and is $$\frac{2\,Wa^3b^3}{3\,EI(l + 2b)^2}$$ |
| $$y = \frac{Wx^2\,(l - x)}{48\,EIl}\,(3l - 2x)$$ | Maximum deflection is at $x = 0.5785\,l$, and is $\dfrac{Wl^3}{185\,EI}$ Deflection at center of span, $\dfrac{Wl^3}{192\,EI}$ Deflection at point of greatest negative stress, at $x = \dfrac{5}{8}\,l$ is $\dfrac{Wl^3}{187\,EI}$ |
| Between $R_1$ and $R$, $$y = \frac{x\,(l_1 - x)}{24\,EI}\left\{ (2l_1 - x)(4R_1 - W_1) - \frac{W_1\,(l_1 - x)^2}{l_1} \right\}$$ Between $R_2$ and $R$, $$y = \frac{u\,(l_2 - u)}{24\,EI}\left\{ (2l_2 - u)(4R_2 - W_2) - \frac{W_2\,(l_2 - u)^2}{l_2} \right\}$$ | This case is so complicated that convenient general expressions for the critical deflections cannot be obtained. |

### Stresses and Deflections in Beams

| Type of Beam | Stresses | |
|---|---|---|
| | General Formula for Stress at any Point | Stresses at Critical Points |
| **Case 23.** — Continuous Beam, with Two Equal Spans, Equal Loads at Center of Each <br><br> *(diagram: B, A, B supports with loads W; spans l/2, l/2; reactions $\frac{5}{16}W$, $\frac{11}{8}W$, $\frac{5}{16}W$)* | Between point A and load, $$s = \frac{W}{16Z}(3l - 11x)$$ Between point B and load, $$s = -\frac{5}{16}\frac{Wv}{Z}$$ | Maximum stress at point A, $\dfrac{3}{16}\dfrac{Wl}{Z}$ <br><br> Stress is zero at $$x = \frac{3}{11}l$$ Greatest negative stress at center of span, $$-\frac{5}{32}\frac{Wl}{Z}$$ |
| **Case 24.** — Continuous Beam, with Two Unequal Spans, Unequal Loads at any Point of Each <br><br> $m = \dfrac{1}{2(l_1+l_2)}\left(\dfrac{W_1 a_1 b_1}{l_1}(l_1+a_1) + \dfrac{W_2 a_2 b_2}{l_2}(l_2+a_2)\right)$ <br><br> 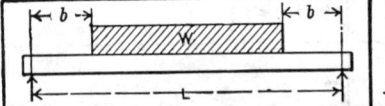 <br><br> $\dfrac{W_1 b_1 - m}{l_1} = r_1 \quad \dfrac{W_1 a_1 + m}{l_1} + \dfrac{W_2 a_2 + m}{l_2} = r \quad \dfrac{W_2 b_2 - m}{l_2} = r_2$ | Between $R_1$ and $W_1$, $$s = -\frac{w r_1}{Z}$$ Between $R$ and $W_1$, $$s = \frac{I}{l_1 Z}[m(l_1 - u) - W_1 a_1 u]$$ Between $R$ and $W_2$, $$s = \frac{I}{l_2 Z}[m(l_2 - x) - W_2 a_2 x]$$ Between $R_2$ and $W_2$, $$s = -\frac{v r_2}{Z}$$ | Stress at load $W_1$, $$-\frac{a_1 r_1}{Z}$$ Stress at support $R$, $$\frac{m}{Z}$$ Stress at load $W_2$, $$-\frac{a_2 r_2}{Z}$$ The greatest of these is the maximum stress. |

**Deflection of Beam Uniformly Loaded for Part of Its Length.** — In the following formulas, lengths are in inches, weights in pounds. $W$ = total load; $L$ = total length between supports; $E$ = modulus of elasticity; $I$ = moment of inertia of beam section; $a$ = fraction of length of beam at each end, that is not loaded = $b \div L$; $f$ = deflection.

$$f = \frac{WL^3}{EI\,384(1 - 2a)}(5 - 24a^2 + 16a^4)$$

The expression for maximum bending moment is: $M_{max.} = \frac{1}{8}WL(1 + 2a)$. These formulas apply to simple beams resting on supports at the ends.

If the formulas are used with metric SI units, $W$ = total load in newtons; $L$ = total length between supports in millimeters; $E$ = modulus of elasticity in newtons per millimeter$^2$; $I$ = moment of inertia of beam section in millimeters$^4$; $a$ = fraction of length of beam at each end, that is not loaded = $b \div L$; and $f$ = deflection in millimeters. The bending moment $M_{max}$ is in newton-millimeters ($\mathbf{N\cdot mm}$). *Note:* A load due to the weight of a mass of $M$ kilograms is $Mg$ newtons, where $g$ = approximately 9.81 meters per second$^2$.

### Stresses and Deflections in Beams

| Deflections (*See footnote at beginning of Table*) | |
|---|---|
| General Formula for Deflections at any Point | Deflections at Critical Points |
| Between point $A$ and load,<br><br>$$y = \frac{Wx^2}{96\,EI}\,(9\,l - 11\,x)$$<br><br>Between point $B$ and load,<br><br>$$y = \frac{Wv}{96\,EI}\,(3\,l^2 - 5\,v^2)$$ | Maximum deflection is at $v = 0.4472\,l$, and is $\dfrac{Wl^3}{107.33\,EI}$<br><br>Deflection at load, $\dfrac{7}{768}\,\dfrac{Wl^3}{EI}$ |
| Between $R_1$ and $W_1$,<br><br>$$y = \frac{w}{6\,EI}\left\{ (l_1 - w)(l_1 + w)\,r_1 - \frac{W_1 b_1^3}{l_1} \right\}$$<br><br>Between $R$ and $W_1$,<br><br>$$y = \frac{u}{6\,EIl_1}\,[\,W_1 a_1 b_1\,(l_1 + a_1)$$<br>$$- W_1 a_1 u^2 - m\,(2\,l_1 - u)(l_1 - u)\,]$$<br><br>Between $R$ and $W_2$,<br><br>$$y = \frac{x}{6\,EIl_2}\,[\,W_2 a_2 b_2\,(l_2 + a_2)$$<br>$$- W_2 a_2 x^2 - m\,(2\,l_2 - x)(l_2 - x)\,]$$<br><br>Between $R_2$ and $W_2$,<br><br>$$y = \frac{v}{6\,EI}\left\{ (l_2 - v)(l_2 + v)\,r_2 - \frac{W_2 b_2^3}{l_2} \right\}$$ | Deflection at load $W_1$.<br><br>$$\frac{a_1 b_1}{6\,EIl_1}\,[\,2\,a_1 b_1 W_1 - m\,(l_1 + a_1)\,]$$<br><br>Deflection at load $W_2$,<br><br>$$\frac{a_2 b_2}{6\,EIl_2}\,[\,2\,a_2 b_2 W_2 - m\,(l_2 + a_2)\,]$$<br><br>This case is so complicated that convenient general expressions for the maximum deflections cannot be obtained. |

**Beams of Uniform Strength Throughout Their Length.** — In nearly all cases, the bending moment in a beam is not uniform throughout its length, but varies. Therefore, a beam of uniform cross-section which is made strong enough at its most strained section, will have an excess of material at every other section. Sometimes it may be desirable to have the cross-section uniform, while in other cases the metal can be more advantageously distributed if the beam is so designed that its cross-section varies from point to point, so that it is at every point just great enough to take care of the bending stresses at that point. A table is given showing beams in which the load is applied in different ways and which are supported by different methods, and the shape of the beam required for uniform strength is indicated. It should be noted that the shape given is the theoretical shape required to resist bending only. It is apparent that sufficient cross-section of beam must also be added either at the points of support (in the case of beams supported at both ends), or at the point of application of the load (in the case of beams loaded at one end), to take care of the vertical shear.

It should be noted that the theoretical shapes of the beams given in the tables on the two following pages are based on the stated assumptions of uniformity of width or depth of cross-section, and unless these are observed in the design, the theoretical outlines do not apply without modifications. For example, in a cantilever with the load at one end, the outline is a parabola only when the width of the

(*Continued on page 420*)

## Beams of Uniform Strength Throughout Their Length

(All loads in pounds, all dimensions in inches.)

| Type of Beam | Description | Formula* |
|---|---|---|
| | Load at one end. Width of beam uniform. Depth of beam decreasing towards loaded end. Outline of beam-shape, parabola with vertex at loaded end. | $P = \dfrac{Sbh^2}{6l}$ |
| | Load at one end. Width of beam uniform. Depth of beam decreasing towards loaded end. Outline of beam, one-half of a parabola with vertex at loaded end. Beam may be reversed so that upper edge is parabolic. | $P = \dfrac{Sbh^2}{6l}$ |
| | Load at one end. Depth of beam uniform. Width of beam decreasing towards loaded end. Outline of beam triangular, with apex at loaded end. | $P = \dfrac{Sbh^2}{6l}$ |
| | Beam of *approximately* uniform strength. Load at one end. Width of beam uniform. Depth of beam decreasing towards loaded end, but not tapering to a sharp point. | $P = \dfrac{Sbh^2}{6l}$ |
| | Uniformly distributed load. Width of beam uniform. Depth of beam decreasing towards outer end. Outline of beam, right-angled triangle. | $P = \dfrac{Sbh^2}{3l}$ |
| | Uniformly distributed load. Depth of beam uniform. Width of beam gradually decreasing towards outer end. Outline of beam is formed by two parabolas which tangent each other at their vertices at the outer end of the beam. | $P = \dfrac{Sbh^2}{3l}$ |

* In the formulas, $P$ = load in pounds; $S$ = safe stress in pounds per square inch; and $a$, $b$, $c$, $h$, and $l$ are in inches. If metric SI units are used, $P$ is in newtons; $S$ = safe stress in N/mm²; and $a$, $b$, $c$, $h$, and $l$ are in millimeters.

## Beams of Uniform Strength Throughout Their Length

| Type of Beam | Description | Formula* |
|---|---|---|
| | Beam supported at both ends. Load concentrated at any point. Depth of beam uniform. Width of beam maximum at point of loading. Outline of beam, two triangles with apexes at points of support. | $P = \dfrac{Sbh^2l}{6\,ac}$ |
| | Beam supported at both ends. Load concentrated at any point. Width of beam uniform. Depth of beam maximum at point of loading. Outline of beam is formed by two parabolas with their vertexes at points of support. | $P = \dfrac{Sbh^2l}{6\,ac}$ |
| | Beam supported at both ends. Load concentrated in the middle. Depth of beam uniform. Width of beam maximum at point of loading. Outline of beam, two triangles with apexes at points of support. | $P = \dfrac{2\,Sbh^2}{3\,l}$ |
| | Beam supported at both ends. Load concentrated at center. Width of beam uniform. Depth of beam maximum at point of loading. Outline of beam, two parabolas with vertices at points of support. | $P = \dfrac{2\,Sbh^2}{3\,l}$ |
| | Beam supported at both ends. Load uniformly distributed. Depth of beam uniform. Width of beam maximum at center. Outline of beam, two parabolas with vertexes at middle of beam. | $P = \dfrac{4\,Sbh^2}{3\,l}$ |
| | Beam supported at both ends. Load uniformly distributed. Width of beam uniform. Depth of beam maximum at center. Outline of beam one-half of an ellipse. | $P = \dfrac{4\,Sbh^2}{3\,l}$ |

\* For details of English and metric SI units used in the formulas, see previous page.

## Rectangular Solid Beams

| Style of Loading and Support | Breadth of Beam, b | Height of Beam, h | Stress in Extreme Fibers, f | Length of Beam, l | Total Load, W |
|---|---|---|---|---|---|
| | Inches | Inches | Lb./sq. in. | Inches | Pounds |
| | Millimeters | Millimeters | N/sq. mm. | Millimeters | Newtons |
| Beam fixed at one end, loaded at the other | $\dfrac{6\,lW}{fh^2}=b$ | $\sqrt{\dfrac{6\,lW}{bf}}=h$ | $\dfrac{6\,lW}{bh^2}=f$ | $\dfrac{bfh^2}{6\,W}=l$ | $\dfrac{bfh^2}{6\,l}=W$ |
| Beam fixed at one end, uniformly loaded | $\dfrac{3\,lW}{fh^2}=b$ | $\sqrt{\dfrac{3\,lW}{bf}}=h$ | $\dfrac{3\,lW}{bh^2}=f$ | $\dfrac{bfh^2}{3\,W}=l$ | $\dfrac{bfh^2}{3\,l}=W$ |
| Beam supported at both ends, single load in middle | $\dfrac{3\,lW}{2\,fh^2}=b$ | $\sqrt{\dfrac{3\,lW}{2\,bf}}=h$ | $\dfrac{3\,lW}{2\,bh^2}=f$ | $\dfrac{2\,bfh^2}{3\,W}=l$ | $\dfrac{2\,bfh^2}{3\,l}=W$ |
| Beam supported at both ends, uniformly loaded | $\dfrac{3\,lW}{4\,fh^2}=b$ | $\sqrt{\dfrac{3\,lW}{4\,bf}}=h$ | $\dfrac{3\,lW}{4\,bh^2}=f$ | $\dfrac{4\,bfh^2}{3\,W}=l$ | $\dfrac{4\,bfh^2}{3\,l}=W$ |
| Beam supported at both ends, single unsymmetrical load | $\dfrac{6\,Wac}{fh^2l}=b$ | $\sqrt{\dfrac{6Wac}{bfl}}=h$ | $\dfrac{6\,Wac}{bh^2l}=f$ | $a+c=l$ | $\dfrac{bh^2fl}{6\,ac}=W$ |
| Beam supported at both ends, two symmetrical loads | $\dfrac{3\,Wa}{fh^2}=b$ | $\sqrt{\dfrac{3\,Wa}{bf}}=h$ | $\dfrac{3\,Wa}{bh^2}=f$ | $l$, any length $\dfrac{bh^2f}{3\,W}=a$ | $\dfrac{bh^2f}{3\,a}=W$ |

## Round Solid Beams

| Style of Loading and Support | Diameter of Beam, $d$ | Stress in Extreme Fibers, $f$ | Length of Beam, $l$ | Total Load, $W$ |
|---|---|---|---|---|
| | Inches | Lb./sq. in. | Inches | Pounds |
| | Millimeters | N/sq. mm. | Millimeters | Newtons |

Beam fixed at one end, loaded at the other

$$\sqrt[3]{\frac{10.18\,lW}{f}}=d \quad\bigg|\quad \frac{10.18\,lW}{d^3}=f \quad\bigg|\quad \frac{d^3 f}{10.18\,W}=l \quad\bigg|\quad \frac{d^3 f}{10.18\,l}=W$$

Beam fixed at one end, uniformly loaded

$$\sqrt[3]{\frac{5.092\,Wl}{f}}=d \quad\bigg|\quad \frac{5.092\,Wl}{d^3}=f \quad\bigg|\quad \frac{d^3 f}{5.092\,W}=l \quad\bigg|\quad \frac{d^3 f}{5.092\,l}=W$$

Beam supported at both ends, single load in middle

$$\sqrt[3]{\frac{2.546\,Wl}{f}}=d \quad\bigg|\quad \frac{2.546\,Wl}{d^3}=f \quad\bigg|\quad \frac{d^3 f}{2.546\,W}=l \quad\bigg|\quad \frac{d^3 f}{2.546\,l}=W$$

Beam supported at both ends, uniformly loaded

$$\sqrt[3]{\frac{1.273\,Wl}{f}}=d \quad\bigg|\quad \frac{1.273\,Wl}{d^3}=f \quad\bigg|\quad \frac{d^3 f}{1.273\,W}=l \quad\bigg|\quad \frac{d^3 f}{1.273\,l}=W$$

Beam supported at both ends, single unsymmetrical load

$$\sqrt[3]{\frac{10.18\,Wac}{fl}}=d \quad\bigg|\quad \frac{10.18\,Wac}{d^3 l}=f \quad\bigg|\quad a+c=l \quad\bigg|\quad \frac{d^3 fl}{10.18\,ac}=W$$

Beam supported at both ends, two symmetrical loads

$$\sqrt[3]{\frac{5.092\,Wa}{f}}=d \quad\bigg|\quad \frac{5.092\,Wa}{d^3}=f \quad\bigg|\quad \begin{array}{c}l,\ \text{any length}\\ \frac{d^3 f}{5.092\,W}=a\end{array} \quad\bigg|\quad \frac{d^3 f}{5.092\,a}=W$$

beam is uniform. It is not correct to use a strictly parabolic shape when the thickness is not uniform, as, for instance, when the beam is made of an I- or T-section. In such cases, some modification may be necessary; but it is evident that whatever the shape adopted, the correct depth of the section can be obtained by an investigation of the bending moment and the shearing load at a number of points, and then a line can be drawn through the points thus ascertained, which will provide for a beam of practically uniform strength whether the cross-section be of uniform width or not.

**Crane Girders with Curved Lower Chords.** — An example of a design which makes use of the principles of beams of uniform strength is found in the ordinary fish-belly type of crane girder. When laying out crane girders, the accompanying tables will be found convenient. The engraving will explain the use of the tables. A crane girder having a span of 61 feet 3 inches has been assumed as an example. The curved part has a span of 60 feet; one-half of this distance, or 30 feet, is divided for the convenience of the templet makers into ten spaces of 3 feet each. The end ordinate, assumed here to be 1 foot, will be found at the extreme left of the tables under the heading H. The lengths of the remaining nine ordinates follow in order. For short spans, say about 30 feet, it is most convenient to divide the base of the curve into five spaces, as it is the usual practice to give ordinates about every 3 feet. In this case, we would use only every other ordinate in the tables, or, beginning with the left-hand column, the ordinates would be as found in the columns headed H, 8, 6, 4 and 2.

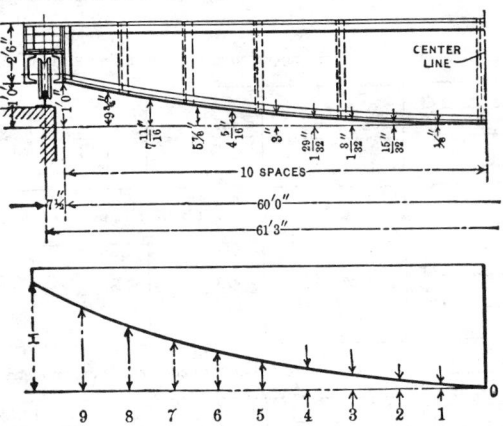

The tables are calculated from the formula: $X = H \times (M^2 \div N^2)$ in which $H$ = end ordinate; $X$ = required ordinate; $N$ = number of equal spaces into which the base line is divided; $M$ = number of spaces from o to the required ordinate. When $N = 10$, as in the case for which the tables are calculated, $N^2 = 100$, and $X = H \times 0.01 M^2$.

Hence, ordinate No. 8 equals $H \times 0.01 \times 64 = 0.64 H$. Ordinate No. 4 equals $H \times 0.01 \times 16 = 0.16 H$.

Opinions vary considerably as to the allowable working stress in crane girders. Many cranes have girders which are designed for a stress of only 8000 pounds per square inch, while in others the stress will be over 14,000 pounds. However, a

### Ordinates of Parabolas for Crane Girder Design — 1

| H (Ft. Ins.) | 9 (Ins.) | 8 (Ins.) | 7 (Ins.) | 6 (Ins.) | 5 (Ins.) | 4 (Ins.) | 3 (Ins.) | 2 (Ins.) | 1 (Ins.) |
|---|---|---|---|---|---|---|---|---|---|
| 6 | $4\frac{7}{8}$ | $3\frac{27}{32}$ | $2\frac{15}{16}$ | $2\frac{5}{32}$ | $1\frac{1}{2}$ | $\frac{31}{32}$ | $\frac{17}{32}$ | $\frac{1}{4}$ | $\frac{1}{16}$ |
| $6\frac14$ | $5\frac{1}{16}$ | $4$ | $3\frac{1}{16}$ | $2\frac{1}{4}$ | $1\frac{9}{16}$ | $1$ | $\frac{9}{16}$ | $\frac{1}{4}$ | $\frac{1}{16}$ |
| $6\frac12$ | $5\frac{1}{4}$ | $4\frac{5}{32}$ | $3\frac{5}{16}$ | $2\frac{11}{32}$ | $1\frac{5}{8}$ | $1\frac{1}{32}$ | $\frac{19}{32}$ | $\frac{1}{4}$ | $\frac{1}{16}$ |
| $6\frac34$ | $5\frac{15}{32}$ | $4\frac{5}{16}$ | $3\frac{5}{16}$ | $2\frac{7}{16}$ | $1\frac{11}{16}$ | $1\frac{3}{32}$ | $\frac{19}{32}$ | $\frac{9}{32}$ | $\frac{1}{16}$ |
| 7 | $5\frac{11}{16}$ | $4\frac{15}{32}$ | $3\frac{7}{16}$ | $2\frac{17}{32}$ | $1\frac{3}{4}$ | $1\frac{1}{8}$ | $\frac{5}{8}$ | $\frac{9}{32}$ | $\frac{1}{16}$ |
| $7\frac14$ | $5\frac{7}{8}$ | $4\frac{21}{32}$ | $3\frac{9}{16}$ | $2\frac{5}{8}$ | $1\frac{13}{16}$ | $1\frac{5}{32}$ | $\frac{21}{32}$ | $\frac{9}{32}$ | $\frac{1}{16}$ |
| $7\frac12$ | $6\frac{1}{16}$ | $4\frac{13}{16}$ | $3\frac{11}{16}$ | $2\frac{11}{16}$ | $1\frac{7}{8}$ | $1\frac{7}{32}$ | $\frac{11}{16}$ | $\frac{5}{16}$ | $\frac{1}{16}$ |
| $7\frac34$ | $6\frac{9}{32}$ | $4\frac{31}{32}$ | $3\frac{13}{16}$ | $2\frac{25}{32}$ | $1\frac{15}{16}$ | $1\frac{1}{4}$ | $\frac{11}{16}$ | $\frac{5}{16}$ | $\frac{1}{16}$ |
| 8 | $6\frac{15}{32}$ | $5\frac{1}{8}$ | $3\frac{15}{16}$ | $2\frac{7}{8}$ | $2$ | $1\frac{9}{32}$ | $\frac{28}{32}$ | $\frac{5}{16}$ | $\frac{8}{32}$ |
| $8\frac14$ | $6\frac{19}{32}$ | $5\frac{9}{32}$ | $4\frac{1}{32}$ | $2\frac{31}{32}$ | $2\frac{1}{16}$ | $1\frac{5}{16}$ | $\frac{3}{4}$ | $\frac{11}{32}$ | $\frac{8}{32}$ |
| $8\frac12$ | $6\frac{7}{8}$ | $5\frac{7}{16}$ | $4\frac{5}{32}$ | $3\frac{1}{16}$ | $2\frac{1}{8}$ | $1\frac{3}{8}$ | $\frac{3}{4}$ | $\frac{11}{32}$ | $\frac{8}{32}$ |
| $8\frac34$ | $7\frac{3}{32}$ | $5\frac{19}{32}$ | $4\frac{9}{32}$ | $3\frac{5}{32}$ | $2\frac{3}{16}$ | $1\frac{13}{32}$ | $\frac{25}{32}$ | $\frac{11}{32}$ | $\frac{8}{32}$ |
| 9 | $7\frac{3}{32}$ | $5\frac{3}{4}$ | $4\frac{13}{32}$ | $3\frac{1}{4}$ | $2\frac{1}{4}$ | $1\frac{7}{16}$ | $\frac{13}{16}$ | $\frac{3}{8}$ | $\frac{8}{32}$ |
| $9\frac14$ | $7\frac{1}{2}$ | $5\frac{15}{16}$ | $4\frac{17}{32}$ | $3\frac{9}{32}$ | $2\frac{5}{16}$ | $1\frac{15}{32}$ | $\frac{27}{32}$ | $\frac{3}{8}$ | $\frac{8}{32}$ |
| $9\frac12$ | $7\frac{11}{16}$ | $6\frac{3}{32}$ | $4\frac{21}{32}$ | $3\frac{7}{16}$ | $2\frac{3}{8}$ | $1\frac{1}{2}$ | $\frac{27}{32}$ | $\frac{3}{8}$ | $\frac{8}{32}$ |
| $9\frac34$ | $7\frac{29}{32}$ | $6\frac{1}{4}$ | $4\frac{25}{32}$ | $3\frac{1}{2}$ | $2\frac{7}{16}$ | $1\frac{9}{16}$ | $\frac{7}{8}$ | $\frac{13}{32}$ | $\frac{8}{32}$ |
| 10 | $8\frac{3}{32}$ | $6\frac{13}{32}$ | $4\frac{29}{32}$ | $3\frac{19}{32}$ | $2\frac{1}{2}$ | $1\frac{19}{32}$ | $\frac{29}{32}$ | $\frac{13}{32}$ | $\frac{8}{32}$ |
| $10\frac14$ | $8\frac{5}{16}$ | $6\frac{9}{16}$ | $5$ | $3\frac{11}{16}$ | $2\frac{9}{16}$ | $1\frac{5}{8}$ | $\frac{15}{16}$ | $\frac{13}{32}$ | $\frac{8}{32}$ |
| $10\frac12$ | $8\frac{1}{2}$ | $6\frac{23}{32}$ | $5\frac{5}{32}$ | $3\frac{25}{32}$ | $2\frac{5}{8}$ | $1\frac{11}{16}$ | $\frac{15}{16}$ | $\frac{13}{32}$ | $\frac{8}{32}$ |
| $10\frac34$ | $8\frac{11}{16}$ | $6\frac{7}{8}$ | $5\frac{1}{4}$ | $3\frac{7}{8}$ | $2\frac{11}{16}$ | $1\frac{23}{32}$ | $\frac{31}{32}$ | $\frac{7}{16}$ | $\frac{8}{32}$ |
| 11 | $8\frac{29}{32}$ | $7\frac{1}{32}$ | $5\frac{13}{32}$ | $3\frac{31}{32}$ | $2\frac{3}{4}$ | $1\frac{3}{4}$ | $1$ | $\frac{7}{16}$ | $\frac{1}{8}$ |
| $11\frac14$ | $9\frac{1}{8}$ | $7\frac{5}{16}$ | $5\frac{1}{2}$ | $4\frac{1}{16}$ | $2\frac{13}{16}$ | $1\frac{13}{16}$ | $1$ | $\frac{7}{16}$ | $\frac{1}{8}$ |
| $11\frac12$ | $9\frac{5}{16}$ | $7\frac{7}{8}$ | $5\frac{5}{8}$ | $4\frac{5}{32}$ | $2\frac{7}{8}$ | $1\frac{27}{32}$ | $1\frac{1}{32}$ | $\frac{15}{32}$ | $\frac{1}{8}$ |
| $11\frac34$ | $9\frac{1}{2}$ | $7\frac{17}{32}$ | $5\frac{3}{4}$ | $4\frac{7}{32}$ | $2\frac{15}{16}$ | $1\frac{7}{8}$ | $1\frac{1}{16}$ | $\frac{15}{32}$ | $\frac{1}{8}$ |
| 12 | $9\frac{3}{4}$ | $7\frac{11}{16}$ | $5\frac{7}{8}$ | $4\frac{5}{16}$ | $3$ | $1\frac{29}{32}$ | $1\frac{3}{32}$ | $\frac{15}{32}$ | $\frac{1}{8}$ |
| 1  $0\frac14$ | $9\frac{15}{16}$ | $7\frac{27}{32}$ | $6$ | $4\frac{13}{32}$ | $3\frac{1}{16}$ | $1\frac{31}{32}$ | $1\frac{3}{32}$ | $\frac{1}{2}$ | $\frac{1}{8}$ |
| 1  $0\frac12$ | $10\frac{1}{8}$ | $8$ | $6\frac{1}{8}$ | $4\frac{1}{2}$ | $3\frac{1}{8}$ | $2$ | $1\frac{1}{8}$ | $\frac{1}{2}$ | $\frac{1}{8}$ |
| 1  $0\frac34$ | $10\frac{5}{16}$ | $8\frac{5}{32}$ | $6\frac{1}{4}$ | $4\frac{19}{32}$ | $3\frac{3}{16}$ | $2\frac{3}{32}$ | $1\frac{1}{8}$ | $\frac{1}{2}$ | $\frac{1}{8}$ |
| 1  1 | $10\frac{17}{32}$ | $8\frac{5}{16}$ | $6\frac{3}{8}$ | $4\frac{11}{16}$ | $3\frac{1}{4}$ | $2\frac{3}{32}$ | $1\frac{5}{32}$ | $\frac{17}{32}$ | $\frac{1}{8}$ |
| 1  $1\frac14$ | $10\frac{3}{4}$ | $8\frac{15}{32}$ | $6\frac{1}{2}$ | $4\frac{25}{32}$ | $3\frac{5}{16}$ | $2\frac{1}{8}$ | $1\frac{3}{16}$ | $\frac{17}{32}$ | $\frac{1}{8}$ |
| 1  $1\frac12$ | $10\frac{15}{16}$ | $8\frac{5}{8}$ | $6\frac{5}{8}$ | $4\frac{7}{8}$ | $3\frac{3}{8}$ | $2\frac{5}{32}$ | $1\frac{7}{32}$ | $\frac{17}{32}$ | $\frac{1}{8}$ |
| 1  $1\frac34$ | $11\frac{1}{8}$ | $8\frac{7}{8}$ | $6\frac{3}{4}$ | $4\frac{15}{16}$ | $3\frac{7}{16}$ | $2\frac{3}{16}$ | $1\frac{1}{4}$ | $\frac{9}{16}$ | $\frac{1}{8}$ |
| 1  2 | $11\frac{11}{32}$ | $8\frac{31}{32}$ | $6\frac{7}{8}$ | $5\frac{1}{32}$ | $3\frac{1}{2}$ | $2\frac{1}{4}$ | $1\frac{1}{4}$ | $\frac{9}{16}$ | $\frac{5}{32}$ |
| 1  $2\frac14$ | $11\frac{17}{32}$ | $9\frac{1}{8}$ | $7$ | $5\frac{1}{8}$ | $3\frac{9}{16}$ | $2\frac{9}{32}$ | $1\frac{9}{32}$ | $\frac{9}{16}$ | $\frac{5}{32}$ |
| 1  $2\frac12$ | $11\frac{3}{4}$ | $9\frac{3}{32}$ | $7\frac{3}{32}$ | $5\frac{1}{8}$ | $3\frac{5}{8}$ | $2\frac{5}{32}$ | $1\frac{5}{16}$ | $\frac{19}{32}$ | $\frac{5}{32}$ |
| 1  $2\frac34$ | $11\frac{15}{16}$ | $9\frac{7}{16}$ | $7\frac{7}{32}$ | $5\frac{5}{16}$ | $3\frac{11}{16}$ | $2\frac{3}{8}$ | $1\frac{5}{16}$ | $\frac{19}{32}$ | $\frac{5}{32}$ |
| 1  3 | $12\frac{5}{32}$ | $9\frac{19}{32}$ | $7\frac{11}{32}$ | $5\frac{13}{32}$ | $3\frac{3}{4}$ | $2\frac{13}{32}$ | $1\frac{11}{32}$ | $\frac{19}{32}$ | $\frac{5}{32}$ |
| 1  $3\frac14$ | $12\frac{11}{32}$ | $9\frac{3}{4}$ | $7\frac{15}{32}$ | $5\frac{1}{2}$ | $3\frac{13}{16}$ | $2\frac{9}{16}$ | $1\frac{3}{8}$ | $\frac{19}{32}$ | $\frac{5}{32}$ |
| 1  $3\frac12$ | $12\frac{9}{16}$ | $9\frac{15}{16}$ | $7\frac{19}{32}$ | $5\frac{19}{32}$ | $3\frac{7}{8}$ | $2\frac{1}{2}$ | $1\frac{13}{32}$ | $\frac{5}{8}$ | $\frac{5}{32}$ |
| 1  $3\frac34$ | $12\frac{3}{4}$ | $10\frac{3}{32}$ | $7\frac{23}{32}$ | $5\frac{21}{32}$ | $3\frac{15}{16}$ | $2\frac{17}{32}$ | $1\frac{13}{32}$ | $\frac{5}{8}$ | $\frac{5}{32}$ |
| 1  4 | $12\frac{31}{32}$ | $10\frac{1}{4}$ | $7\frac{27}{32}$ | $5\frac{3}{4}$ | $4$ | $2\frac{9}{16}$ | $1\frac{7}{16}$ | $\frac{5}{8}$ | $\frac{5}{32}$ |
| 1  $4\frac14$ | $13\frac{5}{32}$ | $10\frac{13}{32}$ | $7\frac{31}{32}$ | $5\frac{27}{32}$ | $4\frac{1}{16}$ | $2\frac{19}{32}$ | $1\frac{15}{32}$ | $\frac{21}{32}$ | $\frac{5}{32}$ |
| 1  $4\frac12$ | $13\frac{3}{8}$ | $10\frac{9}{16}$ | $8\frac{3}{32}$ | $5\frac{15}{16}$ | $4\frac{1}{8}$ | $2\frac{5}{8}$ | $1\frac{1}{2}$ | $\frac{21}{32}$ | $\frac{5}{32}$ |
| 1  $4\frac34$ | $13\frac{9}{16}$ | $10\frac{28}{32}$ | $8\frac{5}{32}$ | $6\frac{1}{32}$ | $4\frac{3}{16}$ | $2\frac{11}{16}$ | $1\frac{1}{2}$ | $\frac{11}{16}$ | $\frac{5}{32}$ |
| 1  5 | $13\frac{25}{32}$ | $10\frac{7}{8}$ | $8\frac{11}{32}$ | $6\frac{1}{8}$ | $4\frac{1}{4}$ | $2\frac{23}{32}$ | $1\frac{17}{32}$ | $\frac{11}{16}$ | $\frac{5}{32}$ |
| 1  $5\frac14$ | $13\frac{31}{32}$ | $11\frac{1}{32}$ | $8\frac{7}{16}$ | $6\frac{7}{32}$ | $4\frac{5}{16}$ | $2\frac{3}{4}$ | $1\frac{9}{16}$ | $\frac{11}{16}$ | $\frac{8}{16}$ |
| 1  $5\frac12$ | $14\frac{3}{16}$ | $11\frac{3}{16}$ | $8\frac{9}{16}$ | $6\frac{5}{16}$ | $4\frac{3}{8}$ | $2\frac{13}{16}$ | $1\frac{9}{16}$ | $\frac{11}{16}$ | $\frac{8}{16}$ |
| 1  $5\frac34$ | $14\frac{3}{8}$ | $11\frac{3}{8}$ | $8\frac{11}{16}$ | $6\frac{13}{32}$ | $4\frac{7}{16}$ | $2\frac{27}{32}$ | $1\frac{19}{32}$ | $\frac{23}{32}$ | $\frac{8}{16}$ |
| 1  6 | $14\frac{19}{32}$ | $11\frac{17}{32}$ | $8\frac{13}{16}$ | $6\frac{15}{32}$ | $4\frac{1}{2}$ | $2\frac{7}{8}$ | $1\frac{5}{8}$ | $\frac{23}{32}$ | $\frac{8}{16}$ |

### Ordinates of Parabolas for Crane Girder Design — 2

| | Ordinates | | | | | | | | |
|---|---|---|---|---|---|---|---|---|---|
| H | 9 | 8 | 7 | 6 | 5 | 4 | 3 | 2 | 1 |
| Ft. Ins. | Ft. Ins. | Ft. Ins. | Ft. Ins. | Ins. | Ins. | Ins. | Ins. | Ins. | Ins. |
| 1 6 | 1 2 19/32 | 11 17/32 | 8 13/16 | 6 15/32 | 4 1/2 | 2 7/8 | 1 5/8 | 23/32 | 3/16 |
| 1 6 1/4 | 1 2 25/32 | 11 11/16 | 8 15/16 | 6 9/16 | 4 9/16 | 2 15/16 | 1 21/32 | 23/32 | 3/16 |
| 1 6 1/2 | 1 3 | 11 27/32 | 9 1/16 | 6 21/32 | 4 5/8 | 2 31/32 | 1 21/32 | 3/4 | 3/16 |
| 1 6 3/4 | 1 3 3/16 | 1 0 | 9 3/16 | 6 3/4 | 4 11/16 | 3 | 1 11/16 | 3/4 | 3/16 |
| 1 7 | 1 3 13/32 | 1 0 5/32 | 9 5/16 | 6 27/32 | 4 3/4 | 3 1/32 | 1 23/32 | 3/4 | 3/16 |
| 1 7 1/4 | 1 3 19/32 | 1 0 5/16 | 9 7/16 | 6 15/16 | 4 13/16 | 3 3/32 | 1 23/32 | 3/4 | 3/16 |
| 1 7 1/2 | 1 3 25/32 | 1 0 15/32 | 9 9/16 | 7 | 4 7/8 | 3 1/8 | 1 3/4 | 25/32 | 3/16 |
| 1 7 3/4 | 1 4 | 1 0 21/32 | 9 11/16 | 7 1/8 | 4 15/16 | 3 5/32 | 1 25/32 | 25/32 | 3/16 |
| 1 8 | 1 4 3/16 | 1 0 13/16 | 9 13/16 | 7 3/16 | 5 | 3 3/16 | 1 13/16 | 13/16 | 3/16 |
| 1 8 1/4 | 1 4 13/32 | 1 0 31/32 | 9 15/16 | 7 9/32 | 5 1/16 | 3 1/4 | 1 13/16 | 13/16 | 3/16 |
| 1 8 1/2 | 1 4 19/32 | 1 1/8 | 10 1/32 | 7 3/8 | 5 1/8 | 3 9/32 | 1 27/32 | 13/16 | 7/32 |
| 1 8 3/4 | 1 4 13/16 | 1 9/32 | 10 5/32 | 7 15/32 | 5 3/16 | 3 5/16 | 1 7/8 | 27/32 | 7/32 |
| 1 9 | 1 5 | 1 7/16 | 10 9/32 | 7 9/16 | 5 1/4 | 3 3/8 | 1 7/8 | 27/32 | 7/32 |
| 1 9 1/4 | 1 5 7/32 | 1 19/32 | 10 13/32 | 7 21/32 | 5 5/16 | 3 13/32 | 1 29/32 | 27/32 | 7/32 |
| 1 9 1/2 | 1 5 13/32 | 1 3/4 | 10 17/32 | 7 3/4 | 5 3/8 | 3 7/16 | 1 15/16 | 7/8 | 7/32 |
| 1 9 3/4 | 1 5 5/8 | 1 15/16 | 10 21/32 | 7 27/32 | 5 7/16 | 3 15/32 | 1 31/32 | 7/8 | 7/32 |
| 1 10 | 1 5 13/16 | 1 2 3/32 | 10 25/32 | 7 15/16 | 5 1/2 | 3 17/32 | 1 31/32 | 7/8 | 7/32 |
| 1 10 1/4 | 1 6 1/32 | 1 2 1/4 | 10 29/32 | 8 | 5 9/16 | 3 9/16 | 2 | 29/32 | 7/32 |
| 1 10 1/2 | 1 6 7/32 | 1 2 7/16 | 11 1/32 | 8 3/32 | 5 5/8 | 3 19/32 | 2 1/32 | 29/32 | 7/32 |
| 1 10 3/4 | 1 6 7/16 | 1 2 9/16 | 11 5/32 | 8 3/16 | 5 11/16 | 3 5/8 | 2 1/16 | 29/32 | 7/32 |
| 1 11 | 1 6 5/8 | 1 2 23/32 | 11 9/32 | 8 9/32 | 5 3/4 | 3 11/16 | 2 1/16 | 15/16 | 7/32 |
| 1 11 1/4 | 1 6 15/16 | 1 2 7/8 | 11 13/32 | 8 3/8 | 5 13/16 | 3 23/32 | 2 3/32 | 15/16 | 7/32 |
| 1 11 1/2 | 1 7 1/32 | 1 3 1/32 | 11 17/32 | 8 15/32 | 5 7/8 | 3 3/4 | 2 1/8 | 15/16 | 1/4 |
| 1 11 3/4 | 1 7 1/4 | 1 3 3/16 | 11 5/8 | 8 9/16 | 5 15/16 | 3 13/16 | 2 1/8 | 31/32 | 1/4 |
| 2 0 | 1 7 7/16 | 1 3 3/8 | 11 3/4 | 8 21/32 | 6 | 3 27/32 | 2 5/32 | 31/32 | 1/4 |
| 2 0 1/4 | 1 7 21/32 | 1 3 17/32 | 11 7/8 | 8 23/32 | 6 1/16 | 3 7/8 | 2 3/16 | 31/32 | 1/4 |
| 2 0 1/2 | 1 7 27/32 | 1 3 11/16 | 1 0 | 8 13/16 | 6 1/8 | 3 15/16 | 2 3/16 | 31/32 | 1/4 |
| 2 0 3/4 | 1 8 1/32 | 1 3 27/32 | 1 1/8 | 8 29/32 | 6 3/16 | 3 31/32 | 2 7/32 | 1 | 1/4 |
| 2 1 | 1 8 1/4 | 1 4 | 1 1/4 | 9 | 6 1/4 | 4 | 2 1/4 | 1 | 1/4 |
| 2 1 1/4 | 1 8 15/32 | 1 4 5/32 | 1 3/8 | 9 3/32 | 6 5/16 | 4 1/32 | 2 9/32 | 1 | 1/4 |
| 2 1 1/2 | 1 8 21/32 | 1 4 5/16 | 1 1/2 | 9 3/16 | 6 3/8 | 4 3/32 | 2 9/32 | 1 1/32 | 1/4 |
| 2 1 3/4 | 1 8 27/32 | 1 4 15/32 | 1 5/8 | 9 9/32 | 6 7/16 | 4 1/8 | 2 5/16 | 1 1/32 | 1/4 |
| 2 2 | 1 9 1/16 | 1 4 21/32 | 1 3/4 | 9 3/8 | 6 1/2 | 4 5/32 | 2 11/32 | 1 1/32 | 1/4 |
| 2 2 1/4 | 1 9 1/4 | 1 4 13/16 | 1 7/8 | 9 15/32 | 6 9/16 | 4 7/32 | 2 3/8 | 1 1/16 | 1/4 |
| 2 2 1/2 | 1 9 15/32 | 1 4 31/32 | 1 1 | 9 9/16 | 6 11/16 | 4 9/32 | 2 13/32 | 1 1/16 | 1/4 |
| 2 2 3/4 | 1 9 21/32 | 1 5 1/8 | 1 1 1/8 | 9 5/8 | 6 11/16 | 4 9/32 | 2 13/32 | 1 1/16 | 9/32 |
| 2 3 | 1 9 7/8 | 1 5 9/32 | 1 1 1/4 | 9 23/32 | 6 3/4 | 4 5/16 | 2 7/16 | 1 3/32 | 9/32 |
| 2 3 1/4 | 1 10 1/16 | 1 5 7/16 | 1 1 11/32 | 9 13/16 | 6 13/16 | 4 3/8 | 2 15/32 | 1 3/32 | 9/32 |
| 2 3 1/2 | 1 10 9/32 | 1 5 19/32 | 1 1 15/32 | 9 29/32 | 6 7/8 | 4 13/32 | 2 15/32 | 1 3/32 | 9/32 |
| 2 3 3/4 | 1 10 15/32 | 1 5 3/4 | 1 1 19/32 | 10 | 6 15/16 | 4 7/16 | 2 1/2 | 1 1/8 | 9/32 |
| 2 4 | 1 10 11/16 | 1 5 15/16 | 1 1 23/32 | 10 3/32 | 7 | 4 15/32 | 2 17/32 | 1 1/8 | 9/32 |
| 2 4 1/4 | 1 10 7/8 | 1 6 3/32 | 1 1 27/32 | 10 3/16 | 7 1/16 | 4 17/32 | 2 17/32 | 1 1/8 | 9/32 |
| 2 4 1/2 | 1 11 3/32 | 1 6 1/4 | 1 1 31/32 | 10 1/4 | 7 1/8 | 4 9/16 | 2 9/16 | 1 5/32 | 9/32 |
| 2 4 3/4 | 1 11 9/32 | 1 6 13/32 | 1 2 3/32 | 10 11/32 | 7 3/16 | 4 19/32 | 2 19/32 | 1 5/32 | 9/32 |
| 2 5 | 1 11 1/2 | 1 6 9/16 | 1 2 7/32 | 10 7/16 | 7 1/4 | 4 21/32 | 2 5/8 | 1 5/32 | 9/32 |
| 2 5 1/4 | 1 11 11/16 | 1 6 23/32 | 1 2 11/32 | 10 17/32 | 7 5/16 | 4 11/16 | 2 5/8 | 1 3/16 | 9/32 |
| 2 5 1/2 | 1 11 29/32 | 1 6 7/8 | 1 2 15/32 | 10 5/8 | 7 3/8 | 4 23/32 | 2 21/32 | 1 3/16 | 5/16 |
| 2 5 3/4 | 2 0 3/32 | 1 7 1/32 | 1 2 19/32 | 10 23/32 | 7 7/16 | 4 3/4 | 2 11/16 | 1 3/16 | 5/16 |
| 2 6 | 2 0 5/16 | 1 7 7/32 | 1 2 23/32 | 10 13/16 | 7 1/2 | 4 13/16 | 2 23/32 | 1 7/32 | 5/16 |

general factor of safety of 5 is the most usual and desirable in crane work, and if that factor of safety is adopted, the working stress should be anywhere from 11,000 to 12,000 pounds per square inch.

**Strength of Channels.** — Experiments on standard channels carried out by Bach (published in 1909) show that the regular bending formula for beams freely supported at their ends and loaded in the center gives too high a value for the strength of structural channels. The experiments show that the amount by which

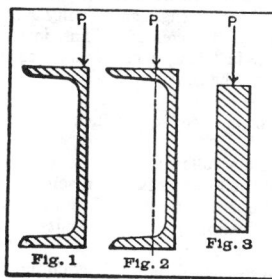

Fig. 1          Fig. 2          Fig. 3

the value obtained from the formula is greater than that obtained by experiments, is, for channels 4¾ inches high, 7 per cent; for channels 8¾ inches high, 18 per cent; and for channels 11¾ inches high, 26 per cent. These values are those found when the load is assumed to be applied in the center line of the web of the channel as shown in Fig. 1. If the load is placed along the line of the vertical neutral axis of the channel as shown in Fig. 2, the permissible load according to the beam formula is 10, 25.5 and 34 per cent greater than that shown by the experiments. These experiments, therefore, indicate that when the usual formulas are employed in calculations, for channels or other structural shapes, a liberal factor of safety should be allowed in order to compensate for the difference of the results given by the formula and those of actual experiments. It should be noted that the formula for bending is fully correct whenever the section of the member is such that the load is fully distributed over the whole sectional area, as in a rectangular section, Fig. 3; but in the case of channels as well as many other structural shapes, the load is not, as a rule, properly distributed over the whole section, but stresses certain portions of the section in a higher degree than others.

**Deflection as a Limiting Factor in Beam Design.** — For some applications, a beam must be stronger than required by the maximum load it is to support, in order to prevent excessive deflection. Since maximum allowable deflections for such cases vary widely for different classes of service, a general formula for determining them cannot be given. When exceptionally stiff girders are required, one rule is to limit the deflection to 1 inch per 100 feet of span; hence, if $l$ = length of span in inches, deflection = $l \div 1200$. According to another formula, deflection limit = $l \div 360$ where beams are adjacent to materials like plaster which would be broken by excessive beam deflection. Some machine parts of the beam type must be very rigid to maintain alignment under load. For example, the deflection of a locomotive guide-bar may be limited to 0.010 inch or less. These examples merely illustrate variations in practice. It is impracticable to give general formulas for determining the allowable deflection in any case, because the allowable amount depends upon the conditions governing each class of work.

*Procedure in Designing for Deflection:* Assume that a deflection equal to $l \div 1200$ is to be the limiting factor in selecting a wide-flange (W-shape) beam having a span length of 144 inches. Supports are at both ends and load at center is 15,000 pounds. Deflection $y$ is to be limited to $144 \div 1200 = 0.12$ inch. According to the formula on page 404 (Case 2), in which $W$ = load on beam in pounds, $l$ = length of span in inches, $E$ = modulus of elasticity of material, $I$ = moment of inertia of cross section.

$$\text{Deflection } y = \frac{Wl^3}{48EI}; \text{ hence } I = \frac{Wl^3}{48yE} = \frac{15,000 \times 144^3}{48 \times 0.12 \times 29,000,000} = 268.1$$

A structural wide-flange beam having a depth of 12 inches and weighing 36 pounds per foot has a moment of inertia $I$ of 281 and a section modulus ($Z$ or $S$) of 46.0 (see table, page 394). Checking now for maximum stress $s$ (Case 2, page 404):

$$s = \frac{Wl}{4Z} = \frac{15,000 \times 144}{4 \times 46.0} = 11,740 \text{ lbs. per sq. in.}$$

Although deflection is the limiting factor in this case, the maximum stress is checked to make sure that it is within the allowable limit. As the limiting deflection is decreased, for a given load and length of span, the beam strength and rigidity must be increased, and, consequently, the maximum stress is decreased. Thus, in the preceding example, if the maximum deflection is 0.08 inch instead of 0.12 inch, then the calculated value for the moment of inertia $I$ will be 402; hence a W 12 × 53 beam having an $I$ value of 426 could be used (nearest value above 402). The maximum stress then would be reduced to 7640 pounds per square inch and the calculated deflection is 0.076 inch.

A similar example using metric SI units is as follows. Assume that a deflection equal to $l \div 1000$ millimeters is to be the limiting factor in selecting a W-beam having a span length of 5 meters. Supports are at both ends and the load at the center is 30 kilonewtons. Deflection $y$ is to be limited to $5000 \div 1000 = 5$ millimeters. The formula on page 404 (Case 2) is applied, and $W$ = load on beam in newtons; $l$ = length of span in mm; $E$ = modulus of elasticity (assume 200,000 N/mm² in this example); and $I$ = moment of inertia of cross-section in millimeters⁴. Thus,

$$\text{Deflection } y = \frac{Wl^3}{48EI};$$

hence

$$I = \frac{Wl^3}{48yE} = \frac{30,000 \times 5000^3}{48 \times 5 \times 200,000} = 78,125,000 \text{ mm}^4$$

Although deflection is the limiting factor in this case, the maximum stress is checked to make sure that it is within the allowable limit, using the formula from page 404 (Case 2):

$$s = \frac{Wl}{4Z}$$

The units of $s$ are newtons per square millimeter; $W$ is the load in newtons; $l$ is the length in mm; and $Z$ = section modulus of the cross-section of the beam = $I \div$ distance from neutral axis to extreme fiber in mm.

**Curved Beams.** — The formula $S = Mc/I$ used to compute stresses due to bending of beams is based on the assumption that the beams are straight before any loads are applied. In the case of beams having initial curvature, however, the stresses may be considerably higher than predicted by the ordinary straight-beam formula since the effect of initial curvature is to shift the neutral axis of a curved member in from the gravity axis toward the center of curvature (the concave side of the beam). This shift in the position of the neutral axis causes an increase in the stress on the concave side of the beam and decreases the stress at the outside fibers.

Hooks, press frames, and other machine members which as a rule have a rather pronounced initial curvature may have a maximum stress at the inside fibers of up to about 3½ times that predicted by the ordinary straight-beam formula.

*Stress Correction Factors for Curved Beams:* A simple method for determining the maximum fiber stress due to bending of curved members consists of (1) calculating the maximum stress using the straight-beam formula $S = Mc/I$; and (2) multiplying the calculated stress by a stress correction factor. The table on page 425 gives

## Values of the Stress Correction Factor $K$ for Various Curved Beam Sections

| Section | $R/c$ | Factor $K$ In-side Fiber | Factor $K$ Out-side Fiber | $y_0$* | Section | $R/c$ | Factor $K$ In-side Fiber | Factor $K$ Out-side Fiber | $y_0$* |
|---|---|---|---|---|---|---|---|---|---|
| (circular section) | 1.2 | 3.41 | .54 | .224 R | (flanged section $4\frac{1}{2}t$, $\frac{3}{2}t$, $4t$) | 1.2 | 3.63 | .58 | .418 R |
| | 1.4 | 2.40 | .60 | .151 R | | 1.4 | 2.54 | .63 | .299 R |
| | 1.6 | 1.96 | .65 | .108 R | | 1.6 | 2.14 | .67 | .229 R |
| | 1.8 | 1.75 | .68 | .084 R | | 1.8 | 1.89 | .70 | .183 R |
| | 2.0 | 1.62 | .71 | .069 R | | 2.0 | 1.73 | .72 | .149 R |
| | 3.0 | 1.33 | .79 | .030 R | | 3.0 | 1.41 | .79 | .069 R |
| | 4.0 | 1.23 | .84 | .016 R | | 4.0 | 1.29 | .83 | .040 R |
| | 6.0 | 1.14 | .89 | .0070 R | | 6.0 | 1.18 | .88 | .018 R |
| | 8.0 | 1.10 | .91 | .0039 R | | 8.0 | 1.13 | .91 | .010 R |
| | 10.0 | 1.08 | .93 | .0025 R | | 10.0 | 1.10 | .92 | .0065 R |
| (rectangular section) | 1.2 | 2.89 | .57 | .305 R | (I-beam section $3t$, $2t$, $6t$) | 1.2 | 3.55 | .67 | .409 R |
| | 1.4 | 2.13 | .63 | .204 R | | 1.4 | 2.48 | .72 | .292 R |
| | 1.6 | 1.79 | .67 | .149 R | | 1.6 | 2.07 | .76 | .224 R |
| | 1.8 | 1.63 | .70 | .112 R | | 1.8 | 1.83 | .78 | .178 R |
| | 2.0 | 1.52 | .73 | .090 R | | 2.0 | 1.69 | .80 | .144 R |
| | 3.0 | 1.30 | .81 | .041 R | | 3.0 | 1.38 | .86 | .067 R |
| | 4.0 | 1.20 | .85 | .021 R | | 4.0 | 1.26 | .89 | .038 R |
| | 6.0 | 1.12 | .90 | .0093 R | | 6.0 | 1.15 | .92 | .018 R |
| | 8.0 | 1.09 | .92 | .0052 R | | 8.0 | 1.10 | .94 | .010 R |
| | 10.0 | 1.07 | .94 | .0033 R | | 10.0 | 1.08 | .95 | .0065 R |
| (trapezoidal section $b$, $2b$) | 1.2 | 3.01 | .54 | .336 R | (I-beam section $4t$, $3t$) | 1.2 | 2.52 | .67 | .408 R |
| | 1.4 | 2.18 | .60 | .229 R | | 1.4 | 1.90 | .71 | .285 R |
| | 1.6 | 1.87 | .65 | .168 R | | 1.6 | 1.63 | .75 | .208 R |
| | 1.8 | 1.69 | .68 | .128 R | | 1.8 | 1.50 | .77 | .160 R |
| | 2.0 | 1.58 | .71 | .102 R | | 2.0 | 1.41 | .79 | .127 R |
| | 3.0 | 1.33 | .80 | .046 R | | 3.0 | 1.23 | .86 | .058 R |
| | 4.0 | 1.23 | .84 | .024 R | | 4.0 | 1.16 | .89 | .030 R |
| | 6.0 | 1.13 | .88 | .011 R | | 6.0 | 1.10 | .92 | .013 R |
| | 8.0 | 1.10 | .91 | .0060 R | | 8.0 | 1.07 | .94 | .0076 R |
| | 10.0 | 1.08 | .93 | .0039 R | | 10.0 | 1.05 | .95 | .0048 R |
| (trapezoidal section $3b$, $2b$) | 1.2 | 3.09 | .56 | .336 R | (hollow circular section $2d$, $d$) | 1.2 | 3.28 | .58 | .269 R |
| | 1.4 | 2.25 | .62 | .229 R | | 1.4 | 2.31 | .64 | .182 R |
| | 1.6 | 1.91 | .66 | .168 R | | 1.6 | 1.89 | .68 | .134 R |
| | 1.8 | 1.73 | .70 | .128 R | | 1.8 | 1.70 | .71 | .104 R |
| | 2.0 | 1.61 | .73 | .102 R | | 2.0 | 1.57 | .73 | .083 R |
| | 3.0 | 1.37 | .81 | .046 R | | 3.0 | 1.31 | .81 | .038 R |
| | 4.0 | 1.26 | .86 | .024 R | | 4.0 | 1.21 | .85 | .020 R |
| | 6.0 | 1.17 | .91 | .011 R | | 6.0 | 1.13 | .90 | .0087 R |
| | 8.0 | 1.13 | .94 | .0060 R | | 8.0 | 1.10 | .92 | .0049 R |
| | 10.0 | 1.11 | .95 | .0039 R | | 10.0 | 1.07 | .93 | .0031 R |
| (trapezoidal section $5b$, $4b$) | 1.2 | 3.14 | .52 | .352 R | (hollow rectangular section $4t$, $2t$) | 1.2 | 2.63 | .68 | .399 R |
| | 1.4 | 2.29 | .54 | .243 R | | 1.4 | 1.97 | .73 | .280 R |
| | 1.6 | 1.93 | .62 | .179 R | | 1.6 | 1.66 | .76 | .205 R |
| | 1.8 | 1.74 | .65 | .138 R | | 1.8 | 1.51 | .78 | .159 R |
| | 2.0 | 1.61 | .68 | .110 R | | 2.0 | 1.43 | .80 | .127 R |
| | 3.0 | 1.34 | .76 | .050 R | | 3.0 | 1.23 | .86 | .058 R |
| | 4.0 | 1.24 | .82 | .028 R | | 4.0 | 1.15 | 89 | .031 R |
| | 6.0 | 1.15 | .87 | .012 R | | 6.0 | 1.09 | .92 | .014 R |
| | 8.0 | 1.12 | .91 | .0060 R | | 8.0 | 1.07 | .94 | .0076 R |
| | 10.0 | 1.10 | .93 | .0039 R | | 10.0 | 1.06 | .95 | .0048 R |
| (triangular section $3b$) | 1.2 | 3.26 | .44 | .361 R | | | | | |
| | 1.4 | 2.39 | .50 | .251 R | | | | | |
| | 1.6 | 1.99 | .54 | .186 R | | | | | |
| | 1.8 | 1.78 | .57 | .144 R | | | | | |
| | 2.0 | 1.66 | .60 | .116 R | | | | | |
| | 3.0 | 1.37 | .70 | .052 R | | | | | |
| | 4.0 | 1.27 | .75 | .029 R | | | | | |
| | 6.0 | 1.16 | .82 | .013 R | | | | | |
| | 8.0 | 1.12 | .86 | .0060 R | | | | | |
| | 10.0 | 1.09 | .88 | .0039 R | | | | | |

*Example:* The fiber stresses of a curved rectangular beam are calculated as 5000 psi using the straight beam formula, $S = Mc/I$. If beam is 8 inches deep and its radius of curvature is 12 inches, what are the true stresses? $R/c = 12/4 = 3$. The factors in the table corresponding to $R/c = 3$ are 0.81 and 1.30. Outside fiber stress = 5000 × 0.81 = 4050 psi; inside fiber stress = 5000 × 1.30 = 6500 psi.

* $y_0$ is the distance from the centroidal axis to the neutral axis of curved beams subjected to pure bending and is measured from the centroidal axis toward the center of curvature.

stress correction factors for some of the common cross-sections and proportions used in the design of curved members.

An example in the application of the method using English units of measurement is given at the bottom of the table. A similar example using metric SI units is as follows: The fiber stresses of a curved rectangular beam are calculated as 40 newtons per millimeter², using the straight beam formula, $S = Mc/I$. If the beam is 150 mm deep and its radius of curvature is 300 mm, what are the true stresses? $R/c = 300/75 = 4$. From the table on page 425, the $K$ factors corresponding to $R/c = 4$ are 1.20 and 0.85. Thus, the inside fiber stress is $40 \times 1.20 = 48$ N/mm² = 48 megapascals; and the outside fiber stress is $40 \times 0.85 = 34$ N/mm² = 34 megapascals.

*Approximate Formula for Stress Correction Factor:* The stress correction factors given in the table on page 425 were determined by Wilson and Quereau and published in the University of Illinois Engineering Experiment Station Circular No. 16, " A Simple Method of Determining Stress in Curved Flexural Members." In this same publication the authors indicate that the following empirical formula may be used to calculate the value of the stress correction factor for the *inside* fibers of sections not covered by the tabular data to within 5 per cent accuracy except in the case of triangular sections where up to 10 per cent deviation may be expected. However, for most engineering calculations, this should prove to be a satisfactory formula for general use in determining the factor for the inside fibers.

$$K = 1.00 + 0.5 \frac{I}{bc^2}\left[\frac{1}{R - c} + \frac{1}{R}\right]$$

(Use 1.05 instead of 0.5 in this formula for circular and elliptical sections.)

> $I$ = Moment of inertia of section about centroidal axis;
> $b$ = maximum width of section;
> $c$ = distance from centroidal axis to inside fiber, i.e., to the extreme fiber nearest the center of curvature;
> $R$ = radius of curvature of centroidal axis of beam.

*Example:* On the accompanying diagram are shown the dimensions of a clamp frame of rectangular cross-section. Determine the maximum stress at points $A$ and $B$ due to a clamping force of 1000 pounds.

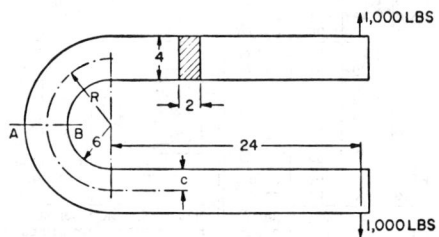

The cross-sectional area = $2 \times 4 = 8$ square inches; the bending moment at section $AB$ is 1000 $(24 + 6 + 2) = 32,000$ inch pounds; the distance from the center of gravity of the section at $AB$ to point $B$ is $c = 2$ inches; and the moment of inertia of the section is, using the formula on page 370, $2 \times (4)^3 \div 12 = 10.666$ inches⁴.

Using the straight-beam formula, page 424, the stress at points $A$ and $B$ due to the bending moment is:

$$S = \frac{Mc}{I} = \frac{32,000 \times 2}{10.666} = 6000 \text{ psi}$$

The stress at $A$ is a compressive stress of 6000 psi while that at $B$ is a tensile stress of 6000 psi.

These values must be corrected to account for the curvature effect. In the table on page 425 for $R/c = (6 + 2)/(2) = 4$, the value of $K$ is found to be 1.20 and 0.85 for points $B$ and $A$ respectively. Thus, the actual stress due to bending at point $B$ is $1.20 \times 6000 = 7200$ psi in tension while the stress at point $A$ is $0.85 \times 6000 = 5100$ psi in compression.

To these stresses at $A$ and $B$ must be added, algebraically, the direct stress at section $AB$ due to the 1000-pound clamping force. The direct stress on section $AB$ will be a tensile stress equal to the clamping force divided by the section area. Thus $1000 \div 8 = 125$ psi in tension.

The maximum unit stress at $A$ is, therefore, $5100 - 125 = 4975$ psi in compression while the maximum unit stress at $B$ is $7200 + 125 = 7325$ psi in tension.

The following is a similar calculation using metric SI units, assuming that it is required to determine the maximum stress at points $A$ and $B$ due to clamping force of 4 kilonewtons acting on the frame. The frame cross-section is 50 by 100 millimeters, the radius $R = 200$ mm, and the length of the straight portions is 600 mm. Thus, the cross-sectional area $= 50 \times 100 = 5000 \text{ mm}^2$; the bending moment at $AB$ is $4000(600 + 200) = 3,200,000$ newton-millimeters; the distance from the center of gravity of the section at $AB$ to point $B$ is $c = 50$ mm; and the moment of inertia of the section is, using the formula on page 370, $50 \times (100)^3 \div 12 = 4,170,000 \text{ mm}^4$.

Using the straight-beam formula, page 424, the stress at points $A$ and $B$ due to the bending moment is:

$$s = \frac{Mc}{I} = \frac{3,200,000 \times 50}{4,170,000}$$

$$= 38.4 \text{ newtons per millimeter}^2 = 38.4 \text{ megapascals.}$$

The stress at $A$ is a compressive stress of 38.4 $\text{N/mm}^2$, while that at $B$ is a tensile stress of 38.4 $\text{N/mm}^2$. These values must be corrected to account for the curvature effect. From the table on page 425, the $K$ factors are 1.20 and 0.85 for points $A$ and $B$ respectively, derived from $R/c = 200/50 = 4$. Thus, the actual stress due to bending at point $B$ is $1.20 \times 38.4 = 46.1 \text{ N/mm}^2$ (46.1 megapascals) in tension; and the stress at point $A$ is $0.85 \times 38.4 = 32.6$ $\text{N/mm}^2$ (32.6 megapascals) in compression.

To these stresses at $A$ and $B$ must be added, algebraically, the direct stress at section $AB$ due to the 4 kN clamping force. The direct stress on section $AB$ will be a tensile stress equal to the clamping force divided by the section area. Thus, $4000/5000 = 0.8 \text{ N/mm}^2$. The maximum unit stress at $A$ is, therefore, $32.61 - 0.8 = 31.8 \text{ N/mm}^2$ (31.8 megapascals) in compression, and the maximum unit stress at $B$ is $46.1 + 0.8 = 46.9 \text{ N/mm}^2$ (46.9 megapascals) in tension.

## Columns

**Strength of Columns or Struts.** — Structural members which are subject to compression may be so long in proportion to the diameter or lateral dimensions that failure may be the result (1) of both compression and bending or (2) of bending or buckling to such a degree that compressive stress may be ignored. In such cases, the *slenderness ratio* is important. This ratio equals the length $l$ of the column in inches divided by the least radius of gyration $r$ of the cross-section. Various formulas have been used for designing columns which are too slender to be designed for compression only.

**Rankine or Gordon Formula.** — This formula is generally applied when slenderness ratios range between 20 and 100, and sometimes for ratios up to 120. The notation, in English and metric SI units of measurement, is given on page 430.

$$p = \frac{S}{1 + K\left(\frac{l}{r}\right)^2} = \text{ultimate load, lbs. per sq. in.}$$

Factor $K$ may be established by tests with a given material and end condition, and for the probable range of $l/r$. If determined by calculation, $K = S/C\pi^2E$. Factor $C$ equals 1 for either rounded or pivoted column ends, 4 for fixed ends, and 1 to 4 for square flat ends. The factors 25,000, 12,500, etc., in the Rankine formulas arranged as on page 430 equal $1/K$, and have been used extensively.

**Straight-line Formula.** — This general type of formula is often used in designing compression members for buildings, bridges, or similar structural work. It is convenient especially in designing a number of columns which are made of the same material but vary in size, assuming that factor $B$ is known. This factor is determined by tests.

$$p = S_y - B\left(\frac{l}{r}\right) = \text{ultimate load, lbs. per sq. in.}$$

$S_y$ equals yield point, lbs. per square inch, and factor $B$ ranges from 50 to 100. Safe unit stress = $p$/factor of safety.

**Formulas of American Railway Engineering Association.** — The formulas which follow apply to structural steel having an ultimate strength of 60,000 to 72,000 pounds per square inch.

For building columns having $l/r$ ratios not greater than 120, allowable unit stress = $17,000 - 0.485\, l^2/r^2$. For columns having $l/r$ ratios greater than 120, allowable unit stress = $\dfrac{18,000}{1 + \dfrac{l^2}{18,000\, r^2}}$

For bridge compression members centrally loaded and with values of $l/r$ not greater than 140:

$$\text{Allowable unit stress, riveted ends} = 15,000 - \frac{1}{4}\frac{l^2}{r^2}$$

$$\text{Allowable unit stress, pin ends} = 15,000 - \frac{1}{3}\frac{l^2}{r^2}$$

**American Institute of Steel Construction.** — For main or secondary compression members with $l/r$ ratios up to 120, safe unit stress = $17,000 - 0.485\, l^2/r^2$. For columns and bracing or other secondary members with $l/r$ ratios above 120,

$$\text{Safe unit stress, psi} = \frac{18,000}{1 + \dfrac{l^2}{18,000\, r^2}} \text{ for bracing and secondary members. For}$$

$$\text{main members, safe unit stress, psi} = \frac{18,000}{1 + \dfrac{l^2}{18,000\, r^2}} \times \left(1.6 - \frac{l/r}{200}\right)$$

*Pipe Columns:* Allowable concentric loads for steel pipe columns based on the above formulas are given in the table on page 431.

**Euler Formula.** — This formula is for columns which are so slender that bending or buckling action predominates and compressive stresses are not taken into account.

$$P = \frac{C\pi^2 IE}{l^2} = \text{total ultimate load, in pounds}$$

The notation, in English and metric SI units of measurement, is given on page 430. Factors $C$ for different end conditions are included in the Euler formulas at the bottom of the table. According to a series of experiments, Euler formulas should be used if the values of $l/r$ exceed the following ratios: Structural steel and flat ends, 195; hinged ends, 155; round ends, 120; cast iron with flat ends, 120; hinged ends, 100; round ends, 75; oak with flat ends, 130. The *critical slenderness ratio* which marks the dividing line between the shorter columns and those slender enough to warrant using the Euler formula, depends upon the column material and its end conditions. If the Euler formula is applied when the slenderness ratio is too small, the *calculated* ultimate strength will exceed the yield point of the material and, obviously, will be incorrect.

**Eccentrically Loaded Columns.** — In the application of the column formulas previously referred to, it is assumed that the action of the load coincides with the axis of the column. If the load is offset relative to the column axis, the column is said to be eccentrically loaded, and its strength is then calculated by using a modification of the Rankine formula, the quantity $cz/r^2$ being added to the denominator, as shown in the table on the next page. This modified formula is applicable to columns having a slenderness ratio varying from 20 or 30 to about 100.

**Machine Elements Subjected to Compressive Loads.** — As in the case of structural compression members, an unbraced machine member that is relatively slender, i.e., its length is more than, say, six times the least dimension perpendicular to its longitudinal axis is usually designed as a column, since failure due to overloading (assuming a compressive load centrally applied in an axial direction) may occur by buckling or a combination of buckling and compression rather than by direct compression alone. In the design of unbraced steel machine "columns" which are to carry compressive loads applied along their longitudinal axes, two formulas are in general use:

(Euler)        $$P_{cr} = \frac{s_y A r^2}{Q} \tag{1}$$

(J. B. Johnson)   $$P_{cr} = A s_y \left(1 - \frac{Q}{4 r^2}\right) \quad \text{where} \quad Q = \frac{s_y l^2}{n\pi^2 E} \tag{2 and 3}$$

In these formulas, $P_{cr}$ = critical load in pounds that would result in failure of the column; $A$ = cross-sectional area, square inches; $s_y$ = yield point of the material, pounds per square inch; $r$ = least radius of gyration of cross-section, inches; $E$ = modulus of elasticity, pounds per square inch; $l$ = column length, inches; and $n$ = a coefficient for end conditions. For both ends fixed, $n = 4$; for one end fixed, one end free, $n = 0.25$; for one end fixed and the other end free but guided, $n = 2$; for round or pinned ends, free but guided, $n = 1$; and for flat ends, $n = 1$ to 4. It should be noted that these values of $n$ represent ideal conditions that are seldom attained in practice; for example, for both ends fixed a value of $n = 3$ to 3.5 may be more realistic than $n = 4$.

If metric SI units are used in these formulas, $P_{cr}$ = critical load in newtons that would result in failure of the column; $A$ = cross-sectional area, square millimeters; $s_y$ = yield point of the material, newtons per square mm; $r$ = least radius of gyration of cross-section, mm; $E$ = modulus of elasticity newtons per square mm; $l$ = column length, mm; and $n$ = a coefficient for end conditions. The coefficients given are valid for calculations in metric units.

### Rankine's and Euler's Formulas for Columns

| Symbol | Quantity | English Unit | Metric SI Units |
|---|---|---|---|
| $p$ | Ultimate unit load | Lbs./sq. in. | Newtons/sq. mm. |
| $P$ | Total ultimate load | Pounds | Newtons |
| $S$ | Ultimate compressive strength of material | Lbs./sq. in. | Newtons/sq. mm. |
| $l$ | Length of column or strut | Inches | Millimeters |
| $r$ | Least radius of gyration | Inches | Millimeters |
| $I$ | Least moment of inertia | Inches$^4$ | Millimeters$^4$ |
| $r^2$ | Moment of inertia/area of section | Inches$^2$ | Millimeters$^2$ |
| $E$ | Modulus of elasticity of material | Lbs./sq. in. | Newtons/sq. mm. |
| $c$ | Distance from neutral axis of cross-section to side under compression | Inches | Millimeters |
| $z$ | Distance from axis of load to axis coinciding with center of gravity of cross-section | Inches | Millimeters |

#### Rankine's Formulas

| Material | Both Ends of Column Fixed | One End Fixed and One End Rounded | Both Ends Rounded |
|---|---|---|---|
| Steel......... | $p = \dfrac{S}{1 + \dfrac{l^2}{25,000\, r^2}}$ | $p = \dfrac{S}{1 + \dfrac{l^2}{12,500\, r^2}}$ | $p = \dfrac{S}{1 + \dfrac{l^2}{6250\, r^2}}$ |
| Cast Iron...... | $p = \dfrac{S}{1 + \dfrac{l^2}{5000\, r^2}}$ | $p = \dfrac{S}{1 + \dfrac{l^2}{2500\, r^2}}$ | $p = \dfrac{S}{1 + \dfrac{l^2}{1250\, r^2}}$ |
| Wrought Iron.. | $p = \dfrac{S}{1 + \dfrac{l^2}{35,000\, r^2}}$ | $p = \dfrac{S}{1 + \dfrac{l^2}{17,500\, r^2}}$ | $p = \dfrac{S}{1 + \dfrac{l^2}{8750\, r^2}}$ |
| Timber........ | $p = \dfrac{S}{1 + \dfrac{l^2}{3000\, r^2}}$ | $p = \dfrac{S}{1 + \dfrac{l^2}{1500\, r^2}}$ | $p = \dfrac{S}{1 + \dfrac{l^2}{750\, r^2}}$ |

#### Formulas Modified for Eccentrically Loaded Columns

| Material * | Both Ends of Column Fixed | One End Fixed and One End Rounded | Both Ends Rounded |
|---|---|---|---|
| Steel.......... | $p = \dfrac{S}{1 + \dfrac{l^2}{25,000\, r^2} + \dfrac{cz}{r^2}}$ | $p = \dfrac{S}{1 + \dfrac{l^2}{12,500\, r^2} + \dfrac{cz}{r^2}}$ | $p = \dfrac{S}{1 + \dfrac{l^2}{6250\, r^2} + \dfrac{cz}{r^2}}$ |

* For other materials such as cast iron, etc., use the Rankine formulas given in the upper table and add to the denominator the quantity $\dfrac{cz}{r^2}$.

#### Euler's Formulas for Slender Columns

| Both Ends of Column Fixed | One End Fixed and One End Rounded | Both Ends Rounded | One End Fixed and One End Free |
|---|---|---|---|
| $P = \dfrac{4\,\pi^2 IE}{l^2}$ | $P = \dfrac{2\,\pi^2 IE}{l^2}$ | $P = \dfrac{\pi^2 IE}{l^2}$ | $P = \dfrac{\pi^2 IE}{4\,l^2}$ |

*Allowable Working Loads for Columns:* To find the total allowable working load for a given section, divide the total ultimate load $P$ (or $p \times$ area), as found by the appropriate formula above, by a suitable factor of safety.

**Allowable Concentric Loads for Steel Pipe Columns†**

| STANDARD STEEL PIPE | | | | | | | | |
|---|---|---|---|---|---|---|---|---|
| | Nominal Diameter of Pipe, Inches | | | | | | | |
| | 12 | 10 | 8 | 6 | 5 | 4 | 3½ | 3 |
| Effective Length (*KL*), Feet* | Wall Thickness of Pipe, Inch | | | | | | | |
| | .375 | .365 | .322 | .280 | .258 | .237 | .226 | .216 |
| | Weight per Foot of Pipe, Pounds | | | | | | | |
| | 49.56 | 40.48 | 28.55 | 18.97 | 14.62 | 10.79 | 9.11 | 7.58 |
| Allowable Concentric Loads in Thousands of Pounds | | | | | | | | |
| 6 | 303 | 246 | 171 | 110 | 83 | 59 | 48 | 38 |
| 7 | 301 | 243 | 168 | 108 | 81 | 57 | 46 | 36 |
| 8 | 299 | 241 | 166 | 106 | 78 | 54 | 44 | 34 |
| 9 | 296 | 238 | 163 | 103 | 76 | 52 | 41 | 31 |
| 10 | 293 | 235 | 161 | 101 | 73 | 49 | 38 | 28 |
| 11 | 291 | 232 | 158 | 98 | 71 | 46 | 35 | 25 |
| 12 | 288 | 229 | 155 | 95 | 68 | 43 | 32 | 22 |
| 13 | 285 | 226 | 152 | 92 | 65 | 40 | 29 | 18.5 |
| 14 | 282 | 223 | 149 | 89 | 61 | 36 | 25 | 15.9 |
| 15 | 278 | 220 | 145 | 86 | 58 | 33 | 22 | 13.9 |
| 16 | 275 | 216 | 142 | 82 | 55 | 29 | 19.4 | 12.2 |
| 17 | 272 | 213 | 138 | 79 | 51 | 26 | 17.1 | 10.8 |
| 18 | 268 | 209 | 135 | 75 | 47 | 23 | 15.3 | 9.6 |
| 19 | 265 | 205 | 131 | 71 | 43 | 21 | 13.7 | 8.6 |
| 20 | 261 | 201 | 127 | 67 | 39 | 18.7 | 12.4 | .... |
| 21 | 257 | 197 | 123 | 63 | 36 | 17.0 | 11.2 | .... |
| 22 | 254 | 193 | 119 | 59 | 32 | 15.5 | 10.2 | .... |
| 23 | 250 | 189 | 115 | 55 | 30 | 14.1 | .... | .... |
| 24 | 246 | 185 | 111 | 51 | 27 | 13.0 | .... | .... |

| EXTRA STRONG STEEL PIPE | | | | | | | | |
|---|---|---|---|---|---|---|---|---|
| | Nominal Diameter of Pipe, Inches | | | | | | | |
| | 12 | 10 | 8 | 6 | 5 | 4 | 3½ | 3 |
| Effective Length (*KL*), Feet* | Wall Thickness of Pipe, Inch | | | | | | | |
| | .500 | .500 | .500 | .432 | .375 | .337 | .318 | .300 |
| | Weight per Foot of Pipe, Pounds | | | | | | | |
| | 65.42 | 54.74 | 43.39 | 28.57 | 20.78 | 14.98 | 12.51 | 10.25 |
| Allowable Concentric Loads in Thousands of Pounds | | | | | | | | |
| 6 | 400 | 332 | 259 | 166 | 118 | 81 | 66 | 52 |
| 7 | 397 | 329 | 255 | 162 | 114 | 78 | 63 | 48 |
| 8 | 394 | 325 | 251 | 159 | 111 | 75 | 59 | 45 |
| 9 | 390 | 321 | 247 | 155 | 107 | 71 | 55 | 41 |
| 10 | 387 | 318 | 243 | 151 | 103 | 67 | 51 | 37 |
| 11 | 383 | 314 | 239 | 146 | 99 | 63 | 47 | 33 |
| 12 | 379 | 309 | 234 | 142 | 95 | 59 | 43 | 28 |
| 13 | 375 | 305 | 229 | 137 | 91 | 54 | 38 | 24 |
| 14 | 371 | 301 | 224 | 132 | 86 | 49 | 33 | 21 |
| 15 | 367 | 296 | 219 | 127 | 81 | 44 | 29 | 17.9 |
| 16 | 363 | 291 | 214 | 122 | 76 | 39 | 25 | 15.7 |
| 17 | 358 | 286 | 209 | 116 | 71 | 34 | 22 | 13.9 |
| 18 | 353 | 281 | 203 | 111 | 65 | 31 | 20 | 12.4 |
| 19 | 349 | 276 | 197 | 105 | 59 | 28 | 18.0 | .... |
| 20 | 344 | 271 | 191 | 99 | 53 | 25 | 16.2 | .... |
| 21 | 339 | 265 | 185 | 93 | 48 | 23 | 14.7 | .... |
| 22 | 334 | 260 | 179 | 86 | 44 | 21 | .... | .... |
| 23 | 328 | 254 | 172 | 79 | 40 | 18.8 | .... | .... |
| 24 | 323 | 248 | 166 | 73 | 37 | 17.3 | .... | .... |

Footnotes appear at bottom of continued table on next page.

**Allowable Concentric Loads for Steel Pipe Columns† (Concluded)**

| Effective Length (KL), Feet* | DOUBLE-EXTRA STRONG STEEL PIPE | | | | | | |
|---|---|---|---|---|---|---|---|
| | Nominal Diameter of Pipe, Inches | | | | | | |
| | 12 | 10 | 8 | 6 | 5 | 4 | 3 |
| | Wall Thickness of Pipe, Inch | | | | | | |
| | 1.000 | 1.000 | .875 | .864 | .750 | .674 | .600 |
| | Weight per Foot of Pipe, Pounds | | | | | | |
| | 125.49 | 104.13 | 72.42 | 53.16 | 38.55 | 27.54 | 18.58 |
| | Allowable Concentric Loads in Thousands of Pounds | | | | | | |
| 6 | 766 | 630 | 431 | 306 | 216 | 147 | 91 |
| 7 | 760 | 623 | 424 | 299 | 209 | 140 | 84 |
| 8 | 753 | 616 | 417 | 292 | 202 | 133 | 77 |
| 9 | 747 | 608 | 410 | 284 | 195 | 126 | 69 |
| 10 | 739 | 601 | 403 | 275 | 187 | 118 | 60 |
| 11 | 732 | 592 | 395 | 266 | 178 | 109 | 51 |
| 12 | 724 | 584 | 387 | 257 | 170 | 100 | 43 |
| 13 | 716 | 575 | 378 | 247 | 160 | 91 | 37 |
| 14 | 707 | 566 | 369 | 237 | 151 | 81 | 32 |
| 15 | 699 | 557 | 360 | 227 | 141 | 70 | 28 |
| 16 | 690 | 547 | 351 | 216 | 130 | 62 | 24 |
| 17 | 681 | 537 | 341 | 205 | 119 | 55 | 21 |
| 18 | 671 | 527 | 331 | 193 | 108 | 49 | .... |
| 19 | 662 | 516 | 321 | 181 | 96 | 44 | .... |
| 20 | 652 | 506 | 310 | 168 | 87 | 40 | .... |
| 21 | 642 | 495 | 299 | 155 | 79 | 36 | .... |
| 22 | 631 | 483 | 288 | 142 | 72 | 33 | .... |
| 23 | 621 | 472 | 276 | 130 | 66 | .... | .... |
| 24 | 610 | 460 | 264 | 119 | 60 | .... | .... |

**EFFECTIVE LENGTH FACTORS (K) FOR VARIOUS COLUMN CONFIGURATIONS**

| | (a) | (b) | (c) | (d) | (e) | (f) |
|---|---|---|---|---|---|---|
| Buckled shape of column is shown by dashed line | | | | | | |
| Theoretical K value | 0.5 | 0.7 | 1.0 | 1.0 | 2.0 | 2.0 |
| Recommended design value when ideal conditions are approximated | 0.65 | 0.80 | 1.2 | 1.0 | 2.10 | 2.0 |

| End condition code | | |
|---|---|---|
| | ⫟ | Rotation fixed and translation fixed |
| | ▽ | Rotation free and translation fixed |
| | ▥ | Rotation fixed and translation free |
| | ○ | Rotation free and translation free |

† Data from American Institute of Steel Construction Manual, Sixth Edition, 1967.
* With respect to radius of gyration. The effective length (KL) is the actual unbraced length, L, in feet, multiplied by the effective length factor (K) which is dependent upon the restraint at the ends of the unbraced length and the means available to resist lateral movements. Determination of K may be made by referring to the last portion of this table.

Loads shown below the heavy horizontal lines are for main members with $Kl/r$ ratios (where $l$ is the actual unbraced length in inches and $r$ the governing radius of gyration in inches) between 120 and 200.

*Application of Euler and Johnson Formulas:* To determine whether the Euler or Johnson formula is applicable in any particular case it is necessary to determine the value of the quantity $Q \div r^2$. If $Q \div r^2$ is greater than 2, then the Euler formula (1) should be used; if $Q \div r^2$ is less than 2, then the J. B. Johnson formula is applicable. Most compression members in machine design are in the range of proportions covered by the Johnson formula. For this reason a good procedure is to design machine elements on the basis of the Johnson formula and then as a check calculate $Q \div r^2$ to determine whether the Johnson formula applies or the Euler formula should have been used.

*Factor of Safety for Machine Columns:* When the conditions of loading and the physical qualities of the material used are accurately known, a factor of safety as low as 1.25 is sometimes used when minimum weight is important. For the usual case, however, a factor of safety of 2 to 2.5 is applied for steady loads. The factor of safety represents the ratio of the critical load $P_{cr}$ to the working load.

*Examples of Compression Member Design:* A rectangular machine member 24 inches long and ½ × 1 inch in cross-section is to carry a compressive load of 4000 pounds along its axis. What is the factor of safety for this load considering that the material is machinery steel having a yield point of 40,000 pounds per square inch, the load is steady, and each end of the rod has a ball connection so that $n = 1$?

From Formula (3)

$$Q = \frac{40,000 \times 24 \times 24}{1 \times 3.1416 \times 3.1416 \times 30,000,000} = 0.0778$$

(The values 40,000 and 30,000,000 were obtained from the table on page 444.)

The radius of gyration $r$ for a rectangular section (page 370) is 0.289 × the dimension in the direction of bending. In columns, bending is most apt to occur in the direction in which the section is the weakest, the ½-inch dimension in this case. Hence, least radius of gyration $r = 0.289 \times \frac{1}{2} = 0.145$ inch.

$$\frac{Q}{r^2} = \frac{0.0778}{(0.145)^2} = 3.70$$

which is more than 2 so that the Euler formula will be used.

$$P_{cr} = \frac{s_y A r^2}{Q} = \frac{40,000 \times \frac{1}{2} \times 1}{3.70}$$

= 5400 pounds so that the factor of safety is 5400 ÷ 4000 = 1.35

In the preceding example, the column formulas were used to check the adequacy of a column of known dimensions. The more usual problem involves determining what the dimensions should be to resist a specified load. As an example, the previous problem can be reworded as follows:

A 24-inch long bar of rectangular cross-section with width $w$ twice its depth $d$ is to carry a load of 4000 pounds. What must the width and depth be if a factor of safety of 1.35 is to be used?

First determine the critical load $P_{cr}$:

$$P_{cr} = \text{working load} \times \text{factor of safety}$$
$$= 4000 \times 1.35 = 5400 \text{ pounds.}$$

Next determine $Q$ which, as before, will be 0.0778.
Assume Formula (2) applies:

$$P_{cr} = A s_y \left( 1 - \frac{Q}{4\,r^2} \right)$$

$$5400 = w \times d \times 40,000 \left( 1 - \frac{0.0778}{4\, r^2} \right) = 2\, d^2 \times 40,000 \left( 1 - \frac{0.01945}{r^2} \right)$$

$$\frac{5400}{40,000 \times 2} = d^2 \left( 1 - \frac{0.01945}{r^2} \right)$$

As mentioned in the previous example the least radius of gyration $r$ of a rectangle is equal to $0.289 \times$ the least dimension, $d$, in this case. Therefore, substituting for $d$ the value $r \div 0.289$,

$$\frac{5400}{40,000 \times 2} = \left( \frac{r}{0.289} \right)^2 \left( 1 - \frac{0.01945}{r^2} \right)$$

$$\frac{5400 \times 0.289 \times 0.289}{40,000 \times 2} = r^2 - 0.01945$$

$$0.005638 = r^2 - 0.01945$$

$$r^2 = 0.0251$$

Checking to determine if $Q \div r^2$ is greater or less than 2,

$$\frac{Q}{r^2} = \frac{0.0778}{0.0251} = 3.1,$$

therefore Formula (1) should have been used to determine $r$ and dimensions $w$ and $d$. Using Formula (1),

$$5400 = \frac{40,000 \times 2\, d^2 \times r^2}{Q} = \frac{40,000 \times 2 \times \left( \dfrac{r}{0.289} \right)^2 r^2}{0.0778}$$

$$r^4 = \frac{5400 \times 0.0778 \times 0.289 \times 0.289}{40,000 \times 2}$$

$$= 0.0004386$$

$$r = 0.145 \quad \text{so that}$$

$$d = \frac{0.145}{0.289} = 0.50 \text{ inch}$$

and $\qquad w = 2\, d = 1$ inch as in the previous example.

## Plates, Shells and Cylinders

**Flat Stayed Surfaces.** — In many cases, large flat areas are held against pressure by stays distributed at regular intervals over the surface. In boiler work, these stays are usually screwed into the plate and the projecting end riveted over to insure steam tightness. The U. S. Board of Supervising Inspectors and the American Boiler Makers Association rules give the following formula for flat stayed surfaces:

$$P = \frac{C \times t^2}{S^2}$$

in which $P$ = pressure in pounds per square inch;

$C$ = a constant which equals 112, for plates $\frac{7}{16}$ inch and under; 120, for plates over $\frac{7}{16}$ inch thick; 140, for plates with stays having a nut and bolt on the inside and outside; and 160, for plates with stays having washers of at least one-half the thickness of the plate, and with a diameter at least one-half of the greatest pitch.

$t$ = thickness of plate in 16ths of an inch (thickness = $\frac{7}{16}$, $t = 7$);

$S$ = greatest pitch of stays in inches.

# STRENGTH OF MATERIALS 435

**Strength and Deflection of Flat Plates.** — In the majority of cases, the formulas used to determine stresses and deflections in flat plates are based on certain assumptions that can be closely approximated in practice. These assumptions are:

1. the thickness of the plate is not greater than one-quarter the least width of the plate;

2. the greatest deflection when the plate is loaded is less than one-half the plate thickness;

3. the maximum tensile stress resulting from the load does not exceed the elastic limit of the material; and

4. all loads are perpendicular to the plane of the plate.

Plates of ductile materials fail when the maximum stress resulting from deflection under load exceeds the yield strength; for brittle materials, failure occurs when the maximum stress reaches the ultimate tensile strength of the material involved.

**Square and Rectangular Flat Plates.** — The formulas that follow give the maximum stress and deflection of flat steel plates supported in various ways and subjected to the loading indicated. These formulas are based upon a modulus of elasticity for steel of 30,000,000 pounds per square inch and a value of Poisson's ratio of 0.3. If the formulas for maximum stress, $S$, are applied without modification to other materials such as cast iron, aluminum, and brass for which the range of Poisson's ratio is about 0.26 to 0.34, the maximum stress calculations will be in error by not more than about 3 per cent. The deflection formulas may also be applied to materials other than steel by substituting in these formulas the appropriate value for $E$, the modulus of elasticity of the material (see pages 444 and 445). The deflections thus obtained will not be in error by more than about 3 per cent.

In the stress and deflection formulas that follow,

$p$ = uniformly distributed load acting on plate, pounds per square inch;
$W$ = total load on plate, pounds; $W = p \times$ area of plate;
$L$ = distance between supports (length of plate), inches. For rectangular plates, $L$ = long side, $l$ = short side;
$t$ = thickness of plate, inches;
$S$ = maximum tensile stress in plate, pounds per square inch;
$d$ = maximum deflection of plate, inches;
$E$ = modulus of elasticity in tension. $E$ = 30,000,000 pounds per square inch for steel.

If metric SI units are used in the formulas, then,

$W$ = total load on plate, newtons;
$L$ = distance between supports (length of plate), millimeters. For rectangular plates, $L$ = long side, $l$ = short side;
$t$ = thickness of plate, millimeters;
$S$ = maximum tensile stress in plate, newtons per mm squared;
$d$ = maximum deflection of plate, mm;
$E$ = modulus of elasticity, newtons per mm squared.

1. Square flat plate supported at top and bottom of all four edges and a uniformly distributed load over the surface of the plate.

$$S = \frac{0.29\,W}{t^2} \quad (1) \qquad d = \frac{0.0443\,WL^2}{Et^3} \quad (2)$$

2. Square flat plate supported at the bottom only of all four edges and a uniformly distributed load over the surface of the plate.

$$S = \frac{0.28\,W}{t^2} \qquad (3) \qquad\qquad d = \frac{0.0443\,WL^2}{Et^3} \qquad (4)$$

**3.** Square flat plate with all edges firmly fixed and a uniformly distributed **load over** the surface of the plate.

$$S = \frac{0.31\,W}{t^2} \qquad (5) \qquad\qquad d = \frac{0.0138\,WL^2}{Et^3} \qquad (6)$$

**4.** Square flat plate with all edges firmly fixed and a uniform load over small circular area at the center. In equations (7) and (9) $r_o$ = radius of area to which load is applied. If $r_o < 1.7t$, use $r_s$ where $r_s = \sqrt{1.6r_o{}^2 + t^2} - 0.675t$.

$$S = \frac{0.62\,W}{t^2}\log_e\left(\frac{L}{2r_o}\right) \qquad (7) \qquad\qquad d = \frac{0.0568\,WL^2}{Et^3} \qquad (8)$$

**5.** Square flat plate with all edges supported above and below, or below only, and a concentrated load at the center. (See Case 4 for definition of $r_o$.)

$$S = \frac{0.62\,W}{t^2}\left[\log_e\left(\frac{L}{2r_o}\right) + 0.577\right] \qquad (9) \qquad\qquad d = \frac{0.1266\,WL^2}{Et^3} \qquad (10)$$

**6.** Rectangular plate with all edges supported at top and bottom and a uniformly distributed load over the surface of the plate.

$$S = \frac{0.75\,W}{t^2\left(\dfrac{L}{l} + 1.61\dfrac{l^2}{L^2}\right)} \qquad (11) \qquad\qquad d = \frac{0.1422\,W}{Et^3\left(\dfrac{L}{l^3} + \dfrac{2.21}{L^2}\right)} \qquad (12)$$

**7.** Rectangular plate with all edges fixed and a uniformly distributed load **over** the surface of the plate.

$$S = \frac{0.5\,W}{t^2\left(\dfrac{L}{l} + \dfrac{0.623\,l^5}{L^5}\right)} \qquad (13) \qquad\qquad d = \frac{0.0284\,W}{Et^3\left(\dfrac{L}{l^3} + \dfrac{1.056\,l^2}{L^4}\right)} \qquad (14)$$

**Circular Flat Plates.** — In the following formulas, $R$ = radius of plate to supporting edge in inches; $W$ = total load in pounds; and other symbols are the same as used for square and rectangular plates.

If metric SI units are used, $R$ = radius of plate to supporting edge in millimeters, and the values of other symbols are the same as those used for square and rectangular plates.

**1.** Edge supported around the circumference and a uniformly distributed load over the surface of the plate.

$$S = \frac{0.39\,W}{t^2} \qquad (1) \qquad\qquad d = \frac{0.221\,WR^2}{Et^3} \qquad (2)$$

**2.** Edge fixed around circumference and a uniformly distributed load over the surface of the plate.

$$S = \frac{0.24\,W}{t^2} \qquad (3) \qquad\qquad d = \frac{0.0543\,WR^2}{Et^3} \qquad (4)$$

3. Edge supported around the circumference and a concentrated load at the center.

$$S = \frac{0.48\,W}{t^2}\left[1 + 1.3 \log_e \frac{R}{0.325\,t} - 0.0185 \frac{t^2}{R^2}\right] \quad (5) \qquad d = \frac{0.55\,W R^2}{E t^3} \quad (6)$$

4. Edge fixed around circumference and a concentrated load at the center.

$$S = \frac{0.62\,W}{t^2}\left[\log_e \frac{R}{0.325\,t} + 0.0264 \frac{t^2}{R^2}\right] \quad (7) \qquad d = \frac{0.22\,W R^2}{E t^3} \quad (8)$$

**Strength of Cylinders Subjected to Internal Pressure.** — In designing a cylinder to withstand internal pressure, the choice of formula to be used depends on (1) the kind of material of which the cylinder is made (whether brittle or ductile); (2) the construction of the cylinder ends (whether open or closed); and (3) whether the cylinder is classed as a thin- or a thick-walled cylinder.

A cylinder is considered to be thin-walled when the ratio of wall thickness to inside diameter is 0.1 or less and thick-walled when this ratio is greater than 0.1. Materials such as cast iron, hard steel, cast aluminum are considered to be brittle materials; low-carbon steel, brass, bronze, etc. are considered to be ductile.

In the formulas that follow, $p$ = internal pressure, pounds per square inch; $D$ = inside diameter of cylinder, inches; $t$ = wall thickness of cylinder, inches; $\mu$ = Poisson's ratio, = 0.3 for steel, 0.26 for cast iron, 0.34 for aluminum and brass; $S$ = allowable tensile stress, pounds per square inch.

Metric SI units can be used in Formulas (1), (3), (4), and (5), where $p$ = internal pressure in newtons per square millimeter; $D$ = inside diameter of cylinder, millimeters; $t$ = wall thickness, mm; $\mu$ = Poisson's ratio, = 0.3 for steel, 0.26 for cast iron, and 0.34 for aluminum and brass; and $S$ = allowable tensile stress, N/mm². For the use of metric SI units in Formula (2), see below.

*Thin-walled cylinders:*

$$t = \frac{Dp}{2S} \tag{1}$$

For low-pressure cylinders of cast iron such as are used for certain engine and press applications, a formula in common use is

$$t = \frac{Dp}{2500} + 0.3 \tag{2}$$

This formula is based on an allowable stress of 1250 pounds per square inch and will give a wall thickness 0.3 inch greater than Formula (1) to allow for variations in metal thickness that may result from the casting process.

If metric SI units are used in Formula (2), $t$ = cylinder wall thickness in millimeters; $D$ = inside diameter of cylinder, mm; and the allowable stress is in newtons per square millimeter. The value of 0.3 inches additional wall thickness is 7.62 mm, and the next highest number in preferred metric basic sizes is 8 mm.

*Thick-walled cylinders of brittle material; ends open or closed:* Lamé's equation is used when cylinders of this type are subjected to internal pressure.

$$t = \frac{D}{2}\left[\sqrt{\frac{S + p}{S - p}} - 1\right] \tag{3}$$

The table of ratios of outside radius to inside radius of thick cylinders on page 439 is for convenience in calculating the dimensions of cylinders under high internal pressure without the use of Formula 3. As an example of the use of the table, assume that a cylinder of 10 inches inside diameter is to withstand a pressure of 2500 pounds per square inch; the material is cast iron and the allowable stress is 6000 pounds per square inch. To solve the problem, locate the allowable stress per square inch in the left-hand column of the table and the working pressure at the top of the columns. Then find the ratio between the outside and inside radii in the body of the table. In this case, the ratio is 1.558, and hence the outside diameter of the cylinder should be 10 × 1.558, or about 15⅝ inches. The thickness of the cylinder wall will therefore be $\frac{15.558 - 10}{2}$ = 2.779 inches.

Unless very high-grade material is used and sound castings assured, cast iron should not be used for pressures exceeding 2000 pounds per square inch. It is well to leave more metal in the bottom of a hydraulic cylinder than is indicated by the results of calculations, because a hole of some size must be cored in the bottom to permit the entrance of a boring bar when finishing the cylinder, and when this hole is subsequently tapped and plugged it often gives trouble if the precaution mentioned is not taken.

For steady or gradually applied stresses, the maximum allowable fiber stress $S$ may be assumed to be from 3500 to 4000 pounds per square inch for cast iron; from 6000 to 7000 pounds per square inch for brass; and 12,000 pounds per square inch for steel castings. For intermittent stresses, such as in cylinders for steam and hydraulic work, 3000 pounds per square inch for cast iron; 5000 pounds per square inch for brass; and 10,000 pounds per square inch for steel castings, is ordinarily used. These values give ample factors of safety.

*Note:* In metric SI units, 1000 pounds per square inch equals 6.895 newtons per square millimeter.

*Thick-walled cylinders of ductile material; closed ends:* Clavarino's equation is used:

$$t = \frac{D}{2}\left[\sqrt{\frac{S + (1 - 2\mu)p}{S - (1 + \mu)p}} - 1\right] \tag{4}$$

*Thick-walled cylinders of ductile material; open ends:* Birnie's equation is used:

$$t = \frac{D}{2}\left[\sqrt{\frac{S + (1 - \mu)p}{S - (1 + \mu)p}} - 1\right] \tag{5}$$

**Spherical Shells Subjected to Internal Pressure.** — Let:

$D$ = internal diameter of shell in inches;
$p$ = internal pressure in pounds per square inch;
$S$ = safe tensile stress per square inch;
$t$ = the thickness of metal in the shell in inches. Then: $t = \frac{pD}{4S}$

If metric SI units are used, then:

$D$ = internal diameter of shell in millimeters;
$p$ = internal pressure in newtons per square millimeter;
$S$ = safe tensile stress in newtons per square millimeter;
$t$ = thickness of metal in the shell in millimeters.

Meters can be used in the formula in place of millimeters, providing the treatment is consistent throughout.

This formula also applies to hemi-spherical shells, such as the hemi-spherical head of a cylindrical container subjected to internal pressure, etc.

Ratio of Outside Radius to Inside Radius, Thick Cylinders

| Allowable Stress in Metal per Sq. In. of Section | Working Pressure in Cylinder, Pounds per Square Inch | | | | | | | | | | | | |
|---|---|---|---|---|---|---|---|---|---|---|---|---|---|
| | 1000 | 1500 | 2000 | 2500 | 3000 | 3500 | 4000 | 4500 | 5000 | 5500 | 6000 | 6500 | 7000 |
| 2,000 | 1.732 | 2.000 | | | | | | | | | | | |
| 2,500 | 1.527 | 1.732 | | | | | | | | | | | |
| 3,000 | 1.414 | 1.581 | 2.236 | | | | | | | | | | |
| 3,500 | 1.341 | 1.483 | 1.915 | 2.449 | | | | | | | | | |
| 4,000 | 1.291 | 1.414 | 1.732 | 2.081 | 2.645 | | | | | | | | |
| 4,500 | 1.253 | 1.362 | 1.612 | 1.871 | 2.236 | 2.828 | | | | | | | |
| 5,000 | 1.224 | 1.322 | 1.527 | 1.732 | 2.000 | 2.380 | 3.000 | | | | | | |
| 5,500 | 1.201 | 1.291 | 1.464 | 1.633 | 1.844 | 2.121 | 2.516 | 3.162 | | | | | |
| 6,000 | 1.183 | 1.264 | 1.414 | 1.558 | 1.732 | 1.949 | 2.236 | 2.645 | 3.316 | | | | |
| 6,500 | | 1.243 | 1.374 | 1.500 | 1.647 | 1.825 | 2.049 | 2.345 | 2.768 | 3.464 | | | |
| 7,000 | | 1.224 | 1.341 | 1.453 | 1.581 | 1.732 | 1.914 | 2.144 | 2.449 | 2.886 | 3.605 | | |
| 7,500 | | 1.209 | 1.314 | 1.414 | 1.527 | 1.658 | 1.813 | 2.000 | 2.236 | 2.549 | 3.000 | 3.741 | |
| 8,000 | | 1.194 | 1.291 | 1.381 | 1.483 | 1.599 | 1.732 | 1.889 | 2.081 | 2.323 | 2.645 | 3.109 | 3.872 |
| 8,500 | | 1.183 | 1.271 | 1.354 | 1.446 | 1.548 | 1.666 | 1.802 | 1.963 | 2.160 | 2.408 | 2.738 | 3.214 |
| 9,000 | | | 1.253 | 1.330 | 1.414 | 1.507 | 1.612 | 1.732 | 1.871 | 2.035 | 2.236 | 2.440 | 2.828 |
| 9,500 | | | 1.235 | 1.306 | 1.386 | 1.472 | 1.566 | 1.673 | 1.795 | 1.936 | 2.104 | 2.369 | 2.569 |
| 10,000 | | | 1.224 | 1.291 | 1.362 | 1.441 | 1.527 | 1.623 | 1.732 | 1.856 | 2.000 | 2.171 | 2.380 |
| 10,500 | | | 1.212 | 1.274 | 1.341 | 1.414 | 1.493 | 1.581 | 1.678 | 1.789 | 1.915 | 2.061 | 2.236 |
| 11,000 | | | 1.201 | 1.260 | 1.322 | 1.390 | 1.464 | 1.544 | 1.633 | 1.732 | 1.844 | 1.972 | 2.121 |
| 11,500 | | | 1.193 | 1.247 | 1.306 | 1.369 | 1.437 | 1.511 | 1.593 | 1.683 | 1.784 | 1.897 | 2.027 |
| 12,000 | | | 1.183 | 1.235 | 1.291 | 1.359 | 1.414 | 1.483 | 1.558 | 1.640 | 1.732 | 1.834 | 1.949 |
| 12,500 | | | | 1.224 | 1.277 | 1.333 | 1.393 | 1.457 | 1.527 | 1.603 | 1.687 | 1.779 | 1.878 |
| 13,000 | | | | 1.215 | 1.264 | 1.318 | 1.374 | 1.434 | 1.500 | 1.570 | 1.647 | 1.732 | 1.825 |
| 13,500 | | | | 1.206 | 1.253 | 1.303 | 1.357 | 1.414 | 1.475 | 1.541 | 1.612 | 1.690 | 1.775 |
| 14,000 | | | | 1.197 | 1.243 | 1.291 | 1.341 | 1.395 | 1.453 | 1.514 | 1.581 | 1.653 | 1.732 |
| 14,500 | | | | 1.189 | 1.233 | 1.279 | 1.327 | 1.378 | 1.432 | 1.490 | 1.553 | 1.620 | 1.693 |
| 15,000 | | | | 1.183 | 1.224 | 1.268 | 1.314 | 1.362 | 1.414 | 1.469 | 1.527 | 1.590 | 1.658 |
| 16,000 | | | | 1.170 | 1.209 | 1.249 | 1.291 | 1.335 | 1.381 | 1.431 | 1.483 | 1.538 | 1.599 |

*Example:* — Find the thickness of metal required in the hemi-spherical end of a cylindrical vessel, 2 feet in diameter, subjected to an internal pressure of 500 pounds per square inch. The material is mild steel and a tensile stress of 10,000 pounds per square inch is allowable.

$$t = \frac{500 \times 2 \times 12}{4 \times 10,000} = 0.3 \text{ inch.}$$

A similar example using metric SI units is as follows: find the thickness of metal required in the hemi-spherical end of a cylindrical vessel, 750 mm in diameter, subjected to an internal pressure of 3 newtons/mm². The material is mild steel and a tensile stress of 70 newtons/mm² is allowable.

$$t = \frac{3 \times 750}{4 \times 70} = 8.04 \text{ mm.}$$

If the radius of curvature of the dome head of a boiler or container subjected to internal pressure is made equal to the diameter of the boiler, the thickness of the cylindrical shell and of the spherical head should be made the same. For example, if a boiler is 3 feet in diameter, the radius of curvature of its head should be made 3 feet, if material of the same thickness is to be used and the stresses are to be equal in both the head and cylindrical portion.

**Collapsing Pressures of Cylinders and Tubes Subjected to External Pressures.** — The following formulas may be used for finding the collapsing pressures of lap-welded Bessemer steel tubes:

$$P = 86,670\,\frac{t}{D} - 1386 \tag{1}$$

$$P = 50,210,000 \left(\frac{t}{D}\right)^3 \tag{2}$$

in which $P$ = collapsing pressure in pounds per square inch; $D$ = outside diameter of tube or cylinder in inches; $t$ = thickness of wall in inches.

Formula (1) is for values of $P$ greater than 580 pounds per square inch, and Formula (2) is for values of $P$ less than 580 pounds per square inch. These formulas are substantially correct for all lengths of pipe greater than six diameters between transverse joints that tend to hold the pipe to a circular form. The pressure $P$ found is the actual collapsing pressure, and a suitable factor of safety must be used. Ordinarily, a factor of safety of 5 is sufficient. In cases where there are repeated fluctuations of the pressure, vibration, shocks and other stresses, a factor of safety of from 6 to 12 should be used.

If metric SI units are used the formulas are:

$$P = 597.6\,\frac{t}{D} - 9.556 \tag{3}$$

$$P = 346,200 \left(\frac{t}{D}\right)^3 \tag{4}$$

where $P$ = collapsing pressure in newtons per square millimeter; $D$ = outside diameter of tube or cylinder in millimeters; and $t$ = thickness of wall in millimeters. Formula (3) is for values of $P$ greater than 4 N/mm², and Formula (4) is for values of $P$ less than 4 N/mm².

The table " Tubes Subjected to External Pressure " is based upon the requirements of the Steam Boat Inspection Service of the Department of Commerce and Labor and gives the permissible working pressures and corresponding minimum thickness of wall for long, plain, lap-welded and seamless steel flues subjected to external pressure only. The table thicknesses have been calculated from the formula:

$$t = \frac{[(F \times p) + 1386] D}{86{,}670}$$

in which $D$ = outside diameter of flue or tube in inches; $t$ = thickness of wall in inches; $p$ = working pressure in pounds per square inch; $F$ = factor of safety. The formula is applicable to working pressures greater than 100 pounds per square inch, to outside diameters from 7 to 18 inches, and to temperatures less than 650° F.

The Formulas (1) and (2) given on the preceding page were determined by Prof. R. T. Stewart, Dean of the Mechanical Engineering Department of the University of Pittsburg, in a series of experiments carried out at the plant of the National Tube Co., McKeesport, Pa.

The apparent fiber stress under which the different tubes failed varied from about 7000 pounds per square inch for the relatively thinnest walls to 35,000 pounds per square inch for the relatively thickest walls. Since the average yield point of the material tested was 37,000 pounds and the tensile strength 58,000 pounds per square inch, it is evident that the strength of a tube subjected to external fluid collapsing pressure is not dependent alone upon the elastic limit or ultimate strength of the material from which it is made.

### Tubes Subjected to External Pressure

| Outside Diameter of Tube, Inches | Working Pressure in Pounds per Square Inch | | | | | | |
|---|---|---|---|---|---|---|---|
| | 100 | 120 | 140 | 160 | 180 | 200 | 220 |
| | Thickness of Tube in Inches.  Safety Factor, 5 | | | | | | |
| 7 | 0.152 | 0.160 | 0.168 | 0.177 | 0.185 | 0.193 | 0.201 |
| 8 | 0.174 | 0.183 | 0.193 | 0.202 | 0.211 | 0.220 | 0.229 |
| 9 | 0.196 | 0.206 | 0.217 | 0.227 | 0.237 | 0.248 | 0.258 |
| 10 | 0.218 | 0.229 | 0.241 | 0.252 | 0.264 | 0.275 | 0.287 |
| 11 | 0.239 | 0.252 | 0.265 | 0.277 | 0.290 | 0.303 | 0.316 |
| 12 | 0.261 | 0.275 | 0.289 | 0.303 | 0.317 | 0.330 | 0.344 |
| 13 | 0.283 | 0.298 | 0.313 | 0.328 | 0.343 | 0.358 | 0.373 |
| 14 | 0.301 | 0.320 | 0.337 | 0.353 | 0.369 | 0.385 | 0.402 |
| 15 | 0.323 | 0.343 | 0.361 | 0.378 | 0.396 | 0.413 | 0.430 |
| 16 | 0.344 | 0.366 | 0.385 | 0.404 | 0.422 | 0.440 | 0.459 |
| 17 | 0.366 | 0.389 | 0.409 | 0.429 | 0.448 | 0.468 | 0.488 |
| 18 | 0.387 | 0.412 | 0.433 | 0.454 | 0.475 | 0.496 | 0.516 |

### Dimensions and Maximum Allowable Pressure of Tubes Subjected to External Pressure

| Outside Diam., Inches | Thickness of Material, Inches | Maximum Pressure Allowed, Pounds | Outside Diam., Inches | Thickness of Material, Inches | Maximum Pressure Allowed, Pounds | Outside Diam., Inches | Thickness of Material, Inches | Maximum Pressure Allowed, Pounds |
|---|---|---|---|---|---|---|---|---|
| 2 | 0.095 | 427 | 3 | 0.109 | 327 | 4 | 0.134 | 303 |
| 2¼ | 0.095 | 380 | 3¼ | 0.120 | 332 | 4½ | 0.134 | 238 |
| 2½ | 0.109 | 392 | 3½ | 0.120 | 308 | 5 | 0.148 | 235 |
| 2¾ | 0.109 | 356 | 3¾ | 0.120 | 282 | 6 | 0.165 | 199 |

## Physical Properties

**Physical Properties of Heat-treated Steels.** — Steels that have been "fully hardened" to the same hardness when quenched will have the same tensile and yield strengths regardless of composition and alloying elements. When the hardness of such a steel is known, it is also possible to predict its reduction of area and tempering temperature. The accompanying charts illustrating these relationships have been prepared by the Society of Automotive Engineers.

Chart 1 gives the range of Brinell hardnesses that could be expected for any particular tensile strength or it may be used to determine the range of tensile strengths that would correspond to any particular hardness. Chart 2 shows the relationship between the tensile strength or hardness and the yield point. The solid-line curve is the normal-expectancy curve. The dotted-line curves give the range of the variation of scatter of the plotted data. Chart 3 shows the relationship that exists between the tensile strength (or hardness) and the reduction of area. The curve to the left represents the alloy steels and that on the right the carbon steels. Both curves are normal-expectancy curves and the extremities of the perpendicular lines going through them represent the variations from the normal-expectancy curves which may be caused by quality differences and by the magnitude of parasitic stresses induced by quenching. Chart 4 shows the relationship between the hardness (or approximately equivalent tensile strength) and the tempering temperature. Three curves are given, one for fully hardened steels with a carbon content between 0.40 and 0.55 per cent, one for fully hardened steels with a carbon content between 0.30 and 0.40 per cent, and one for steels that are not fully hardened.

Referring to Chart 1 it can be seen that for a tensile strength of say 200,000 pounds per square inch the Brinell hardness could range from something in the order of 375 to 425. Taking 400 as the mean hardness value and using Chart 4, it can be seen that the tempering temperature of fully hardened steels of 0.40 to 0.55 per cent carbon content would be 990 degrees F. and that of fully hardened steels of 0.30 to 0.40 per cent carbon would be 870 degrees F. This chart also shows that the tempering temperature for a steel not fully hardened would approach 520 degrees F. A yield point of $0.9 \times 200,000$ or 180,000 pounds per square inch is indicated (Chart 2) for the fully hardened steel with a tensile strength of 200,000 pounds per square inch. Most alloy steels of 200,000 pounds per square inch tensile strength would probably have a reduction in area of close to 44 per cent (Chart 3) but some would have values in the range of 35 to 53 per cent. Carbon steels of the same tensile strength would probably have a reduction in area of close to 24 per cent but could possibly range from 17 to 31 per cent.

Charts 2 and 3 represent steel in the quenched and tempered condition and Chart 1 represents steel in the hardened and tempered, as-rolled, annealed, and normalized conditions. These charts give a good general indication of mechanical properties; however, more exact information when required should be obtained from tests on samples of the individual heats of steel under consideration.

**Strength Data for Ferrous Metals.** — The accompanying Table 1 gives ultimate strengths, yield points and moduli of elasticity for various ferrous metals. Values are given as ranges, minimum values, and average values. Ranges of values are due to differences in size and shape of sections, heat-treatments undergone, and composition in those cases where several slightly different materials are listed under one general classification. The values in the table are meant to serve as a guide in the selection of ferrous materials and should not be used to write specifications. More specific data should be obtained from the supplier.

**Strength Data for Non-ferrous Metals.** — The ultimate tensile, shear, and yield strengths and moduli of elasticity of many non-ferrous metals are given in Table 2. Values for the most part are given in ranges rather than as single values

Physical Property Charts for Heat-Treated Steels (SAE General Information)

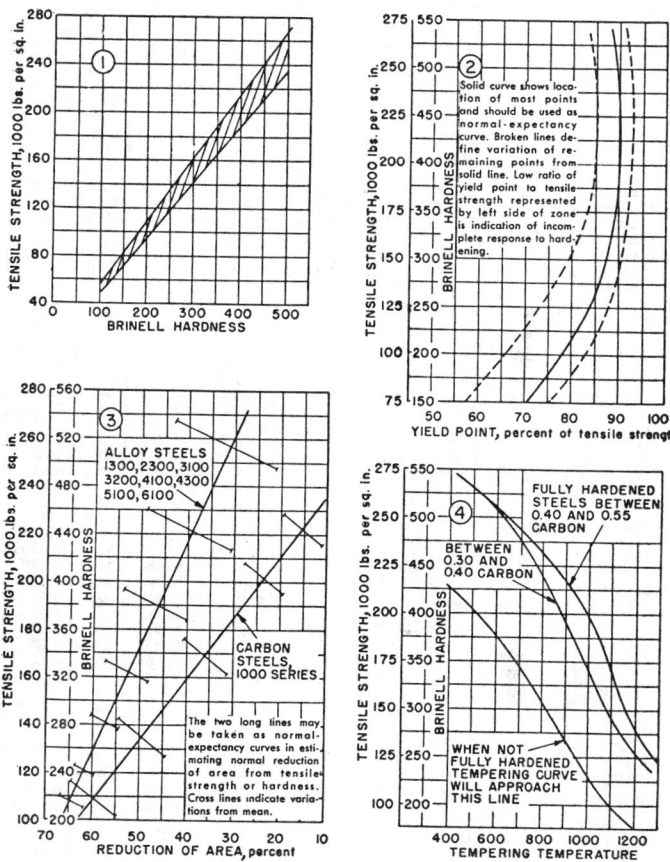

because of differences in composition, forms, sizes and shapes for the aluminum alloys plus differences in heat-treatments undergone for the other non-ferrous metals. The values in the table are meant to serve as a guide, not as specifications. More specific data should be obtained from the supplier.

**Effect of Temperature on Strength and Elasticity of Metals.** — Most ferrous metals have a maximum strength at approximately 400 degrees F. while the strength of non-ferrous alloys is a maximum at about room temperature. The table on page 446 gives general data for variation in metal strength with temperature.

The modulus of elasticity of metals decreases regularly with increasing temperatures above room temperature until at some elevated temperature it falls off rapidly and reaches zero at the melting point.

## Table 1. Strength Data for Iron and Steel

| Material | Ultimate Strength | | | Yield Point, Thousands of Pounds per Square Inch | Modulus of Elasticity | |
| --- | --- | --- | --- | --- | --- | --- |
| | Tension, Thousands of Pounds per Square Inch, $T$ | Compression, in terms of $T$ | Shear, in terms of $T$ | | in Tension, Millions of psi, $E$ | in Shear,[b] in terms of $E$ |
| Cast iron, gray, class 20.. | 20[a] | 3.6 $T$ to 4.4 $T$ | 1.6 $T$ | ..... | 11.6 | 0.40 $E$ |
| class 25 | 25[a] | 3.6 $T$ to 4.4 $T$ | 1.4 $T$ | ..... | 14.2 | 0.40 $E$ |
| class 30 | 30[a] | 3.7 $T$ | 1.4 $T$ | ..... | 14.5 | 0.40 $E$ |
| class 35 | 35[a] | 3.2 $T$ to 3.9 $T$ | 1.4 $T$ | ..... | 16.0 | 0.40 $E$ |
| class 40 | 40[a] | 3.1 $T$ to 3.4 $T$ | 1.3 $T$ | ..... | 17 | 0.40 $E$ |
| class 50 | 50[a] | 3.0 $T$ to 3.4 $T$ | 1.3 $T$ | ..... | 18 | 0.40 $E$ |
| class 60 | 60[a] | 2.8 $T$ | 1.0 $T$ | ..... | 19.9 | 0.40 $E$ |
| malleable | 40 to 100[c] | .... | .... | 30 to 80[c] | 25 | 0.43 $E$ |
| nodular (ductile iron) | 60 to 120[d] | .... | .... | 40 to 90[d] | 23 | |
| Cast steel, carbon | 60 to 100 | $T$ | 0.75 $T$ | 30 to 70 | 30 | 0.38 $E$ |
| low alloy | 70 to 200 | $T$ | 0.75 $T$ | 45 to 170 | 30 | 0.38 $E$ |
| Steel, SAE 950 (low alloy) | 65 to 70 | $T$ | 0.75 $T$ | 45 to 50 | 30 | 0.38 $E$ |
| 1025 (low carbon) | 60 to 103 | $T$ | 0.75 $T$ | 40 to 90 | 30 | 0.38 $E$ |
| 1045 (medium carbon) | 80 to 182 | $T$ | 0.75 $T$ | 50 to 162 | 30 | 0.38 $E$ |
| 1095 (high carbon) | 90 to 213 | $T$ | 0.75 $T$ | 20 to 150 | 30 | 0.39 $E$ |
| 1112 (free cutting)[e] | 60 to 100 | $T$ | 0.75 $T$ | 30 to 95 | 30 | 0.38 $E$ |
| 1212 (free cutting) | 57 to 80 | $T$ | 0.75 $T$ | 25 to 72 | 30 | 0.38 $E$ |
| 1330 (alloy) | 90 to 162 | $T$ | 0.75 $T$ | 27 to 149 | 30 | 0.38 $E$ |
| 2517 (alloy)[e] | 88 to 190 | $T$ | 0.75 $T$ | 60 to 155 | 30 | 0.38 $E$ |
| 3140 (alloy) | 93 to 188 | $T$ | 0.75 $T$ | 62 to 162 | 30 | 0.38 $E$ |
| 3310 (alloy)[e] | 104 to 172 | $T$ | 0.75 $T$ | 56 to 142 | 30 | 0.38 $E$ |
| 4023 (alloy)[e] | 105 to 170 | $T$ | 0.75 $T$ | 60 to 114 | 30 | 0.38 $E$ |
| 4130 (alloy) | 81 to 179 | $T$ | 0.75 $T$ | 46 to 161 | 30 | 0.38 $E$ |
| 4340 (alloy) | 109 to 220 | $T$ | 0.75 $T$ | 68 to 200 | 30 | 0.38 $E$ |
| 4640 (alloy) | 98 to 192 | $T$ | 0.75 $T$ | 62 to 169 | 30 | 0.38 $E$ |
| 4820 (alloy)[e] | 98 to 209 | $T$ | 0.75 $T$ | 68 to 184 | 30 | 0.38 $E$ |
| 5150 (alloy) | 98 to 210 | $T$ | 0.75 $T$ | 51 to 190 | 30 | 0.38 $E$ |
| 52100 (alloy) | 100 to 238 | $T$ | 0.75 $T$ | 81 to 228 | 30 | 0.38 $E$ |
| 6150 (alloy) | 96 to 228 | $T$ | 0.75 $T$ | 59 to 210 | 30 | 0.38 $E$ |
| 8650 (alloy) | 110 to 228 | $T$ | 0.75 $T$ | 69 to 206 | 30 | 0.38 $E$ |
| 8740 (alloy) | 100 to 179 | $T$ | 0.75 $T$ | 60 to 165 | 30 | 0.38 $E$ |
| 9310 (alloy)[e] | 117 to 187 | $T$ | 0.75 $T$ | 63 to 162 | 30 | 0.38 $E$ |
| 9840 (alloy) | 120 to 285 | $T$ | 0.75 $T$ | 45 to 50 | 30 | 0.38 $E$ |
| Steel, stainless, SAE | | | | | | |
| 30302[f] | 85 to 125 | $T$ | ..... | 35 to 95 | 28 | 0.45 $E$ |
| 30321[f] | 85 to 95 | $T$ | ..... | 30 to 60 | 28 | |
| 30347[f] | 90 to 100 | $T$ | ..... | 35 to 65 | 28 | |
| 51420[g] | 95 to 230 | $T$ | ..... | 50 to 195 | 29 | 0.40 $E$ |
| 51430[h] | 75 to 85 | $T$ | ..... | 40 to 70 | 29 | |
| 51446[h] | 80 to 85 | $T$ | ..... | 50 to 70 | 29 | |
| 51501[g] | 70 to 175 | $T$ | ..... | 30 to 135 | 29 | |
| Steel, structural, common | 60 to 75 | $T$ | 0.75 $T$ | 33[a] | 29 | 0.41 $E$ |
| rivet | 52 to 62 | $T$ | 0.75 $T$ | 28[a] | 29 | |
| rivet, high strength | 68 to 82 | $T$ | 0.75 $T$ | 38[a] | 29 | |
| Wrought iron | 34 to 54 | $T$ | 0.83 $T$ | 23 to 32 | 28 | |

[a] Minimum specified value of the American Society of Testing Materials. The specifications for the various materials are as follows: Cast iron, ASTM A48; structural steel for bridges and structures, ASTM A7; structural rivet steel, ASTM A141; high-strength structural rivet steel, ASTM A195.

[b] Synonymous in other literature to the modulus of elasticity in torsion and the modulus of rigidity, $G$.

[c] Range of minimum specified values of the ASTM (ASTM A47, A197, and A220).

[d] Range of minimum specified values of the ASTM (ASTM A339) and the Munitions Board Standards Agency (MIL-I-17166A and MIL-I-11466).

[e] Carburizing grades of steel.

[f] Non-hardenable nickel-chromium and chromium-nickel-manganese steel (austenitic).

[g] Hardenable chromium steel (martensitic).

[h] Non-hardenable chromium steel (ferritic).

Table 2.  Strength Data for Non-Ferrous Metals*

| Material | Ultimate Strength, Thousands of Pounds per Square Inch | | Yield Strength (0.2 per cent offset), Thousands of Pounds per Square Inch | Modulus of Elasticity, Millions of Pounds per Square Inch | |
|---|---|---|---|---|---|
| | in Tension | in Shear | | in Tension, $E$ | in Shear, $G$ |
| Aluminum alloys, cast, | | | | | |
|   sand cast,.............. | 19 to 35 | 14 to 26 | 8 to 25 | 10.3 | ... |
|     heat-treated......... | 20 to 48 | 20 to 34 | 16 to 40 | 10.3 | ... |
|   permanent mold cast, .. | 23 to 35 | 16 to 27 | 9 to 40 | 10.3 | ... |
|     heat-treated......... | 23 to 48 | 15 to 36 | 8.5 to 43 | 10.3 | ... |
|   die-cast............... | 30 to 46 | 19 to 29 | 16 to 27 | 10.3 | ... |
| Aluminum alloys, wrought, | | | | | |
|   annealed.............. | 10 to 42 | 7 to 26 | 4 to 22 | 10.0 to 10.6 | ... |
|   cold-worked.......... | 12 to 63 | 8 to 34 | 11 to 59 | 10.0 to 10.3 | ... |
|   heat-treated........... | 22 to 83 | 14 to 48 | 13 to 73 | 10.0 to 11.4 | ... |
| Aluminum bronze, cast,.. | 62 to 90 | ... | 25 to 37 | 15 to 18 | ... |
|   heat-treated........... | 80 to 110 | ... | 32 to 65 | 15 to 18 | ... |
| Aluminum bronze, wrought, | | | | | |
|   annealed.............. | 55 to 80 | ... | 20 to 40 | 16 to 19 | ... |
|   cold-worked.......... | 71 to 110 | ... | 62 to 66 | 16 to 19 | ... |
|   heat-treated........... | 101 to 151 | ... | 48 to 94 | 16 to 19 | ... |
| Brasses, leaded, cast...... | 32 to 40 | 29 to 31 | 12 to 15 | 12 to 14 | ... |
|   flat products, wrought.. | 46 to 85 | 31 to 45 | 14 to 62 | 14 to 17 | 5.3 to 6.4 |
|   wire, wrought.......... | 50 to 88 | 34 to 46 | .... | 15 | 5.6 |
| Brasses, non-leaded, | | | | | |
|   flat products, wrought.. | 34 to 99 | 28 to 48 | 10 to 65 | 15 to 17 | 5.6 to 6.4 |
|   wire, wrought.......... | 40 to 130 | 29 to 60 | .... | 15 to 17 | 5.6 to 6.4 |
| Copper, wrought, | | | | | |
|   flat products........... | 32 to 57 | 22 to 29 | 10 to 53 | 17 | 6.4 |
|   wire................... | 35 to 66 | 24 to 33 | .... | 17 | 6.4 |
| Inconel, cast............. | 70 to 95 | ... | 30 to 45 | 23 | ... |
|   flat products, wrought.. | 80 to 170 | ... | 30 to 160 | 31 | 11 |
|   wire, wrought.......... | 80 to 185 | ... | 25 to 175 | 31 | 11 |
| Lead.................... | 2.2 to 4.9 | ... | .... | 0.8 to 2.0 | ... |
| Magnesium, cast, | | | | | |
|   sand & permanent mold. | 22 to 40 | 17 to 22 | 12 to 23 | 6.5 | 2.4 |
|   die-cast.............. | 33 | 20 | 22 | 6.5 | 2.4 |
| Magnesium, wrought, | | | | | |
|   sheet and plate........ | 35 to 42 | 21 to 23 | 20 to 32 | 6.5 | 2.4 |
|   bars, rods, and shapes .. | 37 to 55 | 19 to 27 | 26 to 44 | 6.5 | 2.4 |
| Monel, cast............. | 65 to 90 | ... | 32 to 40 | 19 | ... |
|   flat products, wrought. . | 70 to 140 | ... | 25 to 130 | 26 | 9.5 |
|   wire, wrought.......... | 70 to 170 | ... | 25 to 160 | 26 | 9.5 |
| Nickel, cast............. | 45 to 60 | ... | 20 to 30 | 21.5 | ... |
|   flat products, wrought.. | 55 to 130 | ... | 15 to 115 | 30 | 11 |
|   wire, wrought.......... | 50 to 165 | ... | 10 to 155 | 30 | 11 |
| Nickel silver, cast........ | 40 to 50 | ... | 24 to 25 | ... | ... |
|   flat products, wrought.. | 49 to 115 | 41 to 59 | 18 to 90 | 17.5 to 18 | 6.6 to 6.8 |
|   wire, wrought.......... | 50 to 145 | ... | 25 to 90 | 17.5 to 18 | 6.6 to 6.8 |
| Phosphor bronze, wrought, | | | | | |
|   flat products........... | 40 to 128 | ... | 14 to 80 | 15 to 17 | 5.6 to 6.4 |
|   wire.................. | 50 to 147 | ... | 20 to 80 | 16 to 17 | 6 to 6.4 |
| Silicon bronze, wrought, | | | | | |
|   flat products........... | 56 to 110 | 42 to 63 | 21 to 62 | 15 | 5.6 |
|   wire.................. | 50 to 145 | 36 to 70 | 25 to 70 | 15 to 17 | 5.6 to 6.4 |
| Tin bronze, leaded, cast.. | 21 to 38 | 23 to 43 | 15 to 18 | 10 to 14.5 | ... |
| Titanium................ | 50 to 135 | ... | 40 to 120 | 15.0 to 16.5 | ... |
| Zinc, commercial rolled .. | 19.5 to 31 | ... | .... | .... | ... |
| Zirconium............... | 22 to 83 | ... | .... | 9 to 14.5 | 4.8 |

* Consult the index for data on metals not listed and for more data on metals listed.

### Average Ultimate Strength of Common Materials other than Metals
(Pounds per square inch)

| Material | Compression | Tension |
|---|---|---|
| Bricks, best hard........................... | 12,000 | 400 |
| Bricks, light red........................... | 1,000 | 40 |
| Brickwork, common......................... | 1,000 | 50 |
| Brickwork, best............................ | 2,000 | 300 |
| Cement, Portland, one month old............ | 2,000 | 400 |
| Cement, Portland, one year old.............. | 3,000 | 500 |
| Concrete, Portland......................... | 1,000 | 200 |
| Concrete, Portland, one year old............ | 2,000 | 400 |
| Granite................................... | 19,000 | 700 |
| Limestone and sandstone.................... | 9,000 | 300 |
| Trap rock................................. | 20,000 | 800 |
| Slate..................................... | 14,000 | 500 |
| Vulcanized Fiber.......................... | 39,000 | 13,000 |

### Influence of Temperature on the Strength of Metals

| Material | Degrees Fahrenheit | | | | | | | |
|---|---|---|---|---|---|---|---|---|
| | 210 | 400 | 570 | 750 | 930 | 1100 | 1300 | 1475 |
| | Strength in Per Cent of Strength at 70 Degrees F. | | | | | | | |
| Wrought iron ..... | 104 | 112 | 116 | 96 | 76 | 42 | 25 | 15 |
| Cast iron ......... | .... | 100 | 99 | 92 | 76 | 42 | .... | .... |
| Steel castings ..... | 109 | 125 | 121 | 97 | 57 | .... | .... | .... |
| Structural steel.... | 103 | 132 | 122 | 86 | 49 | 28 | .... | .... |
| Copper........... | 95 | 85 | 73 | 59 | 42 | .... | .... | .... |
| Bronze ........... | 101 | 94 | 57 | 26 | 18 | .... | .... | .... |

### Strength of Copper-zinc-tin Alloys
(U. S. Government Tests)

| Copper | Zinc | Tin | Tensile Strength, Lbs. per Sq. In. | Copper | Zinc | Tin | Tensile Strength, Lbs. per Sq. In. | Copper | Zinc | Tin | Tensile Strength, Lbs. per Sq. In. |
|---|---|---|---|---|---|---|---|---|---|---|---|
| 45 | 50 | 5 | 15,000 | 60 | 20 | 20 | 10,000 | 75 | 20 | 5 | 45,000 |
| 50 | 45 | 5 | 50,000 | 65 | 30 | 5 | 50,000 | 75 | 15 | 10 | 45,000 |
| 50 | 40 | 10 | 15,000 | 65 | 25 | 10 | 42,000 | 75 | 10 | 15 | 43,000 |
| 55 | 43 | 2 | 65,000 | 65 | 20 | 15 | 30,000 | 75 | 5 | 20 | 41,000 |
| 55 | 40 | 5 | 62,000 | 65 | 15 | 20 | 18,000 | 80 | 15 | 5 | 45,000 |
| 55 | 35 | 10 | 32,500 | 65 | 10 | 25 | 12,000 | 80 | 10 | 10 | 45,000 |
| 55 | 30 | 15 | 15,000 | 70 | 25 | 5 | 45,000 | 80 | 5 | 15 | 47,500 |
| 60 | 37 | 3 | 60,000 | 70 | 20 | 10 | 44,000 | 85 | 10 | 5 | 43,500 |
| 60 | 35 | 5 | 52,500 | 70 | 15 | 15 | 37,000 | 85 | 5 | 10 | 46,500 |
| 60 | 30 | 10 | 40,000 | 70 | 10 | 20 | 30,000 | 90 | 5 | 5 | 42,000 |

## Permissible Working Stresses for Structural Timbers
(U. S. Government Tests)

| Kind of Timber | Bending, Pounds per Sq. In. | | | Compression, Pounds per Sq. In. | | | |
|---|---|---|---|---|---|---|---|
| | Allowable Stress in Extreme Fiber | | Allowable Horizontal Shear Stress | Allowable Stress Parallel to Grain "Short Columns" | | Allowable Stress Perpendicular to Grain | |
| | Outside Location | Dry Location | All Locations | Outside Location | Dry Location | Outside Location | Inside Location |
| Cedar, western red............... | 800 | 900 | 80 | 700 | 700 | 150 | 200 |
| Cedar, northern white........... | 650 | 750 | 70 | 500 | 550 | 140 | 175 |
| Chestnut......................... | 850 | 950 | 90 | 700 | 800 | 200 | 300 |
| Cypress.......................... | 1100 | 1300 | 100 | 1100 | 1100 | 250 | 350 |
| Douglas fir (No. 1 str'l) *........ | 1400 | 1600 | 100 | 1100 | 1200 | 250 | 350 |
| Douglas fir (No. 2 str'l).......... | 1100 | 1300 | 90 | 900 | 1000 | 225 | 300 |
| Fir, balsam...................... | 750 | 900 | 70 | 600 | 700 | 125 | 150 |
| Gum, red........................ | 900 | 1100 | 100 | 750 | 800 | 200 | 300 |
| Hemlock, western................ | 1100 | 1300 | 75 | 900 | 900 | 225 | 300 |
| Hemlock, eastern................ | 900 | 1000 | 70 | 700 | 700 | 225 | 300 |
| Hickory.......................... | 1500 | 1900 | 140 | 1200 | 1500 | 400 | 600 |
| Maple, sugar or hard............ | 1300 | 1500 | 150 | 1100 | 1200 | 375 | 500 |
| Maple, silver or soft............. | 900 | 1000 | 100 | 700 | 800 | 250 | 350 |
| Oak, white or red................ | 1200 | 1400 | 125 | 900 | 1000 | 375 | 500 |
| Pine, s. yellow (dense) †........ | 1400 | 1600 | 125 | 1100 | 1200 | 250 | 350 |
| Pine, s. yellow (sound).......... | 1100 | 1300 | 105 | 900 | 1000 | 225 | 300 |
| Pine, eastern white.............. | 800 | 900 | 85 | 750 | 750 | 150 | 250 |
| Pine, western white............. | 800 | 900 | 85 | 750 | 750 | 150 | 250 |
| Pine, Norway.................... | 1000 | 1100 | 85 | 800 | 800 | 175 | 300 |
| Redwood......................... | 1000 | 1200 | 70 | 900 | 1000 | 150 | 250 |
| Spruce, red or white............ | 900 | 1100 | 85 | 750 | 800 | 150 | 250 |
| Spruce, Englemann.............. | 650 | 750 | 70 | 550 | 600 | 140 | 175 |

* The strength of large timbers depends chiefly upon the density or weight per cubic foot of the dry wood and upon the character, size, number and location of defects. " Dense " Douglas fir of the " No. 1 structural grade " shows on one end an average of at least six annual rings per inch and at least one-third " summer wood," measured over 3 inches on a line extending from the pith to the corner farthest from the pith when the least dimension of the timber is 5 inches or more. The point where the 3-inch line begins is found by the formula $A = \frac{1}{2}D - 2$, where $A$ = distance in inches from pith to beginning of 3-inch line and $D$ = minimum dimension of timber in inches. The " No. 2 structural grade " for Douglas fir includes timbers not passing the No. 1 grade, because (1) there is less density than required or (2) greater defects than are permitted.

† The term " southern yellow pine " includes the species known heretofore as long-leaf pine, short-leaf pine, loblolly pine, Cuban pine and pond pine. " Dense " southern yellow pine shows on either end an average of at least six annual rings per inch and at least one-third summer wood, or else the greater number of rings shows at least one-third summer wood all as measured over the third, fourth, and fifth inches of a radial line extending from the pith. Wide-ringed material, excluded by this rule, is acceptable, provided the amount of summer wood measured as previously specified is at least one-half. " Sound " southern yellow pine includes pieces without any ring or summer wood requirement.

## Stresses Produced by Shocks

**Stresses in Beams Produced by Shocks.** — Any elastic structure subjected to a shock will deflect until the product of the average resistance, developed by the deflection, and the distance through which it has been overcome, has reached a value equal to the energy of the shock. It follows that for a given shock, the average resisting stresses are inversely proportional to the deflection. If the structure were perfectly rigid, the deflection would be o, and the stress infinite. The effect of a shock is, therefore, to a great extent dependent upon the elastic property (the springiness) of the structure subjected to the impact.

The energy of a body in motion, such as a falling body, may be spent in each of four ways:

1. In deforming the body struck as a whole.
2. In deforming the falling body as a whole.
3. In partial deformation of both bodies on the surface of contact (most of this energy will be transformed into heat).
4. Part of the energy will be taken up by the supports, if these are not perfectly rigid and inelastic.

How much energy is spent in the last three ways it is in most cases difficult to determine, and for this reason it is safest to figure as if the whole amount were spent as in Case 1. In cases where a reliable judgment is possible, as to what percentage of the energy is spent in other ways than the first, a corresponding fraction of the total energy can be assumed as developing stresses in the body subjected to shocks.

One investigation into the stresses produced by shocks led to the following conclusions: 1. A suddenly applied load will produce the same deflection, and, therefore, the same stress as a static load twice as great. 2. The unit stress $p$ (see formulas in the accompanying table) for a given load producing a shock, varies directly as the square root of the modulus of elasticity $E$, and inversely as the square root of the length $L$ of the beam and the area of the section. Thus, for instance, if the sectional area of a beam is increased four times, the unit stress will diminish only by half. This is entirely different from the results produced by static loads where the stress would vary inversely with the area, and within certain limits be practically independent of the modulus of elasticity.

### Stresses Produced in Beams by Shocks

| Method of Support, and Point Struck by Falling Body | Fiber (Unit) Stress $p$ produced by Weight $Q$ Dropped Through a Distance $h$ | Approximate Value of $p$ |
|---|---|---|
| Supported at both ends; struck in center. | $p = \dfrac{QaL}{4\,I}\left(1 + \sqrt{1 + \dfrac{96\,hEI}{QL^3}}\right)$ | $p = a\sqrt{\dfrac{6\,QhE}{LI}}$ |
| Fixed at one end; struck at the other. | $p = \dfrac{QaL}{I}\left(1 + \sqrt{1 + \dfrac{6\,hEI}{QL^3}}\right)$ | $p = a\sqrt{\dfrac{6\,QhE}{LI}}$ |
| Fixed at both ends; struck in center. | $p = \dfrac{QaL}{8\,I}\left(1 + \sqrt{1 + \dfrac{384\,hEI}{QL^3}}\right)$ | $p = a\sqrt{\dfrac{6\,QhE}{LI}}$ |

$I$ = moment of inertia of section; $a$ = distance of extreme fiber from neutral axis; $L$ = length of beam; $E$ = modulus of elasticity.

In the table, the expression for the approximate value of $p$, which is applicable whenever the deflection of the beam is small as compared with the total height $h$ through which the body producing the shock is dropped, is always the same for beams supported at both ends and subjected to shock at *any* point between the supports. In the formulas all dimensions are in inches and weights in pounds.

If metric SI units are used, $p$ is in newtons per square millimeter; $Q$ is in newtons; $E$ = modulus of elasticity in N/mm²; $I$ = moment of inertia of section in millimeters⁴; and $h$, $a$, and $L$ in millimeters. *Note:* If $Q$ is given in kilograms, the value referred to is mass. The weight $Q$ of a mass $M$ kilograms is $Mg$ newtons, where $g$ = approximately 9.81 meters per second².

*Examples of How Formulas for Stresses Produced by Shocks are Derived:* The general formula from which specific formulas for shock stresses in beams, springs, and other machine and structural members are derived is:

$$p = p_s \left( 1 + \sqrt{1 + \frac{2h}{y}} \right) \qquad (1)$$

In this formula, $p$ = stress in pounds per square inch due to shock caused by impact of a moving load; $p_s$ = stress in pounds per square inch resulting when moving load is applied statically; $h$ = distance in inches that load falls before striking beam, spring, or other member; $y$ = deflection, in inches, resulting from static load.

As an example of how Formula (1) may be used to obtain a formula for a specific application, suppose that the load $W$ shown applied to the beam in Case 2 on page 404 were dropped on the beam from a height of $h$ inches instead of being gradually applied (static loading). The maximum stress $p_s$ due to load $W$ for Case 2 is given as $Wl \div 4Z$ and the maximum deflection $y$ is given as $Wl^3 \div 48 EI$. Substituting these values in Formula (1),

$$p = \frac{Wl}{4Z} \left( 1 + \sqrt{1 + \frac{2h}{Wl^3 \div 48 EI}} \right) = \frac{Wl}{4Z} \left( 1 + \sqrt{1 + \frac{96 \, hEI}{Wl^3}} \right) \qquad (2)$$

If in Formula (2) the letter $Q$ is used in place of $W$ and if $Z$, the section modulus, is replaced by its equivalent, $I \div$ distance $a$ from neutral axis to extreme fiber of beam, then Formula (2) becomes the first formula given in the accompanying table "Stresses Produced in Beams by Shocks."

**Stresses in Helical Springs Produced by Shocks.** — A load suddenly applied on a spring will produce the same deflection, and, therefore, also the same unit stress, as a static load twice as great. When the load drops from a height $h$, the stresses are as given in the accompanying table. The approximate values are applicable

### Stresses Produced in Springs by Shocks

| Form of Bar from Which Spring is Made | Fiber (Unit) Stress $f$ Produced by Weight $Q$ Dropped a Height $h$ on a Helical Spring | Approximate Value of $f$ |
|---|---|---|
| Round | $f = \dfrac{8QD}{\pi d^3} \left( 1 + \sqrt{1 + \dfrac{Ghd^4}{4\,QD^3n}} \right)$ | $f = 1.27 \sqrt{\dfrac{QhG}{Dd^2 n}}$ |
| Square | $f = \dfrac{9QD}{4 d^3} \left( 1 + \sqrt{1 + \dfrac{Ghd^4}{0.9\,\pi QD^3n}} \right)$ | $f = 1.34 \sqrt{\dfrac{QhG}{Dd^2 n}}$ |

$G$ = modulus of elasticity for torsion; $d$ = diameter or side of bar; $D$ = mean diameter of spring; $n$ = number of coils in spring.

when the deflection is small as compared with the height $h$. The formulas show that the fiber stress for a given shock will be greater in a spring made from a square bar, than in one made from a round bar, if the diameter of coil be the same, and the side of the square bar equals the diameter of the round bar. It is, therefore, more economical to use round stock for springs which must withstand shocks. This is due to the fact that the deflection for the same fiber stress for a square bar spring is smaller than that for a round bar spring, the ratio being as 4 to 5. The round bar spring is therefore capable of storing more energy than a square bar spring for the same stress.

**Shocks from Bodies in Motion.** — The formulas given can be applied, in general, to shocks from bodies in motion. A body of the weight $W$ moving horizontally with the velocity of $v$ feet per second, has a stored-up energy:

$$A = \frac{1}{2} \times \frac{Wv^2}{g} \text{ foot-pounds, or } \frac{6\,Wv^2}{g} \text{ inch-pounds.}$$

This expression may be substituted for $Qh$ in the tables in the equations for unit stresses containing this quantity, and the stresses produced by the energy of the moving body thereby determined.

The formulas in the tables give the maximum value of the stresses, providing the designer with something definite to guide him even in cases where he may be justified in assuming that only a part of the energy of the shock is taken up by the member under stress.

The formulas can also be applied using metric SI units. The stored-up energy of a body of mass $M$ kilograms moving horizontally with the velocity of $v$ meters per second is:

$$A = \frac{1}{2}Mv^2 \text{ newton-meters.}$$

This expression may be substituted for $Qh$ in the appropriate equations in the tables. For calculation in millimeters, $Qh = 1000\,A$ newton-millimeters.

# SHAFTING

**Torsional Strength of Shafting.** — In the formulas that follow,

$\alpha$ = angular deflection of shaft in degrees;
$c$ = distance from center of gravity to extreme fiber;
$D$ = diameter of shaft in inches;
$G$ = torsional modulus of elasticity = 11,500,000 pounds per square inch for steel;
$J$ = polar moment of inertia of shaft cross-section (see table);
$l$ = length of shaft in inches;
$N$ = angular velocity of shaft in revolutions per minute;
$P$ = power transmitted in horsepower;
$S_s$ = allowable torsional shearing stress in pounds per square inch;
$T$ = torsional or twisting moment in inch-pounds;
$Z_p$ = polar section modulus (see table).

The allowable twisting moment for a shaft of any cross-section such as circular, square, etc. is:

$$T = S_s \times Z_p \tag{1}$$

For a shaft delivering $P$ horsepower at $N$ revolutions per minute the twisting moment $T$ being transmitted is:

$$T = \frac{63,000\,P}{N} \tag{2}$$

The twisting moment $T$ as determined by this formula should be less than the value determined by using Formula (1) if the maximum allowable stress $S_s$ is not to be exceeded.

The diameter of a solid circular shaft required to transmit a given torque $T$ is:

$$D = \sqrt[3]{\frac{5.1\,T}{S_s}}\,; \quad \text{or} \quad D = \sqrt[3]{\frac{321,000\,P}{N S_s}} \qquad \text{(3a) and (3b)}$$

The allowable stresses that are generally used in practice are: 4000 pounds per square inch for main power-transmitting shafts; 6000 pounds per square inch for lineshafts carrying pulleys; and 8500 pounds per square inch for small, short shafts, countershafts, etc. Using these allowable stresses, the horsepower $P$ transmitted by a shaft of diameter $D$, or the diameter $D$ of a shaft to transmit a given horsepower $P$ may be determined from the following formulas:

For main power-transmitting shafts:

$$P = \frac{D^3 N}{80}\,; \quad \text{or} \quad D = \sqrt[3]{\frac{80\,P}{N}} \qquad \text{(4a) and (4b)}$$

For lineshafts carrying pulleys:

$$P = \frac{D^3 N}{53.5}\,; \quad \text{or} \quad D = \sqrt[3]{\frac{53.5\,P}{N}} \qquad \text{(5a) and (5b)}$$

For small, short shafts:

$$P = \frac{D^3 N}{38}\,; \quad \text{or} \quad D = \sqrt[3]{\frac{38\,P}{N}} \qquad \text{(6a) and (6b)}$$

Shafts which are subjected to shocks, sudden starting and stopping, etc., should be given a greater factor of safety resulting in the use of lower allowable stresses than those just mentioned.

*Example:* — What would be the diameter of a lineshaft to transmit 10 horsepower if the shaft makes 150 revolutions per minute? Using Formula (5b),

$$D = \sqrt[3]{\frac{53.5 \times 10}{150}} = 1.53, \quad \text{or, say, } 1\tfrac{9}{16} \text{ inches.}$$

*Example:* — What horsepower would a short shaft, 2 inches in diameter, carrying but two pulleys close to the bearings transmit if the shaft makes 300 revolutions per minute? Using Formula (6a),

$$P = \frac{2^3 \times 300}{38} = 63 \text{ horsepower.}$$

**Torsional Strength of Shafting — Calculations in Metric SI Units.** — The allowable twisting moment for a shaft of any cross-section such as circular, square, etc., can be calculated from:

$$T = S_s \times Z_p \qquad (1')$$

where $T$ = torsional or twisting moment in newton-millimeters; $S_s$ = allowable torsional shearing stress in newtons per square millimeter; and $Z_p$ = polar section modulus in millimeters³.

For a shaft delivering power of $P$ kilowatts at $N$ revolutions per minute, the twisting moment $T$ being transmitted is:

$$T = \frac{9.55 \times 10^6\,P}{N}\,; \quad \text{or} \quad T = \frac{10^6\,P}{\omega} \qquad \text{(2') and (2a')}$$

where $T$ is in newton-millimeters, and $\omega$ = angular velocity in radians per second.

The diameter $D$ of a solid circular shaft required to transmit a given torque $T$ is:

$$D = \sqrt[3]{\frac{5.1\,T}{S_s}}\;; \quad \text{or} \quad D = \sqrt[3]{\frac{48.7 \times 10^6\,P}{NS_s}}\;; \qquad \text{(3a') and (3b')}$$

$$\text{or} \qquad D = \sqrt[3]{\frac{5.1 \times 10^6\,P}{\omega S_s}} \qquad\qquad\qquad \text{(3c')}$$

where $D$ is in millimeters; $T$ is in newton-millimeters; $P$ is power in kilowatts; $N$ = revolutions per minute; $S_s$ = allowable torsional shearing stress in newtons per square millimeter, and $\omega$ = angular velocity in radians per second.

If 28 newtons/mm$^2$ and 59 newtons/mm$^2$ are taken as the generally allowed stresses for main power-transmitting shafts, and small short shafts respectively, then using these allowable stresses, the power $P$ transmitted by a shaft of diameter $D$, or the diameter $D$ of a shaft to transmit a given power $P$ may be determined from the following formulas:

For main power-transmitting shafts:

$$P = \frac{D^3 N}{1.77 \times 10^6}\;; \quad \text{or} \quad D = \sqrt[3]{\frac{1.77 \times 10^6\,P}{N}} \qquad \text{(4a') and (4b')}$$

For small, short shafts:

$$P = \frac{D^3 N}{0.83 \times 10^6}\;; \quad \text{or} \quad D = \sqrt[3]{\frac{0.83 \times 10^6\,P}{N}}\,, \qquad \text{(6a') and (6b')}$$

where $P$ is in kilowatts; $D$ is in millimeters, and $N$ = revolutions per minute.

*Example:* What would be the diameter of a power-transmitting shaft to transmit 150 kW at 500 rpm?

$$D = \sqrt[3]{\frac{1.77 \times 10^6 \times 150}{500}} = 81 \text{ millimeters.}$$

*Example:* What power would a short shaft, 50 millimeters in diameter, transmit at 400 rpm?

$$P = \frac{50^3 \times 400}{0.83 \times 10^6} = 60 \text{ kilowatts.}$$

**Polar Moment of Inertia and Section Modulus.** — The *polar moment of inertia*, $J$, of a cross-section with respect to a polar axis, that is, an axis at right angles to the plane of the cross-section, is defined as the moment of inertia of the cross-section with respect to the point of intersection of the axis and the plane. The polar moment of inertia may be found by taking the sum of the moments of inertia about two perpendicular axes lying in the plane of the cross-section and passing through this point. Thus, for example, the polar moment of inertia of a circular or a square area with respect to a polar axis through the center of gravity is equal to two times the moment of inertia with respect to an axis lying in the plane of the cross-section and passing through the center of gravity.

The polar moment of inertia with respect to a polar axis through the center of gravity is required for problems involving the torsional strength of shafts since this axis is usually the axis about which twisting of the shaft takes place.

The *polar section modulus* (also called section modulus of torsion), $Z_p$, for *circular* sections may be found by dividing the polar moment of inertia, $J$, by the distance $c$ from the center of gravity to the most remote fibre. This method may be used to find the *approximate* value of the polar section modulus of sections that are *nearly* round. For other than circular cross-sections, however, the polar section modulus *does not* equal the polar moment of inertia divided by the distance $c$.

The accompanying table gives formulas for the polar section modulus for several

Polar Moment of Inertia and Polar Section Modulus

| Section | Polar Moment of Inertia $J$ | Polar Section Modulus $Z_p$ |
|---|---|---|
| | $\dfrac{a^4}{6} = 0.1667\,a^4$ | $0.208\,a^3 = 0.074\,d^3$ |
| | $\dfrac{bd\,(b^2 + d^2)}{12}$ | $\dfrac{bd^2}{3 + 1.8\dfrac{d}{b}}$ <br> ($d$ is the shorter side) |
| | $\dfrac{\pi D^4}{32} = 0.098\,D^4$ <br> (see also footnote, p. 385) | $\dfrac{\pi D^3}{16} = 0.196\,D^3$ <br> (see also footnote, p. 385) |
| | $\dfrac{\pi}{32}(D^4 - d^4)$ <br> $= 0.098\,(D^4 - d^4)$ | $\dfrac{\pi}{16}\left(\dfrac{D^4 - d^4}{D}\right)$ <br> $= 0.196\left(\dfrac{D^4 - d^4}{D}\right)$ |
| | $\dfrac{5\sqrt{3}}{8}\,s^4 = 1.0825\,s^4$ <br> $= 0.12\,F^4$ | $0.20\,F^3$ |
| | $\dfrac{\pi D^4}{32} - \dfrac{s^4}{6}$ <br> $= 0.098\,D^4 - 0.167\,s^4$ | $\dfrac{\pi D^3}{16} - \dfrac{s^4}{3\,D}$ <br> $= 0.196\,D^3 - 0.333\,\dfrac{s^4}{D}$ |
| | $\dfrac{\pi D^4}{32} - \dfrac{5\sqrt{3}}{8}\,s^4$ <br> $= 0.098\,D^4 - 1.0825\,s^4$ | $\dfrac{\pi D^3}{16} - \dfrac{5\sqrt{3}\,s^4}{4\,D}$ <br> $= 0.196\,D^3 - 2.165\,\dfrac{s^4}{D}$ |
| | $\dfrac{\sqrt{3}\,s^4}{48} = 0.036\,s^4$ | $\dfrac{s^3}{20} = 0.05\,s^3$ |

different cross-sections. The polar section modulus multiplied by the allowable torsional shearing stress gives the allowable twisting moment to which a shaft may be subjected (see Formula 1).

**Torsional Deflection of Circular Shafts.** — In many cases shafting must be proportioned not only to provide the strength required to transmit a given torque, but also to prevent torsional deflection (twisting) through a greater angle than has been found satisfactory for a given type of service.

For a solid circular shaft the torsional deflection in degrees is given by:

$$\alpha = \frac{584\, Tl}{D^4 G} \tag{7}$$

*Example:* — Find the torsional deflection for a solid steel shaft 4 inches in diameter and 48 inches long, subjected to a twisting moment of 24,000 inch-pounds. By Formula (7),

$$\alpha = \frac{584 \times 24,000 \times 48}{4^4 \times 11,500,000} = 0.23 \text{ degree.}$$

Formula (7) can be used with metric SI units, where $\alpha$ = angular deflection of shaft in degrees; $T$ = torsional moment in newton-millimeters; $l$ = length of shaft in millimeters; $D$ = diameter of shaft in millimeters; and $G$ = torsional modulus of elasticity in newtons per square millimeter.

*Example:* Find the torsional deflection of a solid steel shaft, 100 mm in diameter and 1300 mm long, subjected to a twisting moment of $3 \times 10^6$ newton-millimeters. The torsional modulus of elasticity is 80,000 newtons/mm². By Formula (7)

$$\alpha = \frac{584 \times 3 \times 10^6 \times 1300}{100^4 \times 80,000} = 0.285 \text{ degree.}$$

The diameter of a shaft which is to have a maximum torsional deflection $\alpha$ is given by:

$$D = 4.9 \sqrt[4]{\frac{Tl}{G\alpha}} \tag{8}$$

Formula (8) can be used with metric SI units, where $D$ = diameter of shaft in millimeters; $T$ = torsional moment in newton-millimeters; $l$ = length of shaft in millimeters; $G$ = torsional modulus of elasticity in newtons per square millimeter; and $\alpha$ = angular deflection of shaft in degrees.

According to some authorities, the allowable twist in steel transmission shafting should not exceed 0.08 degree per foot length of the shaft. The diameter $D$ of a shaft which will permit a maximum angular deflection of 0.08 degree per foot of length for a given torque $T$ or for a given horsepower $P$ can be determined from the formulas:

$$D = 0.29 \sqrt[4]{T}; \quad \text{or} \quad D = 4.6 \sqrt[4]{\frac{P}{N}} \tag{9a and 9b}$$

Using metric SI units and assuming an allowable twist in steel transmission shafting of 0.26 degree per meter length, Formulas (9a) and (9b) become:

$$D = 2.26 \sqrt[4]{T}; \quad \text{or} \quad D = 125.7 \sqrt[4]{\frac{P}{N}},$$

where $D$ = diameter of shaft in millimeters; $T$ = torsional moment in newton-millimeters; $P$ = power in kilowatts; and $N$ = revolutions per minute.

Another rule that has been generally used in mill practice limits the deflection to 1 degree in a length equal to 20 times the shaft diameter. For a given torque or horsepower, the diameter of a shaft having this maximum deflection is given by:

$$D = 0.1 \sqrt[3]{T}; \quad \text{or} \quad D = 4.0 \sqrt[3]{\frac{P}{N}} \qquad \text{(10a) and (10b)}$$

*Example:* — Find the diameter of a steel lineshaft to transmit 10 horsepower at 150 revolutions per minute with a torsional deflection not exceeding 0.08 degree per foot of length. By Formula (9b),

$$D = 4.6 \sqrt[4]{\frac{10}{150}} = 2.35 \text{ inches.}$$

This diameter is larger than that obtained for the same horsepower and rpm in the example given for Formula (5b) in which the diameter was calculated for strength considerations only. The usual procedure in the design of shafting which is to have a specified maximum angular deflection is to compute the diameter first by means of Formula (8), (9a), (9b), (10a), or (10b), and then by means of Formula (3a), (3b), (4b), (5b), or (6b), using the larger of the two diameters thus found.

**Linear Deflection of Shafting.** — For steel lineshafting, it is considered good practice to limit the linear deflection to a maximum of 0.010 inch per foot of length. The maximum distance in feet between bearings, for average conditions, in order to avoid excessive linear deflection, is determined by the formulas:

$$L = 8.95 \sqrt[3]{D^2} \text{ for shafting subject to no bending action except its own weight}$$
$$L = 5.2 \sqrt[3]{D^2} \text{ for shafting subject to bending action of pulleys, etc.}$$

in which $D$ = diameter of shaft in inches; $L$ = maximum distance between bearings in feet. Pulleys should be placed as close to the bearings as possible.

**Tables of Horsepower Transmitted by Shafting.** — The accompanying table, "Horsepower Transmitted by Shafting made from Medium Steel" gives the rela-

**Horsepower Transmitted by Shafting made from Medium Steel**

| Diam. of Shaft, Inches | Transmitting Power, but Subject to No Bending Action Except Its Own Weight | | | | | | Transmitting Power, and Subject to Bending Action of Pulleys, Belting, Etc. | | | | | |
|---|---|---|---|---|---|---|---|---|---|---|---|---|
| | Revolutions per Minute | | | | | Max. Distance in Feet Between Bearings | Revolutions per Minute | | | | | Max. Distance in Feet Between Bearings |
| | 100 | 150 | 200 | 250 | 300 | | 100 | 150 | 200 | 250 | 300 | |
| | Horsepower | | | | | | Horsepower | | | | | |
| 1½ | 7 | 10 | 14 | 17 | 20 | 11.7 | 5 | 7 | 10 | 12 | 14 | 6.8 |
| 1⅝ | 9 | 13 | 17 | 21 | 26 | 12.4 | 6 | 9 | 12 | 15 | 18 | 7.2 |
| 1¾ | 11 | 16 | 21 | 26 | 32 | 13.0 | 8 | 11 | 15 | 18 | 22 | 7.5 |
| 1⅞ | 13 | 20 | 26 | 33 | 40 | 13.6 | 9 | 14 | 19 | 23 | 28 | 7.9 |
| 2 | 16 | 24 | 32 | 40 | 48 | 14.2 | 11 | 17 | 23 | 28 | 34 | 8.2 |
| 2⅛ | 19 | 29 | 38 | 48 | 58 | 14.8 | 14 | 21 | 27 | 34 | 42 | 8.6 |
| 2¼ | 23 | 34 | 46 | 57 | 68 | 15.4 | 16 | 24 | 33 | 41 | 48 | 8.9 |
| 2⅜ | 27 | 40 | 54 | 67 | 80 | 16.0 | 19 | 29 | 38 | 48 | 58 | 9.2 |
| 2½ | 31 | 47 | 63 | 78 | 94 | 16.5 | 22 | 33 | 45 | 55 | 66 | 9.6 |
| 2¾ | 42 | 62 | 83 | 102 | 124 | 17.6 | 30 | 44 | 59 | 74 | 89 | 10.2 |
| 3 | 54 | 81 | 108 | 134 | 162 | 18.6 | 39 | 58 | 77 | 96 | 116 | 10.8 |
| 3¼ | 69 | 103 | 137 | 172 | 206 | 19.7 | 49 | 74 | 98 | 123 | 148 | 11.4 |
| 3½ | 86 | 129 | 172 | 215 | 258 | 20.7 | 61 | 92 | 123 | 153 | 184 | 12.0 |
| 3¾ | 105 | 158 | 211 | 264 | 316 | 21.6 | 75 | 113 | 151 | 188 | 226 | 12.5 |
| 4 | 128 | 192 | 256 | 320 | 384 | 22.6 | 91 | 137 | 183 | 228 | 274 | 13.1 |

tion between the diameter of shaft, revolutions per minute, horsepower transmitted, and maximum distance in feet between bearings. Assume, for example, that it is required to find the diameter of a shaft for transmitting 40 horsepower at a speed of 250 revolutions per minute. The shaft is not subjected to any bending action except its own weight. From the table, it is found, by locating "40" in the column under 250 revolutions per minute, that the diameter of the shaft required is 2 inches. The maximum permissible distance between the shaft bearings is slightly more than 14 feet. When the exact horsepower cannot be found in the table, it is advisable to take the nearest larger value listed in the table and find the diameter of shafting required to transmit this horsepower. Tables are also given for the horsepower which can be safely transmitted by cold-rolled and turned steel lineshafting. The cold-rolled steel table on page 458 is carried up to 5 inches only, because this diameter is the largest which is commonly cold-rolled. These tables have been used by the transmission department of the Jones & Laughlin Steel Co., and are based on the assumption that bearings are placed at intervals of from 8 to 10 feet and all pulleys are located as near to the bearings as possible. In these tables, the body part in each gives the number of horsepower to be transmitted. For example, assume that a 3-inch cold-rolled steel lineshaft revolves at a speed of 400 revolutions per minute. Find the power that this shaft can safely transmit. By locating 3 inches in the left-hand column and 400 at the top of the vertical columns, and following the vertical column downward until opposite 3 inches, it is found that under the given conditions 154 horsepower may be safely transmitted.

In general, shafting up to three inches in diameter is almost always made from cold-rolled steel. This shafting is true and straight and needs no turning, but if keyways are cut in the shaft, it must, as a rule, be straightened afterwards, as the cutting of the keyways relieves the tension on the surface of the shaft due to the cold-rolling process. Sizes of shafting from three to five inches in diameter may be either cold-rolled or turned, more frequently the latter, while all larger sizes of shafting must be turned, because cold-rolled shafting is not available in diameters larger than five inches.

**Design of Transmission Shafting.** — The following guidelines for the design of shafting for transmitting a given amount of power under various conditions of loading are based upon formulas given in the former American Standard ASA B17c Code for the Design of Transmission Shafting. These formulas are based on the *maximum-shear theory* of failure which assumes that the elastic limit of a *ductile* ferrous material in shear is practically one-half its elastic limit in tension. This theory agrees, very nearly, with the results of tests on ductile materials and has gained wide acceptance in practice.

The formulas given apply in all cases of shaft design including shafts for special machinery. The limitation of these formulas is that they provide only for the strength of shafting and are not concerned with the torsional or lineal deformation which may, in the case of shafts used in machine design, be the controlling factor (see " Torsional Deflection of Circular Shafts " and " Linear Deflection of Shafting " for deflection considerations). In the formulas that follow,

$B = \sqrt[3]{1 \div (1 - K^4)}$ (see Table 3);

$D$ = outside diameter of shaft in inches;

$D_1$ = inside diameter of a hollow shaft in inches;

$K_m$ = combined shock and fatigue factor to be applied in every case to the computed bending moment (see Table 1);

$K_t$ = combined shock and fatigue factor to be applied in every case to the computed torsional moment (see Table 1);

$M$ = maximum bending moment in inch-pounds;

$N$ = revolutions per minute;

## Horsepower Transmitted by Turned Steel Lineshafting

| Diam. of Shaft | Number of Revolutions per Minute | | | | | | | | | | | | |
|---|---|---|---|---|---|---|---|---|---|---|---|---|---|
| | 100 | 125 | 150 | 175 | 200 | 225 | 250 | 300 | 350 | 400 | 450 | 500 | 600 |
| 1½ | 3.7 | 4.7 | 5.6 | 6.6 | 7.5 | 8.4 | 9.4 | 11.2 | 13.1 | 15.0 | 16.9 | 18.8 | 22 |
| 1⁹⁄₁₆ | 4.2 | 5.3 | 6.4 | 7.4 | 8.5 | 9.5 | 10.6 | 12.7 | 14.8 | 17.0 | 19.0 | 21 | 25 |
| 1⅝ | 4.8 | 5.9 | 7.1 | 8.3 | 9.5 | 10.7 | 11.9 | 14.3 | 16.6 | 19.0 | 21 | 24 | 28 |
| 1¹¹⁄₁₆ | 5.3 | 6.7 | 8.0 | 9.3 | 10.7 | 12.0 | 13.4 | 16.0 | 18.7 | 21 | 24 | 27 | 32 |
| 1¾ | 5.9 | 7.4 | 8.9 | 10.4 | 11.9 | 13.4 | 14.9 | 17.9 | 21 | 24 | 27 | 30 | 36 |
| 1¹³⁄₁₆ | 6.6 | 8.2 | 9.9 | 11.5 | 13.2 | 14.8 | 16.5 | 19.8 | 23 | 26 | 30 | 33 | 40 |
| 1⅞ | 7.3 | 9.1 | 11.0 | 12.8 | 14.7 | 16.5 | 18.3 | 22 | 26 | 29 | 33 | 37 | 44 |
| 1¹⁵⁄₁₆ | 8.1 | 10.0 | 12.1 | 14.1 | 16.1 | 18.2 | 20 | 24 | 28 | 32 | 36 | 40 | 48 |
| 2 | 8.9 | 11.1 | 13.3 | 15.6 | 17.8 | 20 | 22 | 27 | 31 | 35 | 40 | 44 | 53 |
| 2¹⁄₁₆ | 9.8 | 12.3 | 14.7 | 17.2 | 19.6 | 22 | 24 | 29 | 34 | 39 | 44 | 49 | 59 |
| 2⅛ | 10.6 | 13.3 | 16.0 | 18.6 | 21 | 24 | 27 | 32 | 37 | 43 | 48 | 53 | 64 |
| 2³⁄₁₆ | 11.6 | 14.6 | 17.5 | 20.0 | 23 | 26 | 29 | 35 | 41 | 47 | 52 | 58 | 70 |
| 2¼ | 12.6 | 15.8 | 19.0 | 22.0 | 25 | 28 | 32 | 38 | 44 | 51 | 57 | 63 | 76 |
| 2⁵⁄₁₆ | 13.7 | 17.2 | 21 | 24 | 27 | 31 | 34 | 41 | 48 | 55 | 62 | 69 | 82 |
| 2⅜ | 14.9 | 18.6 | 22 | 26 | 30 | 33 | 37 | 45 | 52 | 60 | 67 | 74 | 89 |
| 2⁷⁄₁₆ | 16.0 | 20 | 24 | 28 | 32 | 36 | 40 | 48 | 56 | 64 | 72 | 80 | 96 |
| 2½ | 17.4 | 22 | 26 | 30 | 35 | 39 | 43 | 52 | 61 | 69 | 78 | 87 | 104 |
| 2⁹⁄₁₆ | 18.7 | 23 | 28 | 33 | 37 | 42 | 47 | 56 | 66 | 75 | 84 | 94 | 112 |
| 2⅝ | 20 | 25 | 30 | 35 | 40 | 45 | 50 | 60 | 71 | 80 | 90 | 100 | 120 |
| 2¹¹⁄₁₆ | 21 | 27 | 32 | 38 | 43 | 48 | 54 | 65 | 76 | 86 | 97 | 108 | 129 |
| 2¾ | 23 | 29 | 35 | 40 | 46 | 52 | 58 | 69 | 81 | 92 | 104 | 115 | 138 |
| 2¹³⁄₁₆ | 25 | 31 | 37 | 43 | 49 | 56 | 62 | 74 | 87 | 99 | 111 | 124 | 148 |
| 2⅞ | 26 | 33 | 40 | 46 | 53 | 59 | 66 | 79 | 92 | 105 | 119 | 132 | 158 |
| 2¹⁵⁄₁₆ | 28 | 35 | 42 | 49 | 56 | 63 | 70 | 84 | 99 | 113 | 127 | 141 | 169 |
| 3 | 30 | 37 | 45 | 52 | 60 | 67 | 75 | 90 | 105 | 120 | 135 | 150 | 180 |
| 3⅛ | 34 | 42 | 51 | 59 | 68 | 76 | 85 | 102 | 119 | 136 | 152 | 170 | 203 |
| 3¼ | 38 | 48 | 57 | 67 | 76 | 86 | 95 | 114 | 134 | 153 | 172 | 191 | 229 |
| 3⅜ | 43 | 53 | 64 | 75 | 85 | 96 | 107 | 128 | 150 | 171 | 192 | 213 | 256 |
| 3½ | 48 | 60 | 72 | 83 | 95 | 107 | 119 | 143 | 167 | 190 | 214 | 238 | 286 |
| 3⅝ | 53 | 66 | 79 | 93 | 106 | 119 | 132 | 159 | 185 | 211 | 238 | 265 | 317 |
| 3¾ | 59 | 73 | 88 | 103 | 117 | 132 | 146 | 176 | 205 | 234 | 264 | 293 | 351 |
| 3⅞ | 65 | 81 | 97 | 113 | 129 | 145 | 161 | 194 | 226 | 258 | 291 | 322 | 387 |
| 4 | 71 | 89 | 107 | 125 | 142 | 160 | 178 | 213 | 249 | 284 | 320 | 356 | 427 |
| 4⅛ | 78 | 98 | 117 | 136 | 156 | 176 | 195 | 235 | 273 | 312 | 351 | 390 | 468 |
| 4¼ | 85 | 107 | 128 | 149 | 170 | 192 | 213 | 256 | 298 | 341 | 385 | 426 | 511 |
| 4⅜ | 93 | 116 | 139 | 163 | 186 | 210 | 233 | 279 | 326 | 372 | 419 | 466 | 559 |
| 4½ | 102 | 127 | 152 | 178 | 203 | 228 | 253 | 305 | 356 | 405 | 456 | 507 | 610 |
| 4⅝ | 110 | 138 | 165 | 193 | 220 | 247 | 275 | 330 | 385 | 440 | 495 | 550 | 660 |
| 4¾ | 119 | 149 | 179 | 209 | 238 | 268 | 298 | 357 | 416 | 476 | 537 | 595 | 714 |
| 4⅞ | 129 | 161 | 193 | 226 | 258 | 290 | 322 | 387 | 452 | 516 | 581 | 646 | 775 |
| 5 | 139 | 174 | 208 | 244 | 278 | 313 | 347 | 417 | 486 | 557 | 625 | 695 | 835 |

## Horsepower Transmitted by Cold-rolled Steel Lineshafting

| Diam. of Shaft | Number of Revolutions per Minute | | | | | | | | | | | | |
|---|---|---|---|---|---|---|---|---|---|---|---|---|---|
| | 100 | 125 | 150 | 175 | 200 | 225 | 250 | 300 | 350 | 400 | 450 | 500 | 600 |
| 1½ | 4.8 | 6.0 | 7.2 | 8.4 | 9.6 | 10.8 | 12.0 | 14.4 | 16.9 | 19.2 | 22 | 24 | 29 |
| 1⁹⁄₁₆ | 5.5 | 6.8 | 8.2 | 9.5 | 10.9 | 12.2 | 13.6 | 16.4 | 19.0 | 22 | 25 | 27 | 33 |
| 1⅝ | 6.1 | 7.6 | 9.2 | 10.7 | 12.2 | 13.8 | 15.3 | 18.4 | 21 | 24 | 28 | 31 | 37 |
| 1¹¹⁄₁₆ | 6.9 | 8.6 | 10.3 | 12.0 | 13.7 | 15.4 | 17.1 | 21 | 24 | 27 | 31 | 34 | 41 |
| 1¾ | 7.7 | 9.6 | 11.5 | 13.4 | 15.3 | 17.2 | 19.1 | 23 | 27 | 31 | 34 | 38 | 46 |
| 1¹³⁄₁₆ | 8.5 | 10.6 | 12.7 | 14.8 | 16.9 | 19.0 | 21 | 25 | 30 | 34 | 38 | 42 | 51 |
| 1⅞ | 9.4 | 11.7 | 14.1 | 16.4 | 18.8 | 21 | 23 | 28 | 33 | 38 | 42 | 47 | 57 |
| 1¹⁵⁄₁₆ | 10.4 | 13.0 | 15.6 | 18.2 | 21 | 23 | 26 | 31 | 36 | 42 | 47 | 52 | 62 |
| 2 | 11.4 | 14.3 | 17.2 | 20 | 23 | 26 | 29 | 34 | 40 | 46 | 51 | 57 | 69 |
| 2¹⁄₁₆ | 12.6 | 15.7 | 18.9 | 22 | 25 | 28 | 31 | 38 | 44 | 50 | 56 | 63 | 76 |
| 2⅛ | 13.7 | 17.1 | 21 | 24 | 27 | 31 | 34 | 41 | 48 | 55 | 61 | 68 | 82 |
| 2³⁄₁₆ | 15.0 | 18.7 | 22 | 26 | 30 | 34 | 37 | 45 | 52 | 60 | 67 | 75 | 90 |
| 2¼ | 16.3 | 20 | 24 | 29 | 33 | 37 | 41 | 49 | 57 | 65 | 73 | 81 | 98 |
| 2⁵⁄₁₆ | 17.7 | 22 | 27 | 31 | 35 | 40 | 44 | 53 | 62 | 71 | 80 | 88 | 106 |
| 2⅜ | 19.2 | 24 | 29 | 34 | 38 | 43 | 48 | 57 | 67 | 76 | 86 | 96 | 115 |
| 2⁷⁄₁₆ | 20 | 25 | 30 | 36 | 41 | 46 | 51 | 61 | 72 | 81 | 91 | 102 | 122 |
| 2½ | 22 | 28 | 33 | 39 | 45 | 50 | 56 | 67 | 78 | 89 | 100 | 112 | 133 |
| 2⁹⁄₁₆ | 24 | 30 | 36 | 42 | 48 | 54 | 60 | 72 | 84 | 96 | 108 | 120 | 144 |
| 2⅝ | 26 | 32 | 39 | 45 | 52 | 58 | 64 | 77 | 90 | 104 | 116 | 129 | 155 |
| 2¹¹⁄₁₆ | 28 | 35 | 42 | 48 | 55 | 62 | 69 | 83 | 97 | 111 | 124 | 138 | 166 |
| 2¾ | 30 | 37 | 44 | 52 | 59 | 67 | 74 | 89 | 104 | 119 | 133 | 148 | 178 |
| 2¹³⁄₁₆ | 32 | 40 | 47 | 55 | 63 | 71 | 79 | 95 | 111 | 127 | 143 | 159 | 190 |
| 2⅞ | 34 | 42 | 51 | 59 | 68 | 76 | 85 | 101 | 119 | 135 | 152 | 169 | 203 |
| 2¹⁵⁄₁₆ | 36 | 45 | 54 | 63 | 72 | 81 | 90 | 108 | 127 | 144 | 162 | 181 | 217 |
| 3 | 39 | 48 | 58 | 67 | 77 | 87 | 96 | 116 | 135 | 154 | 173 | 192 | 231 |
| 3⅛ | 44 | 54 | 65 | 76 | 87 | 98 | 109 | 131 | 152 | 174 | 196 | 218 | 261 |
| 3¼ | 49 | 61 | 73 | 86 | 98 | 110 | 122 | 147 | 172 | 196 | 221 | 245 | 294 |
| 3⅜ | 55 | 69 | 83 | 96 | 110 | 124 | 137 | 165 | 192 | 220 | 247 | 275 | 330 |
| 3½ | 61 | 77 | 92 | 107 | 123 | 138 | 153 | 184 | 214 | 245 | 276 | 307 | 367 |
| 3⅝ | 68 | 85 | 102 | 119 | 136 | 153 | 170 | 204 | 238 | 272 | 306 | 340 | 408 |
| 3¾ | 75 | 94 | 113 | 132 | 151 | 170 | 189 | 226 | 264 | 301 | 340 | 377 | 452 |
| 3⅞ | 83 | 104 | 125 | 145 | 166 | 187 | 207 | 249 | 291 | 332 | 379 | 415 | 498 |
| 4 | 92 | 114 | 137 | 160 | 183 | 206 | 229 | 274 | 320 | 366 | 411 | 457 | 549 |
| 4⅛ | 101 | 125 | 150 | 175 | 201 | 226 | 251 | 300 | 351 | 401 | 451 | 501 | 601 |
| 4¼ | 110 | 137 | 164 | 192 | 219 | 246 | 273 | 328 | 383 | 438 | 492 | 547 | 657 |
| 4⅜ | 120 | 150 | 180 | 210 | 239 | 268 | 298 | 358 | 418 | 478 | 538 | 597 | 717 |
| 4½ | 130 | 163 | 195 | 228 | 261 | 293 | 326 | 391 | 455 | 521 | 586 | 651 | 781 |
| 4⅝ | 141 | 177 | 212 | 247 | 283 | 318 | 354 | 425 | 495 | 566 | 636 | 707 | 848 |
| 4¾ | 153 | 191 | 230 | 268 | 307 | 344 | 382 | 459 | 537 | 613 | 688 | 765 | 919 |
| 4⅞ | 166 | 207 | 249 | 290 | 331 | 372 | 413 | 496 | 580 | 662 | 745 | 827 | 994 |
| 5 | 179 | 224 | 268 | 313 | 358 | 402 | 447 | 537 | 625 | 715 | 805 | 895 | 1074 |

**Diameters for Finished Shafting** (former American Standard ASA B17.1)

| Diameters, Inches | | Minus Tolerances, Inches* | Diameters, Inches | | Minus Tolerances, Inches* | Diameters, Inches | | Minus Tolerances, Inches* |
|---|---|---|---|---|---|---|---|---|
| Transmission Shafting | Machinery Shafting | | Transmission Shafting | Machinery Shafting | | Transmission Shafting | Machinery Shafting | |
| | 1/2 | 0.002 | | 1 13/16 | 0.003 | | 3 3/4 | 0.004 |
| | 9/16 | 0.002 | | 1 7/8 | 0.003 | | 3 7/8 | 0.004 |
| | 5/8 | 0.002 | 1 15/16 | 1 15/16 | 0.003 | 3 15/16 | 4 | 0.004 |
| | 11/16 | 0.002 | | 2 | 0.003 | | 4 1/4 | 0.005 |
| | 3/4 | 0.002 | | 2 1/16 | 0.004 | 4 7/16 | 4 1/2 | 0.005 |
| | 13/16 | 0.002 | | 2 1/8 | 0.004 | | 4 3/4 | 0.005 |
| | 7/8 | 0.002 | 2 3/16 | 2 3/16 | 0.004 | 4 15/16 | 5 | 0.005 |
| 15/16 | 15/16 | 0.002 | | 2 1/4 | 0.004 | | 5 1/4 | 0.005 |
| | 1 | 0.002 | | 2 5/16 | 0.004 | 5 7/16 | 5 1/2 | 0.005 |
| | 1 1/16 | 0.003 | | 2 3/8 | 0.004 | | 5 3/4 | 0.005 |
| | 1 1/8 | 0.003 | 2 7/16 | 2 7/16 | 0.004 | 5 15/16 | 6 | 0.005 |
| 1 3/16 | 1 3/16 | 0.003 | | 2 1/2 | 0.004 | | 6 1/4 | 0.006 |
| | 1 1/4 | 0.003 | | 2 5/8 | 0.004 | 6 1/2 | 6 1/2 | 0.006 |
| | 1 5/16 | 0.003 | | 2 3/4 | 0.004 | | 6 3/4 | 0.006 |
| | 1 3/8 | 0.003 | 2 15/16 | 2 7/8 | 0.004 | 7 | 7 | 0.006 |
| 1 7/16 | 1 7/16 | 0.003 | | 3 | 0.004 | | 7 1/4 | 0.006 |
| | 1 1/2 | 0.003 | | 3 1/8 | 0.004 | 7 1/2 | 7 1/2 | 0.006 |
| | 1 9/16 | 0.003 | | 3 1/4 | 0.004 | | 7 3/4 | 0.006 |
| | 1 5/8 | 0.003 | 3 3/16 | 3 3/8 | 0.004 | 8 | 8 | 0.006 |
| 1 11/16 | 1 11/16 | 0.003 | | 3 1/2 | 0.004 | ...... | ...... | ...... |
| | 1 3/4 | 0.003 | | 3 5/8 | 0.004 | ...... | ...... | ...... |

* *Note:* — These tolerances are *negative* or minus and represent the maximum allowable variation *below* the exact nominal size. For instance the maximum diameter of the 1 15/16 inch shaft is 1.938 inch and its minimum allowable diameter is 1.935 inch. Stock lengths of finished transmission shafting shall be: 16, 20 and 24 feet.

$P$ = maximum power to be transmitted by shaft in horsepower;

$p_t$ = maximum allowable shearing stress under combined loading conditions in pounds per square inch (see Table 2);

$S$ = maximum allowable flexural (bending) stress, either in tension or compression in pounds per square inch (see Table 2);

$S_s$ = maximum allowable torsional shearing stress in pounds per square inch (see Table 2); $T$ = maximum torsional moment in inch-pounds;

$V$ = maximum transverse shearing load in pounds.

For shafts subjected to pure torsional loads only,

$$D = B\sqrt[3]{\frac{5.1\,K_t T}{S_s}}\;;\quad \text{or}\quad D = B\sqrt[3]{\frac{321{,}000\,K_t P}{S_s N}} \qquad \text{(1a) and (1b)}$$

For stationary shafts subjected to bending only,

$$D = B\sqrt[3]{\frac{10.2\,K_m M}{S}} \qquad (2)$$

For shafts subjected to combined torsion and bending,

$$D = B\sqrt[3]{\frac{5.1}{p_t}\sqrt{(K_m M)^2 + (K_t T)^2}}\;; \qquad (3a)$$

$$\text{or}\quad D = B\sqrt[3]{\frac{5.1}{p_t}\sqrt{(K_m M)^2 + \left(\frac{63{,}000\,K_t P}{N}\right)^2}} \qquad (3b)$$

Formulas (1a) to (3b) may be used for solid shafts or for hollow shafts. For solid shafts the factor $B$ is equal to 1, whereas for hollow shafts the value of $B$ depends on the value of $K$ which, in turn, depends on the ratio of the inside diameter of the shaft to the outside diameter ($D_1 \div D = K$). Table 3 gives values of $B$ corresponding to various values of $K$.

For short solid shafts subjected only to heavy transverse shear, the diameter of shaft required is:

$$D = \sqrt{\frac{1.7\, V}{S_s}} \qquad (4)$$

Formulas (1a), (2), (3a) and (4), can be used unchanged with metric SI units. Formula (1b) becomes:

$$D = B \sqrt[3]{\frac{48.7\, K_t P}{S_s N}}\; ;$$

and Formula (3b) becomes:

$$D = B \sqrt[3]{\frac{5.1}{P_t} \sqrt{(K_m M)^2 + \left(\frac{9.55\, K_t P}{N}\right)^2}}$$

Throughout the formulas, $D$ = outside diameter of shaft in millimeters; $T$ = maximum torsional moment in newton-millimeters; $S_s$ = maximum allowable torsional shearing stress in newtons per millimeter squared (see Table 2); $P$ = maximum power to be transmitted in milliwatts; $N$ = revolutions per minute; $M$ = maximum bending moment in newton-millimeters; $S$ = maximum allowable flexural (bending) stress, either in tension or compression in newtons per millimeter squared (see Table 2); $p_t$ = maximum allowable shearing stress under combined loading conditions in newtons per millimeter squared; and $V$ = maximum transverse shearing load in kilograms. The factors $K_m$, $K_t$, and $B$ are unchanged, and $D_1$ = the inside diameter of a hollow shaft in millimeters.

Table 1. Recommended Values of the Combined Shock and Fatigue Factors for Various Types of Load

| Type of Load | Stationary Shafts | | Rotating Shafts | |
|---|---|---|---|---|
| | $K_m$ | $K_t$ | $K_m$ | $K_t$ |
| Gradually applied and steady | 1.0 | 1.0 | 1.0 | 1.0 |
| Suddenly applied, minor shocks only | 1.5-2.0 | 1.5-2.0 | 1.5-2.0 | 1.0-1.5 |
| Suddenly applied, heavy shocks | ... | ... | 2.0-3.0 | 1.5-3.0 |

Table 2. Recommended Maximum Allowable Working Stresses for Shafts Under Various Types of Load*

| Material | Type of Load | | |
|---|---|---|---|
| | Simple Bending | Pure Torsion | Combined Stress |
| "Commercial Steel" shafting without keyways | $S = 16{,}000$ | $S_s = 8000$ | $p_t = 8000$ |
| "Commercial Steel" shafting with keyways | $S = 12{,}000$ | $S_s = 6000$ | $p_t = 6000$ |
| Steel purchased under definite physical specs. | Note (a) | Note (b) | Note (b) |

* If the values in the Table are converted to metric SI units, note that 1000 pounds per square inch = 6.895 newtons per square millimeter.

(a) $S$ = 60 per cent of the elastic limit in tension but not more than 36 per cent of the ultimate tensile strength.   (b) $S_s$ and $p_t$ = 30 per cent of the elastic limit in tension but not more than 18 per cent of the ultimate tensile strength.

**Effect of Keyways on Shaft Strength.** — Keyways cut into a shaft reduce its load carrying ability, particularly when impact loads or stress reversals are involved. To insure an adequate factor of safety in the design of a shaft with standard keyway

Table 3. Values of the Factor $B$ Corresponding to Various Values of $K$ for Hollow Shafts*

| $K = \dfrac{D_1}{D} =$ | 0.95 | 0.90 | 0.85 | 0.80 | 0.75 | 0.70 | 0.65 | 0.60 | 0.55 | 0.50 |
|---|---|---|---|---|---|---|---|---|---|---|
| $B = \sqrt[3]{1 \div (1 - K^4)} =$ | 1.75 | 1.43 | 1.28 | 1.19 | 1.14 | 1.10 | 1.07 | 1.05 | 1.03 | 1.02 |

* For solid shafts $B = 1$ since $K = 0$. ($B = \sqrt[3]{1 \div (1 - K^4)} = \sqrt[3]{1 \div (1 - 0)} = 1$).

(width, one-quarter, and depth, one-eighth of shaft diameter), the former Code for Transmission Shafting tentatively recommended that shafts with keyways be designed on the basis of a solid circular shaft using not more than 75 per cent of the working stress recommended for the solid shaft.

**Formula for Shafts of Brittle Materials.** — The preceding formulas are applicable to ductile materials and are based on the maximum-shear theory of failure which assumes that the elastic limit of a *ductile* material in shear is one-half its elastic limit in tension.

Brittle materials are generally stronger in shear than in tension; therefore the maximum-shear theory is not applicable. The *maximum-normal-stress theory* of failure is now generally accepted for the design of shafts made from brittle materials. A material may be considered to be brittle if its elongation in a 2-inch gage length is less than 5 per cent. Materials such as cast iron, hardened tool steel, hard bronze, etc., conform to this rule. The diameter of a shaft made of a brittle material may be determined from the following formula which is based on the maximum-normal-stress theory of failure:

$$D = B \sqrt[3]{\frac{5.1}{S_t} \left[ (K_m M) + \sqrt{(K_m M)^2 + (K_t T)^2} \right]}$$

where $S_t$ is the maximum allowable tensile stress in pounds per square inch and the other quantities are as previously defined.

The formula can be used unchanged with metric SI units, where $D$ = outside diameter of shaft in millimeters; $S_t$ = the maximum allowable tensile stress in newtons per millimeter squared; $M$ = maximum bending moment in newton-millimeters; and $T$ = maximum torsional moment in newton-millimeters. The factors $K_m$, $K_t$, and $B$ are unchanged.

**Critical Speed of Rotating Shafts.** — At certain speeds, a rotating shaft will become dynamically unstable and the resulting vibrations and deflections that occur can result in damage not only to the shaft but to the machine of which it is a part. The speeds at which such dynamic instability occurs are called the critical speeds of the shaft. On pages 345-346 are given formulas for the critical speeds of shafts subject to various conditions of loading and support. A shaft may be safely operated either above or below its critical speed, good practice indicating that the operating speed be at least 20 per cent above or below the critical.

The formulas commonly used to determine critical speeds are sufficiently accurate for general purposes. There are cases, however, where the torque applied to a shaft has an important effect on its critical speed. Investigations have shown that the critical speeds of a uniform shaft are decreased as the applied torque is increased, and that there exist critical torques which will reduce the corresponding critical speed of the shaft to zero. A detailed analysis of the effects of applied torques on critical speeds may be found in a paper, " Critical Speeds of Uniform Shafts under Axial Torque, " by Golomb and Rosenberg presented at the First U. S. National Congress of Applied Mechanics in 1951.

## Table Giving Comparative Torsional Strength and Weight of Hollow and Solid Shafting with Same Outside Diameter

(Upper figures in each line give number of per cent decrease in strength; lower figures give per cent decrease in weight.)

*Example:* — A 4-inch shaft, with a 2-inch hole through it, has a weight 25 per cent less than a solid 4-inch shaft, but its strength is decreased only 6.25 per cent.

| Diam. of Solid and Hollow Shaft, Inches | Diameter of Axial Hole in Hollow Shaft, Inches | | | | | | | | | |
|---|---|---|---|---|---|---|---|---|---|---|
| | 1 | 1¼ | 1½ | 1¾ | 2 | 2½ | 3 | 3½ | 4 | 4½ |
| 1½ | 19.76<br>44.44 | 48.23<br>69.44 | ..... <br>..... | ..... <br>..... | ..... <br>..... | ..... <br>..... | ..... <br>..... | ..... <br>..... | ..... <br>..... | ..... <br>..... |
| 1¾ | 10.67<br>32.66 | 26.04<br>51.02 | 53.98<br>73.49 | ..... <br>..... | ..... <br>..... | ..... <br>..... | ..... <br>..... | ..... <br>..... | ..... <br>..... | ..... <br>..... |
| 2 | 6.25<br>25.00 | 15.26<br>39.07 | 31.65<br>56.25 | 58.62<br>76.54 | ..... <br>..... | ..... <br>..... | ..... <br>..... | ..... <br>..... | ..... <br>..... | ..... <br>..... |
| 2¼ | 3.91<br>19.75 | 9.53<br>30.87 | 19.76<br>44.44 | 36.60<br>60.49 | 62.43<br>79.00 | ..... <br>..... | ..... <br>..... | ..... <br>..... | ..... <br>..... | ..... <br>..... |
| 2½ | 2.56<br>16.00 | 6.25<br>25.00 | 12.96<br>36.00 | 24.01<br>49.00 | 40.96<br>64.00 | ..... <br>..... | ..... <br>..... | ..... <br>..... | ..... <br>..... | ..... <br>..... |
| 2¾ | 1.75<br>13.22 | 4.28<br>20.66 | 8.86<br>29.74 | 16.40<br>40.48 | 27.98<br>52.89 | 68.30<br>82.63 | ..... <br>..... | ..... <br>..... | ..... <br>..... | ..... <br>..... |
| 3 | 1.24<br>11.11 | 3.01<br>17.36 | 6.25<br>25.00 | 11.58<br>34.01 | 19.76<br>44.44 | 48.23<br>69.44 | ..... <br>..... | ..... <br>..... | ..... <br>..... | ..... <br>..... |
| 3¼ | 0.87<br>9.46 | 2.19<br>14.80 | 4.54<br>21.30 | 8.41<br>29.00 | 14.35<br>37.87 | 35.02<br>59.17 | 72.61<br>85.22 | ..... <br>..... | ..... <br>..... | ..... <br>..... |
| 3½ | 0.67<br>8.16 | 1.63<br>12.76 | 3.38<br>18.36 | 6.25<br>25.00 | 10.67<br>32.66 | 26.04<br>51.02 | 53.98<br>73.49 | ..... <br>..... | ..... <br>..... | ..... <br>..... |
| 3¾ | 0.51<br>7.11 | 1.24<br>11.11 | 2.56<br>16.00 | 4.75<br>21.77 | 8.09<br>28.45 | 19.76<br>44.44 | 40.96<br>64.00 | 75.89<br>87.10 | ..... <br>..... | ..... <br>..... |
| 4 | 0.40<br>6.25 | 0.96<br>9.77 | 1.98<br>14.06 | 3.68<br>19.14 | 6.25<br>25.00 | 15.26<br>39.07 | 31.65<br>56.25 | 58.62<br>76.56 | ..... <br>..... | ..... <br>..... |
| 4¼ | 0.31<br>5.54 | 0.74<br>8.65 | 1.56<br>12.45 | 2.89<br>16.95 | 4.91<br>22.15 | 11.99<br>34.61 | 24.83<br>49.85 | 46.00<br>67.83 | 78.47<br>88.59 | ..... <br>..... |
| 4½ | 0.25<br>4.94 | 0.70<br>7.72 | 1.24<br>11.11 | 2.29<br>15.12 | 3.91<br>19.75 | 9.53<br>30.87 | 19.76<br>44.44 | 36.60<br>60.49 | 62.43<br>79.00 | ..... <br>..... |
| 4¾ | 0.20<br>4.43 | 0.50<br>6.93 | 1.00<br>9.97 | 1.85<br>13.57 | 3.15<br>17.73 | 7.68<br>27.70 | 15.92<br>39.90 | 29.48<br>54.29 | 50.29<br>70.91 | 80.56<br>89.75 |
| 5 | 0.16<br>4.00 | 0.40<br>6.25 | 0.81<br>8.10 | 1.51<br>12.25 | 2.56<br>16.00 | 6.25<br>25.00 | 12.96<br>36.00 | 24.01<br>49.00 | 40.96<br>64.00 | 65.61<br>81.00 |
| 5½ | 0.11<br>3.30 | 0.27<br>5.17 | 0.55<br>7.43 | 1.03<br>10.12 | 1.75<br>13.22 | 4.27<br>20.66 | 8.86<br>29.76 | 16.40<br>40.48 | 27.98<br>52.89 | 44.82<br>66.94 |
| 6 | 0.09<br>2.77 | 0.19<br>4.34 | 0.40<br>6.25 | 0.73<br>8.50 | 1.24<br>11.11 | 3.02<br>17.36 | 6.25<br>25.00 | 11.58<br>34.02 | 19.76<br>44.44 | 31.65<br>56.25 |
| 6½ | 0.06<br>2.36 | 0.14<br>3.70 | 0 29<br>5.32 | 0.59<br>7.24 | 0.90<br>9.47 | 2.19<br>14.79 | 4.54<br>21.30 | 8.41<br>28.99 | 14.35<br>37.87 | 23.98<br>47 93 |
| 7 | 0.05<br>2.04 | 0.11<br>3.19 | 0.22<br>4.59 | 0.40<br>6.25 | 0.67<br>8.16 | 1.63<br>12.76 | 3.38<br>18.36 | 6.25<br>25.00 | 10.67<br>32.66 | 17.08<br>41.33 |
| 7½ | 0.04<br>1.77 | 0.08<br>2.77 | 0.16<br>4.00 | 0.30<br>5.44 | 0.51<br>7.11 | 1.24<br>11.11 | 2.56<br>16.00 | 4.75<br>21.77 | 8.09<br>28.45 | 12.96<br>36.00 |
| 8 | 0.03<br>1.56 | 0.06<br>2.44 | 0.13<br>3.51 | 0.23<br>4.78 | 0.40<br>6.25 | 0.96<br>9.77 | 1.98<br>14.06 | 3.68<br>19.14 | 6.25<br>25 00 | 10.02<br>31.64 |

# WIRE AND SHEET-METAL GAGES

The thicknesses of sheet metals and the diameters of wires conform to various gaging systems. These gage sizes are indicated by numbers, and the following tables give the decimal equivalents of the different gage numbers. Much confusion has resulted from the use of gage numbers, and in ordering materials it is preferable to give the exact dimensions in decimal fractions of an inch. While the dimensions thus specified should conform to the gage ordinarily used for a given class of material, any error in the specification due, for example, to the use of a table having "rounded off" or approximate equivalents, will be apparent to the manufacturer at the time the order is placed. Furthermore, the decimal method of indicating wire diameters and sheet metal thicknesses has the advantage of being self-explanatory, whereas arbitrary gage numbers are not. The decimal system of indicating gage sizes is now being used quite generally, and gage numbers are gradually being discarded. Unfortunately, there is considerable variation in the use of different gages. For example, a gage ordinarily used for copper, brass and other non-ferrous materials, may at times be used for steel, and vice versa. The gages specified in the following are the ones ordinarily employed for the materials mentioned, but there are in some cases minor exceptions and variations in the different industries.

**Wire Gages.** — The wire gage system used by practically all of the steel producers in the United States is known by the name Steel Wire Gage or to distinguish it from the Standard Wire Gage (S.W.G.) used in Great Britain it is called the United States Steel Wire Gage. It is the same as the Washburn and Moen, American Steel and Wire Company, and Roebling Wire Gages. The name has the official sanction of the Bureau of Standards at Washington but is not legally effective. The only wire gage which has been recognized in Acts of Congress is the Birmingham Gage (also known as Stub's Iron Wire). The Birmingham Gage is, however, nearly obsolete both in the United States and Great Britain, where it originated. Copper and aluminum wires are specified in decimal fractions. They were formerly universally specified in the United States by the American or Brown & Sharpe Wire Gage. Music spring steel wire, one of the highest quality wires of several types used for mechanical springs, is specified by the piano or music wire gage.

In Great Britain one wire gage has been legalized. This is called the Standard Wire Gage (S.W.G.), formerly called Imperial Wire Gage.

**Gages for Rods.** — Steel wire rod sizes are designated by fractional or decimal parts of an inch and by the gage numbers of the United States Steel Wire Gage. Copper and aluminum rods are specified by decimal fractions and fractions. Drill rod may be specified in decimal fractions but in the carbon and alloy tool steel grades may also be specified in the Stub's Steel Wire Gage and in the high-speed steel drill rod grade may be specified by the Morse Twist Drill Gage (Manufacturers' Standard Gage for Twist Drills). For gage numbers with corresponding decimal equivalents see the tables of American Standard Straight Shank Twist Drills.

**Gages for Wall Thicknesses of Tubing.** — At one time the Birmingham or Stub's Iron Wire Gage was used to specify the wall thickness of the following classes of tubing: Seamless brass, seamless copper, seamless steel, and aluminum. The Brown & Sharpe Wire Gage was used for brazed brass and brazed copper tubing. Wall thicknesses are now specified by decimal parts of an inch but in the case of steel pressure tubes and steel mechanical tubing the wall thickness may be specified by the Birmingham or Stub's Iron Wire Gage. In Great Britain the Standard Wire Gage (S.W.G.) is used to specify the wall thickness of some kinds of steel tubes.

Table 1.    Wire Gages in Approximate Decimals of an Inch

| No. of Wire Gage | American Wire or Brown & Sharpe Gage | Steel Wire Gage (U.S.)* | British Standard Wire Gage (Imperial Wire Gage) | Music or Piano Wire Gage | Birming-ham or Stub's Iron Wire Gage | Stub's Steel Wire Gage | No. of Wire Gage | Stub's Steel Wire Gage |
|---|---|---|---|---|---|---|---|---|
| 7/0 | ... | 0.4900 | 0.5000 | ... | ... | ... | 51 | 0.066 |
| 6/0 | 0.5800 | 0.4615 | 0.4640 | 0.004 | ... | ... | 52 | 0.063 |
| 5/0 | 0.5165 | 0.4305 | 0.4320 | 0.005 | 0.5000 | ... | 53 | 0.058 |
| 4/0 | 0.4600 | 0.3938 | 0.4000 | 0.006 | 0.4540 | ... | 54 | 0.055 |
| 3/0 | 0.4096 | 0.3625 | 0.3720 | 0.007 | 0.4250 | ... | 55 | 0.050 |
| 2/0 | 0.3648 | 0.3310 | 0.3480 | 0.008 | 0.3800 | ... | 56 | 0.045 |
| 1/0 | 0.3249 | 0.3065 | 0.3240 | 0.009 | 0.3400 | ... | 57 | 0.042 |
| 1 | 0.2893 | 0.2830 | 0.3000 | 0.010 | 0.3000 | 0.227 | 58 | 0.041 |
| 2 | 0.2576 | 0.2625 | 0.2760 | 0.011 | 0.2840 | 0.219 | 59 | 0.040 |
| 3 | 0.2294 | 0.2437 | 0.2520 | 0.012 | 0.2590 | 0.212 | 60 | 0.039 |
| 4 | 0.2043 | 0.2253 | 0.2320 | 0.013 | 0.2380 | 0.207 | 61 | 0.038 |
| 5 | 0.1819 | 0.2070 | 0.2120 | 0.014 | 0.2200 | 0.204 | 62 | 0.037 |
| 6 | 0.1620 | 0.1920 | 0.1920 | 0.016 | 0.2030 | 0.201 | 63 | 0.036 |
| 7 | 0.1443 | 0.1770 | 0.1760 | 0.018 | 0.1800 | 0.199 | 64 | 0.035 |
| 8 | 0.1285 | 0.1620 | 0.1600 | 0.020 | 0.1650 | 0.197 | 65 | 0.033 |
| 9 | 0.1144 | 0.1483 | 0.1440 | 0.022 | 0.1480 | 0.194 | 66 | 0.032 |
| 10 | 0.1019 | 0.1350 | 0.1280 | 0.024 | 0.1340 | 0.191 | 67 | 0.031 |
| 11 | 0.0907 | 0.1205 | 0.1160 | 0.026 | 0.1200 | 0.188 | 68 | 0.030 |
| 12 | 0.0808 | 0.1055 | 0.1040 | 0.029 | 0.1090 | 0.185 | 69 | 0.029 |
| 13 | 0.0720 | 0.0915 | 0.0920 | 0.031 | 0.0950 | 0.182 | 70 | 0.027 |
| 14 | 0.0641 | 0.0800 | 0.0800 | 0.033 | 0.0830 | 0.180 | 71 | 0.026 |
| 15 | 0.0571 | 0.0720 | 0.0720 | 0.035 | 0.0720 | 0.178 | 72 | 0.024 |
| 16 | 0.0508 | 0.0625 | 0.0640 | 0.037 | 0.0650 | 0.175 | 73 | 0.023 |
| 17 | 0.0453 | 0.0540 | 0.0560 | 0.039 | 0.0580 | 0.172 | 74 | 0.022 |
| 18 | 0.0403 | 0.0475 | 0.0480 | 0.041 | 0.0490 | 0.168 | 75 | 0.020 |
| 19 | 0.0359 | 0.0410 | 0.0400 | 0.043 | 0.0420 | 0.164 | 76 | 0.018 |
| 20 | 0.0320 | 0.0348 | 0.0360 | 0.045 | 0.0350 | 0.161 | 77 | 0.016 |
| 21 | 0.0285 | 0.0317 | 0.0320 | 0.047 | 0.0320 | 0.157 | 78 | 0.015 |
| 22 | 0.0253 | 0.0286 | 0.0280 | 0.049 | 0.0280 | 0.155 | 79 | 0.014 |
| 23 | 0.0226 | 0.0258 | 0.0240 | 0.051 | 0.0250 | 0.153 | 80 | 0.013 |
| 24 | 0.0201 | 0.0230 | 0.0220 | 0.055 | 0.0220 | 0.151 | ... | ... |
| 25 | 0.0179 | 0.0204 | 0.0200 | 0.059 | 0.0200 | 0.148 | ... | ... |
| 26 | 0.0159 | 0.0181 | 0.0180 | 0.063 | 0.0180 | 0.146 | ... | ... |
| 27 | 0.0142 | 0.0173 | 0.0164 | 0.067 | 0.0160 | 0.143 | ... | ... |
| 28 | 0.0126 | 0.0162 | 0.0148 | 0.071 | 0.0140 | 0.139 | ... | ... |
| 29 | 0.0113 | 0.0150 | 0.0136 | 0.075 | 0.0130 | 0.134 | ... | ... |
| 30 | 0.0100 | 0.0140 | 0.0124 | 0.080 | 0.0120 | 0.127 | ... | ... |
| 31 | 0.00893 | 0.0132 | 0.0116 | 0.085 | 0.0100 | 0.120 | ... | ... |
| 32 | 0.00795 | 0.0128 | 0.0108 | 0.090 | 0.0090 | 0.115 | ... | ... |
| 33 | 0.00708 | 0.0118 | 0.0100 | 0.095 | 0.0080 | 0.112 | ... | ... |
| 34 | 0.00630 | 0.0104 | 0.0092 | 0.100 | 0.0070 | 0.110 | ... | ... |
| 35 | 0.00561 | 0.0095 | 0.0084 | 0.106 | 0.0050 | 0.108 | ... | ... |
| 36 | 0.00500 | 0.0090 | 0.0076 | 0.112 | 0.0040 | 0.106 | ... | ... |
| 37 | 0.00445 | 0.0085 | 0.0068 | 0.118 | ... | 0.103 | ... | ... |
| 38 | 0.00396 | 0.0080 | 0.0060 | 0.124 | ... | 0.101 | ... | ... |
| 39 | 0.00353 | 0.0075 | 0.0052 | 0.130 | ... | 0.099 | ... | ... |
| 40 | 0.00314 | 0.0070 | 0.0048 | 0.138 | ... | 0.097 | ... | ... |
| 41 | 0.00280 | 0.0066 | 0.0044 | 0.146 | ... | 0.095 | ... | ... |
| 42 | 0.00249 | 0.0062 | 0.0040 | 0.154 | ... | 0.092 | ... | ... |
| 43 | 0.00222 | 0.0060 | 0.0036 | 0.162 | ... | 0.088 | ... | ... |
| 44 | 0.00198 | 0.0058 | 0.0032 | 0.170 | ... | 0.085 | ... | ... |
| 45 | 0.00176 | 0.0055 | 0.0028 | 0.180 | ... | 0.081 | ... | ... |
| 46 | 0.00157 | 0.0052 | 0.0024 | ... | ... | 0.079 | ... | ... |
| 47 | 0.00140 | 0.0050 | 0.0020 | ... | ... | 0.077 | ... | ... |
| 48 | 0.00124 | 0.0048 | 0.0016 | ... | ... | 0.075 | ... | ... |
| 49 | 0.00111 | 0.0046 | 0.0012 | ... | ... | 0.072 | ... | ... |
| 50 | 0.00099 | 0.0044 | 0.0010 | ... | ... | 0.069 | ... | ... |

* Also known as Washburn and Moen, American Steel and Wire Co. and Roebling Wire Gages.    A greater selection of sizes is available and is specified by what are known as split gage numbers.    They can be recognized by the ¼ and ½ fractions which follow the gage number; i.e., 4¼, 4½, 4¾.    The decimal equivalents of split gage numbers are in the Steel Products Manual entitled: *Wire and Rods, Carbon Steel* published by the American Iron and Steel Institute, New York, New York.

**Sheet-metal Gages.** — The thicknesses of steel sheets now are based upon a weight of 41.82 pounds per square foot per inch thick. This is known as Manufacturers' Standard Gage for Sheet Steel. (See text in table on page 467.) This gage differs from the older United States Standard Gage for iron and steel sheets and plates, established by Congress in 1893, based upon a weight of 40 pounds per square foot per inch of thickness which is the weight of wrought-iron plate.

Thicknesses of aluminum, copper, and copper-base alloys were formerly designated by the American or Brown & Sharpe Wire Gage but now are specified in decimals or fractions of an inch. Copper and copper-base alloy flat products whose thicknesses are below ¼ inch are specified by the 20-series of American Standard Preferred Numbers given in the American Standard B32.1-1952 entitled Preferred Thicknesses for Uncoated Thin Flat Metals (see accompanying Table 2). The thicknesses in this standard are based on the 20- and 40-series of American Standard Preferred Numbers — A.S.A. Z17.1 (see Handbook page 116) and are applicable to uncoated, thin, flat metals and alloys. Although the table given in the American Standard gives only the 20- and 40-series of numbers the statement is made that when intermediate thicknesses are required, selections should be made from thicknesses based on the 80-series of numbers (see Handbook page 116). Each number of the 20-series is approximately 12 per cent greater than the next smaller one and each number of the 40-series is approximately 6 per cent greater than the next smaller one.

Zinc sheets are more often ordered by specifying decimal thickness although a zinc gage exists and is shown in Table 3.

**Table 2. Preferred Thicknesses for Uncoated Metals and Alloys — Under 0.250 Inch in Thickness (ASA B32.1-1952)***

| Preferred Thickness, Inches | | Preferred Thickness, Inches | | Preferred Thickness, Inches | | Preferred Thickness, Inches | |
|---|---|---|---|---|---|---|---|
| Based on 20-Series | Based on 40-Series | Based on 20-Series | Based on 40-Series | Based on 20-Series | Based on 40-Series | Based on 20-Series | Based on 40-Series |
| ..... | 0.236 | 0.100 | 0.100 | ..... | 0.042 | 0.018 | 0.018 |
| 0.224 | 0.224 | ..... | 0.095 | 0.040 | 0.040 | ..... | 0.017 |
| ..... | 0.212 | 0.090 | 0.090 | ..... | 0.038 | 0.016 | 0.016 |
| 0.200 | 0.200 | ..... | 0.085 | 0.036 | 0.036 | ..... | 0.015 |
| ..... | 0.190 | 0.080 | 0.080 | ..... | 0.034 | 0.014 | 0.014 |
| 0.180 | 0.180 | ..... | 0.075 | 0.032 | 0.032 | ..... | 0.013 |
| ..... | 0.170 | 0.071 | 0.071 | ..... | 0.030 | 0.012 | 0.012 |
| 0.160 | 0.160 | ..... | 0.067 | 0.028 | 0.028 | 0.011 | 0.011 |
| ..... | 0.150 | 0.063 | 0.063 | ..... | 0.026 | 0.010 | 0.010 |
| 0.140 | 0.140 | ..... | 0.060 | 0.025 | 0.025 | 0.009 | 0.009 |
| ..... | 0.132 | 0.056 | 0.056 | ..... | 0.024 | 0.008 | 0.008 |
| 0.125 | 0.125 | ..... | 0.053 | 0.022 | 0.022 | 0.007 | 0.007 |
| ..... | 0.118 | 0.050 | 0.050 | ..... | 0.021 | 0.006 | 0.006 |
| 0.112 | 0.112 | ..... | 0.048 | 0.020 | 0.020 | 0.005 | 0.005 |
| ..... | 0.106 | 0.045 | 0.045 | ..... | 0.019 | 0.004 | 0.004 |

* The American Standard ASA B32.1-1952 lists preferred thicknesses which are based on the 20- and 40-series of preferred numbers and states that those based on the 40-series should provide adequate coverage. However, where intermediate thicknesses are required it states that thicknesses be based on the 80-series of preferred numbers (see Handbook page 116).

Most sheet metal products in Great Britain are specified by the British Standard Wire Gage (Imperial Wire Gage). Black iron and steel sheet and hooping, and galvanized flat and corrugated steel sheet, however, are specified by the Birmingham Gage (B.G.) which was legalized in 1914. This Birmingham Gage should not be confused with the Birmingham or Stub's Iron Wire Gage mentioned previously.

Table 3.   Sheet-Metal Gages in Approximate Decimals of an Inch

| No. of Sheet-Metal Gage | Manufacturers' Standard Gage for Steel* | Birmingham Gage (B.G.) for Sheets, Hoops | Galvanized Sheet Gage | Zinc Gage | No. of Sheet-Metal Gage | Manufacturers' Standard Gage for Steel* | Birmingham Gage (B.G.) for Sheets, Hoops | Galvanized Sheet Gage | Zinc Gage |
|---|---|---|---|---|---|---|---|---|---|
| 15/0 | ... | 1.000 | ... | ... | 20 | 0.0359 | 0.0392 | 0.0396 | 0.070 |
| 14/0 | ... | 0.9583 | ... | ... | 21 | 0.0329 | 0.0349 | 0.0366 | 0.080 |
| 13/0 | ... | 0.9167 | ... | ... | 22 | 0.0299 | 0.03125 | 0.0336 | 0.090 |
| 12/0 | ... | 0.8750 | ... | ... | 23 | 0.0269 | 0.02782 | 0.0306 | 0.100 |
| 11/0 | ... | 0.8333 | ... | ... | 24 | 0.0239 | 0.02476 | 0.0276 | 0.125 |
| 10/0 | ... | 0.7917 | ... | ... | 25 | 0.0209 | 0.02204 | 0.0247 | ... |
| 9/0 | ... | 0.7500 | ... | ... | 26 | 0.0179 | 0.01961 | 0.0217 | ... |
| 8/0 | ... | 0.7083 | ... | ... | 27 | 0.0164 | 0.01745 | 0.0202 | ... |
| 7/0 | ... | 0.6666 | ... | ... | 28 | 0.0149 | 0.01562 | 0.0187 | ... |
| 6/0 | ... | 0.6250 | ... | ... | 29 | 0.0135 | 0.01390 | 0.0172 | ... |
| 5/0 | ... | 0.5883 | ... | ... | 30 | 0.0120 | 0.01230 | 0.0157 | ... |
| 4/0 | ... | 0.5416 | ... | ... | 31 | 0.0105 | 0.01100 | 0.0142 | ... |
| 3/0 | ... | 0.5000 | ... | ... | 32 | 0.0097 | 0.00980 | 0.0134 | ... |
| 2/0 | ... | 0.4452 | ... | ... | 33 | 0.0090 | 0.00870 | ... | ... |
| 1/0 | ... | 0.3964 | ... | ... | 34 | 0.0082 | 0.00770 | ... | ... |
| 1 | ... | 0.3532 | ... | ... | 35 | 0.0075 | 0.00690 | ... | ... |
| 2 | ... | 0.3147 | ... | ... | 36 | 0.0067 | 0.00610 | ... | ... |
| 3 | 0.2391 | 0.2804 | ... | 0.006 | 37 | 0.0064 | 0.00540 | ... | ... |
| 4 | 0.2242 | 0.2500 | ... | 0.008 | 38 | 0.0060 | 0.00480 | ... | ... |
| 5 | 0.2092 | 0.2225 | ... | 0.010 | 39 | ... | 0.00430 | ... | ... |
| 6 | 0.1943 | 0.1981 | ... | 0.012 | 40 | ... | 0.00386 | ... | ... |
| 7 | 0.1793 | 0.1764 | ... | 0.014 | 41 | ... | 0.00343 | ... | ... |
| 8 | 0.1644 | 0.1570 | 0.1681 | 0.016 | 42 | ... | 0.00306 | ... | ... |
| 9 | 0.1495 | 0.1398 | 0.1532 | 0.018 | 43 | ... | 0.00272 | ... | ... |
| 10 | 0.1345 | 0.1250 | 0.1382 | 0.020 | 44 | ... | 0.00242 | ... | ... |
| 11 | 0.1196 | 0.1113 | 0.1233 | 0.024 | 45 | ... | 0.00215 | ... | ... |
| 12 | 0.1046 | 0.0991 | 0.1084 | 0.028 | 46 | ... | 0.00192 | ... | ... |
| 13 | 0.0897 | 0.0882 | 0.0934 | 0.032 | 47 | ... | 0.00170 | ... | ... |
| 14 | 0.0747 | 0.0785 | 0.0785 | 0.036 | 48 | ... | 0.00152 | ... | ... |
| 15 | 0.0673 | 0.0699 | 0.0710 | 0.040 | 49 | ... | 0.00135 | ... | ... |
| 16 | 0.0598 | 0.0625 | 0.0635 | 0.045 | 50 | ... | 0.00120 | ... | ... |
| 17 | 0.0538 | 0.0556 | 0.0575 | 0.050 | 51 | ... | 0.00107 | ... | ... |
| 18 | 0.0478 | 0.0495 | 0.0516 | 0.055 | 52 | ... | 0.00095 | ... | ... |
| 19 | 0.0418 | 0.0440 | 0.0456 | 0.060 | ... | ... | ... | ... | ... |

* For more information and data on Manufacturers' Standard Gage for Steel Sheets see following page.

*The United States Standard Gage* (not shown above) for iron and steel sheets and plates was established by Congress in 1893 and was primarily a *weight* gage rather than a thickness gage. The equivalent thicknesses were derived from the weight of wrought iron. The weight per cubic foot was taken at 480 pounds, thus making the weight of a plate 12 inches square and 1 inch thick, 40 pounds. In converting weight to equivalent thickness, gage tables formerly published contained thicknesses equivalent to the basic weights just mentioned. For example, a No. 3 U. S. gage represents a wrought-iron plate having a weight of 10 pounds per square foot; hence, if the weight per square foot per inch thick is 40 pounds, the plate thickness for a No. 3 gage = 10 ÷ 40 = 0.25 inch, which was the original thickness equivalent for this gage number. Since this and the other thickness equivalents were derived from the weight of wrought iron, they are not correct for steel.

## Manufacturers' Standard Gage for Sheet Steel

*Manufacturers' Standard Gage for Steel Sheets:* Although the basic weight of steel used in the manufacture of steel plate, bars, and other steel products is 40.8 pounds per square foot per inch of thickness, the *Manufacturers' Standard Gage for Steel Sheets* is based on a weight of 41.82 pounds per square foot per inch of thickness. This modified figure provides an adjustment for the variation in thickness from the edges to the center of sheets resulting from the rolling process and also for the shearing tolerances which are on the over side. The thicknesses in the table below are based upon this weight of 41.82 pounds and represent standard mill practice. These nominal thicknesses, however, are subject to tolerances or permissible variations as given in the table on the following pages which include both hot-rolled and cold-rolled sheets.

| Standard Gage No. | Ounces per Square Foot | Pounds per Square Foot | Equivalent Thickness, Inch | Standard Gage No. | Ounces per Square Foot | Pounds per Square Foot | Equivalent Thickness, Inch |
|---|---|---|---|---|---|---|---|
| 3 | 160 | 10.0000 | 0.2391 | 21 | 22 | 1.3750 | .0329 |
| 4 | 150 | 9.3750 | .2242 | 22 | 20 | 1.2500 | .0299 |
| 5 | 140 | 8.7500 | .2092 | 23 | 18 | 1.1250 | .0269 |
| 6 | 130 | 8.1250 | .1943 | 24 | 16 | 1.0000 | .0239 |
| 7 | 120 | 7.5000 | .1793 | 25 | 14 | 0.87500 | .0209 |
| 8 | 110 | 6.8750 | .1644 | 26 | 12 | .75000 | .0179 |
| 9 | 100 | 6.2500 | .1495 | 27 | 11 | .68750 | .0164 |
| 10 | 90 | 5.6250 | .1345 | 28 | 10 | .62500 | .0149 |
| 11 | 80 | 5.0000 | .1196 | 29 | 9 | .56250 | .0135 |
| 12 | 70 | 4.3750 | .1046 | 30 | 8 | .50000 | .0120 |
| 13 | 60 | 3.7500 | .0897 | 31 | 7 | .43750 | .0105 |
| 14 | 50 | 3.1250 | .0747 | 32 | 6.5 | .40625 | .0097 |
| 15 | 45 | 2.8125 | .0673 | 33 | 6 | .37500 | .0090 |
| 16 | 40 | 2.5000 | .0598 | 34 | 5.5 | .34375 | .0082 |
| 17 | 36 | 2.2500 | .0538 | 35 | 5 | .31250 | .0075 |
| 18 | 32 | 2.0000 | .0478 | 36 | 4.5 | .28125 | .0067 |
| 19 | 28 | 1.7500 | .0418 | 37 | 4.25 | .26562 | .0064 |
| 20 | 24 | 1.5000 | .0359 | 38 | 4 | .25000 | .0060 |

## Standard Thickness Tolerances for Steel Sheets — 1
### (American Iron and Steel Institute*)

| COLD ROLLED ALLOY STEEL (For coils or cut lengths)† | | | | | | | | | |
|---|---|---|---|---|---|---|---|---|---|
| | Thickness Range and Plus or Minus Thickness Tolerance | | | | | | | | |
| Width | .2299 .1800 | .1799 .1420 | .1419 .0972 | .0971 .0822 | .0821 .0710 | .0709 .0568 | .0567 .0509 | .0508 .0314 | .0313 .0195 |
| 24 to 32 | .008 | .008 | .007 | .006 | .005 | .005 | .005 | .004 | .003 |
| Over 32 to 40 | .009 | .009 | .008 | .007 | .006 | .005 | .005 | .004 | .003 |
| Over 40 to 48 | .010 | .010 | .009 | .007 | .006 | .005 | .005 | .004 | .003 |
| Over 48 to 60 | .... | .010 | .010 | .008 | .006 | .006 | .005 | .004 | .003 |
| Over 60 to 70 | .... | .011 | .010 | .009 | .007 | .006 | .006 | .005 | .... |
| Over 70 to 80 | .... | .012 | .011 | .009 | .007 | .... | .... | .... | .... |
| Over 80 to 90 | .... | .012 | .012 | .... | .... | .... | .... | .... | .... |
| Over 90 | .... | .012 | .012 | .... | .... | .... | .... | .... | .... |

| HOT ROLLED ALLOY STEEL (Hand Mill Product)† | | | |
|---|---|---|---|
| Thickness | Plus or Minus Tolerance | Thickness | Plus or Minus Tolerance |
| .229 to .188 | .015 | Under .084 to .073 | .007 |
| Under .188 to .146 | .014 | Under .073 to .059 | .006 |
| Under .146 to .131 | .012 | Under .059 to .041 | .005 |
| Under .131 to .115 | .010 | Under .041 to .027 | .004 |
| Under .115 to .099 | .009 | Under .027 to .019 | .003 |
| Under .099 to .084 | .008 | .... | .... |

All dimensions are in inches. See end of table for footnotes.

### Standard Thickness Tolerances for Steel Sheets — 2
(American Iron and Steel Institute*)

#### HOT ROLLED ALLOY STEEL SHEETS
(Continuous Mill Product — Coils or Cut Lengths)†

| Width | Thickness Range and Plus or Minus Thickness Tolerance | | | | |
|---|---|---|---|---|---|
| | .2299 .1800 | .1799 .0972 | .0971 .0822 | .0821 .0710 | .0709 .0568 |
| 24 to 32 | .009 | .008 | .007 | .007 | .006 |
| Over 32 to 40 | .009 | .009 | .008 | .007 | .006 |
| Over 40 to 48 | .010 | .010 | .008 | .007 | .006 |
| Over 48 to 60 | .... | .010 | .008 | .007 | .007 |
| Over 60 to 72 | .... | .011 | .009 | .008 | .007 |
| Over 72 to 80 | .... | .012 | .009 | .008 | .... |
| Over 80 to 90 | .... | .012 | .010 | .... | .... |
| Over 90 | .... | .012 | .... | .... | .... |

#### HOT ROLLED HIGH STRENGTH STEEL SHEETS
(For Coils and Cut Lengths, Including Pickled Sheets)†§

| Width | Thickness Range and Plus or Minus Thickness Tolerance | | | |
|---|---|---|---|---|
| | .2299 .1800 | .1799 .0972 | .0971 .0822 | .0821 .0710 |
| Over 12 to 15 | .007 | .007 | .006 | .006 |
| Over 15 to 20 | .008 | .008 | .007 | .007 |
| Over 20 to 32 | .009 | .008 | .007 | .007 |
| Over 32 to 40 | .009 | .009 | .008 | .007 |
| Over 40 to 48 | .010 | .010 | .008 | .007 |
| Over 48 to 60 | .... | .010 | .008 | .007 |
| Over 60 to 72 | .... | .011 | .009 | .008 |
| Over 72 to 80 | .... | .012 | .009 | .008 |
| Over 80 | .... | .012 | .010 | .... |

#### COLD ROLLED HIGH STRENGTH STEEL SHEETS
(Coils and Cut Lengths)‡

| Width | Thickness Range and Plus or Minus Thickness Tolerance | | | | | | |
|---|---|---|---|---|---|---|---|
| | Over .1419 | .1419 .0972 | .0971 .0710 | .0709 .0568 | .0567 .0389 | .0388 .0195 | .0194 .0142 |
| Over 12 to 15 | .006 | .005 | .005 | .005 | .004 | .003 | .002 |
| Over 15 to 24 | .006 | .006 | .005 | .005 | .004 | .003 | .002 |
| Over 24 to 32 | .007 | .006 | .005 | .005 | .004 | .003 | .002 |
| Over 32 to 40 | .007 | .006 | .006 | .005 | .004 | .003 | .002 |
| Over 40 to 48 | .007 | .006 | .006 | .005 | .004 | .003 | .002 |
| Over 48 to 60 | .008 | .007 | .006 | .005 | .004 | .003 | .002 |
| Over 60 to 72 | .008 | .007 | .007 | .006 | .005 | .004 | .... |
| Over 72 to 80 | .008 | .007 | .007 | .006 | .005 | .004 | .... |
| Over 80 | .009 | .008 | .007 | .006 | .006 | .... | .... |

#### COLD ROLLED HIGH STRENGTH STEEL SHEETS
(Coils and Cut Lengths)‡

| Width | Thickness Range and Plus or Minus Thickness Tolerance | | | | | |
|---|---|---|---|---|---|---|
| | .0821 .0710 | .0709 .0509 | .0508 .0389 | .0388 .0314 | .0313 .0195 | .0194 .0142 |
| 2 to 12 (incl.) | .006 | .005 | .004 | .0035 | .003 | .002 |

All dimensions are in inches.
* Steel Products Manuals — Alloy Steel Sheets and Strip, April 1969; and High Strength Low Alloy Steel and High Strength Intermediate Manganese Steel, October 1967.
† Thickness is measured at any point on the sheet not less than ⅜ inch from a cut edge and not less than ¾ inch from a mill edge.
§ Tolerances listed for this steel do not apply to the uncropped ends of mill edge coils.
‡ Thickness is measured at any point on the sheet not less than ⅜ inch from an edge.

## Standard Thickness Tolerances for Carbon Steel Sheet
(American Iron and Steel Institute*)

### HOT ROLLED SHEET (For coils and cut lengths, including pickled sheets)††

| Width | Thickness range and plus or minus thickness tolerance | | | | | |
|---|---|---|---|---|---|---|
| | .2299 .1800 | .1799 .0972 | .0971 .0710 | .0709 .0568 | .0567 .0509 | .0508 .0449 |
| Over 12 to 20 | .007 | .007 | .006 | .006 | .005 | .005 |
| Over 20 to 40 | .008 | .007 | .007 | .006 | .005 | .005 |
| Over 40 to 48 | .009 | .008 | .007 | .006 | .006 | .005 |
| Over 48 to 60 | .001▲ | .008 | .007 | .007 | .006 | .... |
| Over 60 to 72 | .012▲ | .008 | .008 | .007 | .007 | .... |
| Over 72 | .... | .008 | .008 | .008 | .... | .... |

### HOT ROLLED SHEET (For coils only)††

| Width | Thickness range and plus or minus thickness tolerance | | | |
|---|---|---|---|---|
| | .5000 .3751 | .3750 .3125 | .3124 .2300 | .2299 .1800 |
| Over 12 to 15 | .014 | .012 | .010 | .007 |
| Over 15 to 20 | .015 | .014 | .011 | .007 |
| Over 20 to 32 | .016 | .015 | .012 | .008 |
| Over 32 to 40 | .017 | .016 | .013 | .008 |
| Over 40 to 48 | .018 | .017 | .014 | .009 |
| Over 48 to 60 | .019 | .019 | .015 | .011 |
| Over 60 to 72 | .021 | .021 | .016 | .012 |

### COLD ROLLED SHEET (For coils and cut lengths)‡§

| Width | Thickness range and plus or minus thickness tolerance | | | | | |
|---|---|---|---|---|---|---|
| | .1419 .0972 | .0971 .0710 | .0709 .0568 | .0567 .0389 | .0388 .0195 | .0194 .0142 |
| Over 12 to 15 | .005 | .005 | .005 | .004 | .003 | .002 |
| Over 15 to 72 | .006 | .005 | .005 | .004 | .003 | .002 |
| Over 72 | .007 | .006 | .005 | .004 | .003 | .... |

### COLD ROLLED SHEET (For coils and cut lengths)‡§¶

| Width | Thickness range and plus or minus thickness tolerance | | | |
|---|---|---|---|---|
| | .0821 .0568 | .0567 .0389 | .0388 .0195 | .0194 .0142 |
| 2 to 12 incl. | .005 | .004 | .003 | .002 |

### HOT DIPPED GALVANIZED SHEET (For coils and cut lengths)‡§

| Width | Thickness range and plus or minus thickness tolerance | | | | | |
|---|---|---|---|---|---|---|
| | .1868 .1009 | .1008 .0748 | .0747 .0606 | .0605 .0426 | .0425 .0232 | Under .0232 |
| To 32 | .008 | .007 | .006 | .005 | .004 | .003 |
| Over 32 to 40 | .008 | .008 | .006 | .005 | .004 | .003 |
| Over 40 to 60 | .009 | .008 | .006 | .005 | .004 | .003 |
| Over 60 to 72 | .009 | .008 | .006 | .005 | .004 | .... |

### ELECTROLYTIC ZINC COATED SHEET (For coils and cut lengths)‡§

| Width | Thickness range and plus or minus thickness tolerance | | | | |
|---|---|---|---|---|---|
| | .0971 .0568 | .0567 .0389 | .0388 .0195 | .0194 .0142 | Under .0142 |
| Up to 40 | .005 | .004 | .003 | .002 | .0015 |
| Over 40 to 60 | .005 | .004 | .003 | .002 | .... |

All dimensions are in inches.
* Steel Products Manual — Carbon Sheet Steel, May 1970.
† Thickness is measured at any point across the width not less than ⅜ inch from a cut edge and not less than ¾ inch from a mill edge. Tolerances do not apply to the uncropped ends of mill edge coils.
‡ Regardless of whether total thickness tolerance is specified equally or unequally, plus or minus, the total tolerance should be equal to twice the tabular tolerances.
§ Thickness is measured at any point across the width not less than ⅜ inch from a side edge.
¶ This portion of the table applies to widths produced by slitting from wider sheet.
▲ Coil product only.

**Permissible Variations in Sizes of Cold-finished and Hot-rolled Bars — 1**
(American Iron and Steel Institute*)

HOT ROLLED CARBON STEEL BARS

| Size | Tolerance | | Out-of-Section† | Size | Tolerance | | Out-of-Section† |
|---|---|---|---|---|---|---|---|
| | Plus | Minus | | | Plus | Minus | |
| **Rounds, Squares and Round-Cornered Squares** | | | | | | | |
| To 5/16 | .005 | .005 | .008 | Over 1½ to 2 | 1/64 | 1/64 | .023 |
| Over 5/16 to 7/16 | .006 | .006 | .009 | Over 2 to 2½ | 1/32 | 0 | .023 |
| Over 7/16 to 5/8 | .007 | .007 | .010 | Over 2½ to 3½ | 3/64 | 0 | .035 |
| Over 5/8 to 7/8 | .008 | .008 | .012 | Over 3½ to 4½ | 1/16 | 0 | .046 |
| Over 7/8 to 1 | .009 | .009 | .013 | Over 4½ to 5½ | 5/64 | 0 | .058 |
| Over 1 to 1⅛ | .010 | .010 | .015 | Over 5½ to 6½ | 1/8 | 0 | .070 |
| Over 1⅛ to 1¼ | .011 | .011 | .016 | Over 6½ to 8¼ | 5/32 | 0 | .085 |
| Over 1¼ to 1⅜ | .012 | .012 | .018 | Over 8¼ to 9½ | 3/16 | 0 | .100 |
| Over 1⅜ to 1½ | .014 | .014 | .021 | Over 9½ to 10 | 1/4 | 0 | .120 |
| **Hexagons** | | | | | | | |
| To ½ | .007 | .007 | .011 | Over 1½ to 2 | 1/32 | 1/64 | 1/32 |
| Over ½ to 1 | .010 | .010 | .015 | Over 2 to 2½ | 3/64 | 1/64 | 3/64 |
| Over 1 to 1½ | .021 | .013 | .025 | Over 2½ to 3½ | 1/16 | 1/64 | 1/16 |

COLD FINISHED CARBON STEEL BARS

| Size | Max. % C | | | |
|---|---|---|---|---|
| | Up to .28 | Over .28 to .55 | Up to .55‡ | Over .55§ |
| | Minus Tolerance | | | |
| **Cold Drawn Rounds¶** | | | | |
| To 1½ | .002 | .003 | .004 | .005 |
| Over 1½ to 2½ | .003 | .004 | .005 | .006 |
| Over 2½ to 4 | .004 | .005 | .006 | .007 |
| **Cold Drawn Hexagons¶** | | | | |
| To ¾ | .002 | .003 | .004 | .006 |
| Over ¾ to 1½ | .003 | .004 | .005 | .007 |
| Over 1½ to 2½ | .004 | .005 | .006 | .008 |
| Over 2½ to 3⅛ | .005 | .006 | .007 | .009 |
| **Cold Drawn Squares¶** | | | | |
| To ¾ | .002 | .004 | .005 | .007 |
| Over ¾ to 1½ | .003 | .005 | .006 | .008 |
| Over 1½ to 2½ | .004 | .006 | .007 | .009 |
| Over 2½ to 4 | .006 | .008 | .009 | .011 |

| Width or Diameter | Max. % C | | | |
|---|---|---|---|---|
| | Up to .28 | Over .28 to .55 | Up to .55‡ | Over .55§ |
| | Minus Tolerance | | | |
| **Cold Drawn Flats¶★** | | | | |
| To ¾ | .003 | .004 | .006 | .008 |
| Over ¾ to 1½ | .004 | .005 | .008 | .010 |
| Over 1½ to 3 | .005 | .006 | .010 | .012 |
| Over 3 to 4 | .006 | .008 | .011 | .016 |
| Over 4 to 6 | .008 | .010 | .012 | .020 |
| Over 6 | .013 | .015 | .... | .... |
| **Turned and Polished Rounds¶** | | | | |
| To 1½ | .002 | .003 | .004 | .005 |
| Over 1½ to 2½ | .003 | .004 | .005 | .006 |
| Over 2½ to 4 | .004 | .005 | .006 | .007 |
| Over 4 to 6 | .005 | .006 | .007 | .008 |
| Over 6 to 8 | .006 | .007 | .008 | .009 |
| Over 8 to 9 | .007 | .008 | .009 | .010 |
| Over 9 | .008 | .009 | .010 | .011 |

All dimensions are in inches. * Steel Products Manual — Carbon Steel: Semi-finished For Forging; Hot Rolled and Cold Finished Bars; etc., May 1964. † Means out-of-round, out-of-square or out-of-hexagon. Out-of-round is the difference between the maximum and minimum diameters of the bar, measured at the same cross section. Out-of-square is the difference in the two dimensions at the same cross section of a square bar between opposite faces. Out-of-hexagon is the greatest difference between any two dimensions at the same cross section between opposite faces. ‡ Stress relieved or annealed after cold finishing. § Or all grades quenched and tempered or normalized and tempered before cold finishing. ¶ Values shown include tolerances for bars that have been annealed, spheroidize annealed, normalized, normalized and tempered or quenched and tempered *before* cold finishing. Values do not include tolerances for bars that are spheroidize annealed, normalized, normalized and tempered or quenched and tempered *after* cold finishing. ★ Width governs the tolerances for both width and thickness of flats. For example, when the maximum of carbon range is .28 per cent or less, for a flat 2 inches wide and 1 inch thick, the width tolerance is .005 inch and the thickness tolerance is the same, namely, .005 inch.

**Permissible Variations in Sizes of Cold-finished and Hot-rolled Bars — 2**
(American Iron and Steel Institute*)

## COLD FINISHED ALLOY STEEL BARS

| Size | Max. % C | | | | Width or Diameter | Max. % C | | | |
|---|---|---|---|---|---|---|---|---|---|
| | Up to .28 | Over .28 to .55 | To .55† | Over .55‡ | | Up to .28 | Over .28 to .55 | To .55† | Over .55‡ |
| | Minus Tolerance | | | | | Minus Tolerance | | | |

### Cold Drawn Rounds

| Size | | | | | Cold Drawn Flats§ | | | | |
|---|---|---|---|---|---|---|---|---|---|
| To 1, in coils | .002 | .003 | .004 | .005 | | | | | |
| To 1½ | .003 | .004 | .005 | .006 | To ¾ | .004 | .005 | .007 | .009 |
| Over 1½ to 2½ | .004 | .005 | .006 | .007 | Over ¾ to 1½ | .005 | .006 | .009 | .011 |
| Over 2½ to 4 | .005 | .006 | .007 | .008 | Over 1½ to 3 | .006 | .007 | .011 | .013 |
| | | | | | Over 3 to 4 | .007 | .009 | .012 | .017 |
| **Cold Drawn Hexagons** | | | | | Over 4 to 6 | .009 | .011 | .013 | .021 |
| To ¾ | .003 | .004 | .005 | .007 | Over 6 | .014 | .... | .... | .... |
| Over ¾ to 1½ | .004 | .005 | .006 | .008 | | | | | |
| Over 1½ to 2½ | .005 | .006 | .007 | .009 | | | | | |
| Over 2½ to 3⅛ | .006 | .007 | .008 | .010 | **Turned and Polished Rounds** | | | | |
| Over 3⅛ to 4 | .006 | .... | .... | .... | To 1½ | .003 | .004 | .005 | .006 |
| | | | | | Over 1½ to 2½ | .004 | .005 | .006 | .007 |
| **Cold Drawn Squares** | | | | | Over 2½ to 4 | .005 | .006 | .007 | .008 |
| To ¾ | .003 | .005 | .006 | .008 | Over 4 to 6 | .006 | .007 | .008 | .009 |
| Over ¾ to 1½ | .004 | .006 | .007 | .009 | Over 6 to 8 | .007 | .008 | .009 | .010 |
| Over 1½ to 2½ | .005 | .007 | .008 | .010 | Over 8 to 9 | .008 | .009 | .010 | .011 |
| Over 2½ to 4 | .007 | .009 | .010 | .012 | Over 9 | .009 | .010 | .011 | .012 |
| Over 4 to 5 | .011 | .... | .... | .... | | | | | |

## HOT ROLLED TOOL STEEL BARS¶

| Size | Tolerance | | Size | Tolerance | |
|---|---|---|---|---|---|
| | Minus | Plus | | Minus | Plus |
| To ½ | .005 | .012 | Over 2½ to 3 | .010 | .040 |
| Over ½ to 1 | .005 | .016 | Over 3 to 4 | .012 | .050 |
| Over 1 to 1½ | .006 | .020 | Over 4 to 5½ | .015 | .060 |
| Over 1½ to 2 | .008 | .025 | Over 5½ to 6½ | .018 | .100 |
| Over 2 to 2½ | .010 | .030 | Over 6½ to 8 | .020 | .150 |

## HOT ROLLED TOOL STEEL FLAT BARS

| Width | Thickness Range and Thickness Tolerance | | | | | | | | | | | |
|---|---|---|---|---|---|---|---|---|---|---|---|---|
| | To ¼ | | Over ¼ to ½ | | Over ½ to 1 | | Over 1 to 2 | | Over 2 to 3 | | Over 3 to 4 | |
| | − | + | − | + | − | + | − | + | − | + | − | + |
| To 1 | .006 | .010 | .008 | .012 | .010 | .016 | .... | .... | .... | .... | .... | .... |
| Over 1 to 2 | .006 | .014 | .008 | .016 | .010 | .020 | .020 | .024 | .... | .... | .... | .... |
| Over 2 to 3 | .006 | .018 | .008 | .020 | .010 | .024 | .020 | .027 | .026 | .034 | .... | .... |
| Over 3 to 4 | .008 | .020 | .010 | .022 | .013 | .024 | .024 | .030 | .032 | .042 | .040 | .048 |
| Over 4 to 5 | .010 | .020 | .012 | .024 | .015 | .030 | .027 | .035 | .032 | .042 | .042 | .050 |
| Over 5 to 6 | .012 | .020 | .014 | .024 | .018 | .030 | .030 | .035 | .036 | .046 | .044 | .054 |
| Over 6 to 7 | .014 | .027 | .016 | .032 | .018 | .035 | .030 | .040 | .036 | .048 | .046 | .056 |
| Over 7 to 10 | .018 | .030 | .020 | .035 | .024 | .040 | .035 | .040 | .040 | .054 | .052 | .064 |
| Over 10 to 12 | .020 | .035 | .025 | .040 | .030 | .045 | .040 | .050 | .046 | .060 | .056 | .072 |

All dimensions are in inches.    * Steel Products Manuals — Alloy Steel: Semifinished; Hot Rolled and Cold Finished Bars, Feb. 1964; and Tool Steels, January 1970.    † Stress relieved or annealed after cold finishing.    ‡ With or without stress relieving or annealing after cold finishing.  Also, all carbons, quenched and tempered (heat treated), or normalized and tempered before cold finishing.    § Width governs the tolerances for both width and thickness of flats.  For example: when the maximum of the carbon range is .28 per cent or less, for a flat 2 inches wide and 1 inch thick, the width tolerance is .006 inch and the thickness tolerance is the same, namely, .006 inch.    ¶ For rounds (other than high speed steel free of scale and decarburization), squares, octagons, quarter octagons and hexagons.

**Permissible Variations in Sizes of Tool Steel Bars and Flats**
(American Iron and Steel Institute*)

TOOL STEEL BARS

| Round Bars, High Speed Steels Free of Scale and Decarburization | | | Forged Rounds, Squares, Octagons and Hexagons | | |
|---|---|---|---|---|---|
| Diameter | Tolerance | | Size | Tolerance | |
| | Minus | Plus | | Minus | Plus |
| ¼ to under ⅝ | .0015 | .0015 | Over 1 to 2 | .030 | .060 |
| ⅝ to under 3 1/16 | .000 | .004 | Over 2 to 3 | .030 | .080 |
| 3 1/16 to under 4 1/16 | .000 | .006 | Over 3 to 5 | .060 | .125 |
| 4 1/16 to under 7 | .000 | .031 | Over 5 to 7 | .125 | .187 |
| .... | .... | .... | Over 7 to 9 | .187 | .312 |

Forged Flats

| Width | Thickness Range and Thickness Tolerance | | | | | | | | | |
|---|---|---|---|---|---|---|---|---|---|---|
| | To 1 | | Over 1 to 3 | | Over 3 to 5 | | Over 5 to 7 | | Over 7 to 9 | |
| | Minus | Plus | Minus | Plus | Minus | Plus | Minus | Plus | Minus | Plus |
| Over 1 to 3 | .016 | .031 | .031 | .078 | .... | .... | .... | .... | .... | .... |
| Over 3 to 5 | .031 | .062 | .047 | .094 | .062 | .125 | .... | .... | .... | .... |
| Over 5 to 7 | .047 | .094 | .062 | .125 | .078 | .156 | .125 | .187 | .... | .... |
| Over 7 to 9 | .062 | .125 | .078 | .156 | .094 | .187 | .156 | .219 | .187 | .312 |

All dimensions are in inches.
* Steel Products Manual — Tool Steels, January 1970.

**Standard Stock Sizes of Machined Tool Steel Bars — Flats and Squares**
(U. S. Simplified Practice Recommendation R267-65)

| Thickness, Inches | Width, Inches | | | | | | | | | | | | | | | | | | | |
|---|---|---|---|---|---|---|---|---|---|---|---|---|---|---|---|---|---|---|---|---|
| | ⅞ | 1 | 1⅛ | 1¼ | 1⅜ | 1½ | 1¾ | 2 | 2¼ | 2½ | 2¾ | 3 | 3½ | 4 | 4½ | 5 | 6 | 8 | 10 | 12 |
| ⅞ | X | X | X | X | .. | X | X | X | .. | X | .. | X | X | X | .. | X | X | X | X | X |
| 1 | .. | X | .. | X | .. | X | X | X | X | X | X | X | X | X | X | X | X | X | X | X |
| 1⅛ | .. | .. | X | X | .. | X | X | X | X | X | .. | X | X | X | .. | X | X | X | X | X |
| 1¼ | .. | .. | .. | X | .. | X | X | X | X | X | X | X | X | X | X | X | X | X | X | X |
| 1⅜ | .. | .. | .. | .. | X | X | X | X | X | X | X | X | X | X | .. | X | X | X | X | X |
| 1½ | .. | .. | .. | .. | .. | X | X | X | X | X | X | X | X | X | X | X | X | X | X | X |
| 1¾ | .. | .. | .. | .. | .. | .. | X | X | X | X | X | X | X | X | X | X | X | X | X | X |
| 2 | .. | .. | .. | .. | .. | .. | .. | X | X | X | X | X | X | X | X | X | X | X | X | X |
| 2½ | .. | .. | .. | .. | .. | .. | .. | .. | .. | X | X | X | X | X | X | X | X | X | X | X |
| 3 | .. | .. | .. | .. | .. | .. | .. | .. | .. | .. | .. | X | X | X | X | X | X | X | X | X |
| 4 | .. | .. | .. | .. | .. | .. | .. | .. | .. | .. | .. | .. | .. | X | X | X | X | X | X | X |
| 5 | .. | .. | .. | .. | .. | .. | .. | .. | .. | .. | .. | .. | .. | .. | .. | X | X | X | X | .. |
| 6 | .. | .. | .. | .. | .. | .. | .. | .. | .. | .. | .. | .. | .. | .. | .. | .. | X | X | X | .. |

*Material:* It is recommended that the machine tool steel bars covered by this recommendation shall conform to the chemical and mechanical requirements for the following tool steels as listed in the latest issue of the Tool Steel Manual of the American Iron and Steel Institute: Air-hardening, AISI — A2, A4 and A6; Oil-hardening, AISI — O1, O2 and O6; and High carbon-high chromium, AISI — D2 and D3.

*Dimensional tolerances:* It is recommended that the sizes in this table be finished in conformity with the following oversize and tolerances: Thickness, 0.015 inch oversize, +0.020, −0.000 inch and Width, 0.015 inch oversize, +0.020, −0.000 inch.

*Example:* A 4-inch by 2-inch bar of machined tool steel should measure as follows: 4.015 inch, +0.020 inch, −0.000 inch by 2.015 inch, + 0.020 inch, − 0.000 inch.

*Surface finish:* It is recommended that the surface finish have a roughness height value of 90-110 microinches (see Handbook section Surface Texture). The bars shall be machined on all four sides and be free from surface imperfections such as carburization, decarburization, scale, etc.

## Thickness Tolerances for Aluminum Sheet and Plate* (ANSI H35.2-1967) — I

Specified Width — Tolerance — Plus or Minus

| Specified Thickness | Up thru 18 | Over 18 thru 36 | Over 36 thru 48 | Over 48 thru 54 | Over 54 thru 60 | Over 60 thru 66 | Over 66 thru 72 | Over 72 thru 78 | Over 78 thru 84 | Over 84 thru 90 | Over 90 thru 96 | Over 96 thru 132 | Over 132 thru 144 | Over 144 thru 156 | Over 156 thru 168 |
|---|---|---|---|---|---|---|---|---|---|---|---|---|---|---|---|
| 0.006 to 0.010 | .001 | .0015 | .0025 | .0025 | | | | | | | | | | | |
| 0.011 to 0.017 | .0015 | .0015 | .0025 | .0035 | | | | | | | | | | | |
| 0.018 to 0.028 | .0015 | .002 | .0025 | .0035 | | | | | | | | | | | |
| 0.029 to 0.036 | .002 | .002 | .003 | .004 | | | | | | | | | | | |
| 0.037 to 0.045 | .002 | .0025 | .004 | .004 | .004 | .004 | .004 | | | | | | | | |
| 0.046 to 0.068 | .0025 | .003 | .004 | .005 | .005 | .005 | .005 | .006 | .006 | .007 | .009 | | | | |
| 0.069 to 0.076 | .003 | .003 | .004 | .005 | .006 | .006 | .006 | .007 | .007 | .007 | .011 | | | | |
| 0.077 to 0.096 | .0035 | .0035 | .005 | .005 | .006 | .006 | .006 | .007 | .007 | .008 | .012 | .013 | | | |
| 0.097 to 0.108 | .004 | .004 | .005 | .005 | .006 | .006 | .006 | .007 | .007 | .012 | .012 | .016 | | | |
| 0.109 to 0.125 | .0045 | .0045 | .005 | .005 | .007 | .007 | .007 | .008 | .008 | .012 | .012 | .016 | | | |
| 0.126 to 0.140 | .0045 | .0045 | .005 | .005 | .007 | .010 | .012 | .013 | .014 | .016 | .018 | .020 | | | |
| 0.141 to 0.172 | .006 | .006 | .008 | .008 | .009 | .012 | .014 | .015 | .016 | .016 | .018 | .023 | | | |
| 0.173 to 0.203 | .007 | .007 | .010 | .010 | .011 | .014 | .016 | .017 | .017 | .017 | .019 | .026 | | | |
| 0.204 to 0.249 | .009 | .009 | .011 | .011 | .013 | .016 | .018 | .018 | .018 | .018 | .022 | .028 | | | |
| 0.250 to 0.320 | .013 | .013 | .013 | .013 | .015 | .018 | .020 | .020 | .020 | .020 | .024 | .030 | | | |
| 0.321 to 0.438 | .019 | .019 | .019 | .019 | .020 | .020 | .023 | .023 | .025 | .025 | .025 | .033 | .035 | .042 | .053 |
| 0.439 to 0.625 | .025 | .025 | .025 | .025 | .025 | .025 | .025 | .030 | .030 | .030 | .035 | .035 | .038 | .045 | .057 |
| 0.626 to 0.875 | .030 | .030 | .030 | .030 | .030 | .030 | .030 | .037 | .037 | .037 | .035 | .045 | .043 | .049 | .067 |
| 0.876 to 1.125 | .035 | .035 | .035 | .035 | .035 | .035 | .035 | .045 | .045 | .045 | .045 | .055 | .054 | .059 | .077 |
| 1.126 to 1.375 | .040 | .040 | .040 | .040 | .040 | .040 | .040 | .052 | .052 | .052 | .055 | .065 | .065 | .070 | .088 |
| 1.376 to 1.625 | .045 | .045 | .045 | .045 | .045 | .045 | .045 | .060 | .060 | .060 | .065 | .075 | .075 | .080 | .098 |
| 1.626 to 1.875 | .052 | .052 | .052 | .052 | .052 | .052 | .052 | .070 | .070 | .070 | .075 | .088 | .085 | .090 | .108 |
| 1.876 to 2.250 | .060 | .060 | .060 | .060 | .060 | .060 | .060 | .080 | .080 | .080 | .088 | .100 | | | |
| 2.251 to 2.750 | .075 | .075 | .075 | .075 | .075 | .075 | .075 | .100 | .100 | .100 | .100 | .125 | | | |
| 2.751 to 3.000 | .090 | .090 | .090 | .090 | .090 | .090 | .090 | .120 | .120 | .120 | .125 | .150 | | | |
| 3.001 to 4.000 | .110 | .110 | .110 | .110 | .110 | .110 | .110 | .140 | .140 | .140 | .150 | .160 | | | |
| 4.001 to 5.000 | .125 | .125 | .125 | .125 | .125 | .125 | .125 | .150 | .150 | .150 | .160 | | | | |
| 5.001 to 6.000 | .135 | .135 | .135 | .135 | .135 | .135 | .135 | .160 | .160 | .160 | .170 | | | | |

All dimensions are given in inches.    *For alloys 2014, 2219, 2024, 5052, 3004, 5154, 5454, 5456, 5083, 5086, 6061, 7039, 7075, 7178, 7079; Alclad alloys, 2014, 2219, 2024, 3004, 6061, 7075, 7075 (One Side), 7178, 7079; and Brazing Sheet Nos. 11, 12, 21, 22, 23 and 24.

## Thickness Tolerances for Aluminum Sheet and Plate* (ANSI H35.2-1967) — 2

| Specified Thickness | Specified Width | | | | | | | | | |
|---|---|---|---|---|---|---|---|---|---|---|
| | Up thru 18 | Over 18 thru 36 | Over 36 thru 54 | Over 54 thru 72 | Over 72 thru 90 | Over 90 thru 102 | Over 102 thru 132 | Over 132 thru 144 | Over 144 thru 156 | Over 156 thru 168 |
| | Tolerance — Plus or Minus | | | | | | | | | |
| 0.006 to 0.007 | .001 | .001 | .002 | .... | .... | .... | .... | .... | .... | .... |
| 0.008 to 0.010 | .001 | .0015 | .002 | .... | .... | .... | .... | .... | .... | .... |
| 0.011 to 0.017 | .0015 | .0015 | .002 | .... | .... | .... | .... | .... | .... | .... |
| 0.018 to 0.028 | .0015 | .002 | .0025 | .003 | .004 | .... | .... | .... | .... | .... |
| 0.029 to 0.036 | .002 | .002 | .0025 | .0035 | .005 | .006 | .... | .... | .... | .... |
| 0.037 to 0.045 | .002 | .0025 | .003 | .004 | .005 | .006 | .... | .... | .... | .... |
| 0.046 to 0.068 | .0025 | .003 | .004 | .005 | .006 | .007 | .008 | .... | .... | .... |
| 0.069 to 0.076 | .0025 | .003 | .004 | .006 | .008 | .008 | .009 | .... | .... | .... |
| 0.077 to 0.096 | .003 | .003 | .004 | .006 | .008 | .009 | .010 | .... | .... | .... |
| 0.097 to 0.108 | .0035 | .004 | .005 | .007 | .009 | .009 | .010 | .012 | .... | .... |
| 0.109 to 0.140 | .0045 | .0045 | .005 | .007 | .009 | .010 | .012 | .... | .... | .... |
| 0.141 to 0.172 | .006 | .006 | .008 | .009 | .011 | .012 | .015 | .... | .... | .... |
| 0.173 to 0.203 | .007 | .007 | .009 | .011 | .013 | .015 | .017 | .... | .... | .... |
| 0.204 to 0.249 | .009 | .009 | .011 | .013 | .015 | .017 | .020 | .... | .... | .... |
| 0.250 to 0.320 | .013 | .013 | .013 | .015 | .017 | .020 | .023 | .032 | .040 | .050 |
| 0.321 to 0.438 | .019 | .019 | .019 | .019 | .023 | .026 | .026 | .035 | .043 | .052 |
| 0.439 to 0.625 | .025 | .025 | .025 | .025 | .030 | .035 | .035 | .040 | .046 | .055 |
| 0.626 to 0.875 | .030 | .030 | .030 | .030 | .037 | .045 | .045 | .050 | .056 | .064 |
| 0.876 to 1.125 | .035 | .035 | .035 | .035 | .045 | .055 | .055 | .060 | .066 | .074 |
| 1.126 to 1.375 | .040 | .040 | .040 | .040 | .052 | .065 | .065 | .070 | .075 | .082 |
| 1.376 to 1.625 | .045 | .045 | .045 | .045 | .060 | .075 | .075 | .080 | .085 | .092 |
| 1.626 to 1.875 | .052 | .052 | .052 | .052 | .070 | .088 | .088 | .... | .... | .... |
| 1.876 to 2.250 | .060 | .060 | .060 | .060 | .080 | .100 | .100 | .... | .... | .... |
| 2.251 to 2.750 | .075 | .075 | .075 | .075 | .100 | .125 | .125 | .... | .... | .... |
| 2.751 to 3.000 | .090 | .090 | .090 | .090 | .120 | .150 | .150 | .... | .... | .... |
| 3.001 to 4.000 | .110 | .110 | .110 | .110 | .140 | .160 | .160 | .... | .... | .... |
| 4.001 to 5.000 | .125 | .125 | .125 | .125 | .150 | .160 | .... | .... | .... | .... |
| 5.001 to 6.000 | .135 | .135 | .135 | .135 | .160 | .170 | .... | .... | .... | .... |

All dimensions are given in inches. * For alloys EC, 1100, 1060, 3003, Alclad 3003, 5005, 5050, 5257, 5457, 5557, 5657, 1100 Reflector Sheet, Clad 1100 Reflector Sheet, and Clad 3003 Reflector Sheet.

## Length and Width Tolerances for Aluminum Sheared Flat Sheet and Plate
### (ANSI H35.2-1967)

| Specified Thickness | Specified Length | | | | | | |
|---|---|---|---|---|---|---|---|
| | Up thru 18 | Over 18 thru 48 | Over 48 thru 120 | Over 120 thru 144 | Over 144 thru 180 | Over 180 thru 240 | Over 240 thru 540 |
| | Length Tolerance | | | | | | |
| 0.006 to 0.249 | $\pm\frac{1}{16}$ | $\pm\frac{3}{32}$ | $\pm\frac{1}{8}$ | $\pm\frac{5}{32}$ | $\pm\frac{5}{32}$ | $\pm\frac{1}{4}$ | $\pm\frac{1}{4}$ |
| 0.250 to 0.500 | $+\frac{3}{8}$ | $+\frac{3}{8}$ | $+\frac{3}{8}$ | $+\frac{3}{8}$ | $+\frac{7}{16}$ | $+\frac{7}{16}$ | $+\frac{1}{2}$ |
| 0.501 to 1.000 | $+\frac{1}{2}$ | $+\frac{1}{2}$ | $+\frac{1}{2}$ | $+\frac{1}{2}$ | $+\frac{9}{16}$ | $+\frac{9}{16}$ | $+\frac{5}{8}$ |
| 1.001 to 1.250 | $+\frac{5}{8}$ | $+\frac{5}{8}$ | $+\frac{5}{8}$ | $+\frac{5}{8}$ | $+\frac{3}{4}$ | $+\frac{3}{4}$ | $+1$ |

| Specified Thickness | Specified Width | | | | | | |
|---|---|---|---|---|---|---|---|
| | Up thru 4 | Over 4 thru 18 | Over 18 thru 36 | Over 36 thru 54 | Over 54 thru 72 | Over 72 thru 132 | Over 132 thru 168 |
| | Width Tolerance | | | | | | |
| 0.006 to 0.064 | $\pm\frac{1}{32}$ | $\pm\frac{1}{16}$ | $\pm\frac{3}{32}$ | $\pm\frac{1}{8}$ | $\pm\frac{5}{32}$ | $\pm\frac{3}{16}$ | .... |
| 0.065 to 0.102 | $\pm\frac{1}{16}$ | $\pm\frac{1}{16}$ | $\pm\frac{3}{32}$ | $\pm\frac{1}{8}$ | $\pm\frac{5}{32}$ | $\pm\frac{3}{16}$ | .... |
| 0.103 to 0.249 | $\pm\frac{1}{8}$ | $\pm\frac{3}{32}$ | $\pm\frac{1}{8}$ | $\pm\frac{3}{16}$ | $\pm\frac{3}{16}$ | $\pm\frac{1}{4}$ | .... |
| 0.250 to 0.500 | .... | $+\frac{3}{8}$ | $+\frac{3}{8}$ | $+\frac{3}{8}$ | $+\frac{3}{8}$ | $+\frac{3}{8}$ | $+\frac{1}{2}$ |
| 0.501 to 1.000 | .... | $+\frac{1}{2}$ | $+\frac{1}{2}$ | $+\frac{1}{2}$ | $+\frac{1}{2}$ | $+\frac{1}{2}$ | $+\frac{5}{8}$ |
| 1.001 to 1.250 | .... | $+\frac{5}{8}$ | $+\frac{5}{8}$ | $+\frac{5}{8}$ | $+\frac{5}{8}$ | $+\frac{5}{8}$ | $+\frac{3}{4}$ |

All dimensions are given in inches.

## Tolerances for Aluminum Sheet and Plate (ANSI H35.2-1967)

### SAWED FLAT PLATE

| Specified Thickness | Specified Length | | | | Specified Thickness | Specified Width | | | | |
|---|---|---|---|---|---|---|---|---|---|---|
| | Up thru 10 | Over 10 thru 48 | Over 48 thru 84 | Over 84 | | Up thru 10 | Over 10 thru 48 | Over 48 thru 84 | Over 84 thru 132 | Over 132 thru 168 |
| | Length Tolerance — Plus or Minus | | | | | Width Tolerance — Plus or Minus | | | | |
| 0.250 to 6.000 | 3/32 | 3/16 | 1/4 | 5/16 | 0.250 to 6.000 | 3/32 | 3/16 | 1/4 | 5/16 | 7/16 |

### FLAT SHEET AND PLATE

| Specified Width | Specified Thickness | Specified Length | | | | | | | |
|---|---|---|---|---|---|---|---|---|---|
| | | Up thru 30 | Over 30 thru 60 | Over 60 thru 90 | Over 90 thru 120 | Over 120 thru 150 | Over 150 thru 180 | Over 180 thru 210 | Over 210 thru 240 |
| | | Lateral Bow Tolerance — Allowable Deviation of a Side Edge From a Straight Line | | | | | | | |
| Up thru 4 | 0.006 to 0.125 | 1/16 | 1/4 | 1/2 | 1 | 1 1/2 | 2 | 3 | 4 |
| Over 4 thru 35 | 0.006 to 0.249 | 1/32 | 1/16 | 3/32 | 1/8 | 3/16 | 1 | 1 1/2 | 2 |
| Over 35 | 0.006 to 0.249 | 1/32 | 1/16 | 3/32 | 1/8 | 3/16 | 5/16 | 7/16 | 9/16 |
| Up thru 10 | 0.250 to 6.000 | 1/16 | 1/4 | 1/2 | 1 | 1 1/2 | 2 | 3 | 4* |
| Over 10 thru 18 | 0.250 to 6.000 | 1/32 | 1/16 | 1/8 | 1/4 | 13/32 | 19/32 | 25/32 | 1* |
| Over 18 | 0.250 to 6.000 | 1/32 | 1/16 | 3/32 | 1/8 | 3/16 | 5/16 | 7/16 | 9/16* |

| Specified Thickness | Specified Length | | | | | | |
|---|---|---|---|---|---|---|---|
| | Up thru 18 | Over 18 thru 48 | Over 48 thru 120 | Over 120 thru 144 | Over 144 thru 180 | Over 180 thru 240 | Over 240 thru 540 |
| | Squareness Tolerance — Allowable Difference in Length of Diagonals | | | | | | |
| 0.006 to 0.249 | 1/8 | 3/16 | 1/4 | 5/16 | 5/16 | 1/2 | 1/2 |
| 0.250 to 0.500 | 3/8 | 3/8 | 3/8 | 3/8 | 7/16 | 9/16 | 5/8 |
| 0.501 to 1.000 | 1/2 | 1/2 | 1/2 | 1/2 | 9/16 | 9/16 | 3/4 |
| 1.001 to 6.000 | 5/8 | 5/8 | 5/8 | 5/8 | 3/4 | 3/4 | 1 |

### SAWED OR SHEARED PLATE — H and T Tempers Except T351, T451, T651 and T851†

| Specified Thickness | Transverse Flatness Tolerance — Allowable Deviation from Flat | | |
|---|---|---|---|
| | Widths over 48 thru 72§ | Widths over 24 thru 48 | Widths 24 and less |
| 0.250 to 0.624 | 1/2 | 3/8 | Use short-cycle tolerance |
| 0.625 to 1.500 | 3/8 | 1/4 | |
| 1.501 to 3.000 | 1/4 | 3/16 | |

### SAWED OR SHEARED PLATE — T351, T451, T651 and T851 Tempers

| Specified Thickness | Transverse Flatness Tolerance — Allowable Deviation from Flat | | |
|---|---|---|---|
| | Widths over 48 thru 72§ | Widths over 24 thru 48 | Widths 24 and less |
| 0.250 to 0.624 | 3/8 | 5/16 | Use short-cycle tolerance |
| 0.625 to 1.500 | 5/16 | 1/4 | |
| 1.501 to 3.000 | 3/16 | 3/16 | |
| 3.001 to 6.000 | 1/8 | 1/8 | |

All dimensions in inches.    *Also applicable to any 240-inch increment of longer plate. †The longitudinal flatness tolerance is 1/4 inch in any 6 feet or less in a specified thickness range of from 0.250 to 3.000 inches and the short-cycle flatness tolerance (allowable deviation from flat) is 0.125 inch in a specified thickness range of from 0.250 to 0.624 inch and 0.090 inch in a specified thickness range of from 0.625 to 3.000 inch. The short-cycle flatness tolerance is measured with the plate resting on a flat surface and by use of a frame with rollers mounted on 2-foot centers and a depth gage in the center. (Short-cycle flatness is the flatness over any 2-foot span in any direction.)    ‡The longitudinal flatness tolerance is 3/16 inch in any 6 feet in a specified thickness range of from 0.250 to 3.000 inches. (For lengths over 6 feet, the tolerance is 1/8 inch.) The longitudinal flatness tolerance is 1/8 inch in any 6 feet or less in a specified thickness range of from 3.001 to 6.000 inches. The short-cycle flatness tolerance (allowable deviation from flat) is 0.100 inch in a specified thickness range of from 0.250 to 0.624 inch and 0.075 inch in a specified thickness range of from 0.625 to 6.000 inches.    §For widths over 6 feet, these tolerances apply for any 6 feet of total width.

### Tolerances for Aluminum Wire, Rod and Bar (ANSI H35.2-1967)

#### ROUND WIRE AND ROD

Diameter Tolerance — Plus or Minus Except as Noted
Allowable Deviation from Specified Diameter

| Specified Diameter | Drawn Wire | Cold Finished Rod | Rolled Rod Plus | Rolled Rod Minus |
|---|---|---|---|---|
| Up thru 0.035 | .0005 | .... | .... | .... |
| 0.036 to 0.064 | .001 | .... | .... | .... |
| 0.065 to 0.374 | .0015 | .... | .... | .... |
| 0.375 to 0.500 | .... | .0015 | .... | .... |
| 0.501 to 1.000 | .... | .002 | .... | .... |
| 1.001 to 1.500 | .... | .0025 | .... | .... |
| 1.501 to 2.000 | .... | .004 | .006 | .006 |
| 2.001 to 3.000 | .... | .006 | .008 | .008 |
| 3.001 to 3.499 | .... | .008 | .012 | .012 |
| 3.500 to 5.000 | .... | .012 | .031 | .016 |
| 5.001 to 8.000 | .... | .... | .062 | .031 |

#### REDRAW ROD

Diameter Tolerance — Plus or Minus
Allowable Deviation from Specified Diameter

| Specified Diameter | |
|---|---|
| 0.375 | 0.020 |

#### RECTANGULAR WIRE AND BAR

Tolerance — Plus or Minus
Allowable Deviation from Specified Thickness and Width

| Specified Thickness or Width | Drawn Wire and Cold Finished Bar Thickness | Width | Rolled Bar Thickness | Width |
|---|---|---|---|---|
| Up thru 0.035 | .001 | .... | .... | .... |
| 0.036 to 0.064 | .0015 | .... | .... | .... |
| 0.065 to 0.500 | .002 | .002 | .006 | .... |
| 0.501 to 0.750 | .0025 | .0025 | .008 | .016 |
| 0.751 to 1.000 | .0025 | .0025 | .012 | .016 |
| 1.001 to 1.500 | .003 | .003 | .016 | .016 |
| 1.501 to 2.000 | .005 | .005 | .016 | .031 |
| 2.001 to 3.000 | .008 | .008 | .020 | .031 |
| 3.001 to 4.000 | .... | .010 | .020 | .031 |
| 4.001 to 6.000 | .... | .... | .... | .047 |
| 6.001 to 10.000 | .... | .... | .... | .062 |

#### RIVET AND COLD HEADING WIRE AND ROD

Diameter Tolerance
Allowable Deviation from Specified Diameter

| Specified Diameter | Rivet Wire Plus | Rivet Wire Minus | Rivet Rod Plus | Rivet Rod Minus |
|---|---|---|---|---|
| Up thru 0.061 | .0005 | .0005 | .... | .... |
| 0.062 to 0.123 | .001 | .0005 | .... | .... |
| 0.124 to 0.154 | .001 | .001 | .... | .... |
| 0.155 to 0.374 | .002 | .001 | .... | .... |
| 0.375 to 0.500 | .... | .... | .002 | .001 |
| 0.501 to 1.000 | .... | .... | .003 | .001 |

#### CENTERLESS GROUND ROUND WIRE AND ROD

Diameter Tolerance — Plus or Minus
Allowable Deviation from Specified Diameter

| Specified Diameter | |
|---|---|
| 0.125 to 0.625 | .0005 |
| 0.626 to 1.500 | .0010 |
| 1.501 to 2.000 | .0025 |

#### SQUARE, HEXAGONAL AND OCTAGONAL WIRE AND BAR

Tolerance — Plus or Minus
Allowable Deviation from Specified Distance Across Flats

| Specified Distance Across Flats | Drawn Wire | Cold Finished Bar | Rolled Bar |
|---|---|---|---|
| Up thru 0.035 | .001 | .... | .... |
| 0.036 to 0.064 | .0015 | .... | .... |
| 0.065 to 0.374 | .002 | .... | .... |
| 0.375 to 0.500 | .... | .002 | .... |
| 0.501 to 1.000 | .... | .0025 | .... |
| 1.001 to 1.500 | .... | .003 | .... |
| 1.501 to 2.000 | .... | .005 | .016 |
| 2.001 to 3.000 | .... | .008 | .020 |
| 3.001 to 4.000 | .... | .... | .020 |

#### FLATTENED AND SLIT WIRE

| Specified Thickness | Tolerance + or − Allowable Deviation from Specified Thickness | Specified Width | Tolerance + or − Allowable Deviation from Specified Width |
|---|---|---|---|
| 0.018 to 0.020 | .001 | 0.500 to 0.625 | .0025 |
| 0.021 to 0.060 | .0015 | 0.626 to 1.500 | .004 |
| 0.061 to 0.080 | .002 | 1.501 to 4.750 | .006 |

#### FLATTENED WIRE (Round Edge)*

| Specified Width | Tolerance + or − Allowable Deviation from Specified Width |
|---|---|
| Up thru 0.875 | .007 |
| 0.876 to 2.000 | .010 |

All dimensions are in inches.    *Thickness tolerance is same as for wire at left except that specified thickness range is "Up thru 0.020" for ±.001 tolerance.

# STRENGTH AND PROPERTIES OF
# WIRE ROPE

**Wire Rope Construction.** — Essentially, a wire rope is made up of a number of strands laid helically about a metallic or non-metallic core. Each strand consists of a number of wires also laid helically about a metallic or non-metallic center. Various types of wire rope have been developed to meet a wide range of uses and operating conditions. These types are distinguished by the kind of core; the number of strands; the number, sizes, and arrangement of the wires in each strand; and the way in which the wires and strands are wound or laid about each other. The following descriptive material is based largely on information supplied by the Bethlehem Steel Co.

*Rope Wire Materials:* Materials used in the manufacture of rope wire are, in the order of increasing strength: iron, phosphor bronze, traction steel, plow steel, improved plow steel, and bridge rope steel. Iron wire rope is largely used for low-strength applications such as elevator ropes not used for hoisting, and for stationary guy ropes.

Phosphor bronze wire rope is used occasionally for elevator governor-cable rope and for certain marine applications as life lines, clearing lines, wheel ropes and rigging.

Traction steel wire rope is used primarily as a hoist rope for passenger and freight elevators of the traction drive type, an application for which it was specifically designed.

Ropes made of galvanized wire or wire coated with zinc by the electrodeposition process are used in certain applications where additional protection against rusting is required. As will be noted from the tables of wire-rope sizes and strengths, the breaking strength of galvanized wire rope is 10 per cent less than that of ungalvanized (bright) wire rope. Bethanized (zinc-coated) wire rope can be furnished to bright wire rope strength when so specified.

Galvanized carbon steel, tinned carbon steel, and stainless steel are used for small cords and strands ranging in diameter from 1/64 to 3/8 inch and larger.

Marline clad wire rope has each strand wrapped with a layer of tarred marline. This provides harsh protection for workmen and wear protection for the rope.

*Rope Cores:* Wire-rope cores are made of fiber, cotton, asbestos, polyvinyl plastic, a small wire rope (independent wire-rope core), a multiple-wire strand (wire-strand core) or a cold-drawn wire-wound spring.

*Fiber* (manila or sisal) is the type of core most widely used when loads are not too great. It supports the strands in their relative positions and also acts as a cushion to prevent nicking of the wires lying next to the core.

*Cotton* is used for small ropes such as sash cord and aircraft cord.

*Asbestos cores* can be furnished for certain special operations where the rope is used in oven operations.

*Polyvinyl plastic cores* are offered for use where exposure to moisture, acids, or caustics is excessive.

A *wire strand core*, often referred to as WSC, consists of a multiple-wire strand that may be the same as one of the strands of the rope. It is smoother and more solid than the independent wire rope core and provides a better support for the rope strands.

The *independent wire rope core*, often referred to as IWRC, is a small 6 × 7 wire rope with a wire strand core and is used to provide greater resistance to crushing and distortion of the wire rope. For certain applications it has the advantage over a wire-strand core in that it stretches at a rate closer to that of the rope itself.

Wire ropes with wire-strand cores are, in general, less flexible than wire ropes with independent wire-rope or non-metallic cores.

Ropes with metallic cores are rated 7½ per cent stronger than those with non-metallic cores.

*Wire-Rope Lay:* The lay of a wire rope is the direction of the helical path in which the strands are laid and, similarly, the lay of a strand is the direction of the helical path in which the wires are laid. If the wires in the strand or the strands in the rope form a helix similar to the threads of a right-hand screw, i.e., they wind around to the right, the lay is called right hand and, conversely, if they wind around to the left, the lay is called left hand. In the *regular lay*, the wires in the strands are laid in the opposite direction to the lay of the strands in the rope. In right-regular lay, the strands are laid to the right and the wires to the left. In left-regular lay, the strands are laid to the left, the wires to the right. In *Lang lay*, the wires and strands are laid in the same direction, i.e., in right Lang lay, both the wires and strands are laid to the right and in left Lang lay they are laid to the left.

Alternate lay ropes having alternate right and left laid strands are used to resist distortion and prevent clamp slippage, but because other advantages are missing, have limited use.

The regular lay wire rope is most widely used and right regular lay rope is customarily furnished. Regular lay rope has less tendency to spin or untwist when placed under load and is generally selected where long ropes are employed and the loads handled are frequently removed. Lang lay ropes have greater flexibility than regular lay ropes and are more resistant to abrasion and fatigue.

In preformed wire ropes the wires and strands are preshaped into a helical form so that when laid to form the rope they tend to remain in place. In a non-preformed rope, broken wires tend to "wicker out" or protrude from the rope and strands that are not seized tend to spring apart. Preforming also tends to remove locked-in stresses, lengthen service life, and make the rope easier to handle and to spool.

*Strand Construction:* Various arrangements of wire are used in the construction of wire rope strands. In the simplest arrangement six wires are grouped around a central wire thus making seven wires, all of the same size. Other types of construction known as "filler-wire," Warrington, Seale, etc. make use of wires of different sizes. Their respective patterns of arrangement are shown diagrammatically in the tables of wire rope weights and strengths.

**Specifying Wire Rope.** — In specifying wire rope the following information will be required: length, diameter, number of strands, number of wires in each strand, type of rope construction, grade of steel used in rope, whether preformed or not preformed, type of center, and type of lay. The manufacturer should be consulted in selecting the best type of wire rope for a new application.

**Properties of Wire Rope.** — Important properties of wire rope are strength, wear resistance, flexibility, and resistance to crushing and distortion.

*Strength:* The strength of wire rope depends upon its size, kind of material of which the wires are made and their number, the type of core, and whether the wire is galvanized or not. Strengths of various types and sizes of wire ropes are given in the accompanying tables together with appropriate factors to apply for ropes with steel cores and for galvanized wire ropes.

*Wear Resistance:* Wire rope which in use must pass back and forth over surfaces which subject it to unusual wear or abrasion must be specially constructed to give satisfactory service. Such construction may make use of (1) relatively large outer wires; (2) Lang lay in which wires in each strand are laid in the same direction as the strand; or (3) flattened strands. The object in each case is to provide a greater outside surface area to take the wear or abrasion. From the standpoint of material, improved plow steel has not only the highest tensile strength but also the greatest resistance to abrasion in regularly stocked wire rope.

*Flexibility:* Wire rope which undergoes repeated and severe bending, such as in passing around small sheaves and drums, must have a high degree of flexibility to prevent premature breakage and failure due to fatigue. Greater flexibility in wire rope is obtained by (1) using small wires in larger numbers, (2) using Lang lay, and (3) preforming, that is the wires and strands of the rope are shaped during manufacture to fit the position they will assume in the finished rope.

*Resistance to Crushing and Distortion:* Where wire rope is to be subjected to transverse loads that may crush or distort it, care should be taken to select a type of construction which will stand up under such treatment. Wire rope designed for such conditions may have (1) large outer wires to spread the load per wire over a greater area and (2) an independent wire core or a high-carbon cold-drawn wound spring core.

**Standard Classes of Wire Rope.** — Wire rope is commonly designated by two figures, the first indicating the number of strands and the second, the number of wires per strand, as: 6 × 7, a six-strand rope having seven wires per strand, 8 × 19, an eight-strand rope having 19 wires per strand, etc. When such numbers are used as designations of standard wire rope classes, the second figure in the designation may be purely nominal in that the number of wires per strand for various ropes in the class may be slightly less or slightly more than the nominal as will be seen from the following brief descriptions. (For ropes with a wire strand core, a second group of two numbers may be used to indicate the construction of the wire core, as 1 × 21, 1 × 43, etc.)

*6 × 7 Class (Standard Coarse Laid Rope):* Wire ropes in this class are for use where resistance to wear, as in dragging over the ground or across rollers, is an important requirement. Heavy hauling, rope transmissions, well drilling are common applications. These wire ropes are furnished in right regular lay and occasionally in Lang lay. The cores may be of fiber, independent wire rope, or wire strand. Since this is a relatively stiff type of construction, ropes in this class should be used with large sheaves and drums. Because of the small number of wires, a larger factor of safety may be called for.

As shown in Table 1, this class includes a 6 × 7 construction with fiber core: a 6 × 7 construction with 1 × 7 wire strand core (sometimes called 7 × 7); a 6 × 7 construction with 1 × 19 wire strand core; and a 6 × 7 construction with independent wire rope core.

Two special types of wire rope in this class are: aircraft cord, a 6 × 6 or 7 × 7 Bethanized wire rope of high tensile strength and sash cord, a 6 × 7 iron rope used for a variety of purposes where strength is not an important factor.

*6 × 19 Class (Standard Hoisting Rope):* This is the most popular and widely used class of wire ropes. Ropes in this class are furnished in regular or Lang lay and may be obtained preformed or not preformed. Cores may be of fiber, independent wire rope, or wire strand. As can be seen from Table 2, there are four common types: 6 × 25 filler wire construction with fiber core (not illustrated), independent wire core, or wire strand core (1 × 25 or 1 × 43); 6 × 19 Warrington construction with fiber core; 6 × 21 filler wire construction with fiber core; and 6 × 19, 6 × 21, and 6 × 17 Seale construction with fiber core.

*6 × 37 Class (Extra Flexible Hoisting Rope):* For a given size of rope, the component wires are of smaller diameter than those in the two classes previously described and hence have less resistance to abrasion. Ropes in this class are furnished in regular and Lang lay with fiber core or independent wire rope core, preformed or not preformed.

**Table 1.** Weights and Strengths of 6 × 7 (Standard Coarse Laid) Wire Ropes, Preformed and Not Preformed

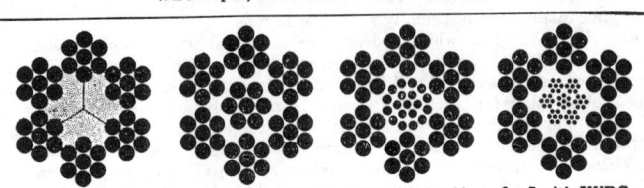

6 x 7 with fiber core    6 x 7 with 1 x 7 WSC    6 x 7 with 1 x 19    6 x 7 with IWRC
         (sometimes called 7 x 7)      WSC

| Diam., Inches | Approx. Weight per Ft., Pounds | Breaking Strength, Tons of 2000 Pounds | | | Diam., Inches | Approx. Weight per Ft., Pounds | Breaking Strength, Tons of 2000 Pounds | | |
|---|---|---|---|---|---|---|---|---|---|
| | | Impr. Plow Steel | Plow Steel | Mild Plow Steel | | | Impr. Plow Steel | Plow Steel | Mild Plow Steel |
| ¼ | 0.094 | 2.64 | 2.30 | 2.00 | ¾ | 0.84 | 22.7 | 19.8 | 17.2 |
| ⁵⁄₁₆ | 0.15 | 4.10 | 3.56 | 3.10 | ⅞ | 1.15 | 30.7 | 26.7 | 23.2 |
| ⅜ | 0.21 | 5.86 | 5.10 | 4.43 | 1 | 1.50 | 39.7 | 34.5 | 30.0 |
| ⁷⁄₁₆ | 0.29 | 7.93 | 6.90 | 6.00 | 1⅛ | 1.90 | 49.8 | 43.3 | 37.7 |
| ½ | 0.38 | 10.3 | 8.96 | 7.79 | 1¼ | 2.34 | 61.0 | 53.0 | 46.1 |
| ⁹⁄₁₆ | 0.48 | 13.0 | 11.3 | 9.82 | 1⅜ | 2.84 | 73.1 | 63.6 | 55.3 |
| ⅝ | 0.59 | 15.9 | 13.9 | 12.0 | 1½ | 3.38 | 86.2 | 75.0 | 65.2 |

For ropes with steel cores, add 7½ per cent to above strengths.
For galvanized ropes, deduct 10 per cent from above strengths.
*Source:* Rope diagrams, Bethlehem Steel Co. All data, U. S. Simplified Practice Recommendation 198-50.

As shown in Table 3, there are four common types: 6 × 29 filler wire construction with fiber core and 6 × 36 filler wire construction with independent wire rope core, a special rope for construction equipment; 6 × 35 (two operations) construction with fiber core and 6 × 41 Warrington Seale construction with fiber core, a standard crane rope in this class of rope construction; 6 × 41 filler wire construction with fiber core or independent wire rope core, a special large shovel rope usually furnished in Lang lay; and 6 × 46 filler wire construction with fiber core or independent wire rope core, a special large shovel and dredge rope.

*8 × 19 Class (Special Flexible Hoisting Rope):* This is a stable smooth-running rope, especially suitable, because of its flexibility, for high speed operation with reverse bends. Ropes in this class are available in regular lay with fiber core.

As shown in Table 4, there are four common types: 8 × 25 filler wire construction, the most flexible but the least wear resistant rope of the four types; Warrington type in 8 × 19 construction, less flexible than the 8 × 25; 8 × 21 filler wire construction, less flexible than the Warrington; and Seale type in 8 × 19 construction, which has the greatest wear resistance of the four types but is also the least flexible.

Also in this class, but not shown in Table 4 are elevator ropes made of traction steel and iron.

*18 × 7 Non-rotating Wire Rope:* This is a rope specially designed for use where a minimum of rotating or spinning is called for, especially in the lifting or lowering

**Table 2.** Weights and Strengths of 6 × 19 (Standard Hoisting) Wire Ropes, Preformed and Not Preformed

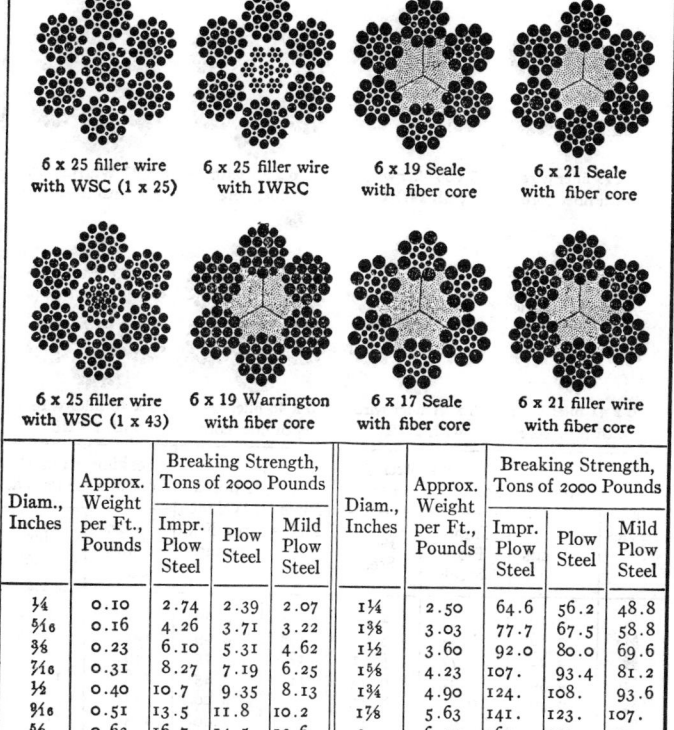

| 6 x 25 filler wire with WSC (1 x 25) | 6 x 25 filler wire with IWRC | 6 x 19 Seale with fiber core | 6 x 21 Seale with fiber core |

| 6 x 25 filler wire with WSC (1 x 43) | 6 x 19 Warrington with fiber core | 6 x 17 Seale with fiber core | 6 x 21 filler wire with fiber core |

| Diam., Inches | Approx. Weight per Ft., Pounds | Breaking Strength, Tons of 2000 Pounds | | | Diam., Inches | Approx. Weight per Ft., Pounds | Breaking Strength, Tons of 2000 Pounds | | |
|---|---|---|---|---|---|---|---|---|---|
| | | Impr. Plow Steel | Plow Steel | Mild Plow Steel | | | Impr. Plow Steel | Plow Steel | Mild Plow Steel |
| ¼ | 0.10 | 2.74 | 2.39 | 2.07 | 1¼ | 2.50 | 64.6 | 56.2 | 48.8 |
| 5⁄16 | 0.16 | 4.26 | 3.71 | 3.22 | 1⅜ | 3.03 | 77.7 | 67.5 | 58.8 |
| ⅜ | 0.23 | 6.10 | 5.31 | 4.62 | 1½ | 3.60 | 92.0 | 80.0 | 69.6 |
| 7⁄16 | 0.31 | 8.27 | 7.19 | 6.25 | 1⅝ | 4.23 | 107. | 93.4 | 81.2 |
| ½ | 0.40 | 10.7 | 9.35 | 8.13 | 1¾ | 4.90 | 124. | 108. | 93.6 |
| 9⁄16 | 0.51 | 13.5 | 11.8 | 10.2 | 1⅞ | 5.63 | 141. | 123. | 107. |
| ⅝ | 0.63 | 16.7 | 14.5 | 12.6 | 2 | 6.40 | 160. | 139. | 121. |
| ¾ | 0.90 | 23.8 | 20.7 | 18.0 | 2⅛ | 7.23 | 179. | 156. | .... |
| ⅞ | 1.23 | 32.2 | 28.0 | 24.3 | 2¼ | 8.10 | 200. | 174. | .... |
| 1 | 1.60 | 41.8 | 36.4 | 31.6 | 2½ | 10.00 | 244. | 212. | .... |
| 1⅛ | 2.03 | 52.6 | 45.7 | 39.8 | 2¾ | 12.10 | 292. | 254. | .... |

The 6 × 25 filler wire with fiber core not illustrated.
For ropes with steel cores, add 7½ per cent to above strengths.
For galvanized ropes, deduct 10 per cent from above strengths.
*Source:* Rope diagrams, Bethlehem Steel Co. All data, U. S. Simplified Practice Recommendation 198–50.

of free loads with a single-part line. It has an inner layer composed of 6 strands of 7 wires each laid in left Lang lay over a fiber core and an outer layer of 12 strands of 7 wires each laid in right regular lay. The combination of opposing lays tends to prevent rotation when the rope is stretched. However, to avoid any tendency to rotate or spin, loads should be kept to at least one-eighth and preferably one-tenth of the

### Table 3. Weights and Strengths of 6 × 37 (Extra Flexible Hoisting) Wire Ropes, Preformed and Not Preformed

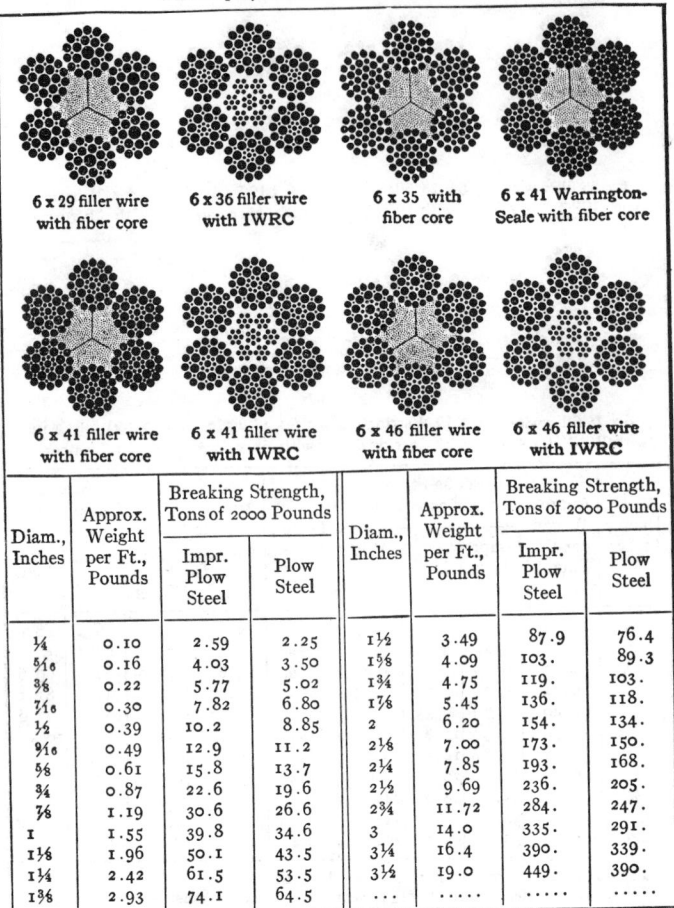

6 x 29 filler wire with fiber core

6 x 36 filler wire with IWRC

6 x 35 with fiber core

6 x 41 Warrington-Seale with fiber core

6 x 41 filler wire with fiber core

6 x 41 filler wire with IWRC

6 x 46 filler wire with fiber core

6 x 46 filler wire with IWRC

| Diam., Inches | Approx. Weight per Ft., Pounds | Breaking Strength, Tons of 2000 Pounds | | Diam., Inches | Approx. Weight per Ft., Pounds | Breaking Strength, Tons of 2000 Pounds | |
|---|---|---|---|---|---|---|---|
| | | Impr. Plow Steel | Plow Steel | | | Impr. Plow Steel | Plow Steel |
| ¼ | 0.10 | 2.59 | 2.25 | 1½ | 3.49 | 87.9 | 76.4 |
| ⁵⁄₁₆ | 0.16 | 4.03 | 3.50 | 1⅝ | 4.09 | 103. | 89.3 |
| ⅜ | 0.22 | 5.77 | 5.02 | 1¾ | 4.75 | 119. | 103. |
| ⁷⁄₁₆ | 0.30 | 7.82 | 6.80 | 1⅞ | 5.45 | 136. | 118. |
| ½ | 0.39 | 10.2 | 8.85 | 2 | 6.20 | 154. | 134. |
| ⁹⁄₁₆ | 0.49 | 12.9 | 11.2 | 2⅛ | 7.00 | 173. | 150. |
| ⅝ | 0.61 | 15.8 | 13.7 | 2¼ | 7.85 | 193. | 168. |
| ¾ | 0.87 | 22.6 | 19.6 | 2½ | 9.69 | 236. | 205. |
| ⅞ | 1.19 | 30.6 | 26.6 | 2¾ | 11.72 | 284. | 247. |
| 1 | 1.55 | 39.8 | 34.6 | 3 | 14.0 | 335. | 291. |
| 1⅛ | 1.96 | 50.1 | 43.5 | 3¼ | 16.4 | 390. | 339. |
| 1¼ | 2.42 | 61.5 | 53.5 | 3½ | 19.0 | 449. | 390. |
| 1⅜ | 2.93 | 74.1 | 64.5 | ... | ..... | ...... | ..... |

For ropes with steel cores, add 7½ per cent to above strengths.
For galvanized ropes, deduct 10 per cent from above strengths.
*Source:* Rope diagrams, Bethlehem Steel Co. All data, U. S. Simplified Practice Recommendation 198–50.

breaking strength of the rope. Weights and strengths are shown in Table 5.

*Flattened Strand Wire Rope:* The wires forming the strands of this type of rope are wound around triangular centers so that a flattened outer surface is provided with a

**Table 4.   Weights and Strengths of 8 × 19 (Special Flexible Hoisting) Wire Ropes, Preformed and Not Preformed**

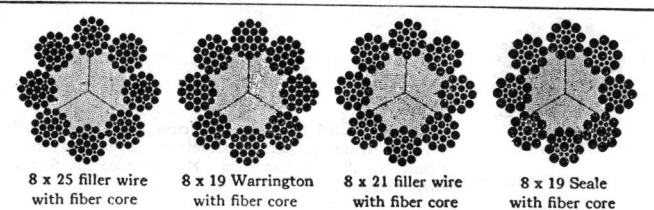

| 8 x 25 filler wire with fiber core | 8 x 19 Warrington with fiber core | 8 x 21 filler wire with fiber core | 8 x 19 Seale with fiber core |
|---|---|---|---|

| Diam., Inches | Approx. Weight per Ft., Pounds | Breaking Strength, Tons of 2000 Lbs. | | Diam., Inches | Approx. Weight per Ft., Pounds | Breaking Strength, Tons of 2000 Lbs. | |
| | | Impr. Plow Steel | Plow Steel | | | Impr. Plow Steel | Plow Steel |
|---|---|---|---|---|---|---|---|
| ¼ | 0.09 | 2.35 | 2.04 | ¾ | 0.82 | 20.5 | 17.8 |
| ⁵⁄₁₆ | 0.14 | 3.65 | 3.18 | ⅞ | 1.11 | 27.7 | 24.1 |
| ⅜ | 0.20 | 5.24 | 4.55 | 1 | 1.45 | 36.0 | 31.3 |
| ⁷⁄₁₆ | 0.28 | 7.09 | 6.17 | 1⅛ | 1.84 | 45.3 | 39.4 |
| ½ | 0.36 | 9.23 | 8.02 | 1¼ | 2.27 | 55.7 | 48.4 |
| ⁹⁄₁₆ | 0.46 | 11.6 | 10.1 | 1⅜ | 2.74 | 67.1 | 58.3 |
| ⅝ | 0.57 | 14.3 | 12.4 | 1½ | 3.26 | 79.4 | 69.1 |

For ropes with steel cores, add 7½ per cent to above strengths.
For galvanized ropes, deduct 10 per cent from above strengths.
*Source:* Rope diagrams, Bethlehem Steel Co.   All data, U. S. Simplified Practice Recommendation 198–50.

greater area than in the regular round rope to withstand severe conditions of abrasion.   The triangular shape of the strands also provides superior resistance to crushing.   Flattened strand wire rope is usually furnished in Lang lay and may be obtained with fiber core or independent wire rope core.   The three types shown in Table 6 are flexible and are designed for hoisting work.

*Flat Wire Rope:* This type of wire rope is made up of a number of four-strand rope units placed side by side and stitched together with soft steel sewing wire.   These four-strand units are alternately right and left lay to resist warping, curling, or rotating in service.   Weights and strengths are shown in Table 7.

**Simplified Practice Recommendations.** — Because the total number of wire rope types is large, manufacturers and users have agreed upon and adopted a U. S. Simplified Practice Recommendation to provide a simplified listing of those kinds and sizes of wire rope which are most commonly used and stocked.   These, then, are the types and sizes which are most generally available.   Other types and sizes for special or limited uses also may be found in individual manufacturer's catalogues.

**Sizes and Strengths of Wire Rope.** — The data shown in Tables 1 through 7 have been taken from U. S. Simplified Practice Recommendation 198–50 but do not include those wire ropes shown in that Simplified Practice Recommendation which are intended primarily for marine use.

**Table 5. Weights and Strengths of Standard 18 × 7 Nonrotating Wire Rope, Preformed and Not Preformed**

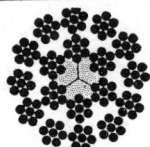

**18 x 7 Non-Rotating Rope**

Recommended Sheave and Drum Diameters: Single layer on drum . . . 36 rope diameters. Multiple layers on drum . . . 48 rope diameters. Mine service . . . 60 rope diameters.

| Diam., Inches | Approx. Weight per Ft., Pounds | Breaking Strength, Tons of 2000 Lbs. | | Diam., Inches | Approx. Weight per Ft., Pounds | Breaking Strength, Tons of 2000 Lbs. | |
|---|---|---|---|---|---|---|---|
| | | Impr. Plow Steel | Plow Steel | | | Impr. Plow Steel | Plow Steel |
| 3/16 | 0.061 | 1.42 | 1.24 | 7/8 | 1.32 | 29.5 | 25.7 |
| 1/4 | 0.108 | 2.51 | 2.18 | 1 | 1.73 | 38.3 | 33.3 |
| 5/16 | 0.169 | 3.90 | 3.39 | 1 1/8 | 2.19 | 48.2 | 41.9 |
| 3/8 | 0.24 | 5.59 | 4.86 | 1 1/4 | 2.70 | 59.2 | 51.5 |
| 7/16 | 0.33 | 7.58 | 6.59 | 1 3/8 | 3.27 | 71.3 | 62.0 |
| 1/2 | 0.43 | 9.85 | 8.57 | 1 1/2 | 3.89 | 84.4 | 73.4 |
| 9/16 | 0.55 | 12.4 | 10.8 | 1 5/8 | 4.57 | 98.4 | 85.6 |
| 5/8 | 0.68 | 15.3 | 13.3 | 1 3/4 | 5.30 | 114. | 98.8 |
| 3/4 | 0.97 | 21.8 | 19.0 | . . . | . . . . | . . . . . | . . . . |

For galvanized ropes, deduct 10 per cent from above strengths.

*Source:* Rope diagrams, sheave and drum diameters, and data for 3/16, 1/4 and 5/16-inch sizes, Bethlehem Steel Co. All other data, U. S. Simplified Practice Recommendation 198–50.

*Wire Rope Diameter:* The diameter of a wire rope is the diameter of that circle which will just enclose it, hence when measuring the diameter with calipers, care must be taken to obtain the largest outside dimension, which is taken across the opposite strands, rather than the smallest dimension which is across opposite "valleys" or "flats." It is standard practice for the nominal diameter to be the minimum with all tolerances taken on the plus side. Limits for diameter as well as for minimum breaking strength and maximum pitch are given in Federal Specification for Wire Rope, RR–R–571a.

*Wire Rope Strengths:* The strength figures shown in the accompanying tables have been obtained by a mathematical derivation based on actual breakage tests of wire rope and represent from 80 to 95 per cent of the total strengths of the individual wires, depending upon the type of rope construction.

**Safe Working Loads and Factors of Safety.** — The maximum load for which a wire rope is to be used should take into account such associated factors as friction, load caused by bending around each sheave, acceleration and deceleration, and, if a long length of rope is to be used for hoisting, the weight of the rope at its maximum extension. The condition of the rope — whether new or old, worn or corroded, and

**Table 6.** Weights and Strengths of Flattened Strand Wire Rope, Preformed and Not Preformed

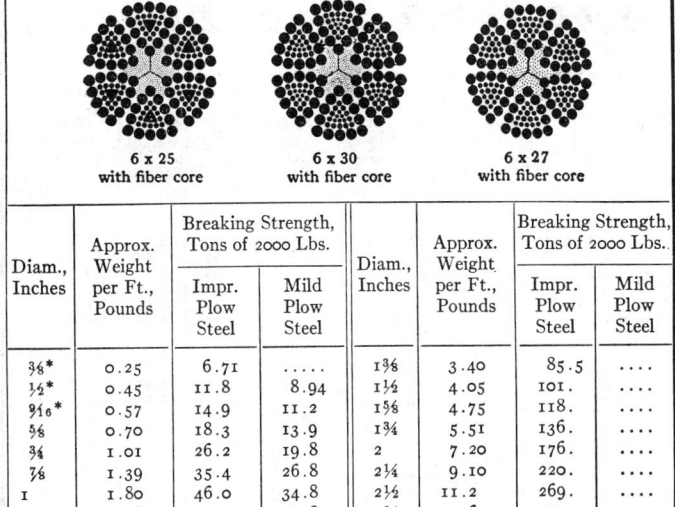

|       | 6 x 25 with fiber core | | 6 x 30 with fiber core | 6 x 27 with fiber core | |

| Diam., Inches | Approx. Weight per Ft., Pounds | Breaking Strength, Tons of 2000 Lbs. | | Diam., Inches | Approx. Weight per Ft., Pounds | Breaking Strength, Tons of 2000 Lbs. | |
| | | Impr. Plow Steel | Mild Plow Steel | | | Impr. Plow Steel | Mild Plow Steel |
| 3/8* | 0.25 | 6.71 | ..... | 1 3/8 | 3.40 | 85.5 | .... |
| 1/2* | 0.45 | 11.8 | 8.94 | 1 1/2 | 4.05 | 101. | .... |
| 9/16* | 0.57 | 14.9 | 11.2 | 1 5/8 | 4.75 | 118. | .... |
| 5/8 | 0.70 | 18.3 | 13.9 | 1 3/4 | 5.51 | 136. | .... |
| 3/4 | 1.01 | 26.2 | 19.8 | 2 | 7.20 | 176. | .... |
| 7/8 | 1.39 | 35.4 | 26.8 | 2 1/4 | 9.10 | 220. | .... |
| 1 | 1.80 | 46.0 | 34.8 | 2 1/2 | 11.2 | 269. | .... |
| 1 1/8 | 2.28 | 57.9 | 43.8 | 2 3/4 | 13.6 | 321. | .... |
| 1 1/4 | 2.81 | 71.0 | 53.7 | ... | ..... | .... | .... |

\* These sizes in Type B only.
Type H is not in U. S. Simplified Practice Recommendation.
*Source:* Rope diagrams, Bethlehem Steel Co. All other data, U. S. Simplified Practice Recommendation 198–50.

type of attachments should also be considered.

Factors of safety for standing rope usually range from 3 to 4; for operating rope, from 5 to 12. Where there is the element of hazard to life or property, higher values are used.

**Installing Wire Rope.** — The main precaution to be taken in removing and installing wire rope is to avoid kinking which greatly lessens its strength and useful life. Thus, it is preferable when removing wire rope from the reel to have the reel with its axis in a horizontal position and, if possible, mounted so that it will revolve and the wire rope taken off straight. If the rope is in a coil, it should be unwound with the coil in a vertical position as by rolling the coil along the ground. Where a drum is to be used, the rope should be run directly onto it from the reel, taking care to see that it is not bent around the drum in a direction opposite to that on the reel, thus causing it to be subject to reverse bending. On flat or smooth-faced drums it is important that the rope be started from the proper end of the drum. A right lay rope that is being overwound on the drum, that is, it passes over the top of the drum as it is wound on, should be started from the right flange of the drum (looking at the drum from the side that the rope is to come) and a left lay rope from the left flange.

### Table 7. Weights and Strengths of Standard Flat Wire Rope, Not Preformed

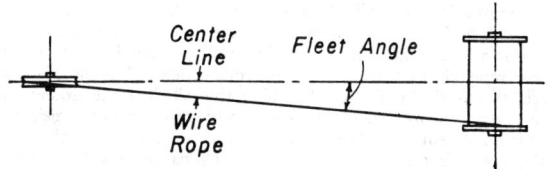

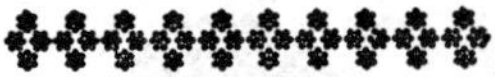

**Flat Wire Rope**

This rope consists of a number of 4-strand rope units placed side by side and stitched together with soft steel sewing wire.

| Width and Thickness, Inches | No. of Ropes | Approx. Weight per Ft., Pounds | Breaking Strength, Tons of 2000 Lbs. | | Width and Thickness, Inches | No. of Ropes | Approx. Weight per Ft., Pounds | Breaking Strength, Tons of 2000 Lbs. | |
|---|---|---|---|---|---|---|---|---|---|
| | | | Plow Steel | Mild Plow Steel | | | | Plow Steel | Mild Plow Steel |
| ¼ × 1½ | 7 | 0.69 | 16.8 | 14.6 | ½ × 4 | 9 | 3.16 | 81.8 | 71.2 |
| ¼ × 2 | 9 | 0.88 | 21.7 | 18.8 | ½ × 4½ | 10 | 3.82 | 90.9 | 79.1 |
| ¼ × 2½ | 11 | 1.15 | 26.5 | 23.0 | ½ × 5 | 12 | 4.16 | 109. | 94.9 |
| ¼ × 3 | 13 | 1.34 | 31.3 | 27.2 | ½ × 5½ | 13 | 4.50 | 118. | 103. |
| | | | | | ½ × 6 | 14 | 4.85 | 127. | 111. |
| 5⁄16 × 1½ | 5 | 0.77 | 18.5 | 16.0 | ½ × 7 | 16 | 5.85 | 145. | 126. |
| 5⁄16 × 2 | 7 | 1.05 | 25.8 | 22.4 | | | | | |
| 5⁄16 × 2½ | 9 | 1.33 | 33.2 | 28.8 | 5⁄8 × 3½ | 6 | 3.40 | 85.8 | 74.6 |
| 5⁄16 × 3 | 11 | 1.61 | 40.5 | 35.3 | 5⁄8 × 4 | 7 | 3.95 | 100. | 87.1 |
| 5⁄16 × 3½ | 13 | 1.89 | 47.9 | 41.7 | 5⁄8 × 4½ | 8 | 4.50 | 114. | 99.5 |
| 5⁄16 × 4 | 15 | 2.17 | 55.3 | 48.1 | 5⁄8 × 5 | 9 | 5.04 | 129. | 112. |
| | | | | | 5⁄8 × 5½ | 10 | 5.59 | 143. | 124. |
| 3⁄8 × 2 | 6 | 1.25 | 31.4 | 27.3 | 5⁄8 × 6 | 11 | 6.14 | 157. | 137. |
| 3⁄8 × 2½ | 8 | 1.64 | 41.8 | 36.4 | 5⁄8 × 7 | 13 | 7.23 | 186. | 162. |
| 3⁄8 × 3 | 9 | 1.84 | 47.1 | 40.9 | 5⁄8 × 8 | 15 | 8.32 | 214. | 186. |
| 3⁄8 × 3½ | 11 | 2.23 | 57.5 | 50.0 | | | | | |
| 3⁄8 × 4 | 12 | 2.44 | 62.7 | 54.6 | ¾ × 5 | 8 | 6.50 | 165. | 143. |
| 3⁄8 × 4½ | 14 | 2.83 | 73.2 | 63.7 | ¾ × 6 | 9 | 7.31 | 185. | 161. |
| 3⁄8 × 5 | 15 | 3.03 | 78.4 | 68.2 | ¾ × 7 | 10 | 8.13 | 206. | 179. |
| 3⁄8 × 5½ | 17 | 3.42 | 88.9 | 77.3 | ¾ × 8 | 11 | 9.70 | 227. | 197. |
| 3⁄8 × 6 | 18 | 3.63 | 94.1 | 81.9 | | | | | |
| | | | | | 7⁄8 × 5 | 7 | 7.50 | 190. | 165. |
| ½ × 2½ | 6 | 2.13 | 54.5 | 47.4 | 7⁄8 × 6 | 8 | 8.56 | 217. | 188. |
| ½ × 3 | 7 | 2.47 | 63.6 | 55.4 | 7⁄8 × 7 | 9 | 9.63 | 244. | 212. |
| ½ × 3½ | 8 | 2.82 | 72.7 | 63.3 | 7⁄8 × 8 | 10 | 10.7 | 271. | 236. |

*Source:* Rope diagram, Bethlehem Steel Co.; all data, U. S. Simplified Practice Recommendation 198-50.

When the rope is underwound on the drum, a right lay rope should be started from the left flange and a left lay rope from the right flange. When this is done, the rope will spool evenly and the turns will lie snugly together.

Center Line    Fleet Angle

Wire Rope

Sheaves and drums should be properly aligned to prevent undue wear. The proper position of the main or lead sheave for the rope as it comes off the drum is governed by what is called the fleet angle or angle between the rope as it stretches from drum to sheave and an imaginary center-line passing through the center of the sheave groove and a point halfway between the ends of the drum. When the rope is at one end of the drum, this angle should not exceed one and a half to two degrees. With the lead sheave mounted with its groove on this center-line, a safe fleet angle is obtained by allowing 30 feet of lead for each two feet of drum width.

*Sheave and Drum Dimensions:* Sheaves and drums should be as large as possible to obtain maximum rope life. However, factors such as the need for lightweight equipment for easy transport and use at high speeds, may call for relatively small sheaves with consequent sacrifice in rope life in the interest of over-all economy. No hard and fast rules can be laid down for any particular rope if the utmost in economical performance is to be obtained. Where maximum rope life is of prime importance, the following recommendations of Federal Specification RR–R–571a for minimum sheave or drum diameters $D$ in terms of rope diameter $d$ will be of interest. For $6 \times 7$ rope (six strands of 7 wires each) $D = 72 d$; for $6 \times 19$ rope, $D = 45 d$; for $6 \times 25$ rope, $D = 45 d$; for $6 \times 29$ rope, $D = 30 d$; for $6 \times 37$ rope, $D = 27 d$; and for $8 \times 19$ rope, $D = 31 d$.

Too small a groove for the rope it is to carry will prevent proper seating of the rope in the bottom of the groove and consequently uneven distribution of load on the rope will result. Too large a groove will not give the rope sufficient side support. Federal specifications RR–R–571a recommend that sheave groove diameters be larger than the nominal rope diameters by the following minimum amounts: For ropes of ¼- to ⁵⁄₁₆-inch diameters, ¹⁄₆₄ inch larger; for ⅜- to ¾-inch diameter ropes, ¹⁄₃₂ inch larger; for ¹³⁄₁₆- to 1⅛-inch diameter ropes, ³⁄₆₄ inch larger; for 1³⁄₁₆- to 1½-inch ropes, ¹⁄₁₆ inch larger; for 1⁹⁄₁₆- to 2¼-inch ropes, ³⁄₃₂ inch larger; and for 2⁵⁄₁₆ and larger diameter ropes, ⅛ inch larger. For new or regrooved sheaves these values should be doubled; in other words for ¼- to ⁵⁄₁₆-inch diameter ropes, the groove diameter should be ¹⁄₃₂ inch larger, etc.

*Drum or Reel Capacity:* The length of wire rope, in feet, that can be spooled onto a drum or reel, is computed by the following formula, where

Table 8. Factors K Used in Calculating Wire Rope Drum and Reel Capacities

| Rope Diam., In. | Factor K | Rope Diam., In. | Factor K | Rope Diam., In. | Factor K |
|---|---|---|---|---|---|
| ³⁄₃₂ | 23.4 | ½ | 0.925 | 1⅜ | 0.127 |
| ⅛ | 13.6 | ⁹⁄₁₆ | 0.741 | 1½ | 0.107 |
| ⁹⁄₆₄ | 10.8 | ⅝ | 0.607 | 1⅝ | 0.0886 |
| ⁵⁄₃₂ | 8.72 | ¹¹⁄₁₆ | 0.506 | 1¾ | 0.0770 |
| ³⁄₁₆ | 6.14 | ¾ | 0.428 | 1⅞ | 0.0675 |
| ⁷⁄₃₂ | 4.59 | ¹³⁄₁₆ | 0.354 | 2 | 0.0597 |
| ¼ | 3.29 | ⅞ | 0.308 | 2⅛ | 0.0532 |
| ⁵⁄₁₆ | 2.21 | 1 | 0.239 | 2¼ | 0.0476 |
| ⅜ | 1.58 | 1⅛ | 0.191 | 2⅜ | 0.0419 |
| ⁷⁄₁₆ | 1.19 | 1¼ | 0.152 | 2½ | 0.0380 |

*Note:* The values of "K" allow for normal oversize of ropes, and the fact that it is practically impossible to "thread-wind" ropes of small diameter. However, the formula is based on uniform rope winding and will not give correct figures if rope is wound non-uniformly on the reel. The amount of tension applied when spooling the rope will also affect the length. The formula is based on the same number of wraps of rope in each layer, which is not strictly correct, but which does not result in appreciable error unless the width (B) of the reel is quite small compared with the flange diameter (H).

$A$ = depth of rope space on drum, inches: $A = (H - D - 2Y) \div 2$
$B$ = width between drum flanges, inches
$D$ = diameter of drum barrel, inches
$H$ = diameter of drum flanges, inches
$K$ = factor from Table 8 for size of line selected
$Y$ = depth not filled on drum or reel where winding is to be less than full capacity
$L$ = length of wire rope on drum or reel, feet.

$$L = (A + D) \times A \times B \times K$$

*Example:* Find the length in feet of $\frac{9}{16}$-inch diameter rope required to fill a drum having the following dimensions: $B$ = 24 inches, $D$ = 18 inches, $H$ = 30 inches,

$A = (30 - 18 - 0) \div 2 = 6$ inches
$L = (6 + 18) \times 6 \times 24 \times 0.741 = 2560.0$ or 2560 feet

The above formula and factors $K$ allow for normal oversize of ropes but will not give correct figures if rope is wound nonuniformly on the reel.

*Load Capacity of Sheave or Drum:* To avoid excessive wear and groove corrugation, the radial pressure exerted by the wire rope on the sheave or drum must be kept within certain maximum limits. The radial pressure of the rope is a function of the rope tension, rope diameter, and tread diameter of the sheave and can be determined by the following equation:

$$P = \frac{2T}{D \times d}$$

where
$P$ = Radial pressure in pounds per square inch (see Table 9)
$T$ = Rope tension in pounds
$D$ = Tread diameter of sheave or drum in inches
$d$ = Rope diameter in inches

According to the Bethlehem Steel Co. the following radial pressures are recommended as maximums according to the material of which the sheave or drum is made:

Table 9.  Maximum Radial Pressures for Drums and Sheaves

| Type of Wire Rope | Drum or Sheave Material | | |
|---|---|---|---|
| | Cast Iron | Cast Steel | Manganese Steel* |
| | Recommended Maximum Radial Pressures, Pounds per Square Inch | | |
| 6 × 7 | 300† | 550† | 1500† |
| 6 × 19 | 500† | 900† | 2500† |
| 6 × 37 | 600 | 1075 | 3000 |
| 6 × 8 Flattened Strand | 450 | 850 | 2200 |
| 6 × 25 Flattened Strand | 800 | 1450 | 4000 |
| 6 × 30 Flattened Strand | 800 | 1450 | 4000 |

* 11 to 13 per cent manganese.
† These values are for regular lay rope. For Lang lay rope these values may be increased by 15 per cent.

*Rope Loads due to Bending:* When a wire rope is bent around a sheave, the resulting bending stress $s_b$ in the outer wire, and equivalent bending load $P_b$ (amount that direct tension load on rope is increased by bending) may be computed by the following formulas: $s_b = Ed_w \div D$; $P_b = s_bA$, where $A = d^2Q$. $E$ is the modulus of elasticity of the wire rope (this varies with the type and condition of rope from 10,000,000 to 14,000,000. An average value of 12,000,000 is frequently used.) $d$ is the diameter of the wire rope, $d_w$ is the diameter of the component wire (for 6 × 7 rope, $d_w = 0.106d$; for 6 × 19 rope, 0.063$d$; for 6 × 37 rope, 0.045$d$; and for 8 × 19 rope, 0.050$d$), $D$ is the pitch diameter of the sheave in inches, $A$ is the metal cross-sectional area of the rope, and $Q$ is a constant, values for which are: 6 × 7 (Fiber Core) rope, 0.380; 6 × 7 (IWRC or WSC), 0.437; 6 × 19 (Fiber Core), 0.405; 6 × 19 (IWRC or WSC), 0.475; 6 × 37 (Fiber Core), 0.400; 6 × 37 (IWRC), 0.470; 8 × 19 (Fiber Core), 0.370; and Flattened Strand Rope, 0.440.

*Example:* Find the bending stress and equivalent bending load due to the bending of a 6 × 19 (Fiber Core) wire rope of ½-inch diameter around a 24-inch pitch diameter sheave.

$$d_w = 0.063 \times 0.5 = 0.0315 \text{ in.}; \quad A = 0.5^2 \times 0.405 = 0.101 \text{ sq. in.}$$
$$s_b = 12,000,000 \times 0.0315 \div 24 = 15,750 \text{ lbs. per sq. in.}$$
$$P_b = 15,750 \times 0.101 = 1590 \text{ lbs.}$$

**Cutting and Seizing of Wire Rope.** — Wire rope can be cut with mechanical wire rope shears, an abrasive wheel, an electric resistance cutter (used for ropes of smaller diameter only), or an acetylene torch. This last method fuses the ends of the wires in the strands. It is important that the rope be seized on either side of where the cut is to be made. Any annealed low carbon steel wire may be used for seizing, the recommended sizes being as follows: For a wire rope diameter of from ¼- to 1⁵⁄₁₆-inch, use a seizing wire of .054-inch (No. 17 Steel Wire Gage); for a rope of 1- to 1⅝-inch diameter, use a .105-inch wire (No. 12); and for rope of 1¾- to 3½-inch diameter, use a .135-inch wire (No. 10). Except for preformed wire ropes, a minimum of two seizings on either side of a cut is recommended. Four seizings should be used on either side of a cut for Lang lay rope, a rope with a steel core, or a non-spinning type of rope.

The following method of seizing is given in Federal specification for wire rope, RR–R–571a. Lay one end of the seizing wire in the groove between two strands of wire rope and wrap the other end tightly in a close helix over the portion in the groove. A seizing iron (round bar ½ to ⅝ inch in diameter by 18 inches long) should be used to wrap the seizing tightly. This bar is placed at right angles to the rope next to the first turn or two of the seizing wire. The seizing wire is brought around the back of the seizing iron so that it now can be wrapped loosely around the wire rope in the opposite direction to that of the seizing coil. As the seizing iron is now rotated around the rope it will carry the seizing wire snugly and tightly into place. When completed, both ends of the seizing should be twisted together tightly.

**Maintenance of Wire Rope.** — Heavy abrasion, overloading, and bending around sheaves or drums which are too small in diameter are the principal reasons for the rapid deterioration of wire rope. Wire rope in use should be inspected periodically for evidence of wear and damage by corrosion. Such inspection should take place at progressively shorter intervals over the useful life of the rope as wear tends to accelerate with use. Where wear is rapid, the outside of a wire rope will show flattened surfaces in a short time.

If there is any hazard involved in the use of the rope, it may be prudent to estimate the remaining strength and service life. This should be done for the weakest point where the most wear or largest number of broken wires are in evidence. One way to arrive at a conclusion is to set an arbitrary number of broken wires in a given strand as an indication that the rope should be removed from service and an ultimate strength test run on the worn sample. The arbitrary figure can then be revised and rechecked until a practical working formula is arrived at. A piece of waste rubbed along the wire rope will help to reveal broken wires. The effects of corrosion are not easy to detect since the exterior wires may appear to be only slightly rusty, while the damaging effects of corrosion may be confined to the hidden inner wires where it cannot be seen. To prevent damage by corrosion, the rope should be kept well lubricated. In some cases zinc coated wire rope may be indicated.

Periodic cleaning of wire rope by using a stiff brush and kerosene or with compressed air or live steam and relubricating will help to lengthen rope life and reduce abrasion and wear on sheaves and drums. Before storing after use, wire rope should be cleaned and lubricated.

**Lubrication of Wire Rope.** — Although wire rope is thoroughly lubricated during manufacture to protect it against corrosion and to reduce friction and wear, this lubrication should be supplemented from time to time. Special lubricants are supplied by wire rope manufacturers. These vary somewhat with the type of rope application and operating condition. Where the preferred lubricant can not be obtained from the wire rope manufacturer, an adhesive type of lubricant similar to that used for open gearing will often be found suitable. At normal temperatures, some wire rope lubricants may be practically solid and will require thinning before application. This may be done by heating to 160 to 200 degrees F. or by diluting with gasoline or some other fluid which will allow the lubricant to penetrate the rope. The lubricant may be painted on the rope or the rope may be passed through a box or tank filled with the lubricant.

**Replacement of Wire Rope.** — When an old wire rope is to be replaced, all drums and sheaves should be examined for wear. All evidence of scoring or imprinting of grooves from previous use should be removed and sheaves with flat spots, defective bearings, and broken flanges, should be repaired or replaced. It will frequently be found that the area of maximum wear is located relatively near one end of the rope. By cutting off that portion, the remainder of the rope may be salvaged for continued use. Sometimes the life of a rope can be increased by simply changing it end for end at about one-half the estimated normal life. The worn sections will then no longer come at points which cause the greatest wear.

**Wire Rope Slings and Fittings.** — A few of the simpler sling arrangements or hitches as they are called, are shown in the accompanying illustration. Normally 6 × 19 Class wire rope is recommended where a diameter in the ¼-inch to 1⅛-inch range is to be used and 6 × 37 Class wire rope where a diameter in the 1¼-inch and larger range is to be used. However, in some cases the 6 × 19 Class may be used even in the larger sizes if resistance to abrasion is of primary importance and the 6 × 37 Class in the smaller sizes if greater flexibility is desired.

The *straight lift hitch*, shown at A, is a straight connector between crane hook and load.

The *basket hitch* may be used with two hooks so that the sides are vertical as shown at B or with a single hook with sides at various angles with the vertical as shown at C, D, and E. As the angle with the vertical increases, a greater tension is placed on the rope so that for any given load, a sling of greater lifting capacity must be used.

The *choker hitch*, shown at F, is widely used for lifting bundles of items such as bars, poles, pipe, and similar objects. The choker hitch holds these items firmly

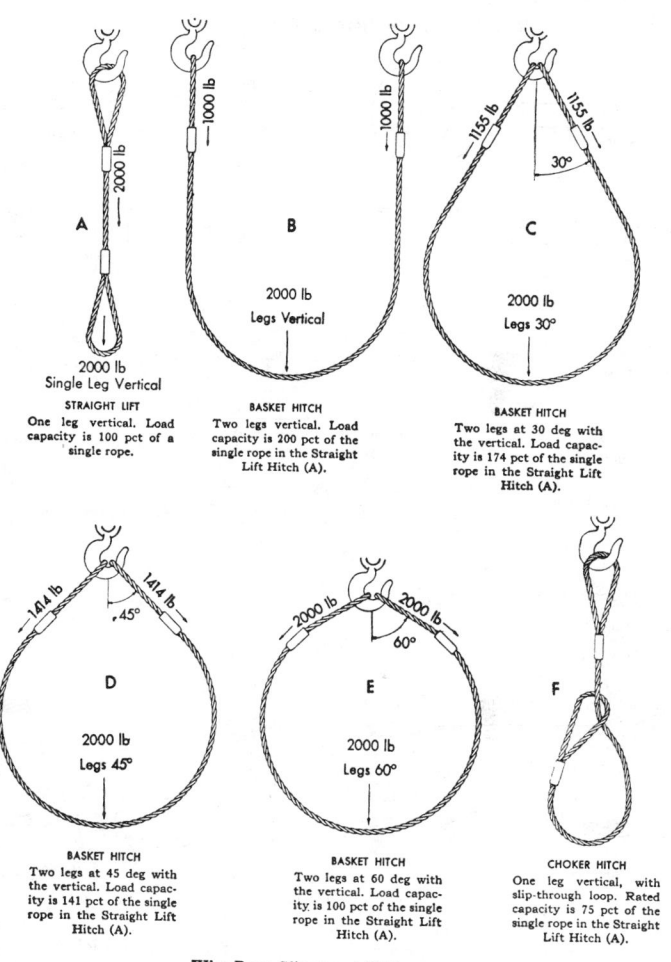

**A**

2000 lb
Single Leg Vertical

**STRAIGHT LIFT**
One leg vertical. Load
capacity is 100 pct of a
single rope.

**B**

2000 lb
Legs Vertical

**BASKET HITCH**
Two legs vertical. Load
capacity is 200 pct of the
single rope in the Straight
Lift Hitch (A).

**C**

30°

2000 lb
Legs 30°

**BASKET HITCH**
Two legs at 30 deg with
the vertical. Load capac-
ity is 174 pct of the single
rope in the Straight Lift
Hitch (A).

**D**

45°

2000 lb
Legs 45°

**BASKET HITCH**
Two legs at 45 deg with
the vertical. Load capac-
ity is 141 pct of the single
rope in the Straight Lift
Hitch (A).

**E**

60°

2000 lb
Legs 60°

**BASKET HITCH**
Two legs at 60 deg with
the vertical. Load capac-
ity is 100 pct of the single
rope in the Straight Lift
Hitch (A).

**F**

**CHOKER HITCH**
One leg vertical, with
slip-through loop. Rated
capacity is 75 pct of the
single rope in the Straight
Lift Hitch (A).

Wire Rope Slings and Fittings

but the load must be balanced so that it rides safely. Since additional stress is
imposed in the rope due to the choking action, the capacity of this type of hitch is
25 per cent less than that of the comparable straight lift. If two choker hitches
are used at an angle, these angles must also be taken into consideration as in the
basket hitches.

## INDUSTRIAL TYPES

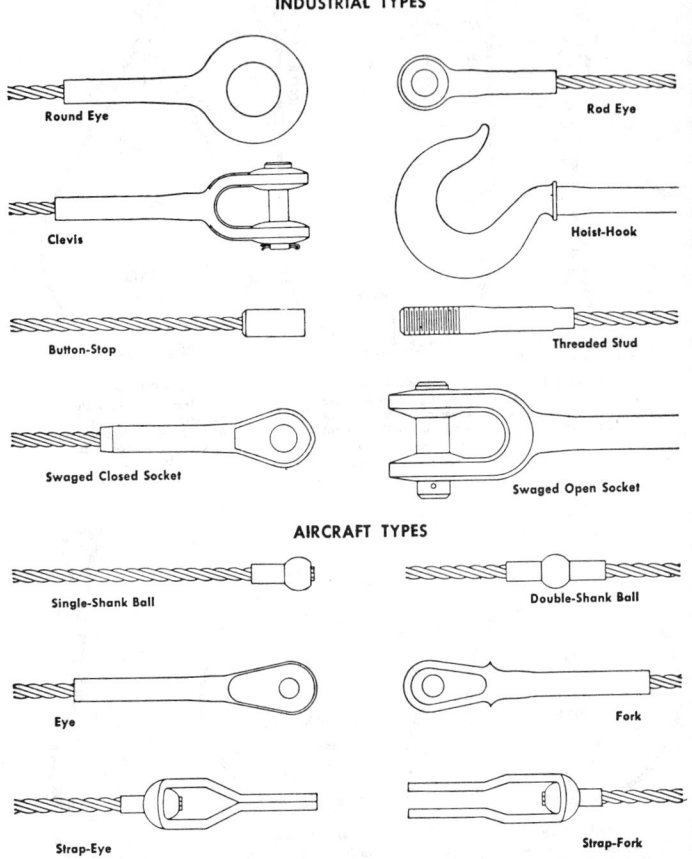

**AIRCRAFT TYPES**

Wire Rope Fittings

**Wire Rope Fittings.** — A wide variety of swaged fittings are available for use with wire rope. A number of industrial and aircraft types are shown in the accompanying illustration. Swaged fittings on wire rope have an efficiency (ability to hold the wire rope) of approximately 100 per cent of the catalogue rope strength. These fittings are applied to the end or body of the wire rope by the application of high pressure through special dies that causes the steel to "flow" around the wires and strands of the rope to form a union that is as strong as the rope itself. The more commonly used type of swaged fittings range from ⅛- to ⅝-inch diameter sizes in the industrial types and from the ¹⁄₁₆- to ⅝-inch sizes in the aircraft types. These fittings are furnished attached to the wire strand, rope, or cable.

**Applying Clips and Attaching Sockets.** — In attaching U-bolt clips for fastening the end of a wire rope to form a loop, it is essential that the saddle or base of the clip bears against the longer or "live" end of the rope loop and the U-bolt against the shorter or "dead" end. The "U" of the clips should never bear against the live end of the rope because the rope may be cut or kinked. A wire-rope thimble should be used in the loop eye of the rope to prevent kinking when rope clips are used. The strength of a clip fastening is usually less than 80 per cent of the strength of the rope. Table 10 gives the proper size, number, and spacing for each size of wire rope.

Table 10. Clips Required for Fastening Wire Rope End

| Rope Diam., In. | U-Bolt Diam., In. | Min. No. of Clips | Clip Spacing, In. | Rope Diam., In. | U-Bolt Diam., In. | Min. No. of Clips | Clip Spacing, In. |
|---|---|---|---|---|---|---|---|
| $3/16$ | $11/32$ | 2 | 3 | $1\frac{1}{8}$ | $1\frac{1}{4}$ | 5 | $9\frac{3}{4}$ |
| $1/4$ | $7/16$ | 2 | $3\frac{1}{4}$ | $1\frac{1}{4}$ | $1\frac{7}{16}$ | 5 | $10\frac{3}{4}$ |
| $5/16$ | $1/2$ | 2 | $3\frac{1}{4}$ | $1\frac{3}{8}$ | $1\frac{1}{2}$ | 6 | $11\frac{1}{2}$ |
| $3/8$ | $9/16$ | 2 | 4 | $1\frac{1}{2}$ | $1\frac{23}{32}$ | 6 | $12\frac{1}{2}$ |
| $7/16$ | $5/8$ | 2 | $4\frac{1}{2}$ | $1\frac{5}{8}$ | $1\frac{3}{4}$ | 6 | $13\frac{1}{4}$ |
| $1/2$ | $11/16$ | 3 | 5 | $1\frac{3}{4}$ | $1\frac{15}{16}$ | 7 | $14\frac{1}{2}$ |
| $5/8$ | $3/4$ | 3 | $5\frac{3}{4}$ | 2 | $2\frac{1}{8}$ | 8 | $16\frac{1}{2}$ |
| $3/4$ | $7/8$ | 4 | $6\frac{3}{4}$ | $2\frac{1}{4}$ | $2\frac{5}{8}$ | 8 | $16\frac{1}{2}$ |
| $7/8$ | 1 | 4 | 8 | $2\frac{1}{2}$ | $2\frac{7}{8}$ | 8 | $17\frac{3}{4}$ |
| 1 | $1\frac{1}{8}$ | 4 | $8\frac{3}{4}$ | ... | .... | ..... | .... |

In attaching commercial sockets of forged steel to wire rope ends, the following procedure is recommended. The wire rope is seized at the end and another seizing is applied at a distance from the end equal to the length of the basket of the socket. As explained in a previous section, soft iron wire is used and particularly for the larger sizes of wire rope, it is important to use a seizing iron to secure a tight winding. For large ropes, the seizing should be several inches long.

The end seizing is now removed and the strands are separated so that the fiber core can be cut back to the next seizing. The individual wires are then untwisted and "broomed out" and for the distance they are to be inserted in the socket are carefully cleaned with benzine, naptha, or unleaded gasoline. The wires are then dipped into commercial muriatic (hydrochloric) acid and left (usually one to three minutes) until the wires are bright and clear or, if zinc coated, until the zinc is removed. After cleaning, the wires are dipped into a hot soda solution (1 pound of soda to 4 gallons of water at 175 degrees F. minimum) to neutralize the acid. The rope is now placed in a vise. A temporary seizing is used to hold the wire ends together until the socket is placed over the rope end. The temporary seizing is then removed and the socket located so that the ends of the wires are about even with the upper end of the basket. The opening around the rope at the bottom of the socket is now sealed with putty or asbestos. A special high grade pure zinc is used to fill the socket. Babbit metal should not be used as it will not hold properly. For proper fluidity and penetration, the zinc is heated to a temperature in the 830- to 900-degree F. range. If a pyrometer is not available to measure the temperature of the molten zinc, a dry soft pine stick dipped into the zinc and quickly withdrawn will show only a slight discoloration and no zinc will adhere to it. If the wood chars, the zinc is too hot. The socket is now permitted to cool and the resulting joint is then ready for use. When properly prepared, its strength should be approximately equal to that of the rope itself.

# SPRINGS <sup>*</sup>

**Introduction.** — Many advances have been made in the spring industry in recent years. For example: developments in materials permit longer fatigue life at higher stresses; simplified design procedures reduce the complexities of design, and improved methods of manufacture help to speed up some of the complicated fabricating procedures and increase production. New types of testing instruments and revised tolerances also permit higher standards of accuracy. Designers should also consider the possibility of using standard springs now available from stock. They can be obtained from spring manufacturing companies located in different areas, and small shipments usually can be made quickly

Designers of springs require information in the following order of precedence to simplify design procedures.

   1. Spring materials and their applications

   2. Allowable spring stresses

   3. Spring design data with tables of spring characteristics, tables of formulas, and tolerances.

Only the more commonly used types of springs are covered in detail. Special types and designs rarely used such as torsion bars, volute springs, Belleville washers, constant force, ring and spiral springs and those made from rectangular wire are only described briefly.

**Notation.** — The following symbols are used in spring equations:

$$AC = \text{Active coils.}$$
$$b = \text{Widest width of rectangular wire, inches.}$$
$$CL = \text{Compressed length, inches.}$$
$$D = \text{Mean coil diameter, inches } = OD - d$$
$$d = \text{Diameter of wire or side of square, inches.}$$
$$E = \text{Modulus of elasticity in tension, pounds per square inch.}$$
$$F = \text{Deflection, for } N \text{ coils, inches.}$$
$$F° = \text{Deflection, for } N \text{ coils, rotary, degrees.}$$
$$f = \text{Deflection, for one active coil.}$$
$$FL = \text{Free length, unloaded spring, inches.}$$
$$G = \text{Modulus of elasticity in torsion, pounds per square inch.}$$
$$IT = \text{Initial tension, pounds.}$$
$$K = \text{Curvature stress correction factor.}$$
$$L = \text{Active length subject to deflection, inches.}$$
$$N = \text{Number of active coils, total.}$$
$$P = \text{Load, pounds.}$$
$$p = \text{pitch, inches.}$$
$$R = \text{Distance from load to central axis, inches.}$$
$$S \text{ or } S_t = \text{Stress, torsional, pounds per square inch.}$$
$$S_b = \text{Stress, bending, pounds per square inch.}$$
$$SH = \text{Solid height.}$$
$$S_{it} = \text{Stress, torsional, due to initial tension, pounds per square inch.}$$
$$T = \text{Torque } = P \times R, \text{ pound-inches.}$$
$$TC = \text{Total coils.}$$
$$t = \text{Thickness, inches.}$$
$$U = \text{Number of revolutions } = F°/360°.$$

*This section was compiled by Harold Carlson, P. E., Consulting Engineer, Lakewood, N. J.

## Spring Materials

The spring materials most commonly used include high-carbon spring steels, alloy spring steels, stainless spring steels, copper-base spring alloys, and nickel-base spring alloys.

**High-Carbon Spring Steels in Wire Form.** — These spring steels are the most commonly used of all spring materials because they are the least expensive, are easily worked, and are readily available. However, they are not satisfactory for springs operating at high or low temperatures or for shock or impact loading. The following wire forms are available:

*Music Wire, ASTM A228* (0.80–0.95 per cent carbon): This is the most widely used of all spring materials for small springs operating at temperatures up to about 250 degrees F. It is tough, has a high tensile strength, and can withstand high stresses under repeated loading. The material is readily available in round form in diameters ranging from 0.005 to 0.125 inch and in some larger sizes up to ¾₆ inch. It is not available with high tensile strengths in square or rectangular sections. Music wire can be plated easily and is obtainable pretinned or preplated with cadmium, but plating after spring manufacture is usually preferred for maximum corrosion resistance.

*Oil-Tempered MB Grade, ASTM A229* (0.60–0.70 per cent carbon): This general-purpose spring steel is commonly used for many types of coil springs where the cost of music wire is prohibitive and in sizes larger than are available in music wire. It is readily available in diameters ranging from 0.125 to 0.500 inch, but both smaller and larger sizes may be obtained. The material should not be used under shock and impact loading conditions, at temperatures above 350 degrees F., or at temperatures in the sub-zero range. Square and rectangular sections of wire are obtainable in fractional sizes. Annealed stock also can be obtained for hardening and tempering after coiling. This material has a heat-treating scale that must be removed before plating.

*Oil-Tempered HB Grade, SAE 1080* (0.75 to 0.85 per cent carbon): This material is similar to the MB Grade except that it has a higher carbon content and a higher tensile strength. It is obtainable in the same sizes and is used for more accurate requirements than the MB Grade, but is not so readily available. In lieu of using this material it may be better to use an alloy spring steel, particularly if a long fatigue life or high endurance properties are needed. Round and square sections are obtainable in the oil-tempered or annealed conditions.

*Hard-Drawn MB Grade, ASTM A227* (0.60–0.70 per cent carbon): This grade is used for general-purpose springs where cost is the most important factor. Although increased use in recent years has resulted in improved quality, it is best not to use it where long life and accuracy of loads and deflections are important. It is available in diameters ranging from 0.031 to 0.500 inch and in some smaller and larger sizes also. The material is available in square sections but at reduced tensile strengths. It is readily plated. Applications should be limited to those in the temperature range of 0 to 250 degrees F.

**High-Carbon Spring Steels in Flat Strip Form.** — Two types of thin, flat, high-carbon spring steel strip are most widely used although several other types are obtainable for specific applications in watches, clocks, and certain instruments. These two compositions are used for over 95 per cent of all such applications. Thin sections of these materials under 0.015 inch having a carbon content of over 0.85 per cent and a hardness of over 47 on the Rockwell C scale are susceptible to hydrogen-embrittlement even though special plating and heating operations are employed. The two types are described as follows:

*Cold-Rolled Spring Steel, Blue-Tempered or Annealed, SAE 1074, also 1064, and 1070* (0.60 to 0.80 per cent carbon): This very popular spring steel is available in thicknesses ranging from 0.005 to 0.062 inch and in some thinner and thicker sections. The material is available in the annealed condition for forming in 4-slide machines and in presses, and can readily be hardened and tempered after forming. It is also available in the heat-treated or blue-tempered condition. The steel is obtainable in several finishes such as straw color, blue color, black, or plain. Hardnesses ranging from 42 to 46 Rockwell C are recommended for spring applications. Uses include spring clips, flat springs, clock springs, and motor, power, and spiral springs.

*Cold-Rolled Spring Steel, Blue-Tempered Clock Steel, SAE 1095* (0.90 to 1.05 per cent carbon): This popular type should be used principally in the blue-tempered condition. Although obtainable in the annealed condition, it does not always harden properly during heat-treatment as it is a "shallow" hardening type. It is used principally in clocks and motor springs. End sections of springs made from this steel are annealed for bending or piercing operations. Hardnesses usually range from 47 to 51 Rockwell C.

Other materials available in strip form and used for flat springs are brass, phosphor-bronze, beryllium-copper, stainless steels, and nickel alloys.

**Alloy Spring Steels.** — These spring steels are used for conditions of high stress, and shock or impact loadings. They can withstand both higher and lower temperatures than the high-carbon steels and are obtainable in either the annealed or pretempered conditions.

*Chromium Vanadium, ASTM A231:* This very popular spring steel is used under conditions involving higher stresses than those for which the high-carbon spring steels are recommended and is also used where good fatigue strength and endurance are needed. It behaves well under shock and impact loading. The material is available in diameters ranging from 0.031 to 0.500 inch and in some larger sizes also. In square sections it is available in fractional sizes. Both the annealed and pretempered types are available in round, square, and rectangular sections. It is used extensively in aircraft-engine valve springs and for springs operating at temperatures up to 425 degrees F.

*Silicon Manganese:* This alloy steel is quite popular in Great Britain. It is less expensive than chromium-vanadium steel and is available in round, square, and rectangular sections in both annealed and pretempered conditions in sizes ranging from 0.031 to 0.500 inch. It was formerly used for knee-action springs in automobiles. It is used in flat leaf springs for trucks and as a substitute for more expensive spring steels.

*Chromium Silicon, ASTM A401:* This alloy is used for highly stressed springs requiring long life and which are subjected to shock loading. It can be heat-treated to higher hardnesses than other spring steels so that high tensile strengths are obtainable. The most popular sizes range from 0.031 to 0.500 inch in diameter. Very rarely are square, flat, or rectangular sections used. Hardnesses ranging from 50 to 53 Rockwell C are quite common and the alloy may be used at temperatures up to 475 degrees F. This material is usually ordered specially for each job.

**Stainless Spring Steels.** — The use of stainless spring steels has increased and several compositions are available all of which may be used for temperatures up to 550 degrees F. They are all corrosion resistant. Only the stainless 18–8 compositions should be used at sub-zero temperatures.

*Stainless Type 302, ASTM A313* (18 per cent chromium, 8 per cent nickel): This is a very popular stainless spring steel because it has the highest tensile strength and quite uniform properties. It is cold-drawn to obtain its mechanical properties and cannot be hardened by heat treatment. This material is nonmagnetic only when fully annealed and becomes slightly magnetic due to the cold-working performed to produce spring properties. It is suitable for use at temperatures up to 550 degrees F. and for sub-zero temperatures. It is very corrosion resistant. The material best exhibits its desirable mechanical properties in diameters ranging from 0.005 to 0.1875 inch although some larger diameters are available. It is also available as hard-rolled flat strip. Square and rectangular sections are available but are infrequently used.

*Stainless Type 304, ASTM A313* (18 per cent chromium, 8 per cent nickel): This material is quite similar to Type 302, but has better bending properties and about 5 per cent lower tensile strength. It is a little easier to draw, due to the slightly lower carbon content.

*Stainless Type 316, ASTM A313* (18 per cent chromium, 12 per cent nickel, 2 per cent molybdenum): This material is quite similar to Type 302 but is slightly more corrosion resistant because of its higher nickel content. Its tensile strength is 10 to 15 per cent lower than Type 302. It is used for aircraft springs.

*Stainless Type 17-7 PH ASTM A313* (17 per cent chromium, 7 per cent nickel): This alloy, which also contains small amounts of aluminum and titanium, is formed in a moderately hard state and then precipitation hardened at relatively low temperatures for several hours to produce tensile strengths nearly comparable to music wire. This material is not readily available in all sizes, and has limited applications due to its high manufacturing cost.

*Stainless Type 414, SAE 51414* (12 per cent chromium, 2 per cent nickel): This alloy has tensile strengths about 15 per cent lower than Type 302 and can be hardened by heat-treatment. For best corrosion resistance it should be highly polished or kept clean. It can be obtained hard drawn in diameters up to 0.1875 inch and is commonly used in flat cold-rolled strip for stampings. The material is not satisfactory for use at low temperatures.

*Stainless Type 420, SAE 51420* (13 per cent chromium): This is the best stainless steel for use in large diameters above 0.1875 inch and is frequently used in smaller sizes. It is formed in the annealed condition and then hardened and tempered. It does not exhibit its stainless properties until after it is hardened. Clean bright surfaces provide the best corrosion resistance, therefore the heat-treating scale must be removed. Bright hardening methods are preferred.

*Stainless Type 431, SAE 51431* (16 per cent chromium, 2 per cent nickel): This spring alloy acquires high tensile properties (nearly the same as music wire) by a combination of heat-treatment to harden the wire plus cold-drawing after heat-treatment. Its corrosion resistance is not equal to Type 302.

**Copper-Base Spring Alloys.** — Copper-base alloys are important spring materials because of their good electrical properties combined with their good resistance to corrosion. Although these materials are more expensive than the high-carbon and the alloy steels, they nevertheless are frequently used in electrical components and in sub-zero temperatures.

*Spring Brass, ASTM B 134* (70 per cent copper, 30 per cent zinc): This material is the least expensive and has the highest electrical conductivity of the copper-base alloys. It has a low tensile strength and poor spring qualities, but is extensively used in flat stampings and where sharp bends are needed. It cannot be hardened by heat-treatment and should not be used at temperatures above 150 degrees F., but is especially good at sub-zero temperatures. Available in round sections and flat strips, this hard-drawn material is usually used in the "spring hard" temper.

*Phosphor Bronze, ASTM B 159* (95 per cent copper, 5 per cent tin): This alloy is the most popular of this group because it combines the best qualities of tensile strength, hardness, electrical conductivity, and corrosion resistance with the least cost. It is more expensive than brass, but can withstand stresses 50 per cent higher.

The material cannot be hardened by heat-treatment. It can be used at temperatures up to 212 degrees F. and at sub-zero temperatures. It is available in round sections and flat strip, usually in the "extra hard" or "spring hard" tempers. It is frequently used for contact fingers in switches because of its low arcing properties. An 8 per cent tin composition is used for flat springs and a superfine grain composition called "Duraflex", has good endurance properties.

*Beryllium Copper, ASTM B 197* (98 per cent copper, 2 per cent beryllium): This alloy can be formed in the annealed condition and then precipitation hardened after forming at temperatures around 600 degrees F, for 2 to 3 hours. This produces a high hardness combined with a high tensile strength. After hardening, the material becomes quite brittle and can withstand very little or no forming. It is the most expensive alloy in the group and heat-treating is expensive due to the need for holding the parts in fixtures to prevent distortion. The principal use of this alloy is for carrying electric current in switches and in electrical components. Flat strip is frequently used for contact fingers.

**Nickel-Base Spring Alloys.** — Nickel-base alloys are corrosion resistant, withstand both elevated and sub-zero temperatures, and their non-magnetic characteristic makes them useful for such applications as gyroscopes, chronoscopes, and indicating instruments. These materials have a high electrical resistance and therefore should not be used for conductors of electrical current.

*Monel*[1] (67 per cent nickel, 30 per cent copper): This material is the least expensive of the nickel-base alloys. It also has the lowest tensile strength but is useful due to its resistance to the corrosive effects of sea water and because it is nearly non-magnetic. The alloy can be subjected to stresses slightly higher than phosphor bronze and nearly as high as beryllium copper. Its high tensile strength and hardness are obtained as a result of cold-drawing and cold-rolling only, since it can not be hardened by heat-treatment. It can be used at temperatures ranging between −100 to +425 degrees F. at normal operating stresses and is available in round wires up to $\frac{3}{16}$ inch in diameter with quite high tensile strengths. Larger diameters and flat strip are available with lower tensile strengths.

*"K" Monel*[1] (66 per cent nickel, 29 per cent copper, 3 per cent aluminum): This material is quite similar to Monel except that the addition of the aluminum makes it a precipitation-hardening alloy. It may be formed in the soft or fairly hard condition and then hardened by a long-time age-hardening heat-treatment to obtain a tensile strength and hardness above Monel and nearly as high as stainless steel. It is used in sizes larger than those usually used with Monel, is non-magnetic and can be used in temperatures ranging from −100 to +450 degrees F. at normal working stresses under 45,000 pounds per square inch.

*Inconel*[1] (78 per cent nickel, 14 per cent chromium, 7 per cent iron): This is one of the most popular of the non-magnetic nickel-base alloys because of its corrosion resistance and because it can be used at temperatures up to 700 degrees F. It is more expensive than stainless steel but less expensive than beryllium copper. Its hardness and tensile strength is higher than that of "K" Monel and is obtained as a result of cold-drawing and cold-rolling only. It cannot be hardened by heat-treatment. Wire diameters up to ¼ inch have the best tensile properties. It is often used in steam valves, regulating valves, and for springs in boilers, compressors, turbines, and jet engines.

*Inconel "X"*[1] (70 per cent nickel, 16 per cent chromium, 7 per cent iron): This material is quite similar to Inconel but the small amounts of titanium, columbium and aluminum in its composition make it a precipitation-hardening alloy. It can be formed in the soft or partially hard condition and then hardened by holding it

[1] Trade name of The International Nickel Company.

at 1200 degrees F. for 4 hours. It is non-magnetic and is used in larger sections than Inconel. This alloy is used at temperatures up to 850 degrees F. and at stresses up to 55,000 pounds per square inch.

*Duranickel*[1] (*"Z" Nickel*) (98 per cent nickel): This alloy is non-magnetic, corrosion resistant, has a high tensile strength and is hardenable by precipitation hardening at 900 degrees F. for 6 hours. It may be used at the same stresses as Inconel but should not be used at temperatures above 500 degrees F.

**Nickel-Base Spring Alloys with Constant Moduli of Elasticity.** — Some special nickel alloys have a constant modulus of elasticity over a wide temperature range. These are especially useful where springs undergo temperature changes and must exhibit uniform spring characteristics. These materials have a low or zero thermoelastic coefficient and therefore do not undergo variations in spring stiffness because of modulus changes due to temperature differentials. They also have low hysteresis and creep values which makes them preferred for use in food-weighing scales, precision instruments, gyroscopes, measuring devices, recording instruments and computing scales where the temperature ranges from −50 to +150 degrees F. These materials are expensive, none being regularly stocked in a wide variety of sizes. They should not be specified without prior discussion with spring manufacturers since some suppliers may not fabricate springs from these alloys because of the special manufacturing processes required. All of these alloys are used in small wire diameters and in thin strip only and are covered by U.S. patents. They are more specifically described as follows:

*Elinvar*[2] (nickel, iron, chromium): This alloy, the first constant-modulus alloy used for hairsprings in watches, is an austenitic alloy hardened only by cold-drawing and cold-rolling. Additions of titanium, tungsten, molybdenum and other alloying elements have brought about improved characteristics and precipitation-hardening abilities. These improved alloys are known by the following trade names: Elinvar Extra, Durinval, Modulvar and Nivarox.

*Ni-Span* C[1] (nickel, iron, chromium, titanium): This very popular constant-modulus alloy is usually formed in the 50 per cent cold-worked condition and precipitation-hardened at 900 degrees F. for 8 hours, although heating to 1250 degrees F. for 3 hours produces hardnesses of 40 to 44 Rockwell C, permitting safe torsional stresses of 60,000 to 80,000 pounds per square inch. This material is ferromagnetic up to 400 degrees F. Above that it becomes non-magnetic.

*Iso-Elastic*[3] (nickel, iron, chromium, molybdenum): This popular alloy is relatively easy to fabricate and is used at safe torsional stresses of 40,000 to 60,000 pounds per square inch and hardnesses of 30 to 36 Rockwell C. It is used principally in dynamometers, instruments, and food-weighing scales.

*Elgiloy*[4] (nickel, iron, chromium, cobalt): This alloy, also known by the trade names 8J Alloy, Durapower, and Cobenium is a nonmagnetic alloy suitable for sub-zero temperatures and temperatures up to about 1000 degrees F., provided that torsional stresses are kept under 75,000 pounds per square inch. It is precipitation-hardened at 900 degrees F. for 8 hours to produce hardnesses of 48 to 50 Rockwell C. The alloy is used in watch and instrument springs.

*Dynavar*[5] (nickel, iron, chromium, cobalt): This alloy is a non-magnetic, corrosion-resistant material suitable for sub-zero temperatures and temperatures up to about 750 degrees F., provided that torsional stresses are kept below 75,000 pounds per square inch. It is precipitation-hardened to produce hardnesses of 48 to 50 Rockwell C and is used in watch and instrument springs.

[1] Trade name of The International Nickel Company. [2] Trade name of Soc. Anon. de Commentry Fourchambault et Decazeville, Paris, France. [3] Trade name of John Chatillon & Sons. [4] Trade name of Elgin National Watch Company. [5] Trade name of Hamilton Watch Company.

## Spring Stresses

**Allowable Working Stresses for Springs.** — The safe working stress for any particular spring depends to a large extent on the following items:

1. Type of spring — whether compression, extension, torsion, etc.;
2. Size of spring — small or large, long or short;
3. Spring material;
4. Size of spring material;
5. Type of service — light, average, or severe;
6. Stress range — low, average, or high;
7. Loading — static, dynamic, or shock;
8. Operating temperature;
9. Design of spring — spring index, sharp bends, hooks.

Consideration should also be given to other factors that affect spring life: corrosion, buckling, friction, and hydrogren embrittlement decrease spring life; manufacturing operations such as high-heat stress-equalizing, presetting, and shot-peening increase spring life.

Item 5, the type of service to which a spring is subjected, is a major factor in determining a safe working stress once consideration has been given to type of spring, kind and size of material, temperature, type of loading, and so on. The types of service are:

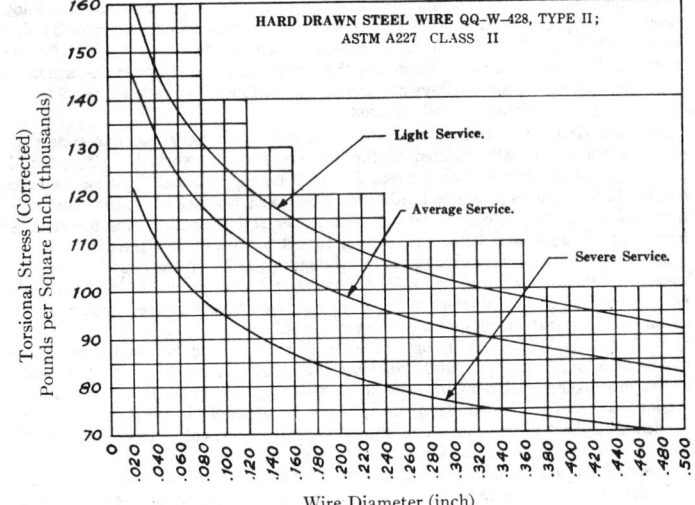

Fig. 1.   Allowable working stresses for compression springs —
Hard drawn steel wire.*

* Although these curves are for compression springs, they may also be used for extension springs; for extension springs, *reduce* the values obtained from the curves by 10 to 15 per cent.

*Light Service.* This includes springs subjected to static loads or small deflections and seldom-used springs such as those in bomb fuses, projectiles, and safety devices. This service is for 1,000 to 10,000 deflections.

*Average Service.* This includes springs in general use in machine tools, mechanical products, and electrical components. Normal frequency of deflections not exceeding 18,000 per hour permit such springs to withstand 100,000 to 1,000,000 deflections.

*Severe Service.* This includes springs subjected to rapid deflections over long periods of time and to shock loading such as in pneumatic hammers, hydraulic controls and valves. This service is for 1,000,000 deflections, and above. Lowering the values 10 per cent permits 10,000,000 deflections.

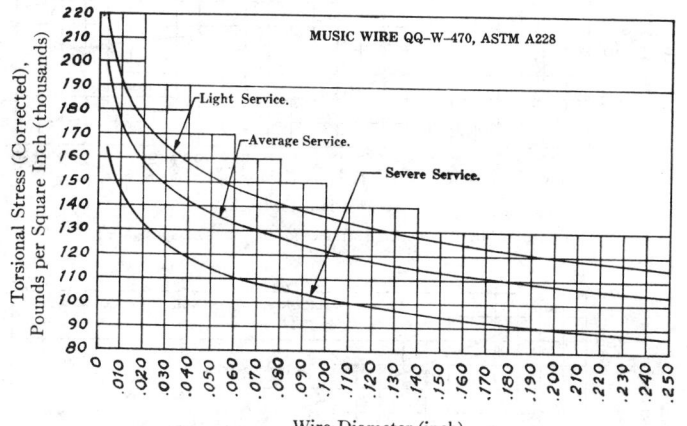

Fig. 2. Allowable working stresses for compression springs —
Music wire.*

Figures 1 through 6 show curves which relate the three types of service conditions to allowable working stresses and wire sizes for compression and extension springs, and safe values are provided. Figures 7 through 10 provide similar information for helical torsion springs. In each chart, the values obtained from the curves may be increased by 20 percent (but not beyond the top curves on the charts if permanent set is to be avoided) for springs that are baked, and shot-peened, and in the case of compression springs, pressed. Springs stressed slightly above the Light Service curves will take a permanent set.

A curvature correction factor is included in all curves, and is used in spring design calculations (see pages 509 and 511). The curves may be used for materials other than those designated in Figs. 1 through 10, by applying multiplication factors as given in Table 1.                                    (*Continued on page 504*)

* Although these curves are for compression springs, they may also be used for extension springs; for extension springs, *reduce* the values obtained from the curves by 10 to 15 per cent.

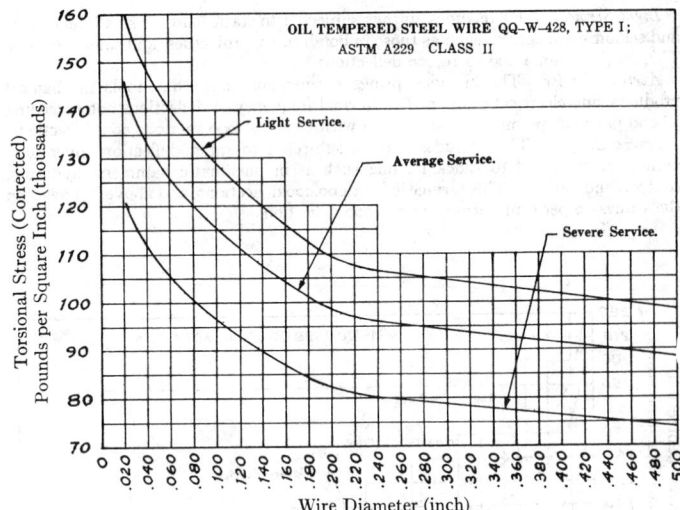

Fig. 3.   Allowable working stresses for compression springs —
Oil tempered steel wire.*

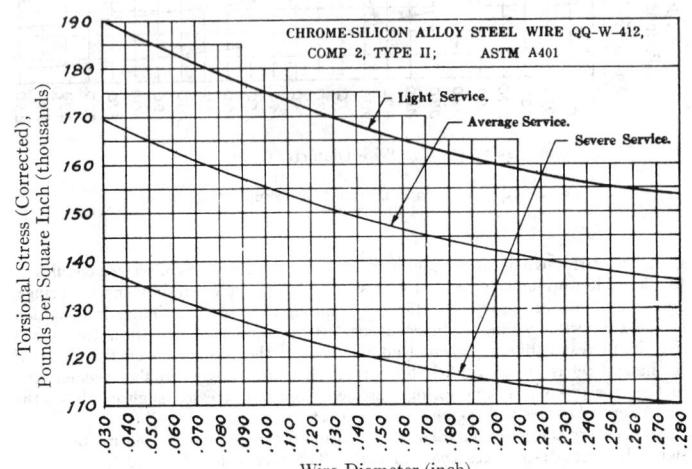

Fig. 4.   Allowable working stresses for compression springs —
Chrome-Silicon alloy steel wire.*

* Although these curves are for compression springs, they may also be used for extension springs; for extension springs, *reduce* the values obtained from the curves by 10 to 15 per cent.

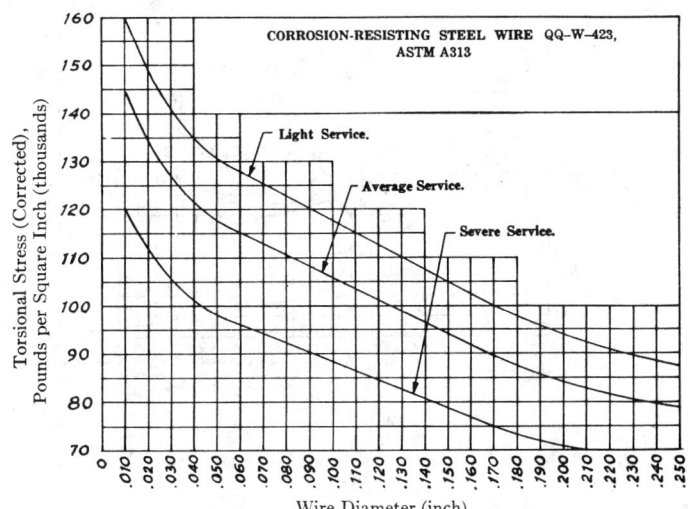

Fig. 5.  Allowable working stresses for compression springs —
Corrosion resisting steel wire.*

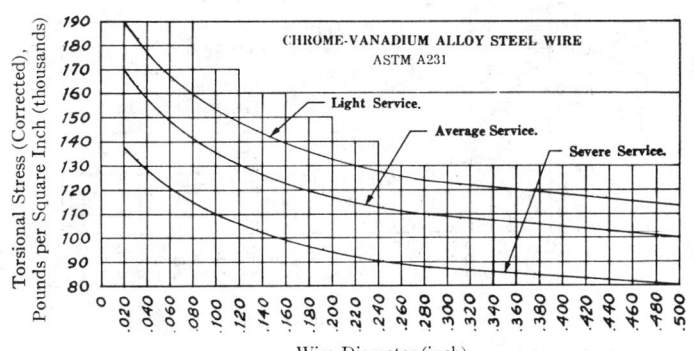

Fig. 6.  Allowable working stresses for compression springs —
Chrome-vanadium alloy steel wire.*

* Although these curves are for compression springs, they may also be used for extension springs; for extension springs, *reduce* the values obtained from the curves by 10 to
15 per cent.

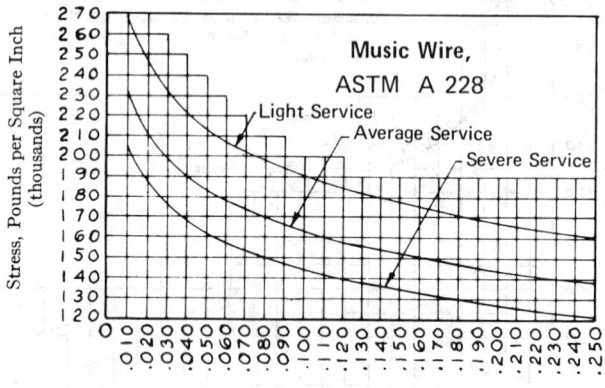

Fig. 7. Recommended design stresses in bending for helical torsion springs —
Round music wire.

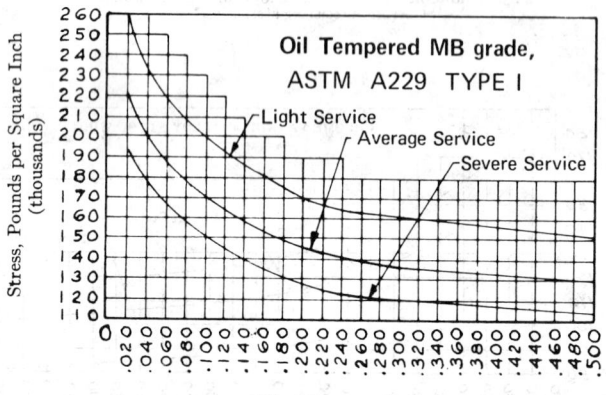

Fig. 8. Recommended design stresses in bending for helical torsion springs —
Oil tempered MB round wire.

**Endurance Limit for Spring Materials.** — When a spring is deflected continually
it will become "tired" and fail at a stress far below its elastic limit. This type of
failure is called *fatigue failure* and usually occurs without warning. *Endurance limit*

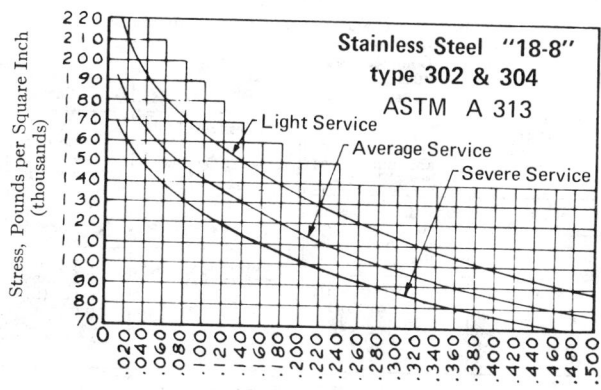

Fig. 9.   Recommended design stresses in bending for helical torsion springs —
Stainless steel round wire.

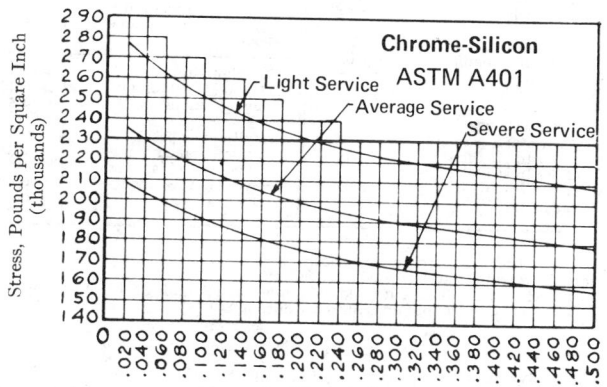

Fig. 10.   Recommended design stresses in bending for helical torsion springs —
Chrome-Silicon round wire.

is the highest stress, or range of stress, in pounds per square inch that can be repeated indefinitely without failure of the spring. Usually ten million cycles of deflection is called "infinite life" and is satisfactory for determining this limit.

## Table 1. Correction Factors for Other Materials*

### Compression and Tension Springs

| Material | Factor | Material | Factor |
|---|---|---|---|
| Silicon-manganese | Multiply the values in the chromium-vanadium curves (Fig. 6) by 0.90 | Stainless Steel, 316 | Multiply the values in the corrosion-resisting steel curves (Fig. 5) by 0.90 |
| Valve-spring quality wire | Use the values in the chromium-vanadium curves (Fig. 6) | Stainless Steel, 431 and 17-7PH | Multiply the values in the music wire curves (Fig. 2) by 0.90 |
| Stainless Steel, 304 and 420 | Multiply the values in the corrosion-resisting steel curves (Fig. 5) by 0.95 | | |

### Helical Torsion Springs

| Material | Factor† | Material | Factor† |
|---|---|---|---|
| Hard Drawn MB | 0.70 | Stainless Steel, 431 | |
| | | Up to $\frac{1}{32}$ inch diameter | 0.80 |
| Stainless Steel, 316 | | Over $\frac{1}{32}$ to $\frac{1}{16}$ inch | 0.85 |
| Up to $\frac{1}{32}$ inch diameter | 0.65 | Over $\frac{1}{16}$ to $\frac{1}{8}$ inch | 0.95 |
| Over $\frac{1}{32}$ to $\frac{3}{16}$ inch | 0.70 | Over $\frac{1}{8}$ inch | 1.00 |
| Over $\frac{3}{16}$ to $\frac{1}{4}$ inch | 0.65 | Chromium-Vanadium | |
| Over $\frac{1}{4}$ inch | 0.50 | Up to $\frac{1}{16}$ inch diameter | 1.05 |
| | | Over $\frac{1}{16}$ inch | 1.10 |
| Stainless Steel, 17-7 PH | | | |
| Up to $\frac{1}{8}$ inch diameter | 1.00 | Phosphor Bronze | |
| Over $\frac{1}{8}$ to $\frac{3}{16}$ inch | 1.07 | Up to $\frac{1}{8}$ inch diameter | 0.45 |
| Over $\frac{3}{16}$ inch | 1.12 | Over $\frac{1}{8}$ inch | 0.55 |
| Stainless Steel, 420 | | Beryllium Copper** | |
| Up to $\frac{1}{32}$ inch diameter | 0.70 | Up to $\frac{1}{32}$ inch diameter | 0.55 |
| Over $\frac{1}{32}$ to $\frac{1}{16}$ inch | 0.75 | Over $\frac{1}{32}$ to $\frac{1}{16}$ inch | 0.60 |
| Over $\frac{1}{16}$ to $\frac{1}{8}$ inch | 0.80 | Over $\frac{1}{16}$ to $\frac{1}{8}$ inch | 0.70 |
| Over $\frac{1}{8}$ to $\frac{3}{16}$ inch | 0.90 | Over $\frac{1}{8}$ inch | 0.80 |
| Over $\frac{3}{16}$ inch | 1.00 | | |

\* For use with design stress curves shown in Figs. 2, 5, 6, and 8.
\*\* Hard drawn and heated treated after coiling.
† Multiply the values in the curves for oil tempered MB grade ASTM A229 Type 1 steel (Fig. 8) by these factors to obtain required values.

For severely worked springs of long life, such as those used in automobile or aircraft engines and in similar applications, it is best to determine the allowable working stresses by referring to the endurance limit curves seen in Fig. 11. These are based principally upon the range or difference between the stress caused by the first or initial load and the stress caused by the final load. Experience with springs designed to stresses within the limits of these curves indicates that they should have infinite or unlimited fatigue life. All values include Wahl curvature correction factor. The stress ranges shown may be increased 20 to 30 per cent for springs that have been properly heated, pressed to remove set, and then shot peened, provided that the increased values are lower than the torsional elastic limit by at least 10 per cent.

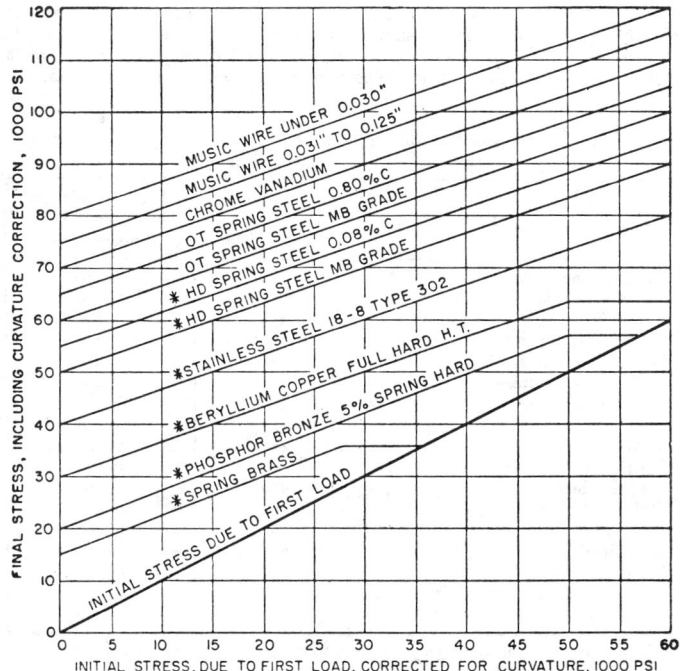

*Notes:* For commercial spring materials with wire diameters up to ¼ inch **except as** noted.

Stress ranges may be increased approximately 30 per cent for properly heated, **preset,** shot-peened springs.

Materials preceded by * are not ordinarily recommended for long continued **service** under severe operating conditions.

Fig. 11. Endurance Limit Curves for Compression Springs

**Working Stresses at Elevated Temperatures.** — Since modulus of elasticity decreases with increase in temperature, springs used at high temperatures exert less load and have larger deflections under load than at room temperature. The torsional modulus of elasticity for steel may be 11,200,000 pounds per square inch at room temperature, but it will drop to 10,600,000 pounds per square inch at 400°F. and will be only 10,000,000 pounds per square inch at 600°F. Also, the elastic limit is reduced, thereby lowering the permissible working stress.

Design stresses should be as low as possible for all springs used at elevated temperatures. In addition, corrosive conditions that usually exist at high temperatures, especially with steam, may require the use of corrosion-resistant material. Table 2 shows the permissible elevated temperatures at which various spring materials may be operated, together with the maximum recommended working stresses at these temperatures. The loss in load at the temperatures shown is less than 5 per cent in 48 hours; however, if the temperatures listed are increased by 20 to 40 degrees, the loss of load may be nearer 10 per cent. Maximum stresses shown in the table are for compression and extension springs and may be increased by 75 per cent for torsion and flat springs. In using the data in Table 2 it should be noted that the values given are for materials in the heat-treated or spring temper condition.

**Table 2.  Recommended Maximum Working Temperatures and Corresponding Maximum Working Stresses for Springs***

| Spring Material | Maximum Working Temperature, Degrees, F. | Maximum Working Stress, Pounds per Square Inch |
|---|---|---|
| Brass Spring Wire | 150 | 30,000 |
| Phosphor Bronze | 225 | 35,000 |
| Music Wire | 250 | 75,000 |
| Beryllium-Copper | 300 | 40,000 |
| Hard Drawn Steel Wire | 325 | 50,000 |
| Carbon Spring Steels | 375 | 55,000 |
| Alloy Spring Steels | 400 | 65,000 |
| Monel | 425 | 40,000 |
| K-Monel | 450 | 45,000 |
| Permanickel† | 500 | 50,000 |
| Stainless Steel 18–8 | 550 | 55,000 |
| Stainless Chromium 431 | 600 | 50,000 |
| Inconel | 700 | 50,000 |
| High Speed Steel | 775 | 70,000 |
| Inconel X | 850 | 55,000 |
| Chromium-Molybdenum-Vanadium | 900 | 55,000 |
| Cobenium, Elgiloy | 1000 | 75,000 |

* Loss of load at temperatures shown is less than 5 per cent in 48 hours.
† Formerly called Z-Nickel, Type B.

## Spring Design Data

This section provides tables of spring characteristics, tables of principal formulas, and other information of a practical nature for designing the more commonly used types of springs.

*Standard wire gages for springs:* Information on wire gages is given in the section beginning on page 471, and gages in decimals of an inch are given in the table on page 472. It should be noted that the range in this table extends from Number 7/0 through Number 80. However, in spring design, the range most commonly used extends only from Gage Number 4/0 through Number 40. When selecting wire use Steel Wire Gage or Washburn and Moen gage for all carbon steels and alloy steels except music wire; use Brown & Sharpe gage for brass and phosphor bronze wire; use Birmingham gage for flat spring steels, and cold rolled strip; and use piano or music wire gage for music wire.

*Spring index:* The spring index is the ratio of the mean coil diameter of a spring to the wire diameter $\left(\dfrac{D}{d}\right)$. This ratio is one of the most important considerations in spring design because the deflection, stress, number of coils, and selection of either annealed or tempered material depend to a considerable extent on this ratio. The best proportioned springs have an index of 7 through 9. Indexes of 4 through 7, and 9 through 16 are often used. Springs with values larger than 16

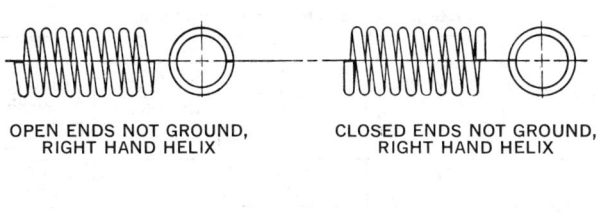

OPEN ENDS NOT GROUND,
RIGHT HAND HELIX

CLOSED ENDS NOT GROUND,
RIGHT HAND HELIX

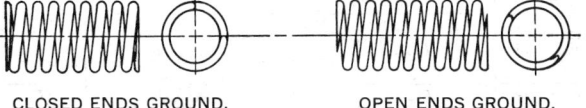

CLOSED ENDS GROUND,
LEFT HAND HELIX

OPEN ENDS GROUND,
LEFT HAND HELIX

Fig. 12.   Types of helical compression spring ends.

require tolerances wider than standard for manufacturing; those with values less than 5 are difficult to coil on automatic coiling machines.

*Direction of helix:* Unless functional requirements call for a definite hand, the helix of compression and extension springs should be specified as optional. When springs are designed to operate, one inside the other, the helices should be opposite hand to prevent intermeshing. For the same reason, a spring which is to operate freely over a threaded member should have a helix of opposite hand to that of the thread. When a spring is to engage with a screw or bolt, it should, of course, have the same helix hand as that of the thread.

**Helical Compression Spring Design.** — After selecting a suitable material and a safe stress value for a given spring, designers should next determine the type of end coil formation best suited for the particular application. Springs with unground ends are less expensive but they do not stand perfectly upright; if this requirement has to be met, closed ground ends are used. Helical compression springs with different types of ends are shown in Fig. 12.

*Spring design formulas:* Table 3 gives formulas for compression spring dimensional characteristics, and Table 4 gives design formulas for compression and extension springs.

*Curvature correction:* In addition to the stress obtained from the formulas for load or deflection, there is a direct shearing stress and an increased stress on the inside of the section due to curvature. Therefore the stress obtained by the usual formulas should be multiplied by a factor $K$ taken from the curve in Fig. 13. The corrected stress thus obtained is used only for comparison with the allowable working stress (fatigue strength) curves to determine if it is a safe stress and should not be used in formulas for deflection. The curvature correction factor $K$ is for compression and extension springs made from round wire. For square wire reduce the $K$ value by approximately 4 percent.

*Design procedure:* The limiting dimensions of a spring are often determined by the available space in the product or assembly in which it is to be used. The loads and deflections on a spring may also be known or can be estimated, but the wire size and number of coils are usually unknown. Design can be carried out with the aid of the tabular data, which is a simple method, or by calculation alone, as follows:

(*Continued on page* 511)

Table 3. Formulas for Compression Springs*

| Feature | Open or Plain (not ground) | Open or Plain (with ends ground) | Squared or Closed (not ground) | Closed and Ground |
|---|---|---|---|---|
| | Formula | | | |
| Pitch $(p)$ | $\dfrac{FL\text{-}d}{N}$ | $\dfrac{FL}{TC}$ | $\dfrac{FL\text{-}3d}{N}$ | $\dfrac{FL\text{-}2d}{N}$ |
| Solid Height $(SH)$ | $(TC+1)d$ | $TC \times d$ | $(TC+1)d$ | $TC \times d$ |
| Number of Active Coils $(N)$ | $N = TC$ or $\dfrac{FL\text{-}d}{p}$ | $N = TC\text{-}1$ or $\dfrac{FL}{p} - 1$ | $N = TC\text{-}2$ or $\dfrac{FL\text{-}3d}{p}$ | $N = TC\text{-}2$ or $\dfrac{FL\text{-}2d}{p}$ |
| Total Coils $(TC)$ | $\dfrac{FL\text{-}d}{p}$ | $\dfrac{FL}{p}$ | $\dfrac{FL\text{-}3d}{p} + 2$ | $\dfrac{FL\text{-}2d}{p} + 2$ |
| Free Length $(FL)$ | $(p \times TC) + d$ | $p \times TC$ | $(p \times N) + 3d$ | $(p \times N) + 2d$ |

Table 4. Formulas for Compression and Extension Springs*

| Feature | Springs made from round wire | Springs made from square wire |
|---|---|---|
| | Formula† | |
| Load $(P)$ Pounds | $\dfrac{0.393\ S\ d^3}{D}$ | $\dfrac{0.416\ S\ d^3}{D}$ |
| | $\dfrac{G\ d^4\ F}{8\ N\ D^3}$ | $\dfrac{G\ d^4\ F}{5.58\ N\ D^3}$ |
| Stress, Torsional $(S)$ Pounds per square inch | $\dfrac{G\ d\ F}{\pi\ N\ D^2}$ | $\dfrac{G\ d\ F}{2.32\ N\ D^2}$ |
| | $\dfrac{P\ D}{0.393\ d^3}$ | $\dfrac{P\ D}{0.416\ d^3}$ |
| Deflection $(F)$ Inch | $\dfrac{8\ P\ N\ D^3}{G\ d^4}$ | $\dfrac{5.58\ P\ N\ D^3}{G\ d^4}$ |
| | $\dfrac{\pi\ S\ N\ D^2}{G\ d}$ | $\dfrac{2.32\ S\ N\ D^2}{G\ d}$ |
| Number of Active Coils $(N)$ | $\dfrac{G\ d^4\ F}{8\ P\ D^3}$ | $\dfrac{G\ d^4\ F}{5.58\ P\ D^3}$ |
| | $\dfrac{G\ d\ F}{\pi\ S\ D^2}$ | $\dfrac{G\ d\ F}{2.32\ S\ D^2}$ |
| Wire Diameter $(d)$ Inch | $\dfrac{\pi\ S\ N\ D^2}{G\ F}$ | $\dfrac{2.32\ S\ N\ D^2}{G\ F}$ |
| | $\sqrt[3]{\dfrac{2.55\ P\ D}{S}}$ | $\sqrt[3]{\dfrac{P\ D}{416\ S}}$ |
| Stress due to Initial Tension $(S_{it})$ | $\dfrac{S}{P} \times IT$ | $\dfrac{S}{P} \times IT$ |

* The symbol notation is given on p. 494.
† Two formulas are given for each feature, and designers can use the one found to be appropriate in a given design instance. The end result from either of any two formulas is the same.

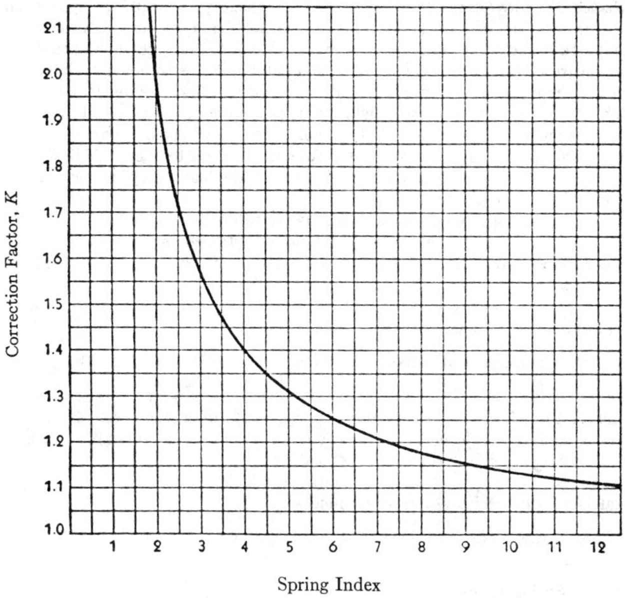

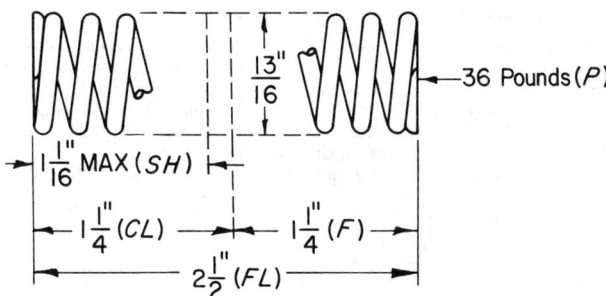

Fig. 13. Compression and extension spring stress correction for curvature.*

*Example:* A compression spring with closed and ground ends is to be made from ASTM A229 high carbon steel wire, as shown in Fig. 14. Determine the wire size and number of coils.

Fig. 14. Compression spring design example.

* For springs made from round wire. For springs made from square wire, reduce the $K$ factor values by approximately 4 per cent.

*Method 1, using table:* Referring to Table 5, page **514**, locate the spring outside diameter ($^{13}/_{16}$ inches, from Fig. 14) in the left-hand column. Note from the drawing that the spring load is 36 pounds. Move to the right in the table to the figure nearest this value, which is 41.7 pounds. This is somewhat above the required value but safe. Immediately above the load value, the deflection $f$ is given, which in this instance is 0.1594 inch. This is the deflection of 1 coil under a load of 41.7 pounds with an uncorrected torsional stress $S$ of 100,000 pounds per square inch. (This is the torsional stress for ASTM A229 oil-tempered MB steel — see page 535). Moving vertically in the table from the load entry, the wire diameter is found to be 0.0915 inch.

The remaining spring design calculations are completed as follows:

*Step 1:* The stress with a load of 36 pounds is obtained as follows: The 36 pound load is 86.3 percent of the 41.7 pound load. Therefore, the stress $S$ at 36 pounds = 0.863 × 100,000 = 86,300 pounds per square inch.

*Step 2:* The 86.3 percent figure is also used to determine the deflection per coil $f$ at 36 pounds load: 0.863 × 0.1594 = 0.1375 inch.

*Step 3:* The number of active coils $AC = \dfrac{F}{f} = \dfrac{1.25}{0.1375} = 9.1$ (say 9)

*Step 4:* Total Coils $TC = AC + 2$ (Table 3) = 9 + 2 = 11

Therefore a quick answer is: 11 coils of 0.0915 inch diameter wire. However, the design procedure should be completed by carrying out these remaining steps:

*Step 5:* Solid Height $SH = TC \times d$ (Table 3)

= 11 × 0.0915 = 1 inch

(Total Deflection = $FL - SH$ = 1.5 inches)

*Step 6:* Stress Solid = $\dfrac{86,300}{1.25} \times 1.5$ = 103,500 pounds per square inch

*Step 7:* Spring Index = $\dfrac{O.D.}{d} - 1 = \dfrac{0.8125}{0.0915} - 1 = 7.9$

*Step 8:* From Fig. 13, the curvature correction factor $K = 1.185$

*Step 9:* Total Stress at 36 pounds load = $S \times K$ = 86,300 × 1.185 = 102,300 pounds per square inch. (This is a safe working stress as it is below the 117,000 pounds per square inch permitted for 0.0915 inch wire shown on the middle curve in Fig. 3.)

*Step 10:* Total Stress at Solid = 103,500 × 1.185 = 122,800 pounds per square inch. (This too, is a safe stress, as it is below the 131,000 pounds per square inch shown on the top curve Fig. 3; therefore the spring will not set.)

*Method 2, using formulas:* The procedure for design using formulas is as follows (the design example is the same as in Method 1, and the spring is shown in Fig. 14):

*Step 1:* Select a safe stress $S$ below the middle fatigue strength curve Fig. 3 for ASTM A229 steel wire, say 90,000 pounds per square inch. Assume a mean diameter $D$ slightly below the $^{13}/_{16}$-inch O.D., say 0.7 inch. Note that the value of $G$ is 11,200,000 pounds per square inch (Table 19).

*Step 2:* A trial wire diameter $d$ and other values are found by formulas from

Table 4 as follows: $d = \sqrt[3]{\dfrac{2.55\,P\,D}{S}} = \sqrt[3]{\dfrac{2.55 \times 36 \times 0.7}{90,000}}$

$$= \sqrt[3]{0.000714} = 0.0894 \text{ inch.}$$

*Note:* Table 20 can be used to avoid solving the cube root.

*Step 3:* From the table on page 464, select the nearest wire gauge size, which is 0.0915 inch diameter. Using this value, the mean diameter $D = {}^{13}/_{16}$ inch − 0.0915 = 0.721 inch. *(Continued on page 517)*

**Table 5. Compression and Extension Spring Deflections***

Wire Size or Washburn and Moen Gauge, and Decimal Equivalent†

Deflection $f$ (inch) per coil, at Load $P$ (pounds)‡

| Outside Diam. Nom. | Dec. | 16 / .0625 | 17 / .054 | 18 / .0475 | 19 / .041 | .038 | .036 | .034 | .032 | .030 | .028 | .026 | .024 | .022 | .020 | .018 | .016 | .014 | .012 | .010 |
|---|---|---|---|---|---|---|---|---|---|---|---|---|---|---|---|---|---|---|---|---|
| 7/64 | .1094 | | | | | | | | | .00589 / 13.83 | .00664 / 10.59 | .00751 / 8.28 | .00853 / 6.36 | .00974 / 4.79 | .01121 / 3.52 | .01302 / 2.51 | .01529 / 1.722 | .01824 / 1.130 | .0222 / .697 | .0277 / .395 |
| 1/8 | .125 | | | | | | .00617 / 20.6 | .00683 / 16.95 | .00758 / 13.83 | .00844 / 11.16 | .00943 / 8.89 | .01058 / 6.97 | .01192 / 5.37 | .01353 / 4.06 | .01548 / 2.99 | .01784 / 2.14 | .0208 / 1.475 | .0247 / .971 | .0299 / .600 | .0371 / .342 |
| 9/64 | .1406 | | | | | .00777 / 21.0 | .00852 / 17.51 | .00937 / 14.47 | .01034 / 11.84 | .01144 / 9.58 | .01271 / 7.66 | .01417 / 6.02 | .01590 / 4.65 | .01794 / 3.53 | .0204 / 2.61 | .0234 / 1.868 | .0272 / 1.291 | .0321 / .852 | .0387 / .528 | .0478 / .301 |
| 5/32 | .1563 | | | | .00909 / 23.5 | .01033 / 18.22 | .01128 / 15.23 | .01234 / 12.62 | .01354 / 10.35 | .01491 / 8.39 | .01649 / 6.72 | .01832 / 5.30 | .0205 / 4.10 | .0230 / 3.11 | .0261 / 2.31 | .0298 / 1.656 | .0345 / 1.146 | .0406 / .758 | .0487 / .470 | .0600 / .268 |
| 11/64 | .1719 | | | .00914 / 33.8 | .01172 / 20.8 | .01324 / 16.09 | .01439 / 13.48 | .01569 / 11.19 | .01716 / 9.19 | .01883 / 7.47 | .0208 / 5.99 | .0230 / 4.73 | .0256 / 3.67 | .0287 / 2.79 | .0324 / 2.07 | .0369 / 1.488 | .0426 / 1.031 | .0500 / .683 | .0598 / .424 | .0735 / .243 |
| 3/16 | .1875 | | .00926 / 46.3 | .01157 / 30.0 | .01468 / 18.47 | .01650 / 14.41 | .01788 / 12.09 | .01944 / 10.05 | .0212 / 8.27 | .0232 / 6.73 | .0255 / 5.40 | .0281 / 4.27 | .0313 / 3.32 | .0349 / 2.53 | .0394 / 1.876 | .0448 / 1.351 | .0516 / .938 | .0603 / .621 | .0720 / .387 | .0884 / .221 |
| 13/64 | .2031 | | .01155 / 41.5 | .01430 / 27.1 | .01798 / 16.69 | .0201 / 13.05 | .0218 / 10.96 | .0236 / 9.13 | .0257 / 7.52 | .0280 / 6.12 | .0307 / 4.92 | .0338 / 3.90 | .0375 / 3.03 | .0418 / 2.31 | .0470 / 1.716 | .0534 / 1.237 | .0614 / .859 | .0717 / .570 | .0854 / .355 | .1046 / .203 |
| 7/32 | .2188 | .01096 / 61.3 | .01411 / 37.5 | .0175 / 24.6 | .0216 / 15.22 | .0241 / 11.92 | .0260 / 10.02 | .0282 / 8.35 | .0306 / 6.88 | .0333 / 5.61 | .0365 / 4.52 | .0401 / 3.58 | .0444 / 2.79 | .0494 / 2.13 | .0555 / 1.580 | .0628 / 1.140 | .0721 / .793 | .0841 / .526 | .1000 / .328 | |
| 15/64 | .2344 | .01326 / 55.8 | .01733 / 34.3 | .0206 / 22.5 | .0256 / 13.99 | .0285 / 10.97 | .0307 / 9.23 | .0331 / 7.70 | .0359 / 6.35 | .0391 / 5.19 | .0427 / 4.18 | .0469 / 3.21 | .0518 / 2.58 | .0575 / 1.969 | .0645 / 1.465 | .0730 / 1.058 | .0836 / .736 | .0974 / .489 | .1156 / .305 | |
| 1/4 | .250 | .01578 / 51.1 | .0216 / 31.6 | .0256 / 20.8 | .0299 / 12.95 | .0332 / 10.17 | .0358 / 8.56 | .0385 / 7.14 | .0417 / 5.90 | .0453 / 4.82 | .0494 / 3.88 | .0541 / 3.08 | .0597 / 2.40 | .0663 / 1.834 | .0742 / 1.366 | .0839 / .987 | .0960 / .687 | .1116 / .457 | | |
| 9/32 | .2813 | .0215 / 43.8 | .0268 / 27.2 | .0323 / 18.01 | .0395 / 11.26 | .0437 / 8.86 | .0469 / 7.47 | .0505 / 6.24 | .0545 / 5.16 | .0591 / 4.24 | .0643 / 3.40 | .0703 / 2.70 | .0774 / 2.11 | .0857 / 1.613 | .0958 / 1.202 | .1080 / .870 | .1234 / .606 | .1432 / .403 | | |
| 5/16 | .3125 | .0281 / 38.3 | .0347 / 23.9 | .0415 / 15.89 | .0504 / 9.97 | .0556 / 7.85 | .0596 / 6.63 | .0640 / 5.54 | .0690 / 4.58 | .0746 / 3.75 | .0811 / 3.03 | .0886 / 2.41 | .0973 / 1.881 | .1076 / 1.440 | .1200 / 1.074 | .1351 / .778 | .1541 / .542 | | | |
| 11/32 | .3438 | .0355 / 34.1 | .0436 / 21.3 | .0518 / 14.21 | .0627 / 8.94 | .0690 / 7.05 | .0733 / 5.95 | .0792 / 4.98 | .0852 / 4.12 | .0921 / 3.40 | .0999 / 2.73 | .1092 / 2.19 | .1196 / 1.697 | .1321 / 1.300 | .1470 / .970 | .1633 / .703 | | | | |
| 3/8 | .375 | .0438 / 30.7 | .0535 / 19.27 | .0634 / 12.85 | .0764 / 8.10 | .0839 / 6.40 | .0895 / 5.40 | .0960 / 4.53 | .1031 / 3.75 | .1113 / 3.07 | .1206 / 2.48 | .1314 / 1.978 | .1440 / 1.546 | .1589 / 1.185 | .1768 / .885 | | | | | |

\* The table is for ASTM A229 oil tempered spring steel with a torsional modulus $G$ of 11,200,000 psi, and an uncorrected torsional stress of 100,000 psi. For other materials use the following factors: stainless steel, multiply by 1.067; phosphor bronze, multiply by 1.24; Monel metal, multiply $f$ by 1.244; beryllium copper, multiply $f$ by 1.725; inconel (non-magnetic), multiply $f$ by 1.045. † Round wire. For square wire, multiply $f$ by .707, and $P$ by 1.2. ‡ The upper figure is the deflection and the lower figure the load as read against each spring size. *Note:* Intermediate values can be obtained within reasonable accuracy by interpolation.

## Table 5. (Continued) Compression and Extension Spring Deflections*

Wire Size or Washburn and Moen Gauge, and Decimal Equivalent

Deflection $f$ (inch) per coil, at Load $P$ (pounds)

| Nom. | Dec. | .026 | .028 | .030 | .032 | .034 | .036 | .038 | 19 / .041 | 18 / .0475 | 17 / .054 | 16 / .0625 | 15 / .072 | 14 / .080 | 13 / .0915 | 3/32 / .0938 | 12 / .1055 | 11 / .1205 | 1/8 / .125 |
|---|---|---|---|---|---|---|---|---|---|---|---|---|---|---|---|---|---|---|---|
| 13/32 | .4063 | .1560 / 1.815 | .1434 / 2.28 | .1324 / 2.82 | .1228 / 3.44 | .1143 / 4.15 | .1068 / 4.95 | .1001 / 5.85 | .0913 / 7.41 | .0760 / 11.73 | .0645 / 17.56 | .0531 / 27.9 | .0436 / 43.9 | .0373 / 61.0 | .0304 / 95.6 | .0292 / 103.7 | .0241 / 153.3 | | |
| 7/16 | .4375 | .1827 / 1.678 | .1680 / 2.11 | .1553 / 2.60 | .1441 / 3.17 | .1343 / 3.82 | .1256 / 4.56 | .1178 / 5.39 | .1075 / 6.82 | .0898 / 10.79 | .0764 / 16.13 | .0631 / 25.6 | .0521 / 40.1 | .0448 / 56.3 | .0370 / 86.9 | .0353 / 94.3 | .0293 / 138.9 | .0234 / 217. | .0219 / 245. |
| 15/32 | .4688 | .212 / 1.559 | .1947 / 1.956 | .1800 / 2.42 | .1673 / 2.94 | .1560 / 3.55 | .1459 / 4.23 | .1370 / 5.00 | .1252 / 6.33 | .1048 / 9.99 | .0894 / 14.91 | .0741 / 23.6 | .0614 / 37.0 | .0530 / 51.7 | .0437 / 79.7 | .0420 / 86.4 | .0351 / 126.9 | .0282 / 197.3 | .0265 / 223. |
| 1/2 | .500 | .243 / 1.456 | .223 / 1.826 | .207 / 2.26 | .1920 / 2.75 | .1792 / 3.31 | .1678 / 3.95 | .1575 / 4.67 | .1441 / 5.90 | .1209 / 9.30 | .1033 / 13.87 | .0859 / 21.9 | .0714 / 34.3 | .0619 / 47.9 | .0512 / 73.6 | .0494 / 80.0 | .0414 / 116.9 | .0335 / 181.1 | .0316 / 205. |
| 17/32 | .5313 | .276 / 1.366 | .254 / 1.713 | .235 / 2.12 | .219 / 2.58 | .204 / 3.10 | .1911 / 3.70 | .1796 / 4.37 | .1645 / 5.52 | .1382 / 8.70 | .1183 / 12.96 | .0987 / 20.5 | .0822 / 31.9 | .0714 / 44.6 | .0593 / 68.4 | .0572 / 74.1 | .0482 / 108.3 | .0393 / 167.3 | .0371 / 188.8 |
| 9/16 | .5625 | | .286 / 1.613 | .265 / 1.991 | .247 / 2.42 | .230 / 2.92 | .216 / 3.48 | .203 / 4.11 | .1861 / 5.19 | .1566 / 8.18 | .1343 / 12.16 | .1122 / 19.17 | .0937 / 29.9 | .0816 / 41.7 | .0680 / 63.9 | .0657 / 69.1 | .0555 / 100.9 | .0455 / 155.5 | .0430 / 175.3 |
| 19/32 | .5938 | | | .297 / 1.880 | .277 / 2.29 | .259 / 2.76 | .242 / 3.28 | .228 / 3.88 | .209 / 4.90 | .1762 / 7.71 | .1514 / 11.46 | .1267 / 18.04 | .1061 / 28.1 | .0926 / 39.1 | .0774 / 60.0 | .0748 / 64.8 | .0634 / 94.4 | .0522 / 145.2 | .0493 / 163.6 |
| 5/8 | .625 | | | .331 / 1.782 | .308 / 2.17 | .288 / 2.61 | .270 / 3.11 | .254 / 3.67 | .233 / 4.63 | .1969 / 7.29 | .1693 / 10.83 | .1420 / 17.04 | .1191 / 26.5 | .1041 / 36.9 | .0873 / 56.4 | .0844 / 61.0 | .0718 / 88.7 | .0593 / 136.2 | .0561 / 153.4 |
| 21/32 | .6563 | | | | .342 / 2.06 | .320 / 2.48 | .300 / 2.95 | .282 / 3.49 | .259 / 4.40 | .219 / 6.92 | .1884 / 10.27 | .1582 / 16.14 | .1330 / 25.1 | .1164 / 34.9 | .0978 / 53.3 | .0946 / 57.6 | .0807 / 83.7 | .0668 / 128.3 | .0634 / 144.3 |
| 11/16 | .6875 | | | | | .352 / 2.36 | .331 / 2.81 | .311 / 3.32 | .286 / 4.19 | .242 / 6.58 | .208 / 9.76 | .1753 / 15.34 | .1476 / 23.8 | .1294 / 33.1 | .1089 / 50.5 | .1054 / 54.6 | .0901 / 79.2 | .0748 / 121.2 | .0710 / 136.3 |
| 23/32 | .7188 | | | | | | .363 / 2.68 | .342 / 3.17 | .314 / 3.99 | .266 / 6.27 | .230 / 9.31 | .1933 / 14.61 | .1630 / 22.7 | .1431 / 31.5 | .1206 / 48.0 | .1168 / 51.9 | .1000 / 75.2 | .0833 / 114.9 | .0791 / 129.2 |
| 3/4 | .750 | | | | | | | .374 / 3.03 | .344 / 3.82 | .291 / 5.99 | .252 / 8.89 | .212 / 13.94 | .1791 / 21.6 | .1574 / 30.0 | .1329 / 45.7 | .1288 / 49.4 | .1105 / 71.5 | .0923 / 109.2 | .0877 / 122.7 |
| 25/32 | .7813 | | | | | | | | .375 / 3.66 | .318 / 5.74 | .275 / 8.50 | .232 / 13.34 | .1960 / 20.7 | .1724 / 28.7 | .1459 / 43.6 | .1413 / 47.1 | .1214 / 68.2 | .1017 / 104.0 | .0967 / 116.9 |
| 13/16 | .8125 | | | | | | | | .407 / 3.51 | .346 / 5.50 | .299 / 8.15 | .253 / 12.78 | .214 / 19.80 | .1881 / 27.5 | .1594 / 41.7 | .1545 / 45.1 | .1329 / 65.2 | .1115 / 99.3 | .1061 / 111.5 |

*Outside Diam. columns shown as Nom. and Dec.*

* The table is for ASTM A229 oil tempered spring steel with a torsional modulus $G$ of 11,200,000 psi, and an uncorrected torsional stress of 100,000 psi. For other materials, and other important footnotes, see page 513.

## Table 5 (Continued). Compression and Extension Spring Deflections*

**Wire Size or Washburn and Moen Gauge, and Decimal Equivalent**

Deflection $f$ (inch) per coil, at Load $P$ (pounds) — each cell shows $f$ / $P$.

| Outside Diam. Nom. | Dec. | 4 (.2253) | 7/32 (.2188) | 5 (.207) | 6 (.192) | 3/16 (.1875) | 7 (.177) | 8 (.162) | 5/32 (.1563) | 9 (.1483) | 10 (.135) | ⅛ (.125) | 11 (.1205) | 12 (.1055) | 3/32 (.0938) | 13 (.0915) | 14 (.080) | 15 (.072) |
|---|---|---|---|---|---|---|---|---|---|---|---|---|---|---|---|---|---|---|
| ⅞ | .875 | .0526 / 691. | .0552 / 626. | .0605 / 521. | .0682 / 407. | .0707 / 377. | .0772 / 312. | .0880 / 234. | .0928 / 209. | .0999 / 176.3 | .1138 / 130.5 | .1262 / 102.3 | .1325 / 91.1 | .1574 / 59.9 | .1825 / 41.5 | .1882 / 39.4 | .222 / 25.3 | .251 / 18.26 |
| 29/32 | .9063 | .0577 / 660. | .0606 / 598. | .0663 / 498. | .0746 / 389. | .0772 / 360. | .0843 / 299. | .0959 / 224. | .1010 / 199.9 | .1087 / 169.0 | .1236 / 125.2 | .1370 / 98.2 | .1438 / 87.5 | .1705 / 57.6 | .1974 / 39.9 | .204 / 36.9 | .239 / 24.3 | .271 / 17.57 |
| 15/16 | .9375 | .0632 / 631. | .0662 / 572. | .0723 / 477. | .0812 / 373. | .0842 / 345. | .0917 / 286. | .1041 / 215. | .1096 / 191.9 | .1178 / 162.3 | .1338 / 120.4 | .1479 / 94.4 | .1554 / 84.1 | .1841 / 55.4 | .213 / 38.4 | .219 / 35.6 | .258 / 23.5 | .292 / 16.94 |
| 31/32 | .9688 | .0688 / 604. | .0721 / 548. | .0786 / 457. | .0882 / 358. | .0913 / 332. | .0994 / 275. | .1127 / 207. | .1183 / 184.5 | .1273 / 156.1 | .1445 / 115.9 | .1598 / 90.9 | .1675 / 81.0 | .1982 / 53.4 | .229 / 37.0 | .236 / 34.3 | .277 / 22.6 | .313 / 16.35 |
| 1 | 1.000 | .0747 / 580. | .0783 / 526. | .0852 / 439. | .0954 / 344. | .0986 / 319. | .1074 / 264. | .1216 / 198.8 | .1278 / 177.6 | .1372 / 150.4 | .1555 / 111.7 | .1718 / 87.6 | .1801 / 78.1 | .213 / 51.5 | .246 / 35.8 | .253 / 33.1 | .297 / 21.9 | .336 / 15.80 |
| 1 1/32 | 1.031 | .0809 / 557. | .0845 / 506. | .0921 / 423. | .1029 / 331. | .1065 / 307. | .1157 / 255. | .1308 / 191.6 | .1374 / 171.3 | .1474 / 145.1 | .1669 / 107.8 | .1843 / 84.6 | .1931 / 75.5 | .228 / 49.8 | .263 / 34.6 | .271 / 32.0 | .317 / 21.1 | .359 / 15.28 |
| 1 1/16 | 1.063 | .0873 / 537. | .0913 / 487. | .0993 / 407. | .1107 / 319. | .1145 / 296. | .1243 / 246. | .1404 / 185.0 | .1474 / 165.4 | .1580 / 140.1 | .1788 / 104.2 | .1972 / 81.8 | .207 / 73.0 | .244 / 48.2 | .281 / 33.5 | .289 / 31.0 | .338 / 20.5 | .382 / 14.80 |
| 1 3/32 | 1.094 | .0939 / 517. | .0982 / 470. | .1066 / 393. | .1188 / 308. | .1229 / 286. | .1332 / 238. | .1503 / 178.8 | .1578 / 159.9 | .1691 / 135.5 | .1910 / 100.8 | .211 / 79.2 | .221 / 70.6 | .260 / 46.7 | .298 / 32.4 | .308 / 30.0 | .360 / 19.83 | .407 / 14.34 |
| 1 ⅛ | 1.125 | .1008 / 499. | .1053 / 454. | .1142 / 379. | .1272 / 298. | .1315 / 276. | .1424 / 230. | .1604 / 173.0 | .1685 / 154.7 | .1804 / 131.2 | .204 / 97.6 | .224 / 76.7 | .235 / 68.4 | .277 / 45.2 | .314 / 31.4 | .328 / 29.1 | .383 / 19.24 | .432 / 13.92 |
| 1 3/16 | 1.188 | .1153 / 467. | .1203 / 424. | .1303 / 355. | .1448 / 279. | .1496 / 259. | .1620 / 215. | .1812 / 162.4 | .1908 / 145.4 | .204 / 123.3 | .231 / 91.7 | .254 / 72.1 | .265 / 64.4 | .311 / 42.6 | .358 / 29.6 | .368 / 27.5 | .431 / 18.15 | .485 / 13.14 |
| 1 ¼ | 1.250 | .1308 / 438. | .1363 / 399. | .1474 / 334. | .1635 / 263. | .1690 / 244. | .1824 / 203. | .205 / 153.1 | .215 / 137.0 | .230 / 116.2 | .258 / 86.6 | .284 / 68.2 | .297 / 60.8 | .349 / 40.3 | .400 / 28.0 | .412 / 26.0 | .480 / 17.19 | .541 / 12.44 |
| 1 5/16 | 1.313 | .1472 / 413. | .1535 / 376. | .1657 / 315. | .1836 / 248. | .1894 / 230. | .205 / 191.6 | .229 / 144.7 | .240 / 129.7 | .256 / 110.1 | .288 / 82.0 | .317 / 64.6 | .331 / 57.7 | .387 / 38.2 | .444 / 26.6 | .457 / 24.6 | .533 / 16.31 | .600 / 11.81 |
| 1 ⅜ | 1.375 | .1650 / 391. | .1713 / 356. | .1848 / 298. | .204 / 235. | .211 / 218. | .227 / 181.7 | .255 / 137.3 | .267 / 123.0 | .285 / 104.4 | .320 / 77.9 | .351 / 61.4 | .367 / 54.8 | .429 / 36.3 | .491 / 25.3 | .506 / 23.4 | .588 / 15.53 | .662 / 11.25 |
| 1 7/16 | 1.438 | .1829 / 371. | .1905 / 337. | .205 / 283. | .227 / 223. | .234 / 207. | .252 / 172.6 | .282 / 130.6 | .295 / 117.0 | .314 / 99.4 | .353 / 74.1 | .387 / 58.4 | .404 / 52.2 | .472 / 34.6 | .540 / 24.1 | .556 / 22.3 | .647 / 14.81 | .727 / 10.73 |

* The table is for ASTM A229 oil tempered spring steel with a torsional modulus $G$ of 11,200,000 psi, and an uncorrected torsional stress of 100,000 psi. For other materials, and other important footnotes, see page 513.

## Table 5 (Concluded). Compression and Extension Spring Deflections*

**Wire Size or Washburn and Moen Gauge, and Decimal Equivalent**

Deflection $f$ (inch) per coil, at Load $P$ (pounds) — each cell shows $f$ / $P$.

| Outside Diam. Nom. | Dec. | 5/16 (.3125) | 0 (.3065) | 9/32 (.2813) | 2 (.2625) | 1/4 (.250) | 3 (.2437) | 4 (.2253) | 7/32 (.2188) | 5 (.207) | 6 (.192) | 3/16 (.1875) | 7 (.177) | 8 (.162) | 5/32 (.1563) | 9 (.1483) | 10 (.135) | 1/8 (.125) | 11 (.1205) |
|---|---|---|---|---|---|---|---|---|---|---|---|---|---|---|---|---|---|---|---|
| 1 1/2 | 1.500 | .1267 / 1008 | .1305 / 947 | .1482 / 717 | .1612 / 574 | .1754 / 499 | .1815 / 452 | .202 / 352 | .210 / 321 | .227 / 269 | .250 / 213 | .258 / 197.1 | .277 / 164.6 | .310 / 124.5 | .324 / 111.8 | .350 / 94.8 | .387 / 70.8 | .424 / 55.8 | .443 / 49.8 |
| 1 5/8 | 1.625 | .1547 / 912 | .1592 / 858 | .1801 / 650 | .1986 / 521 | .212 / 446 | .220 / 411 | .244 / 321 | .254 / 292 | .273 / 246 | .300 / 193.9 | .309 / 180.0 | .332 / 150.3 | .371 / 113.9 | .387 / 102.0 | .413 / 86.7 | .461 / 64.8 | .505 / 51.1 | .527 / 45.7 |
| 1 3/4 | 1.750 | .1856 / 833 | .1908 / 783 | .215 / 595 | .237 / 477 | .253 / 409 | .261 / 377 | .290 / 295 | .301 / 269 | .323 / 226 | .355 / 178.4 | .366 / 165.6 | .392 / 138.5 | .437 / 104.9 | .456 / 94.0 | .485 / 80.0 | .542 / 59.8 | .593 / 47.2 | .619 / 42.2 |
| 1 7/8 | 1.875 | .219 / 767 | .225 / 721 | .253 / 548 | .278 / 440 | .296 / 378 | .306 / 348 | .339 / 272 | .351 / 248 | .377 / 209 | .414 / 165.1 | .426 / 153.4 | .457 / 128.2 | .508 / 97.3 | .530 / 87.2 | .564 / 74.2 | .629 / 55.5 | .687 / 43.8 | .717 / 39.2 |
| 1 15/16 | 1.938 | .237 / 737 | .243 / 693 | .273 / 528 | .300 / 425 | .320 / 364 | .331 / 335 | .365 / 262 | .379 / 239 | .405 / 201 | .446 / 159.2 | .458 / 147.9 | .492 / 123.6 | .546 / 93.8 | .569 / 84.2 | .605 / 71.6 | .676 / 53.6 | .738 / 42.3 | .769 / 37.8 |
| 2 | 2.000 | .256 / 710 | .263 / 668 | .295 / 509 | .323 / 409 | .344 / 351 | .355 / 324 | .392 / 253 | .407 / 231 | .436 / 194.3 | .478 / 153.7 | .492 / 142.8 | .527 / 119.4 | .585 / 90.6 | .610 / 81.3 | .649 / 69.2 | .723 / 51.8 | .789 / 40.9 | .823 / 36.6 |
| 2 1/16 | 2.063 | .275 / 685 | .282 / 644 | .316 / 491 | .346 / 395 | .369 / 339 | .381 / 312 | .421 / 245 | .436 / 223 | .467 / 187.7 | .512 / 148.5 | .526 / 138.1 | .564 / 115.4 | .626 / 87.6 | .652 / 78.7 | .693 / 66.9 | .768 / 50.1 | .843 / 39.6 | .878 / 35.4 |
| 2 1/8 | 2.125 | .295 / 661 | .303 / 622 | .339 / 474 | .371 / 381 | .395 / 327 | .407 / 302 | .449 / 236 | .466 / 216 | .499 / 181.6 | .546 / 143.8 | .562 / 133.6 | .602 / 111.8 | .667 / 84.9 | .696 / 76.1 | .739 / 64.8 | .823 / 48.5 | .898 / 38.3 | .936 / 34.3 |
| 2 3/16 | 2.188 | .316 / 639 | .324 / 601 | .362 / 459 | .396 / 369 | .421 / 317 | .435 / 292 | .479 / 229 | .497 / 209 | .532 / 175.8 | .582 / 139.2 | .598 / 129.5 | .641 / 108.3 | .711 / 82.2 | .740 / 73.8 | .786 / 62.8 | .876 / 47.1 | .955 / 37.2 | .995 / 33.3 |
| 2 1/4 | 2.250 | .337 / 618 | .346 / 582 | .387 / 444 | .423 / 357 | .449 / 307 | .463 / 283 | .511 / 222 | .529 / 202 | .566 / 170.5 | .619 / 135.0 | .637 / 125.5 | .681 / 105.1 | .755 / 79.8 | .787 / 71.6 | .835 / 60.9 | .930 / 45.7 | 1.013 / 36.1 | 1.056 / 32.3 |
| 2 5/16 | 2.313 | .359 / 599 | .368 / 564 | .411 / 430 | .449 / 347 | .478 / 298 | .493 / 275 | .542 / 215 | .562 / 196.3 | .601 / 165.4 | .657 / 131.0 | .676 / 121.8 | .723 / 101.9 | .801 / 77.5 | .834 / 69.5 | .886 / 59.2 | .986 / 44.4 | 1.074 / 35.1 | 1.119 / 31.4 |
| 2 3/8 | 2.375 | .382 / 581 | .392 / 547 | .437 / 417 | .477 / 336 | .507 / 289 | .523 / 267 | .576 / 209 | .596 / 190.7 | .637 / 160.7 | .696 / 127.3 | .716 / 118.3 | .763 / 99.1 | .848 / 75.3 | .884 / 67.6 | .938 / 57.5 | 1.043 / 43.1 | 1.136 / 34.1 | 1.184 / 30.5 |
| 2 7/16 | 2.438 | .405 / 564 | .416 / 531 | .464 / 405 | .506 / 327 | .537 / 281 | .554 / 259 | .609 / 203 | .631 / 185.3 | .674 / 156.1 | .737 / 123.7 | .757 / 115.1 | .810 / 96.3 | .897 / 73.2 | .934 / 65.7 | .991 / 56.0 | 1.102 / 42.0 | 1.201 / 33.2 | .... |
| 2 1/2 | 2.500 | .430 / 548 | .441 / 516 | .491 / 394 | .536 / 317 | .568 / 273 | .586 / 252 | .644 / 197.5 | .667 / 180.2 | .713 / 151.9 | .778 / 120.4 | .800 / 111.6 | .855 / 93.7 | .946 / 71.3 | .986 / 64.0 | 1.046 / 54.5 | 1.162 / 40.9 | 1.266 / 32.3 | .... |

\* The table is for ASTM A229 oil tempered spring steel with a torsional modulus $G$ of 11,200,000 psi, and an uncorrected torsional stress of 100,000 psi. For other materials, and other important footnotes, see page 513.

*Step 4:* The stress $S = \dfrac{P\,D}{0.393\,d^3} = \dfrac{36 \times 0.721}{0.393 \times 0.0915^3} = 86{,}300$ pounds per square inch,

*Step 5:* The number of active coils is $N = \dfrac{G\,d\,F}{\pi S\,D^2}$

$$= \dfrac{11{,}200{,}000 \times 0.0915 \times 1.25}{3.1416 \times 86{,}300 \times 0.721^2} = 9.1 \text{ (say 9)}$$

The answer is the same as before, which is to use 11 total coils of 0.0915 inch diameter wire. The total coils, solid height, etc., are determined in the same manner as in Method 1.

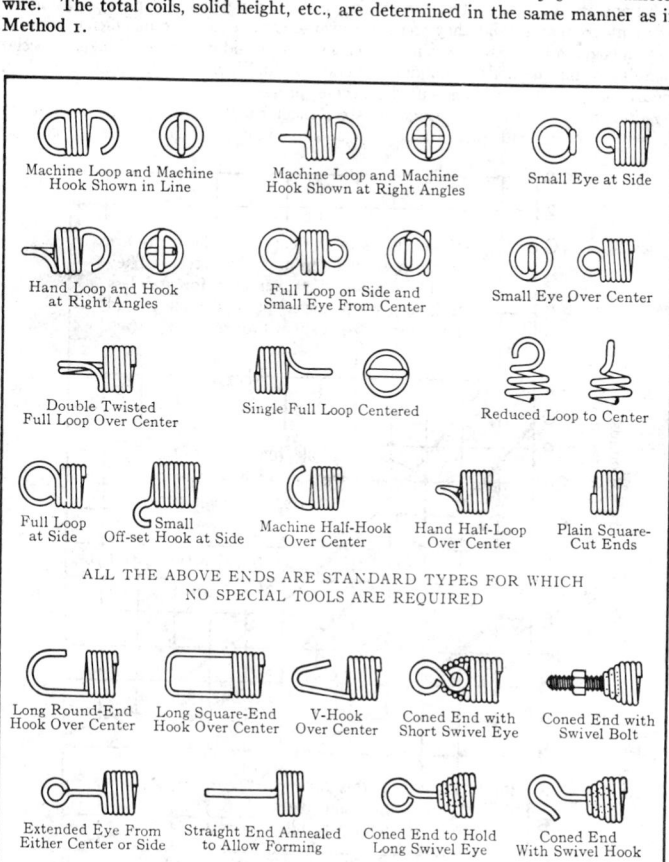

Fig. 15. Types of helical extension spring ends.

**Table of Spring Characteristics.** — Table 5 gives characteristics for compression and extension springs made from ASTM A229 oil-tempered MB spring steel having a torsional modulus of elasticity $G$ of 11,200,000 pounds per square inch, and an uncorrected torsional stress $S$ of 100,000 pounds per square inch. The deflection $f$ for one coil under a load $P$ is shown in the body of the table. The method of using this data is explained in the problems for compression and extension spring design. The table may be used for other materials by applying factors to $f$. The factors are given in a footnote to the table.

**Extension springs.** — These represent about 10 per cent of all springs made by many companies and they frequently cause trouble because insufficient consideration is given to stress due to initial tension, stress and deflection of hooks, special manufacturing methods, secondary operations and overstretching at assembly. Figure 15 shows types of ends used on these springs.

*Initial tension:* In the spring industry, the term "Initial tension" is used to define a force or load, measurable in pounds or ounces, which presses the coils of

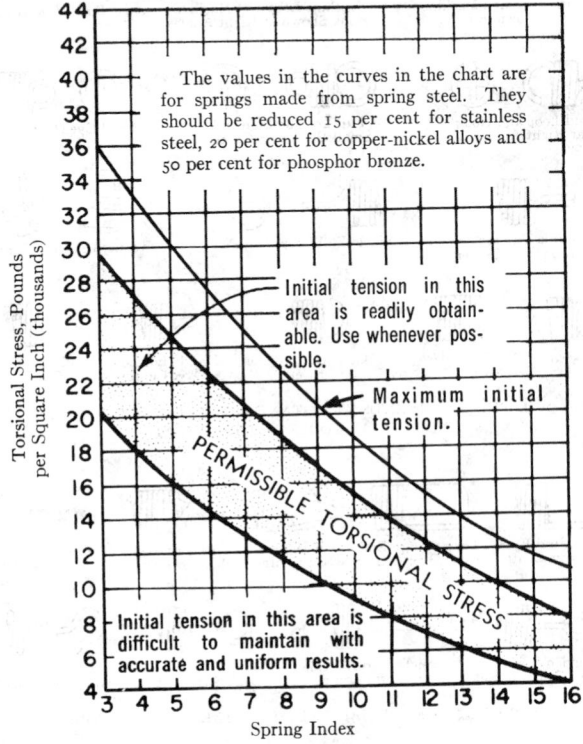

Fig. 16.  Permissible torsional stress caused by initial tension in coiled extension springs for different spring indexes.

a close wound extension spring against one another. This force must be overcome before the coils of a spring begin to open up.

Initial tension is wound into extension springs by bending each coil as it is wound away from its normal plane, thereby producing a slight twist in the wire which causes the coil to spring back tightly against the adjacent coil. Initial tension can be wound into cold-coiled extension springs only. Hot-wound springs and springs made from annealed steel are hardened and tempered after coiling, and therefore initial tension cannot be produced. It is possible to make a spring having initial tension only when a high tensile strength, obtained by cold drawing or by heat-treatment, is possessed by the material as it is being wound into springs. Materials that possess the required characteristics for the manufacture of such springs include hard-drawn wire, music wire, pre-tempered wire, 18-8 stainless steel, phosphor-bronze, and many of the hard-drawn copper-nickel, and nonferrous alloys. Permissible torsional stresses resulting from initial tension for different spring indexes are shown in Fig. 16.

*Hook failure:* The great majority of breakages in extension springs occurs in the hooks. Hooks are subjected to both bending and torsional stresses and have higher stresses than the coils in the spring.

*Stresses in regular hooks:* The calculations for the stresses in hooks are quite complicated and lengthy. Also, the radii of the bends are difficult to determine and frequently vary between specifications and actual production samples. However, regular hooks are more highly stressed than the coils in the body and are subjected to a bending stress at section B (see Fig. 17). The bending stress $S_b$ at section B should be compared with allowable stresses for torsion springs and with the elastic limit of the material in tension (See Fig. 7 through 10).

*Stresses in cross over hooks:* Results of tests on springs having a normal average index show that the cross over hooks last longer than regular hooks. These results may not occur on springs of small index or if the cross over bend is made too sharply.

Inasmuch as both types of hooks have the same bending stress, it would appear that the fatigue life would be the same. However, the large bend radius of the regular hooks causes some torsional stresses to coincide with the bending stresses, thus explaining the earlier breakages. If sharper bends were made on the regular hooks, the life should then be the same as for cross over hooks.

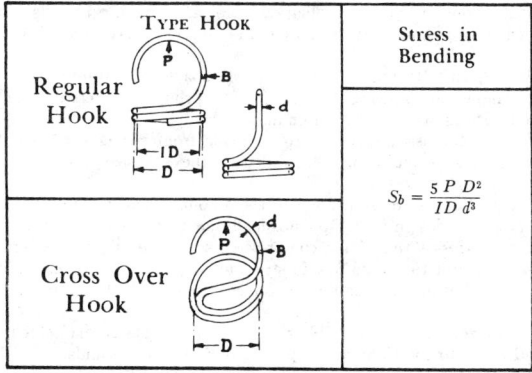

Fig. 17.    Formula for the bending stress at section B on regular and cross-over type hooks.

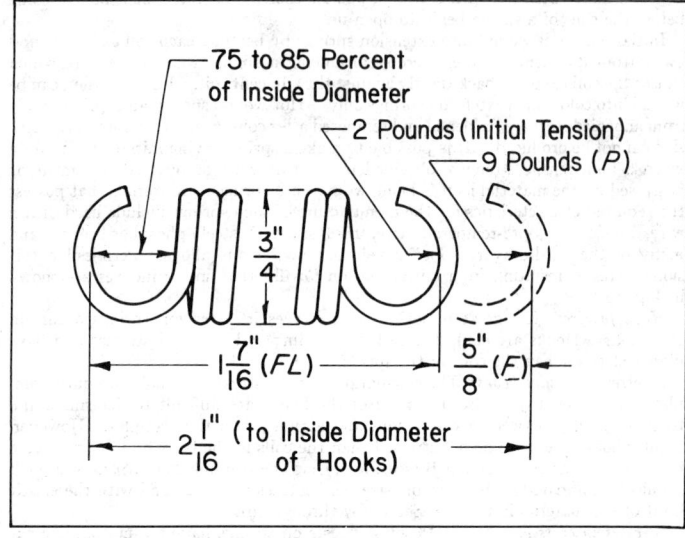

Fig. 18.  Extension spring design example.

*Stresses in half hooks:* The formulas for regular hooks can also be used for half hooks, because the smaller bend radius allows for the increase in strses. It will therefore be observed that half hooks have the same stress in bending as regular hooks.

Frequently overlooked facts by many designers are that one full hook deflects an amount equal to ½ a coil and each half hook deflects an amount equal to ¹⁄₁₀th of a coil. Allowances for these deflections should be made when designing springs. Thus, an extension spring, with regular full hooks and having 10 coils, will have a deflection equal to 11 coils, or 10% more than that which may have been calculated.

**Extension Spring Design.** — The available space in a product or assembly usually determines the limiting dimensions of a spring, but the wire size, number of coils, and initial tension are often unknown.

*Example:* An extension spring is to be made from spring steel ASTM A229, with regular hooks shown in Fig. 18. Calculate the wire size, number of coils and initial tension.

*Note:* Allow about 20 to 25 percent of the 9 pounds load for initial tension, say 2 pounds, and then design for a 7 pounds load (not 9 pounds) at ⅝ inch deflection. Also use lower stresses than for a compression spring to allow for overstretching during assembly and to obtain a safe stress on the hooks. Proceed as for compression springs, but locate a load in the tables somewhat higher than the 9 pounds load.

*Method 1, using table:* From Table 5 locate ¾ inch outside diameter in the left column and move to the right to locate a load $P$ of 13.94 pounds. A deflection $f$ of 0.212 inch appears above this figure. Moving vertically from this position to the top of the column a suitable wire diameter of 0.0625 inch is found.

The remaining design calculations are completed as follows:

*Step 1:* The stress with a load of 7 pounds is obtained as follows: The 7 pounds load is 50.2 percent of the 13.94 pounds load. Therefore, the stress $S$ at 7 pounds = 0.502 percent × 100,000 = 50,200 pounds per square inch.

*Step 2:* The 50.2 percent figure is also used to determine the deflection per coil $f$: 0.502 percent × 0.212 = 0.1062 inch.

*Step 3:* The number of active coils $AC = \dfrac{F}{f} = \dfrac{0.625}{0.1062} = 5.86$ (say 6)

This should be reduced by 1 to allow for deflection of 2 hooks, see notes 1 and 2 that follow these calculations. Therefore, a quick answer is: 5 coils of 0.0625 inch diameter wire. However, the design procedure should be completed by carrying out these remaining steps:

*Step 4:* The body length = $TC + 1 \times d = 5 + 1 \times 0.0625 = 3/8$ inch.

*Step 5:* The length from the body to inside hook

$$= \frac{FL - \text{Body}}{2} = \frac{1.4375 - 0.375}{2} = 0.531 \text{ inch.}$$

Percentage of I.D. $= \dfrac{0.531}{\text{I.D.}} = \dfrac{0.531}{0.625} = 85$ per cent

(This is satisfactory, see Note 3.)

*Step 6:* The spring index $= \dfrac{\text{O.D.}}{d} - 1 = \dfrac{0.75}{0.0625} - 1 = 11$

*Step 7:* The initial tension stress $S_{it} = \dfrac{S \times IT}{P}$

$$= \frac{50,200 \times 2}{7}$$

$$= 14,340 \text{ pounds per square inch.}$$

(Satisfactory, as checked against curve in Fig. 16.)

*Step 8:* The curvature correction factor $K = 1.12$ (Fig. 13.)

*Step 9:* The total stress = 50,200 + 14,340 × 1.12
$$= 72,285 \text{ pounds per square inch.}$$

(This is less than 106,250 pounds per square inch permitted by the middle curve for 0.0625 inch wire in Fig. 3 and therefore is a safe working stress that permits additional deflection as usually necessary for assembly purposes.)

*Step 10:* The large majority of hook breakage is due to high stress in bending and should be checked as follows:

From Fig. 17, stress on hook in bending is:

$$S_b = \frac{5PD^2}{\text{I.D.} \cdot d^3}$$

$$= \frac{5 \times 9 \times 0.6875^2}{0.625 \times 0.0625^3} = 139,200 \text{ pounds per square inch.}$$

(This is less than the top curve value, Fig. 8, for 0.625 inch diameter wire, and is therefore safe. See Note 5.)

*Note:* The following points should be noted when designing extension springs:

1. All coils are active and thus $AC = TC$.

2. Each full hook deflection is approximately equal to ½ coil. Therefore for 2 hooks, reduce the total coils by 1. (Each half hook deflection is nearly equal to 1/10 of a coil.)

3. The distance from the body to the inside of a regular full hook equals 75 to

85 per cent (90 per cent maximum) of the I.D. (For a cross over center hook, it equals the I.D.)

4. Some initial tension should usually be used to hold the spring together. Try not to exceed the maximum curve shown on Fig. 16. Without initial tension, a long spring with many coils will have a different length in the horizontal position than it will when hung vertically.

5. The hooks are stressed in bending, therefore their stress should be less than the maximum bending stress as used for torsion springs — use top fatigue strength curves Figs. 7 through 10.

*Method 2, using formulas:* The sequence of steps for designing extension springs by formulas is similar to that for compression springs. The formulas for this method are given in Table 3.

**Tolerances for Compression and Extension Springs.** — Tolerances for coil diameter, free length, squareness, load, and the angle between loop planes for compression and extension springs are given in Tables 6 through 11. To meet the requirements of load, rate, free length, and solid height, it is necessary to vary the number of coils for compression springs by ± 5 percent. For extension springs, the tolerances on the numbers of coils are: for 3 to 5 coils, ± 20 percent; for 6 to 8 coils, ± 30 percent; for 9 to 12 coils, ± 40 percent. For each additional coil, a further 1½ percent tolerance is added to the extension spring values. Closer

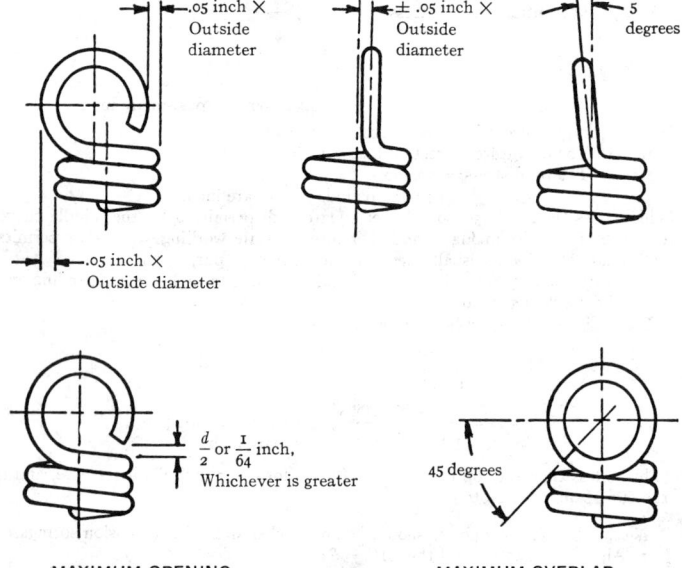

MAXIMUM OPENING FOR CLOSED LOOP      MAXIMUM OVERLAP FOR CLOSED LOOP

Fig. 19. Maximum deviations allowed on ends and variation in alignment of ends (loops) for extension springs.

## Table 6. Compression and Extension Spring Coil Diameter Tolerances*

| Wire Diameter, Inch | Spring Index | | | | | | |
|---|---|---|---|---|---|---|---|
| | 4 | 6 | 8 | 10 | 12 | 14 | 16 |
| | Tolerance, ± inch | | | | | | |
| 0.015 | 0.002 | 0.002 | 0.003 | 0.004 | 0.005 | 0.006 | 0.007 |
| 0.023 | 0.002 | 0.003 | 0.004 | 0.006 | 0.007 | 0.008 | 0.010 |
| 0.035 | 0.002 | 0.004 | 0.006 | 0.007 | 0.009 | 0.011 | 0.013 |
| 0.051 | 0.003 | 0.005 | 0.007 | 0.010 | 0.012 | 0.015 | 0.017 |
| 0.076 | 0.004 | 0.007 | 0.010 | 0.013 | 0.016 | 0.019 | 0.022 |
| 0.114 | 0.006 | 0.009 | 0.013 | 0.018 | 0.021 | 0.025 | 0.029 |
| 0.171 | 0.008 | 0.012 | 0.017 | 0.023 | 0.028 | 0.033 | 0.038 |
| 0.250 | 0.011 | 0.015 | 0.021 | 0.028 | 0.035 | 0.042 | 0.049 |
| 0.375 | 0.016 | 0.020 | 0.026 | 0.037 | 0.046 | 0.054 | 0.064 |
| 0.500 | 0.021 | 0.030 | 0.040 | 0.062 | 0.080 | 0.100 | 0.125 |

## Table 7. Compression Spring Normal Free-Length Tolerances, Squared and Ground Ends*

| Number of Active Coils per Inch | Spring Index | | | | | | |
|---|---|---|---|---|---|---|---|
| | 4 | 6 | 8 | 10 | 12 | 14 | 16 |
| | Tolerance, ± Inch per Inch of Free Length† | | | | | | |
| 0.5 | 0.010 | 0.011 | 0.012 | 0.013 | 0.015 | 0.016 | 0.016 |
| 1 | 0.011 | 0.013 | 0.015 | 0.016 | 0.017 | 0.018 | 0.019 |
| 2 | 0.013 | 0.015 | 0.017 | 0.019 | 0.020 | 0.022 | 0.023 |
| 4 | 0.016 | 0.018 | 0.021 | 0.023 | 0.024 | 0.026 | 0.027 |
| 8 | 0.019 | 0.022 | 0.024 | 0.026 | 0.028 | 0.030 | 0.032 |
| 12 | 0.021 | 0.024 | 0.027 | 0.030 | 0.032 | 0.034 | 0.036 |
| 16 | 0.022 | 0.026 | 0.029 | 0.032 | 0.034 | 0.036 | 0.038 |
| 20 | 0.023 | 0.027 | 0.031 | 0.034 | 0.036 | 0.038 | 0.040 |

## Table 8. Extension Spring Normal Free-Length and End Tolerances*

| FREE-LENGTH TOLERANCES | | END TOLERANCES | |
|---|---|---|---|
| Spring Free-Length (inch) | Tolerance (inch) | Total Number of Coils | Angle Between Loop Planes (degrees) |
| Up to 0.5 | ±0.020 | | |
| Over 0.5 to 1.0 | ±0.030 | | |
| Over 1.0 to 2.0 | ±0.040 | 3 to 6 | ±25 |
| Over 2.0 to 4.0 | ±0.060 | 7 to 9 | ±35 |
| Over 4.0 to 8.0 | ±0.093 | 10 to 12 | ±45 |
| Over 8.0 to 16.0 | ±0.156 | 13 to 16 | ±60 |
| Over 16.0 to 24.0 | ±0.218 | Over 16 | Random |

* Courtesy of the Spring Manufacturers Institute
† For springs less than 0.5 inch long, use the tolerances for 0.5 inch long springs. For springs with unground closed ends, multiply the tolerances by 1.7.

### Table 9. Compression Spring Squareness Tolerances*

| Slenderness Ratio† | Spring Index | | | | | | |
|---|---|---|---|---|---|---|---|
| | 4 | 6 | 8 | 10 | 12 | 14 | 16 |
| | Squareness Tolerance (±degrees) | | | | | | |
| 0.5 | 3.0 | 3.0 | 3.5 | 3.5 | 3.5 | 3.5 | 4.0 |
| 1.0 | 2.5 | 3.0 | 3.0 | 3.0 | 3.0 | 3.5 | 3.5 |
| 1.5 | 2.5 | 2.5 | 2.5 | 3.0 | 3.0 | 3.0 | 3.0 |
| 2.0 | 2.5 | 2.5 | 2.5 | 2.5 | 3.0 | 3.0 | 3.0 |
| 3.0 | 2.0 | 2.5 | 2.5 | 2.5 | 2.5 | 2.5 | 3.0 |
| 4.0 | 2.0 | 2.0 | 2.5 | 2.5 | 2.5 | 2.5 | 2.5 |
| 6.0 | 2.0 | 2.0 | 2.0 | 2.5 | 2.5 | 2.5 | 2.5 |
| 8.0 | 2.0 | 2.0 | 2.0 | 2.0 | 2.5 | 2.5 | 2.5 |
| 10.0 | 2.0 | 2.0 | 2.0 | 2.0 | 2.0 | 2.5 | 2.5 |
| 12.0 | 2.0 | 2.0 | 2.0 | 2.0 | 2.0 | 2.0 | 2.5 |

### Table 10. Compression Spring Normal Load Tolerances

| Length Tolerance, ±inch | Deflection (inch)‡ | | | | | | | |
|---|---|---|---|---|---|---|---|---|
| | 0.05 | 0.10 | 0.15 | 0.20 | 0.25 | 0.30 | 0.40 | 0.50 |
| | Tolerance, ± Per Cent of Load | | | | | | | |
| 0.005 | 12 | 7 | 6 | 5 | .... | .... | .... | .... |
| 0.010 | .... | 12 | 8.5 | 7 | 6.5 | 5.5 | 5 | .... |
| 0.020 | .... | 22 | 15.5 | 12 | 10 | 8.5 | 7 | 6 |
| 0.030 | .... | .... | 22 | 17 | 14 | 12 | 9.5 | 8 |
| 0.040 | .... | .... | .... | 22 | 18 | 15.5 | 12 | 10 |
| 0.050 | .... | .... | .... | .... | 22 | 19 | 14.5 | 12 |
| 0.060 | .... | .... | .... | .... | 25 | 22 | 17 | 14 |
| 0.070 | .... | .... | .... | .... | .... | 25 | 19.5 | 16 |
| 0.080 | .... | .... | .... | .... | .... | .... | 22 | 18 |
| 0.090 | .... | .... | .... | .... | .... | .... | 25 | 20 |
| 0.100 | .... | .... | .... | .... | .... | .... | .... | 22 |
| 0.200 | .... | .... | .... | .... | .... | .... | .... | .... |
| 0.300 | .... | .... | .... | .... | .... | .... | .... | .... |
| 0.400 | .... | .... | .... | .... | .... | .... | .... | .... |
| 0.500 | .... | .... | .... | .... | .... | .... | .... | .... |

| Length Tolerance, ±inch | Deflection (inch)‡ | | | | | | |
|---|---|---|---|---|---|---|---|
| | 0.75 | 1.00 | 1.50 | 2.00 | 3.00 | 4.00 | 6.00 |
| | Tolerance, ± Per Cent of Load | | | | | | |
| 0.005 | .... | .... | .... | .... | .... | .... | .... |
| 0.010 | .... | .... | .... | .... | .... | .... | .... |
| 0.020 | 5 | .... | .... | .... | .... | .... | .... |
| 0.030 | 6 | 5 | .... | .... | .... | .... | .... |
| 0.040 | 7.5 | 6 | 5 | .... | .... | .... | .... |
| 0.050 | 9 | 7 | 5.5 | .... | .... | .... | .... |
| 0.060 | 10 | 8 | 6 | 5 | .... | .... | .... |
| 0.070 | 11 | 9 | 6.5 | 5.5 | .... | .... | .... |
| 0.080 | 12.5 | 10 | 7.5 | 6 | 5 | .... | .... |
| 0.090 | 14 | 11 | 8 | 6 | 5 | .... | .... |
| 0.100 | 15.5 | 12 | 8.5 | 7 | 5.5 | .... | .... |
| 0.200 | .... | 22 | 15.5 | 12 | 8.5 | 7 | 5.5 |
| 0.300 | .... | .... | 22 | 17 | 12 | 9.5 | 7 |
| 0.400 | .... | .... | .... | 21 | 15 | 12 | 8.5 |
| 0.500 | .... | .... | .... | 25 | 18.5 | 14.5 | 10.5 |

* Springs with closed and ground ends, in the free position. Squareness tolerances closer than those shown require special process techniques which increase cost. Springs made from fine wire sizes, and with high spring indices, irregular shapes or long free lengths, require special attention in determining appropriate tolerance and feasibility of grinding ends.

† Slenderness Ratio = $\dfrac{FL}{D}$

‡ From free length to loaded position.

## Table 11.  Extension Spring Normal Load Tolerances

| Spring Index | FL*/F | \multicolumn{11}{c}{Wire Diameter (inch)} |
|---|---|---|

| Spring Index | FL*/F | 0.015 | 0.022 | 0.032 | 0.044 | 0.062 | 0.092 | 0.125 | 0.187 | 0.250 | 0.375 | 0.437 |
|---|---|---|---|---|---|---|---|---|---|---|---|---|
| | | \multicolumn{11}{c}{Tolerance, ± Per Cent of Load} |
| 4 | 12 | 20.0 | 18.5 | 17.6 | 16.9 | 16.2 | 15.5 | 15.0 | 14.3 | 13.8 | 13.0 | 12.6 |
| | 8 | 18.5 | 17.5 | 16.7 | 15.8 | 15.0 | 14.5 | 14.0 | 13.2 | 12.5 | 11.5 | 11.0 |
| | 6 | 16.8 | 16.1 | 15.5 | 14.7 | 13.8 | 13.2 | 12.7 | 11.8 | 11.2 | 9.9 | 9.4 |
| | 4.5 | 15.0 | 14.7 | 14.1 | 13.5 | 12.6 | 12.0 | 11.5 | 10.3 | 9.7 | 8.4 | 7.9 |
| | 2.5 | 13.1 | 12.4 | 12.1 | 11.8 | 10.6 | 10.0 | 9.1 | 8.5 | 8.0 | 6.8 | 6.2 |
| | 1.5 | 10.2 | 9.9 | 9.3 | 8.9 | 8.0 | 7.5 | 7.0 | 6.5 | 6.1 | 5.3 | 4.8 |
| | 0.5 | 6.2 | 5.4 | 4.8 | 4.6 | 4.3 | 4.1 | 4.0 | 3.8 | 3.6 | 3.3 | 3.2 |
| 6 | 12 | 17.0 | 15.5 | 14.6 | 14.1 | 13.5 | 13.1 | 12.7 | 12.0 | 11.5 | 11.2 | 10.7 |
| | 8 | 16.2 | 14.7 | 13.9 | 13.4 | 12.6 | 12.2 | 11.7 | 11.0 | 10.5 | 10.0 | 9.5 |
| | 6 | 15.2 | 14.0 | 12.9 | 12.3 | 11.6 | 10.9 | 10.7 | 10.0 | 9.4 | 8.8 | 8.3 |
| | 4.5 | 13.7 | 12.4 | 11.5 | 11.0 | 10.5 | 10.0 | 9.6 | 9.0 | 8.3 | 7.6 | 7.1 |
| | 2.5 | 11.9 | 10.8 | 10.2 | 9.8 | 9.4 | 9.0 | 8.5 | 7.9 | 7.2 | 6.2 | 6.0 |
| | 1.5 | 9.9 | 9.0 | 8.3 | 7.7 | 7.3 | 7.0 | 6.7 | 6.4 | 6.0 | 4.9 | 4.7 |
| | 0.5 | 6.3 | 5.5 | 4.9 | 4.7 | 4.5 | 4.3 | 4.1 | 4.0 | 3.7 | 3.5 | 3.4 |
| 8 | 12 | 15.8 | 14.3 | 13.1 | 13.0 | 12.1 | 12.0 | 11.5 | 10.8 | 10.2 | 10.0 | 9.5 |
| | 8 | 15.0 | 13.7 | 12.5 | 12.1 | 11.4 | 11.0 | 10.6 | 10.1 | 9.4 | 9.0 | 8.6 |
| | 6 | 14.2 | 13.0 | 11.7 | 11.2 | 10.6 | 10.0 | 9.7 | 9.3 | 8.6 | 8.1 | 7.6 |
| | 4.5 | 12.8 | 11.7 | 10.7 | 10.1 | 9.7 | 9.0 | 8.7 | 8.3 | 7.8 | 7.2 | 6.6 |
| | 2.5 | 11.2 | 10.2 | 9.5 | 8.8 | 8.3 | 7.9 | 7.7 | 7.4 | 6.9 | 6.1 | 5.6 |
| | 1.5 | 9.5 | 8.6 | 7.8 | 7.1 | 6.9 | 6.7 | 6.5 | 6.2 | 5.8 | 4.9 | 4.5 |
| | 0.5 | 6.3 | 5.6 | 5.0 | 4.8 | 4.5 | 4.4 | 4.2 | 4.1 | 3.9 | 3.6 | 3.5 |
| 10 | 12 | 14.8 | 13.3 | 12.0 | 11.9 | 11.1 | 10.9 | 10.5 | 9.9 | 9.3 | 9.2 | 8.8 |
| | 8 | 14.2 | 12.8 | 11.6 | 11.2 | 10.5 | 10.2 | 9.7 | 9.2 | 8.6 | 8.3 | 8.0 |
| | 6 | 13.4 | 12.1 | 10.8 | 10.5 | 9.8 | 9.3 | 8.9 | 8.6 | 8.0 | 7.6 | 7.2 |
| | 4.5 | 12.3 | 10.8 | 10.0 | 9.5 | 9.0 | 8.5 | 8.1 | 7.8 | 7.3 | 6.8 | 6.4 |
| | 2.5 | 10.8 | 9.6 | 9.0 | 8.4 | 8.0 | 7.7 | 7.3 | 7.0 | 6.5 | 5.9 | 5.5 |
| | 1.5 | 9.2 | 8.3 | 7.5 | 6.9 | 6.7 | 6.5 | 6.3 | 6.0 | 5.6 | 5.0 | 4.6 |
| | 0.5 | 6.4 | 5.7 | 5.1 | 4.9 | 4.7 | 4.5 | 4.3 | 4.2 | 4.0 | 3.8 | 3.7 |
| 12 | 12 | 14.0 | 12.3 | 11.1 | 10.8 | 10.1 | 9.8 | 9.5 | 9.0 | 8.5 | 8.2 | 7.9 |
| | 8 | 13.2 | 11.8 | 10.7 | 10.2 | 9.6 | 9.3 | 8.9 | 8.4 | 7.9 | 7.5 | 7.2 |
| | 6 | 12.6 | 11.2 | 10.2 | 9.7 | 9.0 | 8.5 | 8.2 | 7.9 | 7.4 | 6.9 | 6.4 |
| | 4.5 | 11.7 | 10.2 | 9.4 | 9.0 | 8.4 | 8.0 | 7.6 | 7.2 | 6.8 | 6.3 | 5.8 |
| | 2.5 | 10.5 | 9.2 | 8.5 | 8.0 | 7.8 | 7.4 | 7.0 | 6.6 | 6.1 | 5.6 | 5.2 |
| | 1.5 | 8.9 | 8.0 | 7.2 | 6.8 | 6.5 | 6.3 | 6.1 | 5.7 | 5.4 | 4.8 | 4.5 |
| | 0.5 | 6.5 | 5.8 | 5.3 | 5.1 | 4.9 | 4.7 | 4.5 | 4.3 | 4.2 | 4.0 | 3.3 |
| 14 | 12 | 13.1 | 11.3 | 10.2 | 9.7 | 9.1 | 8.8 | 8.4 | 8.1 | 7.6 | 7.2 | 7.0 |
| | 8 | 12.4 | 10.9 | 9.8 | 9.2 | 8.7 | 8.3 | 8.0 | 7.6 | 7.2 | 6.8 | 6.4 |
| | 6 | 11.8 | 10.4 | 9.3 | 8.8 | 8.3 | 7.7 | 7.5 | 7.2 | 6.8 | 6.3 | 5.9 |
| | 4.5 | 11.1 | 9.7 | 8.7 | 8.2 | 7.8 | 7.2 | 7.0 | 6.7 | 6.3 | 5.8 | 5.4 |
| | 2.5 | 10.1 | 8.8 | 8.1 | 7.6 | 7.1 | 6.7 | 6.5 | 6.2 | 5.7 | 5.2 | 5.0 |
| | 1.5 | 8.6 | 7.7 | 7.0 | 6.7 | 6.3 | 6.0 | 5.8 | 5.5 | 5.2 | 4.7 | 4.5 |
| | 0.5 | 6.6 | 5.9 | 5.4 | 5.2 | 5.0 | 4.8 | 4.6 | 4.4 | 4.3 | 4.2 | 4.0 |
| 16 | 12 | 12.3 | 10.3 | 9.2 | 8.6 | 8.1 | 7.7 | 7.4 | 7.2 | 6.8 | 6.3 | 6.1 |
| | 8 | 11.7 | 10.0 | 8.9 | 8.3 | 7.8 | 7.4 | 7.2 | 6.8 | 6.5 | 6.0 | 5.7 |
| | 6 | 11.0 | 9.6 | 8.5 | 8.0 | 7.5 | 7.1 | 6.9 | 6.5 | 6.2 | 5.7 | 5.4 |
| | 4.5 | 10.5 | 9.1 | 8.1 | 7.5 | 7.2 | 6.8 | 6.5 | 6.2 | 5.8 | 5.3 | 5.1 |
| | 2.5 | 9.7 | 8.4 | 7.6 | 7.0 | 6.7 | 6.3 | 6.1 | 5.7 | 5.4 | 4.9 | 4.7 |
| | 1.5 | 8.3 | 7.4 | 6.6 | 6.2 | 6.0 | 5.8 | 5.6 | 5.3 | 5.1 | 4.6 | 4.4 |
| | 0.5 | 6.7 | 5.9 | 5.5 | 5.3 | 5.1 | 5.0 | 4.8 | 4.6 | 4.5 | 4.3 | 4.1 |

* $\frac{FL}{F}$ = the ratio of the spring free length (FL) to the deflection (F).

tolerances on the number of coils for either type of spring, lead to the need for trimming after coiling, and manufacturing time and cost are increased. Figure 19 shows deviations allowed on the ends of extension springs, and variations in end alignments.

**Torsion Spring Design.** — Figure 20 shows the types of ends most commonly used on torsion springs. To produce them requires only limited tooling. The straight torsion end is the least expensive and should be used whenever possible. After determining the spring load or torque required and selecting the end formations, the designer usually estimates suitable space or size limitations. However, the space should be considered approximate until the wire size and number of coils have been determined. The wire size is dependent principally upon the torque. Design data can be carried out with the aid of the tabular data, which is a simple method, or by calculation alone, as shown in the following sections. Also, many other factors affecting the design and operation of torsion springs are covered in the section, Design Recommendations. Design formulas are shown in Table 12.

*Curvature correction:* In addition to the stress obtained from the formulas for load or deflection, there is a direct shearing stress on the inside of the section due to curvature. Therefore, the stress obtained by the usual formulas should be multiplied by the factor $K$ obtained from the curve in Fig. 21. The corrected stress thus obtained is used only for comparison with the allowable working stress (fatigue strength) curves to determine if it is a safe value, and should not be used in the formulas for deflection.

*Torque:* By definition, torque is a force that tends to produce rotation. Torsion springs exert torque in a circular arc and the arms are rotated about the central axis. It should be noted that the stress is in bending, not in torsion. In the spring industry it is customary to specify torque in conjunction with the deflection or with the arms of a spring at a definite position. Formulas for torque are expressed in pound-inches. If ounce-inches are specified, it is necessary to divide this value by 16 in order to use the formulas.

When a load is specified at a distance from a centerline, the torque is, of course, equal to the load multiplied by the distance. The load can be in pounds or ounces with the distances in inches or the load can be in grams or kilograms with the dis-

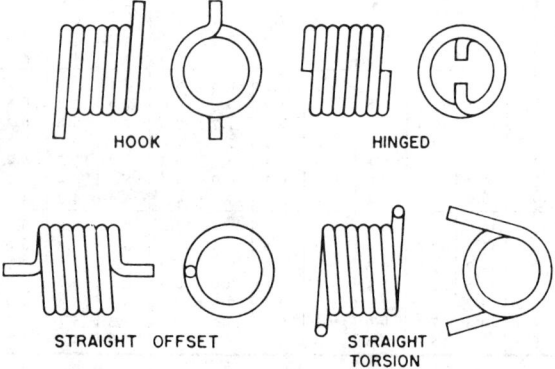

Fig. 20. The most commonly used types of ends for torsion springs.

## Table 12. Formulas for Torsion Springs*

| Feature | Springs made from round wire | Springs made from square wire | Feature | Springs made from round wire | Springs made from square wire |
|---|---|---|---|---|---|
| | Formula† | | | Formula† | |
| $d =$ Wire diameter, Inches | $\sqrt[3]{\dfrac{10.18\,T}{S_b}}$ | $\sqrt[3]{\dfrac{6\,T}{S_b}}$ | $F° =$ Deflection | $\dfrac{392\,S_b\,N\,D}{E\,d}$ | $\dfrac{392\,S_b\,N\,D}{E\,d}$ |
| | $\sqrt[4]{\dfrac{4000\,T\,N\,D}{E\,F°}}$ | $\sqrt[4]{\dfrac{2375\,T\,N\,D}{E\,F°}}$ | | $\dfrac{4000\,T\,N\,D}{E\,d^4}$ | $\dfrac{2375\,T\,N\,D}{E\,d^4}$ |
| $S_b =$ Stress, bending pounds per square inch | $\dfrac{10.18\,T}{d^3}$ | $\dfrac{6\,T}{d^3}$ | $T =$ Torque Inch lbs. (Also $= P \times R$) | $.0982\,S_b\,d^3$ | $.1666\,S_b\,d^3$ |
| $N =$ Active Coils | $\dfrac{E\,d\,F°}{392\,N\,D}$ | $\dfrac{E\,d\,F°}{392\,N\,D}$ | $ID_b =$ Inside Diameter After Deflection, Inches | $\dfrac{E\,d^4\,F°}{4000\,N\,D}$ | $\dfrac{E\,d^4\,F°}{2375\,N\,D}$ |
| | $\dfrac{E\,d\,F°}{392\,S_b\,D}$ | $\dfrac{E\,d\,F°}{392\,S_b\,D}$ | | $\dfrac{N\,(ID\text{ free})}{N+\frac{F°}{360}}$ | $\dfrac{N\,(ID\text{ free})}{N+\frac{F°}{360}}$ |
| | $\dfrac{E\,d^4\,F°}{4000\,T\,D}$ | $\dfrac{E\,d^4\,F°}{2375\,T\,D}$ | .......... | .......... | .......... |

\* The symbol notation is given on page 494.
† Where two formulas are given for one feature, the designer should use the one found to be appropriate in the given design instance. The end result from either of any two formulas is the same.

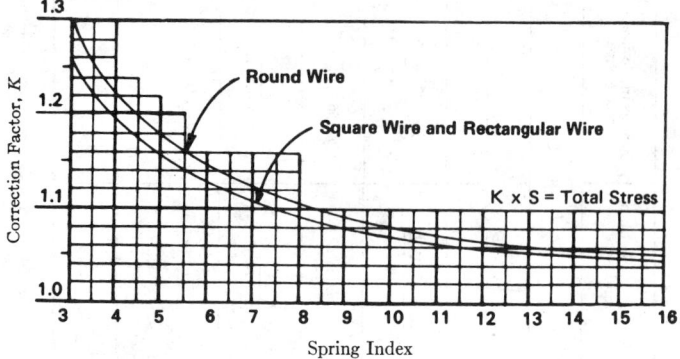

Fig. 21.　Torsion spring stress correction for curvature.

tance in centimeters or millimeters, but to use the design formulas, all values must be converted to pounds and inches.　Design formulas for torque are based on the tangent to the arc of rotation and presume that a rod is used to support the spring. The stress in bending caused by the moment $P \times R$ is identical in magnitude to the torque $T$, provided a rod is used.

Theoretically, it makes no difference how or where the load is applied to the arms of torsion springs.　Thus, in Fig. 22, the loads shown multiplied by their respective distances produce the same torque; i.e., $20 \times 0.5 = 10$ pounds-inches; $10 \times 1 = 10$ pounds-inches; and $5 \times 2 = 10$ pounds-inches.　To further simplify the understanding of torsion spring torque, observe in both Fig. 23 and Fig. 24 that although the turning force is in a circular arc the torque is not equal to $P$ times the radius.　The torque in both cases equals $P \times R$ because the spring rests against the support rod at point $a$.

Torsion spring designs require more effort than other kinds because consideration has to be given to more details such as the proper size of a supporting rod, reduction of the inside diameter, increase in length, deflection of arms, allowance for friction, and method of testing.

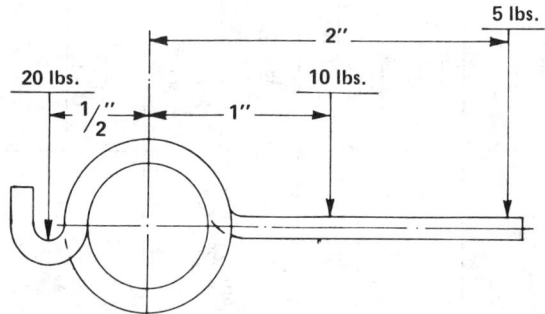

Fig. 22.　Right-hand torsion spring.

*Design example:* What music wire diameter and how many coils are required for the torsion spring shown in Fig. 25, which is to withstand at least 1000 cycles? Also, determine the corrected stress and the reduced inside diameter after deflection.

*Method 1, using table:* From Table 13, page 532, locate the ½ inch inside diameter for the spring in the left-hand column. Move to the right and then vertically to locate a torque value nearest to the required 10 pounds-inches, which is 10.07 pounds-inches. At the top of the same column, the music wire diameter is found, which is Number 31 gauge (0.085 inch). At the bottom of the same column the deflection for one coil is found, which is 15.81 degrees. As a 90-degree deflection is required, the number of coils needed is $\dfrac{90}{15.81}$ = 5.69 (say 5¾ coils).

The spring index $\dfrac{D}{d} = \dfrac{0.500 + 0.085}{0.085}$ = 6.88 and thus the curvature correction factor $K$ from Fig. 24 = 1.13. Therefore the corrected stress equals 167,000 × 1.13 = 188,700 pounds per square inch which is below the Light Service curve (Fig. 7) and therefore should provide a fatigue life of over 1,000 cycles. The reduced inside diameter due to deflection is found from the formula in Table 12:

$$ID_1 = \dfrac{N(ID\ free)}{N + \dfrac{F}{360}} = \dfrac{5.75 \times 0.500}{5.75 + \dfrac{90}{360}} = 0.479\ in.$$

This reduced diameter easily clears a suggested ⅞₆ inch diameter supporting rod: 0.479 − 0.4375 = 0.041 inch clearance, and this also allows for the standard tolerance. The overall length of the spring equals the total number of coils plus one, times the wire diameter. Thus, 6¾ × 0.085 = 0.574 inch. If a small space of about ¹⁄₆₄ in. is allowed between the coils to eliminate coil friction, an overall length of 2¹⁄₃₂ inch results.

Although this completes the design calculations, other tolerances should be applied in accordance with the Torsion Spring Tolerance Tables shown at the end of this section.

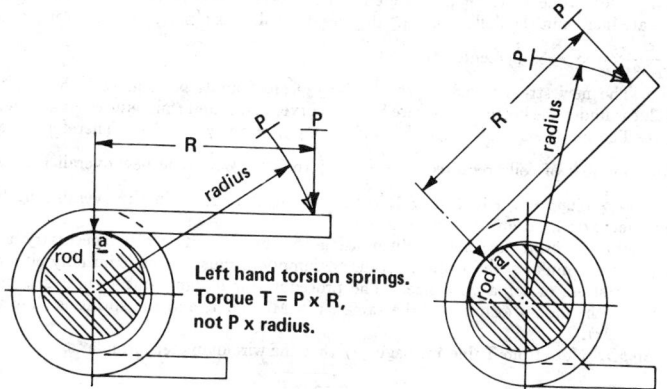

Fig. 23. (*Left*) Left-hand torsion spring. The torque is $T = P \times R$, not $P \times$ radius, because the spring is resting against the support rod at point *a*.

Fig. 24. (*Right*) Left-hand torsion spring. As with the spring in Fig. 23, the torque is $T = P \times R$, not $P \times$ radius, because the support point is at *a*.

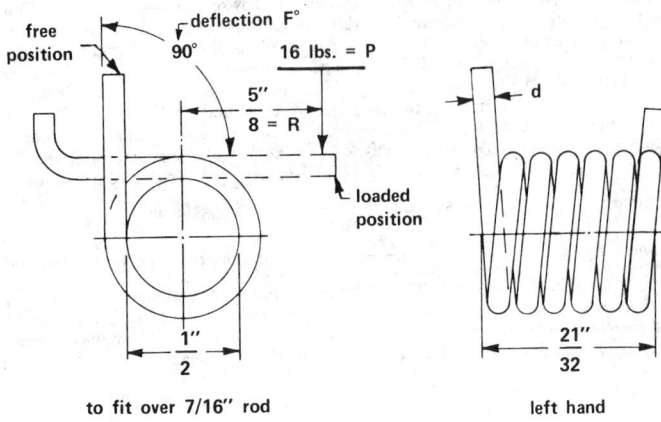

Fig. 25.  Torsion spring design example.  The spring is to be assembled
on a 7/16-inch support rod.

*Longer fatigue life:*  If a longer fatigue life is desired, use a slightly larger wire
diameter.  Usually the next larger gage size is satisfactory.  The larger wire will
reduce the stress and still exert the same torque, but will require more coils and a
longer overall length.

*Percentage method for calculating longer life:*  The spring design can be easily
adjusted for longer life as follows:

1. Select the next larger gage size, which is Number 32 (0.090 inch) from Table
13.  The torque is 11.88 pound-inches, the design stress is 166,000 pounds per
square inch, and the deflection 14.9 degrees per coil.  As a percentage the torque is
$\frac{10}{11.88} \times 100 = 84$ percent.

2. The new stress is $0.84 \times 166,000 = 139,440$ pounds per square inch.  This
value is under the bottom or Severe Service curve, Fig. 7, and thus assures longer life.

3. The new deflection per coil is $0.84 \times 14.97 = 12.57$ degrees.  Therefore, the
total number of coils required $= \frac{90}{12.57} = 7.16$ (say 7⅛).  The new overall length
$= 8\frac{1}{8} \times 0.090 = .73$ inch (say ¾ inch).  A slight increase in the overall length
and new arm location are thus necessary.

*Method 2, using formulas:*  When using this method, it is often necessary to
solve the formulas several times because assumptions must be made initially either
for the stress or for a wire size.  The procedure for design using formulas is as
follows (the design example is the same as in Method 1, and the spring is shown
in Fig. 25):

*Step 1:*  Note from Table 12, page 527 that the wire diameter formula is:

$$d = \sqrt[3]{\frac{10.18T}{S_b}}$$

*Step 2:*  Referring to Fig. 7, select a trial stress, say 150,000 pounds per square inch.

(Continued on page 535)

**Table 13. Torsion Spring Deflections***

**AMW Wire Gauge and Decimal Equivalent†**

| AMW Wire Gauge | 16 | 15 | 14 | 13 | 12 | 11 | 10 | 9 | 8 | 7 | 6 | 5 | 4 | 3 | 2 | 1 | Fractional | Decimal |
|---|---|---|---|---|---|---|---|---|---|---|---|---|---|---|---|---|---|---|
| Decimal Equivalent† | .037 | .035 | .033 | .031 | .029 | .026 | .024 | .022 | .020 | .018 | .016 | .014 | .013 | .012 | .011 | .010 | | |
| Design Stress, pounds per sq. in. (thousands) | 192 | 194 | 196 | 197 | 199 | 202 | 205 | 207 | 210 | 214 | 217 | 221 | 224 | 226 | 229 | 232 | | |
| Torque, pounds-inch | .9550 | .8168 | .6917 | .5763 | .4766 | .3486 | .2783 | .2164 | .1650 | .1226 | .0873 | .0596 | .0483 | .0383 | .0299 | .0228 | | |
| Deflection, degrees per coil | | | | | | | | | | | | | | | | | | |
| | | | | 7.896 | 8.343 | 9.137 | 9.818 | 10.56 | 11.51 | 12.72 | 14.15 | 16.05 | 17.29 | 18.64 | 20.33 | 22.35 | 1/16 | .0625 |
| | | | | 9.215 | 9.768 | 10.75 | 11.59 | 12.52 | 13.69 | 15.19 | 16.96 | 19.32 | 20.86 | 22.55 | 24.66 | 27.17 | 5/64 | .078125 |
| | 9.171 | 9.646 | 10.18 | 10.53 | 11.19 | 12.36 | 13.36 | 14.47 | 15.87 | 17.65 | 19.78 | 22.60 | 24.44 | 26.47 | 28.98 | 31.98 | 3/32 | .09375 |
| | 10.27 | 10.82 | 11.43 | 11.85 | 12.62 | 13.98 | 15.14 | 16.43 | 18.05 | 20.12 | 22.60 | 25.88 | 28.02 | 30.38 | 33.30 | 36.80 | 7/64 | .109375 |
| | 11.36 | 11.99 | 12.68 | 13.17 | 14.04 | 15.59 | 16.91 | 18.38 | 20.23 | 22.59 | 25.41 | 29.16 | 31.60 | 34.29 | 37.62 | 41.62 | 1/8 | .125 |
| | 12.46 | 13.16 | 13.94 | 14.49 | 15.47 | 17.20 | 18.69 | 20.33 | 22.41 | 25.06 | 28.23 | 32.43 | 35.17 | 38.20 | 41.94 | 46.44 | 9/64 | .140625 |
| | 13.56 | 14.33 | 15.19 | 15.81 | 16.89 | 18.82 | 20.46 | 22.29 | 24.59 | 27.53 | 31.04 | 35.71 | 38.75 | 42.11 | 46.27 | 51.25 | 5/32 | .15625 |
| | 15.75 | 16.67 | 17.70 | 18.45 | 19.74 | 22.04 | 24.01 | 26.19 | 28.95 | 32.47 | 36.67 | 42.27 | 45.91 | 49.93 | 54.91 | 60.89 | 3/16 | .1875 |
| | 17.94 | 19.01 | 20.21 | 21.09 | 22.59 | 25.27 | 27.55 | 30.10 | 33.31 | 37.40 | 42.31 | 48.82 | 53.06 | 57.75 | 63.56 | 70.52 | 7/32 | .21875 |
| | 20.13 | 21.35 | 22.72 | 23.73 | 25.44 | 28.49 | 31.10 | 34.01 | 37.67 | 42.34 | 47.94 | 55.38 | 60.22 | 65.57 | 72.20 | 80.15 | 1/4 | .250 |

**AMW Wire Gauge and Decimal Equivalent†**

| AMW Wire Gauge | 31 | 30 | 29 | 28 | 27 | 26 | 25 | 24 | 23 | 22 | 21 | 20 | 19 | 18 | 17 | Fractional | Decimal |
|---|---|---|---|---|---|---|---|---|---|---|---|---|---|---|---|---|---|
| Decimal Equivalent† | .085 | .080 | .075 | .071 | .067 | .063 | .059 | .055 | .051 | .049 | .047 | .045 | .043 | .041 | .039 | | |
| Design Stress, pounds per sq. in. (thousands) | 167 | 169 | 171 | 173 | 174 | 176 | 178 | 180 | 182 | 183 | 184 | 185 | 187 | 188 | 190 | | |
| Torque, pounds-inch | 10.07 | 8.497 | 7.084 | 6.080 | 5.139 | 4.322 | 3.590 | 2.941 | 2.371 | 2.114 | 1.876 | 1.655 | 1.460 | 1.272 | 1.107 | | |
| Deflection, degrees per coil | | | | | | | | | | | | | | | | | |
| | | | | | | | | | | | | | 8.957 | 9.320 | 9.771 | 7/64 | .109375 |
| | | | | | | | | | | 8.784 | 9.102 | 9.447 | 9.876 | 10.29 | 10.80 | 1/8 | .125 |
| | | | | | | | 8.141 | 8.654 | 9.244 | 9.572 | 9.929 | 10.32 | 10.79 | 11.26 | 11.83 | 9/64 | .140625 |
| | | | | | 7.975 | 8.279 | 8.778 | 9.345 | 9.997 | 10.36 | 10.76 | 11.18 | 11.71 | 12.23 | 12.86 | 5/32 | .15625 |
| | 7.364 | 7.772 | 8.232 | 8.663 | 9.091 | 9.459 | 10.05 | 10.73 | 11.50 | 11.94 | 12.41 | 12.92 | 13.55 | 14.16 | 14.92 | 3/16 | .1875 |
| | 8.208 | 8.680 | 9.212 | 9.711 | 10.21 | 10.64 | 11.33 | 12.11 | 13.01 | 13.52 | 14.06 | 14.66 | 15.39 | 16.10 | 16.97 | 7/32 | .21875 |
| | 9.053 | 9.588 | 10.19 | 10.76 | 11.32 | 11.82 | 12.60 | 13.49 | 14.52 | 15.09 | 15.72 | 16.39 | 17.22 | 18.04 | 19.03 | 1/4 | .250 |

* For an example in the use of the table, see page 529. Note: Intermediate values may be interpolated within reasonable accuracy.

† For sizes up to 13 gauge, the table values are for music wire with a modulus E of 29,500,000 psi; for sizes from 14 to 26 gauge, the table values are for music wire with a modulus of 29,000,000 psi; and for sizes from 27 to 31 gauge the values are for music wire with a modulus of 28,000,000 psi.

## Table 13 (Continued). Torsion Spring Deflections

**AMW Wire Gauge and Decimal Equivalent***

| Inside Diam. Fractional | Inside Diam. Decimal | 8 / .020 | 9 / .022 | 10 / .024 | 11 / .026 | 12 / .029 | 13 / .031 | 14 / .033 | 15 / .035 | 16 / .037 | 17 / .039 | 18 / .041 | 19 / .043 | 20 / .045 | 21 / .047 | 22 / .049 | 23 / .051 |
|---|---|---|---|---|---|---|---|---|---|---|---|---|---|---|---|---|---|
| Design Stress, pounds per sq in. (thousands) | | 210 | 207 | 205 | 202 | 199 | 197 | 196 | 194 | 192 | 190 | 188 | 187 | 185 | 184 | 183 | 182 |
| Torque, pounds-inch | | .1650 | .2164 | .2783 | .3486 | .4766 | .5763 | .6917 | .8168 | .9559 | 1.107 | 1.272 | 1.460 | 1.655 | 1.876 | 2.114 | 2.371 |
| **Deflection, degrees per coil** | | | | | | | | | | | | | | | | | |
| 9/32 | .28125 | 42.03 | 37.92 | 34.65 | 31.72 | 28.29 | 26.37 | 25.23 | 23.69 | 22.32 | 21.09 | 19.97 | 19.06 | 18.13 | 17.37 | 16.67 | 16.03 |
| 5/16 | .3125 | 46.39 | 41.82 | 38.19 | 34.95 | 31.14 | 29.01 | 27.74 | 26.04 | 24.51 | 23.15 | 21.91 | 20.90 | 19.87 | 19.02 | 18.25 | 17.53 |
| 11/32 | .34375 | 50.75 | 45.73 | 41.74 | 38.17 | 33.99 | 31.65 | 30.25 | 28.38 | 26.71 | 25.21 | 23.85 | 22.73 | 21.60 | 20.68 | 19.83 | 19.04 |
| 3/8 | .375 | 55.11 | 49.64 | 45.29 | 41.40 | 36.84 | 34.28 | 32.76 | 30.72 | 28.90 | 27.26 | 25.78 | 24.57 | 23.34 | 22.33 | 21.40 | 20.55 |
| 13/32 | .40625 | 59.47 | 53.54 | 48.85 | 44.63 | 39.69 | 36.92 | 35.27 | 33.06 | 31.09 | 29.32 | 27.72 | 26.41 | 25.08 | 23.99 | 22.98 | 22.06 |
| 7/16 | .4375 | 63.83 | 57.45 | 52.38 | 47.85 | 42.54 | 39.56 | 37.77 | 35.40 | 33.28 | 31.38 | 29.66 | 28.25 | 26.81 | 25.64 | 24.56 | 23.56 |
| 15/32 | .46875 | 68.19 | 61.36 | 55.93 | 51.00 | 45.39 | 42.20 | 40.28 | 37.74 | 35.47 | 33.44 | 31.59 | 30.08 | 28.55 | 27.29 | 26.14 | 25.07 |
| 1/2 | .500 | 72.55 | 65.27 | 59.48 | 54.30 | 48.24 | 44.84 | 42.79 | 40.08 | 37.67 | 35.49 | 33.53 | 31.92 | 30.29 | 28.95 | 27.71 | 26.58 |

**AMW Wire Gauge and Decimal Equivalent***

| Inside Diam. Fractional | Inside Diam. Decimal | 24 / .055 | 25 / .059 | 26 / .063 | 27 / .067 | 28 / .071 | 29 / .075 | 30 / .080 | 31 / .085 | 32 / .090 | 33 / .095 | 34 / .100 | 35 / .106 | 36 / .112 | 37 / .118 | 1/8 / .125 |
|---|---|---|---|---|---|---|---|---|---|---|---|---|---|---|---|---|
| Design Stress, pounds per sq in. (thousands) | | 180 | 178 | 176 | 174 | 173 | 171 | 169 | 167 | 166 | 164 | 163 | 161 | 160 | 158 | 156 |
| Torque, pounds-inch | | 2.941 | 3.590 | 4.322 | 5.139 | 6.080 | 7.084 | 8.497 | 10.07 | 11.88 | 13.81 | 16.00 | 18.83 | 22.07 | 25.49 | 29.92 |
| **Deflection, degrees per coil** | | | | | | | | | | | | | | | | |
| 9/32 | .28125 | 14.88 | 13.88 | 13.00 | 12.44 | 11.81 | 11.17 | 10.50 | 9.897 | 9.418 | 8.934 | 8.547 | 8.090 | 7.727 | 7.353 | 6.973 |
| 5/16 | .3125 | 16.26 | 15.15 | 14.18 | 13.56 | 12.85 | 12.15 | 11.40 | 10.74 | 10.21 | 9.676 | 9.248 | 8.743 | 8.341 | 7.929 | 7.510 |
| 11/32 | .34375 | 17.64 | 16.42 | 15.36 | 14.67 | 13.90 | 13.13 | 12.31 | 11.59 | 11.00 | 10.42 | 9.948 | 9.396 | 8.955 | 8.504 | 8.046 |
| 3/8 | .375 | 19.02 | 17.69 | 16.54 | 15.79 | 14.95 | 14.11 | 13.22 | 12.43 | 11.80 | 11.16 | 10.65 | 10.05 | 9.569 | 9.080 | 8.583 |
| 13/32 | .40625 | 20.40 | 18.97 | 17.72 | 16.90 | 15.99 | 15.09 | 14.13 | 13.28 | 12.59 | 11.90 | 11.35 | 10.70 | 10.18 | 9.655 | 9.119 |
| 7/16 | .4375 | 21.79 | 20.25 | 18.90 | 18.02 | 17.04 | 16.07 | 15.04 | 14.12 | 13.38 | 12.64 | 12.05 | 11.35 | 10.80 | 10.23 | 9.655 |
| 15/32 | .46875 | 23.17 | 21.52 | 20.08 | 19.14 | 18.09 | 17.05 | 15.94 | 14.96 | 14.17 | 13.39 | 12.75 | 12.01 | 11.41 | 10.81 | 10.19 |
| 1/2 | .500 | 24.55 | 22.80 | 21.26 | 20.25 | 19.14 | 18.03 | 16.85 | 15.81 | 14.97 | 14.13 | 13.45 | 12.66 | 12.03 | 11.38 | 10.73 |

* For sizes up to 13 gauge, the table values are for music wire with a modulus $E$ of 29,500,000 psi; for sizes from 14 to 26 gauge, the table values are for music wire with a modulus of 29,000,000 psi; and for sizes from 27 gauge to 1/8 inch diameter, the values are for music wire with a modulus of 28,500,000 psi.

## Table 13 (Continued). Torsion Spring Deflections

### AMW Wire Gauge and Decimal Equivalent*

| Inside Diam. Fractional | Inside Diam. Decimal | 16 .037 | 17 .039 | 18 .041 | 19 .043 | 20 .045 | 21 .047 | 22 .049 | 23 .051 | 24 .055 | 25 .059 | 26 .063 | 27 .067 | 28 .071 | 29 .075 | 30 .080 |
|---|---|---|---|---|---|---|---|---|---|---|---|---|---|---|---|---|
| Design Stress, pounds per sq in. (thousands) | | 192 | 190 | 188 | 187 | 185 | 184 | 183 | 182 | 180 | 178 | 176 | 174 | 173 | 171 | 169 |
| Torque, pounds-inch | | .9550 | 1.107 | 1.272 | 1.460 | 1.655 | 1.876 | 2.114 | 2.371 | 2.941 | 3.590 | 4.322 | 5.139 | 6.080 | 7.084 | 8.497 |
| **Deflection, degrees per coil** | | | | | | | | | | | | | | | | |
| 17/32 | .53125 | 39.86 | 37.55 | 35.47 | 33.76 | 32.02 | 30.60 | 29.29 | 28.09 | 25.93 | 24.07 | 22.44 | 21.37 | 20.18 | 19.01 | 17.76 |
| 9/16 | .5625 | 42.05 | 39.61 | 37.40 | 35.59 | 33.76 | 32.25 | 30.87 | 29.59 | 27.32 | 25.35 | 23.62 | 22.49 | 21.23 | 19.99 | 18.67 |
| 19/32 | .59375 | 44.24 | 41.67 | 39.34 | 37.43 | 35.50 | 33.91 | 32.45 | 31.10 | 28.70 | 26.62 | 24.80 | 23.60 | 22.28 | 20.97 | 19.58 |
| 5/8 | .625 | 46.43 | 43.73 | 41.28 | 39.27 | 37.23 | 35.56 | 34.02 | 32.61 | 30.08 | 27.89 | 25.98 | 24.72 | 23.33 | 21.95 | 20.48 |
| 21/32 | .65625 | 48.63 | 45.78 | 43.22 | 41.10 | 38.97 | 37.22 | 35.60 | 34.12 | 31.46 | 29.17 | 27.16 | 25.83 | 24.37 | 22.93 | 21.39 |
| 11/16 | .6875 | 50.82 | 47.84 | 45.15 | 42.94 | 40.71 | 38.87 | 37.18 | 35.60 | 32.85 | 30.44 | 28.34 | 26.95 | 25.42 | 23.91 | 22.30 |
| 23/32 | .71875 | 53.01 | 49.90 | 47.09 | 44.78 | 42.44 | 40.52 | 38.76 | 37.13 | 34.23 | 31.72 | 29.52 | 28.07 | 26.47 | 24.89 | 23.21 |
| 3/4 | .750 | 55.20 | 51.96 | 49.03 | 46.62 | 44.18 | 42.18 | 40.33 | 38.64 | 35.61 | 32.99 | 30.70 | 29.18 | 27.52 | 25.87 | 24.12 |

### Wire Gauge or Size and Decimal Equivalent††

| Inside Diam. Fractional | Inside Diam. Decimal | 31 .085 | 32 .090 | 33 .095 | 34 .100 | 35 .106 | 36 .112 | 37 .118 | 1/8 .125 | 10 .135 | 9 .1483 | 5/32 .1563 | 8 .162 | 7 .177 | 3/16 .1875 | 6 .192 | 5 .207 |
|---|---|---|---|---|---|---|---|---|---|---|---|---|---|---|---|---|---|
| Design Stress, pounds per sq in. (thousands) | | 167 | 166 | 164 | 163 | 161 | 160 | 158 | 156 | 161 | 158 | 156 | 154 | 150 | 149 | 146 | 143 |
| Torque, pounds-inch | | 10.07 | 11.88 | 13.81 | 16.00 | 18.83 | 22.07 | 25.49 | 29.92 | 38.90 | 50.60 | 58.44 | 64.30 | 81.68 | 96.45 | 101.5 | 124.6 |
| **Deflection, degrees per coil** | | | | | | | | | | | | | | | | | |
| 17/32 | .53125 | 16.65 | 15.76 | 14.87 | 14.15 | 13.31 | 12.64 | 11.96 | 10.93 | 10.42 | 9.958 | 9.441 | 9.064 | 8.256 | 7.856 | 7.565 | 7.015 |
| 9/16 | .5625 | 17.50 | 16.55 | 15.61 | 14.85 | 13.97 | 13.25 | 12.53 | 11.44 | 10.87 | 10.42 | 9.870 | 9.473 | 8.620 | 8.198 | 7.891 | 7.312 |
| 19/32 | .59375 | 18.34 | 17.35 | 16.35 | 15.55 | 14.62 | 13.87 | 13.11 | 11.95 | 11.33 | 10.87 | 10.30 | 9.882 | 8.984 | 8.539 | 8.218 | 7.669 |
| 5/8 | .625 | 19.19 | 18.14 | 17.10 | 16.25 | 15.27 | 14.48 | 13.68 | 12.47 | 11.79 | 11.33 | 10.73 | 10.29 | 9.348 | 8.881 | 8.545 | 7.906 |
| 21/32 | .65625 | 20.03 | 18.93 | 17.84 | 16.95 | 15.92 | 15.10 | 14.26 | 12.98 | 12.25 | 11.79 | 11.16 | 10.70 | 9.713 | 9.222 | 8.872 | 8.202 |
| 11/16 | .6875 | 20.88 | 19.72 | 18.58 | 17.65 | 16.58 | 15.71 | 14.83 | 13.49 | 12.71 | 12.25 | 11.59 | 11.11 | 10.08 | 9.564 | 9.199 | 8.499 |
| 23/32 | .71875 | 21.72 | 20.52 | 19.32 | 18.36 | 17.23 | 16.32 | 15.41 | 14.00 | 13.16 | 12.71 | 12.02 | 11.52 | 10.44 | 9.905 | 9.526 | 8.796 |
| 3/4 | .750 | 22.56 | 21.31 | 20.06 | 19.06 | 17.88 | 16.94 | 15.99 | 14.52 | 13.61 | 13.16 | 12.44 | 11.92 | 10.81 | 10.25 | 9.852 | 9.093 |

* For sizes up to 26 gauge, the table values are for music wire with a modulus $E$ of 29,000,000 psi; for sizes from 27 to 1/8 inch diameter the table values are for music wire with a modulus of 28,500,000 psi; for sizes from 1 to 5 gauge are for oil tempered MB with a modulus of 28,500,000 psi.

† Gauges 31 through 37 are AMW gauges. Gauges 10 through 5 are Washburn and Moen.

## Table 13 (Concluded). Torsion Spring Deflections

### AMW Wire Gauge and Decimal Equivalent*

| Inside Diam. Fractional | Inside Diam. Decimal | 24 | 25 | 26 | 27 | 28 | 29 | 30 | 31 | 32 | 33 | 34 | 35 | 36 | 37 | 1/8 |
|---|---|---|---|---|---|---|---|---|---|---|---|---|---|---|---|---|
| | | .055 | .059 | .063 | .067 | .071 | .075 | .080 | .085 | .090 | .095 | .100 | .106 | .112 | .118 | .125 |
| Design Stress, pounds per sq. in. (thousands) | | 180 | 178 | 176 | 174 | 173 | 171 | 169 | 167 | 166 | 164 | 163 | 161 | 160 | 158 | 156 |
| Torque, pounds-inch | | 2.941 | 3.590 | 4.322 | 5.139 | 6.080 | 7.084 | 8.497 | 10.07 | 11.88 | 13.81 | 16.00 | 18.83 | 22.07 | 25.49 | 29.92 |
| **Deflection, degrees per coil** | | | | | | | | | | | | | | | | |
| 13/16 | .8125 | 38.38 | 35.54 | 33.06 | 31.42 | 29.61 | 27.83 | 25.93 | 24.25 | 22.90 | 21.55 | 20.46 | 19.19 | 18.17 | 17.14 | 16.09 |
| 7/8 | .875 | 41.14 | 38.09 | 35.42 | 33.65 | 31.70 | 29.79 | 27.75 | 25.94 | 24.58 | 23.03 | 21.86 | 20.49 | 19.39 | 18.29 | 17.17 |
| 15/16 | .9375 | 43.91 | 40.64 | 37.78 | 35.88 | 33.80 | 31.75 | 29.56 | 27.63 | 26.07 | 24.52 | 23.26 | 21.80 | 20.62 | 19.44 | 18.24 |
| 1 | 1.000 | 46.67 | 43.19 | 40.14 | 38.11 | 35.89 | 33.71 | 31.38 | 29.32 | 27.65 | 26.00 | 24.66 | 23.11 | 21.85 | 20.59 | 19.31 |
| 1 1/16 | 1.0625 | 49.44 | 45.74 | 42.50 | 40.35 | 37.99 | 35.67 | 33.20 | 31.01 | 29.24 | 27.48 | 26.06 | 24.41 | 23.08 | 21.74 | 20.38 |
| 1 1/8 | 1.125 | 52.20 | 48.28 | 44.86 | 42.58 | 40.08 | 37.63 | 35.01 | 32.70 | 30.82 | 28.97 | 27.46 | 25.72 | 24.31 | 22.89 | 21.46 |
| 1 3/16 | 1.1875 | 54.97 | 50.83 | 47.22 | 44.81 | 42.18 | 39.59 | 36.83 | 34.39 | 32.41 | 30.45 | 28.86 | 27.02 | 25.53 | 24.04 | 22.53 |
| 1 1/4 | 1.250 | 57.73 | 53.38 | 49.58 | 47.04 | 44.27 | 41.55 | 38.64 | 36.08 | 33.99 | 31.94 | 30.27 | 28.33 | 26.76 | 25.19 | 23.60 |

### Washburn and Moen Gauge or Size and Decimal Equivalent*

| Inside Diam. Fractional | Inside Diam. Decimal | 10 | 9 | 5/32 | 8 | 7 | 3/16 | 6 | 5 | 7/32 | 4 | 3 | 1/4 | 9/32 | 5/16 | 11/32 | 3/8 |
|---|---|---|---|---|---|---|---|---|---|---|---|---|---|---|---|---|---|
| | | .135 | .1483 | .1563 | .162 | .177 | .1875 | .192 | .207 | .2188 | .2253 | .2437 | .250 | .2813 | .3125 | .3438 | .375 |
| Design Stress, pounds per sq in. (thousands) | | 161 | 158 | 156 | 154 | 150 | 149 | 146 | 143 | 142 | 141 | 140 | 139 | 138 | 137 | 136 | 135 |
| Torque, pounds-inch | | 38.90 | 50.60 | 58.44 | 64.30 | 81.68 | 96.45 | 101.5 | 124.6 | 146.0 | 158.3 | 199.0 | 213.3 | 301.5 | 410.6 | 542.5 | 700.0 |
| **Deflection, degrees per coil** | | | | | | | | | | | | | | | | | |
| 13/16 | .8125 | 15.54 | 14.08 | 13.30 | 12.74 | 11.53 | 10.93 | 10.51 | 9.687 | 9.208 | 8.933 | 8.346 | 8.125 | 7.382 | 6.784 | 6.292 | 5.880 |
| 7/8 | .875 | 16.57 | 15.00 | 14.16 | 13.56 | 12.26 | 11.61 | 11.16 | 10.28 | 9.766 | 9.471 | 8.840 | 8.603 | 7.863 | 7.161 | 6.632 | 6.189 |
| 15/16 | .9375 | 17.59 | 15.92 | 15.02 | 14.38 | 12.99 | 12.30 | 11.81 | 10.87 | 10.32 | 10.01 | 9.333 | 9.081 | 8.225 | 7.537 | 6.972 | 6.499 |
| 1 | 1.000 | 18.62 | 16.83 | 15.88 | 15.19 | 13.72 | 12.98 | 12.47 | 11.47 | 10.88 | 10.55 | 9.827 | 9.559 | 8.647 | 7.914 | 7.312 | 6.808 |
| 1 1/16 | 1.0625 | 19.64 | 17.74 | 16.74 | 16.01 | 14.45 | 13.66 | 13.12 | 12.06 | 11.44 | 11.09 | 10.32 | 10.04 | 9.069 | 8.291 | 7.652 | 7.118 |
| 1 1/8 | 1.125 | 20.67 | 18.66 | 17.59 | 16.83 | 15.18 | 14.35 | 13.77 | 12.66 | 12.00 | 11.62 | 10.81 | 10.52 | 9.491 | 8.668 | 7.993 | 7.427 |
| 1 3/16 | 1.1875 | 21.69 | 19.57 | 18.45 | 17.64 | 15.90 | 15.03 | 14.43 | 13.25 | 12.56 | 12.16 | 11.31 | 10.99 | 9.912 | 9.045 | 8.333 | 7.737 |
| 1 1/4 | 1.250 | 22.72 | 20.49 | 19.31 | 18.46 | 16.63 | 15.71 | 15.08 | 13.84 | 13.11 | 12.70 | 11.80 | 11.47 | 10.33 | 9.422 | 8.673 | 8.046 |

* For sizes up to 26 gauge, the table values are for music wire with a modulus E of 29,000,000 psi; for sizes from 27 to 1/8 inch diameter the table values are for music wire with a modulus of 28,500,000 psi; for sizes from 10 gauge to 3/8 inch diameter, the values are for oil tempered MB with a modulus of 28,500,000 psi.

*Step 3:* Apply the trial stress, and the 10 pound-inches torque value in the wire diameter formula:

$$d = \sqrt[3]{\frac{10.18T}{S_b}} = \sqrt[3]{\frac{10.18 \times 10}{150,000}} = \sqrt[3]{0.000679} = 0.0879 \text{ inch}$$

The nearest gauge sizes are 0.085 and 0.090 inch diameter. *Note:* Table 20, page 541, can be used to avoid solving the cube root.

*Step 4:* Select 0.085 inch wire diameter and solve the equation for the actual stress:

$$S_b = \frac{10.18T}{d^3} = \frac{10.18 \times 10}{0.085^3} = 165,764 \text{ pounds per square inch.}$$

*Step 5:* Calculate the number of coils from the equation, Table 12:

$$N = \frac{E \, d \, F°}{392 S_b D}$$

$$= \frac{28,500,000 \times 0.085 \times 90}{392 \times 165,764 \times 0.585} = 5.73 \text{ (say } 5\tfrac{3}{4})$$

*Step 6:* Calculate the total stress. The spring index is 6.88, and the correction factor $K$ is 1.13, therefore total stress = $165,764 \times 1.13 = 187,313$ pounds per square inch. *Note:* The corrected stress should not be used in any of the formulas as it does not determine the torque or the deflection.

**Table of Torsion Spring Characteristics.** — Table 13 shows design characteristics for the most commonly used torsion springs made from wire of standard gauge sizes. The deflection for one coil at a specified torque and stress is shown in the body of the table. The figures are based on music wire (ASTM A228) and oil-tempered MB grade (ASTM A229), and can be used for several other materials which have a similar values for the modulus of elasticity $E$. However, the design stress may be too high or too low, and the design stress, torque, and deflection per coil should each be multiplied by the appropriate correction factor in Table 14 when using any of the materials given in that table.

**Torsion Spring Design Recommendations.** — The following recommendations should be taken into account when designing torsion springs:

*Hand:* The hand or direction of coiling should be specified and the spring designed so deflection causes the spring to wind up and to have more coils. This increase

Table 14. Correction Factors for Other Materials*

| Material | Factor | Material | Factor |
|---|---|---|---|
| Hard Drawn MB | .75 | Stainless 316 | |
| Chrome-Vanadium | 1.10 | Up to ⅛ inch diameter | .75 |
| Chrome-Silicon | 1.20 | Over ⅛ to ¼ inch diameter | .65 |
| | | Over ¼ inch diameter | .65 |
| Stainless 302 and 304 | | | |
| Up to ⅛ inch diameter | .85 | Stainless 17-7 PH | |
| Over ⅛ to ¼ inch diameter | .75 | Up to ⅛ inch diameter | 1.00 |
| Over ¼ inch diameter | .65 | Over ⅛ to 3⁄16 inch diameter | 1.07 |
| Stainless 420 | .80 | Over 3⁄16 inch diameter | 1.12 |
| Stainless 431 | .85 | | .... |

* For use with values in Table 13. *Note:* The figures in Table 13 are for music wire (ASTM A228) and oil tempered MB grade (ASTM A229) and can be used for several other materials which have a similar modulus of elasticity $E$. However, the design stress may be too high or too low, and therefore the design stress, torque, and deflection per coil should each be multiplied by the appropriate correction factor when using any of the materials given in this table (Table 14).

in coils and overall length should be allowed for during design. Deflecting the spring in an unwinding direction causes higher stresses and may cause early failure. When a spring is sighted down the longitudinal axis, it is "right hand" when the direction of the wire into the spring takes a clockwise direction or if the angle of the coils follows an angle similar to the threads of a standard bolt or screw, otherwise it is "left hand." A spring must be coiled right-handed to engage the threads of a standard machine screw.

*Rods:* Torsion springs should be supported by a rod running through the center whenever possible. If unsupported, or if held by clamps or lugs, the spring will buckle and the torque will be reduced or unusual stresses may occur.

*Diameter Reduction:* The inside diameter reduces during deflection. This should be computed and proper clearance provided over the supporting rod. Also, allowances should be considered for normal spring diameter tolerances.

*Winding:* The coils of a spring may be closely or loosely wound, but they seldom should be wound with the coils pressed tightly together. Tightly wound springs with initial tension on the coils do not deflect uniformly and are difficult to test accurately. A small space between the coils of about 20 to 25 percent of the wire thickness is desirable. Square and rectangular wire sections should be avoided whenever possible as they are difficult to wind, expensive, and are not always readily available.

*Arm Length:* All the wire in a torsion spring is active between the points where the loads are applied. Deflection of long extended arms can be calculated by allowing one third of the arm length, from the point of load contact to the body of the spring, to be converted into coils. However, if the length of arm is equal to or less than one-half the length of one coil, it can be safely neglected in most applications.

*Total Coils:* Torsion springs having less than three coils frequently buckle and are difficult to test accurately. When thirty or more coils are used, light loads will not deflect all the coils simultaneously due to friction with the supporting rod. To facilitate manufacturing it is usually preferable to specify the total number of coils to the nearest fraction in eighths or quarters such as 5⅛, 5¼, 5½, etc.

*Double Torsion:* This design consists of one left-hand-wound series of coils and one series of right-hand-wound coils connected at the center. They are difficult to manufacture and are expensive. It often is better to use two separate springs. For torque and stress calculations, each series is calculated separately as individual springs; then the torque values are added together, but the deflections are not added.

*Bends:* Arms should be kept as straight as possible. Bends are difficult to produce and often are made by secondary operations. They are therefore expensive. Sharp bends raise stresses that cause early failure. Bend radii should be as large as practicable. Hooks tend to open during deflection; their stresses can be calculated by the same procedure as that for tension springs.

*Spring Index:* The spring index must be used with caution. In design formulas it is $\frac{D}{d}$. For shop measurement it is $\frac{O.D.}{d}$. For arbor design it is $\frac{I.D.}{d}$. Conversions are easily performed by either adding or subtracting 1 from $\frac{D}{d}$.

*Proportions:* A spring index between 4 and 14 provides the best proportions. Larger ratios may require more than average tolerances. Ratios of 3 or less, often cannot be coiled on automatic spring coiling machines because of arbor breakage. Also, springs with smaller or larger spring indexes often do not give the results as calculated from the design formulas.

**Torsion Spring Tolerances.** — Torsion springs are coiled in a different manner from other types of coiled springs and therefore different tolerances apply. The

Table 15.  Torsion Spring Tolerance for Angular
Relationship of Ends

| Number of Coils (N) | Spring Index | | | | | | | | |
|---|---|---|---|---|---|---|---|---|---|
| | 4 | 6 | 8 | 10 | 12 | 14 | 16 | 18 | 20 |
| | Free Angle Tolerance, ± degrees | | | | | | | | |
| 1 | 2 | 3 | 3.5 | 4 | 4.5 | 5 | 5.5 | 5.5 | 6 |
| 2 | 4 | 5 | 6 | 7 | 8 | 8.5 | 9 | 9.5 | 10 |
| 3 | 5.5 | 7 | 8 | 9.5 | 10.5 | 11 | 12 | 13 | 14 |
| 4 | 7 | 9 | 10 | 12 | 14 | 15 | 16 | 16.5 | 17 |
| 5 | 8 | 10 | 12 | 14 | 16 | 18 | 20 | 20.5 | 21 |
| 6 | 9.5 | 12 | 14.5 | 16 | 19 | 20.5 | 21 | 22.5 | 24 |
| 8 | 12 | 15 | 18 | 20.5 | 23 | 25 | 27 | 28 | 29 |
| 10 | 14 | 19 | 21 | 24 | 27 | 29 | 31.5 | 32.5 | 34 |
| 15 | 20 | 25 | 28 | 31 | 34 | 36 | 38 | 40 | 42 |
| 20 | 25 | 30 | 34 | 37 | 41 | 44 | 47 | 49 | 51 |
| 25 | 29 | 35 | 40 | 44 | 48 | 52 | 56 | 60 | 63 |
| 30 | 32 | 38 | 44 | 50 | 55 | 60 | 65 | 68 | 70 |
| 50 | 45 | 55 | 63 | 70 | 77 | 84 | 90 | 95 | 100 |

Table 16.  Torsion Spring Coil Diameter Tolerances

| Wire Diameter, Inch | Spring Index | | | | | | |
|---|---|---|---|---|---|---|---|
| | 4 | 6 | 8 | 10 | 12 | 14 | 16 |
| | Coil Diameter Tolerance, ± inch | | | | | | |
| 0.015 | 0.002 | 0.002 | 0.002 | 0.002 | 0.003 | 0.003 | 0.004 |
| 0.023 | 0.002 | 0.002 | 0.002 | 0.003 | 0.004 | 0.005 | 0.006 |
| 0.035 | 0.002 | 0.002 | 0.003 | 0.004 | 0.006 | 0.007 | 0.009 |
| 0.051 | 0.002 | 0.003 | 0.005 | 0.007 | 0.008 | 0.010 | 0.012 |
| 0.076 | 0.003 | 0.005 | 0.007 | 0.009 | 0.012 | 0.015 | 0.018 |
| 0.114 | 0.004 | 0.007 | 0.010 | 0.013 | 0.018 | 0.022 | 0.028 |
| 0.172 | 0.006 | 0.010 | 0.013 | 0.020 | 0.027 | 0.034 | 0.042 |
| 0.250 | 0.008 | 0.014 | 0.022 | 0.030 | 0.040 | 0.050 | 0.060 |

Table 17.  Torsion Spring Tolerance on
Number of Coils

| Number of Coils | Tolerance | Number of Coils | Tolerance |
|---|---|---|---|
| up to    5 | ± 5° | over    10 to    20 | ± 15° |
| over    5 to    10 | ± 10° | over    20 to    40 | ± 30° |

commercial tolerance on loads is ± 10 percent and is specified with reference to the angular deflection. For example: 100 pounds-inches ± 10 percent at 45 degrees deflection. One load specified usually suffices. If two loads and two deflections are specified, the manufacturing and testing times are increased. Smaller tolerances than ± 10 percent require each spring to be individually tested and adjusted, which adds considerably to manufacturing time and cost. Tables 15, 16, and 17 give respectively free angle tolerances, coil diameter tolerances, and tolerances on the number of coils.

**Miscellaneous Springs.** — This section provides information on various springs, some in common use, some less commonly used.

*Conical compression:* These springs taper from top to bottom and are useful where an increasing (instead of a constant) load rate is needed, where solid height must be small, and where vibration must be damped. Conical springs with a uniform pitch are easiest to coil. Load and deflection formulas for compression springs can be used — using the average mean coil diameter, and providing the deflection does not cause the largest active coil to lie against the bottom coil. When this happens, each coil must be calculated separately, using the standard formulas for compression springs.

*Belleville washers:* These washer type springs can sustain relatively large loads with small deflections, but the loads and deflections can be increased by stacking the springs as shown in Fig. 26.

Design data is not given here because the wide variations in ratios of O.D. to I.D., height to thickness, and other factors require too many formulas for convenient use and involve constants obtained from more than 24 curves. It is now practicable to select required sizes from the large stocks carried by several of the larger spring manufacturing companies. Most of these companies also stock curved and wave washers.

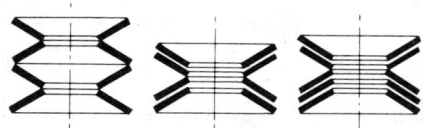

Fig. 26.   Examples of Belleville Spring Combinations.

*Volute springs:* These are often used on army tanks and heavy field artillery, and seldom find additional uses because of their high cost, long production time, difficulties in manufacture and unavailability of a wide range of materials and sizes. Small volute springs are often replaced with standard compression springs.

*Torsion bars:* Although the more simple types are often used on motor cars, the more complicated types with specially forged ends are finding fewer applications as time goes on.

*Constant force springs:* Those made from flat spring steel are finding more applications each year. However, the complicated design procedures can be eliminated by selecting a standard design from thousands now available from several spring manufacturers.

*Spiral, clock, and motor springs:* Although often used in wind-up type motors for toys and other products, they are difficult to design and results cannot be calculated with precise accuracy. However, many useful designs have been developed and are available from spring manufacturing companies.

*Flat springs:* These springs are often used to overcome operating space limitations in various products such as electric switches and relays. Table 18 lists formulas for designing flat springs. They are based on standard beam formulas where the deflection is small.

**Table 18. Formulas for Flat Springs***

| Feature | | | | |
|---|---|---|---|---|
| Deflect. $f$ Inches | $\dfrac{P L^3}{4 E b t^3}$ $\dfrac{S_b L^2}{6 E t}$ | $\dfrac{4 P L^3}{E b t^3}$ $\dfrac{2 S_b L^2}{3 E t}$ | $\dfrac{6 P L^3}{E b t^3}$ $\dfrac{S_b L^2}{E t}$ | $\dfrac{5.22 P L^3}{E b t^3}$ $\dfrac{.87 S_b L^2}{E t}$ |
| Load $P$ Pounds | $\dfrac{2 S_b b t^2}{3 L}$ $\dfrac{4 E b t^3 F}{L^3}$ | $\dfrac{S_b b t^2}{6 L}$ $\dfrac{E b t^3 F}{4 L^3}$ | $\dfrac{S_b b t^2}{6 L}$ $\dfrac{E b t^3 F}{6 L^3}$ | $\dfrac{S_b b t^2}{6 L}$ $\dfrac{E b t^3 F}{5.22 L^3}$ |
| Stress $S_b$ Bending Pounds per sq. inch | $\dfrac{3 P L}{2 b t^2}$ $\dfrac{6 E t F}{L^2}$ | $\dfrac{6 P L}{b t^2}$ $\dfrac{3 E t F}{2 L^2}$ | $\dfrac{6 P L}{b t^2}$ $\dfrac{E t F}{L^2}$ | $\dfrac{6 P L}{b t^2}$ $\dfrac{E t F}{.87 L^2}$ |
| Thickness $t$ Inches | $\dfrac{S_b L^2}{6 E F}$ $\sqrt[3]{\dfrac{P L^3}{4 E b F}}$ | $\dfrac{2 S_b L^2}{3 E F}$ $\sqrt[3]{\dfrac{4 P L^3}{E b F}}$ | $\dfrac{S_b L^2}{E F}$ $\sqrt[3]{\dfrac{6 P L^3}{E b F}}$ | $\dfrac{.87 S_b L^2}{E F}$ $\sqrt[3]{\dfrac{5.22 P L^3}{E b F}}$ |

* Based on standard beam formulas where the deflection is small.
*Note:* Where two formulas are given for one feature, the designer should use the one found to be appropriate for the given design instance. The result from either of any two formulas is the same.

**Moduli of Elasticity of Spring Materials.** — The modulus of elasticity in tension, denoted by the letter $E$, and the modulus of elasticity in torsion, denoted by the letter $G$, are used in formulas relating to spring design. Values of these moduli for various ferrous and nonferrous spring materials are given in Table 19.

**General Heat Treating Information for Springs.** — The following is general information on the heat treatment of springs, and is applicable to pre-tempered or hard-drawn spring materials only.

*Compression springs* are baked after coiling (before setting) to relieve residual stresses and thus permit larger deflections before taking a permanent set.

*Extension springs* also are baked, but heat removes some of the initial tension. Allowance should be made for this loss. Baking at 500 degrees F for 30 minutes removes approximately 50 percent of the initial tension. The shrinkage in diameter however, will slightly increase the load and rate.

*Torsion springs* do not actually require baking because coiling causes residual stresses in a direction that is helpful, but such springs frequently are baked so that jarring or handling will not cause them to lose position of ends.

**Table 19. Moduli of Elasticity in Torsion and Tension of Spring Materials**

| FERROUS MATERIALS | | | NON-FERROUS MATERIALS | | |
|---|---|---|---|---|---|
| Material (Commercial Name) | Modulus of Elasticity, pounds per square inch | | Material (Commercial Name) | Modulus of Elasticity, pounds per square inch | |
| | In Torsion, G | In Tension, E | | In Torsion, G | In Tension, E |
| Hard Drawn MB | | | Spring Brass | | |
| Up to 0.032 inch | 11,700,000 | 28,800,000 | Type 70-30 | 5,000,000 | 15,000,000 |
| 0.033 to 0.063 inch | 11,600,000 | 28,700,000 | Phosphor Bronze | | |
| 0.064 to 0.125 inch | 11,500,000 | 28,600,000 | 5 per cent tin | 6,000,000 | 15,000,000 |
| 0.126 to .625 inch | 11,400,000 | 28,500,000 | | | |
| Music Wire | | | Beryllium Copper | | |
| Up to 0.032 inch | 12,000,000 | 29,500,000 | Cold Drawn 4 Nos. | 7,000,000 | 17,000,000 |
| 0.033 to 0.063 inch | 11,850,000 | 29,000,000 | Pretempered, | | |
| 0.064 to 0.125 inch | 11,750,000 | 28,500,000 | fully hard | 7,250,000 | 19,000,000 |
| 0.126 to 0.250 inch | 11,600,000 | 28,000,000 | Inconel† 600 | 10,500,000 | 31,000,000* |
| Oil Tempered MB | 11,200,000 | 28,500,000 | Inconel† X 750 | 10,500,000 | 31,000,000* |
| Chrome-Vanadium | 11,200,000 | 28,500,000 | Mcnel† 400 | 9,500,000 | 26,000,000 |
| Chrome-Silicon | 11,200,000 | 29,500,000 | Monel† K 500 | 9,500,000 | 26,000,000 |
| Silicon-Manganese | 10,750,000 | 29,000,000 | Duranickel† 300 | 11,000,000 | 30,000,000 |
| Stainless Steel | | | Permanickel† | 11,000,000 | 30,000,000 |
| Types 302, 304, 316 | 10,000,000 | 28,000,000* | Ni Span C† 902 | 10,000,000 | 27,500,000 |
| Type 17-7 PH | 10,500,000 | 29,500,000 | Elgiloy‡ | 12,000,000 | 29,500,000 |
| Type 420 | 11,000,000 | 29,000,000 | Iso-Elastic** | 9,200,000 | 26,000,000 |
| Type 431 | 11,400,000 | 29,500,000 | | | |

\* May be 2,000,000 pounds per square inch less if material is not fully hard.
  † Trade name of the International Nickel Company.   ‡ Trade name of Hamilton Watch Company.   \*\* Trade name of John Chatillon & Sons.
*Note:* Modulus G is used for compression and extension springs; modulus E is used for torsion, flat, and spiral springs.

*Outside diameters shrink* when springs of music wire, pre-tempered MB and other carbon or alloy steels are baked. Baking also slightly increases the free length and these changes cause a little stronger load and increase the rate.

*Outside diameters expand* when springs of stainless steel (18-8) are baked. The free length also reduces slightly and these changes cause a little lighter load and decrease the rate.

*Inconel, Monel, and nickel alloys* do not change much when baked.

*Beryllium-copper shrinks and deforms* when heated. Such springs usually are baked in fixtures or supported on arbors or rods during heating.

*Brass and phosphor bronze* springs should be given a light heat only. Baking above 450 degrees F will soften the material. Do not heat in salt pots.

*Spring brass and phosphor bronze* springs which are not too highly stressed and are not subject to severe operating use may receive adequate stress relieving after coiling by immersing them in boiling water for a period of one hour.

*Position of loops will* change with heat. Parallel hooks may change as much as 45 degrees during baking. Torsion spring arms will alter position considerably. These changes should be allowed for during looping or forming.

*Quick heating* after coiling either in a high temperature salt pot or by passing a spring through a gas flame is not good practice. Samples heated in this way will not conform with production runs that are properly baked. A small, controlled-temperature oven should be used for samples and for small lot orders.

*Plated springs* should always be baked before plating to relieve coiling stresses and again after plating to relieve hydrogen embrittlement.

*Hardness* values fall with high heat — but music wire, hard drawn and stainless steel will increase 2 to 4 points Rockwell C.

Table 20.   Squares, Cubes and Fourth Powers of Wire Diameters

| Steel Wire Gage (U.S.) | Music or Piano Wire Gage | Diameter Inches | Section Area | Square | Cube | Fourth Power |
|---|---|---|---|---|---|---|
| 7-0 | .. | 0.4900 | 0.1886 | 0.24010 | 0.11765 | 0.05765 |
| 6-0 | .. | 0.4615 | 0.1673 | 0.21298 | 0.09829 | 0.04536 |
| 5-0 | .. | 0.4305 | 0.1456 | 0.18533 | 0.07978 | 0.03435 |
| 4-0 | .. | 0.3938 | 0.1218 | 0.15508 | 0.06107 | 0.02405 |
| 3-0 | .. | 0.3625 | 0.1032 | 0.13141 | 0.04763 | 0.01727 |
| 2-0 | .. | 0.331 | 0.0860 | 0.10956 | 0.03626 | 0.01200 |
| 1-0 | .. | 0.3065 | 0.0738 | 0.09394 | 0.02879 | 0.008825 |
| 1 | .. | 0.283 | 0.0629 | 0.08009 | 0.02267 | 0.006414 |
| 2 | .. | 0.2625 | 0.0541 | 0.06891 | 0.01809 | 0.004748 |
| 3 | .. | 0.2437 | 0.0466 | 0.05939 | 0.01447 | 0.003527 |
| 4 | .. | 0.2253 | 0.0399 | 0.05076 | 0.01144 | 0.002577 |
| 5 | .. | 0.207 | 0.0337 | 0.04285 | 0.00887 | 0.001836 |
| 6 | .. | 0.192 | 0.0290 | 0.03686 | 0.00708 | 0.001359 |
| .. | 45 | 0.180 | 0.0254 | 0.03240 | 0.00583 | 0.001050 |
| 7 | .. | 0.177 | 0.0246 | 0.03133 | 0.00555 | 0.000982 |
| .. | 44 | 0.170 | 0.0227 | 0.02890 | 0.00491 | 0.000835 |
| 8 | 43 | 0.162 | 0.0206 | 0.02624 | 0.00425 | 0.000689 |
| .. | 42 | 0.154 | 0.0186 | 0.02372 | 0.00365 | 0.000563 |
| 9 | .. | 0.1483 | 0.0173 | 0.02199 | 0.00326 | 0.000484 |
| .. | 41 | 0.146 | 0.0167 | 0.02132 | 0.00311 | 0.000455 |
| .. | 40 | 0.138 | 0.0150 | 0.01904 | 0.00263 | 0.000363 |
| 10 | .. | 0.135 | 0.0143 | 0.01822 | 0.00246 | 0.000332 |
| .. | 39 | 0.130 | 0.0133 | 0.01690 | 0.00220 | 0.000286 |
| .. | 38 | 0.124 | 0.0121 | 0.01538 | 0.00191 | 0.000237 |
| 11 | .. | 0.1205 | 0.0114 | 0.01452 | 0.00175 | 0.000211 |
| .. | 37 | 0.118 | 0.0109 | 0.01392 | 0.00164 | 0.000194 |
| .. | 36 | 0.112 | 0.0099 | 0.01254 | 0.00140 | 0.000157 |
| .. | 35 | 0.106 | 0.0088 | 0.01124 | 0.00119 | 0.000126 |
| 12 | .. | 0.1055 | 0.0087 | 0.01113 | 0.001174 | 0.0001239 |
| .. | 34 | 0.100 | 0.0078 | 0.0100 | 0.001000 | 0.0001000 |
| .. | 33 | 0.095 | 0.0071 | 0.00902 | 0.000857 | 0.0000815 |
| 13 | .. | 0.0915 | 0.0066 | 0.00837 | 0.000766 | 0.0000701 |
| .. | 32 | 0.090 | 0.0064 | 0.00810 | 0.000729 | 0.0000656 |
| .. | 31 | 0.085 | 0.0057 | 0.00722 | 0.000614 | 0.0000522 |
| 14 | 30 | 0.080 | 0.0050 | 0.0064 | 0.000512 | 0.0000410 |
| .. | 29 | 0.075 | 0.0044 | 0.00562 | 0.000422 | 0.0000316 |
| 15 | .. | 0.072 | 0.0041 | 0.00518 | 0.000373 | 0.0000269 |
| .. | 28 | 0.071 | 0.0040 | 0.00504 | 0.000358 | 0.0000254 |
| .. | 27 | 0.067 | 0.0035 | 0.00449 | 0.000301 | 0.0000202 |
| .. | 26 | 0.063 | 0.0031 | 0.00397 | 0.000250 | 0.0000158 |
| 16 | .. | 0.0625 | 0.0031 | 0.00391 | 0.000244 | 0.0000153 |
| .. | 25 | 0.059 | 0.0027 | 0.00348 | 0.000205 | 0.0000121 |
| .. | 24 | 0.055 | 0.0024 | 0.00302 | 0.000166 | 0.00000915 |
| 17 | .. | 0.054 | 0.0023 | 0.00292 | 0.000157 | 0.00000850 |
| .. | 23 | 0.051 | 0.0020 | 0.00260 | 0.000133 | 0.00000677 |
| .. | 22 | 0.049 | 0.00189 | 0.00240 | 0.000118 | 0.00000576 |
| 18 | .. | 0.0475 | 0.00177 | 0.00226 | 0.000107 | 0.00000509 |
| .. | 21 | 0.047 | 0.00173 | 0.00221 | 0.000104 | 0.00000488 |
| .. | 20 | 0.045 | 0.00159 | 0.00202 | 0.000091 | 0.00000410 |
| .. | 19 | 0.043 | 0.00145 | 0.00185 | 0.0000795 | 0.00000342 |
| 19 | 18 | 0.041 | 0.00132 | 0.00168 | 0.0000689 | 0.00000283 |
| .. | 17 | 0.039 | 0.00119 | 0.00152 | 0.0000593 | 0.00000231 |
| .. | 16 | 0.037 | 0.00108 | 0.00137 | 0.0000507 | 0.00000187 |
| .. | 15 | 0.035 | 0.00096 | 0.00122 | 0.0000429 | 0.00000150 |
| 20 | .. | 0.0348 | 0.00095 | 0.00121 | 0.0000421 | 0.00000147 |
| .. | 14 | 0.033 | 0.00086 | 0.00109 | 0.0000359 | 0.00000119 |
| 21 | .. | 0.0317 | 0.00079 | 0.00100 | 0.0000319 | 0.00000101 |
| .. | 13 | 0.031 | 0.00075 | 0.00096 | 0.0000298 | 0.000000924 |
| .. | 12 | 0.029 | 0.00066 | 0.00084 | 0.0000244 | 0.000000707 |
| 22 | .. | 0.0286 | 0.00064 | 0.00082 | 0.0000234 | 0.000000669 |
| .. | 11 | 0.026 | 0.00053 | 0.00068 | 0.0000176 | 0.000000457 |
| 23 | .. | 0.0258 | 0.00052 | 0.00067 | 0.0000172 | 0.000000443 |
| .. | 10 | 0.024 | 0.00045 | 0.00058 | 0.0000138 | 0.000000332 |
| 24 | .. | 0.023 | 0.00042 | 0.00053 | 0.0000122 | 0.000000280 |
| .. | 9 | 0.022 | 0.00038 | 0.00048 | 0.0000106 | 0.000000234 |

Table 21.　Causes of Spring Failure*

| | Cause | Comments and Recommendations |
|---|---|---|
| **GROUP 1** | High stress | The majority of spring failures are due to high stresses caused by large deflections and high loads. High stresses should be used only for statically loaded springs. Low stresses lengthen fatigue life. |
| | Hydrogen embrittlement | Improper electro-plating methods and acid cleaning of springs, without proper baking treatment, cause spring steels to become brittle, and is a frequent cause of failure. Non-ferrous springs are immune. |
| | Sharp bends and holes | Sharp bends on extension, torsion and flat springs and holes or notches in flat springs, cause high concentration of stress resulting in failure. Bend radii should be as large as possible, and tool marks avoided. |
| | Fatigue | Repeated deflections of springs, especially above 1,000,000 cycles, even with medium stresses, may cause failure. Low stresses should be used if a spring is subject to a very high number of operating cycles conditions. |
| **GROUP 2** | Shock loading | Impact, shock, and rapid loading cause far higher stresses than those computed by the regular spring formulas. High carbon spring steels do not withstand shock loading as well as alloy steels. |
| | Corrosion | Slight rusting or pitting caused by acids, alkali, galvanic corrosion, stress corrosion cracking, or corrosive atmosphere weakens the material and causes higher stresses in the corroded area. |
| | Faulty heat treatment | Keeping spring materials at the hardening temperature for longer periods than necessary causes an undesirable growth in grain structure resulting in brittleness even though the hardness may be correct. |
| | Faulty material | Poor material containing inclusions, seams, slivers, and flat material with rough, slit, or torn edges cause early failure. Overdrawn wire, improper hardness, and poor grain structure also result in early failure. |
| **GROUP 3** | High temperature | High temperatures reduce spring temper (or hardness), lower the modulus of elasticity thereby causing lower loads, reduce the elastic limit and increase corrosion. Corrosion resisting or nickel alloys should be used. |
| | Low temperature | Temperatures below −40 degrees F lessen the ability of carbon steels to withstand shock loads. Carbon steels become brittle at −70 degrees F. Corrosion resisting, nickel or non-ferrous alloys should be used. |
| | Friction | Close fits on rods or in holes result in a wearing away of material and occasional failure. The outside diameters of compression springs expand during deflection but they become smaller on torsion springs. |
| | Other causes | Enlarged hooks on extension springs increase the stress at the bends. Carrying too much electrical current will cause failure. Welding and soldering frequently destroy the spring temper. Tool marks, nicks, and cuts often raise stresses. Deflecting torsion springs outwardly causes high stresses and winding them tightly causes binding on supporting rods. High speed of deflection, vibration and surging due to operation near natural periods of vibration or their harmonics, cause increased stresses. |

* Spring failure may be breakage, high permanent set, or loss of load. The causes are listed in groups in this table. Group 1 covers causes that occur most frequently; Group 2 covers causes that are less frequent; and Group 3 lists causes that occur occasionally.

Table 22. Arbor Diameters for Springs made from Music Wire

| Wire Diam. (inch) | Spring Outside Diameter (inch) | | | | | | | | | | | | |
| | 1/16 | 3/32 | 1/8 | 5/32 | 3/16 | 7/32 | 1/4 | 9/32 | 5/16 | 11/32 | 3/8 | 7/16 | 1/2 |
| | Arbor Diameter (inch) | | | | | | | | | | | | |
|---|---|---|---|---|---|---|---|---|---|---|---|---|---|
| .008 | .039 | .060 | .078 | .093 | .107 | .119 | .129 | ... | ... | .... | ... | .... | |
| .010 | .037 | .060 | .080 | .099 | .115 | .129 | .142 | .154 | .164 | .... | ... | .... | |
| .012 | .034 | .059 | .081 | .101 | .119 | .135 | .150 | .163 | .177 | .189 | .200 | ... | .... |
| .014 | .031 | .057 | .081 | .102 | .121 | .140 | .156 | .172 | .187 | .200 | .213 | .234 | .... |
| .016 | .028 | .055 | .079 | .102 | .123 | .142 | .161 | .178 | .194 | .209 | .224 | .250 | .271 |
| .018 | ... | .053 | .077 | .101 | .124 | .144 | .161 | .182 | .200 | .215 | .231 | .259 | .284 |
| .020 | ... | .049 | .075 | .096 | .123 | .144 | .165 | .184 | .203 | .220 | .237 | .268 | .296 |
| .022 | ... | .046 | .072 | .097 | .122 | .145 | .165 | .186 | .206 | .224 | .242 | .275 | .305 |
| .024 | ... | .043 | .070 | .095 | .120 | .144 | .166 | .187 | .207 | .226 | .245 | .280 | .312 |
| .026 | ... | ... | .067 | .093 | .118 | .143 | .166 | .187 | .208 | .228 | .248 | .285 | .318 |
| .028 | ... | ... | .064 | .091 | .115 | .141 | .165 | .187 | .208 | .229 | .250 | .288 | .323 |
| .030 | ... | ... | .061 | .088 | .113 | .138 | .163 | .187 | .209 | .229 | .251 | .291 | .328 |
| .032 | ... | ... | .057 | .085 | .111 | .136 | .161 | .185 | .209 | .229 | .251 | .292 | .331 |
| .034 | ... | ... | ... | .082 | .109 | .134 | .159 | .184 | .208 | .229 | .251 | .292 | .333 |
| .036 | ... | ... | ... | .078 | .106 | .131 | .156 | .182 | .206 | .229 | .250 | .294 | .333 |
| .038 | ... | ... | ... | .075 | .103 | .129 | .154 | .179 | .205 | .227 | .251 | .293 | .335 |
| .041 | ... | ... | ... | ... | .098 | .125 | .151 | .176 | .201 | .226 | .250 | .294 | .336 |
| .0475 | ... | ... | ... | ... | .087 | .115 | .142 | .168 | .194 | .220 | .244 | .293 | .337 |
| .054 | ... | ... | ... | ... | ... | .103 | .132 | .160 | .187 | .212 | .245 | .287 | .336 |
| .0625 | ... | ... | ... | ... | ... | ... | .108 | .146 | .169 | .201 | .228 | .280 | .330 |
| .072 | ... | ... | ... | ... | ... | ... | ... | .129 | .158 | .186 | .214 | .268 | .319 |
| .080 | ... | ... | ... | ... | ... | ... | ... | ... | .144 | .173 | .201 | .256 | .308 |
| .0915 | ... | ... | ... | ... | ... | ... | ... | ... | ... | ... | .181 | .238 | .293 |
| .1055 | ... | ... | ... | ... | ... | ... | ... | ... | ... | ... | ... | .215 | .271 |
| .1205 | ... | ... | ... | ... | ... | ... | ... | ... | ... | ... | ... | ... | .215 |
| .125 | ... | ... | ... | ... | ... | ... | ... | ... | ... | ... | ... | ... | .239 |

| Wire Diam. (inch) | Spring Outside Diameter (inches) | | | | | | | | | | | | | |
| | 9/16 | 5/8 | 11/16 | 3/4 | 13/16 | 7/8 | 15/16 | 1 | 1 1/8 | 1 1/4 | 1 3/8 | 1 1/2 | 1 3/4 | 2 |
| | Arbor Diameter (inches) | | | | | | | | | | | | | |
|---|---|---|---|---|---|---|---|---|---|---|---|---|---|---|
| .022 | .332 | .357 | .380 | ... | ... | ... | ... | ... | ... | ... | ... | ... | ... | ... |
| .024 | .341 | .367 | .393 | .415 | ... | ... | ... | ... | ... | ... | ... | ... | ... | ... |
| .026 | .350 | .380 | .406 | .430 | ... | ... | ... | ... | ... | ... | ... | ... | ... | ... |
| .028 | .356 | .387 | .416 | .442 | .467 | ... | ... | ... | ... | ... | ... | ... | ... | ... |
| .030 | .362 | .395 | .426 | .453 | .481 | .506 | ... | ... | ... | ... | ... | ... | ... | ... |
| .032 | .367 | .400 | .432 | .462 | .490 | .516 | .540 | ... | ... | ... | ... | ... | ... | ... |
| .034 | .370 | .404 | .437 | .469 | .498 | .526 | .552 | .557 | ... | ... | ... | ... | ... | ... |
| .036 | .372 | .407 | .442 | .474 | .506 | .536 | .562 | .589 | ... | ... | ... | ... | ... | ... |
| .038 | .375 | .412 | .448 | .481 | .512 | .543 | .572 | .600 | .650 | ... | ... | ... | ... | ... |
| .041 | .378 | .416 | .456 | .489 | .522 | .554 | .586 | .615 | .670 | .718 | ... | ... | ... | ... |
| .0475 | .380 | .422 | .464 | .504 | .541 | .576 | .610 | .643 | .706 | .763 | .812 | ... | ... | ... |
| .054 | .381 | .425 | .467 | .509 | .550 | .589 | .625 | .661 | .727 | .792 | .850 | .906 | ... | ... |
| .0625 | .379 | .426 | .468 | .512 | .556 | .597 | .639 | .678 | .753 | .822 | .889 | .951 | 1.06 | 1.17 |
| .072 | .370 | .418 | .466 | .512 | .555 | .599 | .641 | .682 | .765 | .840 | .911 | .980 | 1.11 | 1.22 |
| .080 | .360 | .411 | .461 | .509 | .554 | .599 | .641 | .685 | .772 | .851 | .930 | 1.00 | 1.13 | 1.26 |
| .0915 | .347 | .398 | .448 | .500 | .547 | .597 | .640 | .685 | .776 | .860 | .942 | 1.02 | 1.16 | 1.30 |
| .1055 | .327 | .381 | .433 | .485 | .535 | .586 | .630 | .683 | .775 | .865 | .952 | 1.04 | 1.20 | 1.35 |
| .1205 | .303 | .358 | .414 | .468 | .520 | .571 | .622 | .673 | .772 | .864 | .955 | 1.04 | 1.22 | 1.38 |
| .125 | .295 | .351 | .406 | .461 | .515 | .567 | .617 | .671 | .770 | .864 | .955 | 1.05 | 1.23 | 1.39 |

# FRICTION

Friction is the resistance to motion which takes place when one body is moved upon another, and is generally defined as "that force which acts between two bodies at their surface of contact, so as to resist their sliding on each other." The force of friction, $F$, bears — according to the conditions under which sliding occurs — a certain relation to the pressure between the two bodies; this pressure is called the normal pressure $N$. The relation between force of friction and normal pressure is given by the *coefficient of friction*, generally denoted by the Greek letter $\mu$. Thus:

$$F = \mu \times N, \quad \text{and} \quad \mu = \frac{F}{N}$$

*Example:* — A body weighing 28 pounds rests on a horizontal surface. The force required to keep it in motion along the surface is 7 pounds. Find the coefficient of friction.

$$\mu = \frac{F}{N} = \frac{7}{28} = 0.25$$

If a body is placed on an inclined plane, the friction between the body and the plane will prevent it from sliding down the inclined surface, provided the angle of the plane with the horizontal is not too great. There will be a certain angle, however, at which the body will just barely be able to remain stationary, the frictional resistance being very nearly overcome by the tendency of the body to slide down. This angle is termed the angle of repose, and the tangent of this angle equals the coefficient of friction. The angle of repose is frequently denoted by the Greek letter $\theta$. Thus, $\mu = \tan \theta$.

A greater force is required to start a body from a state of rest than to merely keep it in motion, because the *friction of rest* is greater than the *friction of motion*.

**Laws of Friction.** — The laws of friction for unlubricated or dry surfaces are summarized in the following statements.

1. For low pressures the friction is directly proportional to the normal pressure between the two surfaces. As the pressure increases to a high value the friction does not rise as rapidly; but when the pressure becomes abnormally high, the friction increases at a rapid rate until seizing takes place.

2. The friction both in its total amount and its coefficient is independent of the areas in contact, so long as the total pressure remains the same. This is true for moderate pressures only. For high pressures, this law is modified in the same way as in the first case.

3. At very low velocities the friction is independent of the velocity of rubbing. As the velocities increase, the friction decreases.

*Lubricated Surfaces:* For well lubricated surfaces, the laws of friction are considerably different from those governing dry or poorly lubricated surfaces.

1. The frictional resistance is almost independent of the pressure per square inch, if the surfaces are flooded with oil.

2. The friction varies directly as the speed, at low pressures; but for high pressures the friction is very great at low velocities, approaching a minimum at about two feet per second linear velocity, and afterwards increasing approximately as the square root of the speed.

3. For well lubricated surfaces the frictional resistance depends, to a very great extent, on the temperature, partly because of the change in the viscosity of the oil and partly because, for a journal bearing, the diameter of the bearing increases with the rise of temperature more rapidly than the diameter of the shaft, thus relieving the bearing of side pressure.

4. If the bearing surfaces are flooded with oil, the friction is almost independent of the nature of the material of the surfaces in contact. As the lubrication becomes less ample, the coefficient of friction becomes more dependent upon the material of the surfaces.

**Influence of Friction on the Efficiency of Small Machine Elements.** — Friction between machine parts lowers the efficiency of a machine. In the following are given average values of the efficiency, in per cent, of the most common machine elements when carefully made. Ordinary bearings, 95 to 98; roller bearings, 98; ball bearings, 99; spur gears with cut teeth, including bearings, 99; bevel gears with cut teeth, including bearings, 98; belting, from 96 to 98; high-class silent power transmission chain, 97 to 99; roller chains, 95 to 97.

**Coefficients of Friction.** — Tables 1 and 2 provide representative values of static friction for various combinations of materials with dry (clean, unlubricated) and lubricated surfaces. The static or breakaway friction shown in these tables will generally be higher than the subsequent or sliding friction. Typically, the steel-on-steel static coefficient of 0.8 unlubricated will drop to 0.4 when sliding has been initiated; with oil lubrication, the value will drop from 0.16 to 0.03.

Many factors affect friction, and even slight deviations from normal or test conditions can produce wide variations. Accordingly, when using friction coefficients in design calculations, due allowance or factors of safety should be considered and in critical cases, specific tests conducted to provide actual coefficients for material, geometry, and/or lubricant combinations.

**Table 1.    Coefficients of Static Friction for Steel on Various Materials\***

| Material | Coefficient of Friction, $\mu$ | |
| --- | --- | --- |
|  | Clean | Lubricated |
| Steel | 0.8 | 0.16 |
| Copper-lead alloy | 0.22 | .... |
| Phosphor bronze | 0.35 | .... |
| Aluminum bronze | 0.45 | .... |
| Brass | 0.35 | 0.19 |
| Cast iron | 0.4 | 0.21 |
| Bronze | .... | 0.16 |
| Sintered bronze | .... | 0.13 |
| Hard carbon | 0.14 | 0.11–0.14 |
| Graphite | 0.1 | 0.1 |
| Tungsten carbide | 0.4–0.6 | 0.1–0.2 |
| Plexiglas | 0.4–0.5 | 0.4–0.5 |
| Polystyrene | 0.3–0.35 | 0.3–0.35 |
| Polythene | 0.2 | 0.2 |
| Teflon | 0.04 | 0.04 |

\* With permission from *The Friction and Lubrication of Solids*, Vol. I, by Bowden and Tabor, Clarendon Press, Oxford, 1950.

**Rolling Friction.** — When a body rolls on a surface, the force resisting the motion is termed *rolling friction* or *rolling resistance*. Let $W$ = total weight of rolling body or load on wheel, in pounds; $r$ = radius of wheel, in inches; $f$ = coefficient of rolling resistance, in inches. Then: Resistance to rolling, in pounds = $(W \times f) \div r$.

The coefficient of rolling resistance varies with the conditions. For wood on

Table 2.　Coefficients of Static Friction for Various Material Combinations*

| Material Combination | Coefficient of Friction, $\mu$ | |
| --- | --- | --- |
| | Clean | Lubricated |
| Aluminum — aluminum | 1.35 | 0.30 |
| Cadmium — cadmium | 0.5 | 0.05 |
| Chromium — chromium | 0.41 | 0.34 |
| Copper — copper | 1.0 | 0.08 |
| Iron — iron | 1.0 | 0.15–0.20 |
| Magnesium — magnesium | 0.6 | 0.08 |
| Nickel — nickel | 0.7 | 0.28 |
| Platinum — platinum | 1.2 | 0.25 |
| Silver — silver | 1.4 | 0.55 |
| Zinc — zinc | 0.6 | 0.04 |
| Glass — glass | 0.9–1.0 | 0.1–0.6 |
| Glass — metal | 0.5–0.7 | 0.2–0.3 |
| Diamond — diamond | 0.1 | 0.05–0.1 |
| Diamond — metal | 0.1–0.15 | 0.1 |
| Sapphire — sapphire | 0.2 | 0.2 |
| Hard carbon on carbon | 0.16 | 0.12–0.14 |
| Graphite — graphite (in vacuum) | 0.5–0.8 | .... |
| Graphite — graphite | 0.1 | 0.1 |
| Tungsten carbide — tungsten carbide | 0.2–0.25 | 0.12 |
| Plexiglas — plexiglas | 0.8 | 0.8 |
| Polystyrene — polystyrene | 0.5 | 0.5 |
| Teflon — teflon | 0.04 | 0.04 |
| Nylon — nylon | 0.15–0.25 | .... |
| Solids on rubber | 1–4 | .... |
| Wood on wood (clean) | 0.25–0.5 | .... |
| Wood on wood (wet) | 0.2 | .... |
| Wood on metals (clean) | 0.2–0.6 | .... |
| Wood on metals (wet) | 0.2 | .... |
| Brick on wood | 0.6 | .... |
| Leather on wood | 0.3–0.4 | .... |
| Leather on metal (clean) | 0.6 | .... |
| Leather on metal (wet) | 0.4 | .... |
| Leather on metal (greasy) | 0.2 | .... |
| Brake material on cast iron | 0.4 | .... |
| Brake material on cast iron (wet) | 0.2 | .... |

* With permission from *The Friction and Lubrication of Solids*, Vol. I, by Bowden and Tabor, Clarendon Press, Oxford, 1950.

wood it may be assumed as 0.06 inch; for iron on iron, 0.02 inch; iron on granite, 0.085 inch; iron on asphalt, 0.15 inch; and iron on wood, 0.22 inch.

The coefficient of rolling resistance, $f$, is in inches and is not the same as the ordinary, or sliding coefficient of friction which is a dimensionless ratio between frictional resistance and normal load. As various investigators are not in close agreement on the true values for these coefficients, the foregoing values should only be used for approximate calculation of rolling resistance.

**Lubricants.** — The concept of supporting a load being moved on a friction-reducing film is called lubrication. The material of which the film is made up is defined as a lubricant. Lubricants may be in the form of fluids, such as conventional oils; semisolids, such as grease; and true solids, such as graphite and Teflon. Water, alcohol, and even air have been used as lubricants. Factors affecting the selection of the type of lubricant to be used are load, speed, temperature (both ambient and generated) and environmental contamination such as in the food processing industry. Liquid lubricants usually are matched to the application by proper selection, of viscosity and load-carrying ability. (See "Lubricants and Lubrication," page 665.)

**Wear.** — There is no apparent consistent relationship between friction and wear; friction may be high and wear low, or wear may be high and friction low. The principal types of wear are: adhesive, abrasive, and pitting.

*Adhesive wear:* As two surfaces slide over each other, wear occurs because of the shearing, deformation, and plucking away of material at points of adhesion; these points of adhesion occur at roughness peaks. The amount of wear is generally proportional to the load and to the distance over which the surfaces have slid, and inversely proportional to the hardness of the surface on which wear occurs.

*Abrasive wear:* This occurs when a hard rough surface slides against a softer one, ploughing a series of grooves and removing material; it also occurs when abrasive particles are introduced between sliding surfaces, or when a part is moved through an abrasive medium. Erosion abrasion takes place when abrasive particles impact on surfaces. These particles may be suspended in liquids, carried by air, or flow of their own weight such as sand particles down a chute.

*Pitting:* This is due to the surface fatigue failure of a material as a result of repeated surface or sub-surface stresses that exceed the endurance limit of the material. Pitting is a common mode of failure for gear teeth.

Pitting can be non-progressive or destructive. Non-progressive pitting occurs during the initial operation of a machine because of surface fatigue at roughness peaks. As the peaks wear, the surface becomes smoother and hence the load becomes more uniformly distributed. Pitting stops when the stress becomes less than the endurance limit of the material. Destructive pitting, however, is progressive and continues until the surface disintegrates.

**Designing for Wear Resistance.** — In adhesive wear situations the best results are usually obtained if parts have hard surfaces. Hard surfaces may be obtained by making parts from hard materials or by the application of appropriate surface treatments. Some of the hardest materials are diamond, brittle carbides, and martensitic steels. For less severe applications pearlitic and austenitic steels can be used.

Some common surface treatments of metal surfaces for the prevention of wear are: chromium and nickel plating, flame plating with tungsten carbide, phosphate treatment, carburizing, nitriding, cyaniding, carbonitriding, siliconizing, and chromizing. Aluminum and magnesium can be given hard anodic treatments. For moderate cost applications in which normal wear can be tolerated, low or medium carbon steels can be run against cast iron, brass, bronze, aluminum brass, or nylon. Adequate lubricants should be used. In general it is best if unlike combinations of materials slide against each other. This is often accomplished by employing different surface treatments; for example, having a nitrided steel part run against one that is carburized.

Abrasive wear can be limited in several ways. Surfaces can be given a higher hardness than the abrasive particles. If a circulating lubricant is used, the abrasive particles can be removed by filtering. A combination of a hard and a soft surface can be used so that the abrasive particles can imbed themselves into the softer material. Typical softer materials used for this purpose are babbitt metals, copper-lead alloys, and aluminum alloys.

Pitting is controlled by keeping contact stresses within allowable limits. In the case of gears a hard pinion is often run with a softer gear. As the softer gear wears, there will be better conformity between gear teeth profiles, hence less localized high stresses.

## PLAIN BEARINGS

On the pages that follow are given data and procedures for designing full-film or hydrodynamically lubricated bearings of the journal and thrust types. However, before proceeding to these design methods, it may prove useful to review first those bearing aspects concerning the types of bearings available; lubricants and lubrication methods; hardness and surface finish; machining methods; seals; retainers; and typical length-to-diameter ratios for various applications.

The following paragraphs preceding the design sections provide guidance in these matters and suggest modifications in allowable loads when other than full-film operating conditions exist in a bearing.

**Classes of Plain Bearings.** — Bearings that provide sliding contact between mating surfaces fall into three general classes: *radial bearings* that support rotating shafts or journals; *thrust bearings* that support axial loads on rotating members; and *guide* or *slipper bearings* that guide moving parts in a straight line. Radial sliding bearings, more commonly called sleeve bearings, may be of several types, the most usual being the plain full journal bearing which has 360-degree contact with its mating journal, and the partial journal bearing, which has less than 180-degree contact. This latter type is used when the load direction is constant and has the advantages of simplicity, ease of lubrication, and reduced frictional loss.

The relative motions between the parts of plain bearings may take place: (1) As pure sliding without the benefit of a liquid or gaseous lubricating medium between the moving surfaces such as with the dry operation of nylon or teflon; (2) With hydrodynamic lubrication in which a wedge or film build-up of lubricating medium is produced, with either whole or partial separation of the bearing surfaces; (3) With hydrostatic lubrication in which a lubricating medium is introduced under pressure between the mating surfaces causing a force opposite to the applied load and a lifting or separation of these surfaces; and (4) With a hybrid form or combination of hydrodynamic and hydrostatic lubrication.

Listed below are some of the advantages and disadvantages of sliding contact (plain) bearings as compared with rolling contact (anti-friction) bearings.

*Advantages:* 1. Require less space; 2. Are quieter in operation; 3. Are lower in cost, particularly in high volume production; 4. Have greater rigidity; and 5. Their life is generally not limited by fatigue.

*Disadvantages:* 1. Have higher frictional properties resulting in higher power consumption; 2. Are more susceptible to damage from foreign material in lubrication system; 3. Have more stringent lubrication requirements; and 4. Are more susceptible to damage from interrupted lubrication supply.

**Types of Journal Bearings.** — Many types of journal bearing configurations have been developed; some of these are shown in Fig. 1.

*Circumferential-groove bearings,* (a), have an oil groove extending circumferentially around the bearing. The oil is maintained under pressure in the groove. The groove divides the bearing into two shorter bearings which tend to run at a slightly greater eccentricity. However, the advantage in terms of stability is slight, and this design is most commonly used in reciprocating-load main and connecting-rod bearings because of the uniformity of oil distribution.

Short cylindrical bearings are a better solution than the circumferential-groove bearing for high-speed, low-load service. In many cases, the bearing can be shortened enough to increase the unit loading to a substantial value, causing the shaft to ride at a position of substantial eccentricity in the bearing. Experience has shown that instability rarely results when the shaft eccentricity is greater than 0.6. Very short bearings are not often used for this type of application, because they do not provide a high temporary

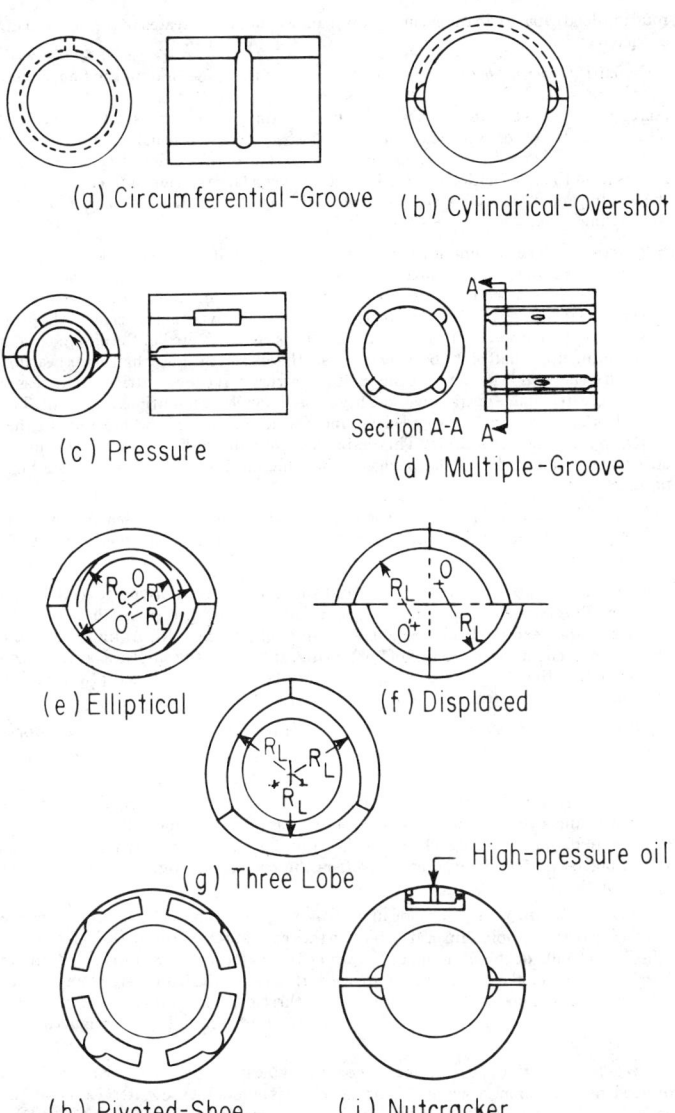

(a) Circumferential-Groove

(b) Cylindrical-Overshot

(c) Pressure

(d) Multiple-Groove

Section A-A

(e) Elliptical

(f) Displaced

(g) Three Lobe

High-pressure oil

(h) Pivoted-Shoe

(i) Nutcracker

Fig. 1. Typical shapes of several types of pressure-fed bearings.

rotating-load capacity in the event some unbalance should be created in the rotor during service.

*Cylindrical-overshot bearings*, (b), are used where surface speeds of 10,000 fpm or more exist, and where additional oil flow is desired to maintain a reasonable bearing temperature. This bearing has a wide circumferential groove extending from one axial oil groove to the other over the upper half of the bearing. Oil is usually admitted to the trailing-edge oil groove. An inlet orifice is used to control the oil flow. Cooler operation results from the elimination of shearing action over a large section of the upper half of the bearing and, to a great extent, from the additional flow of cool oil over the top half of the bearing.

*Pressure bearings*, (c), employ a groove over the top half of the bearing. The groove terminates at a sharp dam about 45 degrees beyond the vertical in the direction of shaft rotation. Oil is pumped into this groove by shear action from the rotation of the shaft and is then stopped by the dam. In high-speed operation, this situation creates a high oil pressure over the upper half of the bearing. The pressure created in the oil groove and surrounding upper half of the bearing increases the load on the lower half of the bearing. This self-generated load increases the shaft eccentricity. If the eccentricity is increased to 0.6 or greater, stable operation under high-speed, low-load conditions can result. The central oil groove can be extended around the lower half of the bearing, further increasing the effective loading. This design has one primary disadvantage: Dirt in the oil will tend to abrade the sharp edge of the dam and impair ability to create high pressures.

*Multiple-groove bearings*, (d), are sometimes used to provide increased oil flow. The interruptions in the oil film also appear to give this bearing some merit as a stable design.

*Elliptical bearings*, (e), are not truly elliptical, but are formed from two sections of a cylinder. This two-piece bearing has a large clearance in the direction of the split and a smaller clearance in the load direction at right angles to the split. At light loads, the shaft runs eccentric to both halves of the bearing, and hence, the elliptical bearing has a higher oil flow than the corresponding cylindrical bearing. Thus, the elliptical bearing will run cooler and will be more stable than a cylindrical bearing.

*Elliptical-overshot bearings*, (not shown), are elliptical bearings in which the upper half is relieved by a wide oil groove connecting the axial oil grooves. They are analogous to cylindrical-overshot bearings.

*Displaced elliptical bearings*, (f), shift the centers of the two bearing arcs in both a horizontal and a vertical direction. This design has greater stiffness than a cylindrical bearing, in both horizontal and vertical directions, with substantially higher oil flow. It has not been extensively used, but offers the prospect of high stability and cool operation.

*Three-lobe bearings*, (g), are made up in cross section of three circular arcs. They are most effective as anti-oilwhip bearings when the centers of curvature of each of the three lobes lie well outside the clearance circle, which the shaft center can describe within the bearing. Three axial oil-feed grooves are used. It is a more difficult design to manufacture, since it is almost necessary to make it in three parts instead of two. The bore is machined with shims between each of the three parts. The shims are removed after manufacture.

*Pivoted-shoe bearings*, (h), are one of the most stable bearings. The bearing surface is divided into three or more segments, each of which is pivoted at the center. In operation, each shoe tilts to form a wedge-shaped oil film, thus creating a force tending to push the shaft toward the center of the bearing. For single-direction rotation, the shoes are sometimes pivoted near one end and forced toward the shaft by springs.

*Nutcracker bearings,* (i), consist of two cylindrical half-bearings. The upper half-bearing is free to move in a vertical direction and is forced toward the shaft by a hydraulic cylinder. External oil pressure may be used to create load on the upper half of the bearing through the hydraulic cylinder. Or, the high-pressure oil may be obtained from the lower half of the bearing by tapping a hole into the high-pressure oil film, thus creating a self-loading bearing. Either type can increase bearing eccentricity to the point where stable operation can be achieved.

**Hydrostatic Bearings.** — Hydrostatic bearings are used when operating conditions require full film lubrication that cannot be developed hydrodynamically. The hydrostatically lubricated bearing, either thrust or radial, is supplied with lubricant under pressure from an external source. Some advantages of the hydrostatic bearing over bearings of other types are: low friction; high load capacity; high reliability; high stiffness; and long life.

Hydrostatic bearings are used successfully in many applications among which are machine tools, rolling mills, and other heavily loaded slow-moving machinery. However, specialized techniques, including a thorough understanding of hydraulic components external to the bearing package is required. The designer is cautioned against use of this type of bearing without a full knowledge of all aspects of the problem. Determination of the operating performance of hydrostatic bearings is a specialized area of the lubrication field and is described in specialized reference books.

**Design.** — The design of a sliding bearing is generally accomplished in one of two ways: (1) a bearing operating under similar conditions is used as a model or basis from which the new bearing is designed; (2) in the absence of any previous experience with similar bearings in similar environments, certain assumptions concerning operating conditions and requirements are made and a tentative design prepared based on general design parameters or rules of thumb. Detailed lubrication analysis is then performed to establish design and operating details and requirements.

**Modes of Bearing Operation.** — The load carrying ability of a sliding bearing depends upon the kind of fluid film which is formed between its moving surfaces. The formation of this film is dependent, in part, on the design of the bearing and, in part, on the speed of rotation. It results in three modes or regions of operation designated as *full film, mixed film* and *boundary* lubrication with effects on bearing friction as shown in Fig. 2.

In terms of physical bearing operation these three modes may be further described as follows:

1. Full film, or hydrodynamic lubrication produces a complete physical separation of the sliding surfaces. This results in low friction and long wear-free service life.

To promote full film lubrication in hydrodynamic operation the following parameters should be satisfied: 1. Lubricant selected has the correct viscosity for the proposed operation; 2. Proper lubricant flow rates are maintained; 3. Proper design methods and considerations have been utilized; and 4. Surface velocity in excess of 25 feet per minute is maintained.

When full film lubrication is achieved, a coefficient of friction between 0.001 and 0.005 can be expected.

2. Boundary lubrication takes place when the sliding surfaces are rubbing together with only an extremely thin film of lubricant present. This type of operation is acceptable only in applications with oscillating or slow rotary motion. In complete boundary lubrication the oscillatory or rotary motion is usually less than 10 feet per minute with resulting coefficients of friction of 0.08 to 0.14. These bearings are usually grease lubricated or periodically oil lubricated.

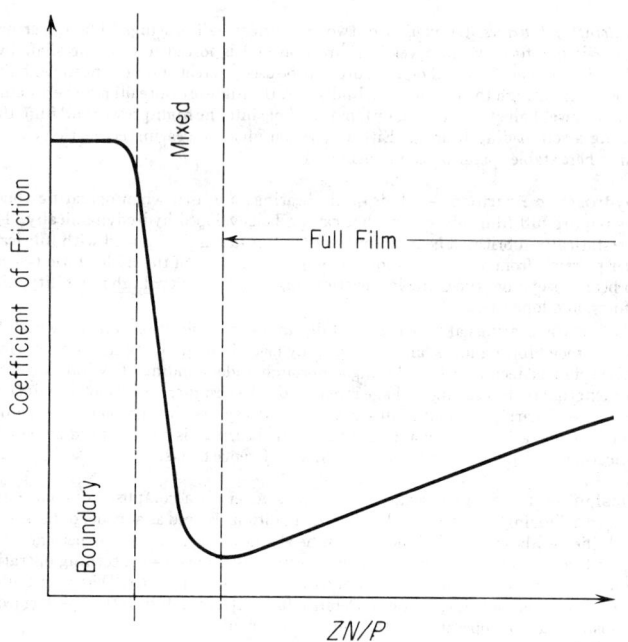

Fig. 2.　Three modes of bearing operation.

3. Mixed-film lubrication is a mode of operation between the full film and boundary modes. With this mode there is a partial separation of the sliding surfaces by the lubricant film; however, as in boundary lubrication, limitations on surface speed and wear will result. With this type of lubrication a surface velocity in excess of 10 feet per minute is required with resulting coefficients of friction of 0.02 to 0.08.

A journal bearing in starting up and accelerating to its operating point passes through all three modes of operation. At rest, the journal and bearing are in contact and thus when starting, the operation is in the boundary lubrication region. As the shaft begins to rotate more rapidly and the hydrodynamic film starts to build up, bearing operation enters the region of mixed-film lubrication. When design speeds and loads are reached, the hydrodynamic action in a properly designed bearing will now promote full film lubrication.

**Methods of Retaining Bearings.** — A number of methods are available to insure that a bearing remain in place within a housing. Which method to use depends upon the particular application but requires first that the unit lends itself to convenient assembly and disassembly; additionally, the bearing wall should be of uniform thickness to avoid introduction of weak points in the construction which may lead to elastic or thermal distortion.

*Press or Shrink Fit:* One common and satisfactory technique for retaining the bearing is to press or shrink the bearing in the housing with an interference fit. This method permits the use of bearings having uniform wall thickness over the entire bearing length.

Standard bushings with finished inside and outside diameters are available in sizes up to approximately 5-inches inside diameter. Stock bushings are commonly provided 0.002- to 0.003-inch over nominal on outside diameter sizes of 3 inches or less. For diameters greater than 3 inches, actual outside diameters are 0.003 to 0.005 inch over nominal. Since these tolerances are built into standard bushings, the amount of press fit is controlled by the housing bore size.

As a result of a press or shrink fit, the bore of the bearing material "closes in" by some amount. In general, this diameter decrease is approximately 70 to 100 per cent of the amount of the interference fit. Any attempt to accurately predict the amount of close-in, in an effort to avoid final clearance machining, should be avoided.

Shrink fits may be accomplished by chilling the bearing in a mixture of dry-ice and alcohol, or in liquid air. These methods are easier than heating the housing and are preferred. Dry ice in alcohol has a temperature of −110 degrees F and liquid air boils at −310 degrees F.

When a bearing is pressed into the housing, the driving force should be uniformly applied to the end of the bearing to avoid upsetting or peening of the bearing. Of equal importance, the mating surfaces must be clean, smoothly finished, and free of machining imperfections.

*Keying Methods:* A variety of methods can be used to fix the position of the bearing with respect to its housing by "keying" the two together. Possible keying methods are

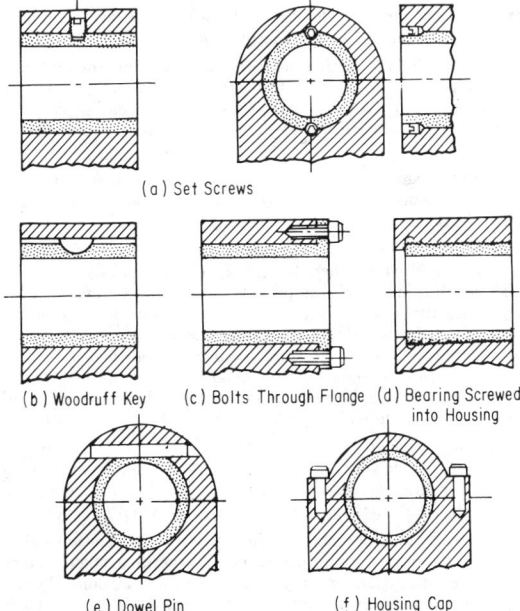

(a) Set Screws

(b) Woodruff Key    (c) Bolts Through Flange    (d) Bearing Screwed into Housing

(e) Dowel Pin    (f) Housing Cap

Fig. 3. Methods of bearing retention.

shown in Fig. 3 including: (a) Set screws; (b) Woodruff keys; (c) Bolted bearing flanges; (d) Threaded bearings; (e) Dowel pins; and (f) Housing caps.

Factors to be considered when selecting one of these methods are:

1. Maintaining uniform wall thickness of the bearing material, if possible, especially in the load-carrying region of the bearing.

2. Providing as much contact area as possible between bearing and housing. Mating surfaces should be clean, smooth, and free from imperfections to facilitate heat transfer.

3. Preventing any local deformation of the bearing that might result from the keying method. Machining after keying is recommended.

4. Considering the possibility of bearing distortion resulting from the effect of temperature·changes on the particular keying method.

**Methods of Sealing.** — In applications where lubricants or process fluids are utilized in operation, provision must be made normally to prevent leakage to other areas. This is accomplished by the use of static and dynamic type sealing devices. In general three terms are used to describe the devices used for sealing:

*Seal:* A means of preventing migration of fluids, gases, or particles across a joint or opening in a container.

*Packing:* A dynamic seal, used where some form of relative motion occurs between rigid members of an assembly.

*Gaskets:* A static seal, used where there is no relative motion between joined parts.

Two major functions must be achieved by all sealing applications: prevent escape of fluid; and prevent migration of foreign matter from the outside.

The first determination in selecting the proper seal is whether the application is static or dynamic. To meet the requirements of a static application, there must be no relative motion betwen the joining parts or between the seal and the mating part. If there is any relative motion, the application must be considered dynamic, and the seal selected accordingly.

Dynamic sealing requires control of fluids leaking between parts with relative motion. Two primary methods are used to this end: positive contact or rubbing seals; and controlled clearance non-contact seals.

*Positive Contact or Rubbing Seals:* These are utilized where positive containment of liquids or gases is required, or where the seal area is continously flooded. If properly selected and applied, contact seals can provide zero leakage for most fluids. However, because they are sensitive to temperature, pressure, and speed, improper application can result in early failure. These seals are applicable to rotating and reciprocating shafts. In many cases, the positive-contact seals are available as off-the-shelf items. In other instances, they are custom-designed to the special demands of a particular application. Custom design is offered by many seal manufacturers and, for extreme cases, probably offers the best solution to the sealing problem.

*Controlled Clearance Non-Contact Seals:* Representative of the controlled-clearance seals, which includes all seals in which there is no rubbing contact between the rotating and stationary members, are throttling bushings and labyrinths. Both of these types operate by fluid-throttling action in narrow annular or radial passages.

Clearance seals are frictionless and very insensitive to temperature and speed. They are chiefly effective as devices for limiting leakage rather than stopping it completely. Although they are employed as primary seals in many applications, the clearance seal also finds use as auxiliary protection in contact-seal applications. These seals are usually designed into the equipment by the designer himself, and they can take on many different forms.

Advantages of this seal are that friction is kept to an absolute minimum and that there is no wear or distortion during the life of the equipment. However, there are two significant disadvantages: The seal has limited use when leakage rates are critical; and it becomes quite costly as the configuration becomes elaborate.

*Static Seals:* Static seals such as gaskets, "O" rings, and molded packings cover very broad ranges of both design and materials. Some of the typical types are listed as follows: 1. Molded packings: A. Lip type, B. Squeeze-molded; 2. Simple compression packings; 3. Diaphragm seals; 4. Non-metallic gaskets; 5. "O" rings; and 6. Metallic gaskets and "O" rings.

Detailed design information for specific products should be obtained directly from manufacturers.

**Hardness and Surface Finish.** — Even in well-lubricated full-film sleeve bearings, momentary contact between journal and bearing may occur under such conditions as starting, stopping, or overloading. In mixed-film and boundary-film lubricated sleeve bearings, continuous metal-to-metal contact occurs. Hence, to allow for any necessary wearing-in, the journal is usually made harder than the bearing material. This allows the effects of scoring or wearing to take place on the bearing, which is more easily replaced, rather than on the more expensive shaft. As a general rule, recommended Brinnell hardness of the journal is at least 100 points harder than the bearing material.

The softer cast bronzes used for bearings are those with high lead content and very little tin. Such bronzes give adequate service in boundary and mixed-film applications where full advantage is taken of their excellent "bearing" characteristics.

High-tin, low-lead content cast bronzes are the harder bronzes and these have high ultimate load-carrying capacity; higher journal hardnesses are required with these bearing bronzes. Aluminum bronze, for example, requires a journal hardness in the range of 550 to 600 Bhn.

In general, harder bearing materials require better alignment and more reliable lubrication to minimize local heat generation if and when the journal touches the shaft. Also, abrasives which find their way into the bearing are a problem for the harder bearing and greater care should be taken to exclude them.

*Surface Finish:* Whether bearing operation is complete boundary, mixed film, or fluid film, surface finish of journal and bearing must receive careful attention. In applications where operation is hydrodynamic or full film, peak surface variations should be less than the expected minimum film thickness; otherwise, peaks on the journal surface will contact peaks on the bearing surface, with resulting high friction and temperature rise. Ranges of surface roughness obtained by various finishing methods are: boring, broaching, and reaming, 32 to 64 microinches, rms; grinding, 16 to 64 microinches, rms; and fine grinding, 4 to 16 microinches, rms.

In general, the better surface finishes are required for full-film bearings operating at high eccentricity ratios since full-film lubrication must be maintained with small clearances, and metal-to-metal contact must be avoided. Also, the harder the material the better the surface finish required. For boundary and mixed-film applications surface finish requirements may be somewhat relaxed since bearing wear-in will in time smooth the surfaces.

Figure 4 is a general guide to the ranges required for bearing and journal surface finishes. Selecting a particular surface finish in each range can be simplified by observing the general rule that smoother finishes are required for the harder materials, for high loads, and for high speeds.

**Machining.** — The methods most commonly used in finishing journal bearing bores are: boring, broaching, reaming, and burnishing.

Boring of journal bearings provides the best concentricity, alignment, and size control; this being the finishing method of choice when close tolerances and clear-

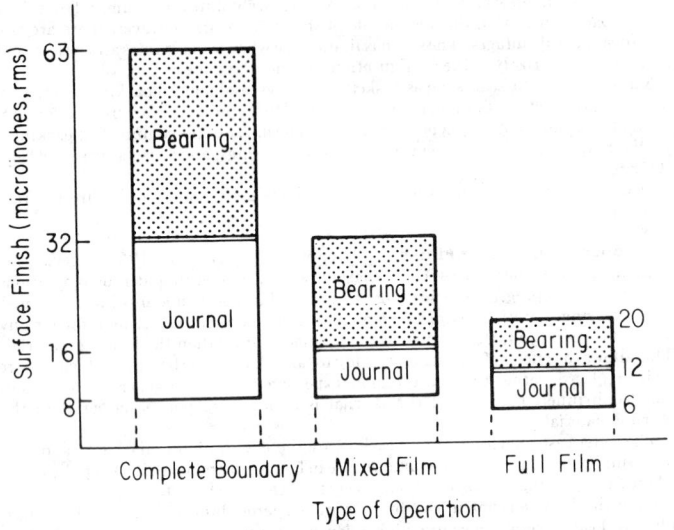

Fig. 4.    Recommended ranges of surface finish for the three types of
sleeve bearing operations.

ances are desirable. Broaching is a rapid finishing method providing good size and
alignment control when adequate piloting is possible. Soft babbitt materials are
particularly compatible with the broaching method. A third finishing method, reaming, facilitates good size and alignment control when piloting is utilized. Reaming can
be accomplished both manually or by machine; the machine method being preferred.
Burnishing is a fast sizing operation which gives good alignment control, but does not
give as good size control as the cutting methods. It is not recommended for soft materials
such as babbitt. Burnishing has an ironing effect which gives added seating of the
bushing outside diameter in the housing bore; consequently, it is often used for this
purpose, especially on a 1/32 inch wall bushing, even if a further sizing operation is to be
used subsequently.

**Methods of Lubrication.** — There are numerous ways to supply lubricant to
bearings. The more common of these are described as follows:

*Pressure lubrication,* in which an abundance of oil is fed to the bearing from a central
groove, single or multiple holes, or axial grooves, is effective and efficient. The moving
oil assists in flushing dirt from the bearing and helps keep the bearing cool. In fact it
removes heat faster than other lubricating methods and, therefore, permits thinner oil
films and unimpaired actual load capacities. The oil supply pressure needed for
bushings carrying the basic load is directly proportional to the shaft speed, but for most
installations 50 psi will be adequate.

*Splash fed* is a term applied to a variety of intermittently lubricated bushings. It
includes everything from bearings spattered with oil from the action of other moving
parts to bearings regularly dipped in oil. Like oil bath lubrication, splash feeding is
practical when the housing can be made oiltight and when the moving parts do not
churn the oil. The fluctuating nature of the load and the intermittent oil supply in

splash fed applications requires the designer to use experience and judgment when determining the probable load capacity of bearings lubricated in this way.

*Oil bath lubrication,* in which the bushing is submerged in oil, is the most reliable of all methods except pressure lubrication. It is practical if the housing can be made oiltight, and if the shaft speed is not so great as to cause excessive churning of the oil.

*Oil ring lubrication,* in which oil is supplied to the bearing by a ring in contact with the shaft will, within reasonable limits, bring enough oil to the bearing to maintain hydrodynamic lubrication. If the shaft speed is too low, little oil will follow the ring to the bearing; and, if the speed is too high, the ring speed will not keep pace with the shaft. Also, a ring revolving at high speed will lose oil by centrifugal force. For best results, the peripheral speed of the shaft should be between 200 and 2000 feet per minute. Safe load to achieve hydrodynamic lubrication should be one-half of that for pressure fed bearings.

*Wick or waste pack lubrication* delivers oil to a bushing by the capillary action of a wick or waste pack; the amount delivered being proportional to the size of the wick or pack. Unless the load is light, hydrodynamic lubrication is doubtful. The safe load, then, to

**Table 1. Oil Viscosity Unit Conversion**

| Convert from | Convert to | | | | |
|---|---|---|---|---|---|
| | Poise (P) | Centipoise (Z) | Reyn ($\mu$) | Stoke (S) | Centistoke ($\nu$) |
| | Multiplying Factors | | | | |
| Poise (P) $\dfrac{\text{dyne-sec}}{\text{cm}^2}$ or $\dfrac{\text{gram mass}}{\text{sec-cm}}$ | 1 | 100 | $1.45 \times 10^{-5}$ | $\dfrac{1}{\rho}$ | $\dfrac{100}{\rho}$ |
| Centipoise (Z) $\dfrac{\text{dyne-sec}}{100\ \text{cm}^2}$ or $\dfrac{\text{gram mass}}{100\ \text{sec-cm}}$ | 0.01 | 1 | $1.45 \times 10^{-7}$ | $\dfrac{0.01}{\rho}$ | $\dfrac{1}{\rho}$ |
| Reyn ($\mu$) $\dfrac{\text{lb force-sec}}{\text{in.}^2}$ | $6.9 \times 10^4$ | $6.9 \times 10^6$ | 1 | $\dfrac{6.9 \times 10^4}{\rho}$ | $\dfrac{6.9 \times 10^6}{\rho}$ |
| Stoke (S) $\dfrac{\text{cm}^2}{\text{sec}}$ | $\rho$ | $100\,\rho$ | $1.45 \times 10^{-5}\,\rho$ | 1 | 100 |
| Centistoke ($\nu$) $\dfrac{\text{cm}^2}{100\ \text{sec}}$ | $0.01\,\rho$ | $\rho$ | $1.45 \times 10^{-7}\,\rho$ | 0.01 | 1 |

$\rho$ = Specific gravity of the oil.

To convert from a value in the "Convert from" column to a value in a "Convert to" column, multiply the "Convert from" column by the figure in the intersecting block, e.g. to change from Centipoise to Reyn, multiply Centipoise value by $1.45 \times 10^{-7}$.

achieve hydrodynamic lubrication, should be one-quarter of that of pressure fed bearings.

*Grease* packed in a cavity surrounding the bushing is less adequate than an oil system, but it has the advantage of being more or less permanent. Although hydrodynamic lubrication is possible under certain very favorable circumstances, boundary lubrication is the usual state.

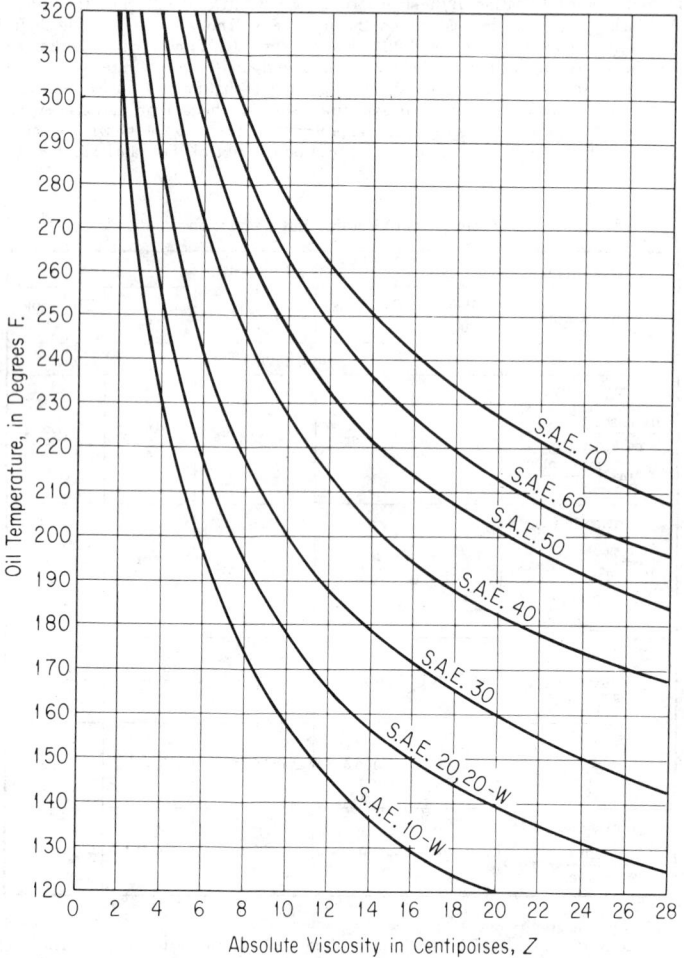

Fig. 5. Viscosity vs. temperature — SAE oils.

**Lubricants.** — The value of an oil as a lubricant depends mainly upon its film-forming capacity; that is, its capability of maintaining a film of oil between the bearing surfaces. The film-forming capacity depends to a large extent on the viscosity of the oil, but this should not be understood to mean that oil of the highest viscosity is in every case the most suitable lubricant. For practical reasons, an oil of the lowest viscosity which will retain an unbroken oil film between the bearing surfaces is the most suitable for purposes of lubrication. This is because a higher viscosity than that necessary to maintain the oil film results in a waste of power due to the expenditure of energy necessary to overcome the internal friction of the oil itself.

Figure 5 provides representative values of viscosity in centipoises for SAE mineral oils. Table 1 is provided as a means of converting viscosities of other units to centipoises.

**Lubricant Selection.** — In selecting lubricants for journal bearing operation several factors must be considered: 1. Type of operation (Full-, mixed-, or boundary film) anticipated; 2. Surface speed; and 3. Bearing loading.

Figure 6 combines these factors and facilitates general selection of the proper lubricant viscosity range.

As an example of using these curves, consider a lightly loaded bearing operating at 2000 rpm. At the bottom of the figure locate 2000 rpm and move vertically to intersect the light-load full-film lubrication curve which would indicate an SAE 5 oil.

As a general rule-of-thumb heavier oils are recommended for high loads and lighter oils for high speeds.

In addition, other than using conventional lubricating oils, journal bearings may be lubricated with either greases or solid lubricants. Some of the reasons for use of these lubricants are to:

  1. Lengthen the period between relubrication;

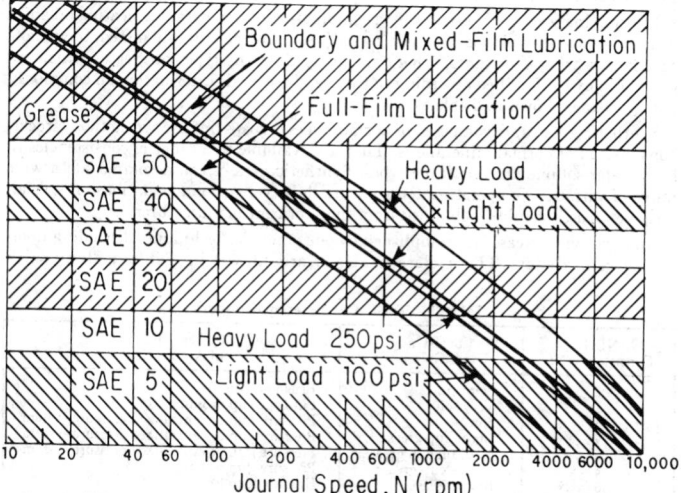

Fig. 6.   Lubricant selection guide.

Table 2.   Commonly Used Greases and Solid Lubricants

| Type | Operating Temperature, Degrees F | Load | Comments |
|---|---|---|---|
| Greases | | | |
| Calcium or lime soap | 160 | Moderate | ... |
| Sodium soap | 300 | Wide | For wide speed range |
| Aluminum soap | 180 | Moderate | ... |
| Lithium soap | 300 | Moderate | Good low temperature |
| Barium soap | 350 | Wide | ... |
| Solid Lubricants | | | |
| Graphite | 1000 | Wide | ... |
| Molybdenum disulphide | −100 to 750 | Wide | ... |

2. Avoid contaminating surrounding equipment or material with "leaking" lubricating oil;

3. Provide effective lubrication under extreme temperature ranges;

4. Provide effective lubrication in the presence of contaminating atmospheres; and

5. Prevent intimate metal-to-metal contact under conditions of high unit pressure which might destroy boundary lubricating films.

*Greases:* Where full-film lubrication is not possible or is impractical for slow-speed fairly high-load applications, greases are widely used as bearing lubricants. Although full-film lubrication with grease is possible, it is not normally considered since an elaborate pumping system is required to continuously supply a prescribed amount of grease to the bearing. Bearings supplied with grease are usually lubricated periodically. Grease lubrication, therefore, implies that the bearing will operate under conditions of complete boundary lubrication and should be designed accordingly.

Lubricating greases are essentially a combination of a mineral lubricating oil and a thickening agent, which is usually a metallic soap. When suitably mixed, they make excellent bearing lubricants. There are many different types of greases which, in general, may be classified according to the soap base used. Information on commonly used greases is shown in Table 2.

Synthetic greases are composed of normal types of soaps but use synthetic hydrocarbons instead of normal mineral oils. They are available in a range of consistencies in both water-soluble and insoluble types. Synthetic greases can accommodate a wide range of variation in operating temperature; however, recommendations on special-purpose greases should be obtained from the lubricant manufacturer.

Application of grease is accomplished by one of several techniques depending upon grease consistency. These classifications are shown in Table 3 along with

Table 3.   NLGI* Consistency Numbers

| NLGI Consistency No. | Consistency of Grease | Typical Method of Application |
|---|---|---|
| 0 | Semifluid | Brush or gun |
| 1 | Very soft | Pin-type cup or gun |
| 2 | Soft | Pressure gun or centralized pressure system |
| 3 | Light cup grease | Pressure gun or centralized pressure system |
| 4 | Medium cup grease | Pressure gun or centralized pressure system |
| 5 | Heavy cup grease | Pressure gun or hand |
| 6 | Block grease | Hand, cut to fit |

*National Lubricating Grease Institute

typical methods of application. Grooves for grease are generally greater in width, up to 1.5 times, than for oil.

Coefficients of friction for grease-lubricated bearings range from 0.08 to 0.16, depending upon consistency of the grease, frequency of lubrication, and type of grease. An average value of 0.12 may be used for design purposes.

*Solid Lubricants:* The need for effective high-temperature lubricants led to the development of several solid lubricants. Essentially, solid lubricants may be described as low-shear-strength solid materials. Their function within a bronze bearing is to act as an intermediary material between sliding surfaces. Since these solids have very low shear strength, they shear more readily than the bearing material and thereby allow relative motion. So long as solid lubricant remains between the moving surfaces, effective lubrication is provided and friction and wear are reduced to acceptable levels.

Solid lubricants provide the most effective boundary films in terms of reduced friction, wear, and transfer of metal from one sliding component to the other. However, there is a significant deterioration in these desirable properties as the operating temperature of the boundary film approaches the melting point of the solid film. At this temperature the friction may increase by a factor of 5 to 10 and the rate of metal transfer may increase by as much as 1000. What occurs is that the molecules of the lubricant lose their orientation to the surface that exists when the lubricant is solid. As the temperature further increases, additional deterioration sets in with the friction increasing by some additional small amount but the transfer of metal accelerates by an additional factor of 20 or more. The final effect of too high temperature is the same as metal-to-metal contact without benefit of lubricant. These changes which are due to the physical state of the lubricant are reversed when cooling takes place.

The effects just described also partially explain why fatty acid lubricants are superior to paraffin base lubricants. The fatty acid lubricants react chemically with the metallic surfaces to form a metallic soap that has a higher melting point than the lubricant itself, the result being that the breakdown temperature of the film, now in the form of a metallic soap is raised so that it acts more like a solid film lubricant than a fluid film lubricant.

## Journal or Sleeve Bearings

Although this type of bearing may take many shapes and forms, there are always three basic components: journal or shaft, bushing or bearing, and lubricant. Figure 7 shows these components with the nomenclature generally used to describe a journal bearing: $W$ = applied load, $N$ = revolution, $e$ = eccentricity of journal center to bearing center, $\theta$ = attitude angle, which is the angle between the applied load and the point of minimum film thickness, $d$ = diameter of the shaft, $c_d$ = bearing clearance, $d + c_d$ = diameter of the bearing and $h_o$ = minimum film thickness.

**Grooving and Oil Feeding.** — Grooving in a journal bearing has two purposes: (1) to establish and maintain an efficient film of lubricant between the bearing moving surfaces and (2) to provide adequate bearing cooling. The obvious and only practical location for introducing lubricant to the bearing is in a region of low pressure. A typical pressure profile of a bearing is shown by Fig. 8. The arrow $W$ shows the applied load. Typical grooving configurations used for journal bearings are shown in Fig. 9.

**Heat Radiating Capacity.** — In a self-contained lubrication system for a journal bearing, the heat generated by bearing friction must be removed to prevent continued temperature rise to an unsatisfactory level. The heat-radiating capacity $H_R$ of the bearing in foot-pounds per minute may be calculated from the formula $H_R = Ld\, Ct_R$ in which $C$ is a constant determined by O. Lasche, and $t_R$ is temperature rise in degrees Fahrenheit. Values for the product $Ct_R$ may be found from

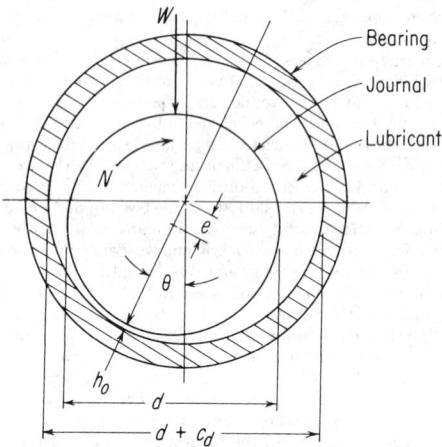

Fig. 7.   Basic components of a journal bearing.

the curves in Fig. 10 for various values of bearing temperature rise $t_R$ and for three operating conditions.   In this equation $L$ = the total length of the bearing in inches and $d$ = the bearing diameter in inches.

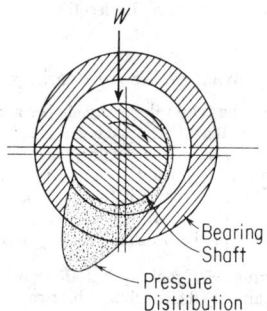

Fig. 8.   Typical pressure profile of journal bearing.

**Journal Bearing Design Notation.** — The symbols used in the following step-by-step procedure for lubrication analysis and design of a plain sleeve or journal bearing are listed below:

$c$ = specific heat of lubricant, Btu/lb/°F
$c_d$ = diametral clearance, inches
$C_n$ = bearing capacity number
$d$ = journal diameter, inches
$e$ = eccentricity, inches

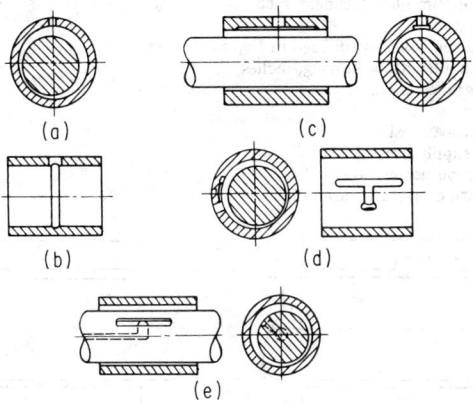

Fig. 9. Types of journal bearing oil grooving: (a) Single inlet hole. (b) Circular groove. (c) Straight axial groove. (d) Straight axial groove with feeder groove. (e) Straight axial groove in shaft.

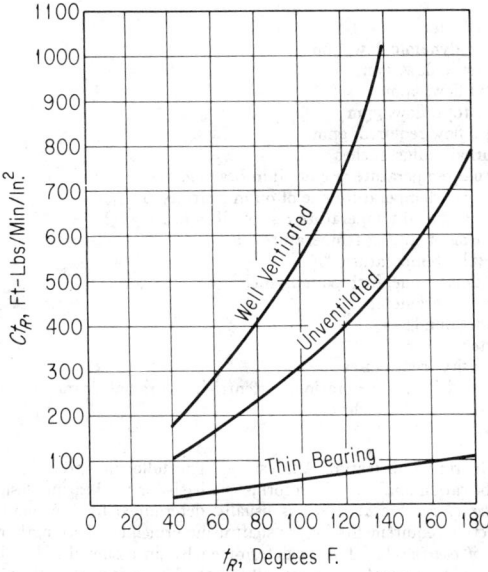

Fig. 10. Heat radiating capacity factor, $Ct_R$, vs. bearing temperature rise, $t_R$ — journal bearings.

$h_o$ = minimum film thickness, inch
$K$ = constants
$l$ = bearing length as defined in Fig. II, inches
$L$ = actual length of bearing, inches
$m$ = clearance modulus
$N$ = rpm
$p_b$ = unit load, psi
$p_s$ = oil supply pressure, psi
$P_f$ = friction horsepower
$P'$ = bearing pressure parameter

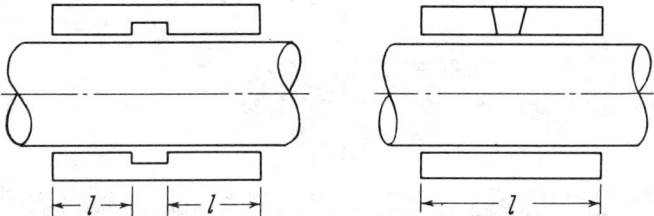

Fig. II. Length, $l$, of bearing for circular groove type (left) and single inlet hole type (right).

$q$ = flow factor
$Q_1$ = hydrodynamic flow, gpm
$Q_2$ = pressure flow, gpm
$Q$ = total flow, gpm
$Q_{new}$ = new total flow, gpm
$Q_R$ = total flow required, gpm
$r$ = journal radius, inches
$\Delta t$ = actual temperature rise of oil in bearing, °F
$\Delta t_a$ = assumed temperature rise of oil in bearing, °F
$\Delta t_{new}$ = new assumed temperature rise of oil in bearing, °F
$t_b$ = bearing operating temperature, °F
$t_{in}$ = oil inlet temperature, °F
$T_f$ = friction torque, inch-pounds/inch
$T'$ = torque parameter
$W$ = load, pounds
$X$ = factor
$Z$ = viscosity, centipoises
$\epsilon$ = eccentricity ratio — ratio of eccentricity to radial clearance
$\alpha$ = oil density, lbs/inch³

**Journal Bearing Lubrication Analysis.** — The following procedure leads to a complete lubrication analysis which forms the basis for the bearing design.

1. *Diameter of bearing d.* This is usually determined by considering strength and/or deflection requirements for the shaft using principles of strength of materials.

2. *Length of bearing L.* This is determined by an assumed $l/d$ ratio in which $l$ may or may not be equal to the overall length, $L$ (See Step 6). Bearing pressure and the possibility of edge loading due to shaft deflection and misalignment are factors to be considered. In general, shaft misalignment resulting from location

tolerances and/or shaft deflections should be maintained below 0.0003 inch per inch of length.

3. *Bearing pressure $p_b$*. The unit load in pounds per square inch is calculated from the formula:

$$p_b = \frac{W}{Kld}$$

where $K$ = 1 for single oil hole
       $K$ = 2 for central groove
      $W$ = load, pounds
       $l$ = bearing length as defined in Fig. 11, inches
      $d$ = journal diameter, inches

Typical unit loads in service are shown in Table 4. These pressures can be used as a safe guide in selection. However, if space limitations impose a higher limit of loading, the complete lubrication analysis and evaluation of material properties will determine acceptability.

4. *Diametral clearance $c_d$*. This is selected on a trial basis from Fig. 12 which shows suggested diametral clearance ranges for various shaft sizes and for two speed ranges. These are *hot* or *operating* clearances so that thermal expansion of journal and bearing to these temperatures must be taken into consideration in establishing machining dimensions. The optimum operating clearance should be determined on the basis of a complete lubrication analysis (See paragraph following Step 23).

5. *Clearance modulus $m$*. This is calculated from the formula:

$$m = \frac{c_d}{d}$$

6. *Length to diameter ratio $l/d$*. This is usually between 1 and 2; however, with the modern trend toward higher speeds and more compact units, lower ratios down to 0.3 are used. In shorter bearings there is a consequent reduction in load carrying capacity due to excessive end or side leakage of lubricant. In longer bearings there may be a tendency towards edge loading. Length $l$ for a single oil feed hole is taken as the total length of the bearing as shown in Fig. 11. For a central oil groove length, $l$ is taken as one-half the total length.

Typical $l/d$ ratio's use for various types of applications are given in Table 5.

**Table 4. Allowable Sleeve Bearing Pressures for Various Classes of Bearings***

| Types of Bearing or Kind of Service | Pressure, Lbs. per Sq. In. | Types of Bearing or Kind of Service | Pressure, Lbs. per Sq. In. |
|---|---|---|---|
| Electric Motor & Generator Bearings (General) | 100–200 | Diesel Engine, Rod | 1000–2000 |
| Turbine & Reduction Gears | 100–250 | Wrist Pins | 1800–2000 |
| Heavy Line Shafting | 100–150 | Automotive, Main Bearings | 500–700 |
| Locomotive Axles | 300–350 | Rod Bearings | 1500–2500 |
| Light Line Shafting | 15–35 | Centrifugal Pumps | 80–100 |
| Diesel Engine, Main | 800–1500 | Aircraft Rod Bearings | 700–3000 |

* These pressures in pounds per square inch of area equal to length times diameter are intended as a general guide only. The allowable unit pressure depends upon operating conditions, especially in regard to lubrication, design of bearings, workmanship, velocity, and nature of load.

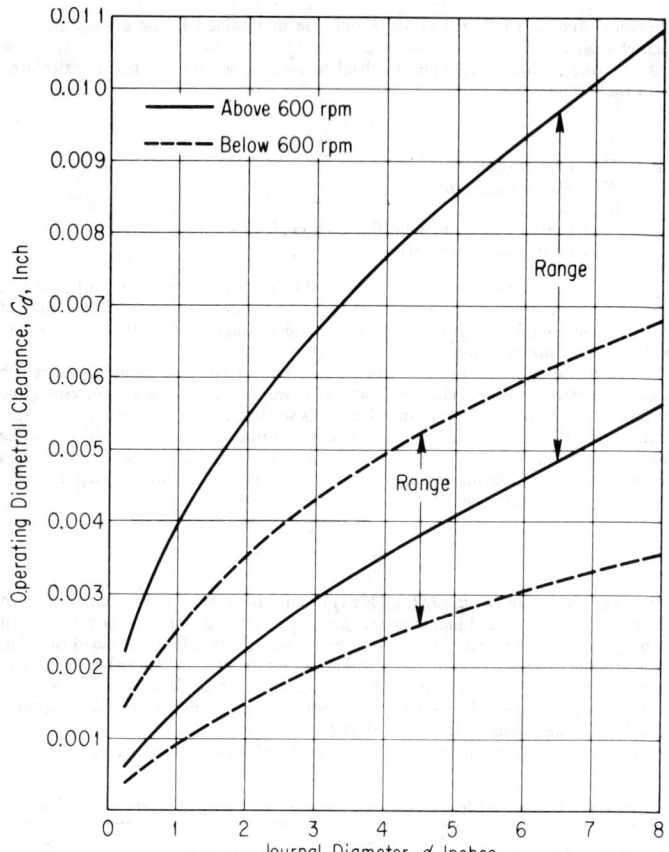

Fig. 12. Operating diametral clearance, $c_d$ vs. journal diameter, $d$.

Table 5.   Representative $l/d$ Ratios

| Type of Service | $l/d$ | Type of Service | $l/d$ |
|---|---|---|---|
| Gasoline and diesel engine | | Light shafting | 2.5 to 3.5 |
| Main bearings and crankpins | 0.3 to 1.0 | Heavy shafting | 2.0 to 3.0 |
| Generators and motors | 1.2 to 2.5 | Steam engine | |
| Turbogenerators | 0.8 to 1.5 | Main bearings | 1.5 to 2.5 |
| Machine tools | 2.0 to 3.0 | Crank and wrist pins | 1.0 to 1.3 |

7. *Assumed operating temperature* $t_b$.   A temperature rise of the lubricant as it passes through the bearing is assumed and the consequent operating temperature in degrees F. is calculated from the formula:

$$t_b = t_{in} + \Delta t_a$$

where $t_{in}$ = inlet temperature of oil in °F

$\Delta t_a$ = assumed temperature rise of·oil in bearing in °F

An initial assumption of 20 °F is usually made.

8. *Viscosity of lubricant Z.* The viscosity in centipoises at the assumed bearing operating temperature is found from the curve in Fig. 5 which shows the viscosity of SAE grade oils versus temperature.

9. *Bearing pressure parameter P'.* This is required to find the eccentricity ratio and is calculated from the formula:

$$P' = \frac{6.9(1000m)^2 p_b}{ZN}$$

where $N$ = rpm

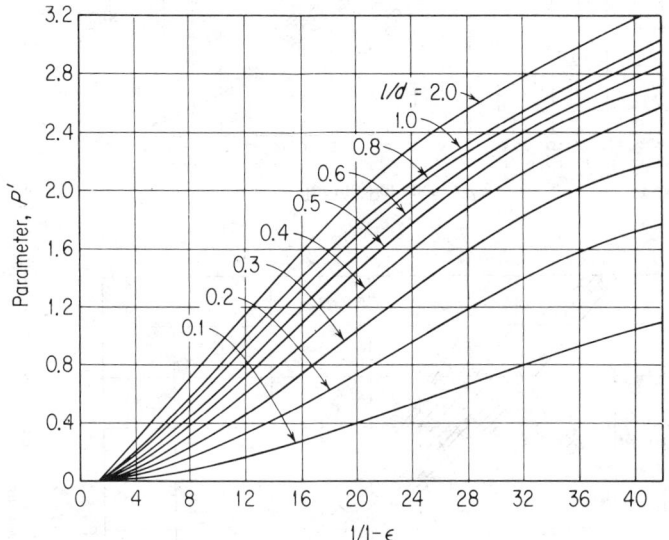

Fig. 13. Bearing parameter, $P'$, vs. eccentricity ratio, $1/(1 - \epsilon)$ — journal bearings.

10. *Eccentricity ratio ϵ.* Using $P'$ and $l/d$, the value of $1/(1 - \epsilon)$ is determined from Fig. 13 and from this ϵ can be determined.

11. *Torque parameter T'.* This is obtained from Fig. 14 or Fig. 15 using $1/(1 - \epsilon)$ and $l/d$.

12. *Friction torque T.* This is calculated from the formula:

$$T = \frac{T' r^2 Z N}{6900(1000m)}$$

where $r$ = journal radius, inches

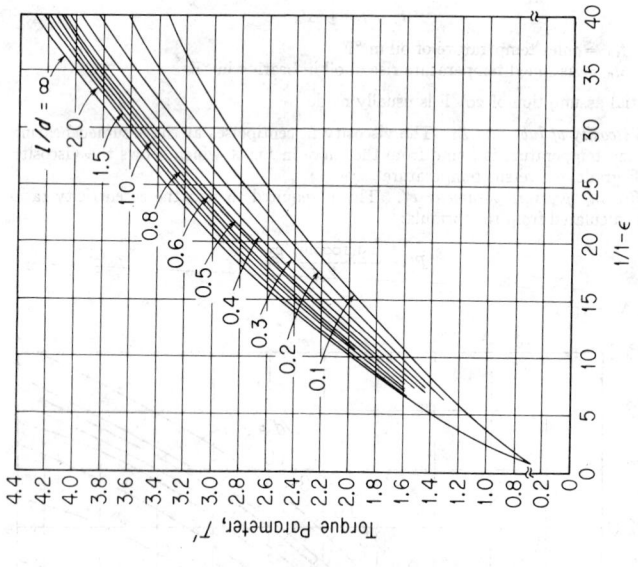

**Fig. 15.** Torque parameter, $T'$, vs. eccentricity ratio, $1/(1 - e)$ — journal bearings.

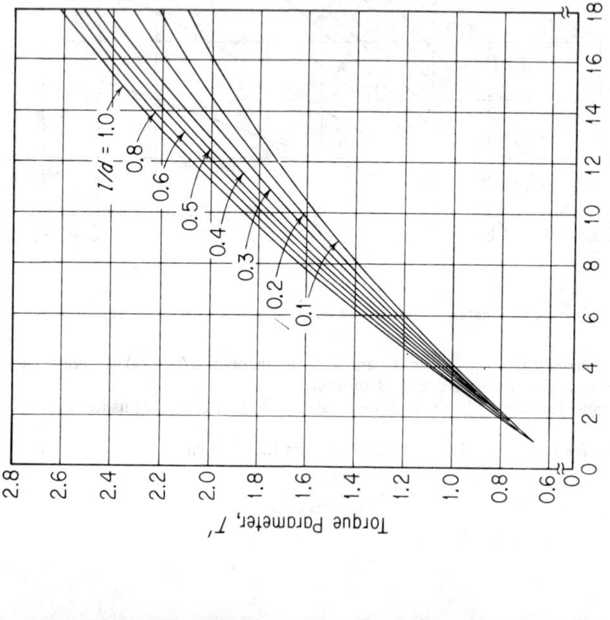

**Fig. 14.** Torque parameter, $T'$, vs. eccentricity ratio, $1/(1 - \epsilon)$ — journal bearings.

**Table 6. X Factor vs. Temperature of Mineral Oils**

| Temperature | X Factor |
|---|---|
| 100 | 12.9 |
| 150 | 12.4 |
| 200 | 12.1 |
| 250 | 11.8 |
| 300 | 11.5 |

13. *Friction horsepower $P_f$.* This is calculated from the formula:

$$P_f = \frac{KTNl}{63,000}$$

Where $K$ = 1 for single oil hole, 2 for central groove

14. *Factor X.* This factor is used in the calculation of the lubricant flow and can either be obtained from Table 6 or calculated from the formula:

$$X = 0.1837/\alpha c$$

where $\alpha$ = oil density in pounds per cubic inch
$c$ = specific heat of lubricant in BTU/lb./°F

15. *Total flow of lubricant required $Q_R$.* This is calculated from the formula:

$$Q_R = \frac{X(P_f)}{\Delta t_a}$$

16. *Bearing capacity number $C_n$.* This is needed to obtain the flow factor and is calculated from the formula:

$$C_n = \left(\frac{l}{d}\right)^2 \Big/ 60 P'$$

17. *Flow factor q.* This is obtained from the curve in Fig. 16.

18. *Actual hydrodynamic flow of lubricant $Q_1$.* This flow in gallons per minute is calculated from the formula:

$$Q_1 = \frac{N l c_d q d}{294}$$

19. *Actual pressure flow of lubricant $Q_2$.* This flow in gallons per minute is calculated from the formula:

$$Q_2 = \frac{K p_s c_d^3 d (1 + 1.5\epsilon^2)}{Zl}$$

where $K$ = 1.64 × 10^5 for single oil hole
$K$ = 2.35 × 10^5 for central groove
$p_s$ = oil supply pressure

20. *Actual total flow of lubricant Q.* This is obtained by adding the hydrodynamic flow and the pressure flow.

$$Q = Q_1 + Q_2$$

21. *Actual bearing temperature rise $\Delta t$.* This temperature rise in degrees F is obtained from the formula:

$$\Delta t = \frac{X(P_f)}{Q}$$

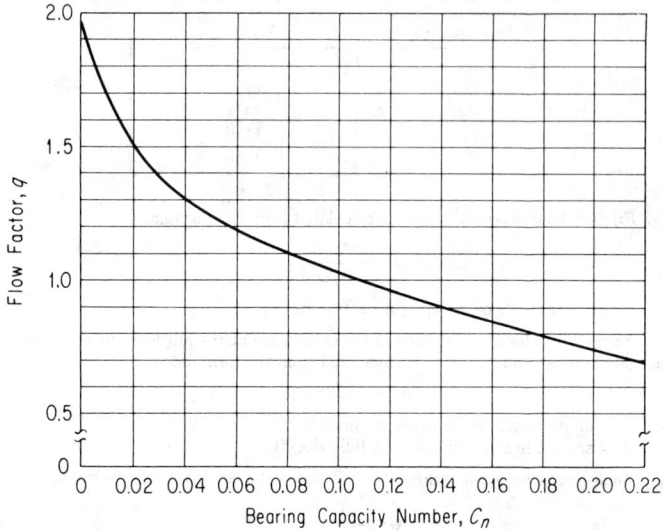

Fig. 16.  Flow factor, $q$, vs. bearing capacity number, $C_n$ — journal bearings.

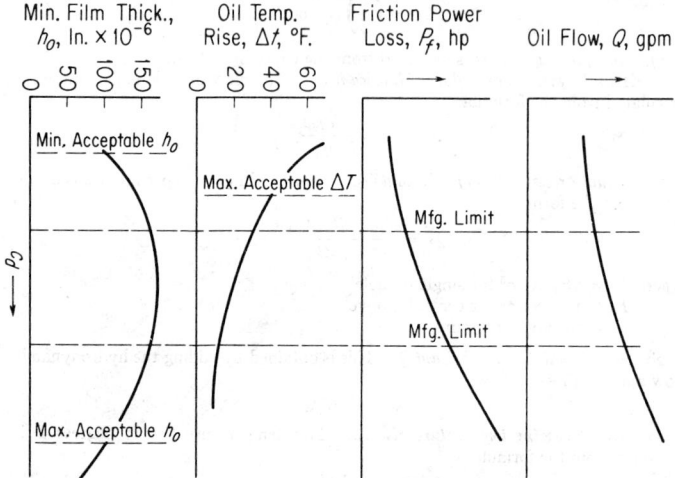

Fig. 17.  Example of lubrication analysis curves for journal bearing.

22. *Comparison of actual and assumed temperature rises.*
At this point if $\Delta t_a$ and $\Delta t$ differ by more than 5 degrees F, Steps 7 through 22 are repeated using a $\Delta t_{new}$ halfway between the former $\Delta t_a$ and $\Delta t$.

23. *Minimum film thickness $h_o$.* When Step 22 has been satisfied, the minimum film thickness in inches is calculated from the formula:

$$h_o = \frac{c_d}{2}(1 - \epsilon)$$

A new diametral clearance $c_d$ is now assumed and Steps 5 through 23 are repeated. When this has been done for a sufficient number of values for $c_d$, the full lubrication study is plotted as shown in Fig. 17. From this chart a working range of diametral clearance can be determined that optimizes film thickness, differential temperature, friction horsepower and oil flow.

**Use of Lubrication Analysis.** — Once the lubrication analysis has been completed and plotted as shown in Fig. 17. the following steps lead to the optimum bearing design, taking into consideration both basic operating requirements and requirements peculiar to the application.

1. Examine the curve (Fig. 17) for minimum film thickness and determine the acceptable range of diametral clearance, $c_d$, based on a minimum of $100 \times 10^{-6}$ inch.

2. Determine the minimum acceptable $c_d$ based on a maximum $\Delta t$ of $40°F$ from the oil temperature rise curve (Fig. 17).

3. If there are no requirements for maintaining low friction horsepower and oil flow, the possible limits of diametral clearance are now defined.

4. The required manufacturing tolerances can now be placed within this band to optimize $h_o$ as shown by Fig. 17.

5. If oil flow and power loss are a consideration, the manufacturing tolerances will then be shifted, within the range permitted by the requirements for $h_o$ and $\Delta t$

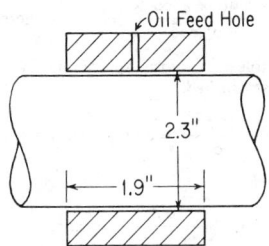

Fig. 18.  Full journal bearing example design.

*Example:* A full journal bearing, Fig. 18, 2.3 inches in diameter and 1.9 inches long is to carry a load of 6000 pounds at 4800 rpm, using SAE 30 oil supplied at 200°F through a single oil hole at 30 psi. Determine the operating characteristics of this bearing as a function of diametral clearance.

1. *Diameter of bearing.*  Given as 2.3 inches.

2. *Length of bearing.*  Given as 1.9 inches.

3. *Bearing pressure.*

$$p_b = \frac{6000}{1 \times 1.9 \times 2.3} = 1372 \text{ lbs. per sq. in.}$$

4. *Diametral clearance.* Assume $c_d$ is equal to 0.003 inch from Fig.12 for first calculation.

5. *Clearance modulus.*

$$m = \frac{0.003}{2.3} = 0.0013 \text{ inch}$$

6. *Length to diameter ratio.*

$$\frac{l}{d} = \frac{1.9}{2.3} = 0.83$$

7. *Assumed operating temperature.* If the temperature rise $\Delta t_a$ is assumed to be 20°F,

$$t_b = 200 + 20 = 220°F$$

8. *Viscosity of lubricant.* From Fig. 5, $Z = 7.7$ centipoises

9. *Bearing pressure parameter.*

$$P' = \frac{6.9 \times 1.3^2 \times 1372}{7.7 \times 4800} = 0.43$$

10. *Eccentricity ratio.* From Fig.13, $\dfrac{1}{1-\epsilon} = 6.8$ and $\epsilon = 0.85$

11. *Torque parameter.* From Fig. 14, $T' = 1.46$

12. *Friction torque.*

$$T_f = \frac{1.46 \times 1.15^2 \times 7.7 \times 4800}{6900 \times 1.3} = 7.96 \text{ inch-pounds per inch}$$

13. *Friction horsepower.*

$$P_f = \frac{1 \times 7.96 \times 4800 \times 1.9}{63,000} = 1.15 \text{ horsepower}$$

14. *Factor X.* From Table 6, $X = 12$, approximately

15. *Total flow of lubricant required.*

$$Q_R = \frac{12 \times 1.15}{20} = 0.69 \text{ gallon per minute}$$

16. *Bearing capacity number*

$$C_n = \frac{0.83^2}{60 \times 0.43} = 0.027$$

17. *Flow factor.* From Fig. 16, $q = 1.43$

18. *Actual hydrodynamic flow of lubricant.*

$$Q_1 = \frac{4800 \times 1.9 \times 0.003 \times 1.43 \times 2.3}{294} = 0.306 \text{ gallon per minute}$$

19. *Actual pressure flow of lubricant.*

$$Q_2 = \frac{1.64 \times 10^5 \times 30 \times 0.003^3 \times 2.3 \times (1 + 1.5 \times 0.85^2)}{7.7 \times 1.9} = 0.044 \text{ gallon per minute}$$

20. *Actual total flow of lubricant.*

$$Q = 0.306 + 0.044 = 0.350 \text{ gallon per minute}$$

21. *Actual bearing temperature rise.*

$$\Delta t = \frac{12 \times 1.15}{0.350} = 39.4°F$$

22. *Comparison of actual and assumed temperature rises.* Since $\Delta t_a$ and $\Delta t$ differ by more than 5°F, a new $\Delta t_a$, midway between these two, of 30°F is assumed and Steps 7 through 22 are repeated.

7a. *Assumed operating temperature.*

$$t_b = 200 + 30 = 230°F$$

8a. *Viscosity of lubricant.* From Fig. 5, $Z = 6.8$ centipoises

9a. *Bearing pressure parameter.*

$$P' = \frac{6.9 \times 1.3^2 \times 1372}{6.8 \times 4800} = 0.49$$

10a. *Eccentricity ratio.* From Fig. 13, $\frac{1}{1-\epsilon} = 7.4$ and $\epsilon = 0.86$

11a. *Torque parameter.* From Fig. 14, $T' = 1.53$

12a. *Friction torque.*

$$T_f = \frac{1.53 \times 1.15^2 \times 6.8 \times 4800}{6900 \times 1.3} = 7.36 \text{ inch-pounds per inch}$$

13a. *Friction horsepower.*

$$P_f = \frac{1 \times 7.36 \times 4800 \times 1.9}{63,000} = 1.07 \text{ horsepower}$$

14a. *Factor X.* From Table 6, $X = 11.9$, approximately

15a. *Total flow of lubricant required.*

$$Q_R = \frac{11.9 \times 1.07}{30} = 0.42 \text{ gallon per minute}$$

16a. *Bearing capacity number.*

$$C_n = \frac{0.83^2}{60 \times 0.49} = 0.023$$

17a. *Flow factor.* From Fig. 16, $q = 1.48$

18a. *Actual hydrodynamic flow of lubricant.*

$$Q_1 = \frac{4800 \times 1.9 \times 0.003 \times 1.48 \times 2.3}{294} = 0.317 \text{ gallon per minute}$$

19a. *Pressure flow.*

$$Q_2 = \frac{1.64 \times 10^5 \times 30 \times 0.003^3 \times 2.3 \times (1 + 1.5 \times 0.86^2)}{6.8 \times 1.9} = 0.050 \text{ gallon per minute}$$

20a. *Actual flow of lubricant.*

$$Q_{new} = 0.317 + 0.050 = 0.367 \text{ gallon per minute}$$

21a. *Actual bearing temperature rise.*

$$\Delta t = \frac{11.9 \times 1.06}{0.367} = 34.4°F$$

22a. *Comparison of actual and assumed temperature rises.* Now $\Delta t$ and $\Delta t_a$ are within 5°F.

23. *Minimum film thickness.*

$$h_o = \frac{0.003}{2}(1 - 0.86) = 0.00021 \text{ inch.}$$

This analysis may now be repeated for other values of $c_d$ determined from Fig. 12 and a complete lubrication analysis performed and plotted as shown in Fig. 17. An operating range for $c_d$ can then be determined to optimize minimum clearance, friction horsepower loss, lubricant flow and temperature rise.

## Thrust Bearings

Thrust bearings, as the name implies, are used to either absorb axial shaft loads or to position shafts. Brief descriptions of the normal designs for these bearings follow with approximate design methods for each. The generally accepted load ranges for these types of bearings are given in Table 7 and the schematic configurations are shown in Fig. 19.

*The parallel or flat plate thrust bearing* is probably the most frequently used type. It is the simplest and lowest in cost of those considered; however, it is also the least capable of absorbing load as can be seen from Table 7. It is most generally used as a positioning device where loads are either light or occasional.

*The step bearing* is, like the parallel plate, also a relatively simple design. This type of bearing will accept the normal range of thrust loads and lends itself to low-cost high-volume production. However, this bearing becomes sensitive to alignment as its size increases.

*The tapered land thrust bearing*, as shown in Table 7, is capable of high load capacity. Where the step bearing is generally used for small sizes, this type can be used in larger sizes. It is, however, more costly to manufacture and does require good alignment as size is increased.

*The tilting pad or Kingsbury thrust bearing* (as it is commonly referred to), is also capable of high thrust capacity. Because of its construction it is more costly, but it has the inherent advantage of being able to absorb significant amounts of misalignment.

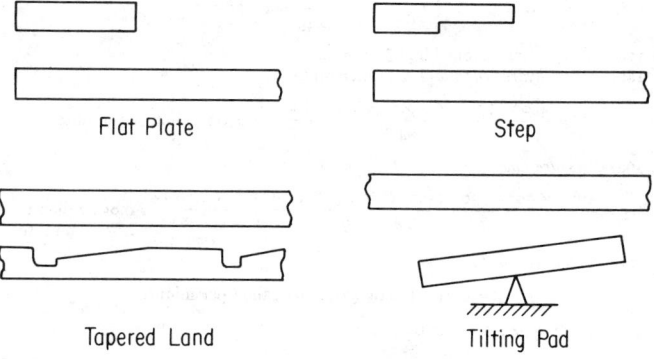

Fig. 19.   Types of thrust bearings.

Table 7.  Thrust Bearing Loads*

| Type | Normal Unit Loads, Lbs. per Sq. In. | Maximum Unit Loads, Lbs. per Sq. In. |
|---|---|---|
| Parallel Surface | <75 | <150 |
| Step | 200 | 500 |
| Tapered Land | 200 | 500 |
| Tilting Pad | 200 | 500 |

**Thrust Bearing Design Notation.** — The symbols used in the design procedures which follow for flat plate, step, tapered land and tilting pad thrust bearings are listed below:

$a$ = radial width of pad, inches
$b$ = circumferential length of pad at pitch line, inches
$b_2$ = pad step length
$B$ = circumference of pitch circle, inches
$c$ = specific heat of oil, BTU/lb/°F
$D$ = diameter, inches
$e$ = depth of step, inch
$f$ = coefficient of friction
$g$ = depth of 45° chamfer, inches
$h$ = film thickness, inch
$i$ = number of pads
$J$ = power loss coefficient
$K$ = film thickness factor
$K_g$ = fraction of circumference occupied by the pads; usually 0.8
$l$ = length of chamfer, inches
$M$ = horsepower per square inch
$N$ = revolutions per minute
$O$ = operating number
$p$ = bearing unit load, psi
$p_s$ = oil supply pressure, psi
$P_f$ = friction horsepower
$Q$ = total flow, gpm
$Q_c$ = required flow per chamfer, gpm
$Q_c^0$ = uncorrected required flow per chamfer, gpm
$Q_F$ = film flow, gpm
$s$ = oil groove width
$\Delta t$ = temperature rise, °F
$U$ = velocity, feet per minute
$V$ = effective width-to-length ratio for one pad
$W$ = applied load, pounds
$Y_G$ = oil flow factor
$Y_L$ = leakage factor
$Y_S$ = shape factor
$Z$ = viscosity, centipoises
$\alpha$ = dimensionless film thickness factor
$\delta$ = taper
$\xi$ = kinetic energy correction factor

*Note:* Subscript 1 denotes inside diameter and subscript 2 denotes outside diameter. Subscript $i$ denotes inlet and subscript $o$ denotes outlet.

* See footnote on page 592.

**Flat Plate Thrust Bearing Design.** — The following steps define the performance of a flat plate thrust bearing, one section of which is shown in Fig. 20. Although each bearing section is wedge shaped, as shown at (2), for the purposes of design calculation it is considered to be a rectangle with a length $b$ equal to the circumferential length along the pitch line of the actual section and a width $a$ equal to the difference in the external and internal radii.

*General Parameters:* (a) From Table 7 the maximum unit load is between 75 and 100 pounds per square inch.   (b) The outside diameter is usually between 1.5 and 2.5 times the inside diameter.

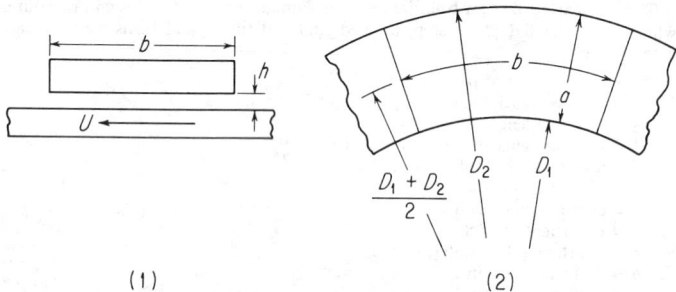

Fig. 20.   Basic elements of flat plate thrust bearing.

1. *Inside diameter, $D_1$.*  This is determined by shaft size and clearance.
2. *Outside diameter, $D_2$.*  This is calculated by the formula:

$$D_2 = \left(\frac{4W}{\pi K_g p} + D_1{}^2\right)^{1/2}$$

where: $W$ = applied load, pounds
$K_g$ = fraction of circumference occupied by pads, usually 0.8
$p$ = bearing unit load, psi

3. *Radial pad width, $a$.*  This is equal to one-half the difference between the inside and outside diameters.

$$a = \frac{D_2 - D_1}{2}$$

4. *Pitch line circumference, $B$.*  This is found from the pitch diameter.

$$B = (\pi)(D_2 - a)$$

5. *Number of pads, $i$.*  Assume an oil groove width, $s$. If the length of pad is assumed to be optimum, i.e., equal to its width,

$$i_{app} = \frac{B}{a + s}$$

Take $i$ as nearest even number.

6. *Length of pad, $b$.*  If number of pads and oil groove width are known

$$b = \frac{B - (i \times s)}{i}$$

7. *Actual unit load, p.* This is calculated in pounds per square inch based on pad dimensions.

$$p = \frac{W}{iab}$$

8. *Pitch line velocity, U.* This is found in feet per minute.

$$U = \frac{BN}{12}$$

where $N$ = rpm

9. *Friction power loss, $P_f$.* This is difficult to calculate for this type of bearing since there is no theoretical method of determining the operating film thickness. However, a good approximation can be made using Fig. 21. From this curve the value of $M$, horsepower loss per square inch of bearing surface, can be obtained. The total power loss $P_f$, is then calculated from

$$P_f = iabM$$

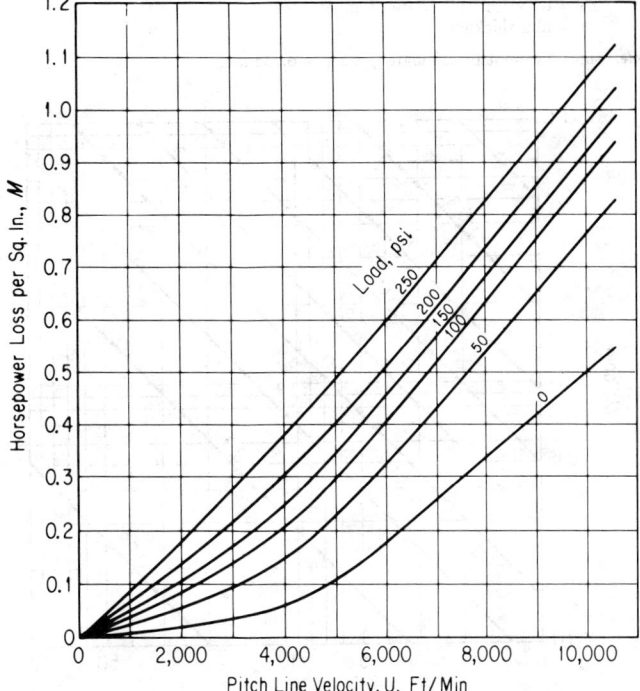

Fig. 21. Friction power loss, $M$, vs. peripheral speed, $U$ — thrust bearings.*

* See footnote on page 592.

10. *Oil flow required, Q.* This may be estimated in gallons per minute for a given temperature rise from:

$$Q = \frac{42.4 P_f}{c\Delta t}$$

where $c$ = specific heat of oil in BTU/lb/°F
$\Delta t$ = temperature rise of the oil in °F

*Note:* A $\Delta t$ of 50°F is an acceptable maximum

Since there is no theoretical method of predicting the minimum film thickness in this type of bearing, only an approximation, based on experience, of the film flow can be made. For this reason and based on practical experience, it is desirable to have a minimum of one-half of the desired oil flow pass through the chamfer.

11. *Film flow, $Q_F$.* This is calculated in gallons per minute from:

$$Q_F = \frac{(1.5)(10^6)iVh^3 p_s}{Z_2}$$

where $V$ = effective width-to-length ratio for one pad $a/b$
$Z_2$ = oil viscosity at outlet temperature
$h$ = film thickness

*Note:* since $h$ cannot be calculated, use $h = 0.002$ inch

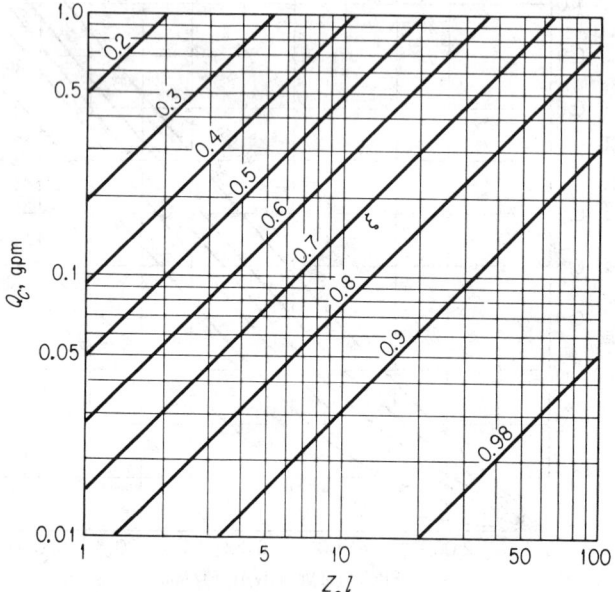

Fig. 22.    Kinetic energy correction factor, $\xi$, — thrust bearings.*

* See footnote on page 592.

12. *Required flow per chamfer, $Q_c$.* This is readily found from the formula:

$$Q_c = \frac{Q}{i}$$

13. *Kinetic energy correction factor, $\xi$.* This is found by assuming a chamfer length $l$ and entering Fig. 22 with a value $Z_2 l$ and $Q_c$.

14. *Uncorrected required flow per chamfer, $Q_c^0$.* This is found from the formula:

$$Q_c^0 = \frac{Q_c}{\xi}$$

15. *Depth of chamfer, $g$.* This is found from the formula:

$$g = \sqrt[4]{\frac{Q_c^0 l Z_2}{4.74 \times 10^4 p_s}}$$

*Example:* Design a flat plate thrust bearing to carry 900 pounds load at 4000 rpm using an SAE 10 oil with a specific heat of 3.5 Btu/lb/°F at 120°F and 30 psi inlet conditions. The shaft is 2 ¾ inches in diameter and the temperature rise is not to exceed 40°F. Figure 23 shows the final design of this bearing.

1. *Inside diameter.* Assumed to be 3 inches to clear shaft.

2. *Outside diameter.* Assuming a unit bearing load of 75 pounds per square inch from Table 7,

$$D_2 = \sqrt{\frac{4 \times 900}{\pi \times 0.8 \times 75} + 3^2} = 5.30 \text{ inches}$$

Use 5½ inches.

3. *Radial pad width.*

$$a = \frac{5.5 - 3}{2} = 1.25 \text{ inches}$$

4. *Pitch line circumference.*

$$B = \pi \times 4.25 = 13.3 \text{ inches}$$

5. *Number of pads.* Assume an oil groove width of ³⁄₁₆ inch. If length of pad is assumed to be equal to width of pad, then

$$i_{app} = \frac{13.3}{1.25 + 0.1875} = 9 +$$

If the number of pads, $i$, is taken as 10, then

6. *Length of pad.* $b = \dfrac{13.3 - (10 \times 0.1875)}{10} = 1.14 \text{ inches}$

7. *Actual unit load.*

$$p = \frac{900}{10 \times 1.25 \times 1.14} = 63 \text{ psi}$$

8. *Pitch line velocity.*

$$U = \frac{13.3 \times 4000}{12} = 4,430 \text{ ft. per min.}$$

9. *Friction power loss.* From Fig. 21, $M = 0.19$

$$P_f = 10 \times 1.25 \times 1.14 \times 0.19 = 2.7 \text{ horsepower}$$

10. *Oil flow required.*

$$Q = \frac{42.4 \times 2.7}{3.5 \times 40} = 0.82 \text{ gallon per minute}$$

(Assuming a temperature rise of 40°F. — the maximum allowable according to the given condition — then the assumed operating temperature will be 120°F + 40°F = 160°F and the oil viscosity $Z_2$ is found from Fig. 5 to be 9.6 centipoises.)

11. *Film flow.*

$$Q_F = \frac{1.5 \times 10^6 \times 10 \times 1 \times (.002)^3 \times 30}{9.6} = 0.038 \text{ gpm}$$

Since this is a very small part of the required flow of 0.82 gpm, the bulk of the flow must be carried through the chamfers.

12. *Required flow per chamfer.* Assume that all of the oil flow is to be carried through the chamfers

$$Q_c = \frac{0.82}{10} = 0.082 \text{ gpm}$$

13. *Kinetic energy correction factor.* If $l$, the length of chamfer is made ⅛ inch, then $Z_2 l = 9.6 \times ⅛ = 1.2$. Entering Fig. 22 with this value and $Q_c = 0.082$,

$$\xi = 0.44$$

14. *Uncorrected required oil flow per chamfer.*

$$Q_c^0 = \frac{0.082}{0.44} = 0.186 \text{ gpm}$$

15. *Depth of chamfer.*

$$g = \sqrt[4]{\frac{0.186 \times 0.125 \times 9.6}{4.74 \times 10^4 \times 30}}$$

$$g = 0.02 \text{ inch}$$

A schematic drawing of this bearing is shown in Fig. 23.

**Step Thrust Bearing Design.** — The following steps define the performance of a step thrust bearing, one section of which is shown in Fig. 24. Although each bearing section is wedge shaped, as shown at (2), for the purposes of design calculation it is considered to be a rectangle with a length $b$ equal to the circumferential length along the pitch line of the actual section and a width $a$ equal to the difference in the external and internal radii.

*General Parameters:* For optimum proportions $a = b$, $b_2 = 1.2b_1$, and $e = 0.7h$.

1. *Internal diameter, $D_1$.* An internal diameter is assumed that is sufficient to clear the shaft.

2. *External diameter, $D_2$.* A unit bearing pressure is assumed from Table 7 and the external diameter is then found from the formula

$$D_2 = \sqrt{\frac{4W}{\pi K_\rho p} + D_1^2}$$

3. *Radial pad width, $a$.* This is equal to the difference between the external and internal radii.

$$a = \frac{D_2 - D_1}{2}$$

4. *Pitch line circumference, $B$.* This is found from the formula

$$B = \frac{\pi(D_1 + D_2)}{2}$$

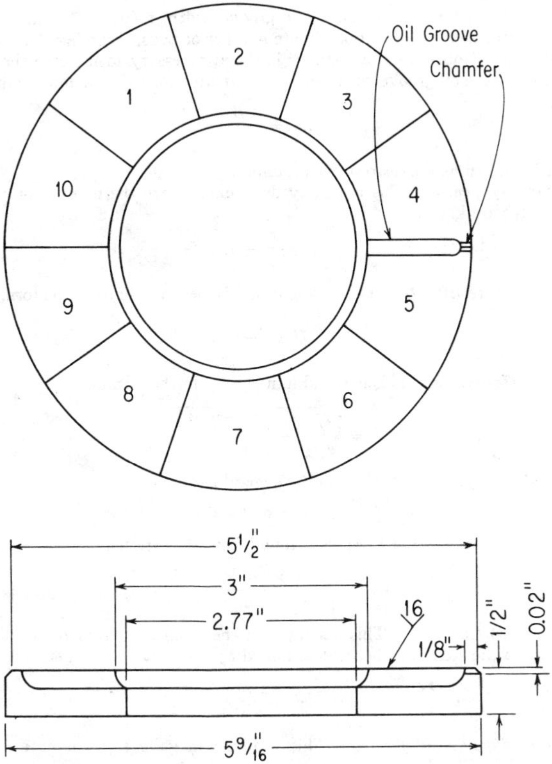

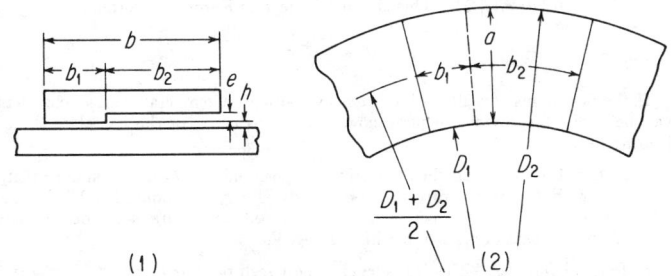

Fig. 23. Flat plate thrust bearing example design.*

Fig. 24. Basic elements of step thrust bearing.*

* See footnote on page 592.

5. *Number of pads, i.* Assume an oil groove width, s, (0.062 inch may be taken as a minimum) and find the approximate number of pads, assuming the pad length is equal to a. Note that if a chamfer is found necessary to increase the oil flow (see Step 13), the oil groove width should be greater than the chamfer width.

$$i_{app} = \frac{B}{a + s}$$

Then *i* is taken as the nearest even number.

6. *Length of pad, b.* This is readily determined since the number of pads and groove width are known.

$$b = \frac{B}{i} - s$$

7. *Pitch line velocity, U.* This is found in feet per minute from the formula

$$U = \frac{BN}{12}$$

8. *Film thickness, h.* This is found in inches from the formula

$$h = \sqrt{\frac{2.09 \times 10^{-9} i a^3 U Z}{W}}$$

9. *Depth of step, e.* According to the general parameter

$$e = 0.7h$$

10. *Friction power loss, $P_f$.* This is found from the formula

$$P_f = \frac{7.35 \times 10^{-13} i a^2 U^2 Z}{h}$$

11. *Pad step length, $b_2$.* This distance, on the pitch line, from the leading edge of the pad to the step in inches is determined by the general parameters

$$b_2 = \frac{1.2b}{2.2}$$

12. *Hydrodynamic oil flow, Q.* This is found in gallons per minute from the formula

$$Q = 6.65 \times 10^{-4} i a h U$$

13. *Temperature rise, $\Delta t$.* This is found in degrees F from the formula

$$\Delta t = \frac{42.4 P_f}{c\,Q}$$

If the flow is insufficient, as indicated by too high a temperature rise, chamfers can be added to provide adequate flow as in Steps 12–15 of the flat plate thrust bearing design.

*Example:* Design a step thrust bearing for positioning a ⅞-inch diameter shaft operating with a 25-pound thrust load and a speed of 5,000 rpm. The lubricating oil has a viscosity of 25 centipoises at the operating temperature of 160 deg. F. and it has a specific heat of 3.4 Btu per lb. per deg. F.

1. *Internal diameter.* This is assumed to be 1 inch to clear the shaft.

2. *External diameter.* Since this is a positioning bearing with low total load, unit load will be negligible and the external diameter is not established by using

the formula given in Step 2 of the procedure but a convenient size is taken to give the desired overall bearing proportions.

$$D_2 = 3 \text{ inches}$$

3. *Radial pad width.*

$$a = \frac{3 - 1}{2} = 1 \text{ inch}$$

4. *Pitch line circumference.*

$$B = \frac{\pi(3 + 1)}{2} = 6.28 \text{ inches}$$

5. *Number of pads.* Assuming a minimum groove width of 0.062 inch

$$i_{app} = \frac{6.28}{1 + 0.062} = 5.9$$

Take $i = 6$

6. *Length of pad.*

$$b = \frac{6.28}{6} - 0.062 = 0.985$$

7. *Pitch line velocity.*

$$U = \frac{6.28 \times 5,000}{12} = 2,620 \text{ fpm}$$

8. *Film thickness.*

$$h = \sqrt{\frac{2.09 \times 10^{-9} \times 6 \times 1^3 \times 2,620 \times 25}{25}} = 0.0057 \text{ inch}$$

9. *Depth of step.*

$$e = 0.7 \times 0.0057 = 0.004 \text{ inch}$$

10. *Power loss.*

$$P_f = \frac{7.35 \times 10^{-13} \times 6 \times 1^2 \times 2,620^2 \times 25}{0.0057} = 0.133 \text{ hp}$$

11. *Pad step length.*

$$b_2 = \frac{1.2 \times 0.985}{2.2} = 0.537 \text{ inch}$$

12. *Total hydrodynamic oil flow.*

$$Q = 6.65 \times 10^{-4} \times 6 \times 1 \times 0.0057 \times 2,620 = 0.060 \text{ gpm}$$

13. *Temperature rise.*

$$\Delta t = \frac{42.4 \times 0.133}{3.4 \times 0.060} = 28°F.$$

**Tapered Land Thrust Bearing Design.** — The following steps define the performance of a tapered land thrust bearing, one section of which is shown in Fig. 25. Although each bearing section is wedge shaped, as shown at (2), for the purposes of design calculation it is considered to be a rectangle with a length $b$ equal to the circumferential length along the pitch line of the actual section and a width $a$ equal to the difference in the external and internal radii.

*General Parameters:* Usually the taper extends to only 80 per cent of the pad length with the remainder being flat, thus: $b_2 = 0.8b$ and $b_1 = 0.2b$.

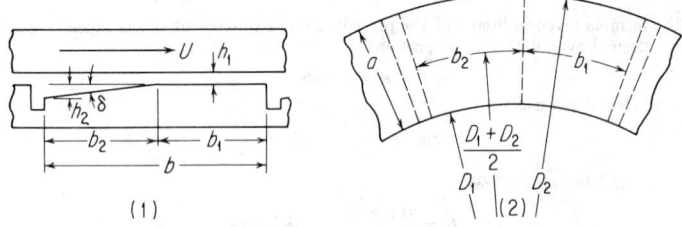

Fig. 25. Basic elements of tapered-land thrust bearing.*

1. *Inside diameter, $D_1$.* This is determined by shaft size and clearance.
2. *Outside diameter, $D_2$.* This is calculated by the formula:

$$D_2 = \left( \frac{4W}{\pi K_g p_a} + D_1{}^2 \right)^{1/2}$$

where: $K_g = 0.8$ or $0.9$ and $W =$ applied load, pounds
$p_a =$ assumed unit load from Table 7, page 575

3. *Radial pad width, a.* This is equal to one half the difference between the inside and outside diameters.

$$a = \frac{D_2 - D_1}{2}$$

4. *Pitch line circumference, B.* This is found from the mean diameter:

$$B = \frac{\pi(D_1 + D_2)}{2}$$

5. *Number of pads, i.* Assume an oil groove width, $s$, and find the approximate number of pads, assuming the pad length is equal to $a$.

$$i_{app} = \frac{B}{a + s}$$

Then $i$ is taken as the nearest even number.

6. *Length of pad, b.* This is readily determined since the number of pads and groove width are known.

$$b = \frac{B - is}{i}$$

7. *Taper values, $\delta_1$ and $\delta_2$.* These can be taken from Table 8.
8. *Actual bearing unit load, p.* This is calculated in pounds per square inch from the formula:

$$p = \frac{W}{iab}$$

9. *Pitch line velocity, U.* This is found in feet per minute at the pitch circle from the formula:

$$U = \frac{BN}{12}$$

where $N =$ rpm

* See footnote on page 592.

Table 8. Taper Values for Tapered Land Thrust Bearing

| Pad Dimensions, Inches | Taper, Inches | |
|---|---|---|
| $a \times b$ | $\delta_1 = h_2 - h_1$ (at ID) | $\delta_2 = h_2 - h_1$ (at OD) |
| ½ × ½ | 0.0025 | 0.0015 |
| 1 × 1 | 0.005 | 0.003 |
| 3 × 3 | 0.007 | 0.004 |
| 7 × 7 | 0.009 | 0.006 |

10. *Oil leakage factor*, $Y_L$. This is found either from Fig. 26 which shows curves for $Y_L$ as functions of the pad width $a$ and length of land $b$ or from the formula:

$$Y_L = \frac{b}{1 + (\pi^2 b^2 / 12 a^2)}$$

11. *Film thickness factor*, $K$. This is calculated using the formula:

$$K = \frac{5.75 \times 10^6 p}{U Y_L Z}$$

12. *Minimum film thickness*, $h$. Using the value of $K$ just determined and the selected taper values $\delta_1$ and $\delta_2$, $h$ is found from Fig. 27. In general, $h$ should be 0.001 inch for small bearings and 0.002 inch for larger and high-speed bearings.

13. *Friction power loss*, $P_f$. Using the film thickness $h$, the coefficient $J$ can be obtained from Fig. 28. The friction loss in horsepower is then calculated from the formula:

$$P_f = 8.79 \times 10^{-13} i a b J U^2 Z$$

14. *Required oil flow*, $Q$. This may be estimated in gallons per minute for a given temperature rise $\Delta t$ from the formula:

$$Q = \frac{42.4 \, P_f}{c \Delta t}$$

where $c$ = specific heat of the oil in Btu/lb/°F

*Note:* A $\Delta t$ of 50°F is an acceptable maximum.

15. *Shape factor*, $Y_S$. This is needed to compute the actual oil flow and is calculated from the formula:

$$Y_S = \frac{8ab}{D_2{}^2 - D_1{}^2}$$

16. *Oil flow factor*, $Y_G$. This is found from Fig. 29 using $Y_S$ and $D_1/D_2$.

17. *Actual oil film flow*, $Q_F$. The amount of oil in gallons per minute that the bearing film will pass is calculated from the formula:

$$Q_F = \frac{8.9 \times 10^{-4} i \delta_2 D_2{}^3 N Y_G Y_S{}^2}{D_2 - D_1}$$

18. If the flow is insufficient, the tapers can be increased or chamfers calculated to provide adequate flow, as in Steps 12–15 of the flat plate thrust bearing design procedure.

*Example:* Design a tapered land thrust bearing for 70,000 pounds at 3600 rpm. The shaft diameter is 6.5 inches. The oil inlet temperature is 110°F @ 20 psi.

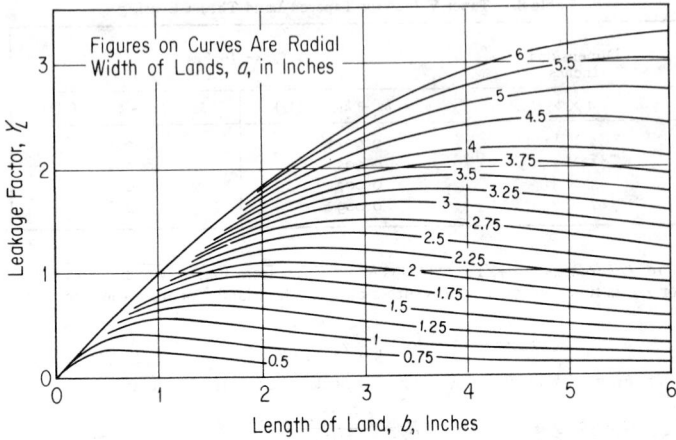

Fig. 26. Leakage factor, $Y_L$, vs. pad dimensions $a$ and $b$ — tapered-land thrust bearings.*

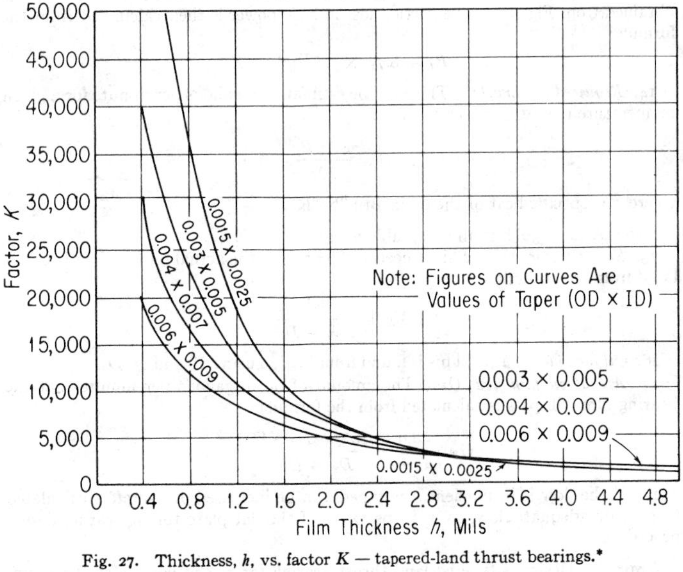

Fig. 27. Thickness, $h$, vs. factor $K$ — tapered-land thrust bearings.*

* See footnote on page 592.

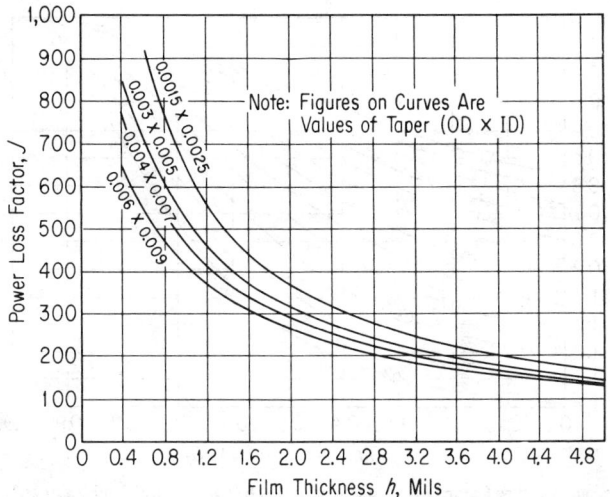

Fig. 28. Power-loss coefficient, $J$, vs. film thickness $h$ — tapered-land thrust bearings.[*]

A maximum temperature rise of 50°F is acceptable and results in a viscosity of 18 centipoises. Use a value of $K_g = 0.9$ and a $c = 3.5$ Btu/lb/°F.

1. *Internal diameter.* Assume $D_1 = 7$ inches to clear shaft.

2. *External diameter.* Assume a unit bearing load $p_a$ of 400 pounds per square inch from Table 7, then

$$D_2 = \sqrt{\frac{4 \times 70{,}000}{3.14 \times 0.9 \times 400} + 7^2} = 17.2 \text{ inches}$$

Round off to 17 inches.

3. *Radial pad width.*

$$a = \frac{17 - 7}{2} = 5 \text{ inches}$$

4. *Pitch line circumference.*

$$B = \frac{3.14(17 + 7)}{2} = 37.7 \text{ inches}$$

5. *Number of pads.* Assume groove width of 0.5 inch, then

$$i_{app} = \frac{37.7}{5 + 0.5} = 6.85$$

Take $i = 6$

6. *Length of pad.*

$$b = \frac{37.7 - 6 \times 0.5}{6} = 5.78 \text{ inches}$$

[*] See footnote on page 592.

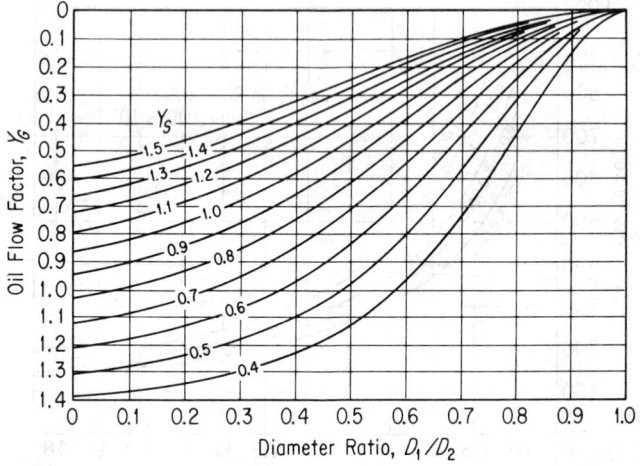

Fig. 29.   Oil-flow factor, $Y_G$, vs. diameter ratio $D_1/D_2$ — tapered-land bearings.*

7. *Taper values.*   Interpolate in Table 8 to obtain

$$\delta_1 = 0.008 \text{ inch and } \delta_2 = 0.005 \text{ inch}$$

8. *Actual bearing unit load.*

$$p = \frac{70,000}{6 \times 5 \times 5.78} = 404 \text{ psi}$$

9. *Pitch line velocity.*

$$U = \frac{37.7 \times 3600}{12} = 11,300 \text{ ft. per min.}$$

10. *Oil leakage factor.*

From Fig. 26, $Y_L = 2.75$

11. *Film thickness factor.*

$$K = \frac{5.75 \times 10^6 \times 404}{11,300 \times 2.75 \times 18} = 4150$$

12. *Minimum film thickness.*

From Fig. 27, $h = 2.2$ mils

13. *Friction power loss.*   From Fig. 28, $J = 260$ then

$$P_f = 8.79 \times 10^{-13} \times 6 \times 5 \times 5.78 \times 260 \times 11,300^2 \times 18 = 91 \text{ hp.}$$

14. *Required oil flow.*

$$Q = \frac{42.4 \times 91}{3.5 \times 50} = 22.0 \text{ gpm}$$

* See footnote on page 592.

15. *Shape factor.*

$$Y_S = \frac{8 \times 5 \times 5.78}{17^2 - 7^2} = 0.684$$

16. *Oil flow factor.*

From Fig. 29, $Y_G = 0.087$

where $D_1/D_2 = 0.41$

17. *Actual oil film flow.*

$$Q_F = \frac{8.9 \times 10^{-4} \times 6 \times 0.005 \times 17^3 \times 3600 \times 0.087 \times (0.684)^2}{17 - 7} = 19.2 \text{ gpm}$$

Since film flow is less than required oil flow, either the tapers can be increased, Step 7, or chamfers calculated, Steps 12–15 of the flat plate thrust bearing design procedure.

**Tilting Pad Thrust Bearing Design.** — The following steps define the performance of a tilting pad thrust bearing, one section of which is shown in Fig. 30. Although each bearing section is wedge shaped, as shown at (2), for the purposes of design calculation it is considered to be a rectangle with a length $b$ equal to the circumferential length along the pitch line of the actual section and a width $a$ equal to the difference in the external and internal radii. The location of the pivot shown in Fig. 30 is optimum. If shaft rotation in both directions is required, however, the pivot must be at the midpoint which results in little or no detrimental effect on the performance.

Fig. 30. Basic elements of tilting pad thrust bearing.*

1. *Inside diameter, $D_1$.* This is determined by shaft size and clearance.

2. *Outside diameter, $D_2$.* This is calculated from the formula:

$$D_2 = \left(\frac{4W}{\pi K_g p} + D_1^2\right)^{\frac{1}{2}}$$

where $W$ = applied load, pounds
$K_g = 0.8$
$p$ = unit load from Table 7

3. *Radial pad width, $a$.* This is equal to one-half the difference between the inside and outside diameters.

$$a = \frac{D_2 - D_1}{2}$$

4. *Pitch line circumference, $B$.* This is found from the mean diameter:

$$B = \pi\left(\frac{D_1 + D_2}{2}\right)$$

* See footnote on page 592.

5. *Number of pads, i.* The number of pads may be estimated from the formula:

$$i = \frac{BK_g}{a}$$

Select the nearest even number.

6. *Length of pad, b.* This can be found from the formula:

$$b \cong \frac{BK_g}{i}$$

7. *Pitch line velocity, U.* This is calculated in feet per minute from the formula:

$$U = \frac{BN}{12}$$

8. *Bearing unit load, p.* This is calculated from the formula:

$$p = \frac{W}{iab}$$

9. *Operating number, O.* This is calculated from the formula:

$$O = \frac{1.45 \times 10^{-7} Z_2 U}{5pb}$$

where $Z_2$ = viscosity of oil at outlet temperature (inlet temperature plus assumed temperature rise through the bearing).

10. *Minimum film thickness, $h_{min}$.* Using the operating number the value of $\alpha$ = dimensionless film thickness is found from Fig. 31. Then the actual minimum film thickness is calculated from the formula:

$$h_{min} = \alpha b$$

In general, this value should be 0.001 inch for small bearings and 0.002 inch for larger and high-speed bearings.

11. *Coefficient of friction, f.* This is found from Fig. 32.

12. *Friction power loss, $P_f$.* This horsepower loss now is calculated by the formula:

$$P_f = \frac{fWU}{33,000}$$

13. *Actual oil flow, Q.* This flow over the pad in gallons per minute is calculated from the formula:

$$Q = 0.0591 \alpha iabU$$

14. *Temperature rise, $\Delta t$.* This is found from the formula:

$$\Delta t = 0.0217 \frac{fp}{\alpha c}$$

where $c$ = specific heat of oil in Btu/lb/°F

If the flow is insufficient, as indicated by too high a temperature rise, chamfers can be added to provide adequate flow as in Steps 12–15 of the flat plate thrust bearing design.

*Example:* Design a tilting pad thrust bearing for 70,000 pounds thrust at 3600 rpm. The shaft diameter is 6.5 inches and a maximum OD of 15 inches is available. The oil inlet temperature is 110°F and the supply pressure is 20 pounds per square inch. A maximum temperature rise of 50°F is acceptable and results in a viscosity of 18 centipoises. Use a value of 3.5 Btu/lb/°F for $c$.

1. *Inside diameter.* Assume $D_1 = 7$ inches to clear shaft.

2. *Outside diameter.* Given maximum $D_2 = 15$ inches

3. *Radial pad width.*

$$a = \frac{15 - 7}{2} = 4 \text{ inches}$$

4. *Pitch line circumference.*

$$B = \pi \left( \frac{7 + 15}{2} \right) = 34.6 \text{ inches}$$

5. *Number of pads.*

$$i = \frac{34.6 \times 0.8}{4} = 6.9$$

Select 6 pads: $i = 6$

6. *Length of pad.*

$$b = \frac{34.6 \times 0.8}{6} = 4.61 \text{ inches}$$

Make $b = 4.75$ inches

7. *Pitch line velocity.*

$$U = \frac{34.6 \times 3600}{12} = 10,400 \text{ ft/min}$$

8. *Bearing unit load.*

$$p = \frac{70,000}{6 \times 4 \times 4.75} = 614 \text{ psi}$$

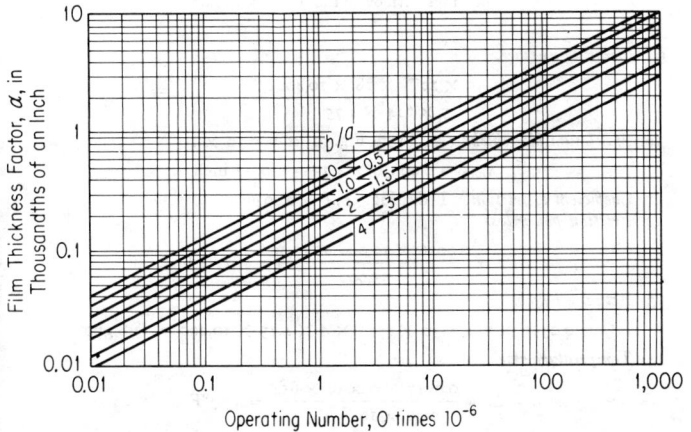

Fig. 31. Dimensionless minimum film thickness, $\alpha$, vs. operating number, $O$ — tilting pad thrust bearings.*

* See footnote on page 592.

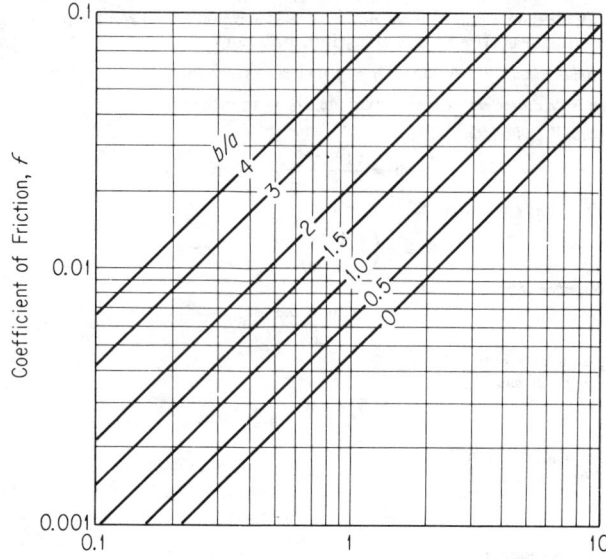

Fig. 32. Coefficient of friction, $f$, vs. dimensionless film thickness $\alpha$ — tilting pad thrust bearings with optimum pivot location.*

9. *Operating number.*

$$O = \frac{1.45 \times 10^{-7} \times 18 \times 10,400}{5 \times 614 \times 4.75} = 1.86 \times 10^{-6}$$

10. *Minimum film thickness.* From Fig. 31, $\alpha = 0.30 \times 10^{-3}$

$$h_{\min} = 0.00030 \times 4.75 = 0.0014 \text{ inch}$$

11. *Coefficient of friction.* From Fig. 32, $f = 0.0036$
12. *Friction power loss.*

$$P_f = \frac{0.0036 \times 70,000 \times 10,400}{33,000} = 79.4 \text{ hp}$$

13. *Actual oil flow.*

$$Q = 0.0591 \times 6 \times 0.30 \times 10^{-3} \times 4 \times 4.75 \times 10,400 = 23.8 \text{ gpm}$$

14. *Temperature rise.*

$$\Delta t = \frac{0.0217 \times 0.0036 \times 614}{0.30 \times 10^{-3} \times 3.5} = 45.7°\text{F}$$

Since this is less than the 50°F which is considered as the acceptable maximum, the design is satisfactory.

* Table 7 and Figs. 21-32 inclusive reproduced with permission from Wilcock and Booser, *Bearing Design and Application*, McGraw Hill Book Co., Copyright © 1957.

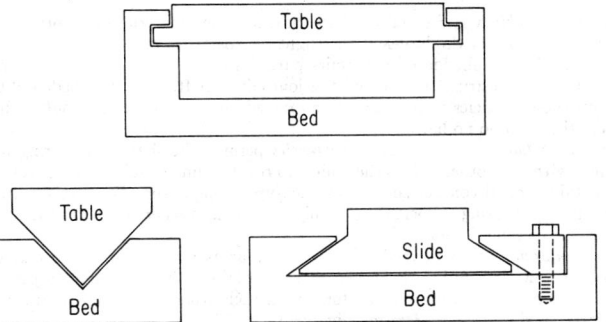

Fig. 33.    Types of guide bearings.

## Guide Bearings

This type of bearing is generally used as a positioning device or as a guide to linear motion such as in machine tools.   Figure 33 shows several examples of this type of bearing.   It is normal for this type of bearing to operate in the boundary lubrication region with either dry, dry film (MoS$_2$ etc.), grease, oil or gaseous lubrication.   In order to improve performance, reduce wear, and increase stability, the use of hydrostatic lubrication is often resorted to.   This type of design provides oil or gas under pressure to pocket designs which provides a film and complete separation of the sliding surfaces.

## Plain Bearing Materials

Materials used for sliding bearings cover a wide range of metals and non-metals. To make the optimum selection requires a complete analysis of the specific application.   The important general categories are: Babbitts, alkali hardened lead, cadmium alloys, copper lead, bronze aluminum, silver, sintered metals, plastic, wood, rubber and carbon graphite.

**Properties of Bearing Materials.** — For a material to be used as a plain bearing it must possess certain physical and chemical properties which permit it to operate properly.   If a material does not possess all of these characteristics to some degree it will not function long as a bearing.   It should be noted, however, that there are few, if any, materials which are outstanding in all of these characteristics.   Therefore, the selection of the optimum bearing material for a given application is at best a compromise to secure the most desirable combination of properties required for that particular application.   The seven properties generally acknowledged to be the most significant are: 1. Fatigue resistance; 2. Embeddability; 3. Compatibility; 4. Conformability; 5. Thermal conductivity; 6. Corrosion resistance; and 7. Load capacity.

These properties are described as follows:

1. *Fatigue resistance* is the ability of the bearing lining material to withstand repeated applications of stress and strain without cracking, flaking or otherwise being destroyed.

2. *Embeddability* is the ability of the bearing lining material to absorb or embed within itself any of the larger of the slight dirt particles present in the lubrication systems. Poor embeddability permits particles circulating around the bearing to score both the bearing surface and the journal or shaft. Good embeddability will permit these particles to be trapped and forced into the surface and out of the way where they can do no harm.

3. *Compatability or anti-scoring tendencies* permits the shaft and bearing to "get along" with each other. It is the ability to resist galling or seizing under conditions of metal-to-metal contact such as at start-up. This is the characteristic which is most truly a bearing property, since only at start up is there contact of the bearing and shaft in good designs.

4. *Conformability* is defined as malleability or as the ability of the material to creep or flow slightly under load, as in the initial stages of running, to permit the shaft and bearing contours to conform with each other or to compensate for non-uniform loading present from misalignment.

5. *High thermal conductivity* is required to absorb and carry away the heat generated in the bearing. This is most important, not in removing frictional heat generated in the oil film, but in preventing seizures due to local hot spots caused by local asperity break-throughs or foreign particles.

6. *Corrosion resistance* is required to resist attack by organic acids that are sometimes formed in oils at operating conditions.

7. *Load capacity or strength* is the ability of the material to withstand the hydro-dynamic pressures exerted upon it during operation.

**Babbitt or White Metal Alloys.** — Many different bearing metal compositions are referred to as babbitt metals. The exact composition of the original babbitt metal is not known; however, the ingredients were probably tin, copper, and antimony in approximately the following percentages: 89.3, 3.6 and 7.1. Tin and lead-base babbitts are probably the best known of all bearing materials. With their excellent embeddability and compatibility characteristics under boundary lubrication, babbitt bearings are used in a wide range of applications including household appliances, automobile and diesel engines, railroad cars, electric motors, generators, steam and gas turbines, and industrial and marine gear units.

Both the Society of Automotive Engineers and American Society for Testing and Materials have classified white metal bearing alloys. Tables 9 and 10 give compositions and properties or characteristics for the two classifications.

In small bushings for fractional-horsepower motors and in automotive engine bearings the babbitt is generally used as a thin coating over a flat steel strip. After forming oil distribution grooves and drilling any required holes, the strip is cut to size, then rolled and shaped into the finished bearing. These are available for shaft diameters from 0.5 to 5 inches. Strip bearings are turned out by the millions yearly from automated factories and offer an excellent combination of low cost with good bearing properties.

For larger bearings in heavy-duty equipment, a thicker babbitt is cast on a rigid backing of steel or cast iron. Chemical and electrolytic cleaning of the bearing shell, thorough rinsing, tinning, and then centrifugal casting of the babbitt are desirable for sound bonding of the babbitt to the bearing shell. After machining, the babbitt layer is usually ⅛ to ¼ inch thick.

Compared to other bearing materials, babbitts generally have lower load-carrying capacity and fatigue strength, are a little higher in cost, and require a more complicated design. Also, their strength decreases rapidly with increasing temperature. These shortcomings can be avoided by using an intermediate layer of high-strength, fatigue resistant material which is placed between a steel backing and a thin babbitt

**Table 9. Bearing and Bushing Alloys — Composition, Forms, Characteristics and Applications** (SAE General Information)

| SAE No. & Alloy Grouping | | Nominal Composition, Per Cent | Form of Use (1), Characteristics (2), and Applications (3) |
|---|---|---|---|
| Sn-Base Alloys | 11 | Sn, 87.5; Sb, 6.75; Cu, 5.75. | (1) Cast on steel, bronze or brass backs, or directly in the bearing housing. (2) Soft, corrosion resistant with fair fatigue resistance. (3) Main and connecting-rod bearings; motor bushings. Operates with either hard or soft journal. |
| | 12 | Sn, 89; Sb, 7.5; Cu, 3.5. | |
| Pb-Base Alloys | 13 | Pb, 84; Sb, 10; Sn, 6. | (1) SAE 13 and 14 are cast on steel, bronze or brass, or in the bearing housing; SAE 15 is cast on steel; and SAE 16 is cast into and on a porous sintered matrix, usually copper-nickel bonded to steel. (2) Soft, moderately fatigue resistant, corrosion resistant. (3) Main and connecting-rod bearings. Operates with hard or soft journal with good finish. |
| | 14 | Pb, 75; Sb, 15; Sn, 10. | |
| | 15 | Pb, 83; Sb, 15; Sn, 1; As, 1. | |
| | 16 | Pb, 92; Sb, 3.5; Sn, 4.5. | |
| Pb-Sn Overlays | 19 | Pb, 90; Sn, 10. | (1) Electrodeposited as a thin layer on copper-lead or silver bearing faces. (2) Soft, corrosion resistant. Bearings so coated run satisfactorily against soft shafts throughout the life of the coating. (3) Heavy-duty, high-speed main and connecting-rod bearings. |
| | 190 | Pb, 93; Sn, 7. | |
| Cu-Pb Alloys | 49 | Cu, 76; Pb, 24. | (1) Cast or sintered on steel back with the exception of SAE 481 which is cast on steel back only. (2) Moderately hard. Somewhat subject to oil corrosion. Some oils minimize this; protection with overlay may be desirable. Fatigue resistance good to fairly good. Listed in order of decreasing hardness and fatigue resistance. (3) Main and connecting-rod bearings. The higher lead alloys can be used unplated against a soft shaft, although an overlay is helpful. The lower lead alloys may be used against a hard shaft, or with an overlay against a soft one. |
| | 48 | Cu, 70; Pb, 30. | |
| | 480 | Cu, 65; Pb, 35. | |
| | 481 | Cu, 60; Pb, 40. | |
| Cu-Pb-Sn Alloys | 482 | Cu, 67; Pb, 28; Sn, 5. | (1) Steel backed and lined with a structure combining sintered copper alloy matrix with corrosion-resistant lead alloy. (2) Moderately hard. Corrosion resistance improved over copper-leads of equal lead content without tin. Fatigue resistance fairly good. Listed in order of decreasing hardness and fatigue resistance. (3) Main and connecting-rod bearings. Generally used without overlay. SAE 484 and 485 may be used with hard or soft shaft, while a hardened or cast shaft is recommended for SAE 482. |
| | 484 | Cu, 55; Pb, 42; Sn, 3. | |
| | 485 | Cu, 46; Pb, 51; Sn, 3. | |
| Al-Base Alloys | 770 | Al, 91.75; Sn, 6.25; Cu, 1; Ni, 1. | (1) SAE 770 cast in permanent molds; work-hardened to improve physical properties. SAE 780 and 782 usually bonded to steel back but is procurable in strip form without steel backing. SAE 781 usually bonded to steel back but can be produced as castings or wrought strip without steel back. (2) Hard, extremely fatigue resistant, resistant to oil corrosion. (3) Main and connecting-rod bearings. Generally used with suitable overlay. SAE 781 and 782 also used for bushings and thrust bearings with or without overlay. |
| | 780 | Al, 91; Sn, 6; Si, 1.5; Cu, 1; Ni, 0.5. | |
| | 781 | Al, 95; Si, 4; Cd, 1. | |
| | 782 | Al, 95; Cu, 1; Ni, 1; Cd, 2. | |
| Other Cu-Base Alloys | 795 | Cu, 90; Zn, 9.5; Sn, 0.5. | (1) Wrought solid bronze. (2) Hard, strong, good fatigue resistance. (3) Intermediate-load oscillating motion such as tie-rods, brake shafts, and so forth. |
| | 791 | Cu, 88; Zn, 4; Sn, 4; Pb, 4. | (1) SAE 791, wrought solid bronze; SAE 793, cast on steel back; SAE 798, sintered on steel back. (2) General-purpose bearing material, good shock and load capacity. Resistant to high temperatures. Hard shaft desirable. Less score resistant than higher lead alloys. (3) Medium to high loads. Transmission bushings and thrust washers. SAE 791 also used for piston-pin and 793 and 798 for chassis bushings. |
| | 793 | Cu, 84; Pb, 8; Sn, 4; Zn, 4. | |
| | 798 | Cu, 84; Pb, 8; Sn, 4; Zn, 4. | |

**Table 9.** (*Continued*)   **Bearing and Bushing Alloys — Composition, Forms, Characteristics and Applications** (SAE General Information)

| SAE No. & Alloy Grouping | Nominal Composition, Per Cent | Form of Use (1), Characteristics (2), and Applications (3) |
|---|---|---|
| Other Cu-Base Alloys | 792 {Cu, 80; Sn, 10; Pb, 10. | (1) SAE 792, cast on steel back. SAE 797, sintered on steel back. (2) Has maximum shock and load carrying capacity of conventional cast bearing alloys; hard, both fatigue and corrosion resistant. Hard shaft desirable. (3) Heavy loads with oscillating or rotating motion. Used for piston pins, steering knuckles, differential axles, thrust washers, and wear plates. |
| | 797 {Cu, 80; Sn, 10; Pb, 10. | |
| | 794 {Cu, 73.5; Pb, 23; Sn, 3.5. | (1) SAE 794, cast on steel back; SAE 799, sintered on steel back. (2) Higher lead content gives improved surface action for higher speeds but results in somewhat less corrosion resistance. (3) Intermediate load application for both oscillating and rotating shafts, that is, rocker-arm bushings, transmissions, and farm implements. |
| | 799 {Cu, 73.5; Pb, 23; Sn, 3.5. | |

**Table 10. White Metal Bearing Alloys — Composition and Properties**
(ASTM B23-49)

| ASTM Alloy Grade Number | Nominal Composition, Per Cent | | | | Compressive Yield Point,[b] psi | | Ultimate Compressive Strength,[c] psi | | Brinell Hardness, 10 mm ball— 500 kg load— 30 seconds | | Melting Point, Deg. F | Proper Pouring Temperature, Deg. F |
|---|---|---|---|---|---|---|---|---|---|---|---|---|
| | Tin | Antimony | Lead | Copper | 68° F | 212° F | 68° F | 212° F | 68° F | 212° F | | |
| 1 | 91.0 | 4.5 | ... | 4.5 | 4400 | 2650 | 12,850 | 6950 | 17.0 | 8.0 | 433 | 825 |
| 2 | 89.0 | 7.5 | ... | 3.5 | 6100 | 3000 | 14,900 | 8700 | 24.5 | 12.0 | 466 | 795 |
| 3 | 83.33 | 8.33 | ... | 8.33 | 6600 | 3150 | 17,600 | 9900 | 27.0 | 14.5 | 464 | 915 |
| 4 | 75.0 | 12.0 | 10.0 | 3.0 | 5550 | 2150 | 16,150 | 6900 | 24.5 | 12.0 | 363 | 710 |
| 5 | 65.0 | 15.0 | 18.0 | 2.0 | 5050 | 2150 | 15,050 | 6750 | 22.5 | 10.0 | 358 | 690 |
| 6 | 20.0 | 15.0 | 63.5 | 1.5 | 3800 | 2050 | 14,550 | 8050 | 21.0 | 10.5 | 358 | 655 |
| 7 | 10.0 | 15.0 | 75.0 | ... | 3550 | 1600 | 15,650 | 6150 | 22.5 | 10.5 | 464 | 640 |
| 8 | 5.0 | 15.0 | 80.0 | ... | 3400 | 1750 | 15,600 | 6150 | 20.0 | 9.5 | 459 | 645 |
| 10 | 2.0 | 15.0 | 83.0 | ... | 3350 | 1850 | 15,450 | 5750 | 17.5 | 9.0 | 468 | 630 |
| 11 | ... | 15.0 | 85.0 | ... | 3050 | 1400 | 12,800 | 5100 | 15.0 | 7.0 | 471 | 630 |
| 12 | ... | 10.0 | 90.0 | ... | 2800 | 1250 | 12,900 | 5100 | 14.5 | 6.5 | 473 | 625 |
| 15[d] | 1.0 | 15.0 | 82.5 | 0.5 | ... | ... | ... | ... | 21.0 | 13.0 | 479 | 662 |
| 16 | 10.0 | 12.5 | 77.0 | 0.5 | ... | ... | ... | ... | 27.5 | 13.6 | 471 | 620 |
| 19 | 5.0 | 9.0 | 86.0 | ... | ... | ... | 15,600 | 6100 | 17.7 | 8.0 | 462 | 620 |

*a* The compression test specimens were cylinders 1.5 inches in length and 0.5 inch in diameter, machined from chill castings 2 inches in length and 0.75 inch in diameter. The Brinell tests were made on the bottom face of parallel machined specimens cast in a 2-inch diameter by 0.625-inch deep steel mold at room temperature.

*b* The values for yield point were taken from stress-strain curves at a deformation of 0.125 per cent reduction of gage length.

*c* The ultimate strength values were taken as the unit load necessary to produce a deformation of 25 per cent of the length of the specimen.

*d* Also nominal arsenic, 1 per cent.

surface layer. Such composite bearings frequently eliminate any need for using alternate materials with poorer bearing characteristics.

Tin babbitt is composed of 80 to 90 per cent tin to which is added about 3 to 8 per cent copper and 4 to 14 per cent antimony. An increase in copper or antimony produces increased hardness and tensile strength and decreased ductility. However, if the percentage of these alloys is increased above those shown in Table 10, the resulting alloy will have decreased fatigue resistance. These alloys have very little tendency to cause wear to their journals because of their ability to embed dirt. They resist the corrosive effects of acids, are not prone to oil-film failure and are easily bonded and cast. Two drawbacks are encountered in their use. They have a low fatigue resistance and their hardness and strength drop appreciably at low temperatures.

Lead babbitt compositions generally range from 10 to 15 per cent antimony and up to 10 per cent tin in combination with the lead. These alloys, like tin-base babbitts, have little tendency to cause wear to their journals, embed dirt well, resist the corrosive effects of acids, are not prone to oil film failure and are easily bonded and cast. Their chief disadvantage when compared to the tin-base alloy is a rather lower strength and a susceptibility to corrosion.

**Cadmium Base.** — Cadmium alloy bearings have a greater resistance to fatigue than babbitt bearings; however, use of these materials is very limited due to their poor corrosion resistance. These alloys contain 1 to 15 per cent nickel, or 0.4 to 0.75 per cent copper, and 0.5 to 2.0 per cent silver. Their prime attribute is their high temperature capability. The load carrying capacity and relative basic bearing properties are shown in Tables 11 and 12.

**Copper Lead.** — Copper lead bearings are a binary mixture of copper and lead containing from 20 to 40 per cent lead. Since the lead is practically insoluble in copper, a cast microstructure consists of lead pockets in a copper matrix. A steel backing is commonly used with this material and high volume is achieved either by continuous casting or by powder metallurgy techniques. This material is very often used with an overplate such as lead tin and lead tin copper to increase basic bearing properties. Tables 11 and 12 provide comparisons of material properties.

The combination of good fatigue strength, high-load capacity, and high temperature performance has resulted in extensive use of this material for heavy-duty main and connecting-rod bearing as well as moderate load and speed applications in turbines and electric motors.

**Leaded Bronze and Tin Bronze.** — Leaded and tin bronzes contain up to 25 per cent lead or approximately 10 per cent tin respectively. Cast leaded bronze bearings offer good compatibility, excellent casting and easy machining character-

Table 11. Properties of Bearing Alloys

| Material | Recommended Shaft Hardness, Brinell | Load Carrying Capacity, psi | Maximum Operating Temp., °F |
|---|---|---|---|
| Tin Base Babbitt | 150 or less | 800–1500 | 300 |
| Lead Base Babbitt | 150 or less | 800–1200 | 300 |
| Cadmium Base | 200–250 | 1200–2000 | 500 |
| Copper Lead | 300 | 1500–2500 | 350 |
| Tin Bronze | 300–400 | 4000+ | 500+ |
| Lead Bronze | 300 | 3000–4500 | 450–500 |
| Aluminum | 300 | 4000+ | 225–300 |
| Silver-Overplated | 300 | 4000+ | 500 |
| Tri Metal-Overplate | 230 or less | 2000–4000+ | 225–300 |

**Table 12.  Bearing Characteristics Ratings**

| Material | Compatibility | Conformability and Embeddability | Corrosion Resistance | Fatigue Strength |
|---|---|---|---|---|
| Tin Base Babbitt | 1 | 1 | 1 | 5 |
| Lead Base Babbitt | 1 | 1 | 3 | 5 |
| Cadmium Base | 1 | 2 | 5 | 4 |
| Copper Lead | 2 | 2 | 5 | 3 |
| Tin Bronze | 3 | 5 | 2 | 1 |
| Lead Bronze | 3 | 4 | 4 | 2 |
| Aluminum | 5 | 3 | 1 | 2 |
| Silver Overplated | 2 | 3 | 1 | 1 |
| Tri Metal-Overplated | 1 | 2 | 2 | 3 |

*Note:* 1 is best; 5 is worst.

istics, low cost, good structural properties and high-load capacity, usefulness as a single material which requires neither a separate overlay nor a steel backing.  Bronzes are available in standard bar stock, sand or permanent molds, investment, centrifugal or continuous casting.  Leaded bronzes have better compatibility than tin bronzes because the spheroids of lead smear over the bearing surface under conditions of inadequate lubrication.  These alloys are generally a first choice at intermediate loads and speeds.  Tables 11 and 12 provide comparisons of basic bearing properties of these materials.

**Aluminum.** — Aluminum bearings are either cast solid aluminum, aluminum with a steel backing or aluminum with a suitable overlay.  The aluminum is usually alloyed with small amounts of tin, silicone, cadmium, nickel, or copper, as shown in Table 9.  An aluminum bearing alloy with 20 to 30 per cent tin alloy and up to 3 per cent copper has shown promise as a substitute for bronzes in some industrial applications.

These bearings are best suited for operation with hard journals.  Due to the high thermal expansion of the metal (which results in diametral contraction when it is confined as a bearing in a rigid housing), large clearances are required, which tend to make the bearing noisy, especially on starting.  Overlays of lead-tin, lead, or lead-tin-copper may be applied to aluminum bearings to facilitate their use with soft shafts.

Aluminum alloys are available with properties specifically designed for bearing applications, such as high load-carrying capacity, fatigue strength and thermal conductivity, in addition to excellent corrosion resistance and low cost.

**Silver.** — Silver bearings were developed for and have an excellent record in heavy duty applications such as aircraft master rod and diesel engine main bearings.  Silver has a higher fatigue rating than any of the other bearing materials; in fact there have been cases where the steel backing used with this material has shown evidences of fatigue before the silver.  The advent of overlays, or more commonly called overplates, made it possible for silver to be used as a bearing material.  Silver by itself does not possess any of the desirable bearing qualities except high fatigue resistance and high thermal conductivity.  The overlays such as lead, lead tin or lead indium improve the embeddability and antiscore properties of silver.  The relative basic properties of this material, when used as an overplate, are shown in Tables 11 and 12.

**Cast Iron.** — Cast Iron is an inexpensive bearing material capable of operation at light loads and low speeds, i.e., to 130 feet per minute and 150 pounds per square inch. These bearings must be well lubricated and have a rather large clearance so as to avoid scoring from particles torn from the cast iron that ride between bearing and journal. A journal hardness of between 150 to 250 Brinell has been found to be best when using cast iron bearings.

**Porous Metals.** — Porous metal self-lubricating bearings are usually made by sintering metals such as plain or leaded bronze, iron, and stainless steel. The sintering produces a spongelike structure capable of absorbing fairly large quantities of oil, usually 10–35 per cent of the total volume. These bearings are used where lubrication supply is difficult, inadequate or infrequent. This type of bearing should be flooded from time to time to resaturate the material. Another use of these materials is to meter a small quantity of oil to the bearings such as in drop feed systems. The general design operating characteristics of this class of materials are shown in Table 13.

Table 13. Application Limits — Semi-Lubricated Sintered-Metal and Nonmetallic Bearings

| Type of Bearing | Load Capacity Psi | Max. Temp. °F | Max. Surface Speed Ft/Min | PV Limit P = Psi Load V = Surface Ft/Min |
|---|---|---|---|---|
| Porous Metals | 4000/8000 | 150 | 1500 | 50,000 |
| Rubber | 50 | 150 | 1000 | 15,000 |
| Graphitic Materials | 600 | 700 | 2500 | 15,000 Dry 150,000 Lubricated |
| Laminated Phenolics | 6000 | 200 | 2500 | 15,000 |
| Nylon | 1000 | 200 | 1000 | 3,000 |
| Wood (Maple & Lignum Vitae) | 2000 | 150 | 2000 | 15,000 |
| TFE | 500 | 500 | 50 | 1,000 |
| Reinforced TFE | 2500 | 500 | 1000 | 10,000 |
| TFE Fabric | 60,000 | 500 | 150 | 25,000 |

The accompanying Table 14 gives the chemical compositions, permissible loads, interference fits, running clearances and size specifications of bronze base and iron base metal powder sintered bearings that are specified in the ASTM specifications for oil impregnated metal powder sintered bearings (B438-70 and B439-70).

**Plastics (Phenolic, Nylon, TFE).** — Plastics are finding increased usage as bearing materials because of their resistance to corrosion, quiet operation, ability to be molded into many configurations, and their excellent compatibility which minimizes or eliminates the need for lubrication. There are many plastics capable of operating as bearings, however, they are mostly phenolic, tetrafluoroethylene (TFE) or polyamide (nylon) resins. The general application limits for these materials are shown in Table 13.

*Laminated Phenolics:* These composite materials consist of cotton fabric, asbestos, or other fillers bonded with phenolic resin. They have excellent compatibility with various fluids as well as strength and shock resistance. However, precautions must be taken to maintain adequate bearing cooling since the thermal conductivity of these materials is low.

*Nylon:* This material has the widest use for small lightly loaded applications. It has low frictional properties and requires no lubrication.

### Table 14. Copper- and Iron-Base Sintered Bearings (Oil Impregnated) —
(ASTM B438-70, B439-70, and Appendices)

#### CHEMICAL REQUIREMENTS

| Alloying Elements[a] | Composition, Per Cent | | | | | |
|---|---|---|---|---|---|---|
| | Copper Base | | Iron Base | | | |
| | Grade 1 — Types 1 & 2 | Grade 2 — Type 1 | Grade | | | |
| | | | 1 | 2 | 3 | 4 |
| Cu | 87.5–90.5 | 82.6–88.5 | ... | ... | 7.0–11.0 | 18.0–22.0 |
| Fe | 1.0 max. | 1.0 max. | 96.25 min. | 95.9 min. | Remainder[d] | Remainder[d] |
| Sn | 9.5–10.5 | 9.5–10.5 | ... | ... | ... | ... |
| Pb | ... | 2.0–4.0 | ... | ... | ... | ... |
| Zn, max. | ... | 0.75 | ... | ... | ... | ... |
| Ni, max. | ... | 0.35 | ... | ... | ... | ... |
| Sb, max. | ... | 0.25 | ... | ... | ... | ... |
| Si, max. | ... | ... | 0.3 | 0.3 | ... | ... |
| Al, max. | ... | ... | 0.2 | 0.2 | ... | ... |
| C, max. | 1.75[c] | 1.75[c] | ... | ... | ... | ... |
| Other, max. | 0.5 | 0.5 | 3.0 | 3.0 | 3.0 | 3.0 |
| Comb. C[b] | ... | ... | 0.25 max. | 0.25–0.60 | ... | ... |

#### PERMISSIBLE LOADS

| Copper-Base Bearings | | | Iron-Base Bearings | | |
|---|---|---|---|---|---|
| Shaft Velocity, fpm | Grade 1-Type 1 | Grade 1-Type 2 Grade 2-Type 1 | Shaft Velocity, fpm | Grade 1, Grade 2 | Grade 3, Grade 4 |
| | Permissible Load, psi | | | Permissible Load, psi | |
| Slow & intermittent | 3200 | 4000 | Slow & intermittent | 3600 | 8000 |
| 25 | 2000 | 2000 | 25 | 1800 | 3000 |
| 50 to 100 incl. | 550 | 500 | 50 to 100 incl. | 450 | 700 |
| Over 100–150 incl. | 365 | 325 | Over 100–150 incl. | 300 | 400 |
| Over 150–200 incl. | 280 | 250 | Over 150–200 incl. | 225 | 300 |
| Over 200 | e | e | Over 200 | e | e |

#### PRESS FITS AND RUNNING CLEARANCES[f]

| Outside Diameter of Bearing | Press Fit | | Shaft Size | Min. Recommended Clearance |
|---|---|---|---|---|
| | Min. | Max. | | |
| Up to 0.760 | .001 | .003 | Up to 0.760 | .0005 |
| 0.761 to 1.510 | .0015 | .004 | 0.761 to 1.510 | .001 |
| 1.511 to 2.510 | .002 | .005 | 1.511 to 2.510 | .0015 |
| 2.511 to 3.010 | .002 | .006 | Over 2.510 | .002 |
| Over 3.010 | .002 | .007 | | |

#### STANDARD COPPER-BASE SLEEVE BEARING SIZE SPECIFICATIONS

| Inside Diam. | | Outside Diam. | | Wall Thickness | Length | Outside Diam. | | Wall Thickness | Length |
|---|---|---|---|---|---|---|---|---|---|
| Fractional | Decimal | Fractional | Decimal | | | Fractional | Decimal | | |
| 1/8 | 0.127 | 3/16 | 0.1905 | 1/32 | 0.250 | 1/4 | 0.253 | 1/16 | 0.250 |
| 5/32 | 0.158 | 1/4 | 0.253 | 3/64 | 0.312 | ... | ... | ... | ... |
| 3/16 | 0.1895 | 1/4 | 0.253 | 1/32 | 0.375 | 5/16 | 0.3155 | 1/16 | 0.375 |
| 1/4 | 0.252 | 3/8 | 0.378 | 1/16 | 0.500 | 7/16 | 0.4405 | 3/32 | 0.500 |
| 5/16 | 0.3145 | 7/16 | 0.4405 | 1/16 | 0.562 | 1/2 | 0.503 | 3/32 | 0.562 |
| 3/8 | 0.377 | 1/2 | 0.503 | 1/16 | 0.625 | 9/16 | 0.5655 | 3/32 | 0.625 |
| 1/2 | 0.502 | 5/8 | 0.628 | 1/16 | 0.750 | 3/4 | 0.753 | 1/8 | 0.750 |
| 5/8 | 0.627 | 3/4 | 0.753 | 1/16 | 0.750 | 7/8 | 0.879 | 1/8 | 0.937 |
| 3/4 | 0.752 | 7/8 | 0.879 | 1/16 | 1.125 | 1 | 1.004 | 1/8 | 1.125 |
| 1 | 1.003 | 1 1/4 | 1.254 | 1/8 | 1.500 | 1 3/8 | 1.379 | 3/16 | 1.500 |
| 1 1/4 | 1.2535 | 1 1/2 | 1.504 | 1/8 | 1.500 | 1 5/8 | 1.630 | 3/16 | 1.875 |
| 1 1/2 | 1.504 | 1 3/4 | 1.755 | 1/8 | 1.500 | 1 7/8 | 1.880 | 3/16 | 2.250 |
| 2 | 2.004 | 2 1/2 | 2.505 | 1/4 | 2.000 | ... | ... | ... | ... |
| 2 1/2 | 2.505 | 3 | 3.006 | 1/4 | 2.500 | ... | ... | ... | ... |
| 3 | 3.006 | 3 1/2 | 3.507 | 1/4 | 3.000 | ... | ... | ... | ... |

*Note.* Footnotes pertaining to the letter references can be found on the bottom of the continued table on the following page.

**Table 14** (Continued). **Copper- and Iron-Base Sintered Bearings (Oil Impregnated) —**
(ASTM B438-70, B439-70, and Appendices)

| COMMERCIAL DIMENSIONAL TOLERANCES | | | | | |
|---|---|---|---|---|---|
| Diameter Tolerance[h] | | | Length Tolerance[h] | | |
| Inside Diameter or Outside Diameter | Total Tolerance for Inside or Outside Diameter | | Length | Total Length Tolerance | |
| | Copper Base | Iron Base | | Copper Base | Iron Base |
| Up to 0.760 | −.001 | −.001 | Up to 1.495 | ±.010 | ±.010 |
| 0.761 to 1.510 | −.001 | −.0015 | 1.496 to 1.990 | ±.015 | ±.015 |
| 1.511 to 2.510 | −.0015 | −.002 | 1.991 to 2.990 | ±.015 | ±.020 |
| 2.511 to 3.010 | −.002 | −.003 | 2.991 to 4.985 | ±.020 | ±.030 |
| 3.011 to 4.010 | −.003 | −.004 | ... | ... | ... |
| 4.011 to 5.010 | −.004 | −.005 | ... | ... | ... |
| 5.011 to 6.010 | −.005 | −.006 | ... | ... | ... |

| Concentricity Tolerance[h,i] | | | | |
|---|---|---|---|---|
| Outside Diameter | Copper Base | | Iron Base | |
| | Wall Thickness, max. | Concentricity Tolerance | Wall Thickness, max. | Concentricity Tolerance |
| Up to 1.010 | Up to 0.255 | .003 | Up to 0.355 | .003 |
| 1.011 to 1.510 | Up to 0.355 | .003 | Up to 0.355 | .003 |
| 1.511 to 2.010 | Up to 0.505 | .004 | Up to 0.505 | .004 |
| 2.011 to 3.010 | Up to 0.760 | .005 | Up to 1.010 | .005 |
| 3.011 to 4.010 | Up to 1.010 | .005 | Up to 1.010 | .005 |
| 4.011 to 5.010 | Up to 1.510 | .006 | Up to 1.510 | .006 |
| 5.011 to 6.010 | Up to 2.010 | .007 | Up to 2.010 | .007 |

| Flange and Thrust Bearings, Diameter and Thickness Tolerances[i] | | | | |
|---|---|---|---|---|
| Diameter Range | Flange Diameter Tolerance | | Flange Thickness Tolerance | |
| | Standard | Special | Standard | Special |
| 0 to 1½ | ±.005 | ±.0025 | ±.005 | ±.0025 |
| Over 1½ to 3 | ±.010 | ±.005 | ±.010 | ±.007 |
| Over 3 to 6 | ±.025 | ±.010 | ±.015 | ±.010 |

| Parallelism on Faces, max. | | | | |
|---|---|---|---|---|
| Diameter Range | Copper Base | | Iron Base | |
| | Standard | Special | Standard | Special |
| 0 to 1½ | .003 | .002 | .005 | .003 |
| Over 1½ to 3 | .004 | .003 | .007 | .005 |
| Over 3 to 6 | .005 | .004 | .010 | .007 |

All dimensions in inches except where otherwise noted.

[a] Abbreviations used for the alloying elements are as follows: Cu, copper; Fe, iron; Sn, tin; Pb, lead; Zn, zinc; Ni, nickel; Sb, antimony; Si, silicon; Al, aluminum; and C, carbon.

[b] Combined carbon (on basis of iron only) may be a metallographic estimate of the carbon in the iron.

[c] Commonly graphite. A minimum of 1.5 per cent of another type of solid lubricant may be substituted when authorized by the purchaser.

[d] Total of iron plus copper shall be 97 per cent, min.

[e] For shaft velocities over 200 fpm the permissible loads may be calculated as follows: $P = 50,000/V$; where $P$ = safe load, psi of projected area, and $V$ = shaft velocity, fpm.

[f] Only *minimum* recommended clearances are listed. It is assumed that ground steel shafting will be used and that all bearings will be oil impregnated.

[g] For some of the inside diameter sizes the standard provides for the selection of two possible sets of outside diameter, wall thickness and length dimensions as shown.

[h] Values given here are intended for copper-base bearings with a 4 to 1 maximum length to inside diameter ratio and a 24 to 1 maximum length to wall thickness ratio; also for iron-base bearings with a 3 to 1 maximum length to inside diameter ratio and a 20 to 1 maximum length to wall thickness ratio. Bearings having greater ratios than these are not covered here.

[i] Total indicator reading.

[j] Standard and special tolerances are specified for diameters, thickness and parallelism. Special tolerances should not be specified unless required since they require additional or secondary operations and, therefore are costlier. Thrust bearings (¼ inch thickness, max.) have a standard thickness tolerance of ±0.005 inch and a special thickness tolerance of ±0.0025 inch for all diameters. Thrust bearing outside diameter tolerances are the same as for flange bearings.

*Teflon:* This material, with its exceptional low coefficient of friction, self lubricating characteristics, resistance to attack by almost any chemicals, and its wide temperature range, is one of the most interesting of the plastics for bearing use. High cost combined with low load capacity cause Teflon to be selected mostly in modified form, where other less expensive materials have proved inadequate for design requirements.

Bearings made of laminated phenolics, nylon or Teflon are all unaffected by acids and alkalies except if highly concentrated and therefore can be used with lubricants containing dilute acids or alkalies. Water is used to lubricate most phenolic laminate bearings but oil, grease, and emulsions of grease and water are also used. Water and oil are used as lubricants for nylon and Teflon bearings. Amost all types of plastic bearings absorb water and oil to some extent. In some the dimensional change caused by the absorption may be as much as three per cent in one direction. This means that bearings have to be treated before use so that proper clearances will be kept. This may be done by boiling in water, for water lubricated bearings. Boiling in water makes bearings swell the maximum amount. Clearances for phenolic bearings are kept at about 0.001 inch per inch of diameter on treated bearings. Partially lubricated or dry nylon bearings are given a clearance of 0.004 to 0.006 inches for a one-inch diameter bearing.

*Wood:* Bearings made from such woods as lignum vitae, rock maple or oak offer self-lubricating properties, low cost and clean operation. However, they have frequently been displaced in recent years by various plastics, rubber and sintered-metal bearings. General applications are shown in Table 10.

*Rubber:* Rubber bearings give excellent performance on propeller shafts and rudders of ships, hydraulic turbines, pumps, sand and gravel washers, dredges and other industrial equipment that handle water or slurries. The resilience of rubber helps to isolate vibration and provide quiet operation, allows running with relatively large clearances and helps to compensate for misalignment. In these bearings a fluted rubber structure is supported by a metal shell. The flutes or scallops in the rubber form a series of grooves through which lubricant or, as generally used, water and foreign material such as sand may pass through the bearing.

**Carbon-Graphite.** — Bearings of molded and machined carbon-graphite are used where regular maintenance and lubrication cannot be given. They are dimensionally stable over a wide range of temperatures, may be lubricated if desired, and are not affected by chemicals. These bearings may be used up to temperatures of 700 to 750 degrees F. in air or 1200 degrees F. in a non-oxidizing atmosphere, and generally are operated at a maximum load of 20 pounds per square inch. In some instances a metal or metal alloy is added to the carbon-graphite composition to improve such properties as compressive strength and density. The temperature limitation depends upon the melting point of the metal or alloy and the maximum load is generally 350 pounds per square inch when used with no lubrication or 600 pounds per square inch when used with lubrication.

Normal running clearances for both types of carbon-graphite bearings used with steel shafts and operating at a temperature of less than 200 degrees F. are as follows: 0.001 inch for bearings of 0.187 to 0.500-inch inside diameter, 0.002 inch for bearings of 0.501 to 1.000-inch inside diameter, 0.003 inch for bearings of 1.001 to 1.250-inch inside diameter, 0.004 inch for bearings of 1.251 to 1.500-inch inside diameter, and 0.005 inch for bearings of 1.501 to 2.000-inch inside diameter. Speeds depend upon too many variables to list specifically so it can only be stated here that high loads require a low number of rpm and low loads permit a high number of rpm. Smooth journals are necessary in these bearings as rough ones tend to abrade the bearings quickly. Cast iron and hard chromium-plate steel shafts of 400 Brinell and over, and phosphor-bronze shafts over 135 Brinell are recommended.

**Babbitting.** — Babbitt metal is extensively used as a lining for bearings, not only for its anti-frictional qualities, but because it is much cheaper than a machined box. Prior to pouring the babbitt, the bearing should be heated to prevent the molten metal from becoming chilled and sluggish. Bronze shells which are to have babbitt linings should first be tinned by immersing in a pot of molten solder. Use solder of " half and half " composition, and zinc chloride as a flux. The shell should be babbitted immediately after tinning. Babbitt should not be used for tinning, because it has a much higher melting point, which makes it difficult to maintain a molten film on the surface to be tinned. Cast-iron shells are rarely, if ever, tinned.

If much work of this kind is being done, babbitting jigs should be made. These are simply fixtures which bear against or fit into any finished surface or hole with which the mandrel must be aligned, and which hold the latter in the correct position while the babbitt is being poured. Whenever practicable, the bearing should be placed in a vertical position while pouring. The ladle should preferably have a rounded spout rather than one which is sharp or broad. A broad, thin stream or one that is intermittent tends to produce porous areas or blow-holes. Putty is preferable to clay for luting or sealing the ends of the bearings, as moisture in the clay tends to produce sputtering.

**Coatings for Babbitting Mandrels.** — For babbitting solid bearings, the surface of the mandrel which comes into contact with the metal, when the latter is poured into the bearing, should be coated with some substance to facilitate the removal of the mandrel from the bearing. One method of coating the mandrel is to hold it near an oil flame so that the smoke will come into contact with it and cover the surface with carbon. Instead of smoking the mandrel, the surface is sometimes covered with a coat of thin white lead. Another method is to wrap a piece of paper about the mandrel. In babbitting two-part bearings a coating or covering for the mandrel will not be necessary.

**Temperature of Molten Babbitt.** — Babbitt metal used for bearings is melted in iron pots or kettles, and the molten metal should be kept at a constant temperature of about 870 degrees F. A constant temperature is very important. The temperature should be increased slowly and the babbitt thoroughly stirred, especially when new babbitt is being melted, or when old babbitt which has been allowed to solidify in the pot is being re-melted. This is necessary in order to prevent certain of the constituent metals from rising to the top and becoming oxidized, as well as to prevent the heavier metals from sinking to the bottom of the pot, thus producing a non-uniform alloy. The babbitt may be melted in the pouring ladle over an open fire. The temperature is about right when a pine stick used for stirring chars but does not ignite. The mandrel temperature is right when water evaporates rapidly from its surface without spluttering.

**Preheating Bearings and Mandrels.** — All iron shells must be preheated to a temperature of from 200 to 300 degrees F. before pouring the babbitt. The higher temperature is preferable, as a rule, except where a lining is being poured in a very heavy shell, when it may be necessary to use the lower temperature to prevent the babbitt from cooling too slowly. The mandrels should be preheated to a temperature of from 200 to 300 degrees F. when pouring babbitt into the shells, but, for bronze shells, a somewhat lower temperature should be used. Oil-holes in the bearing shells are filled with asbestos or wood driven against the mandrel, and the joints are made tight with clay.

**Pouring Babbitt Metal.** — To secure good results, the babbitt must be poured at the correct temperature. If the babbitt is poured at too high a temperature, extreme shrinkage will occur, resulting in porous areas in the lining and in broken anchors; furthermore, the babbitt will be oxidized, softened, and dirty, and its anti-frictional qualities will be lowered. If the babbitt is poured at too low a temperature, a lining having a coarse granular formation is the result. If the shells and mandrels are too cold, blow-holes and similar defects will form, and the lining will shrink away from the shell in cooling. If the temperature of the shell or mandrel is too high, the babbitt will cool too slowly, and the heavier metals will have time to settle, producing a bearing which will be soft at one place and brittle at another.

## BALL, ROLLER, AND NEEDLE BEARINGS (ROLLING CONTACT BEARINGS)

Rolling contact bearings substitute a rolling element, ball or roller, for a hydrodynamic or hydrostatic fluid film to carry an impressed load without wear and with reduced friction. Because of their greatly reduced starting friction, when compared to the conventional journal bearing, they have acquired the common designation of "anti-friction" bearings. Though normally made with hardened rolling elements and races, and usually utilizing a separator to space the rolling elements and reduce friction, many variations are in use throughout the mechanical and electrical industries. The most common anti-friction bearing application is that of the deep-groove ball bearing with ribbon-type separator and sealed-grease lubrication used to support a shaft with radial and thrust loads in rotating equipment. This shielded or sealed bearing has become a standard and commonplace item ordered from a supplier's catalogue in much the same manner as nuts and bolts. Because of the simple design approach and the elimination of a separate lubrication system or device, this bearing is found in as many installations as the wick-fed or impregnated porous plain bushing.

Currently, a number of manufacturers produce a complete range of ball and roller bearings in a fully interchangeable series with standard dimensions, tolerances and fits as specified in Anti-Friction Bearing Manufacturers Association (AFBMA) Standards. Except for deep-groove ball bearings, performance standards are not so well defined and sizing and selection must be done in close conformance with the specific manufacturer's catalogue requirements. In general, desired functional features should be carefully gone over with the vendor's representatives.

Rolling contact bearings are made to high standards of accuracy and with close metallurgical control. Balls and rollers are normally held to diametral tolerances of .0001 inch or less within one bearing and are often used as "gage" blocks in routine toolroom operations. This accuracy is essential to the performance and durability of rolling-contact bearings and also in limiting runout, providing proper radial and axial clearances, and insuring smoothness of operation.

Because of their low friction, both starting and running, rolling-contact bearings are utilized to reduce the complexity of many systems normally functioning with journal bearings. Aside from this advantage and that of precise radial and axial location of rotating elements, however, they also are desirable because of their reduced lubrication requirements and their ability to function during brief interruptions in normal lubrication.

In applying rolling-contact bearings it is well to appreciate that they have a life which is limited by the fatigue life of the material from which they are made and as modified by the lubricant used. In rolling-contact fatigue, precise relationships between life, load, and design characteristics are not predictable, but a statistical function described as the "probability of survival," is used to relate them according to equations recommended by the AFBMA. Deviations from these formulae result when certain extremes in applications such as speed, deflection, temperature, lubrication, and internal geometry must be dealt with.

**Types of Anti-friction Bearings.** — The general types are usually determined by the shape of the rolling element, but many variations have been developed which apply conventional elements in unique ways. Thus it is well to know that special bearings can be procured with races adapted to specific applications, though this is not practical for other than high volume configurations or where the requirements cannot be met in a more economical manner. "Special" races are appreciably more expensive. Quite often, in such situations, races are made to incorporate other functions of the mechanism, or are "submerged" in the surrounding structure, with the rolling elements supported by shaft or housing which has been hardened and finished in a suitable manner. Typical anti-friction bearing types are shown in Fig. 1.

**Types of Ball Bearings.** — Most types of ball bearings originate from three basic designs: the single-row radial, the single-row angular contact and the double-row angular contact.

| BALL BEARINGS, SINGLE ROW, RADIAL CONTACT | | | |
|---|---|---|---|
| Symbol | Description | Symbol | Description |
| BC | Non-filling slot assembly 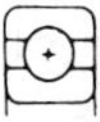 | BH | Non-separable counter-bore assembly 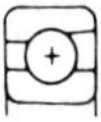 |
| BL | Filling slot assembly 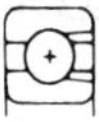 | BM | Separable assembly 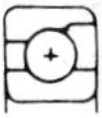 |
| **BALL BEARINGS, SINGLE ROW, ANGULAR CONTACT** | | | |
| Symbol | Description | Symbol | Description |
| BN | Non-separable Nominal contact angle: from above 10° to and including 22°  | BAS | Separable inner ring Nominal contact angle: from above 22° to and including 32°  |
| BNS | Separable outer ring Nominal contact angle: from above 10° to and including 22°  | BT | Non-separable Nominal contact angle: from above 32° to and including 45° 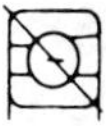 |
| BNT | Separable inner ring Nominal contact angle: from above 10° to and including 22° 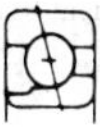 | BY | Two-piece outer ring 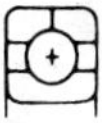 |
| BA | Non-separable Nominal contact angle: from above 22° to and including 32°  | BZ | Two-piece inner ring 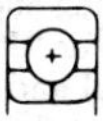 |
| **BALL BEARINGS, SINGLE ROW, RADIAL CONTACT, SPHERICAL OUTSIDE SURFACE** | | | |
| Symbol | Description | Symbol | Description |
| BCA | Non-filling slot assembly 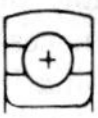 | BLA | Filling slot assembly 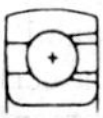 |

Fig. 1. Types of rolling element bearings and their symbols.

| BALL BEARINGS, DOUBLE ROW, RADIAL CONTACT | | | | |
|---|---|---|---|---|
| Symbol | Description | | Symbol | Description |
| BF | Filling slot assembly | 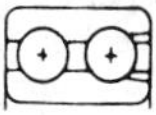 | BHA | Non-separable two-piece outer ring 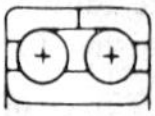 |
| BK | Non-filling slot assembly 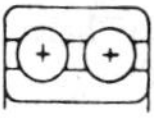 | | | |

| BALL BEARINGS, DOUBLE ROW, ANGULAR CONTACT | | | | |
|---|---|---|---|---|
| Symbol | Description | | Symbol | Description |
| BD | Filling slot assembly Vertex of contact angles inside bearing | 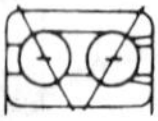 | BG | Non-filling slot assembly Vertex of contact angles outside bearing 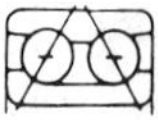 |
| BE | Filling slot assembly Vertex of contact angles outside bearing | 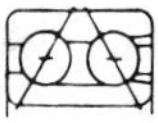 | BAA | Non-separable Vertex of contact angles inside bearing Two-piece outer ring 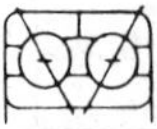 |
| BJ | Non-filling slot assembly Vertex of contact angles inside bearing | 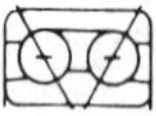 | BVV | Separable Vertex of contact angles outside bearing Two-piece inner ring 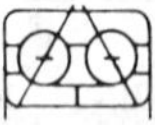 |

| BALL BEARINGS, DOUBLE ROW, SELF-ALIGNING | | |
|---|---|---|
| Symbol | Description | |
| BS | Raceway of outer ring spherical | 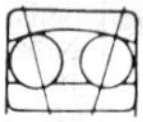 |

Fig. 1 (*Continued*). Types of rolling element bearings and their symbols.

*Single-row Radial, Non-filling Slot:* This is probably the most widely used ball bearing and is employed in many modified forms. It is also known as the "Conrad" type or "Deep-groove" type. It is a symmetrical unit capable of taking combined radial and thrust loads in which the thrust component is relatively high, but is not intended for pure thrust loads, however. Because this type is not self-aligning, accurate alignment between shaft and housing bore is required.

*Single-row Radial, Filling Slot:* This type is designed primarily to carry radial loads. Bearings of this type are assembled with as many balls as can be introduced by eccentric displacement of the rings, as in the non-filling slot type, and then several more balls are inserted through the loading slot, aided by a slight spreading of the rings and heat expansion of the outer ring, if necessary. This type of bearing will take a certain degree of thrust when in combination with a radial load but is not recommended where thrust loads exceed 60 per cent of the radial load.

*Single-row Angular-contact:* This type is designed for combined radial and thrust loads where the thrust component may be large and axial deflection must be confined

| CYLINDRICAL ROLLER BEARINGS, SINGLE ROW, NON-LOCATING TYPE | | | | |
|---|---|---|---|---|
| Symbol | Description | Symbol | Description | |
| RU | Inner ring without ribs Double-ribbed outer ring Inner ring separable 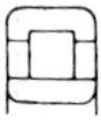 | RNS | Double-ribbed inner ring Outer ring without ribs Outer ring separable Spherical outside surface 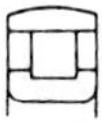 | |
| RUP | Inner ring without ribs Double-ribbed outer ring with one loose rib Both rings separable 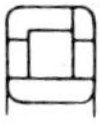 | RAB | Inner ring without ribs Single-ribbed outer ring Both rings separable 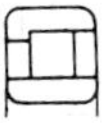 | |
| RUA | Inner ring without ribs Double-ribbed outer ring Inner ring separable Spherical outside surface 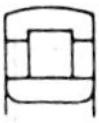 | RM | Inner ring without ribs Rollers located by cage, end-rings or internal snap rings recessed in outer ring Inner ring separable 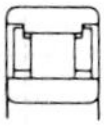 | |
| RN | Double-ribbed inner ring Outer ring without ribs Outer ring separable 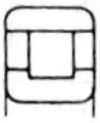 | RNU | Inner ring without ribs Outer ring without ribs Both rings separable 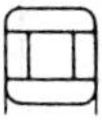 | |

| CYLINDRICAL ROLLER BEARINGS, SINGLE ROW, ONE-DIRECTION-LOCATING TYPE | | | | |
|---|---|---|---|---|
| Symbol | Description | Symbol | Description | |
| RR | Single-ribbed inner ring Outer ring with two internal snap rings Inner ring separable 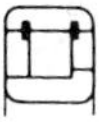 | RF | Double-ribbed inner ring Single-ribbed outer ring Outer ring separable 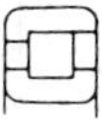 | |
| RJ | Single-ribbed inner ring Double-ribbed outer ring Inner ring separable 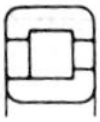 | RS | Single-ribbed inner ring Outer ring with one rib and one internal snap ring Inner ring separable 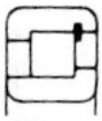 | |
| RJP | Single-ribbed inner ring Double-ribbed outer ring with one loose rib Both rings separable 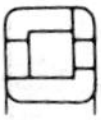 | RAA | Single-ribbed inner ring Single-ribbed outer ring Both rings separable 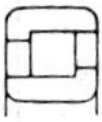 | |

Fig. 1 (*Continued*).  Types of rolling element bearings and their symbols.

within very close limits.   A high shoulder on one side of the outer ring is provided to take the thrust, while the shoulder on the other side is only high enough to make the bearing non-separable.   Except where used for a pure thrust load in one direction, this type is applied either in pairs (duplex) or one at each end of the shaft, opposed.

*Double-row Bearings:* These are, in effect, two single-row angular-contact bearings built as a unit with the internal fit between balls and raceway fixed at the time of

| CYLINDRICAL ROLLER BEARINGS, SINGLE ROW, TWO-DIRECTION-LOCATING TYPE | | | |
|---|---|---|---|
| Symbol | Description | Symbol | Description |
| RK | Double-ribbed inner ring Outer ring with two internal snap rings Non-separable |  RY | Double-ribbed inner ring Outer ring with one rib and one internal snap ring Non-separable 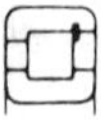 |
| RC | Double-ribbed inner ring Double-ribbed outer ring Non-separable | 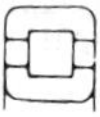 RCS | Double-ribbed inner ring Double-ribbed outer ring Non-separable Spherical outside surface 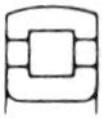 |
| RG | Inner ring, with one rib and one snap ring Double-ribbed outer ring Non-separable | 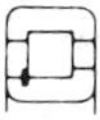 |  |
| RP | Double-ribbed inner ring Double-ribbed outer ring with one loose rib Outer ring separable 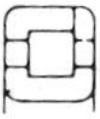 | RT | Double-ribbed inner ring with one loose rib Double-ribbed outer ring Inner ring separable |

CYLINDRICAL ROLLER BEARINGS

| DOUBLE ROW NON-LOCATING TYPE | | DOUBLE ROW TWO-DIRECTION-LOCATING TYPE | |
|---|---|---|---|
| Symbol | Description | Symbol | Description |
| RA | Inner ring without ribs Three integral ribs on outer ring Inner ring separable | RB | Three integral ribs on inner ring Outer ring without ribs, with two internal snap rings Non-separable |
| RD | Three integral ribs on inner ring Outer ring without ribs Outer ring separable | MULTI-ROW NON-LOCATING TYPE | |
| | | Symbol | Description |
| RE | Inner ring without ribs Outer rings without ribs, with two internal snap rings Inner ring separable | RV | Inner ring without ribs Double-ribbed outer ring (loose ribs) Both rings separable |

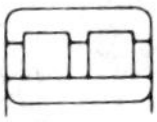

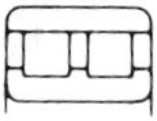

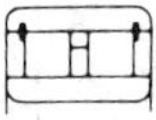

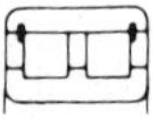

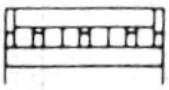

Fig. 1 (*Continued*).  Types of rolling element bearings and their symbols.

bearing assembly.  This fit is therefore not dependent upon mounting methods for internal rigidity.  These bearings usually have a known amount of internal preload built in for maximum resistance to deflection under combined loads with thrust from either direction.  Thus, with balls and races under compression before an external load is applied, due to this internal preload, the bearings are very effective for radial loads where bearing deflection must be minimized.

| SELF-ALIGNING ROLLER BEARINGS, DOUBLE ROW | | | |
|---|---|---|---|
| Symbol | Description | Symbol | Description |
| SD | Three integral ribs on inner ring Raceway of outer ring spherical | SL | Raceway of outer ring spherical Rollers guided by the cage Two integral ribs on inner ring |
| SE | Raceway of outer ring spherical Rollers guided by separate center guide ring in outer ring | SELF-ALIGNING ROLLER BEARINGS, SINGLE ROW | |
| | | Symbol | Description |
| | | SR | Inner ring with ribs Raceway of outer ring spherical Radial contact |
| SW | Raceway of inner ring spherical | SA | Raceway of outer ring spherical Angular contact |
| SC | Raceway of outer ring spherical Rollers guided by separate axially floating guide ring on inner ring | SB | Raceway of inner ring spherical Angular contact |
| THRUST BALL BEARINGS | | THRUST ROLLER BEARINGS | |
| Symbol | Description | Symbol | Description |
| TA | Single direction, grooved raceways, flat seats | TS | Self-aligning, single-direction, flat seats, asymmetrical barrel-shaped rollers |
| TAA | Single row, angular contact where a line through the ball contact points forms an angle from above 45° to and including 75° with a perpendicular to the bearing axis of rotation | TSA | Self-aligning, single-direction, flat seats, symmetrical barrel-shaped rollers |
| | | TP | Single-direction, flat seats, cylindrical rollers |

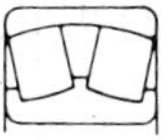

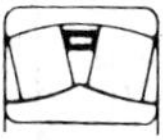

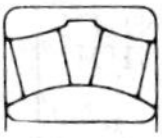

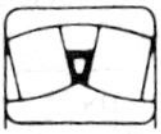

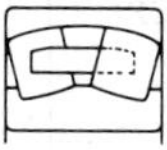

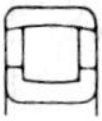

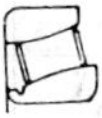

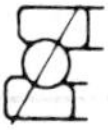

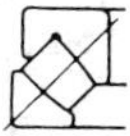

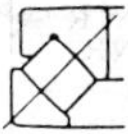

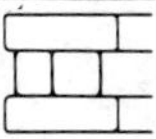

Fig. 1 (*Continued*).   Types of rolling element bearings and their symbols.

*Other Types:* Modifications of these basic types provide arrangements for self-sealing, location by snap ring, shielding, etc., but the fundamentals of mounting are not changed.   A special type is the *self-aligning* ball bearing which can be used to compensate for an appreciable degree of misalignment between shaft and housing due to shaft deflections, mounting inaccuracies, or other causes commonly encountered.   With a single row of balls, alignment is provided by a spherical outer surface on the outer ring; with a double row of balls, alignment is provided by a spherical raceway on the outer ring.   Bearings in the wide series have a considerable amount of thrust capacity.

**Types of Roller Bearings. —** Types of roller bearings are distinguished by the

| NEEDLE ROLLER BEARINGS, DRAWN CUP | | | |
|---|---|---|---|
| Symbol* | Description | Symbol* | Description |
| NIB<br>NB | Needle roller bearing, full complement, drawn cup, without inner ring. | NIYM<br>NYM | Needle roller bearing, full complement, rollers retained by lubricant, drawn cup, closed end, without inner ring. |
| NIBM<br>NBM | Needle roller bearing, full complement, drawn cup, closed end, without inner ring. | NIH<br>NH | Needle roller bearing, with cage, drawn cup, without inner ring. |
| NIY<br>NY | Needle roller bearing, full complement, rollers retained by lubricant, drawn cup, without inner ring. | NIHM<br>NHM | Needle roller bearing, with cage, drawn cup, closed end, without inner ring. |

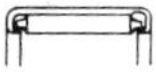

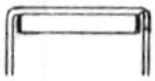

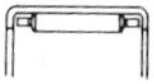

| NEEDLE ROLLER BEARINGS | |
|---|---|
| Symbol* | Description |
| NIA<br>NA | Needle roller bearing, with cage, machined ring, lubrication hole and groove in OD, without inner ring. 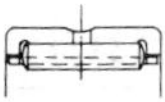 |

| NEEDLE ROLLER AND CAGE ASSEMBLIES | |
|---|---|
| Symbol* | Description |
| NIM<br>NM | Needle roller and cage assembly |

| NEEDLE ROLLER BEARING INNER RINGS | |
|---|---|
| Symbol* | Description |
| NIR<br>NR | Needle roller bearing inner ring, lubrication hole and groove in bore. 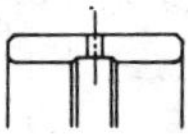 |

Needle Roller Bearings Types NIB, NIY, NIH, and NIA may be used with inch dimensioned inner rings, Type NIR.

* Symbols with I, as NIB, are inch-dimensioned, and those without the I, as NB, are metric dimensioned.

Fig. 1 (*Concluded*).  Types of rolling element bearings and their symbols.

design of rollers and raceways to handle axial, combined axial and thrust, or thrust loads.

*Cylindrical Roller:* These have solid or helically wound hollow cylindrical rollers. The free ring may have a restraining flange to provide some restraint to endwise movement in one direction or may be without a flange so that the bearing rings may be displaced axially with respect to each other.   Either rolls or roller path on the races may be slightly crowned to prevent edge loading under slight shaft misalignment.   Low friction makes this type suitable for relatively high speeds.

*Barrel Roller:* These have rollers that are barrel-shaped and symmetrical.   They are furnished in both single- and double-row mountings.   As with cylindrical roller bearings, the single-row mounting type has a low thrust capacity, but angular mounting of rolls in the double-row type permits its use for combined axial and thrust loads.

*Spherical Roller:* These are usually furnished in a double-row, self-aligning mounting.   Both rows of rollers have a common spherical outer raceway.  The rollers are barrel-shaped with one end smaller than the other to provide a small thrust to keep the rollers in contact with the center guide flange. This type of roller bearing has a high radial and thrust load carrying capacity with the ability to maintain this capacity under some degree of misalignment of shaft and bearing housing.

*Tapered Roller:* In this type, straight tapered rollers are held in accurate alignment by means of a guide flange on the inner ring. The basic characteristic of these bearings is that the apexes of the tapered working surfaces of both rollers and races, if extended, would coincide on the bearing axis. These bearings are separable. They have a high radial and thrust carrying capacity.

**Types of Ball and Roller Thrust Bearings.** — These are designed to take thrust loads alone or, in some cases, in combination with radial loads.

*One-direction Ball Thrust:* These consist of a shaft ring and a flat or spherical housing ring with a single row of balls between. They are capable of carrying pure thrust loads in one direction only. They cannot carry any radial load.

*Two-direction Ball Thrust:* These consist of a shaft ring with a ball groove in either side, two sets of balls, and two housing rings so arranged that thrust loads in either direction can be supported. No radial loads can be carried.

*Spherical Roller Thrust:* This type is similar in design to the radial spherical roller bearing except that it has a much larger contact angle. The rollers are barrel shaped with one end smaller than the other. This type bearing has a very high thrust load carrying capacity and can also carry radial loads.

*Tapered Roller Thrust:* In this type the rollers are straight tapered and several different arrangements of housing and shaft are used.

**Types of Needle Bearings.** — Needle bearings are characterized by their relatively small size rollers, usually not ranging above ¼ inch in diameter, and a relatively high ratio of length to diameter, usually ranging from about 6 to 1 to 10 to 1, although rollers of shorter proportion are often used. Another feature that is characteristic of several types of needle bearings is the absence of a cage or separator for retaining the individual rollers. Needle bearings may be divided into three classes: loose-roller, outer race and retained roller, and non-separable units.

*Loose-roller:* This type of bearing has no integral races or retaining members, the needles being located directly between the shaft and the outer bearing bore. Usually both shaft and outer bore bearing surfaces are hardened and retaining members that have smooth unbroken surfaces are provided to prevent endwise movement. Compactness and high radial load capacity are features of this type.

*Outer Race and Retained Roller:* There are two types of outer race and retained roller bearings. In the *Drawn Shell* type, the needle rollers are enclosed by a hardened shell that acts as a retaining member and also as a hardened outer race. The needles roll directly on the shaft, the bearing surface of which should be hardened. The capacity for given roller length and shaft diameter is about two-thirds of the loose roller type. It is mounted in the housing with a press fit.

In the *Machined Race* type, the outer race consists of a heavy machined member. Various modifications of this type provide heavy ends or faces for end location of the needle rollers, or open end construction with end washers for roller retention, or a cage which maintains alignment of the rollers and is itself held in place by retaining rings. An auxiliary outer member with spherical seat which holds the outer race may be provided for self alignment. This type is applicable where split housings occur or where a press fit of the bearing into the housing is not possible.

*Non-separable:* This type consists of a non-separable unit of outer race, rollers and inner race. It is used where high static or oscillating motion loads are expected as in certain aircraft components, and where both outer and inner races are necessary.

**Special or Unconventional Types.** — Rolling contact bearings have been developed for many highly specialized applications. They may be constructed of non-corrosive materials, non-magnetic materials, plastics, ceramics, and even wood. Though the materials are chosen to adapt more conventional configurations to difficult applications or environments, even greater ingenuity has been applied in utilizing rolling-contact for solving particular problems. Thus, linear or recirculating bearings are available to provide low friction, accurate location, and simplified lubrication features to such applications as machine ways, axial motion devices, jack-screws, steering linkages, collets, and chucks. This type of bearing utilizes the "full-complement" style of loading the rolling elements between "races" or ways without a cage and with each element advancing by the action of "races" in the loaded areas and by contact with the adjacent element in the unloaded areas. The "races" may not be cylindrical or bodies of revolution but plane surfaces, with suitable interruptions to free the rolling elements so that they can then follow a return trough or slot back to the entry-point at the start of the "race" contact area. Combinations of radial and thrust bearings are available for the user with special requirements.

**Plastic Bearings.** — A more recent development has been the use of Acetal Resin rollers and balls in applications where abrasive, corrosive and difficult-to-lubricate conditions exist. Though these bearings do not have the load carrying capacity nor the low friction factor of their hard steel counterparts, they do offer freedom from indentation, wear, and corrosion, while at the same time providing significant weight savings. Of additional value are: (1) their resistance to indentation from shock loads or oscillation and (2) their self-lubricating properties. Usually these bearings are not available in stock, but must be designed and produced in accordance with the data made available by the plastics processor.

**Pillow Block and Flanged Housing Bearings.** — Of great interest to the shop man and particularly adaptable to "line-shafting" applications are a series of ball and roller bearings supplied with their own housings, adapters, and seals. Often called premounted bearings, they come with a wide variety of flange mountings permitting location on faces parallel to or perpendicular to the shaft axis.

Inner races can be mounted directly on ground shafts, or can be adapter-mounted to "drill-rod" or to commercial-shafting. For installations sensitive to unbalance and vibration, the use of accurately ground shaft seats is recommended.

Since most pillow block designs incorporate self-aligning bearing types, they do not require the precision mountings utilized with more normal bearing installations.

**Conventional Bearing Materials.** — Most rolling contact bearings are made with all load carrying members of full hard steel, either through- or case-hardened. For greater reliability this material is controlled and selected for cleanliness and alloying practices in conformity with rigid specifications in order to reduce the incidence of anomalies and inclusions which could limit the useful fatigue life. Magnaflux inspection is employed to insure that elements are free from both material defects and cracks. Likewise, a light etch is employed between rough and finish grinding to insure that burns due to heavy stock removal and associated decarburization will be eliminated from finished pieces.

**Cage Materials.** — Standard bearings are normally made with cages of free-machining brass or low carbon sulphurized steel. In high-speed applications or where lubrication may be intermittent or marginal, special materials may be employed. Iron-silicon-bronze, laminated phenolics, silver-plating over-lays, solid-film baked-on coatings, carbon-graphite inserts, and in extreme cases sintered or even impregnated materials are used in separators.

Commercial bearings usually rely on stamped steel with or without a phosphate treatment; some economical varieties are found with snap-in plastic or metallic cages.

So long as lubrication is adequate and speeds are both reasonable and steady, the materials and design of the cage are of secondary importance when compared with those of the rolling elements and their contacts with the races. In spite of this tolerance however, a good portion of all rolling bearing failures encountered can be traced to cage failures resulting from inadequate lubrication. It can never be over-emphasized that *no bearing can be designed to run continuously without lubrication!*

**Standard Method of Bearing Designation.** — The Anti-Friction Bearing Manufacturers Association has adopted a standard identification code that provides a specific designation for each different ball, roller, and needle bearing. Thus, for any given bearing, a uniform designation is provided for manufacturer and user alike, so that the confusion of different company designations can be avoided.

In this identification code there is a "basic number" for each bearing which consists of three elements: a one- to four-digit number which indicates the size of the bore in numbers of millimeters (metric series); a two- or three-letter symbol which indicates the type of bearing; and a two-digit number which indicates the dimension series to which the bearing belongs.

In addition to this "basic number" other numbers and letters are added to designate type of tolerance, cage, lubrication, fit up, ring modification, the addition of shields, seals, mounting accessories, etc. Thus, a complete designating symbol might be *50BC02JPXE0A10*, for example. The basic number is *50BC02* and the remainder is the supplementary number. For a radial bearing, this latter consists of up to four letters to indicate modification of design, one or two digits to indicate internal fit and tolerances, a letter to indicate lubricants and preservatives, and up to three digits to indicate special requirements.

For a thrust bearing the supplementary number would consist of two letters to indicate modifications of design, one digit to indicate tolerances, one letter to indicate lubricants and preservatives, and up to three digits to indicate special requirements.

For a needle bearing the supplementary number would consist of two letters to indicate cage material or integral seals and one letter to indicate lubricants or preservatives.

*Dimension Series:* Annular ball, cylindrical roller, and self-aligning roller bearings are made in a series of different outside diameters for every given bore diameter and in a series of different widths for every given outside diameter. Thus, each of these bearings belongs to a dimension series which is designated by a two-digit number such as 01, 23, 93, etc. The first digit (8, 0, 1, 2, 3, 4, 5, 6, and 9) indicates the *width series* and the second digit (8, 9, 0, 1, 2, 3, and 4) the *diameter series* to which the bearing belongs. In the case of ball and roller thrust bearings and needle roller bearings, similar types of identification codes are used.

**Bearing Tolerances.** — In order to provide standards of precision for proper application of ball or roller bearings in all types of equipment, four classes of tolerances have been established by the Anti-Friction Bearing Manufacturers Association for ball bearings, two for cylindrical roller bearings and one for spherical roller bearings. These tolerances are given in Tables 1, 2, 3 and 4. They are designated as ABEC-1, ABEC-5, ABEC-7 and ABEC-9 for ball bearings, the ABEC-9 being the most precise, RBEC-1 and RBEC-5 for roller bearings. In general, bearings to specifications closer than ABEC-1 or RBEC-1 are required because of the need for very precise fits on shaft or housing, to reduce eccentricity or runout of shaft or supported part, or to permit operation at very high speeds. All four classes

(Continued on page 618)

**Table 1.   ABEC-1 and RBEC-1 Tolerance Limits for Metric Ball and Roller Bearings** (ANSI/AFBMA Std 20-1977[1])

| Basic Inner Ring Bore Diameter, $d$ | | | | Bore Tolerance Limits,[2] Inch | | | | Radial Runout,[4] $K_i$ Inch |
|---|---|---|---|---|---|---|---|---|
| mm | | Inches | | $d_{mp}$ | | $d_s$[3] | | |
| Over | Incl. | Over | Incl. | Low | High | Low | High | |
| 2.5 | 10 | 0.0984 | 0.3937 | −.0003 | +0 | −.0004 | +.0001 | .0003 |
| 10 | 18 | 0.3937 | 0.7087 | −.0003 | +0 | −.0004 | +.0001 | .0004 |
| 18 | 30 | 0.7087 | 1.1811 | −.0004 | +0 | −.0005 | +.0001 | .0005 |
| 30 | 50 | 1.1811 | 1.9685 | −.0005 | +0 | −.0006 | +.0001 | .0006 |
| 50 | 80 | 1.9685 | 3.1496 | −.0006 | +0 | −.0008 | +.0002 | .0008 |
| 80 | 120 | 3.1496 | 4.7244 | −.0008 | +0 | −.0010 | +.0002 | .0010 |
| 120 | 180 | 4.7244 | 7.0866 | −.0010 | +0 | −.0012 | +.0003 | .0012 |

| Basic Outer Ring Outside Diameter, $D$ | | | | Outside Diameter Tolerance Limits,[5] Inch | | | | | | Radial Runout,[4] $K_e$ Inch |
|---|---|---|---|---|---|---|---|---|---|---|
| mm | | Inches | | $D_{mp}$ | | $D_s$[6] | | $D_s$[7] | | |
| Over | Incl. | Over | Incl. | High | Low | High | Low | High | Low | |
| 6 | 18 | 0.2362 | 0.7087 | +0 | −.0003 | +.0001 | −.0004 | +.0002 | −.0005 | .0006 |
| 18 | 30 | 0.7087 | 1.1811 | +0 | −.0004 | +.0001 | −.0005 | +.0002 | −.0006 | .0006 |
| 30 | 50 | 1.1811 | 1.9685 | +0 | −.0005 | +.0002 | −.0007 | +.0003 | −.0008 | .0008 |
| 50 | 80 | 1.9685 | 3.1496 | +0 | −.0005 | +.0002 | −.0007 | +.0004 | −.0009 | .0010 |
| 80 | 120 | 3.1496 | 4.7244 | +0 | −.0006 | +.0002 | −.0008 | +.0005 | −.0011 | .0014 |
| 120 | 150 | 4.7244 | 5.9055 | +0 | −.0008 | +.0002 | −.0010 | +.0005 | −.0013 | .0016 |
| 150 | 180 | 5.9055 | 7.0866 | +0 | −.0010 | +.0003 | −.0013 | +.0006 | −.0016 | .0018 |
| 180 | 250 | 7.0866 | 9.8425 | +0 | −.0012 | +.0003 | −.0015 | +.0008 | −.0020 | .0020 |
| 250 | 315 | 9.8425 | 12.4015 | +0 | −.0014 | +.0004 | −.0018 | +.0008 | −.0022 | .0024 |
| 315 | 400 | 12.4015 | 15.7480 | +0 | −.0016 | +.0004 | −.0020 | +.0009 | −.0025 | .0028 |

### Width Tolerances

| Normal Single Bearings | | | | Modified Single Bearings[8] | | | |
|---|---|---|---|---|---|---|---|
| Basic Inner Ring Bore, $d$, mm | | Width, $B_s$ Tolerance Limits, Inch | | Basic Inner Ring Bore, $d$, mm | | Width, $B_s$ Tolerance Limits, Inch | |
| Over | Incl. | High | Low | Over | Incl. | High | Low |
| 2.5 | 50 | +0 | −.0050 | 2.5 | 50 | +0 | −.0100 |
| 50 | 80 | +0 | −.0060 | 50 | 80 | +0 | −.0150 |
| 80 | 120 | +0 | −.0080 | 80 | 120 | +0 | −.0150 |
| 120 | 180 | +0 | −.0100 | 120 | 180 | +0 | −.0200 |

For sizes beyond range of this table, see Standard.
[1] Does not cover tapered roller bearings.   [2] The amounts by which the largest and smallest single diameter, $d_s$, of the bore and the mean, $d_{mp}$, of these two vary from the basic.   Bore tolerance limits do not apply to tapered bore inner rings.   [3] $d_s$ applies only to: metric diameter series 0, up to and including $d = 40$ mm and in diameter series 2, up to and including 180 mm.   For larger sizes in series 0 and 2 and all sizes of series 1, 8 and 9, $d_s$ is not restricted.   [4] Total indicator reading.   [5] The amounts by which the largest and smallest outside diameter, $D_s$, and the mean, $D_{mp}$, of these two vary from the basic.   [6] $D_s$ applies only to open bearings of metric diameter series 0 up to and including $D = 80$ mm and diameter series 2, up to and including $D = 315$ mm.   For larger sizes in series 0 and 2 and all sizes of series 1, 8 and 9, $D_s$ is not restricted.   [7] $D_s$ applies only to bearings with shields or seals in metric diameter series 0 up to and including $D = 80$ mm and in diameter series 2 up to and including $D = 315$ mm.   For larger sizes in series 0 and 2 and for all sizes in series 1, 8 and 9, $D_s$ is not restricted.   [8] This refers to a ball bearing on which one or both sides are so modified that two or more bearings can be mounted side by side as a unit.

**Table 2. ABEC-5 and RBEC-5 Tolerance Limits for Metric Ball and Roller Bearings (ANSI/AFBMA Std 20-1977[1])**

## INNER RING

| Inner Ring Bore Basic Diam., d | | | | Bore Tolerance Limits,[2] Inch | | | | | | Radial Run-out $K_i$ | Ref. Side Run-out with Bore | Race-way Run-out with Ref. Side[3] | Width | | | | Width Variation Indiv. Ring |
| --- | --- | --- | --- | --- | --- | --- | --- | --- | --- | --- | --- | --- | --- | --- | --- | --- | --- |
| mm | | Inches | | All Series, $d_{mp}$ | | Diam. Series 4,3,2,1,0 $d_s$ | | Diam. Series 9,8 $d_s$ | | | | | Normal Single Bearings, $B_s$ | | Modified Single Bearings,[4] $B_s$ | | |
| Over | Incl. | Over | Incl. | Low | High | Low | High | Low | High | Max. | Max. | Max. | High | Low | High | Low | Max. |
| 0.6 | 10 | 0.0236 | 0.3937 | −.0002 | +0 | −.0002 | +0 | −.0002 | +0 | .00015 | .0003 | .0003 | +0 | −.0016 | +0 | −.0100 | .0002 |
| 10 | 18 | 0.3937 | 0.7087 | −.0002 | +0 | −.0002 | +0 | −.00025 | +.00005 | .00015 | .0003 | .0003 | +0 | −.0032 | +0 | −.0100 | .0002 |
| 18 | 30 | 0.7087 | 1.1811 | −.0002 | +0 | −.0002 | +0 | −.0003 | +.0001 | .00015 | .0003 | .0003 | +0 | −.0050 | +0 | −.0100 | .0002 |
| 30 | 50 | 1.1811 | 1.9685 | −.0002 | +0 | −.0002 | +0 | −.00035 | +.00015 | .0002 | .0003 | .0003 | +0 | −.0050 | +0 | −.0100 | .0002 |
| 50 | 80 | 1.9685 | 3.1496 | −.0003 | +0 | −.0003 | +0 | −.00045 | +.00015 | .0002 | .0003 | .0003 | +0 | −.0060 | +0 | −.0100 | .0002 |
| 80 | 120 | 3.1496 | 4.7244 | −.0003 | +0 | −.00035 | +.00005 | −.0005 | +.0002 | .00025 | .0003 | .0003 | +0 | −.0080 | +0 | −.0150 | .0003 |
| 120 | 180 | 4.7244 | 7.0866 | −.0004 | +0 | −.0005 | +.0001 | −.0006 | +.0002 | .0003 | .0004 | .0004 | +0 | −.0100 | +0 | −.0150 | .0003 |

## OUTER RING

| Basic Outside Diameter, D | | | | Outside Diameter Tolerance Limits, Inch | | | | | | | | Radial Runout $K_e$ | O.D. Runout with Ref. Side[6] | Race-way Runout with Ref. Side[3] | Width | Width Variation Indiv. Ring |
| --- | --- | --- | --- | --- | --- | --- | --- | --- | --- | --- | --- | --- | --- | --- | --- | --- |
| mm | | Inches | | All Series, $D_{mp}$ | | Open Bearings — Diam. Series 4,3,2,1,0 $D_s$ | | Diam. Series 9,8 $D_s$ | | Diam. Series[5] 4,3,2,1,0 $D_s$ | | | | | Tol. Limits | |
| Over | Incl. | Over | Incl. | High | Low | High | Low | High | Low | High | Low | Max. | Max. | Max. | | Max. |
| 2.5 | 6 | 0.0984 | 0.2362 | +0 | −.0002 | +0 | −.0002 | +0 | −.0002 | +.0001 | −.0003 | .0002 | .0003 | .0003 | Same as those for Inner Ring of the same bearing | .0002 |
| 6 | 18 | 0.2362 | 0.7087 | +0 | −.0002 | +0 | −.0002 | +.00005 | −.00025 | +.0001 | −.0003 | .0002 | .0003 | .0003 | | .0002 |
| 18 | 30 | 0.7087 | 1.1811 | +0 | −.0002 | +0 | −.0002 | +.0001 | −.0003 | +.0001 | −.0003 | .0002 | .0003 | .0003 | | .0002 |
| 30 | 50 | 1.1811 | 1.9685 | +0 | −.0002 | +0 | −.0002 | +.00015 | −.00035 | +.0002 | −.0004 | .0002 | .0003 | .0003 | | .0002 |
| 50 | 80 | 1.9685 | 3.1496 | +0 | −.0003 | +0 | −.0003 | +.0002 | −.00045 | +.0002 | −.0005 | .0003 | .0003 | .0004 | | .0002 |
| 80 | 120 | 3.1496 | 4.7244 | +0 | −.0003 | +.00005 | −.00035 | +.00035 | −.0005 | +.0003 | −.0005 | .0004 | .0003 | .0004 | | .0003 |
| 120 | 150 | 4.7244 | 5.9055 | +0 | −.0004 | +.0001 | −.0005 | +.0006 | −.0006 | +.0003 | −.0007 | .0004 | .0004 | .0005 | | .0003 |
| 150 | 180 | 5.9055 | 7.0866 | +0 | −.0005 | +.0001 | −.0005 | +.0006 | −.0006 | +.0003 | −.0007 | .0005 | .0004 | .0005 | | .0003 |
| 180 | 250 | 7.0866 | 9.8425 | +0 | −.0005 | +.00015 | −.00065 | +.00075 | −.00075 | +.0004 | −.0009 | .0005 | .0004 | .0006 | | .0004 |

All dimensions are in inches unless otherwise indicated. For sizes beyond range of this table see Standard. [1] Does not cover instrument bearings and tapered roller bearings. [2] Bore tolerance limits do not apply to tapered bore inner rings. [3] Does not apply to roller bearings or self-aligning ball bearings. [4] A bearing on which one or both sides are so modified that two or more bearings can be mounted side by side as a unit. [5] For bearings with shields or seals; $D_s$ is not restricted for diameter series 9 and 8. [6] Applies to bearings of width series 1 and narrower.

**Table 3. ABEC-7 Tolerance Limits for Metric Radial and Angular Contact Ball Bearings (ANSI/AFBMA Std 20-1977[1])**

### INNER RING[2]

| Inner Ring Bore Basic Diam., d mm Over | Incl. | Inches Over | Incl. | Bore Tolerance Limits, Inch — All Series $d_{mp}$ Low | High | Diam. Series 4,3,2,1,0 $d_s$ Low | High | Diam. Series 9,8 $d_s$ Low | High | Radial Run-out, $K_i$ Max. | Ref. Side Run-out with Bore Max. | Raceway Run-out with Ref. Side[3] Max. | Normal Single Bearings $B_s$ High | Low | Modified Single Bearings[4] $B_s$ High | Low | Width Variation Indiv. Ring Max. |
|---|---|---|---|---|---|---|---|---|---|---|---|---|---|---|---|---|---|
| 0.6 | 10 | 0.0236 | 0.3937 | −.00015 | +0 | −.00015 | +0 | −.00015 | +0 | .0001 | .0001 | .0001 | +0 | −.0016 | +0 | −.0100 | .0001 |
| 10 | 18 | 0.3937 | 0.7087 | −.00015 | +0 | −.00015 | +0 | −.0002 | +.00005 | .0001 | .0001 | .0001 | +0 | −.0032 | +0 | −.0100 | .0001 |
| 18 | 30 | 0.7087 | 1.1811 | −.00015 | +0 | −.00015 | +0 | −.00025 | +.00025 | .0001 | .00015 | .00015 | +0 | −.0050 | +0 | −.0100 | .0001 |
| 30 | 50 | 1.1811 | 1.9685 | −.0002 | +0 | −.0002 | +0 | −.0003 | +.0001 | .00015 | .00015 | .00015 | +0 | −.0050 | +0 | −.0100 | .0001 |
| 50 | 80 | 1.9685 | 3.1496 | −.0002 | +0 | −.0002 | +0 | −.0003 | +.0001 | .00015 | .00015 | .00015 | +0 | −.0060 | +0 | −.0100 | .00015 |
| 80 | 120 | 3.1496 | 4.7244 | −.00025 | +0 | −.0003 | +.00005 | −.00035 | +.0001 | .0002 | .0002 | .0002 | +0 | −.0080 | +0 | −.0150 | .00015 |
| 120 | 180 | 4.7244 | 7.0866 | −.0003 | +0 | −.00035 | +.00005 | −.00045 | +.00015 | .0003 | .0003 | .0003 | +0 | −.0100 | +0 | −.0150 | .0002 |

### OUTER RING[2]

| Basic Outside Diameter, D mm Over | Incl. | Inches Over | Incl. | Outside Diameter Tolerance Limits, Inch — All Series $D_{mp}$ High | Low | Open Bearings — Diam. Series 4,3,2,1,0 $D_s$ High | Low | Diam. Series[5] 9,8 $D_s$ High | Low | Radial Runout $K_e$ Max. | O.D. Runout with Ref. Side[6] Max. | Raceway Runout with Ref. Side[3] Max. | Width Tol. Limits | Width Variation Indiv. Ring Max. |
|---|---|---|---|---|---|---|---|---|---|---|---|---|---|---|
| 2.5 | 6 | 0.0984 | 0.2362 | +0 | −.0002 | +0 | −.0002 | +0 | −.0002 | .00015 | .00015 | .0002 | Same as those for Inner Ring of the same bearing | .0001 |
| 6 | 18 | 0.2362 | 0.7087 | +0 | −.0002 | +0 | −.0002 | +.00005 | −.00025 | .00015 | .00015 | .0002 | | .0001 |
| 18 | 30 | 0.7087 | 1.1811 | +0 | −.0002 | +0 | −.0002 | +.0001 | −.0003 | .00015 | .00015 | .0002 | | .0001 |
| 30 | 50 | 1.1811 | 1.9685 | +0 | −.0002 | +0 | −.0002 | +.0001 | −.0003 | .0002 | .00015 | .0002 | | .0001 |
| 50 | 80 | 1.9685 | 3.1496 | +0 | −.0002 | +0 | −.0002 | +.0001 | −.0003 | .0002 | .0002 | .0002 | | .0001 |
| 80 | 120 | 3.1496 | 4.7244 | +0 | −.0003 | +.00005 | −.00035 | +.0001 | −.0004 | .0002 | .0002 | .0002 | | .0002 |
| 120 | 150 | 4.7244 | 5.9055 | +0 | −.0004 | +.0001 | −.0005 | +.0003 | −.0006 | .0003 | .0002 | .0003 | | .0002 |
| 150 | 180 | 5.9055 | 7.0866 | +0 | −.0004 | +.0001 | −.0005 | +.0003 | −.0006 | .0003 | .0002 | .0003 | | .0002 |
| 180 | 250 | 7.0866 | 9.8425 | +0 | −.0004 | +.0001 | −.0005 | +.0003 | −.0008 | .0003 | .0003 | .0004 | | .0003 |

All dimensions are in inches unless otherwise indicated. For sizes beyond range of this table see Standard. [1] Does not cover instrument bearings. [2] All tolerance limits are in inches. [3] Does not apply to self-aligning ball bearings. [4] A bearing on which one or both sides are so modified that two or more bearings can be mounted side by side as a unit; $D_s$ is not restricted for dimension series 9 and 8. [5] For bearings with shields or seals. [6] Applies to bearings of width series 1, and narrower.

## Table 4. ABEC-9 Tolerance Limits for Metric Radial and Angular Contact Ball Bearings (ANSI/AFBMA Std 20-1977[1])

### Inner Ring

| Basic Bore Diameter, d | | | | Bore Tolerance Limits, Inch $d_s$ | | Radial Run-out,[2] $K_i$ Inch | Width Variation,[2] Inch | Ref. Side Runout with Bore, Inch | Raceway Runout with Side,[3] Inch |
|---|---|---|---|---|---|---|---|---|---|
| mm | | Inches | | | | | | | |
| Over | Incl. | Over | Incl. | Low | High | | | | |
| 0.6 | 10 | 0.0236 | 0.3937 | −.0001 | +0 | .00005 | .00005 | .00005 | .00005 |
| 10 | 18 | 0.3937 | 0.7087 | −.0001 | +0 | .00005 | .00005 | .00005 | .00005 |
| 18 | 30 | 0.7087 | 1.1811 | −.0001 | +0 | .0001 | .00005 | .00005 | .0001 |
| 30 | 50 | 1.1811 | 1.9685 | −.0001 | +0 | .0001 | .00005 | .00005 | .0001 |
| 50 | 80 | 1.9685 | 3.1496 | −.00015 | +0 | .0001 | .00005 | .00005 | .0001 |
| 80 | 120 | 3.1496 | 4.7244 | −.0002 | +0 | .0001 | .0001 | .0001 | .0001 |
| 120 | 150 | 4.7244 | 5.9055 | −.00025 | +0 | .0001 | .0001 | .0001 | .0001 |
| 150 | 180 | 5.9055 | 7.0866 | −.00025 | +0 | .0002 | .00015 | .00015 | .0002 |

### Outer Ring

| Basic Outside Diameter, D | | | | Outside Diameter Tolerance Limits,[4] Inch $D_s$ | | Radial Run-out,[5] $K_e$ Inch | Width Variations, Inch | Outside Cylin. Surface Runout with Side,[6] Inch | Raceway Runout with Side,[3] Inch |
|---|---|---|---|---|---|---|---|---|---|
| mm | | Inches | | | | | | | |
| Over | Incl. | Over | Incl. | High | Low | | | | |
| 2.5 | 18 | 0.0984 | 0.7087 | +0 | −.0001 | .00005 | .00005 | .00005 | .00005 |
| 18 | 30 | 0.7087 | 1.1811 | +0 | −.00015 | .0001 | .00005 | .00005 | .0001 |
| 30 | 50 | 1.1811 | 1.9685 | +0 | −.00015 | .0001 | .00005 | .00005 | .0001 |
| 50 | 80 | 1.9685 | 3.1496 | +0 | −.00015 | .00015 | .00005 | .00005 | .00015 |
| 80 | 120 | 3.1496 | 4.7244 | +0 | −.0002 | .0002 | .0001 | .0001 | .0002 |
| 120 | 150 | 4.7244 | 5.9055 | +0 | −.0002 | .0002 | .0001 | .0001 | .0002 |
| 150 | 180 | 5.9055 | 7.0866 | +0 | −.00025 | .0002 | .0001 | .0001 | .0002 |
| 180 | 250 | 7.0866 | 9.8425 | +0 | −.0003 | .00025 | .00015 | .00015 | .00025 |
| 250 | 315 | 9.8425 | 12.4015 | +0 | −.0003 | .00025 | .00015 | .00015 | .00025 |

### Width Tolerance Limits

| Normal Single Bearings | | | | Modified Single Bearings[7] | | | |
|---|---|---|---|---|---|---|---|
| Basic Bore Diam., mm | | Width, $B_s$ Tolerance Limits, Inch | | Basic Bore Diam., mm | | Width, $B_s$ Tolerance Limits, Inch | |
| Over | Incl. | High | Low | Over | Incl. | High | Low |
| 0.6 | 10 | +0 | −.0010 | 0.6 | 80 | +0 | −.0100 |
| 10 | 18 | +0 | −.0032 | 80 | 150 | +0 | −.0150 |
| 18 | 50 | +0 | −.0050 | 150 | 250 | +0 | −.0200 |
| 50 | 80 | +0 | −.0060 | .... | .... | .... | .... |
| 80 | 120 | +0 | −.0080 | .... | .... | .... | .... |
| 120 | 150 | +0 | −.0100 | .... | .... | .... | .... |
| 150 | 180 | +0 | −.0120 | .... | .... | .... | .... |

For sizes beyond the range of this table see Standard.
[1] Does not cover instrument bearings. [2] Difference between greatest and smallest radial distance between bore surface and middle of a raceway on outside of the ring. [3] Does not apply to self-aligning ball bearings. "Side" is reference side. [4] These tolerance limits apply before seals or shields are inserted. [5] Difference between greatest and smallest radial distance between outside surface and middle of a raceway on inside of the ring. [6] Applies to bearings of metric width series 1 or narrower. [7] This refers to a ball bearing on which one or both sides are so modified that two or more bearings can be mounted side by side as a unit.

**Table 5.   AFBMA and American National Standard Tolerance Limits for Metric Single Direction Thrust Ball and Roller Bearings** (ANSI/AFBMA Std 20-1977)

| BASIC PLAN METRIC DIMENSIONED† | | | | | | | | | |
|---|---|---|---|---|---|---|---|---|---|
| Basic Bore Diam., d | | Shaft Washer | | Both Indiv. Washers | Bearing Height Tolerance Limits, $H_m$ | | Basic Outside Diameter, D | | Housing Washer |
| | | Bore Tolerance Limits, $d_s$ | | Raceway Runout with Seat Face | | | | | Outside Diam. Tolerance Limits, $D_s$ |
| mm | | Inch | | Inch | Inch | | mm | | Inch |
| Over | Incl. | Low | High | Max. | High | Low | Over | Incl. | High | Low |
| 0 | 18 | −.0003 | +.0002 | .0004 | 0 | −.0030 | 10 | 18 | 0 | −.0004 |
| 18 | 30 | −.0004 | +.0003 | .0004 | 0 | −.0030 | 18 | 30 | 0 | −.0005 |
| 30 | 50 | −.0005 | +.0004 | .0004 | 0 | −.0039 | 30 | 50 | 0 | −.0006 |
| 50 | 80 | −.0006 | +.0005 | .0004 | 0 | −.0049 | 50 | 80 | 0 | −.0007 |
| 80 | 120 | −.0008 | +.0006 | .0006 | 0 | −.0059 | 80 | 120 | 0 | −.0009 |
| 120 | 180 | −.0010 | +.0007 | .0006 | 0 | −.0069 | 120 | 180 | 0 | −.0010 |
| 180 | 250 | −.0012 | +.0009 | .0008 | 0 | −.0079 | 180 | 250 | 0 | −.0012 |
| 250 | 315 | −.0014 | +.0010 | .0010 | 0 | −.0089 | 250 | 315 | 0 | −.0014 |
| 315 | 400 | −.0016 | +.0011 | .0012 | 0 | −.0118 | 315 | 400 | 0 | −.0016 |
| 400 | 500 | −.0018 | +.0013 | .0012 | .... | .... | 400 | 500 | 0 | −.0018 |

† For thrust bearings of types TS and TSA, tolerances in Table 1 apply.
For sizes beyond the range of this table see Standard.

include tolerances for bore, outside diameter, ring width, and radial runouts of inner and outer rings.   ABEC-5, ABEC-7 and ABEC-9 provide added tolerances for parallelism of sides, side runout and groove parallelism with sides.

*Thrust Bearings:* Anti-Friction Bearing Manufacturers Association and American National Standard tolerance limits for metric single direction thrust ball and roller bearings are given in Table 5.   Tolerance limits for single direction thrust ball bearings, inch dimensioned are given in Table 6, and for cylindrical thrust roller bearings, inch dimensioned in Table 7.

There is only one class of tolerance limits established for metric thrust bearings.

*Radial Needle Roller Bearings:* Tolerance limits for needle roller bearings, drawn cup, without inner ring, inch types NIB, NIBM, NIY, NIYM, NIH, and NIHN are given in Table 9 and for metric types NB, NBM, NY, NYM, NH and NHM are given in Table 9. Standard tolerance limits for needle roller bearings, with cage, machined ring, without inner ring, inch type NIA are given in Table 10 and for needle roller bearings inner rings, inch type NIR are given in Table 11.

**Table 6.   Tolerance Limits for Single Direction Ball Thrust Bearings—Inch Design** (ANSI/AFBMA Std 21.2-1977)

| Bore Diameter† d, Inches | | Height Tolerance Limits, Inch | | Outside Diameter D, Inches | | Outside Diameter Tolerance, Inch | |
|---|---|---|---|---|---|---|---|
| Over | Incl. | High | Low | Over | Incl. | High | Low |
| 0 | 1.8125 | +.005 | −.005 | 0 | 5.3125 | +0 | −.002 |
| 1.8125 | 12.0000 | +.010 | −.010 | 5.3125 | 17.3750 | +0 | −.003 |
| 12.0000 | 20.0000 | +.015 | −.015 | 17.3750 | 39.3701 | +0 | −.004 |

† Bore tolerance limits are: For bore diameters over 0 to 6.7500 inches, inclusive, +.005, −0; for bore diameters over 6.7500 to 20.0000 inches, inclusive, +.007, −0.

## Table 7. Tolerance Limits for Cylindrical Roller Thrust Bearings — Inch Design (ANSI/AFBMA Std 21.2-1977)

| Basic Bore Diam., d | | Bore Tolerance Limits | | Height Tolerance Limits | | Basic Outside Diam., D | | Outside Diam. Tolerance Limits | |
|---|---|---|---|---|---|---|---|---|---|
| Over | Incl. | $d_{min}$ | $d_{max}$ | High | Low | Over | Incl. | $D_{max}$ | $D_{min}$ |
| EXTRA LIGHT SERIES — TYPE TP | | | | | | | | | |
| 0 | 0.9375 | +.0040 | +.0060 | +.0050 | -.0050 | 0 | 4.7188 | +0 | -.0030 |
| 0.9375 | 1.9375 | +.0050 | +.0070 | +.0050 | -.0050 | 4.7188 | 5.2188 | +0 | -.0030 |
| 1.9375 | 3.0000 | +.0060 | +.0080 | +.0050 | -.0050 | .... | .... | .... | .... |
| 3.0000 | 3.5000 | +.0080 | +.0100 | +.0100 | -.0100 | .... | .... | .... | .... |

| Basic Bore Diameter, d | | Bore Tolerance Limits | | Basic Outside Diameter,[1] D | | Outside Diam., D Tolerance Limits | | Basic Outside Diameter,[2] $D_1$ | | Outside Diam., $D_1$ Tolerance Limits | |
|---|---|---|---|---|---|---|---|---|---|---|---|
| Over | Incl. | $d_{max}$ | $d_{min}$ | Over | Incl. | $D_{max}$ | $D_{min}$ | Over | Incl. | $D_{max}$ | $D_{min}$ |
| LIGHT SERIES — TYPES TP AND TR | | | | | | | | | | | |
| 0 | 1.1875 | +0 | -.0005 | 0 | 2.8750 | +.0005 | -0 | 0 | 3.0000 | +.0007 | -0 |
| 1.1875 | 1.3750 | +0 | -.0006 | 2.8750 | 3.3750 | +.0007 | -0 | 3.0000 | 3.3750 | +.0009 | -0 |
| 1.3750 | 1.5620 | +0 | -.0007 | 3.3750 | 3.7500 | +.0009 | -0 | 3.3750 | 3.6250 | +.0011 | -0 |
| 1.5620 | 1.7500 | +0 | -.0008 | 3.7500 | 4.1250 | +.0011 | -0 | 3.6250 | 3.8750 | +.0013 | -0 |
| 1.7500 | 1.9370 | +0 | -.0009 | 4.1250 | 4.7180 | +.0013 | -0 | 3.8750 | 4.5312 | +.0015 | -0 |
| 1.9370 | 2.1250 | +0 | -.0010 | 4.7180 | 5.0000 | +.0015 | -0 | 4.5312 | 5.0000 | +.0017 | -0 |
| 2.1250 | 2.5000 | +0 | -.0011 | .... | .... | .... | .... | .... | .... | .... | .... |
| 2.2500 | 3.0000 | +0 | -.0012 | .... | .... | .... | .... | .... | .... | .... | .... |
| 3.0000 | 3.5000 | +0 | -.0013 | .... | .... | .... | .... | .... | .... | .... | .... |
| HEAVY SERIES — TYPES TP AND TR | | | | | | | | | | | |
| 2.0000 | 3.0000 | +0 | -.0010 | 5.0000 | 10.0000 | +.0015 | -0 | 5.0000 | 10.5000 | +.0019 | -0 |
| 3.0000 | 3.5000 | +0 | -.0012 | 10.0000 | 18.0000 | +.0020 | -0 | 10.5000 | 12.7500 | +.0021 | -0 |
| 3.5000 | 9.0000 | +0 | -.0015 | 18.0000 | 26.0000 | +.0025 | -0 | 12.7500 | 17.0000 | +.0023 | -0 |
| 9.0000 | 12.0000 | +0 | -.0018 | 26.0000 | 34.0000 | +.0030 | -0 | 17.0000 | 27.0000 | +.0025 | -0 |
| 12.0000 | 18.0000 | +0 | -.0020 | 34.0000 | 44.0000 | +.0040 | -0 | 27.0000 | 35.0000 | +.0030 | -0 |
| 18.0000 | 22.0000 | +0 | -.0025 | .... | .... | .... | .... | .... | .... | .... | .... |
| 22.0000 | 30.0000 | +0 | -.003 | .... | .... | .... | .... | .... | .... | .... | .... |

### HEIGHT TOLERANCES — TYPES TP AND TR

| Basic Bore Diameter, d | | Height,[3] H Tolerance Limits | | Height,[4] $H_1$ Tolerance Limits | | Basic Bore Diameter, d | | Height,[3] H Tolerance Limits | | Height,[4] $H_1$ Tolerance Limits | |
|---|---|---|---|---|---|---|---|---|---|---|---|
| Over | Incl. | High | Low | High | Low | Over | Incl. | High | Low | High | Low |
| 0 | 2.0000 | +0 | -.0060 | +0 | -.0080 | 6.0000 | 10.0000 | +0 | -.0150 | +0 | -.0200 |
| 2.0000 | 3.0000 | +0 | -.0080 | +0 | -.0100 | 10.0000 | 18.0000 | +0 | -.0200 | +0 | -.0250 |
| 3.0000 | 6.0000 | +0 | -.0100 | +0 | -.0150 | 18.0000 | 30.0000 | +0 | -.0250 | +0 | -.0300 |

All dimensions are in inches.
[1] D is outside diameter over bearing seat.
[2] $D_1$ is outside diameter over aligning washer.
[3] H is height excluding aligning washer.
[4] $H_1$ is height including aligning washer.

**Table 8.   AFBMA and American National Standard Tolerance Limits for Needle Roller Bearings, Drawn Cup, Without Inner Ring — Inch Types NIB, NIBM, NIY, NIYM, NIH, and NIHM (ANSI/AFBMA Std 18.2-1976)**

| Ring Gage Bore Diameter* | | Basic Bore Diameter under Needle Rollers, $F_w$ | | Allowable Deviation from $F_w$* | | Allowable Deviation from Width, $B$ | |
|---|---|---|---|---|---|---|---|
| Basic Outside Diameter, $D$ | Deviation from $D$ | Inch | | Inch | | Inch | |
| Inch | Inch | Over | Incl. | Low | High | High | Low |
| All Diameters | +.0005 | 0.1875 | 1.2500 | +.0015 | +.0024 | +0 | −.0100 |
| | | 1.2500 | 1.3750 | +.0015 | +.0025 | +0 | −.0100 |
| For fitting and mounting practice see Table 17. | | 1.3750 | 1.6250 | +.0015 | +.0026 | +0 | −.0100 |
| * The bore diameter under needle rollers can be measured only when bearing is pressed into a ring gage which rounds and sizes the bearing. | | 1.6250 | 1.8750 | +.0015 | +.0027 | +0 | −.0100 |
| | | 1.8750 | 2.0000 | +.0016 | +.0028 | +0 | −.0100 |
| | | 2.0000 | 2.5000 | +.0016 | +.0030 | +0 | −.0100 |
| | | 2.5000 | 3.5000 | +.0020 | +.0035 | +0 | −.0100 |

**Table 9.   AFBMA and American National Standard Tolerance Limits for Needle Roller Bearings, Drawn Cup, Without Inner Ring — Metric Types NB, NBM, NY, NYM, NH, and NHM (ANSI/AFBMA Std 18.1-1976)**

| Ring Gage Bore Diameter* | | | Basic Bore Diameter under Needle Rollers, $F_w$ | | Allowable Deviation from $F_w$* | | Allowable Deviation from Width, $B$ | |
|---|---|---|---|---|---|---|---|---|
| Basic Outside Diameter, $D$ | | Deviation from $D$ | | | | | | |
| mm | | Micrometers | mm | | Micrometers | | Micrometers | |
| Over | Incl. | | Over | Incl. | Low | High | High | Low |
| 6 | 10 | −16 | 3 | 6 | +10 | +28 | +0 | −250 |
| 10 | 18 | −20 | 6 | 10 | +13 | +31 | +0 | −250 |
| 18 | 30 | −24 | 10 | 18 | +16 | +34 | +0 | −250 |
| 30 | 50 | −28 | 18 | 30 | +20 | +41 | +0 | −250 |
| 50 | 78 | −33 | 30 | 50 | +25 | +50 | +0 | −250 |
| ... | ... | ... | 50 | 70 | +30 | +60 | +0 | −250 |

For fitting and mounting practice, see Table 18.
* The bore diameter under needle rollers can be measured only when bearing is pressed into a ring gage which rounds and sizes the bearing.

**Table 10. AFBMA and American National Standard Tolerance Limits for Needle Roller Bearings, With Cage, Machined Ring, Without Inner Ring — Inch Type NIA** (ANSI/AFBMA Std 18.2-1976)

| Basic Outside Diameter, D | | Allowable Deviation from D of Single Mean Diameter, $D_{mp}$ | | Basic Bore Diameter under Needle Rollers, $F_w$ | | Allowable Deviation from $F_w$ | | Allowable Deviation from Width, B | |
|---|---|---|---|---|---|---|---|---|---|
| Inch | | Inch | | Inch | | Inch | | Inch | |
| Over | Incl. | High | Low | Over | Incl. | Low | High | High | Low |
| 0.7500 | 2.0000 | +0 | −.0005 | 0.3150 | 0.7087 | +.0008 | +.0017 | +0 | −.0050 |
| 2.0000 | 3.2500 | +0 | −.0006 | 0.7087 | 1.1811 | +.0009 | +.0018 | +0 | −.0050 |
| 3.2500 | 4.7500 | +0 | −.0008 | 1.1811 | 1.6535 | +.0010 | +.0019 | +0 | −.0050 |
| 4.7500 | 7.2500 | +0 | −.0010 | 1.6535 | 1.9685 | +.0010 | +.0020 | +0 | −.0050 |
| | | | | 1.9685 | 2.7559 | +.0011 | +.0021 | +0 | −.0050 |
| 7.2500 | 10.2500 | +0 | −.0012 | 2.7559 | 3.1496 | +.0011 | +.0023 | +0 | −.0050 |
| 10.2500 | 11.1250 | +0 | −.0014 | 3.1496 | 4.0157 | +.0012 | +.0024 | +0 | −.0050 |
| .... | .... | ... | .... | 4.0157 | 4.7244 | +.0012 | +.0026 | +0 | −.0050 |
| .... | .... | ... | .... | 4.7244 | 6.2992 | +.0013 | +.0027 | +0 | −.0050 |
| .... | .... | ... | .... | 6.2992 | 7.0866 | +.0013 | +.0029 | +0 | −.0050 |
| .... | .... | ... | .... | 7.0866 | 7.8740 | +.0014 | +.0030 | +0 | −.0050 |
| .... | .... | ... | .... | 7.8740 | 9.2520 | +.0014 | +.0032 | +0 | −.0050 |

All dimensions are in inches.
For fitting and mounting practice, see Table 19.

**Table 11. AFBMA and American National Standard Tolerance Limits for Needle Roller Bearing Inner Rings — Inch Type NIR** (ANSI/AFBMA Std 18.2-1976)

| Basic Outside Diameter, F | | Allowable Deviation from D of Single Mean Diameter, $F_{mp}$ | | Basic Bore Diameter, d | | Allowable Deviation from d of Single Mean Diameter, $d_{mp}$ | | Allowable Deviation from Width, B | |
|---|---|---|---|---|---|---|---|---|---|
| Inch | | Inch | | Inch | | Inch | | Inch | |
| Over | Incl. | High | Low | Over | Incl. | High | Low | High | Low |
| 0.3937 | 0.7087 | −.0005 | −.0009 | 0.3125 | 0.7500 | +0 | −.0004 | +.0100 | +.0050 |
| 0.7087 | 1.0236 | −.0007 | −.0012 | 0.7500 | 2.0000 | +0 | −.0005 | +.0100 | +.0050 |
| 1.0236 | 1.1181 | −.0009 | −.0014 | 2.0000 | 3.2500 | +0 | −.0006 | +.0100 | +.0050 |
| 1.1811 | 1.3780 | −.0009 | −.0015 | 3.2500 | 4.2500 | +0 | −.0008 | +.0100 | +.0050 |
| 1.3780 | 1.9685 | −.0010 | −.0016 | 4.2500 | 4.7500 | +0 | −.0008 | +.0150 | +.0100 |
| 1.9685 | 3.1496 | −.0011 | −.0018 | 4.7500 | 7.2500 | +0 | −.0010 | +.0150 | +.0100 |
| 3.1496 | 3.9370 | −.0013 | −.0022 | 7.2500 | 8.0000 | +0 | −.0012 | +.0150 | +.0100 |
| 3.9370 | 4.7244 | −.0015 | −.0024 | .... | .... | ... | .... | .... | .... |
| 4.7244 | 5.5118 | −.0015 | −.0025 | .... | .... | ... | .... | .... | .... |
| 5.5118 | 7.0866 | −.0017 | −.0027 | .... | .... | ... | .... | .... | .... |
| 7.0866 | 8.2677 | −.0019 | −.0031 | .... | .... | ... | .... | .... | .... |
| 8.2677 | 9.2520 | −.0020 | −.0032 | .... | .... | ... | .... | .... | .... |

All dimensions are in inches.
For fitting and mounting practice, see Table 20.

**Metric Radial Ball and Roller Bearing Shaft and Housing Fits.** — To select the proper fits, it is necessary to consider the type and extent of the load, bearing type, and certain other design and performance requirements.

The required shaft and housing fits are indicated in Tables 12 and 15. The terms "Light," "Normal," and "Heavy" loads refer to radial loads that are generally within the following limits ($C$ being the Basic Load Rating computed in accordance with AFBMA-ANSI Standards):

| Radial Load | Ball Bearings | Roller Bearings |
|---|---|---|
| Light | up to $0.07C$ | up to $0.08C$ |
| Normal | from $0.07C$ to $0.15C$ | from $0.08C$ to $0.18C$ |
| Heavy | over $0.15C$ | over $0.18C$ |

*Shaft Fits:* Table 12 indicates the initial approach to shaft fit selection. Note that for most normal applications where the shaft rotates and the radial load direction is constant, an interference fit should be used. Also, the heavier the load, the greater is the required interference. For stationary shaft conditions and constant radial load direction, the inner ring may be moderately loose on the shaft.

Note that for pure thrust (axial) loading, heavy interference fits are not necessary; only a moderately loose to tight shaft fit is needed.

The upper part of Table 13 shows how the shaft diameters for various ANSI shaft limit classifications deviate from the basic bore diameters.

Table 14 shows the actual diameter limits of bearing bores, and of shaft seats according to ANSI shaft limits g6, h6, h5, etc.

*Housing Fits:* Table 15 indicates the initial approach to housing fit selection. Note that the use of clearance or interference fits is mainly dependent upon which bearing ring rotates in relation to the radial load. For indeterminate or varying load directions, avoid clearance fits. Clearance fits are preferred in axially split housings to avoid distorting bearing outer rings. The extent of the radial load also influences the choice of fit.

The lower part of Table 13 shows how housing bores for various ANSI hole limit classifications deviate from the basic shaft outside diameters.

Table 16 shows the actual diameter limits of bearing outside diameters, and housing seats according to ANSI hole limits G7, H7, H6, etc.

**Design and Installation Considerations.** — Since interference fitting will reduce bearing radial internal clearance, it is recommended that prospective users consult bearing manufacturers to make certain that the required bearings are correctly specified to satisfy all mounting, environmental and other operating conditions and requirements. This is particularly necessary in those cases where heat sources in associated parts may further diminish bearing clearances in operation.

Standard values of radial internal clearances of radial bearings are listed in AFBMA-ANSI Standards.

**Allowance for Axial Displacement.** — Consideration should be given to axial displacement of bearing components due to thermal expansion or contraction of associated parts. Displacement may be accommodated either by the internal construction of the bearing or by allowing one of the bearing rings to be axially displaceable. For unusual applications consult bearing manufacturers.

**Table 12. Selection of Shaft Tolerance Classifications for Metric Radial Ball and Roller Bearings of ABEC-1 and RBEC-1 Tolerance Classes**

| Operating Conditions | Ball Bearings mm | Ball Bearings Inch | Cylindrical Roller Bearings mm | Cylindrical Roller Bearings Inch | Spherical Roller Bearings mm | Spherical Roller Bearings Inch | Tolerance Symbol[1] |
|---|---|---|---|---|---|---|---|
| Inner ring stationary in relation to the direction of the load. — All loads: Inner ring has to be easily displaceable | All diameters | All diameters | All diameters | All diameters | All diameters | All diameters | g6 |
| Inner ring stationary in relation to the direction of the load. — All loads: Inner ring does not have to be easily displaceable | All diameters | All diameters | All diameters | All diameters | All diameters | All diameters | h6 |
| Direction of load indeterminate or the inner ring rotating in relation to the direction of the load. — Radial load: LIGHT | ≤18<br>>18 | ≤0.71<br>>0.71 | ≤40<br>(40)—140<br>(140)—320 | ≤1.57<br>(1.57)—5.52<br>(5.52)—12.6 | ≤40<br>(40)—100<br>(100)—200 | ≤1.57<br>(1.57)—3.94<br>(3.94)—7.88 | h5<br>j6[2]<br>k6[2]<br>m6[2] |
| Radial load: NORMAL | ≤18<br>>18 | ≤0.71<br>>0.71 | ≤40<br>(40)—100<br>(100)—140<br>(140)—320 | ≤1.57<br>(1.57)—3.94<br>(3.94)—5.52<br>(5.52)—12.6 | ≤40<br>(40)—65<br>(65)—100<br>(100)—140<br>(140)—280<br>(280)—500<br>>500 | ≤1.57<br>(1.57)—2.56<br>(2.56)—3.94<br>(3.94)—5.52<br>(5.52)—11.10<br>(11.10)—19.7<br>>19.7 | j5<br>k5<br>m5<br>m6<br>n6<br>p6<br>r6<br>r7 |
| Radial load: HEAVY | (18)—100<br>>100 | (0.71)—3.94<br>>3.94 | ≤40<br>(65)—140<br>(140)—320 | ≤1.57<br>(2.56)—5.52<br>(5.52)—12.6 | ≤40<br>(40)—65<br>(65)—100<br>(100)—140<br>(140)—200<br>>200 | ≤1.57<br>(1.57)—2.56<br>(2.56)—3.94<br>(3.94)—5.52<br>(5.52)—7.88<br>>7.88 | k5<br>m5<br>m6[2]<br>n6[2]<br>p6[2]<br>r6[2]<br>r7[2] |
| Pure Thrust Load | All diams. | All diams. | All diams. | All diams. | All diams. | All diams. | j6 |

[1] For solid steel shafts. For hollow or nonferrous shafts, tighter fits may be needed. [2] When greater accuracy is required use j5, k5 and m5 instead of j6, k6 and m6, respectively. Numerical values are given in Tables 13 and 14.

**Table 13.  AFBMA and American National Standard Shaft Diameter and Housing Bore Tolerance Limits\* (ANSI B3.17-1973)**

### Allowable Deviations of Shaft Diameter from Basic Bore Diameter, Inch

| Inches Over | Inches Incl. | mm Over | mm Incl. | g6 | h6 | h5 | js | j6 | k5 | k6 | m5 | m6 | n6 | p6 | r6 | r7 |
|---|---|---|---|---|---|---|---|---|---|---|---|---|---|---|---|---|
| .1181 | .2362 | 3 | 6 | -.0002/-.0005 | 0/-.0003 | 0/-.0002 | +.0001/-.0001 | +.0003/-.0001 | +.0002/+.0001 | +.0004/+.0001 | | +.0005/+.0002 | | | | |
| .2362 | .3927 | 6 | 10 | -.0002/-.0006 | 0/-.0004 | 0/-.0003 | +.0002/-.0001 | +.0003/-.0001 | +.0003/+.0001 | +.0004/+.0001 | | +.0007/+.0003 | | | | |
| .3937 | .7087 | 10 | 18 | -.0003/-.0007 | 0/-.0004 | 0/-.0003 | +.0002/-.0001 | +.0003/-.0002 | +.0004/+.0001 | +.0005/+.0001 | +.0006/+.0003 | +.0008/+.0003 | | | | |
| .7087 | 1.1811 | 18 | 30 | -.0003/-.0008 | 0/-.0005 | 0/-.0004 | +.0002/-.0002 | +.0004/-.0002 | +.0004/+.0001 | +.0006/+.0001 | +.0007/+.0003 | +.0010/+.0004 | +.0011/+.0006 | | | |
| 1.1811 | 1.9685 | 30 | 50 | -.0004/-.0010 | 0/-.0006 | 0/-.0004 | +.0002/-.0002 | +.0004/-.0003 | +.0005/+.0001 | +.0007/+.0001 | +.0008/+.0004 | +.0012/+.0005 | +.0013/+.0007 | +.0016/+.0010 | | |
| 1.9685 | 3.1496 | 50 | 80 | -.0004/-.0011 | 0/-.0007 | 0/-.0005 | +.0003/-.0003 | +.0005/-.0003 | +.0006/+.0001 | +.0008/+.0001 | +.0010/+.0005 | +.0014/+.0005 | +.0015/+.0008 | +.0021/+.0014 | +.0023/+.0016 | |
| 3.1496 | 4.7244 | 80 | 120 | -.0005/-.0014 | 0/-.0009 | 0/-.0006 | +.0003/-.0003 | +.0005/-.0004 | +.0007/+.0001 | +.0010/+.0001 | +.0011/+.0005 | +.0014/+.0005 | +.0019/+.0010 | +.0025/+.0016 | +.0029/+.0020 | |

### Allowable Deviations of Housing Bore from Basic Outside Diameter of Shaft, Inch

| Inches Over | Inches Incl. | mm Over | mm Incl. | G7 | H6 | H7 | J6 | J7 | K6 | K7 | M6 | M7 | N6 | N7 | P6 | P7 |
|---|---|---|---|---|---|---|---|---|---|---|---|---|---|---|---|---|
| .3937 | .7086 | 10 | 18 | +.0010/+.0003 | +.0004/0 | +.0007/0 | +.0003/-.0002 | +.0004/-.0003 | +.0001/-.0004 | +.0002/-.0006 | -.0002/-.0006 | 0/-.0007 | -.0004/-.0008 | -.0002/-.0009 | -.0006/-.0010 | -.0004/-.0011 |
| .7086 | 1.1811 | 18 | 30 | +.0011/+.0003 | +.0005/0 | +.0008/0 | +.0003/-.0002 | +.0005/-.0003 | +.0001/-.0005 | +.0002/-.0007 | -.0002/-.0007 | 0/-.0008 | -.0005/-.0010 | -.0003/-.0011 | -.0007/-.0012 | -.0005/-.0013 |
| 1.1811 | 1.9685 | 30 | 50 | +.0014/+.0004 | +.0006/0 | +.0010/0 | +.0004/-.0002 | +.0006/-.0004 | +.0001/-.0006 | +.0003/-.0007 | -.0002/-.0008 | 0/-.0010 | -.0005/-.0011 | -.0003/-.0013 | -.0008/-.0014 | -.0006/-.0016 |
| 1.9685 | 3.1496 | 50 | 80 | +.0016/+.0004 | +.0007/0 | +.0012/0 | +.0005/-.0002 | +.0008/-.0004 | +.0002/-.0007 | +.0003/-.0008 | -.0003/-.0010 | 0/-.0012 | -.0006/-.0013 | -.0003/-.0015 | -.0012/-.0019 | -.0009/-.0021 |
| 3.1496 | 4.7244 | 80 | 120 | +.0019/+.0005 | +.0009/0 | +.0014/0 | +.0005/-.0003 | +.0009/-.0005 | +.0002/-.0008 | +.0004/-.0010 | -.0003/-.0012 | 0/-.0014 | -.0007/-.0016 | -.0004/-.0018 | -.0013/-.0022 | -.0011/-.0025 |
| 4.7244 | 7.0866 | 120 | 180 | +.0022/+.0006 | +.0010/0 | +.0016/0 | +.0007/-.0003 | +.0010/-.0006 | +.0002/-.0010 | +.0004/-.0011 | -.0003/-.0013 | 0/-.0016 | -.0009/-.0019 | -.0006/-.0022 | -.0015/-.0025 | -.0012/-.0028 |
| 7.0866 | 9.8425 | 180 | 250 | +.0024/+.0006 | +.0012/0 | +.0018/0 | +.0009/-.0003 | +.0011/-.0007 | +.0002/-.0013 | +.0005/-.0013 | -.0003/-.0015 | 0/-.0018 | -.0010/-.0022 | -.0008/-.0026 | -.0016/-.0028 | -.0014/-.0032 |

\* Based on ANSI B4.1-1967, R1974 Preferred Limits and Fits for Cylindrical Parts.  Symbols g6, h6, etc., are shaft and G7, H7, etc., hole limits designations.  For larger diameters see AFBMA Standard 7.

Table 14. AFBMA and American National Standard Shaft Diameter Limits for Metric Radial Ball and Roller Bearings of ABEC-1 and RBEC-1 Tolerance Classes (ANSI B3.17-1973)

| Mm. | Bearing Bore Diam. Inches | | g6 Shaft Diam., Inch | | h6 Shaft Diam., Inch | | h5 Shaft Diam., Inch | | js Shaft Diam., Inch | | j6 Shaft Diam., Inch | |
|---|---|---|---|---|---|---|---|---|---|---|---|---|
| | Max. | Min. | Max. | Min. | Max. | Min. | Max. | Min. | Max. | Min. | Max. | Min. |
| 4 | .1575 | .1572 | .1573 | .1570 | .1575 | .1572 | .1575 | .1573 | .1576 | .1574 | | |
| 5 | .1969 | .1966 | .1967 | .1964 | .1969 | .1966 | .1969 | .1967 | .1970 | .1968 | | |
| 6 | .2362 | .2359 | .2360 | .2357 | .2362 | .2359 | .2362 | .2360 | .2363 | .2361 | | |
| 7 | .2756 | .2753 | .2754 | .2750 | .2756 | .2752 | .2756 | .2753 | .2758 | .2755 | | |
| 8 | .3150 | .3147 | .3148 | .3144 | .3150 | .3146 | .3150 | .3147 | .3152 | .3149 | | |
| 9 | .3543 | .3540 | .3541 | .3537 | .3543 | .3539 | .3543 | .3540 | .3545 | .3542 | | |
| 10 | .3937 | .3934 | .3935 | .3931 | .3937 | .3933 | .3937 | .3934 | .3939 | .3936 | | |
| 12 | .4724 | .4721 | .4721 | .4717 | .4724 | .4720 | .4724 | .4721 | .4726 | .4723 | .4727 | .4723 |
| 15 | .5906 | .5903 | .5903 | .5899 | .5906 | .5902 | .5906 | .5903 | .5908 | .5905 | .5909 | .5905 |
| 17 | .6693 | .6690 | .6690 | .6686 | .6693 | .6689 | .6693 | .6690 | .6695 | .6692 | .6696 | .6692 |
| 20 | .7874 | .7870 | .7871 | .7866 | .7874 | .7869 | .7874 | .7870 | .7876 | .7872 | .7877 | .7872 |
| 25 | .9843 | .9839 | .9840 | .9835 | .9843 | .9838 | .9843 | .9839 | .9845 | .9841 | .9846 | .9841 |
| 30 | 1.1811 | 1.1807 | 1.1808 | 1.1803 | 1.1811 | 1.1806 | 1.1811 | 1.1807 | 1.1813 | 1.1809 | 1.1814 | 1.1809 |
| 35 | 1.3780 | 1.3775 | 1.3776 | 1.3770 | 1.3780 | 1.3774 | 1.3780 | 1.3776 | 1.3782 | 1.3778 | 1.3784 | 1.3778 |
| 40 | 1.5748 | 1.5743 | 1.5744 | 1.5738 | 1.5748 | 1.5742 | 1.5748 | 1.5744 | 1.5750 | 1.5746 | 1.5752 | 1.5746 |
| 45 | 1.7717 | 1.7712 | 1.7713 | 1.7707 | 1.7717 | 1.7711 | 1.7717 | 1.7713 | 1.7719 | 1.7715 | 1.7721 | 1.7715 |
| 50 | 1.9685 | 1.9680 | 1.9681 | 1.9675 | 1.9685 | 1.9679 | 1.9685 | 1.9681 | 1.9687 | 1.9683 | 1.9689 | 1.9683 |
| 55 | 2.1654 | 2.1648 | 2.1650 | 2.1643 | 2.1654 | 2.1647 | 2.1654 | 2.1649 | 2.1656 | 2.1651 | 2.1658 | 2.1651 |
| 60 | 2.3622 | 2.3616 | 2.3618 | 2.3611 | 2.3622 | 2.3615 | 2.3622 | 2.3617 | 2.3624 | 2.3619 | 2.3626 | 2.3619 |
| 65 | 2.5591 | 2.5585 | 2.5587 | 2.5580 | 2.5591 | 2.5584 | 2.5591 | 2.5586 | 2.5593 | 2.5588 | 2.5595 | 2.5588 |
| 70 | 2.7559 | 2.7553 | 2.7555 | 2.7548 | 2.7559 | 2.7552 | 2.7559 | 2.7554 | 2.7561 | 2.7556 | 2.7563 | 2.7556 |
| 75 | 2.9528 | 2.9522 | 2.9524 | 2.9517 | 2.9528 | 2.9521 | 2.9528 | 2.9523 | 2.9530 | 2.9525 | 2.9532 | 2.9525 |
| 80 | 3.1496 | 3.1490 | 3.1492 | 3.1485 | 3.1496 | 3.1489 | 3.1496 | 3.1491 | 3.1498 | 3.1493 | 3.1500 | 3.1493 |
| 85 | 3.3465 | 3.3457 | 3.3460 | 3.3451 | 3.3465 | 3.3456 | 3.3465 | 3.3459 | 3.3467 | 3.3461 | 3.3470 | 3.3461 |
| 90 | 3.5433 | 3.5425 | 3.5428 | 3.5419 | 3.5433 | 3.5424 | 3.5433 | 3.5427 | 3.5435 | 3.5429 | 3.5438 | 3.5429 |
| 95 | 3.7402 | 3.7394 | 3.7397 | 3.7388 | 3.7402 | 3.7393 | 3.7402 | 3.7396 | 3.7404 | 3.7398 | 3.7407 | 3.7398 |
| 100 | 3.9370 | 3.9362 | 3.9365 | 3.9356 | 3.9370 | 3.9361 | 3.9370 | 3.9364 | 3.9372 | 3.9366 | 3.9375 | 3.9366 |
| 105 | 4.1339 | 4.1331 | 4.1334 | 4.1325 | 4.1339 | 4.1330 | 4.1339 | 4.1333 | 4.1341 | 4.1335 | 4.1344 | 4.1335 |
| 110 | 4.3307 | 4.3299 | 4.3302 | 4.3293 | 4.3307 | 4.3298 | 4.3307 | 4.3301 | 4.3309 | 4.3303 | 4.3312 | 4.3303 |
| 115 | 4.5276 | 4.5268 | 4.5271 | 4.5262 | 4.5276 | 4.5267 | 4.5276 | 4.5270 | 4.5278 | 4.5272 | 4.5281 | 4.5272 |
| 120 | 4.7244 | 4.7236 | 4.7239 | 4.7230 | 4.7244 | 4.7235 | 4.7244 | 4.7238 | 4.7246 | 4.7240 | 4.7249 | 4.7240 |

For larger diameters see AFBMA Standard 7.   See also Tables 12 and 13.

Table 14 (Continued).　AFBMA and American National Standard Shaft Diameter Limits for Metric Radial Ball and Roller Bearings of ABEC-1 and RBEC-1 Tolerance Classes (ANSI B3.17-1973)

| Mm. | Bearing Bore Diam. Inches | | k5 Shaft Diam., Inch | | k6 Shaft Diam., Inch | | m5 Shaft Diam., Inch | | m6 Shaft Diam., Inch | |
|---|---|---|---|---|---|---|---|---|---|---|
| | Max. | Min. | Max. | Min. | Max. | Min. | Max. | Min. | Max. | Min. |
| 12 | .4724 | .4721 | .4728 | .4725 | | | .4730 | .4727 | | |
| 15 | .5906 | .5903 | .5910 | .5907 | | | .5912 | .5909 | | |
| 17 | .6693 | .6690 | .6697 | .6694 | | | .6699 | .6696 | | |
| 20 | .7874 | .7870 | .7879 | .7875 | .7880 | .7875 | .7881 | .7877 | | |
| 25 | .9843 | .9839 | .9848 | .9844 | .9849 | .9844 | .9850 | .9846 | | |
| 30 | 1.1811 | 1.1807 | 1.1816 | 1.1812 | 1.1817 | 1.1812 | 1.1818 | 1.1814 | | |
| 35 | 1.3780 | 1.3775 | 1.3785 | 1.3781 | 1.3787 | 1.3781 | 1.3788 | 1.3784 | 1.3790 | 1.3784 |
| 40 | 1.5748 | 1.5743 | 1.5753 | 1.5749 | 1.5755 | 1.5749 | 1.5756 | 1.5752 | 1.5758 | 1.5752 |
| 45 | 1.7717 | 1.7712 | 1.7722 | 1.7718 | 1.7724 | 1.7718 | 1.7725 | 1.7721 | 1.7727 | 1.7721 |
| 50 | 1.9685 | 1.9680 | 1.9690 | 1.9686 | 1.9692 | 1.9686 | 1.9693 | 1.9689 | 1.9695 | 1.9689 |
| 55 | 2.1654 | 2.1648 | 2.1660 | 2.1655 | 2.1662 | 2.1655 | 2.1664 | 2.1659 | 2.1666 | 2.1659 |
| 60 | 2.3622 | 2.3616 | 2.3628 | 2.3623 | 2.3630 | 2.3623 | 2.3632 | 2.3627 | 2.3634 | 2.3627 |
| 65 | 2.5591 | 2.5585 | 2.5597 | 2.5592 | 2.5599 | 2.5592 | 2.5601 | 2.5596 | 2.5603 | 2.5596 |
| 70 | 2.7559 | 2.7553 | 2.7565 | 2.7560 | 2.7567 | 2.7560 | 2.7569 | 2.7564 | 2.7571 | 2.7564 |
| 75 | 2.9528 | 2.9522 | 2.9534 | 2.9529 | 2.9536 | 2.9529 | 2.9538 | 2.9533 | 2.9540 | 2.9533 |
| 80 | 3.1496 | 3.1490 | 3.1502 | 3.1497 | 3.1504 | 3.1497 | 3.1506 | 3.1501 | 3.1508 | 3.1501 |
| 85 | 3.3465 | 3.3457 | 3.3472 | 3.3466 | 3.3475 | 3.3466 | 3.3476 | 3.3470 | 3.3479 | 3.3470 |
| 90 | 3.5433 | 3.5425 | 3.5440 | 3.5434 | 3.5443 | 3.5434 | 3.5444 | 3.5438 | 3.5447 | 3.5438 |
| 95 | 3.7402 | 3.7394 | 3.7409 | 3.7403 | 3.7412 | 3.7403 | 3.7413 | 3.7407 | 3.7416 | 3.7407 |
| 100 | 3.9370 | 3.9362 | 3.9377 | 3.9371 | 3.9380 | 3.9371 | 3.9381 | 3.9375 | 3.9384 | 3.9375 |
| 105 | 4.1339 | 4.1331 | 4.1346 | 4.1340 | 4.1349 | 4.1340 | 4.1350 | 4.1344 | 4.1353 | 4.1344 |
| 110 | 4.3307 | 4.3299 | 4.3314 | 4.3308 | 4.3317 | 4.3308 | 4.3318 | 4.3312 | 4.3321 | 4.3312 |
| 115 | 4.5276 | 4.5268 | 4.5283 | 4.5277 | 4.5286 | 4.5277 | 4.5287 | 4.5281 | 4.5290 | 4.5281 |
| 120 | 4.7244 | 4.7236 | 4.7251 | 4.7245 | 4.7255 | 4.7245 | 4.7255 | 4.7249 | 4.7258 | 4.7249 |
| 125 | 4.9213 | 4.9203 | 4.9221 | 4.9214 | 4.9224 | 4.9214 | 4.9226 | 4.9219 | 4.9229 | 4.9219 |
| 130 | 5.1181 | 5.1171 | 5.1189 | 5.1182 | 5.1192 | 5.1182 | 5.1194 | 5.1187 | 5.1197 | 5.1187 |
| 140 | 5.5118 | 5.5108 | 5.5126 | 5.5119 | 5.5129 | 5.5119 | 5.5131 | 5.5124 | 5.5134 | 5.5124 |
| 150 | 5.9055 | 5.9045 | 5.9063 | 5.9056 | 5.9066 | 5.9056 | 5.9068 | 5.9061 | 5.9071 | 5.9061 |
| 160 | 6.2992 | 6.2982 | 6.3000 | 6.2993 | 6.3003 | 6.2993 | 6.3005 | 6.2998 | 6.3008 | 6.2998 |
| 170 | 6.6929 | 6.6919 | 6.6937 | 6.6930 | 6.6940 | 6.6930 | 6.6942 | 6.6935 | 6.6945 | 6.6935 |
| 180 | 7.0866 | 7.0856 | 7.0874 | 7.0867 | 7.0877 | 7.0867 | 7.0879 | 7.0872 | 7.0882 | 7.0872 |

For larger diameters see AFBMA Standard 7.　See also Tables 12 and 13.

Table 14 (*Concluded*). AFBMA and American National Standard Shaft Diameter Limits for Metric Radial Ball and Roller Bearings of ABEC-1 and RBEC-1 Tolerance Classes (ANSI B3.17-1973)

| Bearing Bore Diam. | | | n6 Shaft Diam., Inch | | p6 Shaft Diam., Inch | | r6 Shaft Diam., Inch | | r7 Shaft Diam., Inch | |
| Mm. | Inches Max. | Min. | Max. | Min. | Max. | Min. | Max. | Min. | Max. | Min. |
|---|---|---|---|---|---|---|---|---|---|---|
| 35 | 1.3780 | 1.3775 | 1.3793 | 1.3787 | | | | | | |
| 40 | 1.5748 | 1.5743 | 1.5761 | 1.5755 | | | | | | |
| 45 | 1.7717 | 1.7712 | 1.7730 | 1.7724 | | | | | | |
| 50 | 1.9685 | 1.9680 | 1.9698 | 1.9692 | | | | | | |
| 55 | 2.1654 | 2.1648 | 2.1669 | 2.1662 | 2.1675 | 2.1668 | | | | |
| 60 | 2.3622 | 2.3616 | 2.3637 | 2.3630 | 2.3643 | 2.3636 | | | | |
| 65 | 2.5591 | 2.5585 | 2.5606 | 2.5599 | 2.5612 | 2.5605 | | | | |
| 70 | 2.7559 | 2.7553 | 2.7574 | 2.7567 | 2.7580 | 2.7573 | | | | |
| 75 | 2.9528 | 2.9522 | 2.9543 | 2.9536 | 2.9549 | 2.9542 | | | | |
| 80 | 3.1496 | 3.1490 | 3.1511 | 3.1504 | 3.1517 | 3.1510 | | | | |
| 85 | 3.3465 | 3.3457 | 3.3484 | 3.3475 | 3.3490 | 3.3481 | 3.3494 | 3.3485 | | |
| 90 | 3.5433 | 3.5425 | 3.5452 | 3.5443 | 3.5458 | 3.5449 | 3.5462 | 3.5453 | | |
| 95 | 3.7402 | 3.7394 | 3.7421 | 3.7412 | 3.7427 | 3.7418 | 3.7431 | 3.7422 | | |
| 100 | 3.9370 | 3.9362 | 3.9389 | 3.9380 | 3.9395 | 3.9386 | 3.9399 | 3.9390 | | |
| 105 | 4.1339 | 4.1331 | 4.1358 | 4.1349 | 4.1364 | 4.1355 | 4.1368 | 4.1359 | | |
| 110 | 4.3307 | 4.3299 | 4.3326 | 4.3317 | 4.3332 | 4.3323 | 4.3336 | 4.3327 | | |
| 115 | 4.5276 | 4.5268 | 4.5295 | 4.5286 | 4.5301 | 4.5292 | 4.5305 | 4.5296 | | |
| 120 | 4.7244 | 4.7236 | 4.7263 | 4.7254 | 4.7269 | 4.7260 | 4.7273 | 4.7264 | | |
| 125 | 4.9213 | 4.9203 | 4.9235 | 4.9225 | 4.9241 | 4.9231 | 4.9248 | 4.9238 | | |
| 130 | 5.1181 | 5.1171 | 5.1203 | 5.1193 | 5.1209 | 5.1199 | 5.1216 | 5.1206 | | |
| 140 | 5.5118 | 5.5108 | 5.5140 | 5.5130 | 5.5146 | 5.5136 | 5.5153 | 5.5143 | | |
| 150 | 5.9055 | 5.9045 | 5.9077 | 5.9067 | 5.9083 | 5.9073 | 5.9090 | 5.9080 | | |
| 160 | 6.2992 | 6.2982 | 6.3014 | 6.3004 | 6.3020 | 6.3010 | 6.3027 | 6.3017 | | |
| 170 | 6.6929 | 6.6919 | 6.6951 | 6.6941 | 6.6957 | 6.6947 | 6.6964 | 6.6954 | | |
| 180 | 7.0866 | 7.0856 | 7.0888 | 7.0878 | 7.0894 | 7.0884 | 7.0901 | 7.0891 | | |
| 190 | 7.4803 | 7.4791 | 7.4829 | 7.4817 | 7.4835 | 7.4823 | 7.4845 | 7.4833 | 7.4851 | 7.4833 |
| 200 | 7.8740 | 7.8728 | 7.8766 | 7.8754 | 7.8772 | 7.8760 | 7.8782 | 7.8770 | 7.8788 | 7.8770 |
| 220 | 8.6614 | 8.6602 | 8.6640 | 8.6628 | 8.6646 | 8.6634 | 8.6656 | 8.6644 | 8.6662 | 8.6644 |
| 240 | 9.4488 | 9.4476 | 9.4514 | 9.4502 | 9.4520 | 9.4508 | 9.4530 | 9.4518 | 9.4536 | 9.4518 |

For larger diameters see AFBMA Standard 7. See also Tables 12 and 13.

**Table 15. Selection of Housing Tolerance Classifications for Metric Radial Ball and Roller Bearings of ABEC-1 and RBEC-1 Tolerance Classes**

| Rotational Conditions | Loading | Outer Ring Axial Displacement Limitations | Other Conditions | Tolerance Classification* |
|---|---|---|---|---|
| Outer ring stationary in relation to load direction | Light, Normal and Heavy | Outer ring must be easily axially displaceable | Heat input through shaft | G7 |
|  | Light, Normal and Heavy |  | Housing split axially | H7 |
|  | Light, Normal and Heavy |  | Housing not split axially | H6† |
|  | Shock with temporary complete unloading |  | .... | J6† |
| Load direction is indeterminate | Light and Normal | Outer ring need not be axially displaceable |  |  |
|  | Normal and Heavy |  |  | K6† |
|  | Heavy Shock |  | Split housing not recommended | M6† |
| Outer ring rotating in relation to load direction | Light |  |  |  |
|  | Normal and Heavy |  |  | N6† |
|  | Heavy |  | Thin wall housing not split | P6† |

*(The columns "Rotational Conditions," "Loading," "Outer Ring Axial Displacement Limitations," and "Other Conditions" are grouped under the heading "Design and Operating Conditions.")*

* For cast iron or steel housings. For housings of nonferrous alloys tighter fits may be needed.

† Where wider tolerances are permissible, use tolerance classifications P7, N7, M7, K7, J7, and H7, in place of P6, N6, M6, K6, J6, and H6, respectively.

Numerical values are given in Tables 13 and 16.

**Table 16. AFBMA and American National Standard Housing Bore Limits for Metric Radial Ball and Roller Bearings of ABEC-1 and RBEC-1 Tolerance Classes (ANSI B3.17-1973)**

| Bearing Outside Diameter | | | G7 Housing Bore, Inch | | H7 Housing Bore, Inch | | H6 Housing Bore, Inch | | J7 Housing Bore, Inch | | J6 Housing Bore, Inch | | K6 Housing Bore, Inch | |
|---|---|---|---|---|---|---|---|---|---|---|---|---|---|---|
| Mm. | Inches Max. | Inches Min. | Max. | Min. | Max. | Min. | Min. | Max. | Min. | Max. | Min. | Max. | Min. | Max. |
| 16 | .6299 | .6296 | .6309 | .6302 | .6306 | .6299 | .6299 | .6303 | .6296 | .6303 | .6297 | .6301 | .6295 | .6299 |
| 19 | .7480 | .7476 | .7491 | .7483 | .7488 | .7480 | .7480 | .7485 | .7477 | .7485 | .7477 | .7483 | .7475 | .7480 |
| 21 | .8268 | .8264 | .8279 | .8271 | .8276 | .8268 | .8268 | .8273 | .8265 | .8273 | .8266 | .8271 | .8263 | .8268 |
| 22 | .8661 | .8657 | .8672 | .8664 | .8669 | .8661 | .8661 | .8666 | .8658 | .8666 | .8659 | .8664 | .8656 | .8661 |
| 24 | .9449 | .9445 | .9460 | .9452 | .9457 | .9449 | .9449 | .9454 | .9446 | .9454 | .9447 | .9452 | .9444 | .9449 |
| 26 | 1.0236 | 1.0232 | 1.0247 | 1.0239 | 1.0244 | 1.0236 | 1.0236 | 1.0241 | 1.0233 | 1.0241 | 1.0234 | 1.0239 | 1.0231 | 1.0236 |
| 28 | 1.1024 | 1.1020 | 1.1035 | 1.1027 | 1.1032 | 1.1024 | 1.1024 | 1.1029 | 1.1021 | 1.1029 | 1.1022 | 1.1027 | 1.1019 | 1.1024 |
| 30 | 1.1811 | 1.1807 | 1.1822 | 1.1814 | 1.1819 | 1.1811 | 1.1811 | 1.1816 | 1.1808 | 1.1816 | 1.1809 | 1.1814 | 1.1806 | 1.1811 |
| 32 | 1.2598 | 1.2593 | 1.2612 | 1.2602 | 1.2608 | 1.2598 | 1.2598 | 1.2604 | 1.2594 | 1.2604 | 1.2596 | 1.2602 | 1.2593 | 1.2599 |
| 35 | 1.3780 | 1.3775 | 1.3794 | 1.3784 | 1.3790 | 1.3780 | 1.3780 | 1.3786 | 1.3776 | 1.3786 | 1.3778 | 1.3784 | 1.3775 | 1.3781 |
| 37 | 1.4567 | 1.4562 | 1.4581 | 1.4571 | 1.4577 | 1.4567 | 1.4567 | 1.4573 | 1.4563 | 1.4573 | 1.4565 | 1.4571 | 1.4562 | 1.4568 |
| 40 | 1.5748 | 1.5743 | 1.5762 | 1.5752 | 1.5758 | 1.5748 | 1.5748 | 1.5754 | 1.5744 | 1.5754 | 1.5746 | 1.5752 | 1.5743 | 1.5749 |
| 42 | 1.6535 | 1.6530 | 1.6549 | 1.6539 | 1.6545 | 1.6535 | 1.6535 | 1.6541 | 1.6531 | 1.6541 | 1.6533 | 1.6539 | 1.6530 | 1.6536 |
| 47 | 1.8504 | 1.8499 | 1.8518 | 1.8508 | 1.8514 | 1.8504 | 1.8504 | 1.8510 | 1.8500 | 1.8510 | 1.8502 | 1.8508 | 1.8499 | 1.8505 |
| 52 | 2.0472 | 2.0467 | 2.0488 | 2.0476 | 2.0484 | 2.0472 | 2.0472 | 2.0479 | 2.0468 | 2.0480 | 2.0470 | 2.0477 | 2.0466 | 2.0473 |
| 55 | 2.1654 | 2.1649 | 2.1670 | 2.1658 | 2.1666 | 2.1654 | 2.1654 | 2.1661 | 2.1650 | 2.1662 | 2.1652 | 2.1659 | 2.1648 | 2.1655 |
| 58 | 2.2835 | 2.2830 | 2.2851 | 2.2839 | 2.2847 | 2.2835 | 2.2835 | 2.2842 | 2.2831 | 2.2843 | 2.2833 | 2.2840 | 2.2829 | 2.2836 |
| 62 | 2.4409 | 2.4404 | 2.4425 | 2.4413 | 2.4421 | 2.4409 | 2.4409 | 2.4416 | 2.4405 | 2.4417 | 2.4407 | 2.4414 | 2.4403 | 2.4410 |
| 65 | 2.5591 | 2.5586 | 2.5607 | 2.5595 | 2.5603 | 2.5591 | 2.5591 | 2.5598 | 2.5587 | 2.5599 | 2.5589 | 2.5596 | 2.5585 | 2.5592 |
| 68 | 2.6772 | 2.6767 | 2.6788 | 2.6776 | 2.6784 | 2.6772 | 2.6772 | 2.6779 | 2.6768 | 2.6780 | 2.6770 | 2.6777 | 2.6766 | 2.6773 |
| 72 | 2.8346 | 2.8341 | 2.8362 | 2.8350 | 2.8358 | 2.8346 | 2.8346 | 2.8353 | 2.8342 | 2.8354 | 2.8344 | 2.8351 | 2.8340 | 2.8347 |
| 75 | 2.9528 | 2.9523 | 2.9544 | 2.9532 | 2.9540 | 2.9528 | 2.9528 | 2.9535 | 2.9524 | 2.9536 | 2.9526 | 2.9533 | 2.9522 | 2.9529 |
| 78 | 3.0709 | 3.0704 | 3.0725 | 3.0713 | 3.0721 | 3.0709 | 3.0709 | 3.0716 | 3.0705 | 3.0717 | 3.0707 | 3.0714 | 3.0703 | 3.0710 |
| 80 | 3.1496 | 3.1491 | 3.1512 | 3.1500 | 3.1508 | 3.1496 | 3.1496 | 3.1503 | 3.1492 | 3.1504 | 3.1494 | 3.1501 | 3.1490 | 3.1497 |
| 85 | 3.3465 | 3.3459 | 3.3484 | 3.3470 | 3.3479 | 3.3465 | 3.3465 | 3.3474 | 3.3460 | 3.3474 | 3.3463 | 3.3472 | 3.3458 | 3.3467 |
| 90 | 3.5433 | 3.5427 | 3.5452 | 3.5438 | 3.5447 | 3.5433 | 3.5433 | 3.5442 | 3.5428 | 3.5442 | 3.5431 | 3.5440 | 3.5426 | 3.5435 |
| 95 | 3.7402 | 3.7396 | 3.7421 | 3.7407 | 3.7416 | 3.7402 | 3.7402 | 3.7411 | 3.7397 | 3.7411 | 3.7400 | 3.7409 | 3.7395 | 3.7404 |
| 100 | 3.9370 | 3.9364 | 3.9389 | 3.9375 | 3.9384 | 3.9370 | 3.9370 | 3.9379 | 3.9365 | 3.9379 | 3.9368 | 3.9377 | 3.9363 | 3.9372 |
| 105 | 4.1339 | 4.1333 | 4.1358 | 4.1344 | 4.1353 | 4.1339 | 4.1339 | 4.1348 | 4.1334 | 4.1348 | 4.1337 | 4.1346 | 4.1332 | 4.1341 |
| 110 | 4.3307 | 4.3301 | 4.3326 | 4.3312 | 4.3321 | 4.3307 | 4.3307 | 4.3316 | 4.3302 | 4.3316 | 4.3305 | 4.3314 | 4.3300 | 4.3309 |
| 115 | 4.5276 | 4.5270 | 4.5295 | 4.5281 | 4.5290 | 4.5276 | 4.5276 | 4.5285 | 4.5271 | 4.5285 | 4.5274 | 4.5283 | 4.5269 | 4.5278 |
| 120 | 4.7244 | 4.7238 | 4.7263 | 4.7249 | 4.7258 | 4.7244 | 4.7244 | 4.7253 | 4.7239 | 4.7253 | 4.7242 | 4.7251 | 4.7237 | 4.7246 |

For larger diameters see AFBMA Standard 7.   See also Tables 13 and 15.

Table 16 (*Concluded*). AFBMA and American National Standard Housing Bore Limits for Metric Radial Ball and Roller Bearings of ABEC-1 and RBEC-1 Tolerance Classes (ANSI B3.17-1973)

| Bearing Outside Diameter | | | K7 Housing Bore, Inch | | M6 Housing Bore, Inch | | M7 Housing Bore, Inch | | N6 Housing Bore, Inch | | N7 Housing Bore, Inch | | P6 Housing Bore, Inch | | P7 Housing Bore, Inch | |
|---|---|---|---|---|---|---|---|---|---|---|---|---|---|---|---|---|
| Mm | Inches Max | Inches Min | Min | Max | Min | Max | Min | Max | Min | Max | Min | Max | Min | Max | Min | Max |
| 16 | .6299 | .6295 | .6294 | .6301 | .6293 | .6297 | .6292 | .6299 | .6291 | .6295 | .6290 | .6297 | .6289 | .6293 | .6288 | .6295 |
| 19 | .7480 | .7476 | .7474 | .7482 | .7473 | .7478 | .7472 | .7480 | .7470 | .7475 | .7469 | .7477 | .7468 | .7473 | .7467 | .7475 |
| 21 | .8268 | .8264 | .8262 | .8270 | .8261 | .8266 | .8260 | .8268 | .8258 | .8263 | .8257 | .8265 | .8256 | .8261 | .8255 | .8263 |
| 22 | .8661 | .8657 | .8655 | .8663 | .8654 | .8659 | .8653 | .8661 | .8651 | .8656 | .8650 | .8658 | .8649 | .8654 | .8648 | .8656 |
| 24 | .9449 | .9445 | .9443 | .9451 | .9442 | .9447 | .9441 | .9449 | .9439 | .9444 | .9438 | .9446 | .9437 | .9442 | .9436 | .9444 |
| 26 | 1.0236 | 1.0232 | 1.0230 | 1.0238 | 1.0229 | 1.0234 | 1.0228 | 1.0236 | 1.0226 | 1.0231 | 1.0225 | 1.0233 | 1.0224 | 1.0229 | 1.0223 | 1.0231 |
| 28 | 1.1024 | 1.1020 | 1.1018 | 1.1026 | 1.1017 | 1.1022 | 1.1016 | 1.1024 | 1.1014 | 1.1019 | 1.1013 | 1.1021 | 1.1012 | 1.1017 | 1.1011 | 1.1019 |
| 30 | 1.1811 | 1.1807 | 1.1805 | 1.1813 | 1.1804 | 1.1809 | 1.1803 | 1.1811 | 1.1801 | 1.1806 | 1.1800 | 1.1808 | 1.1799 | 1.1804 | 1.1798 | 1.1806 |
| 32 | 1.2598 | 1.2593 | 1.2591 | 1.2601 | 1.2590 | 1.2596 | 1.2588 | 1.2598 | 1.2587 | 1.2593 | 1.2585 | 1.2595 | 1.2584 | 1.2590 | 1.2582 | 1.2592 |
| 35 | 1.3780 | 1.3775 | 1.3773 | 1.3783 | 1.3772 | 1.3778 | 1.3770 | 1.3780 | 1.3769 | 1.3775 | 1.3767 | 1.3777 | 1.3766 | 1.3772 | 1.3764 | 1.3774 |
| 37 | 1.4567 | 1.4562 | 1.4560 | 1.4570 | 1.4559 | 1.4565 | 1.4557 | 1.4567 | 1.4556 | 1.4562 | 1.4554 | 1.4564 | 1.4553 | 1.4559 | 1.4551 | 1.4561 |
| 40 | 1.5748 | 1.5743 | 1.5741 | 1.5751 | 1.5740 | 1.5746 | 1.5738 | 1.5748 | 1.5737 | 1.5743 | 1.5735 | 1.5745 | 1.5734 | 1.5740 | 1.5732 | 1.5742 |
| 42 | 1.6535 | 1.6530 | 1.6528 | 1.6538 | 1.6527 | 1.6533 | 1.6525 | 1.6535 | 1.6524 | 1.6530 | 1.6522 | 1.6532 | 1.6521 | 1.6527 | 1.6519 | 1.6529 |
| 47 | 1.8504 | 1.8499 | 1.8497 | 1.8507 | 1.8496 | 1.8502 | 1.8494 | 1.8504 | 1.8493 | 1.8499 | 1.8491 | 1.8501 | 1.8490 | 1.8496 | 1.8488 | 1.8498 |
| 52 | 2.0472 | 2.0467 | 2.0464 | 2.0476 | 2.0462 | 2.0469 | 2.0460 | 2.0472 | 2.0459 | 2.0466 | 2.0457 | 2.0469 | 2.0453 | 2.0460 | 2.0451 | 2.0463 |
| 55 | 2.1654 | 2.1649 | 2.1646 | 2.1658 | 2.1644 | 2.1651 | 2.1642 | 2.1654 | 2.1641 | 2.1648 | 2.1639 | 2.1651 | 2.1635 | 2.1642 | 2.1633 | 2.1645 |
| 58 | 2.2835 | 2.2830 | 2.2827 | 2.2839 | 2.2825 | 2.2832 | 2.2823 | 2.2835 | 2.2822 | 2.2829 | 2.2820 | 2.2832 | 2.2816 | 2.2823 | 2.2814 | 2.2826 |
| 62 | 2.4409 | 2.4404 | 2.4401 | 2.4413 | 2.4399 | 2.4406 | 2.4397 | 2.4409 | 2.4396 | 2.4403 | 2.4394 | 2.4406 | 2.4390 | 2.4397 | 2.4388 | 2.4400 |
| 65 | 2.5591 | 2.5586 | 2.5583 | 2.5595 | 2.5581 | 2.5588 | 2.5579 | 2.5591 | 2.5578 | 2.5585 | 2.5576 | 2.5588 | 2.5572 | 2.5579 | 2.5570 | 2.5582 |
| 68 | 2.6772 | 2.6767 | 2.6764 | 2.6776 | 2.6762 | 2.6769 | 2.6760 | 2.6772 | 2.6759 | 2.6766 | 2.6757 | 2.6769 | 2.6753 | 2.6760 | 2.6751 | 2.6763 |
| 72 | 2.8346 | 2.8341 | 2.8338 | 2.8350 | 2.8336 | 2.8343 | 2.8334 | 2.8346 | 2.8333 | 2.8340 | 2.8331 | 2.8343 | 2.8327 | 2.8334 | 2.8325 | 2.8337 |
| 75 | 2.9528 | 2.9523 | 2.9520 | 2.9532 | 2.9518 | 2.9525 | 2.9516 | 2.9528 | 2.9515 | 2.9522 | 2.9513 | 2.9525 | 2.9509 | 2.9516 | 2.9507 | 2.9519 |
| 78 | 3.0709 | 3.0704 | 3.0701 | 3.0713 | 3.0699 | 3.0706 | 3.0697 | 3.0709 | 3.0696 | 3.0703 | 3.0694 | 3.0706 | 3.0688 | 3.0697 | 3.0688 | 3.0700 |
| 80 | 3.1496 | 3.1491 | 3.1488 | 3.1500 | 3.1486 | 3.1493 | 3.1484 | 3.1496 | 3.1483 | 3.1490 | 3.1481 | 3.1493 | 3.1477 | 3.1484 | 3.1475 | 3.1487 |
| 85 | 3.3465 | 3.3459 | 3.3455 | 3.3469 | 3.3453 | 3.3462 | 3.3451 | 3.3465 | 3.3449 | 3.3458 | 3.3447 | 3.3461 | 3.3443 | 3.3452 | 3.3440 | 3.3454 |
| 90 | 3.5433 | 3.5427 | 3.5423 | 3.5437 | 3.5421 | 3.5430 | 3.5419 | 3.5433 | 3.5417 | 3.5426 | 3.5415 | 3.5429 | 3.5411 | 3.5420 | 3.5408 | 3.5422 |
| 95 | 3.7402 | 3.7396 | 3.7392 | 3.7406 | 3.7390 | 3.7399 | 3.7388 | 3.7402 | 3.7386 | 3.7395 | 3.7384 | 3.7398 | 3.7380 | 3.7389 | 3.7377 | 3.7391 |
| 100 | 3.9370 | 3.9364 | 3.9360 | 3.9374 | 3.9358 | 3.9367 | 3.9356 | 3.9370 | 3.9354 | 3.9363 | 3.9352 | 3.9366 | 3.9348 | 3.9357 | 3.9345 | 3.9359 |
| 105 | 4.1339 | 4.1333 | 4.1329 | 4.1343 | 4.1327 | 4.1336 | 4.1325 | 4.1339 | 4.1323 | 4.1332 | 4.1321 | 4.1335 | 4.1317 | 4.1326 | 4.1314 | 4.1328 |
| 110 | 4.3307 | 4.3301 | 4.3297 | 4.3311 | 4.3295 | 4.3304 | 4.3293 | 4.3307 | 4.3291 | 4.3300 | 4.3289 | 4.3303 | 4.3285 | 4.3294 | 4.3282 | 4.3296 |
| 115 | 4.5276 | 4.5270 | 4.5266 | 4.5280 | 4.5264 | 4.5273 | 4.5262 | 4.5276 | 4.5260 | 4.5269 | 4.5258 | 4.5272 | 4.5254 | 4.5263 | 4.5251 | 4.5265 |
| 120 | 4.7244 | 4.7238 | 4.7234 | 4.7248 | 4.7232 | 4.7241 | 4.7230 | 4.7244 | 4.7228 | 4.7237 | 4.7226 | 4.7240 | 4.7222 | 4.7231 | 4.7219 | 4.7233 |

For larger diameters see AFBMA Standard 7. See also Tables 13 and 15.

**Needle Roller Bearing Fitting and Mounting Practice.** — The tolerance limits required for shaft and housing seat diameters for needle roller bearings with inner and outer rings as well as limits for raceway diameters where inner and/or outer rings are omitted and rollers operate directly upon these surfaces are given in Tables 17 to 20, inclusive. Unusual design and operating conditions may require a departure from these practices. In such cases, bearing manufacturers should be consulted.

*Needle Roller Bearings, Drawn Cup:* These bearings without inner ring, Types NIB, NB, NIBM, NBM, NIY, NY, NIYM, NYM, NIH, NH, NIHM, NHM, and Inner Rings, Type NIR depend on the housings into which they are pressed for their size and shape. Therefore, the housings must not only have the proper bore dimensions but also must have sufficient strength. Tables 17 and 18, inclusive, show the bore tolerance limits for rigid housings such as those made from cast iron or steel of heavy radial section equal to or greater than the ring gage section given in AFBMA Standard 4 (ANSI B3.4-1971). The bearing manufacturers should be consulted for recommendations if the housings must be of lower strength materials such as aluminum or even of steel of thin radial section. The shape of the housing bores should be such that when the mean bore diameter of a housing is measured in each of several radial planes, the maximum difference between these mean diameters should not exceed 0.0005 inch or one-half the housing bore tolerance limit, if smaller. Also, the radial deviation from circular form should not exceed 0.00025 inch. The housing bore surface finish should not exceed 125 micro-inches arithmetical average.

Most needle roller bearings do not use inner rings, they operate directly on the surfaces of shafts. When shafts are used as inner raceways, they should be made of bearing quality steel hardened to Rockwell C 58 minimum. Tables 17 and 18 show the shaft raceway tolerance limits and Table 20 shows the shaft seat tolerance limits when inner rings are used. However, whether the shaft surfaces are used as inner raceways or as seats for inner rings, the mean outside diameter of the shaft surface in each of several radial planes should be determined. The difference between these mean diameters should not exceed 0.0003 inch or one-half the diameter tolerance limit, if smaller. The radial deviation from circular form should not exceed 0.0001 inch, for diameters up to and including one inch. Above one inch the allowable deviation is 0.0001 times the shaft diameter. The surface finish should not exceed 16 micro-inches arithmetical average. The housing bore and shaft diameter tolerance limits depend upon whether the load rotates relative to the shaft or the housing.

*Needle Roller Bearing With Cage, Machined Ring, Without Inner Ring:* The following covers needle roller bearings Type NIA and inner rings Type NIR. The shape of the housing bores should be such that when the mean bore diameter of a housing is measured in each of several radial planes, the maximum difference between these mean diameters does not exceed 0.0005 inch or one-half the housing bore tolerance limit, if smaller. Also, the radial deviation from circular form should not exceed 0.00025 inch. The housing bore surface finish should not exceed 125 micro-inches arithmetical average. Table 19 shows the housing bore tolerance limits.

When shafts are used as inner raceways their requirements are the same as those given above for Needle Roller Bearings, Drawn Cup. Table 19 shows the shaft raceway tolerance limits and Table 20 shows the shaft seat tolerance limits when inner rings are used.

**Needle Roller and Cage Assemblies, Types NIM and NM.** — For information concerning boundary dimensions, tolerance limits, and fitting and mounting practice, reference should be made to ANSI/AFBMA Std 18.1-1976 and ANSI/AFBMA Std 18.2-1976.

**Table 17. AFBMA and American National Standard Tolerance Limits for Shaft Raceway and Housing Bore Diameters — Needle Roller Bearings, Drawn Cup, Without Inner Ring, Inch Types NIB, NIBM, NIY, NIYM, NIH, and NIHM** (ANSI/AFBMA Std 18.2-1976)

| Basic Bore Diameter under Needle Rollers, $F_w$ | | Shaft Raceway Diameter* Allowable Deviation from $F_w$ | | Basic Outside Diameter, $D$ | | Housing Bore Diameter* Allowable Deviation from $D$ | |
|---|---|---|---|---|---|---|---|
| Inch | | Inch | | Inch | | Inch | |
| Over | Incl. | High | Low | Over | Incl. | Low | High |
| OUTER RING STATIONARY RELATIVE TO LOAD | | | | | | | |
| 0.1875 | 1.8750 | +0 | −.0005 | 0.3750 | 4.0000 | −.0005 | +.0005 |
| 1.8750 | 3.5000 | +0 | −.0006 | .... | .... | .... | .... |
| OUTER RING ROTATING RELATIVE TO LOAD | | | | | | | |
| 0.1875 | 1.8750 | −.0005 | −.0010 | 0.3750 | 4.0000 | −.0010 | +0 |
| 1.8750 | 3.5000 | −.0005 | −.0011 | .... | .... | .... | .... |

For bearing tolerances, see Table 8.
* See text for additional requirements.

**Table 18. AFBMA and American National Standard Tolerance Limits for Shaft Raceway and Housing Bore Diameters — Needle Roller Bearings, Drawn Cup, Without Inner Ring, Metric Types NB, NBM, NY, NYM, NH, and NHM** (ANSI/AFBMA Std 18.1-1976)

| Basic Bore Diameter under Needle Rollers, $F_w$ | | Shaft Raceway Diameter* Allowable Deviation from $F_w$ | | Basic Outside Diameter, $D$ | | Housing Bore Diameter* Allowable Deviation from $D$ | |
|---|---|---|---|---|---|---|---|
| OUTER RING STATIONARY RELATIVE TO LOAD | | | | | | | |
| mm | | Inch | | ANSI h6, Inch | | mm | | Inch | | ANSI N7, Inch | |

| Over | Incl. | Over | Incl. | High | Low | Over | Incl. | Over | Incl. | Low | High |
|---|---|---|---|---|---|---|---|---|---|---|---|
| 3 | 6 | 0.1181 | 0.2362 | +0 | −.0003 | 6 | 10 | 0.2362 | 0.3937 | −.0007 | −.0002 |
| 6 | 10 | 0.2362 | 0.3937 | +0 | −.0004 | 10 | 18 | 0.3937 | 0.7087 | −.0009 | −.0002 |
| 10 | 18 | 0.3937 | 0.7087 | +0 | −.0004 | 18 | 30 | 0.7087 | 1.1811 | −.0011 | −.0003 |
| 18 | 30 | 0.7087 | 1.1811 | +0 | −.0005 | 30 | 50 | 1.1811 | 1.9685 | −.0013 | −.0003 |
| 30 | 50 | 1.1811 | 1.9685 | +0 | −.0006 | 50 | 78 | 1.9685 | 3.0709 | −.0015 | −.0004 |
| 50 | 70 | 1.9685 | 2.7559 | +0 | −.0007 | .... | .... | .... | .... | .... | .... |

| OUTER RING ROTATING RELATIVE TO LOAD | | | | | | | | | | | |
|---|---|---|---|---|---|---|---|---|---|---|---|
| mm | | Inch | | ANSI f6, Inch | | mm | | Inch | | ANSI R7, Inch | |
| Over | Incl. | Over | Incl. | High | Low | Over | Incl. | Over | Incl. | Low | High |
| 3 | 6 | 0.1181 | 0.2362 | −.0004 | −.0007 | 6 | 10 | 0.2362 | 0.3937 | −.0011 | −.0005 |
| 6 | 10 | 0.2362 | 0.3937 | −.0005 | −.0009 | 10 | 18 | 0.3937 | 0.7087 | −.0013 | −.0006 |
| 10 | 18 | 0.3937 | 0.7087 | −.0006 | −.0011 | 18 | 30 | 0.7087 | 1.1811 | −.0016 | −.0008 |
| 18 | 30 | 0.7087 | 1.1811 | −.0008 | −.0013 | 30 | 50 | 1.1811 | 1.9685 | −.0020 | −.0010 |
| 30 | 50 | 1.1811 | 1.9685 | −.0010 | −.0016 | 50 | 65 | 1.9685 | 2.5591 | −.0024 | −.0012 |
| 50 | 70 | 1.9685 | 2.7559 | −.0012 | −.0019 | 65 | 78 | 2.5591 | 3.0709 | −.0024 | −.0013 |

For bearing tolerances, see Table 9.
*See text for additional requirements.

**Table 19. AFBMA and American National Standard Tolerance Limits for Shaft Raceway and Housing Bore Diameters — Needle Roller Bearings, With Cage, Machined Ring, Without Inner Ring, Inch Type NIA**

(ANSI/AFBMA Std 18.2-1976)

| Basic Bore Diameter under Needle Rollers, $F_w$ | | Shaft Raceway Diameter Allowable Deviation from $F_w$ | | Basic Outside Diameter, $D$ | | Housing Bore Diameter* Allowable Deviation from $D$ | |
|---|---|---|---|---|---|---|---|
| OUTER RING STATIONARY RELATIVE TO LOAD | | | | | | | |
| Inch | | ANSI h6, Inch | | Inch | | ANSI H7, Inch | |
| Over | Incl. | High | Low | Over | Incl. | Low | High |
| 0.2362 | 0.7087 | +0 | −.0004 | 0.7087 | 1.1811 | +0 | +.0008 |
| 0.7087 | 1.1811 | +0 | −.0005 | 1.1811 | 1.9685 | +0 | +.0010 |
| 1.1811 | 1.9685 | +0 | −.0006 | 1.9685 | 3.1496 | +0 | +.0012 |
| 1.9685 | 3.1496 | +0 | −.0007 | 3.1496 | 4.7244 | +0 | +.0014 |
| 3.1496 | 4.7244 | +0 | −.0009 | 4.7244 | 7.0866 | +0 | +.0016 |
| 4.7244 | 7.0866 | +0 | −.0010 | 7.0866 | 9.8425 | +0 | +.0018 |
| 7.0866 | 9.2520 | +0 | −.0012 | 9.8425 | 11.2205 | +0 | +.0020 |
| OUTER RING ROTATING RELATIVE TO LOAD | | | | | | | |
| Inch | | ANSI f6, Inch | | Inch | | ANSI N7, Inch | |
| Over | Incl. | High | Low | Over | Incl. | Low | High |
| 0.2362 | 0.3937 | −.0005 | −.0009 | 0.7087 | 1.1811 | −.0011 | −.0003 |
| 0.3937 | 0.7087 | −.0006 | −.0010 | 1.1811 | 1.9685 | −.0013 | −.0003 |
| 0.7087 | 1.1811 | −.0008 | −.0013 | 1.9685 | 3.1496 | −.0015 | −.0003 |
| 1.1811 | 1.9685 | −.0010 | −.0016 | 3.1496 | 4.7244 | −.0019 | −.0005 |
| 1.9685 | 3.1496 | −.0012 | −.0019 | 4.7244 | 7.0866 | −.0022 | −.0006 |
| 3.1496 | 4.7244 | −.0014 | −.0023 | 7.0866 | 9.8425 | −.0024 | −.0006 |
| 4.7244 | 7.0866 | −.0016 | −.0026 | 9.8425 | 11.2205 | −.0026 | −.0006 |
| 7.0866 | 9.2520 | −.0020 | −.0032 | .... | .... | .... | .... |

For bearing tolerances, see Table 10.
* See text for additional requirements.

**Table 20. AFBMA and American National Standard Tolerance Limits for Shaft Diameters — Needle Roller Bearing Inner Rings, Inch Type NIR (Used with Bearings Types NIB, NIY, NIH, NIA)** (ANSI/AFBMA Std 18.2-1976)

| Basic Bore, $d$ | | Shaft Diameter* | | | |
|---|---|---|---|---|---|
| | | Shaft Rotating Relative to Load, Outer Ring Stationary Relative to Load Allowable Deviation from $d$ | | Shaft Stationary Relative to Load, Outer Ring Rotating Relative to Load Allowable Deviation from $d$ | |
| Inch | | ANSI m5, Inch | | ANSI g6, Inch | |
| Over | Incl. | High | Low | High | Low |
| 0.2362 | 0.3937 | +.0005 | +.00025 | −.0002 | −.0006 |
| 0.3937 | 0.7087 | +.0006 | +.0003 | −.00025 | −.00065 |
| 0.7087 | 1.1811 | +.0007 | +.0003 | −.0003 | −.0008 |
| 1.1811 | 1.9685 | +.0008 | +.0004 | −.0004 | −.0010 |
| 1.9685 | 3.1496 | +.0010 | +.0005 | −.0004 | −.0011 |
| 3.1496 | 4.7244 | +.0011 | +.0005 | −.0005 | −.0014 |
| 4.7244 | 7.0866 | +.0013 | +.0006 | −.0006 | −.0016 |
| 7.0866 | 8.0709 | +.0014 | +.0006 | −.0006 | −.0018 |

For inner ring tolerance limits, see Table 11.
* See text for additional requirements.

Table 21. AFBMA Standard Lockwashers (Series W-00) for Ball Bearings and Cylindrical and Spherical Roller Bearings and (Series TW-100) for Tapered Roller Bearings

W-00 THROUGH W-16

TW-100 THROUGH TW-116

SURFACE MUST BE FLAT

| Type W No. | Q | Type TW No. | Q | Tangs | | | Width S | | Key X | | X' | | Bore R | | Diameter | | Diam. Over Tangs, Max. | |
|---|---|---|---|---|---|---|---|---|---|---|---|---|---|---|---|---|---|---|
| | | | | No. | Width* T | Project.* V | Min. | Max. | Min. | Max. | Min. | Max. | Min. | Max. | E | Tol. | B | B' |
| W-00 | .042 | TW-100 | .042 | 9 | .115 | 3/16 | .110 | .120 | .334 | .359 | .334 | .359 | .406 | .421 | 5/8 | +.015 | 7/8 | 57/64 |
| W-01 | .042 | TW-101 | .042 | 9 | .115 | 3/16 | .110 | .120 | .412 | .437 | .412 | .437 | .484 | .499 | 23/32 | +.015 | 1 1/64 | 1 3/32 |
| W-02 | .042 | TW-102 | .042 | 11 | .115 | 3/16 | .110 | .120 | .529 | .554 | .513 | .538 | .601 | .616 | 13/16 | +.015 | 1 5/32 | 1 3/32 |
| W-03 | .042 | TW-103 | .058 | 11 | .115 | 3/16 | .110 | .120 | .607 | .632 | .591 | .616 | .679 | .694 | 15/16 | +.015 | 1 7/32 | 1 9/32 |
| W-04 | .042 | TW-104 | .058 | 11 | .115 | 3/16 | .110 | .120 | .729 | .754 | .713 | .738 | .801 | .816 | 1 1/8 | +.015 | 1 11/32 | 1 17/32 |
| W-05 | .050 | TW-105 | .062 | 13 | .156 | 3/32 | .156 | .176 | .909 | .939 | .897 | .927 | .989 | 1.009 | 1 9/32 | +.015 | 1 17/32 | 1 49/64 |
| W-06 | .050 | TW-106 | .062 | 13 | .156 | 3/32 | .156 | .176 | 1.093 | 1.128 | 1.081 | 1.116 | 1.193 | 1.213 | 1 1/2 | +.015 | 1 49/64 | 2 1/64 |
| | | TW-065 | .062 | 13 | .156 | 3/32 | .156 | .176 | | | 1.221 | 1.256 | 1.333 | 1.353 | 1 11/16 | +.015 | 1 59/64 | 2 5/32 |
| W-07 | .050 | TW-107 | .062 | 15 | .156 | 3/32 | .156 | .176 | 1.296 | 1.331 | 1.284 | 1.319 | 1.396 | 1.416 | 1 11/16 | +.015 | 2 1/4 | 2 1/4 |
| W-08 | .058 | TW-108 | .072 | 15 | .156 | 1/8 | .250 | .290 | 1.475 | 1.510 | 1.461 | 1.496 | 1.583 | 1.603 | 2 | +.030 | 2 15/32 | 2 15/32 |
| W-09 | .058 | TW-109 | .072 | 15 | .219 | 1/8 | .250 | .290 | 1.684 | 1.724 | 1.670 | 1.710 | 1.792 | 1.817 | 2 9/32 | +.030 | 2 47/64 | 2 45/64 |
| W-10 | .058 | TW-110 | .072 | 17 | .219 | 1/8 | .250 | .290 | 1.884 | 1.924 | 1.870 | 1.910 | 1.992 | 2.017 | 2 7/16 | +.030 | 2 59/64 | 2 59/64 |
| W-11 | .063 | TW-111 | .072 | 17 | .219 | 1/8 | .250 | .290 | 2.069 | 2.109 | 2.060 | 2.100 | 2.182 | 2.207 | 2 21/32 | +.030 | 3 7/64 | 3 7/64 |
| W-12 | .063 | TW-112 | .082 | 17 | .219 | 1/8 | .250 | .290 | 2.267 | 2.307 | 2.248 | 2.288 | 2.400 | 2.425 | 2 27/32 | +.030 | 3 3/32 | 3 3/32 |
| W-13 | .063 | TW-113 | .082 | 19 | .219 | 1/8 | .250 | .290 | 2.455 | 2.495 | 2.436 | 2.476 | 2.588 | 2.613 | 3 1/16 | +.030 | 3 7/64 | 3 7/64 |
| W-14 | .063 | TW-114 | .082 | 19 | .219 | 3/16 | .250 | .290 | 2.658 | 2.698 | 2.639 | 2.679 | 2.791 | 2.816 | 3 1/16 | +.030 | 3 43/64 | 3 43/64 |
| W-15 | .072 | TW-115 | .095 | 19 | .313 | 3/16 | .313 | .353 | 2.831 | 2.876 | 2.808 | 2.853 | 2.973 | 3.003 | 3 9/16 | +.030 | 3 59/64 | 3 59/64 |
| W-16 | .072 | TW-116 | .095 | 19 | .313 | 3/16 | .313 | .353 | 3.035 | 3.080 | 3.012 | 3.057 | 3.177 | 3.207 | 3 27/32 | +.030 | 4 3/8 | 4 25/64 |

All dimensions in inches.

* Tolerances: On width T, ±.005 inch for Types W-00 to W-03 and Types TW-100 to TW-103; ±.010 inch for Types W-04 to W-07 and Types TW-104 to TW-107; and ±.015 for Types W-08 to W-16 and Types TW-108 to TW-116. On Projection V, +1/32 inch for Types W-00 to W-13 and Types TW-100 to TW-113, and +1/16 inch on Types W-14 to W-16 and Types TW-114 to TW-116.

For sizes larger than shown, see AFBMA Standard 8.

Table 22. AFBMA Standard Locknuts (Series N-oo) for Ball Bearings and Cylindrical and Spherical Roller Bearings and (Series TN-oo) for Tapered Roller Bearings

Runout and parallelism of faces measured on a tight fitting threaded arbor.

N-oo to N-o6 = .002 Max.
N-o7 to AN-15 = .004 Max.

TN-oo to TAN-15 = .002 Max.

Surface Finish Note

TN-oo to TN-11, 100μ in., max.
TN-12 to TAN-15, 120μ in., max.

SEE FOOTNOTE† · R · r₁ · [60°MAX. 45°MIN.] · 50° · D · E · F · G · H · C

| BB & RB Nut No. | TRB Nut No. | Thds. per Inch | Thread Minor Diam. Min. | Max. | Thread Pitch Diam. Min. | Max. | Thd. Major Diam. Min. | Outside Diam. C (+.005 -.015) | Face Diam. E Min. | Max. | Chamfer J | Diam. F (+.010 -.020) | Width G Min. | Max. | Height H Nominal* | Fillet R | Thickness D Min. | Max. |
|---|---|---|---|---|---|---|---|---|---|---|---|---|---|---|---|---|---|---|
| N-oo | TN-oo | 32 | 0.3572 | 0.3666 | 0.3707 | 0.3733 | 0.391 | 3/4 | .605 | .625 | 1/32 | 5/8 | .120 | .130 | 1/16 | .010 | .209 | .229 |
| N-01 | TN-01 | 32 | 0.4352 | 0.4386 | 0.4487 | 0.4513 | 0.469 | 7/8 | .699 | .719 | 1/32 | 3/4 | .120 | .130 | 1/16 | .010 | .303 | .323 |
| N-02 | TN-02 | 32 | 0.5522 | 0.5556 | 0.5557 | 0.5587 | 0.586 | 1 | .793 | .813 | 1/32 | 13/16 | .120 | .130 | 3/32 | .010 | .303 | .323 |
| N-03 | TN-03 | 32 | 0.6302 | 0.6336 | 0.6437 | 0.6467 | 0.664 | 1 1/16 | .918 | .938 | 1/32 | 15/16 | .120 | .130 | 3/32 | .010 | .334 | .354 |
| N-04 | TN-04 | 32 | 0.7472 | 0.7506 | 0.7607 | 0.7641 | 0.781 | 1 3/16 | 1.105 | 1.125 | 1/32 | 1 1/16 | .178 | .198 | 3/32 | .015 | .365 | .385 |
| N-05 | TN-05 | 32 | 0.9352 | 0.9386 | 0.9487 | 0.9521 | 0.969 | 1 3/8 | 1.261 | 1.281 | 1/32 | 1 3/16 | .178 | .198 | 3/32 | .015 | .396 | .416 |
| N-06 | TN-06 | 18 | 1.1129 | 1.1189 | 1.1309 | 1.1409 | 1.173 | 1 9/16 | 1.480 | 1.500 | 3/64 | 1 7/16 | .178 | .198 | 3/32 | .015 | .396 | .416 |
| N-065 | TN-065 | 18 | 1.2524 | 1.2584 | 1.2764 | 1.2864 | 1.3125 | 1 3/4 | 1.793 | 1.813 | 3/64 | 1 9/16 | .178 | .198 | 3/32 | .015 | .428 | .448 |
| N-07 | TN-07 | 18 | 1.3159 | 1.3219 | 1.3399 | 1.3439 | 1.376 | 2 1/16 | 1.793 | 1.813 | 3/64 | 1 7/8 | .178 | .198 | 3/32 | .015 | .428 | .448 |
| N-08 | TN-08 | 18 | 1.5029 | 1.5089 | 1.5289 | 1.5314 | 1.563 | 2 1/4 | 1.980 | 2.000 | 3/64 | 2 1/16 | .240 | .260 | 3/32 | .020 | .428 | .448 |
| N-09 | TN-09 | 18 | 1.7069 | 1.7129 | 1.7309 | 1.7354 | 1.767 | 2 7/16 | 2.261 | 2.281 | 3/64 | 2 1/2 | .240 | .260 | 3/32 | .020 | .428 | .448 |
| N-10 | TN-10 | 18 | 1.9069 | 1.9129 | 1.9309 | 1.9354 | 1.907 | 2 17/32 | 2.418 | 2.438 | 3/64 | 2 23/32 | .240 | .260 | 3/32 | .020 | .490 | .510 |
| N-11 | TN-11 | 18 | 2.0969 | 2.1029 | 2.1209 | 2.1260 | 2.157 | 2 23/32 | 2.636 | 2.656 | 3/64 | 2 29/32 | .240 | .260 | 1/8 | .020 | .490 | .510 |
| N-12 | TN-12 | 18 | 2.2999 | 2.3059 | 2.3239 | 2.3290 | 2.360 | 3 5/32 | 2.824 | 2.844 | 3/64 | 3 1/8 | .240 | .260 | 1/8 | .020 | .521 | .541 |
| N-13 | TN-13 | 18 | 2.4879 | 2.4939 | 2.5119 | 2.5170 | 2.548 | 3 3/8 | 3.043 | 3.063 | 3/64 | 3 3/8 | .240 | .260 | 1/8 | .020 | .553 | .573 |
| N-14 | TN-14 | 18 | 2.6909 | 2.6969 | 2.7149 | 2.7200 | 2.751 | 3 5/8 | 3.283 | 3.313 | 3/64 | 3 3/8 | .240 | .260 | 1/8 | .020 | .553 | .573 |
| AN-15 | TAN-15 | 12 | 2.8488 | 2.8518 | 2.8789 | 2.8843 | 2.933 | 3 7/8 | 3.533 | 3.563 | 5/64 | 3 5/8 | .360 | .385 | 1/8 | .025 | .584 | .604 |

All dimensions in inches.  * Variation in any one nut ±.010.  † Corner radius r₁ is 1/64 inch greater than chamfer J.
Steels for these locknuts are AISI, C1015, C1020, C1035, C1114, C1115, C1117, C1118, B1111, B1113, B1212, or B1214.
Threads are American National form, Class 3.
For larger sizes see AFBMA Standard 8.

Table 23.  AFBMA Standard for Shafts for Locknuts (Series N-00) for Ball Bearings and Cylindrical and Spherical Roller Bearings

| Thds. per Inch | Thread Major Diam. Max. | Min. | Thread Pitch Diam. Max. | Min. | Thread Minor Diam. Max. | Thread Relief W (+1/64 -0) | Brg. Bore | Diam. K Max. | Thread Length (+1/64 -0) | Keyway Depth L (+1/32 -0) | Keyway Width N (+1/64 -0) | M (+1/64 -0) | Nut No. |
|---|---|---|---|---|---|---|---|---|---|---|---|---|---|
| 32 | 0.391 | 0.3856 | 0.3707 | 0.3681 | 0.3527 | 1/16 | 0.3937 | 5/16 | 9/32 | 1/16 | 1/8 | 3/32 | N-00 |
| 32 | 0.469 | 0.4636 | 0.4487 | 0.4461 | 0.4307 | 1/16 | 0.4724 | 13/32 | 3/8 | 1/16 | 1/8 | 3/32 | N-01 |
| 32 | 0.586 | 0.5806 | 0.5657 | 0.5627 | 0.5477 | 1/16 | 0.5906 | 1/2 | 3/8 | 5/64 | 1/8 | 3/32 | N-02 |
| 32 | 0.664 | 0.6586 | 0.6437 | 0.6407 | 0.6257 | 1/16 | 0.6693 | 9/16 | 13/32 | 5/64 | 1/8 | 3/32 | N-03 |
| 32 | 0.781 | 0.7756 | 0.7607 | 0.7573 | 0.7427 | 1/16 | 0.7874 | 23/32 | 7/16 | 5/64 | 3/16 | 3/32 | N-04 |
| 32 | 0.969 | 0.9636 | 0.9487 | 0.9453 | 0.9307 | 1/16 | 0.9843 | 7/8 | 15/32 | 3/32 | 3/16 | 3/32 | N-05 |
| 18 | 1.173 | 1.1648 | 1.1369 | 1.1329 | 1.1048 | 3/32 | 1.1811 | 1 1/16 | 15/32 | 3/32 | 3/16 | 1/8 | N-06 |
| 18 | 1.376 | 1.3678 | 1.3399 | 1.3359 | 1.3078 | 3/32 | 1.3780 | 1 1/4 | 1/2 | 3/32 | 3/16 | 1/8 | N-07 |
| 18 | 1.563 | 1.5548 | 1.5269 | 1.5224 | 1.4948 | 3/32 | 1.5748 | 1 15/32 | 17/32 | 3/32 | 5/16 | 1/8 | N-08 |
| 18 | 1.767 | 1.7588 | 1.7309 | 1.7264 | 1.6988 | 1/8 | 1.7717 | 1 11/16 | 17/32 | 3/32 | 5/16 | 5/32 | N-09 |
| 18 | 1.967 | 1.9588 | 1.9309 | 1.9264 | 1.9088 | 1/8 | 1.9685 | 1 7/8 | 19/32 | 1/8 | 5/16 | 5/32 | N-10 |
| 18 | 2.157 | 2.1488 | 2.1209 | 2.1158 | 2.0888 | 1/8 | 2.1654 | 2 1/16 | 19/32 | 1/8 | 5/16 | 5/32 | N-11 |
| 18 | 2.360 | 2.3518 | 2.3239 | 2.3188 | 2.2918 | 1/8 | 2.3622 | 2 1/4 | 5/8 | 1/8 | 5/16 | 5/32 | N-12 |
| 18 | 2.548 | 2.5398 | 2.5119 | 2.5068 | 2.4798 | 1/8 | 2.5591 | 2 7/16 | 21/32 | 1/8 | 5/16 | 5/32 | N-13 |
| 18 | 2.751 | 2.7428 | 2.7149 | 2.7098 | 2.6828 | 1/8 | 2.7559 | 2 5/8 | 21/32 | 1/8 | 5/16 | 1/4 | N-14 |
| 12 | 2.933 | 2.9218 | 2.8789 | 2.8735 | 2.8308 | 5/32 | 2.9528 | 2 25/32 | 1 1/16 | 1/8 | 3/8 | 1/4 | AN-15 |
| 12 | 3.137 | 3.1258 | 3.0829 | 3.0770 | 3.0348 | 5/32 | 3.1496 | 3 | 1 1/16 | 1/8 | 3/8 | 1/4 | AN-16 |

All dimensions in inches.
For sizes larger than shown, see AFBMA Standard 8.
Thread relief diameter $A$ = Max. Minor Diam. − 1/64 ± .005 for 32 and 18 threads per inch and Max. Minor Diam. − 1/32 ± .010 for 12 threads per inch.  Threads are American National Form NS, Class 3.  See also footnote to Table 24 for material other than steel.
Length of Bearing Seat $Z$ = Min. Bearing Width − 1/64 ± .010.  Length of Keyway equals $M + P$.

Table 24.   AFBMA Standard for Shafts for Tapered Roller Bearing Locknuts

CLAMPED MOUNTING   ADJUSTABLE MOUNTING

| Thds. per Inch | Thread Major Diam. | | Thread Pitch Diam. | | Thread Minor Diam. Max. | Thread Relief W +1/64 -0 | Diam. K Max. | Thread Length | | Keyway | | | | Nut No. |
|---|---|---|---|---|---|---|---|---|---|---|---|---|---|---|
| | Max. | Min. | Max. | Min. | | | | Y +1/64 -0 | P +1/64 -0 | Depth L +1/64 -0 | Width N +1/64 -0 | Length M +1/64 -0 | U +1/64 -0 | |
| 32 | 0.391 | 0.3856 | 0.3707 | 0.3681 | 0.3527 | 1/16 | 5/16 | 19/32 | 3/8 | 3/32 | 1/16 | 3/32 | 15/32 | TN-00 |
| 32 | 0.409 | 0.4036 | 0.4487 | 0.4461 | 0.4307 | 1/16 | 13/32 | 25/32 | 15/32 | 3/32 | 1/16 | 3/32 | 9/16 | TN-01 |
| 32 | 0.586 | 0.5806 | 0.5057 | 0.5027 | 0.5477 | 1/16 | 1/2 | 13/16 | 1/2 | 3/32 | 1/16 | 3/32 | 19/32 | TN-02 |
| 32 | 0.664 | 0.6586 | 0.6437 | 0.6407 | 0.6257 | 1/16 | 9/16 | 7/8 | 17/32 | 3/32 | 1/16 | 3/32 | 5/8 | TN-03 |
| 32 | 0.781 | 0.7756 | 0.7607 | 0.7573 | 0.7427 | 1/16 | 11/16 | 29/32 | 17/32 | 3/32 | 3/16 | 3/32 | 5/8 | TN-04 |
| 32 | 0.969 | 0.9636 | 0.9487 | 0.9453 | 0.9307 | 1/16 | 7/8 | 1 | 19/32 | 1/8 | 3/16 | 1/8 | 23/32 | TN-05 |
| 18 | 1.173 | 1.1648 | 1.1369 | 1.1329 | 1.1048 | 3/32 | 1 1/16 | 1 | 19/32 | 1/8 | 3/16 | 1/8 | 23/32 | TN-06 |
| 18 | 1.3125 | 1.3043 | 1.2764 | 1.2724 | 1.2443 | 3/32 | 1 3/16 | 1 1/16 | 5/8 | 1/8 | 3/16 | 1/8 | 3/4 | TN-065 |
| 18 | 1.376 | 1.3678 | 1.3399 | 1.3359 | 1.3078 | 3/32 | 1 1/4 | 1 1/16 | 5/8 | 1/8 | 3/16 | 1/8 | 3/4 | TN-07 |
| 18 | 1.563 | 1.5548 | 1.5269 | 1.5224 | 1.4948 | 1/8 | 1 7/16 | 1 1/16 | 5/8 | 1/8 | 3/16 | 1/8 | 3/4 | TN-08 |
| 18 | 1.767 | 1.7588 | 1.7309 | 1.7264 | 1.6988 | 1/8 | 1 21/32 | 1 3/16 | 11/16 | 1/8 | 5/16 | 5/32 | 25/32 | TN-09 |
| 18 | 1.967 | 1.9588 | 1.9309 | 1.9264 | 1.8988 | 1/8 | 1 55/64 | 1 3/16 | 11/16 | 1/8 | 5/16 | 5/32 | 27/32 | TN-10 |
| 18 | 2.157 | 2.1488 | 2.1209 | 2.1158 | 2.0888 | 1/8 | 2 3/64 | 1 9/32 | 3/4 | 5/32 | 5/16 | 5/32 | 27/32 | TN-11 |
| 18 | 2.360 | 2.3518 | 2.3239 | 2.3188 | 2.2918 | 1/8 | 2 17/64 | 1 19/32 | 25/32 | 5/32 | 5/16 | 5/32 | 29/32 | TN-12 |
| 18 | 2.548 | 2.5398 | 2.5119 | 2.5068 | 2.4798 | 1/8 | 2 27/64 | 1 11/32 | 25/32 | 5/32 | 5/16 | 5/32 | 1 5/16 | TN-13 |
| 18 | 2.751 | 2.7428 | 2.7149 | 2.7098 | 2.6828 | 1/8 | 2 5/8 | 1 11/32 | 25/32 | 5/32 | 5/16 | 5/32 | 1 | TN-14 |
| 12 | 2.933 | 2.9218 | 2.8789 | 2.8735 | 2.8308 | 5/32 | 2 25/32 | 1 11/32 | 13/16 | 3/16 | 5/16 | 1/4 | 1 1/32 | TAN-15 |
| 12 | 3.137 | 3.1258 | 3.0829 | 3.0770 | 3.0348 | 5/32 | 3 | 1 11/32 | 13/16 | 3/16 | 5/16 | 1/4 | 1 7/32 | TAN-16 |

All dimensions in inches.

Thread relief diameter A = Max. Minor Diam. − 1/64 ± .005 for 32 and 18 threads per inch and Max. Minor Diam. − 1/32 ± .010 for 12 t.p.i.

This standard is applicable to steel nuts. When either the nut or shaft is stainless steel, aluminum or other material having a tendency to seize, it is recommended that the maximum thread diameters of the shaft both major and pitch be reduced by 20 per cent of the listed pitch diameter tolerances (Max. − Min.). For sizes larger than shown, see AFBMA Standard 8.

**Bearing Mounting Practice.** — Because of their inherent design and material rigidity, rolling contact bearings must be mounted with careful control of their alignment and runout. Medium-speed or slower (400,000 $DN$ values or less where $D$ is the bearing bore in millimeters and $N$ is the bearing speed in revolutions per minute), and medium to light load ($C/P$ values of 7 or greater where $C$ is the bearing specific dynamic capacity in pounds and $P$ is the average bearing load in pounds) applications can endure misalignments equivalent to those acceptable for high-capacity, precision journal bearings utilizing hard bearing materials such as silver, copper-lead, or aluminum. In no case, however, should the maximum shaft deflection exceed .001 inch per inch for well-crowned roller bearings, and .003 inch per inch for deep-groove ball-bearings. Except for self-aligning ball-bearings and spherical or barrel roller bearings, all other types require shaft alignments with deflections no greater than .0002 inch per inch. With preloaded ball bearings, this same limit is recommended as a maximum. Close-clearance tapered bearings or thrust bearings of most types require the same shaft alignment also.

Of major importance for all bearings requiring good reliability, is the location of the races on the shaft and in the housing. Assembly methods must insure: (1) that the faces are square, before the cavity is closed; (2) that the cover face is square to the shoulder and pulled in evenly; and (3) that it will be located by a face parallel to it when finally seated against the housing. These requirements are shown in the accompanying figure. In applications not controlled by automatic tooling with closely controlled fixtures and bolt torquing mechanisms, races should be checked for squareness by sweeping with a dial indicator mounted as shown below. For

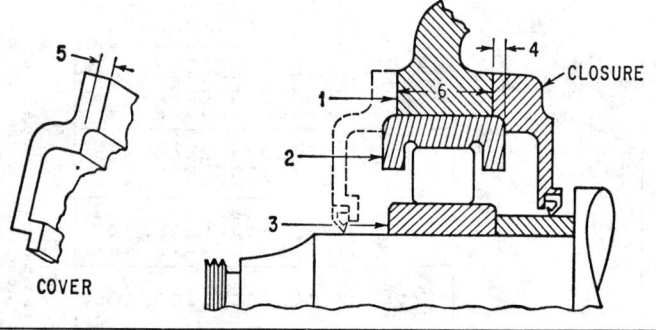

COMMERCIAL APPLICATION ALIGNMENT TOLERANCES

1. HOUSING FACE RUNOUT — Square to shaft center within .0004 inch/inch of radius full indicator reading.

2. OUTER RACE FACE RUNOUT — Square to shaft center within .0004 inch/inch of radius full indicator reading and complementary to the housing runout (not opposed).

3. INNER RACE FACE RUNOUT — Square to shaft center within .0003 inch/inch of radius full indicator reading.

4. and 5. COVER AND CLOSURE MOUNTING FACE PARALLELISM — Parallel within .001.

6. HOUSING MOUNTING FACE PARALLELISM — Parallel within .001.

commercial applications with moderate life and reliability requirements, outer race runouts should be held to .0005 inch per inch of radius and inner race runout to .0004 inch per inch of radius. In preloaded, and precision applications, these tolerances must be cut in half. In regard to the question of alignment, it must be recognized that rolling-contact bearings, being made of fully-hardened steel, do not wear in as may certain journal bearings when carefully applied and initially operated. Likewise, rolling contact bearings absorb relatively little deflection when loaded to $C/P$ values of 6 or less. At such stress levels the rolling element-race deformation is generally not over .0002 inch. Consequently, proper mounting and control of shaft deflections are imperative for reliable bearing performance. Aside from inadequate lubrication, these are the most frequent causes of premature bearing failures.

**Mountings for Precision and Quiet-running Applications.** — In applications of rolling-element bearings where vibration or smoothness of operation is critical, special precautions must be taken to eliminate those conditions which can serve to initiate radial and axial motions. These exciting forces can result in shaft excursions which are in resonance with shaft or housing components over a range of frequencies from well below shaft speed to as much as 100 times above it. The more sensitive the configuration, the greater is the need for precision bearings and mountings to be used.

Precision bearings are normally made to much closer tolerances than standard and therefore benefit from better finishing techniques. Special inspection operations are required, however, to provide races and rolling elements with smoothness and runouts compatible with the needs of the application. Similarly, shafts and housings must be carefully controlled.

Among the important elements to be controlled are shaft, race, and housing roundness; squareness of faces, diameters, shoulders, and rolling paths. Though not readily appreciated, grinding chatter, lobular and compensating out-of-roundness, waviness, and flats of less than .0005 inch deviation from the average or mean diameter can cause significant roughness. To detect these and insure the selection of good pieces, three-point electronic indicator inspection must be made. For ultraprecise or quiet applications, pieces are often checked on a "Talyrond" or a similar continuous recording instrument capable of measuring to within a few millionths of an inch. Though this may seem extreme, it has been found that shaft deformities will be reflected through inner races shrunk onto them. Similarly, tight-fit outer races pick up significant deviations in housings. In many instrument and in missile guidance applications, such deviations and deformities may have to be limited down to less than .00002 inch.

In most of these precision applications, bearings are used with rolling elements controlled to less than 5 millionths of an inch deviation from roundness and within the same range for diameter.

Special attention is required both in housing design and in assembly of the bearing to shaft and housing. Housing response to axial excursions forced by bearing wobble (which in itself is a result of out-of-square mounting) has been found to be a major source of small electric and other rotating equipment noise and howl. Stiffer, more massive housings and careful alignment of bearing races can make significant improvements in applications where noise or vibration has been found to be objectionable.

**Squareness and Alignment.** — In addition to the limits for roundness and wall variation of the races and their supports, squareness of end faces and shoulders must be closely controlled. Tolerances of .0001 inch full indicator reading per inch of diameter are normally required for end faces and shoulders, with appropriately selected limits for fillet eccentricities. The latter must also fall within specified limits for radii tolerances to prevent interference and the resulting cocking of the race. Reference should be made to the bearing dimension tables which list corner radii for typical bearings. Shoulders must also be of a sufficient height to insure proper support for the races, since they are of hardened steel and are less capable of absorbing shock loads and abuse. The general subject of squareness and alignment is of primary importance to the life of rolling element bearings.

The following recommendations for shaft and housing design are given by the New Departure Division of General Motors Corporation:*

"As a rule, there is little trouble experienced with inaccuracies in shafts. Bearings seats and locating shoulders are turned and ground to size with the shaft held on centers and, with ordinary care, there is small chance for serious out-of-roundness or taper. Shaft shoulders should present sufficient surface in contact with the bearing face to assure positive and accurate location.

"Where an undercut must be made for wheel runout in grinding a bearing seat, care should be exercised that no sharp corners are left, for it is at such points that fatigue is most likely to result in shaft breakage. It is best to undercut as little as possible and to have the undercut end in a fillet instead of a sharp corner.

"Where clamping nuts are to be used, it is important to cut the threads as true and square as possible in order to insure even pressure at all points on the bearing inner ring faces when the nuts are set up tight. It is also important not to cut threads so far into the bearing seat as to leave part of the inner ring unsupported or carried on the threads. Excessive deflection is usually the result of improperly designed or undersized machine parts. With a weak shaft, it is possible to seriously affect bearing operation through misalignment due to shaft deflection. Where shafts are comparatively long, the diameter between bearings must be great enough to properly resist bending. In general, the use of more than two bearings on a single shaft should be avoided, owing to the difficulty of securing accurate alignment. With bearings mounted close to each other, this can result in extremely heavy bearing loads.

"Design is as important as careful machining in construction of accurate bearing housings. There should be plenty of metal in the wall sections and large, thin areas should be avoided as much as possible, since they are likely to permit deflection of the boring tool when the housing is being finish-machined.

"Wherever possible, it is best to design a housing so that the radial load placed on the bearing is transmitted as directly as possible to the wall or rib supporting the housing. Diaphragm walls connecting an offset housing to the main wall or side of a machine are apt to deflect unless made thick and well braced.

"When two bearings are to be mounted opposed, but in separate housings, the housings should be so reinforced with fins or webs as to prevent deflection due to the axial load under which the bearings are opposed.

"Where housings are deep and considerable overhang of the boring tool is required, there is a tendency to produce out-of-roundness and taper, unless the tool is very rigid and light finishing cuts are taken. In a too roughly bored housing there is a possibility for the ridges of metal to peen down under load, thus eventually resulting in too loose a fit for the bearing outer ring."

**Soft Metal and Resilient Housings.** — In applications relying on bearing housings made of soft materials (aluminum, magnesium, light sheet metal, etc.) or those which lose their fit because of differential thermal expansion, outer race mounting must be approached in a cautious manner. Of first importance is the

*New Departure Handbook, Vol. II — 1951.

determination of the possible consequences of race loosening and turning. In conjunction with this, the type of loading must be considered for it may serve to magnify the effect of race loosening. It must be remembered that generally, balancing processes do not insure zero unbalance at operating speeds, but rather an "acceptable" maximum. This force exerted by the rotating element on the outer race can initiate a precession which will aggravate the race loosening problem by causing further attrition through wear, pounding, and abrasion. Since this force is generally of an order greater than the friction forces in effect between the outer race, housing, and closures (retaining nuts also), no foolproof method can be recommended for securing outer races in housings which deform significantly under load or after appreciable service wear. Though many such "fixes" are offered, the only sure solution is to press the race into a housing of sufficient stiffness with the heaviest fit consistent with the installed and operating clearances. In many cases, inserts, or liners of cast iron or steel are provided to maintain the desired fit and increase useful life of both bearing and housing.

**Quiet or Vibration-free Mountings.** — In seeming contradiction is the approach to bearing mountings in which all shaft or rotating element vibration must be isolated from the frame, housing, or supporting structure. Here bearing outer races are often supported on elastomeric or metallic springs. Fundamentally, this is an isolation problem and must be approached with caution to insure solution of the primary bearing objective — location and restraint of the rotating body, as well as the reduction or elimination of the dynamic problem. Again, the danger of skidding rolling elements must be considered and reference to the resident engineers or sales engineers of the numerous bearing companies is recommended, as this problem generally develops requirements for special, or non-catalog-type bearings.

**General Mounting Precautions.** — Since the last operations involving the bearing application — mounting and closing — have such important effects on bearing performance, durability, and reliability, it must be cautioned that more bearings are abused or "killed" in this early stage of their life than wear out or "die" under conditions for which they were designed. Hammer and chisel "mechanics" invariably handle bearings as though no blow could be too hard, no dirt too abrasive, and no misalignment of any consequence. Proper tools, fixtures, and techniques are a must for rolling bearing application, and it is the responsibility of the design engineer to provide for this in his design, advisory notes, mounting instructions, and service manuals. Nicks, dents, scores, scratches, corrosion staining, and dirt must be avoided if reliability, long life, and smooth running are to be expected of rolling bearings. All manufacturers have pertinent service instructions available for the bearing user. These should be followed for best performance. In a later section, methods for inspecting bearings and descriptions of most common bearing deficiencies will be given.

**Seating Fits for Bearings.** — Anti-Friction Bearing Manufacturers Association (AFBMA) standard shaft and housing bearing seat tolerances are given in Tables 12 to 16, inclusive.

**Clamping and Retaining Methods.** — Various methods of clamping bearings to prevent axial movement on the shaft are employed, one of the most common being a nut screwed on the end of the shaft and held in place by a tongued lock washer (see Table 21). The shaft thread for the clamping nut (see Table 22) should be cut in accurate relation to bearing seats and shoulders if bearing stresses are to be avoided. The threads used are of American National Form, Class 3; special diameters and data for these are given in Tables 23 and 24. Where somewhat closer than average accuracy is required, the washers and locknut faces may be obtained ground for closer alignment with the threads. For a high degree of accuracy the shaft threads are ground and a more precise clamping means is employed. Where a bearing inner ring is to be clamped it is important to provide a sufficiently high shoulder on the shaft to locate the bearing positively and accurately. In the difference between bearing bore and maximum shaft diameter gives a low shoulder which would enter the corner of the radius of the bearing, a shoulder ring that extends above the shoulder and well into the shaft corner is employed. A shoulder ring with snap wire fitting into a groove in the shaft is sometimes used where no locating shaft shoulder is present. A snap ring fitting into a groove is frequently employed to prevent endwise movement of the bearing away from the locating shoulder where tight clamping is not required. Such a retaining ring should not be used where a slot in the shaft surface might lead to fatigue failure. Snap rings are also used to locate the outer bearing ring in the housing. Dimensions of snap rings used for this latter purpose are given in AFBMA and ANSI standards.

**Bearing Closures.** — Shields, seals, labyrinths, and slingers are employed to retain the lubricant in the bearing and to prevent the entry of dirt, moisture, or other harmful substances. The type selected for a given application depends upon the lubricant, shaft, speed, and the atmospheric conditions in which the unit is to operate. The shields or seals may be located in the bearing itself. Shields differ from seals in that they are attached to one bearing race but there is a definite clearance between the shield and the other, usually the inner, race. When a shielded bearing is placed in a housing in which the grease space has been filled, the bearing in running will tend to expel excess grease past the shields or to accept grease from the housing when the amount in the bearing itself is low.

Seals of leather, rubber, cork, felt, or plastic composition may be used. Since they must bear against the rotating member, excessive pressure should be avoided and some lubricant must be allowed to flow into the area of contact in order to prevent seizing and burning of the seal and scoring of the rotating member. Some seals are made up in the form of cartridges which can be pressed into the end of the bearing housing.

Leather seals may be used over a wide range of speeds. Although lubricant is best retained with a leather cupped inward toward the bearing, this arrangement is not suitable at high speeds due to danger of burning the leather. At high speeds where abrasive dust is present, the seal should be arranged with the leather cupped outward to lead some lubricant into the contact area. Only light pressure of leather against the shaft should be maintained.

**Bearing Fits.** — The slipping or creeping of a bearing ring on a rotating shaft or in a rotating housing occurs when the fit of the ring on the shaft or in the housing is loose. Such slipping or creeping action may cause rapid wear of both shaft and bearing ring when the surfaces are dry and highly loaded. To prevent this action the bearing is customarily mounted with the rotating ring a press fit and the stationary ring a push fit, the tightness or looseness depending upon the service intended. Thus, where shock or vibratory loads are to be encountered, fits should be made somewhat tighter than for ordinary service. The stationary ring, if correctly fitted, is allowed to creep very slowly so that prolonged stressing of one part of the raceway is avoided.

To facilitate the assembly of a bearing on a shaft it may become necessary to expand the inner ring by heating. This should be done in clean oil or in a temperature-controlled furnace at a temperature of between 200 and 250° F. The utmost care must be used to make sure that the temperature does not exceed 250° F. as overheating will tend to reduce the hardness of the rings. Prelubricated bearings should not be mounted by this method.

**Selection of Ball and Roller Bearings.** — As compared with sleeve bearings, ball and roller bearings offer the following advantages: (1) Starting friction is low; (2) Less axial space is required; (3) Relatively accurate shaft alignment can be maintained; (4) Both radial and axial loads can be carried by certain types; (5) Angle of load application is not restricted; (6) Replacement is relatively easy; (7) Comparatively heavy overloads can be carried momentarily; (8) Lubrication is simple; and (9) Design and application can be made with the assistance of bearing supplier engineers.

In selecting a ball or roller bearing for a specific application five choices must be made: (1) the bearing series, (2) the type of bearing, (3) the size of bearing, (4) the method of lubrication, and (5) the type of mounting. Naturally these considerations are modified or affected by the anticipated operating conditions, expected life, cost, and overhaul philosophy. It is well to review the possible history of the bearing and its function in the machine it will be applied to, thus: (1) Will it be expected to endure removal and reapplication? (2) Must it be free from maintenance attention during its useful life? (3) Can wear of the housing or shaft be tolerated during the overhaul period? (4) Must it be adjustable to take up wear, or to change shaft location? (5) How accurately can the load spectrum be estimated? and (6) Will it be relatively free from abuse in operation? Though many cautions could be pointed out, it should always be remembered that inadequate design approaches limit the utilization of rolling element bearings, reduce customer satisfaction, and reduce reliability. Time spent in this stage of design is the most rewarding effort of the bearing engineer, and here again he can depend on the bearing manufacturers' field organization for assistance.

*Type:* Where loads are low, ball bearings are usually less expensive than roller bearings in terms of unit-carrying capacity. Where loads are high, the reverse is usually true.

For a purely radial load, almost any type of radial bearing can be used, the actual choice being determined by other factors. To support a combination of thrust and radial loads, several types of bearings may be considered. If the thrust load component is large, it may be most economical to provide a separate thrust bearing. When a separate thrust bearing cannot be used due to high speed, lack of space, or other factors, the following types may be considered: angular contact ball bearing, deep groove ball bearing without filling slot, tapered roller bearing with steep contact angle, and self-aligning bearing of the wide type. If movement or deflection in an axial direction must be held to a minimum, then a separate thrust bearing or a preloaded bearing capable of taking considerable thrust load is required. To minimize deflection due to a moment in an axial plane, a rigid bearing such as a double row angular contact type with outwardly converging load lines is required. In such cases, the resulting stresses must be taken into consideration in determining the proper size of the bearing.

For shock loads or heavy loads of short duration, roller bearings are usually preferred.

Special bearing designs may be required where accelerations are usually high as in planetary or crank motions.

Where the problem of excessive shaft deflection or misalignment between shaft and housing is present, a self-aligning type of bearing may be a satisfactory solution.

It should be kept in mind that a great deal of difficulty can be avoided if standard types of bearings are used in preference to special designs, wherever possible.

*Size:* The size of bearing required for a given application is determined by the loads that are to be carried and, in some cases, by the amount of rigidity that is necessary to limit deflection to some specified amount.

The forces to which a bearing will be subjected can be calculated by the laws of engineering mechanics from the known loads, power, operating pressure, etc. Where loads are irregular, varying, or of unknown magnitude, it may be difficult to determine the actual forces. In such cases, empirical determination of such forces, based on extensive experience in bearing design, may be needed to attack the problem successfully. Where such experience is lacking, the bearing manufacturer should be consulted or the services of a bearing expert obtained.

If a ball or roller bearing is to be subjected to a combination of radial and thrust loads, an *equivalent radial load* is computed in the case of radial or angular type bearings and an *equivalent thrust load* is computed in the case of thrust bearings.

**Method of Lubrication.** — If speeds are high, relubrication difficult, the shaft angle other than horizontal, the application environment incompatible with normal lubrication, leakage cannot be tolerated; if other elements of the mechanism establish the lubrication requirements, bearing selection must be made with these criteria as controlling influences. Modern bearing types cover a wide selection of lubrication means. Though the most popular type is the "cartridge" type of sealed grease ball bearing, many applications have requirements which dictate against them. Often, operating environments may subject bearings to temperatures too high for seals utilized in the more popular designs. If minute leakage or the accumulation of traces of dirt at seal lips cannot be tolerated by the application (as in baking industry machinery), then the selections of bearings must be made with other sealing and lubrication systems in mind.

High shaft speeds generally dictate bearing selection based on the need for cooling, the suppression of churning or aeration of conventional lubricants, and most important of all, the inherent speed limitations of certain bearing types. An example of the latter is the effect of cage design and of the roller-end thrust-flange contact on the lubrication requirements in commercial taper roller bearings, which limit the speed they can endure and the thrust load they can carry. Reference to the manufacturers' catalog and application-design manuals is recommended before making bearing selections.

**Type of Mounting.** — Many bearing installations are complicated because the best adapted type was not selected. Similarly, performance, reliability, and maintenance operations are restricted because the mounting was not thoroughly considered. There is no universally adaptable bearing for all needs. Careful reviews of the machine requirements should be made before designs are implemented. In many cases complicated machining, redundant shaft and housings, and use of an oversize bearing can be eliminated if the proper bearing in a well-thought-out mounting is chosen.

Advantage should be taken of the many race variations available in "standard" series of bearings. Puller grooves, tapered sleeves, flanged outer races, split races, fully demountable rolling-element and cage assemblies, flexible mountings, hydraulic removal features, relubrication holes and grooves, and many other innovations are available beyond the obvious advantages which are inherent in the basic bearing types.

**Radial and Axial Clearance.** — In designing the bearing mounting, a major consideration is to provide running clearances consistent with the requirements of the application. Race fits must be expected to absorb some of the original bearing clearance so that allowance should be made for approximately 80 per cent of the actual interference showing up in the diameter of the race. This will increase for heavy, stiff housings or for extra light series races shrunk onto solid shafts, while light metal housings (aluminum, magnesium, or sheet metal) and tubular shafts with wall sections less than the race wall thickness will cause a lesser change in the race diameter.

Where the application will impose heat losses through housing or shaft, or where a temperature differential may be expected, allowances must be made in the proper direction to insure proper operating clearance. Some compromises are required in applications where the indicated modification cannot be fully accommodated without endangering the bearing performance at lower speeds, during starting, or under lower temperature conditions than anticipated. Some leeway can be relied on with ball bearings since they can run with moderate preloads (.0005 inch, max.) without affecting bearing life or temperature rise. Roller bearings, however, have a lesser tolerance for preloading, and must be carefully controlled to avoid overheating and resulting self-destruction.

In all critical applications axial and radial clearances should be checked with feeler gages or dial indicators to insure mounted clearances within tolerances established by the design engineer. Since chips, scores, race misalignment, shaft or housing denting, housing distortion, end cover (closure) off-squareness, and mismatch of rotor and housing axial dimensions can rob the bearing of clearance, careful checks of running clearance is recommended.

For precision applications, taper-sleeve mountings, opposed ball or tapered-roller bearings with adjustable or shimmed closures are employed to provide careful control of radial and/or axial clearances. This practice requires skill and experience as well as the initial assistance of the bearing manufacturers' field engineer.

Tapered bore bearings are often used in applications such as these, again requiring careful and well worked-out assembly procedures. They can be assembled on either tapered shafts or on adapter sleeves. Advancement of the inner race over the tapered shaft can be done either by controlled heating (to expand the race as required) or by the use of a hydraulic jack. The adapter sleeve is supplied with a lock-nut which is used to advance the race on the tapered sleeve. With the heavier fits normally required to effect the clearance changes compatible with such mountings, hydraulic removal devices are normally recommended.

For the conventional application, with standard fits, clearances provided in the standard bearing are suitable for normal operation. To insure that the design conditions are "normal," a careful review of the application requirements, environments, operating speed range, anticipated abuses, and design parameters must be made.

**General Bearing Handling Precautions.** — To insure that rolling element bearings are capable of achieving their design life and that they perform without objectionable noise, temperature rise, or shaft excursions, the following precautions are recommended:

1. Use the best bearing available for the application, consistent with the value of the application. Remember, the cost of the best bearing is generally small compared to the replacement costs of the rotating components that can be destroyed if a bearing fails or malfunctions.

2. If questions arise in designing the bearing application, seek out the assistance of the bearing manufacturer's representative.

3. Handle bearings with care, keeping them in the sealed, original container until ready to use.

4. Follow the manufacturer's instructions in handling and assembling the bearings.

5. Work with clean tools, clean dry hands, and in clean surroundings.

6. Do not wash or wipe bearings prior to installation unless special instructions or requirements have been established to do so.

7. Place unwrapped bearings on clean paper and keep them similarly covered until applied, if they cannot be kept in the original container.

8. Don't use wooden mallets, brittle or chipped tools, or dirty fixtures and tools in mounting bearings.

9. Don't spin uncleaned bearings, nor spin *any* bearing with an air blast.

10. Use care not to scratch or nick bearings.

11. Don't strike or press on race flanges.

12. Use adapters for mounting which provide uniform steady pressure rather than hammering on a drift or sleeve.

13. Insure that races are started onto shafts and into housings evenly so as to prevent cocking.

14. Inspect shafts and housings before mounting bearing to insure that proper fits will be maintained.

15. When removing bearings, clean housings, covers, and shafts before exposing the bearings. All dirt can be considered an abrasive, dangerous to the reuse of any rolling bearing.

16. Treat used bearings, which may be reused, as new ones.

17. Protect dismantled bearings from dirt and moisture.

18. Use clean, lint-free rags if bearings are wiped.

19. Wrap bearings in clean, oil-proof paper when not in use.

20. Use clean filtered, water-free Stoddard's solvent or flushing oil to clean bearings.

21. In heating bearings for mounting onto shafts, follow manufacturer's instructions.

22. In assembling bearings onto shafts *never* strike the outer race, or press on it to force the inner race. Apply the pressure on the inner race only. In dismantling follow the same precautions.

23. Do not press, strike, or otherwise force the seal or shield on factory-sealed bearings.

**Bearing Failures, Deficiencies, and Their Origins.** — The general classifications of failures and deficiencies requiring bearing removal are:

(a) Overheating: 1. Inadequate or insufficient lubrication; 2. Excessive lubrication; 3. Grease liquefaction or aeration; 4. Oil foaming; 5. Abrasive or corrosive action due to contaminants in bearing; 6. Distortion of housing due to warping, or out-of-round; 7. Seal rubbing or failure; 8. Inadequate or blocked scavenge oil passages; 9. Inadequate bearing-clearance or bearing-preload; 10. Race turning; 11. Cage wear; and 12. Shaft expansion — loss of bearing or seal clearance.

(b) Vibration: 1. Dirt or chips in bearing; 2. Fatigued race or rolling elements; 3. Race turning; 4. Rotor unbalance; 5. Out-of-round shaft; 6. Race misalignment; 7. Housing resonance; 8. Cage wear; 9. Flats on races or rolling elements; 10. Excessive clearance; 11. Corrosion; 12. False-brinelling or indentation of races; 13. Electrical discharge (similar to corrosion effects); 14. Mixed rolling element diameters; and 15. Out-of-square rolling paths in races.

(c) Turning on shaft: 1. Growth of race due to overheating; 2. Fretting wear; 3. Improper initial fit; 4. Excessive shaft deflection; 5. Initially coarse shaft finish; and 6. Seal rub on inner race.

(d) Binding of the shaft: 1. Lubricant breakdown; 2. Contamination by abrasive or corrosive matter; 3. Housing distortion or out-of-round pinching bearing; 4. Uneven shimming of housing with loss of clearance; 5. Tight rubbing seals; 6. Preloaded bearings; 7. Cocked races; 8. Loss of clearance due to excessive tightening of adapter; 9. Thermal expansion of shaft or housing; and 10. Cage failure.

(e) Noisy bearing: 1. Lubrication breakdown, inadequate lubrication, stiff grease; 2. Contamination; 3. Pinched bearing; 4. Seal rubbing; 5. Loss of clearance and preloading; 6. Bearing slipping on shaft or in housing; 7. Flatted roller or ball; 8. Brinelling due to assembly abuse, handling, or shock loads; 9. Variation in size of rolling elements; 10. Out-of-round or lobular shaft; 11. Housing bore waviness; and 12. Chips or scores under bearing race seat.

(f) Displaced shaft: 1. Bearing wear; 2. Improper housing or closure assembly;

3. Overheated and shifted bearing; 4. Inadequate shaft or housing shoulder; 5. Lubrication and cage failure permitting rolling elements to bunch; 6. Loosened retainer nut or adapter; 7. Excessive heat application in assembling inner race, causing growth and shifting on shaft; and 8. Housing pounding out.

(g) Lubricant leakage: 1. Overfilling of lubricant; 2. Grease churning due to use of too soft a consistency; 3. Grease deterioration due to excessive operating temperature; 4. Operating life longer than grease life (grease breakdown, aeration, and purging); 5. Seal wear; 6. Wrong shaft attitude (bearing seals designed for horizontal mounting only); 7. Seal failure; 8. Clogged breather; 9. Oil foaming due to churning or air flow through housing; 10. Gasket (O-ring) failure or misapplication; 11. Porous housing or closure; and 12. Lubricator set at wrong flow rate.

## Load Ratings and Fatigue Life

**Ball and Roller Bearing Life.** — The performance of ball and roller bearings is a function of many variables. These include the bearing design, the characteristics of the material from which the bearings are made, the way in which they are manufactured, as well as many variables associated with their application. The only sure way to establish the satisfactory operation of a bearing selected for a specific application is by actual performance in the application. As this is often impractical, another basis is required to estimate the suitability of a particular bearing for a given application. Two factors are taken into consideration: the bearing fatigue life, and its ability to withstand static loading.

*Life Criterion:* Even if a ball or roller bearing is properly mounted, adequately lubricated, protected from foreign matter and not subjected to extreme operating conditions, it can ultimately fatigue. Under ideal conditions, the repeated stresses developed in the contact areas between the balls or rollers and the raceways eventually can result in the fatigue of the material which manifests itself with the spalling of the load-carrying surfaces. In most applications the fatigue life is the maximum useful life of a bearing.

*Static Load Criterion:* A static load is a load acting on a non-rotating bearing. Permanent deformations appear in balls or rollers and raceways under a static load of moderate magnitude and increase gradually with increasing load. The permissible static load is, therefore, dependent upon the permissible magnitude of permanent deformation. Experience shows that a total permanent deformation of 0.0001 of the ball or roller diameter, occurring anywhere along the length of contact of the most heavily loaded ball or roller and raceway contact, can be tolerated in most bearing applications without impairment of bearing operation. In certain applications where subsequent rotation of the bearing is slow and where smoothness of operation and friction requirements are not too exacting, a much greater total permanent deformation can be permitted. Conversely, where smoothness or friction requirements are critical, permanent deformation must be reduced.

**Ball Bearing Types Covered.** — AFBMA and American National Standard (ANSI B3.15–1972) sets forth the method of determining ball bearing Rating Life and covers the following types:

1. *Radial, deep groove and angular contact ball bearings* whose inner ring raceways have a cross-sectional radius not larger than 52 percent of the ball diameter and whose outer ring raceways have a cross-sectional radius not larger than 53 percent of the ball diameter.

2. *Radial, self-aligning ball bearings* whose inner ring raceways have cross-sectional radii not larger than 53 percent of the ball diameter.

3. *Thrust ball bearings* whose washer raceways have cross-sectional radii not larger than 54 percent of the ball diameter.

4. *Double row, radial and angular contact ball bearings* and double direction thrust ball bearings are presumed to be symmetrical.

**Limitations for Ball Bearings.** — The following limitations apply:

1. *Truncated contact area.* This standard[1] may not be safely applied to ball bearings subjected to loading which causes the contact area of the ball with the raceway to be truncated by the raceway shoulder. This limitation depends strongly on details of bearing design which are not standardized.

2. *Material.* This standard applies only to ball bearings fabricated from hardened good quality steel.

3. *Types.* The $f_c$ factors specified in the basic load rating formulas are valid only for those ball bearing types specified above.

4. *Lubrication.* The Rating Life calculated according to this standard is based on the assumption that the bearing is adequately lubricated. The determination of adequate lubrication depends upon the bearing application.

5. *Ring support and alignment.* The Rating Life calculated according to this standard assumes that the bearing inner and outer rings are rigidly supported and the inner and outer ring axes are properly aligned.

6. *Internal clearance.* The radial ball bearing Rating Life calculated according to this standard is based on the assumption that only a nominal interior clearance occurs in the mounted bearing at operating speed, load and temperature.

7. *High speed effects.* The Rating Life calculated according to this standard does not account for high speed effects such as ball centrifugal forces and gyroscopic moments. These effects tend to diminish fatigue life. Analytical evaluation of these effects frequently requires the use of high speed digital computation devices and hence is not covered in the standard.

8. *Groove radii.* If groove radii are smaller than those specified in the bearing types covered, the ability of a bearing to resist fatigue is not improved; however, it is diminished by the use of larger radii.

**Calculation of Ball Bearing Rating Life.** — According to the Anti-Friction Bearing Manufacturers Association standards the Rating Life, $L_{10}$ of a group of apparently identical ball bearings is the life in millions of revolutions that 90 percent of the group will complete or exceed. For a single bearing, $L_{10}$ also refers to the life associated with 90 percent reliability.

*Radial and Angular Contact Ball Bearings:* The magnitude of the Rating Life, $L_{10}$, in millions of revolutions, for a radial or angular contact ball bearing application is given by the formula:

$$L_{10} = \left(\frac{C}{P}\right)^3 \tag{1}$$

where $C$ = basic load rating, pounds. See formulas (2) and (3).
$P$ = equivalent radial load, pounds. See formula (4).

For radial and angular contact ball bearings with balls not larger than 1 inch in diameter, $C$ is found by the formula:

$$C = f_c\,(i\cos\alpha)^{0.7}\,Z^{2/3}\,D^{1.8} \tag{2}$$

and with balls larger than 1 inch in diameter $C$ is found by the formula:

$$C = f_c\,(i\cos\alpha)^{0.7}\,Z^{2/3}\,D^{1.4} \tag{3}$$

[1] All references to "standard" are to AFBMA and American National Standard "Load Ratings and Fatigue Life for Ball Bearings" (ANSI B 3.15–1972).

where $f_c$ = a factor which depends on the geometry of the bearing components,
the accuracy to which the various bearings parts are made and the
material. Values of $f_c$ are given in Table 25.

$i$ = number of rows of balls in the bearing

$\alpha$ = nominal contact angle, degrees

$Z$ = number of balls per row in a radial or angular contact ball bearing

$D$ = ball diameter, inches.

The magnitude of the equivalent radial load, $P$, in pounds, for radial and angular
contact ball bearings, under combined constant radial and constant thrust loads is
given by the formula:

$$P = XF_r + YF_a \qquad (4)$$

where $F_r$ = the applied radial load in pounds

$F_a$ = the applied axial load in pounds

$X$ = radial load factor as given in Table 26.

$Y$ = axial load factor as given in Table 26.

Table 25. **Values of $f_c$ for Radial and Angular Contact Ball Bearings**

| $\dfrac{D \cos \alpha}{d_m}$ | Single Row Radial Contact; Single and Double Row Angular Contact, Groove Type* | Double Row Radial Contact, Groove Type | Self-Aligning |
|---|---|---|---|
| | Values of $f_c$ | | |
| 0.05 | 3550 | 3360 | 1310 |
| 0.06 | 3730 | 3530 | 1420 |
| 0.07 | 3880 | 3680 | 1510 |
| 0.08 | 4020 | 3810 | 1600 |
| 0.09 | 4130 | 3900 | 1690 |
| 0.10 | 4220 | 4000 | 1770 |
| 0.12 | 4370 | 4140 | 1940 |
| 0.14 | 4470 | 4230 | 2100 |
| 0.16 | 4530 | 4290 | 2260 |
| 0.18 | 4550 | 4310 | 2410 |
| 0.20 | 4550 | 4310 | 2550 |
| 0.22 | 4530 | 4290 | 2680 |
| 0.24 | 4480 | 4250 | 2790 |
| 0.26 | 4420 | 4190 | 2910 |
| 0.28 | 4340 | 4110 | 3000 |
| 0.30 | 4250 | 4030 | 3060 |
| 0.32 | 4160 | 3950 | 3110 |
| 0.34 | 4050 | 3840 | 3130 |
| 0.36 | 3930 | 3730 | 3140 |
| 0.38 | 3800 | 3610 | 3110 |
| 0.40 | 3670 | 3480 | 3070 |

* a. When calculating the basic load rating for a unit consisting of two similar, single
row, radial contact ball bearings, in a duplex mounting, the pair is considered as one
double row, radial contact ball bearing.

b. When calculating the basic load rating for a unit consisting of two, similar, single
row, angular contact ball bearings in a duplex mounting, "Face-to-Face" or "Back-to-
Back", the pair is considered as one, double row, angular contact ball bearing.

c. When calculating the basic load rating for a unit consisting of two or more similar,
single row, angular contact ball bearings mounted "in Tandem", properly manufactured
and mounted for equal load distribution, the rating of the combination is the number of
bearings to the 0.7 power times the rating of a single row ball bearing. If the unit may be
treated as a number of individually interchangeable single row bearings, this footnote
does not apply.

# 650 BALL AND ROLLER BEARINGS

Table 26. Values of $X$ and $Y$ for Computing Equivalent Radial Load $P$ of Radial and Angular Contact Ball Bearings

| Contact Angle, $\alpha$ | Table Entering Factors | | Single Row[1] Bearings $\frac{F_a}{F_r} > e$ | | Double Row Bearings $\frac{F_a}{F_r} \leqq e$ | | $\frac{F_a}{F_r} > e$ | |
|---|---|---|---|---|---|---|---|---|
| | $F_a/C_o$* | $e$ | $X$ | $Y$ | $X$ | $Y$ | $X$ | $Y$ |
| RADIAL CONTACT GROOVE BEARINGS | | | | | | | | |
| 0° | 0.014 | 0.19 | | 2.30 | | | | 2.30 |
| | 0.028 | 0.22 | | 1.99 | | | | 1.99 |
| | 0.056 | 0.26 | | 1.71 | | | | 1.71 |
| | 0.084 | 0.28 | 0.56 | 1.55 | 1 | 0 | 0.56 | 1.55 |
| | 0.11 | 0.30 | | 1.45 | | | | 1.45 |
| | 0.17 | 0.34 | | 1.31 | | | | 1.31 |
| | 0.28 | 0.38 | | 1.15 | | | | 1.15 |
| | 0.42 | 0.42 | | 1.04 | | | | 1.04 |
| | 0.56 | 0.44 | | 1.00 | | | | 1.00 |
| ANGULAR CONTACT GROOVE BEARINGS | | | | | | | | |
| | $iF_a/C_o$* | $e$ | $X$ | $Y$ | $X$ | $Y$ | $X$ | $Y$ |
| 5° | 0.014 | 0.23 | | | | 2.78 | | 3.74 |
| | 0.028 | 0.26 | For this type use | | | 2.40 | | 3.23 |
| | 0.056 | 0.30 | the $X$, $Y$ and $e$ | | | 2.07 | | 2.78 |
| | 0.085 | 0.34 | values applicable | 1 | | 1.87 | 0.78 | 2.52 |
| | 0.11 | 0.36 | to single row | | | 1.75 | | 2.36 |
| | 0.17 | 0.40 | radial contact | | | 1.58 | | 2.13 |
| | 0.28 | 0.45 | bearings | | | 1.39 | | 1.87 |
| | 0.42 | 0.50 | | | | 1.26 | | 1.69 |
| | 0.56 | 0.52 | | | | 1.21 | | 1.63 |
| 10° | 0.014 | 0.29 | | 1.88 | | 2.18 | | 3.06 |
| | 0.029 | 0.32 | | 1.71 | | 1.98 | | 2.78 |
| | 0.057 | 0.36 | | 1.52 | | 1.76 | | 2.47 |
| | 0.080 | 0.38 | | 1.41 | | 1.63 | | 2.29 |
| | 0.11 | 0.40 | 0.46 | 1.34 | 1 | 1.55 | 0.75 | 2.18 |
| | 0.17 | 0.44 | | 1.23 | | 1.42 | | 2.00 |
| | 0.29 | 0.49 | | 1.10 | | 1.27 | | 1.79 |
| | 0.43 | 0.54 | | 1.01 | | 1.17 | | 1.64 |
| | 0.57 | 0.54 | | 1.00 | | 1.16 | | 1.63 |
| 15° | 0.015 | 0.38 | | 1.47 | | 1.65 | | 2.39 |
| | 0.029 | 0.40 | | 1.40 | | 1.57 | | 2.28 |
| | 0.058 | 0.43 | | 1.30 | | 1.46 | | 2.11 |
| | 0.087 | 0.46 | | 1.23 | | 1.38 | | 2.00 |
| | 0.12 | 0.47 | 0.44 | 1.19 | 1 | 1.34 | 0.72 | 1.93 |
| | 0.17 | 0.50 | | 1.12 | | 1.26 | | 1.82 |
| | 0.29 | 0.55 | | 1.02 | | 1.14 | | 1.66 |
| | 0.44 | 0.56 | | 1.00 | | 1.12 | | 1.63 |
| | 0.58 | 0.56 | | 1.00 | | 1.12 | | 1.63 |
| 20° | ... | 0.57 | 0.43 | 1.00 | 1 | 1.09 | 0.70 | 1.63 |
| 25° | ... | 0.68 | 0.41 | 0.87 | 1 | 0.92 | 0.67 | 1.44 |
| 30° | ... | 0.80 | 0.39 | 0.76 | 1 | 0.78 | 0.63 | 1.24 |
| 35° | ... | 0.95 | 0.37 | 0.66 | 1 | 0.66 | 0.60 | 1.07 |
| 40° | ... | 1.14 | 0.35 | 0.57 | 1 | 0.55 | 0.57 | 0.93 |
| Self-aligning Ball Bearings | 1.5 tan $\alpha$ | 0.40 | 0.4 cot $\alpha$ | 1 | 0.42 cot $\alpha$ | 0.65 | 0.65 cot $\alpha$ | |

Values of $X$, $Y$, and $e$ for contact angles other than shown may be obtained by linear interpolation.

Values of $X$, $Y$, and $e$ shown in this table do not apply to filling slot bearings for applications in which ball-raceway contact areas project substantially into the filling slot under load.

* $C_o$ is the static load rating in pounds of the bearing under consideration and is found by the formula: $C_o = f_o i Z D^2 \cos \alpha$, in which $f_o = 484$ for self-aligning ball bearings and 1780 for radial and angular contact groove ball bearings; $i$ = number of rows of balls in any one bearing; $D$ = ball diameter in inches; and $\alpha$ = contact angle, i.e., nominal angle between line of action of the ball load and a plane perpendicular to the bearing axis.

[1] For single row bearings, when $F_a/F_r \leqq e$, use $X = 1$ and $Y = 0$. Two similar single row angular contact ball bearings mounted "face-to-face" or "back-to-back" are considered as one double row angular contact bearing.

*Thrust Ball Bearings:* The magnitude of the Rating Life, $L_{10}$ in millions of revolutions for a thrust ball bearing application is given by the formula:

$$L_{10} = \left(\frac{C_a}{P_a}\right)^3 \tag{5}$$

where $C_a$ = the basic load rating, pounds. See formulas (6) to (10).
 $P_a$ = equivalent thrust load, pounds. See formula (11).

For single row, single and double direction, thrust ball bearing with balls not larger than 1 inch in diameter, $C_a$ is found by the formulas:

$$\text{for } \alpha = 90 \text{ degrees,} \quad C_a = f_c Z^{2/3} D^{1.8} \tag{6}$$

$$\text{for } \alpha \neq 90 \text{ degrees,} \quad C_a = f_c (\cos \alpha)^{0.7} Z^{2/3} D^{1.8} \tan \alpha \tag{7}$$

and with balls larger than 1 inch in diameter, $C_a$ is found by the formula:

$$\text{for } \alpha = 90 \text{ degrees,} \quad C_a = f_c Z^{2/3} D^{1.4} \tag{8}$$

$$\text{for } \alpha \neq 90 \text{ degrees,} \quad C_a = f_c (\cos \alpha)^{0.7} Z^{2/3} D^{1.4} \tan \alpha \tag{9}$$

where $f_c$ = a factor which depends on the geometry of the bearing components, the accuracy to which the various bearing parts are made, and the material. Values of $f_c$ are given in Table 27.

 $Z$ = number of balls per row in a single row, single direction thrust ball bearing
 $D$ = ball diameter, inches
 $\alpha$ = nominal contact angle, degrees

**Table 27.   Values of $f_c$ for Thrust Ball Bearings**

| $\dfrac{D}{d_m}$ | $\alpha = 90°$ $f_c$ | $\dfrac{D \cos\alpha}{d_m}$ | $\alpha = 45°$ | $\alpha = 60°$ | $\alpha = 75°$ |
|---|---|---|---|---|---|
| | | | | Values of $f_c$ | |
| 0.01 | 2790 | 0.01 | 3200 | 2970 | 2840 |
| 0.02 | 3430 | 0.02 | 3930 | 3650 | 3490 |
| 0.03 | 3880 | 0.03 | 4430 | 4120 | 3930 |
| 0.04 | 4230 | 0.04 | 4810 | 4470 | 4260 |
| 0.05 | 4520 | 0.05 | 5110 | 4760 | 4540 |
| 0.06 | 4780 | 0.06 | 5360 | 4990 | 4760 |
| 0.07 | 5000 | 0.07 | 5580 | 5190 | 4950 |
| 0.08 | 5210 | 0.08 | 5770 | 5360 | 5120 |
| 0.09 | 5390 | 0.09 | 5920 | 5510 | 5250 |
| 0.10 | 5570 | 0.10 | 6050 | 5630 | 5370 |
| 0.12 | 5880 | 0.12 | 6260 | 5830 | .... |
| 0.14 | 6160 | 0.14 | 6390 | 5950 | .... |
| 0.16 | 6410 | 0.16 | 6470 | 6020 | .... |
| 0.18 | 6640 | 0.18 | 6500 | 6050 | .... |
| 0.20 | 6854 | 0.20 | 6490 | 6040 | .... |
| 0.22 | 7060 | 0.22 | 6450 | .... | .... |
| 0.24 | 7240 | 0.24 | 6380 | .... | .... |
| 0.26 | 7410 | 0.26 | 6290 | .... | .... |
| 0.28 | 7600 | 0.28 | 6180 | .... | .... |
| 0.30 | 7750 | 0.30 | 6040 | .... | .... |
| 0.32 | 7900 | .... | .... | .... | .... |
| 0.34 | 8050 | .... | .... | .... | .... |

For thrust ball bearings with two or more rows of similar balls carrying loads in the same direction, the basic load rating, $C_a$, in pounds, is found by the formula:

$$C_a = (Z_1 + Z_2 + \ldots Z_n) \left[ \left( \frac{Z_1}{C_{a1}} \right)^{10/3} + \left( \frac{Z_2}{C_{a2}} \right)^{10/3} + \ldots \left( \frac{Z_n}{C_{an}} \right)^{10/3} \right]^{-0.3} \quad (10)$$

where $Z_1, Z_2 \ldots Z_n$ = number of balls in respective rows of a single-direction multi-row thrust ball bearing.

$C_{a1}, C_{a2} \ldots C_{an}$ = basic load rating per row of a single-direction, multi-row thrust ball bearing, each calculated as a single-row bearing with $Z_1, Z_2 \ldots Z_n$ balls, respectively.

The magnitude of the equivalent thrust load, $P_a$, in pounds for thrust ball bearings with $\alpha \neq 90$ degrees under combined constant thrust and constant radial loads is found by the formula:

$$P_a = XF_r + YF_a \quad (11)$$

where $F_r$ = the applied radial load in pounds
$F_a$ = the applied axial load in pounds
$X$ = radial load factor as given in Table 28.
$Y$ = axial load factor as given in Table 28.

Table 28. Values of $X$ and $Y$ for Computing Equivalent Thrust Load $P_a$ for Thrust Ball Bearings

| Contact Angle $\alpha$ | $e$ | Single Direction Bearings | | Double Direction Bearings | | | |
|---|---|---|---|---|---|---|---|
| | | $\frac{F_a}{F_r} > e$ | | $\frac{F_a}{F_r} \leq e$ | | $\frac{F_a}{F_r} > e$ | |
| | | $X$ | $Y$ | $X$ | $Y$ | $X$ | $Y$ |
| 45° | 1.25 | 0.66 | 1 | 1.18 | 0.59 | 0.66 | 1 |
| 60° | 2.17 | 0.92 | 1 | 1.90 | 0.54 | 0.92 | 1 |
| 75° | 4.67 | 1.66 | 1 | 3.89 | 0.52 | 1.66 | 1 |

For $\alpha = 90°$, $F_r = 0$ and $Y = 1$.

**Roller Bearing Types Covered.** — This standard[1] applies to *cylindrical, tapered and self-aligning radial and thrust roller bearings* and to *needle roller bearings*. These bearings are presumed to be within the size ranges shown in the AFBMA dimensional standards, of good quality and produced in accordance with good manufacturing practice.

Roller bearings vary considerably in design and execution. Since small differences in relative shape of contacting surfaces may account for distinct differences in load carrying ability, this standard does not attempt to cover all design variations, rather it applies to basic roller bearing designs.

**Limitations for Roller Bearings.** — The following limitations apply:

1. *Truncated contact area.* This standard may not be safely applied to roller bearings subjected to application conditions which cause the contact area of the roller with the raceway to be severely truncated by the edge of the raceway or roller.

---

[1] All references to "standard" are to AFBMA and American National Standard "Load Ratings and Fatigue Life for Roller Bearings" (ANSI B3.16-1972).

2. *Stress concentrations.* A cylindrical, tapered or self-aligning roller bearing must be expected to have a basic load rating less than that obtained using a value of $f_c$ taken from Table 29 or Table 31 if, under load a stress concentration is present in some part of the roller-raceway contact. Such stress concentrations occur in the center of nominal point contacts, at the contact extremities for line contacts and at inadequately blended junctions of a rolling surface profile. Stress concentrations can also occur if the rollers are not accurately guided such as in bearings without cages and bearings not having rigid integral flanges. Values of $f_c$ given in Tables 29 and 31 are based upon bearings manufactured to achieve optimized contact. For no bearing type or execution will the factor $f_c$ be greater than that obtained in Tables 29 and 31.

3. *Material.* This standard applies only to roller bearings fabricated from hardened, good quality steel.

4. *Lubrication.* Rating Life calculated according to this standard is based on the assumption that the bearing is adequately lubricated. Determination of adequate lubrication depends upon the bearing application.

5. *Ring support and alignment.* Rating Life calculated according to this standard assumes that the bearing inner and outer rings are rigidly supported, and that the inner and outer ring axes are properly aligned.

6. *Internal clearance.* Radial roller bearing Rating Life calculated according to this standard is based on the assumption that only a nominal internal clearance occurs in the mounted bearing at operating speed, load, and temperature.

7. *High speed effects.* The Rating Life calculated according to this standard does not account for high speed effects such as roller centrifugal forces and gyroscopic moments: These effects tend to diminish fatigue life. Analytical evaluation of these effects frequently requires the use of high speed digital computation devices and hence, cannot be included.

**Calculation of Radial Roller Bearing Rating Life.** — The Rating Life, $L_{10}$ of a group of apparently identical roller bearings is the life in millions of revolutions that 90 percent of the group will complete or exceed. For a single bearing, $L_{10}$ also refers to the life associated with 90 percent reliability.

*Radial Roller Bearings*: The magnitude of the Rating Life, $L_{10}$, in millions of revolutions, for a radial roller bearing application is given by the formula:

$$L_{10} = \left(\frac{C}{P}\right)^{10/3} \qquad (12)$$

where $C$ = the basic load rating in pounds. See formula (13).
      $P$ = equivalent radial load pounds. See formula (14).

For radial roller bearings, $C$ is found by the formula:

$$C = f_c(il_{eff} \cos \alpha)^{7/9} Z^{3/4} D^{29/27} \qquad (13)$$

where $f_c$ = a factor which depends on the geometry of the bearing components, the accuracy to which the various bearing parts are made and the material. Values of $f_c$ are given in Table 29.
      $i$ = number of rows of rollers in the bearing.
      $l_{eff}$ = effective length, inches.
      $\alpha$ = nominal contact angle, degrees.
      $Z$ = number of rollers per row in a radial roller bearing.
      $D$ = roller diameter, inches (mean diameter for a tapered roller, major diameter for a spherical roller).

Table 29.    Values of $f_c$ for Radial Roller Bearings

| $\dfrac{D \cos \alpha}{d_m}$ | $f_c$ | $\dfrac{D \cos \alpha}{d_m}$ | $f_c$ | $\dfrac{D \cos \alpha}{d_m}$ | $f_c$ |
|---|---|---|---|---|---|
| 0.01 | 4700 | 0.11 | 7620 | 0.21 | 7940 |
| 0.02 | 5490 | 0.12 | 7720 | 0.22 | 7920 |
| 0.03 | 5940 | 0.13 | 7800 | 0.23 | 7900 |
| 0.04 | 6340 | 0.14 | 7870 | 0.24 | 7870 |
| 0.05 | 6630 | 0.15 | 7920 | 0.25 | 7800 |
| 0.06 | 6880 | 0.16 | 7970 | 0.26 | 7720 |
| 0.07 | 7130 | 0.17 | 7970 | 0.27 | 7670 |
| 0.08 | 7280 | 0.18 | 7970 | 0.28 | 7620 |
| 0.09 | 7430 | 0.19 | 7970 | 0.29 | 7570 |
| 0.10 | 7520 | 0.20 | 7970 | 0.30 | 7520 |

When rollers are longer than $2.5D$, a reduction in the $f_c$ value must be anticipated. In this case, the bearing manufacturer may be expected to establish load ratings accordingly.

In applications where rollers operate directly on a shaft surface or a housing surface, such a surface must be equivalent in all respects to the raceway it replaces to achieve the basic load rating of the bearing.

When calculating the basic load rating for a unit consisting of two or more similar single-row bearings mounted "in tandem," properly manufactured and mounted for equal load distribution, the rating of the combination is the number of bearings to the 7/9 power times the rating of a single-row bearing. If, for some technical reason, the unit may be treated as a number of individually interchangeable single-row bearings, this consideration does not apply.

The magnitude of the equivalent radial load, $P$, in pounds, for radial roller bearings, under combined constant radial and constant thrust loads is given by the formula:

$$P = XF_r + YF_a \qquad (14)$$

where   $F_r$ = the applied radial load in pounds.  
       $F_a$ = the applied axial load in pounds.  
       $X$ = radial load factor as given in Table 30.  
       $Y$ = axial load factor as given in Table 30.

Table 30.    Values of $X$ and $Y$ for Computing Equivalent  
Load $P_r$ for Self Aligning and Tapered Roller Bearing

| Bearing Type | $\dfrac{F_a}{F_r} \leqq e$ [1] | | $\dfrac{F_a}{F_r} > e$ [1] | |
|---|---|---|---|---|
| | $X$ | $Y$ | $X$ | $Y$ |
| Self Aligning and Tapered Roller Bearings [2] $\alpha \neq 0°$ | Single Row Bearings | | | |
| | 1 | 0 | 0.4 | $0.4 \cot \alpha$ |
| | Double Row Bearings [2] | | | |
| | 1 | $0.45 \cot \alpha$ | 0.67 | $0.67 \cot \alpha$ |

[1] $e = 1.5 \tan \alpha$.  
[2] For $\alpha = 0°$, $F_a = 0$ and $X = 1$.

*Thrust Roller Bearings:* The magnitude of the Rating Life, $L_{10}$, in millions of revolutions for a thrust roller bearing application is given by the formula:

$$L_{10} = \left(\frac{C_a}{P_a}\right)^{10/3} \tag{15}$$

where $C_a$ = basic load rating, pounds.  See formulas (16) to (18).

$P_a$ = static equivalent thrust load, pounds.  See formula (19).

Roller bearings are generally designed to achieve optimized contact; however, they usually support loads other than the loading at which optimized contact is maintained.  The 10/3 exponent in the Rating Life formula was selected to yield satisfactory Rating Life estimates for a broad spectrum from light to heavy loading.  When loading exceeds that which develops optimized contact; e.g., loading greater than $C_a/4$ to $C_a/2$, the user should consult the bearing manufacturer to establish the adequacy of the above Rating Life formula for the particular application.

For single row, single and double direction, thrust roller bearings, the magnitude of the basic load rating, $C_a$, in pounds, is found by the formula:

$$\text{for } \alpha = 90°, \quad C_a = f_c \, l_{eff}^{7/9} Z^{3/4} D^{29/27} \tag{16}$$

$$\text{for } \alpha \neq 90°, \quad C_a = f_c (l_{eff} \cos \alpha)^{7/9} Z^{3/4} D^{29/27} \tan \alpha \tag{17}$$

where  $f_c$ = a factor which depends on the geometry of the bearing components, the accuracy to which the various parts are made, and the material.  Values of $f_c$ are given in Table 31.

$l_{eff}$ = effective length, inches

$Z$ = number of rollers in a single row, single direction, thrust roller bearing

$D$ = roller diameter, inches (mean diameter for a tapered roller, major diameter for a spherical roller)

$\alpha$ = nominal contact angle, degrees

**Table 31.  Values of $f_c$ for Thrust Roller Bearings**

| $\dfrac{D \cos \alpha}{d_m}$ | $45° < \alpha \leq 62°$ | $62° < \alpha \leq 85°$ | $\dfrac{D}{d_m}$ | $\alpha = 90°$ |
|---|---|---|---|---|
| | $f_c$ | | | $f_c$ |
| 0.01 | 10000 | 9480 | 0.01 | 9500 |
| 0.02 | 11400 | 11000 | 0.02 | 11000 |
| 0.03 | 12400 | 12000 | 0.03 | 12100 |
| 0.04 | 13300 | 12700 | 0.04 | 12800 |
| 0.05 | 13800 | 13400 | 0.05 | 13200 |
| 0.06 | 14400 | 13900 | 0.06 | 14100 |
| 0.07 | 14800 | 14300 | 0.07 | 14500 |
| 0.08 | 15200 | 14700 | 0.08 | 15100 |
| 0.09 | 15500 | 14900 | 0.09 | 15400 |
| 0.10 | 15900 | 15200 | 0.10 | 15900 |
| 0.12 | 16100 | 15500 | 0.12 | 16300 |
| 0.14 | 16300 | 15700 | 0.14 | 17000 |
| 0.16 | 16400 | 15800 | 0.16 | 17500 |
| 0.18 | 16400 | 15900 | 0.18 | 18000 |
| 0.20 | 16400 | 15800 | 0.20 | 18500 |
| 0.22 | 16300 | 15700 | 0.22 | 18800 |
| 0.24 | 16200 | 15600 | 0.24 | 19100 |
| 0.26 | 15900 | 15400 | 0.26 | 19600 |
| 0.28 | 15800 | 15200 | 0.28 | 19900 |
| 0.30 | 15600 | 15000 | 0.30 | 20100 |

For thrust roller bearings with two or more rows of rollers carrying loads in the same direction the magnitude of $C_a$ is found by the formula:

$$C_a = (Z_1 l_{eff1} + Z_2 l_{eff2} \ldots Z_n l_{effn}) \left\{ \left[ \frac{Z_1 l_{eff1}}{C_{a1}} \right]^{9/2} + \left[ \frac{Z_2 l_{eff2}}{C_{a2}} \right]^{9/2} + \ldots \right.$$

$$\left. \left[ \frac{Z_n l_{effn}}{C_{an}} \right]^{9/2} \right\}^{-2/9} \tag{18}$$

Where $Z_1, Z_2 \ldots Z_n$ = the number of rollers in respective rows of a single direction, multi-row bearing

$C_{a1}, C_{a2} \ldots C_{an}$ = the basic load rating per row of a single direction, multi-row, thrust roller bearing, each calculated as a single row bearing with $Z_1, Z_2 \ldots Z_n$ rollers respectively.

$l_{eff1}, l_{eff2} \ldots l_{effn}$ = effective length, inches, of rollers in the respective rows.

In applications where rollers operate directly on a surface supplied by the user, such a surface must be equivalent in all respects to the washer raceway it replaces to achieve the basic load rating of the bearing.

In case the bearing is so designed that several rollers are located on a common axis, these rollers are considered as one roller of a length equal to the total effective length of contact of the several rollers. Rollers as defined above, or portions thereof which contact the same washer-raceway area, belong to one row.

When the ratio of the individual roller effective length to the pitch diameter (at which this roller operates) is too large, a reduction in the $f_c$ value must be anticipated due to excessive slip in the roller-raceway contact.

When calculating the basic load rating for a unit consisting of two or more similar single row bearings mounted "in tandem," properly manufactured and mounted for equal load distribution, the rating of the combination is defined by formula (18). If, for some technical reason, the unit may be treated as a unit of individually interchangeable single-row bearings, this consideration does not apply.

The magnitude of the equivalent thrust load, $P_a$, in pounds, for thrust roller bearings with $\alpha$ not equal to 90 degrees under combined constant thrust and constant

Table 32.   Values of $X$ and $Y$ for Computing Equivalent
Thrust Load $P_a$ for Thrust Roller Bearings

| Bearing Type | Single Direction Bearings | | Double Direction Bearings | | | |
|---|---|---|---|---|---|---|
| | $\dfrac{F_a}{F_r} > e^1$ | | $\dfrac{F_a}{F_r} \leqq e^1$ | | $\dfrac{F_a}{F_r} > e^1$ | |
| | $X$ | $Y$ | $X$ | $Y$ | $X$ | $Y$ |
| Self Aligning Tapered Thrust Roller Bearings[2] $\alpha \neq 0$ | $\tan \alpha$ | $1$ | $1.5 \tan \alpha$ | $0.67$ | $\tan \alpha$ | $1$ |

[1] $e = 1.5 \tan \alpha$.
[2] For $\alpha = 90°$, $F_r = 0$ and $Y = 1$.

radial loads is given by the formula:

$$P_a = XF_r + YF_a \qquad (19)$$

where $F_r$ = applied radial load, pounds
$F_a$ = applied axial load, pounds
$X$ = radial load factor as given in Table 32.
$Y$ = axial load factor as given in Table 32.

**Life Adjustment Factors.** — In certain applications of ball or roller bearings it is desirable to specify life for a reliability other than 90 percent. In other cases the bearings may be fabricated from special materials such as a consumable vacuum remelted steel or other steels of exceptionally high quality. Finally, application conditions may indicate other than normal lubrication, load distribution, or temperature. For such conditions a series of life adjustment factors may be applied to the fatigue life formula. This is fully explained in AFBMA and American National Standard "Load Ratings and Fatigue Life for Ball Bearings" (ANSI B3.15-1972) and AFBMA and American National Standard "Load Ratings and Fatigue Life for Roller Bearings" (ANSI B3.16-1972). In addition to consulting these standards it may be advantageous to also obtain information from the bearing manufacturer.

**Life Adjustment Factor for Reliability.** — For certain applications, it is desirable to specify life for a reliability greater than 90 per cent which is the basis of the Rating Life.

To determine the bearing life of ball or roller bearings for reliability greater than 90 per cent, the Rating Life must be adjusted by a factor $a_1$ such that $L_n = a_1 L_{10}$. For a reliability of 95 per cent, designated as $L_5$, the life adjustment factor $a_1$ is 0.62; for 96 per cent, $L_4$, $a_1$ is 0.53; for 97 per cent, $L_3$, $a_1$ is 0.44; for 98 per cent, $L_2$, $a_1$ is 0.33; and for 99 per cent, $L_1$, $a_1$ is 0.21.

**Life Adjustment Factor for Material.** — For certain types of ball or roller bearings which incorporate improved materials and processing, the Rating Life can be adjusted by a factor $a_2$ such that $L_{10}' = a_2 L_{10}$. Factor $a_2$ depends upon steel analysis, metallurgical processes, forming methods, heat treatment, and manufacturing methods in general. Ball and roller bearings fabricated from consumable vacuum remelted steels and certain other special analysis steels, have demonstrated extraordinarily long endurance. These steels are of exceptionally high quality, and bearings fabricated from these are usually considered special manufacture. Generally, $a_2$ values for such steels can be obtained from the bearing manufacturer. However, all of the specified limitations and qualifications for the application of the Rating Life formulas still apply.

**Life Adjustment Factor for Application Condition.** — Application conditions which affect ball or roller bearing life include: 1. lubrication; 2. load distribution (including effects of clearance, misalignment, housing and shaft stiffness, type of loading, and thermal gradients); and 3. temperature. Items 2 and 3 require special analytical and experimental techniques, therefore the user should consult the bearing manufacturer for evaluations and recommendations.

Operating conditions where the factor $a_3$ might be less than 1 include: (a) exceptionally low values of $Nd_m$ (rpm times pitch diameter, in mm); e.g., $Nd_m <$ 10,000; (b) lubricant viscosity at less than 70 SSU for ball bearings and 100 SSU for roller bearings at operating temperature; (c) excessively high operating temperatures. When $a_3$ is less than 1 it may not be assumed that the deficiency in lubrication can be overcome by using an improved steel. When this factor is applied, $L_{10}' = a_3 L_{10}$.

**Calculation of Ball Bearing Static Load Rating.** — The "static load rating" for a radial or angular contact ball bearing is that static radial load which corresponds to a total permanent deformation of ball and raceway at the most heavily stressed contact of 0.0001 of the ball diameter. For a thrust ball bearing, it is that static, centric, thrust load which corresponds to a total permanent deformation of ball and raceway at the most heavily stressed contact of 0.0001 of the ball diameter. Methods for calculation specified in this standard[1] are valid for ball bearings as commonly designed and manufactured of hardened steel.

*Radial and Angular Contact Ball Bearings:* The magnitude of the static load rating, $C_o$ in pounds, for radial and angular contact ball bearings is found by the formula:

$$C_o = f_o \, i \, Z D^2 \cos \alpha \qquad (20)$$

where $f_o$ = a factor which depends on the bearing type, used to determine static load rating. The value of $f_o$ is 484 for self-aligning ball bearings and 1780 for radial and angular contact groove ball bearings.

$i$ = number of rows of balls in bearing

$Z$ = number of balls per row

$D$ = ball diameter, inches

$\alpha$ = nominal contact angle, degrees.

### Table 33. Values of $X_o$ and $Y_o$ for Computing Static Equivalent Load $P_o$ of Ball Bearings

| Contact Angle | Single Row Bearings | | Double Row Bearings | |
|---|---|---|---|---|
| | $X_o$ | $Y_o$ | $X_o$ | $Y_o$ |
| RADIAL CONTACT GROOVE BEARINGS[1,2] | | | | |
| $\alpha = 0°$ | 0.6 | 0.5 | 0.6 | 0.5 |
| ANGULAR CONTACT GROOVE BEARINGS[3] | | | | |
| $\alpha = 20°$ | 0.5 | 0.42 | 1 | 0.84 |
| $\alpha = 25°$ | 0.5 | 0.38 | 1 | 0.76 |
| $\alpha = 30°$ | 0.5 | 0.33 | 1 | 0.66 |
| $\alpha = 35°$ | 0.5 | 0.29 | 1 | 0.58 |
| $\alpha = 40°$ | 0.5 | 0.26 | 1 | 0.52 |
| SELF-ALIGNING BEARINGS | | | | |
| . . . . | 0.5 | 0.22 cot $\alpha$ | 1 | 0.44 cot $\alpha$ |

[1] Permissible maximum value of $F_a/C_o$ depends on the bearing design (groove depth and internal clearance).

[2] $P_o$ is always $\geqq F_r$.

[3] For two similar single row angular contact ball bearings mounted "face-to-face" or "back-to-back" use the values of $X_o$ and $Y_o$ which apply to a double row angular contact ball bearing. For two or more similar, single row angular contact ball bearings mounted "in tandem" use the values of $X_o$ and $Y_o$ which apply to a single row angular contact bearing.

[1] AFBMA and American National Standard "Load Ratings and Fatigue Life for Ball Bearings" (ANSI B3.15-1972).

*Thrust Ball Bearings:* The magnitude of the static load rating, $C_{oa}$, in pounds, for thrust ball bearings is found by the formula:

$$C_{oa} = 7100 \, Z \, D^2 \sin \alpha \qquad (21)$$

where  $Z$ = the number of balls in a single row, single direction thrust ball bearing.
$\quad\quad\;\; D$ = ball diameter, inches
$\quad\quad\;\; \alpha$ = nominal contact angle, degrees.

**Calculation of Ball Bearing Static Equivalent Load.** — The static equivalent load is that calculated, static, radial load which if applied to a radial or angular contact ball bearing, or that calculated, static, centric, thrust load, which if applied to a thrust ball bearing would cause the same total permanent deformation at the most heavily stressed ball and raceway contact as that which occurs under the the actual condition of loading.

*Radial and Angular Contact Ball Bearings:* The magnitude of the static equivalent load, $P_o$, in pounds, for radial and angular contact ball bearings under combined radial and thrust loads is the greater of:

$$P_o = X_o F_r + Y_o F_a \qquad (22)$$
$$P_o = F_r \qquad (23)$$

where  $F_r$ = applied radial load, pounds
$\quad\quad\;\; F_a$ = applied axial load, pounds
$\quad\quad\;\; X_o$ = radial load factor as given in Table 33.
$\quad\quad\;\; Y_o$ = axial load factor as given in Table 33.

*Thrust Ball Bearings:* The magnitude of the static equivalent load, $P_{oa}$, in pounds, for thrust ball bearings with contact angle $\alpha$ not equal to 90 degrees under combined radial and thrust loads is:

$$P_{oa} = F_a + 2.3 F_r \tan \alpha \qquad (24)$$

where  $F_a$ = applied axial load, pounds
$\quad\quad\;\; F_r$ = applied radial load, pounds
$\quad\quad\;\; \alpha$ = nominal contact angle, degrees.

This formula is valid for $F_r \leqq 0.44 F_a \cot \alpha$

**Calculation of Roller Bearing Static Load Rating.** — The static load rating for a radial roller bearing is that static radial load which corresponds to a total permanent deformation of roller and raceway at the most heavily stressed contact of 0.0001 of the roller diameter. For a thrust roller bearing, it is that static, centric, thrust load which corresponds to a total permanent deformation of roller and raceway at the most heavily stressed contact of 0.0001 of the roller diameter. Experience indicates that if optimized contact is achieved in a given application, static load ratings as calculated by this standard are conservative.

*Radial Roller Bearings:* The magnitude of the static load rating, $C_o$, in pounds, for radial roller bearings is found by the formula:

$$C_o = 3130 \, i \, Z \, l_{eff} \, D \cos \alpha \qquad (25)$$

where  $i$ = numbers of rows of rollers in the bearing
$\quad\quad\;\; Z$ = number of rollers per row
$\quad\quad\;\; l_{eff}$ = effective length, inches
$\quad\quad\;\; D$ = roller diameter, inches (mean diameter for a tapered roller, major diameter for a spherical roller)
$\quad\quad\;\; \alpha$ = nominal contact angle, degrees

*Thrust Roller Bearings:* The magnitude of the static load rating, $C_{oa}$, in pounds, for thrust roller bearings is found by the formula:

$$C_{oa} = 14,220 \, Z \, l_{eff} \, D \sin \alpha \qquad (26)$$

where  $Z$ = number of rollers in a single row, single direction, thrust roller bearing
$l_{eff}$ = effective length, inches
$D$ = roller diameter, inches (mean diameter for a tapered roller, major diameter for a spherical roller)
$\alpha$ = nominal contact angle, degrees

In case rollers of different lengths are used to carry load in one direction, $Z \, l_{eff}$ is taken as the total effective length of all the rollers over which the load is distributed.

**Calculation of Roller Bearing Static Equivalent Load.** — The static equivalent load is that calculated, static, radial load which if applied to a radial roller bearing, or that calculated, static, centric, thrust load, which if applied to a thrust roller bearing would cause the same total permanent deformation at the most heavily stressed roller and raceway contact as that which occurs under the actual condition of loading.

*Radial Roller Bearings:* The magnitude of the static equivalent load, $P_o$, in pounds, for radial roller bearings is the greater of

$$P_o = X_o F_r + Y_o F_a \qquad (27)$$

$$P_o = F_r \qquad (28)$$

where  $F_r$ = applied radial load, pounds
$F_a$ = applied axial load, pounds
$X_o$ = radial load factor given in Table 34.
$Y_o$ = axial load factor given in Table 34.

*Thrust Roller Bearings:* The magnitude of the static equivalent load, $P_{oa}$, in pounds, for thrust bearings with contact angle $\alpha$ not equal to 90 degrees, under combined radial and thrust loads is found by the formula:

$$P_{oa} = F_a + 2.3 \, F_r \tan \alpha \qquad (29)$$

where  $F_a$ = applied axial load, pounds
$F_r$ = applied radial load, pounds
$\alpha$ = nominal contact angle, degrees

The accuracy of this formula decreases in the case of single direction bearings when $F_r > 0.44 \, F_a \cot \alpha$

Table 34. **Values of $X_o$ and $Y_o$ for Computing Static Equivalent Load $P_o$ for Self Aligning and Tapered Roller Bearings**

| Bearing Type | Single Row [1] | | Double Row | |
|---|---|---|---|---|
| | $X_o$ | $Y_o$ | $X_o$ | $Y_o$ |
| Self Aligning and Tapered $\alpha \neq 0$ | 0.5 | 0.22 cot $\alpha$ | 1 | 0.44 cot $\alpha$ |

[1] $P_o$ is always $\geqq F_r$.

Table 1. AFBMA Standard Balls for Bearings and Other Purposes

| Grade† | Type of Material* | | | | | | |
|---|---|---|---|---|---|---|---|
| | A | B | C | D†† | E | F | G |
| | Available Size Range (Diameters) | | | | | | |
| 3 | $\frac{1}{32}$ to 1 | ... | ... | ... | ... | ... | ... |
| 5 | $\frac{1}{32}$ to $1\frac{1}{2}$ | $\frac{1}{32}$ to $\frac{3}{4}$ | ... | ... | ... | ... | ... |
| 10 | $\frac{1}{32}$ to $1\frac{1}{2}$ | $\frac{1}{32}$ to $\frac{3}{4}$ | ... | ... | ... | ... | ... |
| 15 | $1\frac{7}{32}$ to $1\frac{1}{2}$ | $\frac{1}{32}$ to $\frac{3}{4}$ | ... | ... | ... | ... | ... |
| 25 | $\frac{1}{32}$ to $1\frac{1}{2}$ | $\frac{1}{32}$ to 1 | ... | ... | ... | ... | ... |
| 50 | $\frac{1}{32}$ to $2\frac{7}{8}$ | $\frac{1}{32}$ to 2 | ... | ... | ... | ... | ... |
| 100 | 3 to $4\frac{1}{2}$ | $\frac{1}{32}$ to $4\frac{1}{2}$ | $\frac{1}{16}$ to $\frac{3}{4}$ | $\frac{1}{16}$ to 1 | ... | ... | ... |
| 200 | $\frac{1}{32}$ to $2\frac{7}{8}$ | $\frac{1}{32}$ to $4\frac{1}{2}$ | $\frac{1}{16}$ to $\frac{3}{4}$ | $\frac{1}{16}$ to 1 | $\frac{1}{4}$ to $1\frac{1}{8}$ | $\frac{1}{16}$ to $\frac{3}{4}$ | $\frac{1}{16}$ to $\frac{3}{4}$ |
| 300 | 3 to $4\frac{1}{2}$ | ... | $\frac{1}{16}$ to $\frac{3}{4}$ | $\frac{1}{16}$ to 1 | ... | ... | ... |
| 500 | $1\frac{11}{32}$ to $4\frac{1}{2}$ | ... | ... | $\frac{1}{16}$ to 1 | ... | $\frac{1}{16}$ to $\frac{3}{4}$ | $\frac{1}{16}$ to $\frac{3}{4}$ |
| 1000 | $\frac{1}{32}$ to $4\frac{1}{2}$ | ... | ... | $\frac{1}{16}$ to 1 | ... | $\frac{1}{16}$ to $\frac{3}{4}$ | $\frac{1}{16}$ to $\frac{3}{4}$ |
| 2000 | ... | ... | ... | $\frac{1}{16}$ to 1 | ... | ... | ... |
| 3000 | ... | ... | ... | $\frac{1}{16}$ to 1 | ... | ... | ... |

| Grade† | Type of Material* | | | | Grade† | Material | |
|---|---|---|---|---|---|---|---|
| | H | I | J | K | | L | M |
| | Available Size Range (Diameters) | | | | | Size Range | |
| 100 | ... | ... | $\frac{1}{16}$ to $\frac{3}{4}$ | ... | 5 | 1mm to $\frac{1}{2}$ | ... |
| 200 | $1\frac{3}{16}$ to 4 | $\frac{1}{16}$ to $1\frac{1}{16}$ | $\frac{1}{16}$ to $1\frac{1}{16}$ | $\frac{1}{16}$ to 1 | 10 | 1mm to $\frac{3}{4}$ | $\frac{1}{32}$ to $\frac{1}{8}$ |
| 500 | ... | $\frac{1}{16}$ to $1\frac{1}{16}$ | ... | ... | 15 | 1mm to 1 | $\frac{1}{32}$ to $\frac{3}{16}$ |
| | | | | | 25 | 1mm to $1\frac{1}{4}$ | $\frac{1}{32}$ to $\frac{1}{4}$ |

All dimensions in inches except where given in millimeters (mm).

* The materials are as follows (letter designations are not part of AFBMA standard):
A — Chrome alloy steel.    H — Aluminum bronze
B — Corrosion resisting hardened steel.    I — Monel metal
C — Corrosion resisting unhardened steel.    J — K-Monel metal
D — Carbon steel.    K — Aluminum
E — Silicon molybdenum steel.    L — Tungsten carbide
F — Brass.    M — Beryllium copper
G — Bronze.

† For grades and tolerances see Table 3.
†† For minimum case depths see Table 5.

Table 2. Ball Hardness Corrections for Curvatures*

| Hardness Reading, Rockwell C | Ball Diameters, Inch | | | | | | |
|---|---|---|---|---|---|---|---|
| | $\frac{1}{4}$ | $\frac{5}{16}$ | $\frac{3}{8}$ | $\frac{1}{2}$ | $\frac{5}{8}$ | $\frac{3}{4}$ | 1 |
| | Correction — Rockwell C | | | | | | |
| 15 | 13.3 | 10.2 | 8.5 | 6.8 | 5.5 | 4.5 | 3.4 |
| 20 | 12.1 | 9.3 | 7.7 | 6.1 | 4.9 | 4.1 | 3.1 |
| 25 | 11.0 | 8.4 | 7.0 | 5.5 | 4.4 | 3.7 | 2.7 |
| 30 | 9.8 | 7.5 | 6.2 | 4.9 | 3.9 | 3.2 | 2.4 |
| 35 | 8.6 | 6.6 | 5.5 | 4.3 | 3.4 | 2.8 | 2.1 |
| 40 | 7.5 | 5.7 | 4.7 | 3.6 | 2.9 | 2.4 | 1.7 |
| 45 | 6.3 | 4.9 | 4.0 | 3.0 | 2.4 | 1.9 | 1.4 |
| 50 | 5.2 | 4.0 | 3.2 | 2.4 | 1.9 | 1.5 | 1.1 |
| 55 | 4.1 | 3.1 | 2.5 | 1.8 | 1.4 | 1.1 | 0.8 |
| 60 | 2.9 | 2.2 | 1.8 | 1.2 | 0.9 | 0.7 | 0.4 |
| 65 | 1.8 | 1.3 | 1.0 | 0.5 | 0.3 | 0.2 | 0.1 |

* Corrections to be added to Rockwell C readings obtained on spherical surfaces of chrome alloy steel, corrosion resisting hardened and unhardened steel, and carbon steel balls. For other ball sizes and hardness readings, interpolate between correction values shown.

Table 3. **AFBMA Standard Balls — Grades and Tolerances**

| Grade | Diameter Tolerance per Ball | "V" Block Out-of-Round in 120° Angle | Diameter Tolerance per Unit Container | Basic Diameter Tolerance | Marking Increments | Maximum Surface Roughness Micro-inch "AA"* |
|---|---|---|---|---|---|---|
| | Inch | Inch | Inch | Inch | Inch | |
| 3 | .000003 | .000003 | .000005 | ±.00003 | .000003 | .5† |
| 5 | .000005 | .000005 | .00001 | ±.00005 | .000005 | .7† |
| 10 | .000010 | .000010 | .00002 | ±.0001 | .000010 | 1.0† |
| 15 | .000015 | .000015 | .00003 | ±.0001 | .000015 | 1.2† |
| 25 | .000025 | .000025 | .00005 | ±.0001 | .000025 | 1.5† |
| 50 | .00005 | .00005 | .0001 | ±.0002 | .00005 | 3.0 |
| 100 | .0001 | .0001 | .0002 | ±.0005 | .0001 | 5.0 |
| 200 | .0002 | .0002 | .0004 | ±.0010 | .0002 | 8.0 |
| 300 | .0003 | .0003 | .0006 | ±.0015 | .0003 | |
| 500 | .0005 | .0005 | .001 | ±.002 | .0005 | |
| 1000 | .001 | .001 | .002 | ±.005 | | Not Applicable |
| 2000 | .002 | .002 | .004 | ±.005 | Not Applicable | |
| 3000 | .003 | .003 | .006 | ±.005 | | |

\* "AA" — Arithmetical average.
† These grades may carry waviness requirements.

Table 4. **AFBMA Standard Balls — Total Hardness Ranges**

| Material | Method* | Rockwell Value |
|---|---|---|
| Steel — | Measured on: | |
|   Chrome alloy |   parallel flats | 60–67 "C" |
|   Corrosion resisting | | |
|     Hardened, 440C | " " | 58–65 "C" |
|     " 440B | " " | 55–62 "C" |
|     Unhardened | " " | 25–39 "C" |
|   Carbon | ball surface | 60 Minimum "C" |
|   Silicon molybdenum | parallel flats | 52–60 "C" |
| Brass | " " | 75–87 "B" |
| Bronze | " " | 75–98 "B" |
| Aluminum bronze | " " | 15–20 "C" |
| Monel metal | " " | 85–95 "B" |
| K-Monel metal | " " | 27 Minimum "C" |
| Aluminum | " " | 54–72 "B" |
| Tungsten carbide | " " | 90.5–91.5 "A" |
| Beryllium copper | " " | 38 Minimum "C" |

\* Rockwell hardness tests shall be conducted in accordance with ASTM Standard E-18. Other hardness measuring methods may be used if properly converted. Correction values for conversion from Rockwell "C" values obtained on the curved surface to Rockwell "C" on parallel flats are given in Table 2.

Table 5. **AFBMA Standard Carbon Steel Balls — Minimum Case Depths**

| Size | Case Depth | Size | Case Depth | Size | Case Depth |
|---|---|---|---|---|---|
| 1/16" | .015" | 5/16" | .045" | 9/16" | .075" |
| 3/32" | .020" | 11/32" | .045" | 5/8" | .075" |
| 1/8" | .025" | 3/8" | .055" | 11/16" | .075" |
| 5/32" | .025" | 13/32" | .055" | 3/4" | .080" |
| 3/16" | .030" | 7/16" | .065" | 13/16" | .080" |
| 7/32" | .035" | 15/32" | .065" | 7/8" | .080" |
| 1/4" | .045" | 1/2" | .070" | 15/16" | .080" |
| 9/32" | .045" | 17/32" | .070" | 1" | .080" |

Table 6. Number of Ferrous Balls per Pound

| Diam., Inches | Chrome Alloy Steel | Corrosion Resisting Steel | | | Silicon Molybdenum Steel | Carbon Steel |
|---|---|---|---|---|---|---|
| | | Hardened | Unhardened | | | |
| | | | AISI 302 | AISI 316 | | |
| 1/32 | 221,138. | 225,928. | ... | ... | ... | ... |
| 1/16 | 27,642.3 | 28.241.0 | 27,352.3 | 27,162.4 | 28,038.6 | 27,545.0 |
| 3/32 | 8,190.32 | 8,367.73 | 8,104.41 | 8,048.13 | 8,307.74 | 8,161.48 |
| 1/8 | 3,455.29 | 3,530.13 | 3,419.04 | 3,395.30 | 3,504.83 | 3,443.12 |
| 5/32 | 1,769.11 | 1,807.43 | 1,750.55 | 1,738.39 | 1,794.47 | 1,762.88 |
| 3/16 | 1,023.79 | 1,045.96 | 1,013.05 | 1,006.01 | 1,038.46 | 1,020.18 |
| 7/32 | 644.719 | 658.684 | 637.956 | 633.526 | 653.962 | 642.449 |
| 1/4 | 431.911 | 441.267 | 427.381 | 424.413 | 438.103 | 430.390 |
| 9/32 | 303.345 | 309.916 | 300.163 | 298.078 | 307.694 | 302.277 |
| 5/16 | 221.138 | 225.928 | 218.819 | 217.299 | 224.309 | 220.360 |
| 11/32 | 166.144 | 169.743 | 164.402 | 163.260 | 168.526 | 165.559 |
| 3/8 | 127.973 | 130.745 | 126.631 | 125.752 | 129.808 | 127.523 |
| 13/32 | 100.654 | 102.835 | 99.5990 | 98.9073 | 102.097 | 100.300 |
| 7/16 | 80.5899 | 82.3355 | 79.7445 | 79.1908 | 81.7453 | 80.3061 |
| 15/32 | 65.5225 | 66.9418 | 64.8352 | 64.3850 | 66.4619 | 65.2918 |
| 1/2 | 53.9889 | 55.1583 | 53.4226 | 53.0516 | 54.7629 | 53.7988 |
| 17/32 | ... | ... | ... | ... | 45.6562 | 44.8524 |
| 9/16 | 37.9181 | 38.7395 | 37.5204 | 37.2598 | 38.4617 | 37.7846 |
| 19/32 | ... | ... | ... | ... | 32.7029 | ... |
| 5/8 | 27.6423 | 28.2410 | 27.3523 | 27.1624 | 28.0386 | 27.5450 |
| 21/32 | ... | ... | ... | ... | 24.2208 | ... |
| 11/16 | 20.7681 | 21.2179 | 20.5502 | 20.4075 | 21.0658 | 20.6949 |
| 23/32 | ... | ... | ... | ... | 18.4358 | ... |
| 3/4 | 15.9967 | 16.3432 | 15.8289 | 15.7190 | 16.2260 | 15.9404 |
| 25/32 | ... | ... | ... | ... | 14.3557 | ... |
| 13/16 | 12.5818 | 12.8543 | ... | ... | 12.7622 | 12.5375 |
| 27/32 | ... | ... | ... | ... | 11.3960 | ... |
| 7/8 | 10.0737 | 10.2919 | ... | ... | 10.2181 | 10.0382 |
| 29/32 | ... | ... | ... | ... | 9.19714 | ... |
| 15/16 | 8.19032 | 8.36773 | ... | ... | 8.30774 | 8.1614 |
| 1 | 6.74861 | 6.89479 | ... | ... | 6.84537 | 6.7248 |
| 1 1/8 | 4.73977 | 4.84243 | ... | ... | 4.80772 | ... |
| 1 1/4 | 3.45529 | 3.53013 | ... | ... | 3.50483 | ... |
| 1 3/8 | 2.59601 | 2.65224 | ... | ... | 2.63323 | ... |
| 1 1/2 | 1.99959 | 2.04290 | ... | ... | 2.02825 | ... |
| 1 5/8 | 1.57273 | 1.60679 | | | | |
| 1 3/4 | 1.25921 | 1.28649 | | | | |
| 1 7/8 | 1.02379 | 1.04596 | | | | |
| 2 | .843577 | .861849 | | | | |
| 2 1/8 | .703296 | .718529 | | | | |
| 2 1/4 | .592471 | .605304 | | | | |
| 2 3/8 | .503760 | .514672 | | | | |
| 2 1/2 | .431911 | .441267 | | | | |
| 2 5/8 | .373101 | .381183 | | | | |
| 2 3/4 | .324501 | .331530 | | | | |
| 2 7/8 | .283988 | .290140 | | | | |
| 3 | .249948 | .255362 | | | | |
| 3 1/8 | .221138 | .225928 | | | | |
| 3 1/4 | .196591 | .200849 | | | | |
| 3 3/8 | .175547 | .179349 | | | | |
| 3 1/2 | .157402 | .160811 | | | | |
| 3 5/8 | .141674 | .144743 | | | | |
| 3 3/4 | .127973 | .130745 | | | | |
| 3 7/8 | .115984 | .118496 | | | | |
| 4 | .105447 | .107731 | | | | |
| 4 1/8 | .096148 | .098231 | | | | |
| 4 1/4 | .087911 | .089816 | | | | |
| 4 3/8 | .080589 | .082335 | | | | |
| 4 1/2 | .074058 | .075663 | | | | |

Densities used in computing various ball quantities per pound:

| Material | Density, lb per cubic inch |
|---|---|
| Chrome alloy steel | .283 |
| Corrosion resisting hardened steel | .277 |
| Corrosion resisting unhardened steel | |
|   AISI Type 302 | .286 |
|   AISI Type 316 | .288 |
| Carbon Steel | .284 |
| Silicon molybdenum steel | .279 |
| †Brass | .306 |
| †Bronze | .304 |
| †Monel metal | .319 |
| †Aluminum bronze | .274 |
| †K–Monel metal | .306 |
| †Aluminum | .101 |
| †Tungsten carbide | .540 |
| †Beryllium copper | .301 |

†Quantities for these materials shown in Table 7.

Table 7.  Number of Nonferrous Balls per Pound

| Diam., Inches | Brass | Bronze | Monel Metal | Aluminum |
|---|---|---|---|---|
| | Number of Balls per Pound* | | | |
| 1/16 | 25,564.6 | 25,732.8 | 24,522.8 | 77,453.2 |
| 3/32 | 7,574.71 | 7,624.54 | 7,266.02 | 22,949.1 |
| 1/8 | 3,195.58 | 3,216.60 | 3,065.35 | 9,681.66 |
| 5/32 | 1,636.13 | 1,646.90 | 1,569.46 | 4,957.01 |
| 3/16 | 946.838 | 953.068 | 908.253 | 2,868.64 |
| 7/32 | 596.260 | 600.182 | 571.961 | 1,806.49 |
| 1/4 | 399.447 | 402.075 | 383.169 | 1,210.20 |
| 9/32 | 280.544 | 282.390 | 269.112 | 849.967 |
| 5/16 | 204.517 | 205.862 | 196.182 | 619.626 |
| 11/32 | 153.656 | 154.667 | 147.394 | 465.534 |
| 3/8 | 118.354 | 119.133 | 113.531 | 358.580 |
| 13/32 | 93.0893 | 93.7017 | 89.2957 | 282.032 |
| 7/16 | 74.5325 | 75.0228 | 71.4951 | 225.811 |
| 15/32 | 60.5976 | 60.9963 | 58.1281 | 183.593 |
| 1/2 | 49.9309 | 50.2594 | 47.8961 | 151.275 |
| 9/16 | 35.0681 | 35.2988 | 33.6390 | 106.245 |
| 5/8 | 25.5646 | 25.7328 | 24.5228 | 77.4532 |
| 11/16 | 19.2071 | 19.3334 | 18.4243 | 58.1918 |
| 3/4 | 14.7943 | 14.8916 | ... | 44.8225 |
| 13/16 | ... | ... | ... | 35.2541 |
| 7/8 | ... | ... | ... | 28.2264 |
| 15/16 | ... | ... | ... | 22.9491 |
| 1 | ... | ... | ... | 18.9094 |

| Diam., Inches† | Beryllium Copper | Tungsten Carbide | K-Monel Metal | Size, Inches | Aluminum Bronze |
|---|---|---|---|---|---|
| | Number of Balls per Pound | | | | No. per Pound |
| 1/16 | 25,989.3 | 14,486.63 | 25,564.6 | 13/16 | 12.9951 |
| 3/32 | 7,700.53 | 4,292.33 | 7,574.71 | 7/8 | 10.4046 |
| 1/8 | 3,248.66 | 1,810.82 | 3,195.58 | 15/16 | 8.45935 |
| 5/32 | 1,663.31 | 927.14 | 1,636.13 | 1 | 6.97028 |
| 3/16 | 962.567 | 536.542 | 946.838 | 1 1/8 | 4.89545 |
| 7/32 | 606.164 | 337.880 | 596.260 | 1 1/4 | 3.56878 |
| 1/4 | 406.083 | 226.353 | 399.447 | 1 3/8 | 2.68128 |
| 9/32 | 285.205 | 158.975 | 280.544 | 1 1/2 | 2.06527 |
| 5/16 | 207.914 | 115.893 | 204.517 | 1 5/8 | 1.62439 |
| 11/32 | 156.209 | 87.0721 | 153.656 | 1 3/4 | 1.30057 |
| 3/8 | 120.320 | 67.0677 | 118.354 | 1 7/8 | 1.05741 |
| 13/32 | 94.6356 | 52.7506 | 93.0893 | 2 | .871286 |
| 7/16 | 75.7706 | 42.2350 | 74.5325 | 2 1/8 | .726397 |
| 15/32 | 61.6043 | 34.3886 | 60.5976 | 2 1/4 | .611932 |
| 1/2 | 50.7603 | 28.2942 | 49.9309 | 2 3/8 | .520307 |
| 17/32 | 42.3192 | 23.5890 | ... | 2 1/2 | .446098 |
| 9/16 | 35.6506 | 19.8719 | 35.0681 | 2 5/8 | .385356 |
| 19/32 | 30.3126 | 16.8965 | ... | 2 3/4 | .335160 |
| 5/8 | 25.9893 | 14.4866 | 25.5646 | 2 7/8 | .293317 |
| 21/32 | 22.4505 | 12.5141 | ... | 3 | .258158 |
| 11/16 | 19.5261 | 10.8840 | 19.2071 | 3 1/8 | .228402 |
| 23/32 | 17.0883 | ... | ... | 3 1/4 | .203048 |
| 3/4 | 15.0401 | 8.3834 | 14.7943 | 3 3/8 | .181313 |
| 25/32 | 13.3065 | ... | ... | 3 1/2 | .162572 |
| 13/16 | 11.8294 | 6.59382 | 11.6361 | 3 5/8 | .146327 |
| 27/32 | 10.5631 | ... | ... | 3 3/4 | .132177 |
| 7/8 | 9.47132 | 5.2793 | 9.31656 | 3 7/8 | .119794 |
| 29/32 | 8.52492 | ... | ... | 4 | .108910 |
| 15/16 | 7.70053 | 4.29233 | 7.57471 | ... | ... |
| 31/32 | 6.97910 | ... | ... | ... | ... |
| 1 | 6.34504 | 3.53677 | 6.24137 | ... | ... |

* For densities used as a basis for calculating quantities see Table 6.
† For sizes over 1 inch, Tungsten Carbide: 1 1/8 inches, 2.48399 balls per pound, 1 1/4, 1.81082; 1 3/8, 1.36050 and 1 1/2, 1.04793.  K-Monel metal: 1 1/8, 4.38351; 1 1/4, 3.19558; 1 3/8, 2.40088; 1 1/2, 1.84929; 1 5/8, 1.45452 and 1 3/4, 1.16457.

## Lubricants and Lubrication

A lubricant is used for one or more of the following purposes: 1. To reduce friction; 2. To prevent wear; 3. To prevent adhesion; 4. To aid in distributing the load; 5. To cool the moving elements; and 6. To prevent corrosion.

The range of materials used as lubricants has been greatly broadened over the years so that in addition to oils and greases many plastics and solids and even gases are now being applied in this role. The only limitations on many of these materials are their ability to replenish themselves, to dissipate frictional heat, their reaction to high environmental temperatures, and their stability in combined environments. Because of the wide selection of lubricating materials available, great care is advisable in choosing the material and the method of application. The following types of lubricants are available: 1. petroleum fluids; 2. synthetic fluids; 3. greases; 4. solid films; 5. working fluids; 6. gases; 7. plastics; 8. animal fat; 9. metallic and mineral films; and 10. vegetable oils.

**Lubricating Oils.** — The most versatile and best-known lubricant is mineral oil. When applied in well-designed applications which provide for the limitations of both mechanical and hydraulic elements, oil is recognized as the most reliable lubricant. Concurrently, it is offered in a wide selection of stocks, carefully developed to meet the requirements of the specific application.

Lubricating oils are seldom marketed without additives blended for a narrow range of applications. Since these "additive-packages" are developed for particular applications it is advisable to consult the sales-engineering representatives of a reputable petroleum company on the proper selection for the conditions under consideration. The following are the most common types of additives: 1. wear preventive; 2. oxidation inhibitor; 3. rust inhibitor; 4. detergent-dispersant; 5. viscosity index improver; 6. defoaming agent; and 7. pour point depressant.

A more recent development in the field of additives, is a series of organic compounds which leave no ash when heated to a temperature high enough to evaporate or burn off the base oil. Initially produced for internal combustion engine applications, they have found ready acceptance in those other applications where metallic or mineral trace elements would promote catalytic, corrosive, deposition, or degradation effects on mechanism materials.

Additives usually are not stable over the entire temperature and shear-rate ranges considered acceptable for the base stock oil application. Because of this, additive type oils must be carefully monitored to insure that they are not continued in service after their principal capabilities have been diminished or depleted. Of primary importance in this regard, is the action of the detergent-dispersant additives which function so well to reduce and control degradation products which would otherwise deposit on the operating parts and oil cavity walls. Since the materials cause the oil to carry a higher than normal amount of the breakdown products in a fine suspension, they may cause an accelerated deposition rate or foaming when they have been depleted or degenerated by thermal or contamination action. In this latter case, the ingestion of water by condensation or leaking can cause markedly harmful effects.

Viscosity index improvers serve to modify oils so that their change in viscosity is reduced over the operating temperature range. These materials may be used to improve both a heavy or a light oil; however, the original stock will tend to revert to its natural state when the additive has been depleted or degraded due to exposure to high temperatures or to the high shear rates normally encountered in the load carrying zones of bearings and gears. In heavy duty installations it is generally advisable to select a heavier or a more highly refined oil (and one that is generally more costly) rather than to rely on a less stable viscosity index improver product.

These oils are generally used in applications where the shear rate is well below 1,000,000 reciprocal seconds as determined in the following:

$$\text{Shear rate (sec.}^{-1}) = \frac{DN}{60t}$$

where $D$ is the journal diameter in inches, $N$ is the journal speed in RPM, and $t$ is the film thickness in inches.

**Types of Oils.** — Aside from being aware of the many additives which can be obtained to satisfy particular application requirements and improve the performance of fluids, the designer must also be acquainted with the wide variety of oils, natural and synthetic, which are available. Each has its own special features which make it suitable for specific applications and which limit its utility in others. Though a complete description of each oil and its application feasibility cannot be given in the following paragraphs, reference to major petroleum and chemical company sales engineers will provide full descriptions and sound recommendations. In many applications, however, it must be accepted that the interrelation of many variables, including shear rate, load, and temperature variations, prohibit precise recommendations or predictions of fluid durability and performance. Thus, prototype and rig testing are often required to insure the final selection of the most satisfactory fluid.

The following table lists the major classifications, and properties of available commercial petroleum oils.

### Properties of Commercial Petroleum Oils*

| Group A | | | | Group B | | | |
|---|---|---|---|---|---|---|---|
| Type | Viscosity, Centistokes | | Density, g/cc at 60°F. | Type | Viscosity, Centistokes | | Density, g/cc at 60°F. |
| | 100°F. | 210°F. | | | 100°F. | 210°F. | |
| SAE 10W | 41 | 6.0 | 0.870 | | 22 | 3.9 | 0.880 |
| SAE 20W | 71 | 8.5 | 0.885 | | 44 | 6.0 | 0.898 |
| SAE 30 | 114 | 11.2 | 0.890 | General Purpose | 66 | 7.0 | 0.915 |
| SAE 40 | 173 | 14.5 | 0.890 | | 110 | 9.9 | 0.915 |
| SAE 50 | 270 | 19.5 | 0.900 | | 200 | 15.5 | 0.890 |
| Group C | | | | Group D | | | |
| SAE 75 | 47 | 7.0 | | Turbine | | | |
| SAE 80 | 69 | 8.0 | | Light | 32 | 5.5 | 0.871 |
| SAE 90 | 285 | 20.5 | 0.930, approx. | Medium | 65 | 8.1 | 0.876 |
| SAE 140 | 725 | 34.0 | | Heavy | 99 | 10.7 | 0.885 |
| SAE 250 | 1,220 | 47.0 | | | | | |
| Group E | | | | Group F | | | |
| Aviation | 5 | 1.5 | 0.858 | Aviation | 76 | 9.3 | 0.875 |
| | 10 | 2.5 | 0.864 | | 268 | 20.0 | 0.891 |
| | | | | | 369 | 25.0 | 0.892 |

* *Applications:*
    Group A. Automotive. With increased additives, diesel and marine reciprocating engines.
    Group B. Gear trains and transmissions. With E. P. additives, hypoid gears.
    Group C. Machine tools and other industrial applications.
    Group D. Marine propulsion and stationary power turbines.
    Group E. Turbojet engines.
    Group F. Reciprocating engines.

**Viscosity.** — As noted above, fluids used as lubricants are generally categorized by their viscosity at 100 and 210 deg. F. Absolute viscosity is defined as a fluid's resistance to shear or motion — its internal friction in other words. This property is described in several ways, but basically it is the force required to move a plane surface of unit area with unit speed parallel to a second plane and at unit distance from it. In the metric system the unit of viscosity is called the "poise" and in the English system is called the "reyn." One reyn is equal to 68,950 poise. One poise is the viscosity of a fluid, such that one dyne force is required to move a surface of one square centimeter with a speed of one centimeter per second, the distance between surfaces being one centimeter. The range of kinematic viscosity for a series of typical fluids is shown in the table on page 666. Kinematic viscosity is related directly to the flow time of a fluid through the viscosimeter capillary. By multiplying the kinematic viscosity by the density of the fluid at the test temperature, one can determine the absolute viscosity. Since, in the metric system, the mass density is equal to the specific gravity, the conversion from kinematic to absolute viscosity is generally made in this system and then converted to English units where required. The densities of typical lubricating fluids with comparable viscosities at 100 deg. F. and 210 deg. F. are shown in this same table.

The following conversion table may be found helpful.

| Multiply | By | To Get |
|---|---|---|
| Centipoises, $Z$, $\dfrac{\text{dyne-sec}}{\text{100 cm}^2}$ | $1.45 \times 10^{-7}$ | Reyns, $\mu$, $\dfrac{\text{lbs. force-sec}}{\text{in.}^2}$ |
| Centistokes, $\nu$, $\dfrac{\text{cm.}^2}{\text{100 sec}}$ | Density in g/cc | Centipoises, $Z$, $\dfrac{\text{dyne-sec}}{\text{100 cm.}^2}$ |
| Saybolt Universal Seconds, $t_s$ | $.22 t_s - \dfrac{180}{t_s}$ | Centistokes, $\nu$, $\dfrac{\text{cm.}^2}{\text{100 sec}}$ |

**Finding Specific Gravity of Oils at Different Temperatures.** — The standard practice in the oil industry is to obtain a measure of specific gravity at 60 deg. F. on an arbitrary scale, in degrees API, as specified by the American Petroleum Institute. As an example, API gravity, $\rho_{API}$, may be expressed as 27.5 degrees at 60 deg. F.

The relation between gravity in API degrees and specific gravity (grams of mass per cubic centimeter) at 60 deg. F., $\rho_{60}$, is:

$$\rho_{60} = \frac{141.5}{131.5 + \rho_{API}}$$

The specific gravity, $\rho_T$, at some other temperature, $T$, is found from the equation:

$$\rho_T = \rho_{60} - 0.00035(T - 60)$$

Normal values of specific gravity for sleeve bearing lubricants range from 0.75 to 0.95 at 60 deg. F. If the API rating is not known, an assumed value of 0.85 may be used.

**Application of Lubricating Oils.** — In the selection and application of lubricating oils careful attention must be given to the temperature in the critical operating area and its effect on oil properties. Analysis of each application should be made with detailed attention given to cooling, friction losses, shear rates, and contaminants.

Many oil selections are found to result in excessive operating temperatures because of a viscosity that is initially too high, which raises the friction losses. As a general rule, the lightest weight oil which can carry the maximum load should be used. Where it is felt that the load carrying capacity is borderline, lubricity improvers may be employed rather than an arbitrarily higher viscosity fluid. It is well to remember that in many mechanisms the thicker fluid may increase friction losses sufficiently to lower the operating viscosity into the range provided by an initially lighter fluid. In such situations also, improved cooling, such as may be accomplished by increasing the oil flow, can improve the fluid properties in the load zone.

Similar improvements can be accomplished in many gear trains and other mechanisms by reducing churning and aeration through improved scavenging, direction of oil jets, and elimination of obstacles to the flow of the fluid. Many devices, such as journal bearings, are extremely sensitive to the effects of cooling flow and can be improved by greater flow rates with a lighter fluid. In other cases it is well to remember that the load carrying capacity of a petroleum oil is affected by pressure, shear rate, and bearing surface finish as well as initial viscosity and therefore these must be considered in the selection of the fluid. Detailed explanation of these factors is not within the scope of this text; however the technical representatives of the petroleum companies can supply practical guides for most applications.

Other factors to consider in the selection of an oil include the following:

1. Compatibility with system materials
2. Water absorption properties
3. Break-in requirements
4. Detergent requirements
5. Corrosion protection
6. Low temperature properties
7. Foaming tendencies
8. Boundary lubrication properties
9. Oxidation resistance (high temperature properties)
10. Viscosity/temperature stability (Viscosity Temperature Index).

Generally, the factors listed above are those which are usually modified by additives as described earlier. Since additives are used in limited amounts in most petroleum products, blended oils are not as durable as the base stock and must therefore be used in carefully worked-out systems. Maintenance procedures must be established to monitor the oil so that it may be replaced when the effect of the additive is noted or expected to degrade. In large systems supervised by a lubricating engineer, sampling and associated laboratory analysis can be relied on, while in customer-maintained systems as in automobiles and reciprocating engines, the design engineer must specify a safe replacement period which takes into account any variation in type of service or utilization.

Some large systems, such as turbine-power units, have complete oil systems which are designed to filter, cool, monitor, meter, and replenish the oil automatically. In such facilities, much larger oil quantities are used and they are maintained by regularly assigned lubricating personnel. Here reliance is placed on conservatively chosen fluids with the expectation that they will endure many months or even years of service.

**Centralized Lubrication Systems.** — Various forms of centralized lubrication systems are used to simplify and render more efficient the task of lubricating machines. In general, a central reservoir provides the supply of oil, which is conveyed to each bearing either through individual lines of tubing or through a single line of tubing that has branches extending to each of the different bearings. Oil is pumped into the lines either manually by a single movement of a lever or handle, or automatically by mechanical drive from some revolving shaft or other part of the machine. In either case, all bearings in the central system are lubricated simultaneously. Centralized force-feed lubrication is adaptable to various classes of machine tools such as lathes, planers, and milling machines and to many other types of machines. It permits the use of a lighter grade of oil, especially where complete coverage of the moving parts is assured.

**Gravity Lubrication Systems.** — Gravity systems of lubrication usually consist of a small number of distributing centers or manifolds from which oil is taken by piping as directly as possible to the various surfaces to be lubricated, each bearing point having its own independent pipe and set of connections. The aim of the gravity system, as of all lubrication systems, is to provide a reliable means of supplying the bearing surfaces with the proper amount of lubricating oil. The means employed to maintain this steady supply of oil include drip feeds, wick feeds, and the wiping type of oiler. Most manifolds are adapted to use either or both drip and wick feeds.

*Drip-feed Lubricators:* A drip feed consists of a simple cup or manifold mounted in a convenient position for filling and connected by a pipe or duct to each bearing to be oiled. The rate of feed in each pipe is regulated by a needle or conical valve. A loose-fitting cover is usually fitted to the manifold in order to prevent cinders or other foreign matter from becoming mixed with the oil. When a cylinder or other chamber operating under pressure is to be lubricated, the oil-cup takes the form of a lubricator having a tight-fitting screw cover and a valve in the oil line. To fill a lubricator of this kind, it is only necessary to close the valve and unscrew the cover.

*Operation of Wick Feeds:* For a wick feed, the siphoning effect of strands of worsted yarn is employed. The worsted wicks give a regular and reliable supply of oil and at the same time act as filters and strainers. A wick composed of the proper number of strands is fitted into each oil-tube. In order to insure using the proper sizes of wicks, a study should be made of the oil requirements of each installation, and the number of strands necessary to meet the demands of bearings at different rates of speed should be determined. When the necessary data have been obtained, a table should be prepared showing the size of wick or the number of strands to be used for each bearing of the machine.

*Oil-conducting Capacity of Wicks:* With the oil level maintained at a point $\frac{3}{8}$ to $\frac{3}{4}$ inch below the top of an oil-tube, each strand of a clean worsted yarn will carry slightly more than one drop of oil a minute. A twenty-four-strand wick feed approximately thirty drops a minute, which is ordinarily sufficient for operating a large bearing at high speed. The wicks should be removed from the oil-tubes when the machinery is idle. If left in place, they will continue to deliver oil to the bearings until the supply in the cup is exhausted, thus wasting a considerable quantity of oil, as well as flooding the bearing. When bearings require an extra supply of oil temporarily, it may be supplied by dipping the wicks or by pouring oil down the tubes from an oil-can or, in the case of drip feeds, by opening the needle valves. When equipment that has remained idle for some time is to be started up, the wicks should be dipped and the moving parts oiled by hand to insure an ample initial supply of oil. The oil should be kept at about the same level in the cup, as otherwise the rate of flow will be affected. Wicks should be lifted periodically to prevent dirt accumulations at the ends from obstructing the flow of oil.

*How Lubricating Wicks are Made:* Wicks for lubricating purposes are made by cutting worsted yarn into lengths about twice the height of the top of the oil-tube above the bottom of the oil-cup, plus 4 inches. Half the required number of strands are then assembled and doubled over a piece of soft copper wire, laid across the middle of the strands. The free ends are then caught together by a small piece of folded sheet lead, and the copper wire twisted together throughout its length.

The lead serves to hold the lower end of the wick in place, and the wire assists in forcing the other end of the wick several inches into the tube. When the wicks are removed, the free end of the copper wire may be hooked over the tube end to indicate which tube the wick belongs to. Dirt from the oil causes the wick to become gummy and to lose its filtering effect. Wicks that have thus become clogged with dirt should be cleaned or replaced by new ones. The cleaning is done by boiling the wicks in soda water and then rinsing them thoroughly to remove all traces of the soda. Oil-pipes are sometimes fitted with openings through which the flow of oil can be observed. In some installations, a short glass tube is substituted for such an opening.

*Wiper-type Lubricating Systems:* Wiper-type lubricators are used for out-of-the-way oscillating parts. A wiper consists of an oil-cup with a central blade or plate extending above the cup, and is attached to a moving part. A strip of fibrous material fed with oil from a source of supply is placed on a stationary part in such a position that the cup in its motion scrapes along the fibrous material and wipes off the oil, which then passes to the bearing surfaces.

Oil manifolds, cups, and pipes should be cleaned occasionally with steam conducted through a hose or with boiling soda water. When soda water is used, the pipes should be disconnected, so that no soda water can reach the bearings.

**Oil Mist Systems.** — A very effective system for both lubricating and cooling many elements which require a limited quantity of fluid is found in a device which generates a mist of oil, separates out the denser and larger (wet) oil particles, and then distributes the mist through a piping or conduit system. The mist is delivered into the bearing, gear, or lubricated element cavity through a condensing or spray nozzle, which also serves to meter the flow. In applications which do not encounter low temperatures or which permit the use of visual devices to monitor the accumulation of solid oil, oil mist devices offer advantages in providing cooling, clean lubricant, pressurized cavities which prevent entrance of contaminants, efficient application of limited lubricant quantities, and near-automatic performance. These devices are supplied with fluid reservoirs holding from a few ounces up to several gallons of oil and with accommodations for either accepting shop air or working from a self-contained compressor powered by electricity. With proper control of the fluid temperature, these units can atomize and dispense most motor and many gear oils.

**Lubricating Greases.** — In many applications, fluid lubricants cannot be used because of the difficulty of retention, relubrication, or the danger of churning. To satisfy these and other requirements such as simplification, greases are applied. These formulations are usually petroleum oils thickened by dispersions of soap, but may consist of synthetic oils with soap or inorganic thickeners, or oil with silaceous dispersions. In all cases, the thickener, which must be carefully prepared and mixed with the fluid, is used to immobilize the oil, serving as a storehouse from which the oil bleeds at a slow rate. Though the thickener very often has lubricating properties itself, the oil bleeding from the bulk of the grease is the determining lubricating function. Thus, it has been shown that when the oil has been depleted to the level of 50 per cent of the total weight of the grease, the lubricating ability of the material is no longer reliable. In some applications requiring an initially softer and wetter material, however, this level may be as high as 60 per cent.

**Grease Consistency Classifications.** — To classify greases as to mobility and oil content, they are divided into Grades by the NLGI (National Lubricating Grease Institute). These grades, ranging from 0, the softest, up through 6, the stiffest, are determined by testing in a penetrometer, with the depth of penetration of a specific cone and weight being the controlling criterion. To insure proper averaging of specimen resistance to the cone, most specifications include a requirement that the specimen be worked in a sieve-like device before being packed into the penetrometer cup for the penetration test. Since many greases exhibit thixotropic properties (they soften with working, as they often do in an application with agitation of the bulk of the grease by the working elements or accelerations), this penetration of the worked specimen should be used as a guide to compare the material to the original manufactured condition of it and other greases, rather than to the exact condition in which it will be found in the application. Conversely, many greases are found to stiffen when exposed to high shear rates at moderate loads as in automatic grease dispensing equipment. The application of a grease, therefore must be determined by a carefully planned cut-and-try procedure.. Most often this is done by the original equipment manufacturer with the aid of the petroleum company representatives, but in many cases it is advisable to include the bearing engineer as well. In this general area it is well to remember that shock loads, axial or thrust movement within or on the grease cavity can cause the grease to contact the moving parts and initiate softening due to the shearing or working thus induced. To limit this action, grease-lubricated bearing assemblies often utilize dams or dividers to keep the bulk of the grease contained and unchanged by this working. Successful application of a grease depends however, on a relatively small amount of mobile lubricant (the oil bled out of the bulk) to replenish that small amount of lubricant in the element to be lubricated. If the space between the bulk of the mobile grease and the bearing is too large, then a critical delay period (which will be regulated by the grease bleed rate and the temperature at which it is held) will ensue before lubricant in the element can be resupplied. Since most lubricants undergo some attrition due to thermal degradation, evaporation, shearing, or decomposition in the bearing area to which applied, this delay can be fatal.

To prevent this from leading to failure, grease is normally applied so that the material in the cavity contacts the bearing in the lower quadrants, insuring that the excess originally packed into it impinges on the material in the reservoir. With the proper selection of a grease which does not slump excessively, and a reservoir construction to prevent churning, the initial action of the bearing when started into operation will be to purge itself of excess grease, and to establish a flow path for bleed oil to enter the bearing. For this purpose, most greases selected will be of a grade 2 or 3 consistency, falling into the "channelling" variety or designation.

**Types of Grease.** — Greases are made with a variety of soaps and are chosen for many particular characteristics. Most popular today, however, are the lithium, or soda-soap grease and the modified-clay thickened materials. For high temperature applications (250 deg. F. and above) certain finely divided dyes and other synthetic thickeners are applied. For all-around use the lithium soap greases are best for moderate temperature applications (up to 225 deg. F.) while a number of soda-soap greases have been found to work well up to 285 deg. F. Since the major suppliers offer a number of different formulations for these temperature ranges it is recommended that the user contact the engineering representatives of a reputable petroleum company before choosing a grease. Greases also vary in volatility and viscosity according to the oil used. Since the former will affect the useful life of the bulk applied to the bearing and the latter will affect the load carrying capacity of the grease, they must both be considered in selecting a grease.

For application to certain gears and slow-speed journal bearings, a variety of greases are thickened with carbon, graphite, molybdenum disulfide, lead, or zinc oxide. Some of these materials are likewise used to inhibit fretting corrosion or wear in sliding or oscillating mechanisms and in screw or thread applications. One material used as a "gear grease" is a residual asphaltic compound which is known as a "Crater Compound." Being extremely stiff and having an extreme temperature-viscosity relationship, its application must also be made with careful consideration of its limitations and only after careful evaluation in the actual application. Its oxidation resistance is limited and its low mobility in winter temperature ranges make it a material to be used with care. However, it is used extensively in the railroad industry and in other applications where containment and application of lubricants is difficult. In such conditions its ability to adhere to gear and chain contact surfaces far outweighs its limitations and in some extremes it is "painted" onto the elements at regular intervals.

**Temperature Effects on Grease Life.** — Since most grease applications are made where long life is important and relubrication is not too practical, operating temperatures must be carefully considered and controlled. Being a hydro-carbon, and normally susceptible to oxidation, grease is subject to the general rule that: Above a critical threshold temperature, each 15- to 18-deg. F. rise in temperature reduces the oxidation life of the lubricant by half. For this reason, it is vital that all elements affecting the operating temperature of the application be considered, correlated, and controlled. With sealed-for-life bearings, in particular, grease life must be determined for representative bearings and limits must be established for all subsequent applications.

Most satisfactory control can be established by measuring bearing temperature rise during a controlled test, at a consistent measuring point or location. Once a base line and limiting range are determined, all deviating bearings should be dismantled, inspected, and reassembled with fresh lubricant for retest. In this manner mavericks or faulty assemblies will be ferreted out and the reliability of the application established. Generally, a well lubricated grease packed bearing will have a temperature rise above ambient, as measured at the outer race, of from 10 to 50 deg. F. In applications where heat is introduced into the bearing through the shaft or housing, a temperature rise must be added to that of the frame or shaft temperature.

In bearing applications care must be taken not to fill the cavity too full. The bearing should have a practical quantity of grease worked into it with the rolling elements thoroughly coated and the cage covered, but the housing (cap and cover) should be no more than 75 per cent filled; with softer greases, this should be no more than 50 per cent. Excessive packing is evidenced by overheating, churning, aerating, and eventual purging with final failure due to insufficient lubrication. In grease lubrication, *never* add a bit more for good luck — hold to the prescribed amount and determine this with care on a number of representative assemblies.

**Relubricating with Grease.** — In some applications, sealed-grease methods are not applicable and addition of grease at regular intervals is required. Where this is recommended by the manufacturer of the equipment, or where the method has been worked out as part of a development program, the procedure must be carefully followed. *First,* use the proper lubricant — the same as recommended by the manufacturer or as originally applied (grease performance can be drastically impaired if contaminated with another lubricant). *Second,* clean the lubrication fitting thoroughly with materials which will not affect the mechanism or penetrate into the grease cavity. *Third,* remove the cap (and if applicable, the drain or purge plug). *Fourth,* clean and inspect the drain or scavenge cavity. *Fifth,* weigh the grease gun or calibrate it to determine delivery rate. *Sixth,* apply the directed quantity or fill until grease is detected coming out the drain or purge hole. *Seventh,* operate the mechanism with the drain open so that excess grease is purged. *Last,* continue to operate the mechanism while determining the temperature rise and insure that it is within limits. Where there is access to a laboratory, samples of the purged material may be analyzed to determine the deterioration of the lubricant and to search for foreign material which may be evidence of contamination or of bearing failure.

Normally, with modern types of grease and bearings, lubrication need only be considered at overhaul periods or over intervals of three to ten years.

**Solid Film Lubricants.** — Solids such as graphite, molybdenum disulfide, polytetrafluoroethylene, lead, babbit, silver, or metallic oxides are used to provide dry film lubrication in high-load, slow-speed or oscillating load conditions. Though most are employed in conjunction with fluid or grease lubricants, they are often applied as the primary or sole lubricant where their inherent limitations are acceptable. Of foremost importance is their inability to carry away heat. Second, they cannot replenish themselves, though they generally do lay down an oriented film on the contacting interface. Third, they are relatively immobile and must be bonded to the substrate by a carrier, by plating, fusing, or by chemical or thermal deposition.

Though these materials do not provide the low coefficient of friction associated with fluid lubrication, they do provide coefficients in the range of 0.4 down to 0.02, depending on the method of application and the material against which they rub. Polytetrafluoroethylene, in normal atmospheres and after establishing a film on both surfaces has been found to exhibit a coefficient of friction down to 0.02. However, this material is subject to cold flow and must be supported by a filler or on a matrix to continue its function. Since it can now be cemented in thin sheets and is often supplied with a fine glass fiber filler, it is practical in a number of installations where the speed and load do not combine to melt the bond or cause the material to sublime.

Bonded films of molybdenum disulfide, using various resins and ceramic combinations as binders, are deposited over phosphate treated steel, aluminum, or other metals with good success. Since its action produces a gradual wear of the lubricant, its life is limited by the thickness which can be applied (not over a thousandth or two in the conventional application). In most applications this is adequate if the material is used to promote break-in, prevent galling or pick-up, and to reduce fretting or abrasion in contacts otherwise impossible to separate.

In all applications of solid film lubricants, the performance of the film is limited by the care and preparation of the surface to which they are applied. If they can't adhere properly, they cannot perform, coming off in flakes and often jamming under flexible components. The best advice is to seek the assistance of the supplier's field engineer and set up a close control of the surface preparation and solid film application procedure. It should be noted that the functions of a good solid film lubricant cannot overcome the need for better surface finishing. Contacting surfaces should be smooth and flat to insure long life and minimum friction forces. Generally, surfaces should be finished to no more than 24 micro-inches AA with waviness no greater than 0.00002 inch.

**Anti-friction Bearing Lubrication.** — The limiting factors in bearing lubrication are the load and the linear velocity of the centers of the balls or rollers. Since these are difficult to evaluate, a speed factor which consists of the inner race bore diameter $\times$ RPM is used as a criterion. This factor will be referred to as $S_i$ where the bore diameter is in inches and $S_m$ where it is in millimeters.

In order to be suitable for use in anti-friction bearings, grease must have the following properties:

1. Freedom from chemically or mechanically active ingredients such as uncombined metals or oxides, and similar mineral or solid contaminants.

2. The slightest possible tendency of change in consistency, such as thickening, separation of oil, evaporation or hardening.

3. A melting point considerably higher than the operating temperatures.

The choice of lubricating oils is easier. They are more uniform in their characteristics and if resistant to oxidation, gumming and evaporation, can be selected primarily with regard to a suitable viscosity.

*Grease Lubrication:* Anti-friction bearings are normally grease lubricated, both because grease is much easier than oil to retain in the housing over a long period and because it acts to some extent as a seal against the entry of dirt and other contaminants into the bearings. For almost all applications, a No. 2 soda-base grease or a mixed-base grease with up to 5 per cent calcium soap to give a smoother consistency, blended with an oil of around 250 to 300 SSU (Saybolt Universal Seconds) at 100 degrees F. is suitable. In cases where speeds are high, say $S_i$ is 5000 or over, a grease made with an oil of about 150 SSU at 100 degrees F. may be more suitable especially if temperatures are also high. In many cases where bearings are exposed to large quantities of water, it has been found that a standard soda-base ball-bearing grease, although classed as water soluble gives better results than water-insoluble types. Greases are available that will give satisfactory lubrication over a temperature range of −40 degrees to +250 degrees F.

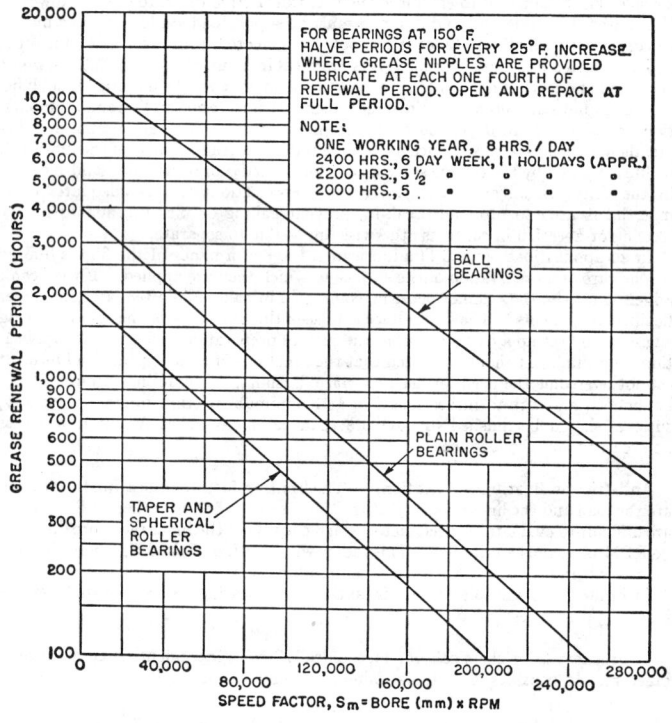

Conservative grease renewal periods will be found in the accompanying chart. Grease should not be allowed to remain in a bearing for longer than 48 months or if the service is very light and temperatures low, 60 months, irrespective of the number of hours' operation during that period as separation of the oil from the soap and oxidation continue whether the bearing is in operation or not.

Before renewing the grease in a hand-packed bearing, the bearing assembly should be removed and washed in clean kerosene, degreasing fluid or other solvent. As soon as the bearing is quite clean it should be washed at once in clean light mineral oil, preferably rust-inhibited. The bearing should *not be spun* before or while it is being oiled. Caustic solutions may be used if the old grease is hard and difficult to remove, but the best method is to soak the bearing for a few hours in light mineral oil, preferably warmed to about 130 degrees F., and then wash in cleaning fluid as described above. The use of chlorinated solvents is best avoided.

When replacing the grease, it should be forced with the fingers between the balls or rollers, dismantling the bearing, if convenient. The available space inside the bearing should be filled completely and the bearing then spun by hand. Any grease thrown out should be wiped off. The space on each side of the bearing in the housing should be not more than half-filled. Too much grease will result in considerable churning, high bearing temperatures and the possibility of early failure. Unlike any other kind of bearing, anti-friction bearings more often give trouble due to over- rather than to under-lubrication.

**Oil Viscosities and Temperature Ranges for Ball Bearing Lubrication***

| Maximum Temperature Range, Degrees F. | Optimum Temperature Range, Degrees F. | Speed Factor, $S_i$† | |
|---|---|---|---|
| | | Under 1000 | Over 1000 |
| | | Viscosity | |
| −40 to +100 | −40 to −10 | 80 to 90 SSU** | 70 to 80 SSU** |
| −10 to +100 | −10 to +30 | 100 to 115 SSU** | 80 to 100 SSU** |
| +30 to +150 | +30 to +150 | SAE 20 | SAE 10 |
| +30 to +200 | +150 to +200 | SAE 40 | SAE 30 |
| +50 to +300 | +200 to +300 | SAE 70 | SAE 60 |

\* Not applicable to air-distributed oil mist lubrication.   \*\* At 100 deg. F.
† Inner race bore diameter (inches) × RPM.

Grease is usually not very suitable for speed factors over 12,000 for $S_i$ or 300,000 for $S_m$ (although successful applications have been made up to an $S_i$ of 50,000) or for temperatures much over 210 degrees F., 300 degrees F. being the extreme practical upper limit, even if synthetics are used. For temperatures above 210 degrees F., the grease renewal periods are very short.

*Oil Lubrication:* Oil lubrication is usually adopted when speeds and temperatures are high or when it is desired to adopt a central oil supply for the machine as a whole. Oil for anti-friction bearing lubrication should be well refined with high film strength and good resistance to oxidation and good corrosion protection. Anti-oxidation additives do no harm but are not really necessary at temperatures below about 200 degrees F. Anti-corrosion additives are always desirable. The accompanying table gives recommended viscosities of oil for ball bearing lubrication other than by an air-distributed oil mist. Within a given temperature and speed range, an oil towards the lighter end of the grade should be used, if convenient, as speeds increase. Roller bearings usually require an oil one grade heavier than do ball bearings for a given speed and temperature range. Cooled oil is sometimes circulated through an anti-friction bearing to carry off excess heat resulting from high speeds and heavy loads.

# CLUTCHES AND COUPLINGS

**Positive Clutches.** — When the driving and driven members of a clutch are connected by the engagement of interlocking teeth or projecting lugs, the clutch is said to be "positive" to distinguish it from the type in which the power is transmitted by frictional contact. The positive clutch is employed when a sudden starting action is not objectionable and when the inertia of the driven parts is relatively small. The various forms of positive clutches differ merely in the angle or shape of the engaging surfaces. The least positive form is one having planes of engagement which incline backward, with respect to the direction of motion. The tendency of such a clutch is to disengage under load, in which case it must be held in

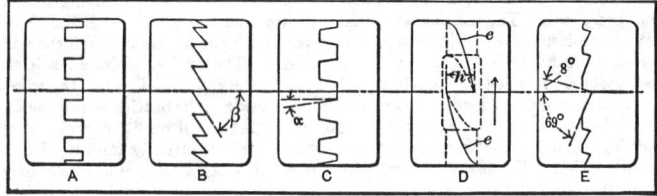

Fig. 1.—Types of Clutch Teeth

position by axial pressure. This pressure may be regulated to perform normal duty, permitting the clutch to slip and disengage when over-loaded. Positive clutches, with the engaging planes parallel to the axis of rotation, are held together to obviate the tendency to jar out of engagement, but they provide no safety feature against over-load. So-called "under-cut" clutches engage more tightly the heavier the load, and are designed to be disengaged only when free from load. The teeth of positive clutches are made in a variety of forms, a few of the more common styles being shown in Fig. 1. Clutch *A* is a straight-toothed type, and *B* has angular

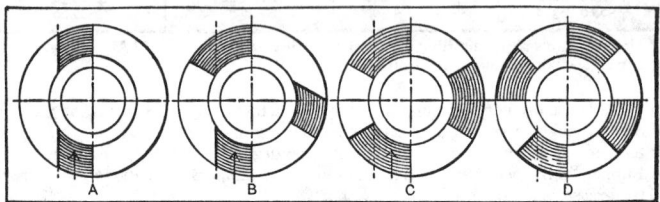

Fig. 2.— Diagrammatical View Showing Method of Cutting Clutch Teeth

or saw-shaped teeth. The driving member of the former can be rotated in either direction; the latter is adapted to the transmission of motion in one direction only, but is more readily engaged. The angle $\beta$ of the cutter for a saw-tooth clutch *B* is ordinarily 60 degrees. Clutch *C* is similar to *A*, except that the sides of the teeth are inclined to facilitate engagement and disengagement. Teeth of this shape are sometimes used when a clutch is required to run in either direction without backlash. Angle $\alpha$ is varied to suit requirements and should not exceed 8 or 9 degrees. The straight-tooth clutch *A* is also modified to make the teeth engage more readily, by rounding the corners of the teeth at the top and bottom. Clutch *D* (commonly called a "spiral-jaw" clutch) differs from *B* in that the surfaces *e* are helicoidal. The driving member of this clutch can only transmit motion in one direction.

Clutches of this type are known as right- and left-hand, the former driving when turning to the right, as indicated by the arrow in the illustration. Clutch $E$ is the form used on the back-shaft of the Brown & Sharpe automatic screw machines. The faces of the teeth are radial and incline at an angle of 8 degrees with the axis, so that the clutch can readily be disengaged. This type of clutch is easily operated, with little jar or noise. The 2-inch diameter size has 10 teeth. Height of working face, ⅛ inch.

**Cutting Clutch Teeth.** — A common method of cutting a straight-tooth clutch is indicated by the diagrams $A$, $B$ and $C$, Fig. 2, which show the first, second and third cuts required for forming the three teeth. The work is held in the chuck of a dividing-head, the latter being set at right angles to the table. A plain milling

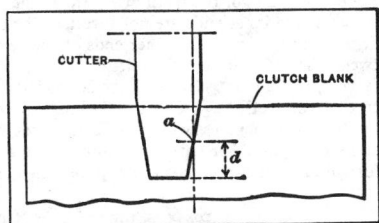

Fig. 3

cutter may be used (unless the corners of the teeth are rounded), the side of the cutter being set to exactly coincide with the center-line. When the number of teeth in the clutch is odd, the cut can be taken clear across the blank as shown, thus finishing the sides of two teeth with one passage of the cutter. When the number of teeth is even, as at $D$, it is necessary to mill all the teeth on one side and then set the cutter for finishing the opposite side. Therefore, clutches of this type com-

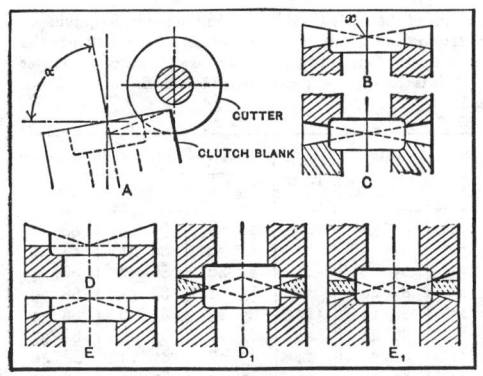

Fig. 4

monly have an odd number of teeth. The maximum width of the cutter depends upon the width of the space at the narrow ends of the teeth If the cutter must be quite narrow in order to pass the narrow ends, some stock may be left in the tooth spaces, which must be removed by a separate cut. If the tooth is of the modified form shown at $C$, Fig. 1, the cutter should be set as indicated in Fig. 3; that is, so that a point $a$ on the cutter at a radial distance $d$ equal to one-half the depth of the clutch teeth lies in a radial plane. When it is important to eliminate all backlash, point $a$ is sometimes located at a radial distance $d$ equal to six-tenths of the depth of the tooth, in order to leave clearance spaces at the bottoms of the teeth; the two clutch members will then fit together tightly. Clutches of this type must be held in mesh.

**Cutting Saw-tooth Clutches.** — When milling clutches having angular teeth as shown at $B$, Fig. 1, the axis of the clutch blank should be inclined a certain angle $\alpha$ from the vertical, as shown at $A$ in Fig. 4. If the teeth were milled with the blank

**vertical,** the tops of the teeth would incline towards the center as at $D$, whereas, if the blank were set to such an angle that the tops of the teeth were square with the axis, the bottoms would incline upwards as at $E$. In either case, the two clutch members would not mesh completely; the engagement of the teeth cut as shown at $D$ and $E$ would be as indicated at $D_1$ and $E_1$ respectively. As will be seen, when the outer points of the teeth at $D_1$ are at the bottom of the grooves in the opposite member, the inner ends are not together, the contact area being represented by the dotted lines. At $E_1$ the inner ends of the teeth strike first and spaces are left between the teeth around the outside of the clutch. To overcome this objectionable feature, the clutch teeth should be cut as indicated at $B$, or so that the bottoms and tops of the teeth have the same inclination, converging at a central point $x$. The teeth of both members will then engage across the entire width as shown at $C$. The angle $\alpha$ required for cutting a clutch as at $B$ can be determined by the following formula in which $\alpha$ equals the required angle, and $N$, the number of teeth;

$$\cos \alpha = \tan \frac{180 \text{ deg.}}{N} \times \cot \text{ cutter angle.}$$

Expressing this formula as a rule: To determine the cosine of angle $\alpha$ (see diagram $A$, Fig. 4) find the tangent of the angle obtained by dividing 180 degrees by the number of teeth, and multiply this tangent by the cotangent of the cutter angle.

These angles for various numbers of teeth and for either a 60-, 70- or 80-degree single-angle cutter are given in the following table:

Angle of Dividing-head for Milling Clutches with Single-angle Cutter

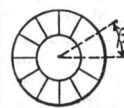 The cosine of the angle to which the milling machine dividing-head is set equals the tangent of one-half of angle $\beta$ between the teeth (see illustration) multiplied by the cotangent of the cutter angle. This is the angle ($a$, Fig. 4) shown by graduations on dividing-head.

| No. of Teeth | Angle of Single-angle Cutter | | | No. of Teeth | Angle of Single-angle Cutter | | |
|---|---|---|---|---|---|---|---|
| | 60° | 70° | 80° | | 60° | 70° | 80° |
| 5 | ....... | ....... | 82° 38′ | 18 | 84° 9′ | 86° 19′ | 88° 13′ |
| 6 | ....... | 77° 52′ | 84 9 | 19 | 84 30 | 86 31 | 88 19 |
| 7 | 73° 50′ | 79 54 | 85 10 | 20 | 84 46 | 86 42 | 88 24 |
| 8 | 76 10 | 81 20 | 85 48 | 21 | 85 1 | 86 51 | 88 29 |
| 9 | 77 52 | 82 23 | 86 19 | 22 | 85 13 | 87 0 | 88 33 |
| 10 | 79 12 | 83 13 | 86 43 | 23 | 85 27 | 87 8 | 88 37 |
| 11 | 80 14 | 83 54 | 87 4 | 24 | 85 38 | 87 15 | 88 40 |
| 12 | 81 6 | 84 24 | 87 18 | 25 | 85 49 | 87 22 | 88 43 |
| 13 | 81 49 | 84 51 | 87 30 | 26 | 85 59 | 87 28 | 88 46 |
| 14 | 82 26 | 85 12 | 87 42 | 27 | 86 8 | 87 34 | 88 50 |
| 15 | 82 57 | 85 34 | 87 51 | 28 | 86 16 | 87 39 | 88 52 |
| 16 | 83 24 | 85 51 | 87 59 | 29 | 86 24 | 87 44 | 88 54 |
| 17 | 83 48 | 86 6 | 88 7 | 30 | 86 31 | 87 48 | 88 56 |

**Friction Clutches.** — Clutches which transmit motion from the driving to the driven member by the friction between the engaging surfaces are built in many different designs, although practically all of them can be classified under four general types, namely, conical clutches; radially-expanding clutches; contracting-

## Angle of Dividing-head for Milling V-shaped Teeth with Double-angle Cutter

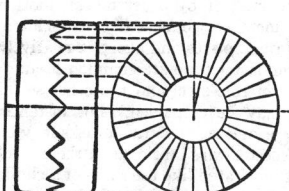

The cosine of the angle to which to set the dividing-head equals the tangent of 90 degrees divided by number of teeth, multiplied by the cotangent of one-half the cutter angle. Thus:

$$\cos \text{ index-head angle} = \tan \frac{90}{N} \times \cot \frac{\text{cutter angle}}{2}$$

This is the angle (a, Fig. 4) shown by graduations on the dividing-head.

| No. of Teeth | Included Angle of Cutter | | No. of Teeth | Included Angle of Cutter | |
|---|---|---|---|---|---|
| | 60° | 90° | | 60° | 90° |
| 10 | 74° 5′ | 80° 53′ | 31 | 84° 57′ | 87° 5′ |
| 11 | 75 35 | 81 53 | 32 | 85 6 | 87 11 |
| 12 | 76 50 | 82 26 | 33 | 85 16 | 87 16 |
| 13 | 77 52 | 83 2 | 34 | 85 25 | 87 21 |
| 14 | 78 45 | 83 32 | 35 | 85 32 | 87 26 |
| 15 | 79 31 | 83 58 | 36 | 85 40 | 87 30 |
| 16 | 80 11 | 84 21 | 37 | 85 47 | 87 34 |
| 17 | 80 46 | 84 41 | 38 | 85 54 | 87 38 |
| 18 | 81 17 | 84 59 | 39 | 86 0 | 87 42 |
| 19 | 81 45 | 85 15 | 40 | 86 6 | 87 45 |
| 20 | 82 10 | 85 29 | 41 | 86 12 | 87 48 |
| 21 | 82 34 | 85 42 | 42 | 86 17 | 87 51 |
| 22 | 82 53 | 85 54 | 43 | 86 22 | 87 54 |
| 23 | 83 12 | 86 5 | 44 | 86 27 | 87 57 |
| 24 | 83 29 | 86 15 | 45 | 86 32 | 88 0 |
| 25 | 83 45 | 86 24 | 46 | 86 37 | 88 3 |
| 26 | 84 1 | 86 32 | 47 | 86 41 | 88 5 |
| 27 | 84 13 | 86 39 | 48 | 86 45 | 88 8 |
| 28 | 84 25 | 86 46 | 49 | 86 49 | 88 10 |
| 29 | 84 37 | 86 53 | 50 | 86 53 | 88 12 |
| 30 | 84 47 | 86 59 | .. | .. .. | .. .. |

The angles given in the table above are applicable to the milling of V-shaped grooves in brackets, etc., which must have toothed surfaces to prevent the two members from turning relative to each other, except when unclamped for angular adjustment.

band clutches; and friction disk clutches in single and multiple types. There are many modifications of these general classes, some of which combine the features of different types. The proportions of various sizes of cone clutches are given in the table " Cast-iron Friction Clutches." The multi-cone friction clutch is a further development of the cone clutch. Instead of having a single cone-shaped surface, there is a series of concentric conical rings which engage annular grooves formed by corresponding rings on the opposite clutch member. The internal-expanding type is provided with shoes which are forced outward against an enclosing drum by the action of levers connecting with a collar free to slide along the shaft. The engaging shoes are commonly lined with wood to increase the coefficient of friction. The well-known Weston disk clutch is based on the principle of multiple-plane friction. It consists of a series of alternating plates or disks so arranged that one set engages with an outside cylindrical case and the other set with the shaft. When these plates are pressed together by spring pressure, or by other means, motion is transmitted from the driving to the driven members connected to the clutch. Some disk clutches have a few rather heavy or thick plates and others a relatively large number of thinner plates. Clutches of the latter type are common in automobile construction. One set of disks may be of soft steel and the other set of phosphor-bronze, or some other combination may be employed. For instance, disks are sometimes provided with cork inserts, or one set or series of disks may be faced with a special friction material such as asbestos-wire fabric, as in the case of "dry plate" clutches, the disks of which are not lubricated like the disks of a clutch having, for example, the steel and phosphor-bronze combination. It is common practice to hold the driving and driven members of friction clutches into engagement by means of spring pressure, although pneumatic and hydraulic pressure is sometimes employed.

### Cast-iron Friction Clutches

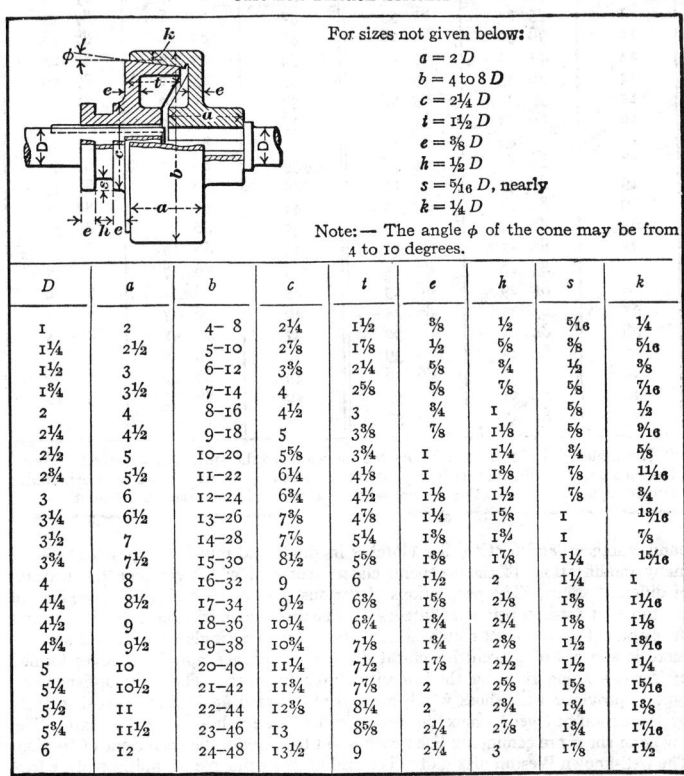

For sizes not given below:

$a = 2 D$
$b = 4 \text{ to } 8 D$
$c = 2\frac{1}{4} D$
$t = 1\frac{1}{2} D$
$e = \frac{3}{8} D$
$h = \frac{1}{2} D$
$s = \frac{5}{16} D$, nearly
$k = \frac{1}{4} D$

Note: — The angle $\phi$ of the cone may be from 4 to 10 degrees.

| D | a | b | c | t | e | h | s | k |
|---|---|---|---|---|---|---|---|---|
| 1 | 2 | 4– 8 | 2¼ | 1½ | ⅜ | ½ | 5/16 | ¼ |
| 1¼ | 2½ | 5–10 | 2⅞ | 1⅞ | ½ | ⅝ | ⅜ | 5/16 |
| 1½ | 3 | 6–12 | 3⅜ | 2¼ | ⅝ | ¾ | ½ | ⅜ |
| 1¾ | 3½ | 7–14 | 4 | 2⅝ | ⅝ | ⅞ | ⅝ | 7/16 |
| 2 | 4 | 8–16 | 4½ | 3 | ¾ | 1 | ⅝ | ½ |
| 2¼ | 4½ | 9–18 | 5 | 3⅜ | ⅞ | 1⅛ | ⅝ | 9/16 |
| 2½ | 5 | 10–20 | 5⅝ | 3¾ | 1 | 1¼ | ¾ | ⅝ |
| 2¾ | 5½ | 11–22 | 6¼ | 4⅛ | 1 | 1⅜ | ⅞ | 11/16 |
| 3 | 6 | 12–24 | 6¾ | 4½ | 1⅛ | 1½ | ⅞ | ¾ |
| 3¼ | 6½ | 13–26 | 7⅜ | 4⅞ | 1¼ | 1⅝ | 1 | 13/16 |
| 3½ | 7 | 14–28 | 7⅞ | 5¼ | 1⅜ | 1¾ | 1 | ⅞ |
| 3¾ | 7½ | 15–30 | 8½ | 5⅝ | 1⅜ | 1⅞ | 1¼ | 15/16 |
| 4 | 8 | 16–32 | 9 | 6 | 1½ | 2 | 1¼ | 1 |
| 4¼ | 8½ | 17–34 | 9½ | 6⅜ | 1⅝ | 2⅛ | 1⅜ | 1 1/16 |
| 4½ | 9 | 18–36 | 10¼ | 6¾ | 1¾ | 2¼ | 1⅜ | 1⅛ |
| 4¾ | 9½ | 19–38 | 10¾ | 7⅛ | 1¾ | 2⅜ | 1½ | 1 3/16 |
| 5 | 10 | 20–40 | 11¼ | 7½ | 1⅞ | 2½ | 1½ | 1¼ |
| 5¼ | 10½ | 21–42 | 11¾ | 7⅞ | 2 | 2⅝ | 1⅝ | 1 5/16 |
| 5½ | 11 | 22–44 | 12⅜ | 8¼ | 2 | 2¾ | 1¾ | 1⅜ |
| 5¾ | 11½ | 23–46 | 13 | 8⅝ | 2¼ | 2⅞ | 1¾ | 1 7/16 |
| 6 | 12 | 24–48 | 13½ | 9 | 2¼ | 3 | 1⅞ | 1½ |

**Power Transmitting Capacity of Friction Clutches.** — When selecting a clutch for a given class of service, it is advisable to consider any overloads that may be encountered and base the power transmitting capacity of the clutch upon such overloads. When the load varies or is subject to frequent release or engagement, the clutch capacity should be greater than the actual amount of power transmitted. If the power is derived from a gas or gasoline engine, the horsepower rating of the clutch should be 75 or 100 per cent greater than that of the engine.

**Formulas for Cone Clutches.** — In cone clutch design, different formulas have been developed for determining the horsepower transmitted. These formulas, at first sight, do not seem to agree, there being a variation due to the fact that in some of the formulas the friction clutch surfaces are assumed to engage without slip, whereas, in others, some allowance is made for slip. The following formulas include both of these conditions:

H.P. = horsepower transmitted;
$N$ = revolutions per minute;
$r$ = mean radius of friction cone, in inches;
$r_1$ = large radius of friction cone, in inches;
$r_2$ = small radius of friction cone, in inches;
$R_1$ = outside radius of leather band, in inches;
$R_2$ = inside radius of leather band, in inches;
$V$ = velocity of a point at distance $r$ from the center, in feet per minute;

$F$ = tangential force acting at radius $r$, in pounds;
$P_n$ = total normal pressure between cone surfaces, in pounds;
$P_s$ = spring pressure, in pounds;
$\alpha$ = angle of clutch surface with axis of shaft = 7 to 13 degrees;
$\beta$ = included angle of clutch leather, when developed, in degrees;
$f$ = coefficient of friction = 0.20 to 0.25 for greasy leather on iron;
$p$ = allowable pressure per square inch of leather band = 7 to 8 pounds;
$W$ = width of clutch leather, in inches.

DEVELOPMENT OF CLUTCH LEATHER

$$R_1 = \frac{r_1}{\sin \alpha} \qquad R_2 = \frac{r_2}{\sin \alpha}$$

$$\beta = \sin \alpha \times 360 \qquad r = \frac{r_1 + r_2}{2}$$

$$V = \frac{2 \pi r N}{12}$$

$$F = \frac{\text{H.P.} \times 33{,}000}{V} \qquad W = \frac{P_n}{2 \pi r p} \qquad \text{H.P.} = \frac{P_n f r N}{63{,}025}$$

For engagement with some slip:

$$P_n = \frac{P_s}{\sin \alpha} \qquad P_s = \frac{\text{H.P.} \times 63{,}025 \sin \alpha}{f r N}$$

For engagement without slip:

$$P_n = \frac{P_s}{\sin \alpha + f \cos \alpha} \qquad P_s = \frac{\text{H.P.} \times 63{,}025 (\sin \alpha + f \cos \alpha)}{f r N}$$

**Angle of Cone.** — If the angle of the conical surface of the cone type of clutch is too small, it may be difficult to release the clutch on account of the wedging effect, whereas, if the angle is too large, excessive spring pressure will be required to prevent slipping. The minimum angle for a leather-faced cone is about 8 or 9 degrees and the maximum angle about 13 degrees. An angle of 12½ degrees appears to be the most common and is generally considered good practice. These angles are given with relation to the clutch axis and are one-half the included angle.

**Power Transmitted by Disk Clutches.** — The approximate amount of power that a disk clutch will transmit may be determined from the following formula, in which $H$ = horsepower transmitted by the clutch: $\mu$ = coefficient of friction; $r$ = mean radius of engaging surfaces; $F$ = axial force in pounds (spring pressure) holding disks in contact; $N$ = number of, frictional surfaces; $S$ = speed of shaft in revolutions per minute:

$$H = \frac{\mu r F N S}{63,000}$$

**Frictional Coefficients for Clutch Calculations.** — While the frictional coefficients used by designers of clutches differ somewhat and depend upon variable factors, the following values may be used in clutch calculations: For greasy leather on cast iron about 0.20 or 0.25, leather on metal that is quite oily 0.15; metal and cork on oily metal 0.32; the same on dry metal 0.35; metal on dry metal 0.15; disk clutches having lubricated surfaces 0.10.

**Magnetic Clutches.** — Clutches of the magnetic type, like other electrical apparatus, are adapted to remote and automatic control. They are especially applicable for high-speed drives; for heavy duty, for use with motors that cannot start heavy loads; and for stopping machinery quickly, in which case a brake is used in combination with the clutch. The Cutler-Hammer magnetic clutch has a field or driving member and an armature or driven member. Each of these parts is carried by a flexible spring steel plate so that when current passes through the winding of the field, the armature is attracted to it and the friction surfaces come into engagement. The turning power of the clutch depends entirely upon the friction surfaces which are held together by magnetic attraction. Current is conducted to the magnetizing winding of the field through two collector rings and graphite brushes. These clutches are operated by direct current. The ratings of some of the different sizes are given in the accompanying table.

### Magnetic Clutch Ratings *

| Nominal Size, Inches | Maximum Speed, R.P.M. | Ratings Type H-30 Clutches | | | Ratings Type H-60 Clutches | | |
|---|---|---|---|---|---|---|---|
| | | Maximum Torque, Lbs. at 1 Ft. Radius | Safe H.P. at 100 R.P.M. | Current Consumption, Watts | Maximum Torque, Lbs. at 1 Ft. Radius | Safe H.P. at 100 R.P.M. | Current Consumption, Watts |
| 10 | 2000 | 89 | 1.1 | 78 | ....... | ....... | ...... |
| 12 | 1680 | 154 | 2.0 | 93 | ....... | ....... | ...... |
| 14 | 1440 | 245 | 3.0 | 115 | 490 | 6 | 130 |
| 16 | 1260 | 366 | 4.5 | 133 | 732 | 9 | 160 |
| 20 | 1000 | 714 | 9.0 | 177 | 1,428 | 18 | 200 |
| 24 | 840 | 1233 | 15.5 | 260 | 2,466 | 31 | 247 |
| 28 | 725 | 1960 | 25.0 | 280 | 3,920 | 49 | 253 |
| 32 | 635 | 2920 | 37.0 | 315 | 5,840 | 74 | 250 |
| 40 | 500 | 5710 | 72.0 | 380 | 10,420 | 132 | 341 |
| 48 | 420 | 9860 | 124.0 | 460 | 19,720 | 250 | 400 |
| 60 | 340 | ....... | ....... | ...... | 38,600 | 485 | 645 |

* Cutler-Hammer Mfg. Co.

**Safety Flange Couplings**

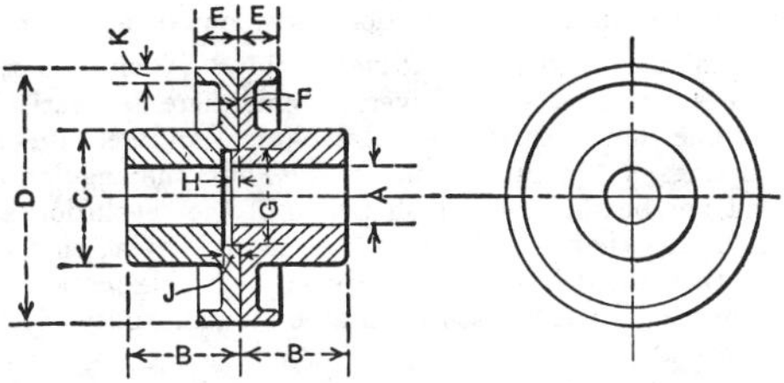


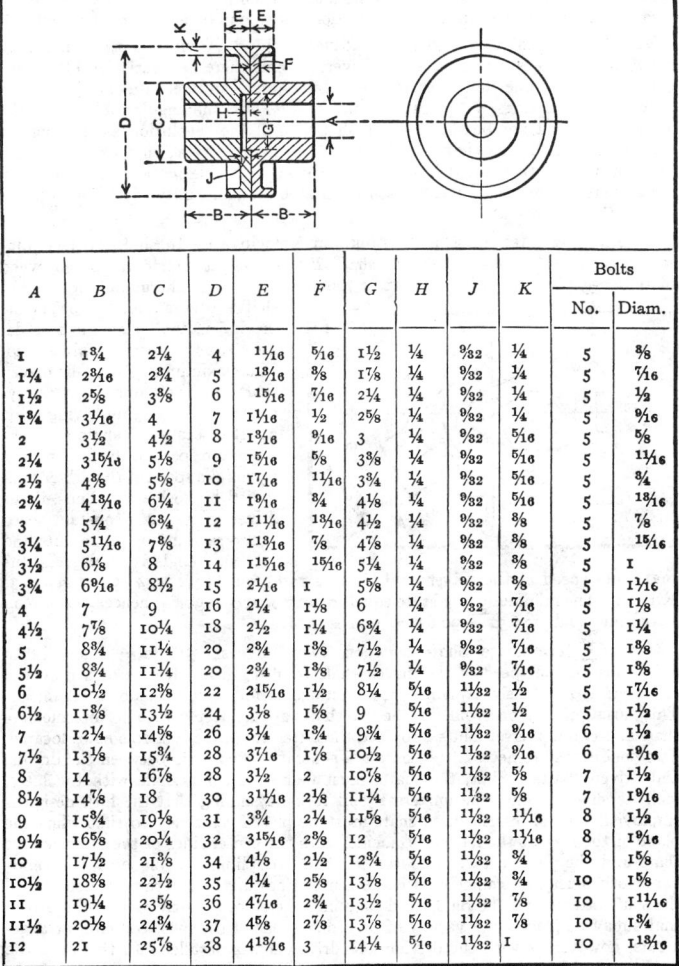

| A | B | C | D | E | F | G | H | J | K | Bolts No. | Bolts Diam. |
|---|---|---|---|---|---|---|---|---|---|---|---|
| 1 | 1 3/4 | 2 1/4 | 4 | 11/16 | 5/16 | 1 1/2 | 1/4 | 9/32 | 1/4 | 5 | 3/8 |
| 1 1/4 | 2 3/16 | 2 3/4 | 5 | 13/16 | 3/8 | 1 7/8 | 1/4 | 9/32 | 1/4 | 5 | 7/16 |
| 1 1/2 | 2 5/8 | 3 3/8 | 6 | 15/16 | 7/16 | 2 1/4 | 1/4 | 9/32 | 1/4 | 5 | 1/2 |
| 1 3/4 | 3 1/16 | 4 | 7 | 1 1/16 | 1/2 | 2 5/8 | 1/4 | 9/32 | 1/4 | 5 | 9/16 |
| 2 | 3 1/2 | 4 1/2 | 8 | 1 3/16 | 9/16 | 3 | 1/4 | 9/32 | 5/16 | 5 | 5/8 |
| 2 1/4 | 3 15/16 | 5 1/8 | 9 | 1 5/16 | 5/8 | 3 3/8 | 1/4 | 9/32 | 5/16 | 5 | 11/16 |
| 2 1/2 | 4 3/8 | 5 5/8 | 10 | 1 7/16 | 11/16 | 3 3/4 | 1/4 | 9/32 | 5/16 | 5 | 3/4 |
| 2 3/4 | 4 13/16 | 6 1/4 | 11 | 1 9/16 | 3/4 | 4 1/8 | 1/4 | 9/32 | 5/16 | 5 | 13/16 |
| 3 | 5 1/4 | 6 3/4 | 12 | 1 11/16 | 13/16 | 4 1/2 | 1/4 | 9/32 | 3/8 | 5 | 7/8 |
| 3 1/4 | 5 11/16 | 7 3/8 | 13 | 1 13/16 | 7/8 | 4 7/8 | 1/4 | 9/32 | 3/8 | 5 | 15/16 |
| 3 1/2 | 6 1/8 | 8 | 14 | 1 15/16 | 15/16 | 5 1/4 | 1/4 | 9/32 | 3/8 | 5 | 1 |
| 3 3/4 | 6 9/16 | 8 1/2 | 15 | 2 1/16 | 1 | 5 5/8 | 1/4 | 9/32 | 3/8 | 5 | 1 1/16 |
| 4 | 7 | 9 | 16 | 2 1/4 | 1 1/8 | 6 | 1/4 | 9/32 | 7/16 | 5 | 1 1/8 |
| 4 1/2 | 7 7/8 | 10 1/4 | 18 | 2 1/2 | 1 1/4 | 6 3/4 | 1/4 | 9/32 | 7/16 | 5 | 1 1/4 |
| 5 | 8 3/4 | 11 1/4 | 20 | 2 3/4 | 1 3/8 | 7 1/2 | 1/4 | 9/32 | 7/16 | 5 | 1 3/8 |
| 5 1/2 | 8 3/4 | 11 1/4 | 20 | 2 3/4 | 1 3/8 | 7 1/2 | 1/4 | 9/32 | 7/16 | 5 | 1 3/8 |
| 6 | 10 1/2 | 12 3/8 | 22 | 2 15/16 | 1 1/2 | 8 1/4 | 5/16 | 11/32 | 1/2 | 5 | 1 7/16 |
| 6 1/2 | 11 3/8 | 13 1/2 | 24 | 3 1/8 | 1 5/8 | 9 | 5/16 | 11/32 | 1/2 | 5 | 1 1/2 |
| 7 | 12 1/4 | 14 5/8 | 26 | 3 1/4 | 1 3/4 | 9 3/4 | 5/16 | 11/32 | 9/16 | 6 | 1 1/2 |
| 7 1/2 | 13 1/8 | 15 3/4 | 28 | 3 7/16 | 1 7/8 | 10 1/2 | 5/16 | 11/32 | 9/16 | 6 | 1 9/16 |
| 8 | 14 | 16 7/8 | 28 | 3 1/2 | 2 | 10 7/8 | 5/16 | 11/32 | 5/8 | 7 | 1 1/2 |
| 8 1/2 | 14 7/8 | 18 | 30 | 3 11/16 | 2 1/8 | 11 1/4 | 5/16 | 11/32 | 5/8 | 7 | 1 9/16 |
| 9 | 15 3/4 | 19 1/8 | 31 | 3 3/4 | 2 1/4 | 11 5/8 | 5/16 | 11/32 | 11/16 | 8 | 1 1/2 |
| 9 1/2 | 16 5/8 | 20 1/4 | 32 | 3 15/16 | 2 3/8 | 12 | 5/16 | 11/32 | 11/16 | 8 | 1 9/16 |
| 10 | 17 1/2 | 21 3/8 | 34 | 4 1/8 | 2 1/2 | 12 3/4 | 5/16 | 11/32 | 3/4 | 8 | 1 5/8 |
| 10 1/2 | 18 3/8 | 22 1/2 | 35 | 4 1/4 | 2 5/8 | 13 1/8 | 5/16 | 11/32 | 3/4 | 10 | 1 5/8 |
| 11 | 19 1/4 | 23 5/8 | 36 | 4 7/16 | 2 3/4 | 13 1/2 | 5/16 | 11/32 | 7/8 | 10 | 1 11/16 |
| 11 1/2 | 20 1/8 | 24 3/4 | 37 | 4 5/8 | 2 7/8 | 13 7/8 | 5/16 | 11/32 | 7/8 | 10 | 1 3/4 |
| 12 | 21 | 25 7/8 | 38 | 4 13/16 | 3 | 14 1/4 | 5/16 | 11/32 | 1 | 10 | 1 13/16 |

**The Universal Joint.** — This form of coupling, originally known as Hooke's coupling, is used for connecting two shafts the axes of which are not in line with each other, but which merely intersect at a point. There are many different designs of universal joints or couplings, which are based on the principle embodied in the original design. One well-known type is shown by the accompanying diagram.

As a rule, a universal joint does not work well if the angle $\alpha$ (see illustration) is more than 45 degrees, and the angle should preferably be limited to about 20 degrees or 25 degrees, excepting when the speed of rotation is slow and little power is transmitted.

**Variation in Angular Velocity of Driven Shaft.** — Owing to the angularity between two shafts connected by a universal joint, there is a variation in the angular velocity of one shaft during a single revolution, and because of this, the use of universal couplings is sometimes prohibited. Thus, the angular velocity of the driven shaft will not be the same at all points of the revolution as the angular velocity of the driving shaft. In other words, if the driving shaft moves with a uniform motion, then the driven shaft will have a variable motion and, therefore, the universal joint should not be used when absolute uniformity of motion is essential for the driven shaft.

**Determining Maximum and Minimum Velocities.** — If shaft $A$ (see diagram) runs at a constant speed, shaft $B$ revolves at maximum speed when shaft $A$ occupies the position shown in the illustration, and the minimum speed of

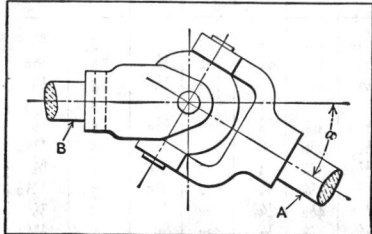

shaft $B$ occurs when the fork of the driving shaft $A$ has turned 90 degrees from the position illustrated. The maximum speed of the driven shaft may be obtained by multiplying the speed of the driving shaft by the secant of angle $\alpha$. The minimum speed of the driven shaft equals the speed of the driver multiplied by cosine $\alpha$. Thus, if the driver rotates at a constant speed of 100 revolutions per minute and the shaft angle is 25 degrees, the maximum speed of the driven shaft is at a rate equal to $1.1034 \times 100 = 110.34$ R.P.M. The minimum speed rate equals $0.9063 \times 100 = 90.63$; hence, the extreme variation equals $110.34 - 90.63 = 19.71$ R.P.M.

**Use of Intermediate Shaft between Two Universal Joints.** — The lack of uniformity in the speed of the driven shaft resulting from the use of a universal coupling, as previously explained, is objectionable for some forms of mechanisms. This variation may be avoided if the two shafts are connected with an intermediate shaft and two universal joints, provided the latter are properly arranged or located. Two conditions are necessary to obtain a constant speed ratio between the driving and driven shafts. First, the shafts must make the same angle with the intermediate shaft; second, the universal joint forks (assuming that the fork design is employed) on the intermediate shaft must be placed relatively so that when the plane of the fork at the left end coincides with the center lines of the intermediate shaft and the shaft attached to the left-hand coupling, the plane of the right-hand fork must also coincide with the center lines of the intermediate shaft and the shaft attached to the right-hand coupling; therefore the driving and the driven shafts may be placed in a variety of positions. One of the most common arrangements, however, is with the driving and driven shafts parallel. In this case, the forks on the intermediate shafts should be placed in the same plane.

This intermediate connecting shaft is frequently made telescoping, and then the driving and driven shafts can be moved independently of each other within certain limits in longitudinal and lateral directions. The telescoping intermediate shaft consists of a rod which enters a sleeve and is provided with a suitable spline, to prevent rotation between the rod and sleeve and permit a sliding movement. This arrangement is applied to various machine tools.

## Proportions of Knuckle Joints

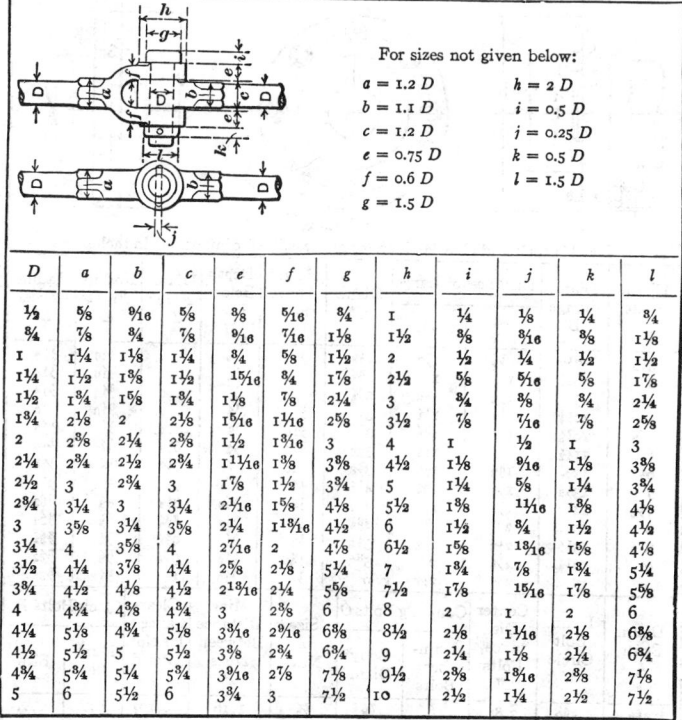

For sizes not given below:

$a = 1.2\,D$  $h = 2\,D$
$b = 1.1\,D$  $i = 0.5\,D$
$c = 1.2\,D$  $j = 0.25\,D$
$e = 0.75\,D$  $k = 0.5\,D$
$f = 0.6\,D$  $l = 1.5\,D$
$g = 1.5\,D$

| D | a | b | c | e | f | g | h | i | j | k | l |
|---|---|---|---|---|---|---|---|---|---|---|---|
| ½ | ⅝ | ⁹⁄₁₆ | ⅝ | ⅜ | ⁵⁄₁₆ | ¾ | 1 | ¼ | ⅛ | ¼ | ¾ |
| ¾ | ⅞ | ¾ | ⅞ | ⁹⁄₁₆ | ⁷⁄₁₆ | 1⅛ | 1½ | ⅜ | ³⁄₁₆ | ⅜ | 1⅛ |
| 1 | 1¼ | 1⅛ | 1¼ | ¾ | ⅝ | 1½ | 2 | ½ | ¼ | ½ | 1½ |
| 1¼ | 1½ | 1⅜ | 1½ | ¹⁵⁄₁₆ | ¾ | 1⅞ | 2½ | ⅝ | ⁵⁄₁₆ | ⅝ | 1⅞ |
| 1½ | 1¾ | 1⅝ | 1¾ | 1⅛ | ⅞ | 2¼ | 3 | ¾ | ⅜ | ¾ | 2¼ |
| 1¾ | 2⅛ | 2 | 2⅛ | 1⁵⁄₁₆ | 1¹⁄₁₆ | 2⅝ | 3½ | ⅞ | ⁷⁄₁₆ | ⅞ | 2⅝ |
| 2 | 2⅜ | 2¼ | 2⅜ | 1½ | 1³⁄₁₆ | 3 | 4 | 1 | ½ | 1 | 3 |
| 2¼ | 2¾ | 2½ | 2¾ | 1¹¹⁄₁₆ | 1⅜ | 3⅜ | 4½ | 1⅛ | ⁹⁄₁₆ | 1⅛ | 3⅜ |
| 2½ | 3 | 2¾ | 3 | 1⅞ | 1½ | 3¾ | 5 | 1¼ | ⅝ | 1¼ | 3¾ |
| 2¾ | 3¼ | 3 | 3¼ | 2¹⁄₁₆ | 1⅝ | 4⅛ | 5½ | 1⅜ | ¹¹⁄₁₆ | 1⅜ | 4⅛ |
| 3 | 3⅝ | 3¼ | 3⅝ | 2¼ | 1¹³⁄₁₆ | 4½ | 6 | 1½ | ¾ | 1½ | 4½ |
| 3¼ | 4 | 3⅝ | 4 | 2⁷⁄₁₆ | 2 | 4⅞ | 6½ | 1⅝ | ¹³⁄₁₆ | 1⅝ | 4⅞ |
| 3½ | 4¼ | 3⅞ | 4¼ | 2⅝ | 2⅛ | 5¼ | 7 | 1¾ | ⅞ | 1¾ | 5¼ |
| 3¾ | 4½ | 4⅛ | 4½ | 2¹³⁄₁₆ | 2¼ | 5⅝ | 7½ | 1⅞ | ¹⁵⁄₁₆ | 1⅞ | 5⅝ |
| 4 | 4¾ | 4⅜ | 4¾ | 3 | 2⅜ | 6 | 8 | 2 | 1 | 2 | 6 |
| 4¼ | 5⅛ | 4¾ | 5⅛ | 3³⁄₁₆ | 2⁹⁄₁₆ | 6⅜ | 8½ | 2⅛ | 1¹⁄₁₆ | 2⅛ | 6⅜ |
| 4½ | 5½ | 5 | 5½ | 3⅜ | 2¾ | 6¾ | 9 | 2¼ | 1⅛ | 2¼ | 6¾ |
| 4¾ | 5¾ | 5¼ | 5¾ | 3⁹⁄₁₆ | 2⅞ | 7⅛ | 9½ | 2⅜ | 1³⁄₁₆ | 2⅜ | 7⅛ |
| 5 | 6 | 5½ | 6 | 3¾ | 3 | 7½ | 10 | 2½ | 1¼ | 2½ | 7½ |

**Flexible Couplings.** — Flexible couplings are used mostly for coupling together electrical machinery or for coupling electrical to other machinery. The general types of flexible couplings include the leather link coupling, the endless belt coupling, and the rubber buffer coupling. The leather link coupling consists of two iron castings with flanges which are connected by leather links and bolts. The bolts are generally six in number and each alternate bolt is tightly fitted in the flange of one casting, but has considerable play in the other. The leather links are placed around pairs of adjacent bolts and provide a slight flexibility for the drive. This coupling is adapted for shafts up to 3½ inches in diameter. The endless belt flexible coupling is adapted for shafts of larger diameter. It consists of two steel rings, one outer and one inner, in which slots are formed and through which two endless leather belts are interwoven. The rubber buffer coupling is formed of two disks; the driving side transmits motion to the driven side by means of studs, bolts or interlocking arms surrounded by heavy rubber bushings which give the necessary flexibility. The "mill type" flexible coupling, which is used chiefly in steel mills, is formed of three steel castings and is adapted to severe service.

**American Standard Shaft Couplings***

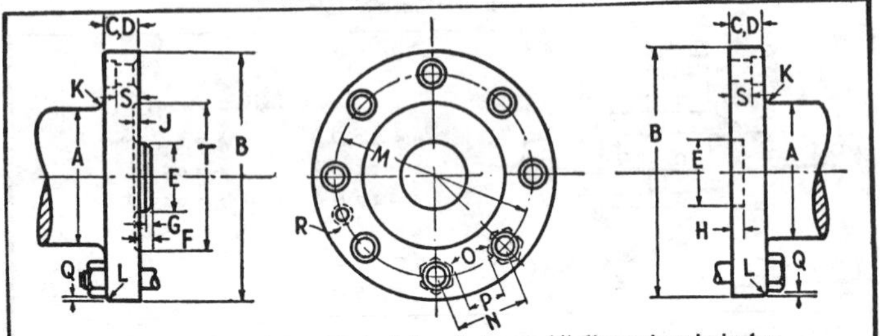


*J* = 1/32 inch and *G* = 1/16 inch for all sizes. All dimensions in inches.

| Diam. Shaft *A* | Diam. Flange *B* | Thickness of Flange *C** | Thickness of Flange *D* | Diam. *E** | Height Above Face *F* | Depth Below Face *H* | Diam. Relief *T* | Radius Fillet *K* | Radius Corner *L** |
|---|---|---|---|---|---|---|---|---|---|
| 3½ | 7½ | ⅞ | ⅞ | 2⅛ | 3/16 | ¼ | ..... | 3/16 | 1/16 |
| 4 | 8½ | 1 | 1 | 2⅜ | 3/16 | ¼ | ..... | 3/16 | 1/16 |
| 4½ | 9 | 1⅛ | 1⅛ | 2¾ | 3/16 | ¼ | ..... | 3/16 | 1/16 |
| 5 | 10¼ | 1¼ | 1¼ | 3 | 3/16 | ¼ | ..... | 3/16 | 1/16 |
| 5½ | 10⅞ | 1⅜ | 1⅜ | 3¼ | 3/16 | ¼ | ..... | ¼ | 1/16 |
| 6 | 11½ | 1½ | 1½ | 3⅝ | 3/16 | ¼ | 6⅝ | ¼ | 1/16 |
| 6½ | 12¾ | 1⅝ | 1⅝ | 3⅞ | 3/16 | ¼ | 7⅛ | ¼ | 1/16 |
| 7 | 13½ | 1¾ | 1¾ | 4¼ | ¼ | 5/16 | 7¾ | ⅜ | 3/32 |
| 7½ | 14 | 1⅞ | 1¾ | 4½ | ¼ | 5/16 | 8¼ | ⅜ | 3/32 |
| 8 | 15 | 2 | 1⅞ | 4¾ | ¼ | 5/16 | 8¾ | ½ | 3/32 |
| 8½ | 15½ | 2⅛ | 1⅞ | 5⅛ | ¼ | 5/16 | 9⅜ | ½ | 3/32 |
| 9 | 16¼ | 2¼ | 2 | 5⅜ | ¼ | 5/16 | 9⅞ | ⅝ | 3/32 |
| 10 | 18 | 2½ | 2¼ | 6 | ¼ | 5/16 | 11 | ¾ | 3/32 |

| Diam. Shaft *A* | Diam. Bolt Circle *M* | Center Distance, Bolts *N* | Coupling Bolts *O* Number Bolts | Coupling Bolts *O* Body Diam. | Size of Nuts | Min. Distance Between Nuts *P* | Clearance for Nut Guard *Q* | Jack Bolts *R* Number | Jack Bolts *R* Size |
|---|---|---|---|---|---|---|---|---|---|
| 3½ | 5⅝ | 2.81 | 6 | ⅞ | ⅞ | 1.16 | 0.107 | ..... | ..... |
| 4 | 6¼ | 3.12 | 6 | ⅞ | ⅞ | 1.47 | 0.107 | ..... | ..... |
| 4½ | 6⅞ | 3.43 | 6 | 1 | 1 | 1.56 | 0.124 | ..... | ..... |
| 5 | 7¾ | 3.87 | 6 | 1¼ | 1¼ | 1.56 | 0.095 | ..... | ..... |
| 5½ | 8⅜ | 4.18 | 6 | 1¼ | 1¼ | 1.87 | 0.095 | ..... | ..... |
| 6 | 9 | 3.44 | 8 | 1¼ | 1¼ | 1.13 | 0.095 | ..... | ..... |
| 6½ | 9¾ | 4.87 | 6 | 1½ | 1½ | 2.13 | 0.128 | ..... | ..... |
| 7 | 10½ | 4.02 | 8 | 1½ | 1½ | 1.28 | 0.129 | 2 | ¾ |
| 7½ | 11 | 4.21 | 8 | 1½ | 1½ | 1.50 | 0.13 | 2 | ¾ |
| 8 | 12 | 4.59 | 8 | 1¾ | 1½ | 1.84 | 0.13 | 2 | ¾ |
| 8½ | 12½ | 4.78 | 8 | 1¾ | 1½ | 2.04 | 0.13 | 2 | ¾ |
| 9 | 13¼ | 4.10 | 10 | 1¾ | 1½ | 1.36 | 0.13 | 2 | ¾ |
| 10 | 14½ | 4.48 | 10 | 2 | 1¾ | 1.29 | 0.16 | 2 | ¾ |

*Integrally forged flange type for hydro-electric units. Complete standard includes sizes up to 40 inches shaft diameter. Flange thicknesses *C* are for shafts subject to bending.

Fitting allowance on dimension *E* for male half-coupling shall be −0.001 inch. Radius *L* omitted when a nut guard is furnished. When outside diameter of nut guard is equal to flange diameter *B* a recess is provided in place of radius *L*. For bolt sizes under 1 inch, use American National Standard, Fine-Thread Series; for sizes 1 inch to 2 inches inclusive, 8 threads per inch; for sizes over 2 inches, 6 threads per inch. Jack bolts furnished with horizontal shafts only and provided in the male half-coupling.

**Double-cone Clamping Coupling**

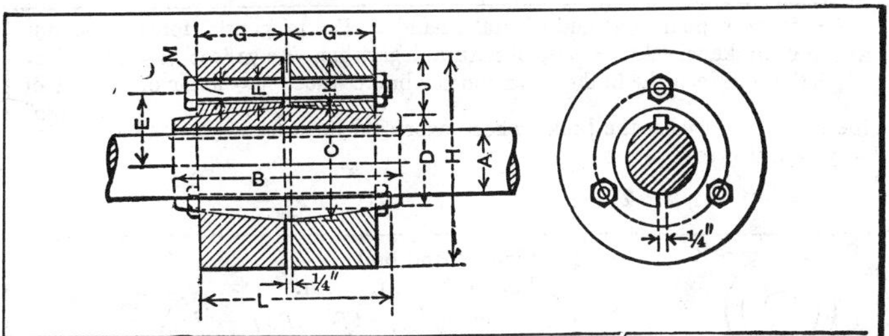


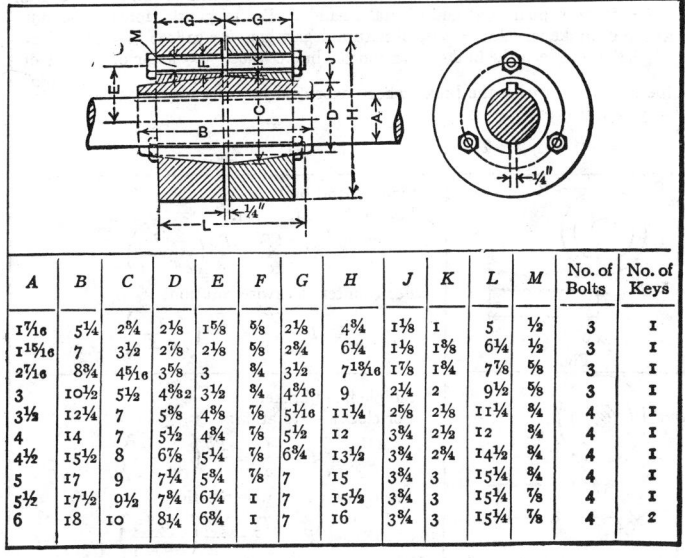

| A | B | C | D | E | F | G | H | J | K | L | M | No. of Bolts | No. of Keys |
|---|---|---|---|---|---|---|---|---|---|---|---|---|---|
| 1 7/16 | 5 1/4 | 2 3/4 | 2 1/8 | 1 5/8 | 5/8 | 2 1/8 | 4 3/4 | 1 1/8 | 1 | 5 | 1/2 | 3 | 1 |
| 1 15/16 | 7 | 3 1/2 | 2 7/8 | 2 1/8 | 5/8 | 2 3/4 | 6 1/4 | 1 1/8 | 1 3/8 | 6 1/4 | 1/2 | 3 | 1 |
| 2 7/16 | 8 3/4 | 4 5/16 | 3 5/8 | 3 | 3/4 | 3 1/2 | 7 13/16 | 1 7/8 | 1 3/4 | 7 7/8 | 5/8 | 3 | 1 |
| 3 | 10 1/2 | 5 1/2 | 4 3/32 | 3 1/2 | 3/4 | 4 3/16 | 9 | 2 1/4 | 2 | 9 1/2 | 5/8 | 3 | 1 |
| 3 1/2 | 12 1/4 | 7 | 5 3/8 | 4 3/8 | 7/8 | 5 1/16 | 11 1/4 | 2 5/8 | 2 1/8 | 11 1/4 | 3/4 | 4 | 1 |
| 4 | 14 | 7 | 5 1/2 | 4 3/4 | 7/8 | 5 1/2 | 12 | 3 3/4 | 2 1/2 | 12 | 3/4 | 4 | 1 |
| 4 1/2 | 15 1/2 | 8 | 6 7/8 | 5 1/4 | 7/8 | 6 3/4 | 13 1/2 | 3 3/4 | 2 3/4 | 14 1/2 | 3/4 | 4 | 1 |
| 5 | 17 | 9 | 7 1/4 | 5 3/4 | 7/8 | 7 | 15 | 3 3/4 | 3 | 15 1/4 | 3/4 | 4 | 1 |
| 5 1/2 | 17 1/2 | 9 1/2 | 7 3/4 | 6 1/4 | 1 | 7 | 15 1/2 | 3 3/4 | 3 | 15 1/4 | 7/8 | 4 | 1 |
| 6 | 18 | 10 | 8 1/4 | 6 3/4 | 1 | 7 | 16 | 3 3/4 | 3 | 15 1/4 | 7/8 | 4 | 2 |

# FRICTION BRAKES

**Formulas for Band Brakes.** — In any band brake, such as shown in Fig. 1, in the tabulation of the formulas, where the brake wheel rotates in a clockwise direction, the tension in that part of the band marked $x$ equals $P\,\dfrac{1}{e^{\mu\theta}-1}$

The tension in that part marked $y$ equals $P\,\dfrac{e^{\mu\theta}}{e^{\mu\theta}-1}$

$P$ = tangential force in pounds at rim of brake wheel;
$e$ = base of natural logarithms = 2.71828;
$\mu$ = coefficient of friction between the brake band and the brake wheel;
$\theta$ = angle of contact of the brake band with the brake wheel expressed in radians (one radian $= \dfrac{180\text{ deg.}}{\pi} = 57.296$ degrees).

For simplicity in the formulas presented, the tensions at $x$ and $y$ (Fig. 1) are denoted by $T_1$ and $T_2$ respectively, for clockwise rotation. When the direction of the rotation is reversed, the tension in $x$ equals $T_2$, and the tension in $y$ equals $T_1$, which is the reverse of the tension in the clockwise direction.

The value of the expression $e^{\mu\theta}$ in these formulas may be most easily solved by means of logarithms. The value of $e^{\mu\theta}$ is found by multiplying the logarithm of $e$ by the product of the numerical values of $\mu$ and $\theta$, and finding the number whose logarithm is equal to the result of this multiplication. The procedure may be best illustrated by an example.

### Formulas for Simple and Differential Band Brakes

$F$ = force in pounds at end of brake handle; $P$ = tangential force in pounds at rim of brake wheel; $e$ = base of natural logarithms = 2.71828; $\mu$ = coefficient of friction between the brake band and the brake wheel; $\theta$ = angle of contact of the brake band with the brake wheel, expressed in radians (one radian = $\dfrac{180°}{\pi}$ = 57.296 degrees).

$$T_1 = P\,\frac{1}{e^{\mu\theta} - 1} \qquad\qquad T_2 = P\,\frac{e^{\mu\theta}}{e^{\mu\theta} - 1}$$

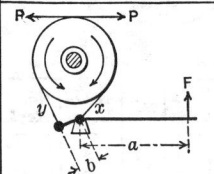

Fig. 1

Simple band brake.
For clockwise rotation:

$$F = \frac{bT_2}{a} = \frac{Pb}{a}\left(\frac{e^{\mu\theta}}{e^{\mu\theta} - 1}\right)$$

For counter clockwise rotation:

$$F = \frac{bT_1}{a} = \frac{Pb}{a}\left(\frac{1}{e^{\mu\theta} - 1}\right)$$

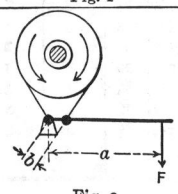

Fig. 2

Simple band brake.
For clockwise rotation:

$$F = \frac{bT_1}{a} = \frac{Pb}{a}\left(\frac{1}{e^{\mu\theta} - 1}\right)$$

For counter clockwise rotation:

$$F = \frac{bT_2}{a} = \frac{Pb}{a}\left(\frac{e^{\mu\theta}}{e^{\mu\theta} - 1}\right)$$

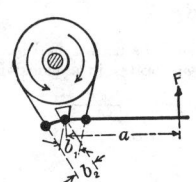

Fig. 3

Differential band brake.
For clockwise rotation:

$$F = \frac{b_2T_2 - b_1T_1}{a} = \frac{P}{a}\left(\frac{b_2e^{\mu\theta} - b_1}{e^{\mu\theta} - 1}\right)$$

For counter clockwise rotation:

$$F = \frac{b_2T_1 - b_1T_2}{a} = \frac{P}{a}\left(\frac{b_2 - b_1e^{\mu\theta}}{e^{\mu\theta} - 1}\right)$$

In this case, if $b_2$ is equal to, or less than, $b_1e^{\mu\theta}$, the force $F$ will be 0 or negative and the band brake works automatically.

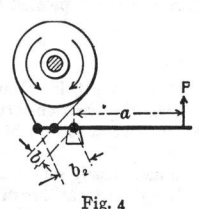

Fig. 4

Differential band brake.
For clockwise rotation:

$$F = \frac{b_2T_2 + b_1T_1}{a} = \frac{P}{a}\left(\frac{b_2e^{\mu\theta} + b_1}{e^{\mu\theta} - 1}\right)$$

For counter clockwise rotation:

$$F = \frac{b_1T_2 + b_2T_1}{a} = \frac{P}{a}\left(\frac{b_1e^{\mu\theta} + b_2}{e^{\mu\theta} - 1}\right)$$

If $b_2 = b_1$, both of the above formulas reduce to $F = \dfrac{Pb_1}{a}\left(\dfrac{e^{\mu\theta} + 1}{e^{\mu\theta} - 1}\right)$. In this case, the same force $F$ is required for rotation in either direction.

In a band brake of the type in Fig. 1, dimension $a = 24$ inches, and $b = 4$ inches; force $P = 100$ pounds; coefficient $\mu = 0.2$, and angle of contact $= 240$ degrees, or

$$\theta = \frac{240}{180} \times \pi = 4.18.$$

The rotation is clockwise. Find force $F$ required.

$$F = \frac{Pb}{a}\left(\frac{e^{\mu\theta}}{e^{\mu\theta}-1}\right) = \frac{100 \times 4}{24}\left(\frac{2.71828^{0.2\times4.18}}{2.71828^{0.2\times4.18}-1}\right)$$

$$= \frac{400}{24} \times \frac{2.71828^{0.836}}{2.71828^{0.836}-1} = 16.66 \times \frac{2.31}{2.31-1} = 29.4.$$

The calculations for determining the value of $e^{\mu\theta}$ are rather cumbersome, and the accompanying table will save calculations.

### Table of Values of $e^{\mu\theta}$

| Proportion of Contact to Whole Circumference, $\frac{\theta}{2\pi}$ | Steel Band on Cast Iron, $\mu = 0.18$ | Leather Belt on | | | |
|---|---|---|---|---|---|
| | | Wood | Cast Iron | | |
| | | Slightly Greasy; $\mu = 0.47$ | Very Greasy; $\mu = 0.12$ | Slightly Greasy; $\mu = 0.28$ | Damp; $\mu = 0.38$ |
| 0.1 | 1.12 | 1.34 | 1.08 | 1.19 | 1.27 |
| 0.2 | 1.25 | 1.81 | 1.16 | 1.42 | 1.61 |
| 0.3 | 1.40 | 2.43 | 1.25 | 1.69 | 2.05 |
| 0.4 | 1.57 | 3.26 | 1.35 | 2.02 | 2.60 |
| 0.425 | 1.62 | 3.51 | 1.38 | 2.11 | 2.76 |
| 0.45 | 1.66 | 3.78 | 1.40 | 2.21 | 2.93 |
| 0.475 | 1.71 | 4.07 | 1.43 | 2.31 | 3.11 |
| 0.5 | 1.76 | 4.38 | 1.46 | 2.41 | 3.30 |
| 0.525 | 1.81 | 4.71 | 1.49 | 2.52 | 3.50 |
| 0.55 | 1.86 | 5.07 | 1.51 | 2.63 | 3.72 |
| 0.6 | 1.97 | 5.88 | 1.57 | 2.81 | 4.19 |
| 0.7 | 2.21 | 7.90 | 1.66 | 3.43 | 5.32 |
| 0.8 | 2.47 | 10.60 | 1.83 | 4.09 | 6.75 |
| 0.9 | 2.77 | 14.30 | 1.97 | 4.87 | 8.57 |
| 1.0 | 3.10 | 19.20 | 2.12 | 5.81 | 10.90 |

**Coefficient of Friction in Brakes.** — The coefficients of friction that may be assumed for friction brake calculations are as follows: Iron on iron, 0.25 to 0.3; leather on iron, 0.3; cork on iron, 0.35. Values somewhat lower than these should be assumed when the velocities exceed 400 feet per minute at the beginning of the braking operation.

For brakes where wooden brake blocks are used on iron drums, poplar has proved the best brake-block material. The best material for the brake drum is wrought iron. Poplar gives a high coefficient of friction, and is little affected by oil. The average coefficient of friction for poplar brake blocks and wrought-iron drums is o.6; for poplar on cast iron, o.35; for oak on wrought iron, o.5; for oak on cast iron, o.3; for beech on wrought iron, o.5; for beech on cast iron, o.3; for elm on wrought iron, o.6; and for elm on cast iron, o.35. The objection to elm is that the friction decreases rapidly if the friction surfaces are oily. The coefficient of friction for elm and wrought iron, if oily, is less than o.4.

### Formulas for Block Brakes

> $F$ = force in pounds at end of brake handle;
> $P$ = tangential force in pounds at rim of brake wheel;
> $\mu$ = coefficient of friction between the brake block and brake wheel.

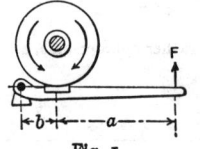

Fig. 1

Block brake.
For rotation in either direction:

$$F = P \frac{b}{a+b} \times \frac{1}{\mu} = \frac{Pb}{a+b}\left(\frac{1}{\mu}\right)$$

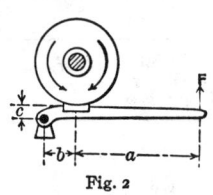

Fig. 2

Block brake.
For clockwise rotation:

$$F = \frac{\dfrac{Pb}{\mu} - Pc}{a+b} = \frac{Pb}{a+b}\left(\frac{1}{\mu} - \frac{c}{b}\right)$$

For counter clockwise rotation:

$$F = \frac{\dfrac{Pb}{\mu} + Pc}{a+b} = \frac{Pb}{a+b}\left(\frac{1}{\mu} + \frac{c}{b}\right)$$

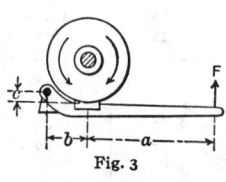

Fig. 3

Block brake.
For clockwise rotation:

$$F = \frac{\dfrac{Pb}{\mu} + Pc}{a+b} = \frac{Pb}{a+b}\left(\frac{1}{\mu} + \frac{c}{b}\right)$$

For counter clockwise rotation:

$$F = \frac{\dfrac{Pb}{\mu} - Pc}{a+b} = \frac{Pb}{a+b}\left(\frac{1}{\mu} - \frac{c}{b}\right)$$

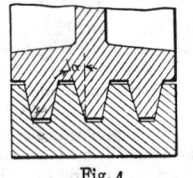

Fig. 4

The brake wheel and friction block of the block brake are often grooved as shown in Fig. 4. In this case, substitute for $\mu$ in the above equations the value $\dfrac{\mu}{\sin\alpha + \mu\cos\alpha}$ where $\alpha$ is one-half the angle included by the faces of the grooves.

**Calculating Horsepower from Dynamometer Tests.** — When a dynamometer is arranged for obtaining the horsepower transmitted by a shaft, as indicated by the diagrammatic view in the accompanying illustration, the horsepower may be obtained by the formula:

$$H.P. = \frac{2\pi LPN}{33,000}$$

in which H.P. = horsepower transmitted; $N$ = number of revolutions per minute; $L$ = distance (as shown in illustration) from center of pulley to point of action of weight $P$, in feet; $P$ = weight hung on brake arm or read on scale.

By adopting a length of brake arm equal to 5 feet 3 inches, the formula may be reduced to the simple form:

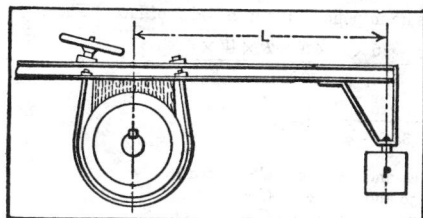

$$H.P. = \frac{NP}{1000}$$

If a length of brake arm equal to 2 feet 7½ inches is adopted as a standard, the formula takes the form:

$$H.P. = \frac{NP}{2000}$$

The *transmission* type of dynamometer measures the power by transmitting it through the mechanism of the dynamometer from the apparatus in which it is generated, or to the apparatus in which it is to be utilized. Dynamometers known as *indicators* operate by simultaneously measuring the pressure and volume of a confined fluid. This type may be used for the measurement of the power generated by steam or gas engines or absorbed by refrigerating machinery, air compressors, or pumps. An electrical dynamometer is for measuring the power of an electric current, based on the mutual action of currents flowing in two coils. It consists principally of one fixed and one movable coil, which, in the normal position, are at right angles to each other. Both coils are connected in series, and, when a current traverses the coils, the fields produced are at right angles; hence, the coils tend to take up a parallel position. The movable coil with an attached pointer will be deflected, the deflection measuring directly the electric current.

## Friction Wheels for Power Transmission

When a rotating member is driven intermittently and the rate of driving does not need to be positive, friction wheels are frequently used, especially when the amount of power to be transmitted is comparatively small. The driven wheels in a pair of friction disks should always be made of a harder material than the driving wheels, so that if the driven wheel should be held stationary by the load, while the driving wheel revolves under its own pressure, a flat spot may not be rapidly worn on the driven wheel. The driven wheels, therefore, are usually made of iron, while the driving wheels are made of or covered with, rubber, paper, leather, wood or fiber. The safe working pressures per inch of contact for various materials are as follows: Straw fiber, 150; leather fiber, 240; tarred fiber, 240; leather, 150; wood, 100 to 150; paper, 150. Coefficients of friction for different combinations of materials are given in the following table. Smaller values should be used for exceptionally high speeds, or when the transmission must be started while under load.

**Horsepower of Friction Wheels.** — Let $D$ = diameter of friction wheel in inches; $N$ = Number of revolutions per minute; $W$ = width of face in inches; $f$ =

### Working Values of Coefficient of Friction

| Materials | Coefficient of Friction | Materials | Coefficient of Friction |
|---|---|---|---|
| Straw fiber and cast iron...... | 0.26 | Tarred fiber and aluminum... | 0.18 |
| Straw fiber and aluminum.... | 0.27 | Leather and cast iron......... | 0.14 |
| Leather fiber and cast iron.... | 0.31 | Leather and aluminum........ | 0.22 |
| Leather fiber and aluminum.. | 0.30 | Leather and typemetal....... | 0.25 |
| Tarred fiber and cast iron..... | 0.15 | Wood and metal.............. | 0.25 |
| Paper and cast iron .......... | 0.20 | | |

coefficient of friction; $P$ = pressure in pounds, per inch width of face. Then:

$$H.P. = \frac{3.1416 \times D \times N \times P \times W \times f}{33,000 \times 12}$$

Assume

$$\frac{3.1416 \times P \times f}{33,000 \times 12} = C;$$

then,

for $P = 100$ and $f = 0.20$, $C = 0.00016$;
for $P = 150$ and $f = 0.20$, $C = 0.00024$;
for $P = 200$ and $f = 0.20$, $C = 0.00032$.

The horsepower transmitted is then:

$$H.P. = D \times N \times W \times C.$$

*Example:* — Find the horsepower transmitted by a pair of friction wheels; the diameter of the driving wheel is 10 inches, and it revolves at 200 revolutions per minute. The width of the wheel is 2 inches. The pressure per inch width of face is 150 pounds, and the coefficient of friction 0.20.

$$H.P. = 10 \times 200 \times 2 \times 0.00024 = 0.96 \text{ horsepower.}$$

### Horsepower Which May be Transmitted by Means of a Clean Paper Friction Wheel of One-inch Face when Run Under a Pressure of 150 Pounds

(Rockwood Mfg. Co.)

| Diameter of Friction | Revolutions per Minute | | | | | | | | | | |
|---|---|---|---|---|---|---|---|---|---|---|---|
| | 25 | 50 | 75 | 100 | 150 | 200 | 300 | 400 | 600 | 800 | 1000 |
| 4 | 0.023 | 0.047 | 0.071 | 0.095 | 0.142 | 0.190 | 0.285 | 0.380 | 0.571 | 0.76 | 0.95 |
| 6 | 0.035 | 0.071 | 0.107 | 0.142 | 0.214 | 0.285 | 0.428 | 0.571 | 0.856 | 1.14 | 1.42 |
| 8 | 0.047 | 0.095 | 0.142 | 0.190 | 0.285 | 0.380 | 0.571 | 0.761 | 1.142 | 1.52 | 1.90 |
| 10 | 0.059 | 0.119 | 0.178 | 0.238 | 0.357 | 0.476 | 0.714 | 0.952 | 1.428 | 1.90 | 2.38 |
| 14 | 0.083 | 0.166 | 0.249 | 0.333 | 0.499 | 0.666 | 0.999 | 1.332 | 1.999 | 2.66 | 3.33 |
| 16 | 0.095 | 0.190 | 0.285 | 0.380 | 0.571 | 0.761 | 1.142 | 1.523 | 2.284 | 3.04 | 3.80 |
| 18 | 0.107 | 0.214 | 0.321 | 0.428 | 0.642 | 0.856 | 1.285 | 1.713 | 2.570 | 3.42 | 4.28 |
| 24 | 0.142 | 0.285 | 0.428 | 0.571 | 0.856 | 1.142 | 1.713 | 2.284 | 3.427 | 4.56 | 5.71 |
| 30 | 0.178 | 0.357 | 0.535 | 0.714 | 1.071 | 1.428 | 2.142 | 2.856 | 4.284 | 5.71 | 7.14 |
| 36 | 0.214 | 0.428 | 0.642 | 0.856 | 1.285 | 1.713 | 2.570 | 3.427 | 5.140 | 6.85 | 8.56 |
| 42 | 0.249 | 0.499 | 0.749 | 0.999 | 1.499 | 1.999 | 2.998 | 3.998 | 5.997 | 7.99 | 9.99 |
| 48 | 0.285 | 0.571 | 0.856 | 1.142 | 1.713 | 2.284 | 3.427 | 4.569 | 6.854 | 9.13 | 11.42 |
| 50 | 0.297 | 0.595 | 0.892 | 1.190 | 1.785 | 2.380 | 3.570 | 4.760 | 7.140 | 9.52 | 11.90 |

# CAMS AND CAM DESIGN

**Classes of Cams.** — Cams may, in general, be divided into two classes: uniform motion cams and uniformly accelerated motion cams. The uniform motion cam moves the follower at the same rate of speed from the beginning to the end of the stroke; but as the movement is started from zero to the full speed of the uniform motion and stops in the same abrupt way, there is a distinct shock at the beginning and end of the stroke, if the movement is at all rapid. In machinery working at a high rate of speed, therefore, it is important that cams are so constructed that sudden shocks are avoided when starting the motion or when reversing the direction of motion of the follower.

The uniformly accelerated motion cam which is shown later, in Fig. 5, is suitable for moderate speeds, but it has the disadvantage of sudden changes in acceleration at the beginning, middle and end of the stroke. A cycloidal motion curve cam produces no abrupt changes in acceleration and is often used in high-speed machinery

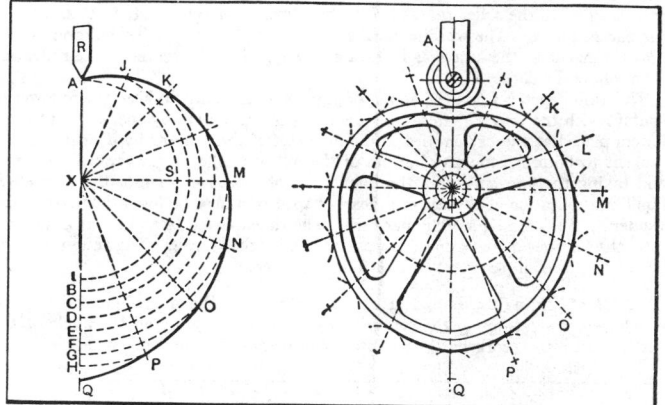

Fig 1                Fig. 2

because it results in low noise, vibration and wear. The cycloidal motion displacement curve is so called because it can be generated from a cycloid which is the locus of a point of a circle rolling on a straight line. Thus, not only a cycloidal motion cam profile but a whole family of similar profiles which are derived from modified cycloids may be generated comparatively easily.*

**Laying-out a Uniform Motion Cam.** — The laying-out of a heart-shaped cam will serve as an illustration of the general method. In Fig. 1, the pointed follower *R* is to be given a reciprocating motion. The throw is assumed to be 1½ inch. Let *X* be the center of the cam. Let *A* be the point at which the follower is at the lower end of the stroke. Draw semi-circle *ASI*, and extend the diameter at the side opposite *A*, a distance *IQ*, equal to the required throw. Divide *IQ* into any number of equal parts, as at *B*, *C*, *D*, etc., and divide the semi-circle by the same number of radii, equally distributed. With *X* as a center and a radius equal to *XB*, describe an arc intersecting *XJ* at *J*. With the same center and a radius equal to *XC*, describe an arc intersecting *XK* at *K*. Continue this process through

* Jensen, P. W., *Cam Design and Manufacture*, Industrial Press Inc., 1965.

the points *D, E, F*, etc., thus obtaining the points *L, M, N*, etc. The latter are points on the required curve. The other half is laid out in the same manner.

The excessive friction of a pointed follower such as that shown at *R* necessitates the use of a follower that will reduce the amount of friction to a minimum. A small roller meets this requirement. If a roller is employed as a follower, the problem of laying out the cam curve becomes modified. A roller traveling along the curve shown in Fig. 1 would not impart to the follower-rod the desired uniform rise and fall. The variation would be but slight, yet sufficient to merit consideration where accuracy is desired.

**Cams with Roller Followers.** — Fig. 2 represents a heart-shaped cam that is, designed for the same movement as the cam, Fig. 1, but it has a roller follower. The curve may be laid out as described in connection with the diagram, Fig. 1, because the curve of Fig. 1 represents the path followed by the center of the roller, Fig. 2. To locate the working surface of a cam having a roller type of follower, draw a series of arcs equal to the roller radius from various points or centers *A, J, K, L*, etc., on the curve, Fig. 1. The working surface of the cam having a roller follower is then drawn tangent to these arcs as illustrated by Fig. 2. The center of the roller will then follow the curve, Fig. 1.

This cam depends upon the action of gravity, or a spring, to keep the follower in contact with the driver. It can be made positive in action by the use of two followers placed at the extremities of the diameter of the cam, or by drawing curves tangent to both the top and bottom of the follower roller in its various positions, and taking the two curves as the boundaries of a groove cut into the metal. A familiar application of the use of a heart-shaped cam may be found in the bobbin-winder of the domestic sewing machine. The thread is fed to and fro at a uniform rate, the follower of the cam acting as a guide for the thread. The action is made positive by the employment of two follower rollers.

**Effect of Changing Location of Cam Roller.** — When the line of motion of a follower passes through the center of rotation of the cam, and the angle of the curve causes it to work hard, the curve may be modified, and the same motion of

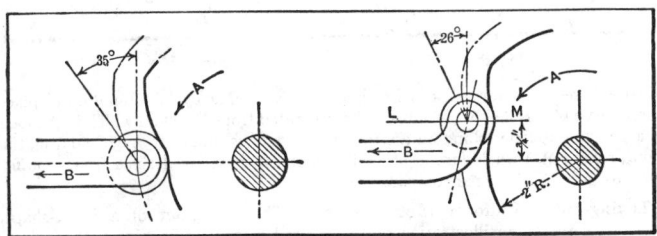

Fig. 3                                        Fig. 4

follower obtained by placing the follower with its line of action parallel to its original position and not passing through the center of the cam. A condition may be assumed, as shown in Fig. 3. Here the cam rotates in the direction indicated by arrow *A*. It moves the follower ¾ inch in the direction indicated by arrow *B*, during a 30-degree angle of motion of the cam-shaft. The angle of the cam presented to the follower at the beginning of the stroke would be 35 degrees, as determined by the tangent to the curve. Should the cam curve work hard at the required speed, the cam would be made of greater diameter, if possible, which would reduce the angle of the cam. The design of the machine, however, might make this change impossible. Another way consists in changing the location of the cam roller. In Fig. 4 all conditions are the same as in Fig. 3, except that the roller has been placed ¾ inch above the line passing through the center of the cam. The center of the roller will now pass along the line *LM*, or parallel to the line of motion in Fig. 3. The angle of the curve presented to the roller in this case is 26 degrees — much less than the angle in Fig. 3 — and the angle decreases as the roller moves away from the center of rotation. There is, of course, a limit to the distance the roller may be changed, for if placed too far away from the center line, the thrust in the direction at right angles to the direction of motion of the follower would be so great as to offset the advantage gained. Even without the aid of an illustration it may be seen that to place the cam roller on the other side of the center would cause the angle of the cam curve to increase, thus making conditions worse. The offset of the roller should be in the direction opposed to the direction of motion of the cam.

**Laying-out a Uniformly Accelerated Motion Cam.** — When a uniformly accelerated motion is used, the distances passed through by the follower during equal periods of time increase uniformly, so that if, for instance, the follower moves a distance equal to 1 length unit during the first second, and 3 during the second, it will move 5 length units during the third second, 7 during the fourth, 9 during

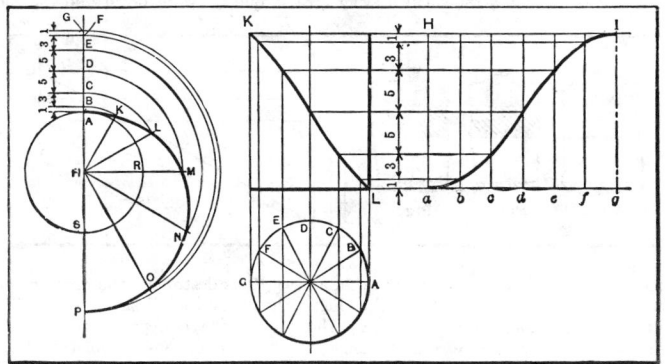

Fig. 5                           Fig. 6

the fifth, and so on. When the motion is retarded, it will move 9, 7, 5, 3 and 1 length units during successive seconds. In Fig. 5 is shown a layout of a uniformly accelerated motion plate cam. The motion of the follower is back and forth from *A* to *G*. The rise takes place during 180 degrees and the fall during an equal period, the two halves of the cam being alike. To construct the cam, divide the half circle *ARS* into six equal parts, and draw radii *HK*, *HL*, etc. Divide *AG* into two equal parts, *AD* and *DG*, and then divide each of these parts into three divisions, the lengths of which are to each other as 1 : 3 : 5. With *H* as a center, draw circular arcs from *B*, *C*, *D*, etc., to *K*, *L*, *M*, etc. The intersections between the circles and the radii are points on the cam curve. If the half circle *ARS* should be divided into eight equal parts instead of six, line *AG* would also have to be divided into eight parts in the proportions 1 : 3 : 5 : 7 : 7 : 5 : 3 : 1. With a cam constructed in this manner, the follower starts at *A* with a velocity of zero; it reaches its maximum velocity at *D*, and at *G*, where the motion is reversed, the velocity is again zero.

**Development and Layout of a Uniformly Accelerated Motion Cylindrical Cam.** — To the right in Fig. 6 is shown the development of a uniformly accelerated motion cam curve laid out on the surface of a cylindrical cam. This development is necessary for finding the projection on the cylindrical surface, as shown at *KL*. To construct the developed curve, first divide the base circle of the cylinder into, say, twelve equal parts. Set off these parts along line *ag*. Only one-half of the layout has been shown, as the other half is constructed in the same manner, except that the curve is here falling instead of rising. Divide line *aH* into the same number of divisions as the half circle, the divisions being in the proportion 1 : 3 : 5 : 5 : 3 : 1. Draw horizontal lines from these division points and vertical lines from *a*, *b*, *c*, etc. The intersections between the two sets of lines are points on the developed cam curve. These points are transferred to the cylindrical surface at the left by projection in the usual manner.

**Shape of Rolls for Cylinder Cams.** — The rolls for cylindrical cams working in a groove in the cam should be conical rather than cylindrical in shape, in order that they may rotate freely and without excessive friction. Fig. 1 shows a straight roll and groove, the action of which is faulty because of the varying surface speed at the top and bottom of the groove. Fig. 2 shows a roll with curved surface. For heavy work, however, the small bearing area is quickly worn down and the roll

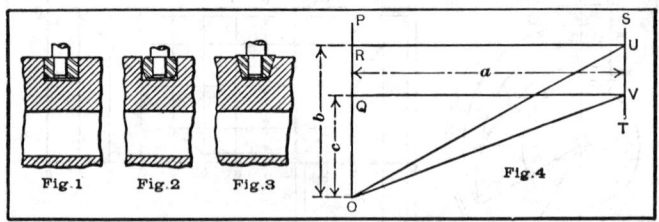

Fig. 1    Fig. 2    Fig. 3              Fig. 4

presses a groove into the side of the cam as well, thus destroying the accuracy of the movement and creating backlash. Fig. 3 shows the conical shape which permits a true rolling action in the groove. The amount of taper depends on the angle of spiral of the cam groove. As this angle, as a rule, is not constant for the whole movement, the roll and groove should be designed to meet the requirements on that section of the cam where the heaviest duty is performed. Frequently the cam groove is of a nearly even spiral angle for a considerable length. The method for determining the angle of the roll and groove to work correctly during the important part of the cycle is as follows:

In Fig. 4, *b* is the circumferential distance on the surface of the cam that includes the section of the groove for which correct rolling action is required. The throw of the cam for this circumferential movement is *a*. Line *OU* is the development of the movement of the cam roll during the given part of the cycle, and *c* is the movement corresponding to *b*, but on a circle the diameter of which is equal to that of the cam at the bottom of the groove. With the same throw *a* as before, the line *OV* will be the development of the cam at the bottom of the groove. *OU* then is the length of the helix traveled by the top of the roll, while *OV* is the travel at the bottom of the groove. If then the top width and bottom width of the groove be made proportional to *OU* and *OV*, the groove will be properly proportioned.

## AFBMA Standard Needle Roller Cam Followers*

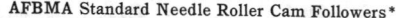

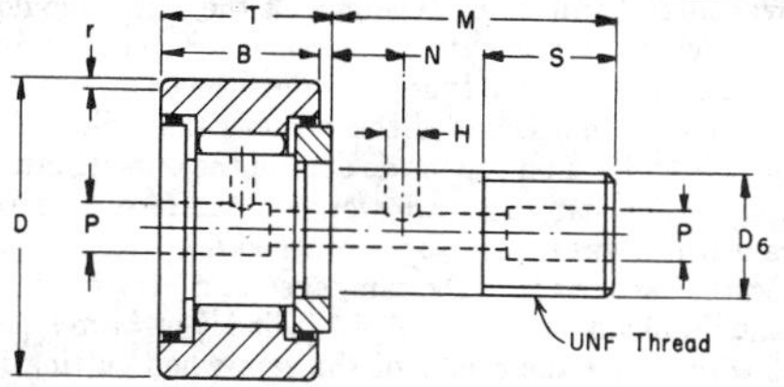

| Code No. | Stud Diam. +0.001 -0.000 $D_6$ | Roller O.D. +0.000 -0.001 $D$ | Overall Width $T$ | Roller Width +0.000 -0.005 $B$ | Stud Length $M$ | Thread Length $S$ | Corner Radius $r$ |
|---|---|---|---|---|---|---|---|
| 001CTA | 0.1900 | 0.5000 | 0.375 | 0.344 | 0.5000 | 0.2500 | 0.01 |
| 003CTA | 0.1900 | 0.5000 | 0.406 | 0.375 | 0.6250 | 0.2500 | 0.01 |
| 005CTA | 0.1900 | 0.5625 | 0.406 | 0.375 | 0.6250 | 0.2500 | 0.01 |
| 009CTA | 0.2500 | 0.6250 | 0.438 | 0.406 | 0.6250 | 0.3125 | 0.02 |
| 013CTA | 0.2500 | 0.6250 | 0.469 | 0.438 | 0.7500 | 0.3125 | 0.02 |
| 015CTA | 0.2500 | 0.6875 | 0.469 | 0.438 | 0.7500 | 0.3125 | 0.02 |
| 021CTA | 0.3750 | 0.7500 | 0.531 | 0.500 | 0.8750 | 0.3750 | 0.03 |
| 027CTA | 0.3750 | 0.8750 | 0.531 | 0.500 | 0.8750 | 0.3750 | 0.03 |
| 031CTA | 0.4375 | 1.0000 | 0.656 | 0.625 | 1.0000 | 0.5000 | 0.05 |
| 035CTA | 0.4375 | 1.1250 | 0.656 | 0.625 | 1.0000 | 0.5000 | 0.05 |
| 045CTA | 0.5000 | 1.2500 | 0.781 | 0.750 | 1.2500 | 0.6250 | 0.07 |
| 047CTA | 0.5000 | 1.3750 | 0.781 | 0.750 | 1.2500 | 0.6250 | 0.07 |
| 051CTA | 0.6250 | 1.5000 | 0.906 | 0.875 | 1.5000 | 0.7500 | 0.09 |
| 057CTA | 0.6250 | 1.6250 | 0.906 | 0.875 | 1.5000 | 0.7500 | 0.09 |
| 065CTA | 0.7500 | 1.7500 | 1.031 | 1.000 | 1.7500 | 0.8750 | 0.10 |
| 069CTA | 0.7500 | 1.8750 | 1.031 | 1.000 | 1.7500 | 0.8750 | 0.10 |
| 075CTA | 0.8750 | 2.0000 | 1.281 | 1.250 | 2.0000 | 1.0000 | 0.12 |
| 079CTA | 0.8750 | 2.2500 | 1.281 | 1.250 | 2.0000 | 1.0000 | 0.12 |
| 085CTA | 1.0000 | 2.5000 | 1.531 | 1.500 | 2.2500 | 1.1250 | 0.15 |
| 093CTA | 1.0000 | 2.7500 | 1.531 | 1.500 | 2.2500 | 1.1250 | 0.15 |
| 101CTA | 1.2500 | 3.0000 | 1.781 | 1.750 | 2.5000 | 1.2500 | 0.18 |
| 109CTA | 1.2500 | 3.2500 | 1.781 | 1.750 | 2.5000 | 1.2500 | 0.18 |

| Code No. | UNF Thread | $N$ | $H$ (Max) | Hole Diameter $P$ | Bore Diameter for Stud Max | Bore Diameter for Stud Min |
|---|---|---|---|---|---|---|
| 001CTA | 10 X 32 | ..... | ..... | 0.1250† | 0.1905 | 0.1900 |
| 003CTA | 10 X 32 | ..... | ..... | 0.1250† | 0.1905 | 0.1900 |
| 005CTA | 10 X 32 | ..... | ..... | 0.1250† | 0.1905 | 0.1900 |
| 009CTA | ¼ X 28 | ..... | ..... | 0.1250† | 0.2505 | 0.2500 |
| 013CTA | ¼ X 28 | ..... | ..... | 0.1250† | 0.2505 | 0.2500 |
| 015CTA | ¼ X 28 | ..... | ..... | 0.1250† | 0.2505 | 0.2500 |
| 021CTA | ⅜ X 24 | 0.2500 | 0.0938 | 0.1875 | 0.3755 | 0.3750 |
| 027CTA | ⅜ X 24 | 0.2500 | 0.0938 | 0.1875 | 0.3755 | 0.3750 |
| 031CTA | ⁷⁄₁₆ X 20 | 0.2500 | 0.1250 | 0.1875 | 0.4380 | 0.4375 |
| 035CTA | ⁷⁄₁₆ X 20 | 0.2500 | 0.1250 | 0.1875 | 0.4380 | 0.4375 |
| 045CTA | ½ X 20 | 0.3125 | 0.1250 | 0.1875 | 0.5005 | 0.5000 |
| 047CTA | ½ X 20 | 0.3125 | 0.1250 | 0.1875 | 0.5005 | 0.5000 |
| 051CTA | ⅝ X 18 | 0.3750 | 0.1562 | 0.1875 | 0.6255 | 0.6250 |
| 057CTA | ⅝ X 18 | 0.3750 | 0.1562 | 0.1875 | 0.6255 | 0.6250 |
| 065CTA | ¾ X 16 | 0.4375 | 0.1562 | 0.1875 | 0.7505 | 0.7500 |
| 069CTA | ¾ X 16 | 0.4375 | 0.1562 | 0.1875 | 0.7505 | 0.7500 |
| 075CTA | ⅞ X 14 | 0.5000 | 0.1875 | 0.1875 | 0.8755 | 0.8750 |
| 079CTA | ⅞ X 14 | 0.5000 | 0.1875 | 0.1875 | 0.8755 | 0.8750 |
| 085CTA | 1 X 14 | 0.5625 | 0.1875 | 0.1875 | 1.0005 | 1.0000 |
| 093CTA | 1 X 14 | 0.5625 | 0.1875 | 0.1875 | 1.0005 | 1.0000 |
| 101CTA | 1¼ X 12 | 0.6250 | 0.2500 | 0.2500 | 1.2505 | 1.2500 |
| 109CTA | 1¼ X 12 | 0.6250 | 0.2500 | 0.2500 | 1.2505 | 1.2500 |

* Anti-Friction Bearing Manufacturers Assn. All dimensions are given in inches.
† Grease fitting hole in head only.
For larger sizes see AFBMA Standards, Section 2.

**Cam Rolls and Roll Studs.** — It is important that the cam roll and roll stud be ground all over after hardening. The end of the roller should be cut back or recessed $\frac{1}{64}$ of an inch or thereabouts on the sides for some distance, beginning at the periphery, so as to avoid undue friction against the collar of the stud or the part in which it is mounted. On account of the warping that takes place in hardening, rolls that are not ground both on the inside and on the outside often will stop under heavy load, until in time flat spots are worn on the face, and then the working surface of the cam will begin to wear or is roughed up. Roll studs that are out of parallel with the working surface of the cam, even to a very small degree, also cause trouble. The same difficulty is met with on cylinder or barrel cams if the milling cutter is set below or above the center of the cam when cutting it. The roll will then bear at one end only at the most important time — when the throw takes place.

There is a great deal of end pressure on the conical rolls used in barrel cams, and this must be taken care of by thrust collars on the studs on which the rolls are mounted, or, better still, by a ball race scored in the collar and the large end of the roll, so as to provide for a ball thrust bearing. The end pressure on a conical roll, however, reduces the side pressure on the stud to a considerable extent, so that the stud may be made shorter or smaller in diameter than when a roll with parallel sides is used.

**Cam Milling.** — Plate cams having a constant rise, such as are used on automatic screw machines, can be cut in a universal milling machine, with the spiral head either in a vertical position or set at an angle $\alpha$, as shown by the illustration. When the spiral head is set vertical, the "lead" of the cam (or its rise for one complete revolution) is the same as the lead for which the machine is geared; but when the spiral head and cutter are inclined, any lead or rise of the cam can be obtained, provided it is less than the lead for which the machine is geared, that is, less than the forward feed of the table for one turn of the spiral-head spindle. The cam lead, then, can be varied within certain limits by simply changing the inclination $\alpha$ of the spiral head and cutter. In the following formulas for determining this angle of inclination, for a given rise of cam and with the machine geared for a certain lead, let

$\alpha$ = angle to which index head and milling attachment are set;

$r$ = rise of cam in given part of circumference;

$R$ = "lead" of cam, or rise if latter were continued at given rate for one complete revolution;

$L$ = spiral lead for which milling machine is geared;

$N$ = part of circumference in which rise is required, expressed as a decimal in hundredths of cam circumference.

$$\sin \alpha = \frac{R}{L}, \text{ and } R = \frac{r}{N}; \text{ hence, } \sin \alpha = \frac{r}{N \times L}$$

For example, suppose a cam is to be milled having a rise of 0.125 inch in 300 degrees or in 0.83 of the circumference, and that the machine is geared for the smallest possible lead, or 0.67 inch; then:

$$\sin \alpha = \frac{r}{N \times L} = \frac{0.125}{0.83 \times 0.67} = 0.2247,$$

which is approximately the sine of 13 degrees. Therefore, to secure a rise of 0.125 inch with the machine geared for 0.67 inch lead, the spiral head is elevated to an angle of 13 degrees and the vertical milling attachment is also swiveled around to locate the cutter in line with the spiral-head spindle, so that the edge of the finished cam will be parallel to its axis of rotation. When there are several lobes on a cam, having different leads, the machine can be geared for a lead somewhat in excess of the

greatest lead on the cam, and then all the lobes can be milled without changing the spiral head gearing, by simply varying the angle of the spiral head and cutter to suit the different cam leads. Whenever possible, it is advisable to mill on the under side of the cam, as there is less interference from chips; moreover, it is easier to see any lines that may be laid out on the cam face. To set the cam for a new cut, it is first

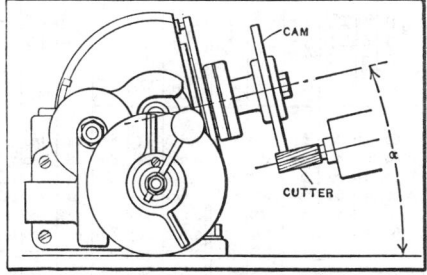

turned back by operating the handle of the table feed screw, after which the index crank is disengaged from the plate and turned the required amount.

The accompanying tables give the combinations of change gears and the angular setting required for cutting a cam of any lead likely to be met with in practice. The figures in the column headed "Lead of Cam," represent the rise for one complete revolution. Set the vertical attachment to the angle given in the table. For the dividing head, subtract the angle in the table from 90 degrees; the difference is the angle to which the spindle must be raised from the horizontal position.

*Example:* — If the angle is 39½ degrees, set the spindle of the vertical attachment 39½ degrees from the vertical. Set the dividing head 50½ degrees from the horizontal position (90 − 39½ = 50½). These tables were compiled by the Cincinnati Milling Machine Co.

## Simple Method for Cutting Uniform Motion Cams.

— Cams are generally laid out with dividers, machined and filed to the line; but for a cam that must

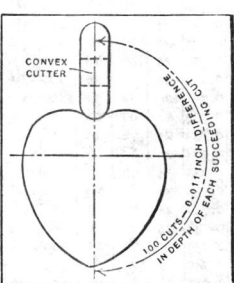

advance a certain number of thousandths per revolution of spindle this method is not accurate. Cams are easily and accurately cut in the following manner. Let it be required to make the heart cam shown in the illustration. The throw of this cam is 1.1 inch. Now, by setting the index on the milling machine to cut 200 teeth and also dividing 1.1 inch by 100, we find that we have 0.011 inch to recede from or advance towards the cam center for each cut across the cam. Placing the cam securely on an arbor, and the latter between the centers of the milling machine, and using a convex cutter set the proper distance from the center of the arbor, make the first cut across the cam. Then, by lowering the milling machine knee 0.011 inch and

turning the index pin the proper number of holes on the index plate, take the next cut and so on. Each cut should be marked on paper so that there will be no mistake as to the number of cuts taken; when 100 cuts have been made, the knee must be raised in order to complete the opposite side of the cam.

This method can also be used to advantage for milling uniform motion cam lobes extending only over a portion of the cam circumference. After the milling has been completed, the surface of the cam must be smoothed off by means of filing.

## Change Gears and Angles for Cam Milling — 1

| Lead of Cam | Gear on Worm | First Intermediate | Second Intermediate | Gear on Screw | Angle | Lead of Cam | Gear on Worm | First Intermediate | Second Intermediate | Gear on Screw | Angle | Lead of Cam | Gear on Worm | First Intermediate | Second Intermediate | Gear on Screw | Angle |
|---|---|---|---|---|---|---|---|---|---|---|---|---|---|---|---|---|---|
| 0.600 | 24 | 86 | 24 | 100 | 26½ | 0.650 | 24 | 86 | 24 | 100 | 14 | 0.700 | 24 | 72 | 24 | 100 | 29 |
| 0.601 | 24 | 86 | 24 | 100 | 26 | 0.651 | 24 | 86 | 28 | 100 | 33½ | 0.701 | 24 | 72 | 24 | 86 | 41 |
| 0.602 | 24 | 86 | 24 | 100 | 26 | 0.652 | 24 | 86 | 24 | 100 | 13½ | 0.702 | 24 | 86 | 28 | 100 | 26 |
| 0.603 | 24 | 86 | 28 | 100 | 39½ | 0.653 | 24 | 86 | 32 | 100 | 43 | 0.703 | 24 | 72 | 24 | 100 | 28½ |
| 0.604 | 24 | 72 | 24 | 100 | 41 | 0.654 | 24 | 86 | 24 | 100 | 12½ | 0.704 | 24 | 86 | 32 | 100 | 38 |
| 0.605 | 24 | 86 | 24 | 100 | 25½ | 0.655 | 24 | 86 | 24 | 100 | 12 | 0.705 | 24 | 86 | 28 | 100 | 25½ |
| 0.606 | 24 | 86 | 28 | 100 | 39 | 0.656 | 24 | 86 | 24 | 100 | 11½ | 0.706 | 24 | 72 | 24 | 100 | 28 |
| 0.607 | 24 | 86 | 24 | 100 | 25 | 0.657 | 24 | 86 | 24 | 100 | 11 | 0.707 | 24 | 72 | 24 | 86 | 40½ |
| 0.608 | 24 | 72 | 24 | 100 | 40½ | 0.658 | 24 | 86 | 32 | 100 | 42½ | 0.708 | 24 | 86 | 28 | 100 | 25 |
| 0.609 | 24 | 86 | 24 | 100 | 24½ | 0.659 | 24 | 72 | 24 | 100 | 34½ | 0.709 | 24 | 72 | 28 | 100 | 40½ |
| 0.610 | 24 | 86 | 24 | 100 | 24½ | 0.660 | 24 | 86 | 24 | 100 | 10 | 0.710 | 24 | 72 | 24 | 100 | 27½ |
| 0.611 | 24 | 86 | 28 | 100 | 38½ | 0.661 | 24 | 86 | 28 | 100 | 32 | 0.711 | 24 | 86 | 28 | 100 | 24½ |
| 0.612 | 24 | 86 | 24 | 100 | 24 | 0.662 | 24 | 86 | 28 | 100 | 32 | 0.712 | 24 | 72 | 24 | 86 | 40 |
| 0.613 | 24 | 72 | 24 | 100 | 40 | 0.663 | 24 | 72 | 24 | 100 | 34 | 0.713 | 24 | 72 | 24 | 100 | 27 |
| 0.614 | 24 | 86 | 24 | 100 | 23½ | 0.664 | 24 | 86 | 32 | 100 | 42 | 0.714 | 24 | 64 | 24 | 100 | 37½ |
| 0.615 | 24 | 86 | 28 | 100 | 38 | 0.665 | 24 | 72 | 28 | 100 | 44½ | 0.715 | 24 | 72 | 28 | 100 | 40 |
| 0.616 | 24 | 86 | 24 | 100 | 23 | 0.666 | 24 | 86 | 28 | 100 | 31½ | 0.716 | 24 | 86 | 28 | 100 | 23½ |
| 0.617 | 24 | 72 | 24 | 100 | 39½ | 0.667 | 24 | 72 | 24 | 100 | 33½ | 0.717 | 24 | 72 | 24 | 86 | 39½ |
| 0.618 | 24 | 72 | 24 | 100 | 39½ | 0.668 | 24 | 64 | 24 | 100 | 42 | 0.718 | 24 | 86 | 32 | 100 | 36½ |
| 0.619 | 24 | 86 | 24 | 86 | 22½ | 0.669 | 24 | 72 | 24 | 86 | 44 | 0.719 | 24 | 86 | 28 | 100 | 23 |
| 0.620 | 24 | 86 | 28 | 100 | 37½ | 0.670 | 24 | 86 | 28 | 100 | 31 | 0.720 | 24 | 72 | 28 | 100 | 39½ |
| 0.621 | 24 | 86 | 24 | 100 | 22 | 0.671 | 24 | 72 | 24 | 100 | 33 | 0.721 | 24 | 86 | 28 | 100 | 22½ |
| 0.622 | 24 | 72 | 24 | 100 | 39 | 0.672 | 24 | 86 | 28 | 100 | 30½ | 0.722 | 24 | 72 | 24 | 100 | 25½ |
| 0.623 | 24 | 86 | 28 | 100 | 21½ | 0.673 | 24 | 86 | 28 | 100 | 30½ | 0.723 | 24 | 64 | 24 | 100 | 36½ |
| 0.624 | 24 | 86 | 28 | 100 | 37 | 0.674 | 24 | 64 | 24 | 100 | 41½ | 0.724 | 24 | 86 | 28 | 100 | 22 |
| 0.625 | 24 | 86 | 24 | 100 | 21 | 0.675 | 24 | 72 | 24 | 100 | 32½ | 0.725 | 24 | 72 | 24 | 100 | 25 |
| 0.626 | 24 | 86 | 32 | 100 | 45½ | 0.676 | 24 | 86 | 28 | 100 | 30 | 0.726 | 24 | 86 | 28 | 100 | 21½ |
| 0.627 | 24 | 86 | 24 | 100 | 20½ | 0.677 | 24 | 72 | 28 | 100 | 43½ | 0.727 | 24 | 86 | 32 | 100 | 35½ |
| 0.628 | 24 | 86 | 28 | 100 | 36½ | 0.678 | 24 | 72 | 24 | 100 | 32 | 0.728 | 24 | 72 | 24 | 100 | 24½ |
| 0.629 | 24 | 86 | 24 | 100 | 20 | 0.679 | 24 | 86 | 32 | 100 | 40½ | 0.729 | 24 | 86 | 28 | 100 | 21 |
| 0.630 | 24 | 72 | 24 | 100 | 38 | 0.680 | 24 | 72 | 24 | 86 | 43 | 0.730 | 24 | 72 | 28 | 100 | 38½ |
| 0.631 | 24 | 86 | 32 | 100 | 45 | 0.681 | 24 | 72 | 24 | 100 | 31½ | 0.731 | 24 | 72 | 24 | 100 | 24 |
| 0.632 | 24 | 86 | 28 | 100 | 36 | 0.682 | 24 | 72 | 28 | 100 | 43 | 0.732 | 24 | 86 | 28 | 100 | 20½ |
| 0.633 | 24 | 86 | 24 | 100 | 19 | 0.683 | 24 | 86 | 28 | 100 | 29 | 0.733 | 24 | 72 | 24 | 86 | 38 |
| 0.634 | 24 | 72 | 24 | 100 | 37½ | 0.684 | 24 | 86 | 32 | 100 | 40 | 0.734 | 24 | 86 | 28 | 100 | 20 |
| 0.635 | 24 | 86 | 24 | 100 | 18½ | 0.685 | 24 | 72 | 24 | 100 | 42½ | 0.735 | 24 | 72 | 28 | 100 | 38 |
| 0.636 | 24 | 86 | 28 | 100 | 35½ | 0.686 | 24 | 86 | 28 | 100 | 28½ | 0.736 | 24 | 86 | 28 | 100 | 19½ |
| 0.637 | 24 | 86 | 32 | 100 | 44½ | 0.687 | 24 | 72 | 28 | 100 | 42½ | 0.737 | 24 | 64 | 24 | 100 | 35 |
| 0.638 | 24 | 72 | 24 | 100 | 37 | 0.688 | 24 | 72 | 28 | 100 | 42½ | 0.738 | 24 | 72 | 24 | 86 | 37½ |
| 0.639 | 24 | 86 | 24 | 100 | 17½ | 0.689 | 24 | 86 | 32 | 100 | 39½ | 0.739 | 24 | 72 | 24 | 100 | 22½ |
| 0.640 | 24 | 86 | 28 | 100 | 35 | 0.690 | 24 | 86 | 28 | 100 | 28 | 0.740 | 24 | 72 | 28 | 100 | 37½ |
| 0.641 | 24 | 86 | 24 | 100 | 17 | 0.691 | 24 | 72 | 24 | 86 | 42 | 0.741 | 24 | 86 | 28 | 100 | 18½ |
| 0.642 | 24 | 86 | 32 | 100 | 44 | 0.692 | 24 | 86 | 28 | 100 | 27½ | 0.742 | 24 | 72 | 24 | 100 | 22 |
| 0.643 | 24 | 86 | 28 | 100 | 34½ | 0.693 | 24 | 72 | 28 | 100 | 42 | 0.743 | 24 | 86 | 28 | 100 | 18 |
| 0.644 | 24 | 86 | 24 | 100 | 16 | 0.694 | 24 | 86 | 32 | 100 | 39 | 0.744 | 24 | 72 | 24 | 100 | 21½ |
| 0.645 | 24 | 86 | 24 | 100 | 15½ | 0.695 | 24 | 64 | 24 | 100 | 39½ | 0.745 | 24 | 72 | 28 | 100 | 37 |
| 0.646 | 24 | 86 | 24 | 100 | 15½ | 0.696 | 24 | 86 | 28 | 100 | 27 | 0.746 | 24 | 64 | 24 | 100 | 34 |
| 0.647 | 24 | 86 | 24 | 100 | 15 | 0.697 | 24 | 72 | 24 | 86 | 41½ | 0.747 | 24 | 86 | 28 | 100 | 17 |
| 0.648 | 24 | 86 | 32 | 100 | 43½ | 0.698 | 24 | 72 | 28 | 100 | 41½ | 0.748 | 24 | 72 | 24 | 86 | 36½ |
| 0.649 | 24 | 86 | 24 | 100 | 14½ | 0.699 | 24 | 86 | 32 | 100 | 38½ | 0.749 | 24 | 86 | 28 | 100 | 16½ |

## Change Gears and Angles for Cam Milling — 2

| Lead of Cam | Gear on Worm | First Intermediate | Second Intermediate | Gear on Screw | Angle |
|---|---|---|---|---|---|
| 0.750 | 24 | 72 | 28 | 100 | 36½ |
| 0.751 | 24 | 86 | 28 | 100 | 16 |
| 0.752 | 24 | 72 | 24 | 100 | 20 |
| 0.753 | 24 | 86 | 32 | 100 | 32½ |
| 0.754 | 24 | 72 | 24 | 100 | 19½ |
| 0.755 | 24 | 72 | 28 | 100 | 36 |
| 0.756 | 24 | 86 | 28 | 100 | 14½ |
| 0.757 | 24 | 72 | 24 | 86 | 35½ |
| 0.758 | 24 | 86 | 28 | 100 | 14 |
| 0.759 | 24 | 64 | 24 | 100 | 32½ |
| 0.760 | 24 | 72 | 28 | 100 | 35½ |
| 0.761 | 24 | 86 | 28 | 100 | 13 |
| 0.762 | 24 | 72 | 24 | 86 | 35 |
| 0.763 | 24 | 72 | 24 | 100 | 17½ |
| 0.764 | 24 | 86 | 28 | 100 | 12 |
| 0.765 | 24 | 72 | 28 | 100 | 17 |
| 0.766 | 24 | 72 | 24 | 86 | 34½ |
| 0.767 | 24 | 72 | 24 | 100 | 16½ |
| 0.768 | 24 | 86 | 28 | 100 | 10½ |
| 0.769 | 24 | 86 | 28 | 100 | 10 |
| 0.770 | 24 | 86 | 32 | 100 | 30½ |
| 0.771 | 24 | 72 | 24 | 86 | 34 |
| 0.772 | 24 | 72 | 24 | 100 | 15 |
| 0.773 | 24 | 86 | 32 | 100 | 30 |
| 0.774 | 24 | 72 | 24 | 100 | 14½ |
| 0.775 | 24 | 64 | 24 | 100 | 30½ |
| 0.776 | 24 | 72 | 24 | 100 | 14 |
| 0.777 | 24 | 86 | 32 | 100 | 29½ |
| 0.778 | 24 | 72 | 28 | 100 | 33½ |
| 0.779 | 24 | 72 | 24 | 100 | 13 |
| 0.780 | 24 | 72 | 24 | 86 | 33 |
| 0.781 | 24 | 72 | 24 | 100 | 12½ |
| 0.782 | 24 | 72 | 24 | 100 | 33 |
| 0.783 | 24 | 64 | 24 | 100 | 29½ |
| 0.784 | 24 | 72 | 24 | 100 | 11½ |
| 0.785 | 24 | 86 | 32 | 100 | 28½ |
| 0.786 | 28 | 86 | 32 | 100 | 41 |
| 0.787 | 24 | 64 | 24 | 100 | 29 |
| 0.788 | 24 | 72 | 24 | 100 | 10 |
| 0.789 | 24 | 72 | 24 | 86 | 32 |
| 0.790 | 24 | 64 | 24 | 86 | 41 |
| 0.791 | 24 | 64 | 24 | 100 | 28½ |
| 0.792 | 24 | 86 | 32 | 100 | 27½ |
| 0.793 | 24 | 72 | 24 | 86 | 31½ |
| 0.794 | 24 | 72 | 28 | 86 | 43 |
| 0.795 | 24 | 72 | 28 | 100 | 31½ |
| 0.796 | 24 | 64 | 24 | 86 | 40½ |
| 0.797 | 24 | 72 | 24 | 86 | 31 |
| 0.798 | 28 | 86 | 32 | 100 | 40 |
| 0.799 | 24 | 72 | 32 | 100 | 41½ |
| 0.800 | 24 | 72 | 28 | 100 | 31 |
| 0.801 | 24 | 72 | 24 | 86 | 30½ |
| 0.802 | 24 | 64 | 24 | 86 | 40 |
| 0.803 | 24 | 86 | 32 | 100 | 26 |
| 0.804 | 28 | 86 | 32 | 100 | 39½ |
| 0.805 | 24 | 72 | 32 | 100 | 41 |
| 0.806 | 24 | 86 | 32 | 100 | 25½ |
| 0.807 | 24 | 64 | 24 | 86 | 39½ |
| 0.808 | 24 | 72 | 28 | 100 | 30 |
| 0.809 | 24 | 64 | 24 | 100 | 26 |
| 0.810 | 28 | 86 | 32 | 100 | 39 |
| 0.811 | 24 | 72 | 32 | 100 | 40½ |
| 0.812 | 24 | 72 | 28 | 100 | 29½ |
| 0.813 | 24 | 72 | 24 | 86 | 29 |
| 0.814 | 24 | 64 | 24 | 86 | 39 |
| 0.815 | 28 | 86 | 32 | 100 | 38½ |
| 0.816 | 24 | 86 | 32 | 100 | 29 |
| 0.817 | 24 | 72 | 24 | 86 | 28½ |
| 0.818 | 24 | 72 | 28 | 86 | 41 |
| 0.819 | 24 | 86 | 32 | 100 | 23½ |
| 0.820 | 24 | 72 | 28 | 100 | 28½ |
| 0.821 | 24 | 72 | 24 | 86 | 28 |
| 0.822 | 24 | 86 | 32 | 100 | 23 |
| 0.823 | 24 | 72 | 32 | 100 | 39½ |
| 0.824 | 24 | 72 | 28 | 100 | 28 |
| 0.825 | 24 | 72 | 24 | 86 | 27½ |
| 0.826 | 28 | 86 | 32 | 100 | 37½ |
| 0.827 | 24 | 72 | 28 | 100 | 27½ |
| 0.828 | 24 | 86 | 32 | 100 | 22 |
| 0.829 | 24 | 72 | 40 | 100 | 42 |
| 0.830 | 24 | 64 | 24 | 86 | 37½ |
| 0.831 | 24 | 72 | 28 | 86 | 40 |
| 0.832 | 24 | 72 | 24 | 86 | 26½ |
| 0.833 | 24 | 56 | 24 | 100 | 36 |
| 0.834 | 24 | 86 | 32 | 100 | 21 |
| 0.835 | 24 | 72 | 32 | 100 | 38½ |
| 0.836 | 24 | 72 | 24 | 86 | 26 |
| 0.837 | 24 | 72 | 28 | 86 | 39½ |
| 0.838 | 24 | 56 | 24 | 100 | 35½ |
| 0.839 | 24 | 86 | 32 | 100 | 20 |
| 0.840 | 24 | 64 | 24 | 100 | 21 |
| 0.841 | 24 | 72 | 24 | 100 | 38 |
| 0.842 | 24 | 86 | 32 | 100 | 19½ |
| 0.843 | 24 | 72 | 28 | 86 | 39 |
| 0.844 | 24 | 86 | 32 | 100 | 19 |
| 0.845 | 24 | 72 | 28 | 100 | 25 |
| 0.846 | 24 | 64 | 24 | 100 | 20 |
| 0.847 | 24 | 86 | 32 | 100 | 18½ |
| 0.848 | 24 | 64 | 24 | 100 | 19½ |
| 0.849 | 24 | 86 | 32 | 100 | 18 |
| 0.850 | 24 | 72 | 24 | 86 | 24 |
| 0.851 | 24 | 64 | 24 | 100 | 19 |
| 0.852 | 24 | 72 | 24 | 86 | 40 |
| 0.853 | 24 | 72 | 24 | 86 | 23½ |
| 0.854 | 24 | 86 | 32 | 100 | 17 |
| 0.855 | 24 | 64 | 28 | 100 | 35½ |
| 0.856 | 24 | 86 | 32 | 100 | 16½ |
| 0.857 | 24 | 64 | 24 | 86 | 35 |
| 0.858 | 24 | 64 | 24 | 86 | 17½ |
| 0.859 | 24 | 72 | 24 | 86 | 22½ |
| 0.860 | 24 | 64 | 28 | 100 | 35 |
| 0.861 | 24 | 72 | 28 | 86 | 37½ |
| 0.862 | 24 | 72 | 28 | 100 | 22½ |
| 0.863 | 24 | 64 | 24 | 100 | 16½ |
| 0.864 | 28 | 86 | 32 | 100 | 34 |
| 0.865 | 24 | 64 | 24 | 100 | 16 |
| 0.866 | 24 | 86 | 32 | 100 | 14 |
| 0.867 | 24 | 64 | 24 | 100 | 15½ |
| 0.868 | 24 | 72 | 28 | 100 | 21½ |
| 0.869 | 24 | 64 | 24 | 100 | 15 |
| 0.870 | 24 | 86 | 32 | 100 | 13 |
| 0.871 | 24 | 64 | 24 | 100 | 14½ |
| 0.872 | 24 | 86 | 32 | 100 | 12½ |
| 0.873 | 24 | 64 | 24 | 100 | 14 |
| 0.874 | 24 | 72 | 24 | 86 | 20 |
| 0.875 | 24 | 72 | 32 | 100 | 11½ |
| 0.876 | 24 | 64 | 28 | 100 | 33½ |
| 0.877 | 24 | 64 | 24 | 100 | 13 |
| 0.878 | 24 | 86 | 32 | 100 | 10½ |
| 0.879 | 24 | 64 | 24 | 100 | 12½ |
| 0.880 | 24 | 64 | 24 | 100 | 12 |
| 0.881 | 24 | 64 | 28 | 100 | 33 |
| 0.882 | 24 | 64 | 24 | 100 | 11½ |
| 0.883 | 24 | 64 | 24 | 86 | 32½ |
| 0.884 | 28 | 86 | 32 | 100 | 32 |
| 0.885 | 24 | 72 | 24 | 100 | 10½ |
| 0.886 | 24 | 64 | 24 | 100 | 10 |
| 0.887 | 24 | 72 | 24 | 86 | 17½ |
| 0.888 | 24 | 64 | 24 | 86 | 32 |
| 0.889 | 24 | 72 | 24 | 86 | 17 |
| 0.890 | 24 | 72 | 28 | 100 | 17½ |
| 0.891 | 24 | 56 | 24 | 100 | 30 |
| 0.892 | 24 | 72 | 24 | 86 | 16½ |
| 0.893 | 28 | 86 | 32 | 100 | 31 |
| 0.894 | 24 | 72 | 24 | 86 | 16 |
| 0.895 | 24 | 64 | 28 | 100 | 31½ |
| 0.896 | 24 | 72 | 24 | 86 | 15½ |
| 0.897 | 24 | 72 | 24 | 100 | 16 |
| 0.898 | 24 | 72 | 24 | 86 | 15 |
| 0.899 | 24 | 72 | 28 | 100 | 15½ |

### Change Gears and Angles for Cam Milling — 3

| Lead of Cam | Gear on Worm | First Intermediate | Second Intermediate | Gear on Screw | Angle |
|---|---|---|---|---|---|
| 0.900 | 24 | 56 | 24 | 100 | 29 |
| 0.901 | 24 | 72 | 28 | 100 | 15 |
| 0.902 | 24 | 72 | 24 | 86 | 14 |
| 0.903 | 24 | 72 | 28 | 100 | 14½ |
| 0.904 | 24 | 72 | 24 | 86 | 13½ |
| 0.905 | 24 | 72 | 28 | 100 | 14 |
| 0.906 | 24 | 72 | 24 | 86 | 13 |
| 0.907 | 24 | 72 | 28 | 100 | 13½ |
| 0.908 | 24 | 72 | 24 | 86 | 12½ |
| 0.909 | 24 | 72 | 28 | 100 | 13 |
| 0.910 | 24 | 72 | 32 | 100 | 31½ |
| 0.911 | 24 | 72 | 28 | 100 | 12½ |
| 0.912 | 24 | 72 | 28 | 100 | 12 |
| 0.913 | 24 | 72 | 24 | 86 | 11 |
| 0.914 | 24 | 72 | 28 | 100 | 11½ |
| 0.915 | 24 | 72 | 32 | 100 | 31 |
| 0.916 | 24 | 72 | 24 | 86 | 10 |
| 0.917 | 24 | 72 | 28 | 100 | 10½ |
| 0.918 | 24 | 64 | 28 | 100 | 29 |
| 0.919 | 24 | 72 | 28 | 100 | 10 |
| 0.920 | 28 | 86 | 32 | 100 | 28 |
| 0.921 | 24 | 56 | 24 | 100 | 26½ |
| 0.922 | 24 | 64 | 28 | 86 | 41 |
| 0.923 | 24 | 64 | 28 | 100 | 28½ |
| 0.924 | 28 | 86 | 32 | 100 | 27½ |
| 0.925 | 24 | 56 | 24 | 100 | 26 |
| 0.926 | 24 | 64 | 32 | 100 | 39½ |
| 0.927 | 24 | 64 | 28 | 100 | 28 |
| 0.928 | 24 | 64 | 28 | 86 | 40½ |
| 0.929 | 24 | 56 | 24 | 100 | 25½ |
| 0.930 | 24 | 72 | 28 | 86 | 31 |
| 0.931 | 24 | 64 | 28 | 100 | 27½ |
| 0.932 | 28 | 72 | 32 | 100 | 41½ |
| 0.933 | 24 | 64 | 24 | 86 | 27 |
| 0.934 | 24 | 86 | 44 | 100 | 40½ |
| 0.935 | 24 | 72 | 28 | 100 | 30½ |
| 0.936 | 24 | 56 | 24 | 100 | 24½ |
| 0.937 | 24 | 64 | 24 | 86 | 26½ |
| 0.938 | 24 | 72 | 32 | 100 | 28½ |
| 0.939 | 24 | 64 | 32 | 100 | 38½ |
| 0.940 | 24 | 56 | 24 | 100 | 24 |
| 0.941 | 24 | 64 | 24 | 86 | 26 |
| 0.942 | 24 | 72 | 32 | 100 | 28 |
| 0.943 | 24 | 72 | 32 | 86 | 40½ |
| 0.944 | 24 | 56 | 24 | 100 | 23½ |
| 0.945 | 24 | 64 | 24 | 86 | 25½ |
| 0.946 | 24 | 72 | 32 | 100 | 27½ |
| 0.947 | 24 | 56 | 24 | 100 | 23 |
| 0.948 | 28 | 86 | 32 | 100 | 24½ |
| 0.949 | 24 | 64 | 24 | 86 | 25 |
| 0.950 | 24 | 72 | 32 | 86 | 40 |
| 0.951 | 24 | 56 | 24 | 100 | 22½ |
| 0.952 | 28 | 86 | 32 | 100 | 24 |
| 0.953 | 24 | 64 | 24 | 86 | 24½ |
| 0.954 | 24 | 56 | 24 | 100 | 22 |
| 0.955 | 24 | 72 | 32 | 100 | 26½ |
| 0.956 | 24 | 64 | 28 | 86 | 38½ |
| 0.957 | 24 | 56 | 24 | 100 | 21½ |
| 0.958 | 24 | 72 | 28 | 86 | 28 |
| 0.959 | 24 | 72 | 32 | 100 | 26 |
| 0.960 | 24 | 64 | 24 | 86 | 23½ |
| 0.961 | 24 | 86 | 44 | 100 | 38½ |
| 0.962 | 24 | 72 | 28 | 86 | 27½ |
| 0.963 | 28 | 86 | 32 | 100 | 22½ |
| 0.964 | 24 | 56 | 24 | 100 | 20½ |
| 0.965 | 24 | 64 | 32 | 100 | 36½ |
| 0.966 | 28 | 86 | 32 | 100 | 22 |
| 0.967 | 24 | 56 | 24 | 100 | 20 |
| 0.968 | 24 | 56 | 24 | 86 | 36 |
| 0.969 | 24 | 64 | 28 | 86 | 37½ |
| 0.970 | 24 | 56 | 24 | 100 | 19½ |
| 0.971 | 24 | 72 | 28 | 86 | 26½ |
| 0.972 | 86 | 44 | 32 | 64 | 6 |
| 0.973 | 24 | 56 | 24 | 100 | 19 |
| 0.974 | 24 | 64 | 24 | 86 | 21½ |
| 0.975 | 24 | 72 | 32 | 100 | 24 |
| 0.976 | 28 | 86 | 32 | 100 | 20½ |
| 0.977 | 24 | 64 | 28 | 100 | 21½ |
| 0.978 | 24 | 56 | 24 | 100 | 18 |
| 0.979 | 28 | 86 | 32 | 100 | 20 |
| 0.980 | 24 | 64 | 28 | 100 | 21 |
| 0.981 | 24 | 64 | 24 | 86 | 20½ |
| 0.982 | 28 | 86 | 32 | 100 | 19½ |
| 0.983 | 24 | 72 | 28 | 86 | 25 |
| 0.984 | 24 | 56 | 24 | 100 | 17 |
| 0.985 | 28 | 86 | 32 | 100 | 19 |
| 0.986 | 24 | 72 | 32 | 100 | 22½ |
| 0.987 | 24 | 64 | 24 | 86 | 19½ |
| 0.988 | 28 | 86 | 32 | 100 | 18½ |
| 0.989 | 24 | 56 | 24 | 100 | 16 |
| 0.990 | 24 | 64 | 24 | 86 | 19 |
| 0.991 | 28 | 86 | 32 | 100 | 18 |
| 0.992 | 24 | 56 | 24 | 100 | 15½ |
| 0.993 | 24 | 64 | 24 | 86 | 18½ |
| 0.994 | 24 | 56 | 24 | 100 | 15 |
| 0.995 | 24 | 72 | 28 | 86 | 23½ |
| 0.996 | 24 | 56 | 24 | 100 | 14½ |
| 0.997 | 24 | 64 | 24 | 86 | 33½ |
| 0.998 | 24 | 56 | 24 | 100 | 14 |
| 0.999 | 28 | 86 | 32 | 100 | 16½ |
| 1.000 | 24 | 86 | 44 | 100 | 35½ |
| 1.001 | 24 | 56 | 24 | 100 | 13½ |
| 1.002 | 28 | 86 | 32 | 100 | 16 |
| 1.003 | 24 | 56 | 24 | 100 | 13 |
| 1.004 | 28 | 86 | 32 | 100 | 15½ |
| 1.005 | 24 | 56 | 24 | 100 | 12½ |
| 1.006 | 24 | 56 | 24 | 100 | 12 |
| 1.007 | 24 | 64 | 24 | 86 | 16 |
| 1.008 | 24 | 56 | 24 | 100 | 11½ |
| 1.009 | 28 | 86 | 32 | 100 | 14½ |
| 1.010 | 24 | 56 | 24 | 100 | 11 |
| 1.011 | 28 | 86 | 32 | 100 | 14 |
| 1.012 | 24 | 56 | 24 | 100 | 10½ |
| 1.013 | 24 | 56 | 24 | 100 | 10 |
| 1.014 | 24 | 64 | 24 | 86 | 14½ |
| 1.015 | 28 | 86 | 32 | 100 | 13 |
| 1.016 | 24 | 64 | 24 | 86 | 14 |
| 1.017 | 28 | 86 | 32 | 100 | 12½ |
| 1.018 | 24 | 64 | 24 | 86 | 13½ |
| 1.019 | 28 | 86 | 32 | 100 | 12 |
| 1.020 | 24 | 64 | 24 | 86 | 13 |
| 1.021 | 28 | 86 | 32 | 100 | 11½ |
| 1.022 | 24 | 64 | 24 | 86 | 12½ |
| 1.023 | 28 | 86 | 32 | 100 | 11 |
| 1.024 | 24 | 64 | 24 | 86 | 12 |
| 1.025 | 24 | 64 | 24 | 28 100 | 12½ |
| 1.026 | 24 | 64 | 24 | 86 | 11½ |
| 1.027 | 24 | 64 | 28 | 100 | 12 |
| 1.028 | 24 | 64 | 24 | 86 | 11 |
| 1.029 | 24 | 64 | 24 | 86 | 10½ |
| 1.030 | 24 | 64 | 24 | 100 | 11 |
| 1.031 | 24 | 64 | 24 | 100 | 11 |
| 1.032 | 24 | 64 | 24 | 100 | 10½ |
| 1.033 | 24 | 72 | 32 | 100 | 14½ |
| 1.034 | 24 | 64 | 28 | 100 | 10 |
| 1.035 | 24 | 72 | 32 | 100 | 14 |
| 1.036 | 24 | 56 | 24 | 86 | 13½ |
| 1.037 | 24 | 72 | 32 | 100 | 13½ |
| 1.038 | 24 | 72 | 28 | 86 | 17 |
| 1.039 | 24 | 64 | 32 | 100 | 30 |
| 1.040 | 24 | 72 | 32 | 100 | 13 |
| 1.041 | 24 | 56 | 24 | 86 | 29½ |
| 1.042 | 24 | 72 | 32 | 100 | 12½ |
| 1.043 | 24 | 72 | 28 | 86 | 16 |
| 1.044 | 24 | 72 | 32 | 100 | 12 |
| 1.045 | 24 | 86 | 40 | 100 | 20½ |
| 1.046 | 24 | 72 | 32 | 100 | 11½ |
| 1.047 | 24 | 72 | 32 | 100 | 11 |
| 1.048 | 24 | 72 | 28 | 86 | 15 |
| 1.049 | 24 | 72 | 32 | 100 | 10½ |

## Change Gears and Angles for Cam Milling — 4

| Lead of Cam | Gear on Worm | First Intermediate | Second Intermediate | Gear on Screw | Angle | Lead of Cam | Gear on Worm | First Intermediate | Second Intermediate | Gear on Screw | Angle | Lead of Cam | Gear on Worm | First Intermediate | Second Intermediate | Gear on Screw | Angle |
|---|---|---|---|---|---|---|---|---|---|---|---|---|---|---|---|---|---|
| 1.050 | 24 | 72 | 28 | 86 | 14½ | 1.100 | 28 | 72 | 32 | 86 | 40½ | 1.150 | 24 | 56 | 24 | 86 | 16 |
| 1.051 | 24 | 72 | 32 | 100 | 10 | 1.101 | 24 | 56 | 24 | 86 | 23 | 1.151 | 24 | 64 | 32 | 100 | 16½ |
| 1.052 | 24 | 86 | 40 | 100 | 19½ | 1.102 | 24 | 64 | 28 | 86 | 25½ | 1.152 | 28 | 86 | 44 | 100 | 36½ |
| 1.053 | 24 | 72 | 28 | 86 | 14 | 1.103 | 28 | 72 | 32 | 100 | 27½ | 1.153 | 24 | 56 | 24 | 86 | 15½ |
| 1.054 | 24 | 72 | 28 | 86 | 14 | 1.104 | 24 | 86 | 44 | 100 | 26 | 1.154 | 24 | 64 | 28 | 86 | 19 |
| 1.055 | 24 | 72 | 28 | 86 | 13½ | 1.105 | 24 | 56 | 24 | 86 | 22½ | 1.155 | 24 | 56 | 24 | 86 | 15 |
| 1.056 | 24 | 56 | 24 | 86 | 28 | 1.106 | 40 | 64 | 24 | 100 | 42½ | 1.156 | 24 | 64 | 32 | 100 | 15½ |
| 1.057 | 24 | 72 | 28 | 86 | 13 | 1.107 | 24 | 64 | 28 | 86 | 25 | 1.157 | 28 | 72 | 32 | 100 | 21½ |
| 1.058 | 24 | 86 | 40 | 100 | 18½ | 1.108 | 24 | 86 | 44 | 100 | 25½ | 1.158 | 24 | 56 | 24 | 86 | 14½ |
| 1.059 | 24 | 72 | 28 | 86 | 12½ | 1.109 | 24 | 56 | 24 | 86 | 22 | 1.159 | 24 | 64 | 32 | 100 | 15 |
| 1.060 | 28 | 86 | 40 | 100 | 35½ | 1.110 | 24 | 72 | 32 | 86 | 26½ | 1.160 | 24 | 56 | 24 | 86 | 14 |
| 1.061 | 24 | 72 | 28 | 86 | 12 | 1.111 | 24 | 64 | 28 | 86 | 24½ | 1.161 | 24 | 64 | 28 | 86 | 18 |
| 1.062 | 24 | 72 | 28 | 86 | 12 | 1.112 | 24 | 72 | 40 | 100 | 33½ | 1.162 | 24 | 64 | 32 | 100 | 14½ |
| 1.063 | 24 | 72 | 28 | 86 | 11½ | 1.113 | 24 | 56 | 24 | 86 | 21½ | 1.163 | 24 | 56 | 24 | 86 | 13½ |
| 1.064 | 24 | 86 | 40 | 100 | 17½ | 1.114 | 24 | 64 | 32 | 86 | 37 | 1.164 | 24 | 64 | 32 | 100 | 14 |
| 1.065 | 24 | 72 | 28 | 86 | 11 | 1.115 | 24 | 64 | 28 | 86 | 24 | 1.165 | 24 | 56 | 24 | 86 | 13 |
| 1.066 | 24 | 56 | 24 | 86 | 27 | 1.116 | 24 | 56 | 24 | 86 | 21 | 1.166 | 24 | 72 | 40 | 100 | 29 |
| 1.067 | 24 | 72 | 28 | 86 | 10½ | 1.117 | 24 | 86 | 44 | 100 | 24½ | 1.167 | 24 | 64 | 32 | 100 | 13½ |
| 1.068 | 24 | 64 | 28 | 86 | 29 | 1.118 | 28 | 72 | 32 | 86 | 26 | 1.168 | 24 | 56 | 24 | 86 | 12½ |
| 1.069 | 24 | 72 | 28 | 86 | 10 | 1.119 | 24 | 72 | 32 | 86 | 25½ | 1.169 | 24 | 64 | 32 | 100 | 13 |
| 1.070 | 24 | 86 | 40 | 100 | 16½ | 1.120 | 24 | 56 | 24 | 86 | 20½ | 1.170 | 24 | 56 | 24 | 86 | 12 |
| 1.071 | 32 | 56 | 24 | 86 | 38½ | 1.121 | 24 | 64 | 32 | 86 | 36½ | 1.171 | 24 | 64 | 28 | 86 | 16½ |
| 1.072 | 28 | 72 | 32 | 100 | 30½ | 1.122 | 24 | 86 | 44 | 100 | 24 | 1.172 | 24 | 56 | 24 | 86 | 11½ |
| 1.073 | 24 | 86 | 40 | 100 | 16 | 1.123 | 28 | 72 | 32 | 100 | 25½ | 1.173 | 28 | 72 | 32 | 100 | 19½ |
| 1.074 | 24 | 64 | 32 | 100 | 26½ | 1.124 | 24 | 56 | 24 | 86 | 20 | 1.174 | 24 | 56 | 24 | 86 | 11 |
| 1.075 | 24 | 86 | 40 | 100 | 15½ | 1.125 | 28 | 64 | 32 | 100 | 36½ | 1.175 | 28 | 86 | 40 | 100 | 25½ |
| 1.076 | 24 | 64 | 32 | 86 | 39½ | 1.126 | 24 | 86 | 44 | 100 | 23½ | 1.176 | 24 | 56 | 24 | 86 | 10½ |
| 1.077 | 28 | 72 | 32 | 100 | 30 | 1.127 | 24 | 56 | 24 | 86 | 19½ | 1.177 | 24 | 64 | 28 | 86 | 15½ |
| 1.078 | 24 | 86 | 40 | 100 | 15 | 1.128 | 24 | 64 | 32 | 100 | 20 | 1.178 | 24 | 56 | 24 | 86 | 10 |
| 1.079 | 24 | 56 | 24 | 86 | 25½ | 1.129 | 24 | 64 | 32 | 86 | 36 | 1.179 | 24 | 64 | 28 | 86 | 15 |
| 1.080 | 24 | 86 | 40 | 100 | 14½ | 1.130 | 24 | 72 | 40 | 100 | 32 | 1.180 | 24 | 64 | 32 | 100 | 10½ |
| 1.081 | 28 | 64 | 32 | 100 | 39½ | 1.131 | 24 | 56 | 24 | 86 | 19 | 1.181 | 32 | 56 | 24 | 100 | 30½ |
| 1.082 | 28 | 86 | 44 | 100 | 41 | 1.132 | 24 | 64 | 28 | 86 | 22 | 1.182 | 24 | 64 | 32 | 100 | 10 |
| 1.083 | 24 | 86 | 40 | 100 | 14 | 1.133 | 24 | 72 | 32 | 86 | 24 | 1.183 | 24 | 86 | 44 | 100 | 15½ |
| 1.084 | 24 | 56 | 24 | 86 | 25 | 1.134 | 24 | 56 | 24 | 86 | 18½ | 1.184 | 24 | 64 | 32 | 100 | 9½ |
| 1.085 | 24 | 86 | 40 | 100 | 13½ | 1.135 | 24 | 64 | 32 | 100 | 19 | 1.185 | 24 | 64 | 28 | 86 | 14 |
| 1.086 | 28 | 86 | 40 | 100 | 33½ | 1.136 | 24 | 64 | 28 | 86 | 21½ | 1.186 | 24 | 86 | 44 | 100 | 15 |
| 1.087 | 24 | 86 | 40 | 100 | 13 | 1.137 | 24 | 56 | 24 | 86 | 18 | 1.187 | 24 | 64 | 28 | 86 | 13½ |
| 1.088 | 24 | 56 | 24 | 86 | 24½ | 1.138 | 24 | 64 | 32 | 100 | 18½ | 1.188 | 24 | 72 | 40 | 100 | 27 |
| 1.089 | 24 | 86 | 40 | 100 | 12½ | 1.139 | 28 | 86 | 40 | 100 | 29 | 1.189 | 24 | 86 | 44 | 100 | 14½ |
| 1.090 | 24 | 72 | 32 | 86 | 28½ | 1.140 | 24 | 64 | 28 | 86 | 21 | 1.190 | 24 | 64 | 28 | 86 | 13 |
| 1.091 | 24 | 86 | 48 | 100 | 35½ | 1.141 | 24 | 56 | 24 | 86 | 17½ | 1.191 | 24 | 86 | 44 | 100 | 14 |
| 1.092 | 24 | 86 | 40 | 100 | 12 | 1.142 | 24 | 64 | 32 | 86 | 35 | 1.192 | 24 | 64 | 28 | 86 | 12½ |
| 1.093 | 24 | 56 | 24 | 86 | 24 | 1.143 | 24 | 86 | 44 | 100 | 21½ | 1.193 | 28 | 72 | 32 | 100 | 16½ |
| 1.094 | 24 | 86 | 40 | 100 | 11½ | 1.144 | 24 | 56 | 24 | 86 | 17 | 1.194 | 24 | 64 | 28 | 86 | 12 |
| 1.095 | 24 | 72 | 32 | 86 | 28 | 1.145 | 28 | 72 | 32 | 100 | 23 | 1.195 | 24 | 72 | 32 | 86 | 15½ |
| 1.096 | 24 | 86 | 40 | 100 | 11 | 1.146 | 24 | 86 | 44 | 100 | 21 | 1.196 | 28 | 72 | 32 | 100 | 16 |
| 1.097 | 24 | 86 | 40 | 100 | 10½ | 1.147 | 24 | 56 | 24 | 86 | 16½ | 1.197 | 24 | 64 | 28 | 86 | 11½ |
| 1.098 | 28 | 72 | 32 | 100 | 28 | 1.148 | 24 | 64 | 32 | 100 | 17 | 1.198 | 24 | 72 | 32 | 86 | 15 |
| 1.099 | 24 | 86 | 40 | 100 | 10 | 1.149 | 28 | 72 | 32 | 100 | 22½ | 1.199 | 24 | 64 | 28 | 86 | 11 |

## Change Gears and Angles for Cam Milling — 5

| Lead of Cam | Gear on Worm | First Intermediate | Second Intermediate | Gear on Screw | Angle | Lead of Cam | Gear on Worm | First Intermediate | Second Intermediate | Gear on Screw | Angle | Lead of Cam | Gear on Worm | First Intermediate | Second Intermediate | Gear on Screw | Angle |
|---|---|---|---|---|---|---|---|---|---|---|---|---|---|---|---|---|---|
| 1.200 | 24 | 72 | 32 | 86 | 14½ | 1.250 | 24 | 64 | 28 | 72 | 31 | 1.300 | 24 | 86 | 48 | 100 | 14 |
| 1.201 | 24 | 64 | 28 | 86 | 10½ | 1.251 | 24 | 86 | 48 | 100 | 21 | 1.301 | 24 | 72 | 40 | 100 | 12½ |
| 1.202 | 24 | 64 | 28 | 86 | 10 | 1.252 | 28 | 86 | 40 | 100 | 16 | 1.302 | 24 | 64 | 32 | 86 | 21 |
| 1.203 | 24 | 86 | 44 | 100 | 11½ | 1.253 | 24 | 72 | 40 | 100 | 20 | 1.303 | 24 | 86 | 48 | 100 | 13½ |
| 1.204 | 28 | 72 | 32 | 100 | 14½ | 1.254 | 24 | 64 | 32 | 86 | 26 | 1.304 | 24 | 72 | 40 | 100 | 12 |
| 1.205 | 24 | 86 | 44 | 100 | 11 | 1.255 | 28 | 86 | 40 | 100 | 15½ | 1.305 | 24 | 64 | 28 | 72 | 26½ |
| 1.206 | 24 | 72 | 32 | 86 | 13½ | 1.256 | 24 | 64 | 28 | 72 | 30½ | 1.306 | 24 | 72 | 40 | 100 | 11½ |
| 1.207 | 24 | 86 | 44 | 100 | 10½ | 1.257 | 24 | 72 | 40 | 100 | 19½ | 1.307 | 32 | 56 | 24 | 100 | 17½ |
| 1.208 | 24 | 72 | 32 | 86 | 13 | 1.258 | 28 | 86 | 40 | 100 | 15 | 1.308 | 24 | 72 | 40 | 100 | 11 |
| 1.209 | 24 | 86 | 44 | 100 | 10 | 1.259 | 24 | 86 | 48 | 100 | 20 | 1.309 | 28 | 86 | 44 | 100 | 24 |
| 1.210 | 28 | 72 | 32 | 100 | 13½ | 1.260 | 28 | 86 | 40 | 100 | 14½ | 1.310 | 24 | 64 | 28 | 72 | 26 |
| 1.211 | 24 | 72 | 32 | 86 | 12½ | 1.261 | 28 | 86 | 40 | 100 | 14½ | 1.311 | 24 | 72 | 40 | 100 | 10½ |
| 1.212 | 28 | 72 | 32 | 100 | 13 | 1.262 | 32 | 56 | 24 | 100 | 23 | 1.312 | 40 | 64 | 24 | 100 | 29 |
| 1.213 | 24 | 72 | 32 | 86 | 12 | 1.263 | 28 | 86 | 40 | 100 | 14 | 1.313 | 24 | 72 | 40 | 100 | 10 |
| 1.214 | 24 | 86 | 48 | 100 | 25 | 1.264 | 24 | 72 | 40 | 100 | 18½ | 1.314 | 28 | 86 | 44 | 100 | 23½ |
| 1.215 | 24 | 72 | 32 | 86 | 11½ | 1.265 | 28 | 86 | 44 | 100 | 28 | 1.315 | 24 | 86 | 48 | 100 | 11 |
| 1.216 | 32 | 56 | 24 | 100 | 27½ | 1.266 | 28 | 86 | 40 | 100 | 13½ | 1.316 | 28 | 64 | 32 | 100 | 20 |
| 1.217 | 24 | 72 | 32 | 86 | 11 | 1.267 | 24 | 86 | 48 | 100 | 19 | 1.317 | 28 | 72 | 32 | 86 | 24½ |
| 1.218 | 24 | 72 | 40 | 100 | 24 | 1.268 | 24 | 72 | 40 | 100 | 18 | 1.318 | 24 | 86 | 48 | 100 | 10½ |
| 1.219 | 24 | 72 | 32 | 86 | 10½ | 1.269 | 28 | 86 | 40 | 100 | 13 | 1.319 | 24 | 64 | 32 | 86 | 19 |
| 1.220 | 28 | 86 | 40 | 100 | 20½ | 1.270 | 24 | 72 | 44 | 100 | 30 | 1.320 | 24 | 86 | 48 | 100 | 10 |
| 1.221 | 24 | 72 | 32 | 86 | 10 | 1.271 | 28 | 86 | 40 | 100 | 12½ | 1.321 | 32 | 56 | 24 | 100 | 15½ |
| 1.222 | 24 | 86 | 40 | 100 | 23½ | 1.272 | 28 | 72 | 32 | 86 | 28½ | 1.322 | 28 | 72 | 32 | 86 | 24 |
| 1.223 | 28 | 72 | 32 | 100 | 10½ | 1.273 | 28 | 86 | 40 | 100 | 12 | 1.323 | 24 | 64 | 32 | 86 | 18½ |
| 1.224 | 28 | 86 | 40 | 100 | 20 | 1.274 | 28 | 86 | 40 | 100 | 12 | 1.324 | 32 | 56 | 24 | 100 | 15 |
| 1.225 | 28 | 72 | 32 | 100 | 10 | 1.275 | 24 | 72 | 40 | 100 | 12 | 1.325 | 28 | 86 | 48 | 100 | 32 |
| 1.226 | 24 | 72 | 48 | 100 | 40 | 1.276 | 28 | 86 | 40 | 100 | 11½ | 1.326 | 32 | 86 | 40 | 100 | 27 |
| 1.227 | 28 | 86 | 40 | 100 | 19½ | 1.277 | 28 | 86 | 44 | 100 | 27 | 1.327 | 32 | 56 | 24 | 100 | 14½ |
| 1.228 | 28 | 86 | 44 | 100 | 31 | 1.278 | 28 | 86 | 40 | 100 | 11 | 1.328 | 28 | 64 | 32 | 100 | 18½ |
| 1.229 | 24 | 86 | 48 | 100 | 23½ | 1.279 | 24 | 64 | 32 | 86 | 23½ | 1.329 | 28 | 86 | 44 | 100 | 22 |
| 1.230 | 28 | 64 | 32 | 100 | 28½ | 1.280 | 28 | 86 | 40 | 100 | 10½ | 1.330 | 32 | 56 | 24 | 100 | 14 |
| 1.231 | 28 | 86 | 40 | 100 | 19 | 1.281 | 24 | 72 | 40 | 100 | 16 | 1.331 | 28 | 64 | 32 | 100 | 18 |
| 1.232 | 24 | 72 | 40 | 100 | 22½ | 1.282 | 28 | 86 | 40 | 100 | 10 | 1.332 | 28 | 64 | 32 | 100 | 18 |
| 1.233 | 32 | 86 | 40 | 100 | 34 | 1.283 | 28 | 72 | 32 | 86 | 27½ | 1.333 | 32 | 56 | 24 | 100 | 13½ |
| 1.234 | 24 | 86 | 48 | 100 | 23 | 1.284 | 24 | 72 | 40 | 100 | 15½ | 1.334 | 24 | 64 | 32 | 86 | 17 |
| 1.235 | 28 | 86 | 40 | 100 | 18½ | 1.285 | 24 | 72 | 40 | 100 | 15½ | 1.335 | 28 | 64 | 32 | 100 | 17½ |
| 1.236 | 24 | 72 | 40 | 100 | 22 | 1.286 | 40 | 64 | 24 | 100 | 31 | 1.336 | 32 | 56 | 24 | 100 | 13 |
| 1.237 | 32 | 56 | 24 | 100 | 25½ | 1.287 | 24 | 72 | 40 | 100 | 15 | 1.337 | 28 | 72 | 32 | 86 | 22½ |
| 1.238 | 28 | 86 | 40 | 100 | 18 | 1.288 | 24 | 72 | 40 | 100 | 15 | 1.338 | 32 | 56 | 24 | 100 | 12½ |
| 1.239 | 24 | 64 | 32 | 72 | 42 | 1.289 | 24 | 64 | 32 | 86 | 22½ | 1.339 | 32 | 56 | 24 | 100 | 12½ |
| 1.240 | 24 | 72 | 40 | 100 | 21½ | 1.290 | 24 | 72 | 40 | 100 | 14½ | 1.340 | 24 | 72 | 44 | 100 | 24 |
| 1.241 | 28 | 86 | 44 | 100 | 30 | 1.291 | 24 | 72 | 40 | 100 | 14½ | 1.341 | 32 | 56 | 24 | 100 | 12 |
| 1.242 | 28 | 86 | 40 | 100 | 17½ | 1.292 | 32 | 56 | 24 | 100 | 19½ | 1.342 | 28 | 64 | 32 | 100 | 16½ |
| 1.243 | 32 | 56 | 24 | 100 | 25 | 1.293 | 24 | 72 | 40 | 100 | 14 | 1.343 | 32 | 56 | 24 | 100 | 11½ |
| 1.244 | 24 | 72 | 40 | 100 | 21 | 1.294 | 24 | 86 | 48 | 100 | 15 | 1.344 | 24 | 64 | 32 | 86 | 15½ |
| 1.245 | 28 | 86 | 40 | 100 | 17 | 1.295 | 28 | 72 | 32 | 86 | 26½ | 1.345 | 24 | 72 | 44 | 100 | 23½ |
| 1.246 | 32 | 72 | 40 | 100 | 45½ | 1.296 | 24 | 72 | 40 | 100 | 13½ | 1.346 | 32 | 56 | 24 | 100 | 11 |
| 1.247 | 24 | 86 | 48 | 100 | 21½ | 1.297 | 24 | 86 | 48 | 100 | 14½ | 1.347 | 24 | 64 | 32 | 86 | 15 |
| 1.248 | 28 | 86 | 40 | 100 | 16½ | 1.298 | 24 | 64 | 32 | 86 | 21½ | 1.348 | 32 | 56 | 24 | 100 | 10½ |
| 1.249 | 24 | 72 | 40 | 100 | 20½ | 1.299 | 24 | 72 | 40 | 100 | 13 | 1.349 | 28 | 64 | 32 | 100 | 15½ |

## Change Gears and Angles for Cam Milling — 6

| Lead of Cam | Gear on Worm | First Intermediate | Second Intermediate | Gear on Screw | Angle | Lead of Cam | Gear on Worm | First Intermediate | Second Intermediate | Gear on Screw | Angle | Lead of Cam | Gear on Worm | First Intermediate | Second Intermediate | Gear on Screw | Angle |
|---|---|---|---|---|---|---|---|---|---|---|---|---|---|---|---|---|---|
| 1.350 | 32 | 56 | 24 | 100 | 10 | 1.400 | 40 | 64 | 24 | 100 | 21 | 1.450 | 32 | 86 | 40 | 100 | 13 |
| 1.351 | 24 | 64 | 32 | 86 | 14½ | 1.401 | 28 | 72 | 32 | 86 | 14½ | 1.451 | 28 | 64 | 32 | 86 | 27 |
| 1.352 | 28 | 64 | 32 | 100 | 15 | 1.402 | 28 | 86 | 44 | 100 | 12 | 1.452 | 40 | 64 | 24 | 100 | 14½ |
| 1.353 | 24 | 72 | 44 | 86 | 37½ | 1.403 | 24 | 72 | 44 | 100 | 17 | 1.453 | 32 | 86 | 40 | 100 | 12½ |
| 1.354 | 24 | 64 | 32 | 86 | 14 | 1.404 | 28 | 72 | 32 | 86 | 14 | 1.454 | 28 | 86 | 48 | 100 | 21½ |
| 1.355 | 28 | 64 | 32 | 100 | 14½ | 1.405 | 24 | 64 | 28 | 72 | 15½ | 1.455 | 32 | 86 | 40 | 100 | 12 |
| 1.356 | 24 | 64 | 32 | 86 | 13½ | 1.406 | 28 | 86 | 44 | 100 | 11 | 1.456 | 24 | 72 | 40 | 86 | 20 |
| 1.357 | 24 | 64 | 28 | 72 | 21½ | 1.407 | 28 | 72 | 32 | 86 | 13½ | 1.457 | 24 | 72 | 40 | 86 | 20 |
| 1.358 | 28 | 64 | 32 | 100 | 14 | 1.408 | 24 | 64 | 28 | 72 | 15 | 1.458 | 32 | 86 | 40 | 100 | 11½ |
| 1.359 | 24 | 64 | 32 | 86 | 13 | 1.409 | 28 | 86 | 44 | 100 | 10½ | 1.459 | 40 | 64 | 24 | 100 | 13½ |
| 1.360 | 28 | 72 | 32 | 86 | 20 | 1.410 | 28 | 72 | 32 | 86 | 13 | 1.460 | 24 | 44 | 32 | 86 | 44 |
| 1.361 | 28 | 64 | 32 | 100 | 13½ | 1.411 | 28 | 86 | 48 | 100 | 25½ | 1.461 | 32 | 86 | 40 | 100 | 11 |
| 1.362 | 24 | 64 | 32 | 86 | 12½ | 1.412 | 24 | 64 | 28 | 72 | 14½ | 1.462 | 40 | 64 | 24 | 100 | 13 |
| 1.363 | 28 | 86 | 44 | 100 | 18 | 1.413 | 28 | 72 | 32 | 86 | 12½ | 1.463 | 32 | 86 | 40 | 100 | 10½ |
| 1.364 | 24 | 64 | 32 | 86 | 12 | 1.414 | 24 | 72 | 44 | 100 | 15½ | 1.464 | 40 | 64 | 24 | 100 | 12½ |
| 1.365 | 24 | 64 | 32 | 86 | 12 | 1.415 | 28 | 72 | 32 | 86 | 12 | 1.465 | 32 | 86 | 40 | 100 | 10 |
| 1.366 | 24 | 64 | 28 | 72 | 20½ | 1.416 | 24 | 64 | 44 | 86 | 42½ | 1.466 | 24 | 72 | 40 | 86 | 19 |
| 1.367 | 24 | 64 | 32 | 86 | 11½ | 1.417 | 24 | 72 | 44 | 100 | 15 | 1.467 | 40 | 64 | 24 | 100 | 12 |
| 1.368 | 28 | 72 | 32 | 86 | 19 | 1.418 | 28 | 72 | 32 | 86 | 11½ | 1.468 | 28 | 64 | 40 | 100 | 33 |
| 1.369 | 24 | 64 | 32 | 86 | 11 | 1.419 | 32 | 86 | 40 | 100 | 17½ | 1.469 | 28 | 86 | 48 | 100 | 20 |
| 1.370 | 28 | 86 | 44 | 100 | 17 | 1.420 | 28 | 72 | 32 | 86 | 11 | 1.470 | 40 | 64 | 24 | 100 | 11½ |
| 1.371 | 24 | 64 | 32 | 86 | 10½ | 1.421 | 24 | 64 | 28 | 72 | 13 | 1.471 | 28 | 72 | 40 | 100 | 19 |
| 1.372 | 24 | 64 | 32 | 86 | 10½ | 1.422 | 40 | 64 | 24 | 100 | 18½ | 1.472 | 40 | 64 | 24 | 100 | 11 |
| 1.373 | 28 | 64 | 44 | 100 | 44½ | 1.423 | 28 | 72 | 32 | 86 | 10½ | 1.473 | 40 | 64 | 24 | 100 | 11 |
| 1.374 | 24 | 64 | 32 | 86 | 10 | 1.424 | 28 | 64 | 32 | 86 | 29 | 1.474 | 24 | 72 | 40 | 86 | 18 |
| 1.375 | 32 | 86 | 40 | 100 | 22½ | 1.425 | 28 | 72 | 32 | 86 | 10 | 1.475 | 40 | 64 | 24 | 100 | 10½ |
| 1.376 | 28 | 72 | 32 | 86 | 18 | 1.426 | 24 | 64 | 28 | 72 | 12 | 1.476 | 28 | 72 | 40 | 100 | 18½ |
| 1.377 | 28 | 64 | 32 | 100 | 10½ | 1.427 | 32 | 86 | 40 | 100 | 16½ | 1.477 | 40 | 64 | 24 | 100 | 10 |
| 1.378 | 28 | 86 | 44 | 100 | 16 | 1.428 | 28 | 86 | 48 | 100 | 24 | 1.478 | 24 | 72 | 40 | 86 | 17½ |
| 1.379 | 28 | 64 | 32 | 100 | 10 | 1.429 | 24 | 64 | 28 | 72 | 11½ | 1.479 | 24 | 64 | 32 | 72 | 27½ |
| 1.380 | 28 | 72 | 32 | 86 | 17½ | 1.430 | 32 | 86 | 40 | 100 | 16 | 1.480 | 28 | 72 | 40 | 100 | 18 |
| 1.381 | 28 | 86 | 44 | 100 | 15½ | 1.431 | 24 | 64 | 28 | 72 | 11 | 1.481 | 28 | 64 | 32 | 86 | 24½ |
| 1.382 | 24 | 64 | 32 | 72 | 34 | 1.432 | 24 | 72 | 44 | 100 | 12½ | 1.482 | 24 | 72 | 40 | 86 | 17 |
| 1.383 | 24 | 64 | 28 | 72 | 18½ | 1.433 | 28 | 86 | 48 | 100 | 23½ | 1.483 | 44 | 64 | 24 | 100 | 26 |
| 1.384 | 28 | 86 | 44 | 100 | 15 | 1.434 | 24 | 64 | 28 | 72 | 10½ | 1.484 | 28 | 72 | 40 | 100 | 17½ |
| 1.385 | 28 | 64 | 44 | 100 | 44 | 1.435 | 24 | 72 | 44 | 100 | 12 | 1.485 | 24 | 64 | 32 | 72 | 27 |
| 1.386 | 40 | 64 | 24 | 100 | 22½ | 1.436 | 24 | 64 | 28 | 72 | 10 | 1.486 | 24 | 72 | 40 | 86 | 16½ |
| 1.387 | 28 | 86 | 44 | 100 | 14½ | 1.437 | 32 | 86 | 40 | 100 | 15 | 1.487 | 28 | 86 | 48 | 100 | 18 |
| 1.388 | 28 | 64 | 32 | 86 | 31½ | 1.438 | 24 | 72 | 44 | 100 | 11½ | 1.488 | 28 | 72 | 40 | 100 | 17 |
| 1.389 | 32 | 86 | 40 | 100 | 21 | 1.439 | 28 | 86 | 48 | 100 | 23 | 1.489 | 24 | 72 | 48 | 100 | 21½ |
| 1.390 | 28 | 86 | 44 | 100 | 14 | 1.440 | 24 | 72 | 44 | 100 | 11 | 1.490 | 24 | 72 | 40 | 86 | 16 |
| 1.391 | 28 | 72 | 32 | 86 | 16 | 1.441 | 32 | 86 | 40 | 100 | 14½ | 1.491 | 28 | 86 | 48 | 100 | 17½ |
| 1.392 | 44 | 64 | 24 | 100 | 32½ | 1.442 | 24 | 72 | 44 | 100 | 10½ | 1.492 | 28 | 72 | 40 | 100 | 16½ |
| 1.393 | 28 | 86 | 44 | 100 | 13½ | 1.443 | 24 | 72 | 44 | 100 | 10½ | 1.493 | 28 | 64 | 32 | 86 | 23½ |
| 1.394 | 28 | 72 | 32 | 86 | 15½ | 1.444 | 32 | 86 | 40 | 100 | 14 | 1.494 | 24 | 72 | 40 | 86 | 15½ |
| 1.395 | 24 | 72 | 44 | 100 | 18 | 1.445 | 24 | 72 | 44 | 100 | 10 | 1.495 | 28 | 86 | 48 | 100 | 17 |
| 1.396 | 28 | 86 | 44 | 100 | 13 | 1.446 | 24 | 72 | 44 | 86 | 32 | 1.496 | 28 | 72 | 40 | 100 | 16 |
| 1.397 | 24 | 64 | 32 | 86 | 35 | 1.447 | 32 | 56 | 40 | 100 | 33½ | 1.497 | 24 | 72 | 40 | 86 | 15 |
| 1.398 | 28 | 72 | 32 | 86 | 15 | 1.448 | 28 | 72 | 40 | 100 | 21½ | 1.498 | 32 | 64 | 40 | 100 | 41½ |
| 1.399 | 28 | 86 | 44 | 100 | 12½ | 1.449 | 40 | 64 | 24 | 100 | 15 | 1.499 | 28 | 72 | 40 | 100 | 15½ |

### Change Gears and Angles for Cam Milling — 7

| Lead of Cam | Gear on Worm | First Intermediate | Second Intermediate | Gear on Screw* | Angle | Lead of Cam | Gear on Worm | First Intermediate | Second Intermediate | Gear on Screw | Angle | Lead of Cam | Gear on Worm | First Intermediate | Second Intermediate | Gear on Screw | Angle |
|---|---|---|---|---|---|---|---|---|---|---|---|---|---|---|---|---|---|
| 1.500 | 28 | 64 | 40 | 100 | 31 | 1.550 | 44 | 64 | 24 | 100 | 20 | 1.600 | 44 | 56 | 24 | 100 | 21 |
| 1.501 | 32 | 86 | 44 | 100 | 23½ | 1.551 | 44 | 64 | 24 | 100 | 20 | 1.601 | 28 | 64 | 32 | 86 | 10½ |
| 1.502 | 28 | 86 | 48 | 100 | 16 | 1.552 | 24 | 72 | 48 | 100 | 14 | 1.602 | 24 | 64 | 32 | 72 | 16 |
| 1.503 | 28 | 72 | 40 | 100 | 15 | 1.553 | 28 | 64 | 32 | 72 | 37 | 1.603 | 28 | 64 | 32 | 86 | 10 |
| 1.504 | 24 | 72 | 40 | 86 | 14 | 1.554 | 24 | 64 | 40 | 86 | 27 | 1.604 | 32 | 86 | 44 | 100 | 11½ |
| 1.505 | 24 | 64 | 32 | 72 | 25½ | 1.555 | 44 | 64 | 24 | 100 | 19½ | 1.605 | 40 | 56 | 24 | 100 | 20½ |
| 1.506 | 28 | 72 | 40 | 100 | 14½ | 1.556 | 24 | 64 | 32 | 72 | 21 | 1.606 | 24 | 64 | 32 | 72 | 15½ |
| 1.507 | 24 | 72 | 40 | 86 | 13½ | 1.557 | 28 | 64 | 32 | 86 | 17 | 1.607 | 32 | 86 | 44 | 100 | 11 |
| 1.508 | 24 | 72 | 48 | 100 | 19½ | 1.558 | 24 | 72 | 44 | 86 | 24 | 1.608 | 44 | 64 | 24 | 100 | 13 |
| 1.509 | 28 | 64 | 32 | 86 | 22 | 1.559 | 24 | 72 | 48 | 100 | 13 | 1.609 | 28 | 72 | 48 | 100 | 30½ |
| 1.510 | 24 | 72 | 40 | 86 | 13 | 1.560 | 44 | 64 | 24 | 100 | 19 | 1.610 | 32 | 86 | 44 | 100 | 10½ |
| 1.511 | 24 | 64 | 32 | 72 | 25 | 1.561 | 28 | 64 | 32 | 86 | 16½ | 1.611 | 44 | 64 | 24 | 100 | 12½ |
| 1.512 | 24 | 64 | 44 | 86 | 38 | 1.562 | 24 | 72 | 48 | 100 | 12½ | 1.612 | 32 | 86 | 44 | 100 | 10 |
| 1.513 | 24 | 72 | 40 | 86 | 12½ | 1.563 | 28 | 72 | 44 | 100 | 24 | 1.613 | 28 | 72 | 44 | 100 | 19½ |
| 1.514 | 24 | 72 | 48 | 86 | 35½ | 1.564 | 24 | 72 | 44 | 86 | 23½ | 1.614 | 44 | 64 | 24 | 100 | 12 |
| 1.515 | 28 | 64 | 32 | 86 | 21½ | 1.565 | 24 | 72 | 48 | 100 | 12 | 1.615 | 24 | 64 | 48 | 86 | 39½ |
| 1.516 | 24 | 72 | 40 | 86 | 12 | 1.566 | 32 | 86 | 44 | 100 | 17 | 1.616 | 40 | 56 | 24 | 100 | 19½ |
| 1.517 | 24 | 72 | 48 | 100 | 18½ | 1.567 | 28 | 64 | 44 | 100 | 35½ | 1.617 | 44 | 64 | 24 | 100 | 11½ |
| 1.518 | 32 | 86 | 44 | 100 | 22 | 1.568 | 24 | 72 | 48 | 100 | 11½ | 1.618 | 24 | 64 | 32 | 72 | 14 |
| 1.519 | 24 | 72 | 40 | 86 | 11½ | 1.569 | 28 | 64 | 32 | 86 | 15½ | 1.619 | 32 | 86 | 48 | 100 | 25 |
| 1.520 | 28 | 86 | 48 | 100 | 13½ | 1.570 | 32 | 72 | 40 | 100 | 28 | 1.620 | 44 | 64 | 24 | 100 | 11 |
| 1.521 | 24 | 72 | 40 | 86 | 11 | 1.571 | 24 | 72 | 48 | 100 | 11 | 1.621 | 24 | 64 | 32 | 72 | 13½ |
| 1.522 | 24 | 72 | 40 | 86 | 11 | 1.572 | 28 | 64 | 32 | 86 | 15 | 1.622 | 44 | 64 | 24 | 100 | 10½ |
| 1.523 | 28 | 86 | 48 | 100 | 13 | 1.573 | 24 | 72 | 48 | 100 | 10½ | 1.623 | 28 | 72 | 44 | 100 | 18½ |
| 1.524 | 24 | 72 | 40 | 86 | 10½ | 1.574 | 32 | 86 | 44 | 100 | 16 | 1.624 | 24 | 64 | 32 | 72 | 13 |
| 1.525 | 28 | 72 | 40 | 100 | 11½ | 1.575 | 24 | 72 | 44 | 86 | 22½ | 1.625 | 44 | 64 | 24 | 100 | 10 |
| 1.526 | 24 | 72 | 40 | 86 | 10 | 1.576 | 24 | 72 | 48 | 100 | 10 | 1.626 | 24 | 72 | 44 | 86 | 17½ |
| 1.527 | 28 | 72 | 40 | 100 | 11 | 1.577 | 32 | 86 | 44 | 100 | 15½ | 1.627 | 24 | 64 | 32 | 72 | 12½ |
| 1.528 | 32 | 86 | 44 | 100 | 21 | 1.578 | 44 | 64 | 24 | 100 | 17 | 1.628 | 24 | 64 | 32 | 72 | 12½ |
| 1.529 | 28 | 86 | 48 | 100 | 12 | 1.579 | 28 | 100 | 56 | 86 | 30 | 1.629 | 32 | 72 | 40 | 86 | 38 |
| 1.530 | 28 | 72 | 40 | 100 | 10½ | 1.580 | 28 | 64 | 32 | 86 | 14 | 1.630 | 24 | 64 | 32 | 72 | 12 |
| 1.531 | 28 | 72 | 44 | 100 | 26½ | 1.581 | 32 | 86 | 44 | 100 | 15 | 1.631 | 24 | 64 | 32 | 72 | 12 |
| 1.532 | 28 | 72 | 40 | 100 | 10 | 1.582 | 44 | 64 | 24 | 100 | 16½ | 1.632 | 28 | 72 | 44 | 100 | 17½ |
| 1.533 | 32 | 86 | 44 | 100 | 20½ | 1.583 | 28 | 64 | 32 | 86 | 13½ | 1.633 | 28 | 72 | 40 | 86 | 25½ |
| 1.534 | 28 | 86 | 48 | 100 | 11 | 1.584 | 40 | 56 | 24 | 100 | 22½ | 1.634 | 24 | 64 | 32 | 72 | 11½ |
| 1.535 | 28 | 64 | 32 | 86 | 19½ | 1.585 | 32 | 86 | 44 | 100 | 14½ | 1.635 | 24 | 72 | 44 | 86 | 16½ |
| 1.536 | 32 | 72 | 40 | 86 | 42 | 1.586 | 28 | 64 | 32 | 86 | 13 | 1.636 | 24 | 64 | 32 | 72 | 11 |
| 1.537 | 28 | 86 | 48 | 100 | 10½ | 1.587 | 24 | 64 | 40 | 86 | 24½ | 1.637 | 32 | 72 | 40 | 100 | 23 |
| 1.538 | 24 | 72 | 48 | 100 | 16 | 1.588 | 32 | 86 | 44 | 100 | 14 | 1.638 | 32 | 86 | 48 | 100 | 23½ |
| 1.539 | 28 | 86 | 48 | 100 | 10 | 1.589 | 28 | 64 | 32 | 86 | 12½ | 1.639 | 24 | 64 | 32 | 72 | 10½ |
| 1.540 | 28 | 100 | 56 | 72 | 45 | 1.590 | 44 | 64 | 24 | 100 | 15½ | 1.640 | 28 | 72 | 40 | 86 | 25 |
| 1.541 | 40 | 56 | 24 | 100 | 26 | 1.591 | 32 | 72 | 40 | 100 | 26½ | 1.641 | 28 | 72 | 44 | 100 | 16½ |
| 1.542 | 24 | 72 | 48 | 100 | 15½ | 1.592 | 28 | 64 | 32 | 86 | 12 | 1.642 | 24 | 64 | 32 | 72 | 10 |
| 1.543 | 32 | 86 | 44 | 100 | 19½ | 1.593 | 24 | 64 | 40 | 86 | 24 | 1.643 | 24 | 72 | 44 | 86 | 15½ |
| 1.544 | 28 | 64 | 32 | 86 | 18½ | 1.594 | 28 | 64 | 32 | 86 | 18½ | 1.644 | 24 | 64 | 40 | 86 | 16½ |
| 1.545 | 24 | 72 | 48 | 100 | 15 | 1.595 | 28 | 64 | 32 | 86 | 11½ | 1.645 | 28 | 72 | 44 | 100 | 16 |
| 1.546 | 24 | 72 | 48 | 100 | 15 | 1.596 | 28 | 64 | 44 | 100 | 34 | 1.646 | 28 | 72 | 40 | 86 | 24½ |
| 1.547 | 40 | 56 | 24 | 100 | 25½ | 1.597 | 44 | 64 | 24 | 100 | 14½ | 1.647 | 24 | 72 | 44 | 86 | 15 |
| 1.548 | 28 | 64 | 32 | 86 | 18 | 1.598 | 28 | 64 | 32 | 86 | 11 | 1.648 | 40 | 56 | 24 | 100 | 16 |
| 1.549 | 24 | 72 | 48 | 100 | 14½ | 1.599 | 24 | 64 | 40 | 86 | 23½ | 1.649 | 28 | 72 | 44 | 100 | 15½ |

## Change Gears and Angles for Cam Milling — 8

| Lead of Cam | Gear on Worm | First Intermediate | Second Intermediate | Gear on Screw | Angle | Lead of Cam | Gear on Worm | First Intermediate | Second Intermediate | Gear on Screw | Angle | Lead of Cam | Gear on Worm | First Intermediate | Second Intermediate | Gear on Screw | Angle |
|---|---|---|---|---|---|---|---|---|---|---|---|---|---|---|---|---|---|
| 1.650 | 28 | 64 | 40 | 100 | 19½ | 1.700 | 32 | 72 | 40 | 100 | 17 | 1.750 | 32 | 86 | 48 | 100 | 11½ |
| 1.651 | 24 | 72 | 44 | 86 | 14½ | 1.701 | 32 | 64 | 40 | 86 | 43 | 1.751 | 32 | 72 | 40 | 100 | 10 |
| 1.652 | 40 | 56 | 24 | 100 | 15½ | 1.702 | 28 | 64 | 40 | 100 | 13½ | 1.752 | 28 | 100 | 56 | 86 | 16 |
| 1.653 | 28 | 72 | 44 | 100 | 15 | 1.703 | 24 | 64 | 40 | 86 | 12½ | 1.753 | 32 | 86 | 48 | 100 | 11 |
| 1.654 | 24 | 72 | 44 | 86 | 14 | 1.704 | 28 | 64 | 40 | 72 | 45½ | 1.754 | 28 | 72 | 48 | 100 | 20 |
| 1.655 | 28 | 64 | 40 | 100 | 19 | 1.705 | 28 | 64 | 40 | 100 | 13 | 1.755 | 28 | 72 | 40 | 86 | 14 |
| 1.656 | 28 | 72 | 44 | 100 | 14½ | 1.706 | 24 | 64 | 40 | 86 | 12 | 1.756 | 32 | 86 | 48 | 100 | 10½ |
| 1.657 | 24 | 72 | 44 | 100 | 14½ | 1.707 | 40 | 86 | 44 | 100 | 33½ | 1.757 | 28 | 100 | 56 | 86 | 15½ |
| 1.658 | 24 | 72 | 44 | 86 | 13½ | 1.708 | 28 | 64 | 40 | 100 | 12½ | 1.758 | 32 | 72 | 44 | 100 | 26 |
| 1.659 | 24 | 64 | 40 | 86 | 18 | 1.709 | 24 | 64 | 40 | 86 | 11½ | 1.759 | 28 | 72 | 48 | 86 | 10 |
| 1.660 | 28 | 72 | 44 | 100 | 14 | 1.710 | 28 | 72 | 40 | 86 | 19 | 1.760 | 28 | 72 | 48 | 100 | 19½ |
| 1.661 | 24 | 72 | 44 | 86 | 13 | 1.711 | 32 | 72 | 44 | 100 | 29 | 1.761 | 28 | 100 | 56 | 86 | 15 |
| 1.662 | 32 | 86 | 48 | 100 | 21½ | 1.712 | 24 | 64 | 40 | 86 | 11 | 1.762 | 28 | 64 | 32 | 72 | 25 |
| 1.663 | 40 | 56 | 24 | 100 | 14 | 1.713 | 32 | 72 | 40 | 100 | 15½ | 1.763 | 28 | 72 | 40 | 86 | 13 |
| 1.664 | 28 | 72 | 44 | 100 | 13½ | 1.714 | 32 | 64 | 40 | 100 | 31 | 1.764 | 24 | 72 | 48 | 86 | 18½ |
| 1.665 | 24 | 72 | 44 | 86 | 12½ | 1.715 | 24 | 64 | 40 | 86 | 10½ | 1.765 | 28 | 100 | 56 | 86 | 14½ |
| 1.666 | 28 | 64 | 32 | 72 | 31 | 1.716 | 28 | 72 | 40 | 86 | 18½ | 1.766 | 28 | 72 | 40 | 86 | 12½ |
| 1.667 | 28 | 72 | 44 | 100 | 13 | 1.717 | 32 | 72 | 40 | 100 | 15 | 1.767 | 44 | 56 | 24 | 100 | 20½ |
| 1.668 | 24 | 72 | 44 | 86 | 12 | 1.718 | 24 | 64 | 40 | 86 | 10 | 1.768 | 32 | 72 | 48 | 100 | 34 |
| 1.669 | 28 | 64 | 40 | 100 | 17½ | 1.719 | 28 | 72 | 48 | 100 | 23 | 1.769 | 28 | 72 | 40 | 86 | 12 |
| 1.670 | 28 | 72 | 44 | 100 | 12½ | 1.720 | 28 | 72 | 40 | 86 | 18 | 1.770 | 28 | 72 | 48 | 100 | 18½ |
| 1.671 | 24 | 72 | 44 | 86 | 11½ | 1.721 | 28 | 64 | 40 | 100 | 10½ | 1.771 | 28 | 72 | 48 | 100 | 18½ |
| 1.672 | 24 | 64 | 40 | 86 | 16½ | 1.722 | 24 | 44 | 32 | 86 | 32 | 1.772 | 44 | 56 | 24 | 100 | 20 |
| 1.673 | 40 | 56 | 24 | 100 | 12½ | 1.723 | 28 | 64 | 40 | 100 | 10 | 1.773 | 28 | 72 | 40 | 86 | 11½ |
| 1.674 | 24 | 72 | 44 | 86 | 11 | 1.724 | 28 | 100 | 56 | 86 | 19 | 1.774 | 24 | 72 | 48 | 86 | 17½ |
| 1.675 | 28 | 64 | 44 | 100 | 29½ | 1.725 | 32 | 72 | 40 | 100 | 14 | 1.775 | 24 | 44 | 32 | 86 | 29 |
| 1.676 | 24 | 72 | 44 | 86 | 10½ | 1.726 | 24 | 56 | 40 | 86 | 30 | 1.776 | 28 | 72 | 40 | 86 | 11 |
| 1.677 | 28 | 72 | 44 | 100 | 11½ | 1.727 | 32 | 72 | 44 | 100 | 28 | 1.777 | 32 | 56 | 40 | 100 | 39 |
| 1.678 | 28 | 64 | 40 | 100 | 16½ | 1.728 | 32 | 72 | 40 | 100 | 13½ | 1.778 | 44 | 56 | 24 | 100 | 19½ |
| 1.679 | 24 | 72 | 44 | 86 | 10 | 1.729 | 32 | 72 | 40 | 100 | 13½ | 1.779 | 28 | 72 | 40 | 86 | 10½ |
| 1.680 | 28 | 72 | 44 | 100 | 11 | 1.730 | 28 | 72 | 40 | 86 | 17 | 1.780 | 28 | 100 | 56 | 86 | 12½ |
| 1.681 | 24 | 64 | 40 | 86 | 15½ | 1.731 | 24 | 72 | 48 | 86 | 21½ | 1.781 | 28 | 72 | 40 | 86 | 10 |
| 1.682 | 28 | 72 | 44 | 100 | 10½ | 1.732 | 32 | 72 | 40 | 100 | 13 | 1.782 | 32 | 64 | 40 | 100 | 27 |
| 1.683 | 40 | 56 | 24 | 100 | 11 | 1.733 | 32 | 86 | 48 | 100 | 14 | 1.783 | 28 | 100 | 56 | 86 | 12 |
| 1.684 | 32 | 86 | 48 | 100 | 19½ | 1.734 | 28 | 72 | 40 | 86 | 16½ | 1.784 | 24 | 56 | 40 | 86 | 26½ |
| 1.685 | 28 | 72 | 44 | 100 | 10 | 1.735 | 28 | 72 | 40 | 86 | 16½ | 1.785 | 28 | 72 | 48 | 100 | 17 |
| 1.686 | 28 | 64 | 40 | 100 | 15½ | 1.736 | 32 | 72 | 40 | 100 | 12½ | 1.786 | 28 | 100 | 56 | 86 | 11½ |
| 1.687 | 32 | 64 | 40 | 100 | 32½ | 1.737 | 32 | 86 | 48 | 100 | 13½ | 1.787 | 32 | 72 | 44 | 100 | 24 |
| 1.688 | 40 | 56 | 24 | 100 | 10 | 1.738 | 32 | 56 | 40 | 100 | 40½ | 1.788 | 24 | 72 | 48 | 86 | 16 |
| 1.689 | 32 | 86 | 48 | 100 | 19 | 1.739 | 32 | 72 | 40 | 100 | 12 | 1.789 | 28 | 100 | 56 | 86 | 11 |
| 1.690 | 28 | 64 | 40 | 100 | 15 | 1.740 | 32 | 86 | 48 | 100 | 13 | 1.790 | 28 | 100 | 56 | 86 | 11 |
| 1.691 | 32 | 72 | 40 | 100 | 18 | 1.741 | 28 | 72 | 44 | 86 | 29 | 1.791 | 24 | 64 | 44 | 86 | 21 |
| 1.692 | 24 | 64 | 40 | 86 | 14 | 1.742 | 32 | 72 | 40 | 86 | 11½ | 1.792 | 28 | 100 | 56 | 86 | 10½ |
| 1.693 | 24 | 72 | 48 | 86 | 24½ | 1.743 | 28 | 72 | 40 | 86 | 15½ | 1.793 | 28 | 100 | 56 | 86 | 10½ |
| 1.694 | 32 | 72 | 44 | 100 | 30 | 1.744 | 32 | 86 | 48 | 100 | 12½ | 1.794 | 44 | 56 | 24 | 100 | 18 |
| 1.695 | 44 | 64 | 40 | 100 | 10 | 1.745 | 32 | 72 | 40 | 100 | 11 | 1.795 | 28 | 100 | 56 | 86 | 10 |
| 1.696 | 24 | 64 | 40 | 86 | 13½ | 1.746 | 24 | 64 | 44 | 86 | 24½ | 1.796 | 28 | 64 | 32 | 72 | 22½ |
| 1.697 | 28 | 72 | 44 | 86 | 31½ | 1.747 | 32 | 86 | 48 | 100 | 12 | 1.797 | 24 | 72 | 48 | 86 | 15 |
| 1.698 | 28 | 64 | 40 | 100 | 14 | 1.748 | 32 | 72 | 40 | 100 | 10½ | 1.798 | 24 | 64 | 44 | 86 | 20½ |
| 1.699 | 24 | 64 | 40 | 86 | 13 | 1.749 | 28 | 72 | 48 | 100 | 20½ | 1.799 | 28 | 72 | 48 | 100 | 15½ |

**Change Gears and Angles for Cam Milling — 9**

| Lead of Cam | Gear on Worm | First Intermediate | Second Intermediate | Gear on Screw | Angle | Lead of Cam | Gear on Worm | First Intermediate | Second Intermediate | Gear on Screw | Angle | Lead of Cam | Gear on Worm | First Intermediate | Second Intermediate | Gear on Screw | Angle |
|---|---|---|---|---|---|---|---|---|---|---|---|---|---|---|---|---|---|
| 1.800 | 32 | 72 | 44 | 100 | 23 | 1.850 | 28 | 64 | 44 | 100 | 16 | 1.900 | 28 | 64 | 40 | 86 | 21 |
| 1.801 | 24 | 72 | 48 | 86 | 14½ | 1.851 | 44 | 56 | 24 | 100 | 11 | 1.901 | 28 | 64 | 32 | 72 | 12 |
| 1.802 | 28 | 64 | 32 | 72 | 22 | 1.852 | 28 | 72 | 44 | 86 | 21½ | 1.902 | 28 | 64 | 32 | 72 | 12 |
| 1.803 | 28 | 72 | 48 | 100 | 15 | 1.853 | 28 | 56 | 32 | 72 | 33½ | 1.903 | 28 | 72 | 44 | 86 | 17 |
| 1.804 | 32 | 72 | 44 | 86 | 37½ | 1.854 | 44 | 56 | 24 | 100 | 10½ | 1.904 | 24 | 64 | 48 | 86 | 24½ |
| 1.805 | 24 | 72 | 48 | 86 | 14 | 1.855 | 28 | 64 | 44 | 100 | 15½ | 1.905 | 28 | 64 | 32 | 72 | 11½ |
| 1.806 | 24 | 56 | 40 | 86 | 25 | 1.856 | 24 | 64 | 40 | 72 | 27 | 1.906 | 32 | 72 | 44 | 100 | 13 |
| 1.807 | 28 | 72 | 48 | 100 | 14½ | 1.857 | 44 | 56 | 24 | 100 | 10 | 1.907 | 32 | 64 | 40 | 100 | 17½ |
| 1.808 | 28 | 72 | 48 | 100 | 14½ | 1.858 | 24 | 64 | 44 | 86 | 14½ | 1.908 | 28 | 64 | 32 | 72 | 11 |
| 1.809 | 24 | 72 | 48 | 86 | 13½ | 1.859 | 28 | 64 | 44 | 100 | 15 | 1.909 | 32 | 72 | 48 | 100 | 26½ |
| 1.810 | 28 | 72 | 48 | 86 | 33½ | 1.860 | 32 | 72 | 44 | 100 | 18 | 1.910 | 32 | 72 | 44 | 100 | 12½ |
| 1.811 | 28 | 72 | 48 | 100 | 14 | 1.861 | 24 | 56 | 40 | 86 | 21 | 1.911 | 24 | 56 | 40 | 86 | 16½ |
| 1.812 | 24 | 72 | 48 | 86 | 13 | 1.862 | 24 | 64 | 44 | 86 | 14 | 1.912 | 28 | 64 | 32 | 72 | 10½ |
| 1.813 | 44 | 56 | 24 | 100 | 16 | 1.863 | 24 | 56 | 40 | 86 | 20½ | 1.913 | 32 | 72 | 44 | 100 | 12 |
| 1.814 | 24 | 64 | 44 | 86 | 19 | 1.864 | 28 | 64 | 44 | 100 | 14½ | 1.914 | 28 | 64 | 32 | 72 | 10 |
| 1.815 | 28 | 72 | 48 | 100 | 13½ | 1.865 | 32 | 72 | 44 | 100 | 17½ | 1.915 | 28 | 64 | 32 | 72 | 10 |
| 1.816 | 24 | 72 | 48 | 86 | 12½ | 1.866 | 24 | 64 | 44 | 86 | 13½ | 1.916 | 24 | 56 | 40 | 86 | 16 |
| 1.817 | 44 | 56 | 24 | 100 | 15½ | 1.867 | 32 | 64 | 40 | 100 | 21 | 1.917 | 32 | 72 | 44 | 100 | 11½ |
| 1.818 | 28 | 72 | 44 | 86 | 24 | 1.868 | 28 | 64 | 44 | 100 | 14 | 1.918 | 28 | 72 | 44 | 86 | 15½ |
| 1.819 | 24 | 72 | 48 | 86 | 12 | 1.869 | 28 | 64 | 32 | 72 | 16 | 1.919 | 24 | 44 | 32 | 86 | 19 |
| 1.820 | 24 | 64 | 44 | 86 | 18½ | 1.870 | 24 | 64 | 44 | 86 | 13 | 1.920 | 32 | 72 | 44 | 100 | 11 |
| 1.821 | 28 | 64 | 32 | 72 | 20½ | 1.871 | 32 | 72 | 44 | 100 | 17 | 1.921 | 24 | 56 | 40 | 86 | 15½ |
| 1.822 | 44 | 56 | 24 | 100 | 15 | 1.872 | 28 | 64 | 44 | 100 | 13½ | 1.922 | 28 | 72 | 44 | 86 | 15 |
| 1.823 | 24 | 72 | 48 | 86 | 11½ | 1.873 | 24 | 64 | 44 | 86 | 12½ | 1.923 | 32 | 72 | 44 | 100 | 10½ |
| 1.824 | 40 | 86 | 44 | 100 | 27 | 1.874 | 24 | 64 | 44 | 86 | 12½ | 1.924 | 28 | 64 | 40 | 86 | 19 |
| 1.825 | 24 | 64 | 44 | 86 | 18 | 1.875 | 32 | 72 | 44 | 100 | 16½ | 1.925 | 24 | 56 | 40 | 86 | 15 |
| 1.826 | 24 | 72 | 48 | 86 | 11 | 1.876 | 28 | 64 | 44 | 100 | 13 | 1.926 | 32 | 72 | 44 | 100 | 10 |
| 1.827 | 28 | 64 | 32 | 72 | 20 | 1.877 | 24 | 64 | 44 | 86 | 12 | 1.927 | 28 | 72 | 44 | 86 | 14½ |
| 1.828 | 24 | 56 | 40 | 86 | 23½ | 1.878 | 28 | 64 | 32 | 72 | 15 | 1.928 | 28 | 64 | 44 | 86 | 30½ |
| 1.829 | 24 | 72 | 48 | 86 | 10½ | 1.879 | 28 | 64 | 44 | 100 | 12½ | 1.929 | 24 | 56 | 40 | 86 | 14½ |
| 1.830 | 28 | 72 | 48 | 100 | 11½ | 1.880 | 24 | 64 | 44 | 86 | 11½ | 1.930 | 24 | 56 | 40 | 86 | 14½ |
| 1.831 | 28 | 64 | 44 | 100 | 18 | 1.881 | 32 | 72 | 40 | 86 | 24½ | 1.931 | 28 | 72 | 44 | 86 | 14 |
| 1.832 | 24 | 72 | 48 | 86 | 10 | 1.882 | 28 | 64 | 32 | 72 | 14½ | 1.932 | 32 | 64 | 40 | 100 | 15 |
| 1.833 | 28 | 72 | 48 | 100 | 11 | 1.883 | 28 | 64 | 44 | 100 | 12 | 1.933 | 48 | 56 | 24 | 100 | 20 |
| 1.834 | 44 | 56 | 24 | 100 | 13½ | 1.884 | 24 | 64 | 44 | 86 | 11 | 1.934 | 24 | 56 | 40 | 86 | 14 |
| 1.835 | 24 | 64 | 44 | 86 | 17 | 1.885 | 32 | 72 | 44 | 100 | 15½ | 1.935 | 28 | 72 | 44 | 86 | 13½ |
| 1.836 | 28 | 72 | 48 | 100 | 10½ | 1.886 | 28 | 64 | 44 | 100 | 11½ | 1.936 | 32 | 64 | 40 | 100 | 14½ |
| 1.837 | 28 | 64 | 40 | 86 | 25½ | 1.887 | 24 | 64 | 44 | 86 | 10½ | 1.937 | 28 | 64 | 48 | 86 | 37½ |
| 1.838 | 44 | 56 | 24 | 100 | 13 | 1.888 | 28 | 72 | 44 | 86 | 18½ | 1.938 | 24 | 56 | 40 | 86 | 13½ |
| 1.839 | 28 | 72 | 48 | 100 | 10 | 1.889 | 32 | 72 | 44 | 100 | 15 | 1.939 | 28 | 72 | 44 | 86 | 13 |
| 1.840 | 24 | 64 | 44 | 86 | 16½ | 1.890 | 24 | 64 | 44 | 86 | 10 | 1.940 | 28 | 64 | 48 | 100 | 22½ |
| 1.841 | 44 | 56 | 24 | 100 | 12½ | 1.891 | 32 | 64 | 40 | 100 | 19 | 1.941 | 32 | 64 | 40 | 100 | 14 |
| 1.842 | 32 | 72 | 40 | 86 | 27 | 1.892 | 32 | 72 | 48 | 100 | 27½ | 1.942 | 24 | 56 | 40 | 86 | 13 |
| 1.843 | 28 | 64 | 48 | 86 | 41 | 1.893 | 28 | 64 | 44 | 100 | 10½ | 1.943 | 28 | 72 | 44 | 86 | 12½ |
| 1.844 | 28 | 64 | 32 | 72 | 18½ | 1.894 | 28 | 64 | 32 | 72 | 13 | 1.944 | 32 | 56 | 40 | 86 | 43 |
| 1.845 | 44 | 56 | 24 | 100 | 12 | 1.895 | 24 | 56 | 40 | 86 | 18 | 1.945 | 48 | 56 | 24 | 100 | 19 |
| 1.846 | 28 | 64 | 44 | 100 | 16½ | 1.896 | 28 | 64 | 44 | 100 | 10 | 1.946 | 28 | 72 | 44 | 86 | 12 |
| 1.847 | 24 | 44 | 32 | 86 | 24½ | 1.897 | 32 | 64 | 40 | 100 | 18½ | 1.947 | 28 | 72 | 44 | 86 | 12 |
| 1.848 | 44 | 56 | 24 | 100 | 11½ | 1.898 | 28 | 64 | 32 | 72 | 12½ | 1.948 | 32 | 72 | 40 | 86 | 19½ |
| 1.849 | 24 | 64 | 44 | 86 | 15½ | 1.899 | 28 | 72 | 48 | 86 | 29 | 1.949 | 24 | 56 | 40 | 86 | 12 |

### Change Gears and Angles for Cam Milling — 10

| Lead of Cam | Gear on Worm | First Intermediate | Second Intermediate | Gear on Screw | Angle | Lead of Cam | Gear on Worm | First Intermediate | Second Intermediate | Gear on Screw | Angle | Lead of Cam | Gear on Worm | First Intermediate | Second Intermediate | Gear on Screw | Angle |
|---|---|---|---|---|---|---|---|---|---|---|---|---|---|---|---|---|---|
| 1.950 | 28 | 72 | 44 | 86 | 11½ | 2.200 | 24 | 56 | 40 | 72 | 22½ | 2.450 | 24 | 64 | 48 | 72 | 11½ |
| 1.955 | 32 | 72 | 48 | 86 | 38 | 2.205 | 48 | 56 | 32 | 100 | 36½ | 2.455 | 40 | 72 | 48 | 100 | 23 |
| 1.960 | 28 | 72 | 44 | 86 | 10 | 2.210 | 48 | 100 | 56 | 86 | 45 | 2.460 | 28 | 64 | 48 | 72 | 32½ |
| 1.965 | 24 | 44 | 32 | 86 | 14½ | 2.215 | 24 | 56 | 40 | 72 | 21½ | 2.465 | 32 | 64 | 44 | 86 | 15½ |
| 1.970 | 32 | 64 | 40 | 100 | 10 | 2.220 | 32 | 72 | 44 | 86 | 12½ | 2.470 | 28 | 40 | 32 | 86 | 18½ |
| 1.975 | 28 | 64 | 40 | 86 | 14 | 2.225 | 28 | 44 | 32 | 86 | 20 | 2.475 | 32 | 64 | 40 | 72 | 27 |
| 1.980 | 28 | 64 | 48 | 100 | 19½ | 2.230 | 32 | 64 | 40 | 86 | 16½ | 2.480 | 44 | 48 | 28 | 100 | 15 |
| 1.985 | 24 | 64 | 40 | 86 | 18½ | 2.235 | 44 | 86 | 48 | 100 | 24½ | 2.485 | 28 | 72 | 56 | 86 | 11 |
| 1.990 | 28 | 64 | 40 | 86 | 12 | 2.240 | 32 | 56 | 40 | 100 | 11½ | 2.490 | 28 | 72 | 56 | 86 | 10½ |
| 1.995 | 40 | 86 | 44 | 100 | 13 | 2.245 | 28 | 64 | 56 | 86 | 38 | 2.495 | 24 | 44 | 40 | 86 | 10½ |
| 2.000 | 48 | 56 | 24 | 100 | 13½ | 2.250 | 24 | 64 | 44 | 72 | 11 | 2.500 | 28 | 64 | 48 | 72 | 31 |
| 2.005 | 28 | 100 | 56 | 72 | 23 | 2.255 | 32 | 64 | 48 | 100 | 20 | 2.505 | 24 | 56 | 44 | 72 | 17 |
| 2.010 | 32 | 72 | 40 | 86 | 13½ | 2.260 | 44 | 56 | 32 | 100 | 26 | 2.510 | 28 | 40 | 44 | 86 | 45½ |
| 2.015 | 40 | 86 | 48 | 100 | 25½ | 2.265 | 28 | 44 | 32 | 86 | 17 | 2.515 | 32 | 64 | 44 | 86 | 10½ |
| 2.020 | 28 | 72 | 48 | 86 | 21½ | 2.270 | 28 | 44 | 32 | 86 | 16½ | 2.520 | 44 | 48 | 28 | 100 | 11 |
| 2.025 | 32 | 72 | 40 | 86 | 11½ | 2.275 | 32 | 64 | 40 | 86 | 12 | 2.525 | 48 | 56 | 32 | 100 | 23 |
| 2.030 | 24 | 64 | 40 | 72 | 13 | 2.280 | 28 | 64 | 44 | 72 | 31½ | 2.530 | 24 | 56 | 44 | 72 | 15 |
| 2.035 | 24 | 64 | 48 | 86 | 13½ | 2.285 | 44 | 86 | 48 | 86 | 21½ | 2.535 | 32 | 56 | 40 | 86 | 17½ |
| 2.040 | 32 | 72 | 48 | 100 | 17 | 2.290 | 24 | 44 | 40 | 86 | 25½ | 2.540 | 32 | 64 | 48 | 86 | 24½ |
| 2.045 | 24 | 64 | 40 | 72 | 11 | 2.295 | 32 | 64 | 48 | 100 | 17 | 2.545 | 32 | 56 | 44 | 86 | 29½ |
| 2.050 | 28 | 64 | 48 | 100 | 12½ | 2.300 | 24 | 56 | 40 | 72 | 15 | 2.550 | 28 | 64 | 44 | 72 | 17½ |
| 2.055 | 24 | 64 | 48 | 86 | 11 | 2.305 | 24 | 56 | 40 | 72 | 14½ | 2.555 | 32 | 56 | 40 | 86 | 16 |
| 2.060 | 32 | 72 | 48 | 100 | 15 | 2.310 | 24 | 56 | 40 | 72 | 14 | 2.560 | 32 | 64 | 48 | 86 | 23½ |
| 2.065 | 28 | 64 | 48 | 100 | 10½ | 2.315 | 24 | 56 | 40 | 72 | 13½ | 2.565 | 28 | 40 | 32 | 86 | 10 |
| 2.070 | 32 | 72 | 48 | 100 | 14 | 2.320 | 28 | 44 | 32 | 86 | 11½ | 2.570 | 44 | 48 | 40 | 100 | 45½ |
| 2.075 | 40 | 44 | 24 | 100 | 18 | 2.325 | 28 | 44 | 32 | 86 | 11 | 2.575 | 24 | 56 | 44 | 72 | 10½ |
| 2.080 | 32 | 64 | 44 | 100 | 19 | 2.330 | 40 | 100 | 56 | 72 | 10½ | 2.580 | 40 | 72 | 56 | 86 | 44½ |
| 2.085 | 24 | 64 | 48 | 72 | 33½ | 2.335 | 28 | 64 | 48 | 86 | 17 | 2.585 | 32 | 56 | 40 | 86 | 13½ |
| 2.090 | 32 | 72 | 48 | 100 | 11½ | 2.340 | 24 | 56 | 48 | 72 | 35 | 2.590 | 32 | 56 | 40 | 86 | 13 |
| 2.095 | 28 | 56 | 32 | 72 | 19½ | 2.345 | 24 | 56 | 40 | 72 | 10 | 2.595 | 44 | 40 | 32 | 100 | 42½ |
| 2.100 | 28 | 44 | 32 | 86 | 27½ | 2.350 | 28 | 64 | 44 | 72 | 28½ | 2.600 | 32 | 56 | 40 | 86 | 12 |
| 2.105 | 40 | 86 | 48 | 100 | 19½ | 2.355 | 44 | 86 | 48 | 100 | 16½ | 2.605 | 32 | 56 | 40 | 86 | 11½ |
| 2.110 | 28 | 64 | 44 | 86 | 19½ | 2.360 | 32 | 64 | 48 | 100 | 10½ | 2.610 | 32 | 64 | 40 | 72 | 20 |
| 2.115 | 28 | 72 | 48 | 86 | 13 | 2.365 | 24 | 56 | 48 | 86 | 8½ | 2.615 | 44 | 48 | 40 | 100 | 44½ |
| 2.120 | 28 | 72 | 48 | 86 | 12½ | 2.370 | 44 | 56 | 32 | 100 | 19½ | 2.620 | 28 | 64 | 44 | 72 | 11½ |
| 2.125 | 32 | 64 | 44 | 100 | 15 | 2.375 | 28 | 64 | 48 | 86 | 13½ | 2.625 | 44 | 56 | 24 | 72 | 27 |
| 2.130 | 28 | 100 | 56 | 72 | 12 | 2.380 | 32 | 100 | 56 | 86 | 17 | 2.630 | 48 | 56 | 32 | 100 | 16½ |
| 2.135 | 28 | 72 | 48 | 86 | 10½ | 2.385 | 32 | 72 | 56 | 86 | 34½ | 2.635 | 40 | 72 | 44 | 86 | 22 |
| 2.140 | 24 | 56 | 40 | 72 | 26 | 2.390 | 28 | 64 | 40 | 72 | 10½ | 2.640 | 48 | 100 | 56 | 72 | 45 |
| 2.145 | 28 | 100 | 56 | 72 | 10 | 2.395 | 40 | 72 | 44 | 100 | 11½ | 2.645 | 24 | 40 | 44 | 86 | 30½ |
| 2.150 | 32 | 72 | 44 | 86 | 19 | 2.400 | 56 | 64 | 32 | 100 | 31 | 2.650 | 40 | 56 | 44 | 86 | 43½ |
| 2.155 | 44 | 56 | 32 | 100 | 31 | 2.405 | 28 | 64 | 48 | 86 | 15 | 2.655 | 56 | 64 | 44 | 100 | 18½ |
| 2.160 | 32 | 64 | 44 | 100 | 11 | 2.410 | 32 | 100 | 56 | 72 | 14½ | 2.660 | 44 | 48 | 40 | 100 | 43½ |
| 2.165 | 28 | 56 | 32 | 72 | 13 | 2.415 | 44 | 86 | 48 | 100 | 10½ | 2.665 | 28 | 64 | 48 | 72 | 24 |
| 2.170 | 32 | 72 | 48 | 86 | 29 | 2.420 | 32 | 100 | 56 | 72 | 13½ | 2.670 | 28 | 44 | 48 | 72 | 41½ |
| 2.175 | 32 | 72 | 44 | 86 | 17 | 2.425 | 32 | 100 | 56 | 72 | 13 | 2.675 | 48 | 64 | 28 | 56 | 44½ |
| 2.180 | 40 | 86 | 48 | 100 | 12½ | 2.430 | 32 | 100 | 56 | 72 | 12½ | 2.680 | 28 | 44 | 48 | 86 | 41 |
| 2.185 | 28 | 56 | 32 | 72 | 10½ | 2.435 | 32 | 72 | 48 | 86 | 11 | 2.685 | 44 | 100 | 56 | 72 | 44 |
| 2.190 | 32 | 56 | 40 | 86 | 34½ | 2.440 | 32 | 72 | 48 | 86 | 10½ | 2.690 | 40 | 64 | 44 | 100 | 12 |
| 2.195 | 28 | 64 | 48 | 86 | 26 | 2.445 | 44 | 56 | 32 | 100 | 13½ | 2.695 | 40 | 64 | 44 | 100 | 11½ |

**Change Gears and Angles for Cam Milling — II**

| Lead of Cam | Gear on Worm | First Intermediate | Second Intermediate | Gear on Screw | Angle | Lead of Cam | Gear on Worm | First Intermediate | Second Intermediate | Gear on Screw | Angle | Lead of Cam | Gear on Worm | First Intermediate | Second Intermediate | Gear on Screw | Angle |
|---|---|---|---|---|---|---|---|---|---|---|---|---|---|---|---|---|---|
| 2.700 | 28 | 44 | 48 | 86 | 40½ | 2.950 | 40 | 64 | 48 | 100 | 10½ | 3.200 | 48 | 100 | 56 | 72 | 31 |
| 2.705 | 56 | 64 | 32 | 100 | 15 | 2.955 | 48 | 64 | 28 | 56 | 38 | 3.205 | 28 | 40 | 44 | 86 | 26½ |
| 2.710 | 40 | 72 | 56 | 86 | 41½ | 2.960 | 24 | 44 | 48 | 72 | 35½ | 3.210 | 24 | 44 | 48 | 72 | 28 |
| 2.715 | 40 | 56 | 44 | 86 | 42 | 2.965 | 32 | 64 | 44 | 72 | 14 | 3.215 | 40 | 64 | 48 | 72 | 39½ |
| 2.720 | 48 | 64 | 28 | 56 | 43½ | 2.970 | 32 | 64 | 48 | 72 | 27 | 3.220 | 56 | 44 | 28 | 86 | 39 |
| 2.725 | 44 | 48 | 40 | 100 | 42 | 2.975 | 40 | 56 | 44 | 86 | 35½ | 3.225 | 24 | 44 | 48 | 72 | 27½ |
| 2.730 | 48 | 100 | 56 | 72 | 43 | 2.980 | 48 | 40 | 28 | 100 | 27½ | 3.230 | 48 | 40 | 28 | 100 | 16 |
| 2.735 | 44 | 40 | 32 | 100 | 39 | 2.985 | 44 | 48 | 40 | 100 | 35½ | 3.235 | 32 | 72 | 64 | 86 | 12 |
| 2.740 | 28 | 44 | 48 | 86 | 39½ | 2.990 | 28 | 48 | 44 | 72 | 33 | 3.240 | 24 | 44 | 48 | 72 | 27 |
| 2.745 | 40 | 72 | 44 | 86 | 15 | 2.995 | 48 | 64 | 28 | 56 | 37 | 3.245 | 28 | 40 | 44 | 86 | 25 |
| 2.750 | 28 | 64 | 48 | 72 | 19½ | 3.000 | 40 | 100 | 56 | 64 | 31 | 3.250 | 44 | 64 | 48 | 100 | 10 |
| 2.755 | 44 | 40 | 32 | 100 | 38½ | 3.005 | 40 | 64 | 48 | 86 | 30½ | 3.255 | 32 | 48 | 40 | 72 | 28½ |
| 2 760 | 28 | 44 | 48 | 86 | 39 | 3.010 | 28 | 56 | 64 | 86 | 36 | 3.260 | 32 | 56 | 44 | 72 | 21 |
| 2.765 | 48 | 64 | 28 | 56 | 42½ | 3.015 | 48 | 64 | 28 | 56 | 36½ | 3.265 | 48 | 100 | 56 | 72 | 29 |
| 2.770 | 28 | 48 | 44 | 72 | 39 | 3.020 | 48 | 100 | 56 | 72 | 36 | 3.270 | 40 | 56 | 44 | 86 | 26½ |
| 2.775 | 40 | 72 | 44 | 86 | 12½ | 3.025 | 40 | 100 | 56 | 72 | 13½ | 3.275 | 44 | 40 | 32 | 100 | 21½ |
| 2.780 | 40 | 72 | 44 | 86 | 12 | 3.030 | 40 | 64 | 44 | 72 | 37½ | 3.280 | 48 | 64 | 28 | 56 | 29 |
| 2.785 | 24 | 44 | 48 | 72 | 40 | 3.035 | 24 | 40 | 48 | 86 | 25 | 3.285 | 32 | 48 | 40 | 72 | 27½ |
| 2.790 | 28 | 48 | 44 | 72 | 38½ | 3.040 | 44 | 48 | 40 | 100 | 34 | 3.290 | 32 | 44 | 40 | 86 | 13½ |
| 2.795 | 32 | 48 | 40 | 72 | 41 | 3.045 | 32 | 64 | 48 | 72 | 24 | 3.295 | 24 | 44 | 48 | 72 | 25 |
| 2.800 | 24 | 56 | 48 | 72 | 11½ | 3.050 | 40 | 56 | 44 | 100 | 14 | 3.300 | 32 | 48 | 40 | 72 | 27 |
| 2.805 | 24 | 56 | 48 | 72 | 11 | 3.055 | 56 | 44 | 28 | 86 | 42½ | 3.305 | 40 | 72 | 56 | 86 | 24 |
| 2.810 | 44 | 56 | 24 | 64 | 17½ | 3.060 | 28 | 44 | 48 | 86 | 30½ | 3.310 | 44 | 48 | 40 | 100 | 25½ |
| 2.815 | 28 | 44 | 40 | 86 | 18 | 3.065 | 40 | 56 | 44 | 86 | 33 | 3.315 | 32 | 48 | 40 | 72 | 26½ |
| 2.820 | 40 | 56 | 44 | 86 | 39½ | 3.070 | 28 | 40 | 44 | 86 | 31 | 3.320 | 28 | 40 | 44 | 86 | 22 |
| 2.825 | 32 | 56 | 44 | 72 | 36 | 3.075 | 44 | 48 | 40 | 100 | 33 | 3.325 | 40 | 56 | 44 | 86 | 24½ |
| 2.830 | 48 | 64 | 28 | 56 | 41 | 3.080 | 40 | 64 | 48 | 86 | 28 | 3.330 | 28 | 56 | 64 | 86 | 26½ |
| 2.835 | 28 | 48 | 40 | 72 | 29 | 3.085 | 28 | 56 | 64 | 86 | 34 | 3.335 | 28 | 64 | 56 | 72 | 11½ |
| 2.840 | 40 | 56 | 44 | 86 | 39 | 3.090 | 48 | 64 | 28 | 56 | 34½ | 3.340 | 40 | 64 | 44 | 72 | 29 |
| 2.845 | 28 | 44 | 40 | 86 | 16 | 3.095 | 48 | 100 | 56 | 72 | 34 | 3.345 | 32 | 44 | 48 | 86 | 34½ |
| 2.850 | 28 | 56 | 64 | 86 | 40 | 3.100 | 24 | 44 | 48 | 72 | 31½ | 3.350 | 44 | 48 | 40 | 100 | 24 |
| 2.855 | 28 | 44 | 48 | 86 | 36½ | 3.105 | 40 | 100 | 56 | 64 | 27½ | 3.355 | 48 | 100 | 56 | 72 | 26 |
| 2.860 | 40 | 56 | 44 | 86 | 38½ | 3.110 | 44 | 48 | 40 | 100 | 32 | 3.360 | 40 | 56 | 48 | 100 | 11½ |
| 2.865 | 24 | 44 | 48 | 72 | 38 | 3.115 | 28 | 48 | 40 | 72 | 16 | 3.365 | 28 | 40 | 44 | 86 | 20 |
| 2.870 | 44 | 48 | 40 | 100 | 38½ | 3.120 | 44 | 64 | 48 | 100 | 19 | 3.370 | 48 | 64 | 28 | 56 | 26 |
| 2.875 | 40 | 64 | 48 | 86 | 34½ | 3.125 | 32 | 56 | 44 | 72 | 26½ | 3.375 | 44 | 48 | 40 | 100 | 23 |
| 2.880 | 48 | 100 | 56 | 72 | 39½ | 3.130 | 32 | 56 | 48 | 86 | 11 | 3.380 | 32 | 56 | 48 | 72 | 27½ |
| 2.885 | 24 | 44 | 48 | 72 | 37½ | 3.135 | 28 | 44 | 48 | 86 | 28 | 3.385 | 48 | 64 | 28 | 56 | 25½ |
| 2.890 | 44 | 48 | 40 | 100 | 38 | 3.140 | 32 | 56 | 48 | 86 | 10 | 3.390 | 48 | 40 | 32 | 100 | 28 |
| 2.895 | 32 | 56 | 44 | 72 | 34 | 3.145 | 48 | 64 | 28 | 56 | 33 | 3.395 | 32 | 56 | 44 | 72 | 13½ |
| 2.900 | 28 | 44 | 40 | 86 | 11½ | 3.150 | 28 | 44 | 48 | 86 | 27½ | 3.400 | 40 | 56 | 44 | 86 | 21½ |
| 2.905 | 40 | 72 | 48 | 86 | 20½ | 3.155 | 28 | 64 | 56 | 72 | 22 | 3.405 | 28 | 44 | 48 | 86 | 16½ |
| 2.910 | 28 | 44 | 40 | 86 | 10½ | 3.160 | 44 | 48 | 40 | 100 | 30½ | 3.410 | 32 | 48 | 40 | 72 | 23 |
| 2.915 | 28 | 40 | 44 | 86 | 35½ | 3.165 | 24 | 44 | 48 | 72 | 29½ | 3.415 | 28 | 40 | 44 | 86 | 17½ |
| 2.920 | 28 | 48 | 44 | 72 | 35 | 3.170 | 28 | 48 | 40 | 72 | 12 | 3.420 | 32 | 40 | 48 | 86 | 40 |
| 2.925 | 40 | 64 | 48 | 86 | 33 | 3.175 | 32 | 48 | 40 | 72 | 31 | 3.425 | 28 | 56 | 64 | 86 | 23 |
| 2.930 | 32 | 64 | 44 | 72 | 16½ | 3.180 | 40 | 56 | 44 | 86 | 29½ | 3.430 | 28 | 44 | 48 | 86 | 15 |
| 2.935 | 48 | 64 | 28 | 56 | 38½ | 3.185 | 40 | 100 | 56 | 64 | 24½ | 3.435 | 44 | 48 | 40 | 100 | 20½ |
| 2.940 | 40 | 64 | 48 | 100 | 11½ | 3.190 | 28 | 56 | 64 | 86 | 31 | 3.440 | 48 | 100 | 56 | 64 | 35 |
| 2.945 | 40 | 72 | 56 | 86 | 35½ | 3.195 | 44 | 72 | 48 | 86 | 20½ | 3.445 | 28 | 44 | 48 | 86 | 14 |

**Definitions of Gear Terms.** — The terms which follow are commonly applied to various classes of gearing.

*Addendum:* Height of tooth above pitch circle or the radial distance between the pitch circle and the top of the tooth (see illustration).

*Approach Ratio:* The ratio of the arc of approach to the arc of action.

*Arc of Action:* Arc of the pitch circle through which a tooth travels from the first point of contact with the mating tooth to the point where contact ceases.

*Arc of Approach:* Arc of the pitch circle through which a tooth travels from the first point of contact with the mating tooth to the pitch point.

*Arc of Recess:* Arc of the pitch circle through which a tooth travels from its contact with the mating tooth at the pitch point to the point where its contact ceases.

*Axial Plane:* In a pair of gears it is the plane that contains the two axes. In a single gear, it may be any plane containing the axis and a given point.

*Backlash:* The amount by which the width of a tooth space exceeds the thickness of the engaging tooth on the pitch circles. As actually indicated by measuring devices, backlash may be determined variously in the transverse, normal or axial planes, and either in the direction of the pitch circles or on the line of action. Such measurements should be converted to corresponding values on transverse pitch circles for general comparison.

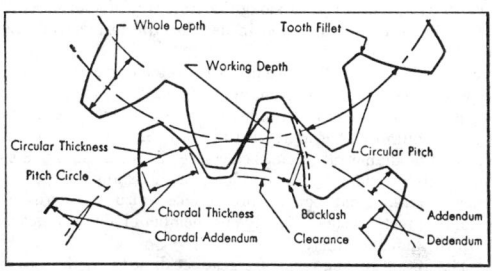

Gear Tooth Parts

*Base Circle:* The circle from which an involute tooth curve is generated or developed.

*Base Helix Angle:* The angle, at the base cylinder of an involute gear, that the tooth makes with the gear axis.

*Base Pitch:* In an involute gear it is the pitch on the base circle or along the line of action. Corresponding sides of involute teeth are parallel curves, and the base pitch is the constant and fundamental distance between them along a common normal in a plane of rotation. The *Normal Base Pitch* is the base pitch in the normal plane, and the *Axial Base Pitch* is the base pitch in the axial plane.

*Center Distance:* The distance between the parallel axes of spur gears and parallel helical gears, or between the crossed axes of crossed helical gears and worm gears. Also, it is the distance between the centers of the pitch circles.

*Central Plane:* In a worm gear this is the plane perpendicular to the gear axis and contains the common perpendicular of the gear and worm axes. In the usual case with the axes at right angles, it contains the worm axis.

*Chordal Addendum:* The height from the top of the tooth to the chord subtending the circular-thickness arc.

*Chordal Thickness:* Length of the chord subtended by the circular thickness arc (the dimension obtained when a gear-tooth caliper is used to measure the thickness at the pitch circle).

*Circular Pitch:* Length of the arc of the pitch circle between the centers or other corresponding points of adjacent teeth (see illustration). *Normal Circular Pitch* is the circular pitch in the normal plane.

*Circular Thickness:* The length of arc between the two sides of a gear tooth, on the pitch circle unless otherwise specified. *Normal Circular Thickness* is the circular thickness in the normal plane.

*Clearance:* The amount by which the dedendum in a given gear exceeds the addendum of its mating gear. It is also the radial distance between the top of a tooth and the bottom of the mating tooth space.

*Contact Diameter:* The smallest diameter on a gear tooth with which the mating gear makes contact.

*Contact Ratio:* The ratio of the arc of action to the circular pitch. It is sometimes thought of as the average number of teeth in contact. For involute gears, the contact ratio is obtained most directly as the ratio of the length of action to the base pitch.

*Contact Stress:* The maximum compressive stress within the contact area between mating gear tooth profiles. It is also called Hertz stress.

*Cycloid:* The curve formed by the path of a point on a circle as it rolls along a straight line. When this circle rolls along the outer side of another circle, the curve is called an *Epicycloid*; when it rolls along the inner side of another circle it is called a *Hypocycloid*. These curves are used in defining the former American Standard composite tooth form.

*Dedendum:* The depth of tooth space below the pitch circle or the radial dimension between the pitch circle and the bottom of the tooth space (see illustration).

*Diametral Pitch:* The ratio of the number of teeth to the number of inches of pitch diameter — equals number of gear teeth to each inch of pitch diameter. *Normal Diametral Pitch* is the diametral pitch as calculated in the normal plane and is equal to the diametral pitch divided by the cosine of the helix angle.

*Effective Face Width:* That portion of the face width that actually comes into contact with mating teeth, as occasionally one member of a pair of gears may have a greater face width than the other.

*Efficiency:* The actual torque ratio of a gear set divided by its gear ratio.

*External Gear:* A gear with teeth on the outer cylindrical surface.

*Face of Tooth:* That surface of the tooth which is between the pitch circle and the top of the tooth.

*Fillet Curve:* The concave portion of the tooth profile where it joins the bottom of the tooth space. The approximate radius of this curve is called the *Fillet Radius*.

*Fillet Stress:* The maximum tensile stress in the gear tooth fillet.

*Flank of Tooth:* That surface which is between the pitch circle and the bottom land. The flank includes the fillet.

*Helical Overlap:* The effective face width of a helical gear divided by the gear axial pitch; also called the *Face Overlap*.

*Helix Angle:* The angle that a helical gear tooth makes with the gear axis at the pitch circle unless otherwise specified.

*Hertz Stress:* See Contact Stress.

*Highest Point of Single Tooth Contact:* The largest diameter on a spur gear at which a single tooth is in contact with the mating gear. Often referred to as HPSTC.

*Internal Diameter:* The diameter of a circle coinciding with the tops of the teeth of an internal gear.

*Internal Gear:* A gear with teeth on the inner cylindrical surface.

*Involute:* The curve formed by the path of a point on a straight line, called the generatrix, as it rolls along a convex base curve. (The base curve is usually a circle.) This curve is generally used as the profile of gear teeth.

*Land:* The *Top Land* is the top surface of a tooth, and the *Bottom Land* is the surface of the gear between the fillets of adjacent teeth.

*Lead:* The distance a helical gear or worm would thread along its axis in one revolution if it were free to move axially.

*Length of Action:* The distance on an involute line of action through which the point of contact moves during the action of the tooth profile.

*Line of Action:* The path of contact in involute gears. It is the straight line passing through the pitch point and tangent to the base circles.

*Lowest Point of Single Tooth Contact:* The smallest diameter on a spur gear at which a single tooth of one gear is in contact with its mating gear. Often referred to as LPSTC. Gear set contact stress is determined with a load placed at this point on the pinion.

*Module:* Ratio of the pitch diameter to the number of teeth. Ordinarily, module is understood to mean ratio of pitch diameter in *millimeters* to the number of teeth. The *English Module* is a ratio of the pitch diameter in inches to the number of teeth.

*Normal Plane:* A plane normal to the tooth surfaces at a point of contact, and perpendicular to the pitch plane.

*Pitch:* The distance between similar, equally-spaced tooth surfaces, in a given direction and along a given curve or line. The single word "pitch" without qualification has been used to designate circular pitch, axial pitch, and diametral pitch, but such confusing usage should be avoided.

*Pitch Circle:* A circle the radius of which is equal to the distance from the gear axis to the pitch point (see illustration).

*Pitch Diameter:* The diameter of the pitch circle. In parallel shaft gears the pitch diameters can be determined directly from the center distance and the numbers of teeth by proportionality. *Operating Pitch Diameter* is the pitch diameter at which the gears operate. *Generating Pitch Diameter* is the pitch diameter at which the gear is generated. In a bevel gear the pitch diameter is understood to be at the outer ends of the teeth unless otherwise specified. (See also reference to standard pitch diameter under *Pressure Angle*.)

*Pitch Plane:* In a pair of gears it is the plane perpendicular to the axial plane and tangent to the pitch surfaces. In a single gear it may be any plane tangent to its pitch surface.

*Pitch Point:* This is the point of tangency of two pitch circles (or of a pitch circle and a pitch line) and is on the line of centers. The pitch point of a tooth profile is at its intersection with the pitch circle.

*Plane of Rotation:* Any plane perpendicular to a gear axis.

*Pressure Angle:* The angle between a tooth profile and a radial line at its pitch point. In involute teeth, pressure angle is often described as the angle between the line of action and the line tangent to the pitch circle. *Standard Pressure Angles* are established in connection with standard gear-tooth proportions. A given pair of involute profiles will transmit smooth motion at the same velocity ratio even when the center distance is changed. Changes in center distance, however, in gear design and gear manufacturing operations, are accompanied by changes in pitch diameter, pitch, and pressure angle. Different values of pitch diameter and pressure angle therefore may occur in the same gear under different conditions. Usually in gear design, and unless otherwise specified, the pressure angle is the *standard pressure angle* at the *standard pitch diameter*, and is standard for the hob or cutter used to generate the teeth. The *Operating Pressure Angle* is determined by the center distance at which a pair of gears operates. The *Generating Pressure Angle* is the angle at the pitch diameter in effect when the gear is generated. Other pressure angles may be considered in gear calculations. In gear cutting tools and cutters, the pressure angle indicates the direction of the cutting edge as referred to some principal direction. In oblique teeth, that is helical, spiral, etc., the pressure angle may be specified in the *transverse, normal,* or *axial* plane. For a spur gear or a straight bevel gear, in which only one direction of cross-section needs to be considered, the general term pressure angle may be used without qualification to indicate transverse pressure angle. In spiral bevel gears, unless otherwise specified, pressure angle means normal pressure angle at the mean cone distance.

*Principal Reference Planes:* These are a pitch plane, axial plane, and transverse plane, all intersecting at a point and mutually perpendicular.

*Rack:* A gear with teeth spaced along a straight line, and suitable for straight-line motion. A *Basic Rack* is one that is adopted as the basis of a system of interchangeable gears. Standard gear-tooth proportions are often illustrated on an outline of the basic rack (see diagrams on page **715**). A *Generating Rack* is a rack outline used to indicate tooth details and dimensions for the design of a required generating tool, such as a hob or gear-shaper cutter.

*Ratio of Gearing:* Ratio of the numbers of teeth on mating gears. Ordinarily the ratio is found by dividing the number of teeth on the larger gear by the number of teeth on the smaller gear or pinion. For example, if the ratio is 2 or "2 to 1," this usually means that the smaller gear or pinion makes two revolutions to one revolution of the larger mating gear.

*Roll Angle:* The angle subtended at the center of a base circle from the origin of an involute to the point of tangency of the generatrix from any point on the same involute. The radian measure of this angle is the tangent of the pressure angle of the point on the involute.

*Root Circle:* A circle coinciding with or tangent to the bottoms of the tooth spaces.

*Root Diameter:* Diameter of the root circle.

*Tangent Plane:* A plane tangent to the tooth surfaces at a point or line of contact.

*Tip Relief:* An arbitrary modification of a tooth profile whereby a small amount of material is removed near the tip of the gear tooth.

*Total Face Width:* The actual width dimension of a gear blank. It may exceed the effective face width, as in the case of double-helical gears where the total face width includes any distance separating the right-hand and left-hand helical teeth.

*Transverse Plane:* A plane perpendicular to the axial plane and to the pitch plane. In gears with parallel axes, the transverse plane and the plane of rotation coincide.

*Trochoid:* The curve formed by the path of a point on the extension of a radius of a circle as it rolls along a curve or line. It is also the curve formed by the path of a point on a perpendicular to a straight line as the straight line rolls along the convex side of a base curve. By the first definition the trochoid is a derivative of the cycloid; by the second definition it is a derivative of the involute.

*True Involute Form Diameter:* The smallest diameter on the tooth at which the involute exists. Usually this is the point of tangency of the involute tooth profile and the fillet curve. This is usually referred to as the *TIF diameter.*

*Undercut:* A condition in generated gear teeth when any part of the fillet curve lies inside of a line drawn tangent to the working profile at its lowest point. Undercut may be deliberately introduced to facilitate finishing operations, as in preshaving.

*Whole Depth:* The total depth of a tooth space, equal to addendum plus dedendum, also equal to working depth plus clearance.

*Working Depth:* The depth of engagement of two gears, that is, the sum of their addendums. The standard working distance is the depth to which a tooth extends into the tooth space of a mating gear when the center distance is standard.

Definitions of gear terms are given in AGMA Standards 112.05, 115.01, and 116.01 entitled "Terms, Definitions, Symbols and Abbreviations," "Reference Information — Basic Gear Geometry," and "Glossary — Terms Used in Gearing," respectively; obtainable from American Gear Manufacturers Assn., 1901 No. Ft. Myer Dr., Arlington, Va. 22209.

**Gear Teeth of Different Diametral Pitch, Full Size**

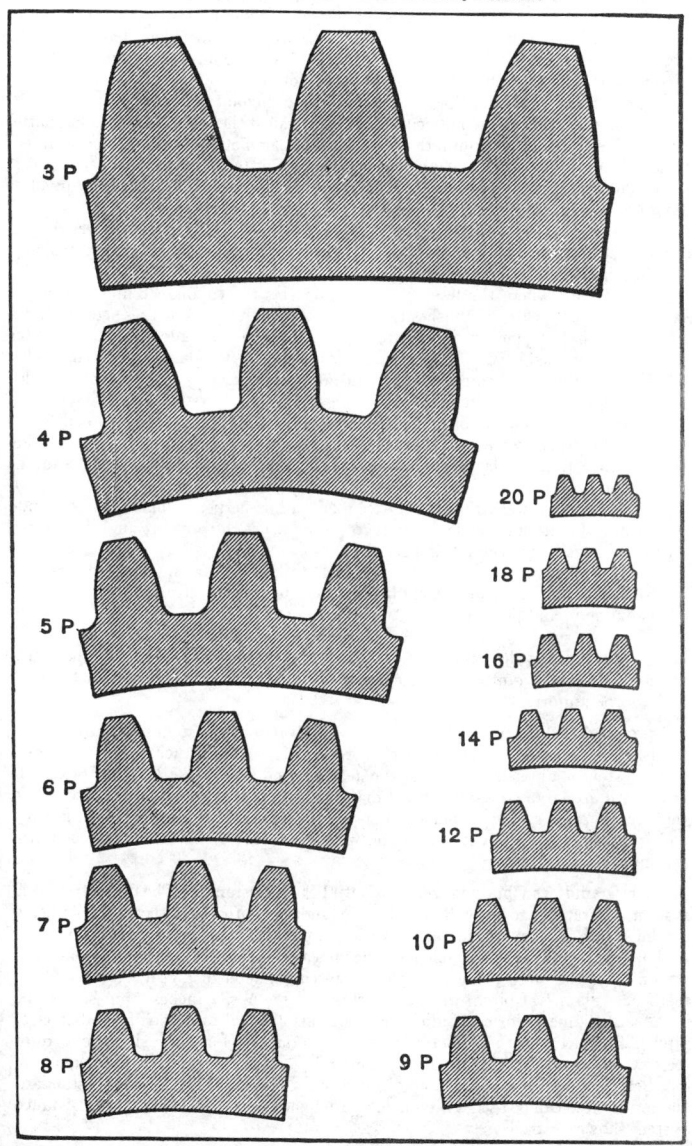

**Properties of the Involute Curve.** — The involute curve is used almost exclusively for gear-tooth profiles, because of the following important properties.

1. The form or shape of an involute curve depends upon the diameter of the base circle from which it is derived. (If a taut line were unwound from the circumference of a circle — the *base circle* of the involute — the end of that line or any point on the unwound portion, would describe an involute curve.)

2. If a gear tooth of involute curvature acts against the involute tooth of a mating gear while rotating at a uniform rate, the angular motion of the driven gear will also be uniform, even though the center-to-center distance is varied.

3. The relative rate of motion between driving and driven gears having involute tooth curves, is established by the diameters of their base circles.

4. Contact between intermeshing involute teeth on a driving and driven gear is along a straight line that is tangent to the two base circles of these gears. This is the *line of action*.

5. The point where the line of action intersects the common center-line of the mating involute gears, establishes the radii of the pitch circles of these gears; hence true pitch circle diameters are affected by a change in the center distance. (Pitch diameters obtained by dividing the number of teeth by the diametral pitch, applies when the center distance equals the total number of teeth on both gears divided by twice the diametral pitch.)

6. The pitch diameters of mating involute gears are directly proportional to the diameters of their respective base circles; thus, if the base circle of one mating gear is three times as large as the other, the pitch circle diameters will be in the same ratio.

7. The angle between the line of action and a line perpendicular to the common center-line of mating gears, is the *pressure angle;* hence the pressure angle is affected by any change in the center distance.

8. When an involute curve acts against a straight line (as in the case of an involute pinion acting against straight-sided rack teeth), the straight line is tangent to the involute and perpendicular to its line of action.

9. The pressure angle, in the case of an involute pinion acting against straight-sided rack teeth, is the angle between the line of action and the line of the rack's motion. If the involute pinion rotates at a uniform rate, movement of the rack will also be uniform.

**Diametral and Circular Pitch Systems.** — Gear tooth system standards are established by specifying the tooth proportions of the basic rack. The diametral pitch system is applied to most of the gearing produced in the United States. If gear teeth are larger than about one diametral pitch, it is common practice to use the circular pitch system. The circular pitch system is also applied to cast gearing and it is commonly used in connection with the design and manufacture of worm gearing.

**Pitch Diameters Obtained with Diametral Pitch System.** — The diametral pitch system is arranged to provide a series of standard tooth sizes, the principle being similar to the standardization of screw thread pitches. Inasmuch as there must be a whole number of teeth on each gear, the increase in pitch diameter per tooth varies according to the pitch. For example, the pitch diameter of a gear having, say, 20 teeth of 4 diametral pitch, will be 5 inches; 21 teeth, 5¼ inches; and so on, the increase in diameter for each additional tooth being equal to ¼ inch for 4 diametral pitch. Similarly, for 2 diametral pitch the variations for successive numbers of teeth would equal ½ inch, and for 10 diametral pitch the variations would equal ¹⁄₁₀ inch, etc. Where a given center distance must be maintained and no standard diametral pitch can be used, gears should be designed to the Variable Center Distance System shown on page 734.

Table 1.  Formulas for Dimensions of Standard Spur Gears

Notation

| | | | |
|---|---|---|---|
| $\phi$ | = Pressure Angle | $D_O$ | = Outside Diameter |
| $a$ | = Addendum | $D_R$ | = Root Diameter |
| $a_G$ | = Addendum of Gear | $F$ | = Face Width |
| $a_P$ | = Addendum of Pinion | $h_k$ | = Working Depth of Tooth |
| $b$ | = Dedendum | $h_t$ | = Whole Depth of Tooth |
| $c$ | = Clearance | $m_G$ | = Gear Ratio |
| $C$ | = Center Distance | $N$ | = Number of Teeth |
| $D$ | = Pitch Diameter | $N_G$ | = Number of Teeth in Gear |
| $D_G$ | = Pitch Diameter of Gear | $N_P$ | = Number of Teeth in Pinion |
| $D_P$ | = Pitch Diameter of Pinion | $p$ | = Circular Pitch |
| $D_B$ | = Base Circle Diameter | $P$ | = Diametral Pitch |

| No. | To Find | Formula | No. | To Find | Formula |
|---|---|---|---|---|---|
| | | General Formulas | | | |
| 1 | Base Circle Diameter | $D_B = D \cos \phi$ | 6a | Number of Teeth | $N = P \times D$ |
| 2a | Circular Pitch | $p = \dfrac{3.1416 D}{N}$ | 6b | Number of Teeth | $N = \dfrac{3.1416 D}{p}$ |
| 2b | Circular Pitch | $p = \dfrac{3.1416}{P}$ | 7a | Outside Diameter (Full-depth Teeth) | $D_O = \dfrac{N+2}{P}$ |
| 3a | Center Distance | $C = \dfrac{N_P(m_G + 1)}{2P}$ | 7b | Outside Diameter (Full-depth Teeth) | $D_O = \dfrac{(N+2)p}{3.1416}$ |
| 3b | Center Distance | $C = \dfrac{D_P + D_G}{2}$ | 8a | Outside Diameter (Amer. Stnd. Stub Teeth) | $D_O = \dfrac{N+1.6}{P}$ |
| 3c | Center Distance | $C = \dfrac{N_G + N_P}{2P}$ | 8b | Outside Diameter (Amer. Stnd. Stub Teeth) | $D_O = \dfrac{(N+1.6)p}{3.1416}$ |
| 3d | Center Distance | $C = \dfrac{(N_G + N_P)p}{6.2832}$ | 9 | Outside Diameter | $D_O = D + 2a$ |
| 4a | Diametral Pitch | $P = \dfrac{3.1416}{p}$ | 10a | Pitch Diameter | $D = \dfrac{N}{P}$ |
| 4b | Diametral Pitch | $P = \dfrac{N}{D}$ | 10b | Pitch Diameter | $D = \dfrac{Np}{3.1416}$ |
| 4c | Diametral Pitch | $P = \dfrac{N_P(m_G + 1)}{2C}$ | 11 | Root Diameter* | $D_R = D - 2b$ |
| 5 | Gear Ratio | $m_G = \dfrac{N_G}{N_P}$ | 12 | Whole Depth | $a + b$ |
| | | | 13 | Working Depth | $a_G + a_P$ |

* See also formulas in table on pages 718 and 720.

**American National Standard Coarse Pitch Spur Gear Tooth Forms.** — The American National Standard (ANSI B6.1-1968) provides tooth proportion information on two involute spur gear forms. These two forms are identical except that one has a pressure angle of 20 degrees and a minimum allowable tooth number of 18 while the other has a pressure angle of 25 degrees and a minimum allowable tooth number of 12. (For pinions with fewer teeth, see tables of tooth proportions for long addendum pinions and their mating short addendum gears on pages 749 to 751.) A gear tooth standard is established by specifying the tooth proportions of the basic rack. Gears made to this standard will thus be conjugate with the specified rack and with each other. The basic rack forms for the 20-degree and 25-degree standard are shown on the following page; basic formulas for these proportions as a function of the gear diametral pitch and also of the circular pitch are given in Table 2. Tooth parts data are given in Table 3.

**Table 2. Formulas for Tooth Parts — American National Standard Coarse Pitch Spur Gear Tooth Forms** (ANSI B6.1-1968)

| To Find | Diametral Pitch, $P$ Known | Circular Pitch, $p$, Known |
|---|---|---|
| 20-DEGREE INVOLUTE FULL-DEPTH TEETH 25-DEGREE INVOLUTE FULL-DEPTH TEETH | | |
| Addendum | $a = 1.000 \div P$ | $a = 0.3183 \times p$ |
| Dedendum (Preferred) (Shaved or Ground Teeth)* | $b = 1.250 \div P$ $b = 1.350 \div P$ | $b = 0.3979 \times p$ $b = 0.4297 \times p$ |
| Working Depth | $h_k = 2.000 \div P$ | $h_k = 0.6366 \times p$ |
| Whole Depth (Preferred) (Shaved or Ground Teeth) | $h_t = 2.250 \div P$ $h_t = 2.350 \div P$ | $h_t = 0.7162 \times p$ $h_t = 0.7480 \times p$ |
| Clearance (Preferred)† (Shaved or Ground Teeth) | $c = 0.250 \div P$ $c = 0.350 \div P$ | $c = 0.0796 \times p$ $c = 0.1114 \times p$ |
| Fillet Radius (Rack)‡ | $r_f = 0.300 \div P$ | $r_f = 0.0955 \times p$ |
| Pitch Diameter | $D = N \div P$ | $D = 0.3183 \times Np$ |
| Outside Diameter | $D_o = (N+2) \div P$ | $D_o = 0.3183 \times (N+2)p$ |
| Root Diameter (Preferred) (Shaved or Ground Teeth) | $D_R = (N-2.5) \div P$ $D_R = (N-2.7) \div P$ | $D_R = 0.3183 \times (N-2.5)p$ $D_R = 0.3183 \times (N-2.7)p$ |
| Circular Thickness — Basic | $t = 1.5708 \div P$ | $t = p \div 2$ |

* When gears are preshave cut on a gear shaper the dedendum will usually need to be increased to $1.40/P$ to allow for the higher fillet trochoid produced by the shaper cutter. This is of particular importance on gears of few teeth or if the gear blank configuration requires the use of a small diameter shaper cutter, in which case the dedendum may need to be increased to as much as $1.45/P$. This should be avoided on highly loaded gears where the consequently reduced $J$ factor will increase gear tooth stress excessively.

† A minimum clearance of $0.157/P$ may be used for the basic 20-degree and 25-degree pressure angle rack in the case of shallow root sections and use of existing hobs or cutters. However, whenever less than standard clearance is used, the location of the TIF diameter should be determined by the method shown on page 755. The TIF diameter must be less than the Contact Diameter determined by the method shown on page 753.

‡ The fillet radius of the basic rack should not exceed $0.235/P$ for a 20-degree pressure angle rack or $0.270/P$ for a 25-degree pressure angle rack for a clearance of $0.157/P$. The basic rack fillet radius must be *reduced* for teeth with a 25-degree pressure angle having a clearance in excess of $0.250/P$.

**American National Standard and Former American Standard Gear Tooth Forms**
(ANSI B6.1-1968 and ASA B6.1-1932)

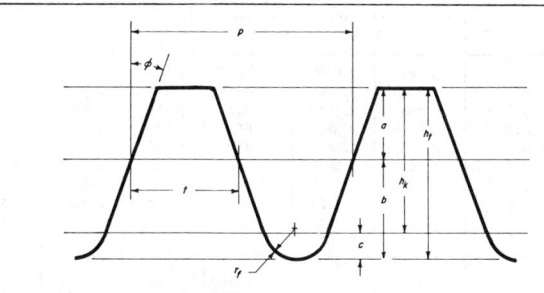

a = addendum        $h_k$ = working depth        $r_f$ = fillet radius of basic rack
b = dedendum        $h_t$ = whole depth          $t$ = circular tooth thickness — basic
c = clearance       $p$ = circular pitch         $\phi$ = pressure angle

**Basic Rack of the 20-Degree and 25-Degree Full-Depth Involute Systems**

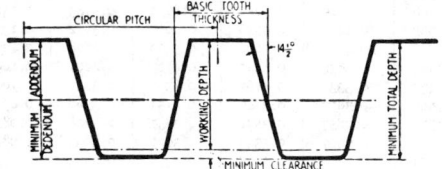

**Basic Rack of the 14 1/2-Degree Full-Depth Involute System**

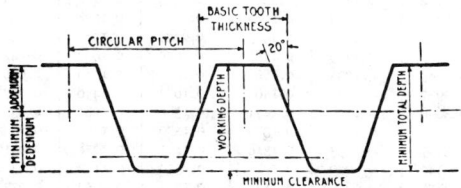

**Basic Rack of the 20-Degree Stub Involute System**

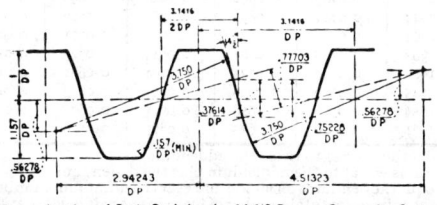

**Approximation of Basic Rack for the 14 1/2-Degree Composite System**

Table 3. Gear Tooth Parts for American National Standard Coarse
Pitch 20- and 25-Degree Pressure Angle Gears

| Diam. Pitch | Circ. Pitch | Stand. Addend.* | Stand. Dedend. | Spec. Dedend.† | Min. Dedend. | Stand. F. Rad. | Min. F. Rad. |
|---|---|---|---|---|---|---|---|
| P | p | a | b | b | b | $r_f$ | $r_f$ |
| 0.3142 | 10. | 3.1831 | 3.9789 | 4.2972 | 3.6828 | 0.9549 | 0.4997 |
| 0.3307 | 9.5 | 3.0239 | 3.7799 | 4.0823 | 3.4987 | 0.9072 | 0.4748 |
| 0.3491 | 9. | 2.8648 | 3.5810 | 3.8675 | 3.3146 | 0.8594 | 0.4498 |
| 0.3696 | 8.5 | 2.7056 | 3.3820 | 3.6526 | 3.1304 | 0.8117 | 0.4248 |
| 0.3927 | 8. | 2.5465 | 3.1831 | 3.4377 | 2.9463 | 0.7639 | 0.3998 |
| 0.4189 | 7.5 | 2.3873 | 2.9842 | 3.2229 | 2.7621 | 0.7162 | 0.3748 |
| 0.4488 | 7. | 2.2282 | 2.7852 | 3.0080 | 2.5780 | 0.6685 | 0.3498 |
| 0.4833 | 6.5 | 2.0690 | 2.5863 | 2.7932 | 2.3938 | 0.6207 | 0.3248 |
| 0.5236 | 6. | 1.9099 | 2.3873 | 2.5783 | 2.2097 | 0.5730 | 0.2998 |
| 0.5712 | 5.5 | 1.7507 | 2.1884 | 2.3635 | 2.0256 | 0.5252 | 0.2749 |
| 0.6283 | 5. | 1.5915 | 1.9894 | 2.1486 | 1.8414 | 0.4775 | 0.2499 |
| 0.6981 | 4.5 | 1.4324 | 1.7905 | 1.9337 | 1.6573 | 0.4297 | 0.2249 |
| 0.7854 | 4. | 1.2732 | 1.5915 | 1.7189 | 1.4731 | 0.3820 | 0.1999 |
| 0.8976 | 3.5 | 1.1141 | 1.3926 | 1.5040 | 1.2890 | 0.3342 | 0.1749 |
| 1. | 3.1416 | 1.0000 | 1.2500 | 1.3500 | 1.1570 | 0.3000 | 0.1570 |
| 1.25 | 2.5133 | 0.8000 | 1.0000 | 1.0800 | 0.9256 | 0.2400 | 0.1256 |
| 1.5 | 2.0944 | 0.6667 | 0.8333 | 0.9000 | 0.7713 | 0.2000 | 0.1047 |
| 1.75 | 1.7952 | 0.5714 | 0.7143 | 0.7714 | 0.6611 | 0.1714 | 0.0897 |
| 2. | 1.5708 | 0.5000 | 0.6250 | 0.6750 | 0.5785 | 0.1500 | 0.0785 |
| 2.25 | 1.3963 | 0.4444 | 0.5556 | 0.6000 | 0.5142 | 0.1333 | 0.0698 |
| 2.5 | 1.2566 | 0.4000 | 0.5000 | 0.5400 | 0.4628 | 0.1200 | 0.0628 |
| 2.75 | 1.1424 | 0.3636 | 0.4545 | 0.4909 | 0.4207 | 0.1091 | 0.0571 |
| 3. | 1.0472 | 0.3333 | 0.4167 | 0.4500 | 0.3857 | 0.1000 | 0.0523 |
| 3.25 | 0.9666 | 0.3077 | 0.3846 | 0.4154 | 0.3560 | 0.0923 | 0.0483 |
| 3.5 | 0.8976 | 0.2857 | 0.3571 | 0.3857 | 0.3306 | 0.0857 | 0.0449 |
| 3.75 | 0.8378 | 0.2667 | 0.3333 | 0.3600 | 0.3085 | 0.0800 | 0.0419 |
| 4. | 0.7854 | 0.2500 | 0.3125 | 0.3375 | 0.2893 | 0.0750 | 0.0392 |
| 4.5 | 0.6981 | 0.2222 | 0.2778 | 0.3000 | 0.2571 | 0.0667 | 0.0349 |
| 5. | 0.6283 | 0.2000 | 0.2500 | 0.2700 | 0.2314 | 0.0600 | 0.0314 |
| 5.5 | 0.5712 | 0.1818 | 0.2273 | 0.2455 | 0.2104 | 0.0545 | 0.0285 |
| 6. | 0.5236 | 0.1667 | 0.2083 | 0.2250 | 0.1928 | 0.0500 | 0.0262 |
| 6.5 | 0.4833 | 0.1538 | 0.1923 | 0.2077 | 0.1780 | 0.0462 | 0.0242 |
| 7. | 0.4488 | 0.1429 | 0.1786 | 0.1929 | 0.1653 | 0.0429 | 0.0224 |
| 7.5 | 0.4189 | 0.1333 | 0.1667 | 0.1800 | 0.1543 | 0.0400 | 0.0209 |
| 8. | 0.3927 | 0.1250 | 0.1563 | 0.1687 | 0.1446 | 0.0375 | 0.0196 |
| 8.5 | 0.3696 | 0.1176 | 0.1471 | 0.1588 | 0.1361 | 0.0353 | 0.0185 |
| 9. | 0.3491 | 0.1111 | 0.1389 | 0.1500 | 0.1286 | 0.0333 | 0.0174 |
| 9.5 | 0.3307 | 0.1053 | 0.1316 | 0.1421 | 0.1218 | 0.0316 | 0.0165 |
| 10. | 0.3142 | 0.1000 | 0.1250 | 0.1350 | 0.1157 | 0.0300 | 0.0157 |
| 11. | 0.2856 | 0.0909 | 0.1136 | 0.1227 | 0.1052 | 0.0273 | 0.0143 |
| 12. | 0.2618 | 0.0833 | 0.1042 | 0.1125 | 0.0964 | 0.0250 | 0.0131 |
| 13. | 0.2417 | 0.0769 | 0.0962 | 0.1038 | 0.0890 | 0.0231 | 0.0121 |
| 14. | 0.2244 | 0.0714 | 0.0893 | 0.0964 | 0.0826 | 0.0214 | 0.0112 |
| 15. | 0.2094 | 0.0667 | 0.0833 | 0.0900 | 0.0771 | 0.0200 | 0.0105 |
| 16. | 0.1963 | 0.0625 | 0.0781 | 0.0844 | 0.0723 | 0.0188 | 0.0098 |
| 17. | 0.1848 | 0.0588 | 0.0735 | 0.0794 | 0.0681 | 0.0176 | 0.0092 |
| 18. | 0.1745 | 0.0556 | 0.0694 | 0.0750 | 0.0643 | 0.0167 | 0.0087 |
| 19. | 0.1653 | 0.0526 | 0.0658 | 0.0711 | 0.0609 | 0.0158 | 0.0083 |
| 20. | 0.1571 | 0.0500 | 0.0625 | 0.0675 | 0.0579 | 0.0150 | 0.0079 |

The working depth is equal to twice the addendum.
The whole depth is equal to the addendum plus the dedendum.
* When using equal addendums on pinion and gear the minimum number of teeth on the pinion is 18 and the minimum total number of teeth in the pair is 36 for 20-degree full depth involute tooth form and 12 and 24, respectively, for 25-degree full depth tooth form.
† The dedendum in this column is used when the gear tooth is shaved. It allows for the higher fillet cut by a protuberance hob.

In recent years the established standard of almost universal use is the ANSI 20-degree standard spur gear form. It provides a gear with good strength and without fillet undercut in pinions of as few as eighteen teeth. Some more recent applications have required a tooth form of even greater strength and fewer teeth than eighteen. This requirement has stimulated the establishment of the ANSI 25-degree standard. This 25-degree form will give greater tooth strength than the 20-degree standard, will provide pinions of as few as twelve teeth without fillet undercut and will provide a lower contact compressive stress for greater gear set surface durability.

**American National Standard Fine Pitch Spur and Helical Gear Tooth Form.** — The American National Standard (ANSI B6.7-1967) provides tooth proportion information on an involute tooth form that is for fine pitch spur or helical gears. Fine pitch is specified as being from 20 diametral pitch to 120 diametral pitch; preferred pitches are 20, 24, 32, 48, 64, 72, 80, 96, and 120. The standard pressure angle is 20 degrees. Basic formulae for the proportions of this standard form are given in Table 4. The basic rack form for this standard is shown on page 719. Tooth, parts data are given in Table 5.

**Other American National Standard Fine Pitch Tooth Forms.** — An appended information sheet in the American National Standard (ANSI B6.7-1967) provides tooth proportion information for two alternate forms of the same proportions as the

**Table 4. Formulas for Tooth Parts — American National Standard Fine Pitch Spur Gear Tooth Forms** (ANSI B6.7-1967)

| 20-DEGREE INVOLUTE FINE-PITCH TEETH* | | |
|---|---|---|
| Addendum | $a = 1.000 \div P$ | $a = 0.3183 \times p$ |
| Dedendum (Preferred) | $b = (1.200 \div P) + 0.002$ | $b = (0.3820 \times p) + 0.002$ |
| (Shaved or Ground Teeth) | $b = (1.350 \div P) + 0.002$ | $b = (0.4297 \times p) + 0.002$ |
| Working Depth | $h_k = 2.000 \div P$ | $h_k = 0.6366 \times p$ |
| Whole Depth (Preferred) | $h_t = (2.200 \div P) + 0.002$ | $h_t = (0.7003 \times p) + 0.002$ |
| (Shaved or Ground Teeth) | $h_t = (2.350 \div P) + 0.002$ | $h_t = (0.7480 \times p) + 0.002$ |
| Clearance (Preferred) | $c = (0.200 \div P) + 0.002$ | $c = (0.0637 \times p) + 0.002$ |
| (Shaved or Ground Teeth) | $c = (0.350 \div P) + 0.002$ | $c = (0.1114 \times p) + 0.002$ |
| Pitch Diameter | $D = N \div P$ | $D = 0.3183 \times Np$ |
| Outside Diameter | $D_o = (N+2) \div P$ | $D_o = 0.3183 \times (N+2)p$ |
| Root Diameter (Preferred) | $D_R = (N-2.2) \div P - 0.004$ | $D_R = 0.3183 \times (N-2.2)p$ |
| (Shaved or Ground Teeth) | $D_R = (N-2.35) \div P - 0.004$ | $D_R = 0.3183 \times (N-2.35)p - 0.004$ |
| Circular Thickness—Basic | $t = 1.5708 \div P$ | $t = p \div 2$ |

* These formulas also apply to the alternate standard 14½-degree and 25-degree involute fine pitch teeth.

20-degree standard form but having 14½-degree and 25-degree pressure angles, respectively. These alternate standard forms do not include the cautionary note limiting their application on new designs as in the alternate forms appended to ANSI B6.1-1968. The 14½-degree form may be preferred in applications that require close control on assembled gear backlash, whereas the 25-degree form may be preferred for ease of production in sintered or precision cast gears and where added tooth strength is required. Formulae for the proportions of the 20-degree fine pitch standard also apply to these alternate standards.

**Other American Spur Gear Standards.** — An appended information sheet in the American National Standard (ANSI B6.1-1968) provides tooth proportion information for three spur gear forms with the notice that they are "not recommended for new designs." These forms are therefore considered to be obsolescent but the information is given on their proportions because they have been used widely in the past. These forms are the 14½-degree full depth form, the 20-degree stub involute form and the 14½-degree composite form which were covered in the former American Standard (ASA B6.1-1932). The basic rack for the 14½-degree

Table 5. American National Standard Fine Pitch Standard Gear Tooth Parts — 14½-, 20-, and 25-Degree Pressure Angles

| Diametral Pitch | Circular Pitch | Circular Thickness | Standard Addend. | Standard Dedend. | Special Dedend. |
|---|---|---|---|---|---|
| $P$ | $p$ | $t$ | $a$ | $b$ | $b$ |
| 20* | .1571 | .0785 | .0500 | .0620 | .0695 |
| 22 | .1428 | .0714 | .0455 | .0565 | .0634 |
| 24* | .1309 | .0654 | .0417 | .0520 | .0582 |
| 26 | .1208 | .0604 | .0385 | .0482 | .0539 |
| 28 | .1122 | .0561 | .0357 | .0449 | .0502 |
| 30 | .1047 | .0524 | .0333 | .0420 | .0470 |
| 32* | .0982 | .0491 | .0313 | .0395 | .0442 |
| 34 | .0924 | .0462 | .0294 | .0373 | .0417 |
| 36 | .0873 | .0436 | .0278 | .0353 | .0395 |
| 38 | .0827 | .0413 | .0263 | .0336 | .0375 |
| 40 | .0785 | .0393 | .0250 | .0320 | .0357 |
| 44 | .0714 | .0357 | .0227 | .0293 | .0327 |
| 48* | .0654 | .0327 | .0208 | .0270 | .0301 |
| 52 | .0604 | .0302 | .0192 | .0251 | .0280 |
| 56 | .0561 | .0280 | .0179 | .0234 | .0261 |
| 60 | .0524 | .0262 | .0167 | .0220 | .0245 |
| 64* | .0491 | .0245 | .0156 | .0207 | .0231 |
| 68 | .0462 | .0231 | .0147 | .0196 | .0219 |
| 72* | .0436 | .0218 | .0139 | .0187 | .0207 |
| 76 | .0413 | .0207 | .0132 | .0178 | .0198 |
| 80* | .0393 | .0196 | .0125 | .0170 | .0189 |
| 88 | .0357 | .0178 | .0114 | .0156 | .0173 |
| 96* | .0327 | .0164 | .0104 | .0145 | .0161 |
| 104 | .0302 | .0151 | .0096 | .0135 | .0150 |
| 112 | .0280 | .0140 | .0089 | .0127 | .0141 |
| 120* | .0262 | .0131 | .0083 | .0120 | .0132 |

The working depth is equal to twice the addendum. The whole depth is equal to the addendum plus the dedendum. For minimum number of teeth see page 749.
* These are standard pitches; others are non-standard.

Table 6.  Formulas for Tooth Parts — Former American Standard Spur Gear
Tooth Forms (ASA B6.1-1932)

| To Find | Diametral Pitch, $P$ Known | Circular Pitch, $p$ Known |
|---|---|---|
| 14½-DEGREE INVOLUTE FULL-DEPTH TEETH 14½-DEGREE COMPOSITE FULL-DEPTH TEETH | | |
| Addendum | $a = 1.000 \div P$ | $a = 0.3183 \times p$ |
| Minimum Dedendum | $b = 1.157 \div P$ | $b = 0.3683 \times p$ |
| Working Depth | $h_k = 2.000 \div P$ | $h_k = 0.6366 \times p$ |
| Minimum Whole Depth | $h_t = 2.157 \div P$ | $h_t = 0.6866 \times p$ |
| Basic Tooth Thickness on Pitch Line | $t = 1.5708 \div P$ | $t = 0.500 \times p$ |
| Minimum Clearance | $c = 0.157 \div P$ | $c = 0.050 \times p$ |
| 20-DEGREE INVOLUTE STUB TEETH | | |
| Addendum | $a = 0.800 \div P$ | $a = 0.2546 \times p$ |
| Minimum Dedendum | $b = 1.000 \div P$ | $b = 0.3183 \times p$ |
| Working Depth | $h_k = 1.600 \div P$ | $h_k = 0.5092 \times p$ |
| Minimum Whole Depth | $h_t = 1.800 \div P$ | $h_t = 0.5729 \times p$ |
| Basic Tooth Thickness on Pitch Line | $t = 1.5708 \div P$ | $t = 0.500 \times p$ |
| Minimum Clearance | $c = 0.200 \div P$ | $c = 0.0637 \times p$ |

*Note:* Radius of fillet equals 1⅓ × clearance for 14½-degree full-depth teeth
and 1½ × clearance for 20-degree full-depth teeth.
*Note:* A suitable working tolerance should be considered in connection with
all minimum recommendations.

full depth form is shown on page 719; basic formulas for these proportions are given
in Table 6  Tooth parts data are given in Tables 7 to 9, inclusive.

**Fellows Stub Tooth.** — The system of stub gear teeth introduced by the Fellows
Gear Shaper Co. is based upon the use of two diametral pitches.  One diametral
pitch, say, 8, is used as the basis for obtaining the dimensions for the addendum and
dedendum, while another diametral pitch, say, 6, is used for obtaining the dimensions
of the thickness of the tooth, the number of teeth, and the pitch diameter.  Teeth
made according to this system are designated as 8/6 pitch, 12/14 pitch, etc., the numerator in this fraction indicating the pitch determining the thickness of the tooth and
the number of teeth, and the denominator, the pitch determining the depth of the
tooth.  The clearance is made greater than in the ordinary gear-tooth system and
equals 0.25 ÷ denominator of the diametral pitch.  The pressure angle is 20 degrees.

This type of stub gear tooth is now used infrequently.  For information as to the
tooth part dimensions see 18th and earlier editions of Machinery's Handbook.

Table 7. Gear Tooth Parts — Former American Standard 14½-degree Involute
Full-depth and 14½-degree Composite Systems* (ASA B6.1-1932)
(Diametral Pitch Gears)

| Diametral Pitch | Equivalent Circular Pitch | Standard Circular Thickness | Standard Addendum | Working Depth of Tooth | Standard Dedendum | Whole Depth of Tooth |
|---|---|---|---|---|---|---|
| P | p | t | a | h_k | b | h_t |
| ½ | 6.2832 | 3.1416 | 2.0000 | 4.0000 | 2.3142 | 4.3142 |
| ¾ | 4.1888 | 2.0944 | 1.3333 | 2.6666 | 1.5428 | 2.8761 |
| 1 | 3.1416 | 1.5708 | 1.0000 | 2.0000 | 1.1571 | 2.1571 |
| 1¼ | 2.5133 | 1.2566 | 0.8000 | 1.6000 | 0.9257 | 1.7257 |
| 1½ | 2.0944 | 1.0472 | 0.6666 | 1.3333 | 0.7714 | 1.4381 |
| 1¾ | 1.7952 | 0.8976 | 0.5714 | 1.1429 | 0.6612 | 1.2326 |
| 2 | 1.5708 | 0.7854 | 0.5000 | 1.0000 | 0.5785 | 1.0785 |
| 2¼ | 1.3963 | 0.6981 | 0.4444 | 0.8888 | 0.5143 | 0.9587 |
| 2½ | 1.2566 | 0.6283 | 0.4000 | 0.8000 | 0.4628 | 0.8628 |
| 2¾ | 1.1424 | 0.5712 | 0.3636 | 0.7273 | 0.4208 | 0.7844 |
| 3 | 1.0472 | 0.5236 | 0.3333 | 0.6666 | 0.3857 | 0.7190 |
| 3½ | 0.8976 | 0.4488 | 0.2857 | 0.5714 | 0.3306 | 0.6163 |
| 4 | 0.7854 | 0.3927 | 0.2500 | 0.5000 | 0.2893 | 0.5393 |
| 5 | 0.6283 | 0.3142 | 0.2000 | 0.4000 | 0.2314 | 0.4314 |
| 6 | 0.5236 | 0.2618 | 0.1666 | 0.3333 | 0.1928 | 0.3595 |
| 7 | 0.4488 | 0.2244 | 0.1429 | 0.2857 | 0.1653 | 0.3081 |
| 8 | 0.3927 | 0.1963 | 0.1250 | 0.2500 | 0.1446 | 0.2696 |
| 9 | 0.3491 | 0.1745 | 0.1111 | 0.2222 | 0.1286 | 0.2397 |
| 10 | 0.3142 | 0.1571 | 0.1000 | 0.2000 | 0.1157 | 0.2157 |
| 11 | 0.2856 | 0.1428 | 0.0909 | 0.1818 | 0.1052 | 0.1961 |
| 12 | 0.2618 | 0.1309 | 0.0833 | 0.1666 | 0.0964 | 0.1798 |
| 13 | 0.2417 | 0.1208 | 0.0769 | 0.1538 | 0.0890 | 0.1659 |
| 14 | 0.2244 | 0.1122 | 0.0714 | 0.1429 | 0.0826 | 0.1541 |
| 15 | 0.2094 | 0.1047 | 0.0666 | 0.1333 | 0.0771 | 0.1438 |
| 16 | 0.1963 | 0.0982 | 0.0625 | 0.1250 | 0.0723 | 0.1348 |
| 17 | 0.1848 | 0.0924 | 0.0588 | 0.1176 | 0.0681 | 0.1269 |
| 18 | 0.1745 | 0.0873 | 0.0555 | 0.1111 | 0.0643 | 0.1198 |
| 19 | 0.1653 | 0.0827 | 0.0526 | 0.1053 | 0.0609 | 0.1135 |
| 20 | 0.1571 | 0.0785 | 0.0500 | 0.1000 | 0.0579 | 0.1079 |
| 22 | 0.1428 | 0.0714 | 0.0455 | 0.0909 | 0.0526 | 0.0980 |
| 24 | 0.1309 | 0.0654 | 0.0417 | 0.0833 | 0.0482 | 0.0898 |
| 26 | 0.1208 | 0.0604 | 0.0385 | 0.0769 | 0.0445 | 0.0829 |
| 28 | 0.1122 | 0.0561 | 0.0357 | 0.0714 | 0.0413 | 0.0770 |
| 30 | 0.1047 | 0.0524 | 0.0333 | 0.0666 | 0.0386 | 0.0719 |
| 32 | 0.0982 | 0.0491 | 0.0312 | 0.0625 | 0.0362 | 0.0674 |
| 34 | 0.0924 | 0.0462 | 0.0294 | 0.0588 | 0.0340 | 0.0634 |
| 36 | 0.0873 | 0.0436 | 0.0278 | 0.0555 | 0.0321 | 0.0599 |
| 38 | 0.0827 | 0.0413 | 0.0263 | 0.0526 | 0.0304 | 0.0568 |
| 40 | 0.0785 | 0.0393 | 0.0250 | 0.0500 | 0.0289 | 0.0539 |
| 42 | 0.0748 | 0.0374 | 0.0238 | 0.0476 | 0.0275 | 0.0514 |
| 44 | 0.0714 | 0.0357 | 0.0227 | 0.0455 | 0.0263 | 0.0490 |
| 46 | 0.0683 | 0.0341 | 0.0217 | 0.0435 | 0.0252 | 0.0469 |
| 48 | 0.0654 | 0.0327 | 0.0208 | 0.0417 | 0.0241 | 0.0449 |
| 50 | 0.0628 | 0.0314 | 0.0200 | 0.0400 | 0.0231 | 0.0431 |

* For 20-degree full-depth teeth of 20 diametral pitch and finer see table on page 722.

**Table 8.** Gear Tooth Parts — Former American Standard 14½-degree Involute Full-depth and 14½-degree Composite Systems* (ASA B6.1-1932)

(Circular Pitch Gears)

| Circular Pitch | Equivalent Diametral Pitch | Standard Circular Thickness | Standard Addendum | Working Depth of Tooth | Standard Dedendum | Whole Depth of Tooth |
|---|---|---|---|---|---|---|
| $p$ | $P$ | $t$ | $a$ | $h_k$ | $b$ | $h_t.$ |
| 4 | 0.7854 | 2.0000 | 1.2732 | 2.5464 | 1.4732 | 2.7464 |
| 3½ | 0.8976 | 1.7500 | 1.1140 | 2.2281 | 1.2890 | 2.4031 |
| 3 | 1.0472 | 1.5000 | 0.9549 | 1.9098 | 1.1049 | 2.0598 |
| 2¾ | 1.1424 | 1.3750 | 0.8753 | 1.7506 | 1.0128 | 1.8881 |
| 2½ | 1.2566 | 1.2500 | 0.7957 | 1.5915 | 0.9207 | 1.7165 |
| 2¼ | 1.3963 | 1.1250 | 0.7162 | 1.4323 | 0.8287 | 1.5448 |
| 2 | 1.5708 | 1.0000 | 0.6366 | 1.2732 | 0.7366 | 1.3732 |
| 1⅞ | 1.6755 | 0.9375 | 0.5968 | 1.1937 | 0.6906 | 1.2874 |
| 1¾ | 1.7952 | 0.8750 | 0.5570 | 1.1141 | 0.6445 | 1.2016 |
| 1⅝ | 1.9333 | 0.8125 | 0.5173 | 1.0345 | 0.5985 | 1.1158 |
| 1½ | 2.0944 | 0.7500 | 0.4775 | 0.9549 | 0.5525 | 1.0299 |
| 1⁷⁄₁₆ | 2.1855 | 0.7187 | 0.4576 | 0.9151 | 0.5294 | 0.9870 |
| 1⅜ | 2.2848 | 0.6875 | 0.4377 | 0.8754 | 0.5064 | 0.9441 |
| 1⁵⁄₁₆ | 2.3936 | 0.6562 | 0.4178 | 0.8356 | 0.4834 | 0.9012 |
| 1¼ | 2.5133 | 0.6250 | 0.3979 | 0.7958 | 0.4604 | 0.8583 |
| 1³⁄₁₆ | 2.6456 | 0.5937 | 0.3780 | 0.7560 | 0.4374 | 0.8154 |
| 1⅛ | 2.7925 | 0.5625 | 0.3581 | 0.7162 | 0.4143 | 0.7724 |
| 1¹⁄₁₆ | 2.9568 | 0.5312 | 0.3382 | 0.6764 | 0.3913 | 0.7295 |
| 1 | 3.1416 | 0.5000 | 0.3183 | 0.6366 | 0.3683 | 0.6866 |
| 15⁄16 | 3.3510 | 0.4687 | 0.2984 | 0.5968 | 0.3453 | 0.6437 |
| ⅞ | 3.5904 | 0.4375 | 0.2785 | 0.5570 | 0.3223 | 0.6007 |
| 13⁄16 | 3.8666 | 0.4062 | 0.2586 | 0.5173 | 0.2993 | 0.5579 |
| ¾ | 4.1888 | 0.3750 | 0.2387 | 0.4775 | 0.2762 | 0.5150 |
| 11⁄16 | 4.5696 | 0.3437 | 0.2189 | 0.4377 | 0.2532 | 0.4720 |
| ⅔ | 4.7124 | 0.3333 | 0.2122 | 0.4244 | 0.2455 | 0.4577 |
| ⅝ | 5.0265 | 0.3125 | 0.1989 | 0.3979 | 0.2301 | 0.4291 |
| 9⁄16 | 5.5851 | 0.2812 | 0.1790 | 0.3581 | 0.2071 | 0.3862 |
| ½ | 6.2832 | 0.2500 | 0.1592 | 0.3183 | 0.1842 | 0.3433 |
| 7⁄16 | 7.1808 | 0.2187 | 0.1393 | 0.2785 | 0.1611 | 0.3003 |
| ⅖ | 7.8540 | 0.2000 | 0.1273 | 0.2546 | 0.1473 | 0.2746 |
| ⅜ | 8.3776 | 0.1875 | 0.1194 | 0.2387 | 0.1381 | 0.2575 |
| ⅓ | 9.4248 | 0.1666 | 0.1061 | 0.2122 | 0.1228 | 0.2289 |
| 5⁄16 | 10.0531 | 0.1562 | 0.0995 | 0.1989 | 0.1151 | 0.2146 |
| 2⁄7 | 10.9956 | 0.1429 | 0.0909 | 0.1819 | 0.1052 | 0.1962 |
| ¼ | 12.5664 | 0.1250 | 0.0796 | 0.1591 | 0.0921 | 0.1716 |
| 2⁄9 | 14.1372 | 0.1111 | 0.0707 | 0.1415 | 0.0818 | 0.1526 |
| ⅕ | 15.7080 | 0.1000 | 0.0637 | 0.1273 | 0.0737 | 0.1373 |
| 3⁄16 | 16.7552 | 0.0937 | 0.0597 | 0.1194 | 0.0690 | 0.1287 |
| ⅙ | 18.8496 | 0.0833 | 0.0531 | 0.1061 | 0.0614 | 0.1144 |
| 1⁄7 | 21.9911 | 0.0714 | 0.0455 | 0.0910 | 0.0526 | 0.0981 |
| ⅛ | 25.1327 | 0.0625 | 0.0398 | 0.0796 | 0.0460 | 0.0858 |
| ⅑ | 28.2743 | 0.0555 | 0.0354 | 0.0707 | 0.0409 | 0.0763 |
| 1⁄10 | 31.4159 | 0.0500 | 0.0318 | 0.0637 | 0.0368 | 0.0687 |
| 1⁄16 | 50.2655 | 0.0312 | 0.0199 | 0.0398 | 0.0230 | 0.0429 |

* For 20-degree full-depth teeth of 20 diametral pitch and finer see table on page 722.

Table 9· Gear Tooth Parts — Former American Standard 20-degree Involute Stub Tooth System (ASA B6.1-1932)

| Diametral Pitch | Circular Pitch | Standard Circular Thickness | Standard Addendum | Working Depth of Tooth | Standard Dedendum | Whole Depth of Tooth |
|---|---|---|---|---|---|---|
| $P$ | $p$ | $t$ | $a$ | $h_k$ | $b$ | $h_t$ |
| ½ | 6.2832 | 3.1416 | 1.6000 | 3.2000 | 2.0000 | 3.6000 |
| ¾ | 4.1888 | 2.0944 | 1.0667 | 2.1334 | 1.3333 | 2.4000 |
| 1 | 3.1416 | 1.5708 | .8000 | 1.6000 | 1.0000 | 1.8000 |
| 1¼ | 2.5133 | 1.2566 | 0.6400 | 1.2800 | 0.8000 | 1.4400 |
| 1½ | 2.0944 | 1.0472 | 0.5333 | 1.0666 | 0.6667 | 1.2000 |
| 1¾ | 1.7952 | 0.8976 | 0.4571 | 0.9142 | 0.5714 | 1.0285 |
| 2 | 1.5708 | 0.7854 | 0.4000 | 0.8000 | 0.5000 | 0.9000 |
| 2¼ | 1.3963 | 0.6981 | 0.3556 | 0.7112 | 0.4444 | 0.8000 |
| 2½ | 1.2566 | 0.6283 | 0.3200 | 0.6400 | 0.4000 | 0.7200 |
| 2¾ | 1.1424 | 0.5712 | 0.2909 | 0.5818 | 0.3636 | 0.6545 |
| 3 | 1.0472 | 0.5236 | 0.2667 | 0.5334 | 0.3333 | 0.6000 |
| 3½ | 0.8976 | 0.4488 | 0.2286 | 0.4572 | 0.2857 | 0.5143 |
| 4 | 0.7854 | 0.3927 | 0.2000 | 0.4000 | 0.2500 | 0.4500 |
| 5 | 0.6283 | 0.3142 | 0.1600 | 0.3200 | 0.2000 | 0.3600 |
| 6 | 0.5236 | 0.2618 | 0.1333 | 0.2666 | 0.1667 | 0.3000 |
| 7 | 0.4488 | 0.2244 | 0.1143 | 0.2286 | 0.1428 | 0.2571 |
| 8 | 0.3927 | 0.1963 | 0.1000 | 0.2000 | 0.1250 | 0.2250 |
| 9 | 0.3491 | 0.1745 | 0.0889 | 0.1778 | 0.1111 | 0.2000 |
| 10 | 0.3142 | 0.1571 | 0.0800 | 0.1600 | 0.1000 | 0.1800 |
| 11 | 0.2856 | 0.1428 | 0.0727 | 0.1454 | 0.0909 | 0.1636 |
| 12 | 0.2618 | 0.1309 | 0.0667 | 0.1334 | 0.0833 | 0.1500 |
| 13 | 0.2417 | 0.1208 | 0.0615 | 0.1230 | 0.0769 | 0.1384 |
| 14 | 0.2244 | 0.1122 | 0.0571 | 0.1142 | 0.0714 | 0.1285 |
| 15 | 0.2094 | 0.1047 | 0.0533 | 0.1066 | 0.0667 | 0.1200 |
| 16 | 0.1963 | 0.0982 | 0.0500 | 0.1000 | 0.0625 | 0.1125 |
| 17 | 0.1848 | 0.0924 | 0.0470 | 0.0940 | 0.0588 | 0.1058 |
| 18 | 0.1745 | 0.0873 | 0.0444 | 0.0888 | 0.0556 | 0.1000 |
| 19 | 0.1653 | 0.0827 | 0.0421 | 0.0842 | 0.0526 | 0.0947 |
| 20 | 0.1571 | 0.0785 | 0.0400 | 0.0800 | 0.0500 | 0.0900 |
| 22 | 0.1428 | 0.0714 | 0.0364 | 0.0728 | 0.0454 | 0.0818 |
| 24 | 0.1309 | 0.0654 | 0.0333 | 0.0666 | 0.0417 | 0.0750 |
| 26 | 0.1208 | 0.0604 | 0.0308 | 0.0616 | 0.0384 | 0.0692 |
| 28 | 0.1122 | 0.0561 | 0.0286 | 0.0572 | 0.0357 | 0.0643 |
| 30 | 0.1047 | 0.0524 | 0.0267 | 0.0534 | 0.0333 | 0.0600 |
| 32 | 0.0982 | 0.0491 | 0.0250 | 0.0500 | 0.0312 | 0.0562 |
| 34 | 0.0924 | 0.0462 | 0.0236 | 0.0472 | 0.0294 | 0.0530 |
| 36 | 0.0873 | 0.0436 | 0.0222 | 0.0444 | 0.0278 | 0.0500 |
| 38 | 0.0827 | 0.0413 | 0.0210 | 0.0420 | 0.0263 | 0.0473 |
| 40 | 0.0785 | 0.0393 | 0.0200 | 0.0400 | 0.0250 | 0.0450 |

**Table 10. Formulas for Outside and Root Diameters of Spur Gears that are Finish-hobbed, Shaped, or Pre-shaved**

## Notation

| | |
|---|---|
| $D$ = Pitch Diameter | $a$ = Standard Addendum |
| $D_O$ = Outside Diameter | $b$ = Standard Minimum Dedendum |
| $D_R$ = Root Diameter | $b_s$ = Standard Dedendum |
| $P$ = Diametral Pitch | $b_{ps}$ = Dedendum for Pre-shaving |

### $14\frac{1}{2}$-, 20-, AND 25-DEGREE INVOLUTE FULL-DEPTH TEETH ($19P$ and coarser)*

$$D_O = D + 2a \ = \frac{N}{P} + \left(2 \times \frac{1}{P}\right)$$

$$D_R = D - 2b \ = \frac{N}{P} - \left(2 \times \frac{1.157}{P}\right) \quad \text{(Hobbed)}[1]$$

$$D_R = D - 2b_s = \frac{N}{P} - \left(2 \times \frac{1.25}{P}\right) \quad \text{(Shaped)}[2]$$

$$D_R = D - 2b_{ps} = \frac{N}{P} - \left(2 \times \frac{1.35}{P}\right) \quad \text{(Pre-shaved)}[3]$$

$$D_R = D - 2b_{ps} = \frac{N}{P} - \left(2 \times \frac{1.40}{P}\right) \quad \text{(Pre-shaved)}[4]$$

### 20-DEGREE INVOLUTE FINE-PITCH FULL-DEPTH TEETH ($20P$ and finer)

$$D_O = D + 2a \ = \frac{N}{P} + \left(2 \times \frac{1}{P}\right)$$

$$D_R = D - 2b \ = \frac{N}{P} - 2\left(\frac{1.2}{P} + 0.002\right) \quad \text{(Hobbed or Shaped)}[5]$$

$$D_R = D - 2b_{ps} = \frac{N}{P} - 2\left(\frac{1.35}{P} + 0.002\right) \quad \text{(Pre-shaved)}[6]$$

### 20-DEGREE INVOLUTE STUB TEETH*

$$D_O = D + 2a = \frac{N}{P} + \left(2 \times \frac{0.8}{P}\right)$$

$$D_R = D - 2b = \frac{N}{P} - \left(2 \times \frac{1}{P}\right) \quad \text{(Hobbed)}$$

$$D_R = D - 2b_{ps} = \frac{N}{P} - \left(2 \times \frac{1.35}{P}\right) \quad \text{(Pre-shaved)}$$

* $14\frac{1}{2}$-degree full-depth and 20-degree stub teeth are not recommended for new designs.

[1] According to ANSI B6.1-1968 a minimum clearance of $0.157/P$ may be used for the basic 20-degree and 25-degree pressure angle rack in the case of shallow root sections and the use of existing hobs and cutters.

[2] According to ANSI B6.1-1968 the preferred clearance is $0.250/P$.

[3] According to ANSI B6.1-1968 the clearance for teeth which are shaved or ground is $0.350/P$.

[4] When gears are preshave cut on a gear shaper the dedendum will usually need to be increased to $1.40/P$ to allow for the higher fillet trochoid produced by the shaper cutter; this is of particular importance on gears of few teeth or if the gear blank configuration requires the use of a small diameter shaper cutter, in which case the dedendum may need to be increased to as much as $1.45/P$. This should be avoided on highly loaded gears where the consequently reduced $J$ factor will increase gear tooth stress excessively.

[5] According to ANSI B6.7-1967 the standard clearance is $0.200/P + 0.002$ (min.).

[6] According to ANSI B6.7-1967 the clearance for shaved or ground teeth is $0.350/P + 0.002$ (min.).

Table 11. Pitch Diameters of Diametral Pitch Gears

| No. of Teeth | Diametral Pitch | | | | | | | | | |
|---|---|---|---|---|---|---|---|---|---|---|
| | 2.5 | 3 | 3.5 | 4 | 4.5 | 5 | 6 | 7 | 8 | 9 |
| | Pitch Diameter | | | | | | | | | |
| 10 | 4.0000 | 3.3333 | 2.8571 | 2.5000 | 2.2222 | 2.0000 | 1.6667 | 1.4286 | 1.2500 | 1.1111 |
| 11 | 4.4000 | 3.6667 | 3.1429 | 2.7500 | 2.4444 | 2.2000 | 1.8333 | 1.5714 | 1.3750 | 1.2222 |
| 12 | 4.8000 | 4.0000 | 3.4286 | 3.0000 | 2.6667 | 2.4000 | 2.0000 | 1.7143 | 1.5000 | 1.3333 |
| 13 | 5.2000 | 4.3333 | 3.7143 | 3.2500 | 2.8889 | 2.6000 | 2.1667 | 1.8571 | 1.6250 | 1.4444 |
| 14 | 5.6000 | 4.6667 | 4.0000 | 3.5000 | 3.1111 | 2.8000 | 2.3333 | 2.0000 | 1.7500 | 1.5556 |
| 15 | 6.0000 | 5.0000 | 4.2857 | 3.7500 | 3.3333 | 3.0000 | 2.5000 | 2.1429 | 1.8750 | 1.6667 |
| 16 | 6.4000 | 5.3333 | 4.5714 | 4.0000 | 3.5556 | 3.2000 | 2.6667 | 2.2857 | 2.0000 | 1.7778 |
| 17 | 6.8000 | 5.6667 | 4.8571 | 4.2500 | 3.7778 | 3.4000 | 2.8333 | 2.4286 | 2.1250 | 1.8889 |
| 18 | 7.2000 | 6.0000 | 5.1429 | 4.5000 | 4.0000 | 3.6000 | 3.0000 | 2.5714 | 2.2500 | 2.0000 |
| 19 | 7.6000 | 6.3333 | 5.4286 | 4.7500 | 4.2222 | 3.8000 | 3.1667 | 2.7143 | 2.3750 | 2.1111 |
| 20 | 8.0000 | 6.6667 | 5.7143 | 5.0000 | 4.4444 | 4.0000 | 3.3333 | 2.8571 | 2.5000 | 2.2222 |
| 21 | 8.4000 | 7.0000 | 6.0000 | 5.2500 | 4.6667 | 4.2000 | 3.5000 | 3.0000 | 2.6250 | 2.3333 |
| 22 | 8.8000 | 7.3333 | 6.2857 | 5.5000 | 4.8889 | 4.4000 | 3.6667 | 3.1429 | 2.7500 | 2.4444 |
| 23 | 9.2000 | 7.6667 | 6.5714 | 5.7500 | 5.1111 | 4.6000 | 3.8333 | 3.2857 | 2.8750 | 2.5556 |
| 24 | 9.6000 | 8.0000 | 6.8571 | 6.0000 | 5.3333 | 4.8000 | 4.0000 | 3.4286 | 3.0000 | 2.6667 |
| 25 | 10.0000 | 8.3333 | 7.1429 | 6.2500 | 5.5556 | 5.0000 | 4.1667 | 3.5714 | 3.1250 | 2.7778 |
| 26 | 10.4000 | 8.6667 | 7.4286 | 6.5000 | 5.7778 | 5.2000 | 4.3333 | 3.7143 | 3.2500 | 2.8889 |
| 27 | 10.8000 | 9.0000 | 7.7143 | 6.7500 | 6.0000 | 5.4000 | 4.5000 | 3.8571 | 3.3750 | 3.0000 |
| 28 | 11.2000 | 9.3333 | 8.0000 | 7.0000 | 6.2222 | 5.6000 | 4.6667 | 4.0000 | 3.5000 | 3.1111 |
| 29 | 11.6000 | 9.6667 | 8.2857 | 7.2500 | 6.4444 | 5.8000 | 4.8333 | 4.1429 | 3.6250 | 3.2222 |
| 30 | 12.0000 | 10.0000 | 8.5714 | 7.5000 | 6.6667 | 6.0000 | 5.0000 | 4.2857 | 3.7500 | 3.3333 |
| 31 | 12.4000 | 10.3333 | 8.8571 | 7.7500 | 6.8889 | 6.2000 | 5.1667 | 4.4286 | 3.8750 | 3.4444 |
| 32 | 12.8000 | 10.6667 | 9.1429 | 8.0000 | 7.1111 | 6.4000 | 5.3333 | 4.5714 | 4.0000 | 3.5556 |
| 33 | 13.2000 | 11.0000 | 9.4286 | 8.2500 | 7.3333 | 6.6000 | 5.5000 | 4.7143 | 4.1250 | 3.6667 |
| 34 | 13.6000 | 11.3333 | 9.7143 | 8.5000 | 7.5556 | 6.8000 | 5.6667 | 4.8571 | 4.2500 | 3.7778 |
| 35 | 14.0000 | 11.6667 | 10.0000 | 8.7500 | 7.7778 | 7.0000 | 5.8333 | 5.0000 | 4.3750 | 3.8889 |
| 36 | 14.4000 | 12.0000 | 10.2857 | 9.0000 | 8.0000 | 7.2000 | 6.0000 | 5.1429 | 4.5000 | 4.0000 |
| 37 | 14.8000 | 12.3333 | 10.5714 | 9.2500 | 8.2222 | 7.4000 | 6.1667 | 5.2857 | 4.6250 | 4.1111 |
| 38 | 15.2000 | 12.6667 | 10.8571 | 9.5000 | 8.4444 | 7.6000 | 6.3333 | 5.4286 | 4.7500 | 4.2222 |
| 39 | 15.6000 | 13.0000 | 11.1429 | 9.7500 | 8.6667 | 7.8000 | 6.5000 | 5.5714 | 4.8750 | 4.3333 |
| 40 | 16.0000 | 13.3333 | 11.4286 | 10.0000 | 8.8889 | 8.0000 | 6.6667 | 5.7143 | 5.0000 | 4.4444 |
| 41 | 16.4000 | 13.6667 | 11.7143 | 10.2500 | 9.1111 | 8.2000 | 6.8333 | 5.8571 | 5.1250 | 4.5556 |
| 42 | 16.8000 | 14.0000 | 12.0000 | 10.5000 | 9.3333 | 8.4000 | 7.0000 | 6.0000 | 5.2500 | 4.6667 |
| 43 | 17.2000 | 14.3333 | 12.2857 | 10.7500 | 9.5556 | 8.6000 | 7.1667 | 6.1429 | 5.3750 | 4.7778 |
| 44 | 17.6000 | 14.6667 | 12.5714 | 11.0000 | 9.7778 | 8.8000 | 7.3333 | 6.2857 | 5.5000 | 4.8889 |
| 45 | 18.0000 | 15.0000 | 12.8571 | 11.2500 | 10.0000 | 9.0000 | 7.5000 | 6.4286 | 5.6250 | 5.0000 |
| 46 | 18.4000 | 15.3333 | 13.1429 | 11.5000 | 10.2222 | 9.2000 | 7.6667 | 6.5714 | 5.7500 | 5.1111 |
| 47 | 18.8000 | 15.6667 | 13.4286 | 11.7500 | 10.4444 | 9.4000 | 7.8333 | 6.7143 | 5.8750 | 5.2222 |
| 48 | 19.2000 | 16.0000 | 13.7143 | 12.0000 | 10.6667 | 9.6000 | 8.0000 | 6.8571 | 6.0000 | 5.3333 |
| 49 | 19.6000 | 16.3333 | 14.0000 | 12.2500 | 10.8889 | 9.8000 | 8.1667 | 7.0000 | 6.1250 | 5.4444 |
| 50 | 20.0000 | 16.6667 | 14.2857 | 12.5000 | 11.1111 | 10.0000 | 8.3333 | 7.1429 | 6.2500 | 5.5556 |
| 51 | 20.4000 | 17.0000 | 14.5714 | 12.7500 | 11.3333 | 10.2000 | 8.5000 | 7.2857 | 6.3750 | 5.6667 |
| 52 | 20.8000 | 17.3333 | 14.8571 | 13.0000 | 11.5556 | 10.4000 | 8.6667 | 7.4286 | 6.5000 | 5.7778 |
| 53 | 21.2000 | 17.6667 | 15.1429 | 13.2500 | 11.7778 | 10.6000 | 8.8333 | 7.5714 | 6.6250 | 5.8889 |
| 54 | 21.6000 | 18.0000 | 15.4286 | 13.5000 | 12.0000 | 10.8000 | 9.0000 | 7.7143 | 6.7500 | 6.0000 |
| 55 | 22.0000 | 18.3333 | 15.7143 | 13.7500 | 12.2222 | 11.0000 | 9.1667 | 7.8571 | 6.8750 | 6.1111 |
| 56 | 22.4000 | 18.6667 | 16.0000 | 14.0000 | 12.4444 | 11.2000 | 9.3333 | 8.0000 | 7.0000 | 6.2222 |
| 57 | 22.8000 | 19.0000 | 16.2857 | 14.2500 | 12.6667 | 11.4000 | 9.5000 | 8.1429 | 7.1250 | 6.3333 |
| 58 | 23.2000 | 19.3333 | 16.5714 | 14.5000 | 12.8889 | 11.6000 | 9.6667 | 8.2857 | 7.2500 | 6.4444 |
| 59 | 23.6000 | 19.6667 | 16.8571 | 14.7500 | 13.1111 | 11.8000 | 9.8333 | 8.4286 | 7.3750 | 6.5556 |
| 60 | 24.0000 | 20.0000 | 17.1429 | 15.0000 | 13.3333 | 12.0000 | 10.0000 | 8.5714 | 7.5000 | 6.6667 |
| 61 | 24.4000 | 20.3333 | 17.4286 | 15.2500 | 13.5556 | 12.2000 | 10.1667 | 8.7143 | 7.6250 | 6.7778 |
| 62 | 24.8000 | 20.6667 | 17.7143 | 15.5000 | 13.7778 | 12.4000 | 10.3333 | 8.8571 | 7.7500 | 6.8889 |
| 63 | 25.2000 | 21.0000 | 18.0000 | 15.7500 | 14.0000 | 12.6000 | 10.5000 | 9.0000 | 7.8750 | 7.0000 |
| 64 | 25.6000 | 21.3333 | 18.2857 | 16.0000 | 14.2222 | 12.8000 | 10.6667 | 9.1429 | 8.0000 | 7.1111 |
| 65 | 26.0000 | 21.6667 | 18.5714 | 16.2500 | 14.4444 | 13.0000 | 10.8333 | 9.2857 | 8.1250 | 7.2222 |
| 66 | 26.4000 | 22.0000 | 18.8571 | 16.5000 | 14.6667 | 13.2000 | 11.0000 | 9.4286 | 8.2500 | 7.3333 |
| 67 | 26.8000 | 22.3333 | 19.1429 | 16.7500 | 14.8889 | 13.4000 | 11.1667 | 9.5714 | 8.3750 | 7.4444 |
| 68 | 27.2000 | 22.6667 | 19.4286 | 17.0000 | 15.1111 | 13.6000 | 11.3333 | 9.7143 | 8.5000 | 7.5556 |
| 69 | 27.6000 | 23.0000 | 19.7143 | 17.2500 | 15.3333 | 13.8000 | 11.5000 | 9.8571 | 8.6250 | 7.6667 |
| 70 | 28.0000 | 23.3333 | 20.0000 | 17.5000 | 15.5556 | 14.0000 | 11.6667 | 10.0000 | 8.7500 | 7.7778 |

Outside diameter of a standard full-depth gear equals the pitch diameter for a gear of the same diametral pitch but with 2 more teeth.

Table 11 (*Continued*). Pitch Diameters of Diametral Pitch Gears

| No. of Teeth | Diametral Pitch | | | | | | | | | |
|---|---|---|---|---|---|---|---|---|---|---|
| | 2.5 | 3 | 3.5 | 4 | 4.5 | 5 | 6 | 7 | 8 | 9 |
| | Pitch Diameter | | | | | | | | | |
| 71 | 28.4000 | 23.6667 | 20.2857 | 17.7500 | 15.7778 | 14.2000 | 11.8333 | 10.1429 | 8.8750 | 7.8889 |
| 72 | 28.8000 | 24.0000 | 20.5714 | 18.0000 | 16.0000 | 14.4000 | 12.0000 | 10.2857 | 9.0000 | 8.0000 |
| 73 | 29.2000 | 24.3333 | 20.8571 | 18.2500 | 16.2222 | 14.6000 | 12.1667 | 10.4286 | 9.1250 | 8.1111 |
| 74 | 29.6000 | 24.6667 | 21.1429 | 18.5000 | 16.4444 | 14.8000 | 12.3333 | 10.5714 | 9.2500 | 8.2222 |
| 75 | 30.0000 | 25.0000 | 21.4286 | 18.7500 | 16.6667 | 15.0000 | 12.5000 | 10.7143 | 9.3750 | 8.3333 |
| 76 | 30.4000 | 25.3333 | 21.7143 | 19.0000 | 16.8889 | 15.2000 | 12.6667 | 10.8571 | 9.5000 | 8.4444 |
| 77 | 30.8000 | 25.6667 | 22.0000 | 19.2500 | 17.1111 | 15.4000 | 12.8333 | 11.0000 | 9.6250 | 8.5556 |
| 78 | 31.2000 | 26.0000 | 22.2857 | 19.5000 | 17.3333 | 15.6000 | 13.0000 | 11.1429 | 9.7500 | 8.6667 |
| 79 | 31.6000 | 26.3333 | 22.5714 | 19.7500 | 17.5556 | 15.8000 | 13.1667 | 11.2857 | 9.8750 | 8.7778 |
| 80 | 32.0000 | 26.6667 | 22.8571 | 20.0000 | 17.7778 | 16.0000 | 13.3333 | 11.4286 | 10.0000 | 8.8889 |
| 81 | 32.4000 | 27.0000 | 23.1429 | 20.2500 | 18.0000 | 16.2000 | 13.5000 | 11.5714 | 10.1250 | 9.0000 |
| 82 | 32.8000 | 27.3333 | 23.4286 | 20.5000 | 18.2222 | 16.4000 | 13.6667 | 11.7143 | 10.2500 | 9.1111 |
| 83 | 33.2000 | 27.6667 | 23.7143 | 20.7500 | 18.4444 | 16.6000 | 13.8333 | 11.8571 | 10.3750 | 9.2222 |
| 84 | 33.6000 | 28.0000 | 24.0000 | 21.0000 | 18.6667 | 16.8000 | 14.0000 | 12.0000 | 10.5000 | 9.3333 |
| 85 | 34.0000 | 28.3333 | 24.2857 | 21.2500 | 18.8889 | 17.0000 | 14.1667 | 12.1429 | 10.6250 | 9.4444 |
| 86 | 34.4000 | 28.6667 | 24.5714 | 21.5000 | 19.1111 | 17.2000 | 14.3333 | 12.2857 | 10.7500 | 9.5556 |
| 87 | 34.8000 | 29.0000 | 24.8571 | 21.7500 | 19.3333 | 17.4000 | 14.5000 | 12.4286 | 10.8750 | 9.6667 |
| 88 | 35.2000 | 29.3333 | 25.1429 | 22.0000 | 19.5556 | 17.6000 | 14.6667 | 12.5714 | 11.0000 | 9.7778 |
| 89 | 35.6000 | 29.6667 | 25.4286 | 22.2500 | 19.7778 | 17.8000 | 14.8333 | 12.7143 | 11.1250 | 9.8889 |
| 90 | 36.0000 | 30.0000 | 25.7143 | 22.5000 | 20.0000 | 18.0000 | 15.0000 | 12.8571 | 11.2500 | 10.0000 |
| 91 | 36.4000 | 30.3333 | 26.0000 | 22.7500 | 20.2222 | 18.2000 | 15.1667 | 13.0000 | 11.3750 | 10.1111 |
| 92 | 36.8000 | 30.6667 | 26.2857 | 23.0000 | 20.4444 | 18.4000 | 15.3333 | 13.1429 | 11.5000 | 10.2222 |
| 93 | 37.2000 | 31.0000 | 26.5714 | 23.2500 | 20.6667 | 18.6000 | 15.5000 | 13.2857 | 11.6250 | 10.3333 |
| 94 | 37.6000 | 31.3333 | 26.8571 | 23.5000 | 20.8889 | 18.8000 | 15.6667 | 13.4286 | 11.7500 | 10.4444 |
| 95 | 38.0000 | 31.6667 | 27.1429 | 23.7500 | 21.1111 | 19.0000 | 15.8333 | 13.5714 | 11.8750 | 10.5556 |
| 96 | 38.4000 | 32.0000 | 27.4286 | 24.0000 | 21.3333 | 19.2000 | 16.0000 | 13.7143 | 12.0000 | 10.6667 |
| 97 | 38.8000 | 32.3333 | 27.7143 | 24.2500 | 21.5556 | 19.4000 | 16.1667 | 13.8571 | 12.1250 | 10.7778 |
| 98 | 39.2000 | 32.6667 | 28.0000 | 24.5000 | 21.7778 | 19.6000 | 16.3333 | 14.0000 | 12.2500 | 10.8889 |
| 99 | 39.6000 | 33.0000 | 28.2857 | 24.7500 | 22.0000 | 19.8000 | 16.5000 | 14.1429 | 12.3750 | 11.0000 |
| 100 | 40.0000 | 33.3333 | 28.5714 | 25.0000 | 22.2222 | 20.0000 | 16.6667 | 14.2857 | 12.5000 | 11.1111 |
| 101 | 40.4000 | 33.6667 | 28.8571 | 25.2500 | 22.4444 | 20.2000 | 16.8333 | 14.4286 | 12.6250 | 11.2222 |
| 102 | 40.8000 | 34.0000 | 29.1429 | 25.5000 | 22.6667 | 20.4000 | 17.0000 | 14.5714 | 12.7500 | 11.3333 |
| 103 | 41.2000 | 34.3333 | 29.4286 | 25.7500 | 22.8889 | 20.6000 | 17.1667 | 14.7143 | 12.8750 | 11.4444 |
| 104 | 41.6000 | 34.6667 | 29.7143 | 26.0000 | 23.1111 | 20.8000 | 17.3333 | 14.8571 | 13.0000 | 11.5556 |
| 105 | 42.0000 | 35.0000 | 30.0000 | 26.2500 | 23.3333 | 21.0000 | 17.5000 | 15.0000 | 13.1250 | 11.6667 |
| 106 | 42.4000 | 35.3333 | 30.2857 | 26.5000 | 23.5556 | 21.2000 | 17.6667 | 15.1429 | 13.2500 | 11.7778 |
| 107 | 42.8000 | 35.6667 | 30.5714 | 26.7500 | 23.7778 | 21.4000 | 17.8333 | 15.2857 | 13.3750 | 11.8889 |
| 108 | 43.2000 | 36.0000 | 30.8571 | 27.0000 | 24.0000 | 21.6000 | 18.0000 | 15.4286 | 13.5000 | 12.0000 |
| 109 | 43.6000 | 36.3333 | 31.1429 | 27.2500 | 24.2222 | 21.8000 | 18.1667 | 15.5714 | 13.6250 | 12.1111 |
| 110 | 44.0000 | 36.6667 | 31.4286 | 27.5000 | 24.4444 | 22.0000 | 18.3333 | 15.7143 | 13.7500 | 12.2222 |
| 111 | 44.4000 | 37.0000 | 31.7143 | 27.7500 | 24.6667 | 22.2000 | 18.5000 | 15.8571 | 13.8750 | 12.3333 |
| 112 | 44.8000 | 37.3333 | 32.0000 | 28.0000 | 24.8889 | 22.4000 | 18.6667 | 16.0000 | 14.0000 | 12.4444 |
| 113 | 45.2000 | 37.6667 | 32.2857 | 28.2500 | 25.1111 | 22.6000 | 18.8333 | 16.1429 | 14.1250 | 12.5556 |
| 114 | 45.6000 | 38.0000 | 32.5714 | 28.5000 | 25.3333 | 22.8000 | 19.0000 | 16.2857 | 14.2500 | 12.6667 |
| 115 | 46.0000 | 38.3333 | 32.8571 | 28.7500 | 25.5556 | 23.0000 | 19.1667 | 16.4286 | 14.3750 | 12.7778 |
| 116 | 46.4000 | 38.6667 | 33.1429 | 29.0000 | 25.7778 | 23.2000 | 19.3333 | 16.5714 | 14.5000 | 12.8889 |
| 117 | 46.8000 | 39.0000 | 33.4286 | 29.2500 | 26.0000 | 23.4000 | 19.5000 | 16.7143 | 14.6250 | 13.0000 |
| 118 | 47.2000 | 39.3333 | 33.7143 | 29.5000 | 26.2222 | 23.6000 | 19.6667 | 16.8571 | 14.7500 | 13.1111 |
| 119 | 47.6000 | 39.6667 | 34.0000 | 29.7500 | 26.4444 | 23.8000 | 19.8333 | 17.0000 | 14.8750 | 13.2222 |
| 120 | 48.0000 | 40.0000 | 34.2857 | 30.0000 | 26.6667 | 24.0000 | 20.0000 | 17.1429 | 15.0000 | 13.3333 |
| 121 | 48.4000 | 40.3333 | 34.5714 | 30.2500 | 26.8889 | 24.2000 | 20.1667 | 17.2857 | 15.1250 | 13.4444 |
| 122 | 48.8000 | 40.6667 | 34.8571 | 30.5000 | 27.1111 | 24.4000 | 20.3333 | 17.4286 | 15.2500 | 13.5556 |
| 123 | 49.2000 | 41.0000 | 35.1429 | 30.7500 | 27.3333 | 24.6000 | 20.5000 | 17.5714 | 15.3750 | 13.6667 |
| 124 | 49.6000 | 41.3333 | 35.4286 | 31.0000 | 27.5556 | 24.8000 | 20.6667 | 17.7143 | 15.5000 | 13.7778 |
| 125 | 50.0000 | 41.6667 | 35.7143 | 31.2500 | 27.7778 | 25.0000 | 20.8333 | 17.8571 | 15.6250 | 13.8889 |
| 126 | 50.4000 | 42.0000 | 36.0000 | 31.5000 | 28.0000 | 25.2000 | 21.0000 | 18.0000 | 15.7500 | 14.0000 |
| 127 | 50.8000 | 42.3333 | 36.2857 | 31.7500 | 28.2222 | 25.4000 | 21.1667 | 18.1429 | 15.8750 | 14.1111 |
| 128 | 51.2000 | 42.6667 | 36.5714 | 32.0000 | 28.4444 | 25.6000 | 21.3333 | 18.2857 | 16.0000 | 14.2222 |
| 129 | 51.6000 | 43.0000 | 36.8571 | 32.2500 | 28.6667 | 25.8000 | 21.5000 | 18.4286 | 16.1250 | 14.3333 |
| 130 | 52.0000 | 43.3333 | 37.1429 | 32.5000 | 28.8889 | 26.0000 | 21.6667 | 18.5714 | 16.2500 | 14.4444 |

Outside diameter of a standard full-depth gear equals the pitch diameter for a gear of the same diametral pitch but with 2 more teeth

Table 11 (*Continued*). Pitch Diameters of Diametral Pitch Gears

| No. of Teeth | Diametral Pitch | | | | | | | | | | |
|---|---|---|---|---|---|---|---|---|---|---|---|
| | 10 | 11 | 12 | 14 | 16 | 18 | 20 | 22 | 24 | 28 | 32 |
| | Pitch Diameter | | | | | | | | | | |
| 10 | 1.0000 | 0.9091 | 0.8333 | 0.7143 | 0.6250 | 0.5556 | 0.5000 | 0.4545 | 0.4167 | 0.3571 | 0.3125 |
| 11 | 1.1000 | 1.0000 | 0.9167 | 0.7857 | 0.6875 | 0.6111 | 0.5500 | 0.5000 | 0.4583 | 0.3929 | 0.3438 |
| 12 | 1.2000 | 1.0909 | 1.0000 | 0.8571 | 0.7500 | 0.6667 | 0.6000 | 0.5455 | 0.5000 | 0.4286 | 0.3750 |
| 13 | 1.3000 | 1.1818 | 1.0833 | 0.9286 | 0.8125 | 0.7222 | 0.6500 | 0.5999 | 0.5417 | 0.4643 | 0.4062 |
| 14 | 1.4000 | 1.2727 | 1.1667 | 1.0000 | 0.8750 | 0.7778 | 0.7000 | 0.6364 | 0.5833 | 0.5000 | 0.4375 |
| 15 | 1.5000 | 1.3636 | 1.2500 | 1.0714 | 0.9375 | 0.8333 | 0.7500 | 0.6818 | 0.6250 | 0.5357 | 0.4688 |
| 16 | 1.6000 | 1.4545 | 1.3333 | 1.1429 | 1.0000 | 0.8889 | 0.8000 | 0.7273 | 0.6667 | 0.5714 | 0.5000 |
| 17 | 1.7000 | 1.5455 | 1.4167 | 1.2143 | 1.0625 | 0.9444 | 0.8500 | 0.7727 | 0.7083 | 0.6071 | 0.5312 |
| 18 | 1.8000 | 1.6364 | 1.5000 | 1.2857 | 1.1250 | 1.0000 | 0.9000 | 0.8182 | 0.7500 | 0.6429 | 0.5625 |
| 19 | 1.9000 | 1.7273 | 1.5833 | 1.3571 | 1.1875 | 1.0556 | 0.9500 | 0.8636 | 0.7917 | 0.6786 | 0.5938 |
| 20 | 2.0000 | 1.8182 | 1.6667 | 1.4286 | 1.2500 | 1.1111 | 1.0000 | 0.9091 | 0.8333 | 0.7143 | 0.6250 |
| 21 | 2.1000 | 1.9091 | 1.7500 | 1.5000 | 1.3125 | 1.1667 | 1.0500 | 0.9545 | 0.8750 | 0.7500 | 0.6562 |
| 22 | 2.2000 | 2.0000 | 1.8333 | 1.5714 | 1.3750 | 1.2222 | 1.1000 | 1.0000 | 0.9167 | 0.7857 | 0.6875 |
| 23 | 2.3000 | 2.0909 | 1.9167 | 1.6429 | 1.4375 | 1.2778 | 1.1500 | 1.0455 | 0.9583 | 0.8214 | 0.7188 |
| 24 | 2.4000 | 2.1818 | 2.0000 | 1.7143 | 1.5000 | 1.3333 | 1.2000 | 1.0909 | 1.0000 | 0.8571 | 0.7500 |
| 25 | 2.5000 | 2.2727 | 2.0833 | 1.7857 | 1.5625 | 1.3889 | 1.2500 | 1.1364 | 1.0417 | 0.8929 | 0.7812 |
| 26 | 2.6000 | 2.3636 | 2.1667 | 1.8571 | 1.6250 | 1.4444 | 1.3000 | 1.1818 | 1.0833 | 0.9286 | 0.8125 |
| 27 | 2.7000 | 2.4545 | 2.2500 | 1.9286 | 1.6875 | 1.5000 | 1.3500 | 1.2273 | 1.1250 | 0.9643 | 0.8438 |
| 28 | 2.8000 | 2.5455 | 2.3333 | 2.0000 | 1.7500 | 1.5556 | 1.4000 | 1.2727 | 1.1667 | 1.0000 | 0.8750 |
| 29 | 2.9000 | 2.6364 | 2.4167 | 2.0714 | 1.8125 | 1.6111 | 1.4500 | 1.3182 | 1.2083 | 1.0357 | 0.9062 |
| 30 | 3.0000 | 2.7273 | 2.5000 | 2.1429 | 1.8750 | 1.6667 | 1.5000 | 1.3636 | 1.2500 | 1.0714 | 0.9375 |
| 31 | 3.1000 | 2.8182 | 2.5833 | 2.2143 | 1.9375 | 1.7222 | 1.5500 | 1.4091 | 1.2917 | 1.1071 | 0.9688 |
| 32 | 3.2000 | 2.9091 | 2.6667 | 2.2857 | 2.0000 | 1.7778 | 1.6000 | 1.4545 | 1.3333 | 1.1429 | 1.0000 |
| 33 | 3.3000 | 3.0000 | 2.7500 | 2.3571 | 2.0625 | 1.8333 | 1.6500 | 1.5000 | 1.3750 | 1.1786 | 1.0312 |
| 34 | 3.4000 | 3.0909 | 2.8333 | 2.4286 | 2.1250 | 1.8889 | 1.7000 | 1.5455 | 1.4167 | 1.2143 | 1.0625 |
| 35 | 3.5000 | 3.1818 | 2.9167 | 2.5000 | 2.1875 | 1.9444 | 1.7500 | 1.5909 | 1.4583 | 1.2500 | 1.0938 |
| 36 | 3.6000 | 3.2727 | 3.0000 | 2.5714 | 2.2500 | 2.0000 | 1.8000 | 1.6364 | 1.5000 | 1.2857 | 1.1250 |
| 37 | 3.7000 | 3.3636 | 3.0833 | 2.6429 | 2.3125 | 2.0556 | 1.8500 | 1.6818 | 1.5417 | 1.3214 | 1.1562 |
| 38 | 3.8000 | 3.4545 | 3.1667 | 2.7143 | 2.3750 | 2.1111 | 1.9000 | 1.7273 | 1.5833 | 1.3571 | 1.1875 |
| 39 | 3.9000 | 3.5455 | 3.2500 | 2.7857 | 2.4375 | 2.1667 | 1.9500 | 1.7727 | 1.6250 | 1.3929 | 1.2188 |
| 40 | 4.0000 | 3.6364 | 3.3333 | 2.8571 | 2.5000 | 2.2222 | 2.0000 | 1.8182 | 1.6667 | 1.4286 | 1.2500 |
| 41 | 4.1000 | 3.7273 | 3.4167 | 2.9286 | 2.5625 | 2.2778 | 2.0500 | 1.8636 | 1.7083 | 1.4643 | 1.2812 |
| 42 | 4.2000 | 3.8182 | 3.5000 | 3.0000 | 2.6250 | 2.3333 | 2.1000 | 1.9091 | 1.7500 | 1.5000 | 1.3125 |
| 43 | 4.3000 | 3.9091 | 3.5833 | 3.0714 | 2.6875 | 2.3889 | 2.1500 | 1.9545 | 1.7917 | 1.5357 | 1.3438 |
| 44 | 4.4000 | 4.0000 | 3.6667 | 3.1429 | 2.7500 | 2.4444 | 2.2000 | 2.0000 | 1.8333 | 1.5714 | 1.3750 |
| 45 | 4.5000 | 4.0909 | 3.7500 | 3.2143 | 2.8125 | 2.5000 | 2.2500 | 2.0455 | 1.8750 | 1.6071 | 1.4062 |
| 46 | 4.6000 | 4.1818 | 3.8333 | 3.2857 | 2.8750 | 2.5556 | 2.3000 | 2.0909 | 1.9167 | 1.6429 | 1.4375 |
| 47 | 4.7000 | 4.2727 | 3.9167 | 3.3571 | 2.9375 | 2.6111 | 2.3500 | 2.1364 | 1.9583 | 1.6786 | 1.4688 |
| 48 | 4.8000 | 4.3636 | 4.0000 | 3.4286 | 3.0000 | 2.6667 | 2.4000 | 2.1818 | 2.0000 | 1.7143 | 1.5000 |
| 49 | 4.9000 | 4.4545 | 4.0833 | 3.5000 | 3.0625 | 2.7222 | 2.4500 | 2.2273 | 2.0417 | 1.7500 | 1.5312 |
| 50 | 5.0000 | 4.5455 | 4.1667 | 3.5714 | 3.1250 | 2.7778 | 2.5000 | 2.2727 | 2.0833 | 1.7857 | 1.5625 |
| 51 | 5.1000 | 4.6364 | 4.2500 | 3.6429 | 3.1875 | 2.8333 | 2.5500 | 2.3182 | 2.1250 | 1.8214 | 1.5938 |
| 52 | 5.2000 | 4.7273 | 4.3333 | 3.7143 | 3.2500 | 2.8889 | 2.6000 | 2.3636 | 2.1667 | 1.8571 | 1.6250 |
| 53 | 5.3000 | 4.8182 | 4.4167 | 3.7857 | 3.3125 | 2.9444 | 2.6500 | 2.4091 | 2.2083 | 1.8929 | 1.6562 |
| 54 | 5.4000 | 4.9091 | 4.5000 | 3.8571 | 3.3750 | 3.0000 | 2.7000 | 2.4545 | 2.2500 | 1.9286 | 1.6875 |
| 55 | 5.5000 | 5.0000 | 4.5833 | 3.9286 | 3.4375 | 3.0556 | 2.7500 | 2.5000 | 2.2917 | 1.9643 | 1.7188 |
| 56 | 5.6000 | 5.0909 | 4.6667 | 4.0000 | 3.5000 | 3.1111 | 2.8000 | 2.5455 | 2.3333 | 2.0000 | 1.7500 |
| 57 | 5.7000 | 5.1818 | 4.7500 | 4.0714 | 3.5625 | 3.1667 | 2.8500 | 2.5909 | 2.3750 | 2.0357 | 1.7812 |
| 58 | 5.8000 | 5.2727 | 4.8333 | 4.1429 | 3.6250 | 3.2222 | 2.9000 | 2.6364 | 2.4167 | 2.0714 | 1.8125 |
| 59 | 5.9000 | 5.3636 | 4.9167 | 4.2143 | 3.6875 | 3.2778 | 2.9500 | 2.6818 | 2.4583 | 2.1071 | 1.8438 |
| 60 | 6.0000 | 5.4545 | 5.0000 | 4.2857 | 3.7500 | 3.3333 | 3.0000 | 2.7273 | 2.5000 | 2.1429 | 1.8750 |
| 61 | 6.1000 | 5.5455 | 5.0833 | 4.3571 | 3.8125 | 3.3889 | 3.0500 | 2.7727 | 2.5417 | 2.1786 | 1.9062 |
| 62 | 6.2000 | 5.6364 | 5.1667 | 4.4286 | 3.8750 | 3.4444 | 3.1000 | 2.8182 | 2.5833 | 2.2143 | 1.9375 |
| 63 | 6.3000 | 5.7273 | 5.2500 | 4.5000 | 3.9375 | 3.5000 | 3.1500 | 2.8636 | 2.6250 | 2.2500 | 1.9688 |
| 64 | 6.4000 | 5.8182 | 5.3333 | 4.5714 | 4.0000 | 3.5556 | 3.2000 | 2.9091 | 2.6667 | 2.2857 | 2.0000 |
| 65 | 6.5000 | 5.9091 | 5.4167 | 4.6429 | 4.0625 | 3.6111 | 3.2500 | 2.9545 | 2.7083 | 2.3214 | 2.0312 |
| 66 | 6.6000 | 6.0000 | 5.5000 | 4.7143 | 4.1250 | 3.6667 | 3.3000 | 3.0000 | 2.7500 | 2.3571 | 2.0625 |
| 67 | 6.7000 | 6.0909 | 5.5833 | 4.7857 | 4.1875 | 3.7222 | 3.3500 | 3.0455 | 2.7917 | 2.3929 | 2.0938 |
| 68 | 6.8000 | 6.1818 | 5.6667 | 4.8571 | 4.2500 | 3.7778 | 3.4000 | 3.0909 | 2.8333 | 2.4286 | 2.1250 |
| 69 | 6.9000 | 6.2727 | 5.7500 | 4.9286 | 4.3125 | 3.8333 | 3.4500 | 3.1364 | 2.8750 | 2.4643 | 2.1562 |
| 70 | 7.0000 | 6.3636 | 5.8333 | 5.0000 | 4.3750 | 3.8889 | 3.5000 | 3.1818 | 2.9167 | 2.5000 | 2.1875 |

Outside diameter of a standard full-depth gear equals the pitch diameter for a gear of the same diametral pitch but with 2 more teeth.

Table 11 (*Continued*).  Pitch Diameters of Diametral Pitch Gears

| No. of Teeth | Diametral Pitch | | | | | | | | | | |
|---|---|---|---|---|---|---|---|---|---|---|---|
| | 10 | 11 | 12 | 14 | 16 | 18 | 20 | 22 | 24 | 28 | 32 |
| | Pitch Diameter | | | | | | | | | | |
| 71 | 7.1000 | 6.4545 | 5.9167 | 5.0714 | 4.4375 | 3.9444 | 3.5500 | 3.2273 | 2.9583 | 2.5357 | 2.2188 |
| 72 | 7.2000 | 6.5455 | 6.0000 | 5.1429 | 4.5000 | 4.0000 | 3.6000 | 3.2727 | 3.0000 | 2.5714 | 2.2500 |
| 73 | 7.3000 | 6.6364 | 6.0833 | 5.2143 | 4.5625 | 4.0556 | 3.6500 | 3.3182 | 3.0417 | 2.6071 | 2.2812 |
| 74 | 7.4000 | 6.7273 | 6.1667 | 5.2857 | 4.6250 | 4.1111 | 3.7000 | 3.3636 | 3.0833 | 2.6429 | 2.3125 |
| 75 | 7.5000 | 6.8182 | 6.2500 | 5.3571 | 4.6875 | 4.1667 | 3.7500 | 3.4091 | 3.1250 | 2.6786 | 2.3438 |
| 76 | 7.6000 | 6.9091 | 6.3333 | 5.4286 | 4.7500 | 4.2222 | 3.8000 | 3.4545 | 3.1667 | 2.7143 | 2.3750 |
| 77 | 7.7000 | 7.0000 | 6.4167 | 5.5000 | 4.8125 | 4.2778 | 3.8500 | 3.5000 | 3.2083 | 2.7500 | 2.4062 |
| 78 | 7.8000 | 7.0909 | 6.5000 | 5.5714 | 4.8750 | 4.3333 | 3.9000 | 3.5455 | 3.2500 | 2.7857 | 2.4375 |
| 79 | 7.9000 | 7.1818 | 6.5833 | 5.6429 | 4.9375 | 4.3889 | 3.9500 | 3.5909 | 3.2917 | 2.8214 | 2.4688 |
| 80 | 8.0000 | 7.2727 | 6.6667 | 5.7143 | 5.0000 | 4.4444 | 4.0000 | 3.6364 | 3.3333 | 2.8571 | 2.5000 |
| 81 | 8.1000 | 7.3636 | 6.7500 | 5.7857 | 5.0625 | 4.5000 | 4.0500 | 3.6818 | 3.3750 | 2.8929 | 2.5312 |
| 82 | 8.2000 | 7.4545 | 6.8333 | 5.8571 | 5.1250 | 4.5556 | 4.1000 | 3.7273 | 3.4167 | 2.9286 | 2.5625 |
| 83 | 8.3000 | 7.5455 | 6.9167 | 5.9286 | 5.1875 | 4.6111 | 4.1500 | 3.7727 | 3.4583 | 2.9643 | 2.5938 |
| 84 | 8.4000 | 7.6364 | 7.0000 | 6.0000 | 5.2500 | 4.6667 | 4.2000 | 3.8182 | 3.5000 | 3.0000 | 2.6250 |
| 85 | 8.5000 | 7.7273 | 7.0833 | 6.0714 | 5.3125 | 4.7222 | 4.2500 | 3.8636 | 3.5417 | 3.0357 | 2.6562 |
| 86 | 8.6000 | 7.8182 | 7.1667 | 6.1429 | 5.3750 | 4.7778 | 4.3000 | 3.9091 | 3.5833 | 3.0714 | 2.6875 |
| 87 | 8.7000 | 7.9091 | 7.2500 | 6.2143 | 5.4375 | 4.8333 | 4.3500 | 3.9545 | 3.6250 | 3.1071 | 2.7188 |
| 88 | 8.8000 | 8.0000 | 7.3333 | 6.2857 | 5.5000 | 4.8889 | 4.4000 | 4.0000 | 3.6667 | 3.1429 | 2.7500 |
| 89 | 8.9000 | 8.0909 | 7.4167 | 6.3571 | 5.5625 | 4.9444 | 4.4500 | 4.0455 | 3.7083 | 3.1786 | 2.7812 |
| 90 | 9.0000 | 8.1818 | 7.5000 | 6.4286 | 5.6250 | 5.0000 | 4.5000 | 4.0909 | 3.7500 | 3.2143 | 2.8125 |
| 91 | 9.1000 | 8.2727 | 7.5833 | 6.5000 | 5.6875 | 5.0556 | 4.5500 | 4.1364 | 3.7917 | 3.2500 | 2.8438 |
| 92 | 9.2000 | 8.3636 | 7.6667 | 6.5714 | 5.7500 | 5.1111 | 4.6000 | 4.1818 | 3.8333 | 3.2857 | 2.8750 |
| 93 | 9.3000 | 8.4545 | 7.7500 | 6.6429 | 5.8125 | 5.1667 | 4.6500 | 4.2273 | 3.8750 | 3.3214 | 2.9062 |
| 94 | 9.4000 | 8.5455 | 7.8333 | 6.7143 | 5.8750 | 5.2222 | 4.7000 | 4.2727 | 3.9167 | 3.3571 | 2.9375 |
| 95 | 9.5000 | 8.6364 | 7.9167 | 6.7857 | 5.9375 | 5.2778 | 4.7500 | 4.3182 | 3.9583 | 3.3929 | 2.9688 |
| 96 | 9.6000 | 8.7273 | 8.0000 | 6.8571 | 6.0000 | 5.3333 | 4.8000 | 4.3636 | 4.0000 | 3.4286 | 3.0000 |
| 97 | 9.7000 | 8.8182 | 8.0833 | 6.9286 | 6.0625 | 5.3889 | 4.8500 | 4.4091 | 4.0417 | 3.4643 | 3.0312 |
| 98 | 9.8000 | 8.9091 | 8.1667 | 7.0000 | 6.1250 | 5.4444 | 4.9000 | 4.4545 | 4.0833 | 3.5000 | 3.0625 |
| 99 | 9.9000 | 9.0000 | 8.2500 | 7.0714 | 6.1875 | 5.5000 | 4.9500 | 4.5000 | 4.1250 | 3.5357 | 3.0938 |
| 100 | 10.0000 | 9.0909 | 8.3333 | 7.1429 | 6.2500 | 5.5556 | 5.0000 | 4.5455 | 4.1667 | 3.5714 | 3.1250 |
| 101 | 10.1000 | 9.1818 | 8.4167 | 7.2143 | 6.3125 | 5.6111 | 5.0500 | 4.5909 | 4.2083 | 3.6071 | 3.1562 |
| 102 | 10.2000 | 9.2727 | 8.5000 | 7.2857 | 6.3750 | 5.6667 | 5.1000 | 4.6364 | 4.2500 | 3.6429 | 3.1875 |
| 103 | 10.3000 | 9.3636 | 8.5833 | 7.3571 | 6.4375 | 5.7222 | 5.1500 | 4.6818 | 4.2917 | 3.6786 | 3.2188 |
| 104 | 10.4000 | 9.4545 | 8.6667 | 7.4286 | 6.5000 | 5.7778 | 5.2000 | 4.7273 | 4.3333 | 3.7143 | 3.2500 |
| 105 | 10.5000 | 9.5455 | 8.7500 | 7.5000 | 6.5625 | 5.8333 | 5.2500 | 4.7727 | 4.3750 | 3.7500 | 3.2812 |
| 106 | 10.6000 | 9.6364 | 8.8333 | 7.5714 | 6.6250 | 5.8889 | 5.3000 | 4.8182 | 4.4167 | 3.7857 | 3.3125 |
| 107 | 10.7000 | 9.7273 | 8.9167 | 7.6429 | 6.6875 | 5.9444 | 5.3500 | 4.8636 | 4.4583 | 3.8214 | 3.3438 |
| 108 | 10.8000 | 9.8182 | 9.0000 | 7.7143 | 6.7500 | 6.0000 | 5.4000 | 4.9091 | 4.5000 | 3.8571 | 3.3750 |
| 109 | 10.9000 | 9.9091 | 9.0833 | 7.7857 | 6.8125 | 6.0556 | 5.4500 | 4.9545 | 4.5417 | 3.8929 | 3.4062 |
| 110 | 11.0000 | 10.0000 | 9.1667 | 7.8571 | 6.8750 | 6.1111 | 5.5000 | 5.0000 | 4.5833 | 3.9286 | 3.4375 |
| 111 | 11.1000 | 10.0909 | 9.2500 | 7.9286 | 6.9375 | 6.1667 | 5.5500 | 5.0455 | 4.6250 | 3.9643 | 3.4688 |
| 112 | 11.2000 | 10.1818 | 9.3333 | 8.0000 | 7.0000 | 6.2222 | 5.6000 | 5.0909 | 4.6667 | 4.0000 | 3.5000 |
| 113 | 11.3000 | 10.2727 | 9.4167 | 8.0714 | 7.0625 | 6.2778 | 5.6500 | 5.1364 | 4.7083 | 4.0357 | 3.5312 |
| 114 | 11.4000 | 10.3636 | 9.5000 | 8.1429 | 7.1250 | 6.3333 | 5.7000 | 5.1818 | 4.7500 | 4.0714 | 3.5625 |
| 115 | 11.5000 | 10.4545 | 9.5833 | 8.2143 | 7.1875 | 6.3889 | 5.7500 | 5.2273 | 4.7917 | 4.1071 | 3.5938 |
| 116 | 11.6000 | 10.5455 | 9.6667 | 8.2857 | 7.2500 | 6.4444 | 5.8000 | 5.2727 | 4.8333 | 4.1429 | 3.6250 |
| 117 | 11.7000 | 10.6364 | 9.7500 | 8.3571 | 7.3125 | 6.5000 | 5.8500 | 5.3182 | 4.8750 | 4.1786 | 3.6562 |
| 118 | 11.8000 | 10.7273 | 9.8333 | 8.4286 | 7.3750 | 6.5556 | 5.9000 | 5.3636 | 4.9167 | 4.2143 | 3.6875 |
| 119 | 11.9000 | 10.8182 | 9.9167 | 8.5000 | 7.4375 | 6.6111 | 5.9500 | 5.4091 | 4.9583 | 4.2500 | 3.7188 |
| 120 | 12.0000 | 10.9091 | 10.0000 | 8.5714 | 7.5000 | 6.6667 | 6.0000 | 5.4545 | 5.0000 | 4.2857 | 3.7500 |
| 121 | 12.1000 | 11.0000 | 10.0833 | 8.6429 | 7.5625 | 6.7222 | 6.0500 | 5.5000 | 5.0417 | 4.3214 | 3.7812 |
| 122 | 12.2000 | 11.0909 | 10.1667 | 8.7143 | 7.6250 | 6.7778 | 6.1000 | 5.5455 | 5.0833 | 4.3571 | 3.8125 |
| 123 | 12.3000 | 11.1818 | 10.2500 | 8.7857 | 7.6875 | 6.8333 | 6.1500 | 5.5909 | 5.1250 | 4.3929 | 3.8438 |
| 124 | 12.4000 | 11.2727 | 10.3333 | 8.8571 | 7.7500 | 6.8889 | 6.2000 | 5.6364 | 5.1667 | 4.4286 | 3.8750 |
| 125 | 12.5000 | 11.3636 | 10.4167 | 8.9286 | 7.8125 | 6.9444 | 6.2500 | 5.6818 | 5.2083 | 4.4643 | 3.9062 |
| 126 | 12.6000 | 11.4545 | 10.5000 | 9.0000 | 7.8750 | 7.0000 | 6.3000 | 5.7273 | 5.2500 | 4.5000 | 3.9375 |
| 127 | 12.7000 | 11.5455 | 10.5833 | 9.0714 | 7.9375 | 7.0556 | 6.3500 | 5.7727 | 5.2917 | 4.5357 | 3.9688 |
| 128 | 12.8000 | 11.6364 | 10.6667 | 9.1429 | 8.0000 | 7.1111 | 6.4000 | 5.8182 | 5.3333 | 4.5714 | 4.0000 |
| 129 | 12.9000 | 11.7273 | 10.7500 | 9.2143 | 8.0625 | 7.1667 | 6.4500 | 5.8636 | 5.3750 | 4.6071 | 4.0312 |
| 130 | 13.0000 | 11.8182 | 10.8333 | 9.2857 | 8.1250 | 7.2222 | 6.5000 | 5.9091 | 5.4167 | 4.6429 | 4.0625 |

Outside diameter of a standard full-depth gear equals the pitch diameter for a gear of the same diametral pitch but with 2 more teeth.

Table 11 (*Continued*). Pitch Diameters of Diametral Pitch Gears

| No. of Teeth | Diametral Pitch | | | | | | | | | | |
|---|---|---|---|---|---|---|---|---|---|---|---|
| | 36 | 40 | 44 | 48 | 56 | 64 | 72 | 80 | 96 | 120 | 128 |
| | Pitch Diameter | | | | | | | | | | |
| 10 | 0.2778 | 0.2500 | 0.2273 | 0.2083 | 0.1786 | 0.1562 | 0.1389 | 0.1250 | 0.1042 | 0.0833 | 0.0781 |
| 11 | 0.3056 | 0.2750 | 0.2500 | 0.2292 | 0.1964 | 0.1719 | 0.1528 | 0.1375 | 0.1146 | 0.0917 | 0.0859 |
| 12 | 0.3333 | 0.3000 | 0.2727 | 0.2500 | 0.2143 | 0.1875 | 0.1667 | 0.1500 | 0.1250 | 0.1000 | 0.0938 |
| 13 | 0.3611 | 0.3250 | 0.2955 | 0.2708 | 0.2321 | 0.2031 | 0.1806 | 0.1625 | 0.1354 | 0.1083 | 0.1016 |
| 14 | 0.3889 | 0.3500 | 0.3182 | 0.2917 | 0.2500 | 0.2188 | 0.1944 | 0.1750 | 0.1458 | 0.1167 | 0.1094 |
| 15 | 0.4167 | 0.3750 | 0.3409 | 0.3125 | 0.2679 | 0.2344 | 0.2083 | 0.1875 | 0.1562 | 0.1250 | 0.1172 |
| 16 | 0.4444 | 0.4000 | 0.3636 | 0.3333 | 0.2857 | 0.2500 | 0.2222 | 0.2000 | 0.1667 | 0.1333 | 0.1250 |
| 17 | 0.4722 | 0.4250 | 0.3864 | 0.3542 | 0.3036 | 0.2656 | 0.2361 | 0.2125 | 0.1771 | 0.1417 | 0.1328 |
| 18 | 0.5000 | 0.4500 | 0.4091 | 0.3750 | 0.3214 | 0.2812 | 0.2500 | 0.2250 | 0.1875 | 0.1500 | 0.1406 |
| 19 | 0.5278 | 0.4750 | 0.4318 | 0.3958 | 0.3393 | 0.2969 | 0.2639 | 0.2375 | 0.1979 | 0.1583 | 0.1484 |
| 20 | 0.5556 | 0.5000 | 0.4545 | 0.4167 | 0.3571 | 0.3125 | 0.2778 | 0.2500 | 0.2083 | 0.1667 | 0.1562 |
| 21 | 0.5833 | 0.5250 | 0.4773 | 0.4375 | 0.3750 | 0.3281 | 0.2917 | 0.2625 | 0.2187 | 0.1750 | 0.1641 |
| 22 | 0.6111 | 0.5500 | 0.5000 | 0.4583 | 0.3929 | 0.3438 | 0.3056 | 0.2750 | 0.2292 | 0.1833 | 0.1719 |
| 23 | 0.6389 | 0.5750 | 0.5227 | 0.4792 | 0.4107 | 0.3594 | 0.3194 | 0.2875 | 0.2396 | 0.1917 | 0.1797 |
| 24 | 0.6667 | 0.6000 | 0.5455 | 0.5000 | 0.4286 | 0.3750 | 0.3333 | 0.3000 | 0.2500 | 0.2000 | 0.1875 |
| 25 | 0.6944 | 0.6250 | 0.5682 | 0.5208 | 0.4464 | 0.3906 | 0.3472 | 0.3125 | 0.2604 | 0.2083 | 0.1953 |
| 26 | 0.7222 | 0.6500 | 0.5909 | 0.5417 | 0.4643 | 0.4062 | 0.3611 | 0.3250 | 0.2708 | 0.2167 | 0.2031 |
| 27 | 0.7500 | 0.6750 | 0.6136 | 0.5625 | 0.4821 | 0.4219 | 0.3750 | 0.3375 | 0.2812 | 0.2250 | 0.2109 |
| 28 | 0.7778 | 0.7000 | 0.6364 | 0.5833 | 0.5000 | 0.4375 | 0.3889 | 0.3500 | 0.2917 | 0.2333 | 0.2188 |
| 29 | 0.8056 | 0.7250 | 0.6591 | 0.6042 | 0.5179 | 0.4531 | 0.4028 | 0.3625 | 0.3021 | 0.2417 | 0.2266 |
| 30 | 0.8333 | 0.7500 | 0.6818 | 0.6250 | 0.5357 | 0.4688 | 0.4167 | 0.3750 | 0.3125 | 0.2500 | 0.2344 |
| 31 | 0.8611 | 0.7750 | 0.7045 | 0.6458 | 0.5536 | 0.4844 | 0.4306 | 0.3875 | 0.3229 | 0.2583 | 0.2422 |
| 32 | 0.8889 | 0.8000 | 0.7273 | 0.6667 | 0.5714 | 0.5000 | 0.4444 | 0.4000 | 0.3333 | 0.2667 | 0.2500 |
| 33 | 0.9167 | 0.8250 | 0.7500 | 0.6875 | 0.5893 | 0.5156 | 0.4583 | 0.4125 | 0.3437 | 0.2750 | 0.2578 |
| 34 | 0.9444 | 0.8500 | 0.7727 | 0.7083 | 0.6071 | 0.5312 | 0.4722 | 0.4250 | 0.3542 | 0.2833 | 0.2656 |
| 35 | 0.9722 | 0.8750 | 0.7955 | 0.7292 | 0.6250 | 0.5469 | 0.4861 | 0.4375 | 0.3646 | 0.2917 | 0.2734 |
| 36 | 1.0000 | 0.9000 | 0.8182 | 0.7500 | 0.6429 | 0.5625 | 0.5000 | 0.4500 | 0.3750 | 0.3000 | 0.2812 |
| 37 | 1.0278 | 0.9250 | 0.8409 | 0.7708 | 0.6607 | 0.5781 | 0.5139 | 0.4625 | 0.3854 | 0.3083 | 0.2891 |
| 38 | 1.0556 | 0.9500 | 0.8636 | 0.7917 | 0.6786 | 0.5938 | 0.5278 | 0.4750 | 0.3958 | 0.3167 | 0.2969 |
| 39 | 1.0833 | 0.9750 | 0.8864 | 0.8125 | 0.6964 | 0.6094 | 0.5417 | 0.4875 | 0.4062 | 0.3250 | 0.3047 |
| 40 | 1.1111 | 1.0000 | 0.9091 | 0.8333 | 0.7143 | 0.6250 | 0.5556 | 0.5000 | 0.4167 | 0.3333 | 0.3125 |
| 41 | 1.1389 | 1.0250 | 0.9318 | 0.8542 | 0.7321 | 0.6406 | 0.5694 | 0.5125 | 0.4271 | 0.3417 | 0.3203 |
| 42 | 1.1667 | 1.0500 | 0.9545 | 0.8750 | 0.7500 | 0.6562 | 0.5833 | 0.5250 | 0.4375 | 0.3500 | 0.3281 |
| 43 | 1.1944 | 1.0750 | 0.9773 | 0.8958 | 0.7679 | 0.6719 | 0.5972 | 0.5375 | 0.4479 | 0.3583 | 0.3359 |
| 44 | 1.2222 | 1.1000 | 1.0000 | 0.9167 | 0.7857 | 0.6875 | 0.6111 | 0.5500 | 0.4583 | 0.3667 | 0.3438 |
| 45 | 1.2500 | 1.1250 | 1.0227 | 0.9375 | 0.8036 | 0.7031 | 0.6250 | 0.5625 | 0.4687 | 0.3750 | 0.3516 |
| 46 | 1.2778 | 1.1500 | 1.0455 | 0.9583 | 0.8214 | 0.7188 | 0.6389 | 0.5750 | 0.4792 | 0.3833 | 0.3594 |
| 47 | 1.3056 | 1.1750 | 1.0682 | 0.9792 | 0.8393 | 0.7344 | 0.6528 | 0.5875 | 0.4896 | 0.3917 | 0.3672 |
| 48 | 1.3333 | 1.2000 | 1.0909 | 1.0000 | 0.8571 | 0.7500 | 0.6667 | 0.6000 | 0.5000 | 0.4000 | 0.3750 |
| 49 | 1.3611 | 1.2250 | 1.1136 | 1.0208 | 0.8750 | 0.7656 | 0.6806 | 0.6125 | 0.5104 | 0.4083 | 0.3828 |
| 50 | 1.3889 | 1.2500 | 1.1364 | 1.0417 | 0.8929 | 0.7812 | 0.6944 | 0.6250 | 0.5208 | 0.4167 | 0.3906 |
| 51 | 1.4167 | 1.2750 | 1.1591 | 1.0625 | 0.9107 | 0.7969 | 0.7083 | 0.6375 | 0.5312 | 0.4250 | 0.3984 |
| 52 | 1.4444 | 1.3000 | 1.1818 | 1.0833 | 0.9286 | 0.8125 | 0.7222 | 0.6500 | 0.5417 | 0.4333 | 0.4062 |
| 53 | 1.4722 | 1.3250 | 1.2045 | 1.1042 | 0.9464 | 0.8281 | 0.7361 | 0.6625 | 0.5521 | 0.4417 | 0.4141 |
| 54 | 1.5000 | 1.3500 | 1.2273 | 1.1250 | 0.9643 | 0.8438 | 0.7500 | 0.6750 | 0.5625 | 0.4500 | 0.4219 |
| 55 | 1.5278 | 1.3750 | 1.2500 | 1.1458 | 0.9821 | 0.8594 | 0.7639 | 0.6875 | 0.5729 | 0.4583 | 0.4297 |
| 56 | 1.5556 | 1.4000 | 1.2727 | 1.1667 | 1.0000 | 0.8750 | 0.7778 | 0.7000 | 0.5833 | 0.4667 | 0.4375 |
| 57 | 1.5833 | 1.4250 | 1.2955 | 1.1875 | 1.0179 | 0.8906 | 0.7917 | 0.7125 | 0.5937 | 0.4750 | 0.4453 |
| 58 | 1.6111 | 1.4500 | 1.3182 | 1.2083 | 1.0357 | 0.9062 | 0.8056 | 0.7250 | 0.6042 | 0.4833 | 0.4531 |
| 59 | 1.6389 | 1.4750 | 1.3409 | 1.2292 | 1.0536 | 0.9219 | 0.8194 | 0.7375 | 0.6146 | 0.4917 | 0.4609 |
| 60 | 1.6667 | 1.5000 | 1.3636 | 1.2500 | 1.0714 | 0.9375 | 0.8333 | 0.7500 | 0.6250 | 0.5000 | 0.4688 |
| 61 | 1.6944 | 1.5250 | 1.3864 | 1.2708 | 1.0893 | 0.9531 | 0.8472 | 0.7625 | 0.6354 | 0.5083 | 0.4766 |
| 62 | 1.7222 | 1.5500 | 1.4091 | 1.2917 | 1.1071 | 0.9688 | 0.8611 | 0.7750 | 0.6458 | 0.5167 | 0.4844 |
| 63 | 1.7500 | 1.5750 | 1.4318 | 1.3125 | 1.1250 | 0.9844 | 0.8750 | 0.7875 | 0.6562 | 0.5250 | 0.4922 |
| 64 | 1.7778 | 1.6000 | 1.4545 | 1.3333 | 1.1429 | 1.0000 | 0.8889 | 0.8000 | 0.6667 | 0.5333 | 0.5000 |
| 65 | 1.8056 | 1.6250 | 1.4773 | 1.3542 | 1.1607 | 1.0156 | 0.9028 | 0.8125 | 0.6771 | 0.5417 | 0.5078 |
| 66 | 1.8333 | 1.6500 | 1.5000 | 1.3750 | 1.1786 | 1.0312 | 0.9167 | 0.8250 | 0.6875 | 0.5500 | 0.5156 |
| 67 | 1.8611 | 1.6750 | 1.5227 | 1.3958 | 1.1964 | 1.0469 | 0.9306 | 0.8375 | 0.6979 | 0.5583 | 0.5234 |
| 68 | 1.8889 | 1.7000 | 1.5455 | 1.4167 | 1.2143 | 1.0625 | 0.9444 | 0.8500 | 0.7083 | 0.5667 | 0.5312 |
| 69 | 1.9167 | 1.7250 | 1.5682 | 1.4375 | 1.2321 | 1.0781 | 0.9583 | 0.8625 | 0.7187 | 0.5750 | 0.5391 |
| 70 | 1.9444 | 1.7500 | 1.5909 | 1.4583 | 1.2500 | 1.0938 | 0.9722 | 0.8750 | 0.7292 | 0.5833 | 0.5469 |

Outside diameter of a standard full-depth gear equals the pitch diameter for a gear of the same diametral pitch but with 2 more teeth.

Table II (*Concluded*). Pitch Diameters of Diametral Pitch Gears

| No. of Teeth | Diametral Pitch | | | | | | | | | | |
|---|---|---|---|---|---|---|---|---|---|---|---|
| | 36 | 40 | 44 | 48 | 56 | 64 | 72 | 80 | 96 | 120 | 128 |
| | Pitch Diameter | | | | | | | | | | |
| 71 | 1.9722 | 1.7750 | 1.6136 | 1.4792 | 1.2679 | 1.1094 | 0.9861 | 0.8875 | 0.7396 | 0.5917 | 0.5547 |
| 72 | 2.0000 | 1.8000 | 1.6364 | 1.5000 | 1.2857 | 1.1250 | 1.0000 | 0.9000 | 0.7500 | 0.6000 | 0.5625 |
| 73 | 2.0278 | 1.8250 | 1.6591 | 1.5208 | 1.3036 | 1.1406 | 1.0139 | 0.9125 | 0.7604 | 0.6083 | 0.5703 |
| 74 | 2.0556 | 1.8500 | 1.6818 | 1.5417 | 1.3214 | 1.1562 | 1.0278 | 0.9250 | 0.7708 | 0.6167 | 0.5781 |
| 75 | 2.0833 | 1.8750 | 1.7045 | 1.5625 | 1.3393 | 1.1719 | 1.0417 | 0.9375 | 0.7812 | 0.6250 | 0.5859 |
| 76 | 2.1111 | 1.9000 | 1.7273 | 1.5833 | 1.3571 | 1.1875 | 1.0556 | 0.9500 | 0.7917 | 0.6333 | 0.5938 |
| 77 | 2.1389 | 1.9250 | 1.7500 | 1.6042 | 1.3750 | 1.2031 | 1.0694 | 0.9625 | 0.8021 | 0.6417 | 0.6016 |
| 78 | 2.1667 | 1.9500 | 1.7727 | 1.6250 | 1.3929 | 1.2188 | 1.0833 | 0.9750 | 0.8125 | 0.6500 | 0.6094 |
| 79 | 2.1944 | 1.9750 | 1.7955 | 1.6458 | 1.4107 | 1.2344 | 1.0972 | 0.9875 | 0.8229 | 0.6583 | 0.6172 |
| 80 | 2.2222 | 2.0000 | 1.8182 | 1.6667 | 1.4286 | 1.2500 | 1.1111 | 1.0000 | 0.8333 | 0.6667 | 0.6250 |
| 81 | 2.2500 | 2.0250 | 1.8409 | 1.6875 | 1.4464 | 1.2656 | 1.1250 | 1.0125 | 0.8437 | 0.6750 | 0.6328 |
| 82 | 2.2778 | 2.0500 | 1.8636 | 1.7083 | 1.4643 | 1.2812 | 1.1389 | 1.0250 | 0.8542 | 0.6833 | 0.6406 |
| 83 | 2.3056 | 2.0750 | 1.8864 | 1.7292 | 1.4821 | 1.2969 | 1.1528 | 1.0375 | 0.8646 | 0.6917 | 0.6484 |
| 84 | 2.3333 | 2.1000 | 1.9091 | 1.7500 | 1.5000 | 1.3125 | 1.1667 | 1.0500 | 0.8750 | 0.7000 | 0.6562 |
| 85 | 2.3611 | 2.1250 | 1.9318 | 1.7708 | 1.5179 | 1.3281 | 1.1806 | 1.0625 | 0.8854 | 0.7083 | 0.6641 |
| 86 | 2.3889 | 2.1500 | 1.9545 | 1.7917 | 1.5357 | 1.3438 | 1.1944 | 1.0750 | 0.8958 | 0.7167 | 0.6719 |
| 87 | 2.4167 | 2.1750 | 1.9773 | 1.8125 | 1.5536 | 1.3594 | 1.2083 | 1.0875 | 0.9062 | 0.7250 | 0.6797 |
| 88 | 2.4444 | 2.2000 | 2.0000 | 1.8333 | 1.5714 | 1.3750 | 1.2222 | 1.1000 | 0.9167 | 0.7333 | 0.6875 |
| 89 | 2.4722 | 2.2250 | 2.0227 | 1.8542 | 1.5893 | 1.3906 | 1.2361 | 1.1125 | 0.9271 | 0.7417 | 0.6953 |
| 90 | 2.5000 | 2.2500 | 2.0455 | 1.8750 | 1.6071 | 1.4062 | 1.2500 | 1.1250 | 0.9375 | 0.7500 | 0.7031 |
| 91 | 2.5278 | 2.2750 | 2.0682 | 1.8958 | 1.6250 | 1.4219 | 1.2639 | 1.1375 | 0.9479 | 0.7583 | 0.7109 |
| 92 | 2.5556 | 2.3000 | 2.0909 | 1.9167 | 1.6429 | 1.4375 | 1.2778 | 1.1500 | 0.9583 | 0.7667 | 0.7188 |
| 93 | 2.5833 | 2.3250 | 2.1136 | 1.9375 | 1.6607 | 1.4531 | 1.2917 | 1.1625 | 0.9687 | 0.7750 | 0.7266 |
| 94 | 2.6111 | 2.3500 | 2.1364 | 1.9583 | 1.6786 | 1.4688 | 1.3056 | 1.1750 | 0.9792 | 0.7833 | 0.7344 |
| 95 | 2.6389 | 2.3750 | 2.1591 | 1.9792 | 1.6964 | 1.4844 | 1.3194 | 1.1875 | 0.9896 | 0.7917 | 0.7422 |
| 96 | 2.6667 | 2.4000 | 2.1818 | 2.0000 | 1.7143 | 1.5000 | 1.3333 | 1.2000 | 1.0000 | 0.8000 | 0.7500 |
| 97 | 2.6944 | 2.4250 | 2.2045 | 2.0208 | 1.7321 | 1.5156 | 1.3472 | 1.2125 | 1.0104 | 0.8083 | 0.7578 |
| 98 | 2.7222 | 2.4500 | 2.2273 | 2.0417 | 1.7500 | 1.5312 | 1.3611 | 1.2250 | 1.0208 | 0.8167 | 0.7656 |
| 99 | 2.7500 | 2.4750 | 2.2500 | 2.0625 | 1.7679 | 1.5469 | 1.3750 | 1.2375 | 1.0312 | 0.8250 | 0.7734 |
| 100 | 2.7778 | 2.5000 | 2.2727 | 2.0833 | 1.7857 | 1.5625 | 1.3889 | 1.2500 | 1.0417 | 0.8333 | 0.7812 |
| 101 | 2.8056 | 2.5250 | 2.2955 | 2.1042 | 1.8036 | 1.5781 | 1.4028 | 1.2625 | 1.0521 | 0.8417 | 0.7891 |
| 102 | 2.8333 | 2.5500 | 2.3182 | 2.1250 | 1.8214 | 1.5938 | 1.4167 | 1.2750 | 1.0625 | 0.8500 | 0.7969 |
| 103 | 2.8611 | 2.5750 | 2.3409 | 2.1458 | 1.8393 | 1.6094 | 1.4306 | 1.2875 | 1.0729 | 0.8583 | 0.8047 |
| 104 | 2.8889 | 2.6000 | 2.3636 | 2.1667 | 1.8571 | 1.6250 | 1.4444 | 1.3000 | 1.0833 | 0.8667 | 0.8125 |
| 105 | 2.9167 | 2.6250 | 2.3864 | 2.1875 | 1.8750 | 1.6406 | 1.4583 | 1.3125 | 1.0937 | 0.8750 | 0.8203 |
| 106 | 2.9444 | 2.6500 | 2.4091 | 2.2083 | 1.8929 | 1.6562 | 1.4722 | 1.3250 | 1.1042 | 0.8833 | 0.8281 |
| 107 | 2.9722 | 2.6750 | 2.4318 | 2.2292 | 1.9107 | 1.6719 | 1.4861 | 1.3375 | 1.1146 | 0.8917 | 0.8359 |
| 108 | 3.0000 | 2.7000 | 2.4545 | 2.2500 | 1.9286 | 1.6875 | 1.5000 | 1.3500 | 1.1250 | 0.9000 | 0.8438 |
| 109 | 3.0278 | 2.7250 | 2.4773 | 2.2708 | 1.9464 | 1.7031 | 1.5139 | 1.3625 | 1.1354 | 0.9083 | 0.8516 |
| 110 | 3.0556 | 2.7500 | 2.5000 | 2.2917 | 1.9643 | 1.7188 | 1.5278 | 1.3750 | 1.1458 | 0.9167 | 0.8594 |
| 111 | 3.0833 | 2.7750 | 2.5227 | 2.3125 | 1.9821 | 1.7344 | 1.5417 | 1.3875 | 1.1562 | 0.9250 | 0.8672 |
| 112 | 3.1111 | 2.8000 | 2.5455 | 2.3333 | 2.0000 | 1.7500 | 1.5556 | 1.4000 | 1.1667 | 0.9333 | 0.8750 |
| 113 | 3.1389 | 2.8250 | 2.5682 | 2.3542 | 2.0179 | 1.7656 | 1.5694 | 1.4125 | 1.1771 | 0.9417 | 0.8828 |
| 114 | 3.1667 | 2.8500 | 2.5909 | 2.3750 | 2.0357 | 1.7812 | 1.5833 | 1.4250 | 1.1875 | 0.9500 | 0.8906 |
| 115 | 3.1944 | 2.8750 | 2.6136 | 2.3958 | 2.0536 | 1.7969 | 1.5972 | 1.4375 | 1.1979 | 0.9583 | 0.8984 |
| 116 | 3.2222 | 2.9000 | 2.6364 | 2.4167 | 2.0714 | 1.8125 | 1.6111 | 1.4500 | 1.2083 | 0.9667 | 0.9062 |
| 117 | 3.2500 | 2.9250 | 2.6591 | 2.4375 | 2.0893 | 1.8281 | 1.6250 | 1.4625 | 1.2187 | 0.9750 | 0.9141 |
| 118 | 3.2778 | 2.9500 | 2.6818 | 2.4583 | 2.1071 | 1.8438 | 1.6389 | 1.4750 | 1.2292 | 0.9833 | 0.9219 |
| 119 | 3.3056 | 2.9750 | 2.7045 | 2.4792 | 2.1250 | 1.8594 | 1.6528 | 1.4875 | 1.2396 | 0.9917 | 0.9297 |
| 120 | 3.3333 | 3.0000 | 2.7273 | 2.5000 | 2.1429 | 1.8750 | 1.6667 | 1.5000 | 1.2500 | 1.0000 | 0.9375 |
| 121 | 3.3611 | 3.0250 | 2.7500 | 2.5208 | 2.1607 | 1.8906 | 1.6806 | 1.5125 | 1.2604 | 1.0083 | 0.9453 |
| 122 | 3.3889 | 3.0500 | 2.7727 | 2.5417 | 2.1786 | 1.9062 | 1.6944 | 1.5250 | 1.2708 | 1.0167 | 0.9531 |
| 123 | 3.4167 | 3.0750 | 2.7955 | 2.5625 | 2.1964 | 1.9219 | 1.7083 | 1.5375 | 1.2812 | 1.0250 | 0.9609 |
| 124 | 3.4444 | 3.1000 | 2.8182 | 2.5833 | 2.2143 | 1.9375 | 1.7222 | 1.5500 | 1.2917 | 1.0333 | 0.9688 |
| 125 | 3.4722 | 3.1250 | 2.8409 | 2.6042 | 2.2321 | 1.9531 | 1.7361 | 1.5625 | 1.3021 | 1.0417 | 0.9766 |
| 126 | 3.5000 | 3.1500 | 2.8636 | 2.6250 | 2.2500 | 1.9688 | 1.7500 | 1.5750 | 1.3125 | 1.0500 | 0.9844 |
| 127 | 3.5278 | 3.1750 | 2.8864 | 2.6458 | 2.2679 | 1.9844 | 1.7639 | 1.5875 | 1.3229 | 1.0583 | 0.9922 |
| 128 | 3.5556 | 3.2000 | 2.9091 | 2.6667 | 2.2857 | 2.0000 | 1.7778 | 1.6000 | 1.3333 | 1.0667 | 1.0000 |
| 129 | 3.5833 | 3.2250 | 2.9318 | 2.6875 | 2.3036 | 2.0156 | 1.7917 | 1.6125 | 1.3437 | 1.0750 | 1.0078 |
| 130 | 3.6111 | 3.2500 | 2.9545 | 2.7083 | 2.3214 | 2.0312 | 1.8056 | 1.6250 | 1.3542 | 1.0833 | 1.0156 |

Outside diameter of a standard full-depth gear equals the pitch diameter for a gear of the same diametral pitch but with 2 more teeth.

**Basic Gear Dimensions.** — The basic dimensions for all involute spur gears may be obtained using the formulas shown in Table 1. This table is used in conjunction with Table 3 to obtain dimensions for coarse pitch gears and Table 5 to obtain dimensions for fine pitch standard spur gears. To obtain the dimensions of gears that are specified at a standard circular pitch, the equivalent diametral pitch is first calculated by using the formula in Table 1. If the required number of teeth in the pinion ($N_P$) is less than the minimum specified in either Table 3 or Table 5, whichever is applicable, the gears must be proportioned by the long and short addendum method shown on page 749.

**Gear Set Center Distance.** — For any set of gears that operate on parallel shafts, the overall size of the gear set is essentially determined by the center distance. This center distance $C$ is a function of the pinion pitch diameter $D_P$ and the gear ratio $m_G$ by this relation:

$$C = \frac{D_P (m_G + 1)}{2}$$

On page 716 it is indicated that the pitch diameter will change in steps; using the above relation it is seen that the center distance will also change in steps. The following Table 12, "Gear Set Center Distance" is given to assist in preliminary gear set sizing. It gives center distances to three significant figures for pinion pitch diameters ranging from one-half to 24 inches and for a gear set ratio of from 1.25 to 10. This table must not be used to obtain the final center distance; for this determination the appropriate formula in Table 1 must be used. This table may also be used to estimate the required size of the space within a housing that is needed for the gear set. The length of this space is:

$$2(C + a)$$

where $a$ = the gear addendum. The width of this space is the large gear outside diameter which may be estimated as:

$$D_P m_G + 2a$$

**Gears for Given Center Distance and Ratio.** — When it is necessary to use a pair of gears of given ratio at a specified center distance $C_1$, it may be found that no gears of standard diametral pitch will satisfy the center distance requirement. In this case, gears of standard diametral pitch $P$ may be redesigned to operate at other than their standard pitch diameter $D$ and standard pressure angle $\phi$. The diametral pitch $P_1$ at which these gears will operate is:

$$P_1 = \frac{N_P + N_G}{2C_1} \tag{1}$$

where $N_P$ = number of teeth in pinion
$N_G$ = number of teeth in gear

and their operating pressure angle $\phi_1$ is:

$$\phi_1 = \arccos\left(\frac{P_1}{P}\cos\phi\right) \tag{2}$$

Thus the pair of gears are cut to a diametral pitch $P$ and a pressure angle $\phi$, however, they operate as standard gears of diametral pitch $P_1$ and pressure angle $\phi_1$.

Table 12. Gear Set Center Distance

| Pitch Dia. | Ratio | | | | | | | | | |
|---|---|---|---|---|---|---|---|---|---|---|
| | 1.25 | 1.5 | 1.75 | 2. | 2.25 | 2.5 | 2.75 | 3. | 3.5 | 4. |
| | Center Distance, Inches | | | | | | | | | |
| 0.5 | 0.563 | 0.625 | 0.688 | 0.75 | 0.813 | 0.875 | 0.938 | 1. | 1.13 | 1.25 |
| 0.6 | 0.675 | 0.75 | 0.825 | 0.9 | 0.975 | 1.05 | 1.12 | 1.2 | 1.35 | 1.5 |
| 0.7 | 0.787 | 0.875 | 0.962 | 1.05 | 1.14 | 1.22 | 1.31 | 1.4 | 1.57 | 1.75 |
| 0.8 | 0.9 | 1. | 1.1 | 1.2 | 1.3 | 1.4 | 1.5 | 1.6 | 1.8 | 2. |
| 0.9 | 1.01 | 1.12 | 1.24 | 1.35 | 1.46 | 1.57 | 1.69 | 1.8 | 2.02 | 2.25 |
| 1. | 1.12 | 1.25 | 1.37 | 1.5 | 1.62 | 1.75 | 1.87 | 2. | 2.25 | 2.5 |
| 1.1 | 1.24 | 1.37 | 1.51 | 1.65 | 1.79 | 1.92 | 2.06 | 2.2 | 2.47 | 2.75 |
| 1.2 | 1.35 | 1.6 | 1.65 | 1.8 | 1.95 | 2.1 | 2.25 | 2.4 | 2.7 | 3. |
| 1.3 | 1.46 | 1.62 | 1.79 | 1.95 | 2.11 | 2.27 | 2.44 | 2.6 | 2.92 | 3.25 |
| 1.4 | 1.57 | 1.75 | 1.92 | 2.1 | 2.27 | 2.45 | 2.62 | 2.8 | 3.15 | 3.5 |
| 1.5 | 1.69 | 1.87 | 2.06 | 2.25 | 2.44 | 2.62 | 2.81 | 3. | 3.37 | 3.75 |
| 1.6 | 1.8 | 2. | 2.2 | 2.4 | 2.6 | 2.8 | 3. | 3.2 | 3.6 | 4. |
| 1.7 | 1.91 | 2.12 | 2.34 | 2.55 | 2.76 | 2.97 | 3.19 | 3.4 | 3.82 | 4.25 |
| 1.8 | 2.02 | 2.25 | 2.47 | 2.7 | 2.92 | 3.15 | 3.37 | 3.6 | 4.05 | 4.5 |
| 1.9 | 2.14 | 2.37 | 2.61 | 2.85 | 3.09 | 3.32 | 3.56 | 3.8 | 4.27 | 4.75 |
| 2. | 2.25 | 2.5 | 2.75 | 3. | 3.25 | 3.5 | 3.75 | 4. | 4.5 | 5. |
| 2.25 | 2.53 | 2.81 | 3.09 | 3.38 | 3.66 | 3.94 | 4.22 | 4.5 | 5.06 | 5.63 |
| 2.5 | 2.81 | 3.13 | 3.44 | 3.75 | 4.06 | 4.38 | 4.69 | 5. | 5.63 | 6.25 |
| 2.75 | 3.09 | 3.44 | 3.78 | 4.13 | 4.47 | 4.81 | 5.16 | 5.5 | 6.19 | 6.88 |
| 3. | 3.38 | 3.75 | 4.13 | 4.5 | 4.88 | 5.25 | 5.63 | 6. | 6.75 | 7.5 |
| 3.25 | 3.66 | 4.06 | 4.47 | 4.88 | 5.28 | 5.69 | 6.09 | 6.5 | 7.31 | 8.13 |
| 3.5 | 3.94 | 4.38 | 4.81 | 5.25 | 5.69 | 6.13 | 6.56 | 7. | 7.88 | 8.75 |
| 3.75 | 4.22 | 4.69 | 5.16 | 5.63 | 6.09 | 6.56 | 7.03 | 7.5 | 8.44 | 9.38 |
| 4. | 4.5 | 5. | 5.5 | 6. | 6.5 | 7. | 7.5 | 8. | 9. | 10. |
| 4.25 | 4.78 | 5.31 | 5.84 | 6.38 | 6.91 | 7.44 | 7.97 | 8.5 | 9.56 | 10.6 |
| 4.5 | 5.06 | 5.63 | 6.19 | 6.75 | 7.31 | 7.88 | 8.44 | 9. | 10.1 | 11.3 |
| 4.75 | 5.34 | 5.94 | 6.53 | 7.13 | 7.72 | 8.31 | 8.91 | 9.5 | 10.7 | 11.9 |
| 5. | 5.63 | 6.25 | 6.88 | 7.5 | 8.13 | 8.75 | 9.38 | 10. | 11.3 | 12.5 |
| 5.25 | 5.91 | 6.56 | 7.22 | 7.88 | 8.53 | 9.19 | 9.84 | 10.5 | 11.8 | 13.1 |
| 5.5 | 6.19 | 6.88 | 7.56 | 8.25 | 8.94 | 9.63 | 10.3 | 11. | 12.4 | 13.8 |
| 5.75 | 6.47 | 7.19 | 7.91 | 8.63 | 9.34 | 10.1 | 10.8 | 11.5 | 12.9 | 14.4 |
| 6. | 6.75 | 7.5 | 8.25 | 9. | 9.75 | 10.5 | 11.3 | 12. | 13.5 | 15. |
| 6.5 | 7.31 | 8.13 | 8.94 | 9.75 | 10.6 | 11.4 | 12.2 | 13. | 14.6 | 16.3 |
| 7. | 7.88 | 8.75 | 9.63 | 10.5 | 11.4 | 12.3 | 13.1 | 14. | 15.8 | 17.5 |
| 7.5 | 8.44 | 9.38 | 10.3 | 11.3 | 12.2 | 13.1 | 14.1 | 15. | 16.9 | 18.8 |
| 8. | 9. | 10. | 11. | 12. | 13. | 14. | 15. | 16. | 18. | 20. |
| 8.5 | 9.56 | 10.6 | 11.7 | 12.8 | 13.8 | 14.9 | 15.9 | 17. | 19.1 | 21.3 |
| 9. | 10.1 | 11.3 | 12.4 | 13.5 | 14.6 | 15.8 | 16.9 | 18. | 20.3 | 22.5 |
| 9.5 | 10.7 | 11.9 | 13.1 | 14.3 | 15.4 | 16.6 | 17.8 | 19. | 21.4 | 23.8 |
| 10. | 11.3 | 12.5 | 13.8 | 15. | 16.3 | 17.5 | 18.8 | 20. | 22.5 | 25. |
| 10.5 | 11.8 | 13.1 | 14.4 | 15.8 | 17.1 | 18.4 | 19.7 | 21. | 23.6 | 26.3 |
| 11. | 12.4 | 13.8 | 15.1 | 16.5 | 17.9 | 19.3 | 20.6 | 22. | 24.8 | 27.5 |
| 11.5 | 12.9 | 14.4 | 15.8 | 17.3 | 18.7 | 20.1 | 21.6 | 23. | 25.9 | 28.8 |
| 12. | 13.5 | 15. | 16.5 | 18. | 19.5 | 21. | 22.5 | 24. | 27. | 30. |
| 13. | 14.6 | 16.3 | 17.9 | 19.5 | 21.1 | 22.8 | 24.4 | 26. | 29.3 | 32.5 |
| 14. | 15.8 | 17.5 | 19.3 | 21. | 22.8 | 24.5 | 26.3 | 28. | 31.5 | 35. |
| 15. | 16.9 | 18.8 | 20.6 | 22.5 | 24.4 | 26.3 | 28.1 | 30. | 33.8 | 37.5 |
| 16. | 18. | 20. | 22. | 24. | 26. | 28. | 30. | 32. | 36. | 40. |
| 17. | 19.1 | 21.3 | 23.4 | 25.5 | 27.6 | 29.8 | 31.9 | 34. | 38.3 | 42.5 |
| 18. | 20.3 | 22.5 | 24.8 | 27. | 29.3 | 31.5 | 33.8 | 36. | 40.5 | 45. |
| 19. | 21.4 | 23.8 | 26.1 | 28.5 | 30.9 | 33.3 | 35.6 | 38. | 42.8 | 47.5 |
| 20. | 22.5 | 25. | 27.5 | 30. | 32.5 | 35. | 37.5 | 40. | 45. | 50. |
| 21. | 23.6 | 26.3 | 28.9 | 31.5 | 34.1 | 36.8 | 39.4 | 42. | 47.3 | 52.5 |
| 22. | 24.8 | 27.5 | 30.3 | 33. | 35.8 | 38.5 | 41.3 | 44. | 49.5 | 55. |
| 23. | 25.9 | 28.8 | 31.6 | 34.5 | 37.4 | 40.3 | 43.1 | 46. | 51.8 | 57.5 |
| 24. | 27. | 30. | 33. | 36. | 39. | 42. | 45. | 48. | 54. | 60. |

Table 12 (*Continued*).    Gear Set Center Distance

| Pitch Dia. | Ratio | | | | | | | | | |
|---|---|---|---|---|---|---|---|---|---|---|
| | 4.5 | 5. | 5.5 | 6. | 6.5 | 7. | 7.5 | 8. | 9. | 10. |
| | Center Distance, Inches | | | | | | | | | |
| 0.5 | 1.38 | 1.5 | 1.63 | 1.75 | 1.88 | 2. | 2.13 | 2.25 | 2.5 | 2.75 |
| 0.6 | 1.65 | 1.8 | 1.95 | 2.1 | 2.25 | 2.4 | 2.55 | 2.7 | 3. | 3.3 |
| 0.7 | 1.92 | 2.1 | 2.27 | 2.45 | 2.62 | 2.8 | 2.97 | 3.15 | 3.5 | 3.85 |
| 0.8 | 2.2 | 2.4 | 2.6 | 2.8 | 3. | 3.2 | 3.4 | 3.6 | 4. | 4.4 |
| 0.9 | 2.47 | 2.7 | 2.92 | 3.15 | 3.37 | 3.6 | 3.82 | 4.05 | 4.5 | 4.95 |
| 1. | 2.75 | 3. | 3.25 | 3.5 | 3.75 | 4. | 4.25 | 4.5 | 5. | 5.5 |
| 1.1 | 3.02 | 3.3 | 3.57 | 3.85 | 4.12 | 4.4 | 4.67 | 4.95 | 5.5 | 6.05 |
| 1.2 | 3.3 | 3.6 | 3.9 | 4.2 | 4.5 | 4.8 | 5.1 | 5.4 | 6. | 6.6 |
| 1.3 | 3.57 | 3.9 | 4.22 | 4.55 | 4.87 | 5.2 | 5.52 | 5.85 | 6.5 | 7.15 |
| 1.4 | 3.85 | 4.2 | 4.55 | 4.9 | 5.25 | 5.6 | 5.95 | 6.3 | 7. | 7.7 |
| 1.5 | 4.12 | 4.5 | 4.87 | 5.25 | 5.62 | 6. | 6.37 | 6.75 | 7.5 | 8.25 |
| 1.6 | 4.4 | 4.8 | 5.2 | 5.6 | 6. | 6.4 | 6.8 | 7.2 | 8. | 8.8 |
| 1.7 | 4.67 | 5.1 | 5.52 | 5.95 | 6.37 | 6.8 | 7.22 | 7.65 | 8.5 | 9.35 |
| 1.8 | 4.95 | 5.4 | 5.85 | 6.3 | 6.75 | 7.2 | 7.65 | 8.1 | 9. | 9.9 |
| 1.9 | 5.22 | 5.7 | 6.17 | 6.65 | 7.12 | 7.6 | 8.07 | 8.55 | 9.5 | 10.4 |
| 2. | 5.5 | 6. | 6.5 | 7. | 7.5 | 8. | 8.5 | 9. | 10. | 11. |
| 2.25 | 6.19 | 6.75 | 7.31 | 7.88 | 8.44 | 9. | 9.56 | 10.1 | 11.3 | 12.4 |
| 2.5 | 6.88 | 7.5 | 8.13 | 8.75 | 9.38 | 10. | 10.6 | 11.3 | 12.5 | 13.8 |
| 2.75 | 7.56 | 8.25 | 8.94 | 9.63 | 10.3 | 11. | 11.7 | 12.4 | 13.8 | 15.1 |
| 3. | 8.25 | 9. | 9.75 | 10.5 | 11.3 | 12. | 12.8 | 13.5 | 15. | 16.5 |
| 3.25 | 8.94 | 9.75 | 10.6 | 11.4 | 12.2 | 13. | 13.8 | 14.6 | 16.3 | 17.9 |
| 3.5 | 9.63 | 10.5 | 11.4 | 12.3 | 13.1 | 14. | 14.9 | 15.8 | 17.5 | 19.3 |
| 3.75 | 10.3 | 11.3 | 12.2 | 13.1 | 14.1 | 15. | 15.9 | 16.9 | 18.8 | 20.6 |
| 4. | 11. | 12. | 13. | 14. | 15. | 16. | 17. | 18. | 20. | 22. |
| 4.25 | 11.7 | 12.8 | 13.8 | 14.9 | 15.9 | 17. | 18.1 | 19.1 | 21.3 | 23.4 |
| 4.5 | 12.4 | 13.5 | 14.6 | 15.8 | 16.9 | 18. | 19.1 | 20.3 | 22.5 | 24.8 |
| 4.75 | 13.1 | 14.3 | 15.4 | 16.6 | 17.8 | 19. | 20.2 | 21.4 | 23.8 | 26.1 |
| 5. | 13.8 | 15. | 16.3 | 17.5 | 18.8 | 20. | 21.3 | 22.5 | 25. | 27.5 |
| 5.25 | 14.4 | 15.8 | 17.1 | 18.4 | 19.7 | 21. | 22.3 | 23.6 | 26.3 | 28.9 |
| 5.5 | 15.1 | 16.5 | 17.9 | 19.3 | 20.6 | 22. | 23.4 | 24.8 | 27.5 | 30.3 |
| 5.75 | 15.8 | 17.3 | 18.7 | 20.1 | 21.6 | 23. | 24.4 | 25.9 | 28.8 | 31.6 |
| 6. | 16.5 | 18. | 19.5 | 21. | 22.5 | 24. | 25.5 | 27. | 30. | 33. |
| 6.5 | 17.9 | 19.5 | 21.1 | 22.8 | 24.4 | 26. | 27.6 | 29.3 | 32.5 | 35.8 |
| 7. | 19.3 | 21. | 22.8 | 24.5 | 26.3 | 28. | 29.8 | 31.5 | 35. | 38.5 |
| 7.5 | 20.6 | 22.5 | 24.4 | 26.3 | 28.1 | 30. | 31.9 | 33.8 | 37.5 | 41.3 |
| 8. | 22. | 24. | 26. | 28. | 30. | 32. | 34. | 36. | 40. | 44. |
| 8.5 | 23.4 | 25.5 | 27.6 | 29.8 | 31.9 | 34. | 36.1 | 38.3 | 42.5 | 46.8 |
| 9. | 24.8 | 27. | 29.3 | 31.5 | 33.8 | 36. | 38.3 | 40.5 | 45. | 49.5 |
| 9.5 | 26.1 | 28.5 | 30.9 | 33.3 | 35.6 | 38. | 40.4 | 42.8 | 47.5 | 52.3 |
| 10. | 27.5 | 30. | 32.5 | 35. | 37.5 | 40. | 42.5 | 45. | 50. | 55. |
| 10.5 | 28.9 | 31.5 | 34.1 | 36.8 | 39.4 | 42. | 44.6 | 47.3 | 52.5 | 57.8 |
| 11. | 30.3 | 33. | 35.8 | 38.5 | 41.3 | 44. | 46.8 | 49.5 | 55. | 60.5 |
| 11.5 | 31.6 | 34.5 | 37.4 | 40.3 | 43.1 | 46. | 48.9 | 51.8 | 57.5 | 63.3 |
| 12. | 33. | 36. | 39. | 42. | 45. | 48. | 51. | 54. | 60. | 66. |
| 13. | 35.8 | 39. | 42.3 | 45.5 | 48.8 | 52. | 55.3 | 58.5 | 65. | 71.5 |
| 14. | 38.5 | 42. | 45.5 | 49. | 52.5 | 56. | 59.5 | 63. | 70. | 77. |
| 15. | 41.3 | 45. | 48.8 | 52.5 | 56.3 | 60. | 63.8 | 67.5 | 75. | 82.5 |
| 16. | 44. | 48. | 52. | 56. | 60. | 64. | 68. | 72. | 80. | 88. |
| 17. | 46.8 | 51. | 55.3 | 59.5 | 63.8 | 68. | 72.3 | 76.5 | 85. | 93.5 |
| 18. | 49.5 | 54. | 58.5 | 63. | 67.5 | 72. | 76.5 | 81. | 90. | 99. |
| 19. | 52.3 | 57. | 61.8 | 66.5 | 71.3 | 76. | 80.8 | 85.5 | 95. | 105. |
| 20. | 55. | 60. | 65. | 70. | 75. | 80. | 85. | 90. | 100. | 110. |
| 21. | 57.8 | 63. | 68.3 | 73.5 | 78.8 | 84. | 89.3 | 94.5 | 105. | 116. |
| 22. | 60.5 | 66. | 71.5 | 77. | 82.5 | 88. | 93.5 | 99. | 110. | 121. |
| 23. | 63.3 | 69. | 74.8 | 80.5 | 86.3 | 92. | 97.8 | 104. | 115. | 127. |
| 24. | 66. | 72. | 78. | 84. | 90. | 96. | 102. | 108. | 120. | 132. |

The pitch $P$ and pressure angle $\phi$ should be chosen so that $\phi_1$ lies between about 18 and 25 degrees.

The operating pitch diameters of the pinion $D_{P_1}$ and of the gear $D_{G_1}$ are:

$$D_{P_1} = \frac{N_P}{P_1} \text{ and } D_{G_1} = \frac{N_G}{P_1}$$ (3a) and (3b)

The base diameters of the pinion $D_{PB_1}$ and of the gear $D_{GB_1}$ are:

$$D_{PB_1} = D_{P_1}\cos\phi_1 \text{ and } D_{GB_1} = D_{G_1}\cos\phi_1$$ (4a) and (4b)

The basic tooth thickness $t_1$, at the operating pitch diameter for both pinion and gear is:

$$t_1 = \frac{1.5708}{P_1}$$ (5)

The root diameters of the pinion $D_{PR_1}$ and gear $D_{GR_1}$ and the corresponding outside diameters $D_{PO_1}$ and $D_{GO_1}$ are not standard because each gear is to be cut with a cutter that is not standard for the operating pitch diameters $D_{P_1}$ and $D_{G_1}$.

The root diameters are:

$$D_{PR_1} = \frac{N_P}{P} - 2b_{P_1} \text{ and } D_{GR_1} = \frac{N_G}{P} = 2b_{G_1}$$ (6a) and (6b)

where: $$b_{P_1} = b_c - \left(\frac{t_{P_2} - \frac{1.5708}{P}}{2\tan\phi}\right)$$ (7a)

and $$b_{G_1} = b_c - \frac{t_{G_2} - \frac{1.5708}{P}}{2\tan\phi}$$ (7b)

where $b_c$ is the hob or cutter addendum for the pinion and gear.

The tooth thicknesses of the pinion $t_{P_2}$ and the gear $t_{G_2}$ are:

$$t_{P_2} = \frac{N_P}{P}\left(\frac{1.5708}{N_P} + \text{inv}\,\phi_1 - \text{inv}\,\phi\right)$$ (8a)

$$t_{G_2} = \frac{N_G}{P}\left(\frac{1.5708}{N_G} + \text{inv}\,\phi_1 - \text{inv}\,\phi\right)$$ (8b)

The outside diameter of the pinion $D_{PO}$ and the gear $D_{GO}$ are:

$$D_{PO} = 2 \times C_1 - D_{GR_1} - 2\left(b_c - \frac{1}{P}\right)$$ (9a)

and $$D_{GO} = 2 \times C_1 = D_{PR_1} - 2\left(b_c - \frac{1}{P}\right)$$ (9b)

*Example:* Design gears of 8 diametral pitch, 20-degree pressure angle and 28 and 88 teeth to operate at 7.50-inch center distance. The gears are to be cut with a hob of 0.169-inch addendum.

$$P_1 = \frac{28 + 88}{2 \times 7.50} = 7.7333$$ (1)

$$\phi_1 = \arc\cos\left(\frac{7.7333}{8} \times 0.93969\right) = 24.719° \tag{2}$$

$$D_{P_1} = \frac{28}{7.7333} = 3.6207 \text{ in.} \tag{3a}$$

and

$$D_{G_1} = \frac{88}{7.7333} = 11.3794 \text{ in.} \tag{3b}$$

$$D_{PB_1} = 3.6207 \times 0.90837 = 3.2889 \text{ in.} \tag{4a}$$

and

$$D_{GB_1} = 11.3794 \times 0.90837 = 10.3367 \text{ in.} \tag{4b}$$

$$t_1 = \frac{1.5708}{7.7333} = 0.20312 \text{ in.} \tag{5}$$

$$D_{PR_1} = \frac{28}{8} - 2 \times 0.1016 = 3.2968 \text{ in.} \tag{6a}$$

and

$$D_{GR_1} = \frac{88}{8} - 2 \times (-0.0428) = 11.0856 \text{ in.} \tag{6b}$$

$$b_{P_1} = 0.169 - \left(\frac{0.2454 - \dfrac{1.5708}{8}}{2 \times 0.36397}\right) = 0.1016 \text{ in.} \tag{7a}$$

$$b_{G_1} = 0.169 - \left(\frac{0.3505 - \dfrac{1.5708}{8}}{2 \times 0.36397}\right) = -0.0428 \text{ in.} \tag{7b}$$

$$t_{P_2} = \frac{28}{8}\left(\frac{1.5708}{28} + 0.028922 - 0.014904\right) = 0.2454 \text{ in.} \tag{8a}$$

$$t_{G_2} = \frac{88}{8}\left(\frac{1.5708}{88} + 0.028922 - 0.014904\right) = 0.3505 \text{ in.} \tag{8b}$$

$$D_{PO_1} = 2 \times 7.50 - 11.0856 - 2\left(0.169 - \frac{1}{8}\right) = 3.8264 \text{ in.} \tag{9a}$$

$$D_{GO_1} = 2 \times 7.50 - 3.2968 - 2\left(0.169 - \frac{1}{8}\right) = 11.6152 \text{ in.} \tag{9b}$$

**Tooth Thickness Allowance for Shaving.** — Proper stock allowance is important for good results in shaving operations. If too much stock is left for shaving, the life of the shaving tool is reduced and, in addition, shaving time is increased. The following figures represent the amount of stock to be left on the teeth for removal by shaving under average conditions: For diametral pitches of 2 to 4, a thickness of 0.003 to 0.004 inch (one-half on each side of the tooth); for 5 to 6 diametral pitch, 0.0025 to 0.0035 inch; for 7 to 10 diametral pitch, 0.002 to 0.003 inch; for 11 to 14 diametral pitch, 0.0015 to 0.0020 inch; for 16 to 18 diametral pitch, 0.001 to 0.002 inch; for 20 to 48 diametral pitch, 0.0005 to 0.0015 inch; and for 52 to 72 diametral pitch, 0.0003 to 0.0007 inch.

The thickness of the gear teeth may be measured in several ways to determine the amount of stock left on the sides of the teeth to be removed by shaving. If it is necessary to measure the tooth thickness during the pre-shaving operation while the gear is in the gear shaper or hobbing machine, a gear tooth caliper or pins would be employed. Caliper methods of measuring gear teeth are explained in detail on pages 742 and 743 for measurements over single teeth and on pages 961 to 963 for measurements over two or more teeth.

When the pre-shaved gear can be removed from the machine for checking, the center distance method may be employed. In this method the pre-shaved gear is meshed without backlash with a gear of standard tooth thickness and the increase in center distance over standard is noted. The amount of total tooth thickness over standard that is left on the pre-shaved gear can then be determined by the formula: $t_2 = 2 \tan \phi \times d$, where: $t_2$ = amount that total thickness of the tooth exceeds the standard thickness, $\phi$ = pressure angle, and $d$ = amount that the center distance between the two gears exceeds the standard center distance.

**Circular Pitch for Given Center Distance and Ratio.** — When it is necessary to use a pair of gears of given ratio at a specified center distance it may be found that no gears of standard diametral pitch will satisfy the center distance requirement. Hence, circular pitch gears may be selected. To find the required circular pitch $p$, when the center distance $C$ and total number of teeth $N$ in both gears are known, use the following formula:

$$p = \frac{C \times 6.2832}{N}$$

*Example:* A pair of gears having a ratio of 3 is to be used at a center distance of 10.230 inches. If one gear has 60 teeth and the other 20, what must be their circular pitch?

$$p = \frac{10.230 \times 6.2832}{60 + 20} = 0.8035 \text{ inch}$$

**Circular Thickness of Tooth when Outside Diameter is Standard.** — For a full-depth or stub tooth gear of standard outside diameter, the tooth thickness on the pitch circle (circular thickness or arc thickness) is found by the following formula:

$$t = \frac{1.5708}{P}$$

where $t$ = circular thickness and $P$ = diametral pitch. In the case of Fellows stub tooth gears the diametral pitch used is the numerator of the pitch fraction (for example, 6 if the pitch is 6/8).

*Example 1:* Find the tooth thickness on the pitch circle of a 14½-degree full-depth tooth of 12 diametral pitch.

$$t = \frac{1.5708}{12} = 0.1309 \text{ inch}$$

*Example 2:* Find the tooth thickness on the tooth circle of a 20-degree full-depth involute tooth having a diametral pitch of 5.

$$t = \frac{1.5708}{5} = 0.31416, \text{ say } 0.3142 \text{ inch}$$

The tooth thickness on the pitch circle can be determined very accurately by means of measurement over wires which are located in tooth spaces that are diametrically opposite or as nearly diametrically opposite as possible. Where measurement over wires is not feasible, the circular or arc tooth thickness can be used in determining the chordal thickness which is the dimension measured with a gear tooth caliper.

**Circular Thickness of Tooth when Outside Diameter has been Enlarged.** — When the outside diameter of a small pinion is not standard but is enlarged to avoid undercut and to improve tooth action, the teeth are located farther out radially relative to the standard pitch diameter and consequently the circular tooth thickness at the standard pitch diameter is increased. To find this increased arc thickness the following formula is used, where $t$ = tooth thickness; $e$ = amount outside diameter is increased over standard; $\phi$ = pressure angle; and $p$ = circular pitch at the standard pitch diameter.

$$t = \frac{p}{2} + e \tan \phi$$

*Example:* The outside diameter of a pinion having 10 teeth of 5 diametral pitch and a pressure angle of 14½ degrees is to be increased by 0.2746 inch. The circular pitch equivalent to 5 diametral pitch is 0.6283 inch. Find the arc tooth thickness at the standard pitch diameter.

$$t = \frac{0.6283}{2} + (0.2746 \times \tan 14\tfrac{1}{2}°)$$

$$t = 0.3142 + (0.2746 \times 0.25862) = 0.3852 \text{ inch}$$

**Circular Thickness of Tooth when Outside Diameter has been Reduced.** — If the outside diameter of a gear is reduced, as is frequently done to maintain the standard center distance when the outside diameter of the mating pinion is increased, the circular thickness of the gear teeth at the standard pitch diameter will be reduced. This decreased circular thickness can be found by the following formula where $t$ = circular thickness at the standard pitch diameter; $e$ = amount outside diameter is reduced under standard; $\phi$ = pressure angle; and $p$ = circular pitch.

$$t = \frac{p}{2} - e \tan \phi$$

*Example:* The outside diameter of a gear having a pressure angle of 14½ degrees is to be reduced by 0.2746 inch or an amount equal to the increase in diameter of its mating pinion. The circular pitch is 0.6283 inch. Determine the circular tooth thickness at the standard pitch diameter.

$$t = \frac{0.6283}{2} - (0.2746 \times \tan 14\tfrac{1}{2}°)$$

$$t = 0.3142 - (0.2746 \times 0.25862) = 0.2432 \text{ inch}$$

**Chordal Thickness of Tooth when Outside Diameter is Standard.** — To find the chordal or straight line thickness of a gear tooth the following formula can be used where $t_c$ = chordal thickness; $D$ = pitch diameter; and $N$ = number of teeth.

$$t_c = D \sin \left( \frac{90°}{N} \right)$$

*Example:* A pinion has 15 teeth of 3 diametral pitch; the pitch diameter is equal to 15 ÷ 3 or 5 inches. Find the chordal thickness at the standard pitch diameter.

$$t_c = 5 \sin \left( \frac{90°}{15} \right) = 5 \sin 6°$$

$$t_c = 5 \times 0.10453 = 0.5226 \text{ inch}$$

**Chordal Thickness of Tooth when Outside Diameter is Special.** — When the outside diameter is larger or smaller than standard the chordal thickness at the standard pitch diameter is found by the following formula where $t_c$ = chordal thickness at the standard pitch diameter $D$; $t$ = circular thickness at the standard pitch diameter of the enlarged pinion or reduced gear being measured.

$$t_c = t - \frac{t^3}{6 \times D^2}$$

*Example 1:* The outside diameter of a pinion having 10 teeth of 5 diametral pitch has been *enlarged* by 0.2746 inch. This enlargement has increased the circular tooth thickness at the standard pitch diameter (as determined by the formula previously given) to 0.3852 inch. Find the equivalent chordal thickness.

$$t_c = 0.3852 - \frac{(0.385)^3}{6 \times (2)^2} = 0.3852 - 0.0024 = 0.3828 \text{ inch.}$$

(The error introduced by rounding the circular thickness to three significant figures before cubing it only affects the fifth decimal place in the result. Values of cubes can be taken from the table beginning on page 2.)

*Example 2:* A gear having 30 teeth is to mesh with the pinion in Example 1 and is *reduced* so that the circular tooth thickness at the standard pitch diameter is 0.2432 inch. Find the equivalent chordal thickness.

$$t_c = 0.2432 - \frac{(0.243)^3}{6 \times (6)^2} = 0.2432 - 0.00007 = 0.2431 \text{ inch.}$$

**Chordal Addendum.** — In measuring the chordal thickness, the vertical scale of a gear tooth caliper is set to the chordal or " corrected " addendum to locate the caliper jaws at the pitch line (see illustration on page 743 ). The simplified formula which follows may be used in determining the chordal addendum either when the addendum is standard for full-depth or stub teeth or when the addendum is either longer or shorter than standard as in case of an enlarged pinion or a gear which is to mesh with an enlarged pinion and has a reduced addendum to maintain the standard center distance. If $a_c$ = chordal addendum; $a$ = addendum; and $t$ = circular thickness of tooth at pitch diameter $D$; then,

$$a_c = a + \frac{t^2}{4D}$$

*Example 1:* The outside diameter of an 8 diametral pitch 14-tooth pinion with 20-degree full-depth teeth is to be increased by using an enlarged addendum of $1.234 \div 8 = 0.1542$ inch (see Table 1 on page 749). The basic tooth thickness of the enlarged pinion is $1.741 \div 8 = 0.2176$ inch. What is the chordal addendum?

$$\text{Chordal addendum} = 0.1542 + \frac{0.2176^2}{4 \, (14 \div 8)}$$
$$= 0.1610 \text{ inch.}$$

*Example 2:* The outside diameter of a 14½-degree pinion having 12 teeth of 2 diametral pitch is to be enlarged 0.624 inch to avoid undercut (see Table 2, on page 750), thus increasing the addendum from 0.5000 to 0.8119 inch and the arc thickness at the pitch line from 0.7854 to 0.9467 inch. Then,

$$\text{Chordal addendum of pinion} = 0.8120 + \frac{0.9467^2}{4 \times (12 \div 2)} = 0.8493 \text{ inch}$$

*Example 3:* The outside diameter of the mating gear for the pinion in Example 2 is to be reduced 0.624 inch. The gear has 60 teeth and the addendum is reduced from 0.5000 to 0.1881 inch (to maintain the standard center distance), thus reducing the arc thickness to 0.6240 inch. Then,

$$\text{Chordal addendum of gear} = 0.1881 + \frac{0.6240^2}{4 \times (60 \div 2)} = 0.1913 \text{ inch}$$

When a gear addendum is reduced as much as the mating pinion addendum is enlarged, the minimum number of gear teeth required to prevent undercutting depends upon the enlargement of the mating pinion. To illustrate, if a 14½-degree pinion with 13 teeth is enlarged 1.185 inches, then the reduced mating gear should have a minimum of 51 teeth to avoid undercut (see Table 2, page 750).

**Table for Chordal Thicknesses and Chordal Addenda of Full-depth Teeth.** — The table on page 744 gives values for chordal thickness and chordal addendum of full-depth spur gear teeth of 1 diametral pitch and from 10 to 156 teeth for gears of standard outside diameter. For any other diametral pitch the values are to be divided by the required pitch.

*Helical Gears:* In applying this table to helical gears, especially when the number of teeth is small and the helix angle large, the equivalent number of teeth $N_e$ for entering the table is found by the formula, $N_e = N \div \cos^3 \psi$, where $N$ is the actual number of teeth in the helical gear and $\psi$ is the helix angle. The values obtained from the table should be divided by the *normal* diametral pitch of the helical gear to get the normal chordal thickness and the normal chordal addendum.

*Example:* Find the normal chordal thickness and the normal chordal addendum of a helical gear having 54 teeth of 6 normal diametral pitch and a helix angle of 45 degrees.

$$N_e = \frac{54}{\cos^3 45^\circ} = \frac{54}{(0.70711)^3} = 153 \text{ teeth}$$

$$\text{Normal chordal thickness} = \frac{1.57077}{6} = 0.26180 \text{ inch}$$

$$\text{Normal chordal addendum} = \frac{1.00405}{6} = 0.16734 \text{ inch}$$

**Tables for Chordal Thicknesses and Chordal Addenda of Milled, Full-depth Teeth.** — Two convenient tables for checking gears with milled, full-depth teeth are given on pages 747 and 748. The first shows chordal thicknesses and chordal addenda for the lowest number of teeth cut by gear cutters Nos. 1 through 8, and for the commonly used diametral pitches. The second gives similar data for commonly used circular pitches. In each case the data shown are accurate for the number of gear teeth indicated, but are approximate for other numbers of teeth within the range of the cutter under which they appear in the table. For the higher diametral pitches and lower circular pitches, the error introduced by using the data for any tooth number within the range of the cutter under which it appears is comparatively small. The chordal thicknesses and chordal addenda for gear cutters Nos. 1 through 8 of the more commonly used diametral and circular pitches can be obtained from the table and formulas on pages 747 and 748.

**Caliper Measurement of Gear Tooth.** — In cutting gear teeth, the general practice is to adjust the cutter or hob until it grazes the outside diameter of the blank; the cutter is then sunk to the total depth of the tooth space plus whatever slight additional amount may be required to provide the necessary play or backlash between the teeth. (For recommendations concerning backlash and excess depth of cut required, see pages 761 to 763.) If the outside diameter of the gear blank is correct, the tooth thickness should also be correct after the cutter has been sunk to the depth required for a given pitch and backlash. However, it is advisable to check the tooth thickness by measuring it, and the vernier gear-tooth caliper (see accompanying illustration) is commonly used in measuring the thickness.

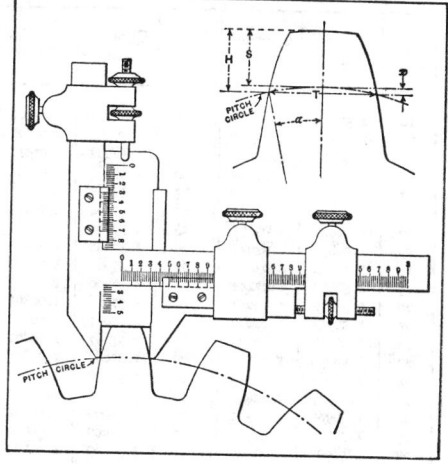

The vertical scale of this caliper is set so that when it rests upon the top of the tooth as shown, the lower ends of the caliper jaws will be at the height of the pitch circle; the horizontal scale then shows the chordal thickness of the tooth at this point. If the gear is being cut on a milling machine or with the type of gear-cutting machine employing a formed milling cutter, the tooth thickness is checked by first taking a trial cut for a short distance at one side of the blank; then the gear blank is indexed for the next space and another cut is taken far enough to mill the full outline of the tooth. The tooth thickness is then measured.

Before the gear-tooth caliper can be used, it is necessary to determine the correct chordal thickness and also the chordal addendum (or "corrected addendum" as it is sometimes called). The vertical scale is set to the chordal addendum, thus locating

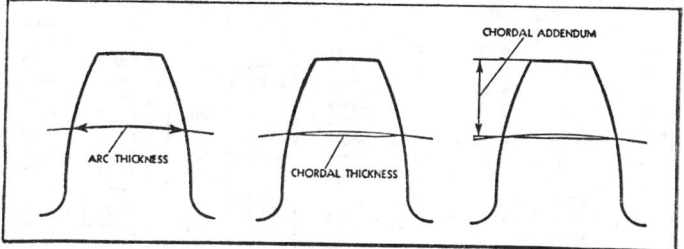

the ends of the jaws at the height of the pitch circle. The rules or formulas to use in determining the chordal thickness and chordal addendum will depend upon the outside diameter of the gear; for example, if the outside diameter of a small pinion is enlarged to avoid undercut and improve the tooth action, this must be taken into account in figuring the chordal thickness and chordal addendum as shown by the accompanying rules. The detail of a gear tooth included with the gear-tooth caliper illustration, represents the chordal thickness $T$, the addendum $S$, and the chordal addendum $H$. For caliper measurements over two or more teeth see pages 961 to 963.

## Chordal Thicknesses and Chordal Addenda of Full-depth Gear Teeth

This table is for spur gears of one diametral pitch. For any other diametral pitch, divide the given value by the required pitch. Table gives the chordal thickness and chordal addendum at the pitch circle when addendum is standard for full-depth teeth. Table is applicable to helical gears as explained on page 742.

| No. of Teeth | Chordal Thickness | Chordal Addend. | No. of Teeth | Chordal Thickness | Chordal Addend. | No. of Teeth | Chordal Thickness | Chordal Addend. |
|---|---|---|---|---|---|---|---|---|
| 10 | 1.56435 | 1.06156 | 59 | 1.57061 | 1.01046 | 108 | 1.57074 | 1.00570 |
| 11 | 1.56546 | 1.05598 | 60 | 1.57062 | 1.01029 | 109 | 1.57075 | 1.00565 |
| 12 | 1.56631 | 1.05133 | 61 | 1.57062 | 1.01011 | 110 | 1.57075 | 1.00560 |
| 13 | 1.56698 | 1.04739 | 62 | 1.57063 | 1.00994 | 111 | 1.57075 | 1.00556 |
| 14 | 1.56752 | 1.04401 | 63 | 1.57063 | 1.00978 | 112 | 1.57075 | 1.00551 |
| 15 | 1.56794 | 1.04109 | 64 | 1.57064 | 1.00963 | 113 | 1.57075 | 1.00546 |
| 16 | 1.56827 | 1.03852 | 65 | 1.57064 | 1.00947 | 114 | 1.57075 | 1.00541 |
| 17 | 1.56856 | 1.03625 | 66 | 1.57065 | 1.00933 | 115 | 1.57075 | 1.00537 |
| 18 | 1.56880 | 1.03425 | 67 | 1.57065 | 1.00920 | 116 | 1.57075 | 1.00533 |
| 19 | 1.56899 | 1.03244 | 68 | 1.57066 | 1.00907 | 117 | 1.57075 | 1.00529 |
| 20 | 1.56918 | 1.03083 | 69 | 1.57066 | 1.00893 | 118 | 1.57075 | 1.00524 |
| 21 | 1.56933 | 1.02936 | 70 | 1.57067 | 1.00880 | 119 | 1.57075 | 1.00519 |
| 22 | 1.56948 | 1.02803 | 71 | 1.57067 | 1.00867 | 120 | 1.57075 | 1.00515 |
| 23 | 1.56956 | 1.02681 | 72 | 1.57067 | 1.00855 | 121 | 1.57075 | 1.00511 |
| 24 | 1.56967 | 1.02569 | 73 | 1.57068 | 1.00843 | 122 | 1.57075 | 1.00507 |
| 25 | 1.56977 | 1.02466 | 74 | 1.57068 | 1.00832 | 123 | 1.57076 | 1.00503 |
| 26 | 1.56986 | 1.02371 | 75 | 1.57068 | 1.00821 | 124 | 1.57076 | 1.00499 |
| 27 | 1.56991 | 1.02284 | 76 | 1.57069 | 1.00810 | 125 | 1.57076 | 1.00495 |
| 28 | 1.56998 | 1.02202 | 77 | 1.57069 | 1.00799 | 126 | 1.57076 | 1.00491 |
| 29 | 1.57003 | 1.02127 | 78 | 1.57069 | 1.00789 | 127 | 1.57076 | 1.00487 |
| 30 | 1.57008 | 1.02055 | 79 | 1.57069 | 1.00780 | 128 | 1.57076 | 1.00483 |
| 31 | 1.57012 | 1.01990 | 80 | 1.57070 | 1.00772 | 129 | 1.57076 | 1.00479 |
| 32 | 1.57016 | 1.01926 | 81 | 1.57070 | 1.00762 | 130 | 1.57076 | 1.00475 |
| 33 | 1.57019 | 1.01869 | 82 | 1.57070 | 1.00752 | 131 | 1.57076 | 1.00472 |
| 34 | 1.57021 | 1.01813 | 83 | 1.57070 | 1.00743 | 132 | 1.57076 | 1.00469 |
| 35 | 1.57025 | 1.01762 | 84 | 1.57071 | 1.00734 | 133 | 1.57076 | 1.00466 |
| 36 | 1.57028 | 1.01714 | 85 | 1.57071 | 1.00725 | 134 | 1.57076 | 1.00462 |
| 37 | 1.57032 | 1.01667 | 86 | 1.57071 | 1.00716 | 135 | 1.57076 | 1.00457 |
| 38 | 1.57035 | 1.01623 | 87 | 1.57071 | 1.00708 | 136 | 1.57076 | 1.00454 |
| 39 | 1.57037 | 1.01582 | 88 | 1.57071 | 1.00700 | 137 | 1.57076 | 1.00451 |
| 40 | 1.57039 | 1.01542 | 89 | 1.57072 | 1.00693 | 138 | 1.57076 | 1.00447 |
| 41 | 1.57041 | 1.01504 | 90 | 1.57072 | 1.00686 | 139 | 1.57076 | 1.00444 |
| 42 | 1.57043 | 1.01471 | 91 | 1.57072 | 1.00679 | 140 | 1.57076 | 1.00441 |
| 43 | 1.57045 | 1.01434 | 92 | 1.57072 | 1.00672 | 141 | 1.57076 | 1.00439 |
| 44 | 1.57047 | 1.01404 | 93 | 1.57072 | 1.00665 | 142 | 1.57076 | 1.00435 |
| 45 | 1.57048 | 1.01370 | 94 | 1.57072 | 1.00658 | 143 | 1.57076 | 1.00432 |
| 46 | 1.57050 | 1.01341 | 95 | 1.57073 | 1.00651 | 144 | 1.57076 | 1.00429 |
| 47 | 1.57051 | 1.01311 | 96 | 1.57073 | 1.00644 | 145 | 1.57077 | 1.00425 |
| 48 | 1.57052 | 1.01285 | 97 | 1.57073 | 1.00637 | 146 | 1.57077 | 1.00422 |
| 49 | 1.57053 | 1.01258 | 98 | 1.57073 | 1.00630 | 147 | 1.57077 | 1.00419 |
| 50 | 1.57054 | 1.01233 | 99 | 1.57073 | 1.00623 | 148 | 1.57077 | 1.00416 |
| 51 | 1.57055 | 1.01209 | 100 | 1.57073 | 1.00617 | 149 | 1.57077 | 1.00413 |
| 52 | 1.57056 | 1.01187 | 101 | 1.57074 | 1.00611 | 150 | 1.57077 | 1.00411 |
| 53 | 1.57057 | 1.01165 | 102 | 1.57074 | 1.00605 | 151 | 1.57077 | 1.00409 |
| 54 | 1.57058 | 1.01143 | 103 | 1.57074 | 1.00599 | 152 | 1.57077 | 1.00407 |
| 55 | 1.57058 | 1.01121 | 104 | 1.57074 | 1.00593 | 153 | 1.57077 | 1.00405 |
| 56 | 1.57059 | 1.01102 | 105 | 1.57074 | 1.00587 | 154 | 1.57077 | 1.00402 |
| 57 | 1.57060 | 1.01083 | 106 | 1.57074 | 1.00581 | 155 | 1.57077 | 1.00400 |
| 58 | 1.57061 | 1.01064 | 107 | 1.57074 | 1.00575 | 156 | 1.57077 | 1.00397 |

## Circular Pitch Gears — Pitch Diameters, Outside Diameters, and Root Diameters

For any particular circular pitch and number of teeth, use the table as shown in the example to find the pitch diameter, outside diameter, and root diameter. *Example:* Pitch diameter for 57 teeth of 6-inch circular pitch=10× pitch diameter given under factor for 5 teeth plus pitch diameter given under factor for 7 teeth. (10 × 9.5493) + 13.3690 = 108.862 inches.

Outside diameter of gear equals pitch diameter plus outside diameter factor from next-to-last column in table = 108.862 + 3.8197 = 112.682 inches.

Root diameter of gear equals pitch diameter minus root diameter factor from last column in table = 108.862 − 4.4194 = 104.443 inches.

| Circular Pitch in Inches | Factor for Number of Teeth | | | | | | | | | Outside Diam. Factor | Root Diameter Factor |
|---|---|---|---|---|---|---|---|---|---|---|---|
| | 1 | 2 | 3 | 4 | 5 | 6 | 7 | 8 | 9 | | |
| | Pitch Diameter Corresponding to Factor for Number of Teeth | | | | | | | | | | |
| 6 | 1.9099 | 3.8197 | 5.7296 | 7.6394 | 9.5493 | 11.4591 | 13.3690 | 15.2788 | 17.1887 | 3.8197 | 4.4194 |
| 5½ | 1.7507 | 3.5014 | 5.2521 | 7.0028 | 8.7535 | 10.5042 | 12.2549 | 14.0056 | 15.7563 | 3.5014 | 4.0511 |
| 5 | 1.5915 | 3.1831 | 4.7746 | 6.3662 | 7.9577 | 9.5493 | 11.1408 | 12.7324 | 14.3239 | 3.1831 | 3.6828 |
| 4½ | 1.4324 | 2.8648 | 4.2972 | 5.7296 | 7.1620 | 8.5943 | 10.0267 | 11.4591 | 12.8915 | 2.8648 | 3.3146 |
| 4 | 1.2732 | 2.5465 | 3.8197 | 5.0929 | 6.3662 | 7.6394 | 8.9127 | 10.1859 | 11.4591 | 2.5465 | 2.9463 |
| 3½ | 1.1141 | 2.2282 | 3.3422 | 4.4563 | 5.5704 | 6.6845 | 7.7986 | 8.9127 | 10.0267 | 2.2282 | 2.5780 |
| 3 | 0.9549 | 1.9099 | 2.8648 | 3.8197 | 4.7746 | 5.7296 | 6.6845 | 7.6394 | 8.5943 | 1.9099 | 2.2097 |
| 2½ | 0.7958 | 1.5915 | 2.3873 | 3.1831 | 3.9789 | 4.7746 | 5.5704 | 6.3662 | 7.1620 | 1.5915 | 1.8414 |
| 2 | 0.6366 | 1.2732 | 1.9099 | 2.5465 | 3.1831 | 3.8197 | 4.4563 | 5.0929 | 5.7296 | 1.2732 | 1.4731 |
| 1⅞ | 0.5968 | 1.1937 | 1.7905 | 2.3873 | 2.9841 | 3.5810 | 4.1778 | 4.7746 | 5.3715 | 1.1937 | 1.3811 |
| 1¾ | 0.5570 | 1.1141 | 1.6711 | 2.2282 | 2.7852 | 3.3422 | 3.8993 | 4.4563 | 5.0134 | 1.1141 | 1.2890 |
| 1⅝ | 0.5173 | 1.0345 | 1.5518 | 2.0690 | 2.5863 | 3.1035 | 3.6208 | 4.1380 | 4.6553 | 1.0345 | 1.1969 |
| 1½ | 0.4775 | 0.9549 | 1.4324 | 1.9099 | 2.3873 | 2.8648 | 3.3422 | 3.8197 | 4.2972 | 0.9549 | 1.1049 |
| 1⁷⁄₁₆ | 0.4576 | 0.9151 | 1.3727 | 1.8303 | 2.2878 | 2.7454 | 3.2030 | 3.6606 | 4.1181 | 0.9151 | 1.0588 |
| 1⅜ | 0.4377 | 0.8754 | 1.3130 | 1.7507 | 2.1884 | 2.6261 | 3.0637 | 3.5014 | 3.9391 | 0.8754 | 1.0128 |
| 1⁵⁄₁₆ | 0.4178 | 0.8356 | 1.2533 | 1.6711 | 2.0889 | 2.5067 | 2.9245 | 3.3422 | 3.7600 | 0.8356 | 0.9667 |
| 1¼ | 0.3979 | 0.7958 | 1.1937 | 1.5915 | 1.9894 | 2.3873 | 2.7852 | 3.1831 | 3.5810 | 0.7958 | 0.9207 |
| 1³⁄₁₆ | 0.3780 | 0.7560 | 1.1340 | 1.5120 | 1.8900 | 2.2680 | 2.6459 | 3.0239 | 3.4019 | 0.7560 | 0.8747 |
| 1⅛ | 0.3581 | 0.7162 | 1.0743 | 1.4324 | 1.7905 | 2.1486 | 2.5067 | 2.8648 | 3.2229 | 0.7162 | 0.8286 |
| 1¹⁄₁₆ | 0.3382 | 0.6764 | 1.0146 | 1.3528 | 1.6910 | 2.0292 | 2.3674 | 2.7056 | 3.0438 | 0.6764 | 0.7826 |
| 1 | 0.3183 | 0.6366 | 0.9549 | 1.2732 | 1.5915 | 1.9099 | 2.2282 | 2.5465 | 2.8648 | 0.6366 | 0.7366 |
| 15⁄16 | 0.2984 | 0.5968 | 0.8952 | 1.1937 | 1.4921 | 1.7905 | 2.0889 | 2.3873 | 2.6857 | 0.5968 | 0.6905 |
| ⅞ | 0.2785 | 0.5570 | 0.8356 | 1.1141 | 1.3926 | 1.6711 | 1.9496 | 2.2282 | 2.5067 | 0.5570 | 0.6445 |
| 13⁄16 | 0.2586 | 0.5173 | 0.7759 | 1.0345 | 1.2931 | 1.5518 | 1.8104 | 2.0690 | 2.3276 | 0.5173 | 0.5985 |
| ¾ | 0.2387 | 0.4775 | 0.7162 | 0.9549 | 1.1937 | 1.4324 | 1.6711 | 1.9099 | 2.1486 | 0.4775 | 0.5524 |
| 11⁄16 | 0.2188 | 0.4377 | 0.6565 | 0.8754 | 1.0942 | 1.3130 | 1.5319 | 1.7507 | 1.9695 | 0.4377 | 0.5064 |
| ⅔ | 0.2122 | 0.4244 | 0.6366 | 0.8488 | 1.0610 | 1.2732 | 1.4854 | 1.6977 | 1.9099 | 0.4244 | 0.4910 |
| ⅝ | 0.1989 | 0.3979 | 0.5968 | 0.7958 | 0.9947 | 1.1937 | 1.3926 | 1.5915 | 1.7905 | 0.3979 | 0.4604 |
| 9⁄16 | 0.1790 | 0.3581 | 0.5371 | 0.7162 | 0.8952 | 1.0743 | 1.2533 | 1.4324 | 1.6114 | 0.3581 | 0.4143 |
| ½ | 0.1592 | 0.3183 | 0.4775 | 0.6366 | 0.7958 | 0.9549 | 1.1141 | 1.2732 | 1.4324 | 0.3183 | 0.3683 |
| 7⁄16 | 0.1393 | 0.2785 | 0.4178 | 0.5570 | 0.6963 | 0.8356 | 0.9748 | 1.1141 | 1.2533 | 0.2785 | 0.3222 |
| ⅜ | 0.1194 | 0.2387 | 0.3581 | 0.4775 | 0.5968 | 0.7162 | 0.8356 | 0.9549 | 1.0743 | 0.2387 | 0.2762 |
| ⅓ | 0.1061 | 0.2122 | 0.3183 | 0.4244 | 0.5305 | 0.6366 | 0.7427 | 0.8488 | 0.9549 | 0.2122 | 0.2455 |
| 5⁄16 | 0.0995 | 0.1989 | 0.2984 | 0.3979 | 0.4974 | 0.5968 | 0.6963 | 0.7958 | 0.8952 | 0.1989 | 0.2302 |
| ¼ | 0.0796 | 0.1592 | 0.2387 | 0.3183 | 0.3979 | 0.4775 | 0.5570 | 0.6366 | 0.7162 | 0.1592 | 0.1841 |
| 3⁄16 | 0.0597 | 0.1194 | 0.1790 | 0.2387 | 0.2984 | 0.3581 | 0.4178 | 0.4775 | 0.5371 | 0.1194 | 0.1381 |
| ⅛ | 0.0398 | 0.0796 | 0.1194 | 0.1592 | 0.1989 | 0.2387 | 0.2785 | 0.3183 | 0.3581 | 0.0796 | 0.0921 |
| 1⁄16 | 0.0199 | 0.0398 | 0.0597 | 0.0796 | 0.0995 | 0.1194 | 0.1393 | 0.1592 | 0.1790 | 0.0398 | 0.0460 |

**Selection of Involute Gear Milling Cutter for a Given Diametral Pitch and Number of Teeth.** — When gear teeth are cut by using formed milling cutters, the cutter must be selected to suit both the pitch and the number of teeth, because the shapes of the tooth spaces vary according to the number of teeth. For instance, the tooth spaces of a small pinion are not of the same shape as the spaces of a large gear of equal pitch. Theoretically, there should be a different formed cutter for every tooth number, but such refinement is unnecessary in practice. The involute formed cutters commonly used are made in series of eight cutters for each diametral pitch (see accompanying table). The shape of each cutter in this series is correct for a certain number of teeth only, but it can be used for other numbers within the limits given. For instance, a No. 6 cutter may be used for gears having from 17 to 20 teeth, but the tooth outline is correct only for 17 teeth or the lowest number in the range, which is also true of the other cutters listed. When this cutter is used for a gear having, say, 19 teeth, too much material is removed from the upper surfaces of the teeth, although the gear meets ordinary requirements. When greater accuracy of tooth shape is desired to insure smoother or quieter operation, an intermediate series of cutters having half-numbers may be used provided the number of gear teeth is between the number listed for the regular cutters (see table).

Involute gear milling cutters are designed to cut a composite tooth form, the center portion being a true involute while the top and bottom portions are cycloidal. This composite form is necessary to prevent tooth interference when milled mating gears are meshed with each other. Because of their composite form, milled gears will not mate satisfactorily enough for high grade work with those of generated, full-involute form. Composite form hobs are available, however, which will produce generated gears that mesh with those cut by gear milling cutters.

In contrast to gears cut by milling, gear teeth cut by a generating process are given the required shape or curvature by the rotation of the gear blank relative to the cutter; consequently, a hob or any generating type of cutter of a given pitch may be used for any number of teeth. In the practical application of the generating principle to gear-hobbing machines, the hob used has cutting edges of the same shape as teeth of a rack of corresponding pitch, except for minor variations such, for example, as increasing the length of the hob teeth to provide for clearance at the bottom of the tooth spaces.

**Series of Involute, Gear Milling Cutters for Each Pitch**

| Number of Cutter | Will cut Gears from | Number of Cutter | Will cut Gears from |
|---|---|---|---|
| 1 | 135 teeth to a rack | 5 | 21 to 25 teeth |
| 2 | 55 to 134 teeth | 6 | 17 to 20 teeth |
| 3 | 35 to 54 teeth | 7 | 14 to 16 teeth |
| 4 | 26 to 34 teeth | 8 | 12 to 13 teeth |

The regular cutters listed above are used ordinarily. The cutters listed below (an intermediate series having half numbers) may be used when greater accuracy of tooth shape is essential in cases where the number of teeth is between the numbers for which the regular cutters are intended.

| Number of Cutter | Will cut Gears from | Number of Cutter | Will cut Gears from |
|---|---|---|---|
| 1½ | 80 to 134 teeth | 5½ | 19 to 20 teeth |
| 2½ | 42 to 54 teeth | 6½ | 15 to 16 teeth |
| 3½ | 30 to 34 teeth | 7½ | 13 teeth |
| 4½ | 23 to 25 teeth | ... | ............ |

### Chordal Thicknesses and Chordal Addenda of Milled, Full-depth Gear Teeth and of Gear Milling Cutters

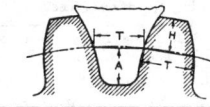

$T$ = chordal thickness of gear tooth and cutter tooth at pitch line;
$H$ = chordal addendum for full-depth gear tooth;
$A$ = chordal addendum of cutter = (2.157 ÷ diametral pitch) − H = (0.6866 × circular pitch) − H.

| Diametral Pitch | Dimension | No. 1 135 Teeth | No. 2 55 Teeth | No. 3 35 Teeth | No. 4 26 Teeth | No. 5 21 Teeth | No. 6 17 Teeth | No. 7 14 Teeth | No. 8 12 Teeth |
|---|---|---|---|---|---|---|---|---|---|
| 1 | T | 1.5707 | 1.5706 | 1.5702 | 1.5698 | 1.5694 | 1.5686 | 1.5675 | 1.5663 |
|   | H | 1.0047 | 1.0112 | 1.0176 | 1.0237 | 1.0294 | 1.0362 | 1.0440 | 1.0514 |
| 1½ | T | 1.0471 | 1.0470 | 1.0468 | 1.0465 | 1.0462 | 1.0457 | 1.0450 | 1.0442 |
|   | H | 0.6698 | 0.6741 | 0.6784 | 0.6824 | 0.6862 | 0.6908 | 0.6960 | 0.7009 |
| 2 | T | 0.7853 | 0.7853 | 0.7851 | 0.7849 | 0.7847 | 0.7843 | 0.7837 | 0.7831 |
|   | H | 0.5023 | 0.5056 | 0.5088 | 0.5118 | 0.5147 | 0.5181 | 0.5220 | 0.5257 |
| 2½ | T | 0.6283 | 0.6282 | 0.6281 | 0.6279 | 0.6277 | 0.6274 | 0.6270 | 0.6265 |
|   | H | 0.4018 | 0.4044 | 0.4070 | 0.4094 | 0.4117 | 0.4144 | 0.4176 | 0.4205 |
| 3 | T | 0.5235 | 0.5235 | 0.5234 | 0.5232 | 0.5231 | 0.5228 | 0.5225 | 0.5221 |
|   | H | 0.3349 | 0.3370 | 0.3392 | 0.3412 | 0.3431 | 0.3454 | 0.3480 | 0.3504 |
| 3½ | T | 0.4487 | 0.4487 | 0.4486 | 0.4485 | 0.4484 | 0.4481 | 0.4478 | 0.4475 |
|   | H | 0.2870 | 0.2889 | 0.2907 | 0.2919 | 0.2935 | 0.2954 | 0.2977 | 0.3004 |
| 4 | T | 0.3926 | 0.3926 | 0.3926 | 0.3924 | 0.3923 | 0.3921 | 0.3919 | 0.3915 |
|   | H | 0.2511 | 0.2528 | 0.2544 | 0.2559 | 0.2573 | 0.2590 | 0.2610 | 0.2628 |
| 5 | T | 0.3141 | 0.3141 | 0.3140 | 0.3139 | 0.3138 | 0.3137 | 0.3135 | 0.3132 |
|   | H | 0.2009 | 0.2022 | 0.2035 | 0.2047 | 0.2058 | 0.2072 | 0.2088 | 0.2102 |
| 6 | T | 0.2618 | 0.2617 | 0.2617 | 0.2616 | 0.2615 | 0.2614 | 0.2612 | 0.2610 |
|   | H | 0.1674 | 0.1685 | 0.1696 | 0.1706 | 0.1715 | 0.1727 | 0.1740 | 0.1752 |
| 7 | T | 0.2244 | 0.2243 | 0.2243 | 0.2242 | 0.2242 | 0.2240 | 0.2239 | 0.2237 |
|   | H | 0.1435 | 0.1444 | 0.1453 | 0.1462 | 0.1470 | 0.1480 | 0.1491 | 0.1502 |
| 8 | T | 0.1963 | 0.1963 | 0.1962 | 0.1962 | 0.1961 | 0.1960 | 0.1959 | 0.1958 |
|   | H | 0.1255 | 0.1264 | 0.1272 | 0.1279 | 0.1286 | 0.1295 | 0.1305 | 0.1314 |
| 9 | T | 0.1745 | 0.1745 | 0.1744 | 0.1744 | 0.1743 | 0.1743 | 0.1741 | 0.1740 |
|   | H | 0.1116 | 0.1123 | 0.1130 | 0.1137 | 0.1143 | 0.1151 | 0.1160 | 0.1168 |
| 10 | T | 0.1570 | 0.1570 | 0.1570 | 0.1569 | 0.1569 | 0.1568 | 0.1567 | 0.1566 |
|   | H | 0.1004 | 0.1011 | 0.1017 | 0.1023 | 0.1029 | 0.1036 | 0.1044 | 0.1051 |
| 11 | T | 0.1428 | 0.1428 | 0.1427 | 0.1427 | 0.1426 | 0.1426 | 0.1425 | 0.1424 |
|   | H | 0.0913 | 0.0919 | 0.0925 | 0.0930 | 0.0935 | 0.0942 | 0.0949 | 0.0955 |
| 12 | T | 0.1309 | 0.1309 | 0.1308 | 0.1308 | 0.1308 | 0.1307 | 0.1306 | 0.1305 |
|   | H | 0.0837 | 0.0842 | 0.0848 | 0.0853 | 0.0857 | 0.0863 | 0.0870 | 0.0876 |
| 14 | T | 0.1122 | 0.1122 | 0.1121 | 0.1121 | 0.1121 | 0.1120 | 0.1119 | 0.1118 |
|   | H | 0.0717 | 0.0722 | 0.0726 | 0.0731 | 0.0735 | 0.0740 | 0.0745 | 0.0751 |
| 16 | T | 0.0981 | 0.0981 | 0.0981 | 0.0981 | 0.0980 | 0.0980 | 0.0979 | 0.0979 |
|   | H | 0.0628 | 0.0632 | 0.0636 | 0.0639 | 0.0643 | 0.0647 | 0.0652 | 0.0657 |
| 18 | T | 0.0872 | 0.0872 | 0.0872 | 0.0872 | 0.0872 | 0.0871 | 0.0870 | 0.0870 |
|   | H | 0.0558 | 0.0561 | 0.0565 | 0.0568 | 0.0571 | 0.0575 | 0.0580 | 0.0584 |
| 20 | T | 0.0785 | 0.0785 | 0.0785 | 0.0785 | 0.0784 | 0.0784 | 0.0783 | 0.0783 |
|   | H | 0.0502 | 0.0505 | 0.0508 | 0.0511 | 0.0514 | 0.0518 | 0.0522 | 0.0525 |

### Chordal Thicknesses and Chordal Addenda of Milled, Full-depth Gear Teeth and of Gear Milling Cutters

| Circular Pitch | Dimension | Number of Gear Cutter, and Corresponding Number of Teeth | | | | | | | |
|---|---|---|---|---|---|---|---|---|---|
| | | No. 1 135 Teeth | No. 2 55 Teeth | No. 3 35 Teeth | No. 4 26 Teeth | No. 5 21 Teeth | No. 6 17 Teeth | No. 7 14 Teeth | No. 8 12 Teeth |
| ¼ | T | 0.1250 | 0.1250 | 0.1249 | 0.1249 | 0.1249 | 0.1248 | 0.1247 | 0.1246 |
| | H | 0.0799 | 0.0804 | 0.0809 | 0.0814 | 0.0819 | 0.0824 | 0.0830 | 0.0836 |
| 5/16 | T | 0.1562 | 0.1562 | 0.1562 | 0.1561 | 0.1561 | 0.1560 | 0.1559 | 0.1558 |
| | H | 0.0999 | 0.1006 | 0.1012 | 0.1018 | 0.1023 | 0.1030 | 0.1038 | 0.1045 |
| 3/8 | T | 0.1875 | 0.1875 | 0.1874 | 0.1873 | 0.1873 | 0.1872 | 0.1871 | 0.1870 |
| | H | 0.1199 | 0.1207 | 0.1214 | 0.1221 | 0.1228 | 0.1236 | 0.1245 | 0.1254 |
| 7/16 | T | 0.2187 | 0.2187 | 0.2186 | 0.2186 | 0.2185 | 0.2184 | 0.2183 | 0.2181 |
| | H | 0.1399 | 0.1408 | 0.1416 | 0.1425 | 0.1433 | 0.1443 | 0.1453 | 0.1464 |
| ½ | T | 0.2500 | 0.2500 | 0.2499 | 0.2498 | 0.2498 | 0.2496 | 0.2495 | 0.2493 |
| | H | 0.1599 | 0.1609 | 0.1619 | 0.1629 | 0.1638 | 0.1649 | 0.1661 | 0.1673 |
| 9/16 | T | 0.2812 | 0.2812 | 0.2811 | 0.2810 | 0.2810 | 0.2808 | 0.2806 | 0.2804 |
| | H | 0.1799 | 0.1810 | 0.1821 | 0.1832 | 0.1842 | 0.1855 | 0.1868 | 0.1882 |
| 5/8 | T | 0.3125 | 0.3125 | 0.3123 | 0.3123 | 0.3122 | 0.3120 | 0.3118 | 0.3116 |
| | H | 0.1998 | 0.2012 | 0.2023 | 0.2036 | 0.2047 | 0.2061 | 0.2076 | 0.2091 |
| 11/16 | T | 0.3437 | 0.3437 | 0.3436 | 0.3435 | 0.3434 | 0.3432 | 0.3430 | 0.3427 |
| | H | 0.2198 | 0.2213 | 0.2226 | 0.2239 | 0.2252 | 0.2267 | 0.2283 | 0.2300 |
| ¾ | T | 0.3750 | 0.3750 | 0.3748 | 0.3747 | 0.3747 | 0.3744 | 0.3742 | 0.3740 |
| | H | 0.2398 | 0.2414 | 0.2428 | 0.2443 | 0.2457 | 0.2473 | 0.2491 | 0.2509 |
| 13/16 | T | 0.4062 | 0.4062 | 0.4060 | 0.4059 | 0.4059 | 0.4056 | 0.4054 | 0.4050 |
| | H | 0.2598 | 0.2615 | 0.2631 | 0.2647 | 0.2661 | 0.2679 | 0.2699 | 0.2718 |
| 7/8 | T | 0.4375 | 0.4375 | 0.4373 | 0.4372 | 0.4371 | 0.4368 | 0.4366 | 0.4362 |
| | H | 0.2798 | 0.2816 | 0.2833 | 0.2850 | 0.2866 | 0.2885 | 0.2906 | 0.2927 |
| 15/16 | T | 0.4687 | 0.4687 | 0.4685 | 0.4684 | 0.4683 | 0.4680 | 0.4678 | 0.4674 |
| | H | 0.2998 | 0.3018 | 0.3035 | 0.3054 | 0.3071 | 0.3092 | 0.3114 | 0.3137 |
| 1 | T | 0.5000 | 0.5000 | 0.4998 | 0.4997 | 0.4996 | 0.4993 | 0.4990 | 0.4986 |
| | H | 0.3198 | 0.3219 | 0.3238 | 0.3258 | 0.3276 | 0.3298 | 0.3322 | 0.3346 |
| 1⅛ | T | 0.5625 | 0.5625 | 0.5623 | 0.5621 | 0.5620 | 0.5617 | 0.5613 | 0.5610 |
| | H | 0.3597 | 0.3621 | 0.3642 | 0.3665 | 0.3685 | 0.3710 | 0.3737 | 0.3764 |
| 1¼ | T | 0.6250 | 0.6250 | 0.6247 | 0.6246 | 0.6245 | 0.6241 | 0.6237 | 0.6232 |
| | H | 0.3997 | 0.4023 | 0.4047 | 0.4072 | 0.4095 | 0.4122 | 0.4152 | 0.4182 |
| 1⅜ | T | 0.6875 | 0.6875 | 0.6872 | 0.6870 | 0.6869 | 0.6865 | 0.6861 | 0.6856 |
| | H | 0.4397 | 0.4426 | 0.4452 | 0.4479 | 0.4504 | 0.4534 | 0.4567 | 0.4600 |
| 1½ | T | 0.7500 | 0.7500 | 0.7497 | 0.7495 | 0.7494 | 0.7489 | 0.7485 | 0.7480 |
| | H | 0.4797 | 0.4828 | 0.4857 | 0.4887 | 0.4914 | 0.4947 | 0.4983 | 0.5019 |
| 1¾ | T | 0.8750 | 0.8750 | 0.8746 | 0.8744 | 0.8743 | 0.8737 | 0.8732 | 0.8726 |
| | H | 0.5596 | 0.5633 | 0.5666 | 0.5701 | 0.5733 | 0.5771 | 0.5813 | 0.5855 |
| 2 | T | 1.0000 | 1.0000 | 0.9996 | 0.9994 | 0.9992 | 0.9986 | 0.9980 | 0.9972 |
| | H | 0.6396 | 0.6438 | 0.6476 | 0.6516 | 0.6552 | 0.6596 | 0.6644 | 0.6692 |
| 2¼ | T | 1.1250 | 1.1250 | 1.1246 | 1.1242 | 1.1240 | 1.1234 | 1.1226 | 1.1220 |
| | H | 0.7195 | 0.7242 | 0.7285 | 0.7330 | 0.7371 | 0.7420 | 0.7474 | 0.7528 |
| 2½ | T | 1.2500 | 1.2500 | 1.2494 | 1.2492 | 1.2490 | 1.2482 | 1.2474 | 1.2464 |
| | H | 0.7995 | 0.8047 | 0.8095 | 0.8145 | 0.8190 | 0.8245 | 0.8305 | 0.8365 |
| 3 | T | 1.5000 | 1.5000 | 1.4994 | 1.4990 | 1.4990 | 1.4978 | 1.4970 | 1.4960 |
| | H | 0.9594 | 0.9657 | 0.9714 | 0.9774 | 0.9828 | 0.9894 | 0.9966 | 1.0038 |

**Increasing Pinion Diameter to Avoid Undercut or Interference.** — On pinions with small numbers of teeth (10 to 17 for 20 degree- and 10 and 11 for 25-degree involute tooth forms) undercutting of the tooth profile or fillet interference with the tip of the mating gear can be avoided by making certain changes from the standard tooth proportions that are specified in Table 3 on page 720. These changes consist essentially in increasing the addendum and hence the outside diameter of the pinion and decreasing the addendum and hence the outside diameter of the mating gear. These changes in outside diameters of pinion and gear do not change the velocity ratio or the procedures in cutting the teeth on a hobbing machine or generating type of shaper or planer.

Data in Table 1 which follows are taken from ANSI Standard B6.1-1968 and show for 20-degree and 25-degree full-depth standard tooth forms, respectively, the addendums and tooth thicknesses for long addendum pinions and their mating short addendum gears when the number of teeth in the pinion is as given.

Similar data for former standard 14½-degree full-depth teeth (20 diametral pitch and coarser) are given in Table 2 and for standard 20-degree fine pitch and alternate standard 14½- and 25-degree fine pitch teeth are given in Table 3.

**Table 1. Addendums and Tooth Thicknesses for Long-Addendum Pinions and their Mating Short-Addendum Gears — 20- and 25-degree Pressure Angles\***
(ANSI-B6.1-1968)

| Number of Teeth in Pinion | Addendum | | Basic Tooth Thickness | | Number of Teeth in Gear |
|---|---|---|---|---|---|
| $N_P$ | Pinion $a_P$ | Gear $a_G$ | Pinion $t_P$ | Gear $t_G$ | $N_G$ (min) |
| 20-DEGREE INVOLUTE FULL DEPTH TOOTH FORM (Less than 20 Diametral Pitch) | | | | | |
| 10 | 1.468 | .532 | 1.912 | 1.230 | 25 |
| 11 | 1.409 | .591 | 1.868 | 1.273 | 24 |
| 12 | 1.351 | .649 | 1.826 | 1.315 | 23 |
| 13 | 1.292 | .708 | 1.783 | 1.358 | 22 |
| 14 | 1.234 | .766 | 1.741 | 1.400 | 21 |
| 15 | 1.175 | .825 | 1.698 | 1.443 | 20 |
| 16 | 1.117 | .883 | 1.656 | 1.486 | 19 |
| 17 | 1.058 | .942 | 1.613 | 1.529 | 18 |
| 25-DEGREE INVOLUTE FULL DEPTH TOOTH FORM (Less than 20 Diametral Pitch) | | | | | |
| 10 | 1.184 | .816 | 1.742 | 1.399 | 15 |
| 11 | 1.095 | .905 | 1.659 | 1.482 | 14 |

\*All values are for 1 diametral pitch. For any other sizes of teeth all linear dimensions should be divided by the diametral pitch. Basic tooth thicknesses do not include an allowance for backlash.

*Example:* A 14-tooth, 20-degree pressure angle pinion of 6 diametral pitch is to be enlarged. What will be the outside diameters of the pinion and a 60-tooth mating gear? If the mating gear is to have the minimum number of teeth to avoid undercut, what will be its outside diameter?

$$D_O \text{ (pinion)} = \frac{N_P}{P} + 2a = \frac{14}{6} + 2\left(\frac{1.234}{6}\right) = 2.745 \text{ inches}$$

$$D_O \text{ (gear)} = \frac{N_G}{P} + 2a = \frac{60}{6} + 2\left(\frac{0.766}{6}\right) = 10.255 \text{ inches}$$

Table 2.   Enlarged Pinion and Reduced Gear Dimensions to Avoid Interference —
14½-degree Involute Full Depth Teeth

| Number of Pinion Teeth | Changes in Pinion and Gear Diameters | Circular Tooth Thickness | | Min. No. of Teeth in Mating Gear | |
|---|---|---|---|---|---|
| | | Pinion | Mating Gear | To avoid Undercut | For full involute Action |
| 10 | 1.3731 | 1.9259 | 1.2157 | 54 | 27 |
| 11 | 1.3104 | 1.9097 | 1.2319 | 53 | 27 |
| 12 | 1.2477 | 1.8935 | 1.2481 | 52 | 28 |
| 13 | 1.1850 | 1.8773 | 1.2643 | 51 | 28 |
| 14 | 1.1223 | 1.8611 | 1.2805 | 50 | 28 |
| 15 | 1.0597 | 1.8449 | 1.2967 | 49 | 28 |
| 16 | 0.9970 | 1.8286 | 1.3130 | 48 | 28 |
| 17 | 0.9343 | 1.8124 | 1.3292 | 47 | 28 |
| 18 | 0.8716 | 1.7962 | 1.3454 | 46 | 28 |
| 19 | 0.8089 | 1.7800 | 1.3616 | 45 | 28 |
| 20 | 0.7462 | 1.7638 | 1.3778 | 44 | 28 |
| 21 | 0.6835 | 1.7476 | 1.3940 | 43 | 28 |
| 22 | 0.6208 | 1.7314 | 1.4102 | 42 | 27 |
| 23 | 0.5581 | 1.7151 | 1.4265 | 41 | 27 |
| 24 | 0.4954 | 1.6989 | 1.4427 | 40 | 27 |
| 25 | 0.4328 | 1.6827 | 1.4589 | 39 | 26 |
| 26 | 0.3701 | 1.6665 | 1.4751 | 38 | 26 |
| 27 | 0.3074 | 1.6503 | 1.4913 | 37 | 26 |
| 28 | 0.2447 | 1.6341 | 1.5075 | 36 | 25 |
| 29 | 0.1820 | 1.6179 | 1.5237 | 35 | 25 |
| 30 | 0.1193 | 1.6017 | 1.5399 | 34 | 24 |
| 31 | 0.0566 | 1.5854 | 1.5562 | 33 | 24 |

All dimensions are given in inches and are for 1 diametral pitch.   For other pitches divide tabular values by desired diametral pitch.

Add to the standard outside diameter of the pinion the amount given in the second column of the table divided by the desired diametral pitch, and, (to maintain standard center distance) subtract the same amount from the outside diameter of the mating gear. Long addendum pinions will mesh with standard gears, but the center distance will be greater than standard.

For a mating gear with minimum number of teeth to avoid undercut:

$$D_O \text{ (gear)} = \frac{N_G}{P} + 2a = \frac{21}{6} + 2\left(\frac{0.766}{6}\right) = 3.755 \text{ inches}$$

**Minimum Number of Teeth to Avoid Undercutting by Hob.** — The data in the above tables give tooth proportions for low numbers of teeth to avoid interference between the gear tooth tip and the pinion tooth flank.   Consideration must also be given to possible undercutting of the pinion tooth flank by the hob used to cut the pinion.   The minimum number of teeth $N_{min}$ of standard proportion that may be cut without undercut is: $N_{min} = 2P \csc^2 \phi \, [a_H - r_t \, (1 - \sin \phi)]$ where: $a_H =$ cutter addendum; $r_t =$ radius at cutter tip or corners; $\phi =$ cutter pressure angle; and $P =$ diametral pitch.

**Gear to Mesh with Enlarged Pinion.** — Data in the fifth column of Table 2 show minimum number of teeth in a mating gear which can be cut with hob or rack type cutter without undercut, when outside diameter of gear has been reduced an

## Table 3.  Dimensions Required when Using Enlarged Fine-pitch Pinions — 20 Diametral Pitch and Finer (ANSI B6.7-1967)

| No. of Teeth n | Outside Diameter | Cir. Tooth Thickness at Standard Pitch Diam. | Decrease in Standard Outside Diam.[1] | Cir. Tooth Thickness at Standard Pitch Diam. | Recommended Minimum No. of Teeth N | Contact Ratio, n Mating with N | Enlarged Pinion Mating with St'd. Gear / Increase over St'd. Center Distance | Two Equal Enlarged Mating Pinions[2] | Contact Ratio of Two Equal Enlarged Mating Pinions |
|---|---|---|---|---|---|---|---|---|---|
| STANDARD 20-DEGREE PRESSURE ANGLE | | | | | | | | | |
| 10 | 12.8302 | 1.8730 | 0.8302 | 1.2686 | 33 | 1.419 | 0.4151 | 0.8302 | 1.135 |
| 11 | 13.7132 | 1.8304 | 0.7132 | 1.3113 | 30 | 1.450 | 0.3566 | 0.7132 | 1.186 |
| 12 | 14.5963 | 1.7878 | 0.5963 | 1.3538 | 27 | 1.473 | 0.2982 | 0.5962 | 1.238 |
| 13 | 15.4793 | 1.7452 | 0.4793 | 1.3964 | 25 | 1.493 | 0.2397 | 0.4794 | 1.290 |
| 14 | 16.3623 | 1.7027 | 0.3623 | 1.4389 | 23 | 1.508 | 0.1812 | 0.3624 | 1.344 |
| 15 | 17.2453 | 1.6601 | 0.2453 | 1.4815 | 21 | 1.516 | 0.1227 | 0.2454 | 1.398 |
| 16 | 18.1284 | 1.6175 | 0.1284 | 1.5241 | 19 | 1.519 | 0.0642 | 0.1284 | 1.436 |
| 17 | 19.0114 | 1.5749 | 0.0114 | 1.5667 | 18 | 1.522 | 0.0057 | 0.0114 | 1.511 |
| ALTERNATE 25-DEGREE PRESSURE ANGLE | | | | | | | | | |
| 8 | 10.5712 | 1.8372 | 0.5712 | 1.3044 | 15 | 1.264 | 0.2856 | 0.5712 | 1.096 |
| 9 | 11.3926 | 1.7539 | 0.3926 | 1.3877 | 14 | 1.286 | 0.1963 | 0.3926 | 1.158 |
| 10 | 12.2140 | 1.6706 | 0.2140 | 1.4710 | 13 | 1.300 | 0.1070 | 0.2140 | 1.222 |
| 11 | 13.0354 | 1.5873 | 0.0354 | 1.5543 | 12 | 1.306 | 0.0177 | 0.0354 | 1.288 |
| ALTERNATE 14½-DEGREE PRESSURE ANGLE | | | | | | | | | |
| 10 | 13.3731 | 1.9259 | 1.3731 | 1.2157 | 54 | 1.831 | 0.6866 | 1.3732 | 1.053 |
| 11 | 14.3104 | 1.9097 | 1.3104 | 1.2319 | 53 | 1.847 | 0.6552 | 1.3104 | 1.088 |
| 12 | 15.2477 | 1.8935 | 1.2477 | 1.2481 | 52 | 1.860 | 0.6239 | 1.2477 | 1.121 |
| 13 | 16.1850 | 1.8773 | 1.1850 | 1.2643 | 51 | 1.873 | 0.5925 | 1.1850 | 1.154 |
| 14 | 17.1223 | 1.8611 | 1.1223 | 1.2805 | 50 | 1.885 | 0.5612 | 1.2223 | 1.186 |
| 15 | 18.0597 | 1.8448 | 1.0597 | 1.2967 | 49 | 1.896 | 0.5299 | 1.0597 | 1.217 |
| 16 | 18.9970 | 1.8286 | 0.9970 | 1.3130 | 48 | 1.906 | 0.4985 | 0.9970 | 1.248 |
| 17 | 19.9343 | 1.8124 | 0.9343 | 1.3292 | 47 | 1.914 | 0.4672 | 0.9343 | 1.278 |
| 18 | 20.8716 | 1.7962 | 0.8716 | 1.3454 | 46 | 1.922 | 0.4358 | 0.8716 | 1.307 |
| 19 | 21.8089 | 1.7800 | 0.8089 | 1.3616 | 45 | 1.929 | 0.4045 | 0.8089 | 1.336 |
| 20 | 22.7462 | 1.7638 | 0.7462 | 1.3778 | 44 | 1.936 | 0.3731 | 0.7462 | 1.364 |
| 21 | 23.6835 | 1.7476 | 0.6835 | 1.3940 | 43 | 1.942 | 0.3418 | 0.6835 | 1.392 |
| 22 | 24.6208 | 1.7314 | 0.6208 | 1.4102 | 42 | 1.948 | 0.3104 | 0.6208 | 1.419 |
| 23 | 25.5581 | 1.7151 | 0.5581 | 1.4265 | 41 | 1.952 | 0.2791 | 0.5581 | 1.446 |
| 24 | 26.4954 | 1.6989 | 0.4954 | 1.4427 | 40 | 1.956 | 0.2477 | 0.4954 | 1.472 |
| 25 | 27.4328 | 1.6827 | 0.4328 | 1.4589 | 39 | 1.960 | 0.2164 | 0.4328 | 1.498 |
| 26 | 28.3701 | 1.6665 | 0.3701 | 1.4751 | 38 | 1.963 | 0.1851 | 0.3701 | 1.524 |
| 27 | 29.3074 | 1.6503 | 0.3074 | 1.4913 | 37 | 1.965 | 0.1537 | 0.3074 | 1.549 |
| 28 | 30.2447 | 1.6341 | 0.2448 | 1.5075 | 36 | 1.967 | 0.1224 | 0.2448 | 1.573 |
| 29 | 31.1820 | 1.6179 | 0.1820 | 1.5237 | 35 | 1.969 | 0.0910 | 0.1820 | 1.598 |
| 30 | 32.1193 | 1.6017 | 0.1193 | 1.5399 | 34 | 1.970 | 0.0597 | 0.1193 | 1.622 |
| 31 | 33.0566 | 1.5854 | 0.0566 | 1.5562 | 33 | 1.971 | 0.0283 | 0.0566 | 1.646 |

All dimensions are given in inches and are for 1 diametral pitch.  For other pitches divide tabulated dimensions by the diametral pitch.

[1] To maintain standard center distance when using an enlarged pinion, the mating gear diameter must be decreased by the amount of the pinion enlargement.

[2] If enlarged mating pinions are of unequal size, the center distance is increased by an amount equal to one-half the sum of their increase over standard outside diameters. Data in this column are not given in the standard.

amount equal to the pinion enlargement to retain the standard center distance. To calculate $N$ for the gear, insert addendum $a$ of enlarged mating pinion in the formula $N = 2a \times \csc^2 \phi$.

*Example:* A gear is to mesh with a 24-tooth pinion of 1 diametral pitch which has been enlarged 0.4954 inch, as shown by the table. The pressure angle is 14½ degrees. Find minimum number of teeth $N$ for reduced gear.

Pinion addendum $= 1 + (0.4954 \div 2) = 1.2477$; hence

$$N = 2 \times 1.2477 \times 15.95 = 39.8 \text{ (use 40)}$$

In the case of 20-degree fine pitch gears with reduced outside diameters, the recommended minimum numbers of teeth given in Table 3, sixth column, are somewhat more than the minimum numbers required to prevent undercutting and are based upon studies made of optimum surface sliding by the American Gear Manufacturers Association Spur and Planetary Committee.

**Standard Center-distance System for Enlarged Pinions.** — In this system, sometimes referred to as " long and short addendums," the center distance is made standard for the numbers of teeth in pinion and gear. The outside diameter of the gear is decreased by the same amount that the outside of the pinion is enlarged. The advantages of this system are: (1) No change in center distance or ratio is required; (2) The operating pressure angle remains standard; and (3) A slightly greater contact ratio is obtained than when the center distance is increased. The disadvantages are (1) The gears as well as the pinion must be changed from standard dimensions; (2) Pinions having fewer than the minimum number of teeth to avoid undercut cannot be satisfactorily meshed together; and (3) In most cases where gear trains include idler gears, the standard center-distance system cannot be used.

**Enlarged Center-distance System for Enlarged Pinions.** — If an enlarged pinion is meshed with another enlarged pinion or with a gear of standard outside diameter, the center distance must be increased. For fine-pitch gears, it is usually satisfactory to increase the center distance by an amount equal to one-half of the enlargements (see eighth column of Table 3). This is an approximation as theoretically there is a slight increase in backlash. The advantages of this system are: (1) Only the pinions need be changed from the standard dimensions; (2) Pinions having fewer than 18 teeth may engage other pinions in this range; (3) The pinion tooth, which is the weaker member, is made stronger by the enlargement; and (4) The tooth contact stress, which controls gear durability, is lowered by being moved away from the pinion base circle. The disadvantages are: (1) Center distances must be enlarged over the standard; (2) The operating pressure angle increases slightly with different combinations of pinions and gears, which is usually not important; and (3) The contact ratio is slightly smaller than that obtained with the standard center-distance system. This consideration is of minor importance as in the worst case the loss is approximately only 6 per cent.

*Enlarged Pinions Meshing without Backlash:* When two enlarged pinions are to mesh without backlash, their center distance will be greater than the standard and less than that for the enlarged center-distance system. This center distance may be calculated by the formulas given in the following section.

**Center Distance at Which Modified Mating Spur Gears Will Mesh with No Backlash.** — When the tooth thickness of one or both of a pair of mating spur gears has been increased or decreased from the standard value ($\pi \div 2P$), the center distance at which they will mesh tightly (without backlash) may be calculated from the following formulas:

$$\text{inv } \phi_1 = \text{inv } \phi + \frac{P(t + T) - \pi}{n + N}$$

$$C = \frac{n + N}{2P}$$

$$C_1 = \frac{\cos \phi}{\cos \phi_1} \times C$$

In these formulas, $P$ = diametral pitch; $n$ = number of teeth in pinion; $N$ = number of teeth in gear; $t$ and $T$ are the actual tooth thicknesses of the pinion and gear, respectively, on their standard pitch circles; inv $\phi$ = involute function of standard pressure angle of gears; $C$ = standard center distance for the gears; $C_1$ = center distance at which the gears mesh without backlash; and inv $\phi_1$ = involute function of operating pressure angle when gears are meshed tightly at center distance $C_1$.

*Example:* Calculate the center distance for no backlash when an enlarged 10-tooth pinion of 100 diametral pitch and 20-degree pressure angle is meshed with a standard 30-tooth gear, the circular thickness of the pinion and gear, respectively, being 0.01873 and 0.015708 inch.

$$\text{inv } \phi_1 = \text{inv } 20° + \frac{100(0.01873 + 0.015708) - \pi}{(10 + 30)}$$

From the table of involute functions, page 198 inv 20-degrees = 0.014904. Therefore,

$$\text{inv } \phi_1 = 0.014904 + \frac{0.34438 - 0.31416}{4} = 0.022459$$

$$\phi_1 = 22°49' \text{ from page 200}$$

$$C = \frac{n + N}{2P} = \frac{10 + 30}{2 \times 100} = 0.2000 \text{ inch}$$

$$C_1 = \frac{\cos 20°}{\cos 22°49'} \times 0.2000 = \frac{0.93969}{0.92175} \times 0.2000 = 0.2039 \text{ inch}$$

**Contact Diameter.** — For two meshing gears it is important to know the contact diameter of each. A first gear with number of teeth, $n$, and outside diameter, $d_o$, meshes at a standard center distance with a second gear with number of teeth, $N$, and outside diameter, $D_o$; both gears have a diametral pitch, $P$, and pressure angle, $\phi$. $a$, $A$, $b$, and $B$ are unnamed angles used only in the calculations. The contact diameter, $d_c$, is found by a three-step calculation that can be done by hand using a trigonometric table and a logarithmic table or a desk calculator. Slide rule calculation is not recommended because it is not accurate enough to give good results. The three-step formulas to find the contact diameter, $d_c$, of the first gear are:

$$\cos A = \frac{N \cos \phi}{D_o \times P} \tag{1}$$

$$\tan b = \tan \phi - \frac{N}{n}(\tan A - \tan \phi) \tag{2}$$

$$d_c = \frac{n \cos \phi}{P \cos b} \tag{3}$$

Similarly the three-step formulas to find the contact diameter, $D_c$, of the second gear are:

$$\cos a = \frac{n \cos \phi}{d_o \times P} \tag{4}$$

$$\tan B = \tan \phi - \frac{n}{N} (\tan a - \tan \phi) \tag{5}$$

$$D_c = \frac{N \cos \phi}{P \cos B} \tag{6}$$

**Contact Ratio.** — The contact ratio of a pair of mating spur gears must be well over 1.0 to assure a smooth transfer of load from one pair of teeth to the next pair as the two gears rotate under load. Because of a reduction in contact ratio due to such factors as tooth deflection, tooth spacing errors, tooth tip breakage, and outside diameter and center distance tolerances, the contact ratio of gears for power transmission as a general rule should not be less than about 1.4. A contact ratio of as low as 1.15 may be used in extreme cases, provided the tolerance effects mentioned above are accounted for in the calculation. The formula for determining the contact ratio, $m_f$, using the nomenclature in the previous section is:

$$m_f = \frac{N}{6.28318} (\tan A - \tan B) \tag{7a}$$

or

$$m_f = \frac{n}{6.28318} (\tan a - \tan b) \tag{7b}$$

Both formulas should give the same answer. It is good practice to use both formulas as a check on the previous calculations.

**Lowest Point of Single Tooth Contact.** — This diameter on the pinion (sometimes referred to as LPSTC) is used to find the maximum contact compressive stress (sometimes called the Hertz Stress) of a pair of mating spur gears. The two-step formulas for determining this pinion diameter, $d_L$, using the same nomenclature as in the previous sections with $c$ and $C$ as unnamed angles used only in the calculations are:

$$\tan c = \tan a - \frac{6.28318}{n} \tag{8}$$

$$d_L = \frac{n \cos \phi}{P \cos c} \tag{9}$$

In some cases it is necessary to have a plot of the compressive stress over the whole cycle of contact; in this case the LPSTC for the gear is required also. The similar two-step formulas for this gear diameter are:

$$\tan C = \tan A - \frac{6.28318}{N} \tag{10}$$

$$D_L = \frac{N \cos \phi}{P \cos C} \tag{11}$$

**Maximum Hob Tip Radius.** — The standard gear tooth proportions given by the formulas in Table 2 on page 718 provide a specified size for the rack fillet radius in the general form of (a constant) × (pitch). For any given standard this constant may vary up to a maximum which it is geometrically impossible to exceed; this maximum constant, $r_c(\max)$, is found by the formula:

$$r_c(\max) = \frac{0.785398 \cos \phi - b \sin \phi}{1 - \sin \phi} \tag{12}$$

where $b$ is the similar constant in the specified formula for the gear dedendum. The hob tip radius of any standard hob to finish cut any standard gear may vary from zero up to this limiting value.

**Undercut Limit for Hobbed Involute Gears.** — It is well to avoid designing and specifying gears that will have a hobbed trochoidal fillet that undercuts the involute gear tooth profile. This should be avoided because it may cause the involute profile to be cut away up to a point above the required contact diameter with the mating gear so that involute action is lost and the contact ratio reduced to a level that may be too low for proper conjugate action. An undercut fillet will also weaken the beam strength and thus raise the fillet tensile stress of the gear tooth. To assure that the hobbed gear tooth will not have an undercut fillet, the following formula must be satisfied:

$$\frac{b - r_c}{\sin \phi} + r_c \leqq 0.5 \, n \sin \phi \tag{13}$$

where $b$ is the dedendum constant; $r_c$ is the hob or rack tip radius constant; $n$ is the number of teeth in the gear; and $\phi$ is the gear and hob pressure angle. If the gear is not standard or the hob does not roll at the gear pitch diameter, this formula can not be applied and the determination of the expected existence of undercut becomes a considerably more complicated procedure.

**Highest Point of Single Tooth Contact.** — This diameter is used to place the maximum operating load for the determination of the gear tooth fillet stress. The two-step formulas for determining this diameter, $d_H$, of the pinion using the same nomenclature as in the previous sections with $d$ and $D$ as unnamed angles used only in the calculations are:

$$\tan d = \tan b + \frac{6.28318}{n} \tag{14}$$

$$d_H = \frac{n \cos \phi}{P \cos d} \tag{15}$$

Similarly for the gear: —

$$\tan D = \tan B + \frac{6.28318}{N} \tag{16}$$

$$D_H = \frac{N \cos \phi}{P \cos D} \tag{17}$$

**True Involute Form Diameter.** — The point on the gear tooth at which the fillet and the involute profile are tangent to each other should be determined to assure that it lies at a smaller diameter than the required contact diameter with the mating gear. If the TIF diameter is larger than the contact diameter, then fillet interference will occur with severe damage to the gear tooth profile and rough action of the gear set. This two-step calculation is made by using the following two formulas with $e$ and $E$ as unnamed angles used only in the calculations:

$$\tan e = \tan \phi - \frac{4}{n}\left(\frac{b - r_c}{\sin 2\phi} + \frac{r_c}{2 \cos \phi}\right) \tag{18}$$

$$d_{TIF} = \frac{n \cos \phi}{P \cos e} \tag{19}$$

As in the previous sections, $\phi$ is the pressure angle of the gear; $n$ is the number of teeth in the pinion; $b$ is the dedendum constant, $r_c$ is the rack or hob tip radius constant, $P$ is the gear diametral pitch and $d_{TIF}$ is the true involute form diameter.

Similarly, for the mating gear:

$$\tan E = \tan \phi - \frac{4}{N}\left(\frac{b - r_c}{\sin 2\phi} + \frac{r_c}{2 \cos \phi}\right) \tag{20}$$

$$D_{TIF} = \frac{N \cos \phi}{P \cos E} \tag{21}$$

Where $N$ is the number of teeth in this mating gear and $D_{TIF}$ is the true involute form diameter.

**Profile Checker Settings.** — The actual tooth profile tolerance will need to be determined on high performance gears that operate either at high unit loads or at high pitch-line velocity. This is done on an involute checker, a machine which requires two settings, the gear base radius and the roll angle in degrees to significant points on the involute. From the smallest diameter outward these significant points are: TIF, Contact Diameter, LPSTC, Pitch Diameter, HPSTC, and Outside Diameter.

The base radius is:

$$R_b = \frac{N \cos \phi}{2P} \tag{22}$$

The roll angle, in degrees, at any point is equal to the tangent of the pressure angle at that point multiplied by 57.2958. The following table shows the tangents to be used at each significant diameter.

| Significant Point on Tooth Profile | Pinion | Gear | For Computation |
|---|---|---|---|
| TIF | $\tan e$ | $\tan E$ | (See Formulas 18 & 20) |
| Contact Diam. | $\tan b$ | $\tan B$ | (See Formulas 2 & 5) |
| LPSTC | $\tan c$ | $\tan C$ | (See Formulas 8 & 10) |
| Pitch Diam. | $\tan \phi$ | $\tan \phi$ | ($\phi$ = Pressure angle) |
| HPSTC | $\tan d$ | $\tan D$ | (See Formulas 14 & 16) |
| Outside Diam. | $\tan a$ | $\tan A$ | (See Formulas 4 & 1) |

*Example:* Find the significant diameters, contact ratio and hob tip radius for a 10-diametral pitch, 23-tooth, 20-degree pressure angle pinion of 2.5-inch outside diameter if it is to mesh with a 31-tooth gear of 3.3-inch outside diameter.

$$\text{Thus: } n = 23 \qquad N = 31$$
$$d_o = 2.5 \qquad D_o = 3.3$$
$$P = 10 \qquad \phi = 20°$$

1. Pinion contact diameter $d_c$:

$$\cos A = \frac{31 \times 0.93969}{3.3 \times 10} \tag{1}$$

$$= 0.88274 \qquad A = 28°1'30''$$

$$\tan b = 0.36397 - \frac{31}{23} \ (0.53227 - 0.36397) \tag{2}$$

$$= 0.13713 \qquad b = 7°48'26''$$

$$d_c = \frac{23 \times 0.93969}{10 \times 0.99073} \tag{3}$$

$$= 2.1815 \text{ inches}$$

2. Gear contact diameter, $D_c$

$$\cos a = \frac{23 \times 0.93963}{2.5 \times 10} \tag{4}$$

$$= 0.86452 \qquad a = 30°10'20''$$

$$\tan B = 0.36397 - \frac{23}{31} \ (0.58136 - 0.36937) \tag{5}$$

$$= 0.20267 \qquad B = 11°27'26''$$

$$D_c = \frac{31 \times 0.93969}{10 \times 0.98000} \tag{6}$$

$$= 2.9725 \text{ inches}$$

3. Contact ratio, $m_f$

$$m_f = \frac{31}{6.28318} \ (0.53227 - 0.20267) \tag{7a}$$

$$= 1.626$$

$$m_f = \frac{23}{6.28318} \ (0.58136 - 0.13713) \tag{7b}$$

$$= 1.626$$

4. Pinion LPSTC, $d_L$

$$\tan c = 0.58136 - \frac{6.28318}{23} \tag{8}$$

$$= 0.30818 \qquad c = 17°7'41''$$

$$d_L = \frac{23 \times 0.93969}{10 \times 0.95565} \tag{9}$$

$$= 2.2616 \text{ inches}$$

5. Gear LPSTC, $D_L$

$$\tan C = 0.53227 - \frac{6.28318}{31} \tag{10}$$

$$= 0.32959 \qquad C = 18°14'30''$$

$$D_L = \frac{31 \times 0.93969}{10 \times 0.94974} \tag{11}$$
$$= 3.0672 \text{ inches}$$

6. Maximum permissible hob tip radius, $r_c$ (max). The dedendum factor is 1.25.

$$r_c \text{ (max)} = \frac{0.785398 \times 0.93969 - 1.25 \times 0.34202}{1 - 0.34202} \tag{12}$$
$$= 0.4719 \text{ inch.}$$

7. If the hop tip radius $r_c$ is 0.30, determine if the pinion involute is undercut.

$$\frac{1.25 - 0.30}{0.34202} + 0.30 \leqq 0.5 \times 23 \times 0.34202 \tag{13}$$

$$3.0776 < 3.9332$$

therefore there is no involute undercut.

8. Pinion HPSTC, $D_H$

$$\tan d = 0.13713 + \frac{6.28318}{23} \tag{14}$$
$$= 0.41031 \qquad d = 22°18'32''$$

$$d_H = \frac{23 \times 0.93969}{10 \times 0.92515} \tag{15}$$
$$= 2.3362 \text{ inches}$$

9. Gear HPSTC, $D_H$

$$\tan D = 0.20267 + \frac{6.28318}{31} \tag{16}$$
$$= 0.40535 \qquad D = 22°3'55''$$

$$D_H = \frac{31 \times 0.93969}{10 \times 0.92676} \tag{17}$$
$$= 3.1433 \text{ inches}$$

10. Pinion TIF diameter, $d_{TIF}$

$$\tan e = 0.36397 - \frac{4}{23}\left(\frac{1.25 - 0.30}{0.64279} + \frac{0.30}{2 \times 0.93969}\right) \tag{18}$$
$$= 0.07917 \qquad e = 4°31'36''$$

$$d_{TIF} = \frac{23 \times 0.93969}{10 \times 0.99688} \tag{19}$$
$$= 2.1681 \text{ inches}$$

11. Gear TIF diameter $D_{TIF}$

$$\tan E = 0.36397 - \frac{4}{31}\left(\frac{1.25 - 0.30}{0.64279} + \frac{0.30}{2 \times 0.93969}\right) \tag{20}$$
$$= 0.15267 \qquad E = 8°40'50''$$

$$D_{TIF} = \frac{31 \times 0.93969}{10 \times 0.98855} = 2.9468 \text{ inches} \tag{21}$$

**Backlash in Gears.** — In general, backlash in gears is play between mating teeth. For purposes of measurement and calculation, backlash is defined as the amount by which a tooth space exceeds the thickness of an engaging tooth. It does not include the effect of center-distance changes of the mountings and variations in bearings. When not otherwise specified, numerical values of backlash are understood to be given on the pitch circles.

The general purpose of backlash is to prevent gears from jamming together and making contact on both sides of their teeth simultaneously. Lack of backlash may cause noise, overloading, overheating of the gears and bearings, and even seizing and failure.

Excessive backlash is objectionable, particularly if the drive is frequently reversing, or if there is an overrunning load as in cam drives. On the other hand, specification of an unnecessarily small amount of backlash allowance will increase the cost of gears, because errors in runout, pitch, profile, and mounting must be held correspondingly smaller. Backlash in no way affects involute action and usually is not detrimental to proper gear action.

**Determining Proper Amount of Backlash.** — In specifying proper backlash and tolerances for a pair of gears, the most important factor is probably the maximum permissible amount of runout in both gear and pinion (or worm). Next are the allowable errors in profile, pitch, tooth thickness, and helix angle. It can be seen that the backlash between a pair of gears will vary as successive teeth make contact because of the effect of composite tooth errors, particularly runout, and errors in the gear mounting centers and bearings.

Other important considerations are speed and space for lubricant film. Slow-moving gears, in general, require the least backlash. Fast-moving fine-pitch gears are usually lubricated with relatively light oil, but if there is insufficient clearance for an oil film, and particularly if oil trapped at the root of the teeth cannot escape, heat and excessive tooth loading will occur.

Heat is a factor because gears may operate warmer than, and therefore expand more than, the housings. The heat may result from oil churning or from frictional losses between the teeth, at bearings or oil seals, or from external causes. Moreover, for the same temperature rise the material of the gears — for example, bronze and aluminum — may expand more than the material of the housings, usually steel or cast iron.

The higher the helix angle or spiral angle, the more transverse backlash is required for a given normal backlash. The transverse backlash is equal to the normal backlash divided by the cosine of the helix angle.

In designs employing normal pressure angles higher than 20 degrees, special consideration must be given to backlash, since more backlash is required on the pitch circles to obtain a given amount of backlash in a direction normal to the tooth profiles.

Errors in boring the gear housings, both in center distance and alignment, are of extreme importance in determining allowance to obtain the backlash desired. The same is true in the mounting of the gears, which is affected by the type and adjustment of bearings, and similar factors. Other influences in backlash specification are heat-treatment subsequent to cutting the teeth, lapping operations, the possible necessity for recutting for any reason, and reduction of tooth thickness through normal wear.

Minimum backlash is necessary for timing, indexing, gun-sighting, and certain instrument gear trains. If the operating speed is very low and the necessary precautions are taken in the manufacture of such gear trains, the backlash may be held to extremely small limits. However, the specification of "zero backlash," so commonly stipulated for gears of this nature, usually involves special and expensive technique, and is difficult to obtain.

**Recommended Backlash.** — In the following tables American Gear Manufacturers Association recommendations for backlash ranges for various kinds of gears are given.* For purposes of measurement and calculation, backlash is defined as the amount by which a tooth space exceeds the thickness of an engaging tooth. When not otherwise specified, numerical values of backlash are understood to be measured at the tightest point of mesh on the pitch circle in a direction normal to the tooth surface when the gears are mounted in their specified position.

*Coarse-Pitch Gears:* Table 1 gives the recommended backlash range for coarse-pitch spur, helical and herringbone gearing. Because backlash for helical and herringbone gears is more conveniently measured in the normal plane, Table 1 has been prepared to show backlash in the normal plane for coarse-pitch helical and herringbone gearing and in the transverse plane for spur gears. To obtain backlash in the transverse plane for helical and herringbone gears, divide the normal plane backlash in Table 1 by the cosine of the helix angle.

The backlash tolerances given in this table contain allowances for gear expansion due to a differential in the operating temperature of the gearing and their supporting structure. The values may be used where the operating temperature is up to 70 degrees F. higher than the ambient temperature. These backlash values will provide proper running clearances for most of the applications listed in Table 9, beginning on page 786.

The following important factors must be considered in establishing backlash tolerances: (a) Center distance tolerance, (b) Parallelism of gear axes, (c) Side runout or wobble, (d) Tooth thickness tolerance, (e) Pitch line runout tolerance, (f) Profile tolerance, (g) Pitch tolerance, (h) Lead tolerance, (i) Types of bearings and subsequent wear, (j) Deflection under load, (k) Gear tooth wear, (l) Pitch line velocity, (m) Lubrication requirements, (n) Thermal expansion of gears and housing.

A tight mesh may result in objectionable gear sound, increased power losses, overheating, rupture of the lubricant film, overloaded bearings and premature gear failure. However, it is recognized that there are some gearing applications where a tight mesh (zero backlash) may be required.

Specifying unnecessarily close backlash tolerances will increase the cost of the gearing. It is obvious from the above summary that the desired amount of backlash is difficult to evaluate. It is, therefore, recommended that when a designer, user or purchaser includes a reference to backlash in a gearing specification and drawing, consultation be arranged with the manufacturer.

*Bevel and Hypoid Gears:* Table 2 gives similar backlash range values for bevel and hypoid gears. These are values based upon average conditions for general purpose gearing, but may require modification to meet specific needs.

Backlash on bevel and hypoid gears can be controlled to some extent by axial adjustment of the gears during assembly. However, due to the fact that actual adjustment of a bevel or hypoid gear in its mounting will alter the amount of backlash, it is imperative that the amount of backlash cut into the gears during manufacture is not excessive. Bevel and hypoid gears must always be capable of operation without interference when adjusted for zero backlash. This requirement is imposed by the fact that a failure of the axial thrust bearing might permit the gears to operate under this condition. Therefore, bevel and hypoid gears should never be designed to operate with normal backlash in excess of $0.080/P$ where $P$ is diametral pitch.

*Fine-Pitch Gears:* Table 3 gives similar backlash range values for fine-pitch spur, helical and herringbone gearing.

* Extracted from Gear Classification Manual, AGMA 390.02 with permission of the publisher, the American Gear Manufacturers Association, 1901 North Fort Myer Drive, Arlington, Va. 22209.

**Table 1. AGMA Recommended Backlash Range for Coarse-Pitch Spur, Helical and Herringbone Gearing**

| Center Distance (Inches) | Normal Diametral Pitches | | | | |
|---|---|---|---|---|---|
| | 0.5-1.99 | 2-3.49 | 3.5-5.99 | 6-9.99 | 10-19.99 |
| | Backlash, Normal Plane, Inches* | | | | |
| Up to 5 | | | | | .005-.015 |
| Over 5 to 10 | | | | .010-.020 | .010-.020 |
| Over 10 to 20 | | | .020-.030 | .015-.025 | .010-.020 |
| Over 20 to 30 | | .030-.040 | .025-.030 | .020-.030 | |
| Over 30 to 40 | .040-.060 | .035-.045 | .030-.040 | .025-.035 | |
| Over 40 to 50 | .050-.070 | .040-.055 | .035-.050 | .030-.040 | |
| Over 50 to 80 | .060-.080 | .045-.065 | .040-.060 | | |
| Over 80 to 100 | .070-.095 | .050-.080 | | | |
| Over 100 to 120 | .080-.110 | | | | |

* Suggested backlash, on nominal centers, measured after rotating to the point of closest engagement. For helical and herringbone gears, divide above values by the cosine of the helix angle to obtain the transverse backlash.

The above backlash tolerances contain allowance for gear expansion due to differential in the operating temperature of the gearing and their supporting structure. The values may be used where the operating temperatures are up to 70 deg F higher than the ambient temperature.

For most of the gearing applications listed in Table 9, the recommended backlash ranges will provide proper running clearance between the engaging teeth of mating gears. Deviation below the minimum or above the maximum values shown, which do not affect operational use of the gearing, should not be cause for rejection.

Definite backlash tolerances on coarse-pitch gearing are to be considered binding on the gear manufacturer only when agreed upon in writing.

Some applications may require less backlash than shown in the above table. In such cases the amount and tolerance should be by agreement between the manufacturer and purchaser.

**Table 2. AGMA Recommended Backlash Range for Bevel and Hypoid Gears\***

| Diametral Pitch | Normal Backlash, Inch | | Diametral Pitch | Normal Backlash, Inch | |
|---|---|---|---|---|---|
| | Quality Numbers 7 through 13 | Quality Numbers 4 through 6 | | Quality Numbers 7 through 13 | Quality Numbers 4 through 6 |
| 1.00 to 1.25 | 0.020-0.030 | 0.045-0.065 | 5.00 to 6.00 | 0.005-0.007 | 0.006-0.013 |
| 1.25 to 1.50 | 0.018-0.026 | 0.035-0.055 | 6.00 to 8.00 | 0.004-0.006 | 0.005-0.010 |
| 1.50 to 1.75 | 0.016-0.022 | 0.025-0.045 | 8.00 to 10.00 | 0.003-0.005 | 0.004-0.008 |
| 1.75 to 2.00 | 0.014-0.018 | 0.020-0.040 | 10.00 to 16.00 | 0.002-0.004 | 0.003-0.005 |
| 2.00 to 2.50 | 0.012-0.016 | 0.020-0.030 | 16.00 to 20.00 | 0.001-0.003 | 0.002-0.004 |
| 2.50 to 3.00 | 0.010-0.013 | 0.015-0.025 | 20 to 50 | 0.000-0.002 | 0.000-0.002 |
| 3.00 to 3.50 | 0.008-0.011 | 0.012-0.022 | 50 to 80 | 0.000-0.001 | 0.000-0.001 |
| 3.50 to 4.00 | 0.007-0.009 | 0.010-0.020 | 80 and finer | 0.000-0.0007 | 0.000-0.0007 |
| 4.00 to 5.00 | 0.006-0.008 | 0.008-0.016 | ............ | ............ | ............ |

* Measured at tightest point of mesh.

**Table 3. AGMA Backlash Allowance and Tolerance for Fine-Pitch Spur, Helical and Herringbone Gearing**

| Backlash Designation | Normal Diametral Pitch Range | Tooth Thinning to Obtain Backlash (Note 1) | | Resulting Approximate Backlash (per Mesh) Normal Plane (Note 2), Inch |
|---|---|---|---|---|
| | | Allowance, per Gear, Inch | Tolerance, per Gear, Inch | |
| A | 20 thru 45 | .002 | 0 to .002 | .004 to .008 |
| | 46 thru 70 | .0015 | 0 to .002 | .003 to .007 |
| | 71 thru 90 | .001 | 0 to .00175 | .002 to .0055 |
| | 91 thru 200 | .00075 | 0 to .00075 | .0015 to .003 |
| B | 20 thru 60 | .001 | 0 to .001 | .002 to .004 |
| | 61 thru 120 | .00075 | 0 to .00075 | .0015 to .003 |
| | 121 thru 200 | .0005 | 0 to .0005 | .001 to .002 |
| C | 20 thru 60 | .0005 | 0 to .0005 | .001 to .002 |
| | 61 thru 120 | .00035 | 0 to .0004 | .0007 to .0015 |
| | 121 thru 200 | .0002 | 0 to .0003 | .0004 to .001 |
| D | 20 thru 60 | .00025 | 0 to .00025 | .0005 to .001 |
| | 61 thru 120 | .0002 | 0 to .0002 | .0004 to .0008 |
| | 121 thru 200 | .0001 | 0 to .0001 | .0002 to .0004 |
| E | 20 thru 60 | | 0 to .00025 | 0 to .0005 |
| | 60 thru 120 | Zero (Note 3) | 0 to .0002 | 0 to .0004 |
| | 121 thru 200 | | 0 to .0001 | 0 to .0002 |

*Backlash* in gears is the play between mating tooth surfaces. For purposes of measurement and calculation, backlash is defined as the amount by which a tooth space exceeds the thickness of an engaging tooth. When not otherwise specified, numerical values of backlash are understood to be measured at the tightest point of mesh on the pitch circle in a direction normal to the tooth surface when the gears are mounted in their specified position.

*Allowance* is the basic amount that a tooth is thinned from basic calculated circular tooth thickness to obtain the required backlash class.

*Tolerance* is the total permissible variation in the circular thickness of the teeth.

*Note 1:* These dimensions are shown primarily for the benefit of the gear manufacturer and represent the amount that the thickness of teeth should be reduced in the pinion and gear below the standard calculated value, to provide for backlash in the mesh. In some cases, particularly with pinions involving small numbers of teeth, it may be desirable to provide for total backlash by thinning the teeth in the gear member only by twice the allowance value shown in column (3). In this case both members will have the tolerance shown in column (4).

In some cases, particularly in meshes with a small total number of teeth, backlash may be achieved by an increase in basic center distance. In such cases, neither member is reduced by the allowance shown in column (3).

*Note 2:* These dimensions indicate the approximate backlash that will occur in a mesh in which each of the mating pairs of gears have the teeth thinned by the amount referred to in Note 1, and are meshed on theoretical centers.

*Note 3:* Backlash in gear sets can also be achieved by increasing the center distance above nominal and using the teeth at standard tooth thickness. Class E backlash designation infers gear sets operating under these conditions.

**Providing Backlash.** — In order to obtain the amount of backlash desired, it is necessary to decrease tooth thickness. However, because of manufacturing and assembling inaccuracies not only in the gears but also in other parts, the allowances made on tooth thickness almost always must exceed the desired amount of backlash. Since the amounts of these allowances depend on the closeness of control exercised on all manufacturing operations, no general recommendations for them can be given.

It is customary to make half of the allowance for backlash on the tooth thickness of each gear of a pair, although there are exceptions. For example, on pinions having very low numbers of teeth it is desirable to provide all of the allowance on the mating gear, so as not to weaken the pinion teeth. In worm gearing, ordinary practice is to provide all of the allowance on the worm which is usually made of a material stronger than that of the worm gear.

In some instances the backlash allowance is provided in the cutter, and the cutter is then operated at the standard tooth depth. In still other cases, backlash is obtained by setting the distance between two tools for cutting the two sides of the teeth, as in straight bevel gears, or by taking side cuts, or by changing the center distance between the gears in their mountings. In spur and helical gearing, backlash allowance is usually obtained by sinking the cutter deeper into the blank than the standard depth. The accompanying table gives the excess depth of cut for various pressure angles.

**Excess Depth of Cut $E$ to Provide Backlash Allowance**

| Distribution of Backlash | Pressure Angle $\phi$, Degrees | | | | |
|---|---|---|---|---|---|
| | $14\frac{1}{2}$ | $17\frac{1}{2}$ | $20°$ | $25°$ | $30°$ |
| Excess Depth of Cut $E$ to Obtain Circular Backlash $B$* | | | | | |
| All on One Gear | $1.93B$ | $1.59B$ | $1.37B$ | $1.07B$ | $0.87B$ |
| One-half on Each Gear | $0.97B$ | $0.79B$ | $0.69B$ | $0.54B$ | $0.43B$ |
| Excess Depth of Cut $E$ to Obtain Backlash $B_b$ Normal to Tooth Profile† | | | | | |
| All on One Gear | $2.00B_b$ | $1.66B_b$ | $1.46B_b$ | $1.18B_b$ | $1.00B_b$ |
| One-half on Each Gear | $1.00B_b$ | $0.83B_b$ | $0.73B_b$ | $0.59B_b$ | $0.50B_b$ |

\* Circular backlash is the amount by which the width of a tooth space is greater than the thickness of the engaging tooth on the pitch circles. As described on pages 759 and 764, this is what is meant by backlash unless otherwise specified.

† Backlash measured normal to the tooth profile by inserting a feeler gage between meshing teeth; to convert to circular backlash, $B = B_b \div \cos \phi$.

**Control of Backlash Allowances in Production.** — Measurement of the tooth thickness of gears is perhaps the simplest way of controlling backlash allowances in production. There are several ways in which this may be done including: (1) chordal thickness measurements as described on page 742; (2) caliper measurements over two or more teeth as described on page 961; and (3) measurements over wires. In this last method, first the theoretical measurement over wires when the backlash allowance is zero is determined by the method described on page 945; then the amount this measurement must be reduced to obtain a desired backlash allowance is taken from the table on page 955.

It should be understood, as explained in the section "Measurement of Backlash," that merely making tooth thickness allowances will not guarantee the amount of backlash in the ready-to-run assembly of two or more gears. Manufacturing limitations will introduce such gear errors as runout, pitch error, profile error, and lead error, and gear-housing errors in both center distance and alignment. All of these make the backlash of the assembled gears different from that indicated by tooth thickness measurements on the individual gears.

**Measurement of Backlash.** — Backlash is commonly measured by holding one gear of a pair stationary and rocking the other back and forth. The movement is registered by a dial indicator having its pointer or finger in a plane of rotation at or near the pitch diameter and in a direction parallel to a tangent to the pitch circle of the moving gear. If the direction of measurement is normal to the teeth, or other than as specified above, it is recommended that readings be converted to the plane of rotation and in a tangent direction at or near the pitch diameter, for purposes of standardization and comparison.

In spur gears, parallel helical gears, and bevel gears, it is immaterial whether the pinion or gear is held stationary for the test. In crossed helical and hypoid gears, readings may vary according to which member is stationary; hence it is customary to hold the pinion stationary and measure on the gear.

In some instances backlash is measured by thickness gages or feelers. A similar method utilizes lead wire inserted between the teeth as they pass through mesh. In both cases it is likewise recommended that readings be converted to the plane of rotation and in a tangent direction at or near the pitch diameter, taking into account the normal pressure angle, and the helix angle or spiral angle of the teeth.

Sometimes backlash in parallel helical or herringbone gears is checked by holding the gear stationary, and axially moving the pinion back and forth, readings being taken on the face or shaft of the pinion, and converted to the plane of rotation by calculation. Another method consists of meshing a pair of gears tightly together on centers and observing the variation from the specified center distance. Such readings should also be converted to the plane of rotation and in a tangent direction at or near the pitch diameter for the reasons previously given.

Measurements of backlash may vary in the same pair of gears, depending on accuracy of manufacturing and assembling. Incorrect tooth profiles will cause a change of backlash at different phases of the tooth action. Eccentricity may cause a substantial difference between maximum and minimum backlash at different positions around the gears. In stating amounts of backlash, it should always be remembered that merely making allowances on tooth thickness does not guarantee the minimum amount of backlash that will exist in actual gears when assembled.

*Fine-pitch Gears:* The measurement of backlash of fine-pitch gears, when assembled, cannot be made in the same manner and by the same techniques employed for gears of coarser pitches. In the very fine pitches it is virtually impossible to use indicating devices for measuring backlash. Sometimes a toolmaker's microscope is used for this purpose to good advantage on very small mechanisms.

Another means of measuring backlash in fine-pitch gears is to attach a beam to one of the shafts and measure the angular displacement in inches when one member is held stationary. The ratio of the length of the beam to the nominal pitch radius of the gear or pinion to which the beam is attached gives the approximate ratio of indicator reading to circular backlash. Because of the limited means of measuring backlash between a pair of fine-pitch gears, gear centers and tooth thickness of the gears when cut must be held to very close limits. Tooth thickness of fine-pitch spur and helical gears can best be checked on a variable center distance fixture using a master gear. When checked in this manner, tooth thickness change = 2 × center distance change × tangent of transverse pressure angle, approximately.

**Control of Backlash in Assemblies.** — Provision is often made for adjusting one gear relative to the other, thereby affording complete control over backlash at initial assembly and throughout the life of the gears. Such practice is most common in bevel gearing. It is fairly common in spur and helical gearing when the application permits slight changes between shaft centers. It is practical in worm gearing only for single thread worms with low lead angles. Otherwise faulty contact results.

Another method of controlling backlash quite common in bevel gears and less common in spur and helical gears is to match the high and low spots of the runout gears of one to one ratio and mark the engaging teeth at the point where the runout of one gear cancels the runout of the mating gear.

**Angular Backlash in Gears.** — When the backlash on the pitch circles of a meshing pair of gears is known, the angular backlash or angular play corresponding to this backlash may be computed from the following formulas.

$$\theta_D = \frac{6875B}{D} \text{ minutes}; \qquad \theta_d = \frac{6875B}{d} \text{ minutes}$$

In these formulas, $B$ = backlash between gears, in inches; $D$ = pitch diameter of larger gear, in inches; $d$ = pitch diameter of smaller gear in inches; $\theta_D$ = angular backlash or angular movement of larger gear in minutes when smaller gear is held fixed and larger gear rocked back and forth; and $\theta_d$ = angular backlash or angular movement of smaller gear in minutes when larger gear is held fixed and smaller gear rocked back and forth.

**Inspection of Gears.** — Perhaps the most widely used method of determining relative accuracy in a gear is to rotate the gear through at least one complete revolution in intimate contact with a master gear of known accuracy. The gear to be tested and the master gear are mounted on a variable-center-distance fixture and the resulting radial displacements or changes in center distance during rotation of the gear are measured by a suitable device. Except for the effect of backlash, this so-called "composite check" approximates the action of the gear under operating conditions and gives the combined effect of the following errors: Runout; pitch error; tooth-thickness variation; profile error; and lateral runout (sometimes called wobble).

*Tooth-to-tooth Composite Error* (Table 5. page 776) is the error which shows up as flicker on the indicator of a variable-center-distance fixture as the gear being tested is rotated from tooth to tooth in intimate contact with the master gear. Such flicker shows the combined or composite effect of |circular pitch error, tooth-thickness variation, and profile error.

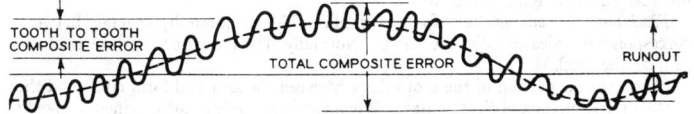

TOOTH TO TOOTH COMPOSITE ERROR

TOTAL COMPOSITE ERROR

RUNOUT

Chart Showing Nature of Composite Errors

*Total Composite Error* (Table 4) is made up of runout, wobble, and the tooth-to-tooth composite error; it is the total center distance displacement read on the indicating device of the testing fixture as shown in the accompanying diagram.

**Pressure for Composite Checking of Fine-pitch Gears.** — In using a variable-center-distance fixture, excessive pressure on fine-pitch gears of narrow face width will result in incorrect readings due to deflection of the teeth. Based on tests, the following checking pressures are recommended for gears of 0.100-inch face width: 20 to 29 diametral pitch, 28 ounces; 30 to 39 pitch, 24 ounces; 40 to 49 pitch, 20 ounces; 50 to 59 pitch, 16 ounces; 60 to 79 pitch, 12 ounces; 80 to 99 pitch, 8 ounces; 100 to 149 pitch, 4 ounces; and 150 and finer pitches, 2 ounces, minimum. These recommended checking pressures are based on the use of antifriction mountings for the movable head of the checking fixture and include the pressure of the indicating device. For face widths less than 0.100-inch, the recommended pressures should be reduced proportionally; for larger widths no increase is necessary although this may be done safely in the proper proportion.

**Gear Selection.** — To aid users and specifiers in the proper selection of gears for various applications, the American Gear Manufacturers Association has prepared a Gear Classification Manual, AGMA 390.03, which covers coarse- and fine-pitch spur, helical, herringbone, bevel and hypoid gears. The data in Tables 1 through 9 and information in the accompanying text have been extracted from this manual with the permission of the publisher, the American Gear Manufacturers Association, 1901 North Fort Myer Drive, Arlington, Va. 22209.

For convenience and simplification, gear selections are made according to assigned AGMA class numbers which combine a dimensional Quality Number, identifying tooth tolerances and a Material Treatment Number. Examples of the way these class numbers are used to denote or specify gears are given in Tables 1a and 1b.

**Gear Tolerance Classifications.** — Table 2 gives the American Gear Manufacturers Association tolerances for runout, pitch, and profile for coarse-pitch, spur, helical and herringbone gears. Table 3 gives lead tolerances; Table 4, total composite tolerances and Table 5, tooth-to-tooth composite tolerances, all for coarse-pitch, spur, helical and herringbone gears. Table 6 gives bevel gear tolerances. Table 7 gives the correlation of Quality Numbers with former fine-pitch classes.

*Axial runout* is the total variation in a direction parallel to the axis of rotation of a reference surface from a surface of revolution. Generally, the reference surface is a plane perpendicular to the axis of rotation.

*Radial runout* is the total variation in a direction perpendicular to the axis of rotation of a reference surface from a surface of revolution. Generally, the reference surface is the pitch surface of the gear. Radial runout includes eccentricity and out-of-roundness, and is approximately equal to twice the eccentricity.

*Runout tolerance* is the total allowable runout. In the case of gear teeth, it is measured by a specified probe such as a cylinder, ball, cone, rack or gear tooth. For spur, helical and herringbone gear teeth, the measurement is made perpendicular to the surface of revolution. On bevel gears, both axial and radial runout are included in the one measurement.

*Tooth-to-tooth spacing tolerance* is the allowable variation in spacing between corresponding sides of adjacent teeth. Normally, the measurements are made at or near the pitch circle.

*Profile* is that portion of the tooth flank between the specified form circle and the outside circle or start of tip chamfer. When the form circle is not specified, it should be that for a meshing rack.

*Profile tolerance* is the allowable deviation of the actual tooth form from the theoretical involute profile with the measurement made in the plane of rotation. Tip relief and any portion of the tooth surface below the active profile is not to be considered.

*Lead* is the axial advance of a helix for one complete turn, as in the threads of cylindrical worms and teeth of helical gears.

*Lead tolerance* as applied to spur, helical and herringbone gears and racks is the allowable deviation across the face width of a tooth surface measured normal to the surface established by the specified lead.

*Composite action* is the variation in center distance when a gear is rolled in tight mesh (double flank contact) with a specified gear.

*Total composite tolerance* is the allowance in center-distance variation in one complete revolution of the gear being inspected. This includes the effects of variations in active profile, lead, pitch, tooth thickness and runout.

*Pitch variation* (formerly pitch error) is the amount by which the circular arc between corresponding points on adjacent teeth differs from the value obtained by dividing the circumference of the corresponding circle by the number of teeth.

*Pitch tolerance* is the allowable amount of pitch variation. In general, measurements are made in a plane of rotation at or near the pitch surface. For bevel and hypoid gears it is customary to make measurements at or near the middle of the face width.

*Spacing variation* (also tooth-to-tooth spacing) is the difference in measured spacing between corresponding sides of adjacent teeth.

*Spacing tolerance* is the allowable amount of spacing variation. In general, measurements are made in a plane of rotation at or near the pitch surface. For bevel and hypoid gears, it is customary to make measurements at or near the middle of the face width.

*Index variation* (formerly accumulated pitch error) is the greatest displacement of any tooth profile from its theoretical position with respect to any other corresponding tooth profile measured at or near the pitch surface.

*Index tolerance* is the allowable amount of index variation. In general, measurements are made in a plane of rotation at or near the pitch surface. For bevel and hypoid gears, it is customary to make the measurements at or near the middle of the face width.

*Tooth-to-tooth composite variation* is the center-distance variation (on bevel and hypoid gears, movement of the pinion in a direction at right angles to the pinion axis) obtained as the gear is rotated from tooth to tooth (360 degrees divided by the number of teeth) in tight mesh with a specified mating gear for one complete revolution of the gear being checked. It includes the effects of variations in circular pitch, tooth thickness and profile on both tooth flanks.

*Tooth-to-tooth composite tolerance* is the allowable amount of tooth-to-tooth variation.

*Total composite variation* is the total variation in center distance (on bevel and hypoid gears, movement of the pinion in a direction at right angles to the pinion axis) obtained when a gear is rotated through one complete revolution in tight mesh with a specified mating gear. This includes the effect of all tooth and runout variations.

*Total composite tolerance* for bevel and hypoid gears is the allowable amount of total composite variation.

**Changes in Gear Tolerances.** — Any changes in gear element tolerances from those indicated in a specified AGMA Quality Number should be by mutual agreement between the manufacturer and the purchaser. It is recognized that there are precision gear applications that require closer tolerances than are indicated in Tables 2, 3 and 4. In such cases the designer or purchaser may wish to consult with the manufacturer to assure that manufacturing facilities and inspection equipment will be available to permit production and inspection to the closer tolerances specified.

In some applications requiring high accuracy gearing it is necessary to match pinion and gear tooth surfaces to obtain specified tooth contact. Matched sets can be provided but usually involve extra cost. In such a case, the manufacturer and purchaser should agree on details of the additional specifications covering how the matching is to be done and checked.

Unusual conditions may require that one or more of the individual element tolerances be of a lower or higher Quality Number than the other element tolerances. In such cases it is possible to modify the AGMA Class Number to include the Quality Number for each gear element tolerance. This should be done only when it results in a more satisfactory gear. Examples of such designations are shown in Table 1.

Certain control gearing applications require gears having a high degree of accuracy in the spacing of teeth. For such applications, a specification of the tooth spacing will be required in addition to the accuracy class specifications shown in Tables 2–4.

(*Continued on page 782*)

**Table 1a.  Designation of Coarse-Pitch Gears and Fine-Pitch Bevel and
Hypoid Gears and Racks by AGMA Class Numbers — Typical Examples**

| When All Gear Element Tolerances are the **Same** Quality Number | | |
|---|---|---|
| Gear Selection Based On | | AGMA Class Number |
| AGMA Quality Number* | AGMA Material Number†<br>Material<br>Treatment<br>Hardness | |
| Runout            8<br>T-T Spacing     8<br>Profile          8<br>Lead             8 | H-14<br>Alloy Steel<br>Quench and Temper<br>285 to 325 Bhn | 8-H-14 |
| Runout            5<br>T-T Spacing     5<br>Profile          5 | IN-3<br>Carbon Steel<br>Induction Harden<br>52 Rc Min. | 5-IN-3 |
| Runout            8<br>T-T Spacing     8<br>Profile          8<br>Lead             8 | NI-7<br>Nodular Iron<br>Quench and Temper<br>270 Bhn Min. | 8-NI-7 |
| Runout           12<br>T-T Spacing    12<br>Profile         12<br>Lead            12 | S-2<br>Stainless Steel<br>No Treatment<br>Hardness as Received | 12-S-2 |
| When All Gear Element Tolerances are **not the Same** Quality Number | | |
| Runout            8<br>T-T Spacing    10<br>Profile          8<br>Lead             8 | CH-15<br>Alloy Steel<br>Carburize<br>60 Rc Min. | 8-CH-15<br>(Except T-T Spacing<br>Quality 10) |
| Runout           11<br>T-T Spacing    11<br>Profile         11<br>Lead            10 | H-5<br>Carbon Steel<br>Quench and Temper<br>285 to 325 Bhn | 11-H-5<br>(Except Lead<br>Quality 10) |
| Runout            8<br>T-T Spacing     9<br>Profile          9<br>Lead             8 | F-12<br>Alloy Steel<br>Flame Harden<br>48 Rc Min. | 8-F-12<br>(Except T-T Spacing<br>and Profile Quality 9) |

\* See Tables 2 and 3.      † See Table 8.

**Table 1b.  Designation of Fine-Pitch Spur, Helical and Herringbone Gears
by AGMA Quality Numbers — Typical Example**

| Gear Selection Based On | | | AGMA Class<br>Number |
|---|---|---|---|
| AGMA Quality<br>Number* | AGMA Backlash**<br>Designation | AGMA Material Number†<br>Material<br>Treatment<br>Hardness | |
| 10 | C | A-4<br>Aluminum Bar 2024-T4<br>Heat Treated<br>120 Bhn (500 Kg) | 10C-A-4 |

\* See Table 4.      \*\* See page 762.      † See Table 8.

**Table 2. AGMA Coarse-Pitch Spur, Helical and Herringbone Gear Tolerances\* — 1**
(In ten-thousandths of an inch)

| AGMA Quality Number | Normal Diametral Pitch | Pitch Diameter, inches | | | | | | | | | |
|---|---|---|---|---|---|---|---|---|---|---|---|
| | | ¾ | 1½ | 3 | 6 | 12 | 25 | 50 | 100 | 200 | 400 |
| | | Runout Tolerance | | | | | | | | | |
| 3 | ½ | .... | .... | .... | .... | 788.2 | 938.6 | 1106.9 | 1305.5 | 1539.6 | 1815.7 |
| | 1 | .... | .... | .... | 477.8 | 563.5 | 671.1 | 791.4 | 933.4 | 1100.8 | 1298.1 |
| | 2 | .... | .... | 289.7 | 341.6 | 402.9 | 479.8 | 565.9 | 667.4 | 787.1 | 928.2 |
| | 4 | .... | .... | 207.1 | 244.3 | 288.1 | 343.1 | 404.6 | 477.2 | 562.7 | 663.7 |
| | 8 | .... | .... | 148.1 | 174.7 | 206.0 | 245.3 | 289.3 | 341.2 | 402.4 | 474.5 |
| 4 | ½ | .... | .... | .... | .... | 563.0 | 670.4 | 790.7 | 932.5 | 1099.7 | 1297.0 |
| | 1 | .... | .... | .... | 341.3 | 402.5 | 479.3 | 565.3 | 666.7 | 786.3 | 927.3 |
| | 2 | .... | .... | 206.9 | 244.0 | 287.8 | 342.7 | 404.2 | 476.7 | 562.2 | 663.0 |
| | 4 | .... | .... | 147.9 | 174.5 | 205.8 | 245.0 | 289.0 | 340.8 | 402.0 | 474.1 |
| | 8 | .... | .... | 105.8 | 124.8 | 147.1 | 175.2 | 206.6 | 243.7 | 287.4 | 338.9 |
| 5 | ½ | .... | .... | .... | .... | 402.1 | 478.9 | 564.8 | 666.1 | 785.5 | 926.4 |
| | 1 | .... | .... | .... | 243.8 | 287.5 | 342.4 | 403.8 | 476.2 | 561.6 | 662.4 |
| | 2 | .... | .... | 147.8 | 174.3 | 205.6 | 244.8 | 288.7 | 340.5 | 401.6 | 473.6 |
| | 4 | .... | 89.6 | 105.7 | 124.6 | 147.0 | 175.0 | 206.4 | 243.5 | 287.1 | 338.6 |
| | 8 | .... | 64.1 | 75.6 | 89.1 | 105.1 | 125.1 | 147.6 | 174.1 | 205.3 | 242.1 |
| 6 | ½ | .... | .... | .... | .... | 287.2 | 342.1 | 403.4 | 475.8 | 561.1 | 661.7 |
| | 1 | .... | .... | .... | 174.1 | 205.4 | 244.6 | 288.4 | 340.2 | 401.2 | 473.1 |
| | 2 | .... | .... | 105.6 | 124.5 | 146.8 | 174.9 | 206.2 | 243.2 | 286.8 | 338.3 |
| | 4 | .... | 64.0 | 75.5 | 89.0 | 105.0 | 125.0 | 147.4 | 173.9 | 205.1 | 241.9 |
| | 8 | 38.8 | 45.8 | 54.0 | 63.6 | 75.1 | 89.4 | 105.4 | 124.3 | 146.6 | 172.9 |
| | 12 | 31.9 | 37.6 | 44.4 | 52.3 | 61.7 | 73.5 | 86.6 | 102.2 | 120.5 | 142.1 |
| | 20 | 24.9 | 29.4 | 34.6 | 40.8 | 48.2 | 57.4 | 67.7 | 79.8 | 94.1 | 111.0 |
| 7 | ½ | .... | .... | .... | .... | 205.2 | 244.3 | 288.1 | 339.8 | 400.8 | 472.7 |
| | 1 | .... | .... | .... | 124.4 | 146.7 | 174.7 | 206.0 | 243.0 | 286.6 | 337.9 |
| | 2 | .... | .... | 75.4 | 88.9 | 104.9 | 124.9 | 147.3 | 173.7 | 204.9 | 241.6 |
| | 4 | .... | 45.7 | 53.9 | 63.6 | 75.0 | 89.3 | 105.3 | 124.2 | 146.5 | 172.8 |
| | 8 | 27.7 | 32.7 | 38.5 | 45.5 | 53.6 | 63.9 | 75.3 | 88.8 | 104.7 | 123.5 |
| | 12 | 22.8 | 26.9 | 31.7 | 37.4 | 44.1 | 52.5 | 61.9 | 73.0 | 86.1 | 101.5 |
| | 20 | 17.8 | 21.0 | 24.7 | 29.2 | 34.4 | 41.0 | 48.3 | 57.0 | 67.2 | 79.3 |

| AGMA Quality Number | Normal Diametral Pitch | Pitch Diameter, inches | | | | | | | | | |
|---|---|---|---|---|---|---|---|---|---|---|---|
| | | ¾ | 1½ | 3 | 6 | 12 | 25 | 50 | 100 | 200 | 400 |
| | | Pitch Tolerance | | | | | | | | | |
| 6 | ½ | .... | .... | .... | .... | 38.4 | 43.7 | 49.4 | 55.9 | 63.2 | 71.4 |
| | 1 | .... | .... | .... | 29.1 | 32.9 | 37.4 | 42.3 | 47.8 | 54.1 | 61.1 |
| | 2 | .... | .... | 22.0 | 24.9 | 28.1 | 32.0 | 36.2 | 41.0 | 46.3 | 52.3 |
| | 4 | .... | 16.7 | 18.9 | 21.3 | 24.1 | 27.4 | 31.0 | 35.1 | 39.6 | 44.8 |
| | 8 | 12.6 | 14.3 | 16.1 | 18.2 | 20.6 | 23.5 | 26.6 | 30.0 | 33.9 | 38.4 |
| | 12 | 11.5 | 13.0 | 14.7 | 16.7 | 18.8 | 21.5 | 24.3 | 27.4 | 31.0 | 35.0 |
| | 20 | 10.3 | 11.6 | 13.1 | 14.9 | 16.8 | 19.1 | 21.6 | 24.5 | 27.6 | 31.3 |
| 7 | ½ | .... | .... | .... | .... | 27.0 | 30.8 | 34.8 | 39.3 | 44.5 | 50.3 |
| | 1 | .... | .... | .... | 20.5 | 23.1 | 26.4 | 29.8 | 33.7 | 38.1 | 43.1 |
| | 2 | .... | .... | 15.5 | 17.5 | 19.8 | 22.6 | 25.5 | 28.8 | 32.6 | 36.9 |
| | 4 | .... | 11.7 | 13.3 | 15.0 | 17.0 | 19.3 | 21.8 | 24.7 | 29.9 | 31.6 |
| | 8 | 8.9 | 10.1 | 11.4 | 12.9 | 14.5 | 16.5 | 18.7 | 21.1 | 23.9 | 27.0 |
| | 12 | 8.1 | 9.2 | 10.4 | 11.7 | 13.3 | 15.1 | 17.1 | 19.3 | 21.8 | 24.7 |
| | 20 | 7.2 | 8.2 | 9.3 | 10.5 | 11.8 | 13.5 | 15.2 | 17.2 | 19.5 | 22.0 |

\* Extracted from AGMA Handbook for Unassembled Gears — Volume 1 — Gear Classifications, Materials, and Inspection (AGMA 390.03), with permission of the publisher, the American Gear Manufacturers Association, 1901 North Fort Myer Drive, Arlington, Va. 22209.

**Table 2. AGMA Coarse-Pitch Spur, Helical and Herringbone Gear Tolerances — 2**
(In ten-thousandths of an inch)

| AGMA Quality Number | Normal Diametral Pitch | Pitch Diameter, inches | | | | | | | | | |
|---|---|---|---|---|---|---|---|---|---|---|---|
| | | ¾ | 1½ | 3 | 6 | 12 | 25 | 50 | 100 | 200 | 400 |
| | | Runout Tolerance | | | | | | | | | |
| 8 | ½ | ..... | ..... | ..... | 88.8 | 146.5 | 174.5 | 205.8 | 242.7 | 286.3 | 337.6 |
| | 1 | ..... | ..... | ..... | 104.8 | 124.8 | 147.2 | 173.6 | 204.7 | 241.4 | |
| | 2 | ..... | ..... | 53.9 | 63.5 | 74.9 | 89.2 | 105.2 | 124.1 | 146.3 | 172.6 |
| | 4 | ..... | 32.7 | 38.5 | 45.4 | 53.6 | 63.8 | 75.2 | 88.7 | 104.6 | 123.4 |
| | 8 | 19.8 | 23.3 | 27.5 | 32.5 | 38.3 | 45.6 | 53.8 | 63.4 | 74.8 | 88.2 |
| | 12 | 16.3 | 19.2 | 22.6 | 26.7 | 31.5 | 37.5 | 44.2 | 52.1 | 61.5 | 72.5 |
| | 20 | 12.7 | 15.0 | 17.7 | 20.8 | 24.6 | 29.3 | 34.5 | 40.7 | 48.0 | 56.6 |
| 9 | ½ | ..... | ..... | ..... | ..... | 104.7 | 124.7 | 147.0 | 173.4 | 204.5 | 241.2 |
| | 1 | ..... | ..... | ..... | 63.5 | 74.8 | 89.1 | 105.1 | 124.0 | 146.2 | 172.4 |
| | 2 | ..... | ..... | 38.5 | 45.4 | 53.5 | 63.7 | 75.2 | 88.6 | 104.5 | 123.3 |
| | 4 | ..... | 23.3 | 27.5 | 32.4 | 38.3 | 45.6 | 53.7 | 63.4 | 74.7 | 88.1 |
| | 8 | 14.1 | 16.7 | 19.7 | 23.2 | 27.4 | 32.6 | 38.4 | 45.3 | 53.4 | 63.0 |
| | 12 | 11.6 | 13.7 | 16.2 | 19.1 | 22.5 | 26.8 | 31.6 | 37.2 | 43.9 | 51.8 |
| | 20 | 9.1 | 10.7 | 12.6 | 14.9 | 17.6 | 20.9 | 24.7 | 29.1 | 34.3 | 40.4 |
| 10 | ½ | ..... | ..... | ..... | ..... | 74.8 | 89.0 | 105.0 | 123.8 | 146.1 | 172.3 |
| | 1 | ..... | ..... | ..... | 45.3 | 53.5 | 63.7 | 75.1 | 88.5 | 104.4 | 123.2 |
| | 2 | ..... | ..... | 27.5 | 32.4 | 38.2 | 45.5 | 53.7 | 63.3 | 74.7 | 88.1 |
| | 4 | ..... | 16.7 | 19.6 | 23.2 | 27.3 | 32.5 | 38.4 | 45.3 | 53.4 | 63.0 |
| | 8 | 10.1 | 11.9 | 14.0 | 16.6 | 19.5 | 23.3 | 27.4 | 32.4 | 38.2 | 45.0 |
| | 12 | 8.3 | 9.8 | 11.5 | 13.6 | 16.1 | 19.1 | 22.6 | 26.6 | 31.4 | 37.0 |
| | 20 | 6.5 | 7.6 | 9.0 | 10.6 | 12.5 | 14.9 | 17.6 | 20.8 | 24.5 | 28.9 |
| 11 | ½ | ..... | ..... | ..... | ..... | 53.4 | 63.6 | 75.0 | 88.5 | 104.3 | 123.0 |
| | 1 | ..... | ..... | ..... | 32.4 | 38.2 | 45.5 | 53.6 | 63.2 | 74.6 | 88.0 |
| | 2 | ..... | ..... | 19.6 | 23.1 | 27.3 | 32.5 | 38.3 | 45.2 | 53.3 | 62.9 |
| | 4 | ..... | 11.9 | 14.0 | 16.6 | 19.5 | 23.2 | 27.4 | 32.3 | 38.1 | 45.0 |
| | 8 | 7.2 | 8.5 | 10.0 | 11.8 | 14.0 | 16.6 | 19.6 | 23.1 | 27.3 | 32.2 |
| | 12 | 5.9 | 7.0 | 8.2 | 9.7 | 11.5 | 13.7 | 16.1 | 19.0 | 22.4 | 26.4 |
| | 20 | 4.6 | 5.5 | 6.4 | 7.6 | 9.0 | 10.7 | 12.6 | 14.8 | 17.5 | 20.6 |
| 12 | ½ | ..... | ..... | ..... | ..... | 38.1 | 45.4 | 53.6 | 63.2 | 74.5 | 87.9 |
| | 1 | ..... | ..... | ..... | 23.1 | 27.3 | 32.5 | 38.3 | 45.2 | 53.3 | 62.8 |
| | 2 | ..... | ..... | 14.0 | 16.5 | 19.5 | 23.2 | 27.4 | 32.3 | 38.1 | 44.9 |
| | 4 | ..... | 8.5 | 10.0 | 11.8 | 13.9 | 16.6 | 19.6 | 23.1 | 27.2 | 32.1 |
| | 8 | 5.2 | 6.1 | 7.2 | 8.5 | 10.0 | 11.9 | 14.0 | 16.5 | 19.5 | 23.0 |
| | 12 | 4.2 | 5.0 | 5.9 | 6.9 | 8.2 | 9.8 | 11.5 | 13.6 | 16.0 | 18.9 |
| | 20 | 3.3 | 3.9 | 4.6 | 5.4 | 6.4 | 7.6 | 9.0 | 10.6 | 12.5 | 14.7 |
| 13 | ½ | ..... | ..... | ..... | ..... | 27.2 | 32.4 | 38.3 | 45.1 | 53.2 | 62.8 |
| | 1 | ..... | ..... | ..... | 16.5 | 19.5 | 23.2 | 27.4 | 32.3 | 38.1 | 44.9 |
| | 2 | ..... | ..... | 10.0 | 11.8 | 13.9 | 16.6 | 19.6 | 23.1 | 27.2 | 32.1 |
| | 4 | ..... | 6.1 | 7.2 | 8.4 | 10.0 | 11.9 | 14.0 | 16.5 | 19.5 | 22.9 |
| | 8 | 3.7 | 4.3 | 5.1 | 6.0 | 7.1 | 8.5 | 10.0 | 11.8 | 13.9 | 16.4 |
| | 12 | 3.0 | 3.6 | 4.2 | 5.0 | 5.9 | 7.0 | 8.2 | 9.7 | 11.4 | 13.5 |
| | 20 | 2.4 | 2.8 | 3.3 | 3.9 | 4.6 | 5.4 | 6.4 | 7.6 | 8.9 | 10.5 |
| 14 | ½ | ..... | ..... | ..... | ..... | 19.5 | 23.2 | 27.3 | 32.2 | 38.0 | 44.8 |
| | 1 | ..... | ..... | ..... | 11.8 | 13.9 | 16.6 | 19.6 | 23.0 | 27.2 | 32.1 |
| | 2 | ..... | ..... | 7.2 | 8.4 | 9.9 | 11.8 | 14.0 | 16.5 | 19.4 | 22.9 |
| | 4 | ..... | 4.3 | 5.1 | 6.0 | 7.1 | 8.5 | 10.0 | 11.8 | 13.9 | 16.4 |
| | 8 | 2.6 | 3.1 | 3.7 | 4.3 | 5.1 | 6.1 | 7.1 | 8.4 | 9.9 | 11.7 |
| | 12 | 2.2 | 2.5 | 3.0 | 3.5 | 4.2 | 5.0 | 5.9 | 6.9 | 8.2 | 9.6 |
| | 20 | 1.7 | 2.0 | 2.3 | 2.8 | 3.3 | 3.9 | 4.6 | 5.4 | 6.4 | 7.5 |
| 15 | ½ | ..... | ..... | ..... | ..... | 13.9 | 16.6 | 19.5 | 23.0 | 27.2 | 32.0 |
| | 1 | ..... | ..... | ..... | 8.4 | 9.9 | 11.8 | 14.0 | 16.5 | 19.4 | 22.9 |
| | 2 | ..... | ..... | 5.1 | 6.0 | 7.1 | 8.5 | 10.0 | 11.8 | 13.9 | 16.4 |
| | 4 | ..... | 3.1 | 3.7 | 4.3 | 5.1 | 6.1 | 7.1 | 8.4 | 9.9 | 11.7 |
| | 8 | 1.9 | 2.2 | 2.6 | 3.1 | 3.6 | 4.3 | 5.1 | 6.0 | 7.1 | 8.4 |
| | 12 | 1.5 | 1.8 | 2.1 | 2.5 | 3.0 | 3.6 | 4.2 | 4.9 | 5.8 | 6.9 |
| | 20 | 1.2 | 1.4 | 1.7 | 2.0 | 2.3 | 2.8 | 3.3 | 3.9 | 4.6 | 5.4 |

**Table 2. AGMA Coarse-Pitch Spur, Helical and Herringbone Gear Tolerances — 3**
(In ten-thousandths of an inch)

| AGMA Quality Number | Normal Diametral Pitch | Pitch Diameter, inches — Pitch Tolerance | | | | | | | | | |
|---|---|---|---|---|---|---|---|---|---|---|---|
| | | ¾ | 1½ | 3 | 6 | 12 | 25 | 50 | 100 | 200 | 400 |
| **8** | ½ | | | | | 19.0 | 21.7 | 24.5 | 27.7 | 31.3 | 35.4 |
| | 1 | | | | 14.4 | 16.3 | 18.6 | 21.0 | 23.7 | 26.8 | 30.3 |
| | 2 | | | 10.9 | 12.3 | 14.0 | 15.9 | 18.0 | 20.3 | 23.0 | 26.0 |
| | 4 | | 8.3 | 9.3 | 10.6 | 11.9 | 13.6 | 15.4 | 17.4 | 19.7 | 22.2 |
| | 8 | 6.3 | 7.1 | 8.0 | 9.0 | 10.2 | 11.7 | 13.2 | 14.9 | 16.8 | 19.0 |
| | 12 | 5.7 | 6.5 | 7.3 | 8.3 | 9.3 | 10.6 | 12.0 | 13.6 | 15.4 | 17.4 |
| | 20 | 5.1 | 5.8 | 6.5 | 7.4 | 8.3 | 9.5 | 10.7 | 12.1 | 13.7 | 15.5 |
| **9** | ½ | | | | | 13.4 | 15.3 | 17.3 | 19.5 | 22.1 | 24.9 |
| | 1 | | | | 10.2 | 11.5 | 13.1 | 14.8 | 16.7 | 18.9 | 21.4 |
| | 2 | | | 7.7 | 8.7 | 9.8 | 11.2 | 12.7 | 14.3 | 16.2 | 18.3 |
| | 4 | | 5.8 | 6.6 | 7.4 | 8.4 | 9.6 | 10.8 | 12.2 | 13.8 | 15.7 |
| | 8 | 4.4 | 5.0 | 5.6 | 6.4 | 7.2 | 8.2 | 9.3 | 10.5 | 11.9 | 13.4 |
| | 12 | 4.0 | 4.6 | 5.1 | 5.8 | 6.6 | 7.5 | 8.5 | 9.6 | 10.8 | 12.2 |
| | 20 | 3.6 | 4.1 | 4.6 | 5.2 | 5.9 | 6.7 | 7.6 | 8.5 | 9.7 | 10.9 |
| **10** | ½ | | | | | 9.4 | 10.8 | 12.2 | 13.7 | 15.5 | 17.6 |
| | 1 | | | | 7.2 | 8.1 | 9.2 | 10.4 | 11.8 | 13.3 | 15.0 |
| | 2 | | | 5.4 | 6.1 | 6.9 | 7.9 | 8.9 | 10.1 | 11.4 | 12.9 |
| | 4 | | 4.1 | 4.6 | 5.2 | 5.9 | 6.7 | 7.6 | 8.6 | 9.8 | 11.0 |
| | 8 | 3.1 | 3.5 | 4.0 | 4.5 | 5.1 | 5.8 | 6.5 | 7.4 | 8.3 | 9.4 |
| | 12 | 2.8 | 3.2 | 3.6 | 4.1 | 4.6 | 5.3 | 6.0 | 6.7 | 7.6 | 8.6 |
| | 20 | 2.5 | 2.9 | 3.2 | 3.7 | 4.1 | 4.7 | 5.3 | 6.0 | 6.8 | 7.7 |
| **11** | ½ | | | | | 6.6 | 7.6 | 8.6 | 9.7 | 10.9 | 12.4 |
| | 1 | | | | 5.0 | 5.7 | 6.5 | 7.3 | 8.3 | 9.4 | 10.6 |
| | 2 | | | 3.8 | 4.3 | 4.9 | 5.6 | 6.3 | 7.1 | 8.0 | 9.1 |
| | 4 | | 2.9 | 3.3 | 3.7 | 4.2 | 4.8 | 5.4 | 6.1 | 6.9 | 7.8 |
| | 8 | 2.2 | 2.5 | 2.8 | 3.2 | 3.6 | 4.1 | 4.6 | 5.2 | 5.9 | 6.6 |
| | 12 | 2.0 | 2.3 | 2.6 | 2.9 | 3.3 | 3.7 | 4.2 | 4.7 | 5.4 | 6.1 |
| | 20 | 1.8 | 2.0 | 2.3 | 2.6 | 2.9 | 3.3 | 3.7 | 4.2 | 4.8 | 5.4 |
| **12** | ½ | | | | | 4.7 | 5.3 | 6.0 | 6.8 | 7.7 | 8.7 |
| | 1 | | | | 3.5 | 4.0 | 4.6 | 5.2 | 5.8 | 6.6 | 7.5 |
| | 2 | | | 2.7 | 3.0 | 3.4 | 3.9 | 4.4 | 5.0 | 5.6 | 6.4 |
| | 4 | | 2.0 | 2.3 | 2.6 | 2.9 | 3.3 | 3.8 | 4.3 | 4.8 | 5.5 |
| | 8 | 1.5 | 1.7 | 2.0 | 2.2 | 2.5 | 2.9 | 3.2 | 3.7 | 4.1 | 4.7 |
| | 12 | 1.4 | 1.6 | 1.8 | 2.0 | 2.3 | 2.6 | 3.0 | 3.3 | 3.8 | 4.3 |
| | 20 | 1.3 | 1.4 | 1.6 | 1.8 | 2.0 | 2.3 | 2.6 | 3.0 | 3.4 | 3.8 |
| **13** | ½ | | | | | 3.3 | 3.8 | 4.2 | 4.8 | 5.4 | 6.1 |
| | 1 | | | | 2.5 | 2.8 | 3.2 | 3.6 | 4.1 | 4.6 | 5.3 |
| | 2 | | | 1.9 | 2.1 | 2.4 | 2.8 | 3.1 | 3.5 | 4.0 | 4.5 |
| | 4 | | 1.4 | 1.6 | 1.8 | 2.1 | 2.4 | 2.7 | 3.0 | 3.4 | 3.8 |
| | 8 | 1.1 | 1.3 | 1.4 | 1.6 | 1.8 | 2.0 | 2.3 | 2.6 | 2.9 | 3.3 |
| | 12 | 1.0 | 1.1 | 1.3 | 1.4 | 1.6 | 1.8 | 2.1 | 2.4 | 2.7 | 3.0 |
| | 20 | 0.9 | 1.0 | 1.1 | 1.3 | 1.4 | 1.6 | 1.9 | 2.1 | 2.4 | 2.7 |
| **14** | ½ | | | | | 2.3 | 2.6 | 3.0 | 3.4 | 3.8 | 4.3 |
| | 1 | | | | 1.8 | 2.0 | 2.3 | 2.6 | 2.9 | 3.3 | 3.7 |
| | 2 | | | 1.3 | 1.5 | 1.7 | 1.9 | 2.2 | 2.5 | 2.8 | 3.2 |
| | 4 | | 1.0 | 1.1 | 1.3 | 1.5 | 1.7 | 1.9 | 2.1 | 2.4 | 2.7 |
| | 8 | 0.8 | 0.9 | 1.0 | 1.1 | 1.2 | 1.4 | 1.6 | 1.8 | 2.1 | 2.3 |
| | 12 | 0.7 | 0.8 | 0.9 | 1.0 | 1.1 | 1.3 | 1.5 | 1.7 | 1.9 | 2.1 |
| | 20 | 0.6 | 0.7 | 0.8 | 0.9 | 1.0 | 1.2 | 1.3 | 1.5 | 1.7 | 1.9 |
| **15** | ½ | | | | | 1.6 | 1.9 | 2.1 | 2.4 | 2.7 | 3.0 |
| | 1 | | | | 1.2 | 1.4 | 1.6 | 1.8 | 2.0 | 2.3 | 2.6 |
| | 2 | | | 0.9 | 1.1 | 1.2 | 1.4 | 1.5 | 1.7 | 2.0 | 2.2 |
| | 4 | | 0.7 | 0.8 | 0.9 | 1.0 | 1.2 | 1.3 | 1.5 | 1.7 | 1.9 |
| | 8 | 0.5 | 0.6 | 0.7 | 0.8 | 0.9 | 1.0 | 1.1 | 1.3 | 1.4 | 1.6 |
| | 12 | 0.5 | 0.6 | 0.6 | 0.7 | 0.8 | 0.9 | 1.0 | 1.2 | 1.3 | 1.5 |
| | 20 | 0.4 | 0.5 | 0.6 | 0.6 | 0.7 | 0.8 | 0.9 | 1.0 | 1.2 | 1.3 |

**Table 2. AGMA Coarse-Pitch Spur, Helical and Herringbone Gear Tolerances — 4**
(In ten-thousandths of an inch)

| AGMA Quality Number | Normal Diametral Pitch | 3/4 | 1½ | 3 | 6 | 12 | 25 | 50 | 100 | 200 | 400 |
|---|---|---|---|---|---|---|---|---|---|---|---|
| | | Profile Tolerance | | | | | | | | | |
| 8 | ½ | .... | .... | .... | .... | 42.6 | 47.7 | 53.1 | 59.1 | 65.7 | 73.1 |
| | 1 | .... | .... | .... | 28.3 | 31.5 | 35.3 | 39.3 | 43.7 | 48.6 | 54.1 |
| | 2 | .... | .... | 18.8 | 21.0 | 23.3 | 26.1 | 29.0 | 32.3 | 36.0 | 40.0 |
| | 4 | .... | 12.5 | 13.9 | 15.5 | 17.2 | 19.3 | 21.5 | 23.9 | 26.6 | 29.6 |
| | 8 | 8.3 | 9.3 | 10.3 | 11.5 | 12.8 | 14.3 | 15.9 | 17.7 | 19.7 | 21.9 |
| | 12 | 7.0 | 7.8 | 8.6 | 9.6 | 10.7 | 12.0 | 13.3 | 14.8 | 16.5 | 18.4 |
| | 20 | 5.6 | 6.2 | 6.9 | 7.7 | 8.6 | 9.6 | 10.7 | 11.9 | 13.2 | 14.7 |
| 9 | ½ | .... | .... | .... | .... | 30.4 | 34.1 | 37.9 | 42.2 | 46.9 | 52.2 |
| | 1 | .... | .... | .... | 20.2 | 22.5 | 25.2 | 28.1 | 31.2 | 34.7 | 38.6 |
| | 2 | .... | .... | 13.5 | 15.0 | 16.7 | 18.6 | 20.7 | 23.1 | 25.7 | 28.6 |
| | 4 | .... | 8.9 | 10.0 | 11.1 | 12.3 | 13.8 | 15.3 | 17.1 | 19.0 | 21.1 |
| | 8 | 5.9 | 6.6 | 7.4 | 8.2 | 9.1 | 10.2 | 11.4 | 12.6 | 14.1 | 15.6 |
| | 12 | 5.0 | 5.5 | 6.2 | 6.9 | 7.6 | 8.6 | 9.5 | 10.6 | 11.8 | 13.1 |
| | 20 | 4.0 | 4.4 | 4.9 | 5.5 | 6.1 | 6.8 | 7.6 | 8.5 | 9.4 | 10.5 |
| 10 | ½ | .... | .... | .... | .... | 21.7 | 24.3 | 27.1 | 30.1 | 33.5 | 37.3 |
| | 1 | .... | .... | .... | 14.5 | 16.1 | 18.0 | 20.0 | 22.3 | 24.8 | 27.6 |
| | 2 | .... | .... | 9.6 | 10.7 | 11.9 | 13.3 | 14.8 | 16.5 | 18.3 | 20.4 |
| | 4 | .... | 6.4 | 7.1 | 7.9 | 8.8 | 9.9 | 11.0 | 12.2 | 13.6 | 15.1 |
| | 8 | 4.2 | 4.7 | 5.3 | 5.9 | 6.5 | 7.3 | 8.1 | 9.0 | 10.0 | 11.2 |
| | 12 | 3.6 | 4.0 | 4.4 | 4.9 | 5.5 | 6.1 | 6.8 | 7.6 | 8.4 | 9.4 |
| | 20 | 2.9 | 3.2 | 3.5 | 3.9 | 4.4 | 4.9 | 5.4 | 6.1 | 6.7 | 7.5 |
| 11 | ½ | .... | .... | .... | .... | 15.5 | 17.4 | 19.3 | 21.5 | 24.0 | 26.7 |
| | 1 | .... | .... | .... | 10.3 | 11.5 | 12.9 | 14.3 | 15.9 | 17.7 | 19.7 |
| | 2 | .... | .... | 6.9 | 7.6 | 8.5 | 9.5 | 10.6 | 11.8 | 13.1 | 14.6 |
| | 4 | .... | 4.6 | 5.1 | 5.6 | 6.3 | 7.0 | 7.8 | 8.7 | 9.7 | 10.8 |
| | 8 | 3.0 | 3.4 | 3.8 | 4.2 | 4.6 | 5.2 | 5.8 | 6.4 | 7.2 | 8.0 |
| | 12 | 2.5 | 2.8 | 3.1 | 3.5 | 3.9 | 4.4 | 4.9 | 5.4 | 6.0 | 6.7 |
| | 20 | 2.0 | 2.3 | 2.5 | 2.8 | 3.1 | 3.5 | 3.9 | 4.3 | 4.8 | 5.4 |
| 12 | ½ | .... | .... | .... | .... | 11.1 | 12.4 | 13.8 | 15.4 | 17.1 | 19.0 |
| | 1 | .... | .... | .... | 7.4 | 8.2 | 9.2 | 10.2 | 11.4 | 12.7 | 14.1 |
| | 2 | .... | .... | 4.9 | 5.5 | 6.1 | 6.8 | 7.6 | 8.4 | 9.4 | 10.4 |
| | 4 | .... | 3.3 | 3.6 | 4.0 | 4.5 | 5.0 | 5.6 | 6.2 | 6.9 | 7.7 |
| | 8 | 2.2 | 2.4 | 2.7 | 3.0 | 3.3 | 3.7 | 4.1 | 4.6 | 5.1 | 5.7 |
| | 12 | 1.8 | 2.0 | 2.2 | 2.5 | 2.8 | 3.1 | 3.5 | 3.9 | 4.3 | 4.8 |
| | 20 | 1.5 | 1.6 | 1.8 | 2.0 | 2.2 | 2.5 | 2.8 | 3.1 | 3.4 | 3.8 |
| 13 | ½ | .... | .... | .... | .... | 7.9 | 8.9 | 9.9 | 11.0 | 12.2 | 13.6 |
| | 1 | .... | .... | .... | 5.3 | 5.9 | 6.6 | 7.3 | 8.1 | 9.0 | 10.1 |
| | 2 | .... | .... | 3.5 | 3.9 | 4.3 | 4.9 | 5.4 | 6.0 | 6.7 | 7.4 |
| | 4 | .... | 2.3 | 2.6 | 2.9 | 3.2 | 3.6 | 4.0 | 4.4 | 4.9 | 5.5 |
| | 8 | 1.5 | 1.7 | 1.9 | 2.1 | 2.4 | 2.7 | 3.0 | 3.3 | 3.7 | 4.1 |
| | 12 | 1.3 | 1.4 | 1.6 | 1.8 | 2.0 | 2.2 | 2.5 | 2.8 | 3.1 | 3.4 |
| | 20 | 1.0 | 1.2 | 1.3 | 1.4 | 1.6 | 1.8 | 2.0 | 2.2 | 2.5 | 2.7 |
| 14 | ½ | .... | .... | .... | .... | 5.7 | 6.3 | 7.1 | 7.8 | 8.7 | 9.7 |
| | 1 | .... | .... | .... | 3.8 | 4.2 | 4.7 | 5.2 | 5.8 | 6.5 | 7.2 |
| | 2 | .... | .... | 2.5 | 2.8 | 3.1 | 3.5 | 3.9 | 4.3 | 4.8 | 5.3 |
| | 4 | .... | 1.7 | 1.9 | 2.1 | 2.3 | 2.6 | 2.9 | 3.2 | 3.5 | 3.9 |
| | 8 | 1.1 | 1.2 | 1.4 | 1.5 | 1.7 | 1.9 | 2.1 | 2.3 | 2.6 | 2.9 |
| | 12 | 0.9 | 1.0 | 1.1 | 1.3 | 1.4 | 1.6 | 1.8 | 2.0 | 2.2 | 2.4 |
| | 20 | 0.7 | 0.8 | 0.9 | 1.0 | 1.1 | 1.3 | 1.4 | 1.6 | 1.8 | 2.0 |
| 15 | ½ | .... | .... | .... | .... | 4.0 | 4.5 | 5.0 | 5.6 | 6.2 | 6.9 |
| | 1 | .... | .... | .... | 2.7 | 3.0 | 3.3 | 3.7 | 4.1 | 4.6 | 5.1 |
| | 2 | .... | .... | 1.8 | 2.0 | 2.2 | 2.5 | 2.8 | 3.1 | 3.4 | 3.8 |
| | 4 | .... | 1.2 | 1.3 | 1.5 | 1.6 | 1.8 | 2.0 | 2.3 | 2.5 | 2.8 |
| | 8 | 0.8 | 0.9 | 1.0 | 1.1 | 1.2 | 1.4 | 1.5 | 1.7 | 1.9 | 2.1 |
| | 12 | 0.7 | 0.7 | 0.8 | 0.9 | 1.0 | 1.1 | 1.3 | 1.4 | 1.6 | 1.7 |
| | 20 | 0.5 | 0.6 | 0.7 | 0.7 | 0.8 | 0.9 | 1.0 | 1.1 | 1.3 | 1.4 |

**Table 3. AGMA Coarse-Pitch Spur, Helical and Herringbone Gear Lead Tolerances***

(In ten-thousandths of an inch)

| AGMA Quality Number | Face Width, inches | | | | |
|---|---|---|---|---|---|
| | Lead Tolerance | | | | |
| | 1 to 2 | 2 | 3 | 4 | 5 |
| 8 | 5 | 8 | 11 | 13 | 16 |
| 9 | 4 | 7 | 9 | 11 | 13 |
| 10 | 3 | 5 | 7 | 9 | 10 |

| AGMA Quality Number | Face Width, inches | | | | |
|---|---|---|---|---|---|
| | Lead Tolerance | | | | |
| | 1 to 2 | 2 | 3 | 4 | 5 |
| 11 | 3 | 4 | 6 | 7 | 8 |
| 12 | 2 | 3 | 5 | 6 | 7 |
| 13 | 2 | 3 | 4 | 4 | 5 |

| AGMA Quality Number | Face Width, inches | | | | |
|---|---|---|---|---|---|
| | Lead Tolerance | | | | |
| | 1 to 2 | 2 | 3 | 4 | 5 |
| 14 | 1 | 2 | 3 | 4 | 4 |
| 15 | 1 | 2 | 2 | 3 | 3 |

* See footnote to table below.

**Table 4. AGMA Coarse-Pitch Spur, Helical and Herringbone Gear Total Composite Tolerances* — 1**

(In ten-thousandths of an inch)

| AGMA Quality Number | Normal Diametral Pitch† | Pitch Diameter, inches | | | | | | | | | | | | | | | | | |
|---|---|---|---|---|---|---|---|---|---|---|---|---|---|---|---|---|---|---|---|
| | | Total Composite Tolerance | | | | | | | | | | | | | | | | | |
| | | 0.040 | 0.063 | 0.100 | 0.160 | 0.250 | 0.400 | 0.630 | 1.0 | 1.6 | 2.5 | 4.0 | 6.3 | 10 | 16 | 25 | 50 | 100 | 200 |
| 5 | 12.0 | | | | | | | 94.8 | 95.0 | 95.4 | 98.4 | 106.4 | 114.7 | 123.8 | 133.8 | 144.0 | 161.5 | 181.2 | |
| | 20.0 | | | | | | 74.3 | 74.3 | 74.3 | 77.0 | 82.5 | 88.6 | 95.0 | 102.0 | 109.6 | 117.3 | 130.5 | | |
| 6 | 0.5 | | | | | | | | | | | | | | 434.3 | 447.5 | 480.3 | 577.7 | 694.8 |
| | 1.0 | | | | | | | | | | | | 268.5 | 273.8 | 282.3 | 293.0 | 346.1 | 408.7 | 482.7 |
| | 2.0 | | | | | | | | | | | 176.3 | 179.4 | 184.4 | 196.9 | 216.8 | 251.9 | 292.6 | 340.0 |
| | 4.0 | | | | | | | | | 118.4 | 119.4 | 121.2 | 123.6 | 135.2 | 148.2 | 161.7 | 185.0 | 211.8 | 242.5 |
| | 8.0 | | | | | | | | 82.6 | 83.1 | 83.9 | 87.9 | 95.2 | 103.3 | 112.2 | 121.3 | 137.1 | 154.8 | |
| | 12.0 | | | | | | | 67.5 | 67.7 | 68.1 | 70.3 | 76.0 | 81.9 | 88.4 | 95.6 | 102.2 | 115.4 | 129.4 | |
| | 20.0 | | | | | | | 53.0 | 53.0 | 55.0 | 58.9 | 63.3 | 67.9 | 72.8 | 78.3 | 83.8 | 93.2 | | |
| 7 | 0.5 | | | | | | | | | | | | | | 300.2 | 313.4 | 343.1 | 412.6 | 496.3 |
| | 1.0 | | | | | | | | | | | | 186.2 | 191.5 | 200.0 | 209.3 | 247.2 | 291.9 | 344.8 |
| | 2.0 | | | | | | | | | | | 123.6 | 126.7 | 131.7 | 140.6 | 154.9 | 179.9 | 209.0 | 242.8 |
| | 4.0 | | | | | | | | | 83.4 | 84.5 | 86.3 | 88.3 | 96.6 | 105.8 | 115.5 | 132.2 | 151.3 | 173.2 |
| | 8.0 | | | | | | | | 58.6 | 59.1 | 59.9 | 62.8 | 68.0 | 73.8 | 80.1 | 86.7 | 97.9 | 110.6 | |
| | 12.0 | | | | | | | 48.0 | 48.3 | 48.6 | 50.2 | 54.3 | 58.5 | 63.2 | 68.3 | 73.5 | 82.4 | 92.4 | |
| | 20.0 | | | | | | 37.9 | 37.9 | 37.9 | 39.3 | 42.1 | 45.2 | 48.5 | 52.0 | 55.9 | 59.9 | 66.6 | | |

* Extracted from AGMA Handbook for Unassembled Gears — Volume 1 — Gear Classifications, Materials, and Inspection (AGMA 390.03), with the permission of the publisher, the American Gear Manufacturers Association, 1901 North Fort Myer Drive, Arlington, Va. 22209.
† For diametral pitches above 20, see above mentioned Handbook.

Table 4. AGMA Coarse-Pitch Spur, Helical and Herringbone Gear Total Composite Tolerances*—2

(In ten-thousandths of an inch)

Total Composite Tolerance

| AGMA Quality Number | Normal Diametral Pitch† | Pitch Diameter, inches | | | | | | | | | | | | | | | | | |
|---|---|---|---|---|---|---|---|---|---|---|---|---|---|---|---|---|---|---|---|
| | | 0.040 | 0.063 | 0.100 | 0.160 | 0.250 | 0.400 | 0.630 | 1.0 | 1.6 | 2.5 | 4.0 | 6.3 | 10 | 16 | 25 | 50 | 100 | 200 |
| 8 | 0.5 | | | | | | | | | | | | | | 204.4 | 217.6 | 245.5 | 294.7 | 354.5 |
| | 1.0 | | | | | | | | | | | | 127.4 | 132.7 | 141.2 | 149.5 | 176.6 | 208.5 | 246.3 |
| | 2.0 | | | | | | | | | | | 86.0 | 89.1 | 94.1 | 100.4 | 110.6 | 128.5 | 149.3 | 173.5 |
| | 4.0 | | | | | | | | | 58.4 | 59.5 | 61.3 | 63.0 | 69.0 | 75.6 | 82.5 | 94.4 | 108.1 | 123.7 |
| | 8.0 | | | | | | | | 41.5 | 42.0 | 42.8 | 44.9 | 48.6 | 52.7 | 57.2 | 61.9 | 69.9 | 79.0 | |
| | 12.0 | | | | | | | 34.1 | 34.4 | 34.7 | 35.9 | 38.8 | 41.8 | 45.1 | 48.8 | 52.5 | 58.9 | 66.0 | |
| | 20.0 | | | | | | 27.1 | 27.1 | 27.1 | 28.1 | 30.1 | 32.3 | 34.6 | 37.2 | 39.9 | 42.8 | 47.6 | | |
| 9 | 0.5 | | | | | | | | | | | | | | 136.0 | 149.1 | 175.0 | 210.5 | 253.2 |
| | 1.0 | | | | | | | | | | | | 85.4 | 90.7 | 99.3 | 106.8 | 126.1 | 148.9 | 175.9 |
| | 2.0 | | | | | | | | | | | 59.1 | 62.2 | 67.2 | 71.7 | 79.0 | 91.8 | 106.7 | 123.9 |
| | 4.0 | | | | | | | | | 40.5 | 41.6 | 43.4 | 45.0 | 49.3 | 54.0 | 58.9 | 67.4 | 77.2 | 88.4 |
| | 8.0 | | | | | | | | 29.2 | 29.8 | 30.6 | 32.0 | 34.7 | 37.0 | 40.9 | 44.2 | 49.9 | 56.4 | |
| | 12.0 | | | | | | | 24.2 | 24.4 | 24.8 | 25.6 | 27.7 | 29.9 | 32.2 | 34.8 | 37.5 | 42.0 | 47.2 | |
| | 20.0 | | | | | | 19.3 | 19.3 | 19.2 | 20.1 | 21.5 | 23.1 | 24.7 | 26.5 | 28.5 | 30.5 | 34.0 | | |
| 10 | 0.5 | | | | | | | | | | | | | | 87.1 | 100.3 | 125.0 | 150.4 | 180.9 |
| | 1.0 | | | | | | | | | | | | 55.5 | 60.7 | 69.3 | 76.3 | 90.1 | 106.4 | 125.6 |
| | 2.0 | | | | | | | | | | | 39.9 | 43.0 | 48.0 | 51.2 | 56.4 | 65.6 | 76.2 | 88.5 |
| | 4.0 | | | | | | | | | 27.8 | 28.9 | 30.7 | 32.2 | 35.2 | 38.6 | 42.1 | 48.2 | 55.1 | 63.1 |
| | 8.0 | | | | | | | | 20.5 | 21.0 | 21.8 | 22.9 | 24.8 | 26.9 | 29.2 | 31.6 | 35.7 | 40.3 | |
| | 12.0 | | | | | | | 17.1 | 17.3 | 17.7 | 18.3 | 19.8 | 21.3 | 23.0 | 24.9 | 26.8 | 30.0 | 33.7 | |
| | 20.0 | | | | | | 13.8 | 13.8 | 13.8 | 14.3 | 15.3 | 16.5 | 17.7 | 19.0 | 20.4 | 21.8 | 24.3 | | |
| 11 | 0.5 | | | | | | | | | | | | | | 52.2 | 65.3 | 89.3 | 107.4 | 129.2 |
| | 1.0 | | | | | | | | | | | | 34.0 | 39.3 | 47.9 | 54.5 | 64.3 | 76.3 | 89.7 |
| | 2.0 | | | | | | | | | | | 26.2 | 29.3 | 34.3 | 36.6 | 40.3 | 46.8 | 54.4 | 63.2 |
| | 4.0 | | | | | | | | | 18.7 | 19.8 | 21.6 | 23.0 | 25.1 | 27.6 | 30.1 | 34.4 | 39.4 | 45.1 |
| | 8.0 | | | | | | | | 14.3 | 14.8 | 15.6 | 16.4 | 17.7 | 19.2 | 20.9 | 22.6 | 25.5 | 28.8 | |
| | 12.0 | | | | | | | 12.0 | 12.3 | 12.6 | 13.1 | 14.1 | 15.2 | 16.4 | 17.8 | 19.1 | 21.5 | 24.1 | |
| | 20.0 | | | | | | 9.9 | 9.9 | 9.9 | 10.2 | 11.0 | 11.8 | 12.6 | 13.5 | 14.6 | 15.6 | 17.3 | | |

*† See footnotes on first page of table.

Table 4. AGMA Coarse-Pitch Spur, Helical and Herringbone Gear Total Composite Tolerances* — 3
(In ten-thousandths of an inch)

| AGMA Quality Number | Normal Diametral Pitch† | Pitch Diameter, inches | | | | | | | | | | | | | | | | | |
|---|---|---|---|---|---|---|---|---|---|---|---|---|---|---|---|---|---|---|---|
| | | 0.040 | 0.063 | 0.100 | 0.160 | 0.250 | 0.400 | 0.630 | 1.0 | 1.6 | 2.5 | 4.0 | 6.3 | 10 | 16 | 25 | 50 | 100 | 200 |
| | | Total Composite Tolerance | | | | | | | | | | | | | | | | | |
| 12 | 0.5 | | | | | | | | | | | | | | 27.2 | 40.4 | 63.8 | 76.7 | 92.3 |
| | 1.0 | | | | | | | | | | | | 18.7 | 24.0 | 32.6 | 38.0 | 46.0 | 54.3 | 64.1 |
| | 2.0 | | | | | | | | | | | 16.4 | 19.5 | 24.5 | 26.1 | 28.8 | 33.5 | 38.9 | 45.2 |
| | 4.0 | | | | | | | | | 12.2 | 13.3 | 15.1 | 16.4 | 18.0 | 19.9 | 21.5 | 24.6 | 28.1 | 32.2 |
| | 8.0 | | | | | | | | 9.8 | 10.3 | 11.1 | 11.7 | 12.6 | 13.7 | 14.9 | 16.1 | 18.2 | 20.6 | |
| | 12.0 | | | | | | | 8.4 | 8.6 | 9.0 | 9.3 | 10.1 | 10.9 | 11.7 | 12.7 | 13.7 | 15.3 | 17.2 | |
| | 20.0 | | | | | | 7.0 | 7.0 | 7.0 | 7.3 | 7.8 | 8.4 | 9.0 | 9.7 | 10.4 | 11.1 | 12.4 | | |
| 13 | 0.5 | | | | | | | | | | | | | | 9.4 | 22.6 | 45.6 | 54.8 | 65.9 |
| | 1.0 | | | | | | | | | | | | 7.8 | 13.1 | 21.6 | 27.8 | 32.8 | 38.8 | 45.8 |
| | 2.0 | | | | | | | | | | | 9.4 | 12.5 | 17.5 | 18.7 | 20.5 | 23.6 | 27.6 | 32.3 |
| | 4.0 | | | | | | | | | 7.5 | 8.6 | 10.4 | 11.7 | 12.8 | 14.6 | 15.3 | 17.6 | 20.1 | 23.0 |
| | 8.0 | | | | | | | | 6.0 | 7.1 | 8.0 | 8.3 | 9.8 | 9.8 | 10.6 | 11.5 | 13.0 | 14.7 | |
| | 12.0 | | | | | | | 5.8 | 6.1 | 6.4 | 6.7 | 7.2 | 7.8 | 8.4 | 9.1 | 9.8 | 10.9 | 12.3 | |
| | 20.0 | | | | | | 5.0 | 5.0 | 5.0 | 5.2 | 5.6 | 6.0 | 6.4 | 6.9 | 7.4 | 8.0 | 8.8 | | |
| 14 | 0.5 | | | | | | | | | | | | | | | 9.9 | 32.5 | 39.1 | 47.1 |
| | 1.0 | | | | | | | | | | | | 7.5 | 5.3 | 13.8 | 19.9 | 23.5 | 27.7 | 32.7 |
| | 2.0 | | | | | | | | | | | 4.4 | 8.4 | 12.5 | 13.3 | 14.7 | 17.1 | 19.8 | 23.0 |
| | 4.0 | | | | | | | | | 4.2 | 5.3 | 7.1 | 6.5 | 9.2 | 10.6 | 11.0 | 12.5 | 14.4 | 16.4 |
| | 8.0 | | | | | | | | 4.3 | 4.9 | 5.7 | 6.0 | 5.6 | 7.0 | 7.6 | 8.2 | 9.3 | 10.5 | |
| | 12.0 | | | | | | | 4.0 | 4.2 | 4.6 | 4.8 | 5.1 | 4.6 | 6.0 | 6.5 | 7.0 | 7.8 | 8.8 | |
| | 20.0 | | | | | | 3.6 | 3.6 | 3.6 | 3.7 | 4.0 | 4.3 | | 4.9 | 5.3 | 5.7 | 6.3 | | |
| 15 | 20.0 | | | | | | 2.6 | 2.6 | 2.6 | 2.7 | 2.9 | 3.1 | 3.3 | 3.5 | 3.8 | 4.1 | 4.5 | | |
| 16 | 20.0 | | | | | | 1.8 | 1.8 | 1.8 | 1.9 | 2.0 | 2.2 | 2.3 | 2.5 | 2.7 | 2.9 | 3.2 | | |

*† See footnotes on first page of table.

**Table 5. AGMA Coarse-Pitch Spur, Helical and Herringbone Gear Tooth-to-Tooth Composite Tolerances\* — I**
(In ten-thousandths of an inch)

*Pitch Diameter, inches — Tooth-to-Tooth Composite Tolerance*

| AGMA Quality Number | Normal Diametral Pitch | 0.04 | 0.06 | 0.10 | 0.16 | 0.25 | 0.40 | 0.63 | 1.0 | 1.6 | 2.5 | 4.0 | 6.3 | 10 | 16 | 25 | 50 | 100 | 200 |
|---|---|---|---|---|---|---|---|---|---|---|---|---|---|---|---|---|---|---|---|
| 5 | 20.0 | | | | | | 44.4 | 39.8 | 35.6 | 33.5 | 33.4 | 33.4 | 33.4 | 33.4 | 33.4 | 33.4 | 33.4 | | |
| 6 | 0.5 | | | | | | | | | | | | | | 76.9 | 69.1 | 57.8 | 57.9 | 57.9 |
|   | 1.0 | | | | | | | | | | | | 68.9 | 61.7 | 55.1 | 49.3 | 49.0 | 49.0 | 49.0 |
|   | 2.0 | | | | | | | | | | | 55.1 | 49.4 | 44.2 | 40.7 | 41.5 | 41.5 | 41.5 | 41.5 |
|   | 4.0 | | | | | | | | | 49.2 | 44.2 | 39.6 | 35.8 | 35.8 | 35.1 | 35.1 | 35.1 | 35.1 | 35.1 |
|   | 8.0 | | | | | | | | 39.5 | 35.3 | 31.7 | 29.6 | 29.7 | 29.7 | 29.7 | 29.7 | 29.7 | 29.7 | |
|   | 12.0 | | | | | | | 36.3 | 32.5 | 29.1 | 27.2 | 27.0 | 27.0 | 27.0 | 27.0 | 27.0 | 27.0 | 27.0 | |
|   | 20.0 | | | | | | 31.7 | 28.4 | 25.5 | 24.0 | 23.9 | 23.9 | 23.9 | 23.9 | 23.9 | 23.9 | 23.9 | | |
| 7 | 0.5 | | | | | | | | | | | | | | 54.9 | 49.3 | 41.3 | 41.3 | 41.3 |
|   | 1.0 | | | | | | | | | | | | 49.2 | 44.6 | 39.4 | 35.2 | 35.0 | 35.0 | 35.0 |
|   | 2.0 | | | | | | | | | | | 39.4 | 35.3 | 31.6 | 29.1 | 29.6 | 29.6 | 29.6 | 29.6 |
|   | 4.0 | | | | | | | | | 35.2 | 31.6 | 28.2 | 25.6 | 25.6 | 25.1 | 25.1 | 25.1 | 25.1 | 25.1 |
|   | 8.0 | | | | | | | | 28.2 | 25.2 | 22.7 | 21.1 | 21.3 | 21.2 | 21.3 | 21.2 | 21.2 | 21.2 | |
|   | 12.0 | | | | | | | 26.0 | 23.2 | 20.8 | 19.3 | 19.3 | 19.3 | 19.3 | 19.3 | 19.3 | 19.3 | 19.3 | |
|   | 20.0 | | | | | | 22.7 | 20.3 | 18.2 | 17.1 | 17.1 | 17.1 | 17.1 | 17.1 | 17.1 | 17.1 | 17.1 | | |
| 8 | 0.5 | | | | | | | | | | | | | | 39.2 | 35.2 | 29.5 | 29.5 | 29.5 |
|   | 1.0 | | | | | | | | | | | | 35.2 | 31.5 | 28.1 | 25.1 | 25.0 | 25.0 | 25.0 |
|   | 2.0 | | | | | | | | | | | 28.1 | 25.2 | 22.6 | 20.8 | 21.2 | 21.2 | 21.2 | 21.2 |
|   | 4.0 | | | | | | | | | 25.1 | 22.6 | 20.2 | 18.3 | 17.9 | 17.9 | 17.9 | 17.9 | 17.9 | 17.9 |
|   | 8.0 | | | | | | | | 20.2 | 18.0 | 16.9 | 15.8 | 15.8 | 15.2 | 15.2 | 15.2 | 15.2 | 15.2 | |
|   | 12.0 | | | | | | | 18.5 | 16.6 | 14.8 | 13.9 | 13.8 | 13.8 | 13.8 | 13.8 | 13.8 | 13.8 | 13.8 | |
|   | 20.0 | | | | | | 16.2 | 14.5 | 13.0 | 12.2 | 12.2 | 12.2 | 12.2 | 12.2 | 12.2 | 12.2 | 12.2 | | |

* Extracted from AGMA Handbook for Unassembled Gears — Volume I — Gear Classifications, Materials, and Inspection (AGMA 390.03), with the permission of the publisher, the American Gear Manufacturers Association, 1901 North Fort Myer Drive, Arlington, Va. 22209.

Table 5. AGMA Coarse-Pitch Spur, Helical and Herringbone Gear Tooth-to-Tooth Composite Tolerances* — 2
(In ten-thousandths of an inch)

| AGMA Quality Number | Normal Diametral Pitch | Pitch Diameter, inches — Tooth-to-Tooth Composite Tolerance | | | | | | | | | | | | | | | | | |
|---|---|---|---|---|---|---|---|---|---|---|---|---|---|---|---|---|---|---|---|
| | | 0.04 | 0.06 | 0.10 | 0.16 | 0.25 | 0.40 | 0.63 | 1.0 | 1.6 | 2.5 | 4.0 | 6.3 | 10 | 16 | 25 | 50 | 100 | 200 |
| 9 | 0.5 | | | | | | | | | | | | 25.1 | 22.5 | 28.0 | 25.2 | 21.1 | 21.1 | 21.1 |
| | 1.0 | | | | | | | | | | | 20.1 | 17.9 | 16.1 | 20.1 | 18.0 | 17.9 | 17.9 | 17.9 |
| | 2.0 | | | | | | | | | 17.9 | 16.1 | 14.4 | 13.0 | 12.8 | 14.8 | 15.1 | 15.1 | 15.1 | 15.1 |
| | 4.0 | | | | | | | | 14.4 | 12.9 | 11.6 | 10.8 | 9.8 | 10.8 | 12.8 | 12.8 | 12.8 | 12.8 | 12.8 |
| | 8.0 | | | | | | | 13.2 | 11.9 | 10.6 | 9.9 | 9.8 | 9.8 | 9.8 | 9.8 | 9.8 | 9.8 | 10.8 | |
| | 20.0 | | | | | | 11.6 | 10.4 | 9.3 | 8.7 | 8.7 | 8.7 | 8.7 | 8.7 | 8.7 | 8.7 | 8.7 | 9.8 | |
| 10 | 0.5 | | | | | | | | | | | | 17.9 | 16.1 | 20.0 | 18.0 | 15.0 | 15.1 | 15.1 |
| | 1.0 | | | | | | | | | | | 14.3 | 12.9 | 11.5 | 14.3 | 12.8 | 12.8 | 12.8 | 12.8 |
| | 2.0 | | | | | | | | | 12.8 | 11.5 | 10.3 | 9.3 | 9.3 | 10.6 | 10.8 | 10.8 | 10.8 | 10.8 |
| | 4.0 | | | | | | | | 10.3 | 9.2 | 8.3 | 7.7 | 7.7 | 7.7 | 9.1 | 9.1 | 9.1 | 9.1 | 9.1 |
| | 8.0 | | | | | | | 9.5 | 8.5 | 7.6 | 7.1 | 7.0 | 7.0 | 7.0 | 7.7 | 7.7 | 7.7 | 7.7 | |
| | 20.0 | | | | | | 8.3 | 7.4 | 6.6 | 6.2 | 6.2 | 6.2 | 6.2 | 6.2 | 6.2 | 6.2 | 6.2 | 7.0 | |
| 11 | 0.5 | | | | | | | | | | | | 12.8 | 11.5 | 14.3 | 12.8 | 10.7 | 10.8 | 10.8 |
| | 1.0 | | | | | | | | | | | 10.2 | 9.2 | 8.2 | 10.2 | 9.2 | 9.1 | 9.1 | 9.1 |
| | 2.0 | | | | | | | | | 9.2 | 8.2 | 7.3 | 6.7 | 6.5 | 7.6 | 7.7 | 7.7 | 7.7 | 7.7 |
| | 4.0 | | | | | | | | 7.3 | 6.6 | 5.9 | 5.5 | 5.5 | 5.5 | 6.5 | 6.5 | 6.5 | 6.5 | 6.5 |
| | 12.0 | | | | | | | 6.8 | 6.0 | 5.4 | 5.1 | 5.0 | 5.0 | 5.0 | 5.5 | 5.5 | 5.5 | 5.5 | |
| | 20.0 | | | | | | 5.9 | 5.3 | 4.7 | 4.5 | 4.4 | 4.4 | 4.4 | 4.4 | 4.4 | 4.4 | 4.4 | 5.0 | |
| 12 | 0.5 | | | | | | | | | | | | 9.6 | 8.2 | 10.2 | 9.2 | 7.7 | 7.7 | 7.7 |
| | 1.0 | | | | | | | | | | | 7.3 | 6.6 | 5.9 | 7.3 | 6.5 | 6.5 | 6.5 | 6.5 |
| | 2.0 | | | | | | | | | 6.5 | 5.9 | 5.2 | 4.8 | 4.7 | 5.4 | 5.5 | 5.5 | 5.5 | 5.5 |
| | 4.0 | | | | | | | | 5.2 | 4.7 | 4.2 | 3.9 | 4.6 | 4.0 | 4.7 | 4.7 | 4.7 | 4.7 | 4.7 |
| | 12.0 | | | | | | | 4.8 | 4.3 | 3.9 | 3.6 | 3.6 | 3.6 | 3.6 | 4.0 | 4.0 | 4.0 | 4.0 | |
| | 20.0 | | | | | | 4.2 | 3.8 | 3.4 | 3.2 | 3.2 | 3.2 | 3.2 | 3.2 | 3.2 | 3.2 | 3.2 | 3.6 | |

* See footnote on first page of table.

**Table 5. AGMA Coarse-Pitch Spur, Helical and Herringbone Gear Tooth-to-Tooth Composite Tolerances\* — 3**
(In ten-thousandths of an inch)

Pitch Diameter, inches — Tooth-to-Tooth Composite Tolerance

| AGMA Quality Number | Normal Diametral Pitch | 0.04 | 0.06 | 0.10 | 0.16 | 0.25 | 0.40 | 0.63 | 1.0 | 1.6 | 2.5 | 4.0 | 6.3 | 10 | 16 | 25 | 50 | 100 | 200 |
|---|---|---|---|---|---|---|---|---|---|---|---|---|---|---|---|---|---|---|---|
| 13 | 0.5 | | | | | | | | | | | | | | 7.3 | 6.6 | 5.5 | 5.5 | 5.5 |
| | 1.0 | | | | | | | | | | | | 6.5 | 5.9 | 5.2 | 4.7 | 4.6 | 4.6 | 4.6 |
| | 2.0 | | | | | | | | | | | 5.2 | 4.7 | 4.3 | 3.9 | 3.9 | 3.9 | 3.9 | 3.9 |
| | 4.0 | | | | | | | | | 4.7 | 4.2 | 3.7 | 3.4 | 3.3 | 3.3 | 3.3 | 3.3 | 3.3 | 3.3 |
| | 8.0 | | | | | | | | 3.7 | 3.3 | 3.0 | 2.8 | 2.8 | 2.8 | 2.8 | 2.8 | 2.8 | 2.8 | |
| | 12.0 | | | | | | | 3.4 | 3.1 | 2.8 | 2.6 | 2.6 | 2.6 | 2.6 | 2.6 | 2.6 | 2.6 | 2.6 | |
| | 20.0 | | | | | | 3.0 | 2.7 | 2.4 | 2.3 | 2.3 | 2.3 | 2.3 | 2.3 | 2.3 | 2.3 | 2.3 | | |
| 14 | 0.5 | | | | | | | | | | | | | | 5.2 | 4.7 | 3.9 | 3.9 | 3.9 |
| | 1.0 | | | | | | | | | | | | 4.7 | 4.2 | 3.7 | 3.3 | 3.3 | 3.3 | 3.3 |
| | 2.0 | | | | | | | | | | | 3.7 | 3.3 | 3.0 | 2.8 | 2.8 | 2.8 | 2.8 | 2.8 |
| | 4.0 | | | | | | | | | 3.3 | 3.0 | 2.7 | 2.4 | 2.4 | 2.4 | 2.4 | 2.4 | 2.4 | 2.4 |
| | 8.0 | | | | | | | | 2.7 | 2.4 | 2.1 | 2.0 | 2.0 | 2.0 | 2.0 | 2.0 | 2.0 | 2.0 | |
| | 12.0 | | | | | | | 2.5 | 2.2 | 2.0 | 1.8 | 1.8 | 1.8 | 1.8 | 1.8 | 1.8 | 1.8 | 1.8 | |
| | 20.0 | | | | | | 2.1 | 1.9 | 1.7 | 1.6 | 1.6 | 1.6 | 1.6 | 1.6 | 1.6 | 1.6 | 1.6 | | |
| 15 | 20.0 | | | | | | 1.5 | 1.4 | 1.2 | 1.2 | 1.2 | 1.2 | 1.2 | 1.2 | 1.2 | 1.2 | 1.2 | | |
| 16 | 20.0 | | | | | | 1.1 | 1.0 | 0.9 | 0.8 | 0.8 | 0.8 | 0.8 | 0.8 | 0.8 | 0.8 | 0.8 | | |

\* See footnote on first page of table.

**Table 6. AGMA Bevel Gear Tolerances — 1**
(In ten-thousandths of an inch)

| AGMA Quality Number | Diametral Pitch | Pitch Diameter (Inches) | | | | | | | | |
|---|---|---|---|---|---|---|---|---|---|---|
| | | ¾ | 1½ | 3 | 5 | 12 | 25 | 50 | 100 | 200 |
| | | Pitch Tolerance | | | | | | | | |
| 6 | ½ | .... | .... | .... | .... | .... | 50 | 55 | 62 | 70 |
| | 1 | .... | .... | .... | .... | 31 | 33 | 38 | 42 | 50 |
| | 2 | .... | .... | .... | 26 | 27 | 29 | 33 | 37 | 42 |
| | 4 | .... | .... | 22 | 22 | 24 | 26 | 28 | 32 | 37 |
| | 8 | .... | 18 | 18 | 19 | 21 | 23 | 25 | 28 | .... |
| | 16–19.99 | 16 | 16 | 16 | 17 | 18 | 19 | 21 | 23 | .... |
| 7 | ½ | .... | .... | .... | .... | .... | 37 | 40 | 45 | 50 |
| | 1 | .... | .... | .... | .... | 23 | 25 | 28 | 32 | 37 |
| | 2 | .... | .... | .... | 19 | 20 | 22 | 24 | 27 | 31 |
| | 4 | .... | .... | 16 | 16 | 17 | 19 | 21 | 24 | 28 |
| | 8 | .... | 13½ | 14 | 14½ | 15½ | 17 | 19 | 21 | .... |
| | 16–19.99 | 11 | 11½ | 12 | 12½ | 13 | 14 | 15½ | 17½ | .... |
| 8 | ½ | .... | .... | .... | .... | .... | 26 | 28 | 31 | .... |
| | 1 | .... | .... | .... | .... | 16 | 18 | 19 | 22 | .... |
| | 2 | .... | .... | .... | 14 | 14 | 15 | 17 | 19 | .... |
| | 4 | .... | .... | 11 | 11 | 12 | 13 | 15 | .... | .... |
| | 8 | .... | 9 | 10 | 10 | 11 | 12 | .... | .... | .... |
| | 16–19.99 | 8 | 8 | 8 | 9 | 9 | .... | .... | .... | .... |
| 9 | ½ | .... | .... | .... | .... | .... | 19 | 20 | 22 | .... |
| | 1 | .... | .... | .... | .... | 11 | 12 | 14 | 16 | .... |
| | 2 | .... | .... | .... | 10 | 10 | 11 | 12 | 14 | .... |
| | 4 | .... | .... | 8 | 8 | 9 | 9½ | 11 | .... | .... |
| | 8 | .... | 7 | 7 | 7½ | 8 | 8½ | .... | .... | .... |
| | 16–19.99 | 6 | 6 | 6 | 6½ | 7 | .... | .... | .... | .... |
| 10 | 1 | .... | .... | .... | .... | 8½ | 9 | .... | .... | .... |
| | 2 | .... | .... | .... | 7 | 7½ | 8 | .... | .... | .... |
| | 4 | .... | .... | 6 | 6 | 6½ | 7 | .... | .... | .... |
| | 8 | .... | 5 | 5 | 5½ | 6 | 6½ | .... | .... | .... |
| | 16–19.99 | 4½ | 4½ | 4½ | 4½ | 5 | .... | .... | .... | .... |
| 11 | 1 | .... | .... | .... | .... | 6 | 6 | .... | .... | .... |
| | 2 | .... | .... | .... | 5 | 5 | 6 | .... | .... | .... |
| | 4 | .... | .... | 4 | 4 | 4½ | 5 | .... | .... | .... |
| | 8 | .... | 3½ | 3½ | 4 | 4 | 4½ | .... | .... | .... |
| | 16–19.99 | 3 | 3 | 3 | 3½ | 3½ | .... | .... | .... | .... |
| 12 | 2 | .... | .... | .... | 3½ | 4 | 4 | .... | .... | .... |
| | 4 | .... | .... | 3 | 3 | 3½ | 3½ | .... | .... | .... |
| | 8 | .... | 2½ | 2½ | 3 | 3 | 3½ | .... | .... | .... |
| | 16–19.99 | 2½ | 2½ | 2½ | 2½ | 2½ | .... | .... | .... | .... |
| 13 | 2 | .... | .... | .... | 2½ | 3 | 3 | .... | .... | .... |
| | 4 | .... | .... | 2 | 2 | 2½ | 2½ | .... | .... | .... |
| | 8 | .... | 2 | 2 | 2 | 2 | 2½ | .... | .... | .... |
| | 16–19.99 | 2 | 2 | 2 | 2 | 2 | .... | .... | .... | .... |

**Table 6.  AGMA Bevel Gear Tolerances — 2**
(In ten-thousandths of an inch)

| AGMA Quality Number | Diametral Pitch | Pitch Diameter (Inches) | | | | | | | | |
| --- | --- | --- | --- | --- | --- | --- | --- | --- | --- | --- |
| | | ¾ | 1½ | 3 | 6 | 12 | 25 | 50 | 100 | 200 |
| | | Runout Tolerance | | | | | | | | |
| 3 | ½ | .... | .... | .... | .... | .... | 770 | 1010 | 1360 | .... |
| | 1 | .... | .... | .... | .... | 540 | 710 | 930 | 1250 | .... |
| | 2 | .... | .... | .... | 382 | 498 | 660 | 860 | 1150 | .... |
| | 4 | .... | .... | 280 | 355 | 460 | 608 | 800 | .... | .... |
| 4 | ½ | .... | .... | .... | .... | .... | 540 | 700 | 940 | .... |
| | 1 | .... | .... | .... | .... | 378 | 496 | 640 | 860 | .... |
| | 2 | .... | .... | .... | 272 | 348 | 452 | 590 | 790 | .... |
| | 4 | .... | .... | 198 | 250 | 320 | 419 | 542 | 720 | .... |
| 5 | ½ | .... | .... | .... | .... | .... | 396 | 510 | 665 | 880 |
| | 1 | .... | .... | .... | .... | 270 | 350 | 450 | 582 | 775 |
| | 2 | .... | .... | .... | 184 | 233 | 302 | 390 | 510 | 680 |
| | 4 | .... | .... | 130 | 160 | 203 | 262 | 340 | 440 | 590 |
| | 8 | .... | 91 | 112 | 140 | 177 | 228 | 290 | 380 | .... |
| 6 | ½ | .... | .... | .... | .... | .... | 280 | 350 | 450 | 600 |
| | 1 | .... | .... | .... | .... | 188 | 235 | 295 | 378 | 508 |
| | 2 | .... | .... | .... | 131 | 160 | 200 | 250 | 322 | 425 |
| | 4 | .... | .... | 92 | 110 | 135 | 170 | 210 | 270 | 360 |
| | 8 | .... | 64 | 76 | 93 | 114 | 143 | 180 | 230 | .... |
| | 16-19.99 | 46 | 55 | 66 | 80 | 98 | 122 | 152 | 193 | .... |
| 7 | ½ | .... | .... | .... | .... | .... | 209 | 260 | 335 | 445 |
| | 1 | .... | .... | .... | .... | 132 | 165 | 205 | 261 | 350 |
| | 2 | .... | .... | .... | 84 | 103 | 130 | 163 | 210 | 280 |
| | 4 | .... | .... | 55 | 67 | 82 | 103 | 130 | 167 | 225 |
| | 8 | .... | 37 | 44 | 54 | 66 | 82 | 103 | 132 | .... |
| | 16-19.99 | 26 | 31 | 37 | 45 | 55 | 69 | 86 | 110 | .... |
| 8 | ½ | .... | .... | .... | .... | .... | 160 | 200 | 255 | .... |
| | 1 | .... | .... | .... | .... | 95 | 115 | 140 | 180 | .... |
| | 2 | .... | .... | .... | 58 | 68 | 82 | 100 | 125 | .... |
| | 4 | .... | .... | 36 | 41 | 47 | 56 | 67 | .... | .... |
| | 8 | .... | 25 | 28 | 32 | 36 | 42 | .... | .... | .... |
| | 16-19.99 | 19 | 21 | 26 | 30 | .... | .... | .... | .... | .... |
| | 20-120 | .... | .... | .... | .... | .... | .... | .... | .... | .... |
| 9 | ½ | .... | .... | .... | .... | .... | 113 | 140 | 180 | .... |
| | 1 | .... | .... | .... | .... | 68 | 81 | 100 | 125 | .... |
| | 2 | .... | .... | .... | 40 | 48 | 58 | 70 | 88 | .... |
| | 4 | .... | .... | 26 | 29 | 34 | 40 | 48 | .... | .... |
| | 8 | .... | 18 | 20 | 22 | 26 | 30 | .... | .... | .... |
| | 16-19.99 | 14 | 15 | 16 | 18 | 21 | .... | .... | .... | .... |
| | 20-120 | .... | .... | .... | .... | .... | .... | .... | .... | .... |
| 10 | 1 | .... | .... | .... | .... | 50 | 58 | .... | .... | .... |
| | 2 | .... | .... | .... | 30 | 34 | 40 | .... | .... | .... |
| | 4 | .... | .... | 18 | 21 | 24 | 28 | .... | .... | .... |
| | 8 | .... | 13 | 14 | 16 | 18 | 21 | .... | .... | .... |
| | 16-19.99 | 10 | 11 | 12 | 13 | 15 | .... | .... | .... | .... |
| | 20-120 | .... | .... | .... | .... | .... | .... | .... | .... | .... |
| 11 | 1 | .... | .... | .... | .... | 34 | 41 | .... | .... | .... |
| | 2 | .... | .... | .... | 21 | 24 | 28 | .... | .... | .... |
| | 4 | .... | .... | 13 | 15 | 17 | 20 | .... | .... | .... |
| | 8 | .... | 9 | 10 | 11 | 13 | 15 | .... | .... | .... |
| | 16-19.99 | 7 | 8 | 8 | 9 | 11 | .... | .... | .... | .... |
| | 20-120 | .... | .... | .... | .... | .... | .... | .... | .... | .... |
| 12 | 2 | .... | .... | .... | 15 | 18 | 21 | .... | .... | .... |
| | 4 | .... | .... | 9 | 11 | 12 | 14 | .... | .... | .... |
| | 8 | .... | 6½ | 7 | 8 | 9 | 11 | .... | .... | .... |
| | 16-19.99 | .... | 5½ | 6 | 7 | 8 | .... | .... | .... | .... |
| | 20-120 | .... | .... | .... | .... | .... | .... | .... | .... | .... |
| 13 | 2 | .... | .... | .... | 11 | 13 | 15 | .... | .... | .... |
| | 4 | .... | .... | 7 | 7½ | 9 | 10½ | .... | .... | .... |
| | 8 | .... | 5 | 5½ | 6 | 7 | 8 | .... | .... | .... |
| | 16-19.99 | 5 | 5 | 5 | 5 | 5½ | .... | .... | .... | .... |
| | 20-120 | .... | .... | .... | .... | .... | .... | .... | .... | .... |

**Table 6.   AGMA Bevel Gear Tolerances — 3**
(In ten-thousandths of an inch)

| AGMA Quality Number | Dia-metral Pitch | Pitch Diameter (Inches) | | | | | | | Pitch Diameter (Inches) | | | | | | |
|---|---|---|---|---|---|---|---|---|---|---|---|---|---|---|---|
| | | 3/16 | 3/8 | 3/4 | 1½ | 3 | 6 | 12 | 3/16 | 3/8 | 3/4 | 1½ | 3 | 6 | 12 |
| | | Tooth to Tooth Composite Tolerance | | | | | | | Total Composite Tolerance | | | | | | |
| 5 | 20–24 | 27 | 27 | 27 | 27 | 27 | 27 | 27 | 52 | 52 | 52 | 52 | 61 | 72 | 72 |
| | 24–32 | 27 | 27 | 27 | 27 | 27 | .. | .. | 52 | 52 | 52 | 52 | 61 | .. | .. |
| | 32–48 | 27 | 27 | 27 | 27 | .. | .. | .. | 52 | 52 | 52 | 52 | .. | .. | .. |
| 6 | 20–32 | 19 | 19 | 19 | 19 | 19 | 19 | 19 | 37 | 37 | 37 | 37 | 44 | 52 | 52 |
| | 32–48 | 19 | 19 | 19 | 19 | 19 | .. | .. | 37 | 37 | 37 | 37 | 44 | .. | .. |
| | 48–64 | 19 | 19 | 19 | 19 | .. | .. | .. | 37 | 37 | 37 | 37 | .. | .. | .. |
| 7 | 20–48 | 14 | 14 | 14 | 14 | 14 | 14 | 14 | 27 | 27 | 27 | 27 | 32 | 37 | 37 |
| | 48–64 | 14 | 14 | 14 | 14 | 14 | .. | .. | 27 | 27 | 27 | 27 | 32 | .. | .. |
| | 64–96 | 14 | 14 | 14 | 14 | .. | .. | .. | 27 | 27 | 27 | 27 | .. | .. | .. |
| 8 | 2 | .. | .. | .. | .. | .. | .. | .. | .. | .. | .. | .. | .. | .. | 80 |
| | 4 | .. | .. | .. | .. | .. | .. | .. | .. | .. | .. | .. | 46 | 52 | 58 |
| | 8 | .. | .. | .. | .. | .. | .. | .. | .. | .. | .. | 35 | 38 | 42 | 46 |
| | 16–19.99 | .. | .. | .. | .. | .. | .. | .. | .. | .. | .. | 27 | 30 | 34 | 37 |
| | 20–64 | 10 | 10 | 10 | 10 | 10 | 10 | 10 | 19 | 19 | 19 | 19 | 23 | 27 | 27 |
| | 64–96 | 10 | 10 | 10 | 10 | 10 | .. | .. | 19 | 19 | 19 | 19 | 23 | .. | .. |
| | 96–120 | 10 | 10 | 10 | 10 | .. | .. | .. | 19 | 19 | 19 | 19 | .. | .. | .. |
| 9 | 2 | .. | .. | .. | .. | .. | .. | .. | .. | .. | .. | .. | .. | .. | 57 |
| | 4 | .. | .. | .. | .. | .. | .. | .. | .. | .. | .. | .. | 33 | 37 | 42 |
| | 8 | .. | .. | .. | .. | .. | .. | .. | .. | .. | .. | 24 | 27 | 30 | 33 |
| | 16–19.99 | .. | .. | .. | .. | .. | .. | .. | .. | .. | .. | 19 | 22 | 24 | 27 |
| | 20–120 | 7 | 7 | 7 | 7 | 7 | 7 | .. | 14 | 14 | 14 | 14 | 16 | 19 | .. |
| 10 | 2 | .. | .. | .. | .. | .. | .. | .. | .. | .. | .. | .. | .. | .. | 40 |
| | 4 | .. | .. | .. | .. | .. | .. | .. | .. | .. | .. | .. | 23 | 26 | 29 |
| | 8 | .. | .. | .. | .. | .. | .. | .. | .. | .. | .. | 17 | 19 | 21 | 24 |
| | 16–19.99 | .. | .. | .. | .. | .. | .. | .. | .. | .. | .. | 14 | 15 | 17 | 19 |
| | 20–120 | 5 | 5 | 5 | 5 | 5 | 5 | .. | 10 | 10 | 10 | 10 | 12 | 14 | .. |
| 11 | 4 | .. | .. | .. | .. | .. | .. | .. | .. | .. | .. | .. | 17 | 18 | 21 |
| | 8 | .. | .. | .. | .. | .. | .. | .. | .. | .. | .. | 13 | 14 | 15 | 17 |
| | 16–19.99 | .. | .. | .. | .. | .. | .. | .. | .. | .. | .. | 10 | 11 | 12 | 14 |
| | 20–120 | 4 | 4 | 4 | 4 | 4 | 4 | .. | 7 | 7 | 7 | 7 | 9 | 10 | .. |
| 12 | 4 | .. | .. | .. | .. | .. | .. | .. | .. | .. | .. | .. | 12 | 13 | 15 |
| | 8 | .. | .. | .. | .. | .. | .. | .. | .. | .. | .. | 9 | 10 | 11 | 12 |
| | 16–19.99 | .. | .. | .. | .. | .. | .. | .. | .. | .. | .. | 7 | 8 | 9 | 10 |
| | 20–120 | 3 | 3 | 3 | 3 | 3 | 3 | .. | 5 | 5 | 5 | 5 | 6 | 7 | .. |
| 13 | 4 | .. | .. | .. | .. | .. | .. | .. | .. | .. | .. | .. | 10 | 11 | 12 |
| | 8 | .. | .. | .. | .. | .. | .. | .. | .. | .. | .. | 8 | 9 | 10 | 11 |
| | 16–19.99 | .. | .. | .. | .. | .. | .. | .. | .. | .. | .. | 6 | 7 | 8 | .. |
| | 20–120 | 2 | 2 | 2 | 2 | 2 | 2 | .. | 4 | 4 | 4 | 4 | 4 | 5 | .. |

**Table 7.**  Correlation of Former Fine-Pitch Spur and Helical Gear Classes
(AGMA 236.04) With the Quality Numbering System of the
Gear Classification Manual (AGMA 390.03)

| Previous AGMA Standard — AGMA 236.04 | Correlation of Quality Numbers of AGMA 390.03 |
|---|---|
| Commercial 1 | 5 or 6 |
| Commercial 2 | 6 or 7 |
| Commercial 3 | 8 |
| Commercial 4 | 9 |
| Precision 1 | 10 or 11 |
| Precision 2 | 12 |
| Precision 3 | 13 or 14 |

**Gear Material and Treatment.** — Table 8 gives AGMA designations for materials and heat treatments for coarse-pitch and fine-pitch gears. Because several of the listed material specifications could be selected for a given gearing application, there are other considerations that will help to determine a proper material selection.

Some of the fundamental factors to consider when making material and treatment selections are as follows:

*a.* The gear designer will have information as to the factors of safety, loading and operating conditions which will dictate the use of carbon steels or alloy steels and what hardness is required.

*b.* For replacement gearing, the life obtained from the previous gearing should be checked. If satisfactory, a similar material may be used for replacement. If longer life is required, the selection of a better material and treatment specification with higher hardness may provide the desired improvement.

*c.* Annealed carbon steels, bar stock, forgings, or castings are usually satisfactory for pinions and gears subjected to uniform or moderate shock loads when the size of the gearing is not an important factor.

*d.* Annealed carbon steel pinions with cast iron gears are sometimes used for the same reason mentioned in item *c.*

*e.* Alloy steel pinions are used for increased loads or where greater life is desired. They may be used with cast iron, or annealed (forged or cast) steel gears, usually where the ratio is about 6 to 1 or higher.

*f.* Alloy steel pinions and gears, heat treated, should be used with the higher hardness ranges when space limitation is a factor, *i.e.*, where smaller centers and face widths may be necessary.

*g.* For steel pinions and gears having ratios of 2 to 1 up to about 8 to 1, with both heat treated before machining and cutting, specify the minimum hardness on the pinion to be 40 Bhn higher than the minimum hardness on the gear. A higher hardness on the pinion than this specification (in relation to the gear hardness) will provide increased wear resistance when the ratio is about 8 to 1 or higher.

*h.* When steel pinions and gears with ratios below 2 to 1 are both heat treated before cutting, they are generally made to the same hardness.

*i.* Steel pinions and gears hardened after cutting to 400 Bhn or higher are generally specified with the same hardness, unless extremely high hardness is desired for the pinion.

*j.* For the convenience of the heat treater, a range of 40 Bhn should be specified. For example, a pinion might be specified with a range from 265 to 305 Bhn and the mating gear 225 to 265 Bhn.

*(Continued on page 785)*

**Table 8. AGMA Materials and Treatments for Coarse-Pitch and Fine-Pitch Gears***

| Designation Number | Material | Treatment | Hardness Range (See Notes 1 and 4) |
|---|---|---|---|
| UC-1 | Carbon Steel | Anneal or as rolled | Equivalent to 179 Bhn for AGMA Durability Rating |
| UA-11 | Alloy Steel | Anneal or as rolled | Equivalent to 179 Bhn for AGMA Durability Rating |
| HC-1 | Carbon Steel | Normalize and Temper or Quench and Temper | 212 to 248 Bhn |
| HC-2 | Carbon Steel | Quench and Temper | 223 to 262 Bhn |
| HC-3 | Carbon Steel | Quench and Temper | 248 to 285 Bhn |
| HC-4 | Carbon Steel | Quench and Temper | 262 to 302 Bhn |
| HC-5 | Carbon Steel | Quench and Temper | 285 to 321 Bhn |
| HC-6 | Carbon Steel | Quench and Temper | 302 to 351 Bhn |
| HA-11 | Alloy Steel | Normalize and Temper or Quench and Temper | 223 to 262 Bhn |
| HA-12 | Alloy Steel | Quench and Temper | 248 to 285 Bhn |
| HA-13 | Alloy Steel | Quench and Temper | 262 to 302 Bhn |
| HA-14 | Alloy Steel | Quench and Temper | 285 to 321 Bhn |
| HA-15 | Alloy Steel | Quench and Temper | 302 to 351 Bhn |
| HA-16 | Alloy Steel | Quench and Temper | 331 to 388 Bhn |
| HA-17 | Alloy Steel | Quench and Temper | 351 to 402 Bhn |
| HA-18 | Alloy Steel | Quench and Temper | 402 to 461 Bhn (42 to 49 Rc) |
| FC-1 | Carbon Steel | Flame Harden | 43 Rc Min. |
| FC-2 | Carbon Steel | Flame Harden | 48 Rc Min. |
| FC-3 | Carbon Steel | Flame Harden | 52 Rc Min. |
| FC-4 | Carbon Steel | Flame Harden | 55 Rc Min. |
| FA-11 | Alloy Steel | Flame Harden | 43 Rc Min. |
| FA-12 | Alloy Steel | Flame Harden | 48 Rc Min. |
| FA-13 | Alloy Steel | Flame Harden | 52 Rc Min. |
| FA-14 | Alloy Steel | Flame Harden | 55 Rc Min. |
| IC-1 | Carbon Steel | Induction Harden | 43 Rc Min. |
| IC-2 | Carbon Steel | Induction Harden | 48 Rc Min. |
| IC-3 | Carbon Steel | Induction Harden | 52 Rc Min. |
| IC-4 | Carbon Steel | Induction Harden | 55 Rc Min. |
| IA-11 | Alloy Steel | Induction Harden | 43 Rc Min. |
| IA-12 | Alloy Steel | Induction Harden | 48 Rc Min. |
| IA-13 | Alloy Steel | Induction Harden | 52 Rc Min. |
| IA-14 | Alloy Steel | Induction Harden | 55 Rc Min. |
| CC-1 | Carbon Steel | Cyanide | 55 Rc Min. |
| CA-11 | Alloy Steel | Cyanide | 55 Rc Min. |
| CN-1 | Carbon Steel | Carbonitride | 55 Rc Min. |
| CN-11 | Alloy Steel | Carbonitride | 55 Rc Min. |
| NA-11 | Alloy Steel (4140-4340-4640) | Nitride | 48 Rc Min. |
| NA-12 | Alloy Steel | Nitride | 64 Rc Min. |
| CH-1 | Carbon Steel | Carburize | 48 Rc Min. |
| CH-2 | Carbon Steel | Carburize | 50 Rc Min. |
| CH-3 | Carbon Steel | Carburize | 55 Rc Min. |
| CH-4 | Carbon Steel | Carburize | 58 Rc Min. |
| CH-5 | Carbon Steel | Carburize | 60 Rc Min. |
| CH-11 | Alloy Steel | Carburize | 48 Rc Min. |
| CH-12 | Alloy Steel | Carburize | 50 Rc Min. |
| CH-13 | Alloy Steel | Carburize | 55 Rc Min. |
| CH-14 | Alloy Steel | Carburize | 58 Rc Min. |
| CH-15 | Alloy Steel | Carburize | 60 Rc Min. |
| CI-20 | Cast Iron | As Required | |
| CI-30 | Cast Iron | As Required | 174 Bhn Min. |
| CI-35 | Cast Iron | As Required | 183 Bhn Min. |

* Extracted from AGMA Handbook for Unassembled Gears—Volume 1—Gear Classifications, Materials, and Inspection (AGMA 390.03), with permission of the publisher, the American Gear Manufacturers Association, 1901 North Fort Myer Drive, Arlington, Va. 22209.

**Table 8.** *(Continued).*  **AGMA Materials and Treatments for Coarse-Pitch and Fine-Pitch Gears**

| Designation Number | Material | Treatment | Hardness Range (See Notes 1 and 4) |
|---|---|---|---|
| CI-40 | Cast Iron | As Required | 202 Bhn Min. |
| CI-50 | Cast Iron | As Required | 217 Bhn Min. |
| CI-60 | Cast Iron | As Required | 223 Bhn Min. |
| NI-1 | Nodular Iron | Anneal | 179 Bhn Min. |
| NI-2 | Nodular Iron | Anneal or Normalize and Temper | 212 Bhn Min. |
| NI-3 | Nodular Iron | Anneal or Normalize and Temper | 223 Bhn Min. |
| NI-4 | Nodular Iron | Anneal or Normalize and Temper | 248 Bhn Min. |
| NI-5 | Nodular Iron | Quench and Temper | 255 Bhn Min. |
| NI-6 | Nodular Iron | Quench and Temper | 262 Bhn Min. |
| NI-7 | Nodular Iron | Quench and Temper | 269 Bhn Min. |
| NI-8 | Nodular Iron | Quench and Temper | 277 Bhn Min. |
| NI-9 | Nodular Iron | Quench and Temper | 285 Bhn Min. |
| NI-10 | Nodular Iron | Quench and Temper | 302 Bhn Min. |
| NI-11 | Nodular Iron | Quench and Temper | 311 Bhn Min. |
| NI-12 | Nodular Iron | Quench and Temper | 331 Bhn Min. |
| NI-13 | Nodular Iron | Quench and Temper | 351 Bhn Min. |
| NI-14 | Nodular Iron | Flame Harden or Induction Harden | 48 Rc Min. (Note 3) |
| SN-1 | Stainless Steel Non Magnetic (300 Series) | None Required | .......... |
| SM-2 | Stainless Steel Magnetic (400 Series) | None Required | .......... |
| SM-3 | Stainless Steel Magnetic (410, 416, 440) | Quench and Temper Induction Harden or Bright Harden | As Specified |
| SM-4 | Stainless Steel Magnetic (440) | Harden and Temper Induction Harden or Bright Harden | As Specified |
| AB-1 | Aluminum Bronze | As Cast | 116 Bhn Min. |
| AB-2 | Aluminum Bronze | As Cast | 116 Bhn Min. |
| AB-2 | Aluminum Bronze | Heat Treated | 121 Bhn Min. |
| AB-3 | Aluminum Bronze | As Cast | 140 Bhn Min. |
| AB-3 | Aluminum Bronze | Heat Treated | 190 Bhn Min. |
| AB-4 | Aluminum Bronze | As Cast | 175 Bhn Min. |
| AB-4 | Aluminum Bronze | Heat Treated | 202 Bhn Min. |
| AB-5 | Aluminum Bronze | Wrought, Heat Treated | 180 Bhn Min. |
| AB-6 | Aluminum Bronze | Wrought, Heat Treated | 180 Bhn Min. |
| MB-1 | Manganese Bronze | As Cast | 85 Bhn (500 Kg) |
| MB-2 | Manganese Bronze | As Cast | 125 Bhn (500 Kg) |
| MB-3 | Manganese Bronze | As Cast | 200 Bhn (500 Kg) |
| MB-4 | Manganese Bronze | As Cast | 210 Bhn (500 Kg) |
| MB-5 | Manganese Bronze | Wrought — Soft | 150 Bhn (500 Kg) |
| MB-6 | Manganese Bronze | Wrought — Half Hard | 190 Bhn (500 Kg) |
| MB-7 | Manganese Bronze | Wrought — Hard | 210 Bhn (500 Kg) |
| BZ-1 | Tin Bronze | As Cast | 70 Bhn (500 Kg) |
| BZ-2 | Tin Bronze | Chill Cast | 80 Bhn (500 Kg) |
| BZ-3 | Tin Bronze | As Cast | 70 Bhn (500 Kg) |

See notes at end of table.

**Table 8.** *(Concluded.)* **AGMA Materials and Treatments for Coarse Pitch and Fine Pitch Gears**

| Desig- nation Number | Material | Treatment | Hardness Range (See Notes 1 and 3) |
|---|---|---|---|
| BZ-4 | Tin Bronze | Chill Cast | 85 Bhn (500 Kg) |
| BZ-5 | Tin Bronze | As Cast | .......... |
| BZ-6 | Tin Bronze | Chill Cast | 95 Bhn (1500 Kg) |
| BZ-7 | Tin Bronze | Centrifugal Cast | .......... |
| BZ-8 | Tin Bronze | Chill Cast | 95 Bhn (1500 Kg) |
| AL-1 | Aluminum 2017 T3 Sheet | Heat Treated | .......... |
| AL-2 | Aluminum 2017 T4 Sheet | Heat Treated | 105 Bhn (500 Kg) |
| AL-3 | Aluminum 2024 T3 Sheet | Heat Treated | 120 Bhn (500 Kg) |
| AL-4 | Aluminum 2024 T4 Sheet | Heat Treated | 120 Bhn (500 Kg) |
| AL-5 | Aluminum 6061 T6 Bar or Sheet | Heat Treated | 95 Bhn (500 Kg) |
| BR-1 | HH Brass | As Rolled | .......... |
| NM | Nonmetallic | .......... | (Note 2) |

*Note 1:* All Brinell hardness numbers are those obtained with a 3000 kg load unless otherwise specified. For routine acceptance tests, the diameter of the impression should be read to 0.05 mm. The range of Brinell hardness should be at least 40 points, up to 285 Bhn and 50 points, over 302 Bhn.

*Note 2:* The use of and specifications for nonmetallic materials shall be established by agreement between gear manufacturer and buyer.

*Note 3:* Where a hardness of Rc Min. is shown, a designer may prefer to provide a range of hardness. In such cases, the range should provide at least 5 points Rc above the minimum indicated in the table, and should be by agreement between the gear manufacturer and the buyer. This includes carbon and alloy steels, for all surface-hardening methods, except nitriding.

For all surface-hardened treatments where Rockwell C measurements cannot be used, the hardnesses as specified are conversion readings in accordance with ASTM specification E140.

See also footnote on first page of table.

*k.* When core hardness is requested, consideration should be given to the section involved.

*l.* Where considerable impact loads exist, the use of alloys, plus a lowering of the hardness to a range of 50 to 56 Rockwell C, is recommended for carburized gears and pinions.

*m.* When accelerated wear is encountered in service, heat treated gearing to provide higher hardness will, in most cases, help to alleviate these conditions. The gear manufacturer should be consulted for appropriate recommendations.

**AGMA Applications and Quality Numbers.** — Table 9 includes a tabulation of many industrial and end use applications for spur, helical, herringbone, bevel and hypoid gearing, racks, and fine-pitch worms and worm gearing. A typical AGMA Quality Number range is shown for each of the many industries and applications. When selecting a Quality Number for an industry or an application which is not shown, use a similar industry or application as a guide.

The AGMA Quality Number shown opposite each item of equipment identifies the quality of gearing generally used. There may be certain designs or operating conditions that would justify specifying gears to a lower or higher Quality Number.

**Table 9.** AGMA Applications and Quality Numbers for Racks and Gears — 1

| Gearing Application | Quality No.* | Gearing Application | Quality No.* |
|---|---|---|---|
| AEROSPACE | | CEMENT INDUSTRY | |
| Actuators | 7–11 | (Continued) | |
| Control Gearing | 7–11 | Overhead Crane | 5–6 |
| Engine Accessories | 10–13 | Pug, Rod and Tube Mills | 5–6 |
| Engine Power | 10–13 | Pulverizer | 5–6 |
| Engine Starting | 10–13 | Raw and Finish Mill | 5–6 |
| Loading Hoist | 7–11 | Rotary Dryer | 5–6 |
| Propeller Feathering | 10–13 | Slurry Agitator | 5–6 |
| Small Engines | 12–13 | CHEWING GUM INDUSTRY | |
| AGRICULTURE | | Chicle Grinder | 6–8 |
| Baler | 3–7 | Coater | 6–8 |
| Beet Harvester | 5–7 | Mixer-Kneader | 6–8 |
| Combine | 5–7 | Molder-Roller | 6–8 |
| Corn Picker | 5–7 | Wrapper | 6–8 |
| Cotton Picker | 5–7 | CHOCOLATE INDUSTRY | |
| Farm Elevator | 3–7 | Glazer, Finisher | 6–8 |
| Field Harvester | 5–7 | Mixer, Mill | 6–8 |
| Peanut Harvester | 3–7 | Molder | 6–8 |
| Potato Digger | 5–7 | Presser, Refiner | 6–8 |
| AIR COMPRESSOR | 10–11 | Tampering | 6–8 |
| AUTOMOTIVE INDUSTRY | 10–11 | Wrapper | 6–8 |
| BALING MACHINE | 5–7 | CLAY WORKING MACHINERY | 5–7 |
| BOTTLING INDUSTRY | | CONSTRUCTION EQUIPMENT | |
| Capping | 6–7 | Backhoe | 6–8 |
| Filling | 6–7 | Cranes | |
| Labeling | 6–7 | Open Gearing | 3–6 |
| Washer, Sterilizer | 6–7 | Enclosed Gearing | 6–8 |
| BREWING INDUSTRY | | Ditch Digger | 3–8 |
| Agitator | 6–8 | Transmission | 6–8 |
| Barrel Washer | 6–8 | Drag Line | 5–8 |
| Cookers | 6–8 | Dumpster | 6–8 |
| Filling Machine | 6–8 | Paver Loader | 3 |
| Mash Tubs | 6–8 | Transmission | 8 |
| Pasteurizer | 6–8 | Mixer | 3–5 |
| Racking Machine | 6–8 | Swing Gear | 3–5 |
| BRICK MAKING MACHINERY | 5–7 | Mixing Bucket | 3 |
| BRIDGE MACHINERY | 5–7 | Shaker | 8 |
| BRIQUETTE MACHINES | 5–7 | Shovels | |
| CEMENT INDUSTRY | | Open Gearing | 3–6 |
| (Quarry Operation) | | Enclosed Gearing | 6–8 |
| Conveyor | 5–6 | Stationary Mixer | |
| Crusher | 5–6 | Transmission | 8 |
| Diesel Electric | | Drum Gears | 3–5 |
| Locomotive | 8–9 | Stone Crusher | |
| Electric Dragline | | Transmission | 8 |
| (cast gear) | 3 | Conveyor | 6 |
| (cut gear) | 6–8 | Truck Mixer | |
| Electric Locomotive | 6–8 | Transfer Case | 9 |
| Electric Shovel | | Drum Gears | 3–5 |
| (cast gear) | 3 | COMMERCIAL METERS | |
| (cut gear) | 6–8 | Gas | 7–9 |
| Elevator | 5–6 | Liquid, Water, Milk | 7–9 |
| Locomotive Crane | | Parking | 7–9 |
| (cast gear) | 3 | COMPUTING AND ACCOUNTING | |
| (cut gear) | 5–6 | MACHINES | |
| (Plant Operation) | | Accounting-Billing | 9–10 |
| Air Separator | 5–6 | Adding Machine-Calculator | 7–9 |
| Ball Mill | 5–7 | Addressograph | 7 |
| Compeb Mill | 5–6 | Bookkeeping | 9–10 |
| Conveyor | 5–6 | Cash Register | 7 |
| Cooler | 5–6 | Comptometer | 6–8 |
| Elevator | 5–6 | Computing | 10–11 |
| Feeder | 5–6 | Data Processing | 7–9 |
| Filter | 5–6 | Dictating Machine | 9 |
| Kiln | 5–6 | Typewriter | 8 |
| Kiln Slurry Agitator | 5–6 | | |

* Quality Numbers are inclusive, from the lowest to highest numbers shown.

**Table 9. AGMA Applications and Quality Numbers for Racks and Gears — 2**

| Gearing Application | Quality No.* | Gearing Application | Quality No.* |
|---|---|---|---|
| CRANES | | FOUNDRY INDUSTRY | |
| Boom Hoist | 5–6 | Conveyor | 5–6 |
| Gantry | 5–6 | Elevator | 5–6 |
| Load Hoist | 5–7 | Ladle | 5–6 |
| Overhead | 5–6 | Molding Machine | 5–6 |
| Ship | 5–7 | Overhead Cranes | 5–6 |
| CRUSHERS | | Sand Mixer | 5–6 |
| Ice, Feed | 6–8 | Sand Slinger | 5–6 |
| Portable and Stationary | 6–8 | Tumbling Mill | 5–6 |
| Rock, Ore, Coal | 6–8 | HOME APPLIANCES | |
| DAIRY INDUSTRY | | Blender | 6–8 |
| Bottle Washer | 6–7 | Mixer | 7–9 |
| Homogenizer | 7–9 | Timer | 8–10 |
| Separator | 7–9 | Washing Machine | 8–10 |
| DISH WASHER | | MACHINE TOOL INDUSTRY | |
| Commercial | 5–7 | Hand Motion (but not Indexing and Positioning) | 6–9 |
| DISTILLERY INDUSTRY | | Power Drives | |
| Agitator | 5–7 | 0–800 FPM | 6–8 |
| Bottle Filler | 5–7 | 800–2000 FPM | 8–10 |
| Conveyor, Elevator | 6–7 | 2000–4000 FPM | 10–12 |
| Grain Pulverizer | 6–8 | Over 4000 FPM | 12 & Up |
| Mash Tub | 5–7 | Indexing and Positioning | |
| Mixer | 5–7 | Approximate Positioning | 6–10 |
| Yeast Tub | 5–7 | Accurate Indexing and Positioning | 12 & Up |
| ELECTRIC FURNACE | | MARINE INDUSTRY | |
| Tilting Gears | 5–7 | Anchor Hoist | 6–8 |
| ELECTRONIC INSTRUMENT CONTROL AND GUIDANCE SYSTEMS | | Cargo Hoist | 7–8 |
| | | Conveyor | 5–7 |
| Accelerometer | 10–12 | Davit Gearing | 5–7 |
| Airborne Temperature Recorder | 12–14 | Elevator | 6–7 |
| | | Small Propulsion | 10–12 |
| Aircraft Instrument | 12 | Steering Gear | 8 |
| Altimeter-Stabilizer | 9–11 | Winch | 5–8 |
| Analog Computer | 10–12 | METAL WORKING | |
| Antenna Assembly | 7–9 | Bending Roll | 5–7 |
| Antiaircraft Detector | 12 | Draw Bench | 6–8 |
| Automatic Pilot | 9–11 | Forge Press | 5–7 |
| Digital Computer | 10–12 | Punch Press | 5–7 |
| Gun Data Computer | 12–14 | Roll Lathe | 5–7 |
| Gyro Caging Mechanism | 10–12 | MINING AND PREPARATION | |
| Gyroscope-Computer | 12–14 | Breaker | 5–6 |
| Pressure Transducer | 12–14 | Car Dump | 5–6 |
| Radar, Sonar, Tuner | 10–12 | Concentrator | 5–6 |
| Recorder, Telemeter | 10–12 | Continuous Miner | 6–7 |
| Servo System Component | 9–11 | Conveyor | 5–7 |
| Sound Detector | 9 | Cutting Machine | 6–10 |
| Transmitter, Receiver | 10–12 | Drag Line | |
| ENGINES | | Open Gearing | 3–6 |
| Combustion | | Enclosed Gearing | 6–8 |
| Engine Accessories | 10–12 | Drills | 5–6 |
| Supercharger | 10–12 | Drier | 5–6 |
| Timing Gearings | 10–12 | Electric Locomotive | 6–8 |
| Transmission | 8–10 | Elevator | 5–6 |
| FARM EQUIPMENT | | Feeder | 6–8 |
| Milking Machine | 6–8 | Flotation | 5–6 |
| Separator | 8–10 | Grizzly | 5–7 |
| Sweeper | 4–6 | Hoists, Skips | 7–8 |
| FLOUR MILL INDUSTRY | | Loader (Underground) | 5–8 |
| Bleacher | 7–8 | Rock Drill | 5–6 |
| Grain Cleaner | 7–8 | Rotary Car Dump | 6–8 |
| Grinder | 7–8 | Screen (Rotary) | 7–8 |
| Hulling | 7–8 | Screen (Shaking) | 7–8 |
| Milling, Scouring | 7–8 | Separator | 5–6 |
| Polisher | 7–8 | | |
| Separator | 7–8 | | |

* Quality Numbers are inclusive, from the lowest to highest numbers shown.

Table 9. AGMA Applications and Quality Numbers for Racks and Gears — 3

| Gearing Application | Quality No.* | Gearing Application | Quality No.* |
|---|---|---|---|
| MINING AND PREPARATION (Continued) | | Mixer, Tuber | 6–8 |
| | | Refiner, Calender | 5–7 |
| Sedimentation | 5–6 | Rubber Mill, Scrap Cutter | 5–7 |
| Shaker | 6–8 | Tire Building | 6–8 |
| Shovel | 3–8 | Tire Chopper | 5–7 |
| Tipple Gearing | 5–7 | Washer, Banbury Mixer | 5–7 |
| Washer | 6–8 | | |
| PAPER AND PULP | | SMALL POWER TOOLS | |
| Bag Machines | 6–8 | Bench Grinder | 6–8 |
| Box Machines | 6–8 | Drills-Saws | 7–9 |
| Building Paper | 6–8 | Hair Clipper | 7–9 |
| Calendar | 6–8 | Hedge Clipper | 7–9 |
| Chipper | 6–8 | Sander, Polisher | 8–10 |
| Coating | 6–8 | Sprayer | 6–8 |
| Envelope Machines | 6–8 | | |
| Food Container | 6–8 | SPACE NAVIGATION | |
| Glazing | 6–8 | Sextant and Star Tracker | 14 & Up |
| Log Conveyor-Elevator | 5–7 | | |
| Mixer, Agitator | 6–8 | STEEL INDUSTRY | |
| Paper Machine | | Miscellaneous Drives | |
| Auxiliary | 8–9 | Bessemer Tilt-Car Dump | 5–6 |
| Main Drive | 10–12 | Coke Pusher, Distributor | 5–6 |
| Press, Couch, Drier Rolls | 6–8 | Conveyor, Door Lift | 5–6 |
| Slitting | 10–12 | Electric Furnace Tilt | 5–6 |
| Steam Drum | 6–8 | Hot Metal Car Tilt | 5–6 |
| Varnishing | 6–8 | Hot Metal Charger | 5–6 |
| Wall Paper Machines | 6–8 | Jib Hoist, Dolomite | |
| PAVING INDUSTRY | | Machine | 5–6 |
| Aggregate Drier | 5–7 | Larry Car, Mud Gun | 5–6 |
| Aggregate Spreader | 5–7 | Mixing Bin, Mixer Tilt | 5–6 |
| Asphalt Mixer | 5–7 | Ore Crusher, Pig Machine | 5–6 |
| Asphalt Spreader | 5–7 | Pulverizer, Quench Car | 5–6 |
| Concrete Batch Mixer | 5–7 | Shaker, Sinter Conveyor | 5–6 |
| PHOTOGRAPHIC EQUIPMENT | | Sinter Machine Skip Hoist | 5–6 |
| Aerial | 10–12 | Slag Crusher, Slag Shovel | 5–6 |
| Commercial | 8–10 | | |
| PRINTING INDUSTRY | | Primary and Secondary Rolling Mill Drives | |
| Press | | Blooming and Plate Mill | 5–6 |
| Book | 9–11 | Heavy Duty Hot Mill Drives | 5–6 |
| Flat | 9–11 | | |
| Magazine | 9–11 | Slabbing and Strip Mill | 5–6 |
| Newspaper | 9–11 | | |
| Roll Reels | 6–7 | Hot Mill Drives | |
| Rotary | 9–11 | Sendzimer-Stekel | 7–8 |
| PUMP INDUSTRY | | Tandem-Temper-Skin | 6–7 |
| Liquid | 10–12 | | |
| Rotary | 6–8 | Cold Mill Drives | |
| Slush-Duplex-Triplex | 6–8 | Bar, Merchant, Rail, Rod | 5–6 |
| Vacuum | 6–8 | Structural, Tube | 5–6 |
| QUARRY INDUSTRY | | Auxiliary and Miscellaneous Drives | |
| Conveyor-Elevator | 6–7 | Annealing Furnace Car | 5–6 |
| Crusher | 5–7 | Bending Roll | 5–6 |
| Rotary Screen | 7–8 | Blooming Mill | |
| RADAR AND MISSILE | | Manipulator | 5–6 |
| Antenna Elevating | 8–10 | Blooming Mill Rack and Pinion | 5–6 |
| Data Gear | 10–12 | | |
| Launch Pad Azimuth | 8 | Blooming Mill Side Guard | 5–6 |
| Ring Gear | 9–12 | Car Haul | 5–6 |
| Rotating Drive | 10–12 | Coil Conveyor | 5–6 |
| RAILROADS | | Coil Dump | 5–6 |
| Construction Hoist | 5–7 | Crop Conveyor | 5–6 |
| Wrecking Crane | 6–8 | Edger Drives | 5–6 |
| RUBBER AND PLASTICS | | Electrolytic Line | 6–7 |
| Boot and Shoe Machines | 6–8 | | |
| Drier, Press | 6–8 | | |
| Extruder, Strainer | 6–8 | | |

* Quality Numbers are inclusive, from the lowest to highest numbers shown.

**Table 9. AGMA Applications and Quality Numbers for Racks and Gears — 4**

| Gearing Application | Quality No.* | Gearing Application | Quality No.* |
|---|---|---|---|
| STEEL INDUSTRY (Continued) | | STEEL INDUSTRY (Continued) | |
| Flange Machine Ingot | | Turbine | 9–10 |
| Buggy | 5–6 | Overhead Cranes | |
| Leveler | 6–7 | Billet Charger, Cold Mill | 5–6 |
| Magazine Pusher | 6–7 | Bucket Handling | 5–6 |
| Mill Shear Drives | 6–7 | Car Repair Shop | 5–6 |
| Mill Table Drives | | Cast House, Coil Storage | 5–6 |
| (under 800 ft/Min) | 5–6 | Charging Machine | 5–6 |
| Mill Table Drives | | Cinder Yard, Hot Top | 5–6 |
| (over 800–1800 ft/Min) | 6–7 | Coal and Ore Bridges | 5–6 |
| Mill Table Drives | | Electric Furnace Charger | 5–6 |
| (over 1800 ft/Min) | 8 | Hot Metal, Ladle | 5–6 |
| Nail and Spike Machine | 5–6 | Hot Mill, Ladle House | 5–6 |
| Piler | 5–6 | Jib Crane, Motor Room | 5–6 |
| Plate Mill Rack and | | Mold Yard, Rod Mill | 5–6 |
| Pinion | 5–6 | Ore Unloader, Stripper | 5–6 |
| Plate Mill Side Guards | 5–6 | Overhead Hoist | 5–6 |
| Plate Turnover | 5–6 | Pickler Building | 5–6 |
| Preheat Furnace Pusher | 5–6 | Pig Machine, Sand House | 5–6 |
| Processor | 6–7 | Portable Hoist | 5–6 |
| Pusher Rack and Pinion | 5–6 | Scale Pit, Shipping | 5–6 |
| Rotary Furnace | 5–6 | Scrap Balers and Shears | 5–6 |
| Shear Depress Table | 5–6 | Scrap Preparation | 5–6 |
| Slab Squeezer | 5–6 | Service Shops | 5–6 |
| Slab Squeezer Rack and | | Skull Cracker | 5–6 |
| Pinion | 5–6 | Slab Handling | 5–6 |
| Slitter, Side Trimmer | 6–7 | | |
| Tension Reel | 6–7 | MISCELLANEOUS | |
| Tilt Table, Upcoiler | 5–6 | Clocks | 6 |
| Transfer Car | 5–6 | Counters | 7–9 |
| Wire Drawing Machine | 6–7 | Fishing Reel | 6 |
| Precision Gear Drives | | Gauges | 8–10 |
| Diesel Electric Gearing | 8–9 | IBM Card Puncher, Sorter | 8 |
| Flying Shear | 9–10 | Metering Pumps | 7–8 |
| Shear Timing Gears | 9–10 | Motion-Picture Equipment | 8 |
| High Speed Reels | 8–9 | Popcorn Machine, Comm. | 6–7 |
| Locomotive Timing | | Pumps | 5–7 |
| Gears | 9–10 | Sewing Machine | 8 |
| Pump Gears | 8–9 | Slicer | 7–8 |
| Tube Reduction Gearing | 8–9 | Vending Machines | 6–7 |

Extracted from AGMA Handbook for Unassembled Gears — Volume 1 — Gear Classifications, Materials, and Inspection (AGMA 390.03), with permission of the publisher, the American Gear Manufacturers Association, 1901 North Fort Myer Drive, Arlington, Va. 22209.
* Quality Numbers are inclusive, from the lowest to highest number shown.

**AGMA Measuring Methods and Practices.** — Process control and inspection procedures for individual element checks of coarse- and fine-pitch, spur, helical, herringbone, bevel and hypoid gears are presented in AGMA Gear Handbook, Volume 1 (AGMA 390.03). It also covers composite inspection of coarse- and fine-pitch, spur, helical, herringbone, bevel and hypoid gears. In addition, backlash and tooth-contact pattern evaluation of bevel and hypoid gearing is discussed.

The quality level of a gear is determined during its manufacture by the specific sequence of steps followed and the degree of care employed at each step. This AGMA Handbook formalizes the long-established concept of process control as a quality determinant. Process control is a method by which gear accuracy is achieved and maintained through control of manufacturing equipment, methods and processes, without resort to inspection of individual elements of every gear produced.

This Gear Handbook is published by the American Gear Manufacturers Association, 1901 North Fort Myer Drive, Arlington, Va. 22209.

**Gear Blanks for Fine-pitch Gears.** — The accuracy to which gears can be produced is considerably affected by the design of the gear blank and the accuracy to which the various surfaces of the blank are machined. The following recommendations should not be regarded as inflexible rules, but rather as minimum average requirements for gear-blank quality compatible with the expected quality class of the finished gear.

*Design of Gear Blanks:* Since the accuracy to which gears can be produced is affected by the design of the blank, the following points of design should be noted: (1) Gears designed with a hole should have the hole large enough to adequately support the blank during the machining of the teeth and yet not so large as to cause distortion; (2) Face widths should be wide enough, in proportion to outside diameters, to avoid springing and to permit obtaining flatness in important surfaces; (3) Short bore lengths should be avoided wherever possible. It is feasible, however, to machine relatively thin blanks in stacks, provided the surfaces are flat and parallel to each other; (4) Where gear blanks with hubs are to be designed, attention should be given to the wall sections of the hubs. Too thin a section will not permit proper clamping of the blank during machining operations and may also affect proper mounting of the gear; and (5) Where pinions or gears integral with their shafts are to be designed, deflection of the shaft can be minimized by having the shaft length and shaft diameter well proportioned to the gear or pinion diameter. The foregoing general principles may also be useful when applied to blanks for coarser pitch gears.

**Specifying Spur and Helical Gear Data on Drawings.** — The data that may be shown on drawings of spur and helical gears falls into three groups: The first group consists of data basic to the design of the gear; the second group consists of data used in manufacturing and inspection; and the third group consists of engineering reference data. The accompanying table may be used as a checklist for the various data which may be placed on gear drawings and the order in which they should appear.

*Explanation of Terms Used in Gear Specifications:*

1. Number of teeth is the number of teeth in 360 deg of gear circumference. In the case of a sector gear, both the actual number of teeth in the sector and the theoretical number of teeth in 360 deg should be given.

2. Diametral pitch is the ratio of the number of teeth in the gear to the number of inches in the standard pitch diameter. It is used in this standard as a nominal specification of tooth size.

2a. Normal diametral pitch is the diametral pitch in the normal plane.

2b. Transverse diametral pitch is the diametral pitch in the transverse plane.

3. Pressure angle is the angle between the gear tooth profile and a radial line at the pitch point. It is used in this standard to specify the pressure angle of the basic rack used in defining the gear tooth profile.

3a. Normal pressure angle is the pressure angle in the normal plane.

3b. Transverse pressure angle is the pressure angle in the transverse plane.

4. Helix angle is the angle between the pitch helix and an element of the pitch cylinder, unless otherwise specified.

4a. Hand of helix is the direction in which the teeth twist as they recede from an observer along the axis. A right hand helix twists clockwise and a left hand helix twists counterclockwise.

5. Standard pitch diameter is the diameter of the pitch circle. It equals the number of teeth divided by the transverse diametral pitch.

6. Tooth form may be specified as standard addendum, long addendum, short addendum, modified involute or special. In case a modified involute or special tooth form is required, a detailed view should be shown on the drawing. If a special tooth form is specified, roll angles must be supplied (see page 756).

7. Addendum is the radial distance between the standard pitch circle and the outside circle. Its actual value is dependent on the specification of outside diameter.

8. Whole depth is the total radial depth of the tooth space. Its actual value is dependent on the specification of outside diameter and root diameter.

9. Maximum calculated circular thickness on the standard pitch circle is the tooth thickness which will provide the desired minimum backlash when the gear is assembled in mesh with its mate on minimum center distance. It is best controlled by testing in tight mesh with a master which integrates all errors in the several teeth in mesh through the arc of action as explained on page 765. It is independent of the effect of runout.

9a. Maximum calculated *normal* circular thickness is the circular tooth thickness in the normal plane which satisfies requirements explained in (9).

10. Gear testing radius is the distance from its axis of rotation to the standard pitch line of a standard master when in intimate contact under recommended pressure on a variable-center-distance running gage. Maximum testing radius should be calculated to provide the maximum circular tooth thickness specified in (9) when checked as explained on page 765. It is affected by the runout of the gear. Tolerance on testing radius must be equal to or greater than the total composite error permitted by the quality class specified in (11).

11. Quality class is specified for convenience when talking or writing about the accuracy of the gear. These classes are explained on page 766.

12. Maximum total composite error.
13. Maximum tooth-to-tooth composite error.
Actual tolerance values (12 and 13) permitted by the quality class (11) are specified in inches to provide machine operator or inspector with tolerances required to inspect the gear.

14. Testing pressure recommendations are given on page 765. Incorrect testing pressure will result in incorrect measurement of testing radius.

15. Master specifications by tool or code number may be required to call for the use of a special master gear when tooth thickness deviates excessively from standard.

16. Measurement over two 0.xxxx diameter pins may be specified to assist manufacturing in determining size at machine for setup only.

17. Outside diameter is usually shown on the drawing of the gear together with other blank dimensions so that it will not be necessary for machine operators to search gear tooth data for this dimension. Since outside diameter is also frequently used in the manufacture and inspection of the teeth, it may be included in the data block with other tooth specifications if preferred. To permit use of topping hobs for cutting gears on which the tooth thickness has been modified from standard, the outside diameter should be related to the specified gear testing radius (10).

18. Maximum root diameter is specified to assure adequate clearance for the outside diameter of the mating gear. This dimension is usually considered acceptable if the gear is checked with a master and meets specifications (10) through (13).

19. Active profile diameter of a gear is the smallest diameter at which the mating gear tooth profile can contact. Because of difficulties involved in checking, this specification is not recommended for gears finer than 48 pitch.

20. Surface roughness on active profile surfaces may be specified in microinches to be checked by instrument up to about 32 pitch, or by visual comparison in the finer pitch ranges. It is very difficult to accurately determine the surface roughness of fine pitch gears. For many commercial applications it is considered acceptable on gears which meet the maximum tooth-to-tooth-error specification (13).

21. Mating gear part number may be shown as a convenient reference. If the gear is used in several applications, all mating gears may be listed but usual practice is to record this information in a reference file.

22. Number of teeth in mating gear. ⎫
23. Minimum operating center distance. ⎭ This information is often specified to eliminate the necessity of getting prints of the mating gear and assemblies for checking the design specifications, interference, backlash, determination of master gear specification, and acceptance or rejection of gears made out of tolerance.

### Data for Spur and Helical Gear Drawings

| Type of Data | Min. Spur Gear Data | Min. Helical Gear Data | Add'l Optional Data | Item Number* | Data* |
|---|---|---|---|---|---|
| Basic Specifications | X | X | | 1 | Number of teeth |
| | X | | | 2 | Diametral pitch |
| | | X | | 2a | Normal diametral pitch |
| | | | X | 2b | Transverse diametral pitch |
| | X | | | 3 | Pressure angle |
| | | X | | 3a | Normal pressure angle |
| | | | X | 3b | Transverse pressure angle |
| | | X | | 4 | Helix angle |
| | | X | | 4a | Hand of helix |
| | X | X | | 5 | Standard pitch diameter |
| | X | X | | 6 | Tooth form |
| | | | X | 7 | Addendum |
| | | | X | 8 | Whole depth |
| | X | | | 9 | Max. calc. circular thickness on std. pitch circle |
| | | X | | 9a | Max. calc. normal circular thickness on std. pitch circle |
| Manufacturing and Inspection | | | X | 10 | Roll angles |
| | X | X | | 11 | A.G.M.A. quality class |
| | X | X | | 12 | Max. total composite error |
| | X | X | | 13 | Max. tooth-to-tooth composite error |
| | | | X | 14 | Testing pressure (Ounces) |
| | X | X | | 15 | Master specification |
| | | | X | 16 | Meas. over two .xxxx dia. pins (For setup only) |
| | X | X | | 17 | Outside diameter (Preferably shown on drawing of gear) |
| | | | X | 18 | Max. root diameter |
| | | | X | 19 | Active profile diameter |
| | | | X | 20 | Surface roughness of active profile |
| Engineering Reference | | | X | 21 | Mating gear part number |
| | | | X | 22 | Number of teeth in mating gear |
| | | | X | 23 | Minimum operating center distance |

* An item-by-item explanation of the terms used in this table is given beginning on page 790.

**Internal Spur Gears.** — An internal gear may be proportioned like a standard spur gear turned "outside in" or with addendum and dedendum in reverse positions; however, to avoid interference or improve the tooth form and action, the internal diameter of the gear should be increased and the outside diameter of the mating pinion is also made larger than the size based upon standard or conventional tooth proportions. The extent of these enlargements will be illustrated by means of examples given in connection with the Rules for Internal Gears. The 20-degree involute full-depth tooth form is recommended for internal gears; the 20-degree stub tooth and the 14½-degree full-depth tooth are also used.

**Methods of Cutting Internal Gears.** — Internal spur gears are cut by methods similar in principle to those employed for external spur gears. They may be cut by one of the following methods: (1) By a generating process, as when using a Fellows gear shaper; (2) by using a formed cutter and milling the teeth; (3) by planing, using a machine of the templet or form-copying type (especially applicable to gears of large pitch); and (4) by using a formed tool which reproduces its shape and is given a planing action either on a slotting or a planing type of machine. Internal gears frequently have a web at one side which limits the amount of clearance space at the ends of the teeth. Such gears may be cut readily on a gear shaper. The most practical method of cutting very large internal gears is on a planer of the form-copying type. A regular spur gear planer is equipped with a special tool-holder for locating the tool in the position required for cutting internal teeth.

**Formed Cutters for Internal Gears.** — When formed cutters are used, a special cutter usually is desirable, because the tooth spaces of an internal gear are not the same shape as the tooth spaces of external gearing having the same pitch and number of teeth. This is due to the fact that an internal gear is a spur gear "turned outside in." According to one rule, the standard No. 1 cutter for external gearing may be used for internal gears of 4 diametral pitch and finer, when there are sixty teeth or more. This No. 1 cutter, as applied to external gearing, is intended for all gears having from 135 teeth to a rack. The finer the pitch and the larger the number of teeth, the better the results obtained with a No. 1 cutter. The standard No. 1 cutter is considered satisfactory for jobbing work, and usually when the number of gears to be cut does not warrant obtaining a special cutter, although the use of the No. 1 cutter is not practicable when the number of teeth in the pinion is large in proportion to the number of teeth in the internal gear.

**Arc Thickness of Internal Gear Tooth.** — *Rule:* If internal diameter of internal gear is enlarged as determined by Rules 1 and 2 for Internal Diameters (see Rules for Internal Gears), the arc tooth thickness at the pitch circle equals 1.3888 divided by the diametral pitch, assuming a pressure angle of 20 degrees.

**Arc Thickness of Pinion Tooth.** — *Rule:* If pinion for internal gear is larger than conventional size (see Outside Diameter of Pinion for Internal Gear, under Rules for Internal Gears), then the arc tooth thickness on pitch circle equals 1.7528 divided by the diametral pitch, assuming a pressure angle of 20 degrees.

*Note:* For chordal thickness and chordal addendum, see rules and formulas for spur gears.

**Relative Sizes of Internal Gear and Pinion.** — If a pinion is too large or too near the size of its mating internal gear, serious interference or modification of the tooth shape may occur.

*Rule:* For internal gears having a 20-degree pressure angle and full-depth teeth, the difference between number of teeth in gear and pinion should not be less than 12. For teeth of stub form, the smallest difference should be 7 or 8 teeth. For a pressure angle of 14½ degrees, the difference in tooth numbers should not be less than 15.

### Rules for Internal Gears — 20-degree Full-depth Teeth

| To Find | Rule |
|---------|------|
| Pitch Diameter | *Rule:* To find the pitch diameter of an internal gear, divide number of internal gear teeth by the diametral pitch. The pitch diameter of mating pinion also equals number of pinion teeth divided by diametral pitch, the same as for external spur gears. |
| Internal Diameter (Enlarged to avoid Interference) | *Rule 1:* For internal gears to mesh with pinions having 16 teeth or more, subtract 1.2 from the number of teeth and divide remainder by diametral pitch.<br>*Example:* An internal gear has 72 teeth of 6 diametral pitch and the mating pinion has 18 teeth; then $$\text{Internal diameter} = \frac{72 - 1.2}{6} = 11.8 \text{ inches}$$ *Rule 2:* If circular pitch is used, subtract 1.2 from the number of internal gear teeth, multiply remainder by the circular pitch, and divide the product by 3.1416. |
| Internal Diameter (Based upon Spur Gear Reversed) | *Rule:* If the internal gear is to be designed to conform to a spur gear turned outside in, then subtract 2 from the number of teeth and divide remainder by the diametral pitch to find internal diameter.<br>*Example:* (Same as Example above.) $$\text{Internal diameter} = \frac{72 - 2}{6} = 11.666 \text{ inches}$$ |
| Outside Diameter of Pinion for Internal Gear | *Note:* If the internal gearing is to be proportioned like standard spur gearing, use the rule or formula previously given for spur gears in determining the outside diameter. The rule and formula following apply to a pinion that is enlarged and intended to mesh with an internal gear enlarged as determined by the preceding Rules 1 and 2 above.<br>*Rule:* For pinions having 16 teeth or more, add 2.5 to the number of pinion teeth and divide by the diametral pitch.<br>*Example 1:* A pinion for driving an internal gear is to have 18 teeth (full depth) of 6 diametral pitch; then $$\text{Outside diameter} = \frac{18 + 2.5}{6} = 3.416 \text{ inches}$$ Using the rule for external spur gears, the outside diameter = 3.333 inches. |
| Center Distance | *Rule:* Subtract the number of pinion teeth from the number of internal gear teeth and divide remainder by two times the diametral pitch. |
| Tooth Thickness | See paragraphs, Arc Thickness of Internal Gear Tooth and Arc Thickness of Pinion Tooth, on page 793. |

**British Standard for Spur and Helical Gears.** — This revised standard No. 436–1940, amended in 1941, 1943, and 1956, applies to machine cut or ground spur gears and to single or double helical gears connecting parallel shafts. Internal as well as external gears are included. The pressure angle is 20 degrees and the working depth equals twice the module (whether English or metric should be stated). The tooth form represents a well-balanced compromise between strength, wear resistance, and quietness of operation. Gears are divided into five general classes.

*Class A1,* Precision Ground Gears (nominal proportions of basic rack tooth for this class are the same as shown by accompanying diagram for circular pitch of 1, except that fillet radius at the root is 0.0938 instead of 0.124 and the dedendum is 0.4583 instead of 0.3979).

*Class A2,* Precision Cut Gears for peripheral speeds above 2000 feet per minute.

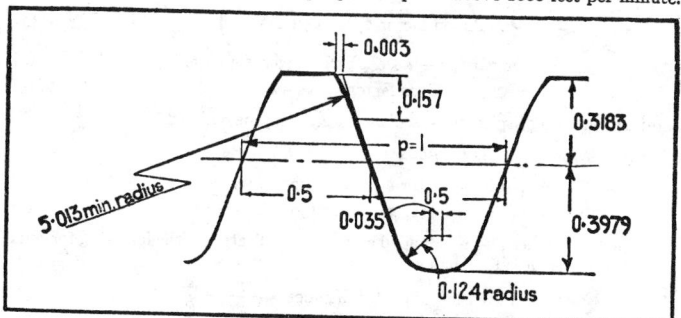

*Class B,* High-class Cut Gears for peripheral speeds between 750 and 3000 feet per minute (nominal proportions of basic rack for Classes A2 and B are shown by accompanying diagram).

*Class C,* Commercial Cut Gears for peripheral speeds below 1200 feet per minute; *Class D,* Large Internal Gears (basic rack for Classes C and D same as diagram excepting tip radius is 4.098, easing 0.006 and its depth, 0.200.

The range of speeds specified for gears of Classes A2, B and C permits considerable overlap between the classes. The notation follows:

$A$ = Gear addendum  
$a$ = Pinion addendum  
$k_p$ = Pinion correction factor  
$k_w$ = Gear correction factor  
$P_n$ = Normal diametral pitch  
$p$ = Circular pitch  

$p_n$ = Normal pitch  
$T$ = Number of gear teeth  
$t$ = Number of pinion teeth  
$\Delta$ = Center distance extension factor  
$\sigma$ = Helix angle  

**Easing or Tip Relief.** — The form is involute excepting for a slight easing at the point. The maximum amount of this easing or tip relief is as follows:

For Classes A1, A2, and B (Precision Ground, Precision Cut and High-class Cut Gears)

$$e = 0.003p \text{ extending } 0.157p \text{ in depth}$$

For Classes C and D (Commercial Cut and Large Internal Gears)

$$e = 0.006p \text{ extending } 0.20p \text{ in depth.}$$

**Helical Gear Teeth.** — The shape and proportions of helical gear teeth in the normal section, corresponds to the basic rack tooth forms, normal pitch being substituted for circular pitch. (Note that $p_n = p \cos \sigma$; also $P_n = \pi \div p_n$.)

**Addendum — Gear and Pinion.** — The *recommended* addendum values vary according to pitch and numbers of teeth in mating gears, in order to obtain full involute action, avoid undercutting in some cases, and obtain better zone and strength factors (factors used in calculating the horsepower rating).

$$\text{Pinion addendum } a = \frac{p_n}{\pi}\,(1 + k_p) = \frac{1 + k_p}{P_n} \qquad (1)$$

$$\text{Gear addendum } A = \frac{p_n}{\pi}\,(1 + k_w) = \frac{1 + k_w}{P_n} \qquad (2)$$

The correction factors $k_p$ and $k_w$ for pinion and gear, are determined as follows:
*When $(t + T)\sec^3 \sigma$ is 60 or greater:*

$$\text{Pinion factor } k_p = 0.4\left(1 - \frac{t}{T}\right) \qquad (3)$$

$$\text{or } 0.02(30 - t\sec^3 \sigma) \text{ whichever is greater} \qquad (4)$$

$$\text{Gear factor } k_w = -k_p \qquad (5)$$

Note: In the case of spur gears, $\sec^3 \sigma = 1$ and may be omitted.

*When $(t + T)\sec^3 \sigma$ is less than 60:*

$$\text{Pinion factor } k_p = 0.02(30 - t\sec^3 \sigma) \qquad (6)$$

$$\text{Gear factor } k_w = 0.02(30 - T\sec^3 \sigma) \qquad (7)$$

The center distance also is extended an amount indicated by the following formula when $(t + T)\sec^3 \sigma$ is less than 60.

$$\text{Extension of center distance} = \frac{\Delta p_n}{\pi} = \frac{\Delta}{P_n} \qquad (8)$$

### Factors used in Center Distance Extension Formula No. 8

| Sum of Correction Factors $k_p + k_w$ | Factor $\Delta$ | Sum of Correction Factors $k_p + k_w$ | Factor $\Delta$ | Sum of Correction Factors $k_p + k_w$ | Factor $\Delta$ | Sum of Correction Factors $k_p + k_w$ | Factor $\Delta$ |
|---|---|---|---|---|---|---|---|
| 0.025 | 0.025 | 0.225 | 0.218 | 0.425 | 0.400 | 0.625 | 0.555 |
| 0.050 | 0.050 | 0.250 | 0.243 | 0.450 | 0.420 | 0.650 | 0.575 |
| 0.075 | 0.075 | 0.275 | 0.267 | 0.475 | 0.444 | 0.675 | 0.588 |
| 0.100 | 0.100 | 0.300 | 0.288 | 0.500 | 0.462 | 0.700 | 0.606 |
| 0.125 | 0.122 | 0.325 | 0.313 | 0.525 | 0.480 | 0.725 | 0.623 |
| 0.150 | 0.146 | 0.350 | 0.332 | 0.550 | 0.500 | 0.750 | 0.636 |
| 0.175 | 0.170 | 0.375 | 0.356 | 0.575 | 0.516 | 0.775 | 0.650 |
| 0.200 | 0.196 | 0.400 | 0.376 | 0.600 | 0.536 | 0.800 | 0.663 |

After finding the sum of $k_p$ and $k_w$, the value of $\Delta$ is obtained either directly, or by interpolation, from the accompanying table (based upon chart in the British standard).

For internal gears (irrespective of numbers of teeth)

$$k_p = 0.4 \quad \text{and} \quad k_w = -k_p \qquad (9)$$

*Outside Diameter of Pinion.* — If number of pinion teeth is such that $t\sec^3 \sigma$ is less than 17, outside diameter is reduced but the pitch diameter and root diameter are not changed.

$$\text{Pinion diam. reduction} = \frac{p_n}{\pi} \times 0.04(17 - t\sec^3 \sigma) = \frac{0.04(17 - t\sec^3 \sigma)}{P_n} \qquad (10)$$

**Example 1.** — Find the addendum values for a pair of spur gears. The pinion has 26 teeth, the gear 73 teeth, and the circular pitch is 0.5 inch. In this case, $t + T = 26 + 73 = 99$. Since this sum is larger than 60, pinion correction factor $k_p$ is determined either by formula (3) or (4), whichever yields the greater value.

Applying formula (3), $k_p = 0.4 \left( 1 - \dfrac{26}{73} \right) = 0.258$.

Applying formula (4), $k_p = 0.02(30 - 26) = 0.08$.

Hence, pinion correction factor $k_p = 0.258$ and gear correction factor $k_w = -0.258$.

Applying formula (1), pinion addendum $a = \dfrac{0.5}{\pi} \times 1.258 = 0.200$ inch.

Applying formula (2), gear addendum $A = \dfrac{0.5}{\pi} \times 0.742 = 0.118$ inch.

Note: The regular unmodified addendum in this case would equal $0.3183 \times 0.5 = 0.159$ inch for pinion and gear.

**Example 2.** — Find the addendum values for a pair of helical gears. The pinion has 11 teeth, the gear 22 teeth. The normal diametral pitch is 4 and the helix angle is $22°30'$ ($\sec^3 22.5° = 1.268$).

First determine whether Formulas (3), (4) and (5) or Formulas (6) and (7) are to be used for finding the pinion and gear correction factors $k_p$ and $k_w$.

$$(t + T) \sec^3 \sigma = (11 + 22) \times 1.268 = 41.8$$

Since 41.8 is less than 60, Formulas (6) and (7) should be used.

$$k_p = 0.02(30 - 11 \times 1.268) = 0.321$$
$$k_w = 0.02(30 - 22 \times 1.268) = 0.042$$

Next, determine the extension of the center distance using Formula (8).

$$k_p + k_w = 0.321 + 0.042 = 0.363$$

Factor $\Delta$ obtained from the accompanying table by interpolation is about 0.345; hence, using Formula (8)

Extension of center distance $= \dfrac{0.345}{4} = 0.086$ inch

Center distance $= \dfrac{t + T}{2P_n \cos \sigma} + 0.086 = \dfrac{11 + 22}{2 \times 4 \times 0.9239} + 0.086 = 4.551$ inches

Gear addendum $= \dfrac{1 + 0.042}{4} = 0.260$ inch

Finally, check to see if the outside diameter of the pinion should be reduced. In this example, $t \sec^3 \sigma = 11 \times 1.268 = 13.948$. Since this is less than 17, the pinion addendum is first obtained by Formula (1) and then it is reduced.

By Formula (1), $a = \dfrac{1 + 0.321}{4} = 0.330$ inch.

By Formula (10), diam. reduction $= \dfrac{0.04(17 - 13.948)}{4} = 0.030$ inch.

Actual pinion addendum $= 0.330 - \dfrac{0.030}{2} = 0.315$ inch

**British Standard Spur and Helical Gears.** — Metric Modules (B.S. 436:Part 2: 1970). — This British Standard is a metric-unit specification for external and internal spur and helical gears for use with parallel shafts. Preferred and second-choice modules are given, and the requirements for basic rack tooth profile, and accuracy are covered. Any of ten different grades of accuracy may be applied to each gear element. Thus, gear requirements are met ranging from coarse commercial to high-speed and high-load precision applications. Tolerances on gear blanks are included in the specification. The Standard is a companion specification to B.S. 436:Part 1:1967, which covers the requirements of spur and helical gears in the inch system.

**Notation.** — To promote international usage of common gear terminology, the terms of draft ISO Recommendation No. 888, 'International vocabulary of gears,' have been adopted, and the notation is derived from ISO Recommendation R701, 'International gear notation, symbols for geometrical data.' The following definitions are given:

$a$ = Center distance
$d$ = Reference circle diameter
  $d_1$  Reference circle diameter, pinion
  $d_2$  Reference circle diameter, wheel
$d_a$ = Tip diameter
  $d_{a1}$  Tip diameter, pinion
  $d_{a2}$  Tip diameter, wheel
$y$ = Center distance modification coefficient
$b$ = Face width
  $b_1$  Face width, pinion
  $b_2$  Face width, wheel
$x$ = Addendum modification coefficient
  $x_1$  Addendum modification coefficient, pinion
  $x_2$  Addendum modification coefficient, wheel
$l$ = Length of arc
$m$ = module
$m_n$ = Normal module
$p_t$ = Transverse pitch
$z$ = Number of teeth
  $z_1$  Number of teeth, pinion
  $z_2$  Number of teeth, wheel
$\beta$ = Helix angle at reference cylinder
$\alpha$ = Pressure angle at reference cylinder
$\alpha_n$ = Normal pressure angle at reference cylinder
$\alpha_t$ = Transverse pressure angle at reference cylinder
$\alpha_{tw}$ = Transverse pressure angle, working

**Basic Rack Tooth Profile.** — The basic rack is generally in agreement with ISO Recommendation R53 'Basic rack of cylindrical gears for general and heavy engineering.' In Fig. 1 is shown the profile of the basic rack for unit normal metric module, and this tooth profile has been adopted for the purposes of the Standard. The values shown are proportions of the module. The pressure angle is 20 degrees.

In practice, the basic rack tooth is usually modified, and the extent of modification shall be in accordance with the following: (a) The total depth may vary within the limits 2.25 to 2.40, which permits an increase in root clearance within the same limits to allow for the use of different manufacturing processes. (b) The root radius may vary within the limits 0.25 to 0.39. (c) Tip relief may be applied within the limits shown in Fig. 1.

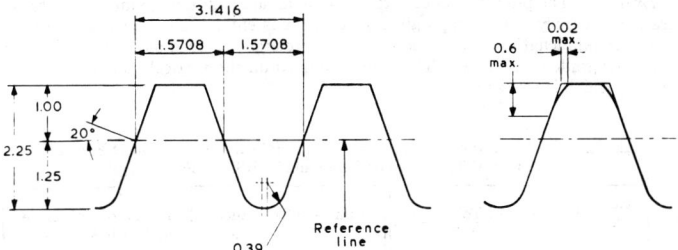

Fig. 1. (Left) British Standard Basic Rack Tooth Profile for Unit Normal Metric Module, and (Right) Limits of Tip Relief (B.S. 436:Part 2:1970)

**Standard Normal Modules.** — The modules, as shown in Table 1, are taken from ISO Recommendation R54 'Modules and diametral pitches of cylindrical gears for general and heavy engineering.' Wherever possible, the preferred modules should be selected, rather than the second choice modules.

**Accuracy Grades.** — The accuracy grades, numbered 3 to 12, are based on the requirements of draft ISO Recommendation No. 1328 'Accuracy of parallel involute gears.' The grade of a gear depends on the limits of tolerance for pitch, tooth profile, and tooth alignment, which are given in Table 2, and shown graphically in the Charts 1, 3 and 4. When these three tolerances are used, the highest grade number selected for any of the three is the grade assigned to the particular gear. The tolerances on the other elements, which are also shown in the table, and in Charts 6, 7, and 8, may be subject to agreement between the purchaser and manufacturer. Generally, the elements of a pair of mating gears are of identical accuracy grades, but different grades may be applied to individual elements by agreement between user and manufacturer. It should be appreciated that the choice of particular element tolerances may depend on the procedure employed for accuracy testing, the choice of disposition of the fundamental tolerance, and assembly requirements, for example.

**Tolerance on Pitch.** — The limits of tolerance on transverse pitch, both adjacent and cumulative, can be obtained from Chart 1 when the length of arc $l$ is known. This arc may be of any selected length in millimeters, less than $\pi d/2$.

Table 1. British Standard Spur and Helical Gears — Standard Normal Metric Modules (B.S. 436:Part 2:1970)

| Preferred Modules* | 1 | 1.25 | 1.5 | 2 | 2.5 | 3 | 4 | 5 | 6 |
|---|---|---|---|---|---|---|---|---|---|
| Second Choice Modules | 1.125 | 1.375 | 1.75 | 2.25 | 2.75 | 3.5 | 4.5 | 5.5 | 7 |
| Preferred Modules* | 8 | 10 | 12 | 16 | 20 | 25 | 32 | 40 | 50 |
| Second Choice Modules | 9 | 11 | 14 | 18 | 22 | 28 | 36 | 45 | ...... |

The values are in millimeters.
* Wherever possible, the preferred modules should be applied rather than those of second choice.

*Example:* The pitch tolerance is required on an arc length of 40 mm for a gear of grade 6 accuracy. The graph given in Chart 1 is entered on the horizontal scale at the 40 mm length position, and the limit of tolerance is read off in relation to the curve for grade 6 accuracy. The figure obtained on the vertical scale is 22 micrometers, which is 0.022 mm.

**Table 2. British Standard Metric Spur and Helical Gears — Basic Formulas for Limits of Tolerance on Elements\*** (B.S. 436:Part 2:1970)

| Gear Accuracy Grade | Limits of Tolerance on Pitch | Limits of Tolerance on Tooth Profile | Limits of Tolerance on Tooth Alignment |
|---|---|---|---|
| 3 | $0.63\sqrt{l} + 1.6$ | $0.16\,\phi_f + 3.15$ | $0.5\,\sqrt{b} + 2.5$ |
| 4 | $1.0\,\sqrt{l} + 2.5$ | $0.25\,\phi_f + 4.0$ | $0.63\sqrt{b} + 3.15$ |
| 5 | $1.6\,\sqrt{l} + 4.0$ | $0.40\,\phi_f + 5.0$ | $0.80\sqrt{b} + 4.0$ |
| 6 | $2.5\,\sqrt{l} + 6.3$ | $0.63\,\phi_f + 6.3$ | $1.0\,\sqrt{b} + 5.0$ |
| 7 | $3.55\sqrt{l} + 9.0$ | $1.0\,\phi_f + 8.0$ | $1.25\sqrt{b} + 6.3$ |
| 8 | $5.0\,\sqrt{l} + 12.5$ | $1.6\,\phi_f + 10.0$ | $2.0\,\sqrt{b} + 10.0$ |
| 9 | $7.1\,\sqrt{l} + 18.0$ | $2.5\,\phi_f + 16.0$ | $3.15\sqrt{b} + 16.0$ |
| 10 | $10.0\,\sqrt{l} + 25.0$ | $4.0\,\phi_f + 25.0$ | $5.0\,\sqrt{b} + 25.0$ |
| 11 | $14.0\,\sqrt{l} + 35.5$ | $6.3\,\phi_f + 40.0$ | $8.0\,\sqrt{b} + 40.0$ |
| 12 | $20.0\,\sqrt{l} + 50.0$ | $10.0\,\phi_f + 63.0$ | $12.5\,\sqrt{b} + 63.0$ |

| Gear Accuracy Grade | Limits of Tolerance on Radial Runout of Teeth | Limits of Tolerance on Tooth-to-Tooth Composite Error | Limits of Tolerance on Total Composite Error |
|---|---|---|---|
| 3 | $0.56\,\phi_p + 7.1$ | $0.32\,\phi_p + 4.0$ | $0.8\,\phi_p + 10$ |
| 4 | $0.90\,\phi_p + 11.2$ | $0.45\,\phi_p + 5.6$ | $1.25\,\phi_p + 16.0$ |
| 5 | $1.40\,\phi_p + 18.0$ | $0.63\,\phi_p + 8.0$ | $2.0\,\phi_p + 25.0$ |
| 6 | $2.24\,\phi_p + 28.0$ | $0.9\,\phi_p + 11.2$ | $3.15\,\phi_p + 40.0$ |
| 7 | $3.15\,\phi_p + 40.0$ | $1.25\,\phi_p + 16.0$ | $4.5\,\phi_p + 56.0$ |
| 8 | $4.0\,\phi_p + 50.0$ | $1.8\,\phi_p + 22.4$ | $5.6\,\phi_p + 71.0$ |
| 9 | $5.0\,\phi_p + 63.0$ | $2.24\,\phi_p + 28.0$ | $7.1\,\phi_p + 90.0$ |
| 10 | $6.3\,\phi_p + 80.0$ | $2.8\,\phi_p + 35.5$ | $9.0\,\phi_p + 112.0$ |
| 11 | $8.0\,\phi_p + 100.0$ | $3.55\,\phi_p + 45.0$ | $11.2\,\phi_p + 140.0$ |
| 12 | $10.0\,\phi_p + 125.0$ | $4.5\,\phi_p + 56.0$ | $14.0\,\phi_p + 180.0$ |

The limits of tolerance are in micro-meters.
\* To simplify application, the limits of tolerance, and tolerance factors are given graphically in Charts 1 through 8.
The values of symbols given in the above formulas are:
    $l$ = any selected length of arc in millimeters, less than $\pi d/2$.
    $\phi_f = m_n + 0.1\,\sqrt{d}$, where $m_n$ = normal module, and $d$ = reference circle diameter in mm.
    $b$ = face width in mm, up to a maximum of 150 mm.
    $\phi_p = m_n + 0.25\,\sqrt{d}$, where $m_n$ = normal module, and $d$ = reference circle diameter in mm.

This tolerance can also be calculated using the appropriate formula given in the pitch tolerance sub-table in Table 2. Thus, for a gear of grade 6 accuracy, the formula is $2.5\,\sqrt{l} + 6.3$. Substituting 40 mm arc length, the calculation is $2.5\,\sqrt{40} + 6.3 = 2.5 \times 6.32 + 6.3 = 22.1$ micro-meters, which rounded down is 0.022 mm.

**Tolerance on Tooth Profile.** — The limits of tolerance on tooth profile may be obtained from Chart 3, when the tolerance factor $\phi_f$ has been arrived at for a particular application using Chart 2.

*Example:* The tolerance on tooth profile is required for a gear of 5 module, with a reference circle diameter of 100 mm. Entering Chart 2 on the vertical scale at 100 mm, the tolerance factor is read off in relation to the curve for 5 module, and the figure is 6, obtained on the horizontal scale. This factor is then applied in Chart 3. If the gear grade accuracy is 9, for example, the graph is entered on the horizontal ordinate at tolerance factor 6, and the profile tolerance is read off in relation to the curve for grade 9. The figure obtained on the vertical ordinate is approximately 31 micro-meters, which is 0.031 mm.

**Chart 1.  British Standard Metric Spur and Helical Gears — Limits of Tolerance on Transverse Pitch (Adjacent and Cumulative)**

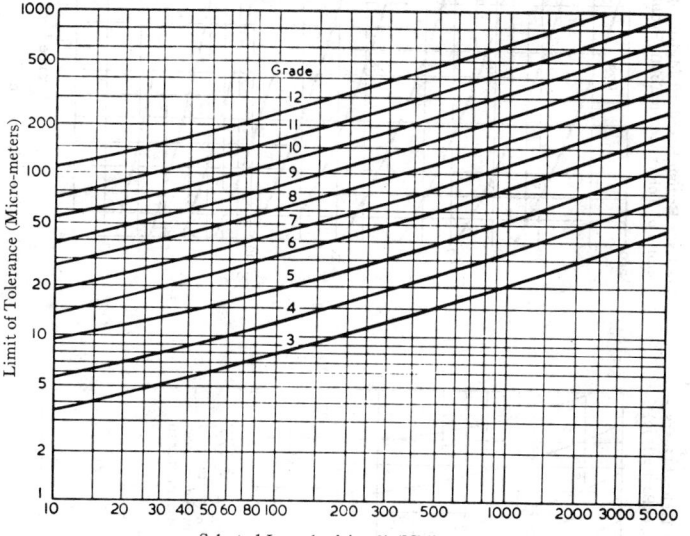

Selected Length of Arc $l^*$ (Millimeters)

\* $l$ = Any selected length of arc in millimeters, less than $\pi d/2$.

The tolerance factor is based on the formula $\phi_f = m_n + 0.1 \sqrt{d}$, where $m_n$ = normal module, and $d$ = reference circle diameter in millimeters. When a particular tolerance factor is known, either by calculation or from Chart 2, the tooth profile tolerance may be calculated as an alternative to obtaining it from Chart 3, using the appropriate formula given in Table 2. Thus, for a gear of grade 9 accuracy, the formula is $2.5\,\phi_f + 16.0$. If the tolerance factor is 6, as in the earlier example, then the calculation is $2.5 \times 6 + 16 = 31$ micrometers, or 0.031 mm.

The spread of the tolerance zone related to the design profile is shown in Fig. 2, left-hand view. Surface irregularities of geometrical form on any one flank of the actual profile, shall be within the tolerance zone contained by the parallel curves A and B. For most applications, positive departure from the design profile should not occur outside the central third of the working depth of the tooth flank as is shown diagrammatically in Fig. 2, right-hand view.

**Chart 2. British Standard Metric Spur and Helical Gears — Tooth Profile Tolerance Factors $\phi_f$** (B.S. 436:Part 2:1970)

Reference Circle Diameter (Millimeters)
2000
1000
500
200
100
50
20
2
3
4
5
10
15
20
Tolerance Factor $\phi_f$
Module $m_n$
2
3
4
5
6
7
8
9
10
11
12
14
16
$\phi_f = m_n + 0.10\sqrt{d}$, where $m_n$ = normal module, and $d$ = reference circle diameter in millimeters.

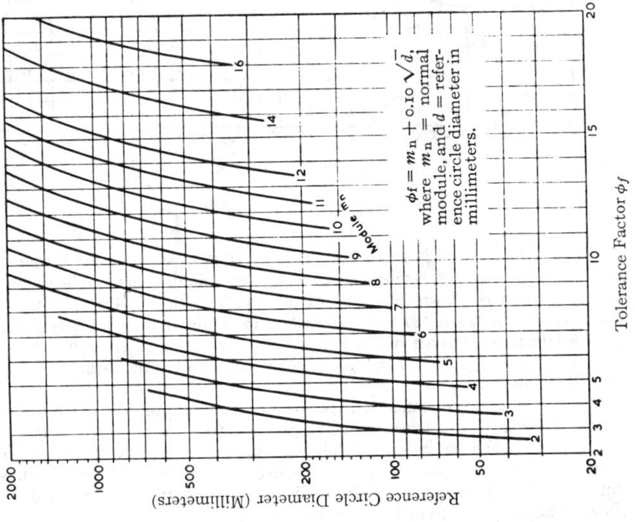

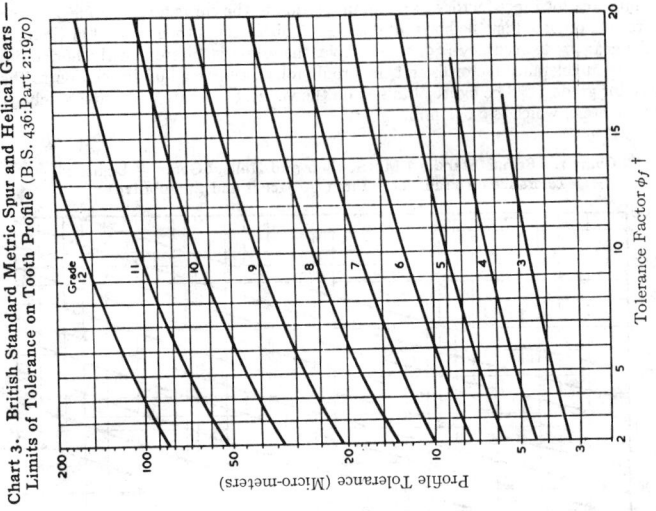

**Chart 3. British Standard Metric Spur and Helical Gears — Limits of Tolerance on Tooth Profile** (B.S. 436:Part 2:1970)

Profile Tolerance (Micro-meters)
200
100
50
20
10
5
3
2
5
10
15
20
Tolerance Factor $\phi_f$ †
Grade
12
11
10
9
8
7
6
5
4
3

† For tolerance factors see Chart 2.

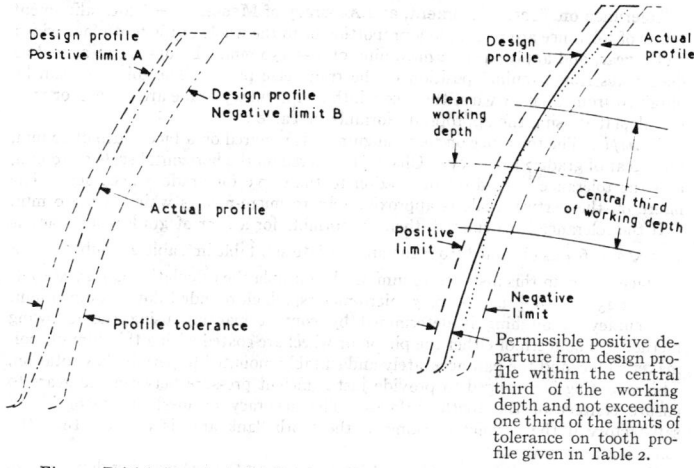

Fig. 2. British Standard Metric Spur and Helical Gears — (Left) Tolerance Zone of Tooth Profile Error, and (Right) Control of Positive Departures from Design Profile (B.S. 436:Part 2:1970)

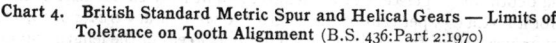

Chart 4. British Standard Metric Spur and Helical Gears — Limits of Tolerance on Tooth Alignment (B.S. 436:Part 2:1970)

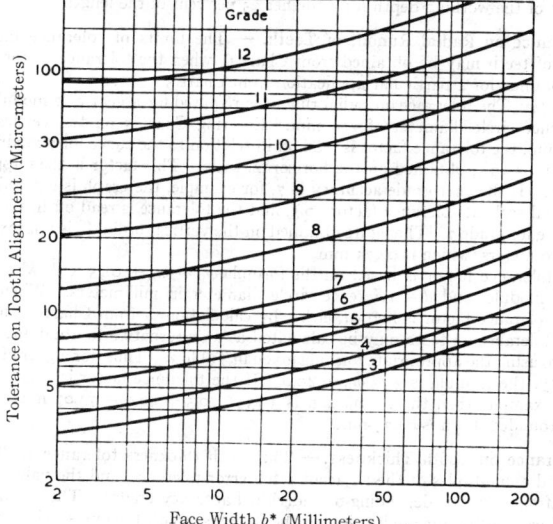

* $b$ = face width up to a maximum of 150 millimeters.

**Tolerance on Tooth Alignment, and Accuracy of Meshing.** — Tooth alignment limits of tolerance are related to a proportion or to the whole of the tooth face width $b$ of a gear, and apply up to a maximum of $b = 150$ mm. Errors are measured as departures from nominal position in the transverse plane. The tolerance can be obtained from Chart 4 when the face width and accuracy grade are known, or may be calculated using the appropriate formula in Table 2.

*Example:* The tolerance on tooth alignment is required on a face width of 10 mm, for a gear of grade 7 accuracy. Chart 4 is entered on the horizontal scale at 10 mm, and the tolerance is read off in relation to the curve for grade 7 accuracy. The figure on the vertical scale is approximately 10 micro-meters which is 0.010 mm.

If the tolerance is calculated, then the formula for a gear of grade 7 accuracy is $1.25 \sqrt{b} + 6.3$ as obtained from the appropriate sub-table in Table 2. Substituting the face width, in this instance 10 mm, in the formula the calculation is $1.25 \sqrt{10} + 6.3 = 1.25 \times 3.16 + 6.3 = 10.25$ micro-meters, which rounded down is 0.010 mm.

Accuracy of meshing is determined by contact marking using the following method. The teeth of either the pinion or wheel are coated with a thin film of tool-maker's blue, and the pair accurately and suitably mounted to permit slow rotation. The mounting is arranged to provide just sufficient pressure between the gears to ensure contact of the tooth surfaces. The accuracy of meshing is considered satisfactory if the contact marking of the tooth flank area is not less than the following percentages:

*Gear accuracy grades 3, 4, and 5:* At least 40 per cent of the working depth for 50 per cent of the length, and at least 20 per cent of the working depth for a further 40 per cent of the length. *Gear accuracy grades 6, 7, and 8:* At least 40 per cent of the working depth for 35 per cent of the length, and at least 20 per cent of the working depth for a further 35 per cent of the length. *Gear accuracy grades 9, 10, 11, and 12:* At least 40 per cent of the working depth for 25 per cent of the length and at least 20 per cent of the working depth for a further 25 per cent of the length.

**Tolerance on Radial Runout of Teeth.** — The limits of tolerance on radial runout of teeth may be obtained from Chart 6, when the tolerance factor $\phi_p$ has been obtained for a particular application from Chart 5.

*Example:* The tolerance on radial runout is required for a gear of 7 module, with a reference circle diameter of 200 mm. Entering Chart 5 on the vertical scale at 200 mm, the tolerance factor is read off in relation to the curve for 7 module, and the figure is 10.5, obtained on the horizontal scale. This factor is then applied in Chart 6. If the gear grade accuracy is 7, for example, the graph is entered on the horizontal scale at tolerance factor 10.5, and the tolerance is read off in relation to the curve for grade 7. The figure obtained on the vertical ordinate is approximately 73 micro-meters, which is 0.073 mm.

The tolerance factor is based on the formula $\phi_p = m_n + 0.25 \sqrt{d}$, where $m_n =$ normal module, and $d =$ reference circle diameter in millimeters. When a particular tolerance factor is known, either by calculation or from Chart 5, the radial runout tolerance on teeth may be calculated as an alternative to obtaining it from Chart 6, using the appropriate formula given in Table 2. Thus, for a gear of grade 7 accuracy, the formula is $3.15 \phi_p + 40.0$. If the tolerance factor is 10.5, as in the earlier example, then the calculation is $3.15 \times 10.5 + 40.0 = 73.07$ micro-meters, which rounded down is 0.073 mm.

**Tolerance on Tooth Thickness.** — The tooth thickness tolerance shall be determined as multiples of the adjacent pitch error tolerance, and the values are obtained from Chart 1, depending on individual accuracy grades. The magnitude of the values selected for application to the designed tooth thickness, will depend on functional considerations, with particular regard to the magnitude of the pitch

**Chart 6. British Standard Metric Spur and Helical Gears — Limits of Tolerance on Radial Runout of Teeth** (B.S. 436:Part 2:1970)

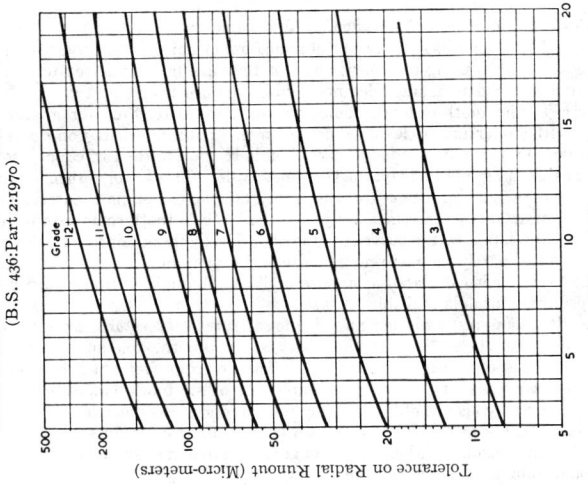

Tolerance Factor $\phi_p$ *

\* For tolerance factors see Chart 5.

**Chart 5. British Standard Metric Spur and Helical Gears — Tolerance Factors $\phi_p$** (B.S.436:Part 2:1970)

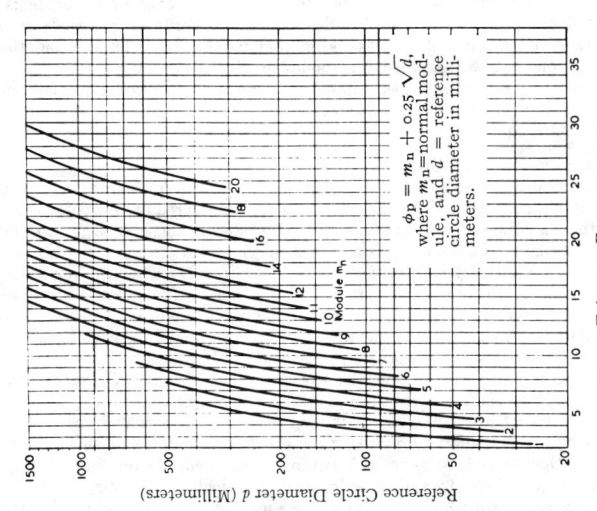

$$\phi_p = m_n + 0.25\ \sqrt{d},$$
where $m_n$ = normal module, and $d$ = reference circle diameter in millimeters.

Tolerance Factor $\phi_p$

error, tooth profile error, and radial runout error, since these features will have a direct effect on variations in tooth thickness around the gear.

**Dual-Flank Composite Tolerance.** — When dual-flank testing is applied, the *tooth-to-tooth composite tolerance* can be obtained from Chart 7. The tolerance factor $\phi_P$ is required when using the chart, and it is the same factor referred to earlier in connection with radial runout tolerance, and is obtained from Chart 5.

*Example:* The tooth-to-tooth composite tolerance is required for a gear of 5 module, with a reference circle diameter of 90 mm. Entering Chart 5 on the vertical scale at 90 mm, the tolerance factor is read off in relation to the curve for 5 module, and the figure is 7.5, obtained on the horizontal scale. This factor is then applied in Chart 7. If the gear grade accuracy is 8, for example, the graph is entered on the horizontal scale at tolerance factor 7.5, and the tooth-to-tooth composite tolerance is read off in relation to the curve for grade 8. The figure obtained on the vertical scale is approximately 36 micro-meters, which is 0.036 mm.

If the tolerance is calculated as an alternative to obtaining it from Chart 7, then the appropriate formula in Table 2 is used.

When dual-flank testing is being applied to measure *total composite error*, the limits of tolerance may be obtained from Chart 8. Again, the tolerance factor $\phi_P$ is required, and is obtained from Chart 5. The procedure for obtaining the final figure is the same as that given in the example for tooth-to-tooth composite tolerance, and the figure may also be obtained using the appropriate formula given in Table 2.

For both *tooth-to-tooth*, and *total composite error*, the total values represent variations in center distances when a product gear is rotated in close mesh with a master gear conforming to BS 3696 'Master Gears.'

**Single-Flank Composite Tolerances.** — When single-flank testing is applied, the tolerance for *tooth-to-tooth composite error* is the sum of the single pitch tolerance and tooth profile tolerance, and the values for addition are obtained from Charts 1 and 3, or by calculation, as described earlier under the appropriate headings. This procedure applies to all gear grade accuracies. For *total composite errors*, the tolerance is the sum of the maximum cumulative pitch tolerance and the tooth profile tolerance. The procedure applies to all gear accuracy grades. The values derived for both *tooth-to-tooth* and *total composite tolerances*, represent errors in angular transmission when a product gear is rotated in mesh with a master gear conforming to BS 3696 'Master gears.'

**Information to be Shown on Drawings.** — Component drawings for spur and helical gears shall be in accordance with BS 308 'Engineering drawing practice,' and the appropriate gear data shall be given covering the requirements for the finished tooth form, dimensions, and accuracy. This information shall be given in tables on the drawing. The essential data that should be given are:

(1) *Manufacturing Information — All Gears:* number of teeth; normal module; basic rack tooth form; axial pitch; tooth profile modification; addendum modification; reference circle diameter; helix angle at reference cylinder (0 degrees for straight spur gears); tooth thickness at reference cylinder; grade of gear; drawing number of mating gear; working center distance; backlash.

*Single Helical Gears:* Hand; lead of tooth helix. *Double Helical Gears:* Hand (in relation to specific part of face width); lead of tooth helix.

(2) *Checking data:* A table containing inspection data shall be provided on the drawing. When information is being inserted in this table, care should be taken to ensure that no confusion arises between the requirements for the accuracy of individual gear elements, and those for dual- and single-flank testing.

(3) *Supplementary and Other Information* may be added to meet particular design, manufacturing, and inspection requirements.

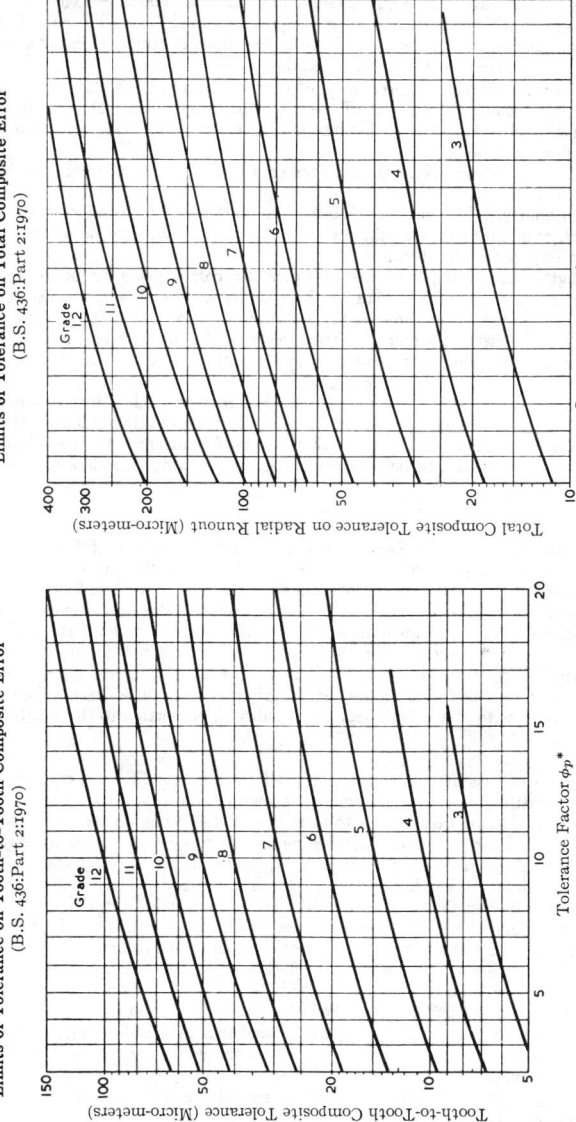

BRITISH METRIC SPUR AND HELICAL GEARS    807

**Chart 8.** British Standard Metric Spur and Helical Gears — Limits of Tolerance on Total Composite Error (B.S. 436:Part 2:1970)

Total Composite Tolerance on Radial Runout (Micro-meters)

Grade 12, 11, 10, 9, 8, 7, 6, 5, 4, 3

Tolerance Factor $\phi_p$ †

† For tolerance factors see Chart 5.

**Chart 7.** British Standard Metric Spur and Helical Gears — Limits of Tolerance on Tooth-to-Tooth Composite Error (B.S. 436:Part 2:1970)

Tooth-to-Tooth Composite Tolerance (Micro-meters)

Grade 12, 11, 10, 9, 8, 7, 6, 5, 4, 3

Tolerance Factor $\phi_p$ *

* For tolerance factors see Chart 5.

**Selecting the Number of Teeth for Gears in Compound Trains.** — The numbers of teeth to provide a required ratio of gearing in a compound gear drive (such as a set of change-gears) may be determined using logarithms as explained in the section, "Tabulated Logarithms of Change-gear Ratios".

Factors other than the required ratio of gearing should also be taken into consideration in choosing the number of teeth for the gears. Some of these factors, listed below, work against each other and it is necessary to use judgment to obtain a satisfactory compromise:

1. Cost. For a given diametral pitch the greater the number of teeth the higher the cost of the gears.

2. Mountability. The gears may be either so large as to interfere with each other or adjacent parts of a machine or, on the other hand, may be too small to make up a required center distance.

3. Strength. Increasing the center distance of a pair of gears by increasing the size of the gears decreases the load on the gear teeth. For low speed drives the loads on the gear teeth are in inverse proportion to the gear sizes. Thus, for example, doubling the sizes of a pair of mating gears results in a one-half reduction in tooth load and doubling of pitch-line velocity with a possible decrease in the dynamic factor (see table 2, page 812).

With regard to the use of intermediate or idler gears, considerations of strength mentioned in (3) do not apply since the loads on the teeth of intermediate gears are the same as on the gears with which they are meshed. Thus, the use of larger intermediate gears should be based on cost and mountability considerations rather than strength.

**Proportions of Cast Spur Gears.** — The table "Dimensions of Spur Gears" gives the formulas for proportioning cast gears with arms of different cross-sections. This table will prove satisfactory for general conditions, but in individual cases modifications must be made to suit particular designs. The oval arm is the one best adapted for small and medium size gears and gives the best appearance. It requires somewhat more metal for the same strength than the designs shown in the lower part of the table, but it is very easily molded. For large size gears, however, arms of the +, T- and H-sections are largely used. In these designs the metal is so distributed as to give a high degree of rigidity in proportion to the weight.

*Cast Teeth.* — Gears having teeth which are formed by molding and casting so as to eliminate cutting are often used where cut gearing is not required. One method of making gears having cast teeth is to form the mold by using a pattern which is a duplicate of the gear required. Another method is to form the mold by using a special machine designed for this purpose. In using such a machine, the molding may be done either by employing a single-tooth hard-wood pattern or a segment-shaped pattern having two or three teeth. The pattern is located radially to suit the gear radius and the teeth impressions are formed around the mold by progressively indexing either the arm which carries the tooth pattern or by indexing the mold itself. The ring of sand inside of the flask is first formed by means of a sweep, and a core box is used for forming the arms of the gear. Molding machines produce accurate molds but the cast gear may be distorted by uneven shrinkage of the arms. In such cases, the rim may be cast separately and then attached to the gear center or spider. Some cast gears are strengthened by shrouding the teeth. Cast teeth are often used where speeds are low and precision gearing is not required. Gear-tooth molding machines have been applied to the molding of spur gears, bevel gears, helical gears and worm gears.

### Dimensions of Spur Gears. (Page 808.)

#### Dimensions of Spur Gears with Oval Arms

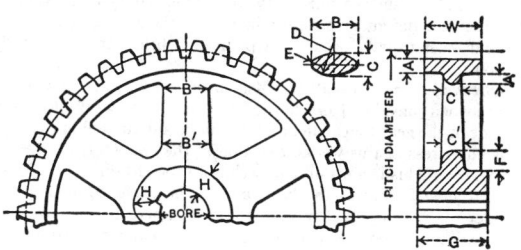

$P$ = diametral pitch, $P'$ = circular pitch.

$A = 1.57 \div P = 0.5 P'$;
$B = 6.28 \div P = 2.0 P'$;
$C = 3.14 \div P = P'$;
$D = 4.71 \div P = 1.5 P'$;
$E = 0.79 \div P = 0.25 P'$;

$F = 2.00 \div P = 0.65 P'$;
$G = W + 0.025$ pitch diameter;
$H = 0.44 \times$ bore;
$B' = B + \frac{3}{4}$ inch per foot;
$C' = C + \frac{3}{4}$ inch per foot.

#### Dimensions of Spur Gears with Ribbed Arms or Arms of H-section

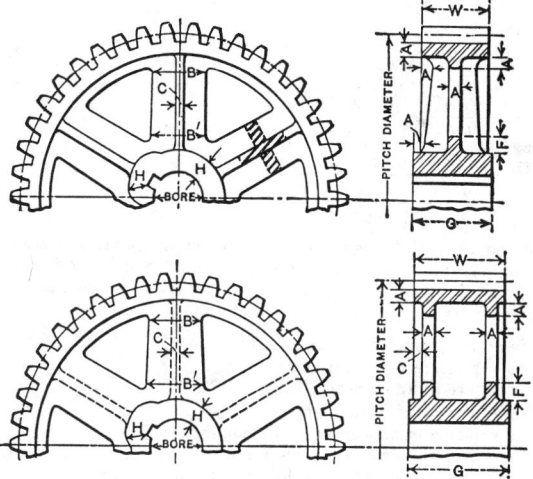

$P$ = diametral pitch, $P'$ = circular pitch.

$A = 1.57 \div P = 0.5 P'$;
$B = 7.85 \div P = 2.5 P'$;
$C = 0.94 \div P = 0.3 P'$;
$F = 2.20 \div P = 0.7 P'$;

$G = W + 0.025$ pitch diameter;
$H = 0.44 \times$ bore;
$B' = B + \frac{3}{4}$ inch per foot.

**Power Transmitting Capacity of Spur Gears.** — If a set of spur gears are made, installed and lubricated properly they normally may be subject to three primary modes of failure:

1. *Pitting:* This is a surface durability or fatigue failure mode that may be anticipated in the gear design by a determination of the gear set contact compressive stress and will most usually occur at a point just below the pitch surface on the driving pinion.

2. *Tooth Breakage:* This is usually a tensile fatigue failure at the weakest section of the gear tooth when considered as a cantilever beam.  The weakest point is normally the tensile side of the gear tooth fillet and it may be anticipated in the gear design by determining the stress at this weakest section of the gear tooth.

3. *Tooth Scoring:* This is a scuffing or welding type of failure.  It is not a fatigue failure but rather a failure of the lubricant in the presence of high tooth sliding velocity under high load.  Well proportioned commercial gears with a pitch line velocity of less than 7,000 feet per minute will normally not score if they have a reasonably good surface finish and are properly lubricated.  If scoring does occur or if it is suspected to be critical in a new high speed design, the scoring temperature index should be determined by the method shown in American Gear Manufacturers Standard AGMA *217.01* or by some similar method.

**Surface Durability Stress.** — The following method of determining the surface durability stress, $s_c$, in pounds per square inch, of a spur gear set is derived from American Gear Manufacturers Standard AGMA *210.02*. The surface compressive stress in a gear set is a maximum at the lowest point of single tooth contact (LPSTC) of the pinion.

It is determined by the following formula:

$$s_c = 3239 \, C \sqrt{\frac{T_P}{F d_p{}^2 I}} \tag{1}$$

$$\text{where:} \quad C = \frac{C_p C_t C_r}{C_l C_h} \sqrt{\frac{C_o C_s C_m C_f}{C_v}} \tag{1a}$$

$$\text{and} \quad I = \frac{\cos^2 \phi}{2 \left( \dfrac{1}{\tan c} + \dfrac{N_P}{N_G \tan D} \right)} \tag{2}$$

(For standard tooth forms the values of the geometry factor $I$ are tabulated in Table 5 on page 813; the interpolation procedure for finding $I$ for values of gear ratio $m_g$ and of number of teeth in pinion that are not tabulated in Table 5, is shown on page 815 in the section Spur Gear Sizing.)

where $d_P$ = pinion pitch diameter in inches

       $\phi$ = pressure angle in degrees

      $F$ = face width in inches

     $N_G$ = number of teeth in gear

     $T_P$ = pinion max. operating torque in lbs.-in.

     $C_p$ = material factor — equals 1.0 for steel pinion and steel gear; for steel and iron use 0.87; and for iron and iron use 0.78

     $C_t$ = temperature factor — equals 1.0 except when gears operate near the tempering temperature of the gear material

     $C_r$ = reliability factor — normally equals 1.0; for very high reliability use 1.25

     $C_l$ = life factor — equals 1.0 for $10^7$ cycles or more; 1.3 for $10^5$ cycles; and 1.5 for $10^4$ cycles

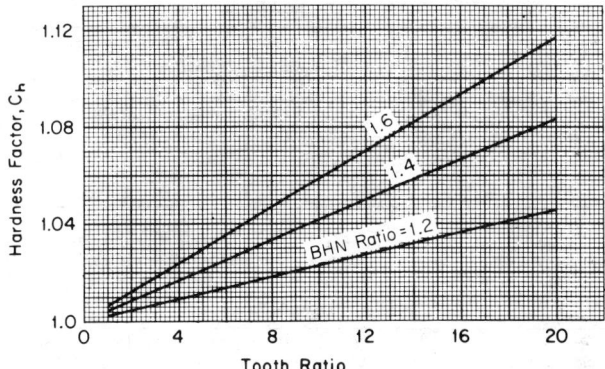

$C_h$ = hardness factor — for a Brinell hardness ratio of pinion to gear of 1.2
  or less, use 1.0, otherwise use value taken from curve in chart shown
  above.

$C_o$ = overload factor — use value from Table 1.

$C_s$ = Size factor, equals 1 provided gear metallurgy and case depth are appropriate.

$C_f$ = surface factor — equals 1.0 for normal surface finish and condition.

$C_v$ = dynamic factor — equals 1.0 for precision gears. For commercial gears
  use value from Table 2.

$C_m$ = Load distribution factor. For normal commercial gearing when the face
  width to pinion pitch diameter ratio is not greater than 2, see Table 3.

tan $c$ = as determined by formula on page 754

tan $D$ = as determined by formula on page 755

Surface durability stress, $s_c$, must not be greater than the limiting values of
allowable contact stress, $s_{ac}$, as given in Table 4.

**Fillet Tensile Stress.** — The following method of determining the fillet tensile
stress of a spur gear is taken from American Gear Manufacturers Standard AGMA
220.02. The fillet tensile stress of a gear is normally determined with the full maximum operating load applied at the highest point of single tooth contact (HPSTC)
diameter; however, if tooth spacing errors are significantly high compared with the
expected full load deflection of the gear tooth mesh, the maximum operating load
should be applied at the gear outside diameter for this stress determination. For all
normal cases using standard gear tooth proportions the fillet tensile stress is highest
on the pinion, this, therefore, is the only one that need be calculated. However, it
can be seen that the comparable stress on the mating gear is inversely proportional to
the value of the $J$ factor of the two gears, all other parameters being equal. The
fillet tensile stress is determined by the formula:

$$s_t = \frac{2K \times T_1 \times n}{F \times d_1^2 \times J} \text{ , where: } K = \frac{K_o \times K_s \times K_m \times K_r \times K_t}{K_v \times K_l} \qquad (3), (4)$$

For definitions of symbols in formulas (3) and (4) see page 813.

**Table 1.  Overload Factor, $C_o$**

| Power Source Shock* | Driven Load Shock | | |
|---|---|---|---|
| | None | Moderate | Heavy |
| | Overload Factor, $C_o$ | | |
| None | 1.00 | 1.25 | 1.75 |
| Light | 1.25 | 1.50 | 2.00 |
| Medium | 1.50 | 1.75 | 2.25 |

Note: For speed increaser drives add $0.01\sqrt{(N/n)}$ to the tabular value.

*For example: No shock — electric motor or turbine drive; Light shock — 4 or more cylinder internal combustion engine drive; and Medium Shock — 1 or 2 cylinder internal combustion engine drive.

**Table 2.  Dynamic Factor $C_v$**

| Pitch Line Velocity, fpm | Dynamic Factor, $C_v$ | Pitch Line Velocity, fpm | Dynamic Factor, $C_v$ |
|---|---|---|---|
| Up to 100 | 0.9 | 701 to 1500 | 0.6 |
| 101 to 300 | 0.8 | 1501 to 3500 | 0.5 |
| 301 to 700 | 0.7 | 3501 to 7000 | 0.4 |

**Table 3.  Load Distribution Factor, $C_m$**

| Face Width, In. | $C_m$ and $K_m$ | Face Width, In. | $C_m$ and $K_m$ | Face Width, In. | $C_m$ and $K_m$ |
|---|---|---|---|---|---|
| Up to 2 | 1.3 | 6 to 9 | 1.5 | 12 to 16 | 1.7 |
| 2 to 6 | 1.4 | 9 to 12 | 1.6 | ... | ... |

**Table 4.  Allowable Contact Stress Number — $s_{ac}$**

| Material | Surface Hardness, Minimum | $s_{ac}$ | Material | Surface Hardness, Minimum | $s_{ac}$ |
|---|---|---|---|---|---|
| | Through Hardened | | Cast Iron | | |
| | 180 Bhn | 85- 95,000 | AGMA Grade 20 | — | 50-60,000 |
| | 240 Bhn | 105-115,000 | AGMA Grade 30 | 175 Bhn | 65-75,000 |
| | | | AGMA Grade 40 | 200 Bhn | 75-85,000 |
| | 300 Bhn | 120-135,000 | Nodular Iron | | 90-100% of the $s_{ac}$ value of steel with the same hardness |
| | 360 Bhn | 145-160,000 | Annealed | 165 Bhn | |
| | | | Normalized | 210 Bhn | |
| | 440 Bhn | 170-190,000 | Oil Quench and Temper | 255 Bhn | |
| Steel | Case Carburized | | | | |
| | 55 $R_c$ | 180-200,000 | Bronze | Tensile Strength psi (Min.) | $s_{ac}$ |
| | 60 $R_c$ | 200-225,000 | Tin Bronze AGMA 2C(10-12% Tin) | 40,000 | 30,000 |
| | Flame or Induction Hardened | | Aluminum Bronze ASTM B 148-52 (Alloy 9C-H.T.) | 90,000 | 65,000 |
| | 50 $R_c$ | 170-190,000 | | | |

Table 5. *J* Tooth Form Factor and *I* Geometry Factor for
American National Standard Involute Gears

| Gear Ratio | No. of Pinion Teeth | 20-degree Pressure Angle | | | 25-degree Pressure Angle | | |
|---|---|---|---|---|---|---|---|
| | | *I* Factor | *J* Factor, Ded. = 1.25* | *J* Factor, Ded. = 1.35* | *I* Factor | *J* Factor, Ded. = 1.25* | *J* Factor, Ded. = 1.35† |
| 1 | 17 | .0755 | .320 | .302 | .0900 | .383 | .353 |
| 1 | 20 | .0774 | .343 | .326 | .0919 | .409 | .379 |
| 1 | 25 | .0789 | .372 | .356 | .0936 | .441 | .410 |
| 1 | 30 | .0795 | .395 | .379 | .0944 | .465 | .435 |
| 1 | 40 | .0800 | .428 | .411 | .0951 | .499 | .468 |
| 1 | 60 | .0802 | .467 | .449 | .0955 | .539 | .505 |
| 1 | 100 | .0803 | .505 | .487 | .0957 | .575 | .540 |
| 1 | 200 | .0803 | .540 | .521 | .0957 | .607 | .570 |
| 3 | 17 | .0983 | .340 | .319 | .1173 | .403 | .369 |
| 3 | 20 | .1037 | .365 | .344 | .1227 | .430 | .396 |
| 3 | 25 | .1089 | .396 | .376 | .1283 | .463 | .429 |
| 3 | 30 | .1121 | .419 | .399 | .1316 | .487 | .452 |
| 3 | 40 | .1154 | .452 | .432 | .1354 | .519 | .485 |
| 3 | 60 | .1180 | .490 | .469 | .1387 | .557 | .520 |
| 3 | 100 | .1195 | .524 | .503 | .1410 | .589 | .553 |
| 3 | 200 | .1202 | .552 | .531 | .1424 | .615 | .576 |
| 10 | 17 | .1128 | .351 | .328 | .1347 | .413 | .377 |
| 10 | 20 | .1204 | .377 | .354 | .1423 | .440 | .404 |
| 10 | 25 | .1282 | .409 | .387 | .1502 | .474 | .437 |
| 10 | 30 | .1328 | .432 | .410 | .1553 | .498 | .461 |
| 10 | 40 | .1379 | .465 | .443 | .1611 | .530 | .492 |
| 10 | 60 | .1420 | .500 | .478 | .1662 | .564 | .526 |
| 10 | 100 | .1444 | .531 | .509 | .1698 | .594 | .555 |
| 10 | 200 | .1456 | .557 | .535 | .1721 | .618 | .579 |

* Rack fillet radius = .30/*P*.   † Rack fillet radius = .21/*P*.

$K_o$ = Overload factor — see Table 1 (same as for $C_o$).

$K_s$ = Size factor — equals 1.0 provided metallurgy and case depth are appropriate.

$K_m$ = Load distribution factor — see Table 3 (same as for $C_m$).

$K_r$ = Reliability factor — normally equals 1.0; for very high reliability use 1.50.

$K_t$ = Temperature factor — equals 1.0 except when gears are operated near the tempering temperature of the gear material.

$K_v$ = Dynamic factor — equals 1.0 for precision gears. For commercial gears use value from Table 2 (same as for $C_v$).

$K_l$ = Life factor — equals 1.0 for $10^7$ cycles or more; for $10^6$ cycles use 1.1.

$J$ = Tooth form factor — for standard gears use value from Table 5.
    ($J$ tooth form factors for values of gear ratio $m_g$ and numbers of teeth in pinion that are not tabulated in Table 5 must be determined by interpolation using the method shown on page 816 in the section Spur Gear Sizing.)

$T_1$ = Pinion maximum operating torque in pound-inches.

$n$ = Number of teeth in the pinion.

$F$ = Gear face width in inches.

$d_1$ = Pinion pitch diameter.

*Example:* Find the surface durability stress and the fillet tensile stress of a 23-tooth, 20-degree AISI standard pinion with a $1.25/P$ dedendum and a diametral pitch of 10 driving a 31-tooth gear. The face width is 1.00 inch; the pinion torque is 500 pounds-inch; the pinion speed is 3600 rpm; and the pinion pitch line velocity is 2000 feet per minute. The gears are made of case carburized steel.

$$n = 23 \qquad\qquad N = 31$$
$$d = 2.3 \qquad\qquad P = 10$$
$$d_o = \frac{n+2}{10} = 2.5 \qquad T_1 = 500$$
$$DF = 1.00 \qquad\qquad V_1 = 2,000$$

From the formulas for significant diameters beginning on page 753:

$$\cos a = \frac{n \cos \phi}{d_o P} = \frac{23 \times 0.93969}{2.5 \times 10} = 0.86451 \qquad (4)$$

$$a = 30°10'24''$$

$$\tan c = \tan a - \frac{6.28318}{n} = 0.58139 - \frac{6.28318}{23} = 0.30821 \qquad (8)$$

$$\tan B = \tan \phi - \frac{n}{N}(\tan a - \tan \phi) \qquad (5)$$

$$= 0.36397 - \frac{23}{31}(0.58139 - 0.36397) = 0.20266$$

$$\tan D = \tan B + \frac{6.28318}{N} = 0.20266 + \frac{6.28318}{31} = 0.40534 \qquad (16)$$

*Surface Durability Stress:*

Based on these data, the following surface durability factors apply: $C_p = 1$, $C_t = 1$, $C_r = 1$, $C_l = 1$, $C_h = 1$, $C_o = 1$. $C_s = 1$, $C_v = 0.55$, $C_m = 1.3$, $C_f = 1$, and

$$I = \frac{0.93969^2}{2\left(\dfrac{1}{0.30821} + \dfrac{23}{31 \times 0.40534}\right)} = 0.087 \qquad (2)$$

$$C = \frac{1 \times 1 \times 1}{1 \times 1}\sqrt{\frac{1 \times 1 \times 1.3 \times 1}{0.55}} = 1.54 \qquad (1a)$$

$$s_c = 3239 \times 1.54 \sqrt{\frac{500}{1 \times (2.3)^2 \times 0.087}} = 164,000 \text{ psi} \qquad (1)$$

*Fillet Tensile Stress:*

Based on the above given data, the following fillet tensile stress factors apply: $K_o = 1$, $K_s = 1$, $K_m = 1.3$, $K_r = 1$, $K_t = 1$, $K_v = 0.55$, and $K_l = 1$.

Using the formula for fillet tensile stress:

For the 23-tooth pinion, $J = 0.369$.

$$K = \frac{1 \times 1 \times 1.3 \times 1 \times 1}{0.55 \times 1} = 2.36 \qquad (4)$$

$$s_t = \frac{2 \times 2.36 \times 500 \times 23}{1 \times 5.3 \times 0.369} = 27,800 \text{ psi.} \qquad (3)$$

The actual tensile stress $s_t$ must not be greater than the allowable tensile stress, $s_{at}$, as given in Table 6.

**Table 6.   Limiting Values of Fillet Tensile Stress, $s_{at}$**

| Material and Brinell Hardness | Allowable Stress, psi | Material and Brinell Hardness | Allowable Stress, psi |
|---|---|---|---|
| Cast Iron — 175 | 8,500 | Steel — 300 | 36,000 |
| Cast Iron — 200 | 13,000 | Steel — 350 | 39,000 |
| Steel, Unhardened | 22,000 | Steel — 400 | 42,000 |
| Steel — 200 | 27,000 | Steel — 450 | 44,000 |
| Steel — 250 | 32,000 | Steel — Carburized | 55,000 to 65,000 |

**Optimum Spur Gear Sizing for Standard Tooth Form.** — The formulas shown in the previous two sections for the determination of surface durability stress $s_c$ and fillet tensile stress $s_t$ may be inverted in such a manner that they can be used to determine the optimum gear size and number of teeth of standard tooth form to be used to satisfy required operating load, stresses, gear ratio, and tooth form. Data that must be given or assumed are:

1. *Gear tooth form:* (a) pressure angle, 20° or 25°; (b) dedendum factor, 1.25 or 1.35; and (c) diametral pitch.

2. *Operating conditions:* (a) maximum pinion torque, inch pounds; (b) gear load factor; and (c) gear reduction ratio.

3. *Gear material hardness:* Brinell hardness number. Material must be steel.

The data to be computed are: (a) gear tooth fillet tensile stress; (b) gear surface durability stress; (c) number of teeth in pinion and gear; and (d) gear face width.

Using these data, complete gear dimensions may be obtained using the formulas in Table 1 on page 717.

The first step in this sizing procedure is to obtain a first estimate of the number of teeth $N_P$ in the pinion using the following formula:

$$\log N_P = \frac{1}{1-q} \log\left(\frac{73000 \times K_m}{6H_b}\right) \qquad (1)$$

where:  $H_b$ = Gear material hardness, Bhn
$K_m$ = Material factor — 0.6 for through-hardening steels
0.9 for case-hardening steels.

$$q = \frac{5}{90} + \frac{1}{7\sqrt[3]{m_g^2}} \qquad (2)$$

where:  $m_g$ = gear reduction ratio; must be greater than 1.

This pinion tooth number $N_P$ and the given gear ratio $m_g$ are used to enter Table 5 on page 813 for the specified tooth form to get values for the $I$ and $J$ factors. In most cases the specified gear ratio and the estimated number of teeth are not tabulated values, thus the $I$ and $J$ factors must be found by interpolation. The interpolation factor $i$ to be used is found by the formula:

$$\text{Factor } i = \frac{h}{m}\left(\frac{m-l}{h-l}\right) \qquad (3)$$

where:  $l$ = tabulated input next lower than the specified value
$m$ = specified input value
$h$ = tabulated input next higher than the specified value

Two interpolation factors are usually required, one is a gear ratio interpolation factor and the other is a pinion tooth number interpolation factor. In later calculations for the $I$ and $J$ factors care must be taken not to confuse one with the other. In the event that the required gear ratio is a tabulated value, no gear ratio interpolation factor is needed; likewise if the estimated pinion tooth number is a tabulated value, no tooth number interpolation factor is needed.

Both the $I$ and $J$ factors are determined in the same manner; the procedure will be explained in terms of $I$. The interpolation formulas are:

$$I = I_l + i(I_h - I_l) \quad \text{and} \quad J = J_l + i(J_h - J_l) \qquad \text{(4a) and (4b)}$$

where:   $I_l$ = the tabulated $I$ for the input next lower than the specified value
         $I_h$ = the tabulated $I$ for the input next higher than the specified value

Each formula is used three times; first, at the tabulated gear ratio lower than the specified value to obtain the $I$ factor for the estimated number of teeth $N_P$ at this tabulated value of gear ratio and second, at the tabulated gear ratio higher than the specified value. Both of these calculations use the tooth number interpolation factor $i_N$. The third interpolation calculation uses the two values of $I$ calculated above to determine the $I$ factor for the specified tooth number at the specified gear ratio; this calculation uses the gear ratio interpolation factor $i_m$.

The allowable fillet tensile stress may be obtained by dividing the appropriate value from Table 6 on page 815 by the gear load factor $K$ (see page 811); or the following formula may be used:

$$\log (s_t) = \frac{3}{4} \log (H_b) - 0.3010 - \log (K) \qquad (5)$$

where:   $s_t$ = Fillet stress in kpsi
         $H_b$ = Gear material hardness in Brinell hardness number
         $K$ = Load factor (see formula 4, page 811)
         (Logarithms are to the base 10)

Note that the stress must be given in thousands of pounds per square inch (kpsi). If the hardness is given in terms of the Rockwell C scale as is often done with case-hardening materials, conversion to Bhn may be done with the conversion table given on page 2130; the required hardness is given in this table in the column labeled 'Tungsten Carbide Ball.' For case-hardening materials, the hardness of the case is to be used in this formula.

To satisfy the requirements of this sizing method, the allowable surface durability stress $s_c$ must be obtained from the following formula:

$$s_c = \sqrt{\frac{s_t \times H_b}{K_m}} \qquad (6)$$

where:   $s_c$ = Durability stress in kpsi
         $s_t$ = Fillet tensile stress in kpsi
         $H_b$ = Material hardness in Bhn
         $K_m$ = Material factor, 0.6 for through-hardening steels, 0.9 for case-hardening steels.

The stress given by this formula is almost the same as that given in Table 4 on page 812, divided by the square root of $K$.

All data needed to size the gear set are now available. To determine the gear size, the following formula is used:

$$FD_P{}^2 = \frac{10.58 T_P}{I s_c{}^2} \qquad (7)$$

where:    $F$ = Effective face width of gear and pinion in inches
$D_P$ = Pinion pitch diameter in inches
$s_c$ = Surface durability stress in kpsi
$I$ = Gear geometry factor
$T_P$ = Pinion maximum operating torque in inch-pounds

and the number of teeth in the pinion $N_P$ is determined by:

$$N_P = \frac{s_t(FD_P{}^2)J_P}{2T_P} \times 10^3 \tag{8}$$

where:    $s_t$ = Fillet tensile stress in kpsi
$(FD_P{}^2)$ = is as determined above
$J_P$ = pinion tooth form factor

This formula will most probably give a result that is not an integer; if so, the actual pinion tooth number must be the next lower integer. The number of teeth $N_G$ in the mating gear is:

$$N_G = N_P m_g \text{ to the nearest integer.} \tag{9}$$

where: $m_g$ = the gear ratio

It is good practice to back check to determine the actual ratio by:

$$m_g = \frac{N_G}{N_P} \tag{10}$$

If this results in an unsatisfactory deviation from the specified gear ratio, the number of teeth in the pinion may be reduced by one or two and a new number is determined for the number of teeth in the mating gear. If this renumbering process is continued excessively, a revised $I$ factor and a revised $FD_P{}^2$ will need to be calculated. As the tooth numbers are reduced, the fillet tensile stress is also reduced, thus a redetermination of $s_t$ by the formula on page 811 need not be made unless the actual reduced value of this stress is needed for some purpose.

If the diametral pitch $P$ has been specified, the pinion pitch diameter $D_P$ may be determined by:

$$D_P = \frac{N_P}{P} \tag{11}$$

If the diametral pitch is not specified, some other criterion is needed to obtain the pinion pitch diameter. This criterion may be the required center distance $C$, in which case the pinion pitch diameter is:

$$D_P = \frac{2C}{m_g + 1} \tag{12}$$

where: $m_g$ = the actual gear ratio.

On the other hand it might be that the gear face width $F_G$ is specified by some ancillary requirement, in which case the pinion pitch diameter is determined by:

$$D_P = \sqrt{\frac{(F_G D_P{}^2)}{F_G}} \tag{13}$$

Note that although $D_P$ is to be determined, $(F_G D_P{}^2)$ has already been calculated.

Or, as is the case with many aerospace applications, it may be required that the smallest practical diameter shall be used; in this case the face width should be no more than 0.7 of the pinion pitch diameter unless special consideration is given to the load patterning of the gear teeth; where the pinion pitch diameter is then determined by:

$$D_P = \sqrt[3]{\frac{(FD_P{}^2)}{0.70}} \qquad (14)$$

The diametral pitch may now be determined by:

$$P = \frac{N_P}{D_P} \qquad (15)$$

and the pitch diameter of the mating gear is:

$$D_G = \frac{N_G}{P} \qquad (16)$$

If the pinion pitch diameter is directly specified because of needed clearance over a bearing, spline, seal, key, or other associated element, then the face width may be directly determined by:

$$F_G = \frac{(FD_P{}^2)}{D_P{}^2} \qquad (17)$$

All data are now at hand for the completion of the required data by use of the formulas in Table 1 (page 717).

A numerical example of this method will illustrate its use; particular attention should be given to the method of interpolation of the tabular values of $I$ and $J$ in Table 5 on page 813.

*Spur Gear Sizing Example:* The following example shows how complete gear dimensions may be calculated using the formulas and procedure just outlined.

*Given data:*

> Pressure angle = 20°
> Dedendum factor $K_o$ = 1.25
> Rack tooth fillet radius $r_f$ = 0.30 in.
> Gear reduction ratio $m_g$ = 2.25
> Pinion torque $T_P$ = 1300 in.-lbs.
> Load factor $K$ = 1.4
> Material hardness $H_b$ = 340 Bhn
> Material factor $K_m$ = 0.60
> Diametral pitch $P$ = 16

*Ratio factor:*

$$q = \frac{5}{90} + \frac{1}{7\sqrt[3]{m_g{}^2}} \qquad (2)$$

$$q = \frac{5}{90} + \frac{1}{7\sqrt[3]{2.25^2}} = 0.139$$

*First estimate of number of teeth in pinion:*

$$\log N_P = \frac{1}{1-q} \log\left(\frac{12.17 \; K_m \times 10^3}{H_b}\right) \qquad (1)$$

$$= \frac{1}{1-0.139} \log\left(\frac{12.17 \times 0.60 \times 10^3}{340}\right)$$

$$= 1.5470$$

$$N_P = 35.2, \text{ nearest integer } 35$$

*Interpolation factor for Table 5* (p. 813):

$$i = \frac{h}{m}\left(\frac{m-l}{h-l}\right) \tag{3}$$

*For tooth number interpolation:* $l = 30$, $m = 35$ and $h = 40$

$$i_N = \frac{40}{35}\left(\frac{35-30}{40-30}\right) = 0.5714$$

*For gear ratio interpolation:* $l = 1$, $m = 2.25$ and $h = 3$

$$i_m = \frac{3}{2.25}\left(\frac{2.25-1}{3-1}\right) = 0.8333$$

*Determination of geometry factor I:*

$$I = I_l + i_N(I_h - I_l) \tag{4a}$$

For $m_g = 1$ and $N_P = 35$, $I_h = 0.0800$, $I_l = 0.0795$ and $i_N = 0.5714$
$$\begin{aligned} I_1 &= 0.0795 + 0.5714\,(0.0800 - 0.0795) \\ &= 0.0798 \end{aligned}$$

For $m_g = 3$ and $N_P = 35$, $I_h = 0.1154$, $I_l = 0.1121$ and $i_N = 0.5714$
$$\begin{aligned} I_2 &= 0.1121 + 0.5714\,(0.1154 - 0.1121) \\ &= 0.1140 \end{aligned}$$

$$I = I_l + i_m(I_h - I_l) \tag{4a}$$

For $m_g = 2.25$ and $N_P = 35$, $I_l = 0.0798$, $I_h = 0.1140$ and $i_m = 0.8333$
$$\begin{aligned} I &= 0.0798 + 0.8333\,(0.1140 - 0.0798) \\ &= 0.1083 \end{aligned}$$

*Determination of form factor J:*

$$J = J_l + i_N(J_h - J_l) \tag{4b}$$

For $m_g = 1$ and $N_P = 35$, $J_l = 0.395$, $J_h = 0.428$ and $i_N = 0.5714$
$$\begin{aligned} J_1 &= 0.395 + 0.5714\,(0.428 - 0.395) \\ &= 0.414 \end{aligned}$$

For $m_g = 3$ and $N_P = 35$, $J_l = 0.419$, $J_h = 0.452$ and $i_N = 0.5714$
$$\begin{aligned} J_2 &= 0.419 + 0.5714\,(0.452 - 0.419) \\ &= 0.438 \end{aligned}$$

$$J = J_l + i_m(J_h - J_l) \tag{4b}$$

For $m_g = 2.25$ and $N_P = 35$, $J_l = 0.414$, $J_h = 0.438$ and $i_m = 0.8333$
$$\begin{aligned} J &= 0.414 + 0.8333\,(0.438 - 0.414) \\ &= 0.434 \end{aligned}$$

*Determination of allowable stresses $s_t$ and $s_c$*

$$\begin{aligned} \log s_t &= \frac{3}{4}\log H_b - 0.301 - \log K \tag{5} \\[4pt] &= \frac{3}{4}\log 340 - 0.301 - \log 1.4 \\[4pt] &= 1.4515 \end{aligned}$$

$$s_t = 28.28, \text{ say } 28 \text{ kpsi}$$

$$s_c = \sqrt{\frac{s_t H_b}{K_m}} \tag{6}$$

$$= \sqrt{\frac{28 \times 340}{0.6}} = \sqrt{15,870}$$

$$s_c = 126 \text{ kpsi}$$

*Determination of Gear Size and Tooth Numbers:*

$$FD_P{}^2 = \frac{10.58 T_P}{I s_c{}^2} \tag{7}$$

$$= \frac{10.58 \times 1300}{0.1083 \times 15,870}$$

$$= 8.00 \text{ in.}^3$$

$$N_P = \frac{s_t (FD_P{}^2) J_P \times 10^3}{2 T_P} \tag{8}$$

$$= \frac{28 \times 8.00 \times 0.434 \times 1000}{2 \times 1300}$$

$$= 37.4, \text{ say, } 37 \text{ teeth}$$

$$N_G = N_P m_g \tag{9}$$

$$= 37 \times 2.25 = 83.25, \text{ say, } 83 \text{ teeth}$$

*Checking on actual ratio:*

$$m_g = \frac{N_G}{N_P} \tag{10}$$

$$m_g = \frac{83}{37} = 2.24$$

*Determination of pinion and gear pitch diameters:*

$$D_P = \frac{N_P}{P} \tag{11}$$

$$= \frac{37}{16} = 2.\overset{.}{3}125 \text{ in., diameter of pinion}$$

$$D_G = \frac{N_G}{P} \tag{11}$$

$$= \frac{83}{16} = 5.1875 \text{ in., diameter of gear}$$

*Determination of face width:*

$$F = \frac{FD_P{}^2}{D_P{}^2} \tag{17}$$

$$= \frac{8.00}{2.3125^2} = 1.50 \text{ in. wide}$$

The designer will need to use judgement in any particular case. This should not be difficult as long as it is remembered that the method described here will determine the maximum tooth numbers and the minimum size ($FD^2$) that must be used to satisfy the specified input criteria. Production cost is usually a function of the amount of metal removal; thus fewer teeth mean increased production cost. An increase in size ($FD^2$) means an increase in gear weight and, in most cases, an increase in gear-box or transmission weight; this, in turn, will mean an increase in production cost.

**British Standard Horsepower Formulas.** — The horsepower formulas which follow are included in the revised British standard specifications No. 436 — 1940, amended in 1941, 1943, and 1956, for machine cut spur gears and also for helical gears used in driving parallel shafts. Two formulas are given. One indicates the horsepower with reference to tooth strength; the other, the horsepower as limited by tooth wear. In deciding upon the horsepower capacity formula, the Committee gave special consideration to the question whether the speed factor should be based on revolutions per minute or pitch-line speed in feet per minute, and the former was adopted. The power capacity of both gear and pinion should be checked (1) for tooth wear, and (2) for tooth strength. The smallest of the four power ratings thus obtained should be used.

In the following formulas $S_c$ = surface stress factor (Table 1); $X_c$ = speed factor for wear (Table 2); $Z$ = zone factor (see chart, page 824); $F$ = face width, inches; $N$ = revolutions per minute; $T$ = number of teeth; $K$ = pitch factor = $P^{0.8}$; $P$ = diametral pitch; $S_b$ = bending stress (Table 1); $X_b$ = speed factor for strength (Table 3); $Y$ = strength factor (see chart, page 825). The wear and strength formulas follow:

$$\text{Horsepower for wear} = \frac{S_c X_c Z F N T}{126{,}000 K P}$$

$$\text{Horsepower for strength} = \frac{S_b X_b Y F N T}{126{,}000 P^2}$$

*Example:* Find the allowable horsepower for spur gears. The pinion and gear speeds are 500 and 100 revolutions per minute; continuous operation 12 hours per day; diametral pitch, 3; face width, 4 inches; pressure angle 20 degrees; pinion, 20 teeth; pinion material, 0.40% carbon steel normalized; gear, 100 teeth; gear material, cast iron of ordinary grade.

Horsepower formulas for wear and for strength are applied to both gear and pinion as shown below and the smallest horsepower rating (approximately 40 in this case) is used.

$$\text{Pinion H.P. for wear} = \frac{1600 \times 0.305 \times 2.20 \times 4 \times 500 \times 20}{126{,}000 \times 2.40 \times 3} = 47$$

$$\text{Gear H.P. for wear} = \frac{1000 \times 0.410 \times 2.20 \times 4 \times 100 \times 100}{126{,}000 \times 2.40 \times 3} = 40$$

$$\text{Pinion H.P. for strength} = \frac{19{,}000 \times 0.305 \times 0.72 \times 4 \times 500 \times 20}{126{,}000 \times 3 \times 3} = 147$$

$$\text{Gear H.P. for strength} = \frac{5800 \times 0.410 \times 0.61 \times 4 \times 100 \times 100}{126{,}000 \times 3 \times 3} = 51$$

Hence, the power rating would be 40 horsepower, assuming sufficient gear usage to warrant a wear rating instead of a tooth breakage rating.

### Table 1. Basic Surface and Bending-Stress Factors of Spur and Helical Gears

Factors for use in British Standard Horsepower Formula

| Type of Material (Numbers in Parentheses Indicate Footnotes) | Minimum Tensile Strength Tons per Sq. In. | Minimum Brinell Hardness Number | Surface Stress Factor $S_c$ | Bending Stress Lb. per Sq. In. $S_b$ |
|---|---|---|---|---|
| Fabric............................ | .. | .. | 560 | 4,500 |
| Cast Iron, Ordinary Grade........... | 12 | 165 | 1,000 | 5,800 |
| "　"　, Medium Grade........... | 16 | 210 | 1,350 | 7,600 |
| "　"　, High Grade, as Cast....... | 22 | 220 | 1,450 | 10,400 |
| Castings, Malleable................. | 20 | 140 | 850 | 11,000 |
| Phosphor Bronze, Sand Cast........ | 12 | 69 | 700 | 7,000 |
| "　"　, Chill Cast......... | 15 | 82 | 850 | 8,500 |
| "　"　, Centrifugally Cast.. | 17 | 90 | 1,000 | 10,000 |
| Cast Steel, 0.35% to 0.45% Carbon... | 35 | 145 | 1,400 | 19,000 |
| "　"　, 0.50% to 0.55% Carbon (1) | 38 | 160 (3) | 3,100 | 13,000 |
| Forged Carbon Steel, 0.15% Carbon (2) | 32 | 140 (4) | 9,000 | 28,000 |
| "　"　"　, 0.40% Carbon (6) | 35 | 145 | 1,400 | 17,000 |
| "　"　"　, 0.40% Carbon (7) | 35 | 145 | 1,600 | 19,000 |
| "　"　"　, 0.40% Carbon (1) | 35 | 145 (8) | 2,800 | 12,000 |
| "　"　"　, 0.40% Carbon (9) | 40 | 175 | 1,800 | 20,000 |
| "　"　"　, 0.40% Carbon(10) | 40 | 175 | 2,000 | 22,000 |
| "　"　"　, 0.55% Carbon (6) | 45 | 200 | 2,000 | 21,600 |
| "　"　"　, 0.55% Carbon (7) | 45 | 200 | 2,300 | 24,000 |
| "　"　"　, 0.55% Carbon (1) | 45 | 200 (11) | 4,000 | 15,000 |
| Nickel Steel, 1% nickel..........(12) | 40 | 175 | 2,000 | 22,000 |
| "　"　, 3% nickel..........(12) | 45 | 200 | 2,300 | 24,000 |
| "　"　, 3% nickel......... (2) | 45 | 200 (13) | 10,200 | 40,000 |
| "　"　, 3½% nickel........ (1) | 55 | 250 (14) | 5,100 | 18,500 |
| "　"　, 3½% nickel........(12) | 55 | 250 | 3,000 | 30,000 |
| "　"　, 3½% nickel........ (2) | 45 | 200 (15) | 10,200 | 40,000 |
| "　"　, 5% nickel.......... (2) | 55 | 250 (16) | 11,200 | 47,000 |
| Nickel-chromium, 1½% Ni, 1% Cr(17) | 55 | 250 | 3,000 | 30,000 |
| "　"　, 1½% Ni, 1% Cr (1) | 55 | 250 (13) | 5,100 | 18,500 |
| "　"　, 1½%..........(17) | 100 | 440 | 5,500 | 40,000 |
| "　"　, 3½%..........(17) | 55 | 250 | 3,000 | 30,000 |
| "　"　, 3½%.......... (1) | 55 | 250 (14) | 5,100 | 18,500 |
| "　"　, 3½%.......... (2) | 55 | 250 (16) | 11,200 | 47,000 |
| Carbon-chromium, 0.55% carbon..(18) | 55 | 250 | 3,000 | 30,000 |
| "　"　, 0.55% carbon..(18) | 65 | 290 | 3,500 | 36,000 |
| "　"　, 0.55% carbon.. (1) | 55 | 250 (14) | 5,100 | 18,500 |

(1) Surface Hardened; (2) Casehardened; (3) Core, 160; case, 530; (4) Hardness of core; (5) Core, 140; case, 640; (6) normalized; for sections thicker than 5 inches; (7) normalized; for sections less than 5 inches thick; (8) Core, 145; case, 460; (9) Heat-treated; for sections thicker than 5 inches; (10) Heat-treated; for sections less 5 inches thick; (11) Core, 200; case, 520; (12) Heat-treated; (13) Core; (14) Core, 250; case, 500; (15) Core, 200; case, 620; (16) Core, 250; case, 600; (17) Oil hardened and tempered to strength given in second col.; (18) Heat-treated to strength given in second column.

Table 2. Speed Factors $X_c$ for Wear

| Rev. per Minute | Running Time — Hours per Day | | | | | | | |
|---|---|---|---|---|---|---|---|---|
| | 1 | 2 | 4 | 6 | 8 | 12 | 18 | 24 |
| | Speed Factors $X_c$ for Wear | | | | | | | |
| 100 | 0.935 | 0.735 | 0.585 | 0.515 | 0.470 | 0.410 | 0.350 | 0.320 |
| 150 | 0.865 | 0.685 | 0.540 | 0.475 | 0.435 | 0.370 | 0.330 | 0.300 |
| 200 | 0.825 | 0.650 | 0.520 | 0.460 | 0.415 | 0.360 | 0.310 | 0.280 |
| 300 | 0.775 | 0.615 | 0.485 | 0.425 | 0.380 | 0.330 | 0.290 | 0.270 |
| 400 | 0.730 | 0.580 | 0.460 | 0.400 | 0.360 | 0.320 | 0.270 | 0.250 |
| 500 | 0.700 | 0.550 | 0.440 | 0.380 | 0.350 | 0.305 | 0.260 | 0.240 |
| 600 | 0.680 | 0.530 | 0.425 | 0.370 | 0.340 | 0.290 | 0.250 | 0.230 |
| 800 | 0.635 | 0.500 | 0.400 | 0.350 | 0.320 | 0.270 | 0.240 | 0.220 |
| 1,000 | 0.610 | 0.480 | 0.380 | 0.335 | 0.305 | 0.260 | 0.230 | 0.210 |
| 1,500 | 0.550 | 0.440 | 0.345 | 0.310 | 0.275 | 0.240 | 0.210 | 0.190 |
| 2,000 | 0.520 | 0.415 | 0.325 | 0.290 | 0.260 | 0.220 | 0.200 | 0.180 |
| 2,500 | 0.480 | 0.380 | 0.305 | 0.265 | 0.240 | 0.210 | 0.185 | 0.165 |
| 3,000 | 0.450 | 0.355 | 0.280 | 0.250 | 0.225 | 0.195 | 0.170 | 0.155 |
| 4,000 | 0.415 | 0.325 | 0.260 | 0.225 | 0.207 | 0.180 | 0.155 | 0.145 |
| 5,000 | 0.380 | 0.305 | 0.240 | 0.210 | 0.190 | 0.165 | 0.145 | 0.132 |
| 6,000 | 0.355 | 0.285 | 0.225 | 0.200 | 0.180 | 0.155 | 0.135 | 0.125 |
| 7,000 | 0.340 | 0.270 | 0.215 | 0.190 | 0.170 | 0.150 | 0.130 | 0.118 |
| 8,000 | 0.325 | 0.260 | 0.205 | 0.180 | 0.165 | 0.142 | 0.125 | 0.113 |
| 9,000 | 0.315 | 0.250 | 0.200 | 0.175 | 0.157 | 0.135 | 0.120 | 0.108 |
| 10,000 | 0.305 | 0.240 | 0.190 | 0.165 | 0.152 | 0.130 | 0.115 | 0.105 |

Table 3. Speed Factors $X_b$ for Strength

| Running Time, Hours per Day | Revolutions per Minute | | | | | | | | | |
|---|---|---|---|---|---|---|---|---|---|---|
| | 100 | 150 | 200 | 300 | 400 | 500 | 600 | 800 | 1000 | 1500 |
| | Speed Factors $X_b$ for Strength | | | | | | | | | |
| 1 | 0.600 | 0.550 | 0.525 | 0.445 | 0.435 | 0.420 | 0.415 | 0.410 | 0.385 | 0.350 |
| 3 | 0.510 | 0.435 | 0.425 | 0.410 | 0.400 | 0.380 | 0.370 | 0.345 | 0.330 | 0.300 |
| 6 | 0.430 | 0.415 | 0.405 | 0.380 | 0.360 | 0.345 | 0.330 | 0.310 | 0.295 | 0.275 |
| 12 | 0.410 | 0.380 | 0.360 | 0.340 | 0.320 | 0.310 | 0.300 | 0.285 | 0.270 | 0.245 |
| 24 | 0.375 | 0.350 | 0.330 | 0.310 | 0.295 | 0.285 | 0.275 | 0.255 | 0.245 | 0.225 |

| Running Time, Hours per Day | Revolutions per Minute | | | | | | | | | |
|---|---|---|---|---|---|---|---|---|---|---|
| | 2000 | 2500 | 3000 | 4000 | 5000 | 6000 | 7000 | 8000 | 9000 | 10,000 |
| | Speed Factors $X_b$ for Strength | | | | | | | | | |
| 1 | 0.325 | 0.305 | 0.285 | 0.260 | 0.240 | 0.225 | 0.215 | 0.208 | 0.200 | 0.192 |
| 3 | 0.285 | 0.260 | 0.245 | 0.225 | 0.208 | 0.195 | 0.185 | 0.178 | 0.170 | 0.165 |
| 6 | 0.255 | 0.235 | 0.220 | 0.200 | 0.185 | 0.175 | 0.165 | 0.160 | 0.153 | 0.148 |
| 12 | 0.230 | 0.215 | 0.200 | 0.182 | 0.168 | 0.158 | 0.150 | 0.145 | 0.140 | 0.135 |
| 24 | 0.210 | 0.195 | 0.180 | 0.165 | 0.152 | 0.143 | 0.138 | 0.130 | 0.126 | 0.120 |

Table 4. Pitch Factors $K$

| Diametral Pitch | Factor $K$ | Diametral Pitch | Factor $K$ | Diametral Pitch | Factor $K$ | Diametral Pitch | Factor $K$ |
|---|---|---|---|---|---|---|---|
| 1 | 1.00 | 2¼ | 1.90 | 4 | 3.05 | 9 | 5.80 |
| 1¼ | 1.20 | 2½ | 2.10 | 5 | 3.65 | 10 | 6.40 |
| 1½ | 1.40 | 2¾ | 2.25 | 6 | 4.25 | 12 | 7.40 |
| 1¾ | 1.55 | 3 | 2.40 | 7 | 4.80 | 14 | 8.30 |
| 2 | 1.75 | 3½ | 2.70 | 8 | 5.40 | 16 | 9.25 |

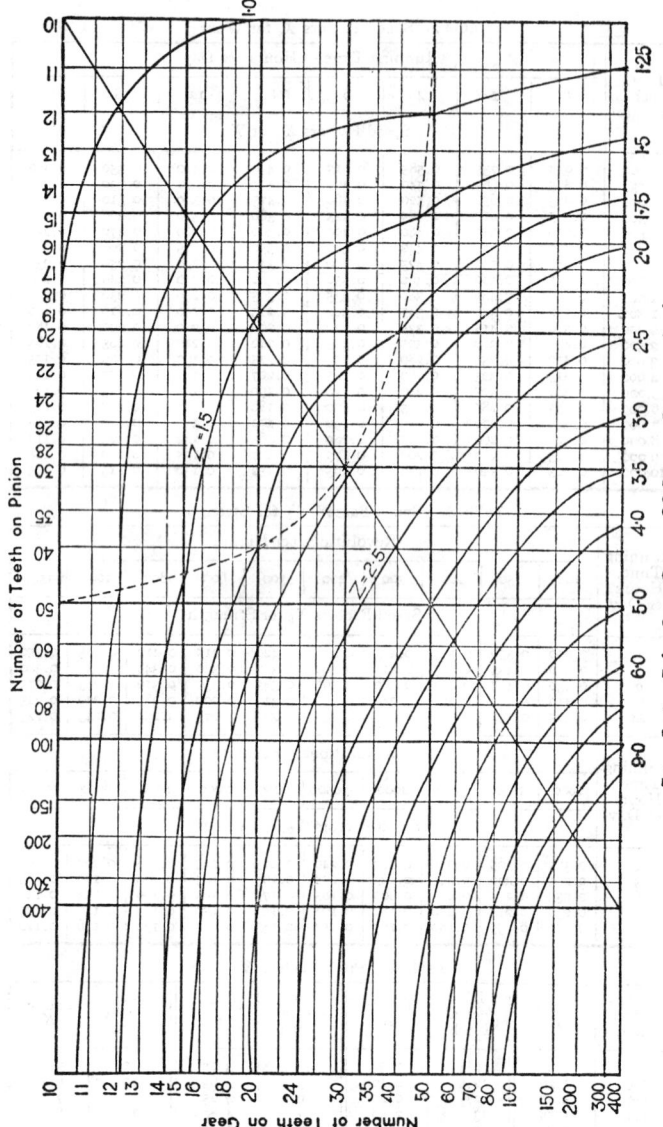

Zone Factor Z for Spur Gears—20-Degree Pressure Angle

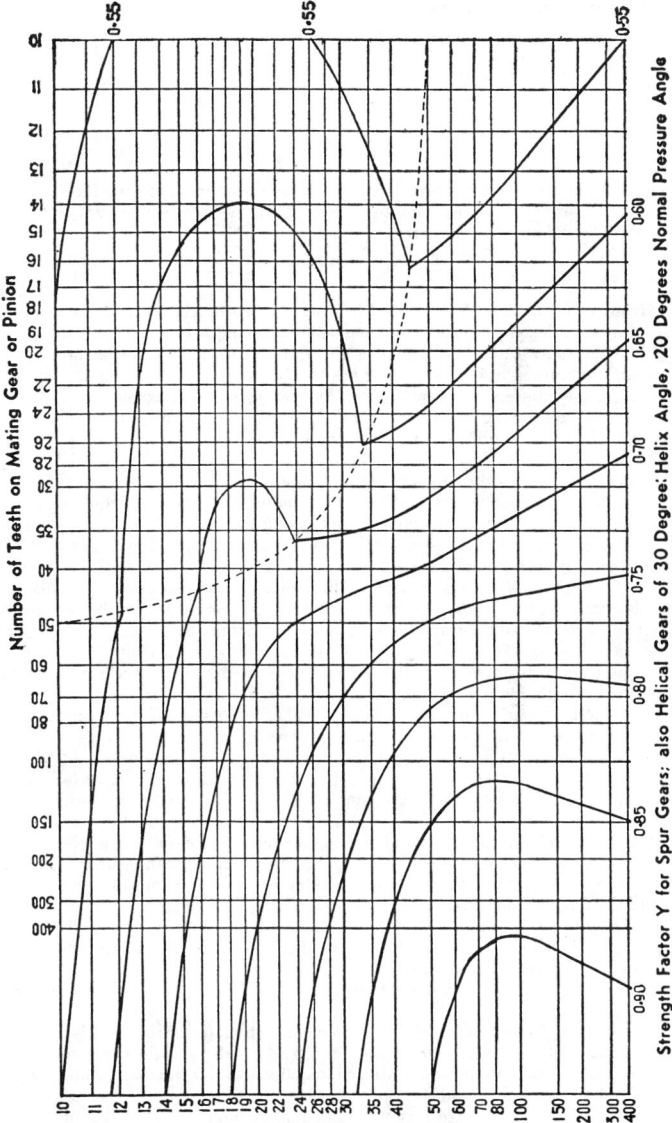

**Spur Gear Tooth and Bearing Loads.** — The tooth load on a gear is the load exerted in a direction normal to the tooth profile. The same load is transmitted to the bearings supporting the gear. To find the bearing load, first calculate the transmitted tangential load, $L$, in pounds:

$$L = \frac{P \times 63,000}{R \times r}$$

where $P$ = horsepower transmitted; $R$ = revolutions per minute of gear; and $r$ = pitch radius of gear, inches. Next, the tooth load, $L_T$ in pounds, which is of the same magnitude as the load on the bearings, is found:

$$L_T = L \div \cos \text{ pressure angle}$$

**Hunting Tooth Gear Ratios.** — When the numbers of teeth in a pair of meshing gears are such that they do not have a common divisor, the ratio of the tooth numbers is said to be a "hunting tooth" or non-factorizing ratio. Each gear in such a pair must make as many revolutions as its mating gear has teeth for a particular tooth on that gear to "hunt" or mesh once with every tooth on the mating gear. For example, with a pair having $^{73}/_{24}$ teeth, a particular tooth on the 73-tooth gear will mesh once with every tooth on the 24-tooth gear when the 73-tooth gear has made 24 revolutions. In the same way, a particular tooth on the 24-tooth gear will mesh once with every tooth on the 73-tooth gear in 73 revolutions of the 24-tooth gear.

When several pairs of gears operating at the same center distance are required to have hunting ratios, this can be accomplished by having the sum of the teeth in each pair equal to a prime number. Thus $^{77}/_{20}$, $^{76}/_{21}$, $^{75}/_{19}$, etc., are all hunting ratios and each pair sum is 97, a prime number.

*Origin of Hunting Tooth Use:* Whenever the number of teeth on a gear is equal to some whole number multiplied by the number of teeth on the mating pinion, then any tooth on the gear meshes with the same tooth on the pinion during each revolution of the gear. For example, with a $^{72}/_{24}$ pair, one tooth on the 24-tooth pinion will always mesh with the same three teeth on the gear. Changing the ratio to $^{73}/_{24}$ produces the hunting tooth condition described previously.

The practice of using "hunting" ratios was common when gears of high accuracy were not available as was the case with the cast iron gears used by millwrights. The theory was that, with a hunting tooth, gear wear would be evenly distributed among all the teeth and they would eventually wear into some indefinite but comparatively true (conjugate) tooth form which operated more quietly and lasted longer than gears of even ratios.

*When to Use Hunting Ratios:* The following conditions favor hunting tooth ratios:

1. Both gears are hardened or hardened and ground. In this case the gears are too hard to wear in much, but use of the hunting tooth prevents high dynamic tooth loads from being confined to a single pair of teeth thus increasing gear life.

2. The gear pair is subject to high cyclic loads such as in a crankshaft. In this case a hunting ratio distributes the high loads to all the teeth evenly and prevents early breakage of certain teeth resulting when integral ratios are used.

*When Not to Use Hunting Ratios:* Hunting ratios should not be used in the following situations:

1. For multiple-start worm drives that require long life. If a hunting ratio is used, the hard material of the worm would continually shave particles from the softer worm gear throughout the life of the drive whereas an integral ratio would bed itself in after a short time, and further wear would be small.

2. When gears are repeatedly assembled and disassembled as in the case of change gears; it is of no importance whether they have a common factor or are hunting ratios.

3. If gears are finished by lapping as in the case of automotive bevel gears, even ratios may be used provided the gears are assembled meshed as they were in the lapping process. If this is not practical, then a hunting ratio is more satisfactory.

## Classification of Gear Steels.

Gear steels may be divided into two general classes — the plain carbon and the alloy steels. Alloy steels are used to some extent in the industrial field, but heat-treated plain carbon steels are far more common. The use of untreated alloy steels for gears is seldom, if ever, justified, and then, only when heat-treating facilities are lacking. The points to be considered in determining whether to use heat-treated plain carbon steels or heat-treated alloy steels are: Does the service condition or design require the superior characteristics of the alloy steels, or, if alloy steels are not required, will the advantages to be derived offset the additional cost? For most applications, plain carbon steels, heat-treated to obtain the best of their qualities for the service intended, are satisfactory and quite economical. The advantages obtained from using heat-treated alloy steels in place of heat-treated plain carbon steels are as follows:

1. Increased surface hardness and depth of hardness penetration for the same carbon content and quench.

2. Ability to obtain the same surface hardness with a less drastic quench and, in the case of some of the alloys, a lower quenching temperature, thus giving less distortion.

3. Increased toughness, as indicated by the higher values of yield point, elongation, and reduction of area.

4. Finer grain size, with the resulting higher impact toughness and increased wear resistance.

5. In the case of some of the alloys, better machining qualities or the possibility of machining at higher hardnesses.

## Use of Casehardening Steels.

Each of the two general classes of gear steels may be further subdivided as follows: (1) Casehardening steels; (2) full-hardening steels; and (3) steels that are heat-treated and drawn to a hardness that will permit machining. The first two — casehardening and full-hardening steels — are interchangeable for some kinds of service, and the choice is often a matter of personal opinion. Casehardening steels with their extremely hard, fine-grained (when properly treated) case and comparatively soft and ductile core are generally used when resistance to wear is desired. Casehardening alloy steels have a fairly tough core, but not as tough as that of the full-hardening steels. In order to realize the greatest benefits from the core properties, casehardened steels should be double-quenched. This is particularly true of the alloy steels, because the benefits derived from their use seldom justify the additional expense, unless the core is refined and toughened by a second quench. The penalty that must be paid for the additional refinement is increased distortion, which may be excessive if the shape or design is not all that it might be.

## Use of "Thru-Hardening" Steels.

Thru-hardening steels are used when great strength, high endurance limit, toughness, and resistance to shock are required. These qualities are governed by the kind of steel and treatment used. Fairly high surface hardnesses are obtainable in this group, though not so high as those of the casehardening steels. For that reason, the resistance to wear is not so great as might be obtained, but when wear resistance combined with great strength and toughness is required, this type of steel is superior to the others. Thru-hardening steels become distorted to some extent when hardened, the amount depending upon the steel and quenching medium used. For that reason, thru-hardening steels are not suitable for high-speed gearing where noise is a factor, or for gearing where

### Steels for Industrial Gearing

| CASE HARDENING STEELS | | | |
|---|---|---|---|
| Material Specification | Hardness | | Typical Heat Treatment, Characteristics and Uses |
| | Case $R_c$ | Core Bhn | |
| AISI 1020 AISI 1116 | 55–60 | 160–230 | Carburize, harden, temper at 350°F. For gears that must be wear resistant. Normalized material is easily machined. Core is ductile but has little strength. |
| AISI 4130 AISI 4140 | 50–55 | 270–370 | Harden, temper at 900°F, Nitride. For parts requiring greater wear resistance than that of thru hardened steels but cannot tolerate the distortion of carburizing. Case is shallow, core is tough. |
| AISI 4615 AISI 4620 AISI 8615 AISI 8620 | 55–60 55–60 | 170–260 200–300 | Carburize, harden, temper at 350°F. For gears requiring high fatigue resistance and strength. The 86xx series has better machinability. The 20 point steels are used for coarser teeth. |
| AISI 9310 | 58–63 | 250–350 | Carburize, harden, temper at 300°F. Primarily for aerospace gears that are highly loaded and operate at high pitch line velocity and for other gears requiring high reliability under extreme operating conditions. This material is not used at high temperature. |
| Nitralloy N and Type 135 Mod. (15-N) | 90–94 | 300–370 | Harden, temper at 1200°F, Nitride. For gears requiring high strength and wear resistance that cannot tolerate the distortion of the carburizing process or that operate at high temperatures. Gear teeth are usually finished before nitriding. Care must be exercised in running nitrided gears together to avoid crazing of case. |
| THRU-HARDENING STEELS | | | |
| AISI 1045 AISI 1140 | 24–40 | .... | Harden and temper to required hardness. Oil quench for lower hardness and water quench for higher hardness. For gears of medium and large size requiring moderate strength and wear resistance. Gears must have consistent, solid sections to withstand quenching. |
| AISI 4140 AISI 4340 | 24–40 | .... | Harden (oil quench), temper to required hardness. For gears requiring high strength and wear resistance, and high shock loading resistance. Use 41xx series for moderate sections and 43xx series for heavy sections. Gears must have consistent, solid sections to withstand quenching. |

accuracy is of paramount importance, except, of course, in cases where grinding of the teeth is practicable. The medium and high-carbon percentages require an oil quench, but a water quench may be necessary for the lower carbon contents, in order to obtain the highest physical properties and hardness. The distortion, however, will be greater with the water quench.

**Heat-Treatment that Permits Machining.** — When the grinding of gear teeth is not practicable and a high degree of accuracy is required, hardened steels may be drawn or tempered to a hardness that will permit the cutting of the teeth. This treatment gives a highly refined structure, great toughness, and, in spite of the low hardness, excellent wearing qualities. The lower strength is somewhat compensated for by the elimination of the increment loads due to the impacts which are caused by inaccuracies. When steels that have a low degree of hardness penetration from surface to core are treated in this manner, the design cannot be based on the physical properties corresponding to the hardness at the surface. Since the physical properties are determined by the hardness, the drop in hardness from surface to core will give lower physical properties at the root of the tooth, where the stress is greatest. The quenching medium may be either oil, water, or brine, depending on the steel used and hardness penetration desired. The amount of distortion, of course, is immaterial, because the machining is done after heat-treating.

**Making Pinion Harder than Gear to Equalize Wear.** — Beneficial results from a wear standpoint are obtained by making the pinion harder than the gear. The pinion, having a lesser number of teeth than the gear, naturally does more work per tooth, and the differential in hardness between the pinion and the gear (the amount being dependent on the ratio) serves to equalize the rate of wear. The harder pinion teeth correct the errors in the gear teeth to some extent by the initial wear and then seem to burnish the teeth of the gear and increase its ability to withstand wear by the greater hardness due to the cold-working of the surface. In applications where the gear ratio is high and there are no severe shock loads, a casehardened pinion running with an oil-treated gear, treated to a Brinell hardness at which the teeth may be cut after treating, is an excellent combination. The pinion, being relatively small, is distorted but little, and distortion in the gear is circumvented by cutting the teeth after treatment.

**Forged and Rolled Carbon Steels for Gears.** — These compositions cover steel for gears in three groups, according to heat treatment, as follows: (a) case-hardened gears, (b) unhardened gears, not heat treated after machining, and (c) hardened and tempered gears.

Forged and rolled carbon gear steels are purchased on the basis of the requirements as to chemical composition specified in Table 1. Class N steel will normally be ordered in ten point carbon ranges within these limits. Requirements as to physical properties have been omitted, but when they are called for the requirements as to carbon shall be omitted. The steels may be made by either or both the open hearth and electric furnace processes.

**Forged and Rolled Alloy Steels for Gears.** — These compositions cover alloy steel for gears, in two classes according to heat treatment, as follows: (a) case-hardened gears, and (b) hardened and tempered gears. Forged and rolled alloy gear steels are purchased on the basis of the requirements as to chemical composition specified in Table 2. Requirements as to physical properties have been omitted. The steel shall be made by either or both the open hearth and electric furnace process.

**Steel Castings for Gears.** — It is recommended that steel castings for cut gears be purchased on the basis of chemical analysis and that only two types of analysis be used, one for case-hardened gears and the other for both untreated gears and those which are to be hardened and tempered. The steel is to be made by the open hearth, crucible or electric furnace processes. The converter process is not recognized. Sufficient risers shall be provided to secure soundness and freedom from undue segregation. Risers shall not be broken off the unannealed castings by force. Where risers are cut off with a torch the cut shall be at least one-half inch above the surface of the castings, and the remaining metal removed by chipping, grinding, or other non-injurious method.

Steel for use in gears shall conform to the requirements, as to chemical composition as indicated in Table 3. All steel castings for gears must be thoroughly normalized or annealed, using such temperature and time as will entirely eliminate the characteristic structure of unannealed castings.

**Effect of Alloying Metals on Gear Steels.** — The effect of the various alloying elements on steel will be summarized in order to assist engineers in deciding upon the particular kind of alloy steel to use for specific purposes. The characteristics, outlined apply only to heat-treated steels. When the effect of the addition of an alloying element is stated, it is understood that reference is made to alloy steels of a given carbon content, compared with a plain carbon steel of the same carbon content.

*Nickel* — The addition of nickel tends to increase the hardness and strength, with but little sacrifice of ductility. The hardness penetration is somewhat greater than that of plain carbon steels. Its use as an alloying element lowers the critical points and produces less distortion, due to the lower quenching temperature. The nickel steels of the case-hardening group carburize more slowly, but the grain growth is less.

*Chromium* — Chromium increases the hardness and strength over that obtained by the use of nickel, though the loss of ductility is greater. Chromium refines the grain and imparts a greater depth of hardness. Chromium steels have a high degree of wear resistance and are easily machined in spite of the fine grain.

*Manganese* — When present in sufficient amounts to warrant the use of the term alloy, the addition of manganese is very effective. It gives greater strength than nickel, and a higher degree of toughness than chromium. Owing to its susceptibility to cold-working, it is likely to flow under severe unit pressures. Up to the present time, it has never been used to any great extent for heat-treated gears, but is now receiving an increasing amount of attention.

*Vanadium* — Vanadium has a similar effect to that of manganese — increasing the hardness, strength, and toughness. The loss of ductility is somewhat more than that due to manganese, but the hardness penetration is greater than for any of the other alloying elements. Owing to the extremely fine-grained structure, the impact strength is high; but vanadium tends to make machining difficult.

*Molybdenum* — Molybdenum has the property of increasing the strength without affecting the ductility. For the same hardness, steels containing molybdenum are more ductile than any other alloy steels, and having nearly the same strength, are tougher; in spite of the increased toughness, the presence of molybdenum does not make machining more difficult. In fact, such steels can be machined at a higher hardness than any of the other alloy steels. The impact strength is nearly as great as that of the vanadium steels.

*Chrome-Nickel Steels* — The combination of the two alloying elements chromium and nickel adds the beneficial qualities of both. The high degree of ductility present in nickel steels is complemented by the high strength, finer grain size, deep hardening and wear-resistant properties imparted by the addition of chromium. The increased toughness makes these steels more difficult to machine than the plain carbon steels, and they are more difficult to heat-treat. The distortion increases with the amount of chromium and nickel.

Table 1. Compositions of Forged and Rolled Carbon Steels for Gears

| Heat-treatment | Class | Carbon | Manganese | Phosphorus | Sulphur |
|---|---|---|---|---|---|
| Case-hardened........ | C | 0.15–0.25 | 0.40–0.70 | 0.045 max | 0.055 max |
| Untreated............ | N | 0.25–0.50 | 0.50–0.80 | 0.045 max | 0.055 max |
| Hardened............ (or untreated) | H | 0.40–0.50 | 0.40–0.70 | 0.045 max | 0.055 max |

Table 2. Compositions of Forged and Rolled Alloy Steels for Gears

| Steel Specification | Chemical Composition* | | | | | |
|---|---|---|---|---|---|---|
| | C | Mn | Si | Ni | Cr | Mo |
| AISI 4130 | .28–.30 | .40– .60 | .20–.35 | .... | .80–1.1 | .15–.25 |
| AISI 4140 | .38–.43 | .75–1.0 | .20–.35 | .... | .80–1.1 | .15–.25 |
| AISI 4340 | .38–.43 | .60– .80 | .20–.35 | 1.65–2.0 | .70– .90 | .20–.30 |
| AISI 4615 | .13–.18 | .45– .65 | .20–.35 | 1.65–2.0 | .... | .20–.30 |
| AISI 4620 | .17–.22 | .46– .65 | .20–.35 | 1.65–2.0 | .... | .20–.30 |
| AISI 8615 | .13–.18 | .70– .90 | .20–.35 | .40– .70 | .40– .60 | .15–.25 |
| AISI 8620 | .18–.23 | .70– .90 | .20–.35 | .40– .70 | .40– .60 | .15–.25 |
| AISI 9310 | .08–.13 | .45– .65 | .20–.35 | 3.0 –3.5 | 1.0 –1.4 | .08–.15 |
| Nitralloy | | | | | | |
| Type N† | .20–.27 | .40– .70 | .20–.40 | 3.2 –3.8 | 1.0 –1.3 | .20–.30 |
| 135 Mod.† | .38–.45 | .40– .70 | .20–.40 | .... | 1.4 –1.8 | .30–.45 |

* C = carbon; Mn = manganese; Si = silicon; Ni = nickel; Cr = chromium, and Mo = Molybdenum.
† Both Nitralloy alloys contain Aluminum .85–1.2%.

Table 3. Compositions of Cast Steels for Gears

| Steel Specification | Chemical Composition* | | |
|---|---|---|---|
| | C | Mn | Si |
| SAE-0022 | .12–.22 | .50–.90 | .60 Max. | May be carburized |
| SAE-0050 | .40–.50 | .50–.90 | .80 Max. | Hardenable 210–250 |

* C = carbon; Mn = manganese; and Si = silicon.

*Chrome-Vanadium Steels* — Chrome-vanadium steels have practically the same tensile properties as the chrome-nickel steels, but the hardening power, impact strength, and wear resistance are increased by the finer grain size. They are difficult to machine and become distorted more easily than the other alloy steels.

*Chrome-Molybdenum Steels* — This group has the same qualities as the straight molybdenum steels, but the hardening depth and wear resistance are increased by the addition of chromium. This steel is very easily heat-treated and machined.

*Nickel-Molybdenum Steels* — Nickel-molybdenum steels have qualities similar to chrome-molybdenum steel. The toughness is said to be greater, but the steel is somewhat more difficult to machine.

**Sintered Materials.** — For high production of low and moderately loaded gears, significant production cost savings may be effected by the use of a sintered metal powder. With this material, the gear is formed in a die under high pressure and then sintered in a furnace. The primary cost saving comes from the great reduction in labor cost of machining the gear teeth and other gear blank surfaces. The volume of production must be high enough to amortize the cost of the die and the gear blank must be of such a configuration that it may be formed and readily ejected from the die.

**Bronze and Brass Gear Castings.** — These specifications cover non-ferrous metals for spur, bevel, and worm gears, bushings and flanges for composition gears. This material shall be purchased on the basis of chemical composition. The alloys may be made by any approved method.

*Spur and Bevel Gears:* For spur and bevel gears, hard cast bronze is recommended (A.S.T.M. B–10–18; S.A.E. No. 62; and the well-known 88–10–2 mixture) with the following limits as to composition: Copper, 86 to 89; tin, 9 to 11; zinc, 1 to 3; lead (max), 0.20; iron (max), 0.06 per cent. Good castings made from this bronze should have the following minimum physical characteristics: Ultimate strength, 30,000 pounds per square inch; yield point, 15,000 pounds per square inch; elongation in 2 inches, 14 per cent.

*Worm Gears:* For bronze worm gears, two alternative analyses of phosphor bronze are recommended, S.A.E. No. 65 and No. 63.

S.A.E. No. 65 (called phosphor gear bronze) has the following composition: Copper, 88 to 90; tin, 10 to 12; phosphorus, 0.1 to 0.3; lead, zinc and impurities (max), 0.5 per cent. Good castings made of this alloy should have the following minimum physical characteristics: Ultimate strength, 35,000 pounds per square inch; yield point, 20,000 pounds per square inch; elongation in 2 inches, 10 per cent.

The composition of S.A.E. No. 63 (called leaded gun metal) follows: Copper, 86 to 89; tin, 9 to 11; lead, 1 to 2.5; phosphorus (max), 0.25; zinc and impurities (max), 0.50 per cent.

Good castings made of this alloy should have the following minimum physical characteristics: Ultimate strength, 30,000 pounds per square inch; yield point, 12,000 pounds per square inch; elongation in 2 inches, 10 per cent.

These alloys, especially No. 65, are adapted to chilling for hardness and refinement of grain. No. 65 is to be preferred for use with worms of great hardness and fine accuracy. No. 63 is to be preferred for use with unhardened worms.

*Gear Bushings:* For bronze bushings for gears, S.A.E. No. 64 is recommended of the following analysis: Copper, 78.5 to 81.5; tin, 9 to 11; lead, 9 to 11; phosphorus, 0.05 to 0.25; zinc (max), 0.75; other impurities (max), 0.25 per cent. Good castings of this alloy should have the following minimum physical characteristics: Ultimate strength, 25,000 pounds per square inch; yield point, 12,000 pounds per square inch; elongation in 2 inches, 8 per cent.

*Flanges for Composition Pinions:* For brass flanges for composition pinions A.S.T.M. B-30-32T, and S.A.E. No. 40 are recommended. This is a good cast red brass of sufficient strength and hardness to take its share of load and wear when the design is such that the flanges mesh with the mating gear. The composition is as follows: Copper, 83 to 86; tin, 4.5 to 5.5; lead, 4.5 to 5.5; zinc, 4.5 to 5.5 iron (max) 0.35; antimony (max), 0.25 per cent; aluminum, none. Good castings made from this alloy should have the following minimum physical characteristics: Ultimate strength, 27,000 pounds per square inch; yield point, 12,000 pounds per square inch; elongation in 2 inches, 16 per cent.

**Materials for Worm Gearing.** — The Hamilton Gear & Machine Co. conducted an extensive series of tests on a variety of materials that might be used for worm gears, to ascertain which material is the most suitable. According to these tests chill-cast nickel-phosphor-bronze ranks first in resistance to wear and deformation. This bronze is composed of approximately 87.5 per cent copper, 11 per cent tin, 1.5 per cent nickel, with from 0.1 to 0.2 per cent phosphorus. The worms used in these tests were made from S.A.E.-2315, 3½ per cent nickel steel, casehardened, ground, and polished. The Shore scleroscope hardness of the worms was between 80 and 90. This nickel alloy steel was adopted after numerous tests of a variety of steels, because it provided the necessary strength, together with the degree of hardness required.

The material that showed up second best in these tests was a No. 65 S.A.E. bronze. Navy bronze (88-10-2) containing 2 per cent of zinc, with no phosphorus, and not chilled, performed satisfactorily at speeds of 600 revolutions per minute, but was not sufficiently strong at lower speeds. Red brass (85-5-5) proved slightly better at from 1500 to 1800 revolutions per minute, but would bend at lower speeds, before it would show actual wear.

**Non-metallic Gearing.** — Non-metallic or composition gearing is used primarily where quietness of operation at high speed is the first consideration. Non-metallic materials are also applied very generally to timing gears and numerous other classes of gearing. Rawhide was used originally for non-metallic gears, but other materials have been introduced which have important advantages. These later materials are sold by different firms under various trade names, such as Micarta, Textolite, Formica, Dilecto, Spauldite, Phenolite, Fibroc, Fabroil, Synthane, Celoron, etc. Most of these gear materials consist of layers of canvas which is impregnated with bakelite and forced together under hydraulic pressure, which, in conjunction with the application of heat, forms a dense rigid mass.

Although phenol resin gears in general are resilient, they are self-supporting and require no side plates or shrouds unless subjected to a heavy starting torque. The phenol resinoid element makes these gears proof against vermin and rodents.

The non-metallic gear materials referred to are generally assumed to have the power-transmitting capacity of cast iron. While the tensile strength may be considerably less than that of cast iron, the resiliency of these materials enables them to withstand impact and abrasion to a degree that might result in excessive wear of cast-iron teeth. Thus in many cases, composition gearing of impregnated canvas has proved to be more durable than cast iron and much more durable than rawhide.

**Application of Non-metallic Gears.** — The most effective field of use for these non-metallic materials is for high-speed duty. At low speeds, where the starting torque may be high, or where the load may fluctuate widely, or when high shock loads may be encountered, these non-metallic materials do not always prove satisfactory. In general, non-metallic materials should not be used for pitch-line velocities below 600 feet per minute.

*Tooth Form:* The best tooth form for non-metallic materials is the 20-degree stub-tooth system. When only a single pair of gears is involved and the center distance can be varied, the best results will be obtained by making the non-metallic driving pinion of all-addendum form, while the driven metal gear is made with standard tooth proportions. Such a drive will carry from 50 to 75 per cent greater loads than one of standard tooth proportions.

*Material for Mating Gear:* For durability under load, the use of hardened steel (over 400 Brinell) for the mating metal gear appears to give the best results. A good second choice for the material of the mating member is cast iron. The use of brass, bronze, or soft steel (under 400 Brinell) as a material for the mating member of phenolic laminated gears leads to excessive abrasive wear.

**Power-transmitting Capacity of Non-metallic Gears.** — The characteristics of gears made of phenolic laminated materials are so different from those of metal gears that they should be considered in a class by themselves. Because of the low modulus of elasticity, most of the effects of small errors in tooth form and spacing are absorbed at the tooth surfaces by the elastic deformation, and have but little effect on the strength of the gears.

If

$S$ = safe working stress for a given velocity

$S_s$ = allowable static stress

$V$ = pitch-line velocity in feet per minute

then, according to the recommended practice of the American Gear Manufacturers' Association,

$$S = S_s \times \left( \frac{150}{200 + V} + 0.25 \right)$$

The value of $S_s$ for phenolic laminated materials is given as 6000 pounds per square inch. The accompanying table gives the safe working stresses $S$ for different pitch-line velocities. When the value of $S$ is known, the horsepower capacity is determined by substituting the value of $S$ for $s_t$ in Equation 3 on page 811 and solving for torque $T_1$. Then H.P. $= T_1 \times$ R.P.M. $\div 63,000$.

Safe Working Stresses for Non-metallic Gears

| Pitch-line Velocity Feet per Minute $V$ | Safe Working Stresses | Pitch-line Velocity Feet per Minute $V$ | Safe Working Stresses | Pitch-line Velocity Feet per Minute $V$ | Safe Working Stresses |
|---|---|---|---|---|---|
| 600 | 2625 | 1800 | 1950 | 4000 | 1714 |
| 700 | 2500 | 2000 | 1909 | 4500 | 1691 |
| 800 | 2400 | 2200 | 1875 | 5000 | 1673 |
| 900 | 2318 | 2400 | 1846 | 5500 | 1653 |
| 1000 | 2250 | 2600 | 1821 | 6000 | 1645 |
| 1200 | 2143 | 2800 | 1800 | 6500 | 1634 |
| 1400 | 2063 | 3000 | 1781 | 7000 | 1622 |
| 1600 | 2000 | 3500 | 1743 | 7500 | 1617 |

The tensile strength of the phenolic laminated materials used for gears, is slightly less than that of cast iron. These materials are far softer than any metal, and the modulus of elasticity is about one-thirtieth that of steel. In other words, if the tooth load on a steel gear which causes a deformation of 0.001 inch were applied to the tooth of a similar gear made of phenolic laminated material, the tooth of the non-metallic gear would be deformed about ½₂ inch. Under these conditions, several things will happen. With all gears, regardless of the theoretical duration of contact, one tooth only will carry the load until the load is sufficient to deform the tooth the amount of the error that may be present. On metal gears, when the tooth has been deformed the amount of the error, the stresses set up in the materials may approach or exceed the elastic limit of the material. Hence for standard tooth forms and those generated from standard basic racks, it is dangerous to calculate their strength as very much greater than that which can safely be carried on a single tooth. On gears made of phenolic laminated materials, on the other hand, the teeth will be deformed the amount of this normal error without setting up any appreciable stresses in the material, so that the load is actually supported by several teeth.

All materials have their own peculiar and distinct characteristics, so that under certain specific conditions, each material has a field of its own where it is superior to any other. Such fields may overlap to some extent, and only in such overlapping fields are different materials directly competitive. For example, steel is more or less ductile, has a high tensile strength, and a high modulus of elasticity. Cast iron, on the other hand, is not ductile, has a low tensile strength, but a high compressive strength, and a low modulus of elasticity. Hence when stiffness and high tensile strength are essential, steel is far superior to cast iron. On the other hand, when these two characteristics are unimportant, but high compressive strength and a moderate amount of elasticity are essential, cast iron is superior to steel.

**Preferred Pitch for Non-metallic Gears.** — The pitch of the gear or pinion should bear a reasonable relation either to the horsepower or speed or to the applied torque, as shown by the accompanying table which conforms to recommended practice of the American Gear Manufacturers' Association. The upper half of this table is based upon horsepower transmitted at a given pitch-line velocity. The lower half gives the torque in pounds-feet or the torque at a 1-foot radius. This torque $T$ for any given horsepower and speed can be obtained from the following formula:

$$T = \frac{5252 \times \text{H.P.}}{\text{R.P.M.}}$$

**Bore Sizes for Non-metallic Gears.** — For plain phenolic laminated pinions, that is, pinions without metal end plates, a drive fit of 0.001 inch per inch of shaft diameter should be used. For shafts above 2.5 inches in diameter, the fit should be constant at 0.0025 to 0.003 inch. When metal reinforcing end plates are used, the drive fit should conform to the same standards as used for metal.

**Preferred Pitches for Non-metallic Gears***

| Diametral Pitch for Given Horsepower and Pitch Line Velocities | | | |
|---|---|---|---|
| Horsepower Transmitted | Pitch Line Velocity up to 1000 Feet per Minute | Pitch Line Velocity from 1000 to 2000 Feet per Minute | Pitch Line Velocity over 2000 Feet per Minute |
| ¼–1 | 8–10 | 10–12 | 12–16 |
| 1–2 | 7–8 | 8–10 | 10–12 |
| 2–3 | 6–7 | 7–8 | 8–10 |
| 3–7½ | 5–6 | 6–7 | 7–8 |
| 7½–10 | 4–5 | 5–6 | 6–7 |
| 10–15 | 3–4 | 4–5 | 5–6 |
| 15–25 | 2½–3 | 3–4 | 4–5 |
| 25–60 | 2–2½ | 2½–3 | 3–4 |
| 60–100 | 1¾–2 | 2–2½ | 2½–3 |
| 100–150 | 1½–1¾ | 1¾–2 | 2–2½ |

| Torque in Pounds-feet for Given Diametral Pitch | | | | | |
|---|---|---|---|---|---|
| Diametral Pitch | Torque in Pounds-feet | | Diametral Pitch | Torque in Pounds-feet | |
| | Minimum | Maximum | | Minimum | Maximum |
| 16 | 1 | 2 | 4 | 50 | 100 |
| 12 | 2 | 4 | 3 | 100 | 200 |
| 10 | 4 | 8 | 2½ | 200 | 450 |
| 8 | 8 | 15 | 2 | 450 | 900 |
| 6 | 15 | 30 | 1½ | 900 | 1800 |
| 5 | 30 | 50 | 1 | 1800 | 3500 |

* These preferred pitches are applicable both to rawhide and the phenolic laminated types of materials.

The root diameter of a pinion of phenolic laminated type should be such that the minimum distance from the edge of the keyway to the root diameter will be at least equal to the depth of tooth.

**Keyway Stresses for Non-metallic Gears.** — The keyway stress should not exceed 3000 pounds per square inch on a plain phenolic laminated gear or pinion. The keyway stress is calculated by the formula:

$$S = \frac{33,000 \times \text{H.P.}}{V \times A}$$

in which

$S$ = unit stress in pounds per square inch;

H.P. = horsepower transmitted;

$V$ = peripheral speed of shaft in feet per minute; and

$A$ = square inch area of keyway in pinion (length $\times$ height).

If the keyway stress formula is expressed in terms of shaft radius $r$ and revolutions per minute, it will read:

$$S = \frac{63,000 \times \text{H.P.}}{\text{R.P.M.} \times r \times A}$$

When the design is such that the keyway stresses exceed 3000 pounds, metal reinforcing end plates may be used. Such end plates should not extend beyond the root diameter of the teeth. The distance from the outer edge of the retaining bolt to the root diameter of the teeth shall not be less than a full tooth depth. The use of drive keys should be avoided, but if required, metal end plates should be used on the pinion to take the wedging action of the key.

For phenolic laminated pinions, the face of the mating gear should be the same or slightly greater than the pinion face.

**Invention of Gear Teeth.** — The invention of gear teeth represents a gradual evolution from gearing of primitive form. The earliest evidence we have of an investigation of the problem of *uniform motion* from toothed gearing and the successful solution of that problem, dates from the time of Olaf Roemer, the celebrated Danish astronomer, who, in the year 1674, proposed the epicycloidal form to obtain uniform motion. Evidently Robert Willis, professor in the University of Cambridge, was the first to make a practical application of the epicycloidal curve so as to provide for an interchangeable series of gears. Willis gives credit to Camus for conceiving the idea of interchangeable gears, but claims for himself its first application. The involute tooth was suggested as a theory by early scientists and mathematicians, but it remained for Willis to present it in a practical form. Perhaps the earliest conception of the application of this form of teeth to gears was by Philippe de Lahire, a Frenchman, who considered it, in theory, equally suitable with the epicycloidal for tooth outlines. This was about 1695 and not long after Roemer had first demonstrated the epicycloidal form. The applicability of the involute had been further elucidated by Leonard Euler, a Swiss mathematician, born at Basel, 1707, who is credited by Willis with being the first to suggest it. Willis devised the Willis odontograph for laying out involute teeth.

A pressure angle of 14½ degrees was selected for three different reasons. First, because the sine of 14½ degrees is nearly ¼, making it convenient in calculation; second, because this angle coincided closely with the pressure angle resulting from the usual construction of epicycloidal gear teeth; third, because the angle of the straight-sided involute rack is the same as the 29-degree worm thread.

## Bevel Gearing

**Types of Bevel Gears.** — Bevel gears are conical gears, that is, gears in the shape of cones, and are used to connect shafts having intersecting axes. Hypoid gears are similar in general form to bevel gears, but operate on axes that are offset. With few exceptions, most bevel gears may be classified as being either of the straight-tooth type or of the curved tooth type. The latter type includes spiral bevels, Zerol bevels, and hypoid gears. The following is a brief description of the distinguishing characteristics of the different types of bevel gears.

*Straight Bevel Gears:* The teeth of this most commonly used type of bevel gear are straight but their sides are tapered so that they would intersect the axis at a common point called the pitch cone apex if extended inward. The face cone elements of most straight bevel gears, however, are now made parallel to the root cone elements of the mating gear to obtain uniform clearance along the length of the teeth. The face cone elements of such gears would, therefore, intersect the axis at a point inside the pitch cone. Straight bevel gears are the easiest to calculate and are economical to produce.

Straight bevel gear teeth may be generated for full length contact or for localized contact. The latter are slightly convex in a lengthwide direction so that some adjustment of the gears during assembly is possible and small displacements due to load deflections can occur without undesirable load concentration on the ends of the teeth. This slight lengthwise rounding of the tooth sides need not be computed in the design but is taken care of automatically in the cutting operation on the newer types of bevel gear generators.

*Zerol\* Bevel Gears:* The teeth of Zerol bevel gears are curved but lie in the same general direction as the teeth of straight bevel gears. They may be thought of as spiral bevel gears of zero spiral angle and are manufactured on the same machines as spiral bevel gears. The face cone elements of Zerol bevel gears do not pass through the pitch cone apex but instead are approximately parallel to the root cone elements of the mating gear to provide uniform tooth clearance. The root cone elements also do not pass through the pitch cone apex because of the manner in which these gears are cut. Zerol bevel gears are used in place of straight bevel gears when generating equipment of the spiral-type but not the straight-type is available, and may be used when hardened bevel gears of high accuracy (produced by grinding) are required.

*Spiral Bevel Gears:* Spiral bevel gears have curved oblique teeth on which contact begins gradually and continues smoothly from end to end. They mesh with a rolling contact similar to straight bevel gears. As a result of their overlapping tooth action, however, spiral bevel gears will transmit motion more smoothly than straight bevel or Zerol bevel gears, reducing noise and vibration which become especially noticeable at high speeds.

One of the advantages associated with spiral bevel gears is the complete control of the localized tooth contact. By making a slight change in the radii of curvature of the mating tooth surfaces, the amount of surface over which tooth contact takes place can be changed to suit the specific requirements of each job. Localized tooth contact promotes smooth, quiet running spiral bevel gears, and permits some mounting deflections without concentrating the load dangerously near either end of the tooth. Permissible deflections established by experience are given under the heading "Mountings for Bevel Gears."

Because their tooth surfaces can be ground, spiral bevel gears have a definite advantage in applications requiring hardened gears of high accuracy. The bottom of the tooth spaces and the tooth profiles may be ground simultaneously, resulting in a smooth blending of the tooth profile, the tooth fillet, and the bottom of the tooth

\* Registered in U.S. Patent Office.

space. This is important from a strength standpoint because it eliminates cutter marks and other surface interruptions which frequently result in stress concentrations.

*Hypoid Gears:* Hypoid gears resemble, in general appearance, spiral bevel gears, except that the axis of the pinion is offset relative to the gear axis. If there is sufficient offset, the shafts may pass one another thus permitting the use of a compact straddle mounting on the gear and pinion. Whereas a spiral bevel pinion has equal pressure angles and symmetrical profile curvatures on both sides of the teeth, a hypoid pinion properly conjugate to a mating gear having equal pressure angles on both sides of the teeth must have non-symmetrical profile curvatures for proper tooth action. In addition, in order to obtain equal arcs of motion for both sides of the teeth, it is necessary to use unequal pressure angles on hypoid pinions. Hypoid gears are usually designed so that the pinion has a larger spiral angle than the gear. The advantage of such design is that the pinion diameter is increased and is stronger than a corresponding spiral bevel pinion. This diameter increment permits the use of comparatively high ratios without the pinion becoming too small to allow a bore or shank of adequate size. The sliding action along the lengthwise direction of their teeth in hypoid gears is a function of the difference in the spiral angles on the gear and pinion. This sliding effect makes such gears even smoother running than spiral bevel gears. Grinding of hypoid gears can be accomplished on the same machines used for grinding spiral bevel and Zerol bevel gears.

**Application of Bevel and Hypoid Gears.** — Bevel and hypoid gears may be used to transmit power between shafts at practically any angle and speed. The particular type of gearing best suited for a specific job, however, depends on the mountings and the operating conditions.

*Straight and Zerol Bevel Gears:* For peripheral speeds up to 1000 feet per minute, where maximum smoothness and quietness are not the primary consideration, straight and Zerol bevel gears are recommended. For such applications, plain bearings may be used for radial and axial loads, although the use of anti-friction bearings is always preferable. Plain bearings permit a more compact and less expensive design; one reason why straight and Zerol bevel gears are much used in differentials. This type of bevel gearing is the simplest to calculate and set up for cutting, and is ideal for small lots where fixed charges must be kept to a minimum.

Zerol bevel gears are recommended for use in place of straight bevel gears: (1) Where hardened gears of high accuracy are required, since Zerol gears may be ground; and (2) when only spiral-type equipment is available for cutting bevel gears.

*Spiral Bevel and Hypoid Gears:* Spiral bevel and hypoid gears are recommended for applications where peripheral speeds exceed 1000 feet per minute or 1000 revolutions per minute, whichever occurs first. In many instances they may be used to advantage at lower speeds, particularly where extreme smoothness and quietness are desired. For peripheral speeds above 8000 feet per minute, ground gears should be used.

For large reduction ratios the use of spiral and hypoid gears will reduce the overall size of the installation since the continuous pitch line contact of these gears makes it practical to obtain smooth performance with a smaller number of teeth in the pinion than is possible with straight or Zerol bevel gears.

Hypoid gears are recommended for industrial applications: (1) When maximum smoothness of operation is desired; (2) for high reduction ratios where compactness of design, smoothness of operation, and maximum pinion strength are important; and (3) for non-intersecting shafts.

Bevel and hypoid gears may be used for both speed-reducing and speed-increasing drives. In speed-increasing drives, however, the ratio should be kept as low as possible and the pinion mounted on anti-friction bearings; otherwise bearing friction will cause the drive to lock.

**Notes on the Design of Bevel Gear Blanks.** — The quality of any finished gear is dependent, to a large degree, on the design and accuracy of the gear blank. A number of factors which affect manufacturing economy as well as performance must be considered.

A gear blank should be designed to avoid localized stresses and serious deflections within itself. Sufficient thickness of metal should be provided under the roots of gear teeth to give them proper support. As a general rule, the amount of metal under the root should equal the whole depth of the tooth; this metal depth should be maintained under the small ends of the teeth as well as under the middle. On webless type ring gears, the minimum stock between the root line and the bottom of tap drill holes should be one-third the tooth depth. For heavily loaded gears, a preliminary analysis of the direction and magnitude of the forces is helpful in the design of both the gear and its mounting. Rigidity is also necessary for proper chucking when cutting the teeth. For this reason, bores, hubs, and other locating surfaces must be in proper proportion to the diameter and pitch of the gear. Small bores, thin webs, or any condition which necessitates excessive overhang in cutting should be avoided.

Other factors to be considered are the ease of machining and, in the case of gears that are to be hardened, proper design to insure the best hardening conditions. It is desirable to provide a locating surface of generous size on the back of gears. This surface should be machined or ground square with the bore and is used both for locating the gear axially in assembly and for holding it when the teeth are cut. The front clamping surface must, of course, be flat and parallel to the back surface. In connection with cutting the teeth on Zerol bevel, spiral bevel, and hypoid gears, clearance must be provided for face-mill type cutters; front and rear hubs should not intersect the extended root line of the gear or they will interfere with the path of the cutter. In addition, there must be enough room in the front of the gear for the clamp nut which holds the gear on the arbor, or in the chuck, while cutting the teeth. The same considerations must be given to straight bevel gears which are to be generated using a circular-type cutter instead of reciprocating tools.

**Mountings for Bevel Gears.** — Rigid mountings should be provided for bevel gears to keep the displacements of the gears under operating loads within recommended limits. In order to align gears properly, care should be exercised to insure accurately machined mountings, properly fitted keys, and couplings that run true and square.

As a result of deflection tests on gears and their mountings, and having observed these same units in service, the Gleason Works recommends that the following allowable deflections be used for gears from 6 to 15 inches in diameter: (1) Neither the pinion nor the gear should lift or depress more than 0.003 inch at the center of the face width; (2) the pinion should not yield axially more than 0.003 inch in either direction; and (3) the gear should not yield axially more than 0.003 inch in either direction on 1 to 1 ratio gears (miter gears), or near miters, or more than 0.010 inch away from the pinion on higher ratios.

When deflections exceed these limits, additional problems are involved in obtaining satisfactory gears. It becomes necessary to narrow and shorten the tooth contacts to suit the more flexible mounting. This decreases the bearing area, raises the unit tooth pressure, and reduces the number of teeth in contact, which results in increased noise and the danger of surface failure as well as tooth breakage.

Spiral bevel and hypoid gears should, in general, be mounted on anti-friction bearings in an oil-tight case. While designs for a given set of conditions may use plain bearings for radial and thrust loads, maintaining gears in satisfactory alignment is usually more easily accomplished with ball or roller bearings.

*Bearing Spacing and Shaft Stiffness:* Bearing spacing and shaft stiffness are extremely important if gear deflections are to be minimized. For both straddle mounted and overhung mounted gears the spread between bearings should never be less than 70 per cent of the pitch diameter of the gear. On overhung mounted gears the spread should be at least 2½ times the overhang and, in addition, the shaft diameter should be equal to or preferably greater than the overhang to provide sufficient shaft stiffness. When two spiral bevel or hypoid gears are mounted on the same shaft, the axial thrust should be taken at one place only and near the gear where the greater thrust is developed. Provision should be made for adjusting both the gear and pinion axially in assembly. Details on how this may be accomplished are given in the Gleason Works booklet, "Assembling Bevel Gears."

**Cutting Bevel Gear Teeth.** — A correctly formed bevel gear tooth has the same sectional shape throughout its length, but on a uniformly diminishing scale from the large to the small end. The only way to obtain this correct form is by using a generating type of bevel gear cutting machine. This accounts, in part, for the extensive use of generating type gear cutting equipment in the production of bevel gears.

Bevel gears too large to be cut by generating equipment (100 inches or over in diameter) may be produced by a form-copying type of gear planer. With this method, a template or former is used to mechanically guide a single cutting tool in the proper path to cut the profile of the teeth. Since the tooth profile produced by this method is dependent on the contour of the template used, it is possible to produce tooth profiles to suit a variety of requirements.

Although generating methods are to be preferred, there are still some cases where straight bevel gears are produced by milling. Milled gears cannot be produced with the accuracy of generated gears and generally are not suitable for use in high-speed applications or where angular motion must be transmitted with a high degree of accuracy. Milled gears are used chiefly as replacement gears in certain applications, and gears which are subsequently to be finished on generating type equipment are sometimes roughed out by milling. Formulas and methods used for the cutting of bevel gears by milling are given in the latter part of this section.

In producing gears by generating methods, the tooth curvature is generated from a straight-sided cutter or tool having an angle equal to the required pressure angle. This tool represents the side of a crown gear tooth. The teeth of a true involute crown gear, however, have sides which are very slightly curved. If the curvature of the cutting tool conforms to that of the involute crown gear, an involute form of bevel gear tooth will be obtained. The use of a straight-sided tool is more practical and results in a very slight change of tooth shape to what is known as the " octoid " form. Both the octoid and involute forms of bevel gear tooth give theoretically correct action.

Bevel gear teeth, like those for spur gears, differ as to pressure angle and tooth proportions. The whole depth and the addendum at the large end of the tooth may be the same as for a spur gear of equal pitch. Most bevel gears, however, both of the straight tooth and spiral-bevel types, have lengthened pinion addendums and shortened gear addendums as in the case of some spur gears, the amount of departure from equal addendums varying with the ratio of gearing. Long addendums on the pinion are used principally to avoid undercut and to increase tooth strength. In addition, where long and short addendums are used, the tooth thickness of the gear is decreased and that of the pinion increased to provide a better balance of strength. See the Gleason Works System; the American Standard for Fine-pitch Straight Bevel Gears; and also the British Standard.

**Nomenclature for Bevel Gears.** — The accompanying diagram illustrates various angles and dimensions referred to in describing bevel gears. In connection with the face angles shown in the diagram, it should be noted that the face cones are made parallel to the root cones of the mating gears to provide uniform clearance along the length of the teeth.

Bevel Gear Nomenclature

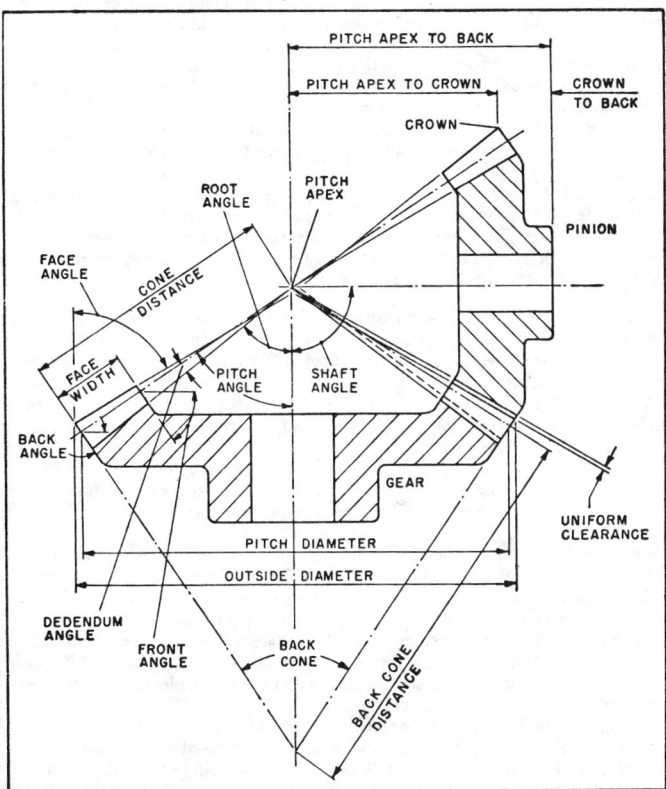

**American Standard System for Straight Bevel Gears (ASA B6.13-1965).** — This system for the design of straight bevel gears is a development from a system originated by the Gleason Works and adopted by the American Gear Manufacturers Association in 1922 and revised in 1971. The general basis of the system and the formulas used to obtain the tooth proportions and blank dimensions are the same as in the revised (1959) Gleason Straight Bevel Gear System which follows:

*Working Depth:* The working depth is equal to 2.000 inches ÷ diametral pitch. The use of stub teeth is not recommended because the reduction in contact increases noise and decreases wear resistance.

*Clearance:* The clearance is equal to (0.188 inches ÷ diametral pitch) + 0.002 inch.

*Pressure Angle:* The basic pressure angle is 20 degrees. In those cases where it is desired to use a 14½-degree pressure angle, the following combinations can be produced without undercut: (a) Those with 29 or more teeth in the pinion; (b) those with 28 teeth in the pinion and 29 or more in the gear; (c) those with 27 teeth in the pinion and 31 or more in the gear; (d) those with 26 teeth in the pinion and 35 or more in the gear; (e) those with 25 teeth in the pinion and 40 or more in the gear; and (f) those with 24 teeth in the pinion and 57 or more in the gear.

*Addendums:* Except when the numbers of teeth in gear and pinion are equal, the pinion has a long addendum and the gear a short addendum. The amount of departure from equal addendums varies with the ratio. Long addendums on the pinion are used principally to avoid undercut and to increase tooth strength.

*Face Angles:* The face cone of the bevel gear is cut so that it will be parallel to the root cone of the mating bevel gear. This puts the face cone apex on the gear axis at a point inside the pitch cone and results in constant clearance between the mating teeth. In addition, it allows the use of larger edge radii on generating tools without fillet interference at the small end, thus increasing tooth strength.

*Tooth Thickness:* In determining tooth thickness to balance the strength of gear and pinion teeth, the load is taken in the position where it is carried entirely on one tooth, this being the worst condition of loading.

**Formulas for Gleason System Straight Bevel Gears.** — The data in Table 2 are used with the formulas in Table 1 to compute the dimensions of Gleason System straight bevel gears with axes at right angles. The following explanatory notes apply to the numbered items in Table 1.

*Number of Teeth (Items 1 and 2):* The Gleason System covers 20-degree pressure angle straight bevel gears for all combinations in which the pinion has 16 or more teeth and combinations in which: (a) the pinion has 15 and the gear 17 or more teeth; (b) the pinion has 14 and the gear 20 or more teeth; and (c) the pinion has 13 and the gear 30 or more teeth. Gears for use in other tooth combinations can, of course, be designed according to the same general principles on which this system is based.

*Face Width (Item 4):* The face width should not exceed one-third of the cone distance (Item 11) or 10 inches ÷ diametral pitch, whichever is smaller. In best design practice the ratio of face width to cone distance will be from 0.25 to 0.3. Increasing the face width over recommended proportions adds strength and durability, but at a rapidly diminishing rate. Extra face width results in manufacturing difficulties by requiring tools of less point width and decreases possible fillet radii. It may even increase the danger of breakage and wear if the load becomes concentrated on the small ends of the teeth.

*Pressure Angle (Item 5):* The standard pressure angle is 20 degrees. However, if a lower pressure angle is preferred, 14½ degrees may be used without undercut for: combinations with 29 or more teeth in the pinion; 28 teeth in the pinion and 29 or more teeth in the gear; 27 teeth in the pinion and 31 or more in the gear; 26 teeth in the pinion and 35 or more in the gear; 25 teeth in the pinion and 40 or more in the gear; and 24 teeth in the pinion and 57 or more in the gear. The 20-degree pressure angle should be used on all aircraft applications where strength is of prime importance, and on all instrument gears which are to run with " zero " backlash.

*Whole Depth (Item 8):* Unless smooth bottoms and fillets are required, it is recommended that gears of 10 diametral pitch and coarser be roughed 0.005 inch deeper than the calculated whole depth so that the finishing tools will not cut on their ends. The calculated whole depth is, of course, used for the finishing operation.

*Dedendum (Item 14):* The value computed in this item is used in subsequent calculations. The actual dedendum, however, will be greater by 0.002 inch.

*Backlash (Item 22):* The table on page 761 gives recommended backlash values when the gear and pinion are assembled and ready to run. Because of manufacturing tolerances and changes caused by heat treatment, it may be necessary in cutting the teeth to allow for more than one-half the tabular value in computing chordal thickness (Item 23) in order to obtain the desired backlash in assembly.

*Tooth Angle (Item 25):* The tooth angle is a machine setting in Gleason two-tool straight bevel gear generators. It is the angle at the machine center between the line to the center of the tooth on the root circle and the line to the point of the tool.

*Limit Point Width (Item 26):* The limit point width is the width of the point of a straight-sided V tool of given pressure angle which will touch both sides and the bottom of a finished tooth space at the small end. The point width of the tools used must be less than the limit point width.

*Tool Advance (Item 27):* The tool advance is a machine setting to extend the tool and cut deeper. Its purpose is to increase the clearance along the full length of the tooth. The 0.002 inch in the formula for clearance is obtained in this way.

**Formulas for Gleason System Angular Straight Bevel Gears.** — Angular bevel gears are those whose shaft angle is other than 90 degrees. The formulas in Table 1 are not directly applicable to angular gears but must be modified as explained in the summary of procedure outlined below in the order of calculation; the item numbers referred to are those in Table 1:

*Items 7, 8, and 9:* Formulas same as given.

*Item 10, Pitch Angles:* There are two cases which must be distinguished.

*Case I* — Shaft Angle ($\Sigma$) less than 90 degrees, for which

$$\tan \gamma = \frac{\sin \Sigma}{\dfrac{N}{n} + \cos \Sigma} \; ; \quad \Gamma = \Sigma - \gamma$$

*Case II* — Shaft Angle ($\Sigma$) greater than 90 degrees, for which

$$\tan \gamma = \frac{\sin (180 - \Sigma)}{\dfrac{N}{n} - \cos (180 - \Sigma)} \; ; \quad \Gamma = \Sigma - \gamma$$

A gear pitch angle ($\Gamma$) greater than 90 degrees indicates an *internal* gear and calculations should be referred to the manufacturer of the generating equipment used to determine whether the gear can be cut.

In either Case I or II,

$$\frac{\sin \gamma}{\sin \Gamma} = \frac{n}{N}$$

*Items 11 and 12:* Formulas same as given.

*Items 13 and 14:* Formulas same as given. However, determination of the tooth proportions requires the use of the equivalent 90-degree bevel gear ratio computed from the formula

$$\text{Equivalent 90-degree ratio } m_{90} = \sqrt{\frac{N \cos \gamma}{n \cos \Gamma}}$$

This equivalent ratio is used to enter Table 2 when determining the gear addendum.

*Items 15, 16, 17, 18, and 19:* Formulas same as given.

*Item 20, Pitch Apex to Crown:* The following formulas must be used:

$$x_O = A_O \cos \gamma - a_P \sin \gamma$$
$$X_O = A_O \cos \Gamma - a_G \sin \Gamma$$

**Table 1. Formulas for 20-degree Straight Bevel Gears — 90-degree Shaft Angle**
(Gleason System and American Standard Straight and Fine-Pitch Straight Systems)

| Given | | | | | |
|---|---|---|---|---|---|
| 1 | Number of Pinion Teeth, $n$ | | 4 | Face Width, $F$ | |
| 2 | Number of Gear Teeth, $N$ | | 5 | Pressure Angle, $\phi = 20°$ | |
| 3 | Diametral Pitch, $P$ | | 6 | Shaft Angle, $\Sigma = 90°$ | |

| No. | Item | To Find — Formula | |
|---|---|---|---|
| | | PINION | GEAR |
| 7 | Working Depth | $h_k = \dfrac{2.000}{P}$ | Same as pinion |
| 8 | Whole Depth | $h_t = \dfrac{2.188}{P} + 0.002$ | Same as pinion |
| 9 | Pitch Diameter | $d = \dfrac{n}{P}$ | $D = \dfrac{N}{P}$ |
| 10 | Pitch Angle | $\gamma = \tan^{-1}\dfrac{n}{N}$ | $\Gamma = 90° - \gamma$ |
| 11 | Cone Distance | $A_O = \dfrac{D}{2\sin\Gamma}$ | Same as pinion |
| 12 | Circular Pitch | $p = \dfrac{3.1416}{P}$ | Same as pinion |
| 13 | Addendum | $a_P = h_k - a_G$ | $a_G = \dfrac{\text{Table 2}}{P}$ |
| 14 | Dedendum (See Note 1) | $b_P = \dfrac{2.188}{P} - a_P$ | $b_G = \dfrac{2.188}{P} - a_G$ |
| 15 | Clearance | $c = h_t - h_k$ | Same as pinion |
| 16 | Dedendum Angle | $\delta_P = \tan^{-1}\dfrac{b_P}{A_O}$ | $\delta_G = \tan^{-1}\dfrac{b_G}{A_O}$ |
| 17 | Face Angle of Blank | $\gamma_O = \gamma + \delta_G$ | $\Gamma_O = \Gamma + \delta_P$ |
| 18 | Root Angle | $\gamma_R = \gamma - \delta_P$ | $\Gamma_R = \Gamma - \delta_G$ |
| 19 | Outside Diameter | $d_O = d + 2a_P\cos\gamma$ | $D_O = D + 2a_G\cos\Gamma$ |
| 20 | Pitch Apex to Crown | $x_O = \dfrac{D}{2} - a_P\sin\gamma$ | $X_O = \dfrac{d}{2} - a_G\sin\Gamma$ |
| 21 | Circular Thickness | $t = p - T$ | $T = \dfrac{p}{2} - (a_P - a_G)\tan\phi - \dfrac{K}{P}$  $K = \dfrac{3}{n} - \left(\dfrac{2.7}{N} + 0.1\right)$ |
| 22 | Backlash (See Note 2) | $B =$ (See Table on page 761) | |
| 23 | Chordal Thickness (See Note 2) | $t_C = t - \dfrac{t^3}{6d^2} - \dfrac{B}{2}$ | $T_C = T - \dfrac{T^3}{6D^2} - \dfrac{B}{2}$ |
| 24 | Chordal Addendum | $a_{CP} = a_P + \dfrac{t^2\cos\gamma}{4d}$ | $a_{CG} = a_G + \dfrac{T^2\cos\Gamma}{4D}$ |
| 25 | Tooth Angle | $\dfrac{3438}{A_O}\left(\dfrac{t}{2} + b_P\tan\phi\right)$ Minutes | Same as pinion |
| 26 | Limit Point Width | $\dfrac{A_O - F}{A_O}(T - 2b_P\tan\phi) - 0.0015$ | Same as pinion |

All linear dimensions are in inches. Calculation of linear dimensions which affect angular values should be carried to 4 or 5 decimal places.

*Note 1:* The actual dedendum will be 0.002-inch greater than calculated due to Item 27.

*Note 2:* The American Standard for Fine-pitch Straight Bevel Gears does not indicate values for $B$ since most fine-pitch bevel gears operate with little backlash.

**Table 2.   Gear Addenda for Generated Straight Bevel Gears***

(For 1 diametral pitch and various gear-to-pinion ratios)

| Ratio = Number of teeth in gear ÷ Number of teeth in pinion | | | | | | | | | | | |
|---|---|---|---|---|---|---|---|---|---|---|---|
| Ratios | | Addendum, Inch | Ratios | | Addendum, Inch | Ratios | | Addendum, Inch | Ratios | | Addendum, Inch |
| From | to | | From | to | | From | to | | From | to | |
| 1.00 | 1.00 | 1.000 | 1.15 | 1.17 | 0.880 | 1.42 | 1.45 | 0.760 | 2.06 | 2.16 | 0.640 |
| 1.00 | 1.02 | 0.990 | 1.17 | 1.19 | 0.870 | 1.45 | 1.48 | 0.750 | 2.16 | 2.27 | 0.630 |
| 1.02 | 1.03 | 0.980 | 1.19 | 1.21 | 0.860 | 1.48 | 1.52 | 0.740 | 2.27 | 2.41 | 0.620 |
| 1.03 | 1.04 | 0.970 | 1.21 | 1.23 | 0.850 | 1.52 | 1.56 | 0.730 | 2.41 | 2.58 | 0.610 |
| 1.04 | 1.05 | 0.960 | 1.23 | 1.25 | 0.840 | 1.56 | 1.60 | 0.720 | 2.58 | 2.78 | 0.600 |
| 1.05 | 1.06 | 0.950 | 1.25 | 1.27 | 0.830 | 1.60 | 1.65 | 0.710 | 2.78 | 3.05 | 0.590 |
| 1.06 | 1.08 | 0.940 | 1.27 | 1.29 | 0.820 | 1.65 | 1.70 | 0.700 | 3.05 | 3.41 | 0.580 |
| 1.08 | 1.09 | 0.930 | 1.29 | 1.31 | 0.810 | 1.70 | 1.76 | 0.690 | 3.41 | 3.94 | 0.570 |
| 1.09 | 1.11 | 0.920 | 1.31 | 1.33 | 0.800 | 1.76 | 1.82 | 0.680 | 3.94 | 4.82 | 0.560 |
| 1.11 | 1.12 | 0.910 | 1.33 | 1.36 | 0.790 | 1.82 | 1.89 | 0.670 | 4.82 | 6.81 | 0.550 |
| 1.12 | 1.14 | 0.900 | 1.36 | 1.39 | 0.780 | 1.89 | 1.97 | 0.660 | 6.81 | ∞ | 0.540 |
| 1.14 | 1.15 | 0.890 | 1.39 | 1.42 | 0.770 | 1.97 | 2.06 | 0.650 | .... | .... | .... |

* Select addendum according to ratio of gearing.   In case of choice, use the *larger* addendum.

As in the original system, long and short addenda have been adopted for all ratios except those with equal numbers of teeth.   A long-addendum pinion and a short-addendum gear have more action in recess than in approach (with pinion driving), have stronger pinion teeth, and can have a lower pressure angle without undercut.

*Item 5, Pressure Angle:* The minimum pressure angle necessary to avoid undercut is computed from the formula:

$$\cos \phi = \sqrt{\frac{\cos \delta \sin (\gamma - \delta)}{\sin \gamma}}$$

The selected pressure angle should be equal to or greater than this calculated value.

*Items 21, 22, 23, 24, 25, 26, and 27:*  Formulas same as given.

*Ratio of Generator Roll Gears:*  Angular gears require special ratio of roll gears for cutting on Gleason generators.   The decimal ratio for the NC/75 ratio gears is found by the formula

$$\text{Decimal ratio of gears} = \frac{A_O P}{37.5}$$

The work roll is then found in the usual manner using the decimal ratio.

**Zerol* Bevel Gears.** — As in the case of straight bevel gears, Zerol gears may be mounted on plain journal bearings to obtain both a smaller and a less-costly assembly.   As with spiral bevel gears, their tooth form can easily be manufactured to have a localized tooth contact that will allow for assembly position tolerance and operating deflection without causing end contact.   These teeth may also be ground if high accuracy is required for reduced noise, or for high pitchline velocity.   The normal limit is a pitchline velocity of 1,000 feet per minute; however, applications have been successful, under special conditions, at speeds up to 15,000 feet per minute. Zerol gears have the same radial and axial loading as straight bevel gears of the same size and proportions.   Therefore, they may be used to supplant a straight bevel gear set where gear performance requires improvement, with no need to also change the gear mounting bearings.

Zerol bevel gears of 20-degree pressure angle should have a pinion with no less than 17 teeth; if a lesser tooth number must be used, the pressure angle may be increased to 22½ or 25 degrees to avoid undercut or involute interference.   In this case the tooth thickness and the dedendum angles must be modified as indicated in AGMA 202.03, or Gleason publication, "Zerol Bevel Gear System."

* Registered in U.S. Patent Office.

Zerol bevel gears of 20-degree pressure angle and 90-degree shaft angle may be proportioned by the use of Table 1, page 844, except that Item 16, dedendum angles $(\delta_P)$ and $(\delta_G)$ must be reduced by:

$$\Delta\delta = \frac{1}{2PA_O}\left(6668 - \frac{300}{F}\sqrt{d \sin \Gamma} - 14P\right) \text{ minutes}$$

**Gleason System Spiral Bevel Gears (1952 Revision).** — The Gleason System provides a basis for designing spiral bevel gears to meet practical operating requirements. The pressure angle used for pinions with small numbers of teeth eliminates undercut in all cases, and the face angles of mating gears are so determined that the clearance is constant along the teeth. The latter decreases the likelihood of fillet interference at the small ends and makes possible the use of slightly larger fillet radii for maximum tooth strength.

Changes incorporated in the 1952 revision are: (1) Adoption of 20 degrees as the basic pressure angle in place of 14½ degrees; (2) inclusion of information regarding angular gears; and (3) revision of tooth thicknesses to give approximately equal stress in gear and pinion.

There are some spiral bevel gear applications and cutting methods which require special tooth designs and for which the system does not apply. These include: (1) Automotive rear-axle drives; (2) Formate gear and pinion; (3) Gears and pinions of 12 diametral pitch and finer, which are usually cut by one of the duplex spread-blade methods; (4) Gear cut spread-blade and pinion cut single-side, with spiral angle less than 20 degrees; and (5) Ratios with fewer than 12 teeth in the pinion.

*Working Depth:* The working depth is 1.700 inches ÷ diametral pitch.

*Clearance:* The clearance is 0.188 inches ÷ diametral pitch.

*Whole Depth:* The whole depth is usually determined as the working depth plus the clearance. It is common practice, however, with gears of 10 diametral pitch and coarser, to rough cut 0.005 inch deeper than the calculated depth in order to avoid having the finishing blades cut on their ends. The calculated whole depth, however, is used in the calculation of tooth proportions and for the finishing operation.

*Pressure Angle:* The basic pressure angle is 20 degrees and permits the following ratios to be cut without undercut: Ratios with 17 or more teeth in the pinion; 19⅛s and higher; 15⁄19 and higher; 14⁄20 and higher; 13⁄22 and higher; and 13⁄26 and higher. However, for those who wish to use a lower pressure angle, the following ratios for 14½-degree and 16-degree pressure angles can be cut without undercut.

*14½-degree Pressure Angle:* Ratios with 28 or more teeth in the pinion; 27⁄29 and higher; 26⁄30 and higher; 25⁄32 and higher; 24⁄33 and higher; 23⁄36 and higher; 23⁄40 and higher; 21⁄42 and higher; 20⁄60 and higher; and 19⁄40 and higher.

*16-degree Pressure Angle:* Ratios with 24 or more teeth in the pinion; 23⁄25 and higher; 23⁄26 and higher; 21⁄27 and higher; 20⁄29 and higher; 19⁄31 and higher; 18⁄36 and higher; 17⁄45 and higher; and 16⁄60 and higher.

*Addendums:* Except where the numbers of teeth are equal, the pinion has a long addendum and the gear a short addendum, the amount of departure from equal addendums varying with the ratio. Long addendums are used on the pinion primarily to avoid undercut and to increase tooth strength.

*Face Angles:* The face cone of a blank is made parallel to the root cone of the mating gear and therefore intersects the gear axis at a point inside the pitch apex. This gives constant clearance along the tooth and allows the use of larger edge radii on generating tools without fillet interference at the small end, thus increasing tooth strength.

*Tooth Thickness:* The tooth thicknesses are proportioned so that the stresses in the gear and pinion will be approximately equal with a left-hand pinion driving clockwise or a right-hand pinion driving counterclockwise. This will give a satisfactory balance of life for gears operating below the endurance limit. If the gears are to operate above the endurance limit, special proportions will be required. Also, on reversible drives which must be designed for optimum load capacity, special proportions will be required. The method of determining the balance of strength for these special cases may be found in the Gleason publication "Strength of Bevel and Hypoid Gears." Determination of tooth thickness measurements will vary with different methods of cutting and will not be given here.

*Spiral Angle:* The system has been designed for a spiral angle of 35 degrees. For smaller spiral angles, undercut may occur and the contact ratio reduced.

*Face Width:* The recommended maximum face width is 0.3 times the cone distance or 10 inches ÷ diametral pitch, whichever is smaller.

*Backlash:* Recommended backlash values for general purpose bevel gearing are given in the table on page 761.

*Standard Cutter Edge Radius:* On cutters for general use, the standard edge radius is made as large as practical for the various jobs for which the cutter may be suitable. Where the maximum fillet radius is required in gears for a specific case, as in aircraft gears, a special determination of the edge radius must be made.

**Formulas for Gleason System Spiral Bevel Gears.** — The data in Table 4 and values from Chart 1 are used in conjunction with the formulas in Table 3 to compute the dimensions of Gleason System spiral bevel gears. These formulas apply only to gears with axes at right angles.

**Formulas for Gleason System Angular Spiral Bevel Gears.** — Angular spiral bevel gears are those whose shaft angle is other than 90 degrees. The formulas in Table 3 are not directly applicable to angular spiral bevel gears but must be modified as explained in the summary of procedure outlined below in the order of calculation; item numbers referred to are those in Table 3.

*Items 7, 8, and 9:* Formulas same as given.

*Item 10, Pitch Angles:* There are two cases which must be distinguished.

*Case I* — Shaft Angle ($\Sigma$) less than 90 degrees, for which

$$\tan \gamma = \frac{\sin \Sigma}{\dfrac{N}{n} + \cos \Sigma} \; ; \quad \Gamma = \Sigma - \gamma$$

*Case II* — Shaft Angle ($\Sigma$) greater than 90 degrees, for which

$$\tan \gamma = \frac{\sin (180 - \Sigma)}{\dfrac{N}{n} - \cos (180 - \Sigma)} \; ; \quad \Gamma = \Sigma - \gamma$$

A gear pitch angle ($\Gamma$) greater than 90 degrees indicates an *internal* gear and calculations should be referred to the manufacturer of the generating equipment used to determine whether the gear can be cut.

In either Case I or II,      $\dfrac{\sin \gamma}{\sin \Gamma} = \dfrac{n}{N}$

*Items 11 and 12:* Formulas same as given.

*Items 13 and 14:* Formulas same as given. However, determination of the tooth proportions requires the use of a modified bevel gear ratio computed from the formula:

$$\frac{n^2}{N} = \frac{n \cos \gamma}{N \cos \Gamma}$$

**Table 3. Formulas for Gleason System 20-degree Pressure Angle, Spiral Bevel Gears — 90-degree Shaft Angle**

| Given | | | | | |
|---|---|---|---|---|---|
| 1 | Number of Pinion Teeth, $n$ | | 4 | Face Width, $F$ | |
| 2 | Number of Gear Teeth, $N$ | | 5 | Pressure Angle, $\phi = 20°$ | |
| 3 | Diametral Pitch, $P$ | | 6 | Shaft Angle, $\Sigma = 90°$ | |

| To Find | | | |
|---|---|---|---|
| No. | Item | Formula | |
| | | PINION | GEAR |
| 7 | Working Depth | $h_k = \dfrac{1.700}{P}$ | Same as pinion |
| 8 | Whole Depth | $h_t = \dfrac{1.888}{P}$ | Same as pinion |
| 9 | Pitch Diameter | $d = \dfrac{n}{P}$ | $D = \dfrac{N}{P}$ |
| 10 | Pitch Angle | $\gamma = \tan^{-1}\dfrac{n}{N}$ | $\Gamma = 90 - \gamma$ |
| 11 | Cone Distance | $A_O = \dfrac{D}{2\sin\Gamma}$ | Same as pinion |
| 12 | Circular Pitch | $p = \dfrac{3.1416}{P}$ | Same as pinion |
| 13 | Addendum | $a_P = h_k - a_G$ | $a_G = \dfrac{1}{P}\left[0.46 + 0.39\left(\dfrac{n}{N}\right)^2\right]$ |
| 14 | Dedendum | $b_P = h_t - a_P$ | $b_G = h_t - a_G$ |
| 15 | Clearance | $c = h_t - h_k$ | Same as pinion |
| 16 | Dedendum Angle | $\delta_P = \tan^{-1}\dfrac{b_P}{A_O}$ | $\delta_G = \tan^{-1}\dfrac{b_G}{A_O}$ |
| 17 | Face Angle of Blank | $\gamma_O = \gamma + \delta_G$ | $\Gamma_O = \Gamma + \delta_P$ |
| 18 | Root Angle | $\gamma_R = \gamma - \delta_P$ | $\Gamma_R = \Gamma - \delta_G$ |
| 19 | Outside Diameter | $d_O = d + 2a_P\cos\gamma$ | $D_O = D + 2a_G\cos\Gamma$ |
| 20 | Pitch Apex to Crown | $x_O = \dfrac{D}{2} - a_P\sin\gamma$ | $X_O = \dfrac{d}{2} - a_G\sin\Gamma$ |
| 21 | Circular Thickness | $t = p - T$ | $T = \dfrac{1}{P}\left[1.5708 - \dfrac{4}{9}(a_P - a_G) - K\right]$<br>For $K$, see p. 860 |
| 22 | Backlash* | $B = $ (See table on page 761) | |

\* When the gear is cut spread-blade, all the backlash is taken from the pinion thickness. When both members are cut single-side, each thickness is reduced by half the backlash.

This equivalent ratio is used when determining the gear addendum.

*Items 15, 16, 17, 18, and 19:* Formulas same as given.

*Item 20, Pitch Apex to Crown:*

$$x_O = A_O \cos \gamma - a_P \sin \gamma$$

$$X_O = A_O \cos \Gamma - a_G \sin \Gamma$$

*Item 5, Pressure Angle:* The minimum pressure angle necessary to avoid undercut can be obtained from Chart 2. The point of intersection on the graph of the dedendum angle and the pitch angle must be on or below the line representing the selected pressure angle if undercut is to be avoided.

*Item 21:* Formulas same as given. However, the value of $K$ from Chart 1 must be determined by using the equivalent 90-degree bevel gear ratio $m_{90}$ which was computed for Item 13, and the number of teeth in the equivalent 90-degree bevel pinion, $n_{90}$, which is computed as follows:

$$n_{90} = \frac{n \sin (\tan^{-1} m_{90})}{\cos \gamma}$$

*Item 22:* Same formulas as given.

**Bevel Gear Mounting.** — To assure successful operation, bevel gears must be rigidly mounted within the housing such that the deflection of one gear relative to the other does not exceed 0.006 inch either axially or tangential to the pitch diameters. Gear mountings that are more flexible will require special treatment in manufacture that will provide a smaller tooth contact area whose position is carefully controlled near the toe end of the face. This will raise the tooth contact stress over the expected value and increase the operating noise level. Satisfactory mounting stiffness is most easily obtained by using large, long-life bearings broadly spaced, and with a toe end bearing on the pinion if at all possible. It is good practice to have the bearing spacing on both gear and pinion at least equal to three-quarters of the gear pitch diameter if at all possible; on high ratio gear sets this will not always be practical on the pinion.

Provision should be made for axially shimming the position of both gears at the time of assembly. The mounting distance of the gear is usually the Pitch Apex to Back (see p. 841) and should be stamped or etched on every bevel gear. At assembly, shims are used to obtain this distance exactly. The gears, when so mounted, should have the proper tooth bearing pattern and backlash. These should be checked; the pattern is checked by bluing the tooth and turning it over under light load while noting the pattern transfer; the backlash is checked with feeler gages.

More complete data is given in "Installation of Bevel Gears" and "Bevel and Hypoid Gear Design" both published by Gleason Works, Rochester, New York.

**Spiral Bevel Gears with Spiral Angle other than 35 Degrees.** — In general, the same tooth proportions used for Gleason System spiral bevel gears may be used for spiral bevel gears with spiral angle of from, say, 20 to 45 degrees. Below 20-degree spiral angle, however, special proportions may be required. In the description of the Gleason System for spiral bevel gears, the reason for stating that the system is based on a 35-degree spiral angle is that the tooth-thickness balance for equal stress in gear and pinion was determined using this spiral angle. This spiral angle is most commonly selected for spiral bevel gears. In general, the higher the spiral angle the greater the face contact ratio becomes and the smoother and quieter the gears will be. The choice of spiral angle has no effect on efficiency, but does affect the thrust loads produced by the gears.

**Chart 1.**   Circular Thickness Factors $K$ for Gleason System Spiral Bevel Gears*

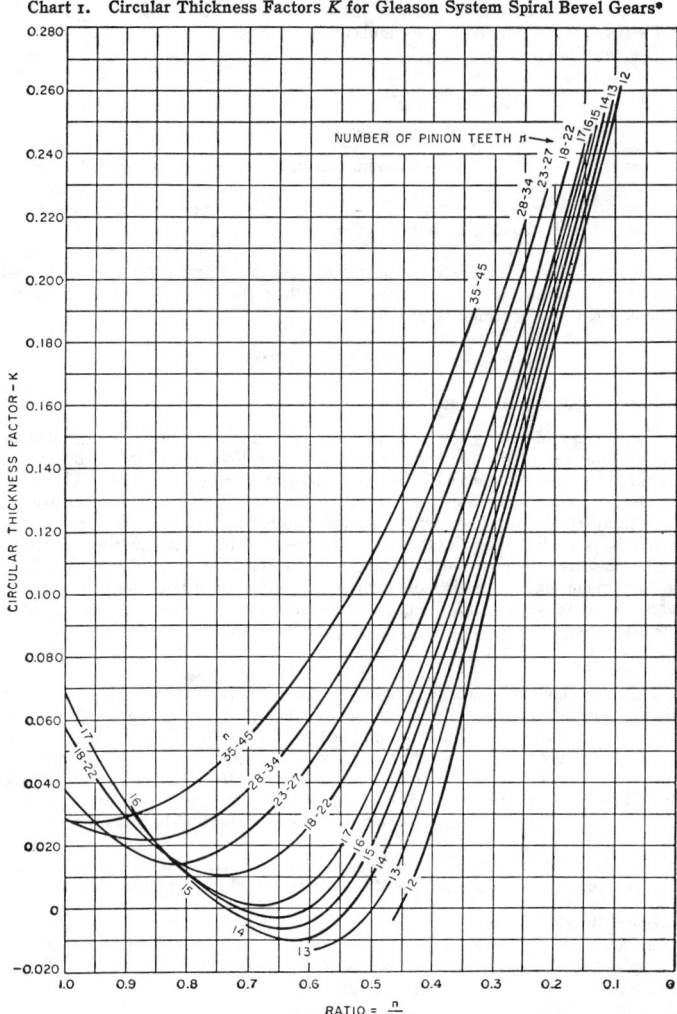

* These circular thickness factors are for spiral bevel gears with 20-degree pressure angle and 35-degree spiral angle for combinations with a left-hand pinion driving clockwise, or a right-hand pinion driving counter-clockwise. (See " Hand of Spiral, Hand of Rotation, and Spiral Angle " for definitions.) When used in the circular thickness formula given in Table 3, the tooth thicknesses of gear and pinion will be proportioned to provide a satisfactory balance of life for gears operating below the endurance limit. For gears operating above the endurance limit, special proportions will be required.

**Chart 2.  Relation between Pressure Angle, Dedendum Angle, and Pitch Angle for No Undercut in Spiral Bevel Gears with 35-degree Spiral Angle***

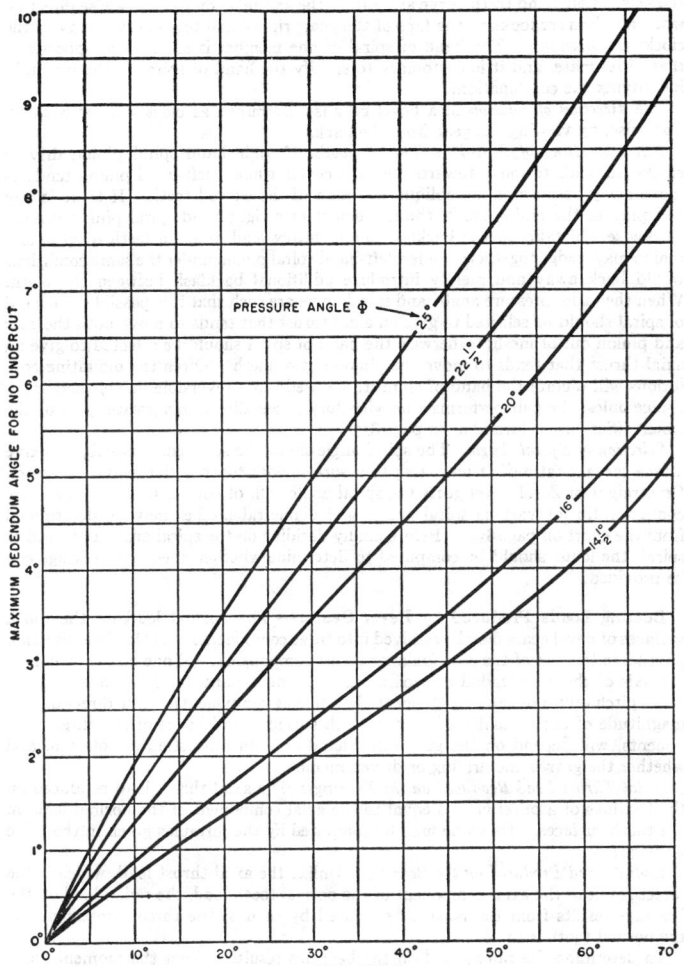

* To determine whether or not a 35-degree spiral angle bevel gear will be undercut, first, locate the point of intersection of the pitch angle and the dedendum angle.  If this point of intersection is on or below the curve representing the selected pressure angle $\phi$, then no undercutting will take place.  If, however, the intersection is above the curve, undercutting will occur and, therefore, a higher pressure angle should be selected or other changes made, such as using larger tooth numbers in the gear and pinion.

**Hand of Spiral, Direction of Rotation, and Spiral Angle.** — The hand of spiral on spiral bevel gears is indicated by the direction in which the teeth curve away from the axis. Left-hand teeth curve away from the axis in a counter-clockwise direction when an observer looks at the face of the gear; right-hand teeth curve away in the clockwise direction. The hand of spiral of one member is always the opposite of that of its mate, and it is customary to specify the hand of spiral of the pinion in identifying the combination.

The *direction of rotation* of a bevel gear is determined as clockwise, or counter-clockwise, by viewing the gear from the back.

*Effect of Hand of Spiral on Thrust Loads:* A right-hand spiral pinion driving clockwise tends to move toward the cone center while a left-hand pinion tends to move away because of the oblique direction of the curved teeth. If there is any end play in the pinion shaft, the movement of a right-hand spiral pinion driving clockwise will take up the backlash under heavy load and the teeth of gear and pinion may wedge together; while a left-hand spiral pinion under the same conditions would back away and merely introduce additional backlash between the teeth. When the ratio, pressure angle, and spiral angle are such that it is possible, the hand of spiral should be selected to give an axial thrust that tends to move both the gear and pinion out of mesh; otherwise the hand of spiral should be selected to give an axial thrust that tends to move the pinion out of mesh. Often the mounting conditions will dictate the hand of spiral to be used; in a reversible drive, there is no choice unless the pair performs a heavier duty in one direction a greater part of the time. (See thrust formulas on page 853.)

*Selection of Spiral Angle:* The spiral angle should be selected, if possible, to give a face contact ratio of at least 1.25 to assure smooth tooth action and quiet gears. On straight or Zerol bevel gears the spiral angle will, of course, be zero. The face contact ratio for various spiral angles and diametral pitches may be determined from the chart on page 859. Before finally deciding on the spiral angle and hand of spiral, the loads should be computed to determine whether adequate bearings can be provided.

**Bearing Loads Produced by Bevel Gears.** — The normal load on the tooth surfaces of bevel gears may be resolved into three components: (1) One in a direction parallel to the axis of the gear (axial or thrust component); (2) one perpendicular to the axis of the gear (radial or separating component); and (3) one tangent to the mean pitch radius of the gear (tangential or driving component). The direction and magnitude of the normal load on the tooth surfaces, and hence of the three components, will depend on the gear ratio, hand of spiral, direction of rotation, and whether the gear is the driving or driven member.

*Axial Thrust Load Produced on the Bearings:* The axial thrust load produced on the bearings of a bevel gear is equal to the axial component of the normal load on the tooth surfaces. Its value may be computed by the formulas given in the table on page 853.

*Radial Load Produced on the Bearings:* Unlike the axial thrust load, which is the direct result of the axial component of the normal tooth load, the radial load on the bearings results from the *moments* produced by each of the three components of the normal tooth load.

To determine the radial load on the bearings resulting from the moments produced by the components, first, compute the magnitudes and directions of these components of the normal tooth load using the formulas in the table on page 853. The values thus obtained, together with the dimensions of the bearing mountings, are then used to determine the radial bearing loads by applying the principle of moments. It should be noted that in computing the moment due to the axial thrust component, this component is considered to act at the mean pitch radius of

**Formulas for Tangential, Axial, and Separating Components of the Normal Tooth Load for Straight and Spiral Bevel Gears**

$W_t$ = tangential load on driving gear at its mean pitch diameter, in pounds; also equal to the tangential load on the driven gear at its mean pitch diameter;
$W_x$ = axial thrust load, in pounds;
$W_s$ = separating component in pounds;
$d_m$ = mean pitch diameter of driving member in inches = $d - F \sin \gamma_d$,

where $d$ = pitch diameter of driving member, $F$ = face width, and $\gamma_d$ = pitch angle;
$n$ = speed of driving member in rpm;
$P$ = horsepower transmitted;
$\phi$ = normal pressure angle;
$\psi$ = spiral angle;
$\gamma_d$ = pitch angle of driving member;
$\gamma_D$ = pitch angle of driven member.

| Component of Normal Tooth Load to be Computed | Hand of Spiral and Direction of Rotation of Driving Member | |
|---|---|---|
| | Right-hand Spiral and Clockwise Rotation, or Left-hand Spiral and Counter-clockwise Rotation | Right-hand Spiral and Counter-clockwise Rotation, or Left-hand Spiral and Clockwise Rotation |
| Tangential, $W_t$ Driving Member and Driven Member | $W_t = \dfrac{126{,}050\,P}{n d_m}$ | |
| *Axial, $W_x$ Driving Member | $W_x = \dfrac{W_t}{\cos\psi}(\tan\phi\sin\gamma_d - \sin\psi\cos\gamma_d)$ | $W_x = \dfrac{W_t}{\cos\psi}(\tan\phi\sin\gamma_d + \sin\psi\cos\gamma_d)$ |
| Driven Member | $W_x = \dfrac{W_t}{\cos\psi}(\tan\phi\sin\gamma_D + \sin\psi\cos\gamma_D)$ | $W_x = \dfrac{W_t}{\cos\psi}(\tan\phi\sin\gamma_D - \sin\psi\cos\gamma_D)$ |
| †Separating, $W_s$ Driving Member | $W_s = \dfrac{W_t}{\cos\psi}(\tan\phi\cos\gamma_d + \sin\psi\sin\gamma_d)$ | $W_s = \dfrac{W_t}{\cos\psi}(\tan\phi\cos\gamma_d - \sin\psi\sin\gamma_d)$ |
| Driven Member | $W_s = \dfrac{W_t}{\cos\psi}(\tan\phi\cos\gamma_D - \sin\psi\sin\gamma_D)$ | $W_s = \dfrac{W_t}{\cos\psi}(\tan\phi\cos\gamma_D + \sin\psi\sin\gamma_D)$ |

* If the computed value of $W_x$ is positive, then the thrust is *away* from the cone center. A negative value indicates the thrust is *toward* the cone center.
† If the computed value of $W_s$ is positive, then the force is *away* from the mating member (separating force). A negative value indicates the *force* is *toward* the mating member (attracting force).

the gear or pinion, that is, at the pitch radius of the mid-face. The mean pitch radius may be computed by subtracting from the pitch radius one-half the product of the face width and the sine of the pitch angle.

**Design of Bevel Gears for Durability and Strength** — The design or selection of a pair of bevel gears to meet a given set of operating conditions is accomplished by a four-step sequence of calculations: (1) Determine the equivalent service horsepower or design load; (2) Select approximate gear sizes by means of a series of design charts; (3) Check the surface durability of the gears selected; and (4) Check the strength of the gears selected. The following information for determining gear ratings is based on methods used by the Gleason Works for general industrial work.

**Equivalent Service Horsepower (Design Load).** — The equivalent service horsepower or design load of a pair of bevel gears is the load for which the gears must be designed. It is computed as follows:

$$\text{Design Load } P_2 = C_s P_1 \quad \text{or,} \quad P_2 = \frac{P_M}{2}, \quad \text{whichever is greater} \quad (1) \text{ and } (2)$$

where $P_1$ = normal operating load in horsepower; $P_M$ = momentary peak load in horsepower; $C_s$ = service factor from Table 1 on page 855.

If the value of $P_M$ to be used in formula (2) is unknown, but the momentary peak pinion torque $T_M$ in inch-pounds is known, then $P_M$ should be computed from the following formula in which $n$ is the pinion speed in revolutions per minute:

$$P_M = \frac{nT_M}{63,025} \qquad (3)$$

*Example:* An electric motor is to deliver 65 horsepower at 1800 rpm to a centrifugal pump which is to run continuously at 600 rpm. The drive is to be through a pair of oil-hardened steel spiral bevel gears at 90-degree shaft angle. The starting torque will be 4000 inch-pounds.

From Formula (3),

$$P_M = \frac{1800 \times 4000}{63,025} = 114 \text{ horsepower}$$

From Formula (1),

$$P_2 = 1.0 \times 65 = 65 \text{ horsepower}$$

From Formula (2),

$$P_2 = \frac{114}{2} = 57 \text{ horsepower}$$

Therefore the value $P_2 = 65$ horsepower is used as the design load in subsequent calculations of the gear sizes.

**Approximate Pinion and Gear Sizes for Design Load.** — The approximate pinion size necessary to transmit design load $P_2$ can be determined from Chart 1 on page 857 by using the gear ratio, and the horsepower to be transmitted per 100 rpm of the pinion computed from the following formula:

$$\text{Horsepower per 100 rpm of pinion } P_{100} = \frac{100\,P_2}{nC_M} \qquad (4)$$

where $C_M$ = material factor from Table 2; $P_2$ = design load from Formula (1) or (2); and $n$ = speed of pinion in rpm. In the previous example, $P_2 = 65$ horsepower; $n = 1800$ rpm; and $C_M$ for an oil-hardened steel gear and pinion from Table 2 is 0.65. Thus, using Formula (4),

$$P_{100} = \frac{100 \times 65}{1800 \times 0.65} = 5.6 \text{ horsepower}$$

and from Chart 1, for a 3:1 ratio, the minimum pinion pitch diameter corresponding to $P_{100} = 5.6$ horsepower is 4.1 inches. The approximate pitch diameter of the gear is equal to the approximate pitch diameter of the pinion multiplied by the gear ratio and in this case is $4.1 \times 3 = 12.3$ inches.

*Number of Teeth:* The relation between pinion size and number of pinion teeth to give a well-balanced design is shown on Chart 2A for spiral bevel gears, and on Chart 2B for straight bevel and Zerol bevel gears. In the previous example the pinion pitch diameter was determined to be approximately 4.1 inches. For a 3:1 gear ratio the number of teeth in the pinion, from Chart 2A, is found to be 15. The number of teeth in the gear is equal to the number of pinion teeth multiplied by the gear ratio and in this case is $15 \times 3 = 45$ teeth. (If hardened gears are to be lapped it is advisable, if possible, to alter the ratio slightly so that the numbers of teeth in gear and pinion do not have a common factor to obtain better results in the lapping process.)

*Diametral Pitch:* The diametral pitch is obtained by dividing the number of teeth in the pinion by the approximate pinion pitch diameter. For the previous example,

$$\text{Diametral Pitch} = \frac{15}{4.1} = 3.7$$

Although it is not necessary to round 3.7 to an integral value, a diametral pitch of

Table 1.  Service Factors $C_s$ for Bevel Gear Drives*

| Character of Power Source | Character of Load on Driven Machine | | |
|---|---|---|---|
| | Uniform | Moderate Shock | Heavy Shock |
| Uniform | 1.00 | 1.25 | 1.75 |
| Light Shock | 1.10 | 1.35 | 1.80 |
| Medium Shock | 1.25 | 1.50 | 1.85 |

* Service factors given are for speed-decreasing drives; for speed-increasing drives, add 0.15 to these factors. Extreme repetitive shock and other applications where exceedingly high energy loads must be absorbed require special consideration and are not covered by the service factors in this table.

4 will be used for convenience in subsequent calculations in this problem. Based on this diametral pitch of 4, the pinion and gear pitch diameters are recomputed:

$$\text{Pitch Diameter of Pinion} = \frac{\text{Number of Pinion Teeth}}{\text{Diametral Pitch}} = \frac{15}{4} = 3.750 \text{ inches}$$

$$\text{Pitch Diameter of Gear} = \frac{\text{Number of Gear Teeth}}{\text{Diametral Pitch}} = \frac{45}{4} = 11.250 \text{ inches}$$

*Face Width:* The face width of the pair of bevel gears can now be determined from the following formula:

$$F = 0.15d\sqrt{1 + \left(\frac{N}{n}\right)^2} \tag{5}$$

where $F$ = face width in inches; $d$ = pitch diameter of pinion in inches; $N$ = number of teeth in gear; and $n$ = number of teeth in pinion. The face width determined by this formula will be equal to 0.3 times the cone distance, which is good design practice. In the previous example, $d$ = 3.750 inches, $N$ = 45, and $n$ = 15. Therefore,

$$F = 0.15 \times 3.750\sqrt{1 + \left(\frac{45}{15}\right)^2} = 1.78, \quad \text{say,} \quad 1\tfrac{3}{4} \text{ inches}$$

*Spiral Angle and Hand of Spiral:* For spiral bevel gears the spiral angle should be sufficient to give a face contact ratio of at least 1.25 to assure smooth tooth action and quiet gears. Design Chart 3 may be used as a guide to assist in the selection of the spiral angle. (See also " Hand of Spiral, Hand of Rotation, and Spiral Angle.") Before making a final decision on the spiral angle, the bearing loads should be computed to determine whether adequate bearings can be provided. (See " Bearing Loads Produced by Bevel Gears.")

In the previous example, the diametral pitch was 4 and the face width of the pinion $1\tfrac{3}{4}$ inches. The product of the diametral pitch and the pinion face width, $4 \times 1\tfrac{3}{4} = 7$, is used to enter Chart 3 to find a spiral angle sufficient to give at least 1.25 face contact ratio. From the chart it is found that spiral angles above 25 degrees will have face contact ratios above 1.25. Therefore a spiral angle of, say, 35 degrees may be selected, provided the gear mountings are rigid and the bearings large enough to absorb the radial and thrust loads associated with this spiral angle. Calculations of the thrust loads may indicate that the spiral angle should not be larger than, say, 30 degrees.

*Summary:* Thus far, the following approximate sizes of gear and pinion have been determined by means of relatively simple formulas and design charts: Teeth in pinion, 15; teeth in gear, 45; face width, 1.75 inches; spiral angle, 35 degrees; pinion pitch diameter, 3.75 inches; and gear pitch diameter, 11.25 inches. The next step is to check this approximate design using the surface durability equation in the following paragraph.

### Table 2.  Material Factors $C_M$ for Surface Durability of Bevel Gears

| Gear | | | Pinion | | | Surface Durability Factor $C_M$ |
|---|---|---|---|---|---|---|
| Material | Hardness | | Material | Hardness | | |
| | Brinell | Rockwell "C" | | Brinell | Rockwell "C" | |
| Cast Iron | ... | ... | Cast Iron | ... | ... | 0.30 |
| Cast Iron | ... | ... | Annealed Steel | 160–200 | ... | 0.30 |
| Cast Iron | ... | ... | Flame-hardened Steel | ... | 50* | 0.40 |
| Cast Iron | ... | ... | Casehardened Steel | ... | 55* | 0.40 |
| Heat-treated Steel | 210–245 | ... | Heat-treated Steel | 245–280 | ... | 0.35 |
| Heat-treated Steel | 250–300 | ... | Casehardened Steel | ... | 55* | 0.50 |
| Oil-hardened Steel | ... | ... | Oil-hardened Steel | ... | ... | 0.65 |
| Flame-hardened Steel | ... | 50* | Flame-hardened Steel | ... | 50* | 1.00 |
| Flame-hardened Steel | ... | 50* | Casehardened Steel | ... | 55* | 1.00 |
| Casehardened Steel | ... | 55* | Casehardened Steel | ... | 55* | 1.00 |

*Minimum values.

**Surface Durability of Bevel Gears.** — Except for cast iron gears, flame-hardened gears, or gears of brittle materials, surface durability is usually the determining factor in tooth failure. Bevel gears are, therefore, usually designed on a surface durability basis and then checked for strength.

The method previously described is a simplified approach of a more detailed method for determining bevel gear sizes for a given application. The gear sizes thus obtained should, therefore, be regarded as tentative until a further check is made using the following formula that takes into account a number of important factors not included in the initial determination.

$$P = FC_MC_sK_3L_T \qquad (6)$$

where  $P$ = maximum equivalent service horsepower. This should be equal or greater than the design horsepower $P_2$ determined by Formula (1) or (2), whichever is greater

$F$ = face width in inches.  (For hypoids, use face width of gear.)

$C_M$ = material factor for surface durability from Table 2.

$C_s$ = factor for contact ratio = $\sqrt{0.4(m_F + m_P)}$, where $m_F$ = face contact ratio from Chart 3. If $m_F$ is greater than 2, use $m_F = 2$ in this formula. For straight bevel and Zerol bevel gears, $m_F = 0$. For hypoids use the average of gear and pinion spiral angle when obtaining $m_F$ from Chart 3. $m_P$ = profile contact ratio from Table 3.

$K_3$ = combined factor for pinion diameter, velocity, and allowable unit load. Values are obtained from Chart 4. For hypoids, the pitch diameter of the corresponding bevel pinion is used to enter the chart.

$L_T$ = lubrication factor = 1 for bevel gears lubricated with mineral oil, or hypoid gear lubricated with extreme pressure lubricant. Use a value of 0.6 for hypoid gears lubricated with mineral oil.

An example using Formula (6) is given on page 858.

**Chart 1.   Horsepower per 100 RPM of the Pinion for Bevel Gears Operating at 90-degree Shaft Angle\***

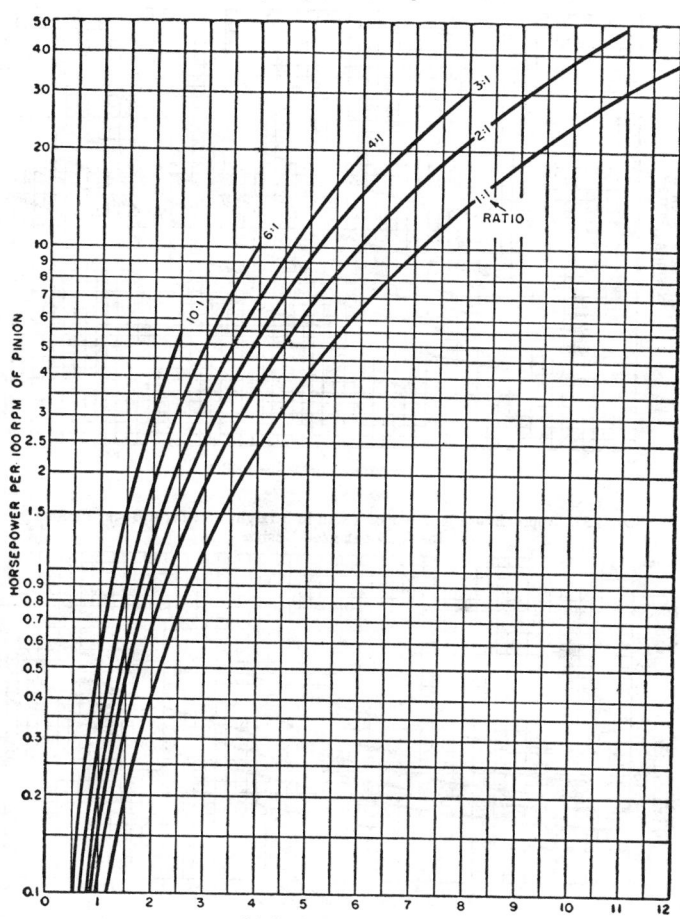

\* This design chart is based on gears having a face width equal to three-tenths of the cone distance and a pitch line velocity of 900 feet per minute for straight and Zerol bevel gears and 2500 feet per minute for spiral bevel gears.  If this chart is used for selecting an approximate pinion size for gears operating at lower speeds than indicated, the size obtained may be somewhat larger than actually needed.

For ratios in between those shown on the curves, interpolation may be employed to select the approximate pinion size required.

### Chart 2A. Approximate Number of Teeth in Spiral Bevel Pinions of 35-degree Spiral Angle for Well-balanced Design

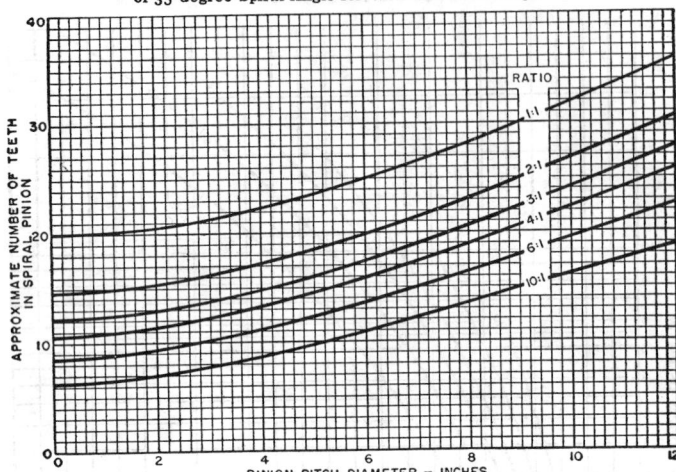

### Chart 2B. Approximate Number of Teeth in Straight and Zerol Bevel Pinions for Well-balanced Design

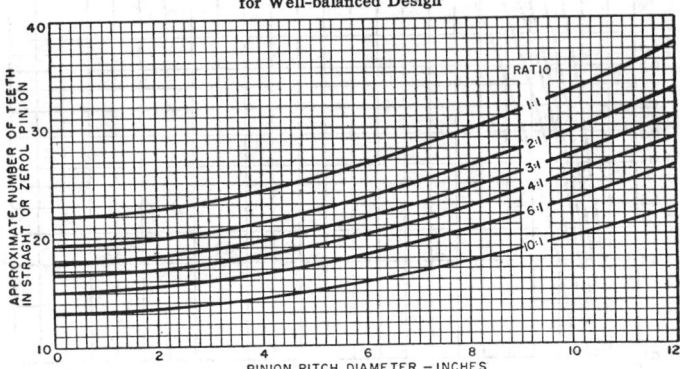

*Example:* Using the pinion and gear sizes previously determined, check the maximum equivalent service horsepower $P$ that the gears can carry according to Formula (6)

$$F = 1.75 \text{ inches}$$
$$C_M = 0.65 \text{ (Table 2)}$$
$$m_F = 1.85 \text{ for 35-degree spiral angle (Chart 3)}$$
$$m_P = 1.22 \text{ for 15 teeth (Table 3)}$$
$$C_s = \sqrt{0.4(1.85 + 1.22)} = 1.11$$

**Chart 3. Spiral Angles Corresponding to Various Face Contact Ratios and Products of Face Width by Diametral Pitch***

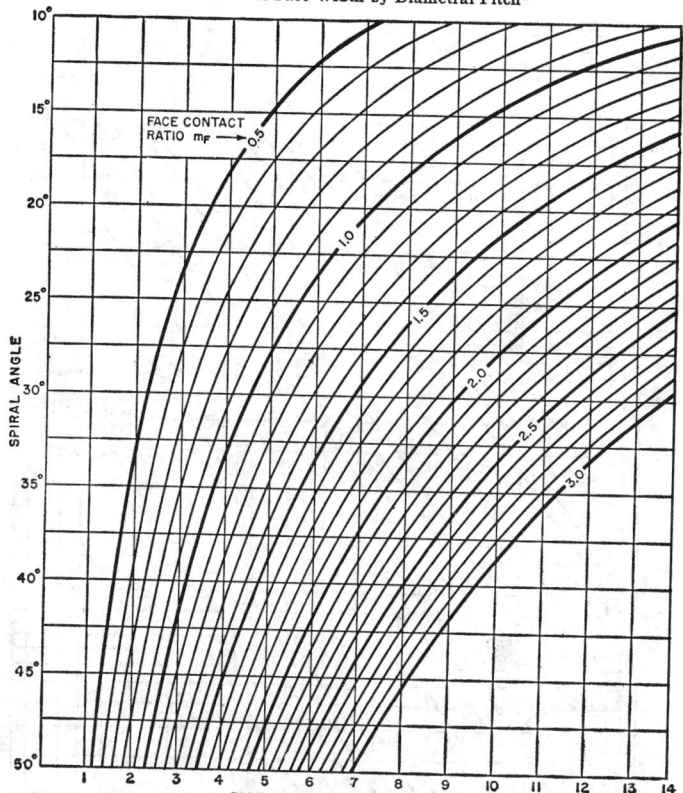

* Chart is used as follows: Assume that it is desired to find the spiral angle corresponding to a face contact ratio of 1.5 and a product of face width by diametral pitch of 8. Move up vertically from the number 8 on the bottom scale until the curved line representing 1.5 face contact ratio is intersected. From the intersection move horizontally left and read the spiral angle of 27 degrees.

Pinion rpm = 1800
Pinion pitch diameter = 3.750 inches
$K_3$ = 50 (Chart 4)
$P = 1.75 \times 0.65 \times 1.11 \times 50 \times 1 = 63$ horsepower maximum

The gears selected by means of the approximate method are, therefore, too small to carry the design load of 65 horsepower. In this case, therefore, it will be necessary to change the proportions somewhat to increase the capacity of the gears by 2 horsepower. One way to accomplish this is to increase the number of teeth

## Chart 4. Combined Durability Factor $K_3$ for Bevel Gears*

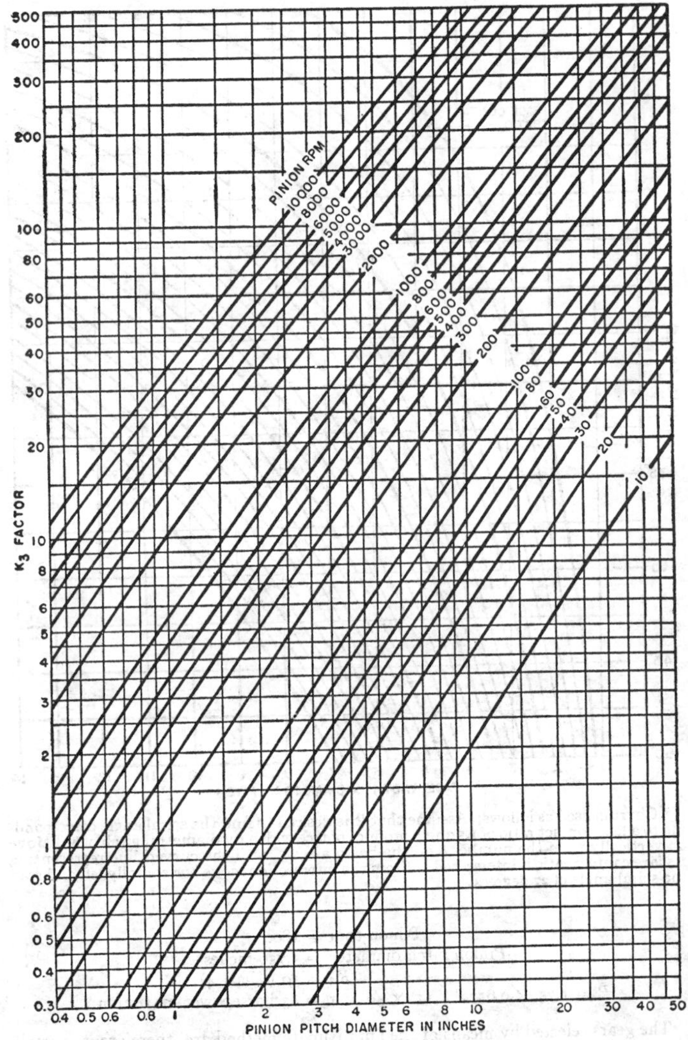

* This combined durability factor, which is used in the surface durability equation on page 856, is based on the pinion pitch diameter, velocity, and allowable unit load.

**Table 3.    Profile Contact Ratios $m_p$ for 20-degree Pressure Angle Bevel and Hypoid Gears**

| Number of Teeth in Pinion | Profile Contact Ratio $m_p$ | | Number of Teeth in Pinion | Profile Contact Ratio $m_p$ | |
|---|---|---|---|---|---|
| | Spiral Bevel and Hypoid Gears | Straight and Zerol Bevel Gears | | Spiral Bevel and Hypoid Gears | Straight and Zerol Bevel Gears |
| 6 | 0.87 | .... | 21 | 1.26 | 1.57 |
| 7 | 0.95 | .... | 22 | 1.27 | 1.58 |
| 8 | 1.02 | .... | 23, 24 | 1.28 | 1.59 |
| 9 | 1.07 | .... | 25 | 1.29 | 1.60 |
| 10 | 1.11 | .... | 26 | 1.29 | 1.61 |
| 11 | 1.15 | .... | 27, 28 | 1.30 | 1.62 |
| 12 | 1.19 | .... | 29 | 1.30 | 1.63 |
| 13 | 1.20 | 1.48 | 30, 31 | 1.31 | 1.64 |
| 14 | 1.21 | 1.49 | 32, 33 | 1.32 | 1.65 |
| 15 | 1.22 | 1.51 | 34, 35 | 1.32 | 1.66 |
| 16 | 1.23 | 1.52 | 36, 37 | 1.33 | 1.67 |
| 17 | 1.24 | 1.53 | 38, 39, 40 | 1.34 | 1.68 |
| 18 | 1.25 | 1.54 | | | |
| 19 | 1.25 | 1.55 | | | |
| 20 | 1.26 | 1.56 | | | |

**Table 4.    Allowable Bending Stress $s'$ for Various Materials Used in Bevel Gears**

| Material | Condition of Material | Heat Treatment | Brinell Hardness Number | Allowable Bending Stress $s'$ |
|---|---|---|---|---|
| Cast Iron | Ordinary | As Cast | .... | 4,600 |
| Cast Iron | High grade | As Cast | 200–300 | 7,000 |
| Steel | Forged or Equivalent | Normalized | 140–180 | 11,000 |
| Steel | Forged or Equivalent | Hardened and Tempered | 180–220 | 13,500 |
| Steel | Forged or Equivalent | Hardened and Tempered | 300–350 | 19,000 |
| Steel | Forged or Equivalent | Flame or Induction Hardened* | 450–550 | 13,500* |
| Steel | Forged or Equivalent | Carburized | 575–625 | 30,000 |

* Unhardened root fillets.

in the pinion slightly.    This will increase the pitch diameter of the pinion and, in proper proportion, the face width recomputed by Formula (5).    Another way would be to use a coarser pitch, say, 3.75, since in bevel gear practice the generating tools are not limited to integral pitches.

Chart 5A.   Geometry Factors $J$ for Straight Bevel Gears with 20-degree
Pressure Angle and 90-degree Shaft Angle

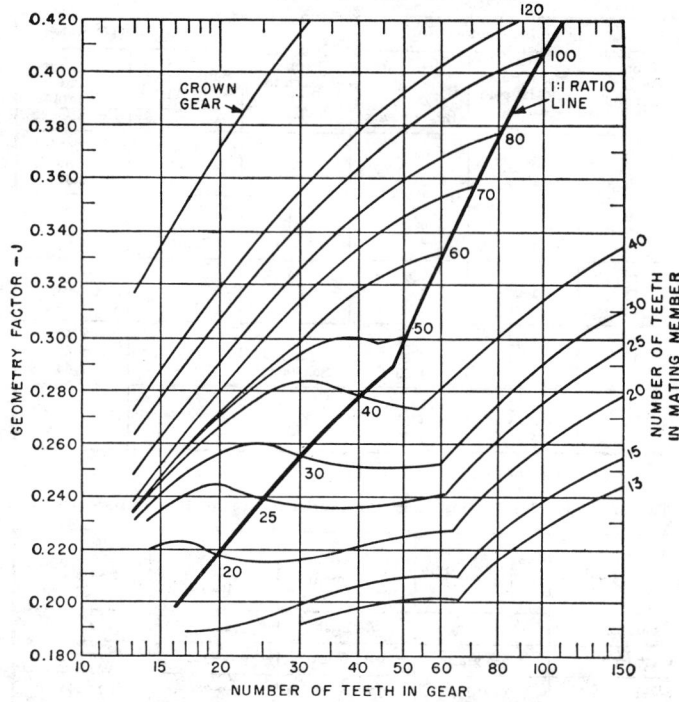

NUMBER OF TEETH IN GEAR
FOR WHICH GEOMETRY FACTOR IS DESIRED

**Strength of Bevel Gears.** — Although surface durability is usually the determining factor in gear tooth failure, there are cases where failure takes place by tooth breakage.  This may be particularly true where loads are high or where brittle materials are used for the gears.

The following formula may be used to check the power which a pair of bevel gears may transmit without failing by tooth breakage:

$$P_3 = \frac{W_t' n d L_f}{126,000} \text{ horsepower} \qquad (7)$$

where   $P_3$ = maximum horsepower.  This must be equal to or greater than the permissible horsepower determined by surface durability requirements, using Formula (6) on page 856.

$W_t' = \dfrac{s'FJ}{K_m P^{0.75}}$ = maximum allowable tangential load in pounds

$s'$ = maximum allowable bending stress in pounds per square inch from Table 4

$F$ = face width in inches (assumed equal on both members)

Chart 5B. Geometry Factors $J$ for Spiral Bevel Gears of 20-degree
Pressure Angle, 35-degree Spiral Angle, and 90-degree Shaft Angle

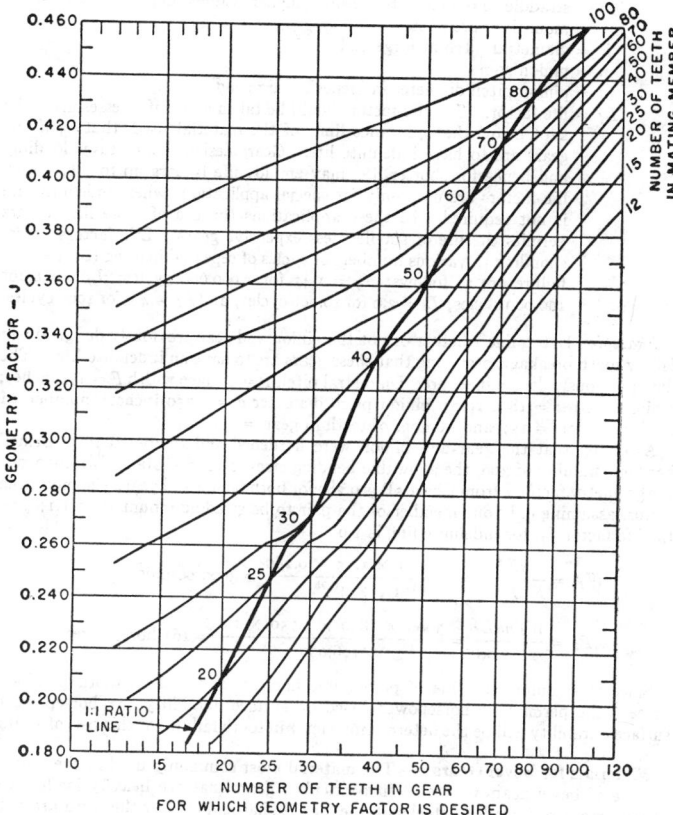

$J$ = geometry factor incorporating stress concentration factor, effective
face width, load location, load distribution, and inertia factor.
Charts 5A and 5B give values for gear and pinion for straight bevel
gears and spiral bevel gears of 35-degree spiral angle with 20-degree
pressure angle, 90-degree shaft angle, and a tool edge radius of 0.240
inch ÷ diametral pitch. Values of $J$ should be reduced by 11 per cent
if tools with standard edge radius of approximately 0.120 inch ÷
diametral pitch are employed.

Because the geometry factor $J$ will be different for the gear and the
mating pinion of straight bevel gears, the smaller of these $J$ values
should be used in computing $W_t'$. For spiral bevel gears, both $J$
values will be the same.

$K_m$ = mounting factor. In general, use a value of 1.0 when both members are straddle mounted and a value of 1.1 when only one member is straddle mounted. Somewhat higher values may be required if mountings are poor.

$P$ = diametral pitch at large end

$n$ = pinion rpm

$d$ = pinion pitch diameter in inches at large end

$L_f$ = life factor. The life factor should be taken as 1.0 if stresses are to be kept below the endurance limit of the material used, that is, if the gears are to have indefinite life. Gears designed for heavy loading, and hence for shorter life, may employ life factors up to 4.6. Such life factors are used only for special applications where indefinite life is not required. In these applications the use of large life factors permits the use of smaller, less expensive gears. Life factors corresponding to various numbers of cycles of repeated loading required for failure are as follows: $L_f$ = 1.45 for 1,000,000 cycles; $L_f$ = 2.0 for 100,000 cycles; $L_f$ = 3.0 for 10,000 cycles; and $L_f$ = 4.6 for 1000 cycles.

*Example:* Determine whether or not the pinion and gear previously designed will fail by tooth breakage, assuming that these gears are to have an indefinite life. The data previously determined, are: Diametral pitch $P = 4$; face width $F = 1\frac{3}{4}$ inches; pinion speed $n = 1800$ rpm; pinion pitch diameter $d = 3.750$ inches; number of teeth in pinion = 15; and number of teeth in gear = 45.

Assuming that the gear and pinion were hardened and tempered to a Brinell hardness number of 300, the allowable bending stress $s'$ from Table 4 is 19,000 psi; the geometry factor $J$ from Chart 5B is 0.280 for both pinion and gear; the mounting factor, assuming only one member of the pair to be straddle mounted, is 1.1; and the life factor $L_f$, for indefinite life, is 1.0.

$$W_t' = \frac{sFJ}{K_m P^{0.75}} = \frac{19,000 \times 1.75 \times 0.280}{1.1 \times 4^{0.75}} = 3000 \text{ pounds}$$

$$P_3 = \frac{W_t' n d L_f}{126,000} = \frac{3000 \times 1800 \times 3.750 \times 1.0}{126,000} = 161 \text{ horsepower}$$

Since this computed value of permissible horsepower based on tooth breakage exceeds the permissible horsepower based on surface durability (63 horsepower), surface durability will be the determining factor in tooth failure for this pair of gears.

**Materials for Bevel Gears.** — The material most commonly used in the manufacture of bevel gears is steel. In particular, gears that are heavily loaded are usually made of a grade of steel suitable for case hardening after the teeth are cut. The table on page 865 lists those steels most frequently specified for bevel gears, and may be used as a guide in selecting the type of steel suitable for a particular application. Gears of small size are usually made from bar stock, while larger gears are machined from forgings. For very large gears, for which forgings cannot be obtained, alloy steel castings are used.

*Forgings:* Where forgings are to be used, any of the commonly accepted forging methods in which the grain flow in the original steel billet follows the direction of the gear axis may be employed. Correct grain flow minimizes hardening distortion.

*Cast Iron:* Bevel gears may be made of high quality cast iron if the loads are relatively light and the size and weight of the gears are not of prime importance.

*Other Gear Materials:* Bevel gears that are lightly loaded and subject to corrosion may be made of brass, bronze, duralumin, or stainless steel. Gears made from plastic or phenolic materials are satisfactory only under very light loads.

**Representative Steels Used for Bevel Gear Applications***

| SAE or AISI No. | Type of Steel | Preliminary Heat Treatment | Brinell Hardness Number | ASTM Grain Size | Remarks |
|---|---|---|---|---|---|
| | | **CARBURIZING STEELS** | | | |
| | | Purchase Specifications | | | |
| 1024 | Manganese | Normalize | | | Low Alloy — oil quench limited to thin sections |
| 2512 | Nickel Alloy | Normalize — Anneal | 163–228 | 5–8 | Aircraft quality |
| 3310 3312X | Nickel-Chromium (Krupp) Nickel-Chromium | Normalize, then heat to 1450°F, cool in furnace. Reheat to 1170°F — cool in air | 163–228 | 5–8 | Used for maximum resistance to wear and fatigue |
| 4028 | Molybdenum | Normalize | 163–217 | | Low Alloy |
| 4615 4620 | Nickel-Molybdenum Nickel-Molybdenum | Normalize — 1700°F-1750°F | 163–217 | 5–8 | Good machining qualities. Well adapted to direct quench — gives tough core with minimum distortion |
| 4815 4820 | Nickel-Molybdenum Nickel-Molybdenum | Normalize | 163–241 | 5–8 | For aircraft and heavily loaded service |
| 5120 | Chromium | Normalize | 163–217 | 5–8 | |
| 8615 8620 8715 8720 | Chromium-Nickel-Molybdenum | Normalize — cool at hammer | 163–217 | 5–8 | Used as an alternate for 4620 |
| | | **OIL HARDENING AND FLAME HARDENING STEELS** | | | |
| 1141 | Sulphurized free cutting carbon steel | Normalize Heat treated | 179–228 255–269 | 5 or Coarser | Free cutting steel used for unhardened gears, oil treated gears, and for gears to be surface hardened where stresses are low |
| 4140 | Chromium-Molybdenum | For oil hardening: Normalize — Anneal | 179–212 | | Used for heat-treated, oil hardened, and surface hardened gears. Machining qualities of 4640 are superior to 4140, and it is the preferred steel for flame hardening |
| 4640 | Nickel-Molybdenum | For surface hardening: Normalize, reheat, quench and draw | 235–269 269–302 302–341 | | |
| 6145 | Chromium-Vanadium | Normalize — reheat, quench, and draw | 235–269 269–302 302–341 | | Fair machining qualities. Used for surface hardened gears when 4640 is not available |
| 8640 8739 | Chromium-Nickel-Molybdenum | Same as for 4640 | | | Used as an alternate for 4640 |
| | | **NITRIDING STEELS** | | | |
| Nitralloy H & G | Special Alloy | Anneal | 163–192 | | Normal hardness range for cutting is 20–28 Rockwell C |

* Any other steels equivalent to those listed in the table may also be used.

**Formulas for Dimensions of Milled Bevel Gears.** — Most bevel gears, as explained on page 840, are produced by generating methods. Even so, there are applications for which it may be desired to cut a pair of mating bevel gears by using rotary formed milling cutters. Examples of such cases include replacement gears for certain types of equipment and gears for use in experimental developments.

The tooth proportions of milled bevel gears differ in some respects from those of generated gears, the principal difference being that for milled bevel gears the tooth thicknesses of pinion and gear are made equal, and the addendum and dedendum of the pinion are respectively the same as those of the gear. The formulas in the accompanying table may be used to calculate the dimensions of milled bevel gears with shafts at a right angle, an acute angle, and an obtuse angle.

In the accompanying diagram and list of notation, the various terms and symbols applied to milled bevel gears are indicated.

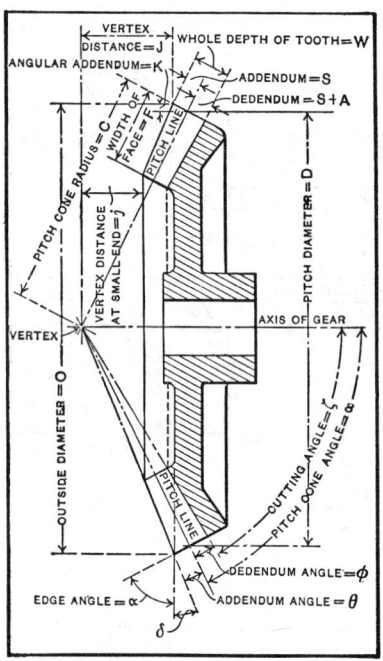

$N$ = number of teeth;
$P$ = diametral pitch;
$p$ = circular pitch;
$\alpha$ = pitch cone angle and edge angle;
$\Sigma$ = angle between shafts;
$D$ = pitch diameter;
$S$ = addendum;
$S+A$ = dedendum ($A$ = clearance);
$W$ = whole depth of tooth;
$T$ = thickness of tooth at pitch line;
$C$ = pitch cone radius;
$F$ = width of face;
$s$ = addendum at small end of tooth;
$t$ = thickness of tooth at pitch line at small end;
$\theta$ = addendum angle;
$\phi$ = dedendum angle;
$\gamma$ = face angle = pitch cone angle + addendum angle;
$\delta$ = angle of compound rest;
$\zeta$ = cutting angle;
$K$ = angular addendum;
$O$ = outside diameter;
$J$ = vertex distance;
$j$ = vertex distance at small end;
$N'$ = number of teeth for which to select cutter.

The Brown & Sharpe Mfg. Co. recommends that the formulas for milled bevel gears be modified to make the clearance at the bottom of the teeth uniform instead of tapering toward the vertex. If this recommendation is followed, then the cutting angle (root angle) should be determined by subtracting the *addendum* angle from the pitch cone angle instead of subtracting the dedendum angle as in the formula given in the table.

## Rules and Formulas for Calculating Dimensions of Milled Bevel Gears

| To Find | | Rule | Formula |
|---|---|---|---|
| Pitch Cone Angle of Pinion | | Divide the sine of the shaft angle by the sum of the cosine of the shaft angle and the quotient obtained by dividing the number of teeth in the gear by the number of teeth in the pinion; this gives the tangent. *Note:* For shaft angles greater than 90° the cosine is negative. | $\tan \alpha_P = \dfrac{\sin \Sigma}{\dfrac{N_G}{N_P} + \cos \Sigma}$ <br> For 90° shaft angle, <br> $\tan \alpha_P = \dfrac{N_P}{N_G}$ |
| Pitch Cone Angle of Gear | | Subtract the pitch cone angle of the pinion from the shaft angle. | $\alpha_G = \Sigma - \alpha_P$ |
| Pitch Diameter | | Divide the number of teeth by the diametral pitch. | $D = N \div P$ |
| These dimensions are the same for both gear and pinion. | Addendum | Divide 1 by the diametral pitch. | $S = 1 \div P$ |
| | Dedendum | Divide 1.157 by the diametral pitch. | $S + A = 1.157 \div P$ |
| | Whole Depth of Tooth | Divide 2.157 by the diametral pitch. | $W = 2.157 \div P$ |
| | Thickness of Tooth at Pitch Line | Divide 1.571 by the diametral pitch. | $T = 1.571 \div P$ |
| | Pitch Cone Radius | Divide the pitch diameter by twice the sine of the pitch cone angle. | $C = \dfrac{D}{2 \times \sin \alpha}$ |
| | Addendum of Small End of Tooth | Subtract the width of face from the pitch cone radius, divide the remainder by the pitch cone radius and multiply by the addendum. | $s = S \times \dfrac{C - F}{C}$ |
| | Thickness of Tooth at Pitch Line at Small End | Subtract the width of face from the pitch cone radius, divide the remainder by the pitch cone radius and multiply by the thickness of the tooth at pitch line. | $t = T \times \dfrac{C - F}{C}$ |
| | Addendum Angle | Divide the addendum by the pitch cone radius to get the tangent. | $\tan \theta = \dfrac{S}{C}$ |
| | Dedendum Angle | Divide the dedendum by the pitch cone radius to get the tangent. | $\tan \phi = \dfrac{S + A}{C}$ |
| | Face Width (Max.) | Divide the pitch cone radius by 3 or divide 8 by the diametral pitch, whichever gives the smaller value. | $F = \dfrac{C}{3}$ or $F = \dfrac{8}{P}$ |
| | Circular Pitch | Divide 3.1416 by the diametral pitch. | $p = 3.1416 \div P$ |
| Face Angle | | Add the addendum angle to the pitch cone angle. | $\gamma = \alpha + \theta$ |
| Compound Rest Angle for Turning Blank | | Subtract both the pitch cone angle and the addendum angle from 90 degrees. | $\delta = 90° - \alpha - \theta$ |
| Cutting Angle | | Subtract the dedendum angle from the pitch cone angle. | $\zeta = \alpha - \phi$ |
| Angular Addendum | | Multiply the addendum by the cosine of the pitch cone angle. | $K = S \times \cos \alpha$ |
| Outside Diameter | | Add twice the angular addendum to the pitch diameter. | $O = D + 2K$ |
| Vertex or Apex Distance | | Multiply one-half the outside diameter by the cotangent of the face angle. | $J = \dfrac{O}{2} \times \cot \gamma$ |
| Vertex Distance at Small End of Tooth | | Subtract the width of face from the pitch cone radius; divide the remainder by the pitch cone radius and multiply by the apex distance. | $j = J \times \dfrac{C - F}{C}$ |
| Number of Teeth for which to Select Cutter | | Divide the number of teeth by the cosine of the pitch cone angle. | $N' = \dfrac{N}{\cos \alpha}$ |

## BEVEL GEARING

### Numbers of Formed Cutters Used to Mill Teeth in Mating Bevel Gear and Pinion with Shafts at Right Angles

(Number of cutter for gear given first, followed by number for pinion. See text, page 870.)

| Number of Teeth in Gear | 12 | 13 | 14 | 15 | 16 | 17 | 18 | 19 | 20 | 21 | 22 | 23 | 24 | 25 | 26 | 27 | 28 |
|---|---|---|---|---|---|---|---|---|---|---|---|---|---|---|---|---|---|
| 12 | 7-7 | | | | | | | | | | | | | | | | |
| 13 | 6-7 | 6-6 | | | | | | | | | | | | | | | |
| 14 | 5-7 | 6-6 | 6-6 | | | | | | | | | | | | | | |
| 15 | 5-7 | 5-6 | 5-6 | 5-5 | | | | | | | | | | | | | |
| 16 | 4-7 | 5-7 | 5-6 | 5-6 | 5-5 | | | | | | | | | | | | |
| 17 | 4-7 | 4-7 | 4-6 | 5-6 | 5-5 | 5-5 | | | | | | | | | | | |
| 18 | 4-7 | 4-7 | 4-6 | 4-6 | 4-5 | 4-5 | 5-5 | | | | | | | | | | |
| 19 | 3-7 | 4-7 | 4-6 | 4-6 | 4-6 | 4-5 | 4-5 | 4-4 | | | | | | | | | |
| 20 | 3-7 | 3-7 | 4-6 | 4-6 | 4-6 | 4-5 | 4-5 | 4-4 | 4-4 | | | | | | | | |
| 21 | 3-8 | 3-7 | 3-7 | 3-6 | 4-6 | 4-5 | 4-5 | 4-5 | 4-4 | 4-4 | | | | | | | |
| 22 | 3-8 | 3-7 | 3-7 | 3-6 | 3-6 | 3-5 | 4-5 | 4-5 | 4-4 | 4-4 | 4-4 | | | | | | |
| 23 | 3-8 | 3-7 | 3-7 | 3-6 | 3-6 | 3-5 | 3-5 | 3-5 | 3-4 | 4-4 | 4-4 | 4-4 | | | | | |
| 24 | 3-8 | 3-7 | 3-7 | 3-6 | 3-6 | 3-6 | 3-5 | 3-5 | 3-4 | 3-4 | 3-4 | 4-4 | 4-4 | | | | |
| 25 | 2-8 | 2-7 | 3-7 | 3-6 | 3-6 | 3-6 | 3-5 | 3-5 | 3-5 | 3-4 | 3-4 | 3-4 | 4-4 | 3-3 | | | |
| 26 | 2-8 | 2-7 | 3-7 | 3-6 | 3-6 | 3-6 | 3-5 | 3-5 | 3-5 | 3-4 | 3-4 | 3-4 | 3-4 | 3-3 | 3-3 | | |
| 27 | 2-8 | 2-7 | 2-7 | 2-6 | 3-6 | 3-6 | 3-5 | 3-5 | 3-5 | 3-5 | 3-4 | 3-4 | 3-4 | 3-4 | 3-4 | 3-3 | 3-3 |
| 28 | 2-8 | 2-7 | 2-7 | 2-6 | 2-6 | 3-6 | 3-5 | 3-5 | 3-5 | 3-5 | 3-4 | 3-4 | 3-4 | 3-4 | 3-4 | 3-3 | 3-3 |
| 29 | 2-8 | 2-7 | 2-7 | 2-7 | 2-6 | 2-6 | 3-5 | 3-5 | 3-5 | 3-5 | 3-4 | 3-4 | 3-4 | 3-4 | 3-4 | 3-3 | 3-3 |
| 30 | 2-8 | 2-7 | 2-7 | 2-7 | 2-6 | 2-6 | 2-5 | 2-5 | 3-5 | 3-5 | 3-4 | 3-4 | 3-4 | 3-4 | 3-4 | 3-3 | 3-3 |
| 31 | 2-8 | 2-7 | 2-7 | 2-7 | 2-6 | 2-6 | 2-6 | 2-5 | 2-5 | 2-5 | 3-4 | 3-4 | 3-4 | 3-4 | 3-4 | 3-3 | 3-3 |
| 32 | 2-8 | 2-7 | 2-7 | 2-7 | 2-6 | 2-6 | 2-6 | 2-5 | 2-5 | 2-5 | 2-4 | 3-4 | 3-4 | 3-4 | 3-4 | 3-3 | 3-3 |
| 33 | 2-8 | 2-8 | 2-7 | 2-7 | 2-6 | 2-6 | 2-6 | 2-5 | 2-5 | 2-5 | 2-4 | 2-4 | 2-4 | 3-4 | 3-4 | 3-4 | 3-3 |
| 34 | 2-8 | 2-8 | 2-7 | 2-7 | 2-6 | 2-6 | 2-6 | 2-5 | 2-5 | 2-5 | 2-4 | 2-4 | 2-4 | 2-4 | 2-4 | 3-4 | 3-3 |
| 35 | 2-8 | 2-8 | 2-7 | 2-7 | 2-6 | 2-6 | 2-6 | 2-5 | 2-5 | 2-5 | 2-4 | 2-4 | 2-4 | 2-4 | 2-4 | 2-4 | 2-3 |
| 36 | 2-8 | 2-8 | 2-7 | 2-7 | 2-6 | 2-6 | 2-6 | 2-5 | 2-5 | 2-5 | 2-5 | 2-4 | 2-4 | 2-4 | 2-4 | 2-4 | 2-3 |
| 37 | 2-8 | 2-8 | 2-7 | 2-7 | 2-6 | 2-6 | 2-6 | 2-5 | 2-5 | 2-5 | 2-5 | 2-4 | 2-4 | 2-4 | 2-4 | 2-4 | 2-3 |
| 38 | 2-8 | 2-8 | 2-7 | 2-7 | 2-6 | 2-6 | 2-6 | 2-5 | 2-5 | 2-5 | 2-5 | 2-4 | 2-4 | 2-4 | 2-4 | 2-4 | 2-4 |
| 39 | 2-8 | 2-8 | 2-7 | 2-7 | 2-6 | 2-6 | 2-6 | 2-5 | 2-5 | 2-5 | 2-5 | 2-4 | 2-4 | 2-4 | 2-4 | 2-4 | 2-4 |
| 40 | 1-8 | 2-8 | 2-7 | 2-7 | 2-6 | 2-6 | 2-6 | 2-6 | 2-5 | 2-5 | 2-5 | 2-4 | 2-4 | 2-4 | 2-4 | 2-4 | 2-4 |
| 41 | 1-8 | 1-8 | 2-7 | 2-7 | 2-6 | 2-6 | 2-6 | 2-6 | 2-5 | 2-5 | 2-5 | 2-4 | 2-4 | 2-4 | 2-4 | 2-4 | 2-4 |
| 42 | 1-8 | 1-8 | 2-7 | 2-7 | 2-6 | 2-6 | 2-6 | 2-6 | 2-5 | 2-5 | 2-5 | 2-5 | 2-4 | 2-4 | 2-4 | 2-4 | 2-4 |
| 43 | 1-8 | 1-8 | 1-7 | 2-7 | 2-6 | 2-6 | 2-6 | 2-6 | 2-5 | 2-5 | 2-5 | 2-5 | 2-4 | 2-4 | 2-4 | 2-4 | 2-4 |
| 44 | 1-8 | 1-8 | 1-7 | 1-7 | 2-6 | 2-6 | 2-6 | 2-6 | 2-5 | 2-5 | 2-5 | 2-5 | 2-4 | 2-4 | 2-4 | 2-4 | 2-4 |
| 45 | 1-8 | 1-8 | 1-7 | 1-7 | 1-6 | 2-6 | 2-6 | 2-6 | 2-5 | 2-5 | 2-5 | 2-5 | 2-4 | 2-4 | 2-4 | 2-4 | 2-4 |
| 46 | 1-8 | 1-8 | 1-7 | 1-7 | 1-7 | 2-6 | 2-6 | 2-6 | 2-5 | 2-5 | 2-5 | 2-5 | 2-4 | 2-4 | 2-4 | 2-4 | 2-4 |
| 47 | 1-8 | 1-8 | 1-7 | 1-7 | 1-7 | 1-6 | 2-6 | 2-6 | 2-5 | 2-5 | 2-5 | 2-5 | 2-4 | 2-4 | 2-4 | 2-4 | 2-4 |
| 48 | 1-8 | 1-8 | 1-7 | 1-7 | 1-7 | 1-6 | 1-6 | 2-6 | 2-5 | 2-5 | 2-5 | 2-5 | 2-4 | 2-4 | 2-4 | 2-4 | 2-4 |
| 49 | 1-8 | 1-8 | 1-7 | 1-7 | 1-7 | 1-6 | 1-6 | 1-6 | 2-5 | 2-5 | 2-5 | 2-5 | 2-4 | 2-4 | 2-4 | 2-4 | 2-4 |
| 50 | 1-8 | 1-8 | 1-7 | 1-7 | 1-7 | 1-6 | 1-6 | 1-6 | 2-5 | 2-5 | 2-5 | 2-5 | 2-4 | 2-4 | 2-4 | 2-4 | 2-4 |
| 51 | 1-8 | 1-8 | 1-7 | 1-7 | 1-7 | 1-6 | 1-6 | 1-6 | 1-5 | 2-5 | 2-5 | 2-5 | 2-4 | 2-4 | 2-4 | 2-4 | 2-4 |
| 52 | 1-8 | 1-8 | 1-7 | 1-7 | 1-7 | 1-6 | 1-6 | 1-6 | 1-5 | 1-5 | 2-5 | 2-5 | 2-4 | 2-4 | 2-4 | 2-4 | 2-4 |
| 53 | 1-8 | 1-8 | 1-7 | 1-7 | 1-7 | 1-6 | 1-6 | 1-6 | 1-5 | 1-5 | 1-5 | 2-5 | 2-4 | 2-4 | 2-4 | 2-4 | 2-4 |
| 54 | 1-8 | 1-8 | 1-7 | 1-7 | 1-7 | 1-6 | 1-6 | 1-6 | 1-5 | 1-5 | 1-5 | 1-5 | 2-4 | 2-4 | 2-4 | 2-4 | 2-4 |
| 55 | 1-8 | 1-8 | 1-7 | 1-7 | 1-7 | 1-6 | 1-6 | 1-6 | 1-5 | 1-5 | 1-5 | 1-5 | 1-4 | 2-4 | 2-4 | 2-4 | 2-4 |

**Numbers of Formed Cutters Used to Mill Teeth in Mating Bevel Gear and Pinion with Shafts at Right Angles** (*Continued*)

(Number of cutter for gear given first, followed by number for pinion. See text, page 870.)

| | | Number of Teeth in Pinion | | | | | | | | | | | | | | | |
|---|---|---|---|---|---|---|---|---|---|---|---|---|---|---|---|---|---|
| | | 12 | 13 | 14 | 15 | 16 | 17 | 18 | 19 | 20 | 21 | 22 | 23 | 24 | 25 | 26 | 27 | 28 |
| Number of Teeth in Gear | 56 | 1-8 | 1-8 | 1-7 | 1-7 | 1-6 | 1-6 | 1-6 | 1-6 | 1-5 | 1-5 | 1-5 | 1-5 | 1-4 | 1-4 | 2-4 | 2-4 | 2-4 |
| | 57 | 1-8 | 1-8 | 1-7 | 1-7 | 1-6 | 1-6 | 1-6 | 1-6 | 1-5 | 1-5 | 1-5 | 1-5 | 1-4 | 1-4 | 1-4 | 2-4 | 2-4 |
| | 58 | 1-8 | 1-8 | 1-7 | 1-7 | 1-6 | 1-6 | 1-6 | 1-6 | 1-6 | 1-5 | 1-5 | 1-5 | 1-5 | 1-4 | 1-4 | 1-4 | 2-4 |
| | 59 | 1-8 | 1-8 | 1-7 | 1-7 | 1-6 | 1-6 | 1-6 | 1-6 | 1-6 | 1-5 | 1-5 | 1-5 | 1-5 | 1-4 | 1-4 | 1-4 | 1-4 |
| | 60 | 1-8 | 1-8 | 1-7 | 1-7 | 1-6 | 1-6 | 1-6 | 1-6 | 1-6 | 1-5 | 1-5 | 1-5 | 1-5 | 1-4 | 1-4 | 1-4 | 1-4 |
| | 61 | 1-8 | 1-8 | 1-7 | 1-7 | 1-6 | 1-6 | 1-6 | 1-6 | 1-6 | 1-5 | 1-5 | 1-5 | 1-5 | 1-4 | 1-4 | 1-4 | 1-4 |
| | 62 | 1-8 | 1-8 | 1-7 | 1-7 | 1-6 | 1-6 | 1-6 | 1-6 | 1-6 | 1-5 | 1-5 | 1-5 | 1-5 | 1-4 | 1-4 | 1-4 | 1-4 |
| | 63 | 1-8 | 1-8 | 1-7 | 1-7 | 1-6 | 1-6 | 1-6 | 1-6 | 1-6 | 1-5 | 1-5 | 1-5 | 1-5 | 1-4 | 1-4 | 1-4 | 1-4 |
| | 64 | 1-8 | 1-8 | 1-7 | 1-7 | 1-6 | 1-6 | 1-6 | 1-6 | 1-6 | 1-5 | 1-5 | 1-5 | 1-5 | 1-4 | 1-4 | 1-4 | 1-4 |
| | 65 | 1-8 | 1-8 | 1-7 | 1-7 | 1-7 | 1-6 | 1-6 | 1-6 | 1-6 | 1-5 | 1-5 | 1-5 | 1-5 | 1-4 | 1-4 | 1-4 | 1-4 |
| | 66 | 1-8 | 1-8 | 1-7 | 1-7 | 1-7 | 1-6 | 1-6 | 1-6 | 1-6 | 1-5 | 1-5 | 1-5 | 1-5 | 1-4 | 1-4 | 1-4 | 1-4 |
| | 67 | 1-8 | 1-8 | 1-7 | 1-7 | 1-7 | 1-6 | 1-6 | 1-6 | 1-6 | 1-5 | 1-5 | 1-5 | 1-5 | 1-4 | 1-4 | 1-4 | 1-4 |
| | 68 | 1-8 | 1-8 | 1-7 | 1-7 | 1-7 | 1-6 | 1-6 | 1-6 | 1-6 | 1-5 | 1-5 | 1-5 | 1-5 | 1-4 | 1-4 | 1-4 | 1-4 |
| | 69 | 1-8 | 1-8 | 1-7 | 1-7 | 1-7 | 1-6 | 1-6 | 1-6 | 1-6 | 1-5 | 1-5 | 1-5 | 1-5 | 1-4 | 1-4 | 1-4 | 1-4 |
| | 70 | 1-8 | 1-8 | 1-7 | 1-7 | 1-7 | 1-6 | 1-6 | 1-6 | 1-6 | 1-5 | 1-5 | 1-5 | 1-5 | 1-4 | 1-4 | 1-4 | 1-4 |
| | 71 | 1-8 | 1-8 | 1-7 | 1-7 | 1-7 | 1-6 | 1-6 | 1-6 | 1-6 | 1-5 | 1-5 | 1-5 | 1-5 | 1-4 | 1-4 | 1-4 | 1-4 |
| | 72 | 1-8 | 1-8 | 1-7 | 1-7 | 1-7 | 1-6 | 1-6 | 1-6 | 1-6 | 1-5 | 1-5 | 1-5 | 1-5 | 1-4 | 1-4 | 1-4 | 1-4 |
| | 73 | 1-8 | 1-8 | 1-7 | 1-7 | 1-7 | 1-6 | 1-6 | 1-6 | 1-6 | 1-5 | 1-5 | 1-5 | 1-5 | 1-4 | 1-4 | 1-4 | 1-4 |
| | 74 | 1-8 | 1-8 | 1-7 | 1-7 | 1-7 | 1-6 | 1-6 | 1-6 | 1-6 | 1-5 | 1-5 | 1-5 | 1-5 | 1-4 | 1-4 | 1-4 | 1-4 |
| | 75 | 1-8 | 1-8 | 1-7 | 1-7 | 1-7 | 1-6 | 1-6 | 1-6 | 1-6 | 1-5 | 1-5 | 1-5 | 1-5 | 1-4 | 1-4 | 1-4 | 1-4 |
| | 76 | 1-8 | 1-8 | 1-7 | 1-7 | 1-7 | 1-6 | 1-6 | 1-6 | 1-6 | 1-5 | 1-5 | 1-5 | 1-5 | 1-4 | 1-4 | 1-4 | 1-4 |
| | 77 | 1-8 | 1-8 | 1-7 | 1-7 | 1-7 | 1-6 | 1-6 | 1-6 | 1-6 | 1-5 | 1-5 | 1-5 | 1-5 | 1-4 | 1-4 | 1-4 | 1-4 |
| | 78 | 1-8 | 1-8 | 1-7 | 1-7 | 1-7 | 1-6 | 1-6 | 1-6 | 1-6 | 1-5 | 1-5 | 1-5 | 1-5 | 1-4 | 1-4 | 1-4 | 1-4 |
| | 79 | 1-8 | 1-8 | 1-7 | 1-7 | 1-7 | 1-6 | 1-6 | 1-6 | 1-6 | 1-5 | 1-5 | 1-5 | 1-5 | 1-4 | 1-4 | 1-4 | 1-4 |
| | 80 | 1-8 | 1-8 | 1-7 | 1-7 | 1-7 | 1-6 | 1-6 | 1-6 | 1-6 | 1-5 | 1-5 | 1-5 | 1-5 | 1-4 | 1-4 | 1-4 | 1-4 |
| | 81 | 1-8 | 1-8 | 1-7 | 1-7 | 1-7 | 1-6 | 1-6 | 1-6 | 1-6 | 1-5 | 1-5 | 1-5 | 1-5 | 1-4 | 1-4 | 1-4 | 1-4 |
| | 82 | 1-8 | 1-8 | 1-7 | 1-7 | 1-7 | 1-6 | 1-6 | 1-6 | 1-6 | 1-5 | 1-5 | 1-5 | 1-5 | 1-4 | 1-4 | 1-4 | 1-4 |
| | 83 | 1-8 | 1-8 | 1-7 | 1-7 | 1-7 | 1-6 | 1-6 | 1-6 | 1-6 | 1-5 | 1-5 | 1-5 | 1-5 | 1-4 | 1-4 | 1-4 | 1-4 |
| | 84 | 1-8 | 1-8 | 1-7 | 1-7 | 1-7 | 1-6 | 1-6 | 1-6 | 1-6 | 1-5 | 1-5 | 1-5 | 1-5 | 1-4 | 1-4 | 1-4 | 1-4 |
| | 85 | 1-8 | 1-8 | 1-7 | 1-7 | 1-7 | 1-6 | 1-6 | 1-6 | 1-6 | 1-5 | 1-5 | 1-5 | 1-5 | 1-4 | 1-4 | 1-4 | 1-4 |
| | 86 | 1-8 | 1-8 | 1-7 | 1-7 | 1-7 | 1-6 | 1-6 | 1-6 | 1-6 | 1-5 | 1-5 | 1-5 | 1-5 | 1-4 | 1-4 | 1-4 | 1-4 |
| | 87 | 1-8 | 1-8 | 1-7 | 1-7 | 1-7 | 1-6 | 1-6 | 1-6 | 1-6 | 1-5 | 1-5 | 1-5 | 1-5 | 1-4 | 1-4 | 1-4 | 1-4 |
| | 88 | 1-8 | 1-8 | 1-7 | 1-7 | 1-7 | 1-6 | 1-6 | 1-6 | 1-6 | 1-5 | 1-5 | 1-5 | 1-5 | 1-4 | 1-4 | 1-4 | 1-4 |
| | 89 | 1-8 | 1-8 | 1-7 | 1-7 | 1-7 | 1-6 | 1-6 | 1-6 | 1-6 | 1-5 | 1-5 | 1-5 | 1-5 | 1-4 | 1-4 | 1-4 | 1-4 |
| | 90 | 1-8 | 1-8 | 1-7 | 1-7 | 1-7 | 1-6 | 1-6 | 1-6 | 1-6 | 1-5 | 1-5 | 1-5 | 1-5 | 1-4 | 1-4 | 1-4 | 1-4 |
| | 91 | 1-8 | 1-8 | 1-7 | 1-7 | 1-7 | 1-6 | 1-6 | 1-6 | 1-6 | 1-5 | 1-5 | 1-5 | 1-5 | 1-4 | 1-4 | 1-4 | 1-4 |
| | 92 | 1-8 | 1-8 | 1-7 | 1-7 | 1-7 | 1-6 | 1-6 | 1-6 | 1-6 | 1-5 | 1-5 | 1-5 | 1-5 | 1-4 | 1-4 | 1-4 | 1-4 |
| | 93 | 1-8 | 1-8 | 1-7 | 1-7 | 1-7 | 1-6 | 1-6 | 1-6 | 1-6 | 1-5 | 1-5 | 1-5 | 1-5 | 1-4 | 1-4 | 1-4 | 1-4 |
| | 94 | 1-8 | 1-8 | 1-7 | 1-7 | 1-7 | 1-6 | 1-6 | 1-6 | 1-6 | 1-5 | 1-5 | 1-5 | 1-5 | 1-4 | 1-4 | 1-4 | 1-4 |
| | 95 | 1-8 | 1-8 | 1-7 | 1-7 | 1-7 | 1-6 | 1-6 | 1-6 | 1-6 | 1-5 | 1-5 | 1-5 | 1-5 | 1-4 | 1-4 | 1-4 | 1-4 |
| | 96 | 1-8 | 1-8 | 1-7 | 1-7 | 1-7 | 1-6 | 1-6 | 1-6 | 1-6 | 1-5 | 1-5 | 1-5 | 1-5 | 1-4 | 1-4 | 1-4 | 1-4 |
| | 97 | 1-8 | 1-8 | 1-7 | 1-7 | 1-7 | 1-6 | 1-6 | 1-6 | 1-6 | 1-5 | 1-5 | 1-5 | 1-5 | 1-4 | 1-4 | 1-4 | 1-4 |
| | 98 | 1-8 | 1-8 | 1-7 | 1-7 | 1-7 | 1-6 | 1-6 | 1-6 | 1-6 | 1-5 | 1-5 | 1-5 | 1-5 | 1-4 | 1-4 | 1-4 | 1-4 |
| | 99 | 1-8 | 1-8 | 1-7 | 1-7 | 1-7 | 1-6 | 1-6 | 1-6 | 1-6 | 1-5 | 1-5 | 1-5 | 1-5 | 1-4 | 1-4 | 1-4 | 1-4 |
| | 100 | 1-8 | 1-8 | 1-7 | 1-7 | 1-7 | 1-6 | 1-6 | 1-6 | 1-6 | 1-5 | 1-5 | 1-5 | 1-5 | 1-4 | 1-4 | 1-4 | 1-4 |

**Selecting Formed Cutters for Milling Bevel Gears.** — For milling 14½-degree pressure angle bevel gears, the standard cutter series furnished by manufacturers of formed milling cutters is commonly used. There are 8 cutters in the series for each diametral pitch to cover the full range from a 12-tooth pinion to a crown gear. The difference between formed cutters used for milling spur gears and those used for bevel gears is that bevel gear cutters are thinner, since they must pass through the narrow tooth space at the small end of the bevel gear; otherwise the shape of the cutter and hence, the cutter number, are the same.

To select the proper number of cutter to be used when a bevel gear is to be milled, it is necessary, first, to compute what is called the " Number of Teeth, $N'$ for which to Select Cutter." This number of teeth can then be used to select the proper number of bevel gear cutter from the spur gear milling cutter table on page 746. The value of $N'$ may be computed using the last formula on page 867.

*Example 1:* What numbers of cutters are required for a pair of bevel gears of 4 diametral pitch and 70 degree shaft angle if the gear has 50 teeth and the pinion 20 teeth?

The pitch cone angle of the pinion is determined by using the first formula on page 867:

$$\tan \alpha_P = \frac{\sin \Sigma}{\dfrac{N_G}{N_P} + \cos \Sigma} = \frac{\sin 70^\circ}{\dfrac{50}{20} + \cos 70^\circ} = 0.33064; \quad \alpha_P = 18^\circ18'$$

The pitch cone angle of the gear is determined from the second formula on page 867:

$$\alpha_G = \Sigma - \alpha_P = 70^\circ - 18^\circ18' = 51^\circ42'$$

The numbers of teeth $N'$ for which to select the cutters for the gear and pinion may now be determined from the last formula on page 867:

$$N' \text{ for the pinion} = \frac{N_P}{\cos \alpha_P} = \frac{20}{\cos 18^\circ18'} = 21.1, \text{ say, 21 teeth}$$

$$N' \text{ for the gear} = \frac{N_G}{\cos \alpha_G} = \frac{50}{\cos 51^\circ42'} = 80.7, \text{ say, 81 teeth}$$

From the table on page 746 the numbers of the cutters for pinion and gear are found to be, respectively, 5 and 2.

*Example 2:* Required the cutters for a pair of bevel gears where the gear has 24 teeth and the pinion 12 teeth. The shaft angle is 90 degrees. As in the first example, the formulas given on page 867 will be used.

$$\tan \alpha_P = N_P \div N_G = 12 \div 24 = 0.5000; \text{ and } \alpha_P = 26^\circ34'$$
$$\alpha_G = \Sigma - \alpha_P = 90^\circ - 26^\circ34' = 63^\circ26'$$

$$N' \text{ for pinion} = 12 \div \cos 26^\circ34' = 13.4, \text{ say, 13 teeth}$$
$$N' \text{ for gear} = 24 \div \cos 63^\circ26' = 53.6, \text{ say, 54 teeth}$$

And from the table on page 746 the cutters for pinion and gear are found to be, respectively, 8 and 3.

**Use of Table for Selecting Formed Cutters for Milling Bevel Gears.** — The table beginning on page 868 gives the numbers of cutters to use for milling various numbers of teeth in the gear and pinion. The table applies only to bevel gears with axes at right angles. Thus, in Example 2 given above, the numbers of the cutters could have been obtained directly by entering the table with the actual numbers of teeth in the gear, 24, and the pinion, 12.

**Offset of Cutter for Milling Bevel Gears.** — When milling bevel gears with a rotary formed cutter, it is necessary to take two cuts through each tooth space with the gear blank slightly off center, first on one side and then on the other, to obtain a tooth of approximately the correct form. The gear blank is also rotated proportionately to obtain the proper tooth thickness at the large and small ends. The amount that the gear blank or cutter should be offset from the central position can be determined quite accurately by the use of the table " Factors for Obtaining Offset for Milling Bevel Gears," in conjunction with the following rule: Find the factor in the table corresponding to the number of cutter used and to the ratio of the pitch cone radius to the face width; then divide this factor by the diametral pitch and subtract the result from half the thickness of the cutter at the pitch line.

### Factors for Obtaining Offset for Milling Bevel Gears

| No. of Cutter | Ratio of Pitch Cone Radius to Width of Face $\left(\dfrac{C}{F}\right)$ | | | | | | | | | | | | |
|---|---|---|---|---|---|---|---|---|---|---|---|---|---|
| | $\dfrac{3}{1}$ | $\dfrac{3\frac{1}{4}}{1}$ | $\dfrac{3\frac{1}{2}}{1}$ | $\dfrac{3\frac{3}{4}}{1}$ | $\dfrac{4}{1}$ | $\dfrac{4\frac{1}{4}}{1}$ | $\dfrac{4\frac{1}{2}}{1}$ | $\dfrac{4\frac{3}{4}}{1}$ | $\dfrac{5}{1}$ | $\dfrac{5\frac{1}{2}}{1}$ | $\dfrac{6}{1}$ | $\dfrac{7}{1}$ | $\dfrac{8}{1}$ |
| 1 | 0.254 | 0.254 | 0.255 | 0.256 | 0.257 | 0.257 | 0.257 | 0.258 | 0.258 | 0.259 | 0.260 | 0.262 | 0.264 |
| 2 | 0.266 | 0.268 | 0.271 | 0.272 | 0.273 | 0.274 | 0.274 | 0.275 | 0.277 | 0.279 | 0.280 | 0.283 | 0.284 |
| 3 | 0.266 | 0.268 | 0.271 | 0.273 | 0.275 | 0.278 | 0.280 | 0.282 | 0.283 | 0.286 | 0.287 | 0.290 | 0.292 |
| 4 | 0.275 | 0.280 | 0.285 | 0.287 | 0.291 | 0.293 | 0.296 | 0.298 | 0.298 | 0.302 | 0.305 | 0.308 | 0.311 |
| 5 | 0.280 | 0.285 | 0.290 | 0.293 | 0.295 | 0.296 | 0.298 | 0.300 | 0.302 | 0.307 | 0.309 | 0.313 | 0.315 |
| 6 | 0.311 | 0.318 | 0.323 | 0.328 | 0.330 | 0.334 | 0.337 | 0.340 | 0.343 | 0.348 | 0.352 | 0.356 | 0.362 |
| 7 | 0.289 | 0.298 | 0.308 | 0.316 | 0.324 | 0.329 | 0.334 | 0.338 | 0.343 | 0.350 | 0.360 | 0.290 | 0.376 |
| 8 | 0.275 | 0.286 | 0.296 | 0.309 | 0.319 | 0.331 | 0.338 | 0.344 | 0.352 | 0.361 | 0.368 | 0.380 | 0.386 |

> **Note.**—For obtaining offset by above table, use formula:
>
> $$\text{Offset} = \frac{T}{2} - \frac{\text{factor from table}}{P}$$
>
> $P$ = diametral pitch of gear to be cut;
>
> $T$ = thickness of cutter used, measured at pitch line.

To illustrate, what would be the amount of offset for a bevel gear having 24 teeth, 6 diametral pitch, 30-degree pitch cone angle and 1¼-inch face or tooth length? In order to obtain a factor from the table, the ratio of the pitch cone radius to the face width must be determined. The pitch cone radius equals the pitch diameter divided by twice the sine of the pitch cone angle = $4 \div (2 \times 0.5) = 4$ inches. As the face width is 1.25, the ratio is $4 \div 1.25$ or about 3¼ to 1. The factor in the table for this ratio is 0.280 with a No. 4 cutter, which would be the cutter number for this particular gear. The thickness of the cutter at the pitch line is measured by using a vernier gear tooth caliper. The depth $S + A$ (see following illustration; $S$ = addendum; $A$ = clearance) at which to take the measurement equals 1.157 divided by the diametral pitch; thus, $1.157 \div 6 = 0.1928$ inch. The cutter thickness at this depth will vary with different cutters and even with the same cutter as it is ground away, because formed bevel gear cutters are commonly provided with side relief. Assuming that the thickness is 0.1745 inch, and substituting the values in the formula given, we have:

$$\text{Offset} = \frac{0.1745}{2} - \frac{0.280}{6} = 0.0406 \text{ inch}$$

**Adjusting the Gear Blank for Milling.** — After the offset is determined, the blank is adjusted laterally this amount, and the tooth spaces are milled around the blank. After having milled one side of each tooth to the proper dimensions, the blank is set over in the opposite direction the same amount from a position central with the cutter, and is rotated to line up the cutter with a tooth space at the small end. A trial cut is then taken, which will leave the tooth being milled a little too thick, provided the cutter is thin enough — as it should be — to pass

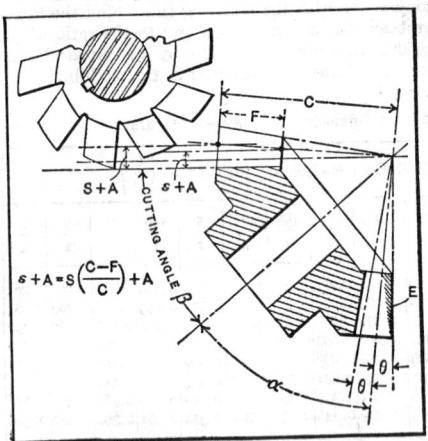

through the small end of the tooth space of the finished gear. This trial tooth is made the proper thickness by rotating the blank toward the cutter. To test the amount of offset measure the tooth thickness (with a vernier caliper) at the large and small ends. The caliper should be set so that the addendum at the small end is in proper proportion to the addendum at the large end; that is, in the ratio, $\dfrac{C-F}{C}$ (see illustration). In taking these measurements, if the thicknesses at both ends (which should be in this same ratio) are too great, rotate the tooth toward the cutter and take trial cuts until the proper thickness at either the large or small end is obtained. If the large end of the tooth is the right thickness and the small end too thick, the blank was offset too much; inversely, if the small end is correct and the large end too thick, the blank was not set enough off center, and, in either case, its position should be changed accordingly. The formula and table previously referred to will enable a properly turned blank to be set accurate enough for general work. The dividing head should be set to the cutting angle $\beta$ (see illustration), which is found by subtracting the addendum angle $\theta$ from the pitch cone angle $\alpha$. After cutting a bevel gear by the method described, the sides of the teeth at the small end should be filed as indicated by the shade lines at $E$; that is, by filing off a triangular area from the point of the tooth at the large end to the point at the small end, thence down to the pitch line and back diagonally to a point at the large end.

**Circular Thickness, Chordal Thickness, and Chordal Addendum of Milled Bevel Gear Teeth.** — In the formulas that follow, $T$ = circular tooth thickness on pitch circle at large end of tooth; $t$ = circular thickness at small end; $T_c$ and $t_c$ = chordal thickness at large and small end respectively; $S_c$ and $s_c$ = chordal addendum at large and small end respectively; $D$ = pitch diameter at large end; and $C, F, P, S, s$ and $\alpha$ are as defined on page 866.

$$T = \frac{1.5708}{P}; \qquad t = \frac{T(C-F)}{C}$$

$$T_c = T - \frac{T^3}{6\,D^2}; \qquad t_c = t - \frac{t^3}{6\,(D - 2\,F\sin\alpha)^2}$$

$$S_c = S + \frac{T^2\cos\alpha}{4\,D}; \qquad s_c = s + \frac{t^2\cos\alpha}{4\,(D - 2\,F\sin\alpha)}$$

**British Standard for Bevel Gears** (B.S. 545: 1949). — This British Standard for machine cut bevel gearing applies to all bevel gears connecting intersecting shafts and includes either straight or curved teeth having a normal pressure angle of 20 degrees at the pitch cone. These gears are divided into three classes:

Class A — Precision gears recommended for peripheral speeds above 2000 feet per minute when transmitting normal loads.

Class B — High-class cut gears suitable for peripheral speeds below 3000 feet per minute when transmitting normal loads.

Class C — Commercial cut gears suitable for peripheral speeds below 1200 feet per minute when transmitting normal loads.

This standard assumes that the gears will be supported on shafts of ample size provided with suitable, effectively-lubricated bearings.

*Form of Tooth:* The normal section of the British Standard basic rack tooth for bevel gears is shown in the accompanying diagram, and corresponds to the developed section of the crown gear on the back cone. The tip easing shown may be provided at the root of the tooth, if preferred, by the manufacturer.

**British Standard Basic Rack for Bevel Gears (for Unit Normal Pitch)**

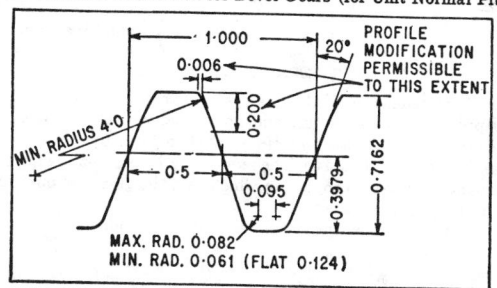

*Cutting Tool:* A cutting tool of form counterpart to a single flank and root fillet of the basic rack may be used to generate single flanks and root fillets of bevel gear teeth of British Standard form. In general, neither the amount of tip easing, nor the fillet radius on the bevel gear tooth will be a constant multiple of the pitch at all distances from the apex.

*Face Width:* The face width should not normally exceed one-third of the pitch cone distance or three times the pitch.

*Tooth Spiral:* The form of tooth spiral is left to the manufacturer as it is dependent on the type of cutter and method of cutting employed. For this reason no formula is given for the spiral angle at different positions along the face, although reference is made both to the spiral angle at midface and to that at the pitch circle. The following notation will be used:

$A$ = gear addendum

$a$ = pinion addendum

$C$ = cone distance = $D \div 2 \sin \gamma_w$

$D$ = gear pitch diameter

$d$ = pinion pitch diameter

$k_p$ = pinion addendum coefficient

$k_w$ = gear addendum coefficient

$P$ = diametral pitch

$P_n$ = normal diametral pitch = $P \sec \sigma_2$

$p$ = circular pitch = $\pi \div P$

$p_n$ = normal pitch = $p \cos \sigma_2$

$T$ = number of gear teeth

$t$ = number of pinion teeth

$\sigma_1$ = spiral angle at midface

$\sigma_2$ = spiral angle at pitch circle

$\gamma_p$ = pitch angle of pinion

$\gamma_w$ = pitch angle of gear

$\psi_n$ = normal pressure angle

$Q$ = interference factor corresponding to various values of $\sigma_1$. The following values are taken from a chart in the Standard: $0°$, 34.2; $5°$, 34; $10°$, 33.3; $20°$, 30.5; $30°$, 26.5; $40°$, 21.7; and $50°$, 16.5.

*Avoiding Combinations that Give Interference:* Where the number of teeth in the two mating gears is small, that is, when ($t \sec \gamma_p + T \sec \gamma_w$) is less than $Q$, interference or undercutting will be present. Use of such gears should be avoided. For straight bevel gears ($t \sec \gamma_p + T \sec \gamma_w$) should not be less than 34.2, and for bevel gears with 30-degree spiral angle, not less than 26.5.

*Addendum:* The addendum shall be determined by the following formulas:

$$\text{Pinion addendum } a = \frac{p_n}{\pi} (1 + k_p) = \frac{1 + k_p}{P_n}$$

$$\text{Gear addendum } A = \frac{p_n}{\pi} (1 + k_w) = \frac{1 + k_w}{P_n}$$

where $k_p$ and $k_w$, the addendum coefficients, are determined by the formulas,

$$k_p = 0.4 \left( 1 - \frac{t \sec \gamma_p}{T \sec \gamma_w} \right)$$

or

$$k_p = \left( 1 - \frac{2t \sec \gamma_p}{Q} \right), \text{ whichever is greater.}$$

Normally,        $k_w = - k_p$.

*Dedendum:* The dedendum of the tooth is made equal to the difference between the whole depth of the tooth and the addendum as determined above.

*Reduced Addendum and Dedendum:* If $t$ is less than $p_n(11.63 \div \sec^2 \sigma_2 \sec \gamma_p)$, the addendum and dedendum should each be reduced by $p_n(0.4 - 0.0344t \sec^2 \sigma_2 \sec \gamma_p)$ unless this quantity exceeds $0.065 \, p_n$, in which case the gears are outside the scope of the Standard.

*Caliper Settings:* Caliper settings shall be obtained as follows:

$$\text{Let } S = \frac{p}{2} + 2 \tan \psi_n \left( A \sec \sigma_2 - \frac{p}{\pi} \right)$$

$$\text{and } s = p - S$$

Then,

$$\text{Caliper thickness setting for gear, } G = \frac{S}{\sec \left( \sigma_2 + \dfrac{S}{2\,C} \right)}$$

$$\text{Caliper thickness setting for pinion, } g = \frac{s}{\sec \left( \sigma_2 + \dfrac{s}{2\,C} \right)}$$

$$\text{Caliper height setting for gear, } H = A + \frac{G^2}{4\,D \sec^2 \sigma_2 \sec \gamma_w}$$

$$\text{Caliper height setting for pinion, } h = a + \frac{g^2}{4\,d \sec^2 \sigma_2 \sec \gamma_p}$$

# Worm Gearing

**Worm Gearing.** — Worm gearing may be divided into two general classes, fine-pitch worm gearing, and coarse-pitch worm gearing. Fine-pitch worm gearing is segregated from coarse-pitch worm gearing for the following reasons:

1. Fine-pitch worms and worm gears are used largely to transmit motion rather than power. Tooth strength except at the coarser end of the fine-pitch range is seldom an important factor; durability and accuracy, as they affect the transmission of uniform angular motion, are of greater importance.

2. Housing constructions and lubricating methods are, in general, quite different for fine-pitch worm gearing.

3. Because of their small size, profile deviations and tooth bearings in fine-pitch worms and worm gears cannot be determined to the same degree of accuracy as can those of coarse pitches.

4. Equipment generally available for cutting fine-pitch worm gears has restrictions which limit the diameter, the lead range, the degree of accuracy attainable, and the kind of tooth bearing obtainable.

5. Special consideration must be given to top lands in fine-pitch hardened worms and worm gear-cutting tools.

6. Interchangeability and high production are important factors in fine-pitch worm gearing; individual matching of the worm to the gear, as often practiced with coarse-pitch precision worms, is impractical in the case of fine-pitch worm drives.

7. The methods used in the production and inspection of fine-pitch worm gears are different, in general, from those used for coarse-pitches.

## American Standard Design for Fine-pitch Worm Gearing (ASA B6.9–1956). —
This standard is intended as a design procedure for fine-pitch worms and worm gears having axes at right angles. It covers cylindrical worms with helical threads, and worm gears hobbed for fully conjugate tooth surfaces. It does not cover helical gears used as worm gears.

*Hobs:* The hob for producing the gear is a duplicate of the mating worm with regard to tooth profile, number of threads, and lead. The hob differs from the worm principally in that the outside diameter of the hob is larger to allow for resharpening and to provide bottom clearance in the worm gear.

*Pitches:* Eight standard axial pitches have been established to provide adequate coverage of the pitch range normally required: 0.030, 0.040, 0.050, 0.065, 0.080, 0.100, 0.130, 0.160 inch.

Axial pitch is used as a basis for this design standard because: (1) Axial pitch establishes lead which is a basic dimension in the production and inspection of worms; (2) the axial pitch of the worm is equal to the circular pitch of the gear in the central plane; (3) only one set of change gears or one master lead cam is required for a given lead, regardless of lead angle, on commonly available worm-producing equipment.

*Lead Angles:* Fifteen standard lead angles have been established to provide adequate coverage: 0.5, 1, 1.5, 2, 3, 4, 5, 7, 9, 11, 14, 17, 21, 25, and 30 degrees.

This series of lead angles has been standardized to: (1) Minimize tooling; (2) permit obtaining geometric similarity between worms of different axial pitch by keeping the same lead angle; and (3) take into account the production distribution found in fine-pitch worm gearing applications. For example, most fine-pitch worms have either one or two threads. This requires smaller increments at the low end of the lead angle series. For the less frequently used thread numbers, proportionately greater increments at the high end of the lead angle series are sufficient.

*Pressure Angle of Worm:* A pressure angle of 20 degrees has been selected as standard for cutters and grinding wheels used to produce worms within the scope of this Standard because it avoids objectionable undercutting regardless of lead angle.

Although the pressure angle of the cutter or grinding wheel used to produce the worm is 20 degrees, the normal pressure angle produced in the worm will actually be slightly greater, and will vary with the worm diameter, lead angle, and diameter of cutter or grinding wheel. A method for calculating the pressure angle change is

**Table 1.   Formulas for Proportions of American Standard Fine-pitch Worms and Worm Gears (ASAB6.9-1956)**

### LETTER SYMBOLS

$p$ = Circular pitch of worm gear = axial pitch of the worm, $p_x$, in the central plane

$p_x$ = Axial pitch of worm

$p_n$ = Normal circular pitch of worm and worm gear = $p_x \cos \lambda = p \cos \psi$

$\lambda$ = Lead angle of worm

$\psi$ = Helix angle of worm gear

$n$ = Number of threads in worm

$N$ = Number of teeth in worm gear = $n m_G$

$m_G$ = Ratio of gearing = $N \div n$

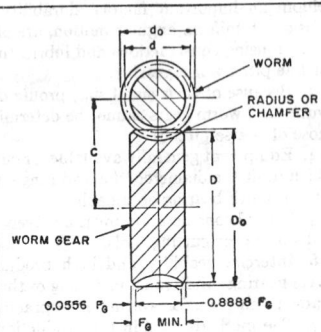

| Item | Formula | Item | Formula |
|------|---------|------|---------|
| **WORM DIMENSIONS** | | **WORM GEAR DIMENSIONS**\*\* | |
| Lead | $l = n p_x$ | Pitch Diameter | $D = N p \div \pi = N p_x \div \pi$ |
| Pitch Diameter | $d = l \div (\pi \tan \lambda)$ | Outside Diameter | $D_o = 2C - d + 2a$ |
| Outside Diameter | $d_o = d + 2a$ | Face Width | $F_{Gmin} = 1.125 \times$ |
| Safe Minimum Length of Threaded Portion of Worm\* | $F_W = \sqrt{D_o{}^2 - D^2}$ | | $\sqrt{(d_o + 2c)^2 - (d_o - 4a)^2}$ |

| **DIMENSIONS FOR BOTH WORM AND WORM GEAR** | | | |
|------|---------|------|---------|
| Addendum | $a = 0.3183 p_n$ | Tooth thickness | $t = 0.5 p_n$ |
| Whole Depth | $h_t = 0.7003 p_n + 0.002$ | Approximate normal pressure angle† | $\phi_n = 20$ degrees |
| Working Depth | $h_k = 0.6366 p_n$ | | |
| Clearance | $c = h_t - h_k$ | Center distance | $C = 0.5\,(d + D)$ |

All dimensions in inches unless otherwise indicated.

\* This formula allows a sufficient length for fine-pitch worms.

\*\* Current practice for fine-pitch worm gearing does not require the use of throated blanks. This results in the much simpler blank shown in the diagram which is quite similar to that for a spur or helical gear. The slight loss in contact resulting from the use of non-throated blanks has little effect on the load-carrying capacity of fine-pitch worm gears.

It is sometimes desirable to use topping hobs for producing worm gears in which the size relation between the outside and pitch diameters must be closely controlled. In such cases the blank is made slightly larger than $D_o$ by an amount (usually from 0.010 to 0.020) depending on the pitch. Topped worm gears will appear to have a small throat which is the result of the hobbing operation. For all intents and purposes, this throating is negligible and a blank so made is not to be considered as being a throated blank.

† As stated in the text on page 875, the actual pressure angle will be slightly greater due to the manufacturing process.

given under the heading " Effect of Production Method on Worm Profile and Pressure Angle."

*Pitch Diameter Range of Worms:* The minimum recommended worm pitch diameter is 0.250 inch and the maximum is 2.000 inches. Pitch diameters for all possible combinations of lead and lead angle, together with the number of threads for each lead, are given in Table 2.

*Tooth Form of Worm and Worm Gear:* The shape of the worm thread in the normal plane is defined as that which is produced by a symmetrical double-conical cutter or grinding wheel having straight elements and an included angle of 40 degrees.

Because worms and worm gears are closely related to their method of manufacture, it is impossible to specify clearly the tooth form of the worm gear without referring to the mating worm. For this reason, worm specifications should include the method of manufacture and the diameter of cutter or grinding wheel used. Similarly, for determining the shape of the generating tool, information about the method of producing the worm threads must be given to the manufacturer if the tools are to be designed correctly.

The worm profile will be a curve that departs from a straight line by varying amounts, depending on the worm diameter, lead angle, and the cutter or grinding wheel diameter. A method for calculating this deviation is given under the heading " Effect of Production Method on Worm Profile and Pressure Angle." The tooth form of the worm gear is understood to be made fully conjugate to the mating worm thread.

**Proportions of Fine-pitch Worms and Worm Gears.** — Hardened worms and cutting tools for worm gears should have adequate top lands. To automatically provide sufficient top lands, regardless of lead angle or axial pitch, the addendum and whole depth proportions of fine-pitch worm gearing are based on the normal circular pitch. Tooth proportions based on normal circular pitch for all combinations of standard axial pitches and lead angles are given in Table 3. Formulas for the proportions of worms and worm gears are given in Table 1.

*Example 1.* Determine the design of a worm and worm gear for a center distance of approximately 3 inches if the ratio is to be 10 to 1; axial pitch, 0.1600 inch; and lead angle, 30 degrees.

From Table 2 it can be determined that there are eight possible worm diameters that will satisfy the given conditions of lead angle and pitch. These worms have from 3 to 10 threads.

To satisfy the 3-inch center distance requirement it is now necessary to determine which of these eight worms, together with its mating worm gear, will come closest to making up this center distance. One way of doing this is as follows:

First use the formula given below to obtain the approximate number of threads necessary. Then from the eight possible worms in Table 2, choose the one whose number of threads is nearest this approximate value:

$$\text{Approximate number of threads needed for required center distance} = \frac{2\pi \times \text{required center distance}}{p_x (\cot \lambda + m_G)}$$

$$\text{Approximate number of threads} = \frac{2 \times 3.1416 \times 3}{0.1600(1.7320 + 10)} = 10.04 \text{ threads}$$

Of the eight possible worms in Table 2, the one having a number of threads nearest this value is the 10-thread worm with a pitch diameter of 0.8821 inch. Since the ratio of gearing is given as 10, $N$ may now be computed as follows:

$$N = 10 \times 10 = 100 \text{ teeth} \qquad \text{(from Table 1)}$$

### Table 2. Pitch Diameters of Fine-pitch Worms for American Standard Combinations of Lead and Lead Angle (ASA B6.9-1956)

| Lead in Inches, $l$ | Number of Threads, $n$ | Lead Angle $\lambda$ in Degrees | | | | | | | |
|---|---|---|---|---|---|---|---|---|---|
| | | 0.5 | 1.0 | 1.5 | 2.0 | 3.0 | 4.0 | 5.0 | 7.0 |
| | | Pitch Diameter $d$ in Inches | | | | | | | |
| 0.030 | 1 | 1.0937 | 0.5472 | 0.3647 | 0.2735 | ...... | ...... | ...... | ...... |
| 0.040 | 1 | 1.4583 | 0.7297 | 0.4863 | 0.3646 | 0.2429 | ...... | ...... | ...... |
| 0.050 | 1 | 1.8228 | 0.9121 | 0.6079 | 0.4558 | 0.3037 | 0.2276 | ...... | ...... |
| 0.060 | 2 | 2.1874 | 1.0945 | 0.7295 | 0.5469 | 0.3644 | 0.2731 | ...... | ...... |
| 0.065 | 1 | ...... | 1.1857 | 0.7903 | 0.5925 | 0.3948 | 0.2959 | 0.2365 | ...... |
| 0.080 | 1,2 | ...... | 1.4593 | 0.9726 | 0.7293 | 0.4859 | 0.3641 | 0.2911 | ...... |
| 0.090 | 3 | ...... | 1.6417 | 1.0942 | 0.8204 | 0.5466 | 0.4097 | 0.3274 | 0.2333 |
| 0.100 | 1,2 | ...... | 1.8242 | 1.2158 | 0.9116 | 0.6073 | 0.4552 | 0.3638 | 0.2592 |
| 0.120 | 3,4 | ...... | 2.1890 | 1.4590 | 1.0939 | 0.7288 | 0.5462 | 0.4366 | 0.3111 |
| 0.130 | 1,2 | ...... | ...... | 1.5805 | 1.1851 | 0.7896 | 0.5917 | 0.4730 | 0.3370 |
| 0.150 | 3,5 | ...... | ...... | 1.8237 | 1.3674 | 0.9110 | 0.6828 | 0.5457 | 0.3889 |
| 0.160 | 1,2,4 | ...... | ...... | 1.9453 | 1.4585 | 0.9718 | 0.7283 | 0.5821 | 0.4148 |
| 0.180 | 6 | ...... | ...... | 2.1884 | 1.6408 | 1.0932 | 0.8193 | 0.6549 | 0.4667 |
| 0.195 | 3 | ...... | ...... | ...... | 1.7776 | 1.1843 | 0.8876 | 0.7095 | 0.5055 |
| 0.200 | 2,4,5 | ...... | ...... | ...... | 1.8232 | 1.2147 | 0.9104 | 0.7276 | 0.5185 |
| 0.210 | 7 | ...... | ...... | ...... | 1.9143 | 1.2754 | 0.9559 | 0.7640 | 0.5444 |
| 0.240 | 3,6,8 | ...... | ...... | ...... | 2.1878 | 1.4576 | 1.0924 | 0.8732 | 0.6222 |
| 0.250 | 5 | | | | | 1.5184 | 1.1380 | 0.9096 | 0.6481 |
| 0.260 | 2,4 | | | | | 1.5791 | 1.1835 | 0.9459 | 0.6741 |
| 0.270 | 9 | | | | | 1.6398 | 1.2290 | 0.9823 | 0.7000 |
| 0.280 | 7 | | | | | 1.7006 | 1.2745 | 1.0187 | 0.7259 |
| 0.300 | 3,6,10 | | | | | 1.8220 | 1.3656 | 1.0915 | 0.7778 |
| 0.320 | 2,4,8 | | | | | 1.9435 | 1.4566 | 1.1642 | 0.8296 |
| 0.325 | 5 | | | | | 1.9739 | 1.4794 | 1.1824 | 0.8426 |
| 0.350 | 7 | | | | | 2.1257 | 1.5932 | 1.2734 | 0.9074 |
| 0.360 | 9 | | | | | ...... | 1.6387 | 1.3098 | 0.9333 |
| 0.390 | 3,6 | | | | | ...... | 1.7752 | 1.4189 | 1.0111 |
| 0.400 | 4,5,8,10 | | | | | ...... | 1.8207 | 1.4553 | 1.0370 |
| 0.450 | 9 | | | | | ...... | 2.0483 | 1.6372 | 1.1666 |
| 0.455 | 7 | | | | | ...... | ...... | 1.6554 | 1.1796 |
| 0.480 | 3,6 | | | | | ...... | ...... | 1.7463 | 1.2444 |
| 0.500 | 5,10 | | | | | ...... | ...... | 1.8191 | 1.2963 |
| 0.520 | 4,8 | | | | | ...... | ...... | 1.8919 | 1.3481 |
| 0.560 | 7 | | | | | ...... | ...... | 2.0374 | 1.4518 |
| 0.585 | 9 | | | | | ...... | ...... | ...... | 1.5166 |
| 0.600 | 6 | | | | | ...... | ...... | ...... | 1.5555 |
| 0.640 | 4,8 | | | | | ...... | ...... | ...... | 1.6592 |
| 0.650 | 5,10 | | | | | ...... | ...... | ...... | 1.6852 |
| 0.700 | 7 | | | | | ...... | ...... | ...... | 1.8148 |
| 0.720 | 9 | | | | | ...... | ...... | ...... | 1.8666 |
| 0.780 | 6 | | | | | ...... | ...... | ...... | 2.0222 |

For each lead shown in the first column, the numbers of threads given in the second column are those corresponding to standard axial pitches.

The standard axial pitch for each pitch diameter shown in the body of the table is obtained by dividing the lead given in column 1 by the number of threads given in column 2. Thus, where more than one number of threads is given in column 2, there will be more than one standard axial pitch for the corresponding pitch diameter.

*Example:* For a lead angle of 2.0 degrees and a lead of 0.240 inch, the corresponding pitch diameter is given in the table as 2.1878 inches.

Since there are three values of $n$ given in column 2, namely, 3, 6, and 8, there will also be three standard axial pitches available for this pitch diameter. These pitches are: $0.240 \div 3 = 0.080$; $0.240 \div 6 = 0.040$; and $0.240 \div 8 = 0.030$.

**Table 2** *(Continued)*. **Pitch Diameters of Fine-pitch Worms for American Standard Combinations of Lead and Lead Angle (ASA B6.9-1956)**

| Lead in Inches, $l$ | Number of Threads, $n$ | Lead Angle λ in Degrees | | | | | | |
|---|---|---|---|---|---|---|---|---|
| | | 9.0 | 11.0 | 14.0 | 17.0 | 21.0 | 25.0 | 30.0 |
| | | Pitch Diameter $d$ in Inches | | | | | | |
| 0.120 | 3,4 | 0.2412 | ...... | ...... | ...... | ...... | ...... | ...... |
| 0.130 | 1,2 | 0.2613 | ...... | ...... | ...... | ...... | ...... | ...... |
| 0.150 | 3,5 | 0.3015 | 0.2456 | ...... | ...... | ...... | ...... | ...... |
| 0.160 | 1,2,4 | 0.3216 | 0.2620 | ...... | ...... | ...... | ...... | ...... |
| 0.180 | 6 | 0.3618 | 0.2948 | ...... | ...... | ...... | ...... | ...... |
| 0.195 | 3 | 0.3919 | 0.3193 | 0.2490 | ...... | ...... | ...... | ...... |
| 0.200 | 2,4,5 | 0.4020 | 0.3275 | 0.2553 | ...... | ...... | ...... | ...... |
| 0.210 | 7 | 0.4221 | 0.3439 | 0.2681 | ...... | ...... | ...... | ...... |
| 0.240 | 3,6,8 | 0.4823 | 0.3930 | 0.3064 | 0.2499 | ...... | ...... | ...... |
| 0.250 | 5 | 0.5024 | 0.4094 | 0.3192 | 0.2603 | ...... | ...... | ...... |
| 0.260 | 2,4 | 0.5225 | 0.4258 | 0.3319 | 0.2707 | ...... | ...... | ...... |
| 0.270 | 9 | 0.5426 | 0.4421 | 0.3447 | 0.2811 | ...... | ...... | ...... |
| 0.280 | 7 | 0.5627 | 0.4585 | 0.3575 | 0.2915 | ...... | ...... | ...... |
| 0.300 | 3,6,10 | 0.6029 | 0.4913 | 0.3830 | 0.3123 | 0.2488 | ...... | ...... |
| 0.320 | 2,4,8 | 0.6431 | 0.5240 | 0.4085 | 0.3332 | 0.2654 | ...... | ...... |
| 0.325 | 5 | 0.6532 | 0.5322 | 0.4149 | 0.3384 | 0.2695 | ...... | ...... |
| 0.350 | 7 | 0.7034 | 0.5731 | 0.4468 | 0.3644 | 0.2902 | ...... | ...... |
| 0.360 | 9 | 0.7235 | 0.5895 | 0.4596 | 0.3748 | 0.2985 | 0.2457 | ...... |
| 0.390 | 3,6 | 0.7838 | 0.6387 | 0.4979 | 0.4060 | 0.3234 | 0.2662 | ...... |
| 0.400 | 4,5,8,10 | 0.8039 | 0.6550 | 0.5107 | 0.4165 | 0.3317 | 0.2730 | ...... |
| 0.450 | 9 | 0.9044 | 0.7369 | 0.5745 | 0.4685 | 0.3732 | 0.3072 | 0.2481 |
| 0.455 | 7 | 0.9144 | 0.7451 | 0.5809 | 0.4737 | 0.3773 | 0.3106 | 0.2509 |
| 0.480 | 3,6 | 0.9647 | 0.7860 | 0.6128 | 0.4998 | 0.3980 | 0.3277 | 0.2646 |
| 0.500 | 5,10 | 1.0049 | 0.8188 | 0.6383 | 0.5206 | 0.4146 | 0.3413 | 0.2757 |
| 0.520 | 4,8 | 1.0451 | 0.8515 | 0.6639 | 0.5414 | 0.4312 | 0.3550 | 0.2867 |
| 0.560 | 7 | 1.1255 | 0.9170 | 0.7149 | 0.5830 | 0.4644 | 0.3823 | 0.3087 |
| 0.585 | 9 | 1.1757 | 0.9580 | 0.7469 | 0.6091 | 0.4851 | 0.3993 | 0.3225 |
| 0.600 | 6 | 1.2059 | 0.9825 | 0.7660 | 0.6247 | 0.4975 | 0.4096 | 0.3308 |
| 0.640 | 4,8 | 1.2863 | 1.0480 | 0.8171 | 0.6663 | 0.5307 | 0.4369 | 0.3529 |
| 0.650 | 5,10 | 1.3064 | 1.0644 | 0.8298 | 0.6767 | 0.5390 | 0.4437 | 0.3584 |
| 0.700 | 7 | 1.4068 | 1.1463 | 0.8937 | 0.7288 | 0.5805 | 0.4778 | 0.3859 |
| 0.720 | 9 | 1.4470 | 1.1791 | 0.9192 | 0.7496 | 0.5971 | 0.4915 | 0.3970 |
| 0.780 | 6 | 1.5676 | 1.2773 | 0.9958 | 0.8121 | 0.6468 | 0.5324 | 0.4300 |
| 0.800 | 5,8,10 | 1.6078 | 1.3101 | 1.0213 | 0.8329 | 0.6634 | 0.5461 | 0.4411 |
| 0.900 | 9 | 1.8088 | 1.4738 | 1.1490 | 0.9370 | 0.7463 | 0.6144 | 0.4962 |
| 0.910 | 7 | 1.8289 | 1.4902 | 1.1618 | 0.9474 | 0.7546 | 0.6212 | 0.5017 |
| 0.960 | 6 | 1.9294 | 1.5721 | 1.2256 | 0.9995 | 0.7961 | 0.6553 | 0.5293 |
| 1.000 | 10 | 2.0098 | 1.6376 | 1.2767 | 1.0412 | 0.8292 | 0.6826 | 0.5513 |
| 1.040 | 8 | ...... | 1.7031 | 1.3277 | 1.0828 | 0.8624 | 0.7099 | 0.5734 |
| 1.120 | 7 | ...... | 1.8341 | 1.4299 | 1.1661 | 0.9287 | 0.7645 | 0.6175 |
| 1.170 | 9 | ...... | 1.9160 | 1.4937 | 1.2181 | 0.9720 | 0.7987 | 0.6451 |
| 1.280 | 8 | ...... | 2.0961 | 1.6341 | 1.3327 | 1.0614 | 0.8738 | 0.7057 |
| 1.300 | 10 | ...... | ...... | 1.6597 | 1.3535 | 1.0780 | 0.8874 | 0.7167 |
| 1.440 | 9 | ...... | ...... | 1.8384 | 1.4993 | 1.1941 | 0.9830 | 0.7939 |
| 1.600 | 10 | ...... | ...... | 2.0427 | 1.6658 | 1.3268 | 1.0922 | 0.8821 |

Other worm and worm gear dimensions may now be calculated using the formulas given in Table 1 or may be taken from the data presented in Tables 2 and 3.

$$l = 1.600 \text{ inches} \qquad\qquad\qquad\qquad\qquad \text{(from Table 2)}$$
$$d = 0.8821 \text{ inch} \qquad\qquad\qquad\qquad\qquad \text{(from Table 2)}$$
$$D = 100 \times 0.1600 \div 3.1416 = 5.0930 \text{ inches} \qquad \text{(from Table 1)}$$
$$C = 0.5(0.8821 + 5.0930) = 2.9876 \text{ inches} \qquad \text{(from Table 1)}$$
$$p_n = 0.1386 \text{ inch} \qquad\qquad\qquad\qquad\qquad \text{(from Table 3)}$$
$$a = 0.0441 \text{ inch} \qquad\qquad\qquad\qquad\qquad \text{(from Table 3)}$$
$$h_t = 0.0990 \text{ inch} \qquad\qquad\qquad\qquad\qquad \text{(from Table 3)}$$
$$h_k = 0.6366 \times 0.1386 = 0.0882 \text{ inch} \qquad \text{(from Table 1)}$$
$$c = 0.0990 - 0.0882 = 0.0108 \text{ inch} \qquad \text{(from Table 1)}$$
$$t = 0.5 \times 0.1386 = 0.0693 \text{ inch} \qquad \text{(from Table 1)}$$
$$d_o = 0.8821 + (2 \times 0.0441) = 0.9703 \text{ inch} \qquad \text{(from Table 1)}$$
$$D_o = (2 \times 2.9876) - 0.8821 + (2 \times 0.0441) = 5.1813 \qquad \text{(from Table 1)}$$
$$F_G = 1.125\sqrt{(0.9703 + 2 \times 0.0108)^2 - (0.9703 - 4 \times 0.0441)^2} = 0.6689 \text{ inch} \qquad \text{(from Table 1)}$$
$$F_W = \sqrt{5.2678^2 - 5.0930^2} = 1.3458 \text{ inches} \qquad \text{(from Table 1)}$$

*Example 2:* Determine the design of a worm and worm gear for a center distance of approximately 0.540 inch if the ratio is to be 50 to 1 and the axial pitch is to be 0.050 inch.

Assume that $n = 1$ (since most fine-pitch worms have either one or two threads). The lead of the worm will then be $np_x = 1 \times 0.050 = 0.050$ inch. From Table 2 it can be determined that there are six possible lead angles and corresponding worm diameters that will satisfy this lead. The approximate lead angle required to meet the conditions of the example can be computed from the following formula:

Cotangent of approximate lead angle =

$$\frac{2\pi \times \text{approximate center distance required}}{\text{assumed number of threads} \times \text{axial pitch}} - m_G$$

Using letter symbols, this formula becomes:

$$\cot\lambda = \frac{2\pi \times C}{n \times p_x} - m_G = \frac{2 \times 3.1416 \times 0.540}{1 \times 0.050} - 50 = 17.859$$

or

$$\lambda = 3° \ 12'$$

Of the six possible worms in Table 2, the one with the 3-degree lead angle is closest to the calculated 3° 12' lead angle. This worm, which has a pitch diameter of 0.3037 inch, is therefore selected.

The remaining worm and worm gear dimensions may now be determined from the data in Tables 2 and 3 and by computation using the formulas given in Table 1.

$$N = 50 \times 1 = 50 \text{ teeth} \qquad\qquad\qquad\qquad \text{(from Table 1)}$$
$$d = 0.3037 \text{ inch} \qquad\qquad\qquad\qquad\qquad \text{(from Table 2)}$$
$$D = 50 \times 0.050 \div 3.1416 = 0.7958 \text{ inch} \qquad \text{(from Table 1)}$$
$$C = 0.5(0.3037 + 0.7958) = 0.5498 \text{ inch} \qquad \text{(from Table 1)}$$
$$p_n = 0.0499 \text{ inch} \qquad\qquad\qquad\qquad\qquad \text{(from Table 3)}$$
$$a = 0.0159 \text{ inch} \qquad\qquad\qquad\qquad\qquad \text{(from Table 3)}$$
$$h_t = 0.0370 \text{ inch} \qquad\qquad\qquad\qquad\qquad \text{(from Table 3)}$$
$$h_k = 0.6366 \times 0.0499 = 0.0318 \text{ inch} \qquad \text{(from Table 1)}$$
$$c = 0.0370 - 0.0318 = 0.0052 \text{ inch} \qquad \text{(from Table 1)}$$

**Table 3. Tooth Proportions of American Standard Fine-pitch Worms and Worm Gears (ASA B6.9-1956)**

| Standard Axial Pitch in Inches, $p_x$ | Tooth Parts* | Lead Angle λ in Degrees | | | | | | | | | | | | | | |
|---|---|---|---|---|---|---|---|---|---|---|---|---|---|---|---|---|
| | | 0.5 | 1 | 1.5 | 2 | 3 | 4 | 5 | 7 | 9 | 11 | 14 | 17 | 21 | 25 | 30 |
| | | Dimensions of Tooth Parts in Inches† | | | | | | | | | | | | | | |
| 0.030 | $a$ | .0095 | .0095 | .0095 | .0095 | .0095 | .0095 | .0095 | .0095 | .0094 | .0094 | .0093 | .0091 | .0089 | .... | .... |
| | $h_t$ | .0229 | .0229 | .0229 | .0229 | .0229 | .0229 | .0229 | .0229 | .0227 | .0227 | .0225 | .0220 | .0216 | .... | .... |
| | $p_n$ | .0300 | .0300 | .0300 | .0300 | .0300 | .0299 | .0299 | .0298 | .0296 | .0294 | .0291 | .0287 | .0280 | .... | .... |
| 0.040 | $a$ | .0127 | .0127 | .0127 | .0127 | .0127 | .0127 | .0127 | .0126 | .0126 | .0125 | .0124 | .0122 | .0119 | .0115 | .... |
| | $h_t$ | .0299 | .0299 | .0299 | .0299 | .0299 | .0299 | .0299 | .0298 | .0297 | .0295 | .0293 | .0288 | .0282 | .0273 | .... |
| | $p_n$ | .0400 | .0400 | .0400 | .0400 | .0399 | .0399 | .0398 | .0397 | .0395 | .0393 | .0388 | .0383 | .0373 | .0363 | .... |
| 0.050 | $a$ | .0159 | .0159 | .0159 | .0159 | .0159 | .0159 | .0159 | .0158 | .0157 | .0156 | .0154 | .0152 | .0149 | .0144 | .0138 |
| | $h_t$ | .0370 | .0370 | .0370 | .0370 | .0370 | .0370 | .0370 | .0368 | .0365 | .0363 | .0359 | .0354 | .0348 | .0337 | .0324 |
| | $p_n$ | .0500 | .0500 | .0500 | .0500 | .0499 | .0499 | .0498 | .0496 | .0494 | .0491 | .0485 | .0478 | .0467 | .0453 | .0433 |
| 0.065 | $a$ | .... | .0207 | .0207 | .0207 | .0207 | .0206 | .0206 | .0205 | .0204 | .0203 | .0201 | .0198 | .0193 | .0188 | .0179 |
| | $h_t$ | .... | .0475 | .0475 | .0475 | .0475 | .0473 | .0473 | .0471 | .0469 | .0467 | .0462 | .0456 | .0445 | .0434 | .0414 |
| | $p_n$ | .... | .0650 | .0650 | .0650 | .0649 | .0648 | .0648 | .0645 | .0642 | .0638 | .0631 | .0622 | .0607 | .0589 | .0563 |
| 0.080 | $a$ | .... | .0255 | .0255 | .0254 | .0254 | .0254 | .0254 | .0253 | .0252 | .0250 | .0247 | .0244 | .0238 | .0231 | .0221 |
| | $h_t$ | .... | .0581 | .0581 | .0579 | .0579 | .0579 | .0579 | .0577 | .0574 | .0570 | .0563 | .0557 | .0544 | .0528 | .0506 |
| | $p_n$ | .... | .0800 | .0800 | .0800 | .0799 | .0798 | .0797 | .0794 | .0790 | .0785 | .0776 | .0765 | .0747 | .0725 | .0693 |
| 0.100 | $a$ | .... | .0318 | .0318 | .0318 | .0318 | .0318 | .0317 | .0316 | .0314 | .0312 | .0309 | .0304 | .0297 | .0288 | .0276 |
| | $h_t$ | .... | .0720 | .0720 | .0720 | .0720 | .0720 | .0717 | .0716 | .0711 | .0706 | .0700 | .0689 | .0673 | .0654 | .0627 |
| | $p_n$ | .... | .1000 | .1000 | .0999 | .0999 | .0998 | .0996 | .0993 | .0988 | .0982 | .0970 | .0956 | .0934 | .0906 | .0866 |
| 0.130 | $a$ | .... | .... | .0414 | .0414 | .0413 | .0413 | .0412 | .0411 | .0409 | .0406 | .0407 | .0396 | .0386 | .0375 | .0358 |
| | $h_t$ | .... | .... | .0931 | .0931 | .0929 | .0929 | .0926 | .0924 | .0920 | .0913 | .0904 | .0891 | .0869 | .0845 | .0808 |
| | $p_n$ | .... | .... | .1300 | .1299 | .1298 | .1297 | .1295 | .1290 | .1284 | .1276 | .1261 | .1243 | .1214 | .1178 | .1126 |
| 0.160 | $a$ | .... | .... | .0509 | .0509 | .0509 | .0508 | .0507 | .0506 | .0503 | .0500 | .0494 | .0487 | .0475 | .0462 | .0441 |
| | $h_t$ | .... | .... | .1140 | .1140 | .1140 | .1138 | .1135 | .1133 | .1127 | .1120 | .1107 | .1091 | .1065 | .1036 | .0990 |
| | $p_n$ | .... | .... | .1599 | .1599 | .1598 | .1596 | .1594 | .1588 | .1580 | .1571 | .1552 | .1530 | .1494 | .1450 | .1386 |

* $a$ = addendum; $h_t$ = whole depth; and $p_n$ = normal circular pitch.
† Tooth proportions are based on the formulas given in Table I.

$$t = 0.5 \times 0.0499 = 0.0250 \text{ inch} \qquad \text{(from Table 1)}$$
$$d_o = 0.3037 + (2 \times 0.0159) = 0.3355 \text{ inch} \qquad \text{(from Table 1)}$$
$$D_o = (2 \times 0.5498) - 0.3037 + (2 \times 0.0159) = 0.8277 \text{ inch} \qquad \text{(from Table 1)}$$
$$F_{G\min} = 1.125\sqrt{(0.3355 + 2 \times 0.0052)^2 - (0.3355 - 4 \times 0.0159)^2}$$
$$= 0.2405 \text{ inch} \qquad \text{(from Table 1)}$$
$$F_W = \sqrt{0.8277^2 - 0.7958^2} = 0.2276 \text{ inch} \qquad \text{(from Table 1)}$$

**Effect of Production Method on Worm Profile and Pressure Angle.** — In worm gearing, tooth bearing is usually used as the means of judging tooth profile accuracy since direct profile measurements on fine-pitch worms or worm gears is not practical. According to American Standard ASA B6.11-1956, Inspection of Fine Pitch Gears, a minimum of 50 per cent initial area of contact is suitable for most fine-pitch worm gearing, although in some cases, such as when the load fluctuates widely, a more restricted initial area of contact may be desirable.

Except where single-pointed lathe tools, end mills, or cutters of special shape are used in the manufacture of worms, the pressure angle and profile produced by the cutter are different from those of the cutter itself. The amounts of these differences depend on several factors, namely, diameter and lead angle of the worm, thickness and depth of the worm thread, and diameter of the cutter or grinding wheel. The accompanying diagram shows the curvature and pressure angle effects produced in the worm by cutters and grinding wheels, and how the amount of variation in worm profile and pressure angle is influenced by the diameter of the cutting tool used.

**Effect of Diameter of Cutting Tool on Profile and Pressure Angle of Worms**

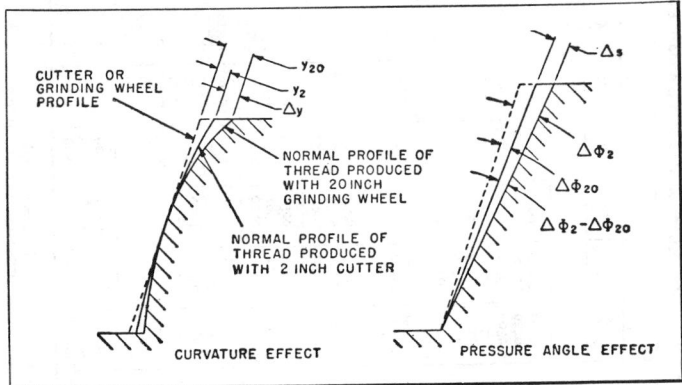

*Calculating Worm Deviations and Pressure Angle Changes:* Included in American Standard ASA B6.9-1956 is an extensive tabulation of profile deviations and pressure angle changes produced by cutters and grinding wheels of 2-inch and 20-inch diameters. These diameters represent the limits of the range commonly used, and the data given are useful in specifying worm profile tolerances. The data also aid in the selection of the method to be used in producing the worm and in specifying the hobs for worm gears. The formulas used to compute the data in the Standard are given here in slightly modified form, and may be used to calculate the profile deviations and pressure angle changes produced in the worm by cutters or grinding wheels.

$$\rho_{ni} = \frac{r \sin \phi_n}{\sin^2 \lambda} \text{ inches} \tag{1}$$

$$\rho_n = \rho_{ni} + \frac{r\rho_{ni}}{R \cos^2 \lambda} \text{ inches} \tag{2}$$

$$\Delta\phi = \frac{5400\, r \sin^3 \lambda}{n(R \cos^2 \lambda + r)} \text{ minutes} \tag{3}$$

$$q = a \sec \phi_n \text{ inches} \tag{4}$$

$$y = \frac{q^2}{2\, \rho_n} \text{ inches} \tag{5}$$

$$s = 0.000582\; q\Delta\phi \text{ inches} \tag{6}$$

$$\Delta y = y_w - y_c \text{ inches} \tag{7}$$

$$\Delta s = s_c - s_w \text{ inches} \tag{8}$$

In these formulas,

$\rho_{ni}$ = radius of curvature of normal thread profile for involute thread;

$r$ = pitch radius of worm;

$\phi_n$ = normal pressure angle of cutter or grinding wheel;

$\lambda$ = lead angle of worm;

$\rho_n$ = radius of curvature of normal thread profile;

$R$ = radius of cutter or grinding wheel;

$\Delta\phi$ = difference between the normal pressure angle of the thread and the normal pressure angle of the cutter or grinding wheel (see diagram). Subscripts $c$ and $w$ are used to denote the cutter and grinding wheel diameters respectively;

$n$ = number of threads in worm;

$a$ = addendum of worm;

$q$ = slant height of worm addendum;

$y$ = amount normal worm profile departs from a straight side (see diagram). Subscripts $c$ and $w$ are used to denote the cutter and grinding wheel diameters, respectively;

$s$ = effect along slant height of worm thread caused by change in pressure angle $\Delta\phi$;

$\Delta y$ = difference in $y$ values of two cutters or grinding wheels of different diameter (see diagram);

$\Delta s$ = effect of $\Delta\phi_c - \Delta\phi_w$ along slant height of thread (see diagram).

*Example 3:* Assuming the worm dimensions are the same as used in Example 1, determine the corrections to be made for two worms, one milled by a 2-inch diameter cutter and the other ground by a 20-inch diameter wheel, both to be assembled with identical worm gears.

To make identical worms when using a 2-inch cutter and a 20-inch wheel, the pressure angle of either the cutter or the wheel must be corrected by an amount corresponding to $\Delta s$ and the profile of the cutter or wheel must be a curve which departs from a straight line by an amount $\Delta y$. The calculations are as follows:

For the 2-inch diameter cutter, using Formulas (1) to (6),

$$\rho_{ni} = \frac{0.4410 \times 0.3420}{0.5000^2} = 0.6033 \text{ inch} \tag{1}$$

$$\rho_n = 0.6033 + \frac{0.4410 \times 0.6033}{1 \times 0.8660^2} = 0.9581 \text{ inch} \tag{2}$$

$$\Delta\phi_c = \frac{5400 \times 0.4410 \times 0.5000^3}{10(1 \times 0.8660^2 + 0.4410)} = 24.99 \text{ minutes} \tag{3}$$

$$q = 0.0441 \times 1.0642 = 0.0469 \text{ inch} \tag{4}$$

$$y_c = \frac{0.0469^2}{2 \times 0.9581} = 0.00115 \text{ inch} \tag{5}$$

$$s_c = 0.000582 \times 0.0469 \times 24.99 = 0.000682 \text{ inch} \tag{6}$$

For the 20-inch diameter wheel, using Formulas (1) to (6)

$$\rho_{ni} = \frac{0.4410 \times 0.3420}{0.5000^2} = 0.6033 \text{ inch} \tag{1}$$

$$\rho_n = 0.6033 + \frac{0.4410 \times 0.6033}{10 \times 0.8660^2} = 0.6387 \text{ inch} \tag{2}$$

$$\Delta\phi_w = \frac{5400 \times 0.4410 \times 0.5000^3}{10(10 \times 0.8660^2 + 0.4410)} = 3.749 \text{ minutes} \tag{3}$$

$$q = 0.0441 \times 1.0642 = 0.0469 \text{ inch} \tag{4}$$

$$y_w = \frac{0.0469^2}{2 \times 0.6387} = 0.00172 \text{ inch} \tag{5}$$

$$s_w = 0.000582 \times 0.0469 \times 3.749 = 0.000102 \text{ inch} \tag{6}$$

Applying formulas (7) and (8):

$$\Delta y = 0.00172 - 0.00115 = 0.00057 \text{ inch} \tag{7}$$

$$\Delta s = 0.000682 - 0.000102 = 0.000580 \text{ inch} \tag{8}$$

Therefore the pressure angle of either the cutter or the wheel must be corrected by an amount corresponding to a $\Delta s$ of 0.000580 inch and the profile of the cutter or wheel must be a curve which departs from a straight line by 0.00057 inch.

**Industrial Worm Gearing.** — The primary considerations in industrial worm gearing are usually: (1) To transmit power efficiently; (2) to transmit power at a considerable reduction in velocity; and (3) to provide a considerable " mechanical advantage " when a given applied force must overcome a comparatively high resisting force. Worm gearing for use in such applications is usually of relatively coarse pitch. The notation below is used in the formulas on the following pages.

$a$ = addendum, worm thread
$A$ = addendum, worm-wheel tooth
$B$ = dedendum, worm-wheel tooth
$b$ = dedendum, worm thread
$C$ = center distance (Fig. 1, p. 888)
$c$ = clearance
$D$ = pitch diameter of worm-wheel
$d$ = pitch diameter of worm
$d_o$ = outside diameter of worm
$D_o$ = outside or over-all diameter of worm-wheel
$D_t$ = throat diameter of worm-wheel
$E$ = efficiency of worm gearing, per cent
$F$ = nominal face width of wheel rim
$F_e$ = effective face width (Fig. 1, p. 888)
$f$ = coefficient of friction
$G$ = length of worm threaded section
$H$ = horsepower rating

$L$ = lead of worm thread = pitch × number of threads or " starts "
$L_a$ = lead angle of worm = helix angle measured from a plane perpendicular to worm axis
$M$ = torque applied to worm-wheel, pound inches
$m$ = module = 0.3183 × axial pitch
$N$ = revolutions per minute of worm-wheel
$n$ = revolutions per minute of worm
$P$ = axial pitch of worm and circular pitch of wheel
$P_n$ = normal pitch of worm
$Q$ = arc length of worm-wheel tooth measured along root
$R$ = ratio of worm gearing = No. of wheel teeth ÷ No. of worm threads.

$S$ = surface stress factor (Table 4)
$S_b$ = bending stress factor, lbs. per sq. in. (Table 4)
$T$ = number of teeth on worm-wheel
$t$ = number of threads or " starts " on worm—2 for double thread, 3 for triple thread, 4 for quadruple thread, etc.
$U$ = radius of worm-wheel throat (Fig. 1)

$V$ = rubbing speed of worm in feet per minute
$W$ = whole tooth depth (worm and worm-wheel)
$X_w$ = speed factor when load rating is limited by *wear* (Fig. 2)
$X_s$ = speed factor when load rating is limited by *strength* (Table 5)
$\phi$ = angle of friction (tan $\phi$ = coefficient of friction)

**Materials for Worm Gearing.** — Worm gears, especially for power transmission, should have steel worms and phosphor bronze wheels. This combination is used extensively. The worms should be hardened and ground to obtain accuracy and a smooth finish.

The phosphor bronze wheels should contain from 10 to 12 per cent of tin. The S.A.E. phosphor gear bronze (No. 65) contains 88–90 % copper, 10–12 % tin, 0.50 % lead, 0.50 % zinc (but with a maximum total lead, zinc and nickel content of 1.0 per cent), phosphorous 0.10–0.30 %, aluminum 0.005 %. The S.A.E. nickel phosphor gear bronze (No. 65 + Ni) contains 87 % copper, 11 % tin, 2 % nickel and 0.2 % phosphorous. Table 4 shows the British Standard wheel and worm materials.

**Single-thread Worms.** — The ratio of the worm speed to the worm-wheel speed may range from 1.5 or even less up to 100 or more. Worm gears having high ratios are not very efficient as transmitters of power; nevertheless high as well as low ratios often are required. Since the ratio equals the number of worm-wheel teeth divided by the number of threads or " starts " on the worm, single-thread worms are used to obtain a high ratio. As a general rule, a ratio of 50 is about the maximum recommended for a single worm and gear combination, although ratios up to 100 or higher are possible. When a high ratio is required, it may be preferable to use, in combination, two sets of worm gears of the multi-thread type in preference to one set of the single-thread type in order to obtain the same total reduction and a higher combined efficiency.

Single-thread worms are comparatively inefficient because of the effect of the low lead angle (see Efficiency of Worm Gearing); consequently, single-thread worms are not used when the primary purpose is to transmit power as efficiently as possible but they may be employed either when a large speed reduction with one set of gearing is necessary, or possibly as a means of adjustment, especially if " mechanical advantage " or self-locking are important factors.

**Multi-thread Worms.** — When worm gearing is designed primarily for transmitting power efficiently, the lead angle of the worm should be as high as is consistent with other requirements and preferably between, say, 25 or 30 and 45 degrees, as indicated by Tables 1 and 2. This means that the worm must be multi-threaded. To obtain a given ratio, some number of worm-wheel teeth divided by some number of worm threads must equal the ratio. Thus, if the ratio is 6, combinations such as the following might be used:

$$\frac{24}{4}, \quad \frac{30}{5}, \quad \frac{36}{6}, \quad \frac{42}{7}$$

The numerators represent the number of worm-wheel teeth and the denominators, the number of worm threads or " starts." The number of worm-wheel teeth may

## Rules and Formulas for Worm Gearing — 1

| No. | To Find | Rule | Formula |
|---|---|---|---|
| 1 | Addendum | Addendum may be affected by lead angle. See paragraph, Addendum and Dedendum. | |
| 2 | Center Distance | Add pitch diameter of worm-wheel to pitch diameter of worm, and divide sum by 2. | $C = \dfrac{(D + d)}{2}$ |
| 3 | | Divide number of worm threads by tangent lead angle, add number of wheel teeth and multiply sum by quotient obtained by dividing pitch by 6.2832. | $C = \dfrac{P}{6.2832}\left(\dfrac{t}{\tan L_a} + T\right)$ |
| 4 | Dedendum | Dedendum may be affected by lead angle. See paragraph, Addendum and Dedendum. | |
| 5 | Clearance | British Standard — multiply cosine lead angle by 0.2 times module. | $c = 0.2\,m\cos L_a$ |
| 6 | Face Width, Worm-wheel | For single and double thread worms, multiply pitch by 2.38 and add 0.25. (Shell type worm.) | $F = 2.38\,P + 0.25$ |
| 7 | | For triple and quadruple thread worm, multiply pitch by 2.15 and add 0.2. (Shell type.) | $F = 2.15\,P + 0.2$ |
| 8 | | When worm threads are integral with shaft, face width of worm-wheel may equal $C^{0.875}$ divided by 3. | $F = \dfrac{C^{0.875}}{3}$ |
| 9 | Lead of Worm Thread | Multiply pitch by number of worm threads or "starts." | $L = tP$ |
| 10 | | Multiply pitch circumference of worm by tangent of lead angle. | $L = \pi d \times \tan L_a$ |
| 11 | | Divide pitch circumference of worm-wheel by ratio. | $L = \pi D \div R$ |
| 12 | Lead Angle, Worm | Divide lead by pitch circumference of worm; quotient is tangent of lead angle. | $\tan L_a = \dfrac{L}{3.1416\,d}$ |
| 13 | Outside Diam., Worm | Add to pitch diameter twice the addendum. See paragraph, Pitch Diameter of Worm; also Addendum and Dedendum. | $d_0 = d + 2a$ |
| 14 | Outside Diam., Worm-wheel | For outside or over-all diameter of worm-wheel, see paragraph, Outside Diameter of Worm-wheel. | |
| 15 | Pitch of Worm and Wheel | Divide lead by number of threads or "starts" on worm = axial pitch of worm and circular pitch of worm-wheel. | $P = \dfrac{L}{t}$ |
| 16 | | Subtract the worm pitch diameter from twice the center distance. Multiply by 3.1416 and divide by number of wheel teeth. | $P = \dfrac{(2C - d) \times 3.1416}{T}$ |

Rules and Formulas for Worm Gearing — 2

| No. | To Find | Rule | Formula |
|-----|---------|------|---------|
| 17 | Pitch of Worm, Normal | Multiply axial pitch by cosine of lead angle to find normal pitch. | $P_n = P \times \cos L_a$ |
| 18 | | Subtract pitch diameter of worm-wheel from twice the center distance. | $d = 2C - D$ |
| 19 | Pitch Diam., Worm | Subtract twice the addendum from outside diameter. See Addendum and Dedendum. | $d = d_0 - 2a$ |
| 20 | | Multiply lead by cotangent lead angle and divide product by 3.1416. | $d = \dfrac{L \times \cot L_a}{3.1416}$ |
| 21 | Pitch Diam., Worm-wheel | Subtract pitch diameter of worm from twice the center distance. | $D = 2C - d$ |
| 22 | | Multiply number of wheel teeth by axial pitch of worm and divide product by 3.1416. | $D = \dfrac{TP}{3.1416}$ |
| 23 | Radius of Rim Corner, Wheel | Multiply pitch by 0.25 | Rad. $= 0.25\,P$ |
| 24 | | British Standard: Radius = 0.5 module. | Rad. $= 0.5\,m$ |
| 25 | Ratio | Divide number of wheel teeth by number of worm threads. | $R = T \div t$ |
| 26 | Rubbing Speed, Ft. per Minute | Divide wheel pitch diameter by ratio; square quotient and add to square of worm pitch diameter; multiply square root of this sum by 0.262 × R.P.M. of worm. | $V = 0.262\,n\sqrt{d^2 + \left(\dfrac{D}{R}\right)^2}$ |
| 27 | | Multiply 0.262 × pitch diameter of worm by R.P.M. of worm; then multiply product by secant of lead angle. | $V = 0.262\,dn \times \sec L_a$ |
| 28 | Throat Diam., Worm-wheel | Add twice the addendum to pitch diameter — see paragraph, Addendum and Dedendum. | $D_t = D + 2A$ |
| 29 | Throat Radius, Worm-wheel | Subtract twice worm addendum from outside radius of worm. | $U = \dfrac{d_0}{2} - 2a$ |
| 30 | Tooth Depth | Whole depth equals addendum + dedendum. See paragraph, Addendum and Dedendum. | $W = a + b$ or $A + B$ |
| 31 | Worm Thread Length | Multiply the number of wheel teeth by 0.02, add 4.5 and multiply sum by pitch. | $G = P(4.5 + 0.02\,T)$ |
| 32 | | British Standard — Subtract square of worm-wheel pitch diameter from square of outside diameter and extract square root of remainder. | $G = \sqrt{D_0^2 - D^2}$ |

not be an exact multiple of the number of threads on a multi-thread worm (as explained later) in order to obtain a " hunting tooth " action.

*Number of Threads or " Starts " on Worm:* The number of threads on the worm ordinarily varies from one to six or eight, depending upon the ratio of the gearing. As the ratio is increased, the number of worm threads is reduced, as a general rule. In some cases, however, the higher of two ratios may also have a larger number of threads. For example, a ratio of $6\frac{1}{5}$ would have 5 threads whereas a ratio of $6\frac{5}{6}$ would have 6 threads. Whenever the ratio is fractional, the number of threads on the worm equals the denominator of the fractional part of the ratio.

**Ratio for Obtaining " Hunting Tooth " Action.** — In designing worm gears having multi-thread worms, it is common practice to select a number of wheel teeth that is not an exact multiple of the number of worm threads. To illustrate, if the desired ratio is about 5 or 6, the actual ratio might be $5\frac{1}{6}$, $5\frac{5}{6}$, $5\frac{2}{7}$, $6\frac{1}{5}$, etc., so that combinations such as $^{31}/_{6}$, $^{35}/_{6}$, $^{37}/_{7}$ or $^{31}/_{5}$ would be obtained. Since the number of wheel teeth and number of worm threads do not have a common divisor, the threads of the worm will mesh with all of the wheel teeth in succession, thus obtaining a "hunting tooth" or self-indexing action. This progressive change will also occur during the worm wheel hobbing operation, and its primary purpose is to produce more accurate worm wheels by uniformly distributing among all of the teeth, any slight errors which might exist in the hob teeth. Another object is to improve the "running-in" action between the hardened and ground worm and the phosphor bronze wheel, but in order to obtain this advantage, the threads on the worm must be accurately or uniformly spaced by precise indexing. With a "hunting tooth ratio," if the thread spacing of a multi-thread worm is inaccurate, load distribution on the threads will be unequal and some threads might not even make contact with the wheel teeth. For this reason, if the indexing is inaccurate, it is preferable to avoid a hunting tooth ratio, but in that case, if the gearing is disassembled, the same worm and gear teeth should be mated when reassembled.

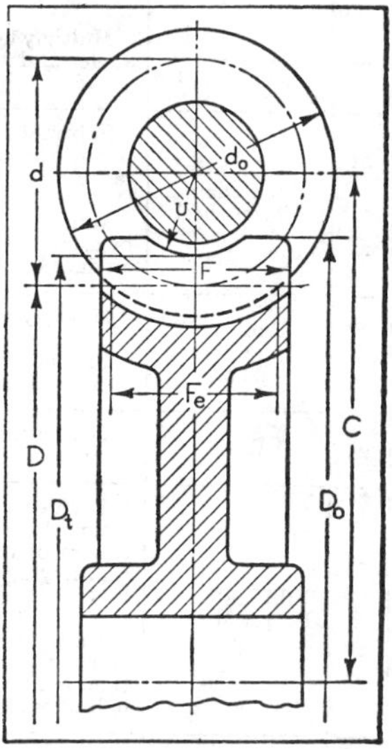


Fig. 1

**Pitch Diameter of Worm.** — The worm must be strong enough to transmit its maximum load without excessive deflection but the diameter should be as small as is consistent with the necessary strength in order to minimize the rubbing speed. It is impracticable to give a rule or formula that is generally applicable, but the following empirical rules are based upon actual practice and may prove useful as a general guide. They apply to casehardened alloy steel worms which are integral with the shaft.

For ratios of 5, 6, or 7, the pitch diameter ranges approximately from 0.38 $C$ when center distance $C$ is 4 inches to 0.33 $C$ when $C$ is 20 inches.

For ratios of 8, 9, 10, the pitch diameter ranges approximately from 0.38 $C$ when center distance $C$ is 4 inches to 0.25 $C$ when $C$ is 30 inches.

For ratios of 10 to 20, the pitch diameter ranges approximately from 0.37 $C$ when center distance $C$ is 4 inches to 0.24 $C$ when $C$ is 30 inches.

For ratios of 20 to 40, the pitch diameter ranges approximately from 0.36 $C$ when center distance $C$ is 4 inches to 0.23 $C$ when $C$ is 30 inches.

According to another empirical formula pitch diameter $d = C^{0.875} \div 2.2$.

**Addendum and Dedendum.** — The following A.G.M.A. formulas are applicable to industrial worm gearing. For single- and double-thread worms, addendum $a = 0.318\,P$ and whole depth $W = 0.686\,P$; for triple and quadruple threads, addendum $a = 0.286\,P$ and whole depth $W = 0.623\,P$.

According to the British standard, $a$ = module $m = 0.3183\,P$; $b = m(2.2 \cos L_a - 1)$; $A = m(2 \cos L_a - 1)$; $B = m(1 + 0.2 \cos L_a)$.

**Outside Diameter of Worm-wheel.** — Practice varies somewhat in determining the outside or over-all diameter of the worm-wheel, as indicated by the following formulas. For usual rim shape, see Fig. 1.

1. For lead angles up to about 15 or 20 degrees, $D_o = D + (3 \times 0.3183\,P)$
2. For lead angles over 20 degrees, $D_o = D + (3 \times 0.3183\,P \times \cos L_a)$
3. For single and double thread, $D_o = D_t + 0.4775\,P$
4. For triple and quadruple thread, $D_o = D_t + 0.3183\,P$
5. According to British standard, $D_o = D + 2A + m$

**Pressure Angles.** — The pressure angle (one-half the included thread angle) ranges from 14½ to 30 degrees. While the practice varies somewhat, the following relationship between lead angle and pressure angle may be used as a general guide.

For lead angles up to about 10 or 12 degrees, pressure angle = 14½ degrees.

For lead angles from 10 to 12 to about 20 or 25 degrees, pressure angle = 20 degrees.

For lead angles from 25 to about 35 degrees, pressure angle = 25 degrees.

For lead angles over 35 degrees, pressure angle = 30 degrees.

In the British Standard specifications, the recommended thread form has a normal pressure angle of 20 degrees.

**Designing Worm Gearing Relative to Center Distance and Ratio.** — In designing worm gears, three general cases or types of problems may be encountered in establishing the proportions of the worm and worm-wheel.

*When Center Distance is Fixed and Ratio may be Varied:* The ratio in this case is nominal and may be varied somewhat to meet other conditions. Assume that the required center distance is 6 inches, the desired ratio is about 7, and the pitch of the worm and wheel is to be approximately 1 inch. Combinations of worm-wheels and worms such as the following might be used in this case:

$$\frac{28}{4}, \quad \frac{35}{5}, \quad \frac{42}{6}, \quad \frac{56}{8}, \text{ etc.}$$

Suppose we select the 28/4 combination for trial but change the number of wheel teeth from 28 to 29 to obtain a self-indexing or "hunting tooth" action. The ratio now equals 29/4 or 7.25. Then, for trial purposes

Pitch diameter $D$ of wheel $= \dfrac{T \times P}{\pi} = \dfrac{29 \times 1}{3.1416} = 9.231$ inches

Pitch diameter $d$ of worm $= 2C - D = 2 \times 6 - 9.231 = 2.769$ inches

Assume that experience, tests, or calculations show that a worm of smaller diameter will have the necessary bending and torsional strength and that a pitch of 1.0625 will be satisfactory. Then the pitch diameter of the worm will be decreased to 2.192 inches and the pitch diameter of the wheel will be increased to 9.808 inches. A check of the lead-angle will show that it equals 31°41′ which is conducive to high efficiency.

*When Ratio is Fixed and Center Distance may be Varied:* Assume that the required ratio is 7¼ and that the center distance may be any value consistent with approved designing practice. This ratio may be obtained with a number of different worm and worm-wheel sizes. For example, in a series of commercial worm gears, the following combinations are employed for gearing having a ratio of 7¼ with center distances varying from 4 to 8.25 inches. The number of worm threads is 4 and the number of teeth on the worm-wheel 29 in all cases.

When $C = 4$ inches, $d = 1.654$; $D = 6.346$; $P = 0.6875$; $L_a = 27°54′$
When $C = 5$ inches, $d = 1.923$; $D = 8.077$; $P = 0.875$; $L_a = 30°5′$
When $C = 6$ inches, $d = 2.192$; $D = 9.808$; $P = 1.0625$; $L_a = 31°41′$
When $C = 7$ inches, $d = 2.461$; $D = 11.539$; $P = 1.25$; $L_a = 32°53′$
When $C = 8.25$ inches, $d = 2.942$; $D = 13.558$; $P = 1.4687$, $L_a = 32°27′$

The horsepower rating increases considerably as the proportions of the worm gearing increase; hence if the gears are intended primarily for power transmission, the general proportions must be selected with reference to the power-transmitting capacity, and, usually the smallest and most compact design that will give satisfactory performance should be selected. The power capacity of the transmission, however, does not depend solely upon the proportions of the worm and worm-wheel. For example, the quality and viscosity of the lubricant is an important factor. The load transmitting capacity of the lubricant may also be increased decidedly when excessive temperature rises are prevented by special means such as forced air cooling. (See " Water and Forced-Air Cooling.")

*When Both Ratio and Center Distance are Fixed:* When both ratio and center distance are fixed, the problem usually is to obtain the *best proportions* of worm and wheel conforming to these fixed values.

*Example:* The required ratio is 6 (6 to 1) and the center distance is fixed at 3.600 inches. Assume that experience or tests show that an axial pitch of 0.50 inch will meet strength requirements. (If normal pitch $P_n$ is given, change to axial pitch ($P = P_n \div \cos L_a$). With a ratio of 6, some of the combinations for trial are:

$$\frac{30}{5} \; ; \; \frac{36}{6} \; ; \; \frac{42}{7} \; .$$

Trial calculations will show that the 36/6 combination gives the best proportions of worm and wheel for the center distance and pitch specified. Thus

$$D = \frac{TP}{\pi} = \frac{36 \times 0.5}{3.1416} = 5.729; \quad d = 2C - D = 2 \times 3.6 - 5.729 = 1.471$$

The lead angle is about 33 degrees. The effect of lead angle on efficiency is dealt with in a following paragraph. The total obtained by adding the number of worm-wheel teeth to the number of worm threads, equals $36 + 6 = 42$ (a total of 40 is a desirable minimum). With the 42/7 combination of the same pitch, the worm would be too small (0.516 inch); and with the 30/5 combination it would be too large (2.426 inches). The present trend in gear designing practice is to use finer pitches than in the past. In the case of worm gears, the pitch may, in certain instances, be changed somewhat either to permit cutting with available equipment or to improve the proportions of worm and wheel.

*When Ratio, Pitch and Lead Angle are Fixed.* Assume that $R = 10$, axial pitch $P = 0.16$ inch, $L_a = 30$ degrees and $C = 3$ inches, approximately.

The first step is to determine for the given ratio, pitch and lead angle, the number of worm threads $t$ which will give a center distance nearest 3 inches.

$$t = \frac{C \times 2\pi}{P \times (\cot L_a + R)} = \frac{3 \times 6.2832}{0.16 \times (1.7320 + 10)} = 10.04$$

The whole number nearest 10.04, or 10, is the required number of worm threads; hence number of teeth on worm-wheel equals $R \times 10 = 100$

$$d = (L \cot L_a) \div \pi = (10 \times 0.16 \times 1.732) \div \pi = 0.8821 \text{ inches}$$

$$D = (TP) \div \pi = (100 \times 0.16) \div \pi = 5.0929 \text{ inches}$$

$$C = (D + d) \div 2 = (5.0929 + 0.8821) \div 2 = 2.9875 \text{ inches}$$

**Efficiency of Worm Gearing.** — The efficiency at a given speed, depends upon the worm lead angle, the workmanship, lubrication, and the general design of the transmission. When worm gearing consists of a hardened and ground worm running with an accurately hobbed wheel properly lubricated, the efficiency depends chiefly upon the lead angle and coefficient of friction between the worm and wheel. In the lower range of lead angles, the efficiency increases considerably as the lead angle increases, as shown by Tables 1 and 2. This increase in efficiency remains practically constant for lead angles between 30 and 45 degrees. Several formulas for obtaining efficiency percentage follow:

*With worm driving:*

$$E = 100 - \frac{R}{2} \text{ (empirical rule)}; \quad E = \frac{100 \times \tan L_a}{\tan (L_a + \phi)}; \quad E = \frac{100 \times L}{L + f\pi d}$$

*With wheel driving:*

$$E = 100 - 2R \text{ (empirical rule)}; \quad E = \frac{100 \times \tan (L_a - \phi)}{\tan L_a}$$

**Table 1.   Efficiency of Worm Gearing for Different Lead Angles and Frictional Coefficients**

| Coefficient of Friction | Lead Angle of Worm | | | | | | | | |
|---|---|---|---|---|---|---|---|---|---|
| | 5 Deg. | 10 Deg. | 15 Deg. | 20 Deg. | 25 Deg. | 30 Deg. | 35 Deg. | 40 Deg. | 45 Deg. |
| 0.01 | 89.7 | 94.5 | 96.1 | 97.0 | 97.4 | 97.7 | 97.9 | 98.0 | 98.0 |
| 0.02 | 81.3 | 89.5 | 92.6 | 94.1 | 95.0 | 95.5 | 95.9 | 96.0 | 96.1 |
| 0.03 | 74.3 | 85.0 | 89.2 | 91.4 | 92.7 | 93.4 | 93.9 | 94.1 | 94.2 |
| 0.04 | 68.4 | 80.9 | 86.1 | 88.8 | 90.4 | 91.4 | 92.0 | 92.2 | 92.3 |
| 0.05 | 63.4 | 77.2 | 83.1 | 86.3 | 88.2 | 89.4 | 90.1 | 90.4 | 90.5 |
| 0.06 | 59.0 | 73.8 | 80.4 | 84.0 | 86.1 | 87.5 | 88.2 | 88.6 | 88.7 |
| 0.07 | 55.2 | 70.7 | 77.8 | 81.7 | 84.1 | 85.6 | 86.4 | 86.9 | 86.9 |
| 0.08 | 51.9 | 67.8 | 75.4 | 79.6 | 82.2 | 83.8 | 84.7 | 85.2 | 85.2 |
| 0.09 | 48.9 | 65.2 | 73.1 | 77.6 | 80.3 | 82.0 | 83.0 | 83.5 | 83.5 |
| 0.10 | 46.3 | 62.7 | 70.9 | 75.6 | 78.5 | 80.3 | 81.4 | 81.9 | 81.8 |

The efficiencies obtained by these formulas and other modifications of them, differ somewhat and do not take into account bearing and oil-churning losses. The efficiency may be improved somewhat after the " running in " period.

**Self-locking or Irreversible Worm Gearing.** — Neglecting friction in the bearings, worm gearing is irreversible when the efficiency is zero or negative, the lead angle being equal to or less than the angle $\phi$ of friction (tan $\phi$ = coefficient of friction). When worm gearing is self-locking or irreversible, this means that the wheel

**Table 2. Efficiency of Worm Gearing for Different Lead Angles and Pitch-line Velocities**

| Velocity at Pitch Line, Feet per Minute | Lead or Helix Angle, Degrees | | | | | |
|---|---|---|---|---|---|---|
| | 5 | 10 | 20 | 30 | 40 | 45 |
| | Efficiency, Per Cent | | | | | |
| 5 | 40 | 56 | 69 | 76 | 79 | 80 |
| 10 | 47 | 62 | 74 | 79 | 82 | 82 |
| 20 | 52 | 67 | 78 | 83 | 85 | 86 |
| 30 | 56 | 71 | 81 | 85 | 87 | 87 |
| 40 | 60 | 74 | 83 | 87 | 88 | 88 |
| 50 | 63 | 76 | 85 | 88 | 89 | 89 |
| 75 | 67 | 80 | 87 | 90 | 90 | 90 |
| 100 | 70 | 82 | 88 | 91 | 91 | 91 |
| 150 | 74 | 84 | 90 | 92 | 92 | 92 |
| 200 | 76 | 85 | 91 | 92 | 92 | 92 |

**Table 3. Coefficients of Friction (f) for Worm Gearing\***

| Rubbing Speed, Ft. per Min. | Coefficients of Friction | Rubbing Speed, Ft. per Min. | Coefficients of Friction | Rubbing Speed, Ft. per Min. | Coefficients of Friction | Rubbing Speed, Ft. per Min. | Coefficients of Friction |
|---|---|---|---|---|---|---|---|
| 30 | 0.073 | 180 | 0.045 | 550 | 0.028 | 1600 | 0.0175 |
| 40 | 0.070 | 190 | 0.044 | 600 | 0.027 | 1700 | 0.0170 |
| 50 | 0.066 | 200 | 0.043 | 650 | 0.026 | 1800 | 0.0165 |
| 60 | 0.062 | 225 | 0.041 | 700 | 0.026 | 1900 | 0.0165 |
| 70 | 0.060 | 250 | 0.040 | 750 | 0.025 | 2000 | 0.0160 |
| 80 | 0.058 | 275 | 0.038 | 800 | 0.024 | 2100 | 0.0160 |
| 90 | 0.056 | 300 | 0.036 | 850 | 0.023 | 2200 | 0.0155 |
| 100 | 0.054 | 325 | 0.035 | 900 | 0.023 | 2300 | 0.0150 |
| 110 | 0.052 | 350 | 0.034 | 950 | 0.022 | 2400 | 0.0150 |
| 120 | 0.051 | 375 | 0.033 | 1000 | 0.022 | 2500 | 0.0150 |
| 130 | 0.050 | 400 | 0.033 | 1100 | 0.021 | 2600 | 0.0145 |
| 140 | 0.049 | 425 | 0.032 | 1200 | 0.020 | 2700 | 0.0145 |
| 150 | 0.048 | 450 | 0.031 | 1300 | 0.019 | 2800 | 0.0140 |
| 160 | 0.047 | 475 | 0.030 | 1400 | 0.019 | 2900 | 0.0140 |
| 170 | 0.046 | 500 | 0.030 | 1500 | 0.018 | 3000 | 0.0140 |

\* These values for different rubbing speeds, are based upon the use of phosphor bronze wheels with case-hardened ground and polished steel worms lubricated with mineral oil.

cannot drive the worm. Since the angle of friction changes rapidly with the rubbing speed, and the static angle of friction may be reduced by external vibration, it is usually impracticable to design irreversible worm gears with any security. If irreversibility is desired it is recommended that some form of brake be employed.

**Worm Gear Operating Temperatures.** — The load capacity of a worm gear lubricant at operating temperature is an important factor in establishing the continuous power-transmitting capacity of the gearing. If the churning or turbulence of the oil generates excessive heat, the viscosity of the lubricant may be reduced below its load-supporting capacity. The temperature measured in the oil sump should not, as a rule, exceed 180 to 200 degrees F. or rise more than 120 to 140 degrees F. above a surrounding air temperature of 60 degrees F. In rear axle motor vehicle transmissions, the maximum operating temperature may be somewhat higher than the figures given and usually is limited to about 220 degrees F.

**Thermal Rating.** — In some cases, especially when the worm speed is comparatively high, the horsepower capacity of worm gearing should be based upon its thermal rating instead of the mechanical rating. To illustrate, worm gearing may have a thermal rating of, say, 60 H.P., and mechanical ratings which are considerably higher than 60 for the higher speed ranges. This means that the gearing is capable of transmitting more than 60 H.P. so far as wear and strength are concerned but not without overheating; hence, in this case a rating of 60 should be considered maximum. Of course, if the power to be transmitted is less than the thermal rating for a given ratio, then the thermal rating may be ignored.

**Water and Forced-Air Cooling.** — One method of increasing the thermal rating of a speed-reducing unit of the worm gear type, is by installing a water-cooling coil through which water is circulated to prevent an excessive rise of the oil temperature. According to one manufacturer, the thermal rating may be increased as much as 35 per cent in this manner. Much larger increases have been obtained by means of a forced air cooling system incorporated in the design of the speed-reducing unit. A fan which is mounted on the worm shaft draws air through a double walled housing, thus maintaining a comparatively low oil bath temperature. A fan cooling system makes it possible to transmit a given amount of power through a worm-gear unit that is much smaller than one not equipped with a fan.

**Horsepower Rating for Worm Gearing.** — According to the British Standard for worm gearing, B. S. 721: 1937, amended in 1941, the permissible load as limited by wear, is determined for both worm and wheel; then the permissible load as limited by strength is determined for both worm and wheel and the lowest of the four load values in pound-inches of torque is used in the horsepower formula, as shown later by an example. The formulas which follow are for determining the "normal rating" which is the safe loading for operating periods of 12 hours per day. The permissible torque in pound-inches as limited by wear is the lower of the two values obtained by Formula (1) when it is applied to both wheel and worm.

$$\text{Limiting } \textit{wear} \text{ load (torque), pound-inches} = 0.18 S X_w F_e D^{1.8} \qquad (1)$$

The permissible torque in pound-inches as limited by strength is the lower of the two values obtained by Formula (2) when it is applied to both wheel and worm.

$$\text{Limiting } \textit{strength} \text{ load (torque), pound-inches} = 0.625 S_b X_s Q m D \cos L_a \qquad (2)$$

$$\text{Horsepower rating} = \frac{MN}{63,000} \text{ where } M \text{ is the smallest of the four values obtained}$$

Table 4.  Bending Stress and Surface Stress Factors for Worm Gears

*Example:*  Wheel is Phosphor Bronze, Centrifugally Cast (Material in Classification A.)  Worm is 3½ per cent Nickel Casehardening Steel (In Classification E.)

*Find Factor S for wheel* under worm material E (upper section of table) and opposite "Phosphor Bronze, Centrifugally Cast." (*S* equals 2000.)

*Find Factor S for worm* under wheel material A (lower section of table) and opposite "3½ per cent Nickel Casehardening Steel." (*S* equals 6500.)

*Find Factor $S_b$ for wheel* in column "Bending Stress Factor" and opposite "Phosphor Bronze, Centrifugally Cast." ($S_b$ equals 10,000.)

*Find Factor $S_b$ for worm* in column "Bending Stress Factor" and opposite "3½ per cent Nickel Casehardening Steel." ($S_b$ equals 40,000.)

| Wheel Materials | | Bending Stress Factor $S_b$ Pounds per Sq. In. | Find surface stress factor $S$ for wheel, under worm material classification letter | | | | |
|---|---|---|---|---|---|---|---|
| | | | A | B | C | D | E |
| A | Phosphor Bronze, Sand Cast | 7,000 | .. | 600* | 600 | 700 | 1400 |
| | Phosphor Bronze, Chill Cast | 8,500 | .. | 800* | 800 | 900 | 1600 |
| | Phosphor Bronze, Centrifugally Cast | 10,000 | .. | 1000* | 1000 | 1100 | 2000 |
| B | Cast Iron (Gray) | 6,000 | 900* | 600* | 600† | 600† | 750† |
| Worm Materials | | Bending Stress Factor $S_b$ Pounds per Sq. In. | Find surface stress factor $S$ for worm, under wheel material classification letter | | | | |
| | | | A | B | C | D | E |
| C | 0.4 Per Cent Carbon Steel, Normalized | 20,000 | 1400 | 900† | .. | .. | .. |
| D | 0.55 Per Cent Carbon Steel, Normalized | 22,000 | 2000 | 1100† | .. | .. | .. |
| E | Low-carbon Casehardening Steel | 27,000 | 6000 | 4000† | .. | .. | .. |
| | 3½ Per Cent Nickel Casehardening Steel | 40,000 | 6500 | 4000† | .. | .. | 2000† |
| | 5 Per Cent Nickel Casehardening Steel | 47,000 | 7000 | 4000† | .. | .. | 2000† |
| | 3½ Per Cent Nickel-chromium Casehardening Steel | 47,000 | 7000 | 4000† | .. | .. | 2000† |
| | High Nickel-chrom. Casehardening Steel | 47,000 | 8000 | 4000† | .. | .. | 2000† |

* Maximum permissable rubbing speeds, 500 feet per minute.
† Should not be used except for hand operated gearing.

by applying Formula (1) to wheel and worm and Formula (2) to wheel and worm. While modern practice is to design gears for power transmission on the basis of wear or durability, it is necessary in many cases to check for strength.

*Example:* Find the normal horsepower rating for worm gearing having a ratio of 15⅓ (or 15⅓ to 1); a center distance of 10¼ inches; worm speed of 680 R.P.M.; wheel speed of about 44.35 R.P.M. The worm has three threads and the wheel, 46 teeth. The worm is made of 3½ per cent nickel steel casehardened and the wheel of phosphor bronze, chill cast. Pitch diameter of worm = 3.112 inches; pitch diameter of wheel = 17.388 inches; lead angle = 20°1′; pitch = 1.1875 inch; and module = 1.1875 × 0.3183 = 0.378; outside diameter of worm = 3.868 inches.

Effective face width $F_e$ (see Fig. 1) which is used in determining wear load, may be obtained readily by measurement of full scale drawing. Assume that $F_e$ = 2.3.

*Rubbing Speed.* — Rubbing speed = 0.262dn sec $L_a$ = 0.262 × 3.112 × 680 × 1.0643 = 590 feet per minute, approximately.

*Wear Load, Wheel.* — Find $S$ in Table 4 opposite " Phosphor Bronze, Chill Cast " and under worm material $E$. $S$ for wheel = 1600.

Find $X_w$ for wheel. Rubbing speed is 590 feet per minute and speed of wheel about 44 R.P.M. Referring to the chart, Fig. 2, the intersection of the vertical line from the rubbing speed and the horizontal line from the wheel speed, is close to the 0.3 curve; hence we may assume that $X_w$ = 0.30, approximately.

Wear load, wheel = 0.18 × 1600 × 0.30 × 2.3 × 17.388$^{1.8}$ = 33,937 pound-inches.

*Wear Load, Worm.* — Find $S$ in Table 4 opposite " 3½ Per Cent Nickel Case-hardened Steel " and under wheel material $A$. $S$ for worm = 6500. Find $X_w$ for worm. Rubbing speed is 590 feet per minute and R.P.M. of worm is 680. The point of intersection on chart (Fig. 2) indicates that $X_w$ = 0.155, approximately.

Wear load, worm = 0.18 × 6500 × 0.155 × 2.3 × 17.388$^{1.8}$ = 71,233 pound-inches.

*Strength Load, Wheel.* — Find $S_b$ in Table 4 opposite " Phosphor Bronze, Chill Cast." $S_b$ = 8500. Find speed factor $X_s$ for wheel in Table 5. This factor for 44.35 R.P.M. is approximately 0.47. The value of $Q$ or arc length of tooth at root, may be determined accurately enough for practical requirements by measuring full scale drawing. Assume in this case that $Q$ = 2.5 inches approximately.

Strength Load, Wheel = 0.625 × 8500 × 0.47 × 2.5 × 0.378 × 17.388 × 0.93959 = 38,550 pound-inches.

*Strength Load, Worm.* — Value of $S_b$ for worm for 3½ per cent nickel casehardened steel is 40,000 as shown by Table 4. Speed factor $X_s$ for worm speed of 680 R.P.M. (see Table 5) is 0.29 approximately.

**Table 5. Speed Factors $X_s$ for Worm Gearing (Strength)**

| R.P.M. | Speed Factor $X_s$ | R.P.M. | Speed Factor $X_s$ | R.P.M. | Speed Factor $X_s$ | R.P.M. | Speed Factor $X_s$ |
|---|---|---|---|---|---|---|---|
| 10 | 0.560 | 90 | 0.420 | 500 | 0.310 | 3000 | 0.200 |
| 15 | 0.540 | 100 | 0.415 | 600 | 0.300 | 3500 | 0.190 |
| 20 | 0.520 | 150 | 0.385 | 700 | 0.290 | 4000 | 0.180 |
| 30 | 0.500 | 200 | 0.365 | 800 | 0.280 | 4500 | 0.175 |
| 40 | 0.480 | 250 | 0.350 | 900 | 0.270 | 5000 | 0.170 |
| 50 | 0.460 | 300 | 0.340 | 1000 | 0.260 | 6000 | 0.160 |
| 60 | 0.450 | 350 | 0.335 | 1500 | 0.240 | 7000 | 0.150 |
| 70 | 0.440 | 400 | 0.330 | 2000 | 0.225 | 8000 | 0.140 |
| 80 | 0.430 | 450 | 0.320 | 2500 | 0.210 | 9000 | 0.135 |

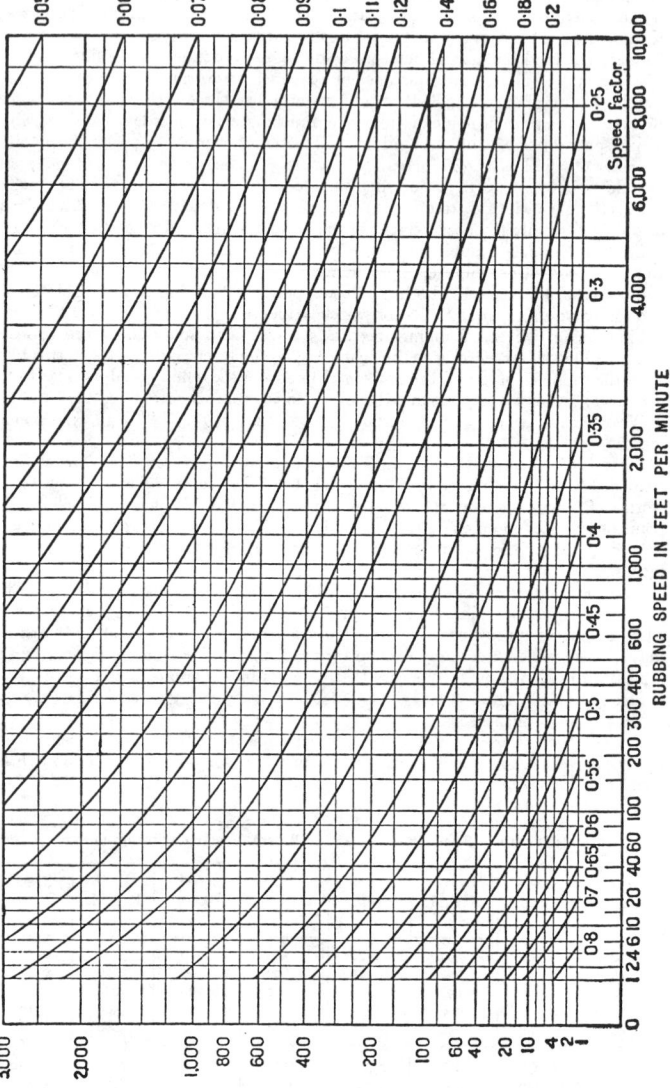

Fig. 2. Speed Factors $X_w$ when Load Rating is Limited by wear.

Strength load, worm = 0.625 × 40,000 × 0.29 × 2.5 × 0.378 × 17.388 × 0.93959 = 111,930 pound-inches.

*Horsepower Rating.* — In the foregoing example, the wear load of 33,937 pound-inches of torque for the wheel is the lowest of the four values; hence

$$\text{Horsepower rating} = \frac{33,937 \times 44.35}{63,000} = 24 \text{ approximately}$$

**Methods of Machining Worm Threads.** — In producing worm threads, the method employed may depend upon quantity required, number of threads on worm or its lead angle, and equipment available. Methods of cutting threads on worms of cylindrical form are described on page 900.

Table 6.  Change Gears for Cutting Diametral Pitch Worms*

| Diametral Pitch to be Cut | Width of Tool Point | Threads per Inch on Lead Screw | | | | | | | |
|---|---|---|---|---|---|---|---|---|---|
| | | 2 | 3 | 4 | 5 | 6 | 7 | 8 | 10 |
| 2 | 0.487 | 22/7 | 33/7 | 44/7 | 55/7 | 66/7 | 77/7 | 88/7 | 110/7 |
| 2¼ | 0.433 | 176/63 | 88/21 | 352/63 | 440/63 | 176/21 | 88/9 | 704/63 | 880/63 |
| 2½ | 0.390 | 88/35 | 132/35 | 176/35 | 44/7 | 264/35 | 44/5 | 352/35 | 88/7 |
| 2¾ | 0.354 | 16/7 | 24/7 | 32/7 | 40/7 | 48/7 | 56/7 | 64/7 | 80/7 |
| 3 | 0.325 | 44/21 | 22/7 | 88/21 | 110/21 | 44/7 | 22/3 | 176/21 | 220/21 |
| 3½ | 0.278 | 88/49 | 132/49 | 176/49 | 220/49 | 264/49 | 44/7 | 352/49 | 440/49 |
| 4 | 0.243 | 11/7 | 33/14 | 22/7 | 55/14 | 33/7 | 11/2 | 44/7 | 55/7 |
| 4½ | 0.217 | 88/63 | 44/21 | 176/63 | 220/63 | 88/21 | 44/9 | 352/63 | 440/63 |
| 5 | 0.195 | 44/35 | 66/35 | 88/35 | 22/7 | 132/35 | 22/5 | 176/35 | 44/7 |
| 6 | 0.162 | 22/21 | 11/7 | 44/21 | 55/21 | 22/7 | 11/3 | 88/21 | 110/21 |
| 7 | 0.139 | 44/49 | 66/49 | 88/49 | 110/49 | 132/49 | 22/7 | 176/49 | 220/49 |
| 8 | 0.122 | 11/14 | 33/28 | 11/7 | 55/28 | 33/14 | 11/4 | 22/7 | 55/14 |
| 9 | 0.108 | 44/63 | 22/21 | 88/63 | 110/63 | 44/21 | 22/9 | 176/63 | 220/63 |
| 10 | 0.097 | 22/35 | 33/35 | 44/35 | 11/7 | 66/35 | 11/5 | 88/35 | 22/7 |
| 11 | 0.088 | 4/7 | 6/7 | 8/7 | 10/7 | 12/7 | 14/7 | 16/7 | 20/7 |
| 12 | 0.081 | 11/21 | 11/14 | 22/21 | 55/42 | 11/7 | 11/6 | 44/21 | 55/21 |
| 14 | 0.069 | 22/49 | 33/49 | 44/49 | 55/49 | 66/49 | 11/7 | 88/49 | 110/49 |
| 16 | 0.061 | 11/28 | 33/56 | 22/28 | 55/56 | 33/28 | 77/56 | 11/7 | 55/28 |
| 18 | 0.054 | 22/63 | 11/21 | 44/63 | 55/63 | 22/21 | 11/9 | 88/63 | 110/63 |
| 20 | 0.049 | 11/35 | 33/70 | 22/35 | 11/14 | 33/35 | 11/10 | 44/35 | 11/7 |
| 22 | 0.044 | 2/7 | 3/7 | 4/7 | 5/7 | 6/7 | 7/7 | 8/7 | 10/7 |
| 24 | 0.040 | 11/42 | 33/84 | 11/21 | 55/84 | 33/42 | 77/84 | 22/21 | 55/42 |
| 26 | 0.037 | 22/91 | 33/91 | 44/91 | 55/91 | 66/91 | 11/13 | 88/91 | 110/91 |
| 28 | 0.035 | 11/49 | 33/98 | 22/49 | 55/98 | 33/49 | 11/14 | 44/49 | 55/49 |
| 30 | 0.032 | 22/105 | 11/35 | 44/105 | 11/21 | 22/35 | 11/15 | 88/105 | 22/21 |
| 32 | 0.030 | 11/56 | 33/112 | 11/28 | 55/112 | 33/56 | 11/16 | 11/14 | 55/56 |
| 40 | 0.024 | 11/70 | 33/140 | 11/35 | 11/28 | 33/70 | 11/20 | 22/35 | 11/14 |
| 48 | 0.020 | 11/84 | 33/168 | 11/42 | 55/168 | 33/84 | 77/168 | 11/21 | 55/84 |

* The ratio of change gears for cutting diametral pitch worms is as 22 times the threads per inch on lead-screw is to 7 times the diametral pitch to be cut. Thus,

$$\frac{22 \times \text{Threads per Inch}}{7 \times \text{Diametral Pitch}} = \text{Ratio of Change Gears}$$

**Table 7. Approximate Worm-thread Lead Angle, in Degrees, for Given Lead and Pitch Diameter**

| Lead of Worm, Inches | Pitch-line Diameter of Worm, Inches | | | | | | | | | | | | | | | | | | | | |
|---|---|---|---|---|---|---|---|---|---|---|---|---|---|---|---|---|---|---|---|---|---|
| | 3½ | 3⅜ | 3¼ | 3⅛ | 3 | 2⅞ | 2¾ | 2⅝ | 2½ | 2⅜ | 2¼ | 2⅛ | 2 | 1⅞ | 1¾ | 1⅝ | 1½ | 1⅜ | 1¼ | 1⅛ | 1 |
| ¼ | 1¼ | 1⅜ | 1⅜ | 1½ | 1½ | 1⅝ | 1⅝ | 1¾ | 1⅞ | 1⅞ | 2 | 2⅛ | 2¼ | 2⅜ | 2⅝ | 2¾ | 3 | 3⅜ | 3⅝ | 4 | 4½ |
| ⅜ | 2 | 2 | 2⅛ | 2¼ | 2¼ | 2⅜ | 2½ | 2⅝ | 2¾ | 2⅞ | 3 | 3¼ | 3⅜ | 3⅝ | 3⅞ | 4¼ | 4½ | 5 | 5½ | 6 | 6¾ |
| ½ | 2⅝ | 2¾ | 2¾ | 2⅞ | 3 | 3⅛ | 3⅜ | 3½ | 3⅝ | 3⅞ | 4 | 4¼ | 4½ | 4⅞ | 5¼ | 5⅝ | 6 | 6⅝ | 7¼ | 8 | 9 |
| ⅝ | 3¼ | 3⅜ | 3½ | 3⅝ | 3¾ | 4 | 4⅛ | 4⅜ | 4½ | 4¾ | 5 | 5⅜ | 5⅝ | 6 | 6½ | 7 | 7½ | 8¼ | 9 | 10 | 11¼ |
| ¾ | 3⅞ | 4 | 4¼ | 4⅜ | 4½ | 4¾ | 5 | 5¼ | 5½ | 5¾ | 6 | 6⅜ | 6¾ | 7¼ | 7¾ | 8⅜ | 9 | 9⅞ | 10¾ | 12 | 13⅜ |
| ⅞ | 4½ | 4¾ | 4⅞ | 5⅛ | 5¼ | 5½ | 5¾ | 6 | 6⅜ | 6¾ | 7 | 7½ | 7⅞ | 8½ | 9 | 9¾ | 10½ | 11½ | 12½ | 13⅞ | 15⅝ |
| 1 | 5¼ | 5⅜ | 5⅝ | 5⅞ | 6 | 6⅜ | 6⅝ | 6⅞ | 7¼ | 7⅝ | 8 | 8½ | 9 | 9⅝ | 10¼ | 11⅛ | 12 | 13 | 14¼ | 15¾ | 17⅝ |
| 1⅛ | 5⅞ | 6 | 6¼ | 6½ | 6¾ | 7⅛ | 7⅜ | 7¾ | 8⅛ | 8⅝ | 9 | 9½ | 10⅛ | 10¾ | 11⅝ | 12⅜ | 13⅜ | 14⅝ | 16 | 17⅝ | 19¾ |
| 1¼ | 6½ | 6¾ | 7 | 7¼ | 7½ | 7⅞ | 8¼ | 8⅝ | 9 | 9½ | 10 | 10⅝ | 11¼ | 12 | 12¾ | 13¾ | 14⅞ | 16⅛ | 17⅝ | 19½ | 21¾ |
| 1⅜ | 7⅛ | 7⅜ | 7⅝ | 8 | 8¼ | 8⅝ | 9 | 9½ | 9⅞ | 10½ | 11 | 11⅝ | 12⅜ | 13⅛ | 14 | 15 | 16¼ | 17⅝ | 19¼ | 21¼ | 23⅝ |
| 1½ | 7¾ | 8 | 8⅜ | 8¾ | 9 | 9⅜ | 9⅞ | 10¼ | 10¾ | 11⅜ | 12 | 12⅝ | 13⅜ | 14¼ | 15¼ | 16⅜ | 17⅝ | 19⅛ | 20⅞ | 23 | 25½ |
| 1⅝ | 8⅜ | 8¾ | 9 | 9⅜ | 9¾ | 10¼ | 10⅝ | 11⅛ | 11¾ | 12¼ | 13 | 13⅝ | 14½ | 15⅜ | 16½ | 17⅝ | 19 | 20½ | 22½ | 24¾ | 27⅜ |
| 1¾ | 9 | 9⅜ | 9¾ | 10⅛ | 10½ | 11 | 11½ | 12 | 12½ | 13¼ | 13⅞ | 14¾ | 15⅝ | 16½ | 17⅝ | 18⅞ | 20⅜ | 22 | 24 | 26⅜ | 29⅛ |
| 1⅞ | 9⅝ | 10 | 10⅜ | 10¾ | 11¼ | 11¾ | 12¼ | 12¾ | 13⅜ | 14⅛ | 14⅞ | 15¾ | 16⅝ | 17⅝ | 18⅞ | 20⅛ | 21¾ | 23½ | 25½ | 28 | 30⅞ |
| 2 | 10¼ | 10⅝ | 11⅛ | 11½ | 12 | 12½ | 13 | 13⅝ | 14¼ | 15 | 15¾ | 16⅝ | 17⅝ | 18¾ | 20 | 21⅜ | 23 | 24⅞ | 27 | 29½ | 32½ |
| 2¼ | 11⅝ | 12 | 12⅜ | 12⅞ | 13⅜ | 14 | 14⅝ | 15¼ | 16 | 16¾ | 17⅝ | 18⅝ | 19¾ | 20⅞ | 22¼ | 23¾ | 25½ | 27½ | 29¾ | 32½ | 35⅝ |
| 2½ | 12¾ | 13¼ | 13¾ | 14¼ | 14⅞ | 15½ | 16⅛ | 16⅞ | 17⅝ | 18½ | 19½ | 20½ | 21¾ | 23 | 24½ | 26⅛ | 28 | 30 | 32½ | 35¼ | 38½ |
| 2¾ | 14 | 14½ | 15 | 15⅝ | 16¼ | 17 | 17⅝ | 18½ | 19¼ | 20¼ | 21¼ | 22⅜ | 23⅝ | 25 | 26⅝ | 28¼ | 30¼ | 32½ | 35 | 37⅞ | 41¼ |
| 3 | 15¼ | 15¾ | 16⅜ | 17 | 17⅝ | 18⅜ | 19⅛ | 20 | 20⅞ | 21⅞ | 23 | 24¼ | 25½ | 27 | 28⅝ | 30⅜ | 32½ | 34¾ | 37⅜ | 40⅜ | 43⅝ |
| 3¼ | 16½ | 17 | 17⅝ | 18¼ | 19 | 19¾ | 20⅝ | 21½ | 22½ | 23½ | 24¾ | 26 | 27⅜ | 28⅞ | 30⅝ | 32½ | 34⅝ | 37 | 39⅝ | 42⅝ | 46 |
| 3½ | 17⅝ | 18¼ | 18⅞ | 19⅝ | 20⅜ | 21⅛ | 22 | 23 | 24 | 25⅛ | 26⅜ | 27⅝ | 29⅛ | 30¾ | 32½ | 34⅜ | 36⅝ | 39 | 41¾ | 45¼ | 48⅛ |
| 3¾ | 18⅞ | 19½ | 20⅛ | 20⅞ | 21¾ | 22½ | 23½ | 24½ | 25½ | 26⅝ | 28 | 29⅜ | 30⅞ | 32½ | 34¼ | 36¼ | 38½ | 41 | 43⅝ | 47½ | 50 |
| 4 | 20 | 20⅝ | 21⅜ | 22⅛ | 23 | 23⅞ | 24⅞ | 25⅞ | 27 | 28¼ | 29½ | 30⅞ | 32½ | 34⅛ | 36 | 38⅛ | 40⅜ | 42¾ | 45½ | 48½ | 51⅞ |
| 4¼ | 21⅛ | 21⅞ | 22⅝ | 23⅜ | 24¼ | 25¼ | 26¼ | 27¼ | 28⅜ | 29⅝ | 31 | 32½ | 34 | 35¾ | 37¾ | 39¾ | 42 | 44½ | 47¼ | 50¼ | 53½ |
| 4½ | 22¼ | 23 | 23¾ | 24⅝ | 25½ | 26½ | 27½ | 28⅝ | 29¾ | 31⅛ | 32½ | 34 | 35⅝ | 37⅜ | 39¼ | 41⅜ | 43⅝ | 46⅛ | 48⅞ | 51⅞ | 55⅛ |
| 4¾ | 23⅜ | 24⅛ | 25 | 25¾ | 26¾ | 27¾ | 28¾ | 30 | 31⅛ | 32½ | 33⅞ | 35⅜ | 37⅛ | 38⅞ | 40¾ | 43 | 45¼ | 47¾ | 50⅜ | 53⅜ | 56½ |
| 5 | 24½ | 25¼ | 26⅛ | 27 | 28 | 29 | 30 | 31¼ | 32½ | 33⅞ | 35¼ | 36⅞ | 38½ | 40⅜ | 42¼ | 44⅜ | 46¾ | 49⅛ | 51⅞ | 54¾ | 57⅞ |
| 5¼ | 25½ | 26⅜ | 27¼ | 28⅛ | 29⅛ | 30⅛ | 31¼ | 32½ | 33¾ | 35⅛ | 36⅝ | 38¼ | 39⅞ | 41¾ | 43⅝ | 45¾ | 48⅛ | 50½ | 53¼ | 56 | 59⅛ |
| 5½ | 26⅝ | 27⅜ | 28¼ | 29¼ | 30⅜ | 31⅜ | 32½ | 33¾ | 35 | 36⅜ | 37⅞ | 39½ | 41¼ | 43 | 45 | 47⅛ | 49⅜ | 51⅞ | 54½ | 57¼ | 60¼ |
| 5¾ | 27⅝ | 28½ | 29⅜ | 30⅜ | 31⅜ | 32½ | 33⅝ | 34⅞ | 36¼ | 37⅝ | 39⅛ | 40¾ | 42½ | 44¼ | 46¼ | 48⅜ | 50⅝ | 53⅛ | 55⅝ | 58⅜ | 61⅜ |
| 6 | 28⅝ | 29½ | 30⅜ | 31⅜ | 32½ | 33⅝ | 34¾ | 36 | 37⅜ | 38¾ | 40⅜ | 42 | 43⅝ | 45½ | 47½ | 49⅝ | 51⅞ | 54¼ | 56¾ | 59½ | 62⅜ |

Table 7. (*Continued*). Approximate Worm-thread Lead Angle, in Degrees, for Given Lead and Pitch Diameter

| Lead of Worm, Inches | Pitch-line Diameter of Worm, Inches | | | | | | | | | | | | | | | | | | | |
|---|---|---|---|---|---|---|---|---|---|---|---|---|---|---|---|---|---|---|---|---|
| | 3⅝ | 3¾ | 3⅞ | 4 | 4⅛ | 4¼ | 4⅜ | 4½ | 4⅝ | 4¾ | 4⅞ | 5 | 5⅛ | 5¼ | 5⅜ | 5½ | 5⅝ | 5¾ | 5⅞ | 6 |
| ¼ | 1¼ | 1¼ | 1¼ | 1¼ | 1¼ | 1¼ | 1 | 1 | 1 | 1 | 1 | 1 | 1 | 1 | 1 | 1 | 1 | 1 | 1 | 1 |
| ⅜ | 2 | 2 | 2 | 1¾ | 1¾ | 1¾ | 1½ | 1½ | 1½ | 1½ | 1½ | 1½ | 1½ | 1½ | 1¼ | 1¼ | 1¼ | 1¼ | 1¼ | 1¼ |
| ½ | 2½ | 2½ | 2½ | 2¼ | 2¼ | 2¼ | 2 | 2 | 2 | 2 | 2 | 2 | 2 | 2 | 1¾ | 1¾ | 1¾ | 1¾ | 1¾ | 1½ |
| ⅝ | 3¼ | 3 | 3 | 2¾ | 2¾ | 2¾ | 2½ | 2½ | 2½ | 2½ | 2½ | 2¼ | 2¼ | 2¼ | 2¼ | 2¼ | 2¼ | 2 | 2 | 2 |
| ¾ | 3¾ | 3¾ | 3½ | 3¼ | 3¼ | 3¼ | 3¼ | 3 | 3 | 3 | 3 | 2¾ | 2¾ | 2¾ | 2¾ | 2½ | 2½ | 2¼ | 2¼ | 2¼ |
| ⅞ | 4½ | 4¼ | 4¼ | 4 | 4 | 4 | 3¾ | 3½ | 3½ | 3½ | 3¼ | 3¼ | 3¼ | 3 | 3 | 3 | 2¾ | 2¾ | 2¾ | 2¾ |
| 1 | 5 | 5 | 4¾ | 4½ | 4½ | 4½ | 4¼ | 4 | 4 | 3¾ | 3¾ | 3¾ | 3½ | 3½ | 3½ | 3¼ | 3¼ | 3¼ | 3¼ | 3 |
| 1⅛ | 5¾ | 5½ | 5½ | 5¼ | 5 | 5 | 4¾ | 4½ | 4½ | 4¼ | 4¼ | 4¼ | 4 | 4 | 3¾ | 3¾ | 3¾ | 3¾ | 3½ | 3½ |
| 1¼ | 6¼ | 6 | 6 | 5¾ | 5½ | 5½ | 5¼ | 5 | 5 | 4¾ | 4¾ | 4½ | 4½ | 4½ | 4¼ | 4¼ | 4¼ | 4 | 4 | 3¾ |
| 1⅜ | 7 | 6¾ | 6½ | 6¼ | 6 | 6 | 5¾ | 5½ | 5½ | 5¼ | 5¼ | 5 | 5 | 4¾ | 4¾ | 4½ | 4½ | 4½ | 4¼ | 4¼ |
| 1½ | 7½ | 7¼ | 7 | 6¾ | 6¾ | 6½ | 6¼ | 6 | 6 | 5¾ | 5½ | 5½ | 5¼ | 5¼ | 5 | 5 | 4¾ | 4¾ | 4¾ | 4½ |
| 1⅝ | 8¼ | 8 | 7¾ | 7½ | 7¼ | 7 | 6¾ | 6½ | 6½ | 6¼ | 6 | 6 | 5¾ | 5¾ | 5½ | 5½ | 5¼ | 5¼ | 5 | 5 |
| 1¾ | 8¾ | 8½ | 8¼ | 8 | 7¾ | 7½ | 7¼ | 7 | 7 | 6¾ | 6½ | 6½ | 6¼ | 6 | 6 | 5¾ | 5¾ | 5½ | 5½ | 5½ |
| 1⅞ | 9½ | 9 | 8¾ | 8½ | 8¼ | 8 | 7¾ | 7½ | 7½ | 7¼ | 7 | 7 | 6¾ | 6½ | 6½ | 6¼ | 6 | 6 | 5¾ | 5¾ |
| 2 | 10 | 9¾ | 9½ | 9 | 8¾ | 8½ | 8½ | 8¼ | 8 | 7¾ | 7½ | 7¼ | 7¼ | 7 | 6¾ | 6¾ | 6½ | 6½ | 6¼ | 6 |
| 2¼ | 11¼ | 11 | 10½ | 10¼ | 10 | 9½ | 9½ | 9 | 9 | 8¾ | 8½ | 8¼ | 8 | 8 | 7¾ | 7½ | 7½ | 7¼ | 7 | 7 |
| 2½ | 12½ | 12 | 11¾ | 11¼ | 11 | 10¾ | 10½ | 10 | 10 | 9½ | 9¼ | 9¼ | 9 | 8¾ | 8½ | 8¼ | 8¼ | 8 | 7¾ | 7½ |
| 2¾ | 13¾ | 13¼ | 12¾ | 12¼ | 12 | 11¾ | 11½ | 11 | 10¾ | 10½ | 10¼ | 10 | 9¾ | 9½ | 9¼ | 9¼ | 9 | 8¾ | 8½ | 8½ |
| 3 | 15 | 14½ | 14 | 13½ | 13 | 13 | 12¼ | 12 | 11½ | 11½ | 11 | 11 | 10¾ | 10½ | 10¼ | 10 | 9¾ | 9½ | 9½ | 9¼ |
| 3¼ | 16 | 15½ | 15 | 14½ | 14¼ | 14 | 13¼ | 13 | 12½ | 12¼ | 12 | 11¾ | 11½ | 11 | 11 | 10¾ | 10½ | 10¼ | 10¼ | 10 |
| 3½ | 17¼ | 16¾ | 16 | 15½ | 15½ | 15 | 14¼ | 14 | 13½ | 13¼ | 13 | 12½ | 12¼ | 12 | 11¾ | 11½ | 11¼ | 11 | 10¾ | 10½ |
| 3¾ | 18½ | 18 | 17¼ | 16½ | 16¼ | 15¾ | 15¼ | 15 | 14½ | 14¼ | 13¾ | 13½ | 13 | 13 | 12½ | 12¼ | 12 | 11¾ | 11½ | 11¼ |
| 4 | 19½ | 19 | 18½ | 17½ | 17¼ | 16¾ | 16¼ | 16 | 15½ | 15 | 14½ | 14¼ | 14 | 13¾ | 13¼ | 13 | 13 | 12½ | 12¼ | 12 |
| 4¼ | 20½ | 20 | 19½ | 18½ | 18½ | 17¾ | 17¼ | 16¾ | 16½ | 16 | 15½ | 15¼ | 15 | 14½ | 14 | 14 | 13¾ | 13½ | 13 | 12¾ |
| 4½ | 21½ | 21 | 20½ | 19¾ | 19½ | 18½ | 18¼ | 17½ | 17¼ | 17 | 16½ | 16 | 15¾ | 15¼ | 15 | 14¾ | 14½ | 14¼ | 13¾ | 13½ |
| 4¾ | 22¾ | 22 | 21½ | 20¾ | 20½ | 19½ | 19¼ | 18½ | 18¼ | 18 | 17¼ | 17 | 16½ | 16 | 16 | 15½ | 15¼ | 15 | 14½ | 14¼ |
| 5 | 23¾ | 23 | 22½ | 21¾ | 21¼ | 20½ | 20 | 19½ | 19 | 18½ | 18¼ | 18 | 17¼ | 17 | 16½ | 16¼ | 16 | 15½ | 15¼ | 15 |
| 5¼ | 24¾ | 24 | 23½ | 22½ | 22 | 21½ | 21 | 20½ | 20 | 19½ | 19 | 18½ | 18¼ | 17¾ | 17½ | 17 | 16¾ | 16¼ | 16 | 15¾ |
| 5½ | 25¾ | 25¼ | 24½ | 23½ | 23 | 22½ | 22 | 21½ | 21 | 20½ | 20 | 19½ | 19 | 18½ | 18¼ | 18 | 17½ | 17 | 16¾ | 16½ |
| 5¾ | 27 | 26¼ | 25½ | 24½ | 24 | 23½ | 23 | 22 | 21¾ | 21 | 20½ | 20¼ | 19¾ | 19½ | 19 | 18½ | 18¼ | 17¾ | 17½ | 17 |
| 6 | 27¾ | 27 | 26½ | 25½ | 25 | 24¼ | 23¾ | 23 | 22½ | 22 | 21½ | 21 | 20½ | 20 | 19½ | 19¼ | 19 | 18½ | 18 | 17¾ |

**1.** *Milling with a Disk-shaped Cutter.* The cutter has straight sides or edges and it is inclined an amount equal to the lead angle to locate the cutting side in alignment with the thread groove. There is a traversing movement per work revolution equal to the required lead. If the worm has two or more threads or starts, these are, of course, milled one at a time and should be uniformly spaced by indexing. *Precise indexing is very important.* A worm- or thread-milling machine is used.

**2.** *Hobbing.* A regular gear-hobbing machine is very efficient, especially for multi-thread worms, because all of the threads are finished simultaneously instead of taking separate cuts and indexing. The hobbing machine is geared with reference to the number of threads on the worm, the procedure being similar to that followed in cutting a helical gear.

**3.** *Generating by using a Worm Thread Generator.* The machine is equipped with a helical type gear shaper cutter, the axis of which is at right angles to the axis of the worm. The cutter generates the thread or threads as it rolls in mesh with the rotating work.

**4.** *Cutting in a Lathe.* One method is to locate the top cutting face of the tool normal or perpendicular to the worm thread. A second method is to locate the top cutting face in the same plane as the axis of the worm. With this method, cutting difficulties are encountered when the lead angles are comparatively large, due to the negative rake on the following side of the tool. As a general rule, the first method is preferable when a lathe must be used.

*Grinding Worm Threads and Hobs.* — A common method of producing worm threads is by milling with a disk-shaped cutter and then grinding after the hardening operation. The milling cutter has straight edges and the grinding wheel straight sides, because of the practical advantage in producing and maintaining these straight edges and sides.

In finishing hardened steel worms by grinding, the straight-sided grinding wheel produces a thread having convex sides and this curvature should be duplicated on the worm-wheel hob. This can be accomplished by finishing both worm and hob with grinding wheels which are maintained within diameter limits established by experience. In the case of multi-thread worms, accuracy of the indexing is essential when grinding both the worm and hob and accurate uniform lead is also important.

**Double-enveloping Worm Gearing.** — Contact between the worm and wheel of the conventional type of worm gearing is theoretically a line contact; however, due to deflection of the materials under load, the line is increased to a narrow band

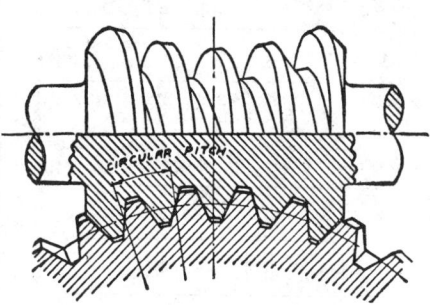

or contact zone. In attempting to produce a double-enveloping type of worm gear (with the worm curved longitudinally to fit the curvature of the wheel as shown by illustration), the problem primarily was that of generating the worm and worm wheel in such a manner as to obtain *area contact* between the engaging teeth. A practical method of obtaining such contact was developed by Samuel I. Cone at the Norfolk Navy Yard, and this is known as "Cone-Drive" worm gearing. The Cone generating method makes it possible to cut the worm and gear without any interference which

DOUBLE-ENVELOPING WORM GEARING 901

Table 1. Input Mechanical Horsepower Ratings of Cone-Drive Worm Gearing*

| Ratio | 100 | 300 | 720 | 870 | 1150 | 1750 |
|---|---|---|---|---|---|---|
| **2-Inch Center Distance** | | | | | | |
| 5:1 | .39 | 1.05 | 2.18 | 2.51 | 3.03 | 3.84 |
| 10:1 | .24 | .66 | 1.40 | 1.62 | 1.97 | 2.52 |
| 15:1 | .16 | .46 | .98 | 1.14 | 1.39 | 1.78 |
| 20:1 | .13 | .34 | .75 | .87 | 1.06 | 1.37 |
| 25:1 | .11 | .28 | .61 | .70 | .86 | 1.10 |
| 30:1 | .08 | .23 | .50 | .59 | .72 | .92 |
| 40:1 | .06 | .17 | .38 | .44 | .54 | .69 |
| 50:1 | .05 | .14 | .30 | .35 | .43 | .56 |
| **2½-Inch Center Distance** | | | | | | |
| 5:1 | .72 | 2.04 | 4.12 | 4.66 | 5.46 | 6.86 |
| 10:1 | .48 | 1.30 | 2.67 | 3.05 | 3.62 | 4.58 |
| 15:1 | .34 | .90 | 1.88 | 2.15 | 2.57 | 3.21 |
| 20:1 | .26 | .69 | 1.44 | 1.65 | 1.97 | 2.47 |
| 25:1 | .20 | .55 | 1.16 | 1.33 | 1.59 | 1.99 |
| 30:1 | .17 | .46 | .97 | 1.11 | 1.33 | 1.67 |
| 40:1 | .13 | .35 | .73 | .83 | 1.00 | 1.27 |
| 50:1 | .11 | .28 | .58 | .67 | .80 | 1.04 |
| **3-Inch Center Distance** | | | | | | |
| 5:1 | 1.31 | 3.60 | 6.97 | 7.87 | 9.02 | 11.30 |
| 10:1 | .87 | 2.31 | 4.65 | 5.26 | 6.17 | 7.72 |
| 15:1 | .61 | 1.62 | 3.27 | 3.72 | 4.37 | 5.47 |
| 20:1 | .47 | 1.23 | 2.51 | 2.86 | 3.36 | 4.22 |
| 25:1 | .37 | .99 | 2.03 | 2.31 | 2.72 | 3.40 |
| 30:1 | .32 | .83 | 1.69 | 1.93 | 2.28 | 2.85 |
| 40:1 | .23 | .62 | 1.28 | 1.45 | 1.71 | 2.14 |
| 50:1 | .20 | .50 | 1.03 | 1.16 | 1.37 | 1.72 |
| 60:1 | .15 | .42 | .85 | .97 | 1.14 | 1.43 |
| **3½-Inch Center Distance** | | | | | | |
| 5:1 | 2.46 | 6.58 | 12.28 | 14.35 | 16.70 | 19.50 |
| 10:1 | 1.62 | 4.24 | 8.23 | 9.18 | 10.70 | 13.10 |
| 15:1 | 1.16 | 2.97 | 5.82 | 6.50 | 7.55 | 9.20 |
| 20:1 | .87 | 2.26 | 4.45 | 5.00 | 5.78 | 7.30 |
| 25:1 | .70 | 1.83 | 3.62 | 4.05 | 4.69 | 5.95 |
| 30:1 | .56 | 1.54 | 3.01 | 3.38 | 3.92 | 5.01 |
| 40:1 | .44 | 1.15 | 2.26 | 2.53 | 2.95 | 3.78 |
| 50:1 | .36 | .93 | 1.81 | 2.04 | 2.38 | 3.00 |
| 60:1 | .29 | .75 | 1.51 | 1.70 | 2.00 | 2.53 |
| **4-Inch Center Distance** | | | | | | |
| 5:1 | 3.62 | 9.30 | 16.75 | 18.50 | 21.80 | 26.65 |
| 10:1 | 2.32 | 6.02 | 11.38 | 12.65 | 14.85 | 18.40 |
| 15:1 | 1.67 | 4.24 | 8.06 | 8.95 | 10.48 | 13.06 |
| 20:1 | 1.24 | 3.24 | 6.17 | 6.88 | 7.97 | 10.09 |
| 25:1 | 1.00 | 2.61 | 5.01 | 5.57 | 6.48 | 8.12 |
| 30:1 | .84 | 2.18 | 4.17 | 4.66 | 5.41 | 6.80 |
| 40:1 | .64 | 1.64 | 3.14 | 3.50 | 4.08 | 5.10 |
| 50:1 | .52 | 1.32 | 2.53 | 2.82 | 3.28 | 4.12 |
| 60:1 | .42 | 1.10 | 2.11 | 2.35 | 2.73 | 3.43 |
| **5-Inch Center Distance** | | | | | | |
| 5:1 | 7.15 | 18.40 | 31.65 | 35.15 | 40.50 | 49.20 |
| 10:1 | 4.63 | 11.87 | 21.13 | 23.40 | 27.17 | 33.45 |
| 15:1 | 3.27 | 8.33 | 15.01 | 16.57 | 19.35 | 23.65 |
| 20:1 | 2.47 | 6.37 | 11.49 | 12.68 | 14.77 | 18.20 |
| 25:1 | 1.99 | 5.15 | 9.32 | 10.30 | 11.97 | 14.75 |
| 30:1 | 1.67 | 4.32 | 7.79 | 8.61 | 10.02 | 12.32 |
| 40:1 | 1.26 | 3.24 | 5.87 | 6.49 | 7.54 | 9.30 |
| 50:1 | 1.01 | 2.60 | 4.72 | 5.20 | 6.07 | 7.47 |
| **5-Inch Center Distance (Continued)** | | | | | | |
| 60:1 | .84 | 2.17 | 3.94 | 4.34 | 5.06 | 6.22 |
| 70:1 | .71 | 1.86 | 3.37 | 3.72 | 4.34 | 5.35 |
| **6-Inch Center Distance** | | | | | | |
| 5:1 | 11.08 | 27.00 | 44.60 | 49.17 | 54.60 | 68.00 |
| 10:1 | 7.08 | 17.86 | 30.40 | 33.70 | 38.65 | 46.60 |
| 15:1 | 4.96 | 12.58 | 21.54 | 23.82 | 27.55 | 33.40 |
| 20:1 | 3.82 | 9.62 | 16.50 | 18.35 | 21.12 | 25.60 |
| 25:1 | 3.07 | 7.79 | 13.38 | 14.96 | 17.16 | 20.78 |
| 30:1 | 2.56 | 6.50 | 11.22 | 12.42 | 14.35 | 17.35 |
| 40:1 | 1.92 | 4.88 | 8.43 | 9.36 | 10.73 | 13.12 |
| 50:1 | 1.55 | 3.92 | 6.77 | 7.52 | 8.66 | 10.53 |
| 60:1 | 1.29 | 3.27 | 5.64 | 6.26 | 7.21 | 8.77 |
| 70:1 | 1.10 | 2.80 | 4.83 | 5.37 | 6.20 | 7.54 |
| **7-Inch Center Distance** | | | | | | |
| 5:1 | 17.40 | 41.50 | 67.30 | 73.80 | 83.50 | 96.40 |
| 10:1 | 11.20 | 27.78 | 46.40 | 51.15 | 58.60 | 68.60 |
| 15:1 | 7.86 | 19.59 | 33.00 | 36.40 | 41.85 | 49.25 |
| 20:1 | 6.01 | 15.02 | 25.30 | 27.97 | 32.17 | 38.00 |
| 25:1 | 4.85 | 12.13 | 20.50 | 22.60 | 26.08 | 30.82 |
| 30:1 | 4.05 | 10.15 | 17.13 | 18.88 | 21.80 | 25.73 |
| 40:1 | 3.06 | 7.64 | 12.87 | 14.28 | 16.38 | 19.45 |
| 50:1 | 2.47 | 6.13 | 10.35 | 11.47 | 13.17 | 15.63 |
| 60:1 | 2.05 | 5.11 | 8.62 | 9.56 | 10.96 | 13.20 |
| 70:1 | 1.75 | 4.37 | 7.38 | 8.20 | 9.42 | 11.70 |
| **8-Inch Center Distance** | | | | | | |
| 5:1 | 25.85 | 59.60 | 95.20 | 104.20 | 116.50 | *134.80* |
| 10:1 | 16.72 | 40.89 | 67.40 | 74.60 | 85.10 | 98.90 |
| 15:1 | 11.75 | 28.97 | 48.10 | 53.05 | 60.95 | 71.20 |
| 20:1 | 8.98 | 22.22 | 36.75 | 40.80 | 46.90 | 54.30 |
| 25:1 | 7.24 | 17.97 | 29.97 | 33.18 | 38.00 | 44.70 |
| 30:1 | 6.05 | 15.05 | 25.20 | 27.70 | 31.77 | 37.20 |
| 40:1 | 4.56 | 11.32 | 18.88 | 20.97 | 24.07 | 27.95 |
| 50:1 | 3.67 | 9.09 | 15.17 | 16.84 | 19.32 | 22.90 |
| 60:1 | 3.03 | 7.59 | 12.68 | 14.03 | 16.08 | 18.70 |
| 70:1 | 2.60 | 6.50 | 10.87 | 12.05 | 13.80 | 16.06 |
| **10-Inch Center Distance** | | | | | | |
| 5:1 | 48.40 | 104.60 | 164.00 | 177.60 | 193.00 | *227.20* |
| 10:1 | 31.47 | 73.43 | 117.20 | 128.90 | 144.80 | 167.18 |
| 15:1 | 22.07 | 51.97 | 80.38 | 92.00 | 103.80 | 119.43 |
| 20:1 | 16.87 | 39.94 | 64.40 | 70.80 | 80.25 | 92.20 |
| 25:1 | 13.64 | 32.33 | 52.30 | 57.35 | 65.10 | 74.87 |
| 30:1 | 11.39 | 27.01 | 43.75 | 47.90 | 54.40 | 62.55 |
| 40:1 | 8.57 | 20.31 | 32.85 | 36.00 | 40.90 | 47.04 |
| 50:1 | 6.89 | 16.38 | 26.58 | 29.15 | 33.10 | 37.99 |
| 60:1 | 5.71 | 13.65 | 22.15 | 24.30 | 27.50 | 31.65 |
| 70:1 | 4.89 | 11.70 | 19.00 | 20.85 | 23.58 | 27.17 |
| **12-Inch Center Distance** | | | | | | |
| 5:1 | 81.50 | 166.30 | 256.00 | 270.00 | *300.50* | *344.00* |
| 10:1 | 53.30 | 117.82 | 186.50 | 200.20 | 222.00 | *258.77* |
| 15:1 | 37.40 | 83.88 | 133.20 | 145.00 | 159.00 | *185.36* |
| 20:1 | 28.68 | 64.34 | 102.00 | 112.00 | 123.20 | *142.40* |
| 25:1 | 23.15 | 52.20 | 82.80 | 90.70 | 99.70 | *115.71* |
| 30:1 | 19.32 | 43.72 | 69.30 | 75.80 | 83.50 | *96.68* |
| 40:1 | 14.58 | 32.87 | 52.10 | 57.00 | 63.00 | *72.71* |
| 50:1 | 11.71 | 26.47 | 42.00 | 45.90 | 50.60 | *58.76* |
| 60:1 | 9.71 | 22.06 | 35.00 | 38.30 | 42.10 | *48.95* |
| 70:1 | 8.32 | 18.91 | 30.00 | 32.80 | 36.20 | *42.02* |

* Horsepower ratings in table are for Class I service: Ratings shown in italics are for force feed lubrication; all others based on splash lubrication.
Other ratios and center distances are available.

would alter the required tooth form. The larger tooth bearing area and multiple tooth contact obtained with this type of worm gearing, increases the load-carrying or horsepower capacity so that as compared with a conventional worm drive a double-enveloping worm drive may be considerably smaller in size. Table 1, which is intended as a general guide, gives input horsepower ratings for Cone-Drive worm gearing for various center distances of from 2 to 12 inches. These ratings were supplied by the Cone-Drive Gear Division of the Michigan Tool Co. and are for Class 1 service. For other classes of service use the appropriate service factors given in the following: *Class 1:* Normal 8-10 hour service, free from recurrent shocks (shock loads that recur at approximately even and frequent intervals). Table 1 applies directly in this classification provided only that the thermal rating is not exceeded; *Class 2:* 8-10 hour service where recurrent shock loading is encountered, or 24-hour service where no shock loading is experienced. Class 1 rating must be reduced by dividing by 1.2 for this service. Thermal rating must not be exceeded; *Class 3:* 24-hour shock load service. Class 1 rating must be reduced by dividing by 1.3 for this service, and thermal rating must not be exceeded; *Class 4:* For intermittent service where worm is operated at 100 revolutions per minute or more, the Class 1 rating must be divided by the factor indicated in Table 2, depending upon the frequency and duration of the load. Thermal rating need not be considered for this class of service; *Class 5:* Low speed service, where the worm speed is less than 100 revolutions per minute, will require output torque ratings in inch-pounds, calculated from the following formula, independent of the nature or duration of the load, except that where such service is intermittent the factors in Table 2 apply as for Class 4.

$$\text{Output Torque} = \text{Input Torque} \times \text{Ratio} \times \text{Efficiency}$$

In this formula, the efficiency, for the purposes of output torque calculations only, is taken as 0.97 for ratios of 6 to 1 or less; for higher ratios, efficiency = 1 − (Ratio ÷ 200).

**Table 2. Service Factors for Class 4 and Intermittent Class 5 Service**

| Multiple Cycles per Hour | 1 Cycle per Hour | 1 Cycle per 2 Hours or More | Service Factor | Service Factor Designation | |
|---|---|---|---|---|---|
| Total Minutes of Operation Allowed | | | | | |
| Per Hour | Per Cycle | Per Cycle | | Class 4 Service | Class 5 Service |
| .. | 5 | 10 | 0.6 | 4/0.6 | 5/0.6 |
| 2 | 10 | 20 | 0.7 | 4/0.7 | 5/0.7 |
| 5 | 15 | 30 | 0.8 | 4/0.8 | 5/0.8 |
| 10 | 20 | 40 | 0.9 | 4/0.9 | 5/0.9 |

**Thermal Rating of Double-enveloping Worm Drives.** — As explained on page 893, the thermal rating and not the mechanical rating may govern the horsepower capacity of worm gearing. For double-enveloping worm drive speed reducers, the following thermal rating formula is designed to limit the input horsepower to such a value that the temperature of the oil in the sump of the reducer will not exceed 100 degrees F. above the ambient temperature:

$$P = AK_t$$

$P$ = thermal horsepower; $A$ = area of housing in square feet; $K_t$ = thermal constant = $\dfrac{f}{25.5(100 - \eta)}$; $f$ = radiation factor = 255 Btu per square foot per hour for 100 degree F. temperature rise; and $\eta$ = efficiency, approximate values for which are as follows for the sets shown in the table on page 901: For ratios of 5:1, $\eta$ = 0.94; for 10:1, 0.93; for 15:1, 0.89; for 20:1, 0.87; for 25:1, 0.85; for 30:1, 0.83; for 40:1, 0.79; for 50:1, 0.73; for 60:1, 0.72; and for 70:1, 0.70.

The oil bath temperature should not exceed 200 degrees F. Where worm speed exceeds 3600 revolutions per minute, or 2000 feet per minute rubbing speed, a force-feed lubrication may be required. Auxiliary cooling by forced air, water coils in sump, or an oil heat exchanger may be provided in a unit where mechanical horsepower rating is in excess of thermal, if full advantage of mechanical capacity is to be realized.

Due to the influence on worm gear speed reducer ratings of such factors as housing shape, surface area, bearing mountings, and lubrication, the preceding thermal rating method is limited to speed reducers meeting the following requirements: (1) The housing must be of the "box" type to minimize oil churning and provide suitable strength to hold the gear set in proper alignment; (2) the housing area should be more than $0.4C^{1.7}$, $C$ being the center distance of the worm and worm-gear set in inches; (3) worm shaft and gear shaft must be mounted on ball or roller bearings of sufficient capacity to have an average life of 25,000 hours at rated load; and (4) worm and gear shafts, mountings, and bolts should be of sufficient size to handle the loads to which the reducer is subjected with adequate safety factors as recommended by American Gear Manufacturers Association or equal.

**Worm Gear Hobs.** — An ideal hob would have exactly the same pitch diameter and lead angle as the worm; repeated sharpening, however, would reduce the hob size because of the form-relieved teeth. Hence, the general practice is to make hobs (especially the radial or in-feed type) "over-size" to provide a grinding allowance and increase the hob life. An over-size hob has a larger pitch diameter and smaller lead angle than the worm, but repeated sharpenings gradually reduce these differences. To compensate for the smaller lead angle of an over-size hob, the hob axis may be set 90-degrees relative to the wheel axis plus the difference between the *lead* angle of the worm at the pitch line, and the *lead* angle of the over-size hob at its pitch line. This angular adjustment is in the direction required to increase the inclination of the worm wheel teeth so that the axis of the assembled worm will be 90 degrees from the wheel axis. ("Lead angle" is measured from a plane perpendicular to worm or hob axis.)

*Hob Diameter Formulas:* If $D$ = pitch diameter of worm; $D_h$ = pitch diameter of hob; $A$ = addendum of worm and worm wheel; $C$ = clearance between worm and worm wheel; and $S$ = increase in hob diameter or "over-size" allowance for sharpening.

Outside diameter $O$ of hob = $D + 2A + 2C + S$
Root diameter of hob = $D - 2A$
Pitch diameter $D_h$ of hob = $O - (2A + 2C)$

*Sharpening Allowance:* Hobs for ordinary commercial work are given the following sharpening allowance, according to the recommended practice of the AGMA: In this formula, $h$ = helix angle of hob at outside diameter measured from axis; $H$ = helix angle of hob at pitch diameter measured from axis.

$$\text{Sharpening allowance} = 0.075 \times \text{normal pitch} \times \left[ \frac{16 - (h - H)}{16} \right] + 0.010$$

**Number of Flutes or Gashes in Hobs.** — For finding the approximate number of flutes in a hob, the following rule may be used: Multiply the diameter of the hob by 3, and divide this product by twice the linear pitch. This rule gives suitable results for hobs for general purposes. Certain modifications, however, are necessary as explained in the following paragraph.

It is important that the number of flutes or gashes in hobs bear a certain relation to the number of threads in the hob and the number of teeth in the worm-wheel to be hobbed. In the first place, avoid having a common factor between the number of threads in the hob and the number of flutes; that is, if the worm is double-threaded, the number of gashes should be, say, 7 or 9, rather than 8. If it is triple-threaded, the number of gashes should be 7 or 11, rather than 6 or 9. The second requirement is to avoid having a common factor between the number of threads in the hob and the number of teeth in the worm-wheel. For example, if the number of teeth in the wheel is 28, it would be best to have the hob triple-threaded, as 3 is not a factor of 28. Again, if there were to be 36 threads in the worm-gear, it would be preferable to have 5 threads in the hob.

The cutter used in gashing hobs should be from ⅛ to ¼ inch thick at the periphery, according to the pitch of the thread of the hob. The width of the gash at the periphery of the hob should be about 0.4 times the pitch of the flutes. The cutter should be sunk into the hob blank so that it reaches from 3⁄16 to ¼ inch below the root of the thread.

**Helical Fluted Hobs.** — Hobs are generally fluted parallel with the axis, but it is obvious that the cutting action will be better if they are fluted on a helix at right angles with the thread helix. The difficulty of relieving the teeth with the ordinary backing-off attachment is the cause for using a flute parallel with the axis. Flutes cut at right angles to the direction of the thread can, however, also be relieved, if the angle of the flutes is slightly modified. In order to relieve hobs with a regular relieving attachment, it is necessary that the number of teeth in one revolution along the thread helix be such that the relieving attachment can be geared to suit it. The following method makes it possible to select an angle of flute that will make the flute come *approximately* at right angles to the thread, and at the same time the angle is so selected that the relieving attachment can be properly geared for relieving the hob.

Let
$C$ = pitch circumference;
$T$ = developed length of thread in one turn;
$N$ = number of teeth in one turn along thread helix;
$F$ = number of flutes;
$\alpha$ = angle of thread helix.

Then (see illustration on the following page):

$C \div F$ = length of each small division on pitch circumference;
$(C \div F) \times \cos \alpha$ = length of division on developed thread;
$C \div \cos \alpha = T$.

Hence
$$\frac{T}{(C \div F) \cos \alpha} = N = \frac{F}{\cos^2 \alpha}$$

Now, if
$\alpha$ = 30 degrees, $N = 1\frac{1}{3} F$;
$\alpha$ = 45 degrees, $N = 2 F$;
$\alpha$ = 60 degrees, $N = 4 F$.

In most cases, however, such simple relations are not obtained. Suppose for example that $F = 7$, and $\alpha = 35$ degrees. Then $N = 10.432$, and no gears could be selected that would relieve this hob. By a very slight change in the helix angle of the flute, however, we can change $N$ to 10 or 10½; in either case we can find suitable gears for the relieving attachment.

The rule for finding the modified helical lead of the flute is: Multiply the lead of the hob by $F$, and divide the product by the difference between the desired values of $N$ and $F$.

Hence, the lead of flute required to make $N = 10$ is:

Lead of hob $\times$ (7 ÷ 3).

To make $N = 10\frac{1}{2}$, we have:

Lead of flute = lead of hob $\times$ (7 ÷ 3.5).

From this the angle of the flute can easily be found.

That the rule given is correct will be understood from the following consideration. Change the angle of the flute helix $\beta$ so that $AG$ contains the required number of parts $N$ desired. Then $EG$ contains $N - F$ parts. But $\cot \beta = BD \div ED$ and by the law of similar triangles,

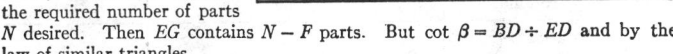

$$BD = \frac{F}{N} \times BG, \text{ and } ED = \frac{N - F}{N} C$$

The lead of the helix of the flute, however, is $C \times \cot \beta$.

Hence, the required lead of the helix of the flute:

$$C \times \cot \beta = \frac{F}{N - F} L$$

This formula makes it possible always to flute hobs so that they can be conveniently relieved, and at the same time have the flutes at approximately right angles to the thread.

## Helical Gearing

**Basic Rules and Formulas for Helical Gear Calculations.** — The ten rules and formulas in the accompanying table may be called the basic rules for helical gear calculations. The following definitions should be clearly understood in order to avoid misunderstandings. The *center angle* of a pair of helical gears is the angle made by the two center lines or axes of the gears. The *tooth angle* is the angle which the direction of the tooth makes with the axis of the gear. The *normal diametral pitch* is the diametral pitch of the cutter used for cutting the teeth in helical gears. In the formulas in the table of "Basic Rules and Formulas for Helical Gear Calculations" the following notation is used:

$P_n$ = normal diametral pitch (pitch of cutter);

$D$ = pitch diameter;

$N$ = number of teeth;

$\alpha$ = helix angle;

$\gamma$ = center angle, or angle between shafts;

$C$ = center distance;

$N'$ = number of teeth for which to select a formed cutter;

$L$ = lead of tooth helix;

$S$ = addendum;

$W$ = whole depth of tooth;

$T_n$ = normal tooth thickness at pitch line;

$O$ = outside diameter.

The rules and formulas are given in the same order as they would ordinarily be used by the designer when calculating a pair of helical gears   The formulas, how-

## Basic Rules and Formulas for Helical Gear Calculations

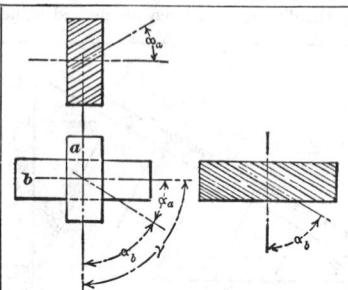

In the formulas, $N$, $\alpha$, etc., are the numbers of teeth, helix angle, etc., for *either* gear or pinion; the notations $N_a$, $N_b$, $\alpha_a$, $\alpha_b$, etc., refer to the teeth or angles in the pinion or gear, respectively, in a pair of gears $a$ and $b$.

| No. | To Find | Rule | Formula |
|---|---|---|---|
| 1 | Relation between Shaft and Tooth Angles. | See rules at bottom of page 909. | |
| 2 | Pitch Diameter. | Divide the number of teeth by the product of the normal pitch and the cosine of the tooth angle. | $D = \dfrac{N}{P_n \cos \alpha}$ |
| 3 | Center Distance. | Add together the pitch diameters of the two gears and divide by 2. | $C = \dfrac{D_a + D_b}{2}$ |
| 4 | Checking Calculations in (2) and (3); for use when angle between shafts is 90 degrees. | To prove the calculations for pitch diameters and center distance, multiply the number of teeth in the first gear by the tangent of the tooth angle of that gear, and add the number of teeth in the second gear to the product; the sum should equal twice the product of the center distance multiplied by the normal diametral pitch, multiplied by the sine of the tooth angle of the first gear. | $N_b + (N_a \times \tan \alpha_a) = 2 CP_n \times \sin \alpha_a$ |
| 5 | Number of Teeth for which to Select Formed Cutter. | Follow procedure outlined under heading "Selecting Cutter for Milling Helical Gears," page 927. | |
| 6 | Lead of Tooth Helix. | Multiply the pitch diameter by 3.1416 times the cotangent of the tooth angle. | $L = \pi D \times \cot \alpha$ |
| 7 | Addendum. | Divide 1 by the normal diametral pitch. | $S = \dfrac{1}{P_n}$ |
| 8 | Whole Depth of Tooth. | Divide 2.157 by the normal diametral pitch.* | $W = \dfrac{2.157}{P_n}$ |
| 9 | Normal Tooth Thickness at Pitch Line. | Divide 1.571 by the normal diametral pitch. | $T_n = \dfrac{1.571}{P_n}$ |
| 10 | Outside Diameter. | Add twice the addendum to the pitch diameter. | $O = D + 2S$ |

*For hobbed 20° pressure angle gears of 20 diametral pitch and finer, $W = (2.200 \div P_n) + 0.002$.

ever, cannot be directly applied to all cases of helical gear problems, and a complete set of formulas for each of the seventeen different cases which are frequently met with is, therefore, given in the following, together with an example for each case. These seventeen cases are:

1. Shafts parallel, ratio 1 ("1 to 1"), center distance approximate.
2. Shafts parallel, ratio 1, center distance exact.
3. Shafts parallel, ratio other than 1, center distance approximate.
4. Shafts parallel, ratio other than 1, center distance exact.
5. Shafts at right angles, ratio 1, center distance approximate.
6. Shafts at right angles, ratio 1, center distance exact.
7. Shafts at right angles, ratio other than 1, center distance approximate.
8A. Shafts at right angles, ratio other than 1, center distance exact.
8B. Shafts at right angles, any ratio, helix angle for minimum center distance.
9. Shafts at 45-degree angle, ratio 1, center distance approximate.
10. Shafts at 45-degree angle, ratio 1, center distance exact.
11. Shafts at 45-degree angle, ratio other than 1, center distance approximate.
12. Shafts at 45-degree angle, ratio other than 1, center distance exact.
13. Shafts at any angle, ratio 1, center distance approximate.
14. Shafts at any angle, ratio 1, center distance exact.
15. Shafts at any angle, ratio other than 1, center distance approximate.
16. Shafts at any angle, ratio other than 1, center distance exact.

**Pitch of Cutter for Helical Gears** — The thickness of the cutter at the pitch line for cutting helical gears should equal one-half the normal circular pitch $n$ (see illustration). If a cutter were used having a thickness, at the pitch line, equal to one-half the circular pitch $P$, as for spur gearing, the spaces between the teeth would be cut too wide, thus producing thin teeth. The normal pitch varies with the angle $\alpha$ of the helix; hence, the helix angle must be considered when selecting a cutter. The cutter should be of the same pitch as the *normal* diametral pitch of the gear. This normal pitch is found by dividing the transverse diametral pitch by the cosine of the helix angle. To illustrate, if the pitch diameter of a helical gear is 6.718 and there are 38 teeth having a helix angle of 45 degrees, the transverse diametral pitch equals 38 ÷ 6.718 = 5.656; then, the normal diametral pitch equals 5.656 divided by the cosine of 45 degrees or 5.656 ÷ 0.707 = 8. A cutter, then, of 8 diametral pitch is the one to use for this particular gear. This same result could

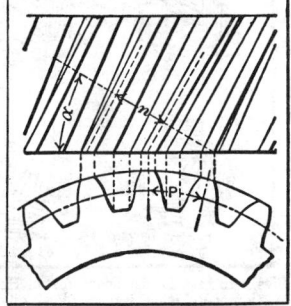

also be obtained as follows: If the circular pitch $P$ is 0.5554, the normal circular pitch $n$ can be found by multiplying the circular pitch $P$ by the cosine of the helix angle. For example, 0.5554 × 0.707 = 0.3927. The normal diametral pitch is then found by dividing 3.1416 by the normal circular pitch. Thus 3.1416 ÷ 0.3927 = 8, which is the diametral pitch of the cutter.

Helical gears should preferably be cut on some type of gear-cutting machine such as a "hobber," or on a shaper or planer of the generating type. Milling machines are used in some shops, especially when a gear-cutting machine is not available. The pitch of the formed cutter used in milling a helical gear, must not only conform to the normal diametral pitch of the gear, but the cutter number must also be determined. See page 927, "Selecting Cutter for Milling Helical Gears."

**Procedure in Calculating Helical Gears.** — One of the first steps necessary in helical gear design is to determine the direction of the thrust, if the thrust is to be taken in one direction only. When the direction of the thrust has been determined and the relative position of the driver and driven gear is known, the direction of helix (right- or left-hand) may be found. The thrust diagrams, Figs. 1 to 28, are used for finding the direction of helix. The arrows at the end bearings of the gears indicate the direction of the reaction against the thrust caused by the tooth pressure. The direction of the thrust depends on the direction of helix, the relative positions of driver and driven gear, and the direction of rotation. If the exact

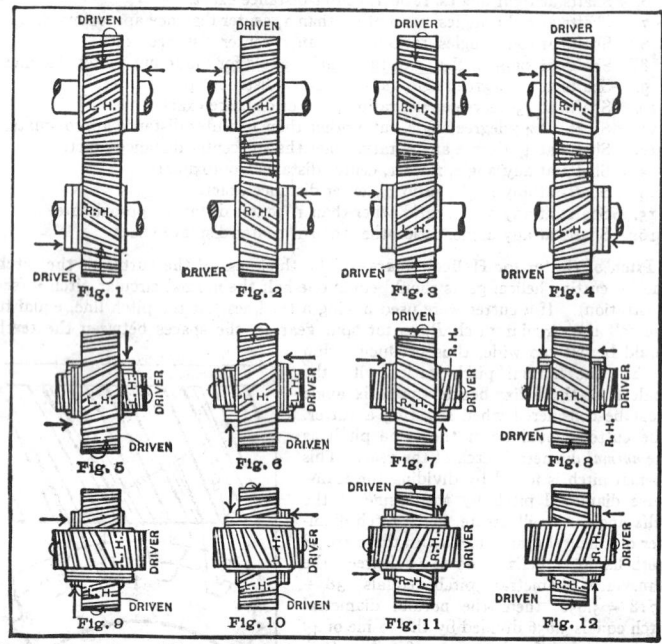

Figs. 1 to 12. Thrust Diagrams for Helical Gears — Direction of Thrust depends upon Direction of Rotation, Relative Position of Driver and Driven Gear, and Direction of Helix

condition with regard to thrust is not found in the diagrams, it may be obtained by changing any one of these three conditions; that is, in Fig. 1 the thrust may be changed to the opposite direction by interchanging driver or driven gear, by reversing the direction of rotation or by changing the direction of helix. Any one of these alterations will produce a thrust in the opposite direction.

The conditions of the design determine the nature of the center distance, whether it must be exact or approximate. The number of teeth in each gear is, of course, determined by the required speed ratios of the shafts. The angle of helix depends upon the conditions of the design and the relative position of the shafts. For parallel shafts the helix angle should not exceed 20 degrees in order to avoid excessive end thrust. In order to obtain smooth running gears, the helix angle should be such that one end of the tooth remains in contact until the opposite end of the following tooth has found a bearing.

As far as the calculations are concerned, the formulas are the same for a 135-degree shaft angle as for an angle of 45 degrees. The following general rule relative to gears having a shaft angle of 45 degrees should be observed: When the helix angle of each gear is less than 45 degrees, then the helix angles are of the same hand and one helix angle is 45 degrees minus the other. When the helix angle of either gear is greater than 45 degrees, then the helix angles are of opposite hand, and the helix angle of one gear is 45 degrees plus the helix angle of the other.

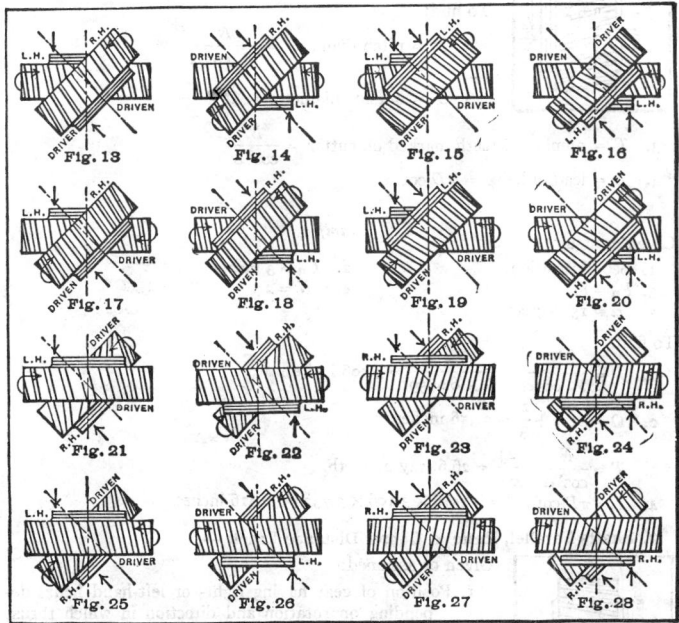

Figs. 13 to 28. Thrust Diagrams for Helical Gears — Direction of Thrust depends upon Direction of Rotation, Relative Position of Driver and Driven Gear, and Direction of Helix

The following rules should be observed for helical gears with shafts at any given angle. If each helix angle is less than the shaft angle, then the sum of the helix angles of the two gears will equal the angle between the shafts, and the helix is of the same hand in both gears; if the helix angle of one of the gears is greater than the shaft angle, then the difference between the helix angles of the two gears will be equal to the shaft angle, and the gears will be of opposite hand.

In some cases of helical gear design, particularly where trial-and-error methods of calculation must be used, it may be of considerable advantage to employ a graphical method for obtaining an approximate solution. (Such a method may be found in *Manual Of Gear Design*, Section Three, by Earle Buckingham.) The approximate solution thus obtained can then be modified to suit the problem at hand by making use of the standard helical gear formulas.

## 1. Shafts Parallel, Ratio 1 (or "1 to 1"), Center Distance Approximate. —

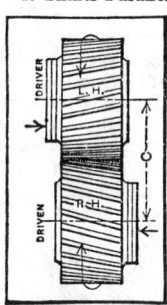

Given or assumed:

1. Hand of helix on driver or driven gear depending on rotation and direction in which thrust is to be received.
2. $C_a$ = approximate center distance.
3. $P_n$ = normal diametral pitch (pitch of cutter).
4. $N$ = number of teeth.
5. $\alpha$ = angle of helix (usually less than 20 degrees for single gears to avoid excessive end thrust).

To find:

1. $D$ = pitch diameter = $\dfrac{N}{P_n \cos \alpha}$

2. $O$ = outside diameter = $D + \dfrac{2}{P_n}$

3. $T$ = number of teeth marked on cutter = $\dfrac{N}{\cos^3 \alpha}$

4. $L$ = lead of helix = $\pi D \cot \alpha$.

Given or assumed:

*Example*

1. See illustration.
2. $C_a = 3$ inches
3. $P_n = 8$.
4. $N = 24$.
5. $\alpha = 15$ degrees.

To find:

1. $D = \dfrac{N}{P_n \cos \alpha} = \dfrac{24}{8 \times 0.9659} = 3.106$ inches.

2. $O = 3.106 + \dfrac{2}{8} = 3.356$ inches.

3. $T = \dfrac{N}{\cos^3 \alpha} = \dfrac{24}{0.9} = 26.6$, say 27 teeth.

4. $L = \pi D \cot 15° = 3.1416 \times 3.106 \times 3.732 = 36.416$ inches.

## 2. Shafts Parallel, Ratio 1, Center Distance Exact. —

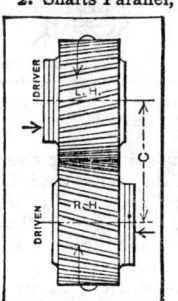

Given or assumed:

1. Position of gear having right- or left-hand helix, depending on rotation and direction in which thrust is to be received.
2. $C$ = exact center distance = pitch diameter $D$.
3. $P_n$ = normal diametral pitch (pitch of cutter).
4. $N$ = number of teeth in each gear.

To find:

1. $\cos \alpha = \dfrac{N}{P_n D}$

2. $O$ = outside diameter = $D + \dfrac{2}{P_n}$

3. $T$ = number of teeth marked on formed milling cutter = $\dfrac{N}{\cos^3 \alpha}$

4. $L$ = lead of helix = $\pi D \cot \alpha$.

$\alpha$ is usually less than 20 degrees to avoid excessive end thrust.

*Example*

**Given or assumed:**

1. See illustration.
2. $C = 3$ inches.
3. $P_n = 8$.
4. $N = 22$.

**To find:**

1. $\cos \alpha = \dfrac{N}{P_n D} = \dfrac{22}{8 \times 3} = 0.9166$, or $\alpha = 23° 34'$.

2. $O = D + \dfrac{2}{P_n} = 3 + \dfrac{2}{8} = 3\frac{1}{4}$ inches.

3. $T = \dfrac{N}{\cos^3 \alpha} = \dfrac{22}{(0.92)^3} = 28.2$, say 28 teeth.

4. $L = \pi D \cot \alpha = 3.1416 \times 3 \times 2.29 = 21.58$ inches.

### 3. Shafts Parallel, Ratio Other Than 1, Center Distance Approximate. —

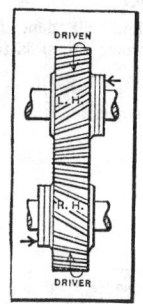

**Given or assumed:**

1. Position of gear having right- or left-hand helix, depending upon rotation and direction in which thrust is to be received.
2. $C_a$ = approximate center distance.
3. $P_n$ = normal diametral pitch.
4. $N$ = number of teeth in large gear.
5. $n$ = number of teeth in small gear.
6. $\alpha$ = angle of helix.

**To find:**

1. $D$ = pitch diameter of large gear = $\dfrac{N}{P_n \cos \alpha}$

2. $d$ = pitch diameter of small gear = $\dfrac{n}{P_n \cos \alpha}$

3. $O$ = outside diameter of large gear = $D + \dfrac{2}{P_n}$

4. $o$ = outside diameter of small gear = $d + \dfrac{2}{P_n}$

5. $T$ = number of teeth marked on formed cutter (large gear) = $\dfrac{N}{\cos^3 \alpha}$

6. $t$ = number of teeth marked on formed cutter (small gear) = $\dfrac{n}{\cos^3 \alpha}$

7. $L$ = lead of helix on large gear = $\pi D \cot \alpha$.
8. $l$ = lead of helix on small gear = $\pi d \cot \alpha$.
9. $C$ = center distance (if not right vary $\alpha$) = $\frac{1}{2}(D + d)$.

*Example*

**Given or assumed:**

1. See illustration.
2. $C_a = 17$ inches.
3. $P_n = 2$.
4. $N = 48$.
5. $n = 20$.
6. $\alpha = 20$ degrees.

**To find:**

1. $D = \dfrac{N}{P_n \cos \alpha} = \dfrac{48}{2 \times 0.9397} = 25.541$ inches.

2. $d = \dfrac{n}{P_n \cos \alpha} = \dfrac{20}{2 \times 0.9397} = 10.642$ inches.

3. $O = D + \dfrac{2}{P_n} = 25.541 + \dfrac{2}{2} = 26.541$ inches.

4. $o = d + \dfrac{2}{P_n} = 10.642 + \dfrac{2}{2} = 11.642$ inches.

5. $T = \dfrac{N}{\cos^3 \alpha} = \dfrac{48}{(0.9397)^3} = 57.8$, say 58 teeth.

6. $t = \dfrac{n}{\cos^3 \alpha} = \dfrac{20}{(0.9397)^3} = 24.1$, say 24 teeth.

7. $L = \pi D \cot \alpha = 3.1416 \times 25.541 \times 2.747 = 220.42$ inches.

8. $l = \pi d \cot \alpha = 3.1416 \times 10.642 \times 2.747 = 91.84$ inches.

9. $C = \frac{1}{2}(D + d) = \frac{1}{2}(25.541 + 10.642) = 18.091$ inches.

### 4. Shafts Parallel, Ratio Other Than 1, Center Distance Exact. —

Given or assumed:

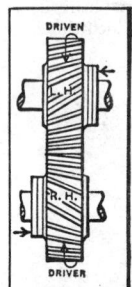

1. Position of gear having right- or left-hand helix, depending upon rotation and direction in which thrust is to be received.
2. $C$ = exact center distance.
3. $P_n$ = normal diametral pitch (pitch of cutter).
4. $N$ = number of teeth in large gear.
5. $n$ = number of teeth in small gear.

To find:

1. $\cos \alpha = \dfrac{N + n}{2 P_n C}$

2. $D$ = pitch diameter of large gear $= \dfrac{N}{P_n \cos \alpha}$

3. $d$ = pitch diameter of small gear $= \dfrac{n}{P_n \cos \alpha}$

4. $O$ = outside diameter of large gear $= D + \dfrac{2}{P_n}$

5. $o$ = outside diameter of small gear $= d + \dfrac{2}{P_n}$

6. $T$ = number of teeth marked on formed milling cutter (large gear) $= \dfrac{N}{\cos^3 \alpha}$

7. $t$ = number of teeth marked on formed milling cutter (small gear) $= \dfrac{n}{\cos^3 \alpha}$

8. $L$ = lead of helix (large gear) $= \pi D \cot \alpha$
9. $l$ = lead of helix (small gear) $= \pi d \cot \alpha$

*Example*

Given or assumed:

1. See illustration. 2. $C = 18.75$ inches. 3. $P_n = 4$. 4. $N = 96$. 5. $n = 48$.

To find:

1. $\cos \alpha = \dfrac{N + n}{2 P_n C} = \dfrac{96 + 48}{2 \times 4 \times 18.75} = 0.96$, or $\alpha = 16° \, 16'$.

2. $D = \dfrac{N}{P_n \cos \alpha} = \dfrac{96}{4 \times 0.96} = 25$ inches.

**3.** $\quad d = \dfrac{n}{P_n \cos \alpha} = \dfrac{48}{4 \times 0.96} = 12.5$ inches.

**4.** $\quad O = D + \dfrac{2}{P_n} = 25 + \dfrac{2}{4} = 25.5$ inches.

**5.** $\quad o = d + \dfrac{2}{P_n} = 12.5 + \dfrac{2}{4} = 13$ inches.

**6.** $\quad T = \dfrac{N}{\cos^3 \alpha} = \dfrac{96}{(0.96)^3} = 108$ teeth.

**7.** $\quad t = \dfrac{n}{\cos^3 \alpha} = \dfrac{48}{(0.96)^3} = 54$ teeth.

**8.** $\quad L = \pi D \cot \alpha = 3.1416 \times 25 \times 3.427 = 269.15$ inches.

**9.** $\quad l = \pi d \cot \alpha = 3.1416 \times 12.5 \times 3.427 = 134.57$ inches.

## 5. Shafts at Right Angles, Ratio 1, Center Distance Approximate. —

When the helix angles are 45 degrees, the gears are exactly alike; when other than 45 degrees, the sum of the helix angles must equal 90 degrees.

Given or assumed:

1. Position of gear having right- or left-hand helix, depending on the rotation and direction in which the thrust is to be received.
2. $C_a$ = approximate center distance.
3. $P_n$ = normal diametral pitch (pitch of cutter).
4. $N$ = number of teeth.
5. $\alpha$ = angle of helix.

To find:

(a) When helix angles are 45 degrees.

    1. $D$ = pitch diameter = $\dfrac{N}{0.70711\, P_n}$    2. $O$ = outside diameter = $D + \dfrac{2}{P_n}$

    3. $T$ = number of teeth marked on formed milling cutter = $\dfrac{N}{0.353}$

    4. $L$ = lead of helix = $\pi D$.      5. $C$ = center distance = $D$.

(b) When helix angles are other than 45 degrees.

    1. $D$ = pitch diameter = $\dfrac{N}{P_n \cos \alpha}$

    2. $T$ = number of teeth marked on formed milling cutter = $\dfrac{N}{\cos^3 \alpha}$

    3. $C$ = center distance = sum of pitch radii.

    4. $L$ = lead of helix = $\pi D \cot \alpha$.

### Example

Given or assumed:

  1. See illustration.      2. $C_a$ = 2.5 inches.      3. $P_n$ = 10.

    4. $N$ = 18 teeth.          5. $\alpha$ = 45 degrees.

To find:

  1. $D = \dfrac{N}{0.70711\, P_n} = \dfrac{18}{0.70711 \times 10} = 2.546$ inches.

  2. $O = D + \dfrac{2}{P_n} = 2.546 + \dfrac{2}{10} = 2.746$ inches.

3. $T = \dfrac{N}{\cos^3 \alpha} = \dfrac{18}{0.353} = 51$ teeth.

4. $L = \pi D \times 1 = 3.1416 \times 2.546 = 7.999$ inches.

## 6. Shafts at Right Angles, Ratio 1, Center Distance Exact. —

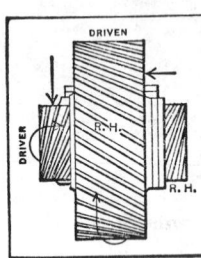

Gears have same direction of helix but probably different pitch diameters and helix angles; the sum of the latter must be 90 degrees. Given or assumed:

1. Position of gear having right- or left-hand helix depending on rotation and direction of thrust.

2. $P_n$ = normal diametral pitch (pitch of cutter).

3. $\phi$ = approximate helix angle of one gear.

4. $C$ = center distance.

5. $N$ = number of teeth = nearest whole number to $CP_n \times \cos \phi$ for approximately 45°; and

$CP_n \times \dfrac{\sin 2\phi}{\sqrt{1 + \sin 2\phi}}$ for any angle.

To find:

1. $\alpha$ = helix angle of one gear

$$\sin 2\,\alpha = \frac{N^2}{2\,C^2 P_n{}^2} \pm \sqrt{\frac{N^2}{C^2 P_n{}^2} + \left(\frac{N^2}{2\,C^2 P_n{}^2}\right)^2}$$

2. $\beta$ = helix angle of other gear = $90° - \alpha$.

3. $D$ = pitch diameter of one gear = $\dfrac{N}{P_n \cos \alpha}$

4. $d$ = pitch diameter of other gear = $\dfrac{N}{P_n \cos \beta}$

5. $O$ = outside diameter of one gear = $D + \dfrac{2}{P_n}$

6. $o$ = outside diameter of other gear = $d + \dfrac{2}{P_n}$

7. $T$ = number of teeth marked on formed cutter for one gear = $\dfrac{N}{\cos^3 \alpha}$

8. $t$ = number of teeth marked on formed cutter for other gear = $\dfrac{N}{\cos^3 \beta}$

9. $L$ = lead of helix for one gear = $\pi D \cot \alpha$.

10. $l$ = lead of helix for other gear = $\pi d \cot \beta$.

*Example*

Given or assumed:

1. See illustration.    2. $P_n = 10$.    3. $\phi = 45$ degrees.    4. $C = 4$ inches.

5. $N = CP_n \cos \phi = 4 \times 10 \times 0.70711 = 28.28$, say 28 teeth.

To find:

1. $\sin 2\,\alpha = 0.98664$, or $\alpha = 40° \ 19'$.    2. $\beta = 90° - \alpha = 49° \ 41'$.

3. $D = \dfrac{N}{P_n \cos \alpha} = \dfrac{28}{10 \times 0.76248} = 3.672$ inches.

4. $d = \dfrac{N}{P_n \cos \beta} = \dfrac{28}{10 \times 0.64701} = 4.328$ inches

5. $O = 3.672 + 0.2 = 3.872$ inches.

6. $o = 4.328 + 0.2 = 4.528$ inches.

7. $T = \dfrac{N}{\cos^3 \alpha} = \dfrac{28}{(0.762)^3} = 63.6$, say 64 teeth.

8. $t = \dfrac{N}{\cos^3 \beta} = \dfrac{28}{(0.647)^3} = 103.8$, say 104 teeth.

9. $L = \pi D \cot \alpha = 3.1416 \times 3.672 \times 1.1787 = 13.597$ inches.

10. $l = \pi d \cot \beta = 3.1416 \times 4.328 \times 0.84841 = 11.536$ inches.

**7. Shafts at Right Angles, Ratio Other Than 1, Center Distance Approx. —**
Sum of helix angles of gear and pinion must equal 90 degrees.

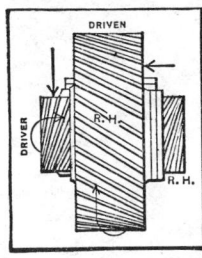

Given or assumed:

1. Position of gear having right- or left-hand **helix**, depending on rotation and direction in which thrust is to be received.
2. $C_a$ = approximate center distance.
3. $P_n$ = normal diametral pitch (pitch of cutter).
4. $R$ = ratio of gear to pinion size.
5. $n$ = number of teeth in pinion $= \dfrac{1.41\, C_a P_n}{R + 1}$ for 45 degrees;  and  $\dfrac{2\, C_a P_n \cos \alpha \cos \beta}{R \cos \beta + \cos \alpha}$ for any angle.
6. $N$ = number of teeth in gear $= nR$.
7. $\alpha$ = angle of helix of gear.
8. $\beta$ = angle of helix of pinion.

To find:

(a) When helix angles are 45 degrees.

1. $D$ = pitch diameter of gear $= \dfrac{N}{0.70711\, P_n}$

2. $d$ = pitch diameter of pinion $= \dfrac{n}{0.70711\, P_n}$

3. $O$ = outside diameter of gear $= D + \dfrac{2}{P_n}$

4. $o$ = outside diameter of pinion $= d + \dfrac{2}{P_n}$

5. $T$ = number of formed cutter (gear) $= \dfrac{N}{0.353}$

6. $t$ = number of formed cutter (pinion) $= \dfrac{n}{0.353}$

7. $L$ = lead of helix of gear $= \pi D$.

8. $l$ = lead of helix of pinion $= \pi d$.

9. $C$ = center distance (exact) $= \dfrac{D + d}{2}$

(b) When helix angles are other than 45 degrees.

1. $D = \dfrac{N}{P_n \cos \alpha}$    2. $d = \dfrac{n}{P_n \cos \beta}$    3. $T = \dfrac{N}{\cos^3 \alpha}$

4.   $t = \dfrac{n}{\cos^3 \beta}$     5.   $L = \pi D \cot \alpha$     6.   $l = \pi d \cot \beta$

*Example*

Given or assumed:

1. See illustration.   2. $C_a = 3.2$ inches.   3. $P_n = 10$.   4. $R = 1.5$.

5.   $n = \dfrac{1.41\, C_a P_n}{R + 1} = \dfrac{1.41 \times 3.2 \times 10}{1.5 + 1} =$ say 18 teeth.

6.   $N = nR = 18 \times 1.5 = 27$ teeth.

7.   $\alpha = 45$ degrees.   8. $\beta = 45$ degrees.

To find:

1.   $D = \dfrac{N}{0.70711\, P_n} = \dfrac{27}{0.70711 \times 10} = 3.818$ inches.

2.   $d = \dfrac{n}{0.70711\, P_n} = \dfrac{18}{0.70711 \times 10} = 2.545$ inches.

3.   $O = D + \dfrac{2}{P_n} = 3.818 + \dfrac{2}{10} = 4.018$ inches.

4.   $o = d + \dfrac{2}{P_n} = 2.545 + \dfrac{2}{10} = 2.745$ inches.

5.   $T = \dfrac{N}{0.353} = \dfrac{27}{0.353} = 76.5$, say 76 teeth.

6.   $t = \dfrac{n}{0.353} = \dfrac{18}{0.353} = 51$ teeth.

7.   $L = \pi D = 3.1416 \times 3.818 = 12$ inches.

8.   $l = \pi d = 3.1416 \times 2.545 = 8$ inches.

9.   $C = \dfrac{D + d}{2} = \dfrac{3.818 + 2.545}{2} = 3.182$ inches.

## 8A. Shafts at Right Angles, Ratios Other Than 1, Center Distance Exact. —

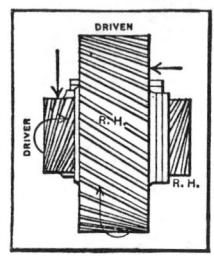

Gears have same direction of helix. The sum of the helix angles will equal 90 degrees.

Given or assumed:

1. Position of gear having right- or left-hand helix depending on rotation and direction in which thrust is to be received.
2. $P_n$ = normal diametral pitch (pitch of cutter).
3. $R$ = ratio of number of teeth in large gear to number of teeth in small gear.
4. $\alpha_a$ = approximate helix angle of large gear.
5. $C$ = exact center distance.

To find:

1. $n$ = number of teeth in small gear nearest

$$\dfrac{2\, C P_n \sin \alpha_a}{1 + R \tan \alpha_a}$$

2. $N$ = number of teeth in large gear = $Rn$.

3. $\alpha$ = exact helix angle of large gear, found by trial from

$$R \sec \alpha + \operatorname{cosec} \alpha = \dfrac{2\, C P_n}{n}$$

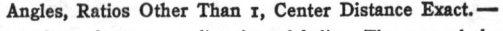

4. $\beta$ = exact helix angle of small gear = $90° - \alpha$.

5. $D$ = pitch diameter of large gear = $N \div P_n \cos \alpha$.

6. $d$ = pitch diameter of small gear = $n \div P_n \cos \beta$.

7. $O$ = outside diameter of large gear = $D + (2 \div P_n)$.

8. $o$ = outside diameter of small gear = $d + (2 \div P_n)$.

9. $N'$ and $n'$ = numbers of teeth marked on cutters for large and small gears (see page 927).

10. $L$ = lead of helix on large gear = $\pi D \cot \alpha$.

11. $l$ = lead of helix on small gear = $\pi d \cot \beta$.

*Example*

Given or assumed:

1. See illustration. 2. $P_n = 8$. 3. $R = 3$. 4. $\alpha_a = 45$ degrees. 5. $C = 10$ in.

To find:

1. $n = \dfrac{2CP_n \sin \alpha_a}{1 + R \tan \alpha_a} = \dfrac{2 \times 10 \times 8 \times 0.70711}{1 + 3} = 28.25$, say 28 teeth.

2. $N = Rn = 3 \times 28 = 84$ teeth.

3. $R \sec \alpha + \operatorname{cosec} \alpha = \dfrac{2CP_n}{n} = \dfrac{2 \times 10 \times 8}{28} = 5.714$, or $\alpha = 46° 6'$.

4. $\beta = 90° - \alpha = 90° - 46° 6' = 43° 54'$.

5. $D = N \div P_n \cos \alpha = 84 \div (8 \times 0.6934) = 15.143$ inches.

6. $d = n \div P_n \cos \beta = 28 \div (8 \times 0.72055) = 4.857$ inches.

7. $O = D + \dfrac{2}{P_n} = 15.143 + 0.25 = 15.393$ inches.

8. $o = d + \dfrac{2}{P_n} = 4.857 + 0.25 = 5.107$ inches.

9. $N' = 275$; $n' = 94$ (see page 927).

10. $L = \pi D \cot \alpha = 3.1416 \times 15.143 \times 0.96232 = 45.78$ inches.

11. $l = \pi d \cot \beta = 3.1416 \times 4.857 \times 1.0392 = 15.857$ inches.

**8B. Shafts at Right Angles, Any Ratio, Helix Angle for Minimum Center Distance.** — Diagram similar to 8A. Gears have same direction of helix. The sum of the helix angles will equal 90 degrees.

For any given ratio of gearing $R$ there is a helix angle $\alpha$ for the larger gear and a helix angle $\beta = 90° - \alpha$ for the smaller gear that will make the center distance $C$ a minimum. Helix angle $\alpha$ is found from the formula $\cot \alpha = \sqrt[3]{R}$. As an example, using the data found in Case 8A, helix angles $\alpha$ and $\beta$ for minimum center distance would be: $\cot \alpha = \sqrt[3]{3} = 1.4422$; $\alpha = 34° 44'$ and $\beta = 90° - 34° 44' = 55° 16'$. Using these helix angles, $D = 12.777$; $d = 6.143$; and $C = 9.460$ from the formulas for $D$ and $d$ given under Case 8A.

**9. Shafts at 45-Degree Angle, Ratio 1, Center Distance Approximate. —**

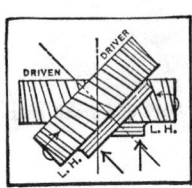

The sum of the helix angles of the two gears equals 45 degrees, and the gears are of the same hand, if each angle is less than 45 degrees. The difference between the helix angles equals 45 degrees, and the gears are of opposite hand, if either angle is greater than 45 degrees.

Given or assumed:

1. Hand of helix, depending on rotation and direction in which thrust is to be received.
2. $C_a$ = approximate center distance.
3. $P_n$ = normal diametral pitch (pitch of cutter).
4. $\alpha$ = angle of helix of driving gear.
5. $\beta$ = angle of helix of driven gear.
6. $N$ = number of teeth nearest $\dfrac{2\,C_a P_n \cos\alpha \cos\beta}{\cos\alpha + \cos\beta}$

To find:

(a) When helix angles are 22½ degrees.

1. $D$ = pitch diameter = $\dfrac{N}{0.9239\,P_n}$   2. $O$ = outside diameter = $D + \dfrac{2}{P_n}$
3. $T$ = number of teeth marked on formed cutter = $N \div 0.788$.
4. $L$ = lead of helix = $7.584\,D$.
5. $C$ = center distance = $D$.

(b) When helix angles are other than 22½ degrees.

1. $D$ = pitch diameter of driver = $\dfrac{N}{P_n \cos\alpha}$

2. $d$ = pitch diameter of driven gear = $\dfrac{N}{P_n \cos\beta}$

3. $O$ = outside diameter of driver = $D + \dfrac{2}{P_n}$

4. $o$ = outside diameter of driven gear = $d + \dfrac{2}{P_n}$

5. $T$ = number of teeth marked on cutter for driver = $N \div \cos^3\alpha$.
6. $t$ = number of teeth marked on cutter for driven gear = $N \div \cos^3\beta$.
7. $L$ = lead of helix for driver = $\pi D \cot\alpha$.
8. $l$ = lead of helix for driven gear = $\pi d \cot\beta$.
9. $C$ = actual center distance = sum of pitch radii.

*Example*

Given or assumed:

1. See illustration.   2. $C_a = 4$ inches.   3. $P_n = 10$.
4 and 5. $\alpha = \beta = 22\tfrac{1}{2}$ deg.   6. $N = 37$.

To find:

1. $D = \dfrac{N}{0.9239\,P_n} = \dfrac{37}{0.9239 \times 10} = 4.005$ inches.

2. $O = D + \dfrac{2}{P_n} = 4.005 + \dfrac{2}{10} = 4.205$ inches.

3. $T = N \div 0.788 = 37 \div 0.788 = 47$ teeth.

4. $L = 7.584 D = 7.584 \times 4.005 = 30.374$ inches.

**10. Shafts at 45-Degree Angle, Ratio 1, Center Distance Exact. —**

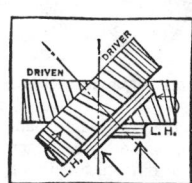

The sum of the helix angles of the two gears equals 45 degrees, and the gears are of the same hand, if each angle is less than 45 degrees. The difference between the helix angles equals 45 degrees, and the gears are of opposite hand, if either angle is greater than 45 degrees.

Given or assumed:

1. Hand of helix, depending on rotation and direction in which thrust is to be received.

2. $P_n$ = normal diametral pitch (pitch of cutter).

3. $C$ = center distance.

4. $\alpha_a$ = approximate helix angle of one gear.

5. $\beta_a$ = approximate helix angle of the other gear.

6. $N$ = number of teeth nearest $\dfrac{2\,CP_n \cos \alpha_a \cos \beta_a}{\cos \alpha_a + \cos \beta_a}$

To find:

1. $\alpha$ and $\beta$ = exact helix angles found by trial from $\sec \alpha + \sec \beta = \dfrac{2\,CP_n}{N}$

2. $D$ = pitch diameter of one gear $= \dfrac{N}{P_n \cos \alpha}$

3. $d$ = pitch diameter of the other gear $= \dfrac{N}{P_n \cos \beta}$

4. $O$ = outside diameter of one gear $= D + \dfrac{2}{P_n}$

5. $o$ = outside diameter of other gear $= d + \dfrac{2}{P_n}$

6. $T$ = number of teeth marked on formed cutter for one gear $= N \div \cos^3 \alpha$.

7. $t$ = number of teeth marked on cutter for other gear $= N \div \cos^3 \beta$.

8. $L$ = lead of helix for one gear $= \pi D \cot \alpha$.

9. $l$ = lead of helix for other gear $= \pi d \cot \beta$.

*Example*

Given or assumed:

1. See illustration. 2. $P_n = 8$. 3. $C = 10$ inches. 4. $\alpha_a = 15°$. 5. $\beta_a = 30°$.

6. $N = \dfrac{2\,CP_n \cos \alpha_a \cos \beta_a}{\cos \alpha_a + \cos \beta_a} = \dfrac{2 \times 10 \times 8 \times 0.96593 \times 0.86603}{0.96593 + 0.86603} = 73$ teeth.

To find:

1. $\alpha$ and $\beta$ from $\sec \alpha + \sec \beta = \dfrac{2\,CP_n}{N} = \dfrac{2 \times 10 \times 8}{73} = 2.1918$; by trial $\alpha$ and $\beta$, respectively, $= 14° \, 44'$ and $30° \, 16'$.

2. $D = \dfrac{N}{P_n \cos \alpha} = \dfrac{73}{8 \times 0.96712} = 9.435$ inches.

3. $d = \dfrac{N}{P_n \cos \beta} = \dfrac{73}{8 \times 0.86369} = 10.565$ inches.

4. $O = D + \dfrac{2}{P_n} = 9.435 + \dfrac{2}{8} = 9.685$ inches.

5.  $o = d + \dfrac{2}{P_n} = 10.565 + \dfrac{2}{8} = 10.815$ inches.

6.  $T = N \div \cos^3 \alpha = 73 \div 0.904 = 81$ teeth.

7.  $t = N \div \cos^3 \beta = 73 \div 0.645 = 113$ teeth.

8.  $L = \pi D \cot \alpha = \pi \times 9.435 \times 3.803 = 112.72$ inches.

9.  $l = \pi d \cot \beta = \pi \times 10.565 \times 1.714 = 56.889$ inches.

**11. Shafts at 45-Degree Angle, Ratio Other Than 1, Center Distance Approximate. —**

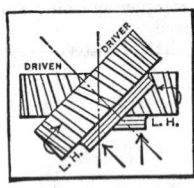

The sum of the helix angles of the two gears equals 45 degrees, and the gears are of the same hand, if each angle is less than 45 degrees. The difference between the helix angles equals 45 degrees, and the gears are of opposite hand, if either angle is greater than 45 degrees.

Given or assumed:

    1. Hand of helix, depending on rotation and direction in which thrust is to be received.

    2. $C_a$ = center distance.

    3. $P_n$ = normal diametral pitch (pitch of cutter).

4.  $R$ = ratio of gear to pinion size, $N \div n$.

5.  $\alpha$ = angle of helix, gear.

6.  $\beta$ = angle of helix, pinion.

7.  $n$ = number of pinion teeth nearest $\dfrac{2\,C_a P_n \cos \alpha \cos \beta}{R \cos \beta + \cos \alpha}$

8.  $N$ = number of gear teeth = $Rn$.

To find:

  (a) When $\alpha = \beta = 22\frac{1}{2}$ degrees.

    1.  $D$ = pitch diameter of gear = $\dfrac{N}{0.9239\,P_n}$

    2.  $d$ = pitch diameter of pinion = $\dfrac{n}{0.9239\,P_n}$

    3.  $O$ = outside diameter of gear = $D + \dfrac{2}{P_n}$

    4.  $o$ = outside diameter of pinion = $d + \dfrac{2}{P_n}$

    5.  $T$ = number of teeth marked on formed cutter for gear = $N \div 0.788$.

    6.  $t$ = number of teeth marked on formed cutter for pinion = $n \div 0.788$.

    7.  $L$ = lead of helix on gear = $7.584\,D$.

    8.  $l$ = lead of helix on pinion = $7.584\,d$.

    9.  $C$ = actual center distance = $\dfrac{D + d}{2}$

  (b) When $\alpha$ and $\beta$ are any angles.

    1.  $D$ = pitch diameter of gear = $\dfrac{N}{P_n \cos \alpha}$

    2.  $d$ = pitch diameter of pinion = $\dfrac{n}{P_n \cos \beta}$

    3.  $O$ = outside diameter of gear = $D + \dfrac{2}{P_n}$

4. $o$ = outside diameter of pinion = $d + \dfrac{2}{P_n}$

5. $T$ = number of teeth marked on formed cutter for gear = $N \div \cos^3 \alpha$.

6. $t$ = number of teeth marked on formed cutter for pinion = $n \div \cos^3 \beta$.

7. $L$ = lead of helix on gear = $\pi D \cot \alpha$.

8. $l$ = lead of helix on pinion = $\pi d \cot \beta$.

9. $C$ = actual center distance = $\dfrac{D+d}{2}$

### Example

Given or assumed:

1. See illustration.    2. $C$ = 12 inches.    3. $P_n$ = 6.    4. $R$ = 3.

5. $\alpha$ = 20 deg.    6. $\beta$ = 25 deg.

7. $n = \dfrac{2\,CP_n \cos \alpha \cos \beta}{R \cos \beta + \cos \alpha} = \dfrac{2 \times 12 \times 6 \times 0.93969 \times 0.90631}{(3 \times 0.90631) + 0.93969}$ = 34 teeth, approx.

8. $N = Rn = 3 \times 34 = 102$ teeth.

To find:

1. $D = \dfrac{N}{P_n \cos \alpha} = \dfrac{102}{6 \times 0.93969} = 18.091$ inches.

2. $d = \dfrac{n}{P_n \cos \beta} = \dfrac{34}{6 \times 0.90631} = 6.252$ inches.

3. $O = D + \dfrac{2}{P_n} = 18.091 + \dfrac{2}{6} = 18.424$ inches.

4. $o = d + \dfrac{2}{P_n} = 6.252 + \dfrac{2}{6} = 6.585$ inches.

5. $T = N \div \cos^3 \alpha = 102 \div 0.83 = 123$ teeth.

6. $t = n \div \cos^3 \beta = 34 \div 0.744 = 46$ teeth.

7. $L = \pi D \cot \alpha = \pi \times 18.091 \times 2.747 = 156.12$ inches.

8. $l = \pi d \cot \beta = \pi \times 6.252 \times 2.145 = 42.13$ inches.

9. $C = \dfrac{D+d}{2} = \dfrac{18.091 + 6.252}{2} = 12.1715$ inches.

**12. Shafts at 45-Degree Angle, Ratio Other Than 1, Center Distance Exact.—**

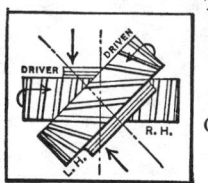

The sum of the helix angles of the two gears equals 45 degrees, and the gears are of the same hand, if each angle is less than 45 degrees. The difference between the helix angles equals 45 degrees, and the gears are of opposite hand, if either angle is greater than 45 degrees.

Given or assumed:

1. Hand of helix, depending on rotation and direction in which thrust is to be received.

2. $P_n$ = normal diametral pitch (pitch of cutter).

3. $R$ = ratio of large to small gear size = $N \div n$.

4. $\alpha_a$ = approximate helix angle of large gear.

5. $\beta_a$ = approximate helix angle of small gear.

6. $C$ = center distance.

7. $n$ = number of teeth in small gear nearest $\dfrac{2\,CP_n \cos \alpha_a \cos \beta_a}{R \cos \beta_a + \cos \alpha_a}$

8. $N$ = number of teeth, large gear = $Rn$.

To find:

1. $\alpha$ and $\beta$, exact helix angles, by trial from $R \sec \alpha + \sec \beta = \dfrac{2\,CP_n}{n}$

2. $D$ = pitch diameter of large gear = $\dfrac{N}{P_n \cos \alpha}$

3. $d$ = pitch diameter of small gear = $\dfrac{n}{P_n \cos \beta}$

4. $O$ = outside diameter of large gear = $D + \dfrac{2}{P_n}$

5. $o$ = outside diameter of small gear = $d + \dfrac{2}{P_n}$

6. $T$ = number of teeth marked on formed cutter for large gear = $N \div \cos^3 \alpha$.

7. $t$ = number of teeth marked on formed cutter for small gear = $n \div \cos^3 \beta$.

8. $L$ = lead of helix for large gear = $\pi D \cot \alpha$.

9. $l$ = lead of helix for small gear = $\pi d \cot \beta$.

### Example

Given or assumed:

1. See illustration.  2. $P_n = 4$.  3. $R = 4$.

4. $\alpha_a = 50$ degrees.  5. $\beta_a = 5$ degrees.  6. $C = 30$ inches.

7. $n = \dfrac{2\,CP_n \cos \alpha_a \cos \beta_a}{R \cos \beta_a + \cos \alpha_a} = \dfrac{2 \times 30 \times 4 \times 0.643 \times 0.996}{(4 \times 0.996) + 0.643} = 33$ teeth.

8. $N = Rn = 4 \times 33 = 132$ teeth.

To find:

1. $\alpha$ and $\beta$ from $R \sec \alpha + \sec \beta = \dfrac{2\,CP_n}{n} = \dfrac{2 \times 30 \times 4}{33} = 7.273$. By trial $\alpha = 50° 21'$, and $\beta = 5° 21'$.

2. $D = \dfrac{N}{P_n \cos \alpha} = \dfrac{132}{4 \times 0.63810} = 51.716$ inches.

3. $d = \dfrac{n}{P_n \cos \beta} = \dfrac{33}{4 \times 0.99564} = 8.286$ inches.

4. $O = D + \dfrac{2}{P_n} = 51.716 + \dfrac{2}{4} = 52.216$ inches.

5. $o = d + \dfrac{2}{P_n} = 8.286 + \dfrac{2}{4} = 8.786$ inches.

6. $T = N \div \cos^3 \alpha = 132 \div 0.26 = 508$ teeth.

7. $t = n \div \cos^3 \beta = 33 \div 0.987 = 33$ teeth.

8. $L = \pi D \cot \alpha = \pi \times 51.716 \times 0.82874 = 134.6$ inches.

9. $l = \pi d \cot \beta = \pi \times 8.286 \times 10.678 = 278$ inches.

### 13. Shafts at any Angle, Ratio 1, Center Distance Approximate. —

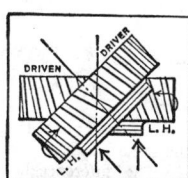

The sum of the helix angles of the two gears equals the shaft angle, and the gears are of the same hand, if each angle is less than the shaft angle. The difference between the helix angles equals the shaft angle, and the gears are of opposite hand, if either angle is greater than the shaft angle.

Given or assumed:

1. Hand of helix, depending on rotation and direction in which thrust is to be received.
2. $C_a$ = approximate center distance.
3. $P_n$ = normal diametral pitch (pitch of cutter).
4. $\alpha$ = angle of helix of one gear.
5. $\beta$ = angle of helix of other gear.
6. $N$ = number of teeth nearest $\dfrac{2\,C_a P_n \cos\alpha \cos\beta}{\cos\alpha + \cos\beta}$

To find:

1. $D$ = pitch diameter of one gear = $\dfrac{N}{P_n \cos\alpha}$

2. $d$ = pitch diameter of other gear = $\dfrac{N}{P_n \cos\beta}$

3. $O$ = outside diameter of one gear = $D + \dfrac{2}{P_n}$

4. $o$ = outside diameter of other gear = $d + \dfrac{2}{P_n}$

5. $T$ = number of teeth marked on formed cutter for one gear = $N \div \cos^3\alpha$.
6. $t$ = number of teeth marked on formed cutter for other gear = $N \div \cos^3\beta$.
7. $L$ = lead of helix for one gear = $\pi D \cot\alpha$.
8. $l$ = lead of helix for other gear = $\pi d \cot\beta$.

9. $C$ = actual center distance = $\dfrac{D+d}{2}$

#### Example

Given or assumed (angle of shafts, 30 degrees):

1. See illustration.   2. $C_a$ = 5 inches.   3. $P_n$ = 10.
4. $\alpha$ = 20 degrees.   5. $\beta$ = 10 degrees.   6. $N$ = 48.

To find:

1. $D = \dfrac{N}{P_n \cos\alpha} = \dfrac{48}{10 \times 0.9397} = 5.108$ inches.

2. $d = \dfrac{N}{P_n \cos\beta} = \dfrac{48}{10 \times 0.9848} = 4.874$ inches.

3. $O = D + \dfrac{2}{P_n} = 5.108 + \dfrac{2}{10} = 5.308$ inches.

4. $o = d + \dfrac{2}{P_n} = 4.874 + \dfrac{2}{10} = 5.074$ inches.

5. $T = N \div \cos^3\alpha = 48 \div 0.83 = 58$ teeth.
6. $t = N \div \cos^3\beta = 48 \div 0.96 = 50$ teeth.

7. $L = \pi D \cot \alpha = \pi \times 5.108 \times 2.747 = 44.08$ inches.
8. $l = \pi d \cot \beta = \pi \times 4.874 \times 5.671 = 86.84$ inches.
9. $C = \dfrac{D + d}{2} = \dfrac{5.108 + 4.874}{2} = 4.991$ inches.

### 14. Shafts at Any Angle, Ratio 1, Center Distance Exact. —

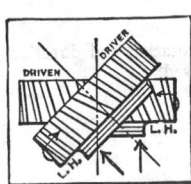

The sum of the helix angles of the two gears equals the shaft angle, and the gears are of the same hand, if each angle is less than the shaft angle. The difference between the helix angles equals the shaft angle, and the gears are of opposite hand, if either angle is greater than the shaft angle.

Given or assumed:

1. Hand of helix, depending on rotation and direction in which thrust is to be received.
2. $C$ = center distance.
3. $P_n$ = normal diametral pitch (pitch of cutter).
4. $\alpha_a$ = approximate helix angle of one gear.
5. $\beta_a$ = approximate helix angle of other gear.
6. $N$ = number of teeth nearest $\dfrac{2\,CP_n \cos \alpha_a \cos \beta_a}{\cos \alpha_a + \cos \beta_a}$

To find:

1. $\alpha$ and $\beta$ = exact helix angles, found by trial from $\sec \alpha + \sec \beta = \dfrac{2\,CP_n}{N}$

2. $D$ = pitch diameter of one gear = $\dfrac{N}{P_n \cos \alpha}$

3. $d$ = pitch diameter of other gear = $\dfrac{N}{P_n \cos \beta}$

4. $O$ = outside diameter of one gear = $D + \dfrac{2}{P_n}$

5. $o$ = outside diameter of other gear = $d + \dfrac{2}{P_n}$

6. $T$ = number of teeth marked on formed cutter for one gear = $N \div \cos^3 \alpha$.
7. $t$ = number of teeth marked on formed cutter for other gear = $N \div \cos^3 \beta$.
8. $L$ = lead of helix for one gear = $\pi D \cot \alpha$.
9. $l$ = lead of helix for other gear = $\pi d \cot \beta$.

#### Example

Given or assumed (angle of shafts, 50 degrees):

1. See illustration.    2. $C = 10$ inches.    3. $P_n = 10$.    4. $\alpha_a = 20$ deg.
5. $\beta_a = 30$ deg.
6. $N = \dfrac{2\,CP_n \cos \alpha_a \cos \beta_a}{\cos \alpha_a + \cos \beta_a} = \dfrac{2 \times 10 \times 10 \times 0.93969 \times 0.86603}{0.93969 + 0.86603} = 90$ teeth.

To find:

1. $\alpha$ and $\beta$ from $\sec \alpha + \sec \beta = \dfrac{2\,CP_n}{N} = \dfrac{2 \times 10 \times 10}{90} = 2.222$. By trial $\alpha$ and $\beta$, respectively, $= 19°\ 20'$ and $30°\ 40'$.

**2.** $D = \dfrac{N}{P_n \cos \alpha} = \dfrac{90}{10 \times 0.94361} = 9.537$ inches.

**3.** $d = \dfrac{N}{P_n \cos \beta} = \dfrac{90}{10 \times 0.86015} = 10.463$ inches.

**4.** $O = D + \dfrac{2}{P_n} = 9.537 + \dfrac{2}{10} = 9.737$ inches.

**5.** $o = d + \dfrac{2}{P_n} = 10.463 + \dfrac{2}{10} = 10.663$ inches.

**6.** $T = N \div \cos^3 \alpha = 90 \div 0.84 = 107$ teeth.

**7.** $t = N \div \cos^3 \beta = 90 \div 0.64 = 141$ teeth.

**8.** $L = \pi D \cot \alpha = \pi \times 9.537 \times 2.85 = 85.39$ inches.

**9.** $l = \pi d \cot \beta = \pi \times 10.463 \times 1.686 = 55.42$ inches.

**15. Shafts at Any Angle, Ratio Other Than 1, Center Distance Approx. —**

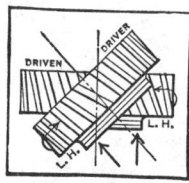

The sum of the helix angles of the two gears equals the shaft angle, and the gears are of the same hand, if each angle is less than the shaft angle. The difference between the helix angles equals the shaft angle, and the gears are of opposite hand, if either angle is greater than the shaft angle.

Given or assumed:

1. Hand of helix, depending on rotation and direction in which thrust is to be received.

2. $C_a$ = center distance.

3. $P_n$ = normal diametral pitch (pitch of cutter).

4. $R$ = ratio of gear to pinion = $N \div n$.

5. $\alpha$ = angle of helix, gear.

6. $\beta$ = angle of helix, pinion.

7. $n$ = number of teeth in pinion nearest $\dfrac{2\,C_a P_n \cos \alpha \cos \beta}{R \cos \beta + \cos \alpha}$ for any angle,

and $\dfrac{2\,C_a P_n \cos \alpha}{R + 1}$ when both angles are equal.

8. $N$ = number of teeth in gear = $Rn$.

To find:

1. $D$ = pitch diameter of gear = $\dfrac{N}{P_n \cos \alpha}$

2. $d$ = pitch diameter of pinion = $\dfrac{n}{P_n \cos \beta}$

3. $O$ = outside diameter of gear = $D + \dfrac{2}{P_n}$

4. $o$ = outside diameter of pinion = $d + \dfrac{2}{P_n}$

5. $T$ = number of teeth marked on cutter for gear = $N \div \cos^3 \alpha$.

6. $t$ = number of teeth marked on cutter for pinion = $n \div \cos^3 \beta$.

7. $L$ = lead of helix on gear = $\pi D \cot \alpha$.

8. $l$ = lead of helix on pinion = $\pi d \cot \beta$.

9. $C$ = actual center distance = $\dfrac{D + d}{2}$

*Example*

**Given or assumed** (angle of shafts, 60 degrees):

1. See illustration.  2. $C_a = 12$ inches.  3. $P_n = 8$.
4. $R = 4$.  5. $\alpha = 30$ degrees.  6. $\beta = 30$ degrees

7. $n = \dfrac{2 C_a P_n \cos \alpha}{R + 1} = \dfrac{2 \times 12 \times 8 \times 0.86603}{4 + 1} = 33$ teeth.

8. $N = 4 \times 33 = 132$ teeth.

**To find:**

1. $D = \dfrac{N}{P_n \cos \alpha} = \dfrac{132}{8 \times 0.86603} = 19.052$ inches.

2. $d = \dfrac{n}{P_n \cos \beta} = \dfrac{33}{8 \times 0.86603} = 4.763$ inches.

3. $O = D + \dfrac{2}{P_n} = 19.052 + \dfrac{2}{8} = 19.302$ inches.

4. $o = d + \dfrac{2}{P_n} = 4.763 + \dfrac{2}{8} = 5.013$ inches.

5. $T = N \div \cos^3 \alpha = 132 \div 0.65 = 203$ teeth.
6. $t = n \div \cos^3 \beta = 33 \div 0.65 = 51$ teeth.
7. $L = \pi D \cot \alpha = \pi \times 19.052 \times 1.732 = 103.66$ inches.
8. $l = \pi d \cot \beta = \pi \times 4.763 \times 1.732 = 25.92$ inches.

9. $C = \dfrac{D + d}{2} = \dfrac{19.052 + 4.763}{2} = 11.9075$ inches.

**16. Shafts at Any Angle, Ratio Other Than 1, Center Distance Exact. —**

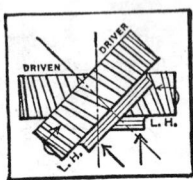

The sum of the helix angles of the two gears equals the shaft angle, and the gears are of the same hand, if each angle is less than the shaft angle. The difference between the helix angles equals the shaft angle, and the gears are of opposite hand, if either angle is greater than the shaft angle.

Given or assumed:

1. Hand of helix, depending on rotation and direction in which thrust is to be received.
2. $C$ = center distance.
3. $P_n$ = normal diametral pitch (pitch of cutter).
4. $\alpha_a$ = approximate helix angle of gear.
5. $\beta_a$ = approximate helix angle of pinion.
6. $R$ = ratio of gear to pinion size = $N \div n$.
7. $n$ = number of pinion teeth nearest $\dfrac{2 C P_n \cos \alpha_a \cos \beta_a}{R \cos \beta_a + \cos \alpha_a}$
8. $N$ = number of gear teeth = $Rn$.

**To find:**

1. $\alpha$ and $\beta$, exact helix angles, found by trial from $R \sec \alpha + \sec \beta = \dfrac{2 C P_n}{n}$

2. $D$ = pitch diameter of gear = $\dfrac{N}{P_n \cos \alpha}$

3. $d$ = pitch diameter of pinion = $\dfrac{n}{P_n \cos \beta}$

4. $O$ = outside diameter of gear = $D + \dfrac{2}{P_n}$

5. $o$ = outside diameter of pinion = $d + \dfrac{2}{P_n}$

6. $N'$ = number of teeth marked on formed cutter for gear (see below).

7. $n'$ = number of teeth marked on formed cutter for pinion (see below).

8. $L$ = lead of helix on gear = $\pi D \cot \alpha$.

9. $l$ = lead of helix on pinion = $\pi d \cot \beta$.

**Selecting Cutter for Milling Helical Gears.** — The proper milling cutter to use for *spur* gears depends upon the pitch of the teeth and also upon the number of teeth as explained on page 746 but a cutter for milling helical gears is not selected with reference to the actual number of teeth in the gear, as in spur gearing, but rather with reference to a calculated number $N'$ that takes into account the effect on the tooth profile of lead angle, normal diametral pitch, and cutter diameter.

In the helical gearing examples on pages 910-927 the number of teeth $N'$ on which to base the selection of the cutter has been determined using the approximate formula $N' = N \div \cos^3 \alpha$ or $N' = N \sec^3 \alpha$, where $N$ = the actual number of teeth in the helical gear and $\alpha$ = the helix angle. However, the use of this formula may,

### Factors for Selecting Cutters for Milling Helical Gears

| Helix Angle, $\alpha$ | $K$ | $K'$ | Helix Angle, $\alpha$ | $K$ | $K'$ | Helix Angle, $\alpha$ | $K$ | $K'$ | Helix Angle, $\alpha$ | $K$ | $K'$ |
|---|---|---|---|---|---|---|---|---|---|---|---|
| 0  | 1.000 | 0     | 16 | 1.127 | 0.082 | 32 | 1.640 | 0.390 | 48 | 3.336 | 1.233 |
| 1  | 1.001 | 0     | 17 | 1.145 | 0.093 | 33 | 1.695 | 0.422 | 49 | 3.540 | 1.323 |
| 2  | 1.002 | 0.001 | 18 | 1.163 | 0.106 | 34 | 1.755 | 0.455 | 50 | 3.767 | 1.420 |
| 3  | 1.004 | 0.003 | 19 | 1.182 | 0.119 | 35 | 1.819 | 0.490 | 51 | 4.012 | 1.525 |
| 4  | 1.007 | 0.005 | 20 | 1.204 | 0.132 | 36 | 1.889 | 0.528 | 52 | 4.284 | 1.638 |
| 5  | 1.011 | 0.008 | 21 | 1.228 | 0.147 | 37 | 1.963 | 0.568 | 53 | 4.586 | 1.761 |
| 6  | 1.016 | 0.011 | 22 | 1.254 | 0.163 | 38 | 2.044 | 0.610 | 54 | 4.925 | 1.894 |
| 7  | 1.022 | 0.015 | 23 | 1.282 | 0.180 | 39 | 2.130 | 0.656 | 55 | 5.295 | 2.039 |
| 8  | 1.030 | 0.020 | 24 | 1.312 | 0.198 | 40 | 2.225 | 0.704 | 56 | 5.710 | 2.198 |
| 9  | 1.038 | 0.025 | 25 | 1.344 | 0.217 | 41 | 2.326 | 0.756 | 57 | 6.190 | 2.371 |
| 10 | 1.047 | 0.031 | 26 | 1.377 | 0.238 | 42 | 2.436 | 0.811 | 58 | 6.720 | 2.561 |
| 11 | 1.057 | 0.038 | 27 | 1.414 | 0.260 | 43 | 2.557 | 0.870 | 59 | 7.321 | 2.770 |
| 12 | 1.068 | 0.045 | 28 | 1.454 | 0.283 | 44 | 2.687 | 0.933 | 60 | 8.000 | 3.000 |
| 13 | 1.080 | 0.053 | 29 | 1.495 | 0.307 | 45 | 2.828 | 1     | 61 | 8.780 | 3.254 |
| 14 | 1.094 | 0.062 | 30 | 1.540 | 0.333 | 46 | 2.983 | 1.072 | 62 | 9.658 | 3.537 |
| 15 | 1.110 | 0.072 | 31 | 1.588 | 0.361 | 47 | 3.152 | 1.150 | 63 | 10.687 | 3.852 |

$K = 1 \div \cos^3 \alpha = \sec^3 \alpha$; $K' = \tan^2 \alpha$

### Outside and Pitch Diameters of Standard Involute-form Milling Cutters*

| Normal Diametral Pitch, $P_n$ | Outside Diam., $D_o$ | Pitch Diam., $D_c$ | $Q = P_n D_c$ | Normal Diametral Pitch, $P_n$ | Outside Diam., $D_o$ | Pitch Diam., $D_c$ | $Q = P_n D_c$ | Normal Diametral Pitch, $P_n$ | Outside Diam., $D_o$ | Pitch Diam., $D_c$ | $Q = P_n D_c$ |
|---|---|---|---|---|---|---|---|---|---|---|---|
| 1    | 8.500 | 6.18 | 6.18  | 6  | 3.125 | 2.76 | 16.56 | 20 | 2.000 | 1.89 | 37.80 |
| 1¼   | 7.750 | 5.70 | 7.12  | 7  | 2.875 | 2.54 | 17.78 | 24 | 1.750 | 1.65 | 39.60 |
| 1½   | 7.000 | 5.46 | 8.19  | 8  | 2.875 | 2.61 | 20.88 | 28 | 1.750 | 1.67 | 46.76 |
| 1¾   | 6.500 | 5.04 | 8.82  | 9  | 2.750 | 2.50 | 22.50 | 32 | 1.750 | 1.68 | 53.76 |
| 2    | 5.750 | 4.60 | 9.20  | 10 | 2.375 | 2.14 | 21.40 | 36 | 1.750 | 1.69 | 60.84 |
| 2½   | 5.750 | 4.83 | 12.08 | 12 | 2.250 | 2.06 | 24.72 | 40 | 1.750 | 1.70 | 68.00 |
| 3    | 4.750 | 3.98 | 11.94 | 14 | 2.125 | 1.96 | 27.44 | 48 | 1.750 | 1.70 | 81.60 |
| 4    | 4.250 | 3.67 | 14.68 | 16 | 2.125 | 1.98 | 31.68 | .. | ..... | .... | ..... |
| 5    | 3.750 | 3.29 | 16.45 | 18 | 2.000 | 1.87 | 33.66 | .. | ..... | .... | ..... |

* The pitch diameters shown in the table are computed from the formula: $D_c = D_o - 2(1.157 \div P_n)$. This same formula may be used to compute the pitch diameter of a non-standard outside diameter cutter when the normal diametral pitch $P_n$ and the outside diameter $D_o$ are known.

where a combination of high helix angle and low tooth number is involved, result in the selection of a higher number of cutter than should actually be used for greatest accuracy. This condition is most likely to occur when the afore-mentioned formula is used to calculate $N'$ for gears of high helix angle and low numbers of teeth.

To avoid the possibility of error in choice of cutter number, the following formula, which gives theoretically correct results for all combinations of helix angle and tooth numbers, is to be preferred:

$$N' = N \sec^3 \alpha + P_n D_c \tan^2 \alpha \qquad (1)$$

where: $N'$ = number of teeth on which to base selection of cutter number from table on page 746; $N$ = actual number of teeth in helical gear; $\alpha$ = helix angle; $P_n$ = normal diametral pitch of gear and cutter; and $D_c$ = pitch diameter of cutter.

To simplify calculations, Formula (1) may be written as follows:

$$N' = NK + QK' \qquad (2)$$

In this formula, $K$, $K'$ and $Q$ are constants obtained from the tables on page 927.

*Example:* Helix angle = 30 degrees; number of teeth in helical gear = 15; and normal diametral pitch = 20. From the tables on page 927 $K$, $K'$, and $Q$ are, respectively, 1.540, 0.333, and 37.80.

$$N' = (15 \times 1.540) + (37.80 \times 0.333) = 23.10 + 12.60$$
$$= 35.70, \text{ say, } 36.$$

Hence, from page 746 select a number 3 cutter. Had the approximate formula been used, then a number 5 cutter would have been selected on the basis of $N' = 23$.

**Milling the Helical Teeth.** — The teeth of a helical gear are proportioned from the normal pitch and not the circular pitch. The whole depth of the tooth can be found by dividing 2.157 by the normal diametral pitch of the gear, which corresponds to the pitch of the cutter. The thickness of the tooth at the pitch line equals 1.571 divided by the normal diametral pitch. After a tooth space has been milled, the cutter should be prevented from dragging through it when being returned for another cut. This can be done by lowering the blank slightly, or by stopping the machine and turning the cutter to such a position that the teeth will not touch the work. If the gear has teeth coarser than 10 or 12 diametral pitch, it is well to take a roughing and a finishing cut. When pressing a helical gear blank on the arbor, it should be remembered that it is more likely to slip when being milled than a spur gear, because the pressure of the cut, being at an angle, tends to rotate the blank on the arbor.

*Angular Position of Table:* When cutting a helical gear on a milling machine, the table is set to the helix angle of the gear. If the lead of the helical gear is known, but not the helix angle, the helix angle is determined by multiplying the pitch diameter of the gear by 3.1416 and dividing this product by the lead; the result is the tangent of the lead angle which may be obtained from trigonometric tables.

**American National Standard Fine-pitch Teeth For Helical Gears.** — The American National Standard ANSI B6.7-1967 provides a 20-degree tooth form for both spur and helical gears of 20 diametrical pitch and finer. Formulas for tooth parts are given in the table on page 721 and data for tooth parts are given on page 722.

The proportions of fine-pitch helical gear teeth are based on the normal diametral pitch and are the same as for fine-pitch spur gears given by the formulas on page 721; the same hobs and cutters are used in their production.

*Increasing Pinion Diameter to Avoid Undercut.* — For 20-degree spur gears the minimum number of teeth which can be generated without undercut is 18 so that pinions having fewer than 18 teeth must be enlarged to avoid undercutting and either the mating gear reduced or the center distance increased by the amounts indicated on pages 749 through 752. As in the case of enlarged spur pinions, when an enlarged helical pinion is used it is necessary either to reduce the diameter of the mating gear or to increase the center distance.

The enlargement of helical pinions to avoid undercut cannot be tabulated as simply as for spur pinions because of the wide range of helix angles used. However, the dimensions of enlarged helical pinions may be calculated by means of the following formulas in which $\phi_n$ = normal pressure angle; $\phi_t$ = transverse pressure angle; $\psi$ = helix angle of pinion; $P_n$ = normal diametral pitch; $P_t$ = transverse diametral pitch; $d$ = pitch diameter of pinion; $d_o$ = outside diameter of enlarged pinion; $K_h$ = enlargement for full-depth pinions of 1 normal diametral pitch; and $n$ = number of teeth in pinion.

$$P_t = P_n \cos \psi \qquad (1)$$

$$d = n \div P_t \qquad (2)$$

$$\tan \phi_t = \tan \phi_n \div \cos \psi \qquad (3)$$

$$K_h = 2 - \frac{n \sin^2 \phi_t}{\cos \psi} \qquad (4)$$

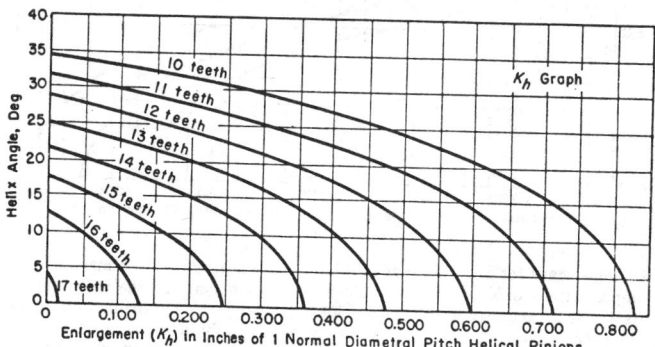

$$d_o = d + \frac{2 + K_h}{P_n} \qquad (5)$$

To eliminate the necessity for making the calculations indicated in Formulas (3) and (4) the accompanying graph may be used to obtain the value of $K_h$ directly for full-depth pinions of 20-degree normal pressure angle.

*Example:* Find the outside diameter of a helical pinion having 12 teeth, 32 normal diametral pitch, 20-degree normal pressure angle, and 18-degree helix angle.

$$P_t = P_n \cos \psi = 32 \cos 18° = 32 \times 0.95106 = 30.4339$$

$$d = n \div P_t = 12 \div 30.4339 = 0.3943 \text{ inch}$$

$$K_h = 0.39 \text{ (from graph)}$$

$$d_o = 0.394 + \frac{2 + 0.39}{32} = 0.469 \text{ inch}$$

It may be noted upon examination of the graph that enlargement of 12 tooth pinions is necessary only when the helix angle is less than about 28 degrees; similarly, for other pinions of 17 teeth or less, the helix angle below which enlargement is necessary is found at the intersection of the curve for the appropriate number of teeth and the left-hand scale.

*Enlarged Helical Pinions of 7, 8, and 9 Teeth:* In a well-designed gear train the minimum number of teeth in the pinions should not be less than 10 teeth. There are, however, special applications where pinions having fewer than 10 teeth must be used because of limited space; the use of such pinions should be restricted to relatively light loads.

The accompanying graph gives the enlargement factor $K_h$ for 7, 8, and 9 tooth helical pinions for use in Formula (5) to determine the enlarged outside diameter of these pinions. The enlarged pinion proportions will be such that there will result minimum practical top land and no undercutting of the tooth profile above the working depth.

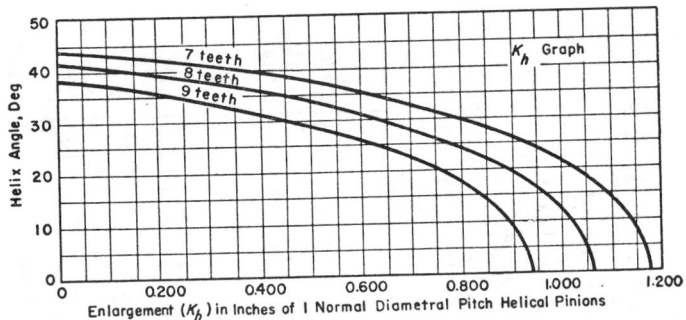

**Change-gears for Helical Gear Hobbing.** — If a gear-hobbing machine is not equipped with a differential, there is a fixed relation between the index and feed gears and it is necessary to compensate for even slight errors in the index gear ratio, to avoid excessive lead errors. This may be done readily (as shown by the example to follow) by modifying the ratio of the feed gears slightly, thus offsetting the index gear error and making very accurate leads possible.

*Machine Without Differential:* The formulas which follow may be applied in computing the index gear ratio.

$R$ = index-gear ratio  
$L$ = lead of gear, inches  
$F$ = feed per gear revolution, inch  
$K$ = machine constant  
$T$ = number of threads on hob  

$N$ = number of teeth on gear  
$P_n$ = normal diametral pitch  
$P_{nc}$ = normal circular pitch  
$A$ = helix angle, relative to axis  
$M$ = feed gear constant  

$$R = \frac{L \div F}{(L \div F) \pm 1} \times \frac{KT}{N} = \frac{L}{L \pm F} \times \frac{KT}{N} = \frac{\text{Driving gear sizes}}{\text{Driven gear sizes}} \qquad (1)$$

Use minus (−) sign in formulas (1) and (2) when gear and hob are the same "hand" and plus (+) sign when they are of opposite hand; when *climb* hobbing is to be used, reverse this rule.

$$R = \frac{KT}{N \pm \dfrac{P_n \times \sin A \times F}{\pi}} = \frac{KT}{N \pm \dfrac{\sin A \times F}{P_{nc}}} \quad (2)$$

$$\text{Ratio of feed gears} = \frac{F}{M}; \quad F = \frac{L(NR - KT)}{NR} \quad (3)$$

$$L = \frac{FNR}{(NR - KT)} = \text{lead obtained with available index and feed gears} \quad (4)$$

*Note:* If gear and hob are of opposite hand, then in Formulas (3) and (4) change $(NR - KT)$ to $(KT - NR)$. This change is also made if gear and hob are of same hand but *climb* hobbing is used.

*Example:* A right-hand helical gear with 48 teeth of 10 normal diametral pitch, has a lead of 44.0894 inches. The feed is to be 0.035 inch, with whatever slight adjustment may be necessary to compensate for the error in available index gears. $K = 30$ and $M = 0.075$. A single-thread right-hand hob is to be used.

$$R = \frac{44.0894}{44.0894 - 0.035} \times \frac{30 \times 1}{48} = 0.62549654;$$

Log 0.62549654 = $\bar{1}$.7962249 (seven-place table).

Since the ratio is less than 1 in this case, eliminate the minus characteristic ($\bar{1}$) by subtracting log of desired ratio, from log of 1 which is 0.00000, thus obtaining log of reciprocal of ratio. Later, the fractions representing gear sizes will be inverted to obtain the actual ratio. If ratio is greater than 1, use log of desired ratio.

$$\begin{array}{r} 0.000000 = \text{Log of 1} \\ -\bar{1}.796225 = \text{Log of ratio desired (rounded to six places)} \\ \hline 0.203775 = \text{Log of reciprocal of ratio} \end{array}$$

*Note that* $0.000000 - \bar{1}.796225 = 0.000000 - (-1 + .796225) = 0.000000 + 1 - .796225 = 0.203775.$

Select trial log from table of logarithms of gear ratios (*see Index*) that is equal approximately to one-half log of reciprocal. Then proceed as follows:

1. Subtract from log of reciprocal, trial log selected from table.
2. Compare difference between logs in step 1, with nearest log in table.
   Repeat step 1 until difference is restricted to at least the fifth place.

Assume that 0.101873 is selected as trial log.

$$\begin{array}{r} 0.203775 = \text{log of reciprocal} \\ -0.101873 = \text{trial log} \\ \hline 0.101902 = \text{difference (not close to any table log)} \end{array}$$

$$\begin{array}{r} 0.203775 = \text{log of reciprocal} \\ -0.101990 = \text{second trial log} \\ \hline 0.101785 = \text{difference (0.101799 nearest log in table)} \end{array}$$

Second trial log 0.101990 = $\log \frac{43}{34}$; log nearest final difference or 0.101799 = $\log \frac{67}{53}$. Inverting, obtainable index ratio = $\frac{34}{43} \times \frac{53}{67} = 0.62547726.$

(If desired index ratio is greater than 1 and log of index ratio is used instead of log of reciprocal of this ratio, the fractions obtained are *not* inverted. The procedure otherwise is the same as here outlined.)

Index ratio error = 0.62549654 − 0.62547726 = 0.00001928.

Now use formula (3) to find slight change required in rate of feed. This change compensates sufficiently for the error in available index gears.

*Change in Feed Rate:* Insert in formula (3) obtainable index ratio.

$$F = \frac{44.0894 \times (48 \times 0.62547726 - 30)}{48 \times 0.62547726} = 0.0336417$$

$$\text{Modified feed gear ratio} = \frac{F}{M} = \frac{0.0336417}{0.075} = 0.448556$$

$$\text{Log } 0.448556 = \bar{1}.651817; \text{ log of reciprocal} = 0.348183$$

To find close approximation to modified feed gear ratio, proceed as in finding suitable gears for index ratio, thus obtaining $\dfrac{106}{71} \times \dfrac{112}{75}$ Inverting, modified feed gear ratio = $\dfrac{71}{106} \times \dfrac{75}{112} = 0.448534$.

Modified feed $F$ = obtainable modified feed ratio $\times M$ = 0.448534 $\times$ 0.075 = 0.03364 inch. If the feed rate is not modified, even a small error in the index gear ratio may result in an excessive lead error.

*Checking Accuracy of Lead:* The modified feed and obtainable index ratio are inserted in formula (4). Desired lead = 44.0894 inches. Lead obtained = 44.087196 inches; hence the computed error = 44.0894 − 44.087196 = 0.002204 inch or about 0.00005 inch per inch of lead.

*Machine with Differential:* If a machine is equipped with a differential, the *lead gears* are computed in order to obtain the required helix angle and lead. The instructions of the hobbing machine manufacturer should be followed in computing the lead gears, because the ratio formula is affected by the location of the differential gears. If these gears are *ahead* of the index gears, the lead gear ratio is not affected by a change in the number of teeth to be cut (see formula 5); hence, the same lead gears are used when, for example, a gear and pinion are cut on the same machine. In the formulas which follow, the notation is the same as previously given, with these exceptions: $R_d$ = lead gear ratio for machine with differential; $P_a$ = axial or linear pitch of helical gear = distance from center of one tooth to center of next tooth measured parallel to gear axis = total lead $L \div$ number of teeth $N$.

$$R_d = \frac{P_a \times T}{K} = \frac{L \times T}{N \times K} = \frac{\pi \times \text{cosec } A \times T}{P_n \times K} = \frac{\text{Driven gear sizes}}{\text{Driving gear sizes}} \quad (5)$$

The number of hob threads $T$ is included in the formula because double-thread hobs are used sometimes, especially for roughing in order to reduce the hobbing time. Lead gears having a ratio sufficiently close to the required ratio, may be determined by using the table of gear ratio logarithms as previously described in connection with the non-differential type of machine. When using a machine equipped with a differential, the effect of a lead-gear ratio error upon the lead of the gear, is small in comparison with the effect of an index gear error when using a non-differential type of machine. The lead obtained with a given or obtainable lead gear ratio may be determined by the following formula: $L = (R_d N K) \div T$. In this formula, $R_d$ represents the ratio obtained with available gears. If the given lead is 44.0894 inches, as in the preceding example, then the desired ratio as obtained with formula (5) would be 0.9185292 if $K = 1$. Assume that the lead gears (selected by using logs of ratios as on page 931) have a ratio of 0.9184704; then this ratio error of 0.0000588 would result in a computed lead error of only 0.000065 inch per inch.

Formula (5), as mentioned, applies to machines having the differential located *ahead* of the index gears. If the differential is located after the index gears, it is necessary to change lead gears whenever the index gears are changed for hobbing a different number of teeth, as indicated by the following formula which gives the lead gear ratio. In this formula, $D$ = pitch diameter.

$$R_d = \frac{L \times T}{K} = \frac{D \times \pi \times T}{K \times \tan A} = \frac{\text{Driven gear sizes}}{\text{Driving gear sizes}} \qquad (6)$$

**General Remarks on Helical Gear Hobbing.** — In cutting teeth having large angles, it is desirable to have the direction of helix of the hob the same as the direction of helix of the gear, or in other words, the gear and the hob of the same "hand." Then the direction of the cut will come against the movement of the blank. At ordinary angles, however, one hob will cut both right- and left-hand gears. In setting up the hobbing machine for helical gears, care should be taken to see that the vertical feed does not trip until the machine has been stopped or the hob has fed down past the finished gear.

## Herringbone Gears

Double helical or herringbone gears are commonly used in parallel-shaft transmissions, especially when a smooth, continuous action (due to the gradual overlapping engagement of the teeth) is essential, as in high-speed drives where the pitch-line velocity may range from about 1000 to 3000 feet per minute in commercial gearing and up to 12,000 feet per minute or higher in more specialized installations. These relatively high speeds are encountered in marine reduction gears, in certain speed-reducing and speed-increasing units, and in various other transmissions, particularly in connection with steam turbine and electric motor drives.

**Causes of Herringbone Gear Failures.** — Where failure occurs in a herringbone gear transmission, it is rarely due to tooth breakage but usually to excessive wear or sub-surface failures, such as pitting and spalling; hence, it is common practice to base the design of such gears upon durability, or upon tooth pressures which are within the allowable limits for wear. In this connection, it seems to have been well established by tests of both spur gears and herringbone gears, that there is a critical surface pressure value for teeth having given physical properties and coefficient of friction. According to these tests, pressures above the critical value result in rapid wear and a short gear life, whereas when pressures are below the critical, wear is negligible. The yield point or endurance limit of the material marks the critical loading point, and in practical designing a reasonable factor of safety would, of course, be employed.

**General Classes of Helical Gear Problems.** — There are two general classes of problems. In one case, the problem is to design gears capable of transmitting a given amount of power at a given speed, safely and without excessive wear; hence, in this case the required proportions must be determined. In the second case, the proportions and speed are known and the power-transmitting capacity is required. The first case is the more difficult and also the more common. In establishing the proportions of the gearing, there are numerous possible combinations of pinion diameter and face width which, theoretically at least, will meet the requirements. The speed of the driver and the ratio of the gearing ordinarily are known.

**A.G.M.A. Horsepower Rating Formula.** — Equation (1) which follows is the standard of the American Gear Manufacturers Association for determining the horsepower ratings of helical and herringbone gears. These ratings for wear or surface durability normally represent tooth loads that are well within the allowable limits for strength. In the equations which follow:

### Table 1.  $F_i$ Factor for Given Face Width

| | | | Face Width ($W$) in Inches | | | | |
|---|---|---|---|---|---|---|---|
| 2 | 4 | 6 | 8 | 10 | 12 | 14 | 16 | 18 |
| $F_i$ factors for high-speed pinions of single, double and triple reduction units (See Note 1) | | | | | | | |
| 1.25 | 2.45 | 3.55 | 4.50 | 5.35 | 6.05 | 6.70 | 7.20 | 7.65 |
| $F_i$ factors for low-speed pinions of double, and intermediate gears of triple reduction units (See Note 2) | | | | | | | |
| 1.70 | 3.25 | 4.70 | 6.00 | 7.10 | 8.10 | 8.95 | 9.65 | 10.20 |
| $F_i$ factors for low-speed pinions of triple reduction units (See Note 3) | | | | | | | |
| 1.85 | 3.50 | 4.95 | 6.30 | 7.50 | 8.50 | 9.45 | 10.20 | 10.80 |

For face widths greater than 18 inches:
$$\begin{cases} \text{Note 1.} & F_i = 0.425 \times W \\ \text{Note 2.} & F_i = 0.570 \times W \\ \text{Note 3.} & F_i = 0.600 \times W \end{cases}$$

$P_w$ = horsepower rating based upon wear or surface durability;

$F_i$ = combined factor relating to face width and pinion location in gearing designed for either single, double, or triple speed reduction (see Table 1);

$K_r$ = combined factor for hardness of pinion and gear, and ratio of gear to pinion size ($K_r$ also takes into account tooth form) — see Table 2;

$D_s$ = combined factor relating to pitch diameter and speed of pinion;

$d$ = pitch diameter of pinion, inches;   $n$ = revolutions per minute of pinion;
$W$ = face width in inches; $V$ = pitch line velocity in feet per minute; $C_v$ = velocity factor. (In equation (3) for trial calculations, the constant 1.5 is an assumed mean value of the reciprocal of $C_v$.)

$$\text{Horsepower rating } P_w = F_i \times K_r \times D_s \tag{1}$$

$$D_s = \frac{d^2 \times C_v \times n}{126,000} \text{ where } C_v = \frac{78}{78 + \sqrt{V}} \tag{2}$$

**Load Capacity of Gearing is Based Upon Size of Pinion.** — In designing herringbone gears it is the general practice except with low ratios, to base the load or power-transmitting capacity upon the pinion size which is assumed to be weaker than the gear and subject to greater wear. In preliminary calculations, one plan which has been applied quite extensively is to assume that the load-carrying capacity of the transmission is directly proportional to the product of the face width and the square of the pinion pitch diameter. The product of the face width and the square of the center distance has also been used as a power capacity factor. A third method which may be utilized in conjunction with the A.G.M.A. horsepower Equation (1) is to use the cube of the pinion diameter in connection with the preliminary calculations for establishing the proportions of gears having a given power capacity. Regardless of the method, the ratio of face width to pinion diameter should agree with established practice.

**Ratio of Face Width to Pinion Diameter.** — The face width is generally established with reference to the pinion diameter. The pinion width must be kept within certain limits to prevent excessive deflection between the supporting bearings. According to some authorities, the face width for ordinary applications should be limited to from 1½ to 2 times the pinion diameter. According to another source, 2 to 2½ times the pinion diameter represents approved practice. In some cases, the ratio of face width to pinion diameter may be as high as 3; hence, it is evident that quite a number of different combinations of pinion width and diameter might be employed in transmitting a given amount of power.

Table 2.   $K_r$ Factor for Given Ratio of Gear to Pinion — External Gears

| Ratio of Gear to Pinion Size | Brinell Hardness of Gear (G) and Pinion (P) (Note: Both gear and pinion are cut after hardening) | | | | | | | |
|---|---|---|---|---|---|---|---|---|
| | (G)180 (P)210 | (G)210 (P)245 | (G)225 (P)265 | (G)245 (P)285 | (G)255 (P)300 | (G)270 (P)315 | (G)285 (P)335 | (G)300 (P)350 |
| | $K_r$ Factor | | | | | | | |
| 1 | 204 | 240 | 261 | 294 | 311 | 339 | 369 | 403 |
| 1.2 | 221 | 262 | 284 | 319 | 338 | 367 | 399 | 436 |
| 1.4 | 236 | 280 | 304 | 341 | 361 | 391 | 425 | 465 |
| 1.6 | 250 | 296 | 322 | 360 | 382 | 413 | 450 | 490 |
| 1.8 | 262 | 310 | 338 | 378 | 400 | 432 | 471 | 513 |
| 2 | 272 | 321 | 350 | 391 | 415 | 450 | 490 | 533 |
| 2.25 | 283 | 334 | 364 | 407 | 431 | 469 | 509 | 555 |
| 2.5 | 291 | 345 | 376 | 420 | 445 | 484 | 526 | 575 |
| 3 | 306 | 361 | 393 | 440 | 467 | 509 | 553 | 604 |
| 3.5 | 318 | 375 | 408 | 457 | 483 | 529 | 575 | 628 |
| 4 | 326 | 385 | 420 | 470 | 497 | 545 | 590 | 644 |
| 4.5 | 333 | 394 | 430 | 481 | 509 | 558 | 603 | 658 |
| 5 | 340 | 401 | 437 | 490 | 518 | 568 | 614 | 669 |
| 6 | 350 | 414 | 449 | 503 | 530 | 581 | 630 | 685 |
| 8 | 361 | 429 | 465 | 521 | 550 | 601 | 654 | 710 |
| 10 | 370 | 439 | 477 | 533 | 563 | 615 | 670 | 727 |
| 15 | 382 | 452 | 492 | 550 | 584 | 635 | 690 | 751 |
| 20 | 388 | 460 | 500 | 559 | 591 | 645 | 700 | 762 |

*Formula for Preliminary Calculations Based upon Cube of Pinion Diameter.* — The following equation will be found convenient to use in preliminary or trial calculations for determining the size of a pinion having a given power-transmitting capacity (constant 1.5 is assumed mean value of $C_v$ reciprocal):

$$d^3 = \frac{1.5 \times P_w \times 126,000}{n \times K_r} \qquad (3)$$

The procedure in using this equation in conjunction with the A.G.M.A. Formula (1) for horsepower rating will be demonstrated by an example.

*Example:* Herringbone gears are to be designed for transmitting about 900 H.P. at 2400 R.P.M. Ratio of gear to pinion size is 4 to 1. Diametral pitch (to be selected) is pitch in transverse plane or plane of rotation. Determine the pinion diameter and face width. Assume for trial purposes that pinion hardness is 265 Brinell.

*Find Factor $K_r$.* — Table 2 shows that $K_r = 420$ for a pinion and gear hardness of 265 and 225 Brinell, respectively, and a ratio of gear to pinion size of 4. A trial pinion pitch diameter will now be determined.

$$d^3 = \frac{1.5 \times 900 \times 126,000}{2400 \times 420} = 169; \quad d = \sqrt[3]{169} = 5.53 \text{ inches}$$

*Selecting Pitch Diameter Corresponding to Standard Diametral Pitch.* — This approximate or trial pitch diameter of 5.53 is next changed to some near value corresponding to a standard diametral pitch and a number of teeth within the usual range for herringbone pinions (ordinarily from 14 to 34 teeth). Since diametral pitch is in the plane of rotation, pitch diameter = number of teeth ÷ diametral pitch; hence spur gear table on page 728 may be used in selecting a suitable diametral pitch and number of teeth. The following combinations have pitch diameters which are close to the trial value of 5.53: for 22 teeth of 4 D.P., pitch diameter = 5.5; for 28 teeth of 5 D.P., pitch diameter = 5.6; for 33 teeth of 6 D.P., pitch diameter = 5.5. Assume that we select for trial 28 teeth of 5 D.P. or a pitch diameter of 5.6 inches. (Note: Since pitch is in plane of rotation, hob or cutter must conform to desired helix angle.)

*Find Trial Factor $F_i$ from Table 1.* — Assume by way of trial that $F_i$ is equal to the pinion diameter, or 5.6 in this case. Table 1 shows that this value lies between $F_i$

Table 3.  Additional $K_r$ Factors — External Gears

| |
|---|
| *Note:* In the following formulas for $K_r$ $$C_r = \frac{\text{gear ratio}}{\text{gear ratio} + 1} \quad \text{where gear ratio} = \frac{\text{No. teeth in gear}}{\text{No. teeth in pinion}}$$ |
| *Gear and Pinion Hardened after Cutting:* — Case-hardened or through-hardened steel: Gear and pinion 575 Brinell, $K_r = 1530 \times 0.9 \times C_r$; gear and pinion 500 Brinell, $K_r = 1350 \times 0.9 \times C_r$; gear 350 Brinell, pinion 450 Brinell, $K_r = 1060 \times 0.9 \times C_r$. Surface-hardened steel: gear and pinion 440 Brinell, $K_r = 0.890 \times 0.9 \times C_r$. |
| *Gear Cut after Hardening, Pinion Hardened after Cutting:* — Case-hardened or through-hardened steel: Gear 335 Brinell, pinion 380 Brinell, $K_r = 973 \times 0.95 \times C_r$; gear 315 Brinell, pinion 360 Brinell, $K_r = 870 \times 0.95 \times C_r$; gear 225 Brinell, pinion 450 Brinell, $K_r = 608 \times 0.95 \times C_r$. |
| *Cast Iron:* — Gear 200 Brinell, pinion 210 Brinell, $K_r = 344 \times C_r$. |
| *Bronze:* — Gear 40,000 pounds per square inch tensile strength, pinion 180 Brinell, $K_r = 274 \times C_r$. |

factors for face widths of 10 and 12 inches. By interpolation, the face width $W$ corresponding to a $F_i$ value of 5.6 is found to be 10.7 inches.

*Find Factor $D_s$.* — This factor relating to pitch diameter and speed of pinion is found by Formula (2). First calculate $V$ and $C_v$.

$$V = \frac{\pi \times d \times n}{12} = \frac{3.1416 \times 5.6 \times 2400}{12} = 3518 \text{ ft. per min.}$$

$$C_v = \frac{78}{78 + \sqrt{3518}} = 0.568; \quad D_s = \frac{5.6^2 \times 0.568 \times 2400}{126,000} = 0.34$$

Inserting these factors $F_i$, $K_r$, and $D_s$ in Equation (1)

$$P_w = 5.6 \times 420 \times 0.34 = 800 \text{ horsepower}$$

*Changing Trial Values to Obtain Given Power Capacity.* — Since the horsepower capacity specified in this case is 900, or 100 more than shown by trial solution, we may either (1) increase face width and factor $F_i$; (2) increase pinion hardness and factor $K_r$; or (3) increase pinion diameter and factor $D_s$. Assume that face width is increased. Then,

$$F_i = 5.6 \times \frac{900}{800} = 6.3$$

If $F_i = 6.3$, then by interpolating in Table 1, we obtain a corresponding face width of 12.8. The ratio of this increased face width to pinion diameter = 12.8 ÷ 5.6 = 2.28, which is within the range of approved practice. When these revised figures are inserted in Equation (1), it will be seen that the capacity has been increased to 900 H.P. Thus,

$$P_w = 6.3 \times 420 \times 0.340 = 900 \text{ horsepower approximately.}$$

**Gear Ratios.** — A single gear train generally is used if the ratio of gear to pinion size is not over 10 or 12. Double or triple reductions would be used for higher ratios.

**Helix Angles.** — For herringbone gears, helix angles usually range from 20 to 45 degrees. Angles of 23 and 30 degrees have been used extensively. The higher angles are for precision gears and comparatively low tooth pressures.

**Pitch of Pinion Teeth.** — Comparatively fine pitches are used for herringbone gears in turbine-driven or other high-speed transmissions. Coarse pitches would not be satisfactory for such applications because of the reduction in contact in the axial plane. Where heavy shock loads are encountered, as in rolling mill drives, for example, large pitches are commonly employed. The number of teeth in a herringbone pinion of given size, does not affect materially the load-carrying capacity, provided the number is somewhere between 14 and 34. According to Farrel-Birmingham Co., the number of teeth should be related to the peripheral velocity as indicated by the following figures which are intended as a general guide. For velocities below 500 feet per minute, 14 to 25 teeth; for velocities between 500 and 1000 feet per minute, 17 to 27 teeth; for velocities higher than 1000 feet per minute, 19 to 33 teeth. Although the number of teeth selected for a given pitch diameter also fixes the pitch, any one of several combinations may meet the requirements.

**Formula for Checking Diametral Pitch.** — According to Buckingham, the pitch should not be finer than indicated by the following equation in which $K$ varies according to pinion hardness as follows: $K = .036$ for 225 Brinell, $.040$ for 245 Brinell; $.045$ for 280 Brinell; $.050$ for 315 Brinell, and $.054$ for 350 Brinell.

Minimum tooth size (diametral pitch) = $(WdnK) \div$ horsepower capacity.

According to this formula, the diametral pitch of the herringbone gear referred to in the preceding example may be as fine as 8.

**Replacement of Spur Gears by Helical Gears.** — If spur gears are to be replaced either by single helical or herringbone gears without changing the center distance the procedure is as follows:

*Rule:* Select a spur gear hob (or cutter) for generating slightly smaller teeth on the helical gearing. For example, if diametral pitch of spur gearing is 6, make normal diametral pitch of herringbone gearing 7. Then, $6 \div 7 =$ cosine of herringbone gear helix angle required to obtain 6 diametral pitch *in plane of rotation*, thus retaining spur gear center distance.

*Note:* If special hob is available having the same diametral pitch *in plane of rotation* as spur gearing to be replaced, merely cut helical or herringbone to whatever helix angle the hob (or other cutter) is intended for.

## Planetary Gearing

Planetary or epicyclic gearing provides means of obtaining a compact design of transmission, with driving and driven shafts in line, and a large speed reduction when required. Typical arrangements of planetary gearing are shown by the following diagrams which are accompanied by speed ratio formulas. When planetary gears are arranged as shown by Figs. 5, 6, 9 and 12, the speed of the follower relative to the driver is increased, whereas Figs. 7, 8, 10 and 11 illustrate speed-reducing mechanisms.

**Direction of Rotation.** — In using the following formulas, if the final result is preceded by a minus sign (negative), this indicates that the driver and follower will rotate in opposite directions; otherwise, both will rotate in the same direction.

**Compound Drive.** — The formulas accompanying Figs. 19 to 22, inclusive, are for obtaining the speed ratios when there are *two* driving members rotating at different speeds. For example, in Fig. 19, the central shaft with its attached link is one driver. The internal gear $z$, instead of being fixed, is also rotated. In Fig. 22, if $z = 24$, $B = 60$ and $S = 3\frac{1}{2}$, with both drivers rotating in the same direction, then $F = 0$, thus indicating, in this case, the point where a larger value of $S$ will reverse follower rotation.

**Planetary Bevel Gears.** — Two forms of planetary gears of the bevel type are shown in Figs. 23 and 24. The planet gear in Fig. 23 rotates about a fixed bevel gear at the center of which is the driven shaft. Fig. 24 illustrates the Humpage reduction gear. This is sometimes referred to as cone-pulley back-gearing because of its use within the cone pulleys of certain types of machine tools.

## Ratios of Planetary or Epicyclic Gearing

$D$ = rotation of *driver* per revolution of follower or driven member.

$F$ = rotation of *follower* or driven member per revolution of driver. (In Figs. 1 to 4, inclusive, $F$ = rotation of planet type follower about its axis.)

$A$ = size of driving gear (use either number of teeth or pitch diameter). Note: When follower derives its motion both from $A$ and from a secondary driving member, $A$ = size of *initial* driving gear, and formula gives speed relationship between $A$ and follower.

$B$ = size of *driven gear or follower* (use either pitch diameter or number of teeth).

$C$ = size of *fixed gear* (use either pitch diameter or number of teeth).

$x$ = size of *planet gear* as shown by diagram (use either pitch diameter or number of teeth).

$y$ = size of *planet gear* as shown by diagram (use either pitch diameter or number of teeth).

$z$ = size of secondary or *auxiliary driving gear*, when follower derives its motion from two driving members.

$S$ = rotation of *secondary driver*, per revolution of *initial* driver. $S$ is negative when secondary and initial drivers rotate in opposite directions. (Formulas in which $S$ is used, give speed relationship between follower and the initial driver.)

Note: In all cases, if $D$ is known, $F = 1 \div D$, or, if $F$ is known, $D = 1 \div F$.

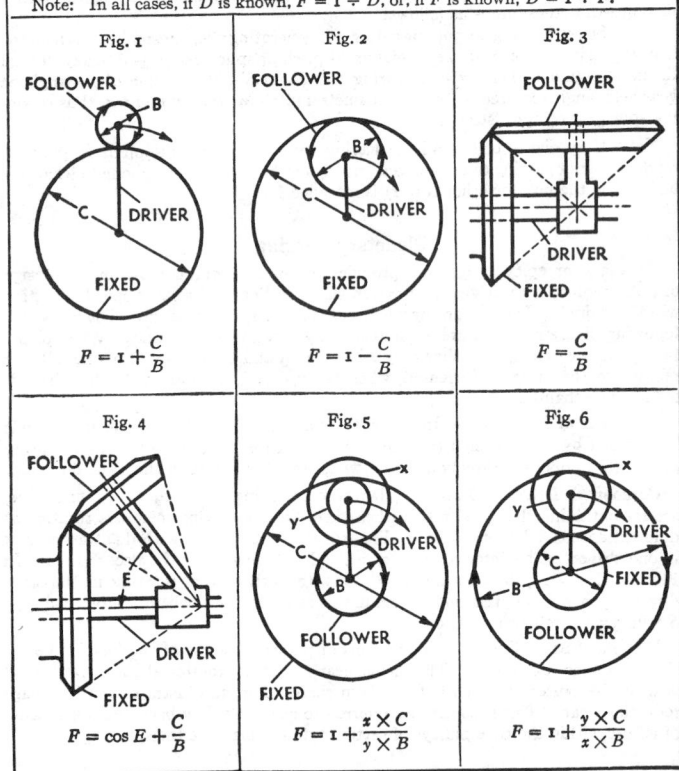

| Fig. 1 | Fig. 2 | Fig. 3 |
|---|---|---|
| $F = 1 + \dfrac{C}{B}$ | $F = 1 - \dfrac{C}{B}$ | $F = \dfrac{C}{B}$ |

| Fig. 4 | Fig. 5 | Fig. 6 |
|---|---|---|
| $F = \cos E + \dfrac{C}{B}$ | $F = 1 + \dfrac{x \times C}{y \times B}$ | $F = 1 + \dfrac{y \times C}{x \times B}$ |

## Ratios of Planetary or Epicyclic Gearing

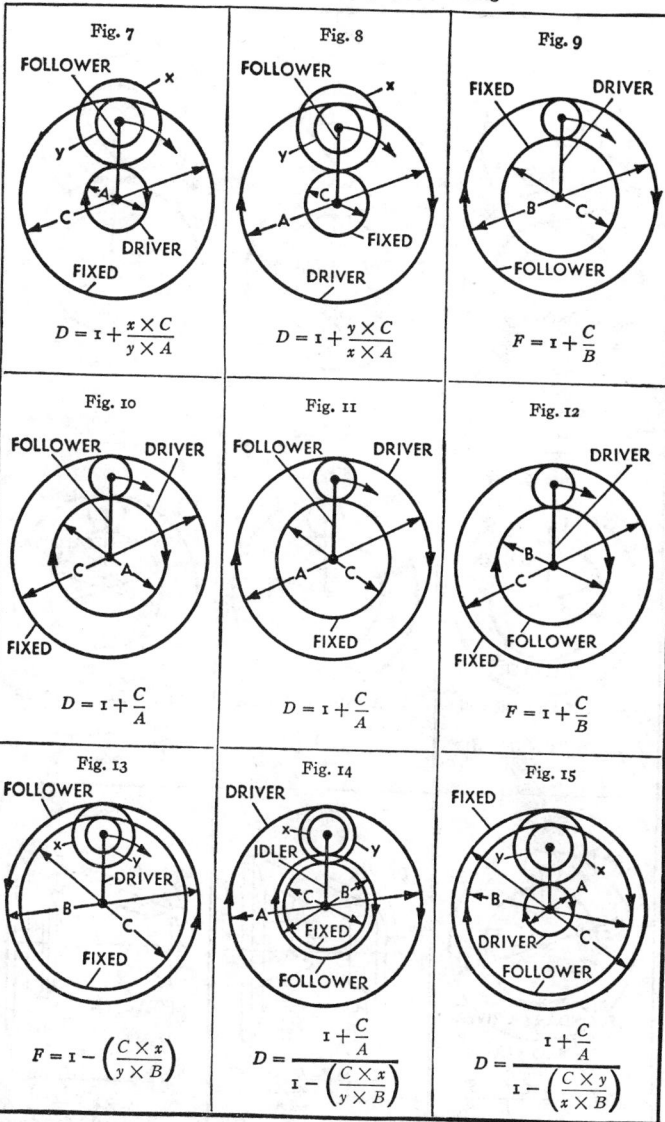

**Fig. 7**

$$D = 1 + \frac{x \times C}{y \times A}$$

**Fig. 8**

$$D = 1 + \frac{y \times C}{x \times A}$$

**Fig. 9**

$$F = 1 + \frac{C}{B}$$

**Fig. 10**

$$D = 1 + \frac{C}{A}$$

**Fig. 11**

$$D = 1 + \frac{C}{A}$$

**Fig. 12**

$$F = 1 + \frac{C}{B}$$

**Fig. 13**

$$F = 1 - \left(\frac{C \times x}{y \times B}\right)$$

**Fig. 14**

$$D = \frac{1 + \dfrac{C}{A}}{1 - \left(\dfrac{C \times x}{y \times B}\right)}$$

**Fig. 15**

$$D = \frac{1 + \dfrac{C}{A}}{1 - \left(\dfrac{C \times y}{x \times B}\right)}$$

### Ratios of Planetary or Epicyclic Gearing

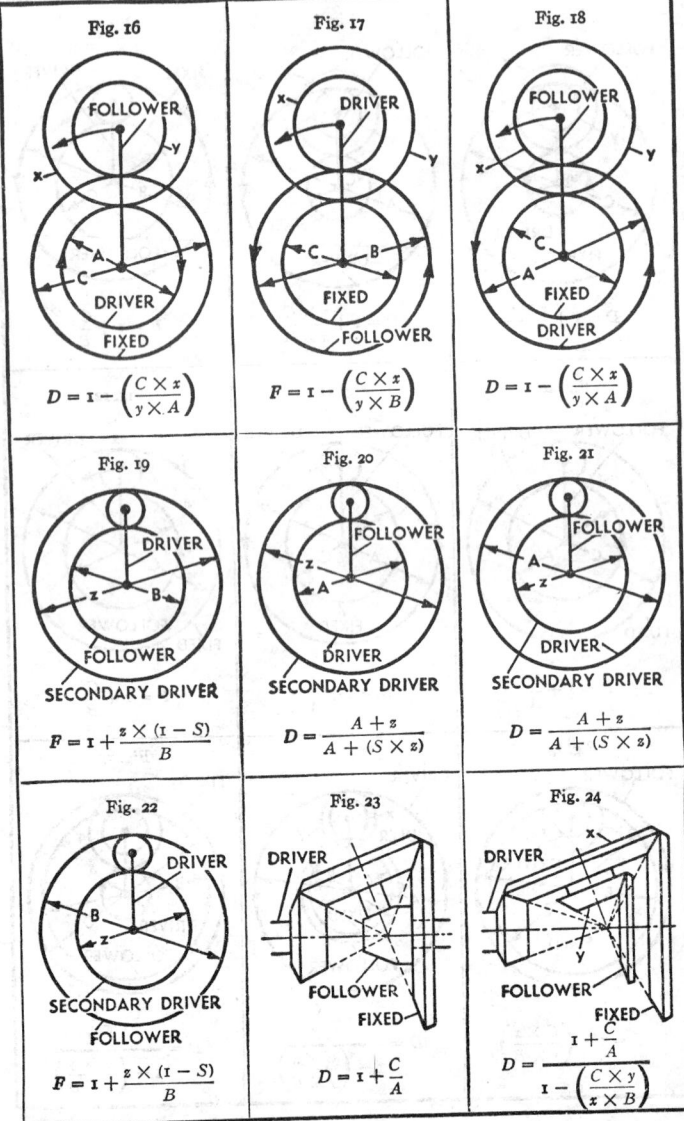

**Fig. 16**

$$D = 1 - \left(\frac{C \times x}{y \times A}\right)$$

**Fig. 17**

$$F = 1 - \left(\frac{C \times x}{y \times B}\right)$$

**Fig. 18**

$$D = 1 - \left(\frac{C \times x}{y \times A}\right)$$

**Fig. 19**

$$F = 1 + \frac{z \times (1 - S)}{B}$$

**Fig. 20**

$$D = \frac{A + z}{A + (S \times z)}$$

**Fig. 21**

$$D = \frac{A + z}{A + (S \times z)}$$

**Fig. 22**

$$F = 1 + \frac{z \times (1 - S)}{B}$$

**Fig. 23**

$$D = 1 + \frac{C}{A}$$

**Fig. 24**

$$D = \frac{1 + \dfrac{C}{A}}{1 - \left(\dfrac{C \times y}{x \times B}\right)}$$

**Gear Design Based upon Module System.** — The *module* of a gear equals the pitch diameter divided by the number of teeth, whereas *diametral pitch* equals the number of teeth divided by the pitch diameter. The module system is in general use in countries which have adopted the metric system; hence the term module is usually understood to mean the pitch diameter *in millimeters* divided by the number of teeth. The module system may, however, also be based upon inch measurements and then it is known as English module to avoid confusion with the metric module. Module is an actual dimension, whereas diametral pitch is only a ratio. Thus, if the pitch diameter of a gear is 50 millimeters and the number of teeth 25, the module is 2 which means that there are 2 millimeters of pitch diameter for each tooth. The table "Tooth Dimensions Based Upon Module System" shows the relation between module, diametral pitch, and circular pitch.

German Standard Tooth Form for Spur and Bevel Gears (DIN — 867)

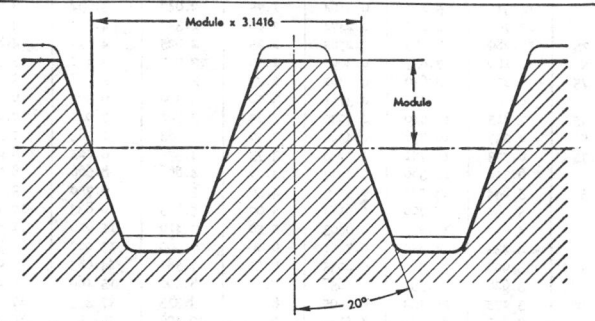

The flanks or sides are straight (involute system) and the pressure angle is 20 degrees. The shape of the root clearance space and the amount of clearance depend upon the method of cutting and special requirements. The amount of clearance may vary from 0.1 × module to 0.3 × module.

| To Find | Module Known | Circular Pitch Known |
|---|---|---|
| Addendum . . . . . . . . . | Equals module | 0.3183 × Circular pitch |
| Dedendum . . . . . . . . . | 1.157 × module*<br>1.167 × module** | 0.3683 × Circular pitch*<br>0.3714 × Circular pitch** |
| Working Depth . . . . . . | 2 × module | 0.6366 × Circular pitch |
| Total Depth . . . . . . . . | 2.157 × module*<br>2.167 × module** | 0.6866 × Circular pitch*<br>0.6898 × Circular pitch** |
| Tooth Thickness on Pitch Line . . . . . . . . . | 1.5708 × module | 0.5 × Circular pitch |

Formulas for dedendum and total depth, marked (*) are used when clearance equals 0.157 × module. Formulas marked (**) are used when clearance equals one-sixth module. It is the common practice among American cutter manufacturers to make the clearance of metric or module cutters equal to 0.157 × module.

## Tooth Dimensions Based Upon Module System

| Module, DIN Standard Series | Equivalent Diametral Pitch | Circular Pitch | | Addendum, Millimeters | Dedendum, Millimeters* | Whole Depth,* Millimeters | Whole Depth,† Millimeters |
|---|---|---|---|---|---|---|---|
| | | Millimeters | Inches | | | | |
| 0.3 | 84.667 | 0.943 | 0.0371 | 0.30 | 0.35 | 0.650 | 0.647 |
| 0.4 | 63.500 | 1.257 | 0.0495 | 0.40 | 0.467 | 0.867 | 0.863 |
| 0.5 | 50.800 | 1.571 | 0.0618 | 0.50 | 0.583 | 1.083 | 1.079 |
| 0.6 | 42.333 | 1.885 | 0.0742 | 0.60 | 0.700 | 1.300 | 1.294 |
| 0.7 | 36.286 | 2.199 | 0.0865 | 0.70 | 0.817 | 1.517 | 1.510 |
| 0.8 | 31.750 | 2.513 | 0.0989 | 0.80 | 0.933 | 1.733 | 1.726 |
| 0.9 | 28.222 | 2.827 | 0.1113 | 0.90 | 1.050 | 1.950 | 1.941 |
| 1 | 25.400 | 3.142 | 0.1237 | 1.00 | 1.167 | 2.167 | 2.157 |
| 1.25 | 20.320 | 3.927 | 0.1546 | 1.25 | 1.458 | 2.708 | 2.697 |
| 1.5 | 16.933 | 4.712 | 0.1855 | 1.50 | 1.750 | 3.250 | 3.236 |
| 1.75 | 14.514 | 5.498 | 0.2164 | 1.75 | 2.042 | 3.792 | 3.774 |
| 2 | 12.700 | 6.283 | 0.2474 | 2.00 | 2.333 | 4.333 | 4.314 |
| 2.25 | 11.289 | 7.069 | 0.2783 | 2.25 | 2.625 | 4.875 | 4.853 |
| 2.5 | 10.160 | 7.854 | 0.3092 | 2.50 | 2.917 | 5.417 | 5.392 |
| 2.75 | 9.236 | 8.639 | 0.3401 | 2.75 | 3.208 | 5.958 | 5.932 |
| 3 | 8.466 | 9.425 | 0.3711 | 3.00 | 3.500 | 6.500 | 6.471 |
| 3.25 | 7.815 | 10.210 | 0.4020 | 3.25 | 3.791 | 7.041 | 7.010 |
| 3.5 | 7.257 | 10.996 | 0.4329 | 3.50 | 4.083 | 7.583 | 7.550 |
| 3.75 | 6.773 | 11.781 | 0.4638 | 3.75 | 4.375 | 8.125 | 8.089 |
| 4 | 6.350 | 12.566 | 0.4947 | 4.00 | 4.666 | 8.666 | 8.628 |
| 4.5 | 5.644 | 14.137 | 0.5566 | 4.50 | 5.25 | 9.750 | 9.707 |
| 5 | 5.080 | 15.708 | 0.6184 | 5.00 | 5.833 | 10.833 | 10.785 |
| 5.5 | 4.618 | 17.279 | 0.6803 | 5.50 | 6.416 | 11.916 | 11.864 |
| 6 | 4.233 | 18.850 | 0.7421 | 6.00 | 7.000 | 13.000 | 12.942 |
| 6.5 | 3.908 | 20.420 | 0.8035 | 6.50 | 7.583 | 14.083 | 14.021 |
| 7 | 3.628 | 21.991 | 0.8658 | 7. | 8.166 | 15.166 | 15.099 |
| 8 | 3.175 | 25.132 | 0.9895 | 8. | 9.333 | 17.333 | 17.256 |
| 9 | 2.822 | 28.274 | 1.1132 | 9. | 10.499 | 19.499 | 19.413 |
| 10 | 2.540 | 31.416 | 1.2368 | 10. | 11.666 | 21.666 | 21.571 |
| 11 | 2.309 | 34.558 | 1.3606 | 11. | 12.833 | 23.833 | 23.728 |
| 12 | 2.117 | 37.699 | 1.4843 | 12. | 14.000 | 26.000 | 25.884 |
| 13 | 1.954 | 40.841 | 1.6079 | 13. | 15.166 | 28.166 | 28.041 |
| 14 | 1.814 | 43.982 | 1.7317 | 14. | 16.332 | 30.332 | 30.198 |
| 15 | 1.693 | 47.124 | 1.8541 | 15. | 17.499 | 32.499 | 32.355 |
| 16 | 1.587 | 50.266 | 1.9790 | 16. | 18.666 | 34.666 | 34.512 |
| 18 | 1.411 | 56.549 | 2.2263 | 18. | 21.000 | 39.000 | 38.826 |
| 20 | 1.270 | 62.832 | 2.4737 | 20. | 23.332 | 43.332 | 43.142 |
| 22 | 1.155 | 69.115 | 2.7210 | 22. | 25.665 | 47.665 | 47.454 |
| 24 | 1.058 | 75.398 | 2.9685 | 24. | 28.000 | 52.000 | 51.768 |
| 27 | 0.941 | 84.823 | 3.339 | 27. | 31.498 | 58.498 | 58.239 |
| 30 | 0.847 | 94.248 | 3.711 | 30. | 35.000 | 65.000 | 64.713 |
| 33 | 0.770 | 103.673 | 4.082 | 33. | 38.498 | 71.498 | 71.181 |
| 36 | 0.706 | 113.097 | 4.453 | 36. | 41.998 | 77.998 | 77.652 |
| 39 | 0.651 | 122.522 | 4.824 | 39. | 45.497 | 84.497 | 84.123 |
| 42 | 0.605 | 131.947 | 5.195 | 42. | 48.997 | 90.997 | 90.594 |
| 45 | 0.564 | 141.372 | 5.566 | 45. | 52.497 | 97.497 | 97.065 |
| 50 | 0.508 | 157.080 | 6.184 | 50. | 58.330 | 108.330 | 107.855 |
| 55 | 0.462 | 172.788 | 6.803 | 55. | 64.163 | 119.163 | 118.635 |
| 60 | 0.423 | 188.496 | 7.421 | 60. | 69.996 | 129.996 | 129.426 |
| 65 | 0.391 | 204.204 | 8.040 | 65. | 75.829 | 140.829 | 140.205 |
| 70 | 0.363 | 219.911 | 8.658 | 70. | 81.662 | 151.662 | 150.997 |
| 75 | 0.339 | 235.619 | 9.276 | 75. | 87.495 | 162.495 | 161.775 |

* Dedendum and total depth when clearance = 0.1666 × module, or one-sixth module.
† Total depth equivalent to American standard full-depth teeth. (Clearance = 0.157 × module.)

### Rules for Module System of Gearing

| To Find | Rule |
|---------|------|
| Metric Module | *Rule 1:* To find the metric module, divide the pitch diameter in millimeters by the number of teeth.<br>*Example 1:* The pitch diameter of a gear is 200 millimeters and the number of teeth, 40; then<br><br>$$\text{module} = \frac{200}{40} = 5$$<br><br>*Rule 2:* Multiply circular pitch in millimeters by 0.3183.<br>*Example 2:* (Same as Example 1. Circular pitch of this gear equals 15.708 millimeters).<br><br>$$\text{module} = 15.708 \times 0.3183 = 5$$<br><br>*Rule 3:* Divide outside diameter in millimeters by the number of teeth plus 2. |
| English Module | *Note:* The module system is usually applied when gear dimensions are expressed in millimeters, but module may also be based upon inch measurements.<br>*Rule:* To find the English module, divide pitch diameter in inches by number of teeth.<br>*Example:* A gear has 48 teeth and a pitch diameter of 12 inches.<br><br>$$\text{module} = \frac{12}{48} = \frac{1}{4} \text{ module or 4 diametral pitch}$$ |
| Metric Module Equivalent to Diametral Pitch | *Rule:* To find the metric module equivalent to a given diametral pitch, divide 25.4 by the diametral pitch.<br>*Example:* Determine metric module equivalent to 10 diametral pitch.<br><br>$$\text{Equivalent module} = \frac{25.4}{10} = 2.54$$<br><br>*Note:* The nearest standard module is 2.5. |
| Diametral Pitch Equivalent to Metric Module | *Rule:* To find the diametral pitch equivalent to a given module, divide 25.4 by the module. (25.4 = number of millimeters per inch.)<br>*Example:* The module is 12; determine equivalent diametral pitch.<br><br>$$\text{Equivalent diametral pitch} = \frac{25.4}{12} = 2.117$$<br><br>*Note:* A diametral pitch of 2 is the nearest *standard* equivalent. |
| Pitch Diameter | *Rule:* Multiply number of teeth by module.<br>*Example:* The metric module is 8 and gear has 40 teeth; then<br><br>$$D = 40 \times 8 = 320 \text{ millimeters} = 12.598 \text{ inches}$$ |
| Outside Diameter | *Rule:* Add 2 to the number of teeth and multiply sum by the module.<br>*Example:* A gear has 40 teeth and module is 6. Find outside or blank diameter.<br><br>$$\text{Outside diameter} = (40 + 2) \times 6 = 252 \text{ millimeters}$$ |

For tooth dimensions, see table Tooth Dimensions Based Upon Module System; also formulas below German Standard Tooth Form.

### Equivalent Diametral Pitches, Circular Pitches, and Metric Modules

Commonly Used Pitches and Modules in Bold Type

| Diametral Pitch | Circular Pitch, Inches | Module Millimeters | Diametral Pitch | Circular Pitch, Inches | Module Millimeters | Diametral Pitch | Circular Pitch, Inches | Module Millimeters |
|---|---|---|---|---|---|---|---|---|
| ½ | 6.2832 | 50.8000 | 2.2848 | 1⅜ | 11.1170 | 10.0531 | 5/16 | 2.5266 |
| 0.5080 | 6.1842 | 50 | 2.3091 | 1.3605 | 11 | 10.1600 | 0.3092 | 2½ |
| 0.5236 | 6 | 48.5104 | 2½ | 1.2566 | 10.1600 | 11 | 0.2856 | 2.3091 |
| 0.5644 | 5.5658 | 45 | 2.5133 | 1¼ | 10.1063 | 12 | 0.2618 | 2.1167 |
| 0.5712 | 5½ | 44.4679 | 2.5400 | 1.2368 | 10 | 12.5664 | ¼ | 2.0213 |
| 0.6283 | 5 | 40.4253 | 2¾ | 1.1424 | 9.2364 | 12.7000 | 0.2474 | 2 |
| 0.6350 | 4.9474 | 40 | 2.7925 | 1⅛ | 9.0957 | 13 | 0.2417 | 1.9538 |
| 0.6981 | 4½ | 36.3828 | 2.8222 | 1.1132 | 9 | 14 | 0.2244 | 1.8143 |
| 0.7257 | 4.3290 | 35 | 3 | 1.0472 | 8.4667 | 15 | 0.2094 | 1.6933 |
| ¾ | 4.1888 | 33.8667 | 3.1416 | 1 | 8.0851 | 16 | 0.1963 | 1.5875 |
| 0.7854 | 4 | 32.3403 | 3.1750 | 0.9895 | 8 | 16.7552 | 3/16 | 1.5160 |
| 0.8378 | 3¾ | 30.3190 | 3.3510 | 15/16 | 7.5797 | 16.9333 | 0.1855 | 1½ |
| 0.8467 | 3.7105 | 30 | 3½ | 0.8976 | 7.2571 | 17 | 0.1848 | 1.4941 |
| 0.8976 | 3½ | 28.2977 | 3.5904 | ⅞ | 7.0744 | 18 | 0.1745 | 1.4111 |
| 0.9666 | 3¼ | 26.2765 | 3.6286 | 0.8658 | 7 | 19 | 0.1653 | 1.3368 |
| 1 | 3.1416 | 25.4000 | 3.8666 | 13/16 | 6.5691 | 20 | 0.1571 | 1.2700 |
| 1.0160 | 3.0921 | 25 | 3.9078 | 0.8040 | 6½ | 22 | 0.1428 | 1.1545 |
| 1.0472 | 3 | 24.2552 | 4 | 0.7854 | 6.3500 | 24 | 0.1309 | 1.0583 |
| 1.1424 | 2¾ | 22.2339 | 4.1888 | ¾ | 6.0638 | 25 | 0.1257 | 1.0160 |
| 1¼ | 2.5133 | 20.3200 | 4.2333 | 0.7421 | 6 | 25.1328 | ⅛ | 1.0106 |
| 1.2566 | 2½ | 20.2127 | 4.5696 | 11/16 | 5.5585 | 25.4000 | 0.1237 | 1 |
| 1.2700 | 2.4737 | 20 | 4.6182 | 0.6803 | 5½ | 26 | 0.1208 | 0.9769 |
| 1.3963 | 2¼ | 18.1914 | 5 | 0.6283 | 5.0800 | 28 | 0.1122 | 0.9071 |
| 1.4111 | 2.2263 | 18 | 5.0265 | ⅝ | 5.0532 | 30 | 0.1047 | 0.8467 |
| 1½ | 2.0944 | 16.9333 | 5.0800 | 0.6184 | 5 | 32 | 0.0982 | 0.7937 |
| 1.5708 | 2 | 16.1701 | 5.5851 | 9/16 | 4.5478 | 34 | 0.0924 | 0.7470 |
| 1.5875 | 1.9790 | 16 | 5.6443 | 0.5566 | 4½ | 36 | 0.0873 | 0.7056 |
| 1.6755 | 1⅞ | 15.1595 | 6 | 0.5236 | 4.2333 | 38 | 0.0827 | 0.6684 |
| 1.6933 | 1.8553 | 15 | 6.2832 | ½ | 4.0425 | 40 | 0.0785 | 0.6350 |
| 1¾ | 1.7952 | 14.5143 | 6.3500 | 0.4947 | 4 | 42 | 0.0748 | 0.6048 |
| 1.7952 | 1¾ | 14.1489 | 7 | 0.4488 | 3.6286 | 44 | 0.0714 | 0.5773 |
| 1.8143 | 1.7316 | 14 | 7.1808 | 7/16 | 3.5372 | 46 | 0.0683 | 0.5522 |
| 1.9333 | 1⅝ | 13.1382 | 7.2571 | 0.4329 | 3½ | 48 | 0.0654 | 0.5292 |
| 1.9538 | 1.6079 | 13 | 8 | 0.3927 | 3.1750 | 50 | 0.0628 | 0.5080 |
| 2 | 1.5708 | 12.7000 | 8.3776 | ⅜ | 3.0319 | 50.2656 | 1/16 | 0.5053 |
| 2.0944 | 1½ | 12.1276 | 8.4667 | 0.3711 | 3 | 50.8000 | 0.0618 | ½ |
| 2.1167 | 1.4842 | 12 | 9 | 0.3491 | 2.8222 | 56 | 0.0561 | 0.4536 |
| 2¼ | 1.3963 | 11.2889 | 10 | 0.3142 | 2.5400 | 60 | 0.0524 | 0.4233 |

The module of a gear is the pitch diameter divided by the number of teeth. The module may be expressed in any units; but when no units are stated, it is understood to be in millimeters. The metric module, therefore, equals the pitch diameter in millimeters divided by the number of teeth. To find the metric module equivalent to a given diametral pitch, divide 25.4 by the diametral pitch. To find the diametral pitch equivalent to a given module, divide 25.4 by the module. (25.4 = number of millimeters per inch.)

## Checking Gear Size by Measurement Over Wires or Pins

The wire or pin method of checking gear sizes is accurate, easily applied, and especially useful in shops with limited inspection equipment. Two cylindrical wires or pins of predetermined diameter are placed in diametrically opposite tooth spaces (see diagram). If the gear has an odd number of teeth, the wires are located as nearly opposite as possible, as shown by the diagram at the right. The over-all measurement $M$ is checked by using any sufficiently accurate method of measurement. The value of measurement $M$ when the pitch diameter is correct can be determined easily and quickly by means of the calculated values in the accompanying tables.

**Measurements for Checking External Spur Gears when Wire Diameter Equals 1.728 Divided by Diametral Pitch.** Tables 1 and 2 give measurements $M$, in inches, for checking the pitch diameters of external spur gears of 1 diametral pitch. For any other diametral pitch, divide the measurement given in the table by whatever diametral pitch is required. The result shows what measurement $M$ should be when the pitch diameter is correct *and there is no allowance for backlash.* The

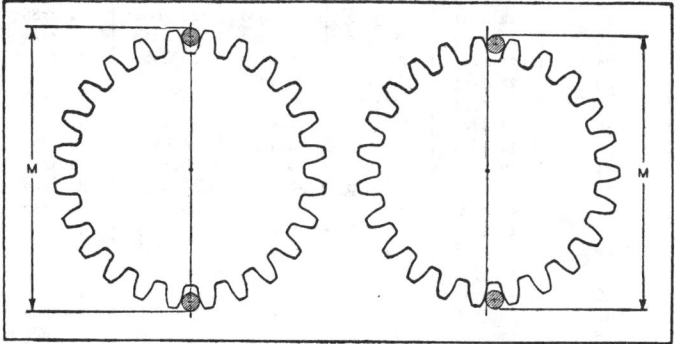

procedure for obtaining a given amount of backlash will be explained later. Tables 1 to 4 inclusive are based upon wire sizes conforming to the Van Keuren standard. For external spur gears the wire size equals 1.728 divided by the diametral pitch. The wire diameters for various diametral pitches will be found in the left-hand section of Table 5.

*Even Number of Teeth:* Table 1 is for even numbers of teeth. To illustrate the use of the table, assume that a spur gear has 32 teeth of 4 diametral pitch and a pressure angle of 20 degrees. Table 1 shows that the measurement for 1 diametral pitch is 34.4130; hence, for 4 diametral pitch, the measurement equals 34.4130 ÷ 4 = 8.6032 inches. This is the actual measurement over the wires when the pitch diameter is correct, provided there is no allowance for backlash. The wire diameter in this case equals 1.728 ÷ 4 = .432 inch (Table 5).

Measurement for even numbers of teeth above 170 and not in the Table 1, may be determined as shown by the following example: Assume that number of teeth = 240 and pressure angle 14½ degrees; then, for 1 diametral pitch, figure at left of decimal point = given No. of teeth + 2 = 240 + 2 = 242. Figure at right of decimal point lies between decimal values given in table for 200 teeth and 3∞ teeth and is obtained by interpolation. Thus, 240 − 200 = 40 (change to .40); .5395 − .5321 = .0074 = difference between decimal values for 300 and 200 teeth;

## Table 1.  Checking External Spur Gear Sizes by Measurement Over Wires

| EVEN NUMBERS OF TEETH | | | | | |
|---|---|---|---|---|---|

Dimensions in table are for 1 diametral pitch and Van Keuren standard wire sizes. For any other diametral pitch, divide dimension in table by given pitch.

$$\text{Wire or pin diameter} = \frac{1.728}{\text{Diametral Pitch}}$$

| No. of Teeth | Pressure Angle | | | | |
|---|---|---|---|---|---|
| | 14½° | 17½° | 20° | 25° | 30° |
| 6 | 8.2846 | 8.2927 | 8.3032 | 8.3340 | 8.3759 |
| 8 | 10.3160 | 10.3196 | 10.3271 | 10.3533 | 10.3919 |
| 10 | 12.3399 | 12.3396 | 12.3445 | 12.3667 | 12.4028 |
| 12 | 14.3590 | 14.3552 | 14.3578 | 14.3768 | 14.4108 |
| 14 | 16.3746 | 16.3671 | 16.3683 | 16.3846 | 16.4169 |
| 16 | 18.3877 | 18.3780 | 18.3768 | 18.3908 | 18.4217 |
| 18 | 20.3989 | 20.3866 | 20.3840 | 20.3959 | 20.4256 |
| 20 | 22.4087 | 22.3940 | 22.3900 | 22.4002 | 22.4288 |
| 22 | 24.4172 | 24.4004 | 24.3952 | 24.4038 | 24.4315 |
| 24 | 26.4247 | 26.4060 | 26.3997 | 26.4069 | 26.4339 |
| 26 | 28.4314 | 28.4110 | 28.4036 | 28.4096 | 28.4358 |
| 28 | 30.4374 | 30.4154 | 30.4071 | 30.4120 | 30.4376 |
| 30 | 32.4429 | 32.4193 | 32.4102 | 32.4141 | 32.4391 |
| 32 | 34.4478 | 34.4228 | 34.4130 | 34.4159 | 34.4405 |
| 34 | 36.4523 | 36.4260 | 36.4155 | 36.4176 | 36.4417 |
| 36 | 38.4565 | 38.4290 | 38.4178 | 38.4191 | 38.4428 |
| 38 | 40.4603 | 40.4317 | 40.4198 | 40.4205 | 40.4438 |
| 40 | 42.4638 | 42.4341 | 42.4217 | 42.4217 | 42.4447 |
| 42 | 44.4671 | 44.4364 | 44.4234 | 44.4228 | 44.4455 |
| 44 | 46.4701 | 46.4385 | 46.4250 | 46.4239 | 46.4463 |
| 46 | 48.4729 | 48.4404 | 48.4265 | 48.4248 | 48.4470 |
| 48 | 50.4756 | 50.4422 | 50.4279 | 50.4257 | 50.4476 |
| 50 | 52.4781 | 52.4439 | 52.4292 | 52.4265 | 52.4482 |
| 52 | 54.4804 | 54.4454 | 54.4304 | 54.4273 | 54.4487 |
| 54 | 56.4826 | 56.4469 | 56.4315 | 56.4280 | 56.4492 |
| 56 | 58.4847 | 58.4483 | 58.4325 | 58.4287 | 58.4497 |
| 58 | 60.4866 | 60.4496 | 60.4335 | 60.4293 | 60.4501 |
| 60 | 62.4884 | 62.4509 | 62.4344 | 62.4299 | 62.4506 |
| 62 | 64.4902 | 64.4520 | 64.4352 | 64.4304 | 64.4510 |
| 64 | 66.4918 | 66.4531 | 66.4361 | 66.4309 | 66.4513 |
| 66 | 68.4933 | 68.4542 | 68.4369 | 68.4314 | 68.4517 |
| 68 | 70.4948 | 70.4552 | 70.4376 | 70.4319 | 70.4520 |
| 70 | 72.4963 | 72.4561 | 72.4383 | 72.4323 | 72.4523 |
| 72 | 74.4977 | 74.4570 | 74.4390 | 74.4327 | 74.4526 |
| 74 | 76.4990 | 76.4578 | 76.4396 | 76.4331 | 76.4529 |
| 76 | 78.5002 | 78.4586 | 78.4402 | 78.4335 | 78.4532 |
| 78 | 80.5014 | 80.4594 | 80.4408 | 80.4339 | 80.4534 |
| 80 | 82.5026 | 82.4601 | 82.4413 | 82.4342 | 82.4536 |
| 82 | 84.5037 | 84.4608 | 84.4418 | 84.4345 | 84.4538 |
| 84 | 86.5047 | 86.4615 | 86.4423 | 86.4348 | 86.4540 |
| 86 | 88.5057 | 88.4621 | 88.4428 | 88.4351 | 88.4542 |
| 88 | 90.5067 | 90.4627 | 90.4433 | 90.4354 | 90.4544 |

Table 1.  Checking External Spur Gear Sizes by Measurement Over Wires

| EVEN NUMBERS OF TEETH | | | | | |
|---|---|---|---|---|---|
| No. of Teeth | 14½° | 17½° | 20° | 25° | 30° |
| 90 | 92.5076 | 92.4633 | 92.4437 | 92.4357 | 92.4546 |
| 92 | 94.5085 | 94.4639 | 94.4441 | 94.4359 | 94.4548 |
| 94 | 96.5094 | 96.4644 | 96.4445 | 96.4362 | 96.4550 |
| 96 | 98.5102 | 98.4649 | 98.4449 | 98.4364 | 98.4552 |
| 98 | 100.5110 | 100.4655 | 100.4453 | 100.4367 | 100.4554 |
| 100 | 102.5118 | 102.4660 | 102.4456 | 102.4369 | 102.4555 |
| 102 | 104.5125 | 104.4665 | 104.4460 | 104.4370 | 104.4557 |
| 104 | 106.5132 | 106.4669 | 106.4463 | 106.4372 | 106.4558 |
| 106 | 108.5139 | 108.4673 | 108.4466 | 108.4374 | 108.4560 |
| 108 | 110.5146 | 110.4678 | 110.4469 | 110.4376 | 110.4561 |
| 110 | 112.5152 | 112.4682 | 112.4472 | 112.4378 | 112.4562 |
| 112 | 114.5159 | 114.4686 | 114.4475 | 114.4380 | 114.4563 |
| 114 | 116.5165 | 116.4690 | 116.4478 | 116.4382 | 116.4564 |
| 116 | 118.5171 | 118.4693 | 118.4481 | 118.4384 | 118.4565 |
| 118 | 120.5177 | 120.4697 | 120.4484 | 120.4385 | 120.4566 |
| 120 | 122.5182 | 122.4701 | 122.4486 | 122.4387 | 122.4567 |
| 122 | 124.5188 | 124.4704 | 124.4489 | 124.4388 | 124.4568 |
| 124 | 126.5193 | 126.4708 | 126.4491 | 126.4390 | 126.4569 |
| 126 | 128.5198 | 128.4711 | 128.4493 | 128.4391 | 128.4570 |
| 128 | 130.5203 | 130.4714 | 130.4496 | 130.4393 | 130.4571 |
| 130 | 132.5208 | 132.4717 | 132.4498 | 132.4394 | 132.4572 |
| 132 | 134.5213 | 134.4720 | 134.4500 | 134.4395 | 134.4573 |
| 134 | 136.5217 | 136.4723 | 136.4502 | 136.4397 | 136.4574 |
| 136 | 138.5221 | 138.4725 | 138.4504 | 138.4398 | 138.4575 |
| 138 | 140.5226 | 140.4728 | 140.4506 | 140.4399 | 140.4576 |
| 140 | 142.5230 | 142.4730 | 142.4508 | 142.4400 | 142.4577 |
| 142 | 144.5234 | 144.4733 | 144.4510 | 144.4401 | 144.4578 |
| 144 | 146.5238 | 146.4736 | 146.4512 | 146.4402 | 146.4578 |
| 146 | 148.5242 | 148.4738 | 148.4513 | 148.4403 | 148.4579 |
| 148 | 150.5246 | 150.4740 | 150.4515 | 150.4404 | 150.4580 |
| 150 | 152.5250 | 152.4742 | 152.4516 | 152.4405 | 152.4580 |
| 152 | 154.5254 | 154.4745 | 154.4518 | 154.4406 | 154.4581 |
| 154 | 156.5257 | 156.4747 | 156.4520 | 156.4407 | 156.4581 |
| 156 | 158.5261 | 158.4749 | 158.4521 | 158.4408 | 158.4582 |
| 158 | 160.5264 | 160.4751 | 160.4523 | 160.4409 | 160.4582 |
| 160 | 162.5267 | 162.4753 | 162.4524 | 162.4410 | 162.4583 |
| 162 | 164.5270 | 164.4755 | 164.4526 | 164.4411 | 164.4584 |
| 164 | 166.5273 | 166.4757 | 166.4527 | 166.4411 | 166.4584 |
| 166 | 168.5276 | 168.4759 | 168.4528 | 168.4412 | 168.4585 |
| 168 | 170.5279 | 170.4760 | 170.4529 | 170.4413 | 170.4585 |
| 170 | 172.5282 | 172.4761 | 172.4531 | 172.4414 | 172.4586 |
| 180 | 182.5297 | 182.4771 | 182.4537 | 182.4418 | 182.4589 |
| 190 | 192.5310 | 192.4780 | 192.4542 | 192.4421 | 192.4591 |
| 200 | 202.5321 | 202.4786 | 202.4548 | 202.4424 | 202.4593 |
| 300 | 302.5395 | 302.4831 | 302.4579 | 302.4443 | 302.4606 |
| 400 | 402.5434 | 402.4854 | 402.4596 | 402.4453 | 402.4613 |
| 500 | 502.5458 | 502.4868 | 502.4606 | 502.4458 | 502.4619 |

## Table 2. Checking External Spur Gear Sizes by Measurement Over Wires

| | | ODD NUMBERS OF TEETH | | | |
|---|---|---|---|---|---|

Dimensions in table are for 1 diametral pitch and Van Keuren standard wire sizes. For any other diametral pitch, divide dimension in table by given pitch.

$$\text{Wire or pin diameter} = \frac{1.728}{\text{Diametral Pitch}}$$

| No. of Teeth | Pressure Angle | | | | |
|---|---|---|---|---|---|
| | 14½° | 17½° | 20° | 25° | 30° |
| 7 | 9.1116 | 9.1172 | 9.1260 | 9.1536 | 9.1928 |
| 9 | 11.1829 | 11.1844 | 11.1905 | 11.2142 | 11.2509 |
| 11 | 13.2317 | 13.2296 | 13.2332 | 13.2536 | 13.2882 |
| 13 | 15.2677 | 15.2617 | 15.2639 | 15.2814 | 15.3142 |
| 15 | 17.2957 | 17.2873 | 17.2871 | 17.3021 | 17.3329 |
| 17 | 19.3182 | 19.3072 | 19.3053 | 19.3181 | 19.3482 |
| 19 | 21.3368 | 21.3233 | 21.3200 | 21.3310 | 21.3600 |
| 21 | 23.3524 | 23.3368 | 23.3321 | 23.3415 | 23.3696 |
| 23 | 25.3658 | 25.3481 | 25.3423 | 25.3502 | 25.3775 |
| 25 | 27.3774 | 27.3579 | 27.3511 | 27.3576 | 27.3842 |
| 27 | 29.3876 | 29.3664 | 29.3586 | 29.3640 | 29.3899 |
| 29 | 31.3966 | 31.3738 | 31.3652 | 31.3695 | 31.3948 |
| 31 | 33.4047 | 33.3804 | 33.3710 | 33.3743 | 33.3991 |
| 33 | 35.4119 | 35.3863 | 35.3761 | 35.3786 | 35.4029 |
| 35 | 37.4185 | 37.3916 | 37.3807 | 37.3824 | 37.4063 |
| 37 | 39.4245 | 39.3964 | 39.3849 | 39.3858 | 39.4094 |
| 39 | 41.4299 | 41.4007 | 41.3886 | 41.3889 | 41.4120 |
| 41 | 43.4348 | 43.4047 | 43.3920 | 43.3917 | 43.4145 |
| 43 | 45.4394 | 45.4083 | 45.3951 | 45.3942 | 45.4168 |
| 45 | 47.4437 | 47.4116 | 47.3980 | 47.3965 | 47.4188 |
| 47 | 49.4477 | 49.4147 | 49.4007 | 49.3986 | 49.4206 |
| 49 | 51.4514 | 51.4175 | 51.4031 | 51.4006 | 51.4223 |
| 51 | 53.4547 | 53.4202 | 53.4053 | 53.4024 | 53.4239 |
| 53 | 55.4579 | 55.4227 | 55.4074 | 55.4041 | 55.4254 |
| 55 | 57.4609 | 57.4249 | 57.4093 | 57.4056 | 57.4267 |
| 57 | 59.4637 | 59.4271 | 59.4111 | 59.4071 | 59.4280 |
| 59 | 61.4664 | 61.4291 | 61.4128 | 61.4084 | 61.4292 |
| 61 | 63.4689 | 63.4310 | 63.4144 | 63.4097 | 63.4303 |
| 63 | 65.4712 | 65.4328 | 65.4159 | 65.4109 | 65.4313 |
| 65 | 67.4734 | 67.4344 | 67.4173 | 67.4120 | 67.4323 |
| 67 | 69.4755 | 69.4360 | 69.4186 | 69.4130 | 69.4332 |
| 69 | 71.4775 | 71.4375 | 71.4198 | 71.4140 | 71.4341 |
| 71 | 73.4795 | 73.4389 | 73.4210 | 73.4150 | 73.4349 |
| 73 | 75.4813 | 75.4403 | 75.4221 | 75.4159 | 75.4357 |
| 75 | 77.4830 | 77.4416 | 77.4232 | 77.4167 | 77.4364 |
| 77 | 79.4847 | 79.4428 | 79.4242 | 79.4175 | 79.4371 |
| 79 | 81.4863 | 81.4440 | 81.4252 | 81.4183 | 81.4378 |
| 81 | 83.4877 | 83.4451 | 83.4262 | 83.4190 | 83.4384 |
| 83 | 85.4892 | 85.4462 | 85.4271 | 85.4196 | 85.4390 |
| 85 | 87.4906 | 87.4472 | 87.4279 | 87.4203 | 87.4395 |
| 87 | 89.4919 | 89.4481 | 89.4287 | 89.4209 | 89.4400 |
| 89 | 91.4932 | 91.4490 | 91.4295 | 91.4215 | 91.4405 |

Table 2.  Checking External Spur Gear Sizes by Measurement Over Wires

| | | ODD NUMBERS OF TEETH | | | |
|---|---|---|---|---|---|
| No. of Teeth | 14½° | 17½° | 20° | 25° | 30° |
| 91 | 93.4944 | 93.4499 | 93.4303 | 93.4221 | 93.4410 |
| 93 | 95.4956 | 95.4508 | 95.4310 | 95.4227 | 95.4415 |
| 95 | 97.4967 | 97.4516 | 97.4317 | 97.4232 | 97.4420 |
| 97 | 99.4978 | 99.4524 | 99.4323 | 99.4237 | 99.4424 |
| 99 | 101.4988 | 101.4532 | 101.4329 | 101.4242 | 101.4428 |
| 101 | 103.4998 | 103.4540 | 103.4335 | 103.4247 | 103.4432 |
| 103 | 105.5008 | 105.4546 | 105.4341 | 105.4252 | 105.4436 |
| 105 | 107.5017 | 107.4553 | 107.4346 | 107.4256 | 107.4440 |
| 107 | 109.5026 | 109.4559 | 109.4352 | 109.4260 | 109.4443 |
| 109 | 111.5035 | 111.4566 | 111.4357 | 111.4264 | 111.4447 |
| 111 | 113.5044 | 113.4572 | 113.4362 | 113.4268 | 113.4450 |
| 113 | 115.5052 | 115.4578 | 115.4367 | 115.4272 | 115.4453 |
| 115 | 117.5060 | 117.4584 | 117.4372 | 117.4275 | 117.4456 |
| 117 | 119.5068 | 119.4589 | 119.4376 | 119.4279 | 119.4459 |
| 119 | 121.5075 | 121.4594 | 121.4380 | 121.4282 | 121.4462 |
| 121 | 123.5082 | 123.4599 | 123.4384 | 123.4285 | 123.4465 |
| 123 | 125.5089 | 125.4604 | 125.4388 | 125.4288 | 125.4468 |
| 125 | 127.5096 | 127.4609 | 127.4392 | 127.4291 | 127.4471 |
| 127 | 129.5103 | 129.4614 | 129.4396 | 129.4294 | 129.4473 |
| 129 | 131.5109 | 131.4619 | 131.4400 | 131.4297 | 131.4476 |
| 131 | 133.5115 | 133.4623 | 133.4404 | 133.4300 | 133.4478 |
| 133 | 135.5121 | 135.4628 | 135.4408 | 135.4302 | 135.4480 |
| 135 | 137.5127 | 137.4632 | 137.4411 | 137.4305 | 137.4483 |
| 137 | 139.5133 | 139.4636 | 139.4414 | 139.4307 | 139.4485 |
| 139 | 141.5139 | 141.4640 | 141.4418 | 141.4310 | 141.4487 |
| 141 | 143.5144 | 143.4644 | 143.4421 | 143.4312 | 143.4489 |
| 143 | 145.5149 | 145.4648 | 145.4424 | 145.4315 | 145.4491 |
| 145 | 147.5154 | 147.4651 | 147.4427 | 147.4317 | 147.4493 |
| 147 | 149.5159 | 149.4655 | 149.4430 | 149.4319 | 149.4495 |
| 149 | 151.5164 | 151.4658 | 151.4433 | 151.4321 | 151.4497 |
| 151 | 153.5169 | 153.4661 | 153.4435 | 153.4323 | 153.4498 |
| 153 | 155.5174 | 155.4665 | 155.4438 | 155.4325 | 155.4500 |
| 155 | 157.5179 | 157.4668 | 157.4440 | 157.4327 | 157.4502 |
| 157 | 159.5183 | 159.4671 | 159.4443 | 159.4329 | 159.4504 |
| 159 | 161.5188 | 161.4674 | 161.4445 | 161.4331 | 161.4505 |
| 161 | 163.5192 | 163.4677 | 163.4448 | 163.4333 | 163.4507 |
| 163 | 165.5196 | 165.4680 | 165.4450 | 165.4335 | 165.4508 |
| 165 | 167.5200 | 167.4683 | 167.4453 | 167.4337 | 167.4510 |
| 167 | 169.5204 | 169.4686 | 169.4455 | 169.4338 | 169.4511 |
| 169 | 171.5208 | 171.4688 | 171.4457 | 171.4340 | 171.4513 |
| 171 | 173.5212 | 173.4691 | 173.4459 | 173.4342 | 173.4514 |
| 181 | 183.5230 | 183.4704 | 183.4469 | 183.4350 | 183.4520 |
| 191 | 193.5246 | 193.4715 | 193.4478 | 193.4357 | 193.4526 |
| 201 | 203.5260 | 203.4725 | 203.4487 | 203.4363 | 203.4532 |
| 301 | 303.5355 | 303.4790 | 303.4538 | 303.4402 | 303.4565 |
| 401 | 403.5404 | 403.4823 | 403.4565 | 403.4422 | 403.4582 |
| 501 | 503.5433 | 503.4843 | 503.4581 | 503.4434 | 503.4592 |

### Table 3. Checking Internal Spur Gear Sizes by Measurement Between Wires

| EVEN NUMBERS OF TEETH |
|---|

Dimensions in table are for 1 diametral pitch and Van Keuren standard wire sizes. For any other diametral pitch, divide dimensions in table by given pitch.

$$\text{Wire or pin diameter} = \frac{1.44}{\text{Diametral Pitch}}$$

| No. of Teeth | Pressure Angle | | | | |
|---|---|---|---|---|---|
| | 14½° | 17½° | 20° | 25° | 30° |
| 10 | 8.8337 | 8.7383 | 8.6617 | 8.5209 | 8.3966 |
| 12 | 10.8394 | 10.7404 | 10.6623 | 10.5210 | 10.3973 |
| 14 | 12.8438 | 12.7419 | 12.6627 | 12.5210 | 12.3978 |
| 16 | 14.8474 | 14.7431 | 14.6630 | 14.5210 | 14.3982 |
| 18 | 16.8504 | 16.7441 | 16.6633 | 16.5210 | 16.3985 |
| 20 | 18.8529 | 18.7449 | 18.6635 | 18.5211 | 18.3987 |
| 22 | 20.8550 | 20.7456 | 20.6636 | 20.5211 | 20.3989 |
| 24 | 22.8569 | 22.7462 | 22.6638 | 22.5211 | 22.3991 |
| 26 | 24.8585 | 24.7467 | 24.6639 | 24.5211 | 24.3992 |
| 28 | 26.8599 | 26.7471 | 26.6640 | 26.5211 | 26.3993 |
| 30 | 28.8612 | 28.7475 | 28.6641 | 28.5211 | 28.3994 |
| 32 | 30.8623 | 30.7478 | 30.6642 | 30.5211 | 30.3995 |
| 34 | 32.8633 | 32.7481 | 32.6641 | 32.5211 | 32.3995 |
| 36 | 34.8642 | 34.7483 | 34.6643 | 34.5212 | 34.3996 |
| 38 | 36.8650 | 36.7486 | 36.6642 | 36.5212 | 36.3996 |
| 40 | 38.8658 | 38.7488 | 38.6644 | 38.5212 | 38.3997 |
| 42 | 40.8665 | 40.7490 | 40.6644 | 40.5212 | 40.3997 |
| 44 | 42.8672 | 42.7492 | 42.6645 | 42.5212 | 42.3998 |
| 46 | 44.8678 | 44.7493 | 44.6645 | 44.5212 | 44.3998 |
| 48 | 46.8683 | 46.7495 | 46.6646 | 46.5212 | 46.3999 |
| 50 | 48.8688 | 48.7496 | 48.6646 | 48.5212 | 48.3999 |
| 52 | 50.8692 | 50.7497 | 50.6646 | 50.5212 | 50.3999 |
| 54 | 52.8697 | 52.7499 | 52.6647 | 52.5212 | 52.4000 |
| 56 | 54.8701 | 54.7500 | 54.6647 | 54.5212 | 54.4000 |
| 58 | 56.8705 | 56.7501 | 56.6648 | 56.5212 | 56.4001 |
| 60 | 58.8709 | 58.7502 | 58.6648 | 58.5212 | 58.4001 |
| 62 | 60.8712 | 60.7503 | 60.6648 | 60.5212 | 60.4001 |
| 64 | 62.8715 | 62.7504 | 62.6648 | 62.5212 | 62.4001 |
| 66 | 64.8718 | 64.7505 | 64.6649 | 64.5212 | 64.4001 |
| 68 | 66.8721 | 66.7505 | 66.6649 | 66.5212 | 66.4001 |
| 70 | 68.8724 | 68.7506 | 68.6649 | 68.5212 | 68.4001 |
| 72 | 70.8727 | 70.7507 | 70.6649 | 70.5212 | 70.4002 |
| 74 | 72.8729 | 72.7507 | 72.6649 | 72.5212 | 72.4002 |
| 76 | 74.8731 | 74.7508 | 74.6649 | 74.5212 | 74.4002 |
| 78 | 76.8734 | 76.7509 | 76.6649 | 76.5212 | 76.4002 |
| 80 | 78.8736 | 78.7509 | 78.6649 | 78.5212 | 78.4002 |
| 82 | 80.8738 | 80.7510 | 80.6649 | 80.5212 | 80.4002 |
| 84 | 82.8740 | 82.7510 | 82.6649 | 82.5212 | 82.4002 |
| 86 | 84.8742 | 84.7511 | 84.6650 | 84.5212 | 84.4002 |
| 88 | 86.8743 | 86.7511 | 86.6650 | 86.5212 | 86.4003 |
| 90 | 88.8745 | 88.7512 | 88.6650 | 88.5212 | 88.4003 |

Table 3.  Checking Internal Spur Gear Sizes by Measurement Between Wires

| No. of Teeth | 14½° | 17½° | 20° | 25° | 30° |
|---|---|---|---|---|---|
| | EVEN NUMBERS OF TEETH | | | | |
| 92 | 90.8747 | 90.7512 | 90.6650 | 90.5212 | 90.4003 |
| 94 | 92.8749 | 92.7513 | 92.6650 | 92.5212 | 92.4003 |
| 96 | 94.8750 | 94.7513 | 94.6650 | 94.5212 | 94.4003 |
| 98 | 96.8752 | 96.7513 | 96.6650 | 96.5212 | 96.4003 |
| 100 | 98.8753 | 98.7514 | 98.6650 | 98.5212 | 98.4003 |
| 102 | 100.8754 | 100.7514 | 100.6650 | 100.5212 | 100.4003 |
| 104 | 102.8756 | 102.7514 | 102.6650 | 102.5212 | 102.4003 |
| 106 | 104.8757 | 104.7515 | 104.6650 | 104.5212 | 104.4003 |
| 108 | 106.8758 | 106.7515 | 106.6650 | 106.5212 | 106.4003 |
| 110 | 108.8759 | 108.7515 | 108.6651 | 108.5212 | 108.4004 |
| 112 | 110.8760 | 110.7516 | 110.6651 | 110.5212 | 110.4004 |
| 114 | 112.8761 | 112.7516 | 112.6651 | 112.5212 | 112.4004 |
| 116 | 114.8762 | 114.7516 | 114.6651 | 114.5212 | 114.4004 |
| 118 | 116.8763 | 116.7516 | 116.6651 | 116.5212 | 116.4004 |
| 120 | 118.8764 | 118.7517 | 118.6651 | 118.5212 | 118.4004 |
| 122 | 120.8765 | 120.7517 | 120.6651 | 120.5212 | 120.4004 |
| 124 | 122.8766 | 122.7517 | 122.6651 | 122.5212 | 122.4004 |
| 126 | 124.8767 | 124.7517 | 124.6651 | 124.5212 | 124.4004 |
| 128 | 126.8768 | 126.7518 | 126.6651 | 126.5212 | 126.4004 |
| 130 | 128.8769 | 128.7518 | 128.6652 | 128.5212 | 128.4004 |
| 132 | 130.8769 | 130.7518 | 130.6652 | 130.5212 | 130.4004 |
| 134 | 132.8770 | 132.7518 | 132.6652 | 132.5212 | 132.4004 |
| 136 | 134.8771 | 134.7519 | 134.6652 | 134.5212 | 134.4004 |
| 138 | 136.8772 | 136.7519 | 136.6652 | 136.5212 | 136.4004 |
| 140 | 138.8773 | 138.7519 | 138.6652 | 138.5212 | 138.4004 |
| 142 | 140.8773 | 140.7519 | 140.6652 | 140.5212 | 140.4004 |
| 144 | 142.8774 | 142.7519 | 142.6652 | 142.5212 | 142.4004 |
| 146 | 144.8774 | 144.7520 | 144.6652 | 144.5212 | 144.4004 |
| 148 | 146.8775 | 146.7520 | 146.6652 | 146.5212 | 146.4004 |
| 150 | 148.8775 | 148.7520 | 148.6652 | 148.5212 | 148.4005 |
| 152 | 150.8776 | 150.7520 | 150.6652 | 150.5212 | 150.4005 |
| 154 | 152.8776 | 152.7520 | 152.6652 | 152.5212 | 152.4005 |
| 156 | 154.8777 | 154.7520 | 154.6652 | 154.5212 | 154.4005 |
| 158 | 156.8778 | 156.7520 | 156.6652 | 156.5212 | 156.4005 |
| 160 | 158.8778 | 158.7520 | 158.6652 | 158.5212 | 158.4005 |
| 162 | 160.8779 | 160.7520 | 160.6652 | 160.5212 | 160.4005 |
| 164 | 162.8779 | 162.7521 | 162.6652 | 162.5212 | 162.4005 |
| 166 | 164.8780 | 164.7521 | 164.6652 | 164.5212 | 164.4005 |
| 168 | 166.8780 | 166.7521 | 166.6652 | 166.5212 | 166.4005 |
| 170 | 168.8781 | 168.7521 | 168.6652 | 168.5212 | 168.4005 |
| 180 | 178.8783 | 178.7522 | 178.6652 | 178.5212 | 178.4005 |
| 190 | 188.8785 | 188.7522 | 188.6652 | 188.5212 | 188.4005 |
| 200 | 198.8788 | 198.7523 | 198.6652 | 198.5212 | 198.4005 |
| 300 | 298.8795 | 298.7525 | 298.6654 | 298.5212 | 298.4005 |
| 400 | 398.8803 | 398.7527 | 398.6654 | 398.5212 | 398.4006 |
| 500 | 498.8810 | 498.7528 | 498.6654 | 498.5212 | 498.4006 |

**Table 4.** Checking Internal Spur Gear Sizes by Measurement Between Wires

ODD NUMBERS OF TEETH

Dimensions in table are for 1 diametral pitch and Van Keuren standard wire sizes. For any other diametral pitch, divide dimensions in table by given pitch.

$$\text{Wire or pin diameter} = \frac{1.44}{\text{Diametral Pitch}}$$

| No. of Teeth | Pressure Angle | | | | |
|---|---|---|---|---|---|
| | 14½° | 17½° | 20° | 25° | 30° |
| 7 | 5.6393 | 5.5537 | 5.4823 | 5.3462 | 5.2232 |
| 9 | 7.6894 | 7.5976 | 7.5230 | 7.3847 | 7.2618 |
| 11 | 9.7219 | 9.6256 | 9.5490 | 9.4094 | 9.2867 |
| 13 | 11.7449 | 11.6451 | 11.5669 | 11.4265 | 11.3040 |
| 15 | 13.7620 | 13.6594 | 13.5801 | 13.4391 | 13.3167 |
| 17 | 15.7752 | 15.6703 | 15.5902 | 15.4487 | 15.3265 |
| 19 | 17.7858 | 17.6790 | 17.5981 | 17.4563 | 17.3343 |
| 21 | 19.7945 | 19.6860 | 19.6045 | 19.4625 | 19.3405 |
| 23 | 21.8017 | 21.6918 | 21.6099 | 21.4676 | 21.3457 |
| 25 | 23.8078 | 23.6967 | 23.6143 | 23.4719 | 23.3501 |
| 27 | 25.8130 | 25.7009 | 25.6181 | 25.4755 | 25.3538 |
| 29 | 27.8176 | 27.7045 | 27.6214 | 27.4787 | 27.3571 |
| 31 | 29.8216 | 29.7076 | 29.6242 | 29.4814 | 29.3599 |
| 33 | 31.8251 | 31.7104 | 31.6267 | 31.4838 | 31.3623 |
| 35 | 33.8282 | 33.7128 | 33.6289 | 33.4860 | 33.3645 |
| 37 | 35.8311 | 35.7150 | 35.6310 | 35.4879 | 35.3665 |
| 39 | 37.8336 | 37.7169 | 37.6327 | 37.4896 | 37.3682 |
| 41 | 39.8359 | 39.7187 | 39.6343 | 39.4911 | 39.3698 |
| 43 | 41.8380 | 41.7203 | 41.6357 | 41.4925 | 41.3712 |
| 45 | 43.8399 | 43.7217 | 43.6371 | 43.4938 | 43.3725 |
| 47 | 45.8416 | 45.7231 | 45.6383 | 45.4950 | 45.3737 |
| 49 | 47.8432 | 47.7243 | 47.6394 | 47.4960 | 47.3748 |
| 51 | 49.8447 | 49.7254 | 49.6404 | 49.4970 | 49.3758 |
| 53 | 51.8461 | 51.7265 | 51.6414 | 51.4979 | 51.3768 |
| 55 | 53.8474 | 53.7274 | 53.6422 | 53.4988 | 53.3776 |
| 57 | 55.8486 | 55.7283 | 55.6431 | 55.4996 | 55.3784 |
| 59 | 57.8497 | 57.7292 | 57.6438 | 57.5003 | 57.3792 |
| 61 | 59.8508 | 59.7300 | 59.6445 | 59.5010 | 59.3799 |
| 63 | 61.8517 | 61.7307 | 61.6452 | 61.5016 | 61.3806 |
| 65 | 63.8526 | 63.7314 | 63.6458 | 63.5022 | 63.3812 |
| 67 | 65.8535 | 65.7320 | 65.6464 | 65.5028 | 65.3818 |
| 69 | 67.8543 | 67.7327 | 67.6469 | 67.5033 | 67.3823 |
| 71 | 69.8551 | 69.7332 | 69.6475 | 69.5038 | 69.3828 |
| 73 | 71.8558 | 71.7338 | 71.6480 | 71.5043 | 71.3833 |
| 75 | 73.8565 | 73.7343 | 73.6484 | 73.5048 | 73.3838 |
| 77 | 75.8572 | 75.7348 | 75.6489 | 75.5052 | 75.3842 |
| 79 | 77.8573 | 77.7352 | 77.6493 | 77.5056 | 77.3846 |
| 81 | 79.8584 | 79.7357 | 79.6497 | 79.5060 | 79.3850 |
| 83 | 81.8590 | 81.7361 | 81.6501 | 81.5064 | 81.3854 |
| 85 | 83.8595 | 83.7365 | 83.6505 | 83.5067 | 83.3858 |
| 87 | 85.8600 | 85.7369 | 85.6508 | 85.5071 | 85.3861 |
| 89 | 87.8605 | 87.7373 | 87.6511 | 87.5074 | 87.3864 |

Table 4. Checking Internal Spur Gear Sizes by Measurement Between Wires

| ODD NUMBERS OF TEETH | | | | | |
|---|---|---|---|---|---|
| No. of Teeth | 14½° | 17½° | 20° | 25° | 30° |
| 91 | 89.8610 | 89.7376 | 89.6514 | 89.5077 | 89.3867 |
| 93 | 91.8614 | 91.7379 | 91.6517 | 91.5080 | 91.3870 |
| 95 | 93.8619 | 93.7383 | 93.6520 | 93.5082 | 93.3873 |
| 97 | 95.8623 | 95.7386 | 95.6523 | 95.5085 | 95.3876 |
| 99 | 97.8627 | 97.7389 | 97.6526 | 97.5088 | 97.3879 |
| 101 | 99.8631 | 99.7391 | 99.6528 | 99.5090 | 99.3881 |
| 103 | 101.8635 | 101.7394 | 101.6531 | 101.5093 | 101.3883 |
| 105 | 103.8638 | 103.7397 | 103.6533 | 103.5095 | 103.3886 |
| 107 | 105.8642 | 105.7399 | 105.6535 | 105.5097 | 105.3888 |
| 109 | 107.8645 | 107.7402 | 107.6537 | 107.5099 | 107.3890 |
| 111 | 109.8648 | 109.7404 | 109.6539 | 109.5101 | 109.3893 |
| 113 | 111.8651 | 111.7406 | 111.6541 | 111.5103 | 111.3895 |
| 115 | 113.8654 | 113.7409 | 113.6543 | 113.5105 | 113.3897 |
| 117 | 115.8657 | 115.7411 | 115.6545 | 115.5107 | 115.3899 |
| 119 | 117.8660 | 117.7413 | 117.6547 | 117.5109 | 117.3900 |
| 121 | 119.8662 | 119.7415 | 119.6548 | 119.5110 | 119.3902 |
| 123 | 121.8663 | 121.7417 | 121.6550 | 121.5112 | 121.3904 |
| 125 | 123.8668 | 123.7418 | 123.6552 | 123.5114 | 123.3905 |
| 127 | 125.8670 | 125.7420 | 125.6554 | 125.5115 | 125.3907 |
| 129 | 127.8672 | 127.7422 | 127.6556 | 127.5117 | 127.3908 |
| 131 | 129.8675 | 129.7424 | 129.6557 | 129.5118 | 129.3910 |
| 133 | 131.8677 | 131.7425 | 131.6559 | 131.5120 | 131.3911 |
| 135 | 133.8679 | 133.7427 | 133.6560 | 133.5121 | 133.3913 |
| 137 | 135.8681 | 135.7428 | 135.6561 | 135.5123 | 135.3914 |
| 139 | 137.8683 | 137.7430 | 137.6563 | 137.5124 | 137.3916 |
| 141 | 139.8685 | 139.7431 | 139.6564 | 139.5125 | 139.3917 |
| 143 | 141.8687 | 141.7433 | 141.6565 | 141.5126 | 141.3918 |
| 145 | 143.8689 | 143.7434 | 143.6566 | 143.5127 | 143.3919 |
| 147 | 145.8691 | 145.7436 | 145.6568 | 145.5128 | 145.3920 |
| 149 | 147.8693 | 147.7437 | 147.6569 | 147.5130 | 147.3922 |
| 151 | 149.8694 | 149.7438 | 149.6570 | 149.5131 | 149.3923 |
| 153 | 151.8696 | 151.7439 | 151.6571 | 151.5132 | 151.3924 |
| 155 | 153.8698 | 153.7441 | 153.6572 | 153.5133 | 153.3925 |
| 157 | 155.8699 | 155.7442 | 155.6573 | 155.5134 | 155.3926 |
| 159 | 157.8701 | 157.7443 | 157.6574 | 157.5135 | 157.3927 |
| 161 | 159.8702 | 159.7444 | 159.6575 | 159.5136 | 159.3928 |
| 163 | 161.8704 | 161.7445 | 161.6576 | 161.5137 | 161.3929 |
| 165 | 163.8705 | 163.7446 | 163.6577 | 163.5138 | 163.3930 |
| 167 | 165.8707 | 165.7447 | 165.6578 | 165.5139 | 165.3931 |
| 169 | 167.8708 | 167.7448 | 167.6579 | 167.5139 | 167.3932 |
| 171 | 169.8710 | 169.7449 | 169.6580 | 169.5140 | 169.3933 |
| 181 | 179.8716 | 179.7453 | 179.6584 | 179.5144 | 179.3937 |
| 191 | 189.8721 | 189.7458 | 189.6588 | 189.5148 | 189.3940 |
| 201 | 199.8727 | 199.7461 | 199.6591 | 199.5151 | 199.3944 |
| 301 | 299.8759 | 299.7485 | 299.6612 | 299.5171 | 299.3965 |
| 401 | 399.8776 | 399.7496 | 399.6623 | 399.5182 | 399.3975 |
| 501 | 499.8786 | 499.7504 | 499.6629 | 499.5188 | 499.3981 |

hence, decimal required = .5321 + (.40 × .0074) = .53506. Total dimension = 242.53506 divided by the diametral pitch required.

*Odd Number of Teeth:* Table 2 is for odd numbers of teeth. Measurement for odd numbers above 171 and not in Table 2, may be determined as shown by the following example: Assume that number of teeth = 335 and pressure angle 20 degrees; then, for 1 diametral pitch, figure at left of decimal point = given No. of teeth + 2 = 335 + 2 = 337. Figure at right of decimal point lies between decimal values given in table for 301 and 401 teeth. Thus, 335 − 301 = 34 (change to .34); .4565 − .4538 = .0027; hence, decimal required = .4538 + (.34 × .0027) = .4547. Total dimension = 337.4547.

### Table 5. Van Keuren Wire Diameters for Gears

| External Gears Wire Diam. = 1.728 ÷ D.P. | | | | Internal Gears Wire Diam. = 1.44 ÷ D.P. | | | |
|---|---|---|---|---|---|---|---|
| D.P. | Diam. | D.P. | Diam. | D.P. | Diam. | D.P. | Diam. |
| 2 | .86400 | 16 | .10800 | 2 | .72000 | 16 | .09000 |
| 2½ | .69120 | 18 | .09600 | 2½ | .57600 | 18 | .08000 |
| 3 | .57600 | 20 | .08640 | 3 | .48000 | 20 | .07200 |
| 4 | .43200 | 22 | .07855 | 4 | .36000 | 22 | .06545 |
| 5 | .34560 | 24 | .07200 | 5 | .28800 | 24 | .06000 |
| 6 | .28800 | 28 | .06171 | 6 | .24000 | 28 | .05143 |
| 7 | .24686 | 32 | .05400 | 7 | .20571 | 32 | .04500 |
| 8 | .21600 | 36 | .04800 | 8 | .18000 | 36 | .04000 |
| 9 | .19200 | 40 | .04320 | 9 | .16000 | 40 | .03600 |
| 10 | .17280 | 48 | .03600 | 10 | .14400 | 48 | .03000 |
| 11 | .15709 | 64 | .02700 | 11 | .13091 | 64 | .02250 |
| 12 | .14400 | 72 | .02400 | 12 | .12000 | 72 | .02000 |
| 14 | .12343 | 80 | .02160 | 14 | .10286 | 80 | .01800 |

### Measurements for Checking Internal Gears when Wire Diameter Equals 1.44 Divided by Diametral Pitch.

Tables 3 and 4 give measurements between wires for checking internal gears of 1 diametral pitch. For any other diametral pitch, divide the measurement given in the table by the diametral pitch required. These measurements are based upon the Van Keuren standard wire size, which, for internal spur gears, equals 1.44 divided by the diametral pitch (see Table 5).

*Even Number of Teeth:* For an even number of teeth above 170 and not in Table 3, proceed as shown by the following example: Assume that the number of teeth = 380 and pressure angle is 14½ degrees; then, for 1 diametral pitch, figure at left of decimal point = given number of teeth − 2 = 380 − 2 = 378. Figure at right of decimal point lies between decimal values given in table for 300 and 400 teeth and is obtained by interpolation. Thus, 380 − 300 = 80 (change to .80); .8803 − .8795 = .0008; hence, decimal required = .8795 + (.80 × .0008) = .88014. Total dimension = 378.88014.

*Odd Number of Teeth:* Table 4 is for internal gears having odd numbers of teeth. For tooth numbers above 171 and not in the table, proceed as shown by the following example: Assume that number of teeth = 337 and pressure angle is 14½ degrees; then, for 1 diametral pitch, figure at left of decimal point = given No. of teeth − 2 = 337 − 2 = 335. Figure at right of decimal point lies between decimal values given in table for 301 and 401 teeth and is obtained by interpolation. Thus, 337 − 301 = 36 (change to .36); .8776 − .8759 = .0017; hence, decimal required = .8759 + (.36 × .0017) = .8765 Total dimension = 335.8765.

**Measurements for Checking External Spur Gears when Wire Diameter Equals 1.68 Divided by Diametral Pitch.** — Tables 7 and 8 give measurements $M$, in inches, for checking the pitch diameters of external spur gears of 1 diametral pitch. For any other diametral pitch, divide the measurement given in the table by whatever diametral pitch is required. The result shows what measurement $M$ should be when the pitch diameter is correct and there is no allowance for backlash. The procedure for checking for a given amount of backlash when the diameter of the measuring wires equals 1.68 divided by the diametral pitch is explained under a subsequent heading. Tables 7 and 8 are based upon wire sizes equal to 1.68 divided by the diametral pitch. The corresponding wire diameters for various diametral pitches are given in Table 9.

To find measurement $M$ of an external spur gear using wire sizes equal to 1.68 inches divided by the diametral pitch, the same method is followed in using Tables 7 and 8 as that outlined for Tables 1 and 2.

**Allowance for Backlash.** — Tables 1, 2, 7 and 8 give measurements over wires when the pitch diameters are correct and there is no allowance for backlash or play between meshing teeth. Backlash is obtained by cutting the teeth somewhat deeper than standard, thus reducing the thickness. Usually, the teeth of both mating gears are reduced in thickness an amount equal to one-half of the total backlash desired. However, if the pinion is small, it is common practice to reduce the gear teeth the full amount of backlash and pinion is made to standard size. The changes in measurements $M$ over wires, for obtaining backlash in external spur gears, are listed in Table 6.

**Table 6.   Backlash Allowances for External and Internal Spur Gears**

*External Gears:* For each 0.001 inch reduction in pitch-line tooth thickness, *reduce* measurement over wires obtained from Tables 1, 2, 7, or 8 by the amount shown below.

*Internal Gears:* For each 0.001 inch reduction in pitch-line tooth thickness, *increase* measurement between wires obtained from Tables 3 or 4 by the amounts shown below. Backlash on pitch line equals double tooth thickness reduction when teeth of *both* mating gears are reduced. If teeth of *one* gear only are reduced, backlash on pitch line equals amount of reduction.

*Example:* For a 30-tooth, 10-diametral pitch, 20-degree pressure angle, external gear the measurement over wires from Table 1 is 32.4102 ÷ 10. For a backlash of 0.002 this measurement must be reduced by 2 × 0.0024 to 3.2362 or (3.2410 − 0.0048).

| No. of Teeth | $14\frac{1}{2}°$ | | $17\frac{1}{2}°$ | | $20°$ | | $25°$ | | $30°$ | |
|---|---|---|---|---|---|---|---|---|---|---|
| | Ext. | Int. | Ext. | Int. | Ext. | Int. | Ext. | Int. | Ext. | Int. |
| 5 | .0019 | .0024 | .0018 | .0024 | .0017 | .0023 | .0015 | .0021 | .0013 | .0019 |
| 10 | .0024 | .0029 | .0022 | .0027 | .0020 | .0026 | .0017 | .0022 | .0015 | .0018 |
| 20 | .0028 | .0032 | .0025 | .0029 | .0023 | .0027 | .0019 | .0022 | .0016 | .0018 |
| 30 | .0030 | .0034 | .0026 | .0030 | .0024 | .0027 | .0020 | .0022 | .0016 | .0018 |
| 40 | .0031 | .0035 | .0027 | .0030 | .0025 | .0027 | .0020 | .0022 | .0017 | .0018 |
| 50 | .0032 | .0036 | .0028 | .0031 | .0025 | .0027 | .0020 | .0022 | .0017 | .0018 |
| 100 | .0035 | .0037 | .0030 | .0031 | .0026 | .0027 | .0021 | .0022 | .0017 | .0017 |
| 200 | .0036 | .0038 | .0031 | .0031 | .0027 | .0027 | .0021 | .0022 | .0017 | .0017 |

**Measurements for Checking External Helical Gears when Wire Diameter Equals 1.68 Inches Divided by Normal Diametral Pitch.** — This method makes use of Table 7 for even tooth external spur gears in checking both even-tooth and odd-tooth helical gears.

A convenient method of checking helical gears is to use three wires or pins held in place between the flat, parallel surfaces of plates. This measurement $M$ between these plates and perpendicular to the gear axis, will be the same for both even and odd teeth numbers because the axial displacement of the wires with the odd numbers of teeth, does not affect the perpendicular measurement between the plates.

## Table 7. Checking External Spur Gear Sizes by Measurement Over Wires

### EVEN NUMBERS OF TEETH

Dimensions in table are for 1 diametral pitch and 1.68-inch series wire sizes (a Van Keuren standard). For any other diametral pitch, divide dimension in table by given pitch.

$$\text{Wire or pin diameter} = \frac{1.68}{\text{Diametral Pitch}}$$

| No. of Teeth | Pressure Angle | | | | |
|---|---|---|---|---|---|
| | 14½° | 17½° | 20° | 25° | 30° |
| 6 | 8.1298 | 8.1442 | 8.1600 | 8.2003 | 8.2504 |
| 8 | 10.1535 | 10.1647 | 10.1783 | 10.2155 | 10.2633 |
| 10 | 12.1712 | 12.1796 | 12.1914 | 12.2260 | 12.2722 |
| 12 | 14.1851 | 14.1910 | 14.2013 | 14.2338 | 14.2785 |
| 14 | 16.1964 | 16.2001 | 16.2091 | 16.2397 | 16.2833 |
| 16 | 18.2058 | 18.2076 | 18.2154 | 18.2445 | 18.2871 |
| 18 | 20.2137 | 20.2138 | 20.2205 | 20.2483 | 20.2902 |
| 20 | 22.2205 | 22.2190 | 22.2249 | 22.2515 | 22.2927 |
| 22 | 24.2265 | 24.2235 | 24.2286 | 24.2542 | 24.2949 |
| 24 | 26.2317 | 26.2275 | 26.2318 | 26.2566 | 26.2967 |
| 26 | 28.2363 | 28.2309 | 28.2346 | 28.2586 | 28.2982 |
| 28 | 30.2404 | 30.2339 | 30.2371 | 30.2603 | 30.2996 |
| 30 | 32.2441 | 32.2367 | 32.2392 | 32.2619 | 32.3008 |
| 32 | 34.2475 | 34.2391 | 34.2412 | 34.2632 | 34.3017 |
| 34 | 36.2505 | 36.2413 | 36.2430 | 36.2644 | 36.3026 |
| 36 | 38.2533 | 38.2433 | 38.2445 | 38.2655 | 38.3035 |
| 38 | 40.2558 | 40.2451 | 40.2460 | 40.2666 | 40.3044 |
| 40 | 42.2582 | 42.2468 | 42.2473 | 42.2675 | 42.3051 |
| 42 | 44.2604 | 44.2483 | 44.2485 | 44.2683 | 44.3057 |
| 44 | 46.2624 | 46.2497 | 46.2496 | 46.2690 | 46.3063 |
| 46 | 48.2642 | 48.2510 | 48.2506 | 48.2697 | 48.3068 |
| 48 | 50.2660 | 50.2522 | 50.2516 | 50.2704 | 50.3073 |
| 50 | 52.2676 | 52.2534 | 52.2525 | 52.2710 | 52.3078 |
| 52 | 54.2691 | 54.2545 | 54.2533 | 54.2716 | 54.3082 |
| 54 | 56.2705 | 56.2555 | 56.2541 | 56.2721 | 56.3086 |
| 56 | 58.2719 | 58.2564 | 58.2548 | 58.2726 | 58.3089 |
| 58 | 60.2731 | 60.2572 | 60.2555 | 60.2730 | 60.3093 |
| 60 | 62.2743 | 62.2580 | 62.2561 | 62.2735 | 62.3096 |
| 62 | 64.2755 | 64.2587 | 64.2567 | 64.2739 | 64.3099 |
| 64 | 66.2765 | 66.2594 | 66.2572 | 66.2742 | 66.3102 |
| 66 | 68.2775 | 68.2601 | 68.2577 | 68.2746 | 68.3104 |
| 68 | 70.2785 | 70.2608 | 70.2582 | 70.2749 | 70.3107 |
| 70 | 72.2794 | 72.2615 | 72.2587 | 72.2752 | 72.3109 |
| 72 | 74.2803 | 74.2620 | 74.2591 | 74.2755 | 74.3111 |
| 74 | 76.2811 | 76.2625 | 76.2596 | 76.2758 | 76.3113 |
| 76 | 78.2819 | 78.2631 | 78.2600 | 78.2761 | 78.3115 |
| 78 | 80.2827 | 80.2636 | 80.2604 | 80.2763 | 80.3117 |
| 80 | 82.2834 | 82.2641 | 82.2607 | 82.2766 | 82.3119 |
| 82 | 84.2841 | 84.2646 | 84.2611 | 84.2768 | 84.3121 |
| 84 | 86.2847 | 86.2650 | 86.2614 | 86.2771 | 86.3123 |
| 86 | 88.2854 | 88.2655 | 88.2617 | 88.2773 | 88.3124 |
| 88 | 90.2860 | 90.2659 | 90.2620 | 90.2775 | 90.3126 |

Table 7. Checking External Spur Gear Sizes by Measurement Over Wires

| EVEN NUMBERS OF TEETH | | | | | |
|---|---|---|---|---|---|
| No. of Teeth | $14\frac{1}{2}°$ | $17\frac{1}{2}°$ | 20° | 25° | 30° |
| 90 | 92.2866 | 92.2662 | 92.2624 | 92.2777 | 92.3127 |
| 92 | 94.2872 | 94.2666 | 94.2627 | 94.2779 | 94.3129 |
| 94 | 96.2877 | 96.2670 | 96.2630 | 96.2780 | 96.3130 |
| 96 | 98.2882 | 98.2673 | 98.2632 | 98.2782 | 98.3131 |
| 98 | 100.2887 | 100.2677 | 100.2635 | 100.2784 | 100.3132 |
| 100 | 102.2892 | 102.2680 | 102.2638 | 102.2785 | 102.3134 |
| 102 | 104.2897 | 104.2683 | 104.2640 | 104.2787 | 104.3135 |
| 104 | 106.2901 | 106.2685 | 106.2642 | 106.2788 | 106.3136 |
| 106 | 108.2905 | 108.2688 | 108.2644 | 108.2789 | 108.3137 |
| 108 | 110.2910 | 110.2691 | 110.2645 | 110.2791 | 110.3138 |
| 110 | 112.2914 | 112.2694 | 112.2647 | 112.2792 | 112.3139 |
| 112 | 114.2918 | 114.2696 | 114.2649 | 114.2793 | 114.3140 |
| 114 | 116.2921 | 116.2699 | 116.2651 | 116.2794 | 116.3141 |
| 116 | 118.2925 | 118.2701 | 118.2653 | 118.2795 | 118.3142 |
| 118 | 120.2929 | 120.2703 | 120.2655 | 120.2797 | 120.3142 |
| 120 | 122.2932 | 122.2706 | 122.2656 | 122.2798 | 122.3143 |
| 122 | 124.2936 | 124.2708 | 124.2658 | 124.2799 | 124.3144 |
| 124 | 126.2939 | 126.2710 | 126.2660 | 126.2800 | 126.3145 |
| 126 | 128.2941 | 128.2712 | 128.2661 | 128.2801 | 128.3146 |
| 128 | 130.2945 | 130.2714 | 130.2663 | 130.2802 | 130.3146 |
| 130 | 132.2948 | 132.2716 | 132.2664 | 132.2803 | 132.3147 |
| 132 | 134.2951 | 134.2718 | 134.2666 | 134.2804 | 134.3147 |
| 134 | 136.2954 | 136.2720 | 136.2667 | 136.2805 | 136.3148 |
| 136 | 138.2957 | 138.2722 | 138.2669 | 138.2806 | 138.3149 |
| 138 | 140.2960 | 140.2724 | 140.2670 | 140.2807 | 140.3149 |
| 140 | 142.2962 | 142.2725 | 142.2671 | 142.2808 | 142.3150 |
| 142 | 144.2965 | 144.2727 | 144.2672 | 144.2808 | 144.3151 |
| 144 | 146.2967 | 146.2729 | 146.2674 | 146.2809 | 146.3151 |
| 146 | 148.2970 | 148.2730 | 148.2675 | 148.2810 | 148.3152 |
| 148 | 150.2972 | 150.2732 | 150.2676 | 150.2811 | 150.3152 |
| 150 | 152.2974 | 152.2733 | 152.2677 | 152.2812 | 152.3153 |
| 152 | 154.2977 | 154.2735 | 154.2678 | 154.2812 | 154.3153 |
| 154 | 156.2979 | 156.2736 | 156.2679 | 156.2813 | 156.3154 |
| 156 | 158.2981 | 158.2737 | 158.2680 | 158.2813 | 158.3155 |
| 158 | 160.2983 | 160.2739 | 160.2681 | 160.2814 | 160.3155 |
| 160 | 162.2985 | 162.2740 | 162.2682 | 162.2815 | 162.3155 |
| 162 | 164.2987 | 164.2741 | 164.2683 | 164.2815 | 164.3156 |
| 164 | 166.2989 | 166.2742 | 166.2684 | 166.2816 | 166.3156 |
| 166 | 168.2990 | 168.2744 | 168.2685 | 168.2816 | 168.3157 |
| 168 | 170.2992 | 170.2745 | 170.2686 | 170.2817 | 170.3157 |
| 170 | 172.2994 | 172.2746 | 172.2687 | 172.2818 | 172.3158 |
| 180 | 182.3003 | 182.2752 | 182.2691 | 182.2820 | 182.3160 |
| 190 | 192.3011 | 192.2757 | 192.2694 | 192.2823 | 192.3161 |
| 200 | 202.3018 | 202.2761 | 202.2698 | 202.2825 | 202.3163 |
| 300 | 302.3063 | 302.2790 | 302.2719 | 302.2839 | 302.3173 |
| 400 | 402.3087 | 402.2804 | 402.2730 | 402.2845 | 402.3178 |
| 500 | 502.3101 | 502.2813 | 502.2736 | 502.2850 | 502.3181 |

## Table 8. Checking External Spur Gear Sizes by Measurement Over Wires

| | ODD NUMBERS OF TEETH | | | | |
|---|---|---|---|---|---|

Dimensions in table are for 1 diametral pitch and 1.68-inch series wire sizes (a VanKeuren standard). For any other diametral pitch, divide dimension in table by given pitch.

$$\text{Wire or pin diameter} = \frac{1.68}{\text{Diametral Pitch}}$$

| No. of Teeth | Pressure Angle | | | | |
|---|---|---|---|---|---|
| | 14½° | 17½° | 20° | 25° | 30° |
| 5 | 6.8485 | 6.8639 | 6.8800 | 6.9202 | 6.9691 |
| 7 | 8.9555 | 8.9679 | 8.9822 | 9.0199 | 9.0675 |
| 9 | 11.0189 | 11.0285 | 11.0410 | 11.0762 | 11.1224 |
| 11 | 13.0615 | 13.0686 | 13.0795 | 13.1126 | 13.1575 |
| 13 | 15.0925 | 15.0973 | 15.1068 | 15.1381 | 15.1819 |
| 15 | 17.1163 | 17.1190 | 17.1273 | 17.1570 | 17.1998 |
| 17 | 19.1351 | 19.1360 | 19.1432 | 19.1716 | 19.2136 |
| 19 | 21.1505 | 21.1498 | 21.1561 | 21.1832 | 21.2245 |
| 21 | 23.1634 | 23.1611 | 23.1665 | 23.1926 | 23.2334 |
| 23 | 25.1743 | 25.1707 | 25.1754 | 25.2005 | 25.2408 |
| 25 | 27.1836 | 27.1788 | 27.1828 | 27.2071 | 27.2469 |
| 27 | 29.1918 | 29.1859 | 29.1892 | 29.2128 | 29.2522 |
| 29 | 31.1990 | 31.1920 | 31.1948 | 31.2177 | 31.2568 |
| 31 | 33.2053 | 33.1974 | 33.1997 | 33.2220 | 33.2607 |
| 33 | 35.2110 | 35.2021 | 35.2041 | 35.2258 | 35.2642 |
| 35 | 37.2161 | 37.2065 | 37.2079 | 37.2292 | 37.2674 |
| 37 | 39.2208 | 39.2104 | 39.2115 | 39.2323 | 39.2702 |
| 39 | 41.2249 | 41.2138 | 41.2147 | 41.2349 | 41.2726 |
| 41 | 43.2287 | 43.2170 | 43.2174 | 43.2374 | 43.2749 |
| 43 | 45.2323 | 45.2199 | 45.2200 | 45.2396 | 45.2769 |
| 45 | 47.2355 | 47.2226 | 47.2224 | 47.2417 | 47.2788 |
| 47 | 49.2385 | 49.2251 | 49.2246 | 49.2435 | 49.2805 |
| 49 | 51.2413 | 51.2273 | 51.2266 | 51.2452 | 51.2820 |
| 51 | 53.2439 | 53.2294 | 53.2284 | 53.2468 | 53.2835 |
| 53 | 55.2463 | 55.2313 | 55.2302 | 55.2483 | 55.2848 |
| 55 | 57.2485 | 57.2331 | 57.2318 | 57.2497 | 57.2861 |
| 57 | 59.2506 | 59.2348 | 59.2333 | 59.2509 | 59.2872 |
| 59 | 61.2526 | 61.2363 | 61.2347 | 61.2521 | 61.2883 |
| 61 | 63.2545 | 63.2378 | 63.2360 | 63.2532 | 63.2893 |
| 63 | 65.2562 | 65.2392 | 65.2372 | 65.2543 | 65.2902 |
| 65 | 67.2579 | 67.2406 | 67.2383 | 67.2553 | 67.2911 |
| 67 | 69.2594 | 69.2419 | 69.2394 | 69.2562 | 69.2920 |
| 69 | 71.2609 | 71.2431 | 71.2405 | 71.2571 | 71.2928 |
| 71 | 73.2623 | 73.2442 | 73.2414 | 73.2579 | 73.2935 |
| 73 | 75.2636 | 75.2452 | 75.2423 | 75.2586 | 75.2942 |
| 75 | 77.2649 | 77.2462 | 77.2432 | 77.2594 | 77.2949 |
| 77 | 79.2661 | 79.2472 | 79.2440 | 79.2601 | 79.2955 |
| 79 | 81.2673 | 81.2481 | 81.2448 | 81.2607 | 81.2961 |
| 81 | 83.2684 | 83.2490 | 83.2456 | 83.2614 | 83.2967 |
| 83 | 85.2694 | 85.2498 | 85.2463 | 85.2620 | 85.2972 |
| 85 | 87.2704 | 87.2506 | 87.2470 | 87.2625 | 87.2977 |
| 87 | 89.2714 | 89.2514 | 89.2476 | 89.2631 | 89.2982 |
| 89 | 91.2723 | 91.2521 | 91.2482 | 91.2636 | 91.2987 |

**Table 8.  Checking External Spur Gear Sizes by Measurement Over Wires**

| No. of Teeth | | | | | |
|---|---|---|---|---|---|
| | | | ODD NUMBERS OF TEETH | | |
| No. of Teeth | 14½° | 17½° | 20° | 25° | 30° |
| 91 | 93.2732 | 93.2528 | 93.2489 | 93.2641 | 93.2991 |
| 93 | 95.2741 | 95.2534 | 95.2495 | 95.2646 | 95.2996 |
| 95 | 97.2749 | 97.2541 | 97.2500 | 97.2650 | 97.3000 |
| 97 | 99.2757 | 99.2547 | 99.2506 | 99.2655 | 99.3004 |
| 99 | 101.2764 | 101.2553 | 101.2511 | 101.2659 | 101.3008 |
| 101 | 103.2771 | 103.2558 | 103.2516 | 103.2663 | 103.3011 |
| 103 | 105.2778 | 105.2563 | 105.2520 | 105.2667 | 105.3015 |
| 105 | 107.2785 | 107.2568 | 107.2525 | 107.2671 | 107.3018 |
| 107 | 109.2791 | 109.2573 | 109.2529 | 109.2674 | 109.3021 |
| 109 | 111.2798 | 111.2578 | 111.2533 | 111.2678 | 111.3024 |
| 111 | 113.2804 | 113.2583 | 113.2537 | 113.2681 | 113.3027 |
| 113 | 115.2809 | 115.2588 | 115.2541 | 115.2684 | 115.3030 |
| 115 | 117.2815 | 117.2592 | 117.2544 | 117.2687 | 117.3033 |
| 117 | 119.2821 | 119.2596 | 119.2548 | 119.2690 | 119.3036 |
| 119 | 121.2826 | 121.2601 | 121.2552 | 121.2693 | 121.3038 |
| 121 | 123.2831 | 123.2605 | 123.2555 | 123.2696 | 123.3041 |
| 123 | 125.2836 | 125.2608 | 125.2558 | 125.2699 | 125.3043 |
| 125 | 127.2841 | 127.2612 | 127.2562 | 127.2702 | 127.3046 |
| 127 | 129.2846 | 129.2615 | 129.2565 | 129.2704 | 129.3048 |
| 129 | 131.2851 | 131.2619 | 131.2568 | 131.2707 | 131.3050 |
| 131 | 133.2855 | 133.2622 | 133.2571 | 133.2709 | 133.3053 |
| 133 | 135.2859 | 135.2626 | 135.2574 | 135.2712 | 135.3055 |
| 135 | 137.2863 | 137.2629 | 137.2577 | 137.2714 | 137.3057 |
| 137 | 139.2867 | 139.2632 | 139.2579 | 139.2716 | 139.3059 |
| 139 | 141.2871 | 141.2635 | 141.2582 | 141.2718 | 141.3060 |
| 141 | 143.2875 | 143.2638 | 143.2584 | 143.2720 | 143.3062 |
| 143 | 145.2879 | 145.2641 | 145.2587 | 145.2722 | 145.3064 |
| 145 | 147.2883 | 147.2644 | 147.2589 | 147.2724 | 147.3066 |
| 147 | 149.2887 | 149.2647 | 149.2591 | 149.2726 | 149.3068 |
| 149 | 151.2890 | 151.2649 | 151.2594 | 151.2728 | 151.3069 |
| 151 | 153.2893 | 153.2652 | 153.2596 | 153.2730 | 153.3071 |
| 153 | 155.2897 | 155.2654 | 155.2598 | 155.2732 | 155.3073 |
| 155 | 157.2900 | 157.2657 | 157.2600 | 157.2733 | 157.3074 |
| 157 | 159.2903 | 159.2659 | 159.2602 | 159.2735 | 159.3076 |
| 159 | 161.2906 | 161.2661 | 161.2604 | 161.2736 | 161.3077 |
| 161 | 163.2909 | 163.2663 | 163.2606 | 163.2738 | 163.3078 |
| 163 | 165.2912 | 165.2665 | 165.2608 | 165.2740 | 165.3080 |
| 165 | 167.2915 | 167.2668 | 167.2610 | 167.2741 | 167.3081 |
| 167 | 169.2917 | 169.2670 | 169.2611 | 169.2743 | 169.3083 |
| 169 | 171.2920 | 171.2672 | 171.2613 | 171.2744 | 171.3084 |
| 171 | 173.2922 | 173.2674 | 173.2615 | 173.2746 | 173.3085 |
| 181 | 183.2936 | 183.2684 | 183.2623 | 183.2752 | 183.3091 |
| 191 | 193.2947 | 193.2692 | 193.2630 | 193.2758 | 193.3097 |
| 201 | 203.2957 | 203.2700 | 203.2636 | 203.2764 | 203.3101 |
| 301 | 303.3022 | 303.2749 | 303.2678 | 303.2798 | 303.3132 |
| 401 | 403.3056 | 403.2774 | 403.2699 | 403.2815 | 403.3147 |
| 501 | 503.3076 | 503.2789 | 503.2711 | 503.2825 | 503.3156 |

Table 9. Wire Diameters for Spur and Helical Gears Based upon 1.68 Constant*

| Diametral or Normal Diametral Pitch | Wire Diameter | Diametral or Normal Diametral Pitch | Wire Diameter | Diametral or Normal Diametral Pitch | Wire Diameter | Diametral or Normal Diametral Pitch | Wire Diameter |
|---|---|---|---|---|---|---|---|
| 2 | .840 | 8 | .210 | 18 | .09333 | 40 | .042 |
| 2½ | .672 | 9 | .18666 | 20 | .084 | 48 | .035 |
| 3 | .560 | 10 | .168 | 22 | .07636 | 64 | .02625 |
| 4 | .420 | 11 | .15273 | 24 | .070 | 72 | .02333 |
| 5 | .336 | 12 | .140 | 28 | .060 | 80 | .021 |
| 6 | .280 | 14 | .120 | 32 | .0525 | .. | ...... |
| 7 | .240 | 16 | .105 | 36 | .04667 | .. | ...... |

* Pin diameter = 1.68 ÷ diametral pitch for spur gears and 1.68 ÷ normal diametral pitch for helical gears.

*Measurement Over Balls:* The measurement can be taken over two balls if they are kept in a plane of the gear's rotation. This can be done by holding them against a surface parallel to the face of the gear. In the case of odd-tooth helical gears, the two balls will not be diametrically opposite each other and hence a correction must be applied to the computed measurement $M$ as described later under "Helical Gear with Odd Number of Teeth."

*Helical Gear with Even Number of Teeth:* First find a number $N_e$, by multiplying the number of teeth $N$ in the helical gear by the secant of the helix angle. Next, multiply the number $N$ of teeth in the helical gear by the cube of the secant of the helix angle, obtaining $N_s$. ($N_e$ is the number of teeth which would be used in selecting a formed cutter in case the teeth are milled. The table on page 927 gives secant cubed values, $K$, for various angles). Take that whole number which is nearest $N_s$ and referring to Table 7 for spur gears with even tooth numbers, find the *decimal* value of the constant for this number of teeth, under the given *normal* pressure angle. Add to this decimal value the sum of $N_e + 2$ and divide the result by the normal diametral pitch to obtain the measurement $M$ over the wires.

*Example:* Assume that a helical gear has 32 teeth of 6 normal diametral pitch, a helix angle of 23 degrees and normal pressure angle of 20 degrees. Determine the measurement $M$ without allowance for backlash. The number $N_e$ is equal to $32 \times \sec 23° = 32 \times 1.0864 = 34.7648$. The number of teeth $N_s$ for which a formed cutter should be selected is equal to $32 \times \sec^3 23° = 32 \times 1.282 = 41.024$. The nearest whole number to $N_s$ is 41 and the decimal part of the constant for 41 teeth must be found from Table 7 by interpolating halfway between the values given for 40 and 42 teeth. The decimal value given for 40 teeth is 0.2473 and for 42 teeth, 0.2485. The value for 41 teeth is halfway between, or 0.2479. Thus, $M$ is computed as follows:

$$M = \frac{34.7648 + 2 + 0.2479}{6} = \frac{37.0127}{6} = 6.1688 \text{ inches}$$

*Helical Gear with Odd Number of Teeth:* The procedure is similar to that just outlined except that a correction is made in the final $M$ value if the measurement is taken over two balls as previously described.

*Example:* Assume that a helical gear has 13 teeth of 8 normal diametral pitch, a helix angle of 45 degrees and a normal pressure angle of 14½ degrees. Wire diameter is 0.210 (see Table 9). To determine the measurement, $M$, without allowance for backlash, the number $N_e$ is equal to $13 \times \sec 45° = 13 \times 1.4142 = 18.3846$. Multiply number of teeth $N$ by cube of secant of helix angle, obtaining $N_s$. Thus

13 × (sec 45°)³ = 13 × 2.828 = 36.764. The nearest whole number to $N_s$ is 37 and the decimal part of the constant for 37 teeth must be found from Table 7 by interpolating halfway between the values given for 36 and 38 teeth. The decimal value given for 36 teeth is 0.2533 and for 38 teeth, 0.2558. The value for 37 teeth is halfway between, or 0.2545. When the measurement is made over three wires and between parallel plates as previously described, $M$ is computed as follows:

$$M = \frac{18.3846 + 2 + 0.2545}{8} = \frac{20.6391}{8} = 2.5799 \text{ inches}$$

When measurement is over two balls, the value of $M$ is corrected for the balls not being diametrically opposite by one-half tooth interval, as follows:

$$M \text{ corrected} = (M - \text{Ball Diam.}) \times \cos\frac{90°}{N} + \text{Ball Diam.}$$

$$M \text{ corrected} = (2.5799 - 0.21) \cos\frac{90°}{13} + 0.21$$

$$= 2.3699 \times 0.99271 + 0.21 = 2.5626$$

**Checking Spur Gear Size by Chordal Measurement Over Two or More Teeth.** — Another method of checking gear sizes, that is generally available, is illustrated by the diagram accompanying Table 10. A vernier caliper is used to measure the distance $M$ over two or more teeth. The diagram illustrates the measurement over two teeth (or with one intervening tooth space), but three or more teeth might be included, depending upon the pitch. The jaws of the caliper are merely held in contact with the sides or profiles of the teeth and perpendicular to the axis of the gear. Measurement $M$ for involute teeth of the correct size, is determined as follows:

*Table for Determining the Chordal Dimension:* Table 10 gives the chordal dimensions for one diametral pitch when measuring over the number of teeth indicated in Table 11. To obtain any chordal dimension, it is simply necessary to divide chord $M$ in the table (opposite the given number of teeth) by the diametral pitch of the gear to be measured and then subtract from the quotient one-half the total backlash between the mating pair of gears. In cases where a small pinion is used with a large gear and all of the backlash is to be obtained by reducing the gear teeth, the total amount of backlash is subtracted from the chordal dimension of the gear and nothing from the chordal dimension of the pinion. The application of the tables will be illustrated by an example.

*Example* — Determine the chordal dimension for checking the size of a gear having 30 teeth of 5 diametral pitch and a pressure angle of 20 degrees. A total backlash of 0.008 inch is to be obtained by reducing equally the teeth of both mating gears. Table 10 shows that the chordal distance for 30 teeth of one diametral pitch and a pressure angle of 20 degrees is 10.7526 inches; one-half of the backlash equals 0.004 inch; hence,

$$\text{Chordal dimension} = \frac{10.7526}{5} - 0.004 = 2.1465 \text{ inches}$$

Table 11 shows that this is the chordal dimension when the vernier caliper spans four teeth, this being the number of teeth to gage over whenever gears of 20-degree pressure angle have any number of teeth from 28 to 36, inclusive.

If it is considered necessary to leave enough stock on the gear teeth for a shaving or finishing cut, this allowance is simply added to the chordal dimension of the finished teeth to obtain the required measurement over the teeth for the roughing

### Table 10. Chordal Measurements over Spur Gear Teeth of 1 Diametral Pitch

Find value of $M$ under pressure angle and opposite number of teeth; divide $M$ by diametral pitch of gear to be measured and then subtract one-half total backlash to obtain a measurement $M$ equivalent to given pitch and backlash. The number of teeth to gage or measure over is shown by Table 11.

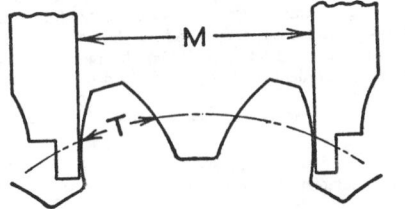

#### Pressure Angle, 14½ Degrees

| Number of Gear Teeth | $M$ in Inches for 1 D. P. | Number of Gear Teeth | $M$ in Inches for 1 D. P. | Number of Gear Teeth | $M$ in Inches for 1 D. P. | Number of Gear Teeth | $M$ in Inches for 1 D. P. |
|---|---|---|---|---|---|---|---|
| 12 | 4.6267 | 37 | 7.8024 | 62 | 14.0197 | 87 | 20.2370 |
| 13 | 4.6321 | 38 | 10.8493 | 63 | 17.0666 | 88 | 23.2838 |
| 14 | 4.6374 | 39 | 10.8547 | 64 | 17.0720 | 89 | 23.2892 |
| 15 | 4.6428 | 40 | 10.8601 | 65 | 17.0773 | 90 | 23.2946 |
| 16 | 4.6482 | 41 | 10.8654 | 66 | 17.0827 | 91 | 23.2999 |
| 17 | 4.6536 | 42 | 10.8708 | 67 | 17.0881 | 92 | 23.3053 |
| 18 | 4.6589 | 43 | 10.8762 | 68 | 17.0934 | 93 | 23.3107 |
| 19 | 7.7058 | 44 | 10.8815 | 69 | 17.0988 | 94 | 23.3160 |
| 20 | 7.7112 | 45 | 10.8869 | 70 | 17.1042 | 95 | 23.3214 |
| 21 | 7.7166 | 46 | 10.8923 | 71 | 17.1095 | 96 | 23.3268 |
| 22 | 7.7219 | 47 | 10.8976 | 72 | 17.1149 | 97 | 23.3322 |
| 23 | 7.7273 | 48 | 10.9030 | 73 | 17.1203 | 98 | 23.3375 |
| 24 | 7.7326 | 49 | 10.9084 | 74 | 17.1256 | 99 | 23.3429 |
| 25 | 7.7380 | 50 | 10.9137 | 75 | 17.1310 | 100 | 23.3483 |
| 26 | 7.7434 | 51 | 13.9606 | 76 | 20.1779 | 101 | 26.3952 |
| 27 | 7.7488 | 52 | 13.9660 | 77 | 20.1833 | 102 | 26.4005 |
| 28 | 7.7541 | 53 | 13.9714 | 78 | 20.1886 | 103 | 26.4059 |
| 29 | 7.7595 | 54 | 13.9767 | 79 | 20.1940 | 104 | 26.4113 |
| 30 | 7.7649 | 55 | 13.9821 | 80 | 20.1994 | 105 | 26.4166 |
| 31 | 7.7702 | 56 | 13.9875 | 81 | 20.2047 | 106 | 26.4220 |
| 32 | 7.7756 | 57 | 13.9929 | 82 | 20.2101 | 107 | 26.4274 |
| 33 | 7.7810 | 58 | 13.9982 | 83 | 20.2155 | 108 | 26.4327 |
| 34 | 7.7863 | 59 | 14.0036 | 84 | 20.2208 | 109 | 26.4381 |
| 35 | 7.7917 | 60 | 14.0090 | 85 | 20.2262 | 110 | 26.4435 |
| 36 | 7.7971 | 61 | 14.0143 | 86 | 20.2316 | ... | ....... |

#### Pressure Angle, 20 Degrees

| Number of Gear Teeth | $M$ in Inches for 1 D. P. | Number of Gear Teeth | $M$ in Inches for 1 D. P. | Number of Gear Teeth | $M$ in Inches for 1 D. P. | Number of Gear Teeth | $M$ in Inches for 1 D. P. |
|---|---|---|---|---|---|---|---|
| 12 | 4.5963 | 30 | 10.7526 | 48 | 16.9090 | 66 | 23.0653 |
| 13 | 4.6103 | 31 | 10.7666 | 49 | 16.9230 | 67 | 23.0793 |
| 14 | 4.6243 | 32 | 10.7806 | 50 | 16.9370 | 68 | 23.0933 |
| 15 | 4.6383 | 33 | 10.7946 | 51 | 16.9510 | 69 | 23.1073 |
| 16 | 4.6523 | 34 | 10.8086 | 52 | 16.9650 | 70 | 23.1214 |
| 17 | 4.6663 | 35 | 10.8226 | 53 | 16.9790 | 71 | 23.1354 |
| 18 | 4.6803 | 36 | 10.8366 | 54 | 16.9930 | 72 | 23.1494 |
| 19 | 7.6464 | 37 | 13.8028 | 55 | 19.9591 | 73 | 26.1155 |
| 20 | 7.6604 | 38 | 13.8168 | 56 | 19.9731 | 74 | 26.1295 |
| 21 | 7.6744 | 39 | 13.8307 | 57 | 19.9872 | 75 | 26.1435 |
| 22 | 7.6884 | 40 | 13.8447 | 58 | 20.0012 | 76 | 26.1575 |
| 23 | 7.7024 | 41 | 13.8587 | 59 | 20.0152 | 77 | 26.1715 |
| 24 | 7.7165 | 42 | 13.8727 | 60 | 20.0292 | 78 | 26.1855 |
| 25 | 7.7305 | 43 | 13.8867 | 61 | 20.0432 | 79 | 26.1995 |
| 26 | 7.7445 | 44 | 13.9007 | 62 | 20.0572 | 80 | 26.2135 |
| 27 | 7.7585 | 45 | 13.9147 | 63 | 20.0712 | 81 | 26.2275 |
| 28 | 10.7246 | 46 | 16.8810 | 64 | 23.0373 | ... | ....... |
| 29 | 10.7386 | 47 | 16.8950 | 65 | 23.0513 | ... | ....... |

**Table 11.  Number of Teeth Included in Chordal Measurement**

This table shows the number of teeth included between the jaws of the vernier caliper in measuring dimension $M$ which is obtained as explained in connection with Table 10.

| Tooth Range for 14½° Pressure Angle | Tooth Range for 20° Pressure Angle | Number of Teeth to Gage Over | Tooth Range for 14½° Pressure Angle | Tooth Range for 20° Pressure Angle | Number of Teeth to Gage Over |
|---|---|---|---|---|---|
| 12 to 18 | 12 to 18 | 2 | 63 to 75 | 46 to 54 | 6 |
| 19 to 37 | 19 to 27 | 3 | 76 to 87 | 55 to 63 | 7 |
| 38 to 50 | 28 to 36 | 4 | 88 to 100 | 64 to 72 | 8 |
| 51 to 62 | 37 to 45 | 5 | 101 to 110 | 73 to 81 | 9 |

operation.  It may be advisable to place this chordal dimension for roughing on the detail drawing.

**Formula for Chordal Dimension $M$.** — The required measurement $M$ over spur gear teeth may be obtained by the following formula in which $R$ = pitch radius of gear, $A$ = pressure angle, $T$ = tooth thickness along pitch circle, $N$ = number of gear teeth, $S$ = number of tooth *spaces* between caliper jaws, $F$ = a factor depending upon the pressure angle = 0.01109 for 14½°; = 0.01973 for 17½°; = 0.0298 for 20°; = 0.04303 for 22½°; = 0.05995 for 25°.  This factor $F$ equals twice the involute function of the pressure angle.

$$M = R \times \cos A \times \left( \frac{T}{R} + \frac{6.2832 \times S}{N} + F \right)$$

*Example* — A spur gear has 30 teeth of 6 diametral pitch and a pressure angle of 14½ degrees.  Determine measurement $M$ over three teeth, there being two intervening tooth spaces.

The pitch radius = 2½ inches, the arc tooth thickness equivalent to 6 diametral pitch is 0.2618 inch (if no allowance is made for backlash) and factor $F$ for 14½ degrees = 0.01109 inch.

$$M = 2.5 \times 0.96815 \times \left( \frac{0.2618}{2.5} + \frac{6.2832 \times 2}{30} + 0.01109 \right) = 1.2941 \text{ inches}$$

**Checking Enlarged Pinions by Measuring Over Pins or Wires.** — When the teeth of small spur gears or pinions would be undercut if generated by an unmodified straight-sided rack cutter or hob, it is common practice to make the outside diameter larger than standard.  The amount of increase in outside diameter varies with the pressure angle and number of teeth, as shown by the table on page 749. In all cases, the teeth are cut to standard depth on a generating type of machine such as a gear hobber or gear shaper; and since the number of teeth and pitch are not changed, the pitch diameter also remains unchanged.  The tooth thickness on the pitch circle, however, is increased and wire sizes suitable for standard gears are not large enough to extend above the tops of these enlarged gears or pinions; hence the Van Keuren wire size recommended for these enlarged pinions equals 1.92 ÷ diametral pitch.  Table 12 gives measurements over wires of this size, for checking full-depth involute gears of 1 diametral pitch.  For any other pitch, merely divide the measurement given in the table by the diametral pitch.  Table 12 applies to pinions which have been enlarged by the same amounts as given in the tables on pages 749 and 750.  These enlarged pinions will mesh with standard gears; but if the standard center distance is to be maintained, reduce the gear diameter below the standard size as much as the pinion diameter is increased.

**Table 12. Checking Enlarged Spur Pinions by Measurement Over Wires**

Measurements over wires are given in table for 1 diametral pitch. For any other diametral pitch, divide measurement in table by given pitch. Wire size equals 1.92 ÷ diametral pitch.

| Number of Teeth | Outside or Major Diameter (Note 1) | Circular Tooth Thickness (Note 2) | Measurement Over Wires | Number of Teeth | Outside or Major Diameter (Note 1) | Circular Tooth Thickness (Note 2) | Measurement Over Wires |
|---|---|---|---|---|---|---|---|
| 14½-degree full-depth involute teeth: | | | | 20-degree full-depth involute teeth: | | | |
| 10 | 13.3731 | 1.9259 | 13.6186 | 10 | 12.8302 | 1.8730 | 13.4408 |
| 11 | 14.3104 | 1.9097 | 14.4966 | 11 | 13.7132 | 1.8304 | 14.2678 |
| 12 | 15.2477 | 1.8935 | 15.6290 | 12 | 14.5963 | 1.7878 | 15.3428 |
| 13 | 16.1850 | 1.8773 | 16.5211 | 13 | 15.4793 | 1.7452 | 16.1807 |
| 14 | 17.1223 | 1.8611 | 17.6244 | 14 | 16.3623 | 1.7027 | 17.2233 |
| 15 | 18.0597 | 1.8449 | 18.5260 | 15 | 17.2453 | 1.6601 | 18.0674 |
| 16 | 18.9970 | 1.8286 | 19.6075 | 16 | 18.1284 | 1.6175 | 19.0851 |
| 17 | 19.9343 | 1.8124 | 20.5156 | 17 | 19.0114 | 1.5749 | 19.9326 |
| 18 | 20.8716 | 1.7962 | 21.5806 | | | | |
| 19 | 21.8089 | 1.7800 | 22.4934 | Note 1: These enlargements, which are to improve the tooth form and avoid undercut, conform to those given in the tables on pages 749 and 750 where data will be found on the minimum number of teeth in the mating gear. | | | |
| 20 | 22.7462 | 1.7638 | 23.5451 | | | | |
| 21 | 23.6835 | 1.7476 | 24.4611 | | | | |
| 22 | 24.6208 | 1.7314 | 25.5018 | | | | |
| 23 | 25.5581 | 1.7151 | 26.4201 | | | | |
| 24 | 26.4954 | 1.6989 | 27.4515 | Note 2: The circular or arc thickness is at the standard pitch diameter. The corresponding chordal thickness may be found as follows: Multiply arc thickness by 90 and then divide product by 3.1416 × pitch radius; find sine of angle thus obtained and multiply it by pitch diameter. | | | |
| 25 | 27.4328 | 1.6827 | 28.3718 | | | | |
| 26 | 28.3701 | 1.6665 | 29.3952 | | | | |
| 27 | 29.3074 | 1.6503 | 30.3168 | | | | |
| 28 | 30.2447 | 1.6341 | 31.3333 | | | | |
| 29 | 31.1820 | 1.6179 | 32.2558 | | | | |
| 30 | 32.1193 | 1.6017 | 33.2661 | | | | |
| 31 | 33.0566 | 1.5854 | 34.1889 | | | | |

**General Formula for Checking External and Internal Spur Gears by Measurement Over Wires.** — The following formulas may be used for pressure angles or wire sizes not covered by the tables. In these formulas, $M$ = measurement *over* wires for external gears or measurement *between* wires for internal gears; $D$ = pitch diameter; $T$ = arc tooth thickness on pitch circle; $W$ = wire diameter; $N$ = number of gear teeth; $A$ = pressure angle of gear; $a$ = angle, the cosine of which is required in Formulas (2) and (3).

First determine the involute function of angle $a$ (inv $a$); then the corresponding angle $a$ is found by referring to the tables of involute functions beginning on page 178,

$$\text{inv } a = \text{inv } A \pm \frac{T}{D} \pm \frac{W}{D \cos A} \mp \frac{\pi}{N} \tag{1}$$

For even numbers of teeth, $$M = \frac{D \cos A}{\cos a} \pm W \tag{2}$$

For odd numbers of teeth, $$M = \left(\frac{D \cos A}{\cos a}\right)\left(\cos \frac{90°}{N}\right) \pm W \tag{3}$$

*Note:* In Formulas (1), (2), and (3), use the upper sign for *external* and the lower sign for *internal* gears wherever a ± or ∓ appears in the formulas.

## Ratchet Gearing

Ratchet gearing may be used to transmit intermittent motion, or its only function may be to prevent the ratchet wheel from rotating backward. Ratchet gearing of this latter form is commonly used in connection with hoisting mechanisms of various kinds, to prevent the hoisting drum or shaft from rotating in a reverse direction under the action of the load.

Ratchet gearing in its simplest form consists of a toothed ratchet wheel *a* (see diagram *A*), and a pawl or detent *b*, and it may be used to transmit intermittent motion or to prevent relative motion between two parts except in one direction. The pawl *b* is pivoted to lever *c* which, when given an oscillating movement, imparts an intermittent rotary movement to ratchet wheel *a*. Diagram *B* illustrates another application of the ordinary ratchet and pawl mechanism. In this instance, the pawl is pivoted to a stationary member and its only function is to prevent the ratchet wheel from rotating backward. With the stationary design, illustrated at *C*, the pawl prevents the ratchet wheel from rotating in either direction, so long as it is in engagement with the wheel.

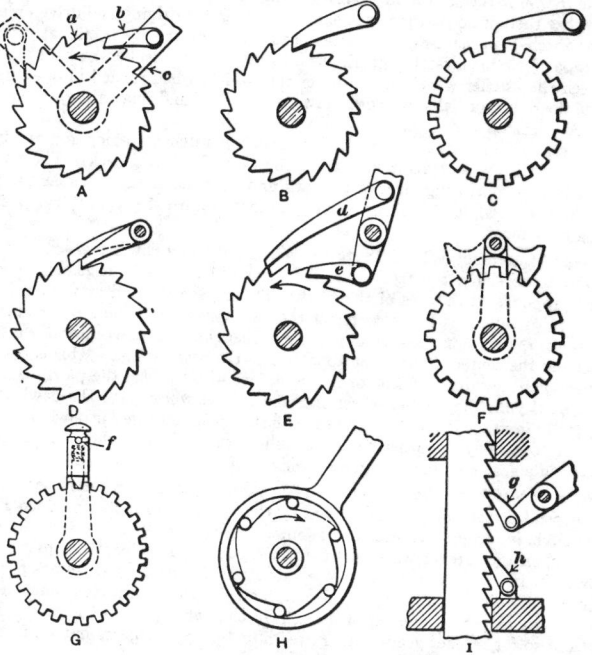

The principle of *multiple-pawl ratchet gearing* is illustrated at *D*, which shows the use of two pawls. One of these pawls is longer than the other, by an amount equal to one-half the pitch of the ratchet-wheel teeth, so that the practical effect is that of reducing the pitch one-half. By placing a number of driving pawls side by side and proportioning their lengths according to the pitch of the teeth, a very fine feed can be obtained with a ratchet wheel of comparatively coarse pitch.

This method of obtaining a fine feed from relatively coarse-pitch ratchets may be preferable to the use of single ratchets of fine pitch which, although providing the feed required, may have considerably weaker teeth.

The type of ratchet gearing shown at *E* is sometimes employed to impart a rotary movement to the ratchet wheel for both the forward and backward motions of the lever to which the two pawls are attached.

A simple form of *reversing ratchet* is illustrated at *F*. The teeth of the wheel are so shaped that either side may be used for driving by simply changing the position of the double-ended pawl, as indicated by the full and dotted lines.

Another form of reversible ratchet gearing for shapers is illustrated at *G*. The pawl, in this case, instead of being a pivoted latch, is in the form of a plunger which is free to move in the direction of its axis, but is normally held into engagement with the ratchet wheel by a small spring. When the pawl is lifted and turned one-half revolution, the driving face then engages the opposite sides of the teeth and the ratchet wheel is given an intermittent rotary motion in the opposite direction.

The *frictional type* of ratchet gearing differs from the designs previously referred to, in that there is no positive engagement between the driving and driven members of the ratchet mechanism, the motion being transmitted by frictional resistance. One type of frictional ratchet gearing is illustrated at *H*. Rollers or balls are placed between the ratchet wheel and an outer ring which, when turned in one direction, causes the rollers or balls to wedge between the wheel and ring as they move up the inclined edges of the teeth.

Diagram *I* illustrates one method of utilizing ratchet gearing for moving the driven member in a straight line, as in the case of a lifting jack. The pawl *g* is pivoted to the operating lever of the jack and does the lifting, whereas the pawl *h* holds the load while the lifting pawl *g* is being returned preparatory to another lifting movement.

**Shape of Ratchet Wheel Teeth.** — When designing ratchet gearing, it is important to so shape the teeth that the pawl will remain in engagement when a load is applied. The faces of the teeth which engage the end of the pawl should be in such relation with the center of the pawl pivot that a line perpendicular to the face of the engaging tooth will pass somewhere between the center of the ratchet wheel and the center of the pivot about which the pawl swings. This is true if the pawl *pushes* the ratchet wheel, or if the ratchet wheel *pushes* the pawl. However, if the pawl *pulls* the ratchet wheel or if the ratchet wheel *pulls* the pawl, the perpendicular from the face of the ratchet teeth should fall outside the pawl pivot center. Ratchet teeth may be either cut by a milling cutter having the correct angle, or hobbed in a gear-hobbing machine by the use of a special hob.

**Pitch of Ratchet Wheel Teeth.** — The pitch of ratchet wheels used for holding suspended loads may be calculated by the following formula, in which $P$ = circular pitch, in inches, measured at the outside circumference; $M$ = turning moment acting upon the ratchet wheel shaft, in inch-pounds; $L$ = length of tooth face, in inches (thickness of ratchet gear); $S$ = safe stress (for steel, 2500 pounds per square inch when subjected to shock, and 4000 pounds per square inch when not subjected to shock); $N$ = number of teeth in ratchet wheel; $F$ = a factor the value of which is 50 for ratchet gears with 12 teeth or less, 35 for gears having from 12 to 20 teeth, and 20 for gears having over 20 teeth:

$$P = \sqrt{\frac{FM}{LSN}}$$

This formula has been used in the calculation of ratchet gears for crane design.

# KEYS AND KEYSEATS

**ANSI Standard Keys and Keyseats.** — American National Standard, B17.1 Keys and Keyseats, based on current industry practice, was approved in 1967, and reaffirmed in 1973. This standard establishes a uniform relationship between shaft sizes and key sizes for parallel and taper keys as shown in Table 2. Other data in this standard are given in Tables 1 and 3-7. The sizes and tolerances shown are for single key applications only.

The following definitions are given in the standard:

*Key:* A demountable machinery part which, when assembled into keyseats, provides a positive means for transmitting torque between the shaft and hub.

*Keyseat:* An axially located rectangular groove in a shaft or hub.

This standard recognizes that there are two classes of stock for parallel keys used by industry. One is a close, plus toleranced key stock and the other is a broad, negative toleranced bar stock. Based on the use of two types of stock, two classes of fit are shown:

*Class 1:* A clearance of metal-to-metal side fit obtained by using bar stock keys and keyseat tolerances as given in Table 4. This is a relatively free fit and applies only to parallel keys.

*Class 2:* A side fit, with possible interference or clearance, obtained by using key stock and keyseat tolerances as given in Table 4. This is a relatively tight fit.

*Class 3:* This is an interference side fit and is not tabulated in Table 4 since the degree of interference has not been standardized. However, it is suggested that the top and bottom fit range given under Class 2 in Table 4, for parallel keys be used.

*Key Size vs. Shaft Diameter:* Shaft diameters are listed in Table 2 for identification of various key sizes and are not intended to establish shaft dimensions, tolerances or selections. For a stepped shaft, the size of a key is determined by the diameter of the shaft at the point of location of the key. Up through 6½-inch diameter shafts square keys are preferred; rectangular keys for larger shafts.

If special considerations dictate the use of a keyseat in the hub shallower than the preferred nominal depth shown, it is recommended that the tabulated preferred nominal standard keyseat be used in the shaft in all cases.

*Keyseat Alignment Tolerances:* A tolerance of 0.010 inch, max is provided for offset (due to parallel displacement of keyseat centerline from centerline of shaft or bore) of keyseats in shaft and bore. The following tolerances for maximum lead (due to angular displacement of keyseat centerline from centerline of shaft or bore and measured at right angles to the shaft or bore centerline) of keyseats in shaft and bore are specified: 0.002 inch for keyseat length up to and including 4 inches; 0.0005 inch per inch of length for keyseat lengths above 4 inches to and including 10 inches; and 0.005 inch for keyseat lengths above 10 inches.

**ANSI Standard Woodruff Keys and Keyseats.** — American National Standard B17.2 was approved in 1967, and reaffirmed in 1978. Data from this standard are shown in Tables 9, 10 and 11.

The following definitions are given in this standard:

*Woodruff Key:* A demountable machinery part which, when assembled into keyseats, provides a positive means for transmitting torque between the shaft and hub.

*Woodruff Key Number:* An identification number by which the size of key may be readily determined.

*Woodruff Keyseat — Shaft:* The circular pocket in which the key is retained.

*Woodruff Keyseat — Hub:* An axially located rectangular groove in a hub. (This has been referred to as a keyway.)

*Woodruff Keyseat Milling Cutter:* An arbor type or shank type milling cutter normally used for milling Woodruff keyseats in shafts.

**Table 1. ANSI Standard Plain and Gib Head Keys** (ANSI B17.1-1967, R1973)

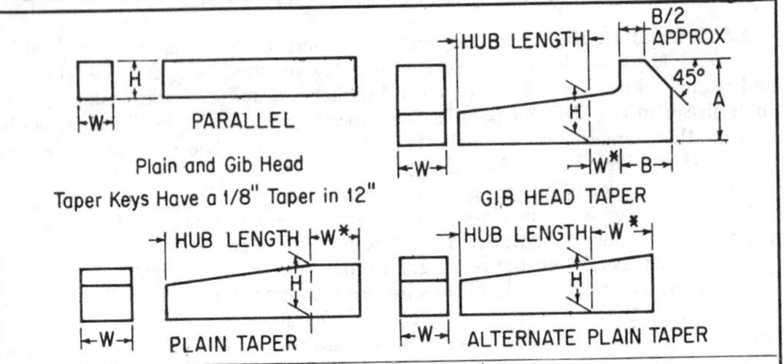


| Key | | | Nominal Key Size Width, W: Over | To (Incl.) | Tolerance Width, W | | Height, H | |
|---|---|---|---|---|---|---|---|---|
| Parallel | Square | Keystock | ... | 1 1/4 | +0.001 | −0.000 | +0.001 | −0.000 |
| | | | 1 1/4 | 3 | +0.002 | −0.000 | +0.002 | −0.000 |
| | | | 3 | 3 1/2 | +0.003 | −0.000 | +0.003 | −0.000 |
| | | Bar Stock | ... | 3/4 | +0.000 | −0.002 | +0.000 | −0.002 |
| | | | 3/4 | 1 1/2 | +0.000 | −0.003 | +0.000 | −0.003 |
| | | | 1 1/2 | 2 1/2 | +0.000 | −0.004 | +0.000 | −0.004 |
| | | | 2 1/2 | 3 1/2 | +0.000 | −0.006 | +0.000 | −0.006 |
| | Rectangular | Keystock | ... | 1 1/4 | +0.001 | −0.000 | +0.005 | −0.005 |
| | | | 1 1/4 | 3 | +0.002 | −0.000 | +0.005 | −0.005 |
| | | | 3 | 7 | +0.003 | −0.000 | +0.005 | −0.005 |
| | | Bar Stock | ... | 3/4 | +0.000 | −0.003 | +0.000 | −0.003 |
| | | | 3/4 | 1 1/2 | +0.000 | −0.004 | +0.000 | −0.004 |
| | | | 1 1/2 | 3 | +0.000 | −0.005 | +0.000 | −0.005 |
| | | | 3 | 4 | +0.000 | −0.006 | +0.000 | −0.006 |
| | | | 4 | 6 | +0.000 | −0.008 | +0.000 | −0.008 |
| | | | 6 | 7 | +0.000 | −0.013 | +0.000 | −0.013 |
| Taper | Plain or Gib Head Square or Rectangular | | ... | 1 1/4 | +0.001 | −0.000 | +0.005 | −0.000 |
| | | | 1 1/4 | 3 | +0.002 | −0.000 | +0.005 | −0.000 |
| | | | 3 | 7 | +0.003 | −0.000 | +0.005 | −0.000 |

GIB HEAD NOMINAL DIMENSIONS

| Nominal Key Size Width, W | Square H | Square A | Square B | Rectangular H | Rectangular A | Rectangular B |
|---|---|---|---|---|---|---|
| 1/8 | 1/8 | 1/4 | 1/4 | 3/32 | 3/16 | 1/8 |
| 3/16 | 3/16 | 5/16 | 5/16 | 1/8 | 1/4 | 1/4 |
| 1/4 | 1/4 | 7/16 | 3/8 | 3/16 | 5/16 | 5/16 |
| 5/16 | 5/16 | 1/2 | 7/16 | 1/4 | 7/16 | 3/8 |
| 3/8 | 3/8 | 5/8 | 1/2 | 1/4 | 7/16 | 3/8 |
| 1/2 | 1/2 | 7/8 | 5/8 | 3/8 | 5/8 | 1/2 |
| 5/8 | 5/8 | 1 | 3/4 | 7/16 | 3/4 | 9/16 |
| 3/4 | 3/4 | 1 1/4 | 7/8 | 1/2 | 7/8 | 5/8 |
| 7/8 | 7/8 | 1 3/8 | 1 | 5/8 | 1 | 3/4 |
| 1 | 1 | 1 5/8 | 1 1/8 | 3/4 | 1 1/4 | 7/8 |
| 1 1/4 | 1 1/4 | 2 | 1 7/16 | 7/8 | 1 3/8 | 1 |
| 1 1/2 | 1 1/2 | 2 3/8 | 1 3/4 | 1 | 1 5/8 | 1 1/8 |
| 1 3/4 | 1 3/4 | 2 3/4 | 2 | 1 1/2 | 2 3/8 | 1 3/4 |
| 2 | 2 | 3 1/2 | 2 1/4 | 1 1/2 | 2 3/8 | 1 3/4 |
| 2 1/2 | 2 1/2 | 4 | 3 | 1 3/4 | 2 3/4 | 2 |
| 3 | 3 | 5 | 3 1/2 | 2 | 3 1/2 | 2 1/4 |
| 3 1/2 | 3 1/2 | 6 | 4 | 2 1/2 | 4 | 3 |
| ... | ... | ... | ... | ... | ... | ... |

All dimensions are given in inches.

* For locating position of dimension $H$. Tolerance does not apply.

For larger sizes the following relationships are suggested as guides for establishing $A$ and $B$: $A = 1.8H$ and $B = 1.2H$.

**Table 2. Key Size Versus Shaft Diameter** (ANSI B17.1-1967, R1973)

| Nominal Shaft Diameter | | Nominal Key Size | | | Nominal Keyseat Depth | |
|---|---|---|---|---|---|---|
| | | Width, W | Height, H | | H/2 | |
| Over | To (Incl.) | | Square | Rectangular | Square | Rectangular |
| 5/16 | 7/16 | 3/32 | 3/32 | .... | 3/64 | .... |
| 7/16 | 9/16 | 1/8 | 1/8 | 3/32 | 1/16 | 3/64 |
| 9/16 | 7/8 | 3/16 | 3/16 | 1/8 | 3/32 | 1/16 |
| 7/8 | 1 1/4 | 1/4 | 1/4 | 3/16 | 1/8 | 3/32 |
| 1 1/4 | 1 3/8 | 5/16 | 5/16 | 1/4 | 5/32 | 1/8 |
| 1 3/8 | 1 3/4 | 3/8 | 3/8 | 1/4 | 3/16 | 1/8 |
| 1 3/4 | 2 1/4 | 1/2 | 1/2 | 3/8 | 1/4 | 3/16 |
| 2 1/4 | 2 3/4 | 5/8 | 5/8 | 7/16 | 5/16 | 7/32 |
| 2 3/4 | 3 1/4 | 3/4 | 3/4 | 1/2 | 3/8 | 1/4 |
| 3 1/4 | 3 3/4 | 7/8 | 7/8 | 5/8 | 7/16 | 5/16 |
| 3 3/4 | 4 1/2 | 1 | 1 | 3/4 | 1/2 | 3/8 |
| 4 1/2 | 5 1/2 | 1 1/4 | 1 1/4 | 7/8 | 5/8 | 7/16 |
| 5 1/2 | 6 1/2 | 1 1/2 | 1 1/2 | 1 | 3/4 | 1/2 |
| 6 1/2 | 7 1/2 | 1 3/4 | 1 3/4 | 1 1/2* | 7/8 | 3/4 |
| 7 1/2 | 9 | 2 | 2 | 1 1/2 | 1 | 3/4 |
| 9 | 11 | 2 1/2 | 2 1/2 | 1 3/4 | 1 1/4 | 7/8 |

All dimensions are given in inches. For larger shaft sizes, see ANSI Standard.
Square keys preferred for shaft diameters above heavy line; rectangular keys, below.
* Some key standards show 1 1/4 inches; preferred height is 1 1/2 inches.

**Table 3. Depth Control Values S and T for Shaft and Hub** (ANSI B17.1-1967, R1973)

| Nominal Shaft Diameter | Parallel and Taper | | Parallel | | Taper | |
|---|---|---|---|---|---|---|
| | Square | Rectangular | Square | Rectangular | Square | Rectangular |
| | S | S | T | T | T | T |
| 1/2 | 0.430 | 0.445 | 0.560 | 0.544 | 0.535 | 0.519 |
| 9/16 | 0.493 | 0.509 | 0.623 | 0.607 | 0.598 | 0.582 |
| 5/8 | 0.517 | 0.548 | 0.709 | 0.678 | 0.684 | 0.653 |
| 11/16 | 0.581 | 0.612 | 0.773 | 0.742 | 0.748 | 0.717 |
| 3/4 | 0.644 | 0.676 | 0.837 | 0.806 | 0.812 | 0.781 |
| 13/16 | 0.708 | 0.739 | 0.900 | 0.869 | 0.875 | 0.844 |
| 7/8 | 0.771 | 0.802 | 0.964 | 0.932 | 0.939 | 0.907 |
| 15/16 | 0.796 | 0.827 | 1.051 | 1.019 | 1.026 | 0.994 |
| 1 | 0.859 | 0.890 | 1.114 | 1.083 | 1.089 | 1.058 |
| 1 1/16 | 0.923 | 0.954 | 1.178 | 1.146 | 1.153 | 1.121 |
| 1 1/8 | 0.986 | 1.017 | 1.241 | 1.210 | 1.216 | 1.185 |
| 1 3/16 | 1.049 | 1.080 | 1.304 | 1.273 | 1.279 | 1.248 |
| 1 1/4 | 1.112 | 1.144 | 1.367 | 1.336 | 1.342 | 1.311 |
| 1 5/16 | 1.137 | 1.169 | 1.455 | 1.424 | 1.430 | 1.311 |
| 1 3/8 | 1.201 | 1.232 | 1.518 | 1.487 | 1.493 | 1.462 |
| 1 7/16 | 1.225 | 1.288 | 1.605 | 1.543 | 1.580 | 1.518 |
| 1 1/2 | 1.289 | 1.351 | 1.669 | 1.606 | 1.644 | 1.581 |
| 1 9/16 | 1.352 | 1.415 | 1.732 | 1.670 | 1.707 | 1.645 |
| 1 5/8 | 1.416 | 1.478 | 1.796 | 1.733 | 1.771 | 1.708 |
| 1 11/16 | 1.479 | 1.541 | 1.859 | 1.796 | 1.834 | 1.771 |
| 1 3/4 | 1.542 | 1.605 | 1.922 | 1.860 | 1.897 | 1.835 |
| 1 13/16 | 1.527 | 1.590 | 2.032 | 1.970 | 2.007 | 1.945 |
| 1 7/8 | 1.591 | 1.654 | 2.096 | 2.034 | 2.071 | 2.009 |
| 1 15/16 | 1.655 | 1.717 | 2.160 | 2.097 | 2.135 | 2.072 |
| 2 | 1.718 | 1.781 | 2.223 | 2.161 | 2.198 | 2.136 |

All dimensions are given in inches. See Table 4 for tolerances.

**Table 3.** *(Concluded).* **Depth Control Values *S* and *T* for Shaft and Hub** (ANSI B17.1-1967, R1973)

| Nominal Shaft Diameter | Parallel and Taper | | Parallel | | Taper | |
|---|---|---|---|---|---|---|
| | Square | Rectangular | Square | Rectangular | Square | Rectangular |
| | *S* | *S* | *T* | *T* | *T* | *T* |
| 2 1/16 | 1.782 | 1.844 | 2.287 | 2.224 | 2.262 | 2.199 |
| 2 1/8 | 1.845 | 1.908 | 2.350 | 2.288 | 2.325 | 2.263 |
| 2 3/16 | 1.909 | 1.971 | 2.414 | 2.351 | 2.389 | 2.326 |
| 2 1/4 | 1.972 | 2.034 | 2.477 | 2.414 | 2.452 | 2.389 |
| 2 5/16 | 1.957 | 2.051 | 2.587 | 2.493 | 2.562 | 2.468 |
| 2 3/8 | 2.021 | 2.114 | 2.651 | 2.557 | 2.626 | 2.532 |
| 2 7/16 | 2.084 | 2.178 | 2.714 | 2.621 | 2.689 | 2.596 |
| 2 1/2 | 2.148 | 2.242 | 2.778 | 2.684 | 2.753 | 2.659 |
| 2 9/16 | 2.211 | 2.305 | 2.841 | 2.748 | 2.816 | 2.723 |
| 2 5/8 | 2.275 | 2.369 | 2.905 | 2.811 | 2.880 | 2.786 |
| 2 11/16 | 2.338 | 2.432 | 2.968 | 2.874 | 2.943 | 2.849 |
| 2 3/4 | 2.402 | 2.495 | 3.032 | 2.938 | 3.007 | 2.913 |
| 2 13/16 | 2.387 | 2.512 | 3.142 | 3.017 | 3.117 | 2.992 |
| 2 7/8 | 2.450 | 2.575 | 3.205 | 3.080 | 3.180 | 3.055 |
| 2 15/16 | 2.514 | 2.639 | 3.269 | 3.144 | 3.244 | 3.119 |
| 3 | 2.577 | 2.702 | 3.332 | 3.207 | 3.307 | 3.182 |
| 3 1/16 | 2.641 | 2.766 | 3.396 | 3.271 | 3.371 | 3.246 |
| 3 1/8 | 2.704 | 2.829 | 3.459 | 3.334 | 3.434 | 3.309 |
| 3 3/16 | 2.768 | 2.893 | 3.523 | 3.398 | 3.498 | 3.373 |
| 3 1/4 | 2.831 | 2.956 | 3.586 | 3.461 | 3.561 | 3.436 |
| 3 5/16 | 2.816 | 2.941 | 3.696 | 3.571 | 3.671 | 3.546 |
| 3 3/8 | 2.880 | 3.005 | 3.760 | 3.635 | 3.735 | 3.610 |
| 3 7/16 | 2.943 | 3.068 | 3.823 | 3.698 | 3.798 | 3.673 |
| 3 1/2 | 3.007 | 3.132 | 3.887 | 3.762 | 3.862 | 3.737 |
| 3 9/16 | 3.070 | 3.195 | 3.950 | 3.825 | 3.925 | 3.800 |
| 3 5/8 | 3.134 | 3.259 | 4.014 | 3.889 | 3.989 | 3.864 |
| 3 11/16 | 3.197 | 3.322 | 4.077 | 3.952 | 4.052 | 3.927 |
| 3 3/4 | 3.261 | 3.386 | 4.141 | 4.016 | 4.116 | 3.991 |
| 3 13/16 | 3.246 | 3.371 | 4.251 | 4.126 | 4.226 | 4.101 |
| 3 7/8 | 3.309 | 3.434 | 4.314 | 4.189 | 4.289 | 4.164 |
| 3 15/16 | 3.373 | 3.498 | 4.378 | 4.253 | 4.353 | 4.228 |
| 4 | 3.436 | 3.561 | 4.441 | 4.316 | 4.416 | 4.291 |
| 4 3/16 | 3.627 | 3.752 | 4.632 | 4.507 | 4.607 | 4.482 |
| 4 1/4 | 3.690 | 3.815 | 4.695 | 4.570 | 4.670 | 4.545 |
| 4 3/8 | 3.817 | 3.942 | 4.822 | 4.697 | 4.797 | 4.672 |
| 4 7/16 | 3.880 | 4.005 | 4.885 | 4.760 | 4.860 | 4.735 |
| 4 1/2 | 3.944 | 4.069 | 4.949 | 4.824 | 4.924 | 4.799 |
| 4 3/4 | 4.041 | 4.229 | 5.296 | 5.109 | 5.271 | 5.084 |
| 4 7/8 | 4.169 | 4.356 | 5.424 | 5.236 | 5.399 | 5.211 |
| 4 15/16 | 4.232 | 4.422 | 5.487 | 5.300 | 5.462 | 5.275 |
| 5 | 4.296 | 4.483 | 5.551 | 5.363 | 5.526 | 5.338 |
| 5 3/16 | 4.486 | 4.674 | 5.741 | 5.554 | 5.716 | 5.529 |
| 5 1/4 | 4.550 | 4.737 | 5.805 | 5.617 | 5.780 | 5.592 |
| 5 7/16 | 4.740 | 4.927 | 5.995 | 5.807 | 5.970 | 5.782 |
| 5 1/2 | 4.803 | 4.991 | 6.058 | 5.871 | 6.033 | 5.846 |
| 5 3/4 | 4.900 | 5.150 | 6.405 | 6.155 | 6.380 | 6.130 |
| 5 15/16 | 5.091 | 5.341 | 6.596 | 6.346 | 6.571 | 6.321 |
| 6 | 5.155 | 5.405 | 6.660 | 6.410 | 6.635 | 6.385 |
| 6 1/4 | 5.409 | 5.659 | 6.914 | 6.664 | 6.889 | 6.639 |
| 6 1/2 | 5.662 | 5.912 | 7.167 | 6.917 | 7.142 | 6.892 |
| 6 3/4 | 5.760 | *5.885 | 7.515 | *7.390 | 7.490 | *7.365 |
| 7 | 6.014 | *6.139 | 7.769 | *7.644 | 7.744 | *7.619 |
| 7 1/4 | 6.268 | *6.393 | 8.023 | *7.898 | 7.998 | *7.873 |
| 7 1/2 | 6.521 | *6.646 | 8.276 | *8.151 | 8.251 | *8.126 |
| 7 3/4 | 6.619 | 6.869 | 8.624 | 8.374 | 8.599 | 8.349 |
| 8 | 6.873 | 7.123 | 8.878 | 8.628 | 8.853 | 8.603 |
| 9 | 7.887 | 8.137 | 9.892 | 9.642 | 9.867 | 9.617 |
| 10 | 8.591 | 8.966 | 11.096 | 10.721 | 11.071 | 10.696 |
| 11 | 9.606 | 9.981 | 12.111 | 11.736 | 12.086 | 11.711 |
| 12 | 10.309 | 10.809 | 13.314 | 12.814 | 13.289 | 12.789 |
| 13 | 11.325 | 11.825 | 14.330 | 13.830 | 14.305 | 13.805 |
| 14 | 12.028 | 12.528 | 15.533 | 15.033 | 15.508 | 15.008 |
| 15 | 13.043 | 13.543 | 16.548 | 16.048 | 16.523 | 16.023 |

All dimensions given in inches. See Table 4 for tolerances.

* 1 3/4 × 1 1/2 inch key.

**Table 4. ANSI Standard Fits for Parallel and Taper Keys** (ANSI B17.1-1967, R1973)

| Type of Key | Key Width | | Side Fit | | | Top and Bottom Fit | | | |
|---|---|---|---|---|---|---|---|---|---|
| | Over | To (Incl.) | Width Tolerance | | Fit Range* | Depth Tolerance | | | Fit Range* |
| | | | Key | Key-Seat | | Key | Shaft Key-Seat | Hub Key-Seat | |
| *Class 1 Fit for Parallel Keys* | | | | | | | | | |
| Square | ... | ½ | +0.000 −0.002 | +0.002 −0.000 | 0.004 CL 0.000 | +0.000 −0.002 | +0.000 −0.015 | +0.010 −0.000 | 0.032 CL 0.005 CL |
| | ½ | ¾ | +0.000 −0.002 | +0.003 −0.000 | 0.005 CL 0.000 | +0.000 −0.002 | +0.000 −0.015 | +0.010 −0.000 | 0.032 CL 0.005 CL |
| | ¾ | 1 | +0.000 −0.003 | +0.003 −0.000 | 0.006 CL 0.000 | +0.000 −0.003 | +0.000 −0.015 | +0.010 −0.000 | 0.033 CL 0.005 CL |
| | 1 | 1½ | +0.000 −0.003 | +0.004 −0.000 | 0.007 CL 0.000 | +0.000 −0.003 | +0.000 −0.015 | +0.010 −0.000 | 0.033 CL 0.005 CL |
| | 1½ | 2½ | +0.000 −0.004 | +0.004 −0.000 | 0.008 CL 0.000 | +0.000 −0.004 | +0.000 −0.015 | +0.010 −0.000 | 0.034 CL 0.005 CL |
| | 2½ | 3½ | +0.000 −0.006 | +0.004 −0.000 | 0.010 CL 0.000 | +0.000 −0.006 | +0.000 −0.015 | +0.010 −0.000 | 0.036 CL 0.005 CL |
| Rectangular | ... | ½ | +0.000 −0.003 | +0.002 −0.000 | 0.005 CL 0.000 | +0.000 −0.003 | +0.000 −0.015 | +0.010 −0.000 | 0.033 CL 0.005 CL |
| | ½ | ¾ | +0.000 −0.003 | +0.003 −0.000 | 0.006 CL 0.000 | +0.000 −0.003 | +0.000 −0.015 | +0.010 −0.000 | 0.033 CL 0.005 CL |
| | ¾ | 1 | +0.000 −0.004 | +0.003 −0.000 | 0.007 CL 0.000 | +0.000 −0.004 | +0.000 −0.015 | +0.010 −0.000 | 0.034 CL 0.005 CL |
| | 1 | 1½ | +0.000 −0.004 | +0.004 −0.000 | 0.008 CL 0.000 | +0.000 −0.004 | +0.000 −0.015 | +0.010 −0.000 | 0.034 CL 0.005 CL |
| | 1½ | 3 | +0.000 −0.005 | +0.004 −0.000 | 0.009 CL 0.000 | +0.000 −0.005 | +0.000 −0.015 | +0.010 −0.000 | 0.035 CL 0.005 CL |
| | 3 | 4 | +0.000 −0.006 | +0.004 −0.000 | 0.010 CL 0.000 | +0.000 −0.006 | +0.000 −0.015 | +0.010 −0.000 | 0.036 CL 0.005 CL |
| | 4 | 6 | +0.000 −0.008 | +0.004 −0.000 | 0.012 CL 0.000 | +0.000 −0.008 | +0.000 −0.015 | +0.010 −0.000 | 0.038 CL 0.005 CL |
| | 6 | 7 | +0.000 −0.013 | +0.004 −0.000 | 0.017 CL 0.000 | +0.000 −0.013 | +0.000 −0.015 | +0.010 −0.000 | 0.043 CL 0.005 CL |
| *Class 2 Fit for Parallel and Taper Keys* | | | | | | | | | |
| Parallel Square | ... | 1¼ | +0.001 −0.000 | +0.002 −0.000 | 0.002 CL 0.001 INT | +0.001 −0.000 | +0.000 −0.015 | +0.010 −0.000 | 0.030 CL 0.004 CL |
| | 1¼ | 3 | +0.002 −0.000 | +0.002 −0.000 | 0.002 CL 0.002 INT | +0.002 −0.000 | +0.000 −0.015 | +0.010 −0.000 | 0.030 CL 0.003 CL |
| | 3 | 3½ | +0.003 −0.000 | +0.002 −0.000 | 0.002 CL 0.003 INT | +0.003 −0.000 | +0.000 −0.015 | +0.010 −0.000 | 0.030 CL 0.002 CL |
| Parallel Rectangular | ... | 1¼ | +0.001 −0.000 | +0.002 −0.000 | 0.002 CL 0.001 INT | +0.005 −0.005 | +0.000 −0.015 | +0.010 −0.000 | 0.035 CL 0.000 CL |
| | 1¼ | 3 | +0.002 −0.000 | +0.002 −0.000 | 0.002 CL 0.002 INT | +0.005 −0.005 | +0.000 −0.015 | +0.010 −0.000 | 0.035 CL 0.000 CL |
| | 3 | 7 | +0.003 −0.000 | +0.002 −0.000 | 0.002 CL 0.003 INT | +0.005 −0.005 | +0.000 −0.015 | +0.010 −0.000 | 0.035 CL 0.000 CL |
| Taper | ... | 1¼ | +0.001 −0.000 | +0.002 −0.000 | 0.002 CL 0.001 INT | +0.005 −0.005 | +0.000 −0.015 | +0.010 −0.000 | 0.005 CL 0.025 INT |
| | 1¼ | 3 | +0.002 −0.000 | +0.002 −0.000 | 0.002 CL 0.002 INT | +0.005 −0.005 | +0.000 −0.015 | +0.010 −0.000 | 0.005 CL 0.025 INT |
| | 3 | § | +0.003 −0.000 | +0.002 −0.000 | 0.002 CL 0.003 INT | +0.005 −0.005 | +0.000 −0.015 | +0.010 −0.000 | 0.005 CL 0.025 INT |

All dimensions are given in inches.  See also text on page 967.
* Limits of variation.  CL = Clearance;  INT = Interference.
§ To (Incl.) 3½-inch Square and 7-inch Rectangular key widths.

**Chamfered Keys and Filleted Keyseats.** — In general practice, chamfered keys and filleted keyseats are not used. However, it is recognized that fillets in keyseats decrease stress concentrations at corners. When used, fillet radii should be as large as possible without causing excessive bearing stresses due to reduced contact area between the key and its mating parts. Keys must be chamfered or rounded to clear fillet radii. Values in Table 5 assume general conditions and should be used only as a guide when critical stresses are encountered.

**Table 5. Suggested Keyseat Fillet Radius and Key Chamfer** (ANSI B17.1-1967, R1973)

| Keyseat Depth, $H/2$ | | Fillet Radius | 45 deg. Chamfer | Keyseat Depth, $H/2$ | | Fillet Radius | 45 deg. Chamfer |
|---|---|---|---|---|---|---|---|
| Over | To (Incl.) | | | Over | To (Incl.) | | |
| 1/8 | 1/4 | 1/32 | 3/64 | 7/8 | 1 1/4 | 3/16 | 7/32 |
| 1/4 | 1/2 | 1/16 | 5/64 | 1 1/4 | 1 3/4 | 1/4 | 9/32 |
| 1/2 | 7/8 | 1/8 | 5/32 | 1 3/4 | 2 1/2 | 3/8 | 13/32 |

All dimensions are given in inches.

**Table 6. ANSI Standard Keyseat Tolerances for Electric Motor and Generator Shaft Extensions** (ANSI B17.1-1967, R1973)

| Keyseat Width | | Width Tolerance | Depth Tolerance |
|---|---|---|---|
| Over | To (Incl.) | | |
| ... | 1/4 | +0.001 −0.001 | +0.000 −0.015 |
| 1/4 | 3/4 | +0.000 −0.002 | +0.000 −0.015 |
| 3/4 | 1 1/4 | +0.000 −0.003 | +0.000 −0.015 |

All dimensions are given in inches.

**Table 7. Set Screws for Use Over Keys\*** (ANSI B17.1-1967, R1973)

| Nom. Shaft Diam. | | Nom. Key Width | Set Screw Diam. | Nom. Shaft Diam. | | Nom. Key Width | Set Screw Diam. |
|---|---|---|---|---|---|---|---|
| Over | To (Incl.) | | | Over | To (Incl.) | | |
| 5/16 | 7/16 | 3/32 | No. 10 | 2 1/4 | 2 3/4 | 5/8 | 1/2 |
| 7/16 | 9/16 | 1/8 | No. 10 | 2 3/4 | 3 1/4 | 3/4 | 5/8 |
| 9/16 | 7/8 | 3/16 | 1/4 | 3 1/4 | 3 3/4 | 7/8 | 3/4 |
| 7/8 | 1 1/4 | 1/4 | 5/16 | 3 3/4 | 4 1/2 | 1 | 3/4 |
| 1 1/4 | 1 3/8 | 5/16 | 3/8 | 4 1/2 | 5 1/2 | 1 1/4 | 7/8 |
| 1 3/8 | 1 3/4 | 3/8 | 3/8 | 5 1/2 | 6 1/2 | 1 1/2 | 1 |
| 1 3/4 | 2 1/4 | 1/2 | 1/2 | ... | ... | ... | ... |

All dimensions are given in inches.
\* These set screw diameter selections are offered as a guide but their use should be dependent upon design considerations.

### Table 8. Finding Depth of Keyseat and Distance from Top of Key to Bottom of Shaft

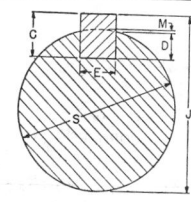

For milling keyseats, the total depth to feed cutter in from outside of shaft to bottom of keyseat is $M + D$, where $D$ is depth of keyseat.

For checking an assembled key and shaft, caliper measurement $J$ between top of key and bottom of shaft is used. $J = S - (M + D) + C$, where $C$ is depth of key. For Woodruff keys, dimensions $C$ and $D$ can be found in Tables 9 to 11. Assuming shaft diameter $S$ is nominal size, the tolerances on dimension $J$ for Woodruff keys in keyslots are +0.000, −0.010 inch.

| Diam. of Shaft S, Inches | Width of Keyseat, E | | | | | | | | | | | | | | |
|---|---|---|---|---|---|---|---|---|---|---|---|---|---|---|---|
| | 1⁄16 | 3⁄32 | 1⁄8 | 5⁄32 | 3⁄16 | 7⁄32 | 1⁄4 | 5⁄16 | 3⁄8 | 7⁄16 | 1⁄2 | 9⁄16 | 5⁄8 | 11⁄16 | 3⁄4 |
| | Dimension M, Inch | | | | | | | | | | | | | | |
| 0.3125 | .0032 | ... | ... | ... | ... | ... | ... | ... | ... | ... | ... | ... | ... | ... | ... |
| 0.3437 | .0029 | .0065 | ... | ... | ... | ... | ... | ... | ... | ... | ... | ... | ... | ... | ... |
| 0.3750 | .0026 | .0060 | .0107 | ... | ... | ... | ... | ... | ... | ... | ... | ... | ... | ... | ... |
| 0.4060 | .0024 | .0055 | .0099 | ... | ... | ... | ... | ... | ... | ... | ... | ... | ... | ... | ... |
| 0.4375 | .0022 | .0051 | .0091 | ... | ... | ... | ... | ... | ... | ... | ... | ... | ... | ... | ... |
| 0.4687 | .0021 | .0047 | .0085 | .0134 | ... | ... | ... | ... | ... | ... | ... | ... | ... | ... | ... |
| 0.5000 | .0020 | .0044 | .0079 | .0125 | ... | ... | ... | ... | ... | ... | ... | ... | ... | ... | ... |
| 0.5625 | ... | .0039 | .0070 | .0111 | .0161 | ... | ... | ... | ... | ... | ... | ... | ... | ... | ... |
| 0.6250 | ... | .0035 | .0063 | .0099 | .0144 | .0198 | ... | ... | ... | ... | ... | ... | ... | ... | ... |
| 0.6875 | ... | .0032 | .0057 | .0090 | .0130 | .0179 | .0235 | ... | ... | ... | ... | ... | ... | ... | ... |
| 0.7500 | ... | .0029 | .0052 | .0082 | .0119 | .0163 | .0214 | .0341 | ... | ... | ... | ... | ... | ... | ... |
| 0.8125 | ... | .0027 | .0048 | .0076 | .0110 | .0150 | .0197 | .0312 | ... | ... | ... | ... | ... | ... | ... |
| 0.8750 | ... | .0025 | .0045 | .0070 | .0102 | .0139 | .0182 | .0288 | ... | ... | ... | ... | ... | ... | ... |
| 0.9375 | ... | ... | .0042 | .0066 | .0095 | .0129 | .0170 | .0268 | .0391 | ... | ... | ... | ... | ... | ... |
| 1.0000 | ... | ... | .0039 | .0061 | .0089 | .0121 | .0159 | .0250 | .0365 | ... | ... | ... | ... | ... | ... |
| 1.0625 | ... | ... | .0037 | .0058 | .0083 | .0114 | .0149 | .0235 | .0342 | ... | ... | ... | ... | ... | ... |
| 1.1250 | ... | ... | .0035 | .0055 | .0079 | .0107 | .0141 | .0221 | .0322 | .0443 | ... | ... | ... | ... | ... |
| 1.1875 | ... | ... | .0033 | .0052 | .0074 | .0102 | .0133 | .0209 | .0304 | .0418 | ... | ... | ... | ... | ... |
| 1.2500 | ... | ... | .0031 | .0049 | .0071 | .0097 | .0126 | .0198 | .0288 | .0395 | ... | ... | ... | ... | ... |
| 1.3750 | ... | ... | ... | .0045 | .0064 | .0088 | .0115 | .0180 | .0261 | .0357 | .0471 | ... | ... | ... | ... |
| 1.5000 | ... | ... | ... | .0041 | .0059 | .0080 | .0105 | .0165 | .0238 | .0326 | .0429 | ... | ... | ... | ... |
| 1.6250 | ... | ... | ... | .0038 | .0054 | .0074 | .0097 | .0152 | .0219 | .0300 | .0394 | .0502 | ... | ... | ... |
| 1.7500 | ... | ... | ... | ... | .0050 | .0069 | .0090 | .0141 | .0203 | .0278 | .0365 | .0464 | ... | ... | ... |
| 1.8750 | ... | ... | ... | ... | .0047 | .0064 | .0084 | .0131 | .0189 | .0259 | .0340 | .0432 | .0536 | ... | ... |
| 2.0000 | ... | ... | ... | ... | .0044 | .0060 | .0078 | .0123 | .0177 | .0242 | .0318 | .0404 | .0501 | ... | ... |
| 2.1250 | ... | ... | ... | ... | ... | .0056 | .0074 | .0116 | .0167 | .0228 | .0298 | .0379 | .0470 | .0572 | .0684 |
| 2.2500 | ... | ... | ... | ... | ... | ... | .0070 | .0109 | .0157 | .0215 | .0281 | .0357 | .0443 | .0538 | .0643 |
| 2.3750 | ... | ... | ... | ... | ... | ... | ... | .0103 | .0149 | .0203 | .0266 | .0338 | .0419 | .0509 | .0608 |
| 2.5000 | ... | ... | ... | ... | ... | ... | ... | ... | .0141 | .0193 | .0253 | .0321 | .0397 | .0482 | .0576 |
| 2.6250 | ... | ... | ... | ... | ... | ... | ... | ... | .0135 | .0184 | .0240 | .0305 | .0377 | .0457 | .0547 |
| 2.7500 | ... | ... | ... | ... | ... | ... | ... | ... | ... | .0175 | .0229 | .0291 | .0360 | .0437 | .0521 |
| 2.8750 | ... | ... | ... | ... | ... | ... | ... | ... | ... | .0168 | .0219 | .0278 | .0344 | .0417 | .0498 |
| 3.0000 | ... | ... | ... | ... | ... | ... | ... | ... | ... | ... | .0210 | .0266 | .0329 | .0399 | .0476 |

**Depths for Milling Keyseats.** — The above table has been compiled to facilitate the accurate milling of keyseats. This table gives the distance $M$ (see illustration accompanying table) between the top of the shaft and a line passing through the upper corners or edges of the keyseat. Dimension $M$ is calculated by the formula: $M = \frac{1}{2}(S - \sqrt{S^2 - E^2})$ where $S$ is diameter of shaft and $E$ is width of keyseat. A simple approximate formula that gives $M$ to within 0.001 inch is: $M = E^2 \div 4S$.

## Table 9. ANSI Standard Woodruff Keys (ANSI B17.2-1967, R1978)

| Key No. | Nominal Key Size $W \times B$ | Actual Length $F$ +0.000 -0.010 | Height of Key | | | | Distance Below Center $E$ |
|---|---|---|---|---|---|---|---|
| | | | $C$ Max. | $C$ Min. | $D$ Max. | $D$ Min. | |
| 202 | 1/16 × 1/4 | 0.248 | 0.109 | 0.104 | 0.109 | 0.104 | 1/64 |
| 202.5 | 1/16 × 5/16 | 0.311 | 0.140 | 0.135 | 0.140 | 0.135 | 1/64 |
| 302.5 | 3/32 × 5/16 | 0.311 | 0.140 | 0.135 | 0.140 | 0.135 | 1/64 |
| 203 | 1/16 × 3/8 | 0.374 | 0.172 | 0.167 | 0.172 | 0.167 | 1/64 |
| 303 | 3/32 × 3/8 | 0.374 | 0.172 | 0.167 | 0.172 | 0.167 | 1/64 |
| 403 | 1/8 × 3/8 | 0.374 | 0.172 | 0.167 | 0.172 | 0.167 | 1/64 |
| 204 | 1/16 × 1/2 | 0.491 | 0.203 | 0.198 | 0.194 | 0.188 | 3/64 |
| 304 | 3/32 × 1/2 | 0.491 | 0.203 | 0.198 | 0.194 | 0.188 | 3/64 |
| 404 | 1/8 × 1/2 | 0.491 | 0.203 | 0.198 | 0.194 | 0.188 | 3/64 |
| 305 | 3/32 × 5/8 | 0.612 | 0.250 | 0.245 | 0.240 | 0.234 | 1/16 |
| 405 | 1/8 × 5/8 | 0.612 | 0.250 | 0.245 | 0.240 | 0.234 | 1/16 |
| 505 | 5/32 × 5/8 | 0.612 | 0.250 | 0.245 | 0.240 | 0.234 | 1/16 |
| 605 | 3/16 × 5/8 | 0.612 | 0.250 | 0.245 | 0.240 | 0.234 | 1/16 |
| 406 | 1/8 × 3/4 | 0.740 | 0.313 | 0.308 | 0.303 | 0.297 | 1/16 |
| 506 | 5/32 × 3/4 | 0.740 | 0.313 | 0.308 | 0.303 | 0.297 | 1/16 |
| 606 | 3/16 × 3/4 | 0.740 | 0.313 | 0.308 | 0.303 | 0.297 | 1/16 |
| 806 | 1/4 × 3/4 | 0.740 | 0.313 | 0.308 | 0.303 | 0.297 | 1/16 |
| 507 | 5/32 × 7/8 | 0.866 | 0.375 | 0.370 | 0.365 | 0.359 | 1/16 |
| 607 | 3/16 × 7/8 | 0.866 | 0.375 | 0.370 | 0.365 | 0.359 | 1/16 |
| 707 | 7/32 × 7/8 | 0.866 | 0.375 | 0.370 | 0.365 | 0.359 | 1/16 |
| 807 | 1/4 × 7/8 | 0.866 | 0.375 | 0.370 | 0.365 | 0.359 | 1/16 |
| 608 | 3/16 × 1 | 0.992 | 0.438 | 0.433 | 0.428 | 0.422 | 1/16 |
| 708 | 7/32 × 1 | 0.992 | 0.438 | 0.433 | 0.428 | 0.422 | 1/16 |
| 808 | 1/4 × 1 | 0.992 | 0.438 | 0.433 | 0.428 | 0.422 | 1/16 |
| 1008 | 5/16 × 1 | 0.992 | 0.438 | 0.433 | 0.428 | 0.422 | 1/16 |
| 1208 | 3/8 × 1 | 0.992 | 0.438 | 0.433 | 0.428 | 0.422 | 1/16 |
| 609 | 3/16 × 1 1/8 | 1.114 | 0.484 | 0.479 | 0.475 | 0.469 | 5/64 |
| 709 | 7/32 × 1 1/8 | 1.114 | 0.484 | 0.479 | 0.475 | 0.469 | 5/64 |
| 809 | 1/4 × 1 1/8 | 1.114 | 0.484 | 0.479 | 0.475 | 0.469 | 5/64 |
| 1009 | 5/16 × 1 1/8 | 1.114 | 0.484 | 0.479 | 0.475 | 0.469 | 5/64 |
| 610 | 3/16 × 1 1/4 | 1.240 | 0.547 | 0.542 | 0.537 | 0.531 | 5/64 |
| 710 | 7/32 × 1 1/4 | 1.240 | 0.547 | 0.542 | 0.537 | 0.531 | 5/64 |
| 810 | 1/4 × 1 1/4 | 1.240 | 0.547 | 0.542 | 0.537 | 0.531 | 5/64 |
| 1010 | 5/16 × 1 1/4 | 1.240 | 0.547 | 0.542 | 0.537 | 0.531 | 5/64 |
| 1210 | 3/8 × 1 1/4 | 1.240 | 0.547 | 0.542 | 0.537 | 0.531 | 5/64 |
| 811 | 1/4 × 1 3/8 | 1.362 | 0.594 | 0.589 | 0.584 | 0.578 | 3/32 |
| 1011 | 5/16 × 1 3/8 | 1.362 | 0.594 | 0.589 | 0.584 | 0.578 | 3/32 |
| 1211 | 3/8 × 1 3/8 | 1.362 | 0.594 | 0.589 | 0.584 | 0.578 | 3/32 |
| 812 | 1/4 × 1 1/2 | 1.484 | 0.641 | 0.636 | 0.631 | 0.625 | 7/64 |
| 1012 | 5/16 × 1 1/2 | 1.484 | 0.641 | 0.636 | 0.631 | 0.625 | 7/64 |
| 1212 | 3/8 × 1 1/2 | 1.484 | 0.641 | 0.636 | 0.631 | 0.625 | 7/64 |

All dimensions are given in inches.

The key numbers indicate nominal key dimensions. The last two digits give the nominal diameter $B$ in eighths of an inch and the digits preceding the last two give the nominal width $W$ in thirty-seconds of an inch.

Table 10. **ANSI Standard Woodruff Keys** (ANSI B17.2-1967, R1978)

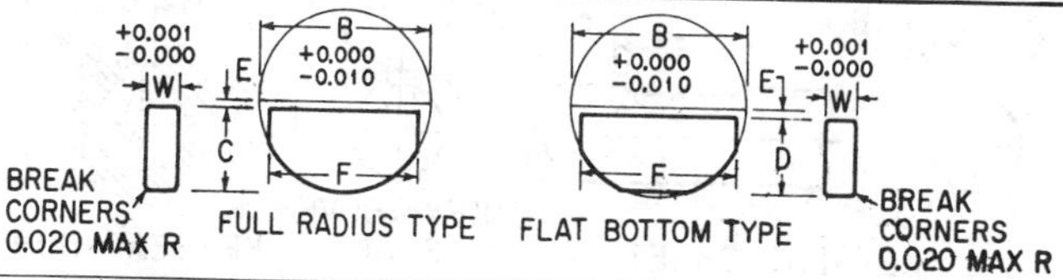


| Key No. | Nominal Key Size $W \times B$ | Actual Length $F$ +0.000 −0.010 | Height of Key C Max. | Height of Key C Min. | Height of Key D Max. | Height of Key D Min. | Distance Below Center $E$ |
|---|---|---|---|---|---|---|---|
| 617-1 | 3/16 × 2 1/8 | 1.380 | 0.406 | 0.401 | 0.396 | 0.390 | 21/32 |
| 817-1 | 1/4 × 2 1/8 | 1.380 | 0.406 | 0.401 | 0.396 | 0.390 | 21/32 |
| 1017-1 | 5/16 × 2 1/8 | 1.380 | 0.406 | 0.401 | 0.396 | 0.390 | 21/32 |
| 1217-1 | 3/8 × 2 1/8 | 1.380 | 0.406 | 0.401 | 0.396 | 0.390 | 21/32 |
| 617 | 3/16 × 2 1/8 | 1.723 | 0.531 | 0.526 | 0.521 | 0.515 | 17/32 |
| 817 | 1/4 × 2 1/8 | 1.723 | 0.531 | 0.526 | 0.521 | 0.515 | 17/32 |
| 1017 | 5/16 × 2 1/8 | 1.723 | 0.531 | 0.526 | 0.521 | 0.515 | 17/32 |
| 1217 | 3/8 × 2 1/8 | 1.723 | 0.531 | 0.526 | 0.521 | 0.515 | 17/32 |
| 822-1 | 1/4 × 2 3/4 | 2.000 | 0.594 | 0.589 | 0.584 | 0.578 | 25/32 |
| 1022-1 | 5/16 × 2 3/4 | 2.000 | 0.594 | 0.589 | 0.584 | 0.578 | 25/32 |
| 1222-1 | 3/8 × 2 3/4 | 2.000 | 0.594 | 0.589 | 0.584 | 0.578 | 25/32 |
| 1422-1 | 7/16 × 2 3/4 | 2.000 | 0.594 | 0.589 | 0.584 | 0.578 | 25/32 |
| 1622-1 | 1/2 × 2 3/4 | 2.000 | 0.594 | 0.589 | 0.584 | 0.578 | 25/32 |
| 822 | 1/4 × 2 3/4 | 2.317 | 0.750 | 0.745 | 0.740 | 0.734 | 5/8 |
| 1022 | 5/16 × 2 3/4 | 2.317 | 0.750 | 0.745 | 0.740 | 0.734 | 5/8 |
| 1222 | 3/8 × 2 3/4 | 2.317 | 0.750 | 0.745 | 0.740 | 0.734 | 5/8 |
| 1422 | 7/16 × 2 3/4 | 2.317 | 0.750 | 0.745 | 0.740 | 0.734 | 5/8 |
| 1622 | 1/2 × 2 3/4 | 2.317 | 0.750 | 0.745 | 0.740 | 0.734 | 5/8 |
| 1228 | 3/8 × 3 1/2 | 2.880 | 0.938 | 0.933 | 0.928 | 0.922 | 13/16 |
| 1428 | 7/16 × 3 1/2 | 2.880 | 0.938 | 0.933 | 0.928 | 0.922 | 13/16 |
| 1628 | 1/2 × 3 1/2 | 2.880 | 0.938 | 0.933 | 0.928 | 0.922 | 13/16 |
| 1828 | 9/16 × 3 1/2 | 2.880 | 0.938 | 0.933 | 0.928 | 0.922 | 13/16 |
| 2028 | 5/8 × 3 1/2 | 2.880 | 0.938 | 0.933 | 0.928 | 0.922 | 13/16 |
| 2228 | 11/16 × 3 1/2 | 2.880 | 0.938 | 0.933 | 0.928 | 0.922 | 13/16 |
| 2428 | 3/4 × 3 1/2 | 2.880 | 0.938 | 0.933 | 0.928 | 0.922 | 13/16 |

All dimensions are given in inches.

The key numbers indicate nominal key dimensions. The last two digits give the nominal diameter $B$ in eighths of an inch and the digits preceding the last two give the nominal width $W$ in thirty-seconds of an inch.

The key numbers with the -1 designation, while representing the nominal key size have a shorter length $F$ and due to a greater distance below center $E$ are less in height than the keys of the same number without the -1 designation.

Table 11.    ANSI Keyseat Dimensions for **Woodruff Keys** (ANSI B17.2-1967, R1978)

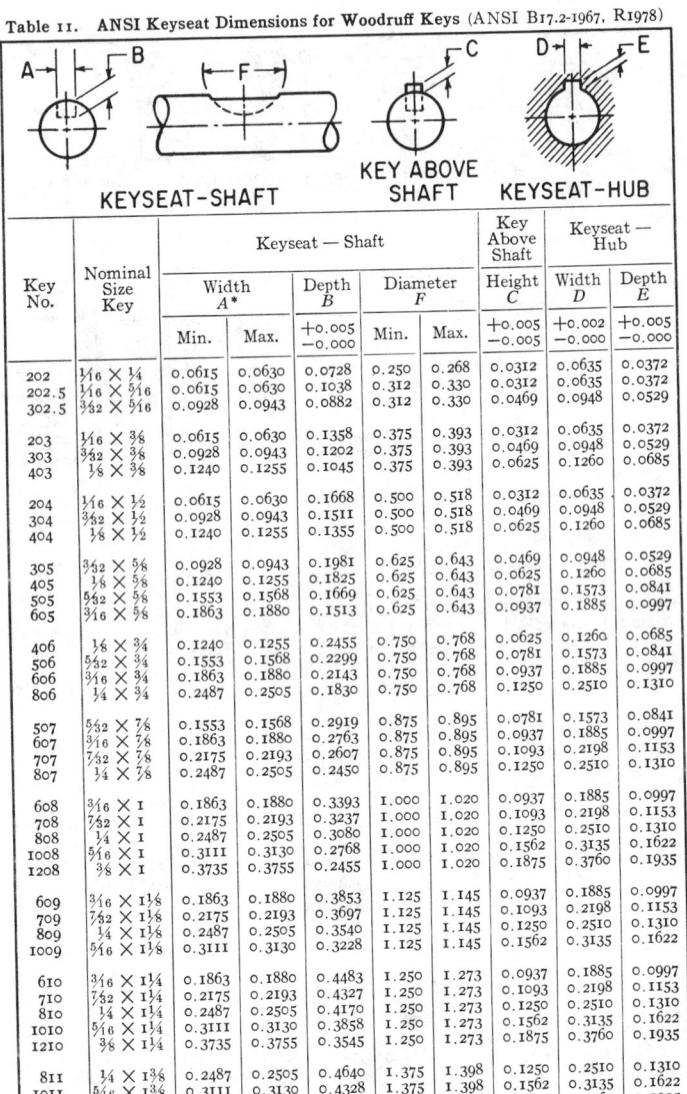

KEYSEAT-SHAFT     KEY ABOVE SHAFT     KEYSEAT-HUB

| Key No. | Nominal Size Key | Keyseat — Shaft | | | | | Key Above Shaft | Keyseat — Hub | |
| | | Width A* | | Depth B | Diameter F | | Height C | Width D | Depth E |
| | | Min. | Max. | +0.005 −0.000 | Min. | Max. | +0.005 −0.005 | +0.002 −0.000 | +0.005 −0.000 |
| 202 | 1⁄16 × 1⁄4 | 0.0615 | 0.0630 | 0.0728 | 0.250 | 0.268 | 0.0312 | 0.0635 | 0.0372 |
| 202.5 | 1⁄16 × 5⁄16 | 0.0615 | 0.0630 | 0.1038 | 0.312 | 0.330 | 0.0312 | 0.0635 | 0.0372 |
| 302.5 | 3⁄32 × 5⁄16 | 0.0928 | 0.0943 | 0.0882 | 0.312 | 0.330 | 0.0469 | 0.0948 | 0.0529 |
| 203 | 1⁄16 × 3⁄8 | 0.0615 | 0.0630 | 0.1358 | 0.375 | 0.393 | 0.0312 | 0.0635 | 0.0372 |
| 303 | 3⁄32 × 3⁄8 | 0.0928 | 0.0943 | 0.1202 | 0.375 | 0.393 | 0.0469 | 0.0948 | 0.0529 |
| 403 | 1⁄8 × 3⁄8 | 0.1240 | 0.1255 | 0.1045 | 0.375 | 0.393 | 0.0625 | 0.1260 | 0.0685 |
| 204 | 1⁄16 × 1⁄2 | 0.0615 | 0.0630 | 0.1668 | 0.500 | 0.518 | 0.0312 | 0.0635 | 0.0372 |
| 304 | 3⁄32 × 1⁄2 | 0.0928 | 0.0943 | 0.1511 | 0.500 | 0.518 | 0.0469 | 0.0948 | 0.0529 |
| 404 | 1⁄8 × 1⁄2 | 0.1240 | 0.1255 | 0.1355 | 0.500 | 0.518 | 0.0625 | 0.1260 | 0.0685 |
| 305 | 3⁄32 × 5⁄8 | 0.0928 | 0.0943 | 0.1981 | 0.625 | 0.643 | 0.0469 | 0.0948 | 0.0529 |
| 405 | 1⁄8 × 5⁄8 | 0.1240 | 0.1255 | 0.1825 | 0.625 | 0.643 | 0.0625 | 0.1260 | 0.0685 |
| 505 | 5⁄32 × 5⁄8 | 0.1553 | 0.1568 | 0.1669 | 0.625 | 0.643 | 0.0781 | 0.1573 | 0.0841 |
| 605 | 3⁄16 × 5⁄8 | 0.1863 | 0.1880 | 0.1513 | 0.625 | 0.643 | 0.0937 | 0.1885 | 0.0997 |
| 406 | 1⁄8 × 3⁄4 | 0.1240 | 0.1255 | 0.2455 | 0.750 | 0.768 | 0.0625 | 0.1260 | 0.0685 |
| 506 | 5⁄32 × 3⁄4 | 0.1553 | 0.1568 | 0.2299 | 0.750 | 0.768 | 0.0781 | 0.1573 | 0.0841 |
| 606 | 3⁄16 × 3⁄4 | 0.1863 | 0.1880 | 0.2143 | 0.750 | 0.768 | 0.0937 | 0.1885 | 0.0997 |
| 806 | 1⁄4 × 3⁄4 | 0.2487 | 0.2505 | 0.1830 | 0.750 | 0.768 | 0.1250 | 0.2510 | 0.1310 |
| 507 | 5⁄32 × 7⁄8 | 0.1553 | 0.1568 | 0.2919 | 0.875 | 0.895 | 0.0781 | 0.1573 | 0.0841 |
| 607 | 3⁄16 × 7⁄8 | 0.1863 | 0.1880 | 0.2763 | 0.875 | 0.895 | 0.0937 | 0.1885 | 0.0997 |
| 707 | 7⁄32 × 7⁄8 | 0.2175 | 0.2193 | 0.2607 | 0.875 | 0.895 | 0.1093 | 0.2198 | 0.1153 |
| 807 | 1⁄4 × 7⁄8 | 0.2487 | 0.2505 | 0.2450 | 0.875 | 0.895 | 0.1250 | 0.2510 | 0.1310 |
| 608 | 3⁄16 × 1 | 0.1863 | 0.1880 | 0.3393 | 1.000 | 1.020 | 0.0937 | 0.1885 | 0.0997 |
| 708 | 7⁄32 × 1 | 0.2175 | 0.2193 | 0.3237 | 1.000 | 1.020 | 0.1093 | 0.2198 | 0.1153 |
| 808 | 1⁄4 × 1 | 0.2487 | 0.2505 | 0.3080 | 1.000 | 1.020 | 0.1250 | 0.2510 | 0.1310 |
| 1008 | 5⁄16 × 1 | 0.3111 | 0.3130 | 0.2768 | 1.000 | 1.020 | 0.1562 | 0.3135 | 0.1622 |
| 1208 | 3⁄8 × 1 | 0.3735 | 0.3755 | 0.2455 | 1.000 | 1.020 | 0.1875 | 0.3760 | 0.1935 |
| 609 | 3⁄16 × 1 1⁄8 | 0.1863 | 0.1880 | 0.3853 | 1.125 | 1.145 | 0.0937 | 0.1885 | 0.0997 |
| 709 | 7⁄32 × 1 1⁄8 | 0.2175 | 0.2193 | 0.3697 | 1.125 | 1.145 | 0.1093 | 0.2198 | 0.1153 |
| 809 | 1⁄4 × 1 1⁄8 | 0.2487 | 0.2505 | 0.3540 | 1.125 | 1.145 | 0.1250 | 0.2510 | 0.1310 |
| 1009 | 5⁄16 × 1 1⁄8 | 0.3111 | 0.3130 | 0.3228 | 1.125 | 1.145 | 0.1562 | 0.3135 | 0.1622 |
| 610 | 3⁄16 × 1 1⁄4 | 0.1863 | 0.1880 | 0.4483 | 1.250 | 1.273 | 0.0937 | 0.1885 | 0.0997 |
| 710 | 7⁄32 × 1 1⁄4 | 0.2175 | 0.2193 | 0.4327 | 1.250 | 1.273 | 0.1093 | 0.2198 | 0.1153 |
| 810 | 1⁄4 × 1 1⁄4 | 0.2487 | 0.2505 | 0.4170 | 1.250 | 1.273 | 0.1250 | 0.2510 | 0.1310 |
| 1010 | 5⁄16 × 1 1⁄4 | 0.3111 | 0.3130 | 0.3858 | 1.250 | 1.273 | 0.1562 | 0.3135 | 0.1622 |
| 1210 | 3⁄8 × 1 1⁄4 | 0.3735 | 0.3755 | 0.3545 | 1.250 | 1.273 | 0.1875 | 0.3760 | 0.1935 |
| 811 | 1⁄4 × 1 3⁄8 | 0.2487 | 0.2505 | 0.4640 | 1.375 | 1.398 | 0.1250 | 0.2510 | 0.1310 |
| 1011 | 5⁄16 × 1 3⁄8 | 0.3111 | 0.3130 | 0.4328 | 1.375 | 1.398 | 0.1562 | 0.3135 | 0.1622 |
| 1211 | 3⁄8 × 1 3⁄8 | 0.3735 | 0.3755 | 0.4015 | 1.375 | 1.398 | 0.1875 | 0.3760 | 0.1935 |

All dimensions are given in inches.
* See footnote at end of table.

**Table 11** *(Concluded)*. **ANSI Standard Keyseat Dimensions for Woodruff Keys**
(ANSI B17.2-1967, R1978)

| Key No. | Nominal Size Key | Keyseat — Shaft | | | | | Key Above Shaft | Keyseat — Hub | |
|---|---|---|---|---|---|---|---|---|---|
| | | Width A* | | Depth B | Diameter F | | Height C | Width D | Depth E |
| | | Min. | Max. | +0.005 −0.000 | Min. | Max. | +0.005 −0.005 | +0.002 −0.000 | +0.005 −0.000 |
| 812 | ¼ × 1½ | 0.2487 | 0.2505 | 0.5110 | 1.500 | 1.523 | 0.1250 | 0.2510 | 0.1310 |
| 1012 | 5⁄16 × 1½ | 0.3111 | 0.3130 | 0.4798 | 1.500 | 1.523 | 0.1562 | 0.3135 | 0.1622 |
| 1212 | 3⁄8 × 1½ | 0.3735 | 0.3755 | 0.4485 | 1.500 | 1.523 | 0.1875 | 0.3760 | 0.1935 |
| 617-1 | 3⁄16 × 2⅛ | 0.1863 | 0.1880 | 0.3073 | 2.125 | 2.160 | 0.0937 | 0.1885 | 0.0997 |
| 817-1 | ¼ × 2⅛ | 0.2487 | 0.2505 | 0.2760 | 2.125 | 2.160 | 0.1250 | 0.2510 | 0.1310 |
| 1017-1 | 5⁄16 × 2⅛ | 0.3111 | 0.3130 | 0.2448 | 2.125 | 2.160 | 0.1562 | 0.3135 | 0.1622 |
| 1217-1 | 3⁄8 × 2⅛ | 0.3735 | 0.3755 | 0.2135 | 2.125 | 2.160 | 0.1875 | 0.3760 | 0.1935 |
| 617 | 3⁄16 × 2⅛ | 0.1863 | 0.1880 | 0.4323 | 2.125 | 2.160 | 0.0937 | 0.1885 | 0.0997 |
| 817 | ¼ × 2⅛ | 0.2487 | 0.2505 | 0.4010 | 2.125 | 2.160 | 0.1250 | 0.2510 | 0.1310 |
| 1017 | 5⁄16 × 2⅛ | 0.3111 | 0.3130 | 0.3698 | 2.125 | 2.160 | 0.1562 | 0.3135 | 0.1622 |
| 1217 | 3⁄8 × 2⅛ | 0.3735 | 0.3755 | 0.3385 | 2.125 | 2.160 | 0.1875 | 0.3760 | 0.1935 |
| 822-1 | ¼ × 2¾ | 0.2487 | 0.2505 | 0.4640 | 2.750 | 2.785 | 0.1250 | 0.2510 | 0.1310 |
| 1022-1 | 5⁄16 × 2¾ | 0.3111 | 0.3130 | 0.4328 | 2.750 | 2.785 | 0.1562 | 0.3135 | 0.1622 |
| 1222-1 | 3⁄8 × 2¾ | 0.3735 | 0.3755 | 0.4015 | 2.750 | 2.785 | 0.1875 | 0.3760 | 0.1935 |
| 1422-1 | 7⁄16 × 2¾ | 0.4360 | 0.4380 | 0.3703 | 2.750 | 2.785 | 0.2187 | 0.4385 | 0.2247 |
| 1622-1 | ½ × 2¾ | 0.4985 | 0.5005 | 0.3390 | 2.750 | 2.785 | 0.2500 | 0.5010 | 0.2560 |
| 822 | ¼ × 2¾ | 0.2487 | 0.2505 | 0.6200 | 2.750 | 2.785 | 0.1250 | 0.2510 | 0.1310 |
| 1022 | 5⁄16 × 2¾ | 0.3111 | 0.3130 | 0.5888 | 2.750 | 2.785 | 0.1562 | 0.3135 | 0.1622 |
| 1222 | 3⁄8 × 2¾ | 0.3735 | 0.3755 | 0.5575 | 2.750 | 2.785 | 0.1875 | 0.3760 | 0.1935 |
| 1422 | 7⁄16 × 2¾ | 0.4360 | 0.4380 | 0.5263 | 2.750 | 2.785 | 0.2187 | 0.4385 | 0.2247 |
| 1622 | ½ × 2¾ | 0.4985 | 0.5005 | 0.4950 | 2.750 | 2.785 | 0.2500 | 0.5010 | 0.2560 |
| 1228 | 3⁄8 × 3½ | 0.3735 | 0.3755 | 0.7455 | 3.500 | 3.535 | 0.1875 | 0.3760 | 0.1935 |
| 1428 | 7⁄16 × 3½ | 0.4360 | 0.4380 | 0.7143 | 3.500 | 3.535 | 0.2187 | 0.4385 | 0.2247 |
| 1628 | ½ × 3½ | 0.4985 | 0.5005 | 0.6830 | 3.500 | 3.535 | 0.2500 | 0.5010 | 0.2560 |
| 1828 | 9⁄16 × 3½ | 0.5610 | 0.5630 | 0.6518 | 3.500 | 3.535 | 0.2812 | 0.5635 | 0.2872 |
| 2028 | 5⁄8 × 3½ | 0.6235 | 0.6255 | 0.6205 | 3.500 | 3.535 | 0.3125 | 0.6260 | 0.3185 |
| 2228 | 11⁄16 × 3½ | 0.6860 | 0.6880 | 0.5893 | 3.500 | 3.535 | 0.3437 | 0.6885 | 0.3497 |
| 2428 | ¾ × 3½ | 0.7485 | 0.7505 | 0.5580 | 3.500 | 3.535 | 0.3750 | 0.7510 | 0.3810 |

All dimensions are given in inches.
* These Width A values were set with the maximum keyseat (shaft) width as that figure which will receive a key with the greatest amount of looseness consistent with assuring the key's sticking in the keyseat (shaft). Minimum keyseat width is that figure permitting the largest shaft distortion acceptable when assembling maximum key in minimum keyseat.
Dimensions A, B, C, D are taken at side intersection.

**Cotters.** — A cotter is a form of key that is used to connect rods, etc., that are subjected either to tension or compression or both, the cotter being subjected to shearing stresses at two transverse cross-sections. When taper cotters are used for drawing and holding parts together, if the cotter is held in place by the friction between the bearing surfaces, the taper should not be too great. Ordinarily a taper varying from ¼ to ½ inch per foot is used for plain cotters. When a set-screw or other device is used to prevent the cotter from backing out of its slot, the taper may vary from 1½ to 2 inches per foot.

**British Standard Metric Keys and Keyways.** — This British Standard (BS 4235:Part 1:1972) covers square and rectangular parallel keys and keyways, and square and rectangular taper keys and keyways. Plain and gib-head taper keys are specified. There are three classes of fit for the square and rectangular parallel keys and keyways, designated free, normal, and close. A *free fit* is applied when the hub of an assembly is required to slide over the key when in use; a *normal fit* is employed when the key is to be inserted in the keyway with the minimum amount of fitting, as may be required in mass production assembly work; and a *close fit* is applied when accurate fitting of the key is required under maximum material conditions, which may involve selection of components.

The Standard does not provide for misalignment or offset greater than can be accommodated within the dimensional tolerances. If an assembly is going to be heavily stressed, a check should be made to ensure that the cumulative effect of misalignment or offset, or both, does not prevent satisfactory bearing on the key. Radii and chamfers are not normally provided on keybar and keys as supplied, but they can be produced during manufacture by agreement between the user and supplier.

Unless otherwise specified, keys in compliance with this Standard are manufactured from steel complying with BS 970 having a tensile strength of not less than 550 MN/m$^2$ in the finished condition. BS 970, Part 1, lists the following steels and maximum section sizes, respectively, that meet this tensile strength requirement: 070M20, 25 $\times$ 14mm; 070M26; 36 $\times$ 20mm; 080M30, 90 $\times$ 45mm; and 080M40, 100 $\times$ 50mm.

At the time of publication of this Standard, the demand for metric keys was not sufficient to enable standard ranges of lengths to be established. The lengths given in the accompanying table are those shown as standard in ISO Recommendations R 773, "Rectangular or Square Parallel Keys and their Corresponding Keyways (Dimensions in millimeters)" and R 774, "Taper Keys and their Corresponding Keyways — with or without Gib Head (Dimensions in millimeters)."

Tables 1 through 4 on the following pages cover the dimensions and tolerances of square and rectangular keys and keyways, and square and rectangular taper keys and keyways.

**British Standard Preferred Lengths of Metric Keys** (BS 4235: Part 1: 1972)

| Length | Type of key | | | | Length | Type of key | | | |
|---|---|---|---|---|---|---|---|---|---|
| | Squ. | Rect. | Squ. taper | Rect. taper | | Squ. | Rect. | Squ. taper | Rect. taper |
| 6 | x | | x | | 63 | x | x | x | x |
| 8 | x | | x | | 70 | x | x | x | x |
| 10 | x | | x | | 80 | | x | | x |
| 12 | x | | x | | 90 | | x | | x |
| 14 | x | | x | | 100 | | x | | x |
| 16 | x | | x | | 110 | | x | | x |
| 18 | x | x | x | x | 125 | | x | | x |
| 20 | x | x | x | x | 140 | | x | | x |
| 22 | x | x | x | x | 160 | | x | | x |
| 25 | x | x | x | x | 180 | | x | | x |
| 28 | x | x | x | x | 200 | | x | | x |
| 32 | x | x | x | x | 220 | | x | | x |
| 36 | x | x | x | x | 250 | | x | | x |
| 40 | x | x | x | x | 280 | | x | | x |
| 45 | x | x | x | x | 320 | | x | | x |
| 50 | x | x | x | x | 360 | | x | | x |
| 56 | x | x | x | x | 400 | | x | | x |

## Table 1. British Standard Metric Keyways for Square and Rectangular Parallel Keys (BS 4235: Part 1: 1972)

Section X-X

Enlarged detail of key and keyways

Keyways for Square Parallel Keys

| Shaft Nominal Diameter d | | Key | Keyway | | | | | | | | | | | |
|---|---|---|---|---|---|---|---|---|---|---|---|---|---|---|
| | | | Width b | | | | | | Depth | | | | Radius r | |
| | | | | | Free Fit | | Normal Fit | | Close Fit | Shaft t1 | | Hub t2 | | | |
| Over | Up to and Incl. | Size, b × h | Nom. | Shaft (H9) | Hub (D10) | Shaft (N9) | Hub (Js9)* | Shaft and Hub (P9) | Nom. | Tol. | Nom. | Tol. | Max. | Min. |
| | | | | Tolerances | | | | | | | | | | |
| 6 | 8 | 2 × 2 | 2 | +0.025 / 0 | +0.060 / +0.020 | −0.004 / −0.029 | +0.012 / −0.012 | −0.006 / −0.031 | 1.2 | +0.1 / 0 | 1 | +0.1 / 0 | 0.16 | 0.08 |
| 8 | 10 | 3 × 3 | 3 | +0.025 / 0 | +0.060 / +0.020 | −0.004 / −0.029 | +0.012 / −0.012 | −0.006 / −0.031 | 1.8 | +0.1 / 0 | 1.4 | +0.1 / 0 | 0.16 | 0.08 |
| 10 | 12 | 4 × 4 | 4 | +0.030 / 0 | +0.078 / +0.030 | 0 / −0.030 | +0.015 / −0.015 | −0.012 / −0.042 | 2.5 | +0.1 / 0 | 1.8 | +0.1 / 0 | 0.16 | 0.08 |
| 12 | 17 | 5 × 5 | 5 | +0.030 / 0 | +0.078 / +0.030 | 0 / −0.030 | +0.015 / −0.015 | −0.012 / −0.042 | 3 | +0.1 / 0 | 2.3 | +0.1 / 0 | 0.25 | 0.16 |
| 17 | 22 | 6 × 6 | 6 | +0.030 / 0 | +0.078 / +0.030 | 0 / −0.030 | +0.015 / −0.015 | −0.012 / −0.042 | 3.5 | +0.1 / 0 | 2.8 | +0.1 / 0 | 0.25 | 0.16 |

All dimensions in millimeters.    * Tolerance limits Js9 are quoted from BS 4500, "ISO Limits and Fits," to three significant figures.

Table 1 (*Continued*). **British Standard Metric Keyways for Square and Rectangular Parallel Keys (BS 4235: Part 1: 1972)**

Keyways for Rectangular Parallel Keys

| Shaft Nom. Dia. *d* Over | Up to and Incl. | Key Size *b × h* | Width *b* Nom. | Free Fit Shaft (H9) | Free Fit Hub (D10) | Normal Fit Shaft (N9) | Normal Fit Hub (Js9)* | Close Fit Shaft and Hub (P9) | Depth Shaft *t₁* Nom. | Shaft *t₁* Tol. | Hub *t₂* Nom. | Hub *t₂* Tol. | Radius *r* Max. | Radius *r* Min. |
|---|---|---|---|---|---|---|---|---|---|---|---|---|---|---|
| 22 | 30 | 8 × 7 | 8 | +0.036 / 0 | +0.098 / +0.040 | 0 / −0.036 | +0.018 / −0.018 | −0.015 / −0.051 | 4 | +0.2 / 0 | 3.3 | +0.2 / 0 | 0.25 | 0.16 |
| 30 | 38 | 10 × 8 | 10 | +0.036 / 0 | +0.098 / +0.040 | 0 / −0.036 | +0.018 / −0.018 | −0.015 / −0.051 | 5 | +0.2 / 0 | 3.3 | +0.2 / 0 | 0.40 | 0.25 |
| 38 | 44 | 12 × 8 | 12 | +0.043 / 0 | +0.120 / +0.050 | 0 / −0.043 | +0.021 / −0.021 | −0.018 / −0.061 | 5 | +0.2 / 0 | 3.3 | +0.2 / 0 | 0.40 | 0.25 |
| 44 | 50 | 14 × 9 | 14 | +0.043 / 0 | +0.120 / +0.050 | 0 / −0.043 | +0.021 / −0.021 | −0.018 / −0.061 | 5.5 | +0.2 / 0 | 3.8 | +0.2 / 0 | 0.40 | 0.25 |
| 50 | 58 | 16 × 10 | 16 | +0.043 / 0 | +0.120 / +0.050 | 0 / −0.043 | +0.021 / −0.021 | −0.018 / −0.061 | 6 | +0.2 / 0 | 4.3 | +0.2 / 0 | 0.40 | 0.25 |
| 58 | 65 | 18 × 11 | 18 | +0.043 / 0 | +0.120 / +0.050 | 0 / −0.043 | +0.021 / −0.021 | −0.018 / −0.061 | 7 | +0.2 / 0 | 4.4 | +0.2 / 0 | 0.40 | 0.25 |
| 65 | 75 | 20 × 12 | 20 | +0.052 / 0 | +0.149 / +0.065 | 0 / −0.052 | +0.026 / −0.026 | −0.022 / −0.074 | 7.5 | +0.2 / 0 | 4.9 | +0.2 / 0 | 0.60 | 0.40 |
| 75 | 85 | 22 × 14 | 22 | +0.052 / 0 | +0.149 / +0.065 | 0 / −0.052 | +0.026 / −0.026 | −0.022 / −0.074 | 9 | +0.2 / 0 | 5.4 | +0.2 / 0 | 0.60 | 0.40 |
| 85 | 95 | 25 × 14 | 25 | +0.052 / 0 | +0.149 / +0.065 | 0 / −0.052 | +0.026 / −0.026 | −0.022 / −0.074 | 9 | +0.2 / 0 | 5.4 | +0.2 / 0 | 0.60 | 0.40 |
| 95 | 110 | 28 × 16 | 28 | +0.052 / 0 | +0.149 / +0.065 | 0 / −0.052 | +0.026 / −0.026 | −0.022 / −0.074 | 10 | +0.2 / 0 | 6.4 | +0.2 / 0 | 0.60 | 0.40 |
| 110 | 130 | 32 × 18 | 32 | +0.062 / 0 | +0.180 / +0.080 | 0 / −0.062 | +0.031 / −0.031 | −0.026 / −0.088 | 11 | +0.2 / 0 | 7.4 | +0.2 / 0 | 1.00 | 0.70 |
| 130 | 150 | 36 × 20 | 36 | +0.062 / 0 | +0.180 / +0.080 | 0 / −0.062 | +0.031 / −0.031 | −0.026 / −0.088 | 12 | +0.3 / 0 | 8.4 | +0.3 / 0 | 1.00 | 0.70 |
| 150 | 170 | 40 × 22 | 40 | +0.062 / 0 | +0.180 / +0.080 | 0 / −0.062 | +0.031 / −0.031 | −0.026 / −0.088 | 13 | +0.3 / 0 | 9.4 | +0.3 / 0 | 1.00 | 0.70 |
| 170 | 200 | 45 × 25 | 45 | +0.062 / 0 | +0.180 / +0.080 | 0 / −0.062 | +0.031 / −0.031 | −0.026 / −0.088 | 15 | +0.3 / 0 | 10.4 | +0.3 / 0 | 1.00 | 0.70 |
| 200 | 230 | 50 × 28 | 50 | +0.062 / 0 | +0.180 / +0.080 | 0 / −0.062 | +0.031 / −0.031 | −0.026 / −0.088 | 17 | +0.3 / 0 | 11.4 | +0.3 / 0 | 1.00 | 0.70 |
| 230 | 260 | 56 × 32 | 56 | +0.074 / 0 | +0.220 / +0.100 | 0 / −0.074 | +0.037 / −0.037 | −0.032 / −0.106 | 20 | +0.3 / 0 | 12.4 | +0.3 / 0 | 1.60 | 1.20 |
| 260 | 290 | 63 × 32 | 63 | +0.074 / 0 | +0.220 / +0.100 | 0 / −0.074 | +0.037 / −0.037 | −0.032 / −0.106 | 20 | +0.3 / 0 | 12.4 | +0.3 / 0 | 1.60 | 1.20 |
| 290 | 330 | 70 × 36 | 70 | +0.074 / 0 | +0.220 / +0.100 | 0 / −0.074 | +0.037 / −0.037 | −0.032 / −0.106 | 22 | +0.3 / 0 | 14.4 | +0.3 / 0 | 1.60 | 1.20 |
| 330 | 380 | 80 × 40 | 80 | +0.074 / 0 | +0.220 / +0.100 | 0 / −0.074 | +0.037 / −0.037 | −0.032 / −0.106 | 25 | +0.3 / 0 | 15.4 | +0.3 / 0 | 1.60 | 1.20 |
| 380 | 440 | 90 × 45 | 90 | +0.087 / 0 | +0.260 / +0.120 | 0 / −0.087 | +0.043 / −0.043 | −0.037 / −0.124 | 28 | +0.3 / 0 | 17.4 | +0.3 / 0 | 2.50 | 2.00 |
| 440 | 500 | 100 × 50 | 100 | +0.087 / 0 | +0.260 / +0.120 | 0 / −0.087 | +0.043 / −0.043 | −0.037 / −0.124 | 31 | +0.3 / 0 | 19.5 | +0.3 / 0 | 2.50 | 2.00 |

All dimensions in millimeters.    * Tolerance limits Js9 are quoted from BS 4500, "ISO Limits and Fits," to three significant figures.

**Table 2. British Standard Metric Keyways for Square and Rectangular Taper Keys**
(BS 4235: Part 1: 1972)

| Shaft | | Key | Keyway | | | | | | |
|---|---|---|---|---|---|---|---|---|---|
| Nominal Diameter $d$ | | Size, $b \times h$ | Width $b$, Shaft and Hub | | Depth | | | | Corner Radius of Keyway |
| | | | | | Shaft $t_1$ | | Hub $t_2$ | | |
| Over | Up to and Incl. | | Nom. | Tol. (D10) | Nom. | Tol. | Nom. | Tol. | Max. | Min. |

| | | Keyways for Square Taper Keys | | | | | | | |
|---|---|---|---|---|---|---|---|---|---|
| 6 | 8 | 2 × 2 | 2 | + 0.060 / + 0.020 | 1.2 | | 0.5 | | 0.16 | 0.08 |
| 8 | 10 | 3 × 3 | 3 | | 1.8 | + 0.1 / 0 | 0.9 | + 0.1 / 0 | 0.16 | 0.08 |
| 10 | 12 | 4 × 4 | 4 | | 2.5 | | 1.2 | | 0.16 | 0.08 |
| 12 | 17 | 5 × 5 | 5 | + 0.078 / + 0.030 | 3 | | 1.7 | | 0.25 | 0.16 |
| 17 | 22 | 6 × 6 | 6 | | 3.5 | + 0.2 / 0 | 2.2 | + 0.2 / 0 | 0.25 | 0.16 |

| | | Keyways for Rectangular Taper Keys | | | | | | | |
|---|---|---|---|---|---|---|---|---|---|
| 22 | 30 | 8 × 7 | 8 | + 0.098 / + 0.040 | 4 | | 2.4 | | 0.25 | 0.16 |
| 30 | 38 | 10 × 8 | 10 | | 5 | | 2.4 | | 0.40 | 0.25 |
| 38 | 44 | 12 × 8 | 12 | | 5 | | 2.4 | | 0.40 | 0.25 |
| 44 | 50 | 14 × 9 | 14 | + 0.120 / + 0.050 | 5.5 | + 0.2 / 0 | 2.9 | + 0.2 / 0 | 0.40 | 0.25 |
| 50 | 58 | 16 × 10 | 16 | | 6 | | 3.4 | | 0.40 | 0.25 |
| 58 | 65 | 18 × 11 | 18 | | 7 | | 3.4 | | 0.40 | 0.25 |
| 65 | 75 | 20 × 12 | 20 | | 7.5 | | 3.9 | | 0.60 | 0.40 |
| 75 | 85 | 22 × 14 | 22 | + 0.149 / + 0.065 | 9 | | 4.4 | | 0.60 | 0.40 |
| 85 | 95 | 25 × 14 | 25 | | 9 | | 4.4 | | 0.60 | 0.40 |
| 95 | 110 | 28 × 16 | 28 | | 10 | | 5.4 | | 0.60 | 0.40 |
| 110 | 130 | 32 × 18 | 32 | | 11 | | 6.4 | | 0.60 | 0.40 |
| 130 | 150 | 36 × 20 | 36 | | 12 | | 7.1 | | 1.00 | 0.70 |
| 150 | 170 | 40 × 22 | 40 | + 0.180 / + 0.080 | 13 | | 8.1 | | 1.00 | 0.70 |
| 170 | 200 | 45 × 25 | 45 | | 15 | | 9.1 | | 1.00 | 0.70 |
| 200 | 230 | 50 × 28 | 50 | | 17 | | 10.1 | | 1.00 | 0.70 |
| 230 | 260 | 56 × 32 | 56 | | 20 | + 0.3 / 0 | 11.1 | + 0.3 / 0 | 1.60 | 1.20 |
| 260 | 290 | 63 × 32 | 63 | + 0.220 / + 0.120 | 20 | | 11.1 | | 1.60 | 1.20 |
| 290 | 330 | 70 × 36 | 70 | | 22 | | 13.1 | | 1.60 | 1.20 |
| 330 | 380 | 80 × 40 | 80 | | 25 | | 14.1 | | 2.50 | 2.00 |
| 380 | 440 | 90 × 45 | 90 | + 0.260 / + 0.120 | 28 | | 16.1 | | 2.50 | 2.00 |
| 440 | 500 | 100 × 50 | 100 | | 31 | | 18.1 | | 2.50 | 2.00 |

## Table 3. British Standard Metric Square and Rectangular Parallel Keys
(BS 4235: Part 1: 1972)

| Width b Nom. | Width b Tol.* | Thickness h Nom. | Thickness h Tol.* | Chamfer s Min. | Chamfer s Max. | Length Range l From | Length Range l To |
|---|---|---|---|---|---|---|---|
| Square Parallel Keys | | | | | | | |
| 2 | 0 / − 0.025 | 2 | 0 / − 0.025 | 0.16 | 0.25 | 6 | 20 |
| 3 | | 3 | | 0.16 | 0.25 | 6 | 36 |
| 4 | | 4 | | 0.16 | 0.25 | 8 | 45 |
| 5 | 0 / − 0.030 | 5 | 0 / − 0.030 | 0.25 | 0.40 | 10 | 56 |
| 6 | | 6 | | 0.25 | 0.40 | 14 | 70 |
| Rectangular Parallel Keys | | | | | | | |
| 8 | 0 / − 0.036 | 7 | | 0.25 | 0.40 | 18 | 90 |
| 10 | | 8 | | 0.40 | 0.60 | 22 | 110 |
| 12 | | 8 | 0 / − 0.090 | 0.40 | 0.60 | 28 | 140 |
| 14 | | 9 | | 0.40 | 0.60 | 36 | 160 |
| 16 | 0 / − 0.043 | 10 | | 0.40 | 0.60 | 45 | 180 |
| 18 | | 11 | | 0.40 | 0.60 | 50 | 200 |
| 20 | | 12 | | 0.60 | 0.80 | 56 | 220 |
| 22 | | 14 | 0 / − 0.110 | 0.60 | 0.80 | 63 | 250 |
| 25 | 0 / − 0.052 | 14 | | 0.60 | 0.80 | 70 | 280 |
| 28 | | 16 | | 0.60 | 0.80 | 80 | 320 |
| 32 | | 18 | | 0.60 | 0.80 | 90 | 360 |
| 36 | | 20 | | 1.00 | 1.20 | 100 | 400 |
| 40 | 0 / − 0.062 | 22 | | 1.00 | 1.20 | ... | ... |
| 45 | | 25 | 0 / − 0.130 | 1.00 | 1.20 | ... | ... |
| 50 | | 28 | | 1.00 | 1.20 | ... | ... |
| 56 | | 32 | | 1.60 | 2.00 | ... | ... |
| 63 | | 32 | | 1.60 | 2.00 | ... | ... |
| 70 | 0 / − 0.074 | 36 | | 1.60 | 2.00 | ... | ... |
| 80 | | 40 | 0 / − 0.160 | 2.50 | 3.00 | ... | ... |
| 90 | 0 / − 0.087 | 45 | | 2.50 | 3.00 | ... | ... |
| 100 | | 50 | | 2.50 | 3.00 | ... | ... |

* The tolerance on the width and thickness of square taper keys is h9, and on the width and thickness of rectangular keys, h9 and h11 respectively, in accordance with ISO metric limits and fits.

## Table 4. British Standard Metric Square and Rectangular Taper Keys
### (BS 4235: Part I: 1972)

Plain key — Form A — Form B — Gib head key Form C — Section X-X — Basic taper 1 in 100

| Width b | | Thickness h | | Chamfer s | | Length Range l | | Gib head h₁ | Radius r |
|---|---|---|---|---|---|---|---|---|---|
| Nom. | Tol.* | Nom. | Tol.* | Min. | Max. | From | To | Nom. | Nom. |
| Square Taper Keys | | | | | | | | | |
| 2 | } 0 −0.025 | 2 | } 0 −0.025 | 0.16 | 0.25 | 6 | 20 | ... | ... |
| 3 | | 3 | | 0.16 | 0.25 | 6 | 36 | ... | ... |
| 4 | } 0 −0.030 | 4 | } 0 −0.030 | 0.16 | 0.25 | 8 | 45 | 7 | 0.25 |
| 5 | | 5 | | 0.25 | 0.40 | 10 | 56 | 8 | 0.25 |
| 6 | | 6 | | 0.25 | 0.40 | 14 | 70 | 10 | 0.25 |
| Rectangular Taper Keys | | | | | | | | | |
| 8 | } 0 −0.036 | 7 | | 0.25 | 0.40 | 18 | 90 | 11 | 1.5 |
| 10 | | 8 | | 0.40 | 0.60 | 22 | 110 | 12 | 1.5 |
| 12 | } 0 −0.043 | 8 | } 0 −0.090 | 0.40 | 0.60 | 28 | 140 | 12 | 1.5 |
| 14 | | 9 | | 0.40 | 0.60 | 36 | 160 | 14 | 1.5 |
| 16 | | 10 | | 0.40 | 0.60 | 45 | 180 | 16 | 3.2 |
| 18 | | 11 | | 0.40 | 0.60 | 50 | 200 | 18 | 3.2 |
| 20 | } 0 −0.052 | 12 | } 0 −0.110 | 0.60 | 0.80 | 56 | 220 | 20 | 3.2 |
| 22 | | 14 | | 0.60 | 0.80 | 63 | 250 | 22 | 3.2 |
| 25 | | 14 | | 0.60 | 0.80 | 70 | 280 | 22 | 3.2 |
| 28 | | 16 | | 0.60 | 0.80 | 80 | 320 | 25 | 3.2 |
| 32 | | 18 | | 0.60 | 0.80 | 90 | 360 | 28 | 6.4 |
| 36 | } 0 −0.062 | 20 | } 0 −0.130 | 1.00 | 1.20 | 100 | 400 | 32 | 6.4 |
| 40 | | 22 | | 1.00 | 1.20 | ... | ... | 36 | 6.4 |
| 45 | | 25 | | 1.00 | 1.20 | ... | ... | 40 | 6.4 |
| 50 | | 28 | | 1.00 | 1.20 | ... | ... | 45 | 6.4 |
| 56 | } 0 −0.074 | 32 | | 1.60 | 2.00 | ... | ... | 50 | 9.5 |
| 63 | | 32 | | 1.60 | 2.00 | ... | ... | 50 | 9.5 |
| 70 | | 36 | } 0 −0.160 | 1.60 | 2.00 | ... | ... | 56 | 9.5 |
| 80 | | 40 | | 2.50 | 3.00 | ... | ... | 63 | 9.5 |
| 90 | } 0 −0.087 | 45 | | 2.50 | 3.00 | ... | ... | 70 | 9.5 |
| 100 | | 50 | | 2.50 | 3.00 | ... | ... | 80 | 9.5 |

* The tolerance on the width and thickness of square taper keys is h9, and on the width and thickness of rectangular taper keys, h9 and h11 respectively, in accordance with ISO metric limits and fits. Does not apply to gib head dimensions.

**British Standard Keys and Keyways.** — Tables 1 through 6 (B.S. 46: Part 1: 1958) provide data for rectangular parallel keys and keyways, square parallel keys and keyways, plain and gib head rectangular taper keys and keyways, plain and gib head square taper keys and keyways, and Woodruff keys and keyways.

*Parallel Keys:* These keys are used for transmitting unidirectional torques in transmissions not subject to heavy starting loads and where periodic withdrawal or sliding of the hub member may be required. In many instances, particularly couplings, a gib-head cannot be accommodated, and there is insufficient room to drift out the key from behind. In these cases it is necessary to withdraw the component over the key and a parallel key is essential. Parallel square and rectangular keys are normally side fitting with top clearance and are usually retained in the shaft rather more securely than in the hub. The rectangular key is the general purpose key for shafts greater than 1 inch in diameter; the square key is intended for use with shafts up to and including 1-inch diameter or for shafts up to 6-inch diameter where it is desirable to have a greater key depth than is provided by rectangular keys. In cases of stepped shafts the larger diameters are usually required by considerations other than torque, e.g. resistance to bending. Where components such as fans, gears, impellers, etc., are attached to the larger shaft diameter, the use of a key smaller than standard for that diameter may be permissible. As this results in unequal disposition of the key in the shaft and its related hub the dimensions $H$ and $h$ must be recalculated to maintain the $T/2$ relationship.

*Taper Keys:* These keys are used for transmitting heavy unidirectional, reversing, or vibrating torques and in applications where periodic withdrawal of the key may be necessary. Taper keys are usually top fitting, but may be top and side fitting where required; in this case the keyway in the hub should have the same width value as the keyway in the shaft. Taper keys of rectangular section are used for general purposes and are of less depth than square keys; square sections are for use with shafts up to and including 1-inch diameter or for shafts up to 6-inch diameter where it is desirable to have greater key depth.

*Woodruff Keys:* These keys are used for light applications or the angular location of associated parts on tapered shaft ends. They are not recommended for other applications, but if so used, corner radii in the shaft and hub keyways are advisable to reduce stress concentration.

**Dimensions and Tolerances for British Parallel and Taper Keys and Keyways.** — Dimensions and tolerances for key and keyway widths given in Tables 1, 2, 3 and 4 are based on the width of key $W$ and provide a fitting allowance. The fitting allowance is designed to permit an interference between the key and the shaft keyway and a slightly easier condition between the key and the hub keyway. In the case of shrink and heavy force fits it may be found necessary to depart from the width and depth tolerances specified. Any variation in the width of the keyway should be such that the greatest width is at the end from which the key enters and any variation in the depth of the keyway should be such that the greatest depth is at the end from which the key enters.

Keys and keybar normally are not chamfered or radiused as supplied, but this may be done at the time of fitting. Radii and chamfers are given in Tables 1, 2, 3 and 4. Corner radii are recommended for keyways to alleviate stress concentration.

**Dimensions and Tolerances of British Woodruff Keys and Keyways.** — Dimensions and tolerances are shown in Table 5. An optional alternative design of Woodruff key which differs from the normal form in its depth is given in the illustration accompanying the table. The method of designating British Woodruff Keys is the same as the American method explained in footnote on page 974.

**Table 1. British Standard Rectangular Parallel Keys, Keyways and Keybars** (B.S. 46: Part 1: 1958)

All dimensions in inches

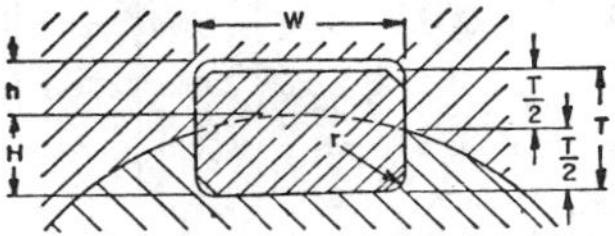


| Diameter of Shaft | | Key | | | | | Keyway in Shaft | | | | Keyway in Hub | | | | Nominal Keyway Radius, $r$* | Keybar | | | |
|---|---|---|---|---|---|---|---|---|---|---|---|---|---|---|---|---|---|---|---|
| Over | Up to and Including | Size, $W \times T$ | Width, $W$ | | Thickness, $T$ | | Width, $W_s$ | | Depth, $H$ | | Width, $W_h$ | | Depth, $h$ | | | Width, $W$ | | Thickness, $T$ | |
| | | | Max. | Min. | Max. | Min. | Min. | Max. | Min. | Max. | Min. | Max. | Min. | Max. | | Max. | Min. | Max. | Min. |
| 1 | 1¼ | 5/16 × ¼ | 0.314 | 0.312 | 0.253 | 0.250 | 0.311 | 0.312 | 0.146 | 0.152 | 0.312 | 0.313 | 0.112 | 0.118 | 0.010 | 0.314 | 0.312 | 0.253 | 0.250 |
| 1¼ | 1½ | ⅜ × ¼ | 0.377 | 0.375 | 0.253 | 0.250 | 0.374 | 0.375 | 0.150 | 0.156 | 0.375 | 0.376 | 0.108 | 0.114 | 0.010 | 0.377 | 0.375 | 0.253 | 0.250 |
| 1½ | 1¾ | 7/16 × 5/16 | 0.440 | 0.438 | 0.315 | 0.312 | 0.437 | 0.438 | 0.186 | 0.192 | 0.438 | 0.439 | 0.135 | 0.141 | 0.020 | 0.440 | 0.438 | 0.315 | 0.312 |
| 1¾ | 2 | ½ × 5/16 | 0.502 | 0.500 | 0.315 | 0.312 | 0.499 | 0.500 | 0.190 | 0.196 | 0.500 | 0.501 | 0.131 | 0.137 | 0.020 | 0.502 | 0.500 | 0.315 | 0.312 |
| 2 | 2½ | ⅝ × 7/16 | 0.627 | 0.625 | 0.441 | 0.438 | 0.624 | 0.625 | 0.260 | 0.266 | 0.625 | 0.626 | 0.185 | 0.191 | 0.020 | 0.627 | 0.625 | 0.441 | 0.438 |
| 2½ | 3 | ¾ × ½ | 0.752 | 0.750 | 0.503 | 0.500 | 0.749 | 0.750 | 0.299 | 0.305 | 0.750 | 0.751 | 0.209 | 0.215 | 0.020 | 0.752 | 0.750 | 0.503 | 0.500 |
| 3 | 3½ | ⅞ × ⅝ | 0.877 | 0.875 | 0.629 | 0.625 | 0.874 | 0.875 | 0.370 | 0.376 | 0.875 | 0.876 | 0.264 | 0.270 | 0.062 | 0.877 | 0.875 | 0.629 | 0.625 |
| 3½ | 4 | 1 × ¾ | 1.003 | 1.000 | 0.754 | 0.750 | 0.999 | 1.000 | 0.441 | 0.447 | 1.000 | 1.001 | 0.318 | 0.324 | 0.062 | 1.003 | 1.000 | 0.754 | 0.750 |
| 4 | 5 | 1¼ × ⅞ | 1.253 | 1.250 | 0.879 | 0.875 | 1.248 | 1.250 | 0.518 | 0.524 | 1.250 | 1.252 | 0.366 | 0.372 | 0.062 | 1.253 | 1.250 | 0.879 | 0.875 |
| 5 | 6 | 1½ × 1 | 1.504 | 1.500 | 1.006 | 1.000 | 1.498 | 1.500 | 0.599 | 0.605 | 1.500 | 1.502 | 0.412 | 0.418 | 0.062 | 1.504 | 1.500 | 1.006 | 1.000 |
| 6 | 7 | 1¾ × 1¼ | 1.754 | 1.750 | 1.256 | 1.250 | 1.748 | 1.750 | 0.740 | 0.746 | 1.750 | 1.752 | 0.526 | 0.532 | 0.125 | | | | |
| 7 | 8 | 2 × 1⅜ | 2.005 | 2.000 | 1.381 | 1.375 | 1.998 | 2.000 | 0.818 | 0.824 | 2.000 | 2.002 | 0.573 | 0.579 | 0.125 | | | | |
| 8 | 9 | 2¼ × 1½ | 2.255 | 2.250 | 1.506 | 1.500 | 2.248 | 2.250 | 0.897 | 0.905 | 2.250 | 2.252 | 0.619 | 0.627 | 0.125 | | | | |
| 9 | 10 | 2½ × 1⅝ | 2.505 | 2.500 | 1.631 | 1.625 | 2.498 | 2.500 | 0.975 | 0.983 | 2.500 | 2.502 | 0.666 | 0.674 | 0.187 | | | | |
| 10 | 11 | 2¾ × 1⅞ | 2.755 | 2.750 | 1.881 | 1.875 | 2.748 | 2.750 | 1.114 | 1.122 | 2.750 | 2.752 | 0.777 | 0.785 | 0.187 | | | | |
| 11 | 12 | 3 × 2 | 3.006 | 3.000 | 2.008 | 2.000 | 2.998 | 3.000 | 1.195 | 1.203 | 3.000 | 3.002 | 0.823 | 0.831 | 0.187 | | | | |
| 12 | 13 | 3¼ × 2⅛ | 3.256 | 3.250 | 2.133 | 2.125 | 3.248 | 3.250 | 1.273 | 1.281 | 3.250 | 3.252 | 0.870 | 0.878 | 0.187 | | | | |
| 13 | 14 | 3½ × 2⅜ | 3.506 | 3.500 | 2.383 | 2.375 | 3.498 | 3.500 | 1.413 | 1.421 | 3.500 | 3.502 | 0.980 | 0.988 | 0.250 | | | | |
| 14 | 15 | 3¾ × 2½ | 3.756 | 3.750 | 2.508 | 2.500 | 3.748 | 3.750 | 1.492 | 1.502 | 3.750 | 3.752 | 1.026 | 1.036 | 0.250 | | | | |
| 15 | 16 | 4 × 2⅝ | 4.008 | 4.000 | 2.633 | 2.625 | 3.998 | 4.000 | 1.571 | 1.581 | 4.000 | 4.002 | 1.072 | 1.082 | 0.250 | | | | |
| 16 | 17 | 4¼ × 2⅞ | 4.258 | 4.250 | 2.883 | 2.875 | 4.248 | 4.250 | 1.711 | 1.721 | 4.250 | 4.252 | 1.182 | 1.192 | 0.312 | | | | |
| 17 | 18 | 4½ × 3 | 4.508 | 4.500 | 3.010 | 3.000 | 4.498 | 4.500 | 1.791 | 1.801 | 4.500 | 4.502 | 1.229 | 1.239 | 0.312 | | | | |
| 18 | 19 | 4¾ × 3⅛ | 4.758 | 4.750 | 3.135 | 3.125 | 4.748 | 4.750 | 1.868 | 1.878 | 4.750 | 4.752 | 1.277 | 1.287 | 0.312 | | | | |
| 19 | 20 | 5 × 3⅜ | 5.008 | 5.000 | 3.385 | 3.375 | 4.998 | 5.000 | 2.010 | 2.020 | 5.000 | 5.002 | 1.385 | 1.395 | 0.312 | | | | |

Bright keybar is not normally available in sections larger than the above.

* The key chamfer shall be the minimum to clear the keyway radius. Nominal values are given.

Table 2.  British Standard Square Parallel Keys, Keyways and Keybars (B.S. 46: Part 1: 1958)

All dimensions in inches

| Diameter of Shaft | | Key | | | Keyway in Shaft | | | | Keyway in Hub | | | | Nominal Keyway Radius, $r^*$ | Bright Keybar | |
| --- | --- | --- | --- | --- | --- | --- | --- | --- | --- | --- | --- | --- | --- | --- | --- |
| | | Size, $W \times T$ | Width, $W$ and Thickness, $T$ | | Width, $W_s$ | | Depth, $H$ | | Width, $W_h$ | | Depth, $h$ | | | Width, $W$ and Thickness, $T$ | |
| Over | Up to and Including | | Max. | Min. | Max. | Min. | Min. | Max. | Max. | Min. | Min. | Max. | | Max. | Min. |
| 1/4 | 1/2 | 1/16×1/16 | 0.127 | 0.125 | 0.125 | 0.124 | 0.072 | 0.078 | 0.126 | 0.125 | 0.060 | 0.066 | 0.010 | 0.127 | 0.125 |
| 1/2 | 3/4 | 3/16×3/16 | 0.190 | 0.188 | 0.188 | 0.187 | 0.107 | 0.113 | 0.189 | 0.188 | 0.088 | 0.094 | 0.010 | 0.190 | 0.188 |
| 3/4 | 1 | 1/4×1/4 | 0.252 | 0.250 | 0.250 | 0.249 | 0.142 | 0.148 | 0.251 | 0.250 | 0.115 | 0.121 | 0.010 | 0.252 | 0.250 |
| 1 | 1 1/4 | 5/16×5/16 | 0.314 | 0.312 | 0.312 | 0.311 | 0.177 | 0.183 | 0.313 | 0.312 | 0.142 | 0.148 | 0.010 | 0.314 | 0.312 |
| 1 1/4 | 1 1/2 | 3/8×3/8 | 0.377 | 0.375 | 0.375 | 0.374 | 0.213 | 0.219 | 0.376 | 0.375 | 0.169 | 0.175 | 0.010 | 0.377 | 0.375 |
| 1 1/2 | 1 3/4 | 7/16×7/16 | 0.440 | 0.438 | 0.438 | 0.437 | 0.248 | 0.254 | 0.439 | 0.438 | 0.197 | 0.203 | 0.020 | 0.440 | 0.438 |
| 1 3/4 | 2 | 1/2×1/2 | 0.502 | 0.500 | 0.500 | 0.499 | 0.283 | 0.289 | 0.501 | 0.500 | 0.224 | 0.230 | 0.020 | 0.502 | 0.500 |
| 2 | 2 1/2 | 5/8×5/8 | 0.627 | 0.625 | 0.625 | 0.624 | 0.354 | 0.360 | 0.626 | 0.625 | 0.278 | 0.284 | 0.020 | 0.627 | 0.625 |
| 2 1/2 | 3 | 3/4×3/4 | 0.752 | 0.750 | 0.750 | 0.749 | 0.424 | 0.430 | 0.751 | 0.750 | 0.333 | 0.339 | 0.020 | 0.752 | 0.750 |
| 3 | 3 1/2 | 7/8×7/8 | 0.877 | 0.875 | 0.875 | 0.874 | 0.495 | 0.501 | 0.876 | 0.875 | 0.387 | 0.393 | 0.062 | 0.877 | 0.875 |
| 3 1/2 | 4 | 1×1 | 1.003 | 1.000 | 1.000 | 0.999 | 0.566 | 0.572 | 1.001 | 1.000 | 0.442 | 0.448 | 0.062 | 1.003 | 1.000 |
| 4 | 5 | 1 1/4×1 1/4 | 1.253 | 1.250 | 1.250 | 1.248 | 0.707 | 0.713 | 1.252 | 1.250 | 0.551 | 0.557 | 0.062 | 1.253 | 1.250 |
| 5 | 6 | 1 1/2×1 1/2 | 1.504 | 1.500 | 1.500 | 1.498 | 0.848 | 0.854 | 1.502 | 1.500 | 0.661 | 0.667 | 0.062 | 1.504 | 1.500 |

* The key chamfer shall be the minimum to clear the keyway radius.  Nominal values are given.

**Table 3. British Standard Rectangular Taper Keys and Keyways, Gib-head and Plain** (B.S. 46: Part 1: 1958)

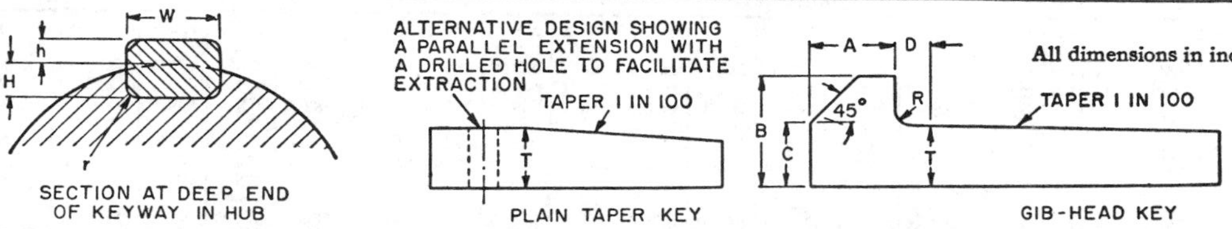
W
h
H
r
SECTION AT DEEP END OF KEYWAY IN HUB
ALTERNATIVE DESIGN SHOWING A PARALLEL EXTENSION WITH A DRILLED HOLE TO FACILITATE EXTRACTION
TAPER 1 IN 100
T
PLAIN TAPER KEY
A
D
45°
R
B
C
TAPER 1 IN 100
T
GIB-HEAD KEY

All dimensions in inches

| Diameter of Shaft | | Key | | | | | Keyway in Shaft and Hub | | | | | | | | Nominal Keyway Radius, r* | Gib-head† | | | | |
|---|---|---|---|---|---|---|---|---|---|---|---|---|---|---|---|---|---|---|---|---|
| Over | Up to and Including | Size, W × T | Width, W | | Thickness, T | | Keyway in Shaft Width, $W_s$ | | Keyway in Hub Width, $W_h$ | | Depth in Shaft, H | | Depth in Hub at Deep End of Keyway, h | | | A | B | C | D | Radius, R |
| | | | Max. | Min. | Max. | Min. | Min. | Max. | Min. | Max. | Min. | Max. | Min. | Max. | | | | | | |
| 1 | 1¼ | 5/16 × ¼ | 0.314 | 0.312 | 0.254 | 0.249 | 0.311 | 0.312 | 0.312 | 0.313 | 0.146 | 0.152 | 0.090 | 0.096 | 0.010 | 3/8 | 7/16 | ¼ | 0.3 | 1/16 |
| 1¼ | 1½ | 3/8 × ¼ | 0.377 | 0.375 | 0.254 | 0.249 | 0.374 | 0.375 | 0.375 | 0.376 | 0.150 | 0.156 | 0.086 | 0.092 | 0.010 | 7/16 | 7/16 | 9/32 | 0.3 | 1/16 |
| 1½ | 1¾ | 7/16 × 5/16 | 0.440 | 0.438 | 0.316 | 0.311 | 0.437 | 0.438 | 0.438 | 0.439 | 0.186 | 0.192 | 0.112 | 0.118 | 0.020 | ½ | 9/16 | 5/16 | 0.4 | 1/16 |
| 1¾ | 2 | ½ × 5/16 | 0.502 | 0.500 | 0.316 | 0.311 | 0.499 | 0.500 | 0.500 | 0.501 | 0.190 | 0.196 | 0.108 | 0.114 | 0.020 | 9/16 | 5/8 | 3/8 | 0.4 | 1/16 |
| 2 | 2½ | 5/8 × 7/16 | 0.627 | 0.625 | 0.442 | 0.437 | 0.624 | 0.625 | 0.625 | 0.626 | 0.260 | 0.266 | 0.162 | 0.168 | 0.020 | 11/16 | ¾ | 7/16 | 0.5 | 1/8 |
| 2½ | 3 | ¾ × ½ | 0.752 | 0.750 | 0.504 | 0.499 | 0.749 | 0.750 | 0.750 | 0.751 | 0.299 | 0.305 | 0.185 | 0.191 | 0.020 | 13/16 | 7/8 | 17/32 | 0.5 | 1/8 |
| 3 | 3½ | 7/8 × 5/8 | 0.877 | 0.875 | 0.630 | 0.624 | 0.874 | 0.875 | 0.875 | 0.876 | 0.370 | 0.376 | 0.239 | 0.245 | 0.062 | 15/16 | 1 | 21/32 | 0.6 | 1/8 |
| 3½ | 4 | 1 × ¾ | 1.003 | 1.000 | 0.755 | 0.749 | 0.999 | 1.000 | 1.000 | 1.001 | 0.441 | 0.447 | 0.293 | 0.299 | 0.062 | 1 1/16 | 1¼ | 23/32 | 0.6 | 1/8 |
| 4 | 5 | 1¼ × 7/8 | 1.253 | 1.250 | 0.880 | 0.874 | 1.248 | 1.250 | 1.250 | 1.252 | 0.518 | 0.524 | 0.340 | 0.346 | 0.062 | 1 5/16 | 1½ | 27/32 | 0.7 | ¼ |
| 5 | 6 | 1½ × 1 | 1.504 | 1.500 | 1.007 | 0.999 | 1.498 | 1.500 | 1.500 | 1.502 | 0.599 | 0.605 | 0.384 | 0.390 | 0.062 | 1 9/16 | 1 5/8 | 1 1/32 | 0.7 | ¼ |
| 6 | 7 | 1¾ × 1¼ | 1.754 | 1.750 | 1.257 | 1.249 | 1.748 | 1.750 | 1.750 | 1.752 | 0.740 | 0.746 | 0.493 | 0.499 | 0.125 | 1 13/16 | 2 | 1 7/32 | 0.8 | ¼ |
| 7 | 8 | 2 × 1 3/8 | 2.005 | 2.000 | 1.382 | 1.374 | 1.998 | 2.000 | 2.000 | 2.002 | 0.818 | 0.824 | 0.539 | 0.545 | 0.125 | 2 1/16 | 2¼ | 1 13/32 | 0.8 | ¼ |
| 8 | 9 | 2¼ × 1½ | 2.255 | 2.250 | 1.509 | 1.499 | 2.248 | 2.250 | 2.250 | 2.252 | 0.897 | 0.905 | 0.581 | 0.589 | 0.125 | 2 5/16 | 2½ | 1 9/16 | 0.9 | 3/8 |
| 9 | 10 | 2½ × 1 5/8 | 2.505 | 2.500 | 1.634 | 1.624 | 2.498 | 2.500 | 2.500 | 2.502 | 0.975 | 0.983 | 0.628 | 0.636 | 0.187 | 2 9/16 | 2¾ | 1 11/16 | 0.9 | 3/8 |
| 10 | 11 | 2¾ × 1 7/8 | 2.755 | 2.750 | 1.884 | 1.874 | 2.748 | 2.750 | 2.750 | 2.752 | 1.114 | 1.122 | 0.738 | 0.746 | 0.187 | 2 13/16 | 3 | 1 15/16 | 1.0 | 3/8 |
| 11 | 12 | 3 × 2 | 3.006 | 3.000 | 2.014 | 1.999 | 2.998 | 3.000 | 3.000 | 3.002 | 1.195 | 1.203 | 0.782 | 0.790 | 0.187 | 3 1/16 | 3¼ | 2 1/16 | 1.0 | 3/8 |

* The key chamfer shall be the minimum to clear the keyway radius. Nominal values shall be given.
† Dimensions *A*, *B*, *C*, *D* and *R* pertain to gib-head keys only.

## Table 4. British Standard Square Taper Keys and Keyways, Gib-head or Plain (B.S. 46: Part 1: 1958)

All dimensions in inches

ALTERNATIVE DESIGN SHOWING A PARALLEL EXTENSION WITH A DRILLED HOLE TO FACILITATE EXTRACTION.

TAPER 1 IN 100 — GIB-HEAD KEY

TAPER 1 IN 100 — PLAIN TAPER KEY

SECTION AT DEEP END OF KEYWAY IN HUB

| Dia. of Shaft Over | Up to and Incl. | Size W×T | Key Width W Max | Key Width W Min | Key Thick. T Max | Key Thick. T Min | Keyway in Shaft Ws Min | Keyway in Shaft Ws Max | Keyway in Hub Wh Min | Keyway in Hub Wh Max | Depth in Shaft H Min | Depth in Shaft H Max | Depth in Hub at Deep End h Min | Depth in Hub at Deep End h Max | Nom. Keyway Radius r* | Gib A | Gib B | Gib H | Gib C | Gib D | Gib Radius R |
|---|---|---|---|---|---|---|---|---|---|---|---|---|---|---|---|---|---|---|---|---|---|
| 1/4 | 1/2 | 1/8×1/8 | 0.127 | 0.125 | 0.129 | 0.124 | 0.124 | 0.125 | 0.125 | 0.126 | 0.072 | 0.078 | 0.039 | 0.045 | 0.010 | 3/16 | 1/4 | 1/8 | 5/32 | 0.1 | 1/32 |
| 1/2 | 3/4 | 3/16×3/16 | 0.190 | 0.188 | 0.192 | 0.187 | 0.187 | 0.188 | 0.188 | 0.189 | 0.107 | 0.113 | 0.067 | 0.073 | 0.010 | 1/4 | 3/8 | 3/16 | 7/32 | 0.2 | 1/32 |
| 3/4 | 1 | 1/4×1/4 | 0.252 | 0.250 | 0.254 | 0.249 | 0.249 | 0.250 | 0.250 | 0.251 | 0.142 | 0.148 | 0.094 | 0.100 | 0.010 | 5/16 | 7/16 | 1/4 | 9/32 | 0.2 | 1/16 |
| 1 | 1 1/4 | 5/16×5/16 | 0.314 | 0.312 | 0.316 | 0.311 | 0.311 | 0.312 | 0.312 | 0.313 | 0.177 | 0.183 | 0.121 | 0.127 | 0.010 | 3/8 | 9/16 | 5/16 | 11/32 | 0.3 | 1/16 |
| 1 1/4 | 1 1/2 | 3/8×3/8 | 0.377 | 0.375 | 0.379 | 0.374 | 0.374 | 0.375 | 0.375 | 0.376 | 0.213 | 0.219 | 0.148 | 0.154 | 0.010 | 7/16 | 5/8 | 3/8 | 13/32 | 0.3 | 1/16 |
| 1 1/2 | 1 3/4 | 7/16×7/16 | 0.440 | 0.438 | 0.442 | 0.437 | 0.437 | 0.438 | 0.438 | 0.439 | 0.248 | 0.254 | 0.175 | 0.181 | 0.020 | 1/2 | 3/4 | 7/16 | 15/32 | 0.4 | 1/16 |
| 1 3/4 | 2 | 1/2×1/2 | 0.502 | 0.500 | 0.504 | 0.499 | 0.499 | 0.500 | 0.500 | 0.501 | 0.283 | 0.289 | 0.202 | 0.208 | 0.020 | 9/16 | 7/8 | 1/2 | 17/32 | 0.4 | 1/16 |
| 2 | 2 1/2 | 5/8×5/8 | 0.627 | 0.625 | 0.630 | 0.624 | 0.624 | 0.625 | 0.625 | 0.626 | 0.354 | 0.360 | 0.256 | 0.262 | 0.020 | 11/16 | 1 1/16 | 5/8 | 21/32 | 0.5 | 1/8 |
| 2 1/2 | 3 | 3/4×3/4 | 0.752 | 0.750 | 0.755 | 0.749 | 0.749 | 0.750 | 0.750 | 0.751 | 0.424 | 0.430 | 0.310 | 0.316 | 0.020 | 13/16 | 1 1/4 | 3/4 | 25/32 | 0.5 | 1/8 |
| 3 | 3 1/2 | 7/8×7/8 | 0.877 | 0.875 | 0.880 | 0.874 | 0.874 | 0.875 | 0.875 | 0.876 | 0.495 | 0.501 | 0.364 | 0.370 | 0.062 | 15/16 | 1 3/8 | 7/8 | 29/32 | 0.6 | 1/8 |
| 3 1/2 | 4 | 1×1 | 1.003 | 1.000 | 1.007 | 0.999 | 0.999 | 1.000 | 1.000 | 1.001 | 0.566 | 0.572 | 0.418 | 0.424 | 0.062 | 1 1/16 | 1 5/8 | 1 | 1 1/16 | 0.6 | 1/8 |
| 4 | 5 | 1 1/4×1 1/4 | 1.253 | 1.250 | 1.257 | 1.249 | 1.248 | 1.250 | 1.250 | 1.252 | 0.707 | 0.713 | 0.526 | 0.532 | 0.062 | 1 5/16 | 2 | 1 1/4 | 1 9/32 | 0.7 | 1/4 |
| 5 | 6 | 1 1/2×1 1/2 | 1.504 | 1.500 | 1.509 | 1.499 | 1.498 | 1.500 | 1.500 | 1.502 | 0.848 | 0.854 | 0.635 | 0.641 | 0.062 | 1 9/16 | 2 1/2 | 1 1/2 | 1 17/32 | 0.7 | 1/4 |

\* The key chamfer shall be the minimum to clear the keyway radius. Nominal values shall be given.

† Dimensions A, B, C, D and R pertain to gib-head keys only.

## Table 5. British Standard Woodruff Keys and Keyways (B.S. 46: Part I: 1958)

All dimensions are in inches

RADIUS 0.005 0.010   VIEW IN DIRECTION OF ARROW X   RADIUS OPTIONAL   ½ D. NOMINAL   OPTIONAL DESIGN

| Key and Cutter No. | Nominal Fractional Size of Key | | Diameter of Key, A | | Depth of Key, B | | Thickness of Key, C | | Keyway Width in Shaft, D | | Width of Keyway in Hub, E | | Keyway Depth in Shaft, F | | Keyway Depth in Hub at Center Line, G | | Depth of Key (Optional Design), H | | Dimension, J |
|---|---|---|---|---|---|---|---|---|---|---|---|---|---|---|---|---|---|---|---|
| | Width | Dia. | Max. | Min. | Max. | Min. | Max. | Min. | Min. | Max. | Min. | Max. | Min. | Max. | Min. | Max. | Max. | Min. | Nom. |
| 203 | 1/16 | 3/8 | 0.375 | 0.370 | .171 | .166 | .063 | .062 | .061 | .063 | .063 | .065 | .135 | .140 | .042 | .047 | .162 | .156 | 1/64 |
| 303 | 3/32 | 3/8 | 0.375 | 0.370 | .171 | .166 | .095 | .094 | .093 | .095 | .095 | .097 | .119 | .124 | .057 | .062 | .162 | .156 | 1/64 |
| 403 | 1/8 | 3/8 | 0.375 | 0.370 | .171 | .166 | .126 | .125 | .124 | .126 | .126 | .128 | .104 | .109 | .073 | .078 | .162 | .156 | 1/64 |
| 204 | 1/16 | 1/2 | 0.500 | 0.490 | .203 | .198 | .063 | .062 | .061 | .063 | .063 | .065 | .167 | .172 | .042 | .047 | .194 | .188 | 3/64 |
| 304 | 3/32 | 1/2 | 0.500 | 0.490 | .203 | .198 | .095 | .094 | .093 | .095 | .095 | .097 | .151 | .156 | .057 | .062 | .194 | .188 | 3/64 |
| 404 | 1/8 | 1/2 | 0.500 | 0.490 | .203 | .198 | .126 | .125 | .124 | .126 | .126 | .128 | .136 | .141 | .073 | .078 | .194 | .188 | 3/64 |
| 305 | 3/32 | 5/8 | 0.625 | 0.615 | .250 | .245 | .095 | .094 | .093 | .095 | .095 | .097 | .198 | .203 | .057 | .062 | .240 | .234 | 1/16 |
| 405 | 1/8 | 5/8 | 0.625 | 0.615 | .250 | .245 | .126 | .125 | .124 | .126 | .126 | .128 | .182 | .187 | .073 | .078 | .240 | .234 | 1/16 |
| 505 | 5/32 | 5/8 | 0.625 | 0.615 | .250 | .245 | .157 | .156 | .155 | .157 | .157 | .159 | .167 | .172 | .089 | .094 | .240 | .234 | 1/16 |
| 406 | 1/8 | 3/4 | 0.750 | 0.740 | .313 | .308 | .126 | .125 | .124 | .126 | .126 | .128 | .246 | .251 | .073 | .078 | .303 | .297 | 1/16 |
| 506 | 5/32 | 3/4 | 0.750 | 0.740 | .313 | .308 | .157 | .156 | .155 | .157 | .157 | .159 | .230 | .235 | .089 | .094 | .303 | .297 | 1/16 |
| 606 | 3/16 | 3/4 | 0.750 | 0.740 | .313 | .308 | .189 | .188 | .187 | .189 | .189 | .191 | .214 | .219 | .104 | .109 | .303 | .297 | 1/16 |
| 507 | 5/32 | 7/8 | 0.875 | 0.865 | .375 | .370 | .157 | .156 | .155 | .157 | .157 | .159 | .292 | .297 | .089 | .094 | .365 | .359 | 1/16 |
| 607 | 3/16 | 7/8 | 0.875 | 0.865 | .375 | .370 | .189 | .188 | .187 | .189 | .189 | .191 | .276 | .281 | .104 | .109 | .365 | .359 | 1/16 |
| 807 | 1/4 | 7/8 | 0.875 | 0.865 | .375 | .370 | .251 | .250 | .249 | .251 | .251 | .253 | .245 | .250 | .104 | .109 | .365 | .359 | 1/16 |
| 608 | 3/16 | 1 | 1.000 | 0.990 | .438 | .433 | .189 | .188 | .187 | .189 | .189 | .191 | .339 | .344 | .136 | .141 | .428 | .422 | 1/16 |
| 808 | 1/4 | 1 | 1.000 | 0.990 | .438 | .433 | .251 | .250 | .249 | .251 | .251 | .253 | .308 | .313 | .136 | .141 | .428 | .422 | 1/16 |
| 1008 | 5/16 | 1 | 1.000 | 0.990 | .438 | .433 | .313 | .312 | .311 | .313 | .313 | .315 | .277 | .282 | .136 | .141 | .428 | .422 | 1/16 |
| 609 | 3/16 | 1 1/8 | 1.125 | 1.115 | .484 | .479 | .189 | .188 | .187 | .189 | .189 | .191 | .385 | .390 | .167 | .172 | .475 | .469 | 5/64 |
| 809 | 1/4 | 1 1/8 | 1.125 | 1.115 | .484 | .479 | .251 | .250 | .249 | .251 | .251 | .253 | .354 | .359 | .167 | .172 | .475 | .469 | 5/64 |
| 1009 | 5/16 | 1 1/8 | 1.125 | 1.115 | .484 | .479 | .313 | .312 | .311 | .313 | .313 | .315 | .323 | .328 | .136 | .141 | .475 | .469 | 5/64 |
| 810 | 1/4 | 1 1/4 | 1.250 | 1.240 | .547 | .542 | .251 | .250 | .249 | .251 | .251 | .253 | .417 | .422 | .167 | .172 | .537 | .531 | 5/64 |
| 1010 | 5/16 | 1 1/4 | 1.250 | 1.240 | .547 | .542 | .313 | .312 | .311 | .313 | .313 | .315 | .386 | .391 | .167 | .172 | .537 | .531 | 5/64 |
| 1210 | 3/8 | 1 1/4 | 1.250 | 1.240 | .547 | .542 | .376 | .375 | .374 | .376 | .376 | .378 | .354 | .359 | .198 | .203 | .537 | .531 | 5/64 |
| 1011 | 5/16 | 1 3/8 | 1.375 | 1.365 | .594 | .589 | .313 | .312 | .311 | .313 | .313 | .315 | .433 | .438 | .198 | .203 | .584 | .578 | 5/64 |
| 1211 | 3/8 | 1 3/8 | 1.375 | 1.365 | .594 | .589 | .376 | .375 | .374 | .376 | .376 | .378 | .402 | .407 | .198 | .203 | .584 | .578 | 5/64 |
| 812 | 1/4 | 1 1/2 | 1.500 | 1.490 | .641 | .636 | .251 | .250 | .249 | .251 | .251 | .253 | .511 | .516 | .167 | .172 | .631 | .625 | 3/32 |
| 1012 | 5/16 | 1 1/2 | 1.500 | 1.490 | .641 | .636 | .313 | .312 | .311 | .313 | .313 | .315 | .480 | .485 | .198 | .203 | .631 | .625 | 7/64 |
| 1212 | 3/8 | 1 1/2 | 1.500 | 1.490 | .641 | .636 | .376 | .375 | .374 | .376 | .376 | .378 | .448 | .453 | .198 | .203 | .631 | .625 | 7/64 |

**Table 6. British Preferred Lengths of Plain (Parallel or Taper) and Gib-head Keys, Rectangular and Square Section (B.S. 46: Part I: 1958 Appendix)**

All dimensions are in inches

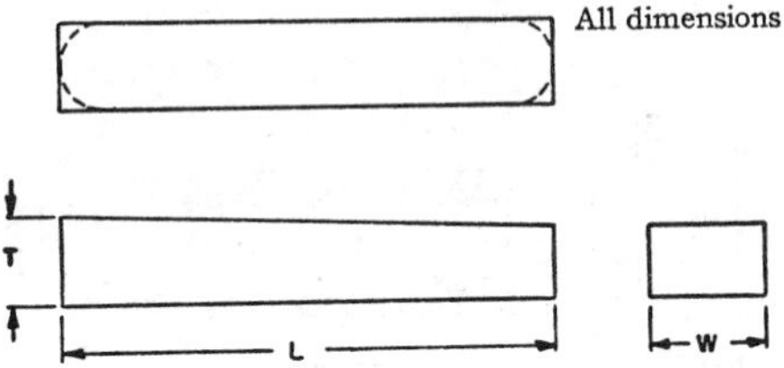

| Plain Key Size $W \times T$ | Overall Length, $L$ | | | | | | | | | | | | | | |
|---|---|---|---|---|---|---|---|---|---|---|---|---|---|---|---|
| | 3/4 | 1 | 1¼ | 1½ | 1¾ | 2 | 2¼ | 2½ | 2¾ | 3 | 3½ | 4 | 4½ | 5 | 6 |
| 1/8 × 1/8 | X | X | | | | | | | | | | | | | |
| 3/16 × 3/16 | X | X | X | X | X | X | | | | | | | | | |
| 1/4 × 1/4 | X | X | X | X | X | X | X | X | X | X | X | | | | |
| 5/16 × 1/4 | X | X | X | X | X | X | X | X | X | X | X | | | | |
| 5/16 × 5/16 | X | X | X | X | X | X | X | X | X | X | | | | | |
| 3/8 × 1/4 | | X | X | X | X | X | X | X | X | X | X | | | | |
| 3/8 × 3/8 | | | X | X | X | X | X | X | X | X | X | X | | | |
| 7/16 × 5/16 | | | | | X | X | X | X | X | X | X | X | X | | |
| 7/16 × 7/16 | | | | | | X | X | X | X | X | X | X | X | | |
| 1/2 × 5/16 | | | | X | X | X | X | X | X | X | X | X | X | X | |
| 1/2 × 1/2 | | | | | | | X | X | X | X | X | X | X | X | |
| 5/8 × 7/16 | | | | | | | | X | X | X | X | X | X | X | |
| 5/8 × 5/8 | | | | | | | | | | X | X | X | X | X | X |
| 3/4 × 1/2 | | | | | | | | | | | X | X | X | X | X |
| 3/4 × 3/4 | | | | | | | | | | | X | X | X | X | X |
| 7/8 × 5/8 | | | | | | | | | | | | | X | X | X |

All dimensions are in inches

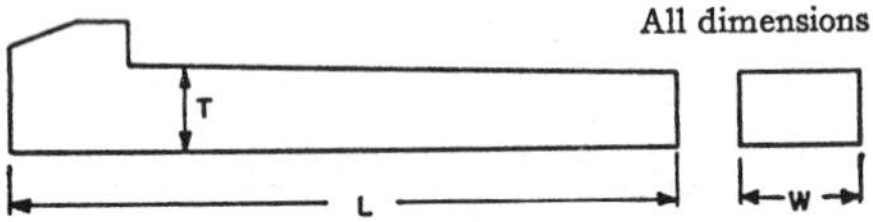

| Gib-head Key Size, $W \times T$ | Overall Length, $L$ | | | | | | | | | | | | | | | | |
|---|---|---|---|---|---|---|---|---|---|---|---|---|---|---|---|---|---|
| | 1½ | 1¾ | 2 | 2¼ | 2½ | 2¾ | 3 | 3½ | 4 | 4½ | 5 | 5½ | 6 | 6½ | 7 | 7½ | 8 |
| 3/16 × 3/16 | X | X | X | X | X | | X | | | | | | | | | | |
| 1/4 × 1/4 | X | X | X | X | X | X | X | X | X | X | | | | | | | |
| 5/16 × 1/4 | | | X | X | X | X | X | X | X | X | | | | | | | |
| 5/16 × 5/16 | | | X | X | X | X | X | X | X | X | X | | | | | | |
| 3/8 × 1/4 | | | X | X | X | X | X | X | X | X | X | | | | | | |
| 3/8 × 3/8 | | | | X | X | X | X | X | X | X | X | X | X | | | | |
| 7/16 × 5/16 | | | | | | | X | X | X | X | X | X | X | X | | | |
| 7/16 × 7/16 | | | | | | | X | X | X | X | X | X | X | X | | | |
| 1/2 × 5/16 | | | | | | X | X | X | X | X | X | X | X | X | X | | |
| 1/2 × 1/2 | | | | | | X | X | X | X | X | X | X | X | X | X | | |
| 5/8 × 7/16 | | | | | | | X | X | X | X | X | X | X | X | X | X | |
| 5/8 × 5/8 | | | | | | | | X | X | X | X | X | X | X | X | X | X |
| 3/4 × 1/2 | | | | | | | | | X | X | X | X | X | X | X | X | X |
| 3/4 × 3/4 | | | | | | | | | X | X | X | X | X | X | X | X | X |
| 7/8 × 5/8 | | | | | | | | | | | X | X | X | X | X | X | X |
| 7/8 × 7/8 | | | | | | | | | | | | | X | X | X | X | X |
| 1 × 3/4 | | | | | | | | | | | | | X | X | X | X | X |
| 1 × 1 | | | | | | | | | | | | | | X | X | X | X |

## SPLINES AND SERRATIONS

A splined shaft is one having a series of parallel keys formed integrally with the shaft and mating with corresponding grooves cut in a hub or fitting; this is in contrast to a shaft having a series of keys or feathers fitted into slots cut into the shaft. This latter construction weakens the shaft to a considerable degree because of the slots cut into it and, as a consequence, reduces its torque-transmitting capacity.

Splined shafts are most generally used in three types of applications: (1) for coupling shafts when relatively heavy torques are to be transmitted without slippage; (2) for transmitting power to slidably-mounted or permanently-fixed gears, pulleys, and other rotating members; and (3) for attaching parts that may require removal for indexing or change in angular position.

Splines having straight-sided teeth have been used in many applications (see SAE Parallel Side Splines for Soft Broached Holes in Fittings); however, the use of splines with teeth of involute profile has steadily increased since (1) involute spline couplings have greater torque-transmitting capacity than any other type; (2) they can be produced by the same techniques and equipment as is used to cut gears; and (3) they have a self-centering action under load even when there is backlash between mating members.

### Involute Splines

**American National Standard Involute Splines.** — These splines or multiple keys are similar in form to internal and external involute gears having a pressure angle (see diagram) of 30 degrees. The general practice is to form the external splines either by hobbing, rolling, or on a gear shaper, and internal splines either by broaching or on a gear shaper. The internal spline is held to basic dimensions and the external spline is varied to control the fit. Involute splines have maximum strength at the base, they can be accurately spaced and are self-centering, thus equalizing the bearing and stresses, and they can be measured and fitted accurately.

In the new standard, ANSI B92.1-1970 most of the features of the 1960 standard are retained; plus the addition of three tolerance classes, for a total of four. The term "involute serration", formerly applied to involute splines with 45-degree pressure angle, has been deleted and the standard now includes involute splines with 30-, 37.5- and 45-degree pressure angles. Tables for these splines have been re-arranged accordingly. The term "serration" will no longer apply to splines covered by this standard.

The revised standard now has only one fit class for all side fit splines; the former Class 2 fit. Class 1 fit has been deleted because of its infrequent use. The major diameter of the flat root side fit spline has been changed and a tolerance applied to include the range of the 1950 and the 1960 standards. The interchangeability limitations with splines made to previous standards are given later in the section entitled "Interchangability."

There have been no tolerance nor fit changes to the major diameter fit section.

This revision recognizes the fact that proper assembly between mating splines is dependent only on the spline being within effective specifications from the tip of the tooth to the form diameter. Therefore, on side fit splines, the internal component major diameter now is shown as a maximum dimension and the external component minor diameter is shown as a minimum dimension. The minimum internal major diameter and the maximum external minor diameter must clear the specified form diameter and thus do not need any additional control.

The spline specification tables now include a greater number of tolerance level selections. These new tolerance classes were added for greater selection to suit end product needs. The selections differ only in the tolerance as applied to space width and tooth thickness. The tolerance class which was used in ASA B5.15-1960 is the basis and is now designated as tolerance Class 5. The new tolerance classes are based on the following formulas:

$$\text{Tolerance Class 4} = \text{Tolerance Class 5} \times 0.71$$
$$\text{Tolerance Class 6} = \text{Tolerance Class 5} \times 1.40$$
$$\text{Tolerance Class 7} = \text{Tolerance Class 5} \times 2.00$$

All dimensions, listed in this standard, are for the finished part. Therefore, any compensation that must be made for operations which take place during processing, such as heat treatment, must be taken into account when selecting the tolerance level for manufacturing.

The standard has the same internal minimum effective space width and external maximum effective tooth thickness for all tolerance classes and has two types of fit. For tooth side fits, the minimum effective space width and the maximum effective tooth thickness are of equal value. This basic concept makes it possible to have interchangeable assembly between mating splines where made to this standard regardless of the tolerance class of the individual members. This allows a tolerance class "mix" of mating members which often is an advantage where one member is considerably less difficult to produce than its mate, and the "average" tolerance applied to the two units is such that it satisfies the design need. For instance, this can be the result of specification of Class 5 tolerance to one member and Class 7 to its mate, thus providing an assembly tolerance in the Class 6 range. The maximum effective tooth thickness is less than the minimum effective space width for major diameter fits to allow for eccentricity variations.

In the event the fit as provided in this standard does not satisfy a particular design need and a specific amount of effective clearance or press fit is desired, the change should be made only to the external spline by a reduction or an increase in effective tooth thickness and a like change in actual tooth thickness. The minimum effective space width, in this standard, is always basic. This basic minimum effective space width should always be retained when special designs are derived from the concept of this standard.

**Terms Applied to Involute Splines.** — The following definitions of involute spline terms, here listed in alphabetical order, are given in the American National Standard:

*Active Spline Length* ($L_a$) is the length of spline which contacts the mating spline. On sliding splines it exceeds the length of engagement.

*Actual Space Width* ($s$) is the circular width on the pitch circle of any single space considering an infinitely thin increment of axial spline length.

*Actual Tooth Thickness* ($t$) is the circular thickness on the pitch circle of any single tooth considering an infinitely thin increment of axial spline length.

*Alignment Variation* is the variation of the effective spline axis with respect to the reference axis (see Fig. 1).

*Base Circle* is the circle from which involute spline tooth profiles are constructed.

*Base Diameter* ($D_b$) is the diameter of the base circle.

*Basic Space Width* is the basic space width for 30-degree pressure angle splines; half the circular pitch. The basic space width for 37.5- and 45-degree pressure angle splines, however, is greater than half the circular pitch. The teeth are proportioned so that the external tooth, at its base, has about the same thickness as the internal tooth at the form diameter. This results in greater minor diameters than those of comparable involute splines of 30-degree pressure angle.

*Circular Pitch* ($p$) is the distance along the pitch circle between corresponding points of adjacent spline teeth.

*Depth of Engagement* is the radial distance from the minor circle of the internal spline to the major circle of the external spline, minus corner clearance and/or chamfer depth.

*Diametral Pitch* (*P*) is the number of spline teeth per inch of pitch diameter. The diametral pitch determines the circular pitch and the basic space width or tooth thickness. In conjunction with the number of teeth, it also determines the pitch diameter. (See also *Pitch*.)

*Effective Clearance* ($c_v$) is the effective space width of the internal spline minus the effective tooth thickness of the mating external spline.

*Effective Space Width,* ($s_v$) of an internal spline is equal to the circular tooth thickness on the pitch circle of an imaginary perfect external spline which would fit the internal spline without looseness or interference considering engagement of the entire axial length of the spline. The minimum effective space width of the internal spline is always basic, as shown in Table 3. Fit variations may be obtained by adjusting the tooth thickness of the external spline.

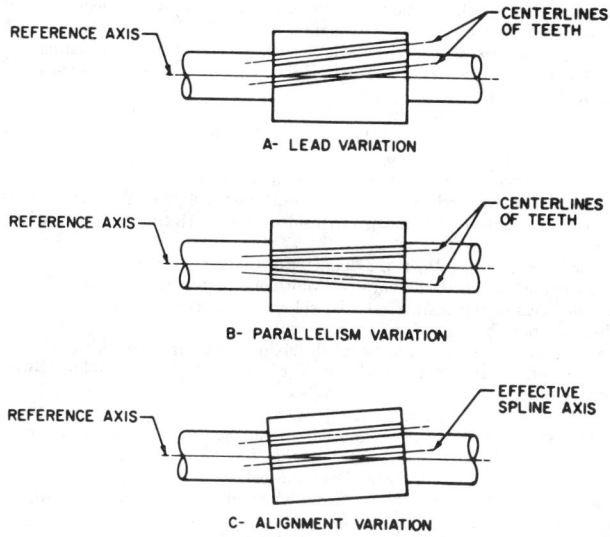

Fig. 1.   Three types of involute spline variations.

*Effective Tooth Thickness* ($t_v$) is the effective tooth thickness of an external spline, equal to the circular space width on the pitch circle of an imaginary perfect internal spline which would fit the external spline without looseness or interference, considering engagement of the entire axial length of the spline.

*Effective Variation* is the accumulated effect of the spline variations on the fit with the mating part.

*External Spline* is a spline formed on the outer surface of a cylinder.

*Fillet* is the concave portion of the tooth profile which joins the sides to the bottom of the space.

*Fillet Root Splines* are those in which a single fillet in the general form of an arc joins the sides of adjacent teeth.

*Flat Root Splines* are those in which fillets join the arcs of major or minor circles to the tooth sides.

*Form Circle* is the circle which defines the deepest points of involute form control of the tooth profile. This circle along with the tooth tip circle (or start of chamfer circle) determines the limits of tooth profile requiring control. It is located near the major circle on the internal spline and near the minor circle on the external spline.

*Form Clearance* ($c_F$) is the radial depth of involute profile beyond the depth of engagement with the mating part. It allows for looseness between mating splines and for eccentricities between the minor circle (internal), the major circle (external), and their respective pitch circles.

*Form Diameter* ($D_{Fe}$, $D_{Fi}$) the diameter of the form circle.

*Internal Spline* is a spline formed on the inner surface of a cylinder.

*Involute Spline* is one having teeth with involute profiles.

*Lead Variation* is the variation of the direction of the spline tooth from its intended direction parallel to the reference axis, also including parallelism and alignment variations (See Fig. 1). *Note*: Straight (nonhelical) splines have an infinite lead.

*Length of Engagement* ($L_q$) is the axial length of contact between mating splines.

*Machining Tolerance* ($m$) is the permissible variation in actual space width or actual tooth thickness.

*Major Circle* is the circle formed by the outermost surface of the spline. It is the outside circle (tooth tip circle) of the external spline or the root circle of the internal spline.

*Major Diameter* ($D_o$, $D_{ri}$) is the diameter of the major circle.

*Minor Circle* is the circle formed by the innermost surface of the spline. It is the root circle of the external spline or the inside circle (tooth tip circle) of the internal spline.

*Minor Diameter* ($D_{re}$, $D_i$) is the diameter of the minor circle.

*Nominal Clearance* is the actual space width of an internal spline minus the actual tooth thickness of the mating external spline. It does not define the fit between mating members, because of the effect of variations.

*Out of Roundness* is the variation of the spline from a true circular configuration.

*Parallelism Variation* is the variation of parallelism of a single spline tooth with respect to any other single spline tooth (See Fig. 1).

*Pitch* ($P/P_s$) is a combination number of a one-to-two ratio indicating the spline proportions; the upper or first number is the diametral pitch, the lower or second number is the stub pitch and denotes, as that fractional part of an inch, the basic radial length of engagement, both above and below the pitch circle.

*Pitch Circle* is the reference circle from which all transverse spline tooth dimensions are constructed.

*Pitch Diameter* ($D$) is the diameter of the pitch circle.

*Pitch Point* is the intersection of the spline tooth profile with the pitch circle.

*Pressure Angle* ($\phi$) is the angle between a line tangent to an involute and a radial line through the point of tangency. Unless otherwise specified, it is the standard pressure angle.

*Profile Variation* is any variation from the specified tooth profile normal to the flank.

*Spline* is a machine element consisting of integral keys (spline teeth) or keyways (spaces) equally spaced around a circle or portion thereof.

*Standard (Main) Pressure Angle*, ($\phi_D$) is the pressure angle at the specified pitch diameter.

*Stub Pitch* ($P_s$) is a number used to denote the radial distance from the pitch circle to the major circle of the external spline and from the pitch circle to the minor circle of the internal spline. The stub pitch for splines in this standard is twice the diametral pitch.

*Total Index Variation* is the greatest difference in any two teeth (adjacent or otherwise) between the actual and the perfect spacing of the tooth profiles.

*Total Tolerance* $(m + \lambda)$ is the machining tolerance plus the variation allowance.

*Variation Allowance* $(\lambda)$ is the permissible effective variation.

Many of these terms and their associated symbols are illustrated in the diagrams in Tables 6 to 10, incl.

**Tooth Proportions.** — There are 17 pitches: 2.5/5, 3/6, 4/8, 5/10, 6/12, 8/16, 10/20, 12/24, 16/32, 20/40, 24/48, 32/64, 40/80, 48/96, 64/128, 80/160, and 128/256. The numerator in this fractional designation is known as the diametral pitch and controls the pitch diameter; the denominator, which is always double the numerator, is known as the stub pitch and controls the tooth depth. For convenience in calculation, only the numerator is used in the formulas given and is designated as $P$. Diametral pitch, as in gears, means the number of teeth per inch of pitch diameter.

Table 1 shows the symbols and Table 2 the formulas for basic tooth dimensions of involute spline teeth of various pitches. Basic dimensions are given in Table 3.

**Table. 1   American National Standard Involute Spline Symbols** (ANSI B92.1-1970)

| | | | |
|---|---|---|---|
| $c_v$ | effective clearance | $M_i$ | measurement between pins, internal spline |
| $c_F$ | form clearance | $N$ | number of teeth |
| $D$ | pitch diameter | $P$ | diametral pitch |
| $D_b$ | base diameter | $P_s$ | stub pitch |
| $D_{ci}$ | pin contact diameter, internal spline | $p$ | circular pitch |
| $D_{ce}$ | pin contact diameter, external spline | $r_f$ | fillet radius |
| | | $s$ | actual space width, circular |
| $D_{Fe}$ | form diameter, external spline | $s_v$ | effective space width, circular |
| $D_{Fi}$ | form diameter, internal spline | $s_c$ | allowable compressive stress, psi |
| $D_i$ | minor diameter, internal spline | $s_s$ | allowable shear stress, psi |
| $D_o$ | major diameter, external spline | $t$ | actual tooth thickness, circular |
| $D_{re}$ | minor diameter, external spline (root) | $t_v$ | effective tooth thickness, circular |
| $D_{ri}$ | major diameter, internal spline (root) | $\lambda$ | variation allowance |
| | | $\epsilon$ | involute roll angle |
| $d_e$ | diameter of measuring pin for external spline | $\phi$ | pressure angle |
| | | $\phi_D$ | standard pressure angle |
| $d_i$ | diameter of measuring pin for internal spline | $\phi_{ci}$ | pressure angle at contact diameter, internal spline |
| $K_e$ | change factor for external spline | $\phi_{ce}$ | pressure angle at contact diameter, external spline |
| $K_i$ | change factor for internal spline | $\phi_i$ | pressure angle at pin center, internal spline |
| $L$ | spline length | | |
| $L_a$ | active spline length | $\phi_e$ | pressure angle at pin center, external spline |
| $L_g$ | length of engagement | | |
| $m$ | machining tolerance | $\phi_F$ | pressure angle at form diameter |
| $M_e$ | measurement over pins, external spline | | |

**Tooth Numbers.** — The American National Standard covers involute splines having tooth numbers ranging from 6 to 60 with a 30- or 37.5-degree pressure angle and from 6 to 100 with a 45-degree pressure angle. In selecting the number of teeth

**Table 2. Formulas for Basic Dimensions (ANSI B92.1-1970)**

| Term | Symbol | Formula — 30 deg $\phi D$<br>Flat Root Side Fit<br>2.5/5-32/64 Pitch | Formula — 30 deg $\phi D$<br>Flat Root Major Dia Fit<br>3/6-16/32 Pitch | Formula — 30 deg $\phi D$<br>Fillet Root Side Fit<br>2.5/5-48/96 Pitch | Formula — 37.5 deg $\phi D$<br>Fillet Root Side Fit<br>2.5/5-48/96 Pitch | Formula — 45 deg $\phi D$<br>Fillet Root Side Fit<br>10/20-128/256 Pitch |
|---|---|---|---|---|---|---|
| Stub Pitch | $P_s$ | $2P$ | $2P$ | $2P$ | $2P$ | $2P$ |
| Pitch Diameter | $D$ | $\dfrac{N}{P}$ | $\dfrac{N}{P}$ | $\dfrac{N}{P}$ | $\dfrac{N}{P}$ | $\dfrac{N}{P}$ |
| Base Diameter | $D_b$ | $D\cos\phi D$ | $D\cos\phi D$ | $D\cos\phi D$ | $D\cos\phi D$ | $D\cos\phi D$ |
| Circular Pitch | $p$ | $\dfrac{\pi}{P}$ | $\dfrac{\pi}{P}$ | $\dfrac{\pi}{P}$ | $\dfrac{\pi}{P}$ | $\dfrac{\pi}{P}$ |
| Minimum Effective Space Width | $s_v$ | $\dfrac{\pi}{2P}$ | $\dfrac{\pi}{2P}$ | $\dfrac{\pi}{2P}$ | $\dfrac{0.5\pi+0.1}{P}$ | $\dfrac{0.5\pi+0.2}{P}$ |
| Major Diameter, Internal | $D_{ri}$ | $\dfrac{N+1.35}{P}$ | $\dfrac{N+1}{P}$ | $\dfrac{N+1.8}{P}$ | $\dfrac{N+1.6}{P}$ | $\dfrac{N+1.4}{P}$ |
| Major Diameter, External | $D_o$ | $\dfrac{N+1}{P}$ | $\dfrac{N+1}{P}$ | $\dfrac{N+1}{P}$ | $\dfrac{N+1}{P}$ | $\dfrac{N+1}{P}$ |
| Minor Diameter, Internal | $D_i$ | $\dfrac{N-1}{P}$ | $\dfrac{N-1}{P}$ | $\dfrac{N-1}{P}$ | $\dfrac{N-0.8}{P}$ | $\dfrac{N-0.6}{P}$ |
| Minor Dia. Ext. — 2.5/5 thru 12/24 pitch | $D_{re}$ | $\dfrac{N-1.35}{P}$ | $\dfrac{N-1.35}{P}$ | $\dfrac{N-1.8}{P}$ | $\dfrac{N-1.3}{P}$ | … |
| Minor Dia. Ext. — 16/32 pitch and finer | | | | $\dfrac{N-2}{P}$ | | |
| Minor Dia. Ext. — 10/20 pitch and finer | | | | | | $\dfrac{N-1}{P}$ |
| Form Diameter, Internal | $D_{Fi}$ | $\dfrac{N+1}{P}+2c_F$ | $\dfrac{N+0.8}{P}-0.004+2c_F$ | $\dfrac{N+1}{P}+2c_F$ | $\dfrac{N+1}{P}+2c_F$ | $\dfrac{N+1}{P}+2c_F$ |
| Form Diameter, External | $D_{Fe}$ | $\dfrac{N-1}{P}-2c_F$ | $\dfrac{N-1}{P}-2c_F$ | $\dfrac{N-1}{P}-2c_F$ | $\dfrac{N-0.8}{P}-2c_F$ | $\dfrac{N-0.6}{P}-2c_F$ |
| Form Clearance (Radial) | $c_F$ | $0.001D$, with max of 0.010, min of 0.002 | | | | |

NOTE: All spline specification table dimensions in the standard are derived from these basic formulas by application of tolerances.

$\pi = 3.1415927$

Table 3.  Basic Dimensions for Involute Splines (ANSIB 92.1-1970)

| Pitch, $P/P_s$ | Circular Pitch, $p$ | Min Effective Space Width (BASIC), $S_v$ min | | | Pitch, $P/P_s$ | Circular Pitch, $p$ | Min Effective Space Width (BASIC), $S_v$ min | | |
|---|---|---|---|---|---|---|---|---|---|
| | | 30 deg $\phi$ | 37.5 deg $\phi$ | 45 deg $\phi$ | | | 30 deg $\phi$ | 37.5 deg $\phi$ | 45 deg $\phi$ |
| 2.5/5 | 1.2566 | 0.6283 | 0.6683 | ... | 20/40 | 0.1571 | 0.0785 | 0.0835 | 0.0885 |
| 3/6 | 1.0472 | 0.5236 | 0.5569 | ... | 24/48 | 0.1309 | 0.0654 | 0.0696 | 0.0738 |
| 4/8 | 0.7854 | 0.3927 | 0.4177 | ... | 32/64 | 0.0982 | 0.0491 | 0.0522 | 0.0553 |
| 5/10 | 0.6283 | 0.3142 | 0.3342 | ... | 40/80 | 0.0785 | 0.0393 | 0.0418 | 0.0443 |
| 6/12 | 0.5236 | 0.2618 | 0.2785 | ... | 48/96 | 0.0654 | 0.0327 | 0.0348 | 0.0369 |
| 8/16 | 0.3927 | 0.1963 | 0.2088 | ... | 64/128 | 0.0491 | ... | ... | 0.0277 |
| 10/20 | 0.3142 | 0.1571 | 0.1671 | 0.1771 | 80/160 | 0.0393 | ... | ... | 0.0221 |
| 12/24 | 0.2618 | 0.1309 | 0.1392 | 0.1476 | 128/256 | 0.0246 | ... | ... | 0.0138 |
| 16/32 | 0.1963 | 0.0982 | 0.1044 | 0.1107 | ... | ... | ... | ... | ... |

for a given spline application, it is well to keep in mind that there are no advantages to be gained by using odd numbers of teeth and that the diameters of splines with odd tooth numbers, particularly internal splines, are troublesome to measure with pins since no two tooth spaces are diametrically opposite each other.

**Types and Classes of Involute Spline Fits.** — Two types of fits are covered by the American National Standard for involute splines, the side fit and the major diameter fit.  Dimensional data for flat root side fit, flat root major diameter fit, and fillet root side fit splines are tabulated in this standard for 30-degree pressure angle splines; but for only the fillet root side fit for 37.5- and 45-degree pressure angle splines.

*Side Fit:* In the side fit, the mating members contact only on the sides of the teeth; major and minor diameters are clearance dimensions.  The tooth sides act as drivers and centralize the mating splines.

*Major Diameter Fit:* Mating parts for this fit contact at the major diameter for centralizing.  The sides of the teeth act as drivers.  The minor diameters are clearance dimensions.

The major diameter fit provides a minimum effective clearance that will allow for contact and location at the major diameter with a minimum amount of location or centralizing effect by the sides of the teeth.  The major diameter fit has only one space width and tooth thickness tolerance which is the same as side fit Class 5.

A fillet root may be specified for an external spline, even though it is otherwise designed to the flat root side fit or major diameter fit standard.  An internal spline with a fillet root can be used only for the side fit.

**Classes of Tolerances.** — This standard includes four classes of tolerances on space width and tooth thickness.  This has been done to provide a range of tolerances for selection to suit a design need.  The classes are variations of the former single tolerance which is now Class 5 and are based on the formulas shown in the footnote of Table 4.  All tolerance classes have the same minimum effective space width and maximum effective tooth thickness limits so that a mix of classes between mating parts is possible.

Table 4. **Maximum Tolerances for Space Width and Tooth Thickness of Tolerance Class 5 Splines\*** (ANSI B92.1-1970)

(Values shown in ten thousandths (20 = 0.0020)

| No. of Teeth | Pitch, $P/P_s$ | | | | | | | |
|---|---|---|---|---|---|---|---|---|
| | 2.5/5 and 3/6 | 4/8 and 5/10 | 6/12 and 8/16 | 10/20 and 12/24 | 16/32 and 20/40 | 24/48 thru 48/96 | 64/128 and 80/160 | 128/256 |
| **N** | Machining Tolerance, $m$ | | | | | | | |
| 10 | 15.8 | 14.5 | 12.5 | 12.0 | 11.7 | 11.7 | 9.6 | 9.5 |
| 20 | 17.6 | 16.0 | 14.0 | 13.0 | 12.4 | 12.4 | 10.2 | 10.0 |
| 30 | 18.4 | 17.5 | 15.5 | 14.0 | 13.1 | 13.1 | 10.8 | 10.5 |
| 40 | 21.8 | 19.0 | 17.0 | 15.0 | 13.8 | 13.8 | 11.4 | — |
| 50 | 23.0 | 20.5 | 18.5 | 16.0 | 14.5 | 14.5 | — | — |
| 60 | 24.8 | 22.0 | 20.0 | 17.0 | 15.2 | 15.2 | — | — |
| 70 | — | — | — | 18.0 | 15.9 | 15.9 | — | — |
| 80 | — | — | — | 19.0 | 16.6 | 16.6 | — | — |
| 90 | — | — | — | 20.0 | 17.3 | 17.3 | — | — |
| 100 | — | — | — | 21.0 | 18.0 | 18.0 | — | — |
| **N** | Variation Allowance, $\lambda$ | | | | | | | |
| 10 | 23.5 | 20.3 | 17.0 | 15.7 | 14.2 | 12.2 | 11.0 | 9.8 |
| 20 | 27.0 | 22.6 | 19.0 | 17.4 | 15.4 | 13.4 | 12.0 | 10.6 |
| 30 | 30.5 | 24.9 | 21.0 | 19.1 | 16.6 | 14.6 | 13.0 | 11.4 |
| 40 | 34.0 | 27.2 | 23.0 | 21.6 | 17.8 | 15.8 | 14.0 | — |
| 50 | 37.5 | 29.5 | 25.0 | 22.5 | 19.0 | 17.0 | — | — |
| 60 | 41.0 | 31.8 | 27.0 | 24.2 | 20.2 | 18.2 | — | — |
| 70 | — | — | — | 25.9 | 21.4 | 19.4 | — | — |
| 80 | — | — | — | 27.6 | 22.6 | 20.6 | — | — |
| 90 | — | — | — | 29.3 | 23.8 | 21.8 | — | — |
| 100 | — | — | — | 31.0 | 25.0 | 23.0 | — | — |
| **N** | Total Index Variation | | | | | | | |
| 10 | 20 | 17 | 15 | 15 | 14 | 12 | 11 | 10 |
| 20 | 24 | 20 | 18 | 17 | 15 | 13 | 12 | 11 |
| 30 | 28 | 22 | 20 | 19 | 16 | 15 | 14 | 13 |
| 40 | 32 | 25 | 22 | 20 | 18 | 16 | 15 | — |
| 50 | 36 | 27 | 25 | 22 | 19 | 17 | — | — |
| 60 | 40 | 30 | 27 | 24 | 20 | 18 | — | — |
| 70 | — | — | — | 26 | 21 | 20 | — | — |
| 80 | — | — | — | 28 | 22 | 21 | — | — |
| 90 | — | — | — | 29 | 24 | 23 | — | — |
| 100 | — | — | — | 31 | 25 | 24 | — | — |
| **N** | Profile Variation | | | | | | | |
| All | +7 −10 | +6 −8 | +5 −7 | +4 −6 | +3 −5 | +2 −4 | +2 −4 | +2 −4 |

| Lead Variation | | | | | | | | | | | | |
|---|---|---|---|---|---|---|---|---|---|---|---|---|
| $L_g$, in. | 0.3 | 0.5 | 1 | 2 | 3 | 4 | 5 | 6 | 7 | 8 | 9 | 10 |
| Variation | 2 | 3 | 4 | 5 | 6 | 7 | 8 | 9 | 10 | 11 | 12 | 13 |

\* For other tolerance classes: Class 4 = 0.71 × Tabulated value
Class 5 = As tabulated in table
Class 6 = 1.40 × Tabulated value
Class 7 = 2.00 × Tabulated value

**Fillets and Chamfers.** — Spline teeth may have either a flat root or a rounded fillet root.

*Flat Root Splines* are suitable for most applications. The fillet which joins the sides to the bottom of the tooth space, if generated, has a varying radius of curvature. Specification of this fillet is usually not required. It is controlled by the form diameter which is the diameter at the deepest point of the desired true involute form (sometimes designated as TIF).

When flat root splines are used for heavily loaded couplings which are not suitable for fillet root spline application, it may be desirable to minimize the stress concentration in the flat root type by specifying an approximate radius for the fillet.

Since internal splines are stronger than external splines because of their broad bases and high pressure angles at the major diameter, broaches for flat root internal splines are normally made with the involute profile extending to the major diameter.

*Fillet Root Splines* are recommended for heavy loads because the larger fillets provided reduce the stress concentrations. The curvature along any generated fillet varies and cannot be specified by a radius of any given value.

External splines may be produced by generating with a pinion type shaper cutter or with a hob, or by cutting with no generating motion using a tool formed to the contour of a tooth space. External splines are also made of cold forming and in these cases are usually of the fillet root design. Internal splines are usually produced by broaching, by form cutting, or by generating with a shaper cutter. Even when full tip radius tools are used, each of these cutting methods produces a fillet contour with individual characteristics. Generated spline fillets are curves related to the prolate epicycloid for external splines and the prolate hypocycloid for internal splines. These fillets have a minimum radius of curvature at the point where the fillet is tangent to the external spline minor diameter circle or the internal spline major diameter circle and a rapidly increasing radius of curvature up to the point where the fillet comes tangent to the involute profile.

*Chamfers and Corner Clearance:* In major diameter fits, it is always necessary to provide corner clearance at the major diameter of the spline coupling. This is usually effected by providing a chamfer on the top corners of the external member. This method may not be possible or feasible because:

(a) If the external member is roll formed by plastic deformation, a chamfer cannot be provided by the process.

(b) A semitopping cutter may not be available.

(c) When cutting external splines with small numbers of teeth, a semitopping cutter may reduce the width of the top land to a prohibitive point.

In such cases the corner clearance can be provided on the internal spline as shown in Fig. 2.

When this option is used, the form diameter may fall in the protuberance area.

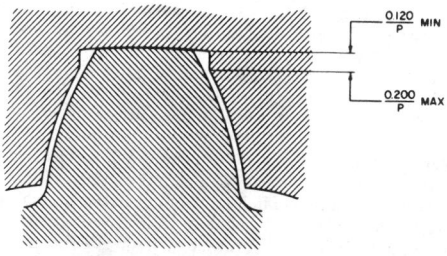

Fig. 2. Internal corner clearance

**Spline Variations.** — The maximum allowable variations for involute splines are listed in Table 4.

*Profile Variation:* The reference profile, from which variations occur, passes through the point which is used to determine the actual space width or tooth thickness. This is either the pitch point or the contact point of the standard measuring pins.

Profile variation is positive in the direction of the space and negative in the direction of the tooth. Profile variations which may occur at any point on the profile for establishing effective fits are shown in Table 4.

*Lead Variations:* The lead tolerance for the total spline length applies also to any portion thereof unless otherwise specified.

*Out of Roundness:* This condition may appear merely as a result of index and profile variations given in Table 4 and requires no further allowance. However, heat treatment and deflection of thin sections may cause out of roundness, which increases index and profile variations. Tolerances for such conditions depend on many variables and are therefore not tabulated. Additional tooth and/or space width tolerance must allow for such conditions.

*Eccentricity:* Eccentricity of major and minor diameters in relation to the effective diameter of side fit splines, should not cause contact beyond the form diameters of the mating splines, even under conditions of maximum effective clearance. This standard does not establish specific tolerances.

Eccentricity of major diameters in relation to the effective diameters of major diameter fit splines should be absorbed within the maximum material limits established by the tolerances on major diameter and effective space width or effective tooth thickness.

If the alignment of mating splines is affected by eccentricity of locating surfaces relative to each other and/or the splines, it may be necessary to decrease the effective and actual tooth thickness of the external splines in order to maintain the desired fit condition. This standard does not include allowances for eccentric location.

**Effect of Spline Variations.** — These can be classified as index variations, profile variations and lead variations.

*Index Variations:* These variations cause the clearance to vary from one set of mating tooth sides to another. Since the fit depends on the areas with minimum clearance, index variations reduce the effective clearance.

*Profile Variations:* Positive profile variations affect the fit by reducing effective clearance. Negative profile variations do not affect the fit but reduce the contact area.

*Lead variations:* These variations will cause clearance variations and therefore reduce the effective clearance.

*Variation Allowance:* The effect of individual spline variations on the fit (effective variation) is less than their total, because areas of more than minimum clearance can be altered without changing the fit. The variation allowance is 60 per cent of the sum of twice the positive profile variation, the total index variation and the lead variation for the length of engagement. The variation allowances in Table 4 are based on a lead variation for an assumed length of engagement equal to one-half the pitch diameter. Adjustment may be required for a greater length of engagement.

**Effective and Actual Dimensions.** — Although each space of an internal spline may have the same width as each tooth of a perfect mating external spline, the two may not fit because of variations of index and profile in the internal spline. In such a case, to allow the perfect external spline to fit in any position, all spaces of the internal spline must be widened by the amount of interference. The resulting width of these tooth spaces is the *actual* space width of the internal spline. The *effective* space width is the tooth thickness of the perfect mating external spline. The same reasoning applied to an external spline which has variations of index and profile when mated with a perfect internal spline leads to the concept of effective tooth thickness which exceeds the actual tooth thickness by the effective variation.

The effective space width of the internal spline minus the effective tooth thickness of the external spline is the effective clearance. This defines the fit of the mating parts. (This is strictly true only if high points of mating parts come into contact.) Positive effective clearance represents looseness or backlash. Negative effective clearance represents tightness or interference.

## Space Width and Tooth Thickness Limits. — The variation of actual space width and actual tooth thickness within the machining tolerance causes corresponding variations of effective dimensions, so that there are four limit dimensions for each component part.

These are shown diagrammatically in Table 5.

The minimum effective space width is always basic. The maximum effective tooth thickness is the same as the minimum effective space width except for the major diameter fit. The major diameter fit maximum effective tooth thickness is less than the minimum effective space width by an amount which allows for eccentricity between the effective spline and the major diameter. The permissible variation of the effective clearance is divided between the internal and external splines to arrive at the maximum effective space width and the minimum effective tooth thickness. Limits of the actual space width and actual tooth thickness are constructed from suitable variation allowances.

## Use of Effective and Actual Dimensions. — Each of the four dimensions for space width and tooth thickness shown in Table 5, has a definite function.

*Minimum Effective Space Width and Maximum Effective Tooth Thickness:* These dimensions control the minimum effective clearance, and must always be specified.

*Minimum Actual Space Width and Maximum Actual Tooth Thickness:* These dimensions cannot be used for acceptance or rejection of parts. If the actual space width is less than the minimum without causing the effective space width to be undersized, or if the actual tooth thickness is more than the maximum without causing the effective tooth thickness to be oversized, the effective variation is less than anticipated; such parts are desirable and not defective. The specification of these actual dimensions as processing reference dimensions is optional. They are also used to analyze undersize effective space width or oversize effective tooth thickness conditions to determine whether or not these conditions are caused by excessive effective variation.

*Maximum Actual Space Width and Minimum Actual Tooth Thickness:* These dimensions control machining tolerance and also limit the effective variation. The spread between these dimensions, reduced by the effective variation of the internal and external spline, is the maximum effective clearance. Where the effective variation obtained in machining is appreciably less than the variation allowance, these dimensions must be adjusted in order to maintain the desired fit.

*Maximum Effective Space Width and Minimum Effective Tooth Thickness:* These dimensions define the maximum effective clearance but they do not limit the effective variation. They may be used, in addition to the maximum actual space width and minimum actual tooth thickness, to prevent the increase of maximum effective clearance due to reduction of effective variations. The notation "inspection optional" may be added where maximum effective clearance is an assembly requirement, but does not need absolute control. It will indicate, without necessarily adding inspection time and equipment, that the actual space width of the internal spline must be held below the maximum, or the actual tooth thickness of the external spline above the minimum, if machining methods result in less than the allowable variations. Where effective variation needs no control or is controlled by laboratory inspection, these limits may be substituted for maximum actual space width and minimum actual tooth thickness.

**Table 5. Specification Guide for Space Width and Tooth Thickness**
(ANSI B92.1-1970)

| Dimension | Disposition of Allowances, Clearances and Tolerances on Part | | Dimensioning Method | | |
|---|---|---|---|---|---|
| | Effective | Actual | Standard | Alternatives A | B |
| Space Width of Internal Spline | | Max | Required | Required | REF |
| | Max | Min | REF | Required | REF |
| | Min | | REF | Required | Required |
| (Basic) | Min / Max | Max | Required | Required | Required |
| Tooth Thickness of External Spline | Min | Max | REF | Required | Required |
| | $c_v$ Min = 0 | | REF | REF | REF |
| | | Min | Required | Required | REF |

*Note:* The minimum effective clearance, $c_v$ min, is greater than zero for major diameter fits. The maximum effective tooth thickness is less than the minimum effective space width for major diameter fits to allow for eccentricity variations.

**Combinations of Involute Spline Types.** — Flat root side fit internal splines may be used with fillet root external splines where the larger radius is desired on the external spline for control of stress concentrations. This combination of fits may also be permitted as a design option by specifying for the minimum root diameter of the external, the value of the minimum root diameter of the fillet root external spline and noting this as "optional root."

A design option may also be permitted to provide either flat root internal or fillet root internal by specifying for the maximum major diameter, the value of the maximum major diameter of the fillet root internal spline and noting this as "optional root."

**Interchangeability.** — Splines made to this standard may interchange with splines made to older standards. Exceptions are listed below.

*External Splines:* These will mate with older internal splines as follows:

| Year | Major Dia. Fit | Flat Root Side Fit | Fillet Root Side Fit |
|---|---|---|---|
| 1946 | Yes | No (A)[c] | No (A) |
| 1950[a] | Yes (B) | Yes (B) | Yes(C) |
| 1950[b] | Yes (B) | No (A) | Yes (C) |
| 1957 SAE | Yes | No (A) | Yes (C) |
| 1960 | Yes | No (A) | Yes (C) |

[a]Full dendendum.
[b]Short dendendum.
[c]For exceptions A, B, C, see paragraph below.

*Internal Splines:* These will mate with older external splines as follows:

| Year | Major Dia. Fit | Flat Root Side Fit | Fillet Root Side Fit |
|------|----------------|--------------------|-----------------------|
| 1946     | No (D)[a]  | No (E) | No (D)  |
| 1950     | Yes (F)    | Yes    | Yes (C) |
| 1957 SAE | Yes (G)    | Yes    | Yes     |
| 1960     | Yes (G)    | Yes    | Yes     |

[a]For exceptions C, D, E, F, G, see paragraph below.

*Exceptions:*

A. The external major diameter, unless chamfered or reduced, may interfere with the internal form diameter on flat root side fit splines. Internal splines made to the 1957 and 1960 standards had the same dimensions as shown for the major diameter fit splines in this standard.

B. For 15 teeth or less, the minor diameter of the internal spline, unless chamfered, will interfere with the form diameter of the external spline.

C. For 9 teeth or less, the minor diameter of the internal spline, unless chamfered, will interfere with form diameter of the external spline.

D. The internal minor diameter, unless chamfered, will interfere with the external form diameter.

E. The internal minor diameter, unless chamfered, will interfere with the external form diameter.

F. For 10 teeth or less, the minimum chamfer on the major diameter of the external spline may not clear the internal form diameter.

G. Depending upon the pitch of the spline, the minimum chamfer on the major diameter may not clear the internal form diameter.

**Drawing Data.** — It is important that uniform specifications be used to show complete information on detail drawings of splines. Much misunderstanding will be avoided by following the suggested arrangement of dimensions and data as given in Tables 6–10, inclusive. The number of x's indicates the number of decimal places normally used. With this tabulated type of spline specifications, it is usually not necessary to show a graphic illustration of the spline teeth.

**Spline Data and Reference Dimensions.** — Spline data are used for engineering and manufacturing purposes. Pitch and pressure angle are not subject to individual inspection.

As used in this standard, *reference* is an added notation or modifier to a dimension, specification, or note when that dimension, specification, or note is:

1. Repeated for drawing clarification.
2. Needed to define a nonfeature datum or basis from which a form or feature is generated.
3. Needed to define a nonfeature dimension from which other specifications or dimensions are developed.
4. Needed to define a nonfeature dimension at which toleranced sizes of a feature are specified.
5. Needed to define a nonfeature dimension from which control tolerances or sizes are developed or added as useful information.

Any dimension, specification, or note that is noted "REF" should not be used as a criterion for part acceptance or rejection.

**Table 6.   Spline Terms, Symbols and Drawing Data, 30-Degree Pressure Angle,
Flat Root Side Fit** (ANSI B92.1-1970)

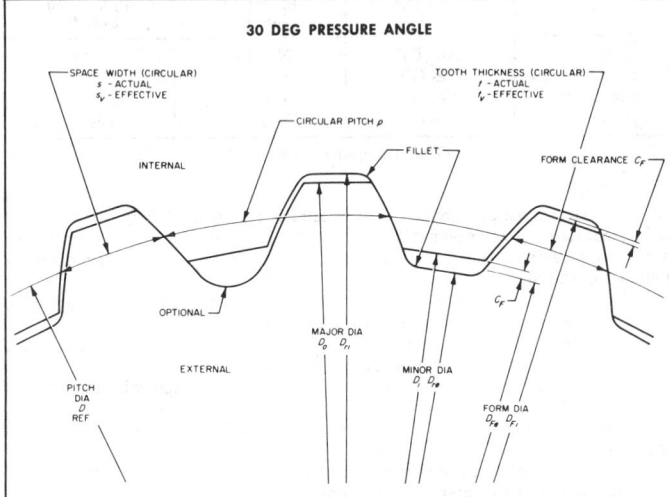

This fit is used in restricted areas (as with tubular parts with wall thickness too small to permit use of fillet roots, and to allow hobbing closer to shoulders, etc.) and for economy (when hobbing, shaping, etc. and shorter broaches for the internal member).

Press fits are not tabulated because their design depends on the degree of tightness desired and must allow for such factors as the shape of the blank, wall thickness, material, hardness, thermal expansion, etc. Close tolerances or selective size grouping may be required to limit fit variations.

**DRAWING DATA**

| INTERNAL INVOLUTE SPLINE DATA | | EXTERNAL INVOLUTE SPLINE DATA | |
|---|---|---|---|
| FLAT ROOT SIDE FIT | | FLAT ROOT SIDE FIT | |
| NUMBER OF TEETH | xx | NUMBER OF TEETH | xx |
| PITCH | xx/xx | PITCH | xx/xx |
| PRESSURE ANGLE | 30° | PRESSURE ANGLE | 30° |
| BASE DIAMETER | x.xxxxxx  REF | BASE DIAMETER | x.xxxxxx  REF |
| PITCH DIAMETER | x.xxxxxx  REF | PITCH DIAMETER | x.xxxxxx  REF |
| MAJOR DIAMETER | x.xxx   max | MAJOR DIAMETER | x.xxx/x.xxx |
| FORM DIAMETER | x.xxx | FORM DIAMETER | x.xxx |
| MINOR DIAMETER | x.xxx/x.xxx | MINOR DIAMETER | x.xxx   min |
| CIRCULAR SPACE WIDTH | | CIRCULAR TOOTH THICKNESS | |
| MAX   ACTUAL | x.xxxx | MAX   EFFECTIVE | x.xxxx |
| MIN   EFFECTIVE | x.xxxx | MIN   ACTUAL | x.xxxx |
| The following information may be added as required: | | The following information may be added as required: | |
| MAX MEASUREMENT BETWEEN PINS | x.xxxx  REF | MIN MEASUREMENT OVER PINS | x.xxxx  REF |
| PIN DIAMETER | x.xxxx | PIN DIAMETER | x.xxxx |

The above drawing data is required for the spline specifications.   The standard system is shown; for alternate systems, see Table 5.   Number of x's indicates number of decimal places normally used.

**Table 7.   Spline Terms, Symbols and Drawing Data, 30-Degree Pressure Angle, Flat Root Major Diameter Fit (ANSI B92.1-1970)**

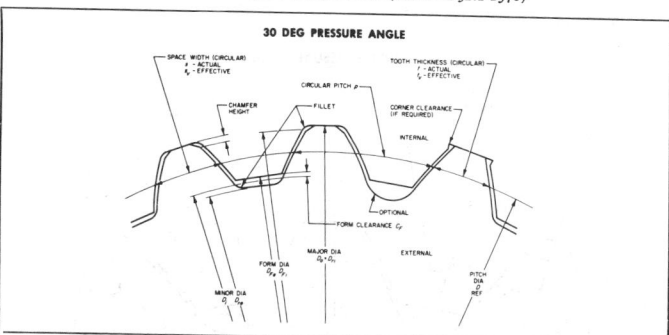

**30 DEG PRESSURE ANGLE**

This fit is used for assemblies where runout of the member having the internal spline must be held to minimum, such as a gear on a shaft as shown below.   In this fit, eccentricity between mating members is limited by the radial major diameter clearance.   Caution should be used in considering this because:

1. When eccentricity between the major circle and the effective spline does not occur, excessive effective clearance may be present.
2. When eccentricity does occur, torque load is borne by only a few teeth.

Caution should also be used in the finer pitches and smaller numbers of teeth to make sure that the major diameter chamfer is not so great that the remaining top land would be so small that the fit would be ineffective.

Press fits are not tabulated because their design depends on the degree of tightness desired and must allow for such factors as the shape of the blank, wall thickness, material, hardness, thermal expansion, etc.   Close tolerances or selective size grouping may be required to limit fit variations.

**DRAWING DATA**

| INTERNAL INVOLUTE SPLINE DATA | | EXTERNAL INVOLUTE SPLINE DATA | |
|---|---|---|---|
| FLAT ROOT MAJOR DIAMETER FIT | | FLAT ROOT MAJOR DIAMETER FIT | |
| NUMBER OF TEETH | xx | NUMBER OF TEETH | xx |
| PITCH | xx/xx | PITCH | xx/xx |
| PRESSURE ANGLE | 30° | PRESSURE ANGLE | 30° |
| BASE DIAMETER | x.xxxxxx  REF | BASE DIAMETER | x.xxxxxx  REF |
| PITCH DIAMETER | x.xxxxxx  REF | PITCH DIAMETER | x.xxxxxx  REF |
| MAJOR DIAMETER | x.xxxx/x.xxxx | MAJOR DIAMETER | x.xxxx/x.xxxx |
| FORM DIAMETER | x.xxx | FORM DIAMETER | x.xxx |
| MINOR DIAMETER | x.xxx/x.xxx | MINOR DIAMETER | x.xxx  min |
| CIRCULAR SPACE WIDTH | | CIRCULAR TOOTH THICKNESS | |
| MAX   ACTUAL | x.xxxx | MAX   EFFECTIVE | x.xxxx |
| MIN   EFFECTIVE | x.xxxx | MIN   ACTUAL | x.xxxx |
| NOTE: THE MAJOR DIAMETER AND EFFECTIVE SPLINE MUST BE CONCENTRIC AT MAXIMUM MATERIAL CONDITIONS. | | NOTE: THE MAJOR DIAMETER AND EFFECTIVE SPLINE MUST BE CONCENTRIC AT MAXIMUM MATERIAL CONDITIONS. | |
| The following information may be added as required: | | The following information may be added as required: | |
| MAX MEASUREMENT BETWEEN PINS | x.xxxx  REF | MIN MEASUREMENT OVER PINS | x.xxxx  REF |
| PIN DIAMETER | x.xxxx | PIN DIAMETER | x.xxxx |
| CORNER CLEARANCE | x.xxx/x.xxx | CHAMFER HEIGHT | x.xxx/x.xxx |

The above drawing data is required for the spline specifications.   The standard system is shown; for alternate systems, see Table 5.   Number of x's indicates number of decimal places normally used.

**Table 8.　Spline Terms, Symbols and Drawing Data, 30-Degree Pressure Angle, Fillet Root Side Fit** (ANSI B92.1-1970)

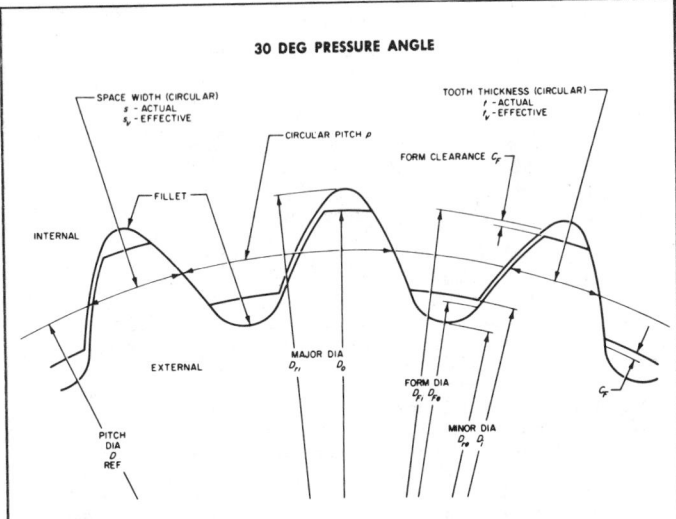

The fillet radius permits heavier loading and effects greater fatigue resistance than flat roots through absence of stress raisers.

Press fits are not tabulated because their design depends on the degree of tightness desired and must allow for such factors as the shape of the blank, wall thickness, material, hardness, thermal expansion, etc. Close tolerances or selective size grouping may be required to limit fit variations.

**DRAWING DATA**

| INTERNAL INVOLUTE SPLINE DATA | | EXTERNAL INVOLUTE SPLINE DATA | |
|---|---|---|---|
| **FILLET ROOT SIDE FIT** | | **FILLET ROOT SIDE FIT** | |
| NUMBER OF TEETH | xx | NUMBER OF TEETH | xx |
| PITCH | xx/xx | PITCH | xx/xx |
| PRESSURE ANGLE | 30° | PRESSURE ANGLE | 30° |
| BASE DIAMETER | x.xxxxxx REF | BASE DIAMETER | x.xxxxxx REF |
| PITCH DIAMETER | x.xxxxxx REF | PITCH DIAMETER | x.xxxxxx REF |
| MAJOR DIAMETER | x.xxx max | MAJOR DIAMETER | x.xxx/x.xxx |
| FORM DIAMETER | x.xxx | FORM DIAMETER | x.xxx |
| MINOR DIAMETER | x.xxx/x.xxx | MINOR DIAMETER | x.xxx min |
| **CIRCULAR SPACE WIDTH** | | **CIRCULAR TOOTH THICKNESS** | |
| MAX ACTUAL | x.xxxx | MAX EFFECTIVE | x.xxxx |
| MIN EFFECTIVE | x.xxxx | MIN ACTUAL | x.xxxx |
| The following information may be added as required: | | The following information may be added as required: | |
| MAX MEASUREMENT BETWEEN PINS | x.xxxx REF | MIN MEASUREMENT OVER PINS | x.xxxx REF |
| PIN DIAMETER | x.xxxx | PIN DIAMETER | x.xxxx |
| FILLET RADIUS MIN | x.xxxx | FILLET RADIUS MIN | x.xxxx |

The above drawing data is required for the spline specifications. The standard system is shown; for alternate systems, see Table 5. Number of x's indicates number of decimal places normally used.

Table 9. Spline Terms, Symbols and Drawing Data, 37.5-Degree Pressure
Angle, Fillet Root Side Fit (ANSI B92.1-1970)

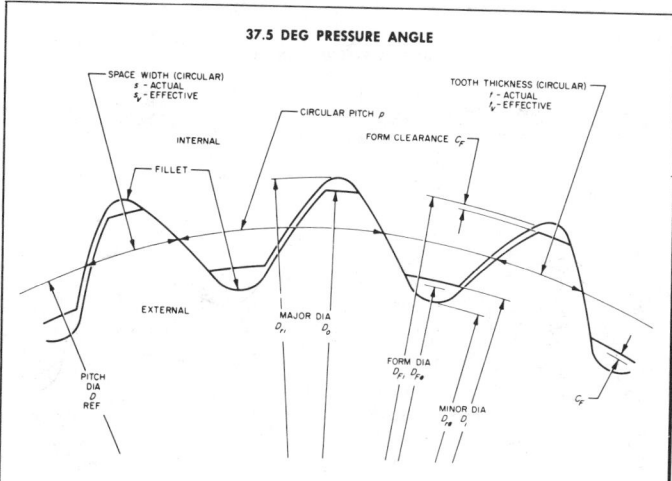

**37.5 DEG PRESSURE ANGLE**

The pressure angle and proportions of this spline are a direct compromise of the 30- and 45-degree pressure angle splines. It is often used on couplings where the external spline is to be cold formed; especially where a 45-degree pressure angle spline will not satisfy functional requirements and the shaft material is above hardness limitation of 30-degree pressure angle cold forming tools.

Press fits are not tabulated because their design depends on the degree of tightness desired and must allow for such factors as the shape of the blank, wall thickness, material, hardness, thermal expansion, etc. Close tolerances or selective size grouping may be required to limit fit variations.

## DRAWING DATA

| INTERNAL INVOLUTE SPLINE DATA | | EXTERNAL INVOLUTE SPLINE DATA | |
|---|---|---|---|
| **FILLET ROOT SIDE FIT** | | **FILLET ROOT SIDE FIT** | |
| NUMBER OF TEETH | xx | NUMBER OF TEETH | xx |
| PITCH | xx/xx | PITCH | xx/xx |
| PRESSURE ANGLE | 37.5° | PRESSURE ANGLE | 37.5° |
| BASE DIAMETER | x.xxxxxx REF | BASE DIAMETER | x.xxxxxx REF |
| PITCH DIAMETER | x.xxxxxx REF | PITCH DIAMETER | x.xxxxxx REF |
| MAJOR DIAMETER | x.xxx max | MAJOR DIAMETER | x.xxx/x.xxx |
| FORM DIAMETER | x.xxx | FORM DIAMETER | x.xxx |
| MINOR DIAMETER | x.xxx/x.xxx | MINOR DIAMETER | x.xxx min |
| **CIRCULAR SPACE WIDTH** | | **CIRCULAR TOOTH THICKNESS** | |
| MAX ACTUAL | x.xxxx | MAX EFFECTIVE | x.xxxx |
| MIN EFFECTIVE | x.xxxx | MIN ACTUAL | x.xxxx |
| The following information may be added as required: | | The following information may be added as required: | |
| MAX MEASUREMENT BETWEEN PINS | x.xxxx REF | MIN MEASUREMENT OVER PINS | x.xxxx REF |
| PIN DIAMETER | x.xxxx | PIN DIAMETER | x.xxxx |

The above drawing data is required for the spline specifications. The standard system is shown; for alternate systems, see Table 5. Number of x's indicates number of decimal places normally used.

**Table 10.   Spline Terms, Symbols and Drawing Data, 45-Degree Pressure Angle,
Fillet Root Side Fit (ANSI B92.1-1970)**

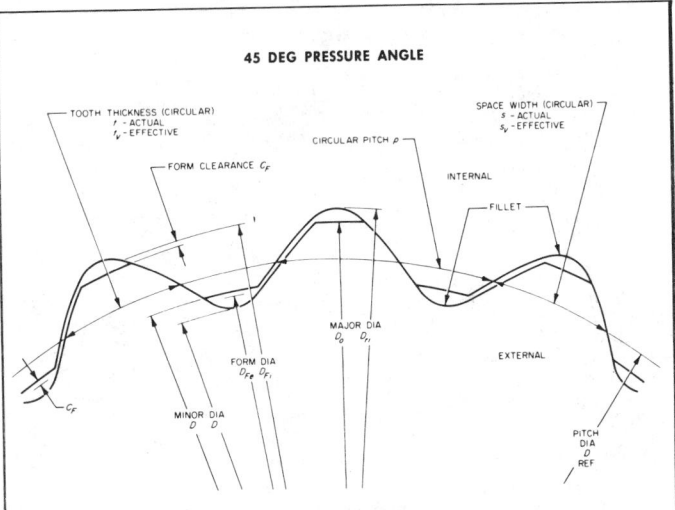

Involute splines with 45-degree pressure angle are used where the toothed
member delivers torque only (does not slide under load) and where wall thicknesses are great enough to resist bursting tendencies.   The 45-degree pressure
angle spline is highly suitable for cold forming.

Press fits are not tabulated because their design depends on the degree of
tightness desired and must allow for such factors as the shape of the blank, wall
thickness, material, hardness, thermal expansion, etc.   Close tolerances or selective size grouping may be required to limit fit variation.

**DRAWING DATA**

| INTERNAL INVOLUTE SPLINE DATA | | EXTERNAL INVOLUTE SPLINE DATA | |
|---|---|---|---|
| **FILLET ROOT SIDE FIT** | | **FILLET ROOT SIDE FIT** | |
| NUMBER OF TEETH | xx | NUMBER OF TEETH | xx |
| PITCH | xx/xx | PITCH | xx/xx |
| PRESSURE ANGLE | 45° | PRESSURE ANGLE | 45° |
| BASE DIAMETER | x.xxxxxx REF | BASE DIAMETER | x.xxxxxx REF |
| PITCH DIAMETER | x.xxxxxx REF | PITCH DIAMETER | x.xxxxxx REF |
| MAJOR DIAMETER | x.xxx max | MAJOR DIAMETER | x.xxx/x.xxx |
| FORM DIAMETER | x.xxx | FORM DIAMETER | x.xxx |
| MINOR DIAMETER | x.xxx/x.xxx | MINOR DIAMETER | x.xxx min |
| **CIRCULAR SPACE WIDTH** | | **CIRCULAR TOOTH THICKNESS** | |
| MAX  ACTUAL | x.xxxx | MAX  EFFECTIVE | x.xxxx |
| MIN  EFFECTIVE | x.xxxx | MIN  ACTUAL | x.xxxx |
| The following information may be added as required: | | The following information may be added as required: | |
| MAX MEASUREMENT BETWEEN PINS | x.xxxx REF | MIN MEASUREMENT OVER PINS | x.xxxx REF |
| PIN DIAMETER | x.xxxx | PIN DIAMETER | x.xxxx |

The above drawing data is required for the spline specifications.   The standard system
is shown; for alternate systems, see Table 5.   Number of x's indicates number of decimal
places normally used.

**Involute Spline Inspection Methods.** — Spline gages are used for routine inspection of production parts.

Analytical inspection, which is the measurement of individual dimensions and variations, may be required:

A. To supplement inspection by gages, for example, where NOT GO composite gages are used in place of NOT GO sector gages and variations must be controlled.

B. To evaluate parts rejected by gages.

C. For prototype parts or short runs where spline gages are not used.

D. To supplement inspection by gages where each individual variation must be restrained from assuming too great a portion of the tolerance between the minimum material actual and the maximum material effective dimensions.

**Inspection with Gages.** — A variety of gages is used in the inspection of involute splines.

*Types of Gages:* A composite spline gage is a gage having a full complement of teeth. A sector spline gage is a gage having two diametrically opposite groups of teeth. A sector plug gage with only two teeth per sector is also known as a "paddle gage." A sector ring gage with only two teeth per sector is also known as a "snap ring gage." A progressive gage is a gage consisting of two or more adjacent sections with different inspection functions. Progressive GO gages are physical combinations of GO gage members which check consecutively first one feature or one group of features, then their relationship to other features. GO and NOT GO gages may also be combined physically to form a progressive gage.

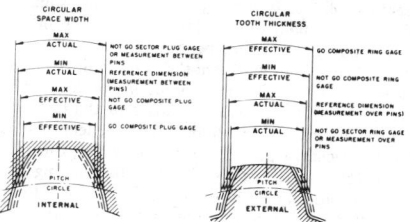

Fig. 3.   Space width and tooth-thickness inspection.

*GO and NOT GO Gages:* GO gages are used to inspect maximum material conditions (maximum external, minimum internal dimensions). They may be used to inspect an individual dimension or the relationship between two or more functional dimensions. They control the minimum looseness or maximum interference.

NOT GO gages are used to inspect minimum material conditions (minimum external, maximum internal dimensions), thereby controlling the maximum looseness or minimum interference. Unless otherwise agreed upon, a product is acceptable only if the NOT GO gage does not enter or go on the part. A NOT GO gage can be used to inspect only one dimension. An attempt at simultaneous NOT GO inspection of more than one dimension could result in failure of such a gage to enter or go on (acceptance of part), even though all but one of the dimensions were outside product limits. In the event all dimensions are outside the limits, their relationship could be such as to allow acceptance.

*Effective and Actual Dimensions:* The effective space width and tooth thickness are inspected by means of an accurate mating member in the form of a composite spline gage.

The actual space width and tooth thickness are inspected with sector plug and ring gages, or by measurements with pins.

**Estimating Key and Spline Sizes and Lengths.** — Figure 1 may be used to estimate the size of American Standard involute splines required to transmit a given torque. It also may be used to find the outside diameter of shafts used with single keys. After the size of the shaft is found, the proportions of the key can be determined from Table 2 on page 969.

Curve A is for flexible splines with teeth hardened to Rockwell C 55–65. For these splines, lengths are generally made equal to or somewhat greater than the pitch diameter for diameters below 1¼ inches; on larger diameters, the length is generally one-third to two-thirds the pitch diameter. Curve A also applies for a single key used as a fixed coupling, the length of the key being one to one and one-quarter times the shaft diameter. The stress in the shaft, neglecting stress concentration at the keyway, is about 7500 pounds per square inch. For the effect of keyways on shaft strength, see page 460.

Curve B represents high-capacity single keys used as fixed couplings for stresses of 9500 pounds per square inch, neglecting stress concentration. Key-length is one to one and one-quarter times shaft diameter and both shaft and key are of moderately hard heat-treated steel. This type of connection is commonly used to key commercial flexible couplings to motor or generator shafts.

Curve C is for multiple-key fixed splines with lengths of three-quarters to one and one-quarter times pitch diameter and shaft hardnesses of 200–300 BHN.

Curve D is for high-capacity splines with lengths one-half to one times the pitch diameter. Hardnesses up to Rockwell C 58 are common and in aircraft applications the shaft is generally hollow to reduce weight.

Curve E represents a solid shaft with 65,000 pounds per square inch shear stress. For hollow shafts with inside diameter equal to three-quarters of the outside diameter the shear stress would be 95,000 pounds per square inch.

*Length of Splines:* Fixed splines with lengths of one-third the pitch diameter will have the same shear strength as the shaft, assuming uniform loading of the teeth; however, errors in spacing of teeth result in only half the teeth being fully loaded. Therefore, for balanced strength of teeth and shaft the length should be two-thirds the pitch diameter. If weight is not important, however, this may be increased to equal the pitch diameter. In the case of flexible splines, long lengths do not contribute to load carrying capacity when there is misalignment to be accommodated. Maximum effective length for flexible splines may be approximated from Fig. 2.

**Formulas for Torque Capacity of Involute Splines.** — The formulas for torque capacity of 30-degree involute splines given in the following paragraphs are derived largely from an article "When Splines Need Stress Control" by D. W. Dudley, *Product Engineering*, Dec. 23, 1957.

In the formulas that follow the symbols used are as defined on page 995 with the following additions: $D_h$ = inside diameter of hollow shaft, inches; $K_a$ = application factor from Table 1; $K_m$ = load distribution factor from Table 2; $K_f$ = fatigue life factor from Table 3; $K_w$ = wear life factor from Table 4; $L_e$ = maximum effective length from Fig. 2, to be used in stress formulas even though the actual length may be greater.

*Definitions:* A *fixed* spline is one which is either shrink fitted or loosely fitted but piloted with rings at each end to prevent rocking of the spline which results in small axial movements that cause wear. A *flexible* spline permits some rocking motion such as occurs when the shafts are not perfectly aligned. This flexing or rocking motion causes axial movement and consequently wear of the teeth. Straight-toothed flexible splines can accommodate only small angular misalignments (less than 1 deg.) before wear becomes a serious problem. For greater amounts of misalignment (up to about 5 deg.), crowned splines are preferable to reduce wear and end-loading of the teeth.

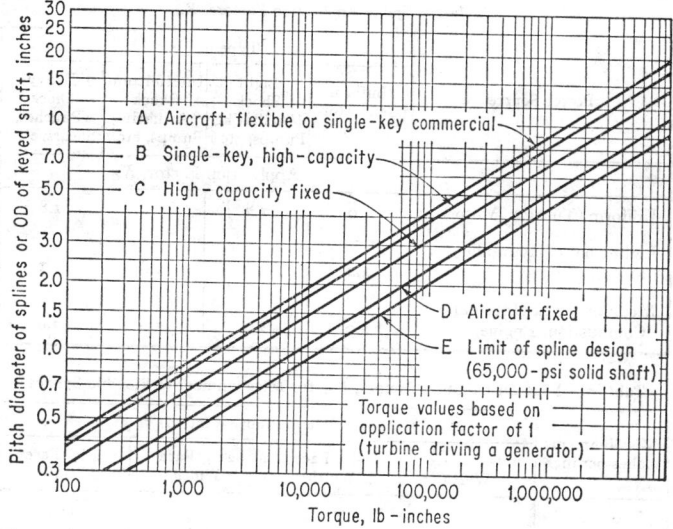

Fig. 1. Chart for estimating involute spline size based on diameter-torque relationships.

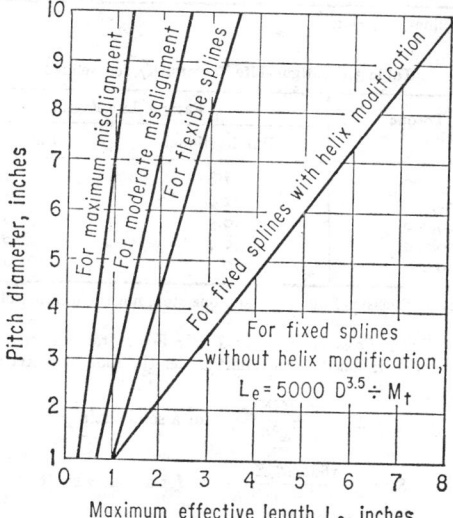

Fig. 2. Maximum effective length for fixed and flexible splines.

Table 1.　Spline Application Factors, $K_a$

| Power Source | Type of Load | | | |
|---|---|---|---|---|
| | Uniform (Generator-Fan) | Light Shock (Oscillating Pumps, etc.) | Intermittent Shock (Actuating Pumps, etc.) | Heavy Shock (Punches, Shears, etc.) |
| | Application Factor, $K_a$ | | | |
| Uniform (Turbine, Motor) | 1.0 | 1.2 | 1.5 | 1.8 |
| Light Shock (Hydraulic Motor) | 1.2 | 1.3 | 1.8 | 2.1 |
| Medium Shock (Internal Combustion, Engine) | 2.0 | 2.2 | 2.4 | 2.8 |

Table 2.　Load Distribution Factors, $K_m$, for Misalignment of Flexible Splines*

| Misalignment, inches per inch | Load Distribution Factor, $K_m$* | | | |
|---|---|---|---|---|
| | ½-in. Face Width | 1-in. Face Width | 2-in. Face Width | 4-in. Face Width |
| 0.001 | 1 | 1 | 1 | 1½ |
| 0.002 | 1 | 1 | 1½ | 2 |
| 0.004 | 1 | 1½ | 2 | 2½ |
| 0.008 | 1½ | 2 | 2½ | 3 |

* For fixed splines, $K_m = 1$.

Table 3.　Fatigue-Life Factors, $K_f$, for Splines

| No. of Torque Cycles* | Fatigue-Life Factor, $K_f$ | |
|---|---|---|
| | Unidirectional | Fully-reversed |
| 1,000 | 1.8 | 1.8 |
| 10,000 | 1.0 | 1.0 |
| 100,000 | 0.5 | 0.4 |
| 1,000,000 | 0.4 | 0.3 |
| 10,000,000 | 0.3 | 0.2 |

* A torque cycle consists of one start and one stop, not the number of revolutions.

*Shear Stress Under Roots of External Teeth:* For a transmitted torque $M_t$, the torsional shear stress induced in the shaft under the root diameter of an external spline is:

$$S_s = \frac{16 M_t K_a}{\pi D_{re}^3 K_f} \qquad \text{for a solid shaft} \qquad (1)$$

$$S_s = \frac{16 M_t D_{re} K_a}{\pi (D_{re}^4 - D_h^4) K_f} \qquad \text{for a hollow shaft} \qquad (2)$$

The computed stress should not exceed the values in Table 5.

Table 4. Wear Life Factors, $K_w$, for Flexible Splines*

| Number of Revolutions of Spline | Life Factor, $K_w$ | Number of Revolutions of Spline | Life Factor, $K_w$ |
|---|---|---|---|
| 10,000 | 4.0 | 100,000,000 | 1.0 |
| 100,000 | 2.8 | 1,000,000,000 | 0.7 |
| 1,000,000 | 2.0 | 10,000,000,000 | 0.5 |
| 10,000,000 | 1.4 | .... | ... |

* Wear life factors, unlike fatigue life factors given in Table 3, are based on the total number of revolutions of the spline, since each revolution of a flexible spline results in a complete cycle of rocking motion which contributes to spline wear.

Table 5. Allowable Shear Stresses for Splines

| Material | Hardness Brinell | Hardness Rockwell C | Max. Allowable Shear Stress, psi |
|---|---|---|---|
| Steel | 160–200 | — | 20,000 |
| Steel | 230–260 | — | 30,000 |
| Steel | 302–351 | 33–38 | 40,000 |
| Surface-hardened Steel | — | 48–53 | 40,000 |
| Case-hardened Steel | — | 58–63 | 50,000 |
| Through-hardened Steel (Aircraft Quality) | — | 42–46 | 45,000 |

*Shear Stress at the Pitch Diameter of Teeth:* The shear stress at the pitch line of the teeth for a transmitted torque $M_t$ is:

$$S_s = \frac{4M_t K_a K_m}{DNL_e t K_f} \tag{3}$$

The factor of 4 in (3) assumes that only half the teeth will carry the load because of spacing errors. For poor manufacturing accuracies, change the factor to 6.

The computed stress should not exceed the values in Table 5.

*Compressive Stresses on Sides of Spline Teeth:* Allowable compressive stresses on splines are very much lower than for gear teeth since non-uniform load distribution and misalignment result in unequal load sharing and end loading of the teeth.

$$\text{For flexible splines, } S_c = \frac{2M_t K_m K_a}{DNL_e h K_w} \tag{4}$$

$$\text{For fixed splines, } S_c = \frac{2M_t K_m K_a}{9DNL_e h K_f} \tag{5}$$

In these formulas, $h$ is the depth of engagement of the teeth, which for flat root splines is $0.9/P$ and for fillet root splines is $1/P$, approximately.

Table 6. Allowable Compressive Stress for Splines

| Material | Hardness | | Max. Allowable Compressive Stress, psi | |
|---|---|---|---|---|
| | Brinell | Rockwell C | Straight | Crowned |
| Steel | 160–200 | — | 1,500 | 6,000 |
| Steel | 230–260 | — | 2,000 | 8,000 |
| Steel | 302–351 | 33–38 | 3,000 | 12,000 |
| Surface-hardened Steel | — | 48–53 | 4,000 | 16,000 |
| Case-hardened Steel | — | 58–63 | 5,000 | 20,000 |

The stresses computed from Formulas (4) and (5) should not exceed the values in Table 6.

*Bursting Stresses on Splines:* Internal splines may burst due to three kinds of tensile stress: (1) tensile stress due to the radial component of the transmitted load; (2) centrifugal tensile stress; and (3) tensile stress due to the tangential force at the pitch line causing bending of the teeth.

$$\text{Radial load tensile stress, } S_1 = \frac{M_t \tan \phi}{\pi D t_w L} \qquad (6)$$

where $t_w$ = wall thickness of internal spline = outside diameter of spline sleeve minus spline major diameter, all divided by 2. $L$ = full length of spline.

$$\text{Centrifugal tensile stress, } S_2 = \frac{1.656 \times (\text{rpm})^2 (D_{oi}^2 + 0.212 D_{ri}^2)}{1,000,000} \qquad (7)$$

Where $D_{oi}$ = outside diameter of spline sleeve.

$$\text{Beam loading tensile stress, } S_3 = \frac{4M_t}{D^2 L_e Y} \qquad (8)$$

In this equation, $Y$ is the Lewis form factor obtained from a tooth layout. For internal splines of 30-deg. pressure angle a value of $Y = 1.5$ is a satisfactory estimate. The factor 4 in (8) assumes that only half the teeth are carrying the load.

Table 7. Allowable Tensile Stresses for Splines

| Material | Hardness | | Max. Allowable Stress, psi |
|---|---|---|---|
| | Brinell | Rockwell C | |
| Steel | 160–200 | — | 22,000 |
| Steel | 230–260 | — | 32,000 |
| Steel | 302–351 | 33–38 | 45,000 |
| Surface-hardened Steel | — | 48–53 | 45,000 |
| Case-hardened Steel | — | 58–63 | 55,000 |
| Through-hardened Steel | — | 42–46 | 50,000 |

The total tensile stress tending to burst the rim of the external member is:

$$S_t = \frac{K_a K_m (S_1 + S_3) + S_2}{K_f} \qquad (9)$$

This computed value should be less than those in Table 7.

**Crowned Splines for Large Misalignments.** — As mentioned on page 1010, crowned splines can accommodate misalignments of up to about 5 degrees. Crowned splines have considerably less capacity than straight splines of the same size if both are operating with precise alignment. However, when large misalignments exist, the crowned spline has greater capacity.

American Standard tooth forms may be used for crowned external members so that they may be mated with straight internal members of Standard form.

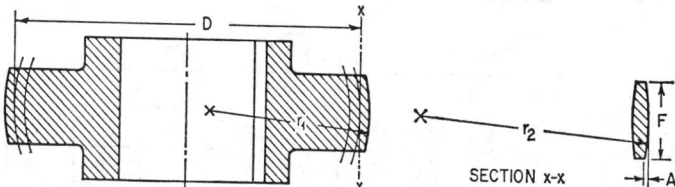

SECTION x-x

The accompanying diagram of a crowned spline shows the radius of the crown $r_1$; the radius of curvature of the crowned tooth, $r_2$; the pitch diameter of the spline, $D$; the face width, $F$; and the relief or crown height $A$ at the ends of the teeth. The crown height $A$ should always be made somewhat greater than one-half the face width multiplied by the tangent of the misalignment angle. For a crown height $A$, the approximate radius of curvature $r_2$, is $F^2 \div 8A$, and $r_1 = r_2 \tan \phi$, where $\phi$ is the pressure angle of the spline.

The compressive stress on the teeth is:

$$S_c = 3190 \sqrt{2T \div DNh r_2}$$

This calculated value should be less than the value in Table 6.

**Fretting Damage to Splines and Other Machine Elements.** — Fretting or fretting corrosion is a type of damage occurring at the interface of two contacting surfaces subject to slippage. On steel, evidence of fretting shows up as debris formed at the interface accompanied by pitting of the metal. This debris is usually red in color and is, in fact, red iron oxide. This characteristic rust-like sludge formation may mistakenly be attributed to rusting of the parts and not to fretting. Fretting damage is not confined to ferrous materials and is observed to occur with aluminum, copper, and other common metals.

Fretting damage may arise in the operation of all machinery subject to vibration and is likely to: (1) destroy close tolerances; (2) clog moving parts; and (3) increase susceptibility to fatigue failure of the parts. Examples of fretting damage are quite commonly found in splines, tie plates, roller shafts in textile machinery; pins in gear trains, suspension springs, connecting rods, and any other elements where relative slippage of contacting parts occurs.

Certain conditions favor the inception of fretting damage: (1) small relative motion of the reciprocating or oscillating type; (2) high contact pressures; (3) oxygen at the contacting surfaces; and (4) lack of means to replenish lubricant at the contacting surfaces. In addition, when similar metals or soft metals constitute the contacting surfaces, fretting is more likely to proceed at a higher rate.

## Table 1. S.A.E. Standard Splined Fittings

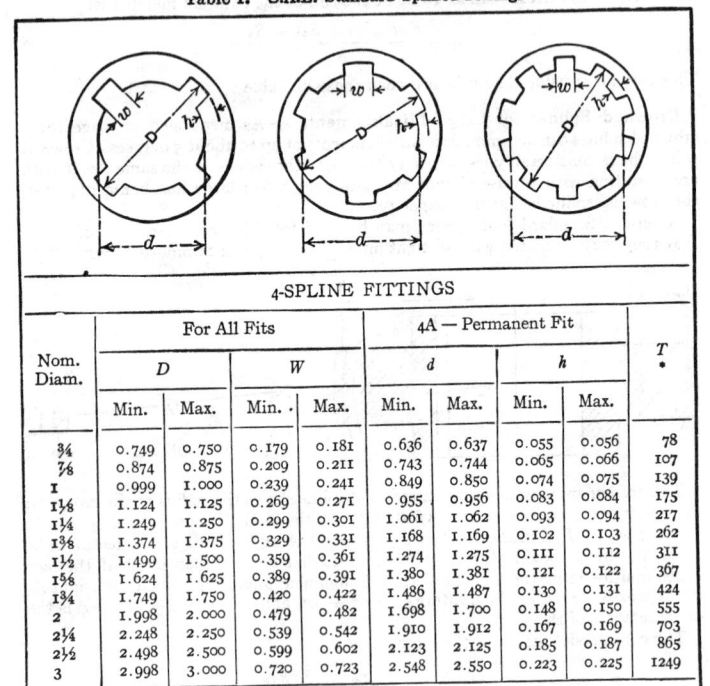

### 4-SPLINE FITTINGS

| Nom. Diam. | For All Fits | | | | 4A — Permanent Fit | | | | T * |
|---|---|---|---|---|---|---|---|---|---|
| | D | | W | | d | | h | | |
| | Min. | Max. | Min. | Max. | Min. | Max. | Min. | Max. | |
| ¾ | 0.749 | 0.750 | 0.179 | 0.181 | 0.636 | 0.637 | 0.055 | 0.056 | 78 |
| ⅞ | 0.874 | 0.875 | 0.209 | 0.211 | 0.743 | 0.744 | 0.065 | 0.066 | 107 |
| I | 0.999 | I.000 | 0.239 | 0.241 | 0.849 | 0.850 | 0.074 | 0.075 | 139 |
| I⅛ | I.124 | I.125 | 0.269 | 0.271 | 0.955 | 0.956 | 0.083 | 0.084 | 175 |
| I¼ | I.249 | I.250 | 0.299 | 0.301 | I.061 | I.062 | 0.093 | 0.094 | 217 |
| I⅜ | I.374 | I.375 | 0.329 | 0.331 | I.168 | I.169 | 0.102 | 0.103 | 262 |
| I½ | I.499 | I.500 | 0.359 | 0.361 | I.274 | I.275 | 0.111 | 0.112 | 311 |
| I⅝ | I.624 | I.625 | 0.389 | 0.391 | I.380 | I.381 | 0.121 | 0.122 | 367 |
| I¾ | I.749 | I.750 | 0.420 | 0.422 | I.486 | I.487 | 0.130 | 0.131 | 424 |
| 2 | I.998 | 2.000 | 0.479 | 0.482 | I.698 | I.700 | 0.148 | 0.150 | 555 |
| 2¼ | 2.248 | 2.250 | 0.539 | 0.542 | I.910 | I.912 | 0.167 | 0.169 | 703 |
| 2½ | 2.498 | 2.500 | 0.599 | 0.602 | 2.123 | 2.125 | 0.185 | 0.187 | 865 |
| 3 | 2.998 | 3.000 | 0.720 | 0.723 | 2.548 | 2.550 | 0.223 | 0.225 | 1249 |

### 4-SPLINE FITTINGS      6-SPLINE FITTINGS

| Nom. Diam. | 4B — To Slide — No Load | | | | | For All Fits | | | |
|---|---|---|---|---|---|---|---|---|---|
| | d | | h | | T * | D | | W | |
| | Min. | Max. | Min. | Max. | | Min. | Max. | Min. | Max. |
| ¾ | 0.561 | 0.562 | 0.093 | 0.094 | 123 | 0.749 | 0.750 | 0.186 | 0.188 |
| ⅞ | 0.655 | 0.656 | 0.108 | 0.109 | 167 | 0.874 | 0.875 | 0.217 | 0.219 |
| I | 0.749 | 0.750 | 0.124 | 0.125 | 219 | 0.999 | I.000 | 0.248 | 0.250 |
| I⅛ | 0.843 | 0.844 | 0.140 | 0.141 | 277 | I.124 | I.125 | 0.279 | 0.281 |
| I¼ | 0.936 | 0.937 | 0.155 | 0.156 | 341 | I.249 | I.250 | 0.311 | 0.313 |
| I⅜ | I.030 | I.031 | 0.171 | 0.172 | 414 | I.374 | I.375 | 0.342 | 0.344 |
| I½ | I.124 | I.125 | 0.186 | 0.187 | 491 | I.499 | I.500 | 0.373 | 0.375 |
| I⅝ | I.218 | I.219 | 0.202 | 0.203 | 577 | I.624 | I.625 | 0.404 | 0.406 |
| I¾ | I.311 | I.312 | 0.218 | 0.219 | 670 | I.749 | I.750 | 0.436 | 0.438 |
| 2 | I.498 | I.500 | 0.248 | 0.250 | 875 | I.998 | 2.000 | 0.497 | 0.500 |
| 2¼ | I.685 | I.687 | 0.279 | 0.281 | 1106 | 2.248 | 2.250 | 0.560 | 0.563 |
| 2½ | I.873 | I.875 | 0.310 | 0.312 | 1365 | 2.498 | 2.500 | 0.622 | 0.625 |
| 3 | 2.248 | 2.250 | 0.373 | 0.375 | 1969 | 2.998 | 3.000 | 0.747 | 0.750 |

\* See note at end of Table 4.

Table 2. S.A.E. Standard Splined Fittings

## 6 – SPLINE FITTINGS

| Nom. Diam. | 6A—Permanent Fit | | | 6B—To Slide—No Load | | | 6C—To Slide Under Load | | |
|---|---|---|---|---|---|---|---|---|---|
| | d | | T * | d | | T * | d | | T * |
| | Min. | Max. | | Min. | Max. | | Min. | Max. | |
| ¾ | 0.674 | 0.675 | 80 | 0.637 | 0.638 | 117 | 0.599 | 0.600 | 152 |
| ⅞ | 0.787 | 0.788 | 109 | 0.743 | 0.744 | 159 | 0.699 | 0.700 | 207 |
| 1 | 0.899 | 0.900 | 143 | 0.849 | 0.850 | 208 | 0.799 | 0.800 | 270 |
| 1⅛ | 1.012 | 1.013 | 180 | 0.955 | 0.956 | 263 | 0.899 | 0.900 | 342 |
| 1¼ | 1.124 | 1.125 | 223 | 1.062 | 1.063 | 325 | 0.999 | 1.000 | 421 |
| 1⅜ | 1.237 | 1.238 | 269 | 1.168 | 1.169 | 393 | 1.099 | 1.100 | 510 |
| 1½ | 1.349 | 1.350 | 321 | 1.274 | 1.275 | 468 | 1.199 | 1.200 | 608 |
| 1⅝ | 1.462 | 1.463 | 376 | 1.380 | 1.381 | 550 | 1.299 | 1.300 | 713 |
| 1¾ | 1.574 | 1.575 | 436 | 1.487 | 1.488 | 637 | 1.399 | 1.400 | 827 |
| 2 | 1.798 | 1.800 | 570 | 1.698 | 1.700 | 833 | 1.598 | 1.600 | 1080 |
| 2¼ | 2.023 | 2.025 | 721 | 1.911 | 1.913 | 1052 | 1.798 | 1.800 | 1367 |
| 2½ | 2.248 | 2.250 | 891 | 2.123 | 2.125 | 1300 | 1.998 | 2.000 | 1688 |
| 3 | 2.698 | 2.700 | 1283 | 2.548 | 2.550 | 1873 | 2.398 | 2.400 | 2430 |

## 10 – SPLINE FITTINGS

| Nom. Diam. | For All Fits | | | | 10A—Permanent Fit | | |
|---|---|---|---|---|---|---|---|
| | D | | W | | d | | T * |
| | Min. | Max. | Min. | Max. | Min. | Max. | |
| ¾ | 0.749 | 0.750 | 0.115 | 0.117 | 0.682 | 0.683 | 120 |
| ⅞ | 0.874 | 0.875 | 0.135 | 0.137 | 0.795 | 0.796 | 165 |
| 1 | 0.999 | 1.000 | 0.154 | 0.156 | 0.909 | 0.910 | 215 |
| 1⅛ | 1.124 | 1.125 | 0.174 | 0.176 | 1.023 | 1.024 | 271 |
| 1¼ | 1.249 | 1.250 | 0.193 | 0.195 | 1.137 | 1.138 | 336 |
| 1⅜ | 1.374 | 1.375 | 0.213 | 0.215 | 1.250 | 1.251 | 406 |
| 1½ | 1.499 | 1.500 | 0.232 | 0.234 | 1.364 | 1.365 | 483 |
| 1⅝ | 1.624 | 1.625 | 0.252 | 0.254 | 1.478 | 1.479 | 566 |
| 1¾ | 1.749 | 1.750 | 0.271 | 0.273 | 1.592 | 1.593 | 658 |
| 2 | 1.998 | 2.000 | 0.309 | 0.312 | 1.818 | 1.820 | 860 |
| 2¼ | 2.248 | 2.250 | 0.348 | 0.351 | 2.046 | 2.048 | 1088 |
| 2½ | 2.498 | 2.500 | 0.387 | 0.390 | 2.273 | 2.275 | 1343 |
| 3 | 2.998 | 3.000 | 0.465 | 0.468 | 2.728 | 2.730 | 1934 |
| 3½ | 3.497 | 3.500 | 0.543 | 0.546 | 3.182 | 3.185 | 2632 |
| 4 | 3.997 | 4.000 | 0.621 | 0.624 | 3.637 | 3.640 | 3438 |
| 4½ | 4.497 | 4.500 | 0.699 | 0.702 | 4.092 | 4.095 | 4351 |
| 5 | 4.997 | 5.000 | 0.777 | 0.780 | 4.547 | 4.550 | 5371 |
| 5½ | 5.497 | 5.500 | 0.855 | 0.858 | 5.002 | 5.005 | 6500 |
| 6 | 5.997 | 6.000 | 0.933 | 0.936 | 5.457 | 5.460 | 7735 |

* See note at end of Table 4.

Table 3. **S.A.E.** Standard Splined Fittings

## 10-SPLINE FITTINGS

| Nom. Diam. | 10B — To Slide — No Load | | | 10C — To Slide Under Load | | |
|---|---|---|---|---|---|---|
| | d | | T • | d | | T • |
| | Min. | Max. | | Min. | Max. | |
| ¾ | 0.644 | 0.645 | 183 | 0.607 | 0.608 | 241 |
| ⅞ | 0.752 | 0.753 | 248 | 0.708 | 0.709 | 329 |
| 1 | 0.859 | 0.860 | 326 | 0.809 | 0.810 | 430 |
| 1⅛ | 0.967 | 0.968 | 412 | 0.910 | 0.911 | 545 |
| 1¼ | 1.074 | 1.075 | 508 | 1.012 | 1.013 | 672 |
| 1⅜ | 1.182 | 1.183 | 614 | 1.113 | 1.114 | 813 |
| 1½ | 1.289 | 1.290 | 732 | 1.214 | 1.215 | 967 |
| 1⅝ | 1.397 | 1.398 | 860 | 1.315 | 1.316 | 1135 |
| 1¾ | 1.504 | 1.505 | 997 | 1.417 | 1.418 | 1316 |
| 2 | 1.718 | 1.720 | 1302 | 1.618 | 1.620 | 1720 |
| 2¼ | 1.933 | 1.935 | 1647 | 1.821 | 1.823 | 2176 |
| 2½ | 2.148 | 2.150 | 2034 | 2.023 | 2.025 | 2688 |
| 3 | 2.578 | 2.580 | 2929 | 2.428 | 2.430 | 3869 |
| 3½ | 3.007 | 3.010 | 3987 | 2.832 | 2.835 | 5266 |
| 4 | 3.437 | 3.440 | 5208 | 3.237 | 3.240 | 6878 |
| 4½ | 3.867 | 3.870 | 6591 | 3.642 | 3.645 | 8705 |
| 5 | 4.297 | 4.300 | 8137 | 4.047 | 4.050 | 10746 |
| 5½ | 4.727 | 4.730 | 9846 | 4.452 | 4.455 | 13003 |
| 6 | 5.157 | 5.160 | 11718 | 4.857 | 4.860 | 15475 |

## 16-SPLINE FITTINGS

| Nom. Diam. | For All Fits | | | | 16A — Permanent Fit | | |
|---|---|---|---|---|---|---|---|
| | D | | W | | d | | T • |
| | Min. | Max. | Min. | Max. | Min. | Max. | |
| 2 | 1.997 | 2.000 | 0.193 | 0.196 | 1.817 | 1.820 | 1375 |
| 2½ | 2.497 | 2.500 | 0.242 | 0.245 | 2.273 | 2.275 | 2149 |
| 3 | 2.997 | 3.000 | 0.291 | 0.294 | 2.727 | 2.730 | 3094 |
| 3½ | 3.497 | 3.500 | 0.340 | 0.343 | 3.182 | 3.185 | 4212 |
| 4 | 3.997 | 4.000 | 0.389 | 0.392 | 3.637 | 3.640 | 5501 |
| 4½ | 4.497 | 4.500 | 0.438 | 0.441 | 4.092 | 4.095 | 6962 |
| 5 | 4.997 | 5.000 | 0.487 | 0.490 | 4.547 | 4.550 | 8595 |
| 5½ | 5.497 | 5.500 | 0.536 | 0.539 | 5.002 | 5.005 | 10395 |
| 6 | 5.997 | 6.000 | 0.585 | 0.588 | 5.457 | 5.460 | 12377 |

• See note at end of Table 4.

Table 4.   S.A.E. Standard Splined Fittings

| 16-SPLINE FITTINGS | | | | | | |
|---|---|---|---|---|---|---|
| Nom. Diam. | 16B — To Slide — No Load | | | 16C — To Slide Under Load | | |
| | d | | T * | d | | T * |
| | Min. | Max. | | Min. | Max. | |
| 2 | 1.717 | 1.720 | 2083 | 1.617 | 1.620 | 2751 |
| 2½ | 2.147 | 2.150 | 3255 | 2.022 | 2.025 | 4299 |
| 3 | 2.577 | 2.580 | 4687 | 2.427 | 2.430 | 6190 |
| 3½ | 3.007 | 3.010 | 6378 | 2.832 | 2.835 | 8426 |
| 4 | 3.437 | 3.440 | 8333 | 3.237 | 3.240 | 11005 |
| 4½ | 3.867 | 3.870 | 10546 | 3.642 | 3.645 | 13928 |
| 5 | 4.297 | 4.300 | 13020 | 4.047 | 4.050 | 17195 |
| 5½ | 4.727 | 4.730 | 15754 | 4.452 | 4.455 | 20806 |
| 6 | 5.157 | 5.160 | 18749 | 4.857 | 4.860 | 24760 |

\* *Torque Capacity of Spline Fittings:* The torque capacities of the different spline fittings are given in the columns headed "T". The torque capacity, per inch of bearing length at 1000 pounds pressure per square inch on the sides of the spline, may be determined by the following formula, in which $T$ = torque capacity in inch-pounds per inch of length, $N$ = number of splines, $R$ = mean radius or radial distance from center of hole to center of spline, $h$ = depth of spline:

$$T = 1000 \, NRh$$

Table 5.   Formulas for Determining Dimensions of S.A.E. Standard Splines

| No. of Splines | W, For All Fits | A Permanent Fit | | B To Slide Without Load | | C To Slide Under Load | |
|---|---|---|---|---|---|---|---|
| | | h | d | h | d | h | d |
| Four | 0.241D[1] | 0.075D | 0.850D | 0.125D | 0.750D | .... | .... |
| Six | 0.250D | 0.050D | 0.900D | 0.075D | 0.850D | 0.100D | 0.800D |
| Ten | 0.156D | 0.045D | 0.910D | 0.070D | 0.860D | 0.095D | 0.810D |
| Sixteen | 0.098D | 0.045D | 0.910D | 0.070D | 0.860D | 0.095D | 0.810D |

These formulas give the maximum dimensions for $W$, $h$ and $d$, as listed in Tables 1 to 4 inclusive. (1) Four splines for fits A and B only.

## S.A.E. Standard Spline Fittings.

The S.A.E. spline fittings (Tables 1 to 4 inclusive) have become an established standard for many applications in the automotive, machine tool, and other industries. The dimensions given, in inches, apply only to soft broached holes. The tolerances given may be readily maintained by usual broaching methods. The tolerances selected for the large and small diameters may depend upon whether the fit between the mating part, as finally made, is on the large or the small diameter. The other diameter, which is designed for clearance, may have a larger manufactured tolerance. If the final fit between the parts is on the sides of the spline only, larger tolerances are permissible for both the large and small diameters. The spline should not be more than 0.006 inch per foot out of parallel with respect to the shaft axis. No allowance is made for corner radii to obtain clearance. Radii at the corners of the spline should not exceed 0.015 inch.

## Proportions of Square Shafts and Fit Allowances

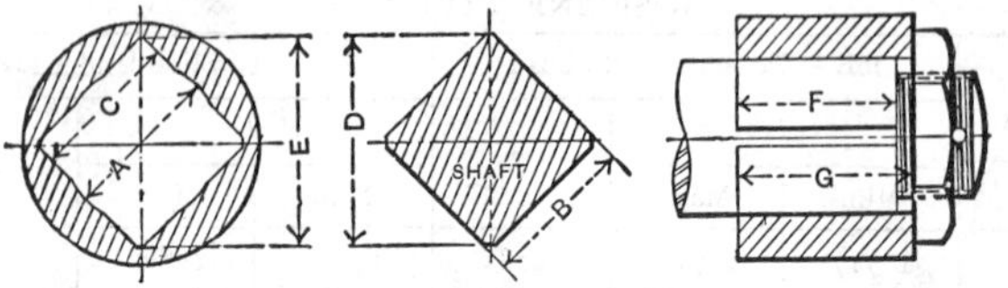


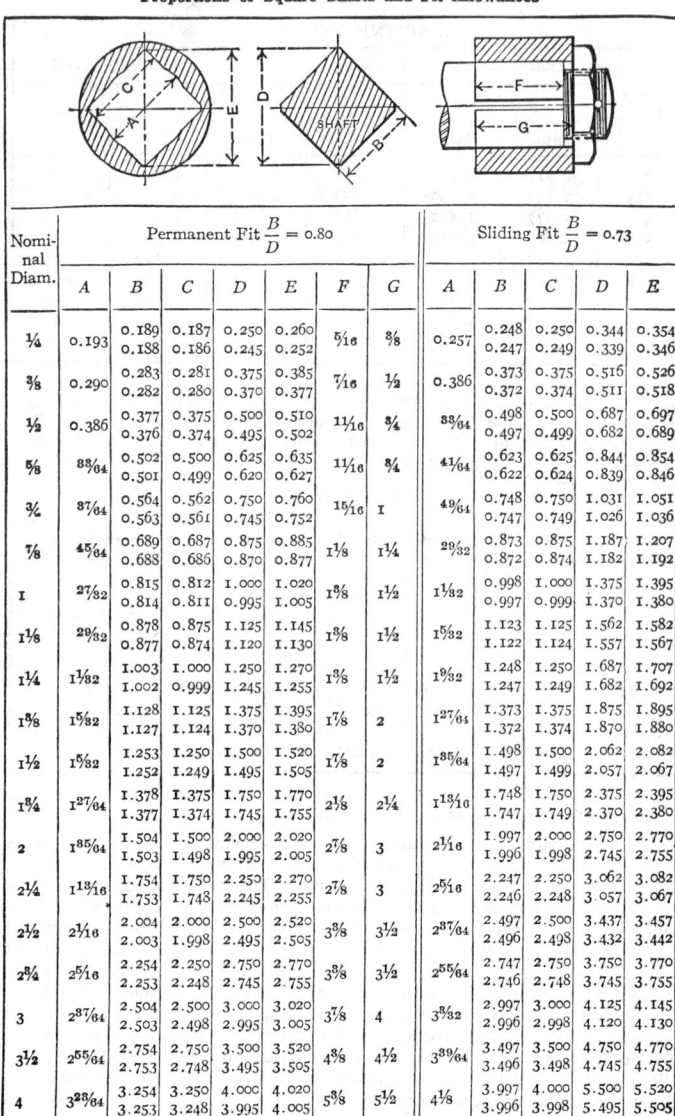

| Nominal Diam. | Permanent Fit $\frac{B}{D} = 0.80$ | | | | | | | Sliding Fit $\frac{B}{D} = 0.73$ | | | | |
|---|---|---|---|---|---|---|---|---|---|---|---|---|
| | *A* | *B* | *C* | *D* | *E* | *F* | *G* | *A* | *B* | *C* | *D* | *E* |
| 1/4 | 0.193 | 0.189<br>0.188 | 0.187<br>0.186 | 0.250<br>0.245 | 0.260<br>0.252 | 5/16 | 3/8 | 0.257 | 0.248<br>0.247 | 0.250<br>0.249 | 0.344<br>0.339 | 0.354<br>0.346 |
| 3/8 | 0.290 | 0.283<br>0.282 | 0.281<br>0.280 | 0.375<br>0.370 | 0.385<br>0.377 | 7/16 | 1/2 | 0.386 | 0.373<br>0.372 | 0.375<br>0.374 | 0.516<br>0.511 | 0.526<br>0.518 |
| 1/2 | 0.386 | 0.377<br>0.376 | 0.375<br>0.374 | 0.500<br>0.495 | 0.510<br>0.502 | 11/16 | 3/4 | 33/64 | 0.498<br>0.497 | 0.500<br>0.499 | 0.687<br>0.682 | 0.697<br>0.689 |
| 5/8 | 33/64 | 0.502<br>0.501 | 0.500<br>0.499 | 0.625<br>0.620 | 0.635<br>0.627 | 11/16 | 3/4 | 41/64 | 0.623<br>0.622 | 0.625<br>0.624 | 0.844<br>0.839 | 0.854<br>0.846 |
| 3/4 | 37/64 | 0.564<br>0.563 | 0.562<br>0.561 | 0.750<br>0.745 | 0.760<br>0.752 | 15/16 | 1 | 49/64 | 0.748<br>0.747 | 0.750<br>0.749 | 1.031<br>1.026 | 1.051<br>1.036 |
| 7/8 | 45/64 | 0.689<br>0.688 | 0.687<br>0.686 | 0.875<br>0.870 | 0.885<br>0.877 | 1 1/8 | 1 1/4 | 29/32 | 0.873<br>0.872 | 0.875<br>0.874 | 1.187<br>1.182 | 1.207<br>1.192 |
| 1 | 27/32 | 0.815<br>0.814 | 0.812<br>0.811 | 1.000<br>0.995 | 1.020<br>1.005 | 1 3/8 | 1 1/2 | 1 1/32 | 0.998<br>0.997 | 1.000<br>0.999 | 1.375<br>1.370 | 1.395<br>1.380 |
| 1 1/8 | 29/32 | 0.878<br>0.877 | 0.875<br>0.874 | 1.125<br>1.120 | 1.145<br>1.130 | 1 3/8 | 1 1/2 | 1 5/32 | 1.123<br>1.122 | 1.125<br>1.124 | 1.562<br>1.557 | 1.582<br>1.567 |
| 1 1/4 | 1 1/32 | 1.003<br>1.002 | 1.000<br>0.999 | 1.250<br>1.245 | 1.270<br>1.255 | 1 3/8 | 1 1/2 | 1 9/32 | 1.248<br>1.247 | 1.250<br>1.249 | 1.687<br>1.682 | 1.707<br>1.692 |
| 1 3/8 | 1 5/32 | 1.128<br>1.127 | 1.125<br>1.124 | 1.375<br>1.370 | 1.395<br>1.380 | 1 7/8 | 2 | 1 27/64 | 1.373<br>1.372 | 1.375<br>1.374 | 1.875<br>1.870 | 1.895<br>1.880 |
| 1 1/2 | 1 5/32 | 1.253<br>1.252 | 1.250<br>1.249 | 1.500<br>1.495 | 1.520<br>1.505 | 1 7/8 | 2 | 1 35/64 | 1.498<br>1.497 | 1.500<br>1.499 | 2.062<br>2.057 | 2.082<br>2.067 |
| 1 3/4 | 1 27/64 | 1.378<br>1.377 | 1.375<br>1.374 | 1.750<br>1.745 | 1.770<br>1.755 | 2 1/8 | 2 1/4 | 1 13/16 | 1.748<br>1.747 | 1.750<br>1.749 | 2.375<br>2.370 | 2.395<br>2.380 |
| 2 | 1 35/64 | 1.504<br>1.503 | 1.500<br>1.498 | 2.000<br>1.995 | 2.020<br>2.005 | 2 7/8 | 3 | 2 1/16 | 1.997<br>1.996 | 2.000<br>1.998 | 2.750<br>2.745 | 2.770<br>2.755 |
| 2 1/4 | 1 13/16 | 1.754<br>1.753 | 1.750<br>1.748 | 2.250<br>2.245 | 2.270<br>2.255 | 2 7/8 | 3 | 2 5/16 | 2.247<br>2.246 | 2.250<br>2.248 | 3.062<br>3.057 | 3.082<br>3.067 |
| 2 1/2 | 2 1/16 | 2.004<br>2.003 | 2.000<br>1.998 | 2.500<br>2.495 | 2.520<br>2.505 | 3 3/8 | 3 1/2 | 2 37/64 | 2.497<br>2.496 | 2.500<br>2.498 | 3.437<br>3.432 | 3.457<br>3.442 |
| 2 3/4 | 2 5/16 | 2.254<br>2.253 | 2.250<br>2.248 | 2.750<br>2.745 | 2.770<br>2.755 | 3 3/8 | 3 1/2 | 2 55/64 | 2.747<br>2.746 | 2.750<br>2.748 | 3.750<br>3.745 | 3.770<br>3.755 |
| 3 | 2 37/64 | 2.504<br>2.503 | 2.500<br>2.498 | 3.000<br>2.995 | 3.020<br>3.005 | 3 7/8 | 4 | 3 3/32 | 2.997<br>2.996 | 3.000<br>2.998 | 4.125<br>4.120 | 4.145<br>4.130 |
| 3 1/2 | 2 55/64 | 2.754<br>2.753 | 2.750<br>2.748 | 3.500<br>3.495 | 3.520<br>3.505 | 4 3/8 | 4 1/2 | 3 39/64 | 3.497<br>3.496 | 3.500<br>3.498 | 4.750<br>4.745 | 4.770<br>4.755 |
| 4 | 3 23/64 | 3.254<br>3.253 | 3.250<br>3.248 | 4.000<br>3.995 | 4.020<br>4.005 | 5 3/8 | 5 1/2 | 4 1/8 | 3.997<br>3.996 | 4.000<br>3.998 | 5.500<br>5.495 | 5.520<br>5.505 |

**Taper Shaft Ends with Slotted Nuts (SAE Standard)**

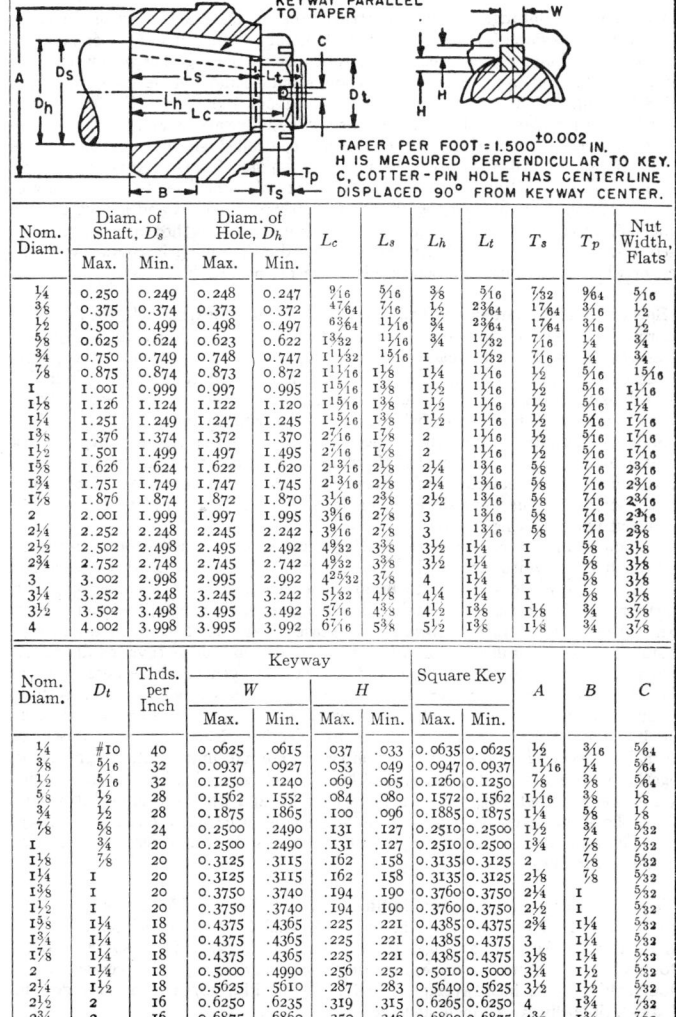

KEYWAY PARALLEL TO TAPER

TAPER PER FOOT = 1.500 +0.002 IN.
H IS MEASURED PERPENDICULAR TO KEY.
C, COTTER-PIN HOLE HAS CENTERLINE DISPLACED 90° FROM KEYWAY CENTER.

| Nom. Diam. | Diam. of Shaft, $D_s$ | | Diam. of Hole, $D_h$ | | $L_c$ | $L_s$ | $L_h$ | $L_t$ | $T_s$ | $T_p$ | Nut Width, Flats |
|---|---|---|---|---|---|---|---|---|---|---|---|
| | Max. | Min. | Max. | Min. | | | | | | | |
| 1/4 | 0.250 | 0.249 | 0.248 | 0.247 | 9/16 | 5/16 | 3/8 | 5/16 | 7/32 | 9/64 | 5/16 |
| 3/8 | 0.375 | 0.374 | 0.373 | 0.372 | 47/64 | 7/16 | 1/2 | 23/64 | 17/64 | 3/16 | 1/2 |
| 1/2 | 0.500 | 0.499 | 0.498 | 0.497 | 63/64 | 11/16 | 3/4 | 23/64 | 17/64 | 3/16 | 1/2 |
| 5/8 | 0.625 | 0.624 | 0.623 | 0.622 | 1 3/32 | 11/16 | 3/4 | 17/32 | 7/16 | 1/4 | 3/4 |
| 3/4 | 0.750 | 0.749 | 0.748 | 0.747 | 1 11/32 | 15/16 | 1 | 17/32 | 7/16 | 1/4 | 3/4 |
| 7/8 | 0.875 | 0.874 | 0.873 | 0.872 | 1 11/16 | 1 1/8 | 1 1/4 | 11/16 | 1/2 | 5/16 | 15/16 |
| 1 | 1.001 | 0.999 | 0.997 | 0.995 | 1 11/16 | 1 1/8 | 1 1/2 | 11/16 | 1/2 | 5/16 | 1 1/16 |
| 1 1/8 | 1.126 | 1.124 | 1.122 | 1.120 | 1 15/16 | 1 3/8 | 1 1/2 | 11/16 | 1/2 | 5/16 | 1 1/4 |
| 1 1/4 | 1.251 | 1.249 | 1.247 | 1.245 | 1 15/16 | 1 3/8 | 1 1/2 | 11/16 | 1/2 | 5/16 | 1 7/16 |
| 1 3/8 | 1.376 | 1.374 | 1.372 | 1.370 | 2 7/16 | 1 7/8 | 2 | 11/16 | 1/2 | 5/16 | 1 7/16 |
| 1 1/2 | 1.501 | 1.499 | 1.497 | 1.495 | 2 7/16 | 1 7/8 | 2 | 11/16 | 1/2 | 5/16 | 1 7/16 |
| 1 5/8 | 1.626 | 1.624 | 1.622 | 1.620 | 2 13/16 | 2 1/8 | 2 1/4 | 13/16 | 5/8 | 7/16 | 2 3/16 |
| 1 3/4 | 1.751 | 1.749 | 1.747 | 1.745 | 2 13/16 | 2 1/8 | 2 1/4 | 13/16 | 5/8 | 7/16 | 2 3/16 |
| 1 7/8 | 1.876 | 1.874 | 1.872 | 1.870 | 3 1/16 | 2 3/8 | 2 1/2 | 13/16 | 5/8 | 7/16 | 2 3/16 |
| 2 | 2.001 | 1.999 | 1.997 | 1.995 | 3 9/16 | 2 7/8 | 3 | 13/16 | 5/8 | 7/16 | 2 3/16 |
| 2 1/4 | 2.252 | 2.248 | 2.245 | 2.242 | 3 9/16 | 2 7/8 | 3 | 13/16 | 5/8 | 7/16 | 2 3/8 |
| 2 1/2 | 2.502 | 2.498 | 2.495 | 2.492 | 4 9/32 | 3 3/8 | 3 1/2 | 1 1/4 | 1 | 5/8 | 3 1/8 |
| 2 3/4 | 2.752 | 2.748 | 2.745 | 2.742 | 4 9/32 | 3 3/8 | 3 1/2 | 1 1/4 | 1 | 5/8 | 3 1/8 |
| 3 | 3.002 | 2.998 | 2.995 | 2.992 | 4 25/32 | 3 7/8 | 4 | 1 1/4 | 1 | 5/8 | 3 1/8 |
| 3 1/4 | 3.252 | 3.248 | 3.245 | 3.242 | 5 5/32 | 4 1/8 | 4 1/4 | 1 1/4 | 1 | 5/8 | 3 1/8 |
| 3 1/2 | 3.502 | 3.498 | 3.495 | 3.492 | 5 7/16 | 4 3/8 | 4 1/2 | 1 3/8 | 1 1/8 | 3/4 | 3 7/8 |
| 4 | 4.002 | 3.998 | 3.995 | 3.992 | 6 7/16 | 5 3/8 | 5 1/2 | 1 3/8 | 1 1/8 | 3/4 | 3 7/8 |

| Nom. Diam. | $D_t$ | Thds. per Inch | Keyway W | | Keyway H | | Square Key | | $A$ | $B$ | $C$ |
|---|---|---|---|---|---|---|---|---|---|---|---|
| | | | Max. | Min. | Max. | Min. | Max. | Min. | | | |
| 1/4 | #10 | 40 | 0.0625 | .0615 | .037 | .033 | 0.0635 | 0.0625 | 1/2 | 3/16 | 5/64 |
| 3/8 | 5/16 | 32 | 0.0937 | .0927 | .053 | .049 | 0.0947 | 0.0937 | 11/16 | 1/4 | 5/64 |
| 1/2 | 9/16 | 32 | 0.1250 | .1240 | .069 | .065 | 0.1260 | 0.1250 | 7/8 | 3/8 | 5/64 |
| 5/8 | 1/2 | 28 | 0.1562 | .1552 | .084 | .080 | 0.1572 | 0.1562 | 1 1/16 | 3/8 | 1/8 |
| 3/4 | 1/2 | 28 | 0.1875 | .1865 | .100 | .096 | 0.1885 | 0.1875 | 1 1/4 | 5/8 | 1/8 |
| 7/8 | 5/8 | 24 | 0.2500 | .2490 | .131 | .127 | 0.2510 | 0.2500 | 1 1/2 | 7/8 | 5/32 |
| 1 | 3/4 | 20 | 0.2500 | .2490 | .131 | .127 | 0.2510 | 0.2500 | 1 3/4 | 7/8 | 5/32 |
| 1 1/8 | 7/8 | 20 | 0.3125 | .3115 | .162 | .158 | 0.3135 | 0.3125 | 2 | 7/8 | 5/32 |
| 1 1/4 | 1 | 20 | 0.3125 | .3115 | .162 | .158 | 0.3135 | 0.3125 | 2 1/8 | 7/8 | 5/32 |
| 1 3/8 | 1 | 20 | 0.3750 | .3740 | .194 | .190 | 0.3760 | 0.3750 | 2 1/4 | 1 | 5/32 |
| 1 1/2 | 1 | 20 | 0.3750 | .3740 | .194 | .190 | 0.3760 | 0.3750 | 2 1/2 | 1 | 5/32 |
| 1 5/8 | 1 1/4 | 18 | 0.4375 | .4365 | .225 | .221 | 0.4385 | 0.4375 | 2 3/4 | 1 1/4 | 5/32 |
| 1 3/4 | 1 1/4 | 18 | 0.4375 | .4365 | .225 | .221 | 0.4385 | 0.4375 | 3 | 1 1/4 | 5/32 |
| 1 7/8 | 1 1/4 | 18 | 0.4375 | .4365 | .225 | .221 | 0.4385 | 0.4375 | 3 1/8 | 1 1/4 | 5/32 |
| 2 | 1 1/4 | 18 | 0.5000 | .4990 | .256 | .252 | 0.5010 | 0.5000 | 3 1/4 | 1 1/2 | 5/32 |
| 2 1/4 | 1 1/2 | 18 | 0.5625 | .5610 | .287 | .283 | 0.5640 | 0.5625 | 3 1/2 | 1 1/2 | 5/32 |
| 2 1/2 | 2 | 16 | 0.6250 | .6235 | .319 | .315 | 0.6265 | 0.6250 | 4 | 1 3/4 | 7/32 |
| 2 3/4 | 2 | 16 | 0.6875 | .6860 | .350 | .346 | 0.6890 | 0.6875 | 4 3/8 | 1 3/4 | 7/32 |
| 3 | 2 | 16 | 0.7500 | .7485 | .381 | .377 | 0.7515 | 0.7500 | 4 3/4 | 2 | 7/32 |
| 3 1/4 | 2 | 16 | 0.7500 | .7485 | .381 | .377 | 0.7515 | 0.7500 | 5 | 2 1/8 | 7/32 |
| 3 1/2 | 2 1/2 | 16 | 0.8750 | .8735 | .444 | .440 | 0.8765 | 0.8750 | 5 1/2 | 2 1/4 | 9/32 |
| 4 | 2 1/2 | 16 | 1.0000 | .9985 | .506 | .502 | 1.0015 | 1.0000 | 6 1/4 | 2 3/4 | 9/32 |

All dimensions in inches except where otherwise noted.

# FLAT BELTS AND PULLEYS

**Flat Leather Belting.** — Three principal types of leather belting produced in the United States are: (1) Oak tanned; (2) Mineral retanned; and (3) Combination oak tanned and mineral retanned. All three types are put together with waterproof cement. The first is for general applications but should not be used where the drive is to operate at an ambient temperature above 120 degrees F. The second type is used for high-speed small-diameter pulley applications and is also suitable for high-speed motor drives and on tension-controlled short-center motor-base applications. It will also resist the mists or vapors of corrosive acids for long periods. The third type is made up of a ply of oak-tanned leather with a ply of mineral retanned leather put together with waterproof cement and is used on step cone pulleys and hard-pull shifting drives. None of these three types should be used, however, when a possibility exists of liquid acid coming in contact with the belt or when the ambient temperature is above 140 degrees F.

**Specifying Flat Leather Belting.** — When specifying a leather belt, the following information should be supplied: (1) Shortest steel tape length around pulleys. If measurement is stated as made with steel tape, manufacturer will make proper deductions for initial stretch. (2) The type, width, and thickness, (3) Whether the belt is to be supplied: (a) made endless at the factory; (b) with laps prepared for cementing on the job, or (c) with ends cut square for lacing.

Slide-rail mounted motors should be moved towards the driven pulley at least three-quarters of the total travel of the slide-rail base before taking measurements. It should be noted that the measurement of an old belt is not a dependable guide for the length of a new one since, being elastic, the new belt will let out somewhat when put under driving tension.

For important drives, where the belting manufacturer is called upon to recommend belt type, width, and thickness, the National Industrial Leather Association lists the following information to be given to the supplier: (1) source of power, whether electric motor, steam engine, diesel engine, or line shaft; if electric motor, then (2) manufacturer, horsepower, rpm, phase, cycles, voltage, current, and type of starting used; (3) full data for both driver and driven pulleys covering diameters, face widths, material, speed, bore, and keyway; (4) general drive data, such as distance between pulley centers and whether tight side of belt is top or bottom; (5) angle of drive center line, whether horizontal vertical, or at a given angle from the horizontal; (6) atmospheric conditions, whether clean, oily, wet, dusty, or normal; (7) type of service, whether temporary, normal, important, or continuous, together with hours per day; (8) type of load, as defined by kind of driven machine, horsepower required, both maximum and average, and whether the load is steady, jerky, shock, or reversing; and (9) whether belt is to be made endless on pulleys or at factory. In addition, a diagram of the drive indicating direction of rotation should be furnished.

**Thicknesses of Flat Leather Belting.** — The following thicknesses have been approved and adopted by the National Industrial Leather Association: Medium Single Ply, 11/64 inch, average; Heavy Single Ply, 13/64 inch, average; Light Double Ply, 18/64 inch, average; Medium Double Ply, 20/64 inch, average; Heavy Double Ply, 23/64 inch, average; Medium Triple Ply, 30/64 inch, average; and Heavy Triple Ply, 34/64 inch, average.

*Measurement:* All of the above thicknesses are average thicknesses and should be determined by measuring 20 coils and dividing the total by the number of coils measured. In rolls containing less than 20 coils, the average thickness should be determined by measuring all of the coils in the roll.

### Table 1.  Belt Capacity Factor, *K* — Flat Leather Belting

Officially adopted by the National Industrial Leather Association

| Belt Speed, Feet per Minute | SINGLE PLY | | DOUBLE PLY | | | TRIPLE PLY | |
|---|---|---|---|---|---|---|---|
| | *11/64″ | *13/64″ | *18/64″ | *20/64″ | *23/64″ | *30/64″ | *34/64″ |
| | Medium | Heavy | Light | Medium | Heavy | Medium | Heavy |
| | Horsepower per Inch of Width, *K* — to be Corrected by Factors from Table 2 | | | | | | |
| 600 | 1.1 | 1.2 | 1.5 | 1.8 | 2.2 | 2.5 | 2.8 |
| 800 | 1.4 | 1.7 | 2.0 | 2.4 | 2.9 | 3.3 | 3.6 |
| 1000 | 1.8 | 2.1 | 2.6 | 3.1 | 3.6 | 4.1 | 4.5 |
| 1200 | 2.1 | 2.5 | 3.1 | 3.7 | 4.3 | 4.9 | 5.4 |
| 1400 | 2.5 | 2.9 | 3.5 | 4.3 | 4.9 | 5.7 | 6.3 |
| 1600 | 2.8 | 3.3 | 4.0 | 4.9 | 5.6 | 6.5 | 7.1 |
| 1800 | 3.2 | 3.7 | 4.5 | 5.4 | 6.2 | 7.3 | 8.0 |
| 2000 | 3.5 | 4.1 | 4.9 | 6.0 | 6.9 | 8.1 | 8.9 |
| 2200 | 3.9 | 4.5 | 5.4 | 6.6 | 7.6 | 8.8 | 9.7 |
| 2400 | 4.2 | 4.9 | 5.9 | 7.1 | 8.2 | 9.5 | 10.5 |
| 2600 | 4.5 | 5.3 | 6.3 | 7.7 | 8.9 | 10.3 | 11.4 |
| 2800 | 4.9 | 5.6 | 6.8 | 8.2 | 9.5 | 11.0 | 12.1 |
| 3000 | 5.2 | 5.9 | 7.2 | 8.7 | 10.0 | 11.6 | 12.8 |
| 3200 | 5.4 | 6.3 | 7.6 | 9.2 | 10.6 | 12.3 | 13.5 |
| 3400 | 5.7 | 6.6 | 7.9 | 9.7 | 11.2 | 12.9 | 14.2 |
| 3600 | 5.9 | 6.9 | 8.3 | 10.1 | 11.7 | 13.4 | 14.8 |
| 3800 | 6.2 | 7.1 | 8.7 | 10.5 | 12.2 | 14.0 | 15.4 |
| 4000 | 6.4 | 7.4 | 9.0 | 10.9 | 12.6 | 14.5 | 16.0 |
| 4200 | 6.7 | 7.7 | 9.3 | 11.3 | 13.0 | 15.0 | 16.5 |
| 4400 | 6.9 | 7.9 | 9.6 | 11.7 | 13.4 | 15.4 | 16.9 |
| 4600 | 7.1 | 8.1 | 9.8 | 12.0 | 13.8 | 15.8 | 17.4 |
| 4800 | 7.2 | 8.3 | 10.1 | 12.3 | 14.1 | 16.2 | 17.8 |
| 5000 | 7.4 | 8.4 | 10.3 | 12.5 | 14.3 | 16.5 | 18.2 |
| 5200 | 7.5 | 8.6 | 10.5 | 12.8 | 14.6 | 16.8 | 18.5 |
| 5400 | 7.6 | 8.7 | 10.6 | 12.9 | 14.8 | 17.1 | 18.8 |
| 5600 | 7.7 | 8.8 | 10.8 | 13.1 | 15.0 | 17.3 | 19.0 |
| 5800 | 7.7 | 8.9 | 10.9 | 13.2 | 15.1 | 17.5 | 19.2 |
| †6000 | 7.8 | 8.9 | 10.9 | 13.2 | 15.2 | 17.6 | 19.3 |
| Belt Speed, fpm | Minimum Allowable Pulley Diameter, Inch, For Belt Thicknesses Listed Above | | | | | | |
| Up to 2500 | 2½ | 3 | 4 | 5¹ | 8¹ | 16² | 20² |
| 2500 to 4000 | 3 | 3½ | 4½ | 6¹ | 9¹ | 18² | 22² |
| 4000 to 6000 | 3½ | 4 | 5 | 7¹ | 10¹ | 20² | 24² |

\* The belt thicknesses are average thicknesses.  See paragraph on Thicknesses of Flat Leather Belting.

† For belt speeds over 6000 feet per minute, consult a leather belting manufacturer.

¹ For belts 8 inches wide and over, add 2 inches to minimum pulley diameter shown.

² For belts 8 inches wide and over, add 4 inches to minimum pulley diameter shown.

*Tolerances:* Allowable tolerances for thicknesses of single and double ply belts are plus or minus 1/64 inch, based on the nominal thickness. At no point shall single-ply belting be more than 3/64 inch thicker or 2/64 inch thinner than the average thickness. For double-ply belting, the variation in thickness shall not be greater than 2/64 inch thicker or thinner than the average.

*Triple-ply Belts:* Most triple-ply belts are constructed for particular drive conditions. The manufacturer should be consulted for specific information concerning thickness and construction.

**Pulleys. —** On step-cone pulleys use a narrow, thick belt rather than a wide, thin belt of the same horsepower capacity. It is good practice to use pulleys with faces from ½ to 2 inches (depending upon their diameters) wider than the belt required for the drive.

**Use with Tension-Controlling Motor Base. —** When used with an electric motor drive, a flat leather belt will give best results if the motor is mounted on some kind of a tension controlling base. Three types are generally available. Two of these are pivoted; one uses the weight of the motor to maintain the proper belt tension; the other utilizes the reaction torque of the motor to accomplish this. The third type has a sliding action and controls the belt tension by means of springs. The effect of all three types is to cause the belt to maintain a uniform pull around and across the pulleys.

**Table 2.   Service Correction Factors *M*, *P* and *F***
**Used in Determining Horsepower Rating**

Select the one appropriate factor from each of the three divisions in this table.

| | | $M$ |
|---|---|---|
| Motor type and Starting Method | Squirrel cage, compensator starting | 1.5 |
| | Squirrel cage, line starting | 2.0 |
| | Slip ring and high starting torque | 2.5 |
| | | $P$ |
| Diameter of Small Pulley | 4 inches and under | 0.5 |
| | 4½ to 8 inches | 0.6 |
| | 9 to 12 inches | 0.7 |
| | 13 to 16 inches | 0.8 |
| | 17 to 30 inches | 0.9 |
| | Over 30 inches | 1.0 |
| | | $F$ |
| Operating Conditions | Oily, wet, or dusty atmosphere | 1.35 |
| | Vertical drives | 1.2 |
| | Jerky loads | 1.2 |
| | Shock and reversing loads | 1.4 |

**Horsepower Ratings. —** In Table 1 are given belt capacity factors for various types and thicknesses of flat leather belting and for various belt speeds in feet per minute. These factors are expressed in terms of horsepower per inch of width and are modified by service correction factors given in Table 2 as shown in the formula below. Where the pulley speed is known in terms of revolutions per minute, the corresponding belt speed in feet per minute can be found from Table 3.

The following formula is used to obtain the horsepower rating, *H*, of flat leather belt:

$$H = \frac{W \times K \times P}{M \times F} \qquad (1)$$

Table 3.  Conversion of Pulley Speeds in Revolutions per Minute into Feet per Minute

| Pulley Diam. in Inches | Revolutions per Minute | | | | | | | | | | | | | | | | | | | | |
|---|---|---|---|---|---|---|---|---|---|---|---|---|---|---|---|---|---|---|---|---|---|
| | 100 | 200 | 300 | 400 | 435 | 490 | 500 | 600 | 690 | 700 | 800 | 900 | 1000 | 1150 | 1200 | 1400 | 1500 | 1600 | 1750 | 1800 | 3600 |
| | Velocity in Feet per Minute* | | | | | | | | | | | | | | | | | | | | |
| 1 | 26 | 52 | 79 | 105 | 114 | 128 | 131 | 157 | 181 | 183 | 209 | 236 | 262 | 301 | 314 | 367 | 393 | 419 | 458 | 471 | 942 |
| 2 | 52 | 105 | 157 | 209 | 228 | 257 | 262 | 314 | 361 | 367 | 419 | 471 | 524 | 602 | 628 | 733 | 785 | 838 | 916 | 942 | 1885 |
| 3 | 79 | 157 | 236 | 314 | 342 | 385 | 393 | 471 | 542 | 550 | 628 | 707 | 785 | 903 | 942 | 1100 | 1178 | 1257 | 1374 | 1414 | 2827 |
| 4 | 105 | 209 | 314 | 419 | 456 | 513 | 524 | 628 | 722 | 733 | 838 | 942 | 1047 | 1204 | 1257 | 1466 | 1571 | 1676 | 1833 | 1885 | 3770 |
| 5 | 131 | 262 | 393 | 524 | 570 | 641 | 654 | 785 | 903 | 916 | 1047 | 1178 | 1309 | 1505 | 1571 | 1833 | 1964 | 2094 | 2291 | 2356 | 4712 |
| 6 | 157 | 314 | 471 | 628 | 683 | 770 | 785 | 942 | 1084 | 1100 | 1257 | 1414 | 1571 | 1806 | 1885 | 2199 | 2356 | 2513 | 2749 | 2827 | 5655 |
| 7 | 183 | 367 | 550 | 733 | 797 | 898 | 916 | 1100 | 1264 | 1283 | 1466 | 1649 | 1833 | 2107 | 2199 | 2566 | 2749 | 2932 | 3207 | 3299 | … |
| 8 | 209 | 419 | 628 | 838 | 911 | 1026 | 1047 | 1257 | 1445 | 1466 | 1676 | 1885 | 2094 | 2409 | 2513 | 2932 | 3142 | 3351 | 3665 | 3770 | … |
| 9 | 236 | 471 | 707 | 942 | 1025 | 1154 | 1178 | 1414 | 1625 | 1649 | 1885 | 2121 | 2356 | 2710 | 2827 | 3299 | 3534 | 3770 | 4123 | 4241 | … |
| 10 | 262 | 524 | 785 | 1048 | 1139 | 1283 | 1309 | 1571 | 1806 | 1833 | 2094 | 2356 | 2618 | 3011 | 3142 | 3665 | 3927 | 4189 | 4582 | 4712 | … |
| 20 | 524 | 1047 | 1571 | 2094 | 2278 | 2566 | 2618 | 3142 | 3612 | 3665 | 4189 | 4712 | 5236 | … | … | … | … | … | … | … | … |
| 30 | 785 | 1571 | 2356 | 3142 | 3416 | 3848 | 3927 | 4712 | 5418 | 5498 | … | … | … | … | … | … | … | … | … | … | … |
| 0.1 | 3 | 5 | 8 | 10 | 11 | 13 | 13 | 16 | 18 | 18 | 21 | 24 | 26 | 30 | 31 | 37 | 39 | 42 | 46 | 47 | 94 |
| 0.2 | 5 | 10 | 16 | 21 | 23 | 26 | 26 | 31 | 36 | 37 | 42 | 47 | 52 | 60 | 63 | 73 | 79 | 84 | 92 | 94 | 188 |
| 0.3 | 8 | 16 | 24 | 31 | 34 | 38 | 39 | 47 | 54 | 55 | 63 | 71 | 79 | 90 | 94 | 110 | 118 | 126 | 137 | 141 | 283 |
| 0.4 | 10 | 21 | 31 | 42 | 46 | 51 | 52 | 63 | 72 | 73 | 84 | 94 | 105 | 120 | 126 | 147 | 157 | 168 | 183 | 188 | 377 |
| 0.5 | 13 | 26 | 39 | 52 | 57 | 64 | 65 | 79 | 90 | 92 | 105 | 118 | 131 | 151 | 157 | 183 | 196 | 209 | 229 | 236 | 471 |
| 0.6 | 16 | 31 | 47 | 63 | 68 | 77 | 78 | 94 | 108 | 110 | 126 | 141 | 157 | 181 | 188 | 220 | 236 | 251 | 275 | 283 | 565 |
| 0.7 | 18 | 37 | 55 | 73 | 80 | 90 | 92 | 110 | 126 | 128 | 147 | 165 | 183 | 211 | 220 | 257 | 275 | 293 | 321 | 330 | 660 |
| 0.8 | 21 | 42 | 63 | 84 | 91 | 103 | 105 | 126 | 144 | 147 | 168 | 189 | 209 | 241 | 251 | 293 | 314 | 335 | 367 | 377 | 754 |
| 0.9 | 24 | 47 | 71 | 94 | 102 | 115 | 118 | 141 | 163 | 165 | 188 | 212 | 236 | 271 | 283 | 330 | 353 | 377 | 412 | 424 | 848 |

* Based on: Velocity in fpm = Diam. in inches × rpm × 0.2618.   For velocities above 6,000 fpm, consult belt manufacturer.
*Example:* Find velocity in fpm of a 28.3-inch diameter pulley rotating at a speed of 600 rpm.   *Solution:* Add the velocities equivalent to the 20-, 8-, and 0.3-inch diameters, thus: V = 3142 + 1257 + 47 = 4446 fpm.

where $W$ = width of belt in inches.
      $K$ = theoretical belt capacity factor taken from Table 1.
      $P$ = correction factor for diameter of smaller pulley taken from Table 2.
      $M$ = correction factor if type of motor and starting method is one of those given in Table 2. If not, use $M = 1$.
      $F$ = special factor if operating condition is one of those shown in Table 2. If not, use $F = 1$.

If the belt is used on an electric motor drive and the horsepower rating of the motor is known at a given speed, then the required width of belt is found from the formula:

$$W = \frac{R \times M \times F}{K \times P} \qquad (2)$$

where $R$ = nameplate horsepower rating of the electric motor.
      $M$, $F$, $K$, and $P$ are as in the previous formula.

*Example:* Find the proper width of flat leather belting for a drive employing a squirrel cage induction motor rated at 15 horsepower at 1750 rpm with across-the-line starting. Pulley diameters are: motor, 8 inches and driven, 16 inches.
    1. For 8-inch pulley running at 1750 rpm, Table 3 gives belt speed of 3665 fpm.
    2. The bottom section of Table 1 shows that for a belt speed of 2500 to 4000 fpm, and an 8-inch pulley, a Medium Double Ply belt is suitable.
    3. Table 1 gives the belt capacity factor for a Medium Double Ply belt running at 3665 fpm as 10.2 (interpolating between 10.1 for 3600 fpm and 10.5 for 3800 fpm).
    4. Table 2 gives the motor correction factor for squirrel cage motor, line starting is 2.0.
    5. The pulley correction factor for 8-inch pulley is given as 0.6 in Table 2.

Hence         $W = \dfrac{R \times M}{K \times P} = \dfrac{15 \times 2.0}{10.2 \times 0.6} = 4.9$ in.

Use a 5-inch wide Medium Double Ply belt.

If the motor was being used on a vertical drive without a tension-controlled motor base, the belt width, as determined, should be multiplied by a factor of 1.2 taken from Table 2. The proper belt for use in this case would be 4.9 × 1.2 or a 6-inch Medium Double Ply belt.

**Speed Limitation.** — The use of flat leather belting for speeds in excess of 6000 feet per minute is not generally recommended. At these speeds the amount of horsepower which the belt can transmit decreases due to increased slippage because of the action of centrifugal force in holding the belt off the pulleys and thus reducing the areas of contact and the contact pressure.

**Installation of Leather Belting.** — An easy method of aligning shafting and pulleys is to first check shafts with a level and then place a taut string across the shafts and check each shaft against it with a large square. The alignment of pulleys can be checked by using a taut string along their edges. If the pulleys are the same width, the string should touch lightly at two opposite points on the rim of each pulley. If the pulleys are of different widths, the distance from the string to the pulley rim at opposite points on its circumference should be the same. If possible, each pulley should be given a half-turn and then rechecked. When pulleys are installed one above the other, a string and plumb-bob can be used to check alignment.

*Belt Tension:* For best results a leather belt should be run with the least tension needed to transmit the load without slipping. If the belt is too slack it will slip, causing its surface to glaze, then crack and peel. If it is too tight, it may put excessive load on the bearings.

Wherever possible, flat leather belts should be operated with the slack side on top. This will provide a greater arc of contact between belt and faces of the pulleys permitting lower tension. On short-center or vertical drives, a tension-controlled motor base should be used.

The maximum pulling power is obtained by running the grain or hair side of the belt next to the pulley faces.

*Endless Belts:* An endless belt should be forced over the pulleys with care to avoid putting a crook in it. Belts, particularly those six inches and wider, should either be made endless on the job as shown in the 19th edition of Machinery's Handbook or slipped on after temporarily shortening the center distance between pulleys. This can be done by moving motor on slide rails, loosening hanger bolts, etc.

*Running Direction:* Care should be taken to have the outside feather edge of the lap face away from the direction in which the belt runs as shown in the accompanying

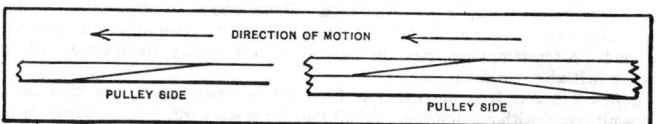

illustration. This tends to protect the outside points of the lap if they should strike guards, guides, or shifters. It also protects the lap from being opened by windage.

*Belt Splicing:* Wherever possible leather belts should be made endless on the job. This avoids possibility of damage to the belt which frequently occurs when an endless belt, that is a tight fit, is forced over the pulleys.

The first step should be to shorten the pulley centers to the minimum. Then, a steel tape measurement is taken around the pulleys. The usual practice is to reduce this measured length by about ⅛ inch per foot to provide adequate belt tension. To the resulting dimension, add the proper allowance as given in Table 4. The result will be the length of the belting to be cut from the roll.

Table 4. Lap and Break Lengths for Splicing Flat Leather Belting

| Single Lap Belt | | | Double Lap Belt | | | |
|---|---|---|---|---|---|---|
| Belt Width | Lap Length | Allowance for Splice | Belt Width | Lap Length | Break Length | Allowance for Splice |
| Under 4 | 4 | 4 | Under 8 | 4 | 8 | 12 |
| 4 to 10 | 5 | 5 | 8 to 12 | 4 | 10 | 14 |
| Over 10 | 6 | 6 | 12 to 18 | 4 | 16 | 20 |
| .... | .. | .. | 18 and up | 4 | 20 | 24 |

All dimensions in inches.

**Maintenance of Leather Belting.** — It is important to establish a system of inspection at regular intervals, at which times the following points should be checked.

1. *Dryness of Belt:* When pulley faces begin to polish, it is a sign that dressing is needed on the belt. Under normal conditions, dress belts every three to six months. Best results are usually obtained by using belt dressing sparingly, but frequently. Use a belt dressing approved by the manufacturer and designed to supply the necessary oils which were lost in use or during cleaning.

2. *Saturation with Oil:* Belts should be kept as clean as possible for best results. Oil or grease thrown from machine bearings will reduce belt life and pulling power. If the leak cannot be stopped at its source, the installation of deflectors or throwing discs will be helpful.

A small amount of oil on a belt can sometimes be removed by ordinary wiping. If this does not do the job, give it a thorough scrubbing with a solution of carbon tetrachloride and naphtha, using a stiff jute brush and working in the direction of lap joints so as not to lift them.

Another method is to remove the belt and soak it for five or six hours in a degreasing solution consisting of one part carbon tetrachloride to three parts naphtha. If carbon tetrachloride is not available, the belt can be soaked in any of the cleaning fluids used by dry cleaning establishments. Due to the fire hazard and toxic effect, the soaking and drying of the belt should be done in the open or where ventilation is good. After removing from the vat, allow belt to dry thoroughly. Always dress a belt after cleaning.

3. *Belt Tension:* It is important to keep the belt tight enough at all times to transmit power without slippage. A belt that is too slack will slip and burn, causing excessive wear. If the belt is too tight, it places undue strain on the bearings and the belt life will be shortened.

4. *Alignment of Shafting and Pulleys:* Belting cannot give good service if the pulleys or shafting are out of alignment. Indications of misalignment are: (a) belt running off the pulley at one side or (b) rubbing or climbing on flanged or step-cone pulleys. A simple test to determine whether the fault is the alignment or a crooked belt, is to turn the belt inside out or end for end. If it still runs to the same side of the pulley as before, the fault is in the alignment, not in the belt.

It is important to check drive alignment at least once a year. In multiple-story buildings, shifting of loads on floors above the shaft may cause it to be distorted or thrown out of alignment.

Some common faults in drive alignment are: (a) Shafts carrying driving and driven pulleys may not be parallel. (b) Shafting may be sprung out of line. (Hangers should always be located near the pulleys, the points of maximum load.) (c) Driving and driven pulleys may be offset. (d) Pulley may be eccentric with shafting.

5. *Laps and Plys:* When cementing laps, if first sizing coat does not dry with a shine, a second coat should be applied and allowed to dry. If sizing coat turns cloudy, apply a second coat and continue brushing until cloudiness disappears. If sizing coat does not dry in thirty minutes, wipe off and apply another sizing coat (this condition caused by an excessive amount of oil in the belt). If cemented laps show signs of opening, re-cement them immediately.

If belt guards, shifters, guides, or pulley flanges rub against edge of belt, laps and plies may open up. This condition should be corrected immediately. A good belt shifter has broad and well-rounded surfaces so as to spread thrust over a large, belt edge area.

Another cause of ply and lap separation is running too thick a belt on a small pulley.

*Shortening of Leather Belts:* When a belt becomes loose, it will slip excessively causing loss of power and undue wear on the belt. A loose belt should be shortened immediately.

Table 5. Trouble-shooting Chart for Flat Leather Transmission Belts

| Trouble | Cause | Remedy |
|---|---|---|
| 1. Belt slips and squeals | (a) Belt too loose<br>(b) Insufficient belt capacity<br>(c) Pulley crown too high, causing increased wear of narrow center section of belt<br>(d) Leather surface too dry and shiny | (a) Increase belt tension<br>(b) Use thicker or wider belt<br>(c) Decrease crown taper to ⅛ inch per foot<br>(d) Apply suitable dressing |
| 2. Excessive belt stretch | Belt capacity too low | Use thicker or wider belt |
| 3. Belt runs crooked | (a) Belt stretched on one side by forcing over pulley<br>(b) Belt ends not squared when joining<br>(c) Belt unevenly stretched by running on misaligned pulleys<br>(d) Loose belt unevenly stretched by running up on flanged or step-cone pulley | (a) (b) (c) (d) Repair damaged belt section or replace belt. Eliminate physical cause when installing |
| 4. Belt runs off pulleys | (a) Misalignment of pulleys or shafting (if belt continues to run off same side when belt is turned end for end)<br>(b) Crooked belt<br>(c) Pulley crown too high | (a) Eliminate cause<br>(b) Repair belt<br>(c) Decrease crown taper to ⅛ inch per foot |
| 5. Belt runs to one side of driven pulley | (a) Belt too slack<br>(b) Load too great<br>(c) Crooked belt (if it runs to opposite side when turned end for end)<br>(d) Misalignment of pulleys or shafting | (a) Increase belt tension<br>(b) Use thicker or wider belt<br>(c) Repair belt<br>(d) Eliminate cause |
| 6. Belt whips and flaps | (a) Pulsating load or power source<br>(b) Shaft, motor, or machine not rigidly supported<br>(c) Lopsided pulley<br>(d) Bent shaft<br>(e) Too much or too little belt tension | (a) (b) (c) (d)<br>(e) Eliminate cause where possible. Try change of speed or addition of flywheel to smooth out load |
| 7. Belt weaves back and forth across pulley | (a) Wobbly pulley<br>(b) High spot on pulley<br>(c) Belt extremely crooked | (a) (b) Correct faulty condition<br>(c) Repair or replace belt |
| 8. Cracked outside ply | (a) Excessive belt tension<br>(b) Pulley diameter too small | (a) Reduce tension<br>(b) Provide proper pulley for belt thickness |
| 9. Cracked inside ply | Burning caused by excessive slip | See Item 1. |
| 10. Peeling grain | (a) Excessive slip<br>(b) Improper belt dressing<br>(c) Chemical fumes or oil | (a) See Item 1.<br>(b) Clean belt with commercial solvent, scrape off any loose grain and use suitable dressing<br>(c) Provide guards if possible, and use type of belt best suited for condition |

Source: National Industrial Leather Association.

**Angular Drive.** — In laying out an angular drive — one in which the pulley shafts are not parallel — there is one fundamental rule to be followed: the belt must leave each pulley in the plane of the pulley toward which it is running. By "plane of the pulley" is meant that plane passing through the center of the face of the pulley and at right angles to its axis or shaft.

If the plane of one pulley of an open belt drive in which the shafts are parallel is rotated through an angle of 90 degrees and the pulleys are placed so that the above fundamental rule will hold, a "quarter-turn drive" will result.

**Quarter-turn Drive.** — To overcome the belt distortion encountered in a quarter-turn drive, and to distribute wear evenly on both of its sides, give one end of the belt a one-half turn (180 degrees) before making the splice. In a single-ply belt, made up in this way, the grain side at the joint is adjacent to the flesh side. This half-turn method of construction is particularly adaptable to double-ply belts, which have the grain exposed on both sides. The turning of the belt side to side also turns it edge to edge, with the result that not only is the wear distributed on the face of the belt, but the tension is also kept equalized on the edges of the belt.

When installing quarter-turn drives, it is particularly advantageous to make the belts endless with clamps and rods right on the job. Fewer shutdowns will result.

**Rubber Belting.** — Rubber belts are used in places exposed to the weather or the action of steam, as they do not absorb moisture or stretch as readily as leather belts, under like conditions. The quality of rubber belting depends on the mixture (containing more or less rubber) that forms the coating, the cotton duck that gives strength to the belt and the method of manufacture. The best grades of rubber belting contain nothing but new rubber; the cheapest grades are composed largely of reclaimed rubber. The weight of the cotton duck is an important consideration. High-grade belts contain what is known as a 32-ounce cotton duck, and the cheaper grades have either a 30-ounce or 28-ounce duck. If the proper weight of duck is used, a 3- or 4-ply rubber belt is equal in strength to a single leather belt; a 5- or 6-ply rubber belt is equal to a double leather belt, and a 7- or 8-ply rubber belt is equal to a triple leather belt.

**Pulleys.** — Flat belt pulleys are usually made of cast iron, fabricated steel, paper, fiber, or various kinds of wood. They may be solid or split and in either case the hub may be split for clamping to the shaft.

Pulley face widths are nominally the same as the widths of the belts they are to carry. Actually, however, the pulley face should be approximately one inch more than the belt width for belts under 12 inches wide, 2 inches more for belts from 12 to 24 inches wide, and 3 inches more for belts over 24 inches in width.

Belts may be made to center themselves on their pulleys by the use of crowned pulleys. The usual figure for the amount of crowning is ⅛ inch per foot of pulley width. Thus, the difference in maximum and minimum radii of a crowned 6-inch wide pulley would be ¹⁄₁₆ inch. Crowned pulleys have a rim section either with a convex curve or a flat V form. Flanges on the sides of flat belt pulleys are in general undesirable as the belt tends to crawl against them. Too much crown is undesirable because of the tendency to "break the belt's back." This is particularly true in the case of riding idlers close to driving pulleys where the curvature of the belt changes rapidly from one pulley to the other. In such cases the idler should under no circumstances be crowned and the adjacent pulley should have very little crown. Pulleys carrying shifting belts are not crowned.

Open belt drives connecting pulleys on short centers with one pulley considerably larger than the other may be unsatisfactory on account of the small angle of wrap on the smaller pulley. This angle may be increased by the use of idler pulleys on one or both sides of the belt.

**Cast-Iron Pulleys.** — Cast-iron pulleys formed of one solid casting may or may not have a split or divided hub. The solid-hub pulley is held to its shaft either by a key, a key and one or two set-screws, or by simply using one or more set-screws without a key as in the case of small pulleys, especially on low-grade machinery where there is little power to transmit. When the hub is split or divided, it is provided with clamping bolts, and when these are tightened, the split hub grips the shaft tightly. In addition to clamping bolts, a key or a key and set-screws may be used. Pulleys of this kind are known as the clamp-hub type.

Most pulleys have six arms. For diameters less than 15 or 20 inches, there may be four arms, and pulleys 5 feet or larger in diameter often have eight arms.

**Split Cast-iron Pulleys.** — The split pulley which is formed of two separate sections bolted together both at the hub and on opposite sides of the rim, can be placed between other pulleys on a shaft without removing either the pulleys or the shaft. These pulleys often have interchangeable hub bushings to fit shafts of different diameter. It is good practice to make pulleys having a face width of 10 inches or over, either of the clamp-hub or split form, because shrinkage strains are either greatly reduced or practically eliminated, and the hub of the pulley can be firmly clamped to a shaft even though the bore is not an accurate fit. If the face width of a cast-iron pulley is greater than from 20 to 24 inches, there should be two sets of arms to provide better support for the rim.

**Wood Pulleys.** — Wood pulleys are not only much lighter than cast-iron pulleys but they are superior as transmitters of power; in fact it is claimed that they will transmit from 35 to 50 per cent more power for the same belt tension. Wood pulleys should not be used where they are exposed to excessive moisture. Ordinarily the rims are built up of segments, and the arrangement of the arms varies on different sizes and makes. Some wood pulleys intended for unusually severe duty have a rim which is joined to an iron center or hub by a solid web of wood. Other pulleys of the iron-center type have cast-iron hubs and arms and a wood rim. Internal shrinkage strains are thus eliminated and the pulleys are adapted to unusually high speeds. Well-seasoned maple is adapted to wood pulleys. Wood bushings are often inserted in the hubs to permit using the pulleys on shafts of different size.

**Steel Pulleys.** — Pulleys formed of sheet steel combine lightness with strength and they are free from the initial stresses which are such an uncertain factor in many cast-iron pulleys. The weight is ordinarily from 45 to 55 per cent less than the weight of a cast-iron pulley of equal power-transmitting capacity, which lessens the weight on the lineshaft and reduces the frictional losses. A series of tests showed that the percentage of slip was from 2.35 to 2.70 per cent less for steel pulleys than for cast-iron pulleys. Steel pulleys are ordinarily of the split type.

**Safe Speeds for Pulleys.** — The maximum safe rim speeds for solid cast-iron pulleys is as a general rule about 5000 feet per minute. If the pulley is split or formed of separate sections which are bolted together at the rim, the maximum speed should be limited to about 55 or 60 per cent of the maximum speed for solid pulleys. While the safe speeds of built-up steel pulleys are subject to some variation on account of differences in design or construction, in general such pulleys may be run at about 6000 feet per minute. The safe speeds recommended for wood pulleys vary considerably according to the type; thus, the maximum speeds recommended may be 5000 feet per minute for some pulleys and 10,000 feet per minute for others of different construction. A pulley having a cast-iron hub and arms, with a wood rim, has been operated under test at a rim speed of five and one-half miles per minute. For additional information on speeds see pages 340 and 341 in the flywheel section.

## Dimensions of Pulleys

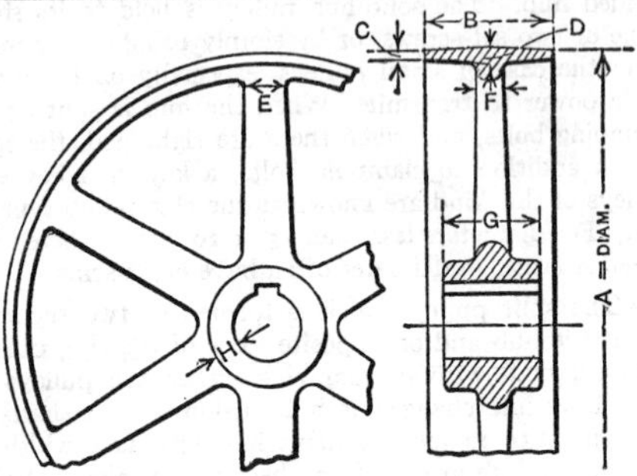


In all cases, the number of arms is 6. The arms increase in size towards the hub, the taper being 1/2 inch per foot. For safe speeds and descriptions of various pulley constructions, see pages 340 and 341.

| Diam. *A* | Face *B* | *C* | *D* | *E* | *F* | *G* | *H* |
|---|---|---|---|---|---|---|---|
| 6 | 4 | 1/8 | 3/16 | 3/4 | 7/16 | 3 | 3/8 |
| 6 | 6 | 1/8 | 3/16 | 3/4 | 7/16 | 3 1/2 | 1/2 |
| 6 | 8 | 1/8 | 3/16 | 3/4 | 7/16 | 3 1/2 | 1/2 |
| 6 | 12 | 1/8 | 3/16 | 3/4 | 7/16 | 4 | 1/2 |
| 8 | 4 | 1/8 | 3/16 | 13/16 | 7/16 | 3 | 3/8 |
| 8 | 6 | 1/8 | 3/16 | 13/16 | 7/16 | 3 1/2 | 1/2 |
| 8 | 8 | 5/32 | 1/4 | 1 1/16 | 9/16 | 4 1/2 | 1/2 |
| 8 | 12 | 5/32 | 1/4 | 1 1/16 | 9/16 | 5 1/2 | 1/2 |
| 10 | 4 | 1/8 | 3/16 | 15/16 | 9/16 | 3 | 1/2 |
| 10 | 6 | 5/32 | 1/4 | 1 1/16 | 9/16 | 3 1/2 | 1/2 |
| 10 | 8 | 5/32 | 1/4 | 1 1/16 | 9/16 | 4 1/2 | 1/2 |
| 10 | 12 | 5/32 | 1/4 | 1 5/16 | 5/8 | 5 1/2 | 5/8 |
| 12 | 4 | 5/32 | 1/4 | 1 | 7/16 | 3 1/4 | 1/2 |
| 12 | 6 | 5/32 | 1/4 | 1 3/4 | 1/2 | 4 | 1/2 |
| 12 | 8 | 5/32 | 1/4 | 1 3/4 | 1/2 | 5 | 5/8 |
| 12 | 12 | 3/16 | 5/16 | 1 1/2 | 3/4 | 6 1/2 | 5/8 |
| 14 | 4 | 5/32 | 1/4 | 1 1/8 | 1/2 | 3 1/2 | 1/2 |
| 14 | 6 | 5/32 | 1/4 | 1 1/8 | 1/2 | 4 1/2 | 5/8 |
| 14 | 8 | 3/16 | 5/16 | 1 5/16 | 9/16 | 5 | 5/8 |
| 14 | 12 | 3/16 | 5/16 | 1 11/16 | 13/16 | 6 1/2 | 5/8 |
| 16 | 4 | 5/32 | 1/4 | 1 3/8 | 9/16 | 3 1/2 | 1/2 |
| 16 | 8 | 3/16 | 5/16 | 1 7/16 | 5/8 | 5 | 5/8 |
| 16 | 12 | 7/32 | 11/32 | 1 7/16 | 5/8 | 6 1/2 | 3/4 |
| 16 | 16 | 7/32 | 11/32 | 1 7/8 | 15/16 | 8 1/4 | 7/8 |
| 18 | 4 | 3/16 | 5/16 | 1 5/16 | 9/16 | 4 | 5/8 |
| 18 | 8 | 7/32 | 11/32 | 1 1/2 | 11/16 | 5 1/2 | 3/4 |
| 18 | 12 | 7/32 | 11/32 | 1 1/2 | 11/16 | 7 1/4 | 7/8 |
| 18 | 20 | 1/4 | 3/8 | 2 1/4 | 1 1/4 | 9 | 7/8 |
| 20 | 4 | 3/16 | 5/16 | 1 3/8 | 5/8 | 4 | 5/8 |
| 20 | 8 | 3/16 | 5/16 | 1 3/8 | 5/8 | 5 | 3/4 |
| 20 | 12 | 7/32 | 11/32 | 1 5/8 | 3/4 | 7 | 3/4 |
| 20 | 20 | 9/32 | 7/16 | 2 1/4 | 1 1/8 | 10 | 1 |
| 22 | 4 | 3/16 | 5/16 | 1 1/2 | 5/8 | 4 | 5/8 |
| 22 | 8 | 3/16 | 5/16 | 1 1/2 | 5/8 | 5 | 3/4 |
| 22 | 12 | 7/32 | 11/32 | 1 3/4 | 13/16 | 6 1/2 | 7/8 |
| 22 | 20 | 9/32 | 7/16 | 2 1/2 | 1 1/4 | 11 | 1 1/8 |
| 24 | 4 | 7/32 | 11/32 | 1 9/16 | 11/16 | 4 | 5/8 |
| 24 | 8 | 7/32 | 11/32 | 1 9/16 | 11/16 | 5 1/2 | 3/4 |

## Rules for Calculating Diameters and Speeds of Pulleys

**Speed of Driven Pulley Required.** — Diameter and speed of driving pulley, and diameter of driven pulley are known. *Rule:* Multiply the diameter of the driving pulley by its speed in revolutions per minute, and divide the product by the diameter of the driven pulley.

*Example:* — If the diameter of the driving pulley is 15 inches and its speed, 180 revolutions per minute, and the diameter of the driven pulley, 9 inches, then the speed of the driven pulley $= \dfrac{15 \times 180}{9} = 300$ revolutions per minute.

**Diameter of Driven Pulley Required.** — Diameter and speed of driving pulley, and revolutions per minute of driven pulley are known. *Rule:* Multiply the diameter of the driving pulley by its speed in revolutions per minute, and divide the product by the required speed of the driven pulley.

*Example:* — If the diameter of the driving pulley is 24 inches and its speed, 100 revolutions per minute, and the driven pulley is to rotate 600 revolutions per minute, then the diameter of the driven pulley $= \dfrac{24 \times 100}{600} = 4$ inches.

**Diameter of Driving Pulley Required.** — Diameter and speed of driven pulley, and speed of driving pulley are known. *Rule:* Multiply the diameter of the driven pulley by its speed in revolutions per minute, and divide the product by the speed of the driving pulley.

*Example:* — If the diameter of the driven pulley is 36 inches and its required speed, 150 revolutions per minute, and the speed of the driving pulley is 600 revolutions per minute, then the diameter of the driving pulley $= \dfrac{36 \times 150}{600} = 9$ inches.

**Speed of Driving Pulley Required.** — Diameters of driving and driven pulleys, and speed of driven pulley are known. *Rule:* Multiply the diameter of the driven pulley by its speed, and divide the product by the diameter of the driving pulley.

*Example:* — If the diameter of driven pulley is 4 inches, its required speed, 800 revolutions per minute, and the diameter of the driver, 26 inches, then the required speed of the driver $= \dfrac{4 \times 800}{26} = 123$ revolutions per minute, approximately.

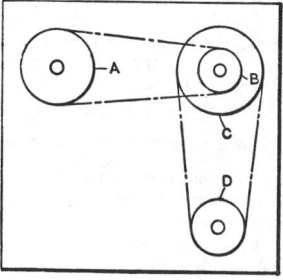

**Speed of Driven Pulley in Compound Drive Required.** — Diameters of pulleys $A$, $B$, $C$ and $D$ (see illustration), and speed of pulley $A$ are known; find speed of pulley $D$. *Rule:* Divide product of diameters of driving pulleys by product of diameters of driven pulleys, and multiply quotient by speed of first driving pulley.

*Example:* — If the diameters of the driving pulleys $A$ and $C$ are 18 and 24 inches; the diameters of the driven pulleys $B$ and $D$, 12 and 13 inches; and the speed of the driver $A$, 260 revolutions per minute; then the speed of the driven pulley $D = \dfrac{18 \times 24}{12 \times 13} \times 260 = 720$ revolutions per minute.

**Pulley Diameters in Compound Drive Required.** — Speeds of driving and driven pulleys are known; find diameters of the four pulleys $A$, $B$, $C$ and $D$. *Rule:* Place the speed of the driving pulley as the numerator of a fraction, and the speed of driven pulley as the denominator, and reduce this fraction to its lowest terms; then resolve both the numerator and denominator into two factors, and multiply each "pair" of factors (a pair being one factor in the numerator and one in the denominator) by a trial number which will give pulleys of suitable diameters.

*Example:* — If the speed of pulley $A$ is 260 revolutions per minute, and the required speed of pulley $D$ is 720 revolutions per minute, find the diameters of the four pulleys. The fraction $\frac{260}{720}$ reduced to its lowest terms is $\frac{13}{36}$, which represents the required speed ratio. Resolve $\frac{13}{36}$ into two factors; $\frac{13}{36} = \frac{1 \times 13}{2 \times 18}$. Multiply by trial numbers 12 and 1:

$$\frac{(1 \times 12) \times (13 \times 1)}{(2 \times 12) \times (18 \times 1)} = \frac{12 \times 13}{24 \times 18}$$

The values 12 and 13 in the numerator represent the diameters of the *driven* pulleys $B$ and $D$ and values 24 and 18 in the denominator, the diameters of the *driving* pulleys.

**Lengths of Open and Crossed Belts.** — The following formulas for determining the lengths of belts on pulleys are accurate enough for practical purposes. No allowance is made for a lap joint. In these formulas, $L$ = length of open belt; $L_c$ = length of cross belt; $D$, $C$ and $R$ = diameter, circumference and radius, respectively, of large pulley; $d$, $c$ and $r$ = diameter, circumference and radius, respectively, of small pulley; $x$ = center-to-center distance.

$$L = \frac{D+d}{2} \times 3\tfrac{1}{7} + 2x; \quad L_c = \frac{C}{2} + \frac{c}{2} + 2\sqrt{x^2 + (R+r)^2}$$

## Rules for Calculating Speeds of Gearing

The relative speeds of shafts connected by spur or bevel gearing can be determined by the foregoing rules for pulley-and-belt drives, provided the *pitch diameter* or the number of teeth in the gear is substituted for the pulley diameter, in each case, as shown by the following examples:

**Speed of Driven Gear Required.** — Number of teeth in driving gear, its speed, and number of teeth in driven gear are known. *Rule:* Multiply the number of teeth in the driving gear by its speed in revolutions per minute, and divide by the number of teeth in the driven gear.

*Example:* — If the driving gear has 20 teeth and rotates 80 revolutions per minute, and the driven gear has 40 teeth, then the speed of the driven gear $= \frac{20 \times 80}{40}$ = 40 revolutions per minute.

If one or more intermediate gears are placed in a direct train between the driving and driven gears, the speed ratio will remain the same.

**Pitch Diameter of Driven Gear Required.** — The pitch diameter of the driving gear, its speed, and speed required for driven gear are known. *Rule:* Multiply the pitch diameter of the driving gear by its speed in revolutions per minute, and divide by the required speed of the driven gear.

*Example:* — If the pitch diameter of the driver is 8 inches, its speed, 75 revolutions per minute, and the speed required for the driven gear, 20 revolutions per minute, then the pitch diameter of the driven gear $= \frac{8 \times 75}{20}$ = 30 inches.

# V-BELTS AND SHEAVES

**V-Belt Drives.** — Belts of the V type, commonly manufactured of fabric, cord, or a combination of these, treated with natural or synthetic rubber compound and vulcanized together, provide a quiet, compact, and resilient form of power transmission. They are used extensively in single and multiple form for automotive, home and commercial equipment and in industrial drives for a wide range of horsepowers extending upwards from fractional values.

The tapered cross-sectional shape of a V-belt causes it to wedge firmly into the sheave groove during operation so that the driving action takes place through the sides of the belt rather than the bottom, which normally is not in contact with the sheave at all.

**Light-duty or Fractional Horsepower V-belts.** — These are typically used as single belts with fractional horsepower motors or engines and are intended for service that is commonly infrequent or intermittent rather than continuous. The Rubber Manufacturers Association with the cooperation of V-belt manufacturers has provided *Standards for Light Duty or Fractional Horsepower V-Belts* in which appear a description of the sizes and load capacities of light-duty V-belts together with sheave dimensions and horsepower ratings.

*Belt Sizes:* Cross-sectional dimensions as given in this standard are shown in Table 1. Standard lengths are given in Table 2. It should be noted that these are *outside* lengths and not pitch lengths as in the case of Multiple V-belts.

*Size Designation:* The size designation of the light-duty belt may show the cross section and the nominal outside length. For example, a 2L belt of 8-inch nominal length would be designated 2L080; a 3L belt of 52-inch nominal length would be designated 3L520; a 4L belt of 100-inch nominal length would be designated 4L1000. Other methods of designation may be used, however.

*Sheave Dimensions:* Dimensions of sheave grooves are given in Table 3. The minimum sheave outside diameters recommended for use with light-duty or fractional horsepower belts are: for 2L belt, 0.8 inch O.D.; 3L belt, 1.5 inch O.D.; 4L belt, 2.5 inch O.D.; and 5L belt, 3.5 inch O.D.

*Speed Ratios:* The sheave diameter used in calculating speed ratios and belt speeds is obtained by subtracting the value "2X" in Table 3 from the effective outside diameter of the sheave also given in this table.

*Belt Length and Center Distance:* The belt length required by a light-duty V-belt drive is computed from the effective *outside* diameters (Table 3) using the following formula:

$$L = 2C + 1.57(D + d) + \frac{(D - d)^2}{4C}$$

where: $L$ = effective outside length of belt in inches
$C$ = distance between centers of sheaves in inches
$D$ = effective outside diameter of large sheave in inches
  = (for flat pulleys) the outside diameter of the pulley plus twice the nominal belt thickness
$d$ = effective outside diameter of small sheave in inches.

If this calculation results in a length which is not standard, the next longer standard length should be used and the necessary correction made in center distance.

If sheave effective diameters and belt length are known, the center distance between sheaves may be calculated as follows:

$$C = \frac{b + \sqrt{b^2 - 32(D - d)^2}}{16}, \text{ where: } b = 4L - 6.28(D + d)$$

Table 1.  Light-duty V-belt Cross-section Dimensions

| Cross Section | 2L* | 3L | 4L | 5L |
|---|---|---|---|---|
| Nominal Top Width | 1/4 | 3/8 | 1/2 | 21/32 |
| Nominal Thickness | 5/32 | 7/32 | 5/16 | 3/8 |

* The 2L cross section is in limited usage and is not made by all manufacturers. All dimensions in inches.

Table 2.  Light-duty V-belt Standard Outside Lengths

| Nom. Outside Length, Inches | 2L | 3L | 4L | 5L | Effect. Outside Length Variations, Inches |
|---|---|---|---|---|---|
| 8 | * | .. | .. | | +1/8, −3/8 |
| 9 | * | .. | .. | | +1/8, −3/8 |
| 10 | * | .. | .. | | +1/8, −3/8 |
| 11 | * | .. | .. | | +1/8, −3/8 |
| 12 | * | .. | .. | | +1/8, −3/8 |
| 13 | * | .. | .. | | +1/8, −3/8 |
| 14 | * | * | .. | | +1/8, −3/8 |
| 15 | * | * | .. | | +1/8, −3/8 |
| 16 | * | * | .. | | +1/8, −3/8 |
| 17 | * | * | .. | | +1/8, −3/8 |
| 18 | * | * | * | | +1/8, −3/8 |
| 19 | * | * | * | | +1/8, −3/8 |
| 20 | * | * | * | | +1/8, −3/8 |
| 21 | | * | * | .. | +1/4, −5/8 |
| 22 | | * | * | .. | +1/4, −5/8 |
| 23 | | * | * | * | +1/4, −5/8 |
| 24 | | * | * | * | +1/4, −5/8 |
| 25 | | * | * | * | +1/4, −5/8 |
| 26 | | * | * | * | +1/4, −5/8 |
| 27 | | * | * | * | +1/4, −5/8 |
| 28 | | * | * | * | +1/4, −5/8 |
| 29 | | * | * | * | +1/4, −5/8 |
| 30 | | * | * | * | +1/4, −5/8 |
| 31 | | * | * | * | +1/4, −5/8 |
| 32 | | * | * | * | +1/4, −5/8 |
| 33 | | * | * | * | +1/4, −5/8 |
| 34 | | * | * | * | +1/4, −5/8 |
| 35 | | * | * | * | +1/4, −5/8 |
| 36 | | * | * | * | +1/4, −5/8 |
| 37 | | * | * | * | +1/4, −5/8 |
| 38 | | * | * | * | +1/4, −5/8 |
| 39 | | * | * | * | +1/4, −5/8 |
| 40 | | * | * | * | +1/4, −5/8 |
| 41 | | * | * | * | +1/4, −5/8 |
| 42 | | * | * | * | +1/4, −5/8 |
| 43 | | * | * | * | +1/4, −5/8 |
| 44 | | * | * | * | +1/4, −5/8 |
| 45 | | * | * | * | +1/4, −5/8 |
| 46 | | * | * | * | +1/4, −5/8 |
| 47 | | * | * | * | +1/4, −5/8 |
| 48 | | * | * | * | +1/4, −5/8 |
| 49 | | * | * | * | +1/4, −5/8 |
| 50 | | * | * | * | +1/4, −5/8 |
| 51 | | .. | * | * | +1/4, −5/8 |
| 52 | | * | * | * | +1/4, −5/8 |
| 53 | | .. | * | * | +1/4, −5/8 |
| 54 | | * | * | * | +1/4, −5/8 |
| 55 | | .. | * | * | +1/4, −5/8 |
| 56 | | * | * | * | +1/4, −5/8 |
| 57 | | .. | * | * | +1/4, −5/8 |
| 58 | | * | * | * | +1/4, −5/8 |
| 59 | | .. | * | * | +1/4, −5/8 |
| 60 | | * | * | * | +1/4, −5/8 |
| 61 | | .. | * | * | +5/16, −11/16 |
| 62 | | .. | * | * | +5/16, −11/16 |
| 63 | | .. | * | * | +5/16, −11/16 |
| 64 | | .. | * | * | +5/16, −11/16 |
| 65 | | .. | * | * | +5/16, −11/16 |
| 66 | | .. | * | * | +5/16, −11/16 |
| 67 | | .. | * | * | +5/16, −11/16 |
| 68 | | .. | * | * | +5/16, −11/16 |
| 69 | | .. | * | * | +5/16, −11/16 |
| 70 | | .. | * | * | +5/16, −11/16 |
| 71 | | .. | * | * | +5/16, −11/16 |
| 72 | | .. | * | * | +5/16, −11/16 |
| 73 | | .. | * | * | +5/16, −11/16 |
| 74 | | .. | * | * | +5/16, −11/16 |
| 75 | | .. | * | * | +5/16, −11/16 |
| 76 | | .. | * | * | +5/16, −11/16 |
| 77 | | .. | * | * | +5/16, −11/16 |
| 78 | | .. | * | * | +5/16, −11/16 |
| 79 | | .. | * | * | +5/16, −11/16 |
| 80 | | .. | * | * | +5/8, −7/8 |
| 82 | | .. | * | * | +5/8, −7/8 |
| 84 | | .. | * | * | +5/8, −7/8 |
| 86 | | .. | * | * | +5/8, −7/8 |
| 88 | | .. | * | * | +5/8, −7/8 |
| 90 | | .. | * | * | +5/8, −7/8 |
| 92 | | .. | * | * | +5/8, −7/8 |
| 94 | | .. | * | * | +5/8, −7/8 |
| 96 | | .. | * | * | +5/8, −7/8 |
| 98 | | .. | * | * | +5/8, −7/8 |
| 100 | | .. | * | * | +5/8, −7/8 |

If not dictated by other considerations, the recommended center distance is as follows: (a) if the speed ratio is less than 3, the recommended center distance is one-half of the sum of the two sheave diameters plus the diameter of the small sheave; (b) if the speed ratio is 3 or more, the recommended center distance is equal to the diameter of the large sheave.

The center distance obtained either from the formula mentioned above or by the rules given are for nominal length belts under operating tension.  Provision must be made to move the sheave centers closer together so that any belt within the length tolerances given in Table 2 can be installed without being pried over the sheave.  Also, the centers must be adjustable beyond the calculated distance to compensate for belt stretch and wear of belt and grooves.  These required installation and take-up allowances in terms of subtractions from or additions to the nominal center distance are given in Table 4.

Table 3.  Light-duty V-belt Sheave Dimensions

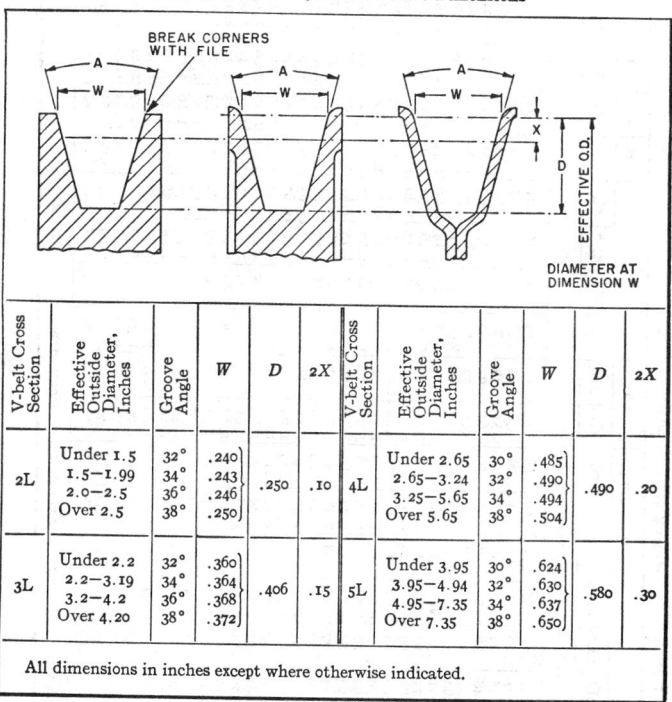

| V-belt Cross Section | Effective Outside Diameter, Inches | Groove Angle | W | D | 2X | V-belt Cross Section | Effective Outside Diameter, Inches | Groove Angle | W | D | 2X |
|---|---|---|---|---|---|---|---|---|---|---|---|
| 2L | Under 1.5<br>1.5–1.99<br>2.0–2.5<br>Over 2.5 | 32°<br>34°<br>36°<br>38° | .240<br>.243<br>.246<br>.250 | .250 | .10 | 4L | Under 2.65<br>2.65–3.24<br>3.25–5.65<br>Over 5.65 | 30°<br>32°<br>34°<br>38° | .485<br>.490<br>.494<br>.504 | .490 | .20 |
| 3L | Under 2.2<br>2.2–3.19<br>3.2–4.2<br>Over 4.20 | 32°<br>34°<br>36°<br>38° | .360<br>.364<br>.368<br>.372 | .406 | .15 | 5L | Under 3.95<br>3.95–4.94<br>4.95–7.35<br>Over 7.35 | 30°<br>32°<br>34°<br>38° | .624<br>.630<br>.637<br>.650 | .580 | .30 |

All dimensions in inches except where otherwise indicated.

Table 4.  Minimum Center Distance Allowances for
Installation and Take-up of Light-duty V-belts

| Lengths | Minimum Allowance Below (−) and Above (+) Standard Center Distance, Inches | | | |
|---|---|---|---|---|
| | 2L | 3L | 4L | 5L |
| 8 to 18 | −⅜,+½ | −⅝,+½ | .......... | .......... |
| 18 to 25 | −⅜,+½ | −⅝,+½ | −¾,+½ | .......... |
| 25 to 38 | .......... | −¾,+½ | −¾,+½ | −1,+½ |
| 38 to 61 | .......... | −¾,+¾ | −⅞,+¾ | −1,+¾ |
| 61 to 80 | .......... | .......... | −1,+1⅛ | −1⅛,+1⅛ |
| 80 to 100, incl. | .......... | .......... | −1⅛,+1½ | −1¼,+1½ |

*Note:* Minus values are for shortening center distance for installation.  Plus values are for lengthening center distance to compensate for stretch and wear.

## Table 5. Horsepower Ratings for Light-duty V-belts

Effective Outside Diameter of Small Sheave, Inches

Horsepower Ratings

| Belt Speed, Ft. per Min. | 1½ | 1¾ | 2 | | 2¼ | | 2½ | | 2¾ | | 3 | | | 3¼ | | 3½ | | 3¾ | | 4 | | 4¼ | 4½ | 4¾ | 5 |
|---|---|---|---|---|---|---|---|---|---|---|---|---|---|---|---|---|---|---|---|---|---|---|---|---|---|
| | 3L | 3L | 3L | 4L | 3L | 4L | 3L | 4L | 3L | 4L | 3L* | 4L | 5L | 4L | 5L | 4L | 5L | 4L | 5L | 4L* | 5L | 5L | 5L | 5L | 5L |
| 200 | .05 | .07 | .08 | .07 | .09 | .09 | .10 | .13 | .10 | .14 | .11 | .16 | .13 | .17 | .16 | .18 | .19 | .19 | .21 | .21 | .24 | .26 | .27 | .29 | .30 |
| 400 | .08 | .12 | .14 | .12 | .16 | .17 | .18 | .23 | .19 | .27 | .20 | .31 | .23 | .34 | .29 | .36 | .35 | .38 | .40 | .40 | .45 | .49 | .52 | .55 | .58 |
| 600 | .11 | .16 | .19 | .15 | .23 | .24 | .25 | .32 | .27 | .38 | .29 | .43 | .31 | .48 | .41 | .51 | .49 | .54 | .57 | .57 | .64 | .69 | .75 | .79 | .83 |
| 800 | .12 | .19 | .24 | .17 | .28 | .30 | .31 | .40 | .33 | .48 | .36 | .54 | .36 | .60 | .50 | .64 | .62 | .69 | .72 | .73 | .80 | .88 | .95 | 1.01 | 1.07 |
| 1000 | .13 | .22 | .28 | .18 | .33 | .33 | .37 | .46 | .40 | .56 | .43 | .65 | .40 | .71 | .57 | .78 | .72 | .83 | .84 | .88 | .95 | 1.05 | 1.14 | 1.21 | 1.28 |
| 1200 | .14 | .24 | .32 | .17 | .38 | .36 | .43 | .51 | .47 | .64 | .50 | .74 | .42 | .83 | .63 | .89 | .80 | .96 | .96 | 1.01 | 1.09 | 1.20 | 1.31 | 1.40 | 1.48 |
| 1400 | .15 | .27 | .35 | .16 | .42 | .39 | .48 | .56 | .52 | .71 | .56 | .82 | .43 | .92 | .67 | 1.01 | .87 | 1.08 | 1.05 | 1.14 | 1.20 | 1.34 | 1.46 | 1.57 | 1.67 |
| 1600 | .15 | .28 | .38 | .14 | .46 | .40 | .52 | .60 | .58 | .77 | .62 | .90 | .42 | 1.01 | .69 | 1.11 | .93 | 1.19 | 1.13 | 1.26 | 1.31 | 1.46 | 1.60 | 1.72 | 1.84 |
| 1800 | .14 | .29 | .41 | .11 | .50 | .40 | .57 | .62 | .63 | .81 | .67 | .96 | .40 | 1.09 | .71 | 1.19 | .97 | 1.28 | 1.20 | 1.37 | 1.40 | 1.57 | 1.73 | 1.87 | 1.99 |
| 2000 | .13 | .30 | .43 | .08 | .53 | .39 | .60 | .64 | .67 | .84 | .72 | 1.02 | .36 | 1.16 | .70 | 1.28 | 1.00 | 1.38 | 1.25 | 1.47 | 1.47 | 1.67 | 1.84 | 1.99 | 2.13 |
| 2200 | .12 | .30 | .44 | .04 | .55 | .37 | .64 | .67 | .71 | .90 | .77 | 1.08 | .31 | 1.24 | .69 | 1.37 | 1.01 | 1.48 | 1.29 | 1.58 | 1.54 | 1.75 | 1.94 | 2.11 | 2.26 |
| 2400 | .10 | .29 | .45 | | .57 | .34 | .66 | .68 | .74 | .92 | .81 | 1.12 | .25 | 1.29 | .66 | 1.43 | 1.00 | 1.55 | 1.33 | 1.66 | 1.62 | 1.81 | 2.02 | 2.21 | 2.37 |
| 2600 | .07 | .28 | .46 | | .58 | .29 | .69 | .66 | .77 | .93 | .84 | 1.16 | .17 | 1.33 | .62 | 1.50 | .97 | 1.63 | 1.31 | 1.75 | 1.63 | 1.85 | 2.09 | 2.29 | 2.47 |
| 2800 | .04 | .26 | .45 | | .59 | .25 | .70 | .65 | .79 | .94 | .87 | 1.18 | .08 | 1.38 | .56 | 1.54 | .93 | 1.68 | 1.31 | 1.81 | 1.64 | 1.91 | 2.15 | 2.36 | 2.56 |
| 3000 | .01 | .23 | .45 | | .60 | .19 | .72 | .63 | .81 | .94 | .89 | 1.19 | | 1.40 | .49 | 1.58 | .87 | 1.74 | 1.27 | 1.87 | 1.63 | 1.93 | 2.19 | 2.42 | 2.63 |
| 3200 | | .20 | .44 | | .59 | .11 | .72 | .60 | .82 | .92 | .91 | 1.20 | | 1.42 | .41 | 1.61 | .79 | 1.78 | 1.22 | 1.92 | 1.60 | 1.94 | 2.21 | 2.46 | 2.68 |
| 3400 | | .16 | .42 | | .59 | .04 | .72 | .55 | .83 | .90 | .92 | 1.19 | | 1.43 | .29 | 1.63 | .69 | 1.80 | 1.15 | 1.96 | 1.54 | 1.92 | 2.22 | 2.48 | 2.72 |
| 3600 | | .12 | .39 | | .57 | | .71 | .50 | .83 | .86 | .92 | 1.16 | | 1.42 | .17 | 1.64 | .57 | 1.82 | 1.05 | 1.98 | 1.47 | 1.89 | 2.21 | 2.48 | 2.73 |
| 3800 | | .06 | .36 | | .54 | | .69 | .43 | .82 | .81 | .91 | 1.13 | | 1.40 | | 1.63 | .43 | 1.83 | .93 | 2.00 | 1.37 | 1.84 | 2.17 | 2.46 | 2.72 |
| 4000 | | | .31 | | .51 | | .67 | .35 | .80 | .76 | .89 | 1.09 | | 1.38 | | 1.61 | .26 | 1.81 | .80 | 2.00 | 1.26 | 1.76 | 2.11 | 2.42 | 2.69 |
| 4200 | | | .26 | | .47 | | .64 | .24 | .77 | .70 | .86 | 1.03 | | 1.32 | | 1.58 | .08 | 1.79 | .64 | 1.98 | 1.12 | 1.65 | 2.03 | 2.35 | 2.64 |
| 4400 | | | .21 | | .42 | | .60 | .14 | .74 | .58 | .82 | .96 | | 1.27 | | 1.53 | | 1.75 | .46 | 1.95 | .97 | 1.55 | 1.93 | 2.27 | 2.57 |
| 4600 | | | .14 | | .37 | | .55 | .01 | .70 | .48 | .77 | .87 | | 1.19 | | 1.46 | | 1.70 | .25 | 1.90 | .78 | 1.41 | 1.80 | 2.16 | 2.48 |
| 4800 | | | .06 | | .30 | | .49 | | .64 | .36 | .72 | .76 | | 1.10 | | 1.39 | | 1.64 | .02 | 1.85 | .58 | 1.25 | 1.66 | 2.03 | 2.36 |
| 5000 | | | | | .22 | | .42 | | .58 | .23 | .65 | .65 | | 1.00 | | 1.30 | | 1.56 | | 1.78 | .34 | 1.06 | 1.49 | 1.88 | 2.22 |
| 5200 | | | | | .14 | | .34 | | .51 | .08 | .58 | .51 | | .88 | | 1.19 | | 1.45 | | 1.70 | .08 | .85 | 1.30 | 1.70 | 2.06 |
| 5400 | | | | | .04 | | .26 | | .43 | | .49 | .36 | | .74 | | 1.07 | | 1.35 | | 1.59 | | .61 | 1.07 | 1.49 | 1.86 |
| 5600 | | | | | | | .16 | | .34 | | .39 | .18 | | .58 | | .91 | | 1.20 | | 1.46 | | .34 | .82 | 1.25 | 1.64 |
| 5800 | | | | | | | .05 | | .24 | | .29 | | | .38 | | .72 | | 1.03 | | 1.29 | | .04 | .54 | .99 | 1.39 |
| 6000 | | | | | | | | | .12 | | | | | .18 | | .54 | | .84 | | 1.11 | | | .23 | .69 | 1.11 |

* These horsepower ratings also hold for this belt size when used with sheaves of larger effective outside diameters.

*Belt Horsepower Capacity:* The horsepower ratings for Light-duty V-belts shown in Table 5 are the basic maximum ratings which take into consideration the degree and rate of belt flexing and the tensile pull, factors common to all V-belt drives. Where the arc of contact on the small sheave is less than 180 degrees as computed by the following formula:

$$\text{Arc of contact} = 180° - \frac{(D-d)60°}{C} \qquad (1)$$

in which $D$ = effective outside diameter of large sheave (Table 3), $d$ = effective outside diameter of smaller sheave (Table 3), and $C$ = center distance of drive, all measured in inches; or where the drive is composed of a small sheave and a large diameter flat pulley (V to flat) in which $D$ is the effective diameter of the flat pulley, the maximum ratings must be reduced by multiplying them by the correction factor for the arc of contact shown in Table 6.

Table 6. Light-duty V-belt Arc of Contact Correction Factors

| Arc of Contact, Deg. | Type of Drive | | Arc of Contact | Type of Drive | | Arc of Contact | Type of Drive | |
|---|---|---|---|---|---|---|---|---|
| | V to V | V to Flat* | | V to V | V to Flat* | | V to V | V to Flat* |
| | Correction Factor | | | Correction Factor | | | Correction Factor | |
| 180 | 1.00 | .75 | 145 | .91 | .83 | 110 | .78 | .78 |
| 175 | .99 | .76 | 140 | .89 | .84 | 105 | .76 | .76 |
| 170 | .98 | .77 | 135 | .88 | .85 | 100 | .74 | .74 |
| 165 | .96 | .78 | 130 | .86 | .86 | 95 | .72 | .72 |
| 160 | .95 | .80 | 125 | .84 | .84 | 90 | .69 | .69 |
| 155 | .94 | .81 | 120 | .82 | .82 | ... | ... | ... |
| 150 | .92 | .82 | 115 | .80 | .80 | ... | ... | ... |

* A V to Flat drive is one comprised of a small sheave and a large diameter flat pulley.

In finding the horsepower capacity required of the drive, it is good practice to start with the horsepower rating of the motor and multiply it by the proper service factor.

For *light service* such as small fans or blowers with light rotors, where the load comes on gradually as they come up to speed and where the final load is relatively steady, a service factor of 1.0 to 1.2 may be applied.

For *medium service* where the driven machine is started or stopped rather frequently or where the starting load is somewhat heavy, as for example, a fan or blower with a large relatively heavy rotor, a compressor with good flywheel effect to smooth out vibrations and with infrequent starting, or light-duty machines used in continuous production service, a service factor of 1.2 to 1.4 may be applied.

For *heavy service* where the machines are subjected to one or more of such severe factors as heavy starting loads, peak or shock loads, reciprocating loads, frequent starting and stopping, or industrial production service (included are refrigerator compressors, air compressors, reciprocating pumps, metalworking and woodworking machinery, stokers, drill presses or grinders, and other machinery subject to similar heavy-duty conditions), a service factor of 1.4 to 1.6 may be applied.

**Grades of Multiple V-Belts.** — Two grades of V-belts for multiple use are recognized in the *Engineering Standards for Multiple V-Belt Drives* sponsored by the Multiple V-Belt Drive and Mechanical Power Transmission Association and the Rubber Manufacturers Association. The *Standard Quality* of V-belt is intended for the great majority of industrial drives with normal loads, speeds, center distances, sheave diameters and operating conditions. Various types of V-belts which fall in the class of *Premium Quality* are designed to meet special drive conditions such as repeated heavy shock loads, pulsating or vibrating loads, substandard sheave diameters, high speed operation, or extremes of temperature, humidity, etc.

**Standard Multiple V-Belt and Sheave Dimensions.** — Five sizes of V-belts are designated in the *Engineering Standards for Multiple V-Belt Drives*. Nominal width and thickness dimensions are as shown in Table 7. However, actual dimensions of V-belts of various manufacturers may vary somewhat from these nominal dimensions. Because of this fact, it is recommended that belts of different makes should never be mixed on the same drive. Standard V-belt *pitch* lengths and permissible pitch-length tolerances are given in Table 8. Groove dimensions and tolerances for multiple V-belt sheaves are given in Table 9.

Table 7.   Standard Multiple V-belt Dimensions

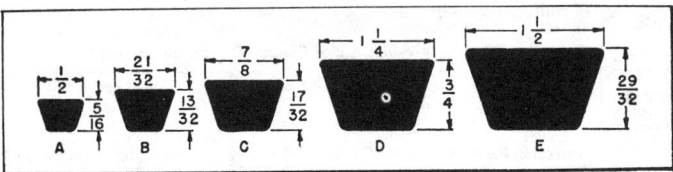

*Measuring a Multiple V-Belt:* The pitch length of a multiple V-belt is determined by placing the belt on a measuring fixture consisting of two equal diameter sheaves having standard dimensions and with a total tension of 50 pounds for an A V-belt, 65 pounds for a B V-belt, 165 pounds for a C V-belt, 300 pounds for a D V-belt, and 400 pounds for an E V-belt. One of the sheaves is fixed in position, while the other is movable along a graduated scale with the specified tension applied to it.

The sheaves should be rotated at least two revolutions to seat the belt properly in the sheave grooves and to equally divide the total tension between the two strands of the belt. The pitch length is the length obtained by adding the pitch circumference of one of the measuring sheaves to twice the measured center distance between them. Deviation of the measured pitch length from the standard pitch length shown in Table 8 should be within the tolerance limits also given in this table.

The grooves of the measuring sheaves should be machined and maintained to the following tolerances: pitch diameter, ±0.002 inch; groove angle, ±0 degrees, 20 minutes; and groove top width, ±0.002 inch.

**Belt Length and Center Distance.** — The relation between center distance and belt *pitch* length is given by the following formula:

$$L = 2C + 1.57(D + d) + \frac{(D - d)^2}{4C} \qquad (2)$$

This formula can be rearranged to solve for center distance, as follows:

$$C = \frac{b + \sqrt{b^2 - 32(D - d)^2}}{16}, \qquad (3)$$

where: $b = 4L - 6.28(D + d)$
$\quad\quad\quad D$ = pitch diameter of large sheave, in inches
$\quad\quad\quad d$ = pitch diameter of small sheave, in inches
$\quad\quad\quad L$ = pitch length of belt in inches
$\quad\quad\quad C$ = center distance in inches.

### Table 8.  Standard Pitch Lengths for Multiple V-belts

| Standard Length Designation* | Standard V-Belt Cross Sections | | | | | Permissible Deviations from Std. Pitch Length, Inches | Matching Limits for One Set†, Inch |
|---|---|---|---|---|---|---|---|
| | A | B | C | D | E | | |
| | Standard Pitch Lengths, Inches | | | | | | |
| 26 | 27.3 | ..... | ..... | ..... | ..... | + .7,− .3 | .10 |
| 31 | 32.3 | ..... | ..... | ..... | ..... | + .7,− .3 | .10 |
| 33 | 34.3 | ..... | ..... | ..... | ..... | + .8,− .4 | .10 |
| 35 | 36.3 | 36.8 | ..... | ..... | ..... | + .8,− .4 | .10 |
| 38 | 39.3 | 39.8 | ..... | ..... | ..... | + .8,− .4 | .10 |
| 42 | 43.3 | 43.8 | ..... | ..... | ..... | + .8,− .4 | .10 |
| 46 | 47.3 | 47.8 | ..... | ..... | ..... | + .8,− .4 | .10 |
| 48 | 49.3 | 49.8 | ..... | ..... | ..... | + .9,− .5 | .10 |
| 51 | 52.3 | 52.8 | 53.9 | ..... | ..... | + .9,− .5 | .10 |
| 53 | 54.3 | 54.8 | ..... | ..... | ..... | + .9,− .5 | .10 |
| 55 | 56.3 | 56.8 | ..... | ..... | ..... | + .9,− .5 | .10 |
| 60 | 61.3 | 61.8 | 62.9 | ..... | ..... | + .9,− .5 | .20 |
| 62 | 63.3 | 63.8 | ..... | ..... | ..... | + .9,− .5 | .20 |
| 64 | 65.3 | 65.8 | ..... | ..... | ..... | + .9,− .5 | .20 |
| 66 | 67.3 | 67.8 | ..... | ..... | ..... | + .9,− .5 | .20 |
| 68 | 69.3 | 69.8 | 70.9 | ..... | ..... | + .9,− .5 | .20 |
| 71 | 72.3 | 72.8 | ..... | ..... | ..... | + .9,− .5 | .20 |
| 75 | 76.3 | 76.8 | 77.9 | ..... | ..... | + .9,− .5 | .20 |
| 78 | 79.3 | 79.8 | ..... | ..... | ..... | +1.0,− .5 | .30 |
| 80 | 81.3 | ..... | ..... | ..... | ..... | +1.0,− .5 | .30 |
| 81 | ..... | 82.8 | 83.9 | ..... | ..... | +1.0,− .5 | .30 |
| 83 | ..... | 84.8 | ..... | ..... | ..... | +1.0,− .5 | .30 |
| 85 | 86.3 | 86.8 | 87.9 | ..... | ..... | +1.0,− .5 | .30 |
| 90 | 91.3 | 91.8 | 92.9 | ..... | ..... | +1.0,− .5 | .30 |
| 96 | 97.3 | ..... | 98.9 | ..... | ..... | +1.0,− .5 | .30 |
| 97 | ..... | 98.8 | ..... | ..... | ..... | +1.0,− .5 | .30 |
| 105 | 106.3 | 106.8 | 107.9 | ..... | ..... | +1.1,− .5 | .40 |
| 112 | 113.3 | 113.8 | 114.9 | ..... | ..... | +1.1,− .5 | .40 |
| 120 | 121.3 | 121.8 | 122.9 | 123.3 | ..... | +1.2,− .5 | .40 |
| 128 | 129.3 | 129.8 | 130.9 | 131.3 | ..... | +1.3,− .6 | .40 |
| 136 | ..... | 137.8 | 138.9 | ..... | ..... | +1.3,− .6 | .40 |
| 144 | ..... | 145.8 | 146.9 | 147.3 | ..... | +1.4,− .6 | .40 |
| 158 | ..... | 159.8 | 160.9 | 161.3 | ..... | +1.5,− .6 | .40 |
| 162 | ..... | ..... | 164.9 | 165.3 | ..... | +1.6,− .6 | .40 |
| 173 | ..... | 174.8 | 175.9 | 176.3 | ..... | +1.7,− .7 | .5c |
| 180 | ..... | 181.8 | 182.9 | 183.3 | 184.5 | +1.7,− .7 | .50 |
| 195 | ..... | 196.8 | 197.9 | 198.3 | 199.5 | +1.8,− .8 | .50 |
| 210 | ..... | 211.8 | 212.9 | 213.3 | 214.5 | +2.0,− .8 | .50 |
| 240 | ..... | 240.3 | 240.9 | 240.8 | 241.0 | +2.2,− .9 | .50 |
| 270 | ..... | 270.3 | 270.9 | 270.8 | 271.0 | +2.4,−1.0 | .50 |
| 300 | ..... | 300.3 | 300.9 | 300.8 | 301.0 | +2.5,−1.2 | .60 |
| 330 | ..... | ..... | 330.9 | 330.8 | 331.0 | +2.5,−1.2 | .60 |
| 360 | ..... | ..... | 360.9 | 360.8 | 361.0 | +2.5,−1.2 | .60 |
| 390 | ..... | ..... | 390.9 | 390.8 | 391.0 | +3.0,−1.5 | .70 |
| 420 | ..... | ..... | 420.9 | 420.8 | 421.0 | +3.5,−2.0 | .70 |
| 480 | ..... | ..... | ..... | 480.8 | 481.0 | +4.0,−2.5 | .70 |
| 540 | ..... | ..... | ..... | 540.8 | 541.0 | +4.5,−3.0 | .70 |
| 600 | ..... | ..... | ..... | 600.8 | 601.0 | +5.0,−3.5 | .70 |
| 660 | ..... | ..... | ..... | 660.8 | 661.0 | +6.0,−4.0 | .70 |

* To specify belt size use the Standard Length Designation prefixed by the letter indicating cross section, for example: B90.

† Maximum allowable difference in actual pitch lengths of longest and shortest V-belts in a given set.

### Table 9. Groove Dimensions and Tolerances for Multiple V-belt Sheaves

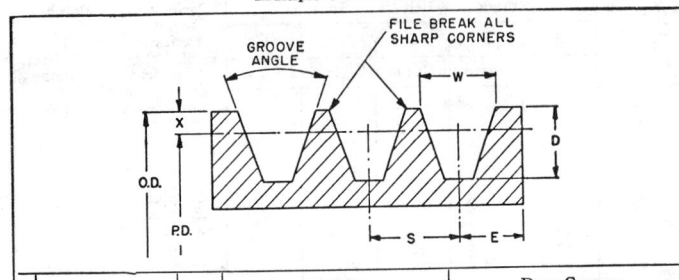

| Belt | Pitch Diameter — Minimum Recommended | Pitch Diameter — Range | Groove Angle ±½° | Standard Groove Dimensions — W (2) | Standard Groove Dimensions — D ±.031 | Standard Groove Dimensions — X .... | Standard Groove Dimensions — S¹ ±.031 | Standard Groove Dimensions — E (3) | Deep Groove Dimensions — W (2) | Deep Groove Dimensions — D ±.031 | Deep Groove Dimensions — X .... | Deep Groove Dimensions — S¹ ±.031 | Deep Groove Dimensions — E (3) |
|---|---|---|---|---|---|---|---|---|---|---|---|---|---|
| A | 3.0 | 2.6 to 5.4 / Over 5.4 | 34° / 38° | .494 / .504 | .490 | .125 | $\frac{5}{8}$ | $\frac{3}{8}$ | .589 / .611 | .645 | .280 | $\frac{3}{4}$ | $\frac{7}{16}$ |
| B | 5.4 | 4.6 to 7.0 / Over 7.0 | 34° / 38° | .637 / .650 | .580 | .175 | $\frac{3}{4}$ | $\frac{1}{2}$ | .747 / .774 | .760 | .355 | $\frac{7}{8}$ | $\frac{9}{16}$ |
| C | 9.0 | 7.0 to 7.99 / 8.0 to 12.0 / Over 12.0 | 34° / 36° / 38° | .879 / .887 / .895 | .780 | .200 | 1 | $1\frac{1}{16}$ | 1.066 / 1.085 / 1.105 | 1.085 | .505 | $1\frac{1}{4}$ | $1\frac{3}{16}$ |
| D | 13.0 | 12.0 to 12.99 / 13.0 to 17.0 / Over 17.0 | 34° / 36° / 38° | 1.259 / 1.271 / 1.283 | 1.050 | .300 | $1\frac{7}{16}$ | $\frac{7}{8}$ | 1.513 / 1.541 / 1.569 | 1.465 | .715 | $1\frac{3}{4}$ | $1\frac{1}{16}$ |
| E | 21.0 | 18.0 to 24.0 / Over 24.0 | 36° / 38° | 1.527 / 1.542 | 1.300 | .400 | $1\frac{3}{4}$ | $1\frac{1}{8}$ | 1.816 / 1.849 | 1.745 | .845 | $2\frac{1}{16}$ | $1\frac{5}{16}$ |

All dimensions in inches except groove angles are in degrees.

¹ Summation of the deviations from $S$ for all grooves in any one sheave shall not exceed ±0.063 inch.

² Tolerances for $W$: for A and B belts are ±.005 inch; for C and D belts, ±.007 inch; and for E belt, ±.010 inch.

³ Tolerances for $E$: for A belts are, +.070, −.000 inch; for B and C belts, +.150, −.000 inch; and for D and E belts, +.250, −.000 inch.

*Outside Diameter Tolerances:* Under 12 inches, ±.020 inch; 12.0 up to 24.0 inches, ±.040 inch; 24 up to 58 inches, .060 inch; 58.0 up to 72.0 inches, ±.120 inch; and for 72 inches and above, ±.250 inch.

*Outside Diameter Eccentricity:* For 10.0-inch pitch diameter and under, .010 inch. Add .0005 inch for each additional inch of pitch diameter up to and including 60.0-inch pitch diameter. Add .001 inch for each additional inch of pitch diameter above 60 inches.

*Side Wobble and Runout:* .001 inch per inch of pitch diameter up to 20 inches. Add .0005 inch for each additional inch of pitch diameter up to and including 60.0 inches. Add .001 inch for each additional inch of pitch diameter above 60.0 inches.

For standard key and keyway dimensions, see pages 968 and 969.

**Installation and Take-up Allowance.** — After calculating a center distance from a standard pitch length, provision should be made for moving the centers together by an amount, as shown by the minus values in Table 10, to permit installing the belts over the sheaves without injury. Also shown in Table 10 is the minimum allowance above the standard center distance (plus values) for which the centers should be adjustable to take up any slack in the belts due to stretch and wear.

Table 10.  Minimum Center Distance Allowances
for Installation and Take-up of Multiple V-belts

| Range of Standard Lengths | Minimum Allowance Below (−) and Above (+) Standard Center Distance | | | | |
|---|---|---|---|---|---|
| | A | B | C | D | E |
| 26 to 38 | −¾, +1 | −1, +1 | ......... | ......... | ......... |
| 38 to 60 | −¾, +1½ | −1, +1½ | −1½, +1½ | ......... | ......... |
| 60 to 90 | −¾, +2 | −1¼, +2 | −1½, +2 | ......... | ......... |
| 90 to 120 | −1, +2½ | −1¼, +2½ | −1½, +2½ | ......... | ......... |
| 120 to 158 | −1, +3 | −1¼, +3 | −1½, +3 | −2, +3 | ......... |
| 158 to 195 | ......... | −1¼, +3½ | −2, +3½ | −2, +3½ | −2½, +3½ |
| 195 to 240 | ......... | −1½, +4 | −2, +4 | −2, +4 | −2½, +4 |
| 240 to 270 | ......... | ......... | −2, +4½ | −2½, +4½ | −2½, +4½ |
| 270 to 330 | ......... | ......... | −2, +5 | −2½, +5 | −3, +5 |
| 330 to 420 | ......... | ......... | −2, +6 | −2½, +6 | −3, +6 |
| 420 and over | ......... | ......... | ......... | −3, * | −3½, * |

All dimensions in inches.
* For this belt size and lengths, 1.5 per cent of belt length above standard center distance for stretch and wear.

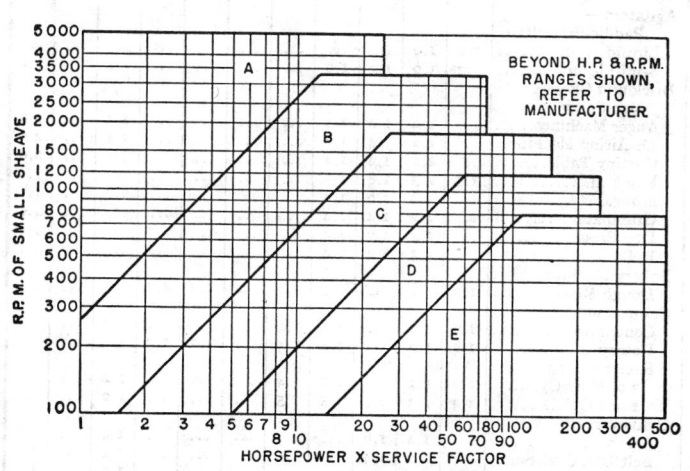

Chart for Selection of V-Belt for Given Drive

**Selection of Multiple V-Belts.** — The chart on page 1043 which appears in Engineering Standards for Multiple V-Belt Drives enables a V-belt of appropriate cross-section to be selected for a given drive if the revolutions per minute of the small sheave, the transmitted horsepower of the driving unit, and the service factor are known. The selection procedure is as follows:

1. Multiply the horsepower to be transmitted by the drive by the proper service factor (Table 11) to obtain the "design horsepower."

2. Enter the chart at the rpm of the small sheave and proceed horizontally to a point in vertical line with the design horsepower.

3. If this point falls in the area marked A, then an A size belt is required or,

Table 11.  Service Factors for Multiple V-belt Applications

| Applications | Electric Motors | | | | | | | | | | |
| | A.C. | | | | | | | | D.C. | | |
| | Squirrel Cage | | | | Syn-chronous | | Single Phase | | | | |
| | Normal Torque Line Start | Normal Torque Compensator Start | High Torque | Wound Rotor (Slip Ring) | Normal Torque | High Torque | Repulsion and Split-Phase | Capacitor | Shunt Wound | Compound Wound | Line Shaft and Clutch Shifting |
| | Service Factors | | | | | | | | | | |
| Agitators — | | | | | | | | | | | |
| Paddle-Propeller | | | | | | | | | | | |
| Liquid | 1.0 | 1.0 | 1.2 | ... | ... | ... | ... | ... | ... | ... | ... |
| Semi-Liquid | 1.2 | 1.0 | 1.4 | 1.2 | ... | ... | ... | ... | ... | ... | ... |
| Brick and Clay Machinery | | | | | | | | | | | |
| Auger Machines | ... | 1.2 | 1.4 | 1.4 | ... | ... | ... | ... | 1.4 | ... | 2.0 |
| De-Airing Machines | ... | 1.2 | 1.4 | 1.4 | ... | ... | ... | ... | 1.4 | ... | 2.0 |
| Cutting Table | ... | 1.2 | 1.4 | 1.4 | ... | ... | ... | ... | ... | ... | 2.0 |
| Pug Mill | 1.5 | 1.3 | 1.8 | 1.5 | ... | ... | ... | ... | ... | ... | ... |
| Mixer | ... | 1.2 | 1.6 | 1.4 | ... | ... | ... | ... | ... | ... | ... |
| Granulator | ... | 1.2 | 1.4 | 1.4 | ... | ... | ... | ... | ... | ... | ... |
| Dry Press | ... | 1.2 | 1.6 | 1.4 | ... | ... | ... | ... | ... | ... | ... |
| Rolls | ... | 1.2 | 1.4 | 1.4 | ... | ... | ... | ... | ... | ... | ... |
| Bakery Machinery | | | | | | | | | | | |
| Dough Mixer | 1.2 | ... | ... | ... | ... | ... | 1.2 | 1.0 | ... | ... | ... |
| Compressors | | | | | | | | | | | |
| Centrifugal | 1.2 | 1.2 | ... | 1.4 | 1.4 | ... | ... | ... | 1.2 | ... | ... |
| Rotary | 1.2 | 1.2 | ... | 1.4 | 1.4 | ... | 1.2 | 1.2 | 1.2 | ... | ... |
| Reciprocating — | | | | | | | | | | | |
| 3 or More Cyl | 1.2 | 1.2 | ... | 1.4 | 1.4 | ... | ... | ... | 1.2 | ... | ... |
| 1 or 2 Cyl | 1.4 | 1.4 | ... | 1.5 | 1.5 | ... | ... | ... | 1.2 | ... | ... |
| Conveyors | | | | | | | | | | | |
| Apron | ... | 1.4 | 1.6 | ... | ... | ... | ... | ... | 1.4 | ... | 1.6 |
| Belt (Ore, Coal, Sand) | ... | 1.2 | 1.4 | ... | ... | ... | ... | ... | 1.2 | ... | 1.4 |
| Belt (Light Package) | ... | 1.0 | 1.1 | ... | ... | ... | ... | ... | 1.0 | ... | 1.2 |

**Table 11.** (*Continued*). Service Factors for Multiple V-Belt Applications

| Applications | Electric Motors | | | | | | | | | | Line Shaft and Clutch Shifting |
| | A.C. | | | | | | | | D.C. | | |
| | Squirrel Cage | | | | Synchronous | | Single Phase | | | | |
| | Normal Torque Line Start | Normal Torque Compensator Start | High Torque | Wound Rotor (Slip Ring) | Normal Torque | High Torque | Repulsion and Split-Phase | Capacitor | Shunt Wound | Compound Wound | |
| | Service Factors | | | | | | | | | | |
|---|---|---|---|---|---|---|---|---|---|---|---|
| Oven | ... | 1.0 | 1.1 | ... | ... | ... | ... | ... | 1.0 | ... | 1.2 |
| Screw | ... | 1.6 | 1.8 | ... | ... | ... | ... | ... | 1.6 | ... | 1.8 |
| Bucket | ... | 1.4 | 1.6 | ... | ... | ... | ... | ... | 1.4 | ... | 1.6 |
| Pan | ... | 1.4 | 1.6 | ... | ... | ... | ... | ... | 1.4 | ... | 1.6 |
| Flight | ... | 1.6 | 1.8 | ... | ... | ... | ... | ... | 1.6 | ... | 1.8 |
| Elevator | ... | 1.4 | 1.6 | ... | ... | ... | ... | ... | 1.4 | ... | 1.6 |
| **Crushing Machinery** | | | | | | | | | | | |
| Jaw Crushers | ... | 1.4 | 1.6 | 1.4 | ... | ... | ... | ... | ... | 1.4 | 1.6 |
| Gyratory Crushers | ... | 1.4 | 1.6 | 1.4 | 1.4 | 1.6 | ... | ... | ... | 1.4 | 1.6 |
| Cone Crushers | ... | 1.4 | 1.6 | 1.4 | ... | ... | ... | ... | ... | 1.6 | 1.6 |
| Crushing Rolls | ... | 1.4 | 1.6 | 1.4 | ... | ... | ... | ... | ... | 1.4 | 1.6 |
| Ball-Pebble and | ... | 1.4 | 1.6 | 1.4 | 1.4 | 1.6 | ... | ... | ... | 1.4 | 1.6 |
| Tube Mills | ... | 1.4 | 1.6 | 1.4 | 1.4 | ... | ... | ... | ... | 1.4 | 1.6 |
| **Fan and Blowers** | | | | | | | | | | | |
| Centrifugal | 1.2 | 1.2 | ... | 1.4 | ... | ... | ... | ... | 1.2 | ... | ... |
| Propeller | 1.4 | 1.4 | 2.0 | 1.6 | ... | 2.0 | ... | ... | 1.4 | ... | ... |
| Induced Draft | 1.2 | 1.2 | ... | 1.4 | ... | ... | ... | ... | 1.4 | ... | ... |
| Positive Blowers | 1.6 | 1.6 | ... | 2.0 | 2.0 | 2.0 | ... | ... | ... | ... | ... |
| Exhausters | 1.2 | 1.2 | ... | 1.4 | ... | ... | ... | ... | 1.4 | ... | 1.5 |
| Line Shafts | 1.4 | 1.4 | ... | 1.4 | 1.4 | 2.0 | 1.4 | 1.4 | 1.4 | 1.4 | 1.6 |
| **Machine Tools** | | | | | | | | | | | |
| Grinders | 1.2 | ... | ... | 1.4 | ... | ... | 1.2 | 1.0 | 1.2 | 1.2 | ... |
| Boring Mills | 1.2 | ... | ... | 1.4 | ... | ... | ... | ... | 1.2 | 1.2 | ... |
| Lathes | 1.0 | ... | ... | 1.2 | ... | ... | 1.0 | 1.0 | 1.0 | 1.0 | ... |
| Milling Machines | 1.2 | ... | ... | 1.4 | ... | ... | ... | ... | 1.2 | 1.2 | ... |
| Screw Machines | 1.0 | ... | ... | 1.0 | ... | ... | 1.0 | 1.0 | 1.0 | 1.0 | ... |
| Cam Cutters | 1.0 | ... | ... | 1.0 | ... | ... | ... | ... | 1.0 | 1.0 | ... |
| Planers | 1.2 | ... | ... | 1.4 | ... | ... | 1.2 | 1.0 | 1.2 | 1.2 | ... |
| Shapers | 1.0 | ... | ... | 1.0 | ... | ... | 1.0 | 1.0 | 1.0 | 1.0 | ... |
| Drill Press | 1.0 | ... | ... | 1.0 | ... | ... | 1.0 | 1.0 | 1.0 | 1.0 | ... |
| Drop Hammers | 1.0 | ... | ... | 1.0 | ... | ... | 1.0 | 1.0 | 1.0 | 1.0 | ... |
| Shears | 1.2 | ... | ... | 1.4 | ... | ... | 1.2 | 1.2 | 1.2 | 1.0 | ... |
| **Mills** | | | | | | | | | | | |
| Pebble | ... | 1.4 | 1.6 | 1.4 | ... | ... | ... | ... | ... | 1.4 | 1.6 |
| Rod | ... | 1.4 | 1.6 | 1.4 | ... | ... | ... | ... | ... | 1.4 | 1.6 |
| Ball | ... | 1.4 | 1.6 | 1.4 | ... | ... | ... | ... | ... | 1.4 | 1.6 |
| Roller Mills | ... | 1.4 | 1.6 | 1.4 | ... | ... | ... | ... | ... | 1.4 | 1.6 |
| Flaking Mills | ... | 1.6 | 1.6 | 1.4 | ... | ... | ... | ... | ... | 1.4 | 1.6 |
| Tumbling Barrels | ... | 1.6 | 1.6 | 1.4 | ... | ... | ... | ... | ... | 1.4 | 1.6 |

Table 11 *(Continued)*. Service Factors for Multiple V-Belt Applications

| Applications | Electric Motors | | | | | | | | | | |
| --- | --- | --- | --- | --- | --- | --- | --- | --- | --- | --- | --- |
| | A.C. | | | | | | | | | D.C. | |
| | Squirrel Cage | | | | Synchronous | | Single Phase | | | | |
| | Normal Torque Line Start | Normal Torque Compensator Start | High Torque | Wound Rotor (Slip Ring) | Normal Torque | High Torque | Repulsion and Split-Phase | Capacitor | Shunt Wound | Compound Wound | Line Shaft and Clutch Shifting |
| | Service Factors | | | | | | | | | | |
| Paper Machinery | | | | | | | | | | | |
| Jordan Engines...... | 1.5 | 1.3 | 1.8 | 1.5 | 1.6 | 1.8 | ... | ... | 1.5 | 1.5 | ... |
| Beaters............. | 1.4 | 1.4 | ... | 1.4 | ... | ... | ... | ... | 1.4 | 1.4 | 1.8 |
| Calenders.......... | 1.2 | 1.2 | ... | 1.2 | ... | ... | ... | ... | 1.2 | 1.2 | ... |
| Agitators.......... | 1.2 | 1.0 | 1.4 | 1.2 | ... | ... | ... | ... | 1.2 | 1.2 | 1.6 |
| Dryers............. | 1.2 | 1.2 | ... | 1.2 | ... | ... | ... | ... | 1.2 | 1.2 | ... |
| Paper Machines...... | 1.4 | 1.4 | ... | 1.5 | ... | ... | ... | ... | 1.5 | 1.5 | 1.6 |
| Pumps | | | | | | | | | | | |
| Centrifugal......... | 1.2 | 1.2 | 1.4 | 1.4 | ... | ... | 1.2 | 1.2 | ... | ... | ... |
| Gear............... | 1.2 | 1.2 | 1.4 | 1.4 | ... | ... | 1.2 | 1.2 | ... | ... | ... |
| Rotary............. | 1.2 | 1.2 | 1.4 | 1.4 | ... | ... | 1.2 | 1.2 | 1.2 | ... | ... |
| Reciprocating — | | | | | | | | | | | |
| 3 or more Cyl...... | 1.2 | 1.2 | ... | 1.4 | 1.4 | 1.6 | ... | ... | ... | ... | ... |
| 1 or 2 Cyl.......... | 1.4 | 1.4 | ... | 1.6 | 1.6 | 1.8 | ... | ... | ... | ... | ... |

similarly, a B, C, D, or E size belt. For example, for a small sheave rotating at 1750 rpm and a design horsepower of 5, a size A belt would be used.

If this point falls near the line of separation between two belt size areas, then both sizes may be considered as suitable for use. For example, a design horsepower of 40 to be transmitted at a small sheave speed of 800 rpm would call for a multiple drive of either C or D size V-belts.

**Horsepower Rating for Multiple V-Belts.** — The following formula and accompanying table of constants may be used to determine the general horsepower rating of a single V-belt.

$$\text{H.P.} = XS^{.91} - \frac{YS}{d_e} - ZS^3 \qquad (4)$$

where, $X$, $Y$, and $Z$ are constants as given in the accompanying table; H.P. = the recommended horsepower which must be multiplied by the appropriate correction factors for length (see Table 12) and arc of contact (see Table 13).

$S$ = belt speed in thousands of feet per minute. This is found by the formula:

$$S = \frac{3.14 \times \text{P.D. (inches)} \times \text{R.P.M.}}{12 \times 1000}$$

$d_e$ = equivalent diameter of small sheave which is equal to pitch diameter (in inches) multiplied by small diameter factor (Table 14). This provides ratings that compensate for the flexing effect of the small and large sheaves of the drive. The maximum value of $d_e$ to be used in the formula is: 5 for A belts; 7 for B belts; 12 for C belts; 17 for D belts; and 28 for E belts.

### Factors $X$, $Y$, and $Z$ for Use in Formula 4

| Factor | Regular Quality Belts | | | | | Premium Quality Belts | | | | |
|---|---|---|---|---|---|---|---|---|---|---|
| | Belt Cross Section | | | | | | | | | |
| | A | B | C | D | E | A | B | C | D | E |
| | Values of X, Y, and Z to be Used in H.P. Formula | | | | | | | | | |
| $X$ | 1.945 | 3.434 | 6.372 | 13.616 | 19.914 | 2.684 | 4.737 | 8.792 | 18.788 | 24.478 |
| $Y$ | 3.801 | 9.830 | 26.948 | 93.899 | 177.74 | 5.326 | 13.962 | 38.819 | 137.70 | 263.04 |
| $Z$ | 0.0136 | 0.0234 | 0.0416 | 0.0848 | 0.1222 | 0.0136 | 0.0234 | 0.0416 | 0.0848 | 0.1222 |

*Example:* Find the horsepower capacity of a standard quality A60 size V-belt for a drive in which the pitch diameter of the small sheave is 3 inches and that of the large, 9 inches.

The small sheave is to rotate at 1750 R.P.M.

1. Find center distance $C$ (Formula 3, page 1040)

$$C = \frac{b + \sqrt{b^2 - 32(D-d)^2}}{16}$$

where $b = 4L - 6.28(D+d)$
$L = 61.3$ inches (Table 8)
$b = 4 \times 61.3 - 6.28(9+3) = 169.8$

$$C = \frac{169.8 + \sqrt{169.8^2 - 32(9-3)^2}}{16} = 21.0 \text{ inches}$$

2. Find arc of contact, $A$ (Formula 5, page 1048)

$$A = 180° - \frac{(D-d)60°}{C} = 180° - \frac{(9-3)60°}{21.0} = 163°$$

3. Find correction factors
Length correction factor = 0.98 (Table 12)
Arc of contact correction factor = 0.96 (Table 13)
Small diameter factor (Speed ratio = $9 \div 3 = 3$) = 1.14 (Table 14)
4. Compute belt speed in thousands of feet per minute:

$$S = \frac{3.14 \times P.D. \times R.P.M.}{12 \times 1000} = \frac{3.14 \times 3 \times 1750}{12 \times 1000} = 1.38$$

5. Compute equivalent diameter of small sheave

$$d_e = 3 \times 1.14 = 3.42 \text{ inches}$$

6. Compute belt H.P. using Formula 4:

$$\text{H.P.} = 1.945 S^{.91} - \frac{3.801 S}{d_e} - .0136 S^3$$

$$= 1.945 \times 1.38^{.91} - \frac{3.801 \times 1.38}{3.42} - .0136 \times 1.38^3$$

$$= 1.945 \times 1.34 - 1.535 - .034 = 1.04$$

7. Apply length and arc of contact correction factors to get horsepower capacity:

$$1.04 \times 0.98 \times 0.96 = 0.98 \text{ H.P.}$$

8. Divide horsepower capacity into horsepower to be transmitted to obtain number of belts required for drive.

### Table 12. Length Correction Factors

| Standard Length Designation | Belt Cross Section A | B | C | Standard Length Designation | Belt Cross Section A | B | C | D | E |
|---|---|---|---|---|---|---|---|---|---|
| | Correction Factor | | | | Correction Factor | | | | |
| 26 | 0.81 | .... | .... | 97 | .... | 1.02 | .... | .... | .... |
| 31 | 0.84 | .... | .... | 105 | 1.10 | 1.04 | 0.94 | .... | .... |
| 33 | 0.86 | .... | .... | 112 | 1.11 | 1.05 | 0.95 | .... | .... |
| 35 | 0.87 | 0.81 | .... | 120 | 1.13 | 1.07 | 0.97 | 0.86 | .... |
| 38 | 0.88 | 0.83 | .... | 128 | 1.14 | 1.08 | 0.98 | 0.87 | .... |
| 42 | 0.90 | 0.85 | .... | 136 | .... | 1.09 | 0.99 | .... | .... |
| 46 | 0.92 | 0.87 | .... | 144 | .... | 1.11 | 1.00 | 0.90 | .... |
| 48 | 0.93 | 0.88 | .... | 158 | .... | 1.13 | 1.02 | 0.92 | .... |
| 51 | 0.94 | 0.89 | 0.80 | 162 | .... | .... | 1.03 | 0.92 | .... |
| 53 | 0.95 | 0.90 | .... | 173 | .... | 1.15 | 1.04 | 0.93 | .... |
| 55 | 0.96 | 0.90 | .... | 180 | .... | 1.16 | 1.05 | 0.94 | 0.91 |
| 60 | 0.98 | 0.92 | 0.82 | 195 | .... | 1.18 | 1.07 | 0.96 | 0.92 |
| 62 | 0.99 | 0.93 | .... | 210 | .... | 1.19 | 1.08 | 0.96 | 0.94 |
| 64 | 0.99 | 0.93 | .... | 240 | .... | 1.22 | 1.11 | 1.00 | 0.96 |
| 66 | 1.00 | 0.94 | .... | 270 | .... | 1.25 | 1.14 | 1.03 | 0.99 |
| 68 | 1.00 | 0.95 | 0.85 | 300 | .... | 1.27 | 1.16 | 1.05 | 1.01 |
| 71 | 1.01 | 0.95 | .... | 330 | .... | .... | 1.19 | 1.07 | 1.03 |
| 75 | 1.02 | 0.97 | 0.87 | 360 | .... | .... | 1.21 | 1.09 | 1.05 |
| 78 | 1.03 | 0.98 | .... | 390 | .... | .... | 1.23 | 1.11 | 1.07 |
| 80 | 1.04 | .... | .... | 420 | .... | .... | 1.24 | 1.12 | 1.09 |
| 81 | .... | 0.98 | 0.89 | 480 | .... | .... | .... | 1.16 | 1.12 |
| 83 | .... | 0.99 | .... | 540 | .... | .... | .... | 1.18 | 1.14 |
| 85 | 1.05 | 0.99 | 0.90 | 600 | .... | .... | .... | 1.20 | 1.17 |
| 90 | 1.06 | 1.00 | 0.91 | 660 | .... | .... | .... | 1.23 | 1.19 |
| 96 | 1.08 | .... | 0.92 | ... | .... | .... | .... | .... | .... |

**Arc of Contact.** — The arc of contact made by the V-belt on the small sheave is of importance when computing the horsepower rating of a V-belt for a given drive. This may be found by the formula:

$$\text{Arc of Contact (degrees)} = 180° - \frac{(D - d)60°}{C} \tag{5}$$

where $D$, $d$ and $C$ are as noted above. Correction factors, for various arcs of contact, used in finding horsepower capacities of mutiple V-belt drives (see example under Horsepower Rating for Multiple V-Belts) are given in Table 13.

### Table 13. Arc of Contact Correction Factors

| Arc of Contact on Small Sheave | Type of Drive V to V | V to Flat* | Arc of Contact on Small Sheave | Type of Drive V to V | V to Flat* |
|---|---|---|---|---|---|
| | Correction Factor | | | Correction Factor | |
| 180° | 1.00 | .75 | 130° | .86 | .86 |
| 170° | .98 | .77 | 120° | .82 | .82 |
| 160° | .95 | .80 | 110° | .78 | .78 |
| 150° | .92 | .82 | 100° | .74 | .74 |
| 140° | .89 | .84 | 90° | .69 | .69 |

* A V-Flat drive is one using a small sheave and a larger diameter flat pulley.

### Table 14. Small Diameter Factors

| Speed Ratio Range | Small Diameter Factor | Speed Ratio Range | Small Diameter Factor | Speed Ratio Range | Small Diameter Factor |
|---|---|---|---|---|---|
| 1.000–1.019 | 1.00 | 1.110–1.142 | 1.05 | 1.341–1.429 | 1.10 |
| 1.020–1.032 | 1.01 | 1.143–1.178 | 1.06 | 1.430–1.562 | 1.11 |
| 1.033–1.055 | 1.02 | 1.179–1.222 | 1.07 | 1.563–1.814 | 1.12 |
| 1.056–1.081 | 1.03 | 1.223–1.274 | 1.08 | 1.815–2.948 | 1.13 |
| 1.082–1.109 | 1.04 | 1.275–1.340 | 1.09 | 2.949 and over | 1.14 |

**Speed of Operation.** — V-belts operate most efficiently at speeds of about 4500 feet per minute. For belt speeds of 5000 feet per minute and more the sheave should be both statically and dynamically balanced. Special design and materials may also be called for and the manufacturer should be consulted. Equivalent belt speed for given sheave pitch diameter and revolutions per minute can be found in Table 3 in Flat Leather Belt section.

**Use of Idlers.** — According to the B. F. Goodrich Company, the most successful drives are those where an idler is not necessary and where proper tension can be had by adjusting the position of either the driver unit or the driven unit. Where these units are not adjustable, a grooved idler can be used on the inside of the drive. Such an idler should have a diameter larger than the recommended minimum sheave diameter for the belt cross section. If the idler has the smallest diameter on the drive, the idler diameter should be used in determining the horsepower per belt. The drive should be designed taking into account the smallest arc of contact whether it be on the driver or on the driven sheaves.

Some fixed-center drives do not leave enough space for a grooved idler inside and for these there is no choice but to install a flat back bend idler. Such idlers are not recommended because they are inefficient and cause trouble. The belts have a tendency to turn over; the belt life is reduced from 20 to 50 per cent; and often a second (grooved) idler must be added to the drive ahead of the flat idler to make it workable. Where a flat back bend idler is used, the belt life will be improved if in the design computations an additional 0.2 service factor (see Table 11) is employed.

**Quarter-turn Drives.** — V-belt quarter-turn drives are used to transmit power from a horizontal shaft to a vertical shaft or vice versa. According to the Engineering Standards for Multiple V-Belt Drives, certain precautions must be taken in setting up this type of drive: (a) Direction of rotation must be such that the tight side of the drive is on the bottom; (b) The axis of the vertical shaft should lie in a plane perpendicular to the horizontal shaft, and intersecting it at the center of the

### Table 15. "Y" Dimensions for Quarter-turn Drives

| Center Distance | 60 | 80 | 100 | 120 | 140 | 160 | 180 | 200 | 220 | 240 |
|---|---|---|---|---|---|---|---|---|---|---|
| "Y" Dimension | 2½ | 2¾ | 3 | 4 | 5¼ | 6½ | 7¾ | 9 | 10½ | 12 |

All dimensions in inches.

face of the sheave on the horizontal shaft; (c) the center of the face of the sheave of the vertical shaft should be below the axis of the horizontal shaft by an amount "Y" which depends on the center distance and is shown in Table 15.

Deep grooved sheaves (see Table 9) should always be used. The drive should have a minimum center distance of $5.5 (D + (N - 1)S + w)$ where, $D$ = pitch diameter of large sheave; $N$ = number of belts; $S$ = deep groove spacing (see Table 9) and $w$ = nominal belt top width (see Table 7).

**Open-end V-Belts.** — V-belts in long lengths which can be cut to the desired length and used with metal V-belt fasteners or connectors are especially designed to permit firm anchorage of these fasteners. The most efficient speed of open-end belts is under 3000 feet per minute and the maximum safe speed is given by one manufacturer as 3500 feet per minute and as 4000 feet per minute by another. Open-end belts are not recommended for either V-flat or quarter-turn drives.

**V-flat Drives.** — When a combination of a V-grooved driver sheave and a flat driven pulley is used, it is called a V-flat drive. V-flat drives are frequently recommended as changeovers from existing flat belt drives where the larger flat pulley can be incorporated into the drive. Although V-flat drives are entirely practical as a means of power transmission, they have definite limitations since the V-belt cannot pull as effectively on a flat pulley as it can in a grooved sheave. V-belts manufacturers recommend that V-flat drives should be used only where the speed ratio is at least 3 to 1 or greater and where the center distance is approximately equal to or slightly less than the diameter of the flat pulley. If the center distance is increased beyond the recommended maximum, or if the speed ratio is reduced below 3 to 1, the efficiency of the V-flat drive drops off sharply. The pulley should be really flat and any crown face should be removed.

The effective pitch diameter of a flat pulley carrying a V-belt is equal to the outside diameter of the pulley plus an amount which depends upon the cross section of the V-belt used. These amounts for a given cross section may vary slightly from one manufacturer to another but the following values may be taken as more or less typical. For an A belt add 0.4 inch; for a B belt, 0.5 inch; for a C belt, 0.7 inch; for a D belt, 0.9 inch; and for an E belt, 1.0 inch.

The preferred minimum face widths of flat pulleys used in V-flat drives is as follows: for an A V-belt, 1¾ inch for one belt and ⅝ inch for each additional belt; for a B V-belt, 2¼ inches for one belt and ¾ inch for each additional belt; for a C V-belt, 2¾ inches for one belt and 1 inch for each additional belt; for a D V-belt, 3¾ inches for one belt and 1⁷⁄₁₆ inch for each additional belt; and for an E V-belt 4¾ inches for one belt and 1¾ inches for each additional belt.

**V-belt Maintenance.** — In obtaining the proper tension on V-belts, it is not necessary to pull them exceedingly taut. They should be tightened only enough to take out slack and undue sag. A good method for checking the proper tension of a V-belt drive is by "striking" the belt with the fist. Slack V-belts feel dead under this test, while properly adjusted V-belts vibrate and feel alive. Another simple test which can be made is to press down firmly on each individual belt in a multi-belt drive. When the top can be depressed so that it is in line with the bottom of other belts on the drive, the correct amount of tension has been applied.

*Destructive Elements:* Belts should be kept clean, free of oil, and protected from sunlight as much as possible. Mineral oil is especially destructive. To clean belts they should be wiped with a dry cloth. The safest way to remove dirt and grime is to wash with soap and water and rinse well. If by accident the belts become grease or oil spattered, remove with carbon tetrachloride. Belt dressing should never be used on a V-belt drive.

**Proper Fit.** — The V-belt should ride in the sheave groove so that the top surface is just above the highest point of the sheave. If the belt rides too high, it loses contact area. A low-riding belt may "bottom" in the sheave groove, reducing the wedging action on the sides and resulting in slipping and burring.

**Double Angle V-Belts.** — For drive applications where power must be transmitted to sheaves from both sides of the belt, a double angle V-belt is available which is hexagonal in cross section. Both sides provide natural V-belt wedging action.

**SAE Standard V-Belts.** — The data for V-belts and pulleys shown in Table 16 cover six sizes of V-belts and the corresponding pulleys for accommodating these. In this standard, as revised in 1954, no change has been made in the 0.380 and the 0.500 inch sizes. The ⅝ inch (28-degree groove) size has been dropped. For the other sizes (1¼₆-, ¾-, ⅞-, and 1-inch), the new dimensions shown in the accompanying table are designed to permit the same actual grooves as in the previous standard.

The diagram and dimensions given permit checking of the pulley by means of measurement over balls or rods of specified diameter placed in the pulley groove as shown.

Standard belt lengths are in increments of one inch, without fractions, up to and including 60 inches. Above 60 inches, the increments are 2 inches without fractions. Standard belt length tolerances are based on the center distance and are as follows: For belt length of 40 inches or less, +⅛, −⁵⁄₃₂ inch; belt lengths over 40 to 50 inches, inclusive, +⅛, −³⁄₁₆ inch; for belt lengths over 50 to 60 inches, inclusive, +⁵⁄₃₂, −⁷⁄₃₂ inch; for belt lengths over 60 to 80 inches, inclusive, +³⁄₁₆, −⁹⁄₃₂ inch; for belt lengths over 80 to 100 inches, inclusive, +⁷⁄₃₂, −¹¹⁄₃₂ inch.

### Table 16.  SAE V-belt and Pulley Dimensions

| SAE Belt and Pulley Size | Nom. Belt Thickness, In. | Nom. Width of Groove, In., $W$ | Pulley Nominal Diameters, In. | Groove Angle, Deg. ±30 Min., $A$ | Ball or Rod Diameter, In., $d$ | Ball Extension | | Min. Depth of Groove, In., $D$ |
|---|---|---|---|---|---|---|---|---|
| | | | | | | $K$ | $2K$ | |
| 0.380 | ⁵⁄₁₆ | 0.380 | 2.75 and up | 36 | 0.3125 | 0.077 | 0.154 | ⁷⁄₁₆ |
| 0.500 | ¹³⁄₃₂ | 0.500 | 3 and up | 36 | 0.4375 | 0.157 | 0.314 | ⁹⁄₁₆ |
| 1¼₆ | ¹³⁄₃₂ | 0.597 | 3 to 4, incl<br>Over 4 to 6, incl<br>Over 6 | 34<br>36<br>38 | 0.500 | 0.129<br>0.140<br>0.151 | 0.258<br>0.280<br>0.302 | ⁹⁄₁₆ |
| ¾ | ⁷⁄₁₆ | 0.660 | 3 to 4, incl<br>Over 4 to 6, incl<br>Over 6 | 34<br>36<br>38 | 0.5625 | 0.164<br>0.176<br>0.187 | 0.328<br>0.352<br>0.374 | ⅝ |
| ⅞ | ½ | 0.785 | 3.5 to 4.5, incl<br>Over 4.5 to 6, incl<br>Over 6 | 34<br>36<br>38 | 0.6875 | 0.236<br>0.248<br>0.260 | 0.472<br>0.496<br>0.520 | 1¼₆ |
| 1 | ⁹⁄₁₆ | 0.910 | 4 to 6, incl<br>Over 6 to 8, incl<br>Over 8 | 34<br>36<br>38 | 0.8125 | 0.308<br>0.321<br>0.333 | 0.616<br>0.642<br>0.666 | 1¹³⁄₁₆ |

# TRANSMISSION CHAINS

In addition to the standard roller and inverted tooth types, a wide variety of drive chains of different construction are available. Such chains are manufactured to various degrees of precision ranging from unfinished castings or forgings to chains having certain machined parts. Practically all of these chains as well as standard roller chains can be equipped with attachments to fit them for conveyor use. A few such types are briefly described in the following paragraphs. Detailed information about them can be obtained from the manufacturers.

*Detachable Chains:* The links of this type of chain, which are identical, are easily detachable. Each has a hook-shaped end in which the bar of the adjacent link articulates. They are available in malleable iron or pressed steel. The chief advantage is the ease with which any link can be removed.

*Cast Roller Chains:* Cast roller chains are constructed, wholly or partly, of cast metal parts and are available in various styles. In general the rollers and side bars are accurately made castings without machine finish. The links are usually connected by means of forged pins secured by nuts or cotters. Such chains are used for slow speeds and moderate loads, or where the precision of standard roller chains is not required.

*Pintle Chains:* Unlike the roller chain, the pintle chain is composed of hollow-cored cylinders cast or forged integrally with two offset side bars and each link identical. The links are joined by pins inserted in holes in the ends of the side bars and through the cored holes in the adjacent links. Lugs prevent turning of the pins in the side bars and thus ensure articulation of the chain between the pin and the cored cylinder.

## Standard Roller Transmission Chains

A roller chain is made up of two kinds of links: roller links and pin links alternately spaced throughout the length of the chain as shown in Table 1.

Roller chains are manufactured in several types, each designed for the particular service required. All roller chains are so constructed that the rollers are evenly spaced throughout the chain. The outstanding advantage of this type of chain is the ability of the rollers to rotate when contacting the teeth of the sprocket. Two arrangements of roller chains are in common use: the single-strand type and the multiple-strand type. In the latter type, two or more chains are joined side by side by means of common pins which maintain the alignment of the rollers in the different strands.

*Standard roller chains* are manufactured to the specifications in the American National Standard for precision power transmission roller chains, attachments, and sprockets (ANSI B29.1-1975) and, where indicated, the data in the subsequent tables have been taken from this standard. These roller chains and sprockets are commonly used for the transmission of power in industrial machinery, machine tools, motor trucks, motorcycles and tractors and similar applications.

*Non-standard roller chains,* developed individually by various manufacturers prior to the adoption of the ANSI standard are similar in form and construction to standard roller chains but do not conform dimensionally to standard chains. Some sizes are still available from the originating manufacturers for replacement on existing equipment. They are not recommended for new installations, since their manufacture is being discontinued as rapidly as possible.

*Standard double-pitch roller chains* are like standard roller chains, except that their link plates have twice the pitch of the corresponding standard-pitch chain. Their design conforms to specifications in the ANSI Standard for double-pitch power transmission roller chains and sprockets (ANSI B29.3-1954, R1974). They are especially useful for low speeds, moderate loads or long center distances.

**Standard Roller Chain Nomenclature, Dimensions and Loads.** — Standard nomenclature for roller chain parts are given in Table 1. Dimensions for Standard Series roller chain are given in Table 2.

*Chain Pitch:* Distance in inches between centers of adjacent joint members. Other dimensions are proportional to the pitch.

*Tolerances for Chain Length:* New chains, under standard measuring load, must not be under-length. Over-length tolerance is 0.001/(pitch in inches)² +0.015 inch per foot. Length measurements are to be taken over a length of at least 12 inches.

*Measuring Load:* This is the load in pounds under which a chain should be measured for length. It is equal to one percent of the ultimate tensile strength, with a minimum of 18 pounds and a maximum of 1000 pounds for both single and multiple strand chain.

*Minimum Ultimate Tensile Strength:* For single-strand chain this is equal to or greater than 12,500 (pitch in inches)² pounds. The minimum tensile strength or breaking strength of a multiple strand chain is equal to that of a single strand chain multiplied by the number of strands.

**Standard Roller Chain Numbers.** — The right-hand figure in the chain number is zero for roller chains of the usual proportions, 1 for a light-weight chain and 5 for a rollerless bushing chain. The numbers to the left of the right-hand figure denote the number of ⅛ inches in the pitch. The letter *H* following the chain number denotes the heavy series; thus the number 80*H* denotes a 1-inch pitch heavy chain. The hyphenated number 2 suffixed to the chain number denotes a double strand, 3 a triple strand, 4 a quadruple strand chain and so on.

*Heavy Series:* These chains, made in ¾-inch and larger pitches, have thicker link plates than those of the regular standard. Their value is only in the acceptance of higher loads at lower speeds.

*Light-weight Machinery Chain:* This chain is designated as No. 41. It is ½ inch pitch; ¼ inch wide; has 0.306-inch diameter rollers and a 0.141-inch pin diameter. The minimum ultimate tensile strength is 1500 pounds.

*Multiple-strand Chain:* This is essentially an assembly of two or more single-strand chains placed side by side with pins that extend through the entire width to maintain alignment of the different strands.

**Types of Sprockets** — Four different designs or types of roller-chain sprockets are shown by the sectional views, Fig. 1. Type *A* is a plain plate; type *B* has a hub on one side only; type *C*, a hub on both sides; and type *D*, a detachable hub. Also used are shear pin and slip clutch sprockets designed to prevent damage to the drive or to other equipment caused by overloads or stalling.

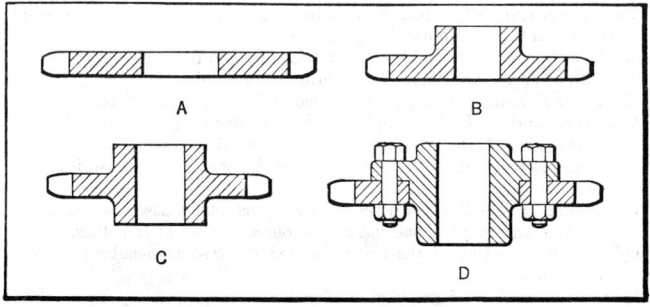

Fig. 1. Types of Sprockets

**Table 1.   ANSI Nomenclature for Roller Chain Parts** (ANSI B29.1-1975)

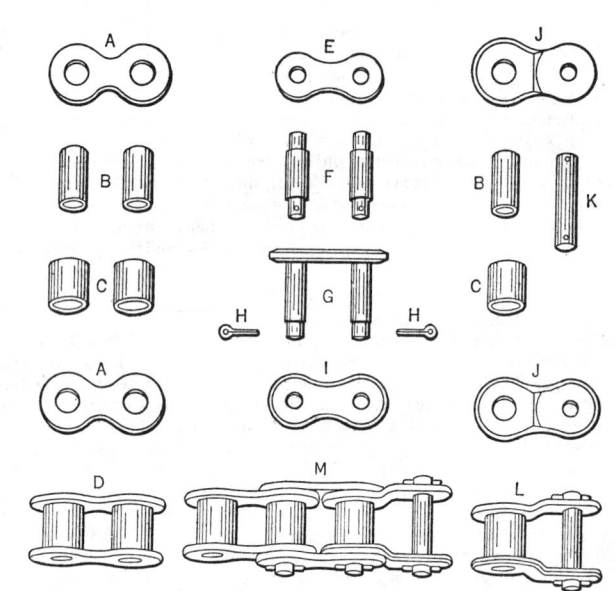

*Roller Link D.* — An inside link consisting of two inside plates, two bushings, and two rollers.

*Pin Link G and E.* — An outside link consisting of two pin-link plates assembled with two pins.

*Inside Plate A.* — One of the plates forming the tension members of a roller link.

*Pin Link Plate E.* — One of the plates forming the tension members of a pin link.

*Pin F.* — A stud articulating within a bushing of an inside link and secured at its ends by the pin-link plates.

*Bushing B.* — A cylindrical bearing in which the pin turns.

*Roller C.* — A ring or thimble which turns over a bushing.

*Assembled Pins G.* — Two pins assembled with one pin-link plate.

*Connecting-Link G and I.* — A pin link having one side plate detachable.

*Connecting-Link Plate I.* — The detachable pin-link plate belonging to a connecting link. It is retained by cotter pins or by a one-piece spring clip (not shown).

*Offset Link L.* — A link consisting of two offset plates assembled with a bushing and roller at one end and an offset link pin at the other.

*Offset Plate J.* — One of the plates forming the tension members of the offset link.

*Offset Link Pin K.* — A pin used in offset links.

Table 2. ANSI Roller Chain Dimensions (ANSI B29.1-1975)

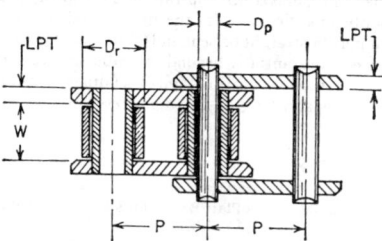

*Roller Diameters* $D_r$ are approximately ⅝ *P*.

The *width W* is defined as the distance between the link plates. It is approximately ⅝ of the chain pitch.

*Pin Diameters* $D_p$ are approximately 5/16 *P* or ½ of the roller diameter.

*Thickness LPT of Inside and Outside Link Plates* for the standard series is approximately ⅛ *P*.

*Thickness of Link Plates* for the heavy series of any pitch is approximately that of the next larger pitch Standard Series chain.

*Maximum Height of Roller Link Plates* = 0.95 Pitch.

*Maximum Height of Pin Link Plates* = 0.82 Pitch.

*Maximum Pin Diameter* = nominal pin diameter + 0.0005 inch.

*Minimum Hole in Bushing* = nominal pin diameter + 0.0015 inch.

*Maximum Width of Roller Link* = nominal width of chain + (2.12 × nominal link plate thickness.)

*Minimum Distance between Pin Link Plates* = maximum width of roller link + 0.005 inch.

| Pitch $P$ | Max. Roller Diameter $D_r$ | Standard Series | | | | | Heavy Series |
|---|---|---|---|---|---|---|---|
| | | Standard Chain No. | Width $W$ | Pin Diameter $D_p$ | Thickness of Link Plates $LPT$ | Measuring Load,† Lb. | Thickness of Link Plates $LPT$ |
| ¼ | *0.130 | 25 | ⅛ | 0.0905 | 0.030 | 18 | ..... |
| ⅜ | *0.200 | 35 | 3/16 | 0.141 | 0.050 | 18 | ..... |
| ½ | 0.306 | 41 | ¼ | 0.141 | 0.050 | 18 | ..... |
| ½ | 0.312 | 40 | 5/16 | 0.156 | 0.060 | 31 | ..... |
| ⅝ | 0.400 | 50 | ⅜ | 0.200 | 0.080 | 49 | ..... |
| ¾ | 0.469 | 60 | ½ | 0.234 | 0.094 | 70 | ..... |
| 1 | 0.625 | 80 | ⅝ | 0.312 | 0.125 | 125 | 0.156 |
| 1¼ | 0.750 | 100 | ¾ | 0.375 | 0.156 | 195 | 0.187 |
| 1½ | 0.875 | 120 | 1 | 0.437 | 0.187 | 281 | 0.219 |
| 1¾ | 1 | 140 | 1 | 0.500 | 0.219 | 383 | 0.250 |
| 2 | 1.125 | 160 | 1¼ | 0.562 | 0.250 | 500 | 0.281 |
| 2¼ | 1.406 | 180 | 1 13/32 | 0.687 | 0.281 | 633 | 0.312 |
| 2½ | 1.562 | 200 | 1½ | 0.781 | 0.312 | 781 | 0.375 |
| 3 | 1.875 | 240 | 1⅞ | 0.937 | 0.375 | 1 000 | 0.500 |

* Bushing diameter. This size chain has no rollers.    † For single-strand chain.

**Attachments.** — Modifications to standard chain components to adapt the chain for use in conveying, elevating, and timing operations are known as "attachments." The components commonly modified are: (1) the link plates, which are provided with extended lugs which may be straight or bent and (2) the chain pins, which are extended in length so as to project substantially beyond the outer surface of the pin link plates. Hole diameters, thicknesses, hole locations and offset dimensions for straight link and bent link plate extensions and lengths and diameters of extended pins are given in Table 3.

### Table 3.  Straight and Bent Link Plate Extensions and Extended Pin Dimensions
(ANSI B29.1-1975)

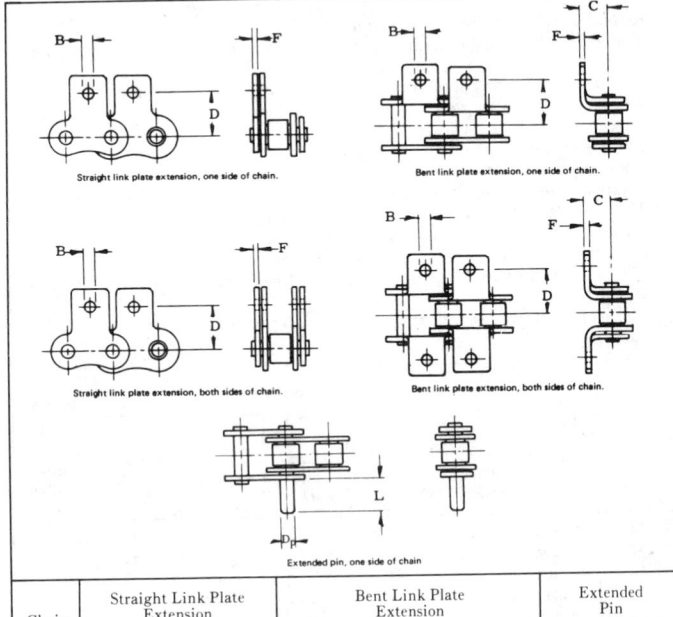

Straight link plate extension, one side of chain.

Bent link plate extension, one side of chain.

Straight link plate extension, both sides of chain.

Bent link plate extension, both sides of chain.

Extended pin, one side of chain

| Chain No. | Straight Link Plate Extension | | | Bent Link Plate Extension | | | | Extended Pin | |
|---|---|---|---|---|---|---|---|---|---|
| | B min. | D | F | B min. | C | D | F | $D_p$ Nominal | L |
| 35 | 0.102 | 0.375 | 0.050 | 0.102 | 0.250 | 0.375 | 0.050 | 0.141 | 0.375 |
| 40 | 0.131 | 0.500 | 0.060 | 0.131 | 0.312 | 0.500 | 0.060 | 0.156 | 0.375 |
| 50 | 0.200 | 0.625 | 0.080 | 0.200 | 0.406 | 0.625 | 0.080 | 0.200 | 0.469 |
| 60 | 0.200 | 0.719 | 0.094 | 0.200 | 0.469 | 0.750 | 0.094 | 0.234 | 0.562 |
| 80 | 0.261 | 0.969 | 0.125 | 0.261 | 0.625 | 1.000 | 0.125 | 0.312 | 0.750 |
| 100 | 0.323 | 1.250 | 0.156 | 0.323 | 0.781 | 1.250 | 0.156 | 0.375 | 0.938 |
| 120 | 0.386 | 1.438 | 0.188 | 0.386 | 0.906 | 1.500 | 0.188 | 0.437 | 1.125 |
| 140 | 0.448 | 1.750 | 0.219 | 0.448 | 1.125 | 1.750 | 0.219 | 0.500 | 1.312 |
| 160 | 0.516 | 2.000 | 0.250 | 0.516 | 1.250 | 2.000 | 0.250 | 0.562 | 1.500 |
| 200 | 0.641 | 2.500 | 0.312 | 0.641 | 1.688 | 2.500 | 0.312 | 0.781 | 1.875 |

All dimensions are in inches.

**Sprocket Classes.** — The American National Standard ANSI B29.1-1975 provides for two classes of sprockets designated as Commercial and Precision. The selection of either is a matter of drive application judgement. The usual moderate to slow speed commercial drive is adequately served by Commercial sprockets. Where extreme high speed in combination with high load is involved, or where the drive involves fixed centers, critical timing, or register problems, or close clearance with outside interference, then the use of Precision sprockets may be more appropriate.

As a general guide, drives requiring Type A or Type B lubrication (see page 1070) would be served by Commercial sprockets. Drives requiring Type C lubrication may require Precision sprockets, the manufacturer should be consulted.

**Keys, Keyways, and Set Screws.** — To secure sprockets to the shaft, both keys and set screws should be used. The key is used to prevent rotation of the sprocket on the shaft. Keys should be fitted carefully in the shaft and sprocket keyways to eliminate all backlash, especially on the fluctuating loads. A set screw should be located over a flat key to secure it against longitudinal displacement.

Where a set screw is to be used with a parallel key, the following sizes are recommended by the American Chain Association. For a sprocket bore and shaft diameter in the range of ½ through ⅞ inch, a ¼-inch set screw; for a range of ¹⁵/₁₆ through 1¾ inches, a ⅜-inch set screw; for a range of 1¹³/₁₆ through 2¼ inches, a ½-inch set screw; for a range of 2⁵/₁₆ through 3¼ inches, a ⅝-inch set screw; for a range of 3⅜ through 4½ inches, a ¾-inch set screw; for a range of 4¾ through 5½ inches, a ⅞-inch set screw; for a range of 5¾ through 7⅜ inches, a 1-inch set screw; and for 7½ through 12½ inches, a 1¼-inch set screw.

**Sprocket Diameters.** — The various diameters of roller chain sprockets are shown in Figure 2. These are defined as follows.

*Pitch Diameter:* The pitch diameter is the diameter of the pitch circle that passes through the centers of the link pins as the chain is wrapped on the sprocket. Since

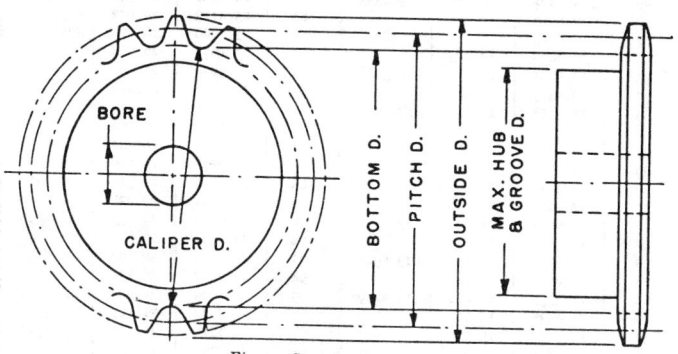

Fig. 2.  Sprocket Diameters

the chain pitch is measured on a straight line between the centers of adjacent pins, the chain pitch lines form a series of chords of the sprocket pitch circle. Sprocket pitch diameters for one-inch pitch and for 9 to 108 teeth are given in Table 4. For lower (5 to 8) or higher (109 to 200) numbers of teeth use the following formula in which $P$ = pitch, $N$ = number of teeth: Pitch Diameter = $P \div \sin(180° \div N)$.

*Bottom Diameter:* The bottom diameter is the diameter of a circle tangent to the curve (called the seating curve) at the bottom of the tooth gap. It equals the pitch diameter minus the diameter of the roller.

**Table 4. ANSI Roller Chain Sprocket Diameters** (ANSI B29.1-1975)

These diameters and caliper factors apply only to chain of 1-inch pitch. For any other pitch, multiply the values given below by the pitch.

     Caliper Diam. (even teeth) = Pitch Diameter − Roller Diam.

     Caliper Diam. (odd teeth) = Caliper factor × Pitch − Roller Diam.

See Table 5 for tolerances on Caliper Diameters.

| No. Teeth* | Pitch Diameter | Outside Diameter | Caliper Factor | No. Teeth* | Pitch Diameter | Outside Diameter | Caliper Factor |
|---|---|---|---|---|---|---|---|
| 9 | 2.9238 | 3.348 | 2.8794 | 59 | 18.7892 | 19.363 | 18.7825 |
| 10 | 3.2361 | 3.678 | | 60 | 19.1073 | 19.681 | |
| 11 | 3.5495 | 4.006 | 3.5133 | 61 | 19.4255 | 20.000 | 19.4190 |
| 12 | 3.8637 | 4.332 | | 62 | 19.7437 | 20.318 | |
| 13 | 4.1786 | 4.657 | 4.1481 | 63 | 20.0618 | 20.637 | 20.0556 |
| 14 | 4.4940 | 4.981 | | 64 | 20.3800 | 20.956 | |
| 15 | 4.8097 | 5.304 | 4.7834 | 65 | 20.6982 | 21.274 | 20.6921 |
| 16 | 5.1258 | 5.627 | | 66 | 21.0164 | 21.593 | |
| 17 | 5.4422 | 5.949 | 5.4190 | 67 | 21.3346 | 21.911 | 21.3287 |
| 18 | 5.7588 | 6.271 | | 68 | 21.6528 | 22.230 | |
| 19 | 6.0755 | 6.593 | 6.0548 | 69 | 21.9710 | 22.548 | 21.9653 |
| 20 | 6.3924 | 6.914 | | 70 | 22.2892 | 22.867 | |
| 21 | 6.7095 | 7.235 | 6.6907 | 71 | 22.6074 | 23.185 | 22.6018 |
| 22 | 7.0267 | 7.555 | | 72 | 22.9256 | 23.504 | |
| 23 | 7.3439 | 7.876 | 7.3268 | 73 | 23.2438 | 23.822 | 23.2384 |
| 24 | 7.6613 | 8.196 | | 74 | 23.5620 | 24.141 | |
| 25 | 7.9787 | 8.516 | 7.9630 | 75 | 23.8802 | 24.459 | 23.8750 |
| 26 | 8.2962 | 8.836 | | 76 | 24.1984 | 24.778 | |
| 27 | 8.6138 | 9.156 | 8.5992 | 77 | 24.5166 | 25.096 | 24.5116 |
| 28 | 8.9314 | 9.475 | | 78 | 24.8349 | 25.415 | |
| 29 | 9.2491 | 9.795 | 9.2355 | 79 | 25.1531 | 25.733 | 25.1481 |
| 30 | 9.5668 | 10.114 | | 80 | 25.4713 | 26.052 | |
| 31 | 9.8845 | 10.434 | 9.8718 | 81 | 25.7896 | 26.370 | 25.7847 |
| 32 | 10.2023 | 10.753 | | 82 | 26.1078 | 26.689 | |
| 33 | 10.5201 | 11.073 | 10.5082 | 83 | 26.4260 | 27.007 | 26.4213 |
| 34 | 10.8379 | 11.392 | | 84 | 26.7443 | 27.326 | |
| 35 | 11.1558 | 11.711 | 11.1446 | 85 | 27.0625 | 27.644 | 27.0579 |
| 36 | 11.4737 | 12.030 | | 86 | 27.3807 | 27.962 | |
| 37 | 11.7916 | 12.349 | 11.7810 | 87 | 27.6990 | 28.281 | 27.6945 |
| 38 | 12.1095 | 12.668 | | 88 | 28.0172 | 28.599 | |
| 39 | 12.4275 | 12.987 | 12.4174 | 89 | 28.3354 | 28.918 | 28.3310 |
| 40 | 12.7455 | 13.306 | | 90 | 28.6537 | 29.236 | |
| 41 | 13.0635 | 13.625 | 13.0539 | 91 | 28.9719 | 29.555 | 28.9676 |
| 42 | 13.3815 | 13.944 | | 92 | 29.2902 | 29.873 | |
| 43 | 13.6995 | 14.263 | 13.6904 | 93 | 29.6084 | 30.192 | 29.6042 |
| 44 | 14.0175 | 14.582 | | 94 | 29.9267 | 30.510 | |
| 45 | 14.3355 | 14.901 | 14.3269 | 95 | 30.2449 | 30.828 | 30.2408 |
| 46 | 14.6536 | 15.219 | | 96 | 30.5632 | 31.147 | |
| 47 | 14.9717 | 15.538 | 14.9634 | 97 | 30.8815 | 31.465 | 30.8774 |
| 48 | 15.2898 | 15.857 | | 98 | 31.1997 | 31.784 | |
| 49 | 15.6079 | 16.176 | 15.5999 | 99 | 31.5180 | 32.102 | 31.5140 |
| 50 | 15.9260 | 16.495 | | 100 | 31.8362 | 32.421 | |
| 51 | 16.2441 | 16.813 | 16.2364 | 101 | 32.1545 | 32.739 | 32.1506 |
| 52 | 16.5622 | 17.132 | | 102 | 32.4727 | 33.057 | |
| 53 | 16.8803 | 17.451 | 16.8729 | 103 | 32.7910 | 33.376 | 32.7872 |
| 54 | 17.1984 | 17.769 | | 104 | 33.1093 | 33.694 | |
| 55 | 17.5165 | 18.088 | 17.5094 | 105 | 33.4275 | 34.013 | 33.4238 |
| 56 | 17.8347 | 18.407 | | 106 | 33.7458 | 34.331 | |
| 57 | 18.1528 | 18.725 | 18.1459 | 107 | 34.0641 | 34.649 | 34.0604 |
| 58 | 18.4710 | 19.044 | | 108 | 34.3823 | 34.968 | |

* For 5-8 and 109-200 teeth see text, pages 1057, 1059.

**Table 5. Minus Tolerances on the Caliper Diameters of Precision Sprockets***
(ANSI B29.1-1975)

| Pitch | Number of Teeth | | | | |
|---|---|---|---|---|---|
| | Up to 16 | 16-24 | 25-35 | 36-48 | 49-63 |
| 1/4 | 0.004 | 0.004 | 0.004 | 0.005 | 0.005 |
| 3/8 | 0.004 | 0.004 | 0.004 | 0.005 | 0.005 |
| 1/2 | 0.004 | 0.005 | 0.0055 | 0.006 | 0.0065 |
| 5/8 | 0.005 | 0.0055 | 0.006 | 0.007 | 0.008 |
| 3/4 | 0.005 | 0.006 | 0.007 | 0.008 | 0.009 |
| 1 | 0.006 | 0.007 | 0.008 | 0.009 | 0.010 |
| 1 1/4 | 0.007 | 0.008 | 0.009 | 0.010 | 0.012 |
| 1 1/2 | 0.007 | 0.009 | 0.0105 | 0.012 | 0.013 |
| 1 3/4 | 0.008 | 0.010 | 0.012 | 0.013 | 0.015 |
| 2 | 0.009 | 0.011 | 0.013 | 0.015 | 0.017 |
| 2 1/4 | 0.010 | 0.012 | 0.014 | 0.016 | 0.018 |
| 2 1/2 | 0.010 | 0.013 | 0.015 | 0.018 | 0.020 |
| 3 | 0.012 | 0.015 | 0.018 | 0.021 | 0.024 |

| Pitch | Number of Teeth | | | | |
|---|---|---|---|---|---|
| | 64-80 | 81-99 | 100-120 | 121-143 | 144 up |
| 1/4 | 0.005 | 0.005 | 0.006 | 0.006 | 0.006 |
| 3/8 | 0.006 | 0.006 | 0.006 | 0.007 | 0.007 |
| 1/2 | 0.007 | 0.0075 | 0.008 | 0.0085 | 0.009 |
| 5/8 | 0.009 | 0.009 | 0.009 | 0.010 | 0.011 |
| 3/4 | 0.010 | 0.010 | 0.011 | 0.012 | 0.013 |
| 1 | 0.011 | 0.012 | 0.013 | 0.014 | 0.015 |
| 1 1/4 | 0.013 | 0.014 | 0.016 | 0.017 | 0.018 |
| 1 1/2 | 0.015 | 0.016 | 0.018 | 0.019 | 0.021 |
| 1 3/4 | 0.017 | 0.019 | 0.020 | 0.022 | 0.024 |
| 2 | 0.019 | 0.021 | 0.023 | 0.025 | 0.027 |
| 2 1/4 | 0.021 | 0.023 | 0.025 | 0.028 | 0.030 |
| 2 1/2 | 0.023 | 0.025 | 0.028 | 0.030 | 0.033 |
| 3 | 0.027 | 0.030 | 0.033 | 0.036 | 0.039 |

* Minus tolerances for Commercial sprockets are twice those shown in this table.

*Caliper Diameter:* The caliper diameter is the same as the bottom diameter for a sprocket with an even number of teeth. For a sprocket with an odd number of teeth, it is defined as the distance from the bottom of one tooth gap to that of the nearest opposite tooth gap. The caliper diameter for an even tooth sprocket is equal to pitch diameter — roller diameter. The caliper diameter for an odd tooth sprocket is equal to caliper factor — roller diameter. Here, the caliper factor = $PD[\cos(90° \div N)]$, where $PD$ = pitch diameter and $N$ = number of teeth. Caliper factors for one-inch pitch and sprockets having 9 to 108 teeth are given in Table 4. For other tooth numbers use above formula. Caliper diameter tolerances are minus only and are equal to $0.002P\sqrt{N} + 0.006$ inch for the Commercial sprockets and $0.001P\sqrt{N} + 0.003$ inch for Precision sprockets. Tolerances are given in Table 5.

*Outside Diameter:* The outside diameter is the diameter over the tips of teeth. Sprocket outside diameters for one-inch pitch and 9 to 108 teeth are given in Table 4. For other tooth numbers the outside diameter may be determined by the following formula in which $O$ = approximate outside diameter; $P$ = pitch of chain; $N$ = number of sprocket teeth: $O = P[0.6 + \cot(180° \div N)]$.

**Table 6. American National Standard Roller Chain Sprocket Flange Thickness and Tooth Section Profile Dimensions** (ANSI B29.1-1975)

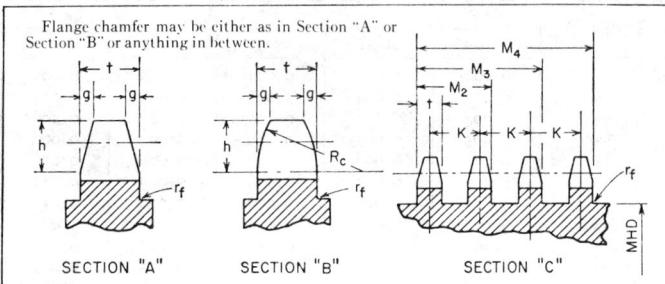

## Sprocket Flange Thickness

| Std. Chain No. | Width of Chain, W | Maximum Sprocket Flange Thickness, t | | | Minus Tolerance on t | | Tolerance on M | | Max. Variation of t on Each Flange | |
|---|---|---|---|---|---|---|---|---|---|---|
| | | Single | Double & Triple | Quad. & Over | Commercial | Precision | Commercial Plus or Minus | Precision Minus Only | Commercial | Precision |
| 25 | 0.125 | 0.110 | 0.106 | 0.096 | 0.021 | 0.007 | 0.007 | 0.007 | 0.021 | 0.004 |
| 35 | 0.188 | 0.169 | 0.163 | 0.150 | 0.027 | 0.008 | 0.008 | 0.008 | 0.027 | 0.004 |
| 41 | 0.250 | 0.226 | ... | ... | 0.032 | 0.009 | ... | ... | 0.032 | 0.004 |
| 40 | 0.312 | 0.284 | 0.275 | 0.256 | 0.035 | 0.009 | 0.009 | 0.009 | 0.035 | 0.004 |
| 50 | 0.375 | 0.343 | 0.332 | 0.310 | 0.036 | 0.010 | 0.010 | 0.010 | 0.036 | 0.005 |
| 60 | 0.500 | 0.459 | 0.444 | 0.418 | 0.036 | 0.011 | 0.011 | 0.011 | 0.036 | 0.006 |
| 80 | 0.625 | 0.575 | 0.556 | 0.526 | 0.040 | 0.012 | 0.012 | 0.012 | 0.040 | 0.006 |
| 100 | 0.750 | 0.692 | 0.669 | 0.633 | 0.046 | 0.014 | 0.014 | 0.014 | 0.046 | 0.007 |
| 120 | 1.000 | 0.924 | 0.894 | 0.848 | 0.057 | 0.016 | 0.016 | 0.016 | 0.057 | 0.008 |
| 140 | 1.000 | 0.924 | 0.894 | 0.848 | 0.057 | 0.016 | 0.016 | 0.016 | 0.057 | 0.008 |
| 160 | 1.250 | 1.156 | 1.119 | 1.063 | 0.062 | 0.018 | 0.018 | 0.018 | 0.062 | 0.009 |
| 180 | 1.406 | 1.302 | 1.259 | 1.198 | 0.068 | 0.020 | 0.020 | 0.020 | 0.068 | 0.010 |
| 200 | 1.500 | 1.389 | 1.344 | 1.278 | 0.072 | 0.021 | 0.021 | 0.021 | 0.072 | 0.010 |
| 240 | 1.875 | 1.738 | 1.682 | 1.602 | 0.087 | 0.025 | 0.025 | 0.025 | 0.087 | 0.012 |

## Sprocket Tooth Section Profile Dimensions

| Stnd. Chain No. | Chain Pitch P | Depth of Chamfer h | Width of Chamfer g | Minimum Radius $R_c$ | Transverse Pitch K | |
|---|---|---|---|---|---|---|
| | | | | | Standard Series | Heavy Series |
| 25 | 1/4 | 1/8 | 1/32 | 0.265 | 0.252 | ..... |
| 35 | 3/8 | 3/16 | 3/64 | 0.398 | 0.399 | ..... |
| 41 | 1/2 | 1/4 | 1/16 | 0.531 | ..... | ..... |
| 40 | 1/2 | 1/4 | 1/16 | 0.531 | 0.566 | ..... |
| 50 | 5/8 | 5/16 | 5/64 | 0.664 | 0.713 | ..... |
| 60 | 3/4 | 3/8 | 3/32 | 0.796 | 0.897 | 1.028 |
| 80 | 1 | 1/2 | 1/8 | 1.062 | 1.153 | 1.283 |
| 100 | 1 1/4 | 5/8 | 5/32 | 1.327 | 1.408 | 1.539 |
| 120 | 1 1/2 | 3/4 | 3/16 | 1.593 | 1.789 | 1.924 |
| 140 | 1 3/4 | 7/8 | 7/32 | 1.858 | 1.924 | 2.055 |
| 160 | 2 | 1 | 1/4 | 2.124 | 2.305 | 2.437 |
| 180 | 2 1/4 | 1 1/8 | 9/32 | 2.392 | 2.592 | 2.723 |
| 200 | 2 1/2 | 1 1/4 | 5/16 | 2.654 | 2.817 | 3.083 |
| 240 | 3 | 1 1/2 | 3/8 | 3.187 | 3.458 | 3.985 |

All dimensions are in inches. $r_f$ max = 0.04 P for max. hub diameter.

## Table 7. Typical Proportions of Single-Strand and Multiple-Strand Cast Roller Chain Sprockets

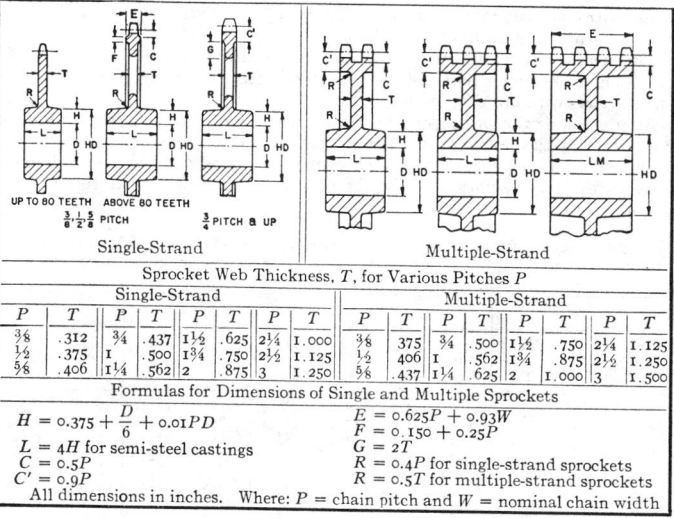

Single-Strand            Multiple-Strand

| Sprocket Web Thickness, $T$, for Various Pitches $P$ | | | | | | | | | | | | | | | | | |
|---|---|---|---|---|---|---|---|---|---|---|---|---|---|---|---|---|---|
| Single-Strand | | | | | | | | Multiple-Strand | | | | | | | | | |
| $P$ | $T$ | $P$ | $T$ | $P$ | $T$ | $P$ | $T$ | $P$ | $T$ | $P$ | $T$ | $P$ | $T$ | $P$ | $T$ | | |
| $\frac{3}{8}$ | .312 | $\frac{3}{4}$ | .437 | $1\frac{1}{2}$ | .625 | $2\frac{1}{4}$ | 1.000 | $\frac{3}{8}$ | 375 | $\frac{3}{4}$ | .500 | $1\frac{1}{2}$ | .750 | $2\frac{1}{4}$ | 1.125 | | |
| $\frac{1}{2}$ | .375 | 1 | .500 | $1\frac{3}{4}$ | .750 | $2\frac{1}{2}$ | 1.125 | $\frac{1}{2}$ | 406 | 1 | .562 | $1\frac{3}{4}$ | .875 | $2\frac{1}{2}$ | 1.250 | | |
| $\frac{5}{8}$ | .406 | $1\frac{1}{4}$ | .562 | 2 | .875 | 3 | 1.250 | $\frac{5}{8}$ | .437 | $1\frac{1}{4}$ | .625 | 2 | 1.000 | 3 | 1.500 | | |

**Formulas for Dimensions of Single and Multiple Sprockets**

$H = 0.375 + \dfrac{D}{6} + 0.01PD$

$L = 4H$ for semi-steel castings

$C = 0.5P$

$C' = 0.9P$

$E = 0.625P + 0.93W$

$F = 0.150 + 0.25P$

$G = 2T$

$R = 0.4P$ for single-strand sprockets

$R = 0.5T$ for multiple-strand sprockets

All dimensions in inches. Where: $P$ = chain pitch and $W$ = nominal chain width

## Table 8. Typical Proportions of Roller Chain Bar-steel Sprockets

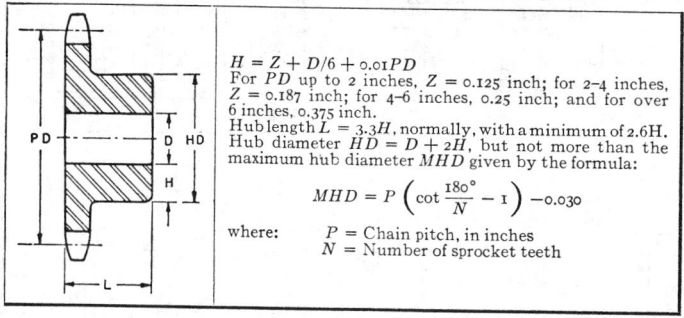

$H = Z + D/6 + 0.01PD$
For $PD$ up to 2 inches, $Z = 0.125$ inch; for 2–4 inches, $Z = 0.187$ inch; for 4–6 inches, $Z = 0.25$ inch; and for over 6 inches, 0.375 inch.
Hub length $L = 3.3H$, normally, with a minimum of $2.6H$.
Hub diameter $HD = D + 2H$, but not more than the maximum hub diameter $MHD$ given by the formula:

$$MHD = P \left( \cot \frac{180°}{N} - 1 \right) - 0.030$$

where:    $P$ = Chain pitch, in inches
          $N$ = Number of sprocket teeth

**Proportions of Sprockets.** — Typical proportions of single-strand and multiple-strand cast roller chain sprockets, as provided by the American Chain Association, are shown in Table 7. Typical proportions of roller chain bar-steel sprockets, also provided by this association, are shown in Table 8.

When sprocket wheels are designed with spokes, the usual assumptions made in order to determine suitable proportions are as follows: (1) That the maximum torque load acting on a sprocket is the chain tensile strength times the sprocket pitch radius; (2) That the torque load is equally divided between the arms by the rim; and (3) That each arm acts as a cantilever beam. The arms are generally elliptical in cross section, with the major axis twice the minor axis.

**Selection of Chain and Sprockets.** — The smallest applicable pitch of roller chain is desirable for quiet operation and high speed. The horsepower capacity varies with the chain pitch as shown in Table 15. However, short pitch with high working load can often be obtained by the use of multiple-strand chain.

The small sprocket selected must be large enough to accommodate the shaft. Table 9 gives maximum bore and hub diameters consistent with commercial practice for sprockets with up to 25 teeth.

After selecting the small sprocket, the number of teeth in the larger sprocket is determined by the desired ratio of the shaft speed. Overemphasis on the exactness in the speed ratio may result in a cumbersome and expensive installation. In most cases, satisfactory operation can be obtained with a minor change in speed of one or both shafts.

**Table 9. Recommended Roller Chain Sprocket Maximum Bore and Hub Diameters***

| No. of Teeth | Roller Chain Pitch 3/8 Max. Bore | 3/8 Max. Hub Dia. | 1/2 Max. Bore | 1/2 Max. Hub Dia. | 5/8 Max. Bore | 5/8 Max. Hub Dia. | 3/4 Max. Bore | 3/4 Max. Hub Dia. | 1 Max. Bore | 1 Max. Hub Dia. |
|---|---|---|---|---|---|---|---|---|---|---|
| 11 | 19/32 | 55/64 | 25/32 | 1 11/64 | 31/32 | 1 15/32 | 1 1/4 | 1 49/64 | 1 5/8 | 2 3/8 |
| 12 | 5/8 | 63/64 | 7/8 | 1 21/64 | 1 5/32 | 1 43/64 | 1 9/32 | 2 1/64 | 1 25/32 | 2 43/64 |
| 13 | 3/4 | 1 7/64 | 1 | 1 1/2 | 1 9/32 | 1 7/8 | 1 1/2 | 2 1/4 | 2 | 3 1/64 |
| 14 | 27/32 | 1 19/64 | 1 5/32 | 1 21/32 | 1 9/16 | 2 9/64 | 1 3/4 | 2 1/2 | 2 9/32 | 3 11/32 |
| 15 | 7/8 | 1 23/64 | 1 1/4 | 1 13/16 | 1 17/32 | 2 9/32 | 1 25/32 | 2 3/4 | 2 13/32 | 3 43/64 |
| 16 | 31/32 | 1 15/32 | 1 9/32 | 1 63/64 | 1 11/16 | 2 31/64 | 1 31/32 | 2 63/64 | 2 23/32 | 3 63/64 |
| 17 | 1 3/32 | 1 19/32 | 1 3/8 | 2 9/64 | 1 25/32 | 2 11/16 | 2 5/32 | 3 7/32 | 2 13/16 | 4 9/16 |
| 18 | 1 7/32 | 1 23/32 | 1 17/32 | 2 19/64 | 1 7/8 | 2 57/64 | 2 9/32 | 3 15/32 | 3 1/8 | 4 41/64 |
| 19 | 1 1/4 | 1 27/32 | 1 11/16 | 2 29/64 | 2 1/16 | 3 5/64 | 2 7/16 | 3 15/64 | 3 5/16 | 4 61/64 |
| 20 | 1 9/32 | 1 61/64 | 1 25/32 | 2 5/8 | 2 1/4 | 3 9/32 | 2 11/16 | 3 63/64 | 3 1/2 | 5 9/32 |
| 21 | 1 9/16 | 2 9/64 | 1 25/32 | 2 25/32 | 2 9/32 | 3 31/64 | 2 13/16 | 4 3/16 | 3 3/4 | 5 19/32 |
| 22 | 1 7/16 | 2 13/64 | 1 15/16 | 2 15/16 | 2 7/16 | 3 11/16 | 3 1/8 | 4 43/64 | 3 7/8 | 5 59/64 |
| 23 | 1 9/16 | 2 9/64 | 2 3/32 | 3 3/32 | 2 5/8 | 3 57/64 | 3 3/8 | 4 43/64 | 4 3/16 | 6 15/64 |
| 24 | 1 11/16 | 2 7/16 | 2 1/4 | 3 17/64 | 2 13/16 | 4 5/64 | 3 1/4 | 4 29/32 | 4 9/16 | 6 9/16 |
| 25 | 1 3/4 | 2 9/16 | 2 9/32 | 3 27/64 | 2 27/32 | 4 9/32 | 3 3/8 | 5 9/32 | 4 11/16 | 6 7/8 |

| No. of Teeth | Roller Chain Pitch 1 1/4 Max. Bore | 1 1/4 Max. Hub Dia. | 1 1/2 Max. Bore | 1 1/2 Max. Hub Dia. | 1 3/4 Max. Bore | 1 3/4 Max. Hub Dia. | 2 Max. Bore | 2 Max. Hub Dia. | 2 1/2 Max. Bore | 2 1/2 Max. Hub Dia. |
|---|---|---|---|---|---|---|---|---|---|---|
| 11 | 1 31/32 | 2 31/32 | 2 5/16 | 3 37/64 | 2 13/16 | 4 11/64 | 3 9/32 | 4 25/32 | 3 15/16 | 5 63/64 |
| 12 | 2 9/32 | 3 3/8 | 2 3/4 | 4 1/16 | 3 1/4 | 4 3/4 | 3 5/8 | 5 27/64 | 4 23/32 | 6 51/64 |
| 13 | 2 17/32 | 3 25/32 | 3 1/16 | 4 35/64 | 3 9/16 | 5 5/16 | 4 1/16 | 6 9/64 | 5 3/32 | 7 39/64 |
| 14 | 2 11/16 | 4 3/16 | 3 9/16 | 5 1/32 | 3 7/8 | 5 7/8 | 4 11/16 | 6 23/32 | 5 23/32 | 8 27/64 |
| 15 | 3 9/32 | 4 19/32 | 3 3/4 | 5 33/64 | 4 5/16 | 6 29/64 | 4 7/8 | 7 3/8 | 6 1/4 | 9 7/32 |
| 16 | 3 9/32 | 5 | 4 | 6 | 4 11/16 | 7 1/64 | 5 1/2 | 8 1/64 | 7 | 10 1/32 |
| 17 | 3 21/32 | 5 13/64 | 4 15/32 | 6 31/64 | 5 1/16 | 7 37/64 | 5 11/16 | 8 21/32 | 7 7/16 | 10 27/32 |
| 18 | 3 25/32 | 5 51/64 | 4 21/32 | 6 31/32 | 5 5/8 | 8 9/64 | 6 1/4 | 9 5/16 | 8 1/8 | 11 41/64 |
| 19 | 4 3/16 | 6 13/64 | 4 15/16 | 7 29/64 | 5 11/16 | 8 45/64 | 6 7/8 | 9 61/64 | 9 | 12 7/16 |
| 20 | 4 19/32 | 6 39/64 | 5 7/16 | 7 15/16 | 6 1/4 | 9 17/64 | 7 | 10 19/32 | 9 3/4 | 13 1/4 |
| 21 | 4 11/16 | 7 | 5 11/16 | 8 27/64 | 6 13/16 | 9 53/64 | 7 3/4 | 11 15/64 | 10 | 14 3/64 |
| 22 | 4 7/8 | 7 13/32 | 5 7/8 | 8 57/64 | 7 1/4 | 10 25/64 | 8 3/8 | 11 7/8 | 10 7/8 | 14 27/32 |
| 23 | 5 3/16 | 7 63/64 | 6 3/8 | 9 3/8 | 7 7/16 | 10 15/16 | 9 | 12 3/64 | 11 5/8 | 15 21/32 |
| 24 | 5 11/16 | 8 13/64 | 6 13/16 | 9 55/64 | 8 | 11 1/2 | 9 5/8 | 13 5/32 | 13 | 16 29/64 |
| 25 | 5 23/32 | 8 39/64 | 7 1/4 | 10 11/32 | 8 9/16 | 12 1/16 | 10 1/4 | 13 51/64 | 13 1/2 | 17 1/4 |

*American Chain Association.
All dimensions in inches.
For standard key dimensions see pages 967 to 969.

**Center Distance between Sprockets.** — The center-to-center distance between sprockets, as a general rule, should not be less than 1½ times the diameter of the larger sprocket and not less than thirty times the pitch nor more than about 50 times the pitch, although much depends upon the speed and other conditions. A center distance equivalent to 80 pitches may be considered an approved maximum. Very long center distances result in catenary tension in the chain. If roller-chain drives are designed correctly, the center-to-center distance for some transmissions may be so short that the sprocket teeth nearly touch each other, assuming that the load is not too great and the number of teeth is not too small. To avoid interference of the sprocket teeth, the center distance must, of course, be somewhat greater than one-half the sum of the outside diameters of the sprockets. The chain should extend around at least 120 degrees of the pinion circumference, and this minimum amount of contact is obtained for all center distances provided the ratio is less than 3½ to 1. Other things being equal, a fairly long chain is recommended in preference to the shortest one allowed by the sprocket diameters, because the rate of chain elongation due to natural wear is inversely proportional to the length, and also because the greater elasticity of the longer strand tends to absorb irregularities of motion and to decrease the effect of shocks.

If possible, the center distance should be adjustable in order to take care of slack due to elongation from wear and this range of adjustment should be at least one and one-half pitches. A little slack is desirable as it allows the chain links to take the best position on the sprocket teeth and reduces the wear on the bearings. Too much sag or an excessive distance between the sprockets may cause the chain to whip up and down — a condition detrimental to smooth running and very destructive to the chain. The sprockets should run in a vertical plane, the sprocket axes being approximately horizontal, unless an idler is used on the slack side to keep the chain in position. The most satisfactory results are obtained when the slack side of the chain is on the bottom.

**Center Distance for a Given Chain Length.** — When the distance between the driving and driven sprockets can be varied to suit the length of the chain, this center distance for a tight chain may be determined by the following formula, in which $c$ = center-to-center distance in inches; $L$ = chain length in pitches; $P$ = pitch of chain; $N$ = number of teeth in large sprocket; $n$ = number of teeth in small sprocket.

$$c = \frac{P}{8}[2L - N - n + \sqrt{(2L - N - n)^2 - 0.810(N - n)^2}]$$

This formula is approximate, but the error is less than the variation in the length of the best chains. The length $L$ in pitches should be an even number for a roller chain, so that the use of an offset connecting link will not be necessary.

**Idler Sprockets.** — When sprockets have a fixed center distance or are non-adjustable, it may be advisable to use an idler sprocket for taking up the slack. The idler should preferably be placed against the slack side between the two strands of the chain. When a sprocket is applied to the tight side of the chain to reduce vibration, it should be on the lower side and so located that the chain will run in a straight line between the two main sprockets. A sprocket will wear excessively if the number of teeth is too small and the speed too high, because there is impact between the teeth and rollers even though the idler carries practically no load.

**Length of Driving Chain.** — The total length of a block chain should be given in multiples of the pitch, whereas for a roller chain, the length should be in multiples of twice the pitch, because the ends must be connected with an outside and inside link. The length of a chain can be calculated accurately enough for ordinary practice by the use of the following formula, in which $L$ = chain length in pitches; $C$ = center distance in pitches; $N$ = number of teeth in large sprocket; $n$ = number of teeth in small sprocket:

$$L = 2C + \frac{N}{2} + \frac{n}{2} + \left(\frac{N-n}{2\pi}\right)^2 \times \frac{1}{C}$$

To the length obtained by this formula, add enough to make a whole number (and for a roller chain, an even number) of pitches. If a roller chain has an odd number of pitches, it will be necessary to use an offset connecting link.

Another formula for obtaining chain length in which $D$ = distance between centers of shafts; $R$ = pitch radius of large sprocket; $r$ = pitch radius of small sprocket; $N$ = number of teeth in large sprocket; $n$ = number of teeth in small sprocket; $P$ = pitch of chain and sprockets; and $l$ = required chain length in inches, is:

$$l = \frac{180° + 2\alpha}{360°} NP + \frac{180° - 2\alpha}{360°} nP + 2D \cos\alpha; \qquad \text{where } \sin\alpha = \frac{R-r}{D}$$

**Cutting Standard Sprocket Tooth Form.** — The proportions and seating curve data for the standard sprocket tooth form for roller chain are given in Table 10. Either formed or generating types of sprocket cutters may be employed.

*Hobs:* Only one hob will be required to cut any number of teeth for a given pitch and roller diameter. All hobs should be marked with pitch and roller diameter to be cut. Formulas and data for standard hob design are given in Table 11.

*Space Cutters:* Five cutters of this type will be required to cut from 7 teeth up for any given roller diameter. The ranges are, respectively, 7–8, 9–11, 12–17, 18–34, and 35 teeth and over. If less than 7 teeth is necessary, special cutters conforming to the required number of teeth should be used.

The regular cutters are based upon an intermediate number of teeth $N_a$ equal to $2N_1N_2 \div (N_1 + N_2)$ in which $N_1$ = minimum number of teeth and $N_2$ = maximum number of teeth for which cutter is intended; but the topping curve radius $F$ (see diagram in Table 12) is designed to produce adequate tooth height on a sprocket of $N_2$ teeth. The values of $N_a$ for the several cutters are, respectively, 7.47, 9.9, 14.07, 23.54, and 56. Formulas and construction data for space cutter layout are given in Table 12 and recommended cutter sizes are given in Table 13.

*Shaper Cutters:* Only one will be required to cut any number of teeth for a given pitch and roller diameter. The manufacturer should be referred to for information concerning the cutter form design to be used.

**Sprocket Manufacture.** — Cast sprockets have cut teeth, and the rim, hub face, and bore are machined. The smaller sprockets are generally cut from steel bar stock, and are finished all over. Sprockets are often made from forgings or forged bars. The extent of finishing depends on the particular specifications that are applicable. Many sprockets are made by welding a steel hub to a steel plate. This process produces a one-piece sprocket of desired proportions and one that can be heat-treated.

**Sprocket Materials.** — For large sprockets, cast iron is commonly used, especially in drives with large speed ratios, since the teeth of the larger sprocket are subjected to fewer chain engagements in a given time. For severe service, cast steel or steel plate is preferred.

The smaller sprockets of a drive are usually made of steel. With this material the body of the sprocket can be heat-treated to produce toughness for shock resistance, and the tooth surfaces can be hardened to resist wear.

Stainless steel or bronze may be used for corrosion resistance and Formica, nylon or other suitable plastic materials for special applications.

**Table 10. ANSI Sprocket Tooth Form for Roller Chain** (ANSI B29.1-1975)

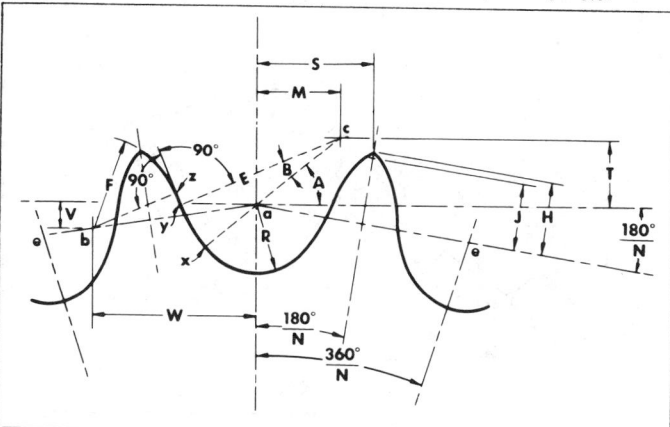

$P$ = pitch $(ae)$; $N$ = number of teeth; $D_r$ = nominal roller diameter
$D_s$ = seating curve diameter = 1.005 $D_r$ + 0.003 (in inches); $R = \frac{1}{2} D_s$
$D_s$ has only plus tolerance
$A = 35° + (60° \div N)$; $B = 18° - (56° \div N)$; $ac = 0.8 D_r$
$M = 0.8 D_r \cos (35° + (60° \div N))$
$T = 0.8 D_r \sin (35° + (60° \div N))$; $E = 1.3025 D_r + 0.0015$ (in inches)
Chord $xy = (2.605 D_r + 0.003) \sin (9° - (28° \div N))$ (in inches)
$yz = D_r [1.4 \sin (17° + (64° \div N)) - 0.8 \sin (18° - (56° \div N))]$
Length of a line between $a$ and $b = 1.4 D_r$
$W = 1.4 D_r \cos (180° \div N)$; $V = 1.4 D_r \sin (180° \div N)$
$F = D_r [0.8 \cos (18° - (56° \div N)) + 1.4 \cos (17° - (64° \div N)) - 1.3025] - 0.0015$ in.
$H = \sqrt{F^2 - (1.4 D_r - 0.5P)^2}$
$S = 0.5P \cos (180° \div N) + H \sin (180° \div N)$
Approximate O.D. of sprocket when $J$ is 0.3$P$ is $P [0.6 + \cot (180° \div N)]$
O.D. of sprocket when tooth is pointed = $P \cot (180° \div N) + \cos (180° \div N)$
$(D_s - D_r) + 2H$
Pressure angle for new chain = $xab = 35° - (120° \div N)$
Minimum pressure angle = $xab - B = 17° - (64° \div N)$; Average pressure angle
= $26° - (92° \div N)$

Seating Curve Data — Inches

| $P$ | $D_r$ | Min. $R$ | Min. $D_s$ | $D_s$ Tol.* | $P$ | $D_r$ | Min. $R$ | Min. $D_s$ | $D_s$ Tol.* |
|---|---|---|---|---|---|---|---|---|---|
| $\frac{1}{4}$ | 0.130 | 0.0670 | 0.134 | 0.0055 | $1\frac{1}{4}$ | 0.750 | 0.3785 | 0.757 | 0.0070 |
| $\frac{3}{8}$ | 0.200 | 0.1020 | 0.204 | 0.0055 | $1\frac{1}{2}$ | 0.875 | 0.4410 | 0.882 | 0.0075 |
| $\frac{1}{2}$ | 0.306 | 0.1585 | 0.317 | 0.0060 | $1\frac{3}{4}$ | 1 | 0.5040 | 1.008 | 0.0080 |
| $\frac{1}{2}$ | 0.312 | 0.1585 | 0.317 | 0.0060 | 2 | 1.125 | 0.5670 | 1.134 | 0.0085 |
| $\frac{5}{8}$ | 0.400 | 0.2025 | 0.405 | 0.0060 | $2\frac{1}{4}$ | 1.406 | 0.7080 | 1.416 | 0.0090 |
| $\frac{3}{4}$ | 0.469 | 0.2370 | 0.474 | 0.0065 | $2\frac{1}{2}$ | 1.562 | 0.7870 | 1.573 | 0.0095 |
| 1 | 0.625 | 0.3155 | 0.631 | 0.0070 | 3 | 1.875 | 0.9435 | 1.887 | 0.0105 |

* Plus tolerance only.

## Table 11. Standard Hob Design for Roller Chain Sprockets

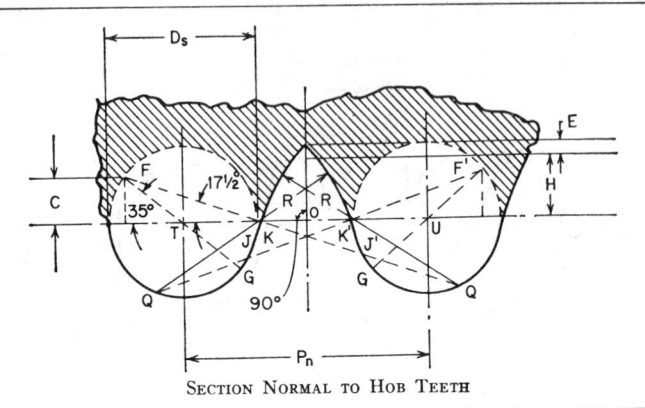

SECTION NORMAL TO HOB TEETH

Hobs designed for a given roller diameter ($D_r$) and chain pitch ($P$) will cut any number of teeth.

$P$ = Pitch of Chain;　$P_n$ = Normal Pitch of Hob = 1.011 $P$ inches

$D_s$ = Minimum Diameter of Seating Curve = 1.005 $D_r$ + 0.003 inches

$F$ = Radius Center for Arc $GK$;　$TO = OU = P_n \div 2$

$H$ = 0.27 $P$;　$E$ = 0.03 $P$ = Radius of Fillet Circle

$Q$ is located on line passing through $F$ and $J$. Point $J$ is intersection of line $XY$ with circle of diameter $D_s$. $R$ is found by trial and the arc of this radius is tangent to arc $KG$ at $K$ and to fillet radius.

$OD$ = Outside Diameter = 1.7 (Bore + $D_r$ + 0.7 $P$) approx.

$D_h$ = Pitch Diameter = $OD - D_s$;　$M$ = Helix Angle;　sin $M$ = $P_n \div \pi D_h$

$L$ = Lead = $P_n \div \cos M$;　$W$ = Width = Not less than 2 × Bore, or 6 $D_r$, or 3.2 $P$

Data for Laying Out Hob Outlines — Inches

| $P$ | $P_n$ | $H$ | $E$ | O.D. | $W$ | Bore | Keyway | No. Gashes |
|---|---|---|---|---|---|---|---|---|
| ¼ | 0.2527 | 0.0675 | 0.0075 | 2⅝ | 2½ | 1.250 | ¼ × ⅛ | 13 |
| ⅜ | 0.379 | 0.101 | 0.012 | 3⅛ | 2½ | 1.250 | ¼ × ⅛ | 13 |
| ½ | 0.506 | 0.135 | 0.015 | 3⅜ | 2½ | 1.250 | ¼ × ⅛ | 12 |
| ⅝ | 0.632 | 0.170 | 0.018 | 3⅝ | 2½ | 1.250 | ¼ × ⅛ | 12 |
| ¾ | 0.759 | 0.202 | 0.023 | 3¾ | 2⅞ | 1.250 | ¼ × ⅛ | 11 |
| 1 | 1.011 | 0.270 | 0.030 | 4⅜ | 3¾ | 1.250 | ¼ × ⅛ | 11 |
| 1¼ | 1.264 | 0.337 | 0.038 | 4¾ | 4½ | 1.250 | ¼ × ⅛ | 10 |
| 1½ | 1.517 | 0.405 | 0.045 | 5⅜ | 5¼ | 1.250 | ¼ × ⅛ | 10 |
| 1¾ | 1.770 | 0.472 | 0.053 | 6⅜ | 6 | 1.500 | ⅜ × 3/16 | 9 |
| 2 | 2.022 | 0.540 | 0.060 | 6⅞ | 6¾ | 1.500 | ⅜ × 3/16 | 9 |
| 2¼ | 2.275 | 0.607 | 0.068 | 8 | 8½ | 1.750 | ⅜ × 3/16 | 8 |
| 2½ | 2.528 | 0.675 | 0.075 | 8⅝ | 9⅜ | 1.750 | ⅜ × 3/16 | 8 |
| 3 | 3.033 | 0.810 | 0.090 | 9¾ | 11¼ | 2.000 | ½ × 3/16 | 8 |

Table 12. Standard Space Cutters for Roller-Chain Sprockets

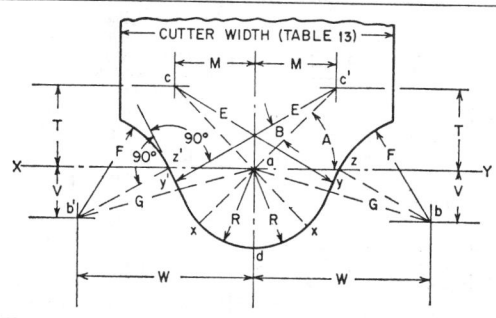

Angle $Yab$ is equal to $180° \div N$ when the cutter is made for a specific number of teeth. For the design of cutters covering a range of teeth, angle $Yab$ was determined by layout to assure chain roller clearance and to avoid pointed teeth on the larger sprockets of each range. It has values as given below for cutters covering the range of teeth shown. The following formulas are for cutters covering the standard ranges of teeth where $N_a$ equals intermediate values given on page 1064.

$$W = 1.4\, D_r \cos Yab; \qquad V = 1.4\, D_r \sin Yab$$

$$yz = D_r \left[ 1.4 \sin \left( 17° + \frac{116°}{N_a} - Yab \right) - 0.8 \sin \left( 18° - \frac{56°}{N_a} \right) \right]$$

$$F = D_r \left[ 0.8 \cos \left( 18° - \frac{56°}{N_a} \right) + 1.4 \cos \left( 17° + \frac{116°}{N_a} - Yab \right) - 1.3025 \right]$$

$$- 0.0015 \text{ in.}$$

For other points, use the value of $N_a$ for $N$ in the standard formulas in Table 10.

Data for Laying Out Space Cutter

| Range of Teeth | M | T | W | V |
|---|---|---|---|---|
| 7–8 | $0.5848\, D_r$ | $0.5459\, D_r$ | $1.2790\, D_r$ | $0.5694\, D_r$ |
| 9–11 | $0.6032\, D_r$ | $0.5255\, D_r$ | $1.3302\, D_r$ | $0.4365\, D_r$ |
| 12–17 | $0.6194\, D_r$ | $0.5063\, D_r$ | $1.3694\, D_r$ | $0.2911\, D_r$ |
| 18–34 | $0.6343\, D_r$ | $0.4875\, D_r$ | $1.3947\, D_r$ | $0.1220\, D_r$ |
| 35 up | $0.6466\, D_r$ | $0.4710\, D_r$ | $1.4000\, D_r$ | $0$ |

| Range of Teeth | F | Chord $xy$ | $yz$ | Angle $Yab$ |
|---|---|---|---|---|
| 7–8 | $0.8686\, D_r - 0.0015$ | $0.2384\, D_r + 0.0003$ | $0.0618\, D_r$ | 24° |
| 9–11 | $0.8554\, D_r - 0.0015$ | $0.2800\, D_r + 0.0003$ | $0.0853\, D_r$ | 18°10′ |
| 12–17 | $0.8364\, D_r - 0.0015$ | $0.3181\, D_r + 0.0004$ | $0.1269\, D_r$ | 12° |
| 18–34 | $0.8073\, D_r - 0.0015$ | $0.3540\, D_r + 0.0004$ | $0.1922\, D_r$ | 5° |
| 35 up | $0.7857\, D_r - 0.0015$ | $0.3850\, D_r + 0.0004$ | $0.2235\, D_r$ | 0° |

$E$ (same for all ranges) $= 1.3025\, D_r + 0.0015$; $G$ (same for all ranges) $= 1.4\, D_r$

**Table 13. Recommended Space Cutter Sizes for Roller-Chain Sprockets**

| Pitch | Roller Diam. | Number of Teeth | | | | | |
|---|---|---|---|---|---|---|---|
| | | 6 | 7–8 | 9–11 | 12–17 | 18–34 | 35 up |
| | | Cutter Diameter (Minimum) | | | | | |
| 1/4 | 0.130 | 2 3/4 | 2 3/4 | 2 3/4 | 2 3/4 | 2 3/4 | 2 3/4 |
| 3/8 | 0.200 | 2 3/4 | 2 3/4 | 2 3/4 | 2 3/4 | 2 3/4 | 2 3/4 |
| 1/2 | 0.312 | 3 | 3 | 3 1/8 | 3 1/8 | 3 1/8 | 3 1/8 |
| 5/8 | 0.400 | 3 1/8 | 3 1/8 | 3 1/4 | 3 1/4 | 3 1/4 | 3 1/4 |
| 3/4 | 0.469 | 3 1/4 | 3 1/4 | 3 3/8 | 3 3/8 | 3 3/8 | 3 3/8 |
| 1 | 0.625 | 3 7/8 | 4 | 4 1/8 | 4 1/8 | 4 1/4 | 4 1/4 |
| 1 1/4 | 0.750 | 4 1/4 | 4 3/8 | 4 1/2 | 4 1/2 | 4 5/8 | 4 5/8 |
| 1 1/2 | 0.875 | 4 3/8 | 4 1/2 | 4 5/8 | 4 5/8 | 4 3/4 | 4 3/4 |
| 1 3/4 | 1.000 | 5 | 5 1/8 | 5 1/4 | 5 3/8 | 5 1/2 | 5 1/2 |
| 2 | 1.125 | 5 3/8 | 5 1/2 | 5 5/8 | 5 3/4 | 5 7/8 | 5 7/8 |
| 2 1/4 | 1.406 | 5 7/8 | 6 | 6 1/4 | 6 3/8 | 6 1/2 | 6 1/2 |
| 2 1/2 | 1.563 | 6 3/8 | 6 5/8 | 6 3/4 | 6 7/8 | 7 | 7 1/8 |
| 3 | 1.875 | 7 1/2 | 7 3/4 | 7 7/8 | 8 | 8 | 8 1/4 |

| Pitch | Roller Diam. | Cutter Width (Minimum) | | | | | |
|---|---|---|---|---|---|---|---|
| 1/4 | 0.130 | 5/16 | 5/16 | 5/16 | 5/16 | 9/32 | 9/32 |
| 3/8 | 0.200 | 15/32 | 15/32 | 15/32 | 7/16 | 7/16 | 13/32 |
| 1/2 | 0.312 | 3/4 | 3/4 | 3/4 | 3/4 | 23/32 | 11/16 |
| 5/8 | 0.400 | 3/4 | 3/4 | 3/4 | 3/4 | 23/32 | 11/16 |
| 3/4 | 0.469 | 29/32 | 29/32 | 29/32 | 7/8 | 27/32 | 13/16 |
| 1 | 0.625 | 1 1/2 | 1 1/2 | 1 15/32 | 1 15/32 | 1 13/32 | 1 11/32 |
| 1 1/4 | 0.750 | 1 13/16 | 1 13/16 | 1 25/32 | 1 3/4 | 1 11/16 | 1 5/8 |
| 1 1/2 | 0.875 | 1 13/16 | 1 13/16 | 1 25/32 | 1 3/4 | 1 11/16 | 1 5/8 |
| 1 3/4 | 1.000 | 2 3/32 | 2 3/32 | 2 1/16 | 2 1/32 | 1 31/32 | 1 7/8 |
| 2 | 1.125 | 2 13/32 | 2 13/32 | 2 3/8 | 2 5/16 | 2 1/4 | 2 5/32 |
| 2 1/4 | 1.406 | 2 11/16 | 2 11/16 | 2 21/32 | 2 19/32 | 2 15/32 | 2 13/32 |
| 2 1/2 | 1.563 | 3 | 3 | 2 15/16 | 2 29/32 | 2 3/4 | 2 11/16 |
| 3 | 1.875 | 3 19/32 | 3 19/32 | 3 17/32 | 3 15/32 | 3 11/32 | 3 7/32 |

Where the same roller diameter is commonly used with chains of two different pitches it is recommended that stock cutters be made wide enough to cut sprockets for both chains.

*Marking of Cutters.* — All cutters are to be marked, giving pitch, roller diameter and range of teeth to be cut.

*Bores for Sprocket Cutters* (recommended practice) are approximately as calculated from the formula:

$$\text{Bore} = 0.7\sqrt{(\text{Width of Cutter} + \text{Roller Diameter} + 0.7\ \text{Pitch})}$$

and are equal to 1 inch for 1/4- through 3/4-inch pitches; 1 1/4 inches for 1- through 1 1/2-inch pitches; 1 1/2 inches for 1 3/4- through 2 1/4-inch pitches; 1 3/4 inches for 2 1/2-inch pitch; and 2 inches for 3-inch pitch.

*Minimum Outside Diameters of Space Cutters* for 35 teeth and over (recommended practice) are approximately as calculated from the formula:

$$\text{Outside Diameter} = 1.2\ (\text{Bore} + \text{Roller Diameter} + 0.7\ \text{Pitch}) + 1\ \text{in.}$$

**Roller Chain Drive Ratings.** — In 1961, under auspices of The American Sprocket Chain Manufacturers Association (now called American Chain Association), a joint research program was begun to study pin-bushing interaction at high speeds and to gain further data on the phenomenon of chain joint galling. The objectives of this program were: 1. To determine how far roller chain could be operated safely into the galling area; 2. To investigate the factors influencing galling; and 3. To learn how to inhibit galling in future chain designs. This study resulted in 1966 in newly revised horsepower ratings of roller chains. Subsequent studies resulted in further revisions which appear in ANSI B29.1-1975. The ratings shown in Table 15 are below the galling range.

The horsepower ratings in Table 14 apply to lubricated, single-pitch, single-strand roller chains, both ANSI Standard and Heavy series. To obtain ratings of multiple-strand chains, a multiple-strand factor is applied. The ratings in Table 14 are based upon: 1. A service factor of 1; 2. A chain length of approximately 100 pitches; 3. Use of recommended lubrication methods; and 4. A drive arrangement where two aligned sprockets are mounted on parallel shafts in a horizontal plane. Under these conditions, approximately 15,000 hours of service life at full load operation may be expected.

Substantial increases in rated speed loads can be utilized, as when a service life of less than 15,000 hours is satisfactory, or when full load operation is encountered only during a portion of the required service life. Chain manufacturers should be consulted for assistance with any special application requirements.

The horsepower ratings shown in Table 15 relate to the speed of the smaller sprocket and drive selections are made on this basis, whether the drive is speed reducing or speed increasing. Drives with more than two sprockets, idlers, composite duty cycles, or other unusual conditions often require special consideration. Where quietness or extra smooth operation are of special importance, a small-pitch chain operating over large diameter sprockets will minimize noise and vibration.

When making drive selection, consideration is given to the loads imposed on the chain by the type of input power and the type of equipment to be driven. Service factors are used to compensate for these loads and the *required* horsepower rating of the chain is determined by the following formula:

$$\text{Required hp Table Rating} = \frac{\text{hp to be Transmitted} \times \text{Service Factor}}{\text{Multiple-Strand Factor}}$$

*Service Factors:* The service factors in Table 14 are for normal chain loading. For unusual or extremely severe operating conditions not shown in this table, it is desirable to use larger service factors.

*Multiple-Strand factors:* The horsepower ratings for multiple-strand chains equal single-strand ratings multiplied by these factors: for two strands, a factor of 1.7; for three strands, 2.5; and for four strands, 3.3

**Table 14. Roller Chain Drive Service Factors**

| Type of Driven Load | Type of Input Power | | |
|---|---|---|---|
| | Internal Combustion Engine with Hydraulic Drive | Electric Motor or Turbine | Internal Combustion Engine with Mechanical Drive |
| Smooth | 1.0 | 1.0 | 1.2 |
| Moderate Shock | 1.2 | 1.3 | 1.4 |
| Heavy Shock | 1.4 | 1.5 | 1.7 |

**Lubrication.** — It has been shown that a separating wedge of fluid lubricant is formed in operating chain joints much like that formed in journal bearings. Therefore, fluid lubricant must be applied to assure an oil supply to the joints and minimize metal-to-metal contact. Lubrication, if supplied in sufficient volume, also provides effective cooling and impact damping at higher speeds. For this reason, it is important that lubrication recommendations be followed. *The ratings in Table 15 apply only to drives lubricated in the manner specified in this table.*

Chain drives should be protected against dirt and moisture and the oil supply kept free of contamination. Periodic oil change is desirable. A good grade of non-detergent petroleum base oil is recommended. Heavy oils and greases are generally too stiff to enter and fill the chain joints. The following lubricant viscosities are recommended: For temperatures of 20° to 40° F, use SAE 20 lubricant; for 40° to 100°, use SAE 30; for 100° to 120°, use SAE 40; and for 120° to 140°, use SAE 50.

There are three basic types of lubrication for roller chain drives. The recommended type shown in Table 15 as Type A, Type B, or Type C is influenced by the chain speed and the amount of power transmitted. These are *minimum* lubrication requirements and the use of a better type (for example, Type C instead of Type B) is acceptable and may be beneficial. Chain life can vary appreciably depending upon the way the drive is lubricated. The better the chain lubrication, the longer the chain life. For this reason, it is important that the lubrication recommendations be followed when using the ratings given in Table 15. The types of lubrication are as follows:

*Type A—Manual or Drip Lubrication:* In manual lubrication, oil is applied copiously with a brush or spout can at least once every eight hours of operation. Volume and frequency should be sufficient to prevent overheating of the chain or discoloration of the chain joints. In drip lubrication, oil drops from a drip lubricator are directed between the link plate edges. The volume and frequency should be sufficient to prevent discoloration of the lubricant in the chain joints. Precaution must be taken against misdirection of the drops by windage.

*Type B—Bath or Disc Lubrication:* In bath lubrication, the lower strand of the chain runs through a sump of oil in the drive housing. The oil level should reach the pitch line of the chain at its lowest point while operating. In disc lubrication, the chain operates above the oil level. The disc picks up oil from the sump and deposits it onto the chain, usually by means of a trough. The diameter of the disc should be such as to produce rim speeds of between 600 and 8000 feet per minute.

*Type C—Oil Stream Lubrication:* The lubricant is usually supplied by a circulating pump capable of supplying each chain drive with a continuous stream of oil. The oil should be applied inside the chain loop evenly across the chain width, and directed at the slack strand.

The chain manufacturer should be consulted when it appears desirable to use a type of lubricant other than that recommended.

**Installation and Alignment.** — Sprockets should have the tooth form, thickness, profile, and diameters conforming to the ANSI B29.1 Standard. For maximum service life small sprockets operating at moderate to high speeds, or near the rated horsepower, should have hardened teeth. Normally, large sprockets should not exceed 120 teeth.

In general a center distance of 30 to 50 chain pitches is most desirable. The distance between sprocket centers should provide at least a 120 degree chain wrap on the smaller sprocket. Drives may be installed with either adjustable or fixed center distances. Adjustable centers simplify the control of chain slack. Sufficient housing clearance must always be provided for the chain slack to obtain full chain life.

Accurate alignment of shafts and sprocket tooth faces provides uniform distribution of the load across the entire chain width and contributes substantially to optimum drive life. Shafting, bearings, and foundations should be suitable to maintain the initial alignment. Periodic maintenance should include an inspection of alignment.

**Table 15. Horsepower Ratings for Roller Chain — 1975**

To properly use this table the following factors must be taken into consideration:

1. Service Factors
2. Multiple Strand Factors
3. Lubrication.

*Service Factors:* See Table 14.

*Multiple Strand Factors:* For two strands, the multiple strand factor is 1.7; for three strands, it is 2.5; and for four strands, it is 3.3.

*Lubrication:* Type A—Manual or Drip Lubrication
Type B—Bath or Disc Lubrication
Type C—Oil Stream Lubrication

Required type of lubrication is indicated at the bottom of each roller chain size section of the table.

For a description of each type of lubrication, see page 1070.

To find the required horsepower table rating, use the following formula:

$$\text{Required hp Table Rating} = \frac{\text{hp to be transmitted} \times \text{Service Factor}}{\text{Multiple Strand Factor}}$$

¼-inch Pitch Standard Single-Strand Roller Chain—No. 25

| No. of Teeth Small Spkt. | Revolutions per Minute—Small Sprocket* | | | | | | | | | | | | |
|---|---|---|---|---|---|---|---|---|---|---|---|---|---|
| | 50 | 100 | 300 | 500 | 700 | 900 | 1200 | 1500 | 1800 | 2100 | 2500 | 3000 | 3500 |
| | Horsepower Rating | | | | | | | | | | | | |
| 9 | 0.02 | 0.04 | 0.12 | 0.18 | 0.25 | 0.31 | 0.41 | 0.50 | 0.58 | 0.67 | 0.79 | 0.93 | 1.06 |
| 10 | 0.03 | 0.05 | 0.13 | 0.21 | 0.28 | 0.35 | 0.45 | 0.56 | 0.65 | 0.75 | 0.88 | 1.04 | 1.19 |
| 11 | 0.03 | 0.05 | 0.14 | 0.23 | 0.31 | 0.39 | 0.50 | 0.62 | 0.73 | 0.83 | 0.98 | 1.15 | 1.32 |
| 12 | 0.03 | 0.06 | 0.16 | 0.25 | 0.34 | 0.43 | 0.55 | 0.68 | 0.80 | 0.92 | 1.07 | 1.26 | 1.45 |
| 13 | 0.04 | 0.06 | 0.17 | 0.27 | 0.37 | 0.47 | 0.60 | 0.74 | 0.87 | 1.00 | 1.17 | 1.38 | 1.58 |
| 14 | 0.04 | 0.07 | 0.19 | 0.30 | 0.40 | 0.50 | 0.65 | 0.80 | 0.94 | 1.08 | 1.27 | 1.49 | 1.71 |
| 15 | 0.04 | 0.08 | 0.20 | 0.32 | 0.43 | 0.54 | 0.70 | 0.86 | 1.01 | 1.17 | 1.36 | 1.61 | 1.85 |
| 16 | 0.04 | 0.08 | 0.22 | 0.34 | 0.47 | 0.58 | 0.76 | 0.92 | 1.09 | 1.25 | 1.46 | 1.72 | 1.98 |
| 17 | 0.05 | 0.09 | 0.23 | 0.37 | 0.50 | 0.62 | 0.81 | 0.99 | 1.16 | 1.33 | 1.56 | 1.84 | 2.11 |
| 18 | 0.05 | 0.09 | 0.25 | 0.39 | 0.53 | 0.66 | 0.86 | 1.05 | 1.24 | 1.42 | 1.66 | 1.96 | 2.25 |
| 19 | 0.05 | 0.10 | 0.26 | 0.41 | 0.56 | 0.70 | 0.91 | 1.11 | 1.31 | 1.50 | 1.76 | 2.07 | 2.38 |
| 20 | 0.06 | 0.10 | 0.28 | 0.44 | 0.59 | 0.74 | 0.96 | 1.17 | 1.38 | 1.59 | 1.86 | 2.19 | 2.52 |
| 21 | 0.06 | 0.11 | 0.29 | 0.46 | 0.62 | 0.78 | 1.01 | 1.24 | 1.46 | 1.68 | 1.96 | 2.31 | 2.66 |
| 22 | 0.06 | 0.11 | 0.31 | 0.48 | 0.66 | 0.82 | 1.07 | 1.30 | 1.53 | 1.76 | 2.06 | 2.43 | 2.79 |
| 23 | 0.06 | 0.12 | 0.32 | 0.51 | 0.69 | 0.86 | 1.12 | 1.37 | 1.61 | 1.85 | 2.16 | 2.55 | 2.93 |
| 24 | 0.07 | 0.13 | 0.34 | 0.53 | 0.72 | 0.90 | 1.17 | 1.43 | 1.69 | 1.94 | 2.27 | 2.67 | 3.07 |
| 25 | 0.07 | 0.13 | 0.35 | 0.56 | 0.75 | 0.94 | 1.22 | 1.50 | 1.76 | 2.02 | 2.37 | 2.79 | 3.21 |
| 26 | 0.07 | 0.14 | 0.37 | 0.58 | 0.79 | 0.98 | 1.28 | 1.56 | 1.84 | 2.11 | 2.47 | 2.91 | 3.34 |
| 28 | 0.08 | 0.15 | 0.40 | 0.63 | 0.85 | 1.07 | 1.38 | 1.69 | 1.99 | 2.29 | 2.68 | 3.15 | 3.62 |
| 30 | 0.08 | 0.16 | 0.43 | 0.68 | 0.92 | 1.15 | 1.49 | 1.82 | 2.15 | 2.46 | 2.88 | 3.40 | 3.90 |
| 32 | 0.09 | 0.17 | 0.46 | 0.73 | 0.98 | 1.23 | 1.60 | 1.95 | 2.30 | 2.64 | 3.09 | 3.64 | 4.18 |
| 35 | 0.10 | 0.19 | 0.51 | 0.80 | 1.08 | 1.36 | 1.76 | 2.15 | 2.53 | 2.91 | 3.41 | 4.01 | 4.61 |
| 40 | 0.12 | 0.22 | 0.58 | 0.92 | 1.25 | 1.57 | 2.03 | 2.48 | 2.93 | 3.36 | 3.93 | 4.64 | 5.32 |
| 45 | 0.13 | 0.25 | 0.66 | 1.05 | 1.42 | 1.78 | 2.31 | 2.82 | 3.32 | 3.82 | 4.47 | 5.26 | 6.05 |
| | Type A | | | | | Type B | | | | | | | |

* For rpm above 3500, see ANSI B29.1-1975.

### Table 15 *(Continued)*. Horsepower Ratings for Roller Chain—1975

**⅜-inch Pitch Standard Single-Strand Roller Chain—No. 35**

| No. of Teeth Small Spkt. | Revolutions per Minute—Small Sprocket* | | | | | | | | | | | | |
|---|---|---|---|---|---|---|---|---|---|---|---|---|
| | 50 | 100 | 300 | 500 | 700 | 900 | 1200 | 1500 | 1800 | 2100 | 2500 | 3000 | 3500 |
| | Horsepower Rating | | | | | | | | | | | | |
| 9 | 0.08 | 0.15 | 0.39 | 0.62 | 0.84 | 1.06 | 1.37 | 1.68 | 1.98 | 2.27 | 2.65 | 2.17 | 1.73 |
| 10 | 0.09 | 0.16 | 0.44 | 0.70 | 0.95 | 1.19 | 1.54 | 1.88 | 2.21 | 2.54 | 2.97 | 2.55 | 2.02 |
| 11 | 0.10 | 0.18 | 0.49 | 0.77 | 1.05 | 1.31 | 1.70 | 2.08 | 2.45 | 2.82 | 3.30 | 2.94 | 2.33 |
| 12 | 0.11 | 0.20 | 0.54 | 0.85 | 1.15 | 1.44 | 1.87 | 2.29 | 2.70 | 3.10 | 3.62 | 3.35 | 2.66 |
| 13 | 0.12 | 0.22 | 0.59 | 0.93 | 1.26 | 1.57 | 2.04 | 2.49 | 2.94 | 3.38 | 3.95 | 3.77 | 3.00 |
| 14 | 0.13 | 0.24 | 0.63 | 1.01 | 1.36 | 1.71 | 2.21 | 2.70 | 3.18 | 3.66 | 4.28 | 4.22 | 3.35 |
| 15 | 0.14 | 0.25 | 0.68 | 1.08 | 1.47 | 1.84 | 2.38 | 2.91 | 3.43 | 3.94 | 4.61 | 4.68 | 3.71 |
| 16 | 0.15 | 0.27 | 0.73 | 1.16 | 1.57 | 1.97 | 2.55 | 3.12 | 3.68 | 4.22 | 4.94 | 5.15 | 4.09 |
| 17 | 0.16 | 0.29 | 0.78 | 1.24 | 1.68 | 2.10 | 2.73 | 3.33 | 3.93 | 4.51 | 5.28 | 5.64 | 4.48 |
| 18 | 0.17 | 0.31 | 0.83 | 1.32 | 1.78 | 2.24 | 2.90 | 3.54 | 4.18 | 4.80 | 5.61 | 6.15 | 4.88 |
| 19 | 0.18 | 0.33 | 0.88 | 1.40 | 1.89 | 2.37 | 3.07 | 3.76 | 4.43 | 5.09 | 5.95 | 6.67 | 5.29 |
| 20 | 0.19 | 0.35 | 0.93 | 1.48 | 2.00 | 2.51 | 3.25 | 3.97 | 4.68 | 5.38 | 6.29 | 7.20 | 5.72 |
| 21 | 0.20 | 0.37 | 0.98 | 1.56 | 2.11 | 2.64 | 3.42 | 4.19 | 4.93 | 5.67 | 6.63 | 7.75 | 6.15 |
| 22 | 0.21 | 0.38 | 1.03 | 1.64 | 2.22 | 2.78 | 3.60 | 4.40 | 5.19 | 5.96 | 6.97 | 8.21 | 6.59 |
| 23 | 0.22 | 0.40 | 1.08 | 1.72 | 2.33 | 2.92 | 3.78 | 4.62 | 5.44 | 6.25 | 7.31 | 8.62 | 7.05 |
| 24 | 0.23 | 0.42 | 1.14 | 1.80 | 2.44 | 3.05 | 3.96 | 4.84 | 5.70 | 6.55 | 7.66 | 9.02 | 7.51 |
| 25 | 0.24 | 0.44 | 1.19 | 1.88 | 2.55 | 3.19 | 4.13 | 5.05 | 5.95 | 6.84 | 8.00 | 9.43 | 7.99 |
| 26 | 0.25 | 0.46 | 1.24 | 1.96 | 2.66 | 3.33 | 4.31 | 5.27 | 6.21 | 7.14 | 8.35 | 9.84 | 8.47 |
| 28 | 0.27 | 0.50 | 1.34 | 2.12 | 2.88 | 3.61 | 4.67 | 5.71 | 6.73 | 7.73 | 9.05 | 10.7 | 9.47 |
| 30 | 0.29 | 0.54 | 1.45 | 2.29 | 3.10 | 3.89 | 5.03 | 6.15 | 7.25 | 8.33 | 9.74 | 11.5 | 10.5 |
| 32 | 0.31 | 0.58 | 1.55 | 2.45 | 3.32 | 4.17 | 5.40 | 6.60 | 7.77 | 8.93 | 10.4 | 12.3 | 11.6 |
| 35 | 0.34 | 0.64 | 1.71 | 2.70 | 3.66 | 4.59 | 5.95 | 7.27 | 8.56 | 9.84 | 11.5 | 13.6 | 13.2 |
| 40 | 0.39 | 0.73 | 1.97 | 3.12 | 4.23 | 5.30 | 6.87 | 8.40 | 9.89 | 11.4 | 13.3 | 15.7 | 16.2 |
| 45 | 0.45 | 0.83 | 2.24 | 3.55 | 4.80 | 6.02 | 7.80 | 9.53 | 11.2 | 12.9 | 15.1 | 17.8 | 19.3 |
| | Type A | | Type B | | | | | | | Type C | | | |

**½-inch Pitch Standard Single-Strand Roller Chain—No. 40**

| No. of Teeth Small Spkt. | Revolutions per Minute—Small Sprocket* | | | | | | | | | | | | |
|---|---|---|---|---|---|---|---|---|---|---|---|---|
| | 50 | 100 | 200 | 300 | 400 | 500 | 700 | 900 | 1000 | 1200 | 1400 | 1600 | 1800 |
| | Horsepower Rating | | | | | | | | | | | | |
| 9 | 0.19 | 0.35 | 0.65 | 0.93 | 1.21 | 1.44 | 2.00 | 2.51 | 2.75 | 3.25 | 3.73 | 4.12 | 3.45 |
| 10 | 0.21 | 0.39 | 0.73 | 1.04 | 1.35 | 1.65 | 2.24 | 2.81 | 3.09 | 3.64 | 4.18 | 4.71 | 4.04 |
| 11 | 0.23 | 0.43 | 0.80 | 1.16 | 1.50 | 1.83 | 2.48 | 3.11 | 3.42 | 4.03 | 4.63 | 5.22 | 4.66 |
| 12 | 0.25 | 0.47 | 0.88 | 1.27 | 1.65 | 2.01 | 2.73 | 3.42 | 3.76 | 4.43 | 5.09 | 5.74 | 5.31 |
| 13 | 0.28 | 0.52 | 0.96 | 1.39 | 1.80 | 2.20 | 2.97 | 3.73 | 4.10 | 4.83 | 5.55 | 6.26 | 5.99 |
| 14 | 0.30 | 0.56 | 1.04 | 1.50 | 1.95 | 2.38 | 3.22 | 4.04 | 4.44 | 5.23 | 6.01 | 6.78 | 6.70 |
| 15 | 0.32 | 0.60 | 1.12 | 1.62 | 2.10 | 2.56 | 3.47 | 4.35 | 4.78 | 5.64 | 6.47 | 7.30 | 7.43 |
| 16 | 0.35 | 0.65 | 1.20 | 1.74 | 2.25 | 2.75 | 3.72 | 4.66 | 5.13 | 6.04 | 6.94 | 7.83 | 8.18 |
| 17 | 0.37 | 0.69 | 1.29 | 1.85 | 2.40 | 2.93 | 3.97 | 4.98 | 5.48 | 6.45 | 7.41 | 8.36 | 8.96 |
| 18 | 0.39 | 0.73 | 1.37 | 1.97 | 2.55 | 3.12 | 4.22 | 5.30 | 5.82 | 6.86 | 7.88 | 8.89 | 9.76 |
| 19 | 0.42 | 0.78 | 1.45 | 2.09 | 2.71 | 3.31 | 4.48 | 5.62 | 6.17 | 7.27 | 8.36 | 9.42 | 10.5 |
| 20 | 0.44 | 0.82 | 1.53 | 2.21 | 2.86 | 3.50 | 4.73 | 5.94 | 6.53 | 7.69 | 8.83 | 9.96 | 11.1 |
| 21 | 0.46 | 0.87 | 1.62 | 2.33 | 3.02 | 3.69 | 4.99 | 6.26 | 6.88 | 8.11 | 9.31 | 10.5 | 11.7 |
| 22 | 0.49 | 0.91 | 1.70 | 2.45 | 3.17 | 3.88 | 5.25 | 6.58 | 7.23 | 8.52 | 9.79 | 11.0 | 12.3 |
| 23 | 0.51 | 0.96 | 1.78 | 2.57 | 3.33 | 4.07 | 5.51 | 6.90 | 7.59 | 8.94 | 10.3 | 11.6 | 12.9 |
| 24 | 0.54 | 1.00 | 1.87 | 2.69 | 3.48 | 4.26 | 5.76 | 7.23 | 7.95 | 9.36 | 10.8 | 12.1 | 13.5 |
| 25 | 0.56 | 1.05 | 1.95 | 2.81 | 3.64 | 4.45 | 6.02 | 7.55 | 8.30 | 9.78 | 11.2 | 12.7 | 14.1 |
| 26 | 0.58 | 1.09 | 2.04 | 2.93 | 3.80 | 4.64 | 6.28 | 7.88 | 8.66 | 10.2 | 11.7 | 13.2 | 14.7 |
| 28 | 0.63 | 1.18 | 2.20 | 3.18 | 4.11 | 5.03 | 6.81 | 8.54 | 9.39 | 11.1 | 12.7 | 14.3 | 15.9 |
| 30 | 0.68 | 1.27 | 2.38 | 3.42 | 4.43 | 5.42 | 7.33 | 9.20 | 10.1 | 11.9 | 13.7 | 15.4 | 17.2 |
| 32 | 0.73 | 1.36 | 2.55 | 3.67 | 4.75 | 5.81 | 7.86 | 9.86 | 10.8 | 12.8 | 14.7 | 16.5 | 18.4 |
| 35 | 0.81 | 1.50 | 2.81 | 4.04 | 5.24 | 6.40 | 8.66 | 10.9 | 11.9 | 14.1 | 16.2 | 18.2 | 20.3 |
| 40 | 0.93 | 1.74 | 3.24 | 4.67 | 6.05 | 7.39 | 10.0 | 12.5 | 13.8 | 16.3 | 18.7 | 21.1 | 23.4 |
| 45 | 1.06 | 1.97 | 3.68 | 5.30 | 6.87 | 8.40 | 11.4 | 14.2 | 15.7 | 18.5 | 21.2 | 23.9 | 26.6 |
| | Type A | | Type B | | | | | | | Type C | | | |

For use of table see page 1071.
*For lower or higher rpms see ANSI B29.1-1975.

**Table 15** *(Continued)*. **Horsepower Ratings for Roller Chain—1975**

½-inch Pitch Light Weight Machinery Roller Chain—No. 41

| No. of Teeth Small Spkt. | Revolutions per Minute—Small Sprocket* | | | | | | | | | | | | |
|---|---|---|---|---|---|---|---|---|---|---|---|---|
| | 10 | 25 | 50 | 100 | 200 | 300 | 400 | 500 | 700 | 900 | 1000 | 1200 | 1400 |
| | Horsepower Rating | | | | | | | | | | | | |
| 9 | 0.02 | 0.05 | 0.10 | 0.19 | 0.36 | 0.51 | 0.66 | 0.81 | 1.10 | 1.38 | 1.52 | 1.27 | 1.01 |
| 10 | 0.03 | 0.06 | 0.11 | 0.21 | 0.40 | 0.57 | 0.74 | 0.91 | 1.23 | 1.54 | 1.70 | 1.49 | 1.18 |
| 11 | 0.03 | 0.07 | 0.13 | 0.24 | 0.44 | 0.64 | 0.82 | 1.01 | 1.37 | 1.71 | 1.88 | 1.71 | 1.36 |
| 12 | 0.03 | 0.07 | 0.14 | 0.26 | 0.49 | 0.70 | 0.91 | 1.11 | 1.50 | 1.88 | 2.07 | 1.95 | 1.55 |
| 13 | 0.04 | 0.08 | 0.15 | 0.28 | 0.53 | 0.76 | 0.99 | 1.21 | 1.63 | 2.05 | 2.25 | 2.20 | 1.75 |
| 14 | 0.04 | 0.09 | 0.16 | 0.31 | 0.57 | 0.83 | 1.07 | 1.31 | 1.77 | 2.22 | 2.44 | 2.46 | 1.95 |
| 15 | 0.04 | 0.09 | 0.18 | 0.33 | 0.62 | 0.89 | 1.15 | 1.41 | 1.91 | 2.39 | 2.63 | 2.73 | 2.17 |
| 16 | 0.04 | 0.10 | 0.19 | 0.36 | 0.66 | 0.95 | 1.24 | 1.51 | 2.05 | 2.57 | 2.82 | 3.01 | 2.39 |
| 17 | 0.05 | 0.11 | 0.20 | 0.38 | 0.71 | 1.02 | 1.32 | 1.61 | 2.18 | 2.74 | 3.01 | 3.29 | 2.61 |
| 18 | 0.05 | 0.12 | 0.22 | 0.40 | 0.75 | 1.08 | 1.40 | 1.72 | 2.32 | 2.91 | 3.20 | 3.59 | 2.85 |
| 19 | 0.05 | 0.12 | 0.23 | 0.43 | 0.80 | 1.15 | 1.49 | 1.82 | 2.46 | 3.09 | 3.40 | 3.89 | 3.09 |
| 20 | 0.06 | 0.13 | 0.24 | 0.45 | 0.84 | 1.21 | 1.57 | 1.92 | 2.60 | 3.26 | 3.59 | 4.20 | 3.33 |
| 21 | 0.06 | 0.14 | 0.26 | 0.48 | 0.89 | 1.28 | 1.66 | 2.03 | 2.74 | 3.44 | 3.78 | 4.46 | 3.59 |
| 22 | 0.06 | 0.14 | 0.27 | 0.50 | 0.93 | 1.35 | 1.74 | 2.13 | 2.89 | 3.62 | 3.98 | 4.69 | 3.85 |
| 23 | 0.06 | 0.15 | 0.28 | 0.53 | 0.98 | 1.41 | 1.83 | 2.24 | 3.03 | 3.80 | 4.17 | 4.92 | 4.11 |
| 24 | 0.07 | 0.16 | 0.29 | 0.55 | 1.03 | 1.48 | 1.92 | 2.34 | 3.17 | 3.97 | 4.37 | 5.15 | 4.38 |
| 25 | 0.07 | 0.17 | 0.31 | 0.57 | 1.07 | 1.55 | 2.00 | 2.45 | 3.31 | 4.15 | 4.57 | 5.38 | 4.66 |
| 26 | 0.07 | 0.17 | 0.32 | 0.60 | 1.12 | 1.61 | 2.09 | 2.55 | 3.46 | 4.33 | 4.76 | 5.61 | 4.94 |
| 28 | 0.08 | 0.19 | 0.35 | 0.65 | 1.21 | 1.75 | 2.26 | 2.77 | 3.74 | 4.69 | 5.16 | 6.08 | 5.52 |
| 30 | 0.08 | 0.20 | 0.38 | 0.70 | 1.31 | 1.88 | 2.44 | 2.98 | 4.03 | 5.06 | 5.56 | 6.55 | 6.13 |
| 32 | 0.09 | 0.22 | 0.40 | 0.75 | 1.40 | 2.02 | 2.61 | 3.20 | 4.33 | 5.42 | 5.96 | 7.03 | 6.75 |
| 35 | 0.10 | 0.24 | 0.44 | 0.83 | 1.54 | 2.22 | 2.88 | 3.52 | 4.76 | 5.97 | 6.57 | 7.74 | 7.72 |
| 40 | 0.12 | 0.27 | 0.51 | 0.96 | 1.78 | 2.57 | 3.33 | 4.07 | 5.50 | 6.90 | 7.59 | 8.94 | 9.43 |
| 45 | 0.14 | 0.31 | 0.58 | 1.08 | 2.02 | 2.92 | 3.78 | 4.62 | 6.25 | 7.84 | 8.62 | 10.2 | 11.3 |
| | Type A | | | Type B | | | | | | | | | T'p C |

⅝-inch Pitch Standard Single-Strand Roller Chain—No. 50

| No. of Teeth Small Spkt. | Revolutions per Minute—Small Sprocket* | | | | | | | | | | | | |
|---|---|---|---|---|---|---|---|---|---|---|---|---|
| | 25 | 50 | 100 | 200 | 300 | 400 | 500 | 700 | 900 | 1000 | 1200 | 1400 | 1600 |
| | Horsepower Rating | | | | | | | | | | | | |
| 9 | 0.19 | 0.36 | 0.67 | 1.26 | 1.81 | 2.35 | 2.87 | 3.89 | 4.88 | 5.36 | 6.32 | 6.02 | 4.92 |
| 10 | 0.22 | 0.41 | 0.76 | 1.41 | 2.03 | 2.63 | 3.22 | 4.36 | 5.46 | 6.01 | 7.08 | 7.05 | 5.77 |
| 11 | 0.24 | 0.45 | 0.84 | 1.56 | 2.25 | 2.92 | 3.57 | 4.83 | 6.06 | 6.66 | 7.85 | 8.13 | 6.65 |
| 12 | 0.26 | 0.49 | 0.92 | 1.72 | 2.47 | 3.21 | 3.92 | 5.31 | 6.65 | 7.31 | 8.62 | 9.26 | 7.58 |
| 13 | 0.29 | 0.54 | 1.00 | 1.87 | 2.70 | 3.50 | 4.27 | 5.78 | 7.25 | 7.97 | 9.40 | 10.4 | 8.55 |
| 14 | 0.31 | 0.58 | 1.09 | 2.03 | 2.92 | 3.79 | 4.63 | 6.27 | 7.86 | 8.64 | 10.2 | 11.7 | 9.55 |
| 15 | 0.34 | 0.63 | 1.17 | 2.19 | 3.15 | 4.08 | 4.99 | 6.75 | 8.47 | 9.31 | 11.0 | 12.6 | 10.6 |
| 16 | 0.36 | 0.67 | 1.26 | 2.34 | 3.38 | 4.37 | 5.35 | 7.24 | 9.08 | 9.98 | 11.8 | 13.5 | 11.7 |
| 17 | 0.39 | 0.72 | 1.34 | 2.50 | 3.61 | 4.67 | 5.71 | 7.73 | 9.69 | 10.7 | 12.6 | 14.4 | 12.8 |
| 18 | 0.41 | 0.76 | 1.43 | 2.66 | 3.83 | 4.97 | 6.07 | 8.22 | 10.3 | 11.3 | 13.4 | 15.3 | 13.9 |
| 19 | 0.43 | 0.81 | 1.51 | 2.82 | 4.07 | 5.27 | 6.44 | 8.72 | 10.9 | 12.0 | 14.2 | 16.3 | 15.1 |
| 20 | 0.46 | 0.86 | 1.60 | 2.98 | 4.30 | 5.57 | 6.80 | 9.21 | 11.5 | 12.7 | 15.0 | 17.2 | 16.3 |
| 21 | 0.48 | 0.90 | 1.69 | 3.14 | 4.53 | 5.87 | 7.17 | 9.71 | 12.2 | 13.4 | 15.8 | 18.1 | 17.6 |
| 22 | 0.51 | 0.95 | 1.77 | 3.31 | 4.76 | 6.17 | 7.54 | 10.2 | 12.8 | 14.1 | 16.6 | 19.1 | 18.8 |
| 23 | 0.53 | 1.00 | 1.86 | 3.47 | 5.00 | 6.47 | 7.91 | 10.7 | 13.4 | 14.8 | 17.4 | 20.0 | 20.1 |
| 24 | 0.56 | 1.04 | 1.95 | 3.63 | 5.23 | 6.78 | 8.29 | 11.2 | 14.1 | 15.5 | 18.2 | 20.9 | 21.4 |
| 25 | 0.58 | 1.09 | 2.03 | 3.80 | 5.47 | 7.08 | 8.66 | 11.7 | 14.7 | 16.2 | 19.0 | 21.9 | 22.8 |
| 26 | 0.61 | 1.14 | 2.12 | 3.96 | 5.70 | 7.39 | 9.03 | 12.2 | 15.3 | 16.9 | 19.9 | 22.8 | 24.2 |
| 28 | 0.66 | 1.23 | 2.30 | 4.29 | 6.18 | 8.01 | 9.79 | 13.2 | 16.6 | 18.3 | 21.5 | 24.7 | 27.0 |
| 30 | 0.71 | 1.33 | 2.48 | 4.62 | 6.66 | 8.63 | 10.5 | 14.3 | 17.9 | 19.7 | 23.2 | 26.6 | 30.0 |
| 32 | 0.76 | 1.42 | 2.66 | 4.96 | 7.14 | 9.25 | 11.3 | 15.3 | 19.2 | 21.1 | 24.9 | 28.6 | 32.2 |
| 35 | 0.84 | 1.57 | 2.93 | 5.46 | 7.86 | 10.2 | 12.5 | 16.9 | 21.1 | 23.2 | 27.4 | 31.5 | 35.5 |
| 40 | 0.97 | 1.81 | 3.38 | 6.31 | 9.08 | 11.8 | 14.4 | 19.5 | 24.4 | 26.8 | 31.6 | 36.3 | 41.0 |
| 45 | 1.10 | 2.06 | 3.84 | 7.16 | 10.3 | 13.4 | 16.3 | 22.1 | 27.7 | 30.5 | 35.9 | 41.3 | 46.5 |
| | Type A | | | Type B | | | | | | Type C | | | |

For use of table see page 1071.
*For lower or higher rpms see ANSI B29.1-1975.

**Table 15** *(Continued)*. **Horsepower Ratings for Roller Chain—1975**

## ¾-inch Pitch Standard Single-Strand Roller Chain—No. 60

| No. of Teeth Small Spkt. | Revolutions per Minute—Small Sprocket* | | | | | | | | | | | | |
|---|---|---|---|---|---|---|---|---|---|---|---|---|
| | 25 | 50 | 100 | 150 | 200 | 300 | 400 | 500 | 600 | 700 | 800 | 900 | 1000 |
| | Horsepower Rating | | | | | | | | | | | | |
| 9 | 0.33 | 0.62 | 1.16 | 1.67 | 2.16 | 3.12 | 4.04 | 4.94 | 5.82 | 6.68 | 7.54 | 8.38 | 9.21 |
| 10 | 0.37 | 0.70 | 1.30 | 1.87 | 2.43 | 3.49 | 4.53 | 5.53 | 6.52 | 7.49 | 8.44 | 9.39 | 10.3 |
| 11 | 0.41 | 0.77 | 1.44 | 2.07 | 2.69 | 3.87 | 5.02 | 6.13 | 7.23 | 8.30 | 9.36 | 10.4 | 11.4 |
| 12 | 0.45 | 0.85 | 1.58 | 2.28 | 2.95 | 4.25 | 5.51 | 6.74 | 7.94 | 9.12 | 10.3 | 11.4 | 12.6 |
| 13 | 0.50 | 0.92 | 1.73 | 2.49 | 3.22 | 4.64 | 6.01 | 7.34 | 8.65 | 9.94 | 11.2 | 12.5 | 13.7 |
| 14 | 0.54 | 1.00 | 1.87 | 2.69 | 3.49 | 5.02 | 6.51 | 7.96 | 9.37 | 10.8 | 12.1 | 13.5 | 14.8 |
| 15 | 0.58 | 1.08 | 2.01 | 2.90 | 3.76 | 5.41 | 7.01 | 8.57 | 10.1 | 11.6 | 13.1 | 14.5 | 16.0 |
| 16 | 0.62 | 1.16 | 2.16 | 3.11 | 4.03 | 5.80 | 7.52 | 9.19 | 10.8 | 12.4 | 14.0 | 15.6 | 17.1 |
| 17 | 0.66 | 1.24 | 2.31 | 3.32 | 4.30 | 6.20 | 8.03 | 9.81 | 11.6 | 13.3 | 15.0 | 16.7 | 18.3 |
| 18 | 0.70 | 1.31 | 2.45 | 3.53 | 4.58 | 6.59 | 8.54 | 10.4 | 12.3 | 14.1 | 15.9 | 17.7 | 19.5 |
| 19 | 0.75 | 1.39 | 2.60 | 3.74 | 4.85 | 6.99 | 9.05 | 11.1 | 13.0 | 15.0 | 16.9 | 18.8 | 20.6 |
| 20 | 0.79 | 1.47 | 2.75 | 3.96 | 5.13 | 7.38 | 9.57 | 11.7 | 13.8 | 15.8 | 17.9 | 19.8 | 21.8 |
| 21 | 0.83 | 1.55 | 2.90 | 4.17 | 5.40 | 7.78 | 10.1 | 12.3 | 14.5 | 16.7 | 18.8 | 20.9 | 23.0 |
| 22 | 0.87 | 1.63 | 3.05 | 4.39 | 5.68 | 8.19 | 10.6 | 13.0 | 15.3 | 17.5 | 19.8 | 22.0 | 24.2 |
| 23 | 0.92 | 1.71 | 3.19 | 4.60 | 5.96 | 8.59 | 11.1 | 13.6 | 16.0 | 18.4 | 20.8 | 23.1 | 25.4 |
| 24 | 0.96 | 1.79 | 3.35 | 4.82 | 6.24 | 8.99 | 11.6 | 14.2 | 16.8 | 19.3 | 21.7 | 24.2 | 26.6 |
| 25 | 1.00 | 1.87 | 3.50 | 5.04 | 6.52 | 9.40 | 12.2 | 14.9 | 17.5 | 20.1 | 22.7 | 25.3 | 27.8 |
| 26 | 1.05 | 1.95 | 3.65 | 5.25 | 6.81 | 9.80 | 12.7 | 15.5 | 18.3 | 21.0 | 23.7 | 26.4 | 29.0 |
| 28 | 1.13 | 2.12 | 3.95 | 5.69 | 7.37 | 10.6 | 13.8 | 16.8 | 19.8 | 22.8 | 25.7 | 28.5 | 31.4 |
| 30 | 1.22 | 2.28 | 4.26 | 6.13 | 7.94 | 11.4 | 14.8 | 18.1 | 21.4 | 24.5 | 27.7 | 30.8 | 33.8 |
| 32 | 1.31 | 2.45 | 4.56 | 6.57 | 8.52 | 12.3 | 15.9 | 19.4 | 22.9 | 26.3 | 29.7 | 33.0 | 36.3 |
| 35 | 1.44 | 2.69 | 5.03 | 7.24 | 9.38 | 13.5 | 17.5 | 21.4 | 25.2 | 29.0 | 32.7 | 36.3 | 39.9 |
| 40 | 1.67 | 3.11 | 5.81 | 8.37 | 10.8 | 15.6 | 20.2 | 24.7 | 29.1 | 33.5 | 37.7 | 42.0 | 46.1 |
| 45 | 1.89 | 3.53 | 6.60 | 9.50 | 12.3 | 17.7 | 23.0 | 28.1 | 33.1 | 38.0 | 42.9 | 47.7 | 52.4 |
| | Type A | | Type B | | | | | | | | Type C | | |

## 1-inch Pitch Standard Single-Strand Roller Chain—No. 80

| No. of Teeth Small Spkt. | Revolutions per Minute—Small Sprocket* | | | | | | | | | | | | |
|---|---|---|---|---|---|---|---|---|---|---|---|---|
| | 25 | 50 | 100 | 150 | 200 | 300 | 400 | 500 | 600 | 700 | 800 | 900 | 1000 |
| | Horsepower Rating | | | | | | | | | | | | |
| 9 | 0.78 | 1.45 | 2.71 | 3.90 | 5.05 | 7.28 | 9.43 | 11.5 | 13.6 | 15.6 | 17.6 | 17.0 | 14.5 |
| 10 | 0.87 | 1.63 | 3.03 | 4.37 | 5.66 | 8.16 | 10.6 | 12.9 | 15.2 | 17.5 | 19.7 | 19.9 | 17.0 |
| 11 | 0.97 | 1.80 | 3.36 | 4.84 | 6.28 | 9.04 | 11.7 | 14.3 | 16.9 | 19.4 | 21.9 | 23.0 | 19.6 |
| 12 | 1.06 | 1.98 | 3.69 | 5.32 | 6.89 | 9.93 | 12.9 | 15.7 | 18.5 | 21.3 | 24.0 | 26.2 | 22.3 |
| 13 | 1.16 | 2.16 | 4.03 | 5.80 | 7.52 | 10.8 | 14.0 | 17.1 | 20.2 | 23.2 | 26.2 | 29.1 | 25.2 |
| 14 | 1.25 | 2.34 | 4.36 | 6.29 | 8.14 | 11.7 | 15.2 | 18.6 | 21.9 | 25.1 | 28.4 | 31.5 | 28.2 |
| 15 | 1.35 | 2.52 | 4.70 | 6.77 | 8.77 | 12.6 | 16.4 | 20.0 | 23.6 | 27.1 | 30.6 | 34.0 | 31.2 |
| 16 | 1.45 | 2.70 | 5.04 | 7.26 | 9.41 | 13.5 | 17.6 | 21.5 | 25.3 | 29.0 | 32.8 | 36.4 | 34.4 |
| 17 | 1.55 | 2.88 | 5.38 | 7.75 | 10.0 | 14.5 | 18.7 | 22.9 | 27.0 | 31.0 | 35.0 | 38.9 | 37.7 |
| 18 | 1.64 | 3.07 | 5.72 | 8.25 | 10.7 | 15.4 | 19.9 | 24.4 | 28.7 | 33.0 | 37.2 | 41.4 | 41.1 |
| 19 | 1.74 | 3.25 | 6.07 | 8.74 | 11.3 | 16.3 | 21.1 | 25.8 | 30.4 | 35.0 | 39.4 | 43.8 | 44.5 |
| 20 | 1.84 | 3.44 | 6.41 | 9.24 | 12.0 | 17.2 | 22.3 | 27.3 | 32.2 | 37.0 | 41.7 | 46.3 | 48.1 |
| 21 | 1.94 | 3.62 | 6.76 | 9.74 | 12.6 | 18.2 | 23.5 | 28.8 | 33.9 | 39.0 | 43.9 | 48.9 | 51.7 |
| 22 | 2.04 | 3.81 | 7.11 | 10.2 | 13.3 | 19.1 | 24.8 | 30.3 | 35.7 | 41.0 | 46.2 | 51.4 | 55.5 |
| 23 | 2.14 | 4.00 | 7.46 | 10.7 | 13.9 | 20.1 | 26.0 | 31.8 | 37.4 | 43.0 | 48.5 | 53.9 | 59.3 |
| 24 | 2.24 | 4.19 | 7.81 | 11.3 | 14.6 | 21.0 | 27.2 | 33.2 | 39.2 | 45.0 | 50.8 | 56.4 | 62.0 |
| 25 | 2.34 | 4.37 | 8.16 | 11.8 | 15.2 | 21.9 | 28.4 | 34.7 | 40.9 | 47.0 | 53.0 | 59.0 | 64.8 |
| 26 | 2.45 | 4.56 | 8.52 | 12.3 | 15.9 | 22.9 | 29.7 | 36.2 | 42.7 | 49.1 | 55.3 | 61.5 | 67.6 |
| 28 | 2.65 | 4.94 | 9.23 | 13.3 | 17.2 | 24.8 | 32.1 | 39.3 | 46.3 | 53.2 | 59.9 | 66.7 | 73.3 |
| 30 | 2.85 | 5.33 | 9.94 | 14.3 | 18.5 | 26.7 | 34.6 | 42.3 | 49.9 | 57.3 | 64.6 | 71.8 | 78.9 |
| 32 | 3.06 | 5.71 | 10.7 | 15.3 | 19.9 | 28.6 | 37.1 | 45.4 | 53.5 | 61.4 | 69.2 | 77.0 | 84.6 |
| 35 | 3.37 | 6.29 | 11.7 | 16.9 | 21.9 | 31.6 | 40.9 | 50.0 | 58.9 | 67.6 | 76.3 | 84.8 | 93.3 |
| 40 | 3.89 | 7.27 | 13.6 | 19.5 | 25.3 | 36.4 | 47.2 | 57.7 | 68.0 | 78.1 | 88.7 | 99.0 | 108 |
| 45 | 4.42 | 8.25 | 15.4 | 22.2 | 28.7 | 41.4 | 53.6 | 65.6 | 77.2 | 88.7 | 100 | 111 | 122 |
| | T'p A | | Type B | | | | | | | Type C | | | |

For use of table see page 1071.
*For lower or higher rpms see ANSI B29.1-1975.

**Table 15** *(Concluded).* **Horsepower Ratings for Roller Chain—1975**

### 1¼-inch Pitch Standard Single-Strand Roller Chain—No. 100

| No. of Teeth Small Spkt. | Revolutions per Minute—Small Sprocket* | | | | | | | | | | | | |
|---|---|---|---|---|---|---|---|---|---|---|---|---|---|
| | 10 | 25 | 50 | 100 | 150 | 200 | 300 | 400 | 500 | 600 | 700 | 800 | 900 |
| | Horsepower Rating | | | | | | | | | | | | |
| 9 | 0.65 | 1.49 | 2.78 | 5.19 | 7.47 | 9.68 | 13.9 | 18.1 | 22.1 | 26.0 | 29.6 | 24.2 | 20.3 |
| 10 | 0.73 | 1.67 | 3.11 | 5.81 | 8.37 | 10.8 | 15.6 | 20.2 | 24.7 | 29.2 | 33.5 | 28.4 | 23.8 |
| 11 | 0.81 | 1.85 | 3.45 | 6.44 | 9.28 | 12.0 | 17.3 | 22.4 | 27.4 | 32.3 | 37.1 | 32.8 | 27.5 |
| 12 | 0.89 | 2.03 | 3.79 | 7.08 | 10.2 | 13.2 | 19.0 | 24.6 | 30.1 | 35.5 | 40.8 | 37.3 | 31.3 |
| 13 | 0.97 | 2.22 | 4.13 | 7.72 | 11.1 | 14.4 | 20.7 | 26.9 | 32.8 | 38.7 | 44.5 | 42.1 | 35.3 |
| 14 | 1.05 | 2.40 | 4.48 | 8.36 | 12.0 | 15.6 | 22.5 | 29.1 | 35.6 | 41.9 | 48.2 | 47.0 | 39.4 |
| 15 | 1.13 | 2.59 | 4.83 | 9.01 | 13.0 | 16.8 | 24.2 | 31.4 | 38.3 | 45.2 | 51.9 | 52.2 | 43.7 |
| 16 | 1.22 | 2.77 | 5.17 | 9.66 | 13.9 | 18.0 | 26.0 | 33.6 | 41.1 | 48.4 | 55.6 | 57.5 | 48.2 |
| 17 | 1.30 | 2.96 | 5.52 | 10.3 | 14.8 | 19.2 | 27.7 | 35.9 | 43.9 | 51.7 | 59.4 | 63.0 | 52.8 |
| 18 | 1.38 | 3.15 | 5.88 | 11.0 | 15.8 | 20.5 | 29.5 | 38.2 | 46.7 | 55.0 | 63.2 | 68.6 | 57.5 |
| 19 | 1.46 | 3.34 | 6.23 | 11.6 | 16.7 | 21.7 | 31.2 | 40.5 | 49.5 | 58.3 | 67.0 | 74.4 | 62.3 |
| 20 | 1.55 | 3.53 | 6.58 | 12.3 | 17.7 | 22.9 | 33.0 | 42.8 | 52.3 | 61.6 | 70.8 | 79.8 | 67.3 |
| 21 | 1.63 | 3.72 | 6.94 | 13.0 | 18.7 | 24.2 | 34.8 | 45.1 | 55.1 | 65.0 | 74.6 | 84.2 | 72.4 |
| 22 | 1.71 | 3.91 | 7.30 | 13.6 | 19.6 | 25.4 | 36.6 | 47.4 | 58.0 | 68.3 | 78.5 | 88.5 | 77.7 |
| 23 | 1.80 | 4.10 | 7.66 | 14.3 | 20.6 | 26.7 | 38.4 | 49.8 | 60.8 | 71.7 | 82.3 | 92.8 | 83.0 |
| 24 | 1.88 | 4.30 | 8.02 | 15.0 | 21.5 | 27.9 | 40.2 | 52.1 | 63.7 | 75.0 | 86.2 | 97.2 | 88.5 |
| 25 | 1.97 | 4.49 | 8.38 | 15.6 | 22.5 | 29.2 | 42.0 | 54.4 | 66.6 | 78.4 | 90.1 | 102 | 94.1 |
| 26 | 2.05 | 4.68 | 8.74 | 16.3 | 23.5 | 30.4 | 43.8 | 56.8 | 69.4 | 81.8 | 94.0 | 106 | 99.8 |
| 28 | 2.22 | 5.07 | 9.47 | 17.7 | 25.5 | 33.0 | 47.5 | 61.5 | 75.2 | 88.6 | 102 | 115 | 112 |
| 30 | 2.40 | 5.47 | 10.2 | 19.0 | 27.4 | 35.5 | 51.2 | 66.3 | 81.0 | 95.5 | 110 | 124 | 124 |
| 32 | 2.57 | 5.86 | 10.9 | 20.4 | 29.4 | 38.1 | 54.9 | 71.1 | 86.9 | 102 | 118 | 133 | 136 |
| 35 | 2.83 | 6.46 | 12.0 | 22.5 | 32.4 | 42.0 | 60.4 | 78.3 | 95.7 | 113 | 130 | 146 | 156 |
| 40 | 3.27 | 7.46 | 13.9 | 26.0 | 37.4 | 48.5 | 69.8 | 90.4 | 111 | 130 | 150 | 169 | 188 |
| 45 | 3.71 | 8.47 | 15.8 | 29.5 | 42.5 | 55.0 | 79.3 | 103 | 126 | 148 | 170 | 192 | 213 |
| | Type A | | Type B | | | | | Type C | | | | | |

### 1½-inch Pitch Standard Single-Strand Roller Chain—No. 120

| No. of Teeth Small Spkt. | Revolutions per Minute—Small Sprocket* | | | | | | | | | | | | |
|---|---|---|---|---|---|---|---|---|---|---|---|---|---|
| | 10 | 25 | 50 | 100 | 150 | 200 | 300 | 400 | 500 | 600 | 700 | 800 | 900 |
| | Horsepower Rating | | | | | | | | | | | | |
| 9 | 1.10 | 2.52 | 4.69 | 8.76 | 12.6 | 16.3 | 23.5 | 30.5 | 37.3 | 43.2 | 34.3 | 28.1 | 23.5 |
| 10 | 1.24 | 2.82 | 5.26 | 9.81 | 14.1 | 18.3 | 26.4 | 34.2 | 41.8 | 49.2 | 40.1 | 32.9 | 27.5 |
| 11 | 1.37 | 3.12 | 5.83 | 10.9 | 15.7 | 20.3 | 29.2 | 37.9 | 46.3 | 54.6 | 46.3 | 37.9 | 31.8 |
| 12 | 1.50 | 3.43 | 6.40 | 11.9 | 17.2 | 22.3 | 32.1 | 41.6 | 50.9 | 59.9 | 52.8 | 43.2 | 36.2 |
| 13 | 1.64 | 3.74 | 6.98 | 13.0 | 18.8 | 24.3 | 35.0 | 45.4 | 55.5 | 65.3 | 59.5 | 48.7 | 40.8 |
| 14 | 1.78 | 4.05 | 7.56 | 14.1 | 20.3 | 26.3 | 37.9 | 49.1 | 60.1 | 70.8 | 66.5 | 54.4 | 45.6 |
| 15 | 1.91 | 4.37 | 8.15 | 15.2 | 21.9 | 28.4 | 40.9 | 53.0 | 64.7 | 76.3 | 73.8 | 60.4 | 50.6 |
| 16 | 2.05 | 4.68 | 8.74 | 16.3 | 23.5 | 30.4 | 43.8 | 56.8 | 69.4 | 81.8 | 81.3 | 66.5 | 55.7 |
| 17 | 2.19 | 5.00 | 9.33 | 17.4 | 25.1 | 32.5 | 46.8 | 60.6 | 74.1 | 87.3 | 89.0 | 72.8 | 61.0 |
| 18 | 2.33 | 5.32 | 9.92 | 18.5 | 26.7 | 34.6 | 49.8 | 64.5 | 78.8 | 92.9 | 97.0 | 79.4 | 66.5 |
| 19 | 2.47 | 5.64 | 10.5 | 19.6 | 28.3 | 36.6 | 52.8 | 68.4 | 83.6 | 98.5 | 105 | 86.1 | 72.1 |
| 20 | 2.61 | 5.96 | 11.1 | 20.7 | 29.9 | 38.7 | 55.8 | 72.2 | 88.3 | 104 | 114 | 92.9 | 77.9 |
| 21 | 2.75 | 6.28 | 11.7 | 21.9 | 31.5 | 40.8 | 58.8 | 76.2 | 93.1 | 110 | 122 | 100 | 83.8 |
| 22 | 2.90 | 6.60 | 12.3 | 23.0 | 33.1 | 42.9 | 61.8 | 80.1 | 97.9 | 115 | 131 | 107 | 89.9 |
| 23 | 3.04 | 6.93 | 12.9 | 24.1 | 34.8 | 45.0 | 64.9 | 84.0 | 103 | 121 | 139 | 115 | 96.1 |
| 24 | 3.18 | 7.25 | 13.5 | 25.3 | 36.4 | 47.1 | 67.9 | 88.0 | 108 | 127 | 146 | 122 | 102 |
| 25 | 3.32 | 7.58 | 14.1 | 26.4 | 38.0 | 49.3 | 71.0 | 91.9 | 112 | 132 | 152 | 130 | 109 |
| 26 | 3.47 | 7.91 | 14.8 | 27.5 | 39.7 | 51.4 | 74.0 | 95.9 | 117 | 138 | 159 | 138 | 115 |
| 28 | 3.76 | 8.57 | 16.0 | 29.8 | 43.0 | 55.7 | 80.2 | 104 | 127 | 150 | 172 | 154 | 129 |
| 30 | 4.05 | 9.23 | 17.2 | 32.1 | 46.3 | 60.0 | 86.4 | 112 | 137 | 161 | 185 | 171 | 143 |
| 32 | 4.34 | 9.90 | 18.5 | 34.5 | 49.6 | 64.3 | 92.6 | 120 | 147 | 173 | 199 | 188 | 158 |
| 35 | 4.78 | 10.9 | 20.3 | 38.0 | 54.7 | 70.9 | 102 | 132 | 162 | 190 | 219 | 215 | 180 |
| 40 | 5.52 | 12.6 | 23.5 | 43.9 | 63.2 | 81.8 | 118 | 153 | 187 | 220 | 253 | ... | ... |
| 45 | 6.27 | 14.3 | 26.7 | 49.8 | 71.7 | 92.9 | 134 | 173 | 212 | 250 | 287 | ... | ... |
| | T'p A | | Type B | | | | | Type C | | | | | |

For use of table see page 1071.
*For higher rpms and larger chain sizes see ANSI B29.1-1975.

**Example of Roller Chain Drive Design Procedure** — The selection of a roller chain and sprockets for a specific design requirement is best accomplished by a systematic step-by-step procedure such as is used in the following example.

*Example:* Select a roller chain drive to transmit 10 horsepower from a counter-shaft to the main shaft of a wire drawing machine. The countershaft is 1 15/16-inches diameter and operates at 1000 rpm. The main shaft is also 1 15/16-inches diameter and must operate between 378 and 382 rpm. Shaft centers, once established, are fixed and by initial calculations must be approximately 22½ inches. The load on the main shaft is uneven and presents "peaks" which places it in the heavy shock load category. The input power is supplied by an electric motor. The driving head is fully enclosed and all parts are lubricated from a central system.

*Step 1. Service Factor:* From Table 14 the service factor for heavy shock load and an electric motor drive is 1.5.

*Step 2. Design Horsepower.* The horsepower upon which the chain selection is based (design horsepower) is equal to the specified horsepower multiplied by the service factor, 10 × 1.5 = 15 hp.

*Step 3. Chain Pitch and Small Sprocket Size for Single-Strand Drive:* In Table 15 under 1000 rpm, a ⅝-inch pitch chain with a 24-tooth sprocket or a ¾-inch pitch chain with a 15-tooth sprocket are possible choices.

*Step 4. Check of Chain Pitch and Sprocket Selection:* From Table 9 it is seen that only the 24-tooth sprocket in Step 3 can be bored to fit the 1 15/16-inch diameter main shaft. In Table 15 a ⅝-pitch chain at a small sprocket speed of 1000 rpm is rated at 15.5 hp for a 24-tooth sprocket.

*Step 5. Selection of Large Sprocket:* Since the driver is to operate at 1000 rpm and the driven at a minimum of 378 rpm, the speed ratio 1000/378 = 2.646. Therefore the large sprocket should have 24 × 2.646 = 63.5 (use 63) teeth.

This combination of 24 and 63 teeth will produce a main drive shaft speed of 381 rpm which is within the limitation of 378 to 382 rpm established in the original specification.

*Step 6. Computation of Chain Length:* Since the 24- and 63-tooth sprockets are to be placed on 22½-inch centers, the chain length is determined from the formula:

$$L = 2C + \frac{N}{2} + \frac{n}{2} + \left(\frac{N-n}{2\pi}\right)^2 \times \frac{1}{C}$$

where $L$ = chain length in pitches; $C$ = shaft center distance in pitches; $N$ = number of teeth in large sprocket; and $n$ = number of teeth in small sprocket.

$$L = 2 \times 36 + \frac{63+24}{2} + \left(\frac{63-24}{6.28}\right)^2 \times \frac{1}{36} = 116.57 \text{ pitches}$$

*Step 7: Correction of Center Distance:* Since the chain is to couple at a whole number of pitches, 116 pitches will be used and the center distance recomputed based on this figure using the formula on page **1063** where $c$ is the center distance in inches and $P$ is the pitch.

$$c = \frac{P}{8}\left(2L - N - n + \sqrt{(2L - N - n)^2 - 0.810(N-n)^2}\right)$$

$$c = \frac{5}{64}\left(2 \times 116 - 63 - 24 + \sqrt{(2 \times 116 - 63 - 24)^2 - 0.810(63-24)^2}\right)$$

$$c = \frac{5}{64}(145 + 140.69) = 22.32 \text{ inches, say } 22\tfrac{3}{8} \text{ inches.}$$

## Silent or Inverted Tooth Chain

Silent or inverted tooth chain consists of a series of toothed links alternately assembled either with pins or with a combination of joint components in such a way that the joints articulate between adjoining pitches. *Side Guide* chain has guide links which engage the sides of the sprocket. *Center Guide* chain has guide links which engage a groove or grooves in the sprocket.

**Characteristics of Silent Chain Drives.** — The silent or "inverted-tooth" driving chain has the following characteristics: The chain passes over the face of the wheel like a belt and the wheel teeth do not project through it; the chain engages the wheel by means of teeth extending across the full width of the under side, with the exception of those chains having a central guide link; the chain teeth and wheel teeth are of such a shape that as the chain pitch increases through wear at the joints, the chain shifts outward upon the teeth, thus engaging the wheel on a pitch circle of increasing diameter; the result of this action is that the pitch of the wheel teeth increases at the same rate as the chain pitch. The accompanying illustration shows an unworn chain to the left, and a worn chain to the right, which

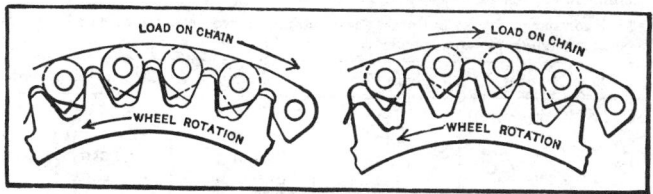

has moved outward as the result of wear. Another distinguishing feature of the silent chain is that the power is transmitted by and to all the teeth in the arc of contact, irrespective of the increasing pitch due to elongation. The links have no sliding action either on or off the teeth, which results in a smooth and practically noiseless action, the chain being originally designed for the transmission of power at higher speeds than are suitable for roller chains. The efficiency of the silent chain itself may be as high as 99 per cent, and for the complete drive, from 96 to 97 per cent, under favorable conditions; from 94 to 96 per cent can be secured with well-designed drives under average conditions.

The life and upkeep of silent chains depend largely upon the design of the entire drive, including the provision for adjustment. If there is much slack, the whipping of the chain will greatly increase the wear, and means of adjustment may double the life of the chain. A slight amount of play is necessary for satisfactory operation. The minimum amount of sag should be about ⅛ inch. Although the silent chain shifts outward from the teeth and adjusts itself for an increase of pitch, it cannot take up the increased pitch in that portion of the chain between the wheels; therefore, the wheel must lag to the extent of the increased pitch in the straight portion of the chain.

**Standard Silent Chain Designation.** — The standard chain number or designation consists of: (1) a two letter symbol SC; (2) one or two numerical digits indicating the pitch in eighths of an inch; and (3) two or three numerical digits indicating the chain width in quarter-inches. Thus, SC302 designates a silent chain of ⅜-inch pitch and ½-inch width, while SC1012 designates a silent chain of 1¼-inch pitch and 3-inch width.

**Silent Chain Links.** — The joint components and link contours vary with each manufacturer's design. As shown in Table 1 minimum crotch height and pitch have been standardized for interchangeability. Chain links are stamped to indicate the pitch as, for example, SC6 or simply 6. Chain link designations are given in Table 1.

**Table 1. American National Standard Silent Chain Links** (ANSI B29.2-1957, R1971)

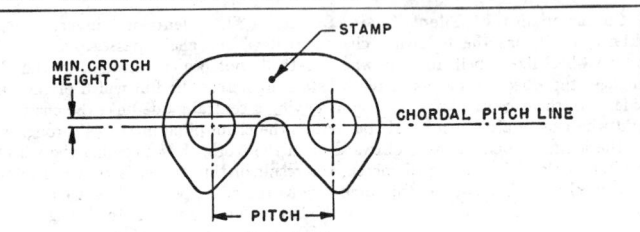

Min. Crotch Height = 0.062 × Chain Pitch.

Link contour may vary but must engage standard sprocket tooth so that joint centers lie on pitch diameter of sprockets.

| Chain Number | | Chain Pitch | Stamp | Crotch Height, Min. |
|---|---|---|---|---|
| SC3 (Width in ¼ in.) | | ⅜ | SC3 or 3 | 0.0232 |
| SC4 | " | ½ | SC4 or 4 | 0.0310 |
| SC5 | " | ⅝ | SC5 or 5 | 0.0388 |
| SC6 | " | ¾ | SC6 or 6 | 0.0465 |
| SC8 | " | 1 | SC8 or 8 | 0.0620 |
| SC10 | " | 1¼ | SC10 or 10 | 0.0775 |
| SC12 | " | 1½ | SC12 or 12 | 0.0930 |
| SC16 | " | 2 | SC16 or 16 | 0.1240 |

**Silent Chain Sprocket Diameters.** — The important sprocket diameters are: (1) outside diameter; (2) pitch diameter; (3) maximum guide groove diameter; and (4) over-pin diameter. These are shown in the diagram in Table 2 and the symbols and formulas for each are also given in this table. Table 3 gives values of outside diameters for sprockets with rounded teeth and with square teeth, pitch diameters, and over-pin diameters for chains of 1-inch pitch and sprockets of various tooth numbers. Values for chains of other pitches (⅜ inch and larger) are found by multiplying the values shown by the pitch. It will be noted that the over-pin diameter is measured over gage pins having a diameter $D_p = 0.625 \times$ chain pitch. Over-pin diameter tolerances are given in Table 4. The diameters of blanks for sprockets with rounded teeth should be 0.020 inch larger than the finished outside diameters shown in the table.

**Silent Chain Sprocket Profiles and Chain Widths.** — Sprocket tooth face profiles for side guide chain, center guide chain and double guide chain are shown in Table 5 together with important dimensions for chains of various pitches and widths. Maximum over-all width $M$ of the three types of chain are also given in this table for various pitches and widths. It should be noted that the sprocket tooth width $W$ for the side guide chain is given in Table 5 for one-half-inch wide chains of ⅜-inch and ½-inch pitches. No values of $W$ for other chain sizes are specified in American National Standard B29.2-1957 (R1971).

Table 2. **ANSI Silent Chain Sprocket Diameters** (ANSI B29.2-1957, R1971)

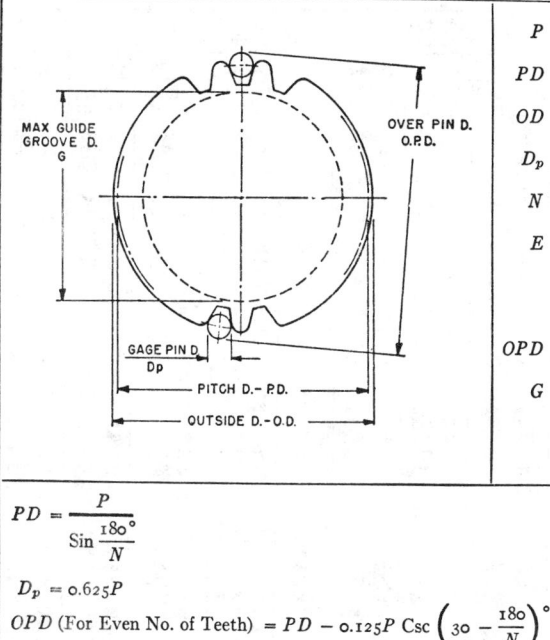

$P$ = Chain Pitch
$PD$ = Pitch Diameter
$OD$ = Outside Diameter
$D_p$ = Gage Pin Diameter
$N$ = Number of Teeth
$E$ = Diameter to Center of Topping Curve
$OPD$ = Over Pin Diameter
$G$ = Max. Guide Groove Diameter

$$PD = \frac{P}{\sin\dfrac{180^\circ}{N}}$$

$$D_p = 0.625P$$

$OPD$ (For Even No. of Teeth) $= PD - 0.125P \operatorname{Csc}\left(30 - \dfrac{180}{N}\right)^\circ + 0.625P$

$OPD$ (For Odd No. of Teeth) =

$$\cos\frac{90^\circ}{N}\left[PD - 0.125P \operatorname{Csc}\left(30 - \frac{180}{N}\right)^\circ\right] + 0.625P$$

$OD$ (For Rounded Teeth) $= P\left(\cot\dfrac{180^\circ}{N} + 0.08\right)$

$OD$ (For Square Teeth) $= 2\sqrt{X^2 + L^2 - 2XL \cos\alpha}$

Where
$$\begin{cases}
X = Y \cos\alpha - \sqrt{(0.15P)^2 - (Y \sin\alpha)^2} \\[2mm]
Y = P(0.500 - 0.375 \sec\alpha)\cot\alpha + 0.11P \\[2mm]
L = Y + \dfrac{E}{2} \text{ (See Table 8 for } E\text{)} \\[2mm]
\alpha = \left(30 - \dfrac{360}{N}\right)^\circ
\end{cases}$$

$G$ (max.) $= P\left(\cot\dfrac{180^\circ}{N} - 1.16\right)$     Tolerance = +0, −1/32 inch.

# SILENT CHAIN

### Table 3. American National Standard Silent Chain Sprocket Diameters
(ANSI B29.2-1957, R1971)

These diameters apply only to chains of 1-inch pitch. For any other pitch (⅜ inch and larger) multiply the values given below by the pitch.

| No. Teeth | Pitch Diameter | Outside Diameter | | Over-Pin Dia.†† | No. Teeth | Pitch Diameter | Outside Diameter | | Over-Pin Dia.†† |
|---|---|---|---|---|---|---|---|---|---|
| | | Rounded Teeth* | Square Teeth† | | | | Rounded Teeth* | Square Teeth† | |
| 17 | 5.442 | 5.429 | 5.298 | 5.669 | 71 | 22.607 | 22.665 | 22.622 | 22.955 |
| 18 | 5.759 | 5.751 | 5.623 | 6.018 | 72 | 22.926 | 22.984 | 22.941 | 23.280 |
| 19 | 6.076 | 6.072 | 5.947 | 6.324 | 73 | 23.244 | 23.302 | 23.259 | 23.593 |
| 20 | 6.393 | 6.393 | 6.271 | 6.669 | 74 | 23.562 | 23.621 | 23.578 | 23.917 |
| 21 | 6.710 | 6.714 | 6.595 | 6.974 | 75 | 23.880 | 23.939 | 23.897 | 24.230 |
| 22 | 7.027 | 7.036 | 6.919 | 7.315 | 76 | 24.198 | 24.257 | 24.216 | 24.553 |
| 23 | 7.344 | 7.356 | 7.243 | 7.621 | 77 | 24.517 | 24.577 | 24.535 | 24.868 |
| 24 | 7.661 | 7.675 | 7.568 | 7.960 | 78 | 24.835 | 24.895 | 24.853 | 25.191 |
| 25 | 7.979 | 7.996 | 7.890 | 8.266 | 79 | 25.153 | 25.213 | 25.172 | 25.504 |
| 26 | 8.296 | 8.315 | 8.213 | 8.602 | 80 | 25.471 | 25.531 | 25.491 | 25.828 |
| 27 | 8.614 | 8.636 | 8.536 | 8.909 | 81 | 25.790 | 25.851 | 25.809 | 26.141 |
| 28 | 8.932 | 8.956 | 8.859 | 9.244 | 82 | 26.108 | 26.169 | 26.128 | 26.465 |
| 29 | 9.249 | 9.275 | 9.181 | 9.551 | 83 | 26.426 | 26.487 | 26.447 | 26.778 |
| 30 | 9.567 | 9.595 | 9.504 | 9.884 | 84 | 26.744 | 26.805 | 26.766 | 27.101 |
| 31 | 9.885 | 9.913 | 9.828 | 10.192 | 85 | 27.063 | 27.125 | 27.084 | 27.415 |
| 32 | 10.202 | 10.233 | 10.150 | 10.524 | 86 | 27.381 | 27.443 | 27.403 | 27.739 |
| 33 | 10.520 | 10.553 | 10.471 | 10.833 | 87 | 27.699 | 27.761 | 27.722 | 28.052 |
| 34 | 10.838 | 10.872 | 10.793 | 11.164 | 88 | 28.017 | 28.079 | 28.040 | 28.375 |
| 35 | 11.156 | 11.191 | 11.115 | 11.472 | 89 | 28.335 | 28.397 | 28.359 | 28.689 |
| 36 | 11.474 | 11.510 | 11.437 | 11.803 | 90 | 28.654 | 28.716 | 28.678 | 29.013 |
| 37 | 11.792 | 11.829 | 11.757 | 12.112 | 91 | 28.972 | 29.035 | 28.997 | 29.327 |
| 38 | 12.110 | 12.149 | 12.077 | 12.442 | 92 | 29.290 | 29.353 | 29.315 | 29.649 |
| 39 | 12.428 | 12.468 | 12.397 | 12.751 | 93 | 29.608 | 29.671 | 29.634 | 29.963 |
| 40 | 12.746 | 12.787 | 12.717 | 13.080 | 94 | 29.926 | 29.989 | 29.953 | 30.285 |
| 41 | 13.064 | 13.106 | 13.037 | 13.390 | 95 | 30.245 | 30.308 | 30.271 | 30.601 |
| 42 | 13.382 | 13.425 | 13.357 | 13.718 | 96 | 30.563 | 30.627 | 30.590 | 30.923 |
| 43 | 13.700 | 13.743 | 13.677 | 14.028 | 97 | 30.881 | 30.945 | 30.909 | 31.237 |
| 44 | 14.018 | 14.062 | 13.997 | 14.356 | 98 | 31.199 | 31.263 | 31.228 | 31.559 |
| 45 | 14.336 | 14.381 | 14.317 | 14.667 | 99 | 31.518 | 31.582 | 31.546 | 31.874 |
| 46 | 14.654 | 14.700 | 14.637 | 14.994 | 100 | 31.836 | 31.900 | 31.865 | 32.196 |
| 47 | 14.972 | 15.018 | 14.957 | 15.305 | 101 | 32.154 | 32.218 | 32.183 | 32.511 |
| 48 | 15.290 | 15.337 | 15.277 | 15.632 | 102 | 32.473 | 32.537 | 32.502 | 32.834 |
| 49 | 15.608 | 15.656 | 15.597 | 15.943 | 103 | 32.791 | 32.856 | 32.820 | 33.148 |
| 50 | 15.926 | 15.975 | 15.917 | 16.270 | 104 | 33.109 | 33.174 | 33.139 | 33.470 |
| 51 | 16.244 | 16.293 | 16.236 | 16.581 | 105 | 33.427 | 33.492 | 33.457 | 33.784 |
| 52 | 16.562 | 16.612 | 16.556 | 16.907 | 106 | 33.746 | 33.811 | 33.776 | 34.107 |
| 53 | 16.880 | 16.930 | 16.876 | 17.218 | 107 | 34.064 | 34.129 | 34.094 | 34.422 |
| 54 | 17.198 | 17.249 | 17.196 | 17.544 | 108 | 34.382 | 34.447 | 34.413 | 34.744 |
| 55 | 17.517 | 17.568 | 17.515 | 17.857 | 109 | 34.701 | 34.767 | 34.731 | 35.059 |
| 56 | 17.835 | 17.887 | 17.834 | 18.183 | 110 | 35.019 | 35.084 | 35.050 | 35.381 |
| 57 | 18.153 | 18.205 | 18.154 | 18.494 | 111 | 35.337 | 35.403 | 35.368 | 35.695 |
| 58 | 18.471 | 18.524 | 18.473 | 18.820 | 112 | 35.655 | 35.721 | 35.687 | 36.017 |
| 59 | 18.789 | 18.842 | 18.793 | 19.131 | 113 | 35.974 | 36.040 | 36.005 | 36.333 |
| 60 | 19.107 | 19.161 | 19.112 | 19.457 | 114 | 36.292 | 36.358 | 36.324 | 36.654 |
| 61 | 19.426 | 19.480 | 19.431 | 19.769 | 115 | 36.610 | 36.676 | 36.642 | 36.969 |
| 62 | 19.744 | 19.799 | 19.750 | 20.095 | 116 | 36.929 | 36.995 | 36.961 | 37.292 |
| 63 | 20.062 | 20.117 | 20.070 | 20.407 | 117 | 37.247 | 37.313 | 37.279 | 37.606 |
| 64 | 20.380 | 20.435 | 20.388 | 20.731 | 118 | 37.565 | 37.632 | 37.598 | 37.928 |
| 65 | 20.698 | 20.754 | 20.708 | 21.044 | 119 | 37.883 | 37.950 | 37.916 | 38.243 |
| 66 | 21.016 | 21.072 | 21.027 | 21.368 | 120 | 38.201 | 38.268 | 38.235 | 38.564 |
| 67 | 21.335 | 21.391 | 21.346 | 21.682 | 121 | 38.519 | 38.586 | 38.553 | 38.879 |
| 68 | 21.653 | 21.710 | 21.665 | 22.006 | 122 | 38.837 | 38.904 | 38.872 | 39.200 |
| 69 | 21.971 | 22.028 | 21.984 | 22.319 | 123 | 39.156 | 39.223 | 39.190 | 39.516 |
| 70 | 22.289 | 22.347 | 22.303 | 22.643 | 124 | 39.475 | 39.542 | 39.508 | 39.839 |

All dimensions in inches.
* Blank diameters are 0.020 inch larger and maximum guide groove diameters $G$ are 1.240 inches smaller than these outside diameters.
† These diameters are maximum; tolerance is +0, −0.50 × pitch, inches.
†† For tolerances on over-pin diameters, see Table 4.
Tolerance for maximum eccentricity (total indicator reading) of pitch diameter with respect to bore is 0.001 × $PD$, but not less than 0.006 nor more than 0.032 inch.

Table 3 (*Continued*). **ANSI Silent Chain Sprocket Diameters**
(ANSI B29.2-1957, R1971)

These diameters apply only to chains of 1-inch pitch. For any other pitch (⅜ inch and larger) multiply the values given below by the pitch.

| No. Teeth | Pitch Diameter | Outside Diameter | | Over-Pin Dia.†† | No. Teeth | Pitch Diameter | Outside Diameter | | Over-Pin Dia.†† |
|---|---|---|---|---|---|---|---|---|---|
| | | Rounded Teeth* | Square Teeth† | | | | Rounded Teeth* | Square Teeth† | |
| 125 | 39.794 | 39.861 | 39.827 | 40.154 | 138 | 43.930 | 43.998 | 43.966 | 44.295 |
| 126 | 40.112 | 40.180 | 40.145 | 40.476 | 139 | 44.249 | 44.317 | 44.284 | 44.611 |
| 127 | 40.430 | 40.497 | 40.464 | 40.790 | 140 | 44.567 | 44.636 | 44.603 | 44.932 |
| 128 | 40.748 | 40.816 | 40.782 | 41.112 | 141 | 44.885 | 44.954 | 44.922 | 45.247 |
| 129 | 41.066 | 41.134 | 41.100 | 41.427 | 142 | 45.203 | 45.271 | 45.240 | 45.568 |
| 130 | 41.384 | 41.452 | 41.419 | 41.748 | 143 | 45.521 | 45.590 | 45.558 | 45.883 |
| 131 | 41.702 | 41.770 | 41.738 | 42.063 | 144 | 45.840 | 45.909 | 45.877 | 46.205 |
| 132 | 42.020 | 42.088 | 42.056 | 42.384 | 145 | 46.158 | 46.227 | 46.195 | 46.520 |
| 133 | 42.338 | 42.406 | 42.374 | 42.699 | 146 | 46.477 | 46.546 | 46.514 | 46.842 |
| 134 | 42.656 | 42.724 | 42.693 | 43.020 | 147 | 46.796 | 46.865 | 46.832 | 47.159 |
| 135 | 42.975 | 43.043 | 43.011 | 43.336 | 148 | 47.114 | 47.183 | 47.151 | 47.479 |
| 136 | 43.293 | 43.362 | 43.329 | 43.657 | 149 | 47.432 | 47.501 | 47.469 | 47.795 |
| 137 | 43.611 | 43.679 | 43.647 | 43.972 | 150 | 47.750 | 47.819 | 47.787 | 48.116 |

All dimensions in inches.
* Blank diameters are 0.020 inch larger and maximum guide groove diameters $G$ are 1.240 inches smaller than these outside diameters.
† These diameters are maximum; tolerance is +0, −0.50 × pitch, inches.
†† For tolerances on over-pin diameters, see Table 4.
Tolerance for maximum eccentricity (total indicator reading) of pitch diameter with respect to bore is 0.001 × $PD$, but not less than 0.006 nor more than 0.032 inch.

Table 4. **Over-Pin Diameter Tolerances for American National Standard Silent Chain Sprocket Measurement** (ANSI B29.2-1957, R1971)

All tolerances are *negative.* Tolerance = $(0.004 + 0.001P\sqrt{N})$, Where $P$ = Chain Pitch, $N$ = Number of Teeth. See Table 3 for over-pin diameters.

| Pitch | Up to 15 | 16-24 | 25-35 | 36-48 | 49-63 | 64-80 | 81-99 | 100-120 | 121-143 | 144 up |
|---|---|---|---|---|---|---|---|---|---|---|
| ⅜ | 0.005 | 0.005 | 0.005 | 0.006 | 0.006 | 0.007 | 0.007 | 0.007 | 0.008 | 0.008 |
| ½ | 0.005 | 0.006 | 0.006 | 0.007 | 0.007 | 0.008 | 0.008 | 0.009 | 0.009 | 0.010 |
| ⅝ | 0.006 | 0.006 | 0.007 | 0.008 | 0.009 | 0.010 | 0.010 | 0.010 | 0.011 | 0.012 |
| ¾ | 0.006 | 0.007 | 0.008 | 0.009 | 0.010 | 0.011 | 0.011 | 0.012 | 0.013 | 0.014 |
| 1 | 0.007 | 0.008 | 0.009 | 0.010 | 0.011 | 0.012 | 0.013 | 0.014 | 0.015 | 0.016 |
| 1¼ | 0.008 | 0.009 | 0.010 | 0.011 | 0.013 | 0.014 | 0.015 | 0.017 | 0.018 | 0.019 |
| 1½ | 0.008 | 0.010 | 0.011 | 0.013 | 0.014 | 0.016 | 0.017 | 0.019 | 0.020 | 0.022 |
| 2 | 0.010 | 0.012 | 0.014 | 0.016 | 0.018 | 0.020 | 0.022 | 0.024 | 0.026 | 0.028 |

Table 5. **American National Standard Silent Chain Widths and Sprocket Face Dimensions** (ANSI B29.2-1957, R1971)

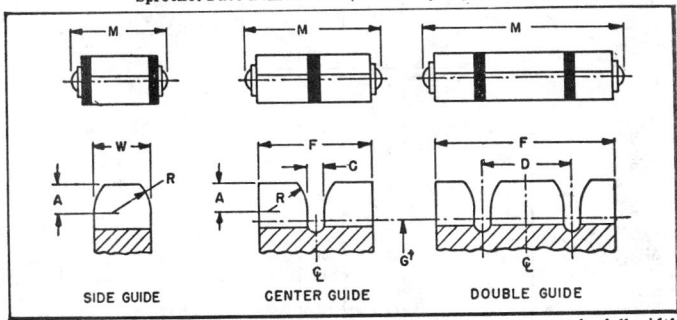

SIDE GUIDE          CENTER GUIDE          DOUBLE GUIDE

† Grooving tool may have either square or round end but groove must be full width down to diameter $G$.

Table 5 (*Continued*).  American National Standard Silent Chain Widths and Sprocket Face Dimensions (ANSI B29.2-1957, R1971)

| Chain No. | Chain Pitch | Type | M Max. | A | C ±0.005 | D ±0.010 | F +0.125 −0.000 | R ±0.003 | W +0.010 −0.000 |
|---|---|---|---|---|---|---|---|---|---|
| SC302 | ⅜ | Side Guide* | 0.594 | 0.133 | ..... | | ..... | 0.200 | 0.410 |
| SC303 | " | Center Guide | 0.844 | " | 0.100 | | 0.750 | " | " |
| SC304 | " | "     " | 1.094 | " | " | | 1.000 | " | ..... |
| SC305 | " | "     " | 1.344 | " | " | | 1.250 | " | |
| SC306 | " | "     " | 1.594 | " | " | | 1.500 | " | |
| SC307 | " | "     " | 1.844 | " | " | | 1.750 | " | |
| SC308 | " | "     " | 2.094 | " | " | | 2.000 | " | |
| SC309 | " | "     " | 2.344 | " | " | | 2.250 | " | |
| SC310 | " | "     " | 2.594 | " | " | ..... | 2.500 | " | |
| SC312 | " | Double Guide | 3.094 | " | " | 1.000 | 3.000 | " | |
| SC316 | " | "     " | 4.094 | " | " | " | 4.000 | " | |
| SC320 | " | "     " | 5.094 | " | " | " | 5.000 | " | |
| SC324 | " | "     " | 6.094 | " | " | " | 6.000 | " | |
| SC402 | ½ | Side Guide* | 0.750 | 0.133 | ..... | | ..... | 0.200 | 0.410 |
| SC403 | " | Center Guide | 0.875 | " | 0.100 | | 0.750 | " | ..... |
| SC404 | " | "     " | 1.125 | " | " | | 1.000 | " | |
| SC405 | " | "     " | 1.375 | " | " | | 1.250 | " | |
| SC406 | " | "     " | 1.625 | " | " | | 1.500 | " | |
| SC407 | " | "     " | 1.875 | " | " | | 1.750 | " | |
| SC408 | " | "     " | 2.125 | " | " | | 2.000 | " | |
| SC409 | " | "     " | 2.375 | " | " | | 2.250 | " | |
| SC410 | " | "     " | 2.625 | " | " | | 2.500 | " | |
| SC411 | " | "     " | 2.875 | " | " | | 2.750 | " | |
| SC412 | " | "     " | 3.125 | " | " | | 3.000 | " | |
| SC414 | " | "     " | 3.625 | " | " | ..... | 3.500 | " | |
| SC416 | " | Double Guide | 4.125 | " | " | 1.000 | 4.000 | " | |
| SC420 | " | "     " | 5.125 | " | " | " | 5.000 | " | |
| SC424 | " | "     " | 6.125 | " | " | " | 6.000 | " | |
| SC432 | " | "     " | 8.125 | " | " | " | 8.000 | " | |
| SC504 | ⅝ | Center Guide | 1.156 | 0.177 | 0.125 | | 1.000 | 0.250 | |
| SC505 | " | "     " | 1.406 | " | " | | 1.250 | " | |
| SC506 | " | "     " | 1.656 | " | " | | 1.500 | " | |
| SC507 | " | "     " | 1.906 | " | " | | 1.750 | " | |
| SC508 | " | "     " | 2.156 | " | " | | 2.000 | " | |
| SC510 | " | "     " | 2.656 | " | " | | 2.500 | " | |
| SC512 | " | "     " | 3.156 | " | " | | 3.000 | " | |
| SC516 | " | "     " | 4.156 | " | " | | 4.000 | " | |
| SC520 | " | Double Guide | 5.156 | " | " | 2.000 | 5.000 | " | |
| SC524 | " | "     " | 6.156 | " | " | " | 6.000 | " | |
| SC528 | " | "     " | 7.156 | " | " | " | 7.000 | " | |
| SC532 | " | "     " | 8.156 | " | " | " | 8.000 | " | |
| SC540 | " | "     " | 10.156 | " | " | " | 10.000 | " | |
| SC604 | ¾ | Center Guide | 1.187 | 0.274 | 0.180 | | 1.000 | 0.360 | |
| SC605 | " | "     " | 1.437 | " | " | | 1.250 | " | |
| SC606 | " | "     " | 1.687 | " | " | | 1.500 | " | |
| SC608 | " | "     " | 2.187 | " | " | | 2.000 | " | |
| SC610 | " | "     " | 2.687 | " | " | | 2.500 | " | |
| SC612 | " | "     " | 3.187 | " | " | | 3.000 | . | |
| SC614 | " | "     " | 3.687 | " | " | | 3.500 | " | |
| SC616 | " | "     " | 4.187 | " | " | | 4.000 | " | |
| SC620 | " | "     " | 5.187 | " | " | | 5.000 | " | |
| SC624 | " | "     " | 6.187 | " | " | ..... | 6.000 | " | |
| SC628 | " | Double Guide | 7.187 | " | " | 4.000 | 7.000 | " | |
| SC632 | " | "     " | 8.187 | " | " | " | 8.000 | " | |
| SC636 | " | "     " | 9.187 | " | " | " | 9.000 | " | |
| SC640 | " | "     " | 10.187 | " | " | " | 10.000 | " | |
| SC648 | " | "     " | 12.187 | " | " | " | 12.000 | " | |

*See footnotes at end of table.

Table 5 (*Continued*).  **American National Standard Silent Chain Widths and Sprocket Face Dimensions** (ANSI B29.2-1957, R1971)

| Chain No. | Chain Pitch | Type | M Max. | A | C ±0.005 | D ±0.010 | F +0.125 -0.000 | R ±0.003 | W +0.010 -0.000 |
|---|---|---|---|---|---|---|---|---|---|
| SC808 | 1 | Center Guide | 2.250 | 0.274 | 0.180 | .... | 2.000 | 0.360 | .... |
| SC810 | " | " " | 2.750 | " | " | .... | 2.500 | " | .... |
| SC812 | " | " " | 3.250 | " | " | .... | 3.000 | " | .... |
| SC816 | " | " " | 4.250 | " | " | .... | 4.000 | " | .... |
| SC820 | " | " " | 5.250 | " | " | .... | 5.000 | " | .... |
| SC824 | " | " " | 6.250 | " | " | .... | 6.000 | " | .... |
| SC828 | " | Double Guide | 7.250 | " | " | 4.000 | 7.000 | " | .... |
| SC832 | " | " " | 8.250 | " | " | " | 8.000 | " | .... |
| SC836 | " | " " | 9.250 | " | " | " | 9.000 | " | .... |
| SC840 | " | " " | 10.250 | " | " | " | 10.000 | " | .... |
| SC848 | " | " " | 12.250 | " | " | " | 12.000 | " | .... |
| SC856 | " | " " | 14.250 | " | " | " | 14.000 | " | .... |
| SC864 | " | " " | 16.250 | " | " | " | 16.000 | " | .... |
| SC1010 | 1¼ | Center Guide | 2.812 | 0.274 | 0.180 | .... | 2.500 | 0.360 | .... |
| SC1012 | " | " " | 3.312 | " | " | .... | 3.000 | " | .... |
| SC1016 | " | " " | 4.312 | " | " | .... | 4.000 | " | .... |
| SC1020 | " | " " | 5.312 | " | " | .... | 5.000 | " | .... |
| SC1024 | " | " " | 6.312 | " | " | .... | 6.000 | " | .... |
| SC1028 | " | " " | 7.312 | " | " | .... | 7.000 | " | .... |
| SC1032 | " | Double Guide | 8.312 | " | " | 4.000 | 8.000 | " | .... |
| SC1036 | " | " " | 9.312 | " | " | " | 9.000 | " | .... |
| SC1040 | " | " " | 10.312 | " | " | " | 10.000 | " | .... |
| SC1048 | " | " " | 12.312 | " | " | " | 12.000 | " | .... |
| SC1056 | " | " " | 14.312 | " | " | " | 14.000 | " | .... |
| SC1064 | " | " " | 16.312 | " | " | " | 16.000 | " | .... |
| SC1072 | " | " " | 18.312 | " | " | " | 18.000 | " | .... |
| SC1080 | " | " " | 20.312 | " | " | " | 20.000 | " | .... |
| SC1212 | 1½ | Center Guide | 3.375 | 0.274 | 0.180 | .... | 3.000 | 0.360 | .... |
| SC1216 | " | " " | 4.375 | " | " | .... | 4.000 | " | .... |
| SC1220 | " | " " | 5.375 | " | " | .... | 5.000 | " | .... |
| SC1224 | " | " " | 6.375 | " | " | .... | 6.000 | " | .... |
| SC1228 | " | " " | 7.375 | " | " | .... | 7.000 | " | .... |
| SC1232 | " | Double Guide | 8.375 | " | " | 4.000 | 8.000 | " | .... |
| SC1236 | " | " " | 9.375 | " | " | " | 9.000 | " | .... |
| SC1240 | " | " " | 10.375 | " | " | " | 10.000 | " | .... |
| SC1248 | " | " " | 12.375 | " | " | " | 12.000 | " | .... |
| SC1256 | " | " " | 14.375 | " | " | " | 14.000 | " | .... |
| SC1264 | " | " " | 16.375 | " | " | " | 16.000 | " | .... |
| SC1272 | " | " " | 18.375 | " | " | " | 18.000 | " | .... |
| SC1280 | " | " " | 20.375 | " | " | " | 20.000 | " | .... |
| SC1288 | " | " " | 22.375 | " | " | " | 22.000 | " | .... |
| SC1296 | " | " " | 24.375 | " | " | " | 24.000 | " | .... |
| SC1616 | 2 | Center Guide | 4.500 | 0.274 | 0.218 | .... | 4.000 | 0.360 | .... |
| SC1620 | " | " " | 5.500 | " | " | .... | 5.000 | " | .... |
| SC1624 | " | " " | 6.500 | " | " | .... | 6.000 | " | .... |
| SC1628 | " | " " | 7.500 | " | " | .... | 7.000 | " | .... |
| SC1632 | " | Double Guide | 8.500 | " | " | 4.000 | 8.000 | " | .... |
| SC1640 | " | " " | 10.500 | " | " | " | 10.000 | " | .... |
| SC1648 | " | " " | 12.500 | " | " | " | 12.000 | " | .... |
| SC1656 | " | " " | 14.500 | " | " | " | 14.000 | " | .... |
| SC1664 | " | " " | 16.500 | " | " | " | 16.000 | " | .... |
| SC1672 | " | " " | 18.500 | " | " | " | 18.000 | " | .... |
| SC1680 | " | " " | 20.500 | " | " | " | 20.000 | " | .... |
| SC1688 | " | " " | 22.500 | " | " | " | 22.000 | " | .... |
| SC1696 | " | " " | 24.500 | " | " | " | 24.000 | " | .... |
| SC16120 | " | " " | 30.500 | " | " | " | 30.000 | " | .... |

* Side Guide chains have single outside guides of same thickness as toothed links.
† Grooving tool may be either square or round end but groove must be full width down to diameter G.  For values of G (max.) see footnote to Table 3.
M = Max. overall width of chain.
The maximum radius over a chain engaged on a sprocket will not exceed the sprocket pitch radius plus 75 percent of the chain pitch.

**Sprocket Hub Dimensions.** — The important hub dimensions are the outside diameter, the bore, and the length. The maximum hub diameter is limited by the need to clear the chain guides and is of particular importance for sprockets with low numbers of teeth. The American National Standard for inverted tooth chains and sprocket teeth ANSI B29.2-1957 (R1971) provides the following formulas for calculating maximum hub diameters, *MHD*.

$$MHD \text{ (for hobbed teeth)} = P\left(\text{Cot}\frac{180°}{N} - 1.33\right)$$

$$MHD \text{ (for straddle cut teeth)} = P\left(\text{Cot}\frac{180°}{N} - 1.25\right)$$

Maximum hub diameters for sprockets with from 17 to 31 teeth are given in Table 6. Maximum hub diameters for other methods of cutting teeth may differ from these values. Recommended maximum bores are given in Table 7.

**Table 6. American National Standard Minimum Hub Diameters for Silent Chain Sprockets (17 to 31 teeth)** (ANSI B29.2-1957, R1971)

Values shown are for 1-inch pitch chain. For other pitches (3/8-inch and larger) multiply the values given by the pitch.

| No. Teeth | Hob Cut | Straddle Cut | No. Teeth | Hob Cut | Straddle Cut | No. Teeth | Hob Cut | Straddle Cut |
|---|---|---|---|---|---|---|---|---|
| | Min. Hub Diam. | | | Min. Hub Diam. | | | Min. Hub Diam. | |
| 17 | 4.019 | 4.099 | 22 | 5.626 | 5.706 | 27 | 7.226 | 7.306 |
| 18 | 4.341 | 4.421 | 23 | 5.946 | 6.026 | 28 | 7.546 | 7.626 |
| 19 | 4.662 | 4.742 | 24 | 6.265 | 6.345 | 29 | 7.865 | 7.945 |
| 20 | 4.983 | 5.063 | 25 | 6.586 | 6.666 | 30 | 8.185 | 8.265 |
| 21 | 5.304 | 5.384 | 26 | 6.905 | 6.985 | 31 | 8.503 | 8.583 |

All dimensions in inches.

Good practice indicates that teeth of sprockets up to and including 31 teeth should have a Rockwell hardness of C50 min.

**Table 7. Recommended Maximum Sprocket Bores for Silent Chains***

| Number of Teeth | Chain Pitch, Inches | | | | | | | |
|---|---|---|---|---|---|---|---|---|
| | 3/8 | 1/2 | 5/8 | 3/4 | 1 | 1 1/4 | 1 1/2 | 2 |
| | Max. Sprocket Bore, Inches | | | | | | | |
| 17 | 1 | 1 3/8 | 1 3/4 | 2 | 2 3/4 | 3 3/8 | 4 1/8 | 5 1/2 |
| 19 | 1 1/4 | 1 5/8 | 2 | 2 3/8 | 3 1/4 | 4 | 4 7/8 | 6 3/4 |
| 21 | 1 3/8 | 1 7/8 | 2 1/4 | 2 3/4 | 3 3/4 | 4 1/2 | 5 1/2 | 7 3/4 |
| 23 | 1 5/8 | 2 1/8 | 2 5/8 | 3 1/4 | 4 3/8 | 5 1/2 | 6 1/2 | 9 |
| 25 | 1 3/4 | 2 3/8 | 3 | 3 5/8 | 4 3/4 | 6 | 7 1/4 | 10 |
| 27 | 2 | 2 5/8 | 3 3/8 | 3 7/8 | 5 3/8 | 6 3/4 | 8 1/8 | 11 1/4 |
| 29 | 2 1/8 | 2 13/16 | 3 5/8 | 4 3/8 | 5 3/4 | 7 3/8 | 9 1/8 | 12 1/2 |
| 31 | 2 5/16 | 3 1/16 | 3 7/8 | 4 5/8 | 6 3/8 | 8 | 9 7/8 | 13 1/2 |

* Association of Roller and Silent Chain Manufacturers.

**Sprocket Design and Tooth Form.** — Except for tooth form, silent chain sprocket design parallels the general design practice of roller chain sprockets as covered in the previous section.

As shown in Table 8, sprockets for American National Standard silent chains have teeth with straight-line working faces. The tops of teeth may be rounded or

**Table 8.   Tooth Form for ANSI Silent Tooth Sprocket** (ANSI B29.2-1957, R1971)

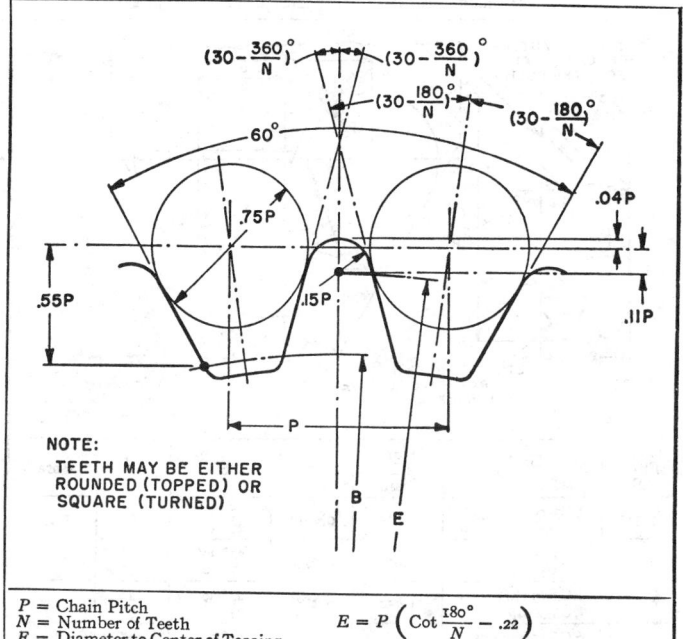

NOTE:
TEETH MAY BE EITHER
ROUNDED (TOPPED) OR
SQUARE (TURNED)

$P$ = Chain Pitch
$N$ = Number of Teeth
$E$ = Diameter to Center of Topping Curve
$B$ = Diameter to Base of Working Face
*Note:* Shape of root line below working face may vary with type of cutter.

$$E = P \left( \text{Cot} \frac{180°}{N} - .22 \right)$$

$$B = P \sqrt{1.515213 + \left( \text{Cot} \frac{180°}{N} - 1.1 \right)^2}$$

square.  Bottom clearance below the working face is not specified but must be sufficient to clear the chain teeth.  The standard tooth form is designed to mesh with link plate contours having an included angle of 60 degrees as shown in the diagram of Table 8.  It will also be seen from this diagram that the angle between the faces of a given tooth $\left( 60° - \frac{720°}{N} \right)$ becomes smaller as the number of teeth decreases so that for a 12-tooth sprocket it will be zero.  In other words, the tooth faces will be parallel.  For smaller tooth numbers, the teeth would be undercut.  For best results, 21 or more teeth are recommended; less than 17 should not be used.

**Cutting Silent Chain Sprocket Teeth.** — Sprocket teeth may be cut either by a straddle cutter or a hob.  Essential dimensions for straddle cutters are given in Table 9 and for hobs in Tables 10 and 11.  American National Standard silent chain hobs are stamped for identification as shown on page 1087.

**Design of Silent Chain Drives.** — The design of silent chain transmissions must be based not only upon the power to be transmitted and the ratio between driving and driven shafts, but also upon such factors as the speed of the faster running shaft, the available space, assuming that it affects the sprocket diameters,

Table 9.　Straddle Cutters for American National Standard
Silent Chain Sprocket Teeth (ANSI B29.2-1957, R1971)

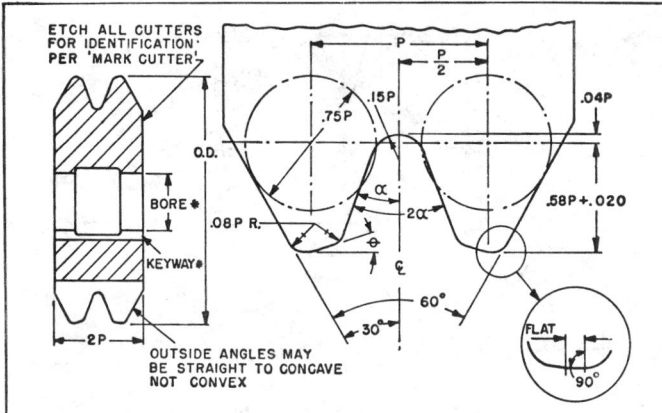

| Chain Pitch $P$ | Mark Cutter† | Out-side Diam. | .75$P$ | $\alpha$ | $\phi$ | Bore* |
|---|---|---|---|---|---|---|
| 0.375 | SC3-15 thru 35<br>SC3-36 up | 3.625 | 0.2813 | 22°-30′<br>27°-30′ | 12°<br>5° | 1.250 |
| 0.500 | SC4-15 thru 35<br>SC4-36 up | 3.875 | 0.3750 | 22°-30′<br>27°-30′ | 12°<br>5° | 1.250 |
| 0.625 | SC5-15 thru 35<br>SC5-36 up | 4.250 | 0.4688 | 22°-30′<br>27°-30′ | 12°<br>5° | 1.250 |
| 0.750 | SC6-15 thru 35<br>SC6-36 up | 4.625 | 0.5625 | 22°-30′<br>27°-30′ | 12°<br>5° | 1.250 |
| 1.000 | SC8-15 thru 35<br>SC8-36 up | 5.250 | 0.7500 | 22°-30′<br>27°-30′ | 12°<br>5° | 1.500 |
| 1.250 | SC10-15 thru 35<br>SC10-36 up | 5.750 | 0.9375 | 22°-30′<br>27°-30′ | 12°<br>5° | 1.500 |
| 1.500 | SC12-15 thru 35<br>SC12-36 up | 6.250 | 1.1250 | 22°-30′<br>27°-30′ | 12°<br>5° | 1.750 |
| 2.000 | SC16-15 thru 35<br>SC16-36 up | 6.500 | 1.5000 | 22°-30′<br>27°-30′ | 12°<br>5° | 1.750 |

* Suggested standard.　Bores other than standard must be specified.
† Range of teeth is indicated in the cutter marking.

the character of the load and certain other factors.　Determining the pitch of the chain and the number of teeth on the smallest sprocket are the important initial steps.　Usually several combinations of pitches and sprocket sizes may be employed for a given installation.　In attempting to select the best combination, it is advisable to consult with the manufacturer of the chain to be used.　Some of the more important fundamental points governing the design of silent chain transmissions will be summarized.

The design of a silent chain drive consists, primarily, of the selection of the chain size, sprockets, determination of chain length, center distance, lubrication method, and arrangement of casings.

**Table 10. Hobs for ANSI Silent Chain Sprocket Teeth** (ANSI B29.2-1957, R1971)

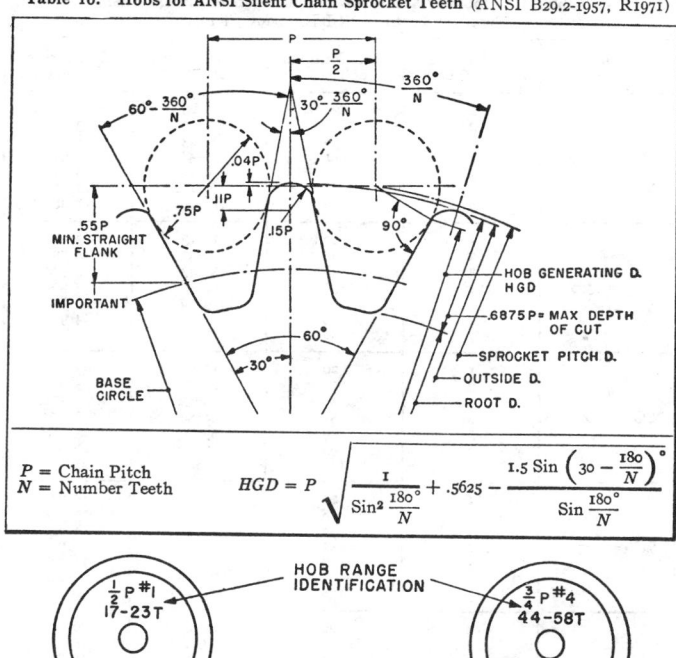

$$P = \text{Chain Pitch}$$
$$N = \text{Number Teeth}$$

$$HGD = P \sqrt{\dfrac{1}{\operatorname{Sin}^2 \dfrac{180°}{N}} + .5625 - \dfrac{1.5 \operatorname{Sin}\left(30 - \dfrac{180}{N}\right)°}{\operatorname{Sin}\dfrac{180°}{N}}}$$

Fig. 1. Identification of American National Standard Inverted Tooth Chain Hobs

**Pitch of Silent Chain.** — The pitch is selected with reference to the speed of the faster running shaft which ordinarily is the driver and holds the smaller sprocket. The following pitches are recommended: for a faster running shaft of 2000 to 5000 rpm, ⅜-inch pitch; for 1500 to 2000 rpm, ½-inch pitch; for 1200 to 1500 rpm, ⅝-inch pitch; for 1000 to 1200 rpm, ¾-inch pitch; for 800 to 1000 rpm, 1-inch pitch; for 650 to 800 rpm, 1¼-inch pitch; for 500 to 600 rpm, 1½-inch pitch; for 300 to 500 rpm, 2-inch pitch; and for below 300 rpm, 2½-inch pitch. As the normal operating speeds increase, the allowable pitch decreases. Recommendations relating to the relationship between pitch and operating speed are intended for normal or average conditions, and the speeds for a given pitch may be exceeded under favorable conditions and may have to be reduced when conditions are unfavorable. In general, smoother or quieter operation will result from using the smallest pitch suitable for a given speed and load. However, a larger pitch which might be applicable under the same conditions, will result in a narrower chain and a less expensive transmission. This usually is true when there is a small speed reduction and comparatively long center distance. If there is a large speed reduction and short center distance, drives having the smaller pitches may be less expensive.

Table 11.  Hobs for American National Standard Silent Chain
Sprocket Teeth (ANSI B29.2-1957, R1971)

| Chain Pitch | Hob Number | Basic Number of Teeth | Tooth Range of Hob | Generating Diameter |
|---|---|---|---|---|
| SC3 = .375 | 1 | 20 | 17-23 | 2.311 |
| | 2 | 28 | 24-32 | 3.247 |
| | 3 | 38 | 33-43 | 4.428 |
| | 4 | 51 | 44-58 | 5.971 |
| | 5 | 69 | 59-79 | 8.114 |
| | 6 | 95 | 80-110 | 11.212 |
| | 7 | 130 | 111-150 | 15.385 |
| SC4 = .500 | 1 | 20 | 17-23 | 3.082 |
| | 2 | 28 | 24-32 | 4.329 |
| | 3 | 38 | 33-43 | 5.904 |
| | 4 | 51 | 44-58 | 7.962 |
| | 5 | 69 | 59-79 | 10.818 |
| | 6 | 95 | 80-110 | 14.950 |
| | 7 | 130 | 111-150 | 20.513 |
| SC5 = .625 | 1 | 20 | 17-23 | 3.852 |
| | 2 | 28 | 24-32 | 5.412 |
| | 3 | 38 | 33-43 | 7.381 |
| | 4 | 51 | 44-58 | 9.952 |
| | 5 | 69 | 59-79 | 13.522 |
| | 6 | 95 | 80-110 | 18.687 |
| | 7 | 130 | 111-150 | 25.641 |
| SC6 = .750 | 1 | 20 | 17-23 | 4.623 |
| | 2 | 28 | 24-32 | 6.494 |
| | 3 | 38 | 33-43 | 8.857 |
| | 4 | 51 | 44-58 | 11.943 |
| | 5 | 69 | 59-79 | 16.227 |
| | 6 | 95 | 80-110 | 22.424 |
| | 7 | 130 | 111-150 | 30.770 |
| SC8 = 1.000 | 1 | 20 | 17-23 | 6.163 |
| | 2 | 28 | 24-32 | 8.659 |
| | 3 | 38 | 33-43 | 11.809 |
| | 4 | 51 | 44-58 | 15.924 |
| | 5 | 69 | 59-79 | 21.636 |
| | 6 | 95 | 80-110 | 29.899 |
| | 7 | 130 | 111-150 | 41.026 |
| SC10 = 1.250 | 1 | 20 | 17-23 | 7.704 |
| | 2 | 28 | 24-32 | 10.823 |
| | 3 | 38 | 33-43 | 14.761 |
| | 4 | 51 | 44-58 | 19.905 |
| | 5 | 69 | 59-79 | 27.045 |
| | 6 | 95 | 80-110 | 37.374 |
| | 7 | 130 | 111-150 | 51.283 |
| SC12 = 1.500 | 1 | 20 | 17-23 | 9.245 |
| | 2 | 28 | 24-32 | 12.988 |
| | 3 | 38 | 33-43 | 17.713 |
| | 4 | 51 | 44-58 | 23.886 |
| | 5 | 69 | 59-79 | 32.454 |
| | 6 | 95 | 80-110 | 44.849 |
| | 7 | 130 | 111-150 | 61.539 |
| SC16 = 2.000 | 1 | 20 | 17-23 | 12.327 |
| | 2 | 28 | 24-32 | 17.317 |
| | 3 | 38 | 33-43 | 23.618 |
| | 4 | 51 | 44-58 | 31.848 |
| | 5 | 69 | 59-79 | 43.272 |
| | 6 | 95 | 80-110 | 59.798 |
| | 7 | 130 | 111-150 | 82.052 |

**Maximum Ratios for Silent Chain Drives.** — The maximum permissible ratios between driving and driven sprockets vary somewhat for different conditions and usually range from 6- or 7-to-1 up to 10-to-1. Some drives have even higher ratios, especially when the operating conditions are exceptionally favorable. When a large speed reduction is necessary, it is preferable as a general rule to use a double reduction or compound type of transmission instead of obtaining the entire reduction with two sprockets. Drives should be so proportioned that the angle between the two strands of a tight chain does not exceed 45 degrees. When the angle is larger, the chain does not have sufficient contact with the driving sprocket.

*Sprocket Size and Chain Speed:* A driving sprocket with not less than 17 teeth is generally recommended. For the driven sprocket, one manufacturer recommends 127 teeth as a maximum limit and less than 100 as preferable. If practicable, the sprocket sizes should be small enough to limit the chain speed to from 1200 to 1400 feet per minute. If the chain speed exceeds these figures, this may indicate that the pitch is too large or that a smaller pitch, and, consequently, a reduction in sprocket diameters (and chain speed) will result in better operating conditions. Both sprockets should preferably have a "hunting tooth ratio" relative to the number of chain links for uniform wear. See "Hunting Tooth Ratios," page 826.

If there is a small reduction in speed between the driving and driven shafts, both sprockets may be made as small as is consistent with satisfactory operation, either to obtain a compact drive or possibly to avoid excessive chain speed in cases where the rotative speed is high for a given horsepower. Under such conditions, one manufacturer recommends driving sprockets ranging from 17 to 30 teeth, and driven sprockets ranging from 19 to 33 teeth. If the number of revolutions per minute is low for a given horsepower and the center distance comparatively long, then the recommended range for driving sprockets is from 23 to 111 teeth, and driven sprockets from 27 to 129 teeth. The preferable range is from 17 to 75 teeth for the driving sprockets, and 19 to 102 teeth for the driven sprockets.

**Center Distance for Silent Chain Drives.** — If the ratio of the drive is small, it is possible to locate the sprockets so close that the teeth just clear; however, as a general rule, the minimum center-to-center distance should equal the sum of the diameters of both sprockets. According to the Whitney Chain & Mfg. Co., if the speed ratio is not over 2½-to-1, the center distance may be equal to one-half the sum of the sprocket diameters plus tooth clearance, providing this distance is not less than the minimum given in Table 12.

Table 12.   Minimum Center Distances for Various Pitches

| Pitch, Inches | ⅜ | ½ | ⅝ | ¾ | 1 | 1¼ | 1½ |
|---|---|---|---|---|---|---|---|
| Minimum Center Distance, Inches | 6 | 9 | 12 | 15 | 21 | 27 | 33 |

If the speed ratio is greater than 2½-to-1, the center distance should not be less than the sum of the sprocket diameters.

When the chain length in pitches is known, the equivalent center distance for a tight chain may be determined by the formula for roller chain found on page 1063.

In selecting chain length, factors determining length should be adjusted so that the use of offset links may be avoided wherever possible. Chain lengths of an uneven number of pitches are also to be avoided.

**Silent Tooth Chain Horsepower Capacity.** — The horsepower ratings given in Table 13 have been established on a life expectancy of approximately 20,000 hours under optimum drive conditions, i.e. for a uniform rate of work where there is relatively little shock or load variation throughout a single revolution of a driven sprocket. Using these horsepower ratings as a basis, engineering judgment should be exercised as to the severity of the operating conditions for the intended installation, taking into consideration both the source of power, the nature of the load, and the resulting effect of inertia, strain, and shock. Thus, for other than optimum drive conditions, the specified horsepower must be multiplied by the applicable service factor to obtain a "design" horsepower value. This is the value used to enter Table 13 to obtain the required size of chain.

*Service Factors:* For a uniform type of load, a service factor of 1.0 for a 10-hour day and 1.3 for a 24-hour day are recommended. For a moderate shock load, service factors of 1.4 for a 10-hour day and 1.7 for a 24-hour day are recommended. For heavy shock loads, service factors of 1.7 for a 10-hour day and 2.0 for a 24-hour day are recommended.

**Installation of Silent Chain Drives.** — In installing chain transmissions of any kind, horizontal drives are those having driving and driven shafts in a horizontal plane. These are always preferable to vertical drives, which have a vertical center line intersecting the driving and driven shafts. If one sprocket must be higher than the other, avoid a vertical drive if possible by so locating the two sprockets that the common center line inclines from the vertical as far as is permitted by other conditions which might govern the installation. If practicable, an adjustment should be provided for the center distance between the driving and driven shafts. Driving motors are often mounted on adjustable base or slide rails to provide this adjustment for the center distance.

*Slack Side of Chain:* As a general rule, the slack strand of a chain should be on the lower side of a horizontal drive. If the drive is not horizontal but angular or at some angle less than 90 degrees from the vertical, the slack should preferably be on that side which causes the strand to curve outward or away from the center line of the driving and driven shafts. Whenever the slack strand is on the upper side of either a horizontal or inclined drive, adjustment for the center distance is especially important to compensate for possible chain elongation.

*Lubrication:* The life of a silent chain subjected to conditions such as are common to automobile drives, depends largely upon the wear of the joints. On account of the high speed and whipping action, it is important to have the chains well oiled. When splash lubrication is employed, the supply pipe should be placed so that the oil will be directed against the inside of the chain.

It is preferable that silent chains be operated in an oil-retaining casing with provisions for lubrication. The viscosity of the oil depends somewhat upon temperature conditions and the chain speed, but in general should be a straight mineral oil of about SAE 30 viscosity. Heavy-bodied oil of about 500 seconds Saybolt may be used on very slow moving chains. Avoid using greases of any kind.

**Double-Flexure Silent Chain.** — In double-flexure chain, the teeth of the link plates project on both sides of the chain and the chain flexes in both directions. This chain is used where the drive arrangements require that sprockets contact both sides of the chain. Neither double-flexure chain or sprockets are covered in American National Standard ANSI B29.2-1957 (R1971).

**Small Pitch Silent Chain.** — A 3⁄16-inch pitch silent chain is covered by American National Standard B29.9-1958 (R1974). This standard covers the sprocket tooth form, standard sprocket diameters, general chain proportions, sprocket face dimensions and horsepower ratings.

## Table 13. Horsepower Ratings for Silent Chain Drives — 1966

The industrial standard horsepower ratings and service factors for silent chain drives as revised by the American Sprocket Chain Manufacturers Association are given below. These ratings are for American Standard silent chain as covered by American Standard ASA B29.2-1957.

These may require modification by using the service factors given on Page 1090. These factors, which apply to typical drives, are intended as a general guide only, and engineering judgment or experience may indicate different modifications to suit the nature of the load.

$$\text{Horsepower capacity of chain per inch of width} = \frac{\text{Rating in Table 13}}{\text{Service factor}}$$

$$\text{Chain width for given total H.P. capacity} = \frac{\text{H.P.} \times \text{Service factor}}{\text{Rating per inch, Table 13}}$$

*Lubrication.* — The horsepower ratings for each pitch listed in Table 1 are divided into three sections by heavy zigzag lines. For drives in the first or left-hand section, bath, splash, oil-cup or brush lubrication may be applied; for drives in the second or middle section, a disk or circulating pump is preferable; for drives in the third or right-hand section, consult the manufacturer's engineering department in regard to proper method of lubrication.

### ⅜″ Pitch — Horsepower per Inch of Chain Width

| No. of Teeth Small Sprkt. | Revolutions per Minute — Small Sprocket | | | | | | | | | | | | |
|---|---|---|---|---|---|---|---|---|---|---|---|---|---|
| | 100 | 500 | 1000 | 1200 | 1500 | 1800 | 2000 | 2500 | 3000 | 3500 | 4000 | 5000 | 6000 |
| *17 | .46 | 2.1 | 4.6 | 4.9 | 5.3 | 6.5 | 6.9 | 7.9 | 8.5 | 8.8 | 8.8 | .. | .. |
| *19 | .53 | 2.5 | 4.8 | 5.4 | 6.5 | 7.4 | 7.9 | 9.1 | 9.9 | 10 | 11 | 9.8 | .. |
| 21 | .58 | 2.8 | 5.1 | 6.0 | 7.3 | 8.3 | 9.0 | 10 | 11 | 12 | 12 | 12 | 10 |
| 23 | .63 | 3.0 | 5.6 | 6.6 | 8.0 | 9.3 | 10 | 12 | 13 | 14 | 14 | 14 | 12 |
| 25 | .69 | 3.3 | 6.1 | 7.3 | 8.8 | 10 | 11 | 13 | 14 | 15 | 15 | 15 | 14 |
| 27 | .74 | 3.5 | 6.8 | 7.9 | 9.5 | 11 | 12 | 14 | 15 | 16 | 18 | 18 | 16 |
| 29 | .80 | 3.8 | 7.3 | 8.5 | 10 | 12 | 13 | 15 | 16 | 18 | 19 | 19 | 18 |
| 31 | .85 | 4.1 | 7.8 | 9.1 | 11 | 13 | 14 | 16 | 18 | 19 | 20 | 20 | 19 |
| 33 | .90 | 4.4 | 8.3 | 9.8 | 12 | 14 | 15 | 18 | 19 | 21 | 21 | 21 | 20 |
| 35 | .96 | 4.6 | 8.8 | 10 | 13 | 15 | 16 | 19 | 20 | 23 | 23 | 23 | 21 |
| 37 | 1.0 | 4.9 | 9.1 | 11 | 14 | 15 | 16 | 20 | 21 | 24 | 24 | 24 | .. |
| 40 | 1.1 | 5.3 | 10 | 12 | 15 | 16 | 18 | 21 | 24 | 25 | 26 | 26 | .. |
| 45 | 1.3 | 6.0 | 11 | 13 | 16 | 19 | 20 | 24 | 26 | 28 | 29 | .. | .. |
| 50 | 1.4 | 6.6 | 13 | 15 | 18 | 20 | 23 | 26 | 29 | 30 | .. | .. | .. |

TYPE I       TYPE II       TYPE III

### ½″ Pitch — Horsepower per Inch of Chain Width

| No. of Teeth Small Sprkt. | Revolutions per Minute — Small Sprocket | | | | | | | | | | |
|---|---|---|---|---|---|---|---|---|---|---|---|
| | 100 | 500 | 700 | 1000 | 1200 | 1800 | 2000 | 2500 | 3000 | 3500 | 4000 |
| *17 | .83 | 3.8 | 5.0 | 6.3 | 7.5 | 10 | 11 | 11 | 11 | 11 | .. |
| *19 | .93 | 3.8 | 5.0 | 7.5 | 8.8 | 11 | 13 | 14 | 14 | 14 | .. |
| 21 | 1.0 | 5.0 | 6.3 | 8.8 | 10 | 14 | 14 | 15 | 16 | 16 | .. |
| 23 | 1.1 | 5.0 | 7.5 | 10 | 11 | 15 | 16 | 18 | 19 | 19 | 18 |
| 25 | 1.2 | 5.0 | 7.5 | 10 | 13 | 16 | 18 | 20 | 21 | 21 | 20 |
| 27 | 1.3 | 6.3 | 8.8 | 11 | 13 | 18 | 19 | 21 | 24 | 24 | 23 |
| 29 | 1.4 | 6.3 | 8.8 | 13 | 14 | 19 | 21 | 24 | 25 | 25 | 25 |
| 31 | 1.5 | 7.5 | 10 | 13 | 15 | 21 | 23 | 25 | 28 | 28 | 28 |
| 33 | 1.6 | 7.5 | 10 | 14 | 16 | 23 | 24 | 28 | 29 | 30 | 29 |
| 35 | 1.8 | 7.5 | 11 | 15 | 18 | 24 | 25 | 29 | 31 | 31 | 30 |
| 37 | 1.9 | 8.8 | 11 | 16 | 19 | 25 | 26 | 30 | 33 | 33 | .. |
| 40 | 2.0 | 8.8 | 13 | 18 | 20 | 28 | 29 | 33 | 35 | 35 | .. |
| 45 | 2.5 | 10 | 14 | 19 | 23 | 30 | 30 | 36 | 39 | .. | .. |
| 50 | 2.5 | 11 | 15 | 21 | 25 | 34 | 36 | 40 | .. | .. | .. |

TYPE I       TYPE II       TYPE III

* For best results, smaller sprocket should have at least 21 teeth.

Table 13 *(Continued)*. Horsepower Ratings for Silent Chain Drives — 1966

**⅝″ Pitch — Horsepower per Inch of Chain Width**

| No. of Teeth Small Sprkt. | Revolutions per Minute — Small Sprocket | | | | | | | | | |
|---|---|---|---|---|---|---|---|---|---|---|
| | 100 | 500 | 700 | 1000 | 1200 | 1800 | 2000 | 2500 | 3000 | 3500 |
| *17 | 1.3 | 6.3 | 7.5 | 10 | 11 | 14 | 15 | 14 | .. | .. |
| *19 | 1.4 | 6.3 | 8.8 | 13 | 14 | 16 | 18 | 18 | .. | .. |
| 21 | 1.6 | 7.5 | 10 | 13 | 15 | 19 | 20 | 20 | 20 | .. |
| 23 | 1.8 | 7.5 | 11 | 15 | 16 | 21 | 23 | 24 | 23 | .. |
| 25 | 1.9 | 8.8 | 11 | 16 | 19 | 24 | 25 | 26 | 26 | 24 |
| 27 | 2.0 | 10 | 13 | 18 | 20 | 26 | 28 | 29 | 29 | 26 |
| 29 | 2.1 | 10 | 14 | 19 | 21 | 28 | 30 | 31 | 31 | 29 |
| 31 | 2.4 | 11 | 15 | 20 | 23 | 30 | 31 | 34 | 34 | 31 |
| 33 | 2.5 | 11 | 16 | 21 | 25 | 33 | 34 | 36 | 36 | 34 |
| 35 | 2.6 | 13 | 16 | 23 | 26 | 34 | 36 | 39 | 39 | 35 |
| 37 | 2.8 | 13 | 18 | 24 | 28 | 36 | 39 | 43 | 41 | .. |
| 40 | 3.0 | 14 | 19 | 26 | 30 | 39 | 41 | 44 | .. | .. |
| 45 | 3.4 | 16 | 21 | 29 | 34 | 44 | 46 | .. | .. | .. |
| 50 | 3.8 | 18 | 24 | 33 | 38 | 48 | 50 | .. | .. | .. |
| | TYPE I | | TYPE II | | | TYPE III | | | | |

**¾″ Pitch — Horsepower per Inch of Chain Width**

| No. of Teeth Small Sprkt. | Revolutions per Minute — Small Sprocket | | | | | | | | |
|---|---|---|---|---|---|---|---|---|---|
| | 100 | 500 | 700 | 1000 | 1200 | 1500 | 1800 | 2000 | 2500 |
| *17 | 1.9 | 8.1 | 11 | 14 | 15 | 16 | 18 | 18 | .. |
| *19 | 2.0 | 9.3 | 13 | 15 | 18 | 20 | 21 | 21 | .. |
| 21 | 2.3 | 10 | 14 | 18 | 20 | 23 | 24 | 25 | 24 |
| 23 | 2.5 | 11 | 15 | 20 | 23 | 25 | 28 | 28 | 28 |
| 25 | 2.8 | 13 | 16 | 21 | 25 | 29 | 31 | 31 | 30 |
| 27 | 2.9 | 14 | 18 | 24 | 28 | 31 | 34 | 35 | 35 |
| 29 | 3.1 | 15 | 20 | 26 | 30 | 34 | 36 | 38 | 38 |
| 31 | 3.4 | 15 | 21 | 28 | 31 | 36 | 40 | 41 | 41 |
| 33 | 3.6 | 16 | 23 | 30 | 34 | 39 | 43 | 44 | 44 |
| 35 | 3.8 | 18 | 24 | 31 | 36 | 41 | 45 | 46 | 46 |
| 37 | 4.0 | 19 | 25 | 34 | 39 | 44 | 48 | 49 | 49 |
| 40 | 4.4 | 20 | 28 | 36 | 41 | 48 | 51 | 53 | 53 |
| 45 | 4.9 | 23 | 30 | 40 | 46 | 53 | 56 | 58 | .. |
| 50 | 5.4 | 25 | 34 | 45 | 51 | 58 | .. | .. | .. |
| | TYPE I | | TYPE II | | | TYPE III | | | |

**1″ Pitch — Horsepower per Inch of Chain Width**

| No. of Teeth Small Sprkt. | Revolutions per Minute — Small Sprocket | | | | | | | | | | |
|---|---|---|---|---|---|---|---|---|---|---|---|
| | 100 | 200 | 300 | 400 | 500 | 700 | 1000 | 1200 | 1500 | 1800 | 2000 |
| *17 | 3.8 | 6.3 | 8.8 | 11 | 14 | 18 | 21 | 23 | .. | .. | .. |
| *19 | 3.8 | 7.5 | 10 | 13 | 15 | 20 | 25 | 26 | 28 | .. | .. |
| 21 | 3.8 | 7.5 | 11 | 15 | 18 | 23 | 29 | 31 | 33 | 33 | .. |
| 23 | 3.8 | 8.8 | 13 | 16 | 19 | 25 | 31 | 35 | 38 | 38 | .. |
| 25 | 5.0 | 8.8 | 14 | 18 | 21 | 28 | 35 | 39 | 41 | 41 | 41 |
| 27 | 5.0 | 10 | 15 | 19 | 24 | 30 | 39 | 43 | 46 | 46 | 45 |
| 29 | 5.0 | 11 | 16 | 20 | 25 | 33 | 41 | 46 | 50 | 51 | 50 |
| 31 | 6.3 | 11 | 16 | 23 | 28 | 35 | 45 | 50 | 54 | 55 | 54 |
| 33 | 6.3 | 13 | 18 | 24 | 29 | 38 | 49 | 54 | 59 | 59 | 58 |
| 35 | 6.3 | 13 | 19 | 25 | 30 | 40 | 51 | 56 | 61 | 63 | 61 |
| 37 | 6.8 | 14 | 20 | 26 | 33 | 43 | 54 | 60 | 65 | 66 | .. |
| 40 | 7.5 | 15 | 23 | 29 | 35 | 45 | 59 | 65 | 70 | .. | .. |
| 45 | 8.8 | 16 | 25 | 31 | 39 | 51 | 65 | 71 | 76 | .. | .. |
| 50 | 10 | 19 | 28 | 35 | 43 | 51 | 71 | 78 | .. | .. | .. |
| | TYPE I | | | TYPE II | | | TYPE III | | | | |

* For best results, smaller sprocket should have at least 21 teeth.

**Table 13** (*Concluded*). **Horsepower Ratings for Silent Chain Drives** — 1966

### 1¼″ Pitch — Horsepower per Inch of Chain Width

| No. of Teeth Small Sprkt. | 100 | 200 | 300 | 400 | 500 | 600 | 700 | 800 | 1000 | 1200 | 1500 |
|---|---|---|---|---|---|---|---|---|---|---|---|
| *19 | 5.6 | 10 | 15 | 20 | 24 | 26 | 29 | 31 | 34 | 35 | .. |
| 21 | 6.3 | 11 | 18 | 23 | 26 | 30 | 33 | 36 | 40 | 41 | .. |
| 23 | 6.9 | 13 | 19 | 24 | 29 | 34 | 36 | 40 | 45 | 46 | 46 |
| 25 | 7.5 | 14 | 20 | 26 | 31 | 36 | 40 | 44 | 50 | 53 | 53 |
| 27 | 8.0 | 15 | 23 | 29 | 35 | 40 | 44 | 49 | 54 | 58 | 58 |
| 29 | 8.6 | 16 | 24 | 31 | 38 | 43 | 48 | 53 | 59 | 63 | 64 |
| 31 | 9.3 | 18 | 26 | 34 | 40 | 46 | 51 | 56 | 64 | 68 | 69 |
| 33 | 9.9 | 18 | 28 | 35 | 43 | 49 | 55 | 60 | 69 | 73 | 74 |
| 35 | 11 | 20 | 29 | 38 | 45 | 53 | 59 | 64 | 73 | 78 | 78 |
| 37 | 11 | 21 | 30 | 40 | 48 | 55 | 63 | 68 | 76 | 81 | .. |
| 40 | 12 | 24 | 34 | 44 | 53 | 60 | 68 | 74 | 83 | 88 | .. |
| 45 | 13 | 26 | 38 | 49 | 59 | 68 | 75 | 81 | 91 | .. | .. |
| 50 | 15 | 29 | 43 | 54 | 65 | 74 | 83 | 90 | 100 | .. | .. |

*Revolutions per Minute — Small Sprocket*

TYPE I          TYPE II          TYPE III

### 1½″ Pitch — Horsepower per Inch of Chain Width

| No. of Teeth Small Sprkt. | 100 | 200 | 300 | 400 | 500 | 600 | 700 | 800 | 900 | 1000 | 1200 |
|---|---|---|---|---|---|---|---|---|---|---|---|
| *19 | 8.0 | 15 | 21 | 28 | 31 | 35 | 39 | 40 | 41 | 43 | .. |
| 21 | 8.8 | 16 | 24 | 30 | 36 | 40 | 44 | 46 | 49 | 49 | .. |
| 23 | 10 | 19 | 26 | 34 | 40 | 45 | 49 | 53 | 55 | 56 | 55 |
| 25 | 10 | 20 | 29 | 38 | 44 | 50 | 55 | 59 | 61 | 65 | 64 |
| 27 | 11 | 23 | 31 | 40 | 48 | 54 | 60 | 64 | 68 | 70 | 70 |
| 29 | 13 | 24 | 34 | 44 | 51 | 59 | 65 | 70 | 74 | 75 | 76 |
| 31 | 14 | 25 | 36 | 46 | 55 | 64 | 70 | 75 | 79 | 81 | 83 |
| 33 | 14 | 28 | 39 | 50 | 59 | 68 | 75 | 80 | 85 | 88 | 89 |
| 35 | 15 | 29 | 41 | 53 | 63 | 71 | 79 | 85 | 90 | 93 | 94 |
| 37 | 16 | 30 | 44 | 59 | 66 | 76 | 84 | 90 | 96 | 99 | .. |
| 40 | 18 | 33 | 48 | 66 | 73 | 83 | 90 | 98 | 105 | .. | .. |
| 45 | 19 | 38 | 54 | 68 | 81 | 93 | 101 | 108 | 113 | .. | .. |
| 50 | 21 | 41 | 59 | 75 | 89 | 101 | 111 | 118 | .. | .. | .. |

*Revolutions per Minute — Small Sprocket*

TYPE I     TYPE II          TYPE III

### 2″ Pitch — Horsepower per Inch of Chain Width

| No. of Teeth Small Sprkt. | 100 | 200 | 300 | 400 | 500 | 600 | 700 | 800 | 900 |
|---|---|---|---|---|---|---|---|---|---|
| *19 | 14 | 26 | 36 | 44 | 50 | 54 | 56 | .. | .. |
| 21 | 16 | 29 | 40 | 50 | 53 | 63 | 65 | .. | .. |
| 23 | 17 | 33 | 45 | 55 | 64 | 70 | 74 | 75 | .. |
| 25 | 18 | 35 | 49 | 61 | 70 | 78 | 83 | 85 | 85 |
| 27 | 20 | 38 | 54 | 66 | 78 | 85 | 91 | 94 | 94 |
| 29 | 21 | 41 | 58 | 73 | 84 | 93 | 99 | 103 | 103 |
| 31 | 23 | 44 | 63 | 78 | 90 | 100 | 106 | 110 | 110 |
| 33 | 25 | 46 | 66 | 83 | 96 | 106 | 114 | 118 | 118 |
| 35 | 26 | 50 | 71 | 88 | 103 | 114 | 121 | 125 | 125 |
| 37 | 28 | 53 | 75 | 93 | 110 | 124 | 128 | 131 | .. |
| 40 | 30 | 58 | 81 | 101 | 118 | 129 | 138 | 141 | .. |
| 45 | 34 | 64 | 90 | 113 | 131 | 144 | 151 | .. | .. |
| 50 | 38 | 71 | 100 | 125 | 144 | 156 | .. | .. | .. |

*Revolutions per Minute — Small Sprocket*

TYPE I     TYPE II          TYPE III

* For best results, smaller sprocket should have at least 21 teeth.

## Sprocket Wheels for Ordinary Link Chain

$$D = \sqrt{\left(\frac{r}{\sin\alpha}\right)^2 + \left(\frac{d}{\cos\alpha}\right)^2} \qquad \alpha = \frac{90}{N} \qquad N = \text{number of teeth.}$$

| No. of Teeth $N =$ | | 7 | 8 | 9 | 10 | 11 | 12 | 13 | 14 | 15 | 16 | 17 | | |
|---|---|---|---|---|---|---|---|---|---|---|---|---|---|---|
| Angle $\alpha =$ | | 12°51′ | 11°15′ | 10°0′ | 9°0′ | 8°11′ | 7°30′ | 6°55′ | 6°25′ | 6°0′ | 5°37′ | 5°17′ | | |
| $d$=Size of Chain | $L$=Length of Link | $W$=Width of Link | | $D$ = Pitch Diameter in Inches | | | | | | | | | $X$ | $y$ |
| $\frac{3}{16}$ | $1\frac{3}{8}$ | $\frac{13}{16}$ | 4.50 | 5.13 | 5.76 | 6.40 | 7.03 | 7.66 | 8.29 | 8.93 | 9.57 | 10.20 | 10.84 | $\frac{1}{16}$ | $\frac{3}{32}$ |
| $\frac{1}{4}$ | $1\frac{1}{2}$ | 1 | 4.50 | 5.13 | 5.76 | 6.40 | 7.03 | 7.66 | 8.29 | 8.93 | 9.57 | 10.20 | 10.84 | $\frac{3}{32}$ | $\frac{3}{32}$ |
| $\frac{5}{16}$ | $1\frac{3}{4}$ | $1\frac{3}{16}$ | 5.06 | 5.77 | 6.48 | 7.18 | 7.91 | 8.62 | 9.33 | 10.05 | 10.76 | 11.47 | 12.19 | $\frac{3}{32}$ | $\frac{3}{32}$ |
| $\frac{3}{8}$ | 2 | $1\frac{3}{8}$ | 5.63 | 6.42 | 7.21 | 8.00 | 8.79 | 9.59 | 10.38 | 11.17 | 11.96 | 12.76 | 13.56 | $\frac{3}{32}$ | $\frac{3}{32}$ |
| $\frac{7}{16}$ | $2\frac{1}{4}$ | $1\frac{9}{16}$ | 6.18 | 7.06 | 7.74 | 8.79 | 9.67 | 10.53 | 11.41 | 12.28 | 13.16 | 14.03 | 14.90 | $\frac{3}{32}$ | $\frac{1}{16}$ |
| $\frac{1}{2}$ | $2\frac{1}{2}$ | $1\frac{3}{4}$ | 6.76 | 7.71 | 8.65 | 9.61 | 10.55 | 11.49 | 12.45 | 13.40 | 14.35 | 15.30 | 16.26 | $\frac{3}{32}$ | $\frac{1}{16}$ |
| $\frac{9}{16}$ | $2\frac{7}{8}$ | $1\frac{15}{16}$ | 7.88 | 8.97 | 10.08 | 11.19 | 12.30 | 13.41 | 14.52 | 15.63 | 16.74 | 17.85 | 18.97 | $\frac{1}{8}$ | $\frac{1}{16}$ |
| $\frac{5}{8}$ | $3\frac{1}{4}$ | $2\frac{1}{8}$ | 9.01 | 10.27 | 11.53 | 12.80 | 14.07 | 15.33 | 16.60 | 17.90 | 19.14 | 20.41 | 21.68 | $\frac{1}{8}$ | $\frac{1}{16}$ |
| $\frac{11}{16}$ | $3\frac{1}{2}$ | $2\frac{5}{16}$ | 9.58 | 10.91 | 12.26 | 13.61 | 14.95 | 16.29 | 17.65 | 18.99 | 20.34 | 21.69 | 23.04 | $\frac{1}{8}$ | $\frac{1}{16}$ |
| $\frac{3}{4}$ | $3\frac{3}{4}$ | $2\frac{1}{2}$ | 10.14 | 11.56 | 12.98 | 14.40 | 15.83 | 17.26 | 18.68 | 20.06 | 21.54 | 22.97 | 24.40 | $\frac{1}{8}$ | $\frac{1}{16}$ |
| $\frac{13}{16}$ | 4 | $2\frac{11}{16}$ | 10.71 | 12.20 | 13.72 | 15.21 | 16.71 | 18.20 | 19.72 | 21.23 | 22.74 | 24.24 | 25.75 | $\frac{1}{8}$ | $\frac{1}{16}$ |
| $\frac{7}{8}$ | $4\frac{1}{4}$ | 3 | 11.27 | 12.85 | 14.43 | 16.01 | 17.55 | 19.17 | 20.76 | 22.35 | 23.93 | 25.52 | 27.11 | $\frac{1}{8}$ | $\frac{1}{16}$ |
| $\frac{15}{16}$ | $4\frac{1}{2}$ | $3\frac{1}{4}$ | 11.84 | 13.50 | 15.15 | 16.81 | 18.47 | 20.13 | 21.80 | 23.46 | 25.13 | 26.80 | 28.47 | $\frac{1}{8}$ | $\frac{1}{16}$ |
| 1 | $4\frac{3}{4}$ | $3\frac{1}{2}$ | 12.40 | 14.13 | 15.87 | 17.61 | 19.35 | 21.09 | 22.84 | 24.58 | 26.33 | 28.08 | 29.83 | $\frac{1}{8}$ | … |

| No. of Teeth $N =$ | | 18 | 19 | 20 | 21 | 22 | 23 | 24 | 25 | 26 | 27 | 28 | $X$ | $y$ |
|---|---|---|---|---|---|---|---|---|---|---|---|---|---|---|
| Angle $\alpha =$ | | 5°0′ | 4°44′ | 4°30′ | 4°17′ | 4°6′ | 3°55′ | 3°45′ | 3°36′ | 3°28′ | 3°20′ | 3°13′ | | |
| $\frac{3}{16}$ | $1\frac{3}{8}$ | $\frac{13}{16}$ | 11.47 | 12.11 | 12.75 | 13.38 | 14.02 | 14.66 | 15.29 | 15.93 | 16.56 | 17.20 | 17.84 | $\frac{1}{16}$ | $\frac{3}{32}$ |
| $\frac{1}{4}$ | $1\frac{1}{2}$ | 1 | 11.47 | 12.11 | 12.75 | 13.38 | 14.02 | 14.66 | 15.29 | 15.93 | 16.56 | 17.20 | 17.84 | $\frac{3}{32}$ | $\frac{3}{32}$ |
| $\frac{5}{16}$ | $1\frac{3}{4}$ | $1\frac{3}{16}$ | 12.91 | 13.62 | 14.34 | 15.05 | 15.77 | 16.49 | 17.20 | 17.92 | 18.62 | 19.34 | 20.06 | $\frac{3}{32}$ | $\frac{3}{32}$ |
| $\frac{3}{8}$ | 2 | $1\frac{3}{8}$ | 14.36 | 15.16 | 15.96 | 16.74 | 17.53 | 18.32 | 19.11 | 19.90 | 20.70 | 21.50 | 22.29 | $\frac{3}{32}$ | $\frac{3}{32}$ |
| $\frac{7}{16}$ | $2\frac{1}{4}$ | $1\frac{9}{16}$ | 15.78 | 16.65 | 17.53 | 18.40 | 19.27 | 20.15 | 21.02 | 21.90 | 22.77 | 23.65 | 24.52 | $\frac{3}{32}$ | $\frac{1}{16}$ |
| $\frac{1}{2}$ | $2\frac{1}{2}$ | $1\frac{3}{4}$ | 17.21 | 18.16 | 19.12 | 20.07 | 21.03 | 21.98 | 22.94 | 23.89 | 24.85 | 25.80 | 26.75 | $\frac{3}{32}$ | $\frac{1}{16}$ |
| $\frac{9}{16}$ | $2\frac{7}{8}$ | $1\frac{15}{16}$ | 20.08 | 21.19 | 22.30 | 23.42 | 24.53 | 25.64 | 26.76 | 27.87 | 28.98 | 30.10 | 31.21 | $\frac{1}{8}$ | $\frac{1}{16}$ |
| $\frac{5}{8}$ | $3\frac{1}{4}$ | $2\frac{1}{8}$ | 22.95 | 24.22 | 25.50 | 26.77 | 28.03 | 29.31 | 30.58 | 31.85 | 33.13 | 34.40 | 35.67 | $\frac{1}{8}$ | $\frac{1}{16}$ |
| $\frac{11}{16}$ | $3\frac{1}{2}$ | $2\frac{5}{16}$ | 24.34 | 25.73 | 27.09 | 28.44 | 29.79 | 31.14 | 32.49 | 33.84 | 35.20 | 36.55 | 37.90 | $\frac{1}{8}$ | $\frac{1}{16}$ |
| $\frac{3}{4}$ | $3\frac{3}{4}$ | $2\frac{1}{2}$ | 25.83 | 27.26 | 28.69 | 30.12 | 31.55 | 32.97 | 34.41 | 35.84 | 37.27 | 38.70 | 40.04 | $\frac{1}{8}$ | $\frac{1}{16}$ |
| $\frac{13}{16}$ | 4 | $2\frac{11}{16}$ | 27.26 | 28.77 | 30.28 | 31.79 | 33.30 | 34.81 | 36.32 | 37.83 | 39.34 | 40.85 | … | $\frac{1}{8}$ | $\frac{1}{16}$ |
| $\frac{7}{8}$ | $4\frac{1}{4}$ | 3 | 28.70 | 30.29 | 31.88 | 33.46 | 35.04 | 36.63 | 38.23 | 39.82 | 41.41 | … | … | $\frac{1}{8}$ | $\frac{1}{16}$ |
| $\frac{15}{16}$ | $4\frac{1}{2}$ | $3\frac{1}{4}$ | 30.14 | 31.80 | 33.46 | 35.13 | 36.83 | 38.48 | 40.15 | … | … | … | … | $\frac{1}{8}$ | $\frac{1}{16}$ |
| 1 | $4\frac{3}{4}$ | $3\frac{1}{2}$ | 31.57 | 33.31 | 35.06 | 36.81 | 38.56 | 40.30 | … | … | … | … | … | $\frac{1}{8}$ | … |

# CRANE CHAIN AND HOOKS

**Material for Crane Chains.** — The best material for crane and hoisting chains is a good grade of wrought iron, in which the percentage of phosphorus, sulphur, silicon and other impurities is comparatively low. The tensile strength of the best grades of wrought iron does not exceed 46,000 pounds per square inch, whereas mild steel with about 0.15 per cent carbon has a tensile strength nearly double this amount. The ductility and toughness of wrought iron, however, is greater than that of ordinary commercial steel, and for this reason it is preferable for chains subjected to heavy intermittent strains, because wrought iron will always give warning by bending or stretching, before breaking. Another important reason for using wrought iron in preference to steel is that a perfect weld can be effected more easily.

**Strength of Chains.** — When calculating the strength of chains, it should be observed that the strength of a link subjected to tensile stresses is not equal to twice the strength of an iron bar of the same diameter as the link stock, but is a certain amount less, owing to the bending action caused by the manner in which the load is applied to the link. The strength is also reduced somewhat by the weld. The following empirical formula is commonly used for calculating the breaking load, in pounds, of wrought-iron crane chains:

$$W = 54,000\,D^2$$

in which $W$ = breaking load in pounds, and $D$ = diameter of bar (in inches) from which links are made. The working load for chains should not exceed one-third the value of $W$, and, in many cases, it is one-fourth or one-fifth of the breaking load. When a chain is wound around a casting and severe bending stresses are introduced, a greater factor of safety should be used.

### Safe Loads in Tons for Ropes, Chains and Cables

| Manila Rope* | | | | Chains | | | | Wire Cable | | | |
|---|---|---|---|---|---|---|---|---|---|---|---|
| Rope Diam. | Single Rope | Two Part | Four Part | Diam. Link Stock | Single Chain | Two Part | Four Part | Cable Diam. | Single Cable | Two Part | Four Part |
| ½ | ⅛ | ¼ | ½ | ¼ | ½ | ⅞ | 1½ | ½ | 1 | 2 | 3½ |
| ⅝ | ¼ | ½ | ¾ | ⅜ | 1 | 1¾ | 3 | ⅝ | 1¾ | 3¼ | 6½ |
| ¾ | ⅜ | ¾ | 1¼ | ½ | 2 | 3½ | 6 | ¾ | 2½ | 4½ | 9 |
| ⅞ | ½ | 1 | 2 | ⅝ | 3 | 5 | 9 | ⅞ | 3¼ | 6 | 12 |
| 1 | ¾ | 1½ | 2½ | ¾ | 5 | 9 | 15 | 1 | 4 | 8 | 16 |
| 1¼ | 1 | 2 | 3 | ⅞ | 6 | 10½ | 18 | 1¼ | 6 | 12 | 24 |
| 1½ | 1¼ | 2½ | 4 | 1 | 8 | 14 | 24 | 1½ | 10 | 19 | 36 |
| 1¾ | 2 | 4 | 6 | 1⅛ | 11 | 19 | 33 | 1¾ | 13 | 25 | 48 |
| 2 | 2½ | 5 | 8 | 1¼ | 13 | 23 | 39 | 2 | 16 | 32 | 60 |
| 2¼ | 3½ | 6½ | 11 | 1½ | 18 | 32 | 54 | ... | ... | ... | ... |
| 2½ | 4½ | 8 | 13 | .... | ... | ... | ... | ... | ... | ... | ... |

\* These figures apply only to a rope in fairly good condition.

**Care of Hoisting and Crane Chains.** — Chains used for hoisting heavy loads are subject to deterioration, both apparent and invisible. The links wear, and repeated loading causes localized deformations to form cracks which spread until the links fail. Chain wear can be reduced by occasional lubrication. The life of a chain can be prolonged by frequent annealing or normalizing unless it has been so highly or frequently stressed that small cracks have formed. If this be the case, annealing or normalizing will not "heal" the material, and the links will eventually fracture. To anneal a chain, heat it to cherry-red and allow it to cool slowly. This should be done every six months, and oftener if the chain is subjected to unusually severe service. Chains should be examined periodically for twists, as a twisted chain will wear rapidly. Any links which have worn excessively should be replaced with new ones, so that every link will do its full share of work during the life of the chain, without exceeding the limit of safety. Chains for hoisting purposes should be made with short links, so that they will wrap closely around the sheaves or drums without bending. The diameter of the winding drums should be not less than 25 or 30 times the diameter of the iron used for the links.

**Studded Chains.** — Tests have demonstrated that the ultimate breaking strength of a chain with studded links is less than that of an unstudded chain. This is probably due to the fact that the open links of an unstudded chain collapse until the sides are approximately parallel, so that the stresses are lower than in the studded links, the sides of which are prevented from collapsing by the studs. The principal function of the stud is to prevent the chain from kinking and catching, so that it will run free from chain lockers, etc. The stud also prevents the chain from becoming rigid under heavy strains.

**Studded Cable Chain**

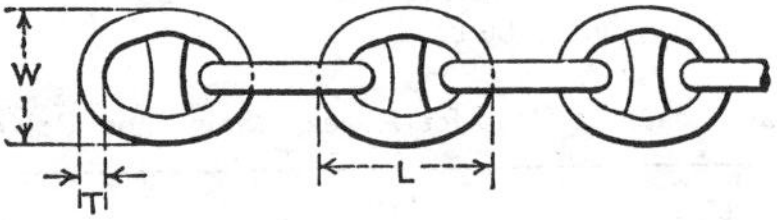


| Size of Chain in Inches | Length of Link in Inches | Width of Link in Inches | Weight per Foot of Chain | Proof Test | Size of Chain in Inches | Length of Link in Inches | Width of Link in Inches | Weight per Foot of Chain | Proof Test |
|---|---|---|---|---|---|---|---|---|---|
| *T* | *L* | *W* | Pounds | Tons | *T* | *L* | *W* | Pounds | Tons |
| 3/4 | 4 3/8 | 2 3/4 | 5.5 | 10.1 | 1 1/2 | 8 1/2 | 5 3/8 | 21.2 | 40.5 |
| 13/16 | 4 3/4 | 3 | 6.3 | 12.0 | 1 9/16 | 8 7/8 | 5 5/8 | 23.8 | 44.0 |
| 7/8 | 5 | 3 1/4 | 8.2 | 13.7 | 1 5/8 | 9 1/4 | 5 7/8 | 25.0 | 47.5 |
| 15/16 | 5 3/8 | 3 1/2 | 9.2 | 15.7 | 1 11/16 | 9 5/8 | 6 | 26.2 | 51.2 |
| 1 | 5 7/8 | 3 3/4 | 10.2 | 18.0 | 1 3/4 | 10 | 6 1/4 | 28.8 | 55.2 |
| 1 1/16 | 6 1/4 | 3 7/8 | 11.5 | 20.3 | 1 7/8 | 10 1/2 | 6 3/4 | 33.8 | 63.3 |
| 1 1/8 | 6 1/2 | 4 1/8 | 12.3 | 22.8 | 1 15/16 | 10 3/4 | 7 | 35.8 | 67.5 |
| 1 3/16 | 6 3/4 | 4 1/4 | 13.5 | 25.5 | 2 | 11 1/8 | 7 1/4 | 38.8 | 72.0 |
| 1 1/4 | 7 1/8 | 4 1/2 | 15.0 | 28.1 | 2 1/16 | 11 1/2 | 7 1/2 | 42.3 | 76.5 |
| 1 5/16 | 7 3/8 | 4 5/8 | 16.2 | 31.0 | 2 1/8 | 12 | 7 3/4 | 46.0 | 81.2 |
| 1 3/8 | 7 3/4 | 4 7/8 | 18.3 | 34.0 | 2 3/16 | 12 1/2 | 8 | 48.3 | 86.1 |
| 1 7/16 | 8 1/8 | 5 1/8 | 18.8 | 37.2 | 2 1/4 | 13 | 8 1/4 | 50.0 | 91.0 |

Note: Safe working loads are one-half of proof test loads.

## Close-link Hoisting, Sling and Crane Chain

| Size | Standard Pitch, P | Average Weight per Foot in Pounds | Outside Length, L | Outside Width, W | Average Safe Working Load in Pounds | Proof Test in Pounds* | Approximate Breaking Strain in Pounds |
|---|---|---|---|---|---|---|---|
| 1/4 | 25/32 | 3/4 | 1 5/16 | 7/8 | 1,200 | 2,500 | 5,000 |
| 5/16 | 27/32 | 1 | 1 1/2 | 1 1/16 | 1,700 | 3,500 | 7,000 |
| 3/8 | 31/32 | 1 1/2 | 1 3/4 | 1 1/4 | 2,500 | 5,000 | 10,000 |
| 7/16 | 1 5/32 | 2 | 2 1/16 | 1 3/8 | 3,500 | 7,000 | 14,000 |
| 1/2 | 1 11/32 | 2 1/2 | 2 3/8 | 1 11/16 | 4,500 | 9,000 | 18,000 |
| 9/16 | 1 15/32 | 3 1/4 | 2 5/8 | 1 7/8 | 5,500 | 11,000 | 22,000 |
| 5/8 | 1 28/32 | 4 | 3 | 2 1/16 | 6,700 | 14,000 | 27,000 |
| 11/16 | 1 18/16 | 5 | 3 1/4 | 2 1/4 | 8,100 | 17,000 | 32,500 |
| 3/4 | 1 15/16 | 6 1/4 | 3 1/2 | 2 1/2 | 10,000 | 20,000 | 40,000 |
| 13/16 | 2 1/16 | 7 | 3 3/4 | 2 11/16 | 10,500 | 23,000 | 42,000 |
| 7/8 | 2 3/16 | 8 | 4 | 2 7/8 | 12,000 | 26,000 | 48,000 |
| 15/16 | 2 7/16 | 9 | 4 3/8 | 3 1/16 | 13,500 | 29,000 | 54,000 |
| 1 | 2 1/2 | 10 | 4 5/8 | 3 1/4 | 15,200 | 32,000 | 61,000 |
| 1 1/16 | 2 5/8 | 12 | 4 7/8 | 3 5/16 | 17,200 | 35,000 | 69,000 |
| 1 1/8 | 2 3/4 | 13 | 5 1/8 | 3 3/4 | 19,500 | 40,000 | 78,000 |
| 1 3/16 | 3 1/16 | 14 1/2 | 5 9/16 | 3 7/8 | 22,000 | 46,000 | 88,000 |
| 1 1/4 | 3 1/8 | 16 | 5 3/4 | 4 1/8 | 23,700 | 51,000 | 95,000 |
| 1 5/16 | 3 3/8 | 17 1/2 | 6 1/8 | 4 1/4 | 26,000 | 54,000 | 104,000 |
| 1 3/8 | 3 9/16 | 19 | 6 7/16 | 4 9/16 | 28,500 | 58,000 | 114,000 |
| 1 7/16 | 3 11/16 | 21 1/2 | 6 11/16 | 4 3/4 | 30,500 | 62,000 | 122,000 |
| 1 1/2 | 3 7/8 | 23 | 7 | 5 | 33,500 | 67,000 | 134,000 |
| 1 9/16 | 4 | 25 | 7 3/8 | 5 5/16 | 35,500 | 70,500 | 142,000 |
| 1 5/8 | 4 1/4 | 28 | 7 3/4 | 5 1/2 | 38,500 | 77,000 | 154,000 |
| 1 11/16 | 4 1/2 | 30 | 8 1/8 | 5 11/16 | 39,500 | 79,000 | 158,000 |
| 1 3/4 | 4 3/4 | 31 | 8 1/2 | 5 7/8 | 41,500 | 83,000 | 166,000 |
| 1 13/16 | 5 | 33 | 8 7/8 | 6 1/16 | 44,500 | 89,000 | 178,000 |
| 1 7/8 | 5 1/4 | 35 | 9 1/4 | 6 3/8 | 47,500 | 95,000 | 190,000 |
| 1 15/16 | 5 1/2 | 38 | 9 5/8 | 6 9/16 | 50,500 | 101,000 | 202,000 |
| 2 | 5 3/4 | 40 | 10 | 6 3/4 | 54,000 | 108,000 | 216,000 |
| 2 1/16 | 6 | 43 | 10 3/8 | 6 15/16 | 57,500 | 115,000 | 230,000 |
| 2 1/8 | 6 1/4 | 47 | 10 3/4 | 7 1/8 | 61,000 | 122,000 | 244,000 |
| 2 3/16 | 6 1/2 | 50 | 11 1/8 | 7 5/16 | 64,500 | 129,000 | 258,000 |
| 2 1/4 | 6 3/4 | 53 | 11 1/2 | 7 5/8 | 68,200 | 136,500 | 273,000 |
| 2 3/8 | 6 7/8 | 58 1/2 | 11 7/8 | 8 | 76,000 | 152,000 | 304,000 |
| 2 1/2 | 7 | 65 | 12 1/4 | 8 3/8 | 84,200 | 168,500 | 337,000 |
| 2 5/8 | 7 1/8 | 70 | 12 5/8 | 8 3/8 | 90,500 | 181,000 | 362,000 |
| 2 3/4 | 7 1/4 | 73 | 13 | 9 1/8 | 96,700 | 193,500 | 387,000 |
| 2 7/8 | 7 1/2 | 76 | 13 1/2 | 9 1/2 | 103,000 | 206,000 | 412,000 |
| 3 | 7 3/4 | 86 | 14 | 9 7/8 | 109,000 | 218,000 | 436,000 |

* Chains tested to U. S. Government and American Bureau of Shipping requirements.

**Safe Loads for Ropes and Chains.** — Safe loads recommended for wire rope or chain slings depends not only upon the strength of the sling but upon the method of applying it to the load, as shown by the accompanying table giving safe loads as prepared originally by the National Founders Association. The loads recommended in this table are more conservative than those usually specified, in order to provide ample allowance for some unobserved weakness in the sling, or the possibility of excessive strains due to misjudgment or accident.

The table in the first column under safe loads, gives the allowable load for a single rope or chain when it is subjected to a straight or downward pull. The three columns which follow give the loads per single rope or chain which can be supported safely by slings. Since the table gives the load *per rope or chain*, a sling arranged in the usual manner and consisting of two ropes or chains, will support safely a load equal to twice the load given in the table. The safe load capacity, however, depends upon the angle between the sling ropes or chains. For example, the safe load recommended for a ⅜-inch plow steel wire rope is 1500 pounds provided the rope is vertical or subjected to a straight downward pull as would be the case when a single rope is used. If a two-rope sling is employed, then the load which can be supported will be reduced an amount depending upon the inclination. To illustrate, if the angle between each side of the sling and the horizontal is 60 degrees, then the allowable load which can be supported per rope is given as 1275 pounds for ⅜-inch plow steel, thus making the total load for a two-rope sling 2550 pounds. When the inclination is reduced to 30 degrees, the allowable load per rope is only 750 pounds, thus making the total load for a two-rope sling equal to 1500 pounds which is the same as that specified for a single rope used vertically.

If the allowable load for a single vertical rope is divided by the cosecant of the angle between one side of a sling and the horizontal, the result will indicate the load reduction required in order to avoid subjecting the inclined rope to a greater stress than when the same rope is in a vertical position. To illustrate, the allowable load for a ¾-inch plow steel rope subjected to a vertical pull is 6000 pounds. If a two-part sling made of this rope is to be used at an angle of 30 degrees, the load reduction equals 6000 ÷ cosecant 30 degrees = 6000 ÷ 2 = 3000 pounds per rope; hence, this sling will support safely a total load of 6000 pounds. The loads as given in the table are well within the ultimate strength of the rope and chain sizes listed to provide an ample factor of safety.

*Protection from Sharp Corners:* When the load to be lifted has sharp corners or edges, as is often the case with castings, and with structural steel and other similar objects, pads or wooden protective pieces should be applied at these corners, to prevent the slings from being abraded or otherwise damaged where they come in contact with the load. This is especially important when the slings consist of wire cable or fiber rope, although it should also be done even when they are made of chain. Wooden corner-pieces are often provided for use in hoisting loads with sharp angles. If pads of burlap or other soft material are used, they should be thick and heavy enough to sustain the pressure, and distribute it over a considerable area, instead of allowing it to be concentrated directly at the edges of the part to be lifted.

**Loads Lifted by Crane Chains.** — To find the approximate weight a chain will lift when rove as a tackle, multiply the safe load given in the table "Close-link Hoisting, Sling and Crane Chain" by the number of parts or chains at the movable block, and subtract one-quarter for frictional resistance. To find the size of chain required for lifting a given weight, divide the weight by the number of chains at the movable block, and add one-third for friction; next find in the column headed "Average Safe Working Load" the corresponding load, and then the corresponding size of chain in the column headed "Size." In case of heavy chain or where chain is unusually long, the weight of the chain itself should also be considered.

## Safe Loads in Pounds for Ropes and Chains

Prepared by National Founders' Association. When handling molten metal, wire ropes and chains should be 25 per cent stronger than indicated in the table.

| Kind of Rope or Chain | Diameter of Rope, or of Rod or Bar for Chain Links, Inch | The safe loads in table are for each *single* rope or chain. When used double or in other multiples, the loads may be increased proportionately.* | | | |
|---|---|---|---|---|---|
| | | Rope or chain Vertical | Sling at 60° | Sling at 45° | Sling at 30° |
| PLOW STEEL | ⅜ | 1,500 | 1,275 | 1,050 | 750 |
| WIRE ROPE | ½ | 2,400 | 2,050 | 1,700 | 1,200 |
| (6 strands | ⅝ | 4,000 | 3,400 | 2,800 | 2,000 |
| of 19 or 37 | ¾ | 6,000 | 5,100 | 4,200 | 3,000 |
| wires.) | ⅞ | 8,000 | 6,800 | 5,600 | 4,000 |
| If crucible | 1 | 10,000 | 8,500 | 7,000 | 5,000 |
| steel rope is | 1⅛ | 13,000 | 11,000 | 9,000 | 6,500 |
| used, reduce | 1¼ | 16,000 | 13,500 | 11,000 | 8,000 |
| loads one-| 1⅜ | 19,000 | 16,000 | 13,000 | 9,500 |
| fifth. | 1½ | 22,000 | 19,000 | 16,000 | 11,000 |
| CRANE | ¼ | 600 | 500 | 425 | 300 |
| CHAIN | ⅜ | 1,200 | 1,025 | 850 | 600 |
| (Best Grade | ½ | 2,400 | 2,050 | 1,700 | 1,200 |
| of Wrought | ⅝ | 4,000 | 3,400 | 2,800 | 2,000 |
| Iron, Hand-| ¾ | 5,500 | 4,700 | 3,900 | 2,750 |
| made, | ⅞ | 7,500 | 6,400 | 5,200 | 3,700 |
| Tested, | 1 | 9,500 | 8,000 | 6,600 | 4,700 |
| Short-linked | 1⅛ | 12,000 | 10,200 | 8,400 | 6,000 |
| chain.) | 1¼ | 15,000 | 12,750 | 10,500 | 7,500 |
| | 1⅜ | 22,000 | 19,000 | 16,000 | 11,000 |
| | ⅜ | 120 | 100 | 85 | 60 |
| | ½ | 250 | 210 | 175 | 125 |
| | ⅝ | 360 | 300 | 250 | 180 |
| | ¾ | 520 | 440 | 360 | 260 |
| MANILA | ⅞ | 620 | 520 | 420 | 300 |
| ROPE | 1 | 750 | 625 | 525 | 375 |
| (Best long | 1⅛ | 1,000 | 850 | 700 | 500 |
| Fiber | 1¼ | 1,200 | 1,025 | 850 | 600 |
| Grade.) | 1½ | 1,600 | 1,350 | 1,100 | 800 |
| | 1¾ | 2,100 | 1,800 | 1,500 | 1,050 |
| | 2 | 2,800 | 2,400 | 2,000 | 1,400 |
| | 2½ | 4,000 | 3,400 | 2,800 | 2,000 |
| | 3 | 6,000 | 5,100 | 4,200 | 3,000 |

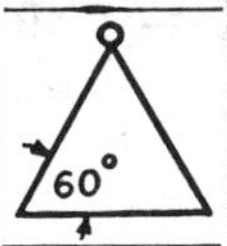

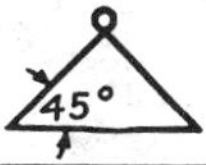

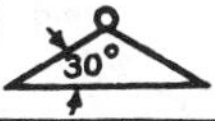

* Note that when the angle between one side of sling and the horizontal, is reduced, the allowable load which may safely be supported per rope or chain, is also reduced. See "Safe Loads for Ropes and Chains."

### Dimensions of Crane Hooks

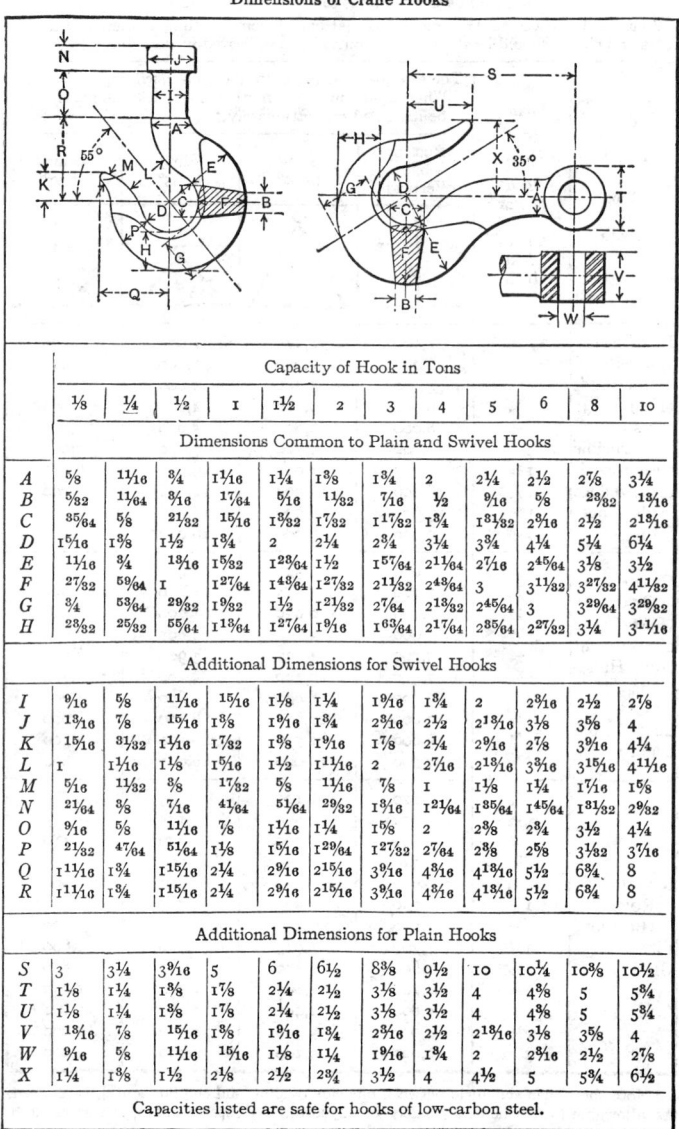

| | Capacity of Hook in Tons | | | | | | | | | | | |
|---|---|---|---|---|---|---|---|---|---|---|---|---|
| | 1/8 | 1/4 | 1/2 | 1 | 1 1/2 | 2 | 3 | 4 | 5 | 6 | 8 | 10 |
| | Dimensions Common to Plain and Swivel Hooks | | | | | | | | | | | |
| A | 5/8 | 11/16 | 3/4 | 1 1/16 | 1 1/4 | 1 3/8 | 1 3/4 | 2 | 2 1/4 | 2 1/2 | 2 7/8 | 3 1/4 |
| B | 5/32 | 11/64 | 3/16 | 17/64 | 5/16 | 11/32 | 7/16 | 1/2 | 9/16 | 5/8 | 28/32 | 13/16 |
| C | 35/64 | 5/8 | 21/32 | 15/16 | 1 3/32 | 1 7/32 | 1 17/32 | 1 3/4 | 1 31/32 | 2 3/16 | 2 1/2 | 2 13/16 |
| D | 1 5/16 | 1 3/8 | 1 1/2 | 1 3/4 | 2 | 2 1/4 | 2 3/4 | 3 1/4 | 3 3/4 | 4 1/4 | 5 1/4 | 6 1/4 |
| E | 11/16 | 3/4 | 13/16 | 15/32 | 1 23/64 | 1 1/2 | 1 57/64 | 2 11/64 | 2 7/16 | | 3 1/8 | 3 1/2 |
| F | 27/32 | 59/64 | 1 | 1 27/64 | 1 48/64 | 1 27/32 | 2 11/32 | 2 48/64 | 3 | 3 11/32 | 3 27/32 | 4 11/32 |
| G | 3/4 | 53/64 | 29/32 | 1 9/32 | 1 1/2 | 1 21/32 | 2 7/64 | 2 13/32 | 2 45/64 | 3 | 3 29/64 | 3 29/32 |
| H | 23/32 | 25/32 | 55/64 | 1 13/64 | 1 27/64 | 1 9/16 | 1 63/64 | 2 17/64 | 2 85/64 | 2 27/32 | 3 1/4 | 3 11/16 |
| | Additional Dimensions for Swivel Hooks | | | | | | | | | | | |
| I | 9/16 | 5/8 | 11/16 | 15/16 | 1 1/8 | 1 1/4 | 1 9/16 | 1 3/4 | 2 | 2 3/16 | 2 1/2 | 2 7/8 |
| J | 13/16 | 7/8 | 15/16 | 1 3/8 | 1 9/16 | 1 3/4 | 2 3/16 | 2 1/2 | 2 13/16 | 3 1/8 | 3 5/8 | 4 |
| K | 15/16 | 31/32 | 1 1/16 | 1 7/32 | 1 3/8 | 1 9/16 | 1 7/8 | 2 1/4 | 2 9/16 | 2 7/8 | 3 9/16 | 4 1/4 |
| L | 1 | 1 1/16 | 1 1/8 | 1 5/16 | 1 1/2 | 1 11/16 | 2 | 2 7/16 | 2 13/16 | 3 3/16 | 3 15/16 | 4 11/16 |
| M | 5/16 | 11/32 | 3/8 | 17/32 | 5/8 | 11/16 | 7/8 | 1 | 1 1/8 | 1 1/4 | 1 7/16 | 1 5/8 |
| N | 21/64 | 3/8 | 7/16 | 41/64 | 51/64 | 29/32 | 1 3/16 | 1 21/64 | 1 85/64 | 1 45/64 | 1 31/32 | 2 9/32 |
| O | 9/16 | 5/8 | 11/16 | 7/8 | 1 1/16 | 1 1/4 | 1 5/8 | 2 | 2 3/8 | 2 3/4 | 3 1/2 | 4 1/4 |
| P | 21/32 | 47/64 | 51/64 | 1 1/8 | 1 5/16 | 1 29/64 | 1 27/32 | 2 7/64 | 2 3/8 | 2 5/8 | 3 1/32 | 3 7/16 |
| Q | 1 11/16 | 1 3/4 | 1 15/16 | 2 1/4 | 2 9/16 | 2 15/16 | 3 9/16 | 4 3/16 | 4 13/16 | 5 1/2 | 6 3/4 | 8 |
| R | 1 11/16 | 1 3/4 | 1 15/16 | 2 1/4 | 2 9/16 | 2 15/16 | 3 9/16 | 4 3/16 | 4 13/16 | 5 1/2 | 6 3/4 | 8 |
| | Additional Dimensions for Plain Hooks | | | | | | | | | | | |
| S | 3 | 3 1/4 | 3 9/16 | 5 | 6 | 6 1/2 | 8 5/8 | 9 1/2 | 10 | 10 1/4 | 10 3/8 | 10 1/2 |
| T | 1 1/8 | 1 1/4 | 1 3/8 | 1 7/8 | 2 1/4 | 2 1/2 | 3 1/8 | 3 1/2 | 4 | 4 3/8 | 5 | 5 3/4 |
| U | 1 1/8 | 1 1/4 | 1 3/8 | 1 7/8 | 2 1/4 | 2 1/2 | 3 1/8 | 3 1/2 | 4 | 4 3/8 | 5 | 5 3/4 |
| V | 13/16 | 7/8 | 15/16 | 1 3/8 | 1 9/16 | 1 3/4 | 2 3/16 | 2 1/2 | 2 13/16 | 3 1/8 | 3 5/8 | 4 |
| W | 9/16 | 5/8 | 11/16 | 15/16 | 1 1/8 | 1 1/4 | 1 9/16 | 1 3/4 | 2 | 2 3/16 | 2 1/2 | 2 7/8 |
| X | 1 1/4 | 1 3/8 | 1 1/2 | 2 1/8 | 2 1/2 | 2 3/4 | 3 1/2 | 4 | 4 1/2 | 5 | 5 3/4 | 6 1/2 |

Capacities listed are safe for hooks of low-carbon steel.

**Dimensions of Chain End-Link and Narrow Shackle — U. S. Navy Standard**

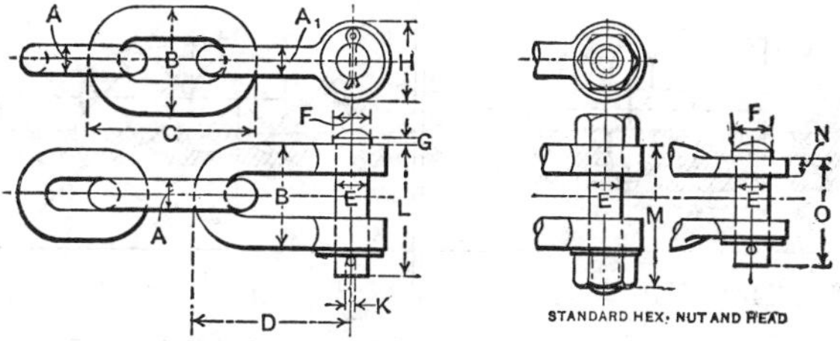


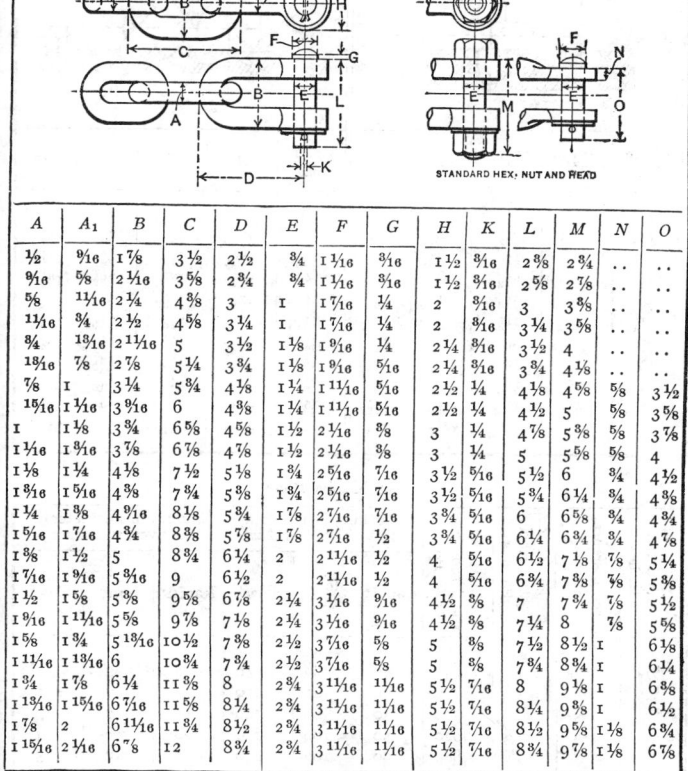

| *A* | *A₁* | *B* | *C* | *D* | *E* | *F* | *G* | *H* | *K* | *L* | *M* | *N* | *O* |
|---|---|---|---|---|---|---|---|---|---|---|---|---|---|
| ½ | 9/16 | 1 7/8 | 3 ½ | 2 ½ | ¾ | 1 1/16 | 3/16 | 1 ½ | 3/16 | 2 3/8 | 2 ¾ | .. | .. |
| 9/16 | 5/8 | 2 1/16 | 3 5/8 | 2 ¾ | ¾ | 1 1/16 | 3/16 | 1 ½ | 3/16 | 2 5/8 | 2 7/8 | .. | .. |
| 5/8 | 11/16 | 2 ¼ | 4 3/8 | 3 | 1 | 1 7/16 | ¼ | 2 | 3/16 | 3 | 3 3/8 | .. | .. |
| 11/16 | ¾ | 2 ½ | 4 5/8 | 3 ¼ | 1 | 1 7/16 | ¼ | 2 | 3/16 | 3 ¼ | 3 5/8 | .. | .. |
| ¾ | 13/16 | 2 11/16 | 5 | 3 ½ | 1 1/8 | 1 9/16 | ¼ | 2 ¼ | 3/16 | 3 ½ | 4 | .. | .. |
| 13/16 | 7/8 | 2 7/8 | 5 ¼ | 3 ¾ | 1 1/8 | 1 9/16 | 5/16 | 2 ¼ | 3/16 | 3 ¾ | 4 1/8 | .. | .. |
| 7/8 | 1 | 3 ¼ | 5 ¾ | 4 1/8 | 1 ¼ | 1 11/16 | 5/16 | 2 ½ | ¼ | 4 1/8 | 4 5/8 | 5/8 | 3 ½ |
| 15/16 | 1 1/16 | 3 9/16 | 6 | 4 3/8 | 1 ¼ | 1 11/16 | 5/16 | 2 ½ | ¼ | 4 ½ | 5 | 5/8 | 3 5/8 |
| 1 | 1 1/8 | 3 ¾ | 6 5/8 | 4 5/8 | 1 ½ | 2 1/16 | 3/8 | 3 | ¼ | 4 7/8 | 5 3/8 | 5/8 | 3 7/8 |
| 1 1/16 | 1 3/16 | 3 7/8 | 6 7/8 | 4 7/8 | 1 ½ | 2 1/16 | 3/8 | 3 | ¼ | 5 | 5 5/8 | 5/8 | 4 |
| 1 1/8 | 1 ¼ | 4 1/8 | 7 ½ | 5 1/8 | 1 ¾ | 2 5/16 | 7/16 | 3 ½ | 5/16 | 5 ½ | 6 | ¾ | 4 ½ |
| 1 3/16 | 1 5/16 | 4 3/8 | 7 ¾ | 5 3/8 | 1 ¾ | 2 5/16 | 7/16 | 3 ½ | 5/16 | 5 ¾ | 6 ¼ | ¾ | 4 3/8 |
| 1 ¼ | 1 3/8 | 4 9/16 | 8 1/8 | 5 ¾ | 1 7/8 | 2 7/16 | 7/16 | 3 ¾ | 5/16 | 6 | 6 5/8 | ¾ | 4 ¾ |
| 1 5/16 | 1 7/16 | 4 ¾ | 8 3/8 | 5 7/8 | 1 7/8 | 2 7/16 | ½ | 3 ¾ | 5/16 | 6 ¼ | 6 ¾ | ¾ | 4 7/8 |
| 1 3/8 | 1 ½ | 5 | 8 ¾ | 6 ¼ | 2 | 2 11/16 | ½ | 4 | 5/16 | 6 ½ | 7 1/8 | 7/8 | 5 ¼ |
| 1 7/16 | 1 9/16 | 5 3/16 | 9 | 6 ½ | 2 | 2 11/16 | ½ | 4 | 5/16 | 6 ¾ | 7 3/8 | 7/8 | 5 3/8 |
| 1 ½ | 1 5/8 | 5 3/8 | 9 5/8 | 6 7/8 | 2 ¼ | 3 1/16 | 9/16 | 4 ½ | 3/8 | 7 | 7 ¾ | 7/8 | 5 ½ |
| 1 9/16 | 1 11/16 | 5 5/8 | 9 7/8 | 7 1/8 | 2 ¼ | 3 1/16 | 9/16 | 4 ½ | 3/8 | 7 ¼ | 8 | 7/8 | 5 5/8 |
| 1 5/8 | 1 ¾ | 5 13/16 | 10 ½ | 7 3/8 | 2 ½ | 3 7/16 | 5/8 | 5 | 3/8 | 7 ½ | 8 ½ | 1 | 6 1/8 |
| 1 11/16 | 1 13/16 | 6 | 10 ¾ | 7 ¾ | 2 ½ | 3 7/16 | 5/8 | 5 | 3/8 | 7 ¾ | 8 ¾ | 1 | 6 ¼ |
| 1 ¾ | 1 7/8 | 6 ¼ | 11 3/8 | 8 | 2 ¾ | 3 11/16 | 11/16 | 5 ½ | 7/16 | 8 | 9 1/8 | 1 | 6 3/8 |
| 1 13/16 | 1 15/16 | 6 7/16 | 11 5/8 | 8 ¼ | 2 ¾ | 3 11/16 | 11/16 | 5 ½ | 7/16 | 8 ¼ | 9 3/8 | 1 | 6 ½ |
| 1 7/8 | 2 | 6 11/16 | 11 ¾ | 8 ½ | 2 ¾ | 3 11/16 | 11/16 | 5 ½ | 7/16 | 8 ½ | 9 5/8 | 1 1/8 | 6 ¾ |
| 1 15/16 | 2 1/16 | 6 7/8 | 12 | 8 ¾ | 2 ¾ | 3 11/16 | 11/16 | 5 ½ | 7/16 | 8 ¾ | 9 7/8 | 1 1/8 | 6 7/8 |

**Winding Drum Score for Wire Rope**

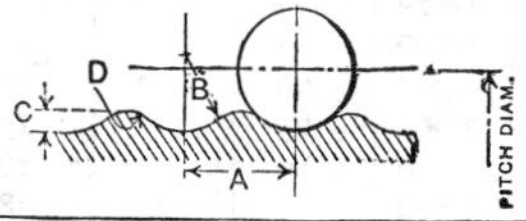


| Rope Diam. | *A* | *B* | *C* | *D* |
|---|---|---|---|---|
| 3/8 | 7/16 | 7/32 | 3/32 | 3/32 |
| 7/16 | ½ | ¼ | 7/64 | 3/32 |
| ½ | 9/16 | 9/32 | 1/8 | 3/32 |
| 9/16 | 5/8 | 5/16 | 9/64 | 1/8 |
| 5/8 | 11/16 | 11/32 | 5/32 | 1/8 |
| 11/16 | ¾ | 3/8 | 11/64 | 1/8 |
| ¾ | 13/16 | 13/32 | 3/16 | 5/32 |
| 13/16 | 7/8 | 7/16 | 13/64 | 5/32 |
| 7/8 | 15/16 | 15/32 | 7/32 | 5/32 |
| 15/16 | 1 | ½ | 15/64 | 3/16 |
| 1 | 1 1/16 | 17/32 | ¼ | 3/16 |

## Winding Drum Scores for Chain

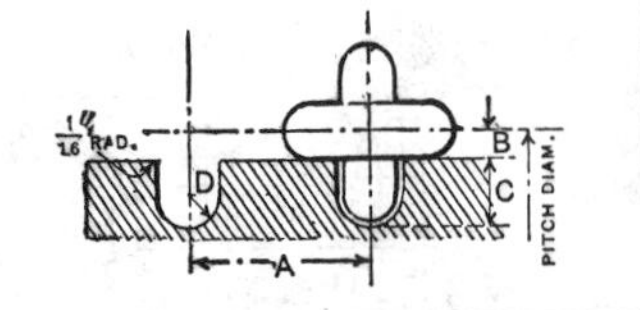


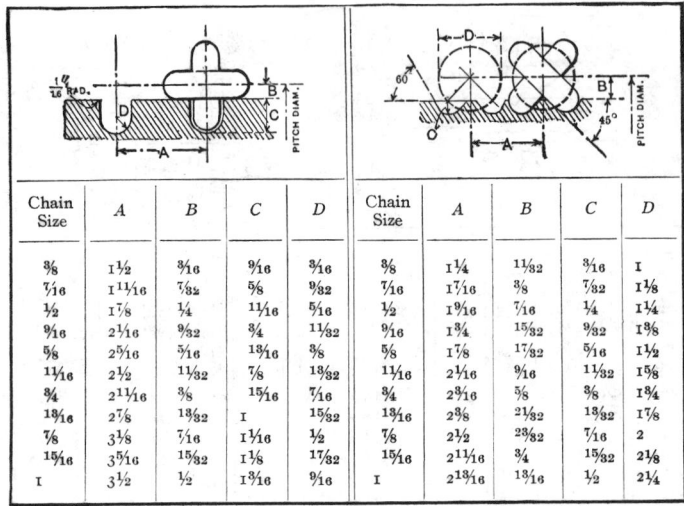

| Chain Size | A | B | C | D |
|---|---|---|---|---|
| 3/8 | 1 1/2 | 3/16 | 9/16 | 3/16 |
| 7/16 | 1 11/16 | 7/32 | 5/8 | 9/32 |
| 1/2 | 1 7/8 | 1/4 | 11/16 | 5/16 |
| 9/16 | 2 1/16 | 9/32 | 3/4 | 11/32 |
| 5/8 | 2 5/16 | 5/16 | 13/16 | 3/8 |
| 11/16 | 2 1/2 | 11/32 | 7/8 | 13/32 |
| 3/4 | 2 11/16 | 3/8 | 15/16 | 7/16 |
| 13/16 | 2 7/8 | 13/32 | 1 | 15/32 |
| 7/8 | 3 1/8 | 7/16 | 1 1/16 | 1/2 |
| 15/16 | 3 5/16 | 15/32 | 1 1/8 | 17/32 |
| 1 | 3 1/2 | 1/2 | 1 3/16 | 9/16 |

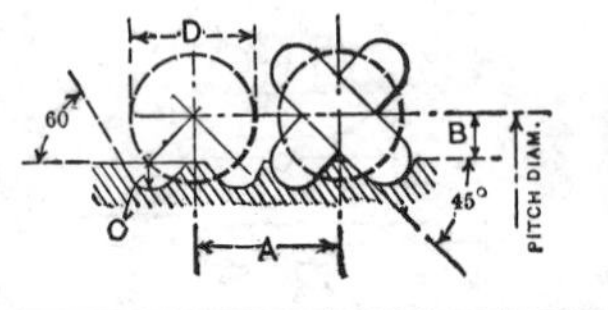


| Chain Size | A | B | C | D |
|---|---|---|---|---|
| 3/8 | 1 1/4 | 11/32 | 3/16 | 1 |
| 7/16 | 1 7/16 | 3/8 | 7/32 | 1 1/8 |
| 1/2 | 1 9/16 | 7/16 | 1/4 | 1 1/4 |
| 9/16 | 1 3/4 | 15/32 | 9/32 | 1 3/8 |
| 5/8 | 1 7/8 | 17/32 | 5/16 | 1 1/2 |
| 11/16 | 2 1/16 | 9/16 | 11/32 | 1 5/8 |
| 3/4 | 2 3/16 | 5/8 | 3/8 | 1 3/4 |
| 13/16 | 2 3/8 | 21/32 | 13/32 | 1 7/8 |
| 7/8 | 2 1/2 | 23/32 | 7/16 | 2 |
| 15/16 | 2 11/16 | 3/4 | 15/32 | 2 1/8 |
| 1 | 2 13/16 | 13/16 | 1/2 | 2 1/4 |

## Strength of Manila Rope

| Approximate Diameter, Inches | Circumference, Inches | Weight of 100 Feet of Rope, Pounds | Ultimate Tensile Strength of Rope, Pounds | Working Load, Total Area of Rope, in Lbs. | | | Minimum Diameter of Sheaves, in Inches | | |
|---|---|---|---|---|---|---|---|---|---|
| | | | | Rapid* | Medium† | Slow† | Rapid* | Medium† | Slow‡ |
| 3/16 | 9/16 | 2 | 230 | 10 | 20 | 40 | 8 | 3 | 1 1/2 |
| 5/16 | 1 | 4 | 630 | 20 | 40 | 100 | 13 | 5 | 2 1/2 |
| 3/8 | 1 1/8 | 5 | 900 | 30 | 60 | 140 | 15 | 6 | 3 |
| 1/2 | 1 1/2 | 7 2/3 | 1,620 | 50 | 100 | 250 | 20 | 8 | 4 |
| 5/8 | 2 | 13 1/3 | 2,880 | 90 | 180 | 450 | 25 | 10 | 5 |
| 13/16 | 2 1/2 | 20 | 4,500 | 150 | 300 | 650 | 35 | 11 | 7 |
| 1 | 3 | 28 1/3 | 6,480 | 200 | 400 | 1,000 | 40 | 12 | 8 |
| 1 1/8 | 3 1/2 | 38 | 8,820 | 250 | 500 | 1,250 | 45 | 13 | 9 |
| 1 5/16 | 4 | 52 | 11,500 | 350 | 700 | 1,700 | 55 | 15 | 11 |
| 1 1/2 | 4 1/2 | 65 | 14,600 | 450 | 900 | 2,200 | 60 | 16 | 12 |
| 1 5/8 | 5 | 80 | 18,000 | 550 | 1100 | 2,600 | 65 | 17 | 13 |
| 2 | 6 | 113 | 25,900 | 750 | 1500 | 3,700 | 80 | 22 | 16 |
| 2 1/4 | 7 | 153 | 35,300 | .... | 2100 | 5,100 | ..... | 24 | 18 |
| 2 5/8 | 8 | 211 | 46,100 | .... | 2700 | 6,600 | ..... | 28 | 21 |
| 3 | 9 | 262 | 58,300 | .... | 3400 | 8,400 | ..... | 32 | 24 |
| 3 1/4 | 10 | 325 | 72,000 | .... | 4200 | 10,300 | ..... | 35 | 26 |

* Speed from 400 to 800 feet per minute. † Hoisting speed from 150 to 300 feet per minute. ‡ Hoisting speed from 50 to 100 feet per minute.

**Drop-forged Weldless Eye-bolts of Carbon Steel**

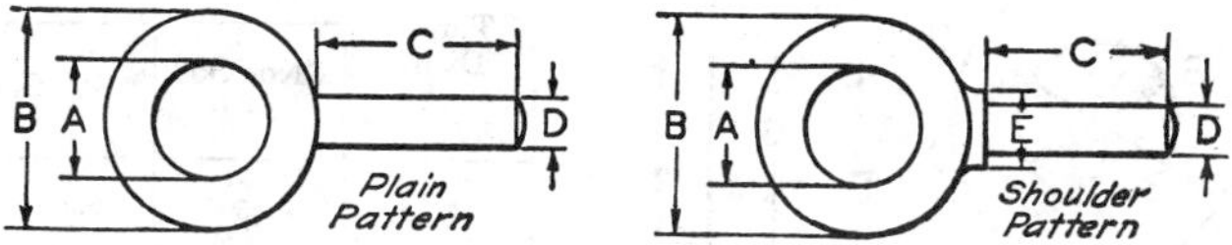


| Plain Pattern — Standard Length † | | | | Plain Pattern — Extra Length | | | | Safe Load, Tons* |
|---|---|---|---|---|---|---|---|---|
| Shank | | Eye Diam. | | Shank | | Eye Diam. | | |
| D | C | A | B | D | C | A | B | |
| 1/4 | 1 3/8 | 1/2 | 7/8 | .... | .... | .... | .... | 0.2 |
| 5/16 | 1 7/8 | 9/16 | 1 1/16 | .... | .... | .... | .... | 0.4 |
| 3/8 | 1 1/8 | 1 1/4 | 2 | .... | .... | .... | .... | 0.7 |
| 7/16 | 1 1/4 | 1 3/8 | 2 1/8 | 3/8 | 4 1/2 | 1 | 1 21/32 | 1. |
| 1/2 | 1 7/16 | 1 1/2 | 2 3/8 | 7/16 | 4 1/2 | 1 3/32 | 1 27/32 | 1.3 |
| 9/16 | 1 5/8 | 1 3/8 | 2 3/8 | 1/2 | 4 1/2 | 1 3/16 | 2 1/16 | 1.5 |
| 5/8 | 1 13/16 | 1 11/16 | 2 15/16 | 9/16 | 4 1/2 | 1 9/32 | 2 9/32 | 2. |
| 3/4 | 2 | 1 13/16 | 3 5/16 | 5/8 | 4 1/2 | 1 3/8 | 2 1/2 | 3. |
| 7/8 | 2 3/16 | 1 15/16 | 3 9/16 | 3/4 | 5 | 1 1/2 | 2 13/16 | 3.5 |
| 1 | 2 3/8 | 2 1/16 | 3 15/16 | 7/8 | 5 | 1 11/16 | 3 1/4 | 4. |
| 1 1/8 | 2 1/2 | 2 3/16 | 4 5/16 | 1 | 5 | 1 13/16 | 3 9/16 | 5. |
| 1 1/4 | 2 3/4 | 2 5/16 | 4 11/16 | 1 1/8 | 5 | 2 | 4 | 7.5 |
| 1 1/2 | 3 1/8 | 2 9/16 | 5 5/16 | 1 1/4 | 6 | 2 3/16 | 4 7/16 | 9. |
| 1 3/4 | 3 1/2 | 3 1/16 | 6 1/16 | 1 1/2 | 6 | 2 1/2 | 5 3/16 | 11. |
| 2 | 4 | 3 1/16 | 6 5/16 | 1 3/4 | 6 | 2 7/8 | 6 1/16 | 13. |
| 2 1/2 | 5 | 4 1/2 | 8 1/2 | 2 | 6 | 3 1/4 | 6 7/8 | 16. |
| 2 3/4 | 4 | 5 | 9 3/4 | 2 1/2 | 6 | 4 | 8 9/16 | 20. |

| Shoulder Pattern — Standard Length ‡ | | | | | Shoulder Pattern — Extra Length | | | | Safe Load, Tons* |
|---|---|---|---|---|---|---|---|---|---|
| Shank | | Eye Diam. | | Diam. E | Shank | | Eye Diam. | | |
| D | C | A | B | | D | C | A | B | |
| 1/4 | 1 | 9/16 | 1 | 1/2 | 1/4 | 3 | 3/4 | 1 3/16 | 0.2 |
| 5/16 | 1 1/8 | 7/8 | 1 7/16 | 9/16 | 5/16 | 4 | 7/8 | 1 7/16 | 0.4 |
| 3/8 | 1 1/4 | 1 1/8 | 1 11/16 | 5/8 | 3/8 | 4 1/2 | 1 | 1 21/32 | 0.7 |
| 7/16 | 1 3/8 | 1 3/32 | 1 27/32 | 13/16 | 7/16 | 4 1/2 | 1 3/32 | 1 27/32 | 1. |
| 1/2 | 1 7/16 | 1 1/4 | 2 3/8 | 7/8 | 1/2 | 4 1/2 | 1 3/16 | 2 1/16 | 1.3 |
| 9/16 | 1 5/8 | 1 9/32 | 2 9/32 | 1 | 9/16 | 4 1/2 | 1 9/32 | 2 9/32 | 1.5 |
| 5/8 | 1 3/4 | 1 3/8 | 2 5/8 | 1 | 5/8 | 4 1/2 | 1 3/8 | 2 1/2 | 2. |
| 3/4 | 2 1/8 | 1 1/2 | 3 | 1 1/8 | 3/4 | 5 | 1 1/2 | 2 13/16 | 3. |
| 7/8 | 2 1/2 | 1 5/8 | 3 1/4 | 1 1/4 | 7/8 | 5 | 1 11/16 | 3 1/4 | 3.5 |
| 1 | 3 1/4 | 1 7/8 | 3 5/8 | 1 1/2 | 1 | 5 | 1 13/16 | 3 9/16 | 4. |
| 1 1/8 | 2 3/4 | 2 | 4 | 1 11/16 | 1 1/8 | 5 | 2 | 4 | 5. |
| 1 1/4 | 3 | 2 3/16 | 4 7/16 | 1 7/8 | 1 1/4 | 6 | 2 3/16 | 4 7/16 | 7.5 |
| 1 1/2 | 3 1/2 | 2 9/16 | 5 5/16 | 2 3/16 | 1 1/2 | 6 | 2 1/2 | 5 3/16 | 9. |
| 1 3/4 | 3 3/4 | 2 7/8 | 6 1/16 | 2 7/16 | 1 3/4 | 6 | 2 7/8 | 6 1/16 | 11. |
| 2 | 4 | 3 1/4 | 6 7/8 | 2 7/8 | 2 | 6 | 3 1/4 | 6 7/8 | 13. |

† The Billings and Spencer Co. standard except for the 2½- and 2¾-inch sizes.
‡ The Billings and Spencer Co. standard except for the 1¾- and 2-inch sizes.
* The ultimate or breaking load is approximately 4 to 5 times the safe working load.

### Eye-Bolts for Motor and Generator Frames, Transformer Covers, etc.

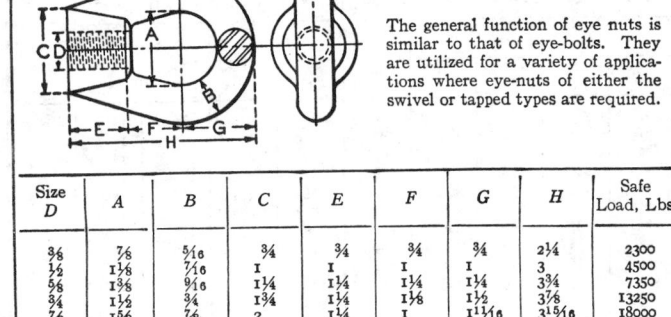

| Thread Diam. | Pitch Diameter | |
|---|---|---|
| | "Not Go" Gage | "Go" Gage |
| 1/2–13 | 0.4500 | 0.4540 |
| 3/4–10 | 0.6850 | 0.6900 |
| 1–8 | 0.9188 | 0.9238 |
| 1 1/4–7 | 1.1572 | 1.1632 |
| 1 1/2–6 | 1.3917 | 1.3977 |
| 1 3/4–5 | 1.6201 | 1.6261 |
| 2–4 1/2 | 1.8557 | ...... |
| 2 1/2–4 | 2.3376 | ...... |

| Shank | | Eye Diam. | | Length E | Thick. F | Width G | Height H | Safe Load, Lbs. |
|---|---|---|---|---|---|---|---|---|
| D | C | A | B | | | | | |
| 1/2 | 1 1/8 | 1 1/2 | 2 3/8 | 3 1/2 | 13/32 | 7/16 | 2 5/8 | 630 |
| 3/4 | 1 9/16 | 1 13/16 | 3 5/8 | 4 7/8 | 5/8 | 3/4 | 3 5/8 | 1500 |
| 1 | 1 15/16 | 2 1/16 | 3 15/16 | 5 7/8 | 13/16 | 15/16 | 4 1/4 | 2760 |
| 1 1/4 | 2 3/4 | 2 9/16 | 4 11/16 | 7 7/16 | 1 | 1 3/16 | 5 5/16 | 4450 |
| 1 1/2 | 3 | 3 | 5 3/4 | 8 3/4 | 1 1/4 | 1 3/8 | 6 1/8 | 6475 |
| 1 3/4 | 3 5/8 | 4 | 7 1/4 | 10 7/8 | 1 1/2 | 1 5/8 | 7 7/8 | 8700 |
| 2 | 3 5/8 | 4 | 7 1/4 | 10 7/8 | 1 1/2 | 1 5/8 | 7 5/8 | 11500 |
| 2 1/2 | 4 1/2 | 4 1/2 | 8 1/2 | 13 | 1 13/16 | 2 | 9 1/4 | 18580 |

Westinghouse Electric & Mfg. Co. standard. Maximum safe loads for direct tension at 5000 pounds per square inch.

### Drop Forged Carbon Steel Eye-Nuts — Billings and Spencer Co. Standard

The general function of eye nuts is similar to that of eye-bolts. They are utilized for a variety of applications where eye-nuts of either the swivel or tapped types are required.

| Size D | A | B | C | E | F | G | H | Safe Load, Lbs. |
|---|---|---|---|---|---|---|---|---|
| 3/8 | 7/8 | 5/16 | 3/4 | 3/4 | 3/4 | 3/4 | 2 1/4 | 2300 |
| 1/2 | 1 1/8 | 7/16 | 1 | 1 | 1 | 1 | 3 | 4500 |
| 5/8 | 1 3/8 | 9/16 | 1 1/4 | 1 1/4 | 1 1/4 | 1 1/4 | 3 3/4 | 7350 |
| 3/4 | 1 1/2 | 3/4 | 1 3/4 | 1 1/4 | 1 1/8 | 1 1/2 | 3 7/8 | 13250 |
| 7/8 | 1 5/8 | 7/8 | 2 | 1 1/4 | 1 | 1 11/16 | 3 15/16 | 18000 |
| 1 | 1 3/4 | 1 | 2 1/4 | 1 5/8 | 1 1/4 | 1 7/8 | 4 3/4 | 23550 |
| 1 1/4 | 2 | 1 1/4 | 2 7/8 | 1 7/8 | 1 3/8 | 2 1/4 | 5 1/2 | 36750 |
| 1 1/2 | 2 1/2 | 1 3/8 | 3 3/8 | 2 1/8 | 1 3/4 | 2 5/8 | 6 1/2 | 44500 |
| 1 3/4 | 3 | 1 1/2 | 3 1/2 | 2 3/8 | 2 | 3 | 7 3/8 | 53100 |
| 2 | 3 1/2 | 1 5/8 | 4 | 3 | 2 1/2 | 3 3/8 | 8 7/8 | 62100 |

Safe load applies to nut only and not to connecting bolt.

# MACHINE DETAILS

## Collar Screws

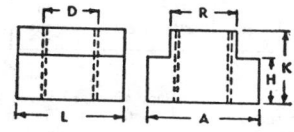

| | Outside Diameter of Screw, D | Number of Threads per Inch | Length of Head, A | Size of Square, B | Thickness of Collar, C | Diameter of Collar, F | Radius of Head, R |
|---|---|---|---|---|---|---|---|
| | 1/8 | 40 | 1/8 | 1/8 | 1/16 | 1/4 | 1/4 |
| | 3/16 | 24 | 3/16 | 3/16 | 5/64 | 11/32 | 3/8 |
| | 1/4 | 20 | 1/4 | 1/4 | 5/64 | 7/16 | 1/2 |
| | 5/16 | 18 | 5/16 | 5/16 | 7/64 | 1/2 | 5/8 |
| | 3/8 | 16 | 3/8 | 3/8 | 1/8 | 5/8 | 3/4 |
| | 7/16 | 14 | 7/16 | 7/16 | 1/8 | 11/16 | 7/8 |
| | 1/2 | 12 or 13 | 1/2 | 1/2 | 3/16 | 13/16 | 1 |
| | 9/16 | 12 | 9/16 | 9/16 | 7/32 | 15/16 | 1 1/8 |
| | 5/8 | 11 | 5/8 | 5/8 | 15/64 | 1 | 1 1/4 |
| | 3/4 | 10 | 3/4 | 3/4 | 9/32 | 1 1/4 | 1 1/2 |

\* On all screws four inches long and under, threads are cut 3/4 of the length L; longer than four inches, threads are cut half of the length L.

## American Standard T-Nuts (ASA B5.1-1949)

Note: No definite provision has been made for the chamfering of corners. Chamfering or rounding is left to manufacturer's discretion.

| Thread Diameter D* | Width of T-slot Throat | Width of Tongue R | | Width of Nut A | | Height of Nut H | | Total Thickness K | Length of Nut L |
|---|---|---|---|---|---|---|---|---|---|
| | | Maximum (Basic) | Minimum | Maximum (Basic) | Minimum | Maximum (Basic) | Minimum | | |
| 1/4 | 11/32 | 0.330 | 0.320 | 9/16 | 17/32 | 3/16 | 11/64 | 9/32 | 9/16 |
| 5/16 | 7/16 | 0.418 | 0.408 | 11/16 | 21/32 | 1/4 | 15/64 | 3/8 | 11/16 |
| 3/8 | 9/16 | 0.543 | 0.533 | 7/8 | 27/32 | 5/16 | 19/64 | 17/32 | 7/8 |
| 1/2 | 11/16 | 0.668 | 0.658 | 1 1/8 | 1 3/32 | 13/32 | 25/64 | 5/8 | 1 1/8 |
| 5/8 | 13/16 | 0.783 | 0.773 | 1 5/16 | 1 9/32 | 17/32 | 1/2 | 25/32 | 1 5/16 |
| 3/4 | 1 1/16 | 1.033 | 1.018 | 1 11/16 | 1 21/32 | 1 1/16 | 1 1/32 | 1 | 1 11/16 |
| 1 | 1 5/16 | 1.273 | 1.258 | 2 1/16 | 2 1/32 | 1 5/16 | 1 9/32 | 1 5/16 | 2 1/16 |
| 1 1/4 | 1 9/16 | 1.523 | 1.508 | 2 1/2 | 2 15/32 | 1 9/16 | 1 17/32 | 1 5/8 | 2 1/2 |

\* Thread diameter in T-nut is made smaller than corresponding T-bolt, to insure full strength of T-nut. A T-nut of given thread diameter requires the next size wider T-slot throat than does a T-bolt of the same thread diameter.

### American Standard T-Slots and T-Bolts (ASA B5.1-1949)

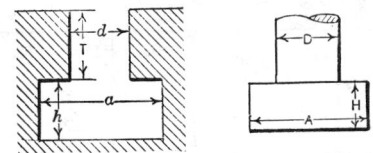

Provision is made in this Standard for optional rounding or breaking of corners in slot and on bolt.

| Diameter of T-bolt* | Width of Throat $d$*† | Depth of Throat $T$ | | Head Space Dimensions and Tolerances | | | | | |
|---|---|---|---|---|---|---|---|---|---|
| | | | | Width $a$ | | | Depth $h$ | | |
| | | Maximum | Minimum | Maximum (Basic) | Tolerance (Minus) | Minimum | Maximum (Basic) | Tolerance (Minus) | Minimum |
| 1/4 | 9/32 | 3/8 | 1/8 | 9/16 | 0.063 | 1/2 | 15/64 | 0.031 | 13/64 |
| 5/16 | 11/32 | 7/16 | 5/32 | 21/32 | 0.063 | 19/32 | 17/64 | 0.031 | 15/64 |
| 3/8 | 7/16 | 9/16 | 7/32 | 25/32 | 0.063 | 23/32 | 21/64 | 0.031 | 19/64 |
| 1/2 | 9/16 | 11/16 | 5/16 | 31/32 | 0.063 | 29/32 | 25/64 | 0.031 | 23/64 |
| 5/8 | 11/16 | 7/8 | 7/16 | 1 1/4 | 0.063 | 1 3/16 | 31/64 | 0.031 | 29/64 |
| 3/4 | 13/16 | 1 1/16 | 9/16 | 1 15/32 | 0.094 | 1 3/8 | 5/8 | 0.031 | 19/32 |
| 1 | 1 1/16 | 1 1/4 | 3/4 | 1 27/32 | 0.094 | 1 3/4 | 53/64 | 0.047 | 25/32 |
| 1 1/4 | 1 5/16 | 1 9/16 | 1 | 2 7/32 | 0.094 | 2 1/8 | 1 3/32 | 0.063 | 1 1/32 |
| 1 1/2 | 1 9/16 | 1 15/16 | 1 1/4 | 2 21/32 | 0.094 | 2 9/16 | 1 11/32 | 0.063 | 1 9/32 |

All dimensions in inches.

\* In addition to the width of throat given, a secondary standard is recognized, having the width of throat the same as the nominal diameter of the T-bolt. This is to provide for the use, during the transition period, of this standard on machine tools where it is already established.

† A tolerance of plus 0.001 is allowed for width of throat when tongues or other parts must fit.

| Diameter of T-bolt $D$ | Threads per Inch | Bolt Head Dimensions and Tolerances | | | | | | |
|---|---|---|---|---|---|---|---|---|
| | | Width across Flats $A$ | | | Width across Corners | Height $H$ | | |
| | | Maximum (Basic) | Tolerance (Minus) | Minimum | | Maximum (Basic) | Tolerance (Minus) | Minimum |
| 1/4 | 20 | 15/32 | 0.031 | 7/16 | 0.663 | 5/32 | 0.016 | 9/64 |
| 5/16 | 18 | 9/16 | 0.031 | 17/32 | 0.796 | 3/16 | 0.016 | 11/64 |
| 3/8 | 16 | 11/16 | 0.031 | 21/32 | 0.972 | 1/4 | 0.016 | 15/64 |
| 1/2 | 13 | 7/8 | 0.031 | 27/32 | 1.238 | 5/16 | 0.016 | 19/64 |
| 5/8 | 11 | 1 1/8 | 0.031 | 1 3/32 | 1.591 | 13/32 | 0.016 | 25/64 |
| 3/4 | 10 | 1 5/16 | 0.031 | 1 9/32 | 1.856 | 17/32 | 0.031 | 1/2 |
| 1 | 8 | 1 11/16 | 0.031 | 1 21/32 | 2.387 | 11/16 | 0.031 | 21/32 |
| 1 1/4 | 7 | 2 1/16 | 0.031 | 2 1/32 | 2.917 | 15/16 | 0.031 | 29/32 |
| 1 1/2 | 6 | 2 1/2 | 0.031 | 2 15/32 | 3.536 | 1 3/16 | 0.031 | 1 9/32 |

## Dimensions of Machine Slides

### Bedded Strips

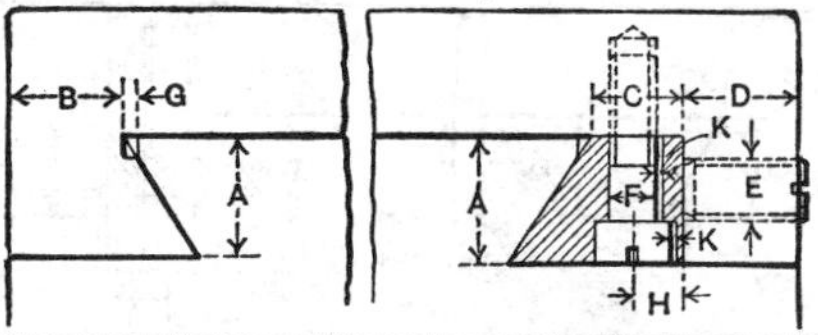

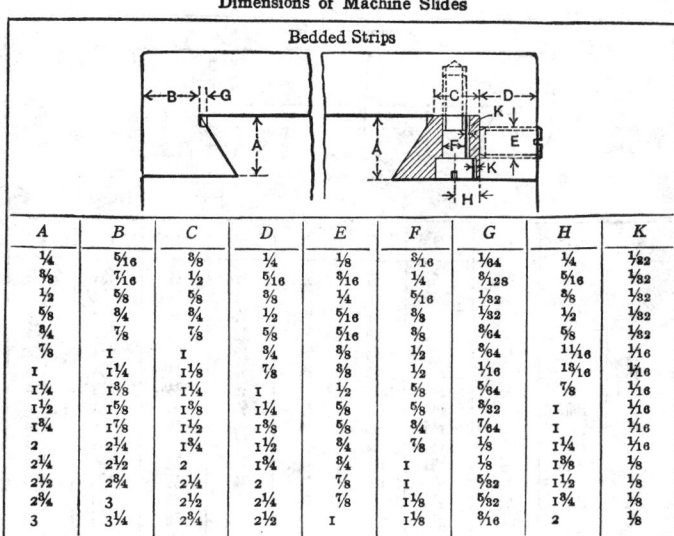

| A | B | C | D | E | F | G | H | K |
|---|---|---|---|---|---|---|---|---|
| 1/4 | 5/16 | 3/8 | 1/4 | 1/8 | 3/16 | 1/64 | 1/4 | 1/32 |
| 3/8 | 7/16 | 1/2 | 5/16 | 3/16 | 1/4 | 3/128 | 5/16 | 1/32 |
| 1/2 | 5/8 | 5/8 | 3/8 | 1/4 | 5/16 | 1/32 | 3/8 | 1/32 |
| 5/8 | 3/4 | 3/4 | 1/2 | 5/16 | 3/8 | 1/32 | 1/2 | 1/32 |
| 3/4 | 7/8 | 7/8 | 5/8 | 5/16 | 3/8 | 3/64 | 5/8 | 1/32 |
| 7/8 | 1 | 1 | 3/4 | 3/8 | 1/2 | 3/64 | 11/16 | 1/16 |
| 1 | 1 1/4 | 1 1/8 | 7/8 | 3/8 | 1/2 | 1/16 | 13/16 | 1/16 |
| 1 1/4 | 1 3/8 | 1 1/4 | 1 | 1/2 | 5/8 | 5/64 | 7/8 | 1/16 |
| 1 1/2 | 1 5/8 | 1 3/8 | 1 1/4 | 5/8 | 5/8 | 3/32 | 1 | 1/16 |
| 1 3/4 | 1 7/8 | 1 1/2 | 1 3/8 | 5/8 | 3/4 | 7/64 | 1 | 1/16 |
| 2 | 2 1/4 | 1 3/4 | 1 1/2 | 3/4 | 7/8 | 1/8 | 1 1/4 | 1/16 |
| 2 1/4 | 2 1/2 | 2 | 1 3/4 | 3/4 | 1 | 1/8 | 1 3/8 | 1/8 |
| 2 1/2 | 2 3/4 | 2 1/4 | 2 | 7/8 | 1 | 5/32 | 1 1/2 | 1/8 |
| 2 3/4 | 3 | 2 1/2 | 2 1/4 | 7/8 | 1 1/8 | 5/32 | 1 3/4 | 1/8 |
| 3 | 3 1/4 | 2 3/4 | 2 1/2 | 1 | 1 1/8 | 3/16 | 2 | 1/8 |

### Square Strips

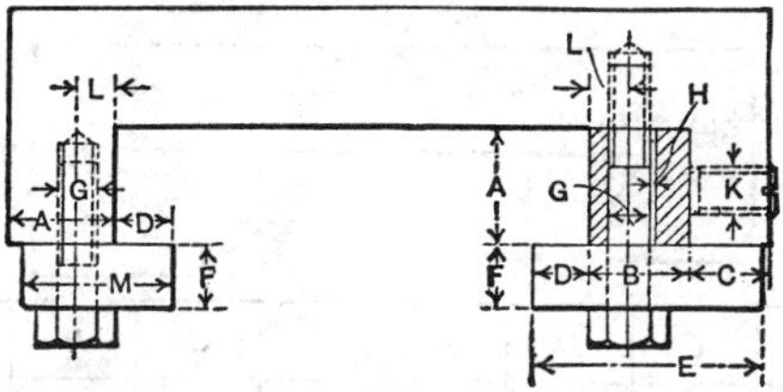

| A | B | C | D | E | F | G | H | K | L | M |
|---|---|---|---|---|---|---|---|---|---|---|
| 1/2 | 5/8 | 3/8 | 3/8 | 1 5/16 | 5/16 | 1/4 | 1/32 | 1/4 | 3/16 | 1 3/16 |
| 5/8 | 3/4 | 1/2 | 1/2 | 1 11/16 | 3/8 | 5/16 | 1/32 | 5/16 | 7/32 | 1 1/16 |
| 3/4 | 7/8 | 5/8 | 5/8 | 2 1/16 | 1/2 | 3/8 | 1/32 | 5/16 | 5/16 | 1 5/16 |
| 7/8 | 1 | 3/4 | 3/4 | 2 7/16 | 5/8 | 1/2 | 1/16 | 3/8 | 3/8 | 1 9/16 |
| 1 | 1 1/8 | 7/8 | 7/8 | 2 3/4 | 3/4 | 1/2 | 1/16 | 3/8 | 7/16 | 1 3/4 |
| 1 1/4 | 1 1/4 | 1 | 1 | 3 1/8 | 7/8 | 5/8 | 1/16 | 1/2 | 7/16 | 2 1/8 |
| 1 1/2 | 1 3/8 | 1 1/8 | 1 1/8 | 3 1/2 | 1 | 5/8 | 1/16 | 1/2 | 1/2 | 2 1/2 |
| 1 3/4 | 1 1/2 | 1 1/4 | 1 1/4 | 3 7/8 | 1 1/8 | 3/4 | 1/16 | 5/8 | 9/16 | 2 7/8 |
| 2 | 1 3/4 | 1 1/2 | 1 1/2 | 4 5/8 | 1 1/4 | 7/8 | 1/16 | 3/4 | 5/8 | 3 3/8 |
| 2 1/4 | 2 | 1 5/8 | 1 5/8 | 5 | 1 3/8 | 7/8 | 1/8 | 3/4 | 11/16 | 3 5/8 |
| 2 1/2 | 2 1/4 | 1 3/4 | 1 3/4 | 5 1/2 | 1 1/2 | 1 | 1/8 | 7/8 | 3/4 | 4 |
| 2 3/4 | 2 1/2 | 1 7/8 | 1 7/8 | 6 | 1 5/8 | 1 | 1/8 | 7/8 | 13/16 | 4 3/8 |
| 3 | 2 3/4 | 2 | 2 | 6 1/2 | 1 3/4 | 1 1/8 | 1/8 | 1 | 7/8 | 4 3/4 |
| 3 1/2 | 3 1/8 | 2 1/4 | 2 1/4 | 7 1/4 | 1 7/8 | 1 1/4 | 1/8 | 1 1/8 | 1 | 5 3/8 |
| 4 | 3 1/2 | 2 1/2 | 2 1/2 | 8 | 2 | 1 1/2 | 1/8 | 1 1/4 | 1 1/8 | 6 |

## Dimensions of Machine Slides

### Overhung Strips

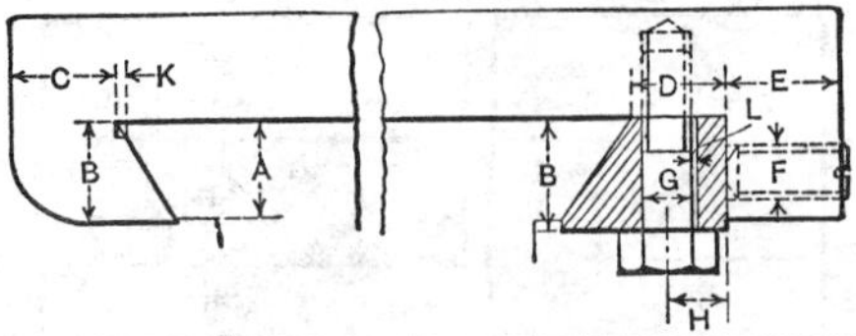


| A | B | C | D | E | F | G | H | K | L |
|---|---|---|---|---|---|---|---|---|---|
| 1/4 | 9/32 | 5/16 | 3/8 | 1/4 | 1/8 | 3/16 | 1/4 | 1/64 | 1/32 |
| 3/8 | 13/32 | 7/16 | 1/2 | 5/16 | 3/16 | 1/4 | 5/16 | 3/128 | 1/32 |
| 1/2 | 9/16 | 5/8 | 5/8 | 3/8 | 1/4 | 5/16 | 3/8 | 1/32 | 1/32 |
| 5/8 | 11/16 | 3/4 | 3/4 | 1/2 | 5/16 | 3/8 | 1/2 | 1/32 | 1/32 |
| 3/4 | 13/16 | 3/4 | 15/16 | 5/8 | 5/16 | 1/2 | 5/8 | 3/64 | 1/32 |
| 7/8 | 15/16 | 1 | 1 | 3/4 | 3/8 | 1/2 | 11/16 | 3/64 | 1/16 |
| 1 | 1 1/8 | 1 1/4 | 1 | 7/8 | 3/8 | 1/2 | 11/16 | 1/16 | 1/16 |
| 1 1/4 | 1 3/8 | 1 1/2 | 1 1/8 | 1 | 1/2 | 5/8 | 3/4 | 5/64 | 1/16 |
| 1 1/2 | 1 5/8 | 1 3/4 | 1 1/4 | 1 1/8 | 1/2 | 5/8 | 7/8 | 3/32 | 1/16 |
| 1 3/4 | 1 7/8 | 2 | 1 1/2 | 1 1/4 | 5/8 | 3/4 | 1 | 3/32 | 1/16 |
| 2 | 2 3/16 | 2 1/4 | 1 3/4 | 1 1/2 | 3/4 | 7/8 | 1 1/4 | 1/8 | 1/16 |
| 2 1/4 | 2 1/2 | 2 1/2 | 2 | 1 5/8 | 3/4 | 7/8 | 1 1/2 | 1/8 | 1/8 |
| 2 1/2 | 2 3/4 | 2 3/4 | 2 1/4 | 1 3/4 | 7/8 | 1 | 1 7/8 | 5/32 | 1/8 |
| 2 3/4 | 3 | 3 | 2 1/2 | 1 7/8 | 7/8 | 1 | 1 7/8 | 5/32 | 1/8 |
| 3 | 3 1/4 | 3 1/4 | 2 3/4 | 2 | 1 | 1 1/8 | 2 | 3/16 | 1/8 |

### Special Strips

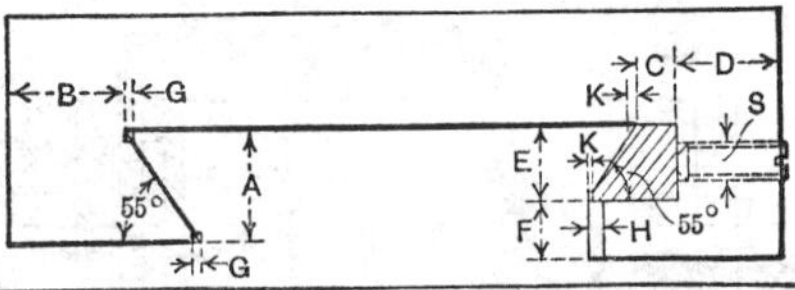


| A | B | C | D | E | F | G | H | K | S |
|---|---|---|---|---|---|---|---|---|---|
| 1 | 1 1/16 | 1/2 | 5/8 | 3/4 | 1/2 | 1/16 | 3/16 | 3/64 | 3/8 |
| 1 1/8 | 1 3/16 | 9/16 | 11/16 | 7/8 | 9/16 | 1/16 | 3/16 | 3/64 | 3/8 |
| 1 1/4 | 1 5/16 | 5/8 | 13/16 | 15/16 | 5/8 | 1/16 | 3/16 | 3/64 | 3/8 |
| 1 3/8 | 1 7/16 | 11/16 | 7/8 | 1 1/16 | 11/16 | 3/32 | 1/4 | 3/64 | 1/2 |
| 1 1/2 | 1 9/16 | 3/4 | 15/16 | 1 1/8 | 3/4 | 3/32 | 1/4 | 3/64 | 1/2 |
| 1 5/8 | 1 11/16 | 13/16 | 1 | 1 1/4 | 13/16 | 3/32 | 1/4 | 3/64 | 1/2 |
| 1 3/4 | 1 13/16 | 7/8 | 1 1/8 | 1 5/16 | 7/8 | 1/8 | 3/8 | 3/32 | 5/8 |
| 1 7/8 | 1 15/16 | 15/16 | 1 3/16 | 1 7/16 | 15/16 | 1/8 | 3/8 | 3/32 | 5/8 |
| 2 | 2 1/8 | 1 | 1 1/4 | 1 1/2 | 1 | 1/8 | 3/8 | 3/32 | 3/4 |
| 2 1/4 | 2 3/8 | 1 1/8 | 1 3/8 | 1 11/16 | 1 1/8 | 1/8 | 1/2 | 3/32 | 3/4 |
| 2 1/2 | 2 5/8 | 1 1/4 | 1 9/16 | 1 7/8 | 1 1/4 | 3/16 | 1/2 | 1/8 | 7/8 |
| 2 3/4 | 2 7/8 | 1 3/8 | 1 3/4 | 2 1/16 | 1 3/8 | 3/16 | 9/16 | 1/8 | 7/8 |
| 3 | 3 3/16 | 1 1/2 | 1 7/8 | 2 1/4 | 1 1/2 | 3/16 | 9/16 | 1/8 | 1 |
| 3 1/4 | 3 7/16 | 1 5/8 | 2 | 2 7/16 | 1 5/8 | 3/16 | 5/8 | 1/8 | 1 1/8 |
| 3 1/2 | 3 11/16 | 1 3/4 | 2 3/16 | 2 5/8 | 1 3/4 | 1/4 | 5/8 | 3/16 | 1 1/4 |
| 3 3/4 | 3 15/16 | 1 7/8 | 2 3/8 | 2 13/16 | 1 7/8 | 1/4 | 3/4 | 3/16 | 1 1/4 |
| 4 | 4 1/4 | 2 | 2 1/2 | 3 | 2 | 1/4 | 3/4 | 3/16 | 1 1/2 |

### Latches for Machine Doors

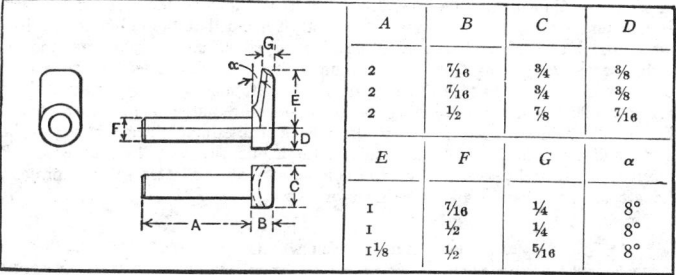

| | A | B | C | D |
|---|---|---|---|---|
| | 2 | $\frac{7}{16}$ | $\frac{3}{4}$ | $\frac{3}{8}$ |
| | 2 | $\frac{7}{16}$ | $\frac{3}{4}$ | $\frac{3}{8}$ |
| | 2 | $\frac{1}{2}$ | $\frac{7}{8}$ | $\frac{7}{16}$ |

| | E | F | G | $\alpha$ |
|---|---|---|---|---|
| | 1 | $\frac{7}{16}$ | $\frac{1}{4}$ | 8° |
| | 1 | $\frac{1}{2}$ | $\frac{1}{4}$ | 8° |
| | $1\frac{1}{8}$ | $\frac{1}{2}$ | $\frac{5}{16}$ | 8° |

### Proportions of Large Handwheels

| Diam. | A | B | C | D | E | F | G |
|---|---|---|---|---|---|---|---|
| 8 | $\frac{3}{4}$ | $\frac{3}{4}$ | $\frac{5}{8}$ | $1\frac{1}{2}$ | $\frac{3}{8}$ | $\frac{5}{16}$ | $1\frac{1}{8}$ |
| 9 | $\frac{13}{16}$ | $\frac{13}{16}$ | $\frac{11}{16}$ | $1\frac{5}{8}$ | $\frac{13}{32}$ | $\frac{11}{32}$ | $1\frac{1}{4}$ |
| 10 | $\frac{7}{8}$ | $\frac{7}{8}$ | $\frac{3}{4}$ | $1\frac{3}{4}$ | $\frac{7}{16}$ | $\frac{3}{8}$ | $1\frac{5}{16}$ |
| 11 | $\frac{15}{16}$ | $\frac{15}{16}$ | $\frac{3}{4}$ | $1\frac{7}{8}$ | $\frac{15}{32}$ | $\frac{3}{8}$ | $1\frac{3}{8}$ |
| 12 | 1 | 1 | $\frac{13}{16}$ | 2 | $\frac{1}{2}$ | $\frac{13}{32}$ | $1\frac{1}{2}$ |
| 13 | $1\frac{1}{16}$ | $1\frac{1}{16}$ | $\frac{13}{16}$ | $2\frac{1}{8}$ | $\frac{17}{32}$ | $\frac{13}{32}$ | $1\frac{5}{8}$ |
| 14 | $1\frac{1}{8}$ | $1\frac{1}{8}$ | $\frac{7}{8}$ | $2\frac{1}{4}$ | $\frac{9}{16}$ | $\frac{7}{16}$ | $1\frac{11}{16}$ |
| 15 | $1\frac{3}{16}$ | $1\frac{3}{16}$ | $\frac{15}{16}$ | $2\frac{3}{8}$ | $\frac{19}{32}$ | $\frac{15}{32}$ | $1\frac{3}{4}$ |
| 16 | $1\frac{1}{4}$ | $1\frac{1}{4}$ | 1 | $2\frac{1}{2}$ | $\frac{5}{8}$ | $\frac{1}{2}$ | $1\frac{7}{8}$ |
| 17 | $1\frac{5}{16}$ | $1\frac{5}{16}$ | 1 | $2\frac{5}{8}$ | $\frac{21}{32}$ | $\frac{1}{2}$ | $1\frac{15}{16}$ |
| 18 | $1\frac{3}{8}$ | $1\frac{3}{8}$ | $1\frac{1}{16}$ | $2\frac{3}{4}$ | $\frac{11}{16}$ | $\frac{17}{32}$ | $2\frac{1}{16}$ |
| 19 | $1\frac{7}{16}$ | $1\frac{7}{16}$ | $1\frac{1}{8}$ | $2\frac{7}{8}$ | $\frac{23}{32}$ | $\frac{9}{16}$ | $2\frac{1}{8}$ |
| 20 | $1\frac{1}{2}$ | $1\frac{1}{2}$ | $1\frac{3}{16}$ | 3 | $\frac{3}{4}$ | $\frac{19}{32}$ | $2\frac{1}{4}$ |
| 21 | $1\frac{9}{16}$ | $1\frac{9}{16}$ | $1\frac{1}{4}$ | $3\frac{1}{8}$ | $\frac{25}{32}$ | $\frac{5}{8}$ | $2\frac{5}{16}$ |
| 22 | $1\frac{5}{8}$ | $1\frac{5}{8}$ | $1\frac{1}{4}$ | $3\frac{1}{4}$ | $\frac{13}{16}$ | $\frac{5}{8}$ | $2\frac{7}{16}$ |
| 23 | $1\frac{11}{16}$ | $1\frac{11}{16}$ | $1\frac{5}{16}$ | $3\frac{3}{8}$ | $\frac{27}{32}$ | $\frac{21}{32}$ | $2\frac{1}{2}$ |
| 24 | $1\frac{3}{4}$ | $1\frac{3}{4}$ | $1\frac{3}{8}$ | $3\frac{1}{2}$ | $\frac{7}{8}$ | $\frac{11}{16}$ | $2\frac{5}{8}$ |
| 27 | $1\frac{15}{16}$ | $1\frac{15}{16}$ | $1\frac{1}{2}$ | $3\frac{7}{8}$ | $\frac{31}{32}$ | $\frac{3}{4}$ | $2\frac{7}{8}$ |
| 30 | $2\frac{1}{8}$ | $2\frac{1}{8}$ | $1\frac{5}{8}$ | $4\frac{1}{4}$ | $1\frac{1}{16}$ | $\frac{13}{16}$ | $3\frac{1}{16}$ |
| 33 | $2\frac{5}{16}$ | $2\frac{5}{16}$ | $1\frac{3}{4}$ | $4\frac{5}{8}$ | $1\frac{5}{32}$ | $\frac{7}{8}$ | $3\frac{7}{16}$ |
| 36 | $2\frac{1}{2}$ | $2\frac{1}{2}$ | 2 | 5 | $1\frac{1}{4}$ | 1 | $3\frac{3}{4}$ |

**Machine Handwheels.** — The accompanying table gives complete dimensions for "dished-arm" machine handwheels. The following remarks relating to hubs, keyways, arms and rims apply to handwheels of the design illustrated. *Hubs.* In Column *D*, the minimum dimension that should be used is given. The hub may be increased in length on side *a*, if necessary. Length *D* is sufficient for Woodruff keys. *Keyway.* This is designed so that either a straight key or a Woodruff key may be used, without changing the dimensions. *Arms.* These incline to form a "dished" wheel. There are two reasons for this: First, it often happens that it is convenient, or necessary, to fasten the handwheel with a nut on the end of the shaft, and if the wheel is dished, there is a recess for this nut, so that the operator does not strike his arm when turning the handwheel with a handle. Second, when casting handwheels with straight arms, the arms often break due to strains, but if dished, the strains are taken up by the arms and hub. The arms are oval in section, as indicated; the taper from $F$ to $G$ is 1 inch to the foot for all sizes. In Column $U$ for a 14-inch handwheel, it will be noted that the radius is less than for a 12-inch size. This is because the 14-inch wheel has six arms, whereas the 12-inch size has four arms. *Rims*. The inner half of the rim is reduced to permit finishing the outside half and to provide an even stopping place for the machined surface. This eliminates the filing that is otherwise required. Note that when a handle is to be used, a boss is cast on the outside of the rim at the end of one of the arms, and a smaller boss on the inside for the handle shank. The counterbore $P$ receives the straight section $Q$ on the handle, so that the latter will not project. Length $R$ on the handle allows 1/16 inch for riveting.

### Machine Handwheels

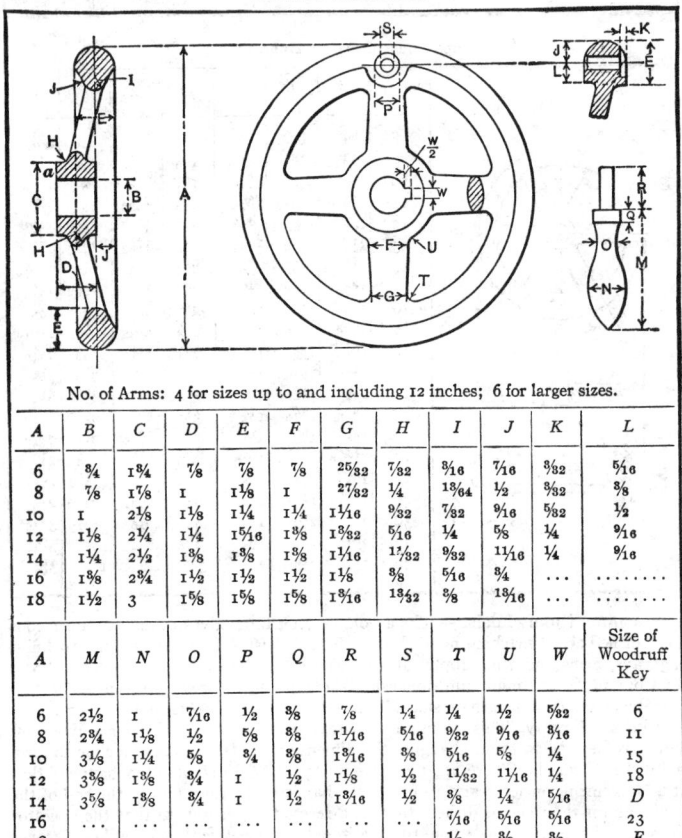

No. of Arms: 4 for sizes up to and including 12 inches; 6 for larger sizes.

| A | B | C | D | E | F | G | H | I | J | K | L |
|---|---|---|---|---|---|---|---|---|---|---|---|
| 6 | 3/4 | 1 3/4 | 7/8 | 7/8 | 7/8 | 25/32 | 7/32 | 3/16 | 7/16 | 3/32 | 5/16 |
| 8 | 7/8 | 1 7/8 | 1 | 1 1/8 | 1 | 27/32 | 1/4 | 13/64 | 1/2 | 3/32 | 3/8 |
| 10 | 1 | 2 1/8 | 1 1/8 | 1 1/4 | 1 1/4 | 1 1/16 | 9/32 | 7/32 | 9/16 | 5/32 | 1/2 |
| 12 | 1 1/8 | 2 1/4 | 1 1/4 | 1 5/16 | 1 3/8 | 1 3/32 | 5/16 | 1/4 | 5/8 | 1/4 | 9/16 |
| 14 | 1 1/4 | 2 1/2 | 1 3/8 | 1 3/8 | 1 3/8 | 1 1/16 | 11/32 | 9/32 | 11/16 | 1/4 | 9/16 |
| 16 | 1 3/8 | 2 3/4 | 1 1/2 | 1 1/2 | 1 1/2 | 1 1/8 | 3/8 | 5/16 | 3/4 | ... | ........ |
| 18 | 1 1/2 | 3 | 1 5/8 | 1 5/8 | 1 5/8 | 1 3/16 | 13/32 | 3/8 | 13/16 | ... | ........ |

| A | M | N | O | P | Q | R | S | T | U | W | Size of Woodruff Key |
|---|---|---|---|---|---|---|---|---|---|---|---|
| 6 | 2 1/2 | 1 | 7/16 | 1/2 | 3/8 | 7/8 | 1/4 | 1/4 | 1/2 | 5/32 | 6 |
| 8 | 2 3/4 | 1 1/8 | 1/2 | 5/8 | 3/8 | 1 1/16 | 5/16 | 9/32 | 9/16 | 3/16 | 11 |
| 10 | 3 1/8 | 1 1/4 | 5/8 | 3/4 | 3/8 | 1 3/16 | 3/8 | 5/16 | 5/8 | 1/4 | 15 |
| 12 | 3 3/8 | 1 3/8 | 3/4 | 1 | 1/2 | 1 1/8 | 1/2 | 11/32 | 11/16 | 1/4 | 18 |
| 14 | 3 5/8 | 1 3/8 | 3/4 | 1 | 1/2 | 1 3/16 | 1/2 | 3/8 | 1/4 | 5/16 | D |
| 16 | ... | ... | ... | ... | ... | ... | ... | 7/16 | 5/16 | 5/16 | 23 |
| 18 | ... | ... | ... | ... | ... | ... | ... | 1/2 | 3/8 | 3/8 | F |

## Handwheels with Straight Arms

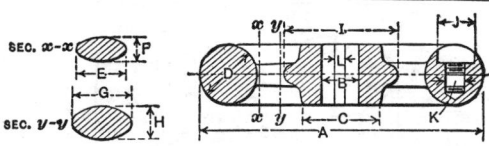

No. of Arms: 4 up to the 10-inch size; 5 in the 11-inch size; 6 in the 12-inch size.

| A | B | C | D | E | F | G | H | I | J | K | L |
|---|---|---|---|---|---|---|---|---|---|---|---|
| 4 | 3/8 | 7/8 | 11/16 | 1/2 | 1/4 | 5/8 | 5/16 | 1 3/8 | 15/32 | 5/16 | 1/16 by 1/32 |
| 5 | 7/16 | 1 | 13/16 | 9/16 | 9/32 | 11/16 | 11/32 | 1 5/8 | 19/32 | 3/8 | 3/32 by 3/64 |
| 6 | 9/16 | 1 3/16 | 15/16 | 5/8 | 5/16 | 3/4 | 3/8 | 1 7/8 | 1/2 | 3/8 | 3/32 by 3/64 |
| 7 | 11/16 | 1 3/8 | 1 | 11/16 | 11/32 | 7/8 | 7/16 | 2 1/8 | 19/32 | 3/8 | 1/8 by 1/16 |
| 8 | 3/4 to 7/8 | 1 1/2 | 1 1/8 | 3/4 | 3/8 | 15/16 | 15/32 | 2 3/8 | 19/32 | 3/8 | 1/8 by 1/16 |
| 9 | 13/16 to 1 | 1 5/8 | 1 3/16 | 13/16 | 13/32 | 1 | 1/2 | 2 5/8 | 19/32 | 3/8 | 1/8 by 1/16 |
| 10 | 7/8 to 1 | 1 3/4 | 1 5/16 | 7/8 | 7/16 | 1 1/8 | 9/16 | 2 7/8 | 21/32 | 7/16 | 3/16 by 3/32 |
| 11 | 15/16 to 1 1/8 | 1 7/8 | 1 3/8 | 15/16 | 15/32 | 1 3/16 | 9/16 | 3 1/8 | 21/32 | 7/16 | 3/16 by 3/32 |
| 12 | 1 to 1 1/4 | 2 | 1 1/2 | 1 | 1/2 | 1 1/4 | 5/8 | 3 3/8 | 3/4 | 1/2 | 1/4 by 1/8 |

### Dimensions of Chains and Rod to Support Counter-weights

| | Size of Chain, A | Diam. of Rod, B | Size of Bolt, C | Diam. of Cotter, E | F | G | Diam. of Hole, H | Max. Safe Load |
|---|---|---|---|---|---|---|---|---|
| | 3/16 | 3/8 | 5/16 | 3/32 | 3/16 | 1 | 3/8 | 350 |
| | 1/4 | 1/2 | 3/8 | 1/8 | 1/4 | 1 1/4 | 7/16 | 650 |
| | 5/16 | 5/8 | 1/2 | 1/8 | 1/4 | 1 1/2 | 9/16 | 1300 |
| | 3/8 | 3/4 | 5/8 | 3/16 | 5/16 | 1 3/4 | 11/16 | 1900 |
| | 1/2 | 7/8 | 3/4 | 3/16 | 5/16 | 2 1/4 | 13/16 | 3300 |
| | 1/2 | 1 | 7/8 | 3/16 | 3/8 | 2 1/2 | 15/16 | 3400 |

**Handwheel Disconnecting Devices.** — A great number of devices for coupling and disconnecting handwheels have been designed. In some designs the hub of the wheel may have two or more teeth on opposite sides of the periphery to engage corresponding slots in a shaft collar or vice versa as shown at *A* in Fig. 1, page 676 The number of slots depends on the number of engaging positions desired. In other designs a spring may be incorporated to keep the teeth disengaged until engaged manually against the pressure of the spring.

Sometimes teeth and slots are provided on both ends of the wheel hub so as to engage different mechanisms when the wheel is moved forward or backward along its axis.

## Machine Handles

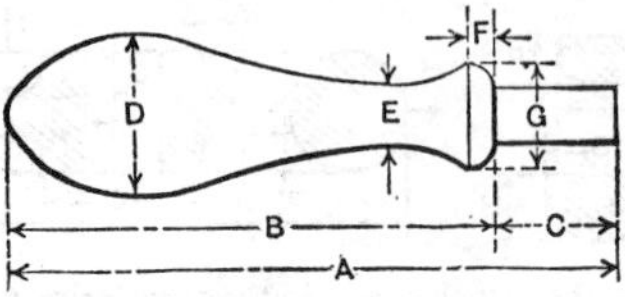


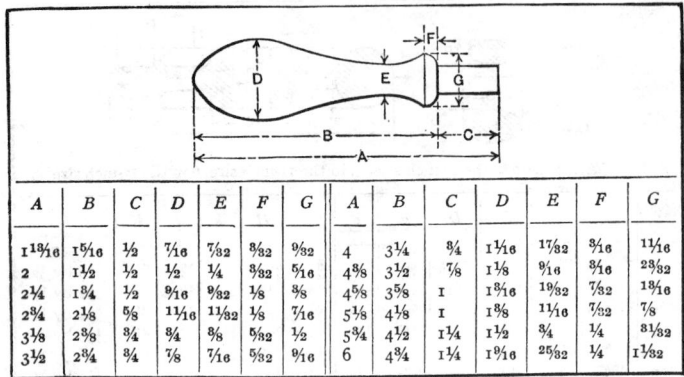

| A | B | C | D | E | F | G | A | B | C | D | E | F | G |
|---|---|---|---|---|---|---|---|---|---|---|---|---|---|
| 1 13/16 | 1 5/16 | 1/2 | 7/16 | 7/32 | 3/32 | 9/32 | 4 | 3 1/4 | 3/4 | 1 1/16 | 17/32 | 3/16 | 11/16 |
| 2 | 1 1/2 | 1/2 | 1/2 | 1/4 | 3/32 | 5/16 | 4 3/8 | 3 1/2 | 7/8 | 1 1/8 | 9/16 | 3/16 | 23/32 |
| 2 1/4 | 1 3/4 | 1/2 | 9/16 | 9/32 | 1/8 | 3/8 | 4 5/8 | 3 5/8 | 1 | 1 3/16 | 19/32 | 7/32 | 13/16 |
| 2 3/4 | 2 1/8 | 5/8 | 11/16 | 11/32 | 1/8 | 7/16 | 5 1/8 | 4 1/8 | 1 | 1 3/8 | 11/16 | 7/32 | 7/8 |
| 3 1/8 | 2 3/8 | 3/4 | 3/4 | 3/8 | 5/32 | 1/2 | 5 3/4 | 4 1/2 | 1 1/4 | 1 1/2 | 3/4 | 1/4 | 31/32 |
| 3 1/2 | 2 3/4 | 3/4 | 7/8 | 7/16 | 5/32 | 9/16 | 6 | 4 3/4 | 1 1/4 | 1 9/16 | 25/32 | 1/4 | 1 1/32 |

## Ball-crank Machine Handles

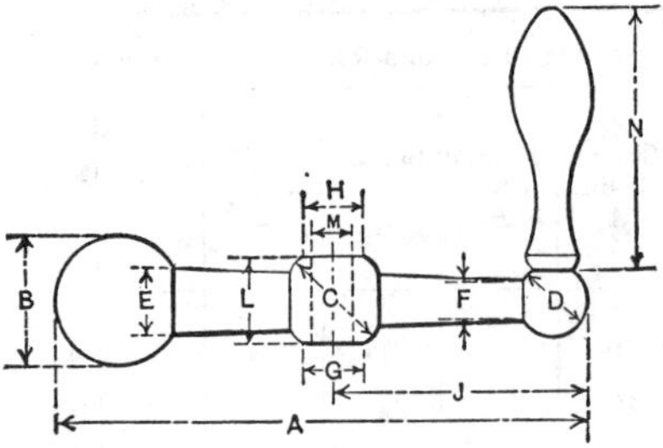


| A | B | C | D | E | F | G | H | J | L | M | N |
|---|---|---|---|---|---|---|---|---|---|---|---|
| 3 | 7/8 | 11/16 | 9/16 | 3/8 | 5/16 | 1/2 | 7/16 | 1 1/2 | 1/2 | 1/4 | 1 1/2 |
| 3 1/4 | 15/16 | 25/32 | 5/8 | 7/16 | 5/16 | 9/16 | 1/2 | 1 5/8 | 9/16 | 5/16 | 1 5/8 |
| 3 1/2 | 1 1/16 | 29/32 | 3/4 | 15/32 | 11/32 | 5/8 | 9/16 | 1 3/4 | 11/16 | 3/8 | 1 3/4 |
| 4 | 1 1/4 | 1 | 3/4 | 17/32 | 13/32 | 3/4 | 11/16 | 2 | 3/4 | 7/16 | 1 7/8 |
| 4 1/2 | 1 5/16 | 1 3/32 | 27/32 | 19/32 | 7/16 | 3/4 | 11/16 | 2 1/4 | 13/16 | 1/2 | 2 |
| 5 | 1 1/2 | 1 5/16 | 1 | 3/4 | 1/2 | 7/8 | 13/16 | 2 1/2 | 1 | 1/2 | 2 3/16 |
| 5 1/2 | 1 1/2 | 1 5/16 | 1 | 3/4 | 1/2 | 7/8 | 13/16 | 2 3/4 | 1 | 1/2 | 2 3/8 |
| 6 | 1 5/8 | 1 3/8 | 1 | 3/4 | 1/2 | 1 | 15/16 | 3 | 1 | 5/8 | 2 9/16 |
| 6 1/2 | 1 5/8 | 1 3/8 | 1 | 3/4 | 1/2 | 1 | 15/16 | 3 1/4 | 1 | 5/8 | 2 3/4 |
| 7 | 1 3/4 | 1 7/16 | 1 | 3/4 | 9/16 | 1 | 15/16 | 3 1/2 | 1 1/16 | 5/8 | 2 15/16 |
| 7 1/2 | 1 3/4 | 1 1/2 | 1 | 3/4 | 9/16 | 1 | 15/16 | 3 3/4 | 1 3/32 | 5/8 | 3 1/8 |
| 8 | 1 3/4 | 1 1/2 | 1 1/16 | 3/4 | 9/16 | 1 | 15/16 | 4 | 1 1/8 | 5/8 | 3 5/16 |
| 8 1/2 | 1 3/4 | 1 9/16 | 1 1/8 | 3/4 | 5/8 | 1 1/8 | 1 1/16 | 4 1/4 | 1 3/16 | 3/4 | 3 1/2 |
| 9 | 1 3/4 | 1 5/8 | 1 3/16 | 3/4 | 5/8 | 1 1/8 | 1 1/16 | 4 1/2 | 1 1/4 | 3/4 | 3 3/4 |

**Two-ball Clamping Levers**

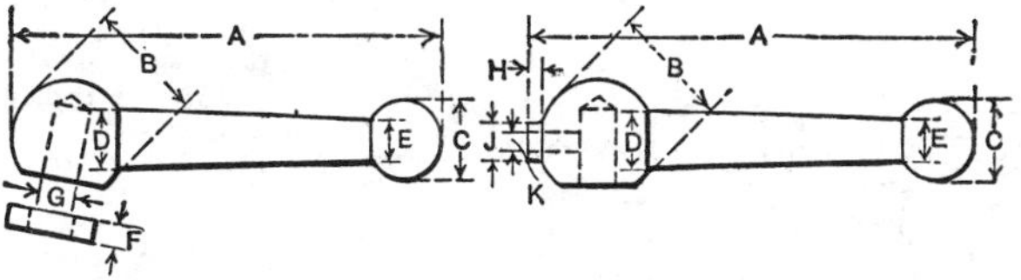


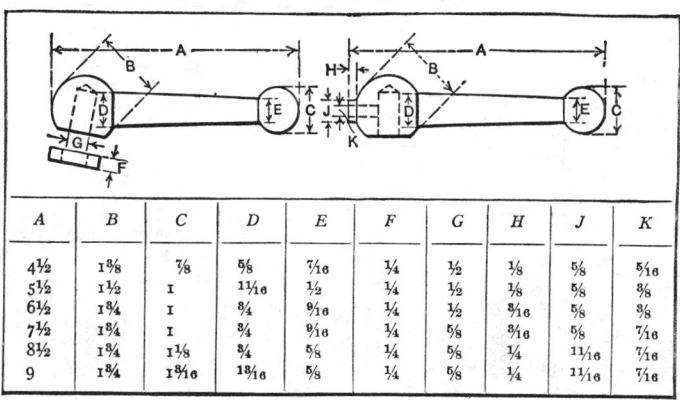

| *A* | *B* | *C* | *D* | *E* | *F* | *G* | *H* | *J* | *K* |
|---|---|---|---|---|---|---|---|---|---|
| 4½ | 1⅜ | ⅞ | ⅝ | 7/16 | ¼ | ½ | ⅛ | ⅝ | 5/16 |
| 5½ | 1½ | 1 | 11/16 | ½ | ¼ | ½ | ⅛ | ⅝ | ⅜ |
| 6½ | 1¾ | 1 | ¾ | 9/16 | ¼ | ½ | 3/16 | ⅝ | ⅜ |
| 7½ | 1¾ | 1 | ¾ | 9/16 | ¼ | ⅝ | 3/16 | ⅝ | 7/16 |
| 8½ | 1¾ | 1⅛ | ¾ | ⅝ | ¼ | ⅝ | ¼ | 11/16 | 7/16 |
| 9 | 1¾ | 1 3/16 | 13/16 | ⅝ | ¼ | ⅝ | ¼ | 11/16 | 7/16 |

**Compound-rest Handles**

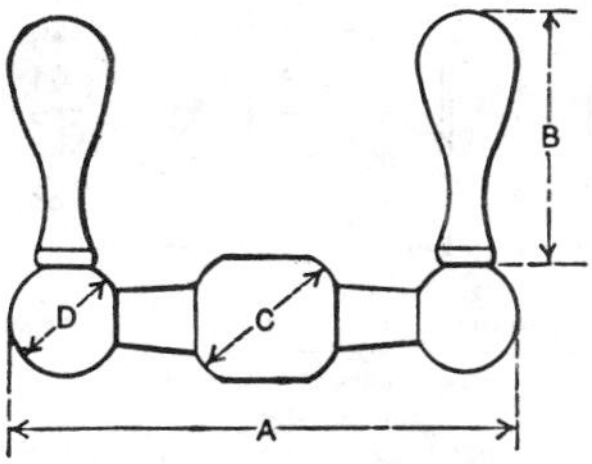


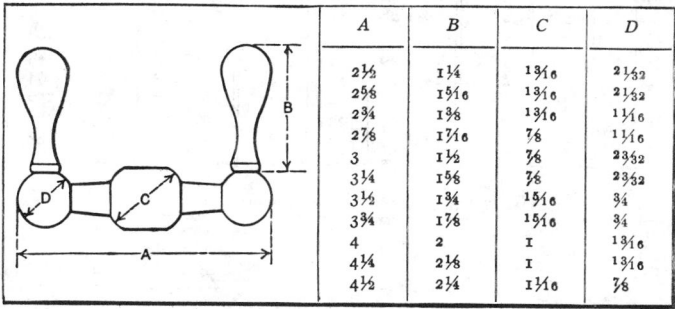

| *A* | *B* | *C* | *D* |
|---|---|---|---|
| 2½ | 1¼ | 13/16 | 21/32 |
| 2⅝ | 1 5/16 | 13/16 | 21/32 |
| 2¾ | 1⅜ | 13/16 | 11/16 |
| 2⅞ | 1 7/16 | ⅞ | 11/16 |
| 3 | 1½ | ⅞ | 23/32 |
| 3¼ | 1⅝ | ⅞ | 23/32 |
| 3½ | 1¾ | 15/16 | ¾ |
| 3¾ | 1⅞ | 15/16 | ¾ |
| 4 | 2 | 1 | 13/16 |
| 4¼ | 2⅛ | 1 | 13/16 |
| 4½ | 2¼ | 1 1/16 | ⅞ |

**Knobs for Machine Doors**

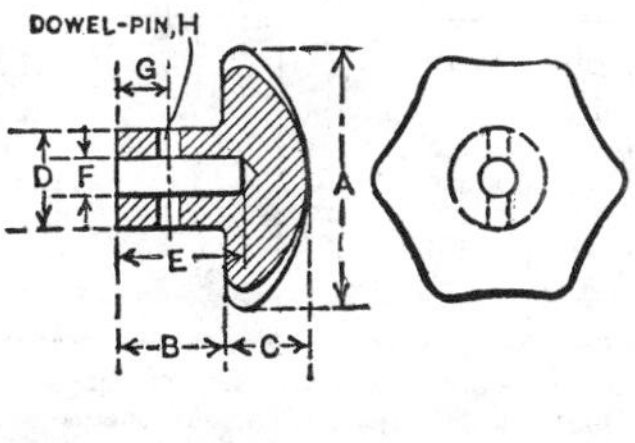


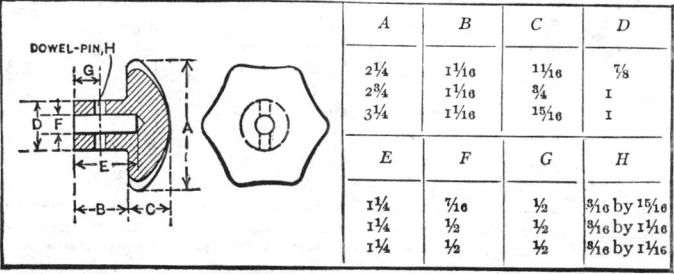

| *A* | *B* | *C* | *D* |
|---|---|---|---|
| 2¼ | 1 1/16 | 11/16 | ⅞ |
| 2¾ | 1 1/16 | ¾ | 1 |
| 3¼ | 1 1/16 | 15/16 | 1 |

| *E* | *F* | *G* | *H* |
|---|---|---|---|
| 1¼ | 7/16 | ½ | 3/16 by 15/16 |
| 1¼ | ½ | ½ | 3/16 by 1 1/16 |
| 1¼ | ½ | ½ | 3/16 by 1 1/16 |

## Dimensions of Turnbuckles

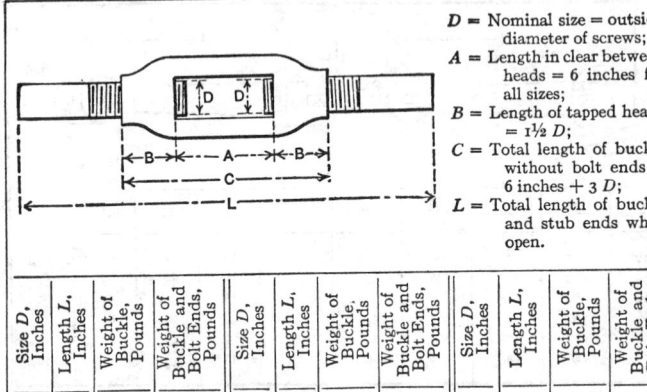

$D$ = Nominal size = outside diameter of screws;
$A$ = Length in clear between heads = 6 inches for all sizes;
$B$ = Length of tapped heads = $1\frac{1}{2}D$;
$C$ = Total length of buckle without bolt ends = 6 inches + 3 $D$;
$L$ = Total length of buckle and stub ends when open.

| Size $D$, Inches | Length $L$, Inches | Weight of Buckle, Pounds | Weight of Buckle and Bolt Ends, Pounds | Size $D$, Inches | Length $L$, Inches | Weight of Buckle, Pounds | Weight of Buckle and Bolt Ends, Pounds | Size $D$, Inches | Length $L$, Inches | Weight of Buckle, Pounds | Weight of Buckle and Bolt Ends, Pounds |
|---|---|---|---|---|---|---|---|---|---|---|---|
| 3/8 | 22 | 1 | 1 1/2 | 1 3/8 | 27 | 7 | 16 | 2 5/8 | 32 | 30 | 70 |
| 7/16 | 22 | 1 | 1 3/4 | 1 1/2 | 27 | 8 | 19 | 2 3/4 | 33 | 33 | 78 |
| 1/2 | 22 | 1 | 2 | 1 5/8 | 28 | 10 | 23 | 2 7/8 | 33 | 36 | 86 |
| 9/16 | 22 | 1 1/4 | 2 1/2 | 1 3/4 | 28 | 11 | 26 | 3 | 34 | 40 | 96 |
| 5/8 | 22 | 1 1/2 | 3 | 1 7/8 | 29 | 12 | 30 | 3 1/8 | 36 | 45 | 108 |
| 3/4 | 23 | 2 | 4 | 2 | 29 | 14 | 35 | 3 1/4 | 36 | 50 | 120 |
| 7/8 | 24 | 3 | 6 | 2 1/8 | 29 | 17 | 41 | 3 3/8 | 37 | 57 | 134 |
| 1 | 25 | 4 | 8 | 2 1/4 | 30 | 20 | 47 | 3 1/2 | 37 | 65 | 150 |
| 1 1/8 | 25 | 5 | 11 | 2 3/8 | 31 | 22 | 53 | 3 3/4 | 39 | 74 | 168 |
| 1 1/4 | 26 | 6 | 13 | 2 1/2 | 32 | 25 | 61 | 4 | 41 | 84 | 188 |

## American National Standard Cotter Pins (ANSI B18.8.1-1972)

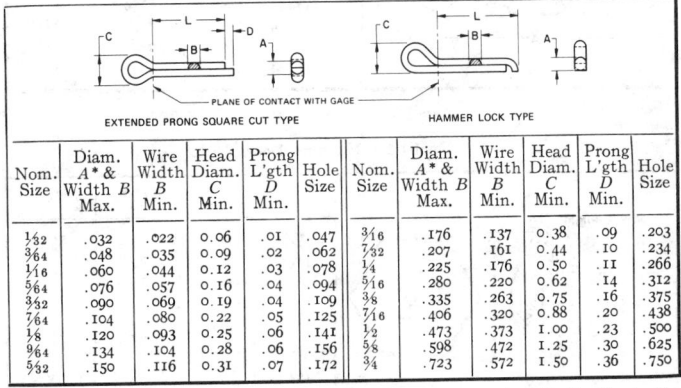

PLANE OF CONTACT WITH GAGE

EXTENDED PRONG SQUARE CUT TYPE                    HAMMER LOCK TYPE

| Nom. Size | Diam. $A^*$ & Width $B$ Max. | Wire Width $B$ Min. | Head Diam. $C$ Min. | Prong L'gth $D$ Min. | Hole Size | Nom. Size | Diam. $A^*$ & Width $B$ Max. | Wire Width $B$ Min. | Head Diam. $C$ Min. | Prong L'gth $D$ Min. | Hole Size |
|---|---|---|---|---|---|---|---|---|---|---|---|
| 1/32 | .032 | .022 | 0.06 | .01 | .047 | 3/16 | .176 | .137 | 0.38 | .09 | .203 |
| 3/64 | .048 | .035 | 0.09 | .02 | .062 | 7/32 | .207 | .161 | 0.44 | .10 | .234 |
| 1/16 | .060 | .044 | 0.12 | .03 | .078 | 1/4 | .225 | .176 | 0.50 | .11 | .266 |
| 5/64 | .076 | .057 | 0.16 | .04 | .094 | 5/16 | .280 | .220 | 0.62 | .14 | .312 |
| 3/32 | .090 | .069 | 0.19 | .04 | .109 | 3/8 | .335 | .263 | 0.75 | .16 | .375 |
| 7/64 | .104 | .080 | 0.22 | .05 | .125 | 7/16 | .406 | .320 | 0.88 | .20 | .438 |
| 1/8 | .120 | .093 | 0.25 | .06 | .141 | 1/2 | .473 | .373 | 1.00 | .23 | .500 |
| 9/64 | .134 | .104 | 0.28 | .06 | .156 | 5/8 | .598 | .472 | 1.25 | .30 | .625 |
| 5/32 | .150 | .116 | 0.31 | .07 | .172 | 3/4 | .723 | .572 | 1.50 | .36 | .750 |

All dimensions are given in inches.    * Tolerances are: —.004 inch for the 1/32- to 3/16-inch sizes, incl.; —.005 inch for the 7/32- to 5/16-inch sizes, incl.; —.006 inch for the 3/8- to 1/2-inch sizes, incl.; and —.008 inch for the 5/8- and 3/4-inch sizes.

American National Standard Clevis Pins (ANSI B18.8.1-1972)

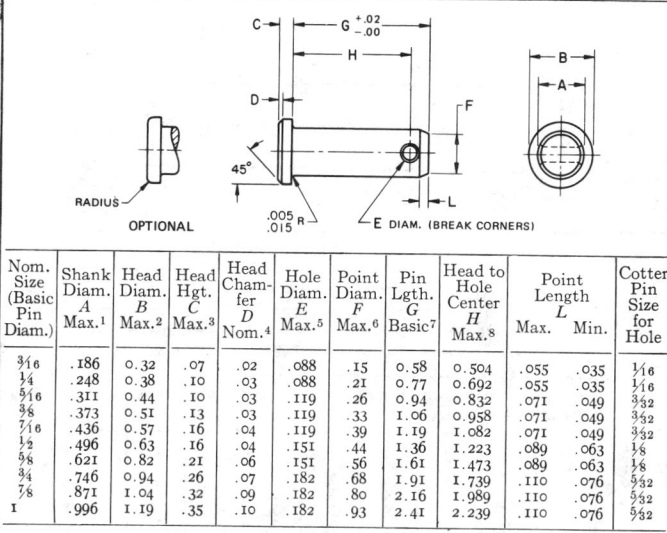

| Nom. Size (Basic Pin Diam.) | Shank Diam. A Max.[1] | Head Diam. B Max.[2] | Head Hgt. C Max.[3] | Head Chamfer D Nom.[4] | Hole Diam. E Max.[5] | Point Diam. F Max.[6] | Pin Lgth. G Basic[7] | Head to Hole Center H Max.[8] | Point Length L Max. | Point Length L Min. | Cotter Pin Size for Hole |
|---|---|---|---|---|---|---|---|---|---|---|---|
| 3/16 | .186 | 0.32 | .07 | .02 | .088 | .15 | 0.58 | 0.504 | .055 | .035 | 1/16 |
| 1/4 | .248 | 0.38 | .10 | .03 | .088 | .21 | 0.77 | 0.692 | .055 | .035 | 1/16 |
| 5/16 | .311 | 0.44 | .10 | .03 | .119 | .26 | 0.94 | 0.832 | .071 | .049 | 3/32 |
| 3/8 | .373 | 0.51 | .13 | .03 | .119 | .33 | 1.06 | 0.958 | .071 | .049 | 3/32 |
| 7/16 | .436 | 0.57 | .16 | .04 | .119 | .39 | 1.19 | 1.082 | .071 | .049 | 3/32 |
| 1/2 | .496 | 0.63 | .16 | .04 | .151 | .44 | 1.36 | 1.223 | .089 | .063 | 1/8 |
| 5/8 | .621 | 0.82 | .21 | .06 | .151 | .56 | 1.61 | 1.473 | .089 | .063 | 1/8 |
| 3/4 | .746 | 0.94 | .26 | .07 | .182 | .68 | 1.91 | 1.739 | .110 | .076 | 5/32 |
| 7/8 | .871 | 1.04 | .32 | .09 | .182 | .80 | 2.16 | 1.989 | .110 | .076 | 5/32 |
| 1 | .996 | 1.19 | .35 | .10 | .182 | .93 | 2.41 | 2.239 | .110 | .076 | 5/32 |

All dimensions are given in inches.   [1] Tolerance is −.005 inch.   [2,3] Tolerance is −.02 inch.   [4] Tolerance is ±.01 inch.   [5] Tolerance is −.015 inch.   [6] Tolerance is −.01 inch.   [7] Lengths tabulated are intended for use with standard clevises, without spacers.  When required, it is recommended that other pin lengths be limited wherever possible to nominal lengths in .06-inch increments.   [8] Tolerance is −.020 inch.

**Dowel-Pins.** — Dowel-pins are used either to retain parts in a fixed position or to preserve alignment.  Under normal conditions a properly fitted dowel-pin is subjected to shearing strain only, and this strain occurs only at the junction of the surfaces of the two parts which are being held by the dowel-pin.  It is seldom necessary to use more than two dowel-pins for holding two pieces together and frequently one is sufficient.  For parts which have to be taken apart frequently, and where driving out of the dowel-pins would tend to wear the holes, and also for very accurately constructed tools and gages which have to be taken apart, or which require to be kept in absolute alignment, the taper dowel-pin is preferable.  As applied to average machine work, the taper dowel-pin is most commonly used but the straight type is given the preference on tool and gage work, except where extreme accuracy is required, or where the tool or gage is to be subjected to rough handling.

The size of the dowel-pin is governed by its application.  For locating nests, gage plates, etc., pins from 1/8 to 3/16 inch in diameter are satisfactory.  For locating dies, the diameter of the dowel-pin should never be less than 1/4 inch; the general rule is to use dowel-pins of the same size as the screws used in fastening the work.  The length of the dowel-pin should be about one and one-half to two times its diameter in each plate or part to be doweled.

When hardened cylindrical dowel-pins are inserted in soft parts, ream the hole about 0.001 inch smaller than the dowel-pin.  If the doweled parts are hardened, grind (or lap) the hole 0.0002 to 0.0003 inch under size.  The hole should be ground or lapped straight or without taper or "bell-mouth."

## American Standard Hardened and Ground Dowel Pins (ASA B5.20-1958)

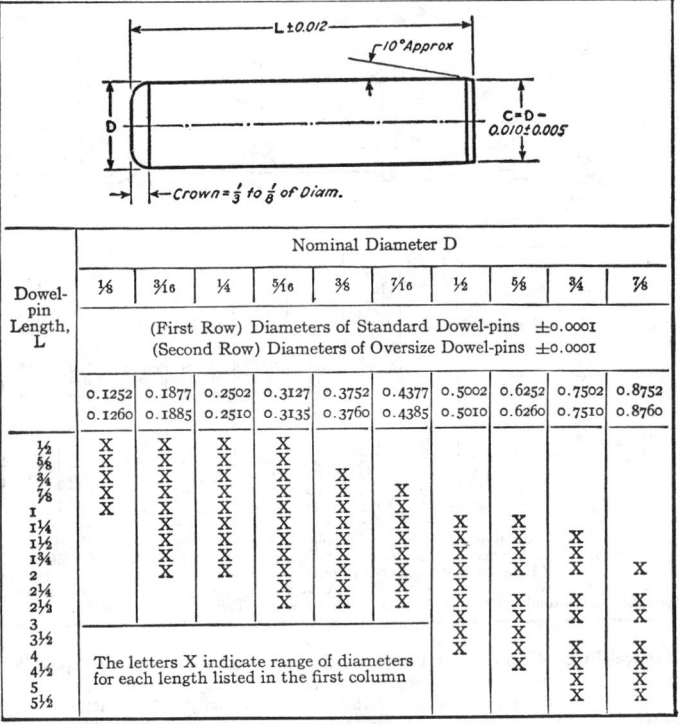

| Dowel-pin Length, L | Nominal Diameter D | | | | | | | | | |
|---|---|---|---|---|---|---|---|---|---|---|
| | ⅛ | 3⁄16 | ¼ | 5⁄16 | ⅜ | 7⁄16 | ½ | ⅝ | ¾ | ⅞ |
| | (First Row) Diameters of Standard Dowel-pins ±0.0001 | | | | | | | | | |
| | (Second Row) Diameters of Oversize Dowel-pins ±0.0001 | | | | | | | | | |
| | 0.1252 | 0.1877 | 0.2502 | 0.3127 | 0.3752 | 0.4377 | 0.5002 | 0.6252 | 0.7502 | 0.8752 |
| | 0.1260 | 0.1885 | 0.2510 | 0.3135 | 0.3760 | 0.4385 | 0.5010 | 0.6260 | 0.7510 | 0.8760 |
| ½ | X | X | X | X | | | | | | |
| ⅝ | X | X | X | X | | | | | | |
| ¾ | X | X | X | X | X | | | | | |
| ⅞ | X | X | X | X | X | X | | | | |
| 1 | | X | X | X | X | X | | | | |
| 1¼ | | X | X | X | X | X | X | X | | |
| 1½ | | X | X | X | X | X | X | X | X | |
| 1¾ | | X | X | X | X | X | X | X | X | |
| 2 | | | X | X | X | X | X | X | X | X |
| 2¼ | | | | X | X | X | X | | | |
| 2½ | | | | X | X | X | X | X | X | X |
| 3 | | | | | | | X | X | X | X |
| 3½ | | | | | | | X | X | X | X |
| 4 | The letters X indicate range of diameters for each length listed in the first column | | | | | | | X | X | X |
| 4½ | | | | | | | | | X | X |
| 5 | | | | | | | | | X | X |
| 5½ | | | | | | | | | X | X |

These dowel-pins are extensively used in the tool and machine industry and a machine reamer of nominal size may be used to produce the holes into which these pins tap or press fit. All dimensions are given in inches.

## American Standard Unhardened Ground Dowel Pins (ASA B5.20-1958)

| Nom. Diam. | Max. Diam. | Min. Diam. | Chamfer Width | Nom. Diam. | Max. Diam. | Min. Diam. | Chamfer Width |
|---|---|---|---|---|---|---|---|
| 0.062 | 0.0600 | 0.0595 | 0.015 | 0.312 | 0.3094 | 0.3089 | 0.030 |
| 0.094 | 0.0912 | 0.0907 | 0.015 | 0.375 | 0.3717 | 0.3712 | 0.030 |
| 0.109 | 0.1068 | 0.1063 | 0.015 | 0.438 | 0.4341 | 0.4336 | 0.030 |
| 0.125 | 0.1223 | 0.1218 | 0.015 | 0.500 | 0.4964 | 0.4959 | 0.030 |
| 0.156 | 0.1535 | 0.1530 | 0.015 | 0.625 | 0.6211 | 0.6206 | 0.045 |
| 0.188 | 0.1847 | 0.1842 | 0.015 | 0.750 | 0.7458 | 0.7453 | 0.045 |
| 0.219 | 0.2159 | 0.2154 | 0.015 | 0.875 | 0.8705 | 0.8700 | 0.060 |
| 0.250 | 0.2470 | 0.2465 | 0.015 | 1.000 | 0.9952 | 0.9947 | 0.060 |

Maximum diameters are graduated from 0.0005 on 1⁄16-inch pins to 0.0028 on 1-inch pins, under the minimum commercial bar stock sizes. All dimensions in inches.

These pins have a 25-degree chamfer on each end. Tolerance on overall length is ±0.012 inch, on chamfer length, ±0.010 inch.

## American Standard Straight Pins (ASA B5.20-1958)

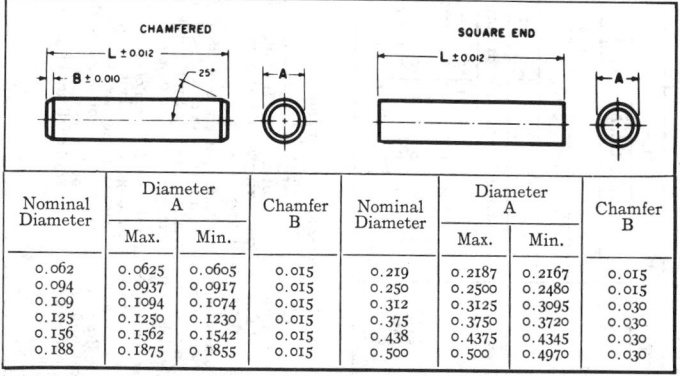

| Nominal Diameter | Diameter A | | Chamfer B | Nominal Diameter | Diameter A | | Chamfer B |
|---|---|---|---|---|---|---|---|
| | Max. | Min. | | | Max. | Min. | |
| 0.062 | 0.0625 | 0.0605 | 0.015 | 0.219 | 0.2187 | 0.2167 | 0.015 |
| 0.094 | 0.0937 | 0.0917 | 0.015 | 0.250 | 0.2500 | 0.2480 | 0.015 |
| 0.109 | 0.1094 | 0.1074 | 0.015 | 0.312 | 0.3125 | 0.3095 | 0.030 |
| 0.125 | 0.1250 | 0.1230 | 0.015 | 0.375 | 0.3750 | 0.3720 | 0.030 |
| 0.156 | 0.1562 | 0.1542 | 0.015 | 0.438 | 0.4375 | 0.4345 | 0.030 |
| 0.188 | 0.1875 | 0.1855 | 0.015 | 0.500 | 0.500 | 0.4970 | 0.030 |

All dimensions are given in inches.

These pins must be straight and free from burrs or any other defects that will affect their serviceability.

**British Standard for Metric Series Dowel Pins.** — Steel parallel dowel pins specified in British Standard 1804:Part 2:1968 are divided into three grades which provide different degrees of pin accuracy.

*Grade 1* is a precision ground pin made from En 32A or En 32B low carbon steel (BS 970) or from high carbon steel to BS 1407 or BS 1423. Pins below 4 mm diameter are unhardened. Those of 4 mm diameter and above are hardened to a minimum of 750 HV 30 in accordance with BS 427, but if they are made from steels to BS 1407 or BS 1423 then the hardness shall be within the range 600 to 700 HV 30, in accordance with BS 427. The values of other hardness scales may be used in accordance with BS 860.

*Grade 2* is a ground pin made from any of the steels used for Grade 1. The pins are normally supplied unhardened, unless a different condition is agreed on between the purchaser and supplier.

*Grade 3* pins are made from En 1A free cutting steel (BS 970) and are supplied with a machined, bright rolled or drawn finish. They are normally supplied unhardened unless a different condition is agreed on between the purchaser and supplier.

Pins of any grade may be made from different steels in accordance with BS 970, by mutual agreement between the purchaser and manufacturer. If steels other than those in the standard range are used, the hardness of the pins shall also be decided on by mutual agreement between purchaser and supplier. As shown in the illustration at the head of the accompanying table, one end of each pin is chamfered to provide a lead. The other end may be similarly chamfered, or domed.

If a dowel pin is driven into a blind hole where no provision is made for releasing air, the worker assembling the pin may be endangered, and damage may be caused to the associated component, or stresses may be set up. The appendix of the Standard describes one method of overcoming this problem, which is the provision of a small flat surface along the length of a pin to permit the release of air.

For purposes of marking, the Standard states that each package or lot of dowel pins shall bear the manufacturer's name or trademark, the BS number, and the grade of pin.

## British Standard Parallel Steel Dowel Pins — Metric Series (BS 1804: Part 2: 1968)

### Limits of Tolerance on Diameter

| Nom. Diam., mm | | Grade* | | |
|---|---|---|---|---|
| | | 1 | 2 | 3 |
| | | Tolerance Zone | | |
| | | m5 | h7 | h11 |
| Over | To & Incl. | Limits of Tolerance, 0.001 mm | | |
| — | 3 | +7 / +2 | 0 / −12† | 0 / −60 |
| 3 | 6 | +9 / +4 | 0 / −12 | 0 / −75 |
| 6 | 10 | +12 / +6 | 0 / −15 | 0 / −90 |
| 10 | 14 | +15 / +7 | 0 / −18 | 0 / −110 |
| 14 | 18 | +15 / +7 | 0 / −18 | 0 / −110 |
| 18 | 24 | +17 / +8 | 0 / −21 | 0 / −130 |
| 24 | 30 | +17 / +8 | 0 / −21 | 0 / −130 |

\* The limits of tolerance for grades 1 and 2 dowel pins have been chosen to provide satisfactory assembly when used in standard reamed holes (H7 and H8 tolerance zones). If the assembly is not satisfactory, BS 1916: Part 1, Limits and Fits for Engineering should be consulted, and a different class of fit chosen.

† This tolerance is larger than that given in BS 1916, and has been included because the use of a closer tolerance would involve precision grinding by the manufacturer, which is uneconomic for a grade 2 dowel pin.

The tolerance limits on the overall length of all grades of dowel pin up to and including 50 mm long are +0.5, −0.0 mm, and for pins over 50 mm long, are +0.8, −0.0 mm. The Standard specifies that the roughness of the cylindrical surface of grades 1 and 2 dowel pins, when assessed in accordance with BS 1134, shall not be greater than 0.4 μm CLA (16 CLA).

Diagram of parallel steel dowel pin showing overall length $L$, nominal diameter $D$, chamfer length $a$, and a 20°–40° chamfer angle.

### Standard Sizes

| | Nominal Diameter D, mm | | | | | | | | | | | | | |
|---|---|---|---|---|---|---|---|---|---|---|---|---|---|---|
| | 1 | 1.5 | 2 | 2.5 | 3 | 4 | 5 | 6 | 8 | 10 | 12 | 16 | 20 | 25 |
| Chamfer a max, mm | 0.3 | 0.3 | 0.3 | 0.4 | 0.45 | 0.6 | 0.75 | 0.9 | 1.2 | 1.5 | 1.8 | 2.5 | 3 | 4 |
| Nom. Length L, mm | | | | | | | | | | | | | | |
| 4 | o | o | o | | | | | | | | | | | |
| 8 | o | o | o | o | o | | | | | | | | | |
| 10 | o | o | o | o | o | o | | | | | | | | |
| 12 | | o | o | o | o | o | o | o | | | | | | |
| 16 | | o | o | o | o | o | o | o | o | | | | | |
| 20 | | | | o | o | o | o | o | o | o | | | | |
| 25 | | | | o | o | o | o | o | o | o | o | | | |
| 30 | | | | | o | o | o | o | o | o | o | o | | |
| 35 | | | | | | o | o | o | o | o | o | o | | |
| 40 | | | | | | | o | o | o | o | o | o | o | |
| 45 | | | | | | | | o | o | o | o | o | o | |
| 50 | | | | | | | | | o | o | o | o | o | o |
| 60 | | | | | | | | | | o | o | o | o | o |
| 70 | | | | | | | | | | o | o | o | o | o |
| 80 | | | | | | | | | | | o | o | o | o |
| 90 | | | | | | | | | | | | o | o | o |
| 100 | | | | | | | | | | | | o | o | o |
| 110 | | | | | | | | | | | | | o | o |
| 120 | | | | | | | | | | | | | o | o |

## American Standard Taper Pins (ASA B5.20-1958)

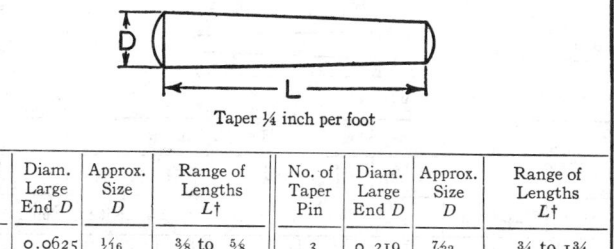

Taper ¼ inch per foot

| No. of Taper Pin | Diam. Large End $D$ | Approx. Size $D$ | Range of Lengths $L\dagger$ | No. of Taper Pin | Diam. Large End $D$ | Approx. Size $D$ | Range of Lengths $L\dagger$ |
|---|---|---|---|---|---|---|---|
| 7/0 | 0.0625 | ¹⁄₁₆ | ⅜ to ⅝ | 3 | 0.219 | ⁷⁄₃₂ | ¾ to 1¾ |
| 6/0 | 0.078 | ⁵⁄₆₄ | ⅜ to ¾ | 4 | 0.250 | ¼ | ¾ to 2 |
| 5/0 | 0.094 | ³⁄₃₂ | ½ to 1 | 5 | 0.289 | ¹⁹⁄₆₄ | 1 to 2¼ |
| 4/0 | 0.109 | ⁷⁄₆₄ | ½ to 1 | 6 | 0.341 | ¹¹⁄₃₂ | 1¼ to 3 |
| 3/0 | 0.125 | ⅛ | ½ to 1 | 7 | 0.409 | ¹³⁄₃₂ | 2 to 3¾ |
| 2/0 | 0.141 | ⁹⁄₆₄ | ½ to 1¼ | 8 | 0.492 | ½ | 2 to 4½ |
| 0 | 0.156 | ⁵⁄₃₂ | ½ to 1¼ | 9 | 0.591 | ¹⁹⁄₃₂ | 2¾ to 5¼ |
| 1 | 0.172 | ¹¹⁄₆₄ | ⅝ to 1¼ | 10 | 0.706 | ⁴⁵⁄₆₄ | 3½ to 6 |
| 2 | 0.193 | ³⁄₁₆ | ¾ to 1½ | | | | |

† These lengths $L$ are suitable for use with the standard reamers listed on page 1654. Longer lengths available for Nos. 1 to 9, incl.

Special sizes No. 11 (0.860), No. 12 (1.032), No. 13 (1.241), and No. 14 (1.523) are also available. Their lengths are special.

Tolerance on diameter is +0.0013 −0.0007 for all sizes.

To find diameter at small end of pin, multiply length $L$ by 0.0208 and subtract product from large end diameter $D$.

### Diameter at Small Ends of Standard Taper Pins

| Pin Length in Inches | Pin Number and Small End Diam. for Given Length | | | | | | | | | | |
|---|---|---|---|---|---|---|---|---|---|---|---|
| | 0 | 1 | 2 | 3 | 4 | 5 | 6 | 7 | 8 | 9 | 10 |
| ¾ | 0.140 | 0.156 | 0.177 | 0.203 | 0.235 | 0.273 | 0.325 | 0.393 | 0.476 | 0.575 | 0.690 |
| 1 | 0.135 | 0.151 | 0.172 | 0.198 | 0.230 | 0.268 | 0.320 | 0.388 | 0.471 | 0.570 | 0.685 |
| 1¼ | 0.130 | 0.146 | 0.167 | 0.192 | 0.224 | 0.263 | 0.315 | 0.382 | 0.466 | 0.565 | 0.680 |
| 1½ | 0.125 | 0.141 | 0.162 | 0.187 | 0.219 | 0.258 | 0.310 | 0.377 | 0.460 | 0.560 | 0.675 |
| 1¾ | 0.120 | 0.136 | 0.157 | 0.182 | 0.214 | 0.252 | 0.305 | 0.372 | 0.455 | 0.554 | 0.669 |
| 2 | 0.114 | 0.130 | 0.151 | 0.177 | 0.209 | 0.247 | 0.299 | 0.367 | 0.450 | 0.549 | 0.664 |
| 2¼ | 0.109 | 0.125 | 0.146 | 0.172 | 0.204 | 0.242 | 0.294 | 0.362 | 0.445 | 0.544 | 0.659 |
| 2½ | 0.104 | 0.120 | 0.141 | 0.166 | 0.198 | 0.237 | 0.289 | 0.356 | 0.440 | 0.539 | 0.654 |
| 2¾ | 0.099 | 0.115 | 0.136 | 0.161 | 0.193 | 0.232 | 0.284 | 0.351 | 0.434 | 0.534 | 0.649 |
| 3 | 0.094 | 0.110 | 0.131 | 0.156 | 0.188 | 0.227 | 0.279 | 0.346 | 0.429 | 0.528 | 0.643 |
| 3¼ | ... | ... | ... | 0.151 | 0.182 | 0.221 | 0.273 | 0.340 | 0.424 | 0.523 | 0.638 |
| 3½ | ... | ... | ... | 0.146 | 0.177 | 0.216 | 0.268 | 0.335 | 0.419 | 0.518 | 0.633 |
| 3¾ | ... | ... | ... | 0.141 | 0.172 | 0.211 | 0.263 | 0.330 | 0.414 | 0.513 | 0.628 |
| 4 | ... | ... | ... | 0.136 | 0.167 | 0.206 | 0.258 | 0.326 | 0.409 | 0.508 | 0.623 |
| 4¼ | ... | ... | ... | 0.131 | 0.162 | 0.201 | 0.253 | 0.321 | 0.403 | 0.502 | 0.617 |
| 4½ | ... | ... | ... | 0.125 | 0.156 | 0.195 | 0.247 | 0.315 | 0.398 | 0.497 | 0.612 |
| 5 | ... | ... | ... | ... | 0.146 | 0.185 | 0.237 | 0.305 | 0.389 | 0.487 | 0.602 |
| 5½ | ... | ... | ... | ... | ... | ... | ... | 0.294 | 0.377 | 0.476 | 0.591 |
| 6 | ... | ... | ... | ... | ... | ... | ... | 0.284 | 0.367 | 0.466 | 0.581 |

## British Standard Solid and Split Taper Pins (B.S. 46: Part 3: 1951)

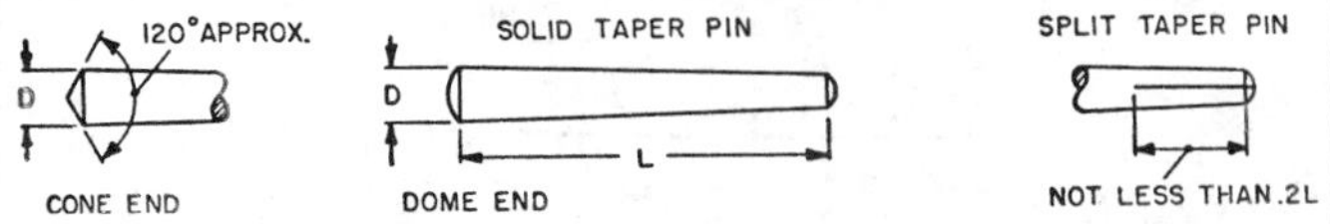

120°APPROX. — CONE END    SOLID TAPER PIN — DOME END    SPLIT TAPER PIN — NOT LESS THAN .2L

| Diam. D at Large End, in. | | | Length L in Inches | | | | | | | | | | | | |
|---|---|---|---|---|---|---|---|---|---|---|---|---|---|---|---|---|
| | | | ¼ | ⅜ | ½ | ⅝ | ¾ | ⅞ | 1 | 1¼ | 1½ | 1¾ | 2 | 2¼ | 2½ | 2¾ |
| Nom. | Max. | Min. | o = Solid taper pins | | | | | | ø = Split taper pins | | | | | | | |
| 1/32* | 0.0312 | 0.0292 | o | o | o | o | | | | | | | | | | |
| 3/64* | 0.0469 | 0.0449 | | o | o | o | o | | | | | | | | | |
| 1/16 | 0.0625 | 0.0605 | | | o | o | oø | oø | oø | oø | o | | | | | |
| 5/64 | 0.0781 | 0.0761 | | | o | o | oø | oø | oø | oø | oø | o | | | | |
| 3/32 | 0.0938 | 0.0918 | | | o | o | oø | oø | oø | oø | oø | oø | o | o | | |
| 7/64* | 0.1094 | 0.1074 | | | | o | o | oø | oø | oø | oø | oø | oø | o | o | |
| 1/8 | 0.1250 | 0.1230 | | | | o | o | oø | oø | oø | oø | oø | oø | oø | o | |
| 9/64* | 0.1406 | 0.1386 | | | | | o | o | oø | oø | oø | oø | oø | oø | oø | o |
| 5/32 | 0.1562 | 0.1542 | | | | | o | o | oø | oø | oø | oø | oø | oø | oø | o |
| 11/64* | 0.1719 | 0.1699 | | | | | | o | o | oø | oø | oø | oø | oø | oø | oø |
| 3/16 | 0.1875 | 0.1855 | | | | | | o | o | oø | oø | oø | oø | oø | oø | oø |
| 7/32 | 0.2188 | 0.2168 | | | | | | | | o | oø | oø | oø | oø | oø | oø |
| 1/4 | 0.2500 | 0.2460 | | | | | | | | o | o | oø | oø | oø | oø | oø |
| 9/32 | 0.2812 | 0.2772 | | | | | | | | | | o | o | o | o | o |
| 5/16 | 0.3125 | 0.3085 | | | | | | | | | | o | o | oø | oø | oø |
| 11/32* | 0.3438 | 0.3398 | | | | | | | | | | | o | o | o | o |
| 3/8 | 0.3750 | 0.3710 | | | | | | | | | | | o | o | o | oø |
| 13/32* | 0.4062 | 0.4022 | | | | | | | | | | | | | | o |

| Nom. | Max. | Min. | 3 | 3¼ | 3½ | 3¾ | 4 | 4½ | 5 | 5½ | 6 | 6½ | 7 | 7½ | 8 |
|---|---|---|---|---|---|---|---|---|---|---|---|---|---|---|---|
| | | | o = Solid taper pins | | | | | | ø = Split taper pins | | | | | | |
| 11/64* | 0.1719 | 0.1699 | o | | | | | | | | | | | | |
| 3/16 | 0.1875 | 0.1855 | oø | | | | | | | | | | | | |
| 7/32 | 0.2188 | 0.2168 | oø | oø | | | | | | | | | | | |
| 1/4 | 0.2500 | 0.2460 | oø | oø | oø | | | | | | | | | | |
| 9/32 | 0.2812 | 0.2772 | o | | o | | | | | | | | | | |
| 5/16 | 0.3125 | 0.3085 | oø | oø | oø | ø | oø | oø | | | | | | | |
| 11/32* | 0.3438 | 0.3398 | o | o | o | | o | o | o | | | | | | |
| 3/8 | 0.3750 | 0.3710 | oø | oø | oø | oø | oø | oø | oø | ø | | | | | |
| 13/32* | 0.4062 | 0.4022 | o | | o | | o | o | o | | | | | | |
| 7/16 | 0.4375 | 0.4335 | oø | oø | oø | oø | oø | oø | oø | ø | oø | ø | | | |
| 1/2 | 0.5000 | 0.4960 | o | o | oø | oø | oø | oø | oø | oø | oø | ø | | | |
| 9/16* | 0.5625 | 0.5585 | | | o | o | oø | oø | oø | oø | oø | ø | ø | ø | |
| 5/8 | 0.6250 | 0.6210 | | | | | o | o | oø | oø | oø | ø | ø | ø | |
| 3/4 | 0.7500 | 0.7460 | | | | | | | o | o | oø | oø | ø | | |
| 7/8 | 0.8750 | 0.8710 | | | | | | | | | o | | o | | |

*Asterisk indicates non-preferred sizes. Taper on diameter is 1 in 48 (¼ inch per foot). Tolerances on lengths are as follows: up to and including 2 inches, ±0.015 inch; above 2 inches and up to and including 6 inches, ±0.040 inch; and above 6 inches, ±0.100 inch. Tolerance on the taper is ±0.0005 inch per inch. These pins come in four Classes: A (mild steel), B (high grade mild steel), C (high tensile steel), and D (stainless steel) with minimum Brinell hardnesses of 121 for the solid pins and 111 for the split pins of Classes A and B and 248 for both the solid and split pins of Classes C and D.

## American Standard Grooved Pins (ASA B5.20-1958)

Diagram labels (Types A, C, F — top): TYPE A, TYPE C, TYPE F; dimensions L ±0.010, A, C, D, E, R, B, 30°

| Nominal Size, In. | 3/64 | 1/16 | 5/64 | 3/32 | 7/64 | 1/8 | 5/32 | 3/16 | 7/32 | 1/4 | 5/16 | 3/8 | 7/16 | 1/2 |
|---|---|---|---|---|---|---|---|---|---|---|---|---|---|---|
| Diameter, $A$, Max. In. | 0.0469 | 0.0625 | 0.0781 | 0.0938 | 0.1094 | 0.1250 | 0.1563 | 0.1875 | 0.2188 | 0.2500 | 0.3125 | 0.3750 | 0.4375 | 0.5000 |
| Diameter, $A$, Min. In. | 0.0459 | 0.0615 | 0.0771 | 0.0928 | 0.1084 | 0.1230 | 0.1543 | 0.1855 | 0.2168 | 0.2480 | 0.3105 | 0.3730 | 0.4355 | 0.4980 |
| Recommended Hole, Max. In. | 0.0478 | 0.0640 | 0.0798 | 0.0956 | 0.1113 | 0.1271 | 0.1587 | 0.1903 | 0.2219 | 0.2534 | 0.3166 | 0.3797 | 0.4428 | 0.5060 |
| Recommended Hole, Min. In. | 0.0465 | 0.0625 | 0.0781 | 0.0938 | 0.1094 | 0.1250 | 0.1563 | 0.1875 | 0.2188 | 0.2500 | 0.3125 | 0.3750 | 0.4375 | 0.5000 |
| Crown Height, $E$, In. | 0.0000 | 0.0065 | 0.0087 | 0.0091 | 0.0110 | 0.0130 | 0.0170 | 0.0180 | 0.0220 | 0.0260 | 0.0340 | 0.0390 | 0.0470 | 0.0520 |
| Radius, $R$, In., ±0.010 | ...... | 3/64 | 3/32 | 1/8 | 9/64 | 5/32 | 3/16 | 1/4 | 9/32 | 5/16 | 3/8 | 15/32 | 17/32 | 9/16 |
| Pilot Length, $C$, In. | ...... | 1/32 | 1/32 | 1/32 | 1/32 | 1/32 | 1/16 | 1/16 | 1/16 | 1/16 | 3/32 | 3/32 | 3/32 | 3/32 |
| Chamfer Length, $D$, In.† | ...... | 1/64 | 1/64 | 1/64 | 1/64 | 1/64 | 1/32 | 1/32 | 1/32 | 1/32 | 3/64 | 3/64 | 3/64 | 3/64 |

Diagram labels (Types B, D, E — bottom): TYPE B, TYPE D, TYPE E; dimensions L ±0.010, L/2, L/4, A, B, E, R

| Nominal Size, In. | 3/64 | 1/16 | 5/64 | 3/32 | 7/64 | 1/8 | 5/32 | 3/16 | 7/32 | 1/4 | 5/16 | 3/8 | 7/16 | 1/2 |
|---|---|---|---|---|---|---|---|---|---|---|---|---|---|---|
| Diameter, $A$, Max. In. | 0.0469 | 0.0625 | 0.0781 | 0.0938 | 0.1094 | 0.1250 | 0.1563 | 0.1875 | 0.2188 | 0.2500 | 0.3125 | 0.3750 | 0.4375 | 0.5000 |
| Diameter, $A$, Min. In. | 0.0459 | 0.0615 | 0.0771 | 0.0928 | 0.1084 | 0.1230 | 0.1543 | 0.1855 | 0.2178 | 0.2480 | 0.3105 | 0.3730 | 0.4355 | 0.4980 |
| Recommended Hole, Max. In. | 0.0478 | 0.0640 | 0.0798 | 0.0956 | 0.1113 | 0.1271 | 0.1587 | 0.1903 | 0.2219 | 0.2534 | 0.3166 | 0.3797 | 0.4428 | 0.5040 |
| Recommended Hole, Min. In. | 0.0465 | 0.0625 | 0.0781 | 0.0938 | 0.1094 | 0.1250 | 0.1563 | 0.1875 | 0.2188 | 0.2500 | 0.3125 | 0.3750 | 0.4375 | 0.5000 |
| Crown Height, $E$, In. | 0.0000 | 0.0065 | 0.0087 | 0.0091 | 0.0110 | 0.0130 | 0.0170 | 0.0180 | 0.0220 | 0.0260 | 0.0340 | 0.0390 | 0.0470 | 0.0520 |
| Radius, $R$, In., ±0.010 | 3/64 | 5/64 | 3/32 | 1/8 | 9/64 | 5/32 | 3/16 | 1/4 | 9/32 | 5/16 | 3/8 | 15/32 | 17/32 | 5/8 |

All dimensions in inches. * Type F is for hopper feeding. † For 1/4-inch size and below, a suitable radius may be substituted by agreement between user and supplier.

**American Standard Grooved Pin Lengths (ASA B5.20-1958)**

| Length, Inches | Nominal Size, Inch | | | | | | | | | | | | | |
|---|---|---|---|---|---|---|---|---|---|---|---|---|---|---|
| | 3/64 | 1/16 | 5/64 | 3/32 | 7/64 | 1/8 | 5/32 | 3/16 | 7/32 | 1/4 | 5/16 | 3/8 | 7/16 | 1/2 |
| 1/4 | X | X | X | X | X | X | | | | | | | | |
| 3/8 | X | X | X | X | X | X | X | X | | | | | | |
| 1/2 | X | X | X | X | X | X | X | X | X | X | | | | |
| 5/8 | X | X | X | X | X | X | X | X | X | X | X | | | |
| 3/4 | | X | X | X | X | X | X | X | X | X | X | X | | |
| 7/8 | | X | X | X | X | X | X | X | X | X | X | X | X | |
| 1 | | X | X | X | X | X | X | X | X | X | X | X | X | X |
| 1 1/4 | | | | X | X | X | X | X | X | X | X | X | X | X |
| 1 1/2 | | | | | | X | X | X | X | X | X | X | X | X |
| 1 3/4 | | | | | | | X | X | X | X | X | X | X | X |
| 2 | | | | | | | X | X | X | X | X | X | X | X |
| 2 1/4 | | | | | | | X | X | X | X | X | X | X | X |
| 2 1/2 | | | | | | | | X | X | X | X | X | X | X |
| 2 3/4 | | | | | | | | X | X | X | X | X | X | X |
| 3 | | | | | | | | X | X | X | X | X | X | X |
| 3 1/4 | | | | | | | | | X | X | X | X | X | X |
| 3 1/2 | | | | | | | | | | X | X | X | X | X |
| 3 3/4 | In 3/64-inch size, Types A, C, and F are available in 1/8- and 3/16-inch lengths. | | | | | | | | | | | X | X | X |
| 4 | | | | | | | | | | | | X | X | X |
| 4 1/4 | | | | | | | | | | | | X | X | X |
| 4 1/2 | | | | | | | | | | | | | X | X |

**Grooved Pins.** — A grooved pin is a cylindrical pin with longitudinal grooves which are rolled or pressed into the cylindrical body to deform the pin stock outward within controlled standard limits. When this type of pin is forced into a drilled hole of the proper diameter, a locking fit is obtained. Only a straight drilled hole is required and no reaming is needed. Application may be by air cylinder or hydraulic press, and hopper feeding is utilized often on large volume applications. Grooved pins are resistant to loosening from shock and vibration and may be removed and reused without appreciably reducing their holding force.

Various types of American Standard grooved pins are shown in the accompanying table together with their dimensions.

*Type A — Half-length Taper Grooved Pin:* This type has full-length tapering grooves. It is used as a connecting and fastening element and has applications similar to those of the conventional taper pins.

*Type B — Half-length Taper Grooved Pin:* This type has tapering grooves extending over one-half the pin length. It is used as a dowel, hinge or linkage bolt for various applications in through drilled holes.

*Type C — Full-length Parallel Grooved Pin:* This type has straight grooves, evenly expanded throughout their length and a short pilot to facilitate insertion. It is especially recommended for applications involving longitudinal stress under severe vibration and shock.

*Type F — Full-length Parallel Grooved Pin:* This type is similar to Type C but is designed for hopper feeding. It has similar applications.

*Type D — Half-length Reverse Taper Grooved Pin:* This type has reverse taper grooves extending over one-half the pin length. It has applications similar to those of Type B for use as a stop, dowel, hinge, or linkage pin in blind holes.

*Type E — Center Grooved Pins:* This type has oval grooves centrally located and extending over one-half the pin length. It is suitable for use as a cross-handle, hinge pin, or fulcrum bolt and is particularly useful for rod assemblies requiring neither head nor additional cotter pins.

## Table 1. American National Standard Metric Tapered Retaining Rings—Basic External Series—3AM1 (ANSI B27.7-1977)

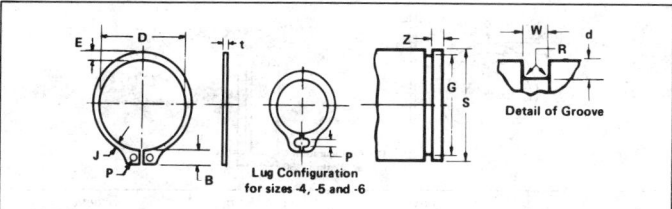

Lug Configuration for sizes -4, -5 and -6

Detail of Groove

| Shaft Diam. | Ring | | Groove | | | | Shaft Diam. | Ring | | Groove | | | |
|---|---|---|---|---|---|---|---|---|---|---|---|---|---|
| | Free Diam. | Thickness | Diam. | Width | Depth | Edge Margin | | Free Diam. | Thickness | Diam. | Width | Depth | Edge Margin |
| S | D | t | G | W | d ref | Z min | S | D | t | G | W | d ref | Z min |
| 4 | 3.60 | 0.25 | 3.80 | 0.32 | 0.1 | 0.3 | 36 | 33.25 | 1.3 | 33.85 | 1.4 | 1.06 | 3.2 |
| 5 | 4.55 | 0.4 | 4.75 | 0.5 | 0.13 | 0.4 | 38 | 35.20 | 1.3 | 35.8 | 1.4 | 1.10 | 3.3 |
| 6 | 5.45 | 0.4 | 5.70 | 0.5 | 0.15 | 0.5 | 40 | 36.75 | 1.6 | 37.7 | 1.75 | 1.15 | 3.4 |
| 7 | 6.35 | 0.6 | 6.60 | 0.7 | 0.20 | 0.6 | 42 | 38.80 | 1.6 | 39.6 | 1.75 | 1.20 | 3.6 |
| 8 | 7.15 | 0.6 | 7.50 | 0.7 | 0.25 | 0.8 | 43 | 39.65 | 1.6 | 40.5 | 1.75 | 1.25 | 3.8 |
| 9 | 8.15 | 0.6 | 8.45 | 0.7 | 0.28 | 0.8 | 45 | 41.60 | 1.6 | 42.4 | 1.75 | 1.30 | 3.9 |
| 10 | 9.00 | 0.6 | 9.40 | 0.7 | 0.30 | 0.9 | 46 | 42.55 | 1.6 | 43.3 | 1.75 | 1.35 | 4.0 |
| 11 | 10.00 | 0.6 | 10.35 | 0.7 | 0.33 | 1.0 | 48 | 44.40 | 1.6 | 45.2 | 1.75 | 1.40 | 4.2 |
| 12 | 10.85 | 0.6 | 11.35 | 0.7 | 0.33 | 1.0 | 50 | 46.20 | 1.6 | 47.2 | 1.75 | 1.40 | 4.2 |
| 13 | 11.90 | 0.9 | 12.30 | 1.0 | 0.35 | 1.0 | 52 | 48.40 | 2.0 | 49.1 | 2.15 | 1.45 | 4.3 |
| 14 | 12.90 | 0.9 | 13.25 | 1.0 | 0.38 | 1.2 | 54 | 49.9 | 2.0 | 51.0 | 2.15 | 1.50 | 4.5 |
| 15 | 13.80 | 0.9 | 14.15 | 1.0 | 0.43 | 1.3 | 55 | 50.6 | 2.0 | 51.8 | 2.15 | 1.60 | 4.8 |
| 16 | 14.70 | 0.9 | 15.10 | 1.0 | 0.45 | 1.4 | 57 | 52.9 | 2.0 | 53.8 | 2.15 | 1.60 | 4.8 |
| 17 | 15.75 | 0.9 | 16.10 | 1.0 | 0.45 | 1.4 | 58 | 53.6 | 2.0 | 54.7 | 2.15 | 1.65 | 4.9 |
| 18 | 16.65 | 1.1 | 17.00 | 1.2 | 0.50 | 1.5 | 60 | 55.8 | 2.0 | 56.7 | 2.15 | 1.65 | 4.9 |
| 19 | 17.60 | 1.1 | 17.95 | 1.2 | 0.53 | 1.6 | 62 | 57.3 | 2.0 | 58.6 | 2.15 | 1.70 | 5.1 |
| 20 | 18.35 | 1.1 | 18.85 | 1.2 | 0.58 | 1.7 | 65 | 60.4 | 2.0 | 61.6 | 2.15 | 1.70 | 5.1 |
| 21 | 19.40 | 1.1 | 19.80 | 1.2 | 0.60 | 1.8 | 68 | 63.1 | 2.0 | 64.5 | 2.15 | 1.75 | 5.3 |
| 22 | 20.30 | 1.1 | 20.70 | 1.2 | 0.65 | 1.9 | 70 | 64.6 | 2.4 | 66.4 | 2.55 | 1.80 | 5.4 |
| 23 | 21.25 | 1.1 | 21.65 | 1.2 | 0.67 | 2.0 | 72 | 66.6 | 2.4 | 68.3 | 2.55 | 1.85 | 5.5 |
| 24 | 22.20 | 1.1 | 22.60 | 1.2 | 0.70 | 2.1 | 75 | 69.0 | 2.4 | 71.2 | 2.55 | 1.90 | 5.7 |
| 25 | 23.10 | 1.1 | 23.50 | 1.2 | 0.75 | 2.3 | 78 | 72.0 | 2.4 | 74.0 | 2.55 | 2.00 | 6.0 |
| 26 | 24.05 | 1.1 | 24.50 | 1.2 | 0.75 | 2.3 | 80 | 74.2 | 2.4 | 75.9 | 2.55 | 2.05 | 6.1 |
| 27 | 24.95 | 1.3 | 25.45 | 1.4 | 0.78 | 2.3 | 82 | 76.4 | 2.4 | 77.8 | 2.55 | 2.10 | 6.3 |
| 28 | 25.80 | 1.3 | 26.40 | 1.4 | 0.80 | 2.4 | 85 | 78.6 | 2.4 | 80.6 | 2.55 | 2.20 | 6.6 |
| 30 | 27.90 | 1.3 | 28.35 | 1.4 | 0.83 | 2.5 | 88 | 81.4 | 2.8 | 83.5 | 2.95 | 2.25 | 6.7 |
| 32 | 29.60 | 1.3 | 30.20 | 1.4 | 0.90 | 2.7 | 90 | 83.2 | 2.8 | 85.4 | 2.95 | 2.30 | 6.9 |
| 34 | 31.40 | 1.3 | 32.00 | 1.4 | 1.00 | 3.0 | 95 | 88.1 | 2.8 | 90.2 | 2.95 | 2.40 | 7.2 |
| 35 | 32.30 | 1.3 | 32.90 | 1.4 | 1.05 | 3.1 | 100 | 92.5 | 2.8 | 95.0 | 2.95 | 2.50 | 7.5 |

All dimensions are in millimeters. Sizes –4, –5, and –6 are available in beryllium copper only.

These rings are designated by series symbol and shaft diameter, thus: for a 4 mm diameter shaft, 3AM1–4; for a 20 mm diameter shaft, 3AM1–20; etc.

*Ring Free Diameter Tolerances:* For ring sizes–4 through–6, +0.05, –0.10 mm; for sizes–7 through –12, +0.05, –0.15 mm; for sizes–13 through–26, +0.15, –0.25 mm; for sizes–27 through–38, +0.25, –0.40 mm; for sizes–40 through–50, +0.35, –0.50 mm; for sizes–52 through–62, +0.35, –0.65 mm; and for sizes–65 through–100, +0.50, –0.75 mm.

*Groove Diameter Tolerances:* For ring sizes–4 through–6, –0.08 mm; for sizes–7 through–10, –0.10 mm; for sizes–11 through–15, –0.12 mm; for sizes–16 through–26, –0.15 mm; for sizes–27 through –36, –0.020 mm; for sizes–38 through–55, –0.30 mm; and for sizes–57 through–100, –0.40 mm.

*Groove Diameter F. I. M. (full indicator movement)* or maximum allowable deviation of concentricity between groove and shaft: For ring sizes–4 through–6, 0.03 mm; for ring sizes–7 through–12, 0.05 mm; for sizes–13 through–28, 0.10 mm; for sizes–30 through–55, 0.15 mm; and for sizes–57 through –100, 0.20 mm.

*Groove Width Tolerances:* For ring sizes–4, +0.05 mm; for sizes–5 and–6, +0.10 mm, for sizes–7 through–38, +0.15 mm; and for sizes–40 through–100, +0.20 mm.

*Groove Maximum Bottom Radii, R:* For ring sizes–4 through–6, none; for sizes–7 through–18, 0.1 mm; for sizes–19 through–30, 0.2 mm; for sizes–32 through–50, 0.3 mm; and for sizes–52 through –100, 0.4 mm. For manufacturing details not shown, including materials, see ANSI B27.7-1977.

**Table 2.  American National Standard Metric Tapered Retaining Rings—Basic Internal Series—3BM1 (ANSI B27.7-1977)**

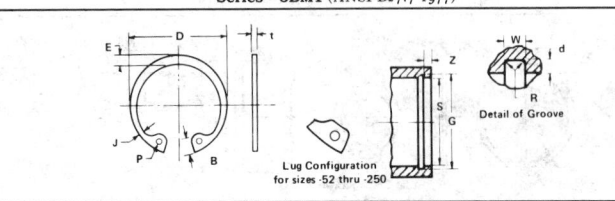

Detail of Groove

Lug Configuration for sizes -52 thru -250

| Shaft Diam. | Ring | | Groove | | | | Shaft Diam. | Ring | | Groove | | | |
|---|---|---|---|---|---|---|---|---|---|---|---|---|---|
| | Free Diam. | Thickness | Diam. | Width | Depth | Edge Margin | | Free Diam. | Thickness | Diam. | Width | Depth | Edge Margin |
| S | D | t | G | W | d ref | Z min | S | D | t | G | W | d ref | Z min |
| 8 | 8.80 | 0.4 | 8.40 | 0.5 | 0.2 | 0.6 | 65 | 72.2 | 2.4 | 69.0 | 2.55 | 2.00 | 6.0 |
| 9 | 10.00 | 0.6 | 9.45 | 0.7 | 0.23 | 0.7 | 68 | 75.7 | 2.4 | 72.2 | 2.55 | 2.10 | 6.3 |
| 10 | 11.10 | 0.6 | 10.50 | 0.7 | 0.25 | 0.8 | 70 | 77.5 | 2.4 | 74.4 | 2.55 | 2.20 | 6.6 |
| 11 | 12.20 | 0.6 | 11.60 | 0.7 | 0.3 | 0.9 | 72 | 79.6 | 2.4 | 76.5 | 2.55 | 2.25 | 6.7 |
| 12 | 13.30 | 0.6 | 12.65 | 0.7 | 0.33 | 1.0 | 75 | 83.3 | 2.4 | 79.7 | 2.55 | 2.35 | 7.1 |
| 13 | 14.25 | 0.9 | 13.70 | 1.0 | 0.35 | 1.1 | 78 | 86.8 | 2.8 | 82.8 | 2.95 | 2.40 | 7.2 |
| 14 | 15.45 | 0.9 | 14.80 | 1.0 | 0.40 | 1.2 | 80 | 89.1 | 2.8 | 85.0 | 2.95 | 2.50 | 7.5 |
| 15 | 16.60 | 0.9 | 15.85 | 1.0 | 0.43 | 1.3 | 82 | 91.1 | 2.8 | 87.2 | 2.95 | 2.60 | 7.8 |
| 16 | 17.70 | 0.9 | 16.90 | 1.0 | 0.45 | 1.4 | 85 | 94.4 | 2.8 | 90.4 | 2.95 | 2.70 | 8.1 |
| 17 | 18.90 | 0.9 | 18.00 | 1.0 | 0.50 | 1.5 | 88 | 97.9 | 2.8 | 93.6 | 2.95 | 2.80 | 8.4 |
| 18 | 20.05 | 0.9 | 19.05 | 1.0 | 0.53 | 1.6 | 90 | 100.0 | 2.80 | 95.7 | 2.95 | 2.85 | 8.6 |
| 19 | 21.10 | 0.9 | 20.10 | 1.0 | 0.55 | 1.7 | 92 | 102.2 | 2.8 | 97.8 | 2.95 | 2.90 | 8.7 |
| 20 | 22.25 | 0.9 | 21.15 | 1.0 | 0.57 | 1.7 | 95 | 105.6 | 2.8 | 101.0 | 2.95 | 3.00 | 9.0 |
| 21 | 23.30 | 0.9 | 22.20 | 1.0 | 0.60 | 1.8 | 98 | 109.0 | 2.8 | 104.2 | 2.95 | 3.10 | 9.3 |
| 22 | 24.40 | 1.1 | 23.30 | 1.2 | 0.65 | 1.9 | 100 | 110.7 | 2.8 | 106.3 | 2.95 | 3.15 | 9.5 |
| 23 | 25.45 | 1.1 | 24.35 | 1.2 | 0.67 | 2.0 | 102 | 112.4 | 2.8 | 108.4 | 2.95 | 3.20 | 9.6 |
| 24 | 26.55 | 1.1 | 25.4 | 1.2 | 0.70 | 2.1 | 105 | 115.8 | 2.8 | 111.5 | 2.95 | 3.25 | 9.8 |
| 25 | 27.75 | 1.1 | 26.6 | 1.2 | 0.80 | 2.4 | 108 | 119.2 | 2.8 | 114.6 | 2.95 | 3.30 | 9.9 |
| 26 | 28.85 | 1.1 | 27.7 | 1.2 | 0.85 | 2.6 | 110 | 120.8 | 2.8 | 116.7 | 2.95 | 3.35 | 10.1 |
| 27 | 29.95 | 1.3 | 28.8 | 1.4 | 0.90 | 2.7 | 115 | 126.0 | 2.8 | 121.9 | 2.95 | 3.45 | 10.4 |
| 28 | 31.10 | 1.3 | 29.8 | 1.4 | 0.90 | 2.7 | 120 | 132.4 | 2.8 | 127.0 | 2.95 | 3.50 | 10.5 |
| 30 | 33.40 | 1.3 | 31.9 | 1.4 | 0.95 | 2.9 | 125 | 137.1 | 2.8 | 132.1 | 2.95 | 3.55 | 10.7 |
| 32 | 35.35 | 1.3 | 33.9 | 1.4 | 0.95 | 2.9 | 130 | 142.5 | 2.8 | 137.2 | 2.95 | 3.60 | 10.8 |
| 34 | 37.75 | 1.3 | 36.1 | 1.4 | 1.05 | 3.2 | 135 | 148.5 | 3.2 | 142.3 | 3.40 | 3.65 | 11.0 |
| 35 | 38.75 | 1.3 | 37.2 | 1.4 | 1.10 | 3.4 | 140 | 154.1 | 3.2 | 147.4 | 3.40 | 3.70 | 11.1 |
| 36 | 40.00 | 1.3 | 38.3 | 1.4 | 1.15 | 3.5 | 145 | 159.5 | 3.2 | 152.5 | 3.40 | 3.75 | 11.3 |
| 37 | 41.05 | 1.3 | 39.3 | 1.4 | 1.15 | 3.5 | 150 | 164.5 | 3.2 | 157.6 | 3.40 | 3.80 | 11.4 |
| 38 | 42.15 | 1.3 | 40.4 | 1.4 | 1.20 | 3.6 | 155 | 168.8 | 3.2 | 162.7 | 3.40 | 3.85 | 11.6 |
| 40 | 44.25 | 1.6 | 42.4 | 1.75 | 1.20 | 3.6 | 160 | 175.1 | 4.0 | 167.8 | 4.25 | 3.90 | 11.7 |
| 42 | 46.60 | 1.6 | 44.5 | 1.75 | 1.25 | 3.7 | 165 | 180.3 | 4.0 | 172.9 | 4.25 | 3.95 | 11.9 |
| 45 | 49.95 | 1.6 | 47.6 | 1.75 | 1.30 | 3.9 | 170 | 185.6 | 4.0 | 178.0 | 4.25 | 4.00 | 12.0 |
| 46 | 51.05 | 1.6 | 48.7 | 1.75 | 1.35 | 4.0 | 175 | 191.3 | 4.0 | 183.2 | 4.25 | 4.10 | 12.3 |
| 47 | 52.15 | 1.6 | 49.8 | 1.75 | 1.40 | 4.2 | 180 | 196.6 | 4.0 | 188.4 | 4.25 | 4.20 | 12.5 |
| 48 | 53.30 | 1.6 | 50.9 | 1.75 | 1.45 | 4.3 | 185 | 202.7 | 4.8 | 193.6 | 5.10 | 4.30 | 12.9 |
| 50 | 55.35 | 1.6 | 53.1 | 1.75 | 1.55 | 4.6 | 190 | 207.7 | 4.8 | 198.8 | 5.10 | 4.40 | 13.2 |
| 52 | 57.90 | 2.0 | 55.3 | 2.15 | 1.65 | 5.0 | 200 | 217.8 | 4.8 | 209.0 | 5.10 | 4.50 | 13.5 |
| 55 | 61.10 | 2.0 | 58.4 | 2.15 | 1.70 | 5.1 | 210 | 230.3 | 4.8 | 219.4 | 5.10 | 4.70 | 14.1 |
| 57 | 63.25 | 2.0 | 60.5 | 2.15 | 1.75 | 5.3 | 220 | 240.5 | 4.8 | 230.0 | 5.10 | 5.00 | 15.0 |
| 58 | 64.4 | 2.0 | 61.6 | 2.15 | 1.80 | 5.4 | 230 | 251.4 | 4.8 | 240.6 | 5.10 | 5.30 | 15.9 |
| 60 | 66.8 | 2.0 | 63.8 | 2.15 | 1.90 | 5.7 | 240 | 262.3 | 4.8 | 251.0 | 5.10 | 5.50 | 16.5 |
| 62 | 68.6 | 2.0 | 65.8 | 2.15 | 1.90 | 5.7 | 250 | 273.3 | 4.8 | 261.4 | 5.10 | 5.70 | 17.1 |
| 63 | 69.9 | 2.0 | 66.9 | 2.15 | 1.95 | 5.9 | ... | ... | ... | ... | ... | ... | ... |

All dimensions are in millimeters.

These rings are designated by series symbol and shaft diameter, thus: for a 9 mm diameter shaft, 3BM1-9; for a 22 mm diameter shaft, 3BM1-22; etc.

*Ring Free Diameter Tolerances:* For ring sizes -8 through -20, +0.25, −0.13 mm; for sizes -21 through -26, +0.40, −0.25 mm; for sizes -27 through -38, +0.65, −0.50 mm; for sizes -40 through-50, +0.90, −0.65 mm; for sizes -52 through-75, +1.00, −0.75 mm; for sizes -78 through -92, +1.40, −1.40

*(Continued on next page)*

**Table 2** *(Continued).* **American National Standard Metric Tapered
Retaining Rings—Basic Internal Series—3BM1** (ANSI B27.7-1977)

mm; for sizes–95 through–155, +1.65, –1.65 mm; for sizes–160 through–180, +2.05, –2.05 mm; and for sizes–185 through–250, +2.30, –2.30 mm.

*Groove Diameter Tolerances:* For ring sizes–8 and –9, +0.06 mm; for sizes–10 through –18, +0.10 mm; for sizes–19 through–28, +0.15 mm; for sizes–30 through–50, +0.20 mm; for sizes–52 through –98, +0.30; for sizes–100 through–160, +0.40 mm; and for sizes–165 through –250, +0.50 mm.

*Groove Diameter F. I. M. (full indicator movement)* or maximum allowable deviation of concentricity between groove and shaft: For ring sizes–8 through–10, 0.03 mm; for sizes–11 through–15, 0.05 mm; for sizes–16 through–25, 0.10 mm; for sizes–26 through–45, 0.15 mm; for sizes–46 through–80, 0.20 mm; for sizes–82 through–150, 0.25 mm; and for sizes–155 through–250, 0.30 mm.

*Groove Width Tolerances:* For ring size–8, +0.10 mm; for sizes–9 through–38, +0.15 mm; for sizes –40 through–130, +0.20 mm; and for sizes–135 through–250, +0.25 mm.

*Groove Maximum Bottom Radii:* For ring sizes–8 through–17, 0.1 mm; for sizes–18 through–30, 0.2 mm; for sizes–32 through–55, 0.3 mm; and for sizes–56 through–250, 0.4 mm.

For manufacturing details not shown, including materials, see ANSI B27.7-1977.

**Table 3. American National Standard Metric Reduced Cross Section
Retaining Rings—E Ring External Series—3CM1** (ANSI B27.7-1977)

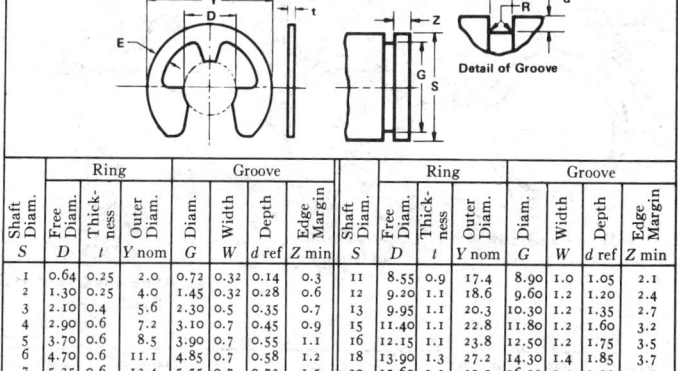

| | Ring | | | Groove | | | | | Ring | | | Groove | | | |
|---|---|---|---|---|---|---|---|---|---|---|---|---|---|---|---|
| Shaft Diam. S | Free Diam. D | Thickness t | Outer Diam. Y nom | Diam. G | Width W | Depth d ref | Edge Margin Z min | Shaft Diam. S | Free Diam. D | Thickness t | Outer Diam. Y nom | Diam. G | Width W | Depth d ref | Edge Margin Z min |
| 1 | 0.64 | 0.25 | 2.0 | 0.72 | 0.32 | 0.14 | 0.3 | 11 | 8.55 | 0.9 | 17.4 | 8.90 | 1.0 | 1.05 | 2.1 |
| 2 | 1.30 | 0.25 | 4.0 | 1.45 | 0.32 | 0.28 | 0.6 | 12 | 9.20 | 1.1 | 18.6 | 9.60 | 1.2 | 1.20 | 2.4 |
| 3 | 2.10 | 0.4 | 5.6 | 2.30 | 0.5 | 0.35 | 0.7 | 13 | 9.95 | 1.1 | 20.3 | 10.30 | 1.2 | 1.35 | 2.7 |
| 4 | 2.90 | 0.6 | 7.2 | 3.10 | 0.7 | 0.45 | 0.9 | 15 | 11.40 | 1.1 | 22.8 | 11.80 | 1.2 | 1.60 | 3.2 |
| 5 | 3.70 | 0.6 | 8.5 | 3.90 | 0.7 | 0.55 | 1.1 | 16 | 12.15 | 1.1 | 23.8 | 12.50 | 1.2 | 1.75 | 3.5 |
| 6 | 4.70 | 0.6 | 11.1 | 4.85 | 0.7 | 0.58 | 1.2 | 18 | 13.90 | 1.3 | 27.2 | 14.30 | 1.4 | 1.85 | 3.7 |
| 7 | 5.25 | 0.6 | 13.4 | 5.55 | 0.7 | 0.73 | 1.5 | 20 | 15.60 | 1.3 | 30.0 | 16.00 | 1.4 | 2.00 | 4.0 |
| 8 | 6.15 | 0.6 | 14.6 | 6.40 | 0.7 | 0.80 | 1.6 | 22 | 17.00 | 1.3 | 33.0 | 17.40 | 1.4 | 2.30 | 4.6 |
| 9 | 6.80 | 0.9 | 15.8 | 7.20 | 1.0 | 0.90 | 1.8 | 25 | 19.50 | 1.3 | 37.1 | 20.00 | 1.4 | 2.50 | 5.0 |
| 10 | 7.60 | 0.9 | 16.8 | 8.00 | 1.0 | 1.00 | 2.0 | ... | ... | ... | ... | ... | ... | ... | ... |

All dimensions are in millimeters. Size–1 is available in beryllium copper only.

These rings are designated by series symbol and shaft diameter, thus: for a 2 mm diameter shaft, 3CM1-2; for a 13 mm shaft, 3CM1-13; etc.

*Ring Free Diameter Tolerances:* For ring sizes–1 through–7, +0.03, –0.08 mm; for sizes–8 through –13, +0.05, –0.10 mm; and for sizes–15 through–25, +0.10, –0.15 mm.

*Groove Diameter Tolerances:* For ring sizes–1 and–2, –0.05 mm; for sizes–3 through–6, –0.08; for sizes–7 through–11, –0.10 mm; for sizes–12 through–18, –0.15 mm; and for sizes–20 through–25, –0.20 mm.

*Groove Diameter F. I. M. (Full Indicator Movement)* or maximum allowable deviation of concentricity between groove and shaft: For ring sizes–1 through–3, 0.04 mm; for–4 through–6, 0.05 mm; for–7 through–10, 0.08 mm; for–11 through–25, 0.10 mm.

*Groove Width Tolerances:* For ring sizes–1 and–2, +0.05 mm; for size–3, +0.10 mm; and for sizes–4 through–25, +0.15 mm.

*Groove Maximum Bottom Radii:* For ring sizes–1 and–2, 0.05 mm; for–3 through–7, 0.15 mm; for–8 through–13, 0.25 mm; and for–15 through–25, 0.4 mm.

For manufacturing details not shown, including materials. See ANSI B27.7-1977.

**Retaining Rings.**—The preceding Table 1 covers Type 3AM1 tapered external retaining rings which are spread over a shaft by means of pliers or a special tool and allowed to relax and seat in a circumferential groove, thereby providing an external protruding shoulder which can be used for locating and retaining a part on a shaft.

Table 2 covers Type 3BM1 tapered internal retaining rings which are compressed into a housing by means of pliers or special tool and allowed to relax and seat in a circumferential groove, thereby providing an internal protruding shoulder which can be used for locating and retaining a part contained inside the housing.

Table 3 covers Type 3CM1 reduced cross section external retaining rings which contain 3 prongs connected by a reduced section bridge to provide greater resilience during installation. They are installed radially, usually by means of an applicator, and provide a high shoulder for abutment by a retained part.

**Table 4.  American National Standard Metric Basic External
Series 3AM1 Retaining Rings—Checking and Performance Data**

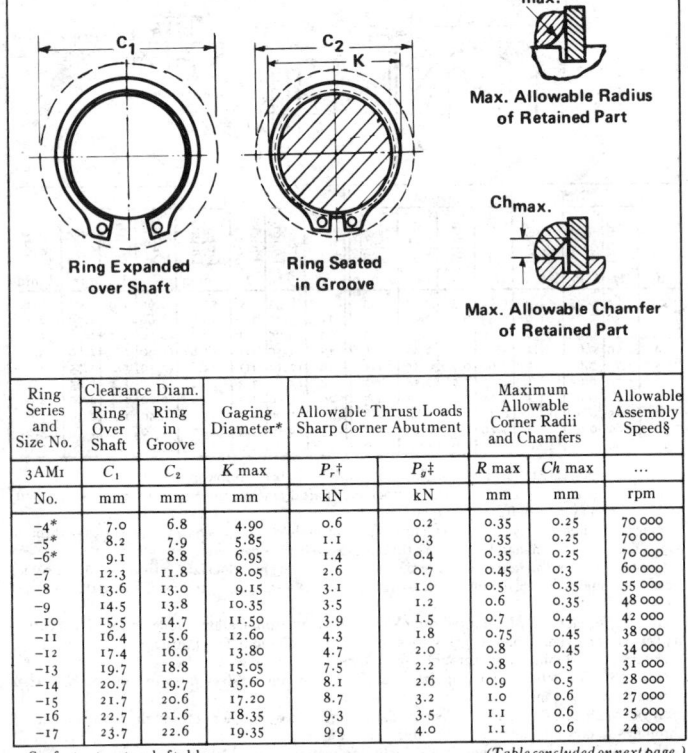

| Ring Series and Size No. | Clearance Diam. | | Gaging Diameter* | Allowable Thrust Loads Sharp Corner Abutment | | Maximum Allowable Corner Radii and Chamfers | | Allowable Assembly Speed§ |
|---|---|---|---|---|---|---|---|---|
| | Ring Over Shaft | Ring in Groove | | | | | | |
| 3AM1 | $C_1$ | $C_2$ | $K$ max | $P_r$† | $P_g$‡ | $R$ max | $Ch$ max | ... |
| No. | mm | mm | mm | kN | kN | mm | mm | rpm |
| −4* | 7.0 | 6.8 | 4.90 | 0.6 | 0.2 | 0.35 | 0.25 | 70 000 |
| −5* | 8.2 | 7.9 | 5.85 | 1.1 | 0.3 | 0.35 | 0.25 | 70 000 |
| −6* | 9.1 | 8.8 | 6.95 | 1.4 | 0.4 | 0.35 | 0.25 | 70 000 |
| −7 | 12.3 | 11.8 | 8.05 | 2.6 | 0.7 | 0.45 | 0.3 | 60 000 |
| −8 | 13.6 | 13.0 | 9.15 | 3.1 | 1.0 | 0.5 | 0.35 | 55 000 |
| −9 | 14.5 | 13.8 | 10.35 | 3.5 | 1.2 | 0.6 | 0.35 | 48 000 |
| −10 | 15.5 | 14.7 | 11.50 | 3.9 | 1.5 | 0.7 | 0.4 | 42 000 |
| −11 | 16.4 | 15.6 | 12.60 | 4.3 | 1.8 | 0.75 | 0.45 | 38 000 |
| −12 | 17.4 | 16.6 | 13.80 | 4.7 | 2.0 | 0.8 | 0.45 | 34 000 |
| −13 | 19.7 | 18.8 | 15.05 | 7.5 | 2.2 | 0.8 | 0.5 | 31 000 |
| −14 | 20.7 | 19.7 | 15.60 | 8.1 | 2.6 | 0.9 | 0.5 | 28 000 |
| −15 | 21.7 | 20.6 | 17.20 | 8.7 | 3.2 | 1.0 | 0.6 | 27 000 |
| −16 | 22.7 | 21.6 | 18.35 | 9.3 | 3.5 | 1.1 | 0.6 | 25 000 |
| −17 | 23.7 | 22.6 | 19.35 | 9.9 | 4.0 | 1.1 | 0.6 | 24 000 |

See footnotes at end of table.                                    (Table concluded on next page.)

Table 4 (Concluded).   **American National Standard Metric Basic External Series 3AM1 Retaining Rings—Checking and Performance Data**

| Ring Series and Size No. | Clearance Diam. | | Gaging Diameter* | Allowable Thrust Loads Sharp Corner Abutment | | Maximum Allowable Corner Radii and Chamfers | | Allowable Assembly Speed§ |
| | Ring Over Shaft | Ring in Groove | | | | | | |
| 3AM1 | $C_1$ | $C_2$ | K max | $P_r$† | $P_g$‡ | R max | Ch max | ... |
| No. | mm | mm | mm | kN | kN | mm | mm | rpm |
| −18 | 26.2 | 25.0 | 20.60 | 16.0 | 4.4 | 1.2 | 0.7 | 23 000 |
| −19 | 27.2 | 25.9 | 21.70 | 16.9 | 4.9 | 1.2 | 0.7 | 21 500 |
| −20 | 28.2 | 26.8 | 22.65 | 17.8 | 5.7 | 1.2 | 0.7 | 20 000 |
| −21 | 29.2 | 27.7 | 23.80 | 18.6 | 6.2 | 1.3 | 0.7 | 19 000 |
| −22 | 30.3 | 28.7 | 24.90 | 19.6 | 7.0 | 1.3 | 0.8 | 18 500 |
| −23 | 31.3 | 29.6 | 26.00 | 20.5 | 7.6 | 1.3 | 0.8 | 18 000 |
| −24 | 34.1 | 32.4 | 27.15 | 21.4 | 8.2 | 1.4 | 0.8 | 17 500 |
| −25 | 35.1 | 33.3 | 28.10 | 22.3 | 9.2 | 1.4 | 0.8 | 17 000 |
| −26 | 36.0 | 34.2 | 29.25 | 23.2 | 9.6 | 1.5 | 0.9 | 16 500 |
| −27 | 37.8 | 35.9 | 30.35 | 28.4 | 10.3 | 1.5 | 0.9 | 16 300 |
| −28 | 38.8 | 36.9 | 31.45 | 28.4 | 11.0 | 1.6 | 1.0 | 15 800 |
| −30 | 40.8 | 38.8 | 33.6 | 31.6 | 12.3 | 1.6 | 1.0 | 15 000 |
| −32 | 42.8 | 40.7 | 35.9 | 33.6 | 14.1 | 1.7 | 1.0 | 14 800 |
| −34 | 44.9 | 42.5 | 37.9 | 36 | 16.7 | 1.7 | 1.1 | 14 000 |
| −35 | 45.9 | 43.4 | 39.0 | 37 | 18.1 | 1.8 | 1.1 | 13 500 |
| −36 | 48.6 | 46.1 | 40.2 | 38 | 18.9 | 1.9 | 1.2 | 13 300 |
| −38 | 50.6 | 48.0 | 42.5 | 40 | 20.5 | 2.0 | 1.2 | 12 700 |
| −40 | 54.0 | 51.3 | 44.5 | 52 | 22.6 | 2.1 | 1.2 | 12 000 |
| −42 | 56.0 | 53.2 | 46.9 | 54 | 24.8 | 2.2 | 1.3 | 11 000 |
| −43 | 57.0 | 54.0 | 47.9 | 55 | 26.4 | 2.3 | 1.4 | 10 800 |
| −45 | 59.0 | 55.9 | 50.0 | 58 | 28.8 | 2.3 | 1.4 | 10 000 |
| −46 | 60.0 | 56.8 | 50.9 | 59 | 30.4 | 2.4 | 1.4 | 9 500 |
| −48 | 62.4 | 59.1 | 53.0 | 62 | 33 | 2.4 | 1.4 | 8 800 |
| −50 | 64.4 | 61.1 | 55.2 | 64 | 35 | 2.4 | 1.4 | 8 000 |
| −52 | 67.6 | 64.1 | 57.4 | 84 | 37 | 2.5 | 1.5 | 7 700 |
| −54 | 69.6 | 66.1 | 59.5 | 87 | 40 | 2.5 | 1.5 | 7 500 |
| −55 | 70.6 | 66.9 | 60.4 | 89 | 44 | 2.5 | 1.5 | 7 400 |
| −57 | 72.6 | 68.9 | 62.7 | 91 | 45 | 2.6 | 1.5 | 7 200 |
| −58 | 73.6 | 69.8 | 63.6 | 93 | 46 | 2.6 | 1.6 | 7 100 |
| −60 | 75.6 | 71.8 | 65.8 | 97 | 49 | 2.6 | 1.6 | 7 000 |
| −62 | 77.6 | 73.6 | 67.9 | 100 | 52 | 2.7 | 1.6 | 6 900 |
| −65 | 80.6 | 76.6 | 71.2 | 105 | 54 | 2.8 | 1.7 | 6 700 |
| −68 | 83.6 | 79.5 | 74.5 | 110 | 58 | 2.9 | 1.7 | 6 500 |
| −70 | 88.1 | 83.9 | 76.4 | 136 | 62 | 2.9 | 1.7 | 6 400 |
| −72 | 90.1 | 85.8 | 78.5 | 140 | 65 | 2.9 | 1.7 | 6 200 |
| −75 | 93.1 | 88.7 | 81.7 | 147 | 69 | 3.0 | 1.8 | 5 900 |
| −78 | 95.4 | 92.1 | 84.6 | 151 | 76 | 3.0 | 1.8 | 5 600 |
| −80 | 97.9 | 93.1 | 87.0 | 155 | 80 | 3.1 | 1.9 | 5 400 |
| −82 | 100.0 | 95.1 | 89.0 | 159 | 84 | 3.2 | 1.9 | 5 200 |
| −85 | 103.0 | 97.9 | 92.1 | 165 | 91 | 3.2 | 1.9 | 5 000 |
| −88 | 107.0 | 100.8 | 95.1 | 199 | 97 | 3.2 | 1.9 | 4 800 |
| −90 | 109.0 | 103.6 | 97.1 | 204 | 101 | 3.2 | 1.9 | 4 500 |
| −95 | 114.0 | 108.6 | 102.7 | 215 | 112 | 3.4 | 2.1 | 4 350 |
| −100 | 119.5 | 113.7 | 108.0 | 227 | 123 | 3.5 | 2.1 | 4 150 |

\* For checking when ring is seated in groove.

† These values apply to rings made from SAE 1060-1090 steels and PH 15-7 Mo stainless steel used on shafts hardened to $R_c$ 50 minimum, with the exception of sizes −4, −5, and −6 which are supplied in beryllium copper only. Values for other sizes made from beryllium copper can be calculated by multiplying the listed values by 0.75. The values listed include a safety factor of 4.

‡ These values are for all standard rings used on low carbon steel shafts. They include a safety factor of 2.

§ These values have been calculated for steel rings.

Maximum allowable assembly loads with R max or Ch max are: For rings sizes −4, 0.2 kN; for sizes −5 and −6, 0.5 kN; for sizes −7 through −12, 2.1 kN; for sizes −13 through −17, 4.0 kN; for sizes −18 through −26, 6.0 kN; for sizes −27 through −38, 8.6 kN; for sizes −40 through −50, 13.2 kN; for sizes −52 through −68, 22.0 kN; for sizes −70 through −85, 32 kN; and for sizes −88 through −100, 47 kN.

*Source:* Appendix to American National Standard ANSI B27.7-1977.

## Table 5. American National Standard Metric Basic Internal Series 3BM1 Retaining Rings—Checking and Performance Data

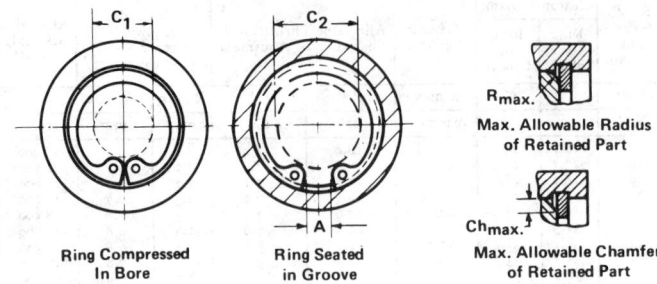

Ring Compressed In Bore — Ring Seated in Groove — $R_{max.}$ Max. Allowable Radius of Retained Part — $Ch_{max.}$ Max. Allowable Chamfer of Retained Part

| Ring Series and Size No. | Clearance Diam. | | Gaging Diameter* | Allowable Thrust Loads Sharp Corner Abutment | | Maximum Allowable Corner Radii and Chamfers | |
|---|---|---|---|---|---|---|---|
| | Ring in Bore | Ring in Groove | | | | | |
| 3BM1 | $C_1$ | $C_2$ | $A$ min | $P_r$† | $P_g$‡ | $R$ max | $Ch$ max |
| No. | mm | mm | mm | kN | kN | mm | mm |
| −8 | 4.4 | 4.8 | 1.40 | 2.4 | 1.0 | 0.4 | 0.3 |
| −9 | 4.6 | 5.0 | 1.50 | 4.4 | 1.2 | 0.5 | 0.35 |
| −10 | 5.5 | 6.0 | 1.85 | 4.9 | 1.5 | 0.5 | 0.35 |
| −11 | 5.7 | 6.3 | 1.95 | 5.4 | 2.0 | 0.6 | 0.4 |
| −12 | 6.7 | 7.3 | 2.25 | 5.8 | 2.4 | 0.6 | 0.4 |
| −13 | 6.8 | 7.5 | 2.35 | 8.9 | 2.6 | 0.7 | 0.5 |
| −14 | 6.9 | 7.7 | 2.65 | 9.7 | 3.2 | 0.7 | 0.5 |
| −15 | 7.9 | 8.7 | 2.80 | 10.4 | 3.7 | 0.7 | 0.5 |
| −16 | 8.8 | 9.7 | 2.80 | 11.0 | 4.2 | 0.7 | 0.5 |
| −17 | 9.8 | 10.8 | 3.35 | 11.7 | 4.9 | 0.75 | 0.6 |
| −18 | 10.3 | 11.3 | 3.40 | 12.3 | 5.5 | 0.75 | 0.6 |
| −19 | 11.4 | 12.5 | 3.40 | 13.1 | 6.0 | 0.8 | 0.65 |
| −20 | 11.6 | 12.7 | 3.8 | 13.7 | 6.6 | 0.9 | 0.7 |
| −21 | 12.6 | 13.8 | 4.2 | 14.5 | 7.3 | 0.9 | 0.7 |
| −22 | 13.5 | 14.8 | 4.3 | 22.5 | 8.3 | 0.9 | 0.7 |
| −23 | 14.5 | 15.9 | 4.9 | 23.5 | 8.9 | 1.0 | 0.8 |
| −24 | 15.5 | 16.9 | 5.2 | 24.8 | 9.7 | 1.0 | 0.8 |
| −25 | 16.5 | 18.1 | 6.0 | 25.7 | 11.6 | 1.0 | 0.8 |
| −26 | 17.5 | 19.2 | 5.7 | 26.8 | 12.7 | 1.2 | 1.0 |
| −27 | 17.4 | 19.2 | 5.9 | 33 | 14.0 | 1.2 | 1.0 |
| −28 | 18.2 | 20.0 | 6.0 | 34 | 14.6 | 1.2 | 1.0 |
| −30 | 20.0 | 21.9 | 6.0 | 37 | 16.5 | 1.2 | 1.0 |
| −32 | 22.0 | 23.9 | 7.3 | 39 | 17.6 | 1.2 | 1.0 |
| −34 | 24.0 | 26.1 | 7.6 | 42 | 20.6 | 1.2 | 1.0 |
| −35 | 25.0 | 27.2 | 8.0 | 43 | 22.3 | 1.2 | 1.0 |
| −36 | 26.0 | 28.3 | 8.3 | 44 | 23.9 | 1.2 | 1.0 |
| −37 | 27.0 | 29.3 | 8.4 | 45 | 24.6 | 1.2 | 1.0 |
| −38 | 28.0 | 30.4 | 8.6 | 46 | 26.4 | 1.2 | 1.0 |
| −40 | 29.2 | 31.6 | 9.7 | 62 | 27.7 | 1.7 | 1.3 |
| −42 | 29.7 | 32.2 | 9.0 | 65 | 30.2 | 1.7 | 1.3 |
| −45 | 32.3 | 34.9 | 9.6 | 69 | 33.8 | 1.7 | 1.3 |
| −46 | 33.3 | 36.0 | 9.7 | 71 | 36 | 1.7 | 1.3 |
| −47 | 34.3 | 37.1 | 10.0 | 72 | 38 | 1.7 | 1.3 |
| −48 | 35.0 | 37.9 | 10.5 | 74 | 40 | 1.7 | 1.3 |
| −50 | 36.9 | 40.0 | 12.1 | 77 | 45 | 1.7 | 1.3 |
| −52 | 38.6 | 41.9 | 11.7 | 99 | 50 | 2.0 | 1.6 |
| −55 | 40.8 | 44.2 | 11.9 | 105 | 54 | 2.0 | 1.6 |
| −57 | 42.2 | 45.7 | 12.5 | 109 | 58 | 2.0 | 1.6 |

See footnotes at end of table.

(Table concluded on next page)

Table 5 *(Concluded)*.   **American National Standard Metric Basic Internal Series 3BM1 Retaining Rings—Checking and Performance Data**

| Ring Series and Size No. | Clearance Diam. | | Gaging Diameter* | Allowable Thrust Loads Sharp Corner Abutment | | Maximum Allowable Corner Radii and Chamfers | |
| | Ring in Bore | Ring in Groove | | | | | |
| 3BM1 | $C_1$ | $C_2$ | $A$ min | $P_r$† | $P_g$‡ | $R$ max | $Ch$ max |
| No. | mm | mm | mm | kN | kN | mm | mm |
| −58 | 43.2 | 46.8 | 13.0 | 111 | 60 | 2.0 | 1.6 |
| −60 | 45.5 | 49.3 | 12.7 | 115 | 66 | 2.0 | 1.6 |
| −62 | 47.0 | 50.8 | 14.0 | 119 | 68 | 2.0 | 1.6 |
| −63 | 47.8 | 51.7 | 14.2 | 120 | 71 | 2.0 | 1.6 |
| −65 | 49.4 | 53.4 | 14.2 | 149 | 75 | 2.0 | 1.6 |
| −68 | 52.0 | 56.2 | 14.4 | 156 | 82 | 2.3 | 1.8 |
| −70 | 53.8 | 58.2 | 16.1 | 161 | 88 | 2.3 | 1.8 |
| −72 | 55.9 | 60.4 | 17.4 | 166 | 93 | 2.3 | 1.8 |
| −75 | 58.2 | 62.9 | 16.8 | 172 | 101 | 2.3 | 1.8 |
| −78 | 61.2 | 66.0 | 17.6 | 209 | 108 | 2.5 | 2.0 |
| −80 | 63.0 | 68.0 | 17.2 | 215 | 115 | 2.5 | 2.0 |
| −82 | 63.5 | 68.7 | 18.8 | 220 | 122 | 2.6 | 2.1 |
| −85 | 66.8 | 72.2 | 19.1 | 228 | 131 | 2.6 | 2.1 |
| −88 | 69.6 | 75.2 | 20.4 | 236 | 141 | 2.8 | 2.2 |
| −90 | 71.6 | 77.3 | 21.4 | 241 | 147 | 2.8 | 2.2 |
| −92 | 73.6 | 79.4 | 22.2 | 247 | 153 | 2.9 | 2.4 |
| −95 | 76.7 | 82.7 | 22.6 | 255 | 164 | 3.0 | 2.5 |
| −98 | 78.3 | 84.5 | 22.6 | 263 | 174 | 3.0 | 2.5 |
| −100 | 80.3 | 86.6 | 24.1 | 269 | 181 | 3.1 | 2.5 |
| −102 | 82.2 | 88.6 | 25.5 | 273 | 187 | 3.2 | 2.6 |
| −105 | 85.1 | 91.6 | 26.0 | 281 | 196 | 3.3 | 2.6 |
| −108 | 88.1 | 94.7 | 26.4 | 290 | 205 | 3.5 | 2.7 |
| −110 | 88.4 | 95.1 | 27.5 | 295 | 212 | 3.6 | 2.8 |
| −115 | 93.2 | 100.1 | 29.4 | 309 | 227 | 3.7 | 2.9 |
| −120 | 98.2 | 105.2 | 27.2 | 321 | 241 | 3.9 | 3.1 |
| −125 | 103.1 | 110.2 | 30.3 | 335 | 255 | 4.0 | 3.2 |
| −130 | 108.0 | 115.2 | 31.0 | 349 | 269 | 4.0 | 3.2 |
| −135 | 110.4 | 117.7 | 30.4 | 415 | 283 | 4.3 | 3.4 |
| −140 | 115.3 | 122.7 | 30.4 | 429 | 298 | 4.3 | 3.4 |
| −145 | 120.4 | 127.9 | 31.6 | 444 | 313 | 4.3 | 3.4 |
| −150 | 125.3 | 132.9 | 33.5 | 460 | 327 | 4.3 | 3.4 |
| −155 | 130.4 | 138.1 | 37.0 | 475 | 343 | 4.3 | 3.4 |
| −160 | 133.8 | 141.6 | 35.0 | 613 | 359 | 4.5 | 3.6 |
| −165 | 138.7 | 146.6 | 33.1 | 632 | 374 | 4.6 | 3.7 |
| −170 | 143.6 | 151.6 | 38.2 | 651 | 390 | 4.6 | 3.7 |
| −175 | 146.0 | 154.2 | 37.7 | 670 | 403 | 4.8 | 3.8 |
| −180 | 151.4 | 159.8 | 39.0 | 690 | 434 | 5.0 | 4.0 |
| −185 | 154.7 | 163.3 | 37.3 | 851 | 457 | 5.1 | 4.1 |
| −190 | 159.5 | 168.3 | 35.0 | 873 | 480 | 5.3 | 4.3 |
| −200 | 169.2 | 178.2 | 43.9 | 919 | 517 | 5.4 | 4.3 |
| −210 | 177.5 | 186.9 | 40.6 | 965 | 566 | 5.8 | 4.6 |
| −220 | 184.1 | 194.1 | 38.3 | 1000 | 608 | 6.1 | 4.9 |
| −230 | 194.0 | 204.6 | 49.0 | 1060 | 686 | 6.3 | 5.1 |
| −240 | 200.4 | 211.4 | 45.4 | 1090 | 725 | 6.6 | 5.3 |
| −250 | 210.0 | 221.4 | 53.0 | 1150 | 808 | 6.7 | 5.4 |

\* For checking when ring is seated in groove.

† These values apply to rings made from SAE 1060-1090 steels and PH 15-7 Mo stainless steel used in bores hardened to $R_c$ 50 minimum. Values for rings made from beryllium copper can be calculated by multiplying the listed values by 0.75. The values listed include a safety factor of 4.

‡ These values are for standard rings used in low carbon steel bores. They include a safety factor of 2.

Maximum allowable assembly loads for $R$ max or $Ch$ max are: for ring size −8, 0.8 kN; for sizes −9 through −12, 2.0 kN; for sizes −13 through −21, 4.0 kN; for sizes −22 through −26, 7.4 kN; for sizes −27 through −38, 10.8 kN; for sizes −40 through −50, 17.4 kN; for sizes −52 through −63, 27.4 kN; for size −65, 42.0 kN; for sizes −68 through −72, 39 kN; for sizes −75 through −130, 54 kN; for sizes −135 through −155, 67 kN; for sizes −160 through −180, 102 kN; and for sizes −185 through −250, 151 kN.

*Source:* Appendix to American National Standard ANSI B27.7-1977.

## Table 6. American National Standard Metric E-Type External Series 3CM1 Retaining Rings—Checking and Performance Data

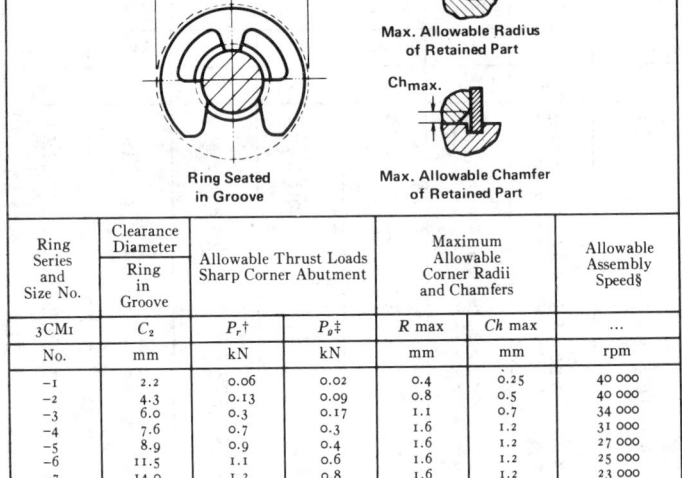

Ring Seated in Groove

Max. Allowable Radius of Retained Part

Max. Allowable Chamfer of Retained Part

| Ring Series and Size No. | Clearance Diameter Ring in Groove | Allowable Thrust Loads Sharp Corner Abutment | | Maximum Allowable Corner Radii and Chamfers | | Allowable Assembly Speed§ |
|---|---|---|---|---|---|---|
| 3CM1 | $C_2$ | $P_r$† | $P_\theta$‡ | $R$ max | $Ch$ max | ... |
| No. | mm | kN | kN | mm | mm | rpm |
| −1 | 2.2 | 0.06 | 0.02 | 0.4 | 0.25 | 40 000 |
| −2 | 4.3 | 0.13 | 0.09 | 0.8 | 0.5 | 40 000 |
| −3 | 6.0 | 0.3 | 0.17 | 1.1 | 0.7 | 34 000 |
| −4 | 7.6 | 0.7 | 0.3 | 1.6 | 1.2 | 31 000 |
| −5 | 8.9 | 0.9 | 0.4 | 1.6 | 1.2 | 27 000 |
| −6 | 11.5 | 1.1 | 0.6 | 1.6 | 1.2 | 25 000 |
| −7 | 14.0 | 1.2 | 0.8 | 1.6 | 1.2 | 23 000 |
| −8 | 15.1 | 1.4 | 1.0 | 1.7 | 1.3 | 21 500 |
| −9 | 16.5 | 3.0 | 1.3 | 1.7 | 1.3 | 19 500 |
| −10 | 17.5 | 3.4 | 1.6 | 1.7 | 1.3 | 18 000 |
| −11 | 18.0 | 3.7 | 1.9 | 1.7 | 1.3 | 16 500 |
| −12 | 19.3 | 4.9 | 2.3 | 1.9 | 1.4 | 15 000 |
| −13 | 21.0 | 5.4 | 2.9 | 2.0 | 1.5 | 13 000 |
| −15 | 23.5 | 6.2 | 4.0 | 2.0 | 1.5 | 11 500 |
| −16 | 24.5 | 6.6 | 4.5 | 2.0 | 1.5 | 10 000 |
| −18 | 27.9 | 8.7 | 5.4 | 2.1 | 1.6 | 9 000 |
| −20 | 30.7 | 9.8 | 6.5 | 2.2 | 1.7 | 8 000 |
| −22 | 33.7 | 10.8 | 8.1 | 2.2 | 1.7 | 7 000 |
| −25 | 37.9 | 12.2 | 10.1 | 2.4 | 1.9 | 5 000 |

† These values apply to rings made from SAE 1060-1090 steels and PH 15-7 Mo stainless steel used on shafts hardened to $R_c$ 50 minimum, with the exception of size −1 which is supplied in beryllium copper only. Values for other sizes made from beryllium copper can be calculated by multiplying the listed values by 0.75. The values listed include a safety factor of 4.
‡ These values apply to all standard rings used on low carbon steel shafts. They include a safety factor of 2.
§ These values have been calculated for steel rings.
Maximum allowable assembly loads with $R$ max or $Ch$ max are as follows:

| Ring Size No. | Maximum Allowable Load, kN | Ring Size No. | Maximum Allowable Load, kN | Ring Size No. | Maximum Allowable Load, kN |
|---|---|---|---|---|---|
| −1 | 0.06 | −8 | 1.4 | −16 | 6.6 |
| −2 | 0.13 | −9 | 3.0 | −18 | 8.7 |
| −3 | 0.3 | −10 | 3.4 | −20 | 9.8 |
| −4 | 0.7 | −11 | 3.7 | −22 | 10.8 |
| −5 | 0.9 | −12 | 4.9 | −25 | 12.2 |
| −6 | 1.1 | −13 | 5.4 | ... | ... |
| −7 | 1.2 | −15 | 6.2 | ... | ... |

*Source:* Appendix to American National Standard ANSI B27.7-1977.

# BOLTS, SCREWS, NUTS, AND WASHERS

Dimensions of bolts, screws, nuts, and washers used in machine construction are given here. For data on thread forms, see "Screw Thread Systems" section.

**American Standard Square and Hexagon Bolts, Screws, and Nuts.** — The 1941 American Standard ASA B18.2 covered head dimensions only. In 1952 and 1955 the Standard was revised to cover the entire product. Some bolt and nut classifications were simplified by elimination or consolidation in agreements reached with the British and Canadians. In 1965 ASA B18.2 was redesignated into two standards: B18.2.1 covering square and hexagon bolts and screws including hexagon cap screws and lag screws and B18.2.2 covering square and hexagon nuts. In B18.2.1-1965, hexagon head cap screws and finished hexagon bolts were consolidated into a single product; heavy semifinished hexagon bolts and heavy finished hexagon bolts were consolidated into a single product; regular semifinished hexagon bolts were eliminated; a new tolerance pattern for all bolts and screws and a positive identification procedure for determining whether an externally threaded product should be designated as a bolt or screw were established. Also included in this standard are heavy hexagon bolts and heavy hexagon structural bolts. In B18.2.2-1965, regular semifinished nuts were discontinued; regular hexagon and heavy hexagon nuts in sizes ¼ through 1 inch, finished hexagon nuts in sizes larger than 1½ inches, washer-faced semifinished style of finished nuts in sizes ⅝-inch and smaller and heavy series nuts in sizes ⁷⁄₁₆-inch and smaller were eliminated.

Further revisions and refinements including the addition of askew head bolts and hex head lag screws and the specifying of countersunk diameters for the various hex nuts and heavy hex nuts resulted in the issuance of the revised standards as ANSI B18.2.1-1972 and ANSI B18.2.2-1972.

**Unified Square and Hexagon Bolts, Screws, and Nuts.** — In certain tables bold-faced items are shown. These are recognized in the Standard as "unified" dimensionally with British and Canadian standards. The other items in the same tables are based on formulas accepted and published by the British for sizes outside the range listed in their standards which, as a matter of information, are B.S. 1768:1963 for Precision (Normal Series) Unified Hexagon Bolts, Screws, Nuts (UNC and UNF Threads); B.S. 1769:1951 for Black (Heavy Series) Unified Hexagon Bolts, etc. and B.S. 2708:1956 for Black (Normal Series) Unified Square and Hexagon Bolts, etc. Tolerances applied to comparable dimensions of American and British Unified bolts and nuts may differ because of rounding off practices and other factors.

**Differentiation between Bolt and Screw.** — A bolt is an externally threaded fastener designed for insertion through holes in assembled parts, and is normally intended to be tightened or released by torquing a nut.

A screw is an externally threaded fastener capable of being inserted into holes in assembled parts, of mating with a preformed internal thread or forming its own thread and of being tightened or released by torquing the head.

An externally threaded fastener which is prevented from being turned during assembly, and which can be tightened or released only by torquing a nut is a *bolt*. (*Example:* round head bolts, track bolts, plow bolts.)

An externally threaded fastener which has a thread form which prohibits assembly with a nut having a straight thread of multiple pitch length is a *screw*. (*Example:* wood screws, tapping screws.)

An externally threaded fastener which must be assembled with a nut to perform its intended service is a *bolt*. (*Example:* heavy hex structural bolt.)

An externally threaded fastener which must be torqued by its head into a tapped or other preformed hole to perform its intended service is a *screw*. (*Example:* square head set screw.)

## American National Standard Square and Hexagon Bolts and Nuts

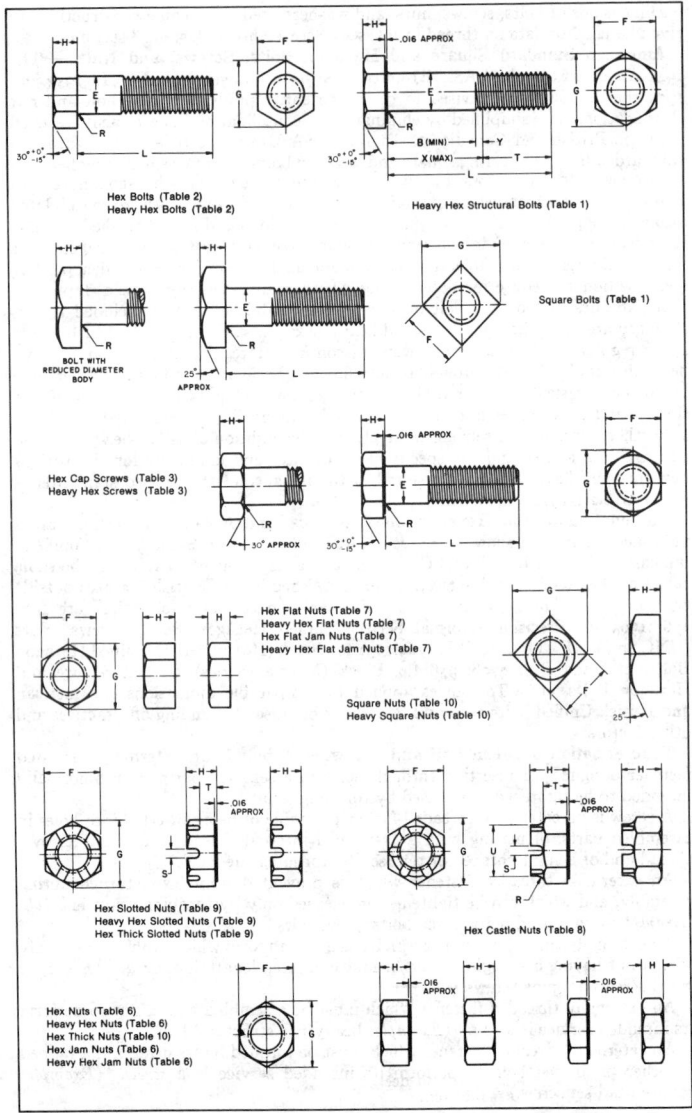

Hex Bolts (Table 2)
Heavy Hex Bolts (Table 2)

Heavy Hex Structural Bolts (Table 1)

BOLT WITH
REDUCED DIAMETER
BODY

Square Bolts (Table 1)

Hex Cap Screws (Table 3)
Heavy Hex Screws (Table 3)

Hex Flat Nuts (Table 7)
Heavy Hex Flat Nuts (Table 7)
Hex Flat Jam Nuts (Table 7)
Heavy Hex Flat Jam Nuts (Table 7)

Square Nuts (Table 10)
Heavy Square Nuts (Table 10)

Hex Slotted Nuts (Table 9)
Heavy Hex Slotted Nuts (Table 9)
Hex Thick Slotted Nuts (Table 9)

Hex Castle Nuts (Table 8)

Hex Nuts (Table 6)
Heavy Hex Nuts (Table 6)
Hex Thick Nuts (Table 10)
Hex Jam Nuts (Table 6)
Heavy Hex Jam Nuts (Table 6)

**Square and Hex Bolts, Screws and Nuts.** — The dimensions for square and hex bolts and screws given in the following tables have been taken from American National Standard ANSI B18.2.1-1972 and for nuts from American National Standard ANSI B18.2.2-1972. Reference should be made to these Standards for information or data not found in the following text and tables:

*Designation:* Bolts and screws should be designated by the following data in the sequence shown: nominal size (fractional and decimal equivalent); threads per inch (omit for lag screws); product length for bolts and screws (fractional or two-place decimal equivalent); product name; material, including specification, where necessary; and protective finish, if required.  Examples: (1) ⅜-16 × 1½ Square Bolt, Steel, Zinc Plated; (2) ½-13 × 3 Hex Cap Screw, SAE Grade 8 Steel; and (3) .75 × 5.00 Hex Lag Screw, Steel.  (4) ½-13 Square Nut, Steel, Zinc Plated; (5) ¾-16 Heavy Hex Nut, SAE Grade 5 Steel; and (6) 1.000-8 Hex Thick Slotted Nut, Corrosion-Resistant Steel.

Table 1.   American National Standard and Unified Standard Square Bolts and American National Standard Heavy Hex Structural Bolts (ANSI B18.2.1-1972)

SQUARE BOLTS

| Nominal Size* or Basic Product Diam. | Body Diam.† E | Width Across Flats F | | | Width Across Corners G | | Height H | | | Radius of Fillet R |
|---|---|---|---|---|---|---|---|---|---|---|
| | Max. | Basic | Max. | Min. | Max. | Min. | Basic | Max. | Min. | Max. |
| ¼ | 0.2500 | ⅜ | 0.375 | 0.362 | 0.530 | 0.498 | 11/64 | 0.188 | 0.156 | 0.03 |
| 5/16 | 0.3125 | 0.324 | ½ | 0.500 | 0.484 | 0.707 | 0.665 | 13/64 | 0.220 | 0.186 | 0.03 |
| ⅜ | 0.3750 | 0.388 | 9/16 | 0.562 | 0.544 | 0.795 | 0.747 | ¼ | 0.268 | 0.232 | 0.03 |
| 7/16 | 0.4375 | 0.452 | ⅝ | 0.625 | 0.603 | 0.884 | 0.828 | 19/64 | 0.316 | 0.278 | 0.03 |
| ½ | 0.5000 | 0.515 | ¾ | 0.750 | 0.725 | 1.061 | 0.995 | 21/64 | 0.348 | 0.308 | 0.03 |
| ⅝ | 0.6250 | 0.642 | 15/16 | 0.938 | 0.906 | 1.326 | 1.244 | 27/64 | 0.444 | 0.400 | 0 06 |
| ¾ | 0.7500 | 0.768 | 1⅛ | 1.125 | 1.088 | 1.591 | 1.494 | ½ | 0.524 | 0.476 | 0 06 |
| ⅞ | 0.8750 | 0.895 | 1 5/16 | 1.312 | 1.269 | 1.856 | 1.742 | 19/32 | 0.620 | 0.568 | 0 06 |
| 1 | 1.0000 | 1.022 | 1½ | 1.500 | 1.450 | 2.121 | 1.991 | 21/32 | 0.684 | 0.628 | 0.09 |
| 1⅛ | 1.1250 | 1.149 | 1 11/16 | 1.688 | 1.631 | 2.386 | 2.239 | ¾ | 0.780 | 0.720 | 0.09 |
| 1¼ | 1.2500 | 1.277 | 1⅞ | 1.875 | 1.812 | 2.652 | 2.489 | 27/32 | 0.876 | 0.812 | 0.09 |
| 1⅜ | 1.3750 | 1.404 | 2 1/16 | 2.062 | 1.994 | 2.917 | 2.738 | 29/32 | 0.940 | 0.872 | 0.09 |
| 1½ | 1.5000 | 1.531 | 2¼ | 2.250 | 2.175 | 3.182 | 2.986 | 1 | 1.036 | 0.964 | 0.09 |

HEAVY HEX STRUCTURAL BOLTS

| Nominal Size* or Basic Product Diam. | Body Diam. E | | Width Across Flats F | | Width Across Corners G | | Height H | | Radius of Fillet R | | Thrd. Lgth. T | Transi- tion Thrd. Y |
|---|---|---|---|---|---|---|---|---|---|---|---|---|
| | Max. | Min. | Max. | Min. | Max. | Min. | Max. | Min. | Max. | Min. | Basic | Max. |
| ½ | 0.5000 | 0.482 | 0.875 | 0.850 | 1.010 | 0.969 | 0.323 | 0.302 | 0.031 | 0.009 | 1.00 | 0.19 |
| ⅝ | 0.6250 | 0.642 | 0.605 | 1.062 | 1.031 | 1.227 | 1.175 | 0.403 | 0.378 | 0.062 | 0.021 | 1.25 | 0.22 |
| ¾ | 0.7500 | 0.768 | 0.729 | 1.250 | 1.212 | 1.443 | 1.383 | 0.483 | 0.455 | 0.062 | 0.021 | 1.38 | 0.25 |
| ⅞ | 0.8750 | 0.895 | 0.852 | 1.438 | 1.394 | 1.660 | 1.589 | 0.563 | 0.531 | 0.062 | 0.031 | 1.50 | 0.28 |
| 1 | 1.0000 | 1.022 | 0.976 | 1.625 | 1.575 | 1.876 | 1.796 | 0.627 | 0.591 | 0.093 | 0.062 | 1.75 | 0.31 |
| 1⅛ | 1.1250 | 1.149 | 1.098 | 1.812 | 1.756 | 2.093 | 2.002 | 0.718 | 0.658 | 0.093 | 0.062 | 2.00 | 0.34 |
| 1¼ | 1.2500 | 1.277 | 1.223 | 2.000 | 1.938 | 2.309 | 2.209 | 0.813 | 0.749 | 0.093 | 0.062 | 2.00 | 0.38 |
| 1⅜ | 1.3750 | 1.404 | 1.345 | 2.188 | 2.119 | 2.526 | 2.416 | 0.878 | 0.810 | 0.093 | 0.062 | 2.25 | 0.44 |
| 1½ | 1.5000 | 1.531 | 1.470 | 2.375 | 2.300 | 2.742 | 2.622 | 0.974 | 0.902 | 0.093 | 0.062 | 2.25 | 0.44 |

All dimensions are in inches.  See diagram on page 1132 .
**Bold type shows bolts unified dimensionally with British and Canadian Standards.**
Threads, when rolled, shall be Unified Coarse, Fine, or 8-thread series (UNRC, UNRF, or 8 UNR Series), Class 2A.  Threads produced by other methods may be Unified Coarse, Fine, or 8-thread series (UNC, UNF, or 8 UN Series), Class 2A.
* Where specifying nominal size in decimals, zeros before the decimal point and in the fourth decimal place are omitted.

**Table 2. American National Standard and Unified Standard Hex and Heavy Hex Bolts** (ANSI B18.2.1-1972)

| Nominal Size* or Basic Diam. | | Body Diam. *E* | Width Across Flats *F* | | | Width Across Corners *G* | | Height *H* | | | Radius of Fillet, *R* | |
|---|---|---|---|---|---|---|---|---|---|---|---|---|
| | | Max. | Basic | Max. | Min. | Max. | Min. | Basic | Max. | Min. | Max. | Min. |
| HEX BOLTS | | | | | | | | | | | | |
| **1/4** | **0.2500** | **0.260** | **7/16** | **0.438** | **0.425** | **0.505** | **0.484** | **11/64** | **0.188** | **0.150** | 0.03 | 0.01 |
| **5/16** | **0.3125** | **0.324** | **1/2** | **0.500** | **0.484** | **0.577** | **0.552** | **7/32** | **0.235** | **0.195** | 0.03 | 0.01 |
| **3/8** | **0.3750** | **0.388** | **9/16** | **0.562** | **0.544** | **0.650** | **0.620** | **1/4** | **0.268** | **0.226** | 0.03 | 0.01 |
| **7/16** | **0.4375** | **0.452** | **5/8** | **0.625** | **0.603** | **0.722** | **0.687** | **19/64** | **0.316** | **0.272** | 0.03 | 0.01 |
| **1/2** | **0.5000** | **0.515** | **3/4** | **0.750** | **0.725** | **0.866** | **0.826** | **11/32** | **0.364** | **0.302** | 0.03 | 0.01 |
| **5/8** | **0.6250** | **0.642** | **15/16** | **0.938** | **0.906** | **1.083** | **1.033** | **27/64** | **0.444** | **0.378** | 0.06 | 0.02 |
| **3/4** | **0.7500** | **0.768** | **1 1/8** | **1.125** | **1.088** | **1.299** | **1.240** | **1/2** | **0.524** | **0.455** | 0.06 | 0.02 |
| **7/8** | **0.8750** | **0.895** | **1 5/16** | **1.312** | **1.269** | **1.516** | **1.447** | **37/64** | **0.604** | **0.531** | 0.06 | 0.02 |
| **1** | **1.0000** | **1.022** | **1 1/2** | **1.500** | **1.450** | **1.732** | **1.653** | **43/64** | **0.700** | **0.591** | 0.09 | 0.03 |
| **1 1/8** | **1.1250** | **1.149** | **1 11/16** | **1.688** | **1.631** | **1.949** | **1.859** | **3/4** | **0.780** | **0.658** | 0.09 | 0.03 |
| **1 1/4** | **1.2500** | **1.277** | **1 7/8** | **1.875** | **1.812** | **2.165** | **2.066** | **27/32** | **0.876** | **0.749** | 0.09 | 0.03 |
| **1 3/8** | **1.3750** | **1.404** | **2 1/16** | **2.062** | **1.994** | **2.382** | **2.273** | **29/32** | **0.940** | **0.810** | 0.09 | 0.03 |
| **1 1/2** | **1.5000** | **1.531** | **2 1/4** | **2.250** | **2.175** | **2.598** | **2.480** | **1** | **1.036** | **0.902** | 0.09 | 0.03 |
| **1 3/4** | **1.7500** | **1.785** | **2 5/8** | **2.625** | **2.538** | **3.031** | **2.893** | **1 5/32** | **1.196** | **1.054** | 0.12 | 0.04 |
| **2** | **2.0000** | **2.039** | **3** | **3.000** | **2.900** | **3.464** | **3.306** | **1 11/32** | **1.388** | **1.175** | 0.12 | 0.04 |
| 2 1/4 | 2.2500 | 2.305 | 3 3/8 | 3.375 | 3.262 | 3.897 | 3.719 | 1 1/2 | 1.548 | 1.327 | 0.19 | 0.06 |
| 2 1/2 | 2.5000 | 2.559 | 3 3/4 | 3.750 | 3.625 | 4.330 | 4.133 | 1 21/32 | 1.708 | 1.479 | 0.19 | 0.06 |
| 2 3/4 | 2.7500 | 2.827 | 4 1/8 | 4.125 | 3.988 | 4.763 | 4.546 | 1 13/16 | 1.869 | 1.632 | 0.19 | 0.06 |
| 3 | 3.0000 | 3.081 | 4 1/2 | 4.500 | 4.350 | 5.196 | 4.959 | 2 | 2.060 | 1.815 | 0.19 | 0.06 |
| 3 1/4 | 3.2500 | 3.335 | 4 7/8 | 4.875 | 4.712 | 5.629 | 5.372 | 2 3/16 | 2.251 | 1.936 | 0.19 | 0.06 |
| 3 1/2 | 3.5000 | 3.589 | 5 1/4 | 5.250 | 5.075 | 6.062 | 5.786 | 2 5/16 | 2.380 | 2.057 | 0.19 | 0.06 |
| 3 3/4 | 3.7500 | 3.858 | 5 5/8 | 5.625 | 5.437 | 6.495 | 6.198 | 2 1/2 | 2.572 | 2.241 | 0.19 | 0.06 |
| 4 | 4.0000 | 4.111 | 6 | 6.000 | 5.800 | 6.928 | 6.612 | 2 11/16 | 2.764 | 2.424 | 0.19 | 0.06 |
| HEAVY HEX BOLTS | | | | | | | | | | | | |
| **1/2** | **0.5000** | **0.515** | **7/8** | **0.875** | **0.850** | **1.010** | **0.969** | 11/32 | 0.364 | 0.302 | 0.03 | 0.01 |
| **5/8** | **0.6250** | **0.642** | **1 1/16** | **1.062** | **1.031** | **1.227** | **1.175** | 27/64 | 0.444 | 0.378 | 0.06 | 0.02 |
| **3/4** | **0.7500** | **0.768** | **1 1/4** | **1.250** | **1.212** | **1.443** | **1.383** | 1/2 | 0.524 | 0.455 | 0.06 | 0.02 |
| **7/8** | **0.8750** | **0.895** | **1 7/16** | **1.438** | **1.394** | **1.660** | **1.589** | 37/64 | 0.604 | 0.531 | 0.06 | 0.02 |
| **1** | **1.0000** | **1.022** | **1 5/8** | **1.625** | **1.575** | **1.876** | **1.796** | 43/64 | 0.700 | 0.591 | 0.09 | 0.03 |
| **1 1/8** | **1.1250** | **1.149** | **1 13/16** | **1.812** | **1.756** | **2.093** | **2.002** | 3/4 | 0.780 | 0.658 | 0.09 | 0.03 |
| **1 1/4** | **1.2500** | **1.277** | **2** | **2.000** | **1.938** | **2.309** | **2.209** | 27/32 | 0.876 | 0.749 | 0.09 | 0.03 |
| **1 3/8** | **1.3750** | **1.404** | **2 3/16** | **2.188** | **2.119** | **2.526** | **2.416** | 29/32 | 0.940 | 0.810 | 0.09 | 0.03 |
| **1 1/2** | **1.5000** | **1.531** | **2 3/8** | **2.375** | **2.300** | **2.742** | **2.622** | 1 | 1.036 | 0.902 | 0.09 | 0.03 |
| **1 3/4** | **1.7500** | **1.785** | **2 3/4** | **2.750** | **2.662** | **3.175** | **3.035** | 1 5/32 | 1.196 | 1.054 | 0.12 | 0.04 |
| **2** | **2.0000** | **2.039** | **3 1/8** | **3.125** | **3.025** | **3.608** | **3.449** | 1 11/32 | 1.388 | 1.175 | 0.12 | 0.04 |
| 2 1/4 | 2.2500 | 2.305 | 3 1/2 | 3.500 | 3.388 | 4.041 | 3.862 | 1 1/2 | 1.548 | 1.327 | 0.19 | 0.06 |
| 2 1/2 | 2.5000 | 2.559 | 3 7/8 | 3.875 | 3.750 | 4.474 | 4.275 | 1 21/32 | 1.708 | 1.479 | 0.19 | 0.06 |
| 2 3/4 | 2.7500 | 2.827 | 4 1/4 | 4.250 | 4.112 | 4.907 | 4.688 | 1 13/16 | 1.869 | 1.632 | 0.19 | 0.06 |
| 3 | 3.0000 | 3.081 | 4 5/8 | 4.625 | 4.475 | 5.340 | 5.102 | 2 | 2.060 | 1.815 | 0.19 | 0.06 |

All dimensions are in inches. See diagram on page 1132.

**Bold type shows bolts unified dimensionally with British and Canadian Standards.**

*Threads:* Threads, when rolled, are Unified Coarse, Fine, or 8-thread series (UNRC, UNRF, or 8 UNR Series), Class 2A. Threads produced by other methods may be Unified Coarse, Fine or 8-thread series (UNC, UNF, or 8 UN Series), Class 2A.

*Body Diameter:* Bolts may be obtained in "reduced diameter body." Where "reduced diameter body" is specified, the body diameter may be reduced to approximately the pitch diameter of the thread. A shoulder of full body diameter under the head may be supplied at the option of the manufacturer.

*Material:* Unless otherwise specified, chemical and mechanical properties of steel bolts conform to ASTM A307, Grade A. Other materials are as agreed upon by manufacturer and purchaser.

* *Nominal Size:* Where specifying nominal size in decimals, zeros preceding the decimal point and in the fourth decimal place are omitted.

Table 3. American National Standard and Unified Standard Heavy Hex Screws and Hex Cap Screws (ANSI B18.2.1-1972)

| Nominal Size* or Basic Diam. | Body Diam.† E | | Width Across Flats F | | | Width Across Corners G | | Height H | | | Fillet | |
|---|---|---|---|---|---|---|---|---|---|---|---|---|
| | Max. | Min. | Basic | Max. | Min. | Max. | Min. | Basic | Max. | Min. | Max. | Min. |
| **HEAVY HEX SCREWS** | | | | | | | | | | | | |
| 1/2 | **0.5000** | **0.5000** | 0.482 | 7/8 | 0.875 | 0.850 | 1.010 | 0.969 | 5/16 | 0.323 | 0.302 | |
| 5/8 | **0.6250** | **0.6250** | 0.605 | 1 1/16 | 1.062 | 1.031 | 1.227 | 1.175 | 25/64 | 0.403 | 0.378 | |
| 3/4 | **0.7500** | **0.7500** | 0.729 | 1 1/4 | 1.250 | 1.212 | 1.443 | 1.383 | 15/32 | 0.483 | 0.455 | |
| 7/8 | **0.8750** | **0.8750** | 0.852 | 1 7/16 | 1.438 | 1.394 | 1.660 | 1.589 | 35/64 | 0.563 | 0.531 | |
| 1 | **1.0000** | **1.0000** | 0.976 | 1 5/8 | 1.625 | 1.575 | 1.876 | 1.796 | 39/64 | 0.627 | 0.591 | |
| 1 1/8 | **1.1250** | **1.1250** | 1.098 | 1 13/16 | 1.812 | 1.756 | 2.093 | 2.002 | 11/16 | 0.718 | 0.658 | |
| 1 1/4 | **1.2500** | **1.2500** | 1.223 | 2 | 2.000 | 1.938 | 2.309 | 2.209 | 25/32 | 0.813 | 0.749 | |
| 1 3/8 | **1.3750** | **1.3750** | 1.345 | 2 3/16 | 2.188 | 2.119 | 2.526 | 2.416 | 27/32 | 0.878 | 0.810 | |
| 1 1/2 | **1.5000** | **1.5000** | 1.470 | 2 3/8 | 2.375 | 2.300 | 2.742 | 2.622 | 15/16 | 0.974 | 0.902 | |
| 1 3/4 | **1.7500** | **1.7500** | 1.716 | 2 3/4 | 2.750 | 2.662 | 3.175 | 3.035 | 1 3/32 | 1.134 | 1.054 | |
| 2 | **2.0000** | **2.0000** | 1.964 | 3 1/8 | 3.125 | 3.025 | 3.608 | 3.449 | 1 7/32 | 1.263 | 1.175 | |
| 2 1/4 | **2.2500** | **2.2500** | 2.214 | 3 1/2 | 3.500 | 3.388 | 4.041 | 3.862 | 1 3/8 | 1.423 | 1.327 | |
| 2 1/2 | **2.5000** | **2.5000** | 2.461 | 3 7/8 | 3.875 | 3.750 | 4.474 | 4.275 | 1 17/32 | 1.583 | 1.479 | |
| 2 3/4 | **2.7500** | **2.7500** | 2.711 | 4 1/4 | 4.250 | 4.112 | 4.907 | 4.688 | 1 11/16 | 1.744 | 1.632 | |
| 3 | **3.0000** | **3.0000** | 2.961 | 4 5/8 | 4.625 | 4.475 | 5.340 | 5.102 | 1 7/8 | 1.935 | 1.815 | |
| **HEX CAP SCREWS (Finished Hex Bolts)** | | | | | | | | | | | | |
| 1/4 | **0.2500** | **0.2500** | 0.2450 | 7/16 | 0.438 | 0.428 | 0.505 | 0.488 | 5/32 | 0.163 | 0.150 | |
| 5/16 | **0.3125** | **0.3125** | 0.3065 | 1/2 | 0.500 | 0.489 | 0.577 | 0.557 | 13/64 | 0.211 | 0.195 | |
| 3/8 | **0.3750** | **0.3750** | 0.3690 | 9/16 | 0.562 | 0.551 | 0.650 | 0.628 | 15/64 | 0.243 | 0.226 | |
| 7/16 | **0.4375** | **0.4375** | 0.4305 | 5/8 | 0.625 | 0.612 | 0.722 | 0.698 | 9/32 | 0.291 | 0.272 | |
| 1/2 | **0.5000** | **0.5000** | 0.4930 | 3/4 | 0.750 | 0.736 | 0.866 | 0.840 | 5/16 | 0.323 | 0.302 | |
| 9/16 | **0.5625** | **0.5625** | 0.5545 | 13/16 | 0.812 | 0.798 | 0.938 | 0.910 | 23/64 | 0.371 | 0.348 | |
| 5/8 | **0.6250** | **0.6250** | 0.6170 | 15/16 | 0.938 | 0.922 | 1.083 | 1.051 | 25/64 | 0.403 | 0.378 | |
| 3/4 | **0.7500** | **0.7500** | 0.7410 | 1 1/8 | 1.125 | 1.100 | 1.299 | 1.254 | 15/32 | 0.483 | 0.455 | |
| 7/8 | **0.8750** | **0.8750** | 0.8660 | 1 5/16 | 1.312 | 1.285 | 1.516 | 1.465 | 35/64 | 0.563 | 0.531 | |
| 1 | **1.0000** | **1.0000** | 0.9900 | 1 1/2 | 1.500 | 1.469 | 1.732 | 1.675 | 39/64 | 0.627 | 0.591 | |
| 1 1/8 | **1.1250** | **1.1250** | 1.1140 | 1 11/16 | 1.688 | 1.631 | 1.949 | 1.859 | 11/16 | 0.718 | 0.658 | |
| 1 1/4 | **1.2500** | **1.2500** | 1.2390 | 1 7/8 | 1.875 | 1.812 | 2.165 | 2.066 | 25/32 | 0.813 | 0.749 | |
| 1 3/8 | **1.3750** | **1.3750** | 1.3630 | 2 1/16 | 2.062 | 1.994 | 2.382 | 2.273 | 27/32 | 0.878 | 0.810 | |
| 1 1/2 | **1.5000** | **1.5000** | 1.4880 | 2 1/4 | 2.250 | 2.175 | 2.598 | 2.480 | 15/16 | 0.974 | 0.902 | |
| 1 3/4 | **1.7500** | **1.7500** | 1.7380 | 2 5/8 | 2.625 | 2.538 | 3.031 | 2.893 | 1 3/32 | 1.134 | 1.054 | |
| 2 | **2.0000** | **2.0000** | 1.9880 | 3 | 3.000 | 2.900 | 3.464 | 3.306 | 1 7/32 | 1.263 | 1.175 | |
| 2 1/4 | **2.2500** | **2.2500** | 2.2380 | 3 3/8 | 3.375 | 3.262 | 3.897 | 3.719 | 1 3/8 | 1.423 | 1.327 | |
| 2 1/2 | **2.5000** | **2.5000** | 2.4880 | 3 3/4 | 3.750 | 3.625 | 4.330 | 4.133 | 1 17/32 | 1.583 | 1.479 | |
| 2 3/4 | **2.7500** | **2.7500** | 2.7380 | 4 1/8 | 4.125 | 3.988 | 4.763 | 4.546 | 1 11/16 | 1.744 | 1.632 | |
| 3 | **3.0000** | **3.0000** | 2.9880 | 4 1/2 | 4.500 | 4.350 | 5.196 | 4.959 | 1 7/8 | 1.935 | 1.815 | |

*Fillet column note (both sections):* For fillet transition diameter, fillet length and radius of fillet, see American National Standard ANSI B18.2.1-1972

All dimensions are in inches. See diagram on page 1132

*Unification:* **Bold type** indicates product features unified dimensionally with British and Canadian standards. Unification of fine thread products is limited to sizes 1 inch and smaller.

*Bearing Surface:* Bearing surface is flat and washer faced or for Hex Cap Screws may have chamfered corners. Diameter of bearing surface is equal to the maximum width across flats within a tolerance of minus 10 per cent.

*Thread Series:* Threads, when rolled, are Unified Coarse, Fine or 8-thread series (UNRC, UNRF, or 8 UNR Series), Class 2A. Threads produced by other methods shall preferably be UNRC, UNRF, or 8 UNR but, at manufacturer's option, may be Unified Coarse, Fine or 8-thread series (UNC, UNF or 8 UN Series), Class 2A.

*Material:* Chemical and mechanical properties of steel screws normally conform to Grades 2, 5, or 8 of SAE J429, ASTM A449 or ASTM A354 Grade BD. Where specified, screws may also be made from brass, bronze, corrosion-resisting steel, aluminum alloy or other materials.

*Nominal Size:* Where specifying nominal size in decimals, zeros preceding the decimal and in the fourth decimal place are omitted.

## Table 4. American National Standard Square Lag Screws (ANSI B18.2.1-1972)

DETAIL OF THREAD

CONE POINT — 60° APPROX

GIMLET POINT — 60° APPROX

25° APPROX

| Nominal Size* or Basic Product Diam. | Body or Shoulder Diam. E Max. | Min. | Width Across Flats F Basic | Max. | Min. | Width Across Corners G Max. | Min. | Height H Basic | Max. | Min. | Shoulder Length S Min. | Radius of Fillet R Max. | Thds. per Inch | Pitch P | Flat at Root B | Depth of Thd. T | Root Diam. D₁ |
|---|---|---|---|---|---|---|---|---|---|---|---|---|---|---|---|---|---|
| No. 10  0.1900 | 0.199 | 0.178 | 9/32 | 0.281 | 0.271 | 0.398 | 0.372 | 1/8 | 0.140 | 0.110 | 0.094 | 0.03 | 11 | 0.091 | 0.039 | 0.035 | 0.120 |
| 1/4  0.2500 | 0.260 | 0.237 | 3/8 | 0.375 | 0.362 | 0.530 | 0.498 | 11/64 | 0.188 | 0.156 | 0.094 | 0.03 | 10 | 0.100 | 0.043 | 0.039 | 0.173 |
| 5/16  0.3125 | 0.324 | 0.298 | 1/2 | 0.500 | 0.484 | 0.707 | 0.665 | 13/64 | 0.220 | 0.186 | 0.125 | 0.03 | 9 | 0.111 | 0.048 | 0.043 | 0.227 |
| 3/8  0.3750 | 0.388 | 0.360 | 9/16 | 0.562 | 0.544 | 0.795 | 0.747 | 1/4 | 0.268 | 0.232 | 0.125 | 0.03 | 7 | 0.143 | 0.062 | 0.055 | 0.265 |
| 7/16  0.4375 | 0.452 | 0.421 | 5/8 | 0.625 | 0.603 | 0.884 | 0.828 | 19/64 | 0.316 | 0.278 | 0.156 | 0.03 | 7 | 0.143 | 0.062 | 0.055 | 0.328 |
| 1/2  0.5000 | 0.515 | 0.482 | 3/4 | 0.750 | 0.725 | 1.061 | 0.995 | 21/64 | 0.348 | 0.308 | 0.156 | 0.03 | 6 | 0.167 | 0.072 | 0.064 | 0.371 |
| 5/8  0.6250 | 0.642 | 0.605 | 15/16 | 0.938 | 0.906 | 1.326 | 1.244 | 27/64 | 0.444 | 0.400 | 0.312 | 0.06 | 5 | 0.200 | 0.086 | 0.077 | 0.471 |
| 3/4  0.7500 | 0.768 | 0.729 | 1 1/8 | 1.125 | 1.088 | 1.591 | 1.494 | 1/2 | 0.524 | 0.476 | 0.375 | 0.06 | 4 1/2 | 0.222 | 0.096 | 0.085 | 0.579 |
| 7/8  0.8750 | 0.895 | 0.852 | 1 5/16 | 1.312 | 1.269 | 1.856 | 1.742 | 19/32 | 0.620 | 0.568 | 0.375 | 0.06 | 4 | 0.250 | 0.108 | 0.096 | 0.683 |
| 1  1.0000 | 1.022 | 0.976 | 1 1/2 | 1.500 | 1.450 | 2.121 | 1.991 | 21/32 | 0.684 | 0.628 | 0.625 | 0.09 | 3 1/2 | 0.286 | 0.123 | 0.110 | 0.780 |
| 1 1/8  1.1250 | 1.149 | 1.098 | 1 11/16 | 1.688 | 1.631 | 2.386 | 2.239 | 3/4 | 0.780 | 0.720 | 0.625 | 0.09 | 3 1/4 | 0.308 | 0.133 | 0.119 | 0.887 |
| 1 1/4  1.2500 | 1.277 | 1.223 | 1 7/8 | 1.875 | 1.812 | 2.652 | 2.489 | 27/32 | 0.876 | 0.812 | 0.625 | 0.09 | 3 1/4 | 0.308 | 0.133 | 0.119 | 1.012 |

All dimensions in inches.  *When specifying decimal nominal size, zeros before decimal point and in fourth decimal place are omitted.

Minimum thread length is ½ length of screw plus 0.50 inch, or 6.00 inches, whichever is shorter. Screws too short for the formula thread length shall be threaded as close to the head as practicable.

Thread formulas: Pitch = 1 ÷ thds. per inch. Flat at root = 0.4305 × pitch. Depth of single thread = 0.385 × pitch.

Table 5. American National Standard Hex Lag Screws (ANSI B18.2.1-1972)

| Nominal Size* or Basic Product Diam. | Body or Shoulder Diam. E Max. | E Min. | Width Across Flats F Basic | F Max. | F Min. | Width Across Corners G Max. | G Min. | Height H Basic | H Max. | H Min. | Shoulder Length S Min. | Radius of Fillet R Max. | Thds. per Inch | Pitch P | Flat at Root B | Depth of Thd. T | Root Diam. D₁ |
|---|---|---|---|---|---|---|---|---|---|---|---|---|---|---|---|---|---|
| No. 10  0.1900 | 0.199 | 0.178 | 9/32 | 0.281 | 0.271 | 0.323 | 0.309 | 1/8 | 0.140 | 0.110 | 0.094 | 0.03 | 11 | 0.091 | 0.039 | 0.035 | 0.120 |
| 1/4  0.2500 | 0.260 | 0.237 | 3/8 | 0.438 | 0.425 | 0.505 | 0.484 | 11/64 | 0.188 | 0.150 | 0.094 | 0.03 | 10 | 0.100 | 0.043 | 0.039 | 0.173 |
| 5/16  0.3125 | 0.324 | 0.298 | 1/2 | 0.500 | 0.484 | 0.577 | 0.552 | 13/64 | 0.235 | 0.195 | 0.125 | 0.03 | 9 | 0.111 | 0.048 | 0.043 | 0.227 |
| 3/8  0.3750 | 0.388 | 0.360 | 9/16 | 0.562 | 0.544 | 0.650 | 0.620 | 1/4 | 0.268 | 0.226 | 0.125 | 0.03 | 7 | 0.143 | 0.062 | 0.055 | 0.265 |
| 7/16  0.4375 | 0.452 | 0.421 | 5/8 | 0.625 | 0.603 | 0.722 | 0.687 | 19/64 | 0.316 | 0.272 | 0.156 | 0.03 | 7 | 0.143 | 0.062 | 0.055 | 0.328 |
| 1/2  0.5000 | 0.515 | 0.482 | 3/4 | 0.750 | 0.725 | 0.866 | 0.826 | 11/32 | 0.364 | 0.302 | 0.156 | 0.03 | 6 | 0.167 | 0.072 | 0.064 | 0.371 |
| 5/8  0.6250 | 0.642 | 0.605 | 15/16 | 0.938 | 0.906 | 1.083 | 1.033 | 27/64 | 0.444 | 0.378 | 0.312 | 0.06 | 5 | 0.200 | 0.086 | 0.077 | 0.471 |
| 3/4  0.7500 | 0.768 | 0.729 | 1 1/8 | 1.125 | 1.088 | 1.299 | 1.240 | 1/2 | 0.524 | 0.455 | 0.375 | 0.06 | 4 1/2 | 0.222 | 0.096 | 0.085 | 0.579 |
| 7/8  0.8750 | 0.895 | 0.852 | 1 5/16 | 1.312 | 1.269 | 1.516 | 1.447 | 37/64 | 0.604 | 0.531 | 0.375 | 0.06 | 4 | 0.250 | 0.108 | 0.096 | 0.683 |
| 1  1.0000 | 1.022 | 0.976 | 1 1/2 | 1.500 | 1.450 | 1.732 | 1.653 | 43/64 | 0.700 | 0.591 | 0.625 | 0.09 | 3 1/2 | 0.286 | 0.123 | 0.110 | 0.780 |
| 1 1/8  1.1250 | 1.149 | 1.098 | 1 11/16 | 1.688 | 1.631 | 1.949 | 1.859 | 3/4 | 0.780 | 0.658 | 0.625 | 0.09 | 3 1/4 | 0.308 | 0.133 | 0.119 | 0.887 |
| 1 1/4  1.2500 | 1.277 | 1.223 | 1 7/8 | 1.875 | 1.812 | 2.165 | 2.066 | 27/32 | 0.876 | 0.749 | 0.625 | 0.09 | 3 1/4 | 0.308 | 0.133 | 0.119 | 1.012 |

All dimensions in inches.    * When specifying decimal nominal size, zeros before decimal point and in fourth decimal place are omitted.
Minimum thread length is 1/2 length of screw plus 0.50 inch, or 6.00 inches, whichever is shorter.    Screws too short for the formula thread length shall be threaded as close to the head as practicable.
Thread formulas: Pitch = 1 ÷ thds. per inch.    Flat at root = 0.4305 × pitch.    Depth of single thread = 0.385 × pitch.

Table 6.  American National Standard and Unified Standard Hex Nuts and
Jam Nuts and Heavy Hex Nuts and Jam Nuts (ANSI B18.2.2-1972)

| Nominal Size* or Basic Major Diam. of Thread | | Width Across Flats F | | | Width Across Corners G | | Thickness, Nuts H | | | Thickness, Jam Nuts H | | |
|---|---|---|---|---|---|---|---|---|---|---|---|---|
| | | Basic | Max. | Min. | Max. | Min. | Basic | Max. | Min. | Basic | Max. | Min. |
| HEX NUTS AND HEX JAM NUTS | | | | | | | | | | | | |
| 1/4 | 0.2500 | 7/16 | 0.438 | 0.428 | 0.505 | 0.488 | 7/32 | 0.226 | 0.212 | 5/32 | 0.163 | 0.150 |
| 5/16 | 0.3125 | 1/2 | 0.500 | 0.489 | 0.577 | 0.557 | 17/64 | 0.273 | 0.258 | 3/16 | 0.195 | 0.180 |
| 3/8 | 0.3750 | 9/16 | 0.562 | 0.551 | 0.650 | 0.628 | 21/64 | 0.337 | 0.320 | 7/32 | 0.227 | 0.210 |
| 7/16 | 0.4375 | 11/16 | 0.688 | 0.675 | 0.794 | 0.768 | 3/8 | 0.385 | 0.365 | 1/4 | 0.260 | 0.240 |
| 1/2 | 0.5000 | 3/4 | 0.750 | 0.736 | 0.866 | 0.840 | 7/16 | 0.448 | 0.427 | 5/16 | 0.323 | 0.302 |
| 9/16 | 0.5625 | 7/8 | 0.875 | 0.861 | 1.010 | 0.982 | 31/64 | 0.496 | 0.473 | 5/16 | 0.324 | 0.301 |
| 5/8 | 0.6250 | 15/16 | 0.938 | 0.922 | 1.083 | 1.051 | 35/64 | 0.559 | 0.535 | 3/8 | 0.387 | 0.363 |
| 3/4 | 0.7500 | 1 1/8 | 1.125 | 1.088 | 1.299 | 1.240 | 41/64 | 0.665 | 0.617 | 27/64 | 0.446 | 0.398 |
| 7/8 | 0.8750 | 1 5/16 | 1.312 | 1.269 | 1.516 | 1.447 | 3/4 | 0.776 | 0.724 | 31/64 | 0.510 | 0.458 |
| 1 | 1.0000 | 1 1/2 | 1.500 | 1.450 | 1.732 | 1.653 | 55/64 | 0.887 | 0.831 | 35/64 | 0.575 | 0.519 |
| 1 1/8 | 1.1250 | 1 11/16 | 1.688 | 1.631 | 1.949 | 1.859 | 31/32 | 0.999 | 0.939 | 39/64 | 0.639 | 0.579 |
| 1 1/4 | 1.2500 | 1 7/8 | 1.875 | 1.812 | 2.165 | 2.066 | 1 1/16 | 1.094 | 1.030 | 23/32 | 0.751 | 0.687 |
| 1 3/8 | 1.3750 | 2 1/16 | 2.062 | 1.994 | 2.382 | 2.273 | 1 11/64 | 1.206 | 1.138 | 25/32 | 0.815 | 0.747 |
| 1 1/2 | 1.5000 | 2 1/4 | 2.250 | 2.175 | 2.598 | 2.480 | 1 9/32 | 1.317 | 1.245 | 27/32 | 0.880 | 0.808 |
| HEAVY HEX NUTS AND HEAVY HEX JAM NUTS | | | | | | | | | | | | |
| 1/4 | 0.2500 | 1/2 | 0.500 | 0.488 | 0.577 | 0.556 | 15/64 | 0.250 | 0.218 | 11/64 | 0.188 | 0.156 |
| 5/16 | 0.3125 | 9/16 | 0.562 | 0.546 | 0.650 | 0.622 | 19/64 | 0.314 | 0.280 | 13/64 | 0.220 | 0.186 |
| 3/8 | 0.3750 | 11/16 | 0.688 | 0.669 | 0.794 | 0.763 | 23/64 | 0.377 | 0.341 | 15/64 | 0.252 | 0.216 |
| 7/16 | 0.4375 | 3/4 | 0.750 | 0.728 | 0.866 | 0.830 | 27/64 | 0.441 | 0.403 | 17/64 | 0.285 | 0.247 |
| 1/2 | 0.5000 | 7/8 | 0.875 | 0.850 | 1.010 | 0.969 | 31/64 | 0.504 | 0.464 | 19/64 | 0.317 | 0.277 |
| 9/16 | 0.5625 | 15/16 | 0.938 | 0.909 | 1.083 | 1.037 | 35/64 | 0.568 | 0.526 | 21/64 | 0.349 | 0.307 |
| 5/8 | 0.6250 | 1 1/16 | 1.062 | 1.031 | 1.227 | 1.175 | 39/64 | 0.631 | 0.587 | 23/64 | 0.381 | 0.337 |
| 3/4 | 0.7500 | 1 1/4 | 1.250 | 1.212 | 1.443 | 1.382 | 47/64 | 0.758 | 0.710 | 27/64 | 0.446 | 0.398 |
| 7/8 | 0.8750 | 1 7/16 | 1.438 | 1.394 | 1.660 | 1.589 | 55/64 | 0.885 | 0.833 | 31/64 | 0.510 | 0.458 |
| 1 | 1.0000 | 1 5/8 | 1.625 | 1.575 | 1.876 | 1.796 | 63/64 | 1.012 | 0.956 | 35/64 | 0.575 | 0.519 |
| 1 1/8 | 1.1250 | 1 13/16 | 1.812 | 1.756 | 2.093 | 2.002 | 1 7/64 | 1.139 | 1.079 | 39/64 | 0.639 | 0.579 |
| 1 1/4 | 1.2500 | 2 | 2.000 | 1.938 | 2.309 | 2.209 | 1 7/32 | 1.251 | 1.187 | 23/32 | 0.751 | 0.687 |
| 1 3/8 | 1.3750 | 2 3/16 | 2.188 | 2.119 | 2.526 | 2.416 | 1 11/32 | 1.378 | 1.310 | 25/32 | 0.815 | 0.747 |
| 1 1/2 | 1.5000 | 2 3/8 | 2.375 | 2.300 | 2.742 | 2.622 | 1 15/32 | 1.505 | 1.433 | 27/32 | 0.880 | 0.808 |
| 1 5/8 | 1.6250 | 2 9/16 | 2.562 | 2.481 | 2.959 | 2.828 | 1 19/32 | 1.632 | 1.556 | 29/32 | 0.944 | 0.868 |
| 1 3/4 | 1.7500 | 2 3/4 | 2.750 | 2.662 | 3.175 | 3.035 | 1 23/32 | 1.759 | 1.679 | 31/32 | 1.009 | 0.929 |
| 1 7/8 | 1.8750 | 2 15/16 | 2.938 | 2.844 | 3.392 | 3.242 | 1 27/32 | 1.886 | 1.802 | 1 1/32 | 1.073 | 0.989 |
| 2 | 2.0000 | 3 1/8 | 3.125 | 3.025 | 3.608 | 3.449 | 1 31/32 | 2.013 | 1.925 | 1 3/32 | 1.138 | 1.050 |
| 2 1/4 | 2.2500 | 3 1/2 | 3.500 | 3.388 | 4.041 | 3.862 | 2 13/64 | 2.251 | 2.155 | 1 13/64 | 1.251 | 1.155 |
| 2 1/2 | 2.5000 | 3 7/8 | 3.875 | 3.750 | 4.474 | 4.275 | 2 29/64 | 2.505 | 2.401 | 1 29/64 | 1.505 | 1.401 |
| 2 3/4 | 2.7500 | 4 1/4 | 4.250 | 4.112 | 4.907 | 4.688 | 2 45/64 | 2.759 | 2.647 | 1 37/64 | 1.634 | 1.522 |
| 3 | 3.0000 | 4 5/8 | 4.625 | 4.475 | 5.340 | 5.102 | 2 61/64 | 3.013 | 2.893 | 1 45/64 | 1.763 | 1.643 |
| 3 1/4 | 3.2500 | 5 | 5.000 | 4.838 | 5.774 | 5.515 | 3 3/16 | 3.252 | 3.124 | 1 13/16 | 1.876 | 1.748 |
| 3 1/2 | 3.5000 | 5 3/8 | 5.375 | 5.200 | 6.207 | 5.928 | 3 7/16 | 3.506 | 3.370 | 1 15/16 | 2.006 | 1.870 |
| 3 3/4 | 3.7500 | 5 3/4 | 5.750 | 5.562 | 6.640 | 6.341 | 3 11/16 | 3.760 | 3.616 | 2 1/16 | 2.134 | 1.990 |
| 4 | 4.0000 | 6 1/8 | 6.125 | 5.925 | 7.073 | 6.755 | 3 15/16 | 4.014 | 3.862 | 2 3/16 | 2.264 | 2.112 |

All dimensions are in inches.  See diagram on page 1132.
Bold dimension shows nuts unified dimensionally with British and Canadian Standards.
Threads are Unified Coarse-, Fine-, or 8-thread series (UNC, UNF or 8UN), Class 2B. Unification of fine-thread nuts is limited to sizes 1 inch and under.
* Where specifying nominal size in decimals, zeros before the decimal point and in the fourth decimal place are omitted.

Table 7. American National Standard and Unified Standard Hex Flat Nuts and Flat Jam Nuts and Heavy Hex Flat Nuts and Flat Jam Nuts (ANSI B18.2.2-1972)

| Nominal Size* or Basic Major Diam. of Thread | | Width Across Flats F | | | Width Across Corners G | | Thickness, Flat Nuts H | | | Thickness, Flat Jam Nuts H | | |
|---|---|---|---|---|---|---|---|---|---|---|---|---|
| | | Basic | Max. | Min. | Max. | Min. | Basic | Max. | Min. | Basic | Max. | Min. |
| HEX FLAT NUTS AND HEX FLAT JAM NUTS | | | | | | | | | | | | |
| 1⅛ | 1.1250 | 1¹¹/₁₆ | 1.688 | 1.631 | 1.949 | 1.859 | 1 | 1.030 | 0.970 | ⅝ | 0.655 | 0.595 |
| 1¼ | 1.2500 | 1⅞ | 1.875 | 1.812 | 2.165 | 2.066 | 1³/₃₂ | 1.126 | 1.062 | ¾ | 0.782 | 0.718 |
| 1⅜ | 1.3750 | 2¹/₁₆ | 2.062 | 1.994 | 2.382 | 2.273 | 1¹³/₆₄ | 1.237 | 1.169 | ¹³/₁₆ | 0.846 | 0.778 |
| 1½ | 1.5000 | 2¼ | 2.250 | 2.175 | 2.598 | 2.480 | 1⁵/₁₆ | 1.348 | 1.276 | ⅞ | 0.911 | 0.839 |
| HEAVY HEX FLAT NUTS AND HEAVY HEX FLAT JAM NUTS | | | | | | | | | | | | |
| 1⅛ | 1.1250 | 1¹³/₁₆ | 1.812 | 1.756 | 2.093 | 2.002 | 1⅛ | 1.155 | 1.079 | ⅝ | 0.655 | 0.579 |
| 1¼ | 1.2500 | 2 | 2.000 | 1.938 | 2.309 | 2.209 | 1¼ | 1.282 | 1.187 | ¾ | 0.782 | 0.687 |
| 1⅜ | 1.3750 | 2³/₁₆ | 2.188 | 2.119 | 2.526 | 2.416 | 1⅜ | 1.409 | 1.310 | ¹³/₁₆ | 0.846 | 0.747 |
| 1½ | 1.5000 | 2⅜ | 2.375 | 2.300 | 2.742 | 2.622 | 1½ | 1.536 | 1.433 | ⅞ | 0.911 | 0.808 |
| 1¾ | 1.7500 | 2¾ | 2.750 | 2.662 | 3.175 | 3.035 | 1¾ | 1.790 | 1.679 | 1 | 1.040 | 0.929 |
| 2 | 2.0000 | 3⅛ | 3.125 | 3.025 | 3.608 | 3.449 | 2 | 2.044 | 1.925 | 1⅛ | 1.169 | 1.050 |
| 2¼ | 2.2500 | 3½ | 3.500 | 3.388 | 4.041 | 3.862 | 2¼ | 2.298 | 2.155 | 1¼ | 1.298 | 1.155 |
| 2½ | 2.5000 | 3⅞ | 3.875 | 3.750 | 4.474 | 4.275 | 2½ | 2.552 | 2.401 | 1½ | 1.552 | 1.401 |
| 2¾ | 2.7500 | 4¼ | 4.250 | 4.112 | 4.907 | 4.688 | 2¾ | 2.806 | 2.647 | 1⅝ | 1.681 | 1.522 |
| 3 | 3.0000 | 4⅝ | 4.625 | 4.475 | 5.340 | 5.102 | 3 | 3.060 | 2.893 | 1¾ | 1.810 | 1.643 |
| 3¼ | 3.2500 | 5 | 5.000 | 4.838 | 5.774 | 5.515 | 3¼ | 3.314 | 3.124 | 1⅞ | 1.939 | 1.748 |
| 3½ | 3.5000 | 5⅜ | 5.375 | 5.200 | 6.207 | 5.928 | 3½ | 3.568 | 3.370 | 2 | 2.068 | 1.870 |
| 3¾ | 3.7500 | 5¾ | 5.750 | 5.562 | 6.640 | 6.341 | 3¾ | 3.822 | 3.616 | 2⅛ | 2.197 | 1.990 |
| 4 | 4.0000 | 6⅛ | 6.125 | 5.925 | 7.073 | 6.755 | 4 | 4.076 | 3.862 | 2¼ | 2.326 | 2.112 |

All dimensions are in inches. See diagram on page 1132.
Bold type are nuts unified dimensionally with British and Canadian Standards.
Threads are Unified Coarse-thread series (UNC), Class 2B.
* Where specifying nominal size in decimals, zeros before the decimal point and in the fourth decimal place are omitted.

Table 8. American National Standard Hex Castle Nuts (ANSI B18.2.2-1972)

| Nominal Size* or Basic Major Diam. of Thread | | Width Across Flats F | | Width Across Corners G | | Thickness H | | Unslotted Thickness T | | Width of Slot S | | Radius of Fillet R | Diam. of Cylindrical Part U |
|---|---|---|---|---|---|---|---|---|---|---|---|---|---|---|
| | | Max. | Min. | Max. | Min. | Max. | Min. | Max. | Min. | Max. | Min. | ±.010 | Min. |
| ¼ | 0.2500 | 0.438 | 0.428 | 0.505 | 0.488 | 0.288 | 0.274 | 0.20 | 0.18 | 0.10 | 0.07 | 0.094 | 0.371 |
| ⁵/₁₆ | 0.3125 | 0.500 | 0.489 | 0.577 | 0.557 | 0.336 | 0.320 | 0.24 | 0.22 | 0.12 | 0.09 | 0.094 | 0.425 |
| ⅜ | 0.3750 | 0.562 | 0.551 | 0.650 | 0.628 | 0.415 | 0.398 | 0.29 | 0.27 | 0.15 | 0.12 | 0.094 | 0.478 |
| ⁷/₁₆ | 0.4375 | 0.688 | 0.675 | 0.794 | 0.768 | 0.463 | 0.444 | 0.31 | 0.29 | 0.15 | 0.12 | 0.094 | 0.582 |
| ½ | 0.5000 | 0.750 | 0.736 | 0.866 | 0.840 | 0.573 | 0.552 | 0.42 | 0.40 | 0.18 | 0.15 | 0.125 | 0.637 |
| ⁹/₁₆ | 0.5625 | 0.875 | 0.861 | 1.010 | 0.982 | 0.621 | 0.598 | 0.43 | 0.41 | 0.18 | 0.15 | 0.156 | 0.744 |
| ⅝ | 0.6250 | 0.938 | 0.922 | 1.083 | 1.051 | 0.731 | 0.706 | 0.51 | 0.49 | 0.24 | 0.18 | 0.156 | 0.797 |
| ¾ | 0.7500 | 1.125 | 1.088 | 1.299 | 1.240 | 0.827 | 0.798 | 0.57 | 0.55 | 0.24 | 0.18 | 0.188 | 0.941 |
| ⅞ | 0.8750 | 1.312 | 1.269 | 1.516 | 1.447 | 0.922 | 0.890 | 0.67 | 0.64 | 0.24 | 0.18 | 0.188 | 1.097 |
| 1 | 1.0000 | 1.500 | 1.450 | 1.732 | 1.653 | 1.018 | 0.982 | 0.73 | 0.70 | 0.30 | 0.24 | 0.188 | 1.254 |
| 1⅛ | 1.1250 | 1.688 | 1.631 | 1.949 | 1.859 | 1.176 | 1.136 | 0.83 | 0.80 | 0.33 | 0.24 | 0.250 | 1.411 |
| 1¼ | 1.2500 | 1.875 | 1.812 | 2.165 | 2.066 | 1.272 | 1.228 | 0.89 | 0.86 | 0.40 | 0.31 | 0.250 | 1.570 |
| 1⅜ | 1.3750 | 2.062 | 1.994 | 2.382 | 2.273 | 1.399 | 1.351 | 1.02 | 0.98 | 0.40 | 0.31 | 0.250 | 1.726 |
| 1½ | 1.5000 | 2.250 | 2.175 | 2.598 | 2.480 | 1.548 | 1.474 | 1.08 | 1.04 | 0.46 | 0.37 | 0.250 | 1.881 |

All dimensions are in inches. See diagram on page 1132.
Threads are Unified Coarse- or Fine-thread series (UNC or UNF), Class 2B.
* Where specifying nominal size in decimals, zeros before the decimal point and in the fourth decimal place are omitted.

Table 9. American National and Unified Standard Hex Slotted Nuts,
Heavy Hex Slotted Nuts and Hex Thick Slotted Nuts (ANSI B18.2.2-1972)

| Nominal Size* or Basic Major Diam. of Thread | | Width Across Flats F | | | Width Across Corners G | | Thickness H | | | Unslotted Thickness T | | Width of Slot S | |
|---|---|---|---|---|---|---|---|---|---|---|---|---|---|
| | | Basic | Max. | Min. | Max. | Min. | Basic | Max. | Min. | Max. | Min. | Max. | Min. |
| **HEX SLOTTED NUTS** | | | | | | | | | | | | | |
| 1/4 | 0.2500 | 7/16 | 0.438 | 0.428 | 0.505 | 0.488 | 7/32 | 0.226 | 0.212 | 0.14 | 0.12 | 0.10 | 0.07 |
| 5/16 | 0.3125 | 1/2 | 0.500 | 0.489 | 0.577 | 0.557 | 17/64 | 0.273 | 0.258 | 0.18 | 0.16 | 0.12 | 0.09 |
| 3/8 | 0.3750 | 9/16 | 0.562 | 0.551 | 0.650 | 0.628 | 21/64 | 0.337 | 0.320 | 0.21 | 0.19 | 0.15 | 0.12 |
| 7/16 | 0.4375 | 11/16 | 0.688 | 0.675 | 0.794 | 0.768 | 3/8 | 0.385 | 0.365 | 0.23 | 0.21 | 0.15 | 0.12 |
| 1/2 | 0.5000 | 3/4 | 0.750 | 0.736 | 0.866 | 0.840 | 7/16 | 0.448 | 0.427 | 0.29 | 0.27 | 0.18 | 0.15 |
| 9/16 | 0.5625 | 7/8 | 0.875 | 0.861 | 1.010 | 0.982 | 31/64 | 0.496 | 0.473 | 0.31 | 0.29 | 0.18 | 0.15 |
| 5/8 | 0.6250 | 15/16 | 0.938 | 0.922 | 1.083 | 1.051 | 35/64 | 0.559 | 0.535 | 0.34 | 0.32 | 0.24 | 0.18 |
| 3/4 | 0.7500 | 1 1/8 | 1.125 | 1.088 | 1.299 | 1.240 | 41/64 | 0.665 | 0.617 | 0.40 | 0.38 | 0.24 | 0.18 |
| 7/8 | 0.8750 | 1 5/16 | 1.312 | 1.269 | 1.516 | 1.447 | 3/4 | 0.776 | 0.724 | 0.52 | 0.49 | 0.24 | 0.18 |
| 1 | 1.0000 | 1 1/2 | 1.500 | 1.450 | 1.732 | 1.653 | 55/64 | 0.887 | 0.831 | 0.59 | 0.56 | 0.30 | 0.24 |
| 1 1/8 | 1.1250 | 1 11/16 | 1.688 | 1.631 | 1.949 | 1.859 | 31/32 | 0.999 | 0.939 | 0.64 | 0.61 | 0.33 | 0.24 |
| 1 1/4 | 1.2500 | 1 7/8 | 1.875 | 1.812 | 2.165 | 2.066 | 1 1/16 | 1.094 | 1.030 | 0.70 | 0.67 | 0.40 | 0.31 |
| 1 3/8 | 1.3750 | 2 1/16 | 2.062 | 1.994 | 2.382 | 2.273 | 1 11/64 | 1.206 | 1.138 | 0.82 | 0.78 | 0.40 | 0.31 |
| 1 1/2 | 1.5000 | 2 1/4 | 2.250 | 2.175 | 2.598 | 2.480 | 1 9/32 | 1.317 | 1.245 | 0.86 | 0.82 | 0.46 | 0.37 |
| **HEAVY HEX SLOTTED NUTS** | | | | | | | | | | | | | |
| 1/4 | 0.2500 | 1/2 | 0.500 | 0.488 | 0.577 | 0.556 | 15/64 | 0.250 | 0.218 | 0.15 | 0.13 | 0.10 | 0.07 |
| 5/16 | 0.3125 | 9/16 | 0.562 | 0.550 | 0.650 | 0.622 | 19/64 | 0.314 | 0.280 | 0.21 | 0.19 | 0.12 | 0.09 |
| 3/8 | 0.3750 | 11/16 | 0.688 | 0.669 | 0.794 | 0.763 | 23/64 | 0.377 | 0.341 | 0.24 | 0.22 | 0.15 | 0.12 |
| 7/16 | 0.4375 | 3/4 | 0.750 | 0.728 | 0.866 | 0.830 | 27/64 | 0.441 | 0.403 | 0.28 | 0.26 | 0.15 | 0.12 |
| 1/2 | 0.5000 | 7/8 | 0.875 | 0.850 | 1.010 | 0.969 | 31/64 | 0.504 | 0.464 | 0.34 | 0.32 | 0.18 | 0.15 |
| 9/16 | 0.5625 | 15/16 | 0.938 | 0.909 | 1.083 | 1.037 | 35/64 | 0.568 | 0.526 | 0.37 | 0.35 | 0.18 | 0.15 |
| 5/8 | 0.6250 | 1 1/16 | 1.062 | 1.031 | 1.227 | 1.175 | 39/64 | 0.631 | 0.587 | 0.40 | 0.38 | 0.24 | 0.18 |
| 3/4 | 0.7500 | 1 1/4 | 1.250 | 1.212 | 1.443 | 1.382 | 47/64 | 0.758 | 0.710 | 0.49 | 0.47 | 0.24 | 0.18 |
| 7/8 | 0.8750 | 1 7/16 | 1.438 | 1.394 | 1.660 | 1.589 | 55/64 | 0.885 | 0.833 | 0.62 | 0.59 | 0.24 | 0.18 |
| 1 | 1.0000 | 1 5/8 | 1.625 | 1.575 | 1.876 | 1.796 | 63/64 | 1.012 | 0.956 | 0.72 | 0.69 | 0.30 | 0.24 |
| 1 1/8 | 1.1250 | 1 13/16 | 1.812 | 1.756 | 2.093 | 2.002 | 1 7/64 | 1.139 | 1.079 | 0.78 | 0.75 | 0.33 | 0.24 |
| 1 1/4 | 1.2500 | 2 | 2.000 | 1.938 | 2.309 | 2.209 | 1 7/32 | 1.251 | 1.187 | 0.86 | 0.83 | 0.40 | 0.31 |
| 1 3/8 | 1.3750 | 2 3/16 | 2.188 | 2.119 | 2.526 | 2.416 | 1 11/32 | 1.378 | 1.310 | 0.99 | 0.95 | 0.40 | 0.31 |
| 1 1/2 | 1.5000 | 2 3/8 | 2.375 | 2.300 | 2.742 | 2.622 | 1 15/32 | 1.505 | 1.433 | 1.05 | 1.01 | 0.46 | 0.43 |
| 1 3/4 | 1.7500 | 2 3/4 | 2.750 | 2.662 | 3.175 | 3.035 | 1 23/32 | 1.759 | 1.679 | 1.24 | 1.20 | 0.52 | 0.43 |
| 2 | 2.0000 | 3 1/8 | 3.125 | 3.025 | 3.608 | 3.449 | 1 31/32 | 2.013 | 1.925 | 1.43 | 1.38 | 0.52 | 0.43 |
| 2 1/4 | 2.2500 | 3 1/2 | 3.500 | 3.388 | 4.041 | 3.862 | 2 13/64 | 2.251 | 2.155 | 1.67 | 1.62 | 0.52 | 0.43 |
| 2 1/2 | 2.5000 | 3 7/8 | 3.875 | 3.750 | 4.474 | 4.275 | 2 29/64 | 2.505 | 2.401 | 1.79 | 1.74 | 0.64 | 0.55 |
| 2 3/4 | 2.7500 | 4 1/4 | 4.250 | 4.112 | 4.907 | 4.688 | 2 45/64 | 2.759 | 2.647 | 2.05 | 1.99 | 0.64 | 0.55 |
| 3 | 3.0000 | 4 5/8 | 4.625 | 4.475 | 5.340 | 5.102 | 2 61/64 | 3.013 | 2.893 | 2.23 | 2.17 | 0.71 | 0.62 |
| 3 1/4 | 3.2500 | 5 | 5.000 | 4.838 | 5.774 | 5.515 | 3 3/16 | 3.252 | 3.124 | 2.47 | 2.41 | 0.71 | 0.62 |
| 3 1/2 | 3.5000 | 5 3/8 | 5.375 | 5.200 | 6.207 | 5.928 | 3 7/16 | 3.506 | 3.370 | 2.72 | 2.65 | 0.71 | 0.62 |
| 3 3/4 | 3.7500 | 5 3/4 | 5.750 | 5.562 | 6.640 | 6.341 | 3 11/16 | 3.760 | 3.616 | 2.97 | 2.90 | 0.71 | 0.62 |
| 4 | 4.0000 | 6 1/8 | 6.125 | 5.925 | 7.073 | 6.755 | 3 15/16 | 4.014 | 3.862 | 3.22 | 3.15 | 0.71 | 0.62 |
| **HEX THICK SLOTTED NUTS** | | | | | | | | | | | | | |
| 1/4 | 0.2500 | 7/16 | 0.438 | 0.428 | 0.505 | 0.488 | 9/32 | 0.288 | 0.274 | 0.20 | 0.18 | 0.10 | 0.07 |
| 5/16 | 0.3125 | 1/2 | 0.500 | 0.489 | 0.577 | 0.557 | 21/64 | 0.336 | 0.320 | 0.24 | 0.22 | 0.12 | 0.09 |
| 3/8 | 0.3750 | 9/16 | 0.562 | 0.551 | 0.650 | 0.628 | 13/32 | 0.415 | 0.398 | 0.29 | 0.27 | 0.15 | 0.12 |
| 7/16 | 0.4375 | 11/16 | 0.688 | 0.675 | 0.794 | 0.768 | 29/64 | 0.463 | 0.444 | 0.31 | 0.29 | 0.15 | 0.12 |
| 1/2 | 0.5000 | 3/4 | 0.750 | 0.736 | 0.866 | 0.840 | 9/16 | 0.573 | 0.552 | 0.42 | 0.40 | 0.18 | 0.15 |
| 9/16 | 0.5625 | 7/8 | 0.875 | 0.861 | 1.010 | 0.982 | 39/64 | 0.621 | 0.598 | 0.43 | 0.41 | 0.18 | 0.15 |
| 5/8 | 0.6250 | 15/16 | 0.938 | 0.922 | 1.083 | 1.051 | 23/32 | 0.731 | 0.706 | 0.51 | 0.49 | 0.24 | 0.18 |
| 3/4 | 0.7500 | 1 1/8 | 1.125 | 1.088 | 1.299 | 1.240 | 13/16 | 0.827 | 0.798 | 0.57 | 0.55 | 0.24 | 0.18 |
| 7/8 | 0.8750 | 1 5/16 | 1.312 | 1.269 | 1.516 | 1.447 | 29/32 | 0.922 | 0.890 | 0.67 | 0.64 | 0.24 | 0.18 |
| 1 | 1.0000 | 1 1/2 | 1.500 | 1.450 | 1.732 | 1.653 | 1 | 1.018 | 0.982 | 0.73 | 0.70 | 0.30 | 0.24 |
| 1 1/8 | 1.1250 | 1 11/16 | 1.688 | 1.631 | 1.949 | 1.859 | 1 5/32 | 1.176 | 1.136 | 0.83 | 0.80 | 0.33 | 0.24 |
| 1 1/4 | 1.2500 | 1 7/8 | 1.875 | 1.812 | 2.165 | 2.066 | 1 1/4 | 1.272 | 1.228 | 0.89 | 0.86 | 0.40 | 0.31 |
| 1 3/8 | 1.3750 | 2 1/16 | 2.062 | 1.994 | 2.382 | 2.273 | 1 3/8 | 1.399 | 1.351 | 1.02 | 0.98 | 0.40 | 0.31 |
| 1 1/2 | 1.5000 | 2 1/4 | 2.250 | 2.175 | 2.598 | 2.480 | 1 1/2 | 1.526 | 1.474 | 1.08 | 1.04 | 0.46 | 0.37 |

All dimensions are in inches. See diagram on page 1132.
Bold type indicates nuts unified dimensionally with British and Canadian Standards.
* See Table 10 footnote.
Threads are Unified Coarse-, Fine-, or 8-thread series (UNC, UNF, or 8UN), Class 2B.
Unification of fine-thread nuts is limited to sizes 1 inch and under.

**Table 10. American National and Unified Standard Square Nuts and Heavy Square Nuts and American National Standard Hex Thick Nuts** (ANSI B18.2.2-1972)

| Nominal Size* or Basic Major Diam. of Thread | | Width Across Flats F | | | Width Across Corners G | | Thickness H | | |
|---|---|---|---|---|---|---|---|---|---|
| | | Basic | Max. | Min. | Max. | Min. | Basic | Max. | Min. |
| SQUARE NUTS† | | | | | | | | | |
| **1/4** | **0.2500** | **7/16** | **0.438** | **0.425** | **0.619** | **0.584** | **7/32** | **0.235** | **0.203** |
| **5/16** | **0.3125** | **9/16** | **0.562** | **0.547** | **0.795** | **0.751** | **17/64** | **0.283** | **0.249** |
| **3/8** | **0.3750** | **5/8** | **0.625** | **0.606** | **0.884** | **0.832** | **21/64** | **0.346** | **0.310** |
| **7/16** | **0.4375** | **3/4** | **0.750** | **0.728** | **1.061** | **1.000** | **3/8** | **0.394** | **0.356** |
| **1/2** | **0.5000** | **13/16** | **0.812** | **0.788** | **1.149** | **1.082** | **7/16** | **0.458** | **0.418** |
| **5/8** | **0.6250** | **1** | **1.000** | **0.969** | **1.414** | **1.330** | **35/64** | **0.569** | **0.525** |
| **3/4** | **0.7500** | **1 1/8** | **1.125** | **1.088** | **1.591** | **1.494** | **21/32** | **0.680** | **0.632** |
| **7/8** | **0.8750** | **1 5/16** | **1.312** | **1.269** | **1.856** | **1.742** | **49/64** | **0.792** | **0.740** |
| **1** | **1.0000** | **1 1/2** | **1.500** | **1.450** | **2.121** | **1.991** | **7/8** | **0.903** | **0.847** |
| **1 1/8** | **1.1250** | **1 11/16** | **1.688** | **1.631** | **2.386** | **2.239** | **1** | **1.030** | **0.970** |
| **1 1/4** | **1.2500** | **1 7/8** | **1.875** | **1.812** | **2.652** | **2.489** | **1 3/32** | **1.126** | **1.062** |
| **1 3/8** | **1.3750** | **2 1/16** | **2.062** | **1.994** | **2.917** | **2.738** | **1 13/64** | **1.237** | **1.169** |
| **1 1/2** | **1.5000** | **2 1/4** | **2.250** | **2.175** | **3.182** | **2.986** | **1 5/16** | **1.348** | **1.276** |
| HEAVY SQUARE NUTS† | | | | | | | | | |
| 1/4 | 0.2500 | 1/2 | 0.500 | 0.488 | 0.707 | 0.670 | 1/4 | 0.266 | 0.218 |
| 5/16 | 0.3125 | 9/16 | 0.562 | 0.546 | 0.795 | 0.750 | 5/16 | 0.330 | 0.280 |
| 3/8 | 0.3750 | 11/16 | 0.688 | 0.669 | 0.973 | 0.919 | 3/8 | 0.393 | 0.341 |
| 7/16 | 0.4375 | 3/4 | 0.750 | 0.728 | 1.060 | 1.000 | 7/16 | 0.456 | 0.403 |
| 1/2 | 0.5000 | 7/8 | 0.875 | 0.850 | 1.237 | 1.167 | 1/2 | 0.520 | 0.464 |
| 5/8 | 0.6250 | 1 1/16 | 1.062 | 1.031 | 1.503 | 1.416 | 5/8 | 0.647 | 0.587 |
| 3/4 | 0.7500 | 1 1/4 | 1.250 | 1.212 | 1.768 | 1.665 | 3/4 | 0.774 | 0.710 |
| 7/8 | 0.8750 | 1 7/16 | 1.438 | 1.394 | 2.033 | 1.914 | 7/8 | 0.901 | 0.833 |
| 1 | 1.0000 | 1 5/8 | 1.625 | 1.575 | 2.298 | 2.162 | 1 | 1.028 | 0.956 |
| 1 1/8 | 1.1250 | 1 13/16 | 1.812 | 1.756 | 2.563 | 2.411 | 1 1/8 | 1.155 | 1.079 |
| 1 1/4 | 1.2500 | 2 | 2.000 | 1.938 | 2.828 | 2.661 | 1 1/4 | 1.282 | 1.187 |
| 1 3/8 | 1.3750 | 2 3/16 | 2.188 | 2.119 | 3.094 | 2.909 | 1 3/8 | 1.409 | 1.310 |
| 1 1/2 | 1.5000 | 2 3/8 | 2.375 | 2.300 | 3.359 | 3.158 | 1 1/2 | 1.536 | 1.433 |
| HEX THICK NUTS‡ | | | | | | | | | |
| 1/4 | 0.2500 | 7/16 | 0.438 | 0.428 | 0.505 | 0.488 | 9/32 | 0.288 | 0.274 |
| 5/16 | 0.3125 | 1/2 | 0.500 | 0.489 | 0.577 | 0.557 | 21/64 | 0.336 | 0.320 |
| 3/8 | 0.3750 | 9/16 | 0.562 | 0.551 | 0.650 | 0.628 | 13/32 | 0.415 | 0.398 |
| 7/16 | 0.4375 | 11/16 | 0.688 | 0.675 | 0.794 | 0.768 | 29/64 | 0.463 | 0.444 |
| 1/2 | 0.5000 | 3/4 | 0.750 | 0.736 | 0.866 | 0.840 | 9/16 | 0.573 | 0.552 |
| 9/16 | 0.5625 | 7/8 | 0.875 | 0.861 | 1.010 | 0.982 | 39/64 | 0.621 | 0.598 |
| 5/8 | 0.6250 | 15/16 | 0.938 | 0.922 | 1.083 | 1.051 | 23/32 | 0.731 | 0.706 |
| 3/4 | 0.7500 | 1 1/8 | 1.125 | 1.088 | 1.299 | 1.240 | 13/16 | 0.827 | 0.798 |
| 7/8 | 0.8750 | 1 5/16 | 1.312 | 1.269 | 1.516 | 1.447 | 29/32 | 0.922 | 0.890 |
| 1 | 1.0000 | 1 1/2 | 1.500 | 1.450 | 1.732 | 1.653 | 1 | 1.018 | 0.982 |
| 1 1/8 | 1.1250 | 1 11/16 | 1.688 | 1.631 | 1.949 | 1.859 | 1 5/32 | 1.176 | 1.136 |
| 1 1/4 | 1.2500 | 1 7/8 | 1.875 | 1.812 | 2.165 | 2.066 | 1 1/4 | 1.272 | 1.228 |
| 1 3/8 | 1.3750 | 2 1/16 | 2.062 | 1.994 | 2.382 | 2.273 | 1 3/8 | 1.399 | 1.351 |
| 1 1/2 | 1.5000 | 2 1/4 | 2.250 | 2.175 | 2.598 | 2.480 | 1 1/2 | 1.526 | 1.474 |

All dimensions are in inches. See diagram on page 1132.

**Bold type indicates nuts unified dimensionally with British and Canadian Standards.**

* Where specifying nominal size in decimals, zeros before the decimal point and in the fourth decimal place are omitted.

† Threads are Unified Coarse-thread series (UNC), Class 2B.

‡ Threads are Unified Coarse-, Fine-, or 8-thread series (UNC, UNF, or 8 UN), Class 2B.

## Low and High Crown (Blind, Acorn) Nuts (SAE Recommended Practice J483a)

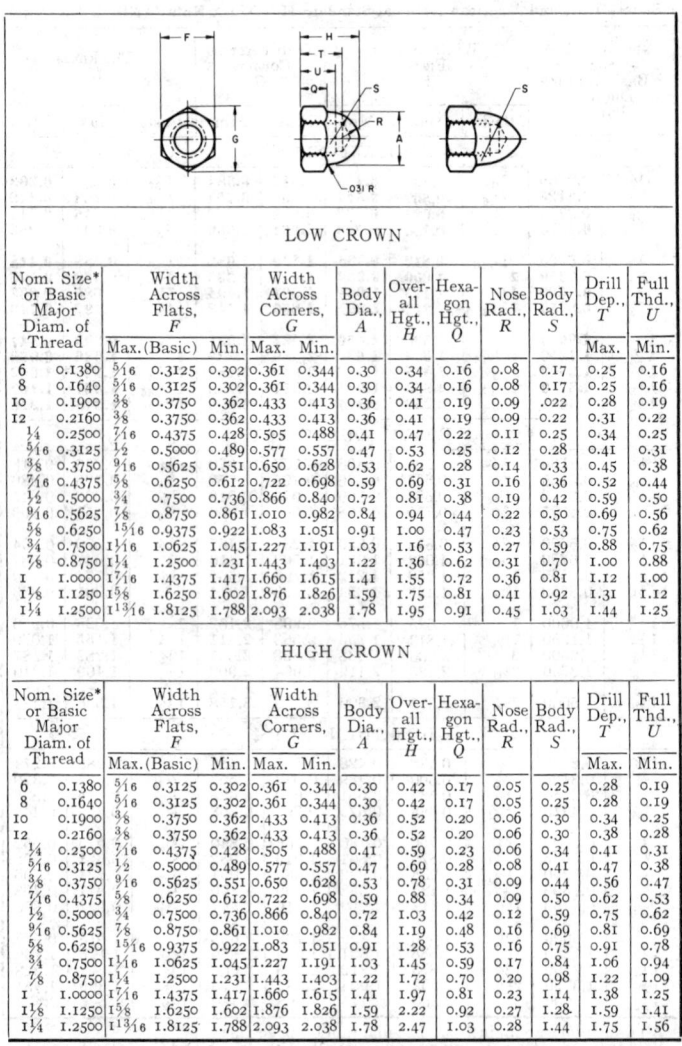

### LOW CROWN

| Nom. Size* or Basic Major Diam. of Thread | | Width Across Flats, F | | Width Across Corners, G | | Body Dia., A | Over-all Hgt., H | Hexagon Hgt., Q | Nose Rad., R | Body Rad., S | Drill Dep., T | Full Thd., U |
|---|---|---|---|---|---|---|---|---|---|---|---|---|
| | | Max.(Basic) | Min. | Max. | Min. | | | | | | Max. | Min. |
| 6 | 0.1380 | 5/16 0.3125 | 0.302 | 0.361 | 0.344 | 0.30 | 0.34 | 0.16 | 0.08 | 0.17 | 0.25 | 0.16 |
| 8 | 0.1640 | 5/16 0.3125 | 0.302 | 0.361 | 0.344 | 0.30 | 0.34 | 0.16 | 0.08 | 0.17 | 0.25 | 0.16 |
| 10 | 0.1900 | 3/8 0.3750 | 0.362 | 0.433 | 0.413 | 0.36 | 0.41 | 0.19 | 0.09 | .022 | 0.28 | 0.19 |
| 12 | 0.2160 | 3/8 0.3750 | 0.362 | 0.433 | 0.413 | 0.36 | 0.41 | 0.19 | 0.09 | 0.22 | 0.31 | 0.22 |
| 1/4 | 0.2500 | 7/16 0.4375 | 0.428 | 0.505 | 0.488 | 0.41 | 0.47 | 0.22 | 0.11 | 0.25 | 0.34 | 0.25 |
| 5/16 | 0.3125 | 1/2 0.5000 | 0.489 | 0.577 | 0.557 | 0.47 | 0.53 | 0.25 | 0.12 | 0.28 | 0.41 | 0.31 |
| 3/8 | 0.3750 | 9/16 0.5625 | 0.551 | 0.650 | 0.628 | 0.53 | 0.62 | 0.28 | 0.14 | 0.33 | 0.45 | 0.38 |
| 7/16 | 0.4375 | 5/8 0.6250 | 0.612 | 0.722 | 0.698 | 0.59 | 0.69 | 0.31 | 0.16 | 0.36 | 0.52 | 0.44 |
| 1/2 | 0.5000 | 3/4 0.7500 | 0.736 | 0.866 | 0.840 | 0.72 | 0.81 | 0.38 | 0.19 | 0.42 | 0.59 | 0.50 |
| 9/16 | 0.5625 | 7/8 0.8750 | 0.861 | 1.010 | 0.982 | 0.84 | 0.94 | 0.44 | 0.22 | 0.50 | 0.69 | 0.56 |
| 5/8 | 0.6250 | 15/16 0.9375 | 0.922 | 1.083 | 1.051 | 0.91 | 1.00 | 0.47 | 0.23 | 0.53 | 0.75 | 0.62 |
| 3/4 | 0.7500 | 1 1/16 1.0625 | 1.045 | 1.227 | 1.191 | 1.03 | 1.16 | 0.53 | 0.27 | 0.59 | 0.88 | 0.75 |
| 7/8 | 0.8750 | 1 1/4 1.2500 | 1.231 | 1.443 | 1.403 | 1.22 | 1.36 | 0.62 | 0.31 | 0.70 | 1.00 | 0.88 |
| 1 | 1.0000 | 1 7/16 1.4375 | 1.417 | 1.660 | 1.615 | 1.41 | 1.55 | 0.72 | 0.36 | 0.81 | 1.12 | 1.00 |
| 1 1/8 | 1.1250 | 1 5/8 1.6250 | 1.602 | 1.876 | 1.826 | 1.59 | 1.75 | 0.81 | 0.41 | 0.92 | 1.31 | 1.12 |
| 1 1/4 | 1.2500 | 1 13/16 1.8125 | 1.788 | 2.093 | 2.038 | 1.78 | 1.95 | 0.91 | 0.45 | 1.03 | 1.44 | 1.25 |

### HIGH CROWN

| Nom. Size* or Basic Major Diam. of Thread | | Width Across Flats, F | | Width Across Corners, G | | Body Dia., A | Over-all Hgt., H | Hexagon Hgt., Q | Nose Rad., R | Body Rad., S | Drill Dep., T | Full Thd., U |
|---|---|---|---|---|---|---|---|---|---|---|---|---|
| | | Max.(Basic) | Min. | Max. | Min. | | | | | | Max. | Min. |
| 6 | 0.1380 | 5/16 0.3125 | 0.302 | 0.361 | 0.344 | 0.30 | 0.42 | 0.17 | 0.05 | 0.25 | 0.28 | 0.19 |
| 8 | 0.1640 | 5/16 0.3125 | 0.302 | 0.361 | 0.344 | 0.30 | 0.42 | 0.17 | 0.05 | 0.25 | 0.28 | 0.19 |
| 10 | 0.1900 | 3/8 0.3750 | 0.362 | 0.433 | 0.413 | 0.36 | 0.52 | 0.20 | 0.06 | 0.30 | 0.34 | 0.25 |
| 12 | 0.2160 | 3/8 0.3750 | 0.362 | 0.433 | 0.413 | 0.36 | 0.52 | 0.20 | 0.06 | 0.30 | 0.38 | 0.28 |
| 1/4 | 0.2500 | 7/16 0.4375 | 0.428 | 0.505 | 0.488 | 0.41 | 0.59 | 0.23 | 0.06 | 0.34 | 0.41 | 0.31 |
| 5/16 | 0.3125 | 1/2 0.5000 | 0.489 | 0.577 | 0.557 | 0.47 | 0.69 | 0.28 | 0.08 | 0.41 | 0.47 | 0.38 |
| 3/8 | 0.3750 | 9/16 0.5625 | 0.551 | 0.650 | 0.628 | 0.53 | 0.78 | 0.31 | 0.09 | 0.44 | 0.56 | 0.47 |
| 7/16 | 0.4375 | 5/8 0.6250 | 0.612 | 0.722 | 0.698 | 0.59 | 0.88 | 0.34 | 0.09 | 0.50 | 0.62 | 0.53 |
| 1/2 | 0.5000 | 3/4 0.7500 | 0.736 | 0.866 | 0.840 | 0.72 | 1.03 | 0.42 | 0.12 | 0.59 | 0.75 | 0.62 |
| 9/16 | 0.5625 | 7/8 0.8750 | 0.861 | 1.010 | 0.982 | 0.84 | 1.19 | 0.48 | 0.16 | 0.69 | 0.81 | 0.69 |
| 5/8 | 0.6250 | 15/16 0.9375 | 0.922 | 1.083 | 1.051 | 0.91 | 1.28 | 0.53 | 0.16 | 0.75 | 0.91 | 0.78 |
| 3/4 | 0.7500 | 1 1/16 1.0625 | 1.045 | 1.227 | 1.191 | 1.03 | 1.45 | 0.59 | 0.17 | 0.84 | 1.06 | 0.94 |
| 7/8 | 0.8750 | 1 1/4 1.2500 | 1.231 | 1.443 | 1.403 | 1.22 | 1.72 | 0.70 | 0.20 | 0.98 | 1.22 | 1.09 |
| 1 | 1.0000 | 1 7/16 1.4375 | 1.417 | 1.660 | 1.615 | 1.41 | 1.97 | 0.81 | 0.23 | 1.14 | 1.38 | 1.25 |
| 1 1/8 | 1.1250 | 1 5/8 1.6250 | 1.602 | 1.876 | 1.826 | 1.59 | 2.22 | 0.92 | 0.27 | 1.28 | 1.59 | 1.41 |
| 1 1/4 | 1.2500 | 1 13/16 1.8125 | 1.788 | 2.093 | 2.038 | 1.78 | 2.47 | 1.03 | 0.28 | 1.44 | 1.75 | 1.56 |

All dimensions are in inches. Threads are Unified Standard Class 2B, UNC or UNF Series. * When specifying a nominal size in decimals, any zero in the fourth decimal place is omitted. *Reprinted with permission. Copyright © 1974, Society of Automotive Engineers, Inc. All rights reserved.*

**Hex High and Hex Slotted High Nuts** (SAE Standard J482a)

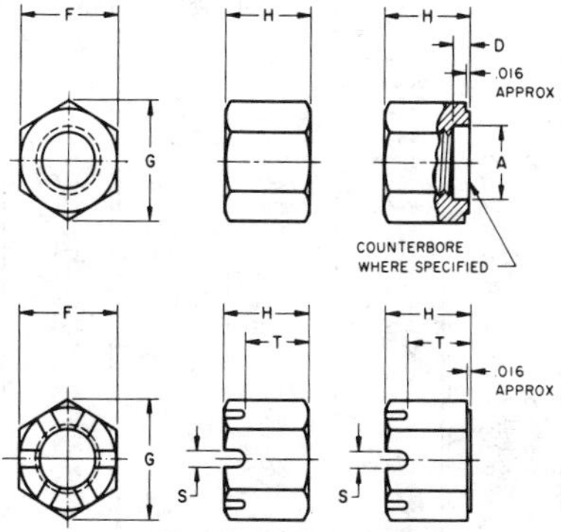


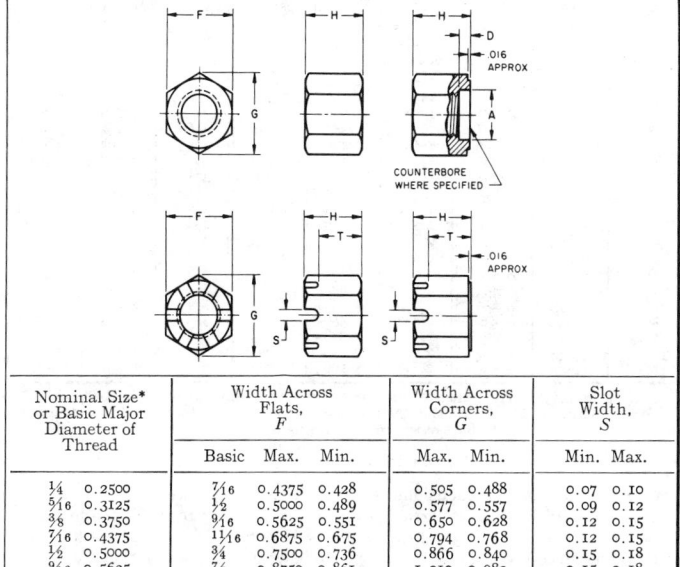

| Nominal Size* or Basic Major Diameter of Thread | | Width Across Flats, *F* | | | Width Across Corners, *G* | | Slot Width, *S* | |
|---|---|---|---|---|---|---|---|---|
| | | Basic | Max. | Min. | Max. | Min. | Min. | Max. |
| 1/4 | 0.2500 | 7/16 | 0.4375 | 0.428 | 0.505 | 0.488 | 0.07 | 0.10 |
| 5/16 | 0.3125 | 1/2 | 0.5000 | 0.489 | 0.577 | 0.557 | 0.09 | 0.12 |
| 3/8 | 0.3750 | 9/16 | 0.5625 | 0.551 | 0.650 | 0.628 | 0.12 | 0.15 |
| 7/16 | 0.4375 | 11/16 | 0.6875 | 0.675 | 0.794 | 0.768 | 0.12 | 0.15 |
| 1/2 | 0.5000 | 3/4 | 0.7500 | 0.736 | 0.866 | 0.840 | 0.15 | 0.18 |
| 9/16 | 0.5625 | 7/8 | 0.8750 | 0.861 | 1.010 | 0.982 | 0.15 | 0.18 |
| 5/8 | 0.6250 | 15/16 | 0.9375 | 0.922 | 1.083 | 1.051 | 0.18 | 0.24 |
| 3/4 | 0.7500 | 1 1/8 | 1.1250 | 1.088 | 1.299 | 1.240 | 0.18 | 0.24 |
| 7/8 | 0.8750 | 1 5/16 | 1.3125 | 1.269 | 1.516 | 1.447 | 0.18 | 0.24 |
| 1 | 1.0000 | 1 1/2 | 1.5000 | 1.450 | 1.732 | 1.653 | 0.24 | 0.30 |
| 1 1/8 | 1.1250 | 1 11/16 | 1.6875 | 1.631 | 1.949 | 1.859 | 0.24 | 0.33 |
| 1 1/4 | 1.2500 | 1 7/8 | 1.8750 | 1.812 | 2.165 | 2.066 | 0.31 | 0.40 |

| Nominal Size* or Basic Major Diameter of Thread | | Thickness, *H* | | | Unslotted Thickness, *T* | | Counterbore (Optional) | |
|---|---|---|---|---|---|---|---|---|
| | | Basic | Max. | Min. | Max. | Min. | Diam., *A* | Depth, *D* |
| 1/4 | 0.2500 | 3/8 | 0.382 | 0.368 | 0.29 | 0.27 | 0.266 | 0.062 |
| 5/16 | 0.3125 | 29/64 | 0.461 | 0.445 | 0.37 | 0.35 | 0.328 | 0.078 |
| 3/8 | 0.3750 | 1/2 | 0.509 | 0.491 | 0.38 | 0.36 | 0.391 | 0.094 |
| 7/16 | 0.4375 | 39/64 | 0.619 | 0.599 | 0.46 | 0.44 | 0.453 | 0.109 |
| 1/2 | 0.5000 | 21/32 | 0.667 | 0.645 | 0.51 | 0.49 | 0.516 | 0.125 |
| 9/16 | 0.5625 | 49/64 | 0.778 | 0.754 | 0.59 | 0.57 | 0.591 | 0.141 |
| 5/8 | 0.6250 | 27/32 | 0.857 | 0.831 | 0.63 | 0.61 | 0.656 | 0.156 |
| 3/4 | 0.7500 | 1 | 1.015 | 0.985 | 0.76 | 0.73 | 0.781 | 0.188 |
| 7/8 | 0.8750 | 1 5/32 | 1.172 | 1.140 | 0.92 | 0.89 | 0.906 | 0.219 |
| 1 | 1.0000 | 1 5/16 | 1.330 | 1.292 | 1.05 | 1.01 | 1.031 | 0.250 |
| 1 1/8 | 1.1250 | 1 1/2 | 1.520 | 1.480 | 1.18 | 1.14 | 1.156 | 0.281 |
| 1 1/4 | 1.2500 | 1 11/16 | 1.710 | 1.666 | 1.34 | 1.29 | 1.281 | 0.312 |

All dimensions are in inches. Threads are Unified Standard Class 2B, UNC or UNF Series. * When specifying a nominal size in decimals, any zero in the fourth decimal place is omitted. *Reprinted with permission. Copyright © 1974, Society of Automotive Engineers, Inc. All rights reserved.*

**Wrench Openings for Nuts** (ANSI B18.2.2-1972 Appendix)

| Max.* Width Across Flats of Nut | Wrench Opening† | | Max.* Width Across Flats of Nut | Wrench Opening† | | Max.* Width Across Flats of Nut | Wrench Opening† | |
|---|---|---|---|---|---|---|---|---|
| | Min. | Max. | | Min. | Max. | | Min. | Max. |
| 5/32 | 0.158 | 0.163 | 1 1/8 | 1.132 | 1.142 | 2 13/16 | 2.827 | 2.845 |
| 3/16 | 0.190 | 0.195 | 1 1/4 | 1.257 | 1.267 | 2 15/16 | 2.954 | 2.973 |
| 7/32 | 0.220 | 0.225 | 1 5/16 | 1.320 | 1.331 | 3 | 3.016 | 3.035 |
| 1/4 | 0.252 | 0.257 | 1 3/8 | 1.383 | 1.394 | 3 1/8 | 3.142 | 3.162 |
| 9/32 | 0.283 | 0.288 | 1 7/16 | 1.446 | 1.457 | 3 3/8 | 3.393 | 3.414 |
| 5/16 | 0.316 | 0.322 | 1 1/2 | 1.508 | 1.520 | 3 1/2 | 3.518 | 3.540 |
| 11/32 | 0.347 | 0.353 | 1 5/8 | 1.634 | 1.646 | 3 3/4 | 3.770 | 3.793 |
| 3/8 | 0.378 | 0.384 | 1 11/16 | 1.696 | 1.708 | 3 7/8 | 3.895 | 3.918 |
| 7/16 | 0.440 | 0.446 | 1 13/16 | 1.822 | 1.835 | 4 1/8 | 4.147 | 4.172 |
| 1/2 | 0.504 | 0.510 | 1 7/8 | 1.885 | 1.898 | 4 1/4 | 4.272 | 4.297 |
| 9/16 | 0.566 | 0.573 | 2 | 2.011 | 2.025 | 4 1/2 | 4.524 | 4.550 |
| 5/8 | 0.629 | 0.636 | 2 1/16 | 2.074 | 2.088 | 4 5/8 | 4.649 | 4.676 |
| 11/16 | 0.692 | 0.699 | 2 3/16 | 2.200 | 2.215 | 4 7/8 | 4.900 | 4.928 |
| 3/4 | 0.755 | 0.763 | 2 1/4 | 2.262 | 2.277 | 5 | 5.026 | 5.055 |
| 13/16 | 0.818 | 0.826 | 2 3/8 | 2.388 | 2.404 | 5 1/4 | 5.277 | 5.307 |
| 7/8 | 0.880 | 0.888 | 2 7/16 | 2.450 | 2.466 | 5 3/8 | 5.403 | 5.434 |
| 15/16 | 0.944 | 0.953 | 2 9/16 | 2.576 | 2.593 | 5 5/8 | 5.654 | 5.686 |
| 1 | 1.006 | 1.015 | 2 5/8 | 2.639 | 2.656 | 5 3/4 | 5.780 | 5.813 |
| 1 1/16 | 1.068 | 1.077 | 2 3/4 | 2.766 | 2.783 | 6 | 6.031 | 6.065 |

All dimensions given in inches.    † Openings for 5/32 to 3/8 widths from old ASA B18.2-1960.
* Wrenches are marked with the "Nominal Size of Wrench" which is equal to the basic or maximum width across flats of the corresponding nut.
Minimum wrench opening equals (1.005W + 0.001). Tolerance on wrench opening equals plus (0.005W + 0.004) from minimum, where W equals nominal size of wrench.

**Wrench Clearance Dimensions.** — Wrench clearances are given in Tables 1 and 2. They are based on a wrench opening corresponding to the dimension across the flats of the fastener. The listed values were obtained from a composite study of the alloy steel wrenches that are commercially available and military specifications. They are suitable for general use as minimum requirements.

**Table 1.   Wrench Clearances for Box Wrench — 12 Point**
(From SAE Aeronautical Drafting Manual)

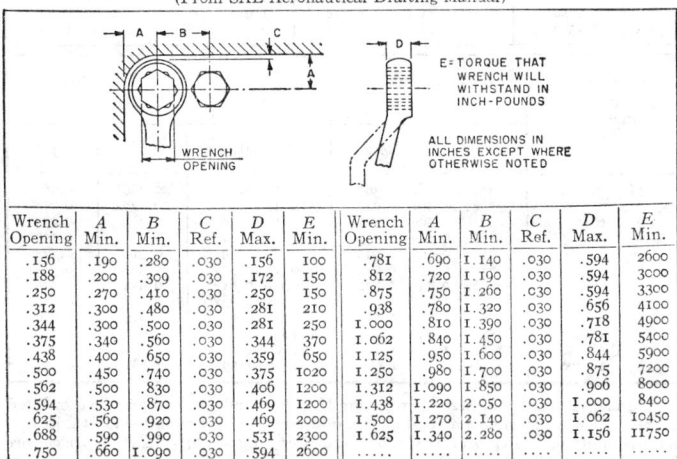

| Wrench Opening | A Min. | B Min. | C Ref. | D Max. | E Min. | Wrench Opening | A Min. | B Min. | C Ref. | D Max. | E Min. |
|---|---|---|---|---|---|---|---|---|---|---|---|
| .156 | .190 | .280 | .030 | .156 | 100 | .781 | .690 | 1.140 | .030 | .594 | 2600 |
| .188 | .200 | .309 | .030 | .172 | 150 | .812 | .720 | 1.190 | .030 | .594 | 3000 |
| .250 | .270 | .410 | .030 | .250 | 150 | .875 | .750 | 1.260 | .030 | .594 | 3300 |
| .312 | .300 | .480 | .030 | .281 | 210 | .938 | .780 | 1.320 | .030 | .656 | 4100 |
| .344 | .300 | .500 | .030 | .281 | 250 | 1.000 | .810 | 1.390 | .030 | .718 | 4900 |
| .375 | .340 | .560 | .030 | .344 | 370 | 1.062 | .840 | 1.450 | .030 | .781 | 5400 |
| .438 | .400 | .650 | .030 | .359 | 650 | 1.125 | .950 | 1.600 | .030 | .844 | 5900 |
| .500 | .450 | .740 | .030 | .375 | 1020 | 1.250 | .980 | 1.700 | .030 | .875 | 7200 |
| .562 | .500 | .830 | .030 | .406 | 1200 | 1.312 | 1.090 | 1.850 | .030 | .906 | 8000 |
| .594 | .530 | .870 | .030 | .469 | 1200 | 1.438 | 1.220 | 2.050 | .030 | 1.000 | 8400 |
| .625 | .560 | .920 | .030 | .469 | 2000 | 1.500 | 1.270 | 2.140 | .030 | 1.062 | 10450 |
| .688 | .590 | .990 | .030 | .531 | 2300 | 1.625 | 1.340 | 2.280 | .030 | 1.156 | 11750 |
| .750 | .660 | 1.090 | .030 | .594 | 2600 | ..... | ..... | ..... | ..... | ..... | ..... |

E = TORQUE THAT WRENCH WILL WITHSTAND IN INCH-POUNDS

ALL DIMENSIONS IN INCHES EXCEPT WHERE OTHERWISE NOTED

WRENCH OPENING

## Table 2. Wrench Clearances for Open End Engineers Wrench 15° and Socket Wrench (Regular Length)
(From SAE Aeronautical Drafting Manual)

Figure legend / diagram labels: H=THICKNESS OF WRENCH HEAD; J=TORQUE THAT WRENCH WILL WITHSTAND IN INCH-POUNDS; WRENCH OPENING; B RADIUS; F RADIUS; OPEN END ENGINEERS WRENCH 15°; ALL DIMENSIONS IN INCHES WHERE OTHERWISE NOTED; SQUARE DRIVE; SOCKET (REGULAR LENGTH); WRENCH OPENING; P=TORQUE THAT WRENCH WILL WITHSTAND IN INCH-POUNDS; *=DOES NOT ALLOW CLEARANCE FOR TORQUE DEVICE.

| Wrench Opening | A Min. | B Max. | C Min. | D Min. | E Min. | F Max. | G Ref. | H Max. | J Min. | K Min. | L Ref. | Q=.250 M Max. | Q=.250 N Max. | Q=.250 P Min. | Q=.375 M Max. | Q=.375 N Max. | Q=.375 P Min. | Q=.500 M Max. | Q=.500 N Max. | Q=.500 P Min. | Q=.750 M Max. | Q=.750 N Max. | Q=.750 P Min. | Wrench Opening |
|---|---|---|---|---|---|---|---|---|---|---|---|---|---|---|---|---|---|---|---|---|---|---|---|---|
| .156 | .220 | .250 | .390 | .166 | .250 | .200 | .030 | .094 | 25 | .370 | .030 | | | 125 | | | | | | | | | | .188 |
| .188 | .250 | .280 | .430 | .190 | .270 | .230 | .030 | .172 | 40 | .470 | .030 | | | 200 | | | | | | | | | | .250 |
| .250 | .280 | .340 | .530 | .270 | .310 | .310 | .030 | .172 | 60 | .470 | .030 | 1.000 | .510 | 300 | | | | | | | | | | .312 |
| .312 | .380 | .470 | .660 | .280 | .390 | .390 | .050 | .203 | 125 | .550 | .030 | 1.000 | .510 | 400 | 1.250 | .690 | 250 | | | | | | | .344 |
| .344 | .420 | .500 | .750 | .340 | .450 | .450 | .050 | .203 | 175 | .580 | .030 | 1.000 | .510 | 450 | 1.250 | .690 | 400 | | | | | | | .375 |
| .375 | .420 | .500 | .780 | .360 | .450 | .520 | .050 | .219 | 250 | .620 | .030 | 1.000 | .519 | 550 | 1.250 | .690 | 675 | | | | | | | .438 |
| .438 | .470 | .590 | .890 | .420 | .520 | .590 | .050 | .250 | 375 | .750 | .030 | 1.000 | .580 | 550 | 1.250 | .690 | 900 | 1.500 | .880 | 1600 | | | | .500 |
| .500 | .520 | .640 | 1.000 | .470 | .580 | .660 | .050 | .266 | 490 | .810 | .030 | 1.000 | .683 | 550 | 1.250 | .690 | 900 | 1.500 | .940 | 1700 | | | | .562 |
| .562 | .590 | .770 | 1.130 | .520 | .660 | .700 | .050 | .297 | 700 | .870 | .030 | 1.000 | .692 | 600 | 1.250 | .880 | 1250 | 1.500 | .940 | 1700 | | | | .594 |
| .594 | .640 | .830 | 1.210 | .530 | .700 | .700 | .050 | .344 | 800 | .920 | .030 | | | | 1.250 | .880 | 1450 | 1.500 | .940 | 2000 | | | | .625 |
| .625 | .640 | .830 | 1.230 | .550 | .700 | .700 | .060 | .344 | 935 | .950 | .030 | | | | 1.250 | .932 | 1750 | 1.500 | .940 | 2000 | | | | .688 |
| .688 | .770 | .920 | 1.470 | .660 | .888 | .800 | .060 | .375 | 1250 | 1.030 | .030 | | | | 1.250 | .963 | 2000 | 1.552 | .970 | 2700 | | | | .750 |
| .750 | .770 | .920 | 1.510 | .670 | .888 | .800 | .060 | .375 | 1500 | 1.120 | .030 | | | | 1.250 | .995 | 2000 | 1.552 | 1.000 | 3000 | | | | .781 |
| .781 | .830 | .950 | 1.550 | .690 | .899 | .840 | .060 | .375 | 1615 | 1.150 | .030 | | | | 1.250 | 1.058 | 2000 | 1.552 | 1.000 | 3600 | | | | .812 |
| .812 | .910 | 1.120 | 1.660 | .720 | .970 | .860 | .060 | .406 | 1710 | 1.200 | .030 | | | | 1.250 | 1.126 | 2000 | 1.552 | 1.065 | 4300 | | | | .875 |
| .875 | .970 | 1.150 | 1.810 | .800 | 1.060 | .910 | .060 | .438 | 2250 | 1.280 | .030 | | | | 1.250 | 1.126 | | 1.562 | 1.130 | 5000 | | | | .938 |
| .938 | .970 | 1.150 | 1.850 | .810 | 1.060 | .950 | .060 | .438 | 2750 | 1.370 | .030 | | | | 1.250 | 1.213 | | 1.625 | 1.130 | 5000 | | | | 1.000 |
| 1.000 | 1.050 | 1.230 | 2.000 | .880 | 1.160 | 1.000 | .060 | .500 | 3250 | 1.470 | .030 | | | | | | | 1.750 | 1.222 | 5000 | | | | 1.062 |
| 1.062 | 1.090 | 1.250 | 2.100 | .970 | 1.200 | 1.200 | .080 | .500 | 3500 | 1.550 | .030 | | | | | | | 1.750 | 1.285 | 5000 | | | | 1.125 |
| 1.125 | 1.140 | 1.370 | 2.210 | 1.000 | 1.270 | 1.230 | .080 | .500 | 4000 | 1.610 | .030 | | | | | | | 1.750 | 1.410 | 5000 | | | | 1.250 |
| 1.250 | 1.270 | 1.420 | 2.440 | 1.080 | 1.390 | 1.310 | .080 | .562 | 5250 | 1.890 | .030 | | | | | | | 1.844 | 1.410 | 5000 | 2.375 | 1.855 | 7250 | 1.312 |
| 1.312 | 1.390 | 1.690 | 2.630 | 1.170 | 1.520 | 1.340 | .080 | .562 | 6000 | 1.980 | .030 | | | | | | | 1.938 | 1.505 | 5000 | 2.500 | 1.920 | 8000 | 1.438 |
| 1.438 | 1.470 | 1.720 | 2.800 | 1.250 | 1.590 | 1.340 | .080 | .641 | 7500 | 2.140 | .030 | | | | | | | 2.000 | 1.507 | 5000 | 2.625 | 2.075 | 9550 | 1.500 |
| 1.500 | 1.470 | 1.720 | 2.840 | 1.270 | 1.590 | 1.450 | .090 | .641 | 8250 | 2.200 | .030 | | | | | | | | 1.723 | 5000 | 2.625 | 2.170 | 10450 | 1.625 |
| 1.625 | 1.560 | 1.880 | 3.100 | 1.380 | 1.750 | 1.560 | .090 | .641 | 9000 | 2.390 | .030 | | | | | | | | | | 2.750 | 2.325 | 11750 | 1.625 |

**American National Standard Round Head and Round Head Square Neck Bolts**
(ANSI B18.5-1971)

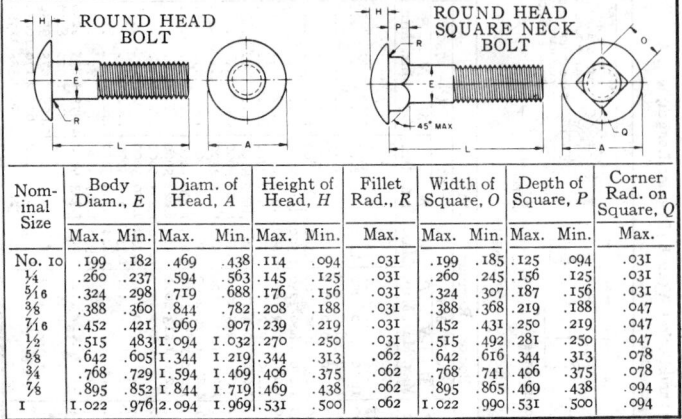

| Nominal Size | Body Diam., E | | Diam. of Head, A | | Height of Head, H | | Fillet Rad., R | Width of Square, O | | Depth of Square, P | | Corner Rad. on Square, Q |
|---|---|---|---|---|---|---|---|---|---|---|---|---|
| | Max. | Min. | Max. | Min. | Max. | Min. | Max. | Max. | Min. | Max. | Min. | Max. |
| No. 10 | .199 | .182 | .469 | .438 | .114 | .094 | .031 | .199 | .185 | .125 | .094 | .031 |
| ¼ | .260 | .237 | .594 | .563 | .145 | .125 | .031 | .260 | .245 | .156 | .125 | .031 |
| ⁵⁄₁₆ | .324 | .298 | .719 | .688 | .176 | .156 | .031 | .324 | .307 | .187 | .156 | .031 |
| ⅜ | .388 | .360 | .844 | .782 | .208 | .188 | .031 | .388 | .368 | .219 | .188 | .047 |
| ⁷⁄₁₆ | .452 | .421 | .969 | .907 | .239 | .219 | .031 | .452 | .431 | .250 | .219 | .047 |
| ½ | .515 | .483 | 1.094 | 1.032 | .270 | .250 | .031 | .515 | .492 | .281 | .250 | .047 |
| ⅝ | .642 | .605 | 1.344 | 1.219 | .344 | .313 | .062 | .642 | .616 | .344 | .313 | .078 |
| ¾ | .768 | .729 | 1.594 | 1.469 | .406 | .375 | .062 | .768 | .741 | .406 | .375 | .078 |
| ⅞ | .895 | .852 | 1.844 | 1.719 | .469 | .438 | .062 | .895 | .865 | .469 | .438 | .094 |
| 1 | 1.022 | .976 | 2.094 | 1.969 | .531 | .500 | .062 | 1.022 | .990 | .531 | .500 | .094 |

All dimensions are in inches unless otherwise specified.

Threads are Unified Standard, Class 2A, UNC Series, in accordance with ANSI B1.1. For threads with additive finish, the maximum diameters shall apply to an unplated or uncoated bolt before plating or coating, whereas the basic diameters (Class 2A maximum diameters plus the allowance) shall apply to a bolt after plating or coating.

Bolts are designated in the sequence shown: nominal size (number, fraction or decimal equivalent); threads per inch; nominal length (fraction or decimal equivalent); product name; material; and protective finish, if required.

    i.e.: ½–13 × 3 Round Head Square Neck Bolt, Steel
    .375–16 × 2.50 Step Bolt, Steel, Zinc Plated

**American National Standard T-Head Bolts** (ANSI B18.5-1971)

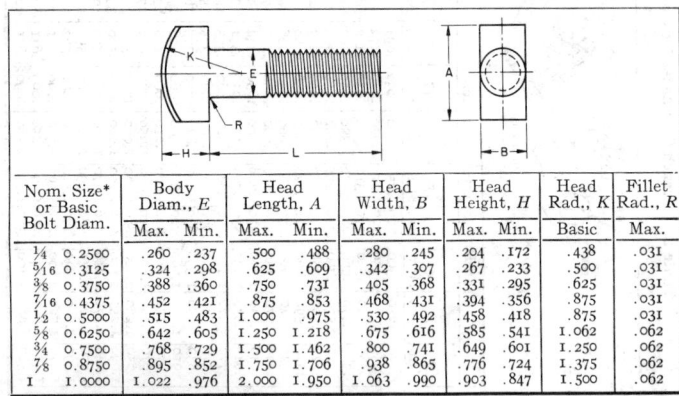

| Nom. Size* or Basic Bolt Diam. | Body Diam., E | | Head Length, A | | Head Width, B | | Head Height, H | | Head Rad., K | Fillet Rad., R |
|---|---|---|---|---|---|---|---|---|---|---|
| | Max. | Min. | Max. | Min. | Max. | Min. | Max. | Min. | Basic | Max. |
| ¼ 0.2500 | .260 | .237 | .500 | .488 | .280 | .245 | .204 | .172 | .438 | .031 |
| ⁵⁄₁₆ 0.3125 | .324 | .298 | .625 | .609 | .342 | .307 | .267 | .233 | .500 | .031 |
| ⅜ 0.3750 | .388 | .360 | .750 | .731 | .405 | .368 | .331 | .295 | .625 | .031 |
| ⁷⁄₁₆ 0.4375 | .452 | .421 | .875 | .853 | .468 | .431 | .394 | .356 | .875 | .031 |
| ½ 0.5000 | .515 | .483 | 1.000 | .975 | .530 | .492 | .458 | .418 | .875 | .031 |
| ⅝ 0.6250 | .642 | .605 | 1.250 | 1.218 | .675 | .616 | .585 | .541 | 1.062 | .062 |
| ¾ 0.7500 | .768 | .729 | 1.500 | 1.462 | .800 | .741 | .649 | .601 | 1.250 | .062 |
| ⅞ 0.8750 | .895 | .852 | 1.750 | 1.706 | .938 | .865 | .776 | .724 | 1.375 | .062 |
| 1 1.0000 | 1.022 | .976 | 2.000 | 1.950 | 1.063 | .990 | .903 | .847 | 1.500 | .062 |

All dimensions are given in inches.
* Where specifying nominal size in decimals, zeros preceding the decimal point and in the fourth decimal place are omitted.

# American National Standard Step and 114 Degree Countersunk Square Neck Bolts (ANSI B18.5-1971)

STEP BOLT

114 DEGREE COUNTERSUNK SQUARE NECK BOLT

| Nominal Size | Step & 114° Countersunk Bolts | | | | | Step Bolts | | | | | | | 114° Countersunk Square Neck Bolts | | | | | | |
|---|---|---|---|---|---|---|---|---|---|---|---|---|---|---|---|---|---|---|---|
| | Body Diam., E | | Width of Square, O | | Corner Rad. on Square, Q | Depth of Square, P | | Diam. of Head, A | | Height of Head, H | | Fillet Radius, R | Depth of Square, P | | Diam. of Head, A | | Flat on Head, F | Height of Head, H | |
| | Max. | Min. | Max. | Min. | Max. | Max. | Min. | Max. | Min. | Max. | Min. | Max. | Max. | Min. | Max. | Min. | Min. | Max. | Min. |
| No. 10 | .199 | .182 | .199 | .185 | .031 | .125 | .094 | .656 | .625 | .114 | .094 | .031 | .125 | .094 | .548 | .500 | .015 | .131 | .112 |
| 1/4 | .260 | .237 | .260 | .245 | .031 | .156 | .125 | .844 | .813 | .145 | .125 | .031 | .156 | .125 | .682 | .625 | .018 | .154 | .135 |
| 5/16 | .324 | .298 | .324 | .307 | .031 | .187 | .156 | 1.031 | 1.000 | .176 | .156 | .031 | .219 | .188 | .821 | .750 | .023 | .184 | .159 |
| 3/8 | .388 | .360 | .388 | .368 | .047 | .219 | .188 | 1.219 | 1.188 | .208 | .188 | .031 | .250 | .219 | .960 | .875 | .027 | .212 | .183 |
| 7/16 | .452 | .421 | .452 | .431 | .047 | .250 | .219 | 1.406 | 1.375 | .239 | .219 | .031 | .281 | .250 | 1.093 | 1.000 | .030 | .235 | .205 |
| 1/2 | .515 | .483 | .515 | .492 | .047 | .281 | .250 | 1.594 | 1.503 | .270 | .250 | .031 | .312 | .281 | 1.233 | 1.125 | .035 | .265 | .229 |
| 5/8* | .642 | .605 | .642 | .616 | .078 | … | … | … | … | … | … | … | .406 | .375 | 1.495 | 1.375 | .038 | .316 | .272 |
| 3/4* | .768 | .729 | .768 | .741 | .078 | … | … | … | … | … | … | … | .500 | .469 | 1.754 | 1.625 | .041 | .368 | .314 |

All dimensions are in inches unless otherwise specified.

\* These sizes pertain to 114 degree countersunk square neck bolts only. Dimensions given in last four columns to the right are for these bolts only. Threads are Unified Standard, Class 2A, UNC Series, in accordance with ANSI B1.1. For threads with additive finish, the maximum diameters shall apply to an unplated or uncoated bolt before plating or coating, whereas the basic diameters (Class 2A maximum diameters plus the allowance) shall apply to a bolt after plating or coating.

Bolts are designated in the sequence shown: nominal size (number, fraction or decimal equivalent); threads per inch; nominal length (fraction or decimal equivalent); product name; material; and protective finish, if required.

i.e.: 1/2-13 × 3 Round Head Square Neck Bolt, Steel

.375-16 × 2.50 Step Bolt, Steel, Zinc Plated

**Lengths of Machine Bolts — Approved by U. S. Department of Commerce; Bureau of Standards; American Institute of Bolt, Nut, and Rivet Manufacturers**

| Lengths, Inches | Diameters, Inch | | | | | | | | | |
|---|---|---|---|---|---|---|---|---|---|---|
| | ¼ | ⁵⁄₁₆ | ⅜ | ⁷⁄₁₆ | ½ | ⅝ | ¾ | ⅞ | 1 | 1⅛ |
| **Square-head Machine Bolts — Stock Sizes (See footnote)** | | | | | | | | | | |
| ½ | ** | ** | | | | | | | | |
| ⅝ | | ** | | | | | | | | |
| ¾ | * | * | * | | | | | | | |
| 1 | * | * | * | | * | * | ** | | | |
| 1¼ | * | * | * | * | * | * | * | | | |
| 1½ | * | * | * | * | * | * | * | ** | | |
| 1¾ | * | * | * | * | * | * | * | ** | | |
| 2 | * | * | * | * | * | * | * | * | ** | |
| 2¼ | * | * | * | ** | * | * | * | * | | |
| 2½ | * | * | * | * | * | * | * | * | * | |
| 2¾ | * | * | * | | * | * | * | * | | |
| 3 | * | * | * | * | * | * | * | * | * | |
| 3¼ | ** | ** | * | * | * | * | * | * | | |
| 3½ | * | * | * | * | * | * | * | * | | |
| 3¾ | | * | * | * | * | * | * | ** | ** | |
| 4 | | * | * | * | * | * | * | * | * | ** |
| 4¼ | | | | | ** | ** | * | ** | | |
| 4½ | * | * | * | | * | * | * | * | * | ** |
| 4¾ | | | | | ** | ** | ** | | | |
| 5 | * | * | * | * | * | * | * | * | * | ** |
| 5½ | * | * | * | * | * | * | * | * | * | ** |
| 6 | * | * | * | * | * | * | * | * | * | * |
| 6½ | ** | ** | * | ** | * | * | * | * | * | |
| 7 | ** | * | * | ** | * | * | * | * | * | ** |
| 7½ | | | * | | * | * | * | * | * | |
| 8 | ** | ** | * | ** | * | * | * | * | * | ** |
| 8½ | | | * | | * | * | * | * | ** | |
| 9 | | | * | ** | * | * | * | * | * | ** |
| 9½ | | | ** | | * | * | * | * | | |
| 10 | | | * | | * | * | * | * | * | |
| 10½ | | | | | * | * | * | | | |
| 11 | | | * | | * | * | * | * | * | ** |
| 11½ | | | | | ** | ** | ** | | | |
| 12 | | | * | | * | * | * | * | | ** |
| 13 | | | | | * | * | * | ** | * | |
| 14 | | | | | * | * | * | * | * | ** |
| 15 | | | | | * | * | * | * | * | |
| 16 | | | | | * | * | * | * | | |
| 17 | | | | | ** | * | * | * | ** | ** |
| 18 | | | | | * | * | * | * | * | * |
| 19 | | | | | ** | ** | * | ** | ** | ** |
| 20 | | | | | * | * | * | * | * | ** |
| 22 | | | | | ** | * | * | ** | ** | |
| 24 | | | | | ** | ** | ** | ** | | |
| **Hexagon-head Machine Bolts — Stock Sizes (See footnote)** | | | | | | | | | | |
| ½ | ** | | | | | | | | | |
| ¾ | * | * | * | | ** | | | | | |
| 1 | * | * | * | | * | | | | | |
| 1¼ | * | * | * | | * | ** | ** | | | |
| 1½ | * | * | * | | * | * | * | | | |
| 1¾ | ** | ** | * | | * | * | * | | | |
| 2 | * | ** | * | | * | * | * | ** | | |
| 2¼ | | | | | ** | ** | ** | | | |
| 2½ | ** | ** | * | | * | ** | * | | | |
| 2¾ | | | | | * | | | | | |
| 3 | | ** | ** | | ** | * | ** | ** | | |
| 3½ | | | ** | | ** | * | ** | ** | | |
| 4 | | | ** | | ** | ** | * | ** | | |
| 4½ | | | | | ** | ** | ** | ** | | |
| 5 | | | | | ** | * | * | ** | | |
| 5½ | | | | | ** | ** | ** | | | |
| 6 | | | | | ** | ** | ** | ** | | |

One asterisk (*) represents stock sizes of maximum demand, and two asterisks (**) stock sizes least frequently used. All other sizes are considered specials.

**Working Strength of Bolts.** — When the nut on a bolt is tightened, an initial tensile load is placed on the bolt which must be taken into account in determining its safe working strength or external load-carrying capacity. The total load on the bolt theoretically varies from a maximum equal to the sum of the initial and external loads (when the bolt is absolutely rigid and the parts held together are elastic) to a minimum equal to either the initial or external loads, whichever is the greater (where the bolt is elastic and the parts held together are absolutely rigid). Since no material is absolutely rigid, in practice the total load values fall somewhere between these maximum and minimum limits, depending upon the relative elasticity of the bolt and joint members.

Some experiments made at Cornell University to determine the initial stress due to tightening nuts on bolts sufficiently to make a packed joint steam tight, showed that experienced machinists tighten nuts with a pull roughly proportional to the bolt diameter. It was also found that the stress due to nut tightening was often sufficient to break a ½-inch bolt, but not larger sizes, assuming that the nut is tightened by an experienced mechanic. It may be concluded, therefore, that bolts smaller than ⅝ inch should not be used for holding cylinder heads or other parts requiring a tight joint. As a result of these tests, the following empirical formula was established for the working strength of bolts used for packed joints or joints where the elasticity of a gasket is greater than the elasticity of the studs or bolts.

$$W = S_t(0.55 \ d^2 - 0.25 \ d)$$

In this formula, $W$ = working strength of bolt or permissible load, in pounds after allowance is made for initial load due to tightening; $S_t$ = allowable working stress in tension, pounds per square inch; and $d$ = nominal outside diameter of stud or bolt, inches. A somewhat more convenient formula, and one which gives approximately the same results, is:

$$W = S_t(A - 0.25 \ d)$$

In this formula, $W$, $S_t$ and $d$ are as previously given, and $A$ = area at the root of the thread, square inches.

*Example.* — What is the working strength of a 1-inch bolt which is screwed up tightly in a packed joint when the allowable working stress is 10,000 psi?

$$W = 10,000(0.55 \times 1 - 0.25 \times 1) = 3000 \text{ pounds approx.}$$

**Formulas for Stress Areas and Length of Engagement of Screw Threads.** — The critical areas of stress of mating screw threads are: (1) The effective cross-sectional area, or tensile stress area, of the external thread; (2) the shear area of the external thread, which depends principally on the minor diameter of the tapped hole; and (3) the shear area of the internal thread, which depends principally on the major diameter of the external thread. The relation of these three stress areas to each other is an important factor in determining how a threaded connection will fail, whether by breakage in the threaded section of the screw (or bolt) or by stripping of either the external or internal thread.

If failure of a threaded assembly should occur, it is preferable for the screw to break rather than have either the external or internal thread strip. In other words, the length of engagement of mating threads should be sufficient to carry the full load necessary to break the screw without the threads stripping.

If mating internal and external threads are manufactured of materials having equal tensile strengths, then to prevent stripping of the external thread, the length of engagement should be not less than that given by Formula 1:

$$L_e = \frac{2 \times A_t}{3.1416 \ K_n\text{max} \left[ \frac{1}{2} + 0.57735n(E_s\text{min} - K_n\text{max}) \right]} \quad (1)$$

In this formula, the factor of 2 means that it is assumed that the area in shear of the screw must be twice the tensile stress area to develop the full strength of the screw (this value is slightly larger than required and thus provides a small factor of safety against stripping); $L_e$ = length of engagement, in inches; $n$ = number of threads per inch; $K_n$max = maximum minor diameter of internal thread; $E_s$min = minimum pitch diameter of external thread for the class of thread specified; and $A_t$ = tensile stress area of screw thread given by Formula (2a) or (2b) or the thread tables on pages 1266 to 1275 for Unified threads which are based on Formula (2a).

For steels of up to 100,000 psi ultimate tensile strength,

$$A_t = 0.7854 \left( D - \frac{0.9743}{n} \right)^2 \tag{2a}$$

For steels of over 100,000 psi ultimate tensile strength,

$$A_t = 3.1416 \left( \frac{E_s\text{min}}{2} - \frac{0.16238}{n} \right)^2 \tag{2b}$$

In these formulas, $D$ = basic major diameter of threads and the other symbols have the same meaning as before.

*Stripping of Internal Thread:* If the internal thread is made of material of lower strength than the external thread, stripping of the internal thread may take place before the screw breaks. To determine whether this condition exists it is necessary to calculate the factor $J$ for the relative strength of the external and internal threads given by Formula (3):

$$J = \frac{A_s \times \text{tensile strength of external thread material}}{A_n \times \text{tensile strength of internal thread material}} \tag{3}$$

If $J$ is less than or equal to 1, the length of engagement determined by Formula (1) is adequate to prevent stripping of the internal thread; if $J$ is greater than 1, the required length of engagement $Q$ to prevent stripping of the internal thread is obtained by multiplying the length of engagement $L_e$, Formula (1), by $J$:

$$Q = JL_e \tag{4}$$

In Formula (3), $A_s$ and $A_n$ are the shear areas of the external and internal threads, respectively, given by Formulas (5) and (6):

$$A_s = 3.1416 \, nL_eK_n\text{max} \left[ \frac{1}{2n} + 0.57735(E_s\text{min} - K_n\text{max}) \right] \tag{5}$$

$$A_n = 3.1416 \, nL_eD_s\text{min} \left[ \frac{1}{2n} + 0.57735(D_s\text{min} - E_n\text{max}) \right] \tag{6}$$

In these formulas, $n$ = threads per inch; $L_e$ = length of engagement from Formula (1); $K_n$max = maximum minor diameter of internal thread; $E_s$min = minimum pitch diameter of the external thread for the class of thread specified; $D_s$min = minimum major diameter of the external thread; and $E_n$max = maximum pitch diameter of internal thread.

**Load to Break Threaded Portion of Screws and Bolts.** — The direct tensile load $P$ to break the threaded portion of a screw or bolt (assuming that no shearing or torsional stresses are acting) can be determined from the following formula:

$$P = SA_t$$

where, $P$ = load in pounds to break screw; $S$ = ultimate tensile strength of material of screw or bolt in pounds per square inch; and $A_t$ = tensile stress area in square inches from Formulas (2a), (2b) or from the screw thread tables.

**U. S. Standard Threads, Bolts and Nuts**

The United States Standard has largely been superseded by the American Standard but the former is sometimes used, especially for screw thread sizes above the present range of the American Standard. For American Standard bolts and nuts, see pages 1131 to 1141; for screw threads, see page 1256.

| Diameter | No. of Threads per Inch | Diameter at Root of Thread | Diameter of Tap Drill | Area in Sq. Inches | | Tensile Strength at Stress of 6000 Pounds per Sq. Inch | Dimensions of Nuts and Bolt Heads | | | | |
|---|---|---|---|---|---|---|---|---|---|---|---|
| | | | | Of Bolt | At Root of Thread | |  |  |  |  |  |
| 1/4 | 20 | 0.185 | 13/64 | 0.049 | 0.026 | 160 | 1/2 | 0.578 | 0.707 | 1/4 | 1/4 |
| 5/16 | 18 | 0.240 | 1/4 | 0.076 | 0.045 | 270 | 19/32 | 0.686 | 0.840 | 5/16 | 19/64 |
| 3/8 | 16 | 0.294 | 5/16 | 0.110 | 0.068 | 410 | 11/16 | 0.794 | 0.972 | 3/8 | 11/32 |
| 7/16 | 14 | 0.345 | 23/64 | 0.150 | 0.093 | 560 | 25/32 | 0.902 | 1.105 | 7/16 | 25/64 |
| 1/2 | 13 | 0.400 | 27/64 | 0.196 | 0.126 | 760 | 7/8 | 1.011 | 1.237 | 1/2 | 7/16 |
| 9/16 | 12 | 0.454 | 15/32 | 0.248 | 0.162 | 1,000 | 31/32 | 1.119 | 1.370 | 9/16 | 31/64 |
| 5/8 | 11 | 0.507 | 17/32 | 0.307 | 0.202 | 1,210 | 1 1/16 | 1.227 | 1.502 | 5/8 | 17/32 |
| 3/4 | 10 | 0.620 | 41/64 | 0.442 | 0.302 | 1,810 | 1 1/4 | 1.444 | 1.768 | 3/4 | 5/8 |
| 7/8 | 9 | 0.731 | 3/4 | 0.601 | 0.419 | 2,520 | 1 7/16 | 1.660 | 2.033 | 7/8 | 23/32 |
| 1 | 8 | 0.838 | 55/64 | 0.785 | 0.551 | 3,300 | 1 5/8 | 1.877 | 2.298 | 1 | 13/16 |
| 1 1/8 | 7 | 0.939 | 31/32 | 0.994 | 0.694 | 4,160 | 1 13/16 | 2.093 | 2.563 | 1 1/8 | 29/32 |
| 1 1/4 | 7 | 1.064 | 1 3/32 | 1.227 | 0.893 | 5,350 | 2 | 2.310 | 2.828 | 1 1/4 | 1 |
| 1 3/8 | 6 | 1.158 | 1 7/32 | 1.485 | 1.057 | 6,340 | 2 3/16 | 2.527 | 3.093 | 1 3/8 | 1 3/32 |
| 1 1/2 | 6 | 1.283 | 1 11/32 | 1.767 | 1.295 | 7,770 | 2 3/8 | 2.743 | 3.358 | 1 1/2 | 1 3/16 |
| 1 5/8 | 5 1/2 | 1.389 | 1 27/64 | 2.074 | 1.515 | 9,090 | 2 9/16 | 2.960 | 3.623 | 1 5/8 | 1 9/32 |
| 1 3/4 | 5 | 1.490 | 1 17/32 | 2.405 | 1.746 | 10,470 | 2 3/4 | 3.176 | 3.889 | 1 3/4 | 1 3/8 |
| 1 7/8 | 5 | 1.615 | 1 21/32 | 2.761 | 2.051 | 12,300 | 2 15/16 | 3.393 | 4.154 | 1 7/8 | 1 15/32 |
| 2 | 4 1/2 | 1.711 | 1 49/64 | 3.142 | 2.302 | 13,800 | 3 1/8 | 3.609 | 4.419 | 2 | 1 9/16 |
| 2 1/4 | 4 1/2 | 1.961 | 2 1/64 | 3.976 | 3.023 | 18,100 | 3 1/2 | 4.043 | 4.949 | 2 1/4 | 1 3/4 |
| 2 1/2 | 4 | 2.175 | 2 13/64 | 4.909 | 3.719 | 22,300 | 3 7/8 | 4.476 | 5.479 | 2 1/2 | 1 15/16 |
| 2 3/4 | 4 | 2.425 | 2 31/64 | 5.940 | 4.620 | 27,700 | 4 1/4 | 4.909 | 6.010 | 2 3/4 | 2 1/8 |
| 3 | 3 1/2 | 2.629 | 2 11/16 | 7.069 | 5.428 | 32,500 | 4 5/8 | 5.342 | 6.540 | 3 | 2 5/16 |
| 3 1/4 | 3 1/2 | 2.879 | 2 15/16 | 8.296 | 6.510 | 39,000 | 5 | 5.775 | 7.070 | 3 1/4 | 2 1/2 |
| 3 1/2 | 3 1/4 | 3.100 | 3 11/64 | 9.621 | 7.548 | 45,300 | 5 3/8 | 6.208 | 7.600 | 3 1/2 | 2 11/16 |
| 3 3/4 | 3 | 3.317 | 3 3/8 | 11.045 | 8.641 | 51,800 | 5 3/4 | 6.641 | 8.131 | 3 3/4 | 2 7/8 |
| 4 | 3 | 3.567 | 3 5/8 | 12.566 | 9.963 | 59,700 | 6 1/8 | 7.074 | 8.661 | 4 | 3 1/16 |
| 4 1/4 | 2 7/8 | 3.798 | 3 27/32 | 14.186 | 11.340 | 68,000 | 6 1/2 | 7.508 | 9.191 | 4 1/4 | 3 1/4 |
| 4 1/2 | 2 3/4 | 4.028 | 4 3/32 | 15.904 | 12.750 | 76,500 | 6 7/8 | 7.941 | 9.721 | 4 1/2 | 3 7/16 |
| 4 3/4 | 2 5/8 | 4.255 | 4 5/16 | 17.721 | 14.215 | 85,500 | 7 1/4 | 8.374 | 10.252 | 4 3/4 | 3 5/8 |
| 5 | 2 1/2 | 4.480 | 4 9/16 | 19.635 | 15.760 | 94,000 | 7 5/8 | 8.807 | 10.782 | 5 | 3 13/16 |
| 5 1/4 | 2 1/2 | 4.730 | 4 13/16 | 21.648 | 17.570 | 105,500 | 8 | 9.240 | 11.312 | 5 1/4 | 4 |
| 5 1/2 | 2 3/8 | 4.953 | 5 1/32 | 23.758 | 19.260 | 116,000 | 8 3/8 | 9.673 | 11.842 | 5 1/2 | 4 3/16 |
| 5 3/4 | 2 3/8 | 5.203 | 5 9/32 | 25.967 | 21.250 | 127,000 | 8 3/4 | 10.106 | 12.373 | 5 3/4 | 4 3/8 |
| 6 | 2 1/4 | 5.423 | 5 1/2 | 28.274 | 23.090 | 138,000 | 9 1/8 | 10.539 | 12.903 | 6 | 4 9/16 |

**British Standard Square and Hexagon Bolts, Screws and Nuts.** — Important dimensions of precision hexagon bolts, screws and nuts (B.S.W. and B.S.F. threads) as covered by British Standard 1083:1965 are given in Tables 1 and 2. The use of fasteners in this standard will decrease as fasteners having Unified inch and ISO metric threads come into increasing use. Dimensions of Unified precision hexagon bolts, screws and nuts (UNC and UNF threads) are given in B.S. 1768:1963; of Unified black hexagon bolts, screws and nuts (UNC and UNF threads) in B.S. 1769:1951; and of Unified black square and hexagon bolts, screws and nuts (UNC and UNF threads) in B.S. 2708:1956. Unified nominal and basic dimensions in these British Standards are the same as the comparable dimensions in the American Standards, but the tolerances applied to these basic dimensions may differ because of rounding-off practices and other factors. For Unified dimensions of square and hexagon bolts and nuts as given in American National Standards B18.2.1 and B18.2.2-1972, see Tables 1 to 3, 6, 7, 9 and 10 on Handbook pages 1133 to 1141. ISO metric precision hexagon bolts, screws and nuts are specified in the British Standard B.S. 3692:1967 (see Handbook pages 1156 to 1162), and ISO metric black hexagon bolts, screws and nuts are covered by British Standard B.S. 4190:1967.

**British Standard Screwed Studs.** — General purpose screwed studs are covered in British Standard 2693: Part 1:1956. The aim in this standard is to provide for a stud having tolerances which would not render it expensive to manufacture and which could be used in association with standard tapped holes for most purposes. Provision has been made for the use of both Unified Fine threads, Unified Coarse threads, British Standard Fine threads, and British Standard Whitworth threads as shown in the table on page 1163.

*Designations:* The *metal end* of the stud is the end which is screwed into the component. The *nut end* is the end of the screw of the stud which is not screwed into the component. The *plain portion* of the stud is the unthreaded length.

*Recommended Fitting Practices for Metal End of Stud:* It is recommended that holes tapped to Class 3B limits (see Handbook pages 1280 to 1302) in accordance with B.S. 1580 Unified screw threads or to Close Class limits (see Handbook page 1348) in accordance with B.S. 84 "Screw Threads of Whitworth Form" as appropriate, be used in association with the metal end of the stud specified in this standard. Where fits are not critical, however, holes may be tapped to Class 2B limits (see tables on Handbook pages 1280 to 1302) in accordance with B.S. 1580 or Normal Class limits (see Handbook page 1348) in accordance with B.S. 84.

It is recommended that the B.A. stud specified in this standard be associated with holes tapped to the limits specified for nuts in B.S. 93, 1919 edition. Where fits for these studs are not critical, holes may be tapped to limits specified for nuts (see Handbook page 1348) in the current edition of B.S. 93.

In general, it will be found that the amount of oversize specified for the studs will produce a satisfactory fit in conjunction with the standard tapping as above. Even when interference is not present, locking will take place on the thread runout which has been carefully controlled for this purpose. Where it is considered essential to assure a true interference fit, higher grade studs should be used. It is recommended that standard studs be used even under special conditions where selective assembly may be necessary.

After several years of use of B.S. 2693: Part 1:1956, it was recognized that it would not meet the requirements of all stud users. The stud tolerances specified could result in clearance or interference fits and, in the former case, locking depended on the run-out threads. Thus, some users felt that true interference fits were essential for their needs. As a result, the British Standards Committee have incorporated the Class 5 interference fit threads specified in American Standard ASA B1.12-1963 into their B.S. 2693: Part 2:1964, "Recommendations for High Grade Studs."

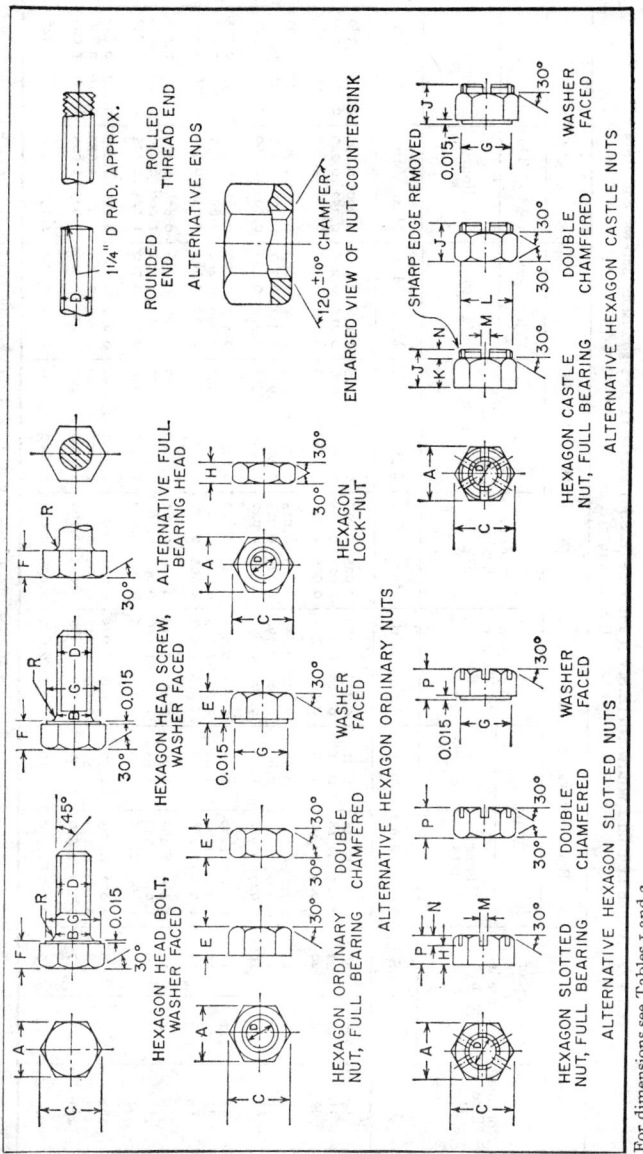

British Standard Whitworth (B.S.W.) and Fine (B.S.F.) Precision Hexagon Bolts, Screws, and Nuts

ROLLED THREAD END

ROUNDED END

ALTERNATIVE ENDS

1¼" D RAD. APPROX.

120 ±10° CHAMFER

ENLARGED VIEW OF NUT COUNTERSINK

WASHER FACED

DOUBLE CHAMFERED

SHARP EDGE REMOVED

HEXAGON CASTLE NUT, FULL BEARING

ALTERNATIVE HEXAGON CASTLE NUTS

ALTERNATIVE FULL BEARING HEAD

HEXAGON LOCK-NUT

HEXAGON HEAD SCREW, WASHER FACED

HEXAGON HEAD BOLT, WASHER FACED

WASHER FACED

DOUBLE CHAMFERED

HEXAGON ORDINARY NUT, FULL BEARING

ALTERNATIVE HEXAGON ORDINARY NUTS

WASHER FACED

DOUBLE CHAMFERED

HEXAGON SLOTTED NUT, FULL BEARING

ALTERNATIVE HEXAGON SLOTTED NUTS

For dimensions see Tables 1 and 2.

## Table 1. British Standard Whitworth (B.S.W.) and Fine (B.S.F.) Precision Hexagon Bolts Screws, and Nuts (B.S. 1083:1965)

| | Number of Threads per Inch | | Bolts, Screws, and Nuts | | | | | | | Bolts and Screws | | | | Nuts | | | |
| | | | Width | | | Diameter of Washer Face G | | Radius Under Head R | | Diameter of Unthreaded Portion of Shank B | | Thickness Head F | | Thickness Ordinary E | | Thickness Lock H | |
| Nominal Size D | B.S.W. | B.S.F. | Across Flats A Max. | Min.† | Across Corners C Max. | Max. | Min. | Max. | Min. | Max. | Min. | Max. | Min. | Max. | Min. | Max. | Min. |
|---|---|---|---|---|---|---|---|---|---|---|---|---|---|---|---|---|---|
| ¼ | 20 | 26 | 0.445 | 0.438 | 0.51 | 0.428 | 0.418 | 0.025 | 0.015 | 0.2500 | 0.2465 | 0.176 | 0.166 | 0.200 | 0.190 | 0.185 | 0.180 |
| 5⁄16 | 18 | 22 | 0.525 | 0.518 | 0.61 | 0.508 | 0.498 | 0.025 | 0.015 | 0.3125 | 0.3090 | 0.218 | 0.208 | 0.250 | 0.240 | 0.210 | 0.200 |
| 3⁄8 | 16 | 20 | 0.600 | 0.592 | 0.69 | 0.582 | 0.572 | 0.025 | 0.015 | 0.3750 | 0.3715 | 0.260 | 0.250 | 0.312 | 0.302 | 0.260 | 0.250 |
| 7⁄16 | 14 | 18 | 0.710 | 0.702 | 0.82 | 0.690 | 0.680 | 0.025 | 0.015 | 0.4375 | 0.4335 | 0.302 | 0.292 | 0.375 | 0.365 | 0.275 | 0.265 |
| ½ | 12 | 16 | 0.820 | 0.812 | 0.95 | 0.800 | 0.790 | 0.025 | 0.015 | 0.5000 | 0.4960 | 0.343 | 0.333 | 0.437 | 0.427 | 0.300 | 0.290 |
| 9⁄16 | 12 | 16 | 0.920 | 0.912 | 1.06 | 0.900 | 0.890 | 0.045 | 0.015 | 0.5625 | 0.5585 | 0.375 | 0.365 | 0.500 | 0.490 | 0.333 | 0.323 |
| 5⁄8 | 11 | 14 | 1.010 | 1.000 | 1.17 | 0.985 | 0.975 | 0.045 | 0.020 | 0.6250 | 0.6190 | 0.417 | 0.407 | 0.562 | 0.552 | 0.375 | 0.365 |
| ¾ | 10 | 12 | 1.200 | 1.190 | 1.39 | 1.175 | 1.165 | 0.045 | 0.020 | 0.7500 | 0.7440 | 0.500 | 0.480 | 0.687 | 0.677 | 0.458 | 0.448 |
| 7⁄8 | 9 | 11 | 1.300 | 1.288 | 1.50 | 1.273 | 1.263 | 0.065 | 0.040 | 0.8750 | 0.8670 | 0.583 | 0.563 | 0.750 | 0.740 | 0.500 | 0.490 |
| 1 | 8 | 10 | 1.480 | 1.468 | 1.71 | 1.453 | 1.443 | 0.095 | 0.060 | 1.0000 | 0.9920 | 0.666 | 0.636 | 0.875 | 0.865 | 0.583 | 0.573 |
| 1⅛ | 7 | 9 | 1.670 | 1.640 | 1.93 | 1.620 | 1.610 | 0.095 | 0.060 | 1.1250 | 1.1170 | 0.750 | 0.710 | 1.000 | 0.990 | 0.666 | 0.656 |
| 1¼ | 7 | 9 | 1.860 | 1.815 | 2.15 | 1.795 | 1.785 | 0.095 | 0.060 | 1.2500 | 1.2420 | 0.830 | 0.790 | 1.125 | 1.105 | 0.750 | 0.730 |
| 1⅜* | | 8 | 2.050 | 2.005 | 2.37 | 1.985 | 1.975 | 0.095 | 0.060 | 1.3750 | 1.3650 | 0.920 | 0.880 | 1.250 | 1.230 | 0.833 | 0.813 |
| 1½ | 6 | 8 | 2.220 | 2.175 | 2.56 | 2.155 | 2.145 | 0.095 | 0.060 | 1.5000 | 1.4900 | 1.000 | 0.960 | 1.375 | 1.355 | 0.916 | 0.896 |
| 1¾ | 5 | 7 | 2.580 | 2.520 | 2.98 | 2.495 | 2.485 | 0.095 | 0.060 | 1.7500 | 1.7400 | 1.170 | 1.110 | 1.625 | 1.605 | 1.083 | 1.063 |
| 2 | 4.5 | 7 | 2.760 | 2.700 | 3.19 | 2.675 | 2.665 | 0.095 | 0.060 | 2.0000 | 1.9900 | 1.330 | 1.270 | 1.750 | 1.730 | 1.166 | 1.146 |

All dimensions in inches except where otherwise noted. * Not standard with B.S.W. thread. † When bolts from ¼ to 1 inch are hot forged, the tolerance on the width across flats shall be two and a half times the tolerance shown in the table and shall be unilaterally minus from maximum size. For dimensional notation, see diagram on page 1153.

Table 2. British Standard Whitworth (B.S.W.) and Fine (B.S.F.) Precision Hexagon Slotted and Castle Nuts (B.S. 1083:1965)

| Nominal Size D | Number of Threads per Inch | | Slotted Nuts | | | | Castle Nuts | | | | | | Slotted and Castle Nuts | | |
|---|---|---|---|---|---|---|---|---|---|---|---|---|---|---|---|
| | | | Thickness P | | Lower Face to Bottom of Slot H | | Total Thickness J | | Lower Face to Bottom of Slot K | | Castellated Portion Diameter L | | Slots Width M | | Depth N |
| | B.S.W. | B.S.F. | Max. | Min. | Max. | Min. | Max. | Min. | Max. | Min. | Max. | Min. | Max. | Min. | Approx. |
| ¼ | 20 | 26 | 0.200 | 0.190 | 0.170 | 0.160 | 0.290 | 0.280 | 0.200 | 0.190 | 0.430 | 0.425 | 0.100 | 0.090 | 0.090 |
| 5⁄16 | 18 | 22 | 0.250 | 0.240 | 0.190 | 0.180 | 0.340 | 0.330 | 0.250 | 0.240 | 0.510 | 0.500 | 0.100 | 0.090 | 0.090 |
| 3⁄8 | 16 | 20 | 0.312 | 0.302 | 0.222 | 0.212 | 0.402 | 0.392 | 0.312 | 0.302 | 0.585 | 0.575 | 0.100 | 0.090 | 0.090 |
| 7⁄16 | 14 | 18 | 0.375 | 0.365 | 0.235 | 0.225 | 0.515 | 0.505 | 0.375 | 0.365 | 0.695 | 0.685 | 0.135 | 0.125 | 0.140 |
| ½ | 12 | 16 | 0.437 | 0.427 | 0.297 | 0.287 | 0.577 | 0.567 | 0.437 | 0.427 | 0.805 | 0.795 | 0.135 | 0.125 | 0.140 |
| 9⁄16 | 12 | 16 | 0.500 | 0.490 | 0.313 | 0.303 | 0.687 | 0.677 | 0.500 | 0.490 | 0.905 | 0.895 | 0.175 | 0.165 | 0.187 |
| 5⁄8 | 11 | 14 | 0.562 | 0.552 | 0.375 | 0.365 | 0.749 | 0.739 | 0.562 | 0.552 | 0.995 | 0.985 | 0.175 | 0.165 | 0.187 |
| ¾ | 10 | 12 | 0.687 | 0.677 | 0.453 | 0.443 | 0.921 | 0.911 | 0.687 | 0.677 | 1.185 | 1.165 | 0.218 | 0.208 | 0.234 |
| 7⁄8 | 9 | 11 | 0.750 | 0.740 | 0.516 | 0.506 | 0.984 | 0.974 | 0.750 | 0.740 | 1.285 | 1.265 | 0.218 | 0.208 | 0.234 |
| 1 | 8 | 10 | 0.875 | 0.865 | 0.595 | 0.585 | 1.155 | 1.145 | 0.875 | 0.865 | 1.465 | 1.445 | 0.260 | 0.250 | 0.280 |
| 1⅛ | 7 | 9 | 1.000 | 0.990 | 0.720 | 0.710 | 1.280 | 1.270 | 1.000 | 0.990 | 1.655 | 1.635 | 0.260 | 0.250 | 0.280 |
| 1¼ | 7 | 9 | 1.125 | 1.105 | 0.797 | 0.777 | 1.453 | 1.433 | 1.125 | 1.105 | 1.845 | 1.825 | 0.300 | 0.290 | 0.328 |
| 1⅜* | | 8 | 1.250 | 1.230 | 0.922 | 0.902 | 1.578 | 1.558 | 1.250 | 1.230 | 2.035 | 2.015 | 0.300 | 0.290 | 0.328 |
| 1½ | 6 | 8 | 1.375 | 1.355 | 1.047 | 1.027 | 1.703 | 1.683 | 1.375 | 1.355 | 2.200 | 2.180 | 0.300 | 0.290 | 0.328 |
| 1¾ | 5 | 7 | 1.625 | 1.605 | 1.250 | 1.230 | 2.000 | 1.980 | 1.625 | 1.605 | 2.555 | 2.535 | 0.343 | 0.333 | 0.375 |
| 2 | 4.5 | 7 | 1.750 | 1.730 | 1.282 | 1.262 | 2.218 | 2.198 | 1.750 | 1.730 | 2.735 | 2.715 | 0.426 | 0.416 | 0.468 |

All dimensions in inches except where otherwise noted. * Not standard with B.S.W. thread. For widths across flats, widths across corners, and diameter of washer face see Table 1. For dimensional notation, see diagram on page 1153.

**British Standard ISO Metric Precision Hexagon Bolts, Screws and Nuts.**
This British Standard (BS 3692:1967) gives the general dimensions and tolerances of precision hexagon bolts, screws and nuts with ISO metric threads in diameters from 1.6 to 68 mm. It is based on the following ISO recommendations and draft recommendations: R 272, R 288, DR 911, DR 947, DR 950, DR 952 and DR 987. Mechanical properties are given only with respect to carbon or alloy steel bolts, screws and nuts, which are not to be used for special applications such as those requiring weldability, corrosion resistance or ability to withstand temperatures above 300°C or below − 50°C. The dimensional requirements of this standard also apply to non-ferrous and stainless steel bolts, screws and nuts.

*Finish:* Finishes may be dull black which results from the heat-treating operation or may be bright finish, the result of bright drawing. Other finishes are possible by mutual agreement between purchaser and producer. It is recommended that reference be made to BS 3382 "Electroplated Coatings on Threaded Components," in this respect.

*General Dimensions:* The bolts, screws and nuts conform to the general dimensions given in Tables 1, 2, 3 and 4.

*Nominal Lengths of Bolts and Screws:* The nominal length of a bolt or screw is the distance from the underside of the head to the extreme end of the shank including any chamfer or radius. Standard nominal lengths and tolerances thereon are given in Table 5.

*Bolt and Screw Ends:* The ends of bolts and screws may be finished with either a 45-degree chamfer to a depth slightly exceeding the depth of thread or a radius approximately equal to 1 ¼ times the nominal diameter of the shank. With rolled threads, the lead formed at the end of the bolt by the thread rolling operation may be regarded as providing the necessary chamfer to the end; the end being reasonably square with the center line of the shank.

*Screw Thread Form:* The form of thread and diameters and associated pitches of standard ISO metric bolts, screws and nuts are in accordance with BS 3643: Part 1, "Thread Data and Standard Thread Series". The screw threads are made to the tolerances for the medium class of fit ($6H/6g$) as specified in BS 3643: Part 2, "Limits and Tolerances for Coarse Pitch Series Threads".

*Length of Thread on Bolts:* The length of thread on bolts is the distance from the end of the bolt (including any chamfer or radius) to the leading face of a screw ring gage which has been screwed as far as possible onto the bolt by hand. Standard thread lengths of bolts are $2d + 6$ mm for a nominal length of bolt up to and including 125 mm, $2d + 12$ mm for a nominal bolt length over 125 mm up to and including 200 mm, and $2d + 25$ mm for a nominal bolt length over 200 mm. Bolts that are too short for minimum thread lengths are threaded as screws and designated as screws. The tolerance on bolt thread lengths are plus two pitches for all diameters.

*Length of Thread on Screws:* Screws are threaded to permit a screw ring gage being screwed by hand to within a distance from the underside of the head not exceeding two and a half times the pitch for diameters up to and including 52 mm and three and a half times the pitch for diameters over 52 mm.

*Angularity and Eccentricity of Bolts, Screws and Nuts:* The axis of the thread of the nut is square to the face of the nut subject to the "squareness tolerance" given in Table 3.

In gaging, the nut is screwed by hand onto a gage, having a truncated taper thread, until the thread of the nut is tight on the thread of the gage. A sleeve sliding on a parallel extension of the gage, which has a face of diameter equal to the minimum distance across the flats of the nut and exactly at 90 degrees to the axis of the gage, is brought into contact with the leading face of the nut. With

*(Continued on page 1161 )*

**Table 1. British Standard ISO Metric Precision Hexagon Bolts, Screws and Nuts**
(BS 3692:1967)

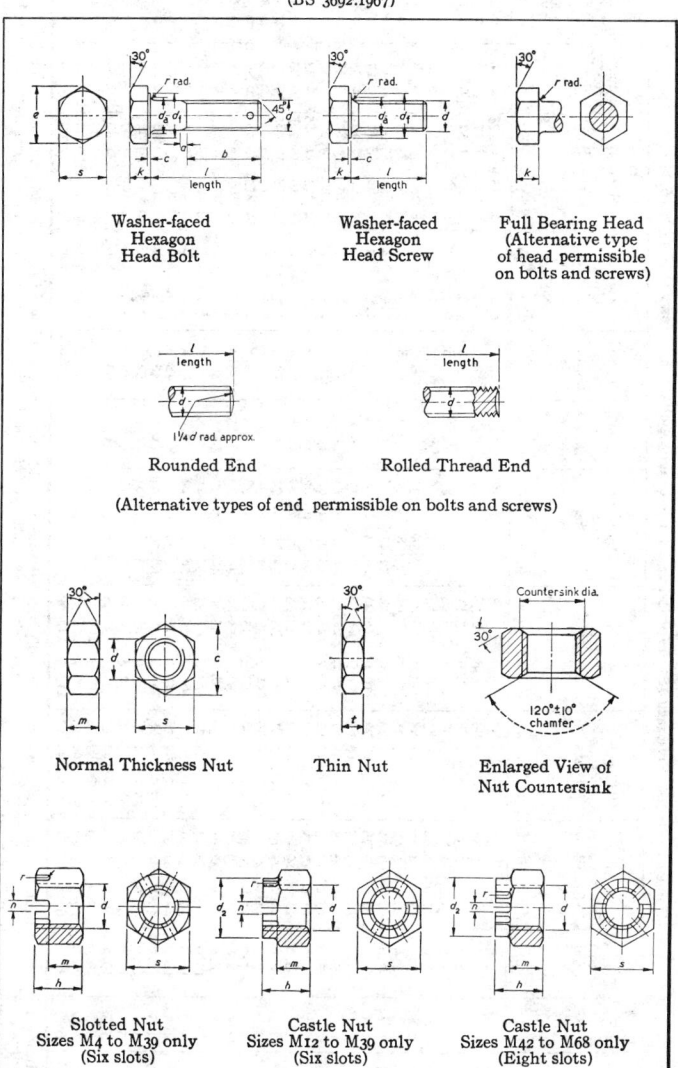

Washer-faced
Hexagon
Head Bolt

Washer-faced
Hexagon
Head Screw

Full Bearing Head
(Alternative type
of head permissible
on bolts and screws)

Rounded End

Rolled Thread End

(Alternative types of end permissible on bolts and screws)

Normal Thickness Nut

Thin Nut

Enlarged View of
Nut Countersink

Slotted Nut
Sizes M4 to M39 only
(Six slots)

Castle Nut
Sizes M12 to M39 only
(Six slots)

Castle Nut
Sizes M42 to M68 only
(Eight slots)

For general dimensions see Tables 2, 3, 4 and 5.

## Table 2.  British Standard ISO Metric Precision Hexagon Bolts and Screws (BS 3692:1967)

| Nom. Size and Thread Diam.* $d$ | Pitch of Thread (Coarse Pitch Series) | Thread Run-out $a$ Max. | Diam. of Unthreaded Shank $d$ Max. | Min. | Width Across Flats $s$ Max. | Min. | Width Across Corners $e$ Max. | Min. | Diam. of Washer Face $d_1$ Max. | Min. | Depth of Washer Face $c$ | Transition Diam.† $d_a$ Max. | Radius Under Head† $r$ Max. | Min. | Height of Head $k$ Max. | Min. | Eccentricity of Head Max. | Eccentricity of Shank and Split Pin Hole to the Thread Max. |
|---|---|---|---|---|---|---|---|---|---|---|---|---|---|---|---|---|---|---|
| M1.6 | 0.35 | 0.8 | 1.6 | 1.46 | 3.2 | 3.08 | 3.7 | 3.48 | | | | 2.0 | 0.2 | 0.1 | 1.225 | 0.975 | 0.18 | 0.14 |
| M2 | 0.4 | 1.0 | 2.0 | 1.86 | 4.0 | 3.88 | 4.6 | 4.38 | | | | 2.6 | 0.3 | 0.1 | 1.525 | 1.275 | 0.18 | 0.14 |
| M2.5 | 0.45 | 1.0 | 2.5 | 2.36 | 5.0 | 4.88 | 5.8 | 5.51 | | | | 3.1 | 0.3 | 0.1 | 1.825 | 1.575 | 0.18 | 0.14 |
| M3 | 0.5 | 1.2 | 3.0 | 2.86 | 5.5 | 5.38 | 6.4 | 6.08 | 5.08 | 4.83 | 0.1 | 3.6 | 0.35 | 0.1 | 2.125 | 1.875 | 0.18 | 0.14 |
| M4 | 0.7 | 1.6 | 4.0 | 3.82 | 7.0 | 6.85 | 8.1 | 7.74 | 6.55 | 6.30 | 0.1 | 4.7 | 0.35 | 0.2 | 2.925 | 2.675 | 0.22 | 0.18 |
| M5 | 0.8 | 2.0 | 5.0 | 4.82 | 8.0 | 7.85 | 9.2 | 8.87 | 7.55 | 7.30 | 0.2 | 5.7 | 0.4 | 0.25 | 3.650 | 3.35 | 0.22 | 0.18 |
| M6 | 1 | 2.5 | 6.0 | 5.82 | 10.0 | 9.78 | 11.5 | 11.05 | 9.48 | 9.23 | 0.3 | 6.8 | 0.4 | 0.4 | 4.15 | 3.85 | 0.22 | 0.18 |
| M8 | 1.25 | 3.0 | 8.0 | 7.78 | 13.0 | 12.73 | 15.0 | 14.38 | 12.43 | 12.18 | 0.4 | 9.2 | 0.6 | 0.4 | 5.65 | 5.35 | 0.27 | 0.22 |
| M10 | 1.5 | 3.5 | 10.0 | 9.78 | 17.0 | 16.73 | 19.6 | 18.90 | 16.43 | 16.18 | 0.4 | 11.2 | 0.6 | 0.6 | 7.18 | 6.82 | 0.27 | 0.22 |
| M12 | 1.75 | 4.0 | 12.0 | 11.73 | 19.0 | 18.67 | 21.9 | 21.10 | 18.37 | 18.12 | 0.4 | 14.2 | 1.1 | 0.6 | 8.18 | 7.82 | 0.33 | 0.27 |
| (M14) | 2 | 5.0 | 14.0 | 13.73 | 22.0 | 21.67 | 25.4 | 24.49 | 21.37 | 21.12 | 0.4 | 16.2 | 1.1 | 0.6 | 9.18 | 8.82 | 0.33 | 0.27 |
| M16 | 2 | 5.0 | 16.0 | 15.73 | 24.0 | 23.67 | 27.7 | 26.75 | 23.27 | 23.02 | 0.4 | 18.2 | 1.1 | 0.6 | 10.18 | 9.82 | 0.33 | 0.27 |
| (M18) | 2.5 | 6.0 | 18.0 | 17.73 | 27.0 | 26.67 | 31.2 | 30.14 | 26.27 | 26.02 | 0.4 | 20.2 | 1.2 | 0.6 | 12.215 | 11.785 | 0.39 | 0.33 |
| M20 | 2.5 | 6.0 | 20.0 | 19.67 | 30.0 | 29.61 | 34.6 | 33.53 | 29.27 | 28.80 | 0.4 | 22.4 | 1.2 | 0.8 | 13.215 | 12.785 | 0.39 | 0.33 |
| (M22) | 2.5 | 6.0 | 22.0 | 21.67 | 32.0 | 31.61 | 36.9 | 35.72 | 31.21 | 30.74 | 0.5 | 24.4 | 1.2 | 0.8 | 14.215 | 13.785 | 0.39 | 0.33 |
| M24 | 3 | 7.0 | 24.0 | 23.67 | 36.0 | 35.38 | 41.6 | 39.98 | 34.98 | 34.51 | 0.5 | 26.4 | 1.7 | 0.8 | 15.215 | 14.785 | 0.39 | 0.33 |
| (M27) | 3 | 7.0 | 27.0 | 26.67 | 41.0 | 40.38 | 47.3 | 45.63 | 39.98 | 39.36 | 0.5 | 30.4 | 1.7 | 1.0 | 17.215 | 16.785 | 0.39 | 0.39 |
| M30 | 3.5 | 8.0 | 30.0 | 29.67 | 46.0 | 45.38 | 53.1 | 51.28 | 44.98 | 44.36 | 0.5 | 33.4 | 1.7 | 1.0 | 19.26 | 18.74 | 0.39 | 0.39 |
| (M33) | 3.5 | 8.0 | 33.0 | 32.61 | 50.0 | 49.38 | 57.7 | 55.80 | 48.98 | 48.36 | 0.5 | 36.4 | 1.7 | 1.0 | 21.26 | 20.74 | 0.46 | 0.39 |
| M36 | 4 | 10.0 | 36.0 | 35.61 | 55.0 | 54.26 | 63.5 | 61.31 | 53.86 | 53.24 | 0.6 | 39.4 | 1.7 | 1.0 | 23.26 | 22.74 | 0.46 | 0.39 |
| (M39) | 4 | 10.0 | 39.0 | 38.61 | 60.0 | 59.26 | 69.3 | 66.96 | 58.86 | 58.24 | 0.6 | 42.4 | 1.7 | 1.0 | 25.26 | 24.74 | 0.46 | 0.39 |
| M42 | 4.5 | 11.0 | 42.0 | 41.61 | 65.0 | 64.26 | 75.1 | 72.61 | 63.76 | 63.04 | 0.6 | 45.6 | 1.8 | 1.2 | 26.26 | 25.74 | 0.46 | 0.39 |
| (M45) | 4.5 | 11.0 | 45.0 | 44.61 | 70.0 | 69.26 | 80.8 | 78.26 | 68.76 | 68.04 | 0.6 | 48.6 | 1.8 | 1.2 | 28.26 | 27.74 | 0.46 | 0.39 |
| M48 | 5 | 12.0 | 48.0 | 47.61 | 75.0 | 74.26 | 86.6 | 83.91 | 73.76 | 73.04 | 0.6 | 52.6 | 1.8 | 1.6 | 30.26 | 29.74 | 0.46 | 0.39 |
| (M52) | 5 | 12.0 | 52.0 | 51.54 | 80.0 | 79.26 | 92.4 | 89.56 | | | | 56.6 | 2.3 | 1.6 | 33.31 | 32.69 | 0.46 | 0.46 |
| M56 | 5.5 | 19.0 | 56.0 | 55.54 | 85.0 | 84.13 | 98.1 | 95.07 | | | | 63.0 | 3.5 | 2.0 | 35.31 | 34.69 | 0.54 | 0.46 |
| (M60) | 5.5 | 19.0 | 60.0 | 59.54 | 90.0 | 89.13 | 103.9 | 100.72 | | | | 67.0 | 3.5 | 2.0 | 38.31 | 37.69 | 0.54 | 0.46 |
| M64 | 6 | 21.0 | 64.0 | 63.54 | 95.0 | 94.13 | 109.7 | 106.37 | | | | 71.0 | 3.5 | 2.0 | 40.31 | 39.69 | 0.54 | 0.46 |
| (M68) | 6 | 21.0 | 68.0 | 67.54 | 100.0 | 99.13 | 115.5 | 112.02 | | | | 75.0 | 3.5 | 2.0 | 43.31 | 42.69 | 0.54 | 0.46 |

All dimensions are in millimeters.  For illustration of bolts and screws see Table 1.

\* Sizes shown in parentheses are non-preferred.

† A true radius is not essential provided that the curve is smooth and lies wholly within the maximum radius, determined from the maximum transitional diameter, and the minimum radius specified.

**Table 3.  British Standard ISO Metric Precision Hexagon Nuts and Thin Nuts (BS 3692:1967)**

| Nominal Size and Thread Diameter* $d$ | Pitch of Thread (Coarse Pitch Series) | Width Across Flats $s$ Max. | Min. | Width Across Corners $e$ Max. | Min. | Thickness of Normal Nut $m$ Max. | Min. | Tolerance on Squareness of Thread to Face of Nut† Max. | Eccentricity of Hexagon Max. | Thickness of Thin Nut $t$ Max. | Min. |
|---|---|---|---|---|---|---|---|---|---|---|---|
| M1.6 | 0.35 | 3.20 | 3.08 | 3.70 | 3.48 | 1.30 | 1.05 | 0.05 | 0.14 | … | … |
| M2 | 0.4 | 4.00 | 3.88 | 4.60 | 4.38 | 1.60 | 1.35 | 0.06 | 0.14 | … | … |
| M2.5 | 0.45 | 5.00 | 4.88 | 5.80 | 5.51 | 2.00 | 1.75 | 0.08 | 0.14 | … | … |
| M3 | 0.5 | 5.50 | 5.38 | 6.40 | 6.08 | 2.40 | 2.15 | 0.09 | 0.14 | … | … |
| M4 | 0.7 | 7.00 | 6.85 | 8.10 | 7.74 | 3.20 | 2.90 | 0.11 | 0.18 | … | … |
| M5 | 0.8 | 8.00 | 7.85 | 9.20 | 8.87 | 4.00 | 3.70 | 0.13 | 0.18 | … | … |
| M6 | 1 | 10.00 | 9.78 | 11.50 | 11.05 | 5.00 | 4.70 | 0.17 | 0.18 | … | … |
| M8 | 1.25 | 13.00 | 12.73 | 15.00 | 14.38 | 6.50 | 6.14 | 0.22 | 0.22 | 5.0 | 4.70 |
| M10 | 1.5 | 17.00 | 16.73 | 19.60 | 18.90 | 8.00 | 7.64 | 0.29 | 0.22 | 6.0 | 5.70 |
| M12 | 1.75 | 19.00 | 18.67 | 21.90 | 21.10 | 10.00 | 9.64 | 0.32 | 0.27 | 7.0 | 6.64 |
| (M14) | 2 | 22.00 | 21.67 | 25.4 | 24.49 | 11.00 | 10.57 | 0.37 | 0.27 | 8.0 | 7.64 |
| M16 | 2 | 24.00 | 23.67 | 27.7 | 26.75 | 13.00 | 12.57 | 0.41 | 0.27 | 8.0 | 7.64 |
| (M18) | 2.5 | 27.00 | 26.67 | 31.20 | 30.14 | 15.00 | 14.57 | 0.46 | 0.27 | 9.0 | 8.64 |
| M20 | 2.5 | 30.00 | 29.67 | 34.60 | 33.53 | 16.00 | 15.57 | 0.51 | 0.33 | 9.0 | 8.64 |
| (M22) | 2.5 | 32.00 | 31.61 | 36.90 | 35.72 | 18.00 | 17.57 | 0.54 | 0.33 | 10.0 | 9.64 |
| M24 | 3 | 36.00 | 35.38 | 41.60 | 39.98 | 19.00 | 18.48 | 0.61 | 0.33 | 10.0 | 9.64 |
| (M27) | 3 | 41.00 | 40.38 | 47.3 | 45.63 | 22.00 | 21.48 | 0.70 | 0.33 | 12.0 | 11.57 |
| M30 | 3.5 | 46.00 | 45.38 | 53.1 | 51.28 | 24.00 | 23.48 | 0.78 | 0.33 | 12.0 | 11.57 |
| (M33) | 3.5 | 50.00 | 49.38 | 57.70 | 55.80 | 26.00 | 25.48 | 0.85 | 0.39 | 14.0 | 13.57 |
| M36 | 4 | 55.00 | 54.26 | 63.50 | 61.31 | 29.00 | 28.48 | 0.94 | 0.39 | 14.0 | 13.57 |
| (M39) | 4 | 60.00 | 59.26 | 69.30 | 66.96 | 31.00 | 30.38 | 1.03 | 0.39 | 16.0 | 15.57 |
| M42 | 4.5 | 65.00 | 64.26 | 75.10 | 72.61 | 34.00 | 33.38 | 1.11 | 0.39 | 16.0 | 15.57 |
| (M45) | 4.5 | 70.00 | 69.26 | 80.80 | 78.26 | 36.00 | 35.38 | 1.20 | 0.39 | 18.0 | 17.57 |
| M48 | 5 | 75.00 | 74.26 | 86.60 | 83.91 | 38.00 | 37.38 | 1.29 | 0.39 | 18.0 | 17.57 |
| (M52) | 5 | 80.00 | 79.26 | 92.40 | 89.56 | 42.00 | 41.38 | 1.37 | 0.46 | 20.0 | 19.48 |
| M56 | 5.5 | 85.00 | 84.13 | 98.10 | 95.07 | 45.00 | 44.38 | 1.46 | 0.46 | … | … |
| (M60) | 5.5 | 90.00 | 89.13 | 103.90 | 100.72 | 48.00 | 47.38 | 1.55 | 0.46 | … | … |
| M64 | 6 | 95.00 | 94.13 | 109.70 | 106.37 | 51.00 | 50.26 | 1.63 | 0.46 | … | … |
| (M68) | 6 | 100.00 | 99.13 | 115.50 | 112.02 | 54.00 | 53.26 | 1.72 | 0.46 | … | … |

All dimensions are in millimeters.  For illustration of hexagon nuts and thin nuts see Table I.  * Sizes shown in parentheses are non-preferred.
† As measured with the nut squareness gage described in the text and illustrated in Appendix A of the Standard and a feeler gage.

Table 4. British Standard ISO Metric Precision Hexagon Slotted Nuts and Castle Nuts (BS 3692:1967)

| Nominal Size and Thread Diameter* d | Width Across Flats s Max. | Min. | Width Across Corners e Max. | Min. | Diameter d₂ Max. | Min. | Thickness h Max. | Min. | Lower Face of Nut to Bottom of Slot m Max. | Min. | Width of Slot n Max. | Min. | Radius (0.25 n) r Min. | Eccentricity of the Slots Max. |
|---|---|---|---|---|---|---|---|---|---|---|---|---|---|---|
| M4 | 7.00 | 6.85 | 8.10 | 7.74 | ... | ... | 5 | 4.70 | 3.2 | 2.90 | 1.45 | 1.2 | 0.3 | 0.18 |
| M5 | 8.00 | 7.85 | 9.20 | 8.87 | ... | ... | 6 | 5.70 | 4.0 | 3.70 | 1.65 | 1.4 | 0.35 | 0.18 |
| M6 | 10.00 | 9.78 | 11.50 | 11.05 | ... | ... | 7.5 | 7.14 | 5 | 4.70 | 2.25 | 2 | 0.5 | 0.18 |
| M8 | 13.00 | 12.73 | 15.00 | 14.38 | ... | ... | 9.5 | 9.14 | 6.5 | 6.14 | 2.75 | 2.5 | 0.625 | 0.22 |
| M10 | 17.00 | 16.73 | 19.60 | 18.90 | ... | ... | 12 | 11.57 | 8 | 7.64 | 3.05 | 2.8 | 0.70 | 0.22 |
| M12 | 19.00 | 18.67 | 21.90 | 21.10 | 17 | 16.57 | 15 | 14.57 | 10 | 9.64 | 3.80 | 3.5 | 0.875 | 0.27 |
| (M14) | 22.00 | 21.67 | 25.4 | 24.49 | 19 | 18.48 | 16 | 15.57 | 11 | 10.57 | 3.80 | 3.5 | 0.875 | 0.27 |
| M16 | 24.00 | 23.67 | 27.7 | 26.75 | 22 | 21.48 | 19 | 18.48 | 13 | 12.57 | 4.80 | 4.5 | 1.125 | 0.27 |
| (M18) | 27.00 | 26.67 | 31.20 | 30.14 | 25 | 24.48 | 21 | 20.48 | 15 | 14.57 | 4.80 | 4.5 | 1.125 | 0.27 |
| M20 | 30.00 | 29.67 | 34.60 | 33.53 | 28 | 27.48 | 22 | 21.48 | 16 | 15.57 | 4.80 | 4.5 | 1.125 | 0.33 |
| (M22) | 32.00 | 31.61 | 36.90 | 35.72 | 30 | 29.48 | 26 | 25.48 | 18 | 17.57 | 5.80 | 5.5 | 1.375 | 0.33 |
| M24 | 36.00 | 35.38 | 41.60 | 39.98 | 34 | 33.38 | 27 | 26.48 | 19 | 18.48 | 5.80 | 5.5 | 1.375 | 0.33 |
| (M27) | 41.00 | 40.38 | 47.3 | 45.63 | 38 | 37.38 | 30 | 29.48 | 22 | 21.48 | 5.80 | 5.5 | 1.375 | 0.33 |
| M30 | 46.00 | 45.38 | 53.1 | 51.28 | 42 | 41.38 | 33 | 32.38 | 24 | 23.48 | 7.36 | 7 | 1.75 | 0.33 |
| M33 | 50.00 | 49.38 | 57.70 | 55.80 | 46 | 45.38 | 35 | 34.38 | 26 | 25.48 | 7.36 | 7 | 1.75 | 0.39 |
| M36 | 55.00 | 54.26 | 63.50 | 61.31 | 50 | 49.38 | 38 | 37.38 | 29 | 28.48 | 7.36 | 7 | 1.75 | 0.39 |
| (M39) | 60.00 | 59.26 | 69.30 | 66.96 | 55 | 54.26 | 40 | 39.38 | 31 | 30.38 | 7.36 | 7 | 1.75 | 0.39 |
| M42 | 65.00 | 64.26 | 75.10 | 72.61 | 58 | 57.26 | 46 | 45.38 | 34 | 33.38 | 9.36 | 9 | 2.25 | 0.39 |
| (M45) | 70.00 | 69.26 | 80.80 | 78.26 | 62 | 61.26 | 48 | 47.38 | 36 | 35.38 | 9.36 | 9 | 2.25 | 0.39 |
| M48 | 75.00 | 74.26 | 86.60 | 83.91 | 65 | 64.26 | 50 | 49.38 | 38 | 37.38 | 9.36 | 9 | 2.25 | 0.39 |
| (M52) | 80.00 | 79.26 | 92.40 | 89.56 | 70 | 69.26 | 54 | 53.26 | 42 | 41.38 | 9.36 | 9 | 2.25 | 0.46 |
| M56 | 85.00 | 84.13 | 98.10 | 95.07 | 75 | 74.26 | 57 | 56.26 | 45 | 44.38 | 9.36 | 9 | 2.25 | 0.46 |
| (M60) | 90.00 | 89.13 | 103.90 | 100.72 | 80 | 79.26 | 63 | 62.26 | 48 | 47.38 | 11.43 | 11 | 2.75 | 0.46 |
| M64 | 95.00 | 94.13 | 109.70 | 106.37 | 85 | 84.13 | 66 | 65.26 | 51 | 50.26 | 11.43 | 11 | 2.75 | 0.46 |
| (M68) | 100.00 | 99.13 | 115.50 | 112.02 | 90 | 89.13 | 69 | 68.26 | 54 | 53.26 | 11.43 | 11 | 2.75 | 0.46 |

All dimensions are in millimeters. For illustration of hexagon slotted nuts and castle nuts see Table 1.    * Sizes shown in parentheses are non-preferred.

**Table 5.  British Standard ISO Metric Bolt and Screw Nominal Lengths**
(BS 3692:1967)

| Nominal Length* $l$ | Tolerance | Nominal Length* $l$ | Tolerance | Nominal Length* $l$ | Tolerance | Nominal Length* $l$ | Tolerance |
|---|---|---|---|---|---|---|---|
| 5 | ±0.24 | 30 | ±0.42 | 90 | ±0.70 | 200 | ±0.925 |
| 6 | ±0.24 | (32) | ±0.50 | (95) | ±0.70 | 220 | ±0.925 |
| (7) | ±0.29 | 35 | ±0.50 | 100 | ±0.70 | 240 | ±0.925 |
| 8 | ±0.29 | (38) | ±0.50 | (105) | ±0.70 | 260 | ±1.05 |
| (9) | ±0.29 | 40 | ±0.50 | 110 | ±0.70 | 280 | ±1.05 |
| 10 | ±0.29 | 45 | ±0.50 | (115) | ±0.70 | 300 | ±1.05 |
| (11) | ±0.35 | 50 | ±0.50 | 120 | ±0.70 | 325 | ±1.15 |
| 12 | ±0.35 | 55 | ±0.60 | (125) | ±0.80 | 350 | ±1.15 |
| 14 | ±0.35 | 60 | ±0.60 | 130 | ±0.80 | 375 | ±1.15 |
| 16 | ±0.35 | 65 | ±0.60 | 140 | ±0.80 | 400 | ±1.15 |
| (18) | ±0.35 | 70 | ±0.60 | 150 | ±0.80 | 425 | ±1.25 |
| 20 | ±0.42 | 75 | ±0.60 | 160 | ±0.80 | 450 | ±1.25 |
| (22) | ±0.42 | 80 | ±0.60 | 170 | ±0.80 | 475 | ±1.25 |
| 25 | ±0.42 | 85 | ±0.70 | 180 | ±0.80 | 500 | ±1.25 |
| (28) | ±0.42 | .... | .... | 190 | ±0.925 | .... | .... |

All dimensions are in millimeters.       *Nominal lengths shown in parentheses are non-preferred.

the sleeve in this position, it should not be possible for a feeler gage of thickness equal to the "squareness tolerance" to enter anywhere between the leading nut face and sleeve face.

The hexagon flats of bolts, screws and nuts are square to the bearing face, and the angularity of the head is within the limits of 90 degrees, plus or minus 1 degree. The eccentricity of the hexagon flats of nuts relative to the thread diameter should not exceed the values given in Table 3 and the eccentricity of the head relative to the width across flats and eccentricity between the shank and thread of bolts and screws should not exceed the values given in Table 2.

*Chamfering, Washer Facing and Countersinking:* Bolt and screw heads have a chamfer of approximately 30 degrees on their upper faces and, at the option of the manufacturer, a washer face or full bearing face on the underside.  Nuts are countersunk at an included angle of 120 degrees plus or minus 10 degrees at both ends of the thread.  The diameter of the countersink should not exceed the nominal major diameter of the thread plus 0.13 mm up to and including 12 mm diameter, and plus 0.25 mm above 12 mm diameter.  This stipulation does not apply to slotted, castle or nuts.

*Strength Grade Designation System for Steel Bolts and Screws:* This Standard includes a strength grade designation system consisting of two figures.  The first figure is one tenth of the minimum tensile strength in kgf/mm², and the second figure is one tenth of the ratio between the minimum yield stress (or stress at permanent set limit, $R_{0.2}$) and the minimum tensile strength, expressed as a percentage.  For example with the strength designation grade 8.8, the first figure 8 represents 1/10 the minimum tensile strength of 80 kgf/mm² and the second figure 8 represents 1/10 the ratio

$$\frac{\text{stress at permanent set limit } R_{0.2}\%}{\text{minimum tensile strength}} = \frac{1}{10} \times \frac{64}{80} \times \frac{100}{1};$$

the numerical values of stress and strength being obtained from the accompanying table.

## Strength Grade Designations of Steel Bolts and Screws

| Strength Grade Designation | 4.6 | 4.8 | 5.6 | 5.8 | 6.6 | 6.8 | 8.8 | 10.9 | 12.9 | 14.9 |
|---|---|---|---|---|---|---|---|---|---|---|
| Tensile Strength ($R_m$), Min. | 40 | 40 | 50 | 50 | 60 | 60 | 80 | 100 | 120 | 140 |
| Yield Stress ($R_e$), Min. | 24 | 32 | 30 | 40 | 36 | 48 | ... | ... | ... | ... |
| Stress at Permanent Set Limit ($R_{0.2}$), Min. | ... | ... | ... | ... | ... | ... | 64 | 90 | 108 | 126 |

All stress and strength values are in kgf/mm² units.

*Strength Grade Designation System for Steel Nuts:* The strength grade designation system for steel nuts is a number which is one-tenth of the specified proof load stress in kgf/mm². The proof load stress corresponds to the minimum tensile strength of the highest grade of bolt or screw with which the nut can be used.

## Strength Grade Designations of Steel Nuts

| Strength Grade Designation | 4 | 5 | 6 | 8 | 12 | 14 |
|---|---|---|---|---|---|---|
| Proof Load Stress (kgf/mm²) | 40 | 50 | 60 | 80 | 120 | 140 |

## Recommended Bolt and Nut Combinations

| Grade of Bolt | 4.6 | 4.8 | 5.6 | 5.8 | 6.6 | 6.8 | 8.8 | 10.9 | 12.9 | 14.9 |
|---|---|---|---|---|---|---|---|---|---|---|
| Recommended Grade of Nut | 4 | 4 | 5 | 5 | 6 | 6 | 8 | 12 | 12 | 14 |

*Note:* Nuts of a higher strength grade may be substituted for nuts of a lower strength grade.

*Marking:* The marking and identification requirements of this Standard are only mandatory for steel bolts, screws and nuts of 6 mm diameter and larger; manufactured to strength grade designations 8.8 (for bolts or screws) and 8 (for nuts) or higher. Bolts and screws are identified as ISO metric by either of the symbols "ISO M" or "M", embossed or indented on top of the head. Nuts may be indented or embossed by alternative methods depending on their method of manufacture.

*Designation:* Bolts 10 mm diameter, 50 mm long manufactured from steel of strength grade 8.8, would be designated:

"Bolts M10 × 50 to BS 3692 — 8.8".

Brass screws 8 mm diameter, 20 mm long would be designated:

"Brass screws M8 × 20 to BS 3692".

Nuts 12 mm diameter, manufactured from steel of strength grade 6, cadmium plated could be designated:

"Nuts M12 to BS 3692 — 6, plated to BS 3382: Part 1".

*Miscellaneous Information:* The Standard also gives mechanical properties of steel bolts, screws and nuts [i.e., tensile strengths; hardnesses (Brinell, Rockwell, Vickers); stresses (yield, proof load); etc.], material and manufacture of steel bolts, screws and nuts; and information on inspection and testing. Appendices to the Standard give information on gaging; chemical composition; testing of mechanical properties; examples of marking of bolts, screws and nuts; and a table of preferred standard sizes of bolts and screws, to name some.

**British Standard General Purpose Studs (B.S. 2693: Part 1:1956)**

## LIMITS FOR END SCREWED INTO COMPONENT (All threads except B.A.)

| Nom. Diam. D | Major Diam. Max. | Threads per Inch | Major Diam. Min. | Effective Diameter Max. | Effective Diameter Min. | Minor Diameter Max. | Minor Diameter Min. | Threads per Inch | Major Diam. Min. | Effective Diameter Max. | Effective Diameter Min. | Minor Diameter Max. | Minor Diameter Min. |
|---|---|---|---|---|---|---|---|---|---|---|---|---|---|
| **UN TH'DS.** | | | | **UNF THREADS** | | | | | | **UNC THREADS** | | | |
| ¼ | .2500 | 28 | .2435 | .2294 | .2265 | .2088 | .2037 | 20 | .2419 | .2201 | .2172 | .1913 | .1849 |
| ⁵⁄₁₆ | .3125 | 24 | .3053 | .2883 | .2852 | .2643 | .2586 | 18 | .3038 | .2793 | .2762 | .2472 | .2402 |
| ⅜ | .3750 | 24 | .3678 | .3510 | .3478 | .3270 | .3211 | 16 | .3656 | .3375 | .3343 | .3014 | .2936 |
| ⁷⁄₁₆ | .4375 | 20 | .4294 | .4084 | .4050 | .3796 | .3729 | 14 | .4272 | .3945 | .3911 | .3533 | .3447 |
| ½ | .5000 | 20 | .4919 | .4712 | .4675 | .4424 | .4356 | 13 | .4891 | .4537 | .4500 | .4093 | .4000 |
| ⁹⁄₁₆ | .5625 | 18 | .5538 | .5302 | .5264 | .4981 | .4907 | 12 | .5511 | .5122 | .5084 | .4641 | .4542 |
| ⅝ | .6250 | 18 | .6163 | .5929 | .5889 | .5608 | .5533 | 11 | .6129 | .5700 | .5660 | .5175 | .5069 |
| ¾ | .7500 | 16 | .7406 | .7137 | .7094 | .6776 | .6693 | 10 | .7371 | .6893 | .6850 | .6316 | .6200 |
| ⅞ | .8750 | 14 | .8647 | .8332 | .8286 | .7920 | .7828 | 9 | .8611 | .8074 | .8028 | .7433 | .7306 |
| 1 | 1.0000 | 12 | .9886 | .9510 | .9459 | .9029 | .8925 | 8 | .9850 | .9239 | .9188 | .8517 | .8376 |
| 1⅛ | 1.1250 | 12 | 1.1136 | 1.0762 | 1.0709 | 1.0281 | 1.0176 | 7 | 1.1086 | 1.0375 | 1.0322 | .9550 | .9393 |
| 1¼ | 1.2500 | 12 | 1.2386 | 1.2014 | 1.1959 | 1.1533 | 1.1427 | 7 | 1.2336 | 1.1627 | 1.1572 | 1.0802 | 1.0644 |
| 1⅜ | 1.3750 | 12 | 1.3636 | 1.3265 | 1.3209 | 1.2784 | 1.2677 | 6 | 1.3568 | 1.2723 | 1.2667 | 1.1761 | 1.1581 |
| 1½ | 1.5000 | 12 | 1.4886 | 1.4517 | 1.4459 | 1.4036 | 1.3928 | 6 | 1.4818 | 1.3975 | 1.3917 | 1.3013 | 1.2832 |
| **B.S. TH'DS.** | | | | **B.S.F. THREADS** | | | | | | **B.S.W. THREADS** | | | |
| ¼ | .2500 | 26 | .2455 | .2280 | .2251 | .2034 | .1984 | 20 | .2452 | .2206 | .2177 | .1886 | .1831 |
| ⁵⁄₁₆ | .3125 | 22 | .3077 | .2863 | .2832 | .2572 | .2517 | 18 | .3073 | .2798 | .2767 | .2442 | .2383 |
| ⅜ | .3750 | 20 | .3699 | .3461 | .3429 | .3141 | .3083 | 16 | .3695 | .3381 | .3349 | .2981 | .2919 |
| ⁷⁄₁₆ | .4375 | 18 | .4320 | .4053 | .4019 | .3697 | .3635 | 14 | .4316 | .3952 | .3918 | .3495 | .3428 |
| ½ | .5000 | 16 | .4942 | .4637 | .4600 | .4237 | .4172 | 12 | .4937 | .4503 | .4466 | .3969 | .3897 |
| ⁹⁄₁₆ | .5625 | 16 | .5566 | .5263 | .5225 | .4863 | .4797 | 12 | .5560 | .5129 | .5091 | .4595 | .4521 |
| ⅝ | .6250 | 14 | .6187 | .5833 | .5793 | .5376 | .5305 | 11 | .6183 | .5708 | .5668 | .5126 | .5050 |
| ¾ | .7500 | 12 | .7432 | .7009 | .6966 | .6475 | .6398 | 10 | .7428 | .6903 | .6860 | .6263 | .6182 |
| ⅞ | .8750 | 11 | .8678 | .8214 | .8168 | .7632 | .7551 | 9 | .8674 | .8085 | .8039 | .7374 | .7288 |
| 1 | 1.0000 | 10 | .9924 | .9411 | .9360 | .8771 | .8686 | 8 | .9920 | .9251 | .9200 | .8451 | .8360 |
| 1⅛ | 1.1250 | 9 | 1.1171 | 1.0592 | 1.0539 | .9881 | .9792 | 7 | 1.1164 | 1.0388 | 1.0335 | .9473 | .9376 |
| 1¼ | 1.2500 | 9 | 1.2419 | 1.1844 | 1.1789 | 1.1133 | 1.1042 | 7 | 1.2413 | 1.1640 | 1.1585 | 1.0725 | 1.0627 |
| 1⅜ | 1.3750 | 8 | 1.3665 | 1.3006 | 1.2950 | 1.2206 | 1.2110 | . . . . | . . . . | . . . . | . . . . | . . . . | . . . . |
| 1½ | 1.5000 | 8 | 1.4913 | 1.4258 | 1.4200 | 1.3458 | 1.3360 | 6 | 1.4906 | 1.3991 | 1.3933 | 1.2924 | 1.2818 |

## LIMITS FOR END SCREWED INTO COMPONENT (B.A. Threads)[1]

| Designation No. | Pitch | Major Diameter Max. | Major Diameter Min. | Effective Diameter Max. | Effective Diameter Min. | Minor Diameter Max. | Minor Diameter Min. |
|---|---|---|---|---|---|---|---|
| 2 | .8100 mm .03189 in. | 4.700 mm .1850 in. | 4.580 mm .1803 in. | 4.275 mm .1683 in. | 4.200 mm .1654 in. | 3.790 mm .1492 in. | 3.620 mm .1425 in. |
| 4 | .6600 mm .02598 in. | 3.600 mm .1417 in. | 3.500 mm .1378 in. | 3.260 mm .1283 in. | 3.190 mm .1256 in. | 2.865 mm .1128 in. | 2.720 mm .1071 in. |

## MINIMUM NOMINAL LENGTHS OF STUDS[2]

| Nom. Stud Diam. | For Thread Length (Component End) of 1D | 1.5D | Nom. Stud Diam. | 1D | 1.5D | Nom. Stud Diam. | 1D | 1.5D |
|---|---|---|---|---|---|---|---|---|
| ¼ | ⅞ | 1 | ⁹⁄₁₆ | 2 | 2⅜ | 1⅛ | 4 | 4⅝ |
| ⁵⁄₁₆ | 1⅛ | 1⅜ | ⅝ | 2¼ | 2⅝ | 1¼ | 4¾ | 5½ |
| ⅜ | 1⅜ | 1⅝ | ¾ | 2⅝ | 3 | 1⅜ | 5 | 5¾ |
| ⁷⁄₁₆ | 1⅝ | 1⅞ | ⅞ | 3⅛ | 3⅝ | 1½ | 5¼ | 6 |
| ½ | 1¾ | 2 | 1 | 3½ | 4 | ... | ... | ... |

All dimensions are in inches except where otherwise noted.

[1] Approximate inch equivalents are shown below the dimensions given in mm.

[2] The standard also gives preferred and standard lengths of studs:

*Preferred* lengths of studs: ⅞, 1, 1⅛, 1¼, 1⅜, 1½, 1¾, 2, 2¼, 2½, 2¾, 3, 3¼, 3½ and for lengths above 3½ the preferred increment is ½.

*Standard* lengths of studs: ⅞, 1, 1⅛, 1¼, 1⅜, 1½, 1⅝, 1¾, 1⅞, 2, 2⅛, 2¼, 2⅜, 2½, 2⅝, 2¾, 2⅞, 3, 3⅛, 3¼, 3⅜, 3½ and for lengths above 3½ the standard increment is ¼.

*Tolerance* for length is ±¹⁄₃₂ for all lengths up to and including 4 and ±¹⁄₁₆ over 4.

### Table 1A.   American National Standard Type A Plain Washers — Preferred Sizes** (ANSI B18.22.1-1965, R1975)

| Nominal Washer Size*** | | Series | Inside Diameter | | | Outside Diameter | | | Thickness | | |
|---|---|---|---|---|---|---|---|---|---|---|---|
| | | | Basic | Tolerance Plus | Tolerance Minus | Basic | Tolerance Plus | Tolerance Minus | Basic | Max. | Min. |
| — | — | | 0.078 | 0.000 | 0.005 | 0.188 | 0.000 | 0.005 | 0.020 | 0.025 | 0.016 |
| — | — | | 0.094 | 0.000 | 0.005 | 0.250 | 0.000 | 0.005 | 0.020 | 0.025 | 0.016 |
| — | — | | 0.125 | 0.008 | 0.005 | 0.312 | 0.008 | 0.005 | 0.032 | 0.040 | 0.025 |
| No. 6 | 0.138 | | 0.156 | 0.008 | 0.005 | 0.375 | 0.015 | 0.005 | 0.049 | 0.065 | 0.036 |
| No. 8 | 0.164 | | 0.188 | 0.008 | 0.005 | 0.438 | 0.015 | 0.005 | 0.049 | 0.065 | 0.036 |
| No. 10 | 0.190 | | 0.219 | 0.008 | 0.005 | 0.500 | 0.015 | 0.005 | 0.049 | 0.065 | 0.036 |
| 3/16 | 0.188 | | 0.250 | 0.015 | 0.005 | 0.562 | 0.015 | 0.005 | 0.049 | 0.065 | 0.036 |
| No. 12 | 0.216 | | 0.250 | 0.015 | 0.005 | 0.562 | 0.015 | 0.005 | 0.065 | 0.080 | 0.051 |
| 1/4 | 0.250 | N | 0.281 | 0.015 | 0.005 | 0.625 | 0.015 | 0.005 | 0.065 | 0.080 | 0.051 |
| 1/4 | 0.250 | W | 0.312 | 0.015 | 0.005 | 0.734* | 0.015 | 0.007 | 0.065 | 0.080 | 0.051 |
| 5/16 | 0.312 | N | 0.344 | 0.015 | 0.005 | 0.688 | 0.015 | 0.007 | 0.065 | 0.080 | 0.051 |
| 5/16 | 0.312 | W | 0.375 | 0.015 | 0.005 | 0.875 | 0.030 | 0.007 | 0.083 | 0.104 | 0.064 |
| 3/8 | 0.375 | N | 0.406 | 0.015 | 0.005 | 0.812 | 0.015 | 0.007 | 0.065 | 0.080 | 0.051 |
| 3/8 | 0.375 | W | 0.438 | 0.015 | 0.005 | 1.000 | 0.030 | 0.007 | 0.083 | 0.104 | 0.064 |
| 7/16 | 0.438 | N | 0.469 | 0.015 | 0.005 | 0.922 | 0.015 | 0.007 | 0.065 | 0.080 | 0.051 |
| 7/16 | 0.438 | W | 0.500 | 0.015 | 0.005 | 1.250 | 0.030 | 0.007 | 0.083 | 0.104 | 0.064 |
| 1/2 | 0.500 | N | 0.531 | 0.015 | 0.005 | 1.062 | 0.030 | 0.007 | 0.095 | 0.121 | 0.074 |
| 1/2 | 0.500 | W | 0.562 | 0.015 | 0.005 | 1.375 | 0.030 | 0.007 | 0.109 | 0.132 | 0.086 |
| 9/16 | 0.562 | N | 0.594 | 0.015 | 0.005 | 1.156* | 0.030 | 0.007 | 0.095 | 0.121 | 0.074 |
| 9/16 | 0.562 | W | 0.625 | 0.015 | 0.005 | 1.469* | 0.030 | 0.007 | 0.109 | 0.132 | 0.086 |
| 5/8 | 0.625 | N | 0.656 | 0.030 | 0.007 | 1.312 | 0.030 | 0.007 | 0.095 | 0.121 | 0.074 |
| 5/8 | 0.625 | W | 0.688 | 0.030 | 0.007 | 1.750 | 0.030 | 0.007 | 0.134 | 0.160 | 0.108 |
| 3/4 | 0.750 | N | 0.812 | 0.030 | 0.007 | 1.469 | 0.030 | 0.007 | 0.134 | 0.160 | 0.108 |
| 3/4 | 0.750 | W | 0.812 | 0.030 | 0.007 | 2.000 | 0.030 | 0.007 | 0.148 | 0.177 | 0.122 |
| 7/8 | 0.875 | N | 0.938 | 0.030 | 0.007 | 1.750 | 0.030 | 0.007 | 0.134 | 0.160 | 0.108 |
| 7/8 | 0.875 | W | 0.938 | 0.030 | 0.007 | 2.250 | 0.030 | 0.007 | 0.165 | 0.192 | 0.136 |
| 1 | 1.000 | N | 1.062 | 0.030 | 0.007 | 2.000 | 0.030 | 0.007 | 0.134 | 0.160 | 0.108 |
| 1 | 1.000 | W | 1.062 | 0.030 | 0.007 | 2.500 | 0.030 | 0.007 | 0.165 | 0.192 | 0.136 |
| 1 1/8 | 1.125 | N | 1.250 | 0.030 | 0.007 | 2.250 | 0.030 | 0.007 | 0.134 | 0.160 | 0.108 |
| 1 1/8 | 1.125 | W | 1.250 | 0.030 | 0.007 | 2.750 | 0.030 | 0.007 | 0.165 | 0.192 | 0.136 |
| 1 1/4 | 1.250 | N | 1.375 | 0.030 | 0.007 | 2.500 | 0.030 | 0.007 | 0.165 | 0.192 | 0.136 |
| 1 1/4 | 1.250 | W | 1.375 | 0.030 | 0.007 | 3.000 | 0.030 | 0.007 | 0.165 | 0.192 | 0.136 |
| 1 3/8 | 1.375 | N | 1.500 | 0.030 | 0.007 | 2.750 | 0.030 | 0.007 | 0.165 | 0.192 | 0.136 |
| 1 3/8 | 1.375 | W | 1.500 | 0.045 | 0.010 | 3.250 | 0.045 | 0.010 | 0.180 | 0.213 | 0.153 |
| 1 1/2 | 1.500 | N | 1.625 | 0.030 | 0.007 | 3.000 | 0.030 | 0.007 | 0.165 | 0.192 | 0.136 |
| 1 1/2 | 1.500 | W | 1.625 | 0.045 | 0.010 | 3.500 | 0.045 | 0.010 | 0.180 | 0.213 | 0.153 |
| 1 5/8 | 1.625 | | 1.750 | 0.045 | 0.010 | 3.750 | 0.045 | 0.010 | 0.180 | 0.213 | 0.153 |
| 1 3/4 | 1.750 | | 1.875 | 0.045 | 0.010 | 4.000 | 0.045 | 0.010 | 0.180 | 0.213 | 0.153 |
| 1 7/8 | 1.875 | | 2.000 | 0.045 | 0.010 | 4.250 | 0.045 | 0.010 | 0.180 | 0.213 | 0.153 |
| 2 | 2.000 | | 2.125 | 0.045 | 0.010 | 4.500 | 0.045 | 0.010 | 0.180 | 0.213 | 0.153 |
| 2 1/4 | 2.250 | | 2.375 | 0.045 | 0.010 | 4.750 | 0.045 | 0.010 | 0.220 | 0.248 | 0.193 |
| 2 1/2 | 2.500 | | 2.625 | 0.045 | 0.010 | 5.000 | 0.045 | 0.010 | 0.238 | 0.280 | 0.210 |
| 2 3/4 | 2.750 | | 2.875 | 0.065 | 0.010 | 5.250 | 0.065 | 0.010 | 0.259 | 0.310 | 0.228 |
| 3 | 3.000 | | 3.125 | 0.065 | 0.010 | 5.500 | 0.065 | 0.010 | 0.284 | 0.327 | 0.249 |

All dimensions are in inches.

* The 0.734-inch, 1.156-inch, and 1.469-inch outside diameters avoid washers which could be used in coin operated devices.

** Preferred sizes are for the most part from series previously designated "Standard Plate" and "SAE." Where common sizes existed in the two series, the SAE size is designated "N" (narrow) and the Standard Plate "W" (wide). These sizes as well as all other sizes of Type A Plain Washers are to be ordered by ID, OD, and thickness dimensions.

*** Nominal washer sizes are intended for use with comparable nominal screw or bolt sizes.

Additional selected sizes of Type A Plain Washers are shown in Table 1B.

Table 1B. American National Standard Type A Plain Washers —
Additional Selected Sizes (ANSI B18.22.1-1965, R1975)

| Inside Diameter | | | Outside Diameter | | | Thickness | | |
|---|---|---|---|---|---|---|---|---|
| Basic | Tolerance | | Basic | Tolerance | | Basic | Max. | Min. |
| | Plus | Minus | | Plus | Minus | | | |
| 0.094 | 0.000 | 0.005 | 0.219 | 0.000 | 0.005 | 0.020 | 0.025 | 0.016 |
| 0.125 | 0.000 | 0.005 | 0.250 | 0.000 | 0.005 | 0.022 | 0.028 | 0.017 |
| 0.156 | 0.008 | 0.005 | 0.312 | 0.008 | 0.005 | 0.035 | 0.048 | 0.027 |
| 0.172 | 0.008 | 0.005 | 0.406 | 0.015 | 0.005 | 0.049 | 0.065 | 0.036 |
| 0.188 | 0.008 | 0.005 | 0.375 | 0.015 | 0.005 | 0.049 | 0.065 | 0.036 |
| 0.203 | 0.008 | 0.005 | 0.469 | 0.015 | 0.005 | 0.049 | 0.065 | 0.036 |
| 0.219 | 0.008 | 0.005 | 0.438 | 0.015 | 0.005 | 0.049 | 0.065 | 0.036 |
| 0.234 | 0.008 | 0.005 | 0.531 | 0.015 | 0.005 | 0.049 | 0.065 | 0.036 |
| 0.250 | 0.015 | 0.005 | 0.500 | 0.015 | 0.005 | 0.049 | 0.065 | 0.036 |
| 0.266 | 0.015 | 0.005 | 0.625 | 0.015 | 0.005 | 0.049 | 0.065 | 0.036 |
| 0.312 | 0.015 | 0.005 | 0.875 | 0.015 | 0.007 | 0.065 | 0.080 | 0.051 |
| 0.375 | 0.015 | 0.005 | 0.734* | 0.015 | 0.007 | 0.065 | 0.080 | 0.051 |
| 0.375 | 0.015 | 0.005 | 1.125 | 0.015 | 0.007 | 0.065 | 0.080 | 0.051 |
| 0.438 | 0.015 | 0.005 | 0.875 | 0.030 | 0.007 | 0.083 | 0.104 | 0.064 |
| 0.438 | 0.015 | 0.005 | 1.375 | 0.030 | 0.007 | 0.083 | 0.104 | 0.064 |
| 0.500 | 0.015 | 0.005 | 1.125 | 0.030 | 0.007 | 0.083 | 0.104 | 0.064 |
| 0.500 | 0.015 | 0.005 | 1.625 | 0.030 | 0.007 | 0.083 | 0.104 | 0.064 |
| 0.562 | 0.015 | 0.005 | 1.250 | 0.030 | 0.007 | 0.109 | 0.132 | 0.086 |
| 0.562 | 0.015 | 0.005 | 1.875 | 0.030 | 0.007 | 0.109 | 0.132 | 0.086 |
| 0.625 | 0.015 | 0.005 | 1.375 | 0.030 | 0.007 | 0.109 | 0.132 | 0.086 |
| 0.625 | 0.015 | 0.005 | 2.125 | 0.030 | 0.007 | 0.134 | 0.160 | 0.108 |
| 0.688 | 0.030 | 0.007 | 1.469* | 0.030 | 0.007 | 0.134 | 0.160 | 0.108 |
| 0.688 | 0.030 | 0.007 | 2.375 | 0.030 | 0.007 | 0.165 | 0.192 | 0.136 |
| 0.812 | 0.030 | 0.007 | 1.750 | 0.030 | 0.007 | 0.148 | 0.177 | 0.122 |
| 0.812 | 0.030 | 0.007 | 2.875 | 0.030 | 0.007 | 0.165 | 0.192 | 0.136 |
| 0.938 | 0.030 | 0.007 | 2.000 | 0.030 | 0.007 | 0.165 | 0.192 | 0.136 |
| 0.938 | 0.030 | 0.007 | 3.375 | 0.045 | 0.010 | 0.180 | 0.213 | 0.153 |
| 1.062 | 0.030 | 0.007 | 2.250 | 0.030 | 0.007 | 0.165 | 0.192 | 0.136 |
| 1.062 | 0.045 | 0.010 | 3.875 | 0.045 | 0.010 | 0.238 | 0.280 | 0.210 |
| 1.250 | 0.030 | 0.007 | 2.500 | 0.030 | 0.007 | 0.165 | 0.192 | 0.136 |
| 1.375 | 0.030 | 0.007 | 2.750 | 0.030 | 0.007 | 0.165 | 0.192 | 0.136 |
| 1.500 | 0.045 | 0.010 | 3.000 | 0.045 | 0.010 | 0.180 | 0.213 | 0.153 |
| 1.625 | 0.045 | 0.010 | 3.250 | 0.045 | 0.010 | 0.180 | 0.213 | 0.153 |
| 1.688 | 0.045 | 0.010 | 3.500 | 0.045 | 0.010 | 0.180 | 0.213 | 0.153 |
| 1.812 | 0.045 | 0.010 | 3.750 | 0.045 | 0.010 | 0.180 | 0.213 | 0.153 |
| 1.938 | 0.045 | 0.010 | 4.000 | 0.045 | 0.010 | 0.180 | 0.213 | 0.153 |
| 2.062 | 0.045 | 0.010 | 4.250 | 0.045 | 0.010 | 0.180 | 0.213 | 0.153 |

All dimensions are in inches.
* The 0.734-inch and 1.469-inch outside diameters avoid washers which could be used in coin operated devices.
The above sizes are to be ordered by ID, OD, and thickness dimensions.
Preferred Sizes of Type A Plain Washers are shown in Table 1A.

**ANSI Standard Plain Washers.** — The Type A plain washers were originally developed in a light, medium, heavy and extra heavy series. These series have been discontinued and the washers are now designated by their nominal dimensions.

The Type B plain washers are available in a narrow, regular and wide series with proportions designed to distribute the load over larger areas of lower strength materials.

Plain washers are made of ferrous or non-ferrous metal, plastic or other material as specified. The tolerances indicated in the tables are intended for metal washers only.

### Table 2. American National Standard Type B Plain Washers
(ANSI B18.22.1-1965, R1975)

| Nominal Washer Size** | | Series† | Inside Diameter | | | Outside Diameter | | | Thickness | | |
|---|---|---|---|---|---|---|---|---|---|---|---|
| | | | Basic | Tolerance | | Basic | Tolerance | | Basic | Max. | Min. |
| | | | | Plus | Minus | | Plus | Minus | | | |
| No. 0 | 0.060 | N | 0.068 | 0.000 | 0.005 | 0.125 | 0.000 | 0.005 | 0.025 | 0.028 | 0.022 |
| | | R | 0.068 | 0.000 | 0.005 | 0.188 | 0.000 | 0.005 | 0.025 | 0.028 | 0.022 |
| | | W | 0.068 | 0.000 | 0.005 | 0.250 | 0.000 | 0.005 | 0.025 | 0.028 | 0.022 |
| No. 1 | 0.073 | N | 0.084 | 0.000 | 0.005 | 0.156 | 0.000 | 0.005 | 0.025 | 0.028 | 0.022 |
| | | R | 0.084 | 0.000 | 0.005 | 0.219 | 0.000 | 0.005 | 0.025 | 0.028 | 0.022 |
| | | W | 0.084 | 0.000 | 0.005 | 0.281 | 0.000 | 0.005 | 0.032 | 0.036 | 0.028 |
| No. 2 | 0.086 | N | 0.094 | 0.000 | 0.005 | 0.188 | 0.000 | 0.005 | 0.025 | 0.028 | 0.022 |
| | | R | 0.094 | 0.000 | 0.005 | 0.250 | 0.000 | 0.005 | 0.032 | 0.036 | 0.028 |
| | | W | 0.094 | 0.000 | 0.005 | 0.344 | 0.000 | 0.005 | 0.032 | 0.036 | 0.028 |
| No. 3 | 0.099 | N | 0.109 | 0.000 | 0.005 | 0.219 | 0.000 | 0.005 | 0.025 | 0.028 | 0.022 |
| | | R | 0.109 | 0.000 | 0.005 | 0.312 | 0.000 | 0.005 | 0.032 | 0.036 | 0.028 |
| | | W | 0.109 | 0.008 | 0.005 | 0.406 | 0.008 | 0.005 | 0.040 | 0.045 | 0.036 |
| No. 4 | 0.112 | N | 0.125 | 0.000 | 0.005 | 0.250 | 0.000 | 0.005 | 0.032 | 0.036 | 0.028 |
| | | R | 0.125 | 0.008 | 0.005 | 0.375 | 0.008 | 0.005 | 0.040 | 0.045 | 0.036 |
| | | W | 0.125 | 0.008 | 0.005 | 0.438 | 0.008 | 0.005 | 0.040 | 0.045 | 0.036 |
| No. 5 | 0.125 | N | 0.141 | 0.000 | 0.005 | 0.281 | 0.000 | 0.005 | 0.032 | 0.036 | 0.028 |
| | | R | 0.141 | 0.008 | 0.005 | 0.406 | 0.008 | 0.005 | 0.040 | 0.045 | 0.036 |
| | | W | 0.141 | 0.008 | 0.005 | 0.500 | 0.008 | 0.005 | 0.040 | 0.045 | 0.036 |
| No. 6 | 0.138 | N | 0.156 | 0.000 | 0.005 | 0.312 | 0.000 | 0.005 | 0.032 | 0.036 | 0.028 |
| | | R | 0.156 | 0.008 | 0.005 | 0.438 | 0.008 | 0.005 | 0.040 | 0.045 | 0.036 |
| | | W | 0.156 | 0.008 | 0.005 | 0.562 | 0.008 | 0.005 | 0.040 | 0.045 | 0.036 |
| No. 8 | 0.164 | N | 0.188 | 0.008 | 0.005 | 0.375 | 0.008 | 0.005 | 0.040 | 0.045 | 0.036 |
| | | R | 0.188 | 0.008 | 0.005 | 0.500 | 0.008 | 0.005 | 0.040 | 0.045 | 0.036 |
| | | W | 0.188 | 0.008 | 0.005 | 0.625 | 0.015 | 0.005 | 0.063 | 0.071 | 0.056 |
| No. 10 | 0.190 | N | 0.203 | 0.008 | 0.005 | 0.406 | 0.008 | 0.005 | 0.040 | 0.045 | 0.036 |
| | | R | 0.203 | 0.008 | 0.005 | 0.562 | 0.008 | 0.005 | 0.040 | 0.045 | 0.036 |
| | | W | 0.203 | 0.008 | 0.005 | 0.734* | 0.015 | 0.007 | 0.063 | 0.071 | 0.056 |
| No. 12 | 0.216 | N | 0.234 | 0.008 | 0.005 | 0.438 | 0.008 | 0.005 | 0.040 | 0.045 | 0.036 |
| | | R | 0.234 | 0.008 | 0.005 | 0.625 | 0.015 | 0.005 | 0.063 | 0.071 | 0.056 |
| | | W | 0.234 | 0.008 | 0.005 | 0.875 | 0.015 | 0.007 | 0.063 | 0.071 | 0.056 |
| ¼ | 0.250 | N | 0.281 | 0.015 | 0.005 | 0.500 | 0.015 | 0.005 | 0.063 | 0.071 | 0.056 |
| | | R | 0.281 | 0.015 | 0.005 | 0.734* | 0.015 | 0.007 | 0.063 | 0.071 | 0.056 |
| | | W | 0.281 | 0.015 | 0.005 | 1.000 | 0.015 | 0.007 | 0.063 | 0.071 | 0.056 |
| ⁵⁄₁₆ | 0.312 | N | 0.344 | 0.015 | 0.005 | 0.625 | 0.015 | 0.005 | 0.063 | 0.071 | 0.056 |
| | | R | 0.344 | 0.015 | 0.005 | 0.875 | 0.015 | 0.007 | 0.063 | 0.071 | 0.056 |
| | | W | 0.344 | 0.015 | 0.005 | 1.125 | 0.015 | 0.007 | 0.063 | 0.071 | 0.056 |
| ⅜ | 0.375 | N | 0.406 | 0.015 | 0.005 | 0.734* | 0.015 | 0.007 | 0.063 | 0.071 | 0.056 |
| | | R | 0.406 | 0.015 | 0.005 | 1.000 | 0.015 | 0.007 | 0.063 | 0.071 | 0.056 |
| | | W | 0.406 | 0.015 | 0.005 | 1.250 | 0.030 | 0.007 | 0.100 | 0.112 | 0.090 |
| ⁷⁄₁₆ | 0.438 | N | 0.469 | 0.015 | 0.005 | 0.875 | 0.015 | 0.007 | 0.063 | 0.071 | 0.056 |
| | | R | 0.469 | 0.015 | 0.005 | 1.125 | 0.015 | 0.007 | 0.063 | 0.071 | 0.056 |
| | | W | 0.469 | 0.015 | 0.005 | 1.469* | 0.030 | 0.007 | 0.100 | 0.112 | 0.090 |

All dimensions are in inches.

   * The 0.734-inch and 1.469-inch outside diameters avoid washers which could be used in coin operated devices.

   ** Nominal washer sizes are intended for use with comparable nominal screw or bolt sizes.

   † N indicates Narrow; R, Regular, and W, Wide Series.

Inside and outside diameters shall be concentric within at least the inside diameter tolerance.

Washers shall be flat within 0.005 inch for basic outside diameters up to and including 0.875 inch, and within 0.010 inch for larger outside diameters.

Table 2 (*Concluded*). American National Standard Type B Plain Washers
(ANSI B18.22.1-1965, R1975).

| Nominal Washer Size** | | Series† | Inside Diameter | | | Outside Diameter | | | Thickness | | |
|---|---|---|---|---|---|---|---|---|---|---|---|
| | | | Basic | Plus | Minus | Basic | Plus | Minus | Basic | Max. | Min. |
| ½ | 0.500 | N | 0.531 | 0.015 | 0.005 | 1.000 | 0.015 | 0.007 | 0.063 | 0.071 | 0.056 |
| | | R | 0.531 | 0.015 | 0.005 | 1.250 | 0.030 | 0.007 | 0.100 | 0.112 | 0.090 |
| | | W | 0.531 | 0.015 | 0.005 | 1.750 | 0.030 | 0.007 | 0.100 | 0.112 | 0.090 |
| 9⁄16 | 0.562 | N | 0.594 | 0.015 | 0.005 | 1.125 | 0.015 | 0.007 | 0.063 | 0.071 | 0.056 |
| | | R | 0.594 | 0.015 | 0.005 | 1.469* | 0.030 | 0.007 | 0.100 | 0.112 | 0.090 |
| | | W | 0.594 | 0.015 | 0.005 | 2.000 | 0.030 | 0.007 | 0.100 | 0.112 | 0.090 |
| ⅝ | 0.625 | N | 0.656 | 0.030 | 0.007 | 1.250 | 0.030 | 0.007 | 0.100 | 0.112 | 0.090 |
| | | R | 0.656 | 0.030 | 0.007 | 1.750 | 0.030 | 0.007 | 0.100 | 0.112 | 0.090 |
| | | W | 0.656 | 0.030 | 0.007 | 2.250 | 0.030 | 0.007 | 0.160 | 0.174 | 0.146 |
| ¾ | 0.750 | N | 0.812 | 0.030 | 0.007 | 1.375 | 0.030 | 0.007 | 0.100 | 0.112 | 0.090 |
| | | R | 0.812 | 0.030 | 0.007 | 2.000 | 0.030 | 0.007 | 0.100 | 0.112 | 0.090 |
| | | W | 0.812 | 0.030 | 0.007 | 2.500 | 0.030 | 0.007 | 0.160 | 0.174 | 0.146 |
| ⅞ | 0.875 | N | 0.938 | 0.030 | 0.007 | 1.469* | 0.030 | 0.007 | 0.100 | 0.112 | 0.090 |
| | | R | 0.938 | 0.030 | 0.007 | 2.250 | 0.030 | 0.007 | 0.160 | 0.174 | 0.146 |
| | | W | 0.938 | 0.030 | 0.007 | 2.750 | 0.030 | 0.007 | 0.160 | 0.174 | 0.146 |
| 1 | 1.000 | N | 1.062 | 0.030 | 0.007 | 1.750 | 0.030 | 0.007 | 0.100 | 0.112 | 0.090 |
| | | R | 1.062 | 0.030 | 0.007 | 2.500 | 0.030 | 0.007 | 0.160 | 0.174 | 0.146 |
| | | W | 1.062 | 0.030 | 0.007 | 3.000 | 0.030 | 0.007 | 0.160 | 0.174 | 0.146 |
| 1⅛ | 1.125 | N | 1.188 | 0.030 | 0.007 | 2.000 | 0.030 | 0.007 | 0.100 | 0.112 | 0.090 |
| | | R | 1.188 | 0.030 | 0.007 | 2.750 | 0.030 | 0.007 | 0.160 | 0.174 | 0.146 |
| | | W | 1.188 | 0.030 | 0.007 | 3.250 | 0.030 | 0.007 | 0.160 | 0.174 | 0.146 |
| 1¼ | 1.250 | N | 1.312 | 0.030 | 0.007 | 2.250 | 0.030 | 0.007 | 0.160 | 0.174 | 0.146 |
| | | R | 1.312 | 0.030 | 0.007 | 3.000 | 0.030 | 0.007 | 0.160 | 0.174 | 0.146 |
| | | W | 1.312 | 0.045 | 0.010 | 3.500 | 0.045 | 0.010 | 0.250 | 0.266 | 0.234 |
| 1⅜ | 1.375 | N | 1.438 | 0.030 | 0.007 | 2.500 | 0.030 | 0.007 | 0.160 | 0.174 | 0.146 |
| | | R | 1.438 | 0.030 | 0.007 | 3.250 | 0.030 | 0.007 | 0.160 | 0.174 | 0.146 |
| | | W | 1.438 | 0.045 | 0.010 | 3.750 | 0.045 | 0.010 | 0.250 | 0.266 | 0.234 |
| 1½ | 1.500 | N | 1.562 | 0.030 | 0.007 | 2.750 | 0.030 | 0.007 | 0.160 | 0.174 | 0.146 |
| | | R | 1.562 | 0.045 | 0.010 | 3.500 | 0.045 | 0.010 | 0.250 | 0.266 | 0.234 |
| | | W | 1.562 | 0.045 | 0.010 | 4.000 | 0.045 | 0.010 | 0.250 | 0.266 | 0.234 |
| 1⅝ | 1.625 | N | 1.750 | 0.030 | 0.007 | 3.000 | 0.030 | 0.007 | 0.160 | 0.174 | 0.146 |
| | | R | 1.750 | 0.045 | 0.010 | 3.750 | 0.045 | 0.010 | 0.250 | 0.266 | 0.234 |
| | | W | 1.750 | 0.045 | 0.010 | 4.250 | 0.045 | 0.010 | 0.250 | 0.266 | 0.234 |
| 1¾ | 1.750 | N | 1.875 | 0.030 | 0.007 | 3.250 | 0.030 | 0.007 | 0.160 | 0.174 | 0.146 |
| | | R | 1.875 | 0.045 | 0.010 | 4.000 | 0.045 | 0.010 | 0.250 | 0.266 | 0.234 |
| | | W | 1.875 | 0.045 | 0.010 | 4.500 | 0.045 | 0.010 | 0.250 | 0.266 | 0.234 |
| 1⅞ | 1.875 | N | 2.000 | 0.045 | 0.010 | 3.500 | 0.045 | 0.010 | 0.250 | 0.266 | 0.234 |
| | | R | 2.000 | 0.045 | 0.010 | 4.250 | 0.045 | 0.010 | 0.250 | 0.266 | 0.234 |
| | | W | 2.000 | 0.045 | 0.010 | 4.750 | 0.045 | 0.010 | 0.250 | 0.266 | 0.234 |
| 2 | 2.000 | N | 2.125 | 0.045 | 0.010 | 3.750 | 0.045 | 0.010 | 0.250 | 0.266 | 0.234 |
| | | R | 2.125 | 0.045 | 0.010 | 4.500 | 0.045 | 0.010 | 0.250 | 0.266 | 0.234 |
| | | W | 2.125 | 0.045 | 0.010 | 5.000 | 0.045 | 0.010 | 0.250 | 0.266 | 0.234 |

All dimensions are in inches.
* The 1.469-inch outside diameter avoids washers which could be used in coin operated devices.
** Nominal washer sizes are intended for use with comparable nominal screw or bolt sizes.
† N indicates Narrow; R, Regular; and W, Wide Series.
Inside and outside diameters shall be concentric within at least the inside diameter tolerance.
Washers shall be flat within 0.005-inch for basic outside diameters up through 0.875-inch and within 0.010 inch for larger outside diameters.
For 2¼-, 2½-, 2¾-, and 3-inch sizes see ANSI B18.22.1-1965, (R1975)

Table 1. American National Standard Helical Spring Lock Washers (ANSI B18.21.1-1972)

ENLARGED SECTION

| Nominal Washer Size | | Inside Diameter, A | | Regular* | | | Heavy† | | | Extra Duty‡ | | |
|---|---|---|---|---|---|---|---|---|---|---|---|---|
| | | Max. | Min. | O.D., B Max.¶ | Section Width, W | Section Thickness, T§ | O.D., B Max.¶ | Section Width, W | Section Thickness, T§ | O.D., B Max.¶ | Section Width, W | Section Thickness, T§ |
| No. 2 | 0.086 | 0.094 | 0.088 | 0.172 | 0.035 | 0.020 | 0.182 | 0.040 | 0.025 | 0.208 | 0.053 | 0.027 |
| No. 3 | 0.099 | 0.107 | 0.101 | 0.195 | 0.040 | 0.025 | 0.209 | 0.047 | 0.031 | 0.239 | 0.062 | 0.034 |
| No. 4 | 0.112 | 0.120 | 0.114 | 0.209 | 0.040 | 0.025 | 0.223 | 0.047 | 0.031 | 0.253 | 0.062 | 0.034 |
| No. 5 | 0.125 | 0.133 | 0.127 | 0.236 | 0.047 | 0.031 | 0.252 | 0.055 | 0.040 | 0.300 | 0.079 | 0.045 |
| No. 6 | 0.138 | 0.148 | 0.141 | 0.250 | 0.047 | 0.031 | 0.266 | 0.055 | 0.040 | 0.314 | 0.079 | 0.045 |
| No. 8 | 0.164 | 0.174 | 0.167 | 0.293 | 0.055 | 0.040 | 0.307 | 0.062 | 0.047 | 0.375 | 0.096 | 0.057 |
| No. 10 | 0.190 | 0.200 | 0.193 | 0.334 | 0.062 | 0.047 | 0.350 | 0.070 | 0.056 | 0.434 | 0.112 | 0.068 |
| No. 12 | 0.216 | 0.227 | 0.220 | 0.377 | 0.070 | 0.056 | 0.391 | 0.077 | 0.063 | 0.497 | 0.130 | 0.080 |
| 1/4 | 0.250 | 0.262 | 0.254 | 0.489 | 0.109 | 0.062 | 0.491 | 0.110 | 0.077 | 0.535 | 0.132 | 0.084 |
| 5/16 | 0.312 | 0.326 | 0.317 | 0.586 | 0.125 | 0.078 | 0.596 | 0.130 | 0.097 | 0.622 | 0.143 | 0.108 |
| 3/8 | 0.375 | 0.390 | 0.380 | 0.683 | 0.141 | 0.094 | 0.691 | 0.145 | 0.115 | 0.741 | 0.170 | 0.123 |
| 7/16 | 0.438 | 0.455 | 0.443 | 0.779 | 0.156 | 0.109 | 0.787 | 0.160 | 0.133 | 0.839 | 0.186 | 0.143 |
| 1/2 | 0.500 | 0.518 | 0.506 | 0.873 | 0.171 | 0.125 | 0.883 | 0.176 | 0.151 | 0.939 | 0.204 | 0.162 |
| 9/16 | 0.562 | 0.582 | 0.570 | 0.971 | 0.188 | 0.141 | 0.981 | 0.193 | 0.170 | 1.041 | 0.223 | 0.182 |
| 5/8 | 0.625 | 0.650 | 0.635 | 1.079 | 0.203 | 0.156 | 1.093 | 0.210 | 0.189 | 1.157 | 0.242 | 0.202 |
| 11/16 | 0.688 | 0.713 | 0.698 | 1.176 | 0.219 | 0.172 | 1.192 | 0.227 | 0.207 | 1.258 | 0.260 | 0.221 |
| 3/4 | 0.750 | 0.775 | 0.760 | 1.271 | 0.234 | 0.188 | 1.291 | 0.244 | 0.226 | 1.361 | 0.279 | 0.241 |
| 13/16 | 0.812 | 0.843 | 0.824 | 1.367 | 0.250 | 0.203 | 1.391 | 0.262 | 0.246 | 1.463 | 0.298 | 0.261 |
| 7/8 | 0.875 | 0.905 | 0.887 | 1.464 | 0.266 | 0.219 | 1.494 | 0.281 | 0.266 | 1.576 | 0.322 | 0.285 |
| 15/16 | 0.938 | 0.970 | 0.950 | 1.560 | 0.281 | 0.234 | 1.594 | 0.298 | 0.284 | 1.688 | 0.345 | 0.308 |
| 1 | 1.000 | 1.042 | 1.017 | 1.661 | 0.297 | 0.250 | 1.705 | 0.319 | 0.306 | 1.799 | 0.366 | 0.330 |
| 1 1/16 | 1.062 | 1.107 | 1.080 | 1.756 | 0.312 | 0.266 | 1.808 | 0.338 | 0.326 | 1.910 | 0.389 | 0.352 |
| 1 1/8 | 1.125 | 1.172 | 1.144 | 1.853 | 0.328 | 0.281 | 1.909 | 0.356 | 0.345 | 2.019 | 0.411 | 0.375 |
| 1 3/16 | 1.188 | 1.237 | 1.208 | 1.950 | 0.344 | 0.297 | 2.008 | 0.373 | 0.364 | 2.124 | 0.432 | 0.396 |
| 1 1/4 | 1.250 | 1.302 | 1.271 | 2.045 | 0.359 | 0.312 | 2.113 | 0.393 | 0.384 | 2.231 | 0.452 | 0.417 |
| 1 5/16 | 1.312 | 1.366 | 1.334 | 2.141 | 0.375 | 0.328 | 2.211 | 0.410 | 0.403 | 2.335 | 0.472 | 0.438 |
| 1 3/8 | 1.375 | 1.432 | 1.398 | 2.239 | 0.391 | 0.344 | 2.311 | 0.427 | 0.422 | 2.439 | 0.491 | 0.458 |
| 1 7/16 | 1.438 | 1.497 | 1.462 | 2.334 | 0.406 | 0.359 | 2.406 | 0.442 | 0.440 | 2.540 | 0.509 | 0.478 |
| 1 1/2 | 1.500 | 1.561 | 1.525 | 2.430 | 0.422 | 0.375 | 2.502 | 0.458 | 0.458 | 2.638 | 0.526 | 0.496 |

All dimensions are given in inches.   * Formerly designated Medium Helical Spring Lock Washers.   † Not recommended for new applications.   ‡ Formerly designated Extra Heavy Helical Spring Lock Washers.   ¶ The maximum outside diameters specified allow for the commercial tolerances on cold-drawn wire.   § $T$ = mean section thickness = $(t_i + t_o) \div 2$

**Table 2.   American National Standard Hi-Collar\* Helical Spring Lock Washers**
(ANSI B18.21.1-1972)

| Nominal Washer Size | Inside Diameter | | Outside Diameter | Washer Section | |
|---|---|---|---|---|---|
| | | | | Width | Thickness |
| | Min. | Max. | Max.\*\* | Min. | Min. |
| No. 4 | 0.112 | 0.114 | 0.120 | 0.173 | 0.022 | 0.022 |
| No. 5 | 0.125 | 0.127 | 0.133 | 0.202 | 0.030 | 0.030 |
| No. 6 | 0.138 | 0.141 | 0.148 | 0.216 | 0.030 | 0.030 |
| No. 8 | 0.164 | 0.167 | 0.174 | 0.267 | 0.042 | 0.047 |
| No. 10 | 0.190 | 0.193 | 0.200 | 0.294 | 0.042 | 0.047 |
| ¼ | 0.250 | 0.254 | 0.262 | 0.365 | 0.047 | 0.078 |
| ⁵⁄₁₆ | 0.312 | 0.317 | 0.326 | 0.460 | 0.062 | 0.093 |
| ⅜ | 0.375 | 0.380 | 0.390 | 0.553 | 0.076 | 0.125 |
| ⁷⁄₁₆ | 0.438 | 0.443 | 0.455 | 0.647 | 0.090 | 0.140 |
| ½ | 0.500 | 0.506 | 0.518 | 0.737 | 0.103 | 0.172 |
| ⅝ | 0.625 | 0.635 | 0.650 | 0.923 | 0.125 | 0.203 |
| ¾ | 0.750 | 0.760 | 0.775 | 1.111 | 0.154 | 0.218 |
| ⅞ | 0.875 | 0.887 | 0.905 | 1.296 | 0.182 | 0.234 |
| 1 | 1.000 | 1.017 | 1.042 | 1.483 | 0.208 | 0.250 |
| 1⅛ | 1.125 | 1.144 | 1.172 | 1.669 | 0.236 | 0.313 |
| 1¼ | 1.250 | 1.271 | 1.302 | 1.799 | 0.236 | 0.313 |
| 1⅜ | 1.375 | 1.398 | 1.432 | 2.041 | 0.292 | 0.375 |
| 1½ | 1.500 | 1.525 | 1.561 | 2.170 | 0.292 | 0.375 |
| 1¾ | 1.750 | 1.775 | 1.811 | 2.602 | 0.383 | 0.469 |
| 2 | 2.000 | 2.025 | 2.061 | 2.852 | 0.383 | 0.469 |
| 2¼ | 2.250 | 2.275 | 2.311 | 3.352 | 0.508 | 0.508 |
| 2½ | 2.500 | 2.525 | 2.561 | 3.602 | 0.508 | 0.508 |
| 2¾ | 2.750 | 2.775 | 2.811 | 4.102 | 0.633 | 0.633 |
| 3 | 3.000 | 3.025 | 3.061 | 4.352 | 0.633 | 0.633 |

\* For use with 1960 Series Socket Head Cap Screws.  See page 1202.
\*\* The maximum outside diameters specified allow for the commercial tolerances on cold-drawn wire.

**American National Standard Helical Spring and Tooth Lock Washers** (ANSI B18.21.1-1972). — This standard covers helical spring lock washers of carbon steel; corrosion resistant steel, Types 302 and 305; aluminum-zinc alloy; phosphor-bronze; silicon-bronze; and K-monel; in various series.  It also covers tooth lock washers of carbon steel having internal teeth, external teeth, and both internal and external teeth, of two constructions, designated as Type A and Type B. These washers are intended for general industrial application.

*Helical spring lock washers:* They have the function of: (1) providing good bolt tension per unit of applied torque for tight assemblies; (2) providing hardened bearing surfaces to create uniform torque control; (3) providing uniform load distribution through controlled radii — section — cut-off; and (4) providing protection against looseness resulting from vibration and corrosion.

Nominal washer sizes are intended for use with comparable nominal screw or bolt sizes.  These washers are designated by the following data in the sequence shown: Product name; nominal size (number, fraction or decimal equivalent); series; material; and protective finish, if required. For example: Helical Spring Lock Washer, .375 Extra Duty, Steel, Phosphate Coated.

Carbon steel helical spring lock washers are available in four series: Regular, heavy, extra duty and hi-collar as given in Tables 1 and 2. All other helical spring lock washers made of other materials are available in the regular series as given in Table 1.

When carbon steel helical spring lock washers are to be hot-dipped galvanized for use with hot-dipped galvanized bolts or screws, they are to be coiled to limits 0.020 inch in excess of those specified in Tables 1 and 2 for minimum inside diameter and maximum outside diameter. Galvanizing on washers under ¼ inch nominal size is not recommended.

*Tooth lock washers:* They serve to lock fasteners, such as bolts and nuts, to the component parts of an assembly, or increase the friction between the fasteners and the assembly. These washers are designated in a manner similar to helical spring lock washers, and are available in carbon steel. Dimensions are given in Tables 3 and 4.

### Table 3. American National Standard Internal-External Tooth Lock Washers
(ANSI B18.21.1-1972)

All dimensions are given in inches except whole numbers under "Size"

TYPE A     TYPE B

| Size | A Inside Diameter Max. | A Min. | B Outside Diameter Max. | B Min. | C Thickness Max. | C Min. |
|---|---|---|---|---|---|---|
| No. 4 | .123 | .115 | .475 | .460 | .021 | .016 |
|  | .123 | .115 | .510 | .495 | .021 | .017 |
|  |  |  | .610 | .580 |  |  |
| No. 6 | .150 | .141 | .510 | .495 | .028 | .023 |
|  | .150 | .141 | .610 | .580 |  |  |
|  |  |  | .690 | .670 |  |  |
| No. 8 | .176 | .168 | .610 | .580 | .034 | .028 |
|  | .176 | .168 | .690 | .670 |  |  |
|  |  |  | .760 | .740 |  |  |
| No. 10 | .204 | .195 | .610 | .580 | .034 | .028 |
|  | .204 | .195 | .690 | .670 | .040 | .032 |
|  |  |  | .760 | .740 |  |  |
|  |  |  | .900 | .880 |  |  |
| No. 12 | .231 | .221 | .690 | .670 | .040 | .032 |
|  | .231 | .221 | .760 | .725 |  |  |
|  |  |  | .900 | .880 |  |  |
|  |  |  | .985 | .965 | .045 | .037 |
| ¼ | .267 | .256 | .760 | .725 | .040 | .032 |
|  | .267 | .256 | .900 | .880 |  |  |
|  |  |  | .985 | .965 | .045 | .037 |
|  |  |  | 1.070 | 1.045 |  |  |

| Size | A Inside Diameter Max. | A Min. | B Outside Diameter Max. | B Min. | C Thickness Max. | C Min. |
|---|---|---|---|---|---|---|
| 5/16 | .332 | .320 | .900 | .865 | .040 | .032 |
|  | .332 | .320 | .985 | .965 | .045 | .037 |
|  | .332 | .320 | 1.070 | 1.045 | .050 | .042 |
|  |  |  | 1.155 | 1.130 |  |  |
| 3/8 | .398 | .384 | .985 | .965 | .045 | .037 |
|  | .398 | .384 | 1.070 | 1.045 | .050 | .042 |
|  |  |  | 1.155 | 1.130 |  |  |
|  |  |  | 1.260 | 1.220 |  |  |
| 7/16 | .464 | .448 | 1.070 | 1.045 | .050 | .042 |
|  | .464 | .448 | 1.155 | 1.130 |  |  |
|  |  |  | 1.260 | 1.220 | .055 | .047 |
|  |  |  | 1.315 | 1.290 |  |  |
| ½ | .530 | .512 | 1.260 | 1.220 | .055 | .047 |
|  | .530 | .512 | 1.315 | 1.290 |  |  |
|  | .530 | .512 | 1.410 | 1.380 | .060 | .052 |
|  |  |  | 1.620 | 1.590 | .067 | .059 |
| 9/16 | .596 | .576 | 1.315 | 1.290 | .055 | .047 |
|  | .596 | .576 | 1.430 | 1.380 | .060 | .052 |
|  | .596 | .576 | 1.620 | 1.590 |  |  |
|  |  |  | 1.830 | 1.797 | .067 | .059 |
| 5/8 | .663 | .640 | 1.410 | 1.380 | .060 | .052 |
|  | .663 | .640 | 1.620 | 1.590 |  |  |
|  |  |  | 1.830 | 1.797 | .067 | .059 |
|  |  |  | 1.975 | 1.935 |  |  |

**Table 4. American National Standard Internal and External Tooth Lock Washers** (ANSI B18.21.1-1972)

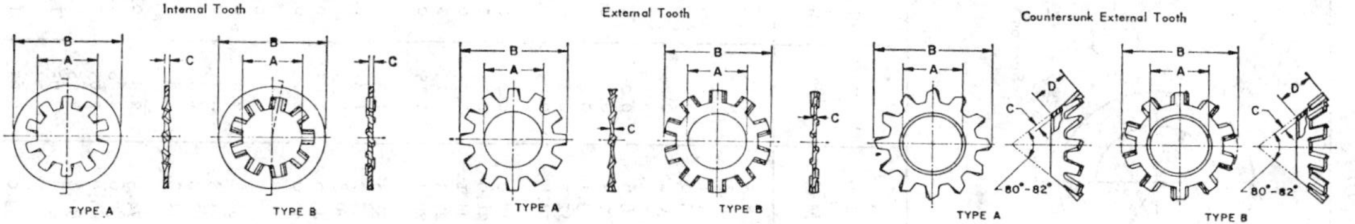


**Internal Tooth Lock Washers**

| | Size | #2 | #3 | #4 | #5 | #6 | #8 | #10 | #12 | ¼ | 5/16 | ⅜ | 7/16 | ½ | 9/16 | ⅝ | 11/16 | ¾ | 13/16 | ⅞ | 1 | 1⅛ | 1¼ |
|---|---|---|---|---|---|---|---|---|---|---|---|---|---|---|---|---|---|---|---|---|---|---|---|
| A | Max | 0.095 | 0.109 | 0.123 | 0.136 | 0.150 | 0.176 | 0.204 | 0.231 | 0.267 | 0.332 | 0.398 | 0.464 | 0.530 | 0.596 | 0.663 | 0.728 | 0.795 | 0.861 | 0.927 | 1.060 | 1.192 | 1.325 |
| | Min | 0.089 | 0.102 | 0.115 | 0.129 | 0.141 | 0.168 | 0.195 | 0.221 | 0.256 | 0.320 | 0.384 | 0.448 | 0.512 | 0.576 | 0.640 | 0.704 | 0.769 | 0.832 | 0.894 | 1.019 | 1.144 | 1.275 |
| B | Max | 0.200 | 0.232 | 0.270 | 0.280 | 0.295 | 0.340 | 0.381 | 0.410 | 0.478 | 0.610 | 0.692 | 0.789 | 0.900 | 0.985 | 1.071 | 1.166 | 1.245 | 1.315 | 1.410 | 1.637 | 1.830 | 1.975 |
| | Min | 0.175 | 0.215 | 0.255 | 0.245 | 0.275 | 0.325 | 0.365 | 0.394 | 0.460 | 0.594 | 0.670 | 0.740 | 0.867 | 0.957 | 1.045 | 1.130 | 1.220 | 1.290 | 1.364 | 1.590 | 1.799 | 1.921 |
| C | Max | 0.015 | 0.019 | 0.019 | 0.021 | 0.021 | 0.023 | 0.025 | 0.025 | 0.028 | 0.034 | 0.040 | 0.040 | 0.045 | 0.045 | 0.050 | 0.050 | 0.055 | 0.055 | 0.060 | 0.067 | 0.067 | 0.067 |
| | Min | 0.010 | 0.012 | 0.015 | 0.017 | 0.017 | 0.018 | 0.020 | 0.020 | 0.023 | 0.028 | 0.032 | 0.032 | 0.037 | 0.037 | 0.042 | 0.042 | 0.047 | 0.047 | 0.052 | 0.059 | 0.059 | 0.059 |

**External Tooth Lock Washers**

| | Size | #2 | #3 | #4 | #5 | #6 | #8 | #10 | #12 | ¼ | 5/16 | ⅜ | 7/16 | ½ | 9/16 | ⅝ | 11/16 | ¾ | 13/16 | ⅞ | 1 | 1⅛ | 1¼ |
|---|---|---|---|---|---|---|---|---|---|---|---|---|---|---|---|---|---|---|---|---|---|---|---|
| A | Max | .... | **0.109** | 0.123 | **0.136** | 0.150 | 0.176 | 0.204 | 0.231 | 0.267 | 0.332 | 0.398 | 0.464 | 0.530 | 0.596 | 0.663 | 0.728 | 0.795 | 0.861 | 0.927 | 1.060 | .... | .... |
| | Min | .... | **0.102** | 0.115 | **0.129** | 0.141 | 0.168 | 0.195 | 0.221 | 0.256 | 0.320 | 0.384 | 0.448 | 0.513 | 0.576 | 0.641 | 0.704 | 0.768 | 0.833 | 0.897 | 1.025 | .... | .... |
| B | Max | .... | **0.235** | 0.260 | **0.285** | 0.320 | 0.381 | 0.410 | 0.475 | 0.510 | 0.610 | 0.694 | 0.760 | 0.900 | 0.985 | 1.070 | 1.155 | 1.260 | 1.315 | 1.410 | 1.620 | .... | .... |
| | Min | .... | **0.220** | 0.245 | **0.270** | 0.305 | 0.365 | 0.395 | 0.460 | 0.494 | 0.588 | 0.670 | 0.740 | 0.880 | 0.960 | 1.045 | 1.130 | 1.220 | 1.290 | 1.380 | 1.590 | .... | .... |
| C | Max | .... | **0.015** | 0.019 | **0.019** | 0.022 | 0.023 | 0.025 | 0.028 | 0.028 | 0.034 | 0.040 | 0.040 | 0.045 | 0.045 | 0.050 | 0.050 | 0.055 | 0.055 | 0.060 | 0.067 | .... | .... |
| | Min | .... | **0.012** | 0.015 | **0.014** | 0.016 | 0.018 | 0.020 | 0.023 | 0.023 | 0.028 | 0.032 | 0.032 | 0.037 | 0.037 | 0.042 | 0.042 | 0.047 | 0.047 | 0.052 | 0.059 | .... | .... |

**Heavy Internal Tooth Lock Washers**

| | Size | ¼ | 5/16 | ⅜ | 7/16 | ½ | 9/16 | ⅝ | ¾ | ⅞ |
|---|---|---|---|---|---|---|---|---|---|---|
| A | Max | 0.267 | 0.332 | 0.398 | 0.464 | 0.530 | 0.596 | 0.663 | 0.795 | 0.927 |
| | Min | 0.256 | 0.320 | 0.384 | 0.448 | 0.512 | 0.576 | 0.640 | 0.768 | 0.894 |
| B | Max | 0.536 | 0.607 | 0.748 | 0.858 | 0.924 | 1.034 | 1.135 | 1.265 | 1.447 |
| | Min | 0.500 | 0.590 | 0.700 | 0.800 | 0.880 | 0.990 | 1.100 | 1.240 | 1.400 |
| C | Max | 0.045 | 0.050 | 0.050 | 0.067 | 0.067 | 0.067 | 0.067 | 0.084 | 0.084 |
| | Min | 0.035 | 0.040 | 0.042 | 0.050 | 0.055 | 0.055 | 0.059 | 0.070 | 0.075 |

**Countersunk External Tooth Lock Washers***

| | Size | #4 | #6 | #8 | #10 | #12 | ¼ | #16 | 5/16 | ⅜ | 7/16 | ½ |
|---|---|---|---|---|---|---|---|---|---|---|---|---|
| A | Max | 0.123 | 0.150 | 0.177 | 0.205 | 0.231 | 0.267 | 0.287 | 0.333 | 0.398 | 0.463 | 0.529 |
| | Min | 0.113 | 0.140 | 0.167 | 0.195 | 0.220 | 0.255 | 0.273 | 0.318 | 0.383 | 0.448 | 0.512 |
| C | Max | 0.019 | 0.021 | 0.021 | 0.025 | 0.025 | 0.025 | 0.028 | 0.028 | 0.034 | 0.045 | 0.045 |
| | Min | 0.015 | 0.017 | 0.017 | 0.020 | 0.020 | 0.020 | 0.023 | 0.023 | 0.028 | 0.037 | 0.037 |
| D | Max | 0.065 | 0.092 | 0.105 | 0.099 | 0.128 | 0.128 | 0.147 | 0.192 | 0.255 | 0.270 | 0.304 |
| | Min | 0.050 | 0.082 | 0.088 | 0.083 | 0.118 | 0.113 | 0.137 | 0.165 | 0.242 | 0.260 | 0.294 |

All dimensions are given in inches. * Starting with #4, approx. O.D.'s. are: 0.213, 0.289, 0.322, 0.354, 0.421, 0.454, 0.505, 0.599, 0.765, 0.867 and 0.976.

## British Standard Single Coil Rectangular Section Spring Washers; Metric Series — Types B and BP (BS 4464: 1969)

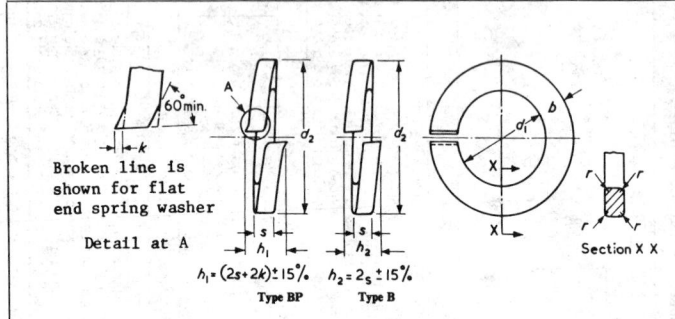

Broken line is shown for flat end spring washer

Detail at A

$h_1 = (2s+2k) \pm 15\%$    $h_2 = 2s \pm 15\%$
Type BP          Type B

Section X X

| Nom. Size & Thread Diam., d | Inside Diam., d₁ | | Width, b | Thickness, s | Outside Diam., d₂ Max | Radius, r Max | k (Type BP Only) |
|---|---|---|---|---|---|---|---|
| | Max | Min | | | | | |
| M1.6 | 1.9 | 1.7 | 0.7 ± 0.1 | 0.4 ± 0.1 | 3.5 | 0.15 | ... |
| M2 | 2.3 | 2.1 | 0.9 ± 0.1 | 0.5 ± 0.1 | 4.3 | 0.15 | ... |
| (M2.2) | 2.5 | 2.3 | 1.0 ± 0.1 | 0.6 ± 0.1 | 4.7 | 0.2 | ... |
| M2.5 | 2.8 | 2.6 | 1.0 ± 0.1 | 0.6 ± 0.1 | 5.0 | 0.2 | ... |
| M3 | 3.3 | 3.1 | 1.3 ± 0.1 | 0.8 ± 0.1 | 6.1 | 0.25 | ... |
| (M3.5) | 3.8 | 3.6 | 1.3 ± 0.1 | 0.8 ± 0.1 | 6.6 | 0.25 | 0.15 |
| M4 | 4.35 | 4.1 | 1.5 ± 0.1 | 0.9 ± 0.1 | 7.55 | 0.3 | 0.15 |
| M5 | 5.35 | 5.1 | 1.8 ± 0.1 | 1.2 ± 0.1 | 9.15 | 0.4 | 0.15 |
| M6 | 6.4 | 6.1 | 2.5 ± 0.15 | 1.6 ± 0.1 | 11.7 | 0.5 | 0.2 |
| M8 | 8.55 | 8.2 | 3 ± 0.15 | 2 ± 0.1 | 14.85 | 0.65 | 0.3 |
| M10 | 10.6 | 10.2 | 3.5 ± 0.2 | 2.2 ± 0.15 | 18.0 | 0.7 | 0.3 |
| M12 | 12.6 | 12.2 | 4 ± 0.2 | 2.5 ± 0.15 | 21.0 | 0.8 | 0.4 |
| (M14) | 14.7 | 14.2 | 4.5 ± 0.2 | 3 ± 0.15 | 24.1 | 1.0 | 0.4 |
| M16 | 16.9 | 16.3 | 5 ± 0.2 | 3.5 ± 0.2 | 27.3 | 1.15 | 0.4 |
| (M18) | 19.0 | 18.3 | 5 ± 0.2 | 3.5 ± 0.2 | 29.4 | 1.15 | 0.4 |
| M20 | 21.1 | 20.3 | 6 ± 0.2 | 4 ± 0.2 | 33.5 | 1.3 | 0.4 |
| (M22) | 23.3 | 22.4 | 6 ± 0.2 | 4 ± 0.2 | 35.7 | 1.3 | 0.4 |
| M24 | 25.3 | 24.4 | 7 ± 0.25 | 5 ± 0.2 | 39.8 | 1.65 | 0.5 |
| (M27) | 28.5 | 27.5 | 7 ± 0.25 | 5 ± 0.2 | 43.0 | 1.65 | 0.5 |
| M30 | 31.5 | 30.5 | 8 ± 0.25 | 6 ± 0.25 | 48.0 | 2.0 | 0.8 |
| (M33) | 34.6 | 33.5 | 10 ± 0.25 | 6 ± 0.25 | 55.1 | 2.0 | 0.8 |
| M36 | 37.6 | 36.5 | 10 ± 0.25 | 6 ± 0.25 | 58.1 | 2.0 | 0.8 |
| (M39) | 40.8 | 39.6 | 10 ± 0.25 | 6 ± 0.25 | 61.3 | 2.0 | 0.8 |
| M42 | 43.8 | 42.6 | 12 ± 0.25 | 7 ± 0.25 | 68.3 | 2.3 | 0.8 |
| (M45) | 46.8 | 45.6 | 12 ± 0.25 | 7 ± 0.25 | 71.3 | 2.3 | 0.8 |
| M48 | 50.0 | 48.8 | 12 ± 0.25 | 7 ± 0.25 | 74.5 | 2.3 | 0.8 |
| (M52) | 54.1 | 52.8 | 14 ± 0.25 | 8 ± 0.25 | 82.6 | 2.65 | 1.0 |
| M56 | 58.1 | 56.8 | 14 ± 0.25 | 8 ± 0.25 | 86.6 | 2.65 | 1.0 |
| (M60) | 62.3 | 60.9 | 14 ± 0.25 | 8 ± 0.25 | 90.8 | 2.65 | 1.0 |
| M64 | 66.3 | 64.9 | 14 ± 0.25 | 8 ± 0.25 | 93.8 | 2.65 | 1.0 |
| (M68) | 70.5 | 69.0 | 14 ± 0.25 | 8 ± 0.25 | 99.0 | 2.65 | 1.0 |

All dimensions are given in millimeters.   Sizes shown in parentheses are non-preferred, and are not usually stock sizes.

WASHERS 1173

British Standard Double Coil Rectangular Section Spring Washers; Metric
Series — Type D (BS 4464: 1969)

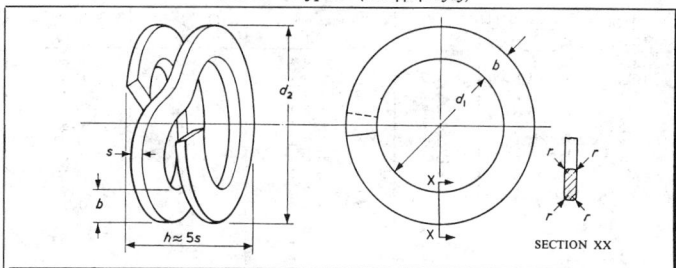

| Nom. | Inside Diam., $d_1$ | | Width, $b$ | Thick-ness, $s$ | O.D., $d_2$ Max | Radius, $r$ Max |
|------|-----|-----|----|----|----|----|
| Size, $d$ | Max | Min | | | | |
| M2 | 2.4 | 2.1 | 0.9 ± 0.1 | 0.5 ± 0.05 | 4.4 | 0.15 |
| (M2.2) | 2.6 | 2.3 | 1.0 ± 0.1 | 0.6 ± 0.05 | 4.8 | 0.2 |
| M2.5 | 2.9 | 2.6 | 1.2 ± 0.1 | 0.7 ± 0.1 | 5.5 | 0.23 |
| M3.0 | 3.6 | 3.3 | 1.2 ± 0.1 | 0.8 ± 0.1 | 6.2 | 0.25 |
| (M3.5) | 4.1 | 3.8 | 1.6 ± 0.1 | 0.8 ± 0.1 | 7.5 | 0.25 |
| M4 | 4.6 | 4.3 | 1.6 ± 0.1 | 0.8 ± 0.1 | 8.0 | 0.25 |
| M5 | 5.6 | 5.3 | 2 ± 0.1 | 0.9 ± 0.1 | 9.8 | 0.3 |
| M6 | 6.6 | 6.3 | 3 ± 0.15 | 1 ± 0.1 | 12.9 | 0.33 |
| M8 | 8.8 | 8.4 | 3 ± 0.15 | 1.2 ± 0.1 | 15.1 | 0.4 |
| M10 | 10.8 | 10.4 | 3.5 ± 0.20 | 1.2 ± 0.1 | 18.2 | 0.4 |
| M12 | 12.8 | 12.4 | 3.5 ± 0.2 | 1.6 ± 0.1 | 20.2 | 0.5 |
| (M14) | 15.0 | 14.5 | 5 ± 0.2 | 1.6 ± 0.1 | 25.4 | 0.5 |
| M16 | 17.0 | 16.5 | 5 ± 0.2 | 2 ± 0.1 | 27.4 | 0.65 |
| (M18) | 19.0 | 18.5 | 5 ± 0.2 | 2 ± 0.1 | 29.4 | 0.65 |
| M20 | 21.5 | 20.8 | 5 ± 0.2 | 2 ± 0.1 | 31.9 | 0.65 |
| (M22) | 23.5 | 22.8 | 6 ± 0.2 | 2.5 ± 0.15 | 35.9 | 0.8 |
| M24 | 26.0 | 25.0 | 6.5 ± 0.2 | 3.25 ± 0.15 | 39.4 | 1.1 |
| (M27) | 29.5 | 28.0 | 7 ± 0.25 | 3.25 ± 0.15 | 44.0 | 1.1 |
| M30 | 33.0 | 31.5 | 8 ± 0.25 | 3.25 ± 0.15 | 49.5 | 1.1 |
| (M33) | 36.0 | 34.5 | 8 ± 0.25 | 3.25 ± 0.15 | 52.5 | 1.1 |
| M36 | 40.0 | 38.0 | 10 ± 0.25 | 3.25 ± 0.15 | 60.5 | 1.1 |
| (M39) | 43.0 | 41.0 | 10 ± 0.25 | 3.25 ± 0.15 | 63.5 | 1.1 |
| M42 | 46.0 | 44.0 | 10 ± 0.25 | 4.5 ± 0.2 | 66.5 | 1.5 |
| M48 | 52.0 | 50.0 | 10 ± 0.25 | 4.5 ± 0.2 | 72.5 | 1.5 |
| M56 | 60.0 | 58.0 | 12 ± 0.25 | 4.5 ± 0.2 | 84.5 | 1.5 |
| M64 | 70.0 | 67.0 | 12 ± 0.25 | 4.5 ± 0.2 | 94.5 | 1.5 |

All dimensions are given in millimeters. Sizes shown in parentheses are non-preferred,
and are not usually stock sizes. The free height of double coil washers before compression is normally approximately five times the thickness but, if required, washers with
other free heights may be obtained by arrangement with manufacturer.

British Standard Single Coil Square Section Spring Washers; Metric Series —
Type A — 1 (BS 4464:1969)

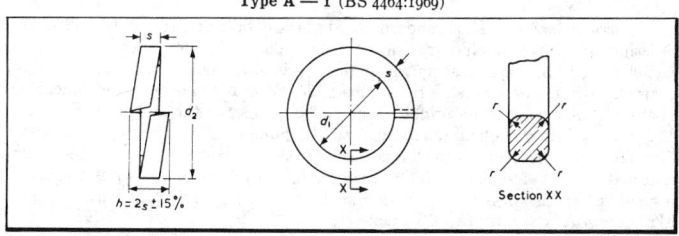

British Standard Single Coil Square Section Spring Washers; Metric Series —
Type A — 2 (BS 4464: 1969)

| Nom. Size, d | Inside Diam., $d_1$ Max | Inside Diam., $d_1$ Min | Thickness & Width, s | O.D., $d_2$ Max | Radius, r Max |
|---|---|---|---|---|---|
| M3 | 3.3 | 3.1 | 1 ± 0.1 | 5.5 | 0.3 |
| (M3.5) | 3.8 | 3.6 | 1 ± 0.1 | 6.0 | 0.3 |
| M4 | 4.35 | 4.1 | 1.2 ± 0.1 | 6.95 | 0.4 |
| M5 | 5.35 | 5.1 | 1.5 ± 0.1 | 8.55 | 0.5 |
| M6 | 6.4 | 6.1 | 1.5 ± 0.1 | 9.6 | 0.5 |
| M8 | 8.55 | 8.2 | 2 ± 0.1 | 12.75 | 0.65 |
| M10 | 10.6 | 10.2 | 2.5 ± 0.15 | 15.9 | 0.8 |
| M12 | 12.6 | 12.2 | 2.5 ± 0.15 | 17.9 | 0.8 |
| (M14) | 14.7 | 14.2 | 3 ± 0.2 | 21.1 | 1.0 |
| M16 | 16.9 | 16.3 | 3.5 ± 0.2 | 24.3 | 1.15 |
| (M18) | 19.0 | 18.3 | 3.5 ± 0.2 | 26.4 | 1.15 |
| M20 | 21.1 | 20.3 | 4.5 ± 0.2 | 30.5 | 1.5 |
| (M22) | 23.3 | 22.4 | 4.5 ± 0.2 | 32.7 | 1.5 |
| M24 | 25.3 | 24.4 | 5 ± 0.2 | 35.7 | 1.65 |
| (M27) | 28.5 | 27.5 | 5 ± 0.2 | 38.9 | 1.65 |
| M30 | 31.5 | 30.5 | 6 ± 0.2 | 43.9 | 2.0 |
| (M33) | 34.6 | 33.5 | 6 ± 0.2 | 47.0 | 2.0 |
| M36 | 37.6 | 36.5 | 7 ± 0.25 | 52.1 | 2.3 |
| (M39) | 40.8 | 39.6 | 7 ± 0.25 | 55.3 | 2.3 |
| M42 | 43.8 | 42.6 | 8 ± 0.25 | 60.3 | 2.65 |
| (M45) | 46.8 | 45.6 | 8 ± 0.25 | 63.3 | 2.65 |
| M48 | 50.0 | 48.8 | 8 ± 0.25 | 66.5 | 2.65 |

All dimensions are given in millimeters.   Sizes shown in parentheses are non-preferred and are not usually stock sizes.

**British Standard for Metric Series Metal Washers.** — BS 4320:1968 specifies bright and black metal washers for general engineering purposes.

*Bright metal washers:* These washers are made from either CS4 cold rolled strip steel (BS 1449:Part 3B) or from CZ 108 brass strip (BS 265), both in the hard condition. However, by mutual agreement between purchaser and supplier, washers may be made available with the material in any other condition, or they may be made from another material, or may be coated with a protective or decorative finish to some appropriate British Standard. Washers are reasonably flat and free from burrs and are normally supplied unchamfered. They may, however, have a 30-degree chamfer on one edge of the external diameter. These washers are made available in two size categories, normal and large diameter, and in two thicknesses, normal (Form A or C), and light (Form B or D). The thickness of a light range washer ranges from ½ to ⅔ the thickness of a normal range washer.

*Black metal washers:* These washers are made from mild steel, and can be supplied in three size categories designated normal, large, and extra large diameters. The normal diameter series are intended for bolts ranging from M5 to M68 (Form E washers); the large diameter series for bolts ranging from M8 to M39 (Form F washers) and the extra large series for bolts from M5 to M39 (Form G washers). A protective finish can be specified by the purchaser in accordance with any appropriate British Standard.

*Washer designations:* The Standard specifies the details that should be given when ordering or placing an inquiry for washers. They are the general description, namely, bright or black washers; the nominal size of the bolt or screw involved, for example, M5; the designated form, for example, Form A or Form E; the dimensions of any chamfer required on bright washers; the number of the Standard (BS 4320), and coating information if required, with the number of the appropriate British Standard and the coating thickness needed. As an example, in the use of this information, the designation for a chamfered, normal diameter series washer of normal range thickness to suit a 12-mm diameter bolt would be: Bright washers M12 (Form A) chamfered to BS 4320.

## British Standard Bright Metal Washers — Metric Series (BS 4320:1968)

### NORMAL DIAMETER SIZES

| Nominal Size of Bolt or Screw | Inside Diameter | | | Outside Diameter | | | Thickness | | | | | |
| | | | | | | | Form A (Normal Range) | | | Form B (Light Range) | | |
| | Nom | Max | Min | Nom | Max | Min | Nom | Max | Min | Nom | Max | Min |
|---|---|---|---|---|---|---|---|---|---|---|---|---|
| M 1.0 | 1.1 | 1.25 | 1.1 | 2.5 | 2.5 | 2.3 | 0.3 | 0.4 | 0.2 | .... | .... | .... |
| M 1.2 | 1.3 | 1.45 | 1.3 | 3.0 | 3.0 | 2.8 | 0.3 | 0.4 | 0.2 | .... | .... | .... |
| (M 1.4) | 1.5 | 1.65 | 1.5 | 3.0 | 3.0 | 2.8 | 0.3 | 0.4 | 0.2 | .... | .... | .... |
| M 1.6 | 1.7 | 1.85 | 1.7 | 4.0 | 4.0 | 3.7 | 0.3 | 0.4 | 0.2 | .... | .... | .... |
| M 2.0 | 2.2 | 2.35 | 2.2 | 5.0 | 5.0 | 4.7 | 0.3 | 0.4 | 0.2 | .... | .... | .... |
| (M 2.2) | 2.4 | 2.55 | 2.4 | 5.0 | 5.0 | 4.7 | 0.5 | 0.6 | 0.4 | .... | .... | .... |
| M 2.5 | 2.7 | 2.85 | 2.7 | 6.5 | 6.5 | 6.2 | 0.5 | 0.6 | 0.4 | .... | .... | .... |
| M 3 | 3.2 | 3.4 | 3.2 | 7 | 7 | 6.7 | 0.5 | 0.6 | 0.4 | .... | .... | .... |
| (M 3.5) | 3.7 | 3.9 | 3.7 | 7 | 7 | 6.7 | 0.5 | 0.6 | 0.4 | .... | .... | .... |
| M 4 | 4.3 | 4.5 | 4.3 | 9 | 9 | 8.7 | 0.8 | 0.9 | 0.7 | .... | .... | .... |
| (M 4.5) | 4.8 | 5.0 | 4.8 | 9 | 9 | 8.7 | 0.8 | 0.9 | 0.7 | .... | .... | .... |
| M 5 | 5.3 | 5.5 | 5.3 | 10 | 10 | 9.7 | 1.0 | 1.1 | 0.9 | .... | .... | .... |
| M 6 | 6.4 | 6.7 | 6.4 | 12.5 | 12.5 | 12.1 | 1.6 | 1.8 | 1.4 | 0.8 | 0.9 | 0.7 |
| (M 7) | 7.4 | 7.7 | 7.4 | 14 | 14 | 13.6 | 1.6 | 1.8 | 1.4 | 0.8 | 0.9 | 0.7 |
| M 8 | 8.4 | 8.7 | 8.4 | 17 | 17 | 16.6 | 1.6 | 1.8 | 1.4 | 1.0 | 1.1 | 0.9 |
| M 10 | 10.5 | 10.9 | 10.5 | 21 | 21 | 20.5 | 2.0 | 2.2 | 1.8 | 1.25 | 1.45 | 1.05 |
| M 12 | 13.0 | 13.4 | 13.0 | 24 | 24 | 23.5 | 2.5 | 2.7 | 2.3 | 1.6 | 1.80 | 1.40 |
| (M 14) | 15.0 | 15.4 | 15.0 | 28 | 28 | 27.5 | 2.5 | 2.7 | 2.3 | 1.6 | 1.8 | 1.4 |
| M 16 | 17.0 | 17.4 | 17.0 | 30 | 30 | 29.5 | 3.0 | 3.3 | 2.7 | 2.0 | 2.2 | 1.8 |
| (M 18) | 19.0 | 19.5 | 19.0 | 34 | 34 | 33.2 | 3.0 | 3.3 | 2.7 | 2.0 | 2.2 | 1.8 |
| M 20 | 21 | 21.5 | 21 | 37 | 37 | 36.2 | 3.0 | 3.3 | 2.7 | 2.0 | 2.2 | 1.8 |
| (M 22) | 23 | 23.5 | 23 | 39 | 39 | 38.2 | 3.0 | 3.3 | 2.7 | 2.0 | 2.2 | 1.8 |
| M 24 | 25 | 25.5 | 25 | 44 | 44 | 43.2 | 4.0 | 4.3 | 3.7 | 2.5 | 2.7 | 2.3 |
| (M 27) | 28 | 28.5 | 28 | 50 | 50 | 49.2 | 4.0 | 4.3 | 3.7 | 2.5 | 2.7 | 2.3 |
| M 30 | 31 | 31.6 | 31 | 56 | 56 | 55.0 | 4.0 | 4.3 | 3.7 | 2.5 | 2.7 | 2.3 |
| (M 33) | 34 | 34.6 | 34 | 60 | 60 | 59.0 | 5.0 | 5.6 | 4.4 | 3.0 | 3.3 | 2.7 |
| M 36 | 37 | 37.6 | 37 | 66 | 66 | 65.0 | 5.0 | 5.6 | 4.4 | 3.0 | 3.3 | 2.7 |
| (M 39) | 40 | 40.6 | 40 | 72 | 72 | 71.0 | 6.0 | 6.6 | 5.4 | 3.0 | 3.3 | 2.7 |

### LARGE DIAMETER SIZES

| Nominal Size of Bolt or Screw | Inside Diameter | | | Outside Diameter | | | Thickness | | | | | |
| | | | | | | | Form C (Normal Range) | | | Form D (Light Range) | | |
| | Nom | Max | Min | Nom | Max | Min | Nom | Max | Min | Nom | Max | Min |
|---|---|---|---|---|---|---|---|---|---|---|---|---|
| M 4 | 4.3 | 4.5 | 4.3 | 10.0 | 10.0 | 9.7 | 0.8 | 0.9 | 0.7 | .... | .... | .... |
| M 5 | 5.3 | 5.5 | 5.3 | 12.5 | 12.5 | 12.1 | 1.0 | 1.1 | 0.9 | .... | .... | .... |
| M 6 | 6.4 | 6.7 | 6.4 | 14 | 14 | 13.6 | 1.6 | 1.8 | 1.4 | 0.8 | 0.9 | 0.7 |
| M 8 | 8.4 | 8.7 | 8.4 | 21 | 21 | 20.5 | 1.6 | 1.8 | 1.4 | 1.0 | 1.1 | 0.9 |
| M 10 | 10.5 | 10.9 | 10.5 | 24 | 24 | 23.5 | 2.0 | 2.2 | 1.8 | 1.25 | 1.45 | 1.05 |
| M 12 | 13.0 | 13.4 | 13.0 | 28 | 28 | 27.5 | 2.5 | 2.7 | 2.3 | 1.6 | 1.8 | 1.4 |
| (M 14) | 15.0 | 15.4 | 15 | 30 | 30 | 29.5 | 2.5 | 2.7 | 2.3 | 1.6 | 1.8 | 1.4 |
| M 16 | 17.0 | 17.4 | 17 | 34 | 34 | 33.2 | 3.0 | 3.3 | 2.7 | 2.0 | 2.2 | 1.8 |
| (M 18) | 19.0 | 19.5 | 19 | 37 | 37 | 36.2 | 3.0 | 3.3 | 2.7 | 2.0 | 2.2 | 1.8 |
| M 20 | 21 | 21.5 | 21 | 39 | 39 | 38.2 | 3.0 | 3.3 | 2.7 | 2.0 | 2.2 | 1.8 |
| (M 22) | 23 | 23.5 | 23 | 44 | 44 | 43.2 | 3.0 | 3.3 | 2.7 | 2.0 | 2.2 | 1.8 |
| M 24 | 25 | 25.5 | 25 | 50 | 50 | 49.2 | 4.0 | 4.3 | 3.7 | 2.5 | 2.7 | 2.3 |
| (M 27) | 28 | 28.5 | 28 | 56 | 56 | 55 | 4.0 | 4.3 | 3.7 | 2.5 | 2.7 | 2.3 |
| M 30 | 31 | 31.6 | 31 | 60 | 60 | 59 | 4.0 | 4.3 | 3.7 | 2.5 | 2.7 | 2.3 |
| (M 33) | 34 | 34.6 | 34 | 66 | 66 | 65 | 5.0 | 5.6 | 4.4 | 3.0 | 3.3 | 2.7 |
| M 36 | 37 | 37.6 | 37 | 72 | 72 | 71 | 5.0 | 5.6 | 4.4 | 3.0 | 3.3 | 2.7 |
| (M 39) | 40 | 40.6 | 40 | 77 | 77 | 76 | 6.0 | 6.6 | 5.4 | 3.0 | 3.3 | 2.7 |

All dimensions are given in millimeters.
Nominal bolt or screw sizes shown in parentheses are non-preferred.

### British Standard Black Metal Washers — Metric Series (BS 4320:1968)

#### NORMAL DIAMETER SIZES (Form E)

| Nom Bolt or Screw Size | Inside Diameter | | | Outside Diameter | | | Thickness | | |
|---|---|---|---|---|---|---|---|---|---|
| | Nom | Max | Min | Nom | Max | Min | Nom | Max | Min |
| M 5 | 5.5 | 5.8 | 5.5 | 10.0 | 10.0 | 9.2 | 1.0 | 1.2 | 0.8 |
| M 6 | 6.6 | 7.0 | 6.6 | 12.5 | 12.5 | 11.7 | 1.6 | 1.9 | 1.3 |
| (M 7) | 7.6 | 8.0 | 7.6 | 14.0 | 14.0 | 13.2 | 1.6 | 1.9 | 1.3 |
| M 8 | 9.0 | 9.4 | 9.0 | 17 | 17 | 16.2 | 1.6 | 1.9 | 1.3 |
| M 10 | 11.0 | 11.5 | 11.0 | 21 | 21 | 20.2 | 2.0 | 2.3 | 1.7 |
| M 12 | 14 | 14.5 | 14 | 24 | 24 | 23.2 | 2.5 | 2.8 | 2.2 |
| (M 14) | 16 | 16.5 | 16 | 28 | 28 | 27.2 | 2.5 | 2.8 | 2.2 |
| M 16 | 18 | 18.5 | 18 | 30 | 30 | 29.2 | 3.0 | 3.6 | 2.4 |
| (M 18) | 20 | 20.6 | 20 | 34 | 34 | 32.8 | 3.0 | 3.6 | 2.4 |
| M 20 | 22 | 22.6 | 22 | 37 | 37 | 35.8 | 3.0 | 3.6 | 2.4 |
| (M 22) | 24 | 24.6 | 24 | 39 | 39 | 37.8 | 3.0 | 3.6 | 2.4 |
| M 24 | 26 | 26.6 | 26 | 44 | 44 | 42.8 | 4 | 4.6 | 3.4 |
| (M 27) | 30 | 30.6 | 30 | 50 | 50 | 48.8 | 4 | 4.6 | 3.4 |
| M 30 | 33 | 33.8 | 33 | 56 | 56 | 54.5 | 4 | 4.6 | 3.4 |
| (M 33) | 36 | 36.8 | 36 | 60 | 60 | 58.5 | 5 | 6.0 | 4.0 |
| M 36 | 39 | 39.8 | 39 | 66 | 66 | 64.5 | 5 | 6.0 | 4.0 |
| (M 39) | 42 | 42.8 | 42 | 72 | 72 | 70.5 | 6 | 7.0 | 5.0 |
| M 42 | 45 | 45.8 | 45 | 78 | 78 | 76.5 | 7 | 8.2 | 5.8 |
| (M 45) | 48 | 48.8 | 48 | 85 | 85 | 83 | 7 | 8.2 | 5.8 |
| M 48 | 52 | 53 | 52 | 92 | 92 | 90 | 8 | 9.2 | 6.8 |
| (M 52) | 56 | 57 | 56 | 98 | 98 | 96 | 8 | 9.2 | 6.8 |
| M 56 | 62 | 63 | 62 | 105 | 105 | 103 | 9 | 10.2 | 7.8 |
| (M 60) | 66 | 67 | 66 | 110 | 110 | 108 | 9 | 10.2 | 7.8 |
| M 64 | 70 | 71 | 70 | 115 | 115 | 113 | 9 | 10.2 | 7.8 |
| (M 68) | 74 | 75 | 74 | 120 | 120 | 118 | 10 | 11.2 | 8.8 |

#### LARGE DIAMETER SIZES (Form F)

| Nom Bolt or Screw Size | Inside Diameter | | | Outside Diameter | | | Thickness | | |
|---|---|---|---|---|---|---|---|---|---|
| | Nom | Max | Min | Nom | Max | Min | Nom | Max | Min |
| M 8 | 9 | 9.4 | 9.0 | 21 | 21 | 20.2 | 1.6 | 1.9 | 1.3 |
| M 10 | 11 | 11.5 | 11 | 24 | 24 | 23.2 | 2 | 2.3 | 1.7 |
| M 12 | 14 | 14.5 | 14 | 28 | 28 | 27.2 | 2.5 | 2.8 | 2.2 |
| (M 14) | 16 | 16.5 | 16 | 30 | 30 | 29.2 | 2.5 | 2.8 | 2.2 |
| M 16 | 18 | 18.5 | 18 | 34 | 34 | 32.8 | 3 | 3.6 | 2.4 |
| (M 18) | 20 | 20.6 | 20 | 37 | 37 | 35.8 | 3 | 3.6 | 2.4 |
| M 20 | 22 | 22.6 | 22 | 39 | 39 | 37.8 | 3 | 3.6 | 2.4 |
| (M 22) | 24 | 24.6 | 24 | 44 | 44 | 42.8 | 3 | 3.6 | 2.4 |
| M 24 | 26 | 26.6 | 26 | 50 | 50 | 48.8 | 4 | 4.6 | 3.4 |
| (M 27) | 30 | 30.6 | 30 | 56 | 56 | 54.5 | 4 | 4.6 | 3.4 |
| M 30 | 33 | 33.8 | 33 | 60 | 60 | 58.5 | 4 | 4.6 | 3.4 |
| (M 33) | 36 | 36.8 | 36 | 66 | 66 | 64.5 | 5 | 6.0 | 4 |
| M 36 | 39 | 39.8 | 39 | 72 | 72 | 70.5 | 5 | 6.0 | 4 |
| (M 39) | 42 | 42.8 | 42 | 77 | 77 | 75.5 | 6 | 7 | 5 |

#### EXTRA LARGE DIAMETER SIZES (Form G)

| Nom Bolt or Screw Size | Inside Diameter | | | Outside Diameter | | | Thickness | | |
|---|---|---|---|---|---|---|---|---|---|
| | Nom | Max | Min | Nom | Max | Min | Nom | Max | Min |
| M 5 | 5.5 | 5.8 | 5.5 | 15 | 15 | 14.2 | 1.6 | 1.9 | 1.3 |
| M 6 | 6.6 | 7.0 | 6.6 | 18 | 18 | 17.2 | 2 | 2.3 | 1.7 |
| (M 7) | 7.6 | 8.0 | 7.6 | 21 | 21 | 20.2 | 2 | 2.3 | 1.7 |
| M 8 | 9 | 9.4 | 9.0 | 24 | 24 | 23.2 | 2 | 2.3 | 1.7 |
| M 10 | 11 | 11.5 | 11.0 | 30 | 30 | 29.2 | 2.5 | 2.8 | 2.2 |
| M 12 | 14 | 14.5 | 14.0 | 36 | 36 | 34.8 | 3 | 3.6 | 2.4 |
| (M 14) | 16 | 16.5 | 16.0 | 42 | 42 | 40.8 | 3 | 3.6 | 2.4 |
| M 16 | 18 | 18.5 | 18 | 48 | 48 | 46.8 | 4 | 4.6 | 3.4 |
| (M 18) | 20 | 20.6 | 20 | 54 | 54 | 52.5 | 4 | 4.6 | 3.4 |
| M 20 | 22 | 22.6 | 22 | 60 | 60 | 58.5 | 5 | 6.0 | 4 |
| (M 22) | 24 | 24.6 | 24 | 66 | 66 | 64.5 | 5 | 6.0 | 4 |
| M 24 | 26 | 26.6 | 26 | 72 | 72 | 70.5 | 6 | 7 | 5 |
| (M 27) | 30 | 30.6 | 30 | 81 | 81 | 79 | 6 | 7 | 5 |
| M 30 | 33 | 33.8 | 33 | 90 | 90 | 88 | 8 | 9.2 | 6.8 |
| (M 33) | 36 | 36.8 | 36 | 99 | 99 | 97 | 8 | 9.2 | 6.8 |
| M 36 | 39 | 39.8 | 39 | 108 | 108 | 106 | 10 | 11.2 | 8.8 |
| (M 39) | 42 | 42.8 | 42 | 117 | 117 | 115 | 10 | 11.2 | 8.8 |

All dimensions are given in millimeters.
Nominal bolt or screw sizes shown in parentheses are non-preferred.

**American National Standard Machine Screws and Machine Screw Nuts.** — This Standard (ANSI B18.6.3) covers both slotted and recessed head machine screws. Dimensions of various types of slotted machine screws, machine screw nuts, and header points are given in Tables 1 through 12. The Standard also covers flat trim head, oval trim head and drilled fillister head machine screws and gives cross recess dimensions and gaging dimensions for all types of machine screw heads.

*Threads:* Except for sizes 0000, 000, and 00, machine screw threads may be either Unified Coarse (UNC) and Fine thread (UNF) Class 2A (see Unified Screw Threads starting on page 1261) or UNRC and UNRF Series, at option of manufacturer. Thread dimensions for sizes 0000, 000, and 00 are given in Table 7 on page 1182.

Threads for hexagon machine screw nuts may be either UNC or UNF, Class 2B, and in square machine screw nuts are UNC Class 2B.

*Length of thread:* Machine screws of sizes No. 5 and smaller with nominal lengths equal to 3 diameters and shorter have full form threads extending to within 1 pitch (thread) of the bearing surface of the head, or closer, if practicable. Nominal lengths greater than 3 diameters, up to and including 1⅛ inch, have full form threads extending to within two pitches (threads) of the bearing surface of the head, or closer, if practicable. Unless otherwise specified, screws of longer nominal length have a minimum length of full form thread of 1.00 inch.

Machine screws of sizes No. 6 and larger with nominal length equal to 3 diameters and shorter have full form threads extending to within 1 pitch (thread) of the bearing surface of the head, or closer, if practicable. Nominal lengths greater than 3 diameters, up to and including 2 inches, have full form threads extending to within 2 pitches (threads) of the bearing surface of the head, or closer, if practicable. Screws of longer nominal length, unless otherwise specified, have a minimum length of full form thread of 1.50 inches.

**Table 1.   American National Standard Square and Hexagon Machine Screw Nuts**
(ANSI B18.6.3-1972, R1977)

| Nom. Size | Basic Diam. | Basic F | Max. F | Min. F | Max. G | Min. G | Max. $G_1$ | Min. $G_1$ | Max. H | Min. H |
|---|---|---|---|---|---|---|---|---|---|---|
| 0 | .0600 | 5⁄32 | .156 | .150 | .221 | .206 | .180 | .171 | .050 | .043 |
| 1 | .0730 | 5⁄32 | .156 | .150 | .221 | .206 | .180 | .171 | .050 | .043 |
| 2 | .0860 | 3⁄16 | .188 | .180 | .265 | .247 | .217 | .205 | .066 | .057 |
| 3 | .0990 | 3⁄16 | .188 | .180 | .265 | .247 | .217 | .205 | .066 | .057 |
| 4 | .1120 | ¼ | .250 | .241 | .354 | .331 | .289 | .275 | .098 | .087 |
| 5 | .1250 | 5⁄16 | .312 | .302 | .442 | .415 | .361 | .344 | .114 | .102 |
| 6 | .1380 | 5⁄16 | .312 | .302 | .442 | .415 | .361 | .344 | .114 | .102 |
| 8 | .1640 | 11⁄32 | .344 | .332 | .486 | .456 | .397 | .378 | .130 | .117 |
| 10 | .1900 | 3⁄8 | .375 | .362 | .530 | .497 | .433 | .413 | .130 | .117 |
| 12 | .2160 | 7⁄16 | .438 | .423 | .619 | .581 | .505 | .482 | .161 | .148 |
| ¼ | .2500 | 7⁄16 | .438 | .423 | .619 | .581 | .505 | .482 | .193 | .178 |
| 5⁄16 | .3125 | 9⁄16 | .562 | .545 | .795 | .748 | .650 | .621 | .225 | .208 |
| 3⁄8 | .3750 | 5⁄8 | .625 | .607 | .884 | .833 | .722 | .692 | .257 | .239 |

All dimensions in inches. Hexagon machine screw nuts have tops flat and chamfered. Diameter of top circle should be the maximum width across flats within a tolerance of minus 15 per cent. Bottoms are flat but may be chamfered if so specified. Square machine screw nuts have tops and bottoms flat without chamfer.

*Diameter of body:* The diameter of machine screw bodies is not less than Class 2A thread minimum pitch diameter nor greater than the basic major diameter of the thread. Cross-recessed trim head machine screws not threaded to the head have an 0.062 in. minimum length shoulder under the head with diameter limits as specified in the dimensional tables in the standard.

*Designation:* Machine screws are designated by the following data in the sequence shown: Nominal size (number, fraction or decimal equivalent); threads per inch; nominal length (fraction or decimal equivalent); product name, including head type and driving provision; header point, if desired; material; and protective finish, if required. For example:

¼ - 20 x 1¼ Slotted Pan Head Machine Screw, Steel, Zinc Plated
6 - 32 x ¾ Type IA Cross Recessed Fillister Head Machine Screw, Brass.

Machine screw nuts are designated by the following data in sequence shown: Nominal size (number, fraction or decimal equivalent); threads per inch; product name; material; and protective finish, if required. For example:

10 - 24 Hexagon Machine Screw Nut, Steel, Zinc Plated
.138 - 32 Square Machine Screw Nut, Brass.

**Table 2. American National Standard Slotted 100-Degree Flat Countersunk Head Machine Screws** (ANSI B18.6.3-1972, R1977)

| Nominal Size[1] or Basic Screw Diam. | | Head Diam., *A* | | Head Height, *H* | Slot Width, *J* | | Slot Depth, *T* | |
|---|---|---|---|---|---|---|---|---|
| | | Max., Edge Sharp | Min., Edge Rounded or Flat | Ref. | Max. | Min. | Max. | Min. |
| 0000 | 0.0210 | .043 | .037 | .009 | .008 | .005 | .008 | .004 |
| 000 | 0.0340 | .064 | .058 | .014 | .012 | .008 | .011 | .007 |
| 00 | 0.0470 | .093 | .085 | .020 | .017 | .010 | .013 | .008 |
| 0 | 0.0600 | .119 | .096 | .026 | .023 | .016 | .013 | .008 |
| 1 | 0.0730 | .146 | .120 | .031 | .026 | .019 | .016 | .010 |
| 2 | 0.0860 | .172 | .143 | .037 | .031 | .023 | .019 | .012 |
| 3 | 0.0990 | .199 | .167 | .043 | .035 | .027 | .022 | .014 |
| 4 | 0.1120 | .225 | .191 | .049 | .039 | .031 | .024 | .017 |
| 6 | 0.1380 | .279 | .238 | .060 | .048 | .039 | .030 | .022 |
| 8 | 0.1640 | .332 | .285 | .072 | .054 | .045 | .036 | .027 |
| 10 | 0.1900 | .385 | .333 | .083 | .060 | .050 | .042 | .031 |
| ¼ | 0.2500 | .507 | .442 | .110 | .075 | .064 | .055 | .042 |
| ⁵⁄₁₆ | 0.3125 | .635 | .556 | .138 | .084 | .072 | .069 | .053 |
| ⅜ | 0.3750 | .762 | .670 | .165 | .094 | .081 | .083 | .065 |

All dimensions are in inches.
[1] When specifying nominal size in decimals, zeros preceding the decimal point and in the fourth decimal place are omitted.

**Table 3.** American National Standard Slotted Flat Countersunk Head and Close Tolerance 100-Degree Flat Countersunk Head Machine Screws
(ANSI B18.6.3-1972, R1977)

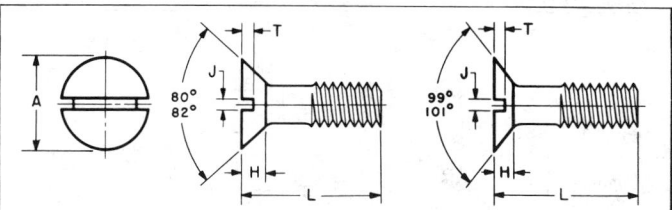

### SLOTTED FLAT COUNTERSUNK HEAD TYPE

| Nominal Size[1] or Basic Screw Diam. | | Max., L[2] | Head Diam., A | | Head Height, H | Slot Width, J | | Slot Depth, T | |
|---|---|---|---|---|---|---|---|---|---|
| | | | Max., Edge Sharp | Min., Edge[3] | Ref. | Max. | Min. | Max. | Min. |
| 0000 | 0.0210 | .... | .043 | .037 | .011 | .008 | .004 | .007 | .003 |
| 000 | 0.0340 | .... | .064 | .058 | .016 | .011 | .007 | .009 | .005 |
| 00 | 0.0470 | .... | .093 | .085 | .028 | .017 | .010 | .014 | .009 |
| 0 | 0.0600 | ⅛ | .119 | .099 | .035 | .023 | .016 | .015 | .010 |
| 1 | 0.0730 | ⅛ | .146 | .123 | .043 | .026 | .019 | .019 | .012 |
| 2 | 0.0860 | ⅛ | .172 | .147 | .051 | .031 | .023 | .023 | .015 |
| 3 | 0.0990 | ⅛ | .199 | .171 | .059 | .035 | .027 | .027 | .017 |
| 4 | 0.1120 | 3⁄16 | .225 | .195 | .067 | .039 | .031 | .030 | .020 |
| 5 | 0.1250 | 3⁄16 | .252 | .220 | .075 | .043 | .035 | .034 | .022 |
| 6 | 0.1380 | 3⁄16 | .279 | .244 | .083 | .048 | .039 | .038 | .024 |
| 8 | 0.1640 | ¼ | .332 | .292 | .100 | .054 | .045 | .045 | .029 |
| 10 | 0.1900 | 5⁄16 | .385 | .340 | .116 | .060 | .050 | .053 | .034 |
| 12 | 0.2160 | ⅜ | .438 | .389 | .132 | .067 | .056 | .060 | .039 |
| ¼ | 0.2500 | 7⁄16 | .507 | .452 | .153 | .075 | .064 | .070 | .046 |
| 5⁄16 | 0.3125 | ½ | .635 | .568 | .191 | .084 | .072 | .088 | .058 |
| ⅜ | 0.3750 | 9⁄16 | .762 | .685 | .230 | .094 | .081 | .106 | .070 |
| 7⁄16 | 0.4375 | ⅝ | .812 | .723 | .223 | .094 | .081 | .103 | .066 |
| ½ | 0.5000 | ¾ | .875 | .775 | .223 | .106 | .091 | .103 | .065 |
| 9⁄16 | 0.5625 | ... | 1.000 | .889 | .260 | .118 | .102 | .120 | .077 |
| ⅝ | 0.6250 | ... | 1.125 | 1.002 | .298 | .133 | .116 | .137 | .088 |
| ¾ | 0.7500 | ... | 1.375 | 1.230 | .372 | .149 | .131 | .171 | .111 |

### CLOSE TOLERANCE 100-DEGREE FLAT COUNTERSUNK HEAD TYPE

| Nominal Size[1] or Basic Screw Diam. | | Head Diameter, A | | Head Height, H | Slot Width, J | | Slot Depth, T | |
|---|---|---|---|---|---|---|---|---|
| | | Max., Edge Sharp | Min., Edge[3] | Ref. | Max. | Min. | Max. | Min. |
| 4 | 0.1120 | .225 | .191 | .049 | .039 | .031 | .024 | .017 |
| 6 | 0.1380 | .279 | .238 | .060 | .048 | .039 | .030 | .022 |
| 8 | 0.1640 | .332 | .285 | .072 | .054 | .045 | .036 | .027 |
| 10 | 0.1900 | .385 | .333 | .083 | .060 | .050 | .042 | .031 |
| ¼ | 0.2500 | .507 | .442 | .110 | .075 | .064 | .055 | .042 |
| 5⁄16 | 0.3125 | .635 | .556 | .138 | .084 | .072 | .069 | .053 |
| ⅜ | 0.3750 | .762 | .670 | .165 | .094 | .081 | .083 | .065 |
| 7⁄16 | 0.4375 | .890 | .783 | .193 | .094 | .081 | .097 | .076 |
| ½ | 0.5000 | 1.017 | .897 | .221 | .106 | .091 | .111 | .088 |
| 9⁄16 | 0.5625 | 1.145 | 1.011 | .249 | .118 | .102 | .125 | .099 |
| ⅝ | 0.6250 | 1.272 | 1.124 | .276 | .133 | .116 | .139 | .111 |

All dimensions are in inches. [1] When specifying nominal size in decimals, zeros preceding the decimal point and in the fourth decimal place are omitted. [2] These lengths or shorter are undercut. [3] May be rounded or flat.

**Table 4. American National Standard Slotted Undercut Flat Countersunk Head and Plain and Slotted Hex Washer Head Machine Screws** (ANSI B18.6.3-1972, R1977)

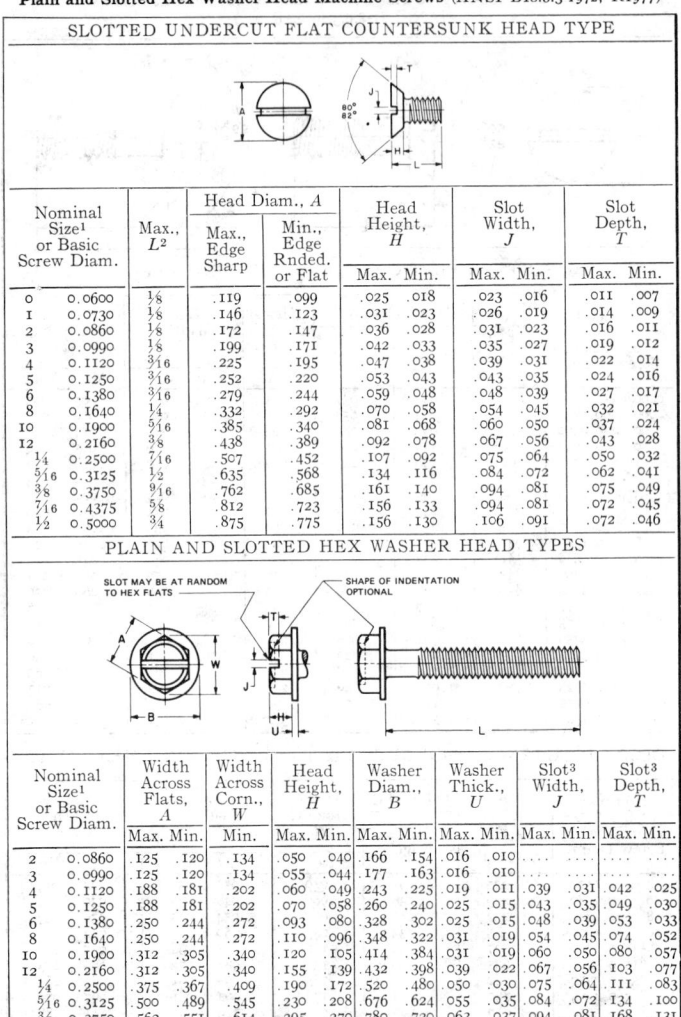

### SLOTTED UNDERCUT FLAT COUNTERSUNK HEAD TYPE

| Nominal Size[1] or Basic Screw Diam. | | Max., $L^2$ | Head Diam., A | | Head Height, H | | Slot Width, J | | Slot Depth, T | |
|---|---|---|---|---|---|---|---|---|---|---|
| | | | Max., Edge Sharp | Min., Edge Rnded. or Flat | Max. | Min. | Max. | Min. | Max. | Min. |
| 0 | 0.0600 | ⅛ | .119 | .099 | .025 | .018 | .023 | .016 | .011 | .007 |
| 1 | 0.0730 | ⅛ | .146 | .123 | .031 | .023 | .026 | .019 | .014 | .009 |
| 2 | 0.0860 | ⅛ | .172 | .147 | .036 | .028 | .031 | .023 | .016 | .011 |
| 3 | 0.0990 | ⅛ | .199 | .171 | .042 | .033 | .035 | .027 | .019 | .012 |
| 4 | 0.1120 | 3⁄16 | .225 | .195 | .047 | .038 | .039 | .031 | .022 | .014 |
| 5 | 0.1250 | 3⁄16 | .252 | .220 | .053 | .043 | .043 | .035 | .024 | .016 |
| 6 | 0.1380 | 3⁄16 | .279 | .244 | .059 | .048 | .048 | .039 | .027 | .017 |
| 8 | 0.1640 | ¼ | .332 | .292 | .070 | .058 | .054 | .045 | .032 | .021 |
| 10 | 0.1900 | 5⁄16 | .385 | .340 | .081 | .068 | .060 | .050 | .037 | .024 |
| 12 | 0.2160 | 3⁄8 | .438 | .389 | .092 | .078 | .067 | .056 | .043 | .028 |
| ¼ | 0.2500 | 7⁄16 | .507 | .452 | .107 | .092 | .075 | .064 | .050 | .032 |
| 5⁄16 | 0.3125 | ½ | .635 | .568 | .134 | .116 | .084 | .072 | .062 | .041 |
| 3⁄8 | 0.3750 | 9⁄16 | .762 | .685 | .161 | .140 | .094 | .081 | .075 | .049 |
| 7⁄16 | 0.4375 | 5⁄8 | .812 | .723 | .156 | .133 | .094 | .081 | .072 | .045 |
| ½ | 0.5000 | ¾ | .875 | .775 | .156 | .130 | .106 | .091 | .072 | .046 |

### PLAIN AND SLOTTED HEX WASHER HEAD TYPES

SLOT MAY BE AT RANDOM TO HEX FLATS — SHAPE OF INDENTATION OPTIONAL

| Nominal Size[1] or Basic Screw Diam. | | Width Across Flats, A | | Width Across Corn., W | Head Height, H | | Washer Diam., B | | Washer Thick., U | | Slot[3] Width, J | | Slot[3] Depth, T | |
|---|---|---|---|---|---|---|---|---|---|---|---|---|---|---|
| | | Max. | Min. | Min. | Max. | Min. | Max. | Min. | Max. | Min. | Max. | Min. | Max. | Min. |
| 2 | 0.0860 | .125 | .120 | .134 | .050 | .040 | .166 | .154 | .016 | .010 | ... | ... | ... | ... |
| 3 | 0.0990 | .125 | .120 | .134 | .055 | .044 | .177 | .163 | .016 | .010 | ... | ... | ... | ... |
| 4 | 0.1120 | .188 | .181 | .202 | .060 | .049 | .243 | .225 | .019 | .011 | .039 | .031 | .042 | .025 |
| 5 | 0.1250 | .188 | .181 | .202 | .070 | .058 | .260 | .240 | .025 | .015 | .043 | .035 | .049 | .030 |
| 6 | 0.1380 | .250 | .244 | .272 | .093 | .080 | .328 | .302 | .025 | .015 | .048 | .039 | .053 | .033 |
| 8 | 0.1640 | .250 | .244 | .272 | .110 | .096 | .348 | .322 | .031 | .019 | .054 | .045 | .074 | .052 |
| 10 | 0.1900 | .312 | .305 | .340 | .120 | .105 | .414 | .384 | .031 | .019 | .060 | .050 | .080 | .057 |
| 12 | 0.2160 | .312 | .305 | .340 | .155 | .139 | .432 | .398 | .039 | .022 | .067 | .056 | .103 | .077 |
| ¼ | 0.2500 | .375 | .367 | .409 | .190 | .172 | .520 | .480 | .050 | .030 | .075 | .064 | .111 | .083 |
| 5⁄16 | 0.3125 | .500 | .489 | .545 | .230 | .208 | .676 | .624 | .055 | .035 | .084 | .072 | .134 | .100 |
| 3⁄8 | 0.3750 | .562 | .551 | .614 | .295 | .270 | .780 | .720 | .063 | .037 | .094 | .081 | .168 | .131 |

All dimensions are in inches.

[1] When specifying nominal size in decimals, zeros preceding the decimal point and in the fourth decimal place are omitted.

[2] These lengths or shorter are undercut.

[3] Unless otherwise specified, hexagon washer head machine screws are not slotted.

Table 5.  American National Standard Slotted Truss Head and Plain and Slotted Hexagon Head Machine Screws (ANSI B18.6.3-1972, R1977)

### SLOTTED TRUSS HEAD TYPE

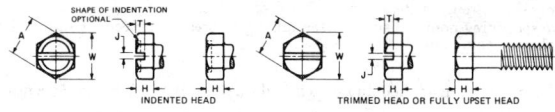

| Nominal Size or Basic Screw Diam. | | Head Diam., A | | Head Height, H | | Head Radius, R | Slot Width, J | | Slot Depth, T | |
|---|---|---|---|---|---|---|---|---|---|---|
| | | Max. | Min. | Max. | Min. | Max. | Max. | Min. | Max. | Min. |
| 0000 | 0.0210 | .049 | .043 | .014 | .010 | .032 | .009 | .005 | .009 | .005 |
| 000 | 0.0340 | .077 | .071 | .022 | .018 | .051 | .013 | .009 | .013 | .009 |
| 00 | 0.0470 | .106 | .098 | .030 | .024 | .070 | .017 | .010 | .018 | .012 |
| 0 | 0.0600 | .131 | .119 | .037 | .029 | .087 | .023 | .016 | .022 | .014 |
| 1 | 0.0730 | .164 | .149 | .045 | .037 | .107 | .026 | .019 | .027 | .018 |
| 2 | 0.0860 | .194 | .180 | .053 | .044 | .129 | .031 | .023 | .031 | .022 |
| 3 | 0.0990 | .226 | .211 | .061 | .051 | .151 | .035 | .027 | .036 | .026 |
| 4 | 0.1120 | .257 | .241 | .069 | .059 | .169 | .039 | .031 | .040 | .030 |
| 5 | 0.1250 | .289 | .272 | .078 | .066 | .191 | .043 | .035 | .045 | .034 |
| 6 | 0.1380 | .321 | .303 | .086 | .074 | .211 | .048 | .039 | .050 | .037 |
| 8 | 0.1640 | .384 | .364 | .102 | .088 | .254 | .054 | .045 | .058 | .045 |
| 10 | 0.1900 | .448 | .425 | .118 | .103 | .283 | .060 | .050 | .068 | .053 |
| 12 | 0.2160 | .511 | .487 | .134 | .118 | .336 | .067 | .056 | .077 | .061 |
| 1/4 | 0.2500 | .573 | .546 | .150 | .133 | .375 | .075 | .064 | .087 | .070 |
| 5/16 | 0.3125 | .698 | .666 | .183 | .162 | .457 | .084 | .072 | .106 | .085 |
| 3/8 | 0.3750 | .823 | .787 | .215 | .191 | .538 | .094 | .081 | .124 | .100 |
| 7/16 | 0.4375 | .948 | .907 | .248 | .221 | .619 | .094 | .081 | .142 | .116 |
| 1/2 | 0.5000 | 1.073 | 1.028 | .280 | .250 | .701 | .106 | .091 | .161 | .131 |
| 9/16 | 0.5625 | 1.198 | 1.149 | .312 | .279 | .783 | .118 | .102 | .179 | .146 |
| 5/8 | 0.6250 | 1.323 | 1.269 | .345 | .309 | .863 | .133 | .116 | .196 | .162 |
| 3/4 | 0.7500 | 1.573 | 1.511 | .410 | .368 | 1.024 | .149 | .131 | .234 | .182 |

### PLAIN AND SLOTTED HEXAGON HEAD TYPES

SHAPE OF INDENTATION OPTIONAL

INDENTED HEAD    TRIMMED HEAD OR FULLY UPSET HEAD

| Nominal Size or Basic Screw Diam. | | Regular Head | | | Large Head | | Head Height, H | | Slot[2] Width, J | | Slot[2] Depth, T | |
|---|---|---|---|---|---|---|---|---|---|---|---|---|
| | | Width Across Flats, A | | Across Corn., W | Width Across Flats, A | | Across Corn., W | | | | | | |
| | | Max. | Min. | Min. | Max. | Min. | Min. | Max. | Min. | Max. | Min. | Max. | Min. |
| 1 | 0.0730 | .125 | .120 | .134 | .... | .... | .... | .044 | .036 | .... | .... | .... | .... |
| 2 | 0.0860 | .125 | .120 | .134 | .... | .... | .... | .050 | .040 | .... | .... | .... | .... |
| 3 | 0.0990 | .188 | .181 | .202 | .... | .... | .... | .055 | .044 | .... | .... | .... | .... |
| 4 | 0.1120 | .188 | .181 | .202 | .219 | .213 | .238 | .060 | .049 | .039 | .031 | .036 | .025 |
| 5 | 0.1250 | .188 | .181 | .202 | .250 | .244 | .272 | .070 | .058 | .043 | .035 | .042 | .030 |
| 6 | 0.1380 | .250 | .244 | .272 | .... | .... | .... | .093 | .080 | .048 | .039 | .046 | .033 |
| 8 | 0.1640 | .250 | .244 | .272 | .312 | .305 | .340 | .110 | .096 | .054 | .045 | .066 | .052 |
| 10 | 0.1900 | .312 | .305 | .340 | .... | .... | .... | .120 | .105 | .060 | .050 | .072 | .057 |
| 12 | 0.2160 | .312 | .305 | .340 | .375 | .367 | .409 | .155 | .139 | .067 | .056 | .093 | .077 |
| 1/4 | 0.2500 | .375 | .367 | .409 | .438 | .428 | .477 | .190 | .172 | .075 | .064 | .101 | .083 |
| 5/16 | 0.3125 | .500 | .489 | .545 | .... | .... | .... | .230 | .208 | .084 | .072 | .122 | .100 |
| 3/8 | 0.3750 | .562 | .551 | .614 | .... | .... | .... | .295 | .270 | .094 | .081 | .156 | .131 |

All dimensions are in inches.  [1] Where specifying nominal size in decimals, zeros preceding decimal points and in the fourth decimal place are omitted.  [2] Unless otherwise specified, hexagon head machine screws are not slotted.

## Table 6. American National Standard Slotted Pan Head Machine Screws
(ANSI B18.6.3-1972, R1977)

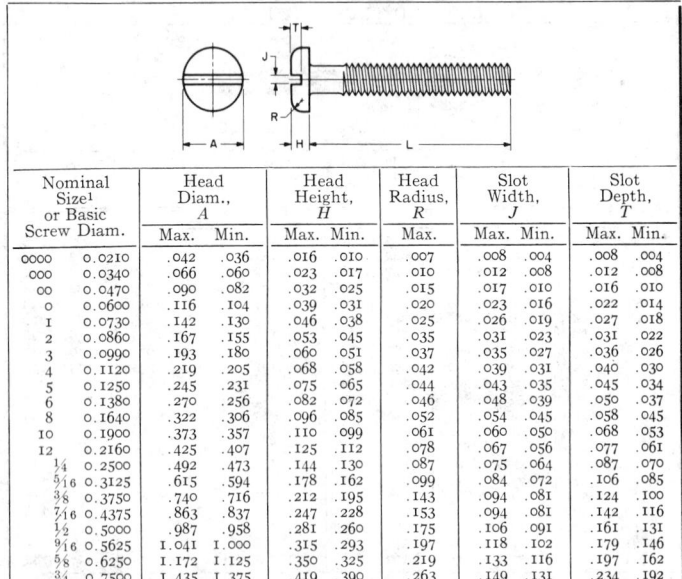

| Nominal Size[1] or Basic Screw Diam. | | Head Diam., A | | Head Height, H | | Head Radius, R | Slot Width, J | | Slot Depth, T | |
|---|---|---|---|---|---|---|---|---|---|---|
| | | Max. | Min. | Max. | Min. | Max. | Max. | Min. | Max. | Min. |
| 0000 | 0.0210 | .042 | .036 | .016 | .010 | .007 | .008 | .004 | .008 | .004 |
| 000 | 0.0340 | .066 | .060 | .023 | .017 | .010 | .012 | .008 | .012 | .008 |
| 00 | 0.0470 | .090 | .082 | .032 | .025 | .015 | .017 | .010 | .016 | .010 |
| 0 | 0.0600 | .116 | .104 | .039 | .031 | .020 | .023 | .016 | .022 | .014 |
| 1 | 0.0730 | .142 | .130 | .046 | .038 | .025 | .026 | .019 | .027 | .018 |
| 2 | 0.0860 | .167 | .155 | .053 | .045 | .035 | .031 | .023 | .031 | .022 |
| 3 | 0.0990 | .193 | .180 | .060 | .051 | .037 | .035 | .027 | .036 | .026 |
| 4 | 0.1120 | .219 | .205 | .068 | .058 | .042 | .039 | .031 | .040 | .030 |
| 5 | 0.1250 | .245 | .231 | .075 | .065 | .044 | .043 | .035 | .045 | .034 |
| 6 | 0.1380 | .270 | .256 | .082 | .072 | .046 | .048 | .039 | .050 | .037 |
| 8 | 0.1640 | .322 | .306 | .096 | .085 | .052 | .054 | .045 | .058 | .045 |
| 10 | 0.1900 | .373 | .357 | .110 | .099 | .061 | .060 | .050 | .068 | .053 |
| 12 | 0.2160 | .425 | .407 | .125 | .112 | .078 | .067 | .056 | .077 | .061 |
| ¼ | 0.2500 | .492 | .473 | .144 | .130 | .087 | .075 | .064 | .087 | .070 |
| ⁵⁄₁₆ | 0.3125 | .615 | .594 | .178 | .162 | .099 | .084 | .072 | .106 | .085 |
| ⅜ | 0.3750 | .740 | .716 | .212 | .195 | .143 | .094 | .081 | .124 | .100 |
| ⁷⁄₁₆ | 0.4375 | .863 | .837 | .247 | .228 | .153 | .094 | .081 | .142 | .116 |
| ½ | 0.5000 | .987 | .958 | .281 | .260 | .175 | .106 | .091 | .161 | .131 |
| ⁹⁄₁₆ | 0.5625 | 1.041 | 1.000 | .315 | .293 | .197 | .118 | .102 | .179 | .146 |
| ⅝ | 0.6250 | 1.172 | 1.125 | .350 | .325 | .219 | .133 | .116 | .197 | .162 |
| ¾ | 0.7500 | 1.435 | 1.375 | .419 | .390 | .263 | .149 | .131 | .234 | .192 |

All dimensions are in inches.

[1] Where specifying nominal size in decimals, zeros preceding decimal and in the fourth decimal place are omitted.

## Table 7. Nos. 0000, 000 and 00 Threads (ANSI B18.6.3, R1977 Appendix)

| Nominal Size[1] and Threads Per Inch | Series Designat. | External[2] | | | | | | Internal[3] | | | |
|---|---|---|---|---|---|---|---|---|---|---|---|
| | | Class | Major Diameter | | Pitch Diameter | | Minor Diam. | Class | Pitch Diameter | | | Major Diam. |
| | | | Max. | Min. | Max. | Min. | Tol. | | Min. | Max. | Tol. | Min. |
| 0000 – 160 or 0.0210 – 160 | NS | 2 | .0210 | .0195 | .0169 | .0158 | .0011 | .0128 | 2 | .0169 | .0181 | .0012 | .0210 |
| 000 – 120 or 0.0340 – 120 | NS | 2 | .0340 | .0325 | .0286 | .0272 | .0014 | .0232 | 2 | .0286 | .0300 | .0014 | .0340 |
| 00 – 90 or 0.0470 – 90 | NS | 2 | .0470 | .0450 | .0398 | .0382 | .0016 | .0326 | 2 | .0398 | .0414 | .0016 | .0470 |
| 00 – 96 or 0.0470 – 96 | NS | 2 | .0470 | .0450 | .0402 | .0386 | .0016 | .0334 | 2 | .0402 | .0418 | .0016 | .0470 |

All dimensions are in inches.

[1] Where specifying nominal size in decimals, zeros preceding decimal and in the fourth decimal place are omitted.

[2] There is no allowance provided on the external threads.

[3] The minor diameter limits for internal threads are not specified, they being determined by the amount of thread engagement necessary to satisfy the strength requirements and tapping performance in the intended application.

Table 8.   American National Standard Slotted Fillister and Slotted Drilled Fillister
Head Machine Screws (ANSI B18.6.3-1972, R1977)

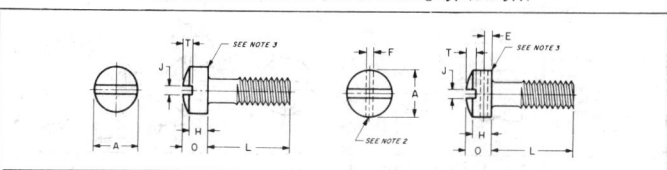

## SLOTTED FILLISTER HEAD TYPE

| Nominal Size[1] or Basic Screw Diam. | | Head Diam., A | | Head Side Height, H | | Total Head Height, O | | Slot Width, J | | Slot Depth, T | |
|---|---|---|---|---|---|---|---|---|---|---|---|
| | | Max. | Min. | Max. | Min. | Max. | Min. | Max. | Min. | Max. | Min. |
| 0000 | 0.0210 | .038 | .032 | .019 | .011 | .025 | .015 | .008 | .004 | .012 | .006 |
| 000 | 0.0340 | .059 | .053 | .029 | .021 | .035 | .027 | .012 | .006 | .017 | .011 |
| 00 | 0.0470 | .082 | .072 | .037 | .028 | .047 | .039 | .017 | .010 | .022 | .015 |
| 0 | 0.0600 | .096 | .083 | .043 | .038 | .055 | .047 | .023 | .016 | .025 | .015 |
| 1 | 0.0730 | .118 | .104 | .053 | .045 | .066 | .058 | .026 | .019 | .031 | .020 |
| 2 | 0.0860 | .140 | .124 | .062 | .053 | .083 | .066 | .031 | .023 | .037 | .025 |
| 3 | 0.0990 | .161 | .145 | .070 | .061 | .095 | .077 | .035 | .027 | .043 | .030 |
| 4 | 0.1120 | .183 | .166 | .079 | .069 | .107 | .088 | .039 | .031 | .048 | .035 |
| 5 | 0.1250 | .205 | .187 | .088 | .078 | .120 | .100 | .043 | .035 | .054 | .040 |
| 6 | 0.1380 | .226 | .208 | .096 | .086 | .132 | .111 | .048 | .039 | .060 | .045 |
| 8 | 0.1640 | .270 | .250 | .113 | .102 | .156 | .133 | .054 | .045 | .071 | .054 |
| 10 | 0.1900 | .313 | .292 | .130 | .118 | .180 | .156 | .060 | .050 | .083 | .064 |
| 12 | 0.2160 | .357 | .334 | .148 | .134 | .205 | .178 | .067 | .056 | .094 | .074 |
| 1/4 | 0.2500 | .414 | .389 | .170 | .155 | .237 | .207 | .075 | .064 | .109 | .087 |
| 5/16 | 0.3125 | .518 | .490 | .211 | .194 | .295 | .262 | .084 | .072 | .137 | .110 |
| 3/8 | 0.3750 | .622 | .590 | .253 | .233 | .355 | .315 | .094 | .081 | .164 | .133 |
| 7/16 | 0.4375 | .625 | .589 | .265 | .242 | .368 | .321 | .094 | .081 | .170 | .135 |
| 1/2 | 0.5000 | .750 | .710 | .297 | .273 | .412 | .362 | .106 | .091 | .190 | .151 |
| 9/16 | 0.5625 | .812 | .768 | .336 | .308 | .466 | .410 | .118 | .102 | .214 | .172 |
| 5/8 | 0.6250 | .875 | .827 | .375 | .345 | .521 | .461 | .133 | .116 | .240 | .193 |
| 3/4 | 0.7500 | 1.000 | .945 | .441 | .406 | .612 | .542 | .149 | .131 | .281 | .226 |

## SLOTTED DRILLED FILLISTER HEAD TYPE

| Nominal Size[1] or Basic Screw Diam. | | Head Diam., A | | Head Side Height, H | | Total Head Height, O | | Slot Width, J | | Slot Depth, T | | Drilled Hole Locat., E | Drilled Hole Diam., F |
|---|---|---|---|---|---|---|---|---|---|---|---|---|---|
| | | Max. | Min. | Max. | Min. | Max. | Min. | Max. | Min. | Max. | Min. | Basic | Basic |
| 2 | 0.0860 | .140 | .124 | .062 | .055 | .083 | .070 | .031 | .023 | .030 | .022 | .026 | .031 |
| 3 | 0.0990 | .161 | .145 | .070 | .064 | .095 | .082 | .035 | .027 | .034 | .026 | .030 | .037 |
| 4 | 0.1120 | .183 | .166 | .079 | .072 | .107 | .094 | .039 | .031 | .038 | .030 | .035 | .037 |
| 5 | 0.1250 | .205 | .187 | .088 | .081 | .120 | .106 | .043 | .033 | .042 | .033 | .038 | .046 |
| 6 | 0.1380 | .226 | .208 | .096 | .089 | .132 | .118 | .048 | .039 | .045 | .035 | .043 | .046 |
| 8 | 0.1640 | .270 | .250 | .113 | .106 | .156 | .141 | .054 | .045 | .065 | .054 | .043 | .046 |
| 10 | 0.1900 | .313 | .292 | .130 | .123 | .180 | .165 | .060 | .050 | .075 | .064 | .043 | .046 |
| 12 | 0.2160 | .357 | .334 | .148 | .139 | .205 | .188 | .067 | .056 | .087 | .074 | .053 | .046 |
| 1/4 | 0.2500 | .414 | .389 | .170 | .161 | .237 | .219 | .075 | .064 | .102 | .087 | .062 | .062 |
| 5/16 | 0.3125 | .518 | .490 | .211 | .201 | .295 | .276 | .084 | .072 | .130 | .110 | .078 | .070 |
| 3/8 | 0.3750 | .622 | .590 | .253 | .242 | .355 | .333 | .094 | .081 | .154 | .134 | .094 | .070 |

All dimensions are in inches.

[1] Where specifying nominal size in decimals, zeros preceding decimal points and in the fourth decimal place are omitted.

[2] Drilled hole shall be approximately perpendicular to the axis of slot and may be permitted to break through bottom of the slot.  Edges of the hole shall be free from burrs.

[3] A slight rounding of the edges at periphery of head is permissible provided the diameter of the bearing circle is equal to no less than 90 per cent of the specified minimum head diameter.

### Table 9. American National Standard Slotted Oval Countersunk Head Machine Screws (ANSI B18.6.3-1972, R1977)

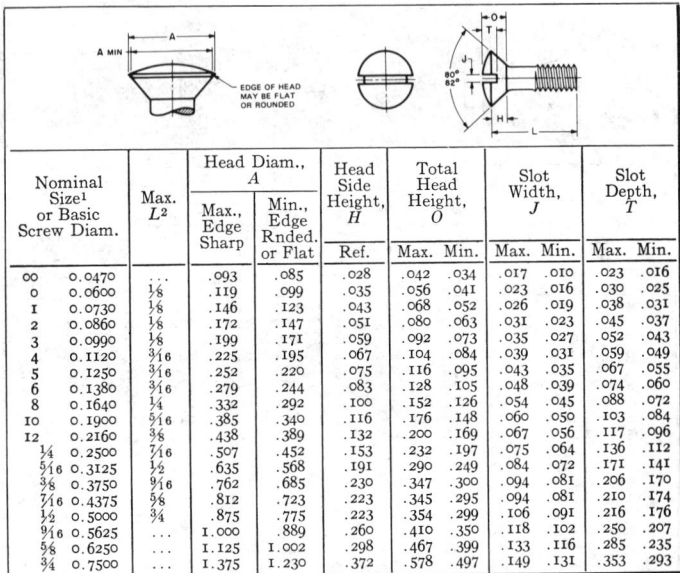

| Nominal Size[1] or Basic Screw Diam. | | Max. L[2] | Head Diam., A | | Head Side Height, H | Total Head Height, O | | Slot Width, J | | Slot Depth, T | |
|---|---|---|---|---|---|---|---|---|---|---|---|
| | | | Max., Edge Sharp | Min., Edge Rnded. or Flat | Ref. | Max. | Min. | Max. | Min. | Max. | Min. |
| 00 | 0.0470 | ... | .093 | .085 | .028 | .042 | .034 | .017 | .010 | .023 | .016 |
| 0 | 0.0600 | 1/8 | .119 | .099 | .035 | .056 | .041 | .023 | .016 | .030 | .025 |
| 1 | 0.0730 | 1/8 | .146 | .123 | .043 | .068 | .052 | .026 | .019 | .038 | .031 |
| 2 | 0.0860 | 1/8 | .172 | .147 | .051 | .080 | .063 | .031 | .023 | .045 | .037 |
| 3 | 0.0990 | 1/8 | .199 | .171 | .059 | .092 | .073 | .035 | .027 | .052 | .043 |
| 4 | 0.1120 | 3/16 | .225 | .195 | .067 | .104 | .084 | .039 | .031 | .059 | .049 |
| 5 | 0.1250 | 3/16 | .252 | .220 | .075 | .116 | .095 | .043 | .035 | .067 | .055 |
| 6 | 0.1380 | 3/16 | .279 | .244 | .083 | .128 | .105 | .048 | .039 | .074 | .060 |
| 8 | 0.1640 | 1/4 | .332 | .292 | .100 | .152 | .126 | .054 | .045 | .088 | .072 |
| 10 | 0.1900 | 5/16 | .385 | .340 | .116 | .176 | .148 | .060 | .050 | .103 | .084 |
| 12 | 0.2160 | 3/8 | .438 | .389 | .132 | .200 | .169 | .067 | .056 | .117 | .096 |
| 1/4 | 0.2500 | 7/16 | .507 | .452 | .153 | .232 | .197 | .075 | .064 | .136 | .112 |
| 5/16 | 0.3125 | 1/2 | .635 | .568 | .191 | .290 | .249 | .084 | .072 | .171 | .141 |
| 3/8 | 0.3750 | 9/16 | .762 | .685 | .230 | .347 | .300 | .094 | .081 | .206 | .170 |
| 7/16 | 0.4375 | 5/8 | .812 | .723 | .223 | .345 | .295 | .094 | .081 | .210 | .174 |
| 1/2 | 0.5000 | 3/4 | .875 | .775 | .223 | .354 | .299 | .106 | .091 | .216 | .176 |
| 9/16 | 0.5625 | ... | 1.000 | .889 | .260 | .410 | .350 | .118 | .102 | .250 | .207 |
| 5/8 | 0.6250 | ... | 1.125 | 1.002 | .298 | .467 | .399 | .133 | .116 | .285 | .235 |
| 3/4 | 0.7500 | ... | 1.375 | 1.230 | .372 | .578 | .497 | .149 | .131 | .353 | .293 |

All dimensions are in inches.   [1] Where specifying nominal size in decimals, zeros preceding decimal points and in the fourth decimal place are omitted.   [2] These lengths or shorter are undercut.

### Table 10. American National Standard Header Points for Machine Screws before Threading (ANSI B18.6.3-1972, R1977)

| Nom. Size | Threads per Inch | Max. P | Min. P | Max. L | Nom. Size | Threads per Inch | Max. P | Min. P | Max. L |
|---|---|---|---|---|---|---|---|---|---|
| | | | | | 10 | 24 | 0.125 | 0.112 | 1 1/4 |
| | | | | | | 32 | 0.138 | 0.124 | 1 1/4 |
| 2 | 56 | 0.057 | 0.050 | 1/2 | 12 | 24 | 0.149 | 0.134 | 1 3/8 |
| | 64 | 0.060 | 0.053 | 1/2 | | 28 | 0.156 | 0.141 | 1 3/8 |
| 4 | 40 | 0.074 | 0.065 | 1/2 | 1/4 | 20 | 0.170 | 0.153 | 1 1/2 |
| | 48 | 0.079 | 0.070 | 1/2 | | 28 | 0.187 | 0.169 | 1 1/2 |
| 5 | 40 | 0.086 | 0.076 | 1/2 | 5/16 | 18 | 0.221 | 0.200 | 1 1/2 |
| | 44 | 0.088 | 0.079 | 1/2 | | 24 | 0.237 | 0.215 | 1 1/2 |
| 6 | 32 | 0.090 | 0.080 | 3/4 | 3/8 | 16 | 0.270 | 0.244 | 1 1/2 |
| | 40 | 0.098 | 0.087 | 3/4 | | 24 | 0.295 | 0.267 | 1 1/2 |
| 8 | 32 | 0.114 | 0.102 | 1 | 7/16 | 14 | 0.316 | 0.287 | 1 1/2 |
| | 36 | 0.118 | 0.106 | 1 | | 20 | 0.342 | 0.310 | 1 1/2 |
| | | | | | 1/2 | 13 | 0.367 | 0.333 | 1 1/2 |
| | | | | | | 20 | 0.399 | 0.362 | 1 1/2 |

All dimensions in inches.   Edges of point may be rounded and end of point need not be flat nor perpendicular to shank.   Machine screws normally have plain sheared ends but when specified may have header points as shown above.

Table 11.   American National Standard Slotted Binding Head and Slotted Undercut
Oval Countersunk Head Machine Screws (ANSI B18.6.3-1972, R1977)

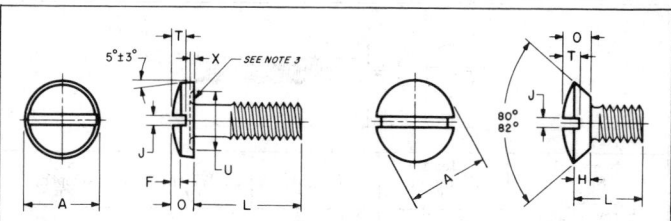

## SLOTTED BINDING HEAD TYPE

| Nominal Size[1] or Basic Screw Diam. | | Head Diam., A | | Total Head Height, O | | Head Oval Height, F | | Slot Width, J | | Slot Depth, T | | Undercut[3] Diam., U | | Undercut[3] Depth, X | |
|---|---|---|---|---|---|---|---|---|---|---|---|---|---|---|---|
| | | Max. | Min. | Max. | Min. | Max. | Min. | Max. | Min. | Max. | Min. | Max. | Min. | Max. | Min. |
| 0000 | 0.0210 | .046 | .040 | .014 | .009 | .006 | .003 | .008 | .004 | .009 | .005 | .... | .... | .... | .... |
| 000 | 0.0340 | .073 | .067 | .021 | .015 | .008 | .005 | .012 | .006 | .013 | .009 | .... | .... | .... | .... |
| 00 | 0.0470 | .098 | .090 | .028 | .023 | .011 | .007 | .017 | .010 | .018 | .012 | .... | .... | .... | .... |
| 0 | 0.0600 | .126 | .119 | .032 | .026 | .012 | .008 | .023 | .016 | .018 | .009 | .098 | .086 | .007 | .002 |
| 1 | 0.0730 | .153 | .145 | .041 | .035 | .015 | .011 | .026 | .019 | .024 | .014 | .120 | .105 | .008 | .003 |
| 2 | 0.0860 | .181 | .171 | .050 | .043 | .018 | .013 | .031 | .023 | .030 | .020 | .141 | .124 | .010 | .005 |
| 3 | 0.0990 | .208 | .197 | .059 | .052 | .022 | .016 | .035 | .027 | .036 | .025 | .162 | .143 | .011 | .006 |
| 4 | 0.1120 | .235 | .223 | .068 | .061 | .025 | .018 | .039 | .031 | .042 | .030 | .184 | .161 | .012 | .007 |
| 5 | 0.1250 | .263 | .249 | .078 | .069 | .029 | .021 | .043 | .035 | .048 | .035 | .205 | .180 | .014 | .009 |
| 6 | 0.1380 | .290 | .275 | .087 | .078 | .032 | .024 | .048 | .039 | .053 | .040 | .226 | .199 | .015 | .010 |
| 8 | 0.1640 | .344 | .326 | .105 | .095 | .039 | .029 | .054 | .045 | .065 | .050 | .269 | .236 | .017 | .012 |
| 10 | 0.1900 | .399 | .378 | .123 | .112 | .045 | .034 | .060 | .050 | .077 | .060 | .312 | .274 | .020 | .015 |
| 12 | 0.2160 | .454 | .430 | .141 | .130 | .052 | .039 | .067 | .056 | .089 | .070 | .354 | .311 | .023 | .018 |
| ¼ | 0.2500 | .525 | .498 | .165 | .152 | .061 | .046 | .075 | .064 | .105 | .084 | .410 | .360 | .026 | .021 |
| 5⁄16 | 0.3125 | .656 | .622 | .209 | .194 | .077 | .059 | .084 | .072 | .134 | .108 | .513 | .450 | .032 | .027 |
| 3⁄8 | 0.3750 | .788 | .746 | .253 | .235 | .094 | .071 | .094 | .081 | .163 | .132 | .615 | .540 | .041 | .034 |

## SLOTTED UNDERCUT OVAL COUNTERSUNK HEAD TYPES

| Nominal Size[1] or Basic Screw Diam. | | Max. L[2] | Head Diam., A | | Head Side Height, H | Total Head Height, O | | Slot Width, J | | Slot Depth, T | |
|---|---|---|---|---|---|---|---|---|---|---|---|
| | | | Max., Edge Sharp | Min., Edge Rnded. or Flat | Ref. | Max. | Min. | Max. | Min. | Max. | Min. |
| 0 | 0.0600 | ⅛ | .119 | .099 | .025 | .046 | .033 | .023 | .016 | .028 | .022 |
| 1 | 0.0730 | ⅛ | .146 | .123 | .031 | .056 | .042 | .026 | .019 | .034 | .027 |
| 2 | 0.0860 | ⅛ | .172 | .147 | .036 | .065 | .050 | .031 | .023 | .040 | .033 |
| 3 | 0.0990 | ⅛ | .199 | .171 | .042 | .075 | .059 | .035 | .027 | .047 | .038 |
| 4 | 0.1120 | 3⁄16 | .225 | .195 | .047 | .084 | .067 | .039 | .031 | .053 | .043 |
| 5 | 0.1250 | 3⁄16 | .252 | .220 | .053 | .094 | .076 | .043 | .035 | .059 | .048 |
| 6 | 0.1380 | 3⁄16 | .279 | .244 | .059 | .104 | .084 | .048 | .039 | .065 | .053 |
| 8 | 0.1640 | ¼ | .332 | .292 | .070 | .123 | .101 | .054 | .045 | .078 | .064 |
| 10 | 0.1900 | 5⁄16 | .385 | .340 | .081 | .142 | .118 | .060 | .050 | .090 | .074 |
| 12 | 0.2160 | ⅜ | .438 | .389 | .092 | .161 | .135 | .067 | .056 | .103 | .085 |
| ¼ | 0.2500 | 7⁄16 | .507 | .452 | .107 | .186 | .158 | .075 | .064 | .119 | .098 |
| 5⁄16 | 0.3125 | ½ | .635 | .568 | .134 | .232 | .198 | .084 | .072 | .149 | .124 |
| ⅜ | 0.3750 | 9⁄16 | .762 | .685 | .161 | .278 | .239 | .094 | .081 | .179 | .149 |
| 7⁄16 | 0.4375 | ⅝ | .812 | .723 | .156 | .279 | .239 | .094 | .081 | .184 | .154 |
| ½ | 0.5000 | ¾ | .875 | .775 | .156 | .288 | .244 | .106 | .091 | .204 | .169 |

All dimensions are in inches.   [1] Where specifying nominal size in decimals, zeros preceding decimal points and in the fourth decimal place are omitted.   [2] These lengths or shorter are undercut.   [3] Unless otherwise specified, slotted binding head machine screws are not undercut.

## Table 12. Slotted Round Head Machine Screws
(ANSI B18.6.3-1972, R1977 Appendix)*

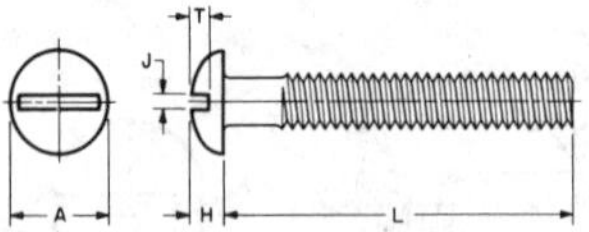

| Nominal Size[1] or Basic Screw Diam. | | Head Diameter, A | | Head Height, H | | Slot Width, J | | Slot Depth, T | |
|---|---|---|---|---|---|---|---|---|---|
| | | Max. | Min. | Max. | Min. | Max. | Min. | Max. | Min. |
| 0000 | 0.0210 | .041 | .035 | .022 | .016 | .008 | .004 | .017 | .013 |
| 000 | 0.0340 | .062 | .056 | .031 | .025 | .012 | .008 | .018 | .012 |
| 00 | 0.0470 | .089 | .080 | .045 | .036 | .017 | .010 | .026 | .018 |
| 0 | 0.0600 | .113 | .099 | .053 | .043 | .023 | .016 | .039 | .029 |
| 1 | 0.0730 | .138 | .122 | .061 | .051 | .026 | .019 | .044 | .033 |
| 2 | 0.0860 | .162 | .146 | .069 | .059 | .031 | .023 | .048 | .037 |
| 3 | 0.0990 | .187 | .169 | .078 | .067 | .035 | .027 | .053 | .040 |
| 4 | 0.1120 | .211 | .193 | .086 | .075 | .039 | .031 | .058 | .044 |
| 5 | 0.1250 | .236 | .217 | .095 | .083 | .043 | .035 | .063 | .047 |
| 6 | 0.1380 | .260 | .240 | .103 | .091 | .048 | .039 | .068 | .051 |
| 8 | 0.1640 | .309 | .287 | .120 | .107 | .054 | .045 | .077 | .058 |
| 10 | 0.1900 | .359 | .334 | .137 | .123 | .060 | .050 | .087 | .065 |
| 12 | 0.2160 | .408 | .382 | .153 | .139 | .067 | .056 | .096 | .073 |
| 1/4 | 0.2500 | .472 | .443 | .175 | .160 | .075 | .064 | .109 | .082 |
| 5/16 | 0.3125 | .590 | .557 | .216 | .198 | .084 | .072 | .132 | .099 |
| 3/8 | 0.3750 | .708 | .670 | .256 | .237 | .094 | .081 | .155 | .117 |
| 7/16 | 0.4375 | .750 | .707 | .328 | .307 | .094 | .081 | .196 | .148 |
| 1/2 | 0.5000 | .813 | .766 | .355 | .332 | .106 | .091 | .211 | .159 |
| 9/16 | 0.5625 | .938 | .887 | .410 | .385 | .118 | .102 | .242 | .183 |
| 5/8 | 0.6250 | 1.000 | .944 | .438 | .411 | .133 | .116 | .258 | .195 |
| 3/4 | 0.7500 | 1.250 | 1.185 | .547 | .516 | .149 | .131 | .320 | .242 |

All dimensions are in inches.
* Not recommended, use Pan Head machine screws.
[1] Where specifying nominal size in decimals, zeros preceding decimal point and in the fourth decimal place are omitted.

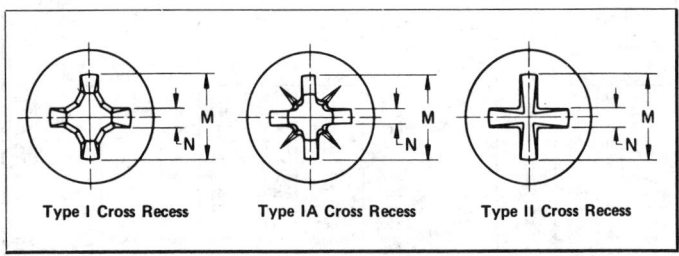

American National Standard Cross Recesses for Machine Screws

**Machine Screw Cross Recesses.** — Three cross recesses, Types I, IA, and II may be used in lieu of slots in machine screw heads. Dimensions for recess diameter $M$, width $N$, and depth $T$ (not shown above) together with recess penetration gaging depths are given in American National Standard ANSI B18.6.3-1972, R1977.

**British Machine Screws.** — Currently (1967) covering fasteners of this type are British Standards B.S. 57:1951, "B.A. Screws, Bolts and Nuts"; B.S. 450:1958, "Machine Screws and Machine Screw Nuts (B.S.W. and B.S.F. Threads)"; B.S. 1981:1953, "Unified Machine Screws and Machine Screw Nuts"; B.S. 2827:1957, "Machine Screw Nuts, Pressed Type (B.A. and Whitworth Form Threads)"; B.S. 3155:1960, "American Machine Screws and Nuts in Sizes Below ¼ inch Diameter"; and B.S. 4183: 1967, "Machine Screws and Machine Screw Nuts, Metric Series." At a conference organized by the British Standards Institution in 1965 at which the major sectors of British industry were represented, a policy statement was approved which urged British firms to regard the traditional screw thread systems — Whitworth, B.A. and B.S.F. — as obsolescent, and to make the internationally-agreed ISO metric thread their first choice (with ISO Unified thread as second choice) for all future designs. It is recognized that some sections of British industry already using ISO inch (Unified) screw threads may find it necessary, for various reasons, to continue with their use for some time; Whitworth and B.A. threads should, however, be superseded by ISO metric threads in preference to an intermediate change to ISO inch threads. This means that eventually fasteners covered by B.S. 57, B.S. 450 and B.S. 2827 would be superseded and replaced by fasteners specified by B.S. 4183.

**British Standard Whitworth (B.S.W.) and Fine (B.S.F.) Machine Screws.** — British Standard B.S. 450:1958 covers machine screws and nuts with British Standard Whitworth and British Standard Fine threads. It covers all of the various heads in common use in both slotted and recessed forms. Head shapes are shown on page 1197 and dimensions on page 1199. It is intended that this standard will eventually be superseded and replaced by B.S. 4183, "Machine Screws and Machine Screw Nuts, Metric Series."

**British Standard Machine Screws and Machine Screw Nuts, Metric Series.** — British Standard B.S. 4183:1967 gives dimensions and tolerances for: countersunk head, raised countersunk head, and cheese head slotted head screws in a diameter range from M1 (1 mm) to M20 (20 mm); pan head slotted head screws in a diameter range from M2.5 (2.5 mm) to M10 (10 mm); countersunk head and raised countersunk head recessed head screws in a diameter range from M2.5 (2.5 mm) to M12 (12 mm); pan head recessed head screws in a diameter range from M2.5 (2.5 mm) to M10 (10 mm); and square and hexagon machine screw nuts in a diameter range from M1.6 (1.6 mm) to M10 (10 mm). Mechanical properties are also specified for steel, brass and aluminum alloy machine screws and machine screw nuts in this standard.

*Material:* The materials from which the screws and nuts are manufactured have a tensile strength not less than the following: steel, 40 kgf/mm² (392 N/mm²); brass, 32 kgf/mm² (314 N/mm²); and aluminum alloy, 32 kgf/mm² (314 N/mm²). The unit, kgf/mm² is in accordance with ISO DR 911 and the unit in parentheses has the relationship, 1 kgf = 9.80665 Newtons. These minimum strengths are applicable to the finished products. Steel machine screws conform to the requirements for strength grade designation 4.8. The strength grade designation system for machine screws consists of two figures, the first is $\frac{1}{10}$ of the minimum tensile strength in kgf/mm², the second is $\frac{1}{10}$ of the ratio between the yield stress and the minimum tensile strength expressed as a percentage: $\frac{1}{10}$ minimum tensile strength of 40 kgf/mm² gives the symbol "4"; $\frac{1}{10}$ ratio $\frac{\text{yield stress}}{\text{minimum tensile strength}} \% = \frac{1}{10} \times$ 32/40 × 100/1 = "8"; giving the strength grade designation "4.8." Multiplication of these two figures gives the minimum yield stress in kgf/mm².

*Coating of Screws and Nuts:* It is recommended that the coating comply with the appropriate part of B.S. 3382 "Electroplated Coatings on Threaded Components."

*Screw Threads:* Screw threads are ISO metric coarse pitch series threads in accordance with B.S. 3643, "ISO Metric Screw Threads," Part 1, "Thread Data and Standard Thread Series." The external threads used for screws conform to tolerance Class 6g limits (medium fit) as given in B.S. 3643, "ISO Metric Screw Threads," Part 2, "Limits and Tolerances for Coarse Pitch Series Threads." The internal threads used for nuts conform to tolerance Class 6H limits (medium fit) as given in B.S. 3643: Part 2.

*Nominal Length of Screws:* For countersunk head screws the nominal length is the distance from the upper surface of the head to the extreme end of the shank, including any chamfer, radius, or cone point. For raised countersunk head screws the nominal length is the distance from the upper surface of the head (excluding the raised portion) to the extreme end of the shank, including any chamfer, radius, or cone point. For pan and cheese head the nominal length is the distance from the underside of the head to the extreme end of the shank, including any chamfer, radius, or cone point. Standard nominal lengths and tolerances are given in Table 5.

*Length of Thread on Screws:* The length of thread is the distance from the end of the screw (including any chamfer, radius, or cone point) to the leading face of a nut without countersink which has been screwed as far as possible onto the screw by hand. The minimum thread length is as given below:

| Nominal Thread Diam. $d$† | M1 | M1.2 | (M1.4) | M1.6 | M2 | (M2.2) | M2.5 | M3 | (M3.5) | M4 |
|---|---|---|---|---|---|---|---|---|---|---|
| Thread Length $b$ (Min.) | * | * | * | 15 | 16 | 17 | 18 | 19 | 20 | 22 |

| Nominal Thread Diam. $d$† | (M4.5) | M5 | M6 | M8 | M10 | M12 | (M14) | M16 | (M18) | M20 |
|---|---|---|---|---|---|---|---|---|---|---|
| Thread Length $b$ (Min.) | 24 | 25 | 28 | 34 | 40 | 46 | 52 | 58 | 64 | 70 |

All dimensions are in millimeters.
† Items shown in parentheses are non-preferred.
* Threaded up to the head.

Screws of nominal thread diameter M1, M1.2 and M1.4 and screws of larger diameters which are too short for the above thread lengths are threaded as far as possible up to the head. In these the length of unthreaded shank under the head does not exceed 1½ pitches for lengths up to twice the diameter and 2 pitches for longer lengths, and is defined as the distance from the leading face of a nut which has been screwed as far as possible onto the screw by hand to: (1) the junction of the basic major diameter and the countersunk portion of the head on countersunk and raised countersunk heads; (2) the underside of the head on other types of heads.

*Diameter of Unthreaded Shank on Screws:* The diameter of the unthreaded portion of the shank on screws is not greater than the basic major diameter of the screw head and not less than the minimum effective diameter of the screw thread. The diameter of the unthreaded portion of shank is closely associated with the method of manufacture; it will generally be nearer the major diameter of the thread for turned screws and nearer the effective diameter for those produced by cold heading.

*Radius Under the Head of Screws:* The radius under the head of pan and cheese head screws runs smoothly into the face of the head and shank without any step or discontinuity. A true radius is not essential providing that the curve is smooth and lies wholly within the maximum radius. Any radius under the head of countersunk head screws runs smoothly into the conical bearing surface of the head and the shank without any step or discontinuity. The radius values given in Tables 1 and 2 are regarded as the maximum where the shank diameter is equal to the major diameter of the thread and minimum where the shank diameter is approximately equal to the effective diameter of the thread.

*Ends of Screws:* When screws are made with rolled threads the "lead" formed by the thread rolling operation is normally regarded as providing the necessary chamfer and no other machining is necessary. The ends of screws with cut threads are normally finished with a chamfer conforming to the dimension in Fig. 1. At the option of the manufacturer, the ends of screws smaller than M6 (6 mm diameter) may be finished with a radius approximately equal to 1¼ times the nominal diameter of the shank. When cone point ends are required they should have the dimensions given in Fig. 1.

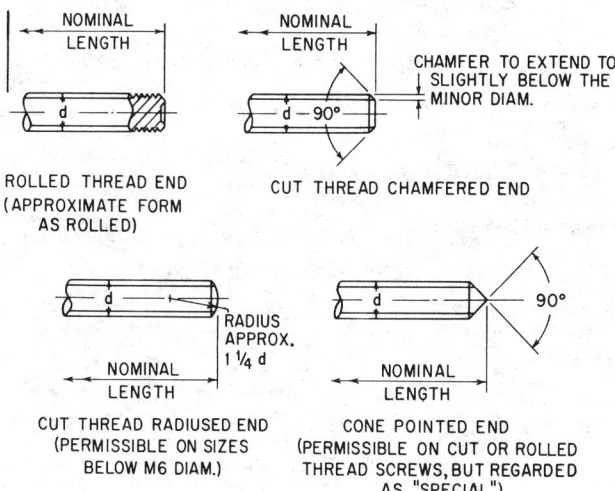

Fig. 1. Alternative Types of End Permissible on Screws

*Dimensions of 90-Degree Countersunk Head Screws:* One of the appendices to this British Standard states that countersunk head screws should fit into the countersunk hole with as great a degree of flushness as possible. To achieve this, it is necessary for the dimensions of both the head of the screw and the countersunk hole to be controlled within prescribed limits. The maximum or design size of the head is controlled by a theoretical diameter to a sharp corner and the minimum head

*Illustrative matter continued on page 1190 and text on page 1196.*

## British Standard Machine Screws and Machine Screw Nuts — Metric Series

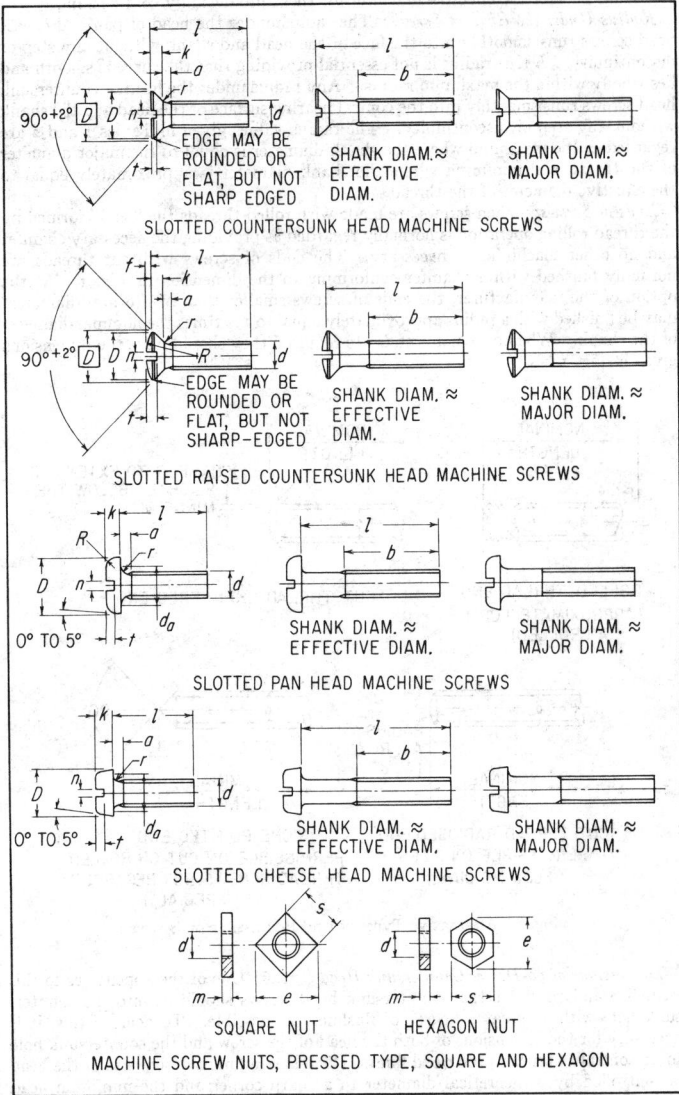

SLOTTED COUNTERSUNK HEAD MACHINE SCREWS

SLOTTED RAISED COUNTERSUNK HEAD MACHINE SCREWS

SLOTTED PAN HEAD MACHINE SCREWS

SLOTTED CHEESE HEAD MACHINE SCREWS

SQUARE NUT     HEXAGON NUT

MACHINE SCREW NUTS, PRESSED TYPE, SQUARE AND HEXAGON

For dimensions see Tables 1 through 5.

Table 1.    British Standard Slotted Countersunk Head Machine Screws — Metric Series (B.S. 4183:1967)

| Nominal Size d* | Head Diameter D Max (Theor. Sharp) 2d | Head Diameter D Min 1.75d | Head Height k Max 0.5d | Head Height k Min 0.45d | Radius r† | Thread Length b Min | Thread Run-out a Max $2\frac{2}{3}$§ | Flushness Tolerance¶ Max | Slot Width n Max | Slot Width n Min | Slot Depth t Max 0.3d | Slot Depth t Min 0.2d |
|---|---|---|---|---|---|---|---|---|---|---|---|---|
| M1 | 2.00 | 1.75 | 0.50 | 0.45 | 0.1 | ‡ | 0.50 | .... | 0.45 | 0.31 | 0.30 | 0.20 |
| M1.2 | 2.40 | 2.10 | 0.60 | 0.54 | 0.1 | ‡ | 0.50 | .... | 0.50 | 0.36 | 0.36 | 0.24 |
| (M1.4) | 2.80 | 2.45 | 0.70 | 0.63 | 0.1 | ‡ | 0.60 | .... | 0.50 | 0.36 | 0.42 | 0.28 |
| M1.6 | 3.20 | 2.80 | 0.80 | 0.72 | 0.1 | 15.0 | 0.70 | .... | 0.60 | 0.46 | 0.48 | 0.32 |
| M2.0 | 4.00 | 3.50 | 1.00 | 0.90 | 0.1 | 16.0 | 0.80 | .... | 0.70 | 0.56 | 0.60 | 0.40 |
| (M2.2) | 4.40 | 3.85 | 1.10 | 0.99 | 0.1 | 17.0 | 0.90 | .... | 0.80 | 0.66 | 0.66 | 0.44 |
| M2.5 | 5.00 | 4.38 | 1.25 | 1.12 | 0.1 | 18.0 | 0.90 | 0.10 | 0.80 | 0.66 | 0.75 | 0.50 |
| M3 | 6.00 | 5.25 | 1.50 | 1.35 | 0.1 | 19.0 | 1.00 | 0.12 | 1.00 | 0.86 | 0.90 | 0.60 |
| (M3.5) | 7.00 | 6.10 | 1.75 | 1.57 | 0.2 | 20.0 | 1.20 | 0.13 | 1.00 | 0.86 | 1.05 | 0.70 |
| M4 | 8.00 | 7.00 | 2.00 | 1.80 | 0.2 | 22.0 | 1.40 | 0.15 | 1.20 | 1.06 | 1.20 | 0.80 |
| (M4.5) | 9.00 | 7.85 | 2.25 | 2.03 | 0.2 | 24.0 | 1.50 | 0.17 | 1.20 | 1.06 | 1.35 | 0.90 |
| M5 | 10.00 | 8.75 | 2.50 | 2.25 | 0.2 | 25.0 | 1.60 | 0.19 | 1.51 | 1.26 | 1.50 | 1.00 |
| M6 | 12.00 | 10.50 | 3.00 | 2.70 | 0.25 | 28.0 | 2.00 | 0.23 | 1.91 | 1.66 | 1.80 | 1.20 |
| M8 | 16.00 | 14.00 | 4.00 | 3.60 | 0.4 | 34.0 | 2.50 | 0.29 | 2.31 | 2.06 | 2.40 | 1.60 |
| M10 | 20.00 | 17.50 | 5.00 | 4.50 | 0.4 | 40.0 | 3.00 | 0.37 | 2.81 | 2.56 | 3.00 | 2.00 |
| M12 | 24.00 | 21.00 | 6.00 | 5.40 | 0.6 | 46.0 | 3.50 | 0.44 | 3.31 | 3.06 | 3.60 | 2.40 |
| (M14) | 28.00 | 24.50 | 7.00 | 6.30 | 0.6 | 52.0 | 4.00 | 0.52 | 3.31 | 3.06 | 4.20 | 2.80 |
| M16 | 32.00 | 28.00 | 8.00 | 7.20 | 0.6 | 58.0 | 4.00 | 0.60 | 4.37 | 4.07 | 4.80 | 3.20 |
| (M18) | 36.00 | 31.50 | 9.00 | 8.10 | 0.6 | 64.0 | 5.00 | 0.67 | 4.37 | 4.07 | 5.40 | 3.60 |
| M20 | 40.00 | 35.00 | 10.00 | 9.00 | 0.8 | 70.0 | 5.00 | 0.75 | 5.37 | 5.07 | 6.00 | 4.00 |

All dimensions are given in millimeters.    For dimensional notation, see diagram on page 1190.    Recessed head screws are also standard and are available.    For dimensions see British Standard.

* Nominal sizes shown in parentheses are non-preferred.
† See Radius Under the Head of Screws description in text.
‡ Threaded up to head.
§ See text following table in Length of Thread on Screws description in text.
¶ See Dimensions of 90-Degree Countersunk Head of Screws description in text.

Table 2. British Standard Slotted Raised Countersunk Head Machine Screws — Metric Series (B.S. 4183:1967)

| Nominal Size d* | Head Diameter D Max. (Theor. Sharp) 2d | Head Diameter D Min. 1.75d | Head Height k Max. 0.5d | Head Height k Min. 0.45d | Radius Under Head r† | Thread Length b Min. | Thread Run-out a Max. 2p§ | Height of Raised Portion f Nom. 0.25d | Head Radius R Nom. | Slot Width n Max. | Slot Width n Min. | Slot Depth t Max. 0.5d | Slot Depth t Min. 0.4d |
|---|---|---|---|---|---|---|---|---|---|---|---|---|---|
| M1 | 2.00 | 1.75 | 0.50 | 0.45 | 0.1 | ‡‡‡‡‡‡ | 0.50 | 0.25 | 2.0 | 0.45 | 0.31 | 0.50 | 0.40 |
| M1.2 | 2.40 | 2.10 | 0.60 | 0.54 | 0.1 | ‡‡‡‡‡‡ | 0.50 | 0.30 | 2.5 | 0.50 | 0.36 | 0.60 | 0.48 |
| (M1.4) | 2.80 | 2.45 | 0.70 | 0.63 | 0.1 | ‡‡‡‡‡‡ | 0.60 | 0.35 | 2.5 | 0.50 | 0.36 | 0.70 | 0.56 |
| M1.6 | 3.20 | 2.80 | 0.80 | 0.72 | 0.1 | 15.0 | 0.70 | 0.40 | 3.0 | 0.60 | 0.46 | 0.80 | 0.64 |
| M2.0 | 4.00 | 3.50 | 1.00 | 0.90 | 0.1 | 16.0 | 0.80 | 0.50 | 4.0 | 0.70 | 0.56 | 1.00 | 0.80 |
| (M2.2) | 4.40 | 3.85 | 1.10 | 0.99 | 0.1 | 17.0 | 0.90 | 0.55 | 4.0 | 0.80 | 0.66 | 1.10 | 0.88 |
| M2.5 | 5.00 | 4.38 | 1.25 | 1.12 | 0.1 | 18.0 | 0.90 | 0.60 | 5.0 | 0.80 | 0.66 | 1.25 | 1.00 |
| M3 | 6.00 | 5.25 | 1.50 | 1.35 | 0.1 | 19.0 | 1.00 | 0.75 | 6.0 | 1.00 | 0.86 | 1.50 | 1.20 |
| (M3.5) | 7.00 | 6.10 | 1.75 | 1.57 | 0.2 | 20.0 | 1.20 | 0.90 | 6.0 | 1.00 | 0.86 | 1.75 | 1.40 |
| M4 | 8.00 | 7.00 | 2.00 | 1.80 | 0.2 | 22.0 | 1.40 | 1.00 | 8.0 | 1.20 | 1.06 | 2.00 | 1.60 |
| (M4.5) | 9.00 | 7.85 | 2.25 | 2.03 | 0.2 | 24.0 | 1.50 | 1.10 | 8.0 | 1.20 | 1.06 | 2.25 | 1.80 |
| M5 | 10.00 | 8.75 | 2.50 | 2.25 | 0.2 | 25.0 | 1.60 | 1.25 | 10.0 | 1.51 | 1.26 | 2.50 | 2.00 |
| M6 | 12.00 | 10.50 | 3.00 | 2.70 | 0.25 | 28.0 | 2.00 | 1.50 | 12.0 | 1.91 | 1.66 | 3.00 | 2.40 |
| M8 | 16.00 | 14.00 | 4.00 | 3.60 | 0.4 | 34.0 | 2.50 | 2.00 | 16.0 | 2.31 | 2.06 | 4.00 | 3.20 |
| M10 | 20.00 | 17.50 | 5.00 | 4.50 | 0.4 | 40.0 | 3.00 | 2.50 | 20.0 | 2.81 | 2.56 | 5.00 | 4.00 |
| M12 | 24.00 | 21.00 | 6.00 | 5.40 | 0.6 | 46.0 | 3.50 | 3.00 | 25.0 | 3.31 | 3.06 | 6.00 | 4.80 |
| (M14) | 28.00 | 24.50 | 7.00 | 6.30 | 0.6 | 52.0 | 4.00 | 3.50 | 25.0 | 3.31 | 3.06 | 7.00 | 5.60 |
| M16 | 32.00 | 28.00 | 8.00 | 7.20 | 0.6 | 58.0 | 4.00 | 4.00 | 32.0 | 4.37 | 4.07 | 8.00 | 6.40 |
| (M18) | 36.00 | 31.50 | 9.00 | 8.10 | 0.6 | 64.0 | 5.00 | 4.50 | 32.0 | 4.37 | 4.07 | 9.00 | 7.20 |
| M20 | 40.00 | 35.00 | 10.00 | 9.00 | 0.8 | 70.0 | 5.00 | 5.00 | 40.0 | 5.37 | 5.07 | 10.00 | 8.00 |

All dimensions are given in millimeters. For dimensional notation see diagram on page 1190. Recessed head screws are also standard and available. For dimensions see British Standard.

* Nominal sizes shown in parentheses are non-preferred.
† See *Radius Under the Head of Screws* description in text.
‡ Threaded up to head.
‡‡ See text following table in *Length of Thread on Screws* description in text.

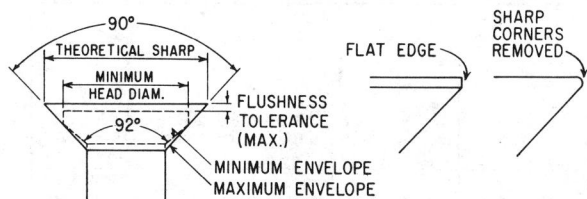

Fig. 2. Head Configuration        Fig. 3. Edge Configuration

Table 3. British Standard Slotted Pan Head Machine Screws —
Metric Series (B.S. 4183:1967)

| Nominal Size d* | Head Diameter D | | Head Height k | | Head Radius R | Radius Under Head r | Transition Diameter $d_a$ |
|---|---|---|---|---|---|---|---|
| | Max. 2d | Min. | Max. 0.6d | Min. | Max. 0.4d | Min. | Max. |
| M2.5 | 5.00 | 4.70 | 1.50 | 1.36 | 1.00 | 0.10 | 3.10 |
| M3 | 6.00 | 5.70 | 1.80 | 1.66 | 1.20 | 0.10 | 3.60 |
| (M3.5) | 7.00 | 6.64 | 2.10 | 1.96 | 1.40 | 0.20 | 4.30 |
| M4 | 8.00 | 7.64 | 2.40 | 2.26 | 1.60 | 0.20 | 4.70 |
| (M4.5) | 9.00 | 8.64 | 2.70 | 2.56 | 1.80 | 0.20 | 5.20 |
| M5 | 10.00 | 9.64 | 3.00 | 2.86 | 2.00 | 0.20 | 5.70 |
| M6 | 12.00 | 11.57 | 3.60 | 3.42 | 2.50 | 0.25 | 6.80 |
| M8 | 16.00 | 15.57 | 4.80 | 4.62 | 3.20 | 0.40 | 9.20 |
| M10 | 20.00 | 19.48 | 6.00 | 5.82 | 4.00 | 0.40 | 11.20 |

| Nominal Size d* | Thread Length b | Thread Run-out a | Slot Width n | | Slot Depth t | |
|---|---|---|---|---|---|---|
| | Min. | Max. 2p† | Max. | Min. | Max. 0.6k | Min. 0.4k |
| M2.5 | 18.00 | 0.90 | 0.80 | 0.66 | 0.90 | 0.60 |
| M3 | 19.00 | 1.00 | 1.00 | 0.86 | 1.08 | 0.72 |
| (M3.5) | 20.00 | 1.20 | 1.00 | 0.86 | 1.26 | 0.84 |
| M4 | 22.00 | 1.40 | 1.20 | 1.06 | 1.44 | 0.96 |
| (M4.5) | 24.00 | 1.50 | 1.20 | 1.06 | 1.62 | 1.08 |
| M5 | 25.00 | 1.60 | 1.51 | 1.26 | 1.80 | 1.20 |
| M6 | 28.00 | 2.00 | 1.91 | 1.66 | 2.16 | 1.44 |
| M8 | 34.00 | 2.50 | 2.31 | 2.06 | 2.88 | 1.92 |
| M10 | 40.00 | 3.00 | 2.81 | 2.56 | 3.60 | 2.40 |

All dimensions are given in millimeters. For dimensional notation, see diagram on page 1190. Recessed head screws are also standard and available. For dimensions see British Standard.
* Nominal sizes shown in parentheses are non-preferred.
† See text following table in *Length of Thread on Screws* description in text.

Table 4. British Standard Slotted Cheese Head Machine Screws — Metric Series (B.S. 4183:1967)

| Nominal Size $d^*$ | Head Diameter $D$ Max. | Head Diameter $D$ Min. | Head Height $k$ Max. | Head Height $k$ Min. | Radius $r$† Min. | Transition Diameter $d_a$ Max. | Thread Length $b$ Min. | Thread Run-out $a$ Max.§ | Slot Width $n$ Max. | Slot Width $n$ Min. | Slot Depth $t$ Max. | Slot Depth $t$ Min. |
|---|---|---|---|---|---|---|---|---|---|---|---|---|
| M1 | 2.00 | 1.75 | 0.70 | 0.56 | 0.10 | 1.30 | † | 0.50 | 0.45 | 0.31 | 0.44 | 0.30 |
| M1.2 | 2.30 | 2.05 | 0.80 | 0.66 | 0.10 | 1.50 | † | 0.50 | 0.50 | 0.36 | 0.49 | 0.35 |
| (M1.4) | 2.60 | 2.35 | 0.90 | 0.76 | 0.10 | 1.70 | † | 0.60 | 0.50 | 0.36 | 0.60 | 0.40 |
| M1.6 | 3.00 | 2.75 | 1.00 | 0.86 | 0.10 | 2.00 | 15.00 | 0.70 | 0.60 | 0.46 | 0.65 | 0.45 |
| M2 | 3.80 | 3.50 | 1.30 | 1.16 | 0.10 | 2.60 | 16.00 | 0.80 | 0.70 | 0.56 | 0.85 | 0.60 |
| (M2.2) | 4.00 | 3.70 | 1.50 | 1.36 | 0.10 | 2.80 | 17.00 | 0.90 | 0.80 | 0.66 | 1.00 | 0.70 |
| M2.5 | 4.50 | 4.20 | 1.60 | 1.46 | 0.10 | 3.10 | 18.00 | 0.90 | 0.80 | 0.66 | 1.00 | 0.70 |
| M3 | 5.50 | 5.20 | 2.00 | 1.86 | 0.10 | 3.60 | 19.00 | 1.00 | 1.00 | 0.86 | 1.30 | 0.90 |
| (M3.5) | 6.00 | 5.70 | 2.40 | 2.26 | 0.10 | 4.10 | 20.00 | 1.20 | 1.00 | 0.86 | 1.40 | 1.00 |
| M4 | 7.00 | 6.64 | 2.60 | 2.46 | 0.20 | 4.70 | 22.00 | 1.40 | 1.20 | 1.06 | 1.60 | 1.20 |
| (M4.5) | 8.00 | 7.64 | 3.10 | 2.92 | 0.20 | 5.20 | 24.00 | 1.50 | 1.20 | 1.06 | 1.80 | 1.40 |
| M5 | 8.50 | 8.14 | 3.30 | 3.12 | 0.20 | 5.70 | 25.00 | 1.60 | 1.51 | 1.26 | 2.00 | 1.50 |
| M6 | 10.00 | 9.64 | 3.90 | 3.72 | 0.25 | 6.80 | 28.00 | 2.00 | 1.91 | 1.66 | 2.30 | 1.80 |
| M8 | 13.00 | 12.57 | 5.00 | 4.82 | 0.40 | 9.20 | 34.00 | 2.50 | 2.31 | 2.06 | 2.80 | 2.30 |
| M10 | 16.00 | 15.57 | 6.00 | 5.82 | 0.40 | 11.20 | 40.00 | 3.00 | 2.81 | 2.56 | 3.20 | 2.70 |
| M12 | 18.00 | 17.57 | 7.00 | 6.78 | 0.60 | 14.20 | 46.00 | 3.50 | 3.31 | 3.06 | 3.80 | 3.20 |
| (M14) | 21.00 | 20.48 | 8.00 | 7.78 | 0.60 | 16.20 | 52.00 | 4.00 | 3.31 | 3.06 | 4.20 | 3.60 |
| M16 | 24.00 | 23.48 | 9.00 | 8.78 | 0.60 | 18.20 | 58.00 | 4.00 | 4.37 | 4.07 | 4.60 | 4.00 |
| (M18) | 27.00 | 26.48 | 10.00 | 9.78 | 0.60 | 20.20 | 64.00 | 5.00 | 4.37 | 4.07 | 5.10 | 4.50 |
| M20 | 30.00 | 29.48 | 11.00 | 10.73 | 0.80 | 22.40 | 70.00 | 5.00 | 5.27 | 5.07 | 5.60 | 5.00 |

All dimensions are given in millimeters. For dimensional notation, see diagram on page 1190.
* Nominal sizes shown in parentheses are non-preferred.
† Threaded up to head.
§ See text following table in *Length of Thread on Screws* description in text.

**Table 5. British Standard Machine Screws and Nuts — Metric Series (B.S. 4183:1967)**

## CONCENTRICITY TOLERANCES

COUNTERSUNK & RAISED COUNTERSUNK HEADS — IT 13 / IT 13 — SLOT TO HEAD / HEAD TO SHANK

PAN & CHEESE HEADS — HEAD TO SHANK

| Nominal Size $d*$ | Head to Shank and Slot to Head (IT 13) — Csk., Raised Csk. and Pan Heads | Head to Shank and Slot to Head (IT 13) — Cheese Heads |
|---|---|---|
| M1 | 0.14 | 0.14 |
| M1.2 | 0.14 | 0.14 |
| (M1.4) | 0.14 | 0.14 |
| M1.6 | 0.18 | 0.14 |
| M2 | 0.18 | 0.14 |
| (M2.2) | 0.18 | 0.18 |
| M2.5 | 0.18 | 0.18 |
| M3 | 0.18 | 0.18 |
| (M3.5) | 0.22 | 0.18 |
| M4 | 0.22 | 0.18 |
| (M4.5) | 0.22 | 0.22 |
| M5 | 0.22 | 0.22 |
| M6 | 0.27 | 0.22 |
| M8 | 0.27 | 0.27 |
| M10 | 0.33 | 0.27 |
| M12 | 0.33 | 0.27 |
| (M14) | 0.33 | 0.33 |
| M16 | 0.39 | 0.33 |
| (M18) | 0.39 | 0.33 |
| M20 | 0.39 | 0.39 |

## NOMINAL LENGTHS AND TOLERANCES ON LENGTH FOR MACHINE SCREWS

| Nominal Length* | Tolerance | Nominal Length* | Tolerance |
|---|---|---|---|
| 1.5 | ±0.12 | 45 | ±0.50 |
| 2 | ±0.12 | 50 | ±0.50 |
| 2.5 | ±0.20 | 55 | ±0.60 |
| 3 | ±0.20 | 60 | ±0.60 |
| 4 | ±0.24 | 65 | ±0.60 |
| 5 | ±0.24 | 70 | ±0.60 |
| 6 | ±0.24 | 75 | ±0.60 |
| (7) | ±0.29 | 80 | ±0.60 |
| 8 | ±0.29 | 85 | ±0.70 |
| (9) | ±0.29 | 90 | ±0.70 |
| 10 | ±0.29 | (95) | ±0.70 |
| (11) | ±0.35 | 100 | ±0.70 |
| 12 | ±0.35 | (105) | ±0.70 |
| 14 | ±0.35 | 110 | ±0.70 |
| 16 | ±0.35 | (115) | ±0.70 |
| (18) | ±0.35 | 120 | ±0.70 |
| 20 | ±0.42 | (125) | ±0.80 |
| (22) | ±0.42 | 130 | ±0.80 |
| 25 | ±0.42 | 140 | ±0.80 |
| (28) | ±0.42 | 150 | ±0.80 |
| 30 | ±0.42 | 160 | ±0.80 |
| (32) | ±0.50 | 170 | ±0.80 |
| 35 | ±0.50 | 180 | ±0.80 |
| (38) | ±0.50 | 190 | ±0.925 |
| 40 | ±0.50 | 200 | ±0.925 |

## DIMENSIONS OF MACHINE SCREW NUTS, PRESSED TYPE, SQUARE AND HEXAGON

| Nominal Size $d*$ | Width Across Flats $s$ Max. | Width Across Flats $s$ Min. | Corners $e$ Square |
|---|---|---|---|
| M1.6 | 3.2 | 3.02 | 4.5 |
| M2 | 4.0 | 3.82 | 5.7 |
| (M2.2) | 4.5 | 4.32 | 6.4 |
| M2.5 | 5.0 | 4.82 | 7.1 |
| M3 | 5.5 | 5.32 | 7.8 |
| (M3.5) | 6.0 | 5.82 | 8.5 |
| M4 | 7.0 | 6.78 | 9.9 |
| M5 | 8.0 | 7.78 | 11.3 |
| M6 | 10.0 | 9.78 | 14.1 |
| M8 | 13.0 | 12.73 | 18.4 |
| M10 | 17.0 | 16.73 | 24.0 |

| Nominal Size $d*$ | Width Across Corners $e$ Hexagon | Thickness $m$ Max. | Thickness $m$ Min. |
|---|---|---|---|
| M1.6 | 3.7 | 1.0 | 0.75 |
| M2 | 4.6 | 1.2 | 0.95 |
| (M2.2) | 5.2 | 1.2 | 0.95 |
| M2.5 | 5.8 | 1.6 | 1.35 |
| M3 | 6.4 | 1.6 | 1.35 |
| (M3.5) | 6.9 | 2.0 | 1.75 |
| M4 | 8.1 | 2.0 | 1.75 |
| M5 | 9.2 | 2.5 | 2.25 |
| M6 | 11.5 | 3.0 | 2.75 |
| M8 | 15.0 | 4.0 | 3.70 |
| M10 | 19.6 | 5.0 | 4.70 |

All dimensions are given in millimeters. For dimensional notation see diagram on page 1190.
* Nominal sizes and lengths shown in parentheses are non-preferred.

angle of 90 degrees. The minimum head size is controlled by a minimum head diameter, the maximum head angle of 92 degrees and a flushness tolerance (see Fig. 2, page 1193). The edge of the head may be flat or rounded, as shown in Fig. 3 on page 1193.

*General Dimensions:* The general dimensions and tolerances for screws and nuts are given in the accompanying tables. Although slotted screw dimensions are given, recessed head screws are also standard and available. Dimensions of recessed head screws are given in B.S. 4183:1967.

**British Unified Machine Screws and Nuts.** — British Standard B.S. 1981:1953 covers certain types of machine screws and machine screw nuts for which agreement has been reached with the United States and Canada as to general dimensions for interchangeability. These types are: countersunk, raised-countersunk, pan, and raised-cheese head screws with slotted or recessed heads; small hexagon head screws; and precision and pressed nuts. All have Unified threads. Head shapes are shown on page 1197 and dimensions are given on page 1198.

*Identification:* As revised by Amendment No. 1 in February 1955, this standard now requires that the above-mentioned screws and nuts which conform to this standard should have a distinguishing feature applied to identify them as Unified. All *recessed head screws* are to be identified as Unified by a groove in the form of four arcs of a circle in the upper surface of the head. All *hexagon head screws* are to be identified as Unified by: (1) a circular recess in the upper surface of the head; or

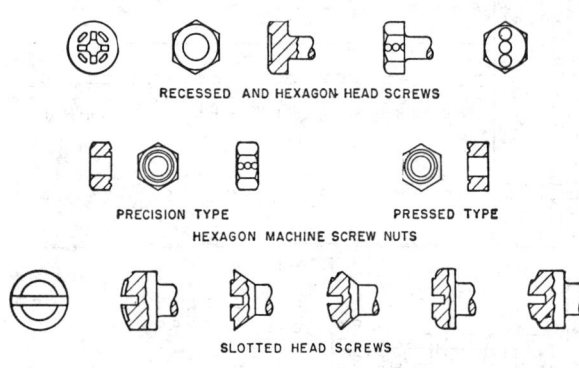

RECESSED AND HEXAGON HEAD SCREWS

PRECISION TYPE            PRESSED TYPE

HEXAGON MACHINE SCREW NUTS

SLOTTED HEAD SCREWS

**Identification Markings for British Standard Unified Machine Screws**

(2) a continuous line of circles indented on one or more of the flats of the hexagon and parallel to the screw axis; or (3) at least two contiguous circles indented on the upper surface of the head. All *machine screw nuts* of the pressed type shall be identified as Unified by means of the application of a groove indented in one face of the nut approximately midway between the major diameter of the thread and flats of the square or hexagon. *Slotted head screws* shall be identified as Unified either by a circular recess or by a circular platform or raised portion on the upper surface of the head. *Machine screw nuts* of the *precision type* shall be identified as Unified by either a groove indented on one face of the front approximately midway between the major diameter of the thread and the flats of the hexagon or a continuous line of circles indented on one or more of the flats of the hexagon and parallel to the nut axis.

**British Standard Machine Screws and Nuts (B.S. 450:1958 and B.S. 1981:1953)**

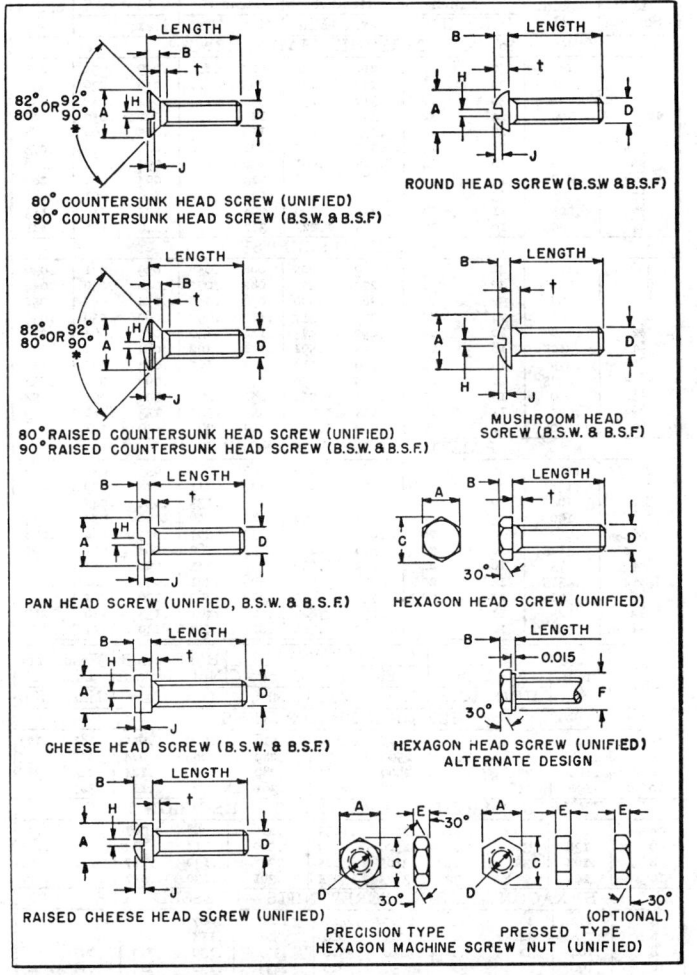

80° COUNTERSUNK HEAD SCREW (UNIFIED)
90° COUNTERSUNK HEAD SCREW (B.S.W. & B.S.F)

ROUND HEAD SCREW (B.S.W & B.S.F)

80° RAISED COUNTERSUNK HEAD SCREW (UNIFIED)
90° RAISED COUNTERSUNK HEAD SCREW (B.S.W & B.S.F.)

MUSHROOM HEAD SCREW (B.S.W. & B.S.F)

PAN HEAD SCREW (UNIFIED, B.S.W. & B.S.F.)

HEXAGON HEAD SCREW (UNIFIED)

CHEESE HEAD SCREW (B.S.W. & B.S.F)

HEXAGON HEAD SCREW (UNIFIED) ALTERNATE DESIGN

RAISED CHEESE HEAD SCREW (UNIFIED)

PRECISION TYPE   PRESSED TYPE
HEXAGON MACHINE SCREW NUT (UNIFIED)

\* Countersinks to suit the screws should have a maximum angle of 80° (Unified) or 90° (B.S.F. and B.S.W.) with a negative tolerance.

† Unified countersunk and raised countersunk head screws 2 inches long and under are threaded right up to the head. Other Unified, B.S.W. and B.S.F. machine screws 2 inches long and under have an unthreaded shank equal to twice the pitch. All Unified, B.S.W. and B.S.F. machine screws longer than 2 inches have a minimum thread length of 1¾ inches.

**British Standard Unified Machine Screws and Machine Screw Nuts (B.S. 1981:1953)[1]**

| Nom. Size of Screw | Basic Diam. D | Threads per Inch | | Diam. of Head A | | Depth of Head B | | Width of Slot H | | Depth of Slot J |
|---|---|---|---|---|---|---|---|---|---|---|
| | | UNC | UNF | Max. | Min. | Max. | Min. | Max. | Min. | |
| **80° COUNTERSUNK HEAD SCREWS[2,3]** | | | | | | | | | | |
| 4 | .112 | 40 | .. | .211 | .194 | .067 | .... | .039 | .031 | .025 |
| 6 | .138 | 32 | .. | .260 | .242 | .083 | .... | .048 | .039 | .031 |
| 8 | .164 | 32 | .. | .310 | .291 | .100 | .... | .054 | .045 | .037 |
| 10 | .190 | 24* | 32 | .359 | .339 | .116 | .... | .060 | .050 | .044 |
| ¼ | .250 | 20 | 28 | .473 | .450 | .153 | .... | .075 | .064 | .058 |
| ⁵⁄₁₆ | .3125 | 18 | 24 | .593 | .565 | .191 | .... | .084 | .072 | .073 |
| ⅜ | .375 | 16 | 24 | .712 | .681 | .230 | .... | .094 | .081 | .086 |
| ⁷⁄₁₆ | .4375 | 14 | 20 | .753 | .719 | .223 | .... | .094 | .081 | .086 |
| ½ | .500 | 13 | 20 | .808 | .770 | .223 | .... | .106 | .091 | .086 |
| ⅝ | .625 | 11 | 18 | 1.041 | .996 | .298 | .... | .133 | .116 | .113 |
| ¾ | .750 | 10 | 16 | 1.275 | 1.223 | .372 | .... | .149 | .131 | .141 |
| **PAN HEAD SCREWS[3]** | | | | | | | | | | |
| 4 | .112 | 40 | .. | .219 | .205 | .068 | .058 | .039 | .031 | .036 |
| 6 | .138 | 32 | .. | .270 | .256 | .082 | .072 | .048 | .039 | .044 |
| 8 | .164 | 32 | .. | .322 | .306 | .096 | .085 | .054 | .045 | .051 |
| 10 | .190 | 24* | 32 | .373 | .357 | .110 | .099 | .060 | .050 | .059 |
| ¼ | .250 | 20 | 28 | .492 | .473† | .144 | .130 | .075 | .064 | .079 |
| ⁵⁄₁₆ | .3125 | 18 | 24 | .615 | .594 | .178 | .162 | .084 | .072 | .101 |
| ⅜ | .375 | 16 | 24 | .740 | .716 | .212 | .195 | .094 | .081 | .122 |
| ⁷⁄₁₆ | .4375 | 14 | 20 | .863 | .838 | .247 | .227 | .094 | .081 | .133 |
| ½ | .500 | 13 | 20 | .987 | .958 | .281 | .260 | .106 | .091 | .152 |
| ⅝ | .625 | 11 | 18 | 1.125 | 1.090 | .350 | .325 | .133 | .116 | .189 |
| ¾ | .750 | 10 | 16 | 1.250 | 1.209 | .419 | .390 | .149 | .131 | .226 |
| **RAISED CHEESE-HEAD SCREWS[3]** | | | | | | | | | | |
| 4 | .112 | 40 | .. | .183 | .166 | .107 | .088 | .039 | .031 | .042 |
| 6 | .138 | 32 | .. | .226 | .208 | .132 | .111 | .048 | .039 | .053 |
| 8 | .164 | 32 | .. | .270 | .250 | .156 | .133 | .054 | .045 | .063 |
| 10 | .190 | 24* | 32 | .313 | .292 | .180 | .156 | .060 | .050 | .074 |
| ¼ | .250 | 20 | 28 | .414 | .389 | .237 | .207 | .075 | .064 | .098 |
| ⁵⁄₁₆ | .3125 | 18 | 24 | .518 | .490 | .295 | .262 | .084 | .072 | .124 |
| ⅜ | .375 | 16 | 24 | .622 | .590 | .355 | .315 | .094 | .081 | .149 |
| ⁷⁄₁₆ | .4375 | 14 | 20 | .625 | .589 | .368 | .321 | .094 | .081 | .153 |
| ½ | .500 | 13 | 20 | .750 | .710 | .412 | .362 | .106 | .091 | .171 |
| ⅝ | .625 | 11 | 18 | .875 | .827 | .521 | .461 | .133 | .116 | .217 |
| ¾ | .750 | 10 | 16 | 1.000 | .945 | .612 | .542 | .149 | .131 | .254 |

| Nom. Size | Basic Diam. D | Threads per Inch | | Width Across | | | | H'd Depth B Nut Thick. E | | Wash. Face Diam. F | |
|---|---|---|---|---|---|---|---|---|---|---|---|
| | | | | Flats A | | Corners C | | | | | |
| | | UNC | UNF | Max. | Min. | Max. | | Max. | Min. | Max. | Min. |
| **HEXAGON HEAD SCREWS** | | | | | | | | | | | |
| 4 | .112 | 40 | .. | .1875 | .1835 | .216 | | .060 | .055 | .183 | .173 |
| 6 | .138 | 32 | .. | .2500 | .2450 | .289 | | .080 | .074 | .245 | .235 |
| 8 | .164 | 32 | .. | .2500 | .2450 | .289 | | .110 | .104 | .245 | .235 |
| 10 | .190 | 24* | 32 | .3125 | .3075 | .361 | | .120 | .113 | .307 | .297 |
| **HEXAGON MACHINE SCREW NUTS — PRECISION TYPE** | | | | | | | | | | | |
| 4 | .112 | 40 | .. | .1875 | .1835 | .216 | | .098 | .087 | .... | .... |
| 6 | .138 | 32 | .. | .2500 | .2450 | .289 | | .114 | .102 | .... | .... |
| 8 | .164 | 32 | .. | .3125 | .3075 | .361 | | .130 | .117 | .... | .... |
| 10 | .190 | 24* | 32 | .3125 | .3075 | .361 | | .130 | .117 | .... | .... |
| **HEXAGON MACHINE SCREW NUTS — PRESSED TYPE** | | | | | | | | | | | |
| 4 | .112 | 40 | .. | .2500 | .2410 | .289 | | .087 | .077 | .... | .... |
| 6 | .138 | 32 | .. | .3125 | .3020 | .361 | | .114 | .102 | .... | .... |
| 8 | .164 | 32 | .. | .3438 | .3320 | .397 | | .130 | .117 | .... | .... |
| 10 | .190 | 24* | 32 | .3750 | .3620 | .433 | | .130 | .117 | .... | .... |
| ¼ | .250 | 20 | 28 | .4375 | .4230 | .505 | | .193 | .178 | .... | .... |
| ⁵⁄₁₆ | .3125 | 18 | 24 | .5625 | .5450 | .649 | | .225 | .208 | .... | .... |
| ⅜ | .375 | 16 | 24 | .6250 | .6070 | .722 | | .257 | .239 | .... | .... |

All dimensions in inches. [1] See page 1197, for a pictorial representation and letter dimensions. [2] All dimensions, except J, given for the No. 4 to ⅜-inch sizes, incl., also apply to all the 80° Raised Countersunk Head Screws given in the Standard. [3] Also available with recessed heads. * Non-preferred. † By arrangement may also be .468.

**British Standard Whitworth (B.S.W.) and Fine (B.S.F.) Machine Screws (B.S. 450:1958)[1]**

| Nom. Size of Screw | Basic Diam. D | Threads per Inch | | Diam. of Head A | | Depth of Head B | | Width of Slot H | | Depth of Slot J |
|---|---|---|---|---|---|---|---|---|---|---|
| | | B.S.W. | B.S.F. | Max. | Min. | Max. | Min. | Max. | Min. | |
| 90° COUNTERSUNK HEAD SCREWS[2,3] | | | | | | | | | | |
| 1/8 | .1250 | 40 | .. | .219 | .201 | .056 | .... | .039 | .032 | .027 |
| 3/16 | .1875 | 24 | 32* | .328 | .307 | .084 | .... | .050 | .042 | .041 |
| 7/32 | .2188 | .. | 28* | .383 | .360 | .098 | .... | .055 | .046 | .048 |
| 1/4 | .2500 | 20 | 26 | .438 | .412 | .113 | .... | .061 | .051 | .055 |
| 5/16 | .3125 | 18 | 22 | .547 | .518 | .141 | .... | .071 | .061 | .069 |
| 3/8 | .3750 | 16 | 20 | .656 | .624 | .169 | .... | .082 | .072 | .083 |
| 7/16 | .4375 | 14 | 18 | .766 | .729 | .197 | .... | .093 | .082 | .097 |
| 1/2 | .5000 | 12 | 16 | .875 | .835 | .225 | .... | .104 | .092 | .111 |
| 9/16 | .5625 | 12* | 16* | .984 | .941 | .253 | .... | .115 | .103 | .125 |
| 5/8 | .6250 | 11 | 14 | 1.094 | 1.046 | .281 | .... | .126 | .113 | .138 |
| 3/4 | .7500 | 10 | 12 | 1.312 | 1.257 | .338 | .... | .148 | .134 | .166 |
| ROUND HEAD SCREWS[3] | | | | | | | | | | |
| 1/8 | .1250 | 40 | .. | .219 | .206 | .087 | .082 | .039 | .032 | .048 |
| 3/16 | .1875 | 24 | 32* | .328 | .312† | .131 | .124 | .050 | .042 | .072 |
| 7/32 | .2188 | .. | 28* | .383 | .365 | .153 | .145 | .055 | .046 | .084 |
| 1/4 | .2500 | 20 | 26 | .438 | .417 | .175 | .165 | .061 | .051 | .096 |
| 5/16 | .3125 | 18 | 22 | .547 | .524 | .219 | .207 | .071 | .061 | .120 |
| 3/8 | .3750 | 16 | 20 | .656 | .629 | .262 | .249 | .082 | .072 | .144 |
| 7/16 | .4375 | 14 | 18 | .766 | .735 | .306 | .291 | .093 | .082 | .168 |
| 1/2 | .5000 | 12 | 16 | .875 | .840 | .350 | .333 | .104 | .092 | .192 |
| 9/16 | .5625 | 12* | 16* | .984 | .946 | .394 | .375 | .115 | .103 | .217 |
| 5/8 | .6250 | 11 | 14 | 1.094 | 1.051 | .437 | .417 | .126 | .113 | .240 |
| 3/4 | .7500 | 10 | 12 | 1.312 | 1.262 | .525 | .500 | .148 | .134 | .288 |
| PAN HEAD SCREWS[3] | | | | | | | | | | |
| 1/8 | .1250 | 40 | .. | .245 | .231 | .075 | .005 | .039 | .032 | .040 |
| 3/16 | .1875 | 24 | 32* | .373 | .375 | .110 | .099 | .050 | .042 | .061 |
| 7/32 | .2188 | .. | 28* | .425 | .407 | .125 | .112 | .055 | .046 | .069 |
| 1/4 | .2500 | 20 | 26 | .492 | .473§ | .144 | .130 | .061 | .051 | .078 |
| 5/16 | .3125 | 18 | 22 | .615 | .594 | .178 | .162 | .071 | .061 | .095 |
| 3/8 | .3750 | 16 | 20 | .740 | .716 | .212 | .195 | .082 | .072 | .112 |
| 7/16 | .4375 | 14 | 18 | .863 | .838 | .247 | .227 | .093 | .082 | .129 |
| 1/2 | .5000 | 12 | 16 | .987 | .958 | .281 | .260 | .104 | .092 | .145 |
| 9/16 | .5625 | 12* | 16* | 1.031 | .999 | .315 | .293 | .115 | .103 | .162 |
| 5/8 | .6250 | 11 | 14 | 1.125 | 1.090 | .350 | .325 | .126 | .113 | .179 |
| 3/4 | .7500 | 10 | 12 | 1.250 | 1.209 | .419 | .390 | .148 | .134 | .213 |
| CHEESE HEAD SCREWS[3] | | | | | | | | | | |
| 1/8 | .1250 | 40 | .. | .188 | .180 | .087 | .082 | .039 | .032 | .039 |
| 3/16 | .1875 | 24 | 32* | .281 | .270 | .131 | .124 | .050 | .042 | .059 |
| 7/32 | .2188 | .. | 28* | .328 | .315 | .153 | .145 | .055 | .046 | .069 |
| 1/4 | .2500 | 20 | 26 | .375 | .360 | .175 | .165 | .061 | .051 | .079 |
| 5/16 | .3125 | 18 | 22 | .469 | .450 | .219 | .207 | .071 | .061 | .098 |
| 3/8 | .3750 | 16 | 20 | .562 | .540 | .262 | .249 | .082 | .072 | .118 |
| 7/16 | .4375 | 14 | 18 | .656 | .630 | .306 | .291 | .093 | .082 | .138 |
| 1/2 | .5000 | 12 | 16 | .750 | .720 | .350 | .333 | .104 | .092 | .157 |
| 9/16 | .5625 | 12* | 16* | .844 | .810 | .394 | .375 | .115 | .103 | .177 |
| 5/8 | .6250 | 11 | 14 | .938 | .900 | .437 | .417 | .126 | .113 | .197 |
| 3/4 | .7500 | 10 | 12 | 1.125 | 1.080 | .525 | .500 | .148 | .134 | .236 |
| MUSHROOM HEAD SCREWS[3] | | | | | | | | | | |
| 1/8 | .1250 | 40 | .. | .289 | .272 | .078 | .066 | .043 | .035 | .040 |
| 3/16 | .1875 | 24 | 32* | .448 | .425 | .118 | .103 | .060 | .050 | .061 |
| 1/4 | .2500 | 20 | 26 | .573 | .546 | .150 | .133 | .075 | .064 | .079 |
| 5/16 | .3125 | 18 | 22 | .698 | .666 | .183 | .162 | .084 | .072 | .096 |
| 3/8 | .3750 | 16 | 20 | .823 | .787 | .215 | .191 | .094 | .081 | .112 |

All dimensions in inches. [1] See diagram on page 1197 for a pictorial representation of screws and letter dimensions. [2] All dimensions, except J, given for the 1/8-through 3/4-inch sizes also apply to all the 90° Raised Countersunk Head Screw dimensions given in the Standard. [3] These screws are also available with recessed heads; dimensions of recess are not given here but may be found in the Standard. * Non-preferred size; avoid use whenever possible. † By arrangement this may also be .309. § By arrangement this may also be .468.

**Slotted Head Cap Screws.** — American National Standard ANSI B18.6.2-1972, R1977 is intended to cover the complete general and dimensional data for the various styles of slotted head cap screws as well as square head and slotted headless set screws (see page 1207). Reference should be made to this Standard for information or data not found in the following text or tables.

*Length of Thread:* The length of complete (full form) thread on cap screws is equal to twice the basic screw diameter plus 0.250 in. with a plus tolerance of 0.188 in. or an amount equal to 2½ times the pitch of the thread, whichever is greater. Cap screws of lengths too short to accommodate the minimum thread length have full form threads extending to within a distance equal to 2½ pitches (threads) of the head.

*Designation:* Slotted head cap screws are designated by the following data in the sequence shown: Nominal size (fraction or decimal equivalent); threads per inch; screw length (fraction or decimal equivalent); product name; material; and protective finish, if required. Examples: ½-13 × 3 Slotted Round Head Cap Screw, SAE Grade 2 Steel, Zinc Plated. .750-16 × 2.25 Slotted Flat Countersunk Head Cap Screw, Corrosion Resistant Steel.

**Table 1. American National Standard Slotted Flat Countersunk Head Cap Screws**
(ANSI B18.6.2-1972, R1977)

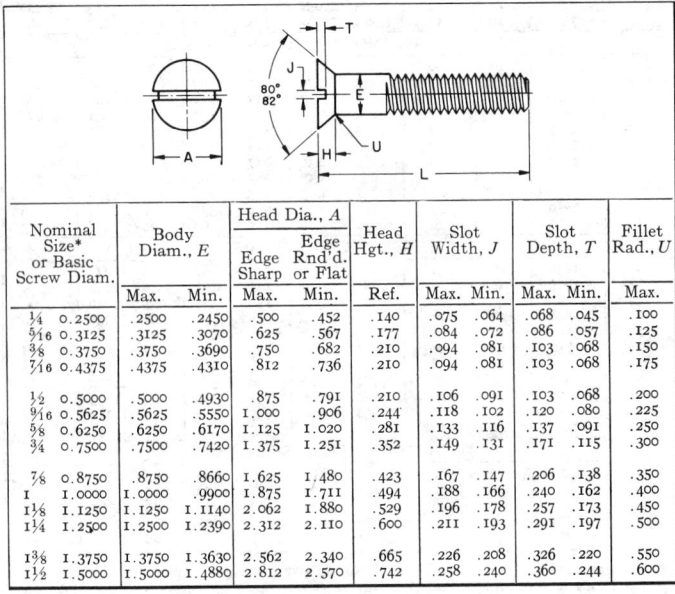

| Nominal Size* or Basic Screw Diam. | Body Diam., E | | Head Dia., A | | Head Hgt., H | Slot Width, J | | Slot Depth, T | | Fillet Rad., U |
|---|---|---|---|---|---|---|---|---|---|---|
| | | | Edge Sharp | Edge Rnd'd. or Flat | | | | | | |
| | Max. | Min. | Max. | Min. | Ref. | Max. | Min. | Max. | Min. | Max. |
| ¼ 0.2500 | .2500 | .2450 | .500 | .452 | .140 | .075 | .064 | .068 | .045 | .100 |
| ⁵⁄₁₆ 0.3125 | .3125 | .3070 | .625 | .567 | .177 | .084 | .072 | .086 | .057 | .125 |
| ⅜ 0.3750 | .3750 | .3690 | .750 | .682 | .210 | .094 | .081 | .103 | .068 | .150 |
| ⁷⁄₁₆ 0.4375 | .4375 | .4310 | .812 | .736 | .210 | .094 | .081 | .103 | .068 | .175 |
| ½ 0.5000 | .5000 | .4930 | .875 | .791 | .210 | .106 | .091 | .103 | .068 | .200 |
| ⁹⁄₁₆ 0.5625 | .5625 | .5550 | 1.000 | .906 | .244 | .118 | .102 | .120 | .080 | .225 |
| ⅝ 0.6250 | .6250 | .6170 | 1.125 | 1.020 | .281 | .133 | .116 | .137 | .091 | .250 |
| ¾ 0.7500 | .7500 | .7420 | 1.375 | 1.251 | .352 | .149 | .131 | .171 | .115 | .300 |
| ⅞ 0.8750 | .8750 | .8660 | 1.625 | 1.480 | .423 | .167 | .147 | .206 | .138 | .350 |
| 1 1.0000 | 1.0000 | .9900 | 1.875 | 1.711 | .494 | .188 | .166 | .240 | .162 | .400 |
| 1⅛ 1.1250 | 1.1250 | 1.1140 | 2.062 | 1.880 | .529 | .196 | .178 | .257 | .173 | .450 |
| 1¼ 1.2500 | 1.2500 | 1.2390 | 2.312 | 2.110 | .600 | .211 | .193 | .291 | .197 | .500 |
| 1⅜ 1.3750 | 1.3750 | 1.3630 | 2.562 | 2.340 | .665 | .226 | .208 | .326 | .220 | .550 |
| 1½ 1.5000 | 1.5000 | 1.4880 | 2.812 | 2.570 | .742 | .258 | .240 | .360 | .244 | .600 |

All dimensions are in inches.
*Threads:* Threads are Unified Standard Class 2A; UNC, UNF and 8 UN Series or UNRC, UNRF and 8 UNR Series.
* When specifying a nominal size in decimals, the zero preceding the decimal point is omitted as is any zero in the fourth decimal place.

### Table 2. American National Standard Slotted Round Head Cap Screws
(ANSI B18.6.2-1972, R1977)

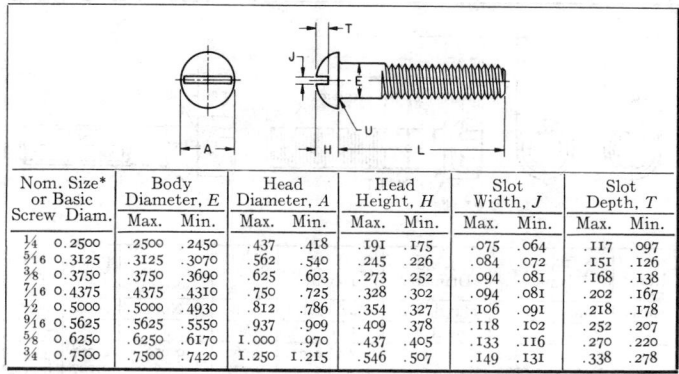

| Nom. Size* or Basic Screw Diam. | | Body Diameter, E | | Head Diameter, A | | Head Height, H | | Slot Width, J | | Slot Depth, T | |
|---|---|---|---|---|---|---|---|---|---|---|---|
| | | Max. | Min. | Max. | Min. | Max. | Min. | Max. | Min. | Max. | Min. |
| ¼ | 0.2500 | .2500 | .2450 | .437 | .418 | .191 | .175 | .075 | .064 | .117 | .097 |
| ⁵⁄₁₆ | 0.3125 | .3125 | .3070 | .562 | .540 | .245 | .226 | .084 | .072 | .151 | .126 |
| ⅜ | 0.3750 | .3750 | .3690 | .625 | .603 | .273 | .252 | .094 | .081 | .168 | .138 |
| ⁷⁄₁₆ | 0.4375 | .4375 | .4310 | .750 | .725 | .328 | .302 | .094 | .081 | .202 | .167 |
| ½ | 0.5000 | .5000 | .4930 | .812 | .786 | .354 | .327 | .106 | .091 | .218 | .178 |
| ⁹⁄₁₆ | 0.5625 | .5625 | .5550 | .937 | .909 | .409 | .378 | .118 | .102 | .252 | .207 |
| ⅝ | 0.6250 | .6250 | .6170 | 1.000 | .970 | .437 | .405 | .133 | .116 | .270 | .220 |
| ¾ | 0.7500 | .7500 | .7420 | 1.250 | 1.215 | .546 | .507 | .149 | .131 | .338 | .278 |

All dimensions are in inches.
*Fillet Radius, U:* For fillet radius see footnote to table below.
*Threads:* Threads are Unified Standard Class 2A; UNC, UNF and 8 UN Series or UNRC, UNRF and 8 UNR Series.
\* When specifying a nominal size in decimals, the zero preceding the decimal point is omitted as is any zero in the fourth decimal place.

### Table 3. American National Standard Slotted Fillister Head Cap Screws
(ANSI B18.6.2-1972, R1977)

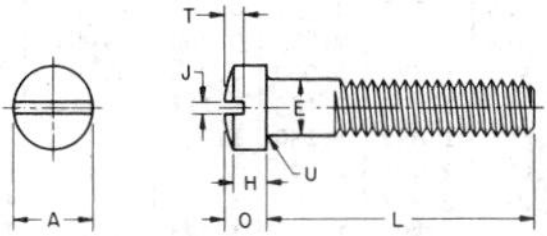

| Nom. Size* or Basic Screw Diam. | | Body Diam., E | | Head Diam., A | | Head Side Height, H | | Total Head Height, O | | Slot Width, J | | Slot Depth, T | |
|---|---|---|---|---|---|---|---|---|---|---|---|---|---|
| | | Max. | Min. | Max. | Min. | Max. | Min. | Max. | Min. | Max. | Min. | Max. | Min. |
| ¼ | 0.2500 | .2500 | .2450 | .375 | .363 | .172 | .157 | .216 | .194 | .075 | .064 | .097 | .077 |
| ⁵⁄₁₆ | 0.3125 | .3125 | .3070 | .437 | .424 | .203 | .186 | .253 | .230 | .084 | .072 | .115 | .090 |
| ⅜ | 0.3750 | .3750 | .3690 | .562 | .547 | .250 | .229 | .314 | .284 | .094 | .081 | .142 | .112 |
| ⁷⁄₁₆ | 0.4375 | .4375 | .4310 | .625 | .608 | .297 | .274 | .368 | .336 | .094 | .081 | .168 | .133 |
| ½ | 0.5000 | .5000 | .4930 | .750 | .731 | .328 | .301 | .413 | .376 | .106 | .091 | .193 | .153 |
| ⁹⁄₁₆ | 0.5625 | .5625 | .5550 | .812 | .792 | .375 | .346 | .467 | .427 | .118 | .102 | .213 | .168 |
| ⅝ | 0.6250 | .6250 | .6170 | .875 | .853 | .422 | .391 | .521 | .478 | .133 | .116 | .239 | .189 |
| ¾ | 0.7500 | .7500 | .7420 | 1.000 | .976 | .500 | .466 | .612 | .566 | .149 | .131 | .283 | .223 |
| ⅞ | 0.8750 | .8750 | .8660 | 1.125 | 1.098 | .594 | .556 | .720 | .668 | .167 | .147 | .334 | .264 |
| 1 | 1.0000 | 1.0000 | .9900 | 1.312 | 1.282 | .656 | .612 | .803 | .743 | .188 | .166 | .371 | .291 |

All dimensions are in inches.
*Fillet Radius, U:* The fillet radius is as follows: For screw sizes ¼ to ⅜ incl., .031 max. and .016 min.; ⁷⁄₁₆ to ⁹⁄₁₆, incl., .047 max., .016 min.; and for ⅝ to 1, incl., .062 max., .031 min.
*Threads:* Threads are Unified Standard Class 2A; UNC, UNF and 8 UN Series or UNRC, UNRF and 8 UNR Series.
\* When specifying a nominal size in decimals, the zero preceding the decimal point is omitted as is any zero in the fourth decimal place.

### Table 1. American National Standard Hexagon and Spline Socket Head Cap Screws (1960 Series) (ANSI B18.3-1976)

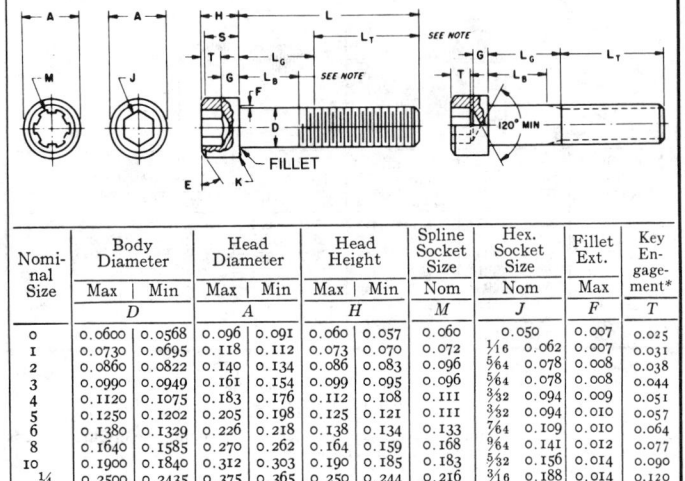

| Nominal Size | Body Diameter Max | Min | Head Diameter Max | Min | Head Height Max | Min | Spline Socket Size Nom | Hex. Socket Size Nom | | Fillet Ext. Max | Key Engagement* |
|---|---|---|---|---|---|---|---|---|---|---|---|
| | D | | A | | H | | M | J | | F | T |
| 0 | 0.0600 | 0.0568 | 0.096 | 0.091 | 0.060 | 0.057 | 0.060 | | 0.050 | 0.007 | 0.025 |
| 1 | 0.0730 | 0.0695 | 0.118 | 0.112 | 0.073 | 0.070 | 0.072 | 1/16 | 0.062 | 0.007 | 0.031 |
| 2 | 0.0860 | 0.0822 | 0.140 | 0.134 | 0.086 | 0.083 | 0.096 | 5/64 | 0.078 | 0.008 | 0.038 |
| 3 | 0.0990 | 0.0949 | 0.161 | 0.154 | 0.099 | 0.095 | 0.096 | 5/64 | 0.078 | 0.008 | 0.044 |
| 4 | 0.1120 | 0.1075 | 0.183 | 0.176 | 0.112 | 0.108 | 0.111 | 3/32 | 0.094 | 0.009 | 0.051 |
| 5 | 0.1250 | 0.1202 | 0.205 | 0.198 | 0.125 | 0.121 | 0.111 | 3/32 | 0.094 | 0.010 | 0.057 |
| 6 | 0.1380 | 0.1329 | 0.226 | 0.218 | 0.138 | 0.134 | 0.133 | 7/64 | 0.109 | 0.010 | 0.064 |
| 8 | 0.1640 | 0.1585 | 0.270 | 0.262 | 0.164 | 0.159 | 0.168 | 9/64 | 0.141 | 0.012 | 0.077 |
| 10 | 0.1900 | 0.1840 | 0.312 | 0.303 | 0.190 | 0.185 | 0.183 | 5/32 | 0.156 | 0.014 | 0.090 |
| 1/4 | 0.2500 | 0.2435 | 0.375 | 0.365 | 0.250 | 0.244 | 0.216 | 3/16 | 0.188 | 0.014 | 0.120 |
| 5/16 | 0.3125 | 0.3053 | 0.469 | 0.457 | 0.312 | 0.306 | 0.291 | 1/4 | 0.250 | 0.017 | 0.151 |
| 3/8 | 0.3750 | 0.3678 | 0.562 | 0.550 | 0.375 | 0.368 | 0.372 | 5/16 | 0.312 | 0.020 | 0.182 |
| 7/16 | 0.4375 | 0.4294 | 0.656 | 0.642 | 0.438 | 0.430 | 0.454 | 3/8 | 0.375 | 0.023 | 0.213 |
| 1/2 | 0.5000 | 0.4919 | 0.750 | 0.735 | 0.500 | 0.492 | 0.454 | 3/8 | 0.375 | 0.026 | 0.245 |
| 5/8 | 0.6250 | 0.6163 | 0.938 | 0.921 | 0.625 | 0.616 | 0.595 | 1/2 | 0.500 | 0.032 | 0.307 |
| 3/4 | 0.7500 | 0.7406 | 1.125 | 1.107 | 0.750 | 0.740 | 0.620 | 5/8 | 0.625 | 0.039 | 0.370 |
| 7/8 | 0.8750 | 0.8647 | 1.312 | 1.293 | 0.875 | 0.864 | 0.698 | 3/4 | 0.750 | 0.044 | 0.432 |
| 1 | 1.0000 | 0.9886 | 1.500 | 1.479 | 1.000 | 0.988 | 0.790 | 3/4 | 0.750 | 0.050 | 0.495 |
| 1 1/8 | 1.1250 | 1.1086 | 1.688 | 1.665 | 1.125 | 1.111 | ..... | 7/8 | 0.875 | 0.055 | 0.557 |
| 1 1/4 | 1.2500 | 1.2336 | 1.875 | 1.852 | 1.250 | 1.236 | ..... | 7/8 | 0.875 | 0.060 | 0.620 |
| 1 3/8 | 1.3750 | 1.3568 | 2.062 | 2.038 | 1.375 | 1.360 | ..... | 1 | 1.000 | 0.065 | 0.682 |
| 1 1/2 | 1.5000 | 1.4818 | 2.250 | 2.224 | 1.500 | 1.485 | ..... | 1 | 1.000 | 0.070 | 0.745 |
| 1 3/4 | 1.7500 | 1.7295 | 2.625 | 2.597 | 1.750 | 1.734 | ..... | 1 1/4 | 1.250 | 0.080 | 0.870 |
| 2 | 2.0000 | 1.9780 | 3.000 | 2.970 | 2.000 | 1.983 | ..... | 1 1/2 | 1.500 | 0.090 | 0.995 |
| 2 1/4 | 2.2500 | 2.2280 | 3.375 | 3.344 | 2.250 | 2.232 | ..... | 1 3/4 | 1.750 | 0.100 | 1.120 |
| 2 1/2 | 2.5000 | 2.4762 | 3.750 | 3.717 | 2.500 | 2.481 | ..... | 1 3/4 | 1.750 | 0.110 | 1.245 |
| 2 3/4 | 2.7500 | 2.7262 | 4.125 | 4.090 | 2.750 | 2.730 | ..... | 2 | 2.000 | 0.120 | 1.370 |
| 3 | 3.0000 | 2.9762 | 4.500 | 4.464 | 3.000 | 2.979 | ..... | 2 1/4 | 2.250 | 0.130 | 1.495 |
| 3 1/4 | 3.2500 | 3.2262 | 4.875 | 4.837 | 3.250 | 3.228 | ..... | 2 1/4 | 2.250 | 0.140 | 1.620 |
| 3 1/2 | 3.5000 | 3.4762 | 5.250 | 5.211 | 3.500 | 3.478 | ..... | 2 3/4 | 2.750 | 0.150 | 1.745 |
| 3 3/4 | 3.7500 | 3.7262 | 5.625 | 5.584 | 3.750 | 3.727 | ..... | 2 3/4 | 2.750 | 0.160 | 1.870 |
| 4 | 4.0000 | 3.9762 | 6.000 | 5.958 | 4.000 | 3.976 | ..... | 3 | 3.000 | 0.170 | 1.995 |

*Key engagement depths are minimum.

All dimensions in inches. The body length $L_B$ of the screw, is the length of the unthreaded cylindrical portion of the shank. The length of thread, $L_T$, is the distance from the extreme point to the last complete (full form) thread. Standard length increments for screw diameters up to 1-inch are $^1\!/_{16}$ inch for lengths $^1\!/_{8}$ through $^1\!/_{4}$ inches, $^1\!/_{8}$ inch for lengths $^1\!/_{4}$ through 1 inch, $^1\!/_{4}$ inch for lengths 1 through 3½ inches, ½ inch for lengths 3½ through 7 inches, 1 inch for lengths 7 through 10 inches and for diameters over 1 inch are ½ inch for lengths 1 through 7 inches, 1 inch for lengths 7 through 10 inches and 2 inches for lengths over 10 inches. Heads may be plain or knurled, and chamfered to an angle $E$ of 30 to 45 degrees with the surface of the flat. The thread conforms to the Unified Standard with radius root, Class 3A, UNRC and UNRF for screw sizes No. o through 1 inch inclusive, Class 2A, UNRC and UNRF for over 1 inch through 1½ inches inclusive, and Class 2A UNRC for sizes larger than 1½ inches. Socket dimensions are given in Table 9. For manufacturing details not shown, including materials, see American National Standard ANSI B18.3-1976.

Table 2.   Drill and Counterbore Sizes For Socket Head Cap Screws (1960 Series)

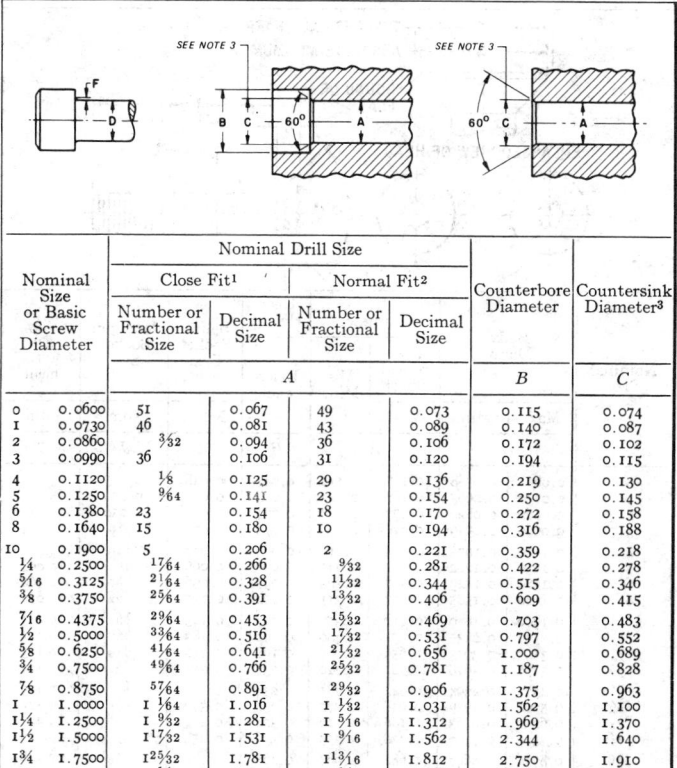

| Nominal Size or Basic Screw Diameter | | Nominal Drill Size | | | | Counterbore Diameter | Countersink Diameter[3] |
|---|---|---|---|---|---|---|---|
| | | Close Fit[1] | | Normal Fit[2] | | | |
| | | Number or Fractional Size | Decimal Size | Number or Fractional Size | Decimal Size | | |
| | | A | | | | B | C |
| 0 | 0.0600 | 51 | 0.067 | 49 | 0.073 | 0.115 | 0.074 |
| 1 | 0.0730 | 46 | 0.081 | 43 | 0.089 | 0.140 | 0.087 |
| 2 | 0.0860 | 3⁄32 | 0.094 | 36 | 0.106 | 0.172 | 0.102 |
| 3 | 0.0990 | 36 | 0.106 | 31 | 0.120 | 0.194 | 0.115 |
| 4 | 0.1120 | 1⁄8 | 0.125 | 29 | 0.136 | 0.219 | 0.130 |
| 5 | 0.1250 | 9⁄64 | 0.141 | 23 | 0.154 | 0.250 | 0.145 |
| 6 | 0.1380 | 23 | 0.154 | 18 | 0.170 | 0.272 | 0.158 |
| 8 | 0.1640 | 15 | 0.180 | 10 | 0.194 | 0.316 | 0.188 |
| 10 | 0.1900 | 5 | 0.206 | 2 | 0.221 | 0.359 | 0.218 |
| 1⁄4 | 0.2500 | 17⁄64 | 0.266 | 9⁄32 | 0.281 | 0.422 | 0.278 |
| 5⁄16 | 0.3125 | 21⁄64 | 0.328 | 11⁄32 | 0.344 | 0.515 | 0.346 |
| 3⁄8 | 0.3750 | 25⁄64 | 0.391 | 13⁄32 | 0.406 | 0.609 | 0.415 |
| 7⁄16 | 0.4375 | 29⁄64 | 0.453 | 15⁄32 | 0.469 | 0.703 | 0.483 |
| 1⁄2 | 0.5000 | 33⁄64 | 0.516 | 17⁄32 | 0.531 | 0.797 | 0.552 |
| 5⁄8 | 0.6250 | 41⁄64 | 0.641 | 21⁄32 | 0.656 | 1.000 | 0.689 |
| 3⁄4 | 0.7500 | 49⁄64 | 0.766 | 25⁄32 | 0.781 | 1.187 | 0.828 |
| 7⁄8 | 0.8750 | 57⁄64 | 0.891 | 29⁄32 | 0.906 | 1.375 | 0.963 |
| 1 | 1.0000 | 1 1⁄64 | 1.016 | 1 1⁄32 | 1.031 | 1.562 | 1.100 |
| 1¼ | 1.2500 | 1 9⁄32 | 1.281 | 1 5⁄16 | 1.312 | 1.969 | 1.370 |
| 1½ | 1.5000 | 1 17⁄32 | 1.531 | 1 9⁄16 | 1.562 | 2.344 | 1.640 |
| 1¾ | 1.7500 | 1 25⁄32 | 1.781 | 1 13⁄16 | 1.812 | 2.750 | 1.910 |
| 2 | 2.0000 | 2 1⁄32 | 2.031 | 2 1⁄16 | 2.062 | 3.125 | 2.180 |

All dimensions in inches.

[1] *Close Fit:* The close fit is normally limited to holes for those lengths of screws which are threaded to the head in assemblies where only one screw is to be used or where two or more screws are to be used and the mating holes are to be produced either at assembly or by matched and coordinated tooling.

[2] *Normal Fit:* The normal fit is intended for screws of relatively long length or for assemblies involving two or more screws where the mating holes are to be produced by conventional tolerancing methods.  It provides for the maximum allowable eccentricity of the longest standard screws and for certain variations in the parts to be fastened, such as: deviations in hole straightness, angularity between the axis of the tapped hole and that of the hole for the shank, differences in center distances of the mating holes, etc.

[3] *Countersink:* It is considered good practice to countersink or break the edges of holes which are smaller than (D Max + 2F Max) in parts having a hardness which approaches, equals or exceeds the screw hardness.  If such holes are not countersunk, the heads of screws may not seat properly or the sharp edges on hole may deform the fillets on screws thereby making them susceptible to fatigue in applications involving dynamic loading. The countersink or corner relief, however, should not be larger than is necessary to insure that the fillet on the screw is cleared.

*Source:* Appendix to American National Standard ANSI B18.3-1976.

### Table 3. American National Standard Hexagon and Spline Socket Flat Head Cap Screws (ANSI B18.3-1976)

| Nominal Size | Body Diam. Max | Body Diam. Min | Head Diameter Theoretical Sharp Max | Head Diameter Abs. Min | Head Height Reference | Spline Socket Size Nom. | Hexagon Socket Size Nom. | Key Engagement Min |
|---|---|---|---|---|---|---|---|---|
| | D | | A | | H | M | J | T |
| 0 | 0.0600 | 0.0568 | 0.138 | 0.117 | 0.044 | 0.048 | 0.035 | 0.025 |
| 1 | 0.0730 | 0.0695 | 0.168 | 0.143 | 0.054 | 0.060 | 0.050 | 0.031 |
| 2 | 0.0860 | 0.0822 | 0.197 | 0.168 | 0.064 | 0.060 | 0.050 | 0.038 |
| 3 | 0.0990 | 0.0949 | 0.226 | 0.193 | 0.073 | 0.072 | 1/16 | 0.044 |
| 4 | 0.1120 | 0.1075 | 0.255 | 0.218 | 0.083 | 0.072 | 1/16 | 0.055 |
| 5 | 0.1250 | 0.1202 | 0.281 | 0.240 | 0.090 | 0.096 | 5/64 | 0.061 |
| 6 | 0.1380 | 0.1329 | 0.307 | 0.263 | 0.097 | 0.096 | 5/64 | 0.066 |
| 8 | 0.1640 | 0.1585 | 0.359 | 0.311 | 0.112 | 0.111 | 3/32 | 0.076 |
| 10 | 0.1900 | 0.1840 | 0.411 | 0.359 | 0.127 | 0.145 | 1/8 | 0.087 |
| 1/4 | 0.2500 | 0.2435 | 0.531 | 0.480 | 0.161 | 0.183 | 5/32 | 0.111 |
| 5/16 | 0.3125 | 0.3053 | 0.656 | 0.600 | 0.198 | 0.216 | 3/16 | 0.135 |
| 3/8 | 0.3750 | 0.3678 | 0.781 | 0.720 | 0.234 | 0.251 | 7/32 | 0.159 |
| 7/16 | 0.4375 | 0.4294 | 0.844 | 0.781 | 0.234 | 0.291 | 1/4 | 0.159 |
| 1/2 | 0.5000 | 0.4919 | 0.938 | 0.872 | 0.251 | 0.372 | 5/16 | 0.172 |
| 5/8 | 0.6250 | 0.6163 | 1.188 | 1.112 | 0.324 | 0.454 | 3/8 | 0.220 |
| 3/4 | 0.7500 | 0.7406 | 1.438 | 1.355 | 0.396 | 0.454 | 1/2 | 0.220 |
| 7/8 | 0.8750 | 0.8647 | 1.688 | 1.604 | 0.468 | ... | 9/16 | 0.248 |
| 1 | 1.0000 | 0.9886 | 1.938 | 1.841 | 0.540 | ... | 5/8 | 0.297 |
| 1 1/8 | 1.1250 | 1.1086 | 2.188 | 2.079 | 0.611 | ... | 3/4 | 0.325 |
| 1 1/4 | 1.2500 | 1.2336 | 2.438 | 2.316 | 0.683 | ... | 7/8 | 0.358 |
| 1 3/8 | 1.3750 | 1.3568 | 2.688 | 2.553 | 0.755 | ... | 7/8 | 0.402 |
| 1 1/2 | 1.5000 | 1.4818 | 2.938 | 2.791 | 0.827 | ... | 1 | 0.435 |

All dimensions in inches.

The body of the screw is the unthreaded cylindrical portion of the shank; the shank being the portion of the screw from the point of juncture of the conical bearing surface and the body to the point of the point. The length of thread $L_T$ is the distance measured from the extreme point to the last complete (full form) thread.

Standard length increments of No. o through 1-inch sizes are as follows: 1/16 inch for nominal screw lengths of 1/8 through 1/4 inch; 1/8 inch for lengths of 1/4 through 1 inch; 1/4 inch for lengths of 1 inch through 3 1/2 inches; 1/2 inch for lengths of 3 1/2 through 7 inches; and 1 inch for lengths of 7 through 10 inches, incl. For screw sizes over 1 inch, length increments are: 1/2 inch for nominal screw lengths of 1 inch through 7 inches; 1 inch for lengths of 7 through 10 inches; and 2 inches for lengths over 10 inches.

Threads shall be Unified external threads with radius root; Class 3A UNRC and UNRF series for sizes No. o through 1 inch and Class 2A UNRC and UNRF series for sizes over 1 inch to 1 1/2 inches, incl.

For manufacturing details not shown, including materials, see American National Standard ANSI B18.3-1976. Socket dimensions are given in Table 9.

**Table 4. American National Standard Hexagon Socket and Spline Socket Button Head Cap Screws*** (ANSI B18.3-1976)

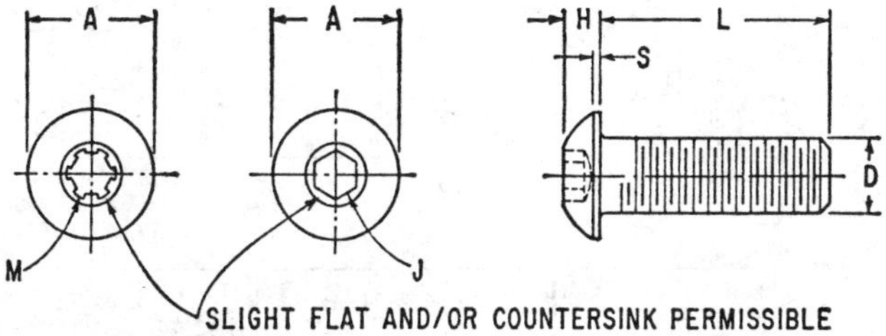
A
A
H
L
S
D
M
J
SLIGHT FLAT AND/OR COUNTERSINK PERMISSIBLE

| Nominal Size | Screw Diameter | Head Diameter | | Head Height | | Head Side Height | Spline Socket Size† | Hexagon Socket Size† | Standard Length |
|---|---|---|---|---|---|---|---|---|---|
| | Basic | Max. | Min. | Max. | Min. | Ref. | Nom. | Nom. | Max. |
| | *D* | *A* | | *H* | | *S* | *M* | *J* | *L* |
| 0 | 0.0600 | 0.114 | 0.104 | 0.032 | 0.026 | 0.010 | 0.048 | 0.035 | ½ |
| 1 | 0.0730 | 0.139 | 0.129 | 0.039 | 0.033 | 0.010 | 0.060 | 0.050 | ½ |
| 2 | 0.0860 | 0.164 | 0.154 | 0.046 | 0.038 | 0.010 | 0.060 | 0.050 | ½ |
| 3 | 0.0990 | 0.188 | 0.176 | 0.052 | 0.044 | 0.010 | 0.072 | 1/16 | ½ |
| 4 | 0.1120 | 0.213 | 0.201 | 0.059 | 0.051 | 0.015 | 0.072 | 1/16 | ½ |
| 5 | 0.1250 | 0.238 | 0.226 | 0.066 | 0.058 | 0.015 | 0.096 | 5/64 | ½ |
| 6 | 0.1380 | 0.262 | 0.250 | 0.073 | 0.063 | 0.015 | 0.096 | 5/64 | 5/8 |
| 8 | 0.1640 | 0.312 | 0.298 | 0.087 | 0.077 | 0.015 | 0.111 | 3/32 | ¾ |
| 10 | 0.1900 | 0.361 | 0.347 | 0.101 | 0.091 | 0.020 | 0.145 | 1/8 | 1 |
| ¼ | 0.2500 | 0.437 | 0.419 | 0.132 | 0.122 | 0.031 | 0.183 | 5/32 | 1 |
| 5/16 | 0.3125 | 0.547 | 0.527 | 0.166 | 0.152 | 0.031 | 0.216 | 3/16 | 1 |
| 3/8 | 0.3750 | 0.656 | 0.636 | 0.199 | 0.185 | 0.031 | 0.251 | 7/32 | 1¼ |
| ½ | 0.5000 | 0.875 | 0.851 | 0.265 | 0.245 | 0.046 | 0.372 | 5/16 | 2 |
| 5/8 | 0.6250 | 1.000 | 0.970 | 0.331 | 0.311 | 0.062 | 0.454 | 3/8 | 2 |

All dimensions in inches.

* These cap screws have been designed and recommended for light fastening applications. They are not suggested for use in critical high-strength applications where socket head cap screws should normally be used.

Standard length increments for socket button head cap screws are as follows: 1/16 inch for nominal screw lengths of 1/8 through ¼ inch, 1/8 inch for nominal screw lengths of ¼ through 1 inch, and ¼ inch for nominal screw lengths of 1 inch through 2 inches. Tolerances on lengths are −0.03 inch for lengths up to 1 inch inclusive. For lengths from 1 through 2½ inches, inclusive, length tolerances are −0.04 inch for sizes No. 0 through 3/8 inch, inclusive and −0.06 inch for sizes ½ and 5/8 inch.

The thread conforms to the Unified standard, Class 3A, with radius root, UNRC and UNRF.

† Socket dimensions are given in Table 9.

For manufacturing details, including materials, not shown, see American National Standard ANSI B18.3-1976.

### Table 5. American National Standard Hexagon Socket Head Shoulder Screws
(ANSI B18.3-1976)

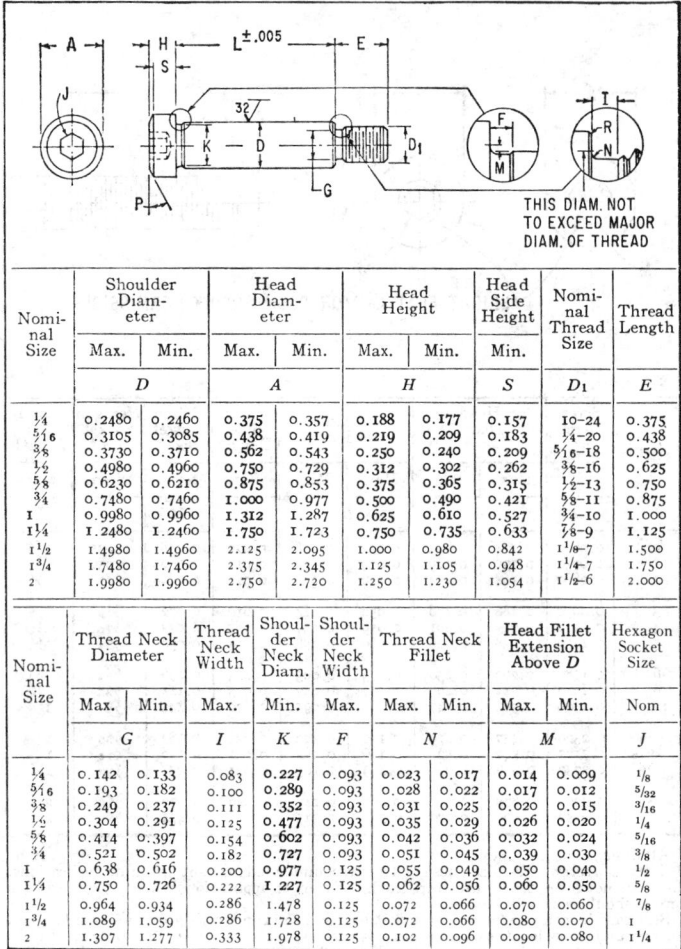

THIS DIAM. NOT
TO EXCEED MAJOR
DIAM. OF THREAD

| Nominal Size | Shoulder Diameter | | Head Diameter | | Head Height | | Head Side Height | Nominal Thread Size | Thread Length |
|---|---|---|---|---|---|---|---|---|---|
| | Max. | Min. | Max. | Min. | Max. | Min. | Min. | | |
| | D | | A | | H | | S | $D_1$ | E |
| ¼ | 0.2480 | 0.2460 | 0.375 | 0.357 | 0.188 | 0.177 | 0.157 | 10–24 | 0.375 |
| ⁵⁄₁₆ | 0.3105 | 0.3085 | 0.438 | 0.419 | 0.219 | 0.209 | 0.183 | ¼–20 | 0.438 |
| ⅜ | 0.3730 | 0.3710 | 0.562 | 0.543 | 0.250 | 0.240 | 0.209 | ⁵⁄₁₆–18 | 0.500 |
| ½ | 0.4980 | 0.4960 | 0.750 | 0.729 | 0.312 | 0.302 | 0.262 | ⅜–16 | 0.625 |
| ⅝ | 0.6230 | 0.6210 | 0.875 | 0.853 | 0.375 | 0.365 | 0.315 | ½–13 | 0.750 |
| ¾ | 0.7480 | 0.7460 | 1.000 | 0.977 | 0.500 | 0.490 | 0.421 | ⅝–11 | 0.875 |
| 1 | 0.9980 | 0.9960 | 1.312 | 1.287 | 0.625 | 0.610 | 0.527 | ¾–10 | 1.000 |
| 1¼ | 1.2480 | 1.2460 | 1.750 | 1.723 | 0.750 | 0.735 | 0.633 | ⅞–9 | 1.125 |
| 1½ | 1.4980 | 1.4960 | 2.125 | 2.095 | 1.000 | 0.980 | 0.842 | 1⅛–7 | 1.500 |
| 1¾ | 1.7480 | 1.7460 | 2.375 | 2.345 | 1.125 | 1.105 | 0.948 | 1¼–7 | 1.750 |
| 2 | 1.9980 | 1.9960 | 2.750 | 2.720 | 1.250 | 1.230 | 1.054 | 1½–6 | 2.000 |

| Nominal Size | Thread Neck Diameter | | Thread Neck Width | Shoulder Neck Diam. | Shoulder Neck Width | Thread Neck Fillet | | Head Fillet Extension Above D | | Hexagon Socket Size |
|---|---|---|---|---|---|---|---|---|---|---|
| | Max. | Min. | Max. | Min. | Max. | Max. | Min. | Max. | Min. | Nom |
| | G | | I | K | F | N | | M | | J |
| ¼ | 0.142 | 0.133 | 0.083 | 0.227 | 0.093 | 0.023 | 0.017 | 0.014 | 0.009 | ⅛ |
| ⁵⁄₁₆ | 0.193 | 0.182 | 0.100 | 0.289 | 0.093 | 0.028 | 0.022 | 0.017 | 0.012 | ⁵⁄₃₂ |
| ⅜ | 0.249 | 0.237 | 0.111 | 0.352 | 0.093 | 0.031 | 0.025 | 0.020 | 0.015 | ³⁄₁₆ |
| ½ | 0.304 | 0.291 | 0.125 | 0.477 | 0.093 | 0.035 | 0.029 | 0.026 | 0.020 | ¼ |
| ⅝ | 0.414 | 0.397 | 0.154 | 0.602 | 0.093 | 0.042 | 0.036 | 0.032 | 0.024 | ⁵⁄₁₆ |
| ¾ | 0.521 | 0.502 | 0.182 | 0.727 | 0.093 | 0.051 | 0.045 | 0.039 | 0.030 | ⅜ |
| 1 | 0.638 | 0.616 | 0.200 | 0.977 | 0.125 | 0.055 | 0.049 | 0.050 | 0.040 | ½ |
| 1¼ | 0.750 | 0.726 | 0.222 | 1.227 | 0.125 | 0.062 | 0.056 | 0.060 | 0.050 | ⅝ |
| 1½ | 0.964 | 0.934 | 0.286 | 1.478 | 0.125 | 0.072 | 0.066 | 0.070 | 0.060 | ⅞ |
| 1¾ | 1.089 | 1.059 | 0.286 | 1.728 | 0.125 | 0.072 | 0.066 | 0.080 | 0.070 | 1 |
| 2 | 1.307 | 1.277 | 0.333 | 1.978 | 0.125 | 0.102 | 0.096 | 0.090 | 0.080 | 1¼ |

All dimensions are in inches. The shoulder refers to the unthreaded portion of the screw. Standard length increments for shoulder screws are as follows: ⅛ inch for nominal screw lengths of ¼ to ¾ inch, ¼ inch for lengths of ¾ to 5 inches and ½ inch for lengths over 5 inches.

The thread conforms to the Unified standard, Class 3A, UNC. Hexagon socket sizes for the respective shoulder screw sizes are the same as for set screws of the same nominal size. Hexagon socket dimensions are given in Table 9.

For manufacturing details, including materials, not shown see American National Standard ANSI B18.3-1976.

## Table 6. American National Standard Slotted Headless Set Screws
(ANSI B18.6.2-1972, R1977)

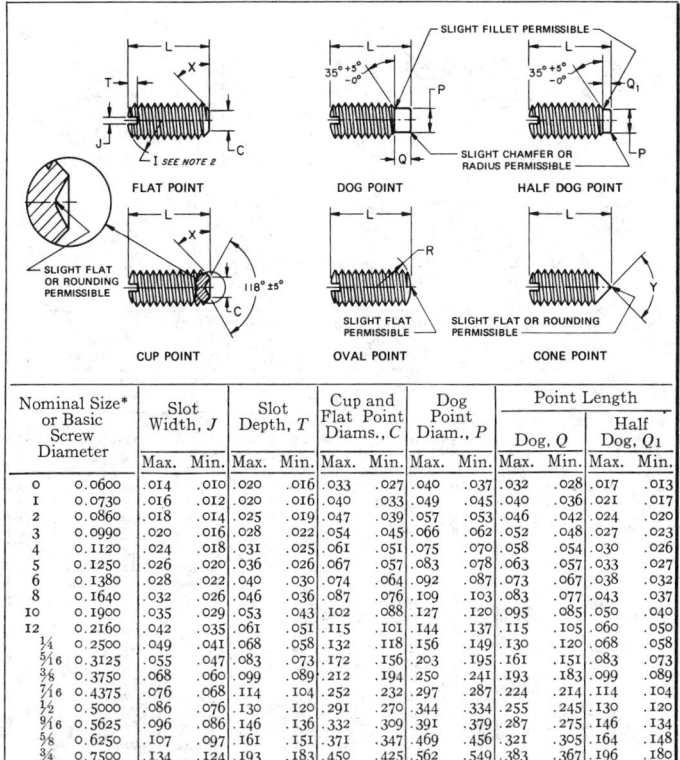

| Nominal Size* or Basic Screw Diameter | | Slot Width, J | | Slot Depth, T | | Cup and Flat Point Diams., C | | Dog Point Diam., P | | Point Length | | | |
|---|---|---|---|---|---|---|---|---|---|---|---|---|---|
| | | | | | | | | | | Dog, Q | | Half Dog, $Q_1$ | |
| | | Max. | Min. | Max. | Min. | Max. | Min. | Max. | Min. | Max. | Min. | Max. | Min. |
| 0 | 0.0600 | .014 | .010 | .020 | .016 | .033 | .027 | .040 | .037 | .032 | .028 | .017 | .013 |
| 1 | 0.0730 | .016 | .012 | .020 | .016 | .040 | .033 | .049 | .045 | .040 | .036 | .021 | .017 |
| 2 | 0.0860 | .018 | .014 | .025 | .019 | .047 | .039 | .057 | .053 | .046 | .042 | .024 | .020 |
| 3 | 0.0990 | .020 | .016 | .028 | .022 | .054 | .045 | .066 | .062 | .052 | .048 | .027 | .023 |
| 4 | 0.1120 | .024 | .018 | .031 | .025 | .061 | .051 | .075 | .070 | .058 | .054 | .030 | .026 |
| 5 | 0.1250 | .026 | .020 | .036 | .026 | .067 | .057 | .083 | .078 | .063 | .057 | .033 | .027 |
| 6 | 0.1380 | .028 | .022 | .040 | .030 | .074 | .064 | .092 | .087 | .073 | .067 | .038 | .032 |
| 8 | 0.1640 | .032 | .026 | .046 | .036 | .087 | .076 | .109 | .103 | .083 | .077 | .043 | .037 |
| 10 | 0.1900 | .035 | .029 | .053 | .043 | .102 | .088 | .127 | .120 | .095 | .085 | .050 | .040 |
| 12 | 0.2160 | .042 | .035 | .061 | .051 | .115 | .101 | .144 | .137 | .115 | .105 | .060 | .050 |
| ¼ | 0.2500 | .049 | .041 | .068 | .058 | .132 | .118 | .156 | .149 | .130 | .120 | .068 | .058 |
| 5⁄16 | 0.3125 | .055 | .047 | .083 | .073 | .172 | .156 | .203 | .195 | .161 | .151 | .083 | .073 |
| ⅜ | 0.3750 | .068 | .060 | .099 | .089 | .212 | .194 | .250 | .241 | .193 | .183 | .099 | .089 |
| 7⁄16 | 0.4375 | .076 | .068 | .114 | .104 | .252 | .232 | .297 | .287 | .224 | .214 | .114 | .104 |
| ½ | 0.5000 | .086 | .076 | .130 | .120 | .291 | .270 | .344 | .334 | .255 | .245 | .130 | .120 |
| 9⁄16 | 0.5625 | .096 | .086 | .146 | .136 | .332 | .309 | .391 | .379 | .287 | .275 | .146 | .134 |
| ⅝ | 0.6250 | .107 | .097 | .161 | .151 | .371 | .347 | .469 | .456 | .321 | .305 | .164 | .148 |
| ¾ | 0.7500 | .134 | .124 | .193 | .183 | .450 | .425 | .562 | .549 | .383 | .367 | .196 | .180 |

All dimensions are in inches.

*Crown Radius, I:* The crown radius has the same value as the basic screw diameter to three decimal places.

*Oval Point Radius, R:* Values of the oval point radius according to nominal screw size are: For a screw size of 0, a radius of .045; 1, .055; 2, .064; 3, .074; 4, .084; 5, .094; 6, .104; 8, .123; 10, .142; 12, .162; ¼, .188; 5⁄16, .234; ⅜, .281; 7⁄16, .328; ½, .375; 9⁄16, .422; ⅝, .469; and for ¾ .562.

*Cone Point Angle, Y:* The cone point angle is 90° ± 2° for the following nominal lengths, or longer, shown according to screw size: For nominal size 0, a length of 5⁄64; 1, 3⁄32; 2, 7⁄64; 3, ⅛; 4, 5⁄32; 5, 3⁄16; 6, 3⁄16; 8, ¼; 10, ¼; 12, 5⁄16; ¼, 5⁄16; 5⁄16, ⅜; ⅜, 7⁄16; 7⁄16, ½; ½, 9⁄16; 9⁄16, ⅝; ⅝, ¾; and for ¾, ⅞. For shorter screws, the cone point angle is 118° ± 2°.

*Point Angle X:* The point angle is 45°, + 5°, − 0°, for screws of nominal lengths, or longer, as given just above for cone point angle, and 30°, min. for shorter screws.

*Threads:* Threads are Unified Standard Class 2A; UNC and UNF Series or UNRC and UNRF Series.

* When specifying a nominal size in decimals a zero preceding the decimal point or any zero in the fourth decimal place is omitted.

## Table 7. American National Standard Hexagon and Spline Socket Set Screws
### (ANSI B18.3-1976)

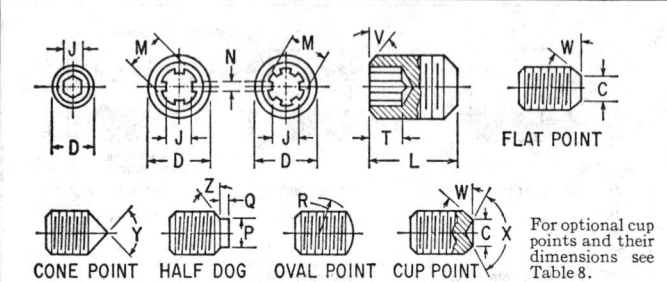

| Nominal Size or Basic Screw Diameter | | Socket Size | | Cup and Flat Point Diameters | | Half Dog Point | | Oval Point Radius | Min. Key Engagement Depth | | Lgth. Limit for Angle |
|---|---|---|---|---|---|---|---|---|---|---|---|
| | | Hex. Nom. | Spl. Nom. | Max. | Min. | Dia. Max. | Lgth. Max. | Basic | Hex. T† | Spl. T† | |
| | | J | M | C | | P | Q | R | | | Y* |
| 0 | 0.0600 | 0.028 | 0.033 | 0.033 | 0.027 | 0.040 | 0.017 | 0.045 | 0.050 | 0.026 | 5/64 |
| 1 | 0.0730 | 0.035 | 0.033 | 0.040 | 0.033 | 0.049 | 0.021 | 0.055 | 0.060 | 0.035 | 3/32 |
| 2 | 0.0860 | 0.035 | 0.048 | 0.047 | 0.039 | 0.057 | 0.024 | 0.064 | 0.060 | 0.040 | 7/64 |
| 3 | 0.0990 | 0.050 | 0.048 | 0.054 | 0.045 | 0.066 | 0.027 | 0.074 | 0.070 | 0.040 | 1/8 |
| 4 | 0.1120 | 0.050 | 0.060 | 0.061 | 0.051 | 0.075 | 0.030 | 0.084 | 0.070 | 0.045 | 5/32 |
| 5 | 0.1250 | 1/16 | 0.072 | 0.067 | 0.057 | 0.083 | 0.033 | 0.094 | 0.080 | 0.055 | 3/16 |
| 6 | 0.1380 | 1/16 | 0.072 | 0.074 | 0.064 | 0.092 | 0.038 | 0.104 | 0.080 | 0.055 | 3/16 |
| 8 | 0.1640 | 5/64 | 0.096 | 0.087 | 0.076 | 0.109 | 0.043 | 0.123 | 0.090 | 0.080 | 1/4 |
| 10 | 0.1900 | 3/32 | 0.111 | 0.102 | 0.088 | 0.127 | 0.049 | 0.142 | 0.100 | 0.080 | 1/4 |
| 1/4 | 0.2500 | 1/8 | 0.145 | 0.132 | 0.118 | 0.156 | 0.067 | 0.188 | 0.125 | 0.125 | 5/16 |
| 5/16 | 0.3125 | 5/32 | 0.183 | 0.172 | 0.156 | 0.203 | 0.082 | 0.234 | 0.156 | 0.156 | 3/8 |
| 3/8 | 0.3750 | 3/16 | 0.216 | 0.212 | 0.194 | 0.250 | 0.099 | 0.281 | 0.188 | 0.188 | 7/16 |
| 7/16 | 0.4375 | 7/32 | 0.251 | 0.252 | 0.232 | 0.297 | 0.114 | 0.328 | 0.219 | 0.219 | 1/2 |
| 1/2 | 0.5000 | 1/4 | 0.291 | 0.291 | 0.270 | 0.344 | 0.130 | 0.375 | 0.250 | 0.250 | 9/16 |
| 5/8 | 0.6250 | 5/16 | 0.372 | 0.371 | 0.347 | 0.469 | 0.164 | 0.469 | 0.312 | 0.312 | 3/4 |
| 3/4 | 0.7500 | 3/8 | 0.454 | 0.450 | 0.425 | 0.562 | 0.196 | 0.562 | 0.375 | 0.375 | 7/8 |
| 7/8 | 0.8750 | 1/2 | 0.595 | 0.530 | 0.502 | 0.656 | 0.227 | 0.656 | 0.500 | 0.500 | 1 |
| 1 | 1.0000 | 9/16 | ... | 0.609 | 0.579 | 0.750 | 0.260 | 0.750 | 0.562 | ... | 1 1/8 |
| 1 1/8 | 1.1250 | 9/16 | ... | 0.689 | 0.655 | 0.844 | 0.291 | 0.844 | 0.562 | ... | 1 1/4 |
| 1 1/4 | 1.2500 | 5/8 | ... | 0.767 | 0.733 | 0.938 | 0.323 | 0.938 | 0.625 | ... | 1 1/2 |
| 1 3/8 | 1.3750 | 5/8 | ... | 0.848 | 0.808 | 1.031 | 0.354 | 1.031 | 0.625 | ... | 1 5/8 |
| 1 1/2 | 1.5000 | 3/4 | ... | 0.926 | 0.886 | 1.125 | 0.385 | 1.125 | 0.750 | ... | 1 3/4 |
| 1 3/4 | 1.7500 | 1 | ... | 1.086 | 1.039 | 1.312 | 0.448 | 1.312 | 1.000 | ... | 2 |
| 2 | 2.0000 | 1 | ... | 1.244 | 1.193 | 1.500 | 0.510 | 1.500 | 1.000 | ... | 2 1/4 |

All dimensions are in inches. The thread conforms to the Unified standard, Class 3A, UNC and UNF series. The socket depth $T$ is included in the Standard and some are shown here. The nominal length $L$ of all socket type set screws is the total or overall length. For Nominal screw lengths of 1/16 through 3/16 inch (0 through 3 sizes incl.) the standard length increment is 1/32 inch; for lengths 1/8 through 1/2 inch the increment is 1/16 inch; for lengths 1/2 through 1 inch the increment is 1/8 inch; for lengths 1 through 2 inches the increment is 1/4 inch; for lengths 2 through 6 inches the increment is 1/2 inch; for lengths 6 inches and longer the increment is 1 inch. Socket dimensions are given in Table 9.

*Cone point angle $Y$ is 90 degrees plus or minus 2 degrees for these nominal lengths or longer and 118 degrees plus or minus 2 degrees for shorter nominal lengths.

†Reference should be made to the Standard for shortest optimum nominal lengths to which the minimum key engagement depths $T$ apply.

Material: High grade alloy steel hardened by quenching from the hardening temperature and tempered to a Rockwell "C" hardness of 45 to 53, or any of the 18-8, AISI 1384 or equivalent types.

Length Tolerance: The allowable tolerance on length $L$ for all set screws of the socket type is ±0.01 inch for set screws up to 5/8 inch long; ±0.02 inch for screws over 5/8 to 2 inches long; ±0.03 inch for screws over 2 to 6 inches long and ±0.06 inch for screws over 6 inches long. Socket dimensions are given in Table 9.

## Table 8. American National Standard Hexagon and Spline Socket Set Screw
Optional Cup Points (ANSI B18.3-1976)

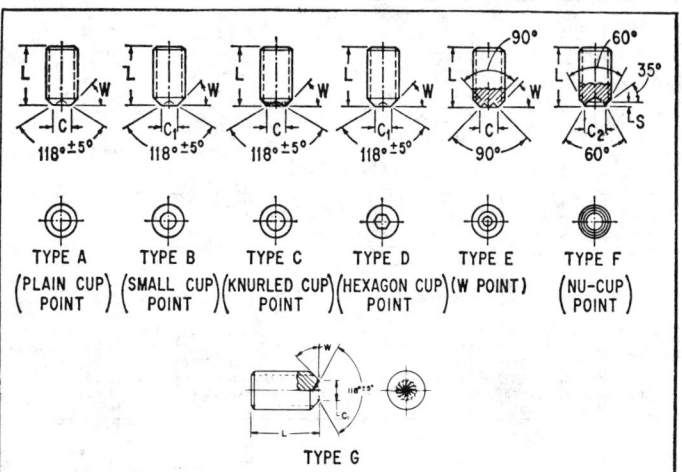

TYPE A
(PLAIN CUP
POINT)

TYPE B
(SMALL CUP
POINT)

TYPE C
(KNURLED CUP
POINT)

TYPE D
(HEXAGON CUP
POINT)

TYPE E
(W POINT)

TYPE F
(NU-CUP
POINT)

TYPE G

| Nom. Size | Point Diam. | | Point Diam. | | Point Diam. | | Point Length | |
|---|---|---|---|---|---|---|---|---|
| | Max. | Min. | Max. | Min. | Max. | Min. | Max. | Min. |
| | C | | C₁ | | C₂ | | S | |
| o | 0.033 | 0.027 | 0.032 | 0.027 | 0.027 | 0.022 | 0.007 | 0.004 |
| 1 | 0.040 | 0.033 | 0.038 | 0.033 | 0.035 | 0.030 | 0.008 | 0.005 |
| 2 | 0.047 | 0.039 | 0.043 | 0.038 | 0.043 | 0.038 | 0.010 | 0.007 |
| 3 | 0.054 | 0.045 | 0.050 | 0.045 | 0.051 | 0.046 | 0.011 | 0.007 |
| 4 | 0.061 | 0.051 | 0.056 | 0.051 | 0.059 | 0.054 | 0.013 | 0.008 |
| 5 | 0.067 | 0.057 | 0.062 | 0.056 | 0.068 | 0.063 | 0.014 | 0.009 |
| 6 | 0.074 | 0.064 | 0.069 | 0.062 | 0.074 | 0.068 | 0.017 | 0.012 |
| 8 | 0.087 | 0.076 | 0.082 | 0.074 | 0.090 | 0.084 | 0.021 | 0.016 |
| 10 | 0.102 | 0.088 | 0.095 | 0.086 | 0.101 | 0.095 | 0.024 | 0.019 |
| ¼ | 0.132 | 0.118 | 0.125 | 0.114 | 0.156 | 0.150 | 0.027 | 0.022 |
| 5⁄16 | 0.172 | 0.156 | 0.156 | 0.144 | 0.190 | 0.185 | 0.038 | 0.033 |
| 3⁄8 | 0.212 | 0.194 | 0.187 | 0.174 | 0.241 | 0.236 | 0.041 | 0.036 |
| 7⁄16 | 0.252 | 0.232 | 0.218 | 0.204 | 0.286 | 0.281 | 0.047 | 0.042 |
| ½ | 0.291 | 0.270 | 0.250 | 0.235 | 0.333 | 0.328 | 0.054 | 0.049 |
| 5⁄8 | 0.371 | 0.347 | 0.312 | 0.295 | 0.425 | 0.420 | 0.067 | 0.062 |
| ¾ | 0.450 | 0.425 | 0.375 | 0.357 | 0.523 | 0.518 | 0.081 | 0.076 |
| 7⁄8 | 0.530 | 0.502 | 0.437 | 0.418 | ... | ... | ... | ... |
| 1 | 0.609 | 0.579 | 0.500 | 0.480 | ... | ... | ... | ... |
| 1⅛ | 0.689 | 0.655 | 0.562 | 0.542 | ... | ... | ... | ... |
| 1¼ | 0.767 | 0.733 | 0.625 | 0.605 | ... | ... | ... | ... |
| 1⅜ | 0.848 | 0.808 | 0.687 | 0.667 | ... | ... | ... | ... |
| 1½ | 0.926 | 0.886 | 0.750 | 0.730 | ... | ... | ... | ... |
| 1¾ | 1.086 | 1.039 | 0.875 | 0.855 | ... | ... | ... | ... |
| 2 | 1.244 | 1.193 | 1.000 | 0.980 | ... | ... | ... | ... |

All dimensions are in inches.
The cup point types shown are those available from various manufacturers.

Table 9. American National Standard Hexagon and Spline Sockets (ANSI B18.3-1976)

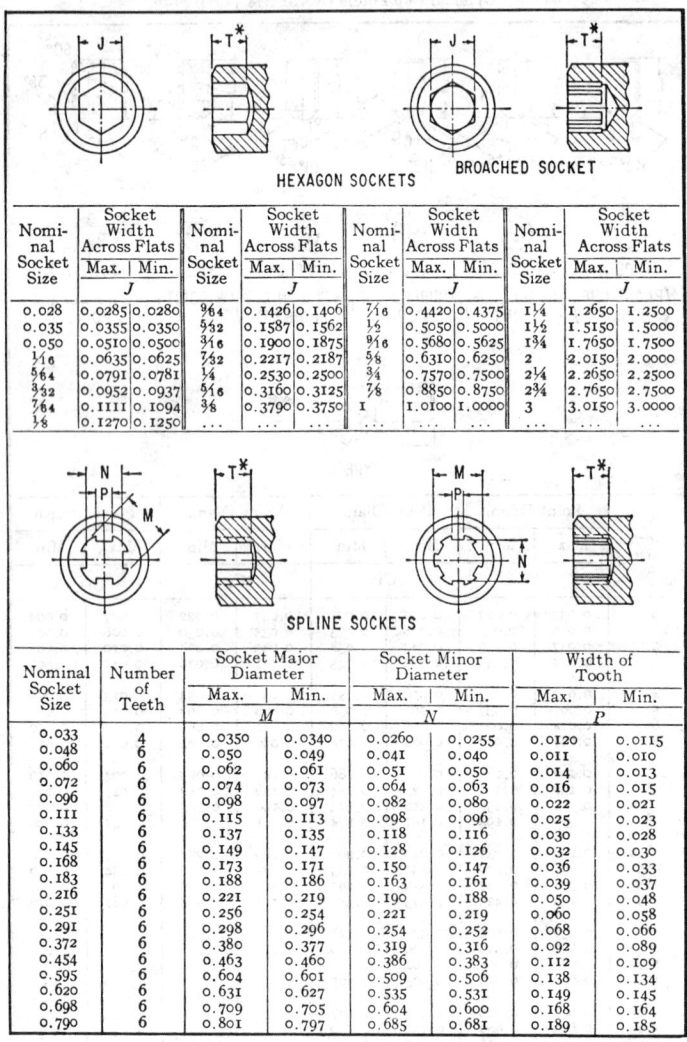

HEXAGON SOCKETS    BROACHED SOCKET

| Nominal Socket Size | Socket Width Across Flats Max. $J$ | Min. | Nominal Socket Size | Socket Width Across Flats Max. $J$ | Min. | Nominal Socket Size | Socket Width Across Flats Max. $J$ | Min. | Nominal Socket Size | Socket Width Across Flats Max. $J$ | Min. |
|---|---|---|---|---|---|---|---|---|---|---|---|
| 0.028 | 0.0285 | 0.0280 | 9/64 | 0.1426 | 0.1406 | 7/16 | 0.4420 | 0.4375 | 1 1/4 | 1.2650 | 1.2500 |
| 0.035 | 0.0355 | 0.0350 | 5/32 | 0.1587 | 0.1562 | 1/2 | 0.5050 | 0.5000 | 1 1/2 | 1.5150 | 1.5000 |
| 0.050 | 0.0510 | 0.0500 | 3/16 | 0.1900 | 0.1875 | 9/16 | 0.5680 | 0.5625 | 1 3/4 | 1.7650 | 1.7500 |
| 1/16 | 0.0635 | 0.0625 | 7/32 | 0.2217 | 0.2187 | 5/8 | 0.6310 | 0.6250 | 2 | 2.0150 | 2.0000 |
| 5/64 | 0.0791 | 0.0781 | 1/4 | 0.2530 | 0.2500 | 3/4 | 0.7570 | 0.7500 | 2 1/4 | 2.2650 | 2.2500 |
| 3/32 | 0.0952 | 0.0937 | 5/16 | 0.3160 | 0.3125 | 7/8 | 0.8850 | 0.8750 | 2 3/4 | 2.7650 | 2.7500 |
| 7/64 | 0.1111 | 0.1094 | 3/8 | 0.3790 | 0.3750 | 1 | 1.0100 | 1.0000 | 3 | 3.0150 | 3.0000 |
| 1/8 | 0.1270 | 0.1250 | ... | ... | ... | ... | ... | ... | ... | ... | ... |

SPLINE SOCKETS

| Nominal Socket Size | Number of Teeth | Socket Major Diameter Max. $M$ | Min. | Socket Minor Diameter Max. $N$ | Min. | Width of Tooth Max. $P$ | Min. |
|---|---|---|---|---|---|---|---|
| 0.033 | 4 | 0.0350 | 0.0340 | 0.0260 | 0.0255 | 0.0120 | 0.0115 |
| 0.048 | 6 | 0.050 | 0.049 | 0.041 | 0.040 | 0.011 | 0.010 |
| 0.060 | 6 | 0.062 | 0.061 | 0.051 | 0.050 | 0.014 | 0.013 |
| 0.072 | 6 | 0.074 | 0.073 | 0.064 | 0.063 | 0.016 | 0.015 |
| 0.096 | 6 | 0.098 | 0.097 | 0.082 | 0.080 | 0.022 | 0.021 |
| 0.111 | 6 | 0.115 | 0.113 | 0.098 | 0.096 | 0.025 | 0.023 |
| 0.133 | 6 | 0.137 | 0.135 | 0.118 | 0.116 | 0.030 | 0.028 |
| 0.145 | 6 | 0.149 | 0.147 | 0.128 | 0.126 | 0.032 | 0.030 |
| 0.168 | 6 | 0.173 | 0.171 | 0.150 | 0.147 | 0.036 | 0.033 |
| 0.183 | 6 | 0.188 | 0.186 | 0.163 | 0.161 | 0.039 | 0.037 |
| 0.216 | 6 | 0.221 | 0.219 | 0.190 | 0.188 | 0.050 | 0.048 |
| 0.251 | 6 | 0.256 | 0.254 | 0.221 | 0.219 | 0.060 | 0.058 |
| 0.291 | 6 | 0.298 | 0.296 | 0.254 | 0.252 | 0.068 | 0.066 |
| 0.372 | 6 | 0.380 | 0.377 | 0.319 | 0.316 | 0.092 | 0.089 |
| 0.454 | 6 | 0.463 | 0.460 | 0.386 | 0.383 | 0.112 | 0.109 |
| 0.595 | 6 | 0.604 | 0.601 | 0.509 | 0.506 | 0.138 | 0.134 |
| 0.620 | 6 | 0.631 | 0.627 | 0.535 | 0.531 | 0.149 | 0.145 |
| 0.698 | 6 | 0.709 | 0.705 | 0.604 | 0.600 | 0.168 | 0.164 |
| 0.790 | 6 | 0.801 | 0.797 | 0.685 | 0.681 | 0.189 | 0.185 |

All dimensions are in inches.

*Socket depths, $T$, for various screw types are given in the standard but are not shown here.

Where sockets are chamfered, the depth of chamfer shall not exceed 10 per cent of the nominal socket size for sizes up to and including 1/16 inch for hexagon sockets and 0.060 for spline sockets, and 7.5 per cent for larger sizes.

## Table 10. American National Standard Square Head Set Screws
(ANSI B18.6.2-1972, R1977)

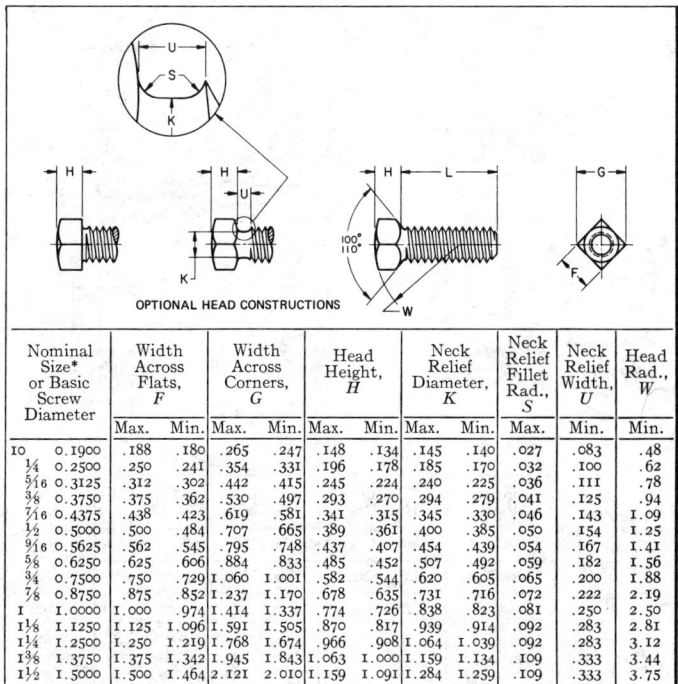

OPTIONAL HEAD CONSTRUCTIONS

| Nominal Size* or Basic Screw Diameter | | Width Across Flats, F | | Width Across Corners, G | | Head Height, H | | Neck Relief Diameter, K | | Neck Relief Fillet Rad., S | Neck Relief Width, U | Head Rad., W |
|---|---|---|---|---|---|---|---|---|---|---|---|---|
| | | Max. | Min. | Max. | Min. | Max. | Min. | Max. | Min. | Max. | Min. | Min. |
| 10 | 0.1900 | .188 | .180 | .265 | .247 | .148 | .134 | .145 | .140 | .027 | .083 | .48 |
| ¼ | 0.2500 | .250 | .241 | .354 | .331 | .196 | .178 | .185 | .170 | .032 | .100 | .62 |
| ⁵⁄₁₆ | 0.3125 | .312 | .302 | .442 | .415 | .245 | .224 | .240 | .225 | .036 | .111 | .78 |
| ⅜ | 0.3750 | .375 | .362 | .530 | .497 | .293 | .270 | .294 | .279 | .041 | .125 | .94 |
| ⁷⁄₁₆ | 0.4375 | .438 | .423 | .619 | .581 | .341 | .315 | .345 | .330 | .046 | .143 | 1.09 |
| ½ | 0.5000 | .500 | .484 | .707 | .665 | .389 | .361 | .400 | .385 | .050 | .154 | 1.25 |
| ⁹⁄₁₆ | 0.5625 | .562 | .545 | .795 | .748 | .437 | .407 | .454 | .439 | .054 | .167 | 1.41 |
| ⅝ | 0.6250 | .625 | .606 | .884 | .833 | .485 | .452 | .507 | .492 | .059 | .182 | 1.56 |
| ¾ | 0.7500 | .750 | .729 | 1.060 | 1.001 | .582 | .544 | .620 | .605 | .065 | .200 | 1.88 |
| ⅞ | 0.8750 | .875 | .852 | 1.237 | 1.170 | .678 | .635 | .731 | .716 | .072 | .222 | 2.19 |
| 1 | 1.0000 | 1.000 | .974 | 1.414 | 1.337 | .774 | .726 | .838 | .823 | .081 | .250 | 2.50 |
| 1⅛ | 1.1250 | 1.125 | 1.096 | 1.591 | 1.505 | .870 | .817 | .939 | .914 | .092 | .283 | 2.81 |
| 1¼ | 1.2500 | 1.250 | 1.219 | 1.768 | 1.674 | .966 | .908 | 1.064 | 1.039 | .092 | .283 | 3.12 |
| 1⅜ | 1.3750 | 1.375 | 1.342 | 1.945 | 1.843 | 1.063 | 1.000 | 1.159 | 1.134 | .109 | .333 | 3.44 |
| 1½ | 1.5000 | 1.500 | 1.464 | 2.121 | 2.010 | 1.159 | 1.091 | 1.284 | 1.259 | .109 | .333 | 3.75 |

All dimensions are in inches.

*Threads:* Threads are Unified Standard Class 2A; UNC, UNF and 8 UN Series or UNRC, UNRF and 8 UNR Series.

*Length of Thread:* Square head set screws have complete (full form) threads extending over that portion of the screw length which is not affected by the point. For the respective constructions, threads extend into the neck relief, to the conical underside of head, or to within one thread (as measured with a thread ring gage) from the flat underside of the head. Threads through angular or crowned portions of points have fully formed roots with partial crests.

*Point Types:* Unless otherwise specified, square head set screws are supplied with cup points. Cup points as furnished by some manufacturers may be externally or internally knurled. Where so specified by the purchaser, screws have cone, dog, half-dog, flat or oval points as given on the following page.

*Designation:* Square head set screws are designated by the following data in the sequence shown: Nominal size (number, fraction or decimal equivalent); threads per inch; screw length (fraction or decimal equivalent); product name; point style; material; and protective finish, if required. Examples: ¼ - 20 x ¾ Square Head Set Screw, Flat Point, Steel, Cadmium Plated. .500 - 13 x 1.25 Square Head Set Screw, Cone Point, Corrosion Resistant Steel.

* When specifying a nominal size in decimals, the zero preceding the decimal point is omitted as is any zero in the fourth decimal place.

**Table 10** (*Continued*). **American National Standard Square Head Set Screws**
(ANSI B18.6.2-1972, R1977)

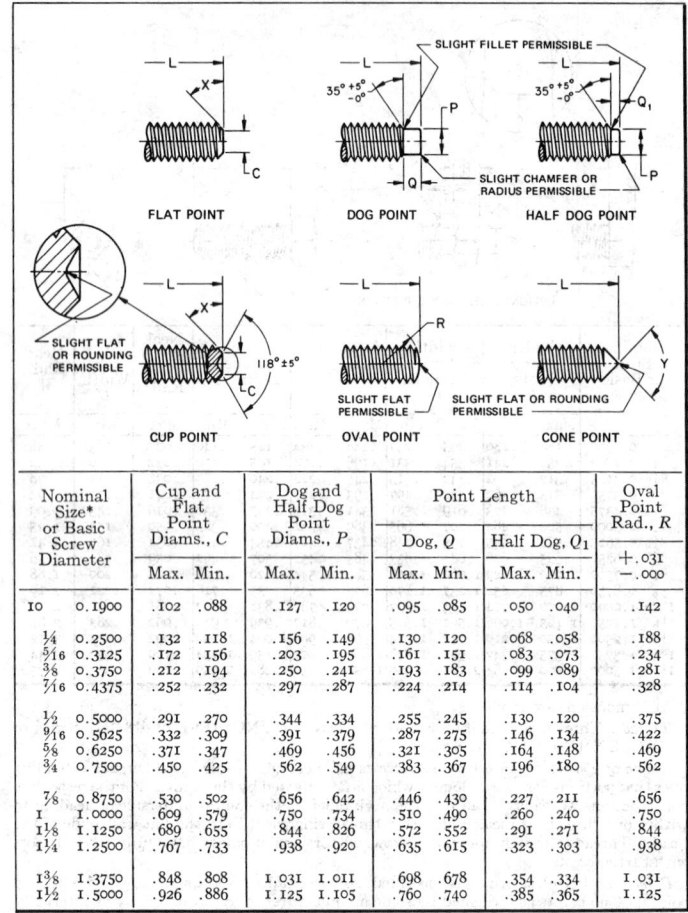

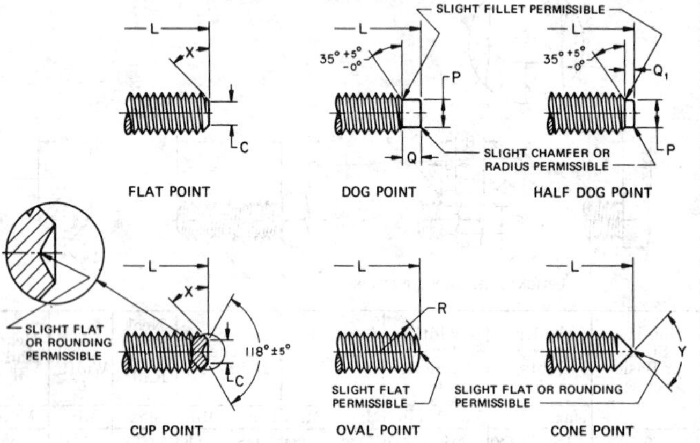
SLIGHT FILLET PERMISSIBLE
L
X
C
FLAT POINT
35° +5° −0°
P
Q
SLIGHT CHAMFER OR RADIUS PERMISSIBLE
DOG POINT
Q1
HALF DOG POINT
SLIGHT FLAT OR ROUNDING PERMISSIBLE
118° ±5°
CUP POINT
R
SLIGHT FLAT PERMISSIBLE
OVAL POINT
SLIGHT FLAT OR ROUNDING PERMISSIBLE
Y
CONE POINT

| Nominal Size* or Basic Screw Diameter | | Cup and Flat Point Diams., *C* | | Dog and Half Dog Point Diams., *P* | | Point Length | | | | Oval Point Rad., *R* |
|---|---|---|---|---|---|---|---|---|---|---|
| | | | | | | Dog, *Q* | | Half Dog, $Q_1$ | | +.031 −.000 |
| | | Max. | Min. | Max. | Min. | Max. | Min. | Max. | Min. | |
| 10 | 0.1900 | .102 | .088 | .127 | .120 | .095 | .085 | .050 | .040 | .142 |
| ¼ | 0.2500 | .132 | .118 | .156 | .149 | .130 | .120 | .068 | .058 | .188 |
| 5⁄16 | 0.3125 | .172 | .156 | .203 | .195 | .161 | .151 | .083 | .073 | .234 |
| ⅜ | 0.3750 | .212 | .194 | .250 | .241 | .193 | .183 | .099 | .089 | .281 |
| 7⁄16 | 0.4375 | .252 | .232 | .297 | .287 | .224 | .214 | .114 | .104 | .328 |
| ½ | 0.5000 | .291 | .270 | .344 | .334 | .255 | .245 | .130 | .120 | .375 |
| 9⁄16 | 0.5625 | .332 | .309 | .391 | .379 | .287 | .275 | .146 | .134 | .422 |
| ⅝ | 0.6250 | .371 | .347 | .469 | .456 | .321 | .305 | .164 | .148 | .469 |
| ¾ | 0.7500 | .450 | .425 | .562 | .549 | .383 | .367 | .196 | .180 | .562 |
| ⅞ | 0.8750 | .530 | .502 | .656 | .642 | .446 | .430 | .227 | .211 | .656 |
| 1 | 1.0000 | .609 | .579 | .750 | .734 | .510 | .490 | .260 | .240 | .750 |
| 1⅛ | 1.1250 | .689 | .655 | .844 | .826 | .572 | .552 | .291 | .271 | .844 |
| 1¼ | 1.2500 | .767 | .733 | .938 | .920 | .635 | .615 | .323 | .303 | .938 |
| 1⅜ | 1.3750 | .848 | .808 | 1.031 | 1.011 | .698 | .678 | .354 | .334 | 1.031 |
| 1½ | 1.5000 | .926 | .886 | 1.125 | 1.105 | .760 | .740 | .385 | .365 | 1.125 |

All dimensions are in inches.

*Cone Point Angle, Y:* For the following nominal lengths, or longer, shown according to nominal size, the cone point angle is 90° ± 2°: For size No. 10, ¼; ¼, 5⁄16; 5⁄16, ⅜; ⅜, 7⁄16; 7⁄16, ½; ½, 9⁄16; 9⁄16, ⅝; ⅝, ¾; ¾, ⅞; ⅞, 1; 1, 1⅛; 1⅛, 1¼; 1¼, 1½; 1⅜, 1⅝; and for 1½, 1¾. For shorter screws the cone point angle is 118° ± 2°.

*Point Angle, X:* The point angle is 45°, + 5°, − 0° for screws of the nominal lengths, or longer, given just above for cone point angle, and 30° min. for shorter lengths.

* When specifying a nominal size in decimals the zero preceding the decimal point is omitted as is any zero in the fourth decimal place.

Table 11. Applicability of Hexagon and Spline Keys and Bits

| Nominal Key or Bit Size | Cap Screws 1960 Series | Flat Countersunk Head Cap Screws | Button Head Cap Screws | Shoulder Screws | Set Screws |
|---|---|---|---|---|---|
| | Nominal Screw Sizes | | | | |
| HEXAGON KEYS AND BITS | | | | | |
| 0.028 | ... | ... | ... | ... | 0 |
| 0.035 | ... | 0 | 0 | ... | 1 & 2 |
| 0.050 | 0 | 1 & 2 | 1 & 2 | ... | 3 & 4 |
| 1/16 0.062 | 1 | 3 & 4 | 3 & 4 | ... | 5 & 6 |
| 5/64 0.078 | 2 & 3 | 5 & 6 | 5 & 6 | ... | 8 |
| 3/32 0.094 | 4 & 5 | 8 | 8 | ... | 10 |
| 7/64 0.109 | 6 | ... | ... | ... | ... |
| 1/8 0.125 | ... | 10 | 10 | 1/4 | 1/4 |
| 9/64 0.141 | 8 | ... | ... | ... | ... |
| 5/32 0.156 | 10 | 1/4 | 1/4 | 5/16 | 5/16 |
| 3/16 0.188 | 1/4 | 5/16 | 5/16 | 3/8 | 3/8 |
| 7/32 0.219 | ... | 3/8 | 3/8 | ... | 7/16 |
| 1/4 0.250 | 5/16 | 7/16 | ... | 1/2 | 1/2 |
| 5/16 0.312 | 3/8 | 1/2 | 1/2 | 5/8 | 5/8 |
| 3/8 0.375 | 7/16 & 1/2 | 5/8 | 5/8 | 3/4 | 3/4 |
| 7/16 0.438 | ... | ... | ... | ... | ... |
| 1/2 0.500 | 5/8 | 3/4 | ... | 1 | 7/8 |
| 9/16 0.562 | ... | 7/8 | ... | ... | 1 & 1 1/8 |
| 5/8 0.625 | 3/4 | 1 | ... | 1 1/4 | 1 1/4 & 1 3/8 |
| 3/4 0.750 | 7/8 & 1 | 1 1/8 | ... | ... | 1 1/2 |
| 7/8 0.875 | 1 1/8 & 1 1/4 | 1 1/4 & 1 3/8 | ... | 1 1/2 | ... |
| 1 1.000 | 1 3/8 & 1 1/2 | 1 1/2 | ... | 1 3/4 | 1 3/4 & 2 |
| 1 1/4 1.250 | 1 3/4 | ... | ... | 2 | ... |
| 1 1/2 1.500 | 2 | ... | ... | ... | ... |
| 1 3/4 1.750 | 2 1/4 & 2 1/2 | ... | ... | ... | ... |
| 2 2.000 | 2 3/4 | ... | ... | ... | ... |
| 2 1/4 2.250 | 3 & 3 1/4 | ... | ... | ... | ... |
| 2 3/4 2.750 | 3 1/2 & 3 3/4 | ... | ... | ... | ... |
| 3 3.000 | 4 | ... | ... | ... | ... |
| SPLINE KEYS AND BITS | | | | | |
| 0.033 | ... | ... | ... | ... | 0 & 1 |
| 0.048 | ... | 0 | 0 | ... | 2 & 3 |
| 0.060 | 0 | 1 & 2 | 1 & 2 | ... | 4 |
| 0.072 | 1 | 3 & 4 | 3 & 4 | ... | 5 & 6 |
| 0.096 | 2 & 3 | 5 & 6 | 5 & 6 | ... | 8 |
| 0.111 | 4 & 5 | 8 | 8 | ... | 10 |
| 0.133 | 6 | ... | ... | ... | ... |
| 0.145 | ... | 10 | 10 | ... | 1/4 |
| 0.168 | 8 | ... | ... | ... | ... |
| 0.183 | 10 | 1/4 | 1/4 | ... | 5/16 |
| 0.216 | 1/4 | 5/16 | 5/16 | ... | 3/8 |
| 0.251 | ... | 3/8 | 3/8 | ... | 7/16 |
| 0.291 | 5/16 | 7/16 | ... | ... | 1/2 |
| 0.372 | 3/8 | 1/2 | 1/2 | ... | 5/8 |
| 0.454 | 7/16 & 1/2 | 5/8 & 3/4 | 5/8 | ... | 3/4 |
| 0.595 | 5/8 | ... | ... | ... | 7/8 |
| 0.620 | 3/4 | ... | ... | ... | ... |
| 0.698 | 7/8 | ... | ... | ... | ... |
| 0.790 | 1 | ... | ... | ... | ... |

*Source:* Appendix to American National Standard ANSI B18.3-1976

**British Standard Hexagon Socket Screws, Metric Series.** — The British Standard specifying hexagon socket screws, metric series is B.S. 4168:1967 "Hexagon Socket Screws and Wrench Keys — Metric Series." Section 1 of this standard gives the dimensions, tolerances and mechanical properties of cap screws, countersunk head screws, button head screws and set screws.

*Material:* Steel hexagon socket screws are manufactured from high grade alloy steel and are hardened in oil and tempered to give certain mechanical properties listed in the standard but not given here. The strength grade designation system for steel hexagon socket cap screws consists of two figures; the first is $\frac{1}{10}$ the minimum tensile strength in kgf/mm², the second is $\frac{1}{10}$ of the ratio between the stress at a permanent set of 0.2 per cent (yield stress) and the minimum tensile strength, written as a percentage; e.g., $\frac{1}{10}$ minimum tensile strength of 120 kgf/mm² gives

symbol "12" and $\frac{1}{10}$ ratio $\dfrac{\text{stress at permanent set 0.2\% (yield stress)}}{\text{minimum tensile strength}} \% = \dfrac{1}{10} \times$

$\dfrac{108}{120} \times \dfrac{100}{1} =$ "9," both when combined giving a strength grade designation = "12.9."

*Finish:* Hexagon socket screws are sound, free from defects and have a blue-black or chemical black oxide finish after heat treatment.

*Head Diameter:* Cap screw heads may be plain or knurled at the option of the manufacturer. For knurled screws the diameter measured over the top of the knurl should not exceed the maximum specified in Table 1 and the diameter of the head before knurling or the unknurled section of head should not be less than the minimum specified in Table 1. For countersunk head screws the maximum sharp values given for 92°/90° countersunk heads in Table 2 are theoretical values only, as it is not practical to make the edge of the head sharp.

*Length of Socket Screws:* The standard nominal lengths and tolerances on lengths for hexagon socket screws are given in Table 3.

*Ends of Socket Screws:* Set screw points are specified in Table 4.

*Screw Threads:* Screw threads are the ISO metric coarse pitch series specified in B.S. 3643, Part 1. Threads are made to the 6g tolerance class as specified in B.S. 3643, Part 2. They may be either cut or rolled at the option of the manufacturer.

*Length of Thread:* In general the length of thread is the distance from the end of the screw, including the chamfer, to the leading face of a screw ring gage which has been screwed as far as possible onto the screw by hand. The length of thread on cap and countersunk head screws is as specified in Table 3. Screws that are too short for the length of thread specified are threaded so that a screw ring gage, being screwed by hand, can come to within a distance from the underside of the head not exceeding 2½ times the pitch. The tolerance on the thread lengths is twice the pitch. The minimum length of thread on button head screws is as specified in Table 3. There is no maximum limit specified. Screws that are too short for the length of thread specified are threaded so that a screw gage, being screwed by hand, can come to within a distance from the underside of the head not exceeding 2½ times the pitch.

*Method of Specifying Socket Screws:* Standard screws are specified by type, nominal diameter of thread, nominal length and the B.S. number; e.g., hexagon socket cap screw, M10 × 50 long, to B.S. 4168.

*Marking:* At the option of the manufacturer hexagon socket cap screws above M5 (5 mm diameter) may have the strength grade designation (12.9) permanently marked, preferably by indenting on the side of the head. The manufacturer's identification may also be shown.

Table 1. British Standard Hexagon Socket Head Cap Screws — Metric Series (B.S. 4168:1967)

| Nominal Size* | Body Diameter, D | | Head Diameter, A | | Head Height, H | | Head Side Height, S | Hexagon Socket Size, J | Key Engagement, K | | Wall Thickness, W | Fillet Radius | | Chamfer or Radius, C |
|---|---|---|---|---|---|---|---|---|---|---|---|---|---|---|
| | | | | | | | | | | | | F | $d_a$ | |
| | Max. | Min. | Max. | Min. | Max. | Min. | Min. | Nom. | Max. | Min. | Min. | Min. | Max. | Max. |
| M3 | 3.00 | 2.86 | 5.50 | 5.20 | 3.00 | 2.86 | 2.70 | 2.50 | 1.70 | 1.30 | 0.96 | 0.10 | 3.60 | 0.125 |
| M4 | 4.00 | 3.82 | 7.00 | 6.64 | 4.00 | 3.82 | 3.60 | 3.00 | 2.40 | 2.00 | 1.28 | 0.20 | 4.70 | 0.125 |
| M5 | 5.00 | 4.82 | 8.50 | 8.14 | 5.00 | 4.82 | 4.50 | 4.00 | 3.10 | 2.70 | 1.60 | 0.20 | 5.70 | 0.125 |
| M6 | 6.00 | 5.82 | 10.00 | 9.64 | 6.00 | 5.82 | 5.40 | 5.00 | 3.78 | 3.30 | 1.92 | 0.25 | 6.80 | 0.200 |
| M8 | 8.00 | 7.78 | 13.00 | 12.57 | 8.00 | 7.78 | 7.20 | 6.00 | 4.78 | 4.30 | 2.56 | 0.40 | 9.20 | 0.200 |
| M10 | 10.00 | 9.78 | 16.00 | 15.57 | 10.00 | 9.78 | 9.00 | 8.00 | 6.25 | 5.50 | 3.20 | 0.40 | 11.20 | 0.200 |
| M12 | 12.00 | 11.73 | 18.00 | 17.57 | 12.00 | 11.73 | 10.80 | 10.00 | 7.50 | 6.60 | 3.84 | 0.60 | 14.20 | 0.250 |
| (M14) | 14.00 | 13.73 | 21.00 | 20.48 | 14.00 | 13.73 | 12.60 | 12.00 | 8.70 | 7.80 | 4.48 | 0.60 | 16.20 | 0.250 |
| M16 | 16.00 | 15.73 | 24.00 | 23.48 | 16.00 | 15.73 | 14.40 | 14.00 | 9.70 | 8.80 | 5.12 | 0.60 | 18.20 | 0.250 |
| (M18) | 18.00 | 17.73 | 27.00 | 26.48 | 18.00 | 17.73 | 16.20 | 14.00 | 10.70 | 9.80 | 5.76 | 0.60 | 20.20 | 0.250 |
| M20 | 20.00 | 19.67 | 30.00 | 29.48 | 20.00 | 19.67 | 18.00 | 17.00 | 11.80 | 10.70 | 6.40 | 0.80 | 22.40 | 0.400 |
| (M22) | 22.00 | 21.67 | 33.00 | 32.38 | 22.00 | 21.67 | 19.80 | 17.00 | 12.40 | 11.30 | 7.04 | 0.80 | 24.40 | 0.400 |
| M24 | 24.00 | 23.67 | 36.00 | 35.38 | 24.00 | 23.67 | 21.60 | 19.00 | 14.00 | 12.90 | 7.68 | 0.80 | 26.40 | 0.400 |

All dimensions are given in millimeters.    * Sizes shown in parentheses are non-preferred.

**Table 2.  British Standard Hexagon Socket Countersunk and
Button Head Screws — Metric Series (B.S. 4168:1967)**

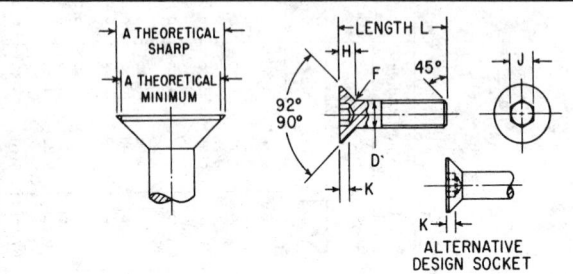

COUNTERSUNK HEAD SCREWS

| Nom. Size* | Body Diameter, D | | Head Diameter, A | | Head Height, H | | Hexagon Socket Size, J | Key Engagement, K | Fillet Radius, F |
|---|---|---|---|---|---|---|---|---|---|
| | Max. | Min. | Theor. Sharp Max. | Absolute Min. | Ref. | Flushness Tolerance | Nom. | Min. | Max. |
| M3 | 3.00 | 2.86 | 6.72 | 5.82 | 1.86 | 0.20 | 2.00 | 1.05 | 0.40 |
| M4 | 4.00 | 3.82 | 8.96 | 7.78 | 2.48 | 0.20 | 2.50 | 1.49 | 0.40 |
| M5 | 5.00 | 4.82 | 11.20 | 9.78 | 3.10 | 0.20 | 3.00 | 1.86 | 0.40 |
| M6 | 6.00 | 5.82 | 13.44 | 11.73 | 3.72 | 0.20 | 4.00 | 2.16 | 0.60 |
| M8 | 8.00 | 7.78 | 17.92 | 15.73 | 4.96 | 0.24 | 5.00 | 2.85 | 0.70 |
| M10 | 10.00 | 9.78 | 22.40 | 19.67 | 6.20 | 0.30 | 6.00 | 3.60 | 0.80 |
| M12 | 12.00 | 11.73 | 26.88 | 23.67 | 7.44 | 0.36 | 8.00 | 4.35 | 1.10 |
| (M14) | 14.00 | 13.73 | 30.24 | 26.67 | 8.12 | 0.40 | 10.00 | 4.65 | 1.10 |
| M16 | 16.00 | 15.73 | 33.60 | 29.67 | 8.80 | 0.45 | 10.00 | 4.89 | 1.10 |
| (M18) | 18.00 | 17.73 | 36.96 | 32.61 | 9.48 | 0.50 | 12.00 | 5.25 | 1.10 |
| M20 | 20.00 | 19.67 | 40.32 | 35.61 | 10.16 | 0.54 | 12.00 | 5.45 | 1.10 |

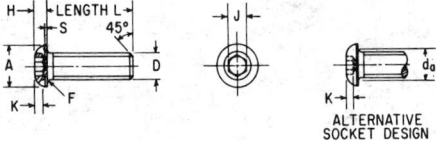

BUTTON HEAD SCREWS

| Nom. Size, D | Head Diameter, A | | Head Height, H | | Head Side Height, S | Hexagon Socket Size, J | Key Engagement, K | Fillet Radius | |
|---|---|---|---|---|---|---|---|---|---|
| | Max. | Min. | Max. | Min. | Ref. | Nom. | Min. | F Min. | $d_a$ Max. |
| M3 | 5.50 | 5.32 | 1.60 | 1.40 | 0.38 | 2.00 | 1.04 | 0.10 | 3.60 |
| M4 | 7.50 | 7.28 | 2.10 | 1.85 | 0.38 | 2.50 | 1.30 | 0.20 | 4.70 |
| M5 | 9.50 | 9.25 | 2.70 | 2.45 | 0.50 | 3.00 | 1.56 | 0.20 | 5.70 |
| M6 | 10.50 | 10.23 | 3.20 | 2.95 | 0.80 | 4.00 | 2.08 | 0.25 | 6.80 |
| M8 | 14.00 | 13.73 | 4.30 | 3.95 | 0.80 | 5.00 | 2.60 | 0.40 | 9.20 |
| M10 | 18.00 | 17.73 | 5.30 | 4.95 | 0.80 | 6.00 | 3.12 | 0.40 | 11.20 |
| M12 | 21.00 | 20.67 | 6.40 | 5.90 | 0.80 | 8.00 | 4.16 | 0.60 | 14.20 |

All dimensions are given in millimeters.
* Sizes shown in parentheses are non-preferred.

**Table 3. British Standard Hexagon Socket Screws — Metric Series** (B.S. 4168:1967)

## DIMENSIONS OF HEXAGON SOCKETS

WIDTH OF OVERCUT (SEE FOOTNOTE*)

LENGTH OF FLAT BROACHED SOCKET

| Nominal Socket Size | Socket Width Across Flats, $J$ | | Nominal Socket Size | Socket Width Across Flats, $J$ | |
|---|---|---|---|---|---|
| | Max. | Min. | | Max. | Min. |
| 1.5 | 1.61 | 1.52 | 6 | 6.15 | 6.03 |
| 2.0 | 2.11 | 2.02 | 8 | 8.19 | 8.04 |
| 2.5 | 2.61 | 2.52 | 10 | 10.19 | 10.04 |
| 3 | 3.11 | 3.02 | 12 | 12.23 | 12.05 |
| 4 | 4.15 | 4.03 | 14 | 14.23 | 14.05 |
| 5 | 5.15 | 5.03 | 17 | 17.23 | 17.05 |
| ... | .... | .... | 19 | 19.275 | 19.065 |

### NOMINAL LENGTHS AND TOLERANCES FOR SOCKET SCREWS†

| Nominal Length | Tolerance | Nominal Length | Tolerance | Nominal Length | Tolerance |
|---|---|---|---|---|---|
| 3 | ±0.20 | (28) | ±0.42 | (95) | ±0.70 |
| 4 | ±0.24 | 30 | ±0.42 | 100 | ±0.70 |
| 5 | ±0.24 | (32) | ±0.50 | (105) | ±0.70 |
| 6 | ±0.24 | 35 | ±0.50 | 110 | ±0.70 |
| (7) | ±0.29 | (38) | ±0.50 | (115) | ±0.70 |
| 8 | ±0.29 | 40 | ±0.50 | 120 | ±0.70 |
| (9) | ±0.29 | 45 | ±0.50 | (125) | ±0.80 |
| 10 | ±0.29 | 50 | ±0.50 | 130 | ±0.80 |
| (11) | ±0.35 | 55 | ±0.60 | 140 | ±0.80 |
| 12 | ±0.35 | 60 | ±0.60 | 150 | ±0.80 |
| 14 | ±0.35 | 65 | ±0.60 | 160 | ±0.80 |
| 16 | ±0.35 | 70 | ±0.60 | 170 | ±0.80 |
| (18) | ±0.35 | 75 | ±0.60 | 180 | ±0.80 |
| 20 | ±0.42 | 80 | ±0.60 | 190 | ±0.925 |
| (22) | ±0.42 | 85 | ±0.70 | 200 | ±0.925 |
| 25 | ±0.42 | 90 | ±0.70 | ... | ...... |

### ASSOCIATION OF THREAD LENGTHS AND NOMINAL THREAD DIAMETERS FOR VARIOUS NOMINAL SCREW LENGTHS‡

| Nominal Thread Diameter, $D$ | Nominal Screw Length | | Nominal Thread Diameter, $D$ | Nominal Screw Length | | |
|---|---|---|---|---|---|---|
| | Up to 125 | Over 125 Up to 200 | | Up to 125 | Over 125 Up to 200 | Above 200 |
| | Length of Thread | | | Length of Thread | | |
| M3 | 12 | .. | M14 | 34 | 40 | .. |
| M4 | 14 | .. | M16 | 38 | 44 | 57 |
| M5 | 16 | .. | M18 | 42 | 48 | 61 |
| M6 | 18 | .. | M20 | 46 | 52 | 65 |
| M8 | 22 | .. | M22 | 50 | 56 | 69 |
| M10 | 26 | 32 | M24 | 54 | 60 | 73 |
| M12 | 30 | 36 | .... | .. | .. | .. |

All dimensions are given in millimeters.

* For broached sockets which are at or near the minimum limit of size across flats the overcut resulting from drilling should not exceed 20 per cent of the length of any flat of the socket.

† The inclusion of dimensional data here is not intended to imply that all of the products described are stock production sizes. Lengths given in parentheses are to be avoided wherever possible.

‡ See *Length of Thread* description in the accompanying text.

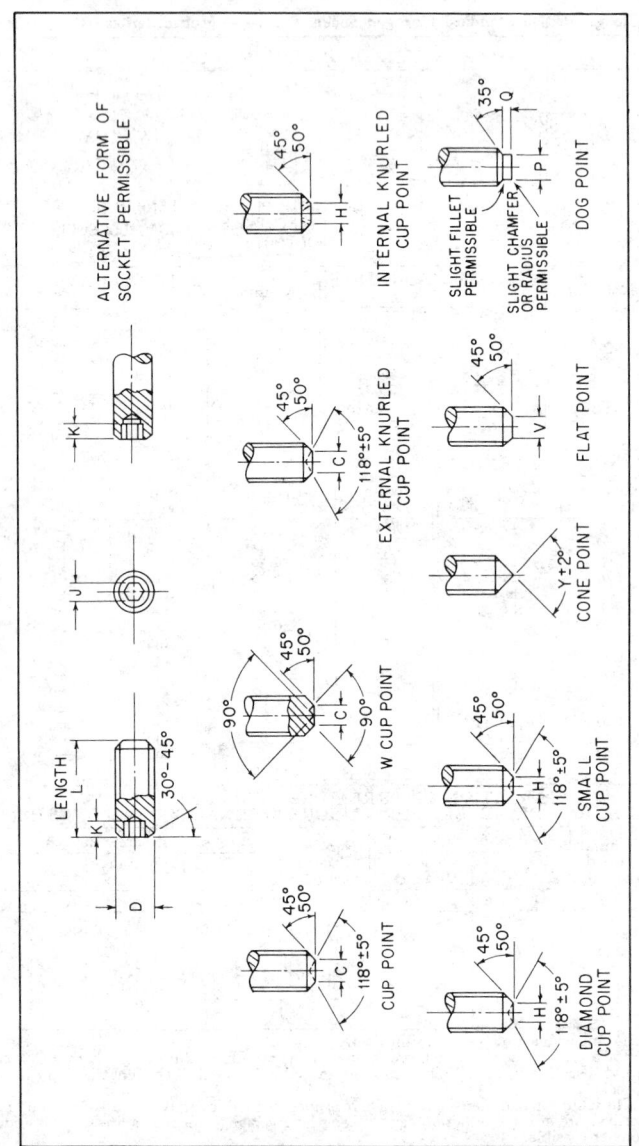

British Standard Hexagon Socket Set Screws — Metric Series

For dimensions see Table 4.

Table 4.  British Standard Hexagon Socket Set Screws — Metric Series (B.S. 4168:1967)

| Nom. Size* D | Hexagon Socket Size, J (Nom.) | Key Engagement K (Min.) | Shortest Nominal Length to Which Dimension K Applies — Cup and Flat Point | Shortest Nominal Length to Which Dimension K Applies — Cone Point | Shortest Nominal Length to Which Dimension K Applies — Dog Point† | W Cup Point, External Knurled Cup Point Diameters, C (Max.) | C (Min.) | Internal Knurled Diamond and Small Cup Point Diameters, H (Max.) | H (Min.) | Cone Point Angle 90° for these Lengths and Over, 118° for Shorter Lengths, Y | Flat Point Diameter, V (Max.) | V (Min.) | Dog Point Diameter, P (Max.) | P (Min.) | Dog Point Length, Q (Max.) | Q (Min.) |
|---|---|---|---|---|---|---|---|---|---|---|---|---|---|---|---|---|
| M3 | 1.50 | 1.20 | 3.00 | 3.00 | 6.00 | 1.40 | 1.00 | 1.50 | 1.10 | 5.00 | 2.00 | 1.60 | 2.00 | 1.86 | 2.40 | 2.25 |
| M4 | 2.00 | 1.60 | 4.00 | 4.00 | 6.00 | 2.00 | 1.60 | 2.00 | 1.60 | 6.00 | 2.60 | 2.20 | 2.50 | 2.36 | 2.80 | 2.60 |
| M5 | 2.50 | 2.00 | 5.00 | 5.00 | 6.00 | 2.50 | 2.10 | 2.50 | 2.10 | 8.00 | 3.40 | 2.92 | 3.50 | 3.32 | 2.80 | 2.60 |
| M6 | 3.00 | 2.40 | 5.00 | 6.00 | 8.00 | 3.00 | 2.60 | 3.00 | 2.60 | 9.00 | 4.00 | 3.52 | 4.50 | 4.32 | 3.35 | 3.15 |
| M8 | 4.00 | 3.20 | 8.00 | 8.00 | 10.00 | 5.00 | 4.52 | 4.00 | 3.52 | 12.00 | 5.50 | 5.02 | 6.00 | 5.82 | 4.70 | 4.50 |
| M10 | 5.00 | 4.00 | 8.00 | 12.00 | 12.00 | 6.00 | 5.52 | 5.00 | 4.52 | 16.00 | 7.00 | 6.42 | 7.00 | 6.78 | 4.90 | 4.70 |
| M12 | 6.00 | 4.80 | 10.00 | 12.00 | 15.00 | 8.00 | 7.42 | 6.00 | 5.52 | 18.00 | 8.50 | 7.92 | 9.00 | 8.78 | 6.40 | 6.20 |
| (M14) | 6.00 | 4.80 | 12.00 | 12.00 | 16.00 | 9.00 | 8.42 | 7.00 | 6.42 | 22.00 | 10.00 | 9.42 | 10.00 | 9.78 | 6.15 | 5.95 |
| M16 | 8.00 | 6.40 | 15.00 | 15.00 | 18.00 | 10.00 | 9.44 | 8.00 | 7.42 | 25.00 | 12.00 | 11.30 | 12.00 | 11.73 | 8.15 | 7.95 |
| (M18) | 10.00 | 8.00 | 18.00 | 18.00 | 20.00 | 12.00 | 11.30 | 9.00 | 8.42 | 28.00 | 13.50 | 12.80 | 13.00 | 12.73 | 7.85 | 7.65 |
| M20 | 10.00 | 8.00 | 18.00 | 18.00 | 22.00 | 14.00 | 13.30 | 10.00 | 9.42 | 30.00 | 15.00 | 14.30 | 15.00 | 14.73 | 7.85 | 7.65 |
| (M22) | 12.00 | 9.60 | 22.00 | 22.00 | 25.00 | 15.00 | 14.30 | 11.00 | 10.30 | 35.00 | 16.50 | 15.80 | 17.00 | 16.73 | 9.90 | 9.65 |
| M24 | 12.00 | 9.60 | 22.00 | 22.00 | 25.00 | 16.00 | 15.30 | 12.00 | 11.30 | 38.00 | 18.00 | 17.30 | 18.00 | 17.73 | 9.65 | 9.40 |

All dimensions are given in millimeters.  For dimensional notation, see diagram on page 1218.
* Sizes shown in parentheses are non-preferred.
† Also denotes minimum length for dog point screws.

## British Standard Whitworth (B.S.W.) and British Standard Fine (B.S.F.)
### Bright Square Head Set-Screws (With Flat Chamfered Ends)

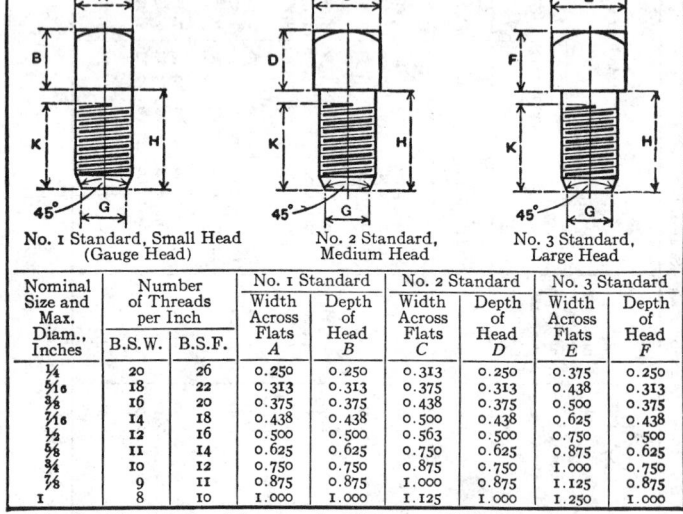

No. 1 Standard, Small Head (Gauge Head)    No. 2 Standard, Medium Head    No. 3 Standard, Large Head

| Nominal Size and Max. Diam., Inches | Number of Threads per Inch | | No. 1 Standard | | No. 2 Standard | | No. 3 Standard | |
|---|---|---|---|---|---|---|---|---|
| | B.S.W. | B.S.F. | Width Across Flats A | Depth of Head B | Width Across Flats C | Depth of Head D | Width Across Flats E | Depth of Head F |
| ¼ | 20 | 26 | 0.250 | 0.250 | 0.313 | 0.250 | 0.375 | 0.250 |
| ⁵⁄₁₆ | 18 | 22 | 0.313 | 0.313 | 0.375 | 0.313 | 0.438 | 0.313 |
| ⅜ | 16 | 20 | 0.375 | 0.375 | 0.438 | 0.375 | 0.500 | 0.375 |
| ⁷⁄₁₆ | 14 | 18 | 0.438 | 0.438 | 0.500 | 0.438 | 0.625 | 0.438 |
| ½ | 12 | 16 | 0.500 | 0.500 | 0.563 | 0.500 | 0.750 | 0.500 |
| ⅝ | 11 | 14 | 0.625 | 0.625 | 0.750 | 0.625 | 0.875 | 0.625 |
| ¾ | 10 | 12 | 0.750 | 0.750 | 0.875 | 0.750 | 1.000 | 0.750 |
| ⅞ | 9 | 11 | 0.875 | 0.875 | 1.000 | 0.875 | 1.125 | 0.875 |
| 1 | 8 | 10 | 1.000 | 1.000 | 1.125 | 1.000 | 1.250 | 1.000 |

* Depth of Head *B*, *D* and *F* same as for Width Across Flats, No. 1 Standard.
Reproduced from Report No. 451-1932 of the British Standards Institution.

**Holding Power of Set-screws.** — While the amount of power a set-screw of given size will transmit without slipping (when used for holding a pulley, gear, or other part from turning relative to a shaft) varies somewhat according to the physical properties of both set-screw and shaft and other variable factors, experiments have shown that the safe holding force in pounds for different diameters of set-screws should be approximately as follows: For ¼-inch diameter set-screws the safe holding force is 100 pounds, for ⅜-inch diameter set-screws the safe holding force is 250 pounds, for ½-inch diameter set-screws the safe holding force is 500 pounds, for ¾-inch diameter set-screws the safe holding force is 1300 pounds, and for 1-inch diameter set-screws the safe holding force is 2500 pounds.

The power or torque that can be safely transmitted by a set-screw may be determined from the formulas, $P = (DNd^{2.3}) \div 50$; or $T = 1250Dd^{2.3}$ in which $P$ is the horsepower transmitted; $T$ is the torque in inch-pounds transmitted; $D$ is the shaft diameter in inches; $N$ is the speed of the shaft in revolutions per minute; and $d$ is the diameter of the set-screw in inches.

*Example:* — How many ½-inch diameter set-screws would be required to transmit 3 horsepower at a shaft speed of 1000 revolutions per minute if the shaft diameter is 1 inch?

Using the first formula given above, the power transmitted by a single ½-inch diameter set-screw is determined: $P = [1 \times 1000 \times (½)^{2.3}] \div 50 = 4.1$ hp. Therefore a single ½-inch diameter set-screw is sufficient.

*Example:* — In the previous example, how many ⅜-inch diameter set-screws would be required? $P = [1 \times 1000 \times (⅜)^{2.3}] \div 50 = 2.1$ hp. Therefore two ⅜-inch diameter set-screws are required.

**ANSI Standard Sheet Metal, Self-Tapping and Metallic Drive Screws.** — Table 1 shows the various types of "self-tapping" screw threads covered by the ANSI Standard ANSI B18.6.4-1966, R1975. Designations of the American National Standards Institute, and the corresponding manufacturers' and federal designations are also shown as well as references to tables where recommended hole sizes for these threads are given. Types A, AB, B, BP and C when turned into a hole of proper size form a thread by a displacing action. Types D, F, G, T, BF and BT when turned into a hole of proper size form a thread by a cutting action. Type U when driven into a hole of proper size forms a series of multiple threads by a displacing action. These screws have the following descriptions and applications:

*Type A:* Spaced-thread screw with gimlet point primarily for use in light sheet metal, resin-impregnated plywood, and asbestos compositions. This type is no longer recommended. Use Type AB.

*Type AB:* Spaced-thread screw with same pitches as Type B but with gimlet point, primarily for similar uses as for Type A.

*Type B:* Spaced-thread screw with a blunt point with pitches generally somewhat finer than Type A. Used for light and heavy sheet metal, non-ferrous castings, plastics, resin-impregnated plywood, and asbestos compositions.

*Type BP:* Spaced-thread screw, the same as Type B but having a cone point. Used primarily in assemblies where holes are misaligned.

*Type C:* Screws having machine screw diameter-pitch combinations with threads approximately American National form and with blunt tapered points. Used where the use of a machine screw thread is preferable to the use of the spaced-thread types of thread forming screws. Also useful when chips from machine screw thread-cutting screws are objectionable. It should be recognized that in specific applications, this type of screw may require extreme driving torques due to long thread engagement or use in hard metals.

*Types F, G, D and T:* Thread-cutting screws with threads approximating machine screw threads, with blunt point, and with tapered entering threads having one or more cutting edges and chip cavities. The tapered threads of the Type F may be complete or incomplete at the producer's option; all other types have incomplete tapered threads. These screws can be used in materials such as aluminum, zinc, and lead die-castings; steel sheets and shapes; cast iron; brass, and plastics.

*Types BF and BT:* Thread-cutting screws with spaced threads as in Type B, with blunt points, and one or more cutting grooves. Used in plastics, die-castings, metal-clad and resin-impregnated plywoods, and asbestos.

*Type U:* Multiple-threaded drive screw with large helix angle, having a pilot, for use in metal and plastics. This screw is forced into the work by pressure and is intended for making permanent fastenings.

**ANSI Standard Head Types for Tapping and Metallic Drive Screws.** — Many of the head types used with "self-tapping" screw threads are similar to the head types of American National Standard machine screws which are shown on pages 1178 through 1186, and include the following:

*Round Head:* The round head has a semi-elliptical top and flat bearing surface. The pan head is now considered preferable.

*Flat Head:* The flat head has a flat top surface and a conical bearing surface with an included angle of approximately 82 degrees or, for another style, 100 degrees.

*Oval Head:* The oval head has a rounded top surface and a conical bearing surface with an included angle of approximately 82 degrees.

*Flat and Oval Trim Heads:* Flat and oval trim heads are similar to the 82-degree flat and oval heads except that the size of head for a given size screw is one or two sizes smaller than the regular flat and oval sizes, and for oval trim heads there is a controlled radius where the top surface meets the conical bearing surface.

*Undercut Flat and Oval Heads:* For short lengths of flat and oval head tapping screws, the heads are undercut to 70 per cent of normal side height to afford greater length of thread on the screws.

*Fillister Head:* The fillister head has a rounded top surface, cylindrical sides and a flat bearing surface.

*Truss Head:* The truss head has a low rounded top surface with a flat bearing surface. For a given screw size, the diameter of the truss head is larger than the diameter of the corresponding round head.

*Pan Head:* The slotted pan head has a flat top surface rounded into cylindrical sides and a flat bearing surface. The recessed pan head has a rounded top and flat bearing surface. This head type is now preferred to the round head.

*Hexagon Head:* The hexagon head has a flat or indented top surface with hexagonal sides and a flat bearing surface.

*Hexagon Washer Head:* The hexagon washer head has an indented top surface and hexagonal sides and a round flat washer bearing surface which projects beyond the hexagon and is formed integrally with the head.

All of the heads are provided with either a slot or one of three types of cross recesses with the exception of the round head which does not have the Type IA cross recess, the oval trim and flat trim heads which are provided with cross recesses only and the hexagon head and hexagon washer heads which may be slotted only if specified.

Table 1.  Type Designations of Sheet Metal Screws, Self-Tapping
Screws and Metallic Drive Screws

| TYPE | DESIGNATION | | | | TYPE | DESIGNATION | | | |
|---|---|---|---|---|---|---|---|---|---|
| | USA STD. | MFG. | FED'R'L. | TABLE NO.* | | USA STD. | MFG. | FED'R'L. | TABLE NO.* |
| | A | A | A | 10 | | G | G | CS ALT. #2 | 15 |
| | AB | AB | AB | 12,13, 14 | | T | 23 | CG | 15 |
| | B | B | B | 12,13, 14 | | BF | BF | BF | 16 |
| | BP | BP | BP | 12,13, 14 | | F | F | CF | 15 |
| | C | C | C | 11 | | BT | 25 | BG | 16 |
| | D | I | CS ALT. #1 | 15 | | U | U | U | 17 |

* Table number refers to table in which recommended hole sizes for a particular thread type is given.

Table 2.   ANSI Standard Threads and Points for Thread
Forming Self-Tapping Screws (ANSI B18.6.4-1966, R1975)

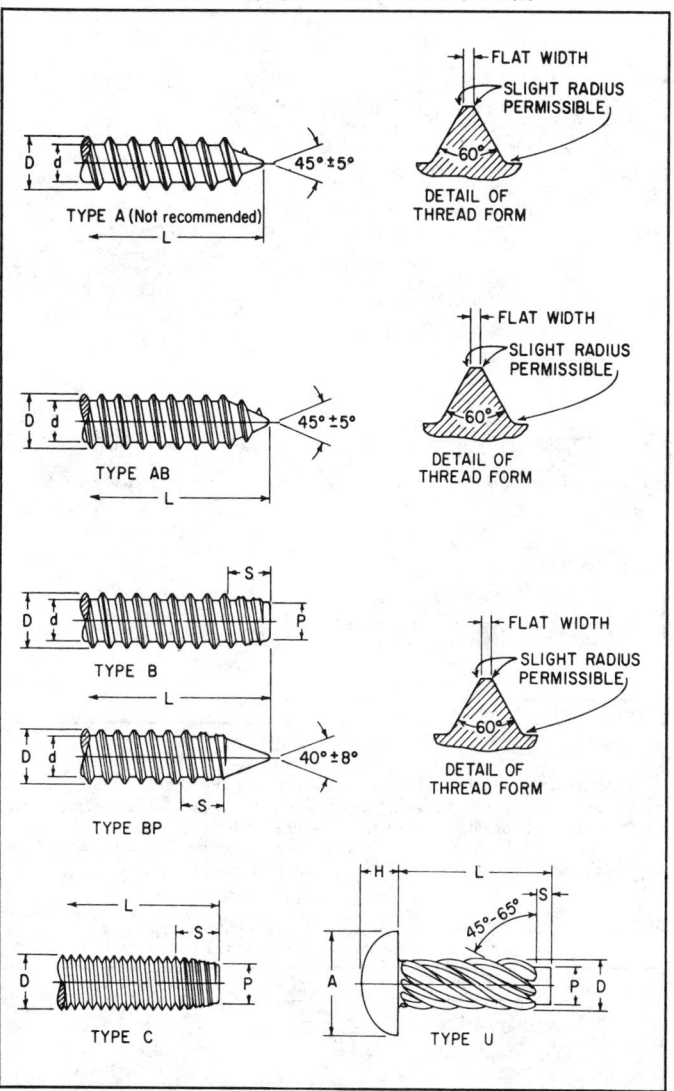

See Tables 5, 6 and 7 for thread data.

Table 3.   ANSI Standard Threads and Points for Thread
Cutting Self-Tapping Screws (ANSI B18.6.4-1966, R1975)

TYPE BF

FLAT WIDTH
SLIGHT RADIUS
PERMISSIBLE
60°
DETAIL OF
THREAD FORM

TYPE BT

TYPE D

TYPE F

TYPE G

TYPE T

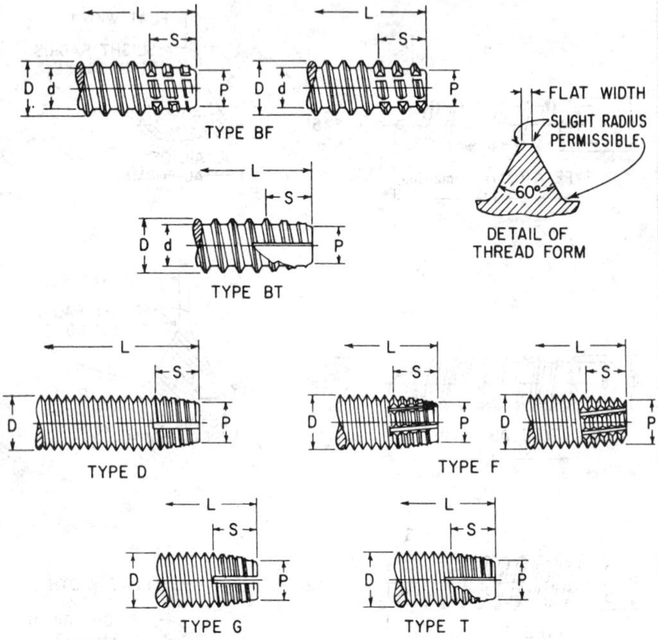

See Tables 8 and 9 for thread data.

**Cross Recesses.** — Type I cross recess has a large center opening, tapered wings, and blunt bottom, with all edges relieved or rounded. Type IA cross recess has a large center opening, wide straight wings, and blunt bottom, with all edges relieved or rounded. Type II consists of two intersecting slots with parallel sides converging to a slightly truncated apex at the bottom of the recess.

Table 4.   ANSI Standard Cross Recesses for Self-Tapping
Screws (ANSI B18.6.4-1966, R1975)

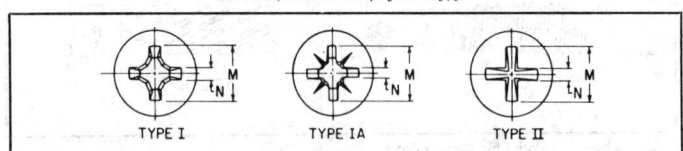

TYPE I      TYPE IA      TYPE II

### Table 5. ANSI Standard Thread and Point Dimensions for Types AB, A and U Thread Forming Tapping Screws (ANSI B18.6.4-1966, R1975)

#### Type AB (Formerly BA)

| Nominal Size or Basic Screw Diameter | Threads per inch | D Major Diameter | | d Minor Diameter | | L Minimum Practical Screw Lengths | |
|---|---|---|---|---|---|---|---|
| | | Max. | Min. | Max. | Min. | 90° Heads | Csk. Heads |
| 0  0.0600 | 48 | 0.060 | 0.057 | 0.036 | 0.033 | 3/32 | 7/64 |
| 1  0.0730 | 42 | 0.075 | 0.072 | 0.049 | 0.046 | 1/8 | 9/64 |
| **2  0.0860** | **32** | **0.088** | **0.084** | **0.064** | **0.060** | 9/64 | 11/64 |
| 3  0.0990 | 28 | 0.101 | 0.097 | 0.075 | 0.071 | 11/64 | 3/16 |
| **4  0.1120** | **24** | **0.114** | **0.110** | **0.086** | **0.082** | 3/16 | 7/32 |
| 5  0.1250 | 20 | 0.130 | 0.126 | 0.094 | 0.090 | 3/16 | 1/4 |
| **6  0.1380** | **20** | **0.139** | **0.135** | **0.104** | **0.099** | 7/32 | 17/64 |
| 7  0.1510 | 19 | 0.154 | 0.149 | 0.115 | 0.109 | 17/64 | 5/16 |
| **8  0.1640** | **18** | **0.166** | **0.161** | **0.122** | **0.116** | 9/32 | 21/64 |
| **10  0.1900** | **16** | **0.189** | **0.183** | **0.141** | **0.135** | 21/64 | 3/8 |
| 12  0.2160 | 14 | 0.215 | 0.209 | 0.164 | 0.157 | 3/8 | 13/32 |
| **1/4  0.2500** | **14** | **0.246** | **0.240** | **0.192** | **0.185** | 13/32 | 15/32 |
| 5/16  0.3125 | 12 | 0.315 | 0.308 | 0.244 | 0.236 | 17/32 | 19/32 |

#### Type A

| Nominal Size or Basic Screw Diameter | Threads per inch | D Major Diameter | | d Minor Diameter | | L† These Lengths or Shorter — Use Type AB | |
|---|---|---|---|---|---|---|---|
| | | Max. | Min. | Max. | Min. | 90° Heads | Csk. Heads |
| 0  0.0600 | 40 | 0.060 | 0.057 | 0.042 | 0.039 | 1/8 | 3/16 |
| 1  0.0730 | 32 | 0.075 | 0.072 | 0.051 | 0.048 | 1/8 | 3/16 |
| 2  0.0860 | 32 | 0.088 | 0.084 | 0.061 | 0.056 | 5/32 | 3/16 |
| 3  0.0990 | 28 | 0.101 | 0.097 | 0.076 | 0.071 | 3/16 | 7/32 |
| 4  0.1120 | 24 | 0.114 | 0.110 | 0.083 | 0.078 | 3/16 | 1/4 |
| 5  0.1250 | 20 | 0.130 | 0.126 | 0.095 | 0.090 | 3/16 | 1/4 |
| 6  0.1380 | 18 | 0.141 | 0.136 | 0.102 | 0.096 | 1/4 | 5/16 |
| 7  0.1510 | 16 | 0.158 | 0.152 | 0.114 | 0.108 | 5/16 | 3/8 |
| 8  0.1640 | 15 | 0.168 | 0.162 | 0.123 | 0.116 | 3/8 | 7/16 |
| 10  0.1900 | 12 | 0.194 | 0.188 | 0.133 | 0.126 | 3/8 | 1/2 |
| 12  0.2160 | 11 | 0.221 | 0.215 | 0.162 | 0.155 | 7/16 | 9/16 |
| 14  0.2420 | 10 | 0.254 | 0.248 | 0.185 | 0.178 | 1/2 | 5/8 |
| 16  0.2680 | 10 | 0.280 | 0.274 | 0.197 | 0.189 | 9/16 | 3/4 |
| 18  0.2940 | 9 | 0.306 | 0.300 | 0.217 | 0.209 | 5/8 | 13/16 |
| 20  0.3200 | 9 | 0.333 | 0.327 | 0.234 | 0.226 | 11/16 | 13/16 |
| 24  0.3720 | 9 | 0.390 | 0.383 | 0.291 | 0.282 | 3/4 | 1 |

#### Type U Metallic Drive Screws

| Nom. Size | No. of Starts | Out. Dia. | | Pilot Dia. | | Nom. Size | No. of Starts | Out. Dia. | | Pilot Dia. | |
|---|---|---|---|---|---|---|---|---|---|---|---|
| | | Max. | Min. | Max. | Min. | | | Max. | Min. | Max. | Min. |
| 00 | 6 | 0.060 | 0.057 | 0.049 | 0.046 | 8 | 8 | 0.167 | 0.162 | 0.136 | 0.132 |
| 0 | 6 | 0.075 | 0.072 | 0.063 | 0.060 | 10 | 8 | 0.182 | 0.177 | 0.150 | 0.146 |
| 2 | 8 | 0.100 | 0.097 | 0.083 | 0.080 | 12 | 8 | 0.212 | 0.206 | 0.177 | 0.173 |
| 4 | 7 | 0.116 | 0.112 | 0.096 | 0.092 | 14 | 9 | 0.242 | 0.236 | 0.202 | 0.198 |
| 6 | 7 | 0.140 | 0.136 | 0.116 | 0.112 | 5/16 | 11 | 0.315 | 0.309 | 0.272 | 0.267 |
| 7 | 8 | 0.154 | 0.150 | 0.126 | 0.122 | 3/8 | 12 | 0.378 | 0.371 | 0.334 | 0.329 |

All dimensions are given in inches.

Sizes shown in bold face type are preferred.  Type A screws no longer recommended.

† For screws of these nominal lengths and shorter use Type AB screws.

For Types A and AB screws no extrusion of excess metal beyond apex of the point resulting from thread rolling is permissible; a slight rounding or truncation of the point is desirable.

The width of flat at crest of thread shall not exceed 0.004 inch for sizes up to and including No. 8, and 0.006 inch for larger sizes of Types A and AB screws.

Table 6.  ANSI Standard Thread and Point Dimensions for Types B and BP Thread Forming Tapping Screws (ANSI B18.6.4-1966. R1975)

| Nominal Size[1] or Basic Screw Diameter | Thds. per inch | D Major Diameter Max. | D Major Diameter Min. | d Minor Diameter Max. | d Minor Diameter Min. | P[2] Point Diameter Max. | P[2] Point Diameter Min. | S[3,4] Point Taper Length — For Short Screws Max. | For Short Screws Min. | For Long Screws Max. | For Long Screws Min. | L — Type B Determinant Length for Point[3] 90° Heads | Csk. Heads | Type B Minimum Practical Screw Lengths 90° Heads | Csk. Heads | Type BP Determinant and Minimum Practical Lengths[4] 90° Heads | Csk. Head |
|---|---|---|---|---|---|---|---|---|---|---|---|---|---|---|---|---|---|
| 0  0.0600 | 48 | 0.060 | 0.057 | 0.036 | 0.033 | 0.031 | 0.027 | 0.042 | 0.031 | 0.052 | 0.042 | 5/64 | 1/8 | 5/64 | 3/32 | 5/32 | 13/64 |
| 1  0.0730 | 42 | 0.075 | 0.072 | 0.049 | 0.046 | 0.044 | 0.040 | 0.048 | 0.036 | 0.060 | 0.048 | 5/64 | 5/32 | 5/64 | 1/8 | 3/16 | 1/4 |
| 2  0.0860 | 32 | 0.088 | 0.084 | 0.064 | 0.060 | 0.058 | 0.054 | 0.062 | 0.047 | 0.078 | 0.062 | 7/64 | 3/16 | 7/64 | 5/32 | 13/64 | 9/32 |
| 3  0.0990 | 28 | 0.101 | 0.097 | 0.075 | 0.071 | 0.068 | 0.063 | 0.071 | 0.054 | 0.089 | 0.071 | 9/64 | 7/32 | 9/64 | 3/16 | 1/4 | 21/64 |
| 4  0.1120 | 24 | 0.114 | 0.110 | 0.086 | 0.082 | 0.079 | 0.074 | 0.083 | 0.063 | 0.104 | 0.083 | 3/16 | 1/4 | 9/64 | 3/16 | 5/16 | 3/8 |
| 5  0.1250 | 20 | 0.130 | 0.126 | 0.094 | 0.090 | 0.087 | 0.082 | 0.100 | 0.075 | 0.125 | 0.100 | 3/16 | 1/4 | 5/32 | 3/16 | 21/64 | 25/64 |
| 6  0.1380 | 20 | 0.139 | 0.135 | 0.104 | 0.099 | 0.095 | 0.089 | 0.100 | 0.075 | 0.125 | 0.100 | 1/4 | 5/16 | 11/64 | 1/4 | 25/64 | 15/32 |
| 7  0.1510 | 19 | 0.154 | 0.149 | 0.115 | 0.109 | 0.105 | 0.099 | 0.105 | 0.079 | 0.132 | 0.105 | 5/16 | 3/8 | 3/16 | 1/4 | 15/32 | 17/32 |
| 8  0.1640 | 18 | 0.166 | 0.161 | 0.122 | 0.116 | 0.112 | 0.106 | 0.111 | 0.083 | 0.139 | 0.111 | 5/16 | 7/16 | 3/16 | 1/4 | 31/64 | 39/64 |
| 10  0.1900 | 16 | 0.189 | 0.183 | 0.141 | 0.135 | 0.130 | 0.123 | 0.125 | 0.094 | 0.156 | 0.125 | 3/8 | 1/2 | 15/64 | 5/16 | 9/16 | 11/16 |
| 12  0.2160 | 14 | 0.215 | 0.209 | 0.164 | 0.157 | 0.152 | 0.145 | 0.143 | 0.107 | 0.179 | 0.143 | 7/16 | 9/16 | 9/32 | 3/8 | 21/32 | 25/32 |
| 1/4  0.2500 | 14 | 0.246 | 0.240 | 0.192 | 0.185 | 0.179 | 0.171 | 0.143 | 0.107 | 0.179 | 0.143 | 1/2 | 5/8 | 9/32 | 3/8 | 3/4 | 7/8 |
| 5/16  0.3125 | 12 | 0.315 | 0.308 | 0.244 | 0.236 | 0.230 | 0.222 | 0.167 | 0.125 | 0.208 | 0.167 | 1/2 | 5/8 | 5/16 | 7/16 | 53/64 | 61/64 |
| 3/8  0.3750 | 12 | 0.380 | 0.371 | 0.309 | 0.299 | 0.293 | 0.285 | 0.167 | 0.125 | 0.208 | 0.167 | 1/2 | 5/8 | 5/16 | 7/16 | 29/32 | 1 1/32 |
| 7/16  0.4375 | 10 | 0.440 | 0.431 | 0.359 | 0.349 | 0.343 | 0.335 | 0.200 | 0.150 | 0.250 | 0.200 | 5/8 | 3/4 | 15/32 | 5/8 | 1 7/64 | 1 5/64 |
| 1/2  0.5000 | 10 | 0.504 | 0.495 | 0.423 | 0.413 | 0.407 | 0.399 | 0.200 | 0.150 | 0.250 | 0.200 | 5/8 | 3/4 | 15/32 | 5/8 | 1 9/16 | 1 9/16 |

All dimensions are given in inches.  See Table 2 for thread diagrams.

Tapered threads shall have unfinished crests.

The width of flat at crest of thread shall not exceed 0.004 inch for sizes up to and including No. 8, and 0.006 inch for larger sizes.

[1] Where specifying nominal size in decimals, zeros in the fourth decimal place shall be omitted.

[2] The tabulated values apply to screw blanks before roll threading.

[3] Type B screws of these nominal lengths and shorter shall have point taper length specified above for short screws.   Longer lengths shall have point taper length specified for long screws.

[4] Type BP screws of these nominal lengths shall have point taper length specified above for short screws.   Longer lengths shall have point taper length specified for long screws.

Table 7. ANSI Standard Thread and Point Dimensions for Type C Thread Forming Tapping Screws (ANSI B18.6.4-1966, R1975)

| Nominal Size¹ or Basic Screw Diameter | | Threads per inch | D Major Diameter | | P² Point Diameter | | S³ Point Taper Length | | | | L Determinant Length for Point Taper³ | | L Minimum Practical Screw Lengths | |
|---|---|---|---|---|---|---|---|---|---|---|---|---|---|---|
| | | | | | | | For Short Screws | | For Long Screws | | | | | |
| | | | Max. | Min. | Max. | Min. | Max. | Min. | Max. | Min. | 90° Heads | Csk. Heads | 90° Heads | Csk. Heads |
| 2 | 0.0860 | 56 | 0.0860 | 0.0820 | 0.067 | 0.061 | 0.062 | 0.045 | 0.080 | 0.062 | 9/64 | 3/16 | 1/8 | 5/32 |
| 2 | 0.0860 | 64 | 0.0860 | 0.0822 | 0.070 | 0.064 | 0.055 | 0.039 | 0.070 | 0.055 | 1/8 | 11/64 | 1/8 | 5/32 |
| 3 | 0.0990 | 48 | 0.0990 | 0.0946 | 0.077 | 0.070 | 0.073 | 0.052 | 0.094 | 0.073 | 11/64 | 7/32 | 1/8 | 3/16 |
| 3 | 0.0990 | 56 | 0.0990 | 0.0950 | 0.080 | 0.074 | 0.062 | 0.045 | 0.080 | 0.062 | 9/64 | 3/16 | 1/8 | 3/16 |
| 4 | 0.1120 | 40 | 0.1120 | 0.1072 | 0.086 | 0.077 | 0.088 | 0.062 | 0.112 | 0.088 | 13/64 | 1/4 | 1/8 | 3/16 |
| 4 | 0.1120 | 48 | 0.1120 | 0.1076 | 0.090 | 0.083 | 0.073 | 0.052 | 0.094 | 0.073 | 13/64 | 7/32 | 1/8 | 3/16 |
| 5 | 0.1250 | 40 | 0.1250 | 0.1202 | 0.099 | 0.090 | 0.088 | 0.062 | 0.112 | 0.088 | 13/64 | 9/32 | 1/8 | 3/16 |
| 5 | 0.1250 | 44 | 0.1250 | 0.1204 | 0.101 | 0.093 | 0.080 | 0.057 | 0.102 | 0.080 | 3/16 | 1/4 | 1/8 | 3/16 |
| 6 | 0.1380 | 32 | 0.1380 | 0.1326 | 0.106 | 0.095 | 0.109 | 0.078 | 0.141 | 0.109 | 3/4 | 5/16 | 3/16 | 1/4 |
| 6 | 0.1380 | 40 | 0.1380 | 0.1332 | 0.112 | 0.103 | 0.088 | 0.062 | 0.112 | 0.088 | 13/64 | 17/64 | 3/16 | 1/4 |
| 8 | 0.1640 | 32 | 0.1640 | 0.1586 | 0.132 | 0.121 | 0.109 | 0.078 | 0.141 | 0.109 | 1/4 | 21/64 | 3/16 | 1/4 |
| 8 | 0.1640 | 36 | 0.1640 | 0.1590 | 0.135 | 0.125 | 0.097 | 0.069 | 0.125 | 0.097 | 9/64 | 19/64 | 3/16 | 1/4 |
| 10 | 0.1900 | 24 | 0.1900 | 0.1834 | 0.147 | 0.133 | 0.146 | 0.104 | 0.188 | 0.146 | 1/32 | 27/64 | 19/64 | 5/16 |
| 10 | 0.1900 | 32 | 0.1900 | 0.1846 | 0.158 | 0.147 | 0.109 | 0.078 | 0.141 | 0.109 | 1/4 | 11/32 | 19/64 | 5/16 |
| 12 | 0.2160 | 24 | 0.2160 | 0.2094 | 0.173 | 0.159 | 0.146 | 0.104 | 0.188 | 0.146 | 11/32 | 7/16 | 17/64 | 3/8 |
| 12 | 0.2160 | 28 | 0.2160 | 0.2098 | 0.179 | 0.167 | 0.125 | 0.089 | 0.161 | 0.125 | 9/64 | 25/64 | 17/64 | 3/8 |
| 1/4 | 0.2500 | 20 | 0.2500 | 0.2428 | 0.198 | 0.181 | 0.175 | 0.125 | 0.225 | 0.175 | 13/32 | 33/64 | 17/64 | 3/8 |
| 1/4 | 0.2500 | 28 | 0.2500 | 0.2438 | 0.213 | 0.201 | 0.125 | 0.089 | 0.161 | 0.125 | 9/64 | 13/32 | 17/64 | 3/8 |
| 5/16 | 0.3125 | 18 | 0.3125 | 0.3043 | 0.255 | 0.236 | 0.194 | 0.139 | 0.250 | 0.194 | 29/64 | 19/32 | 9/32 | 7/16 |
| 5/16 | 0.3125 | 24 | 0.3125 | 0.3059 | 0.269 | 0.255 | 0.146 | 0.104 | 0.188 | 0.146 | 11/32 | 13/32 | 9/32 | 7/16 |
| 3/8 | 0.3750 | 16 | 0.3750 | 0.3660 | 0.310 | 0.289 | 0.219 | 0.156 | 0.281 | 0.219 | 1/32 | 47/64 | 9/32 | 7/16 |
| 3/8 | 0.3750 | 24 | 0.3750 | 0.3684 | 0.332 | 0.318 | 0.146 | 0.104 | 0.188 | 0.146 | 1½/52 | 1/2 | 9/32 | 7/16 |

All dimensions in inches. See Table 2 for thread diagrams.
Tapered threads shall have unfinished crests.
¹ Where specifying nominal size in decimals, zeros in the fourth decimal place shall be omitted.
² The tabulated values apply to screw blanks before roll threading.
³ Screws of these nominal lengths and shorter shall have point taper length specified above for short screws. Longer lengths shall have point taper length specified for long screws. Lengths for 90 deg heads equal 8 times the pitch of the thread, rounded upward to nearest 1/64 inch and for countersunk heads equal 8 times the pitch of the thread plus the maximum head side height (undercut head style), rounded upward to the nearest 1/64 inch.

Table 8. ANSI Standard Thread and Point Dimensions for Types BF and BT* Thread Cutting Tapping Screws (ANSI B18.6.4-1966, R1975)

| Nominal Size[1] or Basic Screw Diameter | Threads per inch | D Major Diameter | | d Minor Diameter | | P[2] Point Diameter | | S[3] Point Taper Length For Short Screws | | For Long Screws | | Determinant Length for Point Taper[3] | | L Minimum Practical Screw Lengths | |
|---|---|---|---|---|---|---|---|---|---|---|---|---|---|---|---|
| | | Max. | Min. | Max. | Min. | Max. | Min. | Max. | Min. | Max. | Min. | 90° Heads | Csk. Heads | 90° Heads | Csk. Heads |
| 0  0.0600 | 48 | 0.060 | 0.057 | 0.036 | 0.033 | 0.031 | 0.027 | 0.042 | 0.031 | 0.052 | 0.042 | 5/64 | 1/8 | 5/64 | 3/32 |
| 1  0.0730 | 42 | 0.075 | 0.072 | 0.049 | 0.046 | 0.044 | 0.040 | 0.048 | 0.036 | 0.060 | 0.048 | 5/64 | 5/32 | 5/64 | 1/8 |
| 2  0.0860 | 32 | 0.088 | 0.084 | 0.064 | 0.060 | 0.058 | 0.054 | 0.062 | 0.047 | 0.078 | 0.062 | 7/64 | 3/16 | 7/64 | 5/32 |
| 3  0.0990 | 28 | 0.101 | 0.097 | 0.075 | 0.071 | 0.068 | 0.063 | 0.071 | 0.054 | 0.089 | 0.071 | 9/64 | 7/32 | 9/64 | 3/16 |
| 4  0.1120 | 24 | 0.114 | 0.110 | 0.086 | 0.082 | 0.079 | 0.074 | 0.083 | 0.063 | 0.104 | 0.083 | 3/16 | 1/4 | 9/64 | 3/16 |
| 5  0.1250 | 20 | 0.130 | 0.126 | 0.094 | 0.090 | 0.087 | 0.082 | 0.100 | 0.075 | 0.125 | 0.100 | 3/16 | 1/4 | 5/32 | 3/16 |
| 6  0.1380 | 20 | 0.139 | 0.135 | 0.104 | 0.099 | 0.095 | 0.089 | 0.100 | 0.075 | 0.125 | 0.100 | 1/4 | 5/16 | 11/64 | 1/4 |
| 7  0.1510 | 19 | 0.154 | 0.149 | 0.115 | 0.109 | 0.105 | 0.099 | 0.105 | 0.079 | 0.132 | 0.105 | 5/16 | 3/8 | 3/16 | 1/4 |
| 8  0.1640 | 18 | 0.166 | 0.161 | 0.122 | 0.116 | 0.112 | 0.106 | 0.111 | 0.083 | 0.139 | 0.111 | 5/16 | 7/16 | 3/16 | 1/4 |
| 10  0.1900 | 16 | 0.189 | 0.183 | 0.141 | 0.135 | 0.130 | 0.123 | 0.125 | 0.094 | 0.156 | 0.125 | 3/8 | 1/2 | 13/64 | 5/16 |
| 12  0.2160 | 14 | 0.215 | 0.209 | 0.164 | 0.157 | 0.152 | 0.145 | 0.143 | 0.107 | 0.179 | 0.143 | 7/16 | 9/16 | 9/32 | 3/8 |
| 1/4  0.2500 | 14 | 0.246 | 0.240 | 0.192 | 0.185 | 0.179 | 0.171 | 0.143 | 0.107 | 0.179 | 0.143 | 1/2 | 5/8 | 9/32 | 3/8 |
| 5/16  0.3125 | 12 | 0.315 | 0.308 | 0.244 | 0.236 | 0.230 | 0.222 | 0.167 | 0.125 | 0.208 | 0.167 | 1/2 | 5/8 | 5/16 | 7/16 |
| 3/8  0.3750 | 12 | 0.380 | 0.371 | 0.309 | 0.299 | 0.293 | 0.285 | 0.167 | 0.125 | 0.208 | 0.167 | 5/8 | 5/8 | 5/16 | 7/16 |
| 7/16  0.4375 | 10 | 0.440 | 0.431 | 0.359 | 0.349 | 0.343 | 0.335 | 0.200 | 0.150 | 0.250 | 0.200 | 5/8 | 3/4 | 15/32 | 5/8 |
| 1/2  0.5000 | 10 | 0.504 | 0.495 | 0.423 | 0.413 | 0.407 | 0.399 | 0.200 | 0.150 | 0.250 | 0.200 | 5/8 | 3/4 | 15/32 | 5/8 |

All dimensions are given in inches. See Table 3 for thread diagrams.

* Otherwise designated "Type 25".

Points of screws shall be tapered and fluted or slotted. Details of taper and flute design shall be optional with manufacturer provided the screws meet the performance requirements. Flutes or slots shall extend through first full form thread except for Type BF screws on which the length of flutes may be one pitch (thread) short of first full form thread.

The width of flat at crest of thread shall not exceed 0.004 inch for sizes up to and including No. 8, and 0.006 inch for larger sizes.

1 Where specifying nominal size in decimals, zeros in the fourth decimal place shall be omitted.

2 The tabulated values apply to screw blanks before roll threading.

3 Screws of these nominal lengths and shorter shall have point taper length specified above for short screws. Longer lengths shall have point taper length specified for long screws.

Table 9. ANSI Standard Thread and Point Dimensions for Types D*, F, and T† Thread Cutting Tapping Screws (ANSI B18.6.4-1966, R1975)

| Nominal Size¹ or Basic Screw Diameter | Threads per inch | D Major Diameter | | P² Point Diameter | | S³ Point Taper Length | | | | L Determinant Length for Point Taper³ | | L Minimum Practical Screw Lengths | |
|---|---|---|---|---|---|---|---|---|---|---|---|---|---|
| | | | | | | For Short Screws | | For Long Screws | | | | | |
| | | Max. | Min. | Max. | Min. | Max. | Min. | Max. | Min. | 90° Heads | Csk. Heads | 90° Heads | Csk. Heads |
| 2 0.0860 | 56 | 0.0860 | 0.0820 | 0.067 | 0.061 | 0.062 | 0.045 | 0.080 | 0.062 | 9/64 | 3/16 | 1/8 | 5/32 |
| 2 0.0860 | 64 | 0.0860 | 0.0822 | 0.070 | 0.064 | 0.055 | 0.039 | 0.070 | 0.055 | 1/8 | 11/64 | 1/8 | 5/32 |
| 3 0.0990 | 48 | 0.0990 | 0.0946 | 0.077 | 0.070 | 0.073 | 0.052 | 0.094 | 0.073 | 1/64 | 7/32 | 1/8 | 3/16 |
| 3 0.0990 | 56 | 0.0990 | 0.0950 | 0.080 | 0.074 | 0.062 | 0.045 | 0.080 | 0.062 | 9/64 | 3/16 | 1/8 | 3/16 |
| 4 0.1120 | 40 | 0.1120 | 0.1072 | 0.086 | 0.077 | 0.088 | 0.062 | 0.112 | 0.088 | 13/64 | 1/4 | 3/16 | 3/16 |
| 4 0.1120 | 48 | 0.1120 | 0.1076 | 0.090 | 0.083 | 0.073 | 0.052 | 0.094 | 0.073 | 13/64 | 7/32 | 1/8 | 3/16 |
| 5 0.1250 | 40 | 0.1250 | 0.1202 | 0.099 | 0.090 | 0.088 | 0.062 | 0.112 | 0.088 | 13/64 | 9/32 | 3/16 | 3/16 |
| 6 0.1380 | 44 | 0.1380 | 0.1326 | 0.106 | 0.093 | 0.080 | 0.057 | 0.102 | 0.080 | 3/16 | 1/4 | 3/16 | 3/16 |
| 6 0.1380 | 32 | 0.1380 | 0.1332 | 0.112 | 0.103 | 0.088 | 0.062 | 0.112 | 0.088 | 9/64 | 5/16 | 3/16 | 1/4 |
| 8 0.1640 | 32 | 0.1640 | 0.1586 | 0.132 | 0.121 | 0.109 | 0.078 | 0.141 | 0.109 | 1/4 | 17/64 | 3/16 | 1/4 |
| 8 0.1640 | 36 | 0.1640 | 0.1590 | 0.135 | 0.125 | 0.097 | 0.069 | 0.125 | 0.097 | 15/64 | 21/64 | 3/16 | 1/4 |
| 10 0.1900 | 24 | 0.1900 | 0.1834 | 0.147 | 0.133 | 0.146 | 0.104 | 0.188 | 0.146 | 11/32 | 19/64 | 15/64 | 5/16 |
| 10 0.1900 | 32 | 0.1900 | 0.1846 | 0.158 | 0.147 | 0.109 | 0.078 | 0.141 | 0.109 | 1/4 | 27/64 | 15/64 | 5/16 |
| 12 0.2160 | 24 | 0.2160 | 0.2094 | 0.173 | 0.159 | 0.146 | 0.104 | 0.188 | 0.146 | 11/32 | 11/32 | 1/4 | 3/8 |
| 12 0.2160 | 28 | 0.2160 | 0.2098 | 0.179 | 0.167 | 0.125 | 0.089 | 0.161 | 0.125 | 19/64 | 25/64 | 1/4 | 3/8 |
| 1/4 0.2500 | 20 | 0.2500 | 0.2428 | 0.198 | 0.181 | 0.175 | 0.125 | 0.225 | 0.175 | 1/4 | 33/64 | 17/64 | 3/8 |
| 1/4 0.2500 | 28 | 0.2500 | 0.2438 | 0.213 | 0.201 | 0.125 | 0.089 | 0.161 | 0.125 | 19/64 | 13/32 | 17/64 | 3/8 |
| 5/16 0.3125 | 18 | 0.3125 | 0.3043 | 0.255 | 0.236 | 0.194 | 0.139 | 0.250 | 0.194 | 29/64 | 19/32 | 9/32 | 7/16 |
| 5/16 0.3125 | 24 | 0.3125 | 0.3059 | 0.269 | 0.255 | 0.146 | 0.104 | 0.188 | 0.146 | 11/32 | 15/32 | 9/32 | 7/16 |
| 3/8 0.3750 | 16 | 0.3750 | 0.3660 | 0.310 | 0.289 | 0.219 | 0.156 | 0.281 | 0.219 | 1/2 | 47/64 | 9/32 | 7/16 |
| 3/8 0.3750 | 24 | 0.3750 | 0.3684 | 0.332 | 0.318 | 0.146 | 0.104 | 0.188 | 0.146 | 17/32 | 1/2 | 9/32 | 7/16 |

All dimensions are given in inches. See Table 3 for thread diagrams.
Points of screws shall be tapered and fluted or slotted. Details of taper and flute design shall be optional with manufacturer provided screws meet the performance requirements. Flutes or slots shall extend through first full form thread except for Type F screws on which the length of flutes may be one pitch (thread) short of first full form thread.

¹ The tabulated values apply to screw blanks before roll threading.
² Where specifying nominal size in decimals, zeros in the fourth decimal place shall be omitted.
³ Screws of these nominal lengths and shorter shall have point taper length specified above for short screws. Longer lengths shall have point taper length specified for long screws. Lengths for 90 deg heads equal 8 times the pitch of the thread, rounded upward to nearest 1/64 inch and for countersunk heads equal 8 times the pitch of the thread plus the maximum head side height (undercut head style), rounded upward to the nearest 1/64 inch.

* Otherwise designated "Type I".   † Otherwise designated "Type 23".

**Table 10. Approximate Hole Sizes for Type A Steel Thread Forming Screws\***

In Steel, Stainless Steel, Monel Metal, Brass, and Aluminum Sheet Metal

| Screw Size | Metal Thickness | Hole Required Pierced or Extruded | Hole Required Drilled or Clean Punched | Drill Size | Screw Size | Metal Thickness | Hole Required Pierced or Extruded | Hole Required Drilled or Clean Punched | Drill Size |
|---|---|---|---|---|---|---|---|---|---|
| 4 | .015 | .... | .086 | 44 | 8 | .024 | .136 | .125 | ⅛ |
| | .018 | .... | .086 | 44 | | .030 | .136 | .125 | ⅛ |
| | .024 | .098 | .094 | 42 | | .036 | .136 | .125 | ⅛ |
| | .030 | .098 | .094 | 42 | | .048 | .136 | .128 | 30 |
| | .036 | .098 | .098 | 40 | 10 | .018 | .... | .136 | 29 |
| 6 | .015 | .... | .104 | 37 | | .024 | .157 | .136 | 29 |
| | .018 | .... | .104 | 37 | | .030 | .157 | .136 | 29 |
| | .024 | .111 | .104 | 37 | | .036 | .157 | .136 | 29 |
| | .030 | .111 | .104 | 37 | | .048 | .157 | .149 | 25 |
| | .036 | .111 | .106 | 36 | 12 | .024 | .... | .161 | 20 |
| 7 | .015 | .... | .116 | 32 | | .030 | .185 | .161 | 20 |
| | .018 | .... | .116 | 32 | | .036 | .185 | .161 | 20 |
| | .024 | .120 | .116 | 32 | | .048 | .185 | .161 | 20 |
| | .030 | .120 | .116 | 32 | 14 | .024 | .... | .185 | 13 |
| | .036 | .120 | .116 | 32 | | .030 | .209 | .189 | 12 |
| | .048 | .120 | .120 | 31 | | .036 | .209 | .191 | 11 |
| 8 | .018 | .... | .125 | ⅛ | | .048 | .209 | .196 | 9 |

| In Plywood (Resin Impregnated) | | | | | | In Asbestos Compositions | | | | | |
|---|---|---|---|---|---|---|---|---|---|---|---|
| Screw Size | Hole Required | Drill Size | Min. Mat'l Thickness | Penetration in Blind Holes Min. | Max. | Screw Size | Hole Required | Drill Size | Min. Mat'l Thickness | Penetration in Blind Holes Min. | Max. |
| 4 | .098 | 40 | 3/16 | 1/4 | 3/4 | 4 | .093 | 42 | 3/16 | 1/4 | 3/4 |
| 6 | .110 | 35 | 3/16 | 1/4 | 3/4 | 6 | .106 | 36 | 3/16 | 1/4 | 3/4 |
| 7 | .128 | 30 | 1/4 | 5/16 | 3/4 | 7 | .125 | ⅛ | 1/4 | 5/16 | 3/4 |
| 8 | .140 | 28 | 1/4 | 5/16 | 3/4 | 8 | .136 | 29 | 1/4 | 5/16 | 3/4 |
| 10 | .169 | 18 | 5/16 | 3/8 | 1 | 10 | .161 | 20 | 5/16 | 3/8 | 1 |
| 12 | .189 | 12 | 5/16 | 3/8 | 1 | 12 | .185 | 13 | 5/16 | 3/8 | 1 |
| 14 | .228 | 1 | 7/16 | 1/2 | 1 | 14 | .213 | 3 | 7/16 | 1/2 | 1 |

See footnote at bottom of Table 11. Type A is not recommended, use Type AB.

**Table 11. Approximate Hole Sizes for Type C Steel Thread Forming Screws\***

In Sheet Steel

| Screw Size | Metal Thickness | Hole Required | Drill Size | Screw Size | Metal Thickness | Hole Required | Drill Size | Screw Size | Metal Thickness | Hole Required | Drill Size |
|---|---|---|---|---|---|---|---|---|---|---|---|
| 4-40 | .037 | .093 | 42 | 10-24 | .037 | .154 | 23 | 1/4-20 | .037 | .221 | 2 |
| | .048 | .093 | 42 | | .048 | .161 | 20 | | .048 | .221 | 2 |
| | .062 | .096 | 41 | | .062 | .166 | 19 | | .062 | .228 | 1 |
| | .075 | .0995 | 39 | | .075 | .1695 | 18 | | .075 | .234 | A |
| | .105 | .101 | 38 | | .105 | .173 | 17 | | .105 | .234 | A |
| | .134 | .101 | 38 | | .134 | .177 | 16 | | .134 | .236 | 6mm |
| 6-32 | .037 | .113 | 33 | 10-32 | .037 | .1695 | 18 | 1/4-28 | .037 | .224 | 5.7mm |
| | .048 | .116 | 32 | | .048 | .1695 | 18 | | .048 | .228 | 1 |
| | .062 | .116 | 32 | | .062 | .1695 | 18 | | .062 | .232 | 5.9mm |
| | .075 | .122 | 3.1mm | | .075 | .173 | 17 | | .075 | .234 | A |
| | .105 | .125 | ⅛ | | .105 | .177 | 16 | | .105 | .238 | B |
| | .134 | .125 | ⅛ | | .134 | .177 | 16 | | .134 | .238 | B |
| 8-32 | .037 | .136 | 29 | 12-24 | .037 | .189 | 12 | 5/16-18 | .037 | .290 | L |
| | .048 | .144 | 27 | | .048 | .1935 | 10 | | .048 | .290 | L |
| | .062 | .144 | 27 | | .062 | .1935 | 10 | | .062 | .290 | L |
| | .075 | .147 | 26 | | .075 | .199 | 8 | | .075 | .295 | M |
| | .105 | .1495 | 25 | | .105 | .199 | 8 | | .105 | .295 | M |
| | .134 | .1495 | 25 | | .134 | .199 | 8 | | .134 | .295 | M |

All dimensions in inches except drill sizes.   \* Since conditions differ widely, it may be necessary to vary the hole size to suit a particular application.

**Table 12.  Approximate Pierced or Extruded Hole Sizes for Types AB, B and BP Steel Thread Forming Screws***

| Screw Size | Metal Thickness | Pierced or Extruded Hole Required | Screw Size | Metal Thickness | Pierced or Extruded Hole Required | Screw Size | Metal Thickness | Pierced or Extruded Hole Required |
|---|---|---|---|---|---|---|---|---|
| In Steel, Stainless Steel, Monel Metal, and Brass Sheet Metal | | | | | | | | |
| 4 | .015 | .086 | 7 | .024 | .120 | 10 | .030 | .157 |
| | .018 | .086 | | .030 | .120 | | .036 | .157 |
| | .024 | .098 | | .036 | .120 | | .048 | .157 |
| | .030 | .098 | | .048 | .120 | 12 | .024 | .185 |
| | .036 | .098 | 8 | .018 | .136 | | .030 | .185 |
| 6 | .015 | .111 | | .024 | .136 | | .036 | .185 |
| | .018 | .111 | | .030 | .136 | | .048 | .185 |
| | .024 | .111 | | .036 | .136 | 1/4 | .030 | .209 |
| | .030 | .111 | | .048 | .136 | | .036 | .209 |
| | .036 | .111 | 10 | .018 | .157 | | .048 | .209 |
| 7 | .018 | .120 | | .024 | .157 | .. | .... | .... |
| In Aluminum Alloy Sheet Metal | | | | | | | | |
| 4 | .024 | .086 | 6 | .048 | .111 | 8 | .036 | .136 |
| | .030 | .086 | 7 | .024 | .120 | | .048 | .136 |
| | .036 | .086 | | .030 | .120 | 10 | .024 | .157 |
| | .048 | .086 | | .036 | .120 | | .030 | .157 |
| 6 | .024 | .111 | | .048 | .120 | | .036 | .157 |
| | .030 | .111 | 8 | .024 | .136 | | .048 | .157 |
| | .036 | .111 | | .030 | .136 | .. | .... | .... |

All dimensions are given in inches except whole number screw sizes.
See footnotes at bottom of page.

**Table 13.  Approximate Drilled Hole Sizes for Types AB, B and BP Steel Thread Forming Screws***

| Screw Size | Hole Required | Drill Size | Min. Mat'l Thickness | Penetration in Blind Holes Min. | Max. | Screw Size | Hole Required | Drill Size | Min. Mat'l Thickness | Penetration in Blind Holes Min. | Max. |
|---|---|---|---|---|---|---|---|---|---|---|---|
| In Plywood (Resin Impregnated) | | | | | | In Asbestos Compositions | | | | | |
| 2 | .073 | 49 | 1/8 | 3/16 | 1/2 | 2 | .076 | 48 | 1/8 | 3/16 | 1/2 |
| 4 | .099 | 39 | 3/16 | 1/4 | 5/8 | 4 | .101 | 38 | 3/16 | 1/4 | 5/8 |
| 6 | .125 | 1/8 | 3/16 | 1/4 | 5/8 | 6 | .120 | 31 | 3/16 | 1/4 | 5/8 |
| 7 | .136 | 29 | 3/16 | 1/4 | 3/4 | 7 | .136 | 29 | 1/4 | 5/16 | 3/4 |
| 8 | .144 | 27 | 3/16 | 1/4 | 3/4 | 8 | .147 | 26 | 5/16 | 3/8 | 3/4 |
| 10 | .173 | 17 | 1/4 | 5/16 | 1 | 10 | .166 | 19 | 5/16 | 3/8 | 1 |
| 12 | .193 | 10 | 5/16 | 3/8 | 1 | 12 | .196 | 9 | 5/16 | 3/8 | 1 |
| 1/4 | .228 | 1 | 5/16 | 3/8 | 1 | 1/4 | .228 | 1 | 7/16 | 1/2 | 1 |
| In Aluminum, Magnesium, Zinc, Brass, and Bronze Castings † | | | | | | In Phenol Formaldehyde Plastics † | | | | | |
| 2 | .078 | 47 | ... | 1/8 | ... | 2 | .078 | 47 | ... | ... | ... |
| 4 | .104 | 37 | ... | 3/16 | ... | 4 | .099 | 39 | ... | ... | ... |
| 6 | .128 | 30 | ... | 1/4 | ... | 6 | .128 | 30 | ... | ... | ... |
| 7 | .144 | 27 | ... | 1/4 | ... | 7 | .136 | 29 | ... | ... | ... |
| 8 | .152 | 24 | ... | 1/4 | ... | 8 | .149 | 25 | ... | ... | ... |
| 10 | .177 | 16 | ... | 1/4 | ... | 10 | .177 | 16 | ... | ... | ... |
| 12 | .199 | 8 | ... | 9/32 | ... | 12 | .199 | 8 | ... | ... | ... |
| 1/4 | .234 | 15/64 | ... | 5/16 | ... | 1/4 | .234 | 15/64 | ... | ... | ... |
| In Cellulose Acetate and Nitrate, and Acrylic and Styrene Resins † | | | | | | | | | | | |
| 2 | .078 | 47 | ... | 3/16 | ... | 8 | .144 | 27 | ... | 5/16 | ... |
| 4 | .093 | 42 | ... | 1/4 | ... | 10 | .169 | 18 | ... | 5/16 | ... |
| 6 | .120 | 31 | ... | 1/4 | ... | 12 | .191 | 11 | ... | 3/8 | ... |
| 7 | .128 | 30 | ... | 1/4 | ... | 1/4 | .221 | 2 | ... | 3/8 | ... |

All dimensions are given in inches except whole number screw and drill sizes.
* Since conditions differ widely, it may be necessary to vary the hole size to suit a particular application.  † Data below apply to Types B and BP only.

#### Table 14. Approximate Drilled or Clean-Punched Hole Sizes for Types AB, B and BP Steel Thread Forming Screws*

| Screw Size | Metal Thickness | Hole Required | Drill Size | Screw Size | Metal Thickness | Hole Required | Drill Size | Screw Size | Metal Thickness | Hole Required | Drill Size |
|---|---|---|---|---|---|---|---|---|---|---|---|
| In Steel, Stainless Steel, Monel Metal, and Brass Sheet Metal |||||||||||| 
| | .015 | .063 | 52 | | .018 | .116 | 32 | 10 | .125 | .169 | 18 |
| | .018 | .063 | 52 | | .024 | .116 | 32 | | .135 | .169 | 18 |
| | .024 | .067 | 51 | | .030 | .116 | 32 | | .164 | .173 | 17 |
| 2 | .030 | .070 | 50 | 7 | .036 | .116 | 32 | | .024 | .166 | 19 |
| | .036 | .073 | 49 | | .048 | .120 | 31 | | .030 | .166 | 19 |
| | .048 | .073 | 49 | | .060 | .128 | 30 | | .036 | .166 | 19 |
| | .060 | .076 | 48 | | .075 | .136 | 29 | | .048 | .169 | 18 |
| | .015 | .086 | 44 | | .105 | .140 | 28 | 12 | .060 | .177 | 16 |
| | .018 | .086 | 44 | | .024 | .125 | ⅛ | | .075 | .182 | 14 |
| | .024 | .089 | 43 | | .030 | .125 | ⅛ | | .105 | .185 | 13 |
| 4 | .030 | .093 | 42 | | .036 | .125 | ⅛ | | .125 | .196 | 9 |
| | .036 | .093 | 42 | 8 | .048 | .128 | 30 | | .135 | .196 | 9 |
| | .048 | .096 | 41 | | .060 | .136 | 29 | | .164 | .201 | 7 |
| | .060 | .099 | 39 | | .075 | .140 | 28 | | .030 | .194 | 10 |
| | .075 | .101 | 38 | | .105 | .149 | 25 | | .036 | .194 | 10 |
| | .015 | .104 | 37 | | .125 | .149 | 25 | | .048 | .194 | 10 |
| | .018 | .104 | 37 | | .135 | .152 | 24 | | .060 | .199 | 8 |
| | .024 | .106 | 36 | | .024 | .144 | 27 | ¼ | .075 | .204 | 6 |
| | .030 | .106 | 36 | | .030 | .144 | 27 | | .105 | .209 | 4 |
| 6 | .036 | .110 | 35 | | .036 | .147 | 26 | | .125 | .228 | 1 |
| | .048 | .111 | 34 | 10 | .048 | .152 | 24 | | .135 | .228 | 1 |
| | .060 | .116 | 32 | | .060 | .152 | 24 | | .164 | .234 | 15/64 |
| | .075 | .120 | 31 | | .075 | .157 | 22 | | .187 | .234 | 15/64 |
| | .105 | .128 | 30 | | .105 | .161 | 20 | | .194 | .234 | 15/64 |
| In Aluminum Alloy Sheet Metal |||||||||||| 
| | .024 | .063 | 52 | | .060 | .120 | 31 | | .164 | .159 | 21 |
| | .030 | .063 | 52 | | .075 | .128 | 30 | 10 | .200 to .375 | .166 | 19 |
| 2 | .036 | .063 | 52 | 7 | .105 | .136 | 29 | | .048 | .161 | 20 |
| | .048 | .067 | 51 | | .128 to .250 | .136 | 29 | | .060 | .166 | 19 |
| | .060 | .070 | 50 | | | | | | .075 | .173 | 17 |
| | .030 | .086 | 44 | | | | | | .105 | .180 | 15 |
| | .036 | .086 | 44 | | .030 | .116 | 32 | 12 | .125 | .182 | 14 |
| 4 | .048 | .086 | 44 | | .036 | .120 | 31 | | .135 | .182 | 14 |
| | .060 | .089 | 43 | | .048 | .128 | 30 | | .164 | .189 | 12 |
| | .075 | .089 | 43 | | .060 | .136 | 29 | | .200 to .375 | .196 | 9 |
| | .105 | .093 | 42 | 8 | .075 | .140 | 28 | | .060 | .199 | 8 |
| | .030 | .104 | 37 | | .105 | .147 | 26 | | .075 | .201 | 7 |
| | .036 | .104 | 37 | | .125 | .147 | 26 | | .105 | .204 | 6 |
| | .048 | .104 | 37 | | .135 | .149 | 25 | | .125 | .209 | 4 |
| 6 | .060 | .106 | 36 | | .162 to .375 | .152 | 24 | ¼ | .135 | .209 | 4 |
| | .075 | .110 | 35 | | | | | | .164 | .213 | 3 |
| | .105 | .111 | 34 | | .036 | .144 | 27 | | .187 | .213 | 3 |
| | .128 to .250 | .120 | 31 | | .048 | .144 | 27 | | .194 | .221 | 2 |
| | | | | | .060 | .144 | 27 | | .200 to .375 | .228 | 1 |
| | .030 | .113 | 33 | 10 | .075 | .147 | 26 | | | | |
| 7 | .036 | .113 | 33 | | .105 | .147 | 26 | | | | |
| | .048 | .116 | 32 | | .125 | .154 | 23 | | | | |
| | | | | | .135 | .154 | 23 | | | | |

All dimensions are given in inches except whole number screw and drill sizes.
* Since conditions differ widely, it may be necessary to vary the hole size to suit a particular application. Hole sizes for metal thicknesses above .075 inch are for Types B and BP only.

Table 15. Approximate Hole Sizes for Types D, F, G, and T Steel Thread Cutting Screws*

| Screw Size | Stock Thickness | | | | | | | | | | |
|---|---|---|---|---|---|---|---|---|---|---|---|
| | .050 | .060 | .083 | .109 | .125 | .140 | 3/16 | 1/4 | 5/16 | 3/8 | 1/2 |
| | Hole Sizes† in Steel | | | | | | | | | | |
| 2-56 | .0730 | .0730 | .0730 | .0730 | .0760 | .0760 | ..... | ..... | ..... | ..... | ..... |
| 3-48 | .0810 | .0810 | .0820 | .0860 | .0860 | .0860 | .0890 | ..... | ..... | ..... | ..... |
| 4-40 | .0890 | .0890 | .0935 | .0960 | .0980 | .0980 | .1015 | ..... | ..... | ..... | ..... |
| 5-40 | .1060 | .1060 | .1060 | .1065 | .1094 | .1100 | .1160 | .1160 | ..... | ..... | ..... |
| 6-32 | .1100 | .1130 | .1160 | .1160 | .1160 | .1200 | .1250 | .1250 | ..... | ..... | ..... |
| 8-32 | .1360 | .1405 | .1405 | .1440 | .1440 | .1470 | .1495 | .1495 | .1495 | ..... | ..... |
| 10-24 | .1520 | .1540 | .1610 | .1610 | .1660 | .1695 | .1730 | .1730 | .1730 | .1730 | ..... |
| 10-32 | .1590 | .1660 | .1660 | .1695 | .1695 | .1695 | .1770 | .1770 | .1770 | .1770 | ..... |
| 12-24 | ..... | .1800 | .1820 | .1875 | .1910 | .1910 | .1990 | .1990 | .1990 | .1990 | .1990 |
| 1/4-20 | ..... | ..... | .2130 | .2188 | .2210 | .2210 | .2280 | .2280 | .2280 | .2280 | .2280 |
| 1/4-28 | ..... | ..... | .2210 | .2280 | .2280 | .2340 | .2344 | .2344 | .2344 | .2344 | .2344 |
| 5/16-18 | ..... | ..... | ..... | .2770 | .2770 | .2813 | .2900 | .2900 | .2900 | .2900 | .2900 |
| 5/16-24 | ..... | ..... | ..... | .2900 | .2900 | .2900 | .2950 | .2950 | .2950 | .2950 | .2950 |
| 3/8-16 | ..... | ..... | ..... | ..... | .3390 | .3390 | .3480 | .3580 | .3580 | .3580 | .3580 |
| 3/8-24 | ..... | ..... | ..... | ..... | .3480 | .3480 | .3580 | .3580 | .3580 | .3580 | .3580 |
| | Hole Sizes† in Aluminum | | | | | | | | | | |
| 2-56 | .0700 | .0730 | .0730 | .0730 | .0730 | .0730 | ..... | ..... | ..... | ..... | ..... |
| 3-48 | .0781 | .0810 | .0820 | .0820 | .0820 | .0860 | .0860 | ..... | ..... | ..... | ..... |
| 4-40 | .0890 | .0890 | .0890 | .0935 | .0935 | .0938 | .0980 | ..... | ..... | ..... | ..... |
| 5-40 | .1015 | .1015 | .1040 | .1040 | .1065 | .1065 | .1100 | .1130 | ..... | ..... | ..... |
| 6-32 | .1094 | .1094 | .1110 | .1130 | .1160 | .1160 | .1200 | .1250 | ..... | ..... | ..... |
| 8-32 | .1360 | .1360 | .1360 | .1405 | .1405 | .1440 | .1470 | .1495 | .1495 | ..... | ..... |
| 10-24 | .1495 | .1520 | .1540 | .1570 | .1590 | .1610 | .1660 | .1719 | .1730 | .1730 | ..... |
| 10-32 | .1610 | .1610 | .1610 | .1660 | .1660 | .1660 | .1719 | .1770 | .1770 | .1770 | ..... |
| 12-24 | ..... | .1770 | .1800 | .1820 | .1850 | .1875 | .1910 | .1990 | .1990 | .1990 | .1990 |
| 1/4-20 | ..... | ..... | .2055 | .2090 | .2130 | .2130 | .2210 | .2280 | .2280 | .2280 | .2280 |
| 1/4-28 | ..... | ..... | .2188 | .2210 | .2210 | .2210 | .2280 | .2344 | .2344 | .2344 | .2344 |
| 5/16-18 | ..... | ..... | ..... | .2660 | .2720 | .2720 | .2810 | .2900 | .2900 | .2900 | .2900 |
| 5/16-24 | ..... | ..... | ..... | .2810 | .2812 | .2812 | .2900 | .2950 | .2950 | .2950 | .2950 |
| 3/8-16 | ..... | ..... | ..... | ..... | .3281 | .3320 | .3390 | .3480 | .3480 | .3480 | .3480 |
| 3/8-24 | ..... | ..... | ..... | ..... | .3438 | .3438 | .3480 | .3580 | .3580 | .3580 | .3580 |
| | Hole Sizes† in Cast Iron | | | | | | | | | | |
| 2-56 | .0760 | .0760 | .0760 | .0781 | .0781 | .0781 | ..... | ..... | ..... | ..... | ..... |
| 3-48 | .0890 | .0890 | .0890 | .0890 | .0890 | .0935 | .0935 | ..... | ..... | ..... | ..... |
| 4-40 | .0995 | .0995 | .1015 | .1015 | .1015 | .1015 | .1040 | ..... | ..... | ..... | ..... |
| 5-40 | .1110 | .1110 | .1130 | .1130 | .1160 | .1160 | .1160 | .1160 | ..... | ..... | ..... |
| 6-32 | .1200 | .1200 | .1250 | .1250 | .1250 | .1250 | .1285 | .1285 | ..... | ..... | ..... |
| 8-32 | .1470 | .1495 | .1495 | .1495 | .1495 | .1495 | .1540 | .1540 | .1540 | ..... | ..... |
| 10-24 | .1695 | .1695 | .1719 | .1730 | .1730 | .1730 | .1770 | .1770 | .1770 | .1770 | ..... |
| 10-32 | .1730 | .1730 | .1770 | .1770 | .1770 | .1770 | .1800 | .1800 | .1800 | .1800 | ..... |
| 12-24 | ..... | .1960 | .1990 | .1990 | .1990 | .1990 | .2031 | .2040 | .2040 | .2040 | .2040 |
| 1/4-20 | ..... | ..... | .2280 | .2280 | .2280 | .2280 | .2344 | .2344 | .2344 | .2344 | .2344 |
| 1/4-28 | ..... | ..... | .2340 | .2344 | .2344 | .2344 | .2380 | .2380 | .2380 | .2380 | .2380 |
| 5/16-18 | ..... | ..... | ..... | .2900 | .2900 | .2900 | .2950 | .2950 | .2950 | .2950 | .2950 |
| 5/16-24 | ..... | ..... | ..... | .2950 | .2950 | .2950 | 3020 | .3020 | .3020 | .3020 | .3020 |
| 3/8-16 | ..... | ..... | ..... | ..... | .3480 | .3480 | .3480 | .3480 | .3480 | .3480 | .3480 |
| 3/8-24 | ..... | ..... | ..... | ..... | .3580 | .3580 | .3580 | .3580 | .3580 | .3580 | .3580 |

All dimensions are given in inches except the whole number screw sizes.
* Since conditions differ widely, it may be necessary to vary the hole size to suit a particular application.
† Hole sizes listed are standard drill sizes.

### Table 15 (Continued). Approximate Hole Sizes for Types D, F, G, and T Steel Thread Cutting Screws*

| Screw Size | Stock Thickness | | | | | | | | | | |
|---|---|---|---|---|---|---|---|---|---|---|---|
| | .050 | .060 | .083 | .109 | .125 | .140 | 3/16 | 1/4 | 5/16 | 3/8 | 1/2 |
| | Hole Sizes† in Zinc and Aluminum Die Castings | | | | | | | | | | |
| 2-56 | .0730 | .0730 | .0760 | .0760 | .0760 | .0760 | ..... | ..... | ..... | ..... | ..... |
| 3-48 | .0820 | .0820 | .0820 | .0860 | .0890 | .0890 | .0890 | ..... | ..... | ..... | ..... |
| 4-40 | .0960 | .0960 | .0960 | .0960 | .0995 | .0995 | .0995 | ..... | ..... | ..... | ..... |
| 5-40 | .1060 | .1060 | .1060 | .1100 | .1100 | .1100 | .1110 | .1130 | ..... | ..... | ..... |
| 6-32 | .1160 | .1200 | .1200 | .1200 | .1200 | .1200 | .1200 | .1200 | .1200 | ..... | ..... |
| 8-32 | .1440 | .1440 | .1440 | .1440 | .1470 | .1470 | .1470 | .1495 | .1495 | ..... | ..... |
| 10-24 | .1610 | .1660 | .1660 | .1660 | .1660 | .1660 | .1660 | .1695 | .1695 | .1719 | ..... |
| 10-32 | .1695 | .1695 | .1719 | .1719 | .1719 | .1719 | .1719 | .1730 | .1730 | .1770 | ..... |
| 12-24 | ..... | .1800 | .1910 | .1910 | .1910 | .1935 | .1935 | .1960 | .1960 | .1990 | .1990 |
| 1/4-20 | ..... | ..... | .2188 | .2210 | .2210 | .2210 | .2210 | .2280 | .2280 | .2280 | .2280 |
| 1/4-28 | ..... | ..... | .2280 | .2280 | .2280 | .2280 | .2280 | .2340 | .2340 | .2344 | .2344 |
| 5/16-18 | ..... | ..... | ..... | .2770 | .2810 | .2810 | .2812 | .2812 | .2900 | .2900 | .2900 |
| 5/16-24 | ..... | ..... | ..... | .2900 | .2900 | .2900 | .2900 | .2900 | .2950 | .2950 | .2950 |
| 3/8-16 | ..... | ..... | ..... | ..... | .3390 | .3390 | .3390 | .3438 | .3438 | .3480 | .3480 |
| 3/8-24 | ..... | ..... | ..... | ..... | .3480 | .3480 | .3480 | .3580 | .3580 | .3580 | .3580 |

| In Phenol Formaldehyde Plastics | | | | | In Acrylic and Other Resins | | | | |
|---|---|---|---|---|---|---|---|---|---|
| Screw Size | Hole Required | Drill Size | Depth of Penetration | | Screw Size | Hole Required | Drill Size | Depth of Penetration | |
| | | | Min. | Max. | | | | Min. | Max. |
| 2-56 | .0781 | 5/64 | 7/32 | 3/8 | 2-56 | .076 | 48 | 7/32 | 3/8 |
| 3-48 | .089 | 43 | 7/32 | 3/8 | 3-48 | .086 | 44 | 7/32 | 3/8 |
| 4-40 | .098 | 40 | 1/4 | 5/16 | 4-40 | .093 | 42 | 1/4 | 5/16 |
| 5-40 | .113 | 33 | 1/4 | 7/16 | 5-40 | .110 | 35 | 1/4 | 7/16 |
| 6-32 | .116 | 32 | 1/4 | 7/16 | 6-32 | .116 | 32 | 1/4 | 1/2 |
| 8-32 | .144 | 27 | 5/16 | 1/2 | 8-32 | .144 | 27 | 5/16 | 1/2 |
| 10-24 | .161 | 20 | 3/8 | 1/2 | 10-24 | .161 | 20 | 3/8 | 1/2 |
| 10-32 | .166 | 19 | 3/8 | 1/2 | 10-32 | .166 | 19 | 3/8 | 1/2 |
| 1/4-20 | .228 | 1 | 3/8 | 5/8 | 1/4-20 | .228 | 1 | 3/8 | 1 |

For footnotes see bottom of table on previous page.

### Table 16. Approximate Hole Sizes for Types BF and BT Steel Thread Cutting Screws*

| Stock Thickness | Screw Size | | | | | | | | | | |
|---|---|---|---|---|---|---|---|---|---|---|---|
| | 2-32 | 3-28 | 4-24 | 5-20 | 6-20 | 8-18 | 10-16 | 12-14 | 1/4-14 | 5/16-12 | 3/8-12 |
| | Hole Sizes† in Zinc and Aluminum Die Castings | | | | | | | | | | |
| .060 | .0730 | .0860 | ..... | ..... | ..... | ..... | ..... | ..... | ..... | ..... | ..... |
| .083 | .0730 | .0860 | ..... | ..... | ..... | ..... | ..... | ..... | ..... | ..... | ..... |
| .109 | .0760 | .0860 | .0980 | .1110 | ..... | ..... | ..... | ..... | ..... | ..... | ..... |
| .125 | .0760 | .0860 | .0995 | .1110 | .1200 | .1490 | .1660 | .1910 | .2210 | .2810 | .344 |
| .140 | .0760 | .0890 | .0995 | .1130 | .1200 | .1490 | .1660 | .1910 | .2210 | .2810 | .344 |
| 3/16 | ..... | .0890 | .0995 | .1130 | .1200 | .1490 | .1660 | .1910 | .2210 | .2810 | .344 |
| 1/4 | ..... | ..... | .1015 | .1160 | .1250 | .1520 | .1695 | .1960 | .2280 | .2810 | .344 |
| 5/16 | ..... | ..... | ..... | ..... | .1250 | .1520 | .1719 | .1960 | .2280 | .2900 | .348 |
| 3/8 | ..... | ..... | ..... | ..... | ..... | ..... | .1719 | .1960 | .2280 | .2900 | .348 |

| In Phenol Formaldehyde Plastics | | | | | In Acrylic and Other Resins | | | | |
|---|---|---|---|---|---|---|---|---|---|
| Screw Size | Hole Required | Drill Size | Depth of Penetration | | Screw Size | Hole Required | Drill Size | Depth of Penetration | |
| | | | Min. | Max. | | | | Min. | Max. |
| 2-32 | .0781 | 5/64 | 3/32 | 1/4 | 2-32 | .076 | 48 | 3/32 | 1/4 |
| 3-28 | .089 | 43 | 1/8 | 5/16 | 3-28 | .089 | 43 | 1/8 | 5/16 |
| 4-24 | .104 | 37 | 1/8 | 5/16 | 4-24 | .0995 | 39 | 1/8 | 5/16 |
| 5-20 | .116 | 32 | 3/16 | 3/8 | 5-20 | .113 | 33 | 3/16 | 3/8 |
| 6-20 | .125 | 1/8 | 3/16 | 3/8 | 6-20 | .120 | 31 | 3/16 | 3/8 |
| 8-18 | .147 | 26 | 1/4 | 1/2 | 8-18 | .144 | 27 | 1/4 | 1/2 |
| 10-16 | .1695 | 18 | 5/16 | 5/8 | 10-16 | .166 | 19 | 5/16 | 5/8 |
| 12-14 | .1935 | 10 | 3/8 | 5/8 | 12-14 | .189 | 12 | 3/8 | 5/8 |
| 1/4-14 | .228 | 1 | 3/8 | 3/4 | 1/4-14 | .221 | 2 | 3/8 | 3/4 |

For footnotes see bottom of table on previous page.

Table 17. Approximate Hole Sizes for Type U Hardened Steel Metallic Drive Screws

| In Ferrous and Non-Ferrous Castings, Sheet Metals, Plastics, Plywood (Resin-Impregnated) and Fiber | | | | | |
|---|---|---|---|---|---|
| Screw Size | Hole Size | Drill Size | Screw Size | Hole Size | Drill Size |
| 00 | .052 | 55 | 8 | .144 | 27 |
| 0 | .067 | 51 | 10 | .161 | 20 |
| 2 | .086 | 44 | 12 | .191 | 11 |
| 4 | .104 | 37 | 14 | .221 | 2 |
| 6 | .120 | 31 | 5⁄16 | .295 | M |
| 7 | .136 | 29 | 3⁄8 | .358 | T |

All dimensions are given in inches except whole number screw and drill sizes and letter drill sizes.

**Self-tapping Thread Inserts.** — Self-tapping screw thread inserts are essentially hard bushings with internal and external threads. The internal threads conform to Unified and American standard classes 2B and 3B, depending on the type of insert used. The external thread has cutting edges on the end that provide the self-tapping feature. These inserts may be used in magnesium, aluminum, cast iron, zinc, plastics, and other materials. Self-tapping inserts are made of case-hardened carbon steel, stainless steel, and brass, the brass type being designed specifically for installation in wood.

**Screw Thread Inserts.** — Screw thread inserts are helically formed coils of diamond-shaped stainless steel or phosphor bronze wire that screw into a threaded hole to form a mating internal thread for a screw or stud. These inserts provide a convenient means of repairing stripped-out threads and are also used to provide stronger threads in soft materials such as aluminum, zinc die castings, wood, magnesium, etc. than can be obtained by direct tapping of the base metal involved.

According to the Heli-Coil Corp., conventional design practice in specifying boss diameters or edge distances can usually be applied since the major diameter of a

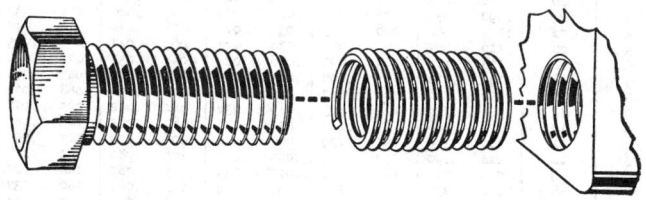

hole tapped to receive a thread insert is not much larger than the major diameter of thread the insert provides. For example, the major diameter of the tapped hole to receive a ¼-28 cap screw is 0.2904 inch while a conventional tapped hole for a ¼-28 cap screw has a major diameter of 0.250 inch.

Screw thread inserts are available in thread sizes from 4-40 to 1½-6 inch National and Unified Coarse Thread Series and in 6-40 to 1½-12 sizes in the fine-thread series. When used in conjunction with appropriate taps and gages, screw thread inserts will meet requirements of 2, 2B, 3 and 3B thread classes.

## American National Standard Flat, Round, and Oval Head Wood Screws
(ANSI B18.6.1-1972, R1977 and Appendix)

| Nominal Size | Threads per Inch | D* Basic Diam. of Screw | J Width of Slot | | A Head Diameter Max., Sharp Edge | A Head Diameter Min., Edge Rounded or Flat | B Head Diameter Max. | B Head Diameter Min. | H Height of Head† Max. | H Height of Head† Min. |
|---|---|---|---|---|---|---|---|---|---|---|
| | | | Max. | Min. | | | | | | |
| 0 | 32 | .060 | .023 | .016 | .119 | .099 | .113 | .099 | .035 | .026 |
| 1 | 28 | .073 | .026 | .019 | .146 | .123 | .138 | .122 | .043 | .033 |
| 2 | 26 | .086 | .031 | .023 | .172 | .147 | .162 | .146 | .051 | .040 |
| 3 | 24 | .099 | .035 | .027 | .199 | .171 | .187 | .169 | .059 | .048 |
| 4 | 22 | .112 | .039 | .031 | .225 | .195 | .211 | .193 | .067 | .055 |
| 5 | 20 | .125 | .043 | .035 | .252 | .220 | .236 | .217 | .075 | .062 |
| 6 | 18 | .138 | .048 | .039 | .279 | .244 | .260 | .240 | .083 | .069 |
| 7 | 16 | .151 | .048 | .039 | .305 | .268 | .285 | .264 | .091 | .076 |
| 8 | 15 | .164 | .054 | .045 | .332 | .292 | .309 | .287 | .100 | .084 |
| 9 | 14 | .177 | .054 | .045 | .358 | .316 | .334 | .311 | .108 | .091 |
| 10 | 13 | .190 | .060 | .050 | .385 | .340 | .359 | .334 | .116 | .098 |
| 12 | 11 | .216 | .067 | .056 | .438 | .389 | .408 | .382 | .132 | .112 |
| 14 | 10 | .242 | .075 | .064 | .491 | .437 | .457 | .429 | .148 | .127 |
| 16 | 9 | .268 | .075 | .064 | .544 | .485 | .506 | .476 | .164 | .141 |
| 18 | 8 | .294 | .084 | .072 | .597 | .534 | .555 | .523 | .180 | .155 |
| 20 | 8 | .320 | .084 | .072 | .650 | .582 | .604 | .570 | .196 | .170 |
| 24 | 7 | .372 | .094 | .081 | .756 | .679 | .702 | .664 | .228 | .198 |

| Nominal Size | Threads per Inch | O Total Height of Head Max. | O Min. | K Height of Head Max. | K Min. | T Depth of Slot Max. | T Min. | U Depth of Slot Max. | U Min. | V Depth of Slot Max. | V Min. |
|---|---|---|---|---|---|---|---|---|---|---|---|
| 0 | 32 | .056 | .041 | .053 | .043 | .015 | .010 | .039 | .029 | .030 | .025 |
| 1 | 28 | .068 | .052 | .061 | .051 | .019 | .012 | .044 | .033 | .038 | .031 |
| 2 | 26 | .080 | .063 | .069 | .059 | .023 | .015 | .048 | .037 | .045 | .037 |
| 3 | 24 | .092 | .073 | .078 | .067 | .027 | .017 | .053 | .040 | .052 | .043 |
| 4 | 22 | .104 | .084 | .086 | .075 | .030 | .020 | .058 | .044 | .059 | .049 |
| 5 | 20 | .116 | .095 | .095 | .083 | .034 | .022 | .063 | .047 | .067 | .055 |
| 6 | 18 | .128 | .105 | .103 | .091 | .038 | .024 | .068 | .051 | .074 | .060 |
| 7 | 16 | .140 | .116 | .111 | .099 | .041 | .027 | .072 | .055 | .081 | .066 |
| 8 | 15 | .152 | .126 | .120 | .107 | .045 | .029 | .077 | .058 | .088 | .072 |
| 9 | 14 | .164 | .137 | .128 | .115 | .049 | .032 | .082 | .062 | .095 | .078 |
| 10 | 13 | .176 | .148 | .137 | .123 | .053 | .034 | .087 | .065 | .103 | .084 |
| 12 | 11 | .200 | .169 | .153 | .139 | .060 | .039 | .096 | .073 | .117 | .096 |
| 14 | 10 | .224 | .190 | .170 | .155 | .068 | .044 | .106 | .080 | .132 | .108 |
| 16 | 9 | .248 | .212 | .187 | .171 | .075 | .049 | .115 | .087 | .146 | .120 |
| 18 | 8 | .272 | .233 | .204 | .187 | .083 | .054 | .125 | .094 | .160 | .132 |
| 20 | 8 | .296 | .254 | .220 | .203 | .090 | .059 | .134 | .101 | .175 | .144 |
| 24 | 7 | .344 | .297 | .254 | .235 | .105 | .069 | .154 | .116 | .204 | .168 |

All dimensions in inches. The edges of flat and oval head screws may be flat or rounded. Wood screws are also available with Types I, IA and II recessed heads. Consult the standard for recessed head dimensions.

* *Diameter Tolerance:* Equals +.004 in. and −.007 in.  *Fillet Under Head:* Has a maximum radius equal to .031 in. for sizes 0 to 4, incl.; .062 in. for sizes 5 to 12, incl.; and .093 in. for sizes 14 to 24, incl. of flat and oval head screws and .016 in. for sizes 0 to 4, incl.; .031 in. for sizes 5 to 12, incl.; and .046 in. for sizes 14 to 24, incl. of round head screws. *Length of Thread:* Approximately equal to ⅔ the screw length. *Threads per Inch:* Maximum permissible variation is ±10 per cent of tabulated number.  † For reference only.

## Standard Wire Nails and Spikes

(Size, Length and Approximate Number to **Pound**)

| Size of Nail | Length, Inches | Common Wire Nails and Brads | | Flooring Brads | | Fence Nails | | Casing, Smooth and Barbed Box | | Finishing Nails | |
|---|---|---|---|---|---|---|---|---|---|---|---|
| | | Gage | No. to Lb. | Gage | No. to Lb. | Gage | No. to Lb. | Gage | No. to Lb. | Gage | No. to Lb. |
| 2 d | 1 | 15 | 876 | ..... | .... | ..... | .... | 15½ | 1010 | 16½ | 1351 |
| 3 d | 1¼ | 14 | 568 | ..... | .... | ..... | .... | 14½ | 635 | 15½ | 807 |
| 4 d | 1½ | 12½ | 316 | ..... | .... | ..... | .... | 14 | 473 | 15 | 584 |
| 5 d | 1¾ | 12½ | 271 | ..... | .... | 10 | 142 | 14 | 406 | 15 | 500 |
| 6 d | 2 | 11½ | 181 | 11 | 157 | 10 | 124 | 12½ | 236 | 13½ | 309 |
| 7 d | 2¼ | 11½ | 161 | 11 | 139 | 9 | 92 | 12½ | 210 | 13 | 238 |
| 8 d | 2½ | 10¼ | 106 | 10 | 99 | 9 | 82 | 11½ | 145 | 12½ | 189 |
| 9 d | 2¾ | 10¼ | 96 | 10 | 90 | 8 | 62 | 11½ | 132 | 12½ | 172 |
| 10 d | 3 | 9 | 69 | 9 | 69 | 7 | 50 | 10½ | 94 | 11½ | 121 |
| 12 d | 3¼ | 9 | 64 | 8 | 54 | 6 | 40 | 10½ | 87 | 11½ | 113 |
| 16 d | 3½ | 8 | 49 | 7 | 43 | 5 | 30 | 10 | 71 | 11 | 90 |
| 20 d | 4 | 6 | 31 | 6 | 31 | 4 | 23 | 9 | 52 | 10 | 62 |
| 30 d | 4½ | 5 | 24 | ..... | .... | ..... | .... | 9 | 46 | ..... | .... |
| 40 d | 5 | 4 | 18 | ..... | .... | ..... | .... | 8 | 35 | ..... | .... |
| 50 d | 5½ | 3 | 16 | ..... | .... | ..... | .... | ..... | .... | ..... | .... |
| 60 d | 6 | 2 | 11 | ..... | .... | ..... | .... | ..... | .... | ..... | .... |

| Size and Length | | Hinge Nails, Heavy | | Hinge Nails, Light | | Clinch Nails | | Barbed Car Nails, Heavy | | Barbed Car Nails, Light | |
|---|---|---|---|---|---|---|---|---|---|---|---|
| 2 d | 1 | ..... | .... | ..... | ..... | 14 | 710 | ..... | .... | ..... | .... |
| 3 d | 1¼ | ..... | .... | ..... | ..... | 13 | 429 | ..... | .... | ..... | .... |
| 4 d | 1½ | 3 | 50 | 6 | 82 | 12 | 274 | 10 | 165 | 12 | 274 |
| 5 d | 1¾ | 3 | 38 | 6 | 62 | 12 | 235 | 9 | 118 | 10 | 142 |
| 6 d | 2 | 3 | 30 | 6 | 50 | 11 | 157 | 9 | 103 | 10 | 124 |
| 7 d | 2¼ | 00 | 12 | 3 | 25 | 11 | 139 | 8 | 76 | 9 | 92 |
| 8 d | 2½ | 00 | 11 | 3 | 23 | 10 | 99 | 8 | 69 | 9 | 82 |
| 9 d | 2¾ | 00 | 10 | 3 | 22 | 10 | 90 | 7 | 54 | 8 | 62 |
| 10 d | 3 | 00 | 9 | 3 | 19 | 9 | 69 | 7 | 50 | 8 | 57 |
| 12 d | 3¼ | ..... | .... | ..... | ..... | 9 | 62 | 6 | 42 | 7 | 50 |
| 16 d | 3½ | ..... | .... | ..... | ..... | 8 | 49 | 6 | 35 | 7 | 43 |
| 20 d | 4 | ..... | .... | ..... | ..... | 7 | 37 | 5 | 26 | 6 | 31 |
| 30 d | 4½ | ..... | .... | ..... | ..... | ..... | .... | 5 | 24 | 6 | 28 |
| 40 d | 5 | ..... | .... | ..... | ..... | ..... | .... | 4 | 18 | 5 | 21 |
| 50 d | 5½ | ..... | .... | ..... | ..... | ..... | .... | 3 | 15 | 4 | 17 |
| 60 d | 6 | ..... | .... | ..... | ..... | ..... | .... | 3 | 13 | 4 | 15 |

| Size and Length | | Boat Nails, Heavy | | Boat Nails, Light | | Slating Nails | |
|---|---|---|---|---|---|---|---|
| 2 d | 1 | ..... | .... | ..... | .... | 12 | 411 |
| 3 d | 1¼ | ..... | .... | ..... | .... | 10½ | 225 |
| 4 d | 1½ | ¼ | 44 | 3/16 | 82 | 10½ | 187 |
| 5 d | 1¾ | ..... | .... | 3/16 | .... | 10 | 142 |
| 6 d | 2 | ¼ | 32 | 3/16 | 62 | 9 | 103 |
| 7 d | 2¼ | ..... | .... | ..... | .... | ..... | .... |
| 8 d | 2½ | ¼ | 26 | 3/16 | 50 | ..... | .... |
| 9 d | 2¾ | ..... | .... | ..... | .... | ..... | .... |
| 10 d | 3 | 3/8 | 14 | ¼ | 22 | ..... | .... |
| 12 d | 3¼ | 3/8 | 13 | ¼ | 20 | ..... | .... |
| 16 d | 3½ | 3/8 | 12 | ¼ | 18 | ..... | .... |
| 20 d | 4 | 3/8 | 10 | ¼ | 16 | ..... | .... |
| 30 d | 4½ | ..... | .... | ..... | .... | ..... | .... |
| 40 d | 5 | ..... | .... | ..... | .... | ..... | .... |
| 50 d | 5½ | ..... | .... | ..... | .... | ..... | .... |
| 60 d | 6 | ..... | .... | ..... | .... | ..... | .... |

Spikes

| Size and Length | | Gage | No. to Lb. |
|---|---|---|---|
| 10 d | 3 | 6 | 41 |
| 12 d | 3¼ | 6 | 38 |
| 16 d | 3½ | 5 | 30 |
| 20 d | 4 | 4 | 23 |
| 30 d | 4½ | 3 | 17 |
| 40 d | 5 | 2 | 13 |
| 50 d | 5½ | 1 | 10 |
| 60 d | 6 | 1 | 8 |
| ..... | | 7 | 0 | 7 |
| ..... | | 8 | 00 | 6 |
| ..... | | 9 | 00 | 5 |
| ..... | | 10 | 3/8 | 4 |
| ..... | | 12 | 3/8 | 3 |

## RIVETS AND RIVETED JOINTS

**Classes and Types of Riveted Joints.** — Riveted joints may be classified by application as: (1) pressure vessel; (2) structural; and (3) machine member. For information and data concerning joints for pressure vessels such as boilers, reference should be made to standard sources such as the ASME Boiler Code. The following sections will cover only structural and machine-member riveted joints.

Basically there are two kinds of riveted joints, the *lap-joint* and the *butt-joint*. In the ordinary *lap-joint* the plates overlap each other and are held together by one or more rows of rivets. In the *butt-joint* the plates being joined are in the same plane and are joined by means of a cover plate or butt strap which is riveted to both plates by one or more rows of rivets. The term *single riveting* means one row of rivets in a lap-joint or one row on each side of a butt-joint; *double riveting* means two rows of rivets in a lap-joint or two rows on each side of the joint in butt riveting. Joints are also triple and quadruple riveted. Lap-joints may also be made with inside or outside cover plates. Types of lap and butt joints are illustrated in the table on pages 1242 and 1243.

**General Considerations of a Riveted Joint.** — Factors to be considered in the design or specification of a riveted joint are: type of joint, spacing of rivets, type and size of rivet, type and size of hole, and rivet material.

*Spacing of Rivets:* The spacing between rivet centers is called *pitch* and between row center lines, *back pitch* or *transverse pitch*. The distance between centers of rivets nearest each other in adjacent rows is called *diagonal pitch*. The distance from the edge of the plate to the center line of the nearest row of rivets is called *margin*.

Examination of a riveted joint made up of several rows of rivets will reveal that after progressing along the joint a given distance the rivet pattern or arrangement is repeated. (For a butt joint, the length of a *repeating section* is usually equal to the *long pitch* or pitch of the rivets in the outer row, i.e. the row farthest from the edge of the joint.) For structural and machine-member joints the proper pitch may be determined by making the tensile strength of the plate over the length of the repeating section, i.e. distance between rivets in the outer row, equal to the total shear strength of the rivets in the repeating section. Minimum pitch and diagonal pitch are also governed by the clearance required for the hold-on (Dolly bar) and rivet set. Dimensions for different sizes of hold-ons and rivet sets are given in the table on page 1248.

When fastening thin plate it is particularly important to maintain accurate spacing to avoid buckling.

*Size and Type of Rivets:* The rivet diameter $d$ commonly falls between $d = 1.2\sqrt{t}$ and $d = 1.4\sqrt{t}$, where $t$ is the thickness of the plate. Dimensions for various types of American Standard large (½-inch diameter and up) rivets and small solid rivets are shown in tables which follow. It may be noted that countersunk heads are not as strong as other types.

*Size and Type of Hole:* Rivet holes may be punched, punched and reamed, or drilled. Rivet holes are usually made $\frac{1}{16}$ inch larger in diameter than the nominal diameter of the rivet although in some classes of work in which the rivet is driven cold, as in automatic machine riveting, the holes are reamed to provide minimum clearance so that the rivet fills the hole completely.

When holes are punched in heavy steel plate, there may be considerable loss of strength unless the holes are reamed to remove the inferior metal immediately surrounding them. This results in the diameter of the punched hole being increased by from $\frac{1}{16}$ to $\frac{1}{8}$ inch. Annealing after punching tends to restore the strength of the plate in the vicinity of the holes.

*Rivet Material:* Rivets for structural and machine member purposes are usually made of wrought iron or soft steel but for aircraft and other applications where light weight or resistance to corrosion is important, copper, aluminum alloy, Monel, Inconel, etc., may be used as rivet material.

**Failure of Riveted Joints.** — Rivets may fail by:

1. Shearing through one cross-section (single shear)
2. Shearing through two cross-sections (double shear)
3. Crushing

Plates may fail by:

4. Shearing along two parallel lines extending from opposite sides of the rivet hole to the edge of the plate
5. Tearing along a single line from middle of rivet hole to edge of plate
6. Crushing
7. Tearing between adjacent rivets (tensile failure) in the same row or in adjacent rows.

Types 4 and 5 failures are caused by rivets being placed too close to the edge of the plate. These types of failure are avoided by placing the center of the rivet at a minimum of one and one-half times the rivet diameter away from the edge.

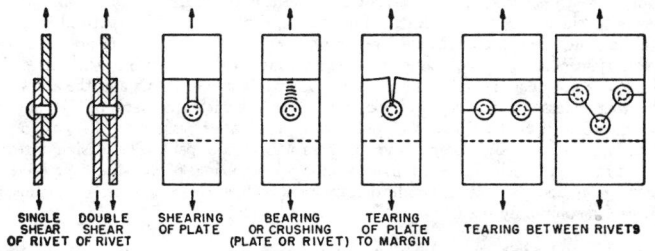

SINGLE SHEAR OF RIVET  DOUBLE SHEAR OF RIVET  SHEARING OF PLATE  BEARING OR CRUSHING (PLATE OR RIVET)  TEARING OF PLATE TO MARGIN  TEARING BETWEEN RIVETS

Types of Rivet and Plate Failure

Failure due to tearing on a diagonal between rivets in adjacent rows when the pitch is four times the rivet diameter or less is avoided by making the transverse pitch one and three-quarters times the rivet diameter.

**Theoretical versus Actual Riveted Joint Failure.** — If it is assumed that the rivets are placed the suggested distance from the edge of the plate and each row the suggested distance from another row, then the failure of a joint is most likely to occur as a result of shear failure of the rivets, bearing failure (crushing) of the plate or rivets, or tensile failure of the plate, alone or in combination depending on the make-up of the joints.

Joint failure in actuality is more complex than this. Rivets do not undergo pure shear especially in lap-joints where rivets are subjected to single shear. The rivet, in this instance, would be subject to a combination of tensile and shearing stresses and it would fail because of combined stresses, not a single stress. Furthermore, the shearing stress is usually considered to be distributed evenly over the cross-section which is also not the case.

Rivets that are usually driven hot contract upon cooling. This contraction in the length of the rivet draws the plates together and sets up a stress in the rivet estimated to be equal in magnitude to the yield point of the rivet steel. The contraction in the diameter of the rivet results in a little clearance between the rivet and the hole in the plate. The tightness in the plates caused by the contraction in length of the rivet gives rise to a condition where quite a sizeable frictional force would have to be overcome before the plates would slip over one another and subject the rivets to a shearing force. It is European practice to design joints for resistance to this slipping. It has been found, however, that the strength-basis designs obtained in American and English practice are not very different from European designs.

**Design of Riveted Joints.** — In the design of riveted joints a simplified treatment is frequently used in which the following assumptions are made:

1. The load is carried equally by the rivets.
2. No combined stresses act on a rivet to cause failure.
3. The shearing stress in a rivet is uniform across the cross-section under question.
4. The load that would cause failure in single shear would have to be doubled to cause failure in double shear.
5. The bearing stress of rivet and plate is distributed equally over the projected area of the rivet.
6. The tensile stress is uniform in the section of metal between the rivets.

**Allowable Stresses.** — The design stresses for riveted joints are usually set by codes, practices, or specifications. The American Institute of Steel Construction issues specifications for the design, fabrication and erection of structural steel for buildings in which the allowable stress permitted in tension for structural steel and rivets is specified at 20,000 pounds per square inch, the allowable bearing stress for rivets is 40,000 psi in double shear and 32,000 psi in single shear, and the allowable shearing stress for rivets is 15,000 psi. The American Society of Mechanical Engineers in its Boiler Code lists the following ultimate stresses: tensile, 55,000 psi; shearing, 44,000 psi; compressive or bearing, 95,000 psi. The design stresses usually used are ⅕th of these: tensile, 11,000 psi; shearing, 8800 psi; compressive or bearing, 19,000 psi. In machine design work values close to these or somewhat lower are commonly used.

**Analysis of Joint Strength.** — The following examples and strength analyses of riveted joints are based upon the six assumptions previously outlined for a simplified treatment.

*Example 1.* Consider a 12-inch section of single-riveted lap-joint made up with plates of ¼-inch thickness and six rivets, ⅝ inch in diameter. Assume that rivet holes are 1/16 inch larger in diameter than the rivets. In this joint, the entire load is transmitted from one plate to the other by means of the rivets. Each plate and the six rivets carries the entire load. The safe tensile load $L$ and the efficiency $\eta$ may be determined in the following way: Design stresses of 8500 psi for shear, 20,000 psi for bearing, and 10,000 psi for tension are arbitrarily assigned and it is assumed that the rivets will not tear or shear through the plate to the edge of the joint.

(a) The safe tensile load $L$ based on single shear of the rivets is equal to the number of rivets $n$ times the cross-sectional area of one rivet $A_r$ times the allowable shearing stress $S_s$ or

$$L = n \times A_r \times S_s$$

$$L = 6 \times \frac{\pi}{4}(0.625)^2 \times 8500$$

$$L = 15,647 \text{ pounds}$$

(b) The safe tensile load $L$ based on bearing stress is equal to the number of rivets $n$ times the projected bearing area of the rivet $A_b$ (diameter times thickness of plate) times the allowable bearing stress $S_c$ or $L = n \times A_b \times S_c = 6 \times (0.625 \times 0.25) \times 20,000 = 18,750$ pounds.

(c) The safe load $L$ based on the tensile stress is equal to the net cross sectional area of the plate between rivet holes $A_p$ times the allowable tensile stress $S_t$ or $L = A_p \times S_t = 0.25[12 - 6(0.625 + 0.0625)] \times 10,000 = 19,688$ pounds.

The safe tensile load for the joint would be the least of the three loads just computed or 15,647 pounds and the efficiency $\eta$ would be equal to this load divided by the tensile strength of the section of plate under consideration, if it were unperforated or $\eta = \dfrac{15,647}{12 \times 0.25 \times 10,000} \times 100 = 52.2$ per cent.

*Example 2.* Under consideration is a 12-inch section of double-riveted butt-joint with main plates ½ inch thick and two cover plates each 5/16 inch thick. There are 3 rivets in the inner row and 2 on the outer and their diameters are 7/8 inch. Assume that the diameter of the rivet holes is 1/16 inch larger than that of the rivets. The rivets are so placed that the main plates will not tear diagonally from one rivet row to the others nor will they tear or fail in shear out to their edges. The safe tensile load $L$ and the efficiency $\eta$ may be determined in the following way: Design stresses of 8500 psi for shear, 20,000 psi for bearing, and 10,000 psi for tension are arbitrarily assigned.

(a) The safe tensile load $L$ based on double shearing of the rivets is equal to the number of rivets $n$ times the number of shearing planes per rivet times the cross sectional area of one rivet $A_r$ times the allowable shearing stress $S_s$ or $L = n \times 2 \times A_r \times S_s = 5 \times 2 \times \dfrac{\pi}{4}(0.875)^2 \times 8500 = 51,112$ pounds.

(b) The safe tensile load $L$ based on bearing stress is equal to the number of rivets $n$ times the projected bearing area of the rivet $A_b$ (diameter times thickness of plate) times the allowable bearing stress $S_c$ or $L = n \times A_b \times S_c = 5 \times (0.875 \times 0.5) \times 20,000 = 43,750$ pounds.

(Cover plates are not considered since their combined thickness is 1/4 inch greater than the main plate thickness.)

(c) The safe tensile load $L$ based on the tensile stress is equal to the net cross sectional area of the plate between the two rivets in the outer row $A_p$ times the allowable tensile stress $S_t$ or $L = A_p \times S_t = 0.5[12 - 2(0.875 + 0.0625)] \times 10,000 = 50,625$ pounds.

In completing the analysis, the sum of the load that would cause tearing between rivets in the three-hole section plus the load carried by the two rivets in the two-hole section is also investigated. The sum is necessary because if the joint is to fail, it must fail at both sections simultaneously. The least safe load that can be carried by the two rivets of the two-hole section is based on the bearing stress (see the foregoing calculations).

(1) The safe tensile load $L$ based on the bearing strength of two rivets of the two-hole section is $L = n \times A_b \times S_c = 2 \times (0.875 \times 0.5) \times 20,000 = 17,500$ pounds.

(2) The safe tensile load $L$ based on the tensile strength of the main plate between holes in the three-hole section is $L = A_p \times S_t = 0.5[12 - 3(0.875 + 0.0625)] \times 10,000 = 45,938$ pounds.

The total safe tensile load based on this combination is $17,500 + 45,938 = 63,438$ pounds which is greater than any of the other results obtained.

The safe tensile load for the joint would be the least of the loads just computed or 43,750 pounds and the efficiency $\eta$ would be equal to this load divided by the tensile strength of the section of plate under consideration, if it were unperforated or $\eta = \dfrac{43,750}{0.5 \times 12 \times 10,000} \times 100 = 72.9$ per cent.

### Analysis of Riveted Joints — 1

A riveted joint may fail by shearing through the rivets (single or double shear), crushing the rivets, tearing the plate between the rivets, crushing the plate or by a combination of two or more of the foregoing causes. Rivets placed too close to the edge of the plate may tear or shear the plate out to the edge but this type of failure is avoided by placing the center of the rivet 1.5 times the rivet diameter away from the edge.

The efficiency of a riveted joint is equal to the strength of the joint divided by the strength of the unriveted plate, expressed as a percentage.

In the following formulas, let,

$d$ = diameter of rivets;          $p$ = pitch of inner row of rivets;
$D$ = diameter of holes;          $P$ = pitch of outer row of rivets;
$t$ = thickness of plate;          $S_s$ = shear stress for rivets;
$t_e$ = thickness of cover plates;          $S_t$ = tensile stress for plates;
$S_c$ = compressive or bearing stress for rivets or plates.

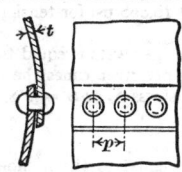

### For Single-riveted Lap-joint

(1) Resistance to shearing one rivet = $\dfrac{\pi d^2}{4} S_s$

(2) Resistance to tearing plate between rivets
= $(p - D)t S_t$

(3) Resistance to crushing rivet or plate
= $d t S_c$

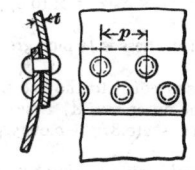

### Double-riveted Lap-joint

(1) Resistance to shearing two rivets = $\dfrac{2\pi d^2}{4} S_s$

(2) Resistance to tearing between two rivets
= $(p - D)t S_t$

(3) Resistance to crushing in front of two rivets = $2 d t S_c$

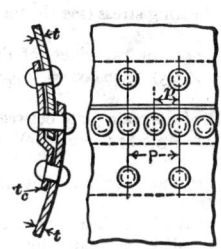

### Single-riveted Lap-joint with Inside Cover Plate

(1) Resistance to tearing between outer row of rivets = $(P - D)t S_t$

(2) Resistance to tearing between inner row of rivets, and shearing outer row of rivets = $(P - 2D)t S_t + \dfrac{\pi d^2}{4} S_s$

(3) Resistance to shearing three rivets = $\dfrac{3\pi d^2}{4} S_s$

(4) Resistance to crushing in front of three rivets = $3 t d S_c$

(5) Resistance to tearing at inner row of rivets, and crushing in front of one rivet in outer row = $(P - 2D)t S_t + t d S_c$

## Analysis of Riveted Joints — 2

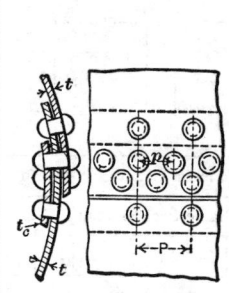

### Double-riveted Lap-joint with Inside Cover Plate

(1) Resistance to tearing at outer row of rivets = $(P - D)tS_t$

(2) Resistance to shearing four rivets = $\dfrac{4\pi d^2}{4} S_s$

(3) Resistance to tearing at inner row and shearing outer row of rivets
$$= (P - 1\tfrac{1}{2}D)tS_t + \frac{\pi d^2}{4} S_s$$

(4) Resistance to crushing in front of four rivets = $4tdS_c$

(5) Resistance to tearing at inner row of rivets, and crushing in front of one rivet = $(P - 1\tfrac{1}{2}D)tS_t + tdS_c$

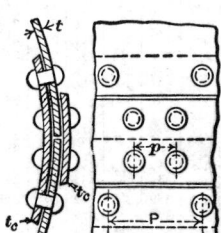

### Double-riveted Butt-joint

(1) Resistance to tearing at outer row of rivets = $(P - D)tS_t$

(2) Resistance to shearing two rivets in double shear and one in single shear = $\dfrac{5\pi d^2}{4} S_s$

(3) Resistance to tearing at inner row of rivets and shearing one rivet of the outer row = $(P - 2D)tS_t + \dfrac{\pi d^2}{4} S_s$

(4) Resistance to crushing in front of three rivets = $3tdS_c$

(5) Resistance to tearing at inner row of rivets, and crushing in front of one rivet in outer row = $(P - 2D)tS_t + tdS_c$

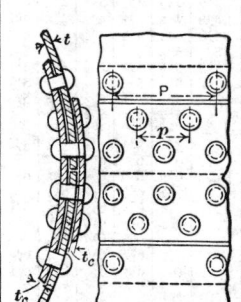

### Triple-riveted Butt-joint

(1) Resistance to tearing at outer row of rivets = $(P - D)tS_t$

(2) Resistance to shearing four rivets in double shear and one in single shear = $\dfrac{9\pi d^2}{4} S_s$

(3) Resistance to tearing at middle row of rivets and shearing one rivet
$$= (P - 2D)tS_t + \frac{\pi d^2}{4} S_s$$

(4) Resistance to crushing in front of four rivets and shearing one rivet = $4dtS_c + \dfrac{\pi d^2}{4} S_s$

(5) Resistance to crushing in front of five rivets = $4dtS_c + dt_cS_c$

Dimensions are usually specified in inches and stresses in pounds per square inch. See page 1240 for a discussion of allowable stresses which may be used in calculating the strengths given by the formulas. The design stresses are usually set by codes, practices, or specifications.

**Rivet Lengths for Forming Round and Countersunk Heads***

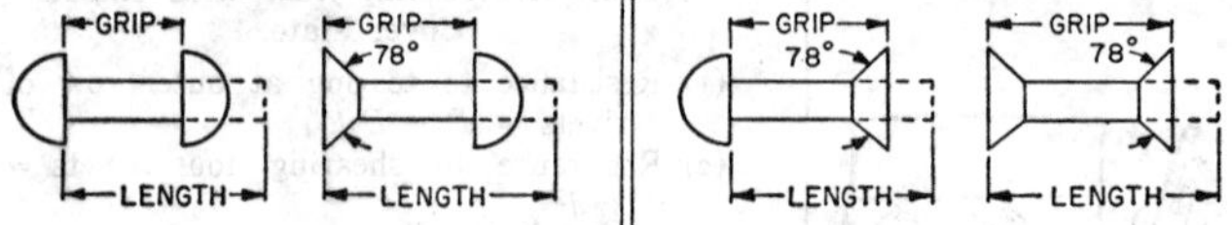


**To Form Round Head**

| Grip in Inches | Diameter of Rivet in Inches: 1/2 | 5/8 | 3/4 | 7/8 | 1 | 1 1/8 | 1 1/4 |
|---|---|---|---|---|---|---|---|
| | Length of Rivet in Inches | | | | | | |
| 1/2 | 1 5/8 | 1 7/8 | 1 7/8 | 2 | 2 1/8 | ... | ... |
| 5/8 | 1 3/4 | 2 | 2 | 2 1/8 | 2 1/4 | ... | ... |
| 3/4 | 1 7/8 | 2 1/8 | 2 1/8 | 2 1/4 | 2 3/8 | ... | ... |
| 7/8 | 2 | 2 1/4 | 2 1/4 | 2 3/8 | 2 1/2 | ... | ... |
| 1 | 2 1/4 | 2 3/8 | 2 3/8 | 2 1/2 | 2 5/8 | 2 3/4 | 2 7/8 |
| 1 1/8 | 2 3/8 | 2 1/2 | 2 1/2 | 2 5/8 | 2 3/4 | 2 7/8 | 3 |
| 1 1/4 | 2 1/2 | 2 5/8 | 2 5/8 | 2 3/4 | 2 7/8 | 3 | 3 1/8 |
| 1 3/8 | 2 5/8 | 2 3/4 | 2 3/4 | 2 7/8 | 3 | 3 1/8 | 3 1/4 |
| 1 1/2 | 2 7/8 | 3 | 3 | 3 1/8 | 3 1/4 | 3 3/8 | 3 1/2 |
| 1 5/8 | 3 | 3 1/8 | 3 1/8 | 3 1/4 | 3 3/8 | 3 1/2 | 3 1/2 |
| 1 3/4 | 3 1/8 | 3 1/4 | 3 1/4 | 3 1/2 | 3 5/8 | 3 3/4 | 3 3/4 |
| 1 7/8 | 3 1/4 | 3 3/8 | 3 3/8 | 3 5/8 | 3 3/4 | 3 7/8 | 3 7/8 |
| 2 | 3 1/2 | 3 1/2 | 3 5/8 | 3 3/4 | 3 7/8 | 4 | 4 |
| 2 1/8 | 3 5/8 | 3 5/8 | 3 3/4 | 3 7/8 | 4 | 4 1/8 | 4 1/8 |
| 2 1/4 | 3 3/4 | 3 7/8 | 3 7/8 | 4 | 4 1/8 | 4 1/4 | 4 1/4 |
| 2 3/8 | 4 | 4 | 4 | 4 1/8 | 4 1/4 | 4 3/8 | 4 3/8 |
| 2 1/2 | 4 1/8 | 4 1/8 | 4 1/8 | 4 1/4 | 4 3/8 | 4 1/2 | 4 1/2 |
| 2 5/8 | 4 1/4 | 4 1/4 | 4 1/4 | 4 3/8 | 4 1/2 | 4 5/8 | 4 5/8 |
| 2 3/4 | 4 3/8 | 4 3/8 | 4 3/8 | 4 1/2 | 4 5/8 | 4 3/4 | 4 3/4 |
| 2 7/8 | 4 5/8 | 4 5/8 | 4 5/8 | 4 5/8 | 4 3/4 | 4 7/8 | 5 |
| 3 | ... | 4 3/4 | 4 3/4 | 4 7/8 | 5 | 5 | 5 1/8 |
| 3 1/8 | ... | 4 7/8 | 4 7/8 | 5 | 5 1/8 | 5 1/4 | 5 1/4 |
| 3 1/4 | ... | 5 | 5 | 5 1/8 | 5 1/4 | 5 3/8 | 5 3/8 |
| 3 3/8 | ... | 5 1/8 | 5 1/8 | 5 1/4 | 5 3/8 | 5 1/2 | 5 1/2 |
| 3 1/2 | ... | 5 3/8 | 5 3/8 | 5 3/8 | 5 1/2 | 5 5/8 | 5 5/8 |
| 3 5/8 | ... | 5 1/2 | 5 1/2 | 5 1/2 | 5 5/8 | 5 3/4 | 5 3/4 |
| 3 3/4 | ... | 5 5/8 | 5 5/8 | 5 5/8 | 5 3/4 | 5 7/8 | 5 7/8 |
| 3 7/8 | ... | 5 3/4 | 5 3/4 | 5 3/4 | 5 7/8 | 6 | 6 |
| 4 | ... | ... | 5 7/8 | 6 | 6 | 6 1/8 | 6 1/4 |
| 4 1/8 | ... | ... | 6 | 6 1/8 | 6 1/4 | 6 3/8 | 6 3/8 |
| 4 1/4 | ... | ... | 6 1/8 | 6 1/4 | 6 1/2 | 6 1/2 | 6 1/2 |
| 4 3/8 | ... | ... | 6 3/8 | 6 1/2 | 6 1/2 | 6 5/8 | 6 5/8 |
| 4 1/2 | ... | ... | 6 1/2 | 6 5/8 | 6 5/8 | 6 3/4 | 6 3/4 |
| 4 5/8 | ... | ... | 6 5/8 | 6 3/4 | 6 3/4 | 6 3/4 | 6 7/8 |
| 4 3/4 | ... | ... | 6 3/4 | 6 7/8 | 6 7/8 | 7 | 7 |
| 4 7/8 | ... | ... | 6 7/8 | 7 | 7 | 7 1/8 | 7 1/8 |
| 5 | ... | ... | ... | 7 1/8 | 7 1/8 | 7 1/4 | 7 1/4 |
| 5 1/8 | ... | ... | ... | 7 1/4 | 7 1/4 | 7 3/8 | 7 3/8 |
| 5 1/4 | ... | ... | ... | 7 3/8 | 7 3/8 | 7 1/2 | 7 1/2 |
| 5 3/8 | ... | ... | ... | 7 5/8 | 7 5/8 | 7 3/4 | 7 5/8 |
| 5 1/2 | ... | ... | ... | 7 3/4 | 7 3/4 | 7 7/8 | 7 7/8 |
| 5 5/8 | ... | ... | ... | 7 7/8 | 7 7/8 | 8 | 8 |
| 5 3/4 | ... | ... | ... | 8 | 8 | 8 1/8 | 8 1/8 |
| 5 7/8 | ... | ... | ... | 8 1/8 | 8 1/8 | 8 1/4 | 8 1/4 |

**To Form Countersunk Head**

| Grip in Inches | Diameter of Rivet in Inches: 1/2 | 5/8 | 3/4 | 7/8 | 1 | 1 1/8 | 1 1/4 |
|---|---|---|---|---|---|---|---|
| | Length of Rivet in Inches | | | | | | |
| 1/2 | 1 | 1 | 1 1/8 | 1 1/4 | 1 1/4 | ... | ... |
| 5/8 | 1 1/8 | 1 1/4 | 1 1/4 | 1 3/8 | 1 3/8 | ... | ... |
| 3/4 | 1 3/8 | 1 3/8 | 1 3/8 | 1 1/2 | 1 1/2 | ... | ... |
| 7/8 | 1 1/2 | 1 1/2 | 1 1/2 | 1 5/8 | 1 5/8 | ... | ... |
| 1 | 1 5/8 | 1 5/8 | 1 5/8 | 1 3/4 | 1 3/4 | 1 7/8 | 1 7/8 |
| 1 1/8 | 1 3/4 | 1 3/4 | 1 7/8 | 1 7/8 | 1 7/8 | 2 | 2 |
| 1 1/4 | 2 | 2 | 2 | 2 | 2 | 2 1/8 | 2 1/8 |
| 1 3/8 | 2 1/8 | 2 1/8 | 2 1/8 | 2 1/4 | 2 1/4 | 2 3/8 | 2 3/8 |
| 1 1/2 | 2 1/4 | 2 1/4 | 2 1/4 | 2 3/8 | 2 3/8 | 2 1/2 | 2 1/2 |
| 1 5/8 | 2 3/8 | 2 3/8 | 2 3/8 | 2 1/2 | 2 1/2 | 2 5/8 | 2 5/8 |
| 1 3/4 | 2 5/8 | 2 5/8 | 2 5/8 | 2 5/8 | 2 5/8 | 2 3/4 | 2 3/4 |
| 1 7/8 | 2 3/4 | 2 3/4 | 2 3/4 | 2 3/4 | 2 3/4 | 2 7/8 | 2 7/8 |
| 2 | 2 7/8 | 2 7/8 | 2 7/8 | 2 7/8 | 2 7/8 | 3 | 3 |
| 2 1/8 | 3 1/8 | 3 | 3 | 3 | 3 | 3 1/8 | 3 1/8 |
| 2 1/4 | 3 1/4 | 3 1/8 | 3 1/8 | 3 1/8 | 3 1/4 | 3 1/4 | 3 1/4 |
| 2 3/8 | 3 3/8 | 3 3/8 | 3 3/8 | 3 3/8 | 3 3/8 | 3 3/8 | 3 3/8 |
| 2 1/2 | 3 1/2 | 3 1/2 | 3 1/2 | 3 1/2 | 3 1/2 | 3 5/8 | 3 5/8 |
| 2 5/8 | 3 3/4 | 3 5/8 | 3 5/8 | 3 5/8 | 3 5/8 | 3 3/4 | 3 3/4 |
| 2 3/4 | 3 7/8 | 3 3/4 | 3 3/4 | 3 3/4 | 3 3/4 | 3 7/8 | 3 7/8 |
| 2 7/8 | 4 | 3 7/8 | 3 7/8 | 3 7/8 | 3 7/8 | 4 | 4 |
| 3 | ... | 4 1/8 | 4 1/8 | 4 1/8 | 4 1/8 | 4 1/8 | 4 1/8 |
| 3 1/8 | ... | 4 1/4 | 4 1/4 | 4 1/4 | 4 1/4 | 4 1/4 | 4 1/4 |
| 3 1/4 | ... | 4 3/8 | 4 3/8 | 4 3/8 | 4 3/8 | 4 3/8 | 4 3/8 |
| 3 3/8 | ... | 4 1/2 | 4 1/2 | 4 1/2 | 4 1/2 | 4 1/2 | 4 1/2 |
| 3 1/2 | ... | 4 5/8 | 4 5/8 | 4 5/8 | 4 5/8 | 4 5/8 | 4 5/8 |
| 3 5/8 | ... | 4 3/4 | 4 3/4 | 4 3/4 | 4 3/4 | 4 7/8 | 4 7/8 |
| 3 3/4 | ... | 5 | 5 | 5 | 5 | 5 | 5 |
| 3 7/8 | ... | 5 1/8 | 5 1/8 | 5 1/8 | 5 1/8 | 5 1/8 | 5 1/8 |
| 4 | ... | ... | 5 1/4 | 5 1/4 | 5 1/4 | 5 1/4 | 5 1/4 |
| 4 1/8 | ... | ... | 5 3/8 | 5 3/8 | 5 3/8 | 5 3/8 | 5 3/8 |
| 4 1/4 | ... | ... | 5 1/2 | 5 1/2 | 5 1/2 | 5 1/2 | 5 1/2 |
| 4 3/8 | ... | ... | 5 5/8 | 5 5/8 | 5 5/8 | 5 5/8 | 5 5/8 |
| 4 1/2 | ... | ... | 5 3/4 | 5 3/4 | 5 3/4 | 5 3/4 | 5 3/4 |
| 4 5/8 | ... | ... | 6 | 6 | 6 | 6 | 6 |
| 4 3/4 | ... | ... | 6 1/8 | 6 1/8 | 6 1/8 | 6 1/8 | 6 1/8 |
| 4 7/8 | ... | ... | 6 1/4 | 6 1/4 | 6 1/4 | 6 1/4 | 6 1/4 |
| 5 | ... | ... | ... | 6 3/8 | 6 3/8 | 6 3/8 | 6 3/8 |
| 5 1/8 | ... | ... | ... | 6 1/2 | 6 1/2 | 6 1/2 | 6 1/2 |
| 5 1/4 | ... | ... | ... | 6 5/8 | 6 5/8 | 6 5/8 | 6 5/8 |
| 5 3/8 | ... | ... | ... | 6 3/4 | 6 3/4 | 6 3/4 | 6 3/4 |
| 5 1/2 | ... | ... | ... | 6 7/8 | 6 7/8 | 6 7/8 | 6 7/8 |
| 5 5/8 | ... | ... | ... | 7 | 7 | 7 | 7 |
| 5 3/4 | ... | ... | ... | 7 1/4 | 7 1/4 | 7 1/4 | 7 1/4 |
| 5 7/8 | ... | ... | ... | 7 3/8 | 7 3/8 | 7 3/8 | 7 3/8 |

* As given by the American Institute of Steel Construction. Values may vary from standard practice of individual fabricators and should be checked against fabricator's standard.

# STANDARD RIVETS

Rivets have been standardized by several standardizing bodies. On the pages which follow will be found standards of the American Standards Association (sponsored by the Society of Automotive Engineers, Inc. and The American Society of Mechanical Engineers) and the British Standards Institution.

**American National Standard Large Rivets.** — The types of rivets covered by this standard (ANSI B18.1.2-1972) are shown on pages 1246 and 1247 It may be noted, however, that when specified, the swell neck included in this standard is applicable to all standard large rivets except the flat countersunk head and oval countersunk head types. Also shown are the hold-on (dolly bar) and rivet set impression dimensions (see page 1248). All standard large rivets have fillets under the head not exceeding an 0.062-inch radius. The length tolerances for these rivets are given as follows: through 6 inches in length, ½ and ⅝-inch diameters, ±0.03 inch; ¾- and ⅞-inch diameters, ±0.06-inch; and 1- through 1¾-inch diameters, ±0.09 inch. For rivets over 6 inches in length, ½- and ⅝-inch diameters, ±0.06 inch; ¾- and ⅞-inch diameters, ±0.12 inch; and 1- through 1¾-inch diameters, ±0.19 inch. Materials suitable for steel and wrought iron rivets are covered in ASTM Specifications A31, A131, A152 and A502.

**American National Standard Small Solid Rivets.** — The types of rivets covered by this standard (ANSI B18.1.1-1972) are shown on pages 1249 to 1251. In addition, the standard gives the dimensions of 60-degree flat countersunk head rivets used to assemble ledger plates and guards for mower cutter bars, but these are not shown. As the heads of standard rivets are not machined or trimmed, the circumference may be somewhat irregular and edges may be rounded or flat. Rivets other than countersunk types are furnished with a definite fillet under the head, whose radius should not exceed 10 per cent of the maximum shank diameter or 0.03 inch, whichever is the smaller. With regard to head dimensions, tolerances shown in the dimensional tables are applicable to rivets produced by the normal cold heading process. Unless otherwise specified, rivets should have plain sheared ends which should be at right angles within 2 degrees to the axis of the rivet and be reasonably flat. When so specified by user, rivets may have the standard header points shown on page 1249. Rivets may be made of ASTM Specification A31, Grade A steel; or may adhere to SAE Recommended Practice, Mechanical and Chemical Requirements for Nonthreaded Fasteners — SAE J430, Grade 0. When so specified, rivets may also be made from other materials.

**British Standard Small Rivets for General Purposes.** — Dimensions of small rivets for general purposes are given in British Standard 641:1951 and are shown in the table on page 1253. In addition, the standard lists the standard lengths of these rivets, gives the dimensions of washers to be used with countersunk head rivets (140°), indicates that the rivets may be made from mild steel, copper, brass and a range of aluminum alloys and pure aluminum specified in B.S. 1473, and gives the dimensions of Coopers' flat head rivets ½ inch in diameter and below, in an appendix. In all types of rivets except those with countersunk heads there is a small radius or chamfer at the junction of the head and the shank.

**British Standard Dimensions of Rivets (½ to 1¾ inch diameter).** — The dimensions of rivets covered in B.S. 275:1927 are given on page 1252 and do not apply to boiler rivets. With regard to this standard the terms "nominal diameter" and "standard diameter" are synonymous. The term "tolerance" refers to the variation from the nominal diameter of the rivet and not to the difference between the diameter under the head and the diameter near the point.

## American National Standard Large Rivets —1 (ANSI B18.1.2-1972)

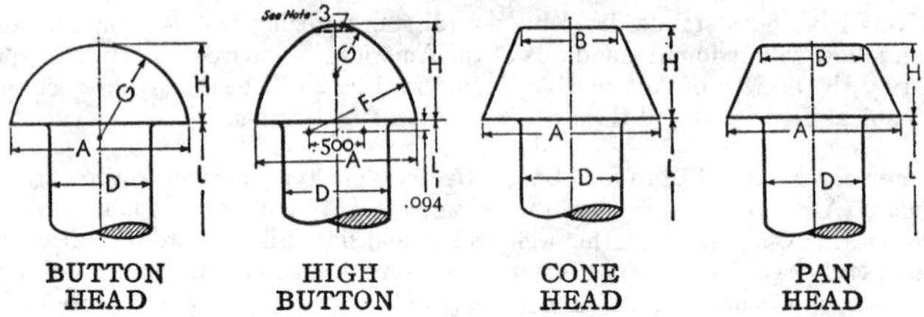

| | BUTTON HEAD | | HIGH BUTTON | | CONE HEAD | | PAN HEAD | |

| Nom. Body Diam. D† | Head Diam. A | | Height H | | Head Diam. A | | Height H | |
|---|---|---|---|---|---|---|---|---|
| | M'f'd Note 1 | Driven Note 2 | M'f'd Note 1 | Driven Note 2 | M'f'd Note 1 | Driven Note 2 | M'f'd Note 1 | Driven Note 2 |
| | BUTTON HEAD | | | | HIGH BUTTON HEAD (ACORN) | | | |
| ½ | 0.875 | 0.922 | 0.375 | 0.344 | 0.781 | 0.875 | 0.500 | 0.375 |
| ⅝ | 1.094 | 1.141 | 0.469 | 0.438 | 0.969 | 1.062 | 0.594 | 0.453 |
| ¾ | 1.312 | 1.375 | 0.562 | 0.516 | 1.156 | 1.250 | 0.688 | 0.531 |
| ⅞ | 1.531 | 1.594 | 0.656 | 0.609 | 1.344 | 1.438 | 0.781 | 0.609 |
| 1 | 1.750 | 1.828 | 0.750 | 0.688 | 1.531 | 1.625 | 0.875 | 0.688 |
| 1⅛ | 1.969 | 2.062 | 0.844 | 0.781 | 1.719 | 1.812 | 0.969 | 0.766 |
| 1¼ | 2.188 | 2.281 | 0.938 | 0.859 | 1.906 | 2.000 | 1.062 | 0.844 |
| 1⅜ | 2.406 | 2.516 | 1.031 | 0.953 | 2.094 | 2.188 | 1.156 | 0.938 |
| 1½ | 2.625 | 2.734 | 1.125 | 1.031 | 2.281 | 2.375 | 1.250 | 1.000 |
| 1⅝ | 2.844 | 2.969 | 1.219 | 1.125 | 2.469 | 2.562 | 1.344 | 1.094 |
| 1¾ | 3.062 | 3.203 | 1.312 | 1.203 | 2.656 | 2.750 | 1.438 | 1.172 |
| | CONE HEAD | | | | PAN HEAD | | | |
| ½ | 0.875 | 0.922 | 0.438 | 0.406 | 0.800 | 0.844 | 0.350 | 0.328 |
| ⅝ | 1.094 | 1.141 | 0.547 | 0.516 | 1.000 | 1.047 | 0.438 | 0.406 |
| ¾ | 1.312 | 1.375 | 0.656 | 0.625 | 1.200 | 1.266 | 0.525 | 0.484 |
| ⅞ | 1.531 | 1.594 | 0.766 | 0.719 | 1.400 | 1.469 | 0.612 | 0.578 |
| 1 | 1.750 | 1.828 | 0.875 | 0.828 | 1.600 | 1.687 | 0.700 | 0.656 |
| 1⅛ | 1.969 | 2.063 | 0.984 | 0.938 | 1.800 | 1.891 | 0.788 | 0.734 |
| 1¼ | 2.188 | 2.281 | 1.094 | 1.031 | 2.000 | 2.094 | 0.875 | 0.812 |
| 1⅜ | 2.406 | 2.516 | 1.203 | 1.141 | 2.200 | 2.312 | 0.962 | 0.906 |
| 1½ | 2.625 | 2.734 | 1.312 | 1.250 | 2.400 | 2.516 | 1.050 | 0.984 |
| 1⅝ | 2.844 | 2.969 | 1.422 | 1.344 | 2.600 | 2.734 | 1.138 | 1.062 |
| 1¾ | 3.062 | 3.203 | 1.531 | 1.453 | 2.800 | 2.938 | 1.225 | 1.141 |

All dimensions are given in inches.

† Tolerance for diameter of body is plus and minus from nominal and for ½-in. size equals +0.020, −0.022; for sizes ⅝ to 1-in., incl., equals +0.030. −0.025; for sizes 1⅛ and 1¼-in. equals +0.035; −0.027; for sizes 1⅜ and 1½-in. equals +0.040, −0.030; for sizes 1⅝ and 1¾-in. equals +0.040, −0.037.

Note 1.    Basic dimensions of head as manufactured.

Note 2.    Dimensions of manufactured head after driving and also of driven head.

Note 3.    Slight flat permissible within the specified head-height tolerance.

The following formulas give the basic dimensions for manufactured shapes: *Button Head*, $A = 1.750D$; $H = 0.750D$; $G = 0.885D$. *High Button Head*, $A = 1.500D + 0.031$; $H = 0.750D + 0.125$; $F = 0.750D + 0.281$; $G = 0.750D − 0.281$. *Cone Head*, $A = 1.750D$; $B = 0.938D$; $H = 0.875D$. *Pan Head*, $A = 1.600D$; $B = 1.000D$; $H = 0.700D$. Length $L$ is measured parallel to the rivet axis, from the extreme end to the bearing surface plane for flat bearing surface head type rivets, or to the intersection of the head top surface with the head diameter for countersunk head type rivets.

## American National Standard Large Rivets — 2 (ANSI B18.1.2-1972)

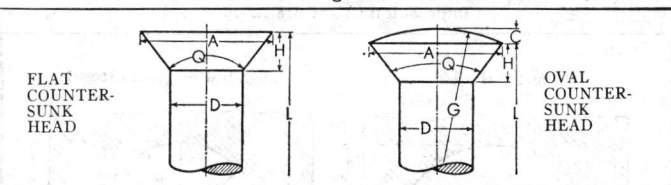

FLAT
COUNTER-
SUNK
HEAD

OVAL
COUNTER-
SUNK
HEAD

| Body Diameter D | | | Head Diam. A | | Head Depth H | Oval Crown Height* C | Oval Crown Radius* G |
|---|---|---|---|---|---|---|---|
| Nominal* | Max. | Min. | Max.† | Min.‡ | Ref. | | |
| FLAT AND OVAL COUNTERSUNK HEAD | | | | | | | |
| ½ 0.500 | 0.520 | 0.478 | 0.936 | 0.872 | 0.260 | 0.095 | 1.125 |
| ⅝ 0.625 | 0.655 | 0.600 | 1.194 | 1.112 | 0.339 | 0.119 | 1.406 |
| ¾ 0.750 | 0.780 | 0.725 | 1.421 | 1.322 | 0.400 | 0.142 | 1.688 |
| ⅞ 0.875 | 0.905 | 0.850 | 1.647 | 1.532 | 0.460 | 0.166 | 1.969 |
| 1 1.000 | 1.030 | 0.975 | 1.873 | 1.745 | 0.520 | 0.190 | 2.250 |
| 1⅛ 1.125 | 1.160 | 1.098 | 2.114 | 1.973 | 0.589 | 0.214 | 2.531 |
| 1¼ 1.250 | 1.285 | 1.223 | 2.340 | 2.199 | 0.650 | 0.238 | 2.812 |
| 1⅜ 1.375 | 1.415 | 1.345 | 2.567 | 2.426 | 0.710 | 0.261 | 3.094 |
| 1½ 1.500 | 1.540 | 1.470 | 2.793 | 2.652 | 0.771 | 0.285 | 3.375 |
| 1⅝ 1.625 | 1.665 | 1.588 | 3.019 | 2.878 | 0.831 | 0.309 | 3.656 |
| 1¾ 1.750 | 1.790 | 1.713 | 3.262 | 3.121 | 0.901 | 0.332 | 3.938 |

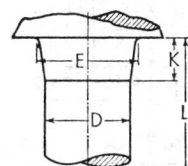

SWELL
NECK

| Body Diameter D | | | Diameter Under Head. E | | Neck Length K* |
|---|---|---|---|---|---|
| Nominal* | Max. | Min. | Max. (Basic) | Min. | |
| SWELL NECK# | | | | | |
| ½ 0.500 | 0.520 | 0.478 | 0.563 | 0.543 | 0.250 |
| ⅝ 0.625 | 0.655 | 0.600 | 0.688 | 0.658 | 0.312 |
| ¾ 0.750 | 0.780 | 0.725 | 0.813 | 0.783 | 0.375 |
| ⅞ 0.875 | 0.905 | 0.850 | 0.938 | 0.908 | 0.438 |
| 1 1.000 | 1.030 | 0.975 | 1.063 | 1.033 | 0.500 |
| 1⅛ 1.125 | 1.160 | 1.098 | 1.188 | 1.153 | 0.562 |
| 1¼ 1.250 | 1.285 | 1.223 | 1.313 | 1.278 | 0.625 |
| 1⅜ 1.375 | 1.415 | 1.345 | 1.438 | 1.398 | 0.688 |
| 1½ 1.500 | 1.540 | 1.470 | 1.563 | 1.523 | 0.750 |
| 1⅝ 1.625 | 1.665 | 1.588 | 1.688 | 1.648 | 0.812 |
| 1¾ 1.750 | 1.790 | 1.713 | 1.813 | 1.773 | 0.875 |

All dimensions are given in inches.
*Basic dimension as manufactured. For tolerances see table on page 1246.
†Sharp edged head.
‡Rounded or flat edged irregularly shaped head (heads are not machined or trimmed).
#The swell neck is applicable to all standard forms of large rivets except the flat countersunk and oval countersunk head types.
The following formulas give basic dimensions for manufactured shapes: *Flat Countersunk Head*, $A = 1.810D$; $H = 1.192$ (Max $A - D$)/2; included angle $Q$ of head = 78 degrees. *Oval Countersunk Head*, $A = 1.810D$; $H = 1.192$ (Max $A - D$)/2; included angle of head = 78 degrees. *Swell Neck*, $E = D + 0.063$; $K = 0.500D$. Length $L$ is measured parallel to the rivet axis, from the extreme end to the bearing surface plane for flat bearing surface head type rivets, or to the intersection of the head top surface with the head diameter for countersunk head type rivets.

**American National Standard Dimensions for Hold-On (Dolly Bar) and Rivet Set Impression** (ANSI B18.1.2-1972)

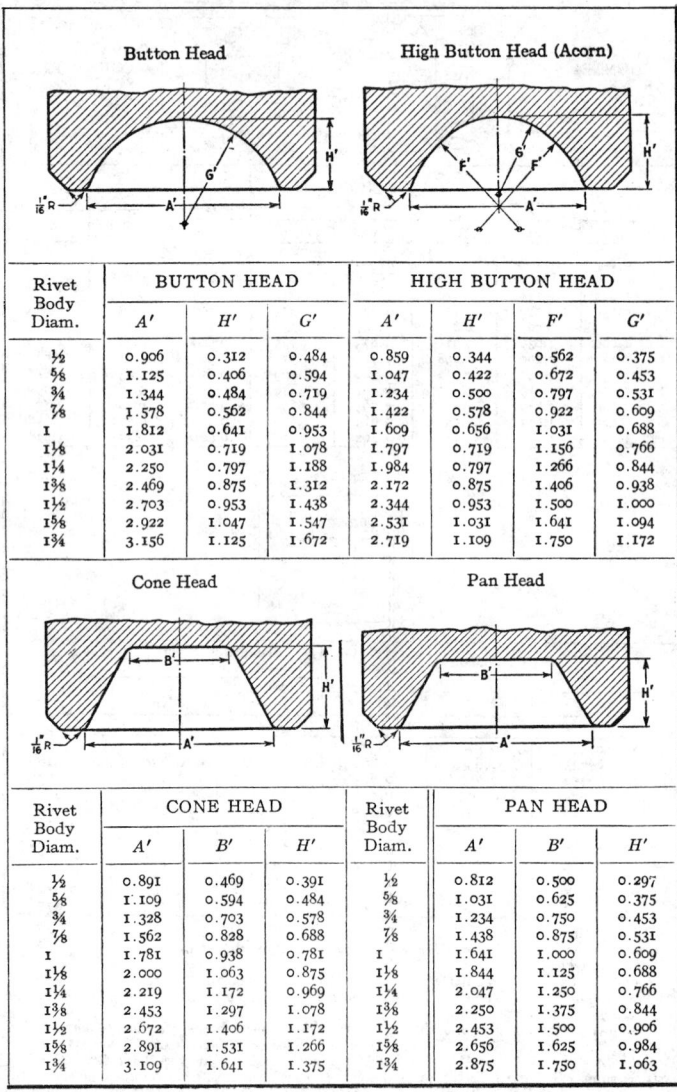

| Rivet Body Diam. | BUTTON HEAD | | | HIGH BUTTON HEAD | | | |
|---|---|---|---|---|---|---|---|
| | $A'$ | $H'$ | $G'$ | $A'$ | $H'$ | $F'$ | $G'$ |
| ½ | 0.906 | 0.312 | 0.484 | 0.859 | 0.344 | 0.562 | 0.375 |
| ⅝ | 1.125 | 0.406 | 0.594 | 1.047 | 0.422 | 0.672 | 0.453 |
| ¾ | 1.344 | 0.484 | 0.719 | 1.234 | 0.500 | 0.797 | 0.531 |
| ⅞ | 1.578 | 0.562 | 0.844 | 1.422 | 0.578 | 0.922 | 0.609 |
| 1 | 1.812 | 0.641 | 0.953 | 1.609 | 0.656 | 1.031 | 0.688 |
| 1⅛ | 2.031 | 0.719 | 1.078 | 1.797 | 0.719 | 1.156 | 0.766 |
| 1¼ | 2.250 | 0.797 | 1.188 | 1.984 | 0.797 | 1.266 | 0.844 |
| 1⅜ | 2.469 | 0.875 | 1.312 | 2.172 | 0.875 | 1.406 | 0.938 |
| 1½ | 2.703 | 0.953 | 1.438 | 2.344 | 0.953 | 1.500 | 1.000 |
| 1⅝ | 2.922 | 1.047 | 1.547 | 2.531 | 1.031 | 1.641 | 1.094 |
| 1¾ | 3.156 | 1.125 | 1.672 | 2.719 | 1.109 | 1.750 | 1.172 |

| Rivet Body Diam. | CONE HEAD | | | Rivet Body Diam. | PAN HEAD | | |
|---|---|---|---|---|---|---|---|
| | $A'$ | $B'$ | $H'$ | | $A'$ | $B'$ | $H'$ |
| ½ | 0.891 | 0.469 | 0.391 | ½ | 0.812 | 0.500 | 0.297 |
| ⅝ | 1.109 | 0.594 | 0.484 | ⅝ | 1.031 | 0.625 | 0.375 |
| ¾ | 1.328 | 0.703 | 0.578 | ¾ | 1.234 | 0.750 | 0.453 |
| ⅞ | 1.562 | 0.828 | 0.688 | ⅞ | 1.438 | 0.875 | 0.531 |
| 1 | 1.781 | 0.938 | 0.781 | 1 | 1.641 | 1.000 | 0.609 |
| 1⅛ | 2.000 | 1.063 | 0.875 | 1⅛ | 1.844 | 1.125 | 0.688 |
| 1¼ | 2.219 | 1.172 | 0.969 | 1¼ | 2.047 | 1.250 | 0.766 |
| 1⅜ | 2.453 | 1.297 | 1.078 | 1⅜ | 2.250 | 1.375 | 0.844 |
| 1½ | 2.672 | 1.406 | 1.172 | 1½ | 2.453 | 1.500 | 0.906 |
| 1⅝ | 2.891 | 1.531 | 1.266 | 1⅝ | 2.656 | 1.625 | 0.984 |
| 1¾ | 3.109 | 1.641 | 1.375 | 1¾ | 2.875 | 1.750 | 1.063 |

All dimensions are given in inches.

**Table 1. American National Standard Small Solid Rivets[1]**
(ANSI B18.1.1-1972 and Appendix)

TRUSS HEAD RIVETS

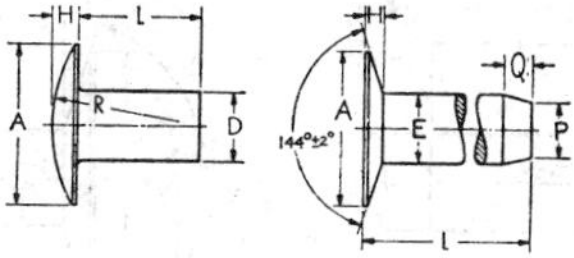


COOPERS RIVETS

TRUSS HEAD RIVETS[2]

| Shank Diam.,[3] D | | Head Dimensions | | | | | Shank Diam.,[3] D | | Head Dimensions | | | | |
|---|---|---|---|---|---|---|---|---|---|---|---|---|---|
| | | Diam., *A* | | Height, *H* | | Rad. *R* | | | Diam., *A* | | Height, *H* | | Rad. *R* |
| Nominal | | Max. | Min. | Max. | Min. | Approx. | Nominal | | Max. | Min. | Max. | Min. | Approx. |
| 3/32 | .094 | .226 | .206 | .038 | .026 | .239 | 9/32 | .281 | 0.661 | .631 | .103 | .085 | 0.706 |
| 1/8 | .125 | .297 | .277 | .048 | .036 | .314 | 5/16 | .312 | 0.732 | .702 | .113 | .095 | 0.784 |
| 5/32 | .156 | .368 | .348 | .059 | .045 | .392 | 11/32 | .344 | 0.806 | .776 | .124 | .104 | 0.862 |
| 3/16 | .188 | .442 | .422 | .069 | .055 | .470 | 3/8 | .375 | 0.878 | .848 | .135 | .115 | 0.942 |
| 7/32 | .219 | .515 | .495 | .080 | .066 | .555 | 13/32 | .406 | 0.949 | .919 | .145 | .123 | 1.028 |
| 1/4 | .250 | .590 | .560 | .091 | .075 | .628 | 7/16 | .438 | 1.020 | .990 | .157 | .135 | 1.098 |

COOPERS RIVETS

| Shank Diameter, *D* | | | Head Diameter, *A* | | Head Height, *H* | | Point Dimensions | | Length, *L* | |
|---|---|---|---|---|---|---|---|---|---|---|
| | | | | | | | Diam., *P* | Length, *Q* | | |
| Size No.[4] | Max. | Min. | Max. | Min. | Max. | Min. | Nom. | Nom. | Max. | Min. |
| 1 lb. | .111 | .105 | .291 | .271 | .045 | .031 | Not Pointed | | .249 | .219 |
| 1¼ lb. | .122 | .116 | .324 | .302 | .050 | .036 | Not Pointed | | .285 | .255 |
| 1½ lb. | .132 | .126 | .324 | .302 | .050 | .036 | Not Pointed | | .285 | .255 |
| 1¾ lb. | .136 | .130 | .324 | .302 | .052 | .034 | Not Pointed | | .318 | .284 |
| 2 lb. | .142 | .136 | .355 | .333 | .056 | .038 | Not Pointed | | .322 | .288 |
| 3 lb. | .158 | .152 | .386 | .364 | .058 | .040 | .123 | .062 | .387 | .353 |
| 4 lb. | .168 | .159 | .388 | .362 | .058 | .040 | .130 | .062 | .418 | .388 |
| 5 lb. | .183 | .174 | .419 | .393 | .063 | .045 | .144 | .062 | .454 | .420 |
| 6 lb. | .206 | .197 | .482 | .456 | .073 | .051 | .160 | .094 | .498 | .457 |
| 7 lb. | .223 | .214 | .513 | .487 | .076 | .054 | .175 | .094 | .561 | .523 |
| 8 lb. | .241 | .232 | .546 | .516 | .081 | .059 | .182 | .094 | .597 | .559 |
| 9 lb. | .248 | .239 | .578 | .548 | .085 | .063 | .197 | .094 | .601 | .563 |
| 10 lb. | .253 | .244 | .578 | .548 | .085 | .063 | .197 | .094 | .632 | .594 |
| 12 lb. | .263 | .251 | .580 | .546 | .086 | .060 | .214 | .094 | .633 | .575 |
| 14 lb. | .275 | .263 | .611 | .577 | .091 | .065 | .223 | .094 | .670 | .612 |
| 16 lb. | .285 | .273 | .611 | .577 | .089 | .063 | .223 | .094 | .699 | .641 |
| 18 lb. | .285 | .273 | .642 | .608 | .108 | .082 | .230 | .125 | .749 | .691 |
| 20 lb. | .316 | .304 | .705 | .671 | .128 | .102 | .250 | .125 | .769 | .711 |
| 3/8 in. | .380 | .365 | .800 | .762 | .136 | .106 | .312 | .125 | .840 | .778 |

*Note:* When specified American National Standard Small Solid Rivets may be obtained with points. Point dimensions for belt and coopers rivets are given in the accompanying tables. Formulas for calculating point dimensions of other rivets are given with the diagram alongside.

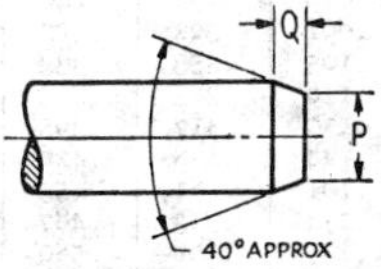


$P = D \times 0.818$
$Q = D \times 0.25$

[1] All dimensions in inches except where otherwise noted. [2] Length tolerance of rivets is plus or minus .016 inch. Approximate proportions of rivets: $A = 2.300 \times D$, $H = 0.330 \times D$, $R = 2.512 \times D$. [3] Tolerances on the nominal shank diameter in inches are given for the following body diameter ranges: 3/32 to 5/32, plus 0.002, minus 0.004; 3/16 to 1/4, plus 0.003, minus 0.006; 9/32 to 11/32, plus 0.004, minus 0.008; and 3/8 to 7/16, plus 0.005, minus 0.010. [4] Size numbers in pounds refer to the approximate weight of 1000 rivets.

## Table 2. American National Standard Small Solid Rivets[1] (ANSI B18.1.1-1972)

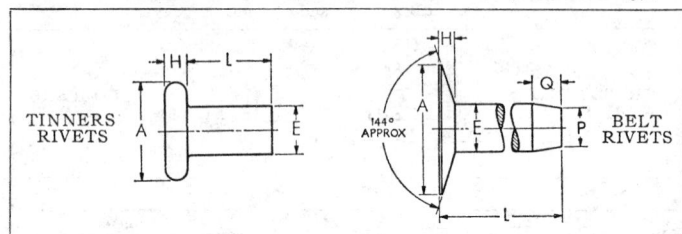

### TINNERS RIVETS

| Size No.[2] | Shank Diameter, E | | Head Diam., A | | Head Height, H | | Length, L | | |
|---|---|---|---|---|---|---|---|---|---|
| | Max. | Min. | Max. | Min. | Max. | Min. | Nom. | Max. | Min. |
| 6 oz. | .081 | .075 | .213 | .193 | .028 | .016 | 1/8 | .135 | .115 |
| 8 oz. | .091 | .085 | .225 | .205 | .036 | .024 | 5/32 | .166 | .146 |
| 10 oz. | .097 | .091 | .250 | .230 | .037 | .025 | 11/64 | .182 | .162 |
| 12 oz. | .107 | .101 | .265 | .245 | .037 | .025 | 3/16 | .198 | .178 |
| 14 oz. | .111 | .105 | .275 | .255 | .038 | .026 | 3/16 | .198 | .178 |
| 1 lb. | .113 | .107 | .285 | .265 | .040 | .028 | 13/64 | .213 | .193 |
| 1¼ lb. | .122 | .116 | .295 | .275 | .045 | .033 | 7/32 | .229 | .209 |
| 1½ lb. | .132 | .126 | .316 | .294 | .046 | .034 | 15/64 | .244 | .224 |
| 1¾ lb. | .136 | .130 | .331 | .309 | .049 | .035 | 1/4 | .260 | .240 |
| 2 lb. | .146 | .140 | .341 | .319 | .050 | .036 | 17/64 | .276 | .256 |
| 2½ lb. | .150 | .144 | .311 | .289 | .069 | .055 | 9/32 | .291 | .271 |
| 3 lb. | .163 | .154 | .329 | .303 | .073 | .059 | 5/16 | .323 | .303 |
| 3½ lb. | .168 | .159 | .348 | .322 | .074 | .060 | 21/64 | .338 | .318 |
| 4 lb. | .179 | .170 | .368 | .342 | .076 | .062 | 11/32 | .354 | .334 |
| 5 lb. | .190 | .181 | .388 | .362 | .084 | .070 | 3/8 | .385 | .365 |
| 6 lb. | .206 | .197 | .419 | .393 | .090 | .076 | 25/64 | .401 | .381 |
| 7 lb. | .223 | .214 | .431 | .405 | .094 | .080 | 13/32 | .416 | .396 |
| 8 lb. | .227 | .218 | .475 | .445 | .101 | .085 | 7/16 | .448 | .428 |
| 9 lb. | .241 | .232 | .490 | .460 | .103 | .087 | 29/64 | .463 | .443 |
| 10 lb. | .241 | .232 | .505 | .475 | .104 | .088 | 15/32 | .479 | .459 |
| 12 lb. | .263 | .251 | .532 | .498 | .108 | .090 | 1/2 | .510 | .490 |
| 14 lb. | .288 | .276 | .577 | .543 | .113 | .095 | 33/64 | .525 | .505 |
| 16 lb. | .304 | .292 | .597 | .563 | .128 | .110 | 17/32 | .541 | .521 |
| 18 lb. | .347 | .335 | .706 | .668 | .156 | .136 | 19/32 | .603 | .583 |

### BELT RIVETS[3]

| Size No.[4] | Shank Diameter, E | | Head Diam., A | | Head Height, H | | Point Dimensions | |
|---|---|---|---|---|---|---|---|---|
| | | | | | | | Diam., P | L'gth., Q |
| | Max. | Min. | Max. | Min. | Max. | Min. | Nominal | Nominal |
| 14 | .085 | .079 | .260 | .240 | .042 | .030 | .065 | .078 |
| 13 | .097 | .091 | .322 | .302 | .051 | .039 | .073 | .078 |
| 12 | .111 | .105 | .353 | .333 | .054 | .040 | .083 | .078 |
| 11 | .122 | .116 | .383 | .363 | .059 | .045 | .097 | .078 |
| 10 | .136 | .130 | .417 | .395 | .065 | .047 | .109 | .094 |
| 9 | .150 | .144 | .448 | .426 | .069 | .051 | .122 | .094 |
| 8 | .167 | .161 | .481 | .455 | .072 | .054 | .135 | .094 |
| 7 | .183 | .174 | .513 | .487 | .075 | .056 | .151 | .125 |
| 6 | .206 | .197 | .606 | .580 | .090 | .068 | .165 | .125 |
| 5 | .223 | .214 | .700 | .674 | .105 | .083 | .185 | .125 |
| 4 | .241 | .232 | .921 | .893 | .138 | .116 | .204 | .141 |

[1] All dimensions in inches.   [2] Size numbers refer to the approximate weight of 1000 rivets.   [3] Length tolerance on belt rivets is plus 0.031 inch, minus 0 inch.   [4] Size number refers to the Stub's iron wire gage number of the stock used in the shank of the rivet.
*Note:* American National Standard Small Solid Rivets may be obtained with or without points. Point proportions are given in the diagram in Table 1.

**Table 3. American National Standard Small Solid Rivets[1] (ANSI B18.1.1-1972 and Appendix)**

Head Dimensions

| Shank Diameter | | | | | FLAT HEAD | | | | FLAT COUNTERSUNK HEAD | | | BUTTON HEAD | | | | | PAN HEAD | | | | | | | |
|---|---|---|---|---|---|---|---|---|---|---|---|---|---|---|---|---|---|---|---|---|---|---|---|---|
| D | | E | | | Diam., A | | Height, H | | Diam., A Sharp | | Height, H Ref. | Diam., A | | Height, H | | Radius, R | Diam., A | | Height, H | | Radii | | | |
| Nominal | | Max. | Min. | | Max. | Min. | Max. | Min. | Max.[2] | Min.[3] | | Max. | Min. | Max. | Min. | Approx. | Max. | Min. | Max. | Min. | R₁ | R₂ | R₃ | |
| | | | | | | | | | | | | | | | | | | | | | Approximate | | |
| 1/16 | .062 | .064 | .059 | | .140 | .120 | .027 | .017 | .118 | .110 | .027 | .122 | .102 | .052 | .042 | .055 | .118 | .098 | .040 | .030 | .019 | .052 | 0.217 | |
| 3/32 | .094 | .096 | .090 | | .200 | .180 | .038 | .026 | .176 | .163 | .040 | .182 | .162 | .077 | .065 | .084 | .173 | .153 | .060 | .048 | .030 | .080 | 0.326 | |
| 1/8 | .125 | .127 | .121 | | .260 | .240 | .048 | .036 | .235 | .217 | .053 | .235 | .215 | .100 | .088 | .111 | .225 | .205 | .078 | .066 | .039 | .106 | 0.429 | |
| 5/32 | .156 | .158 | .152 | | .323 | .301 | .059 | .045 | .293 | .272 | .066 | .290 | .268 | .124 | .110 | .138 | .279 | .257 | .096 | .082 | .049 | .133 | 0.535 | |
| 3/16 | .188 | .191 | .182 | | .387 | .361 | .069 | .055 | .351 | .326 | .079 | .348 | .322 | .147 | .133 | .166 | .334 | .308 | .114 | .100 | .059 | .159 | 0.641 | |
| 7/32 | .219 | .222 | .213 | | .453 | .427 | .080 | .065 | .413 | .384 | .094 | .405 | .379 | .172 | .158 | .195 | .391 | .365 | .133 | .119 | .069 | .186 | 0.754 | |
| 1/4 | .250 | .253 | .244 | | .515 | .485 | .091 | .075 | .469 | .437 | .106 | .460 | .430 | .196 | .180 | .221 | .444 | .414 | .151 | .135 | .079 | .213 | 0.858 | |
| 9/32 | .281 | .285 | .273 | | .579 | .545 | .103 | .085 | .528 | .491 | .119 | .518 | .484 | .220 | .202 | .249 | .499 | .465 | .170 | .152 | .088 | .239 | 0.963 | |
| 5/16 | .312 | .316 | .304 | | .641 | .607 | .113 | .095 | .588 | .547 | .133 | .572 | .538 | .243 | .225 | .276 | .552 | .518 | .187 | .169 | .098 | .266 | 1.070 | |
| 11/32 | .344 | .348 | .336 | | .705 | .667 | .124 | .104 | .646 | .602 | .146 | .630 | .592 | .267 | .247 | .304 | .608 | .570 | .206 | .186 | .108 | .292 | 1.176 | |
| 3/8 | .375 | .380 | .365 | | .769 | .731 | .135 | .115 | .704 | .656 | .159 | .684 | .646 | .291 | .271 | .332 | .663 | .625 | .225 | .205 | .118 | .319 | 1.286 | |
| 13/32 | .406 | .411 | .396 | | .834 | .790 | .146 | .124 | .763 | .710 | .172 | .743 | .699 | .316 | .294 | .358 | .719 | .675 | .243 | .221 | .127 | .345 | 1.392 | |
| 7/16 | .438 | .443 | .428 | | .896 | .852 | .157 | .135 | .823 | .765 | .186 | .798 | .754 | .339 | .317 | .387 | .772 | .728 | .261 | .239 | .137 | .372 | 1.500 | |

[1] All dimensions in inches. Length tolerance of all rivets is plus or minus 0.016 inch. Approximate proportions of rivets: flat head, $A = 2.00 \times D$, $H = 0.33 \ D$; flat countersunk head, $A = 1.850 \times D$, $H = 0.425 \times D$; button head, $A = 1.750 \times D$, $H = 0.750 \times D$, $R = 0.885 \times D$; pan head, $A = 1.720 \times D$, $H = 0.570 \times D$, $R_1 = 0.314 \times D$, $R_2 = 0.850 \times D$, $R_3 = 3.430 \times D$. [2] Tabulated maximum values calculated on basic diameter of rivet and 92° included angle extended to a sharp edge. [3] Minimum of rounded or flat edged irregular shaped head. Rivet heads are not machined or trimmed and the circumference may be irregular and edges rounded or flat. [4] Given for reference purposes only. Variations in this dimension are controlled by the head and shank diameters and the included angle of the head.
Note: ANSI Small Solid Rivets may be obtained with or without points. Point proportions are given in the diagram in Table 1.

### Head Dimensions and Diameters of British Standard Rivets (B.S. 275-1927)
(This standard does not apply to Boiler Rivets)

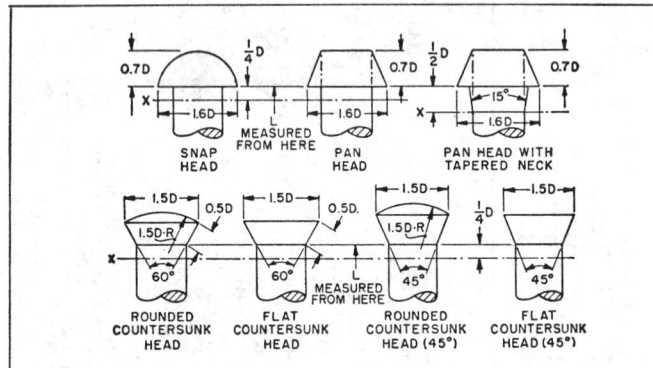

| Nominal Rivet Diameter, D | Shank Diameter† | | | | |
|---|---|---|---|---|---|
| | At Position X§ | | At Position Y§ | | At Position Z§ |
| | Minimum | Maximum | Minimum | Maximum | Minimum |
| ½ | ½ | 17⁄32 | 31⁄64 | ½ | 31⁄64 |
| 9⁄16* | 9⁄16 | 19⁄32 | 35⁄64 | 9⁄16 | 35⁄64 |
| ⅝ | ⅝ | 21⁄32 | 39⁄64 | ⅝ | 39⁄64 |
| 1 1⁄16* | 1 1⁄16 | 23⁄32 | 43⁄64 | 1 1⁄16 | 43⁄64 |
| ¾ | ¾ | 25⁄32 | 47⁄64 | ¾ | 47⁄64 |
| 1 3⁄16* | 1 3⁄16 | 27⁄32 | 51⁄64 | 1 3⁄16 | 51⁄64 |
| ⅞ | ⅞ | 29⁄32 | 55⁄64 | ⅞ | 55⁄64 |
| 1 5⁄16* | 1 5⁄16 | 31⁄32 | 59⁄64 | 1 5⁄16 | 59⁄64 |
| 1 | 1 | 1 1⁄32 | 63⁄64 | 1 | 63⁄64 |
| 1 1⁄16* | 1 1⁄16 | 1 3⁄32 | 1 3⁄64 | 1 1⁄16 | 1 3⁄64 |
| 1⅛ | 1⅛ | 1 5⁄32 | 1 7⁄64 | 1⅛ | 1 7⁄64 |
| 1 3⁄16* | 1 3⁄16 | 1 7⁄32 | 1 11⁄64 | 1 3⁄16 | 1 11⁄64 |
| 1¼ | 1¼ | 1 9⁄32 | 1 15⁄64 | 1¼ | 1 15⁄64 |
| 1 5⁄16* | 1 5⁄16 | 1 11⁄32 | 1 19⁄64 | 1 5⁄16 | 1 19⁄64 |
| 1⅜ | 1⅜ | 1 13⁄32 | 1 23⁄64 | 1⅜ | 1 23⁄64 |
| 1 7⁄16* | 1 7⁄16 | 1 15⁄32 | 1 27⁄64 | 1 7⁄16 | 1 27⁄64 |
| 1½ | 1½ | 1 17⁄32 | 1 31⁄64 | 1½ | 1 31⁄64 |
| 1 9⁄16* | 1 9⁄16 | 1 19⁄32 | 1 35⁄64 | 1 9⁄16 | 1 35⁄64 |
| 1⅝ | 1⅝ | 1 21⁄32 | 1 39⁄64 | 1⅝ | 1 39⁄64 |
| 1 11⁄16* | 1 11⁄16 | 1 23⁄32 | 1 43⁄64 | 1 11⁄16 | 1 43⁄64 |
| 1¾ | 1¾ | 1 25⁄32 | 1 47⁄64 | 1¾ | 1 47⁄64 |

All dimensions that are tabulated are given in inches.

* At the recommendation of the British Standards Institution these sizes are to be dispensed with wherever possible.

† Tolerances of the rivet diameter are as follows: at position X, plus 1⁄32 inch, minus zero; at position Y, plus zero, minus 1⁄64 inch; at position Z, minus 1⁄64 inch but in no case shall the difference between the diameters at positions X and Y exceed 1⁄32 inch, nor shall the diameter of the shank between positions X and Y be less than the minimum diameter specified at position Y.

§ The location of positions Y and Z are as follows: Position Y is located ½D from the end of the rivet for rivet lengths 5 diameters long and under. For longer rivets, position Y is located 4½D from the head of the rivet. Position Z (found only on rivets longer than 5D) is located ½D from the end of the rivet.

## British Standard Small Rivets for General Purposes (B.S. 641:1951)

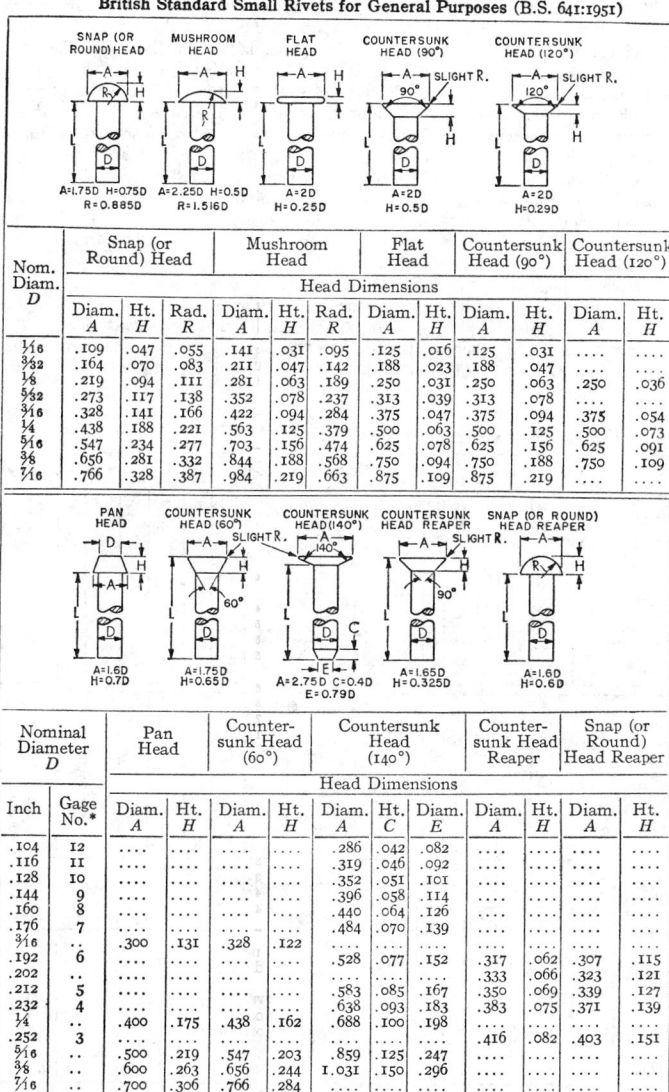

| Nom. Diam. D | Snap (or Round) Head | | | Mushroom Head | | | Flat Head | | Countersunk Head (90°) | | Countersunk Head (120°) | |
|---|---|---|---|---|---|---|---|---|---|---|---|---|
| | Head Dimensions | | | | | | | | | | | |
| | Diam. A | Ht. H | Rad. R | Diam. A | Ht. H | Rad. R | Diam. A | Ht. H | Diam. A | Ht. H | Diam. A | Ht. H |
| 1/16 | .109 | .047 | .055 | .141 | .031 | .095 | .125 | .016 | .125 | .031 | .... | .... |
| 3/32 | .164 | .070 | .083 | .211 | .047 | .142 | .188 | .023 | .188 | .047 | .... | .... |
| 1/8 | .219 | .094 | .111 | .281 | .063 | .189 | .250 | .031 | .250 | .063 | .250 | .036 |
| 5/32 | .273 | .117 | .138 | .352 | .078 | .237 | .313 | .039 | .313 | .078 | .... | .... |
| 3/16 | .328 | .141 | .166 | .422 | .094 | .284 | .375 | .047 | .375 | .094 | .375 | .054 |
| 1/4 | .438 | .188 | .221 | .563 | .125 | .379 | .500 | .063 | .500 | .125 | .500 | .073 |
| 5/16 | .547 | .234 | .277 | .703 | .156 | .474 | .625 | .078 | .625 | .156 | .625 | .091 |
| 3/8 | .656 | .281 | .332 | .844 | .188 | .568 | .750 | .094 | .750 | .188 | .750 | .109 |
| 7/16 | .766 | .328 | .387 | .984 | .219 | .663 | .875 | .109 | .875 | .219 | .... | .... |

| Nominal Diameter D | | Pan Head | | Countersunk Head (60°) | | Countersunk Head (140°) | | | Countersunk Head Reaper | | Snap (or Round) Head Reaper | |
|---|---|---|---|---|---|---|---|---|---|---|---|---|
| | | Head Dimensions | | | | | | | | | | |
| Inch | Gage No.* | Diam. A | Ht. H | Diam. A | Ht. H | Diam. A | Ht. C | Diam. E | Diam. A | Ht. H | Diam. A | Ht. H |
| .104 | 12 | .... | .... | .... | .... | .286 | .042 | .082 | .... | .... | .... | .... |
| .116 | 11 | .... | .... | .... | .... | .319 | .046 | .092 | .... | .... | .... | .... |
| .128 | 10 | .... | .... | .... | .... | .352 | .051 | .101 | .... | .... | .... | .... |
| .144 | 9 | .... | .... | .... | .... | .396 | .058 | .114 | .... | .... | .... | .... |
| .160 | 8 | .... | .... | .... | .... | .440 | .064 | .126 | .... | .... | .... | .... |
| .176 | 7 | .... | .... | .... | .... | .484 | .070 | .139 | .... | .... | .... | .... |
| 3/16 | .. | .300 | .131 | .328 | .122 | .... | .... | .... | .... | .... | .... | .... |
| .192 | 6 | .... | .... | .... | .... | .528 | .077 | .152 | .317 | .062 | .307 | .115 |
| .202 | .. | .... | .... | .... | .... | .... | .... | .... | .333 | .066 | .323 | .121 |
| .212 | 5 | .... | .... | .... | .... | .583 | .085 | .167 | .350 | .069 | .339 | .127 |
| .232 | 4 | .... | .... | .... | .... | .638 | .093 | .183 | .383 | .075 | .371 | .139 |
| 1/4 | .. | .400 | .175 | .438 | .162 | .688 | .100 | .198 | .... | .... | .... | .... |
| .252 | 3 | .... | .... | .... | .... | .... | .... | .... | .416 | .082 | .403 | .151 |
| 5/16 | .. | .500 | .219 | .547 | .203 | .859 | .125 | .247 | .... | .... | .... | .... |
| 3/8 | .. | .600 | .263 | .656 | .244 | 1.031 | .150 | .296 | .... | .... | .... | .... |
| 7/16 | .. | .700 | .306 | .766 | .284 | .... | .... | .... | .... | .... | .... | .... |

All dimensions in inches unless specified otherwise.
* Gage numbers are British Standard Wire Gage (S.W.G.) numbers.

**British Standard Rivets for General Engineering.** — Dimensions in metric units of rivets for general engineering purposes are given in this British Standard, BS 4620:1970, which is based on ISO Recommendation ISO/R 1051. The snap head rivet dimensions of 14 millimeters and above are taken from the German Standard DIN 124, Round Head Rivets for Steel Structures. The shapes of heads have been restricted to those in common use in the United Kingdom. Table 2 shows the rivet dimensions. Table 1 shows a tentative range of preferred nominal lengths as given in an appendix to the Standard. It is stated that these lengths will be reviewed in the light of usage. The rivets are made by cold or hot forging methods from mild steel, copper, brass, pure aluminum, aluminum alloys, or other suitable metal. It is stated that the radius under the head of a rivet shall run smoothly into the face of the head and shank without step or discontinuity.

In this Standard, the following definitions apply: (1) *Nominal diameter.* The diameter of the shank. (2) *Nominal length of rivets other than countersunk or raised countersunk rivets.* The length from the underside of the head to the end of the shank. (3) *Nominal length of countersunk and raised countersunk rivets.* The distance from the periphery of the head to the end of the rivet measured parallel to the axis of the rivet. (4) *Manufactured head.* The head on the rivet as received from the manufacturer.

**Table 1.  Tentative Range of Lengths for Rivets** (Appendix to B.S. 4620:1970)

| Nom. Shank Diam. | 3 | 4 | 5 | 6 | 8 | 10 | 12 | 14 | 16 | (18) | 20 | (22) | 25 | (28) | 30 | (32) | 35 | (38) | 40 | 45 | .. |
|---|---|---|---|---|---|---|---|---|---|---|---|---|---|---|---|---|---|---|---|---|---|
| 1 | X | X | X | X | X | X | X | X | X |  | X |  |  |  |  |  |  |  |  |  |  |
| 1.2 | X | X | X | X | X | X | X | X | X |  | X |  |  |  |  |  |  |  |  |  |  |
| 1.6 | X | X | X | X | X | X | X | X | X | X |  | X | X |  |  |  |  |  |  |  |  |
| 2 | X | X | X | X | X | X | X | X | X |  | X | X | X |  |  |  |  |  |  |  |  |
| 2.5 | X | X | X | X | X | X | X | X | X | X |  | X | X |  |  |  |  |  |  |  |  |
| 3 |  | X | X | X | X | X | X | X | X | X | X | X | X |  |  |  |  |  |  |  |  |
| (3.5) |  |  |  |  |  |  |  |  |  |  |  |  |  |  |  |  |  |  |  |  |  |
| 4 |  |  |  | X | X | X | X | X | X | X |  | X | X |  |  |  |  |  |  |  |  |
| 5 |  |  |  | X | X | X | X | X | X | X | X | X | X |  | X | X | X | X |  | X |  |
| 6 |  |  |  | X | X | X | X | X | X | X | X | X | X |  | X | X | X | X |  | X |  |

| Nom. Shank Diam. | 10 | 12 | 14 | 16 | (18) | 20 | (22) | 25 | (28) | 30 | (32) | 35 | (38) | 40 | 45 | 50 | 55 | 60 | 65 | 70 | 75 |
|---|---|---|---|---|---|---|---|---|---|---|---|---|---|---|---|---|---|---|---|---|---|
| (7) |  |  |  |  |  |  |  |  |  |  |  |  |  |  |  |  |  |  |  |  |  |
| 8 | X | X | X | X | X | X | X | X | X | X |  | X | X |  | X | X |  |  |  |  |  |
| 10 |  |  | X | X | X | X | X | X | X | X |  | X | X |  | X | X |  |  |  |  |  |
| 12 |  |  |  |  |  |  |  | X |  | X |  | X |  | X | X |  | X |  |  |  | X |
| (14) |  |  |  |  |  |  |  |  |  |  |  |  |  |  |  |  |  |  |  |  |  |
| 16 |  |  |  |  |  |  |  |  |  |  |  |  |  | X | X | X | X | X | X |  | X |

| Nom. Shank Diam. | 45 | 50 | 55 | 60 | 65 | 70 | 75 | 80 | 85 | 90 | (95) | 100 | (105) | 110 | (115) | 120 | (125) | 130 | 140 | 150 | 160 |
|---|---|---|---|---|---|---|---|---|---|---|---|---|---|---|---|---|---|---|---|---|---|
| (18) |  |  |  |  |  |  |  |  |  |  |  |  |  |  |  |  |  |  |  |  |  |
| 20 | X |  |  | X |  | X |  | X |  |  |  |  |  |  |  |  |  |  |  |  |  |
| (22) |  |  |  |  |  |  |  |  |  |  |  |  |  |  |  |  |  |  |  |  |  |
| 24 |  |  |  |  | X |  | X |  | X | X |  | X |  |  |  |  |  |  |  |  |  |
| (27) |  |  |  |  |  |  |  |  |  |  |  |  |  |  |  |  |  |  |  |  |  |
| 30 |  |  |  |  |  |  |  |  | X | X |  | X |  | X |  | X |  |  |  |  |  |
| (33) |  |  |  |  |  |  |  |  |  |  |  |  |  |  |  |  |  |  |  |  |  |
| 36 |  |  |  |  |  |  |  |  |  | X |  | X |  | X |  | X |  | X |  |  |  |
| (39) |  |  |  |  |  |  |  |  |  |  |  | X |  | X |  | X |  | X | X | X | X |

All dimensions are in millimetres.

*Note:* Sizes and lengths shown in parenthesis are non-preferred and should be avoided if possible.

**Table 2. British Standard Rivets for General Engineering Purposes** (B.S. 4620:1970)

| 60° Csk. and Raised Csk. Head | 90° Csk. Head | Snap Head | Universal Head | Flat Head |
|---|---|---|---|---|

\* $K = 0.43\,d$ (for ref. only)    † $K = 0.5\,d$ (for ref. only)

### HOT FORGED RIVETS

| Nom. Shank Diam. d | Tol. on Diam. d | 60° Csk. and Raised Csk. Head | | Snap Head | | Universal Head | | | |
|---|---|---|---|---|---|---|---|---|---|
| | | Nom. Diam. D | Height of Raise W | Nom. Diam. D | Nom. Depth K | Nom. Diam. D | Nom. Depth K | Rad. R | Rad. r |
| (14) | | 21 | 2.8 | 22 | 9 | 28 | 5.6 | 42 | 8.4 |
| 16 | ±0.43 | 24 | 3.2 | 25 | 10 | 32 | 6.4 | 48 | 9.6 |
| (18) | | 27 | 3.6 | 28 | 11.5 | 36 | 7.2 | 54 | 11 |
| 20 | | 30 | 4.0 | 32 | 13 | 40 | 8.0 | 60 | 12 |
| (22) | ±0.52 | 33 | 4.4 | 36 | 14 | 44 | 8.8 | 66 | 13 |
| 24 | | 36 | 4.8 | 40 | 16 | 48 | 9.6 | 72 | 14 |
| (27) | | 40 | 5.4 | 43 | 17 | 54 | 10.8 | 81 | 16 |
| 30 | | 45 | 6.0 | 48 | 19 | 60 | 12.0 | 90 | 18 |
| (33) | ±0.62 | 50 | 6.6 | 53 | 21 | 66 | 13.2 | 99 | 20 |
| 36 | | 55 | 7.2 | 58 | 23 | 72 | 14.4 | 108 | 22 |
| (39) | | 59 | 7.8 | 62 | 25 | 78 | 15.6 | 117 | 23 |

### COLD FORGED RIVETS

| Nom. Shank Diam. d | Tol. on Diam. d | 90° Csk. Head | Snap Head | | Universal Head | | | | Flat Head | |
|---|---|---|---|---|---|---|---|---|---|---|
| | | Nom. Diam. D | Nom. Diam. D | Nom. Depth K | Nom. Diam. D | Nom. Depth K | Rad. R | Rad. r | Nom. Diam. D | Nom. Depth K |
| 1 | | 2 | 1.8 | 0.6 | 2 | 0.4 | 3.0 | 0.6 | 2 | 0.25 |
| 1.2 | | 2.4 | 2.1 | 0.7 | 2.4 | 0.5 | 3.6 | 0.7 | 2.4 | 0.3 |
| 1.6 | ±0.07 | 3.2 | 2.8 | 1.0 | 3.2 | 0.6 | 4.8 | 1.0 | 3.2 | 0.4 |
| 2 | | 4 | 3.5 | 1.2 | 4 | 0.8 | 6.0 | 1.2 | 4 | 0.5 |
| 2.5 | | 5 | 4.4 | 1.5 | 5 | 1.0 | 7.5 | 1.5 | 5 | 0.6 |
| 3 | | 6 | 5.3 | 1.8 | 6 | 1.2 | 9.0 | 1.8 | 6 | 0.8 |
| (3.5) | | 7 | 6.1 | 2.1 | 7 | 1.4 | 10.5 | 2.1 | 7 | 0.9 |
| 4 | ±0.09 | 8 | 7 | 2.4 | 8 | 1.6 | 12 | 2.4 | 8 | 1.0 |
| 5 | | 10 | 8.8 | 3.0 | 10 | 2.0 | 15 | 3.0 | 10 | 1.3 |
| 6 | | 12 | 10.5 | 3.6 | 12 | 2.4 | 18 | 3.6 | 12 | 1.5 |
| (7) | | 14 | 12.3 | 4.2 | 14 | 2.8 | 21 | 4.2 | 14 | 1.8 |
| 8 | ±0.11 | 16 | 14 | 4.8 | 16 | 3.2 | 24 | 4.8 | 16 | 2 |
| 10 | | 20 | 18 | 6.0 | 20 | 4.0 | 30 | 6 | 20 | 2.5 |
| 12 | | 24 | 21 | 7.2 | 24 | 4.8 | 36 | 7.2 | ... | ... |
| (14) | ±0.14 | ... | 25 | 8.4 | 28 | 5.6 | 42 | 8.4 | ... | ... |
| 16 | | ... | 28 | 9.6 | 32 | 6.4 | 48 | 9.6 | ... | ... |

All dimensions are in millimetres. Sizes shown in parentheses are non-preferred.

## SCREW THREAD SYSTEMS

**Screw Thread Forms.** — Of the various screw thread forms which have been developed, the most used are those having symmetrical sides inclined at equal angles with a vertical center line through the thread apex. Present-day examples of such threads would include the Unified, the Whitworth and the Acme forms. One of the early forms was the Sharp V which is now used only occasionally. Symmetrical threads are relatively easy to manufacture and inspect and hence are widely used on mass-produced general-purpose threaded fasteners of all types. In addition to general-purpose fastener applications, certain threads are used to repeatedly move or translate machine parts against heavy loads. For these so-called translation threads a stronger form is required. The most widely used translation thread forms are the square, the Acme, and the buttress. Of these, the square thread is the most efficient, but it is also the most difficult to cut owing to its parallel sides and it cannot be adjusted to compensate for wear. Although less efficient, the Acme form of thread has none of the disadvantages of the square form and has the advantage of being somewhat stronger. The buttress form is used for translation of loads in one direction only because of its non-symmetrical form and combines the high efficiency and strength of the square thread with the ease of cutting and adjustment of the Acme thread.

**Sharp V-thread.** — The sides of the thread form an angle of 60 degrees with each other. The top and bottom of the thread are, theoretically, sharp, but in practice it is necessary to make the thread with a slight flat. There is no standard adopted for this flat, but it is usually made about one-twenty-fifth of the pitch. If $p$ = pitch of thread, and $d$ = depth of thread, then:

$$d = p \times \cos 30 \text{ deg.} = 0.866\, p = \frac{0.866}{\text{no. of threads per inch}}$$

Some modified V-threads, for locomotive boiler taps particularly, have a depth of 0.8 × pitch.

**American National and Unified Screw Thread Forms.** — The American National form (formerly known as the United States Standard) was used for many years for most screws, bolts, and miscellaneous threaded products produced in the United States. The American National Standard for Unified Screw Threads now in use includes certain modifications of the former standard as is explained on page 1261. The Basic Profile is shown below and is identical for both UN and UNR screw threads. In this figure $H$ is the height of a sharp V-thread.

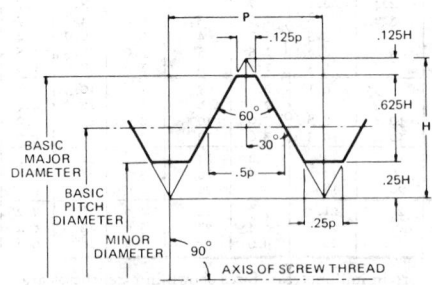

**Definitions of Screw Threads.** — The following definitions are given in American National Standard ANSI B1.7-1977, "Nomenclature, Definitions, and Letter Symbols for Screw Threads," and refer to both straight and taper threads.

*Actual Size:* An actual size is a measured size.

*Allowance:* An allowance is the prescribed difference between the design size and the basic size of a thread.

*Axis of Thread:* The axis of a thread is coincident with the axis of its pitch cylinder or cone.

*Basic Form of Thread:* The basic form of thread is the permanent reference profile from which the design forms for both the external and internal threads are developed.

*Basic Size:* The basic size is that size from which the limits of size are derived by the application of allowances and tolerances.

*Bilateral Tolerance:* This is a tolerance in which variation is permitted in both directions from the specified dimension.

*Black Crest Thread:* This is a thread whose crest displays an unfinished cast, rolled, or forged surface.

*Blunt Start (Blunt End) Thread:* "Blunt start" or "blunt end" designates the removal of the incomplete thread at the end of the thread. This is a feature of threaded parts that are repeatedly assembled by hand, such as hose couplings and thread plug gages, to prevent cutting of hands and crossing of threads. It was formerly known as a Higbee cut.

*Chamfer:* This is a conical surface at the end of a thread.

*Classes of Threads:* These are distinguished from each other by the amounts of tolerance or tolerance and allowance specified.

*Clearance Fit:* This is a fit having limits of size so prescribed that a clearance always results when mating parts are assembled.

*Complete Thread:* The complete (full form) thread is that cross section of a threaded length having full form at crest and root. (See also *Effective Thread* and *Length of Complete Thread.*) *Note:* Formerly in pipe thread terminology this was referred to as "the perfect thread" but that term is no longer considered desirable.

*Crest:* This is that surface of a thread which joins the flanks of the thread and is farthest from the cylinder or cone from which the thread projects.

*Crest Truncation:* This is the radial distance between the sharp crest (crest apex) and the cylinder or cone that would bound the crest.

*Depth of Thread Engagement:* The depth (or height) of thread engagement between two coaxially assembled mating threads is the radial distance by which their thread forms overlap each other.

*Design Size:* This is the basic size with allowance applied, from which the limits of size are derived by the application of a tolerance. If there is no allowance, the design size is the same as the basic size.

*Deviation:* This is a variation from an established dimension, position, standard, or value. (See also *Error.*)

*Dimension:* This is a numerical value expressed in appropriate units of measure and indicated on drawings along with lines, symbols, and notes to define the geometrical characteristic of an object.

*Effective Size:* (See *Virtual Diameter.*)

*Effective Thread:* The effective (or useful) thread includes the complete thread, and those portions of the incomplete thread which are fully formed at the root but not at the crest (in taper pipe threads this includes the so-called black crest threads); thus excluding the vanish thread.

*Element:* Elements of a thread are flank angle, root contour, crest contour, pitch, lead angle, surface finish, major, minor and pitch diameters.

*Error:* This is the algebraic difference between an observed or measured value beyond tolerance limits, and the specified value.

*External Thread:* This is a thread on a cylindrical or conical external surface.

*Fit:* This is the general term used to signify the range of tightness or looseness which results from the application of a specific combination of allowances and tolerances in mating parts.

*Flank:* The flank (or side) of a thread is either surface connecting the crest with the root. The flank surface intersection with an axial plane is theoretically a straight line.

*Flank Angle:* The flank angles are the angles between the individual flanks and the perpendicular to the axis of the thread, measured in an axial plane. A flank angle of a symmetrical thread is commonly termed the half-angle of thread.

*Height of Thread:* The height (or depth) of thread is the distance, measured radially, between the major and minor cylinders or cones, respectively.

*Helix Angle:* On a straight thread, the helix angle is the angle made by the helix of the thread and its relation to the thread axis. On a taper thread, the helix angle at a given axial position is the angle made by the conical spiral of the thread with the axis of the thread. The helix angle is the complement of the lead angle. (See also page 1480 for diagram.)

*Higbee Cut:* (See *Blunt Start (Blunt End) Thread.*)

*Imperfect Thread:* (See *Incomplete Thread.*)

*Included Angle:* This is the angle between the flanks of the thread measured in an axial plane.

*Incomplete Thread:* This is a threaded profile having either crests or roots, or both, not fully formed, resulting from their intersection with the cylindrical or end surface of the work or the vanish cone. It may occur at either end of the thread.

*Interference Fit:* This is a fit having limits of size so prescribed that an interference always results when mating parts are assembled.

*Internal Thread:* This is a thread on a cylindrical or conical internal surface.

*Lead:* When a threaded part is rotated about its axis with respect to a fixed mating thread, the lead is the axial distance moved by the part with relation to the amount of angular rotation. The basic lead is commonly specified as the axial distance moved in one complete rotation. It is necessary to distinguish measurement of lead from measurement of pitch, as uniformity of pitch measurement does not assure uniformity of lead. (Variations in either pitch or lead cause the virtual diameter of the thread to differ from the pitch diameter.)

*Lead Angle:* On a straight thread, the lead angle is the angle made by the helix of the thread at the pitch line with a plane perpendicular to the axis. On a taper thread, the lead angle at a given axial position is the angle made by the conical spiral of the thread with the perpendicular to the axis at the pitch line. (See also page 1480 for diagram.)

*Lead Thread:* This is that portion of the incomplete thread that is fully formed at the root but not fully formed at the crest which occurs at the entering end of either an external or internal thread.

*Left-hand Thread:* A thread is a left-hand thread if, when viewed axially, it winds in a counterclockwise and receding direction. All left-hand threads are designated LH.

*Length of Complete Thread:* This is the axial length of a threaded part where the thread section has full form at both crest and root; that is, the vanish threads are not included. However, on commercial fasteners where there are unfilled crests at the start of rolled threads or a chamfer at the start of a thread not exceeding two pitches in length, this is traditionally included in the specified thread length. (See also *Complete Thread, Lead Thread,* and *Effective Thread.*)

*Length of Thread Engagement:* The length of thread engagement of two mating threads is the axial distance over which the two threads are designed to contact. (See also *Length of Complete Thread.*)

*Limits of Size:* These are the applicable maximum and minimum sizes.

*Major Clearance:* This is the radial distance between the root of the internal thread and the crest of the external thread of the coaxially assembled designed forms of mating threads.

*Major Cone:* This is the cone that would bound the crests of an external taper thread or the roots of an internal taper thread.

*Major Cylinder:* This is the cylinder that would bound the crests of an external straight thread or the roots of an internal straight thread.

*Major Diameter:* On a straight thread the major diameter is that of the major cylinder. On a taper thread the major diameter at a given position on the thread axis is that of the major cone at that position. (See also *Major Cylinder* and *Major Cone.*)

*Maximum Material Condition (MMC):* This is the condition where a feature of size contains the maximum amount of material within the stated limits of size. For example, minimum internal thread size or maximum external thread size.

*Minimum Material Condition (Least Material Condition (LMC)):* This is the condition where a feature of size contains the least amount of material within the stated limits of size. For example, maximum internal thread size or minimum external thread size.

*Minor Clearance:* This is the radial distance between the crest of the internal thread and the root of the external thread of the coaxially assembled design forms of mating threads.

*Minor Cone:* This is the cone that would bound the roots of an external taper thread or the crests of an internal taper thread.

*Minor Cylinder:* This is the cylinder that would bound the roots of an external straight thread or the crests of an internal straight thread.

*Minor Diameter:* On a straight thread the minor diameter is that of the minor cylinder. On a taper thread the minor diameter at a given position on the thread axis is that of the minor cone at that position. (See also *Minor Cylinder* and *Minor Cone.*)

*Multiple-Start Thread:* This is a thread in which the lead is an integral multiple, other than one, of the pitch.

*Nominal Size:* This is the designation used for the purpose of general identification.

*Parallel Thread:* (See *Screw Thread.*)

*Partial Thread:* (See *Vanish Thread.*)

*Pitch:* The pitch of a thread having uniform spacing is the distance measured parallel with its axis between corresponding points on adjacent thread forms in the same axial plane and on the same side of the axis. Pitch is equal to the lead divided by the number of thread starts.

*Pitch Cone:* The pitch cone is one of such apex angle and location of its vertex and axis that its surface would pass through a taper thread in such a manner as to make the widths of the thread ridge and the thread groove equal. It is, therefore, located equidistantly between the sharp major and minor cones of a given thread form. On a theoretically perfect taper thread, these widths are equal to one-half the basic pitch. (See also *Axis of Thread* and *Pitch Diameter.*)

*Pitch Cylinder:* The pitch cylinder is one of such diameter and location of its axis that its surface would pass through a straight thread in such a manner as to make the widths of the thread ridge and groove equal. It is, therefore, located equidistantly between the sharp major and minor cylinders of a given thread form. On a theoretically perfect thread these widths are equal to one-half the basic pitch. (See also *Axis of Thread* and *Pitch Diameter.*)

*Pitch Diameter:* On a straight thread the pitch diameter is the diameter of the pitch cylinder. On a taper thread the pitch diameter at a given position on the thread axis is the diameter of the pitch cone at that position. *Note:* When the crest of a thread is truncated beyond the pitch line, the pitch diameter and pitch cylinder or pitch cone would be based on a theoretical extension of the thread flanks.

*Pitch Line:* This is the generator of the cylinder or cone specified in *Pitch Cylinder* and *Pitch Cone.*

*Right-hand Thread:* A thread is a right-hand thread if, when viewed axially, it winds in a clockwise and receding direction. A thread is considered to be right-hand unless specifically indicated otherwise.

*Root:* A root is that surface of the thread which joins the flanks of adjacent thread forms and is immediately adjacent to the cylinder or cone from which the thread projects.

*Root Truncation:* This is the radial distance between the sharp root (root apex) and the cylinder or cone that would bound the root.

*Runout:* This is a total tolerance used to control the functional relationship of one or more features of a part to a datum axis. There are two types of runout control: (1) circular runout and (2) total runout. Circular runout provides composite control of the circular elements of a surface. Total runout provides composite control of all surface elements.

*Screw Thread:* A screw thread is a ridge, usually of uniform section and produced by forming a groove in the form of a helix on the external or internal surface of a cylinder, or in the form of a conical spiral on the external or internal surface of a cone or frustum of a cone. A screw thread formed on a cylinder is known as a straight or parallel screw thread to distinguish it from a taper screw thread which is formed on a cone or frustum of a cone.

*Sharp Crest (Crest Apex):* This is the apex formed by the intersection of the flanks of a thread when extended, if necessary, beyond the crest.

*Sharp Root (Root Apex):* This is the apex formed by the intersection of the adjacent flanks of adjacent threads when extended, if necessary, beyond the root.

*Standoff:* This is the axial distance between specified reference points on external and internal taper thread members or gages, when assembled with a specified torque or under other specified conditions.

*Straight Thread:* See *Screw Thread.*

*Taper Thread:* See *Screw Thread.*

*Tensile Stress Area:* The tensile stress area of an externally threaded part is the circular cross-sectional area, normal to the axis, of a theoretical circular cylinder which would fail under tension at the same load at which the threaded part fails, if the materials of both were to have the same mechanical properties.

*Thread:* A thread is a portion of a screw thread encompassed by one pitch. On a single-start thread it is equal to one turn. (See also *Threads per Inch* and *Turns per Inch.*)

*Thread Runout:* See *Vanish Thread.*

*Thread Series:* Thread Series are groups of diameter/pitch combinations distinguished from each other by the number of threads per inch applied to specific diameters.

*Thread Shear Area:* The thread shear area of an external thread is the effective area in shear at a specified diameter of the mated internal thread. The thread shear area of an internal thread is the effective area in shear at a specified diameter of the mated external thread. *Note:* The specified diameters are usually the maximum minor diameter of the mated internal thread and the minimum major diameter of the mated external thread, respectively.

*Threads per Inch:* The number of threads per inch is the reciprocal of the pitch in inches.

*Tolerance:* The total amount by which a specific dimension is permitted to vary. The tolerance is the difference between the maximum and minimum limits.

*Tolerance Limit:* This is the variation, positive or negative, by which a size is permitted to depart from the design size.

*Total Thread:* This includes the complete and all of the incomplete thread, thus including the vanish thread and the lead thread.

*Transition Fit:* This is a fit having limits of size so prescribed that either a clearance or an interference may result when mating parts are assembled.

*Turns per Inch:* The number of turns per inch is the reciprocal of the lead in inches.

*Unilateral Tolerance:* A tolerance in which variation is permitted in one direction from the specified dimension.

*Vanish Thread (Partial Thread, Wash-Out Thread,* or *Thread Run-Out):* This is that portion of the incomplete thread which is not fully formed at the root or at crest and root. It is produced by the chamfer at the starting end of the thread forming tool.

*Virtual Diameter:* The virtual diameter of an external or internal thread is the pitch diameter of the enveloping thread of perfect pitch, lead and flank angles, having full depth of engagement but clear at crests and roots, and of a specified length of engagement. It may be derived by adding to the pitch diameter in the case of an external thread, or subtracting from the pitch diameter in the case of an internal thread, the cumulative effects of the deviations from the specified profile, including variations in lead, and flank angle over a specified length of engagement. The effects of taper, out-of-roundness, and surface defects may be positive or negative on either external or internal threads. A perfect external or internal thread gage having a pitch diameter equal to that of the specified material limit, and having a clearance at the crest and root, is the enveloping thread corresponding to that limit. *Note:* This is also called the Functional Diameter, Effective Size, or Virtual Effective Diameter.

*Wash-Out Thread:* See *Vanish Thread.*

## American Standard for Unified Screw Threads. — American Standard B1.1-1949 was the first American standard to cover those Unified Thread Series agreed upon by the United Kingdom, Canada, and the United States to obtain screw thread interchangeability among these three nations. These Unified threads are now the basic American standard for fastening types of screw threads. In relation to previous American practice, Unified threads have substantially the same thread form and are mechanically interchangeable with the former American National threads of the same diameter and pitch. The principal differences between the two systems lie in: (1) the application of allowances; (2) the variation of tolerances with size; (3) difference in amount of pitch diameter tolerance on external and internal threads; and (4) differences in thread designation.

In the Unified system an allowance is provided on both the Classes 1A and 2A external threads whereas in the American National system only the Class 1 external thread has an allowance. Also, in the Unified system the pitch diameter tolerance of an internal thread is 30 per cent greater than that of the external thread, whereas they are equal in the American National system.

*Revised Standard:* The revised screw thread standard ANSI B1.1-1974 is much the same as B1.1-1960 except for several changes in scope. Additions include the UNR thread form and "Unified Screw Threads—Metric Translation" which was formerly B1.1a-1968. Definition of screw thread compatibility criteria is no longer included. In order to emphasize that Unified Screw Threads are based on inch modules, they may be designated as Unified Inch Screw Threads.

Where the letters U, A or B do not appear in the thread designations, the threads conform to the outdated American National screw threads. Data for Classes 2 and 3 of those former screw threads are given, beginning on page 1303, as a matter of reference.

*Advantages of Unified Threads:* The Unified standard is designed to correct certain production difficulties resulting from the former standard. Often, under the old system, the tolerances of the product were practically absorbed by the combined tool and gage tolerances, leaving little for a working tolerance in manufacture. Somewhat greater tolerances are now provided for nut threads. As contrasted with the old "classes of fit" 1, 2, and 3, for each of which the pitch diameter tolerance on the external and internal threads were equal, the Classes 1B, 2B, and 3B (internal) threads in the new standard have, respectively, a 30 per cent larger pitch diameter tolerance than the 1A, 2A, and 3A (external) threads. Relatively more tolerance is provided for fine threads than for coarse threads of the same pitch. In cases where previous tolerances were more liberal than required, they were reduced.

**Thread Form.** — The Design Profiles for Unified screw threads, shown on page 1263, define the maximum material condition for external and internal threads with no allowance and are derived from the Basic Profile, shown on page 1256.

*UN External Screw Threads:* A flat root contour is specified, but it is necessary to provide for some threading tool crest wear, hence a rounded root contour cleared beyond the $0.25p$ flat width of the Basic Profile is optional.

*UNR External Screw Threads:* In order to reduce the rate of threading tool crest wear and to improve fatigue strength of a flat root thread, the Design Profile of the UNR thread has a non-reversing continuous curved root tangent to the thread flanks at the intersection of the minor diameter and thread flanks of the Basic Profile ($0.83333H$, where $H$ is the height of a sharp V-thread). The resulting theoretical radius of curvature is $0.14434p$.

In practice the root contour of the UNR thread may not be definable by a single radius as shown in the Design Profile. In order to allow for this condition and to provide manufacturing tolerance, a minimum radius of $0.10825p$ is specified for any portion of the root contour. The maximum root radius is theoretically limited by the Basic Profile and for other than the basic condition defined by the Design Profile may be greater than $0.14434p$.

*UN and UNR External Screw Threads:* The Design Profiles of both UN and UNR external screw threads have flat crests. However, in practice, product threads are produced with partially or completely rounded crests. A rounded crest tangent at $0.125p$ flat is shown as an option on page 1263.

*UN Internal Screw Thread:* In practice it is necessary to provide for some threading tool crest wear, therefore the root of the Design Profile is rounded and cleared beyond the $0.125p$ flat width of the Basic Profile.

There is no internal UNR screw thread.

**Thread Series.** — Thread series are groups of diameter-pitch combinations distinguished from each other by the number of threads per inch applied to a specific diameter. The various diameter-pitch combinations of eleven standard series are shown in Table 2. The limits of size of threads in the eleven standard series together with certain selected combinations of diameter and pitch, as well as the symbols for designating the various threads, are given in Table 4.

*Coarse-Thread Series:* This series, UNC, is the one most commonly used in the bulk production of bolts, screws, nuts and other general engineering applications. It is also used for threading into lower tensile strength materials such as cast iron, mild steel and softer materials (bronze, brass, aluminum, magnesium and plastics) to obtain the optimum resistance to stripping of the internal thread. It is applicable for rapid assembly or disassembly, or if corrosion or slight damage is possible.

*Fine-Thread Series:* This series, UNF, is suitable for the production of bolts, screws, and nuts and for other applications where the Coarse series is not applicable. External threads of this series have greater tensile stress area than comparable sizes of the Coarse series. The Fine series is suitable when the resistance to stripping

### American National Standard Unified Internal and External Screw Thread Design Forms
(Maximum Material Condition)

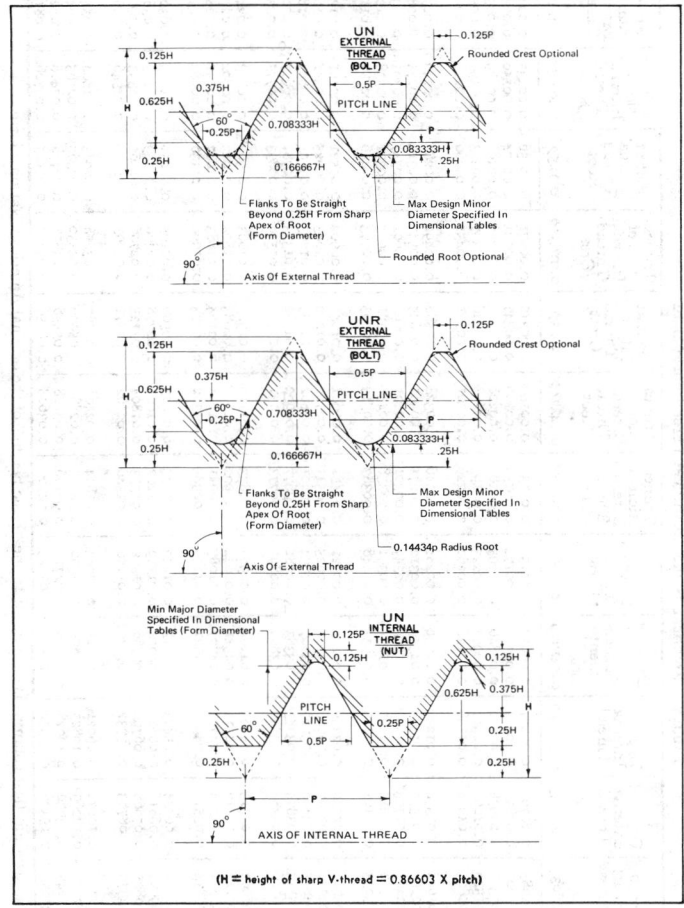

(H = height of sharp V-thread = 0.86603 X pitch)

of both external and mating internal threads equals or exceeds the tensile load carrying capacity of the externally threaded member. It is also used where the length of engagement is short, where a smaller lead angle is desired, or where the wall thickness demands a fine pitch.

*Extra-Fine-Thread Series:* This series, UNEF, is applicable where even finer pitches of threads are desirable, as for short lengths of engagement and for thin-walled tubes, nuts, ferrules, or couplings. It is also generally applicable under the conditions stated above for the fine threads.

Table 1. American Standard Unified Thread Form Data

| Threads per Inch $n$ | Pitch $p$ | Depth of Sharp V-Thread $0.86603p$ | Depth of External Thread $0.61343p$ | Depth of Internal Thread $0.54127p$ | Depth of Thread Engagement $0.54127p$ | Flat at External Thread Crest $0.125p$ | Truncation of External Thread Crest $0.10825p$ | Truncation of External Thread Root $0.14434p$ | Flat at Internal Thread Crest $0.25p$ | Truncation of Internal Thread Crest $0.21651p$ | Flat at Internal Thread Root $0.125p$ | Truncation of Internal Thread Root $0.10825p$ | Addendum of External Thread $0.32476p$ |
|---|---|---|---|---|---|---|---|---|---|---|---|---|---|
| 80 | 0.01250 | 0.01083 | 0.00767 | 0.00677 | 0.00677 | 0.00156 | 0.00135 | 0.00180 | 0.00312 | 0.00271 | 0.00156 | 0.00135 | 0.00406 |
| 72 | 0.01389 | 0.01203 | 0.00852 | 0.00752 | 0.00752 | 0.00174 | 0.00150 | 0.00200 | 0.00347 | 0.00301 | 0.00174 | 0.00150 | 0.00451 |
| 64 | 0.01563 | 0.01353 | 0.00958 | 0.00846 | 0.00846 | 0.00195 | 0.00169 | 0.00226 | 0.00391 | 0.00338 | 0.00195 | 0.00169 | 0.00507 |
| 56 | 0.01786 | 0.01546 | 0.01095 | 0.00967 | 0.00967 | 0.00223 | 0.00193 | 0.00258 | 0.00446 | 0.00387 | 0.00223 | 0.00193 | 0.00580 |
| 48 | 0.02083 | 0.01804 | 0.01278 | 0.01128 | 0.01128 | 0.00260 | 0.00226 | 0.00301 | 0.00521 | 0.00451 | 0.00260 | 0.00226 | 0.00677 |
| 44 | 0.02273 | 0.01968 | 0.01394 | 0.01230 | 0.01230 | 0.00284 | 0.00246 | 0.00328 | 0.00568 | 0.00492 | 0.00284 | 0.00246 | 0.00738 |
| 40 | 0.02500 | 0.02165 | 0.01534 | 0.01353 | 0.01353 | 0.00312 | 0.00271 | 0.00361 | 0.00625 | 0.00541 | 0.00312 | 0.00271 | 0.00812 |
| 36 | 0.02778 | 0.02406 | 0.01704 | 0.01504 | 0.01504 | 0.00347 | 0.00301 | 0.00401 | 0.00694 | 0.00601 | 0.00347 | 0.00301 | 0.00902 |
| 32 | 0.03125 | 0.02706 | 0.01917 | 0.01691 | 0.01691 | 0.00391 | 0.00338 | 0.00451 | 0.00781 | 0.00677 | 0.00391 | 0.00338 | 0.01015 |
| 28 | 0.03571 | 0.03093 | 0.02191 | 0.01933 | 0.01933 | 0.00446 | 0.00387 | 0.00515 | 0.00893 | 0.00773 | 0.00446 | 0.00387 | 0.01160 |
| 24 | 0.04167 | 0.03608 | 0.02556 | 0.02255 | 0.02255 | 0.00521 | 0.00451 | 0.00601 | 0.01042 | 0.00902 | 0.00521 | 0.00451 | 0.01353 |
| 20 | 0.05000 | 0.04330 | 0.03067 | 0.02706 | 0.02706 | 0.00625 | 0.00541 | 0.00722 | 0.01250 | 0.01083 | 0.00625 | 0.00541 | 0.01624 |
| 18 | 0.05556 | 0.04811 | 0.03408 | 0.03007 | 0.03007 | 0.00694 | 0.00601 | 0.00802 | 0.01389 | 0.01203 | 0.00694 | 0.00601 | 0.01804 |
| 16 | 0.06250 | 0.05413 | 0.03834 | 0.03383 | 0.03383 | 0.00781 | 0.00677 | 0.00902 | 0.01562 | 0.01353 | 0.00781 | 0.00677 | 0.02030 |
| 14 | 0.07143 | 0.06186 | 0.04382 | 0.03866 | 0.03866 | 0.00893 | 0.00773 | 0.01031 | 0.01786 | 0.01546 | 0.00893 | 0.00773 | 0.02320 |
| 13 | 0.07692 | 0.06662 | 0.04719 | 0.04164 | 0.04164 | 0.00962 | 0.00833 | 0.01110 | 0.01923 | 0.01665 | 0.00962 | 0.00833 | 0.02498 |
| 12 | 0.08333 | 0.07217 | 0.05112 | 0.04511 | 0.04511 | 0.01042 | 0.00902 | 0.01203 | 0.02083 | 0.01804 | 0.01042 | 0.00902 | 0.02706 |
| 11½ | 0.08696 | 0.07531 | 0.05334 | 0.04707 | 0.04707 | 0.01087 | 0.00941 | 0.01255 | 0.02174 | 0.01883 | 0.01087 | 0.00941 | 0.02824 |
| 11 | 0.09091 | 0.07873 | 0.05577 | 0.04921 | 0.04921 | 0.01136 | 0.00984 | 0.01312 | 0.02273 | 0.01968 | 0.01136 | 0.00984 | 0.02952 |
| 10 | 0.10000 | 0.08660 | 0.06134 | 0.05413 | 0.05413 | 0.01250 | 0.01083 | 0.01443 | 0.02500 | 0.02165 | 0.01250 | 0.01083 | 0.03248 |
| 9 | 0.11111 | 0.09623 | 0.06816 | 0.06014 | 0.06014 | 0.01389 | 0.01203 | 0.01604 | 0.02778 | 0.02406 | 0.01389 | 0.01203 | 0.03608 |
| 8 | 0.12500 | 0.10825 | 0.07668 | 0.06766 | 0.06766 | 0.01562 | 0.01353 | 0.01804 | 0.03125 | 0.02706 | 0.01562 | 0.01353 | 0.04059 |
| 7 | 0.14286 | 0.12372 | 0.08763 | 0.07732 | 0.07732 | 0.01786 | 0.01546 | 0.02062 | 0.03571 | 0.03093 | 0.01786 | 0.01546 | 0.04639 |
| 6 | 0.16667 | 0.14434 | 0.10224 | 0.09021 | 0.09021 | 0.02083 | 0.01804 | 0.02406 | 0.04167 | 0.03608 | 0.02083 | 0.01804 | 0.05413 |
| 5 | 0.20000 | 0.17321 | 0.12269 | 0.10825 | 0.10825 | 0.02500 | 0.02165 | 0.02887 | 0.05000 | 0.04330 | 0.02500 | 0.02165 | 0.06495 |
| 4½ | 0.22222 | 0.19245 | 0.13632 | 0.12028 | 0.12028 | 0.02778 | 0.02406 | 0.03208 | 0.05556 | 0.04811 | 0.02778 | 0.02406 | 0.07217 |
| 4 | 0.25000 | 0.21651 | 0.15336 | 0.13532 | 0.13532 | 0.03125 | 0.02706 | 0.03608 | 0.06250 | 0.05413 | 0.03125 | 0.02706 | 0.08119 |

Twice the external thread addendum (last column) is equivalent to the "basic height" of the original American National form.

**Table 2. Diameter-Pitch Combinations for Standard Series of Unified Threads***

| Sizes† | Basic Major Diam. | Threads per Inch | | | | | | | | | | |
|---|---|---|---|---|---|---|---|---|---|---|---|---|
| | | Series with Graded Pitches | | | Series with Uniform (Constant) Pitches | | | | | | | |
| | | Coarse UNC | Fine[a] UNF | Extra-fine[b] UNEF | 4UN | 6UN | 8UN | 12UN | 16UN | 20UN | 28UN | 32UN |
| 0 | 0.0600 | .. | 80 | Series designation shown indicates the UN thread form, however, the UNR thread form may be specified by substituting UNR in place of UN in all designations for external threads. | | | | | | | | |
| (1) | 0.0730 | 64 | 72 | | | | | | | | | |
| 2 | 0.0860 | 56 | 64 | | | | | | | | | |
| (3) | 0.0990 | 48 | 56 | | | | | | | | | |
| 4 | 0.1120 | 40 | 48 | | | | | | | | | |
| 5 | 0.1250 | 40 | 44 | .. | .. | .. | .. | .. | .. | .. | .. | .. |
| 6 | 0.1380 | 32 | 40 | .. | .. | .. | .. | .. | .. | .. | .. | UNC |
| 8 | 0.1640 | 32 | 36 | .. | .. | .. | .. | .. | .. | .. | .. | UNC |
| 10 | 0.1900 | 24 | 32 | .. | .. | .. | .. | .. | .. | .. | .. | UNF |
| (12) | 0.2160 | 24 | 28 | 32 | .. | .. | .. | .. | .. | .. | UNF | UNEF |
| 1/4 | 0.2500 | 20 | 28 | 32 | .. | .. | .. | .. | .. | UNC | UNF | UNEF |
| 5/16 | 0.3125 | 18 | 24 | 32 | .. | .. | .. | .. | .. | 20 | 28 | UNEF |
| 3/8 | 0.3750 | 16 | 24 | 32 | .. | .. | .. | .. | UNC | 20 | 28 | UNEF |
| 7/16 | 0.4375 | 14 | 20 | 28 | .. | .. | .. | .. | 16 | UNF | UNEF | 32 |
| 1/2 | 0.5000 | 13 | 20 | 28 | .. | .. | .. | .. | 16 | UNF | UNEF | 32 |
| 9/16 | 0.5625 | 12 | 18 | 24 | .. | .. | .. | UNC | 16 | 20 | 28 | 32 |
| 5/8 | 0.6250 | 11 | 18 | 24 | .. | .. | .. | 12 | 16 | 20 | 28 | 32 |
| (11/16) | 0.6875 | .. | .. | 24 | .. | .. | .. | 12 | 16 | 20 | 28 | 32 |
| 3/4 | 0.7500 | 10 | 16 | 20 | .. | .. | .. | 12 | UNF | UNEF | 28 | 32 |
| (13/16) | 0.8125 | .. | .. | 20 | .. | .. | .. | 12 | 16 | UNEF | 28 | 32 |
| 7/8 | 0.8750 | 9 | 14 | 20 | .. | .. | .. | 12 | 16 | UNEF | 28 | 32 |
| (15/16) | 0.9375 | .. | .. | 20 | .. | .. | .. | 12 | 16 | UNEF | 28 | 32 |
| 1 | 1.0000 | 8 | 12 | 20 | .. | .. | UNC | UNF | 16 | UNEF | 28 | 32 |
| (1 1/16) | 1.0625 | .. | .. | 18 | .. | .. | 8 | 12 | 16 | 20 | 28 | .. |
| 1 1/8 | 1.1250 | 7 | 12 | 18 | .. | .. | 8 | UNF | 16 | 20 | 28 | .. |
| (1 3/16) | 1.1875 | .. | .. | 18 | .. | .. | 8 | 12 | 16 | 20 | 28 | .. |
| 1 1/4 | 1.2500 | 7 | 12 | 18 | .. | .. | 8 | UNF | 16 | 20 | 28 | .. |
| (1 5/16) | 1.3125 | .. | .. | 18 | .. | .. | 8 | 12 | 16 | 20 | 28 | .. |
| 1 3/8 | 1.3750 | 6 | 12 | 18 | .. | UNC | 8 | UNF | 16 | 20 | 28 | .. |
| (1 7/16) | 1.4375 | .. | .. | 18 | .. | 6 | 8 | 12 | 16 | 20 | 28 | .. |
| 1 1/2 | 1.5000 | 6 | 12 | 18 | .. | UNC | 8 | UNF | 16 | 20 | 28 | .. |
| (1 9/16) | 1.5625 | .. | .. | 18 | .. | 6 | 8 | 12 | 16 | 20 | .. | .. |
| 1 5/8 | 1.6250 | .. | .. | 18 | .. | 6 | 8 | 12 | 16 | 20 | .. | .. |
| (1 11/16) | 1.6875 | .. | .. | 18 | .. | 6 | 8 | 12 | 16 | 20 | .. | .. |
| 1 3/4 | 1.7500 | 5 | .. | .. | .. | 6 | 8 | 12 | 16 | 20 | .. | .. |
| (1 13/16) | 1.8125 | .. | .. | .. | .. | 6 | 8 | 12 | 16 | 20 | .. | .. |
| 1 7/8 | 1.8750 | .. | .. | .. | .. | 6 | 8 | 12 | 16 | 20 | .. | .. |
| (1 15/16) | 1.9375 | .. | .. | .. | .. | 6 | 8 | 12 | 16 | 20 | .. | .. |
| 2 | 2.0000 | 4 1/2 | .. | .. | .. | 6 | 8 | 12 | 16 | 20 | .. | .. |
| (2 1/8) | 2.1250 | .. | .. | .. | .. | 6 | 8 | 12 | 16 | 20 | .. | .. |
| 2 1/4 | 2.2500 | 4 1/2 | .. | .. | .. | 6 | 8 | 12 | 16 | 20 | .. | .. |
| (2 3/8) | 2.3750 | .. | .. | .. | .. | 6 | 8 | 12 | 16 | 20 | .. | .. |
| 2 1/2 | 2.5000 | 4 | .. | .. | UNC | 6 | 8 | 12 | 16 | 20 | .. | .. |
| (2 5/8) | 2.6250 | .. | .. | .. | 4 | 6 | 8 | 12 | 16 | 20 | .. | .. |
| 2 3/4 | 2.7500 | 4 | .. | .. | UNC | 6 | 8 | 12 | 16 | 20 | .. | .. |
| (2 7/8) | 2.8750 | .. | .. | .. | 4 | 6 | 8 | 12 | 16 | 20 | .. | .. |
| 3 | 3.0000 | 4 | .. | .. | UNC | 6 | 8 | 12 | 16 | 20 | .. | .. |
| (3 1/8) | 3.1250 | .. | .. | .. | 4 | 6 | 8 | 12 | 16 | .. | .. | .. |
| 3 1/4 | 3.2500 | 4 | .. | .. | UNC | 6 | 8 | 12 | 16 | .. | .. | .. |
| (3 3/8) | 3.3750 | .. | .. | .. | 4 | 6 | 8 | 12 | 16 | .. | .. | .. |
| 3 1/2 | 3.5000 | 4 | .. | .. | UNC | 6 | 8 | 12 | 16 | .. | .. | .. |
| (3 5/8) | 3.6250 | .. | .. | .. | 4 | 6 | 8 | 12 | 16 | .. | .. | .. |
| 3 3/4 | 3.7500 | 4 | .. | .. | UNC | 6 | 8 | 12 | 16 | .. | .. | .. |
| (3 7/8) | 3.8750 | .. | .. | .. | 4 | 6 | 8 | 12 | 16 | .. | .. | .. |
| 4 | 4.0000 | 4 | .. | .. | UNC | 6 | 8 | 12 | 16 | .. | .. | .. |

* Dimensions of these Standard Series threads for sizes up to 4 inches are given in Table 4.

† Sizes shown in parentheses are secondary sizes. Primary sizes of 4 1/4, 4 1/2, 4 3/4, 5, 5 1/4, 5 1/2, 5 3/4 and 6 inches also are in the 4, 6, 8, 12, and 16 thread series; secondary sizes of 4 1/8, 4 3/8, 4 5/8, 4 7/8, 5 1/8, 5 3/8, 5 5/8, and 5 7/8 also are in the 4, 6, 8, 12, and 16 thread series.

[a] For diameters over 1 1/2 inches, use 12-thread series.

[b] For diameters over 1 11/16 inches, use 16-thread series.

Table 3a. Coarse-Thread Series, UNC, UNRC, and NC — Basic Dimensions

| Sizes | Basic Major Diam., D | Thds. per Inch, n | Basic Pitch Diam.,[a] E | Minor Diameter[c] | | Lead Angle at Basic P.D. | | Area of Minor Diam. at $D-2h_b$ | Tensile Stress Area[b] |
|---|---|---|---|---|---|---|---|---|---|
| | | | | Ext. Thds., $K_s$ | Int. Thds., $K_n$ | Deg. | Min. | | |
| | Inches | | Inches | Inches | Inches | | | Sq. In. | Sq. In. |
| 1 (.073)* | 0.0730 | 64 | 0.0629 | 0.0538 | 0.0561 | 4 | 31 | 0.00218 | 0.00263 |
| 2 (.086) | 0.0860 | 56 | 0.0744 | 0.0641 | 0.0667 | 4 | 22 | 0.00310 | 0.00370 |
| 3 (.099)* | 0.0990 | 48 | 0.0855 | 0.0734 | 0.0764 | 4 | 26 | 0.00406 | 0.00487 |
| 4 (.112) | 0.1120 | 40 | 0.0958 | 0.0813 | 0.0849 | 4 | 45 | 0.00496 | 0.00604 |
| 5 (.125) | 0.1250 | 40 | 0.1088 | 0.0943 | 0.0979 | 4 | 11 | 0.00672 | 0.00796 |
| 6 (.138) | 0.1380 | 32 | 0.1177 | 0.0997 | 0.1042 | 4 | 50 | 0.00745 | 0.00909 |
| 8 (.164) | 0.1640 | 32 | 0.1437 | 0.1257 | 0.1302 | 3 | 58 | 0.01196 | 0.0140 |
| 10 (.190) | 0.1900 | 24 | 0.1629 | 0.1389 | 0.1449 | 4 | 39 | 0.01450 | 0.0175 |
| 12 (.216) | 0.2160 | 24 | 0.1889 | 0.1649 | 0.1709 | 4 | 1 | 0.0206 | 0.0242 |
| ¼ | 0.2500 | 20 | 0.2175 | 0.1887 | 0.1959 | 4 | 11 | 0.0269 | 0.0318 |
| 5/16 | 0.3125 | 18 | 0.2764 | 0.2443 | 0.2524 | 3 | 40 | 0.0454 | 0.0524 |
| 3/8 | 0.3750 | 16 | 0.3344 | 0.2983 | 0.3073 | 3 | 24 | 0.0678 | 0.0775 |
| 7/16 | 0.4375 | 14 | 0.3911 | 0.3499 | 0.3602 | 3 | 20 | 0.0933 | 0.1063 |
| ½ | 0.5000 | 13 | 0.4500 | 0.4056 | 0.4167 | 3 | 7 | 0.1257 | 0.1419 |
| 9/16 | 0.5625 | 12 | 0.5084 | 0.4603 | 0.4723 | 2 | 59 | 0.162 | 0.182 |
| 5/8 | 0.6250 | 11 | 0.5660 | 0.5135 | 0.5266 | 2 | 56 | 0.202 | 0.226 |
| ¾ | 0.7500 | 10 | 0.6850 | 0.6273 | 0.6417 | 2 | 40 | 0.302 | 0.334 |
| 7/8 | 0.8750 | 9 | 0.8028 | 0.7387 | 0.7547 | 2 | 31 | 0.419 | 0.462 |
| 1 | 1.0000 | 8 | 0.9188 | 0.8466 | 0.8647 | 2 | 29 | 0.551 | 0.606 |
| 1⅛ | 1.1250 | 7 | 1.0322 | 0.9497 | 0.9704 | 2 | 31 | 0.693 | 0.763 |
| 1¼ | 1.2500 | 7 | 1.1572 | 1.0747 | 1.0954 | 2 | 15 | 0.890 | 0.969 |
| 1⅜ | 1.3750 | 6 | 1.2667 | 1.1705 | 1.1946 | 2 | 24 | 1.054 | 1.155 |
| 1½ | 1.5000 | 6 | 1.3917 | 1.2955 | 1.3196 | 2 | 11 | 1.294 | 1.405 |
| 1¾ | 1.7500 | 5 | 1.6201 | 1.5046 | 1.5335 | 2 | 15 | 1.74 | 1.90 |
| 2 | 2.0000 | 4½ | 1.8557 | 1.7274 | 1.7594 | 2 | 11 | 2.30 | 2.50 |
| 2¼ | 2.2500 | 4½ | 2.1057 | 1.9774 | 2.0094 | 1 | 55 | 3.02 | 3.25 |
| 2½ | 2.5000 | 4 | 2.3376 | 2.1933 | 2.2294 | 1 | 57 | 3.72 | 4.00 |
| 2¾ | 2.7500 | 4 | 2.5876 | 2.4433 | 2.4794 | 1 | 46 | 4.62 | 4.93 |
| 3 | 3.0000 | 4 | 2.8376 | 2.6933 | 2.7294 | 1 | 36 | 5.62 | 5.97 |
| 3¼ | 3.2500 | 4 | 3.0876 | 2.9433 | 2.9794 | 1 | 29 | 6.72 | 7.10 |
| 3½ | 3.5000 | 4 | 3.3376 | 3.1933 | 3.2294 | 1 | 22 | 7.92 | 8.33 |
| 3¾ | 3.7500 | 4 | 3.5876 | 3.4133 | 3.4794 | 1 | 16 | 9.21 | 9.66 |
| 4 | 4.0000 | 4 | 3.8376 | 3.6933 | 3.7294 | 1 | 11 | 10.61 | 11.08 |

* Secondary sizes.
a British: Effective Diameter.
b See formula, page 1150.
c Design form. See diagram, page 1263.

*Constant Pitch Series:* The various constant-pitch series, UN, with 4, 6, 8, 12, 16, 20, 28 and 32 threads per inch, given in Table 4, offer a comprehensive range of diameter-pitch combinations for those purposes where the threads in the Coarse, Fine, and Extra-Fine series do not meet the particular requirements of the design.

When selecting threads from these constant-pitch series, preference should be given wherever possible to those tabulated in the 8, 12, or 16 thread series.

*8-Thread Series:* The 8-thread series (8UN) is a uniform-pitch series for large diameters. Although originally intended for high-pressure-joint bolts and nuts, it is now widely used as a substitute for the Coarse-Thread Series for diameters larger than 1 inch.

*12-Thread Series:* The 12-thread series (12UN) is a uniform pitch series for large diameters requiring threads of medium-fine pitch. Although originally intended for boiler practice, it is now used as a continuation of the Fine-Thread Series for diameters larger than 1½ inches.

Table. 3b. Fine-Thread Series, UNF, UNRF, and NF — Basic Dimensions

| Sizes | Basic Major Diam., D | Thds. per Inch, n | Basic Pitch Diam.,a E | Minor Diameterc | | Lead Angle at Basic P.D. | | Area of Minor Diam. at D-2hb | Tensile Stress Areab |
|---|---|---|---|---|---|---|---|---|---|
| | | | | Ext. Thds., Ks | Int. Thds., Kn | Deg. | Min. | | |
| | Inches | | Inches | Inches | Inches | | | Sq. In. | Sq. In. |
| 0 (.060)* | 0.0600 | 80 | 0.0519 | 0.0447 | 0.0465 | 4 | 23 | 0.00151 | 0.00180 |
| 1 (.073)* | 0.0730 | 72 | 0.0640 | 0.0560 | 0.0580 | 3 | 57 | 0.00237 | 0.00278 |
| 2 (.086) | 0.0860 | 64 | 0.0759 | 0.0668 | 0.0691 | 3 | 45 | 0.00339 | 0.00394 |
| 3 (.099)* | 0.0990 | 56 | 0.0874 | 0.0771 | 0.0797 | 3 | 43 | 0.00451 | 0.00523 |
| 4 (.112) | 0.1120 | 48 | 0.0985 | 0.0864 | 0.0894 | 3 | 51 | 0.00566 | 0.00661 |
| 5 (.125) | 0.1250 | 44 | 0.1102 | 0.0971 | 0.1004 | 3 | 45 | 0.00716 | 0.00830 |
| 6 (.138) | 0.1380 | 40 | 0.1218 | 0.1073 | 0.1109 | 3 | 44 | 0.00874 | 0.01015 |
| 8 (.164) | 0.1640 | 36 | 0.1460 | 0.1299 | 0.1339 | 3 | 28 | 0.01285 | 0.01474 |
| 10 (.190) | 0.1900 | 32 | 0.1697 | 0.1517 | 0.1562 | 3 | 21 | 0.0175 | 0.0200 |
| 12 (.216)* | 0.2160 | 28 | 0.1928 | 0.1722 | 0.1773 | 3 | 22 | 0.0226 | 0.0258 |
| 1/4 | 0.2500 | 28 | 0.2268 | 0.2062 | 0.2113 | 2 | 52 | 0.0326 | 0.0364 |
| 5/16 | 0.3125 | 24 | 0.2854 | 0.2614 | 0.2674 | 2 | 40 | 0.0524 | 0.0580 |
| 3/8 | 0.3750 | 24 | 0.3479 | 0.3239 | 0.3299 | 2 | 11 | 0.0809 | 0.0878 |
| 7/16 | 0.4375 | 20 | 0.4050 | 0.3762 | 0.3834 | 2 | 15 | 0.1090 | 0.1187 |
| 1/2 | 0.5000 | 20 | 0.4675 | 0.4387 | 0.4459 | 1 | 57 | 0.1486 | 0.1599 |
| 9/16 | 0.5625 | 18 | 0.5264 | 0.4943 | 0.5024 | 1 | 55 | 0.189 | 0.203 |
| 5/8 | 0.6250 | 18 | 0.5889 | 0.5568 | 0.5649 | 1 | 43 | 0.240 | 0.256 |
| 3/4 | 0.7500 | 16 | 0.7094 | 0.6733 | 0.6823 | 1 | 36 | 0.351 | 0.373 |
| 7/8 | 0.8750 | 14 | 0.8286 | 0.7874 | 0.7977 | 1 | 34 | 0.480 | 0.509 |
| 1 | 1.0000 | 12 | 0.9459 | 0.8978 | 0.9098 | 1 | 36 | 0.625 | 0.663 |
| 1 1/8 | 1.1250 | 12 | 1.0709 | 1.0228 | 1.0348 | 1 | 25 | 0.812 | 0.856 |
| 1 1/4 | 1.2500 | 12 | 1.1959 | 1.1478 | 1.1598 | 1 | 16 | 1.024 | 1.073 |
| 1 3/8 | 1.3750 | 12 | 1.3209 | 1.2728 | 1.2848 | 1 | 9 | 1.260 | 1.315 |
| 1 1/2 | 1.5000 | 12 | 1.4459 | 1.3978 | 1.4098 | 1 | 3 | 1.521 | 1.581 |

* Secondary sizes.
a British: Effective Diameter.
b See formula, page 1150.
c Design form. See diagram, page 1263.

*16-Thread Series:* The 16-thread series (16UN) is a uniform pitch series for large diameters requiring fine-pitch threads. It is suitable for adjusting collars and retaining nuts, and also serves as a continuation of the Extra-fine Thread Series for diameters larger than 1 11/16 inches.

*4-, 6-, 20-, 28-, and 32-Thread Series:* These thread series have been used more or less widely in industry for various applications where the Standard Coarse, Fine or Extra-fine Series were not as applicable. They are now given recognition as Standard Unified Thread Series in a specified selection of diameters for each pitch as shown in Table 2.

Whenever a thread in a constant-pitch series also appears in the UNC, UNF, or UNEF series, the symbols and tolerances for limits of size of UNC, UNF, or UNEF series are applicable, as will be seen in Tables 2 and 4.

*Fine Threads for Thin-Wall Tubing:* Dimensions for a 27-thread series, ranging from 1/4- to 1-inch nominal size, also are included in Table 4. These threads are recommended for general use on thin-wall tubing. The minimum length of complete thread is one-third of the basic major diameter plus 5 threads (+0.185 in.).

*High-Temperature, High-Strength Applications:* For these applications the Coarse Thread Series is recommended in sizes from 1/4 to 1 inch and the 8-thread Series in sizes over 1 inch. Some high-temperature applications involving special physical characteristics or conditions may require modification of dimensions, and it is recommended that when such are necessary they be applied to the external thread.

Table 3c.   Extra-Fine-Thread Series, UNEF, UNREF, and NEF — Basic Dimensions

| Sizes | Basic Major Diam., $D$ | Thds. per Inch, $n$ | Basic Pitch Diam.,[a] $E$ | Minor Diameter[c] | | Lead Angle at Basic P.D. | | Area of Minor Diam. at $D-2h_b$ | Tensile Stress Area[b] |
|---|---|---|---|---|---|---|---|---|---|
| | | | | Ext. Thds., $K_s$ | Int. Thds., $K_n$ | Deg. | Min. | | |
| | Inches | | Inches | Inches | Inches | | | Sq. In. | Sq. In. |
| 12 (.216)* | 0.2160 | 32 | 0.1957 | 0.1777 | 0.1822 | 2 | 55 | 0.0242 | 0.0270 |
| 1/4 | 0.2500 | 32 | 0.2297 | 0.2117 | 0.2162 | 2 | 29 | 0.0344 | 0.0379 |
| 5/16 | 0.3125 | 32 | 0.2922 | 0.2742 | 0.2787 | 1 | 57 | 0.0581 | 0.0625 |
| 3/8 | 0.3750 | 32 | 0.3547 | 0.3367 | 0.3412 | 1 | 36 | 0.0878 | 0.0932 |
| 7/16 | 0.4375 | 28 | 0.4143 | 0.3937 | 0.3988 | 1 | 34 | 0.1201 | 0.1274 |
| 1/2 | 0.5000 | 28 | 0.4768 | 0.4562 | 0.4613 | 1 | 22 | 0.162 | 0.170 |
| 9/16 | 0.5625 | 24 | 0.5354 | 0.5114 | 0.5174 | 1 | 25 | 0.203 | 0.214 |
| 5/8 | 0.6250 | 24 | 0.5979 | 0.5739 | 0.5799 | 1 | 16 | 0.256 | 0.268 |
| 11/16* | 0.6875 | 24 | 0.6604 | 0.6364 | 0.6424 | 1 | 9 | 0.315 | 0.329 |
| 3/4 | 0.7500 | 20 | 0.7175 | 0.6887 | 0.6959 | 1 | 16 | 0.369 | 0.386 |
| 13/16* | 0.8125 | 20 | 0.7800 | 0.7512 | 0.7584 | 1 | 10 | 0.439 | 0.458 |
| 7/8 | 0.8750 | 20 | 0.8425 | 0.8137 | 0.8209 | 1 | 5 | 0.515 | 0.536 |
| 15/16* | 0.9375 | 20 | 0.9050 | 0.8762 | 0.8834 | 1 | 0 | 0.598 | 0.620 |
| 1 | 1.0000 | 20 | 0.9675 | 0.9387 | 0.9459 | 0 | 57 | 0.687 | 0.711 |
| 1 1/16* | 1.0625 | 18 | 1.0264 | 0.9943 | 1.0024 | 0 | 59 | 0.770 | 0.799 |
| 1 1/8 | 1.1250 | 18 | 1.0889 | 1.0568 | 1.0649 | 0 | 56 | 0.871 | 0.901 |
| 1 3/16* | 1.1875 | 18 | 1.1514 | 1.1193 | 1.1274 | 0 | 53 | 0.977 | 1.009 |
| 1 1/4 | 1.2500 | 18 | 1.2139 | 1.1818 | 1.1899 | 0 | 50 | 1.090 | 1.123 |
| 1 5/16* | 1.3125 | 18 | 1.2764 | 1.2443 | 1.2524 | 0 | 48 | 1.208 | 1.244 |
| 1 3/8 | 1.3750 | 18 | 1.3389 | 1.3068 | 1.3149 | 0 | 45 | 1.333 | 1.370 |
| 1 7/16* | 1.4375 | 18 | 1.4014 | 1.3693 | 1.3774 | 0 | 43 | 1.464 | 1.503 |
| 1 1/2 | 1.5000 | 18 | 1.4639 | 1.4318 | 1.4399 | 0 | 42 | 1.60 | 1.64 |
| 1 9/16* | 1.5625 | 18 | 1.5264 | 1.4943 | 1.5024 | 0 | 40 | 1.74 | 1.79 |
| 1 5/8 | 1.6250 | 18 | 1.5889 | 1.5568 | 1.5649 | 0 | 38 | 1.89 | 1.94 |
| 1 11/16* | 1.6875 | 18 | 1.6514 | 1.6193 | 1.6274 | 0 | 37 | 2.05 | 2.10 |

* Secondary sizes.
a British: Effective Diameter.
b See formula, page 1150.
c Design form.  See diagram, page 1263.

*Selected Combinations:* Thread data are tabulated in Table 4 for certain additional selected special combinations of diameter and pitch, with pitch diameter tolerances based on a length of thread engagement of 9 times the pitch. The pitch diameter limits are applicable to a length of engagement of from 5 to 15 times the pitch. (This should not be confused with the lengths of thread on mating parts, as they may exceed the length of engagement by a considerable amount.) Thread symbols are UNS.

*Other Threads of Special Diameters, Pitches, and Lengths of Engagement:* Thread data for special combinations of diameter, pitch, and length of engagement not included in the selected combinations are also given in the Standard but are not given here. Also, when design considerations require non-standard pitches or extreme conditions of engagement not covered by the tables, the allowance and tolerances should be derived from the formulas in the Standard. The thread symbol for such special threads is UNS.

**Thread Classes.** — Thread classes are distinguished from each other by the amounts of tolerance and allowance. Classes identified by a numeral followed by the letters A and B are derived from certain Unified formulas (not shown here) in which the pitch diameter tolerances are based on increments of the basic major (nominal) diameter, the pitch, and the length of engagement. These formulas and the class identification or symbols apply to all of the Unified threads.

Table 3d.   4-Thread Series, 4UN and 4UNR — Basic Dimensions

| Sizes | | Basic Major Diam., D | Basic Pitch Diam.,a E | Minor Diameterc | | Lead Angle λ at Basic P.D. | | Area of Minor Diam. at D-2hb | Tensile Stress Areab |
|---|---|---|---|---|---|---|---|---|---|
| Primary | Secondary | | | Ext. Thds., Ks | Int. Thds., Kn | Deg. | Min. | | |
| Inches | Inches | Inches | Inches | Inches | Inches | | | Sq. In. | Sq. In. |
| 2½† | | 2.5000 | 2.3376 | 2.1933 | 2.2294 | 1 | 57 | 3.72 | 4.00 |
| | 2⅝ | 2.6250 | 2.4626 | 2.3183 | 2.3544 | 1 | 51 | 4.16 | 4.45 |
| 2¾† | | 2.7500 | 2.5876 | 2.4433 | 2.4794 | 1 | 46 | 4.62 | 4.93 |
| | 2⅞ | 2.8750 | 2.7126 | 2.5683 | 2.6044 | 1 | 41 | 5.11 | 5.44 |
| 3† | | 3.0000 | 2.8376 | 2.6933 | 2.7294 | 1 | 36 | 5.62 | 5.97 |
| | 3⅛ | 3.1250 | 2.9626 | 2.8183 | 2.8544 | 1 | 32 | 6.16 | 6.52 |
| 3¼† | | 3.2500 | 3.0876 | 2.9433 | 2.9794 | 1 | 29 | 6.72 | 7.10 |
| | 3⅜ | 3.3750 | 3.2126 | 3.0683 | 3.1044 | 1 | 25 | 7.31 | 7.70 |
| 3½† | | 3.5000 | 3.3376 | 3.1933 | 3.2294 | 1 | 22 | 7.92 | 8.33 |
| | 3⅝ | 3.6250 | 3.4626 | 3.3183 | 3.3544 | 1 | 19 | 8.55 | 9.00 |
| 3¾† | | 3.7500 | 3.5876 | 3.4433 | 3.4794 | 1 | 16 | 9.21 | 9.66 |
| | 3⅞ | 3.8750 | 3.7126 | 3.5683 | 3.6044 | 1 | 14 | 9.90 | 10.36 |
| 4† | | 4.0000 | 3.8376 | 3.6933 | 3.7294 | 1 | 11 | 10.61 | 11.08 |
| | 4⅛ | 4.1250 | 3.9626 | 3.8183 | 3.8544 | 1 | 9 | 11.34 | 11.83 |
| 4¼ | | 4.2500 | 4.0876 | 3.9433 | 3.9794 | 1 | 7 | 12.10 | 12.61 |
| | 4⅜ | 4.3750 | 4.2126 | 4.0683 | 4.1044 | 1 | 5 | 12.88 | 13.41 |
| 4½ | | 4.5000 | 4.3376 | 4.1933 | 4.2294 | 1 | 3 | 13.69 | 14.23 |
| | 4⅝ | 4.6250 | 4.4626 | 4.3183 | 4.3544 | 1 | 1 | 14.52 | 15.1 |
| 4¾ | | 4.7500 | 4.5876 | 4.4433 | 4.4794 | 1 | 0 | 15.4 | 15.9 |
| | 4⅞ | 4.8750 | 4.7126 | 4.5683 | 4.6044 | 0 | 58 | 16.3 | 16.8 |
| 5 | | 5.0000 | 4.8376 | 4.6933 | 4.7294 | 0 | 57 | 17.2 | 17.8 |
| | 5⅛ | 5.1250 | 4.9626 | 4.8183 | 4.8544 | 0 | 55 | 18.1 | 18.7 |
| 5¼ | | 5.2500 | 5.0876 | 4.9433 | 4.9794 | 0 | 54 | 19.1 | 19.7 |
| | 5⅜ | 5.3750 | 5.2126 | 5.0683 | 5.1044 | 0 | 52 | 20.0 | 20.7 |
| 5½ | | 5.5000 | 5.3376 | 5.1933 | 5.2294 | 0 | 51 | 21.0 | 21.7 |
| | 5⅝ | 5.6250 | 5.4626 | 5.3183 | 5.3544 | 0 | 50 | 22.1 | 22.7 |
| 5¾ | | 5.7500 | 5.5876 | 5.4433 | 5.4794 | 0 | 49 | 23.1 | 23.8 |
| | 5⅞ | 5.8750 | 5.7126 | 5.5683 | 5.6044 | 0 | 48 | 24.2 | 24.9 |
| 6 | | 6.0000 | 5.8376 | 5.6933 | 5.7294 | 0 | 47 | 25.3 | 26.0 |

† These are standard sizes of the UNC series.
a British: Effective Diameter.
b See formula, page 1150.
c Design form. See diagram, page 1263.

Classes 1A, 2A, and 3A apply to external threads only, and Classes 1B, 2B, and 3B apply to internal threads only.  The disposition of the tolerances, allowances, and crest clearances for the various classes is illustrated on pages 1276 and 1277

*Classes 2A and 2B:*  Classes 2A and 2B are the most commonly used for general applications, including production of bolts, screws, nuts, and similar fasteners.

The maximum diameters of Class 2A (external) uncoated threads are less than basic by the amount of the allowance.  The allowance minimizes galling and seizing in high-cycle wrench assembly, or it can be used to accommodate plated finishes or other coating.  However, for threads with additive finish, the maximum diameters of Class 2A may be exceeded by the amount of the allowance; for example, the 2A maximum diameters apply to an unplated part or to a part before plating whereas the basic diameters (the 2A maximum diameter plus allowance) apply to a part after plating.  The minimum diameters of Class 2B (internal) threads, whether or not plated or coated, are basic, affording no allowance or clearance in assembly at maximum metal limits.

(Continued on page 1274.)

Table 3e.   6-Thread Series, 6UN and 6UNR — Basic Dimensions

| Sizes | | Basic Major Diam., D | Basic Pitch Diam.,a E | Minor Diameterc | | Lead Angle λ at Basic P.D. | | Area of Minor Diam. at D-2h_b | Tensile Stress Areab |
|---|---|---|---|---|---|---|---|---|---|
| Primary | Secondary | | | Ext. Thds., K_s | Int. Thds., K_n | | | | |
| Inches | Inches | Inches | Inches | Inches | Inches | Deg. | Min. | Sq. In. | Sq. In. |
| 1⅜† | | 1.3750 | 1.2667 | 1.1705 | 1.1946 | 2 | 24 | 1.054 | 1.155 |
| | 1⁷/₁₆ | 1.4375 | 1.3292 | 1.2330 | 1.2571 | 2 | 17 | 1.171 | 1.277 |
| 1½† | | 1.5000 | 1.3917 | 1.2955 | 1.3196 | 2 | 11 | 1.294 | 1.405 |
| | 1⁹/₁₆ | 1.5625 | 1.4542 | 1.3580 | 1.3821 | 2 | 5 | 1.423 | 1.54 |
| 1⅝ | | 1.6250 | 1.5167 | 1.4205 | 1.4446 | 2 | 0 | 1.56 | 1.68 |
| | 1¹¹/₁₆ | 1.6875 | 1.5792 | 1.4830 | 1.5071 | 1 | 55 | 1.70 | 1.83 |
| 1¾ | | 1.7500 | 1.6417 | 1.5455 | 1.5696 | 1 | 51 | 1.85 | 1.98 |
| | 1¹³/₁₆ | 1.8125 | 1.7042 | 1.6080 | 1.6321 | 1 | 47 | 2.00 | 2.14 |
| 1⅞ | | 1.8750 | 1.7667 | 1.6705 | 1.6946 | 1 | 43 | 2.16 | 2.30 |
| | 1¹⁵/₁₆ | 1.9375 | 1.8292 | 1.7330 | 1.7571 | 1 | 40 | 2.33 | 2.47 |
| 2 | | 2.0000 | 1.8917 | 1.7955 | 1.8196 | 1 | 36 | 2.50 | 2.65 |
| | 2⅛ | 2.1250 | 2.0167 | 1.9205 | 1.9446 | 1 | 30 | 2.86 | 3.03 |
| 2¼ | | 2.2500 | 2.1417 | 2.0455 | 2.0696 | 1 | 25 | 3.25 | 3.42 |
| | 2⅜ | 2.3750 | 2.2667 | 2.1705 | 2.1946 | 1 | 20 | 3.66 | 3.85 |
| 2½ | | 2.5000 | 2.3917 | 2.2955 | 2.3196 | 1 | 16 | 4.10 | 4.29 |
| | 2⅝ | 2.6250 | 2.5167 | 2.4205 | 2.4446 | 1 | 12 | 4.56 | 4.76 |
| 2¾ | | 2.7500 | 2.6417 | 2.5455 | 2.5696 | 1 | 9 | 5.04 | 5.26 |
| | 2⅞ | 2.8750 | 2.7667 | 2.6705 | 2.6946 | 1 | 6 | 5.55 | 5.78 |
| 3 | | 3.0000 | 2.8917 | 2.7955 | 2.8196 | 1 | 3 | 6.09 | 6.33 |
| | 3⅛ | 3.1250 | 3.0167 | 2.9205 | 2.9446 | 1 | 0 | 6.64 | 6.89 |
| 3¼ | | 3.2500 | 3.1417 | 3.0455 | 3.0696 | 0 | 58 | 7.23 | 7.49 |
| | 3⅜ | 3.3750 | 3.2667 | 3.1705 | 3.1946 | 0 | 56 | 7.84 | 8.11 |
| 3½ | | 3.5000 | 3.3917 | 3.2955 | 3.3196 | 0 | 54 | 8.47 | 8.75 |
| | 3⅝ | 3.6250 | 3.5167 | 3.4205 | 3.4446 | 0 | 52 | 9.12 | 9.42 |
| 3¾ | | 3.7500 | 3.6417 | 3.5455 | 3.5696 | 0 | 50 | 9.81 | 10.11 |
| | 3⅞ | 3.8750 | 3.7667 | 3.6705 | 3.6946 | 0 | 48 | 10.51 | 10.83 |
| 4 | | 4.0000 | 3.8917 | 3.7955 | 3.8196 | 0 | 47 | 11.24 | 11.57 |
| | 4⅛ | 4.1250 | 4.0167 | 3.9205 | 3.9446 | 0 | 45 | 12.00 | 12.33 |
| 4¼ | | 4.2500 | 4.1417 | 4.0455 | 4.0696 | 0 | 44 | 12.78 | 13.12 |
| | 4⅜ | 4.3750 | 4.2667 | 4.1705 | 4.1946 | 0 | 43 | 13.58 | 13.94 |
| 4½ | | 4.5000 | 4.3917 | 4.2955 | 4.3196 | 0 | 42 | 14.41 | 14.78 |
| | 4⅝ | 4.6250 | 4.5167 | 4.4205 | 4.4446 | 0 | 40 | 15.3 | 15.6 |
| 4¾ | | 4.7500 | 4.6417 | 4.5455 | 4.5696 | 0 | 39 | 16.1 | 16.5 |
| | 4⅞ | 4.8750 | 4.7667 | 4.6705 | 4.6946 | 0 | 38 | 17.0 | 17.5 |
| 5 | | 5.0000 | 4.8917 | 4.7955 | 4.8196 | 0 | 37 | 18.0 | 18.4 |
| | 5⅛ | 5.1250 | 5.0167 | 4.9205 | 4.9446 | 0 | 36 | 18.9 | 19.3 |
| 5¼ | | 5.2500 | 5.1417 | 5.0455 | 5.0696 | 0 | 35 | 19.9 | 20.3 |
| | 5⅜ | 5.3750 | 5.2667 | 5.1705 | 5.1946 | 0 | 35 | 20.9 | 21.3 |
| 5½ | | 5.5000 | 5.3917 | 5.2955 | 5.3196 | 0 | 34 | 21.9 | 22.4 |
| | 5⅝ | 5.6250 | 5.5167 | 5.4205 | 5.4446 | 0 | 33 | 23.0 | 23.4 |
| 5¾ | | 5.7500 | 5.6417 | 5.5455 | 5.5696 | 0 | 32 | 24.0 | 24.5 |
| | 5⅞ | 5.8750 | 5.7667 | 5.6705 | 5.6946 | 0 | 32 | 25.1 | 25.6 |
| 6 | | 6.0000 | 5.8917 | 5.7955 | 5.8196 | 0 | 31 | 26.3 | 26.8 |

† These are standard sizes of the UNC Series.
a British: Effective Diameter.
b See formula, page 1150.
c Design form.  See diagram, page 1263 .

Table 3f.  8-Thread Series, 8UN and 8UNR — Basic Dimensions

| Sizes | | Basic Major Diam., $D$ | Basic Pitch Diam.,$^a$ $E$ | Minor Diameter$^c$ | | Lead Angle λ at Basic P.D. | | Area of Minor Diam. at $D-2h_b$ | Tensile Stress Area$^b$ |
|---|---|---|---|---|---|---|---|---|---|
| Primary | Secondary | | | Ext. Thds., $K_s$ | Int. Thds., $K_n$ | Deg. | Min. | | |
| Inches | Inches | Inches | Inches | Inches | Inches | | | Sq. In. | Sq. In. |
| 1† | | 1.0000 | 0.9188 | 0.8466 | 0.8647 | 2 | 29 | 0.551 | 0.606 |
| | 1¹⁄₁₆ | 1.0625 | 0.9813 | 0.9091 | 0.9272 | 2 | 19 | 0.636 | 0.695 |
| 1⅛ | | 1.1250 | 1.0438 | 0.9716 | 0.9897 | 2 | 11 | 0.728 | 0.790 |
| | 1³⁄₁₆ | 1.1875 | 1.1063 | 1.0341 | 1.0522 | 2 | 4 | 0.825 | 0.892 |
| 1¼ | | 1.2500 | 1.1688 | 1.0966 | 1.1147 | 1 | 57 | 0.929 | 1.000 |
| | 1⁵⁄₁₆ | 1.3125 | 1.2313 | 1.1591 | 1.1772 | 1 | 51 | 1.039 | 1.114 |
| 1⅜ | | 1.3750 | 1.2938 | 1.2216 | 1.2397 | 1 | 46 | 1.155 | 1.233 |
| | 1⁷⁄₁₆ | 1.4375 | 1.3563 | 1.2841 | 1.3022 | 1 | 41 | 1.277 | 1.360 |
| 1½ | | 1.5000 | 1.4188 | 1.3466 | 1.3647 | 1 | 36 | 1.405 | 1.492 |
| | 1⁹⁄₁₆ | 1.5625 | 1.4813 | 1.4091 | 1.4272 | 1 | 32 | 1.54 | 1.63 |
| 1⅝ | | 1.6250 | 1.5438 | 1.4716 | 1.4897 | 1 | 29 | 1.68 | 1.78 |
| | 1¹¹⁄₁₆ | 1.6875 | 1.6063 | 1.5341 | 1.5522 | 1 | 25 | 1.83 | 1.93 |
| 1¾ | | 1.7500 | 1.6688 | 1.5966 | 1.6147 | 1 | 22 | 1.98 | 2.08 |
| | 1¹³⁄₁₆ | 1.8125 | 1.7313 | 1.6591 | 1.6772 | 1 | 19 | 2.14 | 2.25 |
| 1⅞ | | 1.8750 | 1.7938 | 1.7216 | 1.7397 | 1 | 16 | 2.30 | 2.41 |
| | 1¹⁵⁄₁₆ | 1.9375 | 1.8563 | 1.7841 | 1.8022 | 1 | 14 | 2.47 | 2.59 |
| 2 | | 2.0000 | 1.9188 | 1.8466 | 1.8647 | 1 | 11 | 2.65 | 2.77 |
| | 2⅛ | 2.1250 | 2.0438 | 1.9716 | 1.9897 | 1 | 7 | 3.03 | 3.15 |
| 2¼ | | 2.2500 | 2.1688 | 2.0966 | 2.1147 | 1 | 3 | 3.42 | 3.56 |
| | 2⅜ | 2.3750 | 2.2938 | 2.2216 | 2.2397 | 1 | 0 | 3.85 | 3.99 |
| 2½ | | 2.5000 | 2.4188 | 2.3466 | 2.3647 | 0 | 57 | 4.29 | 4.44 |
| | 2⅝ | 2.6250 | 2.5438 | 2.4716 | 2.4897 | 0 | 54 | 4.76 | 4.92 |
| 2¾ | | 2.7500 | 2.6688 | 2.5966 | 2.6147 | 0 | 51 | 5.26 | 5.43 |
| | 2⅞ | 2.8750 | 2.7938 | 2.7216 | 2.7397 | 0 | 49 | 5.78 | 5.95 |
| 3 | | 3.0000 | 2.9188 | 2.8466 | 2.8647 | 0 | 47 | 6.32 | 6.51 |
| | 3⅛ | 3.1250 | 3.0438 | 2.9716 | 2.9897 | 0 | 45 | 6.89 | 7.08 |
| 3¼ | | 3.2500 | 3.1688 | 3.0966 | 3.1147 | 0 | 43 | 7.49 | 7.69 |
| | 3⅜ | 3.3750 | 3.2938 | 3.2216 | 3.2397 | 0 | 42 | 8.11 | 8.31 |
| 3½ | | 3.5000 | 3.4188 | 3.3466 | 3.3647 | 0 | 40 | 8.75 | 8.96 |
| | 3⅝ | 3.6250 | 3.5438 | 3.4716 | 3.4897 | 0 | 39 | 9.42 | 9.64 |
| 3¾ | | 3.7500 | 3.6688 | 3.5966 | 3.6147 | 0 | 37 | 10.11 | 10.34 |
| | 3⅞ | 3.8750 | 3.7938 | 3.7216 | 3.7397 | 0 | 36 | 10.83 | 11.06 |
| 4 | | 4.0000 | 3.9188 | 3.8466 | 3.8647 | 0 | 35 | 11.57 | 11.81 |
| | 4⅛ | 4.1250 | 4.0438 | 3.9716 | 3.9897 | 0 | 34 | 12.34 | 12.59 |
| 4¼ | | 4.2500 | 4.1688 | 4.0966 | 4.1147 | 0 | 33 | 13.12 | 13.38 |
| | 4⅜ | 4.3750 | 4.2938 | 4.2216 | 4.2397 | 0 | 32 | 13.94 | 14.21 |
| 4½ | | 4.5000 | 4.4188 | 4.3466 | 4.3647 | 0 | 31 | 14.78 | 15.1 |
| | 4⅝ | 4.6250 | 4.5438 | 4.4716 | 4.4897 | 0 | 30 | 15.6 | 15.9 |
| 4¾ | | 4.7500 | 4.6688 | 4.5966 | 4.6147 | 0 | 29 | 16.5 | 16.8 |
| | 4⅞ | 4.8750 | 4.7938 | 4.7216 | 4.7397 | 0 | 29 | 17.4 | 17.7 |
| 5 | | 5.0000 | 4.9188 | 4.8466 | 4.8647 | 0 | 28 | 18.4 | 18.7 |
| | 5⅛ | 5.1250 | 5.0438 | 4.9716 | 4.9897 | 0 | 27 | 19.3 | 19.7 |
| 5¼ | | 5.2500 | 5.1688 | 5.0966 | 5.1147 | 0 | 26 | 20.3 | 20.7 |
| | 5⅜ | 5.3750 | 5.2938 | 5.2216 | 5.2397 | 0 | 26 | 21.3 | 21.7 |
| 5½ | | 5.5000 | 5.4188 | 5.3466 | 5.3647 | 0 | 25 | 22.4 | 22.7 |
| | 5⅝ | 5.6250 | 5.5438 | 5.4716 | 5.4897 | 0 | 25 | 23.4 | 23.8 |
| 5¾ | | 5.7500 | 5.6688 | 5.5966 | 5.6147 | 0 | 24 | 24.5 | 24.9 |
| | 5⅞ | 5.8750 | 5.7938 | 5.7216 | 5.7397 | 0 | 24 | 25.6 | 26.0 |
| 6 | | 6.0000 | 5.9188 | 5.8466 | 5.8647 | 0 | 23 | 26.8 | 27.1 |

† This is a standard size of the UNC Series.
$^a$ British: Effective Diameter.
$^b$ See formula, page 1150.
$^c$ Design form.  See diagram, page 1263.

Table 3g.  12-Thread Series, 12UN and 12UNR — Basic Dimensions

| Sizes | | Basic Major Diam., $D$ | Basic Pitch Diam., $E^a$ | Minor Diameter[c] | | Lead Angle λ at Basic P.D. | | Area of Minor Diam. at $D-2h_b$ | Tensile Stress Area[b] |
|---|---|---|---|---|---|---|---|---|---|
| Primary | Secondary | | | Ext. Thds., $K_s$ | Int. Thds., $K_n$ | Deg. | Min. | | |
| Inches | Inches | Inches | Inches | Inches | Inches | | | Sq. In. | Sq. In. |
| 9/16† | | 0.5625 | 0.5084 | 0.4603 | 0.4723 | 2 | 59 | 0.162 | 0.182 |
| | 5/8 | 0.6250 | 0.5709 | 0.5228 | 0.5348 | 2 | 40 | 0.210 | 0.232 |
| | 11/16 | 0.6875 | 0.6334 | 0.5853 | 0.5973 | 2 | 24 | 0.264 | 0.289 |
| 3/4 | | 0.7500 | 0.6959 | 0.6478 | 0.6598 | 2 | 11 | 0.323 | 0.351 |
| | 13/16 | 0.8125 | 0.7584 | 0.7103 | 0.7223 | 2 | 0 | 0.390 | 0.420 |
| 7/8 | | 0.8750 | 0.8209 | 0.7728 | 0.7848 | 1 | 51 | 0.462 | 0.495 |
| | 15/16 | 0.9375 | 0.8834 | 0.8353 | 0.8473 | 1 | 43 | 0.540 | 0.576 |
| 1† | | 1.0000 | 0.9459 | 0.8978 | 0.9098 | 1 | 36 | 0.625 | 0.663 |
| | 1 1/16 | 1.0625 | 1.0084 | 0.9603 | 0.9723 | 1 | 30 | 0.715 | 0.756 |
| 1 1/8† | | 1.1250 | 1.0709 | 1.0228 | 1.0348 | 1 | 25 | 0.812 | 0.856 |
| | 1 3/16 | 1.1875 | 1.1334 | 1.0853 | 1.0973 | 1 | 20 | 0.915 | 0.961 |
| 1 1/4† | | 1.2500 | 1.1959 | 1.1478 | 1.1598 | 1 | 16 | 1.024 | 1.073 |
| | 1 5/16 | 1.3125 | 1.2584 | 1.2103 | 1.2223 | 1 | 12 | 1.139 | 1.191 |
| 1 3/8† | | 1.3750 | 1.3209 | 1.2728 | 1.2848 | 1 | 9 | 1.260 | 1.315 |
| | 1 7/16 | 1.4375 | 1.3834 | 1.3353 | 1.3473 | 1 | 6 | 1.388 | 1.445 |
| 1 1/2† | | 1.5000 | 1.4459 | 1.3978 | 1.4098 | 1 | 3 | 1.52 | 1.58 |
| | 1 9/16 | 1.5625 | 1.5084 | 1.4603 | 1.4723 | 1 | 0 | 1.66 | 1.72 |
| 1 5/8 | | 1.6250 | 1.5709 | 1.5228 | 1.5348 | 0 | 58 | 1.81 | 1.87 |
| | 1 11/16 | 1.6875 | 1.6334 | 1.5853 | 1.5973 | 0 | 56 | 1.96 | 2.03 |
| 1 3/4 | | 1.7500 | 1.6959 | 1.6478 | 1.6598 | 0 | 54 | 2.12 | 2.19 |
| | 1 13/16 | 1.8125 | 1.7584 | 1.7103 | 1.7223 | 0 | 52 | 2.28 | 2.35 |
| 1 7/8 | | 1.8750 | 1.8209 | 1.7728 | 1.7848 | 0 | 50 | 2.45 | 2.53 |
| | 1 15/16 | 1.9375 | 1.8834 | 1.8353 | 1.8473 | 0 | 48 | 2.63 | 2.71 |
| 2 | | 2.0000 | 1.9459 | 1.8978 | 1.9098 | 0 | 47 | 2.81 | 2.89 |
| | 2 1/8 | 2.1250 | 2.0709 | 2.0228 | 2.0348 | 0 | 44 | 3.19 | 3.28 |
| 2 1/4 | | 2.2500 | 2.1959 | 2.1478 | 2.1598 | 0 | 42 | 3.60 | 3.69 |
| | 2 3/8 | 2.3750 | 2.3209 | 2.2728 | 2.2848 | 0 | 39 | 4.04 | 4.13 |
| 2 1/2 | | 2.5000 | 2.4459 | 2.3978 | 2.4098 | 0 | 37 | 4.49 | 4.60 |
| | 2 5/8 | 2.6250 | 2.5709 | 2.5228 | 2.5348 | 0 | 35 | 4.97 | 5.08 |
| 2 3/4 | | 2.7500 | 2.6959 | 2.6478 | 2.6598 | 0 | 34 | 5.48 | 5.59 |
| | 2 7/8 | 2.8750 | 2.8209 | 2.7728 | 2.7848 | 0 | 32 | 6.01 | 6.13 |
| 3 | | 3.0000 | 2.9459 | 2.8978 | 2.9098 | 0 | 31 | 6.57 | 6.69 |
| | 3 1/8 | 3.1250 | 3.0709 | 3.0228 | 3.0348 | 0 | 30 | 7.15 | 7.28 |
| 3 1/4 | | 3.2500 | 3.1959 | 3.1478 | 3.1598 | 0 | 29 | 7.75 | 7.89 |
| | 3 3/8 | 3.3750 | 3.3209 | 3.2728 | 3.2848 | 0 | 27 | 8.38 | 8.52 |
| 3 1/2 | | 3.5000 | 3.4459 | 3.3978 | 3.4098 | 0 | 26 | 9.03 | 9.18 |
| | 3 5/8 | 3.6250 | 3.5709 | 3.5228 | 3.5348 | 0 | 26 | 9.71 | 9.86 |
| 3 3/4 | | 3.7500 | 3.6959 | 3.6478 | 3.6598 | 0 | 25 | 10.42 | 10.57 |
| | 3 7/8 | 3.8750 | 3.8209 | 3.7728 | 3.7848 | 0 | 24 | 11.14 | 11.30 |
| 4 | | 4.0000 | 3.9459 | 3.8978 | 3.9098 | 0 | 23 | 11.90 | 12.06 |
| | 4 1/8 | 4.1250 | 4.0709 | 4.0228 | 4.0348 | 0 | 22 | 12.67 | 12.84 |
| 4 1/4 | | 4.2500 | 4.1959 | 4.1478 | 4.1598 | 0 | 22 | 13.47 | 13.65 |
| | 4 3/8 | 4.3750 | 4.3209 | 4.2728 | 4.2848 | 0 | 21 | 14.30 | 14.48 |
| 4 1/2 | | 4.5000 | 4.4459 | 4.3978 | 4.4098 | 0 | 21 | 15.1 | 15.3 |
| | 4 5/8 | 4.6250 | 4.5709 | 4.5228 | 4.5348 | 0 | 20 | 16.0 | 16.2 |
| 4 3/4 | | 4.7500 | 4.6959 | 4.6478 | 4.6598 | 0 | 19 | 16.9 | 17.1 |
| | 4 7/8 | 4.8750 | 4.8209 | 4.7728 | 4.7848 | 0 | 19 | 17.8 | 18.0 |
| 5 | | 5.0000 | 4.9459 | 4.8978 | 4.9098 | 0 | 18 | 18.8 | 19.0 |
| | 5 1/8 | 5.1250 | 5.0709 | 5.0228 | 5.0348 | 0 | 18 | 19.8 | 20.0 |
| 5 1/4 | | 5.2500 | 5.1959 | 5.1478 | 5.1598 | 0 | 18 | 20.8 | 21.0 |
| | 5 3/8 | 5.3750 | 5.3209 | 5.2728 | 5.2848 | 0 | 17 | 21.8 | 22.0 |
| 5 1/2 | | 5.5000 | 5.4459 | 5.3978 | 5.4098 | 0 | 17 | 22.8 | 23.1 |
| | 5 5/8 | 5.6250 | 5.5709 | 5.5228 | 5.5348 | 0 | 16 | 23.9 | 24.1 |
| 5 3/4 | | 5.7500 | 5.6959 | 5.6478 | 5.6598 | 0 | 16 | 25.0 | 25.2 |
| | 5 7/8 | 5.8750 | 5.8209 | 5.7728 | 5.7848 | 0 | 16 | 26.1 | 26.4 |
| 6 | | 6.0000 | 5.9459 | 5.8978 | 5.9098 | 0 | 15 | 27.3 | 27.5 |

† These are standard sizes of the UNC or UNF Series.
a British: Effective Diameter.
b See formula, page 1150.
c Design form.  See diagram, page 1263.

Table 3h.  16-Thread Series, 16UN and 16UNR — Basic Dimensions

| Sizes Primary (Inches) | Sizes Secondary (Inches) | Basic Major Diam., D (Inches) | Basic Pitch Diam.,[a] E (Inches) | Minor Diameter[c] Ext. Thds., $K_s$ (Inches) | Minor Diameter[c] Int. Thds., $K_n$ (Inches) | Lead Angle λ at Basic P.D. Deg. | Lead Angle λ at Basic P.D. Min. | Area of Minor Diam. at $D-2h_b$ (Sq. In.) | Tensile Stress Area[b] (Sq. In.) |
|---|---|---|---|---|---|---|---|---|---|
| 3/8† | | 0.3750 | 0.3344 | 0.2983 | 0.3073 | 3 | 24 | 0.0678 | 0.0775 |
| 7/16 | | 0.4375 | 0.3969 | 0.3608 | 0.3698 | 2 | 52 | 0.0997 | 0.1114 |
| 1/2 | | 0.5000 | 0.4594 | 0.4233 | 0.4323 | 2 | 29 | 0.1378 | 0.151 |
| 9/16 | | 0.5625 | 0.5219 | 0.4858 | 0.4948 | 2 | 11 | 0.182 | 0.198 |
| 5/8 | | 0.6250 | 0.5844 | 0.5483 | 0.5573 | 1 | 57 | 0.232 | 0.250 |
| | 11/16 | 0.6875 | 0.6469 | 0.6108 | 0.6198 | 1 | 46 | 0.289 | 0.308 |
| 3/4† | | 0.7500 | 0.7094 | 0.6733 | 0.6823 | 1 | 36 | 0.351 | 0.373 |
| | 13/16 | 0.8125 | 0.7719 | 0.7358 | 0.7448 | 1 | 29 | 0.420 | 0.444 |
| 7/8 | | 0.8750 | 0.8344 | 0.7983 | 0.8073 | 1 | 22 | 0.495 | 0.521 |
| | 15/16 | 0.9375 | 0.8969 | 0.8608 | 0.8698 | 1 | 16 | 0.576 | 0.604 |
| 1 | | 1.0000 | 0.9594 | 0.9233 | 0.9323 | 1 | 11 | 0.663 | 0.693 |
| | 1 1/16 | 1.0625 | 1.0219 | 0.9858 | 0.9948 | 1 | 7 | 0.756 | 0.788 |
| 1 1/8 | | 1.1250 | 1.0844 | 1.0483 | 1.0573 | 1 | 3 | 0.856 | 0.889 |
| | 1 3/16 | 1.1875 | 1.1469 | 1.1108 | 1.1198 | 1 | 0 | 0.961 | 0.997 |
| 1 1/4 | | 1.2500 | 1.2094 | 1.1733 | 1.1823 | 0 | 57 | 1.073 | 1.111 |
| | 1 5/16 | 1.3125 | 1.2719 | 1.2358 | 1.2448 | 0 | 54 | 1.191 | 1.230 |
| 1 3/8 | | 1.3750 | 1.3344 | 1.2983 | 1.3073 | 0 | 51 | 1.315 | 1.356 |
| | 1 7/16 | 1.4375 | 1.3969 | 1.3608 | 1.3698 | 0 | 49 | 1.445 | 1.488 |
| 1 1/2 | | 1.5000 | 1.4594 | 1.4233 | 1.4323 | 0 | 47 | 1.58 | 1.63 |
| | 1 9/16 | 1.5625 | 1.5219 | 1.4858 | 1.4948 | 0 | 45 | 1.72 | 1.77 |
| 1 5/8 | | 1.6250 | 1.5844 | 1.5483 | 1.5573 | 0 | 43 | 1.87 | 1.92 |
| | 1 11/16 | 1.6875 | 1.6469 | 1.6108 | 1.6198 | 0 | 42 | 2.03 | 2.08 |
| 1 3/4 | | 1.7500 | 1.7094 | 1.6733 | 1.6823 | 0 | 40 | 2.19 | 2.24 |
| | 1 13/16 | 1.8125 | 1.7719 | 1.7358 | 1.7448 | 0 | 39 | 2.35 | 2.41 |
| 1 7/8 | | 1.8750 | 1.8344 | 1.7983 | 1.8073 | 0 | 37 | 2.53 | 2.58 |
| | 1 15/16 | 1.9375 | 1.8969 | 1.8608 | 1.8698 | 0 | 36 | 2.71 | 2.77 |
| 2 | | 2.0000 | 1.9594 | 1.9233 | 1.9323 | 0 | 35 | 2.89 | 2.95 |
| | 2 1/8 | 2.1250 | 2.0844 | 2.0483 | 2.0573 | 0 | 33 | 3.28 | 3.35 |
| 2 1/4 | | 2.2500 | 2.2094 | 2.1733 | 2.1823 | 0 | 31 | 3.69 | 3.76 |
| | 2 3/8 | 2.3750 | 2.3344 | 2.2983 | 2.3073 | 0 | 29 | 4.13 | 4.21 |
| 2 1/2 | | 2.5000 | 2.4594 | 2.4233 | 2.4323 | 0 | 28 | 4.60 | 4.67 |
| | 2 5/8 | 2.6250 | 2.5844 | 2.5483 | 2.5573 | 0 | 26 | 5.08 | 5.16 |
| 2 3/4 | | 2.7500 | 2.7094 | 2.6733 | 2.6823 | 0 | 25 | 5.59 | 5.68 |
| | 2 7/8 | 2.8750 | 2.8344 | 2.7983 | 2.8073 | 0 | 24 | 6.13 | 6.22 |
| 3 | | 3.0000 | 2.9594 | 2.9233 | 2.9323 | 0 | 23 | 6.69 | 6.78 |
| | 3 1/8 | 3.1250 | 3.0844 | 3.0483 | 3.0573 | 0 | 22 | 7.28 | 7.37 |
| 3 1/4 | | 3.2500 | 3.2094 | 3.1733 | 3.1823 | 0 | 21 | 7.89 | 7.99 |
| | 3 3/8 | 3.3750 | 3.3344 | 3.2983 | 3.3073 | 0 | 21 | 8.52 | 8.63 |
| 3 1/2 | | 3.5000 | 3.4594 | 3.4233 | 3.4323 | 0 | 20 | 9.18 | 9.29 |
| | 3 5/8 | 3.6250 | 3.5844 | 3.5483 | 3.5573 | 0 | 19 | 9.86 | 9.98 |
| 3 3/4 | | 3.7500 | 3.7094 | 3.6733 | 3.6823 | 0 | 18 | 10.57 | 10.69 |
| | 3 7/8 | 3.8750 | 3.8344 | 3.7983 | 3.8073 | 0 | 18 | 11.30 | 11.43 |
| 4 | | 4.0000 | 3.9594 | 3.9233 | 3.9323 | 0 | 17 | 12.06 | 12.19 |
| | 4 1/8 | 4.1250 | 4.0844 | 4.0483 | 4.0573 | 0 | 17 | 12.84 | 12.97 |
| 4 1/4 | | 4.2500 | 4.2094 | 4.1733 | 4.1823 | 0 | 16 | 13.65 | 13.78 |
| | 4 3/8 | 4.3750 | 4.3344 | 4.2983 | 4.3073 | 0 | 16 | 14.48 | 14.62 |
| 4 1/2 | | 4.5000 | 4.4594 | 4.4233 | 4.4323 | 0 | 15 | 15.34 | 15.5 |
| | 4 5/8 | 4.6250 | 4.5844 | 4.5483 | 4.5573 | 0 | 15 | 16.2 | 16.4 |
| 4 3/4 | | 4.7500 | 4.7094 | 4.6733 | 4.6823 | 0 | 15 | 17.1 | 17.3 |
| | 4 7/8 | 4.8750 | 4.8344 | 4.7983 | 4.8073 | 0 | 14 | 18.0 | 18.2 |
| 5 | | 5.0000 | 4.9594 | 4.9233 | 4.9323 | 0 | 14 | 19.0 | 19.2 |
| | 5 1/8 | 5.1250 | 5.0844 | 5.0483 | 5.0573 | 0 | 13 | 20.0 | 20.1 |
| 5 1/4 | | 5.2500 | 5.2094 | 5.1733 | 5.1823 | 0 | 13 | 21.0 | 21.1 |
| | 5 3/8 | 5.3750 | 5.3344 | 5.2983 | 5.3073 | 0 | 13 | 22.0 | 22.2 |
| 5 1/2 | | 5.5000 | 5.4594 | 5.4233 | 5.4323 | 0 | 13 | 23.1 | 23.2 |
| | 5 5/8 | 5.6250 | 5.5844 | 5.5483 | 5.5573 | 0 | 12 | 24.1 | 24.3 |
| 5 3/4 | | 5.7500 | 5.7094 | 5.6733 | 5.6823 | 0 | 12 | 25.2 | 25.4 |
| | 5 7/8 | 5.8750 | 5.8344 | 5.7983 | 5.8073 | 0 | 12 | 26.4 | 26.5 |
| 6 | | 6.0000 | 5.9594 | 5.9233 | 5.9323 | 0 | 11 | 27.5 | 27.7 |

† These are standard sizes of the UNC or UNF Series.
a British: Effective Diameter.
b See formula, page 1150.
c Design form.  See diagram, page 1263.

Table 3i.  20-Thread Series, 20UN and 20UNR — Basic Dimensions

| Sizes | | Basic Major Diam., D | Basic Pitch Diam., E [a] | Minor Diameter [c] | | Lead Angle λ at Basic P.D. | | Area of Minor Diam. at D-2$h_b$ | Tensile Stress Area [b] |
| Primary | Secondary | | | Ext. Thds., $K_s$ | Int. Thds., $K_n$ | | | | |
|---|---|---|---|---|---|---|---|---|---|
| Inches | Inches | Inches | Inches | Inches | Inches | Deg. | Min. | Sq. In. | Sq. In. |
| ¼† | | 0.2500 | 0.2175 | 0.1887 | 0.1959 | 4 | 11 | 0.0269 | 0.0318 |
| 5/16 | | 0.3125 | 0.2800 | 0.2512 | 0.2584 | 3 | 15 | 0.0481 | 0.0547 |
| 3/8 | | 0.3750 | 0.3425 | 0.3137 | 0.3209 | 2 | 40 | 0.0755 | 0.0836 |
| 7/16† | | 0.4375 | 0.4050 | 0.3762 | 0.3834 | 2 | 15 | 0.1090 | 0.1187 |
| ½† | | 0.5000 | 0.4675 | 0.4387 | 0.4459 | 1 | 57 | 0.1486 | 0.160 |
| 9/16 | | 0.5625 | 0.5300 | 0.5012 | 0.5084 | 1 | 43 | 0.194 | 0.207 |
| 5/8 | | 0.6250 | 0.5925 | 0.5637 | 0.5709 | 1 | 32 | 0.246 | 0.261 |
| | 11/16 | 0.6875 | 0.6550 | 0.6262 | 0.6334 | 1 | 24 | 0.304 | 0.320 |
| ¾† | | 0.7500 | 0.7175 | 0.6887 | 0.6959 | 1 | 16 | 0.369 | 0.386 |
| | 13/16 | 0.8125 | 0.7800 | 0.7512 | 0.7584 | 1 | 10 | 0.439 | 0.458 |
| 7/8† | | 0.8750 | 0.8425 | 0.8137 | 0.8209 | 1 | 5 | 0.515 | 0.536 |
| | 15/16† | 0.9375 | 0.9050 | 0.8762 | 0.8834 | 1 | 0 | 0.598 | 0.620 |
| 1† | | 1.0000 | 0.9675 | 0.9387 | 0.9459 | 0 | 57 | 0.687 | 0.711 |
| | 1 1/16 | 1.0625 | 1.0300 | 1.0012 | 1.0084 | 0 | 53 | 0.782 | 0.807 |
| 1⅛ | | 1.1250 | 1.0925 | 1.0637 | 1.0709 | 0 | 50 | 0.882 | 0.910 |
| | 1 3/16 | 1.1875 | 1.1550 | 1.1262 | 1.1334 | 0 | 47 | 0.990 | 1.018 |
| 1¼ | | 1.2500 | 1.2175 | 1.1887 | 1.1959 | 0 | 45 | 1.103 | 1.133 |
| | 1 5/16 | 1.3125 | 1.2800 | 1.2512 | 1.2584 | 0 | 43 | 1.222 | 1.254 |
| 1⅜ | | 1.3750 | 1.3425 | 1.3137 | 1.3209 | 0 | 41 | 1.348 | 1.382 |
| | 1 7/16 | 1.4375 | 1.4050 | 1.3762 | 1.3834 | 0 | 39 | 1.479 | 1.51 |
| 1½ | | 1.5000 | 1.4675 | 1.4387 | 1.4459 | 0 | 37 | 1.62 | 1.65 |
| | 1 9/16 | 1.5625 | 1.5300 | 1.5012 | 1.5084 | 0 | 36 | 1.76 | 1.80 |
| 1⅝ | | 1.6250 | 1.5925 | 1.5637 | 1.5709 | 0 | 34 | 1.91 | 1.95 |
| | 1 11/16 | 1.6875 | 1.6550 | 1.6262 | 1.6334 | 0 | 33 | 2.07 | 2.11 |
| 1¾ | | 1.7500 | 1.7175 | 1.6887 | 1.6959 | 0 | 32 | 2.23 | 2.27 |
| | 1 13/16 | 1.8125 | 1.7800 | 1.7512 | 1.7584 | 0 | 31 | 2.40 | 2.44 |
| 1⅞ | | 1.8750 | 1.8425 | 1.8137 | 1.8209 | 0 | 30 | 2.57 | 2.62 |
| | 1 15/16 | 1.9375 | 1.9050 | 1.8762 | 1.8834 | 0 | 29 | 2.75 | 2.80 |
| 2 | | 2.0000 | 1.9675 | 1.9387 | 1.9459 | 0 | 28 | 2.94 | 2.99 |
| | 2⅛ | 2.1250 | 2.0925 | 2.0637 | 2.0709 | 0 | 26 | 3.33 | 3.39 |
| 2¼ | | 2.2500 | 2.2175 | 2.1887 | 2.1959 | 0 | 25 | 3.75 | 3.81 |
| | 2⅜ | 2.3750 | 2.3425 | 2.3137 | 2.3209 | 0 | 23 | 4.19 | 4.25 |
| 2½ | | 2.5000 | 2.4675 | 2.4387 | 2.4459 | 0 | 22 | 4.66 | 4.72 |
| | 2⅝ | 2.6250 | 2.5925 | 2.5637 | 2.5709 | 0 | 21 | 5.15 | 5.21 |
| 2¾ | | 2.7500 | 2.7175 | 2.6887 | 2.6959 | 0 | 20 | 5.66 | 5.73 |
| | 2⅞ | 2.8750 | 2.8425 | 2.8137 | 2.8209 | 0 | 19 | 6.20 | 6.27 |
| 3 | | 3.0000 | 2.9675 | 2.9387 | 2.9459 | 0 | 18 | 6.77 | 6.84 |

† These are standard sizes of the UNC, UNF, or UNEF Series.
a British: Effective Diameter.
b See formula, page 1150.
c Design form. See diagram, page 1263.

Certain applications require an allowance to permit application of the proper lubricant when making up the assembly, particularly with pressure vessels and steel pipe flanges, fittings, and valves for high-temperature, high-pressure service. For such applications Class 2A, which has an allowance, and Class 2B are recommended, replacing Class 7 which was previously established for such applications but which has been discontinued as an American Standard. In this application, when the thread is coated, the 2A allowance may not be consumed by such coating.

*Classes 3A and 3B:* Classes 3A and 3B may be used if closer tolerances are desired than those provided by Classes 2A and 2B. The maximum diameters of Class 3A

Table 3j. 28-Thread Series, 28UN and 28UNR — Basic Dimensions

| Sizes | | Basic Major Diam., $D$ | Basic Pitch Diam.,[a] $E$ | Minor Diameter[c] | | Lead Angle λ at Basic P.D. | | Area of Minor Diam. at $D-2h_b$ | Tensile Stress Area[b] |
|---|---|---|---|---|---|---|---|---|---|
| Primary | Secondary | | | Ext. Thds., $K_s$ | Int. Thds., $K_n$ | | | | |
| Inches | Inches | Inches | Inches | Inches | Inches | Deg. | Min. | Sq. In. | Sq. In. |
| | 12(.216)† | 0.2160 | 0.1928 | 0.1722 | 0.1773 | 3 | 22 | 0.0226 | 0.0258 |
| 1/4† | | 0.2500 | 0.2268 | 0.2062 | 0.2113 | 2 | 52 | 0.0326 | 0.0364 |
| 5/16 | | 0.3125 | 0.2893 | 0.2687 | 0.2738 | 2 | 15 | 0.0556 | 0.0606 |
| 3/8 | | 0.3750 | 0.3518 | 0.3312 | 0.3363 | 1 | 51 | 0.0848 | 0.0909 |
| 7/16† | | 0.4375 | 0.4143 | 0.3937 | 0.3988 | 1 | 34 | 0.1201 | 0.1274 |
| 1/2† | | 0.5000 | 0.4768 | 0.4562 | 0.4613 | 1 | 22 | 0.162 | 0.170 |
| 9/16 | | 0.5625 | 0.5393 | 0.5187 | 0.5238 | 1 | 12 | 0.209 | 0.219 |
| 5/8 | | 0.6250 | 0.6018 | 0.5812 | 0.5863 | 1 | 5 | 0.263 | 0.274 |
| | 11/16 | 0.6875 | 0.6643 | 0.6437 | 0.6488 | 0 | 59 | 0.323 | 0.335 |
| 3/4 | | 0.7500 | 0.7268 | 0.7062 | 0.7113 | 0 | 54 | 0.389 | 0.402 |
| | 13/16 | 0.8125 | 0.7893 | 0.7687 | 0.7738 | 0 | 50 | 0.461 | 0.475 |
| 7/8 | | 0.8750 | 0.8518 | 0.8312 | 0.8363 | 0 | 46 | 0.539 | 0.554 |
| | 15/16 | 0.9375 | 0.9143 | 0.8937 | 0.8988 | 0 | 43 | 0.624 | 0.640 |
| 1 | | 1.0000 | 0.9768 | 0.9562 | 0.9613 | 0 | 40 | 0.714 | 0.732 |
| | 1 1/16 | 1.0625 | 1.0393 | 1.0187 | 1.0238 | 0 | 38 | 0.811 | 0.830 |
| 1 1/8 | | 1.1250 | 1.1018 | 1.0812 | 1.0863 | 0 | 35 | 0.914 | 0.933 |
| | 1 3/16 | 1.1875 | 1.1643 | 1.1437 | 1.1488 | 0 | 34 | 1.023 | 1.044 |
| 1 1/4 | | 1.2500 | 1.2268 | 1.2062 | 1.2113 | 0 | 32 | 1.138 | 1.160 |
| | 1 5/16 | 1.3125 | 1.2893 | 1.2687 | 1.2738 | 0 | 30 | 1.259 | 1.282 |
| 1 3/8 | | 1.3750 | 1.3518 | 1.3312 | 1.3363 | 0 | 29 | 1.386 | 1.411 |
| | 1 7/16 | 1.4375 | 1.4143 | 1.3937 | 1.3988 | 0 | 28 | 1.52 | 1.55 |
| 1 1/2 | | 1.5000 | 1.4768 | 1.4562 | 1.4613 | 0 | 26 | 1.66 | 1.69 |

† These are standard sizes of the UNF or UNEF Series.
a British: Effective Diameter.
b See formula, page 1150.
c Design form. See diagram, page 1263.

Table 3k. 32-Thread Series, 32UN and 32UNR — Basic Dimensions

| Sizes | | Basic Major Diam., $D$ | Basic Pitch Diam.,[a] $E$ | Minor Diameter[c] | | Lead Angle λ at Basic P.D. | | Area of Minor Diam. at $D-2h_b$ | Tensile Stress Area[b] |
|---|---|---|---|---|---|---|---|---|---|
| Primary | Secondary | | | Ext. Thds., $K_s$ | Int. Thds., $K_n$ | | | | |
| Inches | Inches | Inches | Inches | Inches | Inches | Deg. | Min. | Sq. In. | Sq. In. |
| 6(.138)† | | 0.1380 | 0.1177 | 0.0997 | 0.1042 | 4 | 50 | 0.00745 | 0.00909 |
| 8(.164)† | | 0.1640 | 0.1437 | 0.1257 | 0.1302 | 3 | 58 | 0.01196 | 0.0140 |
| 10(.190)† | | 0.1900 | 0.1697 | 0.1517 | 0.1562 | 3 | 21 | 0.01750 | 0.0200 |
| | 12(.216)† | 0.2160 | 0.1957 | 0.1777 | 0.1822 | 2 | 55 | 0.0242 | 0.0270 |
| 1/4† | | 0.2500 | 0.2297 | 0.2117 | 0.2162 | 2 | 29 | 0.0344 | 0.0379 |
| 5/16† | | 0.3125 | 0.2922 | 0.2742 | 0.2787 | 1 | 57 | 0.0581 | 0.0625 |
| 3/8† | | 0.3750 | 0.3547 | 0.3367 | 0.3412 | 1 | 36 | 0.0878 | 0.0932 |
| 7/16 | | 0.4375 | 0.4172 | 0.3992 | 0.4037 | 1 | 22 | 0.1237 | 0.1301 |
| 1/2 | | 0.5000 | 0.4797 | 0.4617 | 0.4662 | 1 | 11 | 0.166 | 0.173 |
| 9/16 | | 0.5625 | 0.5422 | 0.5242 | 0.5287 | 1 | 3 | 0.214 | 0.222 |
| 5/8 | | 0.6250 | 0.6047 | 0.5867 | 0.5912 | 0 | 57 | 0.268 | 0.278 |
| | 11/16 | 0.6875 | 0.6672 | 0.6492 | 0.6537 | 0 | 51 | 0.329 | 0.339 |
| 3/4 | | 0.7500 | 0.7297 | 0.7117 | 0.7162 | 0 | 47 | 0.395 | 0.407 |
| | 13/16 | 0.8125 | 0.7922 | 0.7742 | 0.7787 | 0 | 43 | 0.468 | 0.480 |
| 7/8 | | 0.8750 | 0.8547 | 0.8367 | 0.8412 | 0 | 40 | 0.547 | 0.560 |
| | 15/16 | 0.9375 | 0.9172 | 0.8992 | 0.9037 | 0 | 37 | 0.632 | 0.646 |
| 1 | | 1.0000 | 0.9797 | 0.9617 | 0.9662 | 0 | 35 | 0.723 | 0.738 |

† These are standard sizes of the UNC, UNF, or UNEF Series.
a British: Effective Diameter.
b See formula, page 1150.
c Design form. See diagram, page 1263.

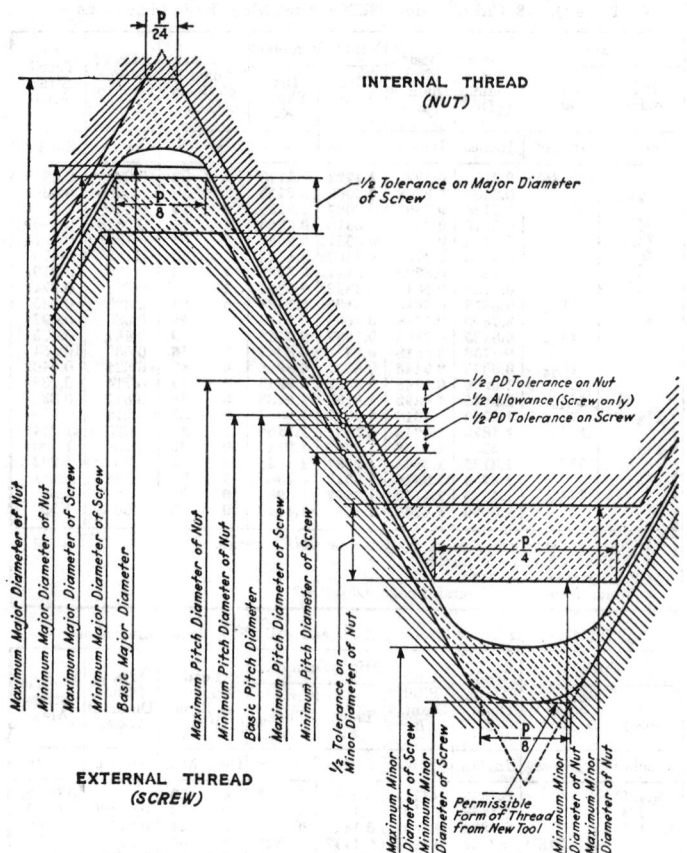

**Limits of Size Showing Tolerances, Allowances (Neutral Space), and Crest Clearances for Unified Classes 1A, 2A, 1B, and 2B**

(external) threads and the minimum diameters of Class 3B (internal) threads, whether or not plated or coated, are basic, affording no allowance or clearance for assembly of maximum metal components.

*Classes 1A and 1B:* Classes 1A and 1B threads replace American National Class 1 for new designs. These classes are intended for ordnance and other special uses. They are used on threaded components where quick and easy assembly is necessary and where a liberal allowance is required to permit ready assembly, even with slightly bruised or dirty threads.

Maximum diameters of Class 1A (external) threads are less than basic by the amount of the same allowance as applied to Class 2A. For the intended applications

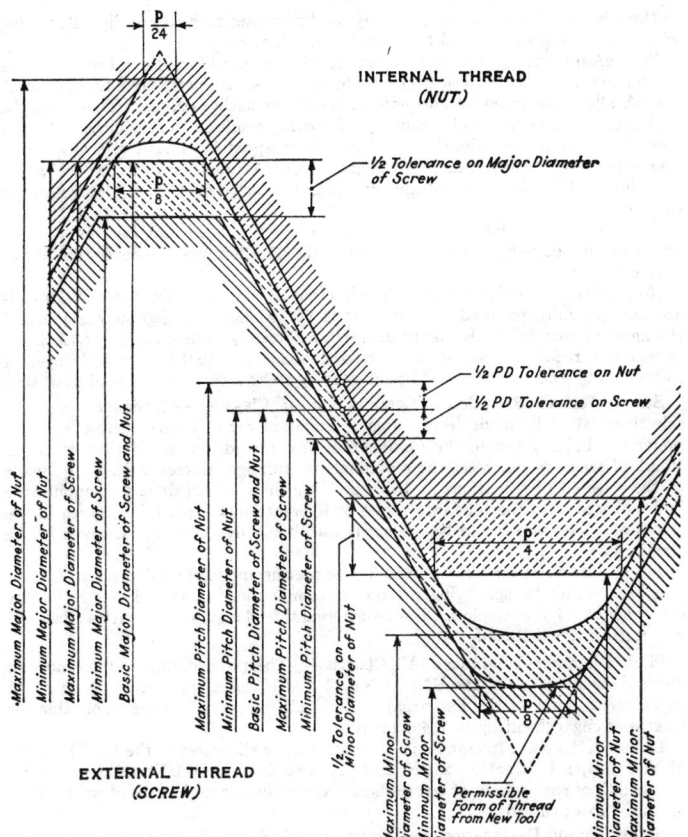

Limits of Size Showing Tolerances and Crest Clearances for Unified Classes 3A and 3B and American National Classes 2 and 3

in American practice the allowance is not available for plating or coating. Where the thread is plated or coated, special provisions are necessary. The minimum diameters of Class 1B (internal) threads, whether or not plated or coated, are basic, affording no allowance or clearance for assembly with maximum metal external thread components having maximum diameters which are basic.

**Coated Threads.** — Although the Standard does not make recommendations for thicknesses of, or specify limits for coatings, it does outline certain principles that will aid mechanical interchangeability if followed whenever conditions permit.

To keep finished threads within the limits of size established in the Standard, external threads should not exceed basic size after plating and internal threads should not be below basic size after plating. This recommendation does not apply to threads coated by certain commonly used processes such as hot-dip galvanizing where it may not be required to maintain these limits.

Class 2A provides both a tolerance and an allowance. Many thread requirements call for coatings such as those deposited by electro-plating processes and, in general, the 2A allowance provides adequate undercut for such coatings. There may be variations in thickness and symmetry of coating resulting from commercial processes but after plating the threads should be accepted by a basic size GO thread ring gage or equivalent functional gage. Class 1A provides an allowance which is maintained for both coated and uncoated product, i.e., it is not available for coating.

Class 3A does not include an allowance so it is suggested that the limits of size before plating be reduced by the amount of the 2A allowance whenever that allowance is adequate.

No provision is made for overcutting internal threads as coatings on such threads are not generally required. Further, it is very difficult to deposit a significant thickness of coating on the flanks of internal threads. Where a specific thickness of coating is required in an internal thread, it is suggested that the thread be overcut so that the thread as coated will be accepted by a GO thread plug gage of basic size.

**Screw Thread Selection — Combination of Classes. —** Whenever possible, selection should be made from Table 2, Standard Series Unified Screw Threads, preference being given to the Coarse- and Fine-thread Series. If threads in the standard series do not meet the requirements of design, reference should be made to the selected combinations in Table 4. The third expedient is to compute the limits of size from the tolerance tables or tolerance increment tables given in the standard. The fourth and last resort is calculation by the formulas given in the standard.

The requirements for screw thread fits for specific applications depend on end use and can be met by specifying the proper combinations of thread classes for the components. For example, a Class 2A external thread may be used with a Class 1B, 2B, or 3B internal thread.

**Pitch Diameter Tolerances, All Classes. —** The pitch diameter tolerances in Table 4 for all classes of the UNC, UNF, and 8UN series are based on a length of engagement equal to the basic major (nominal) diameter and are applicable for lengths of engagement up to 1½ diameters.

The pitch diameter tolerances used in Table 4 for all classes of the UNEF, 4UN, 6UN, 12UN, 16UN, 20UN, 28UN, and 32UN series and the UNS series are based on a length of engagement of 9 threads and are applicable for lengths of engagement of from 5 to 15 threads.

**Screw Thread Designation. —** The basic method of designating a screw thread is used where the standard tolerances or limits of size based on the standard length of engagement are applicable. The designation specifies in sequence the nominal size, number of threads per inch, thread series symbol, and thread class symbol. The nominal size is the basic major diameter and is specified as the fractional diameter, screw number, or their decimal equivalent. Where decimal equivalents are used for size call-out they shall be interpreted as being nominal size designations only and shall have no dimensional significance beyond the fractional size or number designation. The symbol LH is placed after the thread class symbol to indicate a left-hand thread.

*Examples:*

¼–20 UNC–2A or 0.250–20 UNC–2A
10–32 UNF–2A or 0.190–32 UNF–2A
⁷⁄₁₆–20 UNRF–2A or 0.4375–20 UNRF–2A
2–12 UN–2A or 2.000–12 UN–2A
¼–20 UNC–3A–LH or 0.250–20 UNC–3A–LH

For uncoated standard series threads these designations may optionally be supplemented by the addition of the pitch diameter limits of size.

*Example:*   ¼–20 UNC–2A
PD 0.2164–0.2127 (Optional for uncoated threads)

**Designating Coated Threads.** — For coated (or plated) Class 2A external threads, the basic (max) major and basic (max) pitch diameters are given followed by the words AFTER COATING. The major and pitch diameter limits of size before coating are also given followed by the words BEFORE COATING.

*Example:*   ¾–10 UNC–2A
*Major dia 0.7500 max  } AFTER COATING
PD 0.6850 max

†Major dia 0.7482–0.7353  } BEFORE COATING
PD 0.6832–0.6773

Certain applications require an allowance for rapid assembly, to permit application of a proper lubricant, or for residual growth due to high temperature expansion. In such applications where the thread is to be coated and the 2A allowance is not permitted to be consumed by such coating, the thread class symbol is qualified by the addition of the letter G (symbol for allowance) following the class symbol, and the maximum major and maximum pitch diameters are reduced below basic size by the amount of the 2A allowance and followed by the words AFTER COATING. This insures that the allowance is maintained. The major and pitch diameter limits of size before coating are also given followed by SPL and BEFORE COATING. For information concerning the designating of this and other special coating conditions reference should be made to American National Standard ANSI B1.1-1974.

**Designating UNS Threads.** — UNS screw threads which have special combinations of diameter and pitch with tolerance to Unified formulation have the basic form designation set out first followed always by the limits of size.

**Designating Multiple Start Threads.** — If a screw thread is of multiple start it is designated by specifying in sequence the nominal size, pitch (in decimals or threads per inch) and lead (in decimals or fractions).

**Other Special Designations.** — For other special designations including threads with modified limits of size or with special lengths of engagement, reference should be made to American National Standard ANSI B1.1-1974.

**Hole Sizes for Tapping.** — Hole size limits for tapping Classes 1B, 2B, and 3B threads of various lengths of engagement are given in the Tapping Section.

**Internal Thread Minor Diameter Tolerances.** — Internal thread minor diameter tolerances in Table 4 are based on a length of engagement equal to the nominal diameter. For general applications these tolerances are suitable for lengths of engagement up to 1½ diameters. However, some thread applications have lengths of engagement which are greater than 1½ diameters or less than the nominal diameter. For such applications it may be advantageous to increase or decrease the tolerance, respectively, as explained in the Tapping Section.

---

* Major and PD values are equal to basic and correspond to those in Table 4 for Class 3A.
† Major and PD limits are those in Table 4 for Class 2A.

Table 4. Standard Series and Selected Combinations† — Unified Screw Threads

| Nominal Size, Threads per Inch, and Series Designation^f | Class | External^e | | | | | | | Class | Internal^e | | | | |
|---|---|---|---|---|---|---|---|---|---|---|---|---|---|---|
| | | Allowance | Major Diameter Max^b | Major Diameter Min | Major Diameter Min^d | Pitch Diameter Max^b | Pitch Diameter Min | Minor Diameter^a | | Minor Diameter Min | Minor Diameter Max | Pitch Diameter Min | Pitch Diameter Max | Major Diameter Min |
| 0-80 UNF | 2A | 0.0005 | 0.0595 | 0.0563 | — | 0.0514 | 0.0496 | 0.0442 | 2B | 0.0465 | 0.0514 | 0.0519 | 0.0542 | 0.0600 |
| | 3A | 0.0000 | 0.0600 | 0.0568 | — | 0.0519 | 0.0506 | 0.0447 | 3B | 0.0465 | 0.0514 | 0.0519 | 0.0536 | 0.0600 |
| 1-64 UNC | 2A | 0.0006 | 0.0724 | 0.0686 | — | 0.0623 | 0.0603 | 0.0532 | 2B | 0.0561 | 0.0623 | 0.0629 | 0.0655 | 0.0730 |
| | 3A | 0.0000 | 0.0730 | 0.0692 | — | 0.0629 | 0.0614 | 0.0538 | 3B | 0.0561 | 0.0623 | 0.0629 | 0.0648 | 0.0730 |
| 1-72 UNF | 2A | 0.0006 | 0.0724 | 0.0689 | — | 0.0634 | 0.0615 | 0.0554 | 2B | 0.0580 | 0.0635 | 0.0640 | 0.0665 | 0.0730 |
| | 3A | 0.0000 | 0.0730 | 0.0695 | — | 0.0640 | 0.0626 | 0.0560 | 3B | 0.0580 | 0.0635 | 0.0640 | 0.0659 | 0.0730 |
| 2-56 UNC | 2A | 0.0006 | 0.0854 | 0.0813 | — | 0.0738 | 0.0717 | 0.0635 | 2B | 0.0667 | 0.0737 | 0.0744 | 0.0772 | 0.0860 |
| | 3A | 0.0000 | 0.0860 | 0.0819 | — | 0.0744 | 0.0728 | 0.0641 | 3B | 0.0667 | 0.0737 | 0.0744 | 0.0765 | 0.0860 |
| 2-64 UNF | 2A | 0.0006 | 0.0854 | 0.0816 | — | 0.0753 | 0.0733 | 0.0662 | 2B | 0.0691 | 0.0753 | 0.0759 | 0.0786 | 0.0860 |
| | 3A | 0.0000 | 0.0860 | 0.0822 | — | 0.0759 | 0.0744 | 0.0668 | 3B | 0.0691 | 0.0753 | 0.0759 | 0.0779 | 0.0860 |
| 3-48 UNC | 2A | 0.0007 | 0.0983 | 0.0938 | — | 0.0848 | 0.0825 | 0.0727 | 2B | 0.0764 | 0.0845 | 0.0855 | 0.0885 | 0.0990 |
| | 3A | 0.0000 | 0.0990 | 0.0945 | — | 0.0855 | 0.0838 | 0.0734 | 3B | 0.0764 | 0.0845 | 0.0855 | 0.0877 | 0.0990 |
| 3-56 UNF | 2A | 0.0007 | 0.0983 | 0.0942 | — | 0.0867 | 0.0845 | 0.0764 | 2B | 0.0797 | 0.0865 | 0.0874 | 0.0902 | 0.0990 |
| | 3A | 0.0000 | 0.0990 | 0.0949 | — | 0.0874 | 0.0858 | 0.0771 | 3B | 0.0797 | 0.0865 | 0.0874 | 0.0895 | 0.0990 |
| 4-40 UNC | 2A | 0.0008 | 0.1112 | 0.1061 | — | 0.0950 | 0.0925 | 0.0805 | 2B | 0.0849 | 0.0939 | 0.0958 | 0.0991 | 0.1120 |
| | 3A | 0.0000 | 0.1120 | 0.1069 | — | 0.0958 | 0.0939 | 0.0813 | 3B | 0.0849 | 0.0939 | 0.0958 | 0.0982 | 0.1120 |
| 4-48 UNF | 2A | 0.0007 | 0.1113 | 0.1068 | — | 0.0978 | 0.0954 | 0.0857 | 2B | 0.0894 | 0.0968 | 0.0985 | 0.1016 | 0.1120 |
| | 3A | 0.0000 | 0.1120 | 0.1075 | — | 0.0985 | 0.0967 | 0.0864 | 3B | 0.0894 | 0.0968 | 0.0985 | 0.1008 | 0.1120 |
| 5-40 UNC | 2A | 0.0008 | 0.1242 | 0.1191 | — | 0.1080 | 0.1054 | 0.0935 | 2B | 0.0979 | 0.1062 | 0.1088 | 0.1121 | 0.1250 |
| | 3A | 0.0000 | 0.1250 | 0.1199 | — | 0.1088 | 0.1069 | 0.0943 | 3B | 0.0979 | 0.1062 | 0.1088 | 0.1113 | 0.1250 |
| 5-44 UNF | 2A | 0.0007 | 0.1243 | 0.1195 | — | 0.1095 | 0.1070 | 0.0964 | 2B | 0.1004 | 0.1079 | 0.1102 | 0.1134 | 0.1250 |
| | 3A | 0.0000 | 0.1250 | 0.1202 | — | 0.1102 | 0.1083 | 0.0971 | 3B | 0.1004 | 0.1079 | 0.1102 | 0.1126 | 0.1250 |
| 6-32 UNC | 2A | 0.0008 | 0.1372 | 0.1312 | — | 0.1169 | 0.1141 | 0.0989 | 2B | 0.104 | 0.114 | 0.1177 | 0.1214 | 0.1380 |
| | 3A | 0.0000 | 0.1380 | 0.1320 | — | 0.1177 | 0.1156 | 0.0997 | 3B | 0.1040 | 0.1140 | 0.1177 | 0.1204 | 0.1380 |
| 6-40 UNF | 2A | 0.0008 | 0.1372 | 0.1321 | — | 0.1210 | 0.1184 | 0.1065 | 2B | 0.111 | 0.119 | 0.1218 | 0.1252 | 0.1380 |
| | 3A | 0.0000 | 0.1380 | 0.1329 | — | 0.1218 | 0.1198 | 0.1073 | 3B | 0.1110 | 0.1186 | 0.1218 | 0.1243 | 0.1380 |
| 8-32 UNC | 2A | 0.0009 | 0.1631 | 0.1571 | — | 0.1428 | 0.1399 | 0.1248 | 2B | 0.130 | 0.139 | 0.1437 | 0.1475 | 0.1640 |
| | 3A | 0.0000 | 0.1640 | 0.1580 | — | 0.1437 | 0.1415 | 0.1257 | 3B | 0.1300 | 0.1389 | 0.1437 | 0.1465 | 0.1640 |

| Size | Class | Allow. | Maj. Max | Maj. Min | UNR | P.D. Max | P.D. Min | Minor | Class | Minor Min | Minor Max | P.D. Min | P.D. Max | Maj. Min |
|---|---|---|---|---|---|---|---|---|---|---|---|---|---|---|
| 8-36 UNF | 2A | 0.0008 | 0.1632 | 0.1577 | — | 0.1452 | 0.1424 | 0.1291 | 2B | 0.134 | 0.142 | 0.1460 | 0.1496 | 0.1640 |
|  | 3A | 0.0000 | 0.1640 | 0.1585 | — | 0.1460 | 0.1439 | 0.1299 | 3B | 0.1340 | 0.1416 | 0.1460 | 0.1487 | 0.1640 |
| 10-24 UNC | 2A | 0.0010 | 0.1890 | 0.1818 | — | 0.1619 | 0.1586 | 0.1379 | 2B | 0.145 | 0.156 | 0.1629 | 0.1672 | 0.1900 |
|  | 3A | 0.0000 | 0.1900 | 0.1828 | — | 0.1629 | 0.1604 | 0.1389 | 3B | 0.1450 | 0.1555 | 0.1629 | 0.1661 | 0.1900 |
| 10-28 UNS | 2A | 0.0010 | 0.1890 | 0.1825 | — | 0.1658 | 0.1625 | 0.1452 | 2B | 0.151 | 0.160 | 0.1668 | 0.1711 | 0.1900 |
| 10-32 UNF | 2A | 0.0009 | 0.1891 | 0.1831 | — | 0.1688 | 0.1658 | 0.1508 | 2B | 0.156 | 0.164 | 0.1697 | 0.1736 | 0.1900 |
|  | 3A | 0.0000 | 0.1900 | 0.1840 | — | 0.1697 | 0.1674 | 0.1517 | 3B | 0.1560 | 0.1641 | 0.1697 | 0.1726 | 0.1900 |
| 10-36 UNS | 2A | 0.0009 | 0.1891 | 0.1836 | — | 0.1711 | 0.1681 | 0.1550 | 2B | 0.160 | 0.166 | 0.1720 | 0.1759 | 0.1900 |
| 10-40 UNS | 2A | 0.0009 | 0.1891 | 0.1840 | — | 0.1729 | 0.1700 | 0.1584 | 2B | 0.163 | 0.169 | 0.1738 | 0.1775 | 0.1900 |
| 10-48 UNS | 2A | 0.0008 | 0.1892 | 0.1847 | — | 0.1757 | 0.1731 | 0.1636 | 2B | 0.167 | 0.172 | 0.1765 | 0.1799 | 0.1900 |
| 10-56 UNS | 2A | 0.0007 | 0.1893 | 0.1852 | — | 0.1777 | 0.1752 | 0.1674 | 2B | 0.171 | 0.175 | 0.1784 | 0.1816 | 0.1900 |
| 12-24 UNC | 2A | 0.0010 | 0.2150 | 0.2078 | — | 0.1879 | 0.1845 | 0.1639 | 2B | 0.171 | 0.181 | 0.1889 | 0.1933 | 0.2160 |
|  | 3A | 0.0000 | 0.2160 | 0.2088 | — | 0.1889 | 0.1863 | 0.1649 | 3B | 0.1710 | 0.1807 | 0.1889 | 0.1922 | 0.2160 |
| 12-28 UNF | 2A | 0.0010 | 0.2150 | 0.2085 | — | 0.1918 | 0.1886 | 0.1712 | 2B | 0.177 | 0.186 | 0.1928 | 0.1970 | 0.2160 |
|  | 3A | 0.0000 | 0.2160 | 0.2095 | — | 0.1928 | 0.1904 | 0.1722 | 3B | 0.1770 | 0.1857 | 0.1928 | 0.1959 | 0.2160 |
| 12-32 UNEF | 2A | 0.0009 | 0.2151 | 0.2091 | — | 0.1948 | 0.1917 | 0.1768 | 2B | 0.182 | 0.190 | 0.1957 | 0.1998 | 0.2160 |
|  | 3A | 0.0000 | 0.2160 | 0.2100 | — | 0.1957 | 0.1933 | 0.1777 | 3B | 0.1820 | 0.1895 | 0.1957 | 0.1988 | 0.2160 |
| 12-36 UNS | 2A | 0.0009 | 0.2151 | 0.2096 | — | 0.1971 | 0.1941 | 0.1810 | 2B | 0.186 | 0.192 | 0.1980 | 0.2019 | 0.2160 |
| 12-40 UNS | 2A | 0.0009 | 0.2151 | 0.2100 | — | 0.1989 | 0.1960 | 0.1844 | 2B | 0.189 | 0.195 | 0.1998 | 0.2035 | 0.2160 |
| 12-48 UNS | 2A | 0.0008 | 0.2152 | 0.2107 | — | 0.2017 | 0.1991 | 0.1896 | 2B | 0.193 | 0.198 | 0.2025 | 0.2059 | 0.2160 |
| 12-56 UNS | 2A | 0.0007 | 0.2153 | 0.2112 | — | 0.2037 | 0.2012 | 0.1934 | 2B | 0.197 | 0.201 | 0.2044 | 0.2076 | 0.2160 |
| 1/4-20 UNC | 1A | 0.0011 | 0.2489 | 0.2367 | — | 0.2164 | 0.2108 | 0.1876 | 1B | 0.196 | 0.207 | 0.2175 | 0.2248 | 0.2500 |
|  | 2A | 0.0011 | 0.2489 | 0.2408 | 0.2367 | 0.2164 | 0.2127 | 0.1876 | 2B | 0.196 | 0.207 | 0.2175 | 0.2223 | 0.2500 |
|  | 3A | 0.0000 | 0.2500 | 0.2419 | — | 0.2175 | 0.2147 | 0.1887 | 3B | 0.1960 | 0.2067 | 0.2175 | 0.2211 | 0.2500 |
| 1/4-24 UNS | 2A | 0.0011 | 0.2489 | 0.2417 | — | 0.2218 | 0.2181 | 0.1978 | 2B | 0.205 | 0.215 | 0.2229 | 0.2277 | 0.2500 |
| 1/4-27 UNS | 2A | 0.0010 | 0.2490 | 0.2423 | — | 0.2249 | 0.2214 | 0.2036 | 2B | 0.210 | 0.219 | 0.2259 | 0.2304 | 0.2500 |
| 1/4-28 UNF | 1A | 0.0010 | 0.2490 | 0.2392 | — | 0.2258 | 0.2208 | 0.2052 | 1B | 0.211 | 0.220 | 0.2268 | 0.2333 | 0.2500 |
|  | 2A | 0.0010 | 0.2490 | 0.2425 | — | 0.2258 | 0.2225 | 0.2052 | 2B | 0.211 | 0.220 | 0.2268 | 0.2311 | 0.2500 |
|  | 3A | 0.0000 | 0.2500 | 0.2435 | — | 0.2268 | 0.2243 | 0.2062 | 3B | 0.2110 | 0.2190 | 0.2268 | 0.2300 | 0.2500 |
| 1/4-32 UNEF | 2A | 0.0010 | 0.2490 | 0.2430 | — | 0.2287 | 0.2255 | 0.2107 | 2B | 0.216 | 0.224 | 0.2297 | 0.2339 | 0.2500 |
|  | 3A | 0.0000 | 0.2500 | 0.2440 | — | 0.2297 | 0.2273 | 0.2117 | 3B | 0.2160 | 0.2229 | 0.2297 | 0.2328 | 0.2500 |

† Ue UNS threads only if Standard Series do not meet requirements (see page 1278). See footnotes a, b, c, d, e, and f at end of table.

**Table 4 (Continued). Standard Series and Selected Combinations† — Unified Screw Threads**

| Nominal Size, Threads per Inch, and Series Designation^f | External^e | | | | | | | | Internal^e | | | | | |
|---|---|---|---|---|---|---|---|---|---|---|---|---|---|---|
| | Class | Allowance | Major Diameter | | | Pitch Diameter | | Minor Diameter^a | Class | Minor Diameter^e | | Pitch Diameter | | Major Diameter |
| | | | Max^b | Min | Min^d | Max^b | Min | | | Min | Max | Min | Max | Min |
| ¼–36 UNS | 2A | 0.0009 | 0.2491 | 0.2436 | — | 0.2311 | 0.2280 | 0.2150 | 2B | 0.220 | 0.226 | 0.2320 | 0.2360 | 0.2500 |
| ¼–40 UNS | 2A | 0.0009 | 0.2491 | 0.2440 | — | 0.2329 | 0.2300 | 0.2184 | 2B | 0.223 | 0.229 | 0.2338 | 0.2376 | 0.2500 |
| ¼–48 UNS | 2A | 0.0008 | 0.2492 | 0.2447 | — | 0.2357 | 0.2330 | 0.2236 | 2B | 0.227 | 0.232 | 0.2365 | 0.2401 | 0.2500 |
| ¼–56 UNS | 2A | 0.0008 | 0.2492 | 0.2451 | — | 0.2376 | 0.2350 | 0.2273 | 2B | 0.231 | 0.235 | 0.2384 | 0.2417 | 0.2500 |
| 5/16–18 UNC | 1A | 0.0012 | 0.3113 | 0.2982 | — | 0.2752 | 0.2691 | 0.2431 | 1B | 0.252 | 0.265 | 0.2764 | 0.2843 | 0.3125 |
| | 2A | 0.0012 | 0.3113 | 0.3026 | 0.2982 | 0.2752 | 0.2712 | 0.2431 | 2B | 0.252 | 0.265 | 0.2764 | 0.2817 | 0.3125 |
| | 3A | 0.0000 | 0.3125 | 0.3038 | — | 0.2764 | 0.2734 | 0.2443 | 3B | 0.2520 | 0.2630 | 0.2764 | 0.2803 | 0.3125 |
| 5/16–20 UN | 2A | 0.0012 | 0.3113 | 0.3032 | — | 0.2788 | 0.2748 | 0.2500 | 2B | 0.258 | 0.270 | 0.2800 | 0.2852 | 0.3125 |
| | 3A | 0.0000 | 0.3125 | 0.3044 | — | 0.2800 | 0.2770 | 0.2512 | 3B | 0.2580 | 0.2662 | 0.2800 | 0.2839 | 0.3125 |
| 5/16–24 UNF | 1A | 0.0011 | 0.3114 | 0.3006 | — | 0.2843 | 0.2788 | 0.2603 | 1B | 0.267 | 0.277 | 0.2854 | 0.2925 | 0.3125 |
| | 2A | 0.0011 | 0.3114 | 0.3042 | — | 0.2843 | 0.2806 | 0.2603 | 2B | 0.267 | 0.277 | 0.2854 | 0.2902 | 0.3125 |
| | 3A | 0.0000 | 0.3125 | 0.3053 | — | 0.2854 | 0.2827 | 0.2614 | 3B | 0.2670 | 0.2754 | 0.2854 | 0.2890 | 0.3125 |
| 5/16–27 UNS | 2A | 0.0010 | 0.3115 | 0.3048 | — | 0.2874 | 0.2839 | 0.2661 | 2B | 0.272 | 0.281 | 0.2884 | 0.2929 | 0.3125 |
| 5/16–28 UN | 2A | 0.0010 | 0.3115 | 0.3050 | — | 0.2883 | 0.2849 | 0.2677 | 2B | 0.274 | 0.282 | 0.2893 | 0.2937 | 0.3125 |
| | 3A | 0.0000 | 0.3125 | 0.3060 | — | 0.2893 | 0.2867 | 0.2687 | 3B | 0.2740 | 0.2801 | 0.2893 | 0.2926 | 0.3125 |
| 5/16–32 UNEF | 2A | 0.0010 | 0.3115 | 0.3055 | — | 0.2912 | 0.2880 | 0.2732 | 2B | 0.279 | 0.286 | 0.2922 | 0.2964 | 0.3125 |
| | 3A | 0.0000 | 0.3125 | 0.3065 | — | 0.2922 | 0.2898 | 0.2742 | 3B | 0.2790 | 0.2847 | 0.2922 | 0.2953 | 0.3125 |
| 5/16–36 UNS | 2A | 0.0009 | 0.3116 | 0.3061 | — | 0.2936 | 0.2905 | 0.2775 | 2B | 0.282 | 0.289 | 0.2945 | 0.2985 | 0.3125 |
| 5/16–40 UNS | 2A | 0.0009 | 0.3116 | 0.3065 | — | 0.2954 | 0.2925 | 0.2809 | 2B | 0.285 | 0.291 | 0.2963 | 0.3001 | 0.3125 |
| 5/16–48 UNS | 2A | 0.0008 | 0.3117 | 0.3072 | — | 0.2982 | 0.2955 | 0.2861 | 2B | 0.290 | 0.295 | 0.2990 | 0.3026 | 0.3125 |
| 3/8–16 UNC | 1A | 0.0013 | 0.3737 | 0.3595 | — | 0.3331 | 0.3266 | 0.2970 | 1B | 0.307 | 0.321 | 0.3344 | 0.3429 | 0.3750 |
| | 2A | 0.0013 | 0.3737 | 0.3643 | 0.3595 | 0.3331 | 0.3287 | 0.2970 | 2B | 0.307 | 0.321 | 0.3344 | 0.3401 | 0.3750 |
| | 3A | 0.0000 | 0.3750 | 0.3656 | — | 0.3344 | 0.3311 | 0.2983 | 3B | 0.3070 | 0.3182 | 0.3344 | 0.3387 | 0.3750 |
| 3/8–18 UNS | 2A | 0.0013 | 0.3737 | 0.3650 | — | 0.3376 | 0.3333 | 0.3055 | 2B | 0.315 | 0.328 | 0.3389 | 0.3445 | 0.3750 |
| 3/8–20 UN | 2A | 0.0012 | 0.3738 | 0.3657 | — | 0.3413 | 0.3372 | 0.3125 | 2B | 0.321 | 0.332 | 0.3425 | 0.3479 | 0.3750 |
| | 3A | 0.0000 | 0.3750 | 0.3669 | — | 0.3425 | 0.3394 | 0.3137 | 3B | 0.3210 | 0.3287 | 0.3425 | 0.3465 | 0.3750 |
| 3/8–24 UNF | 1A | 0.0011 | 0.3739 | 0.3631 | — | 0.3468 | 0.3411 | 0.3228 | 1B | 0.330 | 0.340 | 0.3479 | 0.3553 | 0.3750 |
| | 2A | 0.0011 | 0.3739 | 0.3667 | — | 0.3468 | 0.3430 | 0.3228 | 2B | 0.330 | 0.340 | 0.3479 | 0.3528 | 0.3750 |

| | | | | | | | | | | | | | | |
|---|---|---|---|---|---|---|---|---|---|---|---|---|---|---|
| 3/8-24 UNF | 3A | 0.0000 | 0.3750 | 0.3678 | — | 0.3479 | 0.3450 | 0.3239 | 3B | 0.3300 | 0.3372 | 0.3479 | 0.3516 | 0.3750 |
| 3/8-27 UNS | 2A | 0.0011 | 0.3739 | 0.3672 | — | 0.3498 | 0.3462 | 0.3285 | 2B | 0.335 | 0.344 | 0.3509 | 0.3556 | 0.3750 |
| 3/8-28 UN | 2A | 0.0011 | 0.3739 | 0.3674 | — | 0.3507 | 0.3471 | 0.3301 | 2B | 0.336 | 0.345 | 0.3518 | 0.3564 | 0.3750 |
| | 3A | 0.0000 | 0.3750 | 0.3685 | — | 0.3518 | 0.3491 | 0.3312 | 3B | 0.3360 | 0.3426 | 0.3518 | 0.3553 | 0.3750 |
| 3/8-32 UNEF | 2A | 0.0010 | 0.3740 | 0.3680 | — | 0.3537 | 0.3503 | 0.3357 | 2B | 0.341 | 0.349 | 0.3547 | 0.3591 | 0.3750 |
| | 3A | 0.0000 | 0.3750 | 0.3690 | — | 0.3547 | 0.3522 | 0.3367 | 3B | 0.3410 | 0.3469 | 0.3547 | 0.3580 | 0.3750 |
| 3/8-36 UNS | 2A | 0.0010 | 0.3740 | 0.3685 | — | 0.3560 | 0.3528 | 0.3399 | 2B | 0.345 | 0.352 | 0.3570 | 0.3612 | 0.3750 |
| 3/8-40 UNS | 2A | 0.0009 | 0.3741 | 0.3690 | — | 0.3579 | 0.3548 | 0.3434 | 2B | 0.348 | 0.354 | 0.3588 | 0.3628 | 0.3750 |
| 0.390-27 UNS | 2A | 0.0011 | 0.3889 | 0.3822 | — | 0.3648 | 0.3612 | 0.3435 | 2B | 0.350 | 0.359 | 0.3659 | 0.3706 | 0.3900 |
| 7/16-14 UNC | 1A | 0.0014 | 0.4361 | 0.4206 | 0.4206 | 0.3897 | 0.3826 | 0.3485 | 1B | 0.360 | 0.376 | 0.3911 | 0.4003 | 0.4375 |
| | 2A | 0.0014 | 0.4361 | 0.4258 | — | 0.3897 | 0.3850 | 0.3485 | 2B | 0.360 | 0.376 | 0.3911 | 0.3972 | 0.4375 |
| | 3A | 0.0000 | 0.4375 | 0.4272 | — | 0.3911 | 0.3876 | 0.3499 | 3B | 0.3600 | 0.3717 | 0.3911 | 0.3957 | 0.4375 |
| 7/16-16 UN | 2A | 0.0014 | 0.4361 | 0.4267 | — | 0.3955 | 0.3909 | 0.3594 | 2B | 0.370 | 0.384 | 0.3969 | 0.4028 | 0.4375 |
| | 3A | 0.0000 | 0.4375 | 0.4281 | — | 0.3969 | 0.3935 | 0.3608 | 3B | 0.3700 | 0.3783 | 0.3969 | 0.4009 | 0.4375 |
| 7/16-18 UNS | 2A | 0.0013 | 0.4362 | 0.4275 | — | 0.4001 | 0.3958 | 0.3680 | 2B | 0.377 | 0.390 | 0.4014 | 0.4070 | 0.4375 |
| 7/16-20 UNF | 1A | 0.0013 | 0.4362 | 0.4240 | — | 0.4037 | 0.3975 | 0.3749 | 1B | 0.383 | 0.395 | 0.4050 | 0.4131 | 0.4375 |
| | 2A | 0.0013 | 0.4362 | 0.4281 | — | 0.4037 | 0.3995 | 0.3749 | 2B | 0.383 | 0.395 | 0.4050 | 0.4104 | 0.4375 |
| | 3A | 0.0000 | 0.4375 | 0.4294 | — | 0.4050 | 0.4019 | 0.3762 | 3B | 0.3830 | 0.3916 | 0.4050 | 0.4091 | 0.4375 |
| 7/16-24 UNS | 2A | 0.0011 | 0.4364 | 0.4292 | — | 0.4093 | 0.4055 | 0.3853 | 2B | 0.392 | 0.402 | 0.4104 | 0.4153 | 0.4375 |
| 7/16-27 UNS | 2A | 0.0011 | 0.4364 | 0.4297 | — | 0.4123 | 0.4087 | 0.3910 | 2B | 0.397 | 0.406 | 0.4134 | 0.4181 | 0.4375 |
| 7/16-28 UNEF | 2A | 0.0011 | 0.4364 | 0.4299 | — | 0.4132 | 0.4096 | 0.3926 | 2B | 0.399 | 0.407 | 0.4143 | 0.4189 | 0.4375 |
| | 3A | 0.0000 | 0.4375 | 0.4310 | — | 0.4143 | 0.4116 | 0.3937 | 3B | 0.3990 | 0.4051 | 0.4143 | 0.4178 | 0.4375 |
| 7/16-32 UN | 2A | 0.0010 | 0.4365 | 0.4305 | — | 0.4162 | 0.4128 | 0.3982 | 2B | 0.404 | 0.411 | 0.4172 | 0.4216 | 0.4375 |
| | 3A | 0.0000 | 0.4375 | 0.4315 | — | 0.4172 | 0.4147 | 0.3992 | 3B | 0.4040 | 0.4094 | 0.4172 | 0.4205 | 0.4375 |
| 1/2-12 UNS | 2A | 0.0016 | 0.4984 | 0.4870 | — | 0.4443 | 0.4389 | 0.3962 | 2B | 0.410 | 0.428 | 0.4459 | 0.4529 | 0.5000 |
| | 3A | 0.0000 | 0.5000 | 0.4886 | — | 0.4459 | 0.4419 | 0.3978 | 3B | 0.4100 | 0.4223 | 0.4459 | 0.4511 | 0.5000 |
| 1/2-13 UNC | 1A | 0.0015 | 0.4985 | 0.4822 | 0.4822 | 0.4485 | 0.4411 | 0.4041 | 1B | 0.417 | 0.434 | 0.4500 | 0.4597 | 0.5000 |
| | 2A | 0.0015 | 0.4985 | 0.4876 | — | 0.4485 | 0.4435 | 0.4041 | 2B | 0.417 | 0.434 | 0.4500 | 0.4565 | 0.5000 |
| | 3A | 0.0000 | 0.5000 | 0.4891 | — | 0.4500 | 0.4463 | 0.4056 | 3B | 0.4170 | 0.4284 | 0.4500 | 0.4548 | 0.5000 |
| 1/2-14 UNS | 2A | 0.0015 | 0.4985 | 0.4882 | — | 0.4521 | 0.4471 | 0.4109 | 2B | 0.423 | 0.438 | 0.4536 | 0.4601 | 0.5000 |

† Use UNS threads only if Standard Series do not meet requirements (see page 1278). See footnotes a, b, c, d, e, and f at end of table.

Table 4 (Continued). Standard Series and Selected Combinations† — Unified Screw Threads

| Nominal Size, Threads per Inch, and Series Designation† | External | | | | | | | | Internal | | | | | |
|---|---|---|---|---|---|---|---|---|---|---|---|---|---|---|
| | Class | Allowance | Major Diameter | | | Pitch Diameter | | Minor Diameter^a | Class | Minor Diameter^e | | Pitch Diameter | | Major Diameter |
| | | | Max^b | Min | Min^d | Max^b | Min | | | Min | Max | Min | Max | Min |
| 1/2–16 UN | 2A | 0.0014 | 0.4986 | 0.4892 | — | 0.4580 | 0.4533 | 0.4219 | 2B | 0.432 | 0.446 | 0.4594 | 0.4655 | 0.5000 |
| | 3A | 0.0000 | 0.5000 | 0.4996 | — | 0.4594 | 0.4559 | 0.4233 | 3B | 0.4320 | 0.4408 | 0.4594 | 0.4640 | 0.5000 |
| 1/2–18 UNS | 2A | 0.0013 | 0.4987 | 0.4900 | — | 0.4626 | 0.4582 | 0.4305 | 2B | 0.440 | 0.453 | 0.4639 | 0.4697 | 0.5000 |
| 1/2–20 UNF | 1A | 0.0013 | 0.4987 | 0.4865 | — | 0.4662 | 0.4598 | 0.4374 | 1B | 0.446 | 0.457 | 0.4675 | 0.4759 | 0.5000 |
| | 2A | 0.0013 | 0.4987 | 0.4906 | — | 0.4662 | 0.4619 | 0.4374 | 2B | 0.446 | 0.457 | 0.4675 | 0.4731 | 0.5000 |
| | 3A | 0.0000 | 0.5000 | 0.4919 | — | 0.4675 | 0.4643 | 0.4387 | 3B | 0.4460 | 0.4537 | 0.4675 | 0.4717 | 0.5000 |
| 1/2–24 UNS | 2A | 0.0012 | 0.4988 | 0.4916 | — | 0.4717 | 0.4678 | 0.4477 | 2B | 0.455 | 0.465 | 0.4729 | 0.4780 | 0.5000 |
| 1/2–27 UNS | 2A | 0.0011 | 0.4989 | 0.4922 | — | 0.4748 | 0.4711 | 0.4535 | 2B | 0.460 | 0.469 | 0.4759 | 0.4807 | 0.5000 |
| 1/2–28 UNEF | 2A | 0.0011 | 0.4989 | 0.4924 | — | 0.4757 | 0.4720 | 0.4551 | 2B | 0.461 | 0.470 | 0.4768 | 0.4816 | 0.5000 |
| | 3A | 0.0000 | 0.5000 | 0.4935 | — | 0.4768 | 0.4740 | 0.4562 | 3B | 0.4610 | 0.4676 | 0.4768 | 0.4804 | 0.5000 |
| 1/2–32 UN | 2A | 0.0010 | 0.4990 | 0.4930 | — | 0.4787 | 0.4752 | 0.4607 | 2B | 0.466 | 0.474 | 0.4797 | 0.4842 | 0.5000 |
| | 3A | 0.0000 | 0.5000 | 0.4940 | — | 0.4797 | 0.4771 | 0.4617 | 3B | 0.4660 | 0.4719 | 0.4797 | 0.4831 | 0.5000 |
| 9/16–12 UNC | 1A | 0.0016 | 0.5609 | 0.5437 | — | 0.5068 | 0.4990 | 0.4587 | 1B | 0.472 | 0.490 | 0.5084 | 0.5186 | 0.5625 |
| | 2A | 0.0016 | 0.5609 | 0.5495 | — | 0.5068 | 0.5016 | 0.4587 | 2B | 0.472 | 0.490 | 0.5084 | 0.5152 | 0.5625 |
| | 3A | 0.0000 | 0.5625 | 0.5511 | 0.5437 | 0.5084 | 0.5045 | 0.4603 | 3B | 0.4720 | 0.4843 | 0.5084 | 0.5135 | 0.5625 |
| 9/16–14 UNS | 2A | 0.0015 | 0.5610 | 0.5507 | — | 0.5146 | 0.5096 | 0.4734 | 2B | 0.485 | 0.501 | 0.5161 | 0.5226 | 0.5625 |
| 9/16–16 UN | 2A | 0.0014 | 0.5611 | 0.5517 | — | 0.5205 | 0.5158 | 0.4844 | 2B | 0.495 | 0.509 | 0.5219 | 0.5280 | 0.5625 |
| | 3A | 0.0000 | 0.5625 | 0.5531 | — | 0.5219 | 0.5184 | 0.4858 | 3B | 0.4950 | 0.5033 | 0.5219 | 0.5265 | 0.5625 |
| 9/16–18 UNF | 1A | 0.0014 | 0.5611 | 0.5480 | — | 0.5250 | 0.5182 | 0.4929 | 1B | 0.502 | 0.515 | 0.5264 | 0.5353 | 0.5625 |
| | 2A | 0.0014 | 0.5611 | 0.5524 | — | 0.5250 | 0.5205 | 0.4929 | 2B | 0.502 | 0.515 | 0.5264 | 0.5323 | 0.5625 |
| | 3A | 0.0000 | 0.5625 | 0.5538 | — | 0.5264 | 0.5230 | 0.4943 | 3B | 0.5020 | 0.5106 | 0.5264 | 0.5308 | 0.5625 |
| 9/16–20 UN | 2A | 0.0013 | 0.5612 | 0.5531 | — | 0.5287 | 0.5245 | 0.4999 | 2B | 0.508 | 0.520 | 0.5300 | 0.5355 | 0.5625 |
| | 3A | 0.0000 | 0.5625 | 0.5544 | — | 0.5300 | 0.5268 | 0.5012 | 3B | 0.5080 | 0.5162 | 0.5300 | 0.5341 | 0.5625 |
| 9/16–24 UNEF | 2A | 0.0012 | 0.5613 | 0.5541 | — | 0.5342 | 0.5303 | 0.5102 | 2B | 0.517 | 0.527 | 0.5354 | 0.5405 | 0.5625 |
| | 3A | 0.0000 | 0.5625 | 0.5553 | — | 0.5354 | 0.5325 | 0.5114 | 3B | 0.5170 | 0.5244 | 0.5354 | 0.5392 | 0.5625 |
| 9/16–27 UNS | 2A | 0.0011 | 0.5614 | 0.5547 | — | 0.5373 | 0.5336 | 0.5160 | 2B | 0.522 | 0.531 | 0.5384 | 0.5432 | 0.5625 |
| 9/16–28 UN | 2A | 0.0011 | 0.5614 | 0.5549 | — | 0.5382 | 0.5345 | 0.5176 | 2B | 0.524 | 0.532 | 0.5393 | 0.5441 | 0.5625 |
| | 3A | 0.0000 | 0.5625 | 0.5560 | — | 0.5393 | 0.5365 | 0.5187 | 3B | 0.5240 | 0.5301 | 0.5393 | 0.5429 | 0.5625 |
| 9/16–32 UN | 2A | 0.0010 | 0.5615 | 0.5555 | — | 0.5412 | 0.5377 | 0.5232 | 2B | 0.529 | 0.536 | 0.5422 | 0.5467 | 0.5625 |
| | 3A | 0.0000 | 0.5625 | 0.5565 | — | 0.5422 | 0.5396 | 0.5242 | 3B | 0.5290 | 0.5344 | 0.5422 | 0.5456 | 0.5625 |

| Nominal Size, Threads per Inch, and Series Designation | Classes | Allowance | Major Dia, Max | Major Dia, Min | UNR Minor Dia, Max, Ext | Pitch Dia, Max | Pitch Dia, Min | Minor Dia, Max, Ext | Classes | Minor Dia, Min | Minor Dia, Max | Pitch Dia, Min | Pitch Dia, Max | Major Dia, Min |
|---|---|---|---|---|---|---|---|---|---|---|---|---|---|---|
| 5/8–11 UNC | 1A | 0.0016 | 0.6234 | 0.6052 | — | 0.5644 | 0.5561 | 0.5119 | 1B | 0.527 | 0.546 | 0.5660 | 0.5767 | 0.6250 |
|  | 2A | 0.0016 | 0.6234 | 0.6113 | — | 0.5644 | 0.5589 | 0.5119 | 2B | 0.527 | 0.546 | 0.5660 | 0.5732 | 0.6250 |
|  | 3A | 0.0000 | 0.6250 | 0.6129 | 0.6052 | 0.5660 | 0.5619 | 0.5135 | 3B | 0.5270 | 0.5391 | 0.5660 | 0.5714 | 0.6250 |
| 5/8–12 UN | 2A | 0.0016 | 0.6234 | 0.6120 | — | 0.5693 | 0.5639 | 0.5212 | 2B | 0.535 | 0.553 | 0.5709 | 0.5780 | 0.6250 |
|  | 3A | 0.0000 | 0.6250 | 0.6136 | — | 0.5709 | 0.5668 | 0.5228 | 3B | 0.5350 | 0.5463 | 0.5709 | 0.5762 | 0.6250 |
| 5/8–14 UNS | 2A | 0.0015 | 0.6235 | 0.6132 | — | 0.5771 | 0.5720 | 0.5359 | 2B | 0.548 | 0.564 | 0.5786 | 0.5852 | 0.6250 |
| 5/8–16 UN | 2A | 0.0014 | 0.6236 | 0.6142 | — | 0.5830 | 0.5782 | 0.5469 | 2B | 0.557 | 0.571 | 0.5844 | 0.5906 | 0.6250 |
|  | 3A | 0.0000 | 0.6250 | 0.6156 | — | 0.5844 | 0.5808 | 0.5483 | 3B | 0.5570 | 0.5658 | 0.5844 | 0.5890 | 0.6250 |
| 5/8–18 UNF | 1A | 0.0014 | 0.6236 | 0.6105 | — | 0.5875 | 0.5805 | 0.5554 | 1B | 0.565 | 0.578 | 0.5889 | 0.5980 | 0.6250 |
|  | 2A | 0.0014 | 0.6236 | 0.6149 | — | 0.5875 | 0.5828 | 0.5554 | 2B | 0.565 | 0.578 | 0.5889 | 0.5949 | 0.6250 |
|  | 3A | 0.0000 | 0.6250 | 0.6163 | — | 0.5889 | 0.5854 | 0.5568 | 3B | 0.5650 | 0.5730 | 0.5889 | 0.5934 | 0.6250 |
| 5/8–20 UN | 2A | 0.0013 | 0.6237 | 0.6156 | — | 0.5912 | 0.5869 | 0.5624 | 2B | 0.571 | 0.582 | 0.5925 | 0.5981 | 0.6250 |
|  | 3A | 0.0000 | 0.6250 | 0.6169 | — | 0.5925 | 0.5893 | 0.5637 | 3B | 0.5710 | 0.5787 | 0.5925 | 0.5967 | 0.6250 |
| 5/8–24 UNEF | 2A | 0.0012 | 0.6238 | 0.6166 | — | 0.5967 | 0.5927 | 0.5727 | 2B | 0.580 | 0.590 | 0.5979 | 0.6031 | 0.6250 |
|  | 3A | 0.0000 | 0.6250 | 0.6178 | — | 0.5979 | 0.5949 | 0.5739 | 3B | 0.5800 | 0.5869 | 0.5979 | 0.6018 | 0.6250 |
| 5/8–27 UNS | 2A | 0.0011 | 0.6239 | 0.6172 | — | 0.5998 | 0.5960 | 0.5785 | 2B | 0.585 | 0.594 | 0.6009 | 0.6059 | 0.6250 |
| 5/8–28 UN | 2A | 0.0011 | 0.6239 | 0.6174 | — | 0.6007 | 0.5969 | 0.5801 | 2B | 0.586 | 0.595 | 0.6018 | 0.6067 | 0.6250 |
|  | 3A | 0.0000 | 0.6250 | 0.6185 | — | 0.6018 | 0.5990 | 0.5812 | 3B | 0.5860 | 0.5936 | 0.6018 | 0.6055 | 0.6250 |
| 5/8–32 UN | 2A | 0.0011 | 0.6239 | 0.6179 | — | 0.6036 | 0.6000 | 0.5856 | 2B | 0.591 | 0.599 | 0.6047 | 0.6093 | 0.6250 |
|  | 3A | 0.0000 | 0.6250 | 0.6190 | — | 0.6047 | 0.6020 | 0.5867 | 3B | 0.5910 | 0.5969 | 0.6047 | 0.6082 | 0.6250 |
| 11/16–12 UN | 2A | 0.0016 | 0.6859 | 0.6745 | — | 0.6318 | 0.6264 | 0.5837 | 2B | 0.597 | 0.615 | 0.6334 | 0.6405 | 0.6875 |
|  | 3A | 0.0000 | 0.6875 | 0.6761 | — | 0.6334 | 0.6293 | 0.5853 | 3B | 0.5970 | 0.6085 | 0.6334 | 0.6387 | 0.6875 |
| 11/16–16 UN | 2A | 0.0014 | 0.6861 | 0.6767 | — | 0.6455 | 0.6407 | 0.6094 | 2B | 0.620 | 0.634 | 0.6469 | 0.6531 | 0.6875 |
|  | 3A | 0.0000 | 0.6875 | 0.6781 | — | 0.6469 | 0.6433 | 0.6108 | 3B | 0.6200 | 0.6283 | 0.6469 | 0.6515 | 0.6875 |
| 11/16–20 UN | 2A | 0.0013 | 0.6862 | 0.6781 | — | 0.6537 | 0.6494 | 0.6249 | 2B | 0.633 | 0.645 | 0.6550 | 0.6606 | 0.6875 |
|  | 3A | 0.0000 | 0.6875 | 0.6794 | — | 0.6550 | 0.6518 | 0.6262 | 3B | 0.6330 | 0.6412 | 0.6550 | 0.6592 | 0.6875 |
| 11/16–24 UNEF | 2A | 0.0012 | 0.6863 | 0.6791 | — | 0.6592 | 0.6552 | 0.6352 | 2B | 0.642 | 0.652 | 0.6604 | 0.6656 | 0.6875 |
|  | 3A | 0.0000 | 0.6875 | 0.6803 | — | 0.6604 | 0.6574 | 0.6364 | 3B | 0.6420 | 0.6494 | 0.6604 | 0.6643 | 0.6875 |
| 11/16–28 UN | 2A | 0.0011 | 0.6864 | 0.6799 | — | 0.6632 | 0.6594 | 0.6426 | 2B | 0.649 | 0.657 | 0.6643 | 0.6692 | 0.6875 |
|  | 3A | 0.0000 | 0.6875 | 0.6810 | — | 0.6643 | 0.6615 | 0.6437 | 3B | 0.6490 | 0.6551 | 0.6643 | 0.6680 | 0.6875 |
| 11/16–32 UN | 2A | 0.0011 | 0.6864 | 0.6804 | — | 0.6661 | 0.6625 | 0.6481 | 2B | 0.654 | 0.661 | 0.6672 | 0.6718 | 0.6875 |
|  | 3A | 0.0000 | 0.6875 | 0.6815 | — | 0.6672 | 0.6645 | 0.6492 | 3B | 0.6540 | 0.6594 | 0.6672 | 0.6707 | 0.6875 |

† Use UNS threads only if Standard Series do not meet requirements (see page 1278). See footnotes a, b, c, d, e, and f at end of table.

Table 4 (*Continued*). Standard Series and Selected Combinations† — Unified Screw Threads

| Nominal Size, Threads per Inch, and Series Designation f | External Class | Allowance | Major Diameter Max b | Major Diameter Min | Major Diameter Min d | Pitch Diameter Max b | Pitch Diameter Min | Minor Diameter a | Internal Class | Minor Diameter Min | Minor Diameter Max e | Pitch Diameter Min | Pitch Diameter Max | Major Diameter Min |
|---|---|---|---|---|---|---|---|---|---|---|---|---|---|---|
| ¾–10 UNC | 1A | 0.0018 | 0.7482 | 0.7288 | — | 0.6832 | 0.6744 | 0.6255 | 1B | 0.642 | 0.663 | 0.6850 | 0.6965 | 0.7500 |
| | 2A | 0.0018 | 0.7482 | 0.7353 | 0.7288 | 0.6832 | 0.6773 | 0.6255 | 2B | 0.642 | 0.663 | 0.6850 | 0.6927 | 0.7500 |
| | 3A | 0.0000 | 0.7500 | 0.7371 | — | 0.6850 | 0.6806 | 0.6273 | 3B | 0.6420 | 0.6545 | 0.6850 | 0.6907 | 0.7500 |
| ¾–12 UN | 2A | 0.0017 | 0.7483 | 0.7369 | — | 0.6942 | 0.6887 | 0.6461 | 2B | 0.660 | 0.678 | 0.6959 | 0.7031 | 0.7500 |
| | 3A | 0.0000 | 0.7500 | 0.7386 | — | 0.6959 | 0.6918 | 0.6478 | 3B | 0.6600 | 0.6707 | 0.6959 | 0.7013 | 0.7500 |
| ¾–14 UNS | 2A | 0.0015 | 0.7485 | 0.7382 | — | 0.7021 | 0.6970 | 0.6609 | 2B | 0.673 | 0.688 | 0.7036 | 0.7103 | 0.7500 |
| ¾–16 UNF | 1A | 0.0015 | 0.7485 | 0.7343 | — | 0.7079 | 0.7004 | 0.6718 | 1B | 0.682 | 0.696 | 0.7094 | 0.7192 | 0.7500 |
| | 2A | 0.0015 | 0.7485 | 0.7391 | — | 0.7079 | 0.7029 | 0.6718 | 2B | 0.682 | 0.696 | 0.7094 | 0.7159 | 0.7500 |
| | 3A | 0.0000 | 0.7500 | 0.7406 | — | 0.7094 | 0.7056 | 0.6733 | 3B | 0.6820 | 0.6908 | 0.7094 | 0.7143 | 0.7500 |
| ¾–18 UNS | 2A | 0.0014 | 0.7486 | 0.7399 | — | 0.7125 | 0.7079 | 0.6804 | 2B | 0.690 | 0.703 | 0.7139 | 0.7199 | 0.7500 |
| ¾–20 UNEF | 2A | 0.0013 | 0.7487 | 0.7406 | — | 0.7162 | 0.7118 | 0.6874 | 2B | 0.696 | 0.707 | 0.7175 | 0.7232 | 0.7500 |
| | 3A | 0.0000 | 0.7500 | 0.7419 | — | 0.7175 | 0.7142 | 0.6887 | 3B | 0.6960 | 0.7037 | 0.7175 | 0.7218 | 0.7500 |
| ¾–24 UNS | 2A | 0.0012 | 0.7488 | 0.7416 | — | 0.7217 | 0.7176 | 0.6977 | 2B | 0.705 | 0.715 | 0.7229 | 0.7282 | 0.7500 |
| ¾–27 UNS | 2A | 0.0012 | 0.7488 | 0.7421 | — | 0.7247 | 0.7208 | 0.7034 | 2B | 0.710 | 0.719 | 0.7259 | 0.7310 | 0.7500 |
| ¾–28 UN | 2A | 0.0012 | 0.7488 | 0.7423 | — | 0.7256 | 0.7218 | 0.7050 | 2B | 0.711 | 0.720 | 0.7268 | 0.7318 | 0.7500 |
| | 3A | 0.0000 | 0.7500 | 0.7435 | — | 0.7268 | 0.7239 | 0.7062 | 3B | 0.7110 | 0.7176 | 0.7268 | 0.7305 | 0.7500 |
| ¾–32 UN | 2A | 0.0011 | 0.7489 | 0.7429 | — | 0.7286 | 0.7250 | 0.7106 | 2B | 0.716 | 0.724 | 0.7297 | 0.7344 | 0.7500 |
| | 3A | 0.0000 | 0.7500 | 0.7440 | — | 0.7297 | 0.7270 | 0.7117 | 3B | 0.7160 | 0.7219 | 0.7297 | 0.7333 | 0.7500 |
| 13/16–12 UN | 2A | 0.0017 | 0.8108 | 0.7994 | — | 0.7567 | 0.7512 | 0.7086 | 2B | 0.722 | 0.740 | 0.7584 | 0.7656 | 0.8125 |
| | 3A | 0.0000 | 0.8125 | 0.8011 | — | 0.7584 | 0.7543 | 0.7103 | 3B | 0.7220 | 0.7329 | 0.7584 | 0.7638 | 0.8125 |
| 13/16–16 UN | 2A | 0.0015 | 0.8110 | 0.8016 | — | 0.7704 | 0.7655 | 0.7343 | 2B | 0.745 | 0.759 | 0.7719 | 0.7782 | 0.8125 |
| | 3A | 0.0000 | 0.8125 | 0.8031 | — | 0.7719 | 0.7683 | 0.7358 | 3B | 0.7450 | 0.7533 | 0.7719 | 0.7766 | 0.8125 |
| 13/16–20 UNEF | 2A | 0.0013 | 0.8112 | 0.8031 | — | 0.7787 | 0.7743 | 0.7498 | 2B | 0.758 | 0.770 | 0.7800 | 0.7857 | 0.8125 |
| | 3A | 0.0000 | 0.8125 | 0.8044 | — | 0.7800 | 0.7767 | 0.7512 | 3B | 0.7580 | 0.7652 | 0.7800 | 0.7843 | 0.8125 |
| 13/16–28 UN | 2A | 0.0012 | 0.8113 | 0.8048 | — | 0.7881 | 0.7843 | 0.7675 | 2B | 0.774 | 0.782 | 0.7893 | 0.7943 | 0.8125 |
| | 3A | 0.0000 | 0.8125 | 0.8060 | — | 0.7893 | 0.7864 | 0.7687 | 3B | 0.7740 | 0.7801 | 0.7893 | 0.7930 | 0.8125 |
| 13/16–32 UN | 2A | 0.0011 | 0.8114 | 0.8054 | — | 0.7911 | 0.7875 | 0.7731 | 2B | 0.779 | 0.786 | 0.7922 | 0.7969 | 0.8125 |
| | 3A | 0.0000 | 0.8125 | 0.8065 | — | 0.7922 | 0.7895 | 0.7742 | 3B | 0.7790 | 0.7844 | 0.7922 | 0.7958 | 0.8125 |

| Nominal Size, Threads per Inch, and Series | Class | | | | | | | Class | | | | | |
|---|---|---|---|---|---|---|---|---|---|---|---|---|---|
| 7/8–9 UNC | 1A | 0.0019 | 0.8731 | — | 0.8009 | 0.7914 | 0.7368 | 1B | 0.755 | 0.778 | 0.8028 | 0.8151 | 0.8750 |
| | 2A | 0.0019 | 0.8731 | 0.8523 | 0.8009 | 0.7946 | 0.7368 | 2B | 0.755 | 0.778 | 0.8028 | 0.8110 | 0.8750 |
| | 3A | 0.0000 | 0.8750 | — | 0.8028 | 0.7981 | 0.7387 | 3B | 0.7550 | 0.7681 | 0.8028 | 0.8089 | 0.8750 |
| 7/8–10 UNS | 2A | 0.0018 | 0.8732 | — | 0.8082 | 0.8022 | 0.7505 | 2B | 0.767 | 0.788 | 0.8100 | 0.8178 | 0.8750 |
| 7/8–12 UN | 2A | 0.0017 | 0.8733 | 0.8619 | 0.8192 | 0.8137 | 0.7711 | 2B | 0.785 | 0.8032 | 0.8209 | 0.8281 | 0.8750 |
| | 3A | 0.0000 | 0.8750 | 0.8636 | 0.8209 | 0.8168 | 0.7728 | 3B | 0.7850 | 0.795 | 0.8209 | 0.8263 | 0.8750 |
| 7/8–14 UNF | 1A | 0.0016 | 0.8734 | 0.8579 | 0.8270 | 0.8189 | 0.7858 | 1B | 0.798 | 0.814 | 0.8286 | 0.8392 | 0.8750 |
| | 2A | 0.0016 | 0.8734 | 0.8631 | 0.8270 | 0.8216 | 0.7858 | 2B | 0.798 | 0.814 | 0.8286 | 0.8356 | 0.8750 |
| | 3A | 0.0000 | 0.8750 | 0.8647 | 0.8286 | 0.8245 | 0.7874 | 3B | 0.7980 | 0.8068 | 0.8286 | 0.8339 | 0.8750 |
| 7/8–16 UN | 2A | 0.0015 | 0.8735 | 0.8641 | 0.8329 | 0.8280 | 0.7968 | 2B | 0.807 | 0.821 | 0.8344 | 0.8407 | 0.8750 |
| | 3A | 0.0000 | 0.8750 | 0.8656 | 0.8344 | 0.8308 | 0.7983 | 3B | 0.8070 | 0.8158 | 0.8344 | 0.8391 | 0.8750 |
| 7/8–18 UNS | 2A | 0.0014 | 0.8736 | 0.8649 | 0.8375 | 0.8329 | 0.8054 | 2B | 0.815 | 0.828 | 0.8389 | 0.8449 | 0.8750 |
| 7/8–20 UNEF | 2A | 0.0013 | 0.8737 | 0.8656 | 0.8412 | 0.8368 | 0.8124 | 2B | 0.821 | 0.832 | 0.8425 | 0.8482 | 0.8750 |
| | 3A | 0.0000 | 0.8750 | 0.8669 | 0.8425 | 0.8392 | 0.8137 | 3B | 0.8210 | 0.8287 | 0.8425 | 0.8468 | 0.8750 |
| 7/8–24 UNS | 2A | 0.0012 | 0.8738 | 0.8666 | 0.8467 | 0.8426 | 0.8227 | 2B | 0.830 | 0.840 | 0.8479 | 0.8532 | 0.8750 |
| 7/8–27 UNS | 2A | 0.0012 | 0.8738 | 0.8671 | 0.8497 | 0.8458 | 0.8284 | 2B | 0.835 | 0.844 | 0.8509 | 0.8560 | 0.8750 |
| 7/8–28 UN | 2A | 0.0012 | 0.8738 | 0.8673 | 0.8506 | 0.8468 | 0.8300 | 2B | 0.836 | 0.845 | 0.8518 | 0.8568 | 0.8750 |
| | 3A | 0.0000 | 0.8750 | 0.8685 | 0.8518 | 0.8489 | 0.8312 | 3B | 0.8360 | 0.8426 | 0.8518 | 0.8555 | 0.8750 |
| 7/8–32 UN | 2A | 0.0011 | 0.8739 | 0.8679 | 0.8536 | 0.8500 | 0.8356 | 2B | 0.841 | 0.849 | 0.8547 | 0.8594 | 0.8750 |
| | 3A | 0.0000 | 0.8750 | 0.8690 | 0.8547 | 0.8520 | 0.8367 | 3B | 0.8410 | 0.8469 | 0.8547 | 0.8583 | 0.8750 |
| 15/16–12 UN | 2A | 0.0017 | 0.9358 | 0.9244 | 0.8817 | 0.8760 | 0.8336 | 2B | 0.847 | 0.865 | 0.8834 | 0.8908 | 0.9375 |
| | 3A | 0.0000 | 0.9375 | 0.9261 | 0.8834 | 0.8793 | 0.8353 | 3B | 0.8470 | 0.8575 | 0.8834 | 0.8889 | 0.9375 |
| 15/16–16 UN | 2A | 0.0015 | 0.9360 | 0.9266 | 0.8954 | 0.8904 | 0.8593 | 2B | 0.870 | 0.884 | 0.8969 | 0.9034 | 0.9375 |
| | 3A | 0.0000 | 0.9375 | 0.9281 | 0.8969 | 0.8932 | 0.8608 | 3B | 0.8700 | 0.8783 | 0.8969 | 0.9018 | 0.9375 |
| 15/16–20 UNEF | 2A | 0.0014 | 0.9361 | 0.9280 | 0.9036 | 0.8991 | 0.8748 | 2B | 0.883 | 0.895 | 0.9050 | 0.9109 | 0.9375 |
| | 3A | 0.0000 | 0.9375 | 0.9294 | 0.9050 | 0.9016 | 0.8762 | 3B | 0.8830 | 0.8912 | 0.9050 | 0.9094 | 0.9375 |
| 15/16–28 UN | 2A | 0.0012 | 0.9363 | 0.9298 | 0.9131 | 0.9091 | 0.8925 | 2B | 0.899 | 0.907 | 0.9143 | 0.9195 | 0.9375 |
| | 3A | 0.0000 | 0.9375 | 0.9310 | 0.9143 | 0.9113 | 0.8937 | 3B | 0.8990 | 0.9061 | 0.9143 | 0.9182 | 0.9375 |
| 15/16–32 UN | 2A | 0.0011 | 0.9364 | 0.9304 | 0.9161 | 0.9123 | 0.8981 | 2B | 0.904 | 0.911 | 0.9172 | 0.9221 | 0.9375 |
| | 3A | 0.0000 | 0.9375 | 0.9315 | 0.9172 | 0.9144 | 0.8992 | 3B | 0.9040 | 0.9094 | 0.9172 | 0.9209 | 0.9375 |
| 1–8 UNC | 1A | 0.0020 | 0.9980 | — | 0.9168 | 0.9067 | 0.8446 | 1B | 0.865 | 0.890 | 0.9188 | 0.9320 | 1.0000 |
| | 2A | 0.0020 | 0.9980 | 0.9755 | 0.9168 | 0.9100 | 0.8446 | 2B | 0.865 | 0.890 | 0.9188 | 0.9276 | 1.0000 |
| | 3A | 0.0000 | 1.0000 | — | 0.9188 | 0.9137 | 0.8466 | 3B | 0.8650 | 0.8797 | 0.9188 | 0.9254 | 1.0000 |

† Use UNS threads only if Standard Series do not meet requirements (see page 1278).      See footnotes a, b, c, d, e, and f at end of table.

**Table 4 (Continued). Standard Series and Selected Combinations† — Unified Screw Threads**

| Nominal Size, Threads per Inch, and Series Designation^f | External Class | Allowance | Major Diameter Max^b | Major Diameter Min | Major Diameter Min^d | Pitch Diameter Max^b | Pitch Diameter Min | Minor Diam^c | Internal Class | Minor Diameter Min | Minor Diameter Max | Pitch Diameter Min | Pitch Diameter Max | Major Diameter Min |
|---|---|---|---|---|---|---|---|---|---|---|---|---|---|---|
| 1-10 UNS | 2A | 0.0018 | 0.9982 | 0.9853 | — | 0.9332 | 0.9270 | 0.8755 | 2B | 0.892 | 0.913 | 0.9350 | 0.9430 | 1.0000 |
| 1-12 UNF | 1A | 0.0018 | 0.9982 | 0.9810 | — | 0.9441 | 0.9353 | 0.8960 | 1B | 0.910 | 0.928 | 0.9459 | 0.9573 | 1.0000 |
|  | 2A | 0.0018 | 0.9982 | 0.9868 | — | 0.9441 | 0.9382 | 0.8960 | 2B | 0.910 | 0.928 | 0.9459 | 0.9535 | 1.0000 |
|  | 3A | 0.0000 | 1.0000 | 0.9886 | — | 0.9459 | 0.9415 | 0.8978 | 3B | 0.9100 | 0.9198 | 0.9459 | 0.9516 | 1.0000 |
| 1-14 UNS | 1A | 0.0017 | 0.9983 | 0.9828 | — | 0.9519 | 0.9435 | 0.9107 | 1B | 0.923 | 0.938 | 0.9536 | 0.9645 | 1.0000 |
|  | 2A | 0.0017 | 0.9983 | 0.9880 | — | 0.9519 | 0.9463 | 0.9107 | 2B | 0.923 | 0.938 | 0.9536 | 0.9609 | 1.0000 |
|  | 3A | 0.0000 | 1.0000 | 0.9897 | — | 0.9536 | 0.9494 | 0.9124 | 3B | 0.9230 | 0.9315 | 0.9536 | 0.9590 | 1.0000 |
| 1-16 UN | 2A | 0.0015 | 0.9985 | 0.9891 | — | 0.9579 | 0.9529 | 0.9218 | 2B | 0.932 | 0.946 | 0.9594 | 0.9659 | 1.0000 |
|  | 3A | 0.0000 | 1.0000 | 0.9906 | — | 0.9594 | 0.9557 | 0.9233 | 3B | 0.9320 | 0.9408 | 0.9594 | 0.9643 | 1.0000 |
| 1-18 UNS | 2A | 0.0014 | 0.9986 | 0.9899 | — | 0.9625 | 0.9578 | 0.9304 | 2B | 0.940 | 0.953 | 0.9639 | 0.9701 | 1.0000 |
| 1-20 UNEF | 2A | 0.0014 | 0.9986 | 0.9905 | — | 0.9661 | 0.9616 | 0.9373 | 2B | 0.946 | 0.957 | 0.9675 | 0.9734 | 1.0000 |
|  | 3A | 0.0000 | 1.0000 | 0.9919 | — | 0.9675 | 0.9641 | 0.9387 | 3B | 0.9460 | 0.9537 | 0.9675 | 0.9719 | 1.0000 |
| 1-24 UNS | 2A | 0.0013 | 0.9987 | 0.9915 | — | 0.9716 | 0.9674 | 0.9476 | 2B | 0.955 | 0.965 | 0.9729 | 0.9784 | 1.0000 |
| 1-27 UNS | 2A | 0.0012 | 0.9988 | 0.9921 | — | 0.9747 | 0.9707 | 0.9534 | 2B | 0.960 | 0.969 | 0.9759 | 0.9811 | 1.0000 |
| 1-28 UN | 2A | 0.0012 | 0.9988 | 0.9923 | — | 0.9756 | 0.9716 | 0.9550 | 2B | 0.961 | 0.970 | 0.9768 | 0.9820 | 1.0000 |
|  | 3A | 0.0000 | 1.0000 | 0.9935 | — | 0.9768 | 0.9738 | 0.9562 | 3B | 0.9610 | 0.9676 | 0.9768 | 0.9807 | 1.0000 |
| 1-32 UN | 2A | 0.0011 | 0.9989 | 0.9929 | — | 0.9786 | 0.9748 | 0.9606 | 2B | 0.966 | 0.974 | 0.9797 | 0.9846 | 1.0000 |
|  | 3A | 0.0000 | 1.0000 | 0.9940 | — | 0.9797 | 0.9769 | 0.9617 | 3B | 0.9660 | 0.9719 | 0.9797 | 0.9834 | 1.0000 |
| 1 1/16-8 UN | 2A | 0.0020 | 1.0605 | 1.0455 | — | 0.9793 | 0.9725 | 0.9071 | 2B | 0.927 | 0.952 | 0.9813 | 0.9902 | 1.0625 |
|  | 3A | 0.0000 | 1.0625 | 1.0475 | — | 0.9813 | 0.9762 | 0.9091 | 3B | 0.9270 | 0.9422 | 0.9813 | 0.9880 | 1.0625 |
| 1 1/16-12 UN | 2A | 0.0017 | 1.0608 | 1.0494 | — | 1.0067 | 1.0010 | 0.9586 | 2B | 0.972 | 0.990 | 1.0084 | 1.0158 | 1.0625 |
|  | 3A | 0.0000 | 1.0625 | 1.0511 | — | 1.0084 | 1.0042 | 0.9603 | 3B | 0.9720 | 0.9823 | 1.0084 | 1.0139 | 1.0625 |
| 1 1/16-16 UN | 2A | 0.0015 | 1.0610 | 1.0516 | — | 1.0204 | 1.0154 | 0.9843 | 2B | 0.995 | 1.009 | 1.0219 | 1.0284 | 1.0625 |
|  | 3A | 0.0000 | 1.0625 | 1.0531 | — | 1.0219 | 1.0182 | 0.9858 | 3B | 0.9950 | 1.0033 | 1.0219 | 1.0268 | 1.0625 |
| 1 1/16-18 UNEF | 2A | 0.0014 | 1.0611 | 1.0524 | — | 1.0250 | 1.0203 | 0.9929 | 2B | 1.002 | 1.015 | 1.0264 | 1.0326 | 1.0625 |
|  | 3A | 0.0000 | 1.0625 | 1.0538 | — | 1.0264 | 1.0228 | 0.9943 | 3B | 1.0020 | 1.0105 | 1.0264 | 1.0310 | 1.0625 |
| 1 1/16-20 UN | 2A | 0.0014 | 1.0611 | 1.0530 | — | 1.0286 | 1.0241 | 0.9998 | 2B | 1.008 | 1.020 | 1.0300 | 1.0359 | 1.0625 |
|  | 3A | 0.0000 | 1.0625 | 1.0544 | — | 1.0300 | 1.0266 | 1.0012 | 3B | 1.0080 | 1.0162 | 1.0300 | 1.0344 | 1.0625 |

| Nominal Size, Threads per Inch, Series | Class | Allowance | Ext Major Max | Ext Major Min | UNR Minor | Ext Pitch Max | Ext Pitch Min | Ext Minor | Class | Int Minor Min | Int Minor Max | Int Pitch Min | Int Pitch Max | Int Major Min |
|---|---|---|---|---|---|---|---|---|---|---|---|---|---|---|
| 1¹/₁₆–28 UN | 2A | 0.0012 | 1.0613 | 1.0548 | — | 1.0381 | 1.0341 | 1.0175 | 2B | 1.024 | 1.032 | 1.0393 | 1.0445 | 1.0625 |
|  | 3A | 0.0000 | 1.0625 | 1.0560 | — | 1.0393 | 1.0363 | 1.0187 | 3B | 1.0240 | 1.0301 | 1.0393 | 1.0432 | 1.0625 |
| 1¹/₈–7 UNC | 1A | 0.0022 | 1.1228 | 1.0932 | — | 1.0300 | 1.0191 | 0.9475 | 1B | 0.970 | 0.998 | 1.0322 | 1.0463 | 1.1250 |
|  | 2A | 0.0022 | 1.1228 | 1.1064 | — | 1.0300 | 1.0228 | 0.9475 | 2B | 0.970 | 0.998 | 1.0322 | 1.0416 | 1.1250 |
|  | 3A | 0.0000 | 1.1250 | 1.1086 | 1.0982 | 1.0322 | 1.0268 | 0.9497 | 3B | 0.9700 | 0.9875 | 1.0322 | 1.0393 | 1.1250 |
| 1¹/₈–8 UN | 2A | 0.0021 | 1.1229 | 1.1079 | — | 1.0417 | 1.0348 | 0.9695 | 2B | 0.990 | 1.015 | 1.0438 | 1.0528 | 1.1250 |
|  | 3A | 0.0000 | 1.1250 | 1.1100 | 1.1004 | 1.0438 | 1.0386 | 0.9716 | 3B | 0.9900 | 1.0047 | 1.0438 | 1.0505 | 1.1250 |
| 1¹/₈–10 UNS | 2A | 0.0018 | 1.1232 | 1.1103 | — | 1.0582 | 1.0520 | 1.0005 | 2B | 1.017 | 1.038 | 1.0600 | 1.0680 | 1.1250 |
| 1¹/₈–12 UNF | 1A | 0.0018 | 1.1232 | 1.1060 | — | 1.0691 | 1.0601 | 1.0210 | 1B | 1.035 | 1.053 | 1.0709 | 1.0826 | 1.1250 |
|  | 2A | 0.0018 | 1.1232 | 1.1118 | — | 1.0691 | 1.0631 | 1.0210 | 2B | 1.035 | 1.053 | 1.0709 | 1.0787 | 1.1250 |
|  | 3A | 0.0000 | 1.1250 | 1.1136 | — | 1.0709 | 1.0664 | 1.0228 | 3B | 1.0350 | 1.0448 | 1.0709 | 1.0768 | 1.1250 |
| 1¹/₈–14 UNS | 2A | 0.0016 | 1.1234 | 1.1131 | — | 1.0770 | 1.0717 | 1.0358 | 2B | 1.048 | 1.064 | 1.0786 | 1.0855 | 1.1250 |
| 1¹/₈–16 UN | 2A | 0.0015 | 1.1235 | 1.1141 | — | 1.0829 | 1.0779 | 1.0468 | 2B | 1.057 | 1.071 | 1.0844 | 1.0909 | 1.1250 |
|  | 3A | 0.0000 | 1.1250 | 1.1156 | — | 1.0844 | 1.0807 | 1.0483 | 3B | 1.0570 | 1.0658 | 1.0844 | 1.0893 | 1.1250 |
| 1¹/₈–18 UNEF | 2A | 0.0014 | 1.1236 | 1.1149 | — | 1.0875 | 1.0828 | 1.0554 | 2B | 1.065 | 1.078 | 1.0889 | 1.0951 | 1.1250 |
|  | 3A | 0.0000 | 1.1250 | 1.1163 | — | 1.0889 | 1.0853 | 1.0568 | 3B | 1.0650 | 1.0730 | 1.0889 | 1.0935 | 1.1250 |
| 1¹/₈–20 UN | 2A | 0.0014 | 1.1236 | 1.1155 | — | 1.0911 | 1.0866 | 1.0623 | 2B | 1.071 | 1.082 | 1.0925 | 1.0984 | 1.1250 |
|  | 3A | 0.0000 | 1.1250 | 1.1169 | — | 1.0925 | 1.0891 | 1.0637 | 3B | 1.0710 | 1.0787 | 1.0925 | 1.0969 | 1.1250 |
| 1¹/₈–24 UNS | 2A | 0.0013 | 1.1237 | 1.1165 | — | 1.0966 | 1.0924 | 1.0726 | 2B | 1.080 | 1.090 | 1.0979 | 1.1034 | 1.1250 |
| 1¹/₈–28 UN | 2A | 0.0012 | 1.1238 | 1.1173 | — | 1.1006 | 1.0966 | 1.0800 | 2B | 1.086 | 1.095 | 1.1018 | 1.1070 | 1.1250 |
|  | 3A | 0.0000 | 1.1250 | 1.1185 | — | 1.1018 | 1.0988 | 1.0812 | 3B | 1.0860 | 1.0926 | 1.1018 | 1.1057 | 1.1250 |
| 1³/₁₆–8 UN | 2A | 0.0021 | 1.1854 | 1.1704 | — | 1.1042 | 1.0972 | 1.0320 | 2B | 1.052 | 1.077 | 1.1063 | 1.1154 | 1.1875 |
|  | 3A | 0.0000 | 1.1875 | 1.1725 | — | 1.1063 | 1.1011 | 1.0341 | 3B | 1.0520 | 1.0672 | 1.1063 | 1.1131 | 1.1875 |
| 1³/₁₆–12 UN | 2A | 0.0017 | 1.1858 | 1.1744 | — | 1.1317 | 1.1259 | 1.0836 | 2B | 1.097 | 1.115 | 1.1334 | 1.1409 | 1.1875 |
|  | 3A | 0.0000 | 1.1875 | 1.1761 | — | 1.1334 | 1.1291 | 1.0853 | 3B | 1.0970 | 1.1073 | 1.1334 | 1.1390 | 1.1875 |
| 1³/₁₆–16 UN | 2A | 0.0015 | 1.1860 | 1.1766 | — | 1.1454 | 1.1403 | 1.1093 | 2B | 1.120 | 1.134 | 1.1469 | 1.1535 | 1.1875 |
|  | 3A | 0.0000 | 1.1875 | 1.1781 | — | 1.1469 | 1.1431 | 1.1108 | 3B | 1.1200 | 1.1283 | 1.1469 | 1.1519 | 1.1875 |
| 1³/₁₆–18 UNEF | 2A | 0.0015 | 1.1860 | 1.1773 | — | 1.1499 | 1.1450 | 1.1178 | 2B | 1.127 | 1.140 | 1.1514 | 1.1577 | 1.1875 |
|  | 3A | 0.0000 | 1.1875 | 1.1788 | — | 1.1514 | 1.1478 | 1.1193 | 3B | 1.1270 | 1.1355 | 1.1514 | 1.1561 | 1.1875 |
| 1³/₁₆–20 UN | 2A | 0.0014 | 1.1861 | 1.1780 | — | 1.1536 | 1.1489 | 1.1248 | 2B | 1.133 | 1.145 | 1.1550 | 1.1611 | 1.1875 |
|  | 3A | 0.0000 | 1.1875 | 1.1794 | — | 1.1550 | 1.1515 | 1.1262 | 3B | 1.1330 | 1.1412 | 1.1550 | 1.1595 | 1.1875 |
| 1³/₁₆–28 UN | 2A | 0.0012 | 1.1863 | 1.1798 | — | 1.1631 | 1.1590 | 1.1425 | 2B | 1.149 | 1.157 | 1.1643 | 1.1696 | 1.1875 |
|  | 3A | 0.0000 | 1.1875 | 1.1810 | — | 1.1643 | 1.1612 | 1.1437 | 3B | 1.1490 | 1.1551 | 1.1643 | 1.1683 | 1.1875 |

† Use UNS threads only if Standard Series do not meet requirements (see page 1278). See footnotes a, b, c, d, e, and f at end of table.

Table 4 (Continued). Standard Series and Selected Combinations† — Unified Screw Threads

| Nominal Size, Threads per Inch, and Series Designation[f] | Class | Allowance | Major Diameter Max[b] | Major Diameter Min | Major Diameter Min[d] | Pitch Diameter Max[b] | Pitch Diameter Min | Minor Diameter[e] | Class | Minor Diameter[e] Min | Minor Diameter[e] Max | Pitch Diameter Min | Pitch Diameter Max | Major Diameter Min |
|---|---|---|---|---|---|---|---|---|---|---|---|---|---|---|
| 1¼-7 UNC | 1A | 0.0022 | 1.2478 | 1.2232 | — | 1.1550 | 1.1439 | 1.0725 | 1B | 1.095 | 1.123 | 1.1572 | 1.1716 | 1.2500 |
|  | 2A | 0.0022 | 1.2478 | 1.2314 | 1.2232 | 1.1550 | 1.1476 | 1.0725 | 2B | 1.095 | 1.123 | 1.1572 | 1.1663 | 1.2500 |
|  | 3A | 0.0000 | 1.2500 | 1.2336 | — | 1.1572 | 1.1517 | 1.0747 | 3B | 1.0950 | 1.1125 | 1.1572 | 1.1644 | 1.2500 |
| 1¼-8 UN | 2A | 0.0021 | 1.2479 | 1.2329 | 1.2254 | 1.1667 | 1.1597 | 1.0945 | 2B | 1.115 | 1.140 | 1.1688 | 1.1780 | 1.2500 |
|  | 3A | 0.0000 | 1.2500 | 1.2350 | — | 1.1688 | 1.1635 | 1.0966 | 3B | 1.1150 | 1.1297 | 1.1688 | 1.1757 | 1.2500 |
| 1¼-10 UNS | 2A | 0.0019 | 1.2481 | 1.2352 | — | 1.1831 | 1.1768 | 1.1254 | 2B | 1.142 | 1.163 | 1.1850 | 1.1932 | 1.2500 |
| 1¼-12 UNF | 1A | 0.0018 | 1.2481 | 1.2310 | — | 1.1941 | 1.1849 | 1.1460 | 1B | 1.160 | 1.178 | 1.1959 | 1.2079 | 1.2500 |
|  | 2A | 0.0018 | 1.2482 | 1.2368 | — | 1.1941 | 1.1879 | 1.1460 | 2B | 1.160 | 1.178 | 1.1959 | 1.2039 | 1.2500 |
|  | 3A | 0.0000 | 1.2500 | 1.2386 | — | 1.1959 | 1.1913 | 1.1478 | 3B | 1.1600 | 1.1698 | 1.1959 | 1.2019 | 1.2500 |
| 1¼-14 UNS | 2A | 0.0016 | 1.2484 | 1.2381 | — | 1.2020 | 1.1966 | 1.1608 | 2B | 1.173 | 1.188 | 1.2036 | 1.2106 | 1.2500 |
| 1¼-16 UN | 2A | 0.0015 | 1.2485 | 1.2391 | — | 1.2079 | 1.2028 | 1.1718 | 2B | 1.182 | 1.196 | 1.2094 | 1.2160 | 1.2500 |
|  | 3A | 0.0000 | 1.2500 | 1.2406 | — | 1.2094 | 1.2056 | 1.1733 | 3B | 1.1820 | 1.1908 | 1.2094 | 1.2144 | 1.2500 |
| 1¼-18 UNEF | 2A | 0.0015 | 1.2485 | 1.2398 | — | 1.2124 | 1.2075 | 1.1803 | 2B | 1.190 | 1.203 | 1.2139 | 1.2202 | 1.2500 |
|  | 3A | 0.0000 | 1.2500 | 1.2413 | — | 1.2139 | 1.2103 | 1.1818 | 3B | 1.1900 | 1.1980 | 1.2139 | 1.2186 | 1.2500 |
| 1¼-20 UN | 2A | 0.0014 | 1.2486 | 1.2405 | — | 1.2161 | 1.2114 | 1.1873 | 2B | 1.196 | 1.207 | 1.2175 | 1.2236 | 1.2500 |
|  | 3A | 0.0000 | 1.2500 | 1.2419 | — | 1.2175 | 1.2140 | 1.1887 | 3B | 1.1960 | 1.2037 | 1.2175 | 1.2220 | 1.2500 |
| 1¼-24 UNS | 2A | 0.0013 | 1.2487 | 1.2415 | — | 1.2216 | 1.2173 | 1.1976 | 2B | 1.205 | 1.215 | 1.2229 | 1.2285 | 1.2500 |
| 1¼-28 UN | 2A | 0.0012 | 1.2488 | 1.2423 | — | 1.2256 | 1.2215 | 1.2050 | 2B | 1.211 | 1.220 | 1.2268 | 1.2321 | 1.2500 |
|  | 3A | 0.0000 | 1.2500 | 1.2435 | — | 1.2268 | 1.2237 | 1.2062 | 3B | 1.2110 | 1.2176 | 1.2268 | 1.2308 | 1.2500 |
| 1⁵⁄₁₆-8 UN | 2A | 0.0021 | 1.3104 | 1.2954 | — | 1.2292 | 1.2221 | 1.1570 | 2B | 1.177 | 1.202 | 1.2313 | 1.2405 | 1.3125 |
|  | 3A | 0.0000 | 1.3125 | 1.2975 | — | 1.2313 | 1.2260 | 1.1591 | 3B | 1.1770 | 1.1922 | 1.2313 | 1.2382 | 1.3125 |
| 1⁵⁄₁₆-12 UN | 2A | 0.0017 | 1.3108 | 1.2994 | — | 1.2567 | 1.2509 | 1.2086 | 2B | 1.222 | 1.240 | 1.2584 | 1.2659 | 1.3125 |
|  | 3A | 0.0000 | 1.3125 | 1.3011 | — | 1.2584 | 1.2541 | 1.2103 | 3B | 1.2220 | 1.2323 | 1.2584 | 1.2640 | 1.3125 |
| 1⁵⁄₁₆-16 UN | 2A | 0.0015 | 1.3110 | 1.3016 | — | 1.2704 | 1.2653 | 1.2343 | 2B | 1.245 | 1.259 | 1.2719 | 1.2785 | 1.3125 |
|  | 3A | 0.0000 | 1.3125 | 1.3031 | — | 1.2719 | 1.2681 | 1.2358 | 3B | 1.2450 | 1.2533 | 1.2719 | 1.2769 | 1.3125 |
| 1⁵⁄₁₆-18 UNEF | 2A | 0.0015 | 1.3110 | 1.3023 | — | 1.2749 | 1.2700 | 1.2428 | 2B | 1.252 | 1.265 | 1.2764 | 1.2827 | 1.3125 |
|  | 3A | 0.0000 | 1.3125 | 1.3038 | — | 1.2764 | 1.2728 | 1.2443 | 3B | 1.2520 | 1.2605 | 1.2764 | 1.2811 | 1.3125 |
| 1⁵⁄₁₆-20 UN | 2A | 0.0014 | 1.3111 | 1.3030 | — | 1.2786 | 1.2739 | 1.2498 | 2B | 1.258 | 1.270 | 1.2800 | 1.2861 | 1.3125 |
|  | 3A | 0.0000 | 1.3125 | 1.3044 | — | 1.2800 | 1.2765 | 1.2512 | 3B | 1.2580 | 1.2662 | 1.2800 | 1.2845 | 1.3125 |

| Nominal Size, Threads per Inch, and Series | Class | Allowance | Major Dia Max | Major Dia Min | | Pitch Dia Max | Pitch Dia Min | Minor Dia | Class | Minor Dia Min | Minor Dia Max | Pitch Dia Min | Pitch Dia Max | Major Dia |
|---|---|---|---|---|---|---|---|---|---|---|---|---|---|---|
| 1 5⁄16-28 UN | 2A | 0.0012 | 1.3113 | 1.3048 | — | 1.2881 | 1.2840 | 1.2675 | 2B | 1.274 | 1.282 | 1.2893 | 1.2946 | 1.3125 |
| | 3A | 0.0000 | 1.3125 | 1.3060 | — | 1.2893 | 1.2862 | 1.2687 | 3B | 1.2740 | 1.2801 | 1.2893 | 1.2933 | 1.3125 |
| 1⅜-6 UNC | 1A | 0.0024 | 1.3726 | 1.3544 | — | 1.2643 | 1.2523 | | 1B | 1.195 | 1.225 | 1.2667 | 1.2822 | 1.3750 |
| | 2A | 0.0024 | 1.3726 | 1.3560 | — | 1.2643 | 1.2563 | 1.1681 | 2B | 1.195 | 1.225 | 1.2667 | 1.2771 | 1.3750 |
| | 3A | 0.0000 | 1.3750 | 1.3568 | 1.3453 | 1.2667 | 1.2607 | 1.1705 | 3B | 1.1950 | 1.2146 | 1.2667 | 1.2745 | 1.3750 |
| 1⅜-8 UN | 2A | 0.0022 | 1.3728 | 1.3578 | — | 1.2916 | 1.2844 | 1.2194 | 2B | 1.240 | 1.265 | 1.2938 | 1.3031 | 1.3750 |
| | 3A | 0.0000 | 1.3750 | 1.3600 | 1.3503 | 1.2938 | 1.2884 | 1.2216 | 3B | 1.2400 | 1.2547 | 1.2938 | 1.3008 | 1.3750 |
| 1⅜-10 UNS | 2A | 0.0019 | 1.3731 | 1.3602 | — | 1.3081 | 1.3018 | 1.2504 | 2B | 1.267 | 1.288 | 1.3100 | 1.3182 | 1.3750 |
| 1⅜-12 UNF | 1A | 0.0019 | 1.3731 | 1.3559 | — | 1.3190 | 1.3096 | | 1B | 1.285 | 1.303 | 1.3209 | 1.3332 | 1.3750 |
| | 2A | 0.0019 | 1.3731 | 1.3617 | — | 1.3190 | 1.3127 | 1.2709 | 2B | 1.285 | 1.303 | 1.3209 | 1.3291 | 1.3750 |
| | 3A | 0.0000 | 1.3750 | 1.3636 | — | 1.3209 | 1.3162 | 1.2728 | 3B | 1.2850 | 1.2948 | 1.3209 | 1.3270 | 1.3750 |
| 1⅜-14 UNS | 2A | 0.0016 | 1.3734 | 1.3631 | — | 1.3270 | 1.3216 | 1.2858 | 2B | 1.298 | 1.314 | 1.3286 | 1.3356 | 1.3750 |
| 1⅜-16 UN | 2A | 0.0015 | 1.3735 | 1.3641 | — | 1.3329 | 1.3278 | 1.2968 | 2B | 1.307 | 1.321 | 1.3344 | 1.3410 | 1.3750 |
| | 3A | 0.0000 | 1.3750 | 1.3656 | — | 1.3344 | 1.3306 | 1.2983 | 3B | 1.3070 | 1.3158 | 1.3344 | 1.3394 | 1.3750 |
| 1⅜-18 UNEF | 2A | 0.0015 | 1.3735 | 1.3648 | — | 1.3374 | 1.3325 | 1.3053 | 2B | 1.315 | 1.328 | 1.3389 | 1.3452 | 1.3750 |
| | 3A | 0.0000 | 1.3750 | 1.3663 | — | 1.3389 | 1.3353 | 1.3068 | 3B | 1.3150 | 1.3230 | 1.3389 | 1.3436 | 1.3750 |
| 1⅜-20 UN | 2A | 0.0014 | 1.3736 | 1.3655 | — | 1.3411 | 1.3364 | 1.3123 | 2B | 1.321 | 1.332 | 1.3425 | 1.3486 | 1.3750 |
| | 3A | 0.0000 | 1.3750 | 1.3669 | — | 1.3425 | 1.3390 | 1.3137 | 3B | 1.3210 | 1.3287 | 1.3425 | 1.3470 | 1.3750 |
| 1⅜-28 UN | 2A | 0.0012 | 1.3738 | 1.3673 | — | 1.3506 | 1.3465 | 1.3300 | 2B | 1.336 | 1.345 | 1.3518 | 1.3571 | 1.3750 |
| | 3A | 0.0000 | 1.3750 | 1.3685 | — | 1.3518 | 1.3487 | 1.3312 | 3B | 1.3360 | 1.3426 | 1.3518 | 1.3558 | 1.3750 |
| 1 7⁄16-6 UN | 2A | 0.0024 | 1.4351 | 1.4169 | — | 1.3268 | 1.3188 | 1.2306 | 2B | 1.257 | 1.288 | 1.3292 | 1.3396 | 1.4375 |
| | 3A | 0.0000 | 1.4375 | 1.4193 | — | 1.3292 | 1.3232 | 1.2330 | 3B | 1.2570 | 1.2771 | 1.3292 | 1.3370 | 1.4375 |
| 1 7⁄16-8 UN | 2A | 0.0022 | 1.4353 | 1.4203 | — | 1.3541 | 1.3469 | 1.2819 | 2B | 1.302 | 1.327 | 1.3563 | 1.3657 | 1.4375 |
| | 3A | 0.0000 | 1.4375 | 1.4225 | — | 1.3563 | 1.3509 | 1.2841 | 3B | 1.3020 | 1.3172 | 1.3563 | 1.3634 | 1.4375 |
| 1 7⁄16-12 UN | 2A | 0.0018 | 1.4357 | 1.4243 | — | 1.3816 | 1.3757 | 1.3335 | 2B | 1.347 | 1.365 | 1.3834 | 1.3910 | 1.4375 |
| | 3A | 0.0000 | 1.4375 | 1.4261 | — | 1.3834 | 1.3790 | 1.3353 | 3B | 1.3470 | 1.3573 | 1.3834 | 1.3891 | 1.4375 |
| 1 7⁄16-16 UN | 2A | 0.0016 | 1.4359 | 1.4265 | — | 1.3953 | 1.3901 | 1.3592 | 2B | 1.370 | 1.384 | 1.3969 | 1.4037 | 1.4375 |
| | 3A | 0.0000 | 1.4375 | 1.4281 | — | 1.3969 | 1.3930 | 1.3608 | 3B | 1.3700 | 1.3783 | 1.3969 | 1.4020 | 1.4375 |
| 1 7⁄16-18 UNEF | 2A | 0.0015 | 1.4360 | 1.4273 | — | 1.3999 | 1.3949 | 1.3678 | 2B | 1.377 | 1.390 | 1.4014 | 1.4079 | 1.4375 |
| | 3A | 0.0000 | 1.4375 | 1.4288 | — | 1.4014 | 1.3977 | 1.3693 | 3B | 1.3770 | 1.3855 | 1.4014 | 1.4062 | 1.4375 |
| 1 7⁄16-20 UN | 2A | 0.0014 | 1.4361 | 1.4280 | — | 1.4036 | 1.3988 | 1.3748 | 2B | 1.383 | 1.395 | 1.4050 | 1.4112 | 1.4375 |
| | 3A | 0.0000 | 1.4375 | 1.4294 | — | 1.4050 | 1.4014 | 1.3762 | 3B | 1.3830 | 1.3912 | 1.4050 | 1.4096 | 1.4375 |
| 1 7⁄16-28 UN | 2A | 0.0013 | 1.4362 | 1.4297 | — | 1.4130 | 1.4088 | 1.3924 | 2B | 1.399 | 1.407 | 1.4143 | 1.4198 | 1.4375 |
| | 3A | 0.0000 | 1.4375 | 1.4310 | — | 1.4143 | 1.4112 | 1.3937 | 3B | 1.3990 | 1.4051 | 1.4143 | 1.4184 | 1.4375 |

† Use UNS threads only if Standard Series do not meet requirements (see page 1278).    See footnotes a, b, c, d, e, and f at end of table.

**Table 4 (Continued). Standard Series and Selected Combinations† — Unified Screw Threads**

| Nominal Size, Threads per Inch, and Series Designation^f | Class | Allowance | Major Diameter Max^b | Major Diameter Min | Major Diameter Min^d | Pitch Diameter Max^b | Pitch Diameter Min | Minor Diam.^c | Class | Minor Diameter Min | Minor Diameter Max | Pitch Diameter Min | Pitch Diameter Max | Major Diameter Min |
|---|---|---|---|---|---|---|---|---|---|---|---|---|---|---|
| 1½–6 UNC | 1A | 0.0024 | 1.4976 | 1.4703 | 1.4703 | 1.3893 | 1.3772 | 1.2931 | 1B | 1.320 | 1.350 | 1.3917 | 1.4075 | 1.5000 |
|  | 2A | 0.0024 | 1.4976 | 1.4794 |  | 1.3893 | 1.3812 | 1.2931 | 2B | 1.320 | 1.350 | 1.3917 | 1.4022 | 1.5000 |
|  | 3A | 0.0000 | 1.5000 | 1.4818 |  | 1.3917 | 1.3856 | 1.2955 | 3B | 1.3200 | 1.3396 | 1.3917 | 1.3996 | 1.5000 |
| 1½–8 UN | 2A | 0.0022 | 1.4978 | 1.4828 | 1.4753 | 1.4166 | 1.4093 | 1.3444 | 2B | 1.365 | 1.390 | 1.4188 | 1.4283 | 1.5000 |
|  | 3A | 0.0000 | 1.5000 | 1.4850 |  | 1.4188 | 1.4133 | 1.3466 | 3B | 1.3650 | 1.3797 | 1.4188 | 1.4259 | 1.5000 |
| 1½–10 UNS | 2A | 0.0019 | 1.4981 | 1.4852 |  | 1.4331 | 1.4267 | 1.3754 | 2B | 1.392 | 1.413 | 1.4350 | 1.4433 | 1.5000 |
| 1½–12 UNF | 1A | 0.0019 | 1.4981 | 1.4809 |  | 1.4440 | 1.4344 | 1.3959 | 1B | 1.410 | 1.428 | 1.4459 | 1.4584 | 1.5000 |
|  | 2A | 0.0019 | 1.4981 | 1.4867 |  | 1.4440 | 1.4376 | 1.3959 | 2B | 1.410 | 1.428 | 1.4459 | 1.4542 | 1.5000 |
|  | 3A | 0.0000 | 1.5000 | 1.4886 |  | 1.4459 | 1.4411 | 1.3978 | 3B | 1.4100 | 1.4198 | 1.4459 | 1.4522 | 1.5000 |
| 1½–14 UNS | 2A | 0.0017 | 1.4983 | 1.4880 |  | 1.4519 | 1.4464 | 1.4107 | 2B | 1.423 | 1.438 | 1.4536 | 1.4608 | 1.5000 |
| 1½–16 UN | 2A | 0.0016 | 1.4984 | 1.4890 |  | 1.4578 | 1.4526 | 1.4217 | 2B | 1.432 | 1.446 | 1.4594 | 1.4662 | 1.5000 |
|  | 3A | 0.0000 | 1.5000 | 1.4906 |  | 1.4594 | 1.4555 | 1.4233 | 3B | 1.4320 | 1.4408 | 1.4594 | 1.4645 | 1.5000 |
| 1½–18 UNEF | 2A | 0.0015 | 1.4985 | 1.4898 |  | 1.4624 | 1.4574 | 1.4303 | 2B | 1.440 | 1.452 | 1.4639 | 1.4704 | 1.5000 |
|  | 3A | 0.0000 | 1.5000 | 1.4913 |  | 1.4639 | 1.4602 | 1.4318 | 3B | 1.4400 | 1.4480 | 1.4639 | 1.4687 | 1.5000 |
| 1½–20 UN | 2A | 0.0014 | 1.4986 | 1.4905 |  | 1.4661 | 1.4613 | 1.4373 | 2B | 1.446 | 1.457 | 1.4675 | 1.4737 | 1.5000 |
|  | 3A | 0.0000 | 1.5000 | 1.4919 |  | 1.4675 | 1.4639 | 1.4387 | 3B | 1.4460 | 1.4537 | 1.4675 | 1.4721 | 1.5000 |
| 1½–24 UNS | 2A | 0.0013 | 1.4987 | 1.4915 |  | 1.4716 | 1.4672 | 1.4476 | 2B | 1.455 | 1.465 | 1.4729 | 1.4787 | 1.5000 |
| 1½–28 UN | 2A | 0.0013 | 1.4987 | 1.4922 |  | 1.4755 | 1.4713 | 1.4549 | 2B | 1.461 | 1.470 | 1.4768 | 1.4823 | 1.5000 |
|  | 3A | 0.0000 | 1.5000 | 1.4935 |  | 1.4768 | 1.4737 | 1.4562 | 3B | 1.4610 | 1.4676 | 1.4768 | 1.4809 | 1.5000 |
| 1⁹⁄₁₆–6 UN | 2A | 0.0024 | 1.5601 | 1.5419 |  | 1.4518 | 1.4436 | 1.3556 | 2B | 1.382 | 1.413 | 1.4542 | **1.4648** | 1.5625 |
|  | 3A | 0.0000 | 1.5625 | 1.5443 |  | 1.4542 | 1.4481 | 1.3580 | 3B | 1.3820 | 1.4021 | 1.4542 | 1.4622 | 1.5625 |
| 1⁹⁄₁₆–8 UN | 2A | 0.0022 | 1.5603 | 1.5453 |  | 1.4791 | 1.4717 | 1.4069 | 2B | 1.427 | 1.452 | 1.4813 | 1.4909 | 1.5625 |
|  | 3A | 0.0000 | 1.5625 | 1.5475 |  | 1.4813 | 1.4758 | 1.4091 | 3B | 1.4270 | 1.4422 | 1.4813 | 1.4885 | 1.5625 |
| 1⁹⁄₁₆–12 UN | 2A | 0.0018 | 1.5607 | 1.5493 |  | 1.5066 | 1.5007 | 1.4585 | 2B | 1.472 | 1.490 | 1.5084 | 1.5160 | 1.5625 |
|  | 3A | 0.0000 | 1.5625 | 1.5511 |  | 1.5084 | 1.5040 | 1.4603 | 3B | 1.4720 | 1.4823 | 1.5084 | 1.5141 | 1.5625 |
| 1⁹⁄₁₆–16 UN | 2A | 0.0016 | 1.5609 | 1.5515 |  | 1.5203 | 1.5151 | 1.4842 | 2B | 1.495 | 1.509 | 1.5219 | 1.5287 | 1.5625 |
|  | 3A | 0.0000 | 1.5625 | 1.5531 |  | 1.5219 | 1.5180 | 1.4858 | 3B | 1.4950 | 1.5033 | 1.5219 | 1.5270 | 1.5625 |
| 1⁹⁄₁₆–18 UNEF | 2A | 0.0015 | 1.5610 | 1.5523 |  | 1.5249 | 1.5199 | 1.4928 | 2B | 1.502 | 1.515 | 1.5264 | 1.5329 | 1.5625 |
|  | 3A | 0.0000 | 1.5625 | 1.5538 |  | 1.5264 | 1.5227 | 1.4943 | 3B | 1.5020 | 1.5105 | 1.5264 | 1.5312 | 1.5625 |

Each thread has two groups of values: external classes (1A/2A/3A) and internal classes (1B/2B/3B). Values shown as the class order given in the "Class" column. ( "—" = not applicable.)

| Identification | Class | Allowance | Major Dia Max | Major Dia Min | UNR Minor Dia Max | Pitch Dia Max | Pitch Dia Min | Minor Dia Max | Minor Dia Min | Minor Dia Max | Pitch Dia Min | Pitch Dia Max | Major Dia Min |
|---|---|---|---|---|---|---|---|---|---|---|---|---|---|
| 1 9/16-20 UN | 2A | 0.0014 | 1.5611 | 1.5530 | — | 1.5286 | 1.5238 | 1.4998 | | | | | |
| | 3A | 0.0000 | 1.5625 | 1.5544 | — | 1.5300 | 1.5264 | 1.5012 | | | | | |
| | 2B | | | | | | | | 1.508 | 1.520 | 1.5300 | 1.5362 | 1.5625 |
| | 3B | | | | | | | | 1.5080 | 1.5162 | 1.5300 | 1.5346 | 1.5625 |
| 1 5/8-6 UN | 2A | 0.0025 | 1.6225 | 1.6043 | — | 1.5142 | 1.5060 | 1.4180 | | | | | |
| | 3A | 0.0000 | 1.6250 | 1.6068 | — | 1.5167 | 1.5105 | 1.4205 | | | | | |
| | 2B | | | | | | | | 1.445 | 1.475 | 1.5167 | 1.5274 | 1.6250 |
| | 3B | | | | | | | | 1.4450 | 1.4646 | 1.5167 | 1.5247 | 1.6250 |
| 1 5/8-8 UN | 2A | 0.0022 | 1.6228 | 1.6078 | — | 1.5416 | 1.5342 | 1.4694 | | | | | |
| | 3A | 0.0000 | 1.6250 | 1.6100 | 1.6003 | 1.5438 | 1.5382 | 1.4716 | | | | | |
| | 2B | | | | | | | | 1.490 | 1.515 | 1.5438 | 1.5535 | 1.6250 |
| | 3B | | | | | | | | 1.4900 | 1.5047 | 1.5438 | 1.5510 | 1.6250 |
| 1 5/8-10 UNS | 2A | 0.0019 | 1.6231 | 1.6102 | — | 1.5581 | 1.5517 | 1.5004 | | | | | |
| | 2B | | | | | | | | 1.517 | 1.538 | 1.5600 | 1.5683 | 1.6250 |
| 1 5/8-12 UN | 2A | 0.0018 | 1.6232 | 1.6118 | — | 1.5691 | 1.5632 | 1.5210 | | | | | |
| | 3A | 0.0000 | 1.6250 | 1.6136 | — | 1.5709 | 1.5665 | 1.5228 | | | | | |
| | 2B | | | | | | | | 1.535 | 1.553 | 1.5709 | 1.5785 | 1.6250 |
| | 3B | | | | | | | | 1.5350 | 1.5448 | 1.5709 | 1.5766 | 1.6250 |
| 1 5/8-14 UNS | 2A | 0.0017 | 1.6233 | 1.6130 | — | 1.5769 | 1.5714 | 1.5357 | | | | | |
| | 2B | | | | | | | | 1.548 | 1.564 | 1.5786 | 1.5858 | 1.6250 |
| 1 5/8-16 UN | 2A | 0.0016 | 1.6234 | 1.6140 | — | 1.5828 | 1.5776 | 1.5467 | | | | | |
| | 3A | 0.0000 | 1.6250 | 1.6156 | — | 1.5844 | 1.5805 | 1.5483 | | | | | |
| | 2B | | | | | | | | 1.557 | 1.571 | 1.5844 | 1.5912 | 1.6250 |
| | 3B | | | | | | | | 1.5570 | 1.5658 | 1.5844 | 1.5895 | 1.6250 |
| 1 5/8-18 UNEF | 2A | 0.0015 | 1.6235 | 1.6148 | — | 1.5874 | 1.5824 | 1.5553 | | | | | |
| | 3A | 0.0000 | 1.6250 | 1.6163 | — | 1.5889 | 1.5852 | 1.5568 | | | | | |
| | 2B | | | | | | | | 1.565 | 1.578 | 1.5889 | 1.5954 | 1.6250 |
| | 3B | | | | | | | | 1.5650 | 1.5730 | 1.5889 | 1.5937 | 1.6250 |
| 1 5/8-20 UN | 2A | 0.0014 | 1.6236 | 1.6155 | — | 1.5911 | 1.5863 | 1.5623 | | | | | |
| | 3A | 0.0000 | 1.6250 | 1.6169 | — | 1.5925 | 1.5889 | 1.5637 | | | | | |
| | 2B | | | | | | | | 1.571 | 1.582 | 1.5925 | 1.5987 | 1.6250 |
| | 3B | | | | | | | | 1.5710 | 1.5787 | 1.5925 | 1.5971 | 1.6250 |
| 1 5/8-24 UNS | 2A | 0.0013 | 1.6237 | 1.6165 | — | 1.5966 | 1.5922 | 1.5726 | | | | | |
| | 2B | | | | | | | | 1.580 | 1.590 | 1.5979 | 1.6037 | 1.6250 |
| 1 11/16-6 UN | 2A | 0.0025 | 1.6850 | 1.6668 | — | 1.5767 | 1.5684 | 1.4805 | | | | | |
| | 3A | 0.0000 | 1.6875 | 1.6693 | — | 1.5792 | 1.5730 | 1.4830 | | | | | |
| | 2B | | | | | | | | 1.507 | 1.538 | 1.5792 | 1.5900 | 1.6875 |
| | 3B | | | | | | | | 1.5070 | 1.5271 | 1.5792 | 1.5873 | 1.6875 |
| 1 11/16-8 UN | 2A | 0.0022 | 1.6853 | 1.6703 | — | 1.6041 | 1.5966 | 1.5319 | | | | | |
| | 3A | 0.0000 | 1.6875 | 1.6725 | — | 1.6063 | 1.6007 | 1.5341 | | | | | |
| | 2B | | | | | | | | 1.552 | 1.577 | 1.6063 | 1.6160 | 1.6875 |
| | 3B | | | | | | | | 1.5520 | 1.5672 | 1.6063 | 1.6136 | 1.6875 |
| 1 11/16-12 UN | 2A | 0.0018 | 1.6857 | 1.6743 | — | 1.6316 | 1.6256 | 1.5835 | | | | | |
| | 3A | 0.0000 | 1.6875 | 1.6761 | — | 1.6334 | 1.6289 | 1.5853 | | | | | |
| | 2B | | | | | | | | 1.597 | 1.615 | 1.6334 | 1.6412 | 1.6875 |
| | 3B | | | | | | | | 1.5970 | 1.6070 | 1.6334 | 1.6392 | 1.6875 |
| 1 11/16-16 UN | 2A | 0.0016 | 1.6859 | 1.6765 | — | 1.6453 | 1.6400 | 1.6092 | | | | | |
| | 3A | 0.0000 | 1.6875 | 1.6781 | — | 1.6469 | 1.6429 | 1.6108 | | | | | |
| | 2B | | | | | | | | 1.620 | 1.634 | 1.6469 | 1.6538 | 1.6875 |
| | 3B | | | | | | | | 1.6200 | 1.6283 | 1.6469 | 1.6521 | 1.6875 |
| 1 11/16-18 UNEF | 2A | 0.0015 | 1.6860 | 1.6773 | — | 1.6499 | 1.6448 | 1.6178 | | | | | |
| | 3A | 0.0000 | 1.6875 | 1.6788 | — | 1.6514 | 1.6476 | 1.6193 | | | | | |
| | 2B | | | | | | | | 1.627 | 1.640 | 1.6514 | 1.6580 | 1.6875 |
| | 3B | | | | | | | | 1.6270 | 1.6355 | 1.6514 | 1.6563 | 1.6875 |
| 1 11/16-20 UN | 2A | 0.0015 | 1.6860 | 1.6779 | — | 1.6535 | 1.6487 | 1.6247 | | | | | |
| | 3A | 0.0000 | 1.6875 | 1.6794 | — | 1.6550 | 1.6514 | 1.6262 | | | | | |
| | 2B | | | | | | | | 1.633 | 1.645 | 1.6550 | 1.6613 | 1.6875 |
| | 3B | | | | | | | | 1.6330 | 1.6412 | 1.6550 | 1.6597 | 1.6875 |
| 1 3/4-5 UNC | 1A | 0.0027 | 1.7473 | 1.7165 | — | 1.6174 | 1.6040 | 1.5019 | | | | | |
| | 2A | 0.0027 | 1.7473 | 1.7268 | 1.7165 | 1.6174 | 1.6085 | 1.5019 | | | | | |
| | 3A | 0.0000 | 1.7500 | 1.7295 | 1.7165 | 1.6201 | 1.6134 | 1.5046 | | | | | |
| | 1B | | | | | | | | 1.534 | 1.568 | 1.6201 | 1.6375 | 1.7500 |
| | 2B | | | | | | | | 1.534 | 1.568 | 1.6201 | 1.6317 | 1.7500 |
| | 3B | | | | | | | | 1.5340 | 1.5575 | 1.6201 | 1.6288 | 1.7500 |

† Use UNS threads only if Standard Series do not meet requirements (see page 1278). See footnotes a, b, c, d, e, and f at end of table.

Table 4 (Continued). Standard Series and Selected Combinations† — Unified Screw Threads

| Nominal Size, Threads per Inch, and Series Designation^f | Class | Allowance | Major Diameter | | | Pitch Diameter | | Minor Diam-eter^a | Class | Minor Diameter^e | | Pitch Diameter | | Major Diameter |
| | | | Max^b | Min | Min^d | Max^b | Min | | | Min | Max | Min | Max | Min |
|---|---|---|---|---|---|---|---|---|---|---|---|---|---|---|
| 1¾–6 UN | 2A | 0.0025 | 1.7475 | 1.7293 | — | 1.6392 | 1.6309 | 1.5430 | 2B | 1.570 | 1.600 | 1.6417 | 1.6525 | 1.7500 |
| | 3A | 0.0000 | 1.7500 | 1.7318 | — | 1.6417 | 1.6354 | 1.5455 | 3B | 1.5700 | 1.5896 | 1.6417 | 1.6498 | 1.7500 |
| 1¾–8 UN | 2A | 0.0023 | 1.7477 | 1.7327 | 1.7252 | 1.6665 | 1.6590 | 1.5943 | 2B | 1.615 | 1.640 | 1.6688 | 1.6786 | 1.7500 |
| | 3A | 0.0000 | 1.7500 | 1.7350 | — | 1.6688 | 1.6632 | 1.5966 | 3B | 1.6150 | 1.6297 | 1.6688 | 1.6762 | 1.7500 |
| 1¾–10 UNS | 2A | 0.0019 | 1.7481 | 1.7352 | — | 1.6831 | 1.6766 | 1.6254 | 2B | 1.642 | 1.663 | 1.6850 | 1.6934 | 1.7500 |
| 1¾–12 UN | 2A | 0.0018 | 1.7482 | 1.7368 | — | 1.6941 | 1.6881 | 1.6460 | 2B | 1.660 | 1.678 | 1.6959 | 1.7037 | 1.7500 |
| | 3A | 0.0000 | 1.7500 | 1.7386 | — | 1.6959 | 1.6914 | 1.6478 | 3B | 1.6600 | 1.6698 | 1.6959 | 1.7017 | 1.7500 |
| 1¾–14 UNS | 2A | 0.0017 | 1.7483 | 1.7380 | — | 1.7019 | 1.6963 | 1.6607 | 2B | 1.673 | 1.688 | 1.7036 | 1.7109 | 1.7500 |
| 1¾–16 UN | 2A | 0.0016 | 1.7484 | 1.7390 | — | 1.7078 | 1.7025 | 1.6717 | 2B | 1.682 | 1.696 | 1.7094 | 1.7163 | 1.7500 |
| | 3A | 0.0000 | 1.7500 | 1.7406 | — | 1.7094 | 1.7054 | 1.6733 | 3B | 1.6820 | 1.6908 | 1.7094 | 1.7146 | 1.7500 |
| 1¾–18 UNS | 2A | 0.0015 | 1.7485 | 1.7398 | — | 1.7124 | 1.7073 | 1.6803 | 2B | 1.690 | 1.703 | 1.7139 | 1.7205 | 1.7500 |
| 1¾–20 UN | 2A | 0.0015 | 1.7485 | 1.7404 | — | 1.7160 | 1.7112 | 1.6872 | 2B | 1.696 | 1.707 | 1.7175 | 1.7238 | 1.7500 |
| | 3A | 0.0000 | 1.7500 | 1.7419 | — | 1.7175 | 1.7139 | 1.6887 | 3B | 1.6960 | 1.7037 | 1.7175 | 1.7222 | 1.7500 |
| 1¹³⁄₁₆–6 UN | 2A | 0.0025 | 1.8100 | 1.7918 | — | 1.7017 | 1.6933 | 1.6055 | 2B | 1.632 | 1.663 | 1.7042 | 1.7151 | 1.8125 |
| | 3A | 0.0000 | 1.8125 | 1.7943 | — | 1.7042 | 1.6979 | 1.6080 | 3B | 1.6320 | 1.6521 | 1.7042 | 1.7124 | 1.8125 |
| 1¹³⁄₁₆–8 UN | 2A | 0.0023 | 1.8102 | 1.7952 | — | 1.7290 | 1.7214 | 1.6568 | 2B | 1.677 | 1.702 | 1.7313 | 1.7412 | 1.8125 |
| | 3A | 0.0000 | 1.8125 | 1.7975 | — | 1.7313 | 1.7256 | 1.6591 | 3B | 1.6770 | 1.6922 | 1.7313 | 1.7387 | 1.8125 |
| 1¹³⁄₁₆–12 UN | 2A | 0.0018 | 1.8107 | 1.7993 | — | 1.7566 | 1.7506 | 1.7085 | 2B | 1.722 | 1.740 | 1.7584 | 1.7662 | 1.8125 |
| | 3A | 0.0000 | 1.8125 | 1.8011 | — | 1.7584 | 1.7539 | 1.7103 | 3B | 1.7220 | 1.7320 | 1.7584 | 1.7642 | 1.8125 |
| 1¹³⁄₁₆–16 UN | 2A | 0.0016 | 1.8109 | 1.8015 | — | 1.7703 | 1.7650 | 1.7342 | 2B | 1.745 | 1.759 | 1.7719 | 1.7788 | 1.8125 |
| | 3A | 0.0000 | 1.8125 | 1.8031 | — | 1.7719 | 1.7679 | 1.7358 | 3B | 1.7450 | 1.7533 | 1.7719 | 1.7771 | 1.8125 |
| 1¹³⁄₁₆–20 UN | 2A | 0.0015 | 1.8110 | 1.8029 | — | 1.7785 | 1.7737 | 1.7497 | 2B | 1.758 | 1.770 | 1.7800 | 1.7863 | 1.8125 |
| | 3A | 0.0000 | 1.8125 | 1.8044 | — | 1.7800 | 1.7764 | 1.7512 | 3B | 1.7580 | 1.7662 | 1.7800 | 1.7847 | 1.8125 |
| 1⅞–6 UN | 2A | 0.0025 | 1.8725 | 1.8543 | — | 1.7642 | 1.7558 | 1.6680 | 2B | 1.695 | 1.725 | 1.7667 | 1.7777 | 1.8750 |
| | 3A | 0.0000 | 1.8750 | 1.8568 | — | 1.7667 | 1.7604 | 1.6705 | 3B | 1.6950 | 1.7146 | 1.7667 | 1.7749 | 1.8750 |
| 1⅞–8 UN | 2A | 0.0023 | 1.8727 | 1.8577 | 1.8502 | 1.7915 | 1.7838 | 1.7193 | 2B | 1.740 | 1.765 | 1.7938 | 1.8038 | 1.8750 |
| | 3A | 0.0000 | 1.8750 | 1.8600 | — | 1.7938 | 1.7881 | 1.7216 | 3B | 1.7400 | 1.7547 | 1.7938 | 1.8013 | 1.8750 |
| 1⅞–10 UNS | 2A | 0.0019 | 1.8731 | 1.8602 | — | 1.8081 | 1.8016 | 1.7504 | 2B | 1.767 | 1.788 | 1.8100 | 1.8184 | 1.8750 |

Standard Series and Selected Combinations — Unified Screw Threads (continued). All dimensions in inches. The columns reproduce, in order, the external-thread data (Allowance, Class, Major Diameter Max and Min, Pitch Diameter Max and Min, UNR Minor Diameter Max) and the internal-thread data (Class, Minor Diameter Min and Max, Pitch Diameter Min and Max, Major Diameter Min).

| Nominal Size, Threads per Inch, and Series Designation | Allowance | Class | Major Dia Max | Major Dia Min | Pitch Dia Max | Pitch Dia Min | UNR Minor Dia Max | Class | Minor Dia Min | Minor Dia Max | Pitch Dia Min | Pitch Dia Max | Major Dia Min |
|---|---|---|---|---|---|---|---|---|---|---|---|---|---|
| 1⅞–12 UN | 0.0018 | 2A | 1.8732 | 1.8618 | 1.8191 | 1.8131 | 1.7710 | 2B | 1.785 | 1.804 | 1.8209 | 1.8287 | 1.8750 |
|  | 0.0000 | 3A | 1.8750 | 1.8636 | 1.8209 | 1.8164 | 1.7728 | 3B | 1.7850 | 1.7948 | 1.8209 | 1.8267 | 1.8750 |
| 1⅞–14 UNS | 0.0017 | 2A | 1.8733 | 1.8630 | 1.8269 | 1.8213 | 1.7857 | 2B | 1.798 | 1.814 | 1.8286 | 1.8359 | 1.8750 |
| 1⅞–16 UN | 0.0016 | 2A | 1.8734 | 1.8640 | 1.8328 | 1.8275 | 1.7967 | 2B | 1.807 | 1.821 | 1.8344 | 1.8413 | 1.8750 |
|  | 0.0000 | 3A | 1.8750 | 1.8656 | 1.8344 | 1.8304 | 1.7983 | 3B | 1.8070 | 1.8158 | 1.8344 | 1.8396 | 1.8750 |
| 1⅞–18 UNS | 0.0015 | 2A | 1.8735 | 1.8648 | 1.8374 | 1.8323 | 1.8053 | 2B | 1.815 | 1.828 | 1.8389 | 1.8455 | 1.8750 |
| 1⅞–20 UN | 0.0015 | 2A | 1.8735 | 1.8654 | 1.8410 | 1.8362 | 1.8122 | 2B | 1.821 | 1.832 | 1.8425 | 1.8488 | 1.8750 |
|  | 0.0000 | 3A | 1.8750 | 1.8669 | 1.8425 | 1.8389 | 1.8137 | 3B | 1.8210 | 1.8287 | 1.8425 | 1.8472 | 1.8750 |
| 1 15/16–6 UN | 0.0026 | 2A | 1.9349 | 1.9167 | 1.8266 | 1.8181 | 1.7304 | 2B | 1.757 | 1.788 | 1.8292 | 1.8403 | 1.9375 |
|  | 0.0000 | 3A | 1.9375 | 1.9193 | 1.8292 | 1.8228 | 1.7330 | 3B | 1.7570 | 1.7771 | 1.8292 | 1.8375 | 1.9375 |
| 1 15/16–8 UN | 0.0023 | 2A | 1.9352 | 1.9202 | 1.8540 | 1.8463 | 1.7818 | 2B | 1.802 | 1.827 | 1.8563 | 1.8663 | 1.9375 |
|  | 0.0000 | 3A | 1.9375 | 1.9225 | 1.8563 | 1.8505 | 1.7841 | 3B | 1.8020 | 1.8172 | 1.8563 | 1.8638 | 1.9375 |
| 1 15/16–12 UN | 0.0018 | 2A | 1.9357 | 1.9243 | 1.8816 | 1.8755 | 1.8335 | 2B | 1.847 | 1.865 | 1.8834 | 1.8913 | 1.9375 |
|  | 0.0000 | 3A | 1.9375 | 1.9261 | 1.8834 | 1.8789 | 1.8353 | 3B | 1.8470 | 1.8570 | 1.8834 | 1.8893 | 1.9375 |
| 1 15/16–16 UN | 0.0016 | 2A | 1.9359 | 1.9265 | 1.8953 | 1.8899 | 1.8592 | 2B | 1.870 | 1.884 | 1.8969 | 1.9039 | 1.9375 |
|  | 0.0000 | 3A | 1.9375 | 1.9281 | 1.8969 | 1.8929 | 1.8608 | 3B | 1.8700 | 1.8783 | 1.8969 | 1.9021 | 1.9375 |
| 1 15/16–20 UN | 0.0015 | 2A | 1.9360 | 1.9279 | 1.9035 | 1.8986 | 1.8747 | 2B | 1.883 | 1.895 | 1.9050 | 1.9114 | 1.9375 |
|  | 0.0000 | 3A | 1.9375 | 1.9294 | 1.9050 | 1.9013 | 1.8762 | 3B | 1.8830 | 1.8912 | 1.9050 | 1.9098 | 1.9375 |
| 2–4½ UNC | 0.0029 | 1A | 1.9971 | 1.9641 | 1.8528 | 1.8385 | 1.7245 | 1B | 1.759 | 1.804 | 1.8557 | 1.8743 | 2.0000 |
|  | 0.0029 | 2A | 1.9971 | 1.9752 | 1.8528 | 1.8433 | 1.7245 | 2B | 1.759 | 1.795 | 1.8557 | 1.8681 | 2.0000 |
|  | 0.0000 | 3A | 2.0000 | 1.9780 | 1.8557 | 1.8486 | 1.7274 | 3B | 1.7590 | 1.786 | 1.8557 | 1.8650 | 2.0000 |
| 2–6 UN | 0.0026 | 2A | 1.9974 | 1.9792 | 1.8891 | 1.8805 | 1.7929 | 2B | 1.820 | 1.850 | 1.8917 | 1.9028 | 2.0000 |
|  | 0.0000 | 3A | 2.0000 | 1.9818 | 1.8917 | 1.8853 | 1.7955 | 3B | 1.8200 | 1.8396 | 1.8917 | 1.9000 | 2.0000 |
| 2–8 UN | 0.0023 | 2A | 1.9977 | 1.9827 | 1.9165 | 1.9087 | 1.8443 | 2B | 1.865 | 1.890 | 1.9188 | 1.9289 | 2.0000 |
|  | 0.0000 | 3A | 2.0000 | 1.9850 | 1.9188 | 1.9130 | 1.8466 | 3B | 1.8650 | 1.8797 | 1.9188 | 1.9264 | 2.0000 |
| 2–10 UNS | 0.0020 | 2A | 1.9980 | 1.9851 | 1.9330 | 1.9266 | 1.8753 | 2B | 1.892 | 1.913 | 1.9350 | 1.9435 | 2.0000 |
| 2–12 UN | 0.0018 | 2A | 1.9982 | 1.9868 | 1.9441 | 1.9380 | 1.8960 | 2B | 1.910 | 1.928 | 1.9459 | 1.9538 | 2.0000 |
|  | 0.0000 | 3A | 2.0000 | 1.9886 | 1.9459 | 1.9414 | 1.8978 | 3B | 1.9100 | 1.9198 | 1.9459 | 1.9518 | 2.0000 |
| 2–14 UNS | 0.0017 | 2A | 1.9983 | 1.9880 | 1.9519 | 1.9462 | 1.9107 | 2B | 1.923 | 1.938 | 1.9536 | 1.9610 | 2.0000 |
| 2–16 UN | 0.0016 | 2A | 1.9984 | 1.9890 | 1.9578 | 1.9524 | 1.9217 | 2B | 1.932 | 1.946 | 1.9594 | 1.9664 | 2.0000 |
|  | 0.0000 | 3A | 2.0000 | 1.9906 | 1.9594 | 1.9554 | 1.9233 | 3B | 1.9320 | 1.9408 | 1.9594 | 1.9646 | 2.0000 |
| 2–18 UNS | 0.0015 | 2A | 1.9985 | 1.9898 | 1.9624 | 1.9573 | 1.9303 | 2B | 1.940 | 1.953 | 1.9639 | 1.9706 | 2.0000 |

† Use UNS threads only if Standard Series do not meet requirements (see page 1278).     See footnotes a, b, c, d, e, and f at end of table.

Table 4 (Continued). Standard Series and Selected Combinations† — Unified Screw Threads

| Nominal Size, Threads per Inch, and Series Designation f | External | | | | | | | | Internal | | | | | |
|---|---|---|---|---|---|---|---|---|---|---|---|---|---|---|
| | Class | Allowance | Major Diameter | | | Pitch Diameter | | Minor Diameter a | Class | Minor Diameter e | | Pitch Diameter | | Major Diameter |
| | | | Max b | Min | Min d | Max b | Min | | | Min | Max | Min | Max | Min |
| 2-20 UN | 2A | 0.0015 | 1.9985 | 1.9904 | — | 1.9660 | 1.9611 | 1.9372 | 2B | 1.946 | 1.957 | 1.9675 | 1.9739 | 2.0000 |
| | 3A | 0.0000 | 2.0000 | 1.9919 | — | 1.9675 | 1.9638 | 1.9387 | 3B | 1.9460 | 1.9537 | 1.9675 | 1.9723 | 2.0000 |
| 2 1/16-16 UNS | 2A | 0.0016 | 2.0609 | 2.0515 | — | 2.0203 | 2.0149 | 1.9842 | 2B | 1.995 | 2.009 | 2.0219 | 2.0289 | 2.0625 |
| | 3A | 0.0000 | 2.0625 | 2.0531 | — | 2.0219 | 2.0179 | 1.9858 | 3B | 1.9950 | 2.0033 | 2.0219 | 2.0271 | 2.0625 |
| 2 1/8-6 UN | 2A | 0.0026 | 2.1224 | 2.1042 | — | 2.0141 | 2.0054 | 1.9179 | 2B | 1.945 | 1.975 | 2.0167 | 2.0280 | 2.1250 |
| | 3A | 0.0000 | 2.1250 | 2.1068 | — | 2.0167 | 2.0102 | 1.9205 | 3B | 1.9450 | 1.9646 | 2.0167 | 2.0251 | 2.1250 |
| 2 1/8-8 UN | 2A | 0.0024 | 2.1226 | 2.1076 | 2.1001 | 2.0414 | 2.0335 | 1.9692 | 2B | 1.990 | 2.015 | 2.0438 | 2.0540 | 2.1250 |
| | 3A | 0.0000 | 2.1250 | 2.1100 | — | 2.0438 | 2.0379 | 1.9716 | 3B | 1.9900 | 2.0047 | 2.0438 | 2.0515 | 2.1250 |
| 2 1/8-12 UN | 2A | 0.0018 | 2.1232 | 2.1118 | — | 2.0691 | 2.0630 | 2.0210 | 2B | 2.035 | 2.053 | 2.0709 | 2.0788 | 2.1250 |
| | 3A | 0.0000 | 2.1250 | 2.1136 | — | 2.0709 | 2.0664 | 2.0228 | 3B | 2.0350 | 2.0448 | 2.0709 | 2.0768 | 2.1250 |
| 2 1/8-16 UN | 2A | 0.0016 | 2.1234 | 2.1140 | — | 2.0828 | 2.0774 | 2.0467 | 2B | 2.057 | 2.071 | 2.0844 | 2.0914 | 2.1250 |
| | 3A | 0.0000 | 2.1250 | 2.1156 | — | 2.0844 | 2.0803 | 2.0483 | 3B | 2.0570 | 2.0658 | 2.0844 | 2.0896 | 2.1250 |
| 2 1/8-20 UN | 2A | 0.0015 | 2.1235 | 2.1154 | — | 2.0910 | 2.0861 | 2.0622 | 2B | 2.071 | 2.082 | 2.0925 | 2.0989 | 2.1250 |
| | 3A | 0.0000 | 2.1250 | 2.1169 | — | 2.0925 | 2.0888 | 2.0637 | 3B | 2.0710 | 2.0787 | 2.0925 | 2.0973 | 2.1250 |
| 2 3/16-16 UNS | 2A | 0.0016 | 2.1859 | 2.1765 | — | 2.1453 | 2.1399 | 2.1092 | 2B | 2.120 | 2.134 | 2.1469 | 2.1539 | 2.1875 |
| | 3A | 0.0000 | 2.1875 | 2.1781 | — | 2.1469 | 2.1428 | 2.1108 | 3B | 2.1200 | 2.1283 | 2.1469 | 2.1521 | 2.1875 |
| 2 1/4-4 1/2 UNC | 1A | 0.0029 | 2.2471 | 2.2141 | 2.2141 | 2.1028 | 2.0882 | 1.9745 | 1B | 2.009 | 2.045 | 2.1057 | 2.1247 | 2.2500 |
| | 2A | 0.0029 | 2.2471 | 2.2251 | — | 2.1028 | 2.0931 | 1.9745 | 2B | 2.009 | 2.045 | 2.1057 | 2.1183 | 2.2500 |
| | 3A | 0.0000 | 2.2500 | 2.2280 | — | 2.1057 | 2.0984 | 1.9774 | 3B | 2.0090 | 2.0361 | 2.1057 | 2.1152 | 2.2500 |
| 2 1/4-6 UN | 2A | 0.0026 | 2.2474 | 2.2292 | — | 2.1391 | 2.1303 | 2.0429 | 2B | 2.070 | 2.100 | 2.1417 | 2.1531 | 2.2500 |
| | 3A | 0.0000 | 2.2500 | 2.2318 | — | 2.1417 | 2.1351 | 2.0455 | 3B | 2.0700 | 2.0896 | 2.1417 | 2.1502 | 2.2500 |
| 2 1/4-8 UN | 2A | 0.0024 | 2.2476 | 2.2326 | 2.2251 | 2.1664 | 2.1584 | 2.0942 | 2B | 2.115 | 2.140 | 2.1688 | 2.1792 | 2.2500 |
| | 3A | 0.0000 | 2.2500 | 2.2350 | — | 2.1688 | 2.1628 | 2.0966 | 3B | 2.1150 | 2.1297 | 2.1688 | 2.1766 | 2.2500 |
| 2 1/4-10 UNS | 2A | 0.0020 | 2.2480 | 2.2351 | — | 2.1830 | 2.1765 | 2.1253 | 2B | 2.142 | 2.163 | 2.1850 | 2.1935 | 2.2500 |
| 2 1/4-12 UN | 2A | 0.0018 | 2.2482 | 2.2368 | — | 2.1941 | 2.1880 | 2.1460 | 2B | 2.160 | 2.178 | 2.1959 | 2.2038 | 2.2500 |
| | 3A | 0.0000 | 2.2500 | 2.2386 | — | 2.1959 | 2.1914 | 2.1478 | 3B | 2.1600 | 2.1698 | 2.1959 | 2.2018 | 2.2500 |
| 2 1/4-14 UNS | 2A | 0.0017 | 2.2483 | 2.2380 | — | 2.2019 | 2.1962 | 2.1607 | 2B | 2.173 | 2.188 | 2.2036 | 2.2110 | 2.2500 |
| 2 1/4-16 UN | 2A | 0.0016 | 2.2484 | 2.2390 | — | 2.2078 | 2.2024 | 2.1717 | 2B | 2.182 | 2.196 | 2.2094 | 2.2164 | 2.2500 |
| | 3A | 0.0000 | 2.2500 | 2.2406 | — | 2.2094 | 2.2053 | 2.1733 | 3B | 2.1820 | 2.1908 | 2.2094 | 2.2146 | 2.2500 |

See footnotes *a*, *b*, *c*, *d*, *e*, and *f* at end of table.

| Nominal Size, Threads per Inch, Series Designation | Class | Allowance | Major Dia Max | Major Dia Min | — | Pitch Dia Max | Pitch Dia Min | UNR Minor Dia Max | Class | Minor Dia Min | Minor Dia Max | Pitch Dia Min | Pitch Dia Max | Major Dia Min |
|---|---|---|---|---|---|---|---|---|---|---|---|---|---|---|
| 2¼–18 UNS | 2A | 0.0015 | 2.2485 | 2.2398 | — | 2.2124 | 2.2073 | 2.1803 | 2B | 2.190 | 2.203 | 2.2139 | 2.2206 | 2.2500 |
| 2¼–20 UN | 2A | 0.0015 | 2.2485 | 2.2404 | — | 2.2160 | 2.2111 | 2.1872 | 2B | 2.196 | 2.207 | 2.2175 | 2.2239 | 2.2500 |
| | 3A | 0.0000 | 2.2500 | 2.2419 | — | 2.2175 | 2.2137 | 2.1887 | 3B | 2.1960 | 2.2037 | 2.2175 | 2.2223 | 2.2500 |
| 2⁵⁄₁₆–16 UNS | 2A | 0.0017 | 2.3108 | 2.3014 | — | 2.2702 | 2.2647 | 2.2341 | 2B | 2.245 | 2.259 | 2.2719 | 2.2791 | 2.3125 |
| | 3A | 0.0000 | 2.3125 | 2.3031 | — | 2.2719 | 2.2678 | 2.2358 | 3B | 2.2450 | 2.2533 | 2.2719 | 2.2773 | 2.3125 |
| 2⅜–6 UN | 2A | 0.0027 | 2.3723 | 2.3541 | — | 2.2640 | 2.2551 | 2.1678 | 2B | 2.195 | 2.226 | 2.2667 | 2.2782 | 2.3750 |
| | 3A | 0.0000 | 2.3750 | 2.3568 | — | 2.2667 | 2.2601 | 2.1705 | 3B | 2.1950 | 2.2146 | 2.2667 | 2.2753 | 2.3750 |
| 2⅜–8 UN | 2A | 0.0024 | 2.3726 | 2.3576 | — | 2.2914 | 2.2833 | 2.2192 | 2B | 2.240 | 2.265 | 2.2938 | 2.3043 | 2.3750 |
| | 3A | 0.0000 | 2.3750 | 2.3600 | — | 2.2938 | 2.2878 | 2.2216 | 3B | 2.2400 | 2.2552 | 2.2938 | 2.3017 | 2.3750 |
| 2⅜–12 UN | 2A | 0.0019 | 2.3731 | 2.3617 | — | 2.3190 | 2.3128 | 2.2709 | 2B | 2.285 | 2.303 | 2.3209 | 2.3290 | 2.3750 |
| | 3A | 0.0000 | 2.3750 | 2.3636 | — | 2.3209 | 2.3163 | 2.2728 | 3B | 2.2850 | 2.2948 | 2.3209 | 2.3269 | 2.3750 |
| 2⅜–16 UN | 2A | 0.0017 | 2.3733 | 2.3639 | — | 2.3327 | 2.3272 | 2.2966 | 2B | 2.307 | 2.321 | 2.3344 | 2.3416 | 2.3750 |
| | 3A | 0.0000 | 2.3750 | 2.3656 | — | 2.3344 | 2.3303 | 2.2983 | 3B | 2.3070 | 2.3158 | 2.3344 | 2.3398 | 2.3750 |
| 2⅜–20 UN | 2A | 0.0015 | 2.3735 | 2.3654 | — | 2.3410 | 2.3359 | 2.3122 | 2B | 2.321 | 2.332 | 2.3425 | 2.3491 | 2.3750 |
| | 3A | 0.0000 | 2.3750 | 2.3669 | — | 2.3425 | 2.3387 | 2.3137 | 3B | 2.3210 | 2.3287 | 2.3425 | 2.3475 | 2.3750 |
| 2⁷⁄₁₆–16 UNS | 2A | 0.0017 | 2.4358 | 2.4264 | — | 2.3952 | 2.3897 | 2.3591 | 2B | 2.370 | 2.384 | 2.3969 | 2.4041 | 2.4375 |
| | 3A | 0.0000 | 2.4375 | 2.4281 | — | 2.3969 | 2.3928 | 2.3608 | 3B | 2.3700 | 2.3783 | 2.3969 | 2.4023 | 2.4375 |
| 2½–4 UNC | 1A | 0.0031 | 2.4969 | 2.4612 | — | 2.3345 | 2.3190 | 2.1902 | 1B | 2.229 | 2.267 | 2.3376 | 2.3578 | 2.5000 |
| | 2A | 0.0031 | 2.4969 | 2.4731 | 2.4612 | 2.3345 | 2.3241 | 2.1902 | 2B | 2.229 | 2.267 | 2.3376 | 2.3578 | 2.5000 |
| | 3A | 0.0000 | 2.5000 | 2.4762 | 2.4751 | 2.3376 | 2.3298 | 2.1933 | 3B | 2.2290 | 2.2594 | 2.3376 | 2.3511 | 2.5000 |
| 2½–6 UN | 2A | 0.0027 | 2.4973 | 2.4791 | — | 2.3890 | 2.3800 | 2.2928 | 2B | 2.320 | 2.350 | 2.3917 | 2.4033 | 2.5000 |
| | 3A | 0.0000 | 2.5000 | 2.4818 | — | 2.3917 | 2.3850 | 2.2955 | 3B | 2.3200 | 2.3396 | 2.3917 | 2.4004 | 2.5000 |
| 2½–8 UN | 2A | 0.0024 | 2.4976 | 2.4826 | — | 2.4164 | 2.4082 | 2.3442 | 2B | 2.365 | 2.390 | 2.4188 | 2.4294 | 2.5000 |
| | 3A | 0.0000 | 2.5000 | 2.4850 | — | 2.4188 | 2.4127 | 2.3466 | 3B | 2.3650 | 2.3797 | 2.4188 | 2.4268 | 2.5000 |
| 2½–10 UNS | 2A | 0.0020 | 2.4980 | 2.4851 | — | 2.4330 | 2.4263 | 2.3753 | 2B | 2.392 | 2.413 | 2.4350 | 2.4437 | 2.5000 |
| 2½–12 UN | 2A | 0.0019 | 2.4981 | 2.4867 | — | 2.4440 | 2.4378 | 2.3959 | 2B | 2.410 | 2.428 | 2.4459 | 2.4540 | 2.5000 |
| | 3A | 0.0000 | 2.5000 | 2.4886 | — | 2.4459 | 2.4413 | 2.3978 | 3B | 2.4100 | 2.4198 | 2.4459 | 2.4519 | 2.5000 |
| 2½–14 UNS | 2A | 0.0017 | 2.4983 | 2.4880 | — | 2.4519 | 2.4461 | 2.4107 | 2B | 2.423 | 2.438 | 2.4536 | 2.4612 | 2.5000 |
| 2½–16 UN | 2A | 0.0017 | 2.4983 | 2.4889 | — | 2.4577 | 2.4522 | 2.4223 | 2B | 2.432 | 2.446 | 2.4594 | 2.4666 | 2.5000 |
| | 3A | 0.0000 | 2.5000 | 2.4906 | — | 2.4594 | 2.4553 | 2.4233 | 3B | 2.4320 | 2.4408 | 2.4594 | 2.4648 | 2.5000 |
| 2½–18 UNS | 2A | 0.0016 | 2.4984 | 2.4897 | — | 2.4623 | 2.4570 | 2.4302 | 2B | 2.440 | 2.453 | 2.4639 | 2.4708 | 2.5000 |
| 2½–20 UN | 2A | 0.0015 | 2.4985 | 2.4904 | — | 2.4660 | 2.4609 | 2.4372 | 2B | 2.446 | 2.457 | 2.4675 | 2.4741 | 2.5000 |
| | 3A | 0.0000 | 2.5000 | 2.4919 | — | 2.4675 | 2.4637 | 2.4387 | 3B | 2.4460 | 2.4537 | 2.4675 | 2.4725 | 2.5000 |

† Use UNS threads only if Standard Series do not meet requirements (see page 1278).

Table 4 (Continued). Standard Series and Selected Combinations† — Unified Screw Threads

| Nominal Size, Threads per Inch, and Series Designation f | Class | Allowance | External Major Diameter Max b | External Major Diameter Min | External Major Diameter Min d | External Pitch Diameter Max b | External Pitch Diameter Min | External Minor Diameter a | Class | Internal Minor Diameter e Min | Internal Minor Diameter e Max | Internal Pitch Diameter Min | Internal Pitch Diameter Max | Internal Major Diameter Min |
|---|---|---|---|---|---|---|---|---|---|---|---|---|---|---|
| 2⅝–6 UN | 2A | 0.0027 | 2.6223 | 2.6041 | — | 2.5140 | 2.5050 | 2.4178 | 2B | 2.445 | 2.475 | 2.5167 | 2.5285 | 2.6250 |
|  | 3A | 0.0000 | 2.6250 | 2.6068 | — | 2.5167 | 2.5099 | 2.4205 | 3B | 2.4450 | 2.4646 | 2.5167 | 2.5255 | 2.6250 |
| 2⅝–8 UN | 2A | 0.0025 | 2.6225 | 2.6075 | — | 2.5413 | 2.5331 | 2.4691 | 2B | 2.490 | 2.515 | 2.5438 | 2.5545 | 2.6250 |
|  | 3A | 0.0000 | 2.6250 | 2.6100 | — | 2.5438 | 2.5376 | 2.4716 | 3B | 2.4900 | 2.5052 | 2.5438 | 2.5518 | 2.6250 |
| 2⅝–12 UN | 2A | 0.0019 | 2.6231 | 2.6117 | — | 2.5690 | 2.5628 | 2.5209 | 2B | 2.535 | 2.553 | 2.5709 | 2.5790 | 2.6250 |
|  | 3A | 0.0000 | 2.6250 | 2.6136 | — | 2.5709 | 2.5663 | 2.5228 | 3B | 2.5350 | 2.5448 | 2.5709 | 2.5769 | 2.6250 |
| 2⅝–16 UN | 2A | 0.0017 | 2.6233 | 2.6139 | — | 2.5827 | 2.5772 | 2.5466 | 2B | 2.557 | 2.571 | 2.5844 | 2.5916 | 2.6250 |
|  | 3A | 0.0000 | 2.6250 | 2.6156 | — | 2.5844 | 2.5803 | 2.5483 | 3B | 2.5570 | 2.5658 | 2.5844 | 2.5898 | 2.6250 |
| 2⅝–20 UN | 2A | 0.0015 | 2.6235 | 2.6154 | — | 2.5910 | 2.5859 | 2.5622 | 2B | 2.571 | 2.582 | 2.5925 | 2.5991 | 2.6250 |
|  | 3A | 0.0000 | 2.6250 | 2.6169 | — | 2.5925 | 2.5887 | 2.5637 | 3B | 2.5710 | 2.5787 | 2.5925 | 2.5975 | 2.6250 |
| 2¾–4 UNC | 1A | 0.0032 | 2.7468 | 2.7111 | 2.7111 | 2.5844 | 2.5686 | 2.4401 | 1B | 2.479 | 2.517 | 2.5876 | 2.6082 | 2.7500 |
|  | 2A | 0.0032 | 2.7468 | 2.7230 | — | 2.5844 | 2.5739 | 2.4401 | 2B | 2.479 | 2.517 | 2.5876 | 2.6013 | 2.7500 |
|  | 3A | 0.0000 | 2.7500 | 2.7262 | — | 2.5876 | 2.5797 | 2.4433 | 3B | 2.4790 | 2.5094 | 2.5876 | 2.5979 | 2.7500 |
| 2¾–6 UN | 2A | 0.0027 | 2.7473 | 2.7291 | — | 2.6390 | 2.6299 | 2.5428 | 2B | 2.570 | 2.600 | 2.6417 | 2.6536 | 2.7500 |
|  | 3A | 0.0000 | 2.7500 | 2.7318 | — | 2.6417 | 2.6349 | 2.5455 | 3B | 2.5700 | 2.5896 | 2.6417 | 2.6506 | 2.7500 |
| 2¾–8 UN | 2A | 0.0025 | 2.7475 | 2.7325 | 2.7250 | 2.6663 | 2.6580 | 2.5941 | 2B | 2.615 | 2.640 | 2.6688 | 2.6796 | 2.7500 |
|  | 3A | 0.0000 | 2.7500 | 2.7351 | — | 2.6688 | 2.6625 | 2.5966 | 3B | 2.6150 | 2.6297 | 2.6688 | 2.6769 | 2.7500 |
| 2¾–10 UNS | 2A | 0.0020 | 2.7480 | 2.7351 | — | 2.6830 | 2.6763 | 2.6253 | 2B | 2.642 | 2.663 | 2.6850 | 2.6937 | 2.7500 |
| 2¾–12 UN | 2A | 0.0019 | 2.7481 | 2.7367 | — | 2.6940 | 2.6878 | 2.6459 | 2B | 2.660 | 2.678 | 2.6959 | 2.7040 | 2.7500 |
|  | 3A | 0.0000 | 2.7500 | 2.7386 | — | 2.6959 | 2.6913 | 2.6478 | 3B | 2.6600 | 2.6698 | 2.6959 | 2.7019 | 2.7500 |
| 2¾–14 UNS | 2A | 0.0017 | 2.7483 | 2.7380 | — | 2.7019 | 2.6961 | 2.6607 | 2B | 2.673 | 2.688 | 2.7036 | 2.7112 | 2.7500 |
| 2¾–16 UN | 2A | 0.0017 | 2.7483 | 2.7389 | — | 2.7077 | 2.7022 | 2.6716 | 2B | 2.682 | 2.696 | 2.7094 | 2.7166 | 2.7500 |
|  | 3A | 0.0000 | 2.7500 | 2.7406 | — | 2.7094 | 2.7053 | 2.6733 | 3B | 2.6820 | 2.6908 | 2.7094 | 2.7148 | 2.7500 |
| 2¾–18 UNS | 2A | 0.0016 | 2.7484 | 2.7397 | — | 2.7123 | 2.7070 | 2.6802 | 2B | 2.690 | 2.703 | 2.7139 | 2.7208 | 2.7500 |
| 2¾–20 UN | 2A | 0.0015 | 2.7485 | 2.7404 | — | 2.7160 | 2.7109 | 2.6872 | 2B | 2.696 | 2.707 | 2.7175 | 2.7241 | 2.7500 |
|  | 3A | 0.0000 | 2.7500 | 2.7419 | — | 2.7175 | 2.7137 | 2.6887 | 3B | 2.6960 | 2.7037 | 2.7175 | 2.7225 | 2.7500 |
| 2⅞–6 UN | 2A | 0.0028 | 2.8722 | 2.8540 | — | 2.7639 | 2.7547 | 2.6677 | 2B | 2.695 | 2.725 | 2.7667 | 2.7787 | 2.8750 |
|  | 3A | 0.0000 | 2.8750 | 2.8568 | — | 2.7667 | 2.7598 | 2.6705 | 3B | 2.6950 | 2.7146 | 2.7667 | 2.7757 | 2.8750 |

| Size | Class | Allowance | Major Dia Max | Major Dia Min | (Ref) | Pitch Dia Max | Pitch Dia Min | Minor Dia | Class | Minor Dia Min | Minor Dia Max | Pitch Dia Min | Pitch Dia Max | Major Dia Min |
|---|---|---|---|---|---|---|---|---|---|---|---|---|---|---|
| 2⅞-8 UN | 2A | 0.0025 | 2.8725 | 2.8575 | — | 2.7913 | 2.7829 | 2.7191 | 2B | 2.740 | 2.765 | 2.7938 | 2.8048 | 2.8750 |
|  | 3A | 0.0000 | 2.8750 | 2.8600 | — | 2.7938 | 2.7875 | 2.7216 | 3B | 2.7400 | 2.7552 | 2.7938 | 2.8020 | 2.8750 |
| 2⅞-12 UN | 2A | 0.0019 | 2.8731 | 2.8617 | — | 2.8190 | 2.8127 | 2.7709 | 2B | 2.785 | 2.803 | 2.8209 | 2.8291 | 2.8750 |
|  | 3A | 0.0000 | 2.8750 | 2.8636 | — | 2.8209 | 2.8162 | 2.7728 | 3B | 2.7850 | 2.7948 | 2.8209 | 2.8271 | 2.8750 |
| 2⅞-16 UN | 2A | 0.0017 | 2.8733 | 2.8639 | — | 2.8327 | 2.8271 | 2.7966 | 2B | 2.807 | 2.821 | 2.8344 | 2.8417 | 2.8750 |
|  | 3A | 0.0000 | 2.8750 | 2.8656 | — | 2.8344 | 2.8302 | 2.7983 | 3B | 2.8070 | 2.8158 | 2.8344 | 2.8399 | 2.8750 |
| 2⅞-20 UN | 2A | 0.0016 | 2.8734 | 2.8653 | — | 2.8409 | 2.8357 | 2.8121 | 2B | 2.821 | 2.832 | 2.8425 | 2.8493 | 2.8750 |
|  | 3A | 0.0000 | 2.8750 | 2.8669 | — | 2.8425 | 2.8386 | 2.8137 | 3B | 2.8210 | 2.8287 | 2.8425 | 2.8476 | 2.8750 |
| 3-4 UNC | 1A | 0.0032 | 2.9968 | 2.9611 | — | 2.8344 | 2.8183 | 2.6901 | 1B | 2.729 | 2.767 | 2.8376 | 2.8585 | 3.0000 |
|  | 2A | 0.0032 | 2.9968 | 2.9730 | — | 2.8344 | 2.8237 | 2.6901 | 2B | 2.729 | 2.767 | 2.8376 | 2.8515 | 3.0000 |
|  | 3A | 0.0000 | 3.0000 | 2.9762 | 2.9611 | 2.8376 | 2.8296 | 2.6933 | 3B | 2.7290 | 2.7594 | 2.8376 | 2.8480 | 3.0000 |
| 3-6 UN | 2A | 0.0028 | 2.9972 | 2.9790 | — | 2.8889 | 2.8796 | 2.7927 | 2B | 2.820 | 2.850 | 2.8917 | 2.9038 | 3.0000 |
|  | 3A | 0.0000 | 3.0000 | 2.9818 | — | 2.8917 | 2.8847 | 2.7955 | 3B | 2.8200 | 2.8396 | 2.8917 | 2.9008 | 3.0000 |
| 3-8 UN | 2A | 0.0026 | 2.9974 | 2.9824 | 2.9749 | 2.9162 | 2.9077 | 2.8440 | 2B | 2.865 | 2.890 | 2.9188 | 2.9299 | 3.0000 |
|  | 3A | 0.0000 | 3.0000 | 2.9850 | — | 2.9188 | 2.9124 | 2.8466 | 3B | 2.8650 | 2.8797 | 2.9188 | 2.9271 | 3.0000 |
| 3-10 UNS | 2A | 0.0020 | 2.9980 | 2.9851 | — | 2.9330 | 2.9262 | 2.8753 | 2B | 2.892 | 2.913 | 2.9350 | 2.9439 | 3.0000 |
| 3-12 UN | 2A | 0.0019 | 2.9981 | 2.9867 | — | 2.9440 | 2.9377 | 2.8959 | 2B | 2.910 | 2.928 | 2.9459 | 2.9541 | 3.0000 |
|  | 3A | 0.0000 | 3.0000 | 2.9886 | — | 2.9459 | 2.9412 | 2.8978 | 3B | 2.9100 | 2.9198 | 2.9459 | 2.9521 | 3.0000 |
| 3-14 UNS | 2A | 0.0018 | 2.9982 | 2.9879 | — | 2.9518 | 2.9459 | 2.9106 | 2B | 2.923 | 2.938 | 2.9536 | 2.9613 | 3.0000 |
| 3-16 UN | 2A | 0.0017 | 2.9983 | 2.9889 | — | 2.9577 | 2.9521 | 2.9216 | 2B | 2.932 | 2.946 | 2.9594 | 2.9667 | 3.0000 |
|  | 3A | 0.0000 | 3.0000 | 2.9906 | — | 2.9594 | 2.9552 | 2.9233 | 3B | 2.9320 | 2.9408 | 2.9594 | 2.9649 | 3.0000 |
| 3-18 UNS | 2A | 0.0016 | 2.9984 | 2.9897 | — | 2.9623 | 2.9569 | 2.9302 | 2B | 2.940 | 2.953 | 2.9639 | 2.9709 | 3.0000 |
| 3-20 UN | 2A | 0.0016 | 2.9984 | 2.9903 | — | 2.9659 | 2.9607 | 2.9371 | 2B | 2.946 | 2.957 | 2.9675 | 2.9743 | 3.0000 |
|  | 3A | 0.0000 | 3.0000 | 2.9919 | — | 2.9675 | 2.9636 | 2.9387 | 3B | 2.9460 | 2.9537 | 2.9675 | 2.9726 | 3.0000 |
| 3⅛-6 UN | 2A | 0.0028 | 3.1222 | 3.1040 | — | 3.0139 | 3.0045 | 2.9177 | 2B | 2.945 | 2.975 | 3.0167 | 3.0289 | 3.1250 |
|  | 3A | 0.0000 | 3.1250 | 3.1068 | — | 3.0167 | 3.0097 | 2.9205 | 3B | 2.9450 | 2.9646 | 3.0167 | 3.0259 | 3.1250 |
| 3⅛-8 UN | 2A | 0.0026 | 3.1224 | 3.1074 | — | 3.0412 | 3.0326 | 2.9690 | 2B | 2.990 | 3.015 | 3.0438 | 3.0550 | 3.1250 |
|  | 3A | 0.0000 | 3.1250 | 3.1100 | — | 3.0438 | 3.0374 | 2.9716 | 3B | 2.9900 | 3.0052 | 3.0438 | 3.0522 | 3.1250 |
| 3⅛-12 UN | 2A | 0.0019 | 3.1231 | 3.1117 | — | 3.0690 | 3.0627 | 3.0209 | 2B | 3.035 | 3.053 | 3.0709 | 3.0791 | 3.1250 |
|  | 3A | 0.0000 | 3.1250 | 3.1136 | — | 3.0709 | 3.0662 | 3.0228 | 3B | 3.0350 | 3.0448 | 3.0709 | 3.0771 | 3.1250 |
| 3⅛-16 UN | 2A | 0.0017 | 3.1233 | 3.1139 | — | 3.0827 | 3.0771 | 3.0466 | 2B | 3.057 | 3.071 | 3.0844 | 3.0917 | 3.1250 |
|  | 3A | 0.0000 | 3.1250 | 3.1156 | — | 3.0844 | 3.0802 | 3.0483 | 3B | 3.0570 | 3.0658 | 3.0844 | 3.0899 | 3.1250 |

† Use UNS threads only if Standard Series do not meet requirements (see page 1278). See footnotes a, b, c, d, e, and f at end of table.

Table 4 (Continued). Standard Series and Selected Combinations† — Unified Screw Threads

| Nominal Size, Threads per Inch, and Series Designation [f] | External Class | External Allowance | External Major Diameter Max [b] | External Major Diameter Min | External Major Diameter Min [d] | External Pitch Diameter Max [b] | External Pitch Diameter Min | External Minor Diameter [c] | Internal Class | Internal Minor Diameter Min [e] | Internal Minor Diameter Max | Internal Pitch Diameter Min | Internal Pitch Diameter Max | Internal Major Diameter Min |
|---|---|---|---|---|---|---|---|---|---|---|---|---|---|---|
| **3¼–4 UNC** | 1A | 0.0033 | 3.2467 | 3.2110 | | 3.0843 | 3.0680 | 2.9400 | 1B | 2.979 | 3.017 | 3.0876 | 3.1088 | 3.2500 |
| | 2A | 0.0033 | 3.2467 | 3.2229 | 3.2110 | 3.0843 | 3.0734 | 2.9400 | 2B | 2.979 | 3.017 | 3.0876 | 3.1017 | 3.2500 |
| | 3A | 0.0000 | 3.2500 | 3.2262 | | 3.0876 | 3.0794 | 2.9433 | 3B | 2.9790 | 3.0094 | 3.0876 | 3.0982 | 3.2500 |
| **3¼–6 UN** | 2A | 0.0028 | 3.2472 | 3.2290 | 3.2249 | 3.1389 | 3.1294 | 3.0427 | 2B | 3.070 | 3.100 | 3.1417 | 3.1540 | 3.2500 |
| | 3A | 0.0000 | 3.2500 | 3.2318 | | 3.1417 | 3.1346 | 3.0455 | 3B | 3.0700 | 3.0896 | 3.1417 | 3.1509 | 3.2500 |
| **3¼–8 UN** | 2A | 0.0026 | 3.2474 | 3.2324 | | 3.1662 | 3.1575 | 3.0940 | 2B | 3.115 | 3.140 | 3.1688 | 3.1801 | 3.2500 |
| | 3A | 0.0000 | 3.2500 | 3.2350 | | 3.1688 | 3.1623 | 3.0966 | 3B | 3.1150 | 3.1297 | 3.1688 | 3.1772 | 3.2500 |
| **3¼–10 UNS** | 2A | 0.0020 | 3.2480 | 3.2351 | | 3.1830 | 3.1762 | 3.1253 | 2B | 3.142 | 3.163 | 3.1850 | 3.1939 | 3.2500 |
| **3¼–12 UN** | 2A | 0.0019 | 3.2481 | 3.2367 | | 3.1940 | 3.1877 | 3.1459 | 2B | 3.160 | 3.178 | 3.1959 | 3.2041 | 3.2500 |
| | 3A | 0.0000 | 3.2500 | 3.2386 | | 3.1959 | 3.1912 | 3.1478 | 3B | 3.1600 | 3.1698 | 3.1959 | 3.2021 | 3.2500 |
| **3¼–14 UNS** | 2A | 0.0018 | 3.2482 | 3.2379 | | 3.2018 | 3.1959 | 3.1606 | 2B | 3.173 | 3.188 | 3.2036 | 3.2113 | 3.2500 |
| **3¼–16 UN** | 2A | 0.0017 | 3.2483 | 3.2389 | | 3.2077 | 3.2021 | 3.1716 | 2B | 3.182 | 3.196 | 3.2094 | 3.2167 | 3.2500 |
| | 3A | 0.0000 | 3.2500 | 3.2406 | | 3.2094 | 3.2052 | 3.1733 | 3B | 3.1820 | 3.1908 | 3.2094 | 3.2149 | 3.2500 |
| **3¼–18 UNS** | 2A | 0.0016 | 3.2484 | 3.2397 | | 3.2123 | 3.2069 | 3.1802 | 2B | 3.190 | 3.203 | 3.2139 | 3.2209 | 3.2500 |
| **3⅜–6 UN** | 2A | 0.0029 | 3.3721 | 3.3539 | | 3.2638 | 3.2543 | 3.1676 | 2B | 3.195 | 3.225 | 3.2667 | 3.2791 | 3.3750 |
| | 3A | 0.0000 | 3.3750 | 3.3568 | | 3.2667 | 3.2595 | 3.1705 | 3B | 3.1950 | 3.2146 | 3.2667 | 3.2760 | 3.3750 |
| **3⅜–8 UN** | 2A | 0.0026 | 3.3724 | 3.3574 | | 3.2912 | 3.2824 | 3.2190 | 2B | 3.240 | 3.265 | 3.2938 | 3.3052 | 3.3750 |
| | 3A | 0.0000 | 3.3750 | 3.3600 | | 3.2938 | 3.2876 | 3.2216 | 3B | 3.2400 | 3.2552 | 3.2938 | 3.3023 | 3.3750 |
| **3⅜–12 UN** | 2A | 0.0019 | 3.3731 | 3.3617 | | 3.3190 | 3.3126 | 3.2709 | 2B | 3.285 | 3.303 | 3.3209 | 3.3293 | 3.3750 |
| | 3A | 0.0000 | 3.3750 | 3.3636 | | 3.3209 | 3.3161 | 3.2728 | 3B | 3.2850 | 3.2948 | 3.3209 | 3.3272 | 3.3750 |
| **3⅜–16 UN** | 2A | 0.0017 | 3.3733 | 3.3639 | | 3.3327 | 3.3269 | 3.2966 | 2B | 3.307 | 3.321 | 3.3344 | 3.3419 | 3.3750 |
| | 3A | 0.0000 | 3.3750 | 3.3656 | | 3.3344 | 3.3301 | 3.2983 | 3B | 3.3070 | 3.3158 | 3.3344 | 3.3400 | 3.3750 |
| **3½–4 UNC** | 1A | 0.0033 | 3.4967 | 3.4610 | | 3.3343 | 3.3177 | 3.1900 | 1B | 3.229 | 3.267 | 3.3376 | 3.3591 | 3.5000 |
| | 2A | 0.0033 | 3.4967 | 3.4729 | 3.4610 | 3.3343 | 3.3233 | 3.1900 | 2B | 3.229 | 3.267 | 3.3376 | 3.3519 | 3.5000 |
| | 3A | 0.0000 | 3.5000 | 3.4762 | | 3.3376 | 3.3293 | 3.1933 | 3B | 3.2290 | 3.2594 | 3.3376 | 3.3484 | 3.5000 |
| **3½–6 UN** | 2A | 0.0029 | 3.4971 | 3.4789 | 3.4749 | 3.3888 | 3.3792 | 3.2926 | 2B | 3.320 | 3.350 | 3.3917 | 3.4042 | 3.5000 |
| | 3A | 0.0000 | 3.5000 | 3.4818 | | 3.3917 | 3.3845 | 3.2955 | 3B | 3.3200 | 3.3396 | 3.3917 | 3.4011 | 3.5000 |
| **3½–8 UN** | 2A | 0.0026 | 3.4974 | 3.4824 | | 3.4162 | 3.4074 | 3.3440 | 2B | 3.365 | 3.390 | 3.4188 | 3.4303 | 3.5000 |
| | 3A | 0.0000 | 3.5000 | 3.4850 | | 3.4188 | 3.4122 | 3.3466 | 3B | 3.3650 | 3.3397 | 3.4188 | 3.4274 | 3.5000 |

| Size | Class | Allowance | Major Dia Max | UNR Minor | Major Dia Min | Pitch Dia Max | Pitch Dia Min | Minor Dia Max | Class | Minor Dia Min | Minor Dia Max | Pitch Dia Min | Pitch Dia Max | Major Dia Min |
|---|---|---|---|---|---|---|---|---|---|---|---|---|---|---|
| 3½–10 UNS | 2A | 0.0021 | 3.4979 | — | 3.4850 | 3.4329 | 3.4260 | 3.3752 | 2B | 3.392 | 3.413 | 3.4350 | 3.4440 | 3.5000 |
| 3½–12 UN | 2A / 3A | 0.0019 / 0.0000 | 3.4981 / 3.5000 | — / — | 3.4867 / 3.4886 | 3.4440 / 3.4459 | 3.4376 / 3.4411 | 3.3959 / 3.3978 | 2B / 3B | 3.410 / 3.4100 | 3.428 / 3.4198 | 3.4459 / 3.4459 | 3.4543 / 3.4522 | 3.5000 / 3.5000 |
| 3½–14 UNS | 2A | 0.0018 | 3.4982 | — | 3.4879 | 3.4518 | 3.4457 | 3.4106 | 2B | 3.423 | 3.438 | 3.4536 | 3.4615 | 3.5000 |
| 3½–16 UN | 2A / 3A | 0.0017 / 0.0000 | 3.4983 / 3.5000 | — / — | 3.4889 / 3.4906 | 3.4577 / 3.4594 | 3.4519 / 3.4551 | 3.4216 / 3.4233 | 2B / 3B | 3.432 / 3.4320 | 3.446 / 3.4408 | 3.4594 / 3.4594 | 3.4669 / 3.4650 | 3.5000 / 3.5000 |
| 3½–18 UNS | 2A | 0.0017 | 3.4983 | — | 3.4896 | 3.4622 | 3.4567 | 3.4301 | 2B | 3.440 | 3.453 | 3.4639 | 3.4711 | 3.5000 |
| 3⅝–6 UN | 2A / 3A | 0.0029 / 0.0000 | 3.6221 / 3.6250 | — / — | 3.6039 / 3.6068 | 3.5138 / 3.5167 | 3.5041 / 3.5094 | 3.4176 / 3.4205 | 2B / 3B | 3.445 / 3.4450 | 3.475 / 3.4646 | 3.5167 / 3.5167 | 3.5293 / 3.5262 | 3.6250 / 3.6250 |
| 3⅝–8 UN | 2A / 3A | 0.0027 / 0.0000 | 3.6223 / 3.6250 | — / — | 3.6073 / 3.6100 | 3.5411 / 3.5438 | 3.5322 / 3.5371 | 3.4689 / 3.4716 | 2B / 3B | 3.490 / 3.4900 | 3.515 / 3.5052 | 3.5438 / 3.5438 | 3.5554 / 3.5525 | 3.6250 / 3.6250 |
| 3⅝–12 UN | 2A / 3A | 0.0019 / 0.0000 | 3.6231 / 3.6250 | — / — | 3.6117 / 3.6136 | 3.5690 / 3.5709 | 3.5626 / 3.5661 | 3.5209 / 3.5228 | 2B / 3B | 3.535 / 3.5350 | 3.553 / 3.5448 | 3.5709 / 3.5709 | 3.5793 / 3.5772 | 3.6250 / 3.6250 |
| 3⅝–16 UN | 2A / 3A | 0.0017 / 0.0000 | 3.6233 / 3.6250 | — / — | 3.6139 / 3.6156 | 3.5827 / 3.5844 | 3.5769 / 3.5801 | 3.5466 / 3.5483 | 2B / 3B | 3.557 / 3.5570 | 3.571 / 3.5658 | 3.5844 / 3.5844 | 3.5919 / 3.5900 | 3.6250 / 3.6250 |
| 3¾–4 UNC | 1A / 2A / 3A | 0.0034 / 0.0034 / 0.0000 | 3.7466 / 3.7466 / 3.7500 | — / — / 3.7109 | 3.7109 / 3.7228 / 3.7262 | 3.5842 / 3.5842 / 3.5876 | 3.5674 / 3.5730 / 3.5792 | 3.4399 / 3.4399 / 3.4433 | 1B / 2B / 3B | 3.479 / 3.479 / 3.4790 | 3.517 / 3.517 / 3.5094 | 3.5876 / 3.5876 / 3.5876 | 3.6094 / 3.6021 / 3.5985 | 3.7500 / 3.7500 / 3.7500 |
| 3¾–6 UN | 2A / 3A | 0.0029 / 0.0000 | 3.7471 / 3.7500 | — / — | 3.7289 / 3.7318 | 3.6388 / 3.6417 | 3.6290 / 3.6344 | 3.5426 / 3.5455 | 2B / 3B | 3.570 / 3.5700 | 3.600 / 3.5896 | 3.6417 / 3.6417 | 3.6544 / 3.6512 | 3.7500 / 3.7500 |
| 3¾–8 UN | 2A / 3A | 0.0027 / 0.0000 | 3.7473 / 3.7500 | — / 3.7248 | 3.7323 / 3.7350 | 3.6661 / 3.6688 | 3.6571 / 3.6621 | 3.5939 / 3.5966 | 2B / 3B | 3.615 / 3.6150 | 3.640 / 3.6297 | 3.6688 / 3.6688 | 3.6805 / 3.6776 | 3.7500 / 3.7500 |
| 3¾–10 UNS | 2A | 0.0021 | 3.7479 | — | 3.7350 | 3.6829 | 3.6760 | 3.6252 | 2B | 3.642 | 3.663 | 3.6850 | 3.6940 | 3.7500 |
| 3¾–12 UN | 2A / 3A | 0.0019 / 0.0000 | 3.7481 / 3.7500 | — / — | 3.7367 / 3.7386 | 3.6940 / 3.6959 | 3.6876 / 3.6911 | 3.6459 / 3.6478 | 2B / 3B | 3.660 / 3.6600 | 3.678 / 3.6698 | 3.6959 / 3.6959 | 3.7043 / 3.7022 | 3.7500 / 3.7500 |
| 3¾–14 UNS | 2A | 0.0018 | 3.7482 | — | 3.7379 | 3.7018 | 3.6957 | 3.6606 | 2B | 3.673 | 3.688 | 3.7036 | 3.7115 | 3.7500 |
| 3¾–16 UN | 2A / 3A | 0.0017 / 0.0000 | 3.7483 / 3.7500 | — / — | 3.7389 / 3.7406 | 3.7077 / 3.7094 | 3.7019 / 3.7051 | 3.6716 / 3.6733 | 2B / 3B | 3.682 / 3.6820 | 3.696 / 3.6908 | 3.7094 / 3.7094 | 3.7169 / 3.7150 | 3.7500 / 3.7500 |
| 3¾–18 UNS | 2A | 0.0017 | 3.7483 | — | 3.7396 | 3.7122 | 3.7067 | 3.6801 | 2B | 3.690 | 3.703 | 3.7139 | 3.7211 | 3.7500 |
| 3⅞–6 UN | 2A / 3A | 0.0030 / 0.0000 | 3.8720 / 3.8750 | — / — | 3.8538 / 3.8568 | 3.7637 / 3.7667 | 3.7538 / 3.7593 | 3.6675 / 3.6705 | 2B / 3B | 3.695 / 3.6950 | 3.725 / 3.7146 | 3.7667 / 3.7667 | 3.7795 / 3.7763 | 3.8750 / 3.8750 |

† Use UNS threads only if Standard Series do not meet requirements (see page 1278).  See footnotes a, b, c, d, e, and f at end of table.

**Table 4 (Continued). Standard Series and Selected Combinations† — Unified Screw Threads**

| Nominal Size, Threads per Inch, and Series Designation^f | Class | Allow-ance | Major Diameter Max^b | Major Diameter Min | Pitch Diameter Max^b | Pitch Diameter Min | Minor Diameter Min^d | Minor Diam-eter^a | Class | Minor Diameter^e Min | Minor Diameter^e Max | Pitch Diameter Min | Pitch Diameter Max | Major Diameter Min |
|---|---|---|---|---|---|---|---|---|---|---|---|---|---|---|
| | | | *External* | | | | | | | *Internal* | | | | |
| 3⅞-8 UN | 2A | 0.0027 | 3.8723 | 3.8573 | 3.7911 | 3.7820 | — | 3.7189 | 2B | 3.740 | 3.765 | 3.7938 | 3.8056 | 3.8750 |
| | 3A | 0.0000 | 3.8750 | 3.8600 | 3.7938 | 3.7870 | — | 3.7216 | 3B | 3.7400 | 3.7552 | 3.7938 | 3.8026 | 3.8750 |
| 3⅞-12 UN | 2A | 0.0020 | 3.8730 | 3.8616 | 3.8189 | 3.8124 | — | 3.7708 | 2B | 3.785 | 3.803 | 3.8209 | 3.8294 | 3.8750 |
| | 3A | 0.0000 | 3.8750 | 3.8636 | 3.8209 | 3.8160 | — | 3.7728 | 3B | 3.7850 | 3.7948 | 3.8209 | 3.8273 | 3.8750 |
| 3⅞-16 UN | 2A | 0.0018 | 3.8732 | 3.8638 | 3.8326 | 3.8267 | — | 3.7965 | 2B | 3.807 | 3.821 | 3.8344 | 3.8420 | 3.8750 |
| | 3A | 0.0000 | 3.8750 | 3.8656 | 3.8344 | 3.8300 | — | 3.7983 | 3B | 3.8070 | 3.8158 | 3.8344 | 3.8401 | 3.8750 |
| 4-4 UNC | 1A | 0.0034 | 3.9966 | 3.9609 | 3.8342 | 3.8172 | 3.9609 | 3.6899 | 1B | 3.729 | 3.767 | 3.8376 | 3.8597 | 4.0000 |
| | 2A | 0.0034 | 3.9966 | 3.9728 | 3.8342 | 3.8229 | — | 3.6899 | 2B | 3.729 | 3.767 | 3.8376 | 3.8523 | 4.0000 |
| | 3A | 0.0000 | 4.0000 | 3.9762 | 3.8376 | 3.8291 | — | 3.6933 | 3B | 3.7290 | 3.7594 | 3.8376 | 3.8487 | 4.0000 |
| 4-6 UN | 2A | 0.0030 | 3.9970 | 3.9788 | 3.8887 | 3.8788 | — | 3.7925 | 2B | 3.820 | 3.850 | 3.8917 | 3.9046 | 4.0000 |
| | 3A | 0.0000 | 4.0000 | 3.9818 | 3.8917 | 3.8843 | — | 3.7955 | 3B | 3.8200 | 3.8396 | 3.8917 | 3.9014 | 4.0000 |
| 4-8 UN | 2A | 0.0027 | 3.9973 | 3.9823 | 3.9161 | 3.9070 | 3.9748 | 3.8439 | 2B | 3.865 | 3.890 | 3.9188 | 3.9307 | 4.0000 |
| | 3A | 0.0000 | 4.0000 | 3.9850 | 3.9188 | 3.9120 | — | 3.8466 | 3B | 3.8650 | 3.8797 | 3.9188 | 3.9277 | 4.0000 |
| 4-10 UNS | 2A | 0.0021 | 3.9979 | 3.9850 | 3.9329 | 3.9259 | — | 3.8752 | 2B | 3.892 | 3.913 | 3.9350 | 3.9441 | 4.0000 |
| 4-12 UN | 2A | 0.0020 | 3.9980 | 3.9866 | 3.9439 | 3.9374 | — | 3.8958 | 2B | 3.910 | 3.928 | 3.9459 | 3.9544 | 4.0000 |
| | 3A | 0.0000 | 4.0000 | 3.9886 | 3.9459 | 3.9410 | — | 3.8978 | 3B | 3.9100 | 3.9198 | 3.9459 | 3.9523 | 4.0000 |
| 4-14 UNS | 2A | 0.0018 | 3.9982 | 3.9879 | 3.9518 | 3.9456 | — | 3.9106 | 2B | 3.923 | 3.938 | 3.9536 | 3.9616 | 4.0000 |
| 4-16 UN | 2A | 0.0018 | 3.9982 | 3.9888 | 3.9576 | 3.9517 | — | 3.9215 | 2B | 3.932 | 3.946 | 3.9594 | 3.9670 | 4.0000 |
| | 3A | 0.0000 | 4.0000 | 3.9906 | 3.9594 | 3.9550 | — | 3.9233 | 3B | 3.9320 | 3.9408 | 3.9594 | 3.9651 | 4.0000 |

† Use UNS threads only if Standard Series do not meet requirements (see page 1278). For sizes above 4 inches see ANSI B1.1-1974.

These dimensions not available at time of this printing.

a See diagrams on pages 1263, 1276, and 1277.
b For Class 2A threads having an additive finish the maximum is increased, by the allowance, to the basic size, the value being the same as for Class 3A.
c Regarding combinations of thread classes, see text on page 1278.
d For unfinished hot-rolled material.
e Revised minor diameter limits Classes 1B and 2B.
f Use UNR designation instead of UN wherever UNR thread form is desired for external use.

Table 5.  Former American Standard Thread Series
Class 2 — External Thread Limits

| No. or Size | Thds. per Inch | Thread Symbol | Major Diameter | | | Pitch Diam. | | Minor Diam. (See p. 1263) |
|---|---|---|---|---|---|---|---|---|
| | | | Max. | Min.[1] | Min.[2] | Max. | Min. | |
| 0 | 80 | NF-2 | 0.0600 | 0.0566 | .... | 0.0519 | 0.0502 | 0.0447 |
| 1 | 64 | NC-2 | 0.0730 | 0.0692 | 0.0678 | 0.0629 | 0.0610 | 0.0538 |
| 1 | 72 | NF-2 | 0.0730 | 0.0694 | .... | 0.0640 | 0.0622 | 0.0560 |
| 2 | 56 | NC-2 | 0.0860 | 0.0820 | 0.0804 | 0.0744 | 0.0724 | 0.0641 |
| 2 | 64 | NF-2 | 0.0860 | 0.0822 | .... | 0.0759 | 0.0740 | 0.0668 |
| 3 | 48 | NC-2 | 0.0990 | 0.0946 | 0.0928 | 0.0855 | 0.0833 | 0.0734 |
| 3 | 56 | NF-2 | 0.0990 | 0.0950 | .... | 0.0874 | 0.0854 | 0.0771 |
| 4 | 40 | NC-2 | 0.1120 | 0.1072 | 0.1052 | 0.0958 | 0.0934 | 0.0813 |
| 4 | 48 | NF-2 | 0.1120 | 0.1076 | .... | 0.0985 | 0.0963 | 0.0864 |
| 5 | 40 | NC-2 | 0.1250 | 0.1202 | 0.1182 | 0.1088 | 0.1064 | 0.0943 |
| 5 | 44 | NF-2 | 0.1250 | 0.1204 | .... | 0.1102 | 0.1079 | 0.0971 |
| 6 | 32 | NC-2 | 0.1380 | 0.1326 | 0.1304 | 0.1177 | 0.1150 | 0.0997 |
| 6 | 40 | NF-2 | 0.1380 | 0.1332 | .... | 0.1218 | 0.1194 | 0.1073 |
| 8 | 32 | NC-2 | 0.1640 | 0.1586 | 0.1564 | 0.1437 | 0.1410 | 0.1257 |
| 8 | 36 | NF-2 | 0.1640 | 0.1590 | .... | 0.1460 | 0.1435 | 0.1299 |
| 10 | 24 | NC-2 | 0.1900 | 0.1834 | 0.1808 | 0.1629 | 0.1596 | 0.1389 |
| 10 | 32 | NF-2 | 0.1900 | 0.1846 | .... | 0.1697 | 0.1670 | 0.1517 |
| 12 | 24 | NC-2 | 0.2160 | 0.2094 | 0.2068 | 0.1889 | 0.1856 | 0.1649 |
| 12 | 28 | NF-2 | 0.2160 | 0.2098 | .... | 0.1928 | 0.1897 | 0.1722 |
| 1/4 | 20 | NC-2 | 0.2500 | 0.2428 | 0.2398 | 0.2175 | 0.2139 | 0.1887 |
| 1/4 | 28 | NF-2 | 0.2500 | 0.2438 | .... | 0.2268 | 0.2237 | 0.2062 |
| 5/16 | 18 | NC-2 | 0.3125 | 0.3043 | 0.3011 | 0.2764 | 0.2723 | 0.2443 |
| 5/16 | 24 | NF-2 | 0.3125 | 0.3059 | .... | 0.2854 | 0.2821 | 0.2614 |
| 3/8 | 16 | NC-2 | 0.3750 | 0.3660 | 0.3624 | 0.3344 | 0.3299 | 0.2983 |
| 3/8 | 24 | NF-2 | 0.3750 | 0.3684 | .... | 0.3479 | 0.3446 | 0.3239 |
| 7/16 | 14 | NC-2 | 0.4375 | 0.4277 | 0.4235 | 0.3911 | 0.3862 | 0.3499 |
| 7/16 | 20 | NF-2 | 0.4375 | 0.4303 | .... | 0.4050 | 0.4014 | 0.3762 |
| 1/2 | 13 | NC-2 | 0.5000 | 0.4896 | 0.4852 | 0.4500 | 0.4448 | 0.4056 |
| 1/2 | 20 | NF-2 | 0.5000 | 0.4928 | .... | 0.4675 | 0.4639 | 0.4387 |
| 9/16 | 12 | NC-2 | 0.5625 | 0.5513 | 0.5467 | 0.5084 | 0.5028 | 0.4603 |
| 9/16 | 18 | NF-2 | 0.5625 | 0.5543 | .... | 0.5264 | 0.5223 | 0.4943 |
| 5/8 | 11 | NC-2 | 0.6250 | 0.6132 | 0.6080 | 0.5660 | 0.5601 | 0.5135 |
| 5/8 | 18 | NF-2 | 0.6250 | 0.6168 | .... | 0.5889 | 0.5848 | 0.5568 |
| 3/4 | 10 | NC-2 | 0.7500 | 0.7372 | 0.7316 | 0.6850 | 0.6786 | 0.6273 |
| 3/4 | 16 | NF-2 | 0.7500 | 0.7410 | .... | 0.7094 | 0.7049 | 0.6733 |
| 7/8 | 9 | NC-2 | 0.8750 | 0.8610 | 0.8550 | 0.8028 | 0.7958 | 0.7387 |
| 7/8 | 14 | NF-2 | 0.8750 | 0.8652 | .... | 0.8286 | 0.8237 | 0.7874 |
| 1 | 8 | NC-2 | 1.0000 | 0.9848 | 0.9778 | 0.9188 | 0.9112 | 0.8466 |
| 1 | 14 | NF-2 | 1.0000 | 0.9902 | .... | 0.9536 | 0.9487 | 0.9124 |

[1] For semi-finished and finished screws and bolts, threaded portion only.

[2] For unfinished hot rolled material, threaded portion only.

A Class 2 external thread need not be combined invariably with a Class 2 internal thread, but may be used with a Class 1B, 2B, 3B, or 3 thread where preferable for a specific application.

Table 5 *(Continued)*. Former American Standard Thread Series
Class 2 — External Thread Limits

| Size | Thds. per Inch | Thread Symbol | Major Diameter Max. | Min.¹ | Min.² | Pitch Diam. Max. | Min. | Minor Diam. (See p. 1263) |
|---|---|---|---|---|---|---|---|---|
| 1 1/8 | 7 | NC-2 | 1.1250 | 1.1080 | 1.1002 | 1.0322 | 1.0237 | 0.9497 |
| 1 1/8 | 8 | N-2 | 1.1250 | 1.1098 | 1.1028 | 1.0438 | 1.0359 | 0.9716 |
| 1 1/8 | 12 | NF-2 | 1.1250 | 1.1138 | .... | 1.0709 | 1.0653 | 1.0228 |
| 1 1/4 | 7 | NC-2 | 1.2500 | 1.2330 | 1.2252 | 1.1572 | 1.1487 | 1.0747 |
| 1 1/4 | 8 | N-2 | 1.2500 | 1.2348 | 1.2278 | 1.1688 | 1.1605 | 1.0966 |
| 1 1/4 | 12 | NF-2 | 1.2500 | 1.2388 | .... | 1.1959 | 1.1903 | 1.1478 |
| 1 3/8 | 6 | NC-2 | 1.3750 | 1.3548 | 1.3460 | 1.2667 | 1.2566 | 1.1705 |
| 1 3/8 | 8 | N-2 | 1.3750 | 1.3598 | 1.3528 | 1.2938 | 1.2852 | 1.2216 |
| 1 3/8 | 12 | NF-2 | 1.3750 | 1.3638 | .... | 1.3209 | 1.3153 | 1.2728 |
| 1 1/2 | 6 | NC-2 | 1.5000 | 1.4798 | 1.4710 | 1.3917 | 1.3816 | 1.2955 |
| 1 1/2 | 8 | N-2 | 1.5000 | 1.4848 | 1.4778 | 1.4188 | 1.4098 | 1.3466 |
| 1 1/2 | 12 | NF-2 | 1.5000 | 1.4888 | .... | 1.4459 | 1.4403 | 1.3978 |
| 1 5/8 | 8 | N-2 | 1.6250 | 1.6098 | 1.6028 | 1.5438 | 1.5345 | 1.4716 |
| 1 3/4 | 5 | NC-2 | 1.7500 | 1.7268 | 1.7162 | 1.6201 | 1.6085 | 1.5046 |
| 1 3/4 | 8 | N-2 | 1.7500 | 1.7348 | 1.7278 | 1.6688 | 1.6591 | 1.5966 |
| 1 7/8 | 8 | N-2 | 1.8750 | 1.8598 | 1.8528 | 1.7938 | 1.7838 | 1.7216 |
| 2 | 4½ | NC-2 | 2.0000 | 1.9746 | 1.9632 | 1.8557 | 1.8430 | 1.7274 |
| 2 | 8 | N-2 | 2.0000 | 1.9848 | 1.9188 | 1.9188 | 1.9084 | 1.8466 |
| 2 1/8 | 8 | N-2 | 2.1250 | 2.1098 | 2.1028 | 2.0438 | 2.0331 | 1.9716 |
| 2 1/4 | 4½ | NC-2 | 2.2500 | 2.2246 | 2.2132 | 2.1057 | 2.0930 | 1.9774 |
| 2 1/4 | 8 | N-2 | 2.2500 | 2.2348 | 2.2278 | 2.1688 | 2.1578 | 2.0966 |
| 2 1/2 | 4 | NC-2 | 2.5000 | 2.4720 | 2.4592 | 2.3376 | 2.3236 | 2.1933 |
| 2 1/2 | 8 | N-2 | 2.5000 | 2.4848 | 2.4778 | 2.4188 | 2.4071 | 2.3466 |
| 2 3/4 | 4 | NC-2 | 2.7500 | 2.7220 | 2.7092 | 2.5876 | 2.5736 | 2.4433 |
| 2 3/4 | 8 | N-2 | 2.7500 | 2.7348 | 2.7278 | 2.6688 | 2.6564 | 2.5966 |
| 3 | 4 | NC-2 | 3.0000 | 2.9720 | 2.9592 | 2.8376 | 2.8236 | 2.6933 |
| 3 | 8 | N-2 | 3.0000 | 2.9848 | 2.9778 | 2.9188 | 2.9058 | 2.8466 |
| 3 1/4 | 4 | NC-2 | 3.2500 | 3.2220 | 3.2092 | 3.0876 | 3.0736 | 2.9433 |
| 3 1/4 | 8 | N-2 | 3.2500 | 3.2348 | 3.2278 | 3.1688 | 3.1556 | 3.0966 |
| 3 1/2 | 4 | NC-2 | 3.5000 | 3.4720 | 3.4592 | 3.3376 | 3.3236 | 3.1933 |
| 3 1/2 | 8 | N-2 | 3.5000 | 3.4848 | 3.4778 | 3.4188 | 3.4055 | 3.3466 |
| 3 3/4 | 4 | NC-2 | 3.7500 | 3.7220 | 3.7092 | 3.5876 | 3.5736 | 3.4433 |
| 3 3/4 | 8 | N-2 | 3.7500 | 3.7348 | 3.7278 | 3.6688 | 3.6554 | 3.5966 |
| 4 | 4 | NC-2 | 4.0000 | 3.9720 | 3.9592 | 3.8376 | 3.8236 | 3.6933 |
| 4 | 8 | N-2 | 4.0000 | 3.9848 | 3.9778 | 3.9188 | 3.9053 | 3.8466 |
| 4 1/4 | 8 | N-2 | 4.2500 | 4.2348 | 4.2278 | 4.1688 | 4.1551 | 4.0966 |
| 4 1/2 | 8 | N-2 | 4.5000 | 4.4848 | 4.4778 | 4.4188 | 4.4050 | 4.3466 |
| 4 3/4 | 8 | N-2 | 4.7500 | 4.7348 | 4.7278 | 4.6688 | 4.6549 | 4.5966 |
| 5 | 8 | N-2 | 5.0000 | 4.9848 | 4.9778 | 4.9188 | 4.9048 | 4.8466 |
| 5 1/4 | 8 | N-2 | 5.2500 | 5.2348 | 5.2278 | 5.1688 | 5.1547 | 5.0966 |
| 5 1/2 | 8 | N-2 | 5.5000 | 5.4848 | 5.4778 | 5.4188 | 5.4046 | 5.3466 |
| 5 3/4 | 8 | N-2 | 5.7500 | 5.7348 | 5.7278 | 5.6688 | 5.6545 | 5.5966 |
| 6 | 8 | N-2 | 6.0000 | 5.9848 | 5.9778 | 5.9188 | 5.9044 | 5.8466 |

¹ For semi-finished and finished screws and bolts, threaded portion only.
² For unfinished hot rolled material, threaded portion only.
A Class 2 external thread need not be combined invariably with a Class 2 internal thread but may be used with a Class 1B, 2B, 3B, or 3 thread where preferable for a specific application.

Table 6.  Former American Standard Thread Series
Class 2 — Internal Thread Limits

| No. or Size | Thds. per Inch | Thd. Sym- bol | Minor Diam. | | Pitch Diam. | | Major Diam. Min. (See p. 1263) |
|---|---|---|---|---|---|---|---|
| | | | Min. | Max. | Min. | Max. | |
| 0 | 80 | NF-2 | 0.0465 | 0.0514 | 0.0519 | 0.0536 | 0.0600 |
| 1 | 64 | NC-2 | 0.0561 | 0.0623 | 0.0629 | 0.0648 | 0.0730 |
| 1 | 72 | NF-2 | 0.0580 | 0.0634 | 0.0640 | 0.0658 | 0.0730 |
| 2 | 56 | NC-2 | 0.0667 | 0.0737 | 0.0744 | 0.0764 | 0.0860 |
| 2 | 64 | NF-2 | 0.0691 | 0.0746 | 0.0759 | 0.0778 | 0.0860 |
| 3 | 48 | NC-2 | 0.0764 | 0.0841 | 0.0855 | 0.0877 | 0.0990 |
| 3 | 56 | NF-2 | 0.0797 | 0.0856 | 0.0874 | 0.0894 | 0.0990 |
| 4 | 40 | NC-2 | 0.0849 | 0.0938 | 0.0958 | 0.0982 | 0.1120 |
| 4 | 48 | NF-2 | 0.0894 | 0.0960 | 0.0985 | 0.1007 | 0.1120 |
| 5 | 40 | NC-2 | 0.0979 | 0.1062 | 0.1088 | 0.1112 | 0.1250 |
| 5 | 44 | NF-2 | 0.1004 | 0.1068 | 0.1102 | 0.1125 | 0.1250 |
| 6 | 32 | NC-2 | 0.1042 | 0.1145 | 0.1177 | 0.1204 | 0.1380 |
| 6 | 40 | NF-2 | 0.1109 | 0.1179 | 0.1218 | 0.1242 | 0.1380 |
| 8 | 32 | NC-2 | 0.1302 | 0.1384 | 0.1437 | 0.1464 | 0.1640 |
| 8 | 36 | NF-2 | 0.1339 | 0.1402 | 0.1460 | 0.1485 | 0.1640 |
| 10 | 24 | NC-2 | 0.1449 | 0.1559 | 0.1629 | 0.1662 | 0.1900 |
| 10 | 32 | NF-2 | 0.1562 | 0.1624 | 0.1697 | 0.1724 | 0.1900 |
| 12 | 24 | NC-2 | 0.1709 | 0.1801 | 0.1889 | 0.1922 | 0.2160 |
| 12 | 28 | NF-2 | 0.1773 | 0.1835 | 0.1928 | 0.1959 | 0.2160 |
| 1/4 | 20 | NC-2 | 0.1959 | 0.2060 | 0.2175 | 0.2211 | 0.2500 |
| 1/4 | 28 | NF-2 | 0.2113 | 0.2173 | 0.2268 | 0.2299 | 0.2500 |
| 5/16 | 18 | NC-2 | 0.2524 | 0.2630 | 0.2764 | 0.2805 | 0.3125 |
| 5/16 | 24 | NF-2 | 0.2674 | 0.2739 | 0.2854 | 0.2887 | 0.3125 |
| 3/8 | 16 | NC-2 | 0.3073 | 0.3184 | 0.3344 | 0.3389 | 0.3750 |
| 3/8 | 24 | NF-2 | 0.3299 | 0.3364 | 0.3479 | 0.3512 | 0.3750 |
| 7/16 | 14 | NC-2 | 0.3602 | 0.3721 | 0.3911 | 0.3960 | 0.4375 |
| 7/16 | 20 | NF-2 | 0.3834 | 0.3906 | 0.4050 | 0.4086 | 0.4375 |
| 1/2 | 13 | NC-2 | 0.4167 | 0.4290 | 0.4500 | 0.4552 | 0.5000 |
| 1/2 | 20 | NF-2 | 0.4459 | 0.4531 | 0.4675 | 0.4711 | 0.5000 |
| 9/16 | 12 | NC-2 | 0.4723 | 0.4850 | 0.5084 | 0.5140 | 0.5625 |
| 9/16 | 18 | NF-2 | 0.5024 | 0.5100 | 0.5264 | 0.5305 | 0.5625 |
| 5/8 | 11 | NC-2 | 0.5266 | 0.5397 | 0.5660 | 0.5719 | 0.6250 |
| 5/8 | 18 | NF-2 | 0.5649 | 0.5725 | 0.5889 | 0.5930 | 0.6250 |
| 3/4 | 10 | NC-2 | 0.6417 | 0.6553 | 0.6850 | 0.6914 | 0.7500 |
| 3/4 | 16 | NF-2 | 0.6823 | 0.6903 | 0.7094 | 0.7139 | 0.7500 |
| 7/8 | 9 | NC-2 | 0.7547 | 0.7689 | 0.8028 | 0.8098 | 0.8750 |
| 7/8 | 14 | NF-2 | 0.7977 | 0.8062 | 0.8286 | 0.8335 | 0.8750 |
| 1 | 8 | NC-2 | 0.8647 | 0.8795 | 0.9188 | 0.9264 | 1.0000 |
| 1 | 14 | NF-2 | 0.9227 | 0.9312 | 0.9536 | 0.9585 | 1.0000 |

A Class 2 internal thread need not be combined invariably with a Class 2 external thread but may be used with a Class 1A, 2A, 3A, or 3 thread where preferable for a specific application.

Class 2 threads have no allowance since the maximum external and minimum internal thread dimensions are equal.

Table 6 (*Continued*).  Former American Standard Thread Series
Class 2 — Internal Thread Limits

| Size | Thds. per Inch | Thd. Symbol | Minor Diam. | | Pitch Diam. | | Major Diam. |
|---|---|---|---|---|---|---|---|
| | | | Min. | Max. | Min. | Max. | Min. (See p. 1263) |
| 1 1/8 | 7 | NC-2 | 0.9704 | 0.9858 | 1.0322 | 1.0407 | 1.1250 |
| 1 1/8 | 8 | N-2 | 0.9897 | 1.0045 | 1.0438 | 1.0517 | 1.1250 |
| 1 1/8 | 12 | NF-2 | 1.0348 | 1.0438 | 1.0709 | 1.0765 | 1.1250 |
| 1 1/4 | 7 | NC-2 | 1.0954 | 1.1108 | 1.1572 | 1.1657 | 1.2500 |
| 1 1/4 | 8 | N-2 | 1.1147 | 1.1295 | 1.1688 | 1.1771 | 1.2500 |
| 1 1/4 | 12 | NF-2 | 1.1598 | 1.1688 | 1.1959 | 1.2015 | 1.2500 |
| 1 3/8 | 6 | NC-2 | 1.1946 | 1.2126 | 1.2667 | 1.2768 | 1.3750 |
| 1 3/8 | 8 | N-2 | 1.2397 | 1.2545 | 1.2938 | 1.3024 | 1.3750 |
| 1 3/8 | 12 | NF-2 | 1.2848 | 1.2938 | 1.3209 | 1.3265 | 1.3750 |
| 1 1/2 | 6 | NC-2 | 1.3196 | 1.3376 | 1.3917 | 1.4018 | 1.5000 |
| 1 1/2 | 8 | N-2 | 1.3647 | 1.3795 | 1.4188 | 1.4278 | 1.5000 |
| 1 1/2 | 12 | NF-2 | 1.4098 | 1.4188 | 1.4459 | 1.4515 | 1.5000 |
| 1 5/8 | 8 | N-2 | 1.4897 | 1.5045 | 1.5438 | 1.5531 | 1.6250 |
| 1 3/4 | 5 | NC-2 | 1.5335 | 1.5551 | 1.6201 | 1.6317 | 1.7500 |
| 1 3/4 | 8 | N-2 | 1.6147 | 1.6295 | 1.6688 | 1.6785 | 1.7500 |
| 1 7/8 | 8 | N-2 | 1.7397 | 1.7545 | 1.7938 | 1.8038 | 1.8750 |
| 2 | 4 1/2 | NC-2 | 1.7594 | 1.7835 | 1.8557 | 1.8684 | 2.0000 |
| 2 | 8 | N-2 | 1.8647 | 1.8795 | 1.9188 | 1.9292 | 2.0000 |
| 2 1/8 | 8 | N-2 | 1.9897 | 2.0045 | 2.0438 | 2.0545 | 2.1250 |
| 2 1/4 | 4 1/2 | NC-2 | 2.0094 | 2.0335 | 2.1057 | 2.1184 | 2.2500 |
| 2 1/4 | 8 | N-2 | 2.1147 | 2.1295 | 2.1688 | 2.1798 | 2.2500 |
| 2 1/2 | 4 | NC-2 | 2.2294 | 2.2564 | 2.3376 | 2.3516 | 2.5000 |
| 2 1/2 | 8 | N-2 | 2.3647 | 2.3795 | 2.4188 | 2.4305 | 2.5000 |
| 2 3/4 | 4 | NC-2 | 2.4794 | 2.5064 | 2.5876 | 2.6016 | 2.7500 |
| 2 3/4 | 8 | N-2 | 2.6147 | 2.6295 | 2.6688 | 2.6812 | 2.7500 |
| 3 | 4 | NC-2 | 2.7294 | 2.7564 | 2.8376 | 2.8516 | 3.0000 |
| 3 | 8 | N-2 | 2.8647 | 2.8795 | 2.9188 | 2.9318 | 3.0000 |
| 3 1/4 | 4 | NC-2 | 2.9794 | 3.0064 | 3.0876 | 3.1016 | 3.2500 |
| 3 1/4 | 8 | N-2 | 3.1147 | 3.1295 | 3.1688 | 3.1820 | 3.2500 |
| 3 1/2 | 4 | NC-2 | 3.2294 | 3.2564 | 3.3376 | 3.3516 | 3.5000 |
| 3 1/2 | 8 | N-2 | 3.3647 | 3.3795 | 3.4188 | 3.4321 | 3.5000 |
| 3 3/4 | 4 | NC-2 | 3.4794 | 3.5064 | 3.5876 | 3.6016 | 3.7500 |
| 3 3/4 | 8 | N-2 | 3.6147 | 3.6295 | 3.6688 | 3.6822 | 3.7500 |
| 4 | 4 | NC-2 | 3.7294 | 3.7564 | 3.8376 | 3.8516 | 4.0000 |
| 4 | 8 | N-2 | 3.8647 | 3.8795 | 3.9188 | 3.9323 | 4.0000 |
| 4 1/4 | 8 | N-2 | 4.1147 | 4.1295 | 4.1688 | 4.1825 | 4.2500 |
| 4 1/2 | 8 | N-2 | 4.3647 | 4.3795 | 4.4188 | 4.4326 | 4.5000 |
| 4 3/4 | 8 | N-2 | 4.6147 | 4.6295 | 4.6688 | 4.6827 | 4.7500 |
| 5 | 8 | N-2 | 4.8647 | 4.8795 | 4.9188 | 4.9328 | 5.0000 |
| 5 1/4 | 8 | N-2 | 5.1147 | 5.1295 | 5.1688 | 5.1829 | 5.2500 |
| 5 1/2 | 8 | N-2 | 5.3647 | 5.3795 | 5.4188 | 5.4330 | 5.5000 |
| 5 3/4 | 8 | N-2 | 5.6147 | 5.6295 | 5.6688 | 5.6831 | 5.7500 |
| 6 | 8 | N-2 | 5.8647 | 5.8795 | 5.9188 | 5.9332 | 6.0000 |

A Class 2 internal thread need not be combined invariably with a Class 2 external thread but may be used with a Class 1A, 2A, 3A, or 3 thread where preferable for a specific application.

Table 7. Former American Standard Thread Series
Class 3 — External Thread Limits

| No. or Size | Thds. per Inch | Thd. Symbol | Major Diam. | | Pitch Diam. | | Minor Diam. (See p. 1263) |
|---|---|---|---|---|---|---|---|
| | | | Max. | Min. | Max. | Min. | |
| 0 | 80 | NF-3 | 0.0600 | 0.0566 | 0.0519 | 0.0506 | 0.0447 |
| 1 | 64 | NC-3 | 0.0730 | 0.0692 | 0.0629 | 0.0615 | 0.0538 |
| 1 | 72 | NF-3 | 0.0730 | 0.0694 | 0.0640 | 0.0627 | 0.0560 |
| 2 | 56 | NC-3 | 0.0860 | 0.0820 | 0.0744 | 0.0729 | 0.0641 |
| 2 | 64 | NF-3 | 0.0860 | 0.0822 | 0.0759 | 0.0745 | 0.0668 |
| 3 | 48 | NC-3 | 0.0990 | 0.0946 | 0.0855 | 0.0839 | 0.0734 |
| 3 | 56 | NF-3 | 0.0990 | 0.0950 | 0.0874 | 0.0859 | 0.0771 |
| 4 | 40 | NC-3 | 0.1120 | 0.1072 | 0.0958 | 0.0941 | 0.0813 |
| 4 | 48 | NF-3 | 0.1120 | 0.1076 | 0.0985 | 0.0969 | 0.0864 |
| 5 | 40 | NC-3 | 0.1250 | 0.1202 | 0.1088 | 0.1071 | 0.0943 |
| 5 | 44 | NF-3 | 0.1250 | 0.1204 | 0.1102 | 0.1086 | 0.0971 |
| 6 | 32 | NC-3 | 0.1380 | 0.1326 | 0.1177 | 0.1158 | 0.0997 |
| 6 | 40 | NF-3 | 0.1380 | 0.1332 | 0.1218 | 0.1201 | 0.1073 |
| 8 | 32 | NC-3 | 0.1640 | 0.1586 | 0.1437 | 0.1418 | 0.1257 |
| 8 | 36 | NF-3 | 0.1640 | 0.1590 | 0.1460 | 0.1442 | 0.1299 |
| 10 | 24 | NC-3 | 0.1900 | 0.1834 | 0.1629 | 0.1605 | 0.1389 |
| 10 | 32 | NF-3 | 0.1900 | 0.1846 | 0.1697 | 0.1678 | 0.1517 |
| 12 | 24 | NC-3 | 0.2160 | 0.2094 | 0.1889 | 0.1865 | 0.1649 |
| 12 | 28 | NF-3 | 0.2160 | 0.2098 | 0.1928 | 0.1906 | 0.1722 |
| 1/4 | 20 | NC-3 | 0.2500 | 0.2428 | 0.2175 | 0.2149 | 0.1887 |
| 1/4 | 28 | NF-3 | 0.2500 | 0.2438 | 0.2268 | 0.2246 | 0.2062 |
| 5/16 | 18 | NC-3 | 0.3125 | 0.3043 | 0.2764 | 0.2734 | 0.2443 |
| 5/16 | 24 | NF-3 | 0.3125 | 0.3059 | 0.2854 | 0.2830 | 0.2614 |
| 3/8 | 16 | NC-3 | 0.3750 | 0.3660 | 0.3344 | 0.3312 | 0.2983 |
| 3/8 | 24 | NF-3 | 0.3750 | 0.3684 | 0.3479 | 0.3455 | 0.3239 |
| 7/16 | 14 | NC-3 | 0.4375 | 0.4277 | 0.3911 | 0.3875 | 0.3499 |
| 7/16 | 20 | NF-3 | 0.4375 | 0.4303 | 0.4050 | 0.4024 | 0.3762 |
| 1/2 | 13 | NC-3 | 0.5000 | 0.4896 | 0.4500 | 0.4463 | 0.4056 |
| 1/2 | 20 | NF-3 | 0.5000 | 0.4928 | 0.4675 | 0.4649 | 0.4387 |
| 9/16 | 12 | NC-3 | 0.5625 | 0.5513 | 0.5084 | 0.5044 | 0.4603 |
| 9/16 | 18 | NF-3 | 0.5625 | 0.5543 | 0.5264 | 0.5234 | 0.4943 |
| 5/8 | 11 | NC-3 | 0.6250 | 0.6132 | 0.5660 | 0.5618 | 0.5135 |
| 5/8 | 18 | NF-3 | 0.6250 | 0.6168 | 0.5889 | 0.5859 | 0.5568 |
| 3/4 | 10 | NC-3 | 0.7500 | 0.7372 | 0.6850 | 0.6805 | 0.6273 |
| 3/4 | 16 | NF-3 | 0.7500 | 0.7410 | 0.7094 | 0.7062 | 0.6733 |
| 7/8 | 9 | NC-3 | 0.8750 | 0.8610 | 0.8028 | 0.7979 | 0.7387 |
| 7/8 | 14 | NF-3 | 0.8750 | 0.8652 | 0.8286 | 0.8250 | 0.7874 |
| 1 | 8 | NC-3 | 1.0000 | 0.9848 | 0.9188 | 0.9134 | 0.8466 |
| 1 | 14 | NF-3 | 1.0000 | 0.9902 | 0.9536 | 0.9500 | 0.9124 |

A Class 3 external thread need not be combined invariably with a Class 3 internal thread but may be used with a Class 1B, 2B, 3B, or 2 thread where preferable for a specific application.

Class 3 threads have no allowance; the maximum external and minimum internal thread dimensions are basic.

Table 7 (*Continued*).   Former American Standard Thread Series
Class 3 — External Thread Limits

| Size | Thds. per Inch | Thd. Symbol | Major Diam. | | Pitch Diam. | | Minor Diam. (See p. 1263) |
|---|---|---|---|---|---|---|---|
| | | | Max. | Min. | Max. | Min. | |
| 1 1/8 | 7 | NC-3 | 1.1250 | 1.1080 | 1.0322 | 1.0263 | 0.9497 |
| 1 1/8 | 8 | N-3 | 1.1250 | 1.1098 | 1.0438 | 1.0383 | 0.9716 |
| 1 1/8 | 12 | NF-3 | 1.1250 | 1.1138 | 1.0709 | 1.0669 | 1.0228 |
| 1 1/4 | 7 | NC-3 | 1.2500 | 1.2330 | 1.1572 | 1.1513 | 1.0747 |
| 1 1/4 | 8 | N-3 | 1.2500 | 1.2348 | 1.1688 | 1.1630 | 1.0966 |
| 1 1/4 | 12 | NF-3 | 1.2500 | 1.2388 | 1.1959 | 1.1919 | 1.1478 |
| 1 3/8 | 6 | NC-3 | 1.3750 | 1.3548 | 1.2667 | 1.2596 | 1.1705 |
| 1 3/8 | 8 | N-3 | 1.3750 | 1.3598 | 1.2938 | 1.2877 | 1.2216 |
| 1 3/8 | 12 | NF-3 | 1.3750 | 1.3638 | 1.3209 | 1.3169 | 1.2728 |
| 1 1/2 | 6 | NC-3 | 1.5000 | 1.4798 | 1.3917 | 1.3846 | 1.2955 |
| 1 1/2 | 8 | N-3 | 1.5000 | 1.4848 | 1.4188 | 1.4125 | 1.3466 |
| 1 1/2 | 12 | NF-3 | 1.5000 | 1.4888 | 1.4459 | 1.4419 | 1.3978 |
| 1 5/8 | 8 | N-3 | 1.6250 | 1.6098 | 1.5438 | 1.5373 | 1.4716 |
| 1 3/4 | 5 | NC-3 | 1.7500 | 1.7268 | 1.6201 | 1.6119 | 1.5046 |
| 1 3/4 | 8 | N-3 | 1.7500 | 1.7348 | 1.6688 | 1.6620 | 1.5966 |
| 1 7/8 | 8 | N-3 | 1.8750 | 1.8598 | 1.7938 | 1.7868 | 1.7216 |
| 2 | 4½ | NC-3 | 2.0000 | 1.9746 | 1.8557 | 1.8468 | 1.7274 |
| 2 | 8 | N-3 | 2.0000 | 1.9848 | 1.9188 | 1.9115 | 1.8466 |
| 2 1/8 | 8 | N-3 | 2.1250 | 2.1098 | 2.0438 | 2.0363 | 1.9716 |
| 2 1/4 | 4½ | NC-3 | 2.2500 | 2.2246 | 2.1057 | 2.0968 | 1.9774 |
| 2 1/4 | 8 | N-3 | 2.2500 | 2.2348 | 2.1688 | 2.1611 | 2.0966 |
| 2 1/2 | 4 | NC-3 | 2.5000 | 2.4720 | 2.3376 | 2.3279 | 2.1933 |
| 2 1/2 | 8 | N-3 | 2.5000 | 2.4848 | 2.4188 | 2.4106 | 2.3466 |
| 2 3/4 | 4 | NC-3 | 2.7500 | 2.7220 | 2.5876 | 2.5779 | 2.4433 |
| 2 3/4 | 8 | N-3 | 2.7500 | 2.7348 | 2.6688 | 2.6601 | 2.5966 |
| 3 | 4 | NC-3 | 3.0000 | 2.9720 | 2.8376 | 2.8279 | 2.6933 |
| 3 | 8 | N-3 | 3.0000 | 2.9848 | 2.9188 | 2.9096 | 2.8466 |
| 3 1/4 | 4 | NC-3 | 3.2500 | 3.2220 | 3.0876 | 3.0779 | 2.9433 |
| 3 1/4 | 8 | N-3 | 3.2500 | 3.2348 | 3.1688 | 3.1595 | 3.0966 |
| 3 1/2 | 4 | NC-3 | 3.5000 | 3.4720 | 3.3376 | 3.3279 | 3.1933 |
| 3 1/2 | 8 | N-3 | 3.5000 | 3.4848 | 3.4188 | 3.4095 | 3.3466 |
| 3 3/4 | 4 | NC-3 | 3.7500 | 3.7220 | 3.5876 | 3.5779 | 3.4433 |
| 3 3/4 | 8 | N-3 | 3.7500 | 3.7348 | 3.6688 | 3.6594 | 3.5966 |
| 4 | 4 | NC-3 | 4.0000 | 3.9720 | 3.8376 | 3.8279 | 3.6933 |
| 4 | 8 | N-3 | 4.0000 | 3.9848 | 3.9188 | 3.9093 | 3.8466 |
| 4 1/4 | 8 | N-3 | 4.2500 | 4.2348 | 4.1688 | 4.1592 | 4.0966 |
| 4 1/2 | 8 | N-3 | 4.5000 | 4.4848 | 4.4188 | 4.4091 | 4.3466 |
| 4 3/4 | 8 | N-3 | 4.7500 | 4.7348 | 4.6688 | 4.6590 | 4.5966 |
| 5 | 8 | N-3 | 5.0000 | 4.9848 | 4.9188 | 4.9089 | 4.8466 |
| 5 1/4 | 8 | N-3 | 5.2500 | 5.2348 | 5.1688 | 5.1589 | 5.0966 |
| 5 1/2 | 8 | N-3 | 5.5000 | 5.4848 | 5.4188 | 5.4088 | 5.3466 |
| 5 3/4 | 8 | N-3 | 5.7500 | 5.7348 | 5.6688 | 5.6587 | 5.5966 |
| 6 | 8 | N-3 | 6.0000 | 5.9848 | 5.9188 | 5.9086 | 5.8466 |

A Class 3 external thread need not be combined invariably with a Class 3 internal thread but may be used with a Class 1B, 2B, 3B, or 2 thread where preferable for a specific application.

Table 8.   Former American Standard Thread Series
Class 3 — Internal Thread Limits

| No. or Size | Thds. per Inch | Thd. Symbol | Minor Diam. | | Pitch Diam. | | Major Diam. Min. (See p.1263) |
|---|---|---|---|---|---|---|---|
| | | | Min. | Max. | Min. | Max. | |
| 0 | 80 | NF-3 | 0.0465 | 0.0514 | 0.0519 | 0.0532 | 0.0600 |
| 1 | 64 | NC-3 | 0.0561 | 0.0623 | 0.0629 | 0.0643 | 0.0730 |
| 1 | 72 | NF-3 | 0.0580 | 0.0634 | 0.0640 | 0.0653 | 0.0730 |
| 2 | 56 | NC-3 | 0.0667 | 0.0737 | 0.0744 | 0.0759 | 0.0860 |
| 2 | 64 | NF-3 | 0.0691 | 0.0746 | 0.0759 | 0.0773 | 0.0860 |
| 3 | 48 | NC-3 | 0.0764 | 0.0841 | 0.0855 | 0.0871 | 0.0990 |
| 3 | 56 | NF-3 | 0.0797 | 0.0856 | 0.0874 | 0.0889 | 0.0990 |
| 4 | 40 | NC-3 | 0.0849 | 0.0938 | 0.0958 | 0.0975 | 0.1120 |
| 4 | 48 | NF-3 | 0.0894 | 0.0960 | 0.0985 | 0.1001 | 0.1120 |
| 5 | 40 | NC-3 | 0.0979 | 0.1062 | 0.1088 | 0.1105 | 0.1250 |
| 5 | 44 | NF-3 | 0.1004 | 0.1068 | 0.1102 | 0.1118 | 0.1250 |
| 6 | 32 | NC-3 | 0.1042 | 0.1145 | 0.1177 | 0.1196 | 0.1380 |
| 6 | 40 | NF-3 | 0.1109 | 0.1179 | 0.1218 | 0.1235 | 0.1380 |
| 8 | 32 | NC-3 | 0.1302 | 0.1384 | 0.1437 | 0.1456 | 0.1640 |
| 8 | 36 | NF-3 | 0.1339 | 0.1402 | 0.1460 | 0.1478 | 0.1640 |
| 10 | 24 | NC-3 | 0.1449 | 0.1559 | 0.1629 | 0.1653 | 0.1900 |
| 10 | 32 | NF-3 | 0.1562 | 0.1624 | 0.1697 | 0.1716 | 0.1900 |
| 12 | 24 | NC-3 | 0.1709 | 0.1801 | 0.1889 | 0.1913 | 0.2160 |
| 12 | 28 | NF-3 | 0.1773 | 0.1835 | 0.1928 | 0.1950 | 0.2160 |
| 1/4 | 20 | NC-3 | 0.1959 | 0.2060 | 0.2175 | 0.2201 | 0.2500 |
| 1/4 | 28 | NF-3 | 0.2113 | 0.2173 | 0.2268 | 0.2290 | 0.2500 |
| 5/16 | 18 | NC-3 | 0.2524 | 0.2630 | 0.2764 | 0.2794 | 0.3125 |
| 5/16 | 24 | NF-3 | 0.2674 | 0.2739 | 0.2854 | 0.2878 | 0.3125 |
| 3/8 | 16 | NC-3 | 0.3073 | 0.3184 | 0.3344 | 0.3376 | 0.3750 |
| 3/8 | 24 | NF-3 | 0.3299 | 0.3364 | 0.3479 | 0.3503 | 0.3750 |
| 7/16 | 14 | NC-3 | 0.3602 | 0.3721 | 0.3911 | 0.3947 | 0.4375 |
| 7/16 | 20 | NF-3 | 0.3834 | 0.3906 | 0.4050 | 0.4076 | 0.4375 |
| 1/2 | 13 | NC-3 | 0.4167 | 0.4290 | 0.4500 | 0.4537 | 0.5000 |
| 1/2 | 20 | NF-3 | 0.4459 | 0.4531 | 0.4675 | 0.4701 | 0.5000 |
| 9/16 | 12 | NC-3 | 0.4723 | 0.4850 | 0.5084 | 0.5124 | 0.5625 |
| 9/16 | 18 | NF-3 | 0.5024 | 0.5100 | 0.5264 | 0.5294 | 0.5625 |
| 5/8 | 11 | NC-3 | 0.5266 | 0.5397 | 0.5660 | 0.5702 | 0.6250 |
| 5/8 | 18 | NF-3 | 0.5649 | 0.5725 | 0.5889 | 0.5919 | 0.6250 |
| 3/4 | 10 | NC-3 | 0.6417 | 0.6553 | 0.6850 | 0.6895 | 0.7500 |
| 3/4 | 16 | NF-3 | 0.6823 | 0.6903 | 0.7094 | 0.7126 | 0.7500 |
| 7/8 | 9 | NC-3 | 0.7547 | 0.7689 | 0.8028 | 0.8077 | 0.8750 |
| 7/8 | 14 | NF-3 | 0.7977 | 0.8062 | 0.8286 | 0.8322 | 0.8750 |
| 1 | 8 | NC-3 | 0.8647 | 0.8795 | 0.9188 | 0.9242 | 1.0000 |
| 1 | 14 | NF-3 | 0.9227 | 0.9312 | 0.9536 | 0.9572 | 1.0000 |

A Class 3 internal thread need not be combined invariably with a Class 3 external thread but may be used with a Class 1A, 2A, 3A, or 2 thread where preferable for a specific application.

Class 3 threads have no allowance since the maximum external and minimum internal thread dimensions are equal.

Table 8 (*Continued*).  Former American Standard Thread Series
Class 3 — Internal Thread Limits

| Size | Thds. per Inch | Thd. Symbol | Minor Diam. | | Pitch Diam. | | Major Diam. Min. (See p. 1263) |
|---|---|---|---|---|---|---|---|
| | | | Min. | Max. | Min. | Max. | |
| 1 1/8 | 7 | NC-3 | 0.9704 | 0.9858 | 1.0322 | 1.0381 | 1.1250 |
| 1 1/8 | 8 | N-3 | 0.9897 | 1.0045 | 1.0438 | 1.0493 | 1.1250 |
| 1 1/8 | 12 | NF-3 | 1.0348 | 1.0438 | 1.0709 | 1.0749 | 1.1250 |
| 1 1/4 | 7 | NC-3 | 1.0954 | 1.1108 | 1.1572 | 1.1631 | 1.2500 |
| 1 1/4 | 8 | N-3 | 1.1147 | 1.1295 | 1.1688 | 1.1746 | 1.2500 |
| 1 1/4 | 12 | NF-3 | 1.1598 | 1.1688 | 1.1959 | 1.1999 | 1.2500 |
| 1 3/8 | 6 | NC-3 | 1.1946 | 1.2126 | 1.2667 | 1.2738 | 1.3750 |
| 1 3/8 | 8 | N-3 | 1.2397 | 1.2545 | 1.2938 | 1.2999 | 1.3750 |
| 1 3/8 | 12 | NF-3 | 1.2848 | 1.2938 | 1.3209 | 1.3249 | 1.3750 |
| 1 1/2 | 6 | NC-3 | 1.3196 | 1.3376 | 1.3917 | 1.3988 | 1.5000 |
| 1 1/2 | 8 | N-3 | 1.3647 | 1.3795 | 1.4188 | 1.4251 | 1.5000 |
| 1 1/2 | 12 | NF-3 | 1.4098 | 1.4188 | 1.4459 | 1.4499 | 1.5000 |
| 1 5/8 | 8 | N-3 | 1.4897 | 1.5045 | 1.5438 | 1.5503 | 1.6250 |
| 1 3/4 | 5 | NC-3 | 1.5335 | 1.5551 | 1.6201 | 1.6283 | 1.7500 |
| 1 3/4 | 8 | N-3 | 1.6147 | 1.6295 | 1.6688 | 1.6756 | 1.7500 |
| 1 7/8 | 8 | N-3 | 1.7397 | 1.7545 | 1.7938 | 1.8008 | 1.8750 |
| 2 | 4½ | NC-3 | 1.7594 | 1.7835 | 1.8557 | 1.8646 | 2.0000 |
| 2 | 8 | N-3 | 1.8647 | 1.8795 | 1.9188 | 1.9261 | 2.0000 |
| 2 1/8 | 8 | N-3 | 1.9897 | 2.0045 | 2.0438 | 2.0513 | 2.1250 |
| 2 1/4 | 4½ | NC-3 | 2.0094 | 2.0335 | 2.1057 | 2.1146 | 2.2500 |
| 2 1/4 | 8 | N-3 | 2.1147 | 2.1295 | 2.1688 | 2.1765 | 2.2500 |
| 2 1/2 | 4 | NC-3 | 2.2294 | 2.2564 | 2.3376 | 2.3473 | 2.5000 |
| 2 1/2 | 8 | N-3 | 2.3647 | 2.3795 | 2.4188 | 2.4270 | 2.5000 |
| 2 3/4 | 4 | NC-3 | 2.4794 | 2.5064 | 2.5876 | 2.5973 | 2.7500 |
| 2 3/4 | 8 | N-3 | 2.6147 | 2.6295 | 2.6688 | 2.6775 | 2.7500 |
| 3 | 4 | NC-3 | 2.7294 | 2.7564 | 2.8376 | 2.8473 | 3.0000 |
| 3 | 8 | N-3 | 2.8647 | 2.8795 | 2.9188 | 2.9280 | 3.0000 |
| 3 1/4 | 4 | NC-3 | 2.9794 | 3.0064 | 3.0876 | 3.0973 | 3.2500 |
| 3 1/4 | 8 | N-3 | 3.1147 | 3.1295 | 3.1688 | 3.1781 | 3.2500 |
| 3 1/2 | 4 | NC-3 | 3.2294 | 3.2564 | 3.3376 | 3.3473 | 3.5000 |
| 3 1/2 | 8 | N-3 | 3.3647 | 3.3795 | 3.4188 | 3.4281 | 3.5000 |
| 3 3/4 | 4 | NC-3 | 3.4794 | 3.5064 | 3.5876 | 3.5973 | 3.7500 |
| 3 3/4 | 8 | N-3 | 3.6147 | 3.6295 | 3.6688 | 3.6782 | 3.7500 |
| 4 | 4 | NC-3 | 3.7294 | 3.7564 | 3.8376 | 3.8473 | 4.0000 |
| 4 | 8 | N-3 | 3.8647 | 3.8795 | 3.9188 | 3.9283 | 4.0000 |
| 4 1/4 | 8 | N-3 | 4.1147 | 4.1295 | 4.1688 | 4.1784 | 4.2500 |
| 4 1/2 | 8 | N-3 | 4.3647 | 4.3795 | 4.4188 | 4.4285 | 4.5000 |
| 4 3/4 | 8 | N-3 | 4.6147 | 4.6295 | 4.6688 | 4.6786 | 4.7500 |
| 5 | 8 | N-3 | 4.8647 | 4.8795 | 4.9188 | 4.9287 | 5.0000 |
| 5 1/4 | 8 | N-3 | 5.1147 | 5.1295 | 5.1688 | 5.1787 | 5.2500 |
| 5 1/2 | 8 | N-3 | 5.3647 | 5.3795 | 5.4188 | 5.4288 | 5.5000 |
| 5 3/4 | 8 | N-3 | 5.6147 | 5.6295 | 5.6688 | 5.6789 | 5.7500 |
| 6 | 8 | N-3 | 5.8647 | 5.8795 | 5.9188 | 5.9290 | 6.0000 |

A Class 3 internal thread need not invariably be combined with a Class 3 external thread but may be used with a Class 1A, 2A, 3A, or 2 thread where preferable for a specific application.

**American Standard for Unified Miniature Screw Threads.** — This American Standard (B1.10-1958) introduces a new thread series to be known as Unified Miniature Screw Threads and intended for general purpose fastening screws and similar uses in watches, instruments, and miniature mechanisms. Use of this series is recommended on all new products in place of the many improvised and unsystematized sizes now in existence which have never achieved broad acceptance nor recognition by standardization bodies. The series covers a diameter range from 0.30 to 1.40 millimeters (0.0118 to 0.0551 inch) and thus supplements the Unified and American thread series which begins at 0.060 inch (number 0 of the machine screw series). It comprises a total of fourteen sizes which, together with their respective pitches, are those endorsed by the American-British-Canadian Conference of April 1955 as the basis for a Unified standard among the inch-using countries, and coincide with the corresponding range of sizes in ISO (International Organization for Standardization) Recommendation No. 68. Additionally, it utilizes thread forms which are compatible in all significant respects with both the Unified and ISO basic thread profiles. Thus, threads in this series are interchangeable with the corresponding sizes in both the American-British-Canadian and ISO standardization programs.

*Basic Form of Thread:* The basic profile by which the design forms of the threads covered by this standard are governed is shown in Table 1. The thread angle is 60 degrees and except for basic height and depth of engagement which are $0.52p$, instead of $0.54127p$, the basic profile for this thread standard is identical with the Unified and American basic thread form. The selection of 0.52 as the exact value of the coefficient for the height of this basic form is based on practical manufacturing considerations and a plan evolved to simplify calculations and achieve more precise agreement between the metric and inch dimensional tables.

Products made to this standard will be interchangeable with products made to other standards which allow a maximum depth of engagement (or combined addendum height) of $0.54127p$. The resulting difference is negligible (only 0.00025 inch for the coarsest pitch) and is completely offset by practical considerations in tapping, since internal thread heights exceeding $0.52p$ are avoided in these (Unified Miniature) small thread sizes in order to escape excessive tap breakage.

*Design Forms of Threads:* The design (maximum material) forms of the external and internal threads are shown in Table 2. These forms are derived from the basic profile shown in Table 1 by the application of clearances for the crests of the addenda at the roots of the mating dedendum forms. Basic and design form dimensions are given in Table 3.

*Nominal Sizes:* The thread sizes comprising this series and their respective pitches are shown in the first two columns of Table 5. The fourteen sizes shown in Table 5 have been systematically distributed to provide a uniformly proportioned selection over the entire range. They are separated alternately into two categories: The sizes shown in bold type are selections made in the interest of simplification and are those to which it is recommended that usage be confined wherever the circumstances of design permit. Where these sizes do not meet requirements the intermediate sizes shown in light type are available.

*Limits of Size:* Formulas used to determine limits of size are given in Table 4; the limits of size are given in Table 5. The diagram on page 1315 illustrates the limits of size and Table 6 gives values for the minimum flat at the root of the external thread shown on the diagram.

*Classes of Threads:* The standard establishes one class of thread with zero allowance on all diameters. When coatings of a measurable thickness are required, they should be included within the maximum material limits of the threads since these limits apply to both coated and uncoated threads.

*Hole Sizes for Tapping:* Suggested hole sizes are given in the Tapping Section.

**Table 1. Unified Miniature Screw Threads — Basic Thread Form**

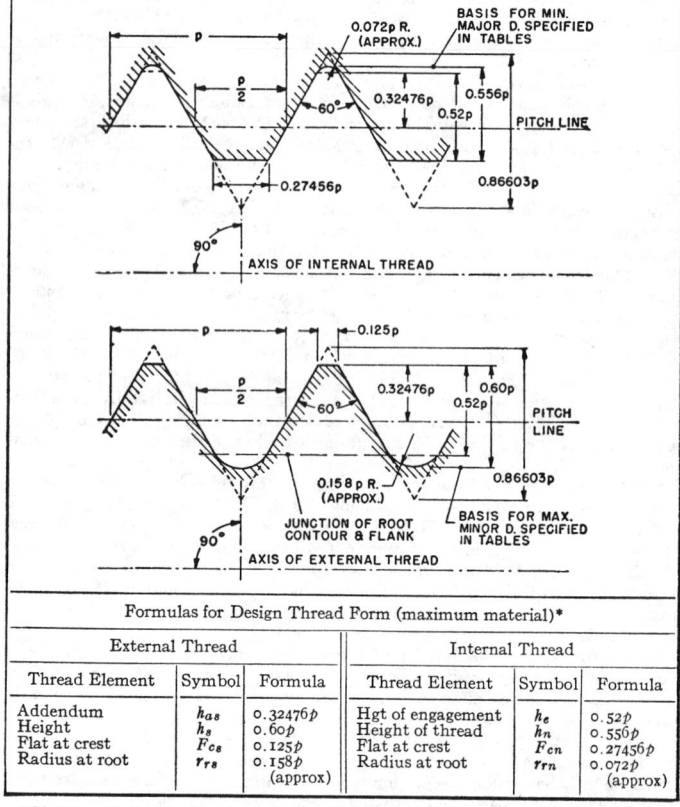

| Formulas for Basic Thread Form* | | |
|---|---|---|
| Thread Element | Symbol | Formula |
| Angle of thread | $2\alpha$ | 60° |
| Half angle of thread | $\alpha$ | 30° |
| Pitch of thread | $p$ | |
| No. of threads per inch | $n$ | $25.4/p$ |
| Height of sharp V thread | $H$ | $0.86603p$ |
| Addendum of basic thread | $h_{ab}$ | $0.32476p$ |
| Height of basic thread | $h_b$ | $0.52p$ |

* Metric units (millimeters) are used in all formulas.

**Table 2. Unified Miniature Screw Threads — Design Thread Form**

| Formulas for Design Thread Form (maximum material)* | | | | | |
|---|---|---|---|---|---|
| External Thread | | | Internal Thread | | |
| Thread Element | Symbol | Formula | Thread Element | Symbol | Formula |
| Addendum | $h_{as}$ | $0.32476p$ | Hgt of engagement | $h_e$ | $0.52p$ |
| Height | $h_s$ | $0.60p$ | Height of thread | $h_n$ | $0.556p$ |
| Flat at crest | $F_{cs}$ | $0.125p$ | Flat at crest | $F_{cn}$ | $0.27456p$ |
| Radius at root | $r_{rs}$ | $0.158p$ (approx) | Radius at root | $r_{rn}$ | $0.072p$ (approx) |

* Metric units (millimeters) are used in all formulas.

**Table 3. Unified Miniature Screw Threads — Basic and Design Form Dimensions**

| | | Basic Thread Form | | | | External Thread Design Form | | | Internal Thread Design Form | |
|---|---|---|---|---|---|---|---|---|---|---|
| Threads per inch $n^*$ | Pitch $p$ | Height of Sharp V $H = 0.86603p$ | Height $h_b = 0.52p$ | Addendum $h_{ab} = h_{aa} = 0.32476p$ | Height $h_s = 0.60p$ | Flat at Crest $F_{cs} = 0.125p$ | Radius at Root $r_{rs} = 0.158p$ | Height $h_n = 0.556p$ | Flat at Crest $F_{cn} = 0.27456p$ | Radius at Root $r_{rn} = 0.072p$ |
| | | | | Millimeter Dimensions | | | | | | |
| ... | .080 | .0693 | .0416 | .0260 | .048 | .0100 | .0126 | 0445 | 0220 | 0058 |
| ... | .090 | .0779 | .0468 | .0292 | .054 | .0112 | .0142 | .0500 | .0247 | .0065 |
| ... | .100 | .0866 | .0520 | .0325 | 060 | .0125 | .0158 | .0556 | .0275 | .0072 |
| ... | .125 | .1083 | .0650 | .0406 | .075 | .0156 | .0198 | .0695 | .0343 | .0090 |
| ... | .150 | .1299 | .0780 | .0487 | .090 | .0188 | .0237 | .0834 | .0412 | .0108 |
| ... | .175 | .1516 | .0910 | .0568 | .105 | .0219 | .0277 | .0973 | 0480 | .0126 |
| ... | .200 | .1732 | .1040 | .0650 | .120 | .0250 | .0316 | .1112 | .0549 | .0144 |
| ... | .225 | .1949 | .1170 | .0731 | .135 | .0281 | .0356 | .1251 | .0618 | .0162 |
| ... | .250 | .2165 | .1300 | .0812 | .150 | .0312 | .0395 | .1390 | .0686 | .0180 |
| ... | .300 | .2598 | .1560 | .0974 | .180 | .0375 | .0474 | .1668 | .0824 | .0216 |
| | | | | Inch Dimensions | | | | | | |
| 317 1/2 | .003150 | .00273 | .00164 | .00102 | .00189 | .00039 | .00050 | .00175 | .00086 | .00023 |
| 282 2/9 | .003543 | .00307 | .00184 | .00115 | .00213 | .00044 | .00056 | .00197 | .00097 | .00026 |
| 254 | .003937 | .00341 | .00205 | .00128 | .00236 | .00049 | .00062 | .00219 | .00108 | .00028 |
| 203 1/5 | .004921 | .00426 | .00256 | .00160 | .00295 | .00062 | .00078 | .00274 | .00135 | .00035 |
| 169 1/3 | .005906 | .00511 | .00307 | .00192 | .00354 | .00074 | .00093 | .00328 | .00162 | .00043 |
| 145 1/7 | .006890 | .00597 | .00358 | .00224 | .00413 | .00086 | .00109 | .00383 | .00189 | .00050 |
| 127 | .007874 | .00682 | .00409 | .00256 | .00472 | .00098 | .00124 | .00438 | .00216 | .00057 |
| 112 8/9 | .008858 | .00767 | .00461 | .00288 | .00531 | .00111 | .00140 | .00493 | .00243 | .00064 |
| 101 3/5 | .009843 | .00852 | .00512 | .00320 | .00591 | .00123 | .00156 | .00547 | .00270 | .00071 |
| 84 2/3 | .011811 | .01023 | .00614 | .00384 | .00709 | .00148 | .00187 | .00657 | .00324 | .00085 |

\* In Tables 5 and 6 these values are shown rounded to the nearest whole number.

**Table 4. Unified Miniature Screw Threads — Formulas for Basic and Design Dimensions and Tolerances**

| FORMULAS FOR BASIC DIMENSIONS | |
|---|---|
| $D$ = Basic Major Diameter and Nominal Size in millimeters; $p$ = Pitch in millimeters; $E$ = Basic Pitch Diameter in millimeters = $D - 0.64952p$; and $K$ = Basic Minor Diameter in millimeters = $D - 1.04p$ | |
| FORMULAS FOR DESIGN DIMENSIONS (MAXIMUM MATERIAL) | |
| External Thread | Internal Thread |
| $D_s$ = Major Diameter = $D$ <br> $E_s$ = Pitch Diameter = $E$ <br> $K_s$ = Minor Diameter = $D - 1.20p$ | $D_n$ = Major Diameter = $D + 0.072p$ <br> $E_n$ = Pitch Diameter = $E$ <br> $K_n$ = Minor Diameter = $K$ |
| FORMULAS FOR TOLERANCES ON DESIGN DIMENSIONS‡ | |
| External Thread (−) | Internal Thread (+) |
| Major Diameter Tol., $0.12p + 0.006$ <br> Pitch Diameter Tol., $0.08p + 0.008$ <br> \* Minor Diameter Tol., $0.16p + 0.008$ | †Major Diameter Tol., $0.168p + 0.008$ <br> Pitch Diameter Tol., $0.08p + 0.008$ <br> Minor Diameter Tol., $0.32p + 0.012$ |

Metric units (millimeters) apply in all formulas. Inch tolerances are not derived by direct conversion of the metric values. They are the differences between the rounded off limits of size in inch units.

\* This tolerance establishes the minimum limit of the minor diameter of the external thread. In practice, this limit is applied to the threading tool and only gaged on the product in confirming new tools. Values for this tolerance are, therefore, not given in Table 5.

† This tolerance establishes the maximum limit of the major diameter of the internal thread. In practice, this limit is applied to the threading tool (tap) and not gaged on the product. Values for this tolerance are, therefore, not given in Table 5.

‡ These tolerances are based on lengths of engagement of ⅔D to 1½D.

## Table 5. Unified Miniature Screw Threads — Limits Of Size and Tolerances

### Metric (mm) values

| Size Designation[a] | Pitch mm | External Major Diam. Max* | External Major Diam. Min | External Pitch Diam. Max* | External Pitch Diam. Min | External Minor Diam. Max[b] | External Minor Diam. Min[c] | Internal Minor Diam. Min* | Internal Minor Diam. Max | Internal Pitch Diam. Min* | Internal Pitch Diam. Max | Internal Major Diam. Min[d] | Internal Major Diam. Max[e] | Lead Angle deg | Lead Angle min | Sectional Area at Minor Diam. at D − 1.28p sq mm |
|---|---|---|---|---|---|---|---|---|---|---|---|---|---|---|---|---|
| **0.30 UNM** | 0.080 | 0.300 | 0.284 | 0.248 | 0.234 | 0.204 | 0.183 | 0.217 | 0.254 | 0.248 | 0.262 | 0.306 | 0.327 | 5 | 52 | 0.0307 |
| **0.35 UNM** | 0.090 | 0.350 | 0.333 | 0.292 | 0.277 | 0.242 | 0.220 | 0.256 | 0.297 | 0.292 | 0.307 | 0.356 | 0.380 | 5 | 37 | 0.0433 |
| **0.40 UNM** | 0.100 | 0.400 | 0.382 | 0.335 | 0.319 | 0.280 | 0.256 | 0.296 | 0.340 | 0.335 | 0.351 | 0.407 | 0.432 | 5 | 26 | 0.0581 |
| 0.45 UNM | 0.100 | 0.450 | 0.432 | 0.385 | 0.369 | 0.330 | 0.306 | 0.346 | 0.390 | 0.385 | 0.401 | 0.457 | 0.482 | 4 | 44 | 0.0814 |
| **0.50 UNM** | 0.125 | 0.500 | 0.479 | 0.419 | 0.401 | 0.350 | 0.322 | 0.370 | 0.422 | 0.419 | 0.437 | 0.509 | 0.538 | 4 | 26 | 0.0908 |
| 0.55 UNM | 0.125 | 0.550 | 0.529 | 0.469 | 0.451 | 0.400 | 0.372 | 0.420 | 0.472 | 0.469 | 0.487 | 0.559 | 0.588 | 4 | 51 | 0.1195 |
| **0.60 UNM** | 0.150 | 0.600 | 0.576 | 0.503 | 0.483 | 0.420 | 0.388 | 0.444 | 0.504 | 0.503 | 0.523 | 0.611 | 0.644 | 5 | 26 | 0.1307 |
| **0.70 UNM** | 0.175 | 0.700 | 0.673 | 0.586 | 0.564 | 0.490 | 0.454 | 0.518 | 0.586 | 0.586 | 0.608 | 0.713 | 0.750 | 5 | 26 | 0.1780 |
| **0.80 UNM** | 0.200 | 0.800 | 0.770 | 0.670 | 0.646 | 0.560 | 0.520 | 0.592 | 0.668 | 0.670 | 0.694 | 0.814 | 0.856 | 5 | 26 | 0.232 |
| 0.90 UNM | 0.225 | 0.900 | 0.867 | 0.754 | 0.728 | 0.630 | 0.586 | 0.666 | 0.750 | 0.754 | 0.780 | 0.916 | 0.962 | 5 | 26 | 0.294 |
| **1.00 UNM** | 0.250 | 1.000 | 0.964 | 0.838 | 0.810 | 0.700 | 0.652 | 0.740 | 0.832 | 0.838 | 0.866 | 1.018 | 1.068 | 5 | 26 | 0.363 |
| 1.10 UNM | 0.250 | 1.100 | 1.064 | 0.938 | 0.910 | 0.800 | 0.752 | 0.840 | 0.932 | 0.938 | 0.966 | 1.118 | 1.168 | 4 | 51 | 0.478 |
| **1.20 UNM** | 0.250 | 1.200 | 1.164 | 1.038 | 1.010 | 0.900 | 0.852 | 0.940 | 1.032 | 1.038 | 1.066 | 1.218 | 1.268 | 4 | 23 | 0.608 |
| **1.40 UNM** | 0.300 | 1.400 | 1.358 | 1.205 | 1.173 | 1.040 | 0.984 | 1.088 | 1.196 | 1.205 | 1.237 | 1.422 | 1.480 | 4 | 32 | 0.811 |

### Inch values

| Size Designation[a] | Pitch Thds. to in. | External Major Diam. Max* | External Major Diam. Min | External Pitch Diam. Max* | External Pitch Diam. Min | External Minor Diam. Max[b] | External Minor Diam. Min[c] | Internal Minor Diam. Min* | Internal Minor Diam. Max | Internal Pitch Diam. Min* | Internal Pitch Diam. Max | Internal Major Diam. Min[d] | Internal Major Diam. Max[e] | Lead Angle deg | Lead Angle min | Sectional Area at Minor Diam. at D − 1.28p sq in |
|---|---|---|---|---|---|---|---|---|---|---|---|---|---|---|---|---|
| **0.30 UNM** | 318 | 0.0118 | 0.0112 | 0.0098 | 0.0092 | 0.0080 | 0.0072 | 0.0085 | 0.0100 | 0.0098 | 0.0104 | 0.0120 | 0.0129 | 5 | 52 | 0.0000475 |
| **0.35 UNM** | 282 | 0.0138 | 0.0131 | 0.0115 | 0.0109 | 0.0095 | 0.0086 | 0.0101 | 0.0117 | 0.0115 | 0.0121 | 0.0140 | 0.0149 | 5 | 37 | 0.0000671 |
| **0.40 UNM** | 254 | 0.0157 | 0.0150 | 0.0132 | 0.0126 | 0.0110 | 0.0101 | 0.0117 | 0.0134 | 0.0132 | 0.0138 | 0.0160 | 0.0170 | 5 | 26 | 0.0000901 |
| 0.45 UNM | 254 | 0.0177 | 0.0170 | 0.0152 | 0.0145 | 0.0130 | 0.0120 | 0.0136 | 0.0154 | 0.0152 | 0.0158 | 0.0180 | 0.0190 | 4 | 44 | 0.0001262 |
| **0.50 UNM** | 203 | 0.0197 | 0.0189 | 0.0165 | 0.0158 | 0.0138 | 0.0127 | 0.0146 | 0.0166 | 0.0165 | 0.0172 | 0.0200 | 0.0212 | 4 | 26 | 0.0001407 |
| 0.55 UNM | 203 | 0.0217 | 0.0208 | 0.0185 | 0.0177 | 0.0157 | 0.0146 | 0.0165 | 0.0186 | 0.0185 | 0.0192 | 0.0220 | 0.0231 | 4 | 51 | 0.0001852 |
| **0.60 UNM** | 169 | 0.0236 | 0.0227 | 0.0198 | 0.0190 | 0.0165 | 0.0153 | 0.0175 | 0.0198 | 0.0198 | 0.0206 | 0.0240 | 0.0254 | 5 | 26 | 0.000203 |
| **0.70 UNM** | 145 | 0.0276 | 0.0265 | 0.0231 | 0.0222 | 0.0193 | 0.0179 | 0.0204 | 0.0231 | 0.0231 | 0.0240 | 0.0281 | 0.0295 | 5 | 26 | 0.000276 |
| **0.80 UNM** | 127 | 0.0315 | 0.0303 | 0.0264 | 0.0254 | 0.0220 | 0.0205 | 0.0233 | 0.0263 | 0.0264 | 0.0273 | 0.0321 | 0.0337 | 5 | 26 | 0.000360 |
| 0.90 UNM | 113 | 0.0354 | 0.0341 | 0.0297 | 0.0287 | 0.0248 | 0.0231 | 0.0262 | 0.0295 | 0.0297 | 0.0307 | 0.0361 | 0.0379 | 5 | 26 | 0.000456 |
| **1.00 UNM** | 102 | 0.0394 | 0.0380 | 0.0330 | 0.0319 | 0.0276 | 0.0257 | 0.0291 | 0.0327 | 0.0330 | 0.0341 | 0.0401 | 0.0420 | 5 | 26 | 0.000563 |
| 1.10 UNM | 102 | 0.0433 | 0.0419 | 0.0369 | 0.0358 | 0.0315 | 0.0296 | 0.0331 | 0.0367 | 0.0369 | 0.0380 | 0.0440 | 0.0460 | 5 | 26 | 0.000741 |
| **1.20 UNM** | 102 | 0.0472 | 0.0458 | 0.0409 | 0.0397 | 0.0354 | 0.0335 | 0.0370 | 0.0406 | 0.0409 | 0.0420 | 0.0480 | 0.0499 | 4 | 23 | 0.000943 |
| **1.40 UNM** | 85 | 0.0551 | 0.0535 | 0.0474 | 0.0462 | 0.0409 | 0.0387 | 0.0428 | 0.0471 | 0.0474 | 0.0487 | 0.0560 | 0.0583 | 4 | 32 | 0.001257 |

* This is also the basic dimension.   [a] Sizes shown in bold type are preferred.   [b] This limit, in conjunction with root form shown in Table 2, is advocated for use when optical projection methods of gaging are employed. For mechanical gaging the minimum minor diameter of the internal thread is applied.   [c] This limit is provided for reference only. In practice, the form of the threading tool is relied upon for this limit.   [d] This limit is provided for reference only, and is not gaged. For gaging, the maximum major diameter of the external thread is applied.

Table 6.   Unified Miniature Screw Threads — Minimum
Root Flats for External Threads

| Pitch | No. of Threads | Thread Height for Min. Flat at Root 0.64p | | Minimum Flat at Root $F_{rs} = 0.136p$ | |
|---|---|---|---|---|---|
| mm | Per Inch | mm | Inch | mm | Inch |
| 0.080 | 318 | 0.0512 | 0.00202 | 0.0109 | 0.00043 |
| 0.090 | 282 | 0.0576 | 0.00227 | 0.0122 | 0.00048 |
| 0.100 | 254 | 0.0640 | 0.00252 | 0.0136 | 0.00054 |
| 0.125 | 203 | 0.0800 | 0.00315 | 0.0170 | 0.00067 |
| 0.150 | 169 | 0.0960 | 0.00378 | 0.0204 | 0.00080 |
| 0.175 | 145 | 0.1120 | 0.00441 | 0.0238 | 0.00094 |
| 0.200 | 127 | 0.1280 | 0.00504 | 0.0272 | 0.00107 |
| 0.225 | 113 | 0.1440 | 0.00567 | 0.0306 | 0.00120 |
| 0.250 | 102 | 0.1600 | 0.00630 | 0.0340 | 0.00134 |
| 0.300 | 85 | 0.1920 | 0.00756 | 0.0408 | 0.00161 |

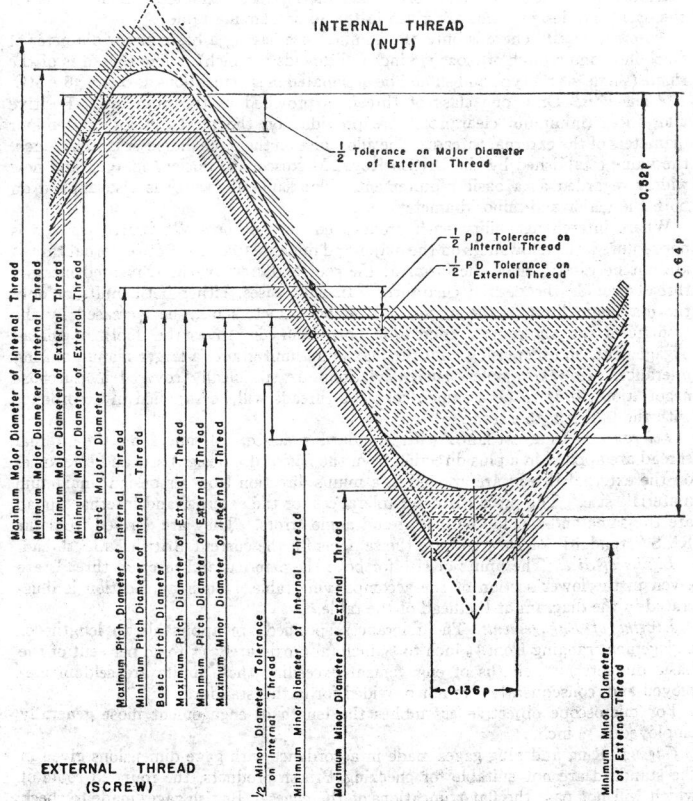

Limits of Size Showing Tolerances and Crest Clearances for UNM Threads

**ANSI Microscope Objective Threads** (ANSI B1.11-1958, R1978). — This standard covers the screw thread used for mounting the objective assembly to the body or lens turret of microscopes. Though utilized principally for microscope objective mountings, this screw thread is recommended also for other optical assemblies of microscopes and associated apparatus, such as photomicrographic equipment. The thread is based upon, and intended to be interchangeable with, the thread introduced and adopted many years ago by the Royal Microscopical Society of Great Britain, generally known as the "RMS thread" and now almost universally accepted as the basic standard for microscope objective mountings (see page 1374).

*Basic and Design Thread Forms:* The basic thread possesses the British Standard Whitworth form, having an included angle of 55 degrees and rounded crests and roots. This same full Whitworth form is also employed as the design, or maximum material, form by the British in the RMS thread. In the American standard, however, the design thread form established in ASA B1.6-1944, American War Standard for Truncated Whitworth Threads, has been adopted. Dimensions for the basic and design forms are given in the accompanying table.

*Nominal Sizes:* There is only one nominal size having a basic major diameter of 0.800 inch and a pitch of 0.027778 inch (36 threads per inch). The thread is of the single (single-start) type and should be designated in specifications as 0.800-36 AMO.

*Allowances:* Only one class of thread is provided in the Standard. Positive allowances (minimum clearances) are provided on the pitch, major, and minor diameters of the external thread. The allowance on the pitch diameter is 0.0018 inch, the value established by the British Royal Microscopical Society in 1924 and now widely regarded as a basic requirement. The same allowance is also applied on both the major and minor diameters.

Where interchangeability with product having full-form Whitworth threads is not required, the allowances on the major and minor diameters of the external thread are not necessary, since the forms at the root and crest of the truncated internal thread provide the desired clearances. In such cases, either both limits or only the maximum limit of the major and minor diameters may be increased by the amount of the allowance. Increasing both limits improves the depth of thread engagement, and increasing only the maximum limit grants a larger manufacturing tolerance. However, unless such deviations are specifically covered in purchase negotiations, it is to be assumed that the threads will be supplied in accordance with the tables in this standard.

*Tolerances:* In accordance with standard practice, tolerances on the internal thread are applied in a plus direction from the basic (also design) size and tolerances on the external thread are applied in a minus direction from its design (maximum material) size. The pitch diameter tolerances for the external and internal thread are the same and include both lead and angle errors. They are derived from the RMS "standard" of 1924 and are the same as for the current British RMS thread.

*Limits of Size:* The limits of size for both the external and internal threads are given in the lower section of the accompanying table. Their application is illustrated in the diagram at the head of the table.

*Lengths of Engagement:* The tolerances specified are applicable to lengths of engagement ranging from ⅛ inch to ⅜ inch (approximately 15 to 50 per cent of the basic diameter). Lengths of engagement exceeding these limits are seldom employed and, consequently, are not provided for in this standard.

For microscope objective assemblies the length of engagement most generally employed is ⅛ inch.

*Gaging:* Ring and plug gages made in accordance with gage dimensions given in the standard are not suitable for checking British products, the rounded roots of which will not pass the flat truncations of the gages. British gages made to check the full Whitworth form will, however, accept American products.

**American National Standard Microscope Objective Thread** (ANSI B1.11-1958, R1978)

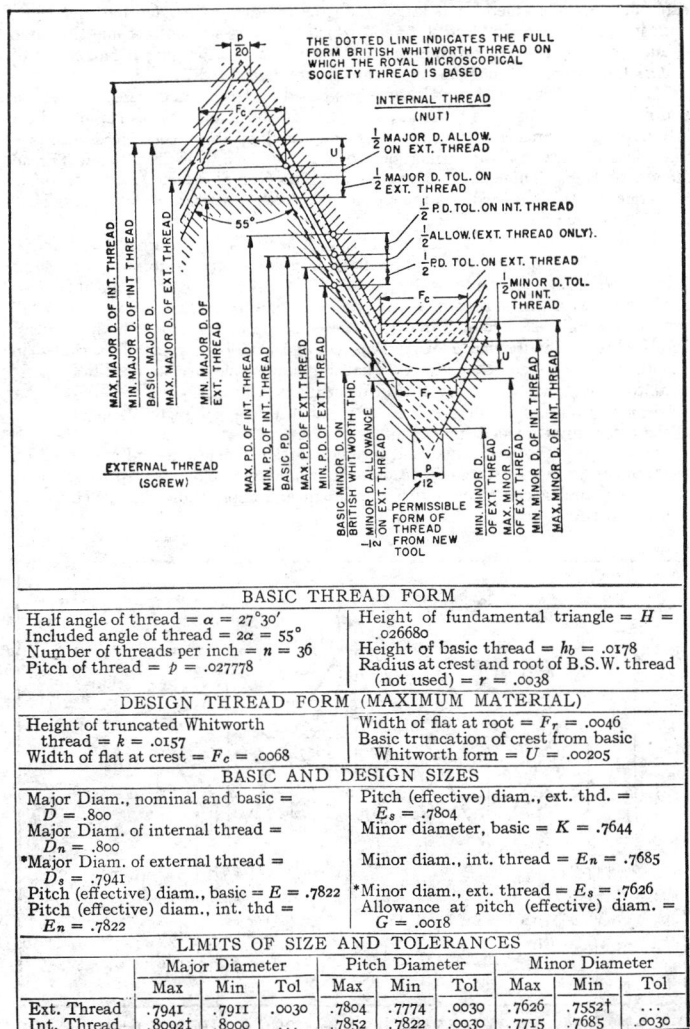

## BASIC THREAD FORM

| | |
|---|---|
| Half angle of thread = $\alpha = 27°30'$ | Height of fundamental triangle = $H = .026680$ |
| Included angle of thread = $2\alpha = 55°$ | Height of basic thread = $h_b = .0178$ |
| Number of threads per inch = $n = 36$ | Radius at crest and root of B.S.W. thread (not used) = $r = .0038$ |
| Pitch of thread = $p = .027778$ | |

## DESIGN THREAD FORM (MAXIMUM MATERIAL)

| | |
|---|---|
| Height of truncated Whitworth thread = $k = .0157$ | Width of flat at root = $F_r = .0046$ |
| Width of flat at crest = $F_c = .0068$ | Basic truncation of crest from basic Whitworth form = $U = .00205$ |

## BASIC AND DESIGN SIZES

| | |
|---|---|
| Major Diam., nominal and basic = $D = .800$ | Pitch (effective) diam., ext. thd. = $E_s = .7804$ |
| Major Diam. of internal thread = $D_n = .800$ | Minor diameter, basic = $K = .7644$ |
| *Major Diam. of external thread = $D_s = .7941$ | Minor diam., int. thread = $E_n = .7685$ |
| Pitch (effective) diam., basic = $E = .7822$ | *Minor diam., ext. thread = $E_s = .7626$ |
| Pitch (effective) diam., int. thd = $E_n = .7822$ | Allowance at pitch (effective) diam. = $G = .0018$ |

## LIMITS OF SIZE AND TOLERANCES

| | Major Diameter | | | Pitch Diameter | | | Minor Diameter | | |
|---|---|---|---|---|---|---|---|---|---|
| | Max | Min | Tol | Max | Min | Tol | Max | Min | Tol |
| Ext. Thread | .7941 | .7911 | .0030 | .7804 | .7774 | .0030 | .7626 | .7552† | ... |
| Int. Thread | .8092‡ | .8000 | ... | .7852 | .7822 | .0030 | .7715 | .7685 | .0030 |

* An allowance equal to that on the pitch diameter is provided for additional clearance.
† Extreme minimum minor diameter produced by a new threading tool having a minimum flat of $p/12$ (= 0.0023 inch). This minimum diameter is not controlled by gages.
‡ Extreme maximum major diameter produced by a new threading tool having a minimum flat of $p/20$ (= 0.0014 inch). This maximum diameter is not controlled by gages.

**Interference-Fit Threads.** — Interference-fit threads are threads in which the externally threaded member is larger than the internally threaded member when both members are in the free state and which, when assembled, become the same size and develop a holding torque through elastic compression, plastic movement of material, or both. By custom, these threads are designated Class 5.

The data in Tables 1, 2, and 3, which is based on ten years of research, testing and field study, represents the first attempt to establish an American standard for interference fit threads that would overcome the difficulties experienced with previous interference fit recommendations such as are given in Federal Screw Thread Handbook H28. These data were adopted as American Standard ASA B1.12-1963. Subsequently the standard was revised and issued as American National Standard ANSI B1.12-1972.

The data in Tables 1, 2, and 3 provide dimensions for external and internal interference-fit (Class 5) threads of modified American National form in the Coarse Thread series, sizes ¼ inch to 1½ inches. It is intended that interference-fit threads conforming with this standard will provide adequate torque conditions which fall within the limits shown in Table 3. These torque limits are from the Federal Screw Thread Handbook H28-1957 Part III and have been generally accepted in service for more than twenty years. The minimum torques are intended to be sufficient to insure that externally threaded members will not loosen in service; the maximum torques establish a ceiling below which seizing, galling or torsional failure of the externally threaded components is unlikely.

Tables 1 and 2 give external and internal thread dimensions and are based on engagement lengths, external thread lengths, and tapping hole depths specified in Table 3 and in compliance with the design and application data given in the following paragraphs.

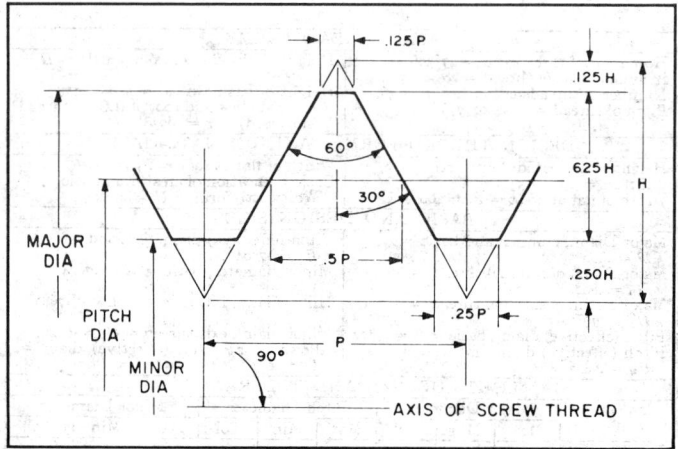

Basic Profile of American National Standard Class 5 Interference Fit Thread

**Design and Application Data for Class 5 Interference-Fit Threads.** — Following are conditions of usage and inspection on which satisfactory application of products made to dimensions in Tables 1, 2, and 3 are based.

*Thread Designations:* The following thread designations provide a means of distinguishing the American Standard Class 5 Threads from the tentative Class 5 and alternate Class 5 threads, specified in Handbook H28. It also distinguishes between external and internal American Standard Class 5 Threads.

Class 5 External Threads are designated as follows:

NC5 HF — For driving in hard ferrous material of hardness over 160 BHN.

NC5 CSF — For driving in copper alloy and soft ferrous material of 160 BHN or less.

NC5 ONF — For driving in other non-ferrous material (non-ferrous materials other than copper alloys), any hardness.

Class 5 Internal Threads are designated as follows:

NC5 IF — Entire ferrous material range.

NC5 INF — Entire non-ferrous material range.

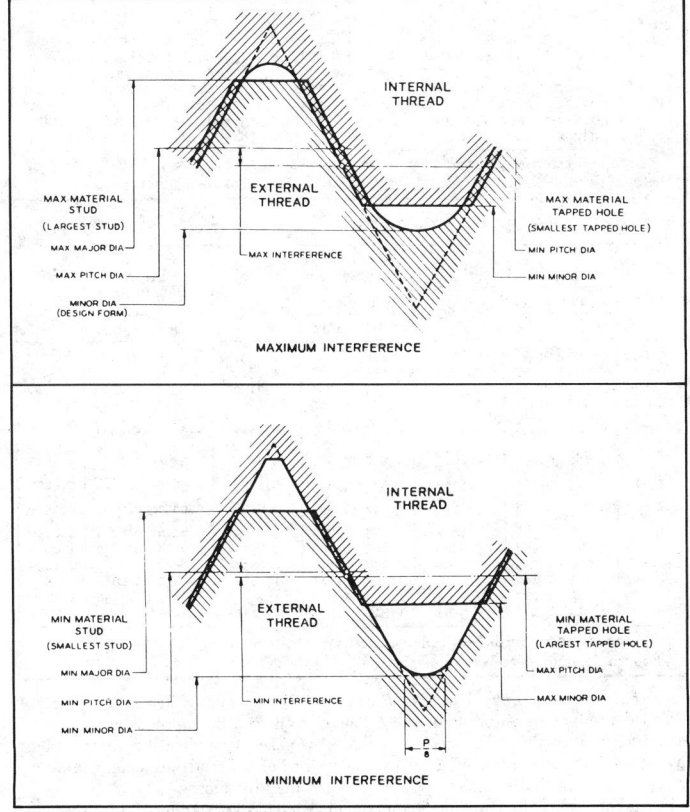

Maximum and Minimum Material Limits for Class 5 Interference-Fit Threads

**Table 1. External Thread Dimensions for Class 5 Interference-Fit Threads***

| Size | Major Diameter, Inches | | | | | | | | Pitch Diameter, Inches | | Minor Diameter, Inches |
|------|------|------|------|------|------|------|------|------|------|------|------|
| | NC5HF for driving in ferrous material with hardness greater than 160 BHN $L_e$=1¼ Diam. | | NC5CSF for driving in brass and ferrous material with hardness equal to or less than 160 BHN $L_e$=1¼ Diam. | | NC5ONF for driving in nonferrous except brass (any hardness) $L_e$=2½ Diam. | | | | | | |
| | Max | Min | Max | Min | Max | Min | Max | Min | Max |
| ¼ –20 | 0.2470 | 0.2408 | 0.2470 | 0.2408 | 0.2470 | 0.2408 | 0.2230 | 0.2204 | 0.1932 |
| ⁵⁄₁₆–18 | 0.3080 | 0.3020 | 0.3090 | 0.3030 | 0.3090 | 0.3030 | 0.2829 | 0.2799 | 0.2508 |
| ⅜ –16 | 0.3690 | 0.3626 | 0.3710 | 0.3646 | 0.3710 | 0.3646 | 0.3414 | 0.3382 | 0.3053 |
| ⁷⁄₁₆–14 | 0.4305 | 0.4233 | 0.4330 | 0.4258 | 0.4330 | 0.4258 | 0.3991 | 0.3955 | 0.3579 |
| ½ –13 | 0.4920 | 0.4846 | 0.4950 | 0.4876 | 0.4950 | 0.4876 | 0.4584 | 0.4547 | 0.4140 |
| ⁹⁄₁₆–12 | 0.5540 | 0.5460 | 0.5580 | 0.5495 | 0.5580 | 0.5495 | 0.5176 | 0.5136 | 0.4695 |
| ⅝ –11 | 0.6140 | 0.6056 | 0.6195 | 0.6111 | 0.6195 | 0.6111 | 0.5758 | 0.5716 | 0.5233 |
| ¾ –10 | 0.7360 | 0.7270 | 0.7440 | 0.7350 | 0.7440 | 0.7350 | 0.6955 | 0.6910 | 0.6378 |
| ⅞ – 9 | 0.8600 | 0.8502 | 0.8685 | 0.8587 | 0.8685 | 0.8587 | 0.8144 | 0.8095 | 0.7503 |
| 1 – 8 | 0.9835 | 0.9727 | 0.9935 | 0.9827 | 0.9935 | 0.9827 | 0.9316 | 0.9262 | 0.8594 |
| 1⅛– 7 | 1.1070 | 1.0952 | 1.1180 | 1.1062 | 1.1180 | 1.1062 | 1.0465 | 1.0406 | 0.9640 |
| 1¼– 7 | 1.232 | 1.220 | 1.2430 | 1.2312 | 1.2430 | 1.2312 | 1.1715 | 1.1656 | 1.0890 |
| 1⅜– 6 | 1.356 | 1.341 | 1.3680 | 1.3538 | 1.3680 | 1.3538 | 1.2839 | 1.2768 | 1.1877 |
| 1½– 6 | 1.481 | 1.467 | 1.4930 | 1.4788 | 1.4930 | 1.4788 | 1.4089 | 1.4018 | 1.3127 |

* Based on externally threaded members being steel ASTM A-325 (SAE Grade 5) or better. For rolled, cut or ground threads. $L_e$ = length of engagement.

**Table 2. Internal Thread Dimensions for Class 5 Interference-Fit Threads**

| Size | NC5IF Ferrous Material | | | NC5INF Nonferrous Material | | | Pitch Diameter | | Major Diam. |
|------|------|------|------|------|------|------|------|------|------|
| | Minor Diam.* | | Tap Drill | Minor Diam.* | | Tap Drill | | | |
| | Min | Max | | Min | Max | | Min | Max | Min |
| ¼ –20 | 0.196 | 0.206 | 0.2031 | 0.196 | 0.206 | 0.2031 | 0.2175 | 0.2201 | 0.2500 |
| ⁵⁄₁₆–18 | 0.252 | 0.263 | 0.2610 | 0.252 | 0.263 | 0.2610 | 0.2764 | 0.2794 | 0.3125 |
| ⅜ –16 | 0.307 | 0.318 | 0.3160 | 0.307 | 0.318 | 0.3160 | 0.3344 | 0.3376 | 0.3750 |
| ⁷⁄₁₆–14 | 0.374 | 0.381 | 0.3750 | 0.360 | 0.372 | 0.3680 | 0.3911 | 0.3947 | 0.4375 |
| ½ –13 | 0.431 | 0.440 | 0.4331 | 0.417 | 0.429 | 0.4219 | 0.4500 | 0.4537 | 0.5000 |
| ⁹⁄₁₆–12 | 0.488 | 0.497 | 0.4921 | 0.472 | 0.485 | 0.4844 | 0.5084 | 0.5124 | 0.5625 |
| ⅝ –11 | 0.544 | 0.554 | 0.5469 | 0.527 | 0.540 | 0.5313 | 0.5660 | 0.5702 | 0.6250 |
| ¾ –10 | 0.667 | 0.678 | 0.6719 | 0.642 | 0.655 | 0.6496 | 0.6850 | 0.6895 | 0.7500 |
| ⅞ – 9 | 0.777 | 0.789 | 0.7812 | 0.755 | 0.769 | 0.7656 | 0.8028 | 0.8077 | 0.8750 |
| 1 – 8 | 0.890 | 0.904 | 0.8906 | 0.865 | 0.880 | 0.8750 | 0.9188 | 0.9242 | 1.0000 |
| 1⅛– 7 | 1.000 | 1.015 | 1.0000 | 0.970 | 0.986 | 0.9844 | 1.0322 | 1.0381 | 1.1250 |
| 1¼– 7 | 1.125 | 1.140 | 1.1250 | 1.095 | 1.111 | 1.1094 | 1.1572 | 1.1631 | 1.2500 |
| 1⅜– 6 | 1.229 | 1.247 | 1.2344 | 1.195 | 1.213 | 1.2031 | 1.2667 | 1.2738 | 1.3750 |
| 1½– 6 | 1.354 | 1.372 | 1.3594 | 1.320 | 1.338 | 1.3281 | 1.3917 | 1.3988 | 1.5000 |

All dimensions are in inches, unless otherwise specified.
* Fourth decimal place is 0 for all sizes.

*Inspection of Externally Threaded Products:* The controlling element for Class 5 threaded products is pitch diameter. This element can be checked by an optical comparator, a thread micrometer, thread snap gages, or indicating thread gages having anvils that are not affected by lead or angle. If studs are zinc, cadmium, or copper plated, limits of size apply to the unplated product.

Points of externally threaded components should be chamfered or otherwise reduced to a diameter below the minimum minor diameter of the thread. The

**Table 3. Torques, Interferences, and Engagement Lengths for Class 5 Interference-Fit Threads**

| Size | Interference on Pitch Diameter | | Engagement Lengths, External Thread Lengths and Tapped Hole Depths* | | | | | | Approx. Torque at Full Engagement of 1-¼D in Ferrous Material | |
|---|---|---|---|---|---|---|---|---|---|---|
| | | | In Brass and Ferrous | | | In Nonferrous Except Brass | | | | |
| | Max | Min | $L_e$ | $T_s$ | $T_h$ min | $L_e$ | $T_s$ | $T_h$ min | Max, Ft-lbs | Min, Ft-lbs |
| ¼ -20 | .0055 | .0003 | 0.312 | 0.375 + .125−0 | 0.375 | 0.625 | 0.688 + .125−0 | 1 1/16 | 12 | 3 |
| 5/16-18 | .0065 | .0005 | 0.391 | 0.469 + .139−0 | 0.469 | 0.781 | 0.859 + .139−0 | 55/64 | 19 | 6 |
| ⅜ -16 | .0070 | .0006 | 0.469 | 0.562 + .156−0 | 0.562 | 0.938 | 1.031 + .156−0 | 1 1/32 | 35 | 10 |
| 7/16-14 | .0080 | .0008 | 0.547 | 0.656 + .179−0 | 0.656 | 1.094 | 1.203 + .179−0 | 1 13/64 | 45 | 15 |
| ½ -13 | .0084 | .0010 | 0.625 | 0.750 + .192−0 | 0.750 | 1.250 | 1.375 + .192−0 | 1 ⅜ | 75 | 20 |
| 9/16-12 | .0092 | .0012 | 0.703 | 0.844 + .208−0 | 0.844 | 1.406 | 1.547 + .208−0 | 1 35/64 | 90 | 30 |
| ⅝ -11 | .0098 | .0014 | 0.781 | 0.938 + .227−0 | 0.938 | 1.562 | 1.719 + .227−0 | 1 23/32 | 120 | 37 |
| ¾ -10 | .0105 | .0015 | 0.938 | 1.125 + .250−0 | 1.125 | 1.875 | 2.062 + .250−0 | 2 1/16 | 190 | 60 |
| ⅞ - 9 | .0116 | .0018 | 1.094 | 1.312 + .278−0 | 1.312 | 2.188 | 2.406 + .278−0 | 2 13/32 | 250 | 90 |
| 1 - 8 | .0128 | .0020 | 1.250 | 1.500 + .312−0 | 1.500 | 2.500 | 2.750 + .312−0 | 2 ¾ | 400 | 125 |
| 1⅛ - 7 | .0143 | .0025 | 1.406 | 1.688 + .357−0 | 1.688 | 2.812 | 3.094 + .357−0 | 3 3/32 | 470 | 155 |
| 1¼ - 7 | .0143 | .0025 | 1.562 | 1.875 + .357−0 | 1.875 | 3.125 | 3.438 + .357−0 | 3 7/16 | 580 | 210 |
| 1⅜ - 6 | .0172 | .0030 | 1.719 | 2.062 + .419−0 | 2.062 | 3.438 | 3.781 + .419−0 | 3 25/32 | 705 | 250 |
| 1½ - 6 | .0172 | .0030 | 1.875 | 2.250 + .419−0 | 2.250 | 3.750 | 4.125 + .419−0 | 4 ⅛ | 840 | 325 |

All dimensions are in inches. * $L_e$ = Length of engagement. $T_s$ = External thread length. $T_h$ = Depth of full form thread in hole.

threads should be free from excessive nicks, burrs, chips, grit or other extraneous material before driving.

*Materials for Externally Threaded Products:* The length of engagement, depth of thread engagement and pitch diameter in Tables 1, 2, and 3 are designed to produce adequate torque conditions when heat-treated medium-carbon steel studs, ASTM A-325 (SAE Grade 5) or better, are used. In many applications, case-carburized and unheat-treated medium-carbon steel products of SAE Grade 4, are satisfactory. SAE Grades 1, 2, 8 and 8.1 may be desirable under certain conditions. This standard is not intended to cover the use of products made of stainless steel, silicon bronze, brass or similar materials. When such materials are used, the tabulated dimensions will probably require adjustment based on pilot experimental work with the materials involved.

*Holes:* GO plain plug and GO thread plug gages should be inserted to full depth in order to detect the effect of excessive drill or tap wear at the bottom of the hole. NOT–GO thread plug gages should enter not more than 1½ threads. Holes must be clean from grit, chips, oil or other extraneous material prior to gaging and before driving studs or screws. Holes should be countersunk to a diameter greater than the major diameter in order to facilitate starting of the externally threaded product and to prevent raising a lip around the hole after driving.

*Lead and Angle Variations:* Angle and lead errors are not normally objectionable since they contribute to interference and this is the purpose of the Class 5 thread. Experience may dictate the need for imposing some limits under certain conditions.

*Lubrication:* For driving in ferrous material, a good lubricant sealer should be used, particularly in the hole. A non-carbonizing type of lubricant (such as a rubber-in-water dispersion) is suggested. The lubricant must be applied to the hole and it may be applied to the male member. In applying it to the hole, care must be taken so that an excess amount of lubricant will not cause the male member to be impeded by hydraulic pressure in a blind hole. Where sealing is involved, the lubricant selected should be insoluble in the medium being sealed.

For driving, in nonferrous material, lubrication may not be needed. Recent British research recommends the use of medium gear oil for driving in aluminum. American research has observed that the minor diameter of lubricated tapped holes in non-ferrous materials may tend to close in, that is, be reduced in driving; whereas with an unlubricated hole the minor diameter may tend to open up in some cases.

*Driving Speed:* This standard makes no recommendation for driving speed. Some opinion has been advanced that careful selection and control of driving speed is desirable to obtain optimum results with various combinations of surface hardness and roughness. Experience with threads made to this standard may indicate what limitations should be placed on driving speeds.

*Relation of Driving Torque to Length of Engagement:* Torques increase directly as the length of engagement. American research indicates that this increase is proportionately more rapid as size increases.

*Breakloose Torques after Reapplication:* The standard does not establish recommended reapplication breakloose torques in cases where repeated usage is involved.

*Assembly Torques for Reapplication:* The standard does not establish assembly torques for reapplication.

**Bottoming and Shouldering of Studs.** — Among the conclusions drawn from stud research is the fact that studs should be driven to a predetermined depth. "Bottoming" or "Shouldering" should be avoided. "Bottoming," which is engagement of the threads of the stud with the imperfect threads at the bottom of a shallow drilled and tapped hole causes the stud to stop suddenly, thus inviting failure in torsional shear. "Shouldering," which is the practice of driving the stud until the thread runout engages with the top threads of the hole, creates radial compressive stresses and upward bulging of the material at the top of the hole. This results in erratic variations in free stud length after driving.

**Extension of the Standard.** — By using the new principles upon which this standard is based, thread sizes may be extended downward. However, adequate data are not now available to permit setting a standard. American research indicates that on smaller sizes the main reliance for producing adequate breakloose torque should be placed on pitch diameter interference and not on increasing the length of engagement.

Although there is some current usage of interference fits on large size threads, adequate data is not now developed to permit setting a standard on larger sizes.

Use of the coarse thread series is urged unless requirements for strength of the male members make a finer pitch necessary. No research data is now available to enable the setting of a trial standard for fine thread products. Indications are, however, that the product of the ratio:

$$\frac{\text{Class 2A UNF PD tolerance}}{\text{Class 2A UNC PD tolerance}}$$

times the coarse thread dimensions given in Tables 1 and 2 will probably work for:

    a. Major diameter tolerance, external threads
    b. Pitch diameter tolerance, external threads
    c. Pitch diameter tolerance, internal threads
    d. Minor diameter tolerance, internal threads
    e. Minimum interference

Similarly, the principles observed in setting the pitch diameter and major diameter limits on the fine series Class 5 external threads above may be followed in deriving the pitch diameter and minor diameter of the fine series internal threads.

**American National Standard Acme Screw Threads.** — This American National Standard ANSI B1.5-1977 is a revision of American Standard ASA B1.5-1973 and provides for two general applications of Acme threads, namely, General Purpose and Centralizing.

The limits and tolerances in this standard relate to single-start Acme threads, and may be used, if considered suitable, for multi-start Acme threads, which provide fast relative traversing motion when this is necessary. For information on additional allowances for multi-start Acme threads, see later section on page 1324.

**General Purpose Acme Threads.** — Three classes of General Purpose threads, 2G, 3G, and 4G, are provided in the standard, each having clearance on all diameters for free movement, and may be used in assemblies with the internal thread rigidly fixed and movement of the external thread in a direction perpendicular to its axis limited by its bearing or bearings. It is suggested that external and internal threads of the same class be used together for general purpose assemblies, Class 2G being the preferred choice. If less backlash or end play is desired, Classes 3G and 4 G are provided.

Where minimal backlash or end play is required, Class 5G is provided. Assemblies of internal and external class 5G threads normally require some fitting-up for satisfactory results. External threads of any class may be assembled with internal threads of any class to provide other degrees of backlash or end play.

*Thread Form:* The accompanying figure shows the thread form of these General Purpose threads, and the formulas accompanying the figure determine their basic dimensions. Table 1 gives the basic dimensions for the most generally used pitches.

*Angle of Thread:* The angle between the sides of the thread, measured in an axial plane, is 29 degrees. The line bisecting this 29-degree angle shall be perpendicular to the axis of the screw thread.

*Thread Series:* A series of diameters and associated pitches is recommended in the Standard as preferred. These diameters and pitches have been chosen to meet present needs with the fewest number of items in order to reduce to a minimum the inventory of both tools and gages. This series of diameters and associated pitches is given in Table 3.

*Chamfers and Fillets:* General Purpose external threads may have the crest corner chamfered to an angle of 45 degrees with the axis to a maximum width of $p/15$, where $p$ is the pitch. This corresponds to a maximum depth of chamfer flat of $0.0945p$.

*Basic Diameters:* The maximum minor diameter of the external thread is basic and is the nominal major diameter for all classes. The minimum pitch diameter of the internal thread is basic and is equal to the basic major diameter minus the basic height of the thread, $h$. The basic minor diameter is the minimum minor diameter of the internal thread. It is equal to the basic major diameter minus twice the basic thread height, $2h$.

*Length of Engagement:* The tolerances specified in this standard are applicable to lengths of engagement not exceeding twice the nominal major diameter.

*Major and Minor Diameter Allowances:* A minimum diametral clearance is provided at the minor diameter of all external threads by establishing the maximum minor diameter 0.020 inch below the basic minor diameter for pitches of 10 threads per inch and coarser, and 0.010 inch for finer pitches. A minimum diametral clearance at the major diameter is obtained by establishing the minimum major diameter of the internal thread 0.020 inch above the basic major diameter for pitches of 10 threads per inch and coarser, and 0.010 inch for finer pitches.

*Major and Minor Diameter Tolerances:* The tolerance on the external thread major diameter is $0.05p$, where $p$ is the pitch, with a minimum of 0.005 inch. The tolerance on the internal thread major diameter is 0.020 inch for 10 threads per inch and coarser and 0.010 for finer pitches. The tolerance on the external thread minor diameter is 1.5 × pitch diameter tolerance. The tolerance on the internal thread minor diameter is $0.05p$ with a minimum of 0.005 inch.

*Pitch Diameter Allowances and Tolerances:* Allowances on the pitch diameter of General Purpose Acme threads are given in Table 4. Pitch diameter tolerances are given in Table 5. The ratios of the pitch diameter tolerances of Classes 2G, 3G, 4G, and 5G General Purpose threads are 3.0, 1.4, 1, and 0.8, respectively.

An increase of 10 per cent in the allowance is recommended for each inch, or fraction thereof, that the length of engagement exceeds two diameters.

*Application of Tolerances:* The tolerances specified are such as to assure interchangeability and maintain a high grade of product. The tolerances on diameters of the internal thread are plus, being applied from minimum sizes to above the minimum sizes. The tolerances on diameters of the external thread are minus, being applied from the maximum sizes to below the maximum sizes. The pitch diameter (otherwise known as thread thickness) tolerances for an external or internal thread of a given class are the same.

*Limiting Dimensions:* Limiting dimensions of General Purpose Acme screw threads in the recommended series are given in Table 2b. These are based on the formulas in Table 2a.

For combinations of pitch and diameter other than those in the recommended series, the formulas in Table 2a and the data in Tables 4 and 5 make it possible to readily determine the limiting dimensions required.

A diagram showing the disposition of allowances, tolerances, and crest clearances for General Purpose Acme threads appears on page 1332.

**Acme Thread Abbreviations.**—The following abbreviations are recommended for use on drawings and in specifications, and on tools and gages:

$$\text{Acme} = \text{Acme threads}$$
$$\text{G} = \text{General Purpose}$$
$$\text{C} = \text{Centralizing}$$
$$\text{p} = \text{pitch}$$
$$\text{L} = \text{lead}$$
$$\text{LH} = \text{left hand}$$

**Designation of General Purpose Acme Threads.** — The following examples listed below are given here to show how these Acme threads are designated on drawings and tools:

1¾–4 Acme– 2G indicates a General Purpose Class 2G Acme thread of 1¾-inch major diameter, 4 threads per inch, single thread, right hand. The same thread, but left hand, is designated 1¾–4 Acme–2G–LH.

2⅞–0.4p–0.8L–Acme–3G indicates a General Purpose Class 3G Acme thread of 2⅞-inch major diameter, pitch 0.4 inch, lead 0.8 inch, double thread, right hand.

**Multiple Start Acme Threads.** — The tabulated diameter-pitch data with allowances and tolerances relates to single-start threads. These data, as tabulated, may be and often are used for two-start Class 2G threads but this usage generally requires reduction of the full working tolerances to provide a greater allowance or clearance zone between the mating threads to assure satisfactory assembly.

When the class of thread requires smaller working tolerances than the 2G class or when threads with 3, 4, or more starts are required, some additional allowances and/or increased tolerances may be needed to insure adequate working tolerances and satisfactory assembly of mating parts.

It is suggested that the allowances shown in Table 4 be used for all external threads and that allowances be applied to internal threads in the following ratios: for two-start threads, 50 per cent of the allowances shown in the third, fourth, and fifth columns of

(*Continued on page* 1330 )

American National Standard General Purpose Acme Screw Thread Form
(ANSI B1.5-1977), and Stub Acme Screw Thread Form (ANSI B1.8-1977)

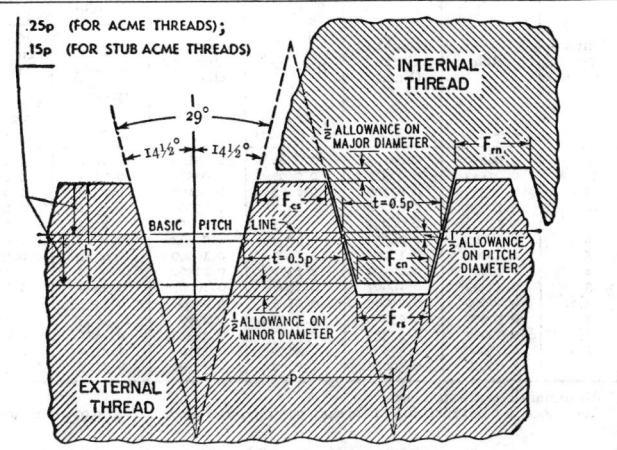

Formulas for Finding Basic Dimensions of General Purpose Acme and Stub Acme Screw Threads*

| General Purpose Acme Threads | Stub Acme Threads |
|---|---|
| Pitch = $p = 1 \div$ No. threads per inch | Pitch = $p = 1 \div$ No. threads per inch |
| Basic thread height $h = 0.5p$ | Basic thread height $h = 0.3p$ |
| Basic thread thickness $t = 0.5p$ | Basic thread thickness $t = 0.5p$ |
| Basic flat at crest $F_{cn} = 0.3707p$ (internal thread) | Basic flat at crest $F_{cn} = 0.4224p$ (internal thread) |
| Basic flat at crest $F_{cs} = 0.3707p - 0.259 \times$ (P.D. allowance on ext. thd.) | Basic flat at crest $F_{cs} = 0.4224p$ (external thread) |
| $F_{rn} = 0.3707p - 0.259 \times$ (major diam. allowance on internal thread) | $F_{rn} = 0.4224p - 0.259 \times$ (major diam. allowance on internal thread) |
| $F_{rs} = 0.3707p - 0.259 \times$ (minor diam. allowance on ext. thread − pitch diam. allowance on ext. thread) | $F_{rs} = 0.4224p - 0.259 \times$ (minor diam. allowance on ext. thread − pitch diam. allowance on ext. thread) |

*Stress Area of General Purpose Acme Threads.* — For computing the tensile strength of the thread section, the minimum stress area based on the mean of the minimum pitch diameter $E_s$ and the minimum minor diameter $K_s$ of the external thread is used:

$$\text{Stress Area} = 3.1416 \left( \frac{E_s + K_s}{4} \right)^2,$$

where $E_s$ and $K_s$ may be computed by Formulas 4 and 6, Table 2a or taken from Table 2b.

*Shear Area of General Purpose Acme Threads.* For computing the shear area per inch length of engagement of the external thread, the maximum minor diameter of the internal thread $K_n$, and the minimum pitch diameter of the external thread $E_s$, Table 2b or Formulas 12 and 4, Table 2a, are used:

$$\text{Shear Area} = 3.1416 K_n \left[ 0.5 + \frac{1}{p} \tan 14\tfrac{1}{2}°(E_s - K_n) \right]$$

* Tables 1 and 6 give basic dimensions for various pitches.

**Table 1. American National Standard General Purpose Acme Screw Thread Form —**
**Basic Dimensions\* (ANSI B1.5-1977)**

| Thds. per Inch | Pitch, | Height of Thread (Basic), | Total Height of Thread, | Thread Thickness (Basic), | Width of Flat | |
|---|---|---|---|---|---|---|
| | | | | | Crest of Internal Thread (Basic), | Root of Internal Thread, |
| $n$ | $p = 1/n$ | $p/2$ | $p/2 + \frac{1}{2}$ allowance† | $p/2$ | $0.3707p$ | $0.3707p - 0.259 \times$ allowance† |
| 16 | 0.06250 | 0.03125 | 0.0362 | 0.03125 | 0.0232 | 0.0206 |
| 14 | 0.07143 | 0.03571 | 0.0407 | 0.03571 | 0.0265 | 0.0239 |
| 12 | 0.08333 | 0.04167 | 0.0467 | 0.04167 | 0.0309 | 0.0283 |
| 10 | 0.10000 | 0.05000 | 0.0600 | 0.05000 | 0.0371 | 0.0319 |
| 8 | 0.12500 | 0.06250 | 0.0725 | 0.06250 | 0.0463 | 0.0411 |
| 6 | 0.16667 | 0.08333 | 0.0933 | 0.08333 | 0.0618 | 0.0566 |
| 5 | 0.20000 | 0.10000 | 0.1100 | 0.10000 | 0.0741 | 0.0689 |
| 4 | 0.25000 | 0.12500 | 0.1350 | 0.12500 | 0.0927 | 0.0875 |
| 3 | 0.33333 | 0.16667 | 0.1767 | 0.16667 | 0.1236 | 0.1184 |
| 2½ | 0.40000 | 0.20000 | 0.2100 | 0.20000 | 0.1483 | 0.1431 |
| 2 | 0.50000 | 0.25000 | 0.2600 | 0.25000 | 0.1853 | 0.1802 |
| 1½ | 0.66667 | 0.33333 | 0.3433 | 0.33333 | 0.2471 | 0.2419 |
| 1⅓ | 0.75000 | 0.37500 | 0.3850 | 0.37500 | 0.2780 | 0.2728 |
| 1 | 1.00000 | 0.50000 | 0.5100 | 0.50000 | 0.3707 | 0.3655 |

\* All dimensions are in inches.
† Allowance is 0.020 inch for 10 threads per inch and coarser, and 0.010 inch for finer threads.

**Table 2a. American National Standard General Purpose Acme Screw Threads —**
**Formulas for Determining Diameters (ANSI B1.5-1977)**

$D$ = Basic Major Diameter and Nominal Size, in Inches.
$p$ = Pitch = 1 ÷ Number of Threads per Inch.
$E$ = Basic Pitch Diameter = $D - 0.5p$
$K$ = Basic Minor Diameter = $D - p$

| No. | EXTERNAL THREADS (SCREWS) |
|---|---|
| 1 | Major Diam., Max. = $D$ |
| 2 | Major Diam., Min. = $D$ *minus* 0.05$p$\* but not less than 0.005. |
| 3 | Pitch Diam., Max. = $E$ *minus* allowance from Table 4. |
| 4 | Pitch Diam., Min. = Pitch Diam., Max. (Formula 3) *minus* tolerance from Table 5. |
| 5 | Minor Diam., Max. = $K$ *minus* 0.020 for 10 threads per inch and coarser and 0.010 for finer pitches. |
| 6 | Minor Diam., Min. = Minor Diam., Max. (Formula 5) *minus* 1.5 × pitch diameter tolerance from Table 5. |
| | INTERNAL THREADS (NUTS) |
| 7 | Major Diam., Min. = $D$ *plus* 0.020 for 10 threads per inch and coarser and 0.010 for finer pitches. |
| 8 | Major Diam., Max. = Major Diam., Min. (Formula 7) *plus* 0.020 for 10 threads per inch and coarser and 0.010 for finer pitches. |
| 9 | Pitch Diam., Min. = $E$ |
| 10 | Pitch Diam., Max. = Pitch Diam., Min. (Formula 9) *plus* tolerance from Table 5. |
| 11 | Minor Diam., Min. = $K$ |
| 12 | Minor Diam., Max. = Minor Diam., Min. (Formula 11) *plus* 0.05$p$\* but not less than 0.005. |

\* If $p$ is between two recommended pitches listed in Table 3, use the coarser of the two pitches in this formula instead of the actual value of $p$.

Table 2b. Limiting Dimensions of American National Standard General Purpose Single Start Acme Screw Threads

| Limiting Diameters | | | Nominal Diameter, D | | | | | | | | | | | |
|---|---|---|---|---|---|---|---|---|---|---|---|---|---|---|
| | | 1/4 | 5/16 | 3/8 | 7/16 | 1/2 | 5/8 | 3/4 | 7/8 | 1 | 1 1/8 | 1 1/4 | 1 3/8 |
| | | Threads per Inch* | | | | | | | | | | | |
| | | 16 | 14 | 12 | 12 | 10 | 8 | 6 | 6 | 5 | 5 | 5 | 4 |
| **External Threads** | | | | | | | | | | | | | |
| Classes 2G, 3G, 4G, and 5G, Major Diameter, Max(D) | Max | 0.2500 | 0.3125 | 0.3750 | 0.4375 | 0.5000 | 0.6250 | 0.7500 | 0.8750 | 1.0000 | 1.1250 | 1.2500 | 1.3750 |
| | Min | 0.2450 | 0.3075 | 0.3700 | 0.4325 | 0.4950 | 0.6188 | 0.7417 | 0.8667 | 0.9900 | 1.1150 | 1.2400 | 1.3625 |
| Classes, 2G, 3G, 4G, and 5G, Minor Diameter | Max | 0.1775 | 0.2311 | 0.2817 | 0.3442 | 0.3800 | 0.4800 | 0.5633 | 0.6883 | 0.7800 | 0.9050 | 1.0300 | 1.1050 |
| Class 2G, Minor Diameter | Min | 0.1618 | 0.2140 | 0.2632 | 0.3253 | 0.3594 | 0.4569 | 0.5372 | 0.6615 | 0.7509 | 0.8753 | 0.9998 | 1.0720 |
| Class 3G, Minor Diameter | Min | 0.1702 | 0.2231 | 0.2730 | 0.3354 | 0.3704 | 0.4692 | 0.5511 | 0.6758 | 0.7664 | 0.8912 | 1.0159 | 1.0896 |
| Class 4G, Minor Diameter | Min | 0.1722 | 0.2254 | 0.2755 | 0.3379 | 0.3731 | 0.4723 | 0.5546 | 0.6794 | 0.7703 | 0.8951 | 1.0199 | 1.0940 |
| Class 5G, Minor Diameter | Min | 0.1733 | 0.2266 | 0.2767 | 0.3391 | 0.3745 | 0.4738 | 0.5563 | 0.6811 | 0.7722 | 0.8971 | 1.0219 | 1.0962 |
| Class 2G, Pitch Diameter | Max | 0.2148 | 0.2728 | 0.3284 | 0.3909 | 0.4443 | 0.5552 | 0.6598 | 0.7842 | 0.8920 | 1.0165 | 1.1411 | 1.2406 |
| | Min | 0.2043 | 0.2614 | 0.3161 | 0.3783 | 0.4306 | 0.5408 | 0.6424 | 0.7663 | 0.8726 | 0.9967 | 1.1210 | 1.2188 |
| Class 3G, Pitch Diameter | Max | 0.2158 | 0.2738 | 0.3296 | 0.3921 | 0.4458 | 0.5578 | 0.6615 | 0.7861 | 0.8940 | 1.0186 | 1.1433 | 1.2430 |
| | Min | 0.2109 | 0.2685 | 0.3238 | 0.3862 | 0.4394 | 0.5506 | 0.6534 | 0.7778 | 0.8849 | 1.0094 | 1.1339 | 1.2327 |
| Class 4G, Pitch Diameter | Max | 0.2168 | 0.2748 | 0.3309 | 0.3934 | 0.4472 | 0.5593 | 0.6632 | 0.7880 | 0.8960 | 1.0208 | 1.1455 | 1.2453 |
| | Min | 0.2133 | 0.2710 | 0.3268 | 0.3892 | 0.4426 | 0.5542 | 0.6574 | 0.7820 | 0.8895 | 1.0142 | 1.1388 | 1.2380 |
| Class 5G, Pitch Diameter | Max | 0.2188 | 0.2768 | 0.3333 | 0.3958 | 0.4500 | 0.5625 | 0.6667 | 0.7917 | 0.9000 | 1.0250 | 1.1500 | 1.2500 |
| | Min | 0.2160 | 0.2738 | 0.3300 | 0.3924 | 0.4463 | 0.5584 | 0.6620 | 0.7869 | 0.8948 | 1.0197 | 1.1446 | 1.2441 |
| **Internal Threads** | | | | | | | | | | | | | |
| Classes 2G, 3G, 4G, and 5G, Major Diameter | Min | 0.2600 | 0.3225 | 0.3850 | 0.4475 | 0.5200 | 0.6450 | 0.7700 | 0.8950 | 1.0200 | 1.1450 | 1.2700 | 1.3950 |
| | Max | 0.2700 | 0.3325 | 0.3950 | 0.4575 | 0.5400 | 0.6650 | 0.7900 | 0.9150 | 1.0400 | 1.1650 | 1.2900 | 1.4150 |
| Classes 2G, 3G, 4G, and 5G, Minor Diameter | Min | 0.1875 | 0.2411 | 0.2917 | 0.3542 | 0.4000 | 0.5000 | 0.5833 | 0.7083 | 0.8000 | 0.9250 | 1.0500 | 1.1250 |
| | Max | 0.1925 | 0.2461 | 0.2967 | 0.3592 | 0.4050 | 0.5062 | 0.5916 | 0.7166 | 0.8100 | 0.9350 | 1.0600 | 1.1375 |
| Class 2G, Pitch Diameter | Min | 0.2188 | 0.2768 | 0.3333 | 0.3958 | 0.4500 | 0.5625 | 0.6667 | 0.7917 | 0.9000 | 1.0250 | 1.1500 | 1.2500 |
| | Max | 0.2293 | 0.2882 | 0.3456 | 0.4084 | 0.4637 | 0.5779 | 0.6841 | 0.8096 | 0.9194 | 1.0448 | 1.1701 | 1.2720 |
| Class 3G, Pitch Diameter | Min | 0.2188 | 0.2768 | 0.3333 | 0.3958 | 0.4500 | 0.5625 | 0.6667 | 0.7917 | 0.9000 | 1.0250 | 1.1500 | 1.2500 |
| | Max | 0.2237 | 0.2821 | 0.3391 | 0.4017 | 0.4564 | 0.5697 | 0.6748 | 0.8000 | 0.9091 | 1.0342 | 1.1594 | 1.2603 |
| Class 4G, Pitch Diameter | Min | 0.2188 | 0.2768 | 0.3333 | 0.3958 | 0.4500 | 0.5625 | 0.6667 | 0.7917 | 0.9000 | 1.0250 | 1.1500 | 1.2500 |
| | Max | 0.2223 | 0.2806 | 0.3374 | 0.4000 | 0.4546 | 0.5676 | 0.6725 | 0.7977 | 0.9065 | 1.0316 | 1.1567 | 1.2573 |
| Class 5G, Pitch Diameter | Min | 0.2188 | 0.2768 | 0.3333 | 0.3958 | 0.4500 | 0.5625 | 0.6667 | 0.7917 | 0.9000 | 1.0250 | 1.1500 | 1.2500 |
| | Max | 0.2216 | 0.2798 | 0.3366 | 0.3992 | 0.4537 | 0.5666 | 0.6714 | 0.7965 | 0.9052 | 1.0303 | 1.1554 | 1.2559 |

* All other dimensions are given in inches.

**Table 2b** *(Continued).*  **Limiting Dimensions of American National Standard General Purpose Single Start Acme Screw Threads**

| Limiting Diameters | | 1½ | 1¾ | 2 | 2¼ | 2½ | 2¾ | 3 | 3½ | 4 | 4½ | 5 |
|---|---|---|---|---|---|---|---|---|---|---|---|---|
| **Nominal Diameter, D** → Threads per Inch* | | 4 | 4 | 4 | 3 | 3 | 3 | 2 | 2 | 2 | 2 | 2 |
| **External Threads** | | | | | | | | | | | | |
| Classes 2G, 3G, 4G, and 5G, Major Diameter | Max (D) | 1.5000 | 1.7500 | 2.0000 | 2.2500 | 2.5000 | 2.7500 | 3.0000 | 3.5000 | 4.0000 | 4.5000 | 5.0000 |
| | Min | 1.4875 | 1.7375 | 1.9875 | 2.2333 | 2.4833 | 2.7333 | 2.9750 | 3.4750 | 3.9750 | 4.4750 | 4.9750 |
| Classes 2G, 3G, 4G, and 5G, Minor Diameter | Max | 1.2300 | 1.4800 | 1.7300 | 1.8967 | 2.1467 | 2.3967 | 2.4800 | 2.9800 | 3.4800 | 3.9800 | 4.4800 |
| Class 2G, Minor Diameter | Min | 1.1965 | 1.4456 | 1.6948 | 1.8572 | 2.1065 | 2.3558 | 2.4336 | 2.9314 | 3.4302 | 3.9291 | 4.4281 |
| Class 3G, Minor Diameter | Min | 1.2144 | 1.4640 | 1.7135 | 1.8783 | 2.1279 | 2.3776 | 2.4579 | 2.9574 | 3.4568 | 3.9563 | 4.4557 |
| Class 4G, Minor Diameter | Min | 1.2189 | 1.4686 | 1.7183 | 1.8835 | 2.1333 | 2.3831 | 2.4642 | 2.9638 | 3.4634 | 3.9631 | 4.4627 |
| Class 5G, Minor Diameter | Min | 1.2210 | 1.4708 | 1.7206 | 1.8862 | 2.1360 | 2.3858 | 2.4674 | 2.9669 | 3.4666 | 3.9663 | 4.4662 |
| Class 2G, Pitch Diameter | Max | 1.3652 | 1.6145 | 1.8637 | 2.0713 | 2.3207 | 2.5700 | 2.7360 | 3.2350 | 3.7340 | 4.2330 | 4.7319 |
| | Min | 1.3429 | 1.5916 | 1.8402 | 2.0450 | 2.2939 | 2.5427 | 2.7044 | 3.2026 | 3.7008 | 4.1991 | 4.6973 |
| Class 3G, Pitch Diameter | Max | 1.3677 | 1.6171 | 1.8665 | 2.0743 | 2.3238 | 2.5734 | 2.7395 | 3.2388 | 3.7380 | 4.2373 | 4.7364 |
| | Min | 1.3573 | 1.6064 | 1.8555 | 2.0620 | 2.3113 | 2.5607 | 2.7248 | 3.2237 | 3.7225 | 4.2215 | 4.7202 |
| Class 4G, Pitch Diameter | Max | 1.3701 | 1.6198 | 1.8693 | 2.0773 | 2.3270 | 2.5767 | 2.7430 | 3.2425 | 3.7420 | 4.2415 | 4.7409 |
| | Min | 1.3627 | 1.6122 | 1.8615 | 2.0685 | 2.3181 | 2.5676 | 2.7325 | 3.2317 | 3.7309 | 4.2302 | 4.7294 |
| Class 5G, Pitch Diameter | Max | 1.3750 | 1.6250 | 1.8750 | 2.0833 | 2.3333 | 2.5833 | 2.7500 | 3.2500 | 3.7500 | 4.2500 | 4.7500 |
| | Min | 1.3690 | 1.6189 | 1.8687 | 2.0763 | 2.3262 | 2.5760 | 2.7416 | 3.2413 | 3.7411 | 4.2409 | 4.7408 |
| **Internal Threads** | | | | | | | | | | | | |
| Classes 2G, 3G, 4G, and 5G, Major Diameter | Min | 1.5200 | 1.7700 | 2.0200 | 2.2700 | 2.5200 | 2.7700 | 3.0200 | 3.5200 | 4.0200 | 4.5200 | 5.0200 |
| | Max | 1.5400 | 1.7900 | 2.0400 | 2.2900 | 2.5400 | 2.7900 | 3.0400 | 3.5400 | 4.0400 | 4.5400 | 5.0400 |
| Classes 2G, 3G, 4G, and 5G, Minor Diameter | Min | 1.2500 | 1.5000 | 1.7500 | 1.9167 | 2.1667 | 2.4167 | 2.5000 | 3.0000 | 3.5000 | 4.0000 | 4.5000 |
| | Max | 1.2625 | 1.5125 | 1.7625 | 1.9334 | 2.1834 | 2.4334 | 2.5250 | 3.0250 | 3.5250 | 4.0250 | 4.5250 |
| Class 2G, Pitch Diameter | Min | 1.3750 | 1.6250 | 1.8750 | 2.0833 | 2.3333 | 2.5833 | 2.7500 | 3.2500 | 3.7500 | 4.2500 | 4.7500 |
| | Max | 1.3973 | 1.6479 | 1.8985 | 2.1096 | 2.3601 | 2.6106 | 2.7816 | 3.2824 | 3.7832 | 4.2839 | 4.7846 |
| Class 3G, Pitch Diameter | Min | 1.3750 | 1.6250 | 1.8750 | 2.0833 | 2.3333 | 2.5833 | 2.7500 | 3.2500 | 3.7500 | 4.2500 | 4.7500 |
| | Max | 1.3854 | 1.6357 | 1.8860 | 2.0956 | 2.3458 | 2.5960 | 2.7647 | 3.2651 | 3.7655 | 4.2658 | 4.7662 |
| Class 4G, Pitch Diameter | Min | 1.3750 | 1.6250 | 1.8750 | 2.0833 | 2.3333 | 2.5833 | 2.7500 | 3.2500 | 3.7500 | 4.2500 | 4.7500 |
| | Max | 1.3824 | 1.6326 | 1.8828 | 2.0921 | 2.3422 | 2.5924 | 2.7605 | 3.2608 | 3.7611 | 4.2613 | 4.7615 |
| Class 5G, Pitch Diameter | Min | 1.3750 | 1.6250 | 1.8750 | 2.0833 | 2.3333 | 2.5833 | 2.7500 | 3.2500 | 3.7500 | 4.2500 | 4.7500 |
| | Max | 1.3810 | 1.6311 | 1.8813 | 2.0903 | 2.3404 | 2.5906 | 2.7584 | 3.2587 | 3.7589 | 4.2591 | 4.7592 |

\* All other dimensions are given in inches.

**Table 3. General Purpose Single Start Acme Screw Thread Data (ANSI B1.5-1977)**

| Identification | | Basic Diameters | | | Thread Data | | | | | | | |
| Nominal Sizes (All Classes) | Threads per Inch.* $n$ | Major Diameter, $D$ | Classes 2G, 3G, 4G & 5G — Pitch Diameter, $E = D - h$ | Minor Diameter, $K = D - 2h$ | Pitch, $p$ | Thickness at Pitch Line, $t = p/2$ | Basic Height of Thread, $h = p/2$ | Basic Width of Flat, $F = 0.3707p$ | Lead Angle at Basic Pitch Diameter* Classes 2G, 3G, 4G & 5G $\lambda$ — Deg | Min | Shear Area† Class 3G | Shear Area‡ Class 3G |
|---|---|---|---|---|---|---|---|---|---|---|---|---|
| 1/4 | 16 | 0.2500 | 0.2188 | 0.1875 | 0.06250 | 0.03125 | 0.03125 | 0.0232 | 5 | 12 | 0.350 | 0.0285 |
| 5/16 | 14 | 0.3125 | 0.2768 | 0.2411 | 0.07143 | 0.03571 | 0.03571 | 0.0265 | 4 | 42 | 0.451 | 0.0474 |
| 3/8 | 12 | 0.3750 | 0.3333 | 0.2917 | 0.08333 | 0.04167 | 0.04167 | 0.0309 | 4 | 33 | 0.545 | 0.0699 |
| 3/8 | 10 | 0.3750 | 0.3250 | 0.2750 | 0.10000 | 0.05000 | 0.05000 | 0.0371 | … | … | … | … |
| 7/16 | 12 | 0.4375 | 0.3958 | 0.3542 | 0.08333 | 0.04167 | 0.04167 | 0.0309 | 3 | 50 | 0.660 | 0.1022 |
| 7/16 | 10 | 0.4375 | 0.3875 | 0.3375 | 0.10000 | 0.05000 | 0.05000 | 0.0371 | … | … | 0.749 | 0.1287 |
| 1/2 | 10 | 0.5000 | 0.4500 | 0.4000 | 0.10000 | 0.05000 | 0.05000 | 0.0371 | 4 | 3 | 0.941 | 0.2043 |
| 5/8 | 8 | 0.6250 | 0.5625 | 0.5000 | 0.12500 | 0.06250 | 0.06250 | 0.0463 | 4 | 33 | 1.108 | 0.2848 |
| 3/4 | 6 | 0.7500 | 0.6667 | 0.5833 | 0.16667 | 0.08333 | 0.08333 | 0.0618 | 4 | 33 | 1.339 | 0.4150 |
| 7/8 | 6 | 0.8750 | 0.7917 | 0.7083 | 0.16667 | 0.08333 | 0.08333 | 0.0618 | 3 | 50 | 1.519 | 0.5354 |
| 1 | 5 | 1.0000 | 0.9000 | 0.8000 | 0.20000 | 0.10000 | 0.10000 | 0.0741 | 4 | 3 | 1.519 | 0.5354 |
| 1 1/8 | 5 | 1.1250 | 1.0250 | 0.9250 | 0.20000 | 0.10000 | 0.10000 | 0.0741 | 3 | 33 | 1.751 | 0.709 |
| 1 1/4 | 5 | 1.2500 | 1.1500 | 1.0500 | 0.20000 | 0.10000 | 0.10000 | 0.0741 | 3 | 10 | 1.983 | 0.907 |
| 1 3/8 | 4 | 1.3750 | 1.2500 | 1.1250 | 0.25000 | 0.12500 | 0.12500 | 0.0927 | 3 | 39 | 2.139 | 1.059 |
| 1 1/2 | 4 | 1.5000 | 1.3750 | 1.2500 | 0.25000 | 0.12500 | 0.12500 | 0.0927 | 3 | 19 | 2.372 | 1.298 |
| 1 3/4 | 4 | 1.7500 | 1.6250 | 1.5000 | 0.25000 | 0.12500 | 0.12500 | 0.0927 | 2 | 48 | 2.837 | 1.851 |
| 2 | 4 | 2.0000 | 1.8750 | 1.7500 | 0.25000 | 0.12500 | 0.12500 | 0.0927 | 2 | 26 | 3.301 | 2.501 |
| 2 1/4 | 3 | 2.2500 | 2.0833 | 1.9167 | 0.33333 | 0.16667 | 0.16667 | 0.1236 | 2 | 55 | 3.643 | 3.049 |
| 2 1/2 | 3 | 2.5000 | 2.3333 | 2.1667 | 0.33333 | 0.16667 | 0.16667 | 0.1236 | 2 | 36 | 4.110 | 3.870 |
| 2 3/4 | 3 | 2.7500 | 2.5833 | 2.4167 | 0.33333 | 0.16667 | 0.16667 | 0.1236 | 2 | 21 | 4.577 | 4.788 |
| 3 | 2 | 3.0000 | 2.7500 | 2.5000 | 0.50000 | 0.25000 | 0.25000 | 0.1853 | 3 | 19 | 4.786 | 5.27 |
| 3 1/2 | 2 | 3.5000 | 3.2500 | 3.0000 | 0.50000 | 0.25000 | 0.25000 | 0.1853 | 3 | 48 | 5.73 | 7.50 |
| 4 | 2 | 4.0000 | 3.7500 | 3.5000 | 0.50000 | 0.25000 | 0.25000 | 0.1853 | 2 | 26 | 6.67 | 10.12 |
| 4 1/2 | 2 | 4.5000 | 4.2500 | 4.0000 | 0.50000 | 0.25000 | 0.25000 | 0.1853 | 2 | 9 | 7.60 | 13.13 |
| 5 | 2 | 5.0000 | 4.7500 | 4.5000 | 0.50000 | 0.25000 | 0.25000 | 0.1853 | 1 | 55 | 8.54 | 16.53 |

* All other dimensions are given in inches.

† Per inch Length of engagement of the external thread in line with the minor diameter crests of the internal thread. Computed from this formula:
Shear Area $= \pi K_n [0.5 + h \tan 14\tfrac{1}{2}° (E_s - K_n)]$. Figures given are the minimum shear area based on max $K_n$ and min $E_s$.

‡ Figures given are the minimum stress area based on the mean of the minimum minor and pitch diameters of the external thread.

Table 4.   American National Standard General Purpose Single Start Acme
Screw Threads—Pitch Diameter Allowances* (ANSI B1.5-1977)

| Nominal Size Range** | | Allowances on External Threads† | | |
|---|---|---|---|---|
| Above | To and Including | Class 2G, $0.008\sqrt{D}$ | Class 3G, $0.006\sqrt{D}$ | Class 4G, $0.004\sqrt{D}$ |
| 0 | 3/16 | 0.0024 | 0.0018 | 0.0012 |
| 3/16 | 5/16 | 0.0040 | 0.0030 | 0.0020 |
| 5/16 | 7/16 | 0.0049 | 0.0037 | 0.0024 |
| 7/16 | 9/16 | 0.0057 | 0.0042 | 0.0028 |
| 9/16 | 11/16 | 0.0063 | 0.0047 | 0.0032 |
| 11/16 | 13/16 | 0.0069 | 0.0052 | 0.0035 |
| 13/16 | 15/16 | 0.0075 | 0.0056 | 0.0037 |
| 15/16 | 11/16 | 0.0080 | 0.0060 | 0.0040 |
| 11/16 | 13/16 | 0.0085 | 0.0064 | 0.0042 |
| 13/16 | 15/16 | 0.0089 | 0.0067 | 0.0045 |
| 15/16 | 17/16 | 0.0094 | 0.0070 | 0.0047 |
| 17/16 | 19/16 | 0.0098 | 0.0073 | 0.0049 |
| 19/16 | 17/8 | 0.0105 | 0.0079 | 0.0052 |
| 17/8 | 21/8 | 0.0113 | 0.0085 | 0.0057 |
| 21/8 | 23/8 | 0.0120 | 0.0090 | 0.0060 |
| 23/8 | 25/8 | 0.0126 | 0.0095 | 0.0063 |
| 25/8 | 27/8 | 0.0133 | 0.0099 | 0.0066 |
| 27/8 | 31/4 | 0.0140 | 0.0105 | 0.0070 |
| 31/4 | 33/4 | 0.0150 | 0.0112 | 0.0075 |
| 33/4 | 41/4 | 0.0160 | 0.0120 | 0.0080 |
| 41/4 | 43/4 | 0.0170 | 0.0127 | 0.0085 |
| 43/4 | 51/2 | 0.0181 | 0.0136 | 0.0091 |

All dimensions in inches.   Class 5G has no allowance on pitch diameter of external
threads.
   * Allowances for Class 2G threads in column 3 also apply to American National Standard
Stub Acme threads (ANSI B1.8-1977).
   ** The values in columns 3 to 5 are to be used for any size within the range shown in
columns 1 and 2.   These values are calculated from the mean of the range.
   It is recommended that the sizes given in Table 3 be used whenever possible.
   † An increase of 10 per cent in the allowance is recommended for each inch, or fraction
thereof, that the length of engagement exceeds two diameters.

Table 4; for three-start threads, 75 per cent of these allowances; and for four-start
threads, 100 per cent of these same values.

These values will provide for a ¼"–16 Acme 2G thread size, 0.002, 0.003, and 0.004
inch additional clearance for 2-, 3-, and 4-start threads, respectively. For a 5"—2 Acme
3G thread size the additional clearances would be 0.0068, 0.0102, and 0.0136 inch,
respectively. GO thread plug gages and taps would be increased by these same values. To
maintain the same working tolerances on multi-start threads, the pitch diameter of the
NOT GO thread plug gage would also be increased by these same values.

For multi-start threads with more than four starts, it is believed that the 100 per cent
allowance provided by the above procedures would be adequate as index spacings
variables would generally be no greater than on a four-start thread.

In general, for multi-start threads of Classes 2G, 3G, and 4G the percentages would be
applied, usually, to allowances for the same class, respectively. However, in cases where
exceptionally good control over lead, angle, and spacing variables would produce close
to theoretical values in the product, it is conceivable that these percentages could be
applied to Class 3G or Class 4G allowances used on Class 2G internally threaded product.
Also these percentages could be applied to Class 4G allowances used on Class 3G
internally threaded product. It is not advocated that any change be made in externally
threaded products.

Designations for gages or tools for internal threads could cover allowance requirements as follows:

GO and NOT GO thread plug gages for: 2⅞–0.4p–0.8L Acme 2G with 4G internal
thread allowance.

**Table 5. American National Standard General Purposes Single Start Acme Screw Threads—Pitch Diameter Tolerances[1,2]** (ANSI B1.5-1977)

(For any particular size of thread, the pitch diameter tolerance is obtained by adding the *diameter increment* from the upper half of the table to the *pitch increment* from the lower half of the table. *Example:* A ¼-16 Acme-2G thread has a pitch diameter tolerance of 0.00300 + 0.00750 = 0.0105 inch.)

| Nom. Dia.,[3] $D$ | Class of Thread | | | | Nom. Dia.,[3] $D$ | Class of Thread | | | |
|---|---|---|---|---|---|---|---|---|---|
| | 2G | 3G | 4G | 5G | | 2G | 3G | 4G | 5G |
| | Diameter Increment | | | | | Diameter Increment | | | |
| | $.006\sqrt{D}$ | $.0028\sqrt{D}$ | $.002\sqrt{D}$ | $.0016\sqrt{D}$ | | $.006\sqrt{D}$ | $.0028\sqrt{D}$ | $.002\sqrt{D}$ | $.0016\sqrt{D}$ |
| ¼ | .00300 | .00140 | .00100 | .00080 | 1½ | .00735 | .00343 | .00245 | .00196 |
| 5/16 | .00335 | .00157 | .00112 | .00089 | 1¾ | .00794 | .00370 | .00265 | .00212 |
| ⅜ | .00367 | .00171 | .00122 | .00098 | 2 | .00849 | .00396 | .00283 | .00226 |
| 7/16 | .00397 | .00185 | .00132 | .00106 | 2¼ | .00900 | .00420 | .00300 | .00240 |
| ½ | .00424 | .00198 | .00141 | .00113 | 2½ | .00949 | .00443 | .00316 | .00253 |
| ⅝ | .00474 | .00221 | .00158 | .00126 | 2¾ | .00995 | .00464 | .00332 | .00265 |
| ¾ | .00520 | .00242 | .00173 | .00139 | 3 | .01039 | .00485 | .00346 | .00277 |
| ⅞ | .00561 | .00262 | .00187 | .00150 | 3½ | .01122 | .00524 | .00374 | .00299 |
| 1 | .00600 | .00280 | .00200 | .00160 | 4 | .01200 | .00560 | .00400 | .00320 |
| 1⅛ | .00636 | .00297 | .00212 | .00170 | 4½ | .01273 | .00594 | .00424 | .00339 |
| 1¼ | .00671 | .00313 | .00224 | .00179 | 5 | .01342 | .00626 | .00447 | .00358 |
| 1⅜ | .00704 | .00328 | .00235 | .00188 | .... | .... | .... | .... | .... |

| Thds. per Inch, $n$ | Class of Thread | | | | Thds. per Inch, $n$ | Class of Thread | | | |
|---|---|---|---|---|---|---|---|---|---|
| | 2G | 3G | 4G | 5G | | 2G | 3G | 4G | 5G |
| | Pitch Increment | | | | | Pitch Increment | | | |
| | $.030\sqrt{1/n}$ | $.014\sqrt{1/n}$ | $.010\sqrt{1/n}$ | $.008\sqrt{1/n}$ | | $.030\sqrt{1/n}$ | $.014\sqrt{1/n}$ | $.010\sqrt{1/n}$ | $.008\sqrt{1/n}$ |
| 16 | .00750 | .00350 | .00250 | .00200 | 4 | .01500 | .00700 | .00500 | .00400 |
| 14 | .00802 | .00374 | .00267 | .00214 | 3 | .01732 | .00808 | .00577 | .00462 |
| 12 | .00866 | .00404 | .00289 | .00231 | 2½ | .01897 | .00885 | .00632 | .00506 |
| 10 | .00949 | .00443 | .00316 | .00253 | 2 | .02121 | .00990 | .00707 | .00566 |
| 8 | .01061 | .00495 | .00354 | .00283 | 1½ | .02449 | .01143 | .00816 | .00653 |
| 6 | .01225 | .00572 | .00408 | .00327 | 1⅓ | .02598 | .01212 | .00866 | .00693 |
| 5 | .01342 | .00626 | .00447 | .00358 | 1 | .03000 | .01400 | .01000 | .00800 |

All dimensions are given in inches.

1 The equivalent tolerance on thread thickness is 0.259 times the pitch diameter tolerance.

2 Columns 2 and 7 of this table also apply to American National Standard Stub Acme threads, ANSI B1.8-1973.

3 For a nominal diameter between any two tabulated nominal diameters, use the diameter increment for the larger of the two tabulated nominal diameters.

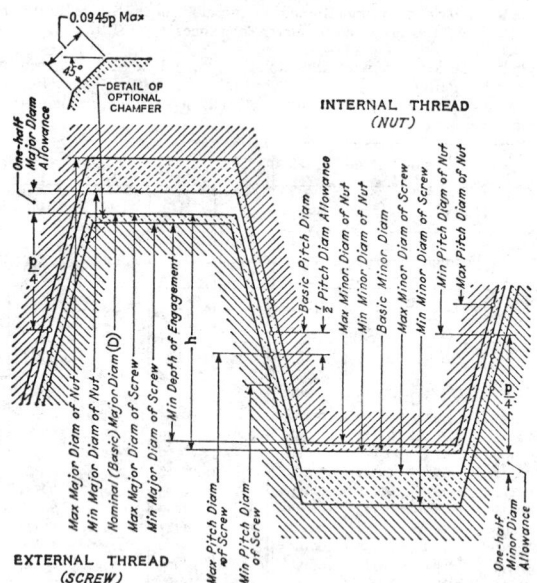

Disposition of Allowances, Tolerances, and Crest Clearances for General
Purpose Single-start Acme Threads (All Classes)

**Centralizing Acme Threads.** — The five classes of Centralizing Acme threads
in American National Standard ANSI B1.5-1977, designated as 2C, 3C, 4C, 5C
and 6C, have limited clearance at the major diameters of internal and external
threads so that a bearing at the major diameters maintains approximate alignment
of the thread axis and prevents wedging on the flanks of the thread. For any
combination of the five classes of threads covered in this standard some end play
or backlash will result. Detailed data on Centralizing Acme threads may be
found in the standard.

**American National Standard Stub Acme Screw Threads.** — This American
National Standard (ANSI B1.8-1977) provides a Stub Acme screw thread for those
unusual applications where, due to mechanical or metallurgical considerations, a
coarse-pitch thread of shallow depth is required. The fit of Stub Acme threads corresponds to the Class 2G General Purpose Acme thread in American National Standard
ANSI B1.5-1977. For a fit having less backlash, the tolerances and allowances for
Classes 3G, 4G and 5G General Purpose Acme threads may be used.

*Thread Form:* The thread form and basic formulas for Stub Acme threads are
given on page **1325** and the basic dimensions in Table 6.

*Allowances and Tolerances:* The major and minor diameter allowances for Stub
Acme threads are the same as those given for General Purpose Acme threads on
page **1323**.

Pitch diameter allowances for Stub Acme threads are the same as for Class 2G

(*Continued on page* 1336)

**Table 6. American National Standard Stub Acme Screw Thread Form —
Basic Dimensions\* (ANSI B1.8-1977)**

| Thds. per Inch | Pitch, | Height of Thread (Basic), | Total Height of Thread, | Thread Thickness (Basic), | Width of Flat | |
|---|---|---|---|---|---|---|
| | | | | | Crest of Internal Thread (Basic), | Root of Internal Thread, |
| | | | $0.3p + \frac{1}{2}$ allowance† | | | $0.4224p - 0.259 \times$ allowance† |
| $n$ | $p = 1/n$ | $0.3p$ | | $p/2$ | $0.4224p$ | |
| 16 | 0.06250 | 0.01875 | 0.0238 | 0.03125 | 0.0264 | 0.0238 |
| 14 | 0.07143 | 0.02143 | 0.0264 | 0.03571 | 0.0302 | 0.0276 |
| 12 | 0.08333 | 0.02500 | 0.0300 | 0.04167 | 0.0352 | 0.0326 |
| 10 | 0.10000 | 0.03000 | 0.0400 | 0.05000 | 0.0422 | 0.0370 |
| 9 | 0.11111 | 0.03333 | 0.0433 | 0.05556 | 0.0469 | 0.0417 |
| 8 | 0.12500 | 0.03750 | 0.0475 | 0.06250 | 0.0528 | 0.0476 |
| 7 | 0.14286 | 0.04285 | 0.0529 | 0.07143 | 0.0603 | 0.0551 |
| 6 | 0.16667 | 0.05000 | 0.0600 | 0.08333 | 0.0704 | 0.0652 |
| 5 | 0.20000 | 0.06000 | 0.0700 | 0.10000 | 0.0845 | 0.0793 |
| 4 | 0.25000 | 0.07500 | 0.0850 | 0.12500 | 0.1056 | 0.1004 |
| 3½ | 0.28571 | 0.08571 | 0.0957 | 0.14286 | 0.1207 | 0.1155 |
| 3 | 0.33333 | 0.10000 | 0.1100 | 0.16667 | 0.1408 | 0.1356 |
| 2½ | 0.40000 | 0.12000 | 0.1300 | 0.20000 | 0.1690 | 0.1638 |
| 2 | 0.50000 | 0.15000 | 0.1600 | 0.25000 | 0.2112 | 0.2060 |
| 1½ | 0.66667 | 0.20000 | 0.2100 | 0.33333 | 0.2816 | 0.2764 |
| 1⅓ | 0.75000 | 0.22500 | 0.2350 | 0.37500 | 0.3168 | 0.3116 |
| 1 | 1.00000 | 0.30000 | 0.3100 | 0.50000 | 0.4224 | 0.4172 |

\* All dimensions in inches. See diagram, page 1325.
† Allowance is 0.020 inch for 10 or less threads per inch and 0.010 inch for more than
10 threads per inch.

**Table 7a. American National Standard Stub Acme Screw Threads — Formulas for
Determining Diameters (ANSI B1.8-1977)**

| | |
|---|---|
| | $D$ = Basic Major Diameter and Nominal Size in Inches |
| | $p$ = Pitch = 1 ÷ Number of Threads per Inch |
| | $E$ = Basic Pitch Diameter = $D - 0.3p$ |
| | $K$ = Basic Minor Diameter = $D - 0.6p$ |
| **No.** | **EXTERNAL THREADS (SCREWS)** |
| 1 | Major Diam., Max. = $D$ |
| 2 | Major Diam., Min. = $D$ *minus* $0.05p$. |
| 3 | Pitch Diam., Max. = $E$ *minus* allowance from column 3, Table 4. |
| 4 | Pitch Diam., Min. = Pitch Diam., Max. (Formula 3) *minus* Class 2G tolerance from Table 5. |
| 5 | Minor Diam., Max. = $K$ *minus* 0.020 for 10 threads per inch and coarser and 0.010 for finer pitches. |
| 6 | Minor Diam., Min. = Minor Diam., Max. (Formula 5) *minus* Class 2G pitch diameter tolerance from Table 5. |
| | **INTERNAL THREADS (NUTS)** |
| 7 | Major Diam., Min. = $D$ *plus* 0.020 for 10 threads per inch and coarser and 0.010 for finer pitches. |
| 8 | Major Diam., Max. = Major Diam., Min. (Formula 7) *plus* Class 2G pitch diameter tolerance from Table 5. |
| 9 | Pitch Diam., Min. = $E$ |
| 10 | Pitch Diam., Max. = Pitch Diam., Min. (Formula 9) *plus* Class 2G tolerance from Table 5. |
| 11 | Minor Diam., Min. = $K$ |
| 12 | Minor Diam., Max. = Minor Diam., Min. (Formula 11) *plus* $0.05p$. |

Table 7b.  Limiting Dimensions for American National Standard Stub Acme Screw Threads (ANSI B1.8-1977)

| Limiting Diameters | | Nominal Diameter (D) | | | | | | | | | | | |
|---|---|---|---|---|---|---|---|---|---|---|---|---|---|
| | | 1/4 | 5/16 | 3/8 | 7/16 | 1/2 | 5/8 | 3/4 | 7/8 | 1 | 1 1/8 | 1 1/4 | 1 3/8 |
| | | Threads per Inch* | | | | | | | | | | | |
| | | 16 | 14 | 12 | 12 | 10 | 8 | 6 | 6 | 5 | 5 | 5 | 4 |
| **External Threads** | | | | | | | | | | | | | |
| Major Diam. {Max(D) | | 0.2500 | 0.3125 | 0.3750 | 0.4375 | 0.5000 | 0.6250 | 0.7500 | 0.8750 | 1.0000 | 1.1250 | 1.2500 | 1.3750 |
| {Min | | 0.2499 | 0.3089 | 0.3708 | 0.4333 | 0.4950 | 0.6188 | 0.7417 | 0.8667 | 0.9900 | 1.1150 | 1.2400 | 1.3625 |
| Pitch Diam. {Max | | 0.2272 | 0.2871 | 0.3451 | 0.4076 | 0.4643 | 0.5812 | 0.6931 | 0.8175 | 0.9320 | 1.0565 | 1.1811 | 1.2906 |
| {Min | | 0.2167 | 0.2757 | 0.3328 | 0.3950 | 0.4506 | 0.5658 | 0.6757 | 0.7996 | 0.9126 | 1.0367 | 1.1610 | 1.2686 |
| Minor Diam. {Max | | 0.2024 | 0.2597 | 0.3150 | 0.3775 | 0.4200 | 0.5300 | 0.6300 | 0.7550 | 0.8600 | 0.9850 | 1.1100 | 1.2050 |
| {Min | | 0.1919 | 0.2483 | 0.3027 | 0.3649 | 0.4063 | 0.5146 | 0.6126 | 0.7371 | 0.8406 | 0.9652 | 1.0899 | 1.1830 |
| **Internal Threads** | | | | | | | | | | | | | |
| Major Diam. {Min | | 0.2600 | 0.3225 | 0.3850 | 0.4475 | 0.5200 | 0.6450 | 0.7700 | 0.8950 | 1.0200 | 1.1450 | 1.2700 | 1.3950 |
| {Max | | 0.2705 | 0.3339 | 0.3973 | 0.4601 | 0.5337 | 0.6604 | 0.7874 | 0.9129 | 1.0394 | 1.1648 | 1.2901 | 1.4170 |
| Pitch Diam. {Min | | 0.2312 | 0.2911 | 0.3500 | 0.4125 | 0.4700 | 0.5875 | 0.7000 | 0.8250 | 0.9400 | 1.0650 | 1.1900 | 1.3000 |
| {Max | | 0.2417 | 0.3025 | 0.3623 | 0.4251 | 0.4837 | 0.6029 | 0.7174 | 0.8429 | 0.9594 | 1.0848 | 1.2101 | 1.3220 |
| Minor Diam. {Min | | 0.2125 | 0.2696 | 0.3250 | 0.3875 | 0.4400 | 0.5500 | 0.6500 | 0.7750 | 0.8800 | 1.0050 | 1.1300 | 1.2250 |
| {Max | | 0.2156 | 0.2732 | 0.3292 | 0.3917 | 0.4450 | 0.5562 | 0.6583 | 0.7833 | 0.8900 | 1.0150 | 1.1400 | 1.2375 |

*All other dimensions are given in inches.

Table 7b (Continued). Limiting Dimensions for American National Standard Stub Acme Screw Threads (ANSI B1.8-1977)

| Limiting Diameters | | 1½ | 1¾ | 2 | 2¼ | 2½ | 2¾ | 3 | 3½ | 4 | 4½ | 5 |
|---|---|---|---|---|---|---|---|---|---|---|---|---|
| | | Nominal Diameter (D) | | | | | | | | | | |
| | | Threads per Inch* | | | | | | | | | | |
| | | 4 | 4 | 4 | 3 | 3 | 3 | 2 | 2 | 2 | 2 | 2 |
| **External Threads** | | | | | | | | | | | | |
| Major Diam. {Max (D) | | 1.5000 | 1.7500 | 2.0000 | 2.2500 | 2.5000 | 2.7500 | 3.0000 | 3.5000 | 4.0000 | 4.5000 | 5.0000 |
| {Min | | 1.4875 | 1.7375 | 1.9875 | 2.2333 | 2.4833 | 2.7333 | 2.9750 | 3.4750 | 3.9750 | 4.4750 | 4.9750 |
| Pitch Diam. {Max | | 1.4152 | 1.6645 | 1.9137 | 2.1380 | 2.3874 | 2.6367 | 2.8360 | 3.3350 | 3.8340 | 4.3330 | 4.8319 |
| {Min | | 1.3929 | 1.6416 | 1.8902 | 2.1117 | 2.3606 | 2.6094 | 2.8044 | 3.3026 | 3.8008 | 4.2991 | 4.7973 |
| Minor Diam. {Max | | 1.3300 | 1.5800 | 1.8300 | 2.0300 | 2.2860 | 2.5300 | 2.6800 | 3.1800 | 3.6800 | 4.1800 | 4.6800 |
| {Min | | 1.3077 | 1.5571 | 1.8065 | 2.0037 | 2.2532 | 2.5027 | 2.6484 | 3.1476 | 3.6468 | 4.1461 | 4.6454 |
| **Internal Threads** | | | | | | | | | | | | |
| Major Diam. {Min | | 1.5200 | 1.7700 | 2.0200 | 2.2700 | 2.5200 | 2.7700 | 3.0200 | 3.5200 | 4.0200 | 4.5200 | 5.0200 |
| {Max | | 1.5423 | 1.7929 | 2.0435 | 2.2963 | 2.5468 | 2.7973 | 3.0516 | 3.5524 | 4.0532 | 4.5539 | 5.0546 |
| Pitch Diam. {Min | | 1.4250 | 1.6750 | 1.9250 | 2.1500 | 2.4000 | 2.6500 | 2.8500 | 3.3500 | 3.8500 | 4.3500 | 4.8500 |
| {Max | | 1.4473 | 1.6979 | 1.9485 | 2.1763 | 2.4268 | 2.6773 | 2.8816 | 3.3824 | 3.8832 | 4.3839 | 4.8846 |
| Minor Diam. {Min | | 1.3500 | 1.6000 | 1.8500 | 2.0500 | 2.3000 | 2.5500 | 2.7000 | 3.2000 | 3.7000 | 4.2000 | 4.7000 |
| {Max | | 1.3625 | 1.6125 | 1.8625 | 2.0667 | 2.3167 | 2.5667 | 2.7250 | 3.2250 | 3.7250 | 4.2250 | 4.7250 |

* All other dimensions are given in inches.

General Purpose Acme threads and are given in column 3 of Table 4. Pitch diameter tolerances for Stub Acme threads are the same as for Class 2G General Purpose Acme threads and are given in columns 2 and 7 of Table 5.

*Limiting Dimensions:* Limiting dimensions of American Standard Stub Acme threads may be determined by using the formulas given in Table 7a, or directly from Table 7b. The diagram below shows the limits of size for Stub Acme threads.

*Thread Series:* The same preferred series of diameters and pitches for General Purpose Acme threads (Table 3) are recommended for Stub Acme threads.

Limits of Size, Allowances, Tolerances, and Crest
Clearances for American National Standard Stub Acme Threads

**Stub Acme Thread Designations.** — The method of designation for Standard Stub Acme threads is illustrated in the following examples: ½–20 Stub Acme indicates a ½-inch major diameter, 20 threads per inch, right hand, single thread, Standard Stub Acme thread. The designation ½–20 Stub Acme–LH indicates the same thread except that it is left hand.

**Alternative Stub Acme Threads.** — Since one Stub Acme thread form may not meet the requirements of all applications, basic data for two of the other commonly used forms is included in the appendix of the American Standard for Stub Acme Threads. These so-called Modified Form 1 and Modified Form 2 threads utilize the same tolerances and allowances as Standard Stub Acme threads and have the same major diameter and basic thread thickness at the pitchline (0.5$p$). The basic height of Form 1 threads, $h$, is 0.375$p$; for Form 2 it is 0.250$p$. The basic width of flat at the crest of the internal thread is 0.4030$p$ for Form 1 and 0.4353$p$ for Form 2.

The pitch diameter and minor diameter for Form 1 threads will be smaller than similar values for the Standard Stub Acme Form and for Form 2 they will be larger owing to the differences in basic thread height $h$. Therefore, in calculating the dimensions of Form 1 and Form 2 threads using Formulas 1 through 12 in Table 7a, it is only necessary to substitute the following values in applying the formulas: For Form 1, $E = D - 0.375p$, $K = D - 0.75p$; for Form 2, $E = D - 0.25p$, $K = D - 0.5p$.

*Thread Designation:* These threads are designated in the same manner as Standard Stub Acme threads except for the insertion of either M1 or M2 after "Acme." Thus, $\frac{1}{2}$–20 Stub Acme M1 for a Form 1 thread; and $\frac{1}{2}$–20 Stub Acme M2 for a Form 2 thread.

**60-Degree Stub Thread.** — Former American Standard B1.3-1941 included a 60-degree stub thread for use where design or operating conditions could be better satisfied by the use of this thread, or other modified threads, than by Acme threads. Data for 60-Degree Stub threads is given in the accompanying table.

60-Degree Stub Thread

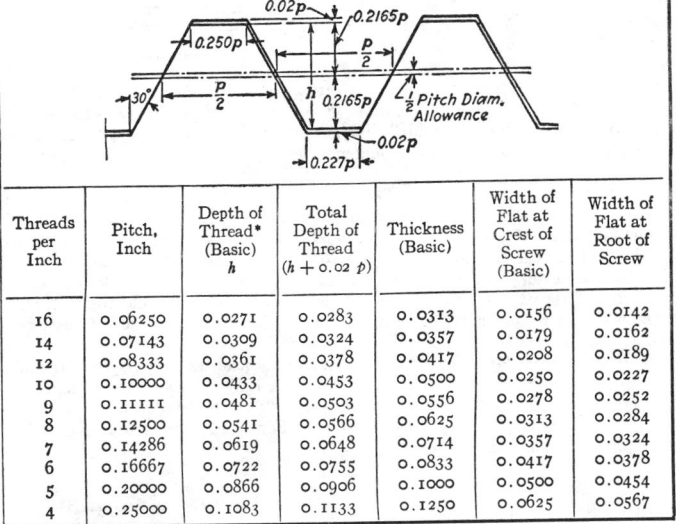

| Threads per Inch | Pitch, Inch | Depth of Thread* (Basic) $h$ | Total Depth of Thread $(h + 0.02\,p)$ | Thickness (Basic) | Width of Flat at Crest of Screw (Basic) | Width of Flat at Root of Screw |
|---|---|---|---|---|---|---|
| 16 | 0.06250 | 0.0271 | 0.0283 | 0.0313 | 0.0156 | 0.0142 |
| 14 | 0.07143 | 0.0309 | 0.0324 | 0.0357 | 0.0179 | 0.0162 |
| 12 | 0.08333 | 0.0361 | 0.0378 | 0.0417 | 0.0208 | 0.0189 |
| 10 | 0.10000 | 0.0433 | 0.0453 | 0.0500 | 0.0250 | 0.0227 |
| 9  | 0.11111 | 0.0481 | 0.0503 | 0.0556 | 0.0278 | 0.0252 |
| 8  | 0.12500 | 0.0541 | 0.0566 | 0.0625 | 0.0313 | 0.0284 |
| 7  | 0.14286 | 0.0619 | 0.0648 | 0.0714 | 0.0357 | 0.0324 |
| 6  | 0.16667 | 0.0722 | 0.0755 | 0.0833 | 0.0417 | 0.0378 |
| 5  | 0.20000 | 0.0866 | 0.0906 | 0.1000 | 0.0500 | 0.0454 |
| 4  | 0.25000 | 0.1083 | 0.1133 | 0.1250 | 0.0625 | 0.0567 |

\* A clearance of at least $0.02 \times$ pitch is added to depth $h$ to produce extra depth, thus avoiding interference with threads of mating part at minor or major diameters.

Basic thread thickness at pitch line = $0.5 \times$ pitch $p$; basic depth $h = 0.433 \times$ pitch; basic width of flat at crest = $0.25 \times$ pitch; width of flat at root of screw thread = $0.227 \times$ pitch; basic pitch diameter = basic major diameter $- 0.433 \times$ pitch; basic minor diameter = basic major diameter $- 0.866 \times$ pitch.

**Square Thread.** — The square form of thread has a pitch about twice that of Unified threads. The thread is difficult to produce economically because of its

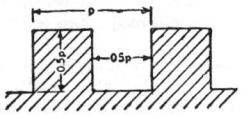

perpendicular sides and has been largely replaced by the Acme form of thread for most applications. Theoretical proportions of a screw are shown in the accompanying diagram. The mating nut has a somewhat larger thread space than the screw to provide a sliding fit.

**10-Degree Modified Square Thread.** — The included angle between the sides of the thread is 10 degrees (see accompanying diagram). The angle of 10 degrees results in a thread which is the practical equivalent of a "square thread," and yet is capable of economical production. Multiple thread milling cutters and ground thread taps should not be specified for modified square threads of the larger lead angles without consulting the cutting tool manufacturer.

*Formulas:* In the following formulas, $D$ = basic major diameter; $E$ = basic pitch diameter; $K$ = basic minor diameter; $p$ = pitch; $h$ = basic depth of thread on screw depth when there is no clearance between root of screw and crest of thread on nut); $t$ = basic thickness of thread at pitch line; $F$ = basic width of flat at crest of screw thread; $G$ = basic width of flat at root of screw thread; $C$ = clearance

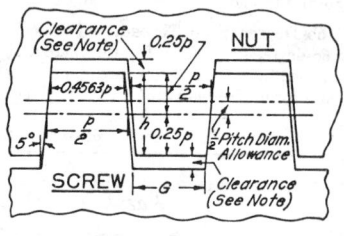

between root of screw and crest of thread on nut: $E = D - 0.5p$; $K = D - p$; $h = 0.5p$ (see Note); $t = 0.5p$; $F = 0.4563p$; $G = 0.4563p - (0.17 \times C)$. *Note:* A clearance should be added to depth $h$ to avoid interference with threads of mating parts at minor or major diameters.

**Threads of Buttress Form.** — The buttress form of thread has certain advantages in applications involving exceptionally high stresses along the thread axis in one direction only. The contacting flank of the thread, which takes the thrust, is referred to as the *pressure flank* and is so nearly perpendicular to the thread axis that the radial component of the thrust is reduced to a minimum. Because of the small radial thrust, this form of thread is particularly applicable where tubular members are screwed together, as in the case of breech mechanisms of large guns and airplane propeller hubs.

Diagram $A$ shows a common form. The front or load-resisting face is perpendicular to the axis of the screw and the thread angle is 45 degrees. According to one rule, the pitch $P = 2 \times$ screw diameter ÷ 15. The thread depth $d$ may equal ¾ × pitch, making the flat $f = \frac{1}{8} \times$ pitch. Sometimes depth $d$ is reduced to ⅔ × pitch, making flat $f = \frac{1}{6} \times$ pitch.

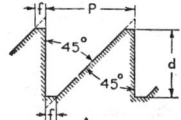
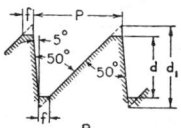
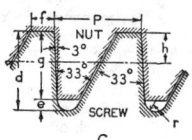

The load-resisting side or flank may be inclined an amount (diagram $B$) ranging usually from 1 to 5 degrees to avoid cutter interference in milling the thread. With an angle of 5 degrees and an included thread angle of 50 degrees, if the width of the flat $f$ at both crest and root equals ⅛ × pitch, then the thread depth equals 0.69 × pitch or ¾ $d_1$.

The saw-tooth form of thread illustrated by diagram $C$ is known in Germany as the "Sägengewinde" and in Italy as the "Fillettatura a dente di Sega." Pitches are standardized from 2 millimeters up to 48 millimeters in the German and Italian specifications. The front face inclines 3 degrees from the perpendicular and the included angle is 33 degrees.

The thread depth $d$ for the screw = 0.86777 × pitch $P$. The thread depth $g$ for the nut = 0.75 × pitch. Dimension $h$ = 0.341 × $P$. The width $f$ of flat at the crest of the thread on the screw = 0.26384 × pitch. Radius $r$ at the root = 0.12427 × pitch. The clearance space $e$ = 0.11777 × pitch.

**American National Standard Buttress Inch Screw Threads.** — The buttress form of thread has certain advantages in applications involving exceptionally high stresses along the thread axis in one direction only. As the thrust side (load flank) of the standard buttress thread is made very nearly perpendicular to the thread axis, the radial component of the thrust is reduced to a minimum. On account of the small radial thrust, the buttress form of thread is particularly applicable when tubular members are screwed together. Examples of actual applications are the breech assemblies of large guns, airplane propeller hubs, and columns for hydraulic presses.

**Table 1. American National Standard Buttress Inch Screw Threads — Basic Dimensions (ANSI B1.9-1973)**

| Thds.* per Inch | Pitch, $p$ | Basic Height of Thread, $h = 0.6p$ | Height of Sharp-V Thread, $H = 0.89064p$ | Crest Trun-cation, $f = 0.14532p$ | Height of Thread, $h_s$ or $h_n = 0.66271p$ | Max. Root Trun-cation,† $s = 0.0826p$ | Max. Root Radius,‡ $r = 0.0714p$ | Width of Flat at Crest, $F = 0.16316p$ |
|---|---|---|---|---|---|---|---|---|
| 20 | 0.0500 | 0.0300 | 0.0445 | 0.0073 | 0.0331 | 0.0041 | 0.0036 | 0.0082 |
| 16 | 0.0625 | 0.0375 | 0.0557 | 0.0091 | 0.0414 | 0.0052 | 0.0045 | 0.0102 |
| 12 | 0.0833 | 0.0500 | 0.0742 | 0.0121 | 0.0552 | 0.0069 | 0.0059 | 0.0136 |
| 10 | 0.1000 | 0.0600 | 0.0891 | 0.0145 | 0.0663 | 0.0083 | 0.0071 | 0.0163 |
| 8 | 0.1250 | 0.0750 | 0.1113 | 0.0182 | 0.0828 | 0.0103 | 0.0089 | 0.0204 |
| 6 | 0.1667 | 0.1000 | 0.1484 | 0.0242 | 0.1105 | 0.0138 | 0.0119 | 0.0271 |
| 5 | 0.2000 | 0.1200 | 0.1781 | 0.0291 | 0.1325 | 0.0165 | 0.0143 | 0.0326 |
| 4 | 0.2500 | 0.1500 | 0.2227 | 0.0363 | 0.1657 | 0.0207 | 0.0179 | 0.0408 |
| 3 | 0.3333 | 0.2000 | 0.2969 | 0.0484 | 0.2209 | 0.0275 | 0.0238 | 0.0543 |
| 2½ | 0.4000 | 0.2400 | 0.3563 | 0.0581 | 0.2651 | 0.0330 | 0.0286 | 0.0653 |
| 2 | 0.5000 | 0.3000 | 0.4453 | 0.0727 | 0.3314 | 0.0413 | 0.0357 | 0.0816 |
| 1½ | 0.6667 | 0.4000 | 0.5938 | 0.0969 | 0.4418 | 0.0551 | 0.0476 | 0.1088 |
| 1¼ | 0.8000 | 0.4800 | 0.7125 | 0.1163 | 0.5302 | 0.0661 | 0.0572 | 0.1305 |
| 1 | 1.0000 | 0.6000 | 0.8906 | 0.1453 | 0.6627 | 0.0826 | 0.0714 | 0.1632 |

* All other dimensions are in inches.
† Minimum root truncation is one-half of maximum.
‡ Minimum root radius is one-half of maximum.

**Table 2. American National Standard Diameter — Pitch Combinations for 7°/45° Buttress Threads (ANSI B1.9-1973)**

| Preferred Nominal Major Diameters, Inches | Threads per Inch* | Preferred Nominal Major Diameters, Inches | Threads per Inch* |
|---|---|---|---|
| 0.5, 0.625, 0.75 | (20, 16, 12) | 4.5, 5, 5.5, 6 | 12, 10, 8, (6, 5, 4), 3 |
| 0.875, 1.0 | (16, 12, 10) | 7, 8, 9, 10 | 10, 8, 6, (5, 4, 3), 2.5, 2 |
| 1.25, 1.375, 1.5 | 16, (12, 10, 8), 6 | 11, 12, 14, 16 | 10, 8, 6, 5, (4, 3, 2.5), 2, 1.5, 1.25 |
| 1.75, 2, 2.25, 2.5 | 16, 12, (10, 8, 6), 5, 4 | 18, 20, 22, 24 | 8, 6, 5, 4, (3, 2.5, 2), 1.5, 1.25, 1 |
| 2.75, 3, 3.5, 4 | 16, 12, 10, (8, 6, 5), 4 | | |

* Preferred pitches are in parentheses.

In selecting the form of thread recommended as standard (ANSI B1.9-1973), manufacture by milling, grinding, rolling, or other suitable means, has been taken into consideration. All dimensions are in inches.

*Form of Thread:* The form of the buttress thread is shown in the accompanying figure and has the following characteristics:

a. A load flank angle, measured in an axial plane, of 7 degrees from the normal to the axis. (*Continued on page* 1342.)

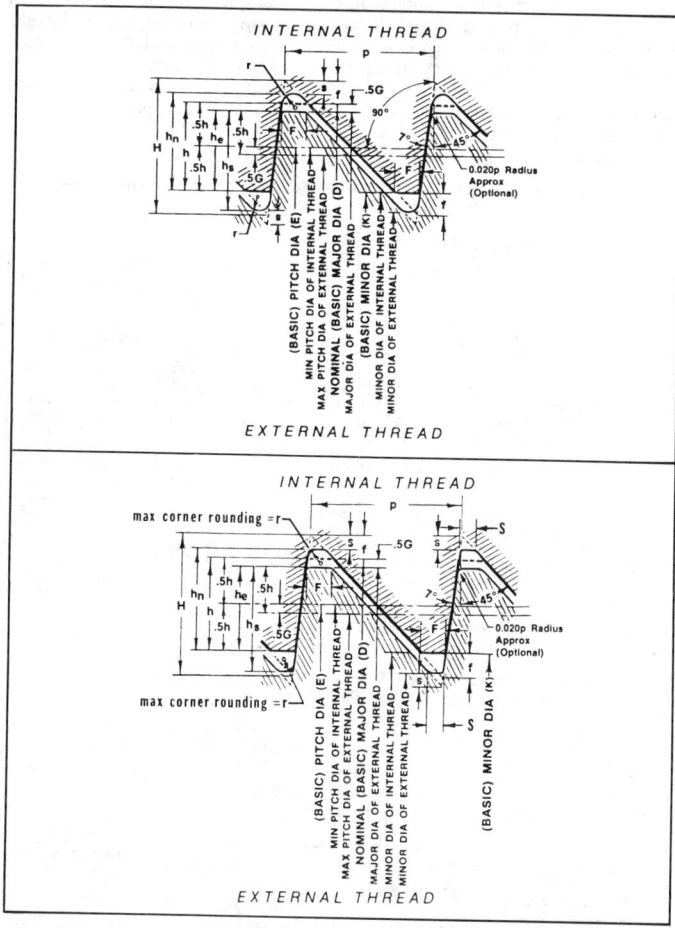

Form of American National Standard 7°/45° Buttress Thread with 0.6p Basic Height of Thread Engagement.    (Top) Round Root.    (Bottom) Flat Root.
Heavy Line Indicates Basic Form

Table 3.  American National Standard Buttress Inch Screw Thread Symbols and Formulas (ANSI B1.9-1973)

| Thread Element | Max. Material (Basic) | Min. Material |
|---|---|---|
| Pitch | $p$ | |
| Height of sharp-V thread | $H = 0.89064p$ | |
| Basic height of thread engagement | $h = 0.6p$ | |
| Root radius (theoretical) (see footnote [a]) | $r = 0.07141p$ | Min. $r = 0.0357p$ |
| Root truncation | $s = 0.0826p$ | Min. $s = 0.5$; Max. $s = 0.0413p$ |
| Root truncation for flat root form | $s = 0.0826p$ | Min. $s = 0.5$; Max. $s = 0.0413p$ |
| Flat width for flat root form | $S = 0.0928p$ | Min. $S = 0.0464p$ |
| Allowance | $G$ (see text) | |
| Height of thread engagement | $h_e = h - 0.5G$ | Min. $h_e$ = Max. $h_e$ − [0.5 tol. on major diam. external thread + 0.5 tol. on minor diam. internal thread]. |
| Crest truncation | $f = 0.14532p$ | |
| Crest width | $F = 0.16316p$ | |
| Major diameter | $D$ | |
| Major diameter of internal thread | $D_n = D + 0.12542p$ | Max. $D_n$ = Max. pitch diam. of internal thread + $0.80803p$ |
| Major diameter of external thread | $D_s = D - G$ | Min. $D_s = D - G - D$ tol. |
| Pitch diameter | $E$ | |
| Pitch diameter of internal thread (see footnote [b]) | $E_n = D - h$ | Max. $E_n = D - h + PD$ tol. |
| Pitch diameter of external thread (see footnote [c]) | $E_s = D - h - G$ | Min. $E_s = D - h - G - PD$ tol. |
| Minor diameter | $K$ | |
| Minor diameter of external thread | $K_s = D - 1.32542p - G$ | Min. $K_s$ = Min. pitch diam. of external thread − $0.80803p$ |
| Minor diameter of internal thread | $K_n = D - 2h$ | Min. $K_n = D - 2h + K$ tol. |
| Height of thread of internal thread | $h_n = 0.66271p$ | |
| Height of thread of external thread | $h_s = 0.66271p$ | |
| Pitch diameter increment for lead | $\Delta E_l$ | |
| Pitch diameter increment for 45° clearance flank angle | $\Delta E_{\alpha_1}$ | |
| Pitch diameter increment for 7° load flank angle | $\Delta E_{\alpha_2}$ | |
| Length of engagement | $L_e$ | |

[a] Unless the flat root form is specified, the rounded root form of the external and internal thread shall be a continuous, smoothly blended curve within the zone defined by $0.07141p$ maximum to $0.0357p$ minimum radius.  The resulting curve shall have no reversals and sudden angular variations, and shall be tangent to the flanks of the thread.  There is, in practice, almost no chance that the rounded thread form will be achieved strictly as basically specified, that is, as a true radius.

[b] The pitch diameter $X$ tolerances for GO and NOT GO threaded plug gages are applied to the internal product limits for $E_n$ and Max. $E_n$.

[c] The pitch diameter $W$ tolerances for GO and NOT GO threaded setting plug gages are applied to the external product limits for $E_s$ and Min. $E_s$.

b. A clearance flank angle, measured in an axial plane, of 45 degrees from the normal to the axis.

c. Equal truncations at the crests of the external and internal threads such that the basic height of thread engagement (assuming no allowance) is equal to 0.6 of the pitch.

d. Equal radii, at the roots of the external and internal basic thread forms tangential to the load flank and the clearance flank. (There is, in practice, almost no chance that the thread forms will be achieved strictly as basically specified, that is, as true radii.) When specified, equal flat roots of the external and internal thread may be supplied.

**Buttress Thread Tolerances.** — Tolerances from basic size on external threads are applied in a minus direction and on internal threads in a plus direction.

*Pitch Diameter Tolerances:* The following formula is used for determining the pitch diameter product tolerance for Class 2 (standard grade) external or internal threads:

$$\text{PD tolerance} = 0.002 \sqrt[3]{D} + 0.00278 \sqrt{L_e} + 0.00854 \sqrt{p}$$

where $D$ = basic major diameter of external thread (assuming no allowance)
$L_e$ = length of engagement
$p$ = pitch of thread

When the length of engagement is taken as $10p$, the formula reduces to:

$$0.002 \sqrt[3]{D} + 0.0173 \sqrt{p}$$

It is to be noted that this formula relates specifically to Class 2 (standard grade) PD tolerances. Class 3 (precision grade) PD tolerances are two-thirds of Class 2 PD tolerances. Pitch diameter tolerances based on this latter formula, for various diameter pitch combinations, are given in Tables 4 and 5.

*Functional Size:* Deviations in lead and flank angle of product threads increase the functional size of an external thread and decrease the functional size of an internal thread by the cumulative effect of the diameter equivalents of these deviations. The functional size of all buttress product threads shall not exceed the maximum-material limit.

*Tolerances on Major Diameter of External Thread and Minor Diameter of Internal Thread:* Unless otherwise specified, these tolerances should be the same as the pitch diameter tolerance for the class used.

*Tolerances on Minor Diameter of External Thread and Major Diameter of Internal Thread:* It will be sufficient in most instances to state only the maximum minor diameter of the external thread and the minimum major diameter of the internal thread without any tolerance. However, the root truncation from a sharp V should not be greater than $0.0826p$ nor less than $0.0413p$.

*Lead and Flank Angle Deviations for Class 2:* The deviations in lead and flank angles may consume the entire tolerance zone between maximum and minimum material product limits given in Table 4.

*Diameter Equivalents for Variations in Lead and Flank Angles for Class 3:* The combined diameter equivalents of variations in lead (including helix deviations), and flank angle for Class 3, shall not exceed 50 percent of the pitch diameter tolerances given in Table 5.

*Tolerances on Taper and Roundness:* There are no requirements for taper and roundness for Class 2 buttress screw threads.

The major and minor diameter of Class 3 buttress thread shall not taper or be out-of-round to the extent that specified limits for major and minor diameter are

Table 4.  American National Standard Class 2 (Standard Grade) Tolerances for Buttress Inch Screw Threads (ANSI B1.9-1973)

| Thds. per Inch | Pitch,* $p$ Inch | Basic Major Diameter, Inch | | | | | | | | | Pitch† Increment, $0.0173\sqrt{p}$ Inch |
|---|---|---|---|---|---|---|---|---|---|---|---|
| | | From 0.5 thru 0.7 | Over 0.7 thru 1.0 | Over 1.0 thru 1.5 | Over 1.5 thru 2.5 | Over 2.5 thru 4 | Over 4 thru 6 | Over 6 thru 10 | Over 10 thru 16 | Over 16 thru 24 | |
| | | Tolerance on Major Diameter of External Thread, Pitch Diameter of External and Internal Threads, and Minor Diameter of Internal Thread, Inch | | | | | | | | | |
| 20 | 0.0500 | .0056 | .... | .... | .... | .... | .... | .... | .... | .... | .00387 |
| 16 | 0.0625 | .0060 | .0062 | .0065 | .0068 | .0073 | .... | .... | .... | .... | .00432 |
| 12 | 0.0833 | .0067 | .0069 | .0071 | .0075 | .0080 | .0084 | .... | .... | .... | .00499 |
| 10 | 0.1000 | .... | .0074 | .0076 | .0080 | .0084 | .0089 | .0095 | .0102 | .... | .00547 |
| 8 | 0.1250 | .... | .... | .0083 | .0086 | .0091 | .0095 | .0101 | .0108 | .0115 | .00612 |
| 6 | 0.1667 | .... | .... | .0092 | .0096 | .0100 | .0105 | .0111 | .0118 | .0125 | .00706 |
| 5 | 0.2000 | .... | .... | .... | .0103 | .0107 | .0112 | .0117 | .0124 | .0132 | .00774 |
| 4 | 0.2500 | .... | .... | .... | .0112 | .0116 | .0121 | .0127 | .0134 | .0141 | .00865 |
| 3 | 0.3333 | .... | .... | .... | .... | .0134 | .0140 | .0147 | .0154 | | .00999 |
| 2.5 | 0.4000 | .... | .... | .... | .... | .... | .0149 | .0156 | .0164 | | .01094 |
| 2.0 | 0.5000 | .... | .... | .... | .... | .... | .0162 | .0169 | .0177 | | .01223 |
| 1.5 | 0.6667 | .... | .... | .... | .... | .... | .... | .0188 | .0196 | | .01413 |
| 1.25 | 0.8000 | .... | .... | .... | .... | .... | .... | .0202 | .0209 | | .01547 |
| 1.0 | 1.0000 | .... | .... | .... | .... | .... | .... | .... | .0227 | | .01730 |
| Diameter Increment,‡ $0.002\sqrt{D}$ | | .00169 | .00189 | .00215 | .00252 | .00296 | .00342 | .00400 | .00470 | .00543 | |

\* For threads with pitches not shown in this table, pitch increment to be used in tolerance formula is to be determined by use of formula P.D. Tolerance = $0.002\sqrt[3]{D} + 0.00278\sqrt{L_e} + 0.00854\sqrt{p}$ where: $D$ = basic major diameter of external thread (assuming no allowance), $L_e$ = length of engagement, and $p$ = pitch of thread.

† When the length of engagement is taken as $10p$, the formula reduces to: $0.002\sqrt[3]{D} + 0.0173\sqrt{p}$.

‡ Diameter $D$, used in diameter increment formula, is based on the average of the range.

Table 5.  American National Standard Class 3 (Precision Grade) Tolerances for Buttress Inch Screw Threads (ANSI B1.9-1973)

| Threads per Inch | Pitch, $p$ Inch | Basic Major Diameter, Inch | | | | | | | | |
|---|---|---|---|---|---|---|---|---|---|---|
| | | From 0.5 thru 0.7 | Over 0.7 thru 1.0 | Over 1.0 thru 1.5 | Over 1.5 thru 2.5 | Over 2.5 thru 4 | Over 4 thru 6 | Over 6 thru 10 | Over 10 thru 16 | Over 16 thru 24 |
| | | Tolerance on Major Diameter of External Thread, Pitch Diameter of External and Internal Threads, and Minor Diameter of Internal Thread, Inch | | | | | | | | |
| 20 | 0.0500 | .0037 | .... | .... | .... | .... | .... | .... | .... | .... |
| 16 | 0.0625 | .0040 | .0042 | .0043 | .0046 | .0049 | .... | .... | .... | .... |
| 12 | 0.0833 | .0044 | .0046 | .0048 | .0050 | .0053 | .0056 | .... | .... | .... |
| 10 | 0.1000 | .... | .0049 | .0051 | .0053 | .0056 | .0059 | .0063 | .0068 | .... |
| 8 | 0.1250 | .... | .... | .0055 | .0058 | .0061 | .0064 | .0067 | .0072 | .0077 |
| 6 | 0.1667 | .... | .... | .0061 | .0064 | .0067 | .0070 | .0074 | .0078 | .0083 |
| 5 | 0.2000 | .... | .... | .... | .0068 | .0071 | .0074 | .0078 | .0083 | .0088 |
| 4 | 0.2500 | .... | .... | .... | .0074 | .0077 | .0080 | .0084 | .0089 | .0094 |
| 3 | 0.3333 | .... | .... | .... | .... | .0089 | .0093 | .0098 | .0103 | |
| 2.5 | 0.4000 | .... | .... | .... | .... | .... | .0100 | .0104 | .0109 | |
| 2.0 | 0.5000 | .... | .... | .... | .... | .... | .0108 | .0113 | .0118 | |
| 1.5 | 0.6667 | .... | .... | .... | .... | .... | .... | .0126 | .0130 | |
| 1.25 | 0.8000 | .... | .... | .... | .... | .... | .... | .0135 | .0139 | |
| 1.0 | 1.0000 | .... | .... | .... | .... | .... | .... | .... | .0152 | |

Table 6.  American National Standard External Thread Allowances for Classes 2 and 3 Buttress Inch Screw Threads (ANSI B1.9-1973)

| Threads per Inch | Pitch, p Inch | Basic Major Diameter, Inch | | | | | | | | |
|---|---|---|---|---|---|---|---|---|---|---|
| | | From 0.5 thru 0.7 | Over 0.7 thru 1.0 | Over 1.0 thru 1.5 | Over 1.5 thru 2.5 | Over 2.5 thru 4 | Over 4 thru 6 | Over 6 thru 10 | Over 10 thru 16 | Over 16 thru 24 |
| | | Allowance on Major, Minor and Pitch Diameters of External Thread, Inch | | | | | | | | |
| 20 | 0.0500 | .0037 | .... | .... | .... | .... | .... | .... | .... | .... |
| 16 | 0.0625 | .0040 | .0042 | .0043 | .0046 | .0049 | .... | .... | .... | .... |
| 12 | 0.0833 | .0044 | .0046 | .0048 | .0050 | .0053 | .0056 | .... | .... | .... |
| 10 | 0.1000 | .... | .0049 | .0051 | .0053 | .0056 | .0059 | .0063 | .0068 | .... |
| 8 | 0.1250 | .... | .... | .0055 | .0058 | .0061 | .0064 | .0067 | .0072 | .0077 |
| 6 | 0.1667 | .... | .... | .0061 | .0064 | .0067 | .0070 | .0074 | .0078 | .0083 |
| 5 | 0.2000 | .... | .... | .... | .0068 | .0071 | .0074 | .0078 | .0083 | .0088 |
| 4 | 0.2500 | .... | .... | .... | .0074 | .0077 | .0080 | .0084 | .0089 | .0094 |
| 3 | 0.3333 | .... | .... | .... | .... | .... | .0089 | .0093 | .0098 | .0103 |
| 2.5 | 0.4000 | .... | .... | .... | .... | .... | .... | .0100 | .0104 | .0109 |
| 2.0 | 0.5000 | .... | .... | .... | .... | .... | .... | .0108 | .0113 | .0118 |
| 1.5 | 0.6667 | .... | .... | .... | .... | .... | .... | .... | .0126 | .0130 |
| 1.25 | 0.8000 | .... | .... | .... | .... | .... | .... | .... | .0135 | .0139 |
| 1.0 | 1.0000 | .... | .... | .... | .... | .... | .... | .... | .... | .0152 |

exceeded.  The taper and out-of-roundness of the pitch diameter for Class 3 buttress threads shall not exceed 50 percent of the pitch diameter tolerances.

**Allowances for Easy Assembly.** — An allowance (clearance) should be provided on all external threads to secure easy assembly of parts.  The amount of the allowance is deducted from the nominal major, pitch and minor diameters of the external thread in order to determine the maximum material condition of the external thread.

The minimum internal thread is basic.

The amount of the allowance is the same for both classes and is equal to the Class 3 pitch diameter tolerance as calculated by the formulas previously given. The allowances for various diameter-pitch combinations are given in Table 6.

**Example Showing Dimensions for a Typical Buttress Thread.** — The dimensions for a 2-inch diameter, 4 threads per inch, Class 2 buttress thread with flank angles of 7 degrees and 45 degrees are:

$h$ = Basic thread height = 0.1500 (Table 1)

$h_s = h_n$ = Height of thread in external and internal threads = 0.1657 (Table 1)

$G$ = Pitch diameter allowance on external thread = 0.0074 (Table 6)

Tolerance on PD of external and internal threads = 0.0112 (Table 4)

Tolerance on major diameter of external thread and minor diameter of internal thread = 0.0112 (Table 4)

*Internal Thread*

Basic Major Diameter = $D$ = 2.0000

Min Major Diameter = $D - 2h + 2h_n$ = 2.0314 (see Table 1)

Min Pitch Diameter = $D - h$ = 1.8500 (see Table 1)

Max Pitch Diameter = $D - h + PD$ Tolerance = 1.8612 (see Table 4)

Min Minor Diameter = $D - 2h$ = 1.7000 (see Table 1)

Max Minor Diameter = $D - 2h$ + Minor Diameter Tolerance
= 1.7112 (see Table 4)

*External Thread*

Max Major Diameter $= D - G = 1.9926$ (see Table 6)
Min Major Diameter $= D - G -$ Major Diameter Tolerance
$= 1.9814$ (see Tables 4 and 6)
Max Pitch Diameter $= D - h - G = 1.8426$ (see Tables 1 and 6)
Min Pitch Diameter $= D - h - G - PD$ Tolerance $= 1.8314$ (see Table 4)
Max Minor Diameter $= D - G - 2h_s = 1.6612$ (see Tables 1 and 6)

**Buttress Thread Designations.** — When only the designation, BUTT is used, the thread is "pull" type buttress (external thread pulls) with the clearance flank leading and the pressure flank 7° following. When the designation, PUSH–BUTT is used, the thread is a push type buttress (external thread pushes) with the load flank 7° leading and the 45° clearance flank following. Whenever possible this description should be confirmed by a simplified view showing thread angles on the drawing of the product that has the buttress thread.

*Standard Buttress Threads:* A buttress thread is considered to be standard when: (a) opposite flank angles are 7° and 45°; (b) basic thread height is 0.6$p$; (c) tolerances and allowances are as shown in Tables 4 through 6; and (d) length of engagement is 10$p$ or less.

*Thread Designation Abbreviations:* In thread designations on drawings, tools, gages, and in specifications, the following abbreviations and letters are to be used:

BUTT          for buttress thread, pull type
PUSH-BUTT     for buttress thread, push type
LH            for left-hand thread (Absence of LH indicates that the thread is
              a right-hand thread.)
P             for pitch
L             for lead
A             for external thread
B             for internal thread

*Note:* Absence of A or B after thread class indicates that designation covers both the external and internal threads.

Le            for length of thread engagement
SPL           for special
FL            for flat root thread
E             for pitch diameter
TPI           for threads per inch
THD           for thread

**Designation Sequence for Buttress Inch Screw Threads.** — When designating single-start standard buttress threads the nominal size is given first, the threads per inch next, then PUSH if the internal member is to push, but nothing if it is to pull, then the class of thread (2 or 3), then whether external (A) or internal (B), then LH if left-hand, but nothing if right-hand, and finally FL if a flat root thread, but nothing if a radiused root thread; thus, 2.5-8 BUTT-2A indicates a 2.5 inch, 8 threads per inch buttress thread, Class 2 external, right-hand, internal member to pull, with radiused root of thread. The designation 2.5-8 PUSH-BUTT-2A-LH-FL signifies a 2.5 inch size, 8 threads per inch buttress thread with internal member to push, Class 2 external, left-hand, and flat root.

A multiple-start standard buttress thread is similarly designated but the pitch is given instead of the threads per inch, this is followed by the lead and the number of starts is indicated in parentheses after the class of thread. Thus, 10-0.25P-0.5L — BUTT-3B (2 start) indicates a 10-inch thread with 4 threads per inch, 0.5 inch lead, buttress form with internal member to pull, Class 3 internal, 2 starts, with radiused root of thread.

**Measurement of Buttress Thread Gages and Product.** — Measuring the pitch diameter of buttress threads presents some difficulty because there is a wide difference between the angle of the load flank and the angle of the clearance flank. The clearance flank of 45° has a greater effect on the pitch diameter measurements than the 7° flank. Therefore, the clearance flank angle on the thread gages should be held to at least as close as the tolerance on the load flank.

For full information on gages and gaging practice for American National Standard buttress threads see American National Standard ANSI B1.9-1973.

### Aero-Thread Formulas and Basic Dimensions

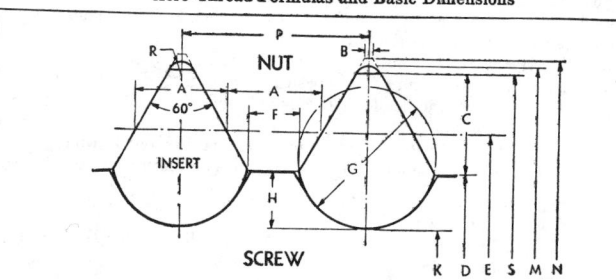

$D$ = major screw diameter and minor diameter of tapped hole (see table); pitch $P = 1 \div$ number of threads per inch (see table); root diameter $K = D - 0.6P = E - P$; tapped hole pitch diameter $E = D + 0.4P$; insert diameter $S = D + 1.05P$; minimum major tap diameter $M = D + 1.122P$; maximum major tap diameter $N = D + 1.194P$; insert engagement $C = 0.525P$; depth $H = 0.3P$; flat $B = P \div 24$; radius $R = 0.072P$; thread form circle diameter $G = 0.75P$; $A = 0.5P$.

| Major Diam. $D$ | Threads per Inch | Pitch Diam. $E$ | Root Diam. $K$ | Major Diam. $S$ Insert | Diam. $G$ Form Circle | Major Tap Diameter | |
|---|---|---|---|---|---|---|---|
| | | | | | | Max. $N$ | Min. $M$ |
| 0.1875 | 24 | 0.2042 | 0.1625 | 0.2313 | 0.0313 | 0.2373 | 0.2343 |
| 0.2500 | 20 | 0.2700 | 0.2200 | 0.3024 | 0.0375 | 0.3098 | 0.3062 |
| 0.3125 | 18 | 0.3347 | 0.2792 | 0.3709 | 0.0417 | 0.3789 | 0.3749 |
| 0.3750 | 16 | 0.4000 | 0.3375 | 0.4406 | 0.0469 | 0.4496 | 0.4452 |
| 0.4375 | 14 | 0.4661 | 0.3946 | 0.5125 | 0.0536 | 0.5228 | 0.5177 |
| 0.5000 | 12 | 0.5333 | 0.4500 | 0.5876 | 0.0625 | 0.5996 | 0.5936 |
| 0.5625 | 12 | 0.5958 | 0.5125 | 0.6500 | 0.0625 | 0.6621 | 0.6561 |
| 0.6250 | 10 | 0.6650 | 0.5650 | 0.7300 | 0.0750 | 0.7444 | 0.7372 |
| 0.6875 | 10 | 0.7275 | 0.6275 | 0.7925 | 0.0750 | 0.8069 | 0.7997 |
| 0.7500 | 9 | 0.7944 | 0.6833 | 0.8666 | 0.0833 | 0.8826 | 0.8746 |
| 0.8750 | 8 | 0.9250 | 0.8000 | 1.0062 | 0.0938 | 1.0242 | 1.0152 |

The " Aero-Thread " is a patented form especially applicable where the internally threaded part is made from soft light materials, such as aluminum or magnesium alloys — as in aircraft construction — and the screw is made from high-strength steel. The spring-shaped insert, usually of phosphor bronze, prevents wear of the light alloy thread especially where frequent bolt removal is necessary. The insert also adjusts itself to bear evenly on all of the thread surfaces.

## Dardelet Self-locking Screw Thread — Standard Series

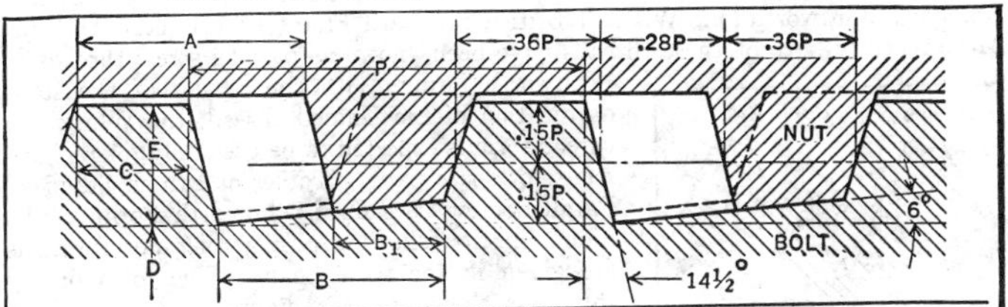

| Nom. Size and Major Dia. | Thds. per Inch | Minor Dia. $D$, Bolt and Nut | External Thread | | | Internal Thread | | |
|---|---|---|---|---|---|---|---|---|
| | | | Thread Depth $E$ | Projected Width $B$ | Width Crest $C$ | Projected Width $B_1$ | Width $A$ | Tap Drill Stock Size |
| ¼ | 16 | 0.2125 | 0.0188 | 0.0361 | 0.0176 | 0.0181 | 0.0351 | No. 5 |
| ⁵⁄₁₆ | 14 | 0.2697 | 0.0214 | 0.0413 | 0.0201 | 0.0207 | 0.0402 | G |
| ⅜ | 12 | 0.3250 | 0.0250 | 0.0482 | 0.0235 | 0.0242 | 0.0468 | O |
| ⁷⁄₁₆ | 11 | 0.3830 | 0.0273 | 0.0526 | 0.0256 | 0.0264 | 0.0511 | ⅜ |
| ½ | 10 | 0.4400 | 0.0300 | 0.0578 | 0.0282 | 0.0290 | 0.0562 | 27⁄64 |
| ⁹⁄₁₆ | 9 | 0.4959 | 0.0333 | 0.0642 | 0.0313 | 0.0322 | 0.0624 | 31⁄64 |
| ⅝ | 8 | 0.5500 | 0.0375 | 0.0723 | 0.0353 | 0.0363 | 0.0703 | 17⁄32 |
| ¾ | 8 | 0.6750 | 0.0375 | 0.0723 | 0.0353 | 0.0363 | 0.0703 | 21⁄32 |
| ⅞ | 7 | 0.7893 | 0.0429 | 0.0826 | 0.0403 | 0.0414 | 0.0804 | 25⁄32 |
| 1 | 6 | 0.9000 | 0.0500 | 0.0964 | 0.0470 | 0.0483 | 0.0937 | 57⁄64 |
| 1⅛ | 5 | 1.0050 | 0.0600 | 0.1156 | 0.0564 | 0.0580 | 0.1125 | 63⁄64 |
| 1¼ | 5 | 1.1300 | 0.0600 | 0.1156 | 0.0564 | 0.0580 | 0.1125 | 1 7⁄64 |
| 1½ | 4 | 1.3500 | 0.0750 | 0.1445 | 0.0705 | 0.0725 | 0.1406 | 1 21⁄64 |
| 1¾ | 4 | 1.6000 | 0.0750 | 0.1445 | 0.0705 | 0.0725 | 0.1406 | 1 37⁄64 |
| 2 | 4 | 1.8500 | 0.0750 | 0.1445 | 0.0705 | 0.0725 | 0.1406 | 1 53⁄64 |
| 2¼ | 4 | 2.1000 | 0.0750 | 0.1445 | 0.0705 | 0.0725 | 0.1406 | 2 3⁄32 |
| 2½ | 4 | 2.3500 | 0.0750 | 0.1445 | 0.0705 | 0.0725 | 0.1406 | 2 11⁄32 |
| 2¾ | 4 | 2.6000 | 0.0750 | 0.1445 | 0.0705 | 0.0725 | 0.1406 | 2 19⁄32 |
| 3 | 4 | 2.8500 | 0.0750 | 0.1445 | 0.0705 | 0.0725 | 0.1406 | 2 27⁄32 |

**Dardelet Thread.** — The Dardelet patented self-locking thread is designed to resist vibrations and remain tight without auxiliary locking devices. The locking surfaces are the tapered root of the bolt thread and the tapered crest of the nut thread. The nut is free to turn until seated tightly against a resisting surface, thus causing it to shift from the free position (indicated by dotted lines) to the locking position. The locking is due to a wedging action between the tapered crest of the nut thread and the tapered root or binding surface of the bolt thread. This self-locking thread is also applied to set-screws and cap-screws. The holes must, of course, be threaded with Dardelet taps. The abutment sides of the Dardelet thread carry the major part of the tensile load. The nut is unlocked simply by turning it backward with a wrench. The Dardelet thread can either be cut or rolled, using standard equipment provided with tools, taps, dies, or rolls made to suit the Dardelet thread profile. The included thread angle is 29 degrees; depth $E = 0.3\ P$; maximum axial movement $= 0.28\ P$. The major internal thread diameter (standard series) equals major external thread diameter plus 0.003 inch except for ¼-inch size which is plus 0.002 inch. The width of both external and internal threads at pitch line equals 0.36 $P$.

**Whitworth Standard Thread Form.** — This thread form is used for the British Standard Whitworth (B.S.W.) and British Standard Fine (B.S.F.) screw threads. More recently both of these threads have been known as parallel screw threads of

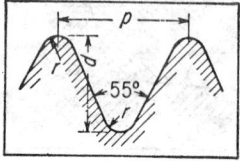

Whitworth form. With the advent of the standardization of the Unified thread, the Whitworth thread form is expected to be used only for replacements or spare parts. Tables of British Standard Parallel Screw Threads of Whitworth Form will be found on the following pages; tolerance formulas are given in the table below. The form of the thread is shown by the diagram. If $p$ = pitch, $d$ = depth of thread, $r$ = radius at crest and root, and $n$ = number of threads per inch, then:

$$d = \tfrac{1}{3}p \times \cot 27°30' = 0.640327p = 0.640327 \div n$$

$$r = 0.137329p = 0.137329 \div n$$

**British Standard Whitworth (B.S.W.) and British Standard Fine (B.S.F.) Threads.** — The B.S.W. is the Coarse Thread series and the B.S.F. is the Fine Thread series of British Standard 84:1956 — Parallel Screw Threads of Whitworth Form. The dimensions given in the table on page 1350 for the major, effective, and minor diameters are, respectively, the maximum limits of these diameters for bolts and the minimum limits for nuts. Formulas for the tolerances on these diameters are given in the table below.

It is recommended that stainless steel bolts of nominal sizes ¾ inch and below should not be made to Close Class limits but rather to Medium or Free Class limits. Nominal sizes above ¾ inch should have maximum and minimum limits which are 0.001 inch smaller than the values obtained from the table.

*Tolerance Classes: Close Class bolts.* Applies to screw threads requiring a fine snug fit, and should be used only for special work where refined accuracy of pitch and thread form are particularly required. *Medium Class bolts and nuts.* Applies to the better class of ordinary interchangeable screw threads. *Free Class bolts.* Applies to the majority of bolts of ordinary commercial quality. *Normal Class nuts.* Applies to ordinary commercial quality nuts; this class is intended for use with Medium or Free Class bolts.

*Allowances:* Only Free Class and Medium Class bolts have an allowance. For nominal sizes of ¾ inch down to ¼ inch, the allowance is 30 per cent of the Medium Class bolt effective diameter tolerance (0.3T); for sizes less than ¼ inch, the allowance for the ¼ inch size applies. Allowances are applied minus from the basic bolt dimensions; the tolerances are then applied to the reduced dimensions.

**Tolerance Formulas for B.S.W. and B.S.F. Threads**

|  | Class or Fit | Tolerance[1] (+ for nuts, − for bolts) | | |
|---|---|---|---|---|
|  |  | Major Diam. | Effective Diam. | Minor Diam. |
| Bolts | Close | $\tfrac{2}{3}T + 0.01\sqrt{p}$ in. | $\tfrac{2}{3}T$ in. | $\tfrac{2}{3}T + 0.013\sqrt{p}$ in. |
|  | Medium | $T + 0.01\sqrt{p}$ in. | $T$ in. | $T + 0.02\ \sqrt{p}$ in. |
|  | Free | $\tfrac{3}{2}T + 0.01\sqrt{p}$ in. | $\tfrac{3}{2}T$ in. | $\tfrac{3}{2}T + 0.02\ \sqrt{p}$ in. |
| Nuts | Close | .... | $\tfrac{2}{3}T$ in. | $\{$ 0.2$p$ + 0.004 in.[2] |
|  | Medium | .... | $T$ in. | $\{$ 0.2$p$ + 0.005 in.[3] |
|  | Normal | .... | $\tfrac{3}{2}T$ in. | $\{$ 0.2$p$ + 0.007 in.[4] |

[1] The symbol, $T = 0.002\sqrt[3]{D} + 0.003\sqrt{L} + 0.005\sqrt{p}$ in which $D$ = major diameter of thread in inches; $L$ = length of engagement in inches; $p$ = pitch in inches. The symbol, $p$, which appears in the table signifies pitch. [2] For 26 threads per inch and finer. [3] For 24 and 22 threads per inch. [4] For 20 threads per inch and coarser.

## Threads of Whitworth Form — Basic Dimensions

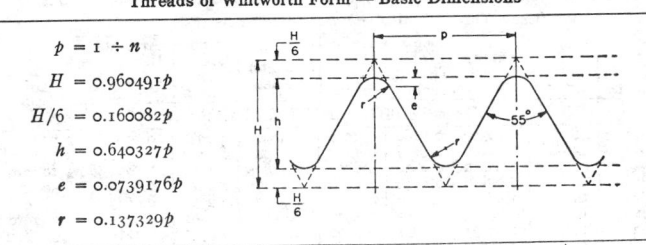

$$p = 1 \div n$$
$$H = 0.960491p$$
$$H/6 = 0.160082p$$
$$h = 0.640327p$$
$$e = 0.0739176p$$
$$r = 0.137329p$$

| Threads per Inch | Pitch | Tri-angular Height | Shorten-ing | Depth of Thread | Depth of Rounding | Radius |
|---|---|---|---|---|---|---|
| $n$ | $p$ | $H$ | $H/6$ | $h$ | $e$ | $r$ |
| 72 | 0.013889 | 0.013340 | 0.002223 | 0.008894 | 0.001027 | 0.001907 |
| 60 | 0.016667 | 0.016009 | 0.002668 | 0.010672 | 0.001232 | 0.002289 |
| 56 | 0.017857 | 0.017151 | 0.002859 | 0.011434 | 0.001320 | 0.002452 |
| 48 | 0.020833 | 0.020010 | 0.003335 | 0.013340 | 0.001540 | 0.002861 |
| 40 | 0.025000 | 0.024012 | 0.004002 | 0.016008 | 0.001848 | 0.003433 |
| 36 | 0.027778 | 0.026680 | 0.004447 | 0.017787 | 0.002053 | 0.003815 |
| 32 | 0.031250 | 0.030015 | 0.005003 | 0.020010 | 0.002310 | 0.004292 |
| 28 | 0.035714 | 0.034303 | 0.005717 | 0.022869 | 0.002640 | 0.004905 |
| 26 | 0.038462 | 0.036942 | 0.006157 | 0.024628 | 0.002843 | 0.005282 |
| 24 | 0.041667 | 0.040020 | 0.006670 | 0.026680 | 0.003080 | 0.005722 |
| 22 | 0.045455 | 0.043659 | 0.007276 | 0.029106 | 0.003360 | 0.006242 |
| 20 | 0.050000 | 0.048025 | 0.008004 | 0.032016 | 0.003696 | 0.006866 |
| 19 | 0.052632 | 0.050553 | 0.008425 | 0.033702 | 0.003890 | 0.007228 |
| 18 | 0.055556 | 0.053361 | 0.008893 | 0.035574 | 0.004107 | 0.007629 |
| 16 | 0.062500 | 0.060031 | 0.010005 | 0.040020 | 0.004620 | 0.008583 |
| 14 | 0.071429 | 0.068607 | 0.011434 | 0.045738 | 0.005280 | 0.009809 |
| 12 | 0.083333 | 0.080041 | 0.013340 | 0.053361 | 0.006160 | 0.011444 |
| 11 | 0.090909 | 0.087317 | 0.014553 | 0.058212 | 0.006720 | 0.012484 |
| 10 | 0.100000 | 0.096049 | 0.016008 | 0.064033 | 0.007392 | 0.013733 |
| 9 | 0.111111 | 0.106721 | 0.017787 | 0.071147 | 0.008213 | 0.015259 |
| 8 | 0.125000 | 0.120061 | 0.020010 | 0.080041 | 0.009240 | 0.017166 |
| 7 | 0.142857 | 0.137213 | 0.022869 | 0.091475 | 0.010560 | 0.019618 |
| 6 | 0.166667 | 0.160082 | 0.026680 | 0.106721 | 0.012320 | 0.022888 |
| 5 | 0.200000 | 0.192098 | 0.032016 | 0.128065 | 0.014784 | 0.027466 |
| 4.5 | 0.222222 | 0.213442 | 0.035574 | 0.142295 | 0.016426 | 0.030518 |
| 4 | 0.250000 | 0.240123 | 0.040020 | 0.160082 | 0.018479 | 0.034332 |
| 3.5 | 0.285714 | 0.274426 | 0.045738 | 0.182951 | 0.021119 | 0.039237 |
| 3.25 | 0.307692 | 0.295536 | 0.049256 | 0.197024 | 0.022744 | 0.042255 |
| 3 | 0.333333 | 0.320164 | 0.053361 | 0.213442 | 0.024639 | 0.045776 |
| 2.875 | 0.347826 | 0.334084 | 0.055681 | 0.222722 | 0.025710 | 0.047767 |
| 2.75 | 0.363636 | 0.349269 | 0.058212 | 0.232846 | 0.026879 | 0.049938 |
| 2.625 | 0.380952 | 0.365901 | 0.060984 | 0.243934 | 0.028159 | 0.052316 |
| 2.5 | 0.400000 | 0.384196 | 0.064033 | 0.256131 | 0.029567 | 0.054932 |

Dimensions are in inches.

## British Standard Whitworth (B.S.W.) and British Standard Fine (B.S.F.)
### Screw Thread Series — Basic Dimensions (B.S. 84:1956)

| Nominal Size, Inches | Threads per Inch | Pitch, Inches | Depth of Thread, Inches | Major Diameter, Inches | Effective Diameter, Inches | Minor Diameter, Inches | Area at Bottom of Thread, Sq. in. | Tap Drill Diam. |
|---|---|---|---|---|---|---|---|---|
| COARSE THREAD SERIES (B.S.W.) | | | | | | | | |
| 1/16* | 40 | 0.02500 | 0.0160 | 0.1250 | 0.1090 | 0.0930 | 0.0068 | 2.55 mm |
| 3/16 | 24 | 0.04167 | 0.0267 | 0.1875 | 0.1608 | 0.1341 | 0.0141 | 3.70 mm |
| 1/4 | 20 | 0.05000 | 0.0320 | 0.2500 | 0.2180 | 0.1860 | 0.0272 | 5.10 mm |
| 5/16 | 18 | 0.05556 | 0.0356 | 0.3125 | 0.2769 | 0.2413 | 0.0457 | 6.50 mm |
| 3/8 | 16 | 0.06250 | 0.0400 | 0.3750 | 0.3350 | 0.2950 | 0.0683 | 5/16 in. |
| 7/16 | 14 | 0.07143 | 0.0457 | 0.4375 | 0.3918 | 0.3461 | 0.0941 | 9.25 mm |
| 1/2 | 12 | 0.08333 | 0.0534 | 0.5000 | 0.4466 | 0.3932 | 0.1214 | 10.50 mm |
| 9/16* | 12 | 0.08333 | 0.0534 | 0.5625 | 0.5091 | 0.4557 | 0.1631 | 12.10 mm |
| 5/8 | 11 | 0.09091 | 0.0582 | 0.6250 | 0.5668 | 0.5086 | 0.2032 | 13.50 mm |
| 11/16* | 11 | 0.09091 | 0.0582 | 0.6875 | 0.6293 | 0.5711 | 0.2562 | .... |
| 3/4 | 10 | 0.10000 | 0.0640 | 0.7500 | 0.6860 | 0.6220 | 0.3039 | 41/64 in. |
| 7/8 | 9 | 0.11111 | 0.0711 | 0.8750 | 0.8039 | 0.7328 | 0.4218 | 19.25 mm |
| 1 | 8 | 0.12500 | 0.0800 | 1.0000 | 0.9200 | 0.8400 | 0.5542 | 22.00 mm |
| 1 1/8 | 7 | 0.14286 | 0.0915 | 1.1250 | 1.0335 | 0.9420 | 0.6969 | 24.75 mm |
| 1 1/4 | 7 | 0.14286 | 0.0915 | 1.2500 | 1.1585 | 1.0670 | 0.8942 | 1 3/32 in. |
| 1 1/2 | 6 | 0.16667 | 0.1067 | 1.5000 | 1.3933 | 1.2866 | 1.3000 | 33.50 mm |
| 1 3/4 | 5 | 0.20000 | 0.1281 | 1.7500 | 1.6219 | 1.4938 | 1.7530 | 39.00 mm |
| 2 | 4.5 | 0.22222 | 0.1423 | 2.0000 | 1.8577 | 1.7154 | 2.3110 | 44.50 mm |
| 2 1/4 | 4 | 0.25000 | 0.1601 | 2.2500 | 2.0899 | 1.9298 | 2.9250 | Tap drill diameters shown in this column are recommended sizes listed in B.S. 1157:1953 and provide from 77 to 87% of full thread. |
| 2 1/2 | 4 | 0.25000 | 0.1601 | 2.5000 | 2.3399 | 2.1798 | 3.7320 | |
| 2 3/4 | 3.5 | 0.28571 | 0.1830 | 2.7500 | 2.5670 | 2.3840 | 4.4640 | |
| 3 | 3.5 | 0.28571 | 0.1830 | 3.0000 | 2.8170 | 2.6340 | 5.4490 | |
| 3 1/4* | 3.25 | 0.30769 | 0.1970 | 3.2500 | 3.0530 | 2.8560 | 6.4060 | |
| 3 1/2 | 3.25 | 0.30769 | 0.1970 | 3.5000 | 3.3030 | 3.1060 | 7.5770 | |
| 3 3/4* | 3 | 0.33333 | 0.2134 | 3.7500 | 3.5366 | 3.3232 | 8.6740 | |
| 4 | 3 | 0.33333 | 0.2134 | 4.0000 | 3.7866 | 3.5732 | 10.0300 | |
| 4 1/2 | 2.875 | 0.34783 | 0.2227 | 4.5000 | 4.2773 | 4.0546 | 12.9100 | |
| 5 | 2.75 | 0.36364 | 0.2328 | 5.0000 | 4.7672 | 4.5344 | 16.1500 | |
| 5 1/2 | 2.625 | 0.38095 | 0.2439 | 5.5000 | 5.2561 | 5.0122 | 19.7300 | |
| 6 | 2.5 | 0.40000 | 0.2561 | 6.0000 | 5.7439 | 5.4878 | 23.6500 | |
| FINE THREAD SERIES (B.S.F.) | | | | | | | | |
| 3/16*† | 32 | 0.03125 | 0.0200 | 0.1875 | 0.1675 | 0.1475 | 0.0171 | 5/32 in. |
| 7/32* | 28 | 0.03571 | 0.0229 | 0.2188 | 0.1959 | 0.1730 | 0.0235 | 4.65 mm |
| 1/4 | 26 | 0.03846 | 0.0246 | 0.2500 | 0.2254 | 0.2008 | 0.0317 | 5.30 mm |
| 9/32* | 26 | 0.03846 | 0.0246 | 0.2812 | 0.2566 | 0.2320 | 0.0423 | .... |
| 5/16 | 22 | 0.04545 | 0.0291 | 0.3125 | 0.2834 | 0.2543 | 0.0508 | 6.75 mm |
| 3/8 | 20 | 0.05000 | 0.0320 | 0.3750 | 0.3430 | 0.3110 | 0.0760 | 8.25 mm |
| 7/16 | 18 | 0.05556 | 0.0356 | 0.4375 | 0.4019 | 0.3663 | 0.1054 | 9.70 mm |
| 1/2 | 16 | 0.06250 | 0.0400 | 0.5000 | 0.4600 | 0.4200 | 0.1385 | 7/16 in. |
| 9/16 | 16 | 0.06250 | 0.0400 | 0.5625 | 0.5225 | 0.4825 | 0.1828 | 1/2 in. |
| 5/8 | 14 | 0.07143 | 0.0457 | 0.6250 | 0.5793 | 0.5336 | 0.2236 | 14.00 mm |
| 11/16* | 14 | 0.07143 | 0.0457 | 0.6875 | 0.6418 | 0.5961 | 0.2791 | .... |
| 3/4 | 12 | 0.08333 | 0.0534 | 0.7500 | 0.6966 | 0.6432 | 0.3249 | 16.75 mm |
| 7/8 | 11 | 0.09091 | 0.0582 | 0.8750 | 0.8168 | 0.7586 | 0.4520 | 25/32 in. |
| 1 | 10 | 0.10000 | 0.0640 | 1.0000 | 0.9360 | 0.8720 | 0.5972 | 22.75 mm |
| 1 1/8 | 9 | 0.11111 | 0.0711 | 1.1250 | 1.0539 | 0.9828 | 0.7586 | 25.50 mm |
| 1 1/4 | 9 | 0.11111 | 0.0711 | 1.2500 | 1.1789 | 1.1078 | 0.9639 | 28.75 mm |
| 1 3/8* | 8 | 0.12500 | 0.0800 | 1.3750 | 1.2950 | 1.2150 | 1.1590 | 31.50 mm |
| 1 1/2 | 8 | 0.12500 | 0.0800 | 1.5000 | 1.4200 | 1.3400 | 1.4100 | 1 23/64 in. |
| 1 5/8* | 8 | 0.12500 | 0.0800 | 1.6250 | 1.5450 | 1.4650 | 1.6860 | Tap drill sizes listed in this column are recommended sizes shown in B.S. 1157:1953 and provide from 78 to 88% of full thread. |
| 1 3/4 | 7 | 0.14286 | 0.0915 | 1.7500 | 1.6585 | 1.5670 | 1.9280 | |
| 2 | 7 | 0.14286 | 0.0915 | 2.0000 | 1.9085 | 1.8170 | 2.5930 | |
| 2 1/4 | 6 | 0.16667 | 0.1067 | 2.2500 | 2.1433 | 2.0366 | 3.2580 | |
| 2 1/2 | 6 | 0.16667 | 0.1067 | 2.5000 | 2.3933 | 2.2866 | 4.1060 | |
| 2 3/4 | 6 | 0.16667 | 0.1067 | 2.7500 | 2.6433 | 2.5366 | 5.0540 | |
| 3 | 5 | 0.20000 | 0.1281 | 3.0000 | 2.8719 | 2.7438 | 5.9130 | |
| 3 1/4 | 5 | 0.20000 | 0.1281 | 3.2500 | 3.1219 | 2.9938 | 7.0390 | |
| 3 1/2 | 4.5 | 0.22222 | 0.1423 | 3.5000 | 3.3577 | 3.2154 | 8.1200 | |
| 3 3/4* | 4.5 | 0.22222 | 0.1423 | 3.7500 | 3.6077 | 3.4654 | 9.4320 | |
| 4 | 4.5 | 0.22222 | 0.1423 | 4.0000 | 3.8577 | 3.7154 | 10.8400 | |
| 4 1/4 | 4 | 0.25000 | 0.1601 | 4.2500 | 4.0899 | 3.9298 | 12.1300 | |

\* To be dispensed with wherever possible.    † The use of 2 B.A. threads is recommended.

**British Association Standard Thread (B.A.)** — This screw thread system is recommended by the British Standards Institution for use in preference to the B.S.W. and B.S.F. systems for all screws smaller than ¼ inch except that the use of the o B.A. thread be discontinued in favor of the ¼ in. B.S.F. It is further recommended that in the selection of sizes preference be given to even numbered B.A. sizes. The thread form is shown by the diagram. It is a symmetrical V-thread,

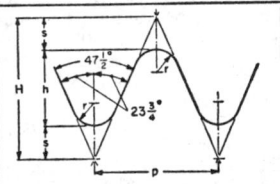

of 47½ degree included angle, having its crests and roots rounded with equal radii, such that the basic depth of the thread is 0.6000 of the pitch. If $p$ = pitch of thread, $H$ = depth of V-thread, $h$ = depth of B.A. thread, $r$ = radius at root and crest of thread, and $s$ = root and crest truncation, then:

$$H = 1.13634 \times p \qquad r = 0.18083 \times p$$

$$h = 0.60000 \times p \qquad s = 0.26817 \times p$$

**British Association Standard Thread (B.A.), Basic Dimensions** (B.S. 93:1951)

| Designation Number | Pitch, mm | Depth of Thread, mm | Bolt and Nut | | | Radius, mm | Threads per Inch (approx.) |
|---|---|---|---|---|---|---|---|
| | | | Major Diameter, mm | Effective Diameter, mm | Minor Diameter, mm | | |
| 0 | 1.0000 | 0.600 | 6.00 | 5.400 | 4.80 | 0.1808 | 25.4 |
| 1 | 0.9000 | 0.540 | 5.30 | 4.760 | 4.22 | 0.1627 | 28.2 |
| 2 | 0.8100 | 0.485 | 4.70 | 4.215 | 3.73 | 0.1465 | 31.4 |
| 3 | 0.7300 | 0.440 | 4.10 | 3.660 | 3.22 | 0.1320 | 34.8 |
| 4 | 0.6600 | 0.395 | 3.60 | 3.205 | 2.81 | 0.1193 | 38.5 |
| 5 | 0.5900 | 0.355 | 3.20 | 2.845 | 2.49 | 0.1067 | 43.0 |
| 6 | 0.5300 | 0.320 | 2.80 | 2.480 | 2.16 | 0.0958 | 47.9 |
| 7 | 0.4800 | 0.290 | 2.50 | 2.210 | 1.92 | 0.0868 | 52.9 |
| 8 | 0.4300 | 0.260 | 2.20 | 1.940 | 1.68 | 0.0778 | 59.1 |
| 9 | 0.3900 | 0.235 | 1.90 | 1.665 | 1.43 | 0.0705 | 65.1 |
| 10 | 0.3500 | 0.210 | 1.70 | 1.490 | 1.28 | 0.0633 | 72.6 |
| 11 | 0.3100 | 0.185 | 1.50 | 1.315 | 1.13 | 0.0561 | 82.0 |
| 12 | 0.2800 | 0.170 | 1.30 | 1.130 | 0.96 | 0.0506 | 90.7 |
| 13 | 0.2500 | 0.150 | 1.20 | 1.050 | 0.90 | 0.0452 | 102 |
| 14 | 0.2300 | 0.140 | 1.00 | 0.860 | 0.72 | 0.0416 | 110 |
| 15 | 0.2100 | 0.125 | 0.90 | 0.775 | 0.65 | 0.0380 | 121 |
| 16 | 0.1900 | 0.115 | 0.79 | 0.675 | 0.56 | 0.0344 | 134 |

*Tolerances and Allowances:* Two classes of bolts and one for nuts are provided: *Close Class bolts* are intended for precision parts subject to stress, no allowance being provided between maximum bolt and minimum nut sizes. *Normal Class bolts* are intended for general commercial production and general engineering use; for sizes o to 10 B.A. an allowance of 0.025 mm is provided.

**Tolerance Formulas for British Association (B.A.) Screw Threads**

| | Class or Fit | Tolerance | | |
|---|---|---|---|---|
| | | Major Diam. | Effective Diam. | Minor Diam. |
| Bolts | Close Class o to 10 B.A. incl. | 0.15$p$ mm | 0.08$p$ + 0.02 mm | 0.16$p$ + 0.04 mm |
| | Normal Class o to 10 B.A. incl. | 0.20$p$ mm | 0.10$p$ + 0.025 mm | 0.20$p$ + 0.05 mm |
| | Normal Class 11 to 16 B.A. incl. | 0.25$p$ mm | 0.10$p$ + 0.025 mm | 0.20$p$ + 0.05 mm |
| Nuts | All Classes | | 0.12$p$ + 0.03 mm | 0.375$p$ mm |

In these formulas, $p$ = pitch in millimeters.

**British Standard Pipe Threads for Non-pressure-tight Joints.** — The threads in BS 2779:1973 — "Specifications for Pipe Threads where Pressure-tight Joints are not Made on the Threads" are Whitworth form parallel fastening threads that are generally used for fastening purposes such as the mechanical assembly of component parts of fittings, cocks and valves. They are not suitable where pressure-tight joints are made on the threads.

The basic Whitworth thread form is shown on page 1349. The crests of the thread may be truncated to certain limits of size given in the Standard except on internal threads, when they are likely to be assembled with external threads conforming to the requirements of BS 21 British Standard Pipe Threads for Pressure-tight Joints (see page 1382).

For external threads two classes of tolerance are provided and for internal, one class. The two classes of tolerance for external threads are Class A and Class B. For economy of manufacture the class B fit should be chosen whenever possible. The class A is reserved for those applications where the closer tolerance is essential. Class A tolerance is an entirely negative value, equivalent to the internal thread tolerance. Class B tolerance is an entirely negative value twice that of class A tolerance. Tables showing limits and dimensions are given in the Standard.

The thread series specified in this Standard shall be designated by the letter "G". A typical reference on a drawing might be "G½", for internal thread; "G ½ A", for external thread, class A; and "G ½ B", for external thread, class B. Where no class reference is stated for external threads, that of class B will be assumed. The designation of truncated threads shall have the addition of the abbreviation "trunc.", ie., G ½ trunc. and G ½ B trunc.

### British Standard Pipe Threads (Non-pressure-tight Joints) — Metric and Inch Basic Sizes* (BS 2779:1973)

| Nominal Size | Threads per Inch† | Depth of Thread | Major Diameter | Pitch Diameter | Minor Diameter |
|---|---|---|---|---|---|
| 1/16 | 28 | 0.581 / *0.0229* | 7.723 / *0.3041* | 7.142 / *0.2812* | 6.561 / *0.2583* |
| 1/8 | 28 | 0.581 / *0.0229* | 9.728 / *0.3830* | 9.147 / *0.3601* | 8.566 / *0.3372* |
| 1/4 | 19 | 0.856 / *0.0337* | 13.157 / *0.5180* | 12.301 / *0.4843* | 11.445 / *0.4506* |
| 3/8 | 19 | 0.856 / *0.0337* | 16.662 / *0.6560* | 15.806 / *0.6223* | 14.950 / *0.5886* |
| 1/2 | 14 | 1.162 / *0.0457* | 20.955 / *0.8250* | 19.793 / *0.7793* | 18.631 / *0.7336* |
| 5/8 | 14 | 1.162 / *0.0457* | 22.911 / *0.9020* | 21.749 / *0.8563* | 20.587 / *0.8106* |
| 3/4 | 14 | 1.162 / *0.0457* | 26.441 / *1.0410* | 25.279 / *0.9953* | 24.117 / *0.9496* |
| 7/8 | 14 | 1.162 / *0.0457* | 30.201 / *1.1890* | 29.039 / *1.1433* | 27.877 / *1.0976* |
| 1 | 11 | 1.479 / *0.0582* | 33.249 / *1.3090* | 31.770 / *1.2508* | 30.291 / *1.1926* |
| 1⅛ | 11 | 1.479 / *0.0582* | 37.897 / *1.4920* | 36.418 / *1.4338* | 34.939 / *1.3756* |
| 1¼ | 11 | 1.479 / *0.0582* | 41.910 / *1.6500* | 40.431 / *1.5918* | 38.952 / *1.5336* |
| 1½ | 11 | 1.479 / *0.0582* | 47.803 / *1.8820* | 46.324 / *1.8238* | 44.845 / *1.7656* |
| 1¾ | 11 | 1.479 / *0.0582* | 53.746 / *2.1160* | 52.267 / *2.0578* | 50.788 / *1.9996* |
| 2 | 11 | 1.479 / *0.0582* | 59.614 / *2.3470* | 58.135 / *2.2888* | 56.656 / *2.2306* |
| 2¼ | 11 | 1.479 / *0.0582* | 65.710 / *2.5870* | 64.231 / *2.5288* | 62.752 / *2.4706* |
| 2½ | 11 | 1.479 / *0.0582* | 75.184 / *2.9600* | 73.705 / *2.9018* | 72.226 / *2.8436* |
| 2¾ | 11 | 1.479 / *0.0582* | 81.534 / *3.2100* | 80.055 / *3.1518* | 78.576 / *3.0936* |
| 3 | 11 | 1.479 / *0.0582* | 87.884 / *3.4600* | 86.405 / *3.4018* | 84.926 / *3.3436* |
| 3½ | 11 | 1.479 / *0.0582* | 100.330 / *3.9500* | 98.851 / *3.8918* | 97.372 / *3.8336* |
| 4 | 11 | 1.479 / *0.0582* | 113.030 / *4.4500* | 111.551 / *4.3918* | 110.072 / *4.3336* |
| 4½ | 11 | 1.479 / *0.0582* | 125.730 / *4.9500* | 124.251 / *4.8918* | 122.772 / *4.8336* |
| 5 | 11 | 1.479 / *0.0582* | 138.430 / *5.4500* | 136.951 / *5.3918* | 135.472 / *5.3336* |
| 5½ | 11 | 1.479 / *0.0582* | 151.130 / *5.9500* | 149.651 / *5.8918* | 148.172 / *5.8336* |
| 6 | 11 | 1.479 / *0.0582* | 163.830 / *6.4500* | 162.351 / *6.3918* | 160.372 / *6.3336* |

* Each basic metric dimension is given in roman figures (nominal sizes excepted) and each basic inch dimension is shown in italics directly beneath it.

† The thread pitches in millimeters are as follows: 0.907 for 28 threads per inch, 1.337 for 19 threads per inch, 1.814 for 14 threads per inch, and 2.309 for 11 threads per inch.

**British Standard Unified Screw Threads of UNJ Basic Profile.** — This British Standard (B.S. 4084:1966) arises from a request originating from within the British aircraft industry and is based upon specifications for Unified screw threads and American military standard MIL-S-8879.

These UNJ threads, having an enlarged root radius, were introduced for applications requiring high fatigue strength where working stress levels are high, in order to minimize size and weight, as in aircraft engines, airframes, missiles, space vehicles and similar designs where size and weight are critical. To meet these requirements, the root radius of external Unified threads is controlled between appreciably enlarged limits, the minor diameter of the mating internal threads being appropriately increased to insure the necessary clearance. The requirement for high strength is further met by restricting the tolerances for UNJ threads to the highest classes, Classes 3A and 3B, of Unified screw threads.

The standard, not described further here, contains both a coarse and a fine pitch series of threads.

**British Standard for ISO Metric Screw Threads.** — The first part of this British Standard (B.S. 3643: Part 1:1963) has been prepared to give particulars of the ISO (International Organization for Standardization) Basic Thread Profile and the standard diameter/pitch combinations of metric screw threads recommended by ISO. Smaller sizes of ISO metric screw threads up to 1.4 mm diameter are covered under Unified Miniature Screw Threads (see Handbook pages 1311 through 1315). It relates to single-start, parallel screw threads from 1.6 to 300 mm diameter having the ISO basic profile for triangular screw threads as given in ISO Recommendation R 68, and metric diameters and pitches as given in ISO Recommendation R 261. Two series of diameters with graded pitches are identified in this standard and are for use with screws, bolts, nuts and other common threaded fasteners. One series has coarse pitches and the other fine pitches.

The second part of this British Standard (B.S. 3643: Part 2:1966) provides information on the ISO metric coarse pitch series only. Limits and tolerances are given for three classes of fit namely "close," "medium" and "free" for a nominal length of engagement, within the range 1 mm to 68 mm diameter. This part of the standard does not apply to screw threads required for pipe joints, interference fit screw threads or screw threads subjected to elevated temperatures.

*Basic Profile:* The ISO basic profile for triangular screw threads is shown in Fig. 1.

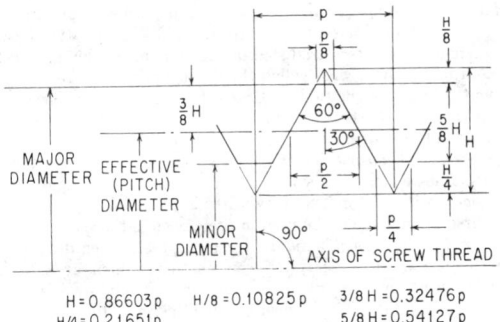

$$H = 0.86603p \qquad H/8 = 0.10825p \qquad 3/8\,H = 0.32476p$$
$$H/4 = 0.21651p \qquad \qquad 5/8\,H = 0.54127p$$

Fig. 1. Basic Form of ISO Metric Thread

NUT (INTERNAL THREAD)

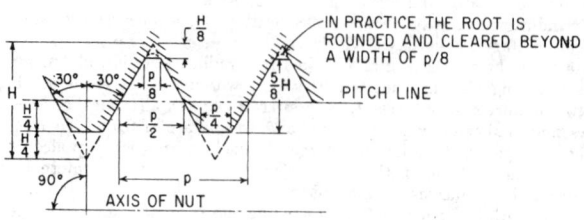

BOLT (EXTERNAL THREAD)

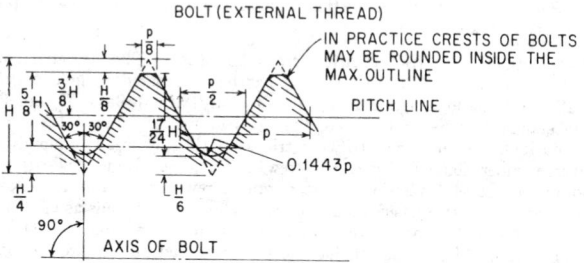

Fig. 2.   Design Forms of Internal and External Threads (Maximum Metal Conditions)

*Design Profiles:* The design profiles for ISO metric internal and external threads are shown in Fig. 2. These represent the profiles of the threads in their maximum metal condition. It may be noted that the root of each thread is deepened so as to clear the basic flat crest of the other thread. The contact between the threads is thus confined to their sloping flanks.

*Basic Numerical Thread Data:* Table 1 gives the basic numerical data for the various standard pitches of ISO metric threads. This table gives the various decimal equivalents to four places to the right of the decimal point, of fractional portions of pitch ($p$) and height ($H$) shown in Figs. 1 and 2 for the various standard pitches.

*Standard Thread Series for General Purposes:* The various standard thread series in the ISO metric system of screw threads are given in Table 2. The basic (nominal) major diameters in millimeters are listed in the first three columns and the various pitches, also in millimeters, with which the diameters may be related are given in the subsequent columns of the table. In selecting diameters, choose for preference from the first column. If those are not suitable for the particular purpose, choose from the second column, or finally from the third column.

*Series for Screws, Bolts and Nuts:* The two standard series of diameters with their related pitches which are recommended for screws, bolts and nuts and other common fasteners with ISO metric screw threads are given in Table 3. Diameters given in the first column should be used in preference to those in the second.

*Special Diameter/Pitch Combinations:* In special cases when design requirements necessitate the use of diameter/pitch combinations which are not included in Table 2, preference should be given the following pitches: 3, 2, 1.5, 1, 0.75, 0.5, 0.35, 0.25, and 0.2 mm. When selecting such pitches, to minimize production difficulties and still comply with specified tolerances it is recommended that diameters larger than the following should not be used with the pitches indicated:

| Pitch (mm) | 0.5 | 0.75 | 1 | 1.5 | 2 | 3 |
|---|---|---|---|---|---|---|
| Max. diameter (mm) | 22 | 33 | 80 | 150 | 200 | 300 |

*Tolerance Zones and Classes of Fit:* Any tolerance zone must be specified both in magnitude and position in relation to the basic size of the fit of which it is a part. The nature of a fit is dependent on both the magnitudes of the tolerances and the positions of the tolerance zones for the two members. The position of a tolerance zone is defined by the distance between the basic size and the nearest end of the tolerance zone. This distance is known as the "fundamental deviation." In the ISO metric screw thread system, fundamental deviations are designated by letters, capitals for internal threads and small letters for external threads. The magnitudes of tolerance zones are designated by tolerance grades (figures). A combination of a tolerance grade (figure) and a fundamental deviation (letter) forms a tolerance class designation, e.g. "6g."

The tolerance grades and fundamental deviations for external threads (bolts) are designated 8g, 6g and 4h and those for internal threads (nuts) are designated 7H, 6H and 5H. The association of 5H internal thread tolerance class with the 4h external thread tolerance class is known as the "close" class of fit, that of the 6H tolerance class with the 6g tolerance class as the "medium" class of fit and that of the 7H with the 8g as the "free" class of fit.

*Limits and Tolerances:* The limits and tolerances for ISO metric coarse pitch series screw threads are given in Tables 4 through 9.

*Application of Classes of Fit:* The "medium" class of fit is appropriate for most general engineering purposes. The minimum clearance associated with this fit assists in free assembly and this minimizes galling and seizing in high speed assembly. The "close" class of fit is applied to threads requiring a closer fit than normally obtained with the "medium" class of fit and should only be used when close accuracy of thread form and pitch is particularly required. Consistent production of threads of this fit demands the use of high quality production equipment and particularly thorough inspection. The "free" class of fit is primarily intended for applications in which quick and easy assembly is needed even when the threads are dirty.

*Coated (Plated) Threads:* Information on the bulk coating (plating) of mass-produced threads is given in B.S. 3382 entitled "Electroplated Coatings on Threaded Components." Normally, in the case of "medium fit" external ISO metric threads (tolerance class 6g) the fundamental deviation is permitted to be absorbed by the thickness of the coating. After coating, such threads should be accepted by a GO screw gage of basic size. The minimum clearance at assembly can, therefore, be zero. The limits of the external threads before coating should be as given in Table 7. After coating, "medium fit" internal ISO metric threads (tolerance class 6H) should be accepted by a GO screw gage of basic size. Thus, "medium fit" internal threads should only be coated if, in manufacture, the tolerance zone has not been fully utilized down to basic size (see Table 6).

*Gaging:* The gaging of ISO metric screw threads should be in accordance with the principles of screw thread gaging given in B.S. 919; "Screw Gauge Limits and Tolerances," Part 2, "Gauges for ISO Metric Screw Threads" if published, otherwise use Part 1, "Gauges for Screw Threads of Unified Form" of the same standard.

*Designation:* ISO metric screw threads are designated according to the following examples: M5×0.8-6H for an internal thread and M8×1.25-6g for an external thread where M denotes the thread system symbol for ISO metric thread, the 5 and 8 denote the nominal size in millimeters, the 0.8 and 1.25 denote the pitch in millimeters, and 6H and 6g denote the thread tolerance class symbol. A fit between a pair of threaded parts is indicated by the internal thread (nut) tolerance class designation followed by the external thread (bolt) class designation, the two separated by a stroke, e.g. M8×1.25-6H/6g or M5×0.8-6H/6g.

Table 1.  British Standard ISO Metric Screw Threads — Thread Data (B.S. 3643: Part 1: 1963)

| Pitch p | p/8 | p/4 | H 0.86603p | H/8 0.10825p | H/6 0.14434p | H/4 0.21651p | 3/8H 0.32476p | 5/8H 0.54127p | 17/24H 0.61344p | 3/4H 0.64952p | 1 1/4H 1.08254p | 1 5/12H 1.22688p |
|---|---|---|---|---|---|---|---|---|---|---|---|---|
| 0.2 | 0.025 | 0.05 | 0.1732 | 0.0216 | 0.0289 | 0.0433 | 0.0650 | 0.1083 | 0.1227 | 0.1299 | 0.2165 | 0.2454 |
| 0.25 | 0.0312 | 0.0625 | 0.2165 | 0.0271 | 0.0361 | 0.0541 | 0.0812 | 0.1353 | 0.1534 | 0.1624 | 0.2706 | 0.3067 |
| 0.35 | 0.0438 | 0.0875 | 0.3031 | 0.0379 | 0.0595 | 0.0758 | 0.1137 | 0.1894 | 0.2147 | 0.2273 | 0.3789 | 0.4294 |
| 0.4 | 0.0500 | 0.1000 | 0.3464 | 0.0433 | 0.0577 | 0.0866 | 0.1299 | 0.2165 | 0.2454 | 0.2598 | 0.4330 | 0.4908 |
| 0.45 | 0.0562 | 0.1125 | 0.3897 | 0.0487 | 0.0650 | 0.0974 | 0.1461 | 0.2436 | 0.2760 | 0.2923 | 0.4871 | 0.5521 |
| 0.5 | 0.0625 | 0.1250 | 0.4330 | 0.0541 | 0.0722 | 0.1083 | 0.1624 | 0.2706 | 0.3067 | 0.3248 | 0.5413 | 0.6134 |
| 0.6 | 0.0750 | 0.1500 | 0.5196 | 0.0650 | 0.0866 | 0.1299 | 0.1949 | 0.3248 | 0.3681 | 0.3897 | 0.6495 | 0.7361 |
| 0.7 | 0.0875 | 0.1750 | 0.6062 | 0.0758 | 0.1010 | 0.1516 | 0.2273 | 0.3789 | 0.4294 | 0.4547 | 0.7578 | 0.8588 |
| 0.75 | 0.0938 | 0.1875 | 0.6495 | 0.0812 | 0.1083 | 0.1624 | 0.2436 | 0.4059 | 0.4601 | 0.4871 | 0.8119 | 0.9202 |
| 0.8 | 0.1000 | 0.2000 | 0.6928 | 0.0866 | 0.1155 | 0.1732 | 0.2598 | 0.4330 | 0.4908 | 0.5196 | 0.8660 | 0.9815 |
| 1 | 0.1250 | 0.2500 | 0.8660 | 0.1082 | 0.1443 | 0.2165 | 0.3248 | 0.5413 | 0.6134 | 0.6495 | 1.0825 | 1.2269 |
| 1.25 | 0.1562 | 0.3125 | 1.0825 | 0.1353 | 0.1804 | 0.2706 | 0.4060 | 0.6766 | 0.7668 | 0.8119 | 1.3532 | 1.5336 |
| 1.5 | 0.1875 | 0.3750 | 1.2990 | 0.1624 | 0.2165 | 0.3248 | 0.4871 | 0.8119 | 0.9202 | 0.9743 | 1.6238 | 1.8403 |
| 1.75 | 0.2188 | 0.4375 | 1.5156 | 0.1894 | 0.2526 | 0.3789 | 0.5683 | 0.9472 | 1.0735 | 1.1367 | 1.8944 | 2.1470 |
| 2 | 0.2500 | 0.5000 | 1.7321 | 0.2165 | 0.2887 | 0.4330 | 0.6495 | 1.0825 | 1.2470 | 1.2990 | 2.1651 | 2.4538 |
| 2.5 | 0.3125 | 0.6250 | 2.1651 | 0.2706 | 0.3608 | 0.5413 | 0.8119 | 1.3532 | 1.5336 | 1.6238 | 2.7064 | 3.0672 |
| 3 | 0.3750 | 0.7500 | 2.5981 | 0.3248 | 0.4330 | 0.6495 | 0.9743 | 1.6238 | 1.8403 | 1.9486 | 3.2476 | 3.6806 |
| 3.5 | 0.4375 | 0.8750 | 3.0311 | 0.3789 | 0.5052 | 0.7578 | 1.1367 | 1.8944 | 2.1470 | 2.2733 | 3.7889 | 4.2941 |
| 4 | 0.5000 | 1.0000 | 3.4641 | 0.4330 | 0.5774 | 0.8660 | 1.2990 | 2.1651 | 2.4538 | 2.5981 | 4.3302 | 4.9075 |
| 4.5 | 0.5625 | 1.1250 | 3.8971 | 0.4871 | 0.6495 | 0.9743 | 1.4614 | 2.4357 | 2.7605 | 2.9228 | 4.8714 | 5.5210 |
| 5 | 0.6250 | 1.2500 | 4.3302 | 0.5412 | 0.7217 | 1.0826 | 1.6238 | 2.7064 | 3.0672 | 3.2476 | 5.4127 | 6.1344 |
| 5.5 | 0.6875 | 1.3750 | 4.7632 | 0.5954 | 0.7939 | 1.1908 | 1.7862 | 2.9770 | 3.3739 | 3.5724 | 5.9540 | 6.7478 |
| 6 | 0.7500 | 1.5000 | 5.1962 | 0.6495 | 0.8660 | 1.2991 | 1.9491 | 3.2476 | 3.6806 | 3.8971 | 6.4952 | 7.3613 |
| 8* | 1.0000 | 2.0000 | 6.9282 | 0.8660 | 1.1547 | 1.7321 | 2.5981 | 4.3302 | 4.9075 | 5.1962 | 8.6603 | 9.8150 |

All dimensions are given in millimeters.
* This pitch is not used in any of the ISO metric standard series.

**Table 2. British Standard ISO Metric Screw Threads — Standard Series**
(B.S. 3643: Part 1: 1963)¶

| Basic Major Diameter | | | Coarse Pitch | Fine Pitch | Basic Major Diameter | | | Coarse Pitch | Fine Pitch |
|---|---|---|---|---|---|---|---|---|---|
| First Choice | Second Choice | Third Choice | | | First Choice | Second Choice | Third Choice | | |
| 1 | 1.1 | ... | 0.25 | ... | ... | ... | 50 | ... | 3, 2, 1.5 |
| 1.2 | ... | ... | 0.25 | ... | ... | ... | 55 | ... | 4, 3, 2, 1.5 |
| ... | 1.4 | ... | 0.3 | ... | 56 | 60 | ... | 5.5 | 4, 3, 2, 1.5 |
| 1.6 | 1.8 | ... | 0.35 | 0.2 | ... | ... | 58 | ... | 4, 3, 2, 1.5 |
| 2 | ... | ... | 0.4 | 0.25 | ... | ... | 62 | ... | 4, 3, 2, 1.5 |
| ... | 2.2 | ... | 0.45 | 0.25 | 64 | 68 | ... | 6 | 4, 3, 2, 1.5 |
| 2.5 | ... | ... | 0.45 | 0.35 | ... | ... | 65 | ... | 4, 3, 2, 1.5 |
| 3 | ... | ... | 0.5 | 0.35 | 72 | ... | 70 | ... | 6, 4, 3, 2, 1.5 |
| ... | 3.5 | ... | 0.6 | 0.35 | ... | ... | 75 | ... | 4, 3, 2, 1.5 |
| 4 | ... | ... | 0.7 | 0.5 | 80 | 76 | ... | ... | 6, 4, 3, 2, 1.5 |
| ... | 4.5 | ... | 0.75 | 0.5 | ... | ... | 78 | ... | 2 |
| 5 | ... | ... | 0.8 | 0.5 | ... | ... | 82 | ... | 2 |
| ... | ... | 5.5 | ... | 0.5 | 90 | 85 | ... | ... | 6, 4, 3, 2 |
| 6 | ... | 7 | 1 | 0.75 | 100 | 95 | ... | ... | 6, 4, 3, 2 |
| 8 | ... | 9 | 1.25 | 1, 0.75 | 110 | 105 | ... | ... | 6, 4, 3, 2 |
| 10 | ... | ... | 1.5 | 1.25, 1, 0.75 | ... | ... | 115 | ... | 6, 4, 3, 2 |
| ... | ... | 11 | 1.5 | 1, 0.75 | 125 | 120 | ... | ... | 6, 4, 3, 2 |
| 12 | ... | ... | 1.75 | 1.5, 1.25, 1 | 140 | 130 | 135 | ... | 6, 4, 3, 2 |
| ... | 14 | ... | 2 | 1.5, 1.25*, 1 | ... | 150 | 145 | ... | 6, 4, 3, 2 |
| ... | ... | 15 | ... | 1.5, 1 | 160 | ... | 155 | ... | 6, 4, 3 |
| 16 | ... | ... | 2 | 1.5, 1 | 180 | 170 | 165 | ... | 6, 4, 3 |
| ... | ... | 17 | ... | 1.5, 1 | ... | ... | 175 | ... | 6, 4, 3 |
| 20 | 18 | ... | 2.5 | 2, 1.5, 1 | ... | 190 | 185 | ... | 6, 4, 3 |
| ... | 22 | ... | 2.5 | 2, 1.5, 1 | 200 | ... | 195 | ... | 6, 4, 3 |
| 24 | ... | ... | 3 | 2, 1.5, 1 | ... | 210 | 205 | ... | 6, 4, 3 |
| ... | ... | 25 | ... | 2, 1.5, 1 | 220 | ... | 215 | ... | 6, 4, 3 |
| ... | ... | 26 | ... | 1.5 | ... | ... | 225 | ... | 6, 4, 3 |
| ... | 27 | ... | 3 | 2, 1.5, 1 | ... | ... | 230 | ... | 6, 4, 3 |
| ... | ... | 28 | ... | 2, 1.5, 1 | ... | ... | 235 | ... | 6, 4, 3 |
| 30 | ... | ... | 3.5 | 3†, 2, 1.5, 1 | 250 | 240 | 245 | ... | 6, 4, 3 |
| ... | ... | 32 | ... | 2, 1.5 | ... | ... | 255 | ... | 6, 4 |
| ... | 33 | ... | 3.5 | 3†, 2, 1.5 | ... | 260 | 265 | ... | 6, 4 |
| ... | ... | 35§ | ... | 1.5 | ... | ... | 270 | ... | 6, 4 |
| 36 | 39 | ... | 4 | 3, 2, 1.5 | 280 | ... | 275 | ... | 6, 4 |
| ... | ... | 38 | ... | 1.5 | ... | ... | 285 | ... | 6, 4 |
| ... | ... | 40 | ... | 3, 2, 1.5 | ... | ... | 290 | ... | 6, 4 |
| 42 | 45 | ... | 4.5 | 4, 3, 2, 1.5 | ... | 300 | 295 | ... | 6, 4 |
| 48 | 52 | ... | 5 | 4, 3, 2, 1.5 | ... | ... | ... | ... | ... |

All dimensions are given in millimeters.
¶ The 1, 1.1, 1.2 and 1.4 basic major diameter sizes are not given in Part 1 of the standard but are given in Part 2.
* The pitch of 1.25 mm for 14 mm diameter is to be used only for spark plugs.
† Avoid these pitches as far as possible.
§ This diameter is to be used for locking nuts for ball bearings.

**Table 3. British Standard ISO Metric Screw Threads — Coarse and Fine Series for Screws, Bolts and Nuts** (B.S. 3643: Part 1: 1963)

| Basic Major Diameter | | Coarse Pitch | Fine Pitch | Basic Major Diameter | | Coarse Pitch | Fine Pitch |
|---|---|---|---|---|---|---|---|
| First Choice | Second Choice | | | First Choice | Second Choice | | |
| 1.6 | 1.8 | 0.35 | ... | 8 | ... | 1.25 | 1 |
| 2 | ... | 0.4 | ... | 10 | ... | 1.5 | 1.25 |
| 2.5 | 2.2 | 0.45 | ... | 12 | ... | 1.75 | 1.25 |
| 3 | ... | 0.5 | ... | 16 | 14 | 2 | 1.5 |
| ... | 3.5 | 0.6 | ... | ... | 18 | 2.5 | 1.5 |
| 4 | ... | 0.7 | ... | 20 | 22 | 2.5 | 1.5 |
| ... | 4.5 | 0.75 | ... | 24 | 27 | 3 | 2 |
| 5 | ... | 0.8 | ... | 30 | 33 | 3.5 | 2 |
| 6 | 7 | 1 | ... | 36 | 39 | 4 | 3 |

All dimensions are given in millimeters.

Table 4. British Standard Coarse Pitch Series ISO Metric Screw Threads —
Limits and Tolerances for Finished Uncoated (Unplated) Internal Threads
(Nuts) — Close Fit (B.S. 3643: Part 2: 1966)

| Thread System Symbol, Nominal Size, Pitch and Tolerance Class | Normal Length of Engagement | | Fundamental Deviation | Major Diameter | Effective Diameter | | | Minor Diameter | | |
|---|---|---|---|---|---|---|---|---|---|---|
| | Over | Up to | | Min.† | Max. | Tol. | Min.‡ | Max. | Tol. | Min. |
| *M1×0.25-5H | 0.6 | 1.8 | 0 | 1.000 | 0.894 | 0.056 | 0.838 | 0.785 | 0.056 | 0.729 |
| M1.1×0.25-5H | 0.6 | 1.8 | 0 | 1.100 | 0.994 | 0.056 | 0.938 | 0.885 | 0.056 | 0.829 |
| *M1.2×0.25-5H | 0.6 | 1.8 | 0 | 1.200 | 1.094 | 0.056 | 1.038 | 0.985 | 0.056 | 0.929 |
| M1.4×0.3-5H | 0.7 | 2.1 | 0 | 1.400 | 1.265 | 0.060 | 1.205 | 1.142 | 0.067 | 1.075 |
| *M1.6×0.35-5H | 0.8 | 2.6 | 0 | 1.600 | 1.440 | 0.067 | 1.373 | 1.301 | 0.080 | 1.221 |
| M1.8×0.35-5H | 0.8 | 2.6 | 0 | 1.800 | 1.640 | 0.067 | 1.573 | 1.501 | 0.080 | 1.421 |
| *M2×0.4-5H | 1 | 3 | 0 | 2.000 | 1.811 | 0.071 | 1.740 | 1.657 | 0.090 | 1.567 |
| M2.2×0.45-5H | 1.3 | 3.8 | 0 | 2.200 | 1.983 | 0.075 | 1.908 | 1.813 | 0.100 | 1.713 |
| *M2.5×0.45-5H | 1.3 | 3.8 | 0 | 2.500 | 2.283 | 0.075 | 2.208 | 2.113 | 0.100 | 2.013 |
| *M3×0.5-5H | 1.5 | 4.5 | 0 | 3.000 | 2.755 | 0.080 | 2.675 | 2.571 | 0.112 | 2.459 |
| M3.5×0.6-5H | 1.7 | 5 | 0 | 3.500 | 3.200 | 0.090 | 3.110 | 2.975 | 0.125 | 2.850 |
| *M4×0.7-5H | 2 | 6 | 0 | 4.000 | 3.640 | 0.095 | 3.545 | 3.382 | 0.140 | 3.242 |
| M4.5×0.75-5H | 2.2 | 6.7 | 0 | 4.500 | 4.108 | 0.095 | 4.013 | 3.838 | 0.150 | 3.688 |
| *M5×0.8-5H | 2.5 | 7.5 | 0 | 5.000 | 4.580 | 0.100 | 4.480 | 4.294 | 0.160 | 4.134 |
| *M6×1-5H | 3 | 9 | 0 | 6.000 | 5.468 | 0.118 | 5.350 | 5.107 | 0.190 | 4.917 |
| M7×1-5H | 3 | 9 | 0 | 7.000 | 6.468 | 0.118 | 6.350 | 6.107 | 0.190 | 5.917 |
| *M8×1.25-5H | 4 | 12 | 0 | 8.000 | 7.313 | 0.125 | 7.188 | 6.859 | 0.212 | 6.647 |
| M9×1.25-5H | 4 | 12 | 0 | 9.000 | 8.313 | 0.125 | 8.188 | 7.859 | 0.212 | 7.647 |
| *M10×1.5-5H | 5 | 15 | 0 | 10.000 | 9.166 | 0.140 | 9.026 | 8.612 | 0.236 | 8.376 |
| M11×1.5-5H | 5 | 15 | 0 | 11.000 | 10.166 | 0.140 | 10.026 | 9.612 | 0.236 | 9.376 |
| *M12×1.75-5H | 6 | 18 | 0 | 12.000 | 11.023 | 0.160 | 10.863 | 10.371 | 0.265 | 10.106 |
| M14×2-5H | 8 | 24 | 0 | 14.000 | 12.871 | 0.170 | 12.701 | 12.135 | 0.300 | 11.835 |
| *M16×2-5H | 8 | 24 | 0 | 16.000 | 14.871 | 0.170 | 14.701 | 14.135 | 0.300 | 13.835 |
| M18×2.5-5H | 10 | 30 | 0 | 18.000 | 16.556 | 0.180 | 16.376 | 15.649 | 0.355 | 15.294 |
| *M20×2.5-5H | 10 | 30 | 0 | 20.000 | 18.556 | 0.180 | 18.376 | 17.649 | 0.355 | 17.294 |
| M22×2.5-5H | 10 | 30 | 0 | 22.000 | 20.556 | 0.180 | 20.376 | 19.644 | 0.355 | 19.294 |
| *M24×3-5H | 12 | 36 | 0 | 24.000 | 22.263 | 0.212 | 22.051 | 21.152 | 0.400 | 20.752 |
| M27×3-5H | 12 | 36 | 0 | 27.000 | 25.263 | 0.212 | 25.051 | 24.152 | 0.400 | 23.752 |
| *M30×3.5-5H | 15 | 45 | 0 | 30.000 | 27.951 | 0.224 | 27.727 | 26.661 | 0.450 | 26.211 |
| M33×3.5-5H | 15 | 45 | 0 | 33.000 | 30.951 | 0.224 | 30.727 | 29.661 | 0.450 | 29.211 |
| *M36×4-5H | 18 | 53 | 0 | 36.000 | 33.638 | 0.236 | 33.402 | 32.145 | 0.475 | 31.670 |
| M39×4-5H | 18 | 53 | 0 | 39.000 | 36.638 | 0.236 | 36.402 | 35.145 | 0.475 | 34.670 |
| *M42×4.5-5H | 21 | 63 | 0 | 42.000 | 39.327 | 0.250 | 39.077 | 37.659 | 0.530 | 37.129 |
| M45×4.5-5H | 21 | 63 | 0 | 45.000 | 42.327 | 0.250 | 42.077 | 40.659 | 0.530 | 40.129 |
| *M48×5-5H | 24 | 71 | 0 | 48.000 | 45.017 | 0.265 | 44.752 | 43.147 | 0.560 | 42.587 |
| M52×5-5H | 24 | 71 | 0 | 52.000 | 49.017 | 0.265 | 48.752 | 47.147 | 0.560 | 46.587 |
| *M56×5.5-5H | 28 | 85 | 0 | 56.000 | 52.708 | 0.280 | 52.428 | 50.646 | 0.600 | 50.046 |
| M60×5.5-5H | 28 | 85 | 0 | 60.000 | 56.708 | 0.280 | 56.428 | 54.646 | 0.600 | 54.046 |
| *M64×6-5H | 32 | 95 | 0 | 64.000 | 60.403 | 0.300 | 60.103 | 58.135 | 0.630 | 57.505 |
| M68×6-5H | 32 | 95 | 0 | 68.000 | 64.403 | 0.300 | 64.103 | 62.135 | 0.630 | 61.505 |

All dimensions are given in millimeters.
* Denotes first choice; other unasterisked items denote second choice with the exception of sizes 7, 9, 11 which denote third choice.
† Also the basic major diameter of this thread system.
‡ Also the basic effective diameter of this thread system.

**Table 5.** British Standard Coarse Pitch Series ISO Metric Screw Threads —
Limits and Tolerances for Finished Uncoated (Unplated) External Threads
(Bolts) — Close Fit (B.S. 3643: Part 2: 1966)¶

| Thread System Symbol, Nominal Size, Pitch, and Tolerance Class | Fundamental Deviation | Major Diameter | | | Effective Diameter | | | Minor Diameter | | |
|---|---|---|---|---|---|---|---|---|---|---|
| | | Max.† | Tol. | Min. | Max.‡ | Tol. | Min. | Max. | Tol. | Min. |
| *M1×0.25-4h | 0 | 1.000 | 0.042 | 0.958 | 0.838 | 0.034 | 0.804 | 0.693 | 0.052 | 0.641 |
| M1.1×0.25-4h | 0 | 1.100 | 0.042 | 1.058 | 0.938 | 0.034 | 0.904 | 0.793 | 0.052 | 0.741 |
| *M1.2×0.25-4h | 0 | 1.200 | 0.042 | 1.158 | 1.038 | 0.034 | 1.004 | 0.893 | 0.052 | 0.841 |
| M1.4×0.3-4h | 0 | 1.400 | 0.048 | 1.352 | 1.205 | 0.036 | 1.169 | 1.032 | 0.058 | 0.974 |
| *M1.6×0.35-4h | 0 | 1.600 | 0.053 | 1.547 | 1.373 | 0.040 | 1.333 | 1.170 | 0.065 | 1.105 |
| M1.8×0.35-4h | 0 | 1.800 | 0.053 | 1.747 | 1.573 | 0.040 | 1.533 | 1.370 | 0.065 | 1.305 |
| *M2×0.4-4h | 0 | 2.000 | 0.060 | 1.940 | 1.740 | 0.042 | 1.698 | 1.509 | 0.071 | 1.438 |
| M2.2×0.45-4h | 0 | 2.200 | 0.063 | 2.137 | 1.908 | 0.045 | 1.863 | 1.648 | 0.077 | 1.571 |
| *M2.5×0.45-4h | 0 | 2.500 | 0.063 | 2.437 | 2.208 | 0.045 | 2.163 | 1.948 | 0.077 | 1.871 |
| *M3×0.5-4h | 0 | 3.000 | 0.067 | 2.933 | 2.675 | 0.048 | 2.627 | 2.387 | 0.084 | 2.303 |
| M3.5×0.6-4h | 0 | 3.500 | 0.080 | 3.420 | 3.110 | 0.053 | 3.047 | 2.764 | 0.096 | 2.668 |
| *M4×0.7-4h | 0 | 4.000 | 0.090 | 3.910 | 3.545 | 0.056 | 3.489 | 3.141 | 0.106 | 3.035 |
| M4.5×0.75-4h | 0 | 4.500 | 0.090 | 4.410 | 4.013 | 0.056 | 3.953 | 3.580 | 0.110 | 3.470 |
| *M5×0.8-4h | 0 | 5.000 | 0.095 | 4.905 | 4.480 | 0.060 | 4.420 | 4.019 | 0.118 | 3.901 |
| *M6×1-4h | 0 | 6.000 | 0.112 | 5.888 | 5.350 | 0.071 | 5.279 | 4.773 | 0.143 | 4.630 |
| M7×1-4h | 0 | 7.000 | 0.112 | 6.888 | 6.350 | 0.071 | 6.279 | 5.773 | 0.143 | 5.630 |
| *M8×1.25-4h | 0 | 8.000 | 0.132 | 7.868 | 7.188 | 0.075 | 7.113 | 6.466 | 0.165 | 6.301 |
| M9×1.25-4h | 0 | 9.000 | 0.132 | 8.868 | 8.188 | 0.075 | 8.113 | 7.466 | 0.165 | 7.301 |
| *M10×1.5-4h | 0 | 10.000 | 0.150 | 9.850 | 9.026 | 0.085 | 8.941 | 8.160 | 0.193 | 7.967 |
| M11×1.5-4h | 0 | 11.000 | 0.150 | 10.850 | 10.026 | 0.085 | 9.941 | 9.160 | 0.193 | 8.967 |
| *M12×1.75-4h | 0 | 12.000 | 0.170 | 11.830 | 10.863 | 0.095 | 10.768 | 9.853 | 0.221 | 9.632 |
| M14×2-4h | 0 | 14.000 | 0.180 | 13.820 | 12.701 | 0.100 | 12.601 | 11.546 | 0.244 | 11.302 |
| *M16×2-4h | 0 | 16.000 | 0.180 | 15.820 | 14.701 | 0.100 | 14.601 | 13.546 | 0.244 | 13.302 |
| M18×2.5-4h | 0 | 18.000 | 0.212 | 17.788 | 16.376 | 0.106 | 16.270 | 14.933 | 0.286 | 14.647 |
| *M20×2.5-4h | 0 | 20.000 | 0.212 | 19.788 | 18.376 | 0.106 | 18.270 | 16.933 | 0.286 | 16.647 |
| M22×2.5-4h | 0 | 22.000 | 0.212 | 21.788 | 20.376 | 0.106 | 20.270 | 18.933 | 0.286 | 18.647 |
| *M24×3-4h | 0 | 24.000 | 0.236 | 23.764 | 22.051 | 0.125 | 21.926 | 20.319 | 0.341 | 19.978 |
| M27×3-4h | 0 | 27.000 | 0.236 | 26.764 | 25.051 | 0.125 | 24.926 | 23.319 | 0.341 | 22.978 |
| *M30×3.5-4h | 0 | 30.000 | 0.265 | 29.735 | 27.727 | 0.132 | 27.595 | 25.706 | 0.384 | 25.322 |
| M33×3.5-4h | 0 | 33.000 | 0.265 | 32.735 | 30.727 | 0.132 | 30.595 | 28.706 | 0.384 | 28.322 |
| *M36×4-4h | 0 | 36.000 | 0.300 | 35.700 | 33.402 | 0.140 | 33.262 | 31.093 | 0.428 | 30.665 |
| M39×4-4h | 0 | 39.000 | 0.300 | 38.700 | 36.402 | 0.140 | 36.262 | 34.093 | 0.428 | 33.665 |
| *M42×4.5-4h | 0 | 42.000 | 0.315 | 41.685 | 39.077 | 0.150 | 38.927 | 36.479 | 0.475 | 36.004 |
| M45×4.5-4h | 0 | 45.000 | 0.315 | 44.685 | 42.077 | 0.150 | 41.927 | 39.479 | 0.475 | 39.004 |
| *M48×5-4h | 0 | 48.000 | 0.335 | 47.665 | 44.753 | 0.160 | 44.592 | 41.866 | 0.521 | 41.345 |
| M52×5-4h | 0 | 52.000 | 0.335 | 51.665 | 48.752 | 0.160 | 48.592 | 45.866 | 0.521 | 45.345 |
| *M56×5.5-4h | 0 | 56.000 | 0.355 | 55.645 | 52.428 | 0.170 | 52.258 | 49.252 | 0.567 | 48.685 |
| M60×5.5-4h | 0 | 60.000 | 0.355 | 59.645 | 56.428 | 0.170 | 56.258 | 53.252 | 0.567 | 52.685 |
| *M64×6-4h | 0 | 64.000 | 0.375 | 63.625 | 60.103 | 0.180 | 59.923 | 56.639 | 0.613 | 56.026 |
| M68×6-4h | 0 | 68.000 | 0.375 | 67.625 | 64.103 | 0.180 | 63.923 | 60.639 | 0.613 | 60.026 |

All dimensions are given in millimeters.
¶ For normal length of engagement see corresponding sizes given in Table 4.
* Denotes first choice; other unasterisked items denote second choice with the exception of sizes 7, 9, 11 which denote third choice.
† Also the basic major diameter of this thread system.
‡ Also the basic effective diameter of this thread system.

Table 6.  British Standard Coarse Pitch Series ISO Metric Screw Threads —
Limits and Tolerances for Finished Uncoated (Unplated) Internal Threads
(Nuts) — Medium Fit (B.S. 3643: Part 2: 1966)

| Thread System Symbol, Nominal Size, Pitch and Tolerance Class | Normal Length of Engagement | | Fundamental Deviation | Major Diameter | Effective Diameter | | | Minor Diameter | | |
|---|---|---|---|---|---|---|---|---|---|---|
| | Over | Up to | | Min.† | Max. | Tol. | Min.‡ | Max. | Tol. | Min. |
| *M1×0.25-6H | 0.6 | 1.8 | ..... | ...... | ....... | ..... | ....... | ....... | ..... | ....... |
| M1.1×0.25-6H | 0.6 | 1.8 | ..... | ...... | ....... | ..... | ....... | ....... | ..... | ....... |
| *M1.2×0.25-6H | 0.6 | 1.8 | ..... | ...... | ....... | ..... | ....... | ....... | ..... | ....... |
| M1.4×0.3-6H | 0.7 | 2.1 | o | 1.400 | 1.280 | 0.075 | 1.205 | 1.160 | 0.085 | 1.075 |
| *M1.6×0.35-6H | 0.8 | 2.6 | o | 1.600 | 1.458 | 0.085 | 1.373 | 1.321 | 0.100 | 1.221 |
| M1.8×0.35-6H | 0.8 | 2.6 | o | 1.800 | 1.650 | 0.085 | 1.573 | 1.521 | 0.100 | 1.421 |
| *M2×0.4-6H | 1 | 3 | o | 2.000 | 1.830 | 0.090 | 1.740 | 1.679 | 0.112 | 1.567 |
| M2.2×0.45-6H | 1.3 | 3.8 | o | 2.200 | 2.003 | 0.095 | 1.908 | 1.838 | 0.125 | 1.713 |
| *M2.5×0.45-6H | 1.3 | 3.8 | o | 2.500 | 2.303 | 0.095 | 2.208 | 2.138 | 0.125 | 2.013 |
| *M3×0.5-6H | 1.5 | 4.5 | o | 3.000 | 2.775 | 0.100 | 2.675 | 2.599 | 0.140 | 2.459 |
| M3.5×0.6-6H | 1.7 | 5 | o | 3.500 | 3.222 | 0.112 | 3.110 | 3.010 | 0.160 | 2.850 |
| *M4×0.7-6H | 2 | 6 | o | 4.000 | 3.663 | 0.118 | 3.545 | 3.422 | 0.180 | 3.242 |
| M4.5×0.75-6H | 2.2 | 6.7 | o | 4.500 | 4.131 | 0.118 | 4.013 | 3.878 | 0.190 | 3.688 |
| *M5×0.8-6H | 2.5 | 7.5 | o | 5.000 | 4.605 | 0.125 | 4.480 | 4.334 | 0.200 | 4.134 |
| *M6×1-6H | 3 | 9 | o | 6.000 | 5.500 | 0.150 | 5.350 | 5.153 | 0.236 | 4.917 |
| M7×1-6H | 3 | 9 | o | 7.000 | 6.500 | 0.150 | 6.350 | 6.153 | 0.236 | 5.917 |
| *M8×1.25-6H | 4 | 12 | o | 8.000 | 7.348 | 0.160 | 7.188 | 6.912 | 0.265 | 6.647 |
| M9×1.25-6H | 4 | 12 | o | 9.000 | 8.348 | 0.160 | 8.188 | 7.912 | 0.265 | 7.647 |
| *M10×1.5-6H | 5 | 15 | o | 10.000 | 9.205 | 0.180 | 9.026 | 8.676 | 0.300 | 8.376 |
| M11×1.5-6H | 5 | 15 | o | 11.000 | 10.206 | 0.180 | 10.026 | 9.676 | 0.300 | 9.376 |
| *M12×1.75-6H | 6 | 18 | o | 12.000 | 11.063 | 0.200 | 10.863 | 10.441 | 0.335 | 10.106 |
| M14×2-6H | 8 | 24 | o | 14.000 | 12.913 | 0.212 | 12.701 | 12.210 | 0.375 | 11.835 |
| *M16×2-6H | 8 | 24 | o | 16.000 | 14.913 | 0.212 | 14.701 | 14.210 | 0.375 | 13.835 |
| M18×2.5-6H | 10 | 30 | o | 18.000 | 16.600 | 0.224 | 16.376 | 15.744 | 0.450 | 15.294 |
| *M20×2.5-6H | 10 | 30 | o | 20.000 | 18.600 | 0.224 | 18.376 | 17.744 | 0.450 | 17.294 |
| M22×2.5-6H | 10 | 30 | o | 22.000 | 20.600 | 0.224 | 20.376 | 19.744 | 0.450 | 19.294 |
| *M24×3-6H | 12 | 36 | o | 24.000 | 22.316 | 0.265 | 22.051 | 21.252 | 0.500 | 20.752 |
| M27×3-6H | 12 | 36 | o | 27.000 | 25.316 | 0.265 | 25.051 | 24.252 | 0.500 | 23.752 |
| *M30×3.5-6H | 15 | 45 | o | 30.000 | 28.007 | 0.280 | 27.727 | 26.771 | 0.560 | 26.211 |
| M33×3.5-6H | 15 | 45 | o | 33.000 | 31.007 | 0.280 | 30.727 | 29.771 | 0.560 | 29.211 |
| *M36×4-6H | 18 | 53 | o | 36.000 | 33.702 | 0.300 | 33.402 | 32.270 | 0.600 | 31.670 |
| M39×4-6H | 18 | 53 | o | 39.000 | 36.702 | 0.300 | 36.402 | 35.270 | 0.600 | 34.670 |
| *M42×4.5-6H | 21 | 63 | o | 42.000 | 39.392 | 0.315 | 39.077 | 37.799 | 0.670 | 37.129 |
| M45×4.5-6H | 21 | 63 | o | 45.000 | 42.392 | 0.315 | 42.077 | 40.799 | 0.670 | 40.129 |
| *M48×5-6H | 24 | 71 | o | 48.000 | 45.087 | 0.335 | 44.752 | 43.297 | 0.710 | 42.587 |
| M52×5-6H | 24 | 71 | o | 52.000 | 49.087 | 0.335 | 48.752 | 47.297 | 0.710 | 46.587 |
| *M56×5.5-6H | 28 | 85 | o | 56.000 | 52.783 | 0.355 | 53.428 | 50.796 | 0.750 | 50.046 |
| M60×5.5-6H | 28 | 85 | o | 60.000 | 56.783 | 0.355 | 56.428 | 54.796 | 0.750 | 54.046 |
| *M64×6-6H | 32 | 95 | o | 64.000 | 60.478 | 0.375 | 60.103 | 58.305 | 0.800 | 57.505 |
| M68×6-6H | 32 | 95 | o | 68.000 | 64.478 | 0.375 | 64.103 | 62.305 | 0.800 | 61.505 |

All dimensions are given in millimeters.
  * Denotes first choice; other unasterisked items denote second choice with exception of sizes 7, 9, 11 which denote third choice.
  † Also the basic major diameter of this thread system.
  ‡ Also the basic effective diameter of this thread system.

Table 7. British Standard Coarse Pitch Series ISO Metric Screw Threads —
Limits and Tolerances for Finished Uncoated (Unplated) External Threads
(Bolts) — Medium Fit (B.S. 3643: Part 2: 1966)†

| Thread System Symbol, Nominal Size, Pitch and Tolerance Class | Fundamental Deviation | Major Diameter | | | Effective Diameter | | | Minor Diameter | | |
|---|---|---|---|---|---|---|---|---|---|---|
| | | Max. | Tol. | Min. | Max. | Tol. | Min. | Max. | Tol. | Min. |
| *M1×0.25-6g | 0.018 | 0.982 | 0.067 | 0.915 | 0.820 | 0.053 | 0.767 | 0.675 | 0.071 | 0.604 |
| M1.1×0.25-6g | 0.018 | 1.082 | 0.067 | 1.015 | 0.920 | 0.053 | 0.867 | 0.775 | 0.071 | 0.704 |
| *M1.2×0.25-6g | 0.018 | 1.182 | 0.067 | 1.115 | 1.020 | 0.053 | 0.967 | 0.875 | 0.071 | 0.804 |
| M1.4×0.3-6g | 0.018 | 1.382 | 0.075 | 1.307 | 1.187 | 0.056 | 1.131 | 1.014 | 0.078 | 0.936 |
| *M1.6×0.35-6g | 0.019 | 1.581 | 0.085 | 1.496 | 1.354 | 0.063 | 1.291 | 1.151 | 0.088 | 1.063 |
| M1.8×0.35-6g | 0.019 | 1.781 | 0.085 | 1.696 | 1.554 | 0.063 | 1.491 | 1.351 | 0.088 | 1.263 |
| *M2×0.4-6g | 0.019 | 1.981 | 0.095 | 1.886 | 1.721 | 0.067 | 1.654 | 1.490 | 0.096 | 1.394 |
| M2.2×0.45-6g | 0.020 | 2.180 | 0.100 | 2.080 | 1.888 | 0.071 | 1.817 | 1.628 | 0.103 | 1.525 |
| *M2.5×0.45-6g | 0.020 | 2.480 | 0.100 | 2.380 | 2.188 | 0.071 | 2.117 | 1.928 | 0.103 | 1.825 |
| *M3×0.5-6g | 0.020 | 2.980 | 0.106 | 2.874 | 2.655 | 0.075 | 2.580 | 2.367 | 0.111 | 2.256 |
| M3.5×0.6-6g | 0.021 | 3.479 | 0.125 | 3.354 | 3.089 | 0.085 | 3.004 | 2.743 | 0.128 | 2.615 |
| *M4×0.7-6g | 0.022 | 3.978 | 0.140 | 3.838 | 3.523 | 0.090 | 3.433 | 3.119 | 0.140 | 2.979 |
| M4.5×0.75-6g | 0.022 | 4.478 | 0.140 | 4.338 | 3.991 | 0.090 | 3.901 | 3.558 | 0.144 | 3.414 |
| *M5×0.8-6g | 0.024 | 4.976 | 0.150 | 4.826 | 4.456 | 0.095 | 4.361 | 3.995 | 0.153 | 3.842 |
| *M6×1-6g | 0.026 | 5.974 | 0.180 | 5.794 | 5.324 | 0.112 | 5.212 | 4.747 | 0.184 | 4.563 |
| M7×1-6g | 0.026 | 6.974 | 0.180 | 6.794 | 6.324 | 0.112 | 6.212 | 5.747 | 0.184 | 5.563 |
| *M8×1.25-6g | 0.028 | 7.972 | 0.212 | 7.760 | 7.160 | 0.118 | 7.042 | 6.438 | 0.208 | 6.230 |
| M9×1.25-6g | 0.028 | 8.972 | 0.212 | 8.760 | 8.160 | 0.118 | 8.042 | 7.438 | 0.208 | 7.230 |
| *M10×1.5-6g | 0.032 | 9.968 | 0.236 | 9.732 | 8.994 | 0.132 | 8.862 | 8.128 | 0.240 | 7.888 |
| M11×1.5-6g | 0.032 | 10.968 | 0.236 | 10.732 | 9.994 | 0.132 | 9.862 | 9.128 | 0.240 | 8.888 |
| *M12×1.75-6g | 0.034 | 11.966 | 0.265 | 11.701 | 10.829 | 0.150 | 10.679 | 9.819 | 0.276 | 9.543 |
| M14×2-6g | 0.038 | 13.962 | 0.280 | 13.682 | 12.663 | 0.160 | 12.503 | 11.508 | 0.304 | 11.204 |
| *M16×2-6g | 0.038 | 15.962 | 0.280 | 15.682 | 14.663 | 0.160 | 14.503 | 13.508 | 0.304 | 13.204 |
| M18×2.5-6g | 0.042 | 17.958 | 0.335 | 17.623 | 16.334 | 0.170 | 16.164 | 14.891 | 0.350 | 14.541 |
| *M20×2.5-6g | 0.042 | 19.958 | 0.335 | 19.623 | 18.334 | 0.170 | 18.164 | 16.891 | 0.350 | 16.541 |
| M22×2.5-6g | 0.042 | 21.958 | 0.335 | 21.623 | 20.334 | 0.170 | 20.164 | 18.891 | 0.350 | 18.541 |
| *M24×3-6g | 0.048 | 23.952 | 0.375 | 23.577 | 22.003 | 0.200 | 21.803 | 20.271 | 0.416 | 19.855 |
| M27×3-6g | 0.048 | 26.952 | 0.375 | 26.577 | 25.003 | 0.200 | 24.803 | 23.271 | 0.416 | 22.855 |
| *M30×3.5-6g | 0.053 | 29.947 | 0.425 | 29.522 | 27.674 | 0.212 | 27.462 | 25.653 | 0.464 | 25.189 |
| M33×3.5-6g | 0.053 | 32.947 | 0.425 | 32.522 | 30.674 | 0.212 | 30.462 | 28.653 | 0.464 | 28.189 |
| *M36×4-6g | 0.060 | 35.940 | 0.475 | 35.465 | 33.342 | 0.224 | 33.118 | 31.033 | 0.512 | 30.521 |
| M39×4-6g | 0.060 | 38.940 | 0.475 | 38.465 | 36.342 | 0.224 | 36.118 | 34.033 | 0.512 | 33.521 |
| *M42×4.5-6g | 0.063 | 41.937 | 0.500 | 41.437 | 39.014 | 0.236 | 38.778 | 36.416 | 0.561 | 35.855 |
| M45×4.5-6g | 0.063 | 44.937 | 0.500 | 44.437 | 42.014 | 0.236 | 41.778 | 39.416 | 0.561 | 38.855 |
| *M48×5-6g | 0.071 | 47.929 | 0.530 | 47.399 | 44.681 | 0.250 | 44.431 | 41.795 | 0.611 | 41.184 |
| M52×5-6g | 0.071 | 51.929 | 0.530 | 51.399 | 48.681 | 0.250 | 48.431 | 45.795 | 0.611 | 45.184 |
| *M56×5.5-6g | 0.075 | 55.925 | 0.560 | 55.365 | 52.353 | 0.265 | 52.088 | 49.177 | 0.662 | 48.515 |
| M60×5.5-6g | 0.075 | 59.925 | 0.560 | 59.365 | 56.353 | 0.265 | 56.088 | 53.177 | 0.662 | 52.515 |
| *M64×6-6g | 0.080 | 63.920 | 0.600 | 63.320 | 60.023 | 0.280 | 59.743 | 56.559 | 0.713 | 55.846 |
| M68×6-6g | 0.080 | 67.920 | 0.600 | 67.320 | 64.023 | 0.280 | 63.743 | 60.559 | 0.713 | 59.846 |

All dimensions are given in millimeters.
* Denotes first choice; other unasterisked items denote second choice with the exception of sizes 7, 9, 11 which denote third choice.
† For normal length of engagement see corresponding sizes given in Table 6.

Table 8. British Standard Coarse Pitch Series ISO Metric Screw Threads — Limits and Tolerances for Finished Uncoated (Unplated) Internal Threads (Nuts) — Free Fit (B.S. 3643: Part 2: 1966)†

| Thread System Symbol, Nominal Size, Pitch and Tolerance Class | Normal Length of Engagement | | Fundamental Deviation | Major Diameter | Effective Diameter | | | Minor Diameter | | |
|---|---|---|---|---|---|---|---|---|---|---|
| | Over | Up to | | Min.† | Max. | Tol. | Min.‡ | Max. | Tol. | Min. |
| *M1×0.25-7H | 0.6 | 1.8 | ..... | ...... | ...... | ..... | ...... | ...... | ..... | ...... |
| M1.1×0.25-7H | 0.6 | 1.8 | ..... | ...... | ...... | ..... | ...... | ...... | ..... | ...... |
| *M1.2×0.25-7H | 0.6 | 1.8 | ..... | ...... | ...... | ..... | ...... | ...... | ..... | ...... |
| M1.4×0.3-7H | 0.7 | 2.1 | ..... | ...... | ...... | ..... | ...... | ...... | ..... | ...... |
| *M1.6×0.35-7H | 0.8 | 2.6 | ..... | ...... | ...... | ..... | ...... | ...... | ..... | ...... |
| M1.8×0.35-7H | 0.8 | 2.6 | ..... | ...... | ...... | ..... | ...... | ...... | ..... | ...... |
| *M2×0.4-7H | 1 | 3 | ..... | ...... | ...... | ..... | ...... | ...... | ..... | ...... |
| M2.2×0.45-7H | 1.3 | 3.8 | ..... | ...... | ...... | ..... | ...... | ...... | ..... | ...... |
| *M2.5×0.45-7H | 1.3 | 3.8 | ..... | ...... | ...... | ..... | ...... | ...... | ..... | ...... |
| *M3×0.5-7H | 1.5 | 4.5 | 0 | 3.000 | 2.800 | 0.125 | 2.675 | 2.639 | 0.180 | 2.459 |
| M3.5×0.6-7H | 1.7 | 5 | 0 | 3.500 | 3.250 | 0.140 | 3.110 | 3.050 | 0.200 | 2.850 |
| *M4×0.7-7H | 2 | 6 | 0 | 4.000 | 3.695 | 0.150 | 3.545 | 3.466 | 0.224 | 3.242 |
| M4.5×0.75-7H | 2.2 | 6.7 | 0 | 4.500 | 4.163 | 0.150 | 4.013 | 3.924 | 0.236 | 3.688 |
| *M5×0.8-7H | 2.5 | 7.5 | 0 | 5.000 | 4.640 | 0.160 | 4.480 | 4.384 | 0.250 | 4.134 |
| *M6×1-7H | 3 | 9 | 0 | 6.000 | 5.540 | 0.190 | 5.350 | 5.217 | 0.300 | 4.917 |
| M7×1-7H | 3 | 9 | 0 | 7.000 | 6.540 | 0.190 | 6.350 | 6.217 | 0.300 | 5.917 |
| *M8×1.25-7H | 4 | 12 | 0 | 8.000 | 7.388 | 0.200 | 7.188 | 6.982 | 0.335 | 6.647 |
| M9×1.25-7H | 4 | 12 | 0 | 9.000 | 8.388 | 0.200 | 8.188 | 7.982 | 0.335 | 7.647 |
| *M10×1.5-7H | 5 | 15 | 0 | 10.000 | 9.250 | 0.224 | 9.026 | 8.751 | 0.375 | 8.376 |
| M11×1.5-7H | 5 | 15 | 0 | 11.000 | 10.250 | 0.224 | 10.026 | 9.751 | 0.375 | 9.376 |
| *M12×1.75-7H | 6 | 18 | 0 | 12.000 | 11.113 | 0.250 | 10.863 | 10.531 | 0.425 | 10.106 |
| M14×2-7H | 8 | 24 | 0 | 14.000 | 12.966 | 0.265 | 12.701 | 12.310 | 0.475 | 11.835 |
| *M16×2-7H | 8 | 24 | 0 | 16.000 | 14.966 | 0.265 | 14.701 | 14.310 | 0.475 | 13.835 |
| M18×2.5-7H | 10 | 30 | 0 | 18.000 | 16.656 | 0.280 | 16.376 | 15.854 | 0.560 | 15.294 |
| *M20×2.5-7H | 10 | 30 | 0 | 20.000 | 18.656 | 0.280 | 18.376 | 17.854 | 0.560 | 17.294 |
| M22×2.5-7H | 10 | 30 | 0 | 22.000 | 20.656 | 0.280 | 20.376 | 19.854 | 0.560 | 19.294 |
| *M24×3-7H | 12 | 36 | 0 | 24.000 | 22.386 | 0.335 | 22.051 | 21.382 | 0.630 | 20.752 |
| M27×3-7H | 12 | 36 | 0 | 27.000 | 25.386 | 0.335 | 25.051 | 24.382 | 0.630 | 23.752 |
| *M30×3.5-7H | 15 | 45 | 0 | 30.000 | 28.082 | 0.355 | 27.727 | 26.921 | 0.710 | 26.211 |
| M33×3.5-7H | 15 | 45 | 0 | 33.000 | 31.082 | 0.355 | 30.727 | 29.921 | 0.710 | 29.211 |
| *M36×4-7H | 18 | 53 | 0 | 36.000 | 33.777 | 0.375 | 33.402 | 32.420 | 0.750 | 31.670 |
| M39×4-7H | 18 | 53 | 0 | 39.000 | 36.777 | 0.375 | 36.402 | 35.420 | 0.750 | 34.670 |
| *M42×4.5-7H | 21 | 63 | 0 | 42.000 | 39.477 | 0.400 | 39.077 | 37.979 | 0.850 | 37.129 |
| M45×4.5-7H | 21 | 63 | 0 | 45.000 | 42.477 | 0.400 | 42.077 | 40.979 | 0.850 | 40.129 |
| *M48×5-7H | 24 | 71 | 0 | 48.000 | 45.177 | 0.425 | 44.752 | 43.487 | 0.900 | 42.587 |
| M52×5-7H | 24 | 71 | 0 | 52.000 | 49.177 | 0.425 | 48.752 | 47.487 | 0.900 | 46.587 |
| *M56×5.5-7H | 28 | 85 | 0 | 56.000 | 52.878 | 0.450 | 52.428 | 50.996 | 0.950 | 50.046 |
| M60×5.5-7H | 28 | 85 | 0 | 60.000 | 56.878 | 0.450 | 56.428 | 54.996 | 0.950 | 54.046 |
| *M64×6-7H | 32 | 95 | 0 | 64.000 | 60.578 | 0.475 | 60.103 | 58.505 | 1.000 | 57.505 |
| M68×6-7H | 32 | 95 | 0 | 68.000 | 64.578 | 0.475 | 64.103 | 62.505 | 1.000 | 61.505 |

All dimensions are given in millimeters.

* Denotes first choice; other unasterisked items denote second choice with exceptions of sizes 7, 9, 11 which denote third choice.

† Also the basic major diameter of this thread system.

‡ Also the basic effective diameter of this thread system.

Table 9. British Standard Coarse Pitch Series ISO Metric Screw Threads —
Limits and Tolerances for Finished Uncoated (Unplated) External Threads
(Bolts) — Free Fit (B.S. 3643: Part 2: 1966)†

| Thread System Symbol, Nominal Size, Pitch and Tolerance Class | Fundamental Deviation | Major Diameter | | | Effective Diameter | | | Minor Diameter | | |
|---|---|---|---|---|---|---|---|---|---|---|
| | | Max. | Tol. | Min. | Max. | Tol. | Min. | Max. | Tol. | Min. |
| *M1×0.25-8g | ..... | ..... | ..... | ..... | ..... | ..... | ..... | ..... | ..... | ..... |
| M1.1×0.25-8g | ..... | ..... | ..... | ..... | ..... | ..... | ..... | ..... | ..... | ..... |
| *M1.2×0.25-8g | ..... | ..... | ..... | ..... | ..... | ..... | ..... | ..... | ..... | ..... |
| M1.4×0.3-8g | ..... | ..... | ..... | ..... | ..... | ..... | ..... | ..... | ..... | ..... |
| *M1.6×0.35-8g | ..... | ..... | ..... | ..... | ..... | ..... | ..... | ..... | ..... | ..... |
| M1.8×0.35-8g | ..... | ..... | ..... | ..... | ..... | ..... | ..... | ..... | ..... | ..... |
| *M2×0.4-8g | ..... | ..... | ..... | ..... | ..... | ..... | ..... | ..... | ..... | ..... |
| M2.2×0.45-8g | ..... | ..... | ..... | ..... | ..... | ..... | ..... | ..... | ..... | ..... |
| *M2.5×0.45-8g | ..... | ..... | ..... | ..... | ..... | ..... | ..... | ..... | ..... | ..... |
| *M3×0.5-8g | ..... | ..... | ..... | ..... | ..... | ..... | ..... | ..... | ..... | ..... |
| M3.5×0.6-8g | ..... | ..... | ..... | ..... | ..... | ..... | ..... | ..... | ..... | ..... |
| *M4×0.7-8g | ..... | ..... | ..... | ..... | ..... | ..... | ..... | ..... | ..... | ..... |
| M4.5×0.75-8g | ..... | ..... | ..... | ..... | ..... | ..... | ..... | ..... | ..... | ..... |
| *M5×0.8-8g | 0.024 | 4.976 | 0.236 | 4.640 | 4.456 | 0.150 | 4.306 | 3.995 | 0.208 | 3.787 |
| *M6×1-8g | 0.026 | 5.974 | 0.280 | 5.694 | 5.324 | 0.180 | 5.144 | 4.747 | 0.252 | 4.495 |
| M7×1-8g | 0.026 | 6.974 | 0.280 | 6.694 | 6.324 | 0.180 | 6.144 | 5.747 | 0.252 | 5.495 |
| *M8×1.25-8g | 0.028 | 7.972 | 0.335 | 7.637 | 7.160 | 0.190 | 6.970 | 6.438 | 0.280 | 6.158 |
| M9×1.25-8g | 0.028 | 8.972 | 0.335 | 8.637 | 8.160 | 0.190 | 7.970 | 7.438 | 0.280 | 7.158 |
| *M10×1.5-8g | 0.032 | 9.968 | 0.375 | 9.593 | 8.994 | 0.212 | 8.782 | 8.128 | 0.320 | 7.808 |
| M11×1.5-8g | 0.032 | 10.968 | 0.375 | 10.593 | 9.994 | 0.212 | 9.782 | 9.128 | 0.320 | 8.808 |
| *M12×1.75-8g | 0.034 | 11.966 | 0.425 | 11.541 | 10.829 | 0.236 | 10.593 | 9.819 | 0.362 | 9.457 |
| M14×2-8g | 0.038 | 13.962 | 0.450 | 13.512 | 12.663 | 0.250 | 12.413 | 11.508 | 0.394 | 11.114 |
| *M16×2-8g | 0.038 | 15.962 | 0.450 | 15.512 | 14.663 | 0.250 | 14.413 | 14.508 | 0.394 | 13.114 |
| M18×2.5-8g | 0.042 | 17.958 | 0.530 | 17.428 | 16.334 | 0.265 | 16.069 | 14.891 | 0.445 | 14.446 |
| *M20×2.5-8g | 0.042 | 19.958 | 0.530 | 19.428 | 18.334 | 0.265 | 18.069 | 16.891 | 0.445 | 16.446 |
| M22×2.5-8g | 0.042 | 21.958 | 0.530 | 21.428 | 20.334 | 0.265 | 20.069 | 18.891 | 0.445 | 18.446 |
| *M24×3-8g | 0.048 | 23.952 | 0.600 | 23.352 | 22.003 | 0.315 | 21.688 | 20.271 | 0.531 | 19.740 |
| M27×3-8g | 0.048 | 26.952 | 0.600 | 26.352 | 25.003 | 0.315 | 24.688 | 23.271 | 0.531 | 22.740 |
| *M30×3.5-8g | 0.053 | 29.947 | 0.670 | 29.277 | 27.674 | 0.335 | 27.339 | 25.653 | 0.587 | 25.066 |
| M33×3.5-8g | 0.053 | 32.947 | 0.670 | 32.277 | 30.674 | 0.335 | 30.339 | 28.653 | 0.587 | 28.066 |
| *M36×4-8g | 0.060 | 35.940 | 0.750 | 35.190 | 33.342 | 0.355 | 32.987 | 31.033 | 0.643 | 30.390 |
| M39×4-8g | 0.060 | 38.940 | 0.750 | 38.190 | 36.342 | 0.355 | 35.987 | 34.033 | 0.643 | 33.390 |
| *M42×4.5-8g | 0.063 | 41.937 | 0.800 | 41.137 | 39.014 | 0.375 | 38.639 | 36.416 | 0.700 | 35.716 |
| M45×4.5-8g | 0.063 | 44.937 | 0.800 | 44.137 | 42.014 | 0.375 | 41.639 | 39.416 | 0.700 | 38.716 |
| *M48×5-8g | 0.071 | 47.929 | 0.850 | 47.079 | 44.681 | 0.400 | 44.201 | 41.795 | 0.761 | 41.034 |
| M52×5-8g | 0.071 | 51.929 | 0.850 | 51.079 | 48.681 | 0.400 | 48.281 | 45.795 | 0.761 | 45.034 |
| *M56×5.5-8g | 0.075 | 55.925 | 0.900 | 55.025 | 52.353 | 0.425 | 51.920 | 49.177 | 0.822 | 48.355 |
| M60×5.5-8g | 0.075 | 59.925 | 0.900 | 59.025 | 56.353 | 0.425 | 55.928 | 53.177 | 0.822 | 52.355 |
| *M64×6-8g | 0.080 | 63.920 | 0.950 | 62.970 | 60.023 | 0.450 | 59.573 | 56.559 | 0.883 | 55.676 |
| M68×6-8g | 0.080 | 67.920 | 0.950 | 66.970 | 64.023 | 0.450 | 63.573 | 60.559 | 0.883 | 59.676 |

All dimensions are given in millimeters.

* Denotes first choice; other unasterisked items denote second choice with the exception of sizes 7, 9, 11 which denote third choice.

† For normal length of engagement see corresponding sizes given in Table 8.

**British Standard Buttress Threads (B.S.1657:1950).** — Specifications for buttress threads in this standard are similar to those in the American Standard except: (1) A basic depth of thread of $0.4p$ is used instead of $0.6p$; (2) Sizes below 1 inch are not included; (3) Tolerances on major and minor diameters are the same as the pitch diameter tolerances, whereas in the American Standard separate tolerances are provided; however, provision is made for smaller major and minor diameter tolerances when crest surfaces of screws or nuts are used as datum surfaces, or when the resulting reduction in depth of engagement must be limited; and (4) Certain combinations of large diameters with fine pitches are provided that are not encouraged in the American Standard.

**International Metric Thread System.** — The Systeme Internationale (S.I.) Thread was adopted at the International Congress for the standardization of screw threads held in Zurich in 1898. The thread form is similar to the American standard (formerly U.S. Standard), excepting the depth which is greater. There is a clearance between the root and mating crest fixed at a maximum of $\frac{1}{16}$ the height of the fundamental triangle or $0.054 \times$ pitch. A rounded root profile is recommended. This system formed the basis of the normal metric series of many European countries.

$$\text{Depth } d = 0.7035\ P \text{ max.; } 0.6855\ P \text{ min.}$$

$$\text{Flat } f = 0.125\ P$$

$$\text{Radius } r = 0.0633\ P \text{ max.; } 0.054\ P \text{ min.}$$

Tap drill diam. = major diam. −pitch

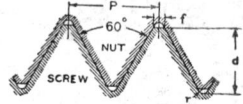

**Löwenherz Thread.** — The Löwenherz thread has flats at the top and bottom the same as the U.S. standard form, but the angle is 53 degrees 8 minutes. The depth equals $0.75 \times$ the pitch, and the width of the flats at the top and bottom is equal to $0.125 \times$ the pitch. This screw thread is based on the metric system and is used for measuring instruments, especially in Germany:

### Löwenherz Thread

| Diameter | | Pitch, Millimeters | Approximate No. of Threads per Inch | Diameter | | Pitch, Millimeters | Approximate No. of Threads per Inch |
|---|---|---|---|---|---|---|---|
| Millimeters | Inches | | | Millimeters | Inches | | |
| 1.0 | 0.0394 | 0.25 | 101.6 | 9.0 | 0.3543 | 1.30 | 19.5 |
| 1.2 | 0.0472 | 0.25 | 101.6 | 10.0 | 0.3937 | 1.40 | 18.1 |
| 1.4 | 0.0551 | 0.30 | 84.7 | 12.0 | 0.4724 | 1.60 | 15.9 |
| 1.7 | 0.0669 | 0.35 | 72.6 | 14.0 | 0.5512 | 1.80 | 14.1 |
| 2.0 | 0.0787 | 0.40 | 63.5 | 16.0 | 0.6299 | 2.00 | 12.7 |
| 2.3 | 0.0905 | 0.40 | 63.5 | 18.0 | 0.7087 | 2.20 | 11.5 |
| 2.6 | 0.1024 | 0.45 | 56.4 | 20.0 | 0.7874 | 2.40 | 10.6 |
| 3.0 | 0.1181 | 0.50 | 50.8 | 22.0 | 0.8661 | 2.80 | 9.1 |
| 3.5 | 0.1378 | 0.60 | 42.3 | 24.0 | 0.9450 | 2.80 | 9.1 |
| 4.0 | 0.1575 | 0.70 | 36.3 | 26.0 | 1.0236 | 3.20 | 7.9 |
| 4.5 | 0.1772 | 0.75 | 33.9 | 28.0 | 1.1024 | 3.20 | 7.9 |
| 5.0 | 0.1968 | 0.80 | 31.7 | 30.0 | 1.1811 | 3.60 | 7.1 |
| 5.5 | 0.2165 | 0.90 | 28.2 | 32.0 | 1.2599 | 3.60 | 7.1 |
| 6.0 | 0.2362 | 1.00 | 25.4 | 36.0 | 1.4173 | 4.00 | 6.4 |
| 7.0 | 0.2756 | 1.10 | 23.1 | 40.0 | 1.5748 | 4.40 | 5.7 |
| 8.0 | 0.3150 | 1.20 | 21.1 | ...... | ...... | ...... | ........ |

**Metric Series Threads — A Comparison of Maximum Metal Dimensions of British (B.S. 1095), French (NF E03-104), German (DIN 13), and Swiss (VSM 12003) Systems**

| Nominal Size and Major Bolt Diam. | Pitch | Pitch Diam. | Bolt Minor Dia. — British | Bolt Minor Dia. — French | Bolt Minor Dia. — German | Bolt Minor Dia. — Swiss | Nut Major Dia. — British & German | Nut Major Dia. — French | Nut Major Dia. — Swiss | Nut Minor Dia. — French, German & Swiss | Nut Minor Dia. — British |
|---|---|---|---|---|---|---|---|---|---|---|---|
| 6 | 1 | 5.350 | 4.863 | 4.59 | 4.700 | 4.60 | 6.000 | 6.108 | 6.100 | 4.700 | 4.863 |
| 7 | 1 | 6.350 | 5.863 | 5.59 | 5.700 | 5.60 | 7.000 | 7.108 | 7.100 | 5.700 | 5.863 |
| 8 | 1.25 | 7.188 | 6.579 | 6.24 | 6.376 | 6.25 | 8.000 | 8.135 | 8.124 | 6.376 | 6.579 |
| 9 | 1.25 | 8.188 | 7.579 | 7.24 | 7.376 | 7.25 | 9.000 | 9.135 | 9.124 | 7.376 | 7.579 |
| 10 | 1.5 | 9.026 | 8.295 | 7.89 | 8.052 | 7.90 | 10.000 | 10.162 | 10.150 | 8.052 | 8.295 |
| 11 | 1.5 | 10.026 | 9.295 | 8.89 | 9.052 | 8.90 | 11.000 | 11.162 | 11.150 | 9.052 | 9.295 |
| 12 | 1.75 | 10.863 | 10.011 | 9.54 | 9.726 | 9.55 | 12.000 | 12.189 | 12.174 | 9.726 | 10.011 |
| 14 | 2 | 12.701 | 11.727 | 11.19 | 11.402 | 11.20 | 14.000 | 14.216 | 14.200 | 11.402 | 11.727 |
| 16 | 2 | 14.701 | 13.727 | 13.19 | 13.402 | 13.20 | 16.000 | 16.216 | 16.200 | 13.402 | 13.727 |
| 18 | 2.5 | 16.376 | 15.158 | 14.48 | 14.752 | 14.50 | 18.000 | 18.270 | 18.250 | 14.752 | 15.158 |
| 20 | 2.5 | 18.376 | 17.158 | 16.48 | 16.752 | 16.50 | 20.000 | 20.270 | 20.250 | 16.752 | 17.158 |
| 22 | 2.5 | 20.376 | 19.158 | 18.48 | 18.752 | 18.50 | 22.000 | 22.270 | 22.250 | 18.752 | 19.158 |
| 24 | 3 | 22.051 | 20.590 | 19.78 | 20.102 | 19.80 | 24.000 | 24.324 | 24.300 | 20.102* | 20.590 |
| 27 | 3 | 25.051 | 23.590 | 22.78 | 23.102 | 22.80 | 27.000 | 27.324 | 27.300 | 23.102† | 23.590 |
| 30 | 3.5 | 27.727 | 26.022 | 25.08 | 25.454 | 25.10 | 30.000 | 30.378 | 30.350 | 25.454 | 26.022 |
| 33 | 3.5 | 30.727 | 29.022 | 28.08 | 28.454 | 28.10 | 33.000 | 33.378 | 33.350 | 28.454 | 29.022 |
| 36 | 4 | 33.402 | 31.453 | 30.37 | 30.804 | 30.40 | 36.000 | 36.432 | 36.400 | 30.804 | 31.453 |
| 39 | 4 | 36.402 | 34.453 | 33.37 | 33.804 | 33.40 | 39.000 | 39.432 | 39.400 | 33.804 | 34.453 |
| 42 | 4.5 | 39.077 | 36.885 | 35.67 | 36.154 | 35.70 | 42.000 | 42.486 | 42.450 | 36.154 | 36.885 |
| 45 | 4.5 | 42.077 | 39.885 | 38.67 | 39.154 | 38.70 | 45.000 | 45.486 | 45.450 | 39.154 | 39.885 |
| 48 | 5 | 44.752 | 42.316 | 40.96 | 41.504 | 41.00 | 48.000 | 48.540 | 48.500 | 41.504 | 42.316 |
| 52 | 5 | 48.752 | 46.316 | 44.96 | 45.504 | 45.00 | 52.000 | 52.540 | 52.500 | 45.504 | 46.316 |
| 56 | 5.5 | 52.428 | 49.748 | 48.26 | 48.856 | 48.30 | 56.000 | 56.594 | 56.550 | 48.856 | 49.748 |
| 60 | 5.5 | 56.428 | 53.748 | 52.26 | 52.856 | 52.30 | 60.000 | 60.594 | 60.550 | 52.856 | 53.748 |

All dimensions shown in mm.
* The value shown is given in the German Standard; the value in the French Standard is 20.002; and in the Swiss Standard, 20.104.
† The value shown is given in the German Standard; the value in the French Standard is 23.002; and in the Swiss Standard, 23.104.

## Trapezoidal Metric Thread — Preferred Basic Sizes (DIN 103)

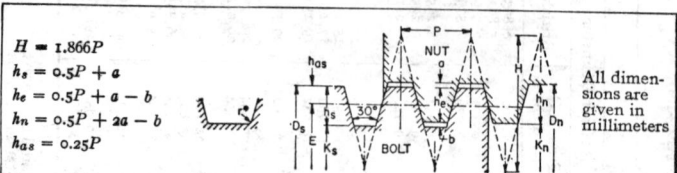

$H = 1.866P$
$h_s = 0.5P + a$
$h_e = 0.5P + a - b$
$h_n = 0.5P + 2a - b$
$h_{as} = 0.25P$

All dimensions are given in millimeters

| Nom. & Major Diam. of Bolt, $D_s$ | Pitch, $P$ | Pitch Diam., $E$ | Depth of Engagement, $h_e$ | Clearance | | Bolt | | Nut | | |
|---|---|---|---|---|---|---|---|---|---|---|
| | | | | $a$ | $b$ | Minor Diam., $K_s$ | Depth of Thread, $h_s$ | Major Diam., $D_n$ | Minor Diam., $K_n$ | Depth of Thread, $h_n$ |
| 10 | 3 | 8.5 | 1.25 | 0.25 | 0.5 | 6.5 | 1.75 | 10.5 | 7.5 | 1.50 |
| 12 | 3 | 10.5 | 1.25 | 0.25 | 0.5 | 8.5 | 1.75 | 12.5 | 9.5 | 1.50 |
| 14 | 4 | 12 | 1.75 | 0.25 | 0.5 | 9.5 | 2.25 | 14.5 | 10.5 | 2.00 |
| 16 | 4 | 14 | 1.75 | 0.25 | 0.5 | 11.5 | 2.25 | 16.5 | 12.5 | 2.00 |
| 18 | 4 | 16 | 1.75 | 0.25 | 0.5 | 13.5 | 2.25 | 18.5 | 14.5 | 2.00 |
| 20 | 4 | 18 | 1.75 | 0.25 | 0.5 | 15.5 | 2.25 | 20.5 | 16.5 | 2.00 |
| 22 | 5 | 19.5 | 2 | 0.25 | 0.75 | 16.5 | 2.75 | 22.5 | 18 | 2.25 |
| 24 | 5 | 21.5 | 2 | 0.25 | 0.75 | 18.5 | 2.75 | 24.5 | 20 | 2.25 |
| 26 | 5 | 23.5 | 2 | 0.25 | 0.75 | 20.5 | 2.75 | 26.5 | 22 | 2.25 |
| 28 | 5 | 25.5 | 2 | 0.25 | 0.75 | 22.5 | 2.75 | 28.5 | 24 | 2.25 |
| 30 | 6 | 27 | 2.5 | 0.25 | 0.75 | 23.5 | 3.25 | 30.5 | 25 | 2.75 |
| 32 | 6 | 29 | 2.5 | 0.25 | 0.75 | 25.5 | 3.25 | 32.5 | 27 | 2.75 |
| 36 | 6 | 33 | 2.5 | 0.25 | 0.75 | 29.5 | 3.25 | 36.5 | 31 | 2.75 |
| 40 | 7 | 36.5 | 3 | 0.25 | 0.75 | 32.5 | 3.75 | 40.5 | 34 | 3.25 |
| 44 | 7 | 40.5 | 3 | 0.25 | 0.75 | 36.5 | 3.75 | 44.5 | 38 | 3.25 |
| 48 | 8 | 44 | 3.5 | 0.25 | 0.75 | 39.5 | 4.25 | 48.5 | 41 | 3.75 |
| 50 | 8 | 46 | 3.5 | 0.25 | 0.75 | 41.5 | 4.25 | 50.5 | 43 | 3.75 |
| 52 | 8 | 48 | 3.5 | 0.25 | 0.75 | 43.5 | 4.25 | 52.5 | 45 | 3.75 |
| 55 | 9 | 50.5 | 4 | 0.25 | 0.75 | 45.5 | 4.75 | 55.5 | 47 | 4.25 |
| 60 | 9 | 55.5 | 4 | 0.25 | 0.75 | 50.5 | 4.75 | 60.5 | 52 | 4.25 |
| 65 | 10 | 60 | 4.5 | 0.25 | 0.75 | 54.5 | 5.25 | 65.5 | 56 | 4.75 |
| 70 | 10 | 65 | 4.5 | 0.25 | 0.75 | 59.5 | 5.25 | 70.5 | 61 | 4.75 |
| 75 | 10 | 70 | 4.5 | 0.25 | 0.75 | 64.5 | 5.25 | 75.5 | 66 | 4.75 |
| 80 | 10 | 75 | 4.5 | 0.25 | 0.75 | 69.5 | 5.25 | 80.5 | 71 | 4.75 |
| 85 | 12 | 79 | 5.5 | 0.25 | 0.75 | 72.5 | 6.25 | 85.5 | 74 | 5.75 |
| 90 | 12 | 84 | 5.5 | 0.25 | 0.75 | 77.5 | 6.25 | 90.5 | 79 | 5.75 |
| 95 | 12 | 89 | 5.5 | 0.25 | 0.75 | 82.5 | 6.25 | 95.5 | 84 | 5.75 |
| 100 | 12 | 94 | 5.5 | 0.25 | 0.75 | 87.5 | 6.25 | 100.5 | 89 | 5.75 |
| 110 | 12 | 104 | 5.5 | 0.25 | 0.75 | 97.5 | 6.25 | 110.5 | 99 | 5.75 |
| 120 | 14 | 113 | 6 | 0.5 | 1.5 | 105 | 7.5 | 121 | 108 | 6.5 |
| 130 | 14 | 123 | 6 | 0.5 | 1.5 | 115 | 7.5 | 131 | 118 | 6.5 |
| 140 | 14 | 133 | 6 | 0.5 | 1.5 | 125 | 7.5 | 141 | 128 | 6.5 |
| 150 | 16 | 142 | 7 | 0.5 | 1.5 | 133 | 8.5 | 151 | 136 | 7.5 |
| 160 | 16 | 152 | 7 | 0.5 | 1.5 | 143 | 8.5 | 161 | 146 | 7.5 |
| 170 | 16 | 162 | 7 | 0.5 | 1.5 | 153 | 8.5 | 171 | 156 | 7.5 |
| 180 | 18 | 171 | 8 | 0.5 | 1.5 | 161 | 9.5 | 181 | 164 | 8.5 |
| 190 | 18 | 181 | 8 | 0.5 | 1.5 | 171 | 9.5 | 191 | 174 | 8.5 |
| 200 | 18 | 191 | 8 | 0.5 | 1.5 | 181 | 9.5 | 201 | 184 | 8.5 |
| 210 | 20 | 200 | 9 | 0.5 | 1.5 | 189 | 10.5 | 211 | 192 | 9.5 |
| 220 | 20 | 210 | 9 | 0.5 | 1.5 | 199 | 10.5 | 221 | 202 | 9.5 |
| 230 | 20 | 220 | 9 | 0.5 | 1.5 | 209 | 10.5 | 231 | 212 | 9.5 |
| 240 | 22 | 229 | 10 | 0.5 | 1.5 | 217 | 11.5 | 241 | 220 | 10.5 |
| 250 | 22 | 239 | 10 | 0.5 | 1.5 | 227 | 11.5 | 251 | 230 | 10.5 |
| 260 | 22 | 249 | 10 | 0.5 | 1.5 | 237 | 11.5 | 261 | 240 | 10.5 |
| 270 | 24 | 258 | 11 | 0.5 | 1.5 | 245 | 12.5 | 271 | 248 | 11.5 |
| 280 | 24 | 268 | 11 | 0.5 | 1.5 | 255 | 12.5 | 281 | 258 | 11.5 |
| 290 | 24 | 278 | 11 | 0.5 | 1.5 | 265 | 12.5 | 291 | 268 | 11.5 |
| 300 | 26 | 287 | 12 | 0.5 | 1.5 | 273 | 13.5 | 301 | 276 | 12.5 |

* Roots are rounded to a radius, $r$ equal to 0.25 mm. for pitches of from 3 to 12 mm. inclusive and 0.5 mm. for pitches of from 14 to 26 mm. inclusive for power transmission.

## S.A.E. Standard Threads for Spark Plugs

| Size Nom. × Pitch | Major Diameter | | Pitch Diameter | | Minor Diameter | |
|---|---|---|---|---|---|---|
| | Max. | Min. | Max. | Min. | Max. | Min. |
| Spark Plug Threads, mm (inches) | | | | | | |
| M18 × 1.5 | 17.955 (0.7069) | 17.803 (0.7009) | 16.980 (0.6685) | 16.853 (0.6635) | 16.053 (0.6320) | ... |
| M14 × 1.25 | 13.868 (0.5460) | 13.741 (0.5410) | 13.104 (0.5159) | 12.997 (0.5117) | 12.339 (0.4858) | ... |
| M12 × 1.25 | 11.862 (0.4670) | 11.735 (0.4620) | 11.100 (0.4370) | 10.998 (0.4330) | 10.211 (0.4020) | ... |
| M10 × 1.0 | 9.974 (0.3927) | 9.794 (0.3856) | 9.324 (0.3671) | 9.212 (0.3627) | 8.747 (0.3444) | ... |
| Tapped Hole Threads, mm (inches) | | | | | | |
| M18 × 1.5 | ... | 18.039 (0.7102) | 17.153 (0.6753) | 17.026 (0.6703) | 16.426 (0.6467) | 16.266 (0.6404) |
| M14 × 1.25 | ... | 14.034 (0.5525) | 13.297 (0.5235) | 13.188 (0.5192) | 12.692 (0.4997) | 12.499 (0.4921) |
| M12 × 1.25 | ... | 11.935 (0.4699) | 11.242 (0.4426) | 11.138 (0.4385) | 10.559 (0.4157) | 10.366 (0.4081) |
| M10 × 1.0 | ... | 10.000 (0.3937) | 9.500 (0.3740) | 9.350 (0.3681) | 9.153 (0.3604) | 9.020 (0.3551) |

In order to keep the wear on the threading tools within permissible limits, the threads in the spark plug GO (ring) gage shall be truncated to the maximum minor diameter of the spark plug, and in the tapped hole GO (plug) gage to the minimum major diameter of the tapped hole. The plain plug gage for checking the minor diameter of the tapped hole shall be the minimum specified. The thread form is that of the ISO metric (see page 1353).

## British Standard for Spark Plugs (BS 45:1972)

This revised British Standard refers solely to spark plugs used in automobiles and industrial spark ignition internal combustion engines. The basic thread form is that of the ISO metric (see page 1353). In assigning tolerances to the threads of the spark plug and the tapped holes, full consideration has been given to the desirability of achieving the closest possible measure of interchangeability between British spark plugs and engines, and those made to the standards of other ISO Member Bodies.

| Basic Thread Dimensions for Spark Plug and Tapped Hole in Cylinder Head | | | | | | | | |
|---|---|---|---|---|---|---|---|---|
| Nom. Size | Pitch | Thread | Major Diam. | | Pitch Diam. | | Minor Diam. | |
| | | | Max. | Min. | Max. | Min. | Max. | Min. |
| 14 | 1.25 | Plug | 13.937 | 13.725 | 13.125 | 12.993 | 12.402 | 12.181 |
| 14 | 1.25 | Hole | * | 14.00 | 13.368 | 13.188 | 12.912 | 12.647 |
| 18 | 1.5 | Plug | 17.933 | 17.697 | 16.959 | 16.819 | 16.092 | 15.845 |
| 18 | 1.5 | Hole | * | 18.00 | 17.216 | 17.026 | 16.676 | 16.376 |

All dimensions are given in millimeters.

*Not specified.

The tolerance grades for finished spark plugs and corresponding tapped holes in the cylinder head are: for 14 mm size, 6e for spark plugs and 6H for tapped holes which gives a minimum clearance of 0.063 mm; and for 18 mm size, 6e for spark plugs and 6H for tapped holes which gives a minimum clearance of 0.067 mm.

These minimum clearances are intended to prevent the possibility of seizure, as a result of combustion deposits on the bare threads, when removing the spark plugs and applies to both ferrous and non-ferrous materials. These clearances are also intended to enable spark plugs with threads in accordance with this standard to be fitted into existing tapped holes.

**Compressed Gas Cylinder Valve Outlet and Inlet Connections.** — Detailed dimensions for all elements of valve outlets and their connections are given in American National Standard ANSI B57.1-1965. Standard outlet connections for the respective gases have been fully defined and are complete in themselves. The relation of one outlet to another is fixed so as to minimize undesirable connections. The threads on the outlet are separated into four basic divisions; internal, external, right-hand, and left-hand. Within each of the four divisions further separation is made by varying the pitch and diameter of the threads. Diameters within each division are so spaced that adjoining sizes either will not enter or will not engage. As far as practicable, the design of connections and assignment of the connections to gases has been made so as to prevent the interchange of connections which may result in a hazard. With the exception of outlets having taper pipe threads which seal at the threads, each outlet provides for screw threads, which do not seal, but merely hold the nipple against its seat. These screw threads have the American National form but are not in the regular series.

**American National Standard Hose Connections for Welding and Cutting Torches**

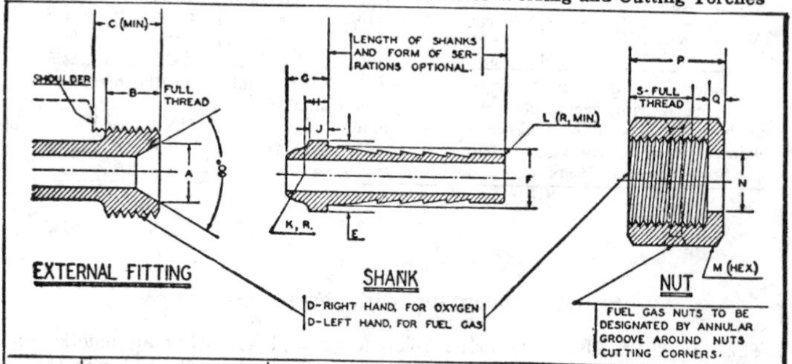

| Class | For Hose Sizes | External Fitting | | | | Shank | |
|---|---|---|---|---|---|---|---|
| | | Seat Diam. A | Thread Length B | Length C | Thread Size | Diam. E | Diam. F |
| A | ³⁄₁₆, ⅛ | 0.250 | ¼ | ⁹⁄₃₂ | ⅜ –24 | 0.326 | 0.248 |
| B | ⅜, ⁵⁄₁₆, ¼, ³⁄₁₆, ⅛ | 0.433 | ⁵⁄₁₆ | ¹³⁄₃₂ | ⁹⁄₁₆–18 | 0.498 | 0.430 |
| C | ½, ⅜, ⁵⁄₁₆, ¼ | 0.625 | ¹¹⁄₁₆ | ²³⁄₃₂ | ⅞ –14 | 0.750 | 0.578 |
| D | ¾, ⅝, ½, ⅜ | 0.954 | ⅞ | 3¹⁄₃₂ | 1¼–12 | 1.136 | 0.875 |

| Shank | | | | | Nut | | | | |
|---|---|---|---|---|---|---|---|---|---|
| Length G | Radius Center H | Length J | Radius K | Radius L | Width Flats M | Diam. N | Length P | Thick. Q | Length Thread S |
| ¼ | 0.182 | ⅛ | 0.099 | ¹⁄₃₂ | ⁷⁄₁₆ | 0.257 | ¹⁵⁄₃₂ | ³⁄₃₂ | ¼ |
| ⁵⁄₁₆ | 0.175 | ⅛ | 0.196 | ³⁄₆₄ | ¹¹⁄₁₆ | 0.4375 | ⅝ | ⅛ | ⁵⁄₁₆ |
| ⁷⁄₁₆ | 0.250 | ³⁄₁₆ | 0.280 | ¹⁄₃₂ | 1⅛ | 0.5937 | 1 | ⁵⁄₃₂ | ¹¹⁄₁₆ |
| ⅝ | 0.327 | ³⁄₁₆ | 0.438 | ³⁄₆₄ | 1½ | 0.9062 | 1¹¹⁄₃₂ | ⁷⁄₃₂ | ¹⁵⁄₁₆ |

Screw threads are American National form, fine-thread series and Class 3 fit. *Right-hand threads are specified for oxygen and left-hand threads for fuel gas.*

Tolerances: Dimension A ±.005 for Classes A to C inclusive, and ±.008 for Class D; dimension E ±.002 for Classes A and B and ±.004 for Classes C and D; dimension F −.005 for Classes A and B and −.010 for Classes C and D; radius H ±.005 for Classes A to C inclusive and ±0.008 for class D; diameter N +.003−000 for Classes A and B; +.006−.003 for Class C and +.006−.002 for Class D.

**ANSI Standard Hose Coupling Screw Threads.** — Threads for hose couplings, valves, and all other fittings used in direct connection with hose intended for domestic, industrial and general service in sizes ½, ⅝, ¾, 1, 1¼, 1½, 2, 2½, 3, 3½, and 4 inches are covered by American National Standard ANSI B2.4-1966 (R1974). These threads are designated as follows:

NH — Standard hose coupling threads of full form as produced by cutting or rolling.

NHR — Standard hose coupling threads for garden hose applications where the design utilizes thin walled material which is formed to the desired thread.

NPSH — Standard straight hose coupling thread series in sizes ½ to 4 inches for joining to American National Standard taper pipe threads using a gasket to seal the joint.

Thread dimensions are given in Table 1 and thread lengths in Table 2.

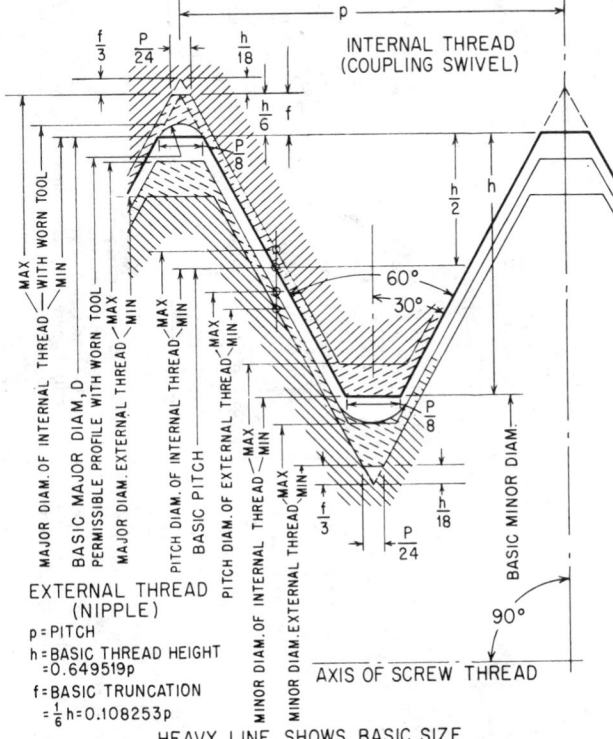

Thread Form for ANSI Standard Hose Coupling Threads, NPSH, NH, and NHR.
Heavy Line Shows Basic Size.

**Table 1. ANSI Standard Hose Coupling Threads for NPSH, NH and NHR Nipples and Coupling Swivels*** (ANSI B2.4-1966, R1974)

| Nom. Size of Hose | Thds. per Inch | Thread Designation | Pitch | Basic Height of Thread | Nipple (External) Thread | | | | | Coupling (Internal) Thread | | | | |
|---|---|---|---|---|---|---|---|---|---|---|---|---|---|---|
| | | | | | Major Diam. | | Pitch Diam. | | Minor Diam. | Minor Diam. | | Pitch Diam. | | Major Diam. |
| | | | | | Max. | Min. | Max. | Min. | Max. | Min. | Max. | Min. | Max. | Min. |
| ½, ⅝, ¾ | 11.5 | .75-11.5NH | .08696 | .05648 | 1.0625 | 1.0455 | 1.0060 | 0.9975 | 0.9495 | 0.9595 | 0.9765 | 1.0160 | 1.0245 | 1.0725 |
| ½, ⅝, ¾ | 11.5 | .75-11.5NHR | .08696 | .05648 | 1.0520 | 1.0350 | 1.0100 | 0.9930 | 0.9495 | 0.9720 | 0.9930 | 1.0160 | 1.0280 | 1.0680 |
| ½ | 14 | .5-14NPSH | .07143 | .04639 | 0.8248 | 0.8108 | 0.7784 | 0.7714 | 0.7320 | 0.7395 | 0.7535 | 0.7859 | 0.7929 | 0.8323 |
| ¾ | 14 | .75-14NPSH | .07143 | .04639 | 1.0353 | 1.0213 | 0.9889 | 0.9819 | 0.9425 | 0.9500 | 0.9640 | 0.9964 | 1.0034 | 1.0428 |
| 1 | 11.5 | 1-11.5NPSH | .08696 | .05648 | 1.2951 | 1.2781 | 1.2386 | 1.2301 | 1.1821 | 1.1921 | 1.2091 | 1.2486 | 1.2571 | 1.3051 |
| 1¼ | 11.5 | 1.25-11.5NPSH | .08696 | .05648 | 1.6399 | 1.6229 | 1.5834 | 1.5749 | 1.5629 | 1.5369 | 1.5539 | 1.5934 | 1.6019 | 1.6499 |
| 1½ | 11.5 | 1.5-11.5NPSH | .08696 | .05648 | 1.8788 | 1.8618 | 1.8223 | 1.8138 | 1.7658 | 1.7758 | 1.7928 | 1.8323 | 1.8408 | 1.8888 |
| 2 | 11.5 | 2-11.5NPSH | .08696 | .05648 | 2.3528 | 2.3358 | 2.2963 | 2.2878 | 2.2398 | 2.2498 | 2.2668 | 2.3063 | 2.3148 | 2.3628 |
| 2½ | 8 | 2.5-8NPSH | .12500 | .08119 | 2.8434 | 2.8212 | 2.7622 | 2.7511 | 2.6810 | 2.6930 | 2.7152 | 2.7742 | 2.7853 | 2.8554 |
| 3 | 8 | 3-8NPSH | .12500 | .08119 | 3.4697 | 3.4475 | 3.3885 | 3.3774 | 3.3073 | 3.3193 | 3.3415 | 3.4005 | 3.4116 | 3.4817 |
| 3½ | 8 | 3.5-8NPSH | .12500 | .08119 | 3.9700 | 3.9478 | 3.8888 | 3.8777 | 3.8076 | 3.8196 | 3.8418 | 3.9008 | 3.9119 | 3.9820 |
| 4 | 8 | 4-8NPSH | .12500 | .08119 | 4.4683 | 4.4461 | 4.3871 | 4.3760 | 4.3059 | 4.3179 | 4.3401 | 4.3991 | 4.4102 | 4.4803 |
| 4 | 6 | 4-6NH(SPL) | .16667 | .10825 | 4.9082 | 4.8722 | 4.7999 | 4.7819 | 4.6916 | 4.7117 | 4.7477 | 4.8200 | 4.8380 | 4.9283 |

All dimensions are given in inches.

* NH and NHR threads are used for garden hose applications. NPSH threads are used for steam, air and all other hose connections to be made up with standard pipe threads. NH (SPL) threads are used for marine applications.

Dimensions given for the maximum minor diameter of the nipple are figured to the intersection of the worn tool arc with a centerline through crest and root. The minimum minor diameter of the nipple shall be that corresponding to a flat at the minor diameter of the minimum nipple equal to $\frac{1}{24}p$, and may be determined by subtracting $0.7939p$ from the minimum pitch diameter of the nipple. (See diagram on p. 1369.)

Dimensions given for the minimum major diameter of the coupling correspond to the basic flat, $\frac{1}{8}p$, and the profile at the major diameter produced by a worn tool must not fall below the basic outline. The maximum major diameter of the coupling shall be that corresponding to a flat at the major diameter of the maximum coupling equal to $\frac{1}{24}p$ and may be determined by adding $0.7939p$ to the maximum pitch diameter of the coupling. (See diagram on p. 1369.)

**Table 2. ANSI Standard Hose Coupling Screw Thread Lengths**
(ANSI B2.4-1966, R1974)

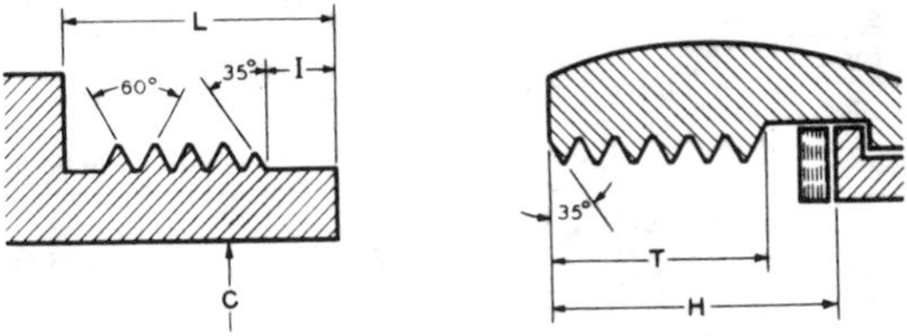


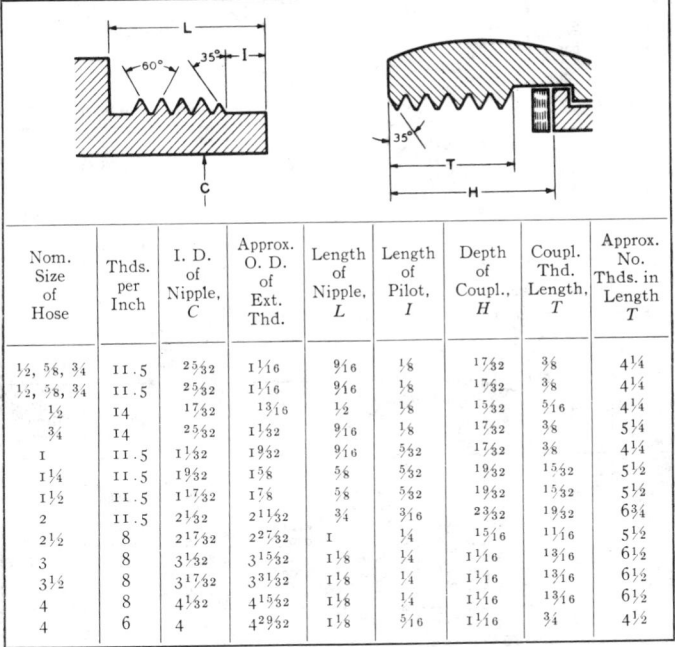

| Nom. Size of Hose | Thds. per Inch | I. D. of Nipple, *C* | Approx. O. D. of Ext. Thd. | Length of Nipple, *L* | Length of Pilot, *I* | Depth of Coupl., *H* | Coupl. Thd. Length, *T* | Approx. No. Thds. in Length *T* |
|---|---|---|---|---|---|---|---|---|
| 1/2, 5/8, 3/4 | 11.5 | 25/32 | 1 1/16 | 9/16 | 1/8 | 17/32 | 3/8 | 4 1/4 |
| 1/2, 5/8, 3/4 | 11.5 | 25/32 | 1 1/16 | 9/16 | 1/8 | 17/32 | 3/8 | 4 1/4 |
| 1/2 | 14 | 17/32 | 13/16 | 1/2 | 1/8 | 15/32 | 5/16 | 4 1/4 |
| 3/4 | 14 | 25/32 | 1 1/32 | 9/16 | 1/8 | 17/32 | 3/8 | 5 1/4 |
| 1 | 11.5 | 1 1/32 | 1 9/32 | 9/16 | 5/32 | 17/32 | 3/8 | 4 1/4 |
| 1 1/4 | 11.5 | 1 9/32 | 1 5/8 | 5/8 | 5/32 | 19/32 | 15/32 | 5 1/2 |
| 1 1/2 | 11.5 | 1 17/32 | 1 7/8 | 5/8 | 5/32 | 19/32 | 15/32 | 5 1/2 |
| 2 | 11.5 | 2 1/32 | 2 11/32 | 3/4 | 3/16 | 23/32 | 19/32 | 6 3/4 |
| 2 1/2 | 8 | 2 17/32 | 2 27/32 | 1 | 1/4 | 15/16 | 11/16 | 5 1/2 |
| 3 | 8 | 3 1/32 | 3 15/32 | 1 1/8 | 1/4 | 1 1/16 | 13/16 | 6 1/2 |
| 3 1/2 | 8 | 3 17/32 | 3 31/32 | 1 1/8 | 1/4 | 1 1/16 | 13/16 | 6 1/2 |
| 4 | 8 | 4 1/32 | 4 15/32 | 1 1/8 | 1/4 | 1 1/16 | 13/16 | 6 1/2 |
| 4 | 6 | 4 | 4 29/32 | 1 1/8 | 5/16 | 1 1/16 | 3/4 | 4 1/2 |

All dimensions are given in inches. For thread designation see Table 1.

**American National Fire Hose Connection Screw Thread.** — This thread is specified in the National Fire Protection Association's Standard NFPA No. 194-1974. It covers the dimensions for screw thread connections for fire hose couplings, suction hose couplings, relay supply hose couplings, fire pump suctions, discharge valves, fire hydrants, nozzles, adaptors, reducers, caps, plugs, wyes, siamese connections, standpipe connections, and sprinkler connections.

*Form of thread:* The basic form of thread is as shown on page 1369. It has an included angle of 60 degrees and is truncated top and botton. The flat at the root and crest of the basic thread form is equal to 1/8 (0.125) times the pitch in inches. The height of the thread is equal to 0.649519 times the pitch. The outer ends of both external and internal threads are terminated by the blunt start or "Higbee Cut" on full thread to avoid crossing and mutilation of thread.

*Thread Designation:* The thread is designated by specifying in sequence the nominal size of the connection, number of threads per inch followed by the thread symbol NH. Thus, .75-8NH indicates a nominal size connection of 0.75 inch diameter with 8 threads per inch.

*Basic Dimensions:* The basic dimensions of the thread are as given in Table 1.

*Thread Limits of Size:* Limits of size for NH external threads are given in Table 2. Limits of size for NH internal threads are given in Table 3.

Table 1.    Basic Dimensions of NH Threads (NFPA 194-1974)

| Nom. Size | Threads per Inch, (tpi) | Thread Designation | Pitch, p | Basic Thread Height, h | Minimum Internal Thread Dimensions | | |
|---|---|---|---|---|---|---|---|
| | | | | | Min. Minor Diam. | Basic Pitch Diam. | Basic Major Diam. |
| ¾ | 8 | .75-8 NH | 0.12500 | 0.08119 | 1.2246 | 1.3058 | 1.3870 |
| 1 | 8 | 1-8 NH | 0.12500 | 0.08119 | 1.2246 | 1.3058 | 1.3870 |
| 1½ | 9 | 1.5-9 NH | 0.11111 | 0.07217 | 1.8577 | 1.9298 | 2.0020 |
| 2½ | 7.5 | 2.5-7.5 NH | 0.13333 | 0.08660 | 2.9104 | 2.9970 | 3.0836 |
| 3 | 6 | 3-6 NH | 0.16667 | 0.10825 | 3.4223 | 3.5306 | 3.6389 |
| 3½ | 6 | 3.5-6 NH | 0.16667 | 0.10825 | 4.0473 | 4.1556 | 4.2639 |
| 4 | 4 | 4-4 NH | 0.25000 | 0.16238 | 4.7111 | 4.8735 | 5.0359 |
| 4½ | 4 | 4.5-4 NH | 0.25000 | 0.16238 | 5.4611 | 5.6235 | 5.7859 |
| 5 | 4 | 5-4 NH | 0.25000 | 0.16238 | 5.9602 | 6.1226 | 6.2850 |
| 6 | 4 | 6-4 NH | 0.25000 | 0.16238 | 6.7252 | 6.8876 | 7.0500 |

| Nom. Size | Threads per Inch, (tpi) | Thread Designation | Pitch, p | External Thread Dimensions (Nipple) | | | |
|---|---|---|---|---|---|---|---|
| | | | | Allowance | Max. Major Diam. | Max. Pitch Diam. | Max. Minor Diam. |
| ¾ | 8 | .75-8 NH | 0.12500 | 0.0120 | 1.3750 | 1.2938 | 1.2126 |
| 1 | 8 | 1-8 NH | 0.12500 | 0.0120 | 1.3750 | 1.2938 | 1.2126 |
| 1½ | 9 | 1.5-9 NH | 0.11111 | 0.0120 | 1.9900 | 1.9178 | 1.8457 |
| 2½ | 7.5 | 2.5-7.5 NH | 0.13333 | 0.0150 | 3.0686 | 2.9820 | 2.8954 |
| 3 | 6 | 3-6 NH | 0.16667 | 0.0150 | 3.6239 | 3.5156 | 3.4073 |
| 3½ | 6 | 3.5-6 NH | 0.16667 | 0.0200 | 4.2439 | 4.1356 | 4.0273 |
| 4 | 4 | 4-4 NH | 0.25000 | 0.0250 | 5.0109 | 4.8485 | 4.6861 |
| 4½ | 4 | 4.5-4 NH | 0.25000 | 0.0250 | 5.7609 | 5.5985 | 5.4361 |
| 5 | 4 | 5-4 NH | 0.25000 | 0.0250 | 6.2600 | 6.0976 | 5.9352 |
| 6 | 4 | 6-4 NH | 0.25000 | 0.0250 | 7.0250 | 6.8626 | 6.7002 |

All dimensions are in inches.

*Tolerances:* The pitch diameter tolerances for a mating external and internal thread are the same. Pitch diameter tolerances include lead and half-angle deviations. Lead deviations consuming one-half of the pitch diameter tolerance are 0.0032 inch for ¾, 1 and 1½-inch sizes; 0.0046 inch for 2½-inch size; 0.0052 inch for 3 and 3½-inch sizes; and 0.0072 inch for 4, 4½, 5, and 6-inch sizes. Half-angle deviations consuming one-half of the pitch diameter tolerance are 1 degree, 42 minutes for ¾, and 1-inch sizes; 1 degree, 54 minutes for 1½-inch size; 2 degrees, 17 minutes for 2½-inch size; 2 degrees, 4 minutes for 3 and 3½-inch sizes; and 1 degree, 55 minutes for 4, 4½, 5, and 6-inch sizes.

Tolerances for the external threads are:

Major diameter tolerance $= 2 \times$ Pitch diameter tolerance
Minor diameter tolerance $=$ Pitch diameter tolerance $+ 2h/9$

The minimum minor diameter of the external thread is such as to result in a flat equal to one-third of the $p/8$ basic flat, or $p/24$, at the root when the pitch diameter of the external thread is at its minimum value. The maximum minor diameter is basic, but may be such as results from the use of a worn or rounded threading tool. This is the maximum minor diameter shown in the figure on page 1369 and is the diameter upon which the minor diameter tolerance formula shown above is based.

# HOSE COUPLING SCREW THREADS 1373

Tolerances for the internal threads are:

Minor diameter tolerance = 2 × Pitch diameter tolerance

The minimum minor diameter of the internal thread is such as to result in a basic flat, $p/8$, at the crest when the pitch diameter of the thread is at its minimum value.

Major diameter tolerance = Pitch diameter tolerance − $2h/9$

*Gages and Gaging:* Full information on gage dimensions and the use of gages in checking the NH thread are given in NFPA Standard No. 194-1974, published by the National Fire Protection Association, 470 Atlantic Avenue, Boston, Mass. 02210.

The information and data taken from this standard are reproduced with the permission of the Association.

**Table 2. Limits of Size and Tolerances for NH External Threads, (Nipples)**
(NFPA 194-1974)

| Nom. Size | Threads per Inch, (tpi) | External Thread (Nipple) | | | | | | |
|---|---|---|---|---|---|---|---|---|
| | | Major Diameter | | | Pitch Diameter | | | Minor* Diam. |
| | | Max. | Min. | Toler. | Max. | Min. | Toler. | Max. |
| ¾ | 8 | 1.3750 | 1.3528 | 0.0222 | 1.2938 | 1.2827 | 0.0111 | 1.2126 |
| 1 | 8 | 1.3750 | 1.3528 | 0.0222 | 1.2938 | 1.2827 | 0.0111 | 1.2126 |
| 1½ | 9 | 1.9900 | 1.9678 | 0.0222 | 1.9178 | 1.9067 | 0.0111 | 1.8457 |
| 2½ | 7.5 | 3.0686 | 3.0366 | 0.0320 | 2.9820 | 2.9660 | 0.0160 | 2.8954 |
| 3 | 6 | 3.6239 | 3.5879 | 0.0360 | 3.5156 | 3.4976 | 0.0180 | 3.4073 |
| 3½ | 6 | 4.2439 | 4.2079 | 0.0360 | 4.1356 | 4.1176 | 0.0180 | 4.0273 |
| 4 | 4 | 5.0109 | 4.9609 | 0.0500 | 4.8485 | 4.8235 | 0.0250 | 4.6861 |
| 4½ | 4 | 5.7609 | 5.7109 | 0.0500 | 5.5985 | 5.5735 | 0.0250 | 5.4361 |
| 5 | 4 | 6.2600 | 6.2100 | 0.0500 | 6.0976 | 6.0726 | 0.0250 | 5.9352 |
| 6 | 4 | 7.0250 | 6.9750 | 0.0500 | 6.8626 | 6.8376 | 0.0250 | 6.7002 |

All dimensions are in inches.
* Dimensions given for the maximum minor diameter of the nipple are figured to the intersection of the worn tool arc with a centerline through crest and root. The minimum minor diameter of the nipple shall be that corresponding to a flat at the minor diameter of the minimum nipple equal to $p/24$ and may be determined by subtracting $11h/9$ (or 0.7939$p$) from the minimum pitch diameter of the nipple.

**Table 3. Limits of Size and Tolerances for NH Internal Threads, (Couplings)**
(NFPA 194-1974)

| Nom. Size | Threads per Inch, (tpi) | Internal Thread (Coupling) | | | | | | |
|---|---|---|---|---|---|---|---|---|
| | | Minor Diameter | | | Pitch Diameter | | | Major* Diam. |
| | | Min. | Max. | Toler. | Min. | Max. | Toler. | Min. |
| ¾ | 8 | 1.2246 | 1.2468 | 0.0222 | 1.3058 | 1.3169 | 0.0111 | 1.3870 |
| 1 | 8 | 1.2246 | 1.2468 | 0.0222 | 1.3058 | 1.3169 | 0.0111 | 1.3870 |
| 1½ | 9 | 1.8577 | 1.8799 | 0.0222 | 1.9298 | 1.9409 | 0.0111 | 2.0020 |
| 2½ | 7.5 | 2.9104 | 2.9424 | 0.0320 | 2.9970 | 3.0130 | 0.0160 | 3.0836 |
| 3 | 6 | 3.4223 | 3.4583 | 0.0360 | 3.5306 | 3.5486 | 0.0180 | 3.6389 |
| 3½ | 6 | 4.0473 | 4.0833 | 0.0360 | 4.1556 | 4.1736 | 0.0180 | 4.2639 |
| 4 | 4 | 4.7111 | 4.7611 | 0.0500 | 4.8735 | 4.8985 | 0.0250 | 5.0359 |
| 4½ | 4 | 5.4611 | 5.5111 | 0.0500 | 5.6235 | 5.6485 | 0.0250 | 5.7859 |
| 5 | 4 | 5.9602 | 6.0102 | 0.0500 | 6.1226 | 6.1476 | 0.0250 | 6.2850 |
| 6 | 4 | 6.7252 | 6.7752 | 0.0500 | 6.8876 | 6.9126 | 0.0250 | 7.0500 |

All dimensions are in inches.
* Dimensions for the minimum major diameter of the coupling correspond to the basic flat ($p/8$), and the profile at the major diameter produced by a worn tool must not fall below the basic outline. The maximum major diameter of the coupling shall be that corresponding to a flat at the major diameter of the maximum coupling equal to $p/24$ and may be determined by adding $11h/9$ (or 0.7939$p$) to the maximum pitch diameter of the coupling.

### Rolled Threads for Screw Shells of Electric Sockets and Lamp Bases — American Standard

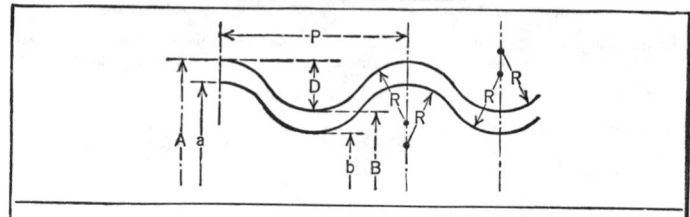

#### Male or Base Screw Shells Before Assembly

| Size | Threads per Inch | Pitch P | Depth of Thread D | Radius Crest Root R | Major Diam. | | Minor Diam. | |
|---|---|---|---|---|---|---|---|---|
| | | | | | Max. A | Min. a | Max. B | Min. b |
| Miniature | 14 | 0.07143 | 0.020 | 0.0210 | 0.375 | 0.370 | 0.335 | 0.330 |
| Candelabra | 10 | 0.10000 | 0.025 | 0.0312 | 0.465 | 0.460 | 0.415 | 0.410 |
| Intermediate | 9 | 0.11111 | 0.027 | 0.0353 | 0.651 | 0.645 | 0.597 | 0.591 |
| Medium | 7 | 0.14286 | 0.033 | 0.0470 | 1.037 | 1.031 | 0.971 | 0.965 |
| Mogul | 4 | 0.25000 | 0.050 | 0.0906 | 1.555 | 1.545 | 1.455 | 1.445 |

#### Socket Screw Shells Before Assembly

| Size | Threads per Inch | Pitch P | Depth of Thread D | Radius Crest Root R | Major Diam. | | Minor Diam. | |
|---|---|---|---|---|---|---|---|---|
| | | | | | Max. A | Min. a | Max. B | Min. b |
| Miniature | 14 | 0.07143 | 0.020 | 0.0210 | 0.3835 | 0.3775 | 0.3435 | 0.3375 |
| Candelabra | 10 | 0.10000 | 0.025 | 0.0312 | 0.476 | 0.470 | 0.426 | 0.420 |
| Intermediate | 9 | 0.11111 | 0.027 | 0.0353 | 0.664 | 0.657 | 0.610 | 0.603 |
| Medium | 7 | 0.14286 | 0.033 | 0.0470 | 1.053 | 1.045 | 0.987 | 0.979 |
| Mogul | 4 | 0.25000 | 0.050 | 0.0906 | 1.577 | 1.565 | 1.477 | 1.465 |

All dimensions given in inches.

*Base Screw Shell Gage Tolerances:* Threaded ring gages — "Go," Max. thread size to minus 0.0003 inch; "Not Go," Min. thread size to plus 0.0003 inch. Plain ring gages — "Go," Max. thread O.D. to minus 0.0002 inch; "Not Go", Min. thread O.D. to plus 0.0002. inch.

*Socket Screw Shell Gages:* Threaded plug gages — "Go," Min. thread size to plus 0.0003 inch; "Not Go," Max. thread size to minus 0.0003 inch. Plain plug gages — "Go," Min. minor diam. to plus 0.0002 inch; "Not Go," Max. minor diam. to minus 0.0002 inch.

*Check Gages for Base Screw Shell Gages:* Threaded plugs for checking threaded ring gages — "Go," Max. thread size to minus 0.0003 inch; "Not Go," Min. thread size to plus 0.0003 inch.

**Instrument Makers' System.** — The standard screw system of the Royal Microscopical Society of London, England, also known as the " Society Thread," is employed for microscope objectives and the nose pieces of the microscope into which these objectives screw. The form of the thread is the standard Whitworth form. The number of threads per inch is 36. The dimensions are as follows:

| | | |
|---|---|---|
| Male thread, outside diam., | max. 0.7982 inch, | min. 0.7952 inch; |
| root diam., | max. 0.7626 inch, | min. 0.7596 inch; |
| Female thread, root of thread, | max. 0.7674 inch, | min. 0.7644 inch; |
| top of thread, | max. 0.8030 inch, | min. 0.8000 inch. |

**Pipe Threads.** — The types of threads used on pipe and pipe fittings may be classed according to their intended use: (1) threads which when assembled with a sealer will produce a pressure-tight joint; (2) threads which when assembled without a sealer will produce a pressure-tight joint; (3) threads which provide free- and loose-fitting mechanical joints without pressure tightness; and (4) threads which produce rigid mechanical joints without pressure tightness. American National Standard pipe threads described in the following paragraphs provide taper and straight pipe threads for use in various combinations and with certain modifications to meet these specific needs.

**American National Standard Taper Pipe Threads.** — The basic dimensions of the ANSI Standard taper pipe thread are given in Table 3.

*Form of Thread:* The angle between the sides of the thread is 60 degrees when measured in an axial plane, and the line bisecting this angle is perpendicular to the axis. The depth of the truncated thread is based on factors entering into the manufacture of cutting tools and the making of tight joints and is given by the formulas in Table 3 or the data in Table 1 obtained from these formulas. While the standard shows flat surfaces at the crest and root of the thread, some rounding may occur in commercial practice, and it is intended that the pipe threads of product shall be acceptable when crest and root of the tools or chasers lie within the limits shown in Table 1.

Table 1.  Limits on Crest and Root of American National Standard External
and Internal Taper Pipe Threads, NPT (ANSI B2.1-1968)

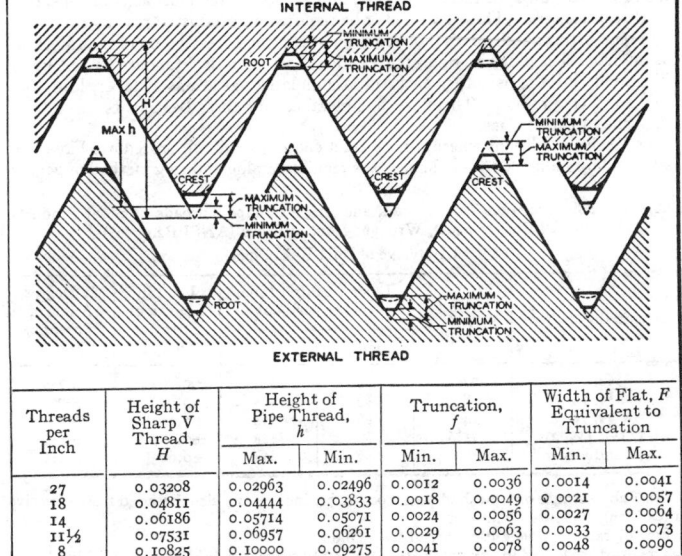

| Threads per Inch | Height of Sharp V Thread, $H$ | Height of Pipe Thread, $h$ | | Truncation, $f$ | | Width of Flat, $F$ Equivalent to Truncation | |
|---|---|---|---|---|---|---|---|
| | | Max. | Min. | Min. | Max. | Min. | Max. |
| 27 | 0.03208 | 0.02963 | 0.02496 | 0.0012 | 0.0036 | 0.0014 | 0.0041 |
| 18 | 0.04811 | 0.04444 | 0.03833 | 0.0018 | 0.0049 | 0.0021 | 0.0057 |
| 14 | 0.06186 | 0.05714 | 0.05071 | 0.0024 | 0.0056 | 0.0027 | 0.0064 |
| 11½ | 0.07531 | 0.06957 | 0.06261 | 0.0029 | 0.0063 | 0.0033 | 0.0073 |
| 8 | 0.10825 | 0.10000 | 0.09275 | 0.0041 | 0.0078 | 0.0048 | 0.0090 |

All dimensions are in inches and are given to four or five decimal places only to avoid errors in computations, not to indicate required precision.

*Pitch Diameter Formulas:* In the following formulas, which apply to the ANSI Standard taper pipe thread, $E_0$ = pitch diameter at end of pipe; $E_1$ = pitch diameter at the large end of the internal thread and also at the gaging notch; $D$ = outside diameter of pipe; $L_1$ = length of hand-tight or normal engagement between external and internal threads; $L_2$ = basic length of effective external taper thread; $p$ = pitch = 1 ÷ number of threads per inch.

$$E_0 = D - (0.05D + 1.1)p$$

$$E_1 = E_0 + 0.0625L_1$$

*Thread Length:* The formula for $L_2$ determines the length of the effective thread and includes approximately two usable threads which are slightly imperfect at the crest. The normal length of engagement $L_1$ between external and internal taper threads, when assembled by hand, is controlled by the use of the gages.

$$L_2 = (0.80D + 6.8)p$$

*Taper:* The taper of the thread is 1 in 16 or 0.75 inch per foot measured on the diameter and along the axis. The corresponding half-angle of taper or angle with the center line is one degree, 47 minutes.

**Tolerances on Thread Elements.** — The maximum allowable variation in the commercial product (manufacturing tolerance) is one turn large or small from the basic dimensions.

The permissible variations in thread elements on steel products and all pipe made of steel, wrought iron, or brass, exclusive of butt-weld pipe, are given in Table 2. This table is a guide for establishing the limits of the thread elements of taps, dies, and thread chasers. These limits may be required on product threads.

On pipe fittings and valves (not steel) for steam pressures 300 pounds and below, it is intended that plug and ring gage practice as set up in the Standard (ANSI B2.1) will provide for a satisfactory check of accumulated variations of taper, lead, and angle in such product. Therefore no tolerances on thread elements have been established for this class.

For service conditions where a more exact check is required, procedures have been developed by industry to supplement the regulation plug and ring method of gaging.

**Table 2. Tolerances on Taper, Lead, and Angle of Pipe Threads of Steel Products and All Pipe of Steel, Wrought-Iron, or Brass** (ANSI B2.1-1968)

(Exclusive of Butt-Weld Pipe)

| Nominal Pipe Size | Threads per Inch | Taper on Pitch Line, Inches per Foot | | Lead in Length of Effective Threads | 60 Degree Angle of Threads, Degrees |
|---|---|---|---|---|---|
| | | Max. | Min. | | |
| 1/16, 1/8 | 27 | 7/8 | 1 1/16 | ±0.003 | ±2 1/2 |
| 1/4, 3/8 | 18 | 7/8 | 1 1/16 | ±0.003 | ±2 |
| 1/2, 3/4 | 14 | 7/8 | 1 1/16 | ±0.003(a) | ±2 |
| 1, 1 1/4, 1 1/2, 2 | 11 1/2 | 7/8 | 1 1/16 | ±0.003(a) | ±1 1/2 |
| 2 1/2 and larger | 8 | 7/8 | 1 1/16 | ±0.003(a) | ±1 1/2 |

(a) The tolerance on lead shall be ±0.003 in. per inch on any size threaded to an effective thread length greater than 1 in.

For tolerances on height of thread see Table 1.

The limits specified in this table are intended to serve as a guide for establishing limits of the thread elements of taps, dies, and thread chasers. These limits may be required on product threads.

**Table 3. Basic Dimensions, American National Standard Taper Pipe Threads,[1] NPT**
(ANSI B2.1-1968)

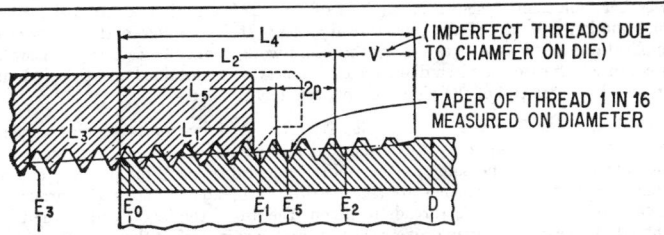

For all dimensions see corresponding reference letters in table.

Angle between sides of thread is 60 degrees. Taper of thread, on diameter, is ¾ inch per foot. Angle of taper with centerline is 1°47′.

The basic maximum thread height, $h$, of the truncated thread is 0.8 × pitch of thread. The crest and root are truncated a minimum of 0.033 × pitch for all pitches. For maximum depth of truncation see Table 1.

| Nominal Pipe Size | Outside Diam. of Pipe, $D$ | Threads per Inch, $n$ | Pitch of Thread, $p$ | Pitch Diameter at Beginning of External Thread, $E_0$ | Handtight Engagement | | Effective Thread, External | |
|---|---|---|---|---|---|---|---|---|
| | | | | | Length,[2] $L_1$ In. | Diam.,[3] $E_1$ | Length,[4] $L_2$ In. | Diam., $E_2$ |
| 1/16 | 0.3125 | 27 | 0.03704 | 0.27118 | 0.160 | 0.28118 | 0.2611 | 0.28750 |
| 1/8 | 0.405 | 27 | 0.03704 | 0.36351 | 0.1615 | 0.37360 | 0.2639 | 0.38000 |
| 1/4 | 0.540 | 18 | 0.05556 | 0.47739 | 0.2278 | 0.49163 | 0.4018 | 0.50250 |
| 3/8 | 0.675 | 18 | 0.05556 | 0.61201 | 0.240 | 0.62701 | 0.4078 | 0.63750 |
| 1/2 | 0.840 | 14 | 0.07143 | 0.75843 | 0.320 | 0.77843 | 0.5337 | 0.79179 |
| 3/4 | 1.050 | 14 | 0.07143 | 0.96768 | 0.339 | 0.98887 | 0.5457 | 1.00179 |
| 1 | 1.315 | 11½ | 0.08696 | 1.21363 | 0.400 | 1.23863 | 0.6828 | 1.25630 |
| 1¼ | 1.660 | 11½ | 0.08696 | 1.55713 | 0.420 | 1.58338 | 0.7068 | 1.60130 |
| 1½ | 1.900 | 11½ | 0.08696 | 1.79609 | 0.420 | 1.82234 | 0.7235 | 1.84130 |
| 2 | 2.375 | 11½ | 0.08696 | 2.26902 | 0.436 | 2.29627 | 0.7565 | 2.31630 |
| 2½ | 2.875 | 8 | 0.12500 | 2.71953 | 0.682 | 2.76216 | 1.1375 | 2.79062 |
| 3 | 3.500 | 8 | 0.12500 | 3.34062 | 0.766 | 3.38850 | 1.2000 | 3.41562 |
| 3½ | 4.000 | 8 | 0.12500 | 3.83750 | 0.821 | 3.88881 | 1.2500 | 3.91562 |
| 4 | 4.500 | 8 | 0.12500 | 4.33438 | 0.844 | 4.38712 | 1.3000 | 4.41562 |
| 5 | 5.563 | 8 | 0.12500 | 5.39073 | 0.937 | 5.44929 | 1.4063 | 5.47862 |
| 6 | 6.625 | 8 | 0.12500 | 6.44609 | 0.958 | 6.50597 | 1.5125 | 6.54062 |
| 8 | 8.625 | 8 | 0.12500 | 8.43359 | 1.063 | 8.50003 | 1.7125 | 8.54062 |
| 10 | 10.750 | 8 | 0.12500 | 10.54531 | 1.210 | 10.62094 | 1.9250 | 10.66562 |
| 12 | 12.750 | 8 | 0.12500 | 12.53281 | 1.360 | 12.61781 | 2.1250 | 12.66562 |
| 14 OD | 14.000 | 8 | 0.12500 | 13.77500 | 1.562 | 13.87262 | 2.2500 | 13.91562 |
| 16 OD | 16.000 | 8 | 0.12500 | 15.76250 | 1.812 | 15.87575 | 2.4500 | 15.91562 |
| 18 OD | 18.000 | 8 | 0.12500 | 17.75000 | 2.000 | 17.87500 | 2.6500 | 17.91562 |
| 20 OD | 20.000 | 8 | 0.12500 | 19.73750 | 2.125 | 19.87031 | 2.8500 | 19.91562 |
| 24 OD | 24.000 | 8 | 0.12500 | 23.71250 | 2.375 | 23.86094 | 3.2500 | 23.91562 |

All dimensions given in inches.

[1] The basic dimensions of the ANSI Standard Taper Pipe Thread are given in inches to four or five decimal places. While this implies a greater degree of precision than is ordinarily attained, these dimensions are the basis of gage dimensions and are so expressed for the purpose of eliminating errors in computations.

[2] Also length of thin ring gage and length from gaging notch to small end of plug gage.

[3] Also pitch diameter at gaging notch (handtight plane).

[4] Also length of plug gage.

**Engagement Between External and Internal Taper Threads.** — The normal length of engagement between external and internal taper threads when screwed together handtight is shown as $L_1$ in Table 3. This length is controlled by the construction and use of the pipe thread gages. It is recognized that in special applications, such as flanges for high-pressure work, longer thread engagement is used, in which case the pitch diameter $E_1$ (Table 3) is maintained and the pitch diameter $E_0$ at the end of the pipe is proportionately smaller.

**Railing Joint Taper Pipe Threads, NPTR.** — Railing joints require a rigid mechanical thread joint with external and internal taper threads. The external thread is basically the same as the ANSI Standard Taper Pipe Thread, except that sizes ½ through 2 inches are shortened by 3 threads and sizes 2½ through 4 inches are shortened by 4 threads to permit the use of the larger end of the pipe thread. A recess in the fitting covers the last scratch or imperfect threads on the pipe.

Table 3 (*Continued*). **Basic Dimensions, American National Standard Taper Pipe Threads, NPT** (ANSI B2.1-1968)

| Nominal Pipe Size | Wrench Makeup Length for Internal Thread | | Vanish Thread, (3.47 thds.), $V$ | Overall Length External Thread, $L_4$ | Nominal Perfect External Threads[5] | | Height of Thread, $h$ | Basic Minor Diam. at Small End of Pipe,[6] $K_0$ |
|---|---|---|---|---|---|---|---|---|
| | Length,[7] $L_3$ | Diam., $E_3$ | | | Length, $L_5$ | Diam., $E_5$ | | |
| 1/16 | 0.1111 | 0.26424 | 0.1285 | 0.3896 | 0.1870 | 0.28287 | 0.02963 | 0.2416 |
| 1/8 | 0.1111 | 0.35656 | 0.1285 | 0.3924 | 0.1898 | 0.37537 | 0.02963 | 0.3339 |
| 1/4 | 0.1667 | 0.46697 | 0.1928 | 0.5946 | 0.2907 | 0.49556 | 0.04444 | 0.4329 |
| 3/8 | 0.1667 | 0.60160 | 0.1928 | 0.6006 | 0.2967 | 0.63056 | 0.04444 | 0.5676 |
| 1/2 | 0.2143 | 0.74504 | 0.2478 | 0.7815 | 0.3909 | 0.78286 | 0.05714 | 0.7013 |
| 3/4 | 0.2143 | 0.95429 | 0.2478 | 0.7935 | 0.4029 | 0.99286 | 0.05714 | 0.9105 |
| 1 | 0.2609 | 1.19733 | 0.3017 | 0.9845 | 0.5089 | 1.24543 | 0.06957 | 1.1441 |
| 1¼ | 0.2609 | 1.54083 | 0.3017 | 1.0085 | 0.5329 | 1.59043 | 0.06957 | 1.4876 |
| 1½ | 0.2609 | 1.77978 | 0.3017 | 1.0252 | 0.5496 | 1.83043 | 0.06957 | 1.7265 |
| 2 | 0.2609 | 2.25272 | 0.3017 | 1.0582 | 0.5826 | 2.30543 | 0.06957 | 2.1995 |
| 2½ | 0.25008 | 2.70391 | 0.4337 | 1.5712 | 0.8875 | 2.77500 | 0.100000 | 2.6195 |
| 3 | 0.25008 | 3.32500 | 0.4337 | 1.6337 | 0.9500 | 3.40000 | 0.100000 | 3.2406 |
| 3½ | 0.2500 | 3.82188 | 0.4337 | 1.6837 | 1.0000 | 3.90000 | 0.100000 | 3.7375 |
| 4 | 0.2500 | 4.31875 | 0.4337 | 1.7337 | 1.0500 | 4.40000 | 0.100000 | 4.2344 |
| 5 | 0.2500 | 5.37511 | 0.4337 | 1.8400 | 1.1563 | 5.46300 | 0.100000 | 5.2907 |
| 6 | 0.2500 | 6.43047 | 0.4337 | 1.9462 | 1.2625 | 6.52500 | 0.100000 | 6.3461 |
| 8 | 0.2500 | 8.41797 | 0.4337 | 2.1462 | 1.4625 | 8.52500 | 0.100000 | 8.3336 |
| 10 | 0.2500 | 10.52969 | 0.4337 | 2.3587 | 1.6750 | 10.65000 | 0.100000 | 10.4453 |
| 12 | 0.2500 | 12.51719 | 0.4337 | 2.5587 | 1.8750 | 12.65000 | 0.100000 | 12.4328 |
| 14 OD | 0.2500 | 13.75938 | 0.4337 | 2.6837 | 2.0000 | 13.90000 | 0.100000 | 13.6750 |
| 16 OD | 0.2500 | 15.74688 | 0.4337 | 2.8837 | 2.2000 | 15.90000 | 0.100000 | 15.6625 |
| 18 OD | 0.2500 | 17.73438 | 0.4337 | 3.0837 | 2.4000 | 17.90000 | 0.100000 | 17.6500 |
| 20 OD | 0.2500 | 19.72188 | 0.4337 | 3.2837 | 2.6000 | 19.90000 | 0.100000 | 19.6375 |
| 24 OD | 0.2500 | 23.69688 | 0.4337 | 3.6837 | 3.0000 | 23.90000 | 0.100000 | 23.6125 |

[5] The length $L_5$ from the end of the pipe determines the plane beyond which the thread form is imperfect at the crest. The next two threads are perfect at the root. At this plane the cone formed by the crests of the thread intersects the cylinder forming the external surface of the pipe. $L_5 = L_2 - 2p$.

[6] Given as information for use in selecting tap drills.

[7] Three threads for 2-inch size and smaller; two threads for larger sizes.

[8] Military Specification MIL—P—7105 gives the wrench makeup as three threads for 3 in. and smaller. The $E_3$ dimensions are then as follows: Size 2½ in., 2.69609 and size 3 in., 3.31719.

Increase in diameter per thread is equal to $0.0625/n$.

Table 4.  American National Standard Internal Threads in Pipe Couplings,
NPSC (ANSI B2.1-1968)

Pressuretight Joints with Lubricant or Sealer

| Nom. Pipe Size | Thds. per Inch | Minor[2] Diam. Min. | Pitch Diameter[1] Min. | Pitch Diameter[1] Max. | Nom. Pipe Size | Thds. per Inch | Minor[2] Diam. Min. | Pitch Diameter[1] Min. | Pitch Diameter[1] Max. |
|---|---|---|---|---|---|---|---|---|---|
| ⅛ | 27 | 0.342 | 0.3701 | 0.3771 | 1½ | 11½ | 1.745 | 1.8142 | 1.8305 |
| ¼ | 18 | 0.440 | 0.4864 | 0.4968 | 2 | 11½ | 2.219 | 2.2881 | 2.3044 |
| ⅜ | 18 | 0.577 | 0.6218 | 0.6322 | 2½ | 8 | 2.650 | 2.7504 | 2.7739 |
| ½ | 14 | 0.715 | 0.7717 | 0.7851 | 3 | 8 | 3.277 | 3.3768 | 3.4002 |
| ¾ | 14 | 0.925 | 0.9822 | 0.9956 | 3½ | 8 | 3.777 | 3.8771 | 3.9005 |
| I | 11½ | 1.161 | 1.2305 | 1.2468 | 4 | 8 | 4.275 | 4.3754 | 4.3988 |
| 1¼ | 11½ | 1.506 | 1.5752 | 1.5915 | .... | .... | .... | .... | .... |

[1] The actual pitch diameter of the straight tapped hole will be slightly smaller than the value given when gaged with a taper plug gage as called for in ANSI B2.1.

[2] As the ANSI Standard Pipe Thread form is maintained, the major and minor diameters of the internal thread vary with the pitch diameter.

**Straight Pipe Threads in Pipe Couplings, NPSC.** — Threads in pipe couplings made in accordance with the ANSI B2.1 specifications are straight (parallel) threads of the same thread form as the ANSI Standard Taper Pipe Thread. They are used to form pressuretight joints when assembled with an ANSI Standard external taper pipe thread and made up with lubricant or sealant. These joints are recommended for comparatively low pressures only.

**Straight Pipe Threads for Mechanical Joints, NPSM, NPSL, and NPSH.** — While external and internal taper pipe threads are recommended for pipe joints in practically every service, there are mechanical joints where straight pipe threads are used to advantage. Three types covered by ANSI B2.1 are:

*Free-Fitting Mechanical Joints for Fixtures (External and Internal), NPSM:* Standard iron, steel, and brass pipe are often used for special applications where there are no internal pressures. Where straight thread joints are required for mechanical assemblies, straight pipe threads are often found more suitable or convenient. Dimensions of these threads are given in Table 5.

*Loose-Fitting Mechanical Joints With Locknuts (External and Internal), NPSL:* This thread is designed to produce a pipe thread having the largest diameter that it is possible to cut on standard pipe. The dimensions of these threads are given in Table 5. It will be noted that the maximum major diameter of the external thread is slightly greater than the nominal outside diameter of the pipe. The normal manufacturer's variation in pipe diameter provides for this increase.

*Loose-Fitting Mechanical Joints for Hose Couplings (External and Internal), NPSH:* Hose coupling joints are ordinarily made with straight internal and external loose-fitting threads. There are several standards of hose threads having various diameters and pitches. One of these is based on the ANSI Standard pipe thread and by the use of this thread series, it is possible to join small hose couplings in sizes ½ to 4 inches, inclusive, to ends of standard pipe having ANSI Standard External Pipe Threads, using a gasket to seal the joints. For the hose coupling thread dimensions see pages 1369 to 1373.

**Thread Designation and Notation.** — American National Standard Pipe Threads are designated by specifying in sequence the nominal size, number of threads per inch, and the symbols for the thread series and form, as: ⅜ — 18 NPT. The symbol designations are as follows: NPT — American National Standard Taper Pipe Thread; NPTR — American National Standard Taper Pipe Thread for Railing Joints; NPSC — American National Standard Straight Pipe Thread for Couplings; NPSM — American National Standard Straight Pipe Thread for Free-fitting Mechanical Joints; NPSL — American National Standard Straight Pipe Thread

### Table 5. American National Standard Straight Pipe Threads for Mechanical Joints, NPSM and NPSL (ANSI B2.2-1968)

| Nominal Pipe Size | Threads per Inch | External Thread | | | | | Internal Thread | | | |
|---|---|---|---|---|---|---|---|---|---|---|
| | | Allowance | Major Diameter | | Pitch Diameter | | Minor Diameter | | Pitch Diameter | |
| | | | Max.[2] | Min. | Max. | Min. | Min.[2] | Max. | Min.[1] | Max. |
| Free-fitting Mechanical Joints for Fixtures — NPSM | | | | | | | | | | |
| 1/8 | 27 | 0.0011 | 0.397 | 0.390 | 0.3725 | 0.3689 | 0.358 | 0.364 | 0.3736 | 0.3783 |
| 1/4 | 18 | 0.0013 | 0.526 | 0.517 | 0.4903 | 0.4859 | 0.468 | 0.481 | 0.4916 | 0.4974 |
| 3/8 | 18 | 0.0014 | 0.662 | 0.653 | 0.6256 | 0.6211 | 0.603 | 0.612 | 0.6270 | 0.6329 |
| 1/2 | 14 | 0.0015 | 0.823 | 0.813 | 0.7769 | 0.7718 | 0.747 | 0.759 | 0.7784 | 0.7851 |
| 3/4 | 14 | 0.0016 | 1.034 | 1.024 | 0.9873 | 0.9820 | 0.958 | 0.970 | 0.9889 | 0.9958 |
| 1 | 11½ | 0.0017 | 1.293 | 1.281 | 1.2369 | 1.2311 | 1.201 | 1.211 | 1.2386 | 1.2462 |
| 1¼ | 11½ | 0.0018 | 1.638 | 1.626 | 1.5816 | 1.5756 | 1.546 | 1.555 | 1.5834 | 1.5912 |
| 1½ | 11½ | 0.0018 | 1.877 | 1.865 | 1.8205 | 1.8144 | 1.785 | 1.794 | 1.8223 | 1.8302 |
| 2 | 11½ | 0.0019 | 2.351 | 2.339 | 2.2944 | 2.2882 | 2.259 | 2.268 | 2.2963 | 2.3044 |
| 2½ | 8 | 0.0022 | 2.841 | 2.826 | 2.7600 | 2.7526 | 2.708 | 2.727 | 2.7622 | 2.7720 |
| 3 | 8 | 0.0023 | 3.467 | 3.452 | 3.3862 | 3.3786 | 3.334 | 3.353 | 3.3885 | 3.3984 |
| 3½ | 8 | 0.0023 | 3.968 | 3.953 | 3.8865 | 3.8788 | 3.835 | 3.848 | 3.8888 | 3.8988 |
| 4 | 8 | 0.0023 | 4.466 | 4.451 | 4.3848 | 4.3771 | 4.333 | 4.346 | 4.3871 | 4.3971 |
| 5 | 8 | 0.0024 | 5.528 | 5.513 | 5.4469 | 5.4390 | 5.395 | 5.408 | 5.4493 | 5.4598 |
| 6 | 8 | 0.0024 | 6.585 | 6.570 | 6.5036 | 6.4955 | 6.452 | 6.464 | 6.5060 | 6.5165 |
| Loose-fitting Mechanical Joints for Locknut Connections — NPSL | | | | | | | | | | |
| 1/8 | 27 | ...... | 0.409 | ..... | 0.3840 | 0.3805 | 0.362 | ..... | 0.3863 | 0.3898 |
| 1/4 | 18 | ...... | 0.541 | ..... | 0.5038 | 0.4986 | 0.470 | ..... | 0.5073 | 0.5125 |
| 3/8 | 18 | ...... | 0.678 | ..... | 0.6409 | 0.6357 | 0.607 | ..... | 0.6444 | 0.6496 |
| 1/2 | 14 | ...... | 0.844 | ..... | 0.7963 | 0.7896 | 0.753 | ..... | 0.8008 | 0.8075 |
| 3/4 | 14 | ...... | 1.054 | ..... | 1.0067 | 1.0000 | 0.964 | ..... | 1.0112 | 1.0179 |
| 1 | 11½ | ...... | 1.318 | ..... | 1.2604 | 1.2523 | 1.208 | ..... | 1.2658 | 1.2739 |
| 1¼ | 11½ | ...... | 1.663 | ..... | 1.6051 | 1.5970 | 1.553 | ..... | 1.6106 | 1.6187 |
| 1½ | 11½ | ...... | 1.902 | ..... | 1.8441 | 1.8360 | 1.792 | ..... | 1.8495 | 1.8576 |
| 2 | 11½ | ...... | 2.376 | ..... | 2.3180 | 2.3099 | 2.265 | ..... | 2.3234 | 2.3315 |
| 2½ | 8 | ...... | 2.877 | ..... | 2.7934 | 2.7817 | 2.718 | ..... | 2.8012 | 2.8129 |
| 3 | 8 | ...... | 3.503 | ..... | 3.4198 | 3.4081 | 3.344 | ..... | 3.4276 | 3.4393 |
| 3½ | 8 | ...... | 4.003 | ..... | 3.9201 | 3.9084 | 3.845 | ..... | 3.9279 | 3.9396 |
| 4 | 8 | ...... | 4.502 | ..... | 4.4184 | 4.4067 | 4.343 | ..... | 4.4262 | 4.4379 |
| 5 | 8 | ...... | 5.564 | ..... | 5.4805 | 5.4688 | 5.405 | ..... | 5.4884 | 5.5001 |
| 6 | 8 | ...... | 6.620 | ..... | 6.5372 | 6.5255 | 6.462 | ..... | 6.5450 | 6.5567 |
| 8 | 8 | ...... | 8.615 | ..... | 8.5313 | 8.5196 | 8.456 | ..... | 8.5391 | 8.5508 |
| 10 | 8 | ...... | 10.735 | ..... | 10.6522 | 10.6405 | 10.577 | ..... | 10.6600 | 10.6717 |
| 12 | 8 | ...... | 12.732 | ..... | 12.6491 | 12.6374 | 12.574 | ..... | 12.6569 | 12.6686 |

*Notes for Free-fitting Fixture Threads:*

[1] This is the same as the pitch diameter at the end of internal thread, $E_1$ *Basic.* (See Table 3.)

The minor diameters of external threads and major diameters of internal threads are those as produced by commercial straight pipe dies and commercial ground straight pipe taps.

The major diameter of the external thread has been calculated on the basis of a truncation of $0.10825p$, and the minor diameter of the internal thread has been calculated on the basis of a truncation of $0.21651p$, to provide no interference at crest and root when product is gaged with gages made in accordance with the Standard.

*Notes for Loose-fitting Locknut Threads:*

[2] As the ANSI Standard Straight Pipe Thread form of thread is maintained, the major and the minor diameters of the internal thread and the major and the minor diameter of the external thread vary with the pitch diameter. The major diameter of the external thread is usually determined by the diameter of the pipe. These theoretical diameters result from adding the depth of the truncated thread ($0.666025 \times p$) to the maximum pitch diameters, and it should be understood that commercial pipe will not always have these maximum major diameters.

The locknut thread is established on the basis of retaining the greatest possible amount of metal thickness between the bottom of the thread and the inside of the pipe.

In order that a locknut may fit loosely on the externally threaded part, an allowance equal to the "increase in pitch diameter per turn" is provided, with a tolerance of 1½ turns for both external and internal threads.

for Loose-fitting Mechanical Joints with Locknuts; and NPSH — American National Standard Straight Pipe Thread for Hose Couplings.

**American National Standard Dryseal Taper Pipe Threads for Pressure-Tight Joints.** — The general form and dimensions of these threads are the same as those of the ANSI Standard taper pipe threads given in Table 3 except for the truncation of crests and roots as given in Table 6. A joint made up with this thread form has no clearance since the flats on the external and internal thread meet, producing metal-to-metal contact or interference at the crest and root of the mating parts that eliminates the need for a sealer. The crest and root of commercially manufactured

Table 6. American National Standard Dryseal Pipe Threads — Limits on Crest and Root Truncation (ANSI B2.2-1968)

| Threads per Inch | Depth of Sharp V Thread Inch | Truncation* | | | | | | | |
| | | Crest | | | | Root | | | |
| | | Maximum | | Minimum | | Maximum | | Minimum | |
| | | Formula | Inch | Formula | Inch | Formula | Inch | Formula | Inch |
| 27 | .03208 | .093$p$ | .0034 | .047$p$ | .0017 | .140$p$ | .0052 | .093$p$ | .0034 |
| 18 | .04811 | .078$p$ | .0043 | .047$p$ | .0026 | .109$p$ | .0061 | .078$p$ | .0043 |
| 14 | .06186 | .060$p$ | .0043 | .036$p$ | .0026 | .085$p$ | .0061 | .060$p$ | .0043 |
| 11½ | .07531 | .060$p$ | .0052 | .040$p$ | .0035 | .090$p$ | .0078 | .060$p$ | .0052 |
| 8 | .10825 | .055$p$ | .0069 | .042$p$ | .0052 | .076$p$ | .0095 | .055$p$ | .0069 |

*The crests of the threads on the plug and ring gages used for gaging Dryseal Threads shall have a minimum truncation of 0.20$p$ and a tolerance of 0.0005 inch for 27 and 18 tpi, 0.0006 inch for 14 and 11.5 tpi, and 0.0007 inch for 8 tpi.

Dryseal taper pipe threads are usually slightly rounded but are acceptable if they lie within the limits set up in Table 6. For purposes of economy and rapid production, a Dryseal joint may be made using a Dryseal external taper pipe thread and a Dryseal internal straight pipe thread.

**Types of Dryseal Pipe Threads and Their Designation.** — American National Standard ANSI B2.2 covers four types of Dryseal threads. They are designated:

NPTF — Dryseal ANSI Standard Pipe Thread
PTF-SAE SHORT — Dryseal SAE Short Taper Pipe Thread
NPSF — Dryseal ANSI Standard Fuel Internal Straight Pipe Thread
NPSI — Dryseal ANSI Standard Intermediate Internal Straight Pipe Thread

The full designation gives in sequence the nominal size, number of threads per inch, form (Dryseal), and symbol, as: ⅛ — 27 DRYSEAL NPTF.

*Type 1, Dryseal ANSI Standard Taper Pipe Thread, NPTF:* This series of threads applies to both external and internal threads of full length and is suitable for pipe joints in practically every type of service. These threads are generally conceded to be superior for strength and seal. Use of the internal tapered thread in hard or brittle materials having thin sections will minimize trouble from fracture.

*Type 2, Dryseal SAE Short Taper Pipe Thread, PTF-SAE SHORT:* External threads of this series conform in all respects with the NPTF threads except that the full thread length has been shortened by eliminating one thread at the small end for increased clearance and economy of material.

Internal threads of this series conform in all respects with NPTF threads except that the full thread length has been shortened by one thread at the large end.

*Type 3, Dryseal ANSI Standard Fuel Internal Straight Pipe Threads, NPSF:* Threads of this series are straight instead of tapered. They are generally used in soft or ductile materials which will adjust at assembly to the taper of external threads but may also be used in hard or brittle materials where the section is heavy.

*Type 4, Dryseal ANSI Standard Intermediate Internal Straight Pipe Threads, NPSI:* Threads of this series are straight instead of tapered. They are generally used in hard or brittle materials where the section is heavy and where there is little expansion at assembly with the external taper threads.

**British Standard Pipe Threads for Pressure-tight Joints.** — The threads in BS 21:1973 — "Specification for Pipe Threads where Pressure-tight Joints are Made on the Threads" are based on the Whitworth thread form and are specified as: (1) *Jointing threads.* These relate to pipe threads for joints made pressure-tight by the mating of the threads; they include taper external threads for assembly with either taper or parallel internal threads (parallel external pipe threads are not suitable as jointing threads). (2) *Longscrew threads.* These relate to parallel external pipe threads used for longscrews (connectors) specified in BS 1387 where a pressure-tight joint is achieved by the compression of a soft material onto the surface of the external thread by tightening a back nut against a socket.

**British Standard External and Internal Pipe Threads (Pressure-tight Joints) — Metric and Inch Dimensions and Limits of Size\* (BS 21:1973)**

| Nominal Size | No. of Threads per Inch† | Basic Diameters at Gage Plane | | | Gage Length | | Number of Useful Threads on Pipe for Basic Gage Length‡ | Tol., + and −, Gage Plane to Face of Int. Taper Thread | Tol., + and − on Diameter of Parallel Int. Threads |
| | | Major | Pitch | Minor | Basic | Tolerance (+ and −) | | | |
|---|---|---|---|---|---|---|---|---|---|
| 1/16 | 28 | 7.723 / *0.304* | 7.142 / *0.2812* | 6.561 / *0.2583* | (4⅜) / *4.0* | (1) / *0.9* | (7⅛) / *6.5* | (1¼) / *1.1* | 0.071 / *0.0028* |
| ⅛ | 28 | 9.728 / *0.383* | 9.147 / *0.3601* | 8.566 / *0.3372* | (4⅜) / *4.0* | (1) / *0.9* | (7⅛) / *6.5* | (1¼) / *1.1* | 0.071 / *0.0028* |
| ¼ | 19 | 13.157 / *0.518* | 12.301 / *0.4843* | 11.445 / *0.4506* | (4½) / *6.0* | (1) / *1.3* | (7¼) / *9.7* | (1¼) / *1.7* | 0.104 / *0.0041* |
| ⅜ | 19 | 16.662 / *0.656* | 15.806 / *0.6223* | 14.950 / *0.5886* | (4¾) / *6.4* | (1) / *1.3* | (7½) / *10.1* | (1¼) / *1.7* | 0.104 / *0.0041* |
| ½ | 14 | 20.955 / *0.825* | 19.793 / *0.7793* | 18.631 / *0.7336* | (4½) / *8.2* | (1) / *1.8* | (7¼) / *13.2* | (1¼) / *2.3* | 0.142 / *0.0056* |
| ¾ | 14 | 26.441 / *1.041* | 25.279 / *0.9953* | 24.117 / *0.9496* | (5¼) / *9.5* | (1) / *1.8* | (8) / *14.5* | (1¼) / *2.3* | 0.142 / *0.0056* |
| 1 | 11 | 33.249 / *1.309* | 31.770 / *1.2508* | 30.291 / *1.1926* | (4½) / *10.4* | (1) / *2.3* | (7¼) / *16.8* | (1¼) / *2.9* | 0.180 / *0.0071* |
| 1¼ | 11 | 41.910 / *1.650* | 40.431 / *1.5918* | 38.952 / *1.5336* | (5½) / *12.7* | (1) / *2.3* | (8¼) / *19.1* | (1¼) / *2.9* | 0.180 / *0.0071* |
| 1½ | 11 | 47.803 / *1.882* | 46.324 / *1.8238* | 44.845 / *1.7656* | (5½) / *12.7* | (1) / *2.3* | (8¼) / *19.1* | (1¼) / *2.9* | 0.180 / *0.0071* |
| 2 | 11 | 59.614 / *2.347* | 58.135 / *2.2888* | 56.656 / *2.2306* | (6⅞) / *15.9* | (1) / *2.3* | (10⅛) / *23.4* | (1¼) / *2.9* | 0.180 / *0.0071* |
| 2½ | 11 | 75.184 / *2.960* | 73.705 / *2.9018* | 72.226 / *2.8436* | (7⁹⁄₁₆) / *17.5* | (1½) / *3.5* | (11⁹⁄₁₆) / *26.7* | (1½) / *3.5* | 0.216 / *0.0085* |
| 3 | 11 | 87.884 / *3.460* | 86.405 / *3.4018* | 84.926 / *3.3436* | (8¹⁵⁄₁₆) / *20.6* | (1½) / *3.5* | (12¹⁵⁄₁₆) / *29.8* | (1½) / *3.5* | 0.216 / *0.0085* |
| 4 | 11 | 113.030 / *4.450* | 111.551 / *4.3918* | 110.072 / *4.3336* | (11) / *25.4* | (1½) / *3.5* | (15½) / *35.8* | (1½) / *3.5* | 0.216 / *0.0085* |
| 5 | 11 | 138.430 / *5.450* | 136.951 / *5.3918* | 135.472 / *5.3336* | (12⅜) / *28.6* | (1½) / *3.5* | (17⅜) / *40.1* | (1½) / *3.5* | 0.216 / *0.0085* |
| 6 | 11 | 163.830 / *6.450* | 162.351 / *6.3918* | 160.872 / *6.3336* | (12⅜) / *28.6* | (1½) / *3.5* | (17⅜) / *40.1* | (1½) / *3.5* | 0.216 / *0.0085* |

\* Each basic metric dimension is given in roman figures (nominal sizes excepted) and each basic inch dimension is shown in italics directly beneath it. Figures in ( ) are numbers of turns of thread with metric linear equivalents given beneath. For basic thread form of parallel threads see page **1349**. Taper of taper thread is 1 in 16 on diameter.

† In the Standard (BS 21:1973) the thread pitches in millimeters are as follows: 0.907 for 28 threads per inch, 1.337 for 19 threads per inch, 1.814 for 14 threads per inch, and 2.309 for 11 threads per inch.

‡ This is the minimum number of useful threads on the pipe for the basic gage length; for the maximum and minimum gage lengths, the minimum numbers of useful threads are, respectively, greater and less by the amount of tolerance in the column to the left. The design of internally threaded parts shall make allowance for receiving pipe ends of up to the minimum number of useful threads corresponding to the maximum gage length; the minimum number of useful *internal* threads shall be no less than 80 per cent of the minimum number of useful external threads for the minimum gage length.

## Measuring Screw Threads

**Pitch and Lead of Screw Threads.** — The *pitch* of a screw thread is the distance from the center of one thread to the center of the next thread. This applies no matter whether the screw has a single, double, triple or quadruple thread. The *lead* of a screw thread is the distance the nut will move forward on the screw if it is turned around one full revolution. In a single-threaded screw, the pitch and lead are equal, because the nut would move forward the distance from one thread to the next, if turned around once. In a double-threaded screw, the nut will move forward two threads, or twice the pitch, so that in this case the lead equals twice the pitch. In a triple-threaded screw, the lead equals three times the pitch, and so on.

The word "pitch" is often, although improperly, used to denote the *number of threads per inch*. Screws are spoken of as having a 12-pitch thread, when twelve threads per inch is what is really meant. The number of threads per inch equals 1 divided by the pitch, or expressed as a formula:

$$\text{Number of threads per inch} = \frac{1}{\text{pitch}}$$

The pitch of a screw equals 1 divided by the number of threads per inch, or:

$$\text{Pitch} = \frac{1}{\text{number of threads per inch}}$$

If the number of threads per inch equals 16, the pitch = 1/16. If the pitch equals 0.05, the number of threads equals 1 ÷ 0.05 = 20. If the pitch is 2/5 inch, the number of threads per inch equals 1 ÷ 2/5 = 2½.

Confusion is often caused by the indefinite designation of multiple-thread screws (double, triple, quadruple, etc.). The expression, "four threads per inch, triple," for example, is not to be recommended. It means that the screw is cut with four triple threads or with twelve threads per inch, if the threads are counted by placing a scale alongside the screw. To cut this screw, the lathe would be geared to cut four threads per inch, but they would be cut only to the depth required for twelve threads per inch. The best expression, when a multiple-thread is to be cut, is to say, in this case, "¼ inch lead, 1/12 inch pitch, triple thread." For single-threaded screws, only the number of threads per inch and the form of the thread are specified. The word "single" is not required.

**Measuring Screw Thread Pitch Diameters by Thread Micrometers.** — As the pitch or angle diameter of a tap or screw is the most important dimension, it is

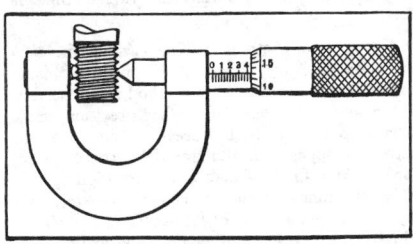

necessary that the pitch diameter of screw threads be measured, in addition to the outside diameter. One method of measuring in the angle of a thread is by means of a special screw thread micrometer, as shown in the accompanying engraving, Fig. 1. The fixed anvil is V-shaped so as to fit over the thread, while the movable point is cone-shaped so as to enable it to enter the space

Fig. 1

between two threads, and at the same time be at liberty to revolve. The contact points are on the sides of the thread, as they necessarily must be in order that the pitch diameter may be determined. The cone-shaped point of the measuring screw is slightly rounded so that it will not bear in the bottom of the thread. There is also sufficient clearance at the bottom of the V-shaped anvil to prevent it from bearing on the top of the thread. The movable point is adapted to measuring all pitches, but the fixed anvil is limited in its capacity. To cover the whole range of pitches, from the finest to the coarsest, a number of fixed anvils are, therefore, required.

To find the theoretical pitch diameter, which is measured by the micrometer, subtract twice the addendum of the thread from the standard outside diameter. The addendum of the thread for the American and other standard threads is given in the section on screw thread systems.

**Ball-point Micrometers.** — If standard plug gages are available, it is not necessary to actually measure the pitch diameter, but merely to compare it with the standard gage. In this case, a ball-point micrometer, as shown in Fig. 2, may be employed. Two types of ball-point micrometers are ordinarily used. One is simply a regular plain micrometer with ball points made to slip over both measuring points. (*See B*, Fig. 2.) This makes a kind of combination plain and ball-point micrometer, the ball points being easily removed. These ball points, however, do not fit solidly on their seats, even if they are split, as shown, and are apt to cause errors in the measurements. The best, and, in the long run, the cheapest, method is to use a regular micrometer arranged as shown at *A*. Drill and ream out both the end of the measuring screw or spindle and the anvil, and fit ball points into them as shown. Care should be taken to have the ball point in the spindle run true. The holes in the micrometer spindle and anvil and the shanks on the points are tapered to insure a good fit. The hole *H* in spindle *G* is provided so that the ball point can be easily driven out when a change for a larger or smaller size of ball point is required.

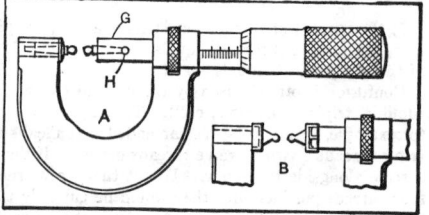

Fig. 2

A ball-point micrometer may be used for comparing the *angle* of a screw thread, with that of a gage. This can be done by using different sizes of ball points, comparing the size first near the root of the thread, then (using a larger ball point) at about the point of the pitch diameter, and finally near the top of the thread (using in the latter case, of course, a much larger ball point). If the gage and thread measurements are the same at each of the three points referred to, this indicates that the thread angle is correct.

**Measuring Screw Threads by Three-wire Method.** — The *effective* or *pitch diameter* of a screw thread may be measured very accurately by means of some form of micrometer and three wires of equal diameter. This method is extensively used in checking the accuracy of threaded plug gages and other precision screw threads. Two of the wires are placed in contact with the thread on one side and the third wire in a position diametrically opposite as illustrated by the diagram, (see Table " Rules and Formulas for Checking Pitch Diameters of Screw Threads ") and the dimension over the wires is determined by means of a micrometer. An ordinary micrometer is commonly used but some form of " floating micrometer " is preferable, especially for measuring thread gages and other precision work. The floating micrometer is mounted upon a compound slide so that it can move freely in directions parallel or at right angles to the axis of the screw, which is held in a horizontal position between adjustable centers. With this arrangement the microm-

eter is held constantly at right angles to the axis of the screw so that only one
wire on each side may be used instead of having two on one side and one on the
other, as is necessary when using an ordinary micrometer. The accuracy of the
pitch diameter may be determined provided the correct micrometer reading for
wires of a given size is known.

**Classes of Formulas for Three-wire Measurement.** — Various formulas have
been established for checking the pitch diameters of screw threads by measurement
over wires of known size. These formulas differ in regard to their simplicity or
complexity and resulting accuracy. They also differ in that some show what
measurement $M$ over the wires should be to obtain a given pitch diameter $E$,
whereas others show the value of the pitch diameter $E$ for a given measurement $M$.

*Formulas for Finding Measurement $M$:* In using a formula for finding the value
of measurement $M$, the required pitch diameter $E$ is inserted in the formula. Then,
in cutting or grinding a screw thread, the *actual* measurement $M$ is made to conform
to the *calculated* value of $M$. Formulas for finding measurement $M$ may be modified
so that the basic major or outside diameter is inserted in the formula instead of the
pitch diameter; however, the pitch diameter type of formula is preferable because
this is a more important dimension than the major diameter.

*Formulas for Finding Pitch Diameters $E$:* Some formulas are arranged to show
the value of the pitch diameter $E$ when measurement $M$ is known. Thus the value
of $M$ is first determined by actual measurement and then it is inserted in the formula
for finding the corresponding pitch diameter $E$. This type of formula is useful for
determining the pitch diameter of an existing thread gage or other screw thread in
connection with inspection work. The formula for finding measurement $M$ is more
convenient to use in the shop or tool-room in cutting or grinding new threads, because the pitch diameter is specified on the drawing and the problem is to find the
value of measurement $M$ for obtaining this pitch diameter.

**General Classes of Screw Thread Profiles.** — Thread profiles may be divided
into three general classes or types as follows:

*Screw Helicoid:* This type is represented by a screw thread having a straight-line
profile in the axial plane. Such a screw thread may be cut in a lathe by using a
straight-sided single-point tool, provided the top surface lies in the axial plane.

*Involute Helicoid:* This type is represented either by a screw thread or a helical
gear tooth having an involute profile in a plane perpendicular to the axis. A rolled
screw thread, theoretically at least, is an exact involute helicoid.

*Intermediate Profiles:* An intermediate profile which lies somewhere between the
screw helicoid and the involute helicoid will be formed on a screw thread either by
milling or grinding with a straight-sided wheel set in alignment with the thread
groove. The resulting form will approach closely the involute helicoid form. In
milling or grinding a thread, the included cutter or wheel angle may either equal the
standard thread angle (which is always measured in the axial plane) or the cutter
or wheel angle may be reduced to approximate, at least, the thread angle in the
normal plane. These variations in practice all affect the three-wire measurement.

**Accuracy of Formulas for Checking Pitch Diameters by Three-wire Method.** —
The exact measurement $M$ for a given pitch diameter depends upon the lead angle,
the thread angle, and the profile or cross-sectional shape of the thread. As pointed
out in the preceding paragraph, the profile depends upon the method of cutting or
forming the thread. In the case of a milled or ground thread, the profile is affected
not only by the cutter or wheel angle but also by the diameter of the cutter or wheel;
hence, because of these variations, an absolutely exact and reasonably simple general formula for measurement $M$ cannot be established; however, if the lead angle

is low, as in the case of a standard single-thread screw, and especially if the thread angle is high like a 60-degree thread, simple formulas which are not arranged to compensate for the lead angle are used ordinarily and meet most practical requirements, particularly in measuring 60-degree threads. If lead angles are large enough to decidedly affect the result, as in the case of most multiple threads (especially Acme or 29-degree worm threads), a formula should be used which compensates for the lead angle sufficiently to obtain the necessary accuracy.

The formulas which follow include (1) a very simple type in which the effect of the lead angle on measurement $M$ is entirely ignored. This simple formula usually is applicable to the measurement of 60-degree single-thread screws, except possibly when gage-making accuracy is required; (2) formulas which do include the effect of the lead angle but, nevertheless, are approximations and not always suitable for the higher lead angles when extreme accuracy is required; (3) formulas for the higher lead angles and the most precise classes of work.

Where approximate formulas are applied consistently in the measurement of both thread plug gages and the threaded "setting plugs" for ring gages, interchangeability might be secured assuming that such approximate formulas were universally employed.

**Wire Sizes for Checking Pitch Diameters of Screw Threads.** — In checking screw threads by the 3-wire method, the general practice is to use measuring wires of the so-called "best size." The "best size" wire is one which contacts at the

### Diameters of Wires for Measuring American Standard and British Standard Whitworth Screw Threads

| Threads per Inch | Pitch, Inch | Wire Diameters for American Standard Threads | | | Wire Diameters for Whitworth Standard Threads | | |
|---|---|---|---|---|---|---|---|
| | | Max. | Min. | Pitch-line Contact | Max. | Min. | Pitch-line Contact |
| 4 | 0.2500 | 0.2250 | 0.1400 | 0.1443 | 0.1900 | 0.1350 | 0.1409 |
| 4½ | 0.2222 | 0.2000 | 0.1244 | 0.1283 | 0.1689 | 0.1200 | 0.1253 |
| 5 | 0.2000 | 0.1800 | 0.1120 | 0.1155 | 0.1520 | 0.1080 | 0.1127 |
| 5½ | 0.1818 | 0.1636 | 0.1018 | 0.1050 | 0.1382 | 0.0982 | 0.1025 |
| 6 | 0.1667 | 0.1500 | 0.0933 | 0.0962 | 0.1267 | 0.0900 | 0.0939 |
| 7 | 0.1428 | 0.1286 | 0.0800 | 0.0825 | 0.1086 | 0.0771 | 0.0805 |
| 8 | 0.1250 | 0.1125 | 0.0700 | 0.0722 | 0.0950 | 0.0675 | 0.0705 |
| 9 | 0.1111 | 0.1000 | 0.0622 | 0.0641 | 0.0844 | 0.0600 | 0.0626 |
| 10 | 0.1000 | 0.0900 | 0.0560 | 0.0577 | 0.0760 | 0.0540 | 0.0564 |
| 11 | 0.0909 | 0.0818 | 0.0509 | 0.0525 | 0.0691 | 0.0491 | 0.0512 |
| 12 | 0.0833 | 0.0750 | 0.0467 | 0.0481 | 0.0633 | 0.0450 | 0.0470 |
| 13 | 0.0769 | 0.0692 | 0.0431 | 0.0444 | 0.0585 | 0.0415 | 0.0434 |
| 14 | 0.0714 | 0.0643 | 0.0400 | 0.0412 | 0.0543 | 0.0386 | 0.0403 |
| 16 | 0.0625 | 0.0562 | 0.0350 | 0.0361 | 0.0475 | 0.0337 | 0.0352 |
| 18 | 0.0555 | 0.0500 | 0.0311 | 0.0321 | 0.0422 | 0.0300 | 0.0313 |
| 20 | 0.0500 | 0.0450 | 0.0280 | 0.0289 | 0.0380 | 0.0270 | 0.0282 |
| 22 | 0.0454 | 0.0409 | 0.0254 | 0.0262 | 0.0345 | 0.0245 | 0.0256 |
| 24 | 0.0417 | 0.0375 | 0.0233 | 0.0240 | 0.0317 | 0.0225 | 0.0235 |
| 28 | 0.0357 | 0.0321 | 0.0200 | 0.0206 | 0.0271 | 0.0193 | 0.0201 |
| 32 | 0.0312 | 0.0281 | 0.0175 | 0.0180 | 0.0237 | 0.0169 | 0.0176 |
| 36 | 0.0278 | 0.0250 | 0.0156 | 0.0160 | 0.0211 | 0.0150 | 0.0156 |
| 40 | 0.0250 | 0.0225 | 0.0140 | 0.0144 | 0.0190 | 0.0135 | 0.0141 |

pitch line or mid-slope of the thread because then the measurement of the pitch diameter is least affected by an error in the thread angle. In the following formula for determining approximately the "best size" wire or the diameter for pitch-line contact, $A$ = one-half included angle of thread in axial plane.

$$\text{Best size wire} = \frac{0.5 \text{ pitch}}{\cos A} = 0.5 \text{ pitch} \times \sec A$$

For 60-degree threads this formula reduces to

$$\text{Best size wire} = 0.57735 \times \text{pitch}$$

These formulas are based upon a thread groove of zero lead angle because ordinary variations in the lead angle have little effect on the wire diameter and it is desirable to use one wire size for a given pitch regardless of the lead angle. A theoretically correct solution for finding the *exact* size for pitch-line contact, involves the use of cumbersome indeterminate equations with solution by successive trials. The accompanying table gives the wire sizes for both American Standard (formerly U. S. Standard) and the Whitworth Standard Threads. The following formulas for determining wire diameters do not give the extreme theoretical limits but the smallest and largest sizes which are practicable. The diameters in the table are based upon these approximate formulas.

American Standard $\begin{cases} \text{Smallest wire diameter} = 0.56 \times \text{pitch} \\ \text{Largest wire diameter} = 0.90 \times \text{pitch} \\ \text{Diameter for pitch-line contact} = 0.57735 \times \text{pitch} \end{cases}$

Whitworth $\begin{cases} \text{Smallest wire diameter} = 0.54 \times \text{pitch} \\ \text{Largest wire diameter} = 0.76 \times \text{pitch} \\ \text{Diameter for pitch-line contact} = 0.56368 \times \text{pitch} \end{cases}$

**Measuring Wire Accuracy.** — A set of three measuring wires should have the same diameter within 0.00002 inch. In order to measure the pitch diameter of a screw-thread gage to an accuracy of 0.0001 inch by means of wires, it is necessary to know the wire diameters to 0.00002 inch. If the diameters of the wires are known only to an accuracy of 0.0001 inch, an accuracy better than 0.0003 inch in the measurement of pitch diameter cannot be expected. The wires should be accurately finished hardened steel cylinders of the maximum possible hardness without being brittle. The hardness should not be less than that corresponding to a Knoop indentation number of 630. A wire of this hardness can be cut with a file only with difficulty. The surface should not be rougher than the equivalent of one measuring 3 microinches deviation from a true cylindrical surface.

**Measuring or Contact Pressure.** — In measuring screw threads or screw thread gages by the 3-wire method, variations in contact pressure will result in different readings. The effect of a variation in contact pressure in measuring threads of fine pitches is indicated by the difference in readings obtained with pressures of 2 pounds and 5 pounds in checking a thread plug gage having 24 threads per inch. The reading over the wires with 5 pounds pressure was 0.00013 inch less than with 2 pounds pressure. For pitches finer than 20 threads per inch a pressure of 16 ounces is recommended by the National Bureau of Standards. For pitches of 20 threads per inch and coarser, a pressure of 2½ pounds is recommended.

In the case of Acme threads, the wire presses against the sides of the thread with a pressure of approximately twice that of the measuring instrument. To limit the tendency of the wires to wedge in between the sides of an Acme thread, it is recommended that pitch diameter measurements on 8 threads per inch and finer be made at 1 pound, and at 2½ pounds for pitches coarser than 8 threads per inch.

**Notation Used in Formulas for Checking Pitch Diameters by Three-wire Method**

> $M$ = dimension over wires
> $D$ = basic major or outside diameter
> $E$ = pitch diameter (basic, maximum, or minimum) for which $M$ is required, or pitch diameter corresponding to measurement $M$
> $T$ = 0.5 pitch $P$ = width of thread in axial plane at diameter $E$
> $T_a$ = arc thickness on pitch cylinder in plane perpendicular to axis
> $P$ = pitch = 1 ÷ number of threads per inch
> $W$ = wire or pin diameter
> $A$ = one-half included thread angle in the axial plane
> $A_n$ = one-half included thread angle in the normal plane or in plane perpendicular to sides of thread = one-half included angle of cutter when thread is milled (tan $A_n$ = tan $A$ × cos $B$). (Note: Included angle of milling cutter or grinding wheel may equal the nominal included angle of thread, or cutter angle may be reduced to whatever normal angle is required to make the thread angle standard in the axial plane. In either case, $A_n$ = one-half cutter angle.)
> $S$ = number of "starts" or threads on a multiple-threaded worm or screw (2 tor double thread, 3 for triple thread, etc.)
> $B$ = lead angle at pitch diameter = helix angle of thread as measured from a plane perpendicular to the axis. Tan $B$ = $L$ ÷ 3.1416$E$
> $L$ = lead of thread = pitch $P$ × number of threads $S$
> $H$ = helix angle at pitch diam. and measured from axis = 90° − $B$ or tan $H$ = cot $B$
> $H_b$ = helix angle at $R_b$ measured from axis
> $R_b$ = radius required in Formulas (4e) and (4)
> $F$ = angle required in Formulas (4b), (4d)( and (4e)
> $G$ = angle required in Formula (4)

**Approximate Three-wire Formulas which do not Compensate for Lead Angle.** — A general formula in which the effect of lead angle is ignored is as follows (see accompanying notation used in formulas):

$$M = E - T \cot A + W (1 + \operatorname{cosec} A) \qquad (1)$$

This formula can be simplified for any given thread angle and pitch. To illustrate, since $T = 0.5\,P$, $M = E - 0.5\,P \cot A + W (1 + 2)$, for a 60-degree thread, such as the American Standard,

$$M = E - 0.86603\,P + 3\,W$$

The accompanying table contains these simplified formulas for different standard threads. Two formulas are given in each case. The upper one includes the standard or basic major diameter and the lower one the pitch diameter or effective diameter. These formulas are sufficiently accurate for practically all checking of standard 60-degree single-thread screws because of the low lead angles which vary from 1° 11′ to 4° 31′ in the American Standard Coarse-thread Series.

**Bureau of Standards General Formula.** — The Formula (2) which follows compensates quite largely for the effect of the lead angle. It is from the National Bureau of Standards Handbook H 28 (1944). The formula, however, as here given has been arranged for finding the value of $M$ (instead of $E$).

$$M = E - T \cot A + W (1 + \operatorname{cosec} A + 0.5 \tan^2 B \cos A \cot A) \qquad (2)$$

The Bureau of Standards uses Formula (2) in preference to Formula (1) when the value of $0.5\,W \tan^2 B \cos A \cot A$ exceeds 0.00015, as in the case of the larger lead angles. If this test is applied to American Standard 60-degree threads it will show that Formula (1) is generally applicable; but in case of 29-degree Acme or worm threads, Formula (2) (or some other which includes effect of lead angle) should be employed.

**Formulas for Checking Pitch Diameters of Screw Threads**

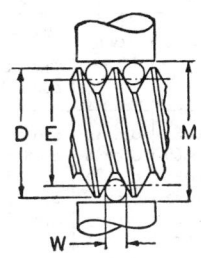

The formulas below do not compensate for the effect of the lead angle upon measurement $M$, but they are sufficiently accurate for checking standard single-thread screws unless exceptional accuracy is required. See accompanying information on effect of lead angle; also matter relating to measuring wire sizes, accuracy required for such wires, and contact or measuring pressure. The approximate best wire size for pitch-line contact may be obtained by the formula

$$W = 0.5 \times \text{pitch} \times \text{sec } \tfrac{1}{2} \text{ included thread angle}$$

For 60-degree threads, $W = 0.57735 \times$ pitch.

| Form of Thread | Formulas for determining measurement $M$ corresponding to correct pitch diameter. Diagram shows dimensions represented by letters in formulas. |
|---|---|
| American (National) Standard | When standard major diameter $D$ is used in formula. $$M = D - (1.5155 \times P) + (3 \times W)$$ When pitch diameter $E$ is used in formula. $$M = E - (0.86603 \times P) + (3 \times W)$$ The American Standard formerly was known as U. S. Standard. |
| British Standard Whitworth | When standard major diameter $D$ is used in formula. $$M = D - (1.6008 \times P) + (3.1657 \times W)$$ When pitch diameter $E$ is used in formula. $$M = E - (0.9605 \times P) + (3.1657 \times W)$$ |
| British Association Standard | When standard major diameter $D$ is used in formula. $$M = D - (1.7363 \times P) + (3.4829 \times W)$$ When pitch diameter $E$ is used in formula. $$M = E - (1.1363 \times P) + (3.4829 \times W)$$ |
| Lowenherz Thread | When standard major diameter $D$ is used in formula. $$M = D - (1.75 \times P) + (3.2359 \times W)$$ When pitch diameter $E$ is used in formula. $$M = E - P + (3.2359 \times W)$$ |
| Sharp V-Thread | When standard major diameter $D$ is used in formula. $$M = D - (1.732 \times P) + (3 \times W)$$ When pitch diameter $E$ is used in formula. $$M = E - (0.86603 \times P) + (3 \times W)$$ |
| International Standard | Use the formula given above for the American (National) Standard. |
| Pipe Thread | See accompanying paragraph on Measuring Taper Screw Threads by Three-wire Method. |
| Acme and Worm Threads | See Buckingham Formulas; also Three-wire Method for Checking Thickness of Acme Threads. |
| Buttress Form of Thread | Different forms of buttress threads are used. See paragraph on wire method applied to buttress threads. |

The wires must be lapped to a uniform diameter and it is very important to insert in the rule or formula the wire diameter as determined by precise means of measurement. Any error will be multiplied. See paragraph on Wire Sizes for Checking Pitch Diameters.

**Values of Constants Used in Formulas for Measuring Pitch Diameters of Screws by the Three-wire System**

| No. of Threads per Inch | V-Thread, 1.732 P | American Standard, 1.5155 P | Whitworth Thread, 1.6008 P | No. of Threads per Inch | V-Thread, 1.732 P | American Standard, 1.5155 P | Whitworth Thread, 1.6008 P |
|---|---|---|---|---|---|---|---|
| 2¼ | 0.7698 | 0.6736 | 0.7115 | 18 | 0.0962 | 0.0842 | 0.0889 |
| 2⅜ | 0.7293 | 0.6381 | 0.6740 | 20 | 0.0866 | 0.0758 | 0.0800 |
| 2½ | 0.6928 | 0.6062 | 0.6403 | 22 | 0.0787 | 0.0689 | 0.0728 |
| 2⅝ | 0.6598 | 0.5773 | 0.6098 | 24 | 0.0722 | 0.0631 | 0.0667 |
| 2¾ | 0.6298 | 0.5511 | 0.5821 | 26 | 0.0666 | 0.0583 | 0.0616 |
| 2⅞ | 0.6025 | 0.5271 | 0.5568 | 28 | 0.0619 | 0.0541 | 0.0572 |
| 3 | 0.5774 | 0.5052 | 0.5336 | 30 | 0.0577 | 0.0505 | 0.0534 |
| 3¼ | 0.5329 | 0.4663 | 0.4926 | 32 | 0.0541 | 0.0474 | 0.0500 |
| 3½ | 0.4949 | 0.4330 | 0.4574 | 34 | 0.0509 | 0.0446 | 0.0471 |
| 4 | 0.4330 | 0.3789 | 0.4002 | 36 | 0.0481 | 0.0421 | 0.0445 |
| 4½ | 0.3849 | 0.3368 | 0.3557 | 38 | 0.0456 | 0.0399 | 0.0421 |
| 5 | 0.3464 | 0.3031 | 0.3202 | 40 | 0.0433 | 0.0379 | 0.0400 |
| 5½ | 0.3149 | 0.2755 | 0.2911 | 42 | 0.0412 | 0.0361 | 0.0381 |
| 6 | 0.2887 | 0.2526 | 0.2668 | 44 | 0.0394 | 0.0344 | 0.0364 |
| 7 | 0.2474 | 0.2165 | 0.2287 | 46 | 0.0377 | 0.0329 | 0.0348 |
| 8 | 0.2165 | 0.1894 | 0.2001 | 48 | 0.0361 | 0.0316 | 0.0334 |
| 9 | 0.1925 | 0.1684 | 0.1779 | 50 | 0.0346 | 0.0303 | 0.0320 |
| 10 | 0.1732 | 0.1515 | 0.1601 | 52 | 0.0333 | 0.0291 | 0.0308 |
| 11 | 0.1575 | 0.1378 | 0.1455 | 56 | 0.0309 | 0.0271 | 0.0286 |
| 12 | 0.1443 | 0.1263 | 0.1334 | 60 | 0.0289 | 0.0253 | 0.0267 |
| 13 | 0.1332 | 0.1166 | 0.1231 | 64 | 0.0271 | 0.0237 | 0.0250 |
| 14 | 0.1237 | 0.1082 | 0.1143 | 68 | 0.0255 | 0.0223 | 0.0235 |
| 15 | 0.1155 | 0.1010 | 0.1067 | 72 | 0.0241 | 0.0210 | 0.0222 |
| 16 | 0.1083 | 0.0947 | 0.1001 | 80 | 0.0217 | 0.0189 | 0.0200 |

**Constants Used for Measuring Pitch Diameters of Metric Screws by the Three-wire System**

| Pitch in Mm. | Pitch in Inches | Metric Thread, 1.5155 P | Pitch in Mm. | Pitch in Inches | Metric Thread, 1.5155 P | Pitch in Mm. | Pitch in Inches | Metric Thread, 1.5155 P |
|---|---|---|---|---|---|---|---|---|
| 0.5 | 0.0197 | 0.0298 | 2.5 | 0.0984 | 0.1492 | 6.0 | 0.2362 | 0.3580 |
| 0.75 | 0.0295 | 0.0447 | 3.0 | 0.1181 | 0.1790 | 6.5 | 0.2559 | 0.3878 |
| 1.0 | 0.0394 | 0.0597 | 3.5 | 0.1378 | 0.2088 | 7.0 | 0.2756 | 0.4177 |
| 1.25 | 0.0492 | 0.0746 | 4.0 | 0.1575 | 0.2387 | 7.5 | 0.2953 | 0.4475 |
| 1.5 | 0.0590 | 0.0895 | 4.5 | 0.1772 | 0.2685 | 8.0 | 0.3150 | 0.4773 |
| 1.75 | 0.0689 | 0.1044 | 5.0 | 0.1969 | 0.2983 | 9.0 | 0.3543 | 0.5370 |
| 2.0 | 0.0787 | 0.1193 | 5.5 | 0.2165 | 0.3282 | 10.0 | 0.3937 | 0.5966 |

This table may be used when a 60-degree metric thread (such as the International Standard or the French Thread) is to be checked, by first converting the metric dimensions into inches. The third column of the table gives the value of 1.5155 × pitch in inches equivalent to the pitch in millimeters given in the first column. The formula for the American Standard thread form is used.

### Why Small Thread Angle Affects Accuracy of Three-wire Measurement. —

In measuring or checking Acme threads, or any others having a comparatively small thread angle $A$, it is particularly important to use a formula which compensates largely, if not entirely, for the effect of the lead angle, especially in all gage and precision work. The effect of the lead angle on the position of the wires and upon the resulting measurement $M$ is much greater in the case of a 29-degree thread than in the case of a higher thread angle such, for example, as a 60-degree thread. This is because the cotangent of the thread angle increases as this angle becomes smaller; consequently, the reduction in the width of the thread groove in the normal plane due to the lead angle causes a wire of given size to rest higher in the groove of a thread having a small thread angle $A$ (like a 29-degree thread) than in the groove with a larger angle (like a 60-degree American Standard).

*Acme Threads:* Three-wire measurements of high accuracy require the use of Formula 4. For most cases, however, Formula 2 or 3 gives satisfactory results. The table on page 1396 lists suitable wire sizes for use in Formulas 2 and 4.

**Dimensions Over Wires of Given Diameter for Checking Screw Threads of American National form (U. S. Standard) and the V-form**

| Diam. of Screw | No. of Threads per Inch | Diam. of Wire used | Dimension over Wires, V-Thread | Dimension over Wires, U. S. Thread | Diam. of Screw | No. of Threads per Inch | Diam. of Wire used | Dimension over Wires, V-Thread | Dimension over Wires, U. S. Thread |
|---|---|---|---|---|---|---|---|---|---|
| ¼ | 18 | 0.035 | 0.2588 | 0.2708 | ⅞ | 8 | 0.090 | 0.9285 | 0.9556 |
| ¼ | 20 | 0.035 | 0.2684 | 0.2792 | ⅞ | 9 | 0.090 | 0.9525 | 0.9766 |
| ¼ | 22 | 0.035 | 0.2763 | 0.2861 | ⅞ | 10 | 0.090 | 0.9718 | 0.9935 |
| ¼ | 24 | 0.035 | 0.2828 | 0.2919 | 15/16 | 8 | 0.090 | 0.9910 | 1.0181 |
| 5/16 | 18 | 0.035 | 0.3213 | 0.3333 | 15/16 | 9 | 0.090 | 1.0150 | 1.0391 |
| 5/16 | 20 | 0.035 | 0.3309 | 0.3417 | 1 | 8 | 0.090 | 1.0535 | 1.0806 |
| 5/16 | 22 | 0.035 | 0.3388 | 0.3486 | 1 | 9 | 0.090 | 1.0775 | 1.1016 |
| 5/16 | 24 | 0.035 | 0.3453 | 0.3544 | 1⅛ | 7 | 0.090 | 1.1476 | 1.1785 |
| ⅜ | 16 | 0.040 | 0.3867 | 0.4003 | 1¼ | 7 | 0.090 | 1.2726 | 1.3035 |
| ⅜ | 18 | 0.040 | 0.3988 | 0.4108 | 1⅜ | 6 | 0.150 | 1.5363 | 1.5724 |
| ⅜ | 20 | 0.040 | 0.4084 | 0.4192 | 1½ | 6 | 0.150 | 1.6613 | 1.6974 |
| 7/16 | 14 | 0.050 | 0.4638 | 0.4793 | 1⅝ | 5½ | 0.150 | 1.7601 | 1.7995 |
| 7/16 | 16 | 0.050 | 0.4792 | 0.4928 | 1¾ | 5 | 0.150 | 1.8536 | 1.8969 |
| ½ | 12 | 0.050 | 0.5057 | 0.5237 | 1⅞ | 5 | 0.150 | 1.9786 | 2.0219 |
| ½ | 13 | 0.050 | 0.5168 | 0.5334 | 2 | 4½ | 0.150 | 2.0651 | 2.1132 |
| ½ | 14 | 0.050 | 0.5263 | 0.5418 | 2¼ | 4½ | 0.150 | 2.3151 | 2.3632 |
| 9/16 | 12 | 0.050 | 0.5682 | 0.5862 | 2½ | 4 | 0.150 | 2.5170 | 2.5711 |
| 9/16 | 14 | 0.050 | 0.5888 | 0.6043 | 2¾ | 4 | 0.150 | 2.7670 | 2.8211 |
| ⅝ | 10 | 0.070 | 0.6618 | 0.6835 | 3 | 3½ | 0.200 | 3.1051 | 3.1670 |
| ⅝ | 11 | 0.070 | 0.6775 | 0.6972 | 3¼ | 3½ | 0.200 | 3.3551 | 3.4170 |
| ⅝ | 12 | 0.070 | 0.6907 | 0.7087 | 3½ | 3¼ | 0.250 | 3.7171 | 3.7837 |
| 11/16 | 10 | 0.070 | 0.7243 | 0.7460 | 3¾ | 3 | 0.250 | 3.9226 | 3.9948 |
| 11/16 | 11 | 0.070 | 0.7400 | 0.7597 | 4 | 3 | 0.250 | 4.1726 | 4.2448 |
| ¾ | 10 | 0.070 | 0.7868 | 0.8085 | 4¼ | 2⅞ | 0.250 | 4.3975 | 4.4729 |
| ¾ | 11 | 0.070 | 0.8025 | 0.8222 | 4½ | 2¾ | 0.250 | 4.6202 | 4.6989 |
| ¾ | 12 | 0.070 | 0.8157 | 0.8337 | 4¾ | 2⅝ | 0.250 | 4.8402 | 4.9227 |
| 13/16 | 9 | 0.070 | 0.8300 | 0.8541 | 5 | 2½ | 0.250 | 5.0572 | 5.1438 |
| 13/16 | 10 | 0.070 | 0.8493 | 0.8710 | .... | .... | ..... | ..... | ..... |

### Table for Measuring Whitworth Standard Threads by the Three-wire Method

| Diam. of Thread | No. of Threads per Inch | Diam. of Wire used | Diam. Measured over Wires | Diam. of Thread | No. of Threads per Inch | Diam. of Wire used | Diam. Measured over Wires |
|---|---|---|---|---|---|---|---|
| 1/8 | 40 | 0.018 | 0.1420 | 2 1/4 | 4 | 0.150 | 2.3247 |
| 3/16 | 24 | 0.030 | 0.2158 | 2 3/8 | 4 | 0.150 | 2.4497 |
| 1/4 | 20 | 0.035 | 0.2808 | 2 1/2 | 4 | 0.150 | 2.5747 |
| 5/16 | 18 | 0.040 | 0.3502 | 2 5/8 | 4 | 0.150 | 2.6997 |
| 3/8 | 16 | 0.040 | 0.4015 | 2 3/4 | 3 1/2 | 0.200 | 2.9257 |
| 7/16 | 14 | 0.050 | 0.4815 | 2 7/8 | 3 1/2 | 0.200 | 3.0507 |
| 1/2 | 12 | 0.050 | 0.5249 | 3 | 3 1/2 | 0.200 | 3.1757 |
| 9/16 | 12 | 0.050 | 0.5874 | 3 1/8 | 3 1/2 | 0.200 | 3.3007 |
| 5/8 | 11 | 0.070 | 0.7011 | 3 1/4 | 3 1/4 | 0.200 | 3.3905 |
| 11/16 | 11 | 0.070 | 0.7636 | 3 3/8 | 3 1/4 | 0.200 | 3.5155 |
| 3/4 | 10 | 0.070 | 0.8115 | 3 1/2 | 3 1/4 | 0.200 | 3.6405 |
| 13/16 | 10 | 0.070 | 0.8740 | 3 5/8 | 3 1/4 | 0.200 | 3.7655 |
| 7/8 | 9 | 0.070 | 0.9187 | 3 3/4 | 3 | 0.200 | 3.8495 |
| 15/16 | 9 | 0.070 | 0.9812 | 3 7/8 | 3 | 0.200 | 3.9745 |
| 1 | 8 | 0.090 | 1.0848 | 4 | 3 | 0.200 | 4.0995 |
| 1 1/16 | 8 | 0.090 | 1.1473 | 4 1/8 | 3 | 0.200 | 4.2245 |
| 1 1/8 | 7 | 0.090 | 1.1812 | 4 1/4 | 2 7/8 | 0.250 | 4.4846 |
| 1 3/16 | 7 | 0.090 | 1.2437 | 4 3/8 | 2 7/8 | 0.250 | 4.6096 |
| 1 1/4 | 7 | 0.090 | 1.3062 | 4 1/2 | 2 7/8 | 0.250 | 4.7346 |
| 1 5/16 | 7 | 0.090 | 1.3687 | 4 5/8 | 2 7/8 | 0.250 | 4.8596 |
| 1 3/8 | 6 | 0.120 | 1.4881 | 4 3/4 | 2 3/4 | 0.250 | 4.9593 |
| 1 7/16 | 6 | 0.120 | 1.5506 | 4 7/8 | 2 3/4 | 0.250 | 5.0843 |
| 1 1/2 | 6 | 0.120 | 1.6131 | 5 | 2 3/4 | 0.250 | 5.2093 |
| 1 9/16 | 6 | 0.120 | 1.6756 | 5 1/8 | 2 3/4 | 0.250 | 5.3343 |
| 1 5/8 | 5 | 0.120 | 1.6847 | 5 1/4 | 2 5/8 | 0.250 | 5.4316 |
| 1 11/16 | 5 | 0.120 | 1.7472 | 5 3/8 | 2 5/8 | 0.250 | 5.5566 |
| 1 3/4 | 5 | 0.120 | 1.8097 | 5 1/2 | 2 5/8 | 0.250 | 5.6816 |
| 1 13/16 | 5 | 0.120 | 1.8722 | 5 5/8 | 2 5/8 | 0.250 | 5.8066 |
| 1 7/8 | 4 1/2 | 0.150 | 1.9942 | 5 3/4 | 2 1/2 | 0.250 | 5.9011 |
| 1 15/16 | 4 1/2 | 0.150 | 2.0567 | 5 7/8 | 2 1/2 | 0.250 | 6.0261 |
| 2 | 4 1/2 | 0.150 | 2.1192 | 6 | 2 1/2 | 0.250 | 6.1511 |
| 2 1/8 | 4 1/2 | 0.150 | 2.2442 | ...... | ...... | ....... | ....... |

**Buckingham Simplified Formula which Includes Effect of Lead Angle.** — The Formula 3 which follows gives very accurate results for the lower lead angles in determining measurement $M$. However, if extreme accuracy is essential, it may be advisable to use the involute helicoid formulas as explained later.

$$M = E + W (1 + \sin A_n) \text{ where } W = \frac{T \times \cos B}{\cos A_n} \qquad \text{(3 and 3a)}$$

Theoretically correct equations for determining measurement $M$ are complex and cumbersome to apply. Formula 3 combines simplicity with a degree of accuracy which meets all but the most exacting requirements, particularly for lead angles below 8 or 10 degrees and the higher thread angles. However, the wire diameter used in Formula (3) must conform to that obtained by Formula (3a) to permit a direct solution or one not involving indeterminate equations and successive trials.

*Application of Buckingham Formula:* In the application of Formula (3) to screw threads or to worms, there are two general cases to be considered.

*Case 1:* The screw thread or worm is to be milled with a cutter having an included angle equal to the nominal or standard thread angle which is assumed to be the angle in the axial plane. For example, a 60-degree cutter is to be used for milling a thread. In this case, the thread angle in the plane of the axis will exceed 60 degrees by an amount increasing with the lead angle. This variation from the standard angle may be of little or no practical importance if the lead angle is small or if the mating nut (or teeth in the case of worm gearing) is formed to suit the thread as milled.

*Case 2:* The screw thread or worm is to be milled with a cutter reduced to whatever normal angle is equivalent to the standard thread angle in the axial plane. For example, a 29-degree Acme thread is to be milled with a cutter having some angle smaller than 29 degrees (the reduction increasing with the lead angle) to make the thread angle standard in the plane of the axis. Theoretically, the milling cutter angle should always be corrected to suit the normal angle; but if the lead angle is small, such correction may be unnecessary.

If the thread is cut in a lathe to the standard angle as measured in the axial plane, Case 2 applies in determining the pin size $W$ and the over-all measurement $M$.

In solving all problems under Case 1, the angle $A_n$ used in Formulas (3) and (3a) equals one-half the included angle of the milling cutter.

When Case 2 applies, the angle $A_n$ for milled threads also equals one-half the included angle of the cutter, but the cutter angle is reduced and is determined as follows:

$$\tan A_n = \tan A \times \cos B$$

The included angle of the cutter or the normal included angle of the thread groove $= 2\, A_n$. Examples 1 and 2 which follow illustrate Cases 1 and 2.

*Example 1 (Case 1):* The example which follows illustrates the case of an Acme screw thread which is milled with a cutter having an included angle of 29 degrees; consequently, the angle of the thread is over 29 degrees in the axial section.

The outside or major diameter is 3 inches; the pitch, ½ inch; the lead, 1 inch; the number of threads or " starts," 2. Find pin size $W$ and measurement $M$.

Pitch diameter $E = 2.75$; $T = 0.25$; $L = 1.0$; $A_n = 14.50°$; $\tan A_n = 0.258618$; $\sin A_n = 0.25038$; $\cos A_n = 0.968148$.

$$\tan B = \frac{1.0}{3.1416 \times 2.75} = 0.115749; \quad B = 6.6025°$$

$$W = \frac{0.25 \times 0.993368}{0.968148} = 0.25651 \text{ inch}$$

$$M = 2.75 + 0.25651 \times (1 + 0.25038) = 3.0707 \text{ inches.}$$

*Note:* This value of $M$ is only 0.0001 inch larger than that obtained by using the very accurate involute helicoid formula (4) referred to on the following page.

*Example 2 (Case 2):* A triple-threaded worm has a pitch diameter of 2.481 inches; pitch of 1.5 inches; lead of 4.5 inches; lead angle of 30 degrees and nominal thread angle of 60 degrees in axial plane. Milling cutter angle is to be reduced. $T = 0.75$ inch; $\cos B = 0.866025$; $\tan A = 0.57735$. Again use Formula (3) to see if it is applicable in this case.

$\text{Tan } A_n = \tan A \times \cos B = 0.57735 \times 0.866025 = 0.5000$; hence $A_n = 26.565°$ making the included cutter angle 53.13°. $\cos A_n = 0.89443$; $\sin A_n = 0.44721$.

$$W = \frac{0.75 \times 0.866025}{0.89443} = 0.72618 \text{ inch.}$$

$$M = 2.481 + 0.72618 \times (1 + 0.44721) = 3.532 \text{ inches.}$$

*Note:* If the value of measurement $M$ is determined by using the following

Formula (4) it will be found that $M = 3.515+$ inches; hence the error equals $3.532 - 3.515 = 0.017$ inch approximately, which indicates that Formula (3) is not accurate enough in this case. The application of this simpler Formula (3) will depend upon the lead angle and thread angle (as previously explained) and also upon the class of work.

**Buckingham Exact Involute Helicoid Formula Applied to Screw Threads.** — When extreme accuracy is required in finding measurement $M$ for obtaining a given pitch diameter, the equations which follow, although somewhat cumbersome to apply, have the merit of providing a direct and very accurate solution; consequently, they are preferable to the indeterminate equations and successive trial solutions heretofore employed when extreme precision is required. These equations are exact for involute helical gears and, consequently, give theoretically correct results when applied to a screw thread of the involute helicoidal form; they also give very close approximations for threads having intermediate profiles.

*Helical Gear Equation Applied to Screw Thread Measurement:* In applying the helical gear equations to a screw thread, use either the axial or normal thread angle and the lead angle of the helix. In order to keep the solution on a practical basis, either thread angle $A$ or $A_n$, as the case may be, is assumed to equal the cutter angle of a milled thread. Actually, the profile of a milled thread will have some curvature in both axial and normal sections; hence angles $A$ and $A_n$ represent the angular approximations of these slightly curved profiles. The equations which follow give the values needed to solve the screw thread problem as a helical gear problem.

$$M = \frac{2\,R_b}{\cos G} + W \qquad (4)$$

$$\tan F = \frac{\tan A}{\tan B} = \frac{\tan A_n}{\sin B}\,; \quad R_b = \frac{E}{2}\cos F \qquad (4a \text{ and } 4b)$$

$$T_a = \frac{T}{\tan B}\,; \quad \tan H_b = \cos F \tan H \qquad (4c \text{ and } 4d)$$

$$\operatorname{inv} G = \frac{T_a}{E} + \operatorname{inv} F + \frac{W}{2\,R_b \cos H_b} - \frac{\pi}{S} \qquad (4e)$$

A table of involute functions is required (see Handbook pages 178 to 222).

*Example 3:* To illustrate the application of Formula (4) and the supplementary formulas, assume that the number of starts $S = 6$; pitch diameter $E = 0.6250$; normal thread angle $A_n = 20°$; lead of thread $L = 0.864$ inch; $T = 0.072$; $W = 0.07013$ inch.

$$\tan B = \frac{L}{\pi E} = \frac{0.864}{1.9635} = 0.44003\,; \quad B = 23.751°$$

Helix angle $H = 90° - 23.751° = 66.249°$

$$\tan F = \frac{\tan A_n}{\sin B} = \frac{0.36397}{0.40276} = 0.90369\,; \quad F = 42.104°$$

$$R_b = \frac{E}{2}\cos F = \frac{0.6250}{2} \times 0.74193 = 0.23185$$

$$T_a = \frac{T}{\tan B} = \frac{0.072}{0.44003} = 0.16362$$

$$\tan H_b = \cos F \tan H = 0.74193 \times 2.27257 = 1.68609\,; \quad H_b = 59.328°$$

The involute function of $G$ is found next by Formula (4e).

$$\text{inv } G = \frac{0.16362}{0.625} + 0.16884 + \frac{0.07013}{2 \times 0.23185 \times 0.51012} - \frac{3.1416}{6} = 0.20351$$

A table of involute functions shows that 44.350° is the angular equivalent of 0.20351; hence $G = 44.350°$.

$$M = \frac{2 R_b}{\cos G} + W = \frac{2 \times 0.23185}{0.71508} + 0.07013 = 0.71859 \text{ inch}$$

**Accuracy of Formulas (3) and (4) Compared.** — With the involute helicoid Formula (4) any wire size which makes contact with the flanks of the thread may be used; however, in the preceding example, the wire diameter $W$ was obtained by Formula (3a) in order to compare Formula (4) with (3). If Example (3) is solved by Formula (3), $M = 0.71912$; hence the difference between the values of $M$ obtained with Formulas (3) and (4) equals $0.71912 - 0.71859 = 0.00053$ inch. The included thread angle in this case is 40 degrees. If Formulas (3) and (4) are applied to a 29-degree thread, the difference in measurements $M$ or the error resulting from the use of Formula (3) will be larger. For example, in case of an Acme thread having a lead angle of about 34 degrees, the difference in values of $M$ obtained by the two formulas equals 0.0008 inch.

**Three-wire Measurement of Acme and Stub Acme Thread Pitch Diameter.** — For single- and multiple-start Acme and Stub Acme threads having lead angles of less than 5 degrees, the approximate three-wire formula given on page 1388 and the best wire size taken from the table on page 1396 may be used.

Multiple-start Acme and Stub Acme threads commonly have a lead angle of greater than 5 degrees. For these, a direct determination of the actual pitch diameter is obtained by using the formula: $E = M - (C + c)$ in conjunction with the table on pages 1397 and 1398. To enter the table, the lead angle $B$ of the thread to be measured must be known. It is found by the formula: $\tan B = L \div 3.1416 E_1$ where $L$ is the lead of the thread and $E_1$ is the nominal pitch diameter. The best wire size is now found by taking the value of $w_1$ as given in the table for lead angle $B$, with interpolation, and dividing it by the number of threads per inch. The value of $(C + c)_1$ given in the table for lead angle $B$ is also divided by the number of threads per inch to get $(C + c)$. Using the best size wires, the actual measurement over wires $M$ is made and the actual pitch diameter $E$ found by using the formula: $E = M - (C + c)$.

*Example:* For a 5 tpi, 4-start Acme thread with a 13.952° lead angle, using three 0.10024-inch wires, $M = 1.1498$ inches, hence $E = 1.1498 - 0.1248 = 1.0250$ inches.

Under certain conditions, a wire may contact one thread flank at two points, in which case it is advisable to substitute balls of the same diameter as the wires.

**Checking Thickness of Acme Screw Threads.** — In some instances it may be preferable to check the thread thickness instead of the pitch diameter, especially if there is a thread thickness tolerance.

A direct method, applicable to the larger pitches, is to use a vernier gear-tooth caliper for measuring the thickness in the *normal* plane of the thread. This measurement, for an American Standard General Purpose Acme thread, should be made at a distance below the *basic* outside diameter equal to $p/4$. The thickness at this basic pitch-line depth and in the axial plane should be $p/2 - 0.259 \times$ the pitch diameter allowance from Table 4 on page 1330 with a tolerance of *minus* $0.259 \times$ the pitch diameter tolerance from Table 5 on page 1331. The thickness in the normal plane or plane of measurement is equal to the thickness in the axial plane multiplied by the cosine of the helix angle. The helix angle may be determined from the formula:

tangent of helix angle = lead of thread $\div 3.1416 \times$ pitch diameter

**Three-Wire Method for Checking Thickness of Acme Threads.** — The application of the 3-wire method of checking the thickness of an Acme screw thread

Wire Sizes for Three-Wire Measurement of Acme Threads with Lead Angles of
Less than 5 Degrees

| Threads per Inch | Best Size | Max. | Min. | Threads per Inch | Best Size | Max. | Min. |
|---|---|---|---|---|---|---|---|
| 1 | 0.51645 | 0.65001 | 0.48726 | 5 | 0.10329 | 0.13000 | 0.09745 |
| 1¼ | 0.38734 | 0.48751 | 0.36545 | 6 | 0.08608 | 0.10834 | 0.08121 |
| 1½ | 0.34430 | 0.43334 | 0.32484 | 8 | 0.06456 | 0.08125 | 0.06091 |
| 2 | 0.25822 | 0.32501 | 0.24363 | 10 | 0.05164 | 0.06500 | 0.04873 |
| 2½ | 0.20658 | 0.26001 | 0.19491 | 12 | 0.04304 | 0.05417 | 0.04061 |
| 3 | 0.17215 | 0.21667 | 0.16242 | 14 | 0.03689 | 0.04643 | 0.03480 |
| 4 | 0.12911 | 0.16250 | 0.12182 | 16 | 0.03228 | 0.04063 | 0.03045 |

Wire sizes are based upon zero helix angle. Best size = 0.51645 × pitch; maximum size = 0.650013 × pitch; minimum size = 0.487263 × pitch.

is included in Report of the National Screw Thread Commission. In applying the 3-wire method for checking thread thickness, the procedure is the same as in checking pitch diameter although a different formula is required. Assume that $D$ = basic major diameter of screw; $M$ = measurement over wires; $W$ = diameter of wires; $S$ = tangent of helix angle at pitch line; $P$ = pitch; $T$ = thread thickness at depth equal to 0.25$P$.

$$T = 1.12931 \times P + 0.25862 \times (M - D) - W \times (1.29152 + 0.48407\, S^2)$$

This formula transposed to show the correct measurement $M$ equivalent to a given required thread thickness is as follows:

$$M = D + \frac{W \times (1.29152 + 0.48407 \times S^2) + T - 1.12931 \times P}{0.25862}$$

*Example:* An Acme General Purpose thread, Class 2G, has a 5-inch basic major diameter, 0.5-inch pitch, and 1-inch lead (double thread). Assume the wire size is 0.258 inch. Determine measurement $M$ for a thread thickness $T$ at the basic pitch line of 0.2454 inch. (This is the maximum thickness at the basic pitch line and equals the basic thickness, 0.5$P$, − 0.259 × allowance from Table 4, page 1330.)

$$M = 5 + \frac{0.258 \times (1.29152 + 0.48407 \times 0.06701^2) + 0.2454 - 1.12931 \times 0.5}{0.25862}$$

    = 5.056 inches

**Testing Angle of Thread by Three-wire Method.** — The error in the angle of a thread may be determined by using sets of wires of two diameters, the measurement over the two sets of wires being followed by calculations to determine the amount of error, assuming that the angle cannot be tested by comparison with a standard plug gage which is known to be correct. The diameter of the small wires for the American standard thread is usually about 0.6 times the pitch and the diameter of the large wires, about 0.9 times the pitch. The total difference between the measurements over the large and small sets of wires is first determined. If the thread is an American standard or any other form having an included angle of 60 degrees, the difference between the two measurements should equal three times the difference between the diameters of the wires used. Thus, if the wires are 0.116 and 0.076 inch in diameter, respectively, the difference equals 0.116 − 0.076 = 0.040 inch. Therefore the difference between the micrometer readings for a standard angle of 60 degrees equals 3 × 0.040 = 0.120 inch in this case. If the

**Best Wire Diameters and Constants for Three-wire Measurement of Acme and Stub Acme Threads with Large Lead Angles — 1-inch Axial Pitch***

| Lead angle, $B$, deg. | 1-start threads | | 2-start threads | | Lead angle, $B$, deg. | 2-start threads | | 3-start threads | |
|---|---|---|---|---|---|---|---|---|---|
| | $w_1$ | $(C+c)_1$ | $w_1$ | $(C+c)_1$ | | $w_1$ | $(C+c)_1$ | $w_1$ | $(C+c)_1$ |
| 5.0 | 0.51450 | 0.64311 | 0.51443 | 0.64290 | 10.0 | 0.50864 | 0.63518 | 0.50847 | 0.63463 |
| 5.1 | 0.51442 | 0.64301 | 0.51435 | 0.64279 | 10.1 | 0.50849 | 0.63498 | 0.50381 | 0.63442 |
| 5.2 | 0.51435 | 0.64291 | 0.51427 | 0.64268 | 10.2 | 0.50834 | 0.63478 | 0.50815 | 0.63420 |
| 5.3 | 0.51427 | 0.64282 | 0.51418 | 0.64256 | 10.3 | 0.50818 | 0.63457 | 0.50800 | 0.63399 |
| 5.4 | 0.51419 | 0.64272 | 0.51410 | 0.64245 | 10.4 | 0.50802 | 0.63436 | 0.50784 | 0.63378 |
| 5.5 | 0.51411 | 0.64261 | 0.51401 | 0.64233 | 10.5 | 0.40786 | 0.63416 | 0.50768 | 0.63356 |
| 5.6 | 0.51403 | 0.64251 | 0.51393 | 0.64221 | 10.6 | 0.50771 | 0.63395 | 0.50751 | 0.63333 |
| 5.7 | 0.51395 | 0.64240 | 0.51384 | 0.64209 | 10.7 | 0.50755 | 0.63375 | 0.50735 | 0.63311 |
| 5.8 | 0.51386 | 0.64229 | 0.51375 | 0.64196 | 10.8 | 0.50739 | 0.53354 | 0.50718 | 0.63288 |
| 5.9 | 0.51377 | 0.64218 | 0.51366 | 0.64184 | 10.9 | 0.50723 | 0.63333 | 0.50701 | 0.63265 |
| 6.0 | 0.51368 | 0.64207 | 0.51356 | 0.64171 | 11.0 | 0.50707 | 0.63313 | 0.50684 | 0.63242 |
| 6.1 | 0.51359 | 0.64195 | 0.51346 | 0.64157 | 11.1 | 0.5069L | 0.63292 | 0.50667 | 0.63219 |
| 6.2 | 0.51350 | 0.64184 | 0.51336 | 0.64144 | 11.2 | 0.50674 | 0.63271 | 0.50649 | 0.63195 |
| 6.3 | 0.51340 | 0.64172 | 0.41327 | 0.64131 | 11.3 | 0.50658 | 0.63250 | 0.50632 | 0.63172 |
| 6.4 | 0.51330 | 0.64160 | 0.51317 | 0.64117 | 11.4 | 0.50641 | 0.63228 | 0.50615 | 0.63149 |
| 6.5 | 0.51320 | 0.64147 | 0.51306 | 0.64103 | 11.5 | 0.50623 | 0.63206 | 0.50597 | 0.63126 |
| 6.6 | 0.51310 | 0.64134 | 0.51296 | 0.64089 | 11.6 | 0.50606 | 0.63184 | 0.50579 | 0.63102 |
| 6.7 | 0.51300 | 0.64122 | 0.51285 | 0.64075 | 11.7 | 0.50589 | 0.63162 | 0.50561 | 0.63078 |
| 6.8 | 0.51290 | 0.64110 | 0.51275 | 0.64061 | 11.8 | 0.50571 | 0.63140 | 0.50544 | 0.63055 |
| 6.9 | 0.51280 | 0.64097 | 0.51264 | 0.64046 | 11.9 | 0.50553 | 0.63117 | 0.50526 | 0.63031 |
| 7.0 | 0.51270 | 0.64085 | 0.51254 | 0.64032 | 12.0 | 0.50535 | 0.63095 | 0.50507 | 0.63006 |
| 7.1 | 0.51259 | 0.64072 | 0.51243 | 0.64017 | 12.1 | 0.50517 | 0.63072 | 0.50488 | 0.62981 |
| 7.2 | 0.51249 | 0.64060 | 0.51232 | 0.64002 | 12.2 | 0.50500 | 0.63050 | 0.50470 | 0.62956 |
| 7.3 | 0.51238 | 0.64047 | 0.51221 | 0.63987 | 12.3 | 0.50482 | 0.63027 | 0.50451 | 0.62931 |
| 7.4 | 0.51227 | 0.64034 | 0.51209 | 0.63972 | 12.4 | 0.50464 | 0.63004 | 0.50432 | 0.62906 |
| 7.5 | 0.51217 | 0.64021 | 0.51198 | 0.63957 | 12.5 | 0.50445 | 0.62981 | 0.50413 | 0.62881 |
| 7.6 | 0.51206 | 0.64008 | 0.51186 | 0.63941 | 12.6 | 0.50427 | 0.62958 | 0.50394 | 0.62856 |
| 7.7 | 0.51196 | 0.63996 | 0.51174 | 0.63925 | 12.7 | 0.50408 | 0.62934 | 0.50375 | 0.62830 |
| 7.8 | 0.51186 | 0.63983 | 0.51162 | 0.63909 | 12.8 | 0.50389 | 0.62911 | 0.50356 | 0.62805 |
| 7.9 | 0.51175 | 0.63970 | 0.51150 | 0.63892 | 12.9 | 0.50371 | 0.62888 | 0.50336 | 0.62779 |
| 8.0 | 0.51164 | 0.63957 | 0.51138 | 0.63876 | 13.0 | 0.50352 | 0.62865 | | |
| 8.1 | 0.51153 | 0.63944 | 0.51125 | 0.63859 | 13.1 | 0.50333 | 0.62841 | | |
| 8.2 | 0.51142 | 0.63930 | 0.51113 | 0.63843 | 13.2 | 0.50313 | 0.62817 | | |
| 8.3 | 0.51130 | 0.63916 | 0.51101 | 0.63827 | 13.3 | 0.50293 | 0.62792 | | |
| 8.4 | 0.51118 | 0.63902 | 0.51088 | 0.63810 | 13.4 | 0.50274 | 0.62778 | | |
| 8.5 | 0.51105 | 0.63887 | 0.51075 | 0.63793 | 13.5 | 0.50254 | 0.62743 | | |
| 8.6 | 0.51093 | 0.63873 | 0.51062 | 0.63775 | 13.6 | 0.50234 | 0.62718 | For these | |
| 8.7 | 0.51081 | 0.63859 | 0.51049 | 0.63758 | 13.7 | 0.50215 | 0.62694 | 3-start | |
| 8.8 | 0.51069 | 0.63845 | 0.51035 | 0.63740 | 13.8 | 0.50195 | 0.62670 | thread | |
| 8.9 | 0.51057 | 0.63831 | 0.51022 | 0.63722 | 13.9 | 0.50175 | 0.62645 | values | |
| 9.0 | 0.51044 | 0.63817 | 0.51008 | 0.63704 | 14.0 | 0.50155 | 0.62621 | see | |
| 9.1 | 0.51032 | 0.63802 | 0.50993 | 0.63685 | 14.1 | 0.50135 | 0.62596 | table | |
| 9.2 | 0.51019 | 0.63788 | 0.50979 | 0.63667 | 14.2 | 0.50115 | 0.62571 | on | |
| 9.3 | 0.51006 | 0.63774 | 0.50965 | 0.63649 | 14.3 | 0.50094 | 0.62546 | following | |
| 9.4 | 0.50993 | 0.63759 | 0.50951 | 0.63630 | 14.4 | 0.50073 | 0.62520 | page. | |
| 9.5 | 0.50981 | 0.63744 | 0.50937 | 0.63612 | 14.5 | 0.50051 | 0.62494 | | |
| 9.6 | 0.50968 | 0.63730 | 0.50922 | 0.63593 | 14.6 | 0.50030 | 0.62468 | | |
| 9.7 | 0.50955 | 0.63715 | 0.50908 | 0.63574 | 14.7 | 0.50009 | 0.62442 | | |
| 9.8 | 0.50941 | 0.63700 | 0.50893 | 0.63555 | 14.8 | 0.49988 | 0.62417 | | |
| 9.9 | 0.50927 | 0.63685 | 0.50879 | 0.63537 | 14.9 | 0.49966 | 0.62391 | | |
| 10.0 | 0.50913 | 0.63670 | 0.50864 | 0.63518 | 15.0 | 0.49945 | 0.62365 | | |

*Courtesy of Van Keuren Co.*

All dimensions are in inches.

* Values given for $w_1$ and $(C+c)_1$ in table are for 1-inch pitch axial threads. For other pitches, divide table values by number of threads per inch.

# 1398 MEASURING SCREW THREADS

**Best Wire Diameters and Constants for Three-wire Measurement of Acme and Stub Acme Threads with Large Lead Angles — 1-inch Axial Pitch***

| Lead angle, $B$, deg. | 3-start threads | | 4-start threads | | Lead angle, $B$, deg. | 3-start threads | | 4-start threads | |
|---|---|---|---|---|---|---|---|---|---|
| | $w_1$ | $(C+c)_1$ | $w_1$ | $(C+c)_1$ | | $w_1$ | $(C+c)_1$ | $w_1$ | $(C+c)_1$ |
| 13.0 | 0.50316 | 0.62752 | 0.50297 | 0.62694 | 18.0 | 0.49154 | 0.61250 | 0.49109 | 0.61109 |
| 13.1 | 0.50295 | 0.62725 | 0.50277 | 0.62667 | 18.1 | 0.49127 | 0.61216 | 0.49082 | 0.61073 |
| 13.2 | 0.50275 | 0.62699 | 0.50256 | 0.62639 | 18.2 | 0.49101 | 0.61182 | 0.49054 | 0.61037 |
| 13.3 | 0.50255 | 0.62672 | 0.50235 | 0.62611 | 18.3 | 0.49074 | 0.61148 | 0.49027 | 0.61001 |
| 13.4 | 0.50235 | 0.62646 | 0.50215 | 0.62583 | 18.4 | 0.49047 | 0.61114 | 0.48999 | 0.60964 |
| 13.5 | 0.50214 | 0.62619 | 0.50194 | 0.62555 | 18.5 | 0.49020 | 0.61080 | 0.48971 | 0.69928 |
| 13.6 | 0.50194 | 0.62592 | 0.50173 | 0.62526 | 18.6 | 0.48992 | 0.61045 | 0.48943 | 0.60981 |
| 13.7 | 0.50173 | 0.62564 | 0.50152 | 0.62498 | 18.7 | 0.48965 | 0.61011 | 0.48915 | 0.60854 |
| 13.8 | 0.50152 | 0.62537 | 0.50131 | 0.62469 | 18.8 | 0.48938 | 0.60976 | 0.48887 | 0.60817 |
| 13.9 | 0.50131 | 0.62509 | 0.50109 | 0.62440 | 18.9 | 0.48910 | 0.60941 | 0.48859 | 0.60780 |
| 14.0 | 0.50110 | 0.62481 | 0.50087 | 0.62411 | 19.0 | 0.48882 | 0.60906 | 0.48830 | 0.60742 |
| 14.1 | 0.50089 | 0.62453 | 0.50065 | 0.62381 | 19.1 | 0.48854 | 0.60871 | 0.48800 | 0.60704 |
| 14.2 | 0.50068 | 0.62425 | 0.50043 | 0.62351 | 19.2 | 0.48825 | 0.60835 | 0.48771 | 0.60666 |
| 14.3 | 0.50046 | 0.62397 | 0.50021 | 0.62321 | 19.3 | 0.48797 | 0.60799 | 0.48742 | 0.60628 |
| 14.4 | 0.50024 | 0.62368 | 0.49999 | 0.62291 | 19.4 | 0.48769 | 0.60764 | 0.48713 | 0.60590 |
| 14.5 | 0.50003 | 0.62340 | 0.49977 | 0.62262 | 19.5 | 0.48741 | 0.60729 | 0.48684 | 0.60552 |
| 14.6 | 0.49981 | 0.62312 | 0.49955 | 0.62232 | 19.6 | 0.48712 | 0.60693 | 0.48655 | 0.60514 |
| 14.7 | 0.49959 | 0.62883 | 0.49932 | 0.62202 | 19.7 | 0.48638 | 0.60657 | 0.48625 | 0.60475 |
| 14.8 | 0.49936 | 0.62253 | 0.49910 | 0.62172 | 19.8 | 0.48655 | 0.60621 | 0.48596 | 0.60437 |
| 14.9 | 0.49914 | 0.62224 | 0.49887 | 0.62141 | 19.9 | 0.48626 | 0.60585 | 0.48566 | 0.60398 |
| 15.0 | 0.49891 | 0.62195 | 0.49864 | 0.62110 | 20.0 | 0.48597 | 0.60549 | 0.48536 | 0.60359 |
| 15.1 | 0.49869 | 0.62166 | 0.49842 | 0.62080 | 20.1 | .... | .... | 0.48506 | 0.60320 |
| 15.2 | 0.49846 | 0.62137 | 0.49819 | 0.62049 | 20.2 | .... | .... | 0.48476 | 0.60281 |
| 15.3 | 0.49824 | 0.62108 | 0.49795 | 0.62017 | 20.3 | .... | .... | 0.48445 | 0.60241 |
| 15.4 | 0.42801 | 0.62078 | 0.49771 | 0.61985 | 20.4 | .... | .... | 0.48415 | 0.60202 |
| 15.5 | 0.49778 | 0.62048 | 0.49747 | 0.61953 | 20.5 | .... | .... | 0.48384 | 0.60162 |
| 15.6 | 0.49754 | 0.62017 | 0.49723 | 0.61921 | 20.6 | .... | .... | 0.48354 | 0.60123 |
| 15.7 | 0.49731 | 0.61987 | 0.49699 | 0.61889 | 20.7 | .... | .... | 0.48323 | 0.60083 |
| 15.8 | 0.49707 | 0.61956 | 0.49675 | 0.61857 | 20.8 | .... | .... | 0.48292 | 0.60042 |
| 15.9 | 0.49683 | 0.61926 | 0.49651 | 0.61825 | 20.9 | .... | .... | 0.48261 | 0.60002 |
| 16.0 | 0.49659 | 0.61895 | 0.49627 | 0.61793 | 21.0 | .... | .... | 0.48230 | 0.59961 |
| 16.1 | 0.49635 | 0.61864 | 0.49602 | 0.61760 | 21.1 | .... | .... | 0.48198 | 0.59920 |
| 16.2 | 0.49611 | 0.61833 | 0.49577 | 0.61727 | 21.2 | .... | .... | 0.48166 | 0.59879 |
| 16.3 | 0.49586 | 0.61801 | 0.49552 | 0.61694 | 21.3 | .... | .... | 0.48134 | 0.59838 |
| 16.4 | 0.49562 | 0.61770 | 0.49527 | 0.61661 | 21.4 | .... | .... | 0.48103 | 0.59797 |
| 16.5 | 0.49537 | 0.61738 | 0.49502 | 0.61628 | 21.5 | .... | .... | 0.48701 | 0.59756 |
| 16.6 | 0.49512 | 0.61706 | 0.49476 | 0.61594 | 21.6 | .... | .... | 0.48040 | 0.59715 |
| 16.7 | 0.49488 | 0.61675 | 0.49451 | 0.61560 | 21.7 | .... | .... | 0.48008 | 0.59674 |
| 16.8 | 0.40463 | 0.61643 | 0.49425 | 0.61526 | 21.8 | .... | .... | 0.47975 | 0.59632 |
| 16.9 | 0.49438 | 0.61611 | 0.49400 | 0.61492 | 21.9 | .... | .... | 0.47943 | 0.59590 |
| 17.0 | 0.49414 | 0.61580 | 0.49375 | 0.61458 | 22.0 | .... | .... | 0.47910 | 0.59548 |
| 17.1 | 0.49389 | 0.61548 | 0.49349 | 0.61424 | 22.1 | .... | .... | 0.47878 | 0.59507 |
| 17.2 | 0.49363 | 0.61515 | 0.49322 | 0.61389 | 22.2 | .... | .... | 0.47845 | 0.59465 |
| 17.3 | 0.49337 | 0.61482 | 0.49296 | 0.61354 | 22.3 | .... | .... | 0.47812 | 0.59422 |
| 17.4 | 0.49311 | 0.61449 | 0.49269 | 0.61319 | 22.4 | .... | .... | 0.47778 | 0.59379 |
| 17.5 | 0.49285 | 0.61416 | 0.49243 | 0.61284 | 22.5 | .... | .... | 0.47745 | 0.59336 |
| 17.6 | 0.49259 | 0.61383 | 0.49217 | 0.61250 | 22.6 | .... | .... | 0.47711 | 0.52993 |
| 17.7 | 0.49233 | 0.61350 | 0.49191 | 0.61215 | 22.7 | .... | .... | 0.47677 | 0.59250 |
| 17.8 | 0.49206 | 0.61316 | 0.49164 | 0.61180 | 22.8 | .... | .... | 0.47643 | 0.59207 |
| 17.9 | 0.49180 | 0.61283 | 0.49137 | 0.61144 | 22.9 | .... | .... | 0.47610 | 0.59164 |
| .... | .... | .... | .... | .... | 23.0 | .... | .... | 0.47577 | 0.59121 |

All dimensions are in inches.  *Courtesy of Van Keuren Co.*

* Values given for $w_1$ and $(C+c)_1$ in table are for 1-inch pitch axial threads. For other pitches, divide table values by number of threads per inch.

angle is incorrect, the amount of error may be determined by the following formula, which applies to any thread regardless of angle:

$$\text{Sin } a = \frac{A}{B - A}$$

In this formula,

$A$ = difference in diameters of the large and small wires used;

$B$ = total difference between the measurements over the large and small wires;

$a$ = one-half the included thread angle.

*Example:* The diameter of the large wires used for testing the angle of a thread is 0.116 inch and of the small wires 0.076 inch. The measurement over the two sets of wires shows a total difference of 0.122 inch instead of the correct difference, 0.120 inch, for a standard angle of 60 degrees when using the sizes of wires mentioned. Therefore the amount of error is determined as follows:

$$\text{Sin } a = \frac{0.040}{0.122 - 0.040} = \frac{0.040}{0.082} = 0.4878$$

By referring to a table of sines it will be seen that this value (0.4878) is the sine of 29 degrees 12 minutes, approximately. Therefore the angle of the thread is 58 degrees, 24 minutes or 1 degree 36 minutes less than the standard angle.

**Measuring Taper Screw Threads by 3-Wire Method.** — When the 3-wire method is used in measuring a taper screw thread the measurement is along a line that is not perpendicular to the axis of the screw thread, the inclination from the perpendicular equalling one-half the included angle of the taper. The formula which follows compensates for this inclination resulting from contact of the measuring instrument surfaces, with two wires on one side and one on the other. The taper thread is measured over the wires in the usual manner excepting that the single wire must be located in the thread at a point where the effective diameter is to be checked (as described more fully later). The formula shows the dimension equivalent to the correct pitch diameter at this given point. The general formula for taper screw threads follows:

$M$ = measurement over the 3 wires; $E$ = pitch diameter; $a$ = one-half the angle of the thread; $N$ = number of threads per inch; $W$ = diameter of wires; $b$ = one-half the angle of taper

$$M = \frac{E - \dfrac{\cot a}{2 N} + W \left(1 + \text{cosec } a\right)}{\text{sec } b}$$

This formula is not theoretically correct but it is, however, accurate for screw threads having tapers of ¾ inch per foot or less. This general formula can be simplified for a given thread angle and taper. The simplified formula following (in which $P$ = pitch) is for an American Standard Pipe Thread:

$$M = \frac{E - (0.86603 \times P) + 3 \times W}{1.00049}$$

Standard pitch diameters for pipe threads will be found in the table "American Standard Pipe Thread." The location of this pitch diameter or distance from the end of the pipe is also shown by the table. In using the formula for finding dimension $M$ over the wires, the single wire is placed in whatever part of the thread groove locates it at the point where the pitch diameter is to be checked. The wire must be accurately located at this point. The other wires are then placed on each side of that thread which is diametrically opposite the single wire. If the pipe thread is straight or without taper,

$$M = E - (0.86603 \times P) + 3 \times W$$

*Application of Formula to Pipe Thread:* To illustrate the use of the formula for taper threads, assume that dimension $M$ is required for an American Standard 3-inch pipe thread gage. The table " American Standard Pipe Thread " shows that the 3-inch size has 8 threads per inch or a pitch of 0.125 inch and a pitch diameter at the gaging notch of 3.3885 inches. Assume that the wire diameter is 0.07217 inch: Then when the pitch diameter is correct

$$M = \frac{3.3885 - (0.86603 \times 0.125) + 3 \times 0.07217}{1.00049} = 3.495 \text{ inch}$$

*Pitch Diameter Equivalent to a Given Measurement Over the Wires:* The formula following may be used to check the pitch diameter at any point along a tapering thread when measurement $M$ over wires of a given diameter is known. In this formula $E$ = the effective or pitch diameter at the position occupied by the single wire. The formula is not theoretically correct but gives very accurate results when applied to tapers of ¾ inch per foot or less.

$$E = 1.00049 \times M + (0.86603 \times P) - 3 \times W$$

*Example:* Measurement $M$ = 3.495 inches at the gaging notch of a 3-inch pipe thread and the wire diameter = 0.07217 inch. Then:

$$E = 1.00049 \times 3.495 + (0.86603 \times 0.125) - 3 \times 0.07217 = 3.3885 \text{ inches}$$

*Pitch Diameter at Any Point Along Taper Screw Thread:* When the pitch diameter in any position along a tapering thread is known, the pitch diameter at any other position may be determined as follows:

Multiply the distance (measured along the axis) between the location of the known pitch diameter and the location of the required pitch diameter, by the taper per inch or by 0.0625 for American Standard Pipe Threads. Add this product to the known diameter, if the required diameter is at a larger part of the taper, or subtract if the required diameter is smaller.

*Example:* The pitch diameter of a 3-inch American Standard Pipe Thread is 3.3885 at the gaging notch. Determine the pitch diameter at the small end. The American Standard Pipe Thread Table shows that the distance between the gaging notch and the small end of a 3-inch pipe is 0.77 inch. Hence the pitch diameter at the small end = 3.3885 − (0.77 × 0.0625) = 3.3404 inches.

**Three-Wire Method Applied to Buttress Threads.** — The angles of buttress threads vary somewhat, especially on the front or load-resisting side. Formula (1) which follows may be applied to any angles required. In this formula $M$ = measurement over wires when *pitch diameter* $E$ is correct; $A$ = included angle of thread and thread groove; $a$ = angle of front face or load-resisting side, measured from a line perpendicular to screw thread axis; $P$ = pitch of thread; $W$ = wire diameter.

$$M = E - \left[ \frac{P}{\tan a + \tan (A - a)} \right] + W \left[ 1 + \cos \left( \frac{A}{2} - a \right) \times \csc \frac{A}{2} \right] \quad (1)$$

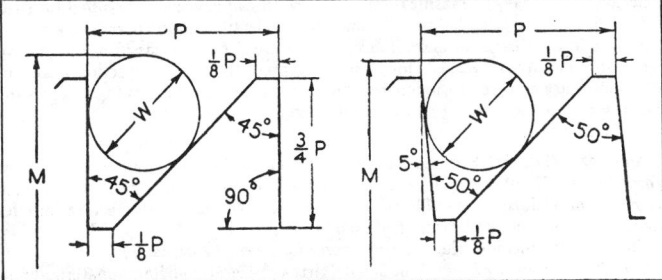

For given angles $A$ and $a$, this general formula may be simplified as shown by Formulas (3) and (4). These simplified formulas contain constants with values depending upon angles $A$ and $a$.

*Wire Diameter:* The wire diameter for obtaining pitch-line contact at the back of a buttress thread may be determined by the following general formula (2):

$$W = P \left( \frac{\cos a}{1 + \cos A} \right) \tag{2}$$

*45-Degree Buttress Thread:* The buttress thread shown by the diagram at the left, has a front or load-resisting side that is perpendicular to the axis of the screw. Measurement $M$ equivalent to a correct pitch diameter $E$ may be determined by formula (3):

$$M = E - P + (W \times 3.4142) \tag{3}$$

Wire diameter $W$ for pitch-line contact at back of thread $= 0.586 \times$ pitch.

*50-Degree Buttress Thread with Front-face Inclination of 5 Degrees:* This buttress thread form is illustrated by the diagram at the right. Measurement $M$ equivalent to the correct pitch diameter $E$ may be determined by formula (4):

$$M = E - (P \times 0.91955) + (W \times 3.2235) \tag{4}$$

Wire diameter $W$ for pitch-line contact at back of thread $= 0.606 \times$ pitch. If the width of flat at crest and root $= \frac{1}{8} \times$ pitch, depth $= 0.69 \times$ pitch.

*American National Standard Buttress Threads (ANSI B1.9-1973):* This buttress screw thread has an included thread angle of 52 degrees and a front face inclination of 7 degrees. Measurements $M$ equivalent to a pitch diameter $E$ may be determined by formula (5):

$$M = E - 0.89064P + 3.15689W + c \tag{5}$$

The wire angle correction factor $c$ is less than 0.0004 inch for recommended combinations of thread diameters and pitches and may be neglected. Use of wire diameter $W = 0.54147P$ is recommended.

**Measurement of Pitch Diameter of Thread Ring Gages.** — The application of direct methods of measurement to determine the pitch diameter of thread ring gages presents serious difficulties, particularly in securing proper contact pressure when a high degree of precision is required. The usual practice is to fit the ring gage to a master setting plug. When the thread ring gage is of correct lead, angle, and thread form, within close limits, this method is quite satisfactory and represents standard American practice. It is the only method available for small sizes of threads. For the larger sizes, various more or less satisfactory methods have been devised, but none of these have found wide application.

**Screw Thread Gage Classification.** — Screw thread gages are classified by their degree of accuracy, that is, by the amount of tolerance afforded the gage manufacturer and the wear allowance, if any. There are also three classifications according to use: (1) Working gages for controlling production; (2) inspection gages for rejection or acceptance of the finished product; (3) reference gages for determining the accuracy of the working and inspection gages.

**American National Standard for Gages and Gaging for Unified Inch Screw Threads** (ANSI B1.2-1974). — This standard covers gaging methods for conformance of Unified Screw Threads and provides the essential specifications for applicable gages required for unified inch screw threads.

The standard includes the following gages for *Product Internal Thread:*

*GO Thread Plug Gage* for functional (virtual) size at maximum-material-limit.

*HI Thread Plug Gage* for HI functional size at minimum-material-limit.

*Indicating Thread Gages* to establish numerical values for determining Functional Differential Reading for use in verifying conformance of thread elements.

*HI Limit Indicating Gages* for HI minimum-material-limit.

*GO and HI Thread Setting Ring Gages* for the two preceding gages described immediately above.

*GO and NOT GO Plain Plug Gages* for minimum and maximum limits of the minor diameter.

It includes the following gages for *Product External Thread:*

*GO Thread Ring Gage* for functional (virtual) size at maximum-material-limit.

*LO Thread Ring Gage* for LO functional size at minimum-material-limit.

*Indicating Thread Gages* to establish numerical values for determining Functional Differential Reading for use in verifying conformance of the thread elements.

*LO Limit Thread Snap or Indicating Gages* for LO minimum-material-limit.

*GO and LO Thread Setting Plug Gages* for the 4 preceding gages described immediately above.

*Plain Gages* for minimum and maximum limits of the major diameter.

The standard lists the following for use of Threaded and Plain Gages for verification of product internal threads:

*Tolerance.* Unless otherwise specified all thread gages which directly check the product thread shall be X tolerance for all classes.

*GO Thread Plug Gages.* GO thread plug gages must enter the full threaded length of the product freely. The GO thread plug gage is a cumulative check of all thread elements except the minor diameter.

*HI Thread Plug Gages.* HI thread plug gages when applied to the product internal thread may engage only the end threads (which may not be representative of the complete thread). Entering threads on product are incomplete and permit gage to start. Starting threads on HI plugs are subject to greater wear than the remaining threads. Such wear in combination with the incomplete product threads permit further entry of the gage. Surveillance facilities ordinarily available in the field are often inadequate for fully determining such gage wear. Also, it is not practical to control nor limit the torque applied by operators, nor that utilized by a specific operator at various times and under varying conditions. For these reasons the following standard practice has been adopted with respect to permissible entry. Threads are acceptable when the HI thread plug gage is applied to the product internal thread: (a) if it does not enter, or (b) if all complete product threads can be entered, provided that a *definite* drag from contact with the product material results on or before the third turn of entry. The gage should not be forced after the drag is definite. Special requirements such as exceptionally thin or ductile material, or a small number of threads, may necessitate modification of this practice.

GO and NOT GO Plain Plug Gages for Minor Diameter of Product Internal Thread. (Recommended in Class Z tolerance.)   GO plain plug gages must completely enter the product internal thread to assure that the minor diameter does not exceed the maximum-material-limit.   NOT GO plain plug gages must not enter the product internal thread to provide adequate assurance that the minor diameter does not exceed the minimum-material-limit.

For Thread Setting Plug Gages the standard gives the following:

GO and LO Truncated Setting Plugs.   W tolerance truncated setting plugs are recommended for setting adjustable thread ring gages to and including 6.25 inches nominal size and may be used for setting thread snap gages and indicating thread gages.   Above 6.25 inches nominal size, the difference in feel between the full form and truncated sections in setting thread ring gages is insignificant, and the basic crest setting plug may be used.

When setting adjustable thread ring gages to size, the truncated portion of the setting plug controls the functional size, and the full form portion assures that adequate clearance is provided at the major diameter of the ring gage.   The full form portion in conjunction with the truncated portion checks — to some degree — the half-angle accuracy of the gage.   The same procedure may be applied to detect uneven angle wear of ring gages in use.

GO and LO Basic-crest (Full Form) Setting Plugs.   W tolerance basic crest setting plugs are frequently used for setting thread snap limit gages and indicating thread gages.   They may also be used for setting large adjustable thread ring gages, especially those above 6.25 inches nominal size.   When they are so used it may be desirable to take a cast of the ring gage thread form to check the half-angle and profile.

GO and NOT GO Plain Plug Acceptance Check Gages for Checking Minor Diameter of Thread Ring Gages.   (Recommended in Class XX tolerance for sizes up to No. 8 (0.164 in.) and in Class X tolerance for larger sizes.)   The GO plain plug gage is made to the minimum minor diameter specified for the thread ring gage (GO or LO), while the NOT GO gage is made to maximum minor diameter specified for the thread ring gage (GO or LO).   After the adjustable thread ring gages have been set to the applicable thread setting plugs, the GO and NOT GO plain plug acceptance check gages are applied to check the minor diameter of the ring gage to assure that it is within the specified limits.   An alternate method for checking minor diameter of thread ring gages is by the use of measuring equipment.

The standard lists the following for use of Threaded and Plain Ring, Snap and Indicating Thread Gages for verification of product external threads:

GO Thread Ring Gages.   GO thread ring gages must be set to the applicable W tolerance setting plugs to assure they are within specified limits.   The product thread must freely enter the GO thread ring gage for the entire length of the threaded portion.   The GO thread ring gage is a cumulative check of all thread elements except the major diameter.

LO Thread Ring Gages.   LO thread ring gages must be set to the applicable W tolerance setting plugs to assure that they are within specified limits.   LO thread ring gages when applied to the product external thread may engage only the end threads (which may not be representative of the complete product thread).

Starting threads on LO rings are subject to greater wear than the remaining threads.   Such wear in combination with the incomplete threads at the end of the product thread permit further entry in the gage.   Surveillance facilities ordinarily available in the field are often inadequate for fully determining such gage wear. Also, it is not practical to control nor limit the torque applied by operators, nor that utilized by a specific operator at various times and under varying conditions.   For these reasons the following standard practice has been adopted with respect to permissible entry.   Threads are acceptable when the LO thread ring gage is applied to the product external thread: (a) if it is not entered, or (b) if all complete product threads can be entered provided that a *definite* drag from contact with the product material results on or before the third turn of entry. The gage should not be forced after the drag is definite. Special requirements such as exceptionally thin or ductile material, or a small number of threads, etc., may necessitate modification of this practice.

*LO Thread Snap Limit Gages or Indicating Thread Gages.* LO thread snap limit gages (or indicating thread gages) check the product external thread LO minimum-material-limit. The gages must be set to the applicable W tolerance setting plugs. The gage is then applied to the product thread at various points around the circumference, and over the entire length of complete product threads. In applying the thread snap limit gage, threads are dimensionally acceptable when the gaging elements do not pass over the product thread or just pass over the product thread with perceptible drag from contact with the product material and the gage. Indicating thread gages provide a numerical value for the product thread size. Product external threads are dimensionally acceptable when the value derived in applying the gage (as described above) is not less than the specified minimum-material-limit.

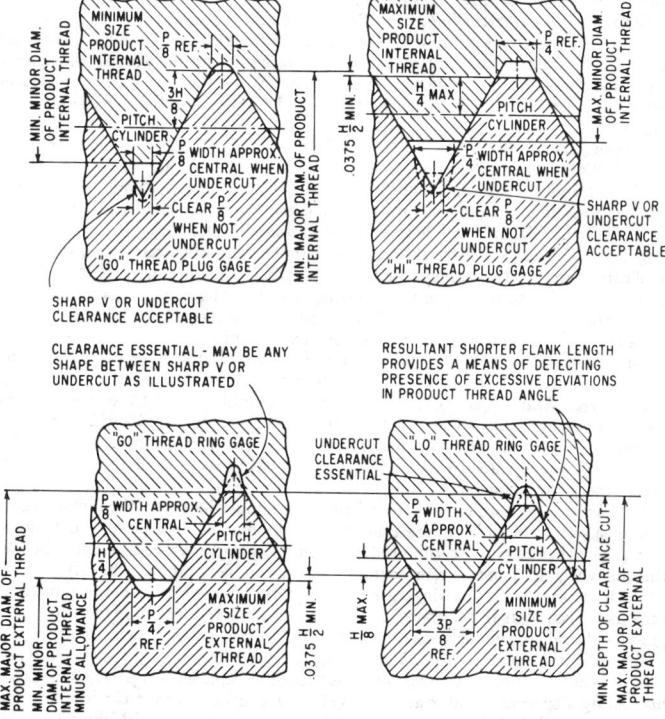

Fig. 1. Thread Forms of Gages for Product Internal and External Threads

Table 1. American National Standard Tolerances for GO, HI, and LO Thread Gages for Unified Inch Screw Threads (ANSI B1.2-1974)

| Thds. per Inch | Tolerance on Lead* | | Tol. on Thread Half-angle (±), minutes | Tol. on Major and Minor Diams. | | | Tolerance on Pitch Diameter | | | | |
|---|---|---|---|---|---|---|---|---|---|---|---|
| | To & incl. ½ in. Diam. | Above ½ in. Diam. | | To & incl. ½ in. Diam. | Above ½ to 4 in. Diam. | Above 4 in. Diam. | To & incl. ½ in. Diam. | Above ½ to 1½ in. Diam. | Above 1½ to 4 in. Diam. | Above 4 to 8 in. Diam. | Above 8 to 12 in. Diam.† |
| **W GAGES** | | | | | | | | | | | |
| 80,72 | .0001 | .00015 | 20 | .0003 | .0003 | ... | .0001 | .00015 | ... | ... | ... |
| 64 | .0001 | .00015 | 20 | .0003 | .0004 | ... | .0001 | .00015 | ... | ... | ... |
| 56 | .0001 | .00015 | 20 | .0003 | .0004 | ... | .0001 | .00015 | .0002 | ... | ... |
| 48 | .0001 | .00015 | 18 | .0003 | .0004 | ... | .0001 | .00015 | .0002 | ... | ... |
| 44,40 | .0001 | .00015 | 15 | .0003 | .0004 | ... | .0001 | .00015 | .0002 | ... | ... |
| 36 | .0001 | .00015 | 12 | .0003 | .0004 | ... | .0001 | .00015 | .0002 | ... | ... |
| 32 | .0001 | .00015 | 12 | .0003 | .0005 | .0007 | .0001 | .00015 | .0002 | .00025 | .0003 |
| 28,27 | .00015 | .00015 | 8 | .0005 | .0005 | .0007 | .0001 | .00015 | .0002 | .00025 | .0003 |
| 24,20 | .00015 | .00015 | 8 | .0005 | .0005 | .0007 | .0001 | .00015 | .0002 | .00025 | .0003 |
| 18 | .00015 | .00015 | 8 | .0005 | .0005 | .0007 | .0001 | .00015 | .0002 | .00025 | .0003 |
| 16 | .00015 | .00015 | 8 | .0006 | .0006 | .0009 | .0001 | .0002 | .00025 | .0003 | .0004 |
| 14,13 | .0002 | .0002 | 6 | .0006 | .0006 | .0009 | .00015 | .0002 | .00025 | .0003 | .0004 |
| 12 | .0002 | .0002 | 6 | .0006 | .0006 | .0009 | .00015 | .0002 | .00025 | .0003 | .0004 |
| 11½ | .0002 | .0002 | 6 | .0006 | .0006 | .0009 | .00015 | .0002 | .00025 | .0003 | .0004 |
| 11 | .0002 | .0002 | 6 | .0006 | .0006 | .0009 | .00015 | .0002 | .00025 | .0003 | .0004 |
| 10 | ... | .00025 | 6 | ... | .0006 | .0009 | ... | .0002 | .00025 | .0003 | .0004 |
| 9 | ... | .00025 | 6 | ... | .0007 | .0011 | ... | .0002 | .00025 | .0003 | .0004 |
| 8 | ... | .00025 | 5 | ... | .0007 | .0011 | ... | .0002 | .00025 | .0003 | .0004 |
| 7 | ... | .0003 | 5 | ... | .0007 | .0011 | ... | .0002 | .00025 | .0003 | .0004 |
| 6 | ... | .0003 | 5 | ... | .0008 | .0013 | ... | .0002 | .00025 | .0003 | .0004 |
| 5 | ... | .0003 | 4 | ... | .0008 | .0013 | ... | ... | .00025 | .0003 | .0004 |
| 4½ | ... | .0003 | 4 | ... | .0008 | .0013 | ... | ... | .00025 | .0003 | .0004 |
| 4 | ... | .0003 | 4 | ... | .0009 | .0015 | ... | ... | .00025 | .0003 | .0004 |
| **X GAGES** | | | | | | | | | | | |
| 80,72 | .0002 | .0002 | 30 | .0003 | .0003 | ... | .0002 | .0002 | ... | ... | ... |
| 64 | .0002 | .0002 | 30 | .0004 | .0004 | ... | .0002 | .0002 | ... | ... | ... |
| 56,48 | .0002 | .0002 | 30 | .0004 | .0004 | ... | .0002 | .0002 | .0003 | ... | ... |
| 44,40 | .0002 | .0002 | 20 | .0004 | .0004 | ... | .0002 | .0002 | .0003 | ... | ... |
| 36 | .0002 | .0002 | 20 | .0004 | .0004 | ... | .0002 | .0002 | .0003 | ... | ... |
| 32,28 | .0003 | .0003 | 15 | .0005 | .0005 | .0007 | .0003 | .0003 | .0004 | .0005 | .0006 |
| 27,24 | .0003 | .0003 | 15 | .0005 | .0005 | .0007 | .0003 | .0003 | .0004 | .0005 | .0006 |
| 20 | .0003 | .0003 | 15 | .0005 | .0005 | .0007 | .0003 | .0003 | .0004 | .0005 | .0006 |
| 18 | .0003 | .0003 | 10 | .0005 | .0005 | .0007 | .0003 | .0003 | .0004 | .0005 | .0006 |
| 16,14 | .0003 | .0003 | 10 | .0006 | .0006 | .0009 | .0003 | .0003 | .0004 | .0006 | .0008 |
| 13,12 | .0003 | .0003 | 10 | .0006 | .0006 | .0009 | .0003 | .0003 | .0004 | .0006 | .0008 |
| 11½ | .0003 | .0003 | 10 | .0006 | .0006 | .0009 | .0003 | .0003 | .0004 | .0006 | .0008 |
| 11,10 | .0003 | .0003 | 10 | .0006 | .0006 | .0009 | .0003 | .0003 | .0004 | .0006 | .0008 |
| 9 | ... | .0003 | 10 | .0007 | .0007 | .0011 | ... | .0003 | .0004 | .0006 | .0008 |
| 8,7 | .0004 | .0004 | 5 | .0007 | .0007 | .0011 | .0004 | .0004 | .0005 | .0006 | .0008 |
| 6 | .0004 | .0004 | 5 | .0008 | .0008 | .0013 | .0004 | .0004 | .0005 | .0006 | .0008 |
| 5,4½ | .0004 | .0004 | 5 | .0008 | .0008 | .0013 | ... | ... | .0005 | .0006 | .0008 |
| 4 | .0004 | .0004 | 5 | .0009 | .0009 | .0015 | ... | ... | .0005 | .0006 | .0008 |

All dimensions are given in inches unless otherwise specified.

* Allowable variation in lead between any two threads not father apart than the length of the standard gage as shown in ANSI B47.1. It has been customary in the past to specify tolerances on lead as plus or minus (±) values. Under the requirement established above, the width of the tolerance zone is the nominal tolerance value specified *regardless of sign*. In view of the preceding, the tolerance symbols, plus or minus (±), should be removed in referencing lead tolerances. The omission of the plus and minus does not change the total tolerance.

† Above 12 inches the tolerance is directly proportional to the tolerance given in this column below, in the ratio of the diameter to 12 inches.

## Table 2. Formulas for Limits of American National Standard Gages for Unified Inch Screw Threads* (ANSI B1.2-1974) — 1

| No. | THREAD GAGES FOR EXTERNAL THREADS |
|---|---|
| 1 | GO Pitch Diameter = Maximum pitch diameter of external thread. Gage tolerance is *minus*. When wear allowance is required, subtract the applicable wear allowance from the maximum pitch diameter and then apply the minus gage tolerance. |
| 2 | GO Minor Diameter = Maximum pitch diameter of external thread minus $H/2$. Gage tolerance is *minus*. |
| 3a | ‡LO Pitch Diameter (for plus tolerance gage) = Minimum pitch diameter of external thread. Gage tolerance is *plus*. |
| 3b | ‡LO Pitch Diameter (for minus tolerance gage) = Minimum pitch diameter of external thread. Gage tolerance is *minus* (optional). |
| 4 | ‡LO Minor Diameter = Minimum pitch diameter of external thread minus $H/4$ but not less than minimum minor diameter of GO thread gage for external thread plus $0.0375H$ ( $= 0.05h_b$). Gage tolerance is *plus*. |

| | PLAIN GAGES FOR MAJOR DIAMETER OF EXTERNAL THREADS |
|---|---|
| 5 | GO = Maximum major diameter of external thread. Gage tolerance is *minus*. |
| 6 | †NOT GO (for semi-finished material) = Minimum major diameter of external thread. Gage tolerance is *plus*. |
| 7 | †NOT GO (for unfinished hot-rolled material) = Minimum major diameter of external thread of hot-rolled material in UNC-2A, and 8UN. Gage tolerance is *plus*. |

| | THREAD GAGES FOR INTERNAL THREADS |
|---|---|
| 8 | GO Major Diameter = Minimum major diameter of internal thread. Gage tolerance is *plus*. |
| 9 | GO Pitch Diameter = Minimum pitch diameter of internal thread. Gage tolerance is *plus*. When wear allowance is required, add the applicable wear allowance to the minimum pitch diameter and then apply the plus gage tolerance. |
| 10 | ‡HI Major Diameter = Maximum pitch diameter of internal thread plus $H/2$, but not to exceed minimum major diameter of GO thread gage for internal thread minus $0.0375H$ ( $= 0.05h_b$). Gage tolerance is *minus*. |
| 11 | ‡HI Pitch Diameter (for minus tolerance gage) = Maximum pitch diameter of internal thread. Gage tolerance is *minus*. |
| 12 | ‡HI Pitch Diameter (for plus tolerance gage) = Maximum pitch diameter of internal thread. Gage tolerance is *plus* (optional). |

| | PLAIN GAGES FOR MINOR DIAMETER OF INTERNAL THREADS |
|---|---|
| 13 | GO = Minimum minor diameter of internal thread. Gage tolerance is *plus*. |
| 14 | †NOT GO = Maximum minor diameter of internal thread. Gage tolerance is *minus*. |

| | TRUNCATED SETTING PLUGS FOR EXTERNAL THREADS |
|---|---|
| 15 | GO Major Diameter (Truncated Portion) = Maximum major diameter of external thread ( = minimum major diameter of full portion of GO setting plug) minus $(0.060\sqrt[3]{p^2} + 0.017p)$. Gage tolerance is *minus*. |
| 16 | GO Major Diameter (Full Portion) = Maximum major diameter of external thread. Gage tolerance is *plus*. |
| 17 | GO Pitch Diameter = Maximum pitch diameter of external thread. Gage tolerance is *minus*. When wear allowance is required, subtract the applicable wear allowance from the maximum pitch diameter and then apply the minus gage tolerance. |
| 18 | ‡LO Major Diameter (Truncated Portion) = Minimum pitch diameter of external thread plus $H/2$. Gage tolerance is *minus*. |

* See data in Screw Thread Systems section for symbols and dimensions of Unified screw threads.

† Plain minimum-material-limit gages retain the term NOT GO as customarily they are not permitted to enter or be entered by acceptable product.

‡ HI is the ANSI designation for a NOT GO gage for internal threads and LO designates a NOT GO gage for external threads. Terms indicate the practice of permissible entry.

*Note:* In the American National Standard ANSI B1.1, it is provided that while the maximum diameters of Class 2A uncoated threads are less than basic by the amount of the allowance, the allowance may be used to accommodate additive finishes. It follows that unless specifically specified otherwise, for threads with additive finish the maximum diameters of Class 2A may be exceeded by the amount of the allowance. In this event GO gages to basic pitch diameter would be applicable. Such gages are made to the same dimensions as listed for Class 3A threads.

**Table 2.** *(Continued).* **Formulas for Limits of American National Standard Gages for Unified Inch Screw Threads\* (ANSI B1.2-1974) — 2**

| No. | TRUNCATED SETTING PLUGS FOR EXTERNAL THREADS |
|-----|-----------------------------------------------|
| 19 | ‡LO Major Diameter (Full Portion) = Maximum major diameter of external thread provided that, after applying the X major diameter tolerance, the maximum major diameter of the gage corresponds to a truncation of not less than o.o67*H* or o.ooo9 inch, whichever is the greater. Gage tolerance is *plus.* |
| 20 | ‡LO Pitch Diameter (for plus tolerance gage) = Minimum pitch diameter of external thread. Gage tolerance is *plus.* |
| 21 | ‡LO Pitch Diameter (for minus tolerance gage) = Minimum pitch diameter of external thread. Gage tolerance is *minus* (optional). |

| No. | BASIC-CREST SETTING PLUGS FOR EXTERNAL THREADS |
|-----|-------------------------------------------------|
| 22 | GO Major Diameter = Maximum major diameter of external thread. Gage tolerance is *plus.* |
| 23 | GO Pitch Diameter = Maximum pitch diameter of external thread. Gage tolerance is *minus.* When wear allowance is required, subtract the applicable wear allowance from the maximum pitch diameter and then apply the minus gage tolerance. |
| 24 | ‡LO Major Diameter = Maximum major diameter of external thread provided that, after applying the X major diameter tolerance, the maximum major diameter of the gage corresponds to a truncation of not less than o.o67*H* or o.ooo9 inch, whichever is the greater. Gage tolerance is *plus.* |
| 25 | LO Pitch Diameter (for plus tolerance gage) = Minimum pitch diameter of external thread. Gage tolerance is *plus.* |
| 26 | ‡LO Pitch Diameter (for minus tolerance gage) = Minimum pitch diameter of external thread. Gage tolerance is *minus* (optional). |

\* See data in Screw Thread Systems section for symbols and dimensions of Unified screw threads.

‡ HI is the ANSI designation for a NOT GO gage for internal threads and LO designates a NOT GO gage for external threads. Terms indicate the practice of permissible entry.

**Table 3.** **American National Standard Tolerances for Plain Gages (ANSI B1.2-1974)**

| Size Range | | Tolerance Class | | | | |
|---|---|---|---|---|---|---|
| Above | To and Including | XX | X | Y | Z | ZZ |
| | | Tolerance | | | | |
| 0.029 | 0.825 | .00002 | .00004 | .00007 | .00010 | .00020 |
| 0.825 | 1.510 | .00003 | .00006 | .00009 | .00012 | .00024 |
| 1.510 | 2.510 | .00004 | .00008 | .00012 | .00016 | .00032 |
| 2.510 | 4.510 | .00005 | .00010 | .00015 | .00020 | .00040 |
| 4.510 | 6.510 | .000065 | .00013 | .00019 | .00025 | .00050 |
| 6.510 | 9.010 | .00008 | .00016 | .00024 | .00032 | .00064 |
| 9.010 | 12.010 | .00010 | .00020 | .00030 | .00040 | .00080 |

All dimensions are given in inches.

*GO and NOT GO Plain Rings and Adjustable Snap Limit and Indicating Gages for Checking Major Diameter of Product External Thread.* (Recommended in Class Z tolerance.) The GO gage must completely receive or pass over the major diameter of the product external thread to assure that the major diameter does not exceed the maximum-material-limit. The NOT GO gage must not pass over the major diameter of the product external thread to assure that the major diameter is not less than the minimum-material-limit.

Limitations concerning the use of gages are given in the standard as follows:

*Product threads* accepted by a gage of one type may be verified by other types. It is possible, however, that parts which are near either rejection limit may be accepted by one type and rejected by another. Also, it is possible for two individual limit gages of the same type to be at the opposite extremes of the gage tolerances permitted, and borderline product threads accepted by one gage could be rejected by another. In such instances, (except when LO limit snap or indicating thread gages

THREAD GAGES

### Table 4.  Constants for Computing Thread Gage Dimensions (ANSI B1.2-1974)

| Threads per Inch | Pitch, $p$ | $.067p$ | $.060\sqrt[3]{p^2}$ | $.017p$ | Height of Sharp V-Thread, $H = .866025p$ | $H/2 = .43301p$ | $.0375H = .05h_b = .03248p$ |
|---|---|---|---|---|---|---|---|
| 80 | .012500 | .00084 | .00323 | .00021 | .010825 | .00541 | .00041 |
| 72 | .013889 | .00093 | .00347 | .00024 | .012028 | .00601 | .00045 |
| 64 | .015625 | .00105 | .00375 | .00027 | .013532 | .00677 | .00051 |
| 56 | .017857 | .00120 | .00410 | .00030 | .015465 | .00773 | .00058 |
| 48 | .020833 | .00140 | .00454 | .00035 | .018042 | .00902 | .00068 |
| 44 | .022727 | .00152 | .00482 | .00039 | .019682 | .00984 | .00074 |
| 40 | .025000 | .00168 | .00513 | .00042 | .021651 | .01083 | .00081 |
| 36 | .027778 | .00186 | .00550 | .00047 | .024056 | .01203 | .00090 |
| 32 | .031250 | .00209 | .00595 | .00053 | .027063 | .01353 | .00101 |
| 28 | .035714 | .00239 | .00651 | .00061 | .030929 | .01546 | .00116 |
| 27 | .037037 | .00248 | .00667 | .00063 | .032075 | .01604 | .00120 |
| 24 | .041667 | .00279 | .00721 | .00071 | .036084 | .01804 | .00135 |
| 20 | .050000 | .00335 | .00814 | .00085 | .043301 | .02165 | .00162 |
| 18 | .055556 | .00372 | .00874 | .00094 | .048113 | .02406 | .00180 |
| 16 | .062500 | .00419 | .00945 | .00106 | .054127 | .02706 | .00203 |
| 14 | .071429 | .00479 | .01033 | .00121 | .061859 | .03093 | .00232 |
| 13 | .076923 | .00515 | .01085 | .00131 | .066617 | .03331 | .00250 |
| 12 | .083333 | .00558 | .01145 | .00142 | .072169 | .03608 | .00271 |
| 11½ | .086957 | .00583 | .01178 | .00148 | .075307 | .03765 | .00282 |
| 11 | .090909 | .00609 | .01213 | .00155 | .078730 | .03936 | .00295 |
| 10 | .100000 | .00670 | .01293 | .00170 | .086603 | .04330 | .00325 |
| 9 | .111111 | .00744 | .01387 | .00189 | .096225 | .04811 | .00361 |
| 8 | .125000 | .00838 | .01500 | .00212 | .108253 | .05413 | .00406 |
| 7 | .142857 | .00957 | .01640 | .00243 | .123718 | .06186 | .00464 |
| 6 | .166667 | .01117 | .01817 | .00283 | .144338 | .07217 | .00541 |
| 5 | .200000 | .01340 | .02052 | .00340 | .173205 | .08660 | .00650 |
| 4½ | .222222 | .01489 | .02201 | .00378 | .192450 | .09623 | .00722 |
| 4 | .250000 | .01675 | .02381 | .00425 | .216506 | .10825 | .00812 |

All dimensions are given in inches unless otherwise specified.

are specified) limit plug and ring thread gages that approximate as closely as practicable the extreme maximum-material-product-limit and minimum-material-product-limit shall be used to determine whether or not the product threads under inspection are within the specified limits of size.

*Large product external and internal threads* above 6.25-inch nominal size may present additional problems for technical and economic reasons.  In these instances verification may be based on use of gages or measurement of thread elements. Various types of gages or measuring devices in addition to those defined in this standard are available and acceptable when properly correlated to this standard. Producer and user should agree on the method and equipment used.

**Thread Forms of Gages.** — Thread forms of gages for product internal and external threads are given in Fig. 1.  The Standard (ANSI B1.2-1974) also gives illustrations of the thread form of truncated thread setting plug gages, the thread forms of basic-crest thread setting plug gages, and an illustration that shows the chip groove and removal of partial thread.

**Thread Gage Tolerances.** — Gage tolerances of thread plug and ring gages and thread setting plugs for Unified screw threads, designated as W and X tolerances, are given in Table 1.  W tolerances represent the highest commercial grade of accuracy and workmanship, and are specified for truncated setting plugs; X tolerances are larger than W tolerances.  Tolerances for plain gages are given in Table 3.

**Formulas for Limits of Gages.** — Formulas for limits of American National Standard Gages for Unified screw threads are given in Table 2.  Some constants which are required to determine gage dimensions are tabulated in Table 4.

# TAPPING AND THREAD CUTTING

**Selection of Taps.** — For most applications, a standard tap supplied by the manufacturer can be used, but some jobs may require special taps. A variety of standard taps can be obtained. In addition to specifying the size of the tap it is necessary to be able to select the one most suitable for the application at hand. The elements of standard taps that are varied are: the number of flutes; the type of flute, whether straight, spiral pointed, or spiral fluted; the chamfer length; the relief of the land, if any; the tool steel used to make the tap; and the surface treatment of the tap. Details regarding the nomenclature of tap elements are given on pages 1687–1710 along with a listing of the standard sizes available.

Factors to consider in selecting a tap include: the method of tapping, by hand or by machine; the material to be tapped and its heat treatment; the length of thread, or depth of the tapped hole; the required tolerance or class of fit; the production requirement and the type of machine to be used. The diameter of the hole must also be considered, although this is usually a matter of design and the specification of the tap drill size.

*Method of Tapping:* The term *hand tap* is used for both hand and machine taps, and almost all taps can be applied to the hand or machine method. While any tap can be used for hand tapping, those having a concentric land without the relief are preferable. In hand tapping the tool is reversed periodically to break the chip, and the heel of the land of a tap with a concentric land (without relief) will cut the chip off cleanly or any portion of it that is attached to the work, whereas a tap with an eccentric or con-eccentric relief may leave a small burr that becomes wedged between the relieved portion of the land and the work. This creates a pressure towards the cutting face of the tap that may cause it to chip; it tends to roughen the threads in the hole, and it increases the overall torque required to turn the tool. When tapping by machine, however, the tap is usually turned only in one direction until the operation is complete, and an eccentric or con-eccentric relief is often an advantage.

*Chamfer Length:* Three types of hand taps, used both for hand and machine tapping, are available, and they are distinguished from each other by the length of chamfer. *Taper taps* have a chamfer angle of 4 degrees which reduces the height about 8-10 teeth; *plug taps* have a 9 degree chamfer angle with 3-5 threads reduced in height; and *bottoming taps* have a 30 degree chamfer angle with 1½ threads reduced in height. Since the teeth that are reduced in height do practically all of the cutting, the chip load or chip thickness per tooth will be least for a taper tap, greater for a plug tap, and greatest for a bottoming tap.

For most through hole tapping applications it is necessary to use only a plug type tap, which is also most suitable for blind holes in which the tap drill hole is deeper than the required thread. If the tap must bottom in a blind hole, the hole is usually threaded first with a plug tap and then finished with a bottoming tap to catch the last threads in the bottom of the hole. Taper taps are used on hard materials having a high tensile strength where the chip load per tooth must be kept to a minimum. However, taper taps should not be used on materials that have a strong tendency to work harden, such as the austenitic stainless steels.

*Spiral Point Taps:* Spiral point taps offer a special advantage when machine tapping through holes in ductile materials because they are designed to handle the long continuous chips that form and which otherwise cause a disposal problem. An angular gash is ground at the point or end of the tap along the face of the chamfered threads or lead teeth of the tap. This gash forms a left-hand helix in the flutes adjacent to the lead teeth which causes the chips to flow ahead of the tap and through the hole. The gash is usually formed to produce a rake angle on the cutting face that increases progressively towards the end of the tool. Since the flutes are used primarily to provide a passage for the cutting fluid, they are usually made narrower and shallower thereby strengthening the tool. For tapping thin workpieces short fluted spiral taps are recommended. They have a spiral point gash along the cutting teeth; the remainder of the threaded portion of the tap has no flute. Most spiral pointed taps are of plug type; however, spiral point bottoming taps are also made.

*Spiral Fluted Taps:* Spiral fluted taps have a helical flute; the helix angle of the flute may be between 15 and 52 degrees and the hand of the helix is the same as that of the threads on the tap. The spiral flute and the axial rake that it forms on the cutting face of the tap combine to induce the chips to flow backward along the helix and out of the hole. Thus, they are ideally suited for tapping blind holes and they are available as plug and bottoming types. A higher helix angle should be specified for tapping very ductile materials; when tapping harder materials chipping at the cutting edge may result and the helix angle must be reduced. For tapping very high strength materials serial taps are recommended having a small helix angle.

Holes having a pronounced interruption such as a groove or a keyway can be tapped with spiral fluted taps on which the helix has the opposite hand as the threads. The land bridges the interruption and allows the tap to cut relatively smoothly.

*Serial Taps and Close Tolerance Threads:* For tapping holes to close tolerances a set of serial taps is used. They are usually available in sets of three: the No. 1 tap is undersize and is the first rougher; the No. 2 tap is of intermediate size and is the second rougher; and the No. 3 tap is used for finishing. The different taps are identified by one, two, and three annular grooves in the shank adjacent to the square. For some applications involving finer pitches only two serial taps are required. Sets are also used to tap hard or tough materials having a high tensile strength, deep blind holes in normal materials, and large coarse threads. A set of more than three taps is sometimes required to produce threads of coarse pitch. Threads to some commercial tolerances, such as American Standard Unified 2B, or ISO Metric 6H, can be produced in one cut using a ground tap; sometimes even closer tolerances can be produced with a single tap. Ground taps are recommended for all close tolerance tapping operations. For much ordinary work, cut taps are satisfactory and more economical than ground taps.

*Tap Steels:* Most taps are made from high speed steel, although carbon tool steel taps are available and in many instances are economical to use. The type of tool steel used is determined by the tap manufacturer and is usually satisfactory when correctly applied except in a few exceptional cases. Typical grades of high speed steel used to make taps are M-1, M-7, and M-10. Carbon tool steel taps are satisfactory where the operating temperature of the tap is low and where a high resistance to abrasion is not required.

*Surface Treatment:* The life of high speed steel taps can sometimes be increased significantly by treating the surface of the tap. A very common treatment is oxide coating, which forms a thin non-metallic oxide coating on the tap that has lubricity and is somewhat porous to absorb and retain oil. This coating reduces the friction between the tap and the work and it makes the surface virtually impervious to rust. It does not increase the hardness of the surface but it significantly reduces or prevents entirely galling, or the tendency of the work material to weld or stick to the cutting edge and to other areas on the tap with which it is in contact. For this reason oxide coated taps are recommended for metals that tend to gall and stick such as non-free cutting low carbon steels and soft copper. It is also useful for tapping other steels having higher strength properties.

Nitriding provides a very hard and wear resistant case on high speed steel. Nitrided taps are especially recommended for tapping plastics; they have also been used successfully on a variety of other materials including high strength high alloy steels. However, some caution must be used in specifying nitrided taps because the nitride case is very brittle and may have a tendency to chip.

Chrome plating has been used to increase the wear resistance of taps but its application has been limited because of the high cost and the danger of hydrogen embrittlement which can cause cracks to form in the tool. A flash plate of about .0001 to .0005 in. in thickness is applied to the tap. Chrome-plated taps have been used successfully to tap a variety of ferrous and nonferrous materials including plastics, hard rubber, mild steel, and tool steel. Other surface treatments that have been used successfully to a limited extent are vapor blasting and liquid honing.

*Rake Angle:* For the majority of applications in both ferrous and nonferrous materials the rake angle machined on the tap by the manufacturer is satisfactory. This angle is approximately 5 to 7 degrees. In some instances it may be desirable to alter the rake angle of the tap to obtain beneficial results and Table 1 provides a guide that can be used. In selecting a rake angle from this table, consideration must be given to the size of the tap and the strength of the land.

**Table 1. Tap Rake Angles for Tapping Different Materials**

| Material | Rake Angle, Degrees | Material | Rake Angle, Degrees |
|---|---|---|---|
| Cast Iron | 0–3 | Aluminum | 10–20 |
| Malleable Iron | 5–8 | Brass | 2–5 |
| Steel | | Naval Brass | 5–8 |
|   AISI 1100 Series | 9–12 | Phosphor Bronze | 9–12 |
|   Low Carbon (up | 9–12 | Tobin Bronze | 5–8 |
|     to .25 per cent) | | Manganese Bronze | 9–12 |
|   Medium Carbon, Annealed | 7–10 | Magnesium | 10–20 |
|     (.30 to .60 per cent) | | Monel | 9–12 |
|   Heat Treated, 225–283 | 0–8 | Copper | 15–18 |
|     Brinell. (.30 to .60 per cent) | | Zinc Die Castings | 12–15 |
|   High Carbon and | 0–5 | Plastic | |
|     High Speed | |   Thermoplastic | 5–8 |
| Stainless | 10–15 |   Thermosetting | 0–3 |
| Titanium | 6–10 | Hard Rubber | 0–3 |

**Cutting Speed.** — The cutting speed for machine tapping is treated in detail on page 1761. It suffices to say here that many variables must be considered in selecting this cutting speed and any tabulation may have to be modified greatly. Where cutting speeds are mentioned in the following section, they are intended only to provide a guideline to show the possible range of speeds that could be used.

**Tapping Specific Materials.** — The work material has a great influence on the ease with which a hole can be tapped. For production work, in many instances, modified taps are recommended; however, for toolroom or short batch work, standard hand taps can be used on most jobs, providing reasonable care is taken when tapping. The following concerns the tapping of metallic materials; information on the tapping of plastics is given on page 1844.

*Low Carbon Steel (Less than 0.15 % C):* These steels are very soft and ductile resulting in a tendency for the work material to tear and to weld to the tap. They produce a continuous chip that is difficult to break and spiral pointed taps are recommended for tapping through holes; for blind holes a spiral fluted tap is recommended. To prevent galling and welding a liberal application of a sulfur base or other suitable cutting fluid is essential and the selection of an oxide coated tap is very helpful.

*Low Carbon Steels (0.15 to 0.30 % C):* The additional carbon in these steels is beneficial as it reduces the tendency to tear and to weld; their machinability is further improved by cold drawing. These steels present no serious problems in tapping provided a suitable cutting fluid is used. An oxide coated tap is recommended, particularly in the lower carbon range.

*Medium Carbon Steels (0.30 to 0.60 % C):* These steels can be tapped without too much difficulty, although a lower cutting speed must be used in machine tapping. The cutting speed is dependent on the carbon content and the heat treatment. Steels that have a higher carbon content must be tapped more slowly, especially if the heat treatment has produced a pearlitic microstructure. The cutting speed and ease of tapping is significantly improved by heat treating to produce a spheroidized microstructure. A suitable cutting fluid must be used.

*High Carbon Steels (More than 0.6 % C):* Usually these materials are tapped in the annealed or normalized condition although sometimes tapping is done after hardening and tempering to a hardness below 55 Rc. Recommendations for tapping after hardening and tempering are given under High Tensile Strength Steels. In the annealed and normalized condition these steels have a higher strength and are more abrasive than steels with a lower carbon content; thus, they are more difficult to tap. The microstructure resulting from the heat treatment has a significant effect on the ease of tapping and the tap life, a spherodite structure being better in this respect than a pearlitic structure. The rake angle of the tap should not exceed 5 degrees and for the harder materials a concentric tap is recommended. The cutting speed is considerably lower for these steels and an activated sulfur-chlorinated cutting fluid is recommended.

*Alloy Steels:* This classification includes a wide variety of steels, each of which may be heat treated to have a wide range of properties. When annealed and normalized they are similar to medium to high carbon steels and usually can be tapped without difficulty, although for some alloy steels a lower tapping speed may be required. Standard taps can be used and for machine tapping a con-eccentric relief may be helpful. A suitable cutting fluid must be used in all cases.

*High Tensile Strength Steels:* Any steel that must be tapped after being heat treated to a hardness range of 40-55 Rc is included in this classification. Low tap life and excessive tap breakage are characteristics of tapping these materials; those that have a high chromium content are particularly troublesome. Best results are obtained with taps that have concentric lands, a rake angle that is at or near zero degrees, and 6 to 8 chamfered threads on the end to reduce the chip load per tooth. The chamfer relief should be kept to a minimum. The load on the tap should be kept to a minimum by every possible means, including using the largest possible tap drill size; keeping the hole depth to a minimum; avoidance of bottoming holes; and, in the larger sizes, using fine instead of coarse pitches. Oxide coated taps are recommended although in some cases a nitrided tap can be used to reduce tap wear. An active sulfur-chlorinated oil is recommended as a cutting fluid and the tapping speed should not exceed about 10 feet per minute.

*Stainless Steels:* Ferritic and martensitic type stainless steels are somewhat like alloy steels that have a high chromium content, and they can be tapped in a similar manner, although a slightly slower cutting speed may have to be used. Standard rake angle oxide coated taps are recommended and a cutting fluid containing molybdenum disulphide is helpful to reduce the friction in tapping. Austenitic stainless steels are very difficult to tap because of their high resistance to cutting and their great tendency to work harden. A work-hardened layer is formed by a cutting edge of the tap and the depth of this layer depends on the severity of the cut and the sharpness of the tool. The next cutting edge must penetrate below the work-hardened layer, if it is to be able to cut. Therefore, the tap must be kept sharp and each succeeding cutting edge on the tool must penetrate below the work-hardened layer formed by the preceding cutting edge. For this reason, a taper tap should not be used, but rather a plug tap having 3-5 chamfered threads. To reduce the rubbing of the lands, an eccentric or con-eccentric relieved land should be used and a 10-15 degree rake angle is recommended. A tough continuous chip is formed that is difficult to break. To control this chip, spiral pointed taps are recommended for through holes and low-helix angle spiral fluted taps for blind holes. An oxide coating on the tap is very helpful and a sulfur-chlorinated mineral lard oil is recommended, although heavy duty soluble oils have also been used successfully.

*Free Cutting Steels:* There is a large number of free-cutting steels, including free cutting stainless steels, which are also called free-machining steels. Sulfur, lead, or phosphorus are added to these steels to improve their machinability. In all cases, the free-machining steels are easier to tap than their counterparts that do not have the free-machining additives. Tool life is usually increased and a somewhat higher cutting speed can be used. The type of tap recommended depends on the particular type of free machining steel and the nature of the tapping operation; in most cases a standard tap can be used.

*High Temperature Alloys:* These are cobalt or nickel base nonferrous alloys that cut like austenitic stainless steel, but are often even more difficult to machine. The recommendations given for austenitic stainless steel also apply to tapping these alloys but the rake angle should be 0 to 10 degrees to strengthen the cutting edge. For most applications a nitrided tap or one made from M41, M42, M43, or M44 steel is recommended. The tapping speed is usually in the range of 5 to 10 feet per minute.

*Titanium and Titanium Alloys:* Titanium and its alloys have a low specific heat and a pronounced tendency to weld on to the tool material; therefore, oxide coated taps are recommended to minimize galling and welding. The rake angle of the tap should be from 6 to 10 degrees. To minimize the contact between the work and the tap an eccentric or con-eccentric relief land should be used. In some cases taps having interrupted threads are helpful. Pure titanium is comparatively easy to tap but the alloys are very difficult. The cutting speed depends on the composition of the alloy and may vary from 40 to 10 feet per minute. Special cutting oils are recommended for tapping titanium.

*Gray Cast Iron:* The microstructure of gray cast iron can vary, even within a single casting, and compositions are used that vary in tensile strength from about 20,000 to 60,000 psi (160 to 250 Bhn). It is seen then that this is not a single material, although in general it is not difficult to tap. The cutting speed may vary from 90 feet per minute for the softer grades to 30 feet per minute for the harder grades. The chip is discontinuous and straight fluted taps should be used for all applications. Oxide coated taps are helpful and in most cases gray cast iron can be tapped dry, although water soluble oils and chemical emulsions are sometimes used.

*Malleable Cast Iron:* Commercial malleable cast irons are also available having a rather wide range of properties, although within a single casting they tend to be quite uniform. They are relatively easy to tap and standard taps can be used. The cutting speed for ferritic cast irons is 60-90 feet per minute, for pearlitic malleable irons 40-50 feet per minute, and for martensitic malleable irons 30-35 feet per minute. A soluble oil cutting fluid is recommended except for martensitic malleable iron where a sulfur base oil may work better.

*Ductile or Nodular Cast Iron:* Several classes of nodular iron are used having a tensile strength varying from 60,000 to 120,000 psi. Moreover, the microstructure in a single casting and in castings produced at different times vary rather widely. The chips are easily controlled but have some tendency to weld to the face and flanks of cutting tools. For this reason oxide coated taps are recommended. The cutting speed may vary from 15 fpm for the harder martensitic ductile irons to 60 fpm for the softer ferritic grades. A suitable cutting fluid should be used.

*Aluminum:* Aluminum and aluminum alloys are relatively soft materials that have little resistance to cutting. The danger in tapping these alloys is that the tap will ream the hole instead of cutting threads, or that it will cut a thread eccentric to the hole. For these reasons, extra care must be taken when aligning the tap and starting the thread. For production tapping a spiral pointed tap is recommended for through holes and a spiral fluted tap for blind holes; preferably these taps should have a 10 to 15 degree rake angle. A lead screw tapping machine is helpful in cutting accurate threads. A heavy duty soluble oil or a light base mineral oil should be used as a cutting fluid.

*Copper Alloys:* Most copper alloys are not difficult to tap, except beryllium copper and a few other hard alloys. Pure copper offers some difficulty because of its ductility and the ductile continuous chip formed, which can be difficult to control. However, with reasonable care and the use of medium heavy duty mineral lard oil it can be tapped successfully. Red brass, yellow brass, and similar alloys containing not more than 35 per cent zinc produce a continuous chip. While straight fluted taps can be used for hand tapping these alloys, machine tapping should be done with spiral pointed or spiral fluted taps for through and blind holes respectively. Naval brass, leaded brass, and .cast brasses produce a discontinuous chip and a straight fluted tap can be used for machine tapping. These alloys do exhibit a tendency to close in on the tap and sometimes an interrupted thread tap is used to reduce the resulting jamming effect. Beryllium copper and the silicon bronzes are the strongest of the copper alloys. Their strength combined with their ability to work harden can cause difficulties in tapping. For these alloys plug type taps should be used and the taps should be kept as sharp as possible. A medium or heavy duty water soluble oil is recommended as a cutting fluid.

**Diameter of Tap Drill.** — Tapping troubles are sometimes caused by tap drills that are too small in diameter. The tap drill should not be smaller than is necessary to give the required strength to the thread as even a very small decrease in the diameter of the drill will increase the torque required and the possibility of broken taps. Tests have shown that any increase in the percentage of full thread over 60 per cent does not significantly increase the strength of the thread. In many cases a 55 to 60 per cent thread is satisfactory, although 75 per cent threads are commonly used to provide an extra measure of safety. The present thread specifications do not always allow the use of the smaller thread depths. In all cases the specification given on a part drawing must be adhered to and these may require smaller minor diameters than might otherwise be recommended.

The depth of the thread in the tapped hole is dependent on the length of thread engagement and on the material from which it is made. In general, when the engagement length is more than one and one-half times the nominal diameter a 50 or 55 per cent thread is satisfactory. Soft ductile materials may permit use of a slightly larger tapping hole than brittle materials such as gray cast iron.

It must be remembered that a twist drill is a roughing tool that may be expected to drill slightly oversize and that some variations in the size of the tapping holes are to be expected. When a closer control of the hole size is required it must be reamed. Reaming is recommended for the larger thread diameters and for some fine pitch threads.

For threads of Unified form (see American Standard Unified Threads, page 1256) the selection of tap drills is covered in the following section, Factors Influencing Minor Diameter Tolerances of Tapped Holes and the hole size limits are given in Table 2. Tables 3 and 4 give tap drill sizes for American National Form threads based on 75 per cent of full thread depth. In the case of smaller-size threads the use of slightly larger drills, if permissible, will reduce tap breakage. The selection of tap drills for these threads also may be based on the hole size limits given in Table 2 for unified threads which take into account lengths of engagement.

The size of the tap drill hole for any desired percentage of full thread depth can be calculated by the formulas below. In these formulas the Per Cent Full Thread is expressed as a decimal; e.g., 75% is expressed as .75. The tap drill size is the size nearest to the calculated hole size.

For American Unified Thread form:

$$\text{Hole Size} = \text{Basic Major Diameter} - \frac{1.08253 \times \text{Per Cent Full Thread}}{\text{Number of Threads per Inch}}$$

For American National Thread form:

$$\text{Hole Size} = \text{Basic Major Diameter} - \frac{1.29904 \times \text{Per Cent Full Thread}}{\text{Number of Threads per Inch}}$$

For ISO Metric threads (all dimensions in millimeters):

$$\text{Hole Size} = \text{Basic Major Diameter} - (1.08253 \times \text{Pitch} \times \text{Per Cent Full Thread})$$

**Factors Influencing Minor Diameter Tolerances of Tapped Holes.** — As stated in the Unified screw thread standard, the principal practical factors which govern minor diameter tolerances of internal threads are tapping difficulties, particularly tap breakage in the small sizes, availability of standard drill sizes in the medium and large sizes, and depth (radial) of engagement. Depth of engagement is related to the stripping strength of the thread assembly, and thus also, to the length of engagement. It also has an influence on the tendency toward disengagement of the threads on one side when assembly is eccentric. The amount of possible eccentricity is one-half of the sum of the pitch diameter allowance and tolerances on both mating threads. For a given pitch, or height of thread, this sum increases with the diameter, and accordingly this factor would require a decrease in minor diameter tolerance with increase in diameter. However, such decrease in tolerance would often require the use of special drill sizes; therefore, to facilitate the use of standard drill sizes, for any given pitch the minor diameter tolerance for Unified thread classes 1B and 2B threads of ¼ inch diameter and larger is constant, in accordance with a formula given in the American Standard for Unified Screw Threads.

*Effect of Length of Engagement on Minor Diameter Tolerances:* There may be applications where the lengths of engagement of mating threads is relatively short or the combination of materials used for mating threads are such that the maximum minor diameter tolerance given in the Standard (based on a length of engagement equal to the nominal diameter) may not provide the desired strength of the fastening. Experience has shown that for lengths of engagement less than ⅔ *D* (the minimum thickness of standard nuts) the minor diameter tolerance may be reduced without causing tapping difficulties. In other applications the length of engagement of mating threads may be long because of design considerations or the combination of materials used for mating threads. As the threads engaged increase in number, a shallower depth of engagement may be permitted and still develop stripping strength greater than the external thread breaking strength. In these cases the maximum tolerance given in the Standard should be increased to reduce the possibility of tapping difficulties. The following paragraphs indicate how the afore-mentioned considerations were taken into account in determining the minor diameter limits for various lengths of engagement given in Table 2.

**Recommended Hole Sizes before Tapping.** — Recommended hole size limits before threading to provide for optimum strength of fastenings and tapping conditions are shown in Table 2 for classes 1B, 2B, and 3B. The hole size limits before threading, and the tolerances between them, are derived from the minimum and maximum minor diameters of the internal thread given in the dimensional tables for Unified threads in the screw thread section using the following rules:

(*Continued on page* 1424)

**Table 2. Recommended Hole Size Limits Before Tapping Unified Threads**

Length of Engagement (D = Nominal Size of Thread) — Recommended Hole Size Limits

First eight data columns are **Classes 1B and 2B**; last eight data columns are **Class 3B**.

| Thread Size | To and Including ½D Min* | Max | Above ½D to ¾D Min | Max | Above ¾D to 1½D Min | Max† | Above 1½D to 3D Min | Max | To and Including ½D Min* | Max | Above ½D to ¾D Min | Max | Above ¾D to 1½D Min | Max† | Above 1½D to 3D Min | Max |
|---|---|---|---|---|---|---|---|---|---|---|---|---|---|---|---|---|
| 0-80 | 0.0465 | 0.0500 | 0.0479 | 0.0514 | 0.0479 | 0.0514 | 0.0479 | 0.0514 | 0.0465 | 0.0500 | 0.0479 | 0.0514 | 0.0479 | 0.0514 | 0.0479 | 0.0514 |
| 1-64 | 0.0561 | 0.0599 | 0.0585 | 0.0623 | 0.0585 | 0.0623 | 0.0585 | 0.0623 | 0.0561 | 0.0599 | 0.0585 | 0.0623 | 0.0585 | 0.0623 | 0.0585 | 0.0623 |
| 1-72 | 0.0580 | 0.0613 | 0.0602 | 0.0635 | 0.0602 | 0.0635 | 0.0602 | 0.0635 | 0.0580 | 0.0613 | 0.0596 | 0.0629 | 0.0602 | 0.0635 | 0.0602 | 0.0635 |
| 2-56 | 0.0667 | 0.0705 | 0.0699 | 0.0737 | 0.0699 | 0.0737 | 0.0699 | 0.0737 | 0.0667 | 0.0705 | 0.0686 | 0.0724 | 0.0699 | 0.0737 | 0.0699 | 0.0737 |
| 2-64 | 0.0691 | 0.0724 | 0.0720 | 0.0753 | 0.0720 | 0.0753 | 0.0720 | 0.0753 | 0.0691 | 0.0724 | 0.0707 | 0.0740 | 0.0720 | 0.0753 | 0.0720 | 0.0753 |
| 3-48 | 0.0764 | 0.0804 | 0.0805 | 0.0845 | 0.0806 | 0.0846 | 0.0806 | 0.0846 | 0.0764 | 0.0804 | 0.0785 | 0.0825 | 0.0805 | 0.0845 | 0.0806 | 0.0846 |
| 3-56 | 0.0797 | 0.0831 | 0.0831 | 0.0865 | 0.0833 | 0.0867 | 0.0833 | 0.0867 | 0.0797 | 0.0831 | 0.0814 | 0.0848 | 0.0831 | 0.0865 | 0.0833 | 0.0867 |
| 4-40 | 0.0849 | 0.0894 | 0.0894 | 0.0939 | 0.0902 | 0.0947 | 0.0902 | 0.0947 | 0.0849 | 0.0894 | 0.0871 | 0.0916 | 0.0894 | 0.0939 | 0.0902 | 0.0947 |
| 4-48 | 0.0894 | 0.0931 | 0.0931 | 0.0968 | 0.0939 | 0.0976 | 0.0939 | 0.0976 | 0.0894 | 0.0931 | 0.0912 | 0.0949 | 0.0931 | 0.0968 | 0.0939 | 0.0976 |
| 5-40 | 0.0979 | 0.1020 | 0.1021 | 0.1062 | 0.1036 | 0.1077 | 0.1036 | 0.1077 | 0.0979 | 0.1020 | 0.1000 | 0.1041 | 0.1021 | 0.1062 | 0.1036 | 0.1077 |
| 5-44 | 0.1004 | 0.1042 | 0.1042 | 0.1079 | 0.1060 | 0.1097 | 0.1060 | 0.1097 | 0.1004 | 0.1042 | 0.1023 | 0.1060 | 0.1042 | 0.1079 | 0.1060 | 0.1097 |
| 6-32 | 0.104 | 0.109 | 0.109 | 0.114 | 0.112 | 0.117 | 0.112 | 0.117 | 0.1040 | 0.1091 | 0.1066 | 0.1115 | 0.1091 | 0.1140 | 0.1115 | 0.1164 |
| 6-40 | 0.111 | 0.115 | 0.115 | 0.119 | 0.117 | 0.121 | 0.117 | 0.121 | 0.1110 | 0.1148 | 0.1128 | 0.1167 | 0.1147 | 0.1186 | 0.1166 | 0.1205 |
| 8-32 | 0.130 | 0.134 | 0.134 | 0.139 | 0.137 | 0.141 | 0.137 | 0.141 | 0.1300 | 0.1345 | 0.1324 | 0.1367 | 0.1346 | 0.1389 | 0.1367 | 0.1410 |
| 8-36 | 0.134 | 0.138 | 0.138 | 0.142 | 0.140 | 0.144 | 0.140 | 0.144 | 0.1340 | 0.1377 | 0.1359 | 0.1397 | 0.1378 | 0.1416 | 0.1397 | 0.1435 |
| 10-24 | 0.145 | 0.150 | 0.150 | 0.156 | 0.152 | 0.159 | 0.152 | 0.159 | 0.1450 | 0.1502 | 0.1475 | 0.1528 | 0.1502 | 0.1555 | 0.1528 | 0.1581 |
| 10-32 | 0.156 | 0.160 | 0.160 | 0.164 | 0.162 | 0.166 | 0.162 | 0.166 | 0.1560 | 0.1601 | 0.1581 | 0.1621 | 0.1601 | 0.1641 | 0.1621 | 0.1661 |
| 12-24 | 0.171 | 0.176 | 0.176 | 0.181 | 0.178 | 0.184 | 0.178 | 0.184 | 0.1710 | 0.1758 | 0.1733 | 0.1782 | 0.1758 | 0.1807 | 0.1782 | 0.1831 |
| 12-28 | 0.177 | 0.182 | 0.182 | 0.186 | 0.184 | 0.188 | 0.184 | 0.188 | 0.1770 | 0.1815 | 0.1794 | 0.1836 | 0.1815 | 0.1857 | 0.1836 | 0.1878 |
| 12-32 | 0.182 | 0.186 | 0.186 | 0.190 | 0.188 | 0.192 | 0.188 | 0.192 | 0.1820 | 0.1858 | 0.1837 | 0.1877 | 0.1855 | 0.1895 | 0.1873 | 0.1913 |
| ¼-20 | 0.196 | 0.202 | 0.202 | 0.207 | 0.204 | 0.210 | 0.204 | 0.210 | 0.1960 | 0.2013 | 0.1986 | 0.2040 | 0.2013 | 0.2067 | 0.2040 | 0.2094 |
| ¼-28 | 0.211 | 0.216 | 0.216 | 0.220 | 0.218 | 0.222 | 0.218 | 0.222 | 0.2110 | 0.2152 | 0.2131 | 0.2171 | 0.2150 | 0.2190 | 0.2169 | 0.2209 |
| ¼-32 | 0.216 | 0.220 | 0.220 | 0.224 | 0.222 | 0.226 | 0.222 | 0.226 | 0.2160 | 0.2196 | 0.2172 | 0.2212 | 0.2189 | 0.2229 | 0.2206 | 0.2246 |
| 5/16-18 | 0.252 | 0.259 | 0.259 | 0.265 | 0.262 | 0.268 | 0.262 | 0.268 | 0.2520 | 0.2577 | 0.2551 | 0.2604 | 0.2577 | 0.2630 | 0.2604 | 0.2657 |
| 5/16-24 | 0.267 | 0.272 | 0.272 | 0.277 | 0.275 | 0.280 | 0.275 | 0.280 | 0.2670 | 0.2714 | 0.2694 | 0.2734 | 0.2714 | 0.2754 | 0.2734 | 0.2774 |
| 5/16-32 | 0.279 | 0.283 | 0.283 | 0.286 | 0.285 | 0.289 | 0.285 | 0.289 | 0.2790 | 0.2817 | 0.2824 | 0.2863 | 0.2837 | 0.2877 | 0.2850 | 0.2890 |
| 3/8-16 | 0.307 | 0.314 | 0.314 | 0.321 | 0.318 | 0.325 | 0.318 | 0.325 | 0.3070 | 0.3127 | 0.3101 | 0.3155 | 0.3128 | 0.3182 | 0.3155 | 0.3209 |
| 3/8-24 | 0.330 | 0.338 | 0.335 | 0.340 | 0.338 | 0.343 | 0.338 | 0.343 | 0.3300 | 0.3336 | 0.3314 | 0.3354 | 0.3332 | 0.3372 | 0.3351 | 0.3391 |
| 3/8-32 | 0.341 | 0.345 | 0.345 | 0.349 | 0.347 | 0.351 | 0.347 | 0.351 | 0.3410 | 0.3441 | 0.3415 | 0.3455 | 0.3429 | 0.3469 | 0.3444 | 0.3484 |

| | 0.3514 | 0.3474 | 0.3501 | 0.3461 | 0.3488 | 0.3449 | 0.3488 | 0.3450 | 0.353 | 0.349 | 0.352 | 0.347 | 0.350 | 0.346 | 0.349 | 0.345 |
|---|---|---|---|---|---|---|---|---|---|---|---|---|---|---|---|---|
| 3⁄16-36 | 0.3514 | 0.3474 | 0.3501 | 0.3461 | 0.3488 | 0.3449 | 0.3488 | 0.3450 | 0.353 | 0.349 | 0.352 | 0.347 | 0.350 | 0.346 | 0.349 | 0.345 |
| 7⁄16-20 | 0.3746 | 0.3688 | 0.3717 | 0.3659 | 0.3688 | 0.3630 | 0.3660 | 0.3600 | 0.380 | 0.372 | 0.376 | 0.368 | 0.372 | 0.364 | 0.368 | 0.360 |
| 7⁄16-28 | 0.3937 | 0.3896 | 0.3916 | 0.3875 | 0.3896 | 0.3855 | 0.3875 | 0.3830 | 0.397 | 0.391 | 0.395 | 0.389 | 0.391 | 0.386 | 0.389 | 0.383 |
| 7⁄16-28 | 0.4067 | 0.4017 | 0.4051 | 0.4011 | 0.4035 | 0.3995 | 0.4020 | 0.3990 | 0.410 | 0.406 | 0.407 | 0.403 | 0.406 | 0.401 | 0.403 | 0.399 |
| 1⁄2-13 | 0.4313 | 0.4255 | 0.4284 | 0.4226 | 0.4254 | 0.4196 | 0.4225 | 0.4170 | 0.438 | 0.430 | 0.434 | 0.426 | 0.430 | 0.421 | 0.426 | 0.417 |
| 1⁄2-12 | 0.4255 | 0.4192 | 0.4223 | 0.4160 | 0.4192 | 0.4129 | 0.4161 | 0.4100 | 0.433 | 0.428 | 0.428 | 0.414 | 0.424 | 0.414 | 0.414 | 0.410 |
| 1⁄2-20 | 0.4556 | 0.4516 | 0.4537 | 0.4497 | 0.4517 | 0.4477 | 0.4498 | 0.4460 | 0.460 | 0.454 | 0.457 | 0.452 | 0.454 | 0.449 | 0.452 | 0.446 |
| 1⁄2-28 | 0.4692 | 0.4652 | 0.4676 | 0.4636 | 0.4660 | 0.4620 | 0.4645 | 0.4610 | 0.472 | 0.468 | 0.470 | 0.466 | 0.468 | 0.463 | 0.467 | 0.461 |
| 9⁄16-12 | 0.4873 | 0.4813 | 0.4843 | 0.4783 | 0.4813 | 0.4753 | 0.4783 | 0.4720 | 0.495 | 0.486 | 0.490 | 0.476 | 0.486 | 0.476 | 0.476 | 0.472 |
| 9⁄16-18 | 0.5127 | 0.5086 | 0.5106 | 0.5065 | 0.5086 | 0.5045 | 0.5065 | 0.5020 | 0.518 | 0.512 | 0.515 | 0.509 | 0.512 | 0.505 | 0.509 | 0.502 |
| 9⁄16-24 | 0.5261 | 0.5221 | 0.5244 | 0.5204 | 0.5226 | 0.5186 | 0.5209 | 0.5170 | 0.530 | 0.525 | 0.527 | 0.522 | 0.525 | 0.520 | 0.522 | 0.517 |
| 9⁄16-28 | 0.5317 | 0.5277 | 0.5301 | 0.5261 | 0.5285 | 0.5245 | 0.5270 | 0.5240 | 0.535 | 0.531 | 0.532 | 0.528 | 0.531 | 0.526 | 0.528 | 0.524 |
| 5⁄8-11 | 0.5422 | 0.5360 | 0.5391 | 0.5329 | 0.5360 | 0.5298 | 0.5328 | 0.5270 | 0.546 | 0.532 | 0.536 | 0.527 | 0.532 | 0.527 | 0.527 | 0.527 |
| 5⁄8-12 | 0.5492 | 0.5434 | 0.5403 | 0.5405 | 0.5435 | 0.5377 | 0.5406 | 0.5350 | 0.553 | 0.544 | 0.549 | 0.541 | 0.544 | 0.540 | 0.544 | 0.535 |
| 5⁄8-18 | 0.5752 | 0.5711 | 0.5730 | 0.5690 | 0.5711 | 0.5670 | 0.5690 | 0.5650 | 0.581 | 0.575 | 0.578 | 0.572 | 0.575 | 0.568 | 0.572 | 0.565 |
| 5⁄8-24 | 0.5886 | 0.5846 | 0.5869 | 0.5829 | 0.5851 | 0.5811 | 0.5834 | 0.5800 | 0.593 | 0.588 | 0.590 | 0.585 | 0.588 | 0.583 | 0.585 | 0.580 |
| 5⁄8-28 | 0.5942 | 0.5902 | 0.5926 | 0.5886 | 0.5910 | 0.5870 | 0.5895 | 0.5860 | 0.597 | 0.593 | 0.595 | 0.591 | 0.593 | 0.588 | 0.591 | 0.586 |
| 11⁄16-12 | 0.6113 | 0.6057 | 0.6085 | 0.6029 | 0.6057 | 0.6001 | 0.6029 | 0.5970 | 0.620 | 0.611 | 0.615 | 0.606 | 0.611 | 0.602 | 0.606 | 0.597 |
| 11⁄16-24 | 0.6511 | 0.6471 | 0.6494 | 0.6454 | 0.6476 | 0.6436 | 0.6459 | 0.6420 | 0.655 | 0.647 | 0.652 | 0.647 | 0.655 | 0.645 | 0.647 | 0.642 |
| 3⁄4-10 | 0.6577 | 0.6513 | 0.6545 | 0.6481 | 0.6513 | 0.6449 | 0.6481 | 0.6420 | 0.668 | 0.658 | 0.658 | 0.653 | 0.668 | 0.647 | 0.658 | 0.642 |
| 3⁄4-12 | 0.6734 | 0.6680 | 0.6707 | 0.6653 | 0.6680 | 0.6626 | 0.6652 | 0.6600 | 0.683 | 0.674 | 0.678 | 0.669 | 0.674 | 0.665 | 0.669 | 0.660 |
| 3⁄4-16 | 0.6929 | 0.6886 | 0.6908 | 0.6865 | 0.6887 | 0.6844 | 0.6866 | 0.6820 | 0.700 | 0.693 | 0.696 | 0.689 | 0.693 | 0.686 | 0.689 | 0.682 |
| 3⁄4-20 | 0.7056 | 0.7016 | 0.7037 | 0.6997 | 0.7017 | 0.6977 | 0.6998 | 0.6960 | 0.710 | 0.704 | 0.707 | 0.702 | 0.704 | 0.699 | 0.702 | 0.696 |
| 3⁄4-28 | 0.7192 | 0.7152 | 0.7176 | 0.7136 | 0.7160 | 0.7120 | 0.7145 | 0.7110 | 0.722 | 0.718 | 0.720 | 0.716 | 0.718 | 0.713 | 0.716 | 0.711 |
| 13⁄16-12 | 0.7356 | 0.7303 | 0.7329 | 0.7276 | 0.7303 | 0.7250 | 0.7276 | 0.7220 | 0.745 | 0.736 | 0.740 | 0.731 | 0.736 | 0.727 | 0.731 | 0.722 |
| 13⁄16-16 | 0.7554 | 0.7511 | 0.7533 | 0.7490 | 0.7512 | 0.7469 | 0.7491 | 0.7450 | 0.763 | 0.756 | 0.759 | 0.752 | 0.756 | 0.749 | 0.752 | 0.745 |
| 13⁄16-20 | 0.7681 | 0.7641 | 0.7662 | 0.7622 | 0.7642 | 0.7602 | 0.7623 | 0.7580 | 0.772 | 0.766 | 0.770 | 0.764 | 0.766 | 0.761 | 0.764 | 0.758 |
| 7⁄8-9 | 0.7714 | 0.7647 | 0.7681 | 0.7614 | 0.7614 | 0.7580 | 0.7614 | 0.7550 | 0.785 | 0.773 | 0.778 | 0.767 | 0.773 | 0.761 | 0.767 | 0.755 |
| 7⁄8-12 | 0.7978 | 0.7926 | 0.7952 | 0.7900 | 0.7926 | 0.7874 | 0.7900 | 0.7850 | 0.808 | 0.799 | 0.803 | 0.794 | 0.799 | 0.790 | 0.794 | 0.785 |
| 7⁄8-14 | 0.8090 | 0.8045 | 0.8068 | 0.8023 | 0.8045 | 0.8000 | 0.8022 | 0.7980 | 0.818 | 0.810 | 0.814 | 0.806 | 0.810 | 0.802 | 0.806 | 0.798 |
| 7⁄8-16 | 0.8179 | 0.8136 | 0.8158 | 0.8115 | 0.8137 | 0.8094 | 0.8116 | 0.8070 | 0.825 | 0.818 | 0.821 | 0.814 | 0.818 | 0.811 | 0.814 | 0.807 |
| 7⁄8-20 | 0.8306 | 0.8266 | 0.8287 | 0.8247 | 0.8267 | 0.8227 | 0.8248 | 0.8210 | 0.835 | 0.829 | 0.832 | 0.827 | 0.829 | 0.824 | 0.827 | 0.821 |
| 7⁄8-28 | 0.8442 | 0.8402 | 0.8426 | 0.8386 | 0.8410 | 0.8370 | 0.8395 | 0.8360 | 0.847 | 0.843 | 0.845 | 0.840 | 0.843 | 0.838 | 0.840 | 0.836 |
| 15⁄16-12 | 0.8601 | 0.8550 | 0.8575 | 0.8524 | 0.8550 | 0.8499 | 0.8524 | 0.8470 | 0.870 | 0.861 | 0.865 | 0.856 | 0.861 | 0.852 | 0.856 | 0.847 |
| 15⁄16-16 | 0.8804 | 0.8761 | 0.8783 | 0.8740 | 0.8762 | 0.8719 | 0.8741 | 0.8700 | 0.888 | 0.881 | 0.884 | 0.877 | 0.881 | 0.874 | 0.877 | 0.870 |
| 15⁄16-20 | 0.8931 | 0.8891 | 0.8912 | 0.8872 | 0.8892 | 0.8852 | 0.8873 | 0.8830 | 0.897 | 0.891 | 0.895 | 0.889 | 0.891 | 0.886 | 0.889 | 0.883 |
| 1-8 | 0.8835 | 0.8760 | 0.8797 | 0.8722 | 0.8759 | 0.8684 | 0.8722 | 0.8650 | 0.896 | 0.884 | 0.890 | 0.878 | 0.884 | 0.865 | 0.878 | 0.865 |
| 1-12 | 0.9223 | 0.9173 | 0.9198 | 0.9148 | 0.9173 | 0.9123 | 0.9148 | 0.9100 | 0.933 | 0.924 | 0.928 | 0.919 | 0.924 | 0.915 | 0.919 | 0.910 |
| 1-14 | 0.9337 | 0.9293 | 0.9315 | 0.9271 | 0.9293 | 0.9249 | 0.9271 | 0.9230 | 0.942 | 0.934 | 0.938 | 0.931 | 0.934 | 0.927 | 0.931 | 0.923 |
| 1-16 | 0.9429 | 0.9386 | 0.9408 | 0.9365 | 0.9387 | 0.9344 | 0.9366 | 0.9330 | 0.950 | 0.943 | 0.946 | 0.939 | 0.943 | 0.936 | 0.939 | 0.932 |
| 1-20 | 0.9556 | 0.9537 | 0.9537 | 0.9497 | 0.9537 | 0.9477 | 0.9498 | 0.9460 | 0.960 | 0.954 | 0.957 | 0.952 | 0.954 | 0.949 | 0.952 | 0.946 |
| 1-28 | 0.9692 | 0.9652 | 0.9676 | 0.9636 | 0.9660 | 0.9620 | 0.9645 | 0.9610 | 0.972 | 0.968 | 0.970 | 0.966 | 0.968 | 0.963 | 0.966 | 0.961 |

*† See footnotes at end of table.

Table 2 (Continued). Recommended Hole Size Limits Before Tapping Unified Threads

Length of Engagement (D = Nominal Size of Thread)

Recommended Hole Size Limits

| Thread Size | **Classes 1B and 2B** To and Including 1/3D — Min* | Max | Above 1/3D to 2/3D — Min | Max† | Above 2/3D to 1 1/2D — Min | Max | **Class 3B** To and Including 1/3D — Min* | Max | Above 1/3D to 2/3D — Min | Max | Above 2/3D to 1 1/2D — Min | Max† | Above 1 1/2D to 3D — Min | Max |
|---|---|---|---|---|---|---|---|---|---|---|---|---|---|---|
| 1 1/16-12 | 0.972 | 0.981 | 0.981 | 0.990 | 0.986 | 0.995 | 0.9720 | 0.9773 | 0.9748 | 0.9798 | 0.9773 | 0.9823 | 0.9798 | 0.9848 |
| 1 1/16-16 | 0.995 | 1.002 | 1.002 | 1.009 | 1.005 | 1.013 | 0.9950 | 0.9991 | 0.9969 | 1.0012 | 0.9990 | 1.0033 | 1.0011 | 1.0054 |
| 1 1/16-18 | 1.002 | 1.009 | 1.009 | 1.015 | 1.012 | 1.018 | 1.0020 | 1.0065 | 1.0044 | 1.0085 | 1.0064 | 1.0105 | 1.0064 | 1.0126 |
| 1 1/8-7 | 0.970 | 0.984 | 0.984 | 0.998 | 0.991 | 1.005 | 0.9700 | 0.9790 | 0.9747 | 0.9833 | 0.9789 | 0.9875 | 0.9832 | 0.9918 |
| 1 1/8-8 | 0.990 | 1.003 | 1.003 | 1.015 | 1.009 | 1.021 | 0.9900 | 0.9972 | 0.9934 | 1.0009 | 0.9972 | 1.0047 | 1.0010 | 1.0085 |
| 1 1/8-12 | 1.035 | 1.044 | 1.044 | 1.053 | 1.049 | 1.058 | 1.0350 | 1.0398 | 1.0373 | 1.0423 | 1.0398 | 1.0448 | 1.0423 | 1.0473 |
| 1 1/8-16 | 1.057 | 1.064 | 1.061 | 1.071 | 1.068 | 1.075 | 1.0570 | 1.0616 | 1.0594 | 1.0637 | 1.0615 | 1.0658 | 1.0636 | 1.0679 |
| 1 1/8-18 | 1.065 | 1.072 | 1.072 | 1.078 | 1.075 | 1.081 | 1.0650 | 1.0690 | 1.0669 | 1.0710 | 1.0689 | 1.0730 | 1.0710 | 1.0751 |
| 1 1/8-20 | 1.071 | 1.077 | 1.077 | 1.082 | 1.079 | 1.085 | 1.0710 | 1.0748 | 1.0727 | 1.0767 | 1.0747 | 1.0787 | 1.0766 | 1.0806 |
| 1 1/8-28 | 1.086 | 1.091 | 1.091 | 1.095 | 1.093 | 1.097 | 1.0860 | 1.0895 | 1.0870 | 1.0910 | 1.0886 | 1.0926 | 1.0902 | 1.0942 |
| 1 3/16-12 | 1.097 | 1.106 | 1.106 | 1.115 | 1.111 | 1.120 | 1.0970 | 1.1023 | 1.0998 | 1.1048 | 1.1023 | 1.1073 | 1.1048 | 1.1098 |
| 1 3/16-16 | 1.120 | 1.127 | 1.124 | 1.131 | 1.131 | 1.138 | 1.1200 | 1.1241 | 1.1219 | 1.1262 | 1.1240 | 1.1283 | 1.1261 | 1.1304 |
| 1 3/16-18 | 1.127 | 1.134 | 1.134 | 1.140 | 1.137 | 1.143 | 1.1270 | 1.1315 | 1.1294 | 1.1335 | 1.1314 | 1.1355 | 1.1335 | 1.1376 |
| 1 1/4-7 | 1.095 | 1.109 | 1.109 | 1.123 | 1.116 | 1.130 | 1.0950 | 1.1040 | 1.0997 | 1.1083 | 1.1039 | 1.1125 | 1.1082 | 1.1168 |
| 1 1/4-8 | 1.115 | 1.128 | 1.128 | 1.140 | 1.134 | 1.146 | 1.1150 | 1.1222 | 1.1184 | 1.1259 | 1.1222 | 1.1297 | 1.1260 | 1.1335 |
| 1 1/4-12 | 1.160 | 1.169 | 1.169 | 1.178 | 1.174 | 1.183 | 1.1600 | 1.1648 | 1.1623 | 1.1673 | 1.1648 | 1.1698 | 1.1673 | 1.1723 |
| 1 1/4-16 | 1.182 | 1.189 | 1.189 | 1.196 | 1.193 | 1.200 | 1.1820 | 1.1866 | 1.1844 | 1.1887 | 1.1865 | 1.1908 | 1.1886 | 1.1929 |
| 1 1/4-18 | 1.190 | 1.197 | 1.197 | 1.203 | 1.200 | 1.206 | 1.1900 | 1.1940 | 1.1919 | 1.1960 | 1.1939 | 1.1980 | 1.1960 | 1.2001 |
| 1 1/4-20 | 1.196 | 1.202 | 1.202 | 1.207 | 1.204 | 1.210 | 1.1960 | 1.1998 | 1.1977 | 1.2017 | 1.1997 | 1.2037 | 1.2016 | 1.2056 |
| 1 5/16-12 | 1.222 | 1.231 | 1.231 | 1.240 | 1.236 | 1.245 | 1.2220 | 1.2273 | 1.2248 | 1.2298 | 1.2273 | 1.2323 | 1.2298 | 1.2348 |
| 1 5/16-16 | 1.245 | 1.252 | 1.249 | 1.256 | 1.256 | 1.263 | 1.2450 | 1.2491 | 1.2469 | 1.2512 | 1.2490 | 1.2533 | 1.2511 | 1.2554 |
| 1 5/16-18 | 1.252 | 1.259 | 1.259 | 1.265 | 1.262 | 1.268 | 1.2520 | 1.2565 | 1.2544 | 1.2585 | 1.2564 | 1.2605 | 1.2585 | 1.2626 |
| 1 3/8-8 | 1.195 | 1.210 | 1.210 | 1.225 | 1.221 | 1.239 | 1.1950 | 1.2046 | 1.1996 | 1.2096 | 1.2046 | 1.2146 | 1.2096 | 1.2196 |
| 1 3/8-12 | 1.240 | 1.253 | 1.253 | 1.265 | 1.259 | 1.271 | 1.2400 | 1.2472 | 1.2434 | 1.2509 | 1.2472 | 1.2547 | 1.2510 | 1.2585 |
| 1 3/8-16 | 1.285 | 1.294 | 1.294 | 1.303 | 1.299 | 1.308 | 1.2850 | 1.2898 | 1.2873 | 1.2923 | 1.2898 | 1.2948 | 1.2923 | 1.2973 |
| 1 3/8-18 | 1.307 | 1.314 | 1.314 | 1.321 | 1.318 | 1.325 | 1.3070 | 1.3116 | 1.3094 | 1.3137 | 1.3115 | 1.3158 | 1.3136 | 1.3179 |
| 1 7/16-12 | 1.315 | 1.322 | 1.322 | 1.328 | 1.325 | 1.331 | 1.3150 | 1.3190 | 1.3169 | 1.3210 | 1.3189 | 1.3230 | 1.3210 | 1.3251 |
| 1 7/16-16 | 1.347 | 1.354 | 1.354 | 1.365 | 1.361 | 1.370 | 1.3470 | 1.3523 | 1.3498 | 1.3548 | 1.3523 | 1.3573 | 1.3548 | 1.3598 |
| 1 7/16-18 | 1.370 | 1.377 | 1.377 | 1.384 | 1.381 | 1.388 | 1.3700 | 1.3741 | 1.3719 | 1.3762 | 1.3740 | 1.3783 | 1.3761 | 1.3804 |
| 1 1/2-6 | 1.377 | 1.384 | 1.384 | 1.390 | 1.387 | 1.393 | 1.3770 | 1.3815 | 1.3794 | 1.3835 | 1.3814 | 1.3855 | 1.3835 | 1.3876 |
| 1 1/2-8 | 1.320 | 1.335 | 1.335 | 1.350 | 1.346 | 1.364 | 1.3200 | 1.3296 | 1.3246 | 1.3346 | 1.3296 | 1.3396 | 1.3346 | 1.3446 |
| 1 1/2-12 | 1.365 | 1.378 | 1.378 | 1.390 | 1.384 | 1.396 | 1.3650 | 1.3722 | 1.3684 | 1.3759 | 1.3722 | 1.3797 | 1.3750 | 1.3835 |

| | | | | | | | | | | | | | | | | |
|---|---|---|---|---|---|---|---|---|---|---|---|---|---|---|---|---|
| 1½-12 | 1.410 | 1.419 | 1.415 | 1.424 | 1.419 | 1.428 | 1.424 | 1.433 | 1.4100 | 1.4148 | 1.4123 | 1.4173 | 1.4148 | 1.4198 | 1.4173 | 1.4223 |
| 1½-16 | 1.432 | 1.439 | 1.436 | 1.443 | 1.439 | 1.446 | 1.443 | 1.450 | 1.4320 | 1.4366 | 1.4344 | 1.4387 | 1.4365 | 1.4408 | 1.4386 | 1.4429 |
| 1½-18 | 1.440 | 1.446 | 1.443 | 1.450 | 1.446 | 1.452 | 1.450 | 1.456 | 1.4400 | 1.4440 | 1.4419 | 1.4460 | 1.4439 | 1.4480 | 1.4460 | 1.4501 |
| 1½-20 | 1.446 | 1.452 | 1.449 | 1.454 | 1.452 | 1.457 | 1.454 | 1.460 | 1.4460 | 1.4498 | 1.4477 | 1.4517 | 1.4497 | 1.4537 | 1.4516 | 1.4556 |
| 1 9/16-16 | 1.495 | 1.502 | 1.499 | 1.506 | 1.502 | 1.509 | 1.506 | 1.513 | 1.4950 | 1.4991 | 1.4969 | 1.5012 | 1.4990 | 1.5033 | 1.5011 | 1.5054 |
| 1 9/16-18 | 1.502 | 1.509 | 1.505 | 1.512 | 1.509 | 1.515 | 1.512 | 1.518 | 1.5020 | 1.5065 | 1.5044 | 1.5085 | 1.5064 | 1.5105 | 1.5085 | 1.5126 |
| 1⅝-8 | 1.490 | 1.498 | 1.494 | 1.509 | 1.498 | 1.515 | 1.509 | 1.521 | 1.4900 | 1.4972 | 1.4934 | 1.5009 | 1.4972 | 1.5047 | 1.5010 | 1.5085 |
| 1⅝-12 | 1.535 | 1.544 | 1.540 | 1.549 | 1.544 | 1.553 | 1.549 | 1.558 | 1.5350 | 1.5398 | 1.5373 | 1.5423 | 1.5398 | 1.5448 | 1.5423 | 1.5473 |
| 1⅝-16 | 1.557 | 1.564 | 1.561 | 1.568 | 1.564 | 1.571 | 1.568 | 1.575 | 1.5570 | 1.5616 | 1.5594 | 1.5637 | 1.5615 | 1.5658 | 1.5636 | 1.5679 |
| 1⅝-18 | 1.565 | 1.572 | 1.568 | 1.575 | 1.572 | 1.578 | 1.575 | 1.581 | 1.5650 | 1.5690 | 1.5669 | 1.5710 | 1.5689 | 1.5730 | 1.5710 | 1.5751 |
| 1 11/16-16 | 1.620 | 1.627 | 1.624 | 1.631 | 1.627 | 1.634 | 1.631 | 1.638 | 1.6200 | 1.6241 | 1.6219 | 1.6262 | 1.6240 | 1.6283 | 1.6261 | 1.6304 |
| 1 11/16-18 | 1.627 | 1.634 | 1.630 | 1.637 | 1.634 | 1.640 | 1.637 | 1.643 | 1.6270 | 1.6315 | 1.6294 | 1.6335 | 1.6314 | 1.6355 | 1.6335 | 1.6376 |
| 1¾-5 | 1.534 | 1.551 | 1.543 | 1.560 | 1.551 | 1.568 | 1.560 | 1.577 | 1.5340 | 1.5455 | 1.5395 | 1.5515 | 1.5455 | 1.5575 | 1.5515 | 1.5635 |
| 1¾-8 | 1.615 | 1.628 | 1.621 | 1.634 | 1.628 | 1.640 | 1.634 | 1.646 | 1.6150 | 1.6222 | 1.6184 | 1.6259 | 1.6222 | 1.6297 | 1.6260 | 1.6335 |
| 1¾-12 | 1.660 | 1.669 | 1.665 | 1.674 | 1.669 | 1.678 | 1.674 | 1.683 | 1.6600 | 1.6648 | 1.6623 | 1.6673 | 1.6648 | 1.6698 | 1.6673 | 1.6723 |
| 1¾-16 | 1.682 | 1.689 | 1.686 | 1.693 | 1.689 | 1.696 | 1.693 | 1.700 | 1.6820 | 1.6866 | 1.6844 | 1.6887 | 1.6865 | 1.6908 | 1.6886 | 1.6929 |
| 1¾-20 | 1.696 | 1.702 | 1.699 | 1.704 | 1.702 | 1.707 | 1.704 | 1.710 | 1.6960 | 1.6998 | 1.6977 | 1.7017 | 1.6997 | 1.7037 | 1.7016 | 1.7056 |
| 1 13/16-16 | 1.745 | 1.752 | 1.749 | 1.756 | 1.752 | 1.759 | 1.756 | 1.763 | 1.7450 | 1.7491 | 1.7469 | 1.7512 | 1.7490 | 1.7533 | 1.7511 | 1.7554 |
| 1⅞-8 | 1.740 | 1.752 | 1.746 | 1.759 | 1.752 | 1.765 | 1.759 | 1.771 | 1.7400 | 1.7472 | 1.7434 | 1.7509 | 1.7472 | 1.7547 | 1.7510 | 1.7585 |
| 1⅞-12 | 1.785 | 1.794 | 1.790 | 1.799 | 1.794 | 1.803 | 1.799 | 1.808 | 1.7850 | 1.7898 | 1.7873 | 1.7923 | 1.7898 | 1.7948 | 1.7923 | 1.7973 |
| 1⅞-16 | 1.807 | 1.814 | 1.810 | 1.818 | 1.814 | 1.821 | 1.818 | 1.825 | 1.8070 | 1.8116 | 1.8094 | 1.8137 | 1.8115 | 1.8158 | 1.8136 | 1.1879 |
| 1 15/16-16 | 1.870 | 1.877 | 1.874 | 1.881 | 1.877 | 1.884 | 1.881 | 1.888 | 1.8700 | 1.8741 | 1.8719 | 1.8762 | 1.8740 | 1.8783 | 1.8761 | 1.8804 |
| 2-4½ | 1.759 | 1.777 | 1.768 | 1.786 | 1.777 | 1.795 | 1.786 | 1.804 | 1.7590 | 1.7727 | 1.7661 | 1.7794 | 1.7728 | 1.7861 | 1.7794 | 1.7927 |
| 2-8 | 1.865 | 1.878 | 1.871 | 1.884 | 1.878 | 1.890 | 1.884 | 1.896 | 1.8650 | 1.8722 | 1.8684 | 1.8759 | 1.8722 | 1.8797 | 1.8760 | 1.8835 |
| 2-12 | 1.910 | 1.919 | 1.915 | 1.924 | 1.919 | 1.928 | 1.924 | 1.933 | 1.9100 | 1.9148 | 1.9123 | 1.9173 | 1.9148 | 1.9198 | 1.9173 | 1.9223 |
| 2-16 | 1.932 | 1.939 | 1.936 | 1.943 | 1.939 | 1.946 | 1.943 | 1.950 | 1.9320 | 1.9366 | 1.9344 | 1.9387 | 1.9365 | 1.9408 | 1.9386 | 1.9429 |
| 2-20 | 1.946 | 1.952 | 1.949 | 1.954 | 1.952 | 1.957 | 1.954 | 1.960 | 1.9460 | 1.9498 | 1.9477 | 1.9517 | 1.9497 | 1.9537 | 1.9516 | 1.9556 |
| 2 1/16-16 | 1.995 | 2.002 | 2.000 | 2.006 | 2.002 | 2.009 | 2.006 | 2.012 | 1.9950 | 1.9991 | 1.9969 | 2.0012 | 1.9990 | 2.0033 | 2.0011 | 2.0054 |
| 2⅛-8 | 1.990 | 2.003 | 1.996 | 2.009 | 2.003 | 2.015 | 2.009 | 2.021 | 1.9900 | 1.9972 | 1.9934 | 2.0009 | 1.9972 | 2.0047 | 2.0010 | 2.0085 |
| 2⅛-12 | 2.035 | 2.044 | 2.040 | 2.049 | 2.044 | 2.053 | 2.049 | 2.058 | 2.0350 | 2.0398 | 2.0373 | 2.0423 | 2.0398 | 2.0448 | 2.0423 | 2.0473 |
| 2⅛-16 | 2.057 | 2.064 | 2.061 | 2.068 | 2.064 | 2.071 | 2.068 | 2.075 | 2.0570 | 2.0616 | 2.0594 | 2.0637 | 2.0615 | 2.0658 | 2.0636 | 2.0679 |
| 2 3/16-16 | 2.120 | 2.127 | 2.124 | 2.131 | 2.127 | 2.134 | 2.131 | 2.138 | 2.1200 | 2.1241 | 2.1219 | 2.1262 | 2.1240 | 2.1283 | 2.1261 | 2.1304 |
| 2¼-4½ | 2.009 | 2.027 | 2.018 | 2.036 | 2.027 | 2.045 | 2.036 | 2.054 | 2.0090 | 2.0227 | 2.0161 | 2.0294 | 2.0228 | 2.0361 | 2.0294 | 2.0427 |
| 2¼-8 | 2.115 | 2.128 | 2.121 | 2.134 | 2.128 | 2.140 | 2.134 | 2.146 | 2.1150 | 2.1222 | 2.1184 | 2.1259 | 2.1222 | 2.1297 | 2.1260 | 2.1335 |
| 2¼-12 | 2.160 | 2.169 | 2.165 | 2.174 | 2.169 | 2.178 | 2.174 | 2.182 | 2.1600 | 2.1648 | 2.1623 | 2.1673 | 2.1648 | 2.1698 | 2.1673 | 2.1723 |
| 2¼-16 | 2.182 | 2.189 | 2.186 | 2.193 | 2.189 | 2.196 | 2.193 | 2.200 | 2.1820 | 2.1866 | 2.1844 | 2.1887 | 2.1865 | 2.1908 | 2.1886 | 2.1929 |
| 2¼-20 | 2.196 | 2.202 | 2.199 | 2.204 | 2.202 | 2.207 | 2.204 | 2.210 | 2.1960 | 2.1998 | 2.1977 | 2.2017 | 2.1997 | 2.2037 | 2.2016 | 2.2056 |
| 2 5/16-16 | 2.245 | 2.252 | 2.249 | 2.256 | 2.252 | 2.259 | 2.256 | 2.263 | 2.2450 | 2.2491 | 2.2469 | 2.2512 | 2.2490 | 2.2533 | 2.2511 | 2.2554 |
| 2⅜-12 | 2.285 | 2.294 | 2.290 | 2.299 | 2.294 | 2.303 | 2.299 | 2.308 | 2.2850 | 2.2898 | 2.2873 | 2.2923 | 2.2898 | 2.2948 | 2.2923 | 2.2973 |
| 2⅜-16 | 2.307 | 2.314 | 2.311 | 2.318 | 2.314 | 2.321 | 2.318 | 2.325 | 2.3070 | 2.3116 | 2.3094 | 2.3137 | 2.3115 | 2.3158 | 2.3136 | 2.3179 |
| 2 7/16-16 | 2.370 | 2.377 | 2.374 | 2.381 | 2.377 | 2.384 | 2.381 | 2.388 | 2.3700 | 2.3741 | 2.3719 | 2.3762 | 2.3740 | 2.3783 | 2.3761 | 2.3804 |
| 2½-4 | 2.229 | 2.248 | 2.238 | 2.258 | 2.248 | 2.267 | 2.258 | 2.277 | 2.2290 | 2.2444 | 2.2369 | 2.2519 | 2.2444 | 2.2594 | 2.2519 | 2.2669 |

*† See footnotes at end of table.

Table 2 (Continued).  Recommended Hole Size Limits Before Tapping Unified Threads

Length of Engagement (D = Nominal Size of Thread)

Recommended Hole Size Limits

| Thread Size | Classes 1B and 2B | | | | | | | | Class 3B | | | | | | | |
|---|---|---|---|---|---|---|---|---|---|---|---|---|---|---|---|---|
| | To and Including ⅓D | | Above ⅓D to ⅔D | | Above ⅔D to 1½D | | Above 1½D to 3D | | To and Including ⅓D | | Above ⅓D to ⅔D | | Above ⅔D to 1½D | | Above 1½D to 3D | |
| | Min* | Max | Min | Max | Min | Maxt | Min | Max | Min* | Max | Min | Max | Min | Maxt | Min | Max |
| 2½-8 | 2.365 | 2.378 | 2.371 | 2.384 | 2.378 | 2.390 | 2.384 | 2.396 | 2.3650 | 2.3722 | 2.3684 | 2.3759 | 2.3722 | 2.3797 | 2.3760 | 2.3835 |
| 2½-12 | 2.410 | 2.419 | 2.415 | 2.424 | 2.419 | 2.428 | 2.424 | 2.433 | 2.4100 | 2.4148 | 2.4123 | 2.4173 | 2.4148 | 2.4198 | 2.4173 | 2.4223 |
| 2½-16 | 2.432 | 2.439 | 2.436 | 2.443 | 2.439 | 2.446 | 2.443 | 2.450 | 2.4320 | 2.4366 | 2.4344 | 2.4387 | 2.4365 | 2.4408 | 2.4386 | 2.4429 |
| 2½-20 | 2.446 | 2.452 | 2.449 | 2.454 | 2.452 | 2.457 | 2.454 | 2.460 | 2.4460 | 2.4498 | 2.4478 | 2.4517 | 2.4497 | 2.4537 | 2.4516 | 2.4556 |
| 2⅝-12 | 2.535 | 2.544 | 2.540 | 2.549 | 2.544 | 2.553 | 2.549 | 2.558 | 2.5350 | 2.5398 | 2.5373 | 2.5423 | 2.5398 | 2.5448 | 2.5423 | 2.5473 |
| 2⅝-16 | 2.557 | 2.564 | 2.561 | 2.568 | 2.564 | 2.571 | 2.568 | 2.575 | 2.5570 | 2.5616 | 2.5594 | 2.5637 | 2.5615 | 2.5658 | 2.5636 | 2.5679 |
| 2¾-4 | 2.479 | 2.498 | 2.489 | 2.508 | 2.498 | 2.517 | 2.508 | 2.527 | 2.4790 | 2.4944 | 2.4869 | 2.5019 | 2.4944 | 2.5094 | 2.5019 | 2.5169 |
| 2¾-8 | 2.615 | 2.628 | 2.621 | 2.634 | 2.628 | 2.640 | 2.634 | 2.646 | 2.6150 | 2.6222 | 2.6184 | 2.6259 | 2.6222 | 2.6297 | 2.6260 | 2.6335 |
| 2¾-12 | 2.660 | 2.669 | 2.665 | 2.674 | 2.669 | 2.678 | 2.674 | 2.683 | 2.6600 | 2.6648 | 2.6623 | 2.6673 | 2.6648 | 2.6698 | 2.6673 | 2.6723 |
| 2¾-16 | 2.682 | 2.689 | 2.686 | 2.693 | 2.689 | 2.696 | 2.693 | 2.700 | 2.6820 | 2.6866 | 2.6844 | 2.6887 | 2.6865 | 2.6908 | 2.6886 | 2.6929 |
| 2⅞-12 | 2.785 | 2.794 | 2.790 | 2.799 | 2.794 | 2.803 | 2.799 | 2.808 | 2.7850 | 2.7898 | 2.7873 | 2.7923 | 2.7898 | 2.7948 | 2.7923 | 2.7973 |
| 2⅞-16 | 2.807 | 2.814 | 2.811 | 2.818 | 2.814 | 2.821 | 2.818 | 2.825 | 2.8070 | 2.8116 | 2.8094 | 2.8137 | 2.8115 | 2.8158 | 2.8136 | 2.8179 |
| 3-4 | 2.729 | 2.748 | 2.739 | 2.758 | 2.748 | 2.767 | 2.758 | 2.777 | 2.7290 | 2.7444 | 2.7369 | 2.7519 | 2.7444 | 2.7594 | 2.7519 | 2.7669 |
| 3-8 | 2.865 | 2.878 | 2.871 | 2.884 | 2.878 | 2.890 | 2.884 | 2.896 | 2.8650 | 2.8722 | 2.8684 | 2.8759 | 2.8722 | 2.8797 | 2.8760 | 2.8835 |
| 3-12 | 2.910 | 2.919 | 2.915 | 2.924 | 2.919 | 2.928 | 2.924 | 2.933 | 2.9100 | 2.9148 | 2.9123 | 2.9173 | 2.9148 | 2.9198 | 2.9173 | 2.9223 |
| 3-16 | 2.932 | 2.939 | 2.936 | 2.943 | 2.939 | 2.946 | 2.943 | 2.950 | 2.9320 | 2.9366 | 2.9344 | 2.9387 | 2.9365 | 2.9408 | 2.9386 | 2.9429 |
| 3⅛-12 | 3.035 | 3.044 | 3.040 | 3.049 | 3.044 | 3.053 | 3.049 | 3.058 | 3.0350 | 3.0398 | 3.0373 | 3.0423 | 3.0398 | 3.0448 | 3.0423 | 3.0473 |
| 3⅛-16 | 3.057 | 3.064 | 3.061 | 3.068 | 3.064 | 3.071 | 3.068 | 3.075 | 3.0570 | 3.0616 | 3.0594 | 3.0637 | 3.0615 | 3.0658 | 3.0636 | 3.0679 |
| 3¼-4 | 2.979 | 2.998 | 2.989 | 3.008 | 2.998 | 3.017 | 3.008 | 3.027 | 2.9790 | 2.9944 | 2.9869 | 3.0019 | 2.9944 | 3.0094 | 3.0019 | 3.0169 |
| 3¼-8 | 3.115 | 3.128 | 3.121 | 3.134 | 3.128 | 3.140 | 3.134 | 3.146 | 3.1150 | 3.1222 | 3.1184 | 3.1259 | 3.1222 | 3.1297 | 3.1260 | 3.1335 |
| 3¼-12 | 3.160 | 3.169 | 3.165 | 3.174 | 3.169 | 3.178 | 3.174 | 3.183 | 3.1600 | 3.1648 | 3.1623 | 3.1673 | 3.1648 | 3.1698 | 3.1673 | 3.1723 |
| 3¼-16 | 3.182 | 3.189 | 3.186 | 3.193 | 3.189 | 3.196 | 3.193 | 3.200 | 3.1820 | 3.1866 | 3.1844 | 3.1887 | 3.1865 | 3.1908 | 3.1886 | 3.1929 |
| 3⅜-12 | 3.285 | 3.294 | 3.290 | 3.299 | 3.294 | 3.303 | 3.299 | 3.308 | 3.2850 | 3.2898 | 3.2873 | 3.2923 | 3.2898 | 3.2948 | 3.2923 | 3.2973 |
| 3⅜-16 | 3.307 | 3.314 | 3.311 | 3.317 | 3.314 | 3.321 | 3.317 | 3.325 | 3.3070 | 3.3116 | 3.3094 | 3.3137 | 3.3115 | 3.3158 | 3.3136 | 3.3179 |
| 3½-4 | 3.229 | 3.248 | 3.239 | 3.258 | 3.248 | 3.267 | 3.258 | 3.277 | 3.2290 | 3.2444 | 3.2369 | 3.2519 | 3.2444 | 3.2594 | 3.2519 | 3.2669 |
| 3½-8 | 3.365 | 3.378 | 3.371 | 3.384 | 3.378 | 3.390 | 3.384 | 3.396 | 3.3650 | 3.3722 | 3.3684 | 3.3759 | 3.3722 | 3.3797 | 3.3760 | 3.3835 |
| 3½-12 | 3.410 | 3.419 | 3.415 | 3.424 | 3.419 | 3.428 | 3.424 | 3.433 | 3.4100 | 3.4148 | 3.4123 | 3.4173 | 3.4148 | 3.4198 | 3.4173 | 3.4223 |
| 3½-16 | 3.432 | 3.439 | 3.436 | 3.443 | 3.439 | 3.446 | 3.443 | 3.450 | 3.4320 | 3.4366 | 3.4344 | 3.4387 | 3.4365 | 3.4408 | 3.4386 | 3.4429 |
| 3⅝-12 | 3.535 | 3.544 | 3.540 | 3.549 | 3.544 | 3.553 | 3.549 | 3.558 | 3.5350 | 3.5398 | 3.5373 | 3.5423 | 3.5398 | 3.5448 | 3.5423 | 3.5473 |
| 3⅝-16 | 3.557 | 3.564 | 3.561 | 3.568 | 3.564 | 3.571 | 3.568 | 3.575 | 3.5570 | 3.5616 | 3.5594 | 3.5637 | 3.5615 | 3.5658 | 3.5636 | 3.5679 |
| 3¾-4 | 3.479 | 3.498 | 3.489 | 3.508 | 3.498 | 3.517 | 3.508 | 3.527 | 3.4790 | 3.4944 | 3.4869 | 3.5019 | 3.4944 | 3.5094 | 3.5019 | 3.5169 |
| 3¾-8 | 3.615 | 3.628 | 3.621 | 3.634 | 3.628 | 3.640 | 3.634 | 3.646 | 3.6150 | 3.6222 | 3.6184 | 3.6259 | 3.6222 | 3.6297 | 3.6260 | 3.6335 |

| Size | C1 | C2 | C3 | C4 | C5 | C6 | C7 | C8 | C9 | C10 | C11 | C12 | C13 | C14 | C15 | C16 |
|---|---|---|---|---|---|---|---|---|---|---|---|---|---|---|---|---|
| 3¾-12 | 3.6723 | 3.6673 | 3.6698 | 3.6648 | 3.6673 | 3.6623 | 3.6648 | 3.6600 | 3.683 | 3.674 | 3.678 | 3.669 | 3.674 | 3.665 | 3.669 | 3.660 |
| 3¾-16 | 3.6929 | 3.6886 | 3.6908 | 3.6865 | 3.6887 | 3.6844 | 3.6866 | 3.6820 | 3.700 | 3.693 | 3.696 | 3.689 | 3.693 | 3.686 | 3.689 | 3.682 |
| 3⅞-12 | 3.7973 | 3.7923 | 3.7948 | 3.7898 | 3.7923 | 3.7873 | 3.7898 | 3.7850 | 3.808 | 3.799 | 3.803 | 3.794 | 3.799 | 3.790 | 3.794 | 3.785 |
| 3⅞-16 | 3.8179 | 3.8136 | 3.8158 | 3.8115 | 3.8137 | 3.8094 | 3.8116 | 3.8070 | 3.825 | 3.818 | 3.821 | 3.814 | 3.818 | 3.811 | 3.814 | 3.807 |
| 4-4 | 3.7669 | 3.7519 | 3.7594 | 3.7444 | 3.7519 | 3.7369 | 3.7444 | 3.7290 | 3.777 | 3.758 | 3.767 | 3.748 | 3.758 | 3.739 | 3.748 | 3.729 |
| 4-8 | 3.8835 | 3.8760 | 3.8797 | 3.8722 | 3.8759 | 3.8684 | 3.8722 | 3.8650 | 3.896 | 3.884 | 3.890 | 3.878 | 3.884 | 3.871 | 3.878 | 3.865 |
| 4-12 | 3.9223 | 3.9173 | 3.9198 | 3.9148 | 3.9173 | 3.9123 | 3.9148 | 3.9100 | 3.933 | 3.924 | 3.928 | 3.919 | 3.924 | 3.915 | 3.919 | 3.910 |
| 4-16 | 3.9429 | 3.9386 | 3.9408 | 3.9365 | 3.9387 | 3.9344 | 3.9366 | 3.9320 | 3.950 | 3.943 | 3.946 | 3.939 | 3.943 | 3.936 | 3.939 | 3.932 |
| 4¼-4 | 4.0169 | 4.0019 | 4.0094 | 3.9944 | 4.0019 | 3.9869 | 3.9944 | 3.9790 | 4.027 | 4.008 | 4.017 | 3.998 | 4.008 | 3.989 | 3.998 | 3.979 |
| 4¼-8 | 4.1335 | 4.1260 | 4.1297 | 4.1222 | 4.1259 | 4.1184 | 4.1222 | 4.1150 | 4.146 | 4.134 | 4.140 | 4.128 | 4.134 | 4.121 | 4.128 | 4.115 |
| 4¼-12 | 4.1723 | 4.1673 | 4.1698 | 4.1648 | 4.1673 | 4.1623 | 4.1648 | 4.1600 | 4.183 | 4.174 | 4.178 | 4.169 | 4.174 | 4.165 | 4.169 | 4.160 |
| 4¼-16 | 4.1929 | 4.1886 | 4.1908 | 4.1865 | 4.1887 | 4.1844 | 4.1866 | 4.1820 | 4.200 | 4.193 | 4.196 | 4.189 | 4.193 | 4.186 | 4.189 | 4.182 |
| 4½-4 | 4.2669 | 4.2519 | 4.2594 | 4.2444 | 4.2519 | 4.2369 | 4.2444 | 4.2290 | 4.277 | 4.258 | 4.267 | 4.248 | 4.258 | 4.239 | 4.248 | 4.229 |
| 4½-8 | 4.3835 | 4.3760 | 4.3797 | 4.3722 | 4.3759 | 4.3684 | 4.3722 | 4.3650 | 4.396 | 4.384 | 4.390 | 4.378 | 4.384 | 4.371 | 4.378 | 4.365 |
| 4½-12 | 4.4223 | 4.4173 | 4.4198 | 4.4148 | 4.4173 | 4.4123 | 4.4148 | 4.4100 | 4.433 | 4.424 | 4.428 | 4.419 | 4.424 | 4.419 | 4.419 | 4.410 |
| 4½-16 | 4.4429 | 4.4386 | 4.4408 | 4.4365 | 4.4387 | 4.4344 | 4.4366 | 4.4320 | 4.455 | 4.444 | 4.446 | 4.439 | 4.444 | 4.437 | 4.439 | 4.432 |
| 4¾-8 | 4.6335 | 4.6260 | 4.6297 | 4.6222 | 4.6259 | 4.6184 | 4.6222 | 4.6150 | 4.646 | 4.646 | 4.640 | 4.628 | 4.646 | 4.621 | 4.628 | 4.615 |
| 4¾-12 | 4.6723 | 4.6673 | 4.6698 | 4.6648 | 4.6673 | 4.6623 | 4.6648 | 4.6600 | 4.683 | 4.674 | 4.678 | 4.669 | 4.674 | 4.665 | 4.669 | 4.660 |
| 4¾-16 | 4.6929 | 4.6886 | 4.6908 | 4.6865 | 4.6887 | 4.6844 | 4.6866 | 4.6820 | 4.700 | 4.693 | 4.696 | 4.689 | 4.693 | 4.686 | 4.689 | 4.682 |
| 5-8 | 4.8835 | 4.8760 | 4.8797 | 4.8722 | 4.8759 | 4.8684 | 4.8722 | 4.8650 | 4.896 | 4.884 | 4.890 | 4.878 | 4.884 | 4.871 | 4.878 | 4.865 |
| 5-12 | 4.9223 | 4.9173 | 4.9198 | 4.9148 | 4.9173 | 4.9123 | 4.9148 | 4.9100 | 4.933 | 4.924 | 4.928 | 4.919 | 4.924 | 4.915 | 4.919 | 4.910 |
| 5-16 | 4.9429 | 4.9386 | 4.9408 | 4.9365 | 4.9387 | 4.9344 | 4.9366 | 4.9320 | 4.950 | 4.943 | 4.946 | 4.939 | 4.943 | 4.936 | 4.939 | 4.932 |
| 5¼-8 | 5.1335 | 5.1260 | 5.1297 | 5.1222 | 5.1259 | 5.1184 | 5.1222 | 5.1150 | 5.146 | 5.134 | 5.140 | 5.128 | 5.134 | 5.121 | 5.128 | 5.115 |
| 5¼-12 | 5.1723 | 5.1673 | 5.1698 | 5.1648 | 5.1673 | 5.1623 | 5.1648 | 5.1600 | 5.183 | 5.174 | 5.178 | 5.169 | 5.174 | 5.165 | 5.169 | 5.160 |
| 5¼-16 | 5.1929 | 5.1886 | 5.1908 | 5.1865 | 5.1887 | 5.1844 | 5.1866 | 5.1820 | 5.200 | 5.193 | 5.196 | 5.189 | 5.193 | 5.186 | 5.189 | 5.182 |
| 5½-8 | 5.3835 | 5.3760 | 5.3797 | 5.3722 | 5.3759 | 5.3684 | 5.3722 | 5.3650 | 5.396 | 5.384 | 5.390 | 5.378 | 5.384 | 5.371 | 5.378 | 5.365 |
| 5½-12 | 5.4223 | 5.4173 | 5.4198 | 5.4148 | 5.4173 | 5.4123 | 5.4148 | 5.4100 | 5.433 | 5.424 | 5.428 | 5.419 | 5.424 | 5.415 | 5.419 | 5.410 |
| 5½-16 | 5.4429 | 5.4386 | 5.4408 | 5.4365 | 5.4387 | 5.4344 | 5.4366 | 5.4320 | 5.450 | 5.442 | 5.446 | 5.439 | 5.442 | 5.436 | 5.439 | 5.432 |
| 5¾-8 | 5.6335 | 5.6260 | 5.6297 | 5.6222 | 5.6259 | 5.6184 | 5.6222 | 5.6150 | 5.646 | 5.634 | 5.640 | 5.628 | 5.634 | 5.621 | 5.628 | 5.615 |
| 5¾-12 | 5.6723 | 5.6673 | 5.6698 | 5.6648 | 5.6673 | 5.6623 | 5.6648 | 5.6600 | 5.683 | 5.674 | 5.678 | 5.669 | 5.674 | 5.665 | 5.669 | 5.660 |
| 5¾-16 | 5.6929 | 5.6886 | 5.6908 | 5.6865 | 5.6887 | 5.6844 | 5.6866 | 5.6820 | 5.700 | 5.693 | 5.696 | 5.689 | 5.693 | 5.686 | 5.689 | 5.682 |
| 6-8 | 5.8835 | 5.8760 | 5.8797 | 5.8722 | 5.8759 | 5.8684 | 5.8722 | 5.8650 | 5.896 | 5.884 | 5.890 | 5.878 | 5.884 | 5.871 | 5.878 | 5.865 |
| 6-12 | 5.9223 | 5.9173 | 5.9198 | 5.9148 | 5.9173 | 5.9123 | 5.9148 | 5.9100 | 5.933 | 5.924 | 5.928 | 5.919 | 5.924 | 5.915 | 5.919 | 5.910 |
| 6-16 | 5.9429 | 5.9386 | 5.9408 | 5.9365 | 5.9387 | 5.9344 | 5.9366 | 5.9320 | 5.950 | 5.943 | 5.946 | 5.939 | 5.943 | 5.935 | 5.939 | 5.932 |

All dimensions are in inches.

For basis of recommended hole size limits see accompanying text.

As an aid in selecting suitable drills, see the listing of American Standard drill sizes in the twist drill section. For amount of expected drill over-size, see page 1669.

* This is the minimum minor diameter specified in the thread tables, page 1280.
† This is the maximum minor diameter specified in the thread tables, page 1280.

## Table 3. Tap Drill Sizes for Threads of American National Form

| Screw Thread | | Commercial Tap Drills* | | Screw Thread | | Commercial Tap Drills* | |
|---|---|---|---|---|---|---|---|
| Outside Diam. Pitch | Root Diam. | Size or Number | Decimal Equiv. | Outside Diam. Pitch | Root Diam. | Size or Number | Decimal Equiv. |
| 1⁄16-64 | 0.0422 | 3⁄64 | 0.0469 | 27 | 0.4519 | 15⁄32 | 0.4687 |
| 72 | 0.0445 | 3⁄64 | 0.0469 | 9⁄16-12 | 0.4542 | 31⁄64 | 0.4844 |
| 5⁄64-60 | 0.0563 | 1⁄16 | 0.0625 | 18 | 0.4903 | 33⁄64 | 0.5156 |
| 72 | 0.0601 | 52 | 0.0635 | 27 | 0.5144 | 17⁄32 | 0.5312 |
| 3⁄32-48 | 0.0667 | 49 | 0.0730 | 5⁄8-11 | 0.5069 | 17⁄32 | 0.5312 |
| 50 | 0.0678 | 49 | 0.0730 | 12 | 0.5168 | 35⁄64 | 0.5469 |
| 7⁄64-48 | 0.0823 | 43 | 0.0890 | 18 | 0.5528 | 37⁄64 | 0.5781 |
| 1⁄8-32 | 0.0844 | 3⁄32 | 0.0937 | 27 | 0.5769 | 19⁄32 | 0.5937 |
| 40 | 0.0925 | 38 | 0.1015 | 11⁄16-16 | 0.5694 | 19⁄32 | 0.5937 |
| 9⁄64-40 | 0.1081 | 32 | 0.1160 | 16 | 0.6063 | 5⁄8 | 0.6250 |
| 5⁄32-32 | 0.1157 | 1⁄8 | 0.1250 | 3⁄4-10 | 0.6201 | 21⁄32 | 0.6562 |
| 36 | 0.1202 | 30 | 0.1285 | 12 | 0.6418 | 43⁄64 | 0.6719 |
| 11⁄64-32 | 0.1313 | 9⁄64 | 0.1406 | 16 | 0.6688 | 11⁄16 | 0.6875 |
| 3⁄16-24 | 0.1334 | 26 | 0.1470 | 27 | 0.7019 | 23⁄32 | 0.7187 |
| 32 | 0.1469 | 22 | 0.1570 | 13⁄16-10 | 0.6826 | 23⁄32 | 0.7187 |
| 13⁄64-24 | 0.1490 | 20 | 0.1610 | 7⁄8- 9 | 0.7307 | 49⁄64 | 0.7656 |
| 7⁄32-24 | 0.1646 | 16 | 0.1770 | 12 | 0.7668 | 51⁄64 | 0.7969 |
| 32 | 0.1782 | 12 | 0.1890 | 14 | 0.7822 | 13⁄16 | 0.8125 |
| 15⁄64-24 | 0.1806 | 10 | 0.1935 | 18 | 0.8028 | 53⁄64 | 0.8281 |
| 1⁄4-20 | 0.1850 | 7 | 0.2010 | 27 | 0.8269 | 27⁄32 | 0.8437 |
| 24 | 0.1959 | 4 | 0.2090 | 15⁄16- 9 | 0.7932 | 53⁄64 | 0.8281 |
| 27 | 0.2019 | 3 | 0.2130 | 1 - 8 | 0.8376 | 7⁄8 | 0.8750 |
| 28 | 0.2036 | 3 | 0.2130 | 12 | 0.8918 | 59⁄64 | 0.9219 |
| 32 | 0.2094 | 7⁄32 | 0.2187 | 14 | 0.9072 | 15⁄16 | 0.9375 |
| 5⁄16-18 | 0.2403 | F | 0.2570 | 27 | 0.9519 | 31⁄32 | 0.9687 |
| 20 | 0.2476 | 17⁄64 | 0.2656 | 1 1⁄8- 7 | 0.9394 | 63⁄64 | 0.9844 |
| 24 | 0.2584 | I | 0.2720 | 12 | 1.0168 | 1 3⁄64 | 1.0469 |
| 27 | 0.2644 | J | 0.2770 | 1 1⁄4- 7 | 1.0644 | 1 7⁄64 | 1.1094 |
| 32 | 0.2719 | 9⁄32 | 0.2812 | 12 | 1.1418 | 1 11⁄64 | 1.1719 |
| 3⁄8-16 | 0.2938 | 5⁄16 | 0.3125 | 1 3⁄8- 6 | 1.1585 | 1 7⁄32 | 1.2187 |
| 20 | 0.3100 | 21⁄64 | 0.3281 | 12 | 1.2668 | 1 19⁄64 | 1.2969 |
| 24 | 0.3209 | Q | 0.3320 | 1 1⁄2- 6 | 1.2835 | 1 11⁄32 | 1.3437 |
| 27 | 0.3269 | R | 0.3390 | 12 | 1.3918 | 1 27⁄64 | 1.4219 |
| 7⁄16-14 | 0.3447 | U | 0.3680 | 1 5⁄8- 5 1⁄2 | 1.3888 | 1 29⁄64 | 1.4531 |
| 20 | 0.3726 | 25⁄64 | 0.3906 | 1 3⁄4- 5 | 1.4902 | 1 9⁄16 | 1.5625 |
| 24 | 0.3834 | X | 0.3970 | 1 7⁄8- 5 | 1.6152 | 1 11⁄16 | 1.6875 |
| 27 | 0.3894 | Y | 0.4040 | 2 - 4 1⁄2 | 1.7113 | 1 25⁄32 | 1.7812 |
| 1⁄2-12 | 0.3918 | 27⁄64 | 0.4219 | 2 1⁄8- 4 1⁄2 | 1.8363 | 1 29⁄32 | 1.9062 |
| 13 | 0.4001 | 27⁄64 | 0.4219 | 2 1⁄4- 4 1⁄2 | 1.9613 | 2 1⁄32 | 2.0312 |
| 20 | 0.4351 | 29⁄64 | 0.4531 | 2 3⁄8- 4 | 2.0502 | 2 1⁄8 | 2.1250 |
| 24 | 0.4459 | 29⁄64 | 0.4531 | 2 1⁄2- 4 | 2.1752 | 2 1⁄4 | 2.2500 |

* These tap drill diameters allow approximately 75 per cent of a full thread. For small thread sizes the use of larger drills will reduce tap breakage.

Table 4. Tap Drills and Clearance Drills for Machine Screws
with American National Thread Form

| Size of Screw | | No. of Threads per Inch | Tap Drills | | Clearance Hole Drills | | | |
|---|---|---|---|---|---|---|---|---|
| | | | | | Close Fit | | Free Fit | |
| No. or Diam. | Decimal Equiv. | | Drill Size | Decimal Equiv. | Drill Size | Decimal Equiv. | Drill Size | Decimal Equiv. |
| o | .060 | 80 | ³⁄₆₄ | .0469 | 52 | .0635 | 50 | .0700 |
| 1 | .073 | 64 | 53 | .0595 | 48 | .0760 | 46 | .0810 |
| | | 72 | 53 | .0595 | | | | |
| 2 | .086 | 56 | 50 | .0700 | 43 | .0890 | 41 | .0960 |
| | | 64 | 50 | .0700 | | | | |
| 3 | .099 | 48 | 47 | .0785 | 37 | .1040 | 35 | .1100 |
| | | 56 | 45 | .0820 | | | | |
| 4 | .112 | 36* | 44 | .0860 | 32 | .1160 | 30 | .1285 |
| | | 40 | 43 | .0890 | | | | |
| | | 48 | 42 | .0935 | | | | |
| 5 | .125 | 40 | 38 | .1015 | 30 | .1285 | 29 | .1360 |
| | | 44 | 37 | .1040 | | | | |
| 6 | .138 | 32 | 36 | .1065 | 27 | .1440 | 25 | .1495 |
| | | 40 | 33 | .1130 | | | | |
| 8 | .164 | 32 | 29 | .1360 | 18 | .1695 | 16 | .1770 |
| | | 36 | 29 | .1360 | | | | |
| 10 | .190 | 24 | 25 | .1495 | 9 | .1960 | 7 | .2010 |
| | | 32 | 21 | .1590 | | | | |
| 12 | .216 | 24 | 16 | .1770 | 2 | .2210 | 1 | .2280 |
| | | 28 | 14 | .1820 | | | | |
| 14 | .242 | 20* | 10 | .1935 | D | .2460 | F | .2570 |
| | | 24* | 7 | .2010 | | | | |
| ¼ | .250 | 20 | 7 | .2010 | F | .2570 | H | .2660 |
| | | 28 | 3 | .2130 | | | | |
| ⁵⁄₁₆ | .3125 | 18 | F | .2570 | P | .3230 | Q | .3320 |
| | | 24 | I | .2720 | | | | |
| ⅜ | .375 | 16 | ⁵⁄₁₆ | .3125 | W | .3860 | X | .3970 |
| | | 24 | Q | .3320 | | | | |
| ⁷⁄₁₆ | .4375 | 14 | U | .3680 | ²⁹⁄₆₄ | .4531 | ¹⁵⁄₃₂ | .4687 |
| | | 20 | ²⁵⁄₆₄ | .3906 | | | | |
| ½ | .500 | 13 | ²⁷⁄₆₄ | .4219 | ³³⁄₆₄ | .5156 | ¹⁷⁄₃₂ | .5312 |
| | | 20 | ²⁹⁄₆₄ | .4531 | | | | |

* Screws marked with asterisk (*) are not in the American Standard but are from the former A.S.M.E. Standard.

1. For lengths of engagement in the range to and including ⅓D, where D equals nominal diameter, the minimum hole size will be equal to the minimum minor diameter of the internal thread and the maximum hole size will be larger by one-half the minor diameter tolerance.

2. For the range from ⅓D to ⅔D, the minimum and maximum hole sizes will each be one quarter of the minor diameter tolerance larger than the corresponding limits for the length of engagement to and including ⅓D.

3. For the range from ⅔D to 1½D the minimum hole size will be larger than the minimum minor diameter of the internal thread by one-half the minor diameter tolerance and the maximum hole size will be equal to the maximum minor diameter.

4. For the range from 1½D to 3D the minimum and maximum hole sizes will each be one-quarter of the minor diameter tolerance of the internal thread larger than the corresponding limits for the ⅔D to 1½D length of engagement.

From the foregoing it will be seen that the difference between limits in each range is the same and equal to one-half of the minor diameter tolerance given in the Unified screw thread dimensional tables. This is a general rule, except that the minimum differences for sizes below ¼ inch are equal to the minor diameter tolerances calculated on the basis of lengths of engagement to and including ⅓D. Also, for lengths of engagement greater than ⅓D and for sizes ¼ inch and larger the values are adjusted so that the difference between limits is never less than 0.004 inch.

For diameter-pitch combinations other than those given in Table 2, the foregoing rules should be applied to the tolerances given in the dimensional tables in the screw thread section or the tolerances derived from the formulas given in the Standard to determine the hole size limits.

*Selection of Tap Drills:* In selecting standard drills to produce holes within the limits given in Table 2 it should be recognized that drills have a tendency to cut oversize. The material on page 1669 may be used as a guide to the expected amount of oversize.

**Hole Sizes for Tapping Unified Miniature Screw Threads.** — Table 5 indicates the hole size limits recommended for tapping. These limits are derived from the internal thread minor diameter limits given in the American Standard for Unified Miniature Screw Threads (ASA B1.10-1958) and are disposed so as to provide the optimum conditions for tapping. The maximum limits are based on providing a functionally adequate fastening for the most common applications, where the material of the externally threaded member is of a strength essentially equal to or greater than that of its mating part. In applications where, because of considerations other than the fastening, the screw is made of an appreciably weaker material, the use of smaller hole sizes is usually necessary to extend thread engagement to a greater depth on the external thread. Recommended minimum hole sizes are greater than the minimum limits of the minor diameters to allow for the spin-up developed in tapping.

In selecting drills to produce holes within the limits given in Table 5 it should be recognized that drills have a tendency to cut oversize. The material on page 1669 may be used as a guide to the expected amount of oversize.

**British Standard Tapping Drill Sizes for Screw and Pipe Threads.** —British Standard BS 1157:1975 provides recommendations for tapping drill sizes for use with fluted taps for various ISO metric, Unified, British Standard fine, British Association, and British Standard Whitworth screw threads as well as British Standard parallel and taper pipe threads.

In the accompanying Table 6, recommended and alternative drill sizes are given for producing holes for ISO metric coarse pitch series threads. These coarse pitch threads are suitable for the large majority of general-purpose applications, and the limits and

Table 5. Unified Miniature Screw Threads —
Recommended Hole Size Limits Before Tapping*

| Thread Size | | Internal Threads | | Lengths of Engagement | | | | | |
| | | | | To and including ⅔D | | Above ⅔D to 1½D | | Above 1½D to 3D | |
| Designation | Pitch | Minor Diameter Limits | | Recommended Hole Size Limits | | | | | |
| | | Min | Max | Min | Max | Min | Max | Min | Max |
| | mm | mm | mm | mm | mm | mm | mm | mm | mm |
| **0.30 UNM** | 0.080 | 0.217 | 0.254 | 0.226 | 0.240 | 0.236 | 0.254 | 0.245 | 0.264 |
| 0.35 UNM | 0.090 | 0.256 | 0.297 | 0.267 | 0.282 | 0.277 | 0.297 | 0.287 | 0.307 |
| **0.40 UNM** | 0.100 | 0.296 | 0.340 | 0.307 | 0.324 | 0.318 | 0.340 | 0.329 | 0.351 |
| 0.45 UNM | 0.100 | 0.346 | 0.390 | 0.357 | 0.374 | 0.368 | 0.390 | 0.379 | 0.401 |
| **0.50 UNM** | 0.125 | 0.370 | 0.422 | 0.383 | 0.402 | 0.396 | 0.422 | 0.409 | 0.435 |
| 0.55 UNM | 0.125 | 0.420 | 0.472 | 0.433 | 0.452 | 0.446 | 0.472 | 0.459 | 0.485 |
| **0.60 UNM** | 0.150 | 0.444 | 0.504 | 0.459 | 0.482 | 0.474 | 0.504 | 0.489 | 0.519 |
| 0.70 UNM | 0.175 | 0.518 | 0.586 | 0.535 | 0.560 | 0.552 | 0.586 | 0.569 | 0.603 |
| **0.80 UNM** | 0.200 | 0.592 | 0.668 | 0.611 | 0.640 | 0.630 | 0.668 | 0.649 | 0.687 |
| 0.90 UNM | 0.225 | 0.666 | 0.750 | 0.687 | 0.718 | 0.708 | 0.750 | 0.729 | 0.771 |
| **1.00 UNM** | 0.250 | 0.740 | 0.832 | 0.763 | 0.798 | 0.786 | 0.832 | 0.809 | 0.855 |
| 1.10 UNM | 0.250 | 0.840 | 0.932 | 0.863 | 0.898 | 0.886 | 0.932 | 0.909 | 0.955 |
| **1.20 UNM** | 0.250 | 0.940 | 1.032 | 0.963 | 0.998 | 0.986 | 1.032 | 1.009 | 1.055 |
| 1.40 UNM | 0.300 | 1.088 | 1.196 | 1.115 | 1.156 | 1.142 | 1.196 | 1.169 | 1.223 |
| Designation | Thds. per Inch | inch | inch | inch | inch | inch | inch | inch | inch |
| **0.30 UNM** | 318 | 0.0085 | 0.0100 | 0.0089 | 0.0095 | 0.0093 | 0.0100 | 0.0096 | 0.0104 |
| 0.35 UNM | 282 | 0.0101 | 0.0117 | 0.0105 | 0.0111 | 0.0109 | 0.0117 | 0.0113 | 0.0121 |
| **0.40 UNM** | 254 | 0.0117 | 0.0134 | 0.0121 | 0.0127 | 0.0125 | 0.0134 | 0.0130 | 0.0138 |
| 0.45 UNM | 254 | 0.0136 | 0.0154 | 0.0141 | 0.0147 | 0.0145 | 0.0154 | 0.0149 | 0.0158 |
| **0.50 UNM** | 203 | 0.0146 | 0.0166 | 0.0150 | 0.0158 | 0.0156 | 0.0166 | 0.0161 | 0.0171 |
| 0.55 UNM | 203 | 0.0165 | 0.0186 | 0.0170 | 0.0178 | 0.0176 | 0.0186 | 0.0181 | 0.0191 |
| **0.60 UNM** | 169 | 0.0175 | 0.0198 | 0.0181 | 0.0190 | 0.0187 | 0.0198 | 0.0193 | 0.0204 |
| 0.70 UNM | 145 | 0.0204 | 0.0231 | 0.0211 | 0.0221 | 0.0217 | 0.0231 | 0.0224 | 0.0237 |
| **0.80 UNM** | 127 | 0.0233 | 0.0263 | 0.0241 | 0.0252 | 0.0248 | 0.0263 | 0.0256 | 0.0270 |
| 0.90 UNM | 113 | 0.0262 | 0.0295 | 0.0270 | 0.0283 | 0.0279 | 0.0295 | 0.0287 | 0.0304 |
| **1.00 UNM** | 102 | 0.0291 | 0.0327 | 0.0300 | 0.0314 | 0.0309 | 0.0327 | 0.0319 | 0.0337 |
| 1.10 UNM | 102 | 0.0331 | 0.0367 | 0.0340 | 0.0354 | 0.0349 | 0.0367 | 0.0358 | 0.0376 |
| **1.20 UNM** | 102 | 0.0370 | 0.0406 | 0.0379 | 0.0393 | 0.0388 | 0.0406 | 0.0397 | 0.0415 |
| 1.40 UNM | 85 | 0.0428 | 0.0471 | 0.0439 | 0.0455 | 0.0450 | 0.0471 | 0.0460 | 0.0481 |

* As an aid in selecting suitable drills, see the listing of American Standard drill sizes in the twist drill section. Thread sizes in heavy type are preferred sizes.

tolerances for internal coarse threads are given on pages 1358, 1360 and 1362. It should be noted that Table 6 is for fluted taps only since a fluteless tap will require for the same screw thread a different size of twist drill than will a fluted tap. When tapped, holes produced with drills of the recommended sizes provide for a theoretical radial engagement with the external thread of about 81 per cent in most cases. Holes produced with drills of the alternative sizes provide for a theoretical radial engagement with the external thread of about 70 to 75 per cent. In some cases, as indicated in Table 6, the alternative drill sizes are suitable only for medium (6H) or free (7H) thread tolerance classes.

When relatively soft material is being tapped, there is a tendency for the metal to be squeezed down towards the root of the tap thread, and in such instances, the minor diameter of the tapped hole may become smaller than the diameter of the drill employed. Users may wish to choose different tapping drill sizes to overcome this problem or for special purposes, and reference can be made to the pages mentioned above to obtain the minor diameter limits for internal coarse pitch series threads.

Reference should be made to this standard (BS 1157:1975) for recommended tapping hole sizes for other types of British Standard screw threads and pipe threads.

Table 6. British Standard Tapping Drill Sizes for ISO Metric Coarse Pitch Series Threads (BS 1157:1975)

| Nom. Size and Thread Diam. | Standard Drill Sizes § | | | | Nom. Size and Thread Diam. | Standard Drill Sizes § | | | |
|---|---|---|---|---|---|---|---|---|---|
| | Recommended | | Alternative | | | Recommended | | Alternative | |
| | Size | Theoretical Radial Engagement with Ext. Thread (Per Cent) | Size | Theoretical Radial Engagement with Ext. Thread (Per Cent) | | Size | Theoretical Radial Engagement with Ext. Thread (Per Cent) | Size | Theoretical Radial Engagement with Ext. Thread (Per Cent) |
| M 1 | 0.75 | 81.5 | 0.78 | 71.7 | M 12 | 10.20 | 83.7 | 10.40 | 74.5* |
| M 1.1 | 0.85 | 81.5 | 0.88 | 71.7 | M 14 | 12.00 | 81.5 | 12.20 | 73.4* |
| M 1.2 | 0.95 | 81.5 | 0.98 | 71.7 | M 16 | 14.00 | 81.5 | 14.25 | 71.3† |
| M 1.4 | 1.10 | 81.5 | 1.15 | 67.9 | M 18 | 15.50 | 81.5 | 15.75 | 73.4† |
| M 1.6 | 1.25 | 81.5 | 1.30 | 69.9 | M 20 | 17.50 | 81.5 | 17.75 | 73.4† |
| M 1.8 | 1.45 | 81.5 | 1.50 | 69.9 | M 22 | 19.50 | 81.5 | 19.75 | 73.4† |
| M 2 | 1.60 | 81.5 | 1.65 | 71.3 | M 24 | 21.00 | 81.5 | 21.25 | 74.7* |
| M 2.2 | 1.75 | 81.5 | 1.80 | 72.5 | M 27 | 24.00 | 81.5 | 24.25 | 74.7* |
| M 2.5 | 2.05 | 81.5 | 2.10 | 72.5 | M 30 | 26.50 | 81.5 | 26.75 | 75.7* |
| M 3 | 2.50 | 81.5 | 2.55 | 73.4 | M 33 | 29.50 | 81.5 | 29.75 | 75.7* |
| M 3.5 | 2.90 | 81.5 | 2.95 | 74.7 | M 36 | 32.00 | 81.5 | ... | ... |
| M 4 | 3.30 | 81.5 | 3.40 | 69.9* | M 39 | 35.00 | 81.5 | ... | ... |
| M 4.5 | 3.70 | 86.8 | 3.80 | 76.1 | M 42 | 37.50 | 81.5 | ... | ... |
| M 5 | 4.20 | 81.5 | 4.30 | 71.3* | M 45 | 40.50 | 81.5 | ... | ... |
| M 6 | 5.00 | 81.5 | 5.10 | 73.4 | M 48 | 43.00 | 81.5 | ... | ... |
| M 7 | 6.00 | 81.5 | 6.10 | 73.4 | M 52 | 47.00 | 81.5 | ... | ... |
| M 8 | 6.80 | 78.5 | 6.90 | 71.7* | M 56 | 50.50 | 81.5 | ... | ... |
| M 9 | 7.80 | 78.5 | 7.90 | 71.7* | M 60 | 54.50 | 81.5 | ... | ... |
| M 10 | 8.50 | 81.5 | 8.60 | 76.1 | M 64 | 58.00 | 81.5 | ... | ... |
| M 11 | 9.50 | 81.5 | 9.60 | 76.1 | M 68 | 62.00 | 81.5 | ... | ... |

§ These tapping drill sizes are for fluted taps only.
* For tolerance class 6H and 7H threads only.
† For tolerance class 7H threads only.

**British Standard Clearance Holes for Metric Bolts and Screws.** — The dimensions of the clearance holes specified in this British Standard (BS 4186:1967) have been chosen in such a way as to require the use of the minimum number of drills. The recommendations cover three series of clearance holes, namely, close fit (H 12), medium fit (H 13), and free fit (H 14) and are suitable for use with bolts and screws specified in the following metric British Standards: BS 3692, ISO metric precision hexagon bolts, screws, and nuts; BS 4168, Hexagon socket screws and wrench keys; BS 4183, Machine screws and machine screw nuts; and BS 4190, ISO metric black hexagon bolts, screws, and nuts. The sizes are in accordance with those given in ISO Recommendation R273, and the range has been extended up to 150 millimeters diameter in accordance with an addendum to that recommendation. The selection of clearance holes sizes to suit particular design requirements can of course be dependent upon many variable factors. It is however felt that the medium fit series should suit the majority of general purpose applications. In the Standard, limiting dimensions are given in a table which is included for reference purposes only, for use in instances where it may be desirable to specify tolerances.

To avoid any risk of interference with the radius under the head of bolts and screws, it is necessary to countersink slightly all recommended clearance holes in the close and medium fit series. Dimensional details for the radius under the head of fasteners made according to BS 3692 are given on page 1158; those for fasteners to BS 4168 are given on page 1215; those to BS 4183 are given on pages 1191 to 1194.

Table 7. British Standard Metric Bolt and Screw Clearance Holes (BS 4186: 1967)

| Nominal Thread Diameter | Clearance Hole Sizes | | | Nominal Thread Diameter | Clearance Hole Sizes | | |
|---|---|---|---|---|---|---|---|
| | Close Fit Series | Medium Fit Series | Free Fit Series | | Close Fit Series | Medium Fit Series | Free Fit Series |
| 1.6 | 1.7 | 1.8 | 2.0 | 52.0 | 54.0 | 56.0 | 62.0 |
| 2.0 | 2.2 | 2.4 | 2.6 | 56.0 | 58.0 | 62.0 | 66.0 |
| 2.5 | 2.7 | 2.9 | 3.1 | 60.0 | 62.0 | 66.0 | 70.0 |
| 3.0 | 3.2 | 3.4 | 3.6 | 64.0 | 66.0 | 70.0 | 74.0 |
| 4.0 | 4.3 | 4.5 | 4.8 | 68.0 | 70.0 | 74.0 | 78.0 |
| 5.0 | 5.3 | 5.5 | 5.8 | 72.0 | 74.0 | 78.0 | 82.0 |
| 6.0 | 6.4 | 6.6 | 7.0 | 76.0 | 78.0 | 82.0 | 86.0 |
| 7.0 | 7.4 | 7.6 | 8.0 | 80.0 | 82.0 | 86.0 | 91.0 |
| 8.0 | 8.4 | 9.0 | 10.0 | 85.0 | 87.0 | 91.0 | 96.0 |
| 10.0 | 10.5 | 11.0 | 12.0 | 90.0 | 93.0 | 96.0 | 101.0 |
| 12.0 | 13.0 | 14.0 | 15.0 | 95.0 | 98.0 | 101.0 | 107.0 |
| 14.0 | 15.0 | 16.0 | 17.0 | 100.0 | 104.0 | 107.0 | 112.0 |
| 16.0 | 17.0 | 18.0 | 19.0 | 105.0 | 109.0 | 112.0 | 117.0 |
| 18.0 | 19.0 | 20.0 | 21.0 | 110.0 | 114.0 | 117.0 | 122.0 |
| 20.0 | 21.0 | 22.0 | 24.0 | 115.0 | 119.0 | 122.0 | 127.0 |
| 22.0 | 23.0 | 24.0 | 26.0 | 120.0 | 124.0 | 127.0 | 132.0 |
| 24.0 | 25.0 | 26.0 | 28.0 | 125.0 | 129.0 | 132.0 | 137.0 |
| 27.0 | 28.0 | 30.0 | 32.0 | 130.0 | 134.0 | 137.0 | 144.0 |
| 30.0 | 31.0 | 33.0 | 35.0 | 140.0 | 144.0 | 147.0 | 155.0 |
| 33.0 | 34.0 | 36.0 | 38.0 | 150.0 | 155.0 | 158.0 | 165.0 |
| 36.0 | 37.0 | 39.0 | 42.0 | .... | .... | .... | .... |
| 39.0 | 40.0 | 42.0 | 45.0 | .... | .... | .... | .... |
| 42.0 | 43.0 | 45.0 | 48.0 | .... | .... | .... | .... |
| 45.0 | 46.0 | 48.0 | 52.0 | .... | .... | .... | .... |
| 48.0 | 50.0 | 52.0 | 56.0 | .... | .... | .... | .... |

All dimensions are given in millimeters.

**Cold Form Tapping.** — Cold form taps do not have cutting edges or conventional flutes; the threads on the tap form the threads in the hole by displacing the metal in an extrusion or swaging process. The threads thus produced are stronger than conventionally cut threads because the grains in the metal are unbroken and the displaced metal is work hardened. The surface of the thread is burnished and has an excellent finish. Although chip problems are eliminated, cold form tapping does displace the metal surrounding the hole and countersinking or chamfering before tapping is recommended. Cold form tapping is not recommended if the wall thickness of the hole is less than two-thirds of the nominal diameter of the thread. If possible, blind holes should be drilled deep enough to permit a cold form tap having a four thread lead to be used as this will require less torque, produce less burr surrounding the hole, and give a greater tool life.

The operation requires 0 to 50 per cent more torque than conventional tapping, and the cold form tap will pick up its own lead when entering the hole; thus, conventional tapping machines and tapping heads can be used. Another advantage is the better tool life obtained. The best results are obtained by using a good lubricating oil instead of a conventional cutting oil.

## Table 8. Theoretical and Tap Drill or Core Hole Sizes for Cold Form Tapping Unified Threads

| Tap Size | Threads Per Inch | Percentage of Full Thread | | | | | | | | |
|---|---|---|---|---|---|---|---|---|---|---|
| | | 75 | | | 65 | | | 55 | | |
| | | Theor. Hole Size | Nearest Drill Size | Dec. Equiv. | Theor. Hole Size | Nearest Drill Size | Dec. Equiv. | Theor. Hole Size | Nearest Drill Size | Dec. Equiv. |
| 0 | 80 | .0536 | 1.35 mm | .0531 | .0545 | .... | .... | .0554 | 54 | .055 |
| 1 | 64 | .0650 | 1.65 mm | .0650 | .0661 | .... | .... | .0672 | 51 | .0670 |
| | 72 | .0659 | 1.65 mm | .0650 | .0669 | 1.7 mm | .0669 | .0679 | 51 | .0670 |
| 2 | 56 | .0769 | 1.95 mm | .0768 | .0781 | 5⁄64 | .0781 | .0794 | 2.0 mm | .0787 |
| | 64 | .0780 | 5⁄64 | .0781 | .0791 | 2.0 mm | .0787 | .0802 | .... | .... |
| 3 | 48 | .0884 | 2.25 mm | .0886 | .0898 | 43 | .089 | .0913 | 2.3 mm | .0906 |
| | 56 | .0889 | 43 | .089 | .0911 | 2.3 mm | .0906 | .0924 | 2.35 mm | .0925 |
| 4 | 40 | .0993 | 2.5 mm | .0984 | .1010 | 39 | .0995 | .1028 | 2.6 mm | .1024 |
| | 48 | .0104 | 38 | .1015 | .1028 | 2.6 mm | .1024 | .1043 | 37 | .1040 |
| 5 | 40 | .1123 | 34 | .1110 | .1140 | 33 | .113 | .1158 | 32 | .1160 |
| | 44 | .1134 | 33 | .113 | .1150 | 2.9 mm | .1142 | .1166 | 32 | .... |
| 6 | 32 | .1221 | 3.1 mm | .1220 | .1243 | .... | .... | .1264 | 3.2 mm | .1260 |
| | 40 | .1253 | 1⁄8 | .1250 | .1270 | 3.2 mm | .1260 | .1288 | 30 | .1285 |
| 8 | 32 | .1481 | 3.75 mm | .1476 | .1503 | 25 | .1495 | .1524 | 24 | .1520 |
| | 36 | .1498 | 25 | .1495 | .1518 | 24 | .1520 | .1537 | 3.9 mm | .1535 |
| 10 | 24 | .1688 | .... | .... | .1717 | 11⁄64 | .1719 | .1746 | 17 | .1730 |
| | 32 | .1741 | 17 | .1730 | .1763 | .... | .... | .1784 | 4.5 mm | .1772 |
| 12 | 24 | .1948 | 10 | .1935 | .1977 | 5.0 mm | .1968 | .2006 | 5.1 mm | .2008 |
| | 28 | .1978 | 5.0 mm | .1968 | .2003 | 8 | .1990 | .2028 | .... | .... |
| ¼ | 20 | .2245 | 5.7 mm | .2244 | .2280 | 1 | .2280 | .2315 | .... | .... |
| | 28 | .2318 | .... | .... | .2243 | A | .2340 | .2368 | 6.0 mm | .2362 |
| 5⁄16 | 18 | .2842 | 7.2 mm | .2835 | .2879 | 7.3 mm | .2874 | .2917 | 7.4 mm | .2913 |
| | 24 | .2912 | 7.4 mm | .2913 | .2941 | M | .2950 | .2969 | 19⁄64 | .2969 |
| 3⁄8 | 16 | .3431 | 11⁄32 | .3437 | .3474 | S | .3480 | .3516 | .... | .... |
| | 24 | .3537 | 9.0 mm | .3543 | .3566 | .... | .... | .3594 | 23⁄64 | .3594 |
| 7⁄16 | 14 | .4011 | .... | .... | .4059 | 13⁄32 | .4062 | .4108 | .... | .... |
| | 20 | .4120 | Z | .413 | .4154 | .... | .... | .4188 | .... | .... |
| ½ | 13 | .4608 | .... | .... | .4660 | .... | .... | .4712 | 12 mm | .4724 |
| | 20 | .4745 | .... | .... | .4779 | .... | .... | .4813 | .... | .... |
| 9⁄16 | 12 | .5200 | .... | .... | .5257 | .... | .... | .5313 | 17⁄32 | .5312 |
| | 18 | .5342 | 13.5 mm | .5315 | .5380 | .... | .... | .5417 | .... | .... |
| 5⁄8 | 11 | .5787 | 37⁄64 | .5781 | .5848 | .... | .... | .5910 | 15 mm | .5906 |
| | 18 | .5976 | 19⁄32 | .5937 | .6004 | .... | .... | .6042 | .... | .... |
| ¾ | 10 | .6990 | .... | .... | .7058 | 45⁄64 | .7031 | .7126 | .... | .... |
| | 16 | .7181 | 23⁄32 | .7187 | .7224 | .... | .... | .7266 | .... | .... |

The method can be applied only to relatively ductile metals, such as low carbon steel, leaded steels, austenitic stainless steels, wrought aluminum, low silicon aluminum die casting alloys, zinc die casting alloys, magnesium, copper, and ductile copper alloys. A higher than normal tapping speed can be used, sometimes by as much as 100 per cent.

Conventional tap drill sizes should not be used for cold form tapping because the metal is displaced to form the thread. Since the cold formed thread is stronger than the conventionally tapped thread, the thread height can be reduced to 60 per cent without a loss of strength; however, the use of a 65 per cent thread is strongly recommended. The following formula is used to calculate the theoretical hole size for cold form tapping:

$$\text{Theoretical Hole Size} = \text{Basic Tap O.D.} - \frac{.0068 \times \text{Per Cent of Full Thread}}{\text{Threads per inch}}$$

The theoretical hole size and the tap drill sizes for American Unified threads are given in Table 8, and Table 9 lists drills for ISO Metric threads. Sharp drills should be used to prevent cold working the walls of the hole, especially on metals that are prone to work hardening. Such damage may cause the torque to increase, possibly stopping the machine or breaking the tap. On materials that can be die cast, cold form tapping can be done in cored holes provided the correct core pin size is used. Since the core pins are slightly tapered, the theoretical hole size should be at the point on the pin where this point is equal to one-half of the required engagement length of the thread in the hole. The core pins should be designed to form a chamfer on the hole to accept the vertical extrusion.

**Table 9. Tap Drill or Core Hole Sizes† for Cold Form Tapping ISO Threads**

| Nominal Size of Tap | Pitch | Recommended Tap Drill Size | Nominal Size of Tap | Pitch | Recommended Tap Drill Size |
|---|---|---|---|---|---|
| 1.6 mm | 0.35 mm | 1.45 mm | 4.0 mm | 0.70 mm | 3.7 mm |
| 1.8 mm | 0.35 mm | 1.65 mm | 4.5 mm | 0.75 mm | 4.2 mm* |
| 2.0 mm | 0.40 mm | 1.8 mm | 5.0 mm | 0.80 mm | 4.6 mm |
| 2.2 mm | 0.45 mm | 2.0 mm | 6.0 mm | 1.00 mm | 5.6 mm* |
| 2.5 mm | 0.45 mm | 2.3 mm | 7.0 mm | 1.00 mm | 6.5 mm |
| 3.0 mm | 0.50 mm | 2.8 mm* | 8.0 mm | 1.25 mm | 7.4 mm |
| 3.5 mm | 0.60 mm | 3.2 mm | 10.0 mm | 1.50 mm | 9.3 mm |

† The sizes are calculated to provide 60 to 75 per cent of full thread.
* These diameters are the nearest stocked drill sizes and not the theoretical hole size, and may not produce 60 to 75 per cent full thread.

**Removing a Broken Tap.**—Broken taps can be removed by electrodischarge machining (EDM), and this method is recommended when available. When an EDM machine is not available, broken taps may be removed by using a tap extractor, which has fingers that enter the flutes of the tap; the tap is backed out of the hole by turning the extractor with a wrench. Sometimes the injection of a small amount of a proprietary solvent into the hole will be helpful. A solvent can be made by diluting about one part nitric acid with five parts water. The action of the proprietary solvent or the diluted nitric acid on the steel loosens the tap so that it can be removed with pliers or with a tap extractor. The hole should be washed out afterwards so that the acid will not continue to work on the part.

Another method is to add, by electric arc welding, additional metal to the shank of the broken tap, above the level of the hole. Care must be taken to prevent depositing metal on the threads in the tapped hole. After the shank has been built up, the head of a bolt or a nut is welded to it and then the tap may be backed out.

### Tap Drills for Pipe Taps *

| Size of Tap | Drills for Briggs Pipe Taps | Drills for Whitworth Pipe Taps | Size of Tap | Drills for Briggs Pipe Taps | Drills for Whitworth Pipe Taps | Size of Tap | Drills for Briggs Pipe Taps | Drills for Whitworth Pipe Taps |
|---|---|---|---|---|---|---|---|---|
| 1/8 | 11/32 | 5/16 | 1 1/4 | 1 1/2 | 1 15/32 | 3 1/4 | . . . . . | 3 1/2 |
| 1/4 | 7/16 | 27/64 | 1 1/2 | 1 23/32 | 1 25/32 | 3 1/2 | 3 3/4 | 3 3/4 |
| 3/8 | 19/32 | 9/16 | 1 3/4 | . . . . . | 1 15/16 | 3 3/4 | . . . . . | 4 |
| 1/2 | 23/32 | 11/16 | 2 | 2 3/16 | 2 9/32 | 4 | 4 1/4 | 4 1/4 |
| 5/8 | . . . . . | 25/32 | 2 1/4 | . . . . . | 2 13/32 | 4 1/2 | 4 3/4 | 4 3/4 |
| 3/4 | 15/16 | 29/32 | 2 1/2 | 2 5/8 | 2 25/32 | 5 | 5 5/16 | 5 1/4 |
| 7/8 | . . . . . | 1 1/16 | 2 3/4 | . . . . . | 3 1/32 | 5 1/2 | . . . . . | 5 3/4 |
| 1 | 1 5/32 | 1 1/8 | 3 | 3 1/4 | 3 9/32 | 6 | 6 5/8 | 6 1/4 |

* To secure the best results, the hole should be reamed before tapping with a reamer having a taper of 3/4 inch per foot.

**Power for Pipe Taps.** — The power required for driving pipe taps is given in the following table, which includes nominal pipe tap sizes from 2 to 8 inches.

The holes to be tapped were reamed with standard pipe tap reamers before tapping. The horsepower recorded was read off just before the tap was reversed.

### Power Required for Pipe Taps

| Nominal Tap Size | Rev. per Min. | Net H.P. | Thickness of Metal | Nominal Tap Size | Rev. per Min. | Net H.P. | Thickness of Metal |
|---|---|---|---|---|---|---|---|
| 2 | 40 | 4.24 | 1 1/8 | 3 1/2 | 25.6 | 7.20 | 1 3/4 |
| 2 1/2 | 40 | 5.15 | 1 1/8 | 4 | 18 | 6.60 | 2 |
| *2 1/2 | 38.5 | 9.14 | 1 1/8 | 5 | 18 | 7.70 | 2 |
| 3 | 40 | 5.75 | 1 1/8 | 6 | 17.8 | 8.80 | 2 |
| *3 | 38.5 | 9.70 | 1 1/8 | 8 | 14 | 7.96 | 2 1/2 |

* Tapping steel casting; other tests in cast iron.

The table gives the net horsepower, deductions being made for the power required to run the machine without a load. The material tapped was cast iron, except in two instances, where steel casting was tapped. It will be seen that nearly double the power is required for tapping steel casting. The power varies, of course, with the conditions. More power than that indicated in the table will be required if the cast iron is of a harder quality or if the taps are not properly relieved. The taps used in these experiments were of the inserted-blade type, the blades being made of high-speed steel.

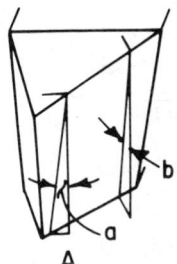

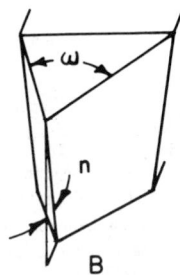

A             B

**Relief Angles for Single-Point Thread Cutting Tools.** — The surface finish on threads cut with single-point thread cutting tools is influenced by the relief angles on the tools. The leading and trailing cutting edges that form the sides of the thread, and the cutting edge at the nose of the tool must all be provided with an adequate amount of relief. Moreover, it is recommended that the effective relief angle, $a_e$, for all of these cutting edges be made equal, although the practice in some shops is to use slightly less relief at the trailing cutting edge. While too much relief may weaken the cutting edge, causing it to chip, an inadequate amount of relief will result in rough threads and in a shortened tool life. Other factors that influence the finish produced on threads include the following: the work material; the cutting speed; the cutting fluid used; the method used to cut the thread; and, the condition of the cutting edge.

Relief angles on single-point thread cutting tools are often specified on the basis of experience. While this method may give satisfactory results in many instances, better results can usually be obtained by calculating these angles, using the formulas provided further on. When special high helix angle threads are to be cut, the magnitude of the relief angles should always be calculated. These calculations are based on the effective relief angle, $a_e$; this is the angle between the flank of the tool and the sloping sides of the thread, measured in a direction parallel to the axis of the thread. Recommended values of this angle are 8 to 14 degrees for high speed steel tools, and 5 to 10 degrees for cemented carbide tools. The larger values are recommended for cutting threads on soft and gummy materials, and the smaller values are for the harder materials, which inherently take a better surface finish. Harder materials also require more support below the cutting edges, which is provided by using a smaller relief angle. These values are recommended for the relief angle below the cutting edge at the nose without any further modification. The angles below the leading and trailing side cutting edges are modified, using the formulas provided. The angles $b$ and $b'$ are the relief angles actually ground on the tool below the leading and trailing side cutting edges respectively; they are measured perpendicular to the side cutting edges. When designing or grinding the thread cutting tool, it is sometimes helpful to know the magnitude of the angle, $n$, for which a formula is provided. This angle would occur only in the event that the tool were ground to a sharp point. It is the angle of the edge formed by the intersection of the flank surfaces.

$$\tan \phi = \frac{\text{lead of thread}}{\pi K}$$

$$\tan \phi' = \frac{\text{lead of thread}}{\pi D}$$

$$a = a_e + \phi$$

$$a' = a_e - \phi'$$

$$\tan b = \tan a \cos (\tfrac{1}{2}\omega)$$

$$\tan b' = \tan a' \cos (\tfrac{1}{2}\omega)$$

$$\tan n = \frac{\tan a - \tan a'}{2 \tan (\tfrac{1}{2}\omega)}$$

Where: $\phi$ = Helix angle of thread at minor diameter

$\phi'$ = Helix angle of thread at major diameter

$K$ = Minor diameter of thread

$D$ = Major diameter of thread

$a$ = Side relief angle parallel to thread axis at leading edge of tool

$a'$ = Side relief angle parallel to thread axis at trailing edge of tool

$a_e$ = Effective relief angle

$b$ = Side relief angle perpendicular to leading edge of tool

$b'$ = Side relief angle perpendicular to trailing edge of tool

$\omega$ = Included angle of thread cutting tool

$n$ = Nose angle resulting from intersection of flank surfaces

*Example:* Calculate the relief angles and the nose angle, $n$, for a single-point thread cutting tool which is to be used to cut a 1-inch diameter, 5 threads-per-inch, double Acme thread. The lead of this thread is $2 \times .200 = .400$ inch. The included angle, $\omega$, of this thread is 29 degrees, the minor diameter, $K$, is .780 inch, and the effective relief angle, $a_e$, below all cutting edges, is to be 10 degrees.

$$\tan \phi = \frac{\text{Lead of thread}}{\pi K} = \frac{.400}{\pi \times .780}$$

$$\phi = 9.27° \ (9°16')$$

$$\tan \phi' = \frac{\text{Lead of thread}}{\pi D} = \frac{.400}{\pi \times 1.000}$$

$$\phi' = 7.26° \ (7°15')$$

$$a = a_e + \phi = 10° + 9.27° = 19.27°$$

$$a' = a_e - \phi' = 10° - 7.26° = 2.74°$$

$$\tan b = \tan a \cos \tfrac{1}{2}\omega = \tan 19.27 \cos 14.5$$

$$b = 18.70° \ (18°42')$$

$$\tan b' = \tan a' \cos \tfrac{1}{2}\omega = \tan 2.74 \cos 14.5$$

$$b' = 2.65° \ (2°39')$$

$$\tan n = \frac{\tan a - \tan a'}{2 \tan (\tfrac{1}{2}\omega)} = \frac{\tan 19.27 - \tan 2.74}{2 \tan 14.5}$$

$$n = 30.26° \ (30°16')$$

## Helix Angles of Screw Threads — Based Upon Pitch Diameters — I

**Threads per Inch**

Helix Angles (Degrees and Minutes) Based Upon Pitch Diameters of Screw Threads of American Standard or U. S. Form.

| Outside Diameter, Inches | 32 | 28 | 26 | 24 | 22 | 20 | 18 | 16 | 14 | 13 | 12 | 11½ | 11 |
|---|---|---|---|---|---|---|---|---|---|---|---|---|---|
| 3/16 | 3°-24' | 3°-57' | 4°-19' | 4°-44' | 5°-14' | 5°-52' | 6°-40' | 7°-43' | .... | .... | .... | .... | .... |
| 1/4 | 2°-29' | 2°-52' | 3°-7' | 3°-24' | 3°-45' | 4°-11' | 4°-44' | 5°-26' | 6°-22' | 6°-59' | .... | .... | .... |
| 5/16 | 1°-57' | 2°-15' | 2°-26' | 2°-40' | 2°-56' | 3°-15' | 3°-40' | 4°-11' | 4°-53' | 5°-20' | 5°-52' | .... | .... |
| 3/8 | 1°-36' | 1°-51' | 2°-00' | 2°-11' | 2°-24' | 2°-40' | 2°-59' | 3°-24' | 3°-57' | 4°-19' | 4°-44' | .... | .... |
| 7/16 | 1°-22' | 1°-34' | 1°-42' | 1°-51' | 2°-2' | 2°-15' | 2°-31' | 2°-52' | 3°-20' | 3°-37' | 3°-57' | .... | 4°-22' |
| 1/2 | 1°-11' | 1°-22' | 1°-29' | 1°-36' | 1°-46' | 1°-57' | 2°-11' | 2°-29' | 2°-52' | 3°-7' | 3°-24' | .... | 3°-45' |
| 9/16 | 1°-3' | 1°-12' | 1°-18' | 1°-25' | 1°-33' | 1°-43' | 1°-55' | 2°-11' | 2°-31' | 2°-44' | 2°-59' | .... | 3°-17' |
| 5/8 | 0°-57' | 1°-5' | 1°-10' | 1°-16' | 1°-23' | 1°-32' | 1°-43' | 1°-57' | 2°-15' | 2°-26' | 2°-40' | .... | 2°-56' |
| 3/4 | 0°-47' | 0°-54' | 0°-58' | 1°-3' | 1°-9' | 1°-16' | 1°-25' | 1°-36' | 1°-51' | 2°-00' | 2°-11' | 2°-17' | 2°-24' |
| 7/8 | 0°-40' | 0°-46' | 0°-50' | 0°-54' | 0°-59' | 1°-5' | 1°-12' | 1°-22' | 1°-34' | 1°-42' | 1°-51' | 1°-56' | 2°-2' |
| 1 | 0°-35' | 0°-40' | 0°-43' | 0°-47' | 0°-51' | 0°-57' | 1°-3' | 1°-11' | 1°-22' | 1°-29' | 1°-36' | 1°-41' | 1°-46' |
| 1⅛ | 0°-31' | 0°-35' | 0°-38' | 0°-42' | 0°-45' | 0°-50' | 0°-56' | 1°-3' | 1°-12' | 1°-18' | 1°-25' | 1°-29' | 1°-33' |
| 1¼ | 0°-28' | 0°-32' | 0°-34' | 0°-37' | 0°-41' | 0°-45' | 0°-50' | 0°-57' | 1°-5' | 1°-10' | 1°-16' | 1°-20' | 1°-23' |
| 1⅜ | 0°-25' | 0°-29' | 0°-31' | 0°-34' | 0°-37' | 0°-41' | 0°-45' | 0°-51' | 0°-59' | 1°-4' | 1°-9' | 1°-12' | 1°-16' |
| 1½ | 0°-23' | 0°-26' | 0°-29' | 0°-31' | 0°-34' | 0°-37' | 0°-42' | 0°-47' | 0°-54' | 0°-58' | 1°-3' | 1°-6' | 1°-9' |
| 1⅝ | 0°-21' | 0°-24' | 0°-26' | 0°-29' | 0°-31' | 0°-34' | 0°-38' | 0°-43' | 0°-49' | 0°-53' | 0°-58' | 1°-1' | 1°-3' |
| 1¾ | 0°-20' | 0°-23' | 0°-24' | 0°-26' | 0°-29' | 0°-32' | 0°-35' | 0°-40' | 0°-46' | 0°-49' | 0°-54' | 0°-56' | 0°-59' |
| 1⅞ | 0°-18' | 0°-21' | 0°-23' | 0°-25' | 0°-27' | 0°-28' | 0°-31' | 0°-37' | 0°-43' | 0°-46' | 0°-50' | 0°-52' | 0°-55' |
| 2 | 0°-17' | 0°-20' | 0°-21' | 0°-23' | 0°-25' | 0°-25' | 0°-27' | 0°-35' | 0°-40' | 0°-43' | 0°-47' | 0°-49' | 0°-51' |
| 2¼ | 0°-15' | 0°-18' | 0°-19' | 0°-21' | 0°-22' | 0°-22' | 0°-25' | 0°-31' | 0°-35' | 0°-38' | 0°-42' | 0°-43' | 0°-45' |
| 2½ | .... | 0°-16' | 0°-17' | 0°-18' | 0°-20' | 0°-20' | 0°-22' | 0°-28' | 0°-32' | 0°-34' | 0°-37' | 0°-39' | 0°-41' |
| 2¾ | .... | .... | 0°-15' | 0°-17' | 0°-18' | 0°-18' | 0°-21' | 0°-25' | 0°-29' | 0°-31' | 0°-34' | 0°-35' | 0°-37' |
| 3 | .... | .... | .... | 0°-15' | 0°-17' | 0°-17' | 0°-19' | 0°-23' | 0°-26' | 0°-29' | 0°-31' | 0°-32' | 0°-34' |
| 3¼ | .... | .... | .... | 0°-14' | 0°-15' | 0°-16' | 0°-18' | 0°-20' | 0°-24' | 0°-26' | 0°-29' | 0°-30' | 0°-31' |
| 3½ | .... | .... | .... | 0°-13' | 0°-14' | 0°-15' | 0°-16' | 0°-18' | 0°-23' | 0°-24' | 0°-26' | 0°-28' | 0°-29' |
| 3¾ | .... | .... | .... | 0°-12' | 0°-13' | 0°-14' | 0°-15' | 0°-17' | 0°-21' | 0°-23' | 0°-25' | 0°-26' | 0°-27' |
| 4 | .... | .... | .... | 0°-11' | 0°-12' | .... | .... | .... | 0°-20' | 0°-21' | 0°-23' | 0°-24' | 0°-25' |

Helix Angles of Screw Threads — Based Upon Pitch Diameters — 2

Helix Angles (Degrees and Minutes) Based Upon Pitch Diameters of Screw Threads of American Standard or U. S. Form.*

| Outside Diameter, Inches | 2 | 3 | 3¼ | 3½ | 4 | 4½ | 5 | 5½ | 6 | 7 | 8 | 9 | 10 |
|---|---|---|---|---|---|---|---|---|---|---|---|---|---|
| | | | | | | | | | | | | Threads per Inch | |
| 7/16 | … | … | … | … | … | … | … | … | … | … | … | … | 4°53' |
| 1/2 | … | … | … | … | … | … | … | … | … | … | 5°26' | 4°44' | 4°11' |
| 9/16 | … | … | … | … | … | … | … | … | … | … | 4°44' | 4°8' | 3°40' |
| 5/8 | … | … | … | … | … | … | … | … | … | 4°53' | 4°11' | 3°40' | 3°15' |
| 3/4 | … | … | … | … | … | … | … | … | 4°44' | 3°58' | 3°24' | 2°59' | 2°40' |
| 7/8 | … | … | … | … | … | … | 4°53' | 4°22' | 3°58' | 3°20' | 2°52' | 2°31' | 2°15' |
| 1 | … | … | … | … | 5°26' | 4°44' | 4°11' | 3°45' | 3°24' | 2°52' | 2°29' | 2°11' | 1°57' |
| 1⅛ | … | … | … | 5°32' | 4°44' | 4°7' | 3°40' | 3°17' | 2°59' | 2°31' | 2°11' | 1°55' | 1°43' |
| 1¼ | … | 5°52' | 5°20' | 4°53' | 4°11' | 3°40' | 3°15' | 2°56' | 2°40' | 2°15' | 1°57' | 1°43' | 1°32' |
| 1⅜ | … | 5°14' | 4°46' | 4°22' | 3°45' | 3°17' | 2°56' | 2°38' | 2°24' | 2°2' | 1°46' | 1°33' | 1°24' |
| 1½ | … | 4°44' | 4°19' | 3°57' | 3°24' | 2°59' | 2°40' | 2°24' | 2°11' | 1°51' | 1°36' | 1°25' | 1°16' |
| 1⅝ | 6°59' | 4°18' | 3°56' | 3°37' | 3°7' | 2°44' | 2°26' | 2°12' | 2°00' | 1°42' | 1°29' | 1°18' | 1°10' |
| 1¾ | 6°22' | 3°57' | 3°37' | 3°20' | 2°52' | 2°31' | 2°15' | 2°2' | 1°51' | 1°34' | 1°22' | 1°12' | 1°5' |
| 1⅞ | 5°52' | 3°40' | 3°21' | 3°5' | 2°40' | 2°20' | 2°5' | 1°53' | 1°43' | 1°28' | 1°16' | 1°7' | 1°00' |
| 2 | 5°26' | 3°24' | 3°7' | 2°52' | 2°29' | 2°11' | 1°57' | 1°46' | 1°36' | 1°22' | 1°11' | 1°3' | 0°57' |
| 2¼ | 4°44' | 2°59' | 2°44' | 2°31' | 2°11' | 1°55' | 1°43' | 1°33' | 1°25' | 1°12' | 1°3' | 0°56' | 0°50' |
| 2½ | 4°11' | 2°40' | 2°26' | 2°15' | 1°57' | 1°43' | 1°32' | 1°24' | 1°16' | 1°5' | 0°57' | 0°50' | 0°45' |
| 2¾ | 3°45' | 2°24' | 2°12' | 2°2' | 1°46' | 1°33' | 1°24' | 1°16' | 1°9' | 0°59' | 0°51' | 0°45' | 0°41' |
| 3 | 3°24' | 2°11' | 2°00' | 1°51' | 1°36' | 1°25' | 1°16' | 1°9' | 1°3' | 0°54' | 0°47' | 0°42' | 0°37' |
| 3¼ | 3°7' | 2°00' | 1°50' | 1°42' | 1°29' | 1°18' | 1°10' | 1°4' | 0°58' | 0°49' | 0°43' | 0°38' | 0°34' |
| 3½ | 2°52' | 1°51' | 1°42' | 1°34' | 1°22' | 1°12' | 1°5' | 0°59' | 0°54' | 0°46' | 0°40' | 0°35' | 0°32' |
| 3¾ | 2°40' | 1°43' | 1°35' | 1°28' | 1°16' | 1°7' | 1°00' | 0°55' | 0°50' | 0°43' | 0°37' | 0°33' | 0°30' |
| 4 | 2°29' | 1°36' | 1°29' | 1°22' | 1°11' | 1°3' | 0°57' | 0°51' | 0°47' | 0°40' | 0°35' | 0°31' | 0°28' |

* To find the tangent of the helix angle equivalent to a given pitch diameter, divide the lead of the thread by 3.1416 × pitch diameter. Pitch diameter = outside diameter − depth of thread.

## Lathe Change Gears

**Change Gears for Thread Cutting.** — To determine the change gears to use for cutting a thread of given pitch, first find what number of threads per inch will be cut when gears of the same size are placed on the lead-screw and spindle stud, either by actual trial or by referring to the index plate; then multiply this number, called the "lathe screw constant," by some trial number to obtain the number of teeth in the gear for the spindle stud, and multiply the threads per inch to be cut by the *same* trial number to obtain the number of teeth in the gear for the lead-screw. Expressing this rule as a formula:

$$\frac{\text{Trial number} \times \text{lathe screw constant}}{\text{Trial number} \times \text{threads per inch to be cut}} = \frac{\text{teeth in gear on spindle stud}}{\text{teeth in gear on lead-screw}}$$

For example, suppose the available change gears supplied with the lathe have 24, 28, 32, 36 teeth, etc., the number increasing by four up to one hundred, and that 10 threads per inch are to be cut in a lathe having a lathe screw constant of 6; then, if the screw constant is written as the numerator, and the number of threads per inch to be cut, as the denominator of a fraction, and both numerator and denominator are multiplied by some trial number, say 4, it is found that gears having 24 and 40 teeth can be used. Thus:

$$\frac{6}{10} = \frac{6 \times 4}{10 \times 4} = \frac{24}{40}$$

The 24-tooth gear goes on the spindle stud and the 40-tooth gear on the lead-screw.

The lathe screw constant is, of course, equal to the number of threads per inch on the lead-screw, provided the spindle stud and spindle are geared in the ratio of 1 to 1, which, however, is not always the case.

**Compound Gearing.** — To find the change gears used in compound gearing, place the screw constant as the numerator, and the number of threads per inch to be cut as the denominator of a fraction; resolve both numerator and denominator into two factors each, and multiply each "pair" of factors by the same number, until values are obtained representing suitable numbers of teeth for the change gears. (One factor in the numerator and one in the denominator make a "pair" of factors.)

*Example:* — $1\frac{3}{4}$ threads per inch are to be cut in a lathe having a screw constant of 8; the available gears have 24, 28, 32, 36, 40 teeth, etc., increasing by four up to one hundred. Following the rule:

$$\frac{8}{1\frac{3}{4}} = \frac{2 \times 4}{1 \times 1\frac{3}{4}} = \frac{(2 \times 36) \times (4 \times 16)}{(1 \times 36) \times (1\frac{3}{4} \times 16)} = \frac{72 \times 64}{36 \times 28}$$

The gears having 72 and 64 teeth are the *driving* gears and those with 36 and 28 teeth are the *driven* gears.

**Fractional Threads.** — Sometimes the lead of a thread is given as a fraction of an inch instead of stating the number of threads per inch. For example, a thread may be required to be cut, having $\frac{3}{8}$ inch lead. The expression "$\frac{3}{8}$ inch lead" should first be transformed to "number of threads per inch." The number of threads per inch (the thread being single) equals:

$$\frac{1}{\frac{3}{8}} = 1 \div \frac{3}{8} = \frac{8}{3} = 2\frac{2}{3}$$

To find the change gears to cut $2\frac{2}{3}$ threads per inch in a lathe having a screw constant 8 and change gears running from 24 to 100 teeth, increasing by 4, proceed as below:

$$\frac{8}{2\frac{2}{3}} = \frac{2 \times 4}{1 \times 2\frac{2}{3}} = \frac{(2 \times 36) \times (4 \times 24)}{(1 \times 36) \times (2\frac{2}{3} \times 24)} = \frac{72 \times 96}{36 \times 64}$$

**Change Gears for Metric Pitches.** — When screws are cut in accordance with the metric system, it is the usual practice to give the lead of the thread in millimeters, instead of the number of threads per unit of measurement. To find the change gears for cutting metric threads, when using a lathe having an English lead-screw, first determine the number of threads per inch corresponding to the given lead in millimeters. Suppose a thread of 3 millimeters lead is to be cut in a lathe having an English lead-screw and a screw constant of 6. As there are 25.4 millimeters per inch, the number of threads per inch will equal $25.4 \div 3$. Place the screw constant as the numerator, and the number of threads per inch to be cut as the denominator:

$$\frac{6}{\frac{25.4}{3}} = 6 \div \frac{25.4}{3} = \frac{6 \times 3}{25.4}$$

The numerator and denominator of this fractional expression of the change gear ratio is next multiplied by some trial number to determine the size of the gears. The first whole number by which 25.4 can be multiplied so as to get a whole number as the result is 5. Thus, $25.4 \times 5 = 127$. Hence, one gear having 127 teeth is always used when cutting metric threads with an English lead-screw. The other gear required in this case has 90 teeth. Thus:

$$\frac{6 \times 3 \times 5}{25.4 \times 5} = \frac{90}{127}$$

Therefore, the following rule can be used to find the change gears for cutting metric pitches with an English lead-screw:

*Rule:* Place the lathe screw constant multiplied by the lead of the required thread in millimeters multiplied by 5, as the numerator of the fraction, and 127 as the denominator. The product of the numbers in the numerator equals the number of teeth for the spindle-stud gear, and 127 is the number of teeth for the lead-screw gear.

If the lathe has a metric pitch lead-screw, and a screw having a given number of threads per inch is to be cut, first find the "metric screw constant" of the lathe or the lead of thread in millimeters that would be cut with change gears of equal size on the lead-screw and spindle stud; then the method of determining the change gears is simply the reverse of the one already explained for cutting a metric thread with an English lead-screw.

*Rule:* To find the change gears for cutting English threads with a metric lead-screw, place 127 in the numerator and the threads per inch to be cut, multiplied by the metric screw constant multiplied by 5, in the denominator; 127 is the number of teeth on the spindle-stud gear and the product of the numbers in the denominator equals the number of teeth in the lead-screw gear.

**Threads per Inch Obtained with a Given Combination of Gears.** — To determine the number of threads per inch that will be obtained with a given combination of gearing, multiply the lathe screw constant by the number of teeth in the *driven* gear (or by the product of the numbers of teeth in both driven gears of compound gearing), and divide the product thus obtained by the number of teeth in the *driving* gear (or by the product of the two driving gears of a compound train). The quotient equals the number of threads per inch.

**Change Gears for Fractional Ratios.** — When gear ratios cannot be expressed exactly in whole numbers which are within the range of ordinary gearing, the combination of gearing required for the fractional ratio may be determined quite easily, in some cases, by the "cancellation method." To illustrate this method, assume that the speeds of two gears are to be in the ratio of 3.423 to 1. The number 3.423 is first changed to $\frac{3423}{1000}$ to clear it of decimals. Then, in order to secure a fraction that can be reduced, 3423 is changed to 3420;

$$\frac{3420}{1000} = \frac{342}{100} = \frac{3 \times 2 \times 57}{2 \times 50} = \frac{3 \times 57}{1 \times 50}$$

Then, multiplying $\frac{3}{1}$ by some trial number, say, 24, the following gear combination is obtained:

$$\frac{72}{24} \times \frac{57}{50} = \frac{4104}{1200} = \frac{3.42}{1}$$

As the desired ratio is 3.423 to 1, there is an error of 0.003. When the ratios are comparatively simple, the cancellation method is not difficult and is frequently used; but by the logarithmic method to be described, more accurate results are possible in most cases.

**Modifying the Quick Change Gearbox Output.** — On most modern lathes the gear train connecting the headstock spindle with the lead screw contains a quick change gearbox. Instead of using different change gears, it is only necessary to position the handles of the gearbox to adjust the speed ratio between the spindle and the lead screw in preparation for cutting a thread. There are, however, occasions when a thread must be cut for which there is no quick change gearbox setting. In this event, it is necessary to modify the normal, or standard, gear ratio between the spindle and the gearbox by installing modifying change gears to replace the standard gears normally used. Metric and other odd pitch threads can be cut on lathes that have an inch thread lead screw and a quick change gearbox having only settings for inch threads by using modifying change gears in the gear train. Likewise, inch threads and other odd pitch threads can be cut on metric lead screw lathes having a gearbox on which only metric thread setting can be made. Modifying change gears also can be used for cutting odd pitch threads on lathes having a quick change gearbox that has both inch and metric thread settings.

The sizes of the modifying change gears can be calculated by formulas, which are given further on; they depend on the thread to be cut and on the setting of the quick change gearbox. Many different sets of gears can be found for each thread to be cut. It is recommended that in each case several calculations be made in order to find the set of gears that is most suitable for installation on the lathe. The modifying change gear formulas that follow are based on the type of lead screw; i.e., whether the lead screw has inch or metric threads.

*Metric Threads on Inch Lead Screw Lathes:* A 127-tooth translating gear must be used in the modifying change gear train in order to be able to cut metric threads on inch lead screw lathes. The formula for calculating the modifying change gears is:

$$\frac{5 \times \text{gearbox setting in thds/in.} \times \text{pitch in mm to be cut}}{127} = \frac{\text{driving gears}}{\text{driven gears}}$$

Equal numbers, called trial numbers, are added to the numerator and to the denominator of this formula in order to find the gears. If suitable gears cannot be found with one set, then another set of equal trial numbers is used. (Since these numbers are equal, such as 15/15 or 24/24, they are equal to the number one when thought of as a fraction; their inclusion has the effect of multiplying the formula by one, which does not change its value.) It is necessary to select the gearbox setting in threads per inch which must be used to cut the metric thread when using the gears calculated by the formula. One method is to select a quick change gearbox setting that is close to the actual number of metric threads in a one-inch length, called the equivalent threads per inch, which can be calculated by the following formula: Equivalent thds/in. = 25.4 ÷ pitch in mm to be cut.

*Example:* Select the quick change gearbox setting and calculate the modifying change gears required to set up a lathe having an inch-thread lead screw in order to cut an M12 × 1.75 metric thread.

$$\text{Equivalent thds/in.} = \frac{25.4}{\text{pitch in mm to be cut}} = \frac{25.4}{1.75} = 14.5 \ (\text{Use 14 thds/in.})$$

$$\frac{5 \times \text{gearbox setting in thds/in.} \times \text{pitch in mm to be cut}}{127} = \frac{5 \times 14 \times 1.75}{127}$$

$$= \frac{(24) \times 5 \times 14 \times 1.75}{(24) \times 127} = \frac{(5 \times 14) \times (24 \times 1.75)}{24 \times 127}$$

$$= \frac{70 \times 42}{24 \times 127} = \frac{\text{driving gears}}{\text{driven gears}}$$

*Odd Inch Pitch Threads:* The calculation of the modifying change gears used for cutting odd pitch threads that are specified by their pitch in inches involves the sizes of the standard gears, which can be found by counting their teeth. Standard gears are those used to enable the lathe to cut the thread for which the gearbox setting is made; they are the gears that are normally used. The threads on worms used with worm gears are among the odd pitch threads that can be cut by this method. As before, it is usually advisable to calculate the actual number of threads per inch of the odd pitch thread and to select a gearbox setting that is close to this value. The following formula is used to calculate the modifying change gears to cut odd inch pitch threads:

$$\frac{\text{standard driving gear} \times \text{pitch to be cut in inches} \times \text{gearbox setting in thds/in.}}{\text{standard driven gear}}$$

$$= \frac{\text{driving gears}}{\text{driven gears}} \ .$$

*Example:* Select the quick change gearbox setting and calculate the modifying change gears required to cut a thread having a pitch equal to .195 inch. The standard driving and driven gears both have 48 teeth. To find equivalent threads per inch:

$$\text{thds/in.} = \frac{1}{\text{pitch}} = \frac{1}{.195} = 5.13 \ (\text{Use 5 thds/in.})$$

$$\frac{\text{standard driving gear} \times \text{pitch to be cut in inches} \times \text{gearbox setting in thds/in.}}{\text{standard driven gear}}$$

$$= \frac{48 \times .195 \times 5}{48} = \frac{(1000) \times .195 \times 5}{(1000)} = \frac{195 \times 5}{500 \times 2} = \frac{39 \times 5}{100 \times 2} = \frac{39 \times 5 \times (8)}{50 \times 2 \times 2 \times (8)}$$

$$= \frac{39 \times 40}{50 \times 32} = \frac{\text{driving gears}}{\text{driven gears}}$$

It will be noted that in the second step above 1000/1000 has been substituted for 48/48. This substitution does not change the ratio. The reason why this substitution was made is that 1000 × .195 = 195, a whole number. Actually 200/200 might have been substituted since 200 × .195 = 39, also a whole number.

The procedure for calculating the modifying gears using the following formulas is the same as illustrated by the two previous examples.

*Odd Threads per Inch on Inch Lead Screw Lathes:*

$$\frac{\text{standard driving gear} \times \text{gearbox setting in thds/in.}}{\text{standard driven gear} \times \text{thds/in. to be cut}} = \frac{\text{driving gears}}{\text{driven gears}}$$

*Inch Threads on Metric Lead Screw Lathes:*

$$\frac{127}{5 \times \text{gearbox setting in mm pitch} \times \text{thds/in. to be cut}} = \frac{\text{driving gears}}{\text{driven gears}}$$

*Odd Metric Pitch Threads on Metric Lead Screw Lathes::*

$$\frac{\text{standard driving gear} \times \text{mm pitch to be cut}}{\text{standard driven gear} \times \text{gearbox setting in mm pitch}} = \frac{\text{driving gears}}{\text{driven gears}}$$

**Tabulated Logarithms of Change-gear Ratios.** — Change-gear problems can be solved readily by the use of the accompanying tables which contain the six-place logarithms of the ratios of all gear combinations between 16 and 120 teeth, inclusive, excepting the 1 to 1 ratios. To illustrate how these logarithms of ratios were obtained, take as an example gears having 72 and 41 teeth, respectively; the ratio equals 72 divided by 41, and to divide by means of logarithms, the logarithm of one number is subtracted from the logarithm of the other, thus:

$$\log 72 = 1.857333$$
$$\log 41 = \underline{1.612784}$$
$$\text{ratio log} = 0.244549$$

The logarithms for ratios of pairs of gears having between 16 and 120 teeth have been arranged in numerical order in the tables. In a number of cases, more than one combination gives the same logarithm, so that the different pairs of gears that equal the logarithm have been repeated. In some simple cases, only the ratio has been given in order to shorten the table; for instance, all the gear combinations that equal a 2 to 1 ratio have been omitted and only the ratio is given.

There are nearly 5000 different ratios represented in the gear tables between the extremes 1.0084 + to 1 (120:119) and 7½ to 1 (120:16). As the sum of any two two-gear logarithms equals a four-gear logarithm, the tables represent over 12,000,000 four-gear combinations; and by using three pairs of gears in a train, there are over 20,000,000,000 six-gear combinations available.

**Driving and Driven Gears.** — Insofar as the use of the gear logarithm tables is concerned, it is immaterial which is the driver and which is the driven gear, and by comparing the gears selected with the ratio, no confusion should result. For example, gears for the ratio 7.32 : 4.17 are selected in the same manner as gears for the ratio 4.17 : 7.32, each ratio being the reciprocal of the other so that the first ratio can be obtained from the second, and vice versa, by merely inverting the ratio. Thus, the logarithm of the smaller number is *always* subtracted from the logarithm of the larger to get the logarithm of the ratio, or its reciprocal, as the case may be, since the table is set up to correspond with this order.

The speeds of the driving and driven gears in a four-gear train are related to the numbers of teeth as follows:

$$\frac{\text{Speed of Driving Shaft}}{\text{Speed of Driven Shaft}} = \frac{\text{Teeth in 1st Driven Gear} \times \text{Teeth in 2nd Driven Gear}}{\text{Teeth in 1st Driving Gear} \times \text{Teeth in 2nd Driving Gear}}$$

**Solving Two-gear Change-gear Problems by Use of Logarithms.** — Suppose that two gears having the ratio 3.423 : 1 are desired. Log 3.423 = 0.534407. From the table, the logarithm nearest to this is log 89 : 26 = 0.534417; therefore, the gears having 89 and 26 teeth are the nearest to the ratio 3.423 to 1, and as 89 ÷ 26 = 3.423077, the ratio error is 0.000077.

When solving gear problems, the ratio should be reduced to terms of 1. For example, what two gears will have teeth numbers in the ratio of 7.182 to 3.902? $\frac{7.182}{3.902} = \frac{1.84059}{1}$ ; the log of 1.84059 is 0.264958. From the table, 81 : 44 = log 0.265032. As 81 ÷ 44 = 1.84091, the error is 0.00032.

A more rapid solution of the same problem makes use of logarithms:

$$\begin{aligned} \log 7.182 &= 0.856245 \\ \log 3.902 &= 0.591287 \\ \log \text{ of ratio} &= 0.264958 \text{ is closest to log 81 : 44 from table.} \end{aligned}$$

**Finding Four-gear Ratios.** — When four gears must be used, the gear logarithms make it possible to obtain results of high accuracy quickly and with minimum effort. For example, suppose it is desired to find four gears that will have numbers of teeth in a ratio of 2.105399 to 1. Log 2.105399 = 0.323334. To keep the ratio about equal in each pair of gears, select from the table that set of gears the logarithm of which is equal to about one-half the ratio logarithm, as log 57 : 37 = 0.187673. By subtracting this from the log of 2.105399, 0.323334, the other logarithm is found to be 0.135561. From the table, log 41 : 30 = 0.135663. The result obtained is: $\frac{57}{37} \times \frac{41}{30} = \frac{2337}{1110} = 2.105405$. The error in the ratio is 0.000006.

In case no combination can be found that nearly equals the logarithm of the ratio, a suitable four-gear combination may be found by a slightly different procedure. For example, what gears will have teeth in the ratio of 595 to 594? As before, take the logarithms from a six-place table,

$$\begin{aligned} \log 595 &= 2.774517 \\ \log 594 &= 2.773786 \\ \log \text{ of ratio} &= 0.000731 \end{aligned}$$

Next, from the table of logarithms for gear ratios, select any ratio, say log 72 : 70 = 0.012235, and add the logarithm of the ratio 595 : 594, or 0.000731; the sum is 0.012966. Select the logarithm nearest the sum from the table; this is found to be log 68 : 66 = 0.012965. Now by inverting the first pair selected, 72 : 70, and multiplying this inverted pair by the ratio just determined, 68 : 66, the desired ratio will be obtained. Thus, inverting 72 : 70 gives 70 : 72. Multiplying by 68 : 66 gives $\frac{70}{72} \times \frac{68}{66} = \frac{4760}{4752} = \frac{595}{594}$.

**Lathe Change-gears.** — For calculating the change-gears to cut any pitch on a lathe, the "constant" of the machine must be known. For any lathe, $C : L =$ driver : driven gear, in which $C$ = constant of machine and $L$ = thds. per inch.

For example, what change-gears are required to cut 1.7345 threads per inch on a lathe having a constant of 4?

$$C : L = 4 : 1.7345$$
$$\log 4 = 0.602060$$
$$\log 1.7345 = 0.239174$$
$$\text{ratio log} = 0.362886$$

From the table,  $\log 113 : 49 = 0.362882$

$$\text{error in log of ratio} = 0.000004$$

Therefore, the driver has 113 teeth, and the driven gear, 49 teeth.

**Relieving Helical Fluted Hobs.** — The problem of relieving hobs that have been fluted at right angles to the thread is an example of the special application of the gear logarithms to difficult problems. The usual method is to alter the angle of the helical flutes to agree with previously calculated change-gears. The ratio between the hob and relieving attachment cam is expressed by the following terms:

$$\frac{N}{\cos^2\alpha} : C = \text{drivers : driven gears}$$

$$\text{Tan } \alpha = \frac{P}{H_c}$$

in which
- $N$ = number of flutes in hob;
- $\alpha$ = helix angle of thread from plane perpendicular to axis;
- $C$ = constant of relieving attachment;
- $P$ = axial lead of hob;
- $H_c$ = hob pitch circumference, or 3.1416 × pitch diameter.

The constant of a relieving attachment can be found on its index-plate, and is determined by the number of flutes that require equal gears on the change-gear studs. This will vary with different makes of lathes.

*Example.* — What four change-gears must be used to relieve a helical fluted worm-gear hob, 24 diametral pitch, sextuple thread, 13 degrees, 41 minutes helix angle of thread, with eleven helical flutes, assuming that a relieving attachment having a constant of 4 is to be used?

$$\text{Cos 13 degrees, 41 minutes} = 0.97162$$
$$0.97162^2 = 0.944045$$
$$\frac{N \div \cos^2\alpha}{C} = \frac{N}{C \times \cos^2\alpha}; \quad \frac{11}{4 \times 0.944045} = \frac{11}{3.776}$$
$$\log 11 = 1.04139$$
$$\log 3.776 = 0.57703$$
$$\log \text{ratio} = 0.46436$$
$$\text{From tables, } \log 67 : 39 = 0.23501$$
$$\text{Subtracting from log ratio} = 0.22935$$
$$\text{From table, } \log 78 : 46 = 0.22933$$

Therefore, the gears are $\dfrac{67}{39} \times \dfrac{78}{46} = \dfrac{\text{drivers}}{\text{driven}}$

The ratio of these gears equals 2.913 which is the ratio represented by 11 ÷ 3.776. In relieving hobs for spur gears, the *normal* pitch (or lead of a single-threaded hob) should equal the circular pitch of the gear, and the *axial* pitch = normal pitch ÷ cos $\alpha$. Sine $\alpha$ = normal pitch ÷ $H_c$.

## Logarithms of Gear Ratios from 1.0084+ to 7.5

| Numbers of Teeth | Logarithm of Ratio | Numbers of Teeth | Logarithm of Ratio | Numbers of Teeth | Logarithm of Ratio | Numbers of Teeth | Logarithm of Ratio |
|---|---|---|---|---|---|---|---|
| 120 : 16 | 0.875061 | 99 : 16 | 0.791515 | 101 : 18 | 0.749049 | 115 : 22 | 0.718275 |
| 119 : 16 | .871427 | 105 : 17 | .790740 | 112 : 20 | .748188 | 94 : 18 | .717855 |
| 118 : 16 | .867762 | 111 : 18 | .790051 | 95 : 17 | .747275 | 120 : 23 | .717453 |
| 117 : 16 | .864065 | 98 : 16 | .787106 | 106 : 19 | .746552 | 99 : 19 | .716882 |
| 116 : 16 | .860338 | 104 : 17 | .786584 | 117 : 21 | .745967 | 109 : 21 | .715207 |
| 115 : 16 | .856578 | 110 : 18 | .786120 | 89 : 16 | .745270 | 83 : 16 | .714958 |
| 114 : 16 | .852785 | 116 : 19 | .785704 | 100 : 18 | .744728 | 114 : 22 | .714482 |
| 113 : 16 | .848958 | 97 : 16 | .782652 | 111 : 20 | .744293 | 88 : 17 | .714034 |
| 120 : 17 | 0.848732 | 103 : 17 | 0.782388 | 94 : 17 | 0.742680 | 119 : 23 | 0.713819 |
| 119 : 17 | .845098 | 109 : 18 | .782154 | 105 : 19 | .742436 | 93 : 18 | .713210 |
| 112 : 16 | .845098 | 115 : 19 | .781944 | 116 : 21 | .742239 | 98 : 19 | .712473 |
| 118 : 17 | .841433 | 114 : 19 | .778151 | 110 : 20 | .740363 | 103 : 20 | .711807 |
| 111 : 16 | .841203 | 108 : 18 | .778151 | 99 : 18 | .740363 | 108 : 21 | .711205 |
| 117 : 17 | .837737 | 102 : 17 | .778151 | 88 : 16 | .740363 | 113 : 22 | .710656 |
| 110 : 16 | .837272 | 96 : 16 | .778151 | 115 : 21 | .738479 | 118 : 23 | .710154 |
| 116 : 17 | .834009 | 119 : 20 | .774517 | 104 : 19 | .738280 | 82 : 16 | .709694 |
| 109 : 16 | 0.833307 | 113 : 19 | 0.774325 | 93 : 17 | 0.738034 | 87 : 17 | 0.709070 |
| 115 : 17 | .830249 | 107 : 18 | .774111 | 120 : 22 | .736759 | 92 : 18 | .708515 |
| 108 : 16 | .829304 | 101 : 17 | .773873 | 109 : 20 | .736397 | 97 : 19 | .708018 |
| 114 : 17 | .826456 | 95 : 16 | .773604 | 98 : 18 | .735954 | 102 : 20 | .707570 |
| 107 : 16 | .825264 | 112 : 19 | .770464 | 87 : 16 | .735400 | 107 : 21 | .707165 |
| 120 : 18 | .823909 | 106 : 18 | .770033 | 114 : 21 | .734686 | 112 : 22 | .706795 |
| 113 : 17 | .822629 | 100 : 17 | .769551 | 103 : 19 | .734084 | 117 : 23 | .706458 |
| 106 : 16 | .821186 | 94 : 16 | .769008 | 92 : 17 | .733339 | 81 : 16 | .704365 |
| 119 : 18 | 0.820275 | 117 : 20 | 0.767156 | 119 : 22 | 0.733124 | 86 : 17 | 0.704050 |
| 112 : 17 | .818769 | 111 : 19 | .766570 | 108 : 20 | .732394 | 91 : 18 | .703769 |
| 105 : 16 | .817069 | 105 : 18 | .765917 | 97 : 18 | .731499 | 96 : 19 | .703518 |
| 118 : 18 | .816609 | 99 : 17 | .765186 | 113 : 21 | .730859 | 101 : 20 | .703291 |
| 111 : 17 | .814874 | 93 : 16 | .764363 | 86 : 16 | .730379 | 106 : 21 | .703087 |
| 117 : 18 | .812913 | 116 : 20 | .763429 | 102 : 19 | .729847 | 111 : 22 | .702900 |
| 104' : 16 | .812913 | 110 : 19 | .762639 | 118 : 22 | .729459 | 116 : 23 | .702730 |
| 110 : 17 | .810944 | 104 : 18 | .761761 | 91 : 17 | .728593 | 120 : 24 | .698970 |
| 116 : 18 | 0.809186 | 98 : 17 | 0.760777 | 107 : 20 | 0.728354 | 115 : 23 | 0.698970 |
| 103 : 16 | .808717 | 115 : 20 | .759668 | 112 : 21 | .726999 | 110 : 22 | .698970 |
| 109 : 17 | .806978 | 92 : 16 | .759668 | 96 : 18 | .726999 | 105 : 21 | .698970 |
| 115 : 18 | .805425 | 109 : 19 | .758673 | 117 : 22 | .725763 | 100 : 20 | .698970 |
| 102 : 16 | .804480 | 103 : 18 | .757565 | 101 : 19 | .725568 | 95 : 19 | .698970 |
| 108 : 17 | .802975 | 120 : 21 | .756962 | 85 : 16 | .725299 | 90 : 18 | .698970 |
| 114 : 18 | .801632 | 114 : 20 | .755875 | 106 : 20 | .724276 | 85 : 17 | .698970 |
| 120 : 19 | .800428 | 91 : 16 | .754921 | 90 : 17 | .723794 | 80 : 16 | .698970 |
| 101 : 16 | 0.800201 | 108 : 19 | 0.754670 | 111 : 21 | 0.723104 | 119 : 24 | 0.695336 |
| 107 : 17 | .798936 | 119 : 21 | .753328 | 95 : 18 | .722451 | 114 : 23 | .695171 |
| 113 : 18 | .797806 | 102 : 18 | .753328 | 116 : 22 | .722035 | 109 : 22 | .695004 |
| 119 : 19 | .796793 | 113 : 20 | .752048 | 100 : 19 | .721246 | 104 : 21 | .694814 |
| 100 : 16 | .795880 | 96 : 16 | .751822 | 105 : 20 | .720159 | 99 : 20 | .694605 |
| 106 : 17 | .794857 | 107 : 19 | .750630 | 84 : 16 | .720159 | 94 : 19 | .694374 |
| 112 : 18 | .793946 | 90 : 16 | .750123 | 110 : 21 | .719173 | 89 : 18 | 0.694118 |
| 118 : 19 | .793128 | 118 : 21 | .749663 | 89 : 17 | .718941 | 84 : 17 | .693830 |

Logarithms of Gear Ratios from 1.0084+ to 7.5

| Numbers of Teeth | Logarithm of Ratio | Numbers of Teeth | Logarithm of Ratio | Numbers of Teeth | Logarithm of Ratio | Numbers of Teeth | Logarithm of Ratio |
|---|---|---|---|---|---|---|---|
| 79 : 16 | 0.693507 | 103 : 22 | 0.670415 | 76 : 17 | 0.650365 | 90 : 21 | 0.632023 |
| 118 : 24 | .691671 | 117 : 25 | .670246 | 116 : 26 | .649485 | 107 : 25 | .631444 |
| 113 : 23 | .691351 | 112 : 24 | .669007 | 107 : 24 | .649173 | 77 : 18 | .631218 |
| 108 : 22 | .691001 | 98 : 21 | .669007 | 98 : 22 | .648803 | 94 : 22 | .630705 |
| 103 : 21 | .690618 | 84 : 18 | .669007 | 89 : 20 | .648360 | 111 : 26 | .630350 |
| 98 : 20 | .690196 | 107 : 23 | .667656 | 80 : 18 | .647817 | 81 : 19 | .629731 |
| 93 : 19 | .689729 | 93 : 20 | .667453 | 120 : 27 | .647817 | 98 : 23 | .629498 |
| 88 : 18 | .689210 | 79 : 17 | .667178 | 111 : 25 | .647383 | 115 : 27 | .629334 |
| 83 : 17 | 0.688629 | 116 : 25 | 0.666518 | 71 : 16 | 0.647138 | 119 : 28 | 0.628389 |
| 117 : 24 | .687975 | 102 : 22 | .666178 | 102 : 23 | .646872 | 85 : 20 | .628389 |
| 78 : 16 | .687975 | 88 : 19 | .665730 | 93 : 21 | .646264 | 68 : 16 | .628389 |
| 112 : 23 | .687490 | 111 : 24 | .665112 | 115 : 26 | .645725 | 106 : 25 | .627366 |
| 107 : 22 | .686961 | 74 : 16 | .665112 | 84 : 19 | .645526 | 89 : 21 | .627171 |
| 102 : 21 | .686381 | 97 : 21 | .664552 | 106 : 24 | .645095 | 72 : 17 | .626884 |
| 97 : 20 | .685742 | 120 : 26 | .664208 | 75 : 17 | .644612 | 110 : 26 | .626419 |
| 92 : 19 | .685034 | 83 : 18 | .663806 | 97 : 22 | .644349 | 93 : 22 | .626060 |
| 116 : 24 | 0.684247 | 106 : 23 | 0.663578 | 119 : 27 | 0.644183 | 114 : 27 | 0.625541 |
| 87 : 18 | .684247 | 115 : 25 | .662758 | 110 : 25 | .643453 | 76 : 18 | .625541 |
| 111 : 23 | .683595 | 92 : 20 | .662758 | 88 : 20 | .643453 | 97 : 23 | .625044 |
| 82 : 17 | .683365 | 101 : 22 | .661899 | 101 : 23 | .642594 | 118 : 28 | .624724 |
| 106 : 22 | .682883 | 78 : 17 | .661645 | 79 : 18 | .642355 | 80 : 19 | .624336 |
| 77 : 16 | .682371 | 110 : 24 | .661182 | 114 : 26 | .641932 | 101 : 24 | .624110 |
| 101 : 21 | .682102 | 87 : 19 | .660766 | 92 : 21 | .641569 | 105 : 25 | .623249 |
| 120 : 25 | .681241 | 119 : 26 | .660574 | 105 : 24 | .640978 | 84 : 20 | .623249 |
| 96 : 20 | 0.681241 | 96 : 21 | 0.660052 | 70 : 16 | 0.640978 | 109 : 26 | 0.622453 |
| 115 : 24 | .680487 | 105 : 23 | .659462 | 118 : 27 | .640518 | 88 : 21 | .622263 |
| 91 : 19 | .680288 | 73 : 16 | .659203 | 83 : 19 | .640324 | 67 : 16 | .621955 |
| 110 : 23 | .679665 | 114 : 25 | .658965 | 96 : 22 | .639849 | 113 : 27 | .621715 |
| 86 : 18 | .679226 | 82 : 18 | .658541 | 109 : 25 | .639487 | 92 : 22 | .621365 |
| 105 : 22 | .678767 | 91 : 20 | .658011 | 74 : 17 | .638783 | 117 : 28 | .621028 |
| 81 : 17 | .678036 | 100 : 22 | .657577 | 87 : 20 | .638489 | 71 : 17 | .620809 |
| 100 : 21 | .677781 | 109 : 24 | .657215 | 100 : 23 | .638272 | 96 : 23 | .620543 |
| 119 : 25 | 0.677607 | 118 : 26 | 0.656909 | 113 : 26 | 0.638105 | 100 : 24 | 0.619789 |
| 114 : 24 | .676694 | 77 : 17 | .656042 | 117 : 27 | .636822 | 75 : 18 | .619789 |
| 95 : 20 | .676694 | 86 : 19 | .655745 | 104 : 24 | .636822 | 104 : 25 | .619093 |
| 76 : 16 | .676694 | 95 : 21 | .655504 | 91 : 21 | .636822 | 79 : 19 | .618874 |
| 109 : 23 | .675699 | 104 : 23 | .655306 | 78 : 18 | .636822 | 108 : 26 | .618451 |
| 90 : 19 | .675489 | 113 : 25 | .655138 | 108 : 25 | .635484 | 83 : 20 | .618048 |
| 104 : 22 | .674611 | 117 : 26 | .653213 | 95 : 22 | .635301 | 112 : 27 | .617854 |
| 85 : 18 | .674146 | 108 : 24 | .653213 | 82 : 19 | .635060 | 116 : 28 | .617300 |
| 118 : 25 | 0.673942 | 99 : 22 | 0.653213 | 69 : 16 | 0.634729 | 87 : 21 | 0.617300 |
| 99 : 21 | .673416 | 90 : 20 | .653213 | 112 : 26 | .634245 | 120 : 29 | .616783 |
| 113 : 24 | .672867 | 81 : 18 | .653213 | 99 : 23 | .633907 | 91 : 22 | .616619 |
| 80 : 17 | .672641 | 72 : 16 | .653213 | 86 : 20 | .633469 | 95 : 23 | .615996 |
| 94 : 20 | .672098 | 112 : 25 | .651278 | 116 : 27 | .633094 | 99 : 24 | .615424 |
| 108 : 23 | .671696 | 103 : 23 | .651109 | 73 : 17 | .632874 | 66 : 16 | .615424 |
| 75 : 16 | .670941 | 94 : 21 | .650909 | 103 : 24 | .632626 | 103 : 25 | .614897 |
| 89 : 19 | .670636 | 85 : 19 | .650665 | 120 : 28 | .632023 | 70 : 17 | .614649 |

## Logarithms of Gear Ratios from 1.0084+ to 7.5

| Numbers of Teeth | Logarithm of Ratio | Numbers of Teeth | Logarithm of Ratio | Numbers of Teeth | Logarithm of Ratio | Numbers of Teeth | Logarithm of Ratio |
|---|---|---|---|---|---|---|---|
| 107 : 26 | 0.614411 | 67 : 17 | 0.595626 | 99 : 26 | 0.580662 | 81 : 22 | 0.566062 |
| 111 : 27 | .613959 | 63 : 16 | .595221 | 118 : 31 | .580520 | 92 : 25 | .565848 |
| 74 : 18 | .613959 | 118 : 30 | .594761 | 114 : 30 | .579784 | 103 : 28 | .565679 |
| 115 : 28 | .613540 | 114 : 29 | .594507 | 95 : 25 | .579784 | 114 : 31 | .565543 |
| 78 : 19 | .613341 | 110 : 28 | .594235 | 76 : 20 | .579784 | 88 : 24 | .564271 |
| 119 : 29 | .613149 | 106 : 27 | .593942 | 110 : 29 | .578995 | 110 : 30 | .564271 |
| 82 : 20 | .612784 | 102 : 26 | .593627 | 91 : 24 | .578830 | 99 : 27 | .564271 |
| 86 : 21 | .612279 | 98 : 25 | .593286 | 72 : 19 | .578579 | 77 : 21 | .564271 |
| 90 : 22 | 0.611820 | 94 : 24 | 0.592917 | 106 : 28 | 0.578148 | 66 : 18 | 0.564271 |
| 94 : 23 | .611400 | 90 : 23 | .592515 | 87 : 23 | .577792 | 117 : 32 | .563036 |
| 98 : 24 | .611015 | 86 : 22 | .592076 | 102 : 27 | .577236 | 106 : 29 | .562908 |
| 102 : 25 | .610660 | 82 : 21 | .591595 | 68 : 18 | .577236 | 95 : 26 | .562750 |
| 106 : 26 | .610333 | 117 : 30 | .591065 | 117 : 31 | .576824 | 84 : 23 | .562552 |
| 110 : 27 | .610029 | 78 : 20 | .591065 | 83 : 22 | .576655 | 73 : 20 | .562293 |
| 114 : 28 | .609747 | 113 : 29 | .590680 | 98 : 26 | .576253 | 62 : 17 | .561943 |
| 118 : 29 | .609484 | 74 : 19 | .590478 | 113 : 30 | .575957 | 113 : 31 | .561717 |
| 65 : 16 | 0.608793 | 109 : 28 | 0.590269 | 64 : 17 | 0.575731 | 102 : 28 | 0.561442 |
| 69 : 17 | .608400 | 105 : 27 | .589826 | 79 : 21 | .575408 | 91 : 25 | .561101 |
| 73 : 18 | .608050 | 70 : 18 | .589826 | 94 : 25 | .575188 | 120 : 33 | .560667 |
| 77 : 19 | .607737 | 101 : 26 | .589348 | 109 : 29 | .575029 | 80 : 22 | .560667 |
| 81 : 20 | .607455 | 66 : 17 | .589105 | 120 : 32 | .574031 | 109 : 30 | .560305 |
| 85 : 21 | .607200 | 97 : 25 | .588832 | 105 : 28 | .574031 | 69 : 19 | .560096 |
| 89 : 22 | .606967 | 62 : 16 | .588272 | 90 : 24 | .574031 | 98 : 27 | .559862 |
| 93 : 23 | .606755 | 93 : 24 | .588272 | 75 : 20 | .574031 | 116 : 32 | .559308 |
| 97 : 24 | 0.606561 | 120 : 31 | 0.587820 | 60 : 16 | 0.574031 | 87 : 24 | 0.559308 |
| 101 : 25 | .606381 | 89 : 23 | .587662 | 116 : 31 | .573096 | 105 : 29 | .558791 |
| 105 : 26 | .606216 | 116 : 30 | .587337 | 86 : 23 | .572770 | 76 : 21 | .558594 |
| 109 : 27 | .606063 | 85 : 22 | .586996 | 71 : 19 | .572505 | 94 : 26 | .558155 |
| 113 : 28 | .605920 | 112 : 29 | .586820 | 112 : 30 | .572097 | 112 : 31 | .557856 |
| 117 : 29 | .605788 | 108 : 28 | .586266 | 97 : 26 | .571798 | 65 : 18 | .557641 |
| 64 : 16 or any 4 to 1 ratio | 0.602060 | 81 : 21 | .586266 | 82 : 22 | .571391 | 83 : 23 | .557350 |
|  |  | 104 : 27 | .585670 | 108 : 29 | .571026 | 101 : 28 | .557163 |
|  |  | 77 : 20 | 0.585461 | 67 : 18 | 0.570802 | 119 : 33 | 0.557033 |
| 119 : 30 | .598426 | 100 : 26 | .585027 | 93 : 25 | .570543 | 108 : 30 | .556303 |
| 115 : 29 | .598300 | 73 : 19 | .584569 | 119 : 32 | .570397 | 90 : 25 | .556303 |
| 111 : 28 | .598165 | 96: 25 | .584331 | 104 : 28 | .569875 | 72 : 20 | .556303 |
| 107 : 27 | .598020 | 119 : 31 | .584185 | 78 : 21 | .569875 | 115 : 32 | .555548 |
| 103 : 26 | .597864 | 115 : 30 | .583577 | 115 : 31 | .569336 | 97 : 27 | .555408 |
| 99 : 25 | 0.597695 | 92 : 24 | .583577 | 89 : 24 | .569179 | 79 : 22 | .555204 |
| 95 : 24 | .597512 | 69 : 18 | .583577 | 63 : 17 | .568892 | 61 : 17 | .554881 |
| 91 : 23 | .597314 | 111 : 29 | 0.582925 | 100 : 27 | 0.568636 | 104 : 29 | 0.554635 |
| 87 : 22 | .597097 | 88: 23 | .582755 | 111 : 30 | .568202 | 86 : 24 | .554287 |
| 83 : 21 | .596859 | 65 : 17 | .582465 | 74 : 20 | .568202 | 111 : 31 | .553961 |
| 79 : 20 | .596597 | 107 : 28 | .582226 | 85 : 23 | .567691 | 68 : 19 | .553755 |
| 75 : 19 | .596308 | 84 : 22 | .581857 | 96 : 26 | .567298 | 93 : 26 | .553510 |
| 71 : 18 | .595986 | 103 : 27 | .581473 | 107 : 29 | .566986 | 118 : 33 | .553368 |
|  |  | 61 : 16 | .581210 | 118 : 32 | .566732 | 100 : 28 | .552842 |
|  |  | 80 : 21 | .580871 | 70 : 19 | .566344 | 75 : 21 | .552842 |

Logarithms of Gear Ratios from 1.0084+ to 7.5

| Numbers of Teeth | Logarithm of Ratio | Numbers of Teeth | Logarithm of Ratio | Numbers of Teeth | Logarithm of Ratio | Numbers of Teeth | Logarithm of Ratio |
|---|---|---|---|---|---|---|---|
| 107 : 30 | 0.552263 | 76 : 22 | 0.538391 | 104 : 31 | 0.525672 | 101 : 31 | 0.512960 |
| 82 : 23 | .552086 | 107 : 31 | .538022 | 114 : 34 | .525426 | 114 : 35 | .512837 |
| 114 : 32 | .551755 | 69 : 20 | .537819 | 67 : 20 | .525045 | 117 : 36 | .511883 |
| 89 : 25 | .551450 | 100 : 29 | .537602 | 77 : 23 | .524763 | 104 : 32 | .511883 |
| 96 : 27 | .550907 | 93 : 27 | .537119 | 87 : 26 | .524546 | 91 : 28 | .511883 |
| 64 : 18 | .550907 | 62 : 18 | .537119 | 97 : 29 | .524374 | 65 : 20 | .511883 |
| 103 : 29 | .550439 | 117 : 34 | .536707 | 107 : 32 | .524234 | 120 : 37 | .510980 |
| 71 : 20 | .550228 | 86 : 25 | .536559 | 117 : 35 | .524118 | 107 : 33 | .510870 |
| 110 : 31 | 0.550031 | 110 : 32 | 0.536243 | 120 : 36 | 0.522879 | 94 : 29 | 0.510730 |
| 117 : 33 | .549672 | 55 : 16 | .536243 | 110 : 33 | .522879 | 81 : 25 | .510545 |
| 78 : 22 | .549672 | 79 : 23 | .535899 | 100 : 30 | .522879 | 68 : 21 | .510290 |
| 85 : 24 | .549208 | 103 : 30 | .535716 | 90 : 27 | .522879 | 110 : 34 | .509914 |
| 92 : 26 | .548815 | 120 : 35 | .535113 | 80 : 24 | .522879 | 97 : 30 | .509650 |
| 99 : 28 | .548477 | 96 : 28 | .535113 | 70 : 21 | .522879 | 84 : 26 | .509306 |
| 106 : 30 | .548185 | 72 : 21 | .535113 | 60 : 18 | .522879 | 113 : 35 | .509010 |
| 113 : 32 | .547928 | 113 : 33 | .534565 | 113 : 34 | .521600 | 71 : 22 | .508836 |
| 120 : 34 | 0.547702 | 89 : 26 | 0.534417 | 103 : 31 | 0.521476 | 100 : 31 | 0.508638 |
| 67 : 19 | .547321 | 65 : 19 | .534160 | 93 : 28 | .521325 | 116 : 36 | .508156 |
| 74 : 21 | .547012 | 106 : 31 | .533944 | 83 : 25 | .521138 | 87 : 27 | .508156 |
| 81 : 23 | .546757 | 82 : 24 | .533603 | 73 : 22 | .520900 | 103 : 32 | .507687 |
| 88 : 25 | .546543 | 99 : 29 | .533237 | 63 : 19 | .520587 | 74 : 23 | .507504 |
| 95 : 27 | .546360 | 116 : 34 | .532980 | 116 : 35 | .520390 | 119 : 37 | .507345 |
| 102 : 29 | .546202 | 75 : 22 | .532639 | 106 : 32 | .520156 | 90 : 28 | .507085 |
| 109 : 31 | .546065 | 92 : 27 | .532424 | 96 : 29 | .519873 | 106 : 33 | .506792 |
| 119 : 34 | 0.544068 | 109 : 32 | 0.532277 | 86 : 26 | 0.519525 | 61 : 19 | 0.506576 |
| 112 : 32 | .544068 | 119 : 35 | .531479 | 119 : 36 | .519245 | 77 : 24 | .506280 |
| 105 : 30 | .544068 | 102 : 30 | .531479 | 76 : 23 | .519086 | 93 : 29 | .506085 |
| 98 : 28 | .544068 | 85 : 25 | .531479 | 109 : 33 | .518913 | 109 : 34 | .505948 |
| 91 : 26 | .544068 | 68 : 20 | .531479 | 99 : 30 | .518514 | 112 : 35 | .505150 |
| 84 : 24 | .544068 | 112 : 33 | .530704 | 89 : 27 | .518026 | 96 : 30 | .505150 |
| 77 : 22 | .544068 | 95 : 28 | .530566 | 112 : 34 | .517739 | 80 : 25 | .505150 |
| 70 : 20 | .544068 | 78 : 23 | .530367 | 79 : 24 | .517416 | 64 : 20 | .505150 |
| 63 : 18 | 0.544068 | 61 : 18 | 0.530057 | 102 : 31 | 0.517239 | 115 : 36 | 0.504395 |
| 56 : 16 | .544068 | 105 : 31 | .529828 | 115 : 35 | .516630 | 99 : 31 | .504274 |
| 115 : 33 | .542184 | 88 : 26 | .529509 | 69 : 21 | .516630 | 83 : 26 | .504105 |
| 108 : 31 | .542062 | 115 : 34 | .529219 | 92 : 28 | .516629 | 67 : 21 | .503856 |
| 101 : 29 | .541923 | 71 : 21 | .529039 | 105 : 32 | .516039 | 118 : 37 | .503680 |
| 87 : 25 | .541579 | 98 : 29 | .528828 | 82 : 25 | .515874 | 102 : 32 | .503450 |
| 80 : 23 | .541362 | 108 : 32 | .528274 | 118 : 36 | .515580 | 86 : 27 | .503135 |
| 73 : 21 | .541104 | 81 : 24 | .528274 | 95 : 29 | .515326 | 105 : 33 | .502675 |
| 66 : 19 | 0.540790 | 118 : 35 | 0.527814 | 108 : 33 | 0.514910 | 70 : 22 | 0.502675 |
| 118 : 34 | .540403 | 91 : 27 | .527678 | 72 : 22 | .514910 | 89 : 28 | .502232 |
| 111 : 32 | .540173 | 64 : 19 | .527426 | 85 : 26 | .514446 | 108 : 34 | .501945 |
| 104 : 30 | .539912 | 101 : 30 | .527200 | 98 : 30 | .514105 | 73 : 23 | .501595 |
| 97 : 28 | .539614 | 111 : 33 | .526809 | 111 : 34 | .513844 | 92 : 29 | .501390 |
| 90 : 26 | .539269 | 74 : 22 | .526809 | 62 : 19 | .513638 | 111 : 35 | .501255 |
| 83 : 24 | .538867 | 84 : 25 | .526339 | 75 : 23 | .513334 | 95 : 30 | .500602 |
| 114 : 33 | .538391 | 94 : 28 | .525970 | 88 : 27 | .513119 | 76 : 24 | .500602 |

## Logarithms of Gear Ratios from 1.0084+ to 7.5

| Numbers of Teeth | Logarithm of Ratio | Numbers of Teeth | Logarithm of Ratio | Numbers of Teeth | Logarithm of Ratio | Numbers of Teeth | Logarithm of Ratio |
|---|---|---|---|---|---|---|---|
| 117 : 37 | 0.499984 | 120 : 39 | 0.488117 | 92 : 31 | 0.472426 | 81 : 28 | 0.461327 |
| 98 : 31 | .499864 | 80 : 26 | .488117 | 89 : 30 | .472269 | 107 : 37 | .461182 |
| 79 : 25 | .499687 | 83 : 27 | .487714 | 86 : 29 | .472101 | 104 : 36 | .460731 |
| 120 : 38 | .499398 | 86 : 28 | .487341 | 83 : 28 | .471920 | 78 : 27 | .460731 |
| 101 : 32 | .499171 | 89 : 29 | .486992 | 80 : 27 | .471726 | 101 : 35 | .460253 |
| 82 : 26 | .498841 | 92 : 30 | .486667 | 77 : 26 | .471517 | 75 : 26 | .460088 |
| 104 : 33 | .498519 | 95 : 31 | .486362 | 74 : 25 | .471291 | 98 : 34 | .459747 |
| 63 : 20 | .498311 | 98 : 32 | .486076 | 71 : 24 | .471047 | 72 : 25 | .459393 |
| 85 : 27 | 0.498055 | 101 : 33 | 0.485808 | 68 : 23 | 0.470781 | 95 : 33 | 0.459210 |
| 107 : 34 | .497905 | 104 : 34 | .485554 | 65 : 22 | .470491 | 118 : 41 | .459098 |
| 110 : 35 | .497325 | 107 : 35 | .485316 | 62 : 21 | .470172 | 115 : 40 | .458638 |
| 88 : 28 | .497325 | 110 : 36 | .485090 | 118 : 40 | .469822 | 92 : 32 | .458638 |
| 113 : 36 | .496776 | 113 : 37 | .484877 | 115 : 39 | .469633 | 69 : 24 | .458638 |
| 91 : 29 | .496643 | 116 : 38 | .484674 | 112 : 38 | .469434 | 112 : 39 | .458153 |
| 69 : 22 | .496426 | 119 : 39 | .484482 | 109 : 37 | .469225 | 89 : 31 | .458028 |
| 116 : 37 | .496256 | 61 : 20 | .484300 | 106 : 36 | .469003 | 66 : 23 | .457816 |
| 94 : 30 | 0.496007 | 64 : 21 | 0.483961 | 103 : 35 | 0.468769 | 109 : 38 | 0.457643 |
| 119 : 38 | .495763 | 67 : 22 | .483652 | 100 : 34 | .468521 | 86 : 30 | .457377 |
| 72 : 23 | .495605 | 70 : 23 | .483370 | 97 : 33 | .468258 | 106 : 37 | .457104 |
| 97 : 31 | .495410 | 73 : 24 | .483112 | 94 : 32 | .467978 | 63 : 22 | .456918 |
| 100 : 32 | .494850 | 76 : 25 | .482874 | 91 : 31 | .467680 | 83 : 29 | .456680 |
| 75 : 24 | .494850 | 79 : 26 | .482654 | 88 : 30 | .467361 | 103 : 36 | .456535 |
| 103 : 33 | .494323 | 82 : 27 | .482450 | 85 : 29 | .467021 | 120 : 42 | .455932 |
| 78 : 25 | .494155 | 85 : 28 | .482261 | 82 : 28 | .466656 | 100 : 35 | .455932 |
| 106 : 34 | 0.493827 | 88 : 29 | 0.482085 | 120 : 41 | 0.466397 | 80 : 28 | 0.455932 |
| 81 : 26 | .493511 | 91 : 30 | .481920 | 79 : 27 | .466263 | 60 : 21 | .455932 |
| 109 : 35 | .493359 | 94 : 31 | .481766 | 117 : 40 | .466126 | 117 : 41 | .455402 |
| 112 : 36 | .492916 | 97 : 32 | .481622 | 114 : 39 | .465840 | 97 : 34 | .455293 |
| 84 : 27 | .492916 | 100 : 33 | .481486 | 76 : 26 | .465840 | 77 : 27 | .455127 |
| 115 : 37 | .492496 | 103 : 34 | .481358 | 111 : 38 | .465539 | 114 : 40 | .454845 |
| 87 : 28 | .492361 | 106 : 35 | .481237 | 73 : 25 | .465383 | 94 : 33 | .454614 |
| 118 : 38 | .492098 | 109 : 36 | .481124 | 108 : 37 | .465222 | 111 : 39 | .454258 |
| 90 : 29 | 0.491845 | 112 : 37 | 0.481016 | 70 : 24 | 0.464887 | 74 : 26 | 0.454258 |
| 93 : 30 | .491362 | 115 : 38 | .480914 | 102 : 35 | .464532 | 91 : 32 | .453891 |
| 62 : 20 | .491362 | 118 : 39 | .480817 | 67 : 23 | .464347 | 71 : 25 | .453318 |
| 96 : 31 | .490910 | 48 : 16 | | 99 : 34 | .464156 | 88 : 31 | .453121 |
| 65 : 21 | .490694 | or any | .477121 | 96 : 33 | .463757 | 105 : 37 | .452988 |
| 99 : 32 | .490485 | 3 to 1 | | 64 : 22 | .463757 | 119 : 42 | .452298 |
| 102 : 33 | .490086 | ratio | | 93 : 32 | .463333 | 102 : 36 | .452298 |
| 68 : 22 | .490086 | 119 : 40 | .473487 | 61 : 21 | .463111 | 85 : 30 | .452298 |
| 105 : 34 | 0.489710 | 116 : 39 | 0.473393 | 90 : 31 | 0.462881 | 68 : 24 | 0.452298 |
| 71 : 23 | .489531 | 113 : 38 | .473295 | 119 : 41 | .462763 | 116 : 41 | .451674 |
| 108 : 35 | .489356 | 110 : 37 | .473191 | 116 : 40 | .462398 | 99 : 35 | .451567 |
| 111 : 36 | .489021 | 107 : 36 | .473081 | 87 : 30 | .462398 | 82 : 29 | .451416 |
| 74 : 24 | .489021 | 104 : 35 | .472965 | 58 : 20 | .462398 | 65 : 23 | .451186 |
| 114 : 37 | .488703 | 101 : 34 | .472843 | 113 : 39 | .462013 | 113 : 40 | .451018 |
| 77 : 25 | .488551 | 98 : 33 | .472712 | 84 : 29 | .461881 | 96 : 34 | .450792 |
| 117 : 38 | .488402 | 95 : 32 | .472574 | 110 : 38 | .461609 | 79 : 28 | .450469 |

Logarithms of Gear Ratios from 1.0084+ to 7.5

| Numbers of Teeth | Logarithm of Ratio | Numbers of Teeth | Logarithm of Ratio | Numbers of Teeth | Logarithm of Ratio | Numbers of Teeth | Logarithm of Ratio |
|---|---|---|---|---|---|---|---|
| 110 : 39 | 0.450328 | 77 : 28 | 0.439333 | 94 : 35 | 0.429060 | 84 : 32 | 0.419129 |
| 93 : 33 | .449969 | 66 : 24 | .439333 | 102 : 38 | .428817 | 118 : 45 | .418670 |
| 62 : 22 | .449969 | 55 : 20 | .439333 | 110 : 41 | .428609 | 97 : 37 | .418570 |
| 107 : 38 | .449600 | 118 : 43 | .438414 | 59 : 22 | .428429 | 76 : 29 | .418416 |
| 76 : 27 | .449450 | 107 : 39 | .438319 | 67 : 25 | .428135 | 55 : 21 | .418143 |
| 90 : 32 | .449092 | 96 : 35 | .438203 | 75 : 28 | .427903 | 89 : 34 | .417911 |
| 104 : 37 | .448832 | 85 : 31 | .438057 | 83 : 31 | .427716 | 102 : 39 | .417536 |
| 118 : 42 | .448633 | 74 : 27 | .437868 | 91 : 34 | .427563 | 68 : 26 | .417536 |
| 73 : 26 | 0.448350 | 63 : 23 | .437613 | 99 : 37 | 0.427434 | 115 : 44 | 0.417245 |
| 87 : 31 | .448158 | 115 : 42 | 0.437449 | 107 : 40 | .427324 | 81 : 31 | .417123 |
| 101 : 36 | .448019 | 104 : 38 | .437250 | 115 : 43 | .427229 | 94 : 36 | .416825 |
| 115 : 41 | .447914 | 93 : 34 | .437004 | 120 : 45 | .425969 | 107 : 41 | .416600 |
| 98 : 35 | .447158 | 82 : 30 | .436693 | 112 : 42 | .425969 | 60 : 23 | .416424 |
| 84 : 30 | .447158 | 112 : 41 | .436434 | 104 : 39 | .425969 | 73 : 28 | .416165 |
| 112 : 40 | .447158 | 71 : 26 | .436285 | 96 : 36 | .425969 | 86 : 33 | .415985 |
| 70 : 25 | .447158 | 101 : 37 | .436120 | 88 : 33 | .425969 | 99 : 38 | .415852 |
| 109 : 39 | 0.446362 | 90 : 33 | .435729 | 80 : 30 | 0.425969 | 112 : 43 | 0.415750 |
| 95 : 34 | .446245 | 60 : 22 | 0.435729 | 72 : 27 | .425969 | 117 : 45 | .414973 |
| 81 : 29 | .446087 | 109 : 40 | .435367 | 64 : 24 | .425969 | 91 : 35 | .414973 |
| 67 : 24 | .445864 | 79 : 29 | .435229 | 56 : 21 | .425969 | 78 : 30 | .414973 |
| 120 : 43 | .445713 | 98 : 36 | .434924 | 117 : 44 | .424733 | 52 : 20 | .414973 |
| 106 : 38 | .445522 | 117 : 43 | .434717 | 109 : 41 | .424643 | 109 : 42 | .414177 |
| 92 : 33 | .445274 | 68 : 25 | .434569 | 101 : 38 | .424538 | 96 : 37 | .414070 |
| 117 : 42 | .444937 | 87 : 32 | .434369 | 93 : 35 | .424415 | 83 : 32 | .413928 |
| 78 : 28 | 0.444937 | 106 : 39 | .434241 | 85 : 32 | 0.424269 | 70 : 27 | 0.413734 |
| 103 : 37 | .444636 | 114 : 42 | 0.433656 | 77 : 29 | .424093 | 57 : 22 | .413452 |
| 64 : 23 | .444452 | 95 : 35 | .433656 | 69 : 26 | .423876 | 101 : 39 | .413257 |
| 89 : 32 | .444240 | 76 : 28 | .433656 | 61 : 23 | .423602 | 88 : 34 | .413004 |
| 114 : 41 | .444121 | 57 : 21 | .433656 | 114 : 43 | .423436 | 119 : 46 | .412789 |
| 100 : 36 | .443698 | 103 : 38 | .433054 | 106 : 40 | .423246 | 75 : 29 | .412663 |
| 75 : 27 | .443698 | 84 : 31 | .432918 | 98 : 37 | .423024 | 106 : 41 | .412522 |
| 111 : 40 | .443263 | 65 : 24 | .432702 | 90 : 34 | .422764 | 62 : 24 | .412180 |
| 86 : 31 | 0.443137 | 111 : 41 | .432539 | 82 : 31 | 0.422452 | 93 : 36 | 0.412180 |
| 61 : 22 | .442907 | 92 : 34 | 0.432309 | 119 : 45 | .422335 | 111 : 43 | .411855 |
| 97 : 35 | .442704 | 119 : 44 | .432094 | 111 : 42 | .422074 | 80 : 31 | .411728 |
| 72 : 26 | .442359 | 73 : 27 | .431959 | 74 : 28 | .422074 | 98 : 38 | .411443 |
| 108 : 39 | .442359 | 100 : 37 | .431798 | 103 : 39 | .421773 | 116 : 45 | .411246 |
| 119 : 43 | .442079 | 108 : 40 | .431364 | 95 : 36 | .421421 | 67 : 26 | .411102 |
| 83 : 30 | .441957 | 81 : 30 | .431364 | 87 : 33 | .421005 | 85 : 33 | .410905 |
| 94 : 34 | .441649 | 54 : 20 | .431364 | 58 : 22 | .421005 | 103 : 40 | .410777 |
| 105 : 38 | 0.441406 | 116 : 43 | .430990 | 108 : 41 | 0.420640 | 90 : 35 | 0.410175 |
| 58 : 21 | .441209 | 89 : 33 | 0.430876 | 79 : 30 | .420506 | 72 : 30 | .410175 |
| 69 : 25 | .440909 | 62 : 23 | .430664 | 100 : 38 | .420216 | 54 : 21 | .410175 |
| 80 : 29 | .440692 | 97 : 36 | .430469 | 71 : 27 | .419893 | 113 : 44 | .409626 |
| 91 : 33 | .440528 | 105 : 39 | .430125 | 92 : 35 | .419720 | 95 : 37 | .409522 |
| 102 : 37 | .440399 | 70 : 26 | .430125 | 113 : 43 | .419610 | 77 : 30 | .409369 |
| 113 : 41 | .440295 | 113 : 42 | .429829 | 105 : 40 | .419130 | 59 : 23 | .409124 |
| 99 : 36 | .439333 | 78 : 29 | .429697 | 63 : 24 | .419130 | 100 : 39 | .408935 |
| 88 : 32 | .439333 | 86 : 32 | .429349 | ...... | ...... | ...... | ...... |

### Logarithms of Gear Ratios from 1.0084+ to 7.5

| Numbers of Teeth | Logarithm of Ratio | Numbers of Teeth | Logarithm of Ratio | Numbers of Teeth | Logarithm of Ratio | Numbers of Teeth | Logarithm of Ratio |
|---|---|---|---|---|---|---|---|
| 82 : 32 | 0.408664 | 40 : 16 | | 110 : 45 | 0.388180 | 48 : 20 | 0.380211 |
| 105 : 41 | .408405 | or any | 0.397940 | 88 : 36 | .388180 | 115 : 48 | .379457 |
| 64 : 25 | .408240 | 2½ to 1 | | 66 : 27 | .388180 | 103 : 43 | .379369 |
| 87 : 34 | .408040 | ratio | | 105 : 43 | .387721 | 91 : 38 | .379258 |
| 110 : 43 | .407924 | 117 : 47 | .396088 | 83 : 34 | .387599 | 79 : 33 | .379113 |
| 115 : 45 | .407485 | 112 : 45 | .396006 | 61 : 25 | .387390 | 67 : 28 | .378916 |
| 92 : 36 | .407485 | 107 : 43 | .395915 | 100 : 41 | .387216 | 55 : 23 | .378635 |
| 69 : 27 | .407485 | 102 : 41 | .395816 | 117 : 48 | .386945 | 98 : 41 | .378442 |
| 120 : 47 | 0.407083 | 97 : 39 | 0.395707 | 78 : 32 | 0.386945 | 86 : 36 | 0.378196 |
| 97 : 38 | .406988 | 92 : 37 | .395586 | 95 : 39 | .386659 | 117 : 49 | .377990 |
| 74 : 29 | .406834 | 87 : 35 | .395451 | 112 : 46 | .386460 | 74 : 31 | .377870 |
| 51 : 20 | .406540 | 82 : 33 | .395300 | 56 : 23 | .386460 | 105 : 44 | .377737 |
| 79 : 31 | .406265 | 77 : 31 | .395129 | 73 : 30 | .386202 | 93 : 39 | .377418 |
| 107 : 42 | .406135 | 72 : 29 | .394935 | 90 : 37 | .386041 | 112 : 47 | .377418 |
| 84 : 33 | .405765 | 67 : 27 | .394711 | 107 : 44 | .385931 | 81 : 34 | .377006 |
| 56 : 22 | .405765 | 62 : 25 | .394452 | 102 : 42 | .385351 | | |
| 117 : 46 | 0.405428 | 119 : 48 | 0.394306 | 119 : 49 | 0.385351 | 100 : 42 | 0.376751 |
| 89 : 35 | .405322 | 57 : 23 | .394147 | 85 : 35 | .385351 | 50 : 21 | .376751 |
| 61 : 24 | .405119 | 109 : 44 | .393974 | 68 : 28 | .385351 | 119 : 50 | .376577 |
| 94 : 37 | .404926 | 52 : 21 | .393784 | 51 : 21 | .385351 | 69 : 29 | .376451 |
| 99 : 39 | .404571 | 99 : 40 | .393575 | 114 : 47 | .384807 | 88 : 37 | .376281 |
| 66 : 26 | .404571 | 94 : 38 | .393344 | 97 : 40 | .384712 | 107 : 45 | .376171 |
| 104 : 41 | .404249 | 89 : 36 | .393088 | 80 : 33 | .384576 | 95 : 40 | .375664 |
| 71 : 28 | .404100 | 84 : 34 | .392800 | 63 : 26 | .384367 | 76 : 32 | .375664 |
| 109 : 43 | 0.403958 | 79 : 32 | 0.392477 | 109 : 45 | 0.384214 | 57 : 24 | 0.375664 |
| 114 : 45 | .403692 | 116 : 47 | .392360 | 92 : 38 | .384004 | 102 : 43 | .375132 |
| 76 : 30 | .403692 | 111 : 45 | .392111 | 75 : 31 | .383700 | 83 : 35 | .375010 |
| 119 : 47 | .403449 | 74 : 30 | .392111 | 104 : 43 | .383565 | 64 : 27 | .374816 |
| 81 : 32 | .403335 | 106 : 43 | .391837 | 87 : 36 | .383217 | 109 : 46 | .374669 |
| 86 : 34 | .403020 | 69 : 28 | .391691 | 58 : 24 | .383217 | 90 : 38 | .374459 |
| 91 : 36 | .402739 | 101 : 41 | .391538 | 99 : 41 | .382851 | 116 : 49 | .374262 |
| 96 : 38 | .402488 | 96 : 39 | .391207 | 70 : 29 | .382700 | 71 : 30 | .374137 |
| 101 : 40 | 0.402261 | 64 : 26 | 0.391207 | 111 : 46 | 0.382565 | 97 : 41 | 0.373988 |
| 53 : 21 | .402057 | 91 : 37 | .390840 | 82 : 34 | .382335 | 78 : 33 | .373581 |
| 111 : 44 | .401870 | 59 : 24 | .390641 | 94 : 39 | .382063 | 52 : 22 | .373581 |
| 58 : 23 | .401700 | 86 : 35 | .390431 | 53 : 22 | .381853 | 111 : 47 | .373225 |
| 63 : 25 | .401401 | 113 : 46 | .390321 | 118 : 49 | .381686 | 85 : 36 | .373116 |
| 68 : 27 | .401145 | 108 : 44 | .389971 | 65 : 27 | .381550 | 59 : 25 | .372912 |
| 73 : 29 | .400925 | 81 : 33 | .389971 | 77 : 32 | .381341 | 92 : 39 | .372723 |
| 78 : 31 | .400733 | 54 : 22 | .389971 | 89 : 37 | .381188 | 66 : 28 | .372386 |
| 83 : 33 | 0.400564 | 103 : 42 | 0.389588 | 101 : 42 | 0.381072 | 99 : 42 | 0.372386 |
| 88 : 35 | .400415 | 76 : 31 | .389452 | 113 : 47 | .380981 | 106 : 45 | .372093 |
| 93 : 37 | .400281 | 98 : 40 | .389166 | 120 : 50 | .380211 | 73 : 31 | .371961 |
| 98 : 39 | .400162 | 49 : 20 | .389166 | 108 : 45 | .380211 | 113 : 48 | .371837 |
| 103 : 41 | .400053 | 120 : 49 | .388985 | 96 : 40 | .380211 | 120 : 51 | .371611 |
| 108 : 43 | .399955 | 71 : 29 | .388860 | 84 : 35 | .380211 | 80 : 34 | .371611 |
| 113 : 45 | .399866 | 93 : 38 | .388699 | 72 : 30 | .380211 | 87 : 37 | .371318 |
| 118 : 47 | .399784 | 115 : 47 | .388600 | 60 : 25 | .380211 | 94 : 40 | .371068 |

**Logarithms of Gear Ratios from 1.0084+ to 7.5**

| Numbers of Teeth | Logarithm of Ratio | Numbers of Teeth | Logarithm of Ratio | Numbers of Teeth | Logarithm of Ratio | Numbers of Teeth | Logarithm of Ratio |
|---|---|---|---|---|---|---|---|
| 47 : 20 | 0.371068 | 115 : 50 | 0.361728 | 70 : 31 | 0.353736 | 62 : 28 | 0.345234 |
| 101 : 43 | .370853 | 92 : 40 | .361728 | 79 : 35 | .353559 | 104 : 47 | .344935 |
| 54 : 23 | .370666 | 69 : 30 | .361728 | 88 : 39 | .353418 | 73 : 33 | .344809 |
| 115 : 49 | .370502 | 46 : 20 | .361728 | 97 : 43 | .353303 | 115 : 52 | .344695 |
| 61 : 26 | .370357 | 108 : 47 | .361326 | 106 : 47 | .353208 | 84 : 38 | .344496 |
| 68 : 29 | .370111 | 85 : 37 | .361217 | 115 : 51 | .353128 | 95 : 43 | .344255 |
| 75 : 32 | .369911 | 62 : 27 | .361028 | 117 : 52 | .352183 | 53 : 24 | .344065 |
| 82 : 35 | .369746 | 101 : 44 | .360869 | 108 : 48 | .352183 | 117 : 53 | .343910 |
| 89 : 38 | 0.369606 | 117 : 51 | 0.360616 | 99 : 44 | 0.352183 | 64 : 29 | 0.343782 |
| 96 : 41 | .369487 | 78 : 34 | .360616 | 90 : 40 | .352183 | 75 : 34 | .343582 |
| 103 : 44 | .369385 | 94 : 41 | .360344 | 81 : 36 | .352183 | 86 : 39 | .343434 |
| 110 : 47 | .369295 | 55 : 24 | .360152 | 72 : 32 | .352183 | 97 : 44 | .343319 |
| 117 : 50 | .369216 | 71 : 31 | .359897 | 63 : 28 | .352183 | 108 : 49 | .343228 |
| 119 : 51 | .367977 | 87 : 38 | .359736 | 45 : 20 | .352183 | 119 : 54 | .343153 |
| 105 : 45 | .367977 | 103 : 45 | .359625 | 119 : 53 | .351271 | 99 : 45 | .342423 |
| 98 : 42 | .367977 | 119 : 52 | .359544 | 110 : 49 | .351197 | 88 : 40 | .342423 |
| 91 : 39 | 0.367977 | 112 : 49 | 0.359022 | 101 : 45 | 0.351109 | 77 : 35 | 0.342423 |
| 84 : 36 | .367977 | 96 : 42 | .359022 | 92 : 41 | .351004 | 66 : 30 | .342423 |
| 77 : 33 | .367977 | 80 : 35 | .359022 | 83 : 37 | .350876 | 55 : 25 | .342423 |
| 70 : 30 | .367977 | 64 : 28 | .359022 | 74 : 33 | .350718 | 112 : 51 | .341648 |
| 63 : 27 | .367977 | 48 : 21 | .359022 | 65 : 29 | .350515 | 101 : 46 | .341564 |
| 56 : 24 | .367977 | 105 : 46 | .358432 | 56 : 25 | .350248 | 90 : 41 | .341459 |
| 49 : 21 | .367977 | 89 : 39 | .358325 | 103 : 46 | .350079 | 79 : 36 | .341325 |
| 114 : 49 | .366709 | 73 : 32 | .358173 | 94 : 42 | .349879 | 57 : 26 | .340902 |
| 107 : 46 | 0.366626 | 57 : 25 | 0.357935 | 85 : 38 | 0.349635 | 103 : 47 | 0.340739 |
| 100 : 43 | .366532 | 98 : 43 | .357758 | 114 : 51 | .349335 | 92 : 42 | .340539 |
| 93 : 40 | .366423 | 82 : 36 | .357511 | 76 : 34 | .349335 | 46 : 21 | .340539 |
| 86 : 37 | .366297 | 107 : 47 | .357286 | 105 : 47 | .349091 | 81 : 37 | .340283 |
| 79 : 34 | .366148 | 66 : 29 | .357146 | 67 : 30 | .348954 | 116 : 53 | .340182 |
| 72 : 31 | .365971 | 91 : 40 | .356981 | 96 : 43 | .348803 | 105 : 48 | .339948 |
| 65 : 28 | .365755 | 116 : 51 | .356888 | 87 : 39 | .348455 | 70 : 32 | .339948 |
| 58 : 25 | .365488 | 75 : 33 | .356547 | 58 : 26 | .348455 | 94 : 43 | .339659 |
| 109 : 47 | 0.365329 | 50 : 22 | 0.356547 | 107 : 48 | 0.348143 | 118 : 54 | 0.339488 |
| 102 : 44 | .365148 | 100 : 44 | .356547 | 78 : 35 | .348027 | 59 : 27 | .339488 |
| 95 : 41 | .364940 | 109 : 48 | .356185 | 98 : 44 | .347773 | 83 : 38 | .339295 |
| 88 : 38 | .364700 | 84 : 37 | .356078 | 49 : 22 | .347773 | 107 : 49 | .339188 |
| 81 : 35 | .364417 | 59 : 26 | .355879 | 118 : 53 | .347606 | 120 : 55 | .338819 |
| 118 : 51 | .364312 | 93 : 41 | .355699 | 69 : 31 | .347487 | 96 : 44 | .338819 |
| 111 : 48 | .364082 | 102 : 45 | .355388 | 89 : 40 | .347330 | 72 : 33 | .338819 |
| 74 : 32 | .364082 | 68 : 30 | .355388 | 109 : 49 | .347230 | 48 : 22 | .338819 |
| 104 : 45 | 0.363821 | 111 : 49 | 0.355127 | 100 : 45 | 0.346788 | 109 : 50 | 0.338457 |
| 67 : 29 | .363677 | 77 : 34 | .355012 | 80 : 36 | .346788 | 85 : 39 | .338354 |
| 90 : 39 | .363178 | 120 : 53 | .354905 | 60 : 27 | .346788 | 61 : 28 | .338172 |
| 60 : 26 | .363178 | 86 : 38 | .354715 | 111 : 50 | .346353 | 98 : 45 | .338014 |
| 113 : 49 | .362882 | 95 : 42 | .354474 | 91 : 41 | .346258 | 111 : 51 | .337753 |
| 53 : 23 | .362548 | 52 : 23 | .354276 | 51 : 23 | .345842 | 74 : 34 | .337753 |
| 76 : 33 | .362300 | 113 : 50 | .354108 | 113 : 51 | .345508 | 87 : 40 | .337459 |
| 99 : 43 | .362167 | 61 : 27 | .353966 | 93 : 42 | .345234 | 100 : 46 | .337242 |

Logarithms of Gear Ratios from 1.0084+ to 7.5

| Numbers of Teeth | Logarithm of Ratio | Numbers of Teeth | Logarithm of Ratio | Numbers of Teeth | Logarithm of Ratio | Numbers of Teeth | Logarithm of Ratio |
|---|---|---|---|---|---|---|---|
| 113 : 52 | 0.337075 | 111 : 52 | 0.329320 | 88 : 42 | 0.321233 | 107 : 52 | 0.313381 |
| 63 : 29 | .336943 | 96 : 45 | .329059 | 44 : 21 | .321233 | 72 : 35 | .313265 |
| 76 : 35 | .336746 | 64 : 30 | .329059 | 111 : 53 | .321047 | 109 : 53 | .313151 |
| 89 : 41 | .336606 | 113 : 53 | .328803 | 67 : 32 | .320925 | 74 : 36 | .312929 |
| 102 : 47 | .336502 | 81 : 38 | .328701 | 90 : 43 | .320774 | 111 : 54 | .312929 |
| 115 : 53 | .336422 | 98 : 46 | .328468 | 113 : 54 | .320685 | 113 : 55 | .312716 |
| 117 : 54 | .335792 | 115 : 54 | .328304 | 115 : 55 | .320335 | 76 : 37 | .312612 |
| 91 : 42 | .335792 | 66 : 31 | .328182 | 92 : 44 | .320335 | 115 : 56 | .312510 |
| 78 : 36 | 0.335792 | 83 : 39 | 0.328014 | 69 : 33 | 0.320335 | 117 : 57 | 0.312311 |
| 65 : 30 | .335792 | 100 : 47 | .327902 | 46 : 22 | .320335 | 78 : 38 | .312311 |
| 52 : 24 | .335792 | 117 : 55 | .327823 | 117 : 56 | .319998 | 119 : 58 | .312119 |
| 119 : 55 | .335184 | 119 : 56 | .327359 | 71 : 34 | .319779 | 80 : 39 | .312025 |
| 106 : 49 | .335110 | 85 : 40 | .327359 | 119 : 57 | .319672 | 82 : 40 | .311754 |
| 93 : 43 | .335014 | 68 : 32 | .327359 | 96 : 46 | .319513 | 41 : 20 | .311754 |
| 80 : 37 | .334888 | 51 : 24 | .327359 | 73 : 35 | .319255 | 84 : 41 | .311495 |
| 67 : 31 | .334713 | 104 : 49 | .326837 | 98 : 47 | .319128 | 86 : 42 | .311249 |
| 54 : 25 | 0.334454 | 87 : 41 | 0.326735 | 100 : 48 | 0.318759 | 43 : 21 | 0.311249 |
| 95 : 44 | .334271 | 70 : 33 | .326584 | 75 : 36 | .318759 | 88 : 43 | .311014 |
| 82 : 38 | .334030 | 89 : 42 | .326141 | 102 : 49 | .318404 | 90 : 44 | .310790 |
| 110 : 51 | .333823 | 108 : 51 | .325854 | 77 : 37 | .318289 | 45 : 22 | .310790 |
| 69 : 32 | .333699 | 72 : 34 | .325854 | 52 : 25 | .318063 | 92 : 45 | .310575 |
| 97 : 45 | .333559 | 91 : 43 | .325573 | 79 : 38 | .317844 | 94 : 46 | .310370 |
| 84 : 39 | .333215 | 55 : 26 | .325389 | 106 : 51 | .317736 | 96 : 47 | .310173 |
| 56 : 26 | .333215 | 74 : 35 | .325164 | 81 : 39 | .317420 | 98 : 48 | .309985 |
| 99 : 46 | 0.332877 | 93 : 44 | 0.325030 | 110 : 53 | 0.317117 | 100 : 49 | 0.309804 |
| 71 : 33 | .332744 | 112 : 53 | .324942 | 83 : 40 | .317018 | 102 : 50 | .309630 |
| 114 : 53 | .332629 | 95 : 45 | .324511 | 112 : 54 | .316824 | 51 : 25 | .309630 |
| 86 : 40 | .332439 | 76 : 36 | .324511 | 85 : 41 | .316635 | 104 : 51 | .309463 |
| 43 : 20 | .332439 | 57 : 27 | .324511 | 114 : 55 | .316542 | 53 : 26 | .309303 |
| 101 : 47 | .332224 | 116 : 55 | .324095 | 87 : 42 | .316270 | 108 : 53 | .309148 |
| 58 : 27 | .332064 | 97 : 46 | .324014 | 58 : 28 | .316270 | 55 : 27 | .308999 |
| 73 : 34 | .331844 | 78 : 37 | .323893 | 118 : 57 | .316007 | 112 : 55 | .308855 |
| 88 : 41 | 0.331699 | 59 : 28 | 0.323694 | 89 : 43 | 0.315922 | 57 : 28 | 0.308717 |
| 103 : 48 | .331596 | 99 : 47 | .323537 | 120 : 58 | .315753 | 116 : 57 | .308583 |
| 118 : 55 | .331519 | 120 : 57 | .323306 | 60 : 29 | .315753 | 59 : 29 | .308454 |
| 105 : 49 | .330993 | 80 : 38 | .323306 | 91 : 44 | .315589 | 118 : 58 | .308454 |
| 90 : 42 | .330993 | 101 : 48 | .323080 | 93 : 45 | .315270 | 120 : 59 | .308329 |
| 75 : 35 | .330993 | 61 : 29 | .322932 | 62 : 30 | .315270 | 61 : 30 | .308209 |
| 60 : 28 | .330993 | 82 : 39 | .322749 | 95 : 46 | .314966 | 63 : 31 | .307979 |
| 45 : 21 | .330993 | 103 : 49 | .322641 | 64 : 31 | .314818 | 65 : 32 | .307763 |
| 107 : 50 | 0.330414 | 105 : 50 | 0.322219 | 97 : 47 | 0.314674 | 67 : 33 | 0.307561 |
| 92 : 43 | .330319 | 84 : 40 | .322219 | 99 : 48 | .314394 | 69 : 34 | .307370 |
| 77 : 36 | .330188 | 63 : 30 | .322219 | 66 : 32 | .314394 | 71 : 35 | .307190 |
| 62 : 29 | .329994 | 42 : 20 | .322219 | 101 : 49 | .314125 | 73 : 36 | .307020 |
| 109 : 51 | .329855 | 107 : 51 | .321814 | 68 : 33 | .313995 | 75 : 37 | .306860 |
| 94 : 44 | .329675 | 86 : 41 | .321715 | 103 : 50 | .313867 | 77 : 38 | .306707 |
| 47 : 22 | .329675 | 65 : 31 | .321552 | 105 : 51 | .313619 | 79 : 39 | .306563 |
| 79 : 37 | .329425 | 109 : 52 | .321423 | 70 : 34 | .313619 | 81 : 40 | .306425 |

Logarithms of Gear Ratios from 1.0084+ to 7.5

| Numbers of Teeth | Logarithm of Ratio | Numbers of Teeth | Logarithm of Ratio | Numbers of Teeth | Logarithm of Ratio | Numbers of Teeth | Logarithm of Ratio |
|---|---|---|---|---|---|---|---|
| 83 : 41 | 0.306294 | 71 : 36 | 0.294956 | 93 : 48 | 0.287242 | 99 : 52 | 0.279632 |
| 85 : 42 | .306170 | 69 : 35 | .294781 | 62 : 32 | .287242 | 59 : 31 | .279490 |
| 87 : 43 | .306051 | 67 : 34 | .294596 | 91 : 47 | .286944 | 78 : 41 | .279311 |
| 89 : 44 | .305937 | 65 : 33 | .294400 | 120 : 62 | .286790 | 97 : 51 | .279202 |
| 91 : 45 | .305829 | 63 : 32 | .294191 | 89 : 46 | .286632 | 116 : 61 | .279128 |
| 93 : 46 | .305725 | 61 : 31 | .293968 | 118 : 61 | .286552 | 95 : 50 | .278754 |
| 95 : 47 | .305626 | 120 : 61 | .293851 | 116 : 60 | .286307 | 76 : 40 | .278754 |
| 97 : 48 | .305531 | 59 : 30 | .293731 | 58 : 30 | .286307 | 57 : 30 | .278754 |
| 99 : 49 | 0.305439 | 116 : 59 | 0.293606 | 114 : 59 | 0.286053 | 38 : 20 | 0.278754 |
| 101 : 50 | .305351 | 57 : 29 | .293477 | 85 : 44 | .285966 | 112 : 59 | .278366 |
| 103 : 51 | .305267 | 112 : 57 | .293343 | 56 : 29 | .285790 | 93 : 49 | .278287 |
| 105 : 52 | .305186 | 55 : 28 | .293205 | 83 : 43 | .285610 | 74 : 39 | .278167 |
| 107 : 53 | .305108 | 108 : 55 | .293061 | 110 : 57 | .285518 | 55 : 29 | .277965 |
| 109 : 54 | .305033 | 53 : 27 | .292912 | 81 : 42 | .285236 | 91 : 48 | .277800 |
| 111 : 55 | .304960 | 104 : 53 | .292757 | 54 : 28 | .285235 | 108 : 57 | .277549 |
| 113 : 56 | .304890 | 51 : 26 | .292597 | 106 : 55 | .284943 | 72 : 38 | .277549 |
| 115 : 57 | 0.304823 | 100 : 51 | 0.292430 | 79 : 41 | 0.284843 | 89 : 47 | 0.277292 |
| 117 : 58 | .304758 | 98 : 50 | .292256 | 52 : 27 | .284640 | 53 : 28 | .277118 |
| 119 : 59 | .304695 | 96 : 49 | .292075 | 77 : 40 | .284431 | 87 : 46 | .276762 |
| 32 : 16 | | 94 : 48 | .291887 | 102 : 53 | .284324 | 104 : 55 | .276671 |
| or | | 92 : 47 | .291690 | 100 : 52 | .283997 | 119 : 63 | .276207 |
| any | .301030 | 90 : 46 | .291485 | 75 : 39 | .283997 | 85 : 45 | .276207 |
| 2 to 1 | | 88 : 45 | .291270 | 98 : 51 | .283656 | 68 : 36 | .276206 |
| ratio | | 86 : 44 | .291046 | 73 : 38 | .283540 | 117 : 62 | .275794 |
| 119 : 60 | 0.297396 | 84 : 43 | 0.290811 | 96 : 50 | 0.283301 | 100 : 53 | 0.275724 |
| 117 : 59 | .297334 | 82 : 42 | .290565 | 119 : 62 | .283155 | 83 : 44 | .275625 |
| 115 : 58 | .297270 | 41 : 21 | .290565 | 71 : 37 | .283057 | 66 : 35 | .275476 |
| 113 : 57 | .297204 | 80 : 41 | .290306 | 94 : 49 | .282932 | 115 : 61 | .275368 |
| 111 : 56 | .297135 | 119 : 61 | .290217 | 117 : 61 | .282856 | 98 : 52 | .275223 |
| 109 : 55 | .297064 | 78 : 40 | .290035 | 92 : 48 | .282547 | 81 : 43 | .275017 |
| 107 : 54 | .296990 | 39 : 20 | .290035 | 69 : 36 | .282547 | 113 : 60 | .274927 |
| 105 : 53 | .296913 | 115 : 59 | .289846 | 113 : 59 | .282226 | 96 : 51 | .274701 |
| 103 : 52 | 0.296833 | 76 : 39 | 0.289749 | 90 : 47 | 0.282145 | 64 : 34 | 0.274701 |
| 101 : 51 | .296751 | 113 : 58 | .289650 | 67 : 35 | .282007 | 111 : 59 | .274471 |
| 99 : 50 | .296665 | 111 : 57 | .289448 | 111 : 58 | .281895 | 79 : 42 | .274378 |
| 97 : 49 | .296576 | 74 : 38 | .289448 | 88 : 46 | .281725 | 94 : 50 | .274158 |
| 95 : 48 | .296482 | 109 : 56 | .289239 | 109 : 57 | .281552 | 109 : 58 | .273999 |
| 93 : 47 | .296385 | 72 : 37 | .289131 | 65 : 34 | .281435 | 62 : 33 | .273878 |
| 91 : 46 | .296284 | 107 : 55 | .289021 | 86 : 45 | .281286 | 77 : 41 | .273707 |
| 89 : 45 | .296178 | 70 : 36 | .288796 | 107 : 56 | .281196 | 92 : 49 | .273592 |
| 87 : 44 | 0.296067 | 103 : 53 | 0.288561 | 84 : 44 | 0.280827 | 107 : 57 | 0.273509 |
| 85 : 43 | .295950 | 68 : 35 | .288441 | 63 : 33 | .280827 | 120 : 64 | .273001 |
| 83 : 42 | .295829 | 101 : 52 | .288318 | 103 : 54 | .280443 | 105 : 56 | .273001 |
| 81 : 41 | .295701 | 66 : 34 | .288065 | 82 : 43 | .280345 | 90 : 48 | .273001 |
| 79 : 40 | .295567 | 99 : 51 | .288065 | 61 : 32 | .280180 | 75 : 40 | .273001 |
| 77 : 39 | .295426 | 97 : 50 | .287802 | 101 : 53 | .280046 | 60 : 32 | .273001 |
| 75 : 38 | .295278 | 64 : 33 | .287666 | 80 : 42 | .279841 | 118 : 63 | .272542 |
| 73 : 37 | .295121 | 95 : 49 | .287528 | 40 : 21 | .279841 | 103 : 55 | .272475 |

### Logarithms of Gear Ratios from 1.0084+ to 7.5

| Numbers of Teeth | Logarithm of Ratio | Numbers of Teeth | Logarithm of Ratio | Numbers of Teeth | Logarithm of Ratio | Numbers of Teeth | Logarithm of Ratio |
|---|---|---|---|---|---|---|---|
| 88 : 47 | 0.272385 | 94 : 51 | 0.265558 | 87 : 48 | 0.258278 | 116 : 65 | 0.251545 |
| 73 : 39 | .272258 | 70 : 38 | .265314 | 58 : 32 | .258278 | 91 : 51 | .251471 |
| 58 : 31 | .272066 | 116 : 63 | .265118 | 96 : 53 | .257995 | 66 : 37 | .251342 |
| 101 : 54 | .271928 | 81 : 44 | .265032 | 67 : 37 | .257873 | 107 : 60 | .251233 |
| 86 : 46 | .271741 | 92 : 50 | .264818 | 105 : 58 | .257761 | 82 : 46 | .251056 |
| 114 : 61 | .271575 | 103 : 56 | .264649 | 38 : 21 | .257564 | 98 : 55 | .250863 |
| 99 : 53 | .271360 | 68 : 37 | .264307 | 76 : 42 | .257564 | 57 : 32 | .250725 |
| 112 : 60 | .271067 | 79 : 43 | .264159 | 85 : 47 | .257321 | 73 : 41 | .250539 |
| 84 : 45 | 0.271067 | 90 : 49 | 0.264046 | 94 : 52 | 0.257125 | 89 : 50 | 0.250420 |
| 56 : 30 | .271067 | 101 : 55 | .263959 | 103 : 57 | .256962 | 105 : 59 | .250337 |
| 97 : 52 | .270768 | 112 : 61 | .263888 | 112 : 62 | .256826 | 112 : 63 | .249878 |
| 69 : 37 | .270647 | 110 : 60 | .263241 | 56 : 31 | .256826 | 96 : 54 | .249878 |
| 110 : 59 | .270541 | 99 : 54 | .263241 | 65 : 36 | .256611 | 80 : 45 | .249878 |
| 82 : 44 | .270361 | 88 : 48 | .263241 | 74 : 41 | .256448 | 64 : 36 | .249878 |
| 41 : 22 | .270361 | 77 : 42 | .263241 | 83 : 46 | .256320 | 119 : 67 | .249472 |
| 95 : 51 | .270153 | 66 : 36 | .263241 | 92 : 51 | .256218 | 103 : 58 | .249409 |
| 54 : 29 | 0.269996 | 55 : 30 | 0.263241 | 101 : 56 | 0.256133 | 87 : 49 | 0.249323 |
| 67 : 36 | .269772 | 119 : 65 | .262634 | 110 : 61 | .256063 | 71 : 40 | .249198 |
| 80 : 43 | .269622 | 108 : 59 | .262572 | 119 : 66 | .256003 | 110 : 62 | .249001 |
| 93 : 50 | .269513 | 97 : 53 | .262496 | 99 : 55 | .255273 | 55 : 31 | .249001 |
| 106 : 57 | .269431 | 86 : 47 | .262401 | 90 : 50 | .255273 | 94 : 53 | .248852 |
| 119 : 64 | .269367 | 75 : 41 | .262277 | 81 : 45 | .255273 | 117 : 66 | .248642 |
| 117 : 63 | .268845 | 64 : 35 | .262112 | 72 : 40 | .255273 | 78 : 44 | .248642 |
| 91 : 49 | .268845 | 53 : 29 | .261878 | 63 : 35 | .255273 | 101 : 57 | .248447 |
| 78 : 42 | 0.268845 | 95 : 52 | 0.261720 | 54 : 30 | 0.255273 | 62 : 35 | 0.248324 |
| 65 : 35 | .268845 | 84 : 46 | .261522 | 115 : 64 | .254518 | 85 : 48 | .248178 |
| 52 : 28 | .268845 | 115 : 63 | .261357 | 106 : 59 | .254454 | 108 : 61 | .248094 |
| 39 : 21 | .268845 | 73 : 40 | .261262 | 97 : 54 | .254378 | 115 : 65 | .247785 |
| 115 : 62 | .268306 | 104 : 57 | .261158 | 88 : 49 | .254287 | 92 : 52 | .247785 |
| 102 : 55 | .268238 | 62 : 34 | .260913 | 79 : 44 | .254174 | 69 : 39 | .247785 |
| 89 : 48 | .268149 | 93 : 51 | .260912 | 70 : 39 | .254033 | 99 : 56 | .247447 |
| 76 : 41 | .268030 | 113 : 62 | .260687 | 61 : 34 | .253851 | 76 : 43 | .247345 |
| 63 : 34 | 0.267862 | 82 : 45 | 0.260601 | 113 : 63 | 0.253738 | 106 : 60 | 0.247155 |
| 113 : 61 | .267749 | 51 : 28 | .260412 | 52 : 29 | .253605 | 53 : 30 | .247155 |
| 100 : 54 | .267606 | 71 : 39 | .260194 | 95 : 53 | .253448 | 83 : 47 | .246980 |
| 87 : 47 | .267421 | 91 : 50 | .260071 | 86 : 48 | .253257 | 113 : 64 | .246898 |
| 111 : 60 | .267172 | 111 : 61 | .259993 | 120 : 67 | .253106 | 120 : 68 | .246672 |
| 74 : 40 | .267172 | 100 : 55 | .259637 | 77 : 43 | .253022 | 90 : 51 | .246672 |
| 37 : 20 | .267172 | 80 : 44 | .259637 | 111 : 62 | .252931 | 60 : 34 | .246672 |
| 98 : 53 | .266950 | 40 : 22 | .259637 | 102 : 57 | .252725 | 97 : 55 | .246409 |
| 61 : 33 | 0.266816 | 109 : 60 | 0.259275 | 68 : 38 | 0.252725 | 67 : 38 | 0.246291 |
| 85 : 46 | .266661 | 89 : 49 | .259194 | 93 : 52 | .252480 | 104 : 59 | .246181 |
| 109 : 59 | .266575 | 69 : 38 | .259066 | 118 : 66 | .252338 | 111 : 63 | .245982 |
| 96 : 52 | .266268 | 118 : 65 | .258969 | 59 : 33 | .252338 | 74 : 42 | .245982 |
| 72 : 39 | .266268 | 98 : 54 | .258832 | 84 : 47 | .252181 | 37 : 21 | .245982 |
| 107 : 58 | .265956 | 78 : 43 | .258626 | 109 : 61 | .252097 | 118 : 67 | .245807 |
| 118 : 64 | .265702 | 107 : 59 | .258532 | 100 : 56 | .251812 | 81 : 46 | .245727 |
| 59 : 32 | .265702 | 116 : 64 | .258278 | 75 : 42 | .251812 | 88 : 50 | .245513 |

Logarithms of Gear Ratios from 1.0084+ to 7.5

| Numbers of Teeth | Logarithm of Ratio | Numbers of Teeth | Logarithm of Ratio | Numbers of Teeth | Logarithm of Ratio | Numbers of Teeth | Logarithm of Ratio |
|---|---|---|---|---|---|---|---|
| 95 : 54 | 0.245330 | 52 : 30 | 0.238882 | 111 : 65 | 0.232410 | 111 : 66 | 0.225779 |
| 51 : 29 | .245172 | 97 : 56 | .238584 | 70 : 41 | .232314 | 74 : 44 | .225779 |
| 109 : 62 | .245035 | 71 : 41 | .238474 | 87 : 51 | .231949 | 116 : 69 | .225609 |
| 58 : 33 | .244914 | 116 : 67 | .238383 | 58 : 34 | .231949 | 79 : 47 | .225529 |
| 65 : 37 | .244712 | 90 : 52 | .238239 | 104 : 61 | .231704 | 84 : 50 | .225309 |
| 72 : 41 | .244549 | 109 : 63 | .238086 | 75 : 44 | .231608 | 89 : 53 | .225114 |
| 79 : 45 | .244415 | 64 : 37 | .237978 | 92 : 54 | .231394 | 94 : 56 | .224940 |
| 86 : 49 | .244302 | 83 : 48 | .237837 | 109 : 64 | .231247 | 99 : 59 | .224783 |
| 93 : 53 | 0.244207 | 102 : 59 | 0.237748 | 63 : 37 | 0.231139 | 52 : 31 | 0.224642 |
| 100 : 57 | .244125 | 95 : 55 | .237361 | 80 : 47 | .230992 | 109 : 65 | .224513 |
| 107 : 61 | .244054 | 76 : 44 | .237361 | 97 : 57 | .230897 | 114 : 68 | .224396 |
| 114 : 65 | .243992 | 57 : 33 | .237361 | 114 : 67 | .230830 | 57 : 34 | .224396 |
| 119 : 68 | .243038 | 38 : 22 | .237361 | 85 : 50 | .230449 | 119 : 71 | .224289 |
| 112 : 64 | .243038 | 107 : 62 | .236992 | 68 : 40 | .230449 | 62 : 37 | .224190 |
| 98 : 56 | .243038 | 88 : 51 | .236913 | 51 : 30 | .230449 | 67 : 40 | .224015 |
| 91 : 52 | .243038 | 69 : 40 | .236789 | 107 : 63 | .230043 | 72 : 43 | .223864 |
| 84 : 48 | 0.243038 | 119 : 69 | 0.236698 | 90 : 53 | 0.229967 | 77 : 46 | 0.223733 |
| 77 : 44 | .243038 | 100 : 58 | .236572 | 73 : 43 | .229854 | 82 : 49 | .223618 |
| 70 : 40 | .243038 | 50 : 29 | .236572 | 112 : 66 | .229674 | 87 : 52 | .223516 |
| 63 : 36 | .243038 | 81 : 47 | .236387 | 56 : 33 | .229674 | 92 : 55 | .223425 |
| 35 : 20 | .243038 | 112 : 65 | .236305 | 95 : 56 | .229536 | 97 : 58 | .223344 |
| 117 : 67 | .242111 | 62 : 36 | .236089 | 117 : 69 | .229337 | 102 : 61 | .223270 |
| 110 : 63 | .242052 | 93 : 54 | .236089 | 78 : 46 | .229337 | 107 : 64 | .223204 |
| 103 : 59 | .241985 | 105 : 61 | .235860 | 100 : 59 | .229148 | 112 : 67 | .223143 |
| 96 : 55 | 0.241909 | 74 : 43 | 0.235763 | 61 : 36 | 0.229027 | 117 : 70 | 0.223088 |
| 89 : 51 | .241820 | 117 : 68 | .235677 | 83 : 49 | .228882 | 100 : 60 | .221849 |
| 82 : 47 | .241716 | 86 : 50 | .235529 | 105 : 62 | .228798 | 95 : 57 | .221849 |
| 75 : 43 | .241593 | 98 : 57 | .235351 | 110 : 65 | .228479 | 90 : 54 | .221849 |
| 68 : 39 | .241444 | 110 : 64 | .235213 | 88 : 52 | .228479 | 85 : 51 | .221849 |
| 61 : 35 | .241262 | 55 : 32 | .235213 | 66 : 39 | .228479 | 80 : 48 | .221849 |
| 115 : 66 | .241154 | 67 : 39 | .235010 | 115 : 68 | .228187 | 75 : 45 | .221849 |
| 108 : 62 | .241032 | 79 : 46 | .234869 | 93 : 55 | .228120 | 70 : 42 | .221849 |
| 54 : 31 | 0.241032 | 91 : 53 | 0.234766 | 71 : 42 | 0.228009 | 65 : 39 | 0.221849 |
| 101 : 58 | .240893 | 103 : 60 | .234686 | 120 : 71 | .227923 | 60 : 36 | .221849 |
| 94 : 54 | .240734 | 115 : 67 | .234623 | 98 : 58 | .227798 | 55 : 33 | .221849 |
| 87 : 50 | .240549 | 96 : 56 | .234083 | 76 : 45 | .227601 | 50 : 30 | .221849 |
| 120 : 69 | .240332 | 84 : 49 | .234083 | 103 : 61 | .227507 | 118 : 71 | .220624 |
| 80 : 46 | .240332 | 72 : 42 | .234083 | 81 : 48 | .227244 | 113 : 68 | .220570 |
| 113 : 65 | .240165 | 60 : 35 | .234083 | 54 : 32 | .227244 | 108 : 65 | .220510 |
| 73 : 42 | .240074 | 113 : 66 | .233535 | 113 : 67 | .227004 | 103 : 62 | .220446 |
| 106 : 61 | 0.239976 | 101 : 59 | 0.233469 | 86 : 51 | 0.226928 | 98 : 59 | 0.220374 |
| 99 : 57 | .239760 | 89 : 52 | .233388 | 59 : 35 | .226784 | 93 : 56 | .220295 |
| 66 : 38 | .239760 | 77 : 45 | .233278 | 91 : 54 | .226648 | 88 : 53 | .220207 |
| 92 : 53 | .239512 | 65 : 38 | .233130 | 64 : 38 | .226396 | 83 : 50 | .220109 |
| 59 : 34 | .239373 | 118 : 69 | .233033 | 96 : 57 | .226396 | 78 : 47 | .219997 |
| 85 : 49 | .239223 | 53 : 31 | .232914 | 101 : 60 | .226170 | 73 : 44 | .219870 |
| 111 : 64 | .239143 | 94 : 55 | .232765 | 69 : 41 | .226065 | 68 : 41 | .219725 |
| 78 : 45 | .238882 | 82 : 48 | .232573 | 106 : 63 | .225965 | 63 : 38 | .219557 |

## Logarithms of Gear Ratios from 1.0084+ to 7.5

| Numbers of Teeth | Logarithm of Ratio | Numbers of Teeth | Logarithm of Ratio | Numbers of Teeth | Logarithm of Ratio | Numbers of Teeth | Logarithm of Ratio |
|---|---|---|---|---|---|---|---|
| 58 : 35 | 0.219360 | 116 : 71 | 0.213200 | 29 : 18 | 0.207126 | 81 : 51 | 0.200915 |
| 111 : 67 | .219248 | 98 : 60 | .213075 | 95 : 59 | .206872 | 54 : 34 | .200915 |
| 53 : 32 | .219126 | 49 : 30 | .213075 | 66 : 41 | .206760 | 100 : 63 | .200660 |
| 101 : 61 | .218992 | 80 : 49 | .212894 | 103 : 64 | .206657 | 73 : 46 | .200565 |
| 96 : 58 | .218843 | 111 : 68 | .212814 | 111 : 69 | .206474 | 119 : 75 | .200486 |
| 91 : 55 | .218679 | 93 : 57 | .212608 | 74 : 46 | .206474 | 92 : 58 | .200360 |
| 86 : 52 | .218495 | 62 : 38 | .212608 | 119 : 74 | .206315 | 111 : 70 | .200225 |
| 81 : 49 | .218289 | 106 : 65 | .212393 | 82 : 51 | .206244 | 65 : 41 | .200130 |
| 119 : 72 | 0.218215 | 75 : 46 | 0.212304 | 90 : 56 | 0.206055 | 84 : 53 | 0.200003 |
| 76 : 46 | .218056 | 119 : 73 | .212224 | 98 : 61 | .205896 | 103 : 65 | .199924 |
| 109 : 66 | .217883 | 88 : 54 | .212089 | 53 : 33 | .205762 | 114 : 72 | .199572 |
| 71 : 43 | .217790 | 101 : 62 | .211930 | 114 : 71 | .205647 | 95 : 60 | .199572 |
| 104 : 63 | .217693 | 57 : 35 | .211807 | 61 : 38 | .205546 | 76 : 48 | .199572 |
| 66 : 40 | .217484 | 70 : 43 | .211630 | 69 : 43 | .205380 | 57 : 36 | .199572 |
| 94 : 57 | .217253 | 96 : 59 | .211419 | 77 : 48 | .205250 | 106 : 67 | .199231 |
| 61 : 37 | .217128 | 109 : 67 | .211352 | 85 : 53 | .205143 | 87 : 55 | .199157 |
| 89 : 54 | 0.216996 | 117 : 72 | 0.210853 | 93 : 58 | 0.205055 | 68 : 43 | 0.199040 |
| 117 : 71 | .216928 | 91 : 56 | .210853 | 101 : 63 | .204981 | 117 : 74 | .198954 |
| 112 : 68 | .216709 | 78 : 48 | .210853 | 109 : 68 | .204918 | 98 : 62 | .198834 |
| 84 : 51 | .216709 | 65 : 40 | .210853 | 117 : 73 | .204863 | 49 : 31 | .198834 |
| 56 : 34 | .216709 | 112 : 69 | .210369 | 96 : 60 | .204120 | 79 : 50 | .198657 |
| 107 : 65 | .216470 | 99 : 61 | .210305 | 88 : 55 | .204120 | 109 : 69 | .198577 |
| 79 : 48 | .216386 | 86 : 53 | .210223 | 80 : 50 | .204120 | 90 : 57 | .198368 |
| 102 : 62 | .216209 | 73 : 45 | .210110 | 72 : 45 | .204120 | 60 : 38 | .198368 |
| 51 : 31 | 0.216209 | 120 : 74 | 0.209950 | 64 : 40 | 0.204120 | 101 : 64 | 0.198141 |
| 74 : 45 | .216019 | 60 : 37 | .209950 | 56 : 35 | .204120 | 71 : 45 | .198046 |
| 97 : 59 | .215920 | 107 : 66 | .209840 | 32 : 20 | .204120 | 112 : 71 | .197960 |
| 120 : 73 | .215858 | 94 : 58 | .209700 | 115 : 72 | .203365 | 82 : 52 | .197811 |
| 115 : 70 | .215600 | 81 : 50 | .209515 | 107 : 67 | .203309 | 93 : 59 | .197631 |
| 92 : 56 | .215600 | 115 : 71 | .209440 | 99 : 62 | .203244 | 52 : 33 | .197489 |
| 69 : 42 | .215600 | 68 : 42 | .209260 | 91 : 57 | .203167 | 115 : 73 | .197375 |
| 110 : 67 | .215318 | 34 : 21 | .209260 | 83 : 52 | .203075 | 63 : 40 | .197281 |
| 87 : 53 | 0.215243 | 89 : 55 | 0.209027 | 75 : 47 | 0.202963 | 74 : 47 | 0.197134 |
| 64 : 39 | .215115 | 55 : 34 | .208884 | 67 : 42 | .202826 | 85 : 54 | .197025 |
| 105 : 64 | .215009 | 76 : 47 | .208716 | 59 : 37 | .202650 | 96 : 61 | .196941 |
| 82 : 50 | .214844 | 97 : 60 | .208620 | 110 : 69 | .202544 | 107 : 68 | .196875 |
| 100 : 61 | .214670 | 118 : 73 | .208559 | 51 : 32 | .202420 | 118 : 75 | .196821 |
| 118 : 72 | .214550 | 84 : 52 | .208276 | 94 : 59 | .202276 | 110 : 70 | .196295 |
| 59 : 36 | .214550 | 63 : 39 | .208276 | 86 : 54 | .202105 | 99 : 63 | .196295 |
| 77 : 47 | .214393 | 113 : 70 | .207980 | 78 : 49 | .201899 | 88 : 56 | .196295 |
| 95 : 58 | 0.214296 | 92 : 57 | 0.207913 | 113 : 71 | 0.201820 | 77 : 49 | 0.196295 |
| 113 : 69 | .214229 | 71 : 44 | .207806 | 105 : 66 | .201645 | 66 : 42 | .196295 |
| 90 : 55 | .213880 | 100 : 62 | .207608 | 70 : 44 | .201645 | 55 : 35 | .196295 |
| 72 : 44 | .213880 | 50 : 31 | .207608 | 97 : 61 | .201442 | 113 : 72 | .195746 |
| 54 : 33 | .213880 | 79 : 49 | .207431 | 62 : 39 | .201327 | 91 : 58 | .195613 |
| 36 : 22 | .213880 | 108 : 67 | .207349 | 89 : 56 | .201202 | 102 : 65 | .195687 |
| 103 : 63 | .213497 | 87 : 54 | .207126 | 116 : 73 | .201135 | 69 : 44 | .195396 |
| 67 : 41 | .213291 | 58 : 36 | .207126 | 108 : 68 | .200915 | 58 : 37 | .195226 |

Logarithms of Gear Ratios from 1.0084+ to 7.5

| Numbers of Teeth | Logarithm of Ratio | Numbers of Teeth | Logarithm of Ratio | Numbers of Teeth | Logarithm of Ratio | Numbers of Teeth | Logarithm of Ratio |
|---|---|---|---|---|---|---|---|
| 105 : 67 | 0.195115 | 99 : 64 | 0.189455 | 110 : 72 | 0.184060 | 71 : 47 | 0.179160 |
| 94 : 60 | .194977 | 116 : 75 | .189397 | 55 : 36 | .184060 | 74 : 49 | .179036 |
| 47 : 30 | .194977 | 119 : 77 | .189056 | 84 : 55 | .183917 | 77 : 51 | .178921 |
| 83 : 53 | .194802 | 102 : 66 | .189056 | 113 : 74 | .183847 | 80 : 53 | .178814 |
| 119 : 76 | .194733 | 85 : 55 | .189056 | 116 : 76 | .183644 | 83 : 55 | .178715 |
| 108 : 69 | .194575 | 68 : 44 | .189056 | 87 : 57 | .183644 | 86 : 57 | .178624 |
| 97 : 62 | .194380 | 51 : 33 | .189056 | 58 : 38 | .183644 | 89 : 59 | .178538 |
| 61 : 39 | .194265 | 34 : 22 | .189056 | 119 : 78 | .183452 | 92 : 61 | .178458 |
| 86 : 55 | 0.194136 | 105 : 68 | 0.188680 | 90 : 59 | 0.183391 | 95 : 63 | 0.178383 |
| 111 : 71 | .194065 | 88 : 57 | .188608 | 61 : 40 | .183270 | 98 : 65 | .178312 |
| 100 : 64 | .193820 | 71 : 46 | .188501 | 93 : 61 | .183153 | 101 : 67 | .178247 |
| 75 : 48 | .193820 | 54 : 35 | .188326 | 96 : 63 | .182931 | 104 : 69 | .178184 |
| 50 : 32 | .193820 | 91 : 59 | .188190 | 64 : 42 | .182931 | 107 : 71 | .178126 |
| 114 : 73 | .193582 | 111 : 72 | .187991 | 32 : 21 | .182931 | 110 : 73 | .178070 |
| 89 : 57 | .193515 | 74 : 48 | .187991 | 99 : 65 | .182722 | 113 : 75 | .178017 |
| 64 : 41 | .193396 | 94 : 61 | .187798 | 67 : 44 | .182622 | 116 : 77 | .177967 |
| 103 : 66 | 0.193293 | 57 : 37 | 0.187673 | 102 : 67 | 0.182525 | 119 : 79 | 0.177920 |
| 117 : 75 | .193125 | 77 : 50 | .187521 | 105 : 69 | .182340 | 24 : 16 | |
| 78 : 50 | .193125 | 97 : 63 | .187431 | 70 : 46 | .182340 | or | |
| 92 : 59 | .192936 | 117 : 76 | .187372 | 108 : 71 | .182166 | any | .176091 |
| 53 : 34 | .192797 | 100 : 65 | .187087 | 73 : 48 | .182082 | 3 to 2 | |
| 120 : 77 | .192691 | 80 : 52 | .187087 | 111 : 73 | .182000 | ratio | |
| 67 : 43 | .192606 | 60 : 39 | .187087 | 114.75 | .181844 | 118 : 79 | .174255 |
| 81 : 52 | .192482 | 103 : 67 | .186762 | 76 : 50 | .181844 | 115 : 77 | .174207 |
| 95 : 61 | 0.192394 | 83 : 54 | 0.186684 | 117 : 77 | 0.181695 | 112 : 75 | 0.174157 |
| 109 : 70 | .192329 | 63 : 41 | .186557 | 79 : 52 | .181624 | 109 : 73 | .174104 |
| 112 : 72 | .191886 | 106 : 69 | .186457 | 120 : 79 | .181554 | 106 : 71 | .174048 |
| 98 : 63 | .191886 | 86 : 56 | .186311 | 82 : 54 | .181420 | 103 : 69 | .173988 |
| 84 : 54 | .191886 | 109 : 71 | .186168 | 85 : 56 | .181231 | 100 : 67 | .173925 |
| 70 : 45 | .191886 | 66 : 43 | .186075 | 88 : 58 | .181055 | 97 : 65 | .173858 |
| 56 : 36 | .191886 | 89 : 58 | .185962 | 91 : 60 | .180890 | 94 : 63 | .173787 |
| 115 : 74 | .191466 | 112 : 73 | .185895 | 94 : 62 | .180736 | 91 : 61 | .173712 |
| 101 : 65 | 0.191408 | 115 : 75 | 0.185637 | 47 : 31 | 0.180736 | 88 : 59 | 0.173631 |
| 87 : 56 | .191331 | 92 : 60 | .185637 | 97 : 64 | .180592 | 85 : 57 | .173544 |
| 73 : 47 | .191225 | 69 : 45 | .185637 | 100 : 66 | .180456 | 82 : 55 | .173451 |
| 59 : 38 | .191068 | 46 : 30 | .185637 | 50 : 33 | .180456 | 79 : 53 | .173351 |
| 104 : 67 | .190959 | 118 : 77 | .185391 | 103 : 68 | .180328 | 76 : 51 | .173243 |
| 90 : 58 | .190815 | 95 : 62 | .185332 | 106 : 70 | .180208 | 73 : 49 | .173127 |
| 76 : 49 | .190618 | 72 : 47 | .185235 | 53 : 35 | .180208 | 70 : 47 | .173000 |
| 107 : 69 | .190535 | 98 : 64 | .185046 | 109 : 72 | .180094 | 67 : 45 | .172862 |
| 93 : 60 | 0.190332 | 49 : 32 | 0.185046 | 112 : 74 | 0.179986 | 64 : 43 | 0.172712 |
| 62 : 40 | .190332 | 75 : 49 | .184865 | 56 : 37 | .179986 | 61 : 41 | .172546 |
| 110 : 71 | .190134 | 101 : 66 | .184778 | 115 : 76 | .179884 | 119 : 80 | .172457 |
| 79 : 51 | .190057 | 104 : 68 | .184524 | 118 : 78 | .179787 | 58 : 39 | .172363 |
| 96 : 62 | .189880 | 78 : 51 | .184524 | 59 : 39 | .179787 | 113 : 76 | .172265 |
| 113 : 73 | .189756 | 52 : 34 | .184524 | 62 : 41 | .179608 | 55 : 37 | .172161 |
| 65 : 42 | .189664 | 107 : 70 | .184286 | 65 : 43 | .179445 | 107 : 72 | .172051 |
| 82 : 53 | .189538 | 81 : 53 | .184209 | 68 : 45 | .179296 | 104 : 70 | .171935 |

### Logarithms of Gear Ratios from 1.0084+ to 7.5

| Numbers of Teeth | Logarithm of Ratio | Numbers of Teeth | Logarithm of Ratio | Numbers of Teeth | Logarithm of Ratio | Numbers of Teeth | Logarithm of Ratio |
|---|---|---|---|---|---|---|---|
| 52 : 35 | 0.171935 | 47 : 32 | 0.166948 | 77 : 53 | 0.162215 | 102 : 71 | 0.157342 |
| 101 : 68 | .171813 | 116 : 79 | .166831 | 61 : 42 | .162081 | 79 : 55 | .157264 |
| 98 : 66 | .171682 | 69 : 47 | .166751 | 106 : 73 | .161983 | 112 : 78 | .157123 |
| 49 : 33 | .171682 | 91 : 62 | .166650 | 90 : 62 | .161851 | 56 : 39 | .157123 |
| 95 : 64 | .171544 | 113 : 77 | .166588 | 45 : 31 | .161851 | 89 : 62 | .156998 |
| 92 : 62 | .171396 | 110 : 75 | .166331 | 119 : 82 | .161733 | 99 : 69 | .156786 |
| 46 : 31 | .171396 | 88 : 60 | .166331 | 74 : 51 | .161662 | 66 : 46 | .156786 |
| 89 : 60 | .171239 | 66 : 45 | .166331 | 103 : 71 | .161579 | 109 : 76 | .156613 |
| 86 : 58 | 0.171071 | 44 : 30 | 0.166331 | 116 : 80 | 0.161368 | 76 : 53 | 0.156538 |
| 83 : 56 | .170890 | 107 : 73 | .166061 | 87 : 60 | .161368 | 119 : 83 | .156469 |
| 120 : 81 | .170696 | 85 : 58 | .165991 | 58 : 40 | .161368 | 86 : 60 | .156347 |
| 80 : 54 | .170696 | 63 : 43 | .165872 | 100 : 69 | .161151 | 43 : 30 | .156347 |
| 117 : 79 | .170559 | 104 : 71 | .165775 | 71 : 49 | .161062 | 96 : 67 | .156196 |
| 77 : 52 | .170487 | 82 : 56 | .165626 | 113 : 78 | .160984 | 106 : 74 | .156074 |
| 114 : 77 | .170414 | 101 : 69 | .165472 | 84 : 58 | .160851 | 53 : 37 | .156074 |
| 111 : 75 | .170262 | 120 : 82 | .165367 | 97 : 67 | .160697 | 116 : 81 | .155973 |
| 74 : 50 | 0.170262 | 60 : 41 | 0.165367 | 110 : 76 | 0.160579 | 63 : 44 | 0.155888 |
| 108 : 73 | .170101 | 79 : 54 | .165233 | 55 : 38 | .160579 | 73 : 51 | .155753 |
| 71 : 48 | .170017 | 98 : 67 | .165151 | 68 : 47 | .160411 | 93 : 65 | .155570 |
| 105 : 71 | .169931 | 117 : 80 | .165096 | 81 : 56 | .160297 | 103 : 72 | .155505 |
| 68 : 46 | .169751 | 114 : 78 | .164810 | 94 : 65 | .160215 | 113 : 79 | .155451 |
| 99 : 67 | .169560 | 95 : 65 | .164810 | 107 : 74 | .160152 | 120 : 84 | .154902 |
| 65 : 44 | .169461 | 76 : 52 | .164810 | 120 : 83 | .160103 | 110 : 84 | .154902 |
| 96 : 65 | .169358 | 57 : 39 | .164810 | 117 : 81 | .159701 | 100 : 70 | .154902 |
| 93 : 63 | 0.169142 | 111 : 76 | 0.164509 | 104 : 72 | 0.159701 | 90 : 63 | 0.154902 |
| 62 : 42 | .169142 | 92 : 63 | .164447 | 91 : 63 | .159701 | 80 : 56 | .154902 |
| 31 : 21 | .169142 | 73 : 50 | .164353 | 78 : 54 | .159701 | 70 : 49 | .154902 |
| 90 : 61 | .168913 | 108 : 74 | .164192 | 65 : 45 | .159701 | 60 : 42 | .154902 |
| 59 : 40 | .168792 | 54 : 37 | .164192 | 52 : 36 | .159701 | 30 : 21 | .154902 |
| 118 : 80 | .168792 | 89 : 61 | .164060 | 114 : 79 | .159278 | 117 : 82 | .154372 |
| 87 : 59 | .168667 | 105 : 72 | .163857 | 101 : 70 | .159223 | 107 : 75 | .154323 |
| 115 : 78 | .168603 | 70 : 48 | .163857 | 88 : 61 | .159153 | 97 : 68 | .154263 |
| 112 : 76 | 0.168404 | 86 : 59 | 0.163647 | 75 : 52 | 0.159058 | 87 : 61 | 0.154190 |
| 84 : 57 | .168404 | 102 : 70 | .163502 | 62 : 43 | .158923 | 77 : 54 | .154097 |
| 56 : 38 | .168404 | 51 : 35 | .163502 | 111 : 77 | .158832 | 67 : 47 | .153977 |
| 109 : 74 | .168196 | 118 : 81 | .163397 | 98 : 68 | .158717 | 114 : 80 | .153815 |
| 81 : 55 | .168122 | 67 : 46 | .163317 | 49 : 34 | .158717 | 57 : 40 | .153815 |
| 106 : 72 | .167973 | 83 : 57 | .163203 | 85 : 59 | .158567 | 104 : 73 | .153710 |
| 53 : 36 | .167973 | 99 : 68 | .163126 | 108 : 75 | .158363 | 94 : 66 | .153584 |
| 78 : 53 | .167819 | 115 : 79 | .163071 | 72 : 50 | .158363 | 47 : 33 | .153584 |
| 103 : 70 | 0.167739 | 112 : 77 | 0.162727 | 95 : 66 | 0.158180 | 84 : 59 | 0.153427 |
| 100 : 68 | .167491 | 96 : 66 | .162727 | 59 : 41 | .158068 | 111 : 78 | .153228 |
| 75 : 51 | .167491 | 80 : 55 | .162727 | 82 : 57 | .157939 | 74 : 52 | .153228 |
| 50 : 34 | .167491 | 64 : 44 | .162727 | 105 : 73 | .157866 | 101 : 71 | .153063 |
| 97 : 66 | .167228 | 48 : 33 | .162727 | 115 : 80 | .157608 | 64 : 45 | .152968 |
| 72 : 49 | .167136 | 32 : 22 | .162727 | 92 : 64 | .157608 | 91 : 64 | .152861 |
| 119 : 81 | .167062 | 109 : 75 | .162365 | 69 : 48 | .157608 | 118 : 83 | .152804 |
| 94 : 64 | .166948 | 93 : 64 | .162303 | 46 : 32 | .157608 | 108 : 76 | .152610 |

## Logarithms of Gear Ratios from 1.0084+ to 7.5

| Numbers of Teeth | Logarithm of Ratio | Numbers of Teeth | Logarithm of Ratio | Numbers of Teeth | Logarithm of Ratio | Numbers of Teeth | Logarithm of Ratio |
|---|---|---|---|---|---|---|---|
| 81 : 57 | 0.152610 | 45 : 32 | 0.148063 | 114 : 82 | 0.143091 | 106 : 77 | 0.138815 |
| 54 : 38 | .152610 | 97 : 69 | .147923 | 57 : 41 | .143091 | 117 : 85 | .138767 |
| 98 : 69 | .152377 | 104 : 74 | .147802 | 82 : 59 | .142962 | 110 : 80 | .138303 |
| 71 : 50 | .152288 | 52 : 37 | .147802 | 107 : 77 | .142893 | 99 : 72 | .138303 |
| 115 : 81 | .152213 | 111 : 79 | .147696 | 100 : 72 | .142668 | 88 : 64 | .138303 |
| 88 : 62 | .152091 | 118 : 84 | .147603 | 75 : 54 | .142668 | 77 : 56 | .138303 |
| 44 : 31 | .152091 | 66 : 47 | .147446 | 50 : 36 | .142668 | 66 : 48 | .138303 |
| 105 : 74 | .151958 | 73 : 52 | .147320 | 118 : 85 | .142463 | 55 : 40 | .138303 |
| 61 : 43 | 0.151861 | 80 : 57 | 0.147215 | 93 : 67 | 0.142408 | 44 : 32 | 0.138303 |
| 78 : 55 | .151732 | 87 : 62 | .147128 | 68 : 49 | .142313 | 114 : 83 | .137827 |
| 95 : 67 | .151649 | 94 : 67 | .147053 | 111 : 80 | .142233 | 103 : 75 | .137776 |
| 112 : 79 | .151591 | 101 : 72 | .146989 | 86 : 62 | .142107 | 92 : 67 | .137713 |
| 119 : 84 | .151268 | 108 : 77 | .146933 | 43 : 31 | .142107 | 81 : 59 | .137633 |
| 102 : 72 | .151268 | 115 : 82 | .146884 | 104 : 75 | .141972 | 70 : 51 | .137528 |
| 85 : 60 | .151268 | 98 : 70 | .146128 | 61 : 44 | .141877 | 118 : 86 | .137384 |
| 68 : 48 | .151268 | 91 : 65 | .146128 | 79 : 57 | .141752 | 59 : 43 | .137384 |
| 51 : 36 | 0.151268 | 84 : 60 | 0.146128 | 97 : 70 | 0.141674 | 107 : 78 | 0.137289 |
| 109 : 77 | .150936 | 77 : 55 | .146128 | 115 : 83 | .141620 | 96 : 70 | .137173 |
| 92 : 65 | .150874 | 70 : 50 | .146128 | 108 : 78 | .141329 | 48 : 35 | .137173 |
| 75 : 53 | .150785 | 63 : 45 | .146128 | 90 : 65 | .141329 | 85 : 62 | .137027 |
| 116 : 82 | .150644 | 56 : 40 | .146128 | 72 : 52 | .141329 | 111 : 81 | .136838 |
| 58 : 41 | .150644 | 49 : 35 | .146128 | 54 : 39 | .141329 | 74 : 54 | .136838 |
| 99 : 70 | .150537 | 42 : 30 | .146128 | 119 : 86 | .141049 | 100 : 73 | .136677 |
| 82 : 58 | .150386 | 35 : 25 | .146128 | 101 : 73 | .140999 | 63 : 46 | .136583 |
| 106 : 75 | 0.150245 | 116 : 83 | 0.145380 | 83 : 60 | 0.140927 | 89 : 65 | 0.136477 |
| 65 : 46 | .150156 | 109 : 78 | .145332 | 65 : 47 | .140816 | 115 : 84 | .136419 |
| 89 : 63 | .150050 | 102 : 73 | .145277 | 112 : 81 | .140733 | 104 : 76 | .136220 |
| 113 : 80 | .149988 | 88 : 63 | .145142 | 94 : 68 | .140619 | 78 : 57 | .136220 |
| 120 : 85 | .149762 | 81 : 58 | .145057 | 47 : 34 | .140619 | 52 : 38 | .136220 |
| 96 : 68 | .149762 | 74 : 53 | .144956 | 76 : 55 | .140451 | 119 : 87 | .136028 |
| 72 : 51 | .149762 | 67 : 48 | .144834 | 105 : 76 | .140376 | 93 : 68 | .135974 |
| 48 : 34 | .149762 | 120 : 86 | .144683 | 116 : 84 | .140179 | 67 : 49 | .135879 |
| 103 : 73 | 0.149514 | 60 : 43 | 0.144683 | 87 : 63 | 0.140179 | 108 : 79 | 0.135797 |
| 79 : 56 | .149439 | 113 : 81 | .144593 | 58 : 42 | .140179 | 82 : 60 | .135663 |
| 110 : 78 | .149298 | 106 : 76 | .144492 | 29 : 21 | .140179 | 41 : 30 | .135663 |
| 55 : 39 | .149298 | 53 : 38 | .144492 | 98 : 71 | .139968 | 97 : 71 | .135513 |
| 86 : 61 | .149169 | 99 : 71 | .144377 | 69 : 50 | .139879 | 112 : 82 | .135404 |
| 117 : 83 | .149108 | 92 : 66 | .144244 | 109 : 79 | .139799 | 56 : 41 | .135404 |
| 93 : 66 | .148939 | 46 : 33 | .144244 | 120 : 87 | .139662 | 71 : 52 | .135255 |
| 62 : 44 | .148939 | 85 : 61 | .144089 | 80 : 58 | .139662 | 86 : 63 | .135158 |
| 31 : 22 | 0.148939 | 117 : 84 | 0.143907 | 91 : 66 | 0.139498 | 101 : 74 | 0.135090 |
| 100 : 71 | .148742 | 78 : 56 | .143907 | 102 : 74 | .139369 | 116 : 85 | .135039 |
| 69 : 49 | .148653 | 110 : 79 | .143766 | 51 : 37 | .139369 | 105 : 77 | .134699 |
| 107 : 76 | .148570 | 71 : 51 | .143688 | 113 : 82 | .139265 | 90 : 66 | .134699 |
| 114 : 81 | .148420 | 103 : 74 | .143606 | 62 : 45 | .139179 | 75 : 55 | .134699 |
| 76 : 54 | .148420 | 96 : 69 | .143422 | 73 : 53 | .139047 | 60 : 44 | .134699 |
| 83 : 59 | .148226 | 64 : 46 | .143422 | 84 : 61 | .138950 | 45 : 33 | .134699 |
| 90 : 64 | .148063 | 89 : 64 | .143210 | 95 : 69 | .138875 | 30 : 22 | .134699 |

## Logarithms of Gear Ratios from 1.0084+ to 7.5

| Numbers of Teeth | Logarithm of Ratio | Numbers of Teeth | Logarithm of Ratio | Numbers of Teeth | Logarithm of Ratio | Numbers of Teeth | Logarithm of Ratio |
|---|---|---|---|---|---|---|---|
| 109 : 80 | 0.134337 | 58 : 43 | 0.129960 | 104 : 78 | 0.124939 | 95 : 72 | 0.120391 |
| 94 : 69 | .134279 | 89 : 66 | .129846 | 88 : 66 | .124939 | 62 : 47 | .120294 |
| 79 : 58 | .134199 | 120 : 89 | .129791 | 84 : 63 | .124939 | 91 : 69 | .120192 |
| 64 : 47 | .134082 | 93 : 69 | .129634 | 64 : 48 | .124939 | 120 : 91 | .120140 |
| 113 : 83 | .134000 | 62 : 46 | .129634 | 60 : 45 | .124939 | 116 : 88 | .119975 |
| 98 : 72 | .133894 | 97 : 72 | .129439 | 44 : 33 | .124939 | 87 : 66 | .119975 |
| 49 : 36 | .133894 | 66 : 49 | .129348 | 40 : 30 | .124939 | 58 : 44 | .119975 |
| 83 : 61 | .133748 | 101 : 75 | .129260 | 28 : 21 | .124939 | 29 : 22 | .119975 |
| 117 : 86 | 0.133687 | 105 : 78 | 0.129095 | 117 : 88 | 0.123703 | 112 : 85 | 0.119799 |
| 102 : 75 | .133539 | 70 : 52 | .129095 | 113 : 85 | .123660 | 83 : 63 | .119738 |
| 68 : 50 | .133539 | 109 : 81 | .128942 | 109 : 82 | .123613 | 108 : 82 | .119610 |
| 87 : 64 | .133339 | 74 : 55 | .128869 | 105 : 79 | .123562 | 54 : 41 | .119610 |
| 106 : 78 | .133211 | 113 : 84 | .128799 | 101 : 76 | .123508 | 79 : 60 | .119476 |
| 72 : 53 | .133057 | 117 : 87 | .128667 | 97 : 73 | .123449 | 104 : 79 | .119406 |
| 91 : 67 | .132967 | 78 : 58 | .128667 | 93 : 70 | .123385 | 100 : 76 | .119186 |
| 110 : 81 | .132908 | 82 : 61 | .128484 | 89 : 67 | .123315 | 75 : 57 | .119186 |
| 114 : 84 | 0.132626 | 86 : 64 | 0.128319 | 85 : 64 | 0.123238 | 50 : 38 | 0.119186 |
| 95 : 70 | .132626 | 43 : 32 | .128319 | 81 : 61 | .123155 | 96 : 73 | .118948 |
| 76 : 56 | .132626 | 90 : 67 | .128168 | 77 : 58 | .123063 | 71 : 54 | .118865 |
| 57 : 42 | .132626 | 94 : 70 | .128030 | 73 : 55 | .122960 | 117 : 89 | .118796 |
| 118 : 87 | .132363 | 47 : 35 | .128030 | 69 : 52 | .122846 | 92 : 70 | .118690 |
| 99 : 73 | .132312 | 98 : 73 | .127903 | 65 : 49 | .122717 | 46 : 35 | .118690 |
| 80 : 59 | .132238 | 102 : 76 | .127787 | 61 : 46 | .122572 | 113 : 86 | .118580 |
| 61 : 45 | .132117 | 51 : 38 | .127787 | 118 : 89 | .122492 | 67 : 51 | .118505 |
| 103 : 76 | 0.132024 | 106 : 79 | 0.127679 | 114 : 86 | 0.122406 | 88 : 67 | 0.118408 |
| 84 : 62 | .131888 | 110 : 82 | .127579 | 57 : 43 | .122406 | 109 : 83 | .118348 |
| 42 : 31 | .131888 | 55 : 41 | .127579 | 110 : 83 | .122315 | 84 : 64 | .118099 |
| 107 : 79 | .131757 | 114 : 85 | .127486 | 53 : 40 | .122216 | 63 : 48 | .118099 |
| 65 : 48 | .131672 | 118 : 88 | .127399 | 106 : 80 | .122216 | 42 : 32 | .118099 |
| 88 : 65 | .131570 | 59 : 44 | .127399 | 102 : 77 | .122110 | 105 : 80 | .118090 |
| 111 : 82 | .131509 | 63 : 47 | .127243 | 98 : 74 | .121994 | 101 : 77 | .117831 |
| 115 : 85 | .131279 | 67 : 50 | .127105 | 49 : 37 | .121994 | 80 : 61 | .117760 |
| 92 : 68 | 0.131279 | 71 : 53 | 0.126982 | 94 : 71 | 0.121870 | 118 : 90 | .117640 |
| 69 : 51 | .131279 | 75 : 56 | .126873 | 90 : 68 | .121734 | 59 : 45 | 0.117640 |
| 46 : 34 | .131279 | 79 : 59 | .126775 | 45 : 34 | .121734 | 97 : 74 | .117540 |
| 119 : 88 | .131064 | 83 : 62 | .126686 | 86 : 65 | .121585 | 114 : 87 | .117386 |
| 96 : 71 | .131013 | 87 : 65 | .126606 | 82 : 62 | .121422 | 76 : 58 | .117386 |
| 73 : 54 | .130929 | 91 : 68 | .126533 | 41 : 31 | .121422 | 93 : 71 | .117225 |
| 100 : 74 | .130768 | 95 : 71 | .126465 | 119 : 90 | .121305 | 110 : 84 | .117113 |
| 50 : 37 | .130768 | 99 : 74 | .126404 | 78 : 59 | .121243 | 55 : 42 | .117113 |
| 77 : 57 | 0.130616 | 103 : 77 | 0.126347 | 115 : 87 | 0.121179 | 72 : 55 | 0.116970 |
| 104 : 77 | .130543 | 107 : 80 | .126294 | 111 : 84 | .121044 | 89 : 68 | .116881 |
| 81 : 60 | .130334 | 111 : 83 | .126245 | 74 : 56 | .121044 | 106 : 81 | .116821 |
| 54 : 40 | .130334 | 115 : 86 | .126199 | 107 : 81 | .120899 | 119 : 91 | .116506 |
| 27 : 20 | .130334 | 119 : 89 | .126157 | 70 : 53 | .120822 | 102 : 78 | .116506 |
| 112 : 83 | .130140 | 120 : 90 | .124939 | 103 : 78 | .120743 | 85 : 65 | .116506 |
| 85 : 63 | .130078 | 112 : 84 | .124939 | 99 : 75 | .120574 | 68 : 52 | .116506 |
| 116 : 86 | .129960 | 108 : 81 | .124939 | 66 : 50 | .120574 | 51 : 39 | .116506 |

Logarithms of Gear Ratios from 1.0084+ to 7.5

| Numbers of Teeth | Logarithm of Ratio | Numbers of Teeth | Logarithm of Ratio | Numbers of Teeth | Logarithm of Ratio | Numbers of Teeth | Logarithm of Ratio |
|---|---|---|---|---|---|---|---|
| 34 : 26 | 0.116506 | 101 : 78 | 0.112227 | 118 : 92 | 0.108094 | 75 : 59 | 0.104209 |
| 115 : 88 | .116215 | 110 : 85 | .111974 | 59 : 46 | .108094 | 61 : 48 | .104089 |
| 98 : 75 | .116165 | 88 : 68 | .111974 | 109 : 85 | .108008 | 108 : 85 | .104005 |
| 81 : 62 | .116093 | 44 : 34 | .111974 | 100 : 78 | .107905 | 94 : 74 | .103896 |
| 64 : 49 | .115984 | 66 : 51 | .111974 | 50 : 39 | .107905 | 47 : 37 | .103896 |
| 111 : 85 | .115904 | 119 : 92 | .111759 | 91 : 71 | .107783 | 80 : 63 | .103750 |
| 94 : 72 | .115795 | 97 : 75 | .111710 | 82 : 64 | .107634 | 113 : 89 | .103688 |
| 47 : 36 | .115795 | 75 : 58 | .111633 | 41 : 32 | .107634 | 99 : 78 | .103541 |
| 77 : 59 | 0.115639 | 106 : 82 | 0.111492 | 114 : 89 | 0.107515 | 66 : 52 | 0.103541 |
| 107 : 82 | .115570 | 53 : 41 | .111492 | 73 : 57 | .107448 | 118 : 93 | .103399 |
| 120 : 92 | .115393 | 84 : 65 | .111366 | 105 : 82 | .107375 | 85 : 67 | .103344 |
| 90 : 69 | .115393 | 115 : 89 | .111308 | 96 : 75 | .107210 | 104 : 82 | .103219 |
| 60 : 46 | .115393 | 93 : 72 | .111151 | 64 : 50 | .107210 | 52 : 41 | .103219 |
| 30 : 23 | .115393 | 62 : 48 | .111151 | 119 : 93 | .107064 | 90 : 71 | .102984 |
| 103 : 79 | .115210 | 102 : 79 | .110973 | 87 : 68 | .107010 | 109 : 86 | .102928 |
| 73 : 56 | .115135 | 71 : 55 | .110896 | 110 : 86 | .106894 | 95 : 75 | .102662 |
| 116 : 89 | 0.115068 | 111 : 86 | 0.110825 | 55 : 43 | 0.106894 | 76 : 60 | 0.102662 |
| 86 : 66 | .114955 | 120 : 93 | .110698 | 78 : 61 | .106765 | 57 : 45 | .102662 |
| 43 : 33 | .114955 | 80 : 62 | .110698 | 101 : 79 | .106694 | 38 : 30 | .102662 |
| 99 : 76 | .114822 | 40 : 31 | .110698 | 115 : 90 | .106455 | 119 : 94 | .102419 |
| 112 : 86 | .114720 | 89 : 69 | .110541 | 92 : 72 | .106455 | 100 : 79 | .102373 |
| 56 : 43 | .114720 | 98 : 76 | .110413 | 69 : 54 | .106455 | 81 : 64 | .102305 |
| 69 : 53 | .114573 | 49 : 38 | .110413 | 46 : 36 | .106455 | 62 : 49 | .102196 |
| 82 : 63 | .114473 | 107 : 83 | .110306 | 106 : 83 | .106228 | 105 : 83 | .102111 |
| 95 : 73 | 0.114401 | 116 : 90 | 0.110216 | 83 : 65 | 0.106165 | 86 : 68 | 0.101990 |
| 108 : 83 | .114346 | 58 : 45 | .110216 | 120 : 94 | .106053 | 43 : 34 | .101990 |
| 117 : 90 | .113943 | 67 : 52 | .110072 | 60 : 47 | .106053 | 110 : 87 | .101873 |
| 104 : 80 | .113943 | 76 : 59 | .109962 | 97 : 76 | .105958 | 67 : 53 | .101799 |
| 91 : 70 | .113943 | 85 : 66 | .109875 | 111 : 87 | .105804 | 91 : 72 | .101709 |
| 78 : 60 | .113943 | 94 : 73 | .109805 | 74 : 58 | .105804 | 115 : 91 | .101656 |
| 65 : 50 | .113943 | 103 : 80 | .109747 | 88 : 69 | .105634 | 120 : 95 | .101458 |
| 52 : 40 | .113943 | 112 : 87 | .109699 | 102 : 80 | .105510 | 96 : 76 | .101458 |
| 39 : 30 | 0.113943 | 117 : 91 | 0.109145 | 51 : 40 | 0.105510 | 72 : 57 | 0.101458 |
| 26 : 20 | .113943 | 108 : 84 | .109145 | 116 : 91 | .105417 | 48 : 38 | .101458 |
| 113 : 87 | .113559 | 99 : 77 | .109145 | 65 : 51 | .105343 | 101 : 80 | .101231 |
| 100 : 77 | .113509 | 90 : 70 | .109145 | 79 : 62 | .105235 | 77 : 61 | .101161 |
| 87 : 67 | .113445 | 81 : 63 | .109145 | 93 : 73 | .105160 | 106 : 84 | .101027 |
| 74 : 57 | .113357 | 72 : 56 | .109145 | 107 : 84 | .105105 | 53 : 42 | .101027 |
| 61 : 47 | .113232 | 63 : 49 | .109145 | 112 : 88 | .104735 | 82 : 65 | .100901 |
| 109 : 84 | .113147 | 54 : 42 | .109145 | 98 : 77 | .104735 | 111 : 88 | .100840 |
| 96 : 74 | 0.113040 | 45 : 35 | 0.109145 | 84 : 66 | 0.104735 | 116 : 92 | 0.100670 |
| 48 : 37 | .113040 | 27 : 21 | .109145 | 70 : 55 | .104735 | 87 : 69 | .100670 |
| 83 : 64 | .112898 | 113 : 88 | .108596 | 56 : 44 | .104735 | 58 : 46 | .100670 |
| 118 : 91 | .112841 | 104 : 81 | .108548 | 42 : 33 | .104735 | 29 : 23 | .100670 |
| 70 : 54 | .112704 | 95 : 74 | .108492 | 28 : 22 | .104735 | 92 : 73 | .100465 |
| 92 : 71 | .112530 | 86 : 67 | .108424 | 117 : 92 | .104398 | 63 : 50 | .100371 |
| 57 : 44 | .112422 | 77 : 60 | .108339 | 103 : 81 | .104352 | 97 : 77 | .100281 |
| 79 : 61 | .112297 | 68 : 53 | .108233 | 89 : 70 | .104292 | 102 : 81 | .100115 |

Logarithms of Gear Ratios from 1.0084+ to 7.5

| Numbers of Teeth | Logarithm of Ratio | Numbers of Teeth | Logarithm of Ratio | Numbers of Teeth | Logarithm of Ratio | Numbers of Teeth | Logarithm of Ratio |
|---|---|---|---|---|---|---|---|
| 68 : 54 | 0.100115 | 117 : 94 | 0.095058 | 74 : 60 | 0.091080 | 88 : 72 | 0.087150 |
| 34 : 27 | .100115 | 56 : 45 | .094976 | 37 : 30 | .091080 | 77 : 63 | .087150 |
| 107 : 85 | .099964 | 107 : 86 | .094885 | 90 : 73 | .090920 | 66 : 54 | .087150 |
| 73 : 58 | .099894 | 102 : 82 | .094786 | 106 : 86 | .090807 | 55 : 45 | .087150 |
| 112 : 89 | .099828 | 51 : 41 | .094786 | 53 : 43 | .090807 | 44 : 36 | .087150 |
| 78 : 62 | .099703 | 97 : 78 | .094677 | 69 : 56 | .090661 | 116 : 95 | .086734 |
| 39 : 31 | .099703 | 92 : 74 | .094556 | 85 : 69 | .090570 | 105 : 86 | .086691 |
| 83 : 66 | .099534 | 46 : 37 | .094556 | 101 : 82 | .090508 | 94 : 77 | .086637 |
| 88 : 70 | 0.099385 | 87 : 70 | 0.094421 | 117 : 95 | 0.090462 | 83 : 68 | 0.086569 |
| 44 : 35 | .099385 | 82 : 66 | .094270 | 112 : 91 | .090177 | 72 : 59 | .086481 |
| 93 : 74 | .099251 | 41 : 33 | .094270 | 96 : 78 | .090177 | 61 : 50 | .086360 |
| 98 : 78 | .099132 | 118 : 95 | .094158 | 80 : 65 | .090177 | 111 : 91 | .086282 |
| 49 : 39 | .099132 | 77 : 62 | .094099 | 64 : 52 | .090177 | 100 : 82 | .086186 |
| 103 : 82 | .099023 | 113 : 91 | .094037 | 48 : 39 | .090177 | 50 : 41 | .086186 |
| 108 : 86 | .098925 | 108 : 87 | .093905 | 107 : 87 | .089865 | 89 : 73 | .086067 |
| 54 : 43 | .098925 | 72 : 58 | .093905 | 91 : 74 | .089810 | 117 : 96 | .085915 |
| 113 : 90 | 0.098836 | 103 : 83 | 0.093759 | 75 : 61 | 0.089732 | 78 : 64 | 0.085915 |
| 118 : 94 | .098754 | 67 : 54 | .093681 | 118 : 96 | .089611 | 39 : 32 | .085915 |
| 59 : 47 | .098754 | 98 : 79 | .093599 | 59 : 48 | .089611 | 106 : 87 | .085787 |
| 64 : 51 | .098610 | 93 : 75 | .093422 | 102 : 83 | .089522 | 67 : 55 | .085712 |
| 69 : 55 | .098486 | 62 : 50 | .093422 | 86 : 70 | .089401 | 95 : 78 | .085629 |
| 74 : 59 | .098380 | 119 : 96 | .093276 | 43 : 35 | .089401 | 112 : 92 | .085430 |
| 79 : 63 | .098287 | 88 : 71 | .093224 | 113 : 92 | .089291 | 84 : 69 | .085430 |
| 84 : 67 | .098205 | 114 : 92 | .093117 | 70 : 57 | .089223 | 56 : 46 | .085430 |
| 89 : 71 | 0.098132 | 57 : 46 | 0.093117 | 97 : 79 | 0.089145 | 28 : 23 | 0.085430 |
| 94 : 75 | .098067 | 83 : 67 | .093003 | 108 : 88 | .088941 | 101 : 83 | .085243 |
| 99 : 79 | .098008 | 109 : 88 | .092944 | 81 : 66 | .088941 | 118 : 97 | .085110 |
| 104 : 83 | .097955 | 104 : 84 | .092754 | 54 : 44 | .088941 | 90 : 74 | .085011 |
| 109 : 87 | .097907 | 78 : 63 | .092754 | 27 : 22 | .088941 | 45 : 37 | .085011 |
| 114 : 91 | .097864 | 52 : 42 | .092754 | 119 : 97 | .088775 | 107 : 88 | .084901 |
| 119 : 95 | .097823 | 26 : 21 | .092754 | 65 : 53 | .088638 | 62 : 51 | .084822 |
| 20 : 16 | | 99 : 80 | .092545 | 103 : 84 | .088558 | 79 : 65 | .084714 |
| or | | 73 : 59 | 0.092471 | 114 : 93 | 0.088422 | 96 : 79 | 0.084644 |
| any | 0.096910 | 120 : 97 | .092410 | 76 : 62 | .088422 | 113 : 93 | .084596 |
| 5 to 4 | | 94 : 76 | .092314 | 38 : 31 | .088422 | 119 : 98 | .084321 |
| ratio | | 47 : 38 | .092314 | 87 : 71 | .088261 | 102 : 84 | .084321 |
| 116 : 93 | .095975 | 115 : 93 | .092215 | 98 : 80 | .088136 | 85 : 70 | .084321 |
| 111 : 89 | .095933 | 68 : 55 | .092146 | 49 : 40 | .088136 | 68 : 56 | .084321 |
| 106 : 85 | .095887 | 89 : 72 | .092058 | 109 : 89 | .088037 | 51 : 42 | .084321 |
| 101 : 81 | .095836 | 110 : 89 | .092003 | 120 : 98 | .087955 | 108 : 89 | .084034 |
| 96 : 77 | 0.095781 | 105 : 85 | 0.091770 | 60 : 49 | 0.087955 | 91 : 75 | 0.083980 |
| 91 : 73 | .095719 | 84 : 68 | .091770 | 71 : 58 | .087830 | 74 : 61 | .083902 |
| 86 : 69 | .095649 | 63 : 51 | .091770 | 82 : 67 | .087739 | 114 : 94 | .083777 |
| 81 : 65 | .095572 | 42 : 34 | .091770 | 93 : 76 | .087669 | 57 : 47 | .083777 |
| 76 : 61 | .095484 | 116 : 94 | .091330 | 104 : 85 | .087614 | 97 : 80 | .083682 |
| 71 : 57 | .095383 | 58 : 47 | .091330 | 115 : 94 | .087570 | 120 : 99 | .083546 |
| 66 : 53 | .095268 | 95 : 77 | 0.091233 | 110 : 90 | .087150 | 80 : 66 | .083546 |
| 61 : 49 | .095134 | 111 : 90 | .091080 | 99 : 81 | .087150 | 40 : 33 | .083546 |

Logarithms of Gear Ratios from 1.0084+ to 7.5

| Numbers of Teeth | Logarithm of Ratio | Numbers of Teeth | Logarithm of Ratio | Numbers of Teeth | Logarithm of Ratio | Numbers of Teeth | Logarithm of Ratio |
|---|---|---|---|---|---|---|---|
| 103 : 85 | 0.083418 | 115 : 96 | 0.078427 | 114 : 96 | 0.074634 | 73 : 62 | 0.070931 |
| 63 : 52 | .083337 | 109 : 91 | .078385 | 95 : 80 | .074634 | 93 : 79 | .070856 |
| 86 : 71 | .083240 | 103 : 86 | .078339 | 76 : 64 | .074634 | 113 : 96 | .070808 |
| 109 : 90 | .083184 | 97 : 81 | .078287 | 57 : 48 | .074634 | 100 : 85 | .070581 |
| 115 : 95 | .082974 | 91 : 76 | .078228 | 38 : 32 | .074634 | 80 : 68 | .070581 |
| 92 : 76 | .082974 | 85 : 71 | .078161 | 108 : 91 | .074382 | 60 : 51 | .070581 |
| 69 : 57 | .082974 | 79 : 66 | .078083 | 89 : 75 | .074329 | 40 : 34 | .070581 |
| 46 : 38 | .082974 | 73 : 61 | .077993 | 70 : 59 | .074246 | 107 : 91 | .070342 |
| 98 : 81 | 0.082741 | 67 : 56 | 0.077887 | 102 : 86 | 0.074102 | 87 : 74 | 0.070288 |
| 75 : 62 | .082670 | 61 : 51 | .077760 | 51 : 43 | .074102 | 67 : 57 | .070200 |
| 52 : 43 | .082535 | 116 : 97 | .077686 | 83 : 70 | .073980 | 114 : 97 | .070133 |
| 81 : 67 | .082410 | 55 : 46 | .077605 | 115 : 97 | .073927 | 94 : 80 | .070038 |
| 110 : 91 | .082351 | 104 : 87 | .077514 | 96 : 81 | .073786 | 47 : 40 | .070038 |
| 116 : 96 | .082187 | 98 : 82 | .077412 | 64 : 54 | .073786 | 74 : 63 | .069891 |
| 87 : 72 | .082187 | 49 : 41 | .077412 | 109 : 92 | .073639 | 101 : 86 | .069823 |
| 58 : 48 | .082187 | 92 : 77 | .077297 | 77 : 65 | .073577 | 81 : 69 | .069636 |
| 29 : 24 | 0.082187 | 86 : 72 | 0.077166 | 90 : 76 | 0.073429 | 54 : 46 | 0.069636 |
| 93 : 77 | .081992 | 43 : 36 | .077166 | 45 : 38 | .073429 | 115 : 98 | .069472 |
| 64 : 53 | .081904 | 80 : 67 | .077015 | 103 : 87 | .073318 | 88 : 75 | .069421 |
| 99 : 82 | .081821 | 117 : 98 | .076960 | 116 : 98 | .073232 | 61 : 52 | .069327 |
| 105 : 87 | .081670 | 111 : 93 | .076840 | 58 : 49 | .073232 | 95 : 81 | .069239 |
| 70 : 58 | .081670 | 74 : 62 | .076840 | 71 : 60 | .073107 | 68 : 58 | .069081 |
| 76 : 63 | .081473 | 37 : 31 | .076840 | 84 : 71 | .073021 | 109 : 93 | .068944 |
| 117 : 97 | .081414 | 105 : 88 | .076707 | 97 : 82 | .072958 | 75 : 64 | .068881 |
| 82 : 68 | 0.081305 | 68 : 57 | 0.076634 | 110 : 93 | 0.072910 | 116 : 99 | 0.068823 |
| 41 : 34 | .081305 | 99 : 83 | .076557 | 117 : 99 | .072551 | 82 : 70 | .068716 |
| 88 : 73 | .081160 | 93 : 78 | .076388 | 104 : 88 | .072551 | 41 : 35 | .068716 |
| 94 : 78 | .081033 | 62 : 52 | .076388 | 91 : 77 | .072551 | 89 : 76 | .068576 |
| 47 : 39 | .081033 | 118 : 99 | .076247 | 78 : 66 | .072551 | 96 : 82 | .068457 |
| 100 : 83 | .080922 | 87 : 73 | .076196 | 65 : 55 | .072551 | 48 : 41 | .068457 |
| 106 : 88 | .080823 | 112 : 94 | .076090 | 39 : 33 | .072551 | 103 : 88 | .068355 |
| 53 : 44 | .080823 | 56 : 47 | .076090 | 52 : 44 | .072551 | 55 : 47 | .068265 |
| 118 : 98 | 0.080656 | 81 : 68 | 0.075976 | 111 : 94 | 0.072195 | 117 : 100 | 0.068186 |
| 59 : 49 | .080656 | 106 : 89 | .075916 | 98 : 83 | .072148 | 62 : 53 | .068116 |
| 65 : 54 | .080520 | 100 : 84 | .075721 | 85 : 72 | .072086 | 69 : 59 | .067997 |
| 71 : 59 | .080406 | 75 : 63 | .075721 | 72 : 61 | .072003 | 76 : 65 | .067900 |
| 77 : 64 | .080311 | 50 : 42 | .075721 | 118 : 100 | .071882 | 83 : 71 | .067820 |
| 83 : 69 | .080229 | 25 : 21 | .075721 | 59 : 50 | .071882 | 90 : 77 | .067752 |
| 89 : 74 | .080158 | 119 : 100 | 0.075547 | 105 : 89 | .071797 | 97 : 83 | .067694 |
| 95 : 79 | .080097 | 94 : 79 | .075501 | 92 : 78 | .071693 | 104 : 89 | .067643 |
| 107 : 89 | 0.079994 | 69 : 58 | 0.075421 | 46 : 39 | .071693 | 111 : 95 | 0.067599 |
| 113 : 94 | .079951 | 113 : 95 | .075355 | 79 : 67 | 0.071552 | 98 : 84 | .066947 |
| 119 : 99 | .079912 | 88 : 74 | .075251 | 112 : 95 | .071494 | 91 : 78 | .066947 |
| 30 : 25 |  | 44 : 37 | .075251 | 99 : 84 | .071356 | 84 : 72 | .066947 |
| or |  | 107 : 90 | .075141 | 66 : 56 | .071356 | 77 : 66 | .066947 |
| any | .079181 | 63 : 53 | .075065 | 86 : 73 | .071176 | 70 : 60 | .066947 |
| 6 to 5 |  | 82 : 69 | .074965 | 106 : 90 | .071063 | 63 : 54 | .066947 |
| ratio |  | 101 : 85 | .074903 | 53 : 45 | .071063 | 56 : 48 | .066947 |

# LOGARITHMS OF GEAR RATIOS

### Logarithms of Gear Ratios from 1.0084+ to 7.5

| Numbers of Teeth | Logarithm of Ratio | Numbers of Teeth | Logarithm of Ratio | Numbers of Teeth | Logarithm of Ratio | Numbers of Teeth | Logarithm of Ratio |
|---|---|---|---|---|---|---|---|
| 49 : 42 | 0.066947 | 75 : 65 | 0.062148 | 113 : 99 | 0.057443 | 69 : 61 | 0.053519 |
| 42 : 36 | .066947 | 60 : 52 | .062148 | 105 : 92 | .057402 | 95 : 84 | .053444 |
| 35 : 30 | .066947 | 45 : 39 | .062148 | 97 : 85 | .057353 | 78 : 69 | .053246 |
| 113 : 97 | .066307 | 113 : 98 | .061852 | 89 : 78 | .057295 | 113 : 100 | .053246 |
| 106 : 91 | .066265 | 98 : 85 | .061807 | 81 : 71 | .057227 | 87 : 77 | .053078 |
| 99 : 85 | .066216 | 83 : 72 | .061746 | 73 : 64 | .057143 | 61 : 54 | .052936 |
| 92 : 79 | .066161 | 68 : 59 | .061657 | 65 : 57 | .057039 | 96 : 85 | .052852 |
| 85 : 73 | .066096 | 53 : 46 | .061518 | 57 : 50 | .056905 | 70 : 62 | 0.052706 |
| 78 : 67 | 0.066020 | 91 : 79 | 0.061414 | 106 : 93 | 0.056823 | 35 : 31 | .052706 |
| 71 : 61 | .065929 | 76 : 66 | .061270 | 98 : 86 | .056728 | 79 : 70 | .052529 |
| 64 : 55 | .065817 | 38 : 33 | .061270 | 49 : 43 | .056728 | 88 : 78 | .052388 |
| 57 : 49 | .065679 | 99 : 86 | .061137 | 90 : 79 | .056615 | 44 : 39 | .052388 |
| 107 : 92 | .065596 | 61 : 53 | .061054 | 82 : 72 | .056481 | 97 : 86 | .052273 |
| 100 : 86 | .065502 | 84 : 73 | .060956 | 41 : 36 | .056481 | 53 : 47 | .052178 |
| 50 : 43 | .065502 | 107 : 93 | .060901 | 74 : 65 | .056318 | 62 : 55 | .052029 |
| 93 : 80 | .065393 | 92 : 80 | .060698 | 107 : 94 | .056256 | 71 : 63 | 0.051918 |
| 86 : 74 | 0.065267 | 69 : 60 | 0.060698 | 99 : 87 | 0.056116 | 80 : 71 | .051832 |
| 43 : 37 | .065267 | 46 : 40 | .060698 | 66 : 58 | .056116 | 89 : 79 | .051763 |
| 79 : 68 | .065118 | 100 : 87 | .060481 | 91 : 80 | .055951 | 98 : 87 | .051707 |
| 115 : 99 | .065063 | 77 : 67 | .060416 | 58 : 51 | .055858 | 107 : 95 | .051660 |
| 72 : 62 | .064941 | 54 : 47 | .060296 | 83 : 73 | .055755 | 99 : 88 | .051153 |
| 36 : 31 | .064941 | 85 : 74 | .060187 | 108 : 95 | .055700 | 90 : 80 | .051153 |
| 101 : 87 | .064802 | 93 : 81 | .059998 | 100 : 88 | .055517 | 81 : 72 | .051153 |
| 65 : 56 | .064725 | 62 : 54 | .059998 | 75 : 66 | .055517 | 72 : 64 | 0.051153 |
| 94 : 81 | 0.064643 | 101 : 88 | 0.059839 | 50 : 44 | 0.055517 | 63 : 56 | .051153 |
| 87 : 75 | .064458 | 70 : 61 | .059768 | 92 : 81 | .055303 | 54 : 48 | .051153 |
| 109 : 94 | .064299 | 109 : 95 | .059703 | 67 : 59 | .055223 | 45 : 40 | .051153 |
| 80 : 69 | .064241 | 78 : 68 | .059586 | 109 : 96 | .055155 | 36 : 32 | .051153 |
| 102 : 88 | .064118 | 39 : 34 | .059586 | 84 : 74 | .055048 | 109 : 97 | .050655 |
| 51 : 44 | .064118 | 86 : 75 | .059437 | 42 : 37 | .055048 | 100 : 89 | .050610 |
| 73 : 63 | .063982 | 47 : 41 | .059314 | 101 : 89 | .054931 | 91 : 81 | .050556 |
| 95 : 82 | .063910 | 94 : 82 | .059314 | 59 : 52 | .054849 | 82 : 73 | 0.050491 |
| 88 : 76 | 0.063669 | 102 : 89 | 0.059210 | 76 : 67 | 0.054739 | 73 : 65 | .050410 |
| 66 : 57 | .063669 | 55 : 48 | .059122 | 93 : 82 | .054669 | 64 : 57 | .050305 |
| 44 : 38 | .063669 | 63 : 55 | .058978 | 110 : 97 | .054621 | 55 : 49 | .050167 |
| 103 : 89 | .063447 | 71 : 62 | .058867 | 102 : 90 | .054358 | 101 : 90 | .050079 |
| 81 : 70 | .063387 | 79 : 69 | .058778 | 85 : 75 | .054358 | 92 : 82 | .049974 |
| 59 : 51 | .063282 | 87 : 76 | .058706 | 68 : 60 | .054358 | 46 : 41 | .049974 |
| 96 : 83 | .063193 | 95 : 83 | .058646 | 51 : 45 | .054358 | 83 : 74 | .049846 |
| 74 : 64 | .063052 | 103 : 90 | .058595 | 34 : 30 | .054358 | 74 : 66 | 0.049688 |
| 37 : 32 | 0.063052 | 96 : 84 | 0.057992 | 111 : 98 | 0.054097 | 37 : 33 | .049688 |
| 89 : 77 | .062899 | 88 : 77 | .057992 | 94 : 83 | .054050 | 102 : 91 | .049558 |
| 52 : 45 | .062791 | 80 : 70 | .057992 | 77 : 68 | .053982 | 65 : 58 | .049485 |
| 67 : 58 | .062647 | 72 : 63 | .057992 | 60 : 53 | .053875 | 93 : 83 | .049405 |
| 82 : 71 | .062556 | 64 : 56 | .057992 | 103 : 91 | .053796 | 56 : 50 | .049218 |
| 97 : 84 | .062493 | 56 : 49 | .057992 | 86 : 76 | .053685 | 84 : 75 | .049218 |
| 112 : 97 | .062446 | 48 : 42 | .057992 | 43 : 38 | .053685 | 103 : 92 | .049049 |
| 90 : 78 | .062148 | 40 : 35 | .057992 | 112 : 99 | .053583 | | |

Logarithms of Gear Ratios from 1.0084+ to 7.5

| Numbers of Teeth | Logarithm of Ratio | Numbers of Teeth | Logarithm of Ratio | Numbers of Teeth | Logarithm of Ratio | Numbers of Teeth | Logarithm of Ratio |
|---|---|---|---|---|---|---|---|
| 75 : 67 | 0.048987 | 41 : 37 | 0.044582 | 57 : 52 | 0.039872 | 38 : 35 | 0.035716 |
| 94 : 84 | .048849 | 72 : 65 | .044419 | 80 : 73 | .039767 | 89 : 82 | .035576 |
| 47 : 42 | .048849 | 103 : 93 | .044354 | 103 : 94 | .039709 | 51 : 47 | .035472 |
| 66 : 59 | .048692 | 93 : 84 | .044204 | 92 : 84 | .039509 | 64 : 59 | .035328 |
| 85 : 76 | .048605 | 62 : 56 | .044204 | 69 : 63 | .039509 | 77 : 71 | .035232 |
| 104 : 93 | .048550 | 83 : 75 | .044017 | 46 : 42 | .039509 | 90 : 83 | .035164 |
| 95 : 85 | .048305 | 52 : 47 | .043905 | 104 : 95 | .039310 | 103 : 95 | .035114 |
| 76 : 68 | .048305 | 73 : 66 | .043779 | 81 : 74 | .039253 | 91 : 84 | .034762 |
| 57 : 51 | 0.048305 | 94 : 85 | 0.043709 | 58 : 53 | 0.039152 | 78 : 72 | 0.034762 |
| 38 : 34 | .048305 | 84 : 76 | .043466 | 93 : 85 | .039064 | 65 : 60 | .034762 |
| 105 : 94 | .048061 | 63 : 57 | .043466 | 105 : 96 | .038918 | 52 : 48 | .034762 |
| 86 : 77 | .048008 | 42 : 38 | .043466 | 70 : 64 | .038918 | 39 : 36 | .034762 |
| 67 : 60 | .047924 | 95 : 86 | .043225 | 35 : 32 | .038918 | 105 : 97 | .034418 |
| 96 : 86 | .047773 | 74 : 67 | .043157 | 82 : 75 | .038753 | 92 : 85 | .034369 |
| 48 : 43 | .047773 | 53 : 48 | .043035 | 94 : 86 | .038629 | 79 : 73 | .034304 |
| 77 : 69 | .047642 | 85 : 77 | .042928 | 47 : 43 | .038629 | 66 : 61 | .034214 |
| 106 : 95 | 0.047582 | 96 : 87 | 0.042752 | 106 : 97 | 0.038534 | 53 : 49 | 0.034080 |
| 87 : 78 | .047425 | 64 : 58 | .042752 | 59 : 54 | .038458 | 93 : 86 | .033984 |
| 58 : 52 | .047425 | 107 : 97 | .042612 | 71 : 65 | .038345 | 80 : 74 | .033858 |
| 29 : 26 | .047425 | 75 : 68 | .042552 | 83 : 76 | .038265 | 40 : 37 | .033858 |
| 97 : 87 | .047252 | 86 : 78 | .042404 | 95 : 87 | .038204 | 107 : 99 | .033749 |
| 39 : 35 | .046997 | 43 : 39 | .042404 | 107 : 98 | .038158 | 67 : 62 | .033683 |
| 88 : 79 | .046856 | 97 : 88 | .042289 | 96 : 88 | .037789 | 94 : 87 | .033609 |
| 98 : 88 | .046743 | 54 : 49 | .042198 | 84 : 77 | 0.037789 | 81 : 75 | .033424 |
| 49 : 44 | 0.046743 | 65 : 59 | 0.042061 | 72 : 66 | .037789 | 54 : 50 | 0.033424 |
| 108 : 97 | .046652 | 76 : 69 | .041965 | 60 : 55 | .037789 | 95 : 88 | .033241 |
| 59 : 53 | .046576 | 87 : 79 | .041892 | 48 : 44 | .037789 | 68 : 63 | .033168 |
| 69 : 62 | .046457 | 98 : 89 | .041836 | 36 : 33 | .037789 | 82 : 76 | .033000 |
| 79 : 71 | .046369 | 109 : 99 | .041791 | 109 : 100 | .037427 | 41 : 38 | .033000 |
| 89 : 80 | .046300 | 99 : 90 | .041393 | 97 : 89 | .037382 | 96 : 89 | .032881 |
| 99 : 89 | .046245 | 88 : 80 | .041393 | 85 : 78 | .037324 | 55 : 51 | .032793 |
| 109 : 98 | .046200 | 77 : 70 | .041393 | 73 : 67 | .037248 | 69 : 64 | .032669 |
| 100 : 90 | 0.045758 | 66 : 60 | 0.041393 | 61 : 56 | 0.037142 | 83 : 77 | 0.032587 |
| 90 : 81 | .045758 | 55 : 50 | .041393 | 98 : 90 | .036984 | 97 : 90 | .032529 |
| 80 : 72 | .045758 | 44 : 40 | .041393 | 49 : 45 | .036984 | 98 : 91 | .032185 |
| 70 : 63 | .045758 | 33 : 30 | .041393 | 86 : 79 | .036871 | 84 : 78 | .032185 |
| 60 : 54 | .045758 | 100 : 91 | .040959 | 74 : 68 | .036723 | 70 : 65 | .032185 |
| 50 : 45 | .045758 | 89 : 81 | .040905 | 37 : 34 | .036723 | 56 : 52 | .032185 |
| 40 : 36 | .045758 | 78 : 71 | .040836 | 99 : 91 | .036594 | 42 : 39 | .032185 |
| 111 : 100 | .045323 | 67 : 61 | .040745 | 62 : 57 | .036517 | 99 : 92 | .031847 |
| 101 : 91 | 0.045280 | 56 : 51 | 0.040618 | 87 : 80 | 0.036429 | 85 : 79 | 0.031792 |
| 91 : 82 | .045228 | 101 : 92 | .040534 | 100 : 92 | .036212 | 71 : 66 | .031714 |
| 81 : 73 | .045162 | 90 : 82 | .040429 | 75 : 69 | .036212 | 57 : 53 | .031599 |
| 71 : 64 | .045078 | 45 : 41 | .040429 | 50 : 46 | .036212 | 100 : 93 | .031517 |
| 61 : 55 | .044967 | 79 : 72 | .040295 | 88 : 81 | .035998 | 86 : 80 | .031409 |
| 51 : 46 | .044812 | 68 : 62 | .040117 | 63 : 58 | .035913 | 43 : 40 | .031409 |
| 92 : 83 | .044710 | 34 : 31 | .040117 | 101 : 93 | .035839 | 72 : 67 | .031258 |
| 82 : 74 | .044582 | 91 : 83 | .039963 | 76 : 70 | .035716 | 101 : 94 | .031194 |

## Logarithms of Gear Ratios from 1.0084+ to 7.5

| Numbers of Teeth | Logarithm of Ratio | Numbers of Teeth | Logarithm of Ratio | Numbers of Teeth | Logarithm of Ratio | Numbers of Teeth | Logarithm of Ratio |
|---|---|---|---|---|---|---|---|
| 87 : 81 | 0.031034 | 50 : 47 | 0.026872 | 80 : 76 | 0.022276 | 100 : 96 | 0.017729 |
| 58 : 54 | .031034 | 67 : 63 | .026734 | 60 : 57 | .022276 | 75 : 72 | .017729 |
| 102 : 95 | .030877 | 84 : 79 | .026652 | 40 : 38 | .022276 | 50 : 48 | .017729 |
| 73 : 68 | .030814 | 101 : 95 | .026598 | 20 : 19 | .022276 | 101 : 97 | .017550 |
| 88 : 82 | .030669 | 85 : 80 | .026329 | 101 : 96 | .022050 | 76 : 73 | .017491 |
| 44 : 41 | .030669 | 68 : 64 | .026329 | 81 : 77 | .021994 | 51 : 49 | .017374 |
| 103 : 96 | .030566 | 51 : 48 | .026329 | 61 : 58 | .021902 | 77 : 74 | .017259 |
| 59 : 55 | .030489 | 34 : 32 | .026329 | 103 : 98 | .021611 | 103 : 99 | .017202 |
| 74 : 69 | 0.030383 | 103 : 97 | 0.026066 | 62 : 59 | 0.021540 | 78 : 75 | 0.017033 |
| 89 : 83 | .030312 | 86 : 81 | .026014 | 83 : 79 | .021451 | 52 : 50 | .017033 |
| 104 : 97 | .030262 | 69 : 65 | .025936 | 104 : 99 | .021398 | 79 : 76 | .016814 |
| 105 : 98 | .029963 | 52 : 49 | .025807 | 84 : 80 | .021189 | 53 : 51 | .016706 |
| 90 : 84 | .029963 | 87 : 82 | .025705 | 63 : 60 | .021189 | 80 : 77 | .016599 |
| 75 : 70 | .029963 | 70 : 66 | .025554 | 42 : 40 | .021189 | 81 : 78 | .016391 |
| 60 : 56 | .029963 | 35 : 33 | .025554 | 85 : 81 | .020934 | 54 : 52 | .016391 |
| 45 : 42 | .029963 | 88 : 83 | .025405 | 64 : 61 | .020850 | 82 : 79 | .016187 |
| 106 : 99 | 0.029671 | 53 : 50 | 0.025306 | 86 : 82 | 0.020685 | 55 : 53 | 0.016087 |
| 91 : 85 | .029623 | 71 : 67 | .025184 | 43 : 41 | .020685 | 83 : 80 | .015988 |
| 76 : 71 | .029555 | 89 : 84 | .025111 | 65 : 62 | .020522 | 84 : 81 | .015794 |
| 61 : 57 | .029455 | 90 : 85 | .024824 | 87 : 83 | .020441 | 56 : 54 | .015794 |
| 107 : 100 | .029384 | 72 : 68 | .024824 | 88 : 84 | .020203 | 85 : 82 | .015605 |
| 92 : 86 | .029289 | 54 : 51 | .024824 | 66 : 63 | .020203 | 57 : 55 | .015512 |
| 46 : 43 | .029289 | 36 : 34 | .024824 | 44 : 42 | .020203 | 86 : 83 | .015420 |
| 77 : 72 | .029158 | 91 : 86 | .024543 | 89 : 85 | .019971 | 87 : 84 | .015240 |
| 93 : 87 | 0.028964 | 73 : 69 | 0.024474 | 67 : 64 | 0.019895 | 58 : 56 | 0.015240 |
| 62 : 58 | .028964 | 55 : 52 | .024359 | 90 : 86 | .019744 | 88 : 85 | .015064 |
| 78 : 73 | .028772 | 92 : 87 | .024269 | 45 : 43 | .019744 | 59 : 57 | .014977 |
| 47 : 44 | .028645 | 74 : 70 | .024134 | 68 : 65 | .019596 | 89 : 86 | .014892 |
| 94 : 88 | .028645 | 37 : 35 | .024134 | 91 : 87 | .019522 | 90 : 87 | .014723 |
| 63 : 59 | .028489 | 93 : 88 | .024000 | 92 : 88 | .019305 | 60 : 58 | .014723 |
| 79 : 74 | .028395 | 56 : 53 | .023912 | 69 : 66 | .019305 | 91 : 88 | .014559 |
| 95 : 89 | .028334 | 75 : 71 | .023803 | 46 : 44 | .019305 | 61 : 59 | .014478 |
| 96 : 90 | 0.028029 | 94 : 89 | 0.023738 | 93 : 89 | 0.019093 | 92 : 89 | 0.014398 |
| 80 : 75 | .028029 | 95 : 90 | .023481 | 70 : 67 | .019023 | 93 : 90 | .014240 |
| 64 : 60 | .028029 | 76 : 72 | .023481 | 94 : 90 | .018885 | 62 : 60 | .014240 |
| 48 : 45 | .028029 | 57 : 54 | .023481 | 47 : 45 | .018885 | 31 : 30 | .014240 |
| 32 : 30 | .028029 | 38 : 36 | .023481 | 71 : 68 | .018749 | 94 : 91 | .014087 |
| 97 : 91 | .027730 | 96 : 91 | .023230 | 95 : 91 | .018682 | 63 : 61 | .014011 |
| 81 : 76 | .027671 | 77 : 73 | .023168 | 96 : 92 | .018483 | 95 : 92 | .013936 |
| 65 : 61 | .027584 | 58 : 55 | .023065 | 72 : 69 | .018483 | 96 : 93 | .013788 |
| 98 : 92 | 0.027438 | 97 : 92 | 0.022984 | 48 : 46 | 0.018483 | 64 : 62 | 0.013788 |
| 49 : 46 | .027438 | 78 : 74 | .022863 | 24 : 23 | .018483 | 32 : 31 | .013788 |
| 82 : 77 | .027323 | 39 : 37 | .022863 | 97 : 93 | .018289 | 97 : 94 | .013644 |
| 99 : 93 | .027152 | 98 : 93 | .022743 | 73 : 70 | .018225 | 65 : 63 | .013573 |
| 66 : 62 | .027152 | 59 : 56 | .022664 | 98 : 94 | .018098 | 98 : 95 | .013503 |
| 33 : 31 | .027152 | 79 : 75 | .022566 | 49 : 47 | .018098 | 99 : 96 | .013364 |
| 83 : 78 | .026984 | 99 : 94 | .022507 | 74 : 71 | .017973 | 66 : 64 | .013364 |
| 100 : 94 | .026872 | 100 : 95 | .022276 | 99 : 95 | .017912 | 33 : 32 | .013364 |

## Logarithms of Gear Ratios from 1.0084+ to 7.5

| Numbers of Teeth | Logarithm of Ratio | Numbers of Teeth | Logarithm of Ratio | Numbers of Teeth | Logarithm of Ratio | Numbers of Teeth | Logarithm of Ratio |
|---|---|---|---|---|---|---|---|
| 100 : 97 | 0.013228 | 87 : 85 | 0.010100 | 61 : 60 | 0.007179 | 93 : 92 | 0.004695 |
| 67 : 65 | .013161 | 88 : 86 | .009984 | 62 : 61 | .007062 | 94 : 93 | .004645 |
| 101 : 98 | .013095 | 44 : 43 | .009984 | 63 : 62 | .006949 | 95 : 94 | .004596 |
| 68 : 66 | .012965 | 89 : 87 | .009871 | 64 : 63 | .006840 | 96 : 95 | .004548 |
| 34 : 33 | .012965 | 90 : 88 | .009760 | 65 : 64 | .006733 | 97 : 96 | .004501 |
| 103 : 100 | .012837 | 45 : 44 | .009760 | 66 : 65 | .006631 | 98 : 97 | .004454 |
| 69 : 67 | .012774 | 91 : 89 | .009651 | 67 : 66 | .006531 | 99 : 98 | .004409 |
| 70 : 68 | .012589 | 92 : 90 | .009545 | 68 : 67 | .006434 | 100 : 99 | .004365 |
| 35 : 34 | 0.012589 | 46 : 45 | 0.009545 | 69 : 68 | 0.006340 | 101 : 100 | 0.004321 |
| 72 : 70 | .012235 | 93 : 91 | .009442 | 70 : 69 | .006249 | 102 : 101 | .004279 |
| 36 : 35 | .012235 | 94 : 92 | .009340 | 71 : 70 | .006160 | 103 : 102 | .004237 |
| 73 : 71 | .012065 | 47 : 46 | .009340 | 72 : 71 | .006074 | 104 : 103 | .004196 |
| 74 : 72 | .011899 | 95 : 93 | .009241 | 73 : 72 | .005990 | 105 : 104 | .004156 |
| 37 : 36 | .011899 | 96 : 94 | .009143 | 74 : 73 | .005909 | 106 : 105 | .004117 |
| 75 : 73 | .011738 | 48 : 47 | .009143 | 75 : 74 | .005830 | 107 : 106 | .004078 |
| 76 : 74 | .011582 | 97 : 95 | .009048 | 76 : 75 | .005752 | 108 : 107 | .004040 |
| 38 : 37 | 0.011582 | 98 : 96 | 0.008955 | 77 : 76 | 0.005677 | 109 : 108 | 0.004002 |
| 77 : 75 | .011429 | 49 : 48 | .008955 | 78 : 77 | .005604 | 110 : 109 | .003967 |
| 78 : 76 | .011281 | 99 : 97 | .008864 | 79 : 78 | .005533 | 111 : 110 | .003930 |
| 39 : 38 | .011281 | 100 : 98 | .008774 | 80 : 79 | .005463 | 112 : 111 | .003895 |
| 79 : 77 | .011136 | 50 : 49 | .008774 | 81 : 80 | .005395 | 113 : 112 | .003860 |
| 80 : 78 | .010995 | 101 : 99 | .008686 | 82 : 81 | .005329 | 114 : 113 | .003827 |
| 40 : 39 | .010995 | 51 : 50 | .008600 | 83 : 82 | .005264 | 115 : 114 | .003793 |
| 81 : 79 | .010858 | 52 : 51 | .008433 | 84 : 83 | .005201 | 116 : 115 | .003760 |
| 82 : 80 | 0.010724 | 53 : 52 | 0.008273 | 85 : 84 | 0.005140 | 117 : 116 | 0.003728 |
| 41 : 40 | .010724 | 54 : 53 | .008118 | 86 : 85 | .005080 | 118 : 117 | .003696 |
| 83 : 81 | .010593 | 55 : 54 | .007969 | 87 : 86 | .005021 | 119 : 118 | .003665 |
| 84 : 82 | .010465 | 56 : 55 | .007825 | 88 : 87 | .004963 | 120 : 119 | .003634 |
| 42 : 41 | .010465 | 57 : 56 | .007687 | 89 : 88 | .004907 | ...... | ...... |
| 85 : 83 | .010341 | 58 : 57 | .007553 | 90 : 89 | .004853 | ...... | ...... |
| 86 : 84 | .010219 | 59 : 58 | .007424 | 91 : 90 | .004799 | ...... | ...... |
| 43 : 42 | .010219 | 60 : 59 | .007299 | 92 : 91 | .004746 | ...... | ...... |

*Example:* A driven shaft is to rotate 6.9078 revolutions while the driving shaft rotates 1.3961 revolutions. Determine sizes of driving and driven gears.

$$\frac{\text{Driving gear size}}{\text{Driven gear size}} = \frac{\text{Driven gear speed}}{\text{Driving gear speed}} = \frac{6.9078}{1.3961}$$

Log 6.9078 = 0.8393398

Log 1.3961 = 0.1449165

$$0.6944233 = \log \frac{\text{Driving gear size}}{\text{Driven gear size}}$$

The nearest logarithm in table is 0.694374 which is the log for ratio 94/19. The driving gear is to have 94 teeth and the driven gear 19 teeth.

Actual revolutions of driven shaft (while driving shaft makes 1.3961 turns) = 1.3961 × 94/19 = 6.9070 or a difference from the desired rotation of 0.0008 revolution. Should a more exact solution be required, some form of compound gearing might be used. (See also examples preceding table.)

# THREAD ROLLING

Screw threads may be formed by rolling either by using some type of thread-rolling machine or by equipping an automatic screw machine or turret lathe with a suitable threading roll. If a thread-rolling machine is used, the unthreaded screw, bolt or other "blank," is placed (either automatically or by hand) between dies having thread-shaped ridges which sink into the blank, and by displacing the metal, form a thread of the required shape and pitch. The thread-rolling process is applied where bolts, screws, studs, threaded rods, etc., are required in large quantities. Screw threads that are within the range of the rolling process may be produced more rapidly by this method than in any other way. The rolled thread, due to the cold-working action of the dies, is 10 to 20 per cent stronger than a cut or ground thread, and the increase may be much higher when tested for fatigue resistance. Another advantage of the rolling process is that no stock is wasted in forming the thread, and the surface of a rolled thread is harder than that of a cut thread, thus increasing wear resistance.

**Thread-Rolling Machine of Flat-die Type.** — One type of machine which is used extensively is equipped with a pair of flat or straight dies  One die is stationary and the other has a reciprocating movement when the machine is in use. The ridges on these dies, which form the screw thread, incline at an angle equal to the helix angle of the thread. In making dies for precision thread rolling, the threads may be formed either by milling and grinding after heat-treatment, or by grinding "from the solid" after heat-treating. A vitrified wheel is used. The thread is formed in one passage of the work, which is inserted at one end of the dies, either by hand or automatically, and then rolls between the die faces until it is ejected at the opposite end. The relation between the position of the dies and a screw thread being rolled is such that the top of the thread-shaped ridge of one die, at the point of contact with the screw thread, is directly opposite the bottom of the thread groove in the other die at the point of contact. Some form of mechanism insures starting the blank at the right time and also square with the dies.

**Thread-Rolling Machine of Cylindrical Die Type.** — With machines of this type, the blank is threaded while being rolled between two or three cylindrical dies (depending upon type of machine) which are pressed into the blank at a rate of penetration adjusted to the hardness of the material, or wall thickness in the case of threading operations on tubing or hollow parts. The dies have ground or ground and lapped threads and a pitch diameter that is a multiple of the pitch diameter of the thread to be rolled. As the dies are much larger in diameter than the work, a multiple thread is required to obtain the same lead angle as that of the work. The thread may be formed in one die revolution or even less, or several revolutions may be required (as in rolling hard materials) to obtain a gradual rate of penetration equivalent to that obtained with flat or straight dies if extended to a length of possibly 15 or 20 feet. Provisions for accurately adjusting or matching the thread rolls to bring them into proper alignment with each other, are important features of these machines.

*Two-Roll Type of Machine:* With a two-roll type of machine, the work is rotated between two horizontal power-driven threading rolls and is supported by a hardened rest bar on the lower side. One roll is fed inward by hydraulic pressure to a depth that is governed automatically.

*Three-Roll Type of Machine:* With this machine the blank to be threaded is held in a "floating position" while being rolled between three cylindrical dies which, through toggle arms, are moved inward at a predetermined rate of penetration until the required pitch diameter is obtained. The die movement is governed by a cam driven through change gears selected to give the required cycle of squeeze, dwell and release.

**Rate of Production.** — Production rates in thread rolling depend upon the type of machine, the size of both machine and work, and whether the parts to be threaded are inserted by hand or automatically. A reciprocating flat die type of machine, applied to ordinary steels, may thread 30 or 40 parts per minute in diameters ranging from about 5/8 to 1 1/8 inch, and 150 to 175 per minute in machine screw sizes from No. 10 (.190) to No. 6 (.138). In the case of heat-treated alloy steels in the usual hardness range of 26 to 32 Rockwell C, the production may be 30 or 40 per minute or less. With a cylindrical die type of machine, which is designed primarily for precision work and hard metals, 10 to 30 parts per minute are common production rates, the amount depending upon the hardness of material and allowable rate of die penetration per work revolution. These production rates are intended as a general guide only. The diameters of rolled threads usually range from the smallest machine screw sizes up to 1 or 1 1/2 inches, depending upon the type and size of machine.

**Precision Thread Rolling.** — Both flat and cylindrical dies are used in aeronautical and other plants for precision work. With accurate dies and blank diameters held to close limits, it is practicable to produce rolled threads for American Standard Class 3 and Class 4 fits. The blank sizing may be by centerless grinding or by means of a die in conjunction with the heading operations. The blank should be round, and, as a general rule, the diameter tolerance should not exceed 1/2 to 3/4 the pitch diameter tolerance. The blank diameter should range from the correct size (which is close to the pitch diameter, but should be determined by actual trial), down to the allowable minimum, the tolerance being minus to insure a correct pitch diameter, even though the major diameter may vary slightly. Precision thread rolling has become an important method of threading alloy steel studs and other threaded parts, especially in aeronautical work where precision and high-fatigue resistance are required. Micrometer screws are also an outstanding example of precision thread rolling. This process has also been applied in tap making, although it is the general practice to finish rolled taps by grinding when the Class 3 and Class 4 fits are required.

**Steels for Thread Rolling.** — Steels vary from soft low-carbon types for ordinary screws and bolts, to nickel, nickel-chromium and molybdenum steels for aircraft studs, bolts, etc., or for any work requiring exceptional strength and fatigue resistance. Typical SAE alloy steels are No. 2330, 3135, 3140, 4027, 4042, 4640 and 6160. The hardness of these steels after heat-treatment usually ranges from 26 to 32 Rockwell C, with tensile strengths varying from 130,000 to 150,000 pounds per square inch. While harder materials might be rolled, grinding is more practicable when the hardness exceeds 40 Rockwell C. Thread rolling is applicable not only to a wide range of steels but for non-ferrous materials, especially if there is difficulty in cutting due to "tearing" the threads.

**Diameter of Blank for Thread Rolling.** — The diameter of the screw blank or cylindrical part upon which a thread is to be rolled should be less than the outside screw diameter by an amount that will just compensate for the metal that is displaced and raised above the original surface by the rolling process. The increase in diameter is approximately equal to the depth of one thread. While there are rules and formulas for determining blank diameters, it may be necessary to make slight changes in the calculated size in order to secure a well-formed thread. The blank diameter should be verified by trial, especially when rolling accurate screw threads. Some stock offers greater resistance to displacement than other stock, owing to the greater hardness or tenacity of the metal. The following figures may prove useful in establishing trial sizes. The blank diameters for screws varying from 1/4 to 1/2 inch are from 0.002 to 0.0025 inch larger than the pitch diameter, and for screws varying from 1/2 to 1 inch or larger, the blank diameters are from 0.0025 to

0.003 inch larger than the pitch diameter. Blanks which are slightly less than the pitch diameter are intended for bolts, screws, etc., which are to have a comparatively free fit. Blanks for this class of work may vary from 0.002 to 0.003 inch less than the pitch diameter for screw thread sizes varying from ¼ to ½ inch, and from 0.003 to 0.005 inch less than the pitch diameter for sizes above ½ inch. If the screw threads are smaller than ¼ inch, the blanks are usually from 0.001 to 0.0015 inch less than the pitch diameter for ordinary grades of work.

**Thread Rolling in Automatic Screw Machines.** — Screw threads are sometimes rolled in automatic screw machines and turret lathes when the thread is behind a shoulder so that it cannot be cut with a die. In such cases, the advantage of rolling the thread is that a second operation is avoided. A circular roll is used for rolling threads in screw machines. The roll may be presented to the work either in a tangential direction or radially, either method producing a satisfactory thread. In the former case, the roll gradually comes into contact with the periphery of the work and completes the thread as it passes across the surface to be threaded. When the roll is held in a radial position, it is simply forced against one side until a complete thread is formed. The method of applying the roll may depend upon the relation between the threading operation and other machining operations. Thread rolling in automatic screw machines is generally applied only to brass and other relatively soft metals, owing to the difficulty of rolling threads in steel. Thread rolls made of chrome-nickel steel containing from 0.15 to 0.20 per cent of carbon have given fairly good results, however, when applied to steel. A 3 per cent nickel steel containing about 0.12 per cent carbon has also proved satisfactory for threading brass.

**Factors Governing the Diameter of Thread Rolling.** — The threading roll used in screw machines may be about the same diameter as the screw thread, but for sizes smaller than, say, ¾ inch, the roll diameter is some multiple of the thread diameter minus a slight amount to obtain a better rolling action. When the diameters of the thread and roll are practically the same, a single-threaded roll is used to form a single thread on the screw. If the diameter of the roll is made double that of the screw, in order to avoid using a small roll, then the roll must have a double thread. If the thread roll is three times the size of the screw thread, a triple thread is used, and so on. These multiple threads are necessary when the roll diameter is some multiple of the work, in order to obtain corresponding helix angles on the roll and work.

**Diameter of Threading Roll.** — The pitch diameter of a threading roll having a single thread, is slightly less than the pitch diameter of the screw thread to be rolled, and in the case of multiple-thread rolls, the pitch diameter is not an exact multiple of the screw thread pitch diameter but is also reduced somewhat. The amount of reduction recommended by one screw machine manufacturer is given by the formula shown at the end of this paragraph. A description of the terms used in the formula is given as follows: $D$ = pitch diameter of threading roll, $d$ = pitch diameter of screw thread, $N$ = number of single threads or "starts" on the roll (this number is selected with reference to diameter of roll desired), $T$ = single depth of thread:

$$D = N\left(d - \frac{T}{2}\right) - T$$

*Example:* Find by using above formula, the pitch diameter of a double-thread roll for rolling a ½-inch American standard screw thread. Pitch diameter $d = 0.4500$ inch and thread depth $T = 0.0499$ inch.

$$D = 2\left(0.4500 - \frac{0.0499}{2}\right) - 0.0499 = 0.8001 \text{ inch}$$

**Kind of Thread on Roll and Its Shape.** — The thread (or threads) on the roll should be left hand for rolling a right-hand thread, and *vice versa*. The roll should be wide enough to overlap the part to be threaded, provided there are clearance spaces at the ends, which should be formed if possible. The thread on the roll should be sharp on top for rolling an American (National) standard form of thread, so that less pressure will be required to displace the metal when rolling the thread. The bottom of the thread groove on the roll may also be left sharp or it may have a flat. If the bottom is sharp, the roll is sunk only far enough into the blank to form a thread having a flat top, assuming that the thread is the American form. The number of threads on the roll (whether double, triple, quadruple, etc.) is selected, as a rule, so that the diameter of the thread roll will be somewhere between 1¼ and 2¼ inches. In making a thread roll, the ends are beveled at an angle of 45 degrees, to prevent the threads on the ends of the roll from chipping. Precautions should be taken in hardening, because if the sharp edges are burnt, the roll will be useless. Thread rolls, as a rule, are lapped after hardening. This is done by holding them on an arbor in the lathe and using emery and oil on a piece of hard wood. A thread roll, to give good results, should fit closely in the holder. If the roll is made to fit loosely, it will mar the threads.

**Application of Thread Roll.** — The shape of the work, and the character of the operations necessary to produce it, govern, to a large extent, the method employed in applying the thread roll. Some of the points to consider are as follows: 1. Diameter of the part to be threaded. 2. Location of the part to be threaded. 3. Length of the part to be threaded. 4. Relation that the thread rolling operation bears to the other operations. 5. Shape of the part to be threaded, whether straight, tapered or otherwise. 6. Method of applying the support. When the diameter to be rolled is much smaller than the diameter of the shoulder preceding it, a cross-slide knurl-holder should be used. If the part to be threaded is not behind a shoulder, a holder on the swing principle should be used. When the work is long (greater in length than two-and-one-half times its diameter) a swing roll-holder should be employed, carrying a support. When the work can be cut off after the thread is rolled, a cross-slide roll-holder should be used. The method of applying the support to the work also governs to some extent the method of applying the thread roll. When no other tool is working at the same time as the thread roll, and when there is freedom from chips, the roll can be held more rigidly by passing it under instead of over the work. When passing the roll over the work, it has a tendency to raise the cross-slide. Where the part to be threaded is tapered, the roll can best be presented to the work by holding it in a cross-slide roll-holder.

**Speeds and Feeds for Thread Rolling.** — When the thread roll is made from high-carbon steel and used on brass, a surface speed as high as 200 feet per minute can be used. Better results, however, are obtained by using a lower speed than this. When the roll is held in a holder attached to the cross-slide, and is presented either tangentially or radially to the work, a considerably higher speed can be used than if it is held in a swing tool. This is due to the lack of rigidity in a holder of the swing type. The feeds to be used when a cross-slide roll-holder is used are given in the upper half of the table "Feeds for Thread Rolling"; the lower half of the table gives the feeds for thread rolling with swing tools. These feeds are applicable for rolling threads without a support, when the root diameter of the blank is not less than five times the double depth of the thread. When the root diameter is less than this, a support should be used. A support should also be used when the width of the roll is more than two-and-one-half times the smallest diameter of the piece to be rolled, irrespective of the pitch of the thread. When the smallest diameter of the piece to be rolled is much less than the root diameter of the thread, the smallest diameter should be taken as the deciding factor for the feed to be used.

## Feeds for Thread Rolling

### Number of Threads per Inch

#### Cross-slide Holders — Feed per Revolution in Inches

| 14 | 18 | 20 | 22 | 24 | 28 | 32 | 36 | 40 | 44 | 48 | 56 | 64 | 72 | Root Diam. of Blank |
|---|---|---|---|---|---|---|---|---|---|---|---|---|---|---|
| | | | | | | 0.0010 | 0.0015 | 0.0020 | 0.0025 | 0.0030 | 0.0035 | 0.0040 | 0.0045 | 1/8 |
| | | | | | 0.0005 | 0.0015 | 0.0020 | 0.0025 | 0.0030 | 0.0035 | 0.0040 | 0.0045 | 0.0050 | 3/16 |
| | | | 0.0005 | 0.0005 | 0.0010 | 0.0020 | 0.0025 | 0.0030 | 0.0035 | 0.0040 | 0.0045 | 0.0050 | 0.0055 | 1/4 |
| | 0.0005 | 0.0005 | 0.0010 | 0.0010 | 0.0015 | 0.0025 | 0.0030 | 0.0035 | 0.0040 | 0.0045 | 0.0050 | 0.0055 | 0.0060 | 5/16 |
| 0.0005 | 0.0010 | 0.0010 | 0.0015 | 0.0015 | 0.0020 | 0.0030 | 0.0035 | 0.0040 | 0.0045 | 0.0050 | 0.0055 | 0.0060 | 0.0065 | 3/8 |
| 0.0010 | 0.0015 | 0.0015 | 0.0020 | 0.0020 | 0.0025 | 0.0035 | 0.0040 | 0.0045 | 0.0050 | 0.0055 | 0.0060 | 0.0065 | 0.0070 | 7/16 |
| 0.0015 | 0.0020 | 0.0020 | 0.0025 | 0.0025 | 0.0030 | 0.0040 | 0.0045 | 0.0050 | 0.0055 | 0.0060 | 0.0065 | 0.0070 | 0.0075 | 1/2 |
| 0.0020 | 0.0025 | 0.0025 | 0.0030 | 0.0030 | 0.0035 | 0.0045 | 0.0050 | 0.0055 | 0.0060 | 0.0065 | 0.0070 | 0.0075 | 0.0080 | 5/8 |
| 0.0025 | 0.0030 | 0.0030 | 0.0035 | 0.0035 | 0.0040 | 0.0050 | 0.0055 | 0.0060 | 0.0065 | 0.0070 | 0.0075 | 0.0080 | 0.0085 | 3/4 |
| 0.0030 | 0.0035 | 0.0035 | 0.0040 | 0.0040 | 0.0045 | 0.0055 | 0.0060 | 0.0065 | 0.0070 | 0.0075 | 0.0080 | 0.0085 | 0.0090 | 7/8 |
| 0.0035 | 0.0040 | 0.0040 | 0.0045 | 0.0045 | 0.0050 | 0.0060 | 0.0065 | 0.0070 | 0.0075 | 0.0080 | 0.0085 | 0.0090 | 0.0095 | 1 |

#### Swing Holders — Feed per Revolution in Inches

| 14 | 18 | 20 | 22 | 24 | 28 | 32 | 36 | 40 | 44 | 48 | 56 | 64 | 72 | Root Diam. |
|---|---|---|---|---|---|---|---|---|---|---|---|---|---|---|
| | | | | | | | | | 0.0005 | 0.0010 | 0.0015 | 0.0020 | 0.0025 | 1/8 |
| | | | | | | | | 0.0005 | 0.0008 | 0.0015 | 0.0020 | 0.0025 | 0.0028 | 3/16 |
| | | | | | 0.0005 | 0.0005 | 0.0005 | 0.0010 | 0.0010 | 0.0020 | 0.0025 | 0.0030 | 0.0030 | 1/4 |
| | | | | 0.0005 | 0.0010 | 0.0010 | 0.0010 | 0.0015 | 0.0015 | 0.0025 | 0.0030 | 0.0035 | 0.0035 | 5/16 |
| | 0.0005 | 0.0005 | 0.0005 | 0.0010 | 0.0015 | 0.0015 | 0.0015 | 0.0020 | 0.0020 | 0.0030 | 0.0035 | 0.0040 | 0.0040 | 3/8 |
| | 0.0010 | 0.0010 | 0.0010 | 0.0015 | 0.0020 | 0.0020 | 0.0020 | 0.0025 | 0.0030 | 0.0035 | 0.0040 | 0.0045 | 0.0045 | 7/16 |
| 0.0005 | 0.0015 | 0.0015 | 0.0015 | 0.0020 | 0.0025 | 0.0025 | 0.0025 | 0.0030 | 0.0035 | 0.0043 | 0.0045 | 0.0048 | 0.0048 | 1/2 |
| 0.0010 | 0.0018 | 0.0020 | 0.0020 | 0.0025 | 0.0028 | 0.0030 | 0.0030 | 0.0035 | 0.0040 | 0.0043 | 0.0048 | 0.0050 | 0.0050 | 5/8 |
| 0.0013 | 0.0020 | 0.0022 | 0.0025 | 0.0028 | 0.0030 | 0.0035 | 0.0035 | 0.0040 | 0.0043 | 0.0045 | 0.0050 | 0.0052 | 0.0055 | 3/4 |
| 0.0015 | 0.0022 | 0.0025 | 0.0028 | 0.0030 | 0.0032 | 0.0038 | 0.0040 | 0.0043 | 0.0045 | 0.0048 | 0.0052 | 0.0055 | 0.0058 | 7/8 |
| 0.0018 | 0.0025 | 0.0028 | 0.0030 | 0.0032 | 0.0035 | 0.0040 | 0.0043 | 0.0047 | 0.0048 | 0.0050 | 0.0054 | 0.0058 | 0.0060 | 1 |

# THREAD GRINDING

Thread grinding is employed for precision tool and gage work and also in producing certain classes of threaded parts. Thread grinding may be utilized (1) because of the accuracy and finish obtained, (2) hardness of material to be threaded, (3) economy in grinding certain classes of screw threads when using modern machines, wheels, and thread-grinding oils. In some cases pre-cut threads are finished by grinding; but usually, threads are ground "from the solid," being formed entirely by the grinding process. Examples of work include thread gages and taps of steel and tungsten carbide, hobs, worms, lead-screws, adjusting or traversing screws, alloy steel studs, etc. Grinding is applied to external, internal, straight, and tapering threads, and to various thread forms.

**Accuracy Obtainable by Thread Grinding.** — With single-edge or single-ribbed wheels it is possible to grind threads on gages to a degree of accuracy that requires but very little lapping to produce a so-called "master" thread gage. As far as lead is concerned, some thread grinding machine manufacturers guarantee to hold the lead within 0.0001 inch per inch of thread; and while it is not guaranteed that a higher degree of accuracy for lead is obtainable, it is known that threads have been ground to closer tolerances than this on the lead. Pitch diameter accuracies for either Class 3 or Class 4 fits are obtainable according to the grinding method used; with single-edge wheels, the thread angle can be ground to an accuracy of within two or three minutes in half the angle.

**Wheels for Thread Grinding.** — The wheels used for steel have an aluminous abrasive and, ordinarily, either a resinoid bond or a vitrified bond. The general rule is to use resinoid wheels when extreme tolerances are not required, and it is desirable to form the thread with a minimum number of passes, as in grinding threaded machine parts, such as studs, adjusting screws which are not calibrated, and for some classes of taps. *Resinoid Wheels*, as a rule, will hold a fine edge longer than a vitrified wheel but they are more flexible and, consequently, less suitable for accurate work, especially when there is lateral grinding pressure that causes wheel deflection. *Vitrified wheels* are utilized for obtaining extreme accuracy in thread form and lead because they are very rigid and not easily deflected by side pressure in grinding. This rigidity is especially important in grinding pre-cut threads on such work as gages, taps and lead-screws. The progressive lead errors in long lead-screws, for example, might cause an increasing lateral pressure that would deflect a resinoid wheel. Vitrified wheels are also recommended for internal grinding.

*Diamond Wheels:* Diamond wheels set in a rubber or plastic bond are also used for thread grinding, especially for grinding threads in carbide materials and in other hardened alloys. Thread grinding is now being done successfully on a commercial basis on both taps and gages made from carbides. Gear hobs made from carbides have also been tested with successful results. Diamond wheels are dressed by means of silicon-carbide grinding wheels which travel past the diamond-wheel thread form at the angle required for the flanks of the thread to be ground. The action of the dressing wheels is, perhaps, best described as a "scrubbing" of the bond which holds the diamond grits. Obviously, the silicon-carbide wheels do not dress the diamonds, but they loosen the bond until the diamonds not wanted drop out.

**Thread Grinding with Single-Edge Wheel.** — With this type of wheel, the edge is trued to the cross-sectional shape of the thread groove. The wheel, when new, may have a diameter of 18 or 20 inches and, when grinding a thread, the wheel is inclined to align it with the thread groove. On some machines, lead variations are obtained by means of change-gears which transmit motion from the work-driving spindle to the lead-screw. Other machines are so designed that a lead-screw is selected to suit the lead of thread to be ground and transmits motion directly to the work-driving spindle.

**Wheels with Edges for Roughing and Finishing.** — The "three-ribbed" type of wheel has a roughing edge or rib which removes about two-thirds of the metal. This is followed by an intermediate rib which leaves about 0.005 inch for the third or finishing rib. The accuracy obtained with this triple-edge type compares with

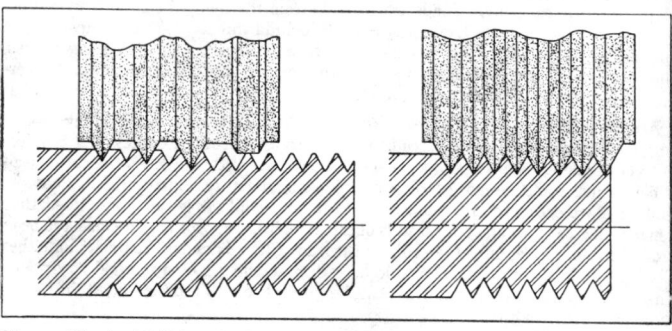

Fig. 1.  Wheel with Edges for Roughing     Fig. 2.  Multi-ribbed Type of Thread-
         and Finishing                              grinding Wheel

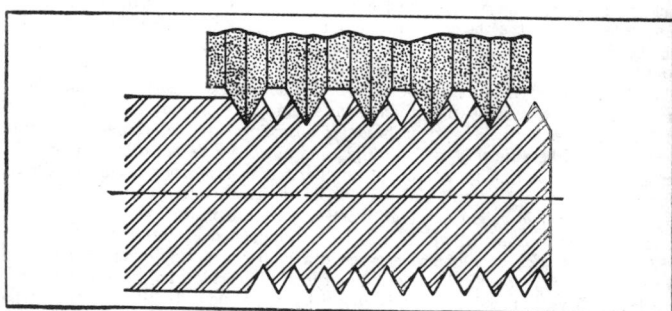

Fig. 3.  Alternate-ribbed Wheel for Grinding the Finer Pitches

that of a single-edge wheel, which means that it may be used for the greatest accuracy obtainable in thread grinding. When the accuracy required makes it necessary, this wheel can be inclined to the helix angle of the thread, the same as is the single-edge wheel.

The three-ribbed wheel is recommended not only for precision work but for grinding threads which are too long for the multi-ribbed wheel referred to later. It is also well adapted to tap grinding, because it is possible to dress a portion of the wheel adjacent to the finish rib for the purpose of grinding the outside diameter of the thread, as indicated in Fig. 1. Furthermore, the wheel can be dressed for grinding or relieving both crests and flanks at the same time.

**Multi-ribbed Wheels.** — This type of wheel is employed when rapid production is more important than extreme accuracy, which means that it is intended primarily for the grinding of duplicate parts in manufacturing. A wheel 1¼ to 2 inches wide has formed upon its face a series of annular thread-shaped ridges (see Fig. 2); hence, if the length of the thread is not greater than the wheel width, a thread may be ground in one work revolution plus about one-half revolution for feeding in and withdrawing the wheel. The principle of operation is the same as that of thread milling with a multiple type cutter. This type of wheel is not inclined to the lead angle. To obtain a Class 3 fit, the lead angle should not exceed 4 degrees.

It is not practicable, to use this form of wheel on thread pitches where the root is less than 0.007 inch wide, because of difficulties in wheel dressing. When this method can be applied, it is the fastest means known of producing threads in hardened materials. It is not recommended, however, that thread gages, taps, and work of this character be ground with multi-ribbed wheels. The single-ribbed wheel has a definite field for accurate, small-lot production.

It is necessary, in multi-ribbed grinding, to use more horsepower than is required for single-ribbed wheel grinding. Coarse threads, in particular, may require a wheel motor with two or three times more horsepower than would be necessary for grinding with a single-ribbed wheel.

**Alternate-ribbed Wheel for Fine Pitches.** — The spacing of ribs on this type of wheel (Fig. 3) equals twice the pitch, so that during the first revolution every other thread groove section is being ground; consequently, about two and one-half work revolutions are required for grinding a complete thread, but the better distribution of cooling oil and resulting increase in work speeds makes this wheel very efficient. This alternate-type of wheel is adapted for grinding threads of fine pitch. Since these wheels cannot be tipped to the helix angle of the thread, they are not recommended for anything closer than Class 3 fits. The "three-ribbed" wheels referred to in a previous paragraph are also made in the alternate type for the finer pitches.

**Grinding Threads "from the Solid."** — The process of forming threads entirely by grinding, or without preliminary cutting, is applied both in the manufacture of certain classes of threaded parts and also in the production of precision tools, such as taps and thread gages. For example, in airplane engine manufacture, certain parts are heat-treated and then the threads are ground "from the solid," thus eliminating distortion and also minute cracks formerly found at the roots of threads which were cut and then hardened. In some cases steel threads of coarse pitch, which are surface hardened, may be rough threaded by cutting, then hardened and finally corrected by grinding. Many ground thread taps are produced by grinding from the solid after heat-treatment. By hardening high-speed steel taps before the thread is formed, there are no narrow or delicate crests to interfere with the application of the high temperature required for uniform hardness and the best steel structure.

**Number of Wheel Passes.** — The number of cuts or passes for grinding from the solid depends upon the type of wheel and accuracy required. In general, threads of 12 or 14 per inch and finer may be ground in one pass of a single-edge wheel unless the "unwrapped" thread length is much greater than normal. Unwrapped length = pitch circumference × total number of thread turns, approximately. For example, a thread gage 1¼ inches long with 24 threads per inch would have an unwrapped length equal to 30 × pitch circumference. (If more convenient, outside circumference may be used instead of pitch circumference.) Assume that there are 6 or 7 feet of unwrapped length on a screw thread having 12 threads per inch. In this case, one pass might be sufficient for a Class 3 fit, whereas two passes might be recommended for a Class 4 fit. When two passes are required, too deep a roughing cut may break down the narrow edge of the wheel. To prevent this, try a roughing cut depth equal to about two-thirds the total thread depth, thus leaving one-third for the finishing cut.

**Wheel and Work Rotation.** — When a screw thread, on the side being ground, is moving *upward* or *against* the grinding wheel rotation, less heat is generated and the grinding operation is more efficient than when wheel and work are moving in the same direction on the grinding side; however, to avoid running a machine idle during its return stroke, many screw threads are ground during both the forward and return traversing movements, by reversing the work rotation at the end of the forward stroke. For this reason, thread grinders generally are equipped so that both forward and return work speeds may be changed; they may also be designed to accelerate the return movement when grinding in one direction only.

**Wheel Speeds.** — Wheel speeds should always be limited to the maximum specified on the wheel by the manufacturer. According to the American National Standard Safety Code, resinoid and vitrified wheels are limited to 12,000 surface feet per minute; however, according to Norton Co., the most efficient speeds are from 9,000 to 10,500 for resinoid wheels and 7,500 to 9,500 for vitrified wheels. Only tested wheels recommended by the wheel manufacturer should be used. After a suitable surface speed has been established, it should be maintained by increasing the R.P.M. of the wheel, as the latter is reduced in diameter by wear.

Since thread grinding wheels work close to the limit of their stock-removing capacity, some adjustment of the wheel or work speed may be required to get the best results. If the wheel speed is too slow for a given job and excessive heat is generated, try an increase in speed, assuming that such increase is within the safety limits. If the wheel is too soft and the edge wears excessively, again an increase in wheel speed will give the effect of a harder wheel and result in better form-retaining qualities.

**Work Speeds.** — The work speed usually ranges from 3 to 10 feet per minute. In grinding with a comparatively heavy feed, and a minimum number of passes, the speed may not exceed 2½ or 3 feet per minute. If very light feeds are employed as in grinding hardened high-speed steel, the work speed may be much higher than 3 feet per minute and should be determined by test. If excessive heat is generated by removing stock too rapidly, a work speed reduction is one remedy. If a wheel is working below its normal capacity, an increase in work speed would prevent dulling of the grains and reduce the tendency to heat or "burn" the work. An increase in work speed and reduction in feed may also be employed to prevent burning while grinding hardened steel.

**Truing Grinding Wheels.** — Thread grinding wheels are trued both to maintain the required thread form and also an efficient grinding surface. Thread grinders ordinarily are equipped with precision truing devices which function automatically. One type automatically dresses the wheel and also compensates for the slight amount removed in dressing, thus automatically maintaining size control of the work. While truing the wheel, a small amount of grinding oil should be used to reduce diamond wear. Light truing cuts are advisable, especially in truing resinoid wheels which may be deflected by excessive truing pressure. A master former for controlling the path followed by the truing diamond may require a modified profile to prevent distortion of the thread form, especially when the lead angles are comparatively large. Such modification usually is not required for 60-degree threads when the pitches for a given diameter are standard because then the resulting lead angles are less than 4½ degrees. In grinding Acme threads or 29-degree worm threads having lead angles greater than 4 or 5 degrees, modified formers may be required to prevent a bulge in the thread profile. The highest point of this bulge is approximately at the pitch line. A bulge of about 0.001 inch may be within allowable limits on some commercial worms but precision worms for gear hobbers, etc., require straight flanks in the axial plane.

*Crushing Method:* Thread grinding wheels are also dressed or formed by the crushing method, which is used in connection with some types of thread grinding machines. When this method is used, the annular ridge or ridges on the wheel are formed by a hardened steel cylindrical dresser or crusher. The crusher has a series of smooth annular ridges which are shaped and spaced like the thread that is to be ground. During the wheel dressing operation, the crusher is positively driven instead of the grinding wheel, and the ridges on the wheel face are formed by the rotating crusher being forced inward.

**Wheel Hardness or Grade.** — Wheel hardness or grade selection is based upon a compromise between efficient cutting and durability of the grinding edge. Grade selection depends on the bond and the character of the work. The following general recommendations are based upon Norton grading.

Vitrified wheels usually range from J to M, and resinoid wheels from R to U. For heat-treated screws or studs and the American Standard Thread, try the following. For 8 to 12 threads per inch, grade S resinoid wheel; for 14 to 20 threads per inch, grade T resinoid; for 24 threads per inch and finer, grades T or U resinoid. For high-speed steel taps 4 to 12 threads per inch, grade J vitrified or S resinoid; 14 to 20 threads per inch, grade K vitrified or T resinoid; 24 to 36 threads per inch, grade M vitrified or T resinoid.

**Grain Size.** — A thread grinding wheel usually operates close to its maximum stock-removing capacity, and the narrow edge which forms the root of the thread is the most vulnerable part. In grain selection, the general rule is to use the coarsest grained wheel that will hold its form while grinding a reasonable amount of work. Pitch of thread and quality of finish are two governing factors. Thus, to obtain an exceptionally fine finish, the grain size might be smaller than is needed to retain the edge profile. The usual grain sizes range from 120 to 150. For heat-treated screws and studs with American Standard Threads, 100 to 180 is the usual range. For precision screw threads of very fine pitch, the grain size may range from 220 to 320. For high-speed steel taps, the usual range is from 150 to 180 for American Standard Threads, and from 80 to 150 for pre-cut Acme threads.

**Thread Grinding by Centerless Method.** — Screw threads may be ground from the solid by the centerless method. A centerless thread grinder is similar in its operating principle to a centerless grinder designed for general work, in that it has a grinding wheel, a regulating or feed wheel (with speed adjustments), and a work-rest. Adjustments are provided to accommodate work of different sizes and for varying the rates of feed. The grinding wheel is a multi-ribbed type, being a series of annular ridges across the face. These ridges conform in pitch and profile with the thread to be ground. The grinding wheel is inclined to suit the helix or lead angle of the thread. In grinding threads on such work as socket type set-screws, the blanks are fed automatically and passed between the grinding and regulating wheels in a continuous stream. To illustrate production possibilities, hardened socket set-screws of ¼-20 size may be ground from the solid at the rate of 60 to 70 per minute and with the wheel operating continuously for 8 hours without redressing. The lead errors of centerless ground screw threads may be limited to 0.0005 inch per inch or even less by reducing the production rate. The pitch diameter tolerances are within 0.0002 to 0.0003 inch of the basic size. The grain size for the wheel is selected with reference to the pitch of the thread, the following sizes being recommended: For 11 to 13 threads per inch, 150; for 16 threads per inch, 180; for 18 to 20 threads per inch, 220; for 24 to 28 threads per inch, 320; for 40 threads per inch, 400.

# THREAD MILLING

*Single-cutter Method:* Whenever a single cutter is used, the axis of the cutter is inclined an amount equal to the lead angle of the screw thread, in order to locate the cutter in line with the thread groove at the point where the cutting action takes place. Tangent of lead angle = lead of screw thread ÷ pitch circumference of screw.

The helical thread groove is generated in practically the same way as when a lathe is used. The single cutter process is especially applicable to the milling of large screw threads of coarse pitch, and either single or multiple threads.

The cutter should revolve as fast as possible without dulling the cutting edges excessively, in order to mill a smooth thread and prevent the unevenness that would result with a slow-moving cutter, on account of the tooth spaces. As the cutter rotates, the part on which a thread is to be milled is also revolved, but at a very slow rate (a few inches per minute), since this rotation of the work is practically a feeding movement. The cutter is ordinarily set to the full depth of the thread groove and finishes a single thread in one passage, although deep threads of coarse pitch may require two or even three cuts. For fine pitches and short threads, the multiple-cutter method (described in the next paragraph) usually is preferable, because it is more rapid. The milling of taper screw threads may be done on a single-cutter type of machine by traversing the cutter laterally as it feeds along in a lengthwise direction, the same as when using a taper attachment on a lathe.

*Multiple-cutter Method:* The multiple cutter for thread milling is practically a series of single cutters, although formed of one solid piece of steel, at least so far as the cutter proper is concerned. The rows of teeth do not lie in a helical path, like the teeth of a hob or tap, but they are annular or without lead. If the cutter had helical teeth the same as a gear hob, it would have to be geared to revolve in a certain fixed ratio with the screw being milled, but a cutter having annular teeth may rotate at any desired cutting speed, while the screw blank is rotated slowly to provide a suitable rate of feed. (The multiple cutters used for thread milling are frequently called "hobs," but the term hob should be applied only to cutters having a helical row of teeth like a gear-cutting hob.)

The object in using a multiple cutter instead of a single cutter is to finish a screw thread complete in approximately one revolution of the work, a slight amount of over-travel being allowed to insure milling the thread to the full depth where the end of cut joins the starting point. The cutter which is at least one or two threads or pitches wider than the thread to be milled, is fed in to the full thread depth and then either the cutter or screw blank is moved in a lengthwise direction a distance equal to the lead of the thread during one revolution of the work.

The multiple cutter is used for milling comparatively short threads and usually medium or fine pitches. The accompanying illustration shows typical examples of external and internal work for which the multiple-cutter type of thread milling has proved very efficient, although its usefulness is not confined to shoulder work and "blind" holes.

In using multiple cutters either for internal or external thread milling, the axis of the cutter is set parallel with the axis of the work, instead of inclining the cutter to suit the lead angle of the thread, as when using a single cutter. Theoretically, this is not the correct position for a cutter, since each cutting edge is revolving in a plane at right angles to the screw's axis while milling a thread groove of helical form. However, as a general rule, interference between the cutter and the thread, does not result a decided change in the standard thread form. Usually the defect is very slight and may be disregarded except when milling threads which incline considerably relative to the axis like a thread of multiple form and large lead angle. Multiple cutters are suitable for external threads having lead angles under 3½ degrees and for internal threads having lead angles under 2½ degrees. Threads

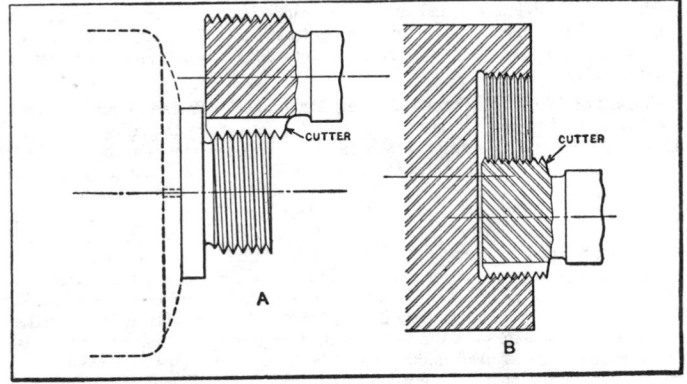

Examples of External and Internal Thread Milling with
a Multiple Type of Cutter

which have steeper sides or smaller included angles than the American Standard
or Whitworth forms should ordinarily be milled with a single cutter, assuming that
the milling process is preferable to other methods. For instance, in milling an Acme
thread which has an included angle between the sides of 29 degrees, there might be
considerable interference if a multiple cutter were used, unless the screw thread
diameter were large enough in proportion to the pitch to prevent such interference.
If an attempt were made to mill a square thread with a multiple cutter, the results
would be unsatisfactory owing to the interference.

Interference between the cutter and work is more pronounced when milling internal threads, because the cutter does not clear itself so well. Experiments have
shown that multiple cutters for internal work should preferably not exceed one-
third the diameter of the hole to be threaded. A cutter that is one-quarter the
diameter of the thread will do very satisfactory work. It is preferable to use as
small a cutter as practicable, either for internal or external work, not only to avoid
interference, but to reduce the strain on the driving mechanism. Some thread milling cutters, known as "topping cutters," are made for milling the outside diameter
of the thread as well as the angular sides and root.

*Planetary Method:* The planetary method of thread milling is similar in principle
to planetary milling. The part to be threaded is held stationary and the thread
milling cutter, while revolving about its own axis, is given a planetary movement
around the work in order to mill the thread in one planetary revolution. The machine spindle and the cutter which is held by it is moved longitudinally for thread
milling, an amount equal to the thread lead during one planetary revolution. This
operation is applicable to both internal and external threads. For the latter operation, the thread milling cutter surrounds the work. This thread milling is frequently accompanied by milling operations on other adjoining surfaces. For example, a planetary type of machine may be used for milling a screw thread and a
concentric cylindrical surface simultaneously. When the milling operation begins,
the eccentrically mounted cutter-spindle feeds the cutter into the right depth and
then the planetary movement begins, thus milling the thread and the cylindrical
surface. Thin sharp starting edges are eliminated on threads milled by the planetary method and the thread begins with a smooth gradual approach. One design of

machine will mill internal and external threads simultaneously. These threads may be of the same hand or one may be right hand and the other left hand. The threads may also be either of the same pitch or of a different pitch, and either straight or tapered.

**Classes of Work for Thread Milling Machines.** — Thread milling machines are used in preference to lathes or taps and dies for certain threading operations. There are four general reasons why a thread milling machine may be preferred: (1) Because the pitch of the thread is too coarse for cutting with a die; (2) because the milling process is more efficient than using a single-point tool in a lathe; (3) to secure a smoother and more accurate thread than would be obtained with a tap or die; (4) because the thread is so located relative to a shoulder or other surface that the milling method is superior, if not the only practicable way. A thread milling machine having a single cutter is especially adapted for coarse pitches, multiple-threaded screws, or any form or size of thread requiring the removal of a relatively large amount of metal, particularly if the pitch of the thread is large in proportion to the screw diameter, since the torsional strain due to the milling process is relatively small. While thread milling has little, if any, advantage over the lathe in regard to accuracy of lead, it gives a higher rate of production, and a thread is usually finished by means of a single passage of the cutter. The multiple-cutter type of thread milling machine frequently comes into competition with dies and taps, and especially self-opening dies and collapsing taps. The use of a multiple cutter is desirable when a thread must be cut close to a shoulder or to the bottom of a shallow recess, although the usefulness of the multiple cutter is not confined to shoulder work and "blind" holes.

**Maximum Pitches of Die-cut Threads.** — Dies of special design could be constructed for practically any pitch, if the screw blank were strong enough to resist the cutting strains and the size and cost of the die were immaterial; but, as a general rule, when the pitch is coarser than four or five threads per inch, the difficulty of cutting threads with dies increases rapidly, although in a few cases some dies are used successfully on screw threads having two or three threads per inch or less. Much depends upon the design of the die, the finish or smoothness required, and the relation between the pitch of the thread and the diameter of the screw. When the screw diameter is relatively small in proportion to the pitch, there may be considerable distortion due to the twisting strains set up when the thread is being cut. If the number of threads per inch is only one or two less than the standard number for a given diameter, a screw blank ordinarily will be strong enough to permit the use of a die.

**Changing Pitch of Screw Thread Slightly.** — A very slight change in the pitch of a screw thread may be necessary as, for example, when the pitch of a tap is increased a small amount to compensate for shrinkage in hardening. One method of obtaining slight variations in pitch is by means of a taper attachment. This attachment is set at an angle and the work is located at the same angle by adjusting the tailstock center. The result is that the tool follows an angular path relative to the movement of the carriage and, consequently, the pitch of the thread is increased slightly, the amount depending upon the angle to which the work and taper attachment are set. The cosine of this angle, for obtaining a given increase in pitch, equals the standard pitch (which would be obtained with the lathe used in the regular way) divided by the increased pitch necessary to compensate for shrinkage.

*Example:* — If the pitch of a ¾-inch American standard screw is to be increased from 0.100 to 0.1005, the cosine of the angle to which the taper attachment and work should be set is found as follows:

$$\text{Cosine of required angle} = \frac{0.100}{0.1005} = 0.9950$$

which is the cosine of 5 degrees 45 minutes, nearly.

## CHANGE GEARS FOR HELICAL MILLING

**Lead of a Milling Machine.** — If gears with an equal number of teeth are placed on the table feed-screw and the worm-gear stud, then the *lead of the milling machine* is the distance the table will travel while the index spindle makes one complete revolution. This distance is a constant used in figuring the change gears.

The lead of a helix or "spiral" is the distance, measured along the axis of the work, in which the helix makes one full turn around the work. The lead of the milling machine may, therefore, also be expressed as the lead of the helix that will be cut when gears with an equal number of teeth are placed on the feed-screw and the worm-gear stud, and an idler of suitable size is interposed between the gears.

*Rule:* To find the lead of a milling machine, place equal gears on the worm-gear stud and on the feed-screw, and multiply the number of revolutions made by the feed-screw to produce one revolution of the index head spindle, by the lead of the thread on the feed-screw. Expressing the rule given as a formula:

$$\frac{\text{lead of milling}}{\text{machine}} = \frac{\text{rev. of feed-screw for one}}{\text{revolution of index spindle}} \times \frac{\text{lead of}}{\text{feed-screw.}}$$
$$\text{with equal gears}$$

Assume that it is necessary to make 40 revolutions of the feed-screw to turn the index head spindle one complete revolution, when the gears are equal, and that the lead of the thread on the feed-screw of the milling machine is ¼ inch; then the lead of the machine equals 40 × ¼ inch = 10 inches.

**Change Gears for Helical Milling.** — To find the change gears to be used in the compound train of gears for helical milling, place the lead of the helix to be cut in the numerator and the lead of the milling machine in the denominator of a fraction; divide numerator and denominator into two factors each; and multiply each "pair" of factors by the *same* number until suitable numbers of teeth for the change gears are obtained. (One factor in the numerator and one in the denominator are considered as one "pair" in this calculation.)

*Example:* Assume that the lead of a machine is 10 inches, and that a helix having a 48-inch lead is to be cut. Following the method explained:

$$\frac{48}{10} = \frac{6 \times 8}{2 \times 5} = \frac{(6 \times 12) \times (8 \times 8)}{(2 \times 12) \times (5 \times 8)} = \frac{72 \times 64}{24 \times 40}$$

The gear having 72 teeth is placed on the worm-gear stud and meshes with the 24-tooth gear on the intermediate stud. On the same intermediate stud is then placed the gear having 64 teeth, which is driven by the gear having 40 teeth placed on the feed-screw. This makes the gears having 72 and 64 teeth the driven gears, and the gears having 24 and 40 teeth the driving gears. In general, for compound gearing, the following formula may be used:

$$\frac{\text{lead of helix to be cut}}{\text{lead of machine}} = \frac{\text{product of driven gears}}{\text{product of driving gears}}$$

**Short-lead Milling.** — If lead to be milled is exceptionally short, the drive may be direct from table feed-screw to dividing head spindle to avoid excessive load on feed-screw and change-gears. If table feed-screw has 4 threads per inch (usual standard), then

$$\text{Change-gear ratio} = \frac{\text{Lead to be milled}}{0.25} = \frac{\text{Driven gears}}{\text{Driving gears}}$$

For indexing, number of teeth on spindle change-gear should be some multiple number of divisions required, to permit indexing by disengaging and turning the gear.

**Helix.** — A helix is a curve generated by a point moving about a cylindrical surface (real or imaginary) at a constant rate in the direction of the cylinder's axis. The curvature of a screw thread is one common example of a helical curve.

*Lead of Helix:* The lead of a helix is the distance that it advances in an axial direction, in one complete turn about the cylindrical surface. To illustrate, the lead of a screw thread equals the distance that a thread advances in one turn; it also equals the distance that a nut would advance in one turn.

*Development of Helix:* If one turn of a helical curve were unrolled onto a plane surface (as shown by diagram), the helix would become a straight line forming the hypotenuse of a right angle triangle. The length of one side of this triangle would equal the circumference of the cylinder with which the helix coincides, and the length of the other side of the triangle would equal the lead of the helix.

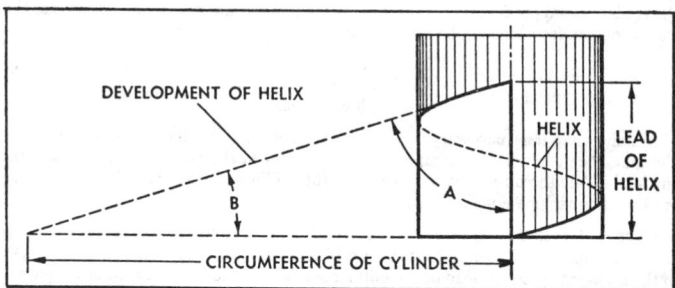

**Helix Angles.** — The triangular development of a helix has one angle $A$ subtended by the circumference of the cylinder, and another angle $B$ subtended by the lead of the helix. The term "helix angle" indicates angle $A$ in some cases and angle $B$ in others. For example, the helix angle of a helical gear, according to the general usage of the term, is always angle $A$, because this is the angle used in helical gear-designing formulas. Helix angle $A$ would also be applied in milling the helical teeth of cutters, reamers, etc. Angle $A$ of a gear or cutter tooth is a measure of its inclination relative to the axis of the gear or cutter.

*Lead Angle:* Angle $B$ is applied to screw threads and worm threads. This angle $B$ is a measure of the inclination of a screw thread from a plane that is perpendicular to the screw thread axis. Angle $B$ is often called the "lead angle" because it is subtended by the lead of the thread, and to distinguish it from the term "helix angle" as applied to helical gears.

*Finding Helix Angle of Helical Gear:* A helical gear tooth has an infinite number of helix angles, but the angle at the pitch diameter or mid-working depth is the one required in gear designing and gear cutting. This angle $A$, relative to the axis of the gear, is found as follows:

$$\text{Cot helix angle} = \frac{\text{Lead of gear tooth}}{3.1416 \times \text{pitch diameter of gear}}$$

*Finding Lead Angle of Screw Thread:* The lead or helix angle at the pitch diameter of a screw thread usually is required when, for example, a thread milling cutter must be aligned with the thread. This angle measured from a plane perpendicular to the screw thread axis, is found as follows:

$$\text{Cot helix angle} = \frac{3.1416 \times \text{pitch diameter of screw thread}}{\text{Lead of screw thread}}$$

Change Gears for Different Leads — 0.670 Inch to 2.658 Inches

| Lead in Inches | Driven Gear on Worm | Driver First Gear on Stud | Driven Second Gear on Stud | Driver Gear on Screw | Lead in Inches | Driven Gear on Worm | Driver First Gear on Stud | Driven Second Gear on Stud | Driver Gear on Screw | Lead in Inches | Driven Gear on Worm | Driver First Gear on Stud | Driven Second Gear on Stud | Driver Gear on Screw |
|---|---|---|---|---|---|---|---|---|---|---|---|---|---|---|
| 0.670 | 24 | 86 | 24 | 100 | 1.711 | 28 | 72 | 44 | 100 | 2.182 | 24 | 44 | 40 | 100 |
| 0.781 | 24 | 86 | 28 | 100 | 1.714 | 24 | 56 | 40 | 100 | 2.188 | 24 | 48 | 28 | 64 |
| 0.800 | 24 | 72 | 24 | 100 | 1.744 | 24 | 64 | 40 | 86 | 2.193 | 24 | 56 | 44 | 86 |
| 0.893 | 24 | 86 | 32 | 100 | 1.745 | 24 | 44 | 32 | 100 | 2.200 | 24 | 44 | 44 | 100 |
| 0.930 | 24 | 72 | 24 | 86 | 1.750 | 28 | 64 | 40 | 100 | 2.222 | 24 | 48 | 32 | 72 |
| 1.029 | 24 | 56 | 24 | 100 | 1.776 | 24 | 44 | 28 | 86 | 2.233 | 40 | 86 | 48 | 100 |
| 1.042 | 28 | 86 | 32 | 100 | 1.778 | 24 | 72 | 40 | 100 | 2.238 | 28 | 64 | 44 | 86 |
| 1.047 | 24 | 64 | 24 | 86 | 1.786 | 24 | 86 | 64 | 100 | 2.240 | 28 | 40 | 32 | 100 |
| 1.050 | 24 | 64 | 28 | 100 | 1.800 | 24 | 64 | 48 | 100 | 2.250 | 24 | 40 | 24 | 64 |
| 1.067 | 24 | 72 | 32 | 100 | 1.809 | 28 | 72 | 40 | 86 | 2.274 | 32 | 72 | 44 | 86 |
| 1.085 | 24 | 72 | 28 | 86 | 1.818 | 24 | 44 | 24 | 72 | 2.286 | 32 | 56 | 40 | 100 |
| 1.116 | 24 | 86 | 40 | 100 | 1.823 | 28 | 86 | 56 | 100 | 2.292 | 24 | 64 | 44 | 72 |
| 1.196 | 24 | 56 | 24 | 86 | 1.860 | 28 | 56 | 32 | 86 | 2.326 | 32 | 64 | 40 | 86 |
| 1.200 | 24 | 48 | 24 | 100 | 1.861 | 24 | 72 | 48 | 86 | 2.333 | 28 | 48 | 40 | 100 |
| 1.221 | 24 | 64 | 28 | 86 | 1.867 | 28 | 48 | 32 | 100 | 2.338 | 24 | 44 | 24 | 56 |
| 1.228 | 24 | 86 | 44 | 100 | 1.875 | 24 | 48 | 24 | 64 | 2.344 | 28 | 86 | 72 | 100 |
| 1.240 | 24 | 72 | 32 | 86 | 1.886 | 24 | 56 | 44 | 86 | 2.368 | 28 | 44 | 32 | 86 |
| 1.250 | 24 | 64 | 24 | 72 | 1.905 | 24 | 56 | 32 | 72 | 2.381 | 32 | 86 | 64 | 100 |
| 1.302 | 28 | 86 | 40 | 100 | 1.919 | 24 | 64 | 44 | 86 | 2.386 | 24 | 44 | 28 | 64 |
| 1.309 | 24 | 44 | 24 | 100 | 1.920 | 24 | 40 | 32 | 100 | 2.392 | 24 | 56 | 48 | 86 |
| 1.333 | 24 | 72 | 40 | 100 | 1.925 | 28 | 64 | 44 | 100 | 2.400 | 28 | 56 | 48 | 100 |
| 1.340 | 24 | 86 | 48 | 100 | 1.944 | 24 | 48 | 28 | 72 | 2.424 | 24 | 44 | 32 | 72 |
| 1.371 | 24 | 56 | 32 | 100 | 1.954 | 24 | 40 | 28 | 86 | 2.431 | 28 | 64 | 40 | 72 |
| 1.395 | 24 | 48 | 24 | 86 | 1.956 | 32 | 72 | 44 | 100 | 2.442 | 24 | 32 | 28 | 86 |
| 1.400 | 24 | 48 | 28 | 100 | 1.990 | 28 | 72 | 44 | 86 | 2.445 | 40 | 72 | 44 | 100 |
| 1.429 | 24 | 56 | 24 | 72 | 1.993 | 24 | 56 | 40 | 86 | 2.450 | 28 | 64 | 56 | 100 |
| 1.440 | 24 | 40 | 24 | 100 | 2.000 | 24 | 40 | 24 | 72 | 2.456 | 44 | 86 | 48 | 100 |
| 1.458 | 24 | 64 | 28 | 72 | 2.009 | 24 | 86 | 72 | 100 | 2.481 | 32 | 72 | 48 | 86 |
| 1.467 | 24 | 72 | 44 | 100 | 2.030 | 24 | 44 | 32 | 86 | 2.489 | 32 | 72 | 56 | 100 |
| 1.488 | 32 | 86 | 40 | 100 | 2.035 | 28 | 64 | 40 | 86 | 2.500 | 24 | 48 | 28 | 56 |
| 1.500 | 24 | 64 | 40 | 100 | 2.036 | 28 | 44 | 32 | 100 | 2.514 | 32 | 56 | 44 | 100 |
| 1.522 | 24 | 44 | 24 | 86 | 2.045 | 24 | 44 | 24 | 64 | 2.532 | 28 | 72 | 56 | 86 |
| 1.550 | 24 | 72 | 40 | 86 | 2.047 | 40 | 86 | 44 | 100 | 2.537 | 24 | 44 | 40 | 86 |
| 1.563 | 24 | 86 | 56 | 100 | 2.057 | 24 | 28 | 24 | 100 | 2.546 | 28 | 44 | 40 | 100 |
| 1.595 | 24 | 56 | 32 | 86 | 2.067 | 32 | 72 | 40 | 86 | 2.558 | 32 | 64 | 44 | 86 |
| 1 600 | 24 | 48 | 32 | 100 | 2.083 | 24 | 64 | 40 | 72 | 2.567 | 28 | 48 | 44 | 100 |
| 1.607 | 24 | 56 | 24 | 64 | 2.084 | 28 | 86 | 64 | 100 | 2.571 | 24 | 40 | 24 | 56 |
| 1.628 | 24 | 48 | 28 | 86 | 2.093 | 24 | 64 | 48 | 86 | 2.593 | 28 | 48 | 32 | 72 |
| 1.637 | 32 | 86 | 44 | 100 | 2.100 | 24 | 64 | 56 | 100 | 2.605 | 28 | 40 | 32 | 86 |
| 1.650 | 24 | 64 | 44 | 100 | 2.121 | 24 | 44 | 28 | 72 | 2.618 | 24 | 44 | 48 | 100 |
| 1.667 | 24 | 56 | 28 | 72 | 2.133 | 24 | 72 | 64 | 100 | 2.619 | 24 | 56 | 44 | 72 |
| 1.674 | 24 | 40 | 24 | 86 | 2.143 | 24 | 56 | 32 | 64 | 2.625 | 24 | 40 | 28 | 64 |
| 1.680 | 24 | 40 | 28 | 100 | 2.171 | 24 | 72 | 56 | 86 | 2.640 | 24 | 40 | 44 | 100 |
| 1.706 | 24 | 72 | 44 | 86 | 2.178 | 28 | 72 | 56 | 100 | 2.658 | 32 | 56 | 40 | 86 |

## Change Gears for Different Leads — 2.667 Inches to 4.040 Inches

| Lead in Inches | Driven | Driver | Driven | Driver | Lead in Inches | Driven | Driver | Driven | Driver | Lead in Inches | Driven | Driver | Driven | Driver |
|---|---|---|---|---|---|---|---|---|---|---|---|---|---|---|
| | Gear on Worm | First Gear on Stud | Second Gear on Stud | Gear on Screw | | Gear on Worm | First Gear on Stud | Second Gear on Stud | Gear on Screw | | Gear on Worm | First Gear on Stud | Second Gear on Stud | Gear on Screw |
| 2.667 | 40 | 72 | 48 | 100 | 3.140 | 24 | 86 | 72 | 64 | 3.588 | 72 | 56 | 24 | 86 |
| 2.674 | 28 | 64 | 44 | 72 | 3.143 | 40 | 56 | 44 | 100 | 3.600 | 72 | 48 | 24 | 100 |
| 2.678 | 24 | 56 | 40 | 64 | 3.150 | 28 | 100 | 72 | 64 | 3.618 | 56 | 72 | 40 | 86 |
| 2.679 | 32 | 86 | 72 | 100 | 3.175 | 32 | 56 | 40 | 72 | 3.636 | 24 | 44 | 32 | 48 |
| 2.700 | 24 | 64 | 72 | 100 | 3.182 | 28 | 44 | 32 | 64 | 3.637 | 48 | 44 | 24 | 72 |
| 2.713 | 28 | 48 | 40 | 86 | 3.189 | 32 | 56 | 48 | 86 | 3.646 | 40 | 48 | 28 | 64 |
| 2.727 | 24 | 44 | 32 | 64 | 3.190 | 24 | 86 | 64 | 56 | 3.655 | 40 | 56 | 44 | 86 |
| 2.743 | 24 | 56 | 64 | 100 | 3.198 | 40 | 64 | 44 | 86 | 3.657 | 64 | 56 | 32 | 100 |
| 2.750 | 40 | 64 | 44 | 100 | 3.200 | 28 | 100 | 64 | 56 | 3.663 | 72 | 64 | 28 | 86 |
| 2.778 | 32 | 64 | 40 | 72 | 3.214 | 24 | 56 | 48 | 64 | 3.667 | 40 | 48 | 44 | 100 |
| 2.791 | 28 | 56 | 48 | 86 | 3.225 | 24 | 100 | 86 | 64 | 3.673 | 24 | 28 | 24 | 56 |
| 2.800 | 24 | 24 | 28 | 100 | 3.241 | 28 | 48 | 40 | 72 | 3.684 | 44 | 86 | 72 | 100 |
| 2.812 | 24 | 32 | 24 | 64 | 3.256 | 24 | 24 | 28 | 86 | 3.686 | 86 | 56 | 24 | 100 |
| 2.828 | 28 | 44 | 32 | 72 | 3.267 | 28 | 48 | 56 | 100 | 3.704 | 32 | 48 | 40 | 72 |
| 2.843 | 40 | 72 | 44 | 86 | 3.273 | 24 | 40 | 24 | 44 | 3.721 | 24 | 24 | 32 | 86 |
| 2.845 | 32 | 72 | 64 | 100 | 3.275 | 44 | 86 | 64 | 100 | 3.733 | 48 | 72 | 56 | 100 |
| 2.849 | 28 | 64 | 56 | 86 | 3.281 | 24 | 32 | 28 | 64 | 3.750 | 24 | 32 | 24 | 48 |
| 2.857 | 24 | 48 | 32 | 56 | 3.300 | 44 | 64 | 48 | 100 | 3.763 | 86 | 64 | 28 | 100 |
| 2.865 | 44 | 86 | 56 | 100 | 3.308 | 32 | 72 | 64 | 86 | 3.771 | 44 | 56 | 48 | 100 |
| 2.867 | 86 | 72 | 24 | 100 | 3.333 | 32 | 64 | 48 | 72 | 3.772 | 24 | 28 | 44 | 100 |
| 2.880 | 24 | 40 | 48 | 100 | 3.345 | 28 | 100 | 86 | 72 | 3.799 | 56 | 48 | 28 | 86 |
| 2.894 | 28 | 72 | 64 | 86 | 3.349 | 40 | 86 | 72 | 100 | 3.809 | 24 | 28 | 32 | 72 |
| 2.909 | 32 | 44 | 40 | 100 | 3.360 | 56 | 40 | 24 | 100 | 3.810 | 64 | 56 | 24 | 72 |
| 2.917 | 24 | 64 | 56 | 72 | 3.383 | 32 | 44 | 40 | 86 | 3.818 | 24 | 40 | 28 | 44 |
| 2.924 | 32 | 56 | 44 | 86 | 3.403 | 28 | 64 | 56 | 72 | 3.819 | 40 | 64 | 44 | 72 |
| 2.933 | 44 | 72 | 48 | 100 | 3.409 | 24 | 44 | 40 | 64 | 3.822 | 86 | 72 | 32 | 100 |
| 2.934 | 32 | 48 | 44 | 100 | 3.411 | 32 | 48 | 44 | 86 | 3.837 | 24 | 32 | 44 | 86 |
| 2.946 | 24 | 56 | 44 | 64 | 3.422 | 44 | 72 | 56 | 100 | 3.840 | 64 | 40 | 24 | 100 |
| 2.960 | 28 | 44 | 40 | 86 | 3.428 | 24 | 40 | 32 | 56 | 3.850 | 44 | 64 | 56 | 100 |
| 2.977 | 40 | 86 | 64 | 100 | 3.429 | 40 | 28 | 24 | 100 | 3.876 | 24 | 72 | 100 | 86 |
| 2.984 | 28 | 48 | 44 | 86 | 3.438 | 24 | 48 | 44 | 64 | 3.889 | 32 | 64 | 56 | 72 |
| 3.000 | 24 | 40 | 28 | 56 | 3.488 | 40 | 64 | 48 | 86 | 3.896 | 24 | 44 | 40 | 56 |
| 3.030 | 24 | 44 | 40 | 72 | 3.491 | 64 | 44 | 24 | 100 | 3.907 | 56 | 40 | 24 | 86 |
| 3.044 | 24 | 44 | 48 | 86 | 3.492 | 32 | 56 | 44 | 72 | 3.911 | 44 | 72 | 64 | 100 |
| 3.055 | 28 | 44 | 48 | 100 | 3.500 | 40 | 64 | 56 | 100 | 3.920 | 28 | 40 | 56 | 100 |
| 3.056 | 32 | 64 | 44 | 72 | 3.520 | 32 | 40 | 44 | 100 | 3.927 | 72 | 44 | 24 | 100 |
| 3.070 | 24 | 40 | 44 | 86 | 3.535 | 28 | 44 | 40 | 72 | 3.929 | 32 | 56 | 44 | 64 |
| 3.080 | 28 | 40 | 44 | 100 | 3.552 | 56 | 44 | 24 | 86 | 3.977 | 28 | 44 | 40 | 64 |
| 3.086 | 24 | 56 | 72 | 100 | 3.556 | 40 | 72 | 64 | 100 | 3.979 | 44 | 72 | 56 | 86 |
| 3.101 | 40 | 72 | 48 | 86 | 3.564 | 56 | 44 | 28 | 100 | 3.987 | 24 | 28 | 40 | 86 |
| 3.111 | 28 | 40 | 32 | 72 | 3.565 | 28 | 48 | 44 | 72 | 4.000 | 24 | 40 | 32 | 48 |
| 3.117 | 24 | 44 | 32 | 56 | 3.571 | 24 | 48 | 40 | 56 | 4.011 | 28 | 48 | 44 | 64 |
| 3.125 | 28 | 56 | 40 | 64 | 3.572 | 48 | 86 | 64 | 100 | 4.019 | 72 | 86 | 48 | 100 |
| 3.126 | 48 | 86 | 56 | 100 | 3.582 | 44 | 40 | 28 | 86 | 4.040 | 32 | 44 | 40 | 72 |

HELICAL MILLING 1483

## Change Gears for Different Leads — 4.059 Inches to 5.568 Inches

| Lead in Inches | Driven Gear on Worm | Driver First Gear on Stud | Driven Second Gear on Stud | Driver Gear on Screw | Lead in Inches | Driven Gear on Worm | Driver First Gear on Stud | Driven Second Gear on Stud | Driver Gear on Screw | Lead in Inches | Driven Gear on Worm | Driver First Gear on Stud | Driven Second Gear on Stud | Driver Gear on Screw |
|---|---|---|---|---|---|---|---|---|---|---|---|---|---|---|
| 4.059 | 32 | 44 | 48 | 86 | 4.567 | 72 | 44 | 24 | 86 | 5.105 | 28 | 48 | 56 | 64 |
| 4.060 | 64 | 44 | 24 | 86 | 4.572 | 40 | 56 | 64 | 100 | 5.116 | 44 | 24 | 24 | 86 |
| 4.070 | 28 | 32 | 40 | 86 | 4.582 | 72 | 44 | 28 | 100 | 5.119 | 86 | 56 | 24 | 72 |
| 4.073 | 64 | 44 | 28 | 100 | 4.583 | 44 | 64 | 48 | 72 | 5.120 | 64 | 40 | 32 | 100 |
| 4.074 | 32 | 48 | 44 | 72 | 4.584 | 32 | 48 | 44 | 64 | 5.133 | 56 | 48 | 44 | 100 |
| 4.091 | 24 | 44 | 48 | 64 | 4.651 | 40 | 24 | 24 | 86 | 5.134 | 44 | 24 | 28 | 100 |
| 4.093 | 32 | 40 | 44 | 86 | 4.655 | 64 | 44 | 32 | 100 | 5.142 | 72 | 56 | 40 | 100 |
| 4.114 | 48 | 28 | 24 | 100 | 4.667 | 28 | 40 | 32 | 48 | 5.143 | 24 | 28 | 24 | 40 |
| 4.125 | 24 | 40 | 44 | 64 | 4.675 | 24 | 28 | 24 | 44 | 5.156 | 44 | 32 | 24 | 64 |
| 4.135 | 40 | 72 | 64 | 86 | 4.687 | 40 | 32 | 24 | 64 | 5.160 | 86 | 40 | 24 | 100 |
| 4.144 | 56 | 44 | 28 | 86 | 4.688 | 56 | 86 | 72 | 100 | 5.168 | 100 | 72 | 32 | 86 |
| 4.167 | 28 | 48 | 40 | 56 | 4.691 | 86 | 44 | 24 | 100 | 5.185 | 28 | 24 | 32 | 72 |
| 4.186 | 72 | 64 | 32 | 86 | 4.714 | 44 | 40 | 24 | 56 | 5.186 | 64 | 48 | 28 | 72 |
| 4.200 | 48 | 64 | 56 | 100 | 4.736 | 64 | 44 | 28 | 86 | 5.195 | 32 | 44 | 40 | 56 |
| 4.242 | 28 | 44 | 32 | 48 | 4.762 | 40 | 28 | 24 | 72 | 5.209 | 100 | 64 | 24 | 72 |
| 4.253 | 64 | 56 | 32 | 86 | 4.773 | 24 | 32 | 28 | 44 | 5.210 | 64 | 40 | 28 | 86 |
| 4.264 | 40 | 48 | 44 | 86 | 4.778 | 86 | 72 | 40 | 100 | 5.226 | 86 | 64 | 28 | 72 |
| 4.267 | 64 | 48 | 32 | 100 | 4.784 | 72 | 56 | 32 | 86 | 5.233 | 72 | 64 | 40 | 86 |
| 4.278 | 28 | 40 | 44 | 72 | 4.785 | 48 | 28 | 24 | 86 | 5.236 | 72 | 44 | 32 | 100 |
| 4.286 | 24 | 28 | 24 | 48 | 4.800 | 48 | 24 | 24 | 100 | 5.238 | 44 | 28 | 24 | 72 |
| 4.300 | 86 | 56 | 28 | 100 | 4.813 | 44 | 40 | 28 | 64 | 5.250 | 24 | 32 | 28 | 40 |
| 4.320 | 72 | 40 | 24 | 100 | 4.821 | 72 | 56 | 24 | 64 | 5.256 | 86 | 72 | 44 | 100 |
| 4.341 | 48 | 72 | 56 | 86 | 4.849 | 32 | 44 | 48 | 72 | 5.280 | 48 | 40 | 44 | 100 |
| 4.342 | 64 | 48 | 28 | 86 | 4.861 | 40 | 32 | 28 | 72 | 5.303 | 28 | 44 | 40 | 48 |
| 4.361 | 100 | 64 | 24 | 86 | 4.884 | 48 | 64 | 56 | 86 | 5.316 | 40 | 28 | 32 | 86 |
| 4.363 | 24 | 40 | 32 | 44 | 4.889 | 32 | 40 | 44 | 72 | 5.328 | 72 | 44 | 28 | 86 |
| 4.364 | 40 | 44 | 48 | 100 | 4.898 | 24 | 28 | 32 | 56 | 5.333 | 40 | 24 | 32 | 100 |
| 4.365 | 40 | 56 | 44 | 72 | 4.900 | 56 | 32 | 28 | 100 | 5.347 | 44 | 64 | 56 | 72 |
| 4.375 | 24 | 24 | 28 | 64 | 4.911 | 40 | 56 | 44 | 64 | 5.348 | 44 | 32 | 28 | 72 |
| 4.386 | 24 | 28 | 44 | 86 | 4.914 | 86 | 56 | 32 | 100 | 5.357 | 40 | 28 | 24 | 64 |
| 4.400 | 24 | 24 | 44 | 100 | 4.950 | 56 | 44 | 28 | 72 | 5.358 | 64 | 86 | 72 | 100 |
| 4.444 | 64 | 56 | 28 | 72 | 4.961 | 64 | 48 | 32 | 86 | 5.375 | 86 | 64 | 40 | 100 |
| 4.465 | 64 | 40 | 24 | 86 | 4.978 | 56 | 72 | 64 | 100 | 5.400 | 72 | 32 | 24 | 100 |
| 4.466 | 48 | 40 | 32 | 86 | 4.984 | 100 | 56 | 24 | 86 | 5.413 | 64 | 44 | 32 | 86 |
| 4.477 | 44 | 32 | 28 | 86 | 5.000 | 24 | 24 | 28 | 86 | 5.426 | 40 | 24 | 28 | 86 |
| 4.479 | 86 | 64 | 24 | 72 | 5.017 | 86 | 48 | 28 | 100 | 5.427 | 40 | 48 | 56 | 86 |
| 4.480 | 56 | 40 | 32 | 100 | 5.023 | 72 | 40 | 24 | 86 | 5.444 | 56 | 40 | 28 | 72 |
| 4.500 | 72 | 64 | 40 | 100 | 5.029 | 44 | 28 | 32 | 100 | 5.455 | 48 | 44 | 28 | 56 |
| 4.522 | 100 | 72 | 28 | 86 | 5.040 | 72 | 40 | 28 | 100 | 5.469 | 40 | 32 | 28 | 64 |
| 4.537 | 56 | 48 | 28 | 72 | 5.074 | 40 | 44 | 48 | 86 | 5.473 | 86 | 44 | 28 | 100 |
| 4.545 | 24 | 44 | 40 | 48 | 5.080 | 64 | 56 | 32 | 72 | 5.486 | 64 | 28 | 24 | 100 |
| 4.546 | 28 | 44 | 40 | 56 | 5.088 | 100 | 64 | 28 | 86 | 5.500 | 44 | 40 | 24 | 48 |
| 4.548 | 44 | 72 | 64 | 86 | 5.091 | 56 | 44 | 40 | 100 | 5.556 | 40 | 24 | 24 | 72 |
| 4.558 | 56 | 40 | 28 | 86 | 5.093 | 40 | 48 | 44 | 72 | 5.568 | 56 | 44 | 28 | 64 |

## Change Gears for Different Leads — 5.581 Inches to 7.500 Inches

| Lead in Inches | Driven — Gear on Worm | Driver — First Gear on Stud | Driven — Second Gear on Stud | Driver — Gear on Screw | Lead in Inches | Driven — Gear on Worm | Driver — First Gear on Stud | Driven — Second Gear on Stud | Driver — Gear on Screw | Lead in Inches | Driven — Gear on Worm | Driver — First Gear on Stud | Driven — Second Gear on Stud | Driver — Gear on Screw |
|---|---|---|---|---|---|---|---|---|---|---|---|---|---|---|
| 5.581 | 64 | 32 | 24 | 86 | 6.172 | 72 | 28 | 24 | 100 | 6.825 | 86 | 56 | 32 | 72 |
| 5.582 | 48 | 24 | 24 | 86 | 6.202 | 40 | 24 | 32 | 86 | 6.857 | 32 | 28 | 24 | 40 |
| 5.600 | 56 | 24 | 24 | 100 | 6.222 | 64 | 40 | 28 | 72 | 6.875 | 44 | 24 | 24 | 64 |
| 5.625 | 48 | 32 | 24 | 64 | 6.234 | 32 | 28 | 24 | 44 | 6.880 | 86 | 40 | 32 | 100 |
| 5.657 | 56 | 44 | 32 | 72 | 6.250 | 24 | 24 | 40 | 64 | 6.944 | 100 | 48 | 24 | 72 |
| 5.698 | 56 | 32 | 28 | 86 | 6.255 | 86 | 44 | 32 | 100 | 6.945 | 100 | 56 | 28 | 72 |
| 5.714 | 48 | 28 | 24 | 72 | 6.279 | 72 | 64 | 48 | 86 | 6.968 | 86 | 48 | 28 | 72 |
| 5.730 | 40 | 48 | 44 | 64 | 6.286 | 44 | 40 | 32 | 56 | 6.977 | 48 | 32 | 40 | 86 |
| 5.733 | 86 | 48 | 32 | 100 | 6.300 | 72 | 32 | 28 | 100 | 6.982 | 64 | 44 | 48 | 100 |
| 5.756 | 72 | 64 | 44 | 86 | 6.343 | 100 | 44 | 24 | 86 | 6.984 | 44 | 28 | 32 | 72 |
| 5.759 | 86 | 56 | 24 | 64 | 6.350 | 86 | 40 | 32 | 72 | 7.000 | 28 | 24 | 24 | 40 |
| 5.760 | 72 | 40 | 32 | 100 | 6.364 | 56 | 44 | 24 | 48 | 7.013 | 72 | 44 | 24 | 56 |
| 5.788 | 64 | 72 | 56 | 86 | 6.379 | 64 | 28 | 24 | 86 | 7.040 | 64 | 40 | 44 | 100 |
| 5.814 | 100 | 64 | 32 | 86 | 6.396 | 64 | 32 | 40 | 86 | 7.071 | 56 | 44 | 40 | 72 |
| 5.818 | 64 | 44 | 40 | 100 | 6.400 | 64 | 24 | 24 | 100 | 7.104 | 56 | 44 | 48 | 86 |
| 5.833 | 28 | 24 | 24 | 48 | 6.417 | 44 | 40 | 28 | 48 | 7.106 | 100 | 72 | 44 | 86 |
| 5.847 | 64 | 56 | 44 | 86 | 6.429 | 24 | 28 | 24 | 32 | 7.111 | 64 | 40 | 32 | 72 |
| 5.848 | 44 | 28 | 32 | 86 | 6.450 | 86 | 64 | 48 | 100 | 7.130 | 44 | 24 | 28 | 72 |
| 5.861 | 72 | 40 | 28 | 86 | 6.460 | 100 | 72 | 40 | 86 | 7.143 | 40 | 28 | 32 | 64 |
| 5.867 | 44 | 24 | 32 | 100 | 6.465 | 64 | 44 | 32 | 72 | 7.159 | 72 | 44 | 28 | 64 |
| 5.893 | 44 | 32 | 24 | 56 | 6.482 | 56 | 48 | 40 | 72 | 7.163 | 56 | 40 | 44 | 86 |
| 5.912 | 86 | 64 | 44 | 100 | 6.512 | 56 | 24 | 24 | 86 | 7.167 | 86 | 40 | 24 | 72 |
| 5.920 | 56 | 44 | 40 | 86 | 6.515 | 86 | 44 | 24 | 72 | 7.176 | 72 | 28 | 24 | 86 |
| 5.926 | 64 | 48 | 32 | 72 | 6.534 | 56 | 24 | 28 | 100 | 7.200 | 72 | 24 | 24 | 100 |
| 5.952 | 100 | 56 | 24 | 72 | 6.545 | 48 | 40 | 24 | 44 | 7.268 | 100 | 64 | 40 | 86 |
| 5.954 | 64 | 40 | 32 | 86 | 6.548 | 44 | 48 | 40 | 56 | 7.272 | 64 | 44 | 28 | 56 |
| 5.969 | 44 | 24 | 28 | 86 | 6.563 | 56 | 32 | 24 | 64 | 7.273 | 32 | 24 | 24 | 44 |
| 5.972 | 86 | 48 | 24 | 72 | 6.578 | 72 | 56 | 44 | 86 | 7.292 | 56 | 48 | 40 | 64 |
| 5.980 | 72 | 56 | 40 | 86 | 6.600 | 48 | 32 | 44 | 100 | 7.310 | 44 | 28 | 40 | 86 |
| 6.000 | 48 | 40 | 28 | 56 | 6.645 | 100 | 56 | 32 | 86 | 7.314 | 64 | 28 | 32 | 100 |
| 6.016 | 44 | 32 | 28 | 64 | 6.667 | 64 | 48 | 28 | 56 | 7.326 | 72 | 32 | 28 | 86 |
| 6.020 | 86 | 40 | 28 | 100 | 6.689 | 86 | 72 | 56 | 100 | 7.330 | 86 | 44 | 24 | 64 |
| 6.061 | 40 | 44 | 32 | 48 | 6.697 | 100 | 56 | 24 | 64 | 7.333 | 44 | 24 | 40 | 100 |
| 6.077 | 100 | 64 | 28 | 72 | 6.698 | 72 | 40 | 32 | 86 | 7.334 | 44 | 40 | 32 | 48 |
| 6.089 | 72 | 44 | 32 | 86 | 6.719 | 86 | 48 | 24 | 64 | 7.347 | 48 | 28 | 24 | 56 |
| 6.109 | 56 | 44 | 48 | 100 | 6.720 | 56 | 40 | 48 | 100 | 7.371 | 86 | 56 | 48 | 100 |
| 6.112 | 24 | 24 | 44 | 72 | 6.735 | 44 | 28 | 24 | 56 | 7.372 | 86 | 28 | 24 | 100 |
| 6.122 | 40 | 28 | 24 | 56 | 6.750 | 72 | 40 | 24 | 64 | 7.400 | 100 | 44 | 28 | 86 |
| 6.125 | 56 | 40 | 28 | 64 | 6.757 | 86 | 56 | 44 | 100 | 7.408 | 40 | 24 | 32 | 72 |
| 6.137 | 72 | 44 | 24 | 64 | 6.766 | 64 | 44 | 40 | 86 | 7.424 | 56 | 44 | 28 | 48 |
| 6.140 | 48 | 40 | 44 | 86 | 6.784 | 100 | 48 | 28 | 86 | 7.442 | 64 | 24 | 24 | 86 |
| 6.143 | 86 | 56 | 40 | 100 | 6.806 | 56 | 32 | 28 | 72 | 7.465 | 86 | 64 | 40 | 72 |
| 6.160 | 56 | 40 | 44 | 100 | 6.818 | 40 | 32 | 24 | 44 | 7.467 | 64 | 24 | 28 | 100 |
| 6.171 | 72 | 56 | 48 | 100 | 6.822 | 44 | 24 | 32 | 86 | 7.500 | 48 | 24 | 24 | 64 |

## Change Gears for Different Leads — 7.525 Inches to 9.598 Inches

| Lead in Inches | Driven (Gear on Worm) | Driver (First Gear on Stud) | Driven (Second Gear on Stud) | Driver (Gear on Screw) | Lead in Inches | Driven (Gear on Worm) | Driver (First Gear on Stud) | Driven (Second Gear on Stud) | Driver (Gear on Screw) | Lead in Inches | Driven (Gear on Worm) | Driver (First Gear on Stud) | Driven (Second Gear on Stud) | Driver (Gear on Screw) |
|---|---|---|---|---|---|---|---|---|---|---|---|---|---|---|
| 7.525 | 86 | 32 | 28 | 100 | 8.140 | 56 | 32 | 40 | 86 | 8.800 | 48 | 24 | 44 | 100 |
| 7.543 | 48 | 28 | 44 | 100 | 8.145 | 64 | 44 | 56 | 100 | 8.838 | 100 | 44 | 28 | 72 |
| 7.576 | 100 | 44 | 24 | 72 | 8.148 | 64 | 48 | 44 | 72 | 8.839 | 72 | 56 | 44 | 64 |
| 7.597 | 56 | 24 | 28 | 86 | 8.149 | 44 | 24 | 32 | 72 | 8.909 | 56 | 40 | 28 | 44 |
| 7.601 | 86 | 44 | 28 | 72 | 8.163 | 40 | 28 | 32 | 56 | 8.929 | 100 | 48 | 24 | 56 |
| 7.611 | 72 | 44 | 40 | 86 | 8.167 | 56 | 40 | 28 | 48 | 8.930 | 64 | 40 | 48 | 86 |
| 7.619 | 64 | 48 | 32 | 56 | 8.182 | 48 | 32 | 24 | 44 | 8.953 | 56 | 32 | 44 | 86 |
| 7.620 | 64 | 28 | 24 | 72 | 8.186 | 64 | 40 | 44 | 86 | 8.959 | 86 | 48 | 28 | 56 |
| 7.636 | 56 | 40 | 24 | 44 | 8.212 | 86 | 64 | 44 | 72 | 8.960 | 64 | 40 | 56 | 100 |
| 7.639 | 44 | 32 | 40 | 72 | 8.229 | 72 | 28 | 32 | 100 | 8.980 | 44 | 28 | 32 | 56 |
| 7.644 | 86 | 72 | 64 | 100 | 8.250 | 44 | 32 | 24 | 40 | 9.000 | 48 | 32 | 24 | 40 |
| 7.657 | 56 | 32 | 28 | 64 | 8.306 | 100 | 56 | 40 | 86 | 9.044 | 100 | 72 | 56 | 86 |
| 7.674 | 72 | 48 | 44 | 86 | 8.312 | 64 | 44 | 32 | 56 | 9.074 | 56 | 24 | 28 | 72 |
| 7.675 | 48 | 32 | 44 | 86 | 8.333 | 40 | 24 | 24 | 48 | 9.091 | 40 | 24 | 24 | 44 |
| 7.679 | 86 | 48 | 24 | 56 | 8.334 | 40 | 24 | 28 | 56 | 9.115 | 100 | 48 | 28 | 64 |
| 7.680 | 64 | 40 | 48 | 100 | 8.361 | 86 | 40 | 28 | 72 | 9.134 | 72 | 44 | 48 | 86 |
| 7.700 | 56 | 32 | 44 | 100 | 8.372 | 72 | 24 | 24 | 86 | 9.137 | 100 | 56 | 44 | 86 |
| 7.714 | 72 | 40 | 24 | 56 | 8.377 | 86 | 44 | 24 | 56 | 9.143 | 64 | 40 | 32 | 56 |
| 7.752 | 100 | 48 | 32 | 86 | 8.400 | 72 | 24 | 28 | 100 | 9.164 | 72 | 44 | 56 | 100 |
| 7.778 | 32 | 24 | 28 | 48 | 8.437 | 72 | 32 | 24 | 64 | 9.167 | 44 | 24 | 24 | 48 |
| 7.792 | 40 | 28 | 24 | 44 | 8.457 | 100 | 44 | 32 | 86 | 9.210 | 72 | 40 | 44 | 86 |
| 7.813 | 100 | 48 | 24 | 64 | 8.484 | 32 | 24 | 28 | 44 | 9.214 | 86 | 40 | 24 | 56 |
| 7.815 | 56 | 40 | 48 | 86 | 8.485 | 64 | 44 | 28 | 48 | 9.260 | 100 | 48 | 32 | 72 |
| 7.818 | 86 | 44 | 40 | 100 | 8.485 | 56 | 44 | 32 | 48 | 9.302 | 48 | 24 | 40 | 86 |
| 7.838 | 86 | 48 | 28 | 64 | 8.506 | 64 | 28 | 32 | 86 | 9.303 | 56 | 28 | 40 | 86 |
| 7.855 | 72 | 44 | 48 | 100 | 8.523 | 100 | 44 | 24 | 64 | 9.333 | 64 | 40 | 28 | 48 |
| 7.857 | 44 | 24 | 24 | 56 | 8.527 | 44 | 24 | 40 | 86 | 9.334 | 32 | 24 | 28 | 40 |
| 7.872 | 44 | 28 | 32 | 64 | 8.532 | 86 | 56 | 40 | 72 | 9.351 | 48 | 28 | 24 | 44 |
| 7.875 | 72 | 40 | 28 | 64 | 8.534 | 64 | 24 | 32 | 100 | 9.375 | 48 | 32 | 40 | 64 |
| 7.883 | 86 | 48 | 44 | 100 | 8.552 | 86 | 44 | 28 | 64 | 9.382 | 86 | 44 | 48 | 100 |
| 7.920 | 72 | 40 | 44 | 100 | 8.556 | 56 | 40 | 44 | 72 | 9.385 | 86 | 56 | 44 | 72 |
| 7.936 | 100 | 56 | 32 | 72 | 8.572 | 64 | 32 | 24 | 56 | 9.406 | 86 | 40 | 28 | 64 |
| 7.954 | 40 | 32 | 28 | 44 | 8.572 | 48 | 24 | 24 | 56 | 9.428 | 44 | 28 | 24 | 40 |
| 7.955 | 56 | 44 | 40 | 64 | 8.594 | 44 | 32 | 40 | 64 | 9.429 | 48 | 40 | 44 | 56 |
| 7.963 | 86 | 48 | 32 | 72 | 8.600 | 86 | 24 | 24 | 100 | 9.460 | 86 | 40 | 44 | 100 |
| 7.974 | 48 | 28 | 40 | 86 | 8.640 | 72 | 40 | 48 | 100 | 9.472 | 64 | 44 | 56 | 86 |
| 7.994 | 100 | 64 | 44 | 86 | 8.681 | 100 | 64 | 40 | 72 | 9.524 | 40 | 28 | 32 | 48 |
| 8.000 | 64 | 32 | 40 | 100 | 8.682 | 64 | 24 | 28 | 86 | 9.545 | 72 | 44 | 28 | 48 |
| 8.021 | 44 | 32 | 28 | 48 | 8.687 | 86 | 44 | 32 | 72 | 9.546 | 56 | 32 | 24 | 44 |
| 8.035 | 72 | 56 | 40 | 64 | 8.721 | 100 | 32 | 24 | 86 | 9.547 | 56 | 44 | 48 | 64 |
| 8.063 | 86 | 40 | 24 | 64 | 8.727 | 44 | 40 | 32 | 44 | 9.549 | 100 | 64 | 44 | 72 |
| 8.081 | 64 | 44 | 40 | 72 | 8.730 | 44 | 28 | 40 | 72 | 9.556 | 86 | 40 | 32 | 72 |
| 8.102 | 100 | 48 | 28 | 72 | 8.750 | 28 | 24 | 24 | 32 | 9.569 | 72 | 28 | 32 | 86 |
| 8.119 | 64 | 44 | 48 | 86 | 8.772 | 48 | 28 | 44 | 86 | 9.598 | 86 | 56 | 40 | 64 |

## Change Gears for Different Leads — 9.600 Inches to 12.375 Inches

| Lead in Inches | Driven — Gear on Worm | Driver — First Gear on Stud | Driven — Second Gear on Stud | Driver — Gear on Screw | Lead in Inches | Driven — Gear on Worm | Driver — First Gear on Stud | Driven — Second Gear on Stud | Driver — Gear on Screw | Lead in Inches | Driven — Gear on Worm | Driver — First Gear on Stud | Driven — Second Gear on Stud | Driver — Gear on Screw |
|---|---|---|---|---|---|---|---|---|---|---|---|---|---|---|
| 9.600 | 72 | 24 | 32 | 100 | 10.370 | 64 | 24 | 28 | 72 | 11.314 | 72 | 28 | 44 | 100 |
| 9.625 | 44 | 32 | 28 | 40 | 10.371 | 64 | 48 | 56 | 72 | 11.363 | 100 | 44 | 24 | 48 |
| 9.643 | 72 | 32 | 24 | 56 | 10.390 | 40 | 28 | 32 | 44 | 11.401 | 86 | 44 | 28 | 48 |
| 9.675 | 86 | 64 | 72 | 100 | 10.417 | 100 | 32 | 24 | 72 | 11.429 | 32 | 24 | 24 | 28 |
| 9.690 | 100 | 48 | 40 | 86 | 10.419 | 64 | 40 | 56 | 86 | 11.454 | 72 | 40 | 28 | 44 |
| 9.697 | 64 | 48 | 32 | 44 | 10.451 | 86 | 32 | 28 | 72 | 11.459 | 44 | 24 | 40 | 64 |
| 9.723 | 40 | 24 | 28 | 48 | 10.467 | 72 | 32 | 40 | 86 | 11.467 | 86 | 24 | 32 | 100 |
| 9.741 | 100 | 44 | 24 | 56 | 10.473 | 72 | 44 | 64 | 100 | 11.512 | 72 | 32 | 44 | 86 |
| 9.768 | 72 | 48 | 56 | 86 | 10.476 | 44 | 24 | 32 | 56 | 11.518 | 86 | 28 | 24 | 64 |
| 9.773 | 86 | 44 | 48 | 100 | 10.477 | 48 | 28 | 44 | 72 | 11.520 | 72 | 40 | 64 | 100 |
| 9.778 | 64 | 40 | 44 | 72 | 10.500 | 56 | 32 | 24 | 40 | 11.574 | 100 | 48 | 40 | 72 |
| 9.796 | 64 | 28 | 24 | 56 | 10.558 | 86 | 56 | 44 | 64 | 11.629 | 100 | 24 | 24 | 86 |
| 9.818 | 72 | 40 | 24 | 44 | 10.571 | 100 | 44 | 40 | 86 | 11.638 | 64 | 40 | 32 | 44 |
| 9.822 | 44 | 32 | 40 | 56 | 10.606 | 56 | 44 | 40 | 48 | 11.667 | 56 | 24 | 24 | 48 |
| 9.828 | 86 | 28 | 32 | 100 | 10.631 | 64 | 28 | 40 | 86 | 11.688 | 72 | 44 | 40 | 56 |
| 9.844 | 72 | 32 | 28 | 64 | 10.655 | 72 | 44 | 56 | 86 | 11.695 | 64 | 28 | 44 | 86 |
| 9.900 | 72 | 32 | 44 | 100 | 10.659 | 100 | 48 | 44 | 86 | 11.719 | 100 | 32 | 24 | 64 |
| 9.921 | 100 | 56 | 40 | 72 | 10.667 | 64 | 40 | 48 | 72 | 11.721 | 72 | 40 | 56 | 86 |
| 9.923 | 64 | 24 | 32 | 86 | 10.694 | 44 | 24 | 28 | 48 | 11.728 | 86 | 40 | 24 | 44 |
| 9.943 | 100 | 44 | 28 | 64 | 10.713 | 40 | 28 | 24 | 32 | 11.733 | 64 | 24 | 44 | 100 |
| 9.954 | 86 | 48 | 40 | 72 | 10.714 | 48 | 32 | 40 | 56 | 11.757 | 86 | 32 | 28 | 64 |
| 9.967 | 100 | 56 | 48 | 86 | 10.750 | 86 | 40 | 24 | 48 | 11.785 | 72 | 48 | 44 | 56 |
| 9.968 | 100 | 28 | 24 | 86 | 10.800 | 72 | 32 | 48 | 100 | 11.786 | 44 | 28 | 24 | 32 |
| 10.000 | 56 | 28 | 24 | 48 | 10.853 | 56 | 24 | 40 | 86 | 11.825 | 86 | 32 | 24 | 100 |
| 10.033 | 86 | 24 | 28 | 100 | 10.859 | 86 | 44 | 40 | 72 | 11.905 | 100 | 28 | 24 | 72 |
| 10.046 | 72 | 40 | 48 | 86 | 10.909 | 72 | 44 | 32 | 48 | 11.938 | 56 | 24 | 44 | 86 |
| 10.057 | 64 | 28 | 44 | 100 | 10.913 | 100 | 56 | 44 | 72 | 11.944 | 86 | 24 | 24 | 72 |
| 10.078 | 86 | 32 | 24 | 64 | 10.937 | 56 | 32 | 40 | 64 | 11.960 | 72 | 28 | 40 | 86 |
| 10.080 | 72 | 40 | 56 | 100 | 10.945 | 86 | 44 | 56 | 100 | 12.000 | 48 | 24 | 24 | 40 |
| 10.101 | 100 | 44 | 32 | 72 | 10.949 | 86 | 48 | 44 | 100 | 12.031 | 56 | 32 | 44 | 64 |
| 10.159 | 64 | 28 | 32 | 72 | 10.972 | 64 | 28 | 48 | 100 | 12.040 | 86 | 40 | 56 | 100 |
| 10.175 | 100 | 32 | 28 | 86 | 11.000 | 44 | 24 | 24 | 40 | 12.121 | 40 | 24 | 32 | 44 |
| 10.182 | 64 | 40 | 28 | 44 | 11.021 | 72 | 28 | 24 | 56 | 12.153 | 100 | 32 | 28 | 72 |
| 10.186 | 44 | 24 | 40 | 72 | 11.057 | 86 | 56 | 72 | 100 | 12.178 | 72 | 44 | 64 | 86 |
| 10.209 | 56 | 24 | 28 | 64 | 11.111 | 40 | 24 | 32 | 48 | 12.216 | 86 | 44 | 40 | 64 |
| 10.228 | 72 | 44 | 40 | 64 | 11.137 | 56 | 32 | 28 | 44 | 12.222 | 44 | 24 | 32 | 48 |
| 10.233 | 48 | 24 | 44 | 86 | 11.160 | 100 | 56 | 40 | 64 | 12.245 | 48 | 28 | 40 | 56 |
| 10.238 | 86 | 28 | 24 | 72 | 11.163 | 72 | 24 | 32 | 86 | 12.250 | 56 | 32 | 28 | 40 |
| 10.267 | 56 | 24 | 44 | 100 | 11.169 | 86 | 44 | 32 | 56 | 12.272 | 72 | 32 | 24 | 44 |
| 10.286 | 48 | 28 | 24 | 40 | 11.198 | 86 | 48 | 40 | 64 | 12.277 | 100 | 56 | 44 | 64 |
| 10.312 | 48 | 32 | 44 | 64 | 11.200 | 56 | 24 | 48 | 100 | 12.286 | 86 | 28 | 40 | 100 |
| 10.313 | 72 | 48 | 44 | 64 | 11.225 | 44 | 28 | 40 | 56 | 12.318 | 86 | 48 | 44 | 64 |
| 10.320 | 86 | 40 | 48 | 100 | 11.250 | 72 | 24 | 24 | 64 | 12.343 | 72 | 28 | 48 | 100 |
| 10.336 | 100 | 72 | 64 | 86 | 11.313 | 64 | 44 | 56 | 72 | 12.375 | 72 | 40 | 44 | 64 |

## Change Gears for Different Leads — 12.403 Inches to 16.000 Inches

| Lead in Inches | Driven (Gear on Worm) | Driver (First Gear on Stud) | Driven (Second Gear on Stud) | Driver (Gear on Screw) | Lead in Inches | Driven (Gear on Worm) | Driver (First Gear on Stud) | Driven (Second Gear on Stud) | Driver (Gear on Screw) | Lead in Inches | Driven (Gear on Worm) | Driver (First Gear on Stud) | Driven (Second Gear on Stud) | Driver (Gear on Screw) |
|---|---|---|---|---|---|---|---|---|---|---|---|---|---|---|
| 12.403 | 64 | 24 | 40 | 86 | 13.438 | 86 | 24 | 24 | 64 | 14.668 | 44 | 24 | 32 | 40 |
| 12.444 | 64 | 40 | 56 | 72 | 13.469 | 48 | 28 | 44 | 56 | 14.694 | 72 | 28 | 32 | 56 |
| 12.468 | 64 | 28 | 24 | 44 | 13.500 | 72 | 32 | 24 | 40 | 14.743 | 86 | 28 | 48 | 100 |
| 12.500 | 40 | 24 | 24 | 32 | 13.514 | 86 | 28 | 44 | 100 | 14.780 | 86 | 40 | 44 | 64 |
| 12.542 | 86 | 40 | 28 | 48 | 13.566 | 100 | 24 | 28 | 86 | 14.800 | 100 | 44 | 56 | 86 |
| 12.508 | 86 | 44 | 64 | 100 | 13.611 | 56 | 24 | 28 | 48 | 14.815 | 64 | 24 | 40 | 72 |
| 12.558 | 72 | 32 | 48 | 86 | 13.636 | 48 | 32 | 40 | 44 | 14.849 | 56 | 24 | 28 | 44 |
| 12.571 | 64 | 40 | 44 | 56 | 13.643 | 64 | 24 | 44 | 86 | 14.880 | 100 | 48 | 40 | 56 |
| 12.572 | 44 | 28 | 32 | 40 | 13.650 | 86 | 28 | 32 | 72 | 14.884 | 64 | 28 | 56 | 86 |
| 12.600 | 72 | 32 | 56 | 100 | 13.672 | 100 | 32 | 28 | 64 | 14.931 | 86 | 32 | 40 | 72 |
| 12.627 | 100 | 44 | 40 | 72 | 13.682 | 86 | 40 | 28 | 44 | 14.933 | 64 | 24 | 56 | 100 |
| 12.686 | 100 | 44 | 48 | 86 | 13.713 | 64 | 40 | 48 | 56 | 14.950 | 100 | 56 | 72 | 86 |
| 12.698 | 64 | 28 | 40 | 72 | 13.715 | 64 | 28 | 24 | 40 | 15.000 | 48 | 24 | 24 | 32 |
| 12.727 | 64 | 32 | 28 | 44 | 13.750 | 44 | 24 | 24 | 32 | 15.050 | 86 | 32 | 56 | 100 |
| 12.728 | 56 | 24 | 24 | 44 | 13.760 | 86 | 40 | 64 | 100 | 15.150 | 100 | 44 | 32 | 48 |
| 12.732 | 100 | 48 | 44 | 72 | 13.889 | 100 | 24 | 24 | 72 | 15.151 | 100 | 44 | 48 | 72 |
| 12.758 | 64 | 28 | 48 | 86 | 13.933 | 86 | 48 | 56 | 72 | 15.202 | 86 | 44 | 56 | 72 |
| 12.791 | 100 | 40 | 44 | 86 | 13.935 | 86 | 24 | 28 | 72 | 15.238 | 64 | 28 | 48 | 72 |
| 12.798 | 86 | 48 | 40 | 56 | 13.953 | 72 | 24 | 40 | 86 | 15.239 | 64 | 28 | 32 | 48 |
| 12.800 | 64 | 28 | 56 | 100 | 13.960 | 86 | 44 | 40 | 56 | 15.272 | 56 | 40 | 48 | 44 |
| 12.834 | 56 | 40 | 44 | 48 | 13.968 | 64 | 28 | 44 | 72 | 15.278 | 44 | 24 | 40 | 48 |
| 12.857 | 72 | 28 | 32 | 64 | 14.000 | 56 | 24 | 24 | 40 | 15.279 | 100 | 40 | 44 | 72 |
| 12.858 | 48 | 28 | 24 | 32 | 14.025 | 72 | 44 | 48 | 56 | 15.306 | 100 | 28 | 24 | 56 |
| 12.900 | 86 | 32 | 48 | 100 | 14.026 | 72 | 28 | 24 | 44 | 15.349 | 72 | 24 | 44 | 86 |
| 12.963 | 56 | 24 | 40 | 72 | 14.063 | 72 | 32 | 40 | 64 | 15.357 | 86 | 28 | 24 | 48 |
| 12.987 | 100 | 44 | 32 | 56 | 14.071 | 86 | 44 | 72 | 100 | 15.429 | 72 | 40 | 48 | 56 |
| 13.020 | 100 | 48 | 40 | 64 | 14.078 | 86 | 48 | 44 | 56 | 15.469 | 72 | 32 | 44 | 64 |
| 13.024 | 56 | 24 | 48 | 86 | 14.142 | 72 | 40 | 44 | 56 | 15.480 | 86 | 40 | 72 | 100 |
| 13.030 | 86 | 44 | 32 | 48 | 14.204 | 100 | 44 | 40 | 64 | 15.504 | 100 | 48 | 64 | 86 |
| 13.062 | 64 | 28 | 32 | 56 | 14.260 | 56 | 24 | 44 | 72 | 15.556 | 64 | 32 | 56 | 72 |
| 13.082 | 100 | 64 | 72 | 86 | 14.286 | 40 | 24 | 24 | 28 | 15.584 | 48 | 28 | 40 | 44 |
| 13.090 | 72 | 40 | 32 | 44 | 14.318 | 72 | 32 | 28 | 44 | 15.625 | 100 | 24 | 24 | 64 |
| 13.096 | 44 | 28 | 40 | 48 | 14.319 | 72 | 44 | 56 | 64 | 15.636 | 86 | 40 | 32 | 44 |
| 13.125 | 72 | 32 | 28 | 48 | 14.322 | 100 | 48 | 44 | 64 | 15.677 | 86 | 32 | 28 | 48 |
| 13.139 | 86 | 40 | 44 | 72 | 14.333 | 86 | 40 | 32 | 48 | 15.714 | 44 | 24 | 24 | 28 |
| 13.157 | 72 | 28 | 44 | 86 | 14.352 | 72 | 28 | 48 | 86 | 15.750 | 72 | 32 | 28 | 40 |
| 13.163 | 86 | 28 | 24 | 56 | 14.400 | 72 | 24 | 48 | 100 | 15.767 | 86 | 24 | 44 | 100 |
| 13.200 | 72 | 24 | 44 | 100 | 14.536 | 100 | 32 | 40 | 86 | 15.873 | 100 | 56 | 64 | 72 |
| 13.258 | 100 | 44 | 28 | 48 | 14.545 | 64 | 24 | 24 | 44 | 15.874 | 100 | 28 | 32 | 72 |
| 13.289 | 100 | 28 | 32 | 86 | 14.583 | 56 | 32 | 40 | 48 | 15.909 | 100 | 40 | 28 | 44 |
| 13.333 | 64 | 24 | 24 | 48 | 14.584 | 40 | 24 | 28 | 32 | 15.925 | 86 | 48 | 64 | 72 |
| 13.393 | 100 | 56 | 48 | 64 | 14.651 | 72 | 32 | 56 | 86 | 15.926 | 86 | 24 | 32 | 72 |
| 13.396 | 72 | 40 | 64 | 86 | 14.659 | 86 | 44 | 48 | 64 | 15.989 | 100 | 32 | 44 | 86 |
| 13.437 | 86 | 32 | 28 | 56 | 14.667 | 64 | 40 | 44 | 48 | 16.000 | 64 | 24 | 24 | 40 |

## Change Gears for Different Leads — 16.042 Inches to 21.39 Inches

| Lead in Inches | Driven (Gear on Worm) | Driver (First Gear on Stud) | Driven (Second Gear on Stud) | Driver (Gear on Screw) | Lead in Inches | Driven (Gear on Worm) | Driver (First Gear on Stud) | Driven (Second Gear on Stud) | Driver (Gear on Screw) | Lead in Inches | Driven (Gear on Worm) | Driver (First Gear on Stud) | Driven (Second Gear on Stud) | Driver (Gear on Screw) |
|---|---|---|---|---|---|---|---|---|---|---|---|---|---|---|
| 16.042 | 56 | 24 | 44 | 64 | 17.442 | 100 | 32 | 48 | 86 | 19.350 | 86 | 32 | 72 | 100 |
| 16.043 | 44 | 24 | 28 | 32 | 17.454 | 64 | 40 | 48 | 44 | 19.380 | 100 | 24 | 40 | 86 |
| 16.071 | 72 | 32 | 40 | 56 | 17.500 | 56 | 24 | 24 | 32 | 19.394 | 64 | 24 | 32 | 44 |
| 16.125 | 86 | 32 | 24 | 40 | 17.550 | 86 | 28 | 32 | 56 | 19.444 | 40 | 24 | 28 | 24 |
| 16.204 | 100 | 24 | 28 | 72 | 17.677 | 100 | 44 | 56 | 72 | 19.480 | 100 | 28 | 24 | 44 |
| 16.233 | 100 | 44 | 40 | 56 | 17.679 | 72 | 32 | 44 | 56 | 19.531 | 100 | 32 | 40 | 64 |
| 16.280 | 100 | 40 | 56 | 86 | 17.778 | 64 | 24 | 32 | 48 | 19.535 | 72 | 24 | 56 | 86 |
| 16.288 | 86 | 44 | 40 | 48 | 17.858 | 100 | 24 | 24 | 56 | 19.545 | 86 | 24 | 24 | 44 |
| 16.296 | 64 | 24 | 44 | 72 | 17.917 | 86 | 24 | 32 | 64 | 19.590 | 64 | 28 | 48 | 56 |
| 16.327 | 64 | 28 | 40 | 56 | 17.918 | 86 | 24 | 48 | 44 | 19.635 | 72 | 40 | 48 | 44 |
| 16.333 | 56 | 24 | 28 | 40 | 17.959 | 64 | 28 | 44 | 56 | 19.642 | 100 | 40 | 44 | 56 |
| 16.364 | 72 | 24 | 24 | 44 | 18.000 | 72 | 24 | 24 | 40 | 19.643 | 44 | 28 | 40 | 32 |
| 16.370 | 100 | 48 | 44 | 56 | 18.181 | 56 | 28 | 40 | 44 | 19.656 | 86 | 28 | 64 | 100 |
| 16.423 | 86 | 32 | 44 | 72 | 18.182 | 48 | 24 | 40 | 44 | 19.687 | 72 | 32 | 56 | 64 |
| 16.456 | 72 | 28 | 64 | 100 | 18.229 | 100 | 32 | 28 | 48 | 19.710 | 86 | 40 | 44 | 48 |
| 16.500 | 72 | 40 | 44 | 48 | 18.273 | 100 | 28 | 44 | 86 | 19.840 | 100 | 28 | 40 | 72 |
| 16.612 | 100 | 28 | 40 | 86 | 18.285 | 64 | 28 | 32 | 40 | 19.886 | 100 | 44 | 56 | 64 |
| 16.623 | 64 | 28 | 32 | 44 | 18.333 | 56 | 28 | 44 | 48 | 19.887 | 100 | 32 | 28 | 44 |
| 16.667 | 56 | 28 | 40 | 48 | 18.367 | 72 | 28 | 40 | 56 | 19.908 | 86 | 24 | 40 | 72 |
| 16.722 | 86 | 40 | 56 | 72 | 18.428 | 86 | 28 | 24 | 40 | 19.934 | 100 | 28 | 48 | 86 |
| 16.744 | 72 | 24 | 48 | 86 | 18.476 | 86 | 32 | 44 | 64 | 20.00 | 72 | 24 | 32 | 48 |
| 16.752 | 86 | 44 | 48 | 56 | 18.519 | 100 | 24 | 32 | 72 | 20.07 | 86 | 24 | 56 | 100 |
| 16.753 | 86 | 28 | 24 | 44 | 18.605 | 100 | 40 | 64 | 86 | 20.09 | 100 | 56 | 72 | 64 |
| 16.797 | 86 | 32 | 40 | 64 | 18.663 | 100 | 64 | 86 | 72 | 20.16 | 86 | 48 | 72 | 64 |
| 16.800 | 72 | 24 | 56 | 100 | 18.667 | 64 | 24 | 28 | 40 | 20.20 | 100 | 44 | 64 | 72 |
| 16.875 | 72 | 32 | 48 | 64 | 18.700 | 72 | 44 | 64 | 56 | 20.35 | 100 | 32 | 56 | 86 |
| 16.892 | 86 | 40 | 44 | 56 | 18.750 | 100 | 32 | 24 | 40 | 20.36 | 64 | 40 | 56 | 44 |
| 16.914 | 100 | 44 | 64 | 86 | 18.750 | 72 | 32 | 40 | 48 | 20.41 | 100 | 28 | 32 | 56 |
| 16.969 | 64 | 44 | 56 | 48 | 18.770 | 86 | 28 | 44 | 72 | 20.42 | 56 | 24 | 28 | 32 |
| 16.970 | 64 | 24 | 28 | 44 | 18.812 | 86 | 32 | 28 | 40 | 20.45 | 72 | 32 | 40 | 44 |
| 17.045 | 100 | 32 | 24 | 44 | 18.858 | 48 | 28 | 44 | 40 | 20.48 | 86 | 48 | 64 | 56 |
| 17.046 | 100 | 44 | 48 | 64 | 18.939 | 100 | 44 | 40 | 48 | 20.57 | 72 | 40 | 64 | 56 |
| 17.062 | 86 | 28 | 40 | 72 | 19.029 | 100 | 44 | 72 | 86 | 20.63 | 72 | 32 | 44 | 48 |
| 17.101 | 86 | 44 | 56 | 64 | 19.048 | 86 | 24 | 32 | 28 | 20.74 | 86 | 24 | 56 | 72 |
| 17.102 | 86 | 32 | 28 | 44 | 19.090 | 56 | 32 | 48 | 44 | 20.78 | 64 | 28 | 40 | 44 |
| 17.141 | 64 | 32 | 48 | 56 | 19.091 | 72 | 24 | 28 | 44 | 20.83 | 100 | 32 | 48 | 72 |
| 17.143 | 64 | 28 | 24 | 32 | 19.096 | 100 | 32 | 44 | 72 | 20.90 | 86 | 32 | 56 | 72 |
| 17.144 | 48 | 24 | 24 | 28 | 19.111 | 86 | 40 | 64 | 72 | 20.93 | 100 | 40 | 72 | 86 |
| 17.188 | 100 | 40 | 44 | 64 | 19.136 | 72 | 28 | 64 | 86 | 20.95 | 64 | 28 | 44 | 48 |
| 17.200 | 86 | 32 | 64 | 100 | 19.197 | 86 | 32 | 40 | 56 | 21.00 | 56 | 32 | 48 | 40 |
| 17.275 | 86 | 56 | 72 | 64 | 19.200 | 72 | 24 | 64 | 100 | 21.12 | 86 | 32 | 44 | 56 |
| 17.361 | 100 | 32 | 40 | 72 | 19.250 | 56 | 32 | 44 | 40 | 21.32 | 100 | 24 | 44 | 86 |
| 17.364 | 64 | 24 | 56 | 86 | 19.285 | 72 | 32 | 48 | 56 | 21.33 | 100 | 56 | 86 | 72 |
| 17.373 | 86 | 44 | 64 | 72 | 19.286 | 72 | 28 | 24 | 32 | 21.39 | 44 | 24 | 28 | 24 |

## Change Gears for Different Leads — 21.43 Inches to 32.09 Inches

| Lead in Inches | Driven — Gear on Worm | Driver — First Gear on Stud | Driven — Second Gear on Stud | Driver — Gear on Screw | Lead in Inches | Driven — Gear on Worm | Driver — First Gear on Stud | Driven — Second Gear on Stud | Driver — Gear on Screw | Lead in Inches | Driven — Gear on Worm | Driver — First Gear on Stud | Driven — Second Gear on Stud | Driver — Gear on Screw |
|---|---|---|---|---|---|---|---|---|---|---|---|---|---|---|
| 21.43 | 100 | 40 | 48 | 56 | 24.88 | 100 | 72 | 86 | 48 | 28.05 | 72 | 28 | 48 | 44 |
| 21.48 | 100 | 32 | 44 | 64 | 24.93 | 64 | 28 | 48 | 44 | 28.06 | 100 | 28 | 44 | 56 |
| 21.50 | 86 | 24 | 24 | 40 | 25.00 | 72 | 24 | 40 | 48 | 28.13 | 100 | 40 | 72 | 64 |
| 21.82 | 72 | 44 | 64 | 48 | 25.08 | 86 | 24 | 28 | 40 | 28.15 | 86 | 28 | 44 | 48 |
| 21.88 | 100 | 40 | 56 | 64 | 25.09 | 86 | 40 | 56 | 48 | 28.29 | 72 | 28 | 44 | 40 |
| 21.90 | 86 | 24 | 44 | 72 | 25.13 | 86 | 44 | 72 | 56 | 28.41 | 100 | 32 | 40 | 44 |
| 21.94 | 86 | 28 | 40 | 56 | 25.14 | 86 | 28 | 44 | 40 | 28.57 | 100 | 56 | 64 | 40 |
| 21.99 | 86 | 44 | 72 | 64 | 25.45 | 64 | 44 | 56 | 32 | 28.64 | 72 | 44 | 56 | 32 |
| 22.00 | 64 | 32 | 44 | 40 | 25.46 | 100 | 24 | 44 | 72 | 28.65 | 100 | 32 | 44 | 48 |
| 22.04 | 72 | 28 | 48 | 56 | 25.51 | 100 | 28 | 40 | 56 | 28.67 | 86 | 40 | 64 | 48 |
| 22.11 | 86 | 28 | 72 | 100 | 25.57 | 100 | 64 | 72 | 44 | 29.09 | 64 | 24 | 48 | 44 |
| 22.22 | 100 | 40 | 64 | 72 | 25.60 | 86 | 28 | 40 | 48 | 29.17 | 100 | 40 | 56 | 48 |
| 22.34 | 86 | 44 | 64 | 56 | 25.67 | 56 | 24 | 44 | 40 | 29.22 | 100 | 56 | 72 | 44 |
| 22.40 | 86 | 32 | 40 | 48 | 25.71 | 72 | 24 | 48 | 56 | 29.32 | 86 | 48 | 72 | 44 |
| 22.50 | 72 | 24 | 48 | 64 | 25.72 | 72 | 24 | 24 | 28 | 29.34 | 64 | 24 | 44 | 40 |
| 22.73 | 100 | 24 | 24 | 44 | 25.80 | 86 | 24 | 72 | 100 | 29.39 | 72 | 28 | 64 | 56 |
| 22.80 | 86 | 48 | 56 | 44 | 25.97 | 100 | 44 | 64 | 56 | 29.56 | 86 | 32 | 44 | 40 |
| 22.86 | 64 | 24 | 24 | 28 | 26.04 | 100 | 32 | 40 | 48 | 29.76 | 100 | 28 | 40 | 48 |
| 22.91 | 72 | 44 | 56 | 40 | 26.06 | 86 | 44 | 64 | 48 | 29.86 | 100 | 40 | 86 | 72 |
| 22.92 | 100 | 40 | 44 | 48 | 26.16 | 100 | 32 | 72 | 86 | 29.90 | 100 | 28 | 72 | 86 |
| 22.93 | 86 | 24 | 64 | 100 | 26.18 | 72 | 40 | 64 | 44 | 30.00 | 56 | 28 | 48 | 32 |
| 23.04 | 86 | 56 | 72 | 48 | 26.19 | 44 | 24 | 40 | 28 | 30.23 | 86 | 32 | 72 | 64 |
| 23.14 | 100 | 24 | 40 | 72 | 26.25 | 72 | 32 | 56 | 48 | 30.30 | 100 | 48 | 64 | 44 |
| 23.26 | 100 | 32 | 64 | 86 | 26.33 | 86 | 28 | 48 | 56 | 30.48 | 64 | 24 | 32 | 28 |
| 23.33 | 64 | 32 | 56 | 48 | 26.52 | 100 | 44 | 56 | 48 | 30.54 | 100 | 44 | 86 | 64 |
| 23.38 | 72 | 28 | 40 | 44 | 26.58 | 100 | 28 | 64 | 86 | 30.56 | 44 | 24 | 40 | 24 |
| 23.44 | 100 | 48 | 72 | 64 | 26.67 | 64 | 28 | 56 | 48 | 30.61 | 100 | 28 | 48 | 56 |
| 23.45 | 86 | 40 | 48 | 44 | 26.79 | 100 | 48 | 72 | 56 | 30.71 | 86 | 24 | 48 | 56 |
| 23.52 | 86 | 32 | 56 | 64 | 26.88 | 86 | 28 | 56 | 64 | 30.72 | 86 | 24 | 24 | 28 |
| 23.57 | 72 | 28 | 44 | 48 | 27.00 | 72 | 32 | 48 | 40 | 30.86 | 72 | 28 | 48 | 40 |
| 23.81 | 100 | 48 | 64 | 56 | 27.13 | 100 | 24 | 56 | 86 | 31.01 | 100 | 24 | 64 | 86 |
| 23.89 | 86 | 32 | 64 | 72 | 27.15 | 100 | 44 | 86 | 72 | 31.11 | 64 | 24 | 56 | 48 |
| 24.00 | 64 | 40 | 72 | 48 | 27.22 | 56 | 24 | 28 | 24 | 31.25 | 100 | 28 | 56 | 64 |
| 24.13 | 86 | 28 | 44 | 56 | 27.27 | 100 | 40 | 48 | 44 | 31.27 | 86 | 40 | 64 | 44 |
| 24.19 | 86 | 40 | 72 | 64 | 27.30 | 86 | 28 | 64 | 72 | 31.35 | 86 | 32 | 56 | 48 |
| 24.24 | 64 | 24 | 40 | 44 | 27.34 | 100 | 32 | 56 | 64 | 31.36 | 86 | 24 | 28 | 32 |
| 24.31 | 100 | 32 | 56 | 72 | 27.36 | 86 | 40 | 56 | 44 | 31.43 | 64 | 28 | 44 | 32 |
| 24.43 | 86 | 32 | 40 | 44 | 27.43 | 64 | 28 | 48 | 40 | 31.50 | 72 | 32 | 56 | 40 |
| 24.44 | 44 | 24 | 32 | 24 | 27.50 | 56 | 32 | 44 | 28 | 31.75 | 100 | 72 | 64 | 28 |
| 24.54 | 72 | 32 | 48 | 44 | 27.64 | 86 | 40 | 72 | 56 | 31.82 | 100 | 44 | 56 | 40 |
| 24.55 | 100 | 32 | 44 | 56 | 27.78 | 100 | 32 | 64 | 72 | 31.85 | 86 | 24 | 64 | 72 |
| 24.57 | 86 | 40 | 64 | 56 | 27.87 | 86 | 24 | 56 | 72 | 31.99 | 100 | 56 | 86 | 48 |
| 24.64 | 86 | 24 | 44 | 64 | 27.92 | 86 | 28 | 40 | 44 | 32.00 | 64 | 28 | 56 | 40 |
| 24.75 | 72 | 32 | 44 | 40 | 28.00 | 100 | 64 | 86 | 48 | 32.09 | 56 | 24 | 44 | 32 |

# HELICAL MILLING

## Change Gears for Different Leads — 32.14 Inches to 60.00 Inches

| Lead in Inches | Driven / Gear on Worm | Driver / First Gear on Stud | Driven / Second Gear on Stud | Driver / Gear on Screw | Lead in Inches | Driven / Gear on Worm | Driver / First Gear on Stud | Driven / Second Gear on Stud | Driver / Gear on Screw | Lead in Inches | Driven / Gear on Worm | Driver / First Gear on Stud | Driven / Second Gear on Stud | Driver / Gear on Screw |
|---|---|---|---|---|---|---|---|---|---|---|---|---|---|---|
| 32.14 | 100 | 56 | 72 | 40 | 38.20 | 100 | 24 | 44 | 48 | 46.07 | 86 | 28 | 72 | 48 |
| 32.25 | 86 | 48 | 72 | 40 | 38.39 | 100 | 40 | 86 | 56 | 46.67 | 64 | 24 | 56 | 32 |
| 32.41 | 100 | 24 | 56 | 72 | 38.57 | 72 | 28 | 48 | 32 | 46.88 | 100 | 32 | 72 | 48 |
| 32.47 | 100 | 28 | 40 | 44 | 38.89 | 56 | 24 | 40 | 24 | 47.15 | 72 | 24 | 44 | 28 |
| 32.58 | 86 | 24 | 40 | 44 | 38.96 | 100 | 28 | 44 | 48 | 47.62 | 100 | 28 | 64 | 48 |
| 32.73 | 72 | 32 | 64 | 44 | 39.09 | 86 | 32 | 64 | 44 | 47.78 | 86 | 24 | 64 | 48 |
| 32.74 | 100 | 28 | 44 | 48 | 39.29 | 100 | 28 | 44 | 40 | 47.99 | 100 | 32 | 86 | 56 |
| 32.85 | 86 | 24 | 44 | 48 | 39.42 | 86 | 24 | 44 | 40 | 48.00 | 72 | 24 | 64 | 40 |
| 33.00 | 72 | 24 | 44 | 40 | 39.49 | 86 | 28 | 72 | 56 | 48.38 | 86 | 32 | 72 | 40 |
| 33.33 | 100 | 24 | 32 | 40 | 39.77 | 100 | 32 | 56 | 44 | 48.61 | 100 | 24 | 56 | 48 |
| 33.51 | 86 | 28 | 48 | 44 | 40.00 | 72 | 24 | 64 | 48 | 48.86 | 100 | 40 | 86 | 44 |
| 33.59 | 100 | 64 | 86 | 40 | 40.18 | 100 | 32 | 72 | 56 | 48.89 | 64 | 24 | 44 | 24 |
| 33.79 | 86 | 28 | 44 | 40 | 40.31 | 86 | 32 | 72 | 48 | 49.11 | 100 | 28 | 44 | 32 |
| 33.94 | 64 | 24 | 56 | 44 | 40.72 | 100 | 44 | 86 | 48 | 49.14 | 86 | 28 | 64 | 40 |
| 34.09 | 100 | 48 | 72 | 44 | 40.82 | 100 | 28 | 64 | 56 | 49.27 | 86 | 24 | 44 | 32 |
| 34.20 | 86 | 44 | 56 | 32 | 40.91 | 100 | 40 | 72 | 44 | 49.77 | 100 | 24 | 86 | 72 |
| 34.29 | 72 | 48 | 64 | 28 | 40.95 | 86 | 28 | 64 | 48 | 50.00 | 100 | 28 | 56 | 40 |
| 34.38 | 100 | 32 | 44 | 40 | 40.96 | 86 | 24 | 32 | 28 | 50.17 | 86 | 24 | 56 | 40 |
| 34.55 | 86 | 32 | 72 | 56 | 41.14 | 72 | 28 | 64 | 40 | 50.26 | 86 | 28 | 72 | 44 |
| 34.72 | 100 | 24 | 40 | 48 | 41.25 | 72 | 24 | 44 | 32 | 51.14 | 100 | 32 | 72 | 44 |
| 34.88 | 100 | 24 | 72 | 86 | 41.67 | 100 | 32 | 64 | 48 | 51.19 | 86 | 24 | 40 | 28 |
| 34.90 | 100 | 56 | 86 | 44 | 41.81 | 86 | 24 | 56 | 48 | 51.43 | 72 | 28 | 64 | 32 |
| 35.00 | 72 | 24 | 56 | 48 | 41.91 | 64 | 24 | 44 | 28 | 51.95 | 100 | 28 | 64 | 44 |
| 35.10 | 86 | 28 | 64 | 56 | 41.99 | 100 | 32 | 86 | 64 | 52.12 | 86 | 24 | 64 | 44 |
| 35.16 | 100 | 32 | 72 | 64 | 42.00 | 72 | 24 | 56 | 40 | 52.50 | 72 | 24 | 56 | 32 |
| 35.18 | 86 | 44 | 72 | 40 | 42.23 | 86 | 28 | 44 | 32 | 53.03 | 100 | 24 | 56 | 44 |
| 35.36 | 72 | 32 | 44 | 28 | 42.66 | 100 | 28 | 86 | 72 | 53.33 | 64 | 24 | 56 | 28 |
| 35.56 | 64 | 24 | 32 | 24 | 42.78 | 56 | 24 | 44 | 24 | 53.57 | 100 | 28 | 72 | 48 |
| 35.71 | 100 | 32 | 64 | 56 | 42.86 | 100 | 28 | 48 | 40 | 53.75 | 86 | 24 | 48 | 32 |
| 35.72 | 100 | 24 | 24 | 28 | 43.00 | 86 | 32 | 64 | 40 | 54.85 | 100 | 28 | 86 | 56 |
| 35.83 | 86 | 32 | 64 | 48 | 43.64 | 72 | 24 | 64 | 44 | 55.00 | 72 | 24 | 44 | 24 |
| 36.00 | 72 | 32 | 64 | 40 | 43.75 | 100 | 32 | 56 | 40 | 55.28 | 86 | 28 | 72 | 40 |
| 36.36 | 100 | 44 | 64 | 40 | 43.98 | 86 | 32 | 72 | 44 | 55.56 | 100 | 24 | 32 | 24 |
| 36.46 | 100 | 48 | 56 | 32 | 44.44 | 64 | 24 | 40 | 24 | 55.99 | 100 | 24 | 86 | 64 |
| 36.67 | 48 | 24 | 44 | 24 | 44.64 | 100 | 28 | 40 | 32 | 56.25 | 100 | 32 | 72 | 40 |
| 36.86 | 86 | 28 | 48 | 40 | 44.68 | 86 | 28 | 64 | 44 | 56.31 | 86 | 24 | 44 | 28 |
| 37.04 | 100 | 24 | 64 | 72 | 44.79 | 100 | 40 | 86 | 48 | 57.14 | 100 | 28 | 64 | 40 |
| 37.33 | 100 | 32 | 86 | 72 | 45.00 | 72 | 28 | 56 | 32 | 57.30 | 100 | 24 | 44 | 32 |
| 37.40 | 72 | 28 | 64 | 44 | 45.45 | 100 | 32 | 64 | 44 | 57.33 | 86 | 24 | 64 | 40 |
| 37.50 | 100 | 48 | 72 | 40 | 45.46 | 100 | 28 | 56 | 44 | 58.33 | 100 | 24 | 56 | 40 |
| 37.63 | 86 | 32 | 56 | 40 | 45.61 | 86 | 24 | 56 | 44 | 58.44 | 100 | 28 | 72 | 44 |
| 37.88 | 100 | 24 | 40 | 44 | 45.72 | 64 | 24 | 48 | 28 | 58.64 | 86 | 24 | 72 | 44 |
| 38.10 | 64 | 24 | 40 | 28 | 45.84 | 100 | 24 | 44 | 40 | 59.53 | 100 | 24 | 40 | 28 |
| 38.18 | 72 | 24 | 56 | 44 | 45.92 | 100 | 28 | 72 | 56 | 60.00 | 72 | 24 | 64 | 32 |

## Lead of Helix for Given Helix Angle Relative to Axis, When Diameter = 1.

| Deg. | o' | 6' | 12' | 18' | 24' | 30' | 36' | 42' | 48' | 54' | 60' |
|---|---|---|---|---|---|---|---|---|---|---|---|
| 0 | Infin. | 1800.001 | 899.997 | 599.994 | 449.993 | 359.992 | 299.990 | 257.130 | 224.986 | 199.983 | 179.982 |
| 1 | 179.982 | 163.616 | 149.978 | 138.438 | 128.545 | 119.973 | 112.471 | 105.851 | 99.967 | 94.702 | 89.964 |
| 2 | 89.964 | 85.676 | 81.778 | 78.219 | 74.956 | 71.954 | 69.183 | 66.617 | 64.235 | 62.016 | 59.945 |
| 3 | 59.945 | 58.008 | 56.191 | 54.485 | 52.879 | 51.365 | 49.934 | 48.581 | 47.299 | 46.082 | 44.927 |
| 4 | 44.927 | 43.827 | 42.780 | 41.782 | 40.830 | 39.918 | 39.046 | 38.212 | 37.412 | 36.645 | 35.909 |
| 5 | 35.909 | 35.201 | 34.520 | 33.866 | 33.235 | 32.627 | 32.040 | 31.475 | 30.928 | 30.400 | 29.890 |
| 6 | 29.890 | 29.397 | 28.919 | 28.456 | 28.008 | 27.573 | 27.152 | 26.743 | 26.346 | 25.961 | 25.586 |
| 7 | 25.586 | 25.222 | 24.868 | 24.524 | 24.189 | 23.863 | 23.545 | 23.236 | 22.934 | 22.640 | 22.354 |
| 8 | 22.354 | 22.074 | 21.801 | 21.535 | 21.275 | 21.021 | 20.773 | 20.530 | 20.293 | 20.062 | 19.835 |
| 9 | 19.835 | 19.614 | 19.397 | 19.185 | 18.977 | 18.773 | 18.574 | 18.379 | 18.188 | 18.000 | 17.817 |
| 10 | 17.817 | 17.637 | 17.460 | 17.287 | 17.117 | 16.950 | 16.787 | 16.626 | 16.469 | 16.314 | 16.162 |
| 11 | 16.162 | 16.013 | 15.866 | 15.722 | 15.581 | 15.441 | 15.305 | 15.170 | 15.038 | 14.908 | 14.780 |
| 12 | 14.780 | 14.654 | 14.530 | 14.409 | 14.289 | 14.171 | 14.055 | 13.940 | 13.828 | 13.717 | 13.608 |
| 13 | 13.608 | 13.500 | 13.394 | 13.290 | 13.187 | 13.086 | 12.986 | 12.887 | 12.790 | 12.695 | 12.600 |
| 14 | 12.600 | 12.507 | 12.415 | 12.325 | 12.237 | 12.148 | 12.061 | 11.975 | 11.890 | 11.807 | 11.725 |
| 15 | 11.725 | 11.643 | 11.563 | 11.484 | 11.405 | 11.328 | 11.252 | 11.177 | 11.102 | 11.029 | 10.956 |
| 16 | 10.956 | 10.884 | 10.813 | 10.743 | 10.674 | 10.606 | 10.538 | 10.471 | 10.405 | 10.340 | 10.276 |
| 17 | 10.276 | 10.212 | 10.149 | 10.086 | 10.025 | 9.964 | 9.904 | 9.844 | 9.785 | 9.727 | 9.669 |
| 18 | 9.669 | 9.612 | 9.555 | 9.499 | 9.444 | 9.389 | 9.335 | 9.281 | 9.228 | 9.176 | 9.124 |
| 19 | 9.124 | 9.072 | 9.021 | 8.971 | 8.921 | 8.872 | 8.823 | 8.774 | 8.726 | 8.679 | 8.631 |
| 20 | 8.631 | 8.585 | 8.539 | 8.493 | 8.447 | 8.403 | 8.358 | 8.314 | 8.270 | 8.227 | 8.184 |
| 21 | 8.184 | 8.142 | 8.099 | 8.058 | 8.016 | 7.975 | 7.935 | 7.894 | 7.855 | 7.815 | 7.776 |
| 22 | 7.776 | 7.737 | 7.698 | 7.660 | 7.622 | 7.584 | 7.547 | 7.510 | 7.474 | 7.437 | 7.401 |
| 23 | 7.401 | 7.365 | 7.330 | 7.295 | 7.260 | 7.225 | 7.191 | 7.157 | 7.123 | 7.089 | 7.056 |
| 24 | 7.056 | 7.023 | 6.990 | 6.958 | 6.926 | 6.894 | 6.862 | 6.830 | 6.799 | 6.768 | 6.737 |
| 25 | 6.737 | 6.707 | 6.676 | 6.646 | 6.617 | 6.586 | 6.557 | 6.528 | 6.499 | 6.470 | 6.441 |
| 26 | 6.441 | 6.413 | 6.385 | 6.357 | 6.329 | 6.300 | 6.274 | 6.246 | 6.219 | 6.192 | 6.166 |
| 27 | 6.166 | 6.139 | 6.113 | 6.087 | 6.061 | 6.035 | 6.009 | 5.984 | 5.959 | 5.933 | 5.908 |
| 28 | 5.908 | 5.884 | 5.859 | 5.835 | 5.810 | 5.786 | 5.762 | 5.738 | 5.715 | 5.691 | 5.668 |
| 29 | 5.668 | 5.644 | 5.621 | 5.598 | 5.575 | 5.553 | 5.530 | 5.508 | 5.486 | 5.463 | 5.441 |

## Lead of Helix for Given Helix Angle Relative to Axis, When Diameter = 1

| Deg. | 0' | 6' | 12' | 18' | 24' | 30' | 36' | 42' | 48' | 54' | 60' |
|------|------|------|------|------|------|------|------|------|------|------|------|
| 30 | 5.441 | 5.420 | 5.398 | 5.376 | 5.355 | 5.333 | 5.312 | 5.291 | 5.270 | 5.249 | 5.228 |
| 31 | 5.228 | 5.208 | 5.187 | 5.167 | 5.147 | 5.127 | 5.107 | 5.087 | 5.067 | 5.047 | 5.028 |
| 32 | 5.028 | 5.008 | 4.989 | 4.969 | 4.950 | 4.931 | 4.912 | 4.894 | 4.875 | 4.856 | 4.838 |
| 33 | 4.838 | 4.819 | 4.801 | 4.783 | 4.764 | 4.746 | 4.728 | 4.711 | 4.693 | 4.675 | 4.658 |
| 34 | 4.658 | 4.640 | 4.623 | 4.605 | 4.588 | 4.571 | 4.554 | 4.537 | 4.520 | 4.503 | 4.487 |
| 35 | 4.487 | 4.470 | 4.453 | 4.437 | 4.421 | 4.404 | 4.388 | 4.372 | 4.356 | 4.340 | 4.324 |
| 36 | 4.324 | 4.308 | 4.292 | 4.277 | 4.261 | 4.246 | 4.230 | 4.215 | 4.199 | 4.184 | 4.169 |
| 37 | 4.169 | 4.154 | 4.139 | 4.124 | 4.109 | 4.094 | 4.079 | 4.065 | 4.050 | 4.036 | 4.021 |
| 38 | 4.021 | 4.007 | 3.992 | 3.978 | 3.964 | 3.950 | 3.935 | 3.921 | 3.907 | 3.893 | 3.880 |
| 39 | 3.880 | 3.866 | 3.852 | 3.838 | 3.825 | 3.811 | 3.798 | 3.784 | 3.771 | 3.757 | 3.744 |
| 40 | 3.744 | 3.731 | 3.718 | 3.704 | 3.691 | 3.678 | 3.665 | 3.652 | 3.640 | 3.627 | 3.614 |
| 41 | 3.614 | 3.601 | 3.589 | 3.576 | 3.563 | 3.551 | 3.538 | 3.526 | 3.514 | 3.501 | 3.489 |
| 42 | 3.489 | 3.477 | 3.465 | 3.453 | 3.440 | 3.428 | 3.416 | 3.405 | 3.393 | 3.381 | 3.369 |
| 43 | 3.369 | 3.358 | 3.346 | 3.334 | 3.322 | 3.311 | 3.299 | 3.287 | 3.276 | 3.265 | 3.253 |
| 44 | 3.253 | 3.242 | 3.231 | 3.219 | 3.208 | 3.197 | 3.186 | 3.175 | 3.164 | 3.153 | 3.142 |
| 45 | 3.142 | 3.131 | 3.120 | 3.109 | 3.098 | 3.087 | 3.076 | 3.066 | 3.055 | 3.044 | 3.034 |
| 46 | 3.034 | 3.023 | 3.013 | 3.002 | 2.992 | 2.981 | 2.971 | 2.960 | 2.950 | 2.940 | 2.930 |
| 47 | 2.930 | 2.919 | 2.909 | 2.899 | 2.889 | 2.879 | 2.869 | 2.859 | 2.849 | 2.839 | 2.829 |
| 48 | 2.829 | 2.819 | 2.809 | 2.799 | 2.789 | 2.779 | 2.770 | 2.760 | 2.750 | 2.741 | 2.731 |
| 49 | 2.731 | 2.721 | 2.712 | 2.702 | 2.693 | 2.683 | 2.674 | 2.664 | 2.655 | 2.645 | 2.636 |
| 50 | 2.636 | 2.627 | 2.617 | 2.608 | 2.599 | 2.590 | 2.581 | 2.571 | 2.562 | 2.553 | 2.544 |
| 51 | 2.544 | 2.535 | 2.526 | 2.517 | 2.508 | 2.499 | 2.490 | 2.481 | 2.472 | 2.463 | 2.454 |
| 52 | 2.454 | 2.446 | 2.437 | 2.428 | 2.419 | 2.411 | 2.402 | 2.393 | 2.385 | 2.376 | 2.367 |
| 53 | 2.367 | 2.359 | 2.350 | 2.342 | 2.333 | 2.325 | 2.316 | 2.308 | 2.299 | 2.291 | 2.282 |
| 54 | 2.282 | 2.274 | 2.266 | 2.257 | 2.249 | 2.241 | 2.233 | 2.224 | 2.216 | 2.208 | 2.200 |
| 55 | 2.200 | 2.192 | 2.183 | 2.175 | 2.167 | 2.159 | 2.151 | 2.143 | 2.135 | 2.127 | 2.119 |
| 56 | 2.119 | 2.111 | 2.103 | 2.095 | 2.087 | 2.079 | 2.072 | 2.064 | 2.056 | 2.048 | 2.040 |
| 57 | 2.040 | 2.032 | 2.025 | 2.017 | 2.009 | 2.001 | 1.994 | 1.986 | 1.978 | 1.971 | 1.963 |
| 58 | 1.963 | 1.955 | 1.948 | 1.940 | 1.933 | 1.925 | 1.918 | 1.910 | 1.903 | 1.895 | 1.888 |
| 59 | 1.888 | 1.880 | 1.873 | 1.865 | 1.858 | 1.851 | 1.843 | 1.836 | 1.828 | 1.821 | 1.814 |

## Lead of Helix for Given Helix Angle Relative to Axis, When Diameter = 1

| Deg. | 0′ | 6′ | 12′ | 18′ | 24′ | 30′ | 36′ | 42′ | 48′ | 54′ | 60′ |
|---|---|---|---|---|---|---|---|---|---|---|---|
| 60 | 1.814 | 1.806 | 1.799 | 1.792 | 1.785 | 1.777 | 1.770 | 1.763 | 1.756 | 1.749 | 1.741 |
| 61 | 1.741 | 1.734 | 1.727 | 1.720 | 1.713 | 1.706 | 1.699 | 1.692 | 1.685 | 1.677 | 1.670 |
| 62 | 1.670 | 1.663 | 1.656 | 1.649 | 1.642 | 1.635 | 1.628 | 1.621 | 1.615 | 1.608 | 1.601 |
| 63 | 1.601 | 1.594 | 1.587 | 1.580 | 1.573 | 1.566 | 1.559 | 1.553 | 1.546 | 1.539 | 1.532 |
| 64 | 1.532 | 1.525 | 1.519 | 1.512 | 1.505 | 1.498 | 1.492 | 1.485 | 1.478 | 1.472 | 1.465 |
| 65 | 1.465 | 1.458 | 1.452 | 1.445 | 1.438 | 1.432 | 1.425 | 1.418 | 1.412 | 1.405 | 1.399 |
| 66 | 1.399 | 1.392 | 1.386 | 1.379 | 1.372 | 1.366 | 1.359 | 1.353 | 1.346 | 1.340 | 1.334 |
| 67 | 1.334 | 1.327 | 1.321 | 1.314 | 1.308 | 1.301 | 1.295 | 1.288 | 1.282 | 1.276 | 1.269 |
| 68 | 1.269 | 1.263 | 1.257 | 1.250 | 1.244 | 1.237 | 1.231 | 1.225 | 1.219 | 1.212 | 1.206 |
| 69 | 1.206 | 1.200 | 1.193 | 1.187 | 1.181 | 1.175 | 1.168 | 1.162 | 1.156 | 1.150 | 1.143 |
| 70 | 1.143 | 1.137 | 1.131 | 1.125 | 1.119 | 1.112 | 1.106 | 1.100 | 1.094 | 1.088 | 1.082 |
| 71 | 1.082 | 1.076 | 1.069 | 1.063 | 1.057 | 1.051 | 1.045 | 1.039 | 1.033 | 1.027 | 1.021 |
| 72 | 1.021 | 1.015 | 1.009 | 1.003 | 0.997 | 0.991 | 0.985 | 0.978 | 0.972 | 0.966 | 0.960 |
| 73 | 0.960 | 0.954 | 0.948 | 0.943 | 0.937 | 0.931 | 0.925 | 0.919 | 0.913 | 0.907 | 0.901 |
| 74 | 0.901 | 0.895 | 0.889 | 0.883 | 0.877 | 0.871 | 0.865 | 0.859 | 0.854 | 0.848 | 0.842 |
| 75 | 0.842 | 0.836 | 0.830 | 0.824 | 0.818 | 0.812 | 0.807 | 0.801 | 0.795 | 0.789 | 0.783 |
| 76 | 0.783 | 0.777 | 0.772 | 0.766 | 0.760 | 0.754 | 0.748 | 0.743 | 0.737 | 0.731 | 0.725 |
| 77 | 0.725 | 0.720 | 0.714 | 0.708 | 0.702 | 0.696 | 0.691 | 0.685 | 0.679 | 0.673 | 0.668 |
| 78 | 0.668 | 0.662 | 0.656 | 0.651 | 0.645 | 0.639 | 0.633 | 0.628 | 0.622 | 0.616 | 0.611 |
| 79 | 0.611 | 0.605 | 0.599 | 0.594 | 0.588 | 0.582 | 0.577 | 0.571 | 0.565 | 0.560 | 0.554 |
| 80 | 0.554 | 0.548 | 0.543 | 0.537 | 0.531 | 0.526 | 0.520 | 0.514 | 0.509 | 0.503 | 0.498 |
| 81 | 0.498 | 0.492 | 0.486 | 0.481 | 0.475 | 0.469 | 0.464 | 0.458 | 0.453 | 0.447 | 0.441 |
| 82 | 0.441 | 0.436 | 0.430 | 0.425 | 0.419 | 0.414 | 0.408 | 0.402 | 0.397 | 0.391 | 0.386 |
| 83 | 0.386 | 0.380 | 0.375 | 0.369 | 0.363 | 0.358 | 0.352 | 0.347 | 0.341 | 0.336 | 0.330 |
| 84 | 0.330 | 0.325 | 0.319 | 0.314 | 0.308 | 0.302 | 0.297 | 0.291 | 0.286 | 0.280 | 0.275 |
| 85 | 0.275 | 0.269 | 0.264 | 0.258 | 0.253 | 0.247 | 0.242 | 0.236 | 0.231 | 0.225 | 0.220 |
| 86 | 0.220 | 0.214 | 0.209 | 0.203 | 0.198 | 0.192 | 0.187 | 0.181 | 0.176 | 0.170 | 0.165 |
| 87 | 0.165 | 0.159 | 0.154 | 0.148 | 0.143 | 0.137 | 0.132 | 0.126 | 0.121 | 0.115 | 0.110 |
| 88 | 0.110 | 0.104 | 0.099 | 0.093 | 0.088 | 0.082 | 0.077 | 0.071 | 0.066 | 0.060 | 0.055 |
| 89 | 0.055 | 0.049 | 0.044 | 0.038 | 0.033 | 0.027 | 0.022 | 0.016 | 0.011 | 0.005 | 0.000 |

**Leads, Change Gears and Angles for Helical Milling**

| Lead of Helix, Inches | Gear on Worm | First Gear on Stud | Second Gear on Stud | Gear on Screw | Diameter of Work, Inches | | | | | | | | | |
|---|---|---|---|---|---|---|---|---|---|---|---|---|---|---|
| | | | | | ⅛ | ¼ | ⅜ | ½ | ⅝ | ¾ | ⅞ | 1 | 1¼ | 1½ |
| 0.67 | 24 | 86 | 24 | 100 | 30¼ | ... | ... | ... | Approximate Angles for Milling Machine Table | | | | | |
| 0.78 | 24 | 86 | 28 | 100 | 26 | 44½ | ... | ... | | | | | | |
| 0.89 | 24 | 86 | 32 | 100 | 23½ | 41 | ... | ... | | | | | | |
| 1.12 | 24 | 86 | 40 | 100 | 19 | 34½ | ... | ... | | | | | | |
| 1.34 | 24 | 86 | 48 | 100 | 16 | 30¼ | 41½ | ... | ... | ... | ... | ... | ... | ... |
| 1.46 | 24 | 64 | 28 | 72 | 14¾ | 28 | 38½ | ... | ... | ... | ... | ... | ... | ... |
| 1.56 | 24 | 86 | 56 | 100 | 13¾ | 26½ | 37 | ... | ... | ... | ... | ... | ... | ... |
| 1.67 | 24 | 64 | 32 | 72 | 12¾ | 25 | 34¾ | 43¼ | ... | ... | ... | ... | ... | ... |
| 1.94 | 32 | 64 | 28 | 72 | 11¼ | 21¾ | 31 | 39 | 45 | ... | ... | ... | ... | ... |
| 2.08 | 24 | 64 | 40 | 72 | 10¼ | 20½ | 29½ | 37 | 43¼ | ... | ... | ... | ... | ... |
| 2.22 | 32 | 56 | 28 | 72 | 9¾ | 19¼ | 27½ | 35 | 41¼ | ... | ... | ... | ... | ... |
| 2.50 | 24 | 64 | 48 | 72 | 8¾ | 17 | 25 | 32 | 38 | 43¼ | ... | ... | ... | ... |
| 2.78 | 40 | 56 | 28 | 72 | 8 | 15½ | 23 | 29½ | 35¼ | 40½ | 44¾ | ... | ... | ... |
| 2.92 | 24 | 64 | 56 | 72 | 7½ | 15 | 21¾ | 28¼ | 34 | 39 | 43¼ | ... | ... | ... |
| 3.24 | 40 | 48 | 28 | 72 | 6¾ | 13¼ | 19¾ | 25¾ | 31¼ | 36 | 40½ | 44¼ | ... | ... |
| 3.70 | 40 | 48 | 32 | 72 | 6 | 11¾ | 17½ | 23 | 28 | 32½ | 36½ | 40½ | ... | ... |
| 3.89 | 56 | 48 | 24 | 72 | 5½ | 11¼ | 16¾ | 22 | 26¾ | 31¼ | 35¼ | 39 | ... | ... |
| 4.17 | 40 | 72 | 48 | 64 | 5¼ | 10½ | 15¾ | 20½ | 25¼ | 29½ | 33½ | 37 | 43¼ | ... |
| 4.46 | 48 | 40 | 32 | 86 | 4¾ | 9¾ | 14¾ | 19¼ | 23¾ | 27¾ | 31½ | 35 | 41½ | ... |
| 4.86 | 40 | 64 | 56 | 72 | 4½ | 9 | 13½ | 17¾ | 22 | 25¾ | 29½ | 33 | 39 | 44¼ |
| 5.33 | 48 | 40 | 32 | 72 | 4 | 8¼ | 12¼ | 16½ | 20¼ | 23¾ | 27¼ | 30½ | 36½ | 41½ |
| 5.44 | 56 | 40 | 28 | 72 | 4 | 8 | 12 | 16 | 20 | 23½ | 26¾ | 30 | 36 | 41 |
| 6.12 | 56 | 40 | 28 | 64 | 3½ | 7¼ | 11 | 14½ | 17¾ | 21 | 24¼ | 27 | 33 | 37¾ |
| 6.22 | 56 | 40 | 32 | 72 | 3½ | 7 | 10¾ | 14¼ | 17½ | 20¾ | 23¾ | 26¾ | 32½ | 37¼ |
| 6.48 | 56 | 48 | 40 | 72 | 3¼ | 6¾ | 10¼ | 13½ | 16¾ | 20 | 23 | 25¾ | 31½ | 36¼ |
| 6.67 | 64 | 48 | 28 | 56 | 3¼ | 6½ | 10 | 13¼ | 16½ | 19½ | 22½ | 25¼ | 30¾ | 35¼ |
| 7.29 | 56 | 48 | 40 | 64 | 3 | 6¼ | 9¼ | 12¼ | 15 | 18 | 20½ | 23½ | 28½ | 33 |
| 7.41 | 64 | 48 | 40 | 72 | 3 | 6 | 9 | 12 | 14¾ | 17¾ | 20¼ | 22¾ | 28¼ | 32½ |
| 7.62 | 64 | 48 | 32 | 56 | 2¾ | 5¾ | 8¾ | 11½ | 14½ | 17¼ | 19¾ | 22¼ | 27½ | 32 |
| 8.33 | 48 | 32 | 40 | 72 | 2½ | 5¼ | 8 | 10½ | 13¼ | 15¾ | 18¼ | 20½ | 25½ | 29½ |
| 8.95 | 86 | 48 | 28 | 56 | 2½ | 5 | 7½ | 10 | 12½ | 14¾ | 17 | 19¼ | 24 | 28 |
| 9.33 | 56 | 40 | 48 | 72 | 2¼ | 4¾ | 7¼ | 9½ | 11¾ | 14 | 16¼ | 18½ | 23 | 27 |
| 9.52 | 64 | 48 | 40 | 56 | 2¼ | 4½ | 7 | 9¼ | 11½ | 13¾ | 16 | 18¼ | 22½ | 26½ |
| 10.29 | 72 | 40 | 32 | 56 | 2 | 4¼ | 6½ | 8¾ | 10¾ | 12¾ | 15 | 17¼ | 21 | 24¾ |
| 10.37 | 64 | 48 | 56 | 72 | 2 | 4¼ | 6½ | 8½ | 10½ | 12¾ | 14¾ | 17 | 20¾ | 24½ |
| 10.50 | 48 | 40 | 56 | 64 | 2 | 4¼ | 6¼ | 8½ | 10½ | 12½ | 14½ | 16¾ | 20½ | 24¼ |
| 10.67 | 64 | 40 | 48 | 72 | 2 | 4 | 6¼ | 8¼ | 10¼ | 12¼ | 14¼ | 16½ | 20¼ | 24 |
| 10.94 | 56 | 32 | 40 | 64 | 2 | 4 | 6 | 8¼ | 10¼ | 12 | 14 | 16¼ | 20 | 23½ |
| 11.11 | 64 | 32 | 40 | 72 | 2 | 4 | 6 | 8 | 10 | 11¾ | 13¾ | 16 | 19¾ | 23 |
| 11.66 | 56 | 32 | 48 | 72 | 1¾ | 3¾ | 5¾ | 7½ | 9½ | 11¼ | 13¼ | 15¼ | 18¾ | 22 |
| 12.00 | 72 | 40 | 32 | 48 | 1¾ | 3¾ | 5½ | 7¼ | 9¼ | 11 | 12¾ | 15 | 18¼ | 21½ |
| 13.12 | 56 | 32 | 48 | 64 | 1½ | 3½ | 5¼ | 6¾ | 8½ | 10¼ | 11¾ | 13½ | 16¾ | 20 |
| 13.33 | 56 | 28 | 48 | 72 | 1½ | 3¼ | 5 | 6½ | 8¼ | 10 | 11½ | 13¼ | 16½ | 19½ |
| 13.71 | 64 | 40 | 48 | 56 | 1½ | 3¼ | 4¾ | 6½ | 8 | 9¾ | 11¼ | 13 | 16 | 19 |
| 15.24 | 64 | 28 | 48 | 72 | 1½ | 3 | 4½ | 5¾ | 7¼ | 8¾ | 10¼ | 11¾ | 14½ | 17¼ |
| 15.56 | 64 | 32 | 56 | 72 | 1¼ | 2¾ | 4¼ | 5¾ | 7¼ | 8¾ | 10 | 11½ | 14¼ | 17 |
| 15.75 | 56 | 64 | 72 | 40 | 1¼ | 2¾ | 4¼ | 5½ | 7 | 8½ | 9¾ | 11¼ | 14 | 16¾ |
| 16.87 | 72 | 32 | 48 | 64 | 1¼ | 2½ | 4 | 5¼ | 6¾ | 7¾ | 9¼ | 10½ | 13¼ | 15¾ |
| 17.14 | 64 | 32 | 48 | 56 | 1¼ | 2½ | 4 | 5¼ | 6½ | 7¾ | 9 | 10¼ | 13 | 15½ |
| 18.75 | 72 | 32 | 40 | 48 | 1 | 2¼ | 3½ | 4¾ | 6 | 7¼ | 8¼ | 9½ | 12 | 14¼ |
| 19.29 | 72 | 32 | 48 | 56 | 1 | 2¼ | 3½ | 4½ | 5¾ | 7 | 8 | 9¼ | 11½ | 13¾ |
| 19.59 | 64 | 28 | 48 | 56 | 1 | 2¼ | 3¼ | 4½ | 5¾ | 6¾ | 8 | 9¼ | 11½ | 13½ |
| 19.69 | 72 | 32 | 56 | 64 | 1 | 2¼ | 3¼ | 4½ | 5¾ | 6¾ | 8 | 9 | 11½ | 13½ |
| 21.43 | 72 | 24 | 40 | 56 | 1 | 2 | 3¼ | 4¼ | 5¼ | 6¼ | 7½ | 8½ | 10½ | 12½ |
| 22.50 | 72 | 28 | 56 | 64 | 1 | 2 | 3 | 4 | 5 | 6 | 7 | 8 | 10 | 12 |
| 23.33 | 64 | 32 | 56 | 48 | 1 | 2 | 3 | 4 | 5 | 5¾ | 6¾ | 7¾ | 9¾ | 11½ |
| 26.25 | 72 | 24 | 56 | 64 | 1 | 1¾ | 2¾ | 3½ | 4¼ | 5 | 6 | 7 | 8½ | 10¼ |
| 26.67 | 64 | 28 | 56 | 48 | ¾ | 1¾ | 2¾ | 3½ | 4¼ | 5 | 6 | 6¾ | 8½ | 10 |
| 28.00 | 64 | 32 | 56 | 40 | ¾ | 1¾ | 2½ | 3¼ | 4 | 4¾ | 5¾ | 6½ | 8 | 9½ |
| 30.86 | 72 | 28 | 48 | 40 | ¾ | 1½ | 2¼ | 3 | 3¾ | 4½ | 5 | 5¾ | 7¼ | 8¾ |

**Leads, Change Gears and Angles for Helical Milling**

| Lead of Helix, Inches | Gear on Worm | First Gear on Stud | Second Gear on Stud | Gear on Screw | Diameter of Work, Inches | | | | | | | | | |
|---|---|---|---|---|---|---|---|---|---|---|---|---|---|---|
| | | | | | 1¾ | 2 | 2¼ | 2½ | 2¾ | 3 | 3¼ | 3½ | 3¾ | 4 |
| 6.12 | 56 | 40 | 28 | 64 | 42 | ... | ... | ... | Approximate Angles for Milling Machine Table | | | | | |
| 6.22 | 56 | 40 | 32 | 72 | 41½ | ... | ... | ... | | | | | | |
| 6.48 | 56 | 48 | 40 | 72 | 40¼ | 44¼ | ... | ... | | | | | | |
| 6.67 | 64 | 48 | 28 | 56 | 39½ | 43½ | ... | ... | ... | ... | ... | ... | ... | ... |
| 7.29 | 56 | 48 | 40 | 64 | 37 | 41 | 44¼ | ... | ... | ... | ... | ... | ... | ... |
| 7.41 | 64 | 48 | 40 | 72 | 36½ | 40¼ | 43¾ | ... | ... | ... | ... | ... | ... | ... |
| 7.62 | 64 | 48 | 32 | 56 | 36 | 39½ | 43 | ... | ... | ... | ... | ... | ... | ... |
| 8.33 | 48 | 32 | 40 | 72 | 33½ | 37 | 40½ | 43½ | ... | ... | ... | ... | ... | ... |
| 8.95 | 86 | 48 | 28 | 56 | 31¾ | 35¼ | 38½ | 41¼ | 44 | ... | ... | ... | ... | ... |
| 9.33 | 56 | 40 | 48 | 72 | 30½ | 34 | 37¼ | 40¼ | 43 | ... | ... | ... | ... | ... |
| 9.52 | 64 | 48 | 40 | 56 | 30 | 33½ | 36½ | 39½ | 42¼ | 45 | ... | ... | ... | ... |
| 10.29 | 72 | 40 | 32 | 56 | 28¼ | 31½ | 34½ | 37½ | 40 | 42½ | 45 | ... | ... | ... |
| 10.37 | 64 | 48 | 56 | 72 | 28 | 31¼ | 34¼ | 37¼ | 39¾ | 42¼ | 44¾ | ... | ... | ... |
| 10.50 | 48 | 40 | 56 | 64 | 27¾ | 31 | 34 | 36¾ | 39½ | 42 | 44¼ | ... | ... | ... |
| 10.67 | 64 | 40 | 48 | 72 | 27¼ | 30½ | 33½ | 36½ | 39 | 41½ | 43¾ | ... | ... | ... |
| 10.94 | 56 | 32 | 40 | 64 | 26¾ | 30 | 33 | 35¾ | 38¼ | 40¾ | 43 | ... | ... | ... |
| 11.11 | 64 | 32 | 40 | 72 | 26½ | 29½ | 32½ | 35¼ | 38 | 40¼ | 42½ | 44¾ | ... | ... |
| 11.66 | 56 | 32 | 48 | 72 | 25¼ | 28½ | 31¼ | 34 | 36½ | 39 | 41¼ | 43½ | ... | ... |
| 12.00 | 72 | 40 | 32 | 48 | 24¾ | 27¾ | 30½ | 33¼ | 35¾ | 38 | 40¼ | 42½ | 44¾ | ... |
| 13.12 | 56 | 32 | 48 | 64 | 22¾ | 25¾ | 28¼ | 31 | 33¼ | 35¾ | 37¾ | 40 | 42 | 43¾ |
| 13.33 | 56 | 28 | 48 | 72 | 22½ | 25½ | 28 | 30½ | 33 | 35¼ | 37½ | 39½ | 41½ | 43¼ |
| 13.71 | 64 | 40 | 48 | 56 | 22 | 24¾ | 27¼ | 30 | 32¼ | 34½ | 36½ | 38¾ | 40¾ | 42½ |
| 15.24 | 64 | 28 | 48 | 72 | 20 | 22½ | 25 | 27¼ | 29½ | 31¾ | 34 | 35¾ | 37¾ | 39½ |
| 15.56 | 64 | 32 | 56 | 72 | 19½ | 22 | 24½ | 27 | 29 | 31¼ | 33¼ | 35¼ | 37 | 39 |
| 15.75 | 56 | 64 | 72 | 40 | 19¼ | 21¾ | 24¼ | 26½ | 28¾ | 31 | 33 | 35 | 36¾ | 38½ |
| 16.87 | 72 | 32 | 48 | 64 | 18¼ | 20½ | 22¾ | 25 | 27 | 29¼ | 31¼ | 33¼ | 35 | 36½ |
| 17.14 | 64 | 32 | 48 | 56 | 17¾ | 20¼ | 22¼ | 24¾ | 26¾ | 29 | 30¾ | 32¾ | 34½ | 36 |
| 18.75 | 72 | 32 | 40 | 48 | 16¼ | 18½ | 20¾ | 22¾ | 25 | 26¾ | 28½ | 30¼ | 32 | 33¾ |
| 19.29 | 72 | 32 | 48 | 56 | 16 | 18¼ | 20¼ | 22¼ | 24 | 26 | 28 | 29¾ | 31½ | 33 |
| 19.59 | 64 | 28 | 48 | 56 | 15¾ | 18 | 20 | 22 | 23¾ | 25¾ | 27½ | 29¼ | 31 | 32¾ |
| 19.69 | 72 | 32 | 56 | 64 | 15¾ | 17¾ | 20 | 21¾ | 23¾ | 25½ | 27½ | 29¼ | 31 | 32½ |
| 21.43 | 72 | 24 | 40 | 56 | 14½ | 16½ | 18½ | 20¼ | 22 | 23¾ | 25½ | 27¼ | 29 | 30¼ |
| 22.50 | 72 | 28 | 56 | 64 | 13¾ | 15¾ | 17½ | 19¼ | 21 | 22¾ | 24½ | 26 | 27¾ | 29¼ |
| 23.33 | 64 | 32 | 56 | 48 | 13¼ | 15¼ | 17 | 18¾ | 20¼ | 22 | 23½ | 25¼ | 27 | 28¼ |
| 26.25 | 72 | 24 | 56 | 64 | 12 | 13½ | 15 | 16¾ | 18¼ | 19¾ | 21¼ | 22¾ | 24¼ | 25½ |
| 26.67 | 64 | 28 | 56 | 48 | 11¾ | 13¼ | 14¾ | 16½ | 18 | 19½ | 21 | 22¼ | 23¾ | 25¼ |
| 28.00 | 64 | 32 | 56 | 40 | 11¼ | 12¾ | 14¼ | 15¾ | 17¼ | 18¾ | 20 | 21½ | 22¾ | 24 |
| 30.86 | 72 | 28 | 48 | 40 | 10 | 11½ | 13 | 14¼ | 15½ | 17 | 18½ | 19½ | 21 | 22 |
| 31.50 | 72 | 32 | 56 | 40 | 10 | 11¼ | 12¾ | 14 | 15¼ | 16½ | 18 | 19¼ | 20½ | 21¾ |
| 36.00 | 72 | 32 | 64 | 40 | 8¾ | 10 | 11 | 12¼ | 13½ | 14¾ | 16 | 17 | 18¼ | 19¼ |
| 41.14 | 72 | 28 | 64 | 40 | 7¾ | 8¾ | 9¾ | 10¾ | 11¾ | 13 | 14 | 15 | 16 | 17 |
| 45.00 | 72 | 28 | 56 | 32 | 7 | 8 | 9 | 10 | 11 | 11¾ | 12¾ | 13¾ | 14¾ | 15½ |
| 48.00 | 72 | 24 | 64 | 40 | 6½ | 7½ | 8½ | 9¼ | 10¼ | 11¼ | 12 | 13 | 13¾ | 14½ |
| 51.43 | 72 | 28 | 64 | 32 | 6 | 7 | 7¾ | 8¾ | 9½ | 10½ | 11¼ | 12 | 12¾ | 13¾ |
| 60.00 | 72 | 24 | 64 | 32 | 5¼ | 6 | 6¾ | 7½ | 8¼ | 9 | 9½ | 10¼ | 11 | 11¾ |
| 68.57 | 72 | 24 | 64 | 28 | 4¼ | 5¼ | 5¾ | 6½ | 7¼ | 8 | 8½ | 9 | 9¾ | 10¼ |

**Helix Angle for Given Lead and Diameter.** — The table on this and the preceding page gives helix angles (relative to axis) equivalent to a range of leads and diameters. The expression " Diameter of Work " at the top of the table might mean pitch diameter or outside diameter, depending upon the class of work. Assume, for example, that a plain milling cutter 4 inches in diameter is to have helical teeth and a helix angle of about 25 degrees is desired. The table shows that this angle will be obtained approximately by using change-gears which will give a lead of 26.67 inches. As the outside diameter of the cutter is 4 inches, the helix angle of 25¼ degrees is at the tops of the teeth. The angles listed for different diameters are used in setting the table of a milling machine. In milling a right-hand helix (or cutter teeth which turn to the right as seen from the end of the cutter), swivel the right-hand end of the machine table toward the rear, and, inversely, for a left-hand helix, swivel the left-hand end of the table toward the rear. The angles in the table are based upon the following formula:

$$\frac{\text{Cot helix angle}}{\text{relative to axis}} = \frac{\text{Lead of helix}}{3.1416 \times \text{diameter}}$$

**Lead of Helix for Given Angle.** — The lead of a helix or "spiral" for given angles measured with the axis of the work is given in the table, pages 1491–1493 for a diameter of 1. For other diameters, lead equals the value found in the table multiplied by the given diameter. Suppose the angle is 55 degrees, and the diameter 5 inches, what would be the lead? By referring to the table (Part 2), it is found that the lead for a diameter of 1 and an angle of 55 degrees 0 minutes equals 2.200. Multiply this value by 5; 5 × 2.200 = 11 inches, which is the required lead. If the lead and diameter are given, and the angle is wanted, divide the given lead by the given diameter, thus obtaining the lead for a diameter equal to 1; then find the angle corresponding to this lead in the table. If the lead and angle are given, and the diameter is wanted, divide the lead by the value in the table for the angle.

# SIMPLE, COMPOUND, DIFFERENTIAL AND BLOCK INDEXING

**Simple Indexing.** — A general rule for determining the number of turns the crank of a dividing head must make, to obtain a given number of divisions, is as follows: Divide the number of turns required for one revolution of the dividing-head spindle by the number of divisions into which the periphery of the work is to be divided.

*Example:* — If 40 turns of the index crank are required for one revolution of the spindle, and 12 divisions are required, the number of turns of the index crank for each indexing would equal 40 ÷ 12 = 3⅓ turns.

**Compound Indexing.** — This method is sometimes used to obtain divisions which are beyond the range of those secured by the simple method. The crank is first turned a definite amount in the regular way, and then the index plate is also turned either in the same or opposite direction, in order to locate the index crank in the proper position. Thus, there are two separate movements which are, in reality, two simple indexing operations. The following rule is for determining what circles of holes can be used for indexing by the compound method.

*Rule:* Resolve into its factors the number of divisions required; then choose at random two circles of holes, subtract one from the other, and factor the difference; place the two sets of factors thus obtained above a horizontal line. Next factor the number of turns of the crank required for one revolution of the spindle, and also the number of holes in each of the chosen circles; place the three sets of factors thus obtained below the horizontal line. If all the factors *above* the line can be canceled by those below, the two circles chosen will give the required number of divisions; if not, other circles must be chosen and another trial made.

*Example:* — Assume that 69 divisions are required, and that circles having 33 and 23 holes are chosen for the first trial. Then, by applying the foregoing rule, it is found that all the factors above the line cancel:

$$\frac{3 \times 23 \times 2 \times 5}{2 \times 2 \times 2 \times 5 \times 3 \times 11 \times 23} = \frac{1}{2 \times 2 \times 11}$$

## Compound Indexing *

| No. of Divisions | Indexing Movements | No. of Times Around | No. of Divisions | Indexing Movements | No. of Times Around | No. of Divisions | Indexing Movements | No. of Times Around |
|---|---|---|---|---|---|---|---|---|
| 51 | $8\frac{1}{47} - 1\frac{24}{49}$ | 11 | 133 | $3\frac{23}{29} - 1\frac{6}{23}$ | 11 | 198* | $\frac{3}{67} + \frac{3}{63}$ | ... |
| 53 | $6\frac{4}{47} - \frac{9}{49}$ | 9 | 134 | $3\frac{27}{47} + 1\frac{5}{49}$ | 13 | 199 | $2\frac{13}{41} - \frac{5}{49}$ | 11 |
| 57 | $4\frac{9}{47} + \frac{3}{49}$ | 7 | 137 | $3\frac{17}{43} - \frac{9}{49}$ | 11 | 201 | $2\frac{15}{47} + 1\frac{9}{49}$ | 13 |
| 59 | $7\frac{10}{47} + 1\frac{24}{49}$ | 11 | 138* | $1\frac{1}{23} - \frac{1}{23}$ | ... | 202 | $3\frac{10}{41} + \frac{9}{49}$ | 17 |
| 61 | $3\frac{42}{47} + \frac{2}{49}$ | 6 | 139 | $2\frac{25}{57} + 2\frac{4}{49}$ | 11 | 203 | $1\frac{23}{59} + \frac{9}{49}$ | 9 |
| 63 | $4\frac{19}{49} + 1\frac{45}{49}$ | 8 | 141 | $1\frac{32}{49} + 2\frac{3}{49}$ | 8 | 204 | $2\frac{29}{41} + \frac{3}{49}$ | 13 |
| 67 | $2\frac{27}{41} + 1\frac{9}{49}$ | 5 | 142 | $4\frac{1}{47} + 1\frac{9}{49}$ | 15 | 206 | $2\frac{35}{59} + \frac{3}{49}$ | 15 |
| 69* | $2\frac{1}{23} - 1\frac{1}{23}$ | ... | 143 | $1\frac{36}{47} - 1\frac{8}{49}$ | 5 | 207 | $3\frac{3}{41} - 2\frac{4}{49}$ | 14 |
| 71 | $3\frac{34}{41} - 2\frac{34}{49}$ | 6 | 146 | $2\frac{3}{57} - \frac{8}{49}$ | 7 | 208 | $1\frac{19}{47} + 1\frac{6}{49}$ | 9 |
| 73 | $6\frac{28}{47} - \frac{1}{49}$ | 12 | 147* | $1\frac{3}{49} - \frac{3}{49}$ | ... | 209 | $\frac{8}{49} + \frac{9}{41}$ | 2 |
| 77* | $\frac{9}{21} + \frac{3}{63}$ | ... | 149 | $3\frac{5}{43} - \frac{8}{49}$ | 11 | 211 | $1\frac{28}{59} + 1\frac{8}{49}$ | 11 |
| 79 | $2\frac{42}{43} + \frac{3}{49}$ | 6 | 151 | $1\frac{42}{43} - \frac{9}{49}$ | 7 | 212 | $3\frac{5}{47} + \frac{9}{49}$ | 17 |
| 81 | $5\frac{3}{41} - \frac{9}{49}$ | 10 | 153 | $2\frac{45}{47} - \frac{8}{49}$ | 11 | 213 | $1\frac{18}{59} + \frac{2}{49}$ | 8 |
| 83 | $3\frac{45}{47} - \frac{9}{49}$ | 8 | 154* | $\frac{9}{61} - \frac{4}{23}$ | ... | 214 | $3\frac{9}{47} - 1\frac{9}{49}$ | 15 |
| 87* | $2\frac{3}{59} - 1\frac{1}{23}$ | ... | 157 | $2\frac{23}{31} + \frac{3}{63}$ | 11 | 217 | $2\frac{3}{43} + 1\frac{6}{49}$ | 13 |
| 89 | $3\frac{28}{39} - \frac{9}{49}$ | 8 | 158 | $5\frac{5}{43} - 1\frac{5}{49}$ | 19 | 218 | $1\frac{25}{47} - \frac{9}{49}$ | 7 |
| 91* | $\frac{9}{49} + 1\frac{5}{49}$ | ... | 159 | $2\frac{7}{57} + 1\frac{6}{49}$ | 10 | 219 | $3\frac{29}{43} - 1\frac{9}{49}$ | 19 |
| 93* | $\frac{3}{21} + 1\frac{1}{23}$ | ... | 161 | $2\frac{10}{59} - \frac{1}{49}$ | 9 | 221 | $1\frac{5}{47} - \frac{1}{49}$ | 6 |
| 96* | $\frac{3}{18} + \frac{5}{20}$ | ... | 162 | $1\frac{39}{59} - \frac{3}{49}$ | 7 | 222 | $2\frac{8}{43} - 1\frac{9}{49}$ | 11 |
| 97 | $4\frac{27}{41} - \frac{9}{49}$ | 11 | 163 | $3\frac{7}{57} - 2\frac{4}{49}$ | 11 | 223 | $2\frac{29}{43} + 1\frac{3}{49}$ | 16 |
| 99* | $1\frac{5}{27} - \frac{5}{23}$ | ... | 166 | $1\frac{19}{43} + 1\frac{24}{49}$ | 7 | 224 | $2\frac{6}{23} + \frac{7}{63}$ | 12 |
| 101 | $4\frac{32}{43} - 1\frac{9}{49}$ | 11 | 167 | $2\frac{1}{59} + \frac{4}{49}$ | 9 | 225* | $\frac{5}{18} - \frac{5}{20}$ | ... |
| 102 | $4\frac{17}{43} - \frac{8}{49}$ | 11 | 169 | $1\frac{32}{57} + 1\frac{3}{49}$ | 9 | 226 | $1\frac{38}{59} + 1\frac{9}{49}$ | 13 |
| 103 | $1\frac{8}{43} + 1\frac{8}{49}$ | 4 | 171 | $1\frac{29}{47} + \frac{1}{49}$ | 7 | 227 | $3\frac{3}{43} + \frac{9}{49}$ | 18 |
| 106 | $2\frac{38}{41} + 2\frac{3}{49}$ | 9 | 173 | $1\frac{7}{43} + 1\frac{1}{49}$ | 6 | 228 | $2\frac{9}{41} - 1\frac{3}{49}$ | 11 |
| 107 | $2\frac{21}{31} - \frac{2}{23}$ | 7 | 174* | $1\frac{1}{23} - \frac{3}{29}$ | ... | 229 | $2\frac{19}{41} - 1\frac{9}{49}$ | 12 |
| 109 | $2\frac{19}{59} + \frac{4}{49}$ | 7 | 175 | $1\frac{4}{51} + \frac{8}{63}$ | 6 | 231* | $\frac{3}{21} + \frac{1}{53}$ | ... |
| 111 | $3\frac{29}{47} + 1\frac{7}{49}$ | 11 | 176 | $1\frac{4}{43} + 1\frac{3}{49}$ | 7 | 233 | $1\frac{39}{47} + \frac{9}{49}$ | 11 |
| 112 | $4\frac{10}{21} - 1\frac{3}{23}$ | 11 | 177 | $2\frac{19}{47} + \frac{4}{49}$ | 11 | 234 | $2\frac{21}{59} + \frac{9}{63}$ | 17 |
| 113 | $3\frac{29}{47} - 1\frac{8}{49}$ | 9 | 178 | $3\frac{28}{47} + 1\frac{1}{49}$ | 17 | 236 | $2\frac{39}{43} + \frac{9}{49}$ | 17 |
| 114 | $1\frac{35}{57} + 2\frac{5}{49}$ | 7 | 179 | $2\frac{34}{47} - 1\frac{3}{49}$ | 11 | 237 | $2\frac{12}{47} - \frac{3}{49}$ | 13 |
| 117 | $7\frac{1}{47} - \frac{9}{49}$ | 20 | 181 | $2\frac{8}{43} + 1\frac{24}{49}$ | 11 | 238 | $2\frac{3}{61} + 1\frac{5}{63}$ | 15 |
| 118 | $1\frac{8}{59} + 2\frac{2}{49}$ | 5 | 182* | $\frac{3}{59} + \frac{7}{49}$ | ... | 239 | $1\frac{23}{43} + 1\frac{5}{49}$ | 11 |
| 119 | $3\frac{5}{23} - 1\frac{6}{23}$ | 8 | 183 | $1\frac{24}{41} + \frac{8}{49}$ | 8 | 241 | $1\frac{1}{41} + 2\frac{3}{49}$ | 9 |
| 121 | $1\frac{4}{47} - 1\frac{5}{49}$ | 3 | 186* | $1\frac{7}{51} - 1\frac{1}{23}$ | ... | 242 | $2\frac{23}{41} - \frac{5}{49}$ | 15 |
| 122 | $3\frac{41}{43} - 1\frac{7}{49}$ | 11 | 187 | $1\frac{20}{47} + 1\frac{9}{49}$ | 8 | 243 | $1\frac{29}{41} - \frac{3}{49}$ | 10 |
| 123 | $1\frac{12}{43} + 1\frac{7}{49}$ | 5 | 189 | $2\frac{26}{41} - 1\frac{9}{49}$ | 11 | 244 | $2\frac{15}{41} + 1\frac{9}{63}$ | 17 |
| 125 | $2\frac{33}{41} - 1\frac{2}{49}$ | 8 | 191 | $1\frac{38}{47} + 1\frac{5}{49}$ | 10 | 246 | $1\frac{9}{43} - 1\frac{6}{49}$ | 5 |
| 126 | $3\frac{10}{19} - \frac{7}{20}$ | 11 | 192 | $2\frac{23}{41} - 1\frac{2}{49}$ | 11 | 247 | $2\frac{15}{43} - \frac{5}{49}$ | 14 |
| 127 | $2\frac{23}{39} + 1\frac{24}{49}$ | 9 | 193 | $1\frac{5}{57} - 1\frac{5}{49}$ | 4 | 249 | $3\frac{5}{43} - \frac{2}{49}$ | 19 |
| 129 | $5\frac{24}{41} + 1\frac{5}{49}$ | 19 | 194 | $2\frac{23}{57} - 1\frac{9}{49}$ | 11 | 250 | $2\frac{9}{57} - \frac{8}{49}$ | 13 |
| 131 | $2\frac{40}{43} + 2\frac{1}{49}$ | 11 | 197 | $1\frac{39}{43} + 1\frac{6}{49}$ | 11 | ... | ......... | ... |

* The indexing movements are exact for the divisions marked with an asterisk (*); the errors of the other divisions are so slight as to be negligible for all ordinary classes of work, such as gear-cutting, etc.

This shows that these circles can be used. The factors 2, 2 and 11 remain uncanceled below the line. The amount the crank and index plate must be moved in their respective circles is next determined by multiplying together all these uncanceled factors. Thus $2 \times 2 \times 11 = 44$. This means that we can index $\frac{1}{8\frac{1}{8}}$ revolution by turning the crank forward 44 holes in the 23-hole circle, and the index plate backward 44 holes in the 33-hole circle. The movement could also be forward 44 holes in the 33-hole circle and backward 44 holes in the 23-hole circle, without affecting the result. The movements obtained by the foregoing rule are expressed in compound indexing tables in the form of fractions, as, for example: $+\frac{44}{23} - \frac{44}{33}$. The numerators represent the number of holes indexed and the denominators the circles used; the $+$ and $-$ signs show that the movements of the crank and index plate are opposite in direction. These fractions can often be reduced and simplified, so that it will not be necessary to move so many holes, by adding some number to them algebraically. The number is chosen by trial, and its sign should be opposite that of the fraction to which it is added. Suppose, for example, a fraction is added representing one complete turn, to each of the fractions referred to; then there will be a movement of 21 holes in the 23-hole circle, and a movement of 11 holes in the opposite direction, in the 33-hole circle.

**Differential Indexing.** — This method is the same, in principle, as compound indexing, but differs from the latter in that the index plate is rotated by suitable gearing which connects it to the spiral-head spindle. This rotation or differential motion of the index plate takes place when the crank is turned, the plate moving either in the same direction as the crank or opposite to it, as may be required. The result is that the *actual* movement of the crank, at every indexing, is either greater or less than its movement with relation to the index plate. The differential method makes it possible to obtain almost any division, by using only one circle of holes for that division and turning the index crank in one direction, the same as for plain indexing. The gears to use for moving the index plate the required amount (when gears are required) are shown by the tables, "Simple and Differential Indexing." This table shows what divisions can be obtained by plain indexing, and also when it is necessary to use gears and the differential system. For example, if 50 divisions are required, the 20-hole index circle is used and the crank is moved 16 holes, but no gears are required. For 51 divisions, a 24-tooth gear is placed on the worm-shaft and a 48-tooth gear is mounted on the spindle. These two gears are connected by two idler gears having 24 and 44 teeth, respectively. To illustrate the principle of differential indexing, suppose a dividing head is to be geared for 271 divisions. The table calls for a gear on the worm-shaft having 56 teeth; a spindle gear with 72 teeth; and a 24-toothed idler which serves to rotate the index plate in the same direction as the crank. The sector should be set for giving the crank a movement of 3 holes in the 21-hole circle. If the spindle and index plate were not connected through gearing, 280 divisions would be obtained by successively moving the crank 3 holes in the 21-hole circle, but the gears cause the index plate to turn in the same direction as the crank at such a rate that, when 271 indexings have been made, the work is turned one complete revolution; therefore, we have 271 divisions instead of 280, the number being reduced because the total movement of the crank, for each indexing, is equal to its movement relative to the index plate, *plus* the movement of the plate itself when (as in this case) the crank and plate rotate in the same direction. If they were rotated in opposite directions, the crank would have a total movement equal to the amount it turned relative to the plate, *minus* the plate's movement. Sometimes it is necessary to use compound gearing, in order to move the index plate the required amount for each turn of the crank. The differential method cannot be used in connection with helical or spiral milling, because the spiral head is then geared to the lead-screw of the machine.

**To Find Ratio of Gearing for Differential Indexing.** — To find the gearing ratio for differential indexing, first select some approximate number $A$ of divisions either greater or less than the required number $N$. To illustrate, if the required number $N$ is 67, the approximate number $A$ might be 70; then if 40 turns of the index crank are required for 1 revolution of the spindle,

$$\text{Gearing ratio } R = (A - N) \times \frac{40}{A}$$

If the approximate number $A$ is less than $N$, the formula is the same as above except that $A - N$ is replaced by $N - A$.

*Example:* Find the gearing ratio and indexing movement for 67 divisions. If $A = 70$,

$$\text{Gearing ratio} = (70 - 67) \times \frac{40}{70} = \frac{12}{7} = \frac{\text{Gear on spindle (driver)}}{\text{Gear on worm (driven)}}$$

The fraction $\frac{12}{7}$ is raised to obtain a numerator and denominator equivalent to available gears. For example, $\frac{12}{7} = \frac{48}{28}$.

Various combinations of gearing and index circles are possible for a given number of divisions. The index movements and gear combinations in the accompanying table apply to a given series of index circles and gear-tooth numbers. The approximate number $A$ upon which any combination is based may be determined by dividing 40 by the fraction representing the indexing movement. For example, the approximate number used for 109 divisions equals $40 \div 1\frac{6}{18}$ or $40 \times 1\frac{6}{6} = 106\frac{2}{3}$. If this approximate number is inserted in the preceding formula, it will be found that the gear ratio is $\frac{7}{6}$ as shown in the table.

**Second Method of Determining Gear Ratio.** — In illustrating a somewhat different method of determining the gear ratio, 67 divisions will again be used. If 70 is selected as the approximate number, then $\frac{40}{70} = \frac{4}{7}$ or $1\frac{2}{7}$ turn of the index crank will be required. If the crank is indexed four-sevenths of a turn sixty-seven times, it will make $\frac{4}{7} \times 67 = 38\frac{2}{7}$ revolutions. This is $1\frac{5}{7}$ turns less than the forty required for one revolution of the work (indicating that the gearing should be arranged to rotate the index plate in the same direction as the index crank to increase the indexing movement); hence the gear ratio = $1\frac{5}{7} = \frac{12}{7}$.

**To Find the Indexing Movement.** — The indexing movement is represented by the fraction $\frac{40}{A}$. For example, if 70 is the approximate number $A$ used in calculating the gear ratio for 67 divisions, then, to find the required movement of the index crank, reduce $\frac{40}{70}$ to any fraction of equal value and having as denominator any number equal to the number of holes available in an index circle. To illustrate,

$$\frac{40}{70} = \frac{4}{7} = \frac{12}{21} = \frac{\text{number of holes indexed}}{\text{number of holes in index circle}}$$

**Use of Idler Gears.** — In differential indexing, idler gears are used (1) to rotate the index plate in the same direction as the index crank, thus *increasing* the actual indexing movement, or (2) to rotate the index plate in the opposite direction, thus *reducing* the actual indexing movement.

*Case 1:* If the approximate number $A$ is *greater* than the actual number of divisions $N$, simple gearing will require one idler, and compound gearing no idler. Index plate and crank rotate in the same direction.

*Case 2:* If the approximate number *A* is *less* than the actual number of divisions *N*, simple gearing requires two idlers, and compound gearing one idler. Index plate and crank rotate in opposite directions.

**When Compound Gearing Is Required.** — In some cases, as will be noted by referring to the table, it is necessary to use a train of four gears in order to obtain the required ratio with gear-tooth numbers in the available series.

*Example:* Find the gear combination and indexing movement for 99 divisions, assuming that an approximate number *A* of 100 is used.

$$\text{Ratio} = (100 - 99) \times \frac{40}{100} = \frac{4}{10} = \frac{4 \times 1}{5 \times 2} = \frac{32}{40} \times \frac{28}{56}$$

These final numbers conform to available gear sizes. The gears having 32 and 28 teeth are the drivers (gear on spindle and first gear on stud), and gears having 40 and 56 teeth are driven (second gear on stud and gear on worm). The indexing movement is represented by the fraction $\frac{40}{100}$ which is reduced to $\frac{8}{20}$, the 20-hole index circle being used in this case.

*Example:* Determine the gear combination to use for indexing 53 divisions. If 56 is used as an approximate number (possibly after one or more trial solutions to find an approximate number and resulting gear ratio coinciding with available gears).

$$\text{Gearing ratio} = (56 - 53) \times \frac{40}{56} = \frac{15}{7} = \frac{3 \times 5}{1 \times 7} = \frac{72 \times 40}{24 \times 56}$$

The tooth numbers above the line represent *gear on spindle* and *first gear on stud.* The numbers below the line represent *second gear on stud* and *gear on worm.*

$$\text{Indexing movement} = \frac{40}{56} = \frac{5}{7} = \frac{5 \times 7}{7 \times 7} = \frac{35 \text{ holes}}{49\text{-hole circle}}$$

In setting sector arms, do not count the hole containing the index crank pin.

**To Check the Number of Divisions Obtained with a Given Gear Ratio and Index Movement.** — Invert the fraction representing the indexing movement and let *C* equal this inverted fraction. *R* = gearing ratio.

*Case 1:* If simple gearing is used with one idler or compound gearing with no idler,

$$\text{Number of divisions } N = 40\,C - RC$$

*Case 2:* If simple gearing is used with two idlers or compound gearing with one idler,

$$\text{Number of divisions } N = 40\,C + RC$$

*Example:* The gear ratio is $\frac{12}{7}$; there is simple gearing and one idler (Case 1), and the indexing movement is $\frac{16}{24}$, making the inverted fraction $C = \frac{24}{16} = 1\frac{1}{2}$; find the number of divisions *N*

$$N = (40 \times 1\tfrac{1}{2}) - (\tfrac{12}{7} \times 1\tfrac{1}{2}) = 70 - \tfrac{24}{7} = 67$$

*Example:* The gear ratio is $\frac{7}{8}$; two idlers are used with simple gearing (Case 2) and the indexing movement is 6 holes in the 16-hole circle. Then

$$N = (40 \times \tfrac{16}{6}) + (\tfrac{7}{8} \times \tfrac{16}{6}) = 109$$

## Simple and Differential Indexing — Brown & Sharpe Milling Machines

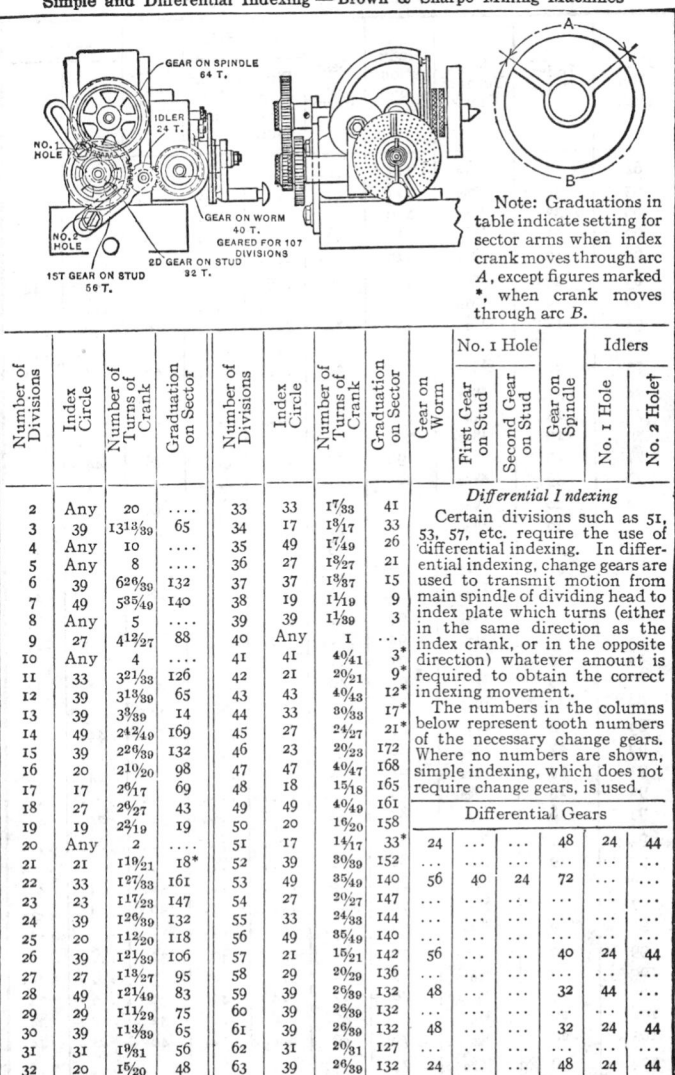

Note: Graduations in table indicate setting for sector arms when index crank moves through arc A, except figures marked *, when crank moves through arc B.

| Number of Divisions | Index Circle | Number of Turns of Crank | Graduation on Sector |
|---|---|---|---|
| 2 | Any | 20 | .... |
| 3 | 39 | $13\frac{13}{39}$ | 65 |
| 4 | Any | 10 | .... |
| 5 | Any | 8 | .... |
| 6 | 39 | $6\frac{26}{39}$ | 132 |
| 7 | 49 | $5\frac{35}{49}$ | 140 |
| 8 | Any | 5 | .... |
| 9 | 27 | $4\frac{12}{27}$ | 88 |
| 10 | Any | 4 | .... |
| 11 | 33 | $3\frac{21}{33}$ | 126 |
| 12 | 39 | $3\frac{13}{39}$ | 65 |
| 13 | 39 | $3\frac{3}{39}$ | 14 |
| 14 | 49 | $2\frac{42}{49}$ | 169 |
| 15 | 39 | $2\frac{26}{39}$ | 132 |
| 16 | 20 | $2\frac{10}{20}$ | 98 |
| 17 | 17 | $2\frac{6}{17}$ | 69 |
| 18 | 27 | $2\frac{6}{27}$ | 43 |
| 19 | 19 | $2\frac{2}{19}$ | 19 |
| 20 | Any | 2 | .... |
| 21 | 21 | $1\frac{19}{21}$ | 18* |
| 22 | 33 | $1\frac{27}{33}$ | 161 |
| 23 | 23 | $1\frac{17}{23}$ | 147 |
| 24 | 39 | $1\frac{26}{39}$ | 132 |
| 25 | 20 | $1\frac{12}{20}$ | 118 |
| 26 | 39 | $1\frac{21}{39}$ | 106 |
| 27 | 27 | $1\frac{13}{27}$ | 95 |
| 28 | 49 | $1\frac{21}{49}$ | 83 |
| 29 | 29 | $1\frac{11}{29}$ | 75 |
| 30 | 39 | $1\frac{13}{39}$ | 65 |
| 31 | 31 | $1\frac{9}{31}$ | 56 |
| 32 | 20 | $1\frac{5}{20}$ | 48 |

| Number of Divisions | Index Circle | Number of Turns of Crank | Graduation on Sector | Gear on Worm | No. 1 Hole First Gear on Stud | No. 1 Hole Second Gear on Stud | Gear on Spindle | Idlers No. 1 Hole | Idlers No. 2 Hole† |
|---|---|---|---|---|---|---|---|---|---|
| 33 | 33 | $1\frac{7}{33}$ | 41 | | | | | | |
| 34 | 17 | $1\frac{8}{17}$ | 33 | | | | | | |
| 35 | 49 | $1\frac{7}{49}$ | 26 | | | | | | |
| 36 | 27 | $1\frac{8}{27}$ | 21 | | | | | | |
| 37 | 37 | $1\frac{8}{37}$ | 15 | | | | | | |
| 38 | 19 | $1\frac{1}{19}$ | 9 | | | | | | |
| 39 | 39 | $1\frac{1}{39}$ | 3 | | | | | | |
| 40 | Any | 1 | ... | | | | | | |
| 41 | 41 | $\frac{40}{41}$ | 3* | | | | | | |
| 42 | 21 | $\frac{20}{21}$ | 9* | | | | | | |
| 43 | 43 | $\frac{40}{43}$ | 12* | | | | | | |
| 44 | 33 | $\frac{30}{33}$ | 17* | | | | | | |
| 45 | 27 | $\frac{24}{27}$ | 21* | | | | | | |
| 46 | 23 | $\frac{20}{23}$ | 172 | | | | | | |
| 47 | 47 | $\frac{40}{47}$ | 168 | | | | | | |
| 48 | 18 | $\frac{15}{18}$ | 165 | | | | | | |
| 49 | 49 | $\frac{40}{49}$ | 161 | | | | | | |
| 50 | 20 | $\frac{16}{20}$ | 158 | | | | | | |
| 51 | 17 | $\frac{14}{17}$ | 33* | 24 | ... | ... | ... | 48 | 24 | 44 |
| 52 | 39 | $\frac{30}{39}$ | 152 | ... | ... | ... | ... | ... | ... |
| 53 | 49 | $\frac{35}{49}$ | 140 | 56 | 40 | 24 | 72 | ... | ... |
| 54 | 27 | $\frac{20}{27}$ | 147 | ... | ... | ... | ... | ... | ... |
| 55 | 33 | $\frac{24}{33}$ | 144 | ... | ... | ... | ... | ... | ... |
| 56 | 49 | $\frac{35}{49}$ | 140 | ... | ... | ... | ... | ... | ... |
| 57 | 21 | $\frac{15}{21}$ | 142 | 56 | ... | ... | 40 | 24 | 44 |
| 58 | 29 | $\frac{20}{29}$ | 136 | ... | ... | ... | ... | ... | ... |
| 59 | 39 | $\frac{26}{39}$ | 132 | 48 | ... | ... | 32 | 44 | ... |
| 60 | 39 | $\frac{26}{39}$ | 132 | ... | ... | ... | ... | ... | ... |
| 61 | 39 | $\frac{26}{39}$ | 132 | 48 | ... | ... | 32 | 24 | 44 |
| 62 | 31 | $\frac{20}{31}$ | 127 | ... | ... | ... | ... | ... | ... |
| 63 | 39 | $\frac{26}{39}$ | 132 | 24 | ... | ... | 48 | 24 | 44 |

### Differential Indexing

Certain divisions such as 51, 53, 57, etc. require the use of differential indexing. In differential indexing, change gears are used to transmit motion from main spindle of dividing head to index plate which turns (either in the same direction as the index crank, or in the opposite direction) whatever amount is required to obtain the correct indexing movement.

The numbers in the columns below represent tooth numbers of the necessary change gears. Where no numbers are shown, simple indexing, which does not require change gears, is used.

**Differential Gears**

† On Nos. 1, 1½ and 2 machines, No. 2 hole is in machine table. On Nos. 3 and 4 machines, No. 2 hole is in head.

## Simple and Differential Indexing

| No. of Divisions | Index Circle | No. of Turns of Crank | Graduation on Sector | Gear on Worm | No. 1 Hole First Gear on Stud | No. 1 Hole Second Gear on Stud | Gear on Spindle | Idlers No. 1 Hole | Idlers No. 2 Hole |
|---|---|---|---|---|---|---|---|---|---|
| 64 | 16 | $1\frac{9}{16}$ | 123 | .... | .... | .... | .... | .... | .... |
| 65 | 39 | $2\frac{4}{39}$ | 121 | .... | .... | .... | .... | .... | .... |
| 66 | 33 | $2\frac{0}{33}$ | 120 | .... | .... | .... | .... | .... | .... |
| 67 | 21 | $1\frac{2}{21}$ | 113 | 28 | .... | .... | 48 | 44 | .... |
| 68 | 17 | $1\frac{0}{17}$ | 116 | .... | .... | .... | .... | .... | .... |
| 69 | 20 | $1\frac{2}{20}$ | 118 | 40 | .... | .... | 56 | 24 | 44 |
| 70 | 49 | $2\frac{8}{49}$ | 112 | .... | .... | .... | .... | .... | .... |
| 71 | 18 | $1\frac{0}{18}$ | 109 | 72 | .... | .... | 40 | 24 | .... |
| 72 | 27 | $1\frac{5}{27}$ | 110 | .... | .... | .... | .... | .... | .... |
| 73 | 21 | $1\frac{2}{21}$ | 113 | 28 | .... | .... | 48 | 24 | 44 |
| 74 | 37 | $2\frac{0}{37}$ | 107 | .... | .... | .... | .... | .... | .... |
| 75 | 15 | $\frac{8}{15}$ | 105 | .... | .... | .... | .... | .... | .... |
| 76 | 19 | $1\frac{0}{19}$ | 103 | .... | .... | .... | .... | .... | .... |
| 77 | 20 | $1\frac{0}{20}$ | 98 | 32 | .... | .... | 48 | 44 | .... |
| 78 | 39 | $2\frac{0}{39}$ | 101 | .... | .... | .... | .... | .... | .... |
| 79 | 20 | $1\frac{0}{20}$ | 98 | 48 | .... | .... | 24 | 44 | .... |
| 80 | 20 | $1\frac{0}{20}$ | 98 | .... | .... | .... | .... | .... | .... |
| 81 | 20 | $1\frac{0}{20}$ | 98 | 48 | .... | .... | 24 | 24 | 44 |
| 82 | 41 | $2\frac{0}{41}$ | 96 | .... | .... | .... | .... | .... | .... |
| 83 | 20 | $1\frac{0}{20}$ | 98 | 32 | .... | .... | 48 | 24 | 44 |
| 84 | 21 | $1\frac{0}{21}$ | 94 | .... | .... | .... | .... | .... | .... |
| 85 | 17 | $\frac{8}{17}$ | 92 | .... | .... | .... | .... | .... | .... |
| 86 | 43 | $2\frac{0}{43}$ | 91 | .... | .... | .... | .... | .... | .... |
| 87 | 15 | $\frac{7}{15}$ | 92 | 40 | .... | .... | 24 | 24 | 44 |
| 88 | 33 | $1\frac{5}{33}$ | 89 | .... | .... | .... | .... | .... | .... |
| 89 | 18 | $\frac{8}{18}$ | 87 | 72 | .... | .... | 32 | 44 | .... |
| 90 | 27 | $1\frac{2}{27}$ | 88 | .... | .... | .... | .... | .... | .... |
| 91 | 39 | $1\frac{8}{39}$ | 91 | 24 | .... | .... | 48 | 24 | 44 |
| 92 | 23 | $1\frac{0}{23}$ | 86 | .... | .... | .... | .... | .... | .... |
| 93 | 18 | $\frac{8}{18}$ | 87 | 24 | .... | .... | 32 | 24 | 44 |
| 94 | 47 | $2\frac{0}{47}$ | 83 | .... | .... | .... | .... | .... | .... |
| 95 | 19 | $\frac{8}{19}$ | 82 | .... | .... | .... | .... | .... | .... |
| 96 | 21 | $\frac{9}{21}$ | 85 | 28 | .... | .... | 32 | 24 | 44 |
| 97 | 20 | $\frac{8}{20}$ | 78 | 40 | .... | .... | 48 | 44 | .... |
| 98 | 49 | $2\frac{0}{49}$ | 79 | .... | .... | .... | .... | .... | .... |
| 99 | 20 | $\frac{8}{20}$ | 78 | 56 | 28 | 40 | 32 | .... | .... |
| 100 | 20 | $\frac{8}{20}$ | 78 | .... | .... | .... | .... | .... | .... |
| 101 | 20 | $\frac{8}{20}$ | 78 | 72 | 24 | 40 | 48 | .... | 24 |
| 102 | 20 | $\frac{8}{20}$ | 78 | 40 | .... | .... | 32 | 24 | 44 |
| 103 | 20 | $\frac{8}{20}$ | 78 | 40 | .... | .... | 48 | 24 | 44 |
| 104 | 39 | $1\frac{5}{39}$ | 75 | .... | .... | .... | .... | .... | .... |
| 105 | 21 | $\frac{8}{21}$ | 75 | .... | .... | .... | .... | .... | .... |
| 106 | 43 | $1\frac{6}{43}$ | 73 | 86 | 24 | 24 | 48 | .... | .... |

### Simple and Differential Indexing

| No. of Divisions | Index Circle | No. of Turns of Crank | Graduation on Sector | Gear on Worm | No. 1 Hole | | Gear on Spindle | Idlers | |
|---|---|---|---|---|---|---|---|---|---|
| | | | | | First Gear on Stud | Second Gear on Stud | | No. 1 Hole | No. 2 Hole |
| 107 | 20 | 8/20 | 78 | 40 | 56 | 32 | 64 | .... | 24 |
| 108 | 27 | 10/27 | 73 | .... | .... | .... | .... | .... | .... |
| 109 | 16 | 9/16 | 73 | 32 | .... | .... | 28 | 24 | 44 |
| 110 | 33 | 12/33 | 71 | .... | .... | .... | .... | .... | .... |
| 111 | 39 | 18/39 | 65 | 24 | .... | .... | 72 | 32 | .... |
| 112 | 39 | 18/39 | 65 | 24 | .... | .... | 64 | 44 | .... |
| 113 | 39 | 18/39 | 65 | 24 | .... | .... | 56 | 44 | .... |
| 114 | 39 | 18/39 | 65 | 24 | .... | .... | 48 | 44 | .... |
| 115 | 23 | 8/23 | 68 | .... | .... | .... | .... | .... | .... |
| 116 | 29 | 10/29 | 68 | .... | .... | .... | .... | .... | .... |
| 117 | 39 | 18/39 | 65 | 24 | .... | .... | 24 | 56 | .... |
| 118 | 39 | 18/39 | 65 | 48 | .... | .... | 32 | 44 | .... |
| 119 | 39 | 18/39 | 65 | 72 | .... | .... | 24 | 44 | .... |
| 120 | 39 | 18/39 | 65 | .... | .... | .... | .... | .... | .... |
| 121 | 39 | 18/39 | 65 | 72 | .... | .... | 24 | 24 | 44 |
| 122 | 39 | 18/39 | 65 | 48 | .... | .... | 32 | 24 | 44 |
| 123 | 39 | 18/39 | 65 | 24 | .... | .... | 24 | 24 | 44 |
| 124 | 31 | 10/31 | 63 | .... | .... | .... | .... | .... | .... |
| 125 | 39 | 13/39 | 65 | 24 | .... | .... | 40 | 24 | 44 |
| 126 | 39 | 18/39 | 65 | 24 | .... | .... | 48 | 24 | 44 |
| 127 | 39 | 18/39 | 65 | 24 | .... | .... | 56 | 24 | 44 |
| 128 | 16 | 5/16 | 61 | .... | .... | .... | .... | .... | .... |
| 129 | 39 | 18/39 | 65 | 24 | .... | .... | 72 | 24 | 44 |
| 130 | 39 | 12/39 | 60 | .... | .... | .... | .... | .... | .... |
| 131 | 20 | 9/20 | 58 | 40 | .... | .... | 28 | 44 | .... |
| 132 | 33 | 10/33 | 59 | .... | .... | .... | .... | .... | .... |
| 133 | 21 | 9/21 | 56 | 24 | .... | .... | 48 | 44 | .... |
| 134 | 21 | 9/21 | 56 | 28 | .... | .... | 48 | 44 | .... |
| 135 | 27 | 8/27 | 58 | .... | .... | .... | .... | .... | .... |
| 136 | 17 | 5/17 | 57 | .... | .... | .... | .... | .... | .... |
| 137 | 21 | 9/21 | 56 | 28 | .... | .... | 24 | 56 | .... |
| 138 | 21 | 9/21 | 56 | 56 | .... | .... | 32 | 44 | .... |
| 139 | 21 | 9/21 | 56 | 56 | 32 | 48 | 24 | .... | .... |
| 140 | 49 | 14/49 | 55 | .... | .... | .... | .... | .... | .... |
| 141 | 18 | 5/18 | 54 | 48 | .... | .... | 40 | 44 | .... |
| 142 | 21 | 9/21 | 56 | 56 | .... | .... | 32 | 24 | 44 |
| 143 | 21 | 9/21 | 56 | 28 | .... | .... | 24 | 24 | 44 |
| 144 | 18 | 5/18 | 54 | .... | .... | .... | .... | .... | .... |
| 145 | 29 | 8/29 | 54 | .... | .... | .... | .... | .... | .... |
| 146 | 21 | 9/21 | 56 | 28 | .... | .... | 48 | 24 | 44 |
| 147 | 21 | 9/21 | 56 | 24 | .... | .... | 48 | 24 | 44 |
| 148 | 37 | 10/37 | 53 | .... | .... | .... | .... | .... | .... |
| 149 | 21 | 9/21 | 56 | 28 | .... | .... | 72 | 24 | 44 |

### Simple and Differential Indexing

| No of Divisions | Index Circle | No. of Turns of Crank | Graduation on Sector | Gear on Worm | No. 1 Hole First Gear on Stud | No. 1 Hole Second Gear on Stud | Gear on Spindle | Idlers No. 1 Hole | Idlers No. 2 Hole |
|---|---|---|---|---|---|---|---|---|---|
| 150 | 15 | 4/15 | 52 | .... | .... | .... | .... | .... | .... |
| 151 | 20 | 5/20 | 48 | 32 | .... | .... | 72 | 44 | .... |
| 152 | 19 | 5/19 | 51 | .... | .... | .... | .... | .... | .... |
| 153 | 20 | 5/20 | 48 | 32 | .... | .... | 56 | 44 | .... |
| 154 | 20 | 5/20 | 48 | 32 | .... | .... | 48 | 44 | .... |
| 155 | 31 | 8/31 | 50 | .... | .... | .... | .... | .... | .... |
| 156 | 39 | 10/39 | 50 | .... | .... | .... | .... | .... | .... |
| 157 | 20 | 5/20 | 48 | 32 | .... | .... | 24 | 56 | .... |
| 158 | 20 | 5/20 | 48 | 48 | .... | .... | 24 | 44 | .... |
| 159 | 20 | 5/20 | 48 | 64 | 32 | 56 | 28 | .... | .... |
| 160 | 20 | 5/20 | 48 | .... | .... | .... | .... | .... | .... |
| 161 | 20 | 5/20 | 48 | 64 | 32 | 56 | 28 | .... | 24 |
| 162 | 20 | 5/20 | 48 | 48 | .... | .... | 24 | 24 | 44 |
| 163 | 20 | 5/20 | 48 | 32 | .... | .... | 24 | 24 | 44 |
| 164 | 41 | 10/41 | 47 | .... | .... | .... | .... | .... | .... |
| 165 | 33 | 8/33 | 47 | .... | .... | .... | .... | .... | .... |
| 166 | 20 | 5/20 | 48 | 32 | .... | .... | 48 | 24 | 44 |
| 167 | 20 | 5/20 | 48 | 32 | .... | .... | 56 | 24 | 44 |
| 168 | 21 | 5/21 | 47 | .... | .... | .... | .... | .... | .... |
| 169 | 20 | 5/20 | 48 | 32 | .... | .... | 72 | 24 | 44 |
| 170 | 17 | 4/17 | 45 | .... | .... | .... | .... | .... | .... |
| 171 | 21 | 5/21 | 47 | 56 | .... | .... | 40 | 24 | 44 |
| 172 | 43 | 10/43 | 44 | .... | .... | .... | .... | .... | .... |
| 173 | 18 | 4/18 | 43 | 72 | 56 | 32 | 64 | .... | .... |
| 174 | 18 | 4/18 | 43 | 24 | .... | .... | 32 | 56 | .... |
| 175 | 18 | 4/18 | 43 | 72 | 40 | 32 | 64 | .... | .... |
| 176 | 18 | 4/18 | 43 | 72 | 24 | 24 | 64 | .... | .... |
| 177 | 18 | 4/18 | 43 | 72 | .... | .... | 48 | 24 | .... |
| 178 | 18 | 4/18 | 43 | 72 | .... | .... | 32 | 44 | .... |
| 179 | 18 | 4/18 | 43 | 72 | 24 | 48 | 32 | .... | .... |
| 180 | 18 | 4/18 | 43 | .... | .... | .... | .... | .... | .... |
| 181 | 18 | 4/18 | 43 | 72 | 24 | 48 | 32 | .... | 24 |
| 182 | 18 | 4/18 | 43 | 72 | .... | .... | 32 | 24 | 44 |
| 183 | 18 | 4/18 | 43 | 48 | .... | .... | 32 | 24 | 44 |
| 184 | 23 | 5/23 | 42 | .... | .... | .... | .... | .... | .... |
| 185 | 37 | 8/37 | 42 | .... | .... | .... | .... | .... | .... |
| 186 | 18 | 4/18 | 43 | 48 | .... | .... | 64 | 24 | 44 |
| 187 | 18 | 4/18 | 43 | 72 | 48 | 24 | 56 | .... | 24 |
| 188 | 47 | 10/47 | 40 | .... | .... | .... | .... | .... | .... |
| 189 | 18 | 4/18 | 43 | 32 | .... | .... | 64 | 24 | 44 |
| 190 | 19 | 4/19 | 40 | .... | .... | .... | .... | .... | .... |
| 191 | 20 | 4/20 | 38 | 40 | .... | .... | 72 | 24 | .... |
| 192 | 20 | 4/20 | 38 | 40 | .... | .... | 64 | 44 | .... |

### Simple and Differential Indexing

| No. of Divisions | Index Circle | No. of Turns of Crank | Graduation on Sector | Gear on Worm | No. 1 Hole | | Gear on Spindle | Idlers | |
| --- | --- | --- | --- | --- | --- | --- | --- | --- | --- |
| | | | | | First Gear on Stud | Second Gear on Stud | | No. 1 Hole | No. 2 Hole |
| 193 | 20 | 4/20 | 38 | 40 | .... | .... | 56 | 44 | .... |
| 194 | 20 | 4/20 | 38 | 40 | .... | .... | 48 | 44 | .... |
| 195 | 39 | 8/39 | 39 | .... | .... | .... | .... | .... | .... |
| 196 | 49 | 10/49 | 38 | .... | .... | .... | .... | .... | .... |
| 197 | 20 | 4/20 | 38 | 40 | .... | .... | 24 | 56 | .... |
| 198 | 20 | 4/20 | 38 | 56 | 28 | 40 | 32 | .... | .... |
| 199 | 20 | 4/20 | 38 | 100 | 40 | 64 | 32 | .... | .... |
| 200 | 20 | 4/20 | 38 | .... | .... | .... | .... | .... | .... |
| 201 | 20 | 4/20 | 38 | 72 | 24 | 40 | 24 | .... | 24 |
| 202 | 20 | 4/20 | 38 | 72 | 24 | 40 | 48 | .... | 24 |
| 203 | 20 | 4/20 | 38 | 40 | .... | .... | 24 | 24 | 44 |
| 204 | 20 | 4/20 | 38 | 40 | .... | .... | 32 | 24 | 44 |
| 205 | 41 | 8/41 | 37 | .... | .... | .... | .... | .... | .... |
| 206 | 20 | 4/20 | 38 | 40 | .... | .... | 48 | 24 | 44 |
| 207 | 20 | 4/20 | 38 | 40 | .... | .... | 56 | 24 | 44 |
| 208 | 20 | 4/20 | 38 | 40 | .... | .... | 64 | 24 | 44 |
| 209 | 20 | 4/20 | 38 | 40 | .... | .... | 72 | 24 | 44 |
| 210 | 21 | 4/21 | 37 | .... | .... | .... | .... | .... | .... |
| 211 | 16 | 3/16 | 36 | 64 | .... | .... | 28 | 44 | .... |
| 212 | 43 | 8/43 | 35 | 86 | 24 | 24 | 48 | .... | .... |
| 213 | 27 | 5/27 | 36 | 72 | .... | .... | 40 | 44 | .... |
| 214 | 20 | 4/20 | 38 | 40 | 56 | 32 | 64 | .... | 24 |
| 215 | 43 | 8/43 | 35 | .... | .... | .... | .... | .... | .... |
| 216 | 27 | 5/27 | 36 | .... | .... | .... | .... | .... | .... |
| 217 | 21 | 4/21 | 37 | 48 | .... | .... | 64 | 24 | 44 |
| 218 | 16 | 3/16 | 36 | 64 | .... | .... | 56 | 24 | 44 |
| 219 | 21 | 4/21 | 37 | 28 | .... | .... | 48 | 24 | 44 |
| 220 | 33 | 9/33 | 35 | .... | .... | .... | .... | .... | .... |
| 221 | 17 | 3/17 | 33 | 24 | .... | .... | 24 | 56 | .... |
| 222 | 18 | 3/18 | 32 | 24 | .... | .... | 72 | 44 | .... |
| 223 | 43 | 8/43 | 35 | 86 | 48 | 24 | 64 | .... | 24 |
| 224 | 18 | 3/18 | 32 | 24 | .... | .... | 64 | 44 | .... |
| 225 | 27 | 5/27 | 36 | 24 | .... | .... | 40 | 24 | 44 |
| 226 | 18 | 3/18 | 32 | 24 | .... | .... | 56 | 44 | .... |
| 227 | 49 | 8/49 | 30 | 56 | 64 | 28 | 72 | .... | .... |
| 228 | 18 | 3/18 | 32 | 24 | .... | .... | 48 | 44 | .... |
| 229 | 18 | 3/18 | 32 | 24 | .... | .... | 44 | 48 | .... |
| 230 | 23 | 4/23 | 34 | .... | .... | .... | .... | .... | .... |
| 231 | 18 | 3/18 | 32 | 32 | .... | .... | 48 | 44 | .... |
| 232 | 29 | 5/29 | 33 | .... | .... | .... | .... | .... | .... |
| 233 | 18 | 3/18 | 32 | 48 | .... | .... | 56 | 44 | .... |
| 234 | 18 | 3/18 | 32 | 24 | .... | .... | 24 | 56 | .... |
| 235 | 47 | 8/47 | 32 | .... | .... | .... | .... | .... | .... |

## Simple and Differential Indexing

| No. of Divisions | Index Circle | No. of Turns of Crank | Graduation on Sector | Gear on Worm | No. 1 Hole First Gear on Stud | No. 1 Hole Second Gear on Stud | Gear on Spindle | Idlers No. 1 Hole | Idlers No. 2 Hole |
|---|---|---|---|---|---|---|---|---|---|
| 236 | 18 | 8/18 | 32 | 48 | .... | .... | 32 | 44 | .... |
| 237 | 18 | 8/18 | 32 | 48 | .... | .... | 24 | 44 | .... |
| 238 | 18 | 8/18 | 32 | 72 | .... | .... | 24 | 44 | .... |
| 239 | 18 | 8/18 | 32 | 72 | 24 | 64 | 32 | .... | .... |
| 240 | 18 | 8/18 | 32 | .... | .... | .... | .... | .... | .... |
| 241 | 18 | 8/18 | 32 | 72 | 24 | 64 | 32 | .... | 24 |
| 242 | 18 | 8/18 | 32 | 72 | .... | .... | 24 | 24 | 44 |
| 243 | 18 | 8/18 | 32 | 64 | .... | .... | 32 | 24 | 44 |
| 244 | 18 | 8/18 | 32 | 48 | .... | .... | 32 | 24 | 44 |
| 245 | 49 | 8/49 | 30 | .... | .... | .... | .... | .... | .... |
| 246 | 18 | 8/18 | 32 | 24 | .... | .... | 24 | 24 | 44 |
| 247 | 18 | 8/18 | 32 | 48 | .... | .... | 56 | 24 | 44 |
| 248 | 31 | 5/31 | 31 | .... | .... | .... | ... | .... | .... |
| 249 | 18 | 8/18 | 32 | 32 | .... | .... | 48 | 24 | 44 |
| 250 | 18 | 8/18 | 32 | 24 | .... | .... | 40 | 24 | 44 |
| 251 | 18 | 8/18 | 32 | 48 | 44 | 32 | 64 | .... | 24 |
| 252 | 18 | 8/18 | 32 | 24 | .... | .... | 48 | 24 | 44 |
| 253 | 33 | 5/33 | 29 | 24 | .... | .... | 40 | 56 | .... |
| 254 | 18 | 8/18 | 32 | 24 | .... | .... | 56 | 24 | 44 |
| 255 | 18 | 8/18 | 32 | 48 | 40 | 24 | 72 | .... | 24 |
| 256 | 18 | 8/18 | 32 | 24 | .... | .... | 64 | 24 | 44 |
| 257 | 49 | 8/49 | 30 | 56 | 48 | 28 | 64 | .... | 24 |
| 258 | 43 | 7/43 | 31 | 32 | .... | .... | 64 | 24 | 44 |
| 259 | 21 | 8/21 | 28 | 24 | .... | .... | 72 | 44 | .... |
| 260 | 39 | 6/39 | 29 | .... | .... | .... | .... | .... | .... |
| 261 | 29 | 4/29 | 26 | 48 | 64 | 24 | 72 | .... | .... |
| 262 | 20 | 3/20 | 28 | 40 | .... | .... | 28 | 44 | .... |
| 263 | 49 | 8/49 | 30 | 56 | 64 | 28 | 72 | .... | 24 |
| 264 | 33 | 5/33 | 29 | .... | .... | .... | .... | .... | .... |
| 265 | 21 | 8/21 | 28 | 56 | 40 | 24 | 72 | .... | .... |
| 266 | 21 | 8/21 | 28 | 32 | .... | .... | 64 | 44 | .... |
| 267 | 27 | 4/27 | 28 | 72 | .... | .... | 32 | 44 | .... |
| 268 | 21 | 8/21 | 28 | 28 | .... | .... | 48 | 44 | .... |
| 269 | 20 | 3/20 | 28 | 64 | 32 | 40 | 28 | .... | 24 |
| 270 | 27 | 4/27 | 28 | .... | .... | .... | .... | .... | .... |
| 271 | 21 | 8/21 | 28 | 56 | 24 | 24 | 72 | .... | .... |
| 272 | 21 | 8/21 | 28 | 56 | .... | .... | 64 | 24 | .... |
| 273 | 21 | 8/21 | 28 | 24 | .... | .... | 24 | 56 | .... |
| 274 | 21 | 8/21 | 28 | 56 | .... | .... | 48 | 44 | .... |
| 275 | 21 | 5/21 | 28 | 56 | .... | .... | 40 | 44 | .... |
| 276 | 21 | 8/21 | 28 | 56 | .... | .... | 32 | 44 | .... |
| 277 | 21 | 8/21 | 28 | 56 | .... | .... | 24 | 44 | .... |
| 278 | 21 | 8/21 | 28 | 56 | 32 | 48 | 24 | .... | .... |

## Simple and Differential Indexing

| No. of Divisions | Index Circle | No. of Turns of Crank | Graduation on Sector | Gear on Worm | No. 1 Hole | | Gear on Spindle | Idlers | |
|---|---|---|---|---|---|---|---|---|---|
| | | | | | First Gear on Stud | Second Gear on Stud | | No. 1 Hole | No. 2 Hole |
| 279 | 27 | 4/27 | 28 | 24 | .... | .... | 32 | 24 | 44 |
| 280 | 49 | 7/49 | 26 | .... | .... | .... | .... | .... | .... |
| 281 | 21 | 8/21 | 28 | 72 | 24 | 56 | 24 | .... | 24 |
| 282 | 43 | 9/43 | 26 | 86 | 24 | 24 | 56 | .... | .... |
| 283 | 21 | 8/21 | 28 | 56 | .... | .... | 24 | 24 | 44 |
| 284 | 21 | 8/21 | 28 | 56 | .... | .... | 32 | 24 | 44 |
| 285 | 21 | 8/21 | 28 | 56 | .... | .... | 40 | 24 | 44 |
| 286 | 21 | 8/21 | 28 | 56 | .... | .... | 48 | 24 | 44 |
| 287 | 21 | 8/21 | 28 | 24 | .... | .... | 24 | 24 | 44 |
| 288 | 21 | 8/21 | 28 | 28 | .... | .... | 32 | 24 | 44 |
| 289 | 21 | 8/21 | 28 | 56 | 24 | 24 | 72 | .... | 24 |
| 290 | 29 | 4/29 | 26 | .... | .... | .... | .... | .... | .... |
| 291 | 15 | 2/15 | 25 | 40 | .... | .... | 48 | 44 | .... |
| 292 | 21 | 8/21 | 28 | 28 | .... | .... | 48 | 24 | 44 |
| 293 | 15 | 2/15 | 25 | 48 | 32 | 40 | 56 | .... | .... |
| 294 | 21 | 8/21 | 28 | 24 | .... | .... | 48 | 24 | 44 |
| 295 | 15 | 2/15 | 25 | 48 | .... | .... | 32 | 44 | .... |
| 296 | 37 | 5/37 | 26 | .... | .... | .... | .... | .... | .... |
| 297 | 33 | 4/33 | 23 | 28 | 48 | 24 | 56 | .... | .... |
| 298 | 21 | 8/21 | 28 | 28 | .... | .... | 72 | 24 | 44 |
| 299 | 23 | 8/23 | 25 | 24 | .... | .... | 24 | 56 | .... |
| 300 | 15 | 2/15 | 25 | .... | .... | .... | .... | .... | .... |
| 301 | 43 | 9/43 | 26 | 24 | .... | .... | 48 | 24 | 44 |
| 302 | 16 | 2/16 | 24 | 32 | .... | .... | 72 | 24 | .... |
| 303 | 15 | 2/15 | 25 | 72 | 24 | 40 | 48 | .... | 24 |
| 304 | 16 | 2/16 | 24 | 24 | .... | .... | 48 | 44 | .... |
| 305 | 15 | 2/15 | 25 | 48 | .... | .... | 32 | 24 | 44 |
| 306 | 15 | 2/15 | 25 | 40 | .... | .... | 32 | 24 | 44 |
| 307 | 15 | 2/15 | 25 | 72 | 48 | 40 | 56 | .... | 24 |
| 308 | 16 | 2/16 | 24 | 32 | .... | .... | 48 | 44 | .... |
| 309 | 15 | 2/15 | 25 | 40 | .... | .... | 48 | 24 | 44 |
| 310 | 31 | 4/31 | 24 | .... | .... | .... | .... | .... | .... |
| 311 | 16 | 2/16 | 24 | 64 | 24 | 24 | 72 | .... | .... |
| 312 | 39 | 5/39 | 24 | .... | .... | .... | .... | .... | .... |
| 313 | 16 | 2/16 | 24 | 32 | .... | .... | 28 | 56 | .... |
| 314 | 16 | 2/16 | 24 | 32 | .... | .... | 24 | 56 | .... |
| 315 | 16 | 2/16 | 24 | 64 | .... | .... | 40 | 24 | .... |
| 316 | 16 | 2/16 | 24 | 64 | .... | .... | 32 | 44 | .... |
| 317 | 16 | 2/16 | 24 | 64 | .... | .... | 24 | 44 | .... |
| 318 | 16 | 2/16 | 24 | 56 | 28 | 48 | 24 | .... | .... |
| 319 | 29 | 4/29 | 26 | 48 | 64 | 24 | 72 | .... | 24 |
| 320 | 16 | 2/16 | 24 | .... | .... | .... | .... | .... | .... |
| 321 | 16 | 2/16 | 24 | 72 | 24 | 64 | 24 | .... | 24 |

## Simple and Differential Indexing

| No. of Divisions | Index Circle | No. of Turns of Crank | Graduation on Sector | Gear on Worm | First Gear on Stud | Second Gear on Stud | Gear on Spindle | Idlers No. 1 Hole | Idlers No. 2 Hole |
|---|---|---|---|---|---|---|---|---|---|
| 322 | 23 | 8/23 | 25 | 32 | .... | .... | 64 | 24 | 44 |
| 323 | 16 | 7/16 | 24 | 64 | .... | .... | 24 | 24 | 44 |
| 324 | 16 | 7/16 | 24 | 64 | .... | .... | 32 | 24 | 44 |
| 325 | 16 | 7/16 | 24 | 64 | .... | .... | 40 | 24 | 44 |
| 326 | 16 | 7/16 | 24 | 32 | .... | .... | 24 | 24 | 44 |
| 327 | 16 | 7/16 | 24 | 32 | .... | .... | 28 | 24 | 44 |
| 328 | 41 | 5/41 | 23 | .... | .... | .... | .... | .... | .... |
| 329 | 16 | 7/16 | 24 | 64 | 24 | 24 | 72 | .... | 24 |
| 330 | 33 | 4/33 | 23 | .... | .... | .... | .... | .... | .... |
| 331 | 16 | 7/16 | 24 | 64 | 44 | 24 | 48 | .... | 24 |
| 332 | 16 | 7/16 | 24 | 32 | .... | .... | 48 | 24 | 44 |
| 333 | 18 | 7/18 | 21 | 24 | .... | .... | 72 | 44 | 44 |
| 334 | 16 | 7/16 | 24 | 32 | .... | .... | 56 | 24 | 44 |
| 335 | 33 | 4/33 | 23 | 72 | 48 | 44 | 40 | .... | 24 |
| 336 | 16 | 7/16 | 24 | 32 | .... | .... | 64 | 24 | 44 |
| 337 | 43 | 5/43 | 21 | 86 | 40 | 32 | 56 | .... | .... |
| 338 | 16 | 7/16 | 24 | 32 | .... | .... | 72 | 24 | 44 |
| 339 | 18 | 7/18 | 21 | 24 | .... | .... | 56 | 44 | .... |
| 340 | 17 | 7/17 | 22 | .... | .... | .... | .... | .... | .... |
| 341 | 43 | 5/43 | 21 | 86 | 24 | 32 | 40 | .... | .... |
| 342 | 18 | 7/18 | 21 | 32 | .... | .... | 64 | 44 | .... |
| 343 | 15 | 7/15 | 25 | 40 | 64 | 24 | 86 | .... | 24 |
| 344 | 43 | 5/43 | 21 | .... | .... | .... | .... | .... | .... |
| 345 | 18 | 7/18 | 21 | 24 | .... | .... | 40 | 56 | .... |
| 346 | 18 | 7/18 | 21 | 72 | 56 | 32 | 64 | .... | .... |
| 347 | 43 | 5/43 | 21 | 86 | 24 | 32 | 40 | .... | 24 |
| 348 | 18 | 7/18 | 21 | 24 | .... | .... | 32 | 56 | .... |
| 349 | 18 | 7/18 | 21 | 72 | 44 | 24 | 48 | .... | .... |
| 350 | 18 | 7/18 | 21 | 72 | 40 | 32 | 64 | .... | .... |
| 351 | 18 | 7/18 | 21 | 24 | .... | .... | 24 | 56 | .... |
| 352 | 18 | 7/18 | 21 | 72 | 24 | 24 | 64 | .... | .... |
| 353 | 18 | 7/18 | 21 | 72 | 24 | 24 | 56 | .... | .... |
| 354 | 18 | 7/18 | 21 | 72 | .... | .... | 48 | 24 | .... |
| 355 | 18 | 7/18 | 21 | 72 | .... | .... | 40 | 24 | .... |
| 356 | 18 | 7/18 | 21 | 72 | .... | .... | 32 | 24 | .... |
| 357 | 18 | 7/18 | 21 | 72 | .... | .... | 24 | 44 | .... |
| 358 | 18 | 7/18 | 21 | 72 | 32 | 48 | 24 | .... | .... |
| 359 | 43 | 5/43 | 21 | 86 | 48 | 32 | 100 | .... | 24 |
| 360 | 18 | 7/18 | 21 | .... | .... | .... | .... | .... | .... |
| 361 | 19 | 7/19 | 19 | 32 | .... | .... | 64 | 44 | .... |
| 362 | 18 | 7/18 | 21 | 72 | 28 | 56 | 32 | .... | 24 |
| 363 | 18 | 7/18 | 21 | 72 | .... | .... | 24 | 24 | 44 |
| 364 | 18 | 7/18 | 21 | 72 | .... | .... | 32 | 24 | 44 |

## Indexing Movements for Standard Index Plate — Cincinnati Milling Machine

The standard index plate indexes all numbers up to and including 60; all even numbers and those divisible by 5 up to 120; and all divisions listed below up to 400. This plate is drilled on both sides, and has holes as follows:

First side: 24, 25, 28, 30, 34, 37, 38, 39, 41, 42, 43.
Second side: 46, 47, 49, 51, 53, 54, 57, 58, 59, 62, 66.

| No. of Divisions | Circle | Turns | Holes | No. of Divisions | Circle | Holes | No. of Divisions | Circle | Holes | No. of Divisions | Circle | Holes |
|---|---|---|---|---|---|---|---|---|---|---|---|---|
| 2 | Any | 20 | ... | 44 | 66 | 60 | 104 | 39 | 15 | 205 | 41 | 8 |
| 3 | 24 | 13 | 8 | 45 | 54 | 48 | 105 | 42 | 16 | 210 | 42 | 8 |
| 4 | Any | 10 | ... | 46 | 46 | 40 | 106 | 53 | 20 | 212 | 53 | 10 |
| 5 | Any | 8 | ... | 47 | 47 | 40 | 108 | 54 | 20 | 215 | 43 | 8 |
| 6 | 24 | 6 | 16 | 48 | 24 | 20 | 110 | 66 | 24 | 216 | 54 | 10 |
| 7 | 28 | 5 | 20 | 49 | 49 | 40 | 112 | 28 | 10 | 220 | 66 | 12 |
| 8 | Any | 5 | ... | 50 | 25 | 20 | 114 | 57 | 20 | 224 | 28 | 5 |
| 9 | 54 | 4 | 24 | 51 | 51 | 40 | 115 | 46 | 16 | 228 | 57 | 10 |
| 10 | Any | 4 | ... | 52 | 39 | 30 | 116 | 58 | 20 | 230 | 46 | 8 |
| 11 | 66 | 3 | 42 | 53 | 53 | 40 | 118 | 59 | 20 | 232 | 58 | 10 |
| 12 | 24 | 3 | 8 | 54 | 54 | 40 | 120 | 66 | 22 | 235 | 47 | 8 |
| 13 | 39 | 3 | 3 | 55 | 66 | 48 | 124 | 62 | 20 | 236 | 59 | 10 |
| 14 | 49 | 2 | 42 | 56 | 28 | 20 | 125 | 25 | 8 | 240 | 66 | 11 |
| 15 | 24 | 2 | 16 | 57 | 57 | 40 | 130 | 39 | 12 | 245 | 49 | 8 |
| 16 | 24 | 2 | 12 | 58 | 58 | 40 | 132 | 66 | 20 | 248 | 62 | 10 |
| 17 | 34 | 2 | 12 | 59 | 59 | 40 | 135 | 54 | 16 | 250 | 25 | 4 |
| 18 | 54 | 2 | 12 | 60 | 42 | 28 | 136 | 34 | 10 | 255 | 51 | 8 |
| 19 | 38 | 2 | 4 | 62 | 62 | 40 | 140 | 28 | 8 | 260 | 39 | 6 |
| 20 | Any | 2 | ... | 64 | 24 | 15 | 144 | 54 | 15 | 264 | 66 | 10 |
| 21 | 42 | 1 | 38 | 65 | 39 | 24 | 145 | 58 | 16 | 270 | 54 | 8 |
| 22 | 66 | 1 | 54 | 66 | 66 | 40 | 148 | 37 | 10 | 272 | 34 | 5 |
| 23 | 46 | 1 | 34 | 68 | 34 | 20 | 150 | 30 | 8 | 280 | 28 | 4 |
| 24 | 24 | 1 | 16 | 70 | 28 | 16 | 152 | 38 | 10 | 290 | 58 | 8 |
| 25 | 25 | 1 | 15 | 72 | 54 | 30 | 155 | 62 | 16 | 296 | 37 | 5 |
| 26 | 39 | 1 | 21 | 74 | 37 | 20 | 156 | 39 | 10 | 300 | 30 | 4 |
| 27 | 54 | 1 | 26 | 75 | 30 | 16 | 160 | 28 | 7 | 304 | 38 | 5 |
| 28 | 42 | 1 | 18 | 76 | 38 | 20 | 164 | 41 | 10 | 310 | 62 | 8 |
| 29 | 58 | 1 | 22 | 78 | 39 | 20 | 165 | 66 | 16 | 312 | 39 | 5 |
| 30 | 24 | 1 | 8 | 80 | 34 | 17 | 168 | 42 | 10 | 320 | 24 | 3 |
| 31 | 62 | 1 | 18 | 82 | 41 | 20 | 170 | 34 | 8 | 328 | 41 | 5 |
| 32 | 28 | 1 | 7 | 84 | 42 | 20 | 172 | 43 | 10 | 330 | 66 | 8 |
| 33 | 66 | 1 | 14 | 85 | 34 | 16 | 176 | 66 | 15 | 336 | 42 | 5 |
| 34 | 34 | 1 | 6 | 86 | 43 | 20 | 180 | 54 | 12 | 340 | 34 | 4 |
| 35 | 28 | 1 | 4 | 88 | 66 | 30 | 184 | 46 | 10 | 344 | 43 | 5 |
| 36 | 54 | 1 | 6 | 90 | 54 | 24 | 185 | 37 | 8 | 360 | 54 | 6 |
| 37 | 37 | 1 | 3 | 92 | 46 | 20 | 188 | 47 | 10 | 368 | 46 | 5 |
| 38 | 38 | 1 | 2 | 94 | 47 | 20 | 190 | 38 | 8 | 370 | 37 | 4 |
| 39 | 39 | 1 | 1 | 95 | 38 | 16 | 192 | 24 | 5 | 376 | 47 | 5 |
| 40 | Any | 1 | ... | 96 | 24 | 10 | 195 | 39 | 8 | 380 | 38 | 4 |
| 41 | 41 | ... | 40 | 98 | 49 | 20 | 196 | 49 | 10 | 390 | 39 | 4 |
| 42 | 42 | ... | 40 | 100 | 25 | 10 | 200 | 30 | 6 | 392 | 49 | 5 |
| 43 | 43 | ... | 40 | 102 | 51 | 20 | 204 | 51 | 10 | 400 | 30 | 3 |

## Indexing Movements for High Numbers — Cincinnati Milling Machine

This set of 3 index plates indexes all numbers up to and including 200; all even numbers and those divisible by 5 up to and including 400. The plates are drilled on each side, making six sides A, B, C, D, E and F.

*Example:* — It is required to index 35 divisions. The preferred side is F, since this requires the least number of holes; but should one of plates D, A or E be in place, either can be used, thus avoiding the changing of plates.

| No. of Divisions | Side | Circle | Turns | Holes | No. of Divisions | Side | Circle | Turns | Holes | No. of Divisions | Side | Circle | Turns | Holes |
|---|---|---|---|---|---|---|---|---|---|---|---|---|---|---|
| 2 | Any | Any | 20 | .... | 15 | C | 93 | 2 | 62 | 28 | D | 77 | 1 | 33 |
| 3 | A | 30 | 13 | 10 | 15 | F | 159 | 2 | 106 | 28 | A | 91 | 1 | 39 |
| 3 | B | 36 | 13 | 12 | 16 | E | 26 | 2 | 13 | 29 | E | 87 | 1 | 33 |
| 3 | E | 42 | 13 | 14 | 16 | F | 28 | 2 | 14 | 30 | A | 30 | 1 | 10 |
| 3 | C | 93 | 13 | 31 | 16 | A | 30 | 2 | 15 | 30 | B | 36 | 1 | 12 |
| 3 | F | 159 | 13 | 53 | 16 | D | 32 | 2 | 16 | 30 | E | 42 | 1 | 14 |
| 4 | Any | Any | 10 | .... | 16 | C | 34 | 2 | 17 | 30 | C | 93 | 1 | 31 |
| 5 | Any | Any | 8 | .... | 16 | B | 36 | 2 | 18 | 30 | F | 159 | 1 | 53 |
| 6 | A | 30 | 6 | 20 | 17 | C | 34 | 2 | 12 | 31 | C | 93 | 1 | 27 |
| 6 | B | 36 | 6 | 24 | 17 | E | 119 | 2 | 42 | 32 | A | 28 | 1 | 7 |
| 6 | E | 42 | 6 | 28 | 17 | C | 153 | 2 | 54 | 32 | D | 32 | 1 | 8 |
| 6 | C | 93 | 6 | 62 | 17 | F | 187 | 2 | 66 | 32 | B | 36 | 1 | 9 |
| 6 | F | 159 | 6 | 106 | 18 | B | 36 | 2 | 8 | 32 | A | 48 | 1 | 12 |
| 7 | F | 28 | 5 | 20 | 18 | A | 99 | 2 | 22 | 33 | A | 99 | 1 | 21 |
| 7 | E | 42 | 5 | 30 | 18 | C | 153 | 2 | 34 | 34 | C | 34 | 1 | 6 |
| 7 | D | 77 | 5 | 55 | 19 | F | 38 | 2 | 4 | 34 | E | 119 | 1 | 21 |
| 7 | A | 91 | 5 | 65 | 19 | E | 133 | 2 | 14 | 34 | F | 187 | 1 | 33 |
| 8 | Any | Any | 5 | .... | 19 | A | 171 | 2 | 18 | 35 | F | 28 | 1 | 4 |
| 9 | B | 36 | 4 | 16 | 20 | Any | Any | 2 | .... | 35 | D | 77 | 1 | 11 |
| 9 | A | 99 | 4 | 44 | 21 | E | 42 | 1 | 38 | 35 | A | 91 | 1 | 13 |
| 9 | C | 153 | 4 | 68 | 21 | A | 147 | 1 | 133 | 35 | E | 119 | 1 | 17 |
| 10 | Any | Any | 4 | .... | 22 | D | 44 | 1 | 36 | 36 | B | 36 | 1 | 4 |
| 11 | D | 44 | 3 | 28 | 22 | A | 99 | 1 | 81 | 36 | A | 99 | 1 | 11 |
| 11 | A | 99 | 3 | 63 | 22 | F | 143 | 1 | 117 | 36 | C | 153 | 1 | 17 |
| 11 | F | 143 | 3 | 91 | 23 | C | 46 | 1 | 34 | 37 | B | 111 | 1 | 9 |
| 12 | A | 30 | 3 | 10 | 23 | A | 69 | 1 | 51 | 38 | F | 38 | 1 | 2 |
| 12 | B | 36 | 3 | 12 | 23 | E | 161 | 1 | 119 | 38 | E | 133 | 1 | 7 |
| 12 | E | 42 | 3 | 14 | 24 | A | 30 | 1 | 20 | 38 | A | 171 | 1 | 9 |
| 12 | C | 93 | 3 | 31 | 24 | B | 36 | 1 | 24 | 39 | A | 117 | 1 | 3 |
| 12 | F | 159 | 3 | 53 | 24 | E | 42 | 1 | 28 | 40 | Any | Any | 1 | .... |
| 13 | E | 26 | 3 | 2 | 24 | C | 93 | 1 | -62 | 41 | C | 123 | .... | 120 |
| 13 | A | 91 | 3 | 7 | 24 | F | 159 | 1 | 106 | 42 | E | 42 | .... | 40 |
| 13 | F | 143 | 3 | 11 | 25 | A | 30 | 1 | 18 | 42 | A | 147 | .... | 140 |
| 13 | B | 169 | 3 | 13 | 25 | E | 175 | 1 | 105 | 43 | A | 129 | .... | 120 |
| 14 | F | 28 | 2 | 24 | 26 | F | 26 | 1 | 14 | 44 | D | 44 | .... | 40 |
| 14 | E | 42 | 2 | 36 | 26 | A | 91 | 1 | 49 | 44 | A | 99 | .... | 90 |
| 14 | D | 77 | 2 | 66 | 26 | B | 169 | 1 | 91 | 44 | F | 143 | .... | 130 |
| 14 | A | 91 | 2 | 78 | 27 | B | 81 | 1 | 39 | 45 | B | 36 | .... | 32 |
| 15 | A | 30 | 2 | 20 | 27 | A | 189 | 1 | 91 | 45 | A | 99 | .... | 88 |
| 15 | B | 36 | 2 | 24 | 28 | F | 28 | 1 | 12 | 45 | C | 153 | .... | 136 |
| 15 | E | 42 | 2 | 28 | 28 | E | 42 | 1 | 18 | 46 | C | 46 | .... | 40 |

Indexing Movements for High Numbers — Cincinnati Milling Machine

| No. of Divisions | Side | Circle | Holes | No. of Divisions | Side | Circle | Holes | No. of Divisions | Side | Circle | Holes |
|---|---|---|---|---|---|---|---|---|---|---|---|
| 46 | A | 69 | 60 | 70 | E | 119 | 68 | 96 | B | 36 | 15 |
| 46 | E | 161 | 140 | 71 | F | 71 | 40 | 96 | A | 48 | 20 |
| 47 | B | 141 | 120 | 72 | B | 36 | 20 | 97 | B | 97 | 40 |
| 48 | A | 30 | 25 | 72 | A | 117 | 65 | 98 | A | 147 | 60 |
| 48 | B | 36 | 30 | 72 | C | 153 | 85 | 99 | A | 99 | 40 |
| 49 | A | 147 | 120 | 73 | E | 73 | 40 | 100 | A | 30 | 12 |
| 50 | A | 30 | 24 | 74 | B | 111 | 60 | 100 | E | 175 | 70 |
| 50 | E | 175 | 140 | 75 | A | 30 | 16 | 101 | F | 101 | 40 |
| 51 | C | 153 | 120 | 76 | F | 38 | 20 | 102 | C | 153 | 60 |
| 52 | E | 26 | 20 | 76 | E | 133 | 70 | 103 | E | 103 | 40 |
| 52 | A | 91 | 70 | 76 | A | 171 | 90 | 104 | E | 26 | 10 |
| 52 | F | 143 | 110 | 77 | D | 77 | 40 | 104 | A | 91 | 35 |
| 52 | B | 169 | 130 | 78 | A | 117 | 60 | 104 | F | 143 | 55 |
| 53 | F | 159 | 120 | 79 | C | 79 | 40 | 104 | B | 169 | 65 |
| 54 | B | 81 | 60 | 80 | E | 26 | 13 | 105 | E | 42 | 16 |
| 54 | A | 189 | 140 | 80 | F | 28 | 14 | 105 | A | 147 | 56 |
| 55 | D | 44 | 32 | 80 | A | 30 | 15 | 106 | F | 159 | 60 |
| 55 | F | 143 | 104 | 80 | D | 32 | 16 | 107 | D | 107 | 40 |
| 56 | F | 28 | 20 | 80 | C | 34 | 17 | 108 | B | 81 | 30 |
| 56 | E | 42 | 30 | 80 | B | 36 | 18 | 108 | A | 189 | 70 |
| 56 | D | 77 | 55 | 80 | E | 42 | 21 | 109 | C | 109 | 40 |
| 56 | A | 91 | 65 | 81 | B | 81 | 40 | 110 | D | 44 | 16 |
| 57 | A | 171 | 120 | 82 | C | 123 | 60 | 110 | A | 99 | 36 |
| 58 | E | 87 | 60 | 83 | F | 83 | 40 | 110 | F | 143 | 52 |
| 59 | A | 177 | 120 | 84 | E | 42 | 20 | 111 | B | 111 | 40 |
| 60 | A | 30 | 20 | 84 | A | 147 | 70 | 112 | F | 28 | 10 |
| 60 | B | 36 | 24 | 85 | C | 34 | 16 | 112 | E | 42 | 15 |
| 60 | E | 42 | 28 | 85 | E | 119 | 56 | 113 | F | 113 | 40 |
| 60 | F | 159 | 106 | 85 | F | 187 | 88 | 114 | A | 171 | 60 |
| 61 | B | 183 | 120 | 86 | A | 129 | 60 | 115 | C | 46 | 16 |
| 62 | C | 93 | 60 | 87 | E | 87 | 40 | 115 | A | 69 | 24 |
| 63 | A | 189 | 120 | 88 | D | 44 | 20 | 115 | E | 161 | 56 |
| 64 | D | 32 | 20 | 88 | A | 99 | 45 | 116 | E | 87 | 30 |
| 64 | A | 48 | 30 | 88 | F | 143 | 65 | 117 | A | 117 | 40 |
| 65 | E | 26 | 16 | 89 | D | 89 | 40 | 118 | A | 177 | 60 |
| 65 | A | 91 | 56 | 90 | B | 36 | 16 | 119 | E | 119 | .40 |
| 65 | F | 143 | 88 | 90 | A | 99 | 44 | 120 | A | 30 | 10 |
| 65 | B | 169 | 104 | 90 | C | 153 | 68 | 120 | B | 36 | 12 |
| 66 | A | 99 | 60 | 91 | A | 91 | 40 | 120 | E | 42 | 14 |
| 67 | B | 67 | 40 | 92 | C | 46 | 20 | 120 | C | 93 | 31 |
| 68 | C | 34 | 20 | 92 | A | 69 | 30 | 120 | F | 159 | 53 |
| 68 | E | 119 | 70 | 92 | E | 161 | 70 | 121 | D | 121 | 40 |
| 68 | F | 187 | 110 | 93 | C | 93 | 40 | 122 | B | 183 | 60 |
| 69 | A | 69 | 40 | 94 | B | 141 | 60 | 123 | C | 123 | 40 |
| 70 | F | 28 | 16 | 95 | F | 38 | 16 | 124 | C | 93 | 30 |
| 70 | D | 42 | 24 | 95 | E | 133 | 56 | 125 | E | 175 | 56 |
| 70 | A | 91 | 52 | 95 | A | 171 | 72 | 126 | A | 189 | 60 |

## Indexing Movements for High Numbers — Cincinnati Milling Machine

| No. of Divisions | Side | Circle | Holes | No. of Divisions | Side | Circle | Holes | No. of Divisions | Side | Circle | Holes |
|---|---|---|---|---|---|---|---|---|---|---|---|
| 127 | B | 127 | 40 | 160 | A | 48 | 12 | 198 | A | 99 | 20 |
| 128 | D | 32 | 10 | 161 | E | 161 | 40 | 199 | B | 199 | 40 |
| 128 | A | 48 | 15 | 162 | B | 81 | 20 | 200 | A | 30 | 6 |
| 129 | A | 129 | 40 | 163 | D | 163 | 40 | 200 | E | 175 | 35 |
| 130 | E | 26 | 8 | 164 | C | 123 | 30 | 202 | F | 101 | 20 |
| 130 | A | 91 | 28 | 165 | A | 99 | 24 | 204 | C | 153 | 30 |
| 130 | F | 143 | 44 | 166 | F | 83 | 20 | 205 | C | 123 | 24 |
| 130 | B | 169 | 52 | 167 | C | 167 | 40 | 206 | E | 103 | 20 |
| 131 | F | 131 | 40 | 168 | E | 42 | 10 | 208 | E | 26 | 5 |
| 132 | A | 99 | 30 | 168 | A | 147 | 35 | 210 | E | 42 | 8 |
| 133 | E | 133 | 40 | 169 | B | 169 | 40 | 210 | A | 147 | 28 |
| 134 | B | 67 | 20 | 170 | C | 34 | 8 | 212 | F | 159 | 30 |
| 135 | B | 81 | 24 | 170 | E | 119 | 28 | 214 | D | 107 | 20 |
| 135 | A | 189 | 56 | 170 | F | 187 | 44 | 215 | A | 129 | 24 |
| 136 | C | 34 | 10 | 171 | A | 171 | 40 | 216 | B | 81 | 15 |
| 136 | E | 119 | 35 | 172 | A | 129 | 30 | 216 | A | 189 | 35 |
| 137 | D | 137 | 40 | 173 | F | 173 | 40 | 218 | C | 109 | 20 |
| 138 | A | 69 | 20 | 174 | E | 87 | 20 | 220 | D | 44 | 8 |
| 139 | C | 139 | 40 | 175 | E | 175 | 40 | 220 | A | 99 | 18 |
| 140 | F | 28 | 8 | 176 | D | 44 | 10 | 220 | F | 143 | 26 |
| 140 | E | 42 | 12 | 177 | A | 177 | 40 | 222 | B | 111 | 20 |
| 140 | D | 77 | 22 | 178 | D | 89 | 20 | 224 | F | 28 | 5 |
| 140 | A | 91 | 26 | 179 | D | 179 | 40 | 226 | F | 113 | 20 |
| 141 | B | 141 | 40 | 180 | B | 36 | 8 | 228 | A | 171 | 30 |
| 142 | F | 71 | 20 | 180 | A | 99 | 22 | 230 | C | 46 | 8 |
| 143 | F | 143 | 40 | 180 | C | 153 | 34 | 230 | A | 69 | 12 |
| 144 | B | 36 | 10 | 181 | C | 181 | 40 | 230 | E | 161 | 28 |
| 145 | E | 87 | 24 | 182 | A | 91 | 20 | 232 | E | 87 | 15 |
| 146 | E | 73 | 20 | 183 | B | 183 | 40 | 234 | A | 117 | 20 |
| 147 | A | 147 | 40 | 184 | C | 46 | 10 | 235 | B | 141 | 24 |
| 148 | B | 111 | 30 | 184 | A | 69 | 15 | 236 | A | 177 | 30 |
| 149 | E | 149 | 40 | 184 | E | 161 | 35 | 238 | E | 119 | 20 |
| 150 | A | 30 | 8 | 185 | B | 111 | 24 | 240 | A | 30 | 5 |
| 151 | D | 151 | 40 | 186 | C | 93 | 20 | 240 | B | 36 | 6 |
| 152 | F | 38 | 10 | 187 | F | 187 | 40 | 240 | E | 42 | 7 |
| 152 | E | 133 | 35 | 188 | B | 141 | 30 | 240 | A | 48 | 8 |
| 152 | A | 171 | 45 | 189 | A | 189 | 40 | 242 | D | 121 | 20 |
| 153 | C | 153 | 40 | 190 | F | 38 | 8 | 244 | B | 183 | 30 |
| 154 | D | 77 | 20 | 190 | E | 133 | 28 | 245 | A | 147 | 24 |
| 155 | C | 93 | 24 | 190 | A | 171 | 40 | 246 | C | 123 | 20 |
| 156 | A | 117 | 30 | 191 | E | 191 | 40 | 248 | C | 93 | 15 |
| 157 | B | 157 | 40 | 192 | A | 48 | 10 | 250 | E | 175 | 28 |
| 158 | C | 79 | 20 | 193 | D | 193 | 40 | 252 | A | 189 | 30 |
| 159 | F | 159 | 40 | 194 | B | 97 | 20 | 254 | B | 127 | 20 |
| 160 | F | 28 | 7 | 195 | A | 117 | 24 | 255 | C | 153 | 24 |
| 160 | D | 32 | 8 | 196 | A | 147 | 30 | 256 | D | 32 | 5 |
| 160 | B | 36 | 9 | 197 | C | 197 | 40 | 258 | A | 129 | 20 |

Indexing Movements for High Numbers — Cincinnati Milling Machine

| No. of Divisions | Side | Circle | Holes | No. of Divisions | Side | Circle | Holes | No. of Divisions | Side | Circle | Holes |
|---|---|---|---|---|---|---|---|---|---|---|---|
| 260 | E | 26 | 4 | 304 | F | 38 | 5 | 354 | A | 177 | 20 |
| 260 | A | 91 | 14 | 305 | B | 183 | 24 | 355 | F | 71 | 8 |
| 260 | F | 143 | 22 | 306 | C | 153 | 20 | 356 | D | 89 | 10 |
| 260 | B | 169 | 26 | 308 | D | 77 | 10 | 358 | D | 179 | 20 |
| 262 | F | 131 | 20 | 310 | C | 93 | 12 | 360 | B | 36 | 4 |
| 264 | A | 99 | 15 | 312 | A | 117 | 15 | 360 | A | 99 | 11 |
| 265 | F | 159 | 24 | 314 | B | 157 | 20 | 360 | C | 153 | 17 |
| 266 | E | 133 | 20 | 315 | A | 189 | 24 | 362 | C | 181 | 20 |
| 268 | B | 67 | 10 | 316 | C | 79 | 10 | 364 | A | 91 | 10 |
| 270 | B | 81 | 12 | 318 | F | 159 | 20 | 365 | E | 73 | 8 |
| 270 | A | 189 | 28 | 320 | D | 32 | 4 | 366 | B | 183 | 20 |
| 272 | C | 34 | 5 | 320 | A | 48 | 6 | 368 | C | 46 | 5 |
| 274 | D | 137 | 20 | 322 | E | 161 | 20 | 370 | B | 111 | 12 |
| 276 | A | 69 | 10 | 324 | B | 81 | 10 | 372 | C | 93 | 10 |
| 278 | C | 139 | 20 | 326 | D | 163 | 20 | 374 | F | 187 | 20 |
| 280 | F | 28 | 4 | 328 | C | 123 | 15 | 376 | B | 141 | 15 |
| 280 | E | 42 | 6 | 330 | A | 99 | 12 | 378 | A | 189 | 20 |
| 280 | D | 77 | 11 | 332 | F | 83 | 10 | 380 | F | 38 | 4 |
| 280 | A | 91 | 13 | 334 | C | 167 | 20 | 380 | E | 133 | 14 |
| 282 | B | 141 | 20 | 335 | B | 67 | 8 | 380 | A | 171 | 18 |
| 284 | F | 71 | 10 | 336 | E | 42 | 5 | 382 | C | 191 | 20 |
| 285 | A | 171 | 24 | 338 | B | 169 | 20 | 384 | A | 48 | 5 |
| 286 | F | 143 | 20 | 340 | C | 34 | 4 | 385 | D | 77 | 8 |
| 288 | B | 36 | 5 | 340 | E | 119 | 14 | 386 | D | 193 | 20 |
| 290 | E | 87 | 12 | 340 | F | 187 | 22 | 388 | B | 97 | 10 |
| 292 | E | 73 | 10 | 342 | A | 171 | 20 | 390 | A | 117 | 12 |
| 294 | A | 147 | 20 | 344 | A | 129 | 15 | 392 | A | 147 | 15 |
| 295 | A | 177 | 24 | 345 | A | 69 | 8 | 394 | C | 197 | 20 |
| 296 | B | 111 | 15 | 346 | F | 173 | 20 | 395 | C | 79 | 8 |
| 298 | E | 149 | 20 | 348 | E | 87 | 10 | 396 | A | 99 | 10 |
| 300 | A | 30 | 4 | 350 | E | 175 | 20 | 398 | B | 199 | 20 |
| 302 | D | 151 | 20 | 352 | D | 44 | 5 | 400 | A | 30 | 3 |

**Angular Indexing.** — With the ordinary indexing head, in which 40 turns of the index crank are required for one revolution of the work, one turn of the index crank equals 9 degrees. Hence, when one complete turn of the index crank equals 9 degrees, two holes in the 18-hole circle, or 3 holes in the 27-hole circle, must correspond to one degree. The first principle or rule for indexing for angles is therefore that two holes in the 18-hole circle or 3 holes in the 27-hole circle equals a movement of one degree of the index head spindle and the work.

Assume that an indexing movement of 35 degrees is required. One complete turn of the index crank equals 9 degrees; therefore, first divide the number of degrees for which to index, by 9, in order to find how many complete turns the index crank should make. The number of degrees left to turn after having completed the full turns are indexed by taking two holes in the 18-hole circle for each degree. In this case, $\dfrac{35}{9}$ = 3⅞, which indicates that the index crank must be turned three full revolutions, and then 8 degrees more are indexed by moving 16 holes in the 18-hole circle.

To index for 11½ degrees, for example, first turn the index crank one revolution, this being a 9-degree movement. Then to index 2½ degrees, move the index crank 5 holes in the 18-hole circle (4 holes for the two whole degrees and one hole for the ½ degree equals the total movement of 5 holes).

Below is shown how this calculation may be carried out to plainly indicate the movement required for this angle:

11½ deg. = 9 deg. + 2 deg. + ½ deg.

1 turn + 4 holes + 1 hole in the 18-hole circle.

Should it be required to index only ⅓ degree, this may be done by using the 27-hole circle. In this circle a three-hole movement equals one degree, and a one-hole movement in that circle thus equals ⅓ degree, or 20 minutes. Assume that it is required to index the work through an angle of 48 degrees 40 minutes. Below is plainly shown how this calculation may be carried out:

48 deg. 40 min. = 45 deg. + 3 deg. + 40 min.

5 turns + 9 holes + 2 holes in the 27-hole circle.

### Angular Values of One-Hole Moves — B. & S. Index Plates

| | |
|---|---|
| 15-hole circle = 36 minutes | 29-hole circle = 18.621 minutes |
| 16-hole circle = 33.750 minutes | 31-hole circle = 17.419 minutes |
| 17-hole circle = 31.765 minutes | 33-hole circle = 16.364 minutes |
| 18-hole circle = 30 minutes | 37-hole circle = 14.595 minutes |
| 19-hole circle = 28.421 minutes | 39-hole circle = 13.846 minutes |
| 20-hole circle = 27 minutes | 41-hole circle = 13.171 minutes |
| 21-hole circle = 25.714 minutes | 43-hole circle = 12.558 minutes |
| 23-hole circle = 23.478 minutes | 47-hole circle = 11.489 minutes |
| 27-hole circle = 20 minutes | 49-hole circle = 11.020 minutes |

**Approximate Indexing for Angles.**— The following general rule for *approximate* indexing of small angles is applicable to any index head requiring 40 revolutions of the index crank for one revolution of the work.

*Rule:* Divide 540 by the total number of minutes to be indexed. If the quotient is approximately equal to the number of holes in any index circle available, the angular movement is obtained by moving the crank one hole in this index circle; but if the quotient is not approximately equal, multiply it by any trial number which will give a product equal to the number of holes in an available index circle and move the index crank as many holes as are indicated by the trial number. (If the quotient of 540 divided by the total number of minutes is greater than the number of holes in any of the index circles, it is not possible to obtain the required movement for the angle by simple indexing.)

*Example:* — Assume that it is required to index to an angle of 2 degrees 46 minutes. Changing this to minutes gives a total of 166 minutes. Dividing 540 by 166 we have 540 ÷ 166 = 3.253. This quotient is next multiplied by some trial number to obtain a product which equals the number of holes in an available index circle. Multiplying by 12, we have 3.253 × 12 = 39.036. Therefore, for indexing 2 degrees 46 minutes, the 39-hole circle can be used and the index crank would be moved 12 holes.

**Tables for Angular Indexing.** — The table, "Angular Indexing," gives the number of turns of the index crank for indexing various angles. In the column headed, "Turns of Index Crank," the whole number (where given) indicates the number of full revolutions; the numerator of the fraction, the number of holes additional; and the denominator, the number of holes in the index circle to be used. The angular movement obtained for a movement of one hole, in·various index plates is given in the table, "Angular Values of One-Hole Moves."

### Angular Indexing

| Angle in Degs. | Turns of Index Crank | Angle in Degs. | Turns of Index Crank | Angle in Degs. | Turns of Index Crank | Angle in Degs. | Turns of Index Crank | Angle in Degs. | Turns of Index Crank |
|---|---|---|---|---|---|---|---|---|---|
| 1 | 2/18 | 10 | 1 2/18 | 19 | 2 2/18 | 28 | 3 2/18 | 37 | 4 2/18 |
| 1⅓ | 4/27 | 10⅓ | 1 4/27 | 19⅓ | 2 4/27 | 28⅓ | 3 4/27 | 37⅓ | 4 4/27 |
| 1½ | 3/18 | 10½ | 1 3/18 | 19½ | 2 3/18 | 28½ | 3 3/18 | 37½ | 4 3/18 |
| 1⅔ | 5/27 | 10⅔ | 1 5/27 | 19⅔ | 2 5/27 | 28⅔ | 3 5/27 | 37⅔ | 4 5/27 |
| 2 | 4/18 | 11 | 1 4/18 | 20 | 2 4/18 | 29 | 3 4/18 | 38 | 4 4/18 |
| 2⅓ | 7/27 | 11⅓ | 1 7/27 | 20⅓ | 2 7/27 | 29⅓ | 3 7/27 | 38⅓ | 4 7/27 |
| 2½ | 5/18 | 11½ | 1 5/18 | 20½ | 2 5/18 | 29½ | 3 5/18 | 38½ | 4 5/18 |
| 2⅔ | 8/27 | 11⅔ | 1 8/27 | 20⅔ | 2 8/27 | 29⅔ | 3 8/27 | 38⅔ | 4 8/27 |
| 3 | 6/18 | 12 | 1 6/18 | 21 | 2 6/18 | 30 | 3 6/18 | 39 | 4 6/18 |
| 3⅓ | 10/27 | 12⅓ | 1 10/27 | 21⅓ | 2 10/27 | 30⅓ | 3 10/27 | 39⅓ | 4 10/27 |
| 3½ | 7/18 | 12½ | 1 7/18 | 21½ | 2 7/18 | 30½ | 3 7/18 | 39½ | 4 7/18 |
| 3⅔ | 11/27 | 12⅔ | 1 11/27 | 21⅔ | 2 11/27 | 30⅔ | 3 11/27 | 39⅔ | 4 11/27 |
| 4 | 8/18 | 13 | 1 8/18 | 22 | 2 8/18 | 31 | 3 8/18 | 40 | 4 8/18 |
| 4⅓ | 13/27 | 13⅓ | 1 13/27 | 22⅓ | 2 13/27 | 31⅓ | 3 13/27 | 40⅓ | 4 13/27 |
| 4½ | 9/18 | 13½ | 1 9/18 | 22½ | 2 9/18 | 31½ | 3 9/18 | 40½ | 4 9/18 |
| 4⅔ | 14/27 | 13⅔ | 1 14/27 | 22⅔ | 2 14/27 | 31⅔ | 3 14/27 | 40⅔ | 4 14/27 |
| 5 | 10/18 | 14 | 1 10/18 | 23 | 2 10/18 | 32 | 3 10/18 | 41 | 4 10/18 |
| 5⅓ | 16/27 | 14⅓ | 1 16/27 | 23⅓ | 2 16/27 | 32⅓ | 3 16/27 | 41⅓ | 4 16/27 |
| 5½ | 11/18 | 14½ | 1 11/18 | 23½ | 2 11/18 | 32½ | 3 11/18 | 41½ | 4 11/18 |
| 5⅔ | 17/27 | 14⅔ | 1 17/27 | 23⅔ | 2 17/27 | 32⅔ | 3 17/27 | 41⅔ | 4 17/27 |
| 6 | 12/18 | 15 | 1 12/18 | 24 | 2 12/18 | 33 | 3 12/18 | 42 | 4 12/18 |
| 6⅓ | 19/27 | 15⅓ | 1 19/27 | 24⅓ | 2 19/27 | 33⅓ | 3 19/27 | 42⅓ | 4 19/27 |
| 6½ | 13/18 | 15½ | 1 13/18 | 24½ | 2 13/18 | 33½ | 3 13/18 | 42½ | 4 13/18 |
| 6⅔ | 20/27 | 15⅔ | 1 20/27 | 24⅔ | 2 20/27 | 33⅔ | 3 20/27 | 42⅔ | 4 20/27 |
| 7 | 14/18 | 16 | 1 14/18 | 25 | 2 14/18 | 34 | 3 14/18 | 43 | 4 14/18 |
| 7⅓ | 22/27 | 16⅓ | 1 22/27 | 25⅓ | 2 22/27 | 34⅓ | 3 22/27 | 43⅓ | 4 22/27 |
| 7½ | 15/18 | 16½ | 1 15/18 | 25½ | 2 15/18 | 34½ | 3 15/18 | 43½ | 4 15/18 |
| 7⅔ | 23/27 | 16⅔ | 1 23/27 | 25⅔ | 2 23/27 | 34⅔ | 3 23/27 | 43⅔ | 4 23/27 |
| 8 | 16/18 | 17 | 1 16/18 | 26 | 2 16/18 | 35 | 3 16/18 | 44 | 4 16/18 |
| 8⅓ | 25/27 | 17⅓ | 1 25/27 | 26⅓ | 2 25/27 | 35⅓ | 3 25/27 | 44⅓ | 4 25/27 |
| 8½ | 17/18 | 17½ | 1 17/18 | 26½ | 2 17/18 | 35½ | 3 17/18 | 44½ | 4 17/18 |
| 8⅔ | 26/27 | 17⅔ | 1 26/27 | 26⅔ | 2 26/27 | 35⅔ | 3 26/27 | 44⅔ | 4 26/27 |
| 9 | 1 | 18 | 2 | 27 | 3 | 36 | 4 | 45 | 5 |
| 9⅓ | 1 1/27 | 18⅓ | 2 1/27 | 27⅓ | 3 1/27 | 36⅓ | 4 1/27 | 45⅓ | 5 1/27 |
| 9½ | 1 1/18 | 18½ | 2 1/18 | 27½ | 3 1/18 | 36½ | 4 1/18 | 45½ | 5 1/18 |
| 9⅔ | 1 2/27 | 18⅔ | 2 2/27 | 27⅔ | 3 2/27 | 36⅔ | 4 2/27 | 45⅔ | 5 2/27 |

### Accurate Angular Indexing Movements — 1 *

| Fractional Indexing Movement | B. & S., Becker, Hendey, K. & T. and Rockford | Cincinnati and LeBlond * | Fractional Indexing Movement | B. & S., Becker, Hendey, K. & T. and Rockford | Cincinnati and LeBlond * | Fractional Indexing Movement | B. & S., Becker, Hendey, K. & T. and Rockford | Cincinnati and LeBlond * |
|---|---|---|---|---|---|---|---|---|
| 0.0152 | ..... | 1/66 | 0.0541 | 2/37 | 2/37 | 0.1000 | 2/20 | 3/30 |
| 0.0161 | ..... | 1/62 | 0.0556 | 1/18 | 3/54 | 0.1017 | ..... | 6/59 |
| 0.0169 | ..... | 1/59 | 0.0566 | ..... | 3/53 | 0.1020 | 5/49 | 5/49 |
| 0.0172 | ..... | 1/58 | 0.0588 | 1/17 | 2/34 | 0.1026 | 4/39 | 4/39 |
| 0.0175 | ..... | 1/57 | 0.0588 | ..... | 3/51 | 0.1034 | 3/29 | 6/58 |
| 0.0185 | ..... | 1/54 | 0.0606 | 2/33 | 4/66 | 0.1053 | 2/19 | 4/38 |
| 0.0189 | ..... | 1/53 | 0.0612 | 3/49 | 3/49 | 0.1053 | ..... | 6/57 |
| 0.0196 | ..... | 1/51 | 0.0625 | 1/16 | ..... | 0.1061 | ..... | 7/66 |
| 0.0204 | 1/49 | 1/49 | 0.0638 | 3/47 | 3/47 | 0.1064 | 5/47 | 5/47 |
| 0.0213 | 1/47 | 1/47 | 0.0645 | 2/31 | 4/62 | 0.1071 | ..... | 3/28 |
| 0.0217 | ..... | 1/46 | 0.0652 | ..... | 3/46 | 0.1081 | 4/37 | 4/37 |
| 0.0233 | 1/43 | 1/43 | 0.0667 | ..... | 2/30 | 0.1087 | ..... | 5/46 |
| 0.0238 | ..... | 1/42 | 0.0678 | ..... | 4/59 | 0.1111 | 2/18 | ..... |
| 0.0244 | 1/41 | 1/41 | 0.0690 | 2/29 | 4/58 | 0.1111 | 3/27 | 6/54 |
| 0.0256 | 1/39 | 1/39 | 0.0698 | 3/43 | 3/43 | 0.1129 | ..... | 7/62 |
| 0.0263 | ..... | 1/38 | 0.0702 | ..... | 4/57 | 0.1132 | ..... | 6/53 |
| 0.0270 | 1/37 | 1/37 | 0.0714 | ..... | 2/28 | 0.1163 | 5/43 | 5/43 |
| 0.0294 | ..... | 1/34 | 0.0714 | ..... | 3/42 | 0.1176 | 3/17 | 4/34 |
| 0.0303 | 1/33 | 2/66 | 0.0732 | 3/41 | 3/41 | 0.1176 | ..... | 6/51 |
| 0.0323 | 1/31 | 2/62 | 0.0741 | 2/27 | 4/54 | 0.1186 | ..... | 7/59 |
| 0.0333 | ..... | 1/30 | 0.0755 | ..... | 4/53 | 0.1190 | ..... | 5/42 |
| 0.0338 | ..... | 2/59 | 0.0758 | ..... | 5/66 | 0.1200 | ..... | 3/25 |
| 0.0345 | 1/29 | 2/58 | 0.0769 | 3/39 | 3/39 | 0.1207 | ..... | 7/58 |
| 0.0351 | ..... | 2/57 | 0.0784 | ..... | 4/51 | 0.1212 | 4/33 | 8/66 |
| 0.0357 | ..... | 1/28 | 0.0789 | ..... | 3/38 | 0.1220 | 5/41 | 5/41 |
| 0.0370 | 1/27 | 2/54 | 0.0800 | ..... | 2/25 | 0.1224 | 6/49 | 6/49 |
| 0.0377 | ..... | 2/53 | 0.0806 | ..... | 5/62 | 0.1228 | ..... | 7/57 |
| 0.0392 | ..... | 2/51 | 0.0811 | 3/37 | 3/37 | 0.1250 | 2/16 | 3/24 |
| 0.0400 | ..... | 1/25 | 0.0816 | 4/49 | 4/49 | 0.1277 | 6/47 | 6/47 |
| 0.0408 | 2/49 | 2/49 | 0.0833 | ..... | 2/24 | 0.1282 | 5/39 | 5/39 |
| 0.0417 | ..... | 1/24 | 0.0847 | ..... | 5/59 | 0.1290 | 4/31 | 8/62 |
| 0.0426 | 2/47 | 2/47 | 0.0851 | 4/47 | 4/47 | 0.1296 | ..... | 7/54 |
| 0.0435 | 1/23 | 2/46 | 0.0862 | ..... | 5/58 | 0.1304 | 3/23 | 6/46 |
| 0.0454 | ..... | 3/66 | 0.0870 | 2/23 | 4/46 | 0.1316 | ..... | 5/38 |
| 0.0465 | 2/43 | 2/43 | 0.0877 | ..... | 5/57 | 0.1321 | ..... | 7/53 |
| 0.0476 | 1/21 | 2/42 | 0.0882 | ..... | 3/34 | 0.1333 | 2/15 | 4/30 |
| 0.0484 | ..... | 3/62 | 0.0909 | 3/33 | 6/66 | 0.1351 | 5/37 | 5/37 |
| 0.0488 | 2/41 | 2/41 | 0.0926 | ..... | 5/54 | 0.1356 | ..... | 8/59 |
| 0.0500 | 1/20 | ..... | 0.0930 | 4/43 | 4/43 | 0.1364 | ..... | 9/66 |
| 0.0508 | ..... | 3/59 | 0.0943 | ..... | 5/53 | 0.1372 | ..... | 7/51 |
| 0.0513 | 2/39 | 2/39 | 0.0952 | 2/21 | 4/42 | 0.1379 | 4/29 | 8/58 |
| 0.0517 | ..... | 3/58 | 0.0968 | 3/31 | 6/62 | 0.1395 | 6/43 | 6/43 |
| 0.0526 | 1/19 | 2/38 | 0.0976 | 4/41 | 4/41 | 0.1404 | ..... | 8/57 |
| 0.0526 | ..... | 3/57 | 0.0980 | ..... | 5/51 | 0.1429 | ..... | 4/28 |

* See explanatory note below Table 8.

Accurate Angular Indexing Movements — 2 *

| Fractional Indexing Movement | B. & S., Becker, Hendey, K. & T. and Rockford | Cincinnati and LeBlond * | Fractional Indexing Movement | B. & S., Becker, Hendey, K. & T. and Rockford | Cincinnati and LeBlond * | Fractional Indexing Movement | B. & S., Becker, Hendey, K. & T. and Rockford | Cincinnati and LeBlond * |
|---|---|---|---|---|---|---|---|---|
| 0.1429 | 3/21 | 9/42 | 0.1864 | .... | 11/59 | 0.2308 | 9/39 | 9/39 |
| 0.1429 | 7/49 | 7/49 | 0.1875 | 3/16 | .... | 0.2326 | 10/43 | 10/43 |
| 0.1452 | .... | 9/62 | 0.1887 | .... | 10/53 | 0.2333 | .... | 7/30 |
| 0.1463 | 9/41 | 9/41 | 0.1892 | 7/37 | 7/37 | 0.2340 | 11/47 | 11/47 |
| 0.1471 | .... | 5/34 | 0.1897 | .... | 11/58 | 0.2353 | 4/17 | 8/34 |
| 0.1481 | 4/27 | 8/54 | 0.1905 | 4/21 | 8/42 | 0.2353 | .... | 12/51 |
| 0.1489 | 7/47 | 7/47 | 0.1915 | 9/47 | 9/47 | 0.2368 | .... | 9/38 |
| 0.1500 | 3/20 | .... | 0.1930 | .... | 11/57 | 0.2373 | .... | 14/59 |
| 0.1509 | .... | 8/53 | 0.1935 | 6/31 | 12/62 | 0.2381 | 5/21 | 10/42 |
| 0.1515 | 5/33 | 10/66 | 0.1951 | 8/41 | 8/41 | 0.2391 | .... | 11/46 |
| 0.1522 | .... | 7/46 | 0.1957 | .... | 9/46 | 0.2400 | .... | 6/25 |
| 0.1525 | .... | 9/59 | 0.1961 | .... | 10/51 | 0.2407 | .... | 13/54 |
| 0.1538 | 6/39 | 9/59 | 0.1970 | .... | 13/66 | 0.2414 | 7/29 | 14/58 |
| 0.1552 | .... | 9/58 | 0.2000 | 3/15 | 5/25 | 0.2419 | .... | 15/62 |
| 0.1569 | .... | 8/51 | 0.2000 | 4/20 | 9/45 | 0.2424 | 8/33 | 16/66 |
| 0.1579 | 3/19 | 9/58 | 0.2034 | .... | 12/59 | 0.2432 | 9/37 | 9/37 |
| 0.1579 | .... | 9/57 | 0.2037 | .... | 11/54 | 0.2439 | 10/41 | 10/41 |
| 0.1600 | .... | 4/25 | 0.2041 | 10/49 | 10/49 | 0.2449 | 12/49 | 12/49 |
| 0.1613 | 5/31 | 10/62 | 0.2051 | 8/39 | 8/39 | 0.2453 | .... | 13/53 |
| 0.1622 | 9/37 | 9/37 | 0.2059 | .... | 7/34 | 0.2456 | .... | 14/57 |
| 0.1628 | 7/43 | 7/43 | 0.2069 | 6/29 | 12/58 | 0.2500 | 4/16 | 6/24 |
| 0.1633 | 8/49 | 8/49 | 0.2075 | .... | 11/53 | 0.2500 | 5/20 | 7/28 |
| 0.1667 | 3/18 | 11/66 | 0.2083 | .... | 5/24 | 0.2542 | .... | 15/59 |
| 0.1667 | .... | 9/54 | 0.2093 | 9/43 | 9/43 | 0.2549 | .... | 13/51 |
| 0.1667 | .... | 7/42 | 0.2097 | .... | 13/62 | 0.2553 | 12/47 | 12/47 |
| 0.1667 | .... | 5/30 | 0.2105 | 4/19 | 8/38 | 0.2558 | 11/43 | 11/43 |
| 0.1667 | .... | 4/24 | 0.2105 | .... | 12/57 | 0.2564 | 10/39 | 10/39 |
| 0.1695 | .... | 10/59 | 0.2121 | 7/33 | 14/66 | 0.2576 | .... | 17/66 |
| 0.1698 | .... | 9/53 | 0.2128 | 10/47 | 10/47 | 0.2581 | 8/31 | 16/62 |
| 0.1702 | 8/47 | 8/47 | 0.2143 | .... | 6/28 | 0.2586 | .... | 15/58 |
| 0.1707 | 7/41 | 7/41 | 0.2143 | .... | 9/42 | 0.2593 | 7/27 | 14/54 |
| 0.1724 | 5/29 | 10/58 | 0.2157 | .... | 11/51 | 0.2609 | 9/23 | 12/46 |
| 0.1739 | 4/23 | 8/46 | 0.2162 | 8/37 | 8/37 | 0.2619 | .... | 11/42 |
| 0.1754 | .... | 10/57 | 0.2174 | 5/23 | 10/46 | 0.2632 | 5/19 | 10/38 |
| 0.1765 | 3/17 | 6/34 | 0.2195 | 9/41 | 9/41 | 0.2632 | .... | 15/57 |
| 0.1765 | .... | 9/51 | 0.2203 | .... | 13/59 | 0.2642 | .... | 14/53 |
| 0.1774 | .... | 11/62 | 0.2222 | 4/18 | .... | 0.2647 | .... | 9/34 |
| 0.1786 | .... | 5/28 | 0.2222 | 6/27 | 12/54 | 0.2653 | 13/49 | 13/49 |
| 0.1795 | 7/39 | 7/39 | 0.2241 | .... | 13/58 | 0.2667 | 4/15 | 8/30 |
| 0.1818 | 6/33 | 12/66 | 0.2245 | 11/49 | 11/49 | 0.2683 | 11/41 | 11/41 |
| 0.1839 | 9/49 | 9/49 | 0.2258 | 7/31 | 14/62 | 0.2703 | 10/37 | 10/37 |
| 0.1842 | .... | 7/38 | 0.2264 | .... | 12/53 | 0.2712 | .... | 16/59 |
| 0.1852 | 5/27 | 10/54 | 0.2273 | .... | 15/66 | 0.2727 | 9/33 | 15/66 |
| 0.1860 | 8/43 | 8/43 | 0.2281 | .... | 13/57 | 0.2742 | .... | 17/62 |

\* See explanatory note below Table 8.

## Accurate Angular Indexing Movements — 3 *

| Fractional Indexing Movement | B. & S., Becker, Hendey, K. & T. and Rockford | Cincinnati and LeBlond * | Fractional Indexing Movement | B. & S., Becker, Hendey, K. & T. and Rockford | Cincinnati and LeBlond * | Fractional Indexing Movement | B. & S., Becker, Hendey, K. & T. and Rockford | Cincinnati and LeBlond * |
|---|---|---|---|---|---|---|---|---|
| 0.2745 | ..... | 14/51 | 0.3191 | 15/47 | 15/47 | 0.3617 | 17/47 | 17/47 |
| 0.2759 | 8/29 | 16/58 | 0.3200 | ..... | 8/25 | 0.3621 | ..... | 21/58 |
| 0.2766 | 13/47 | 13/47 | 0.3208 | ..... | 17/53 | 0.3636 | 12/33 | 24/66 |
| 0.2778 | 5/18 | 15/54 | 0.3214 | ..... | 9/28 | 0.3659 | 15/41 | 15/41 |
| 0.2791 | 12/43 | 12/43 | 0.3220 | ..... | 19/59 | 0.3667 | ..... | 11/30 |
| 0.2800 | ..... | 7/25 | 0.3226 | 10/31 | 20/62 | 0.3673 | 18/49 | 18/49 |
| 0.2807 | ..... | 16/57 | 0.3235 | ..... | 11/34 | 0.3684 | 7/19 | 14/38 |
| 0.2821 | 11/39 | 11/39 | 0.3243 | 12/37 | 12/37 | 0.3684 | ..... | 21/57 |
| 0.2826 | ..... | 13/46 | 0.3256 | 14/43 | 14/43 | 0.3696 | ..... | 17/46 |
| 0.2830 | ..... | 15/53 | 0.3261 | ..... | 15/46 | 0.3704 | 10/27 | 20/54 |
| 0.2857 | ..... | 8/28 | 0.3265 | 16/49 | 16/49 | 0.3710 | ..... | 23/62 |
| 0.2857 | 14/49 | 14/49 | 0.3276 | ..... | 19/58 | 0.3721 | 16/43 | 16/43 |
| 0.2857 | 6/21 | 12/42 | 0.3333 | 6/18 | 8/24 | 0.3725 | ..... | 19/51 |
| 0.2879 | ..... | 19/66 | 0.3333 | 5/15 | 10/30 | 0.3729 | ..... | 22/59 |
| 0.2881 | ..... | 17/59 | 0.3333 | 13/39 | 13/39 | 0.3750 | 6/16 | 9/24 |
| 0.2895 | ..... | 11/38 | 0.3333 | 7/21 | 14/42 | 0.3774 | ..... | 20/53 |
| 0.2903 | 9/31 | 18/62 | 0.3333 | ..... | 17/51 | 0.3784 | 14/37 | 14/37 |
| 0.2917 | ..... | 7/24 | 0.3333 | 9/27 | 18/54 | 0.3788 | ..... | 25/66 |
| 0.2927 | 12/41 | 12/41 | 0.3333 | ..... | 19/57 | 0.3793 | 11/29 | 22/58 |
| 0.2931 | ..... | 17/58 | 0.3333 | 11/33 | 22/66 | 0.3810 | 8/21 | 16/42 |
| 0.2941 | ..... | 15/51 | 0.3387 | ..... | 21/62 | 0.3824 | ..... | 13/34 |
| 0.2941 | 5/17 | 10/34 | 0.3390 | ..... | 20/59 | 0.3830 | 18/47 | 18/47 |
| 0.2963 | 8/27 | 16/54 | 0.3396 | ..... | 18/53 | 0.3846 | 15/39 | 15/39 |
| 0.2973 | 11/37 | 11/37 | 0.3404 | 16/47 | 16/47 | 0.3860 | ..... | 22/57 |
| 0.2979 | 14/47 | 14/47 | 0.3415 | 14/41 | 14/41 | 0.3871 | 12/31 | 24/62 |
| 0.2982 | ..... | 17/57 | 0.3421 | ..... | 13/38 | 0.3878 | 19/49 | 19/49 |
| 0.3000 | 6/20 | 9/30 | 0.3448 | 10/29 | 20/58 | 0.3889 | 7/18 | 21/54 |
| 0.3019 | ..... | 16/53 | 0.3469 | 17/49 | 17/49 | 0.3898 | ..... | 23/59 |
| 0.3023 | 13/43 | 13/43 | 0.3478 | 8/23 | 16/46 | 0.3902 | 16/41 | 16/41 |
| 0.3030 | 10/33 | 20/66 | 0.3485 | ..... | 23/66 | 0.3913 | 9/23 | 18/46 |
| 0.3043 | 7/23 | 14/46 | 0.3488 | 15/43 | 15/43 | 0.3922 | ..... | 20/51 |
| 0.3051 | ..... | 18/59 | 0.3500 | 7/20 | ..... | 0.3929 | ..... | 11/28 |
| 0.3061 | 15/49 | 15/49 | 0.3509 | ..... | 20/57 | 0.3939 | 13/33 | 26/66 |
| 0.3065 | ..... | 19/62 | 0.3514 | 13/37 | 13/37 | 0.3947 | ..... | 15/38 |
| 0.3077 | 12/39 | 12/39 | 0.3519 | ..... | 19/54 | 0.3953 | 17/43 | 17/43 |
| 0.3095 | ..... | 13/42 | 0.3529 | 6/17 | 12/34 | 0.3962 | ..... | 21/53 |
| 0.3103 | 9/29 | 18/58 | 0.3529 | ..... | 18/51 | 0.3966 | ..... | 23/58 |
| 0.3125 | 5/16 | ..... | 0.3548 | 11/31 | 22/62 | 0.4000 | 6/15 | 10/25 |
| 0.3137 | ..... | 16/51 | 0.3559 | ..... | 21/59 | 0.4000 | 8/20 | 12/30 |
| 0.3148 | ..... | 17/54 | 0.3571 | ..... | 10/28 | 0.4032 | ..... | 25/62 |
| 0.3158 | 6/19 | 12/38 | 0.3571 | ..... | 15/42 | 0.4035 | ..... | 23/57 |
| 0.3158 | ..... | 18/57 | 0.3585 | ..... | 19/53 | 0.4043 | 19/47 | 19/47 |
| 0.3171 | 13/41 | 13/41 | 0.3590 | 14/39 | 14/39 | 0.4048 | ..... | 17/42 |
| 0.3182 | ..... | 21/66 | 0.3600 | ..... | 9/25 | 0.4054 | 15/37 | 15/37 |

* See explanatory note below Table 8.

### Accurate Angular Indexing Movements — 4 *

| Fractional Indexing Movement | B. & S., Becker, Hendey, K. & T. and Rockford | Cincinnati and LeBlond * | Fractional Indexing Movement | B. & S., Becker, Hendey, K. & T. and Rockford | Cincinnati and LeBlond * | Fractional Indexing Movement | B. & S., Becker, Hendey, K. & T. and Rockford | Cincinnati and LeBlond * |
|---|---|---|---|---|---|---|---|---|
| 0.4068 | ..... | $2\frac{4}{69}$ | 0.4490 | $2\frac{2}{49}$ | $2\frac{2}{49}$ | 0.4912 | ..... | $2\frac{8}{57}$ |
| 0.4074 | $1\frac{1}{27}$ | $2\frac{4}{64}$ | 0.4500 | $\frac{9}{20}$ | ..... | 0.4915 | ..... | $2\frac{9}{59}$ |
| 0.4082 | $2\frac{4}{9}$ | $2\frac{4}{9}$ | 0.4510 | ..... | $2\frac{3}{51}$ | 0.5000 | $\frac{8}{16}$ | $1\frac{2}{24}$ |
| 0.4091 | ..... | $2\frac{7}{66}$ | 0.4516 | $1\frac{4}{31}$ | $2\frac{8}{62}$ | 0.5000 | $\frac{9}{18}$ | $1\frac{4}{28}$ |
| 0.4103 | $1\frac{6}{39}$ | $1\frac{6}{39}$ | 0.4524 | ..... | $1\frac{9}{42}$ | 0.5000 | $1\frac{0}{20}$ | $1\frac{5}{30}$ |
| 0.4118 | $\frac{7}{17}$ | $1\frac{4}{54}$ | 0.4528 | ..... | $2\frac{4}{53}$ | 0.5000 | ..... | $1\frac{7}{34}$ |
| 0.4118 | ..... | $2\frac{1}{51}$ | 0.4545 | $1\frac{5}{33}$ | $3\frac{9}{66}$ | 0.5000 | ..... | $1\frac{9}{38}$ |
| 0.4130 | ..... | $1\frac{9}{46}$ | 0.4561 | ..... | $2\frac{9}{57}$ | 0.5000 | ..... | $2\frac{1}{42}$ |
| 0.4138 | $1\frac{2}{29}$ | $2\frac{4}{58}$ | 0.4565 | ..... | $2\frac{1}{46}$ | 0.5000 | ..... | $2\frac{3}{46}$ |
| 0.4146 | $1\frac{7}{41}$ | $1\frac{7}{41}$ | 0.4576 | ..... | $2\frac{7}{59}$ | 0.5000 | ..... | $2\frac{7}{54}$ |
| 0.4151 | ..... | $2\frac{2}{53}$ | 0.4583 | ..... | $1\frac{1}{24}$ | 0.5000 | ..... | $2\frac{9}{58}$ |
| 0.4167 | ..... | $1\frac{9}{64}$ | 0.4595 | $1\frac{7}{47}$ | $1\frac{7}{47}$ | 0.5000 | ..... | $3\frac{1}{62}$ |
| 0.4186 | $1\frac{8}{43}$ | $1\frac{8}{43}$ | 0.4615 | $1\frac{8}{39}$ | $1\frac{8}{39}$ | 0.5000 | ..... | $3\frac{3}{66}$ |
| 0.4194 | $1\frac{3}{31}$ | $2\frac{9}{62}$ | 0.4630 | ..... | $2\frac{5}{54}$ | 0.5085 | ..... | $3\frac{9}{59}$ |
| 0.4211 | $\frac{8}{19}$ | $1\frac{6}{38}$ | 0.4634 | $1\frac{9}{41}$ | $1\frac{9}{41}$ | 0.5088 | ..... | $2\frac{9}{57}$ |
| 0.4211 | ..... | $2\frac{4}{67}$ | 0.4643 | ..... | $1\frac{3}{28}$ | 0.5094 | ..... | $2\frac{7}{53}$ |
| 0.4237 | ..... | $2\frac{5}{69}$ | 0.4651 | $2\frac{0}{43}$ | $2\frac{0}{43}$ | 0.5098 | ..... | $2\frac{6}{51}$ |
| 0.4242 | $1\frac{4}{33}$ | $2\frac{8}{66}$ | 0.4655 | ..... | $2\frac{7}{58}$ | 0.5102 | $2\frac{5}{49}$ | $2\frac{5}{49}$ |
| 0.4255 | $2\frac{9}{47}$ | $2\frac{9}{47}$ | 0.4667 | $\frac{7}{15}$ | $1\frac{4}{30}$ | 0.5106 | $2\frac{4}{47}$ | $2\frac{4}{47}$ |
| 0.4259 | ..... | $2\frac{3}{54}$ | 0.4677 | ..... | $2\frac{9}{62}$ | 0.5116 | $2\frac{2}{43}$ | $2\frac{2}{43}$ |
| 0.4286 | ..... | $1\frac{2}{28}$ | 0.4681 | $2\frac{2}{47}$ | $2\frac{2}{47}$ | 0.5122 | $2\frac{1}{41}$ | $2\frac{1}{41}$ |
| 0.4286 | $\frac{9}{21}$ | $1\frac{8}{42}$ | 0.4694 | $2\frac{3}{49}$ | $2\frac{3}{49}$ | 0.5128 | $2\frac{0}{39}$ | $2\frac{0}{39}$ |
| 0.4286 | $2\frac{1}{49}$ | $2\frac{1}{49}$ | 0.4697 | ..... | $3\frac{1}{66}$ | 0.5135 | $1\frac{9}{37}$ | $1\frac{9}{37}$ |
| 0.4310 | ..... | $2\frac{5}{58}$ | 0.4706 | $\frac{8}{17}$ | $1\frac{9}{34}$ | 0.5152 | $1\frac{7}{33}$ | $3\frac{4}{66}$ |
| 0.4314 | ..... | $2\frac{2}{51}$ | 0.4706 | ..... | $2\frac{4}{51}$ | 0.5161 | $1\frac{6}{31}$ | $3\frac{2}{62}$ |
| 0.4324 | $1\frac{6}{47}$ | $1\frac{6}{47}$ | 0.4717 | ..... | $2\frac{5}{53}$ | 0.5172 | $1\frac{5}{29}$ | $3\frac{0}{58}$ |
| 0.4333 | ..... | $1\frac{3}{30}$ | 0.4737 | $\frac{9}{19}$ | $2\frac{7}{57}$ | 0.5185 | $1\frac{4}{27}$ | $2\frac{8}{54}$ |
| 0.4340 | ..... | $2\frac{3}{53}$ | 0.4746 | ..... | $2\frac{8}{59}$ | 0.5200 | ..... | $1\frac{3}{25}$ |
| 0.4348 | $1\frac{9}{23}$ | $2\frac{9}{46}$ | 0.4762 | $1\frac{9}{21}$ | $2\frac{9}{42}$ | 0.5217 | $1\frac{2}{23}$ | $2\frac{4}{46}$ |
| 0.4355 | ..... | $2\frac{7}{62}$ | 0.4783 | $1\frac{1}{23}$ | $2\frac{3}{46}$ | 0.5238 | $1\frac{1}{21}$ | $2\frac{2}{42}$ |
| 0.4359 | $1\frac{7}{39}$ | $1\frac{7}{39}$ | 0.4800 | ..... | $1\frac{2}{25}$ | 0.5254 | ..... | $3\frac{1}{59}$ |
| 0.4375 | $\frac{7}{16}$ | ..... | 0.4814 | ..... | $2\frac{9}{54}$ | 0.5263 | $1\frac{0}{19}$ | $2\frac{9}{58}$ |
| 0.4386 | ..... | $2\frac{5}{57}$ | 0.4815 | $1\frac{3}{27}$ | ..... | 0.5263 | ..... | $3\frac{0}{57}$ |
| 0.4390 | $1\frac{8}{41}$ | $1\frac{8}{41}$ | 0.4828 | $1\frac{4}{29}$ | $2\frac{8}{58}$ | 0.5283 | ..... | $2\frac{8}{53}$ |
| 0.4394 | ..... | $2\frac{9}{66}$ | 0.4839 | $1\frac{5}{31}$ | $3\frac{9}{62}$ | 0.5294 | $\frac{9}{17}$ | $1\frac{8}{34}$ |
| 0.4400 | ..... | $1\frac{1}{25}$ | 0.4848 | $1\frac{5}{33}$ | $3\frac{2}{66}$ | 0.5294 | ..... | $2\frac{7}{51}$ |
| 0.4407 | ..... | $2\frac{9}{69}$ | 0.4865 | $1\frac{8}{37}$ | $1\frac{8}{37}$ | 0.5303 | ..... | $3\frac{5}{66}$ |
| 0.4412 | ..... | $1\frac{5}{34}$ | 0.4872 | $1\frac{9}{39}$ | $1\frac{9}{39}$ | 0.5306 | $2\frac{0}{49}$ | $2\frac{0}{49}$ |
| 0.4419 | $1\frac{9}{43}$ | $1\frac{9}{43}$ | 0.4878 | $2\frac{0}{41}$ | $2\frac{0}{41}$ | 0.5319 | $2\frac{5}{47}$ | $2\frac{5}{47}$ |
| 0.4444 | $\frac{8}{18}$ | ..... | 0.4884 | $2\frac{1}{43}$ | $2\frac{1}{43}$ | 0.5323 | ..... | $3\frac{3}{62}$ |
| 0.4444 | $1\frac{2}{27}$ | $2\frac{4}{54}$ | 0.4894 | $2\frac{3}{47}$ | $2\frac{3}{47}$ | 0.5333 | $\frac{8}{15}$ | $1\frac{9}{30}$ |
| 0.4468 | $2\frac{1}{47}$ | $2\frac{1}{47}$ | 0.4898 | $2\frac{4}{49}$ | $2\frac{4}{49}$ | 0.5345 | ..... | $3\frac{1}{58}$ |
| 0.4474 | ..... | $1\frac{7}{38}$ | 0.4902 | ..... | $2\frac{5}{51}$ | 0.5349 | $2\frac{3}{43}$ | $2\frac{3}{43}$ |
| 0.4483 | $1\frac{3}{29}$ | $2\frac{9}{58}$ | 0.4906 | ..... | $2\frac{6}{53}$ | 0.5357 | ..... | $1\frac{5}{28}$ |

* See explanatory note below Table 8.

## Accurate Angular Indexing Movements — 5 *

| Fractional Indexing Movement | B. & S., Becker, Hendey, K. & T. and Rockford | Cincinnati and LeBlond * | Fractional Indexing Movement | B. & S., Becker, Hendey, K. & T. and Rockford | Cincinnati and LeBlond * | Fractional Indexing Movement | B. & S., Becker, Hendey, K. & T. and Rockford | Cincinnati and LeBlond * |
|---|---|---|---|---|---|---|---|---|
| 0.5366 | $\frac{22}{41}$ | $\frac{22}{41}$ | 0.5789 | ..... | $\frac{33}{57}$ | 0.6250 | $\frac{10}{16}$ | $\frac{15}{24}$ |
| 0.5370 | ..... | $\frac{29}{54}$ | 0.5806 | $\frac{18}{31}$ | $\frac{36}{62}$ | 0.6271 | ..... | $\frac{37}{59}$ |
| 0.5385 | $\frac{21}{39}$ | $\frac{21}{39}$ | 0.5814 | $\frac{25}{43}$ | $\frac{25}{43}$ | 0.6275 | ..... | $\frac{32}{51}$ |
| 0.5405 | $\frac{20}{37}$ | $\frac{20}{37}$ | 0.5833 | ..... | $\frac{14}{24}$ | 0.6279 | $\frac{27}{43}$ | $\frac{27}{43}$ |
| 0.5417 | ..... | $\frac{13}{24}$ | 0.5849 | ..... | $\frac{31}{53}$ | 0.6290 | ..... | $\frac{39}{62}$ |
| 0.5424 | ..... | $\frac{32}{59}$ | 0.5854 | $\frac{24}{41}$ | $\frac{24}{41}$ | 0.6296 | $\frac{17}{27}$ | $\frac{34}{54}$ |
| 0.5435 | ..... | $\frac{25}{46}$ | 0.5862 | $\frac{17}{29}$ | $\frac{34}{58}$ | 0.6304 | ..... | $\frac{29}{46}$ |
| 0.5439 | ..... | $\frac{31}{57}$ | 0.5870 | ..... | $\frac{27}{46}$ | 0.6316 | $\frac{12}{19}$ | $\frac{24}{38}$ |
| 0.5455 | $\frac{18}{33}$ | $\frac{36}{66}$ | 0.5882 | $\frac{10}{17}$ | $\frac{20}{34}$ | 0.6316 | ..... | $\frac{36}{57}$ |
| 0.5472 | ..... | $\frac{29}{53}$ | 0.5882 | ..... | $\frac{30}{51}$ | 0.6327 | $\frac{31}{49}$ | $\frac{31}{49}$ |
| 0.5476 | ..... | $\frac{23}{42}$ | 0.5897 | $\frac{23}{39}$ | $\frac{23}{39}$ | 0.6333 | ..... | $\frac{19}{30}$ |
| 0.5484 | $\frac{17}{31}$ | $\frac{34}{62}$ | 0.5909 | ..... | $\frac{39}{66}$ | 0.6341 | $\frac{26}{41}$ | $\frac{26}{41}$ |
| 0.5490 | ..... | $\frac{28}{51}$ | 0.5918 | $\frac{29}{49}$ | $\frac{29}{49}$ | 0.6364 | $\frac{21}{33}$ | $\frac{42}{66}$ |
| 0.5500 | $\frac{11}{20}$ | ..... | 0.5926 | $\frac{16}{27}$ | $\frac{32}{54}$ | 0.6379 | ..... | $\frac{37}{58}$ |
| 0.5510 | $\frac{27}{49}$ | $\frac{27}{49}$ | 0.5932 | ..... | $\frac{35}{59}$ | 0.6383 | $\frac{30}{47}$ | $\frac{30}{47}$ |
| 0.5517 | $\frac{16}{29}$ | $\frac{32}{58}$ | 0.5946 | $\frac{22}{37}$ | $\frac{22}{37}$ | 0.6400 | ..... | $\frac{16}{25}$ |
| 0.5526 | ..... | $\frac{21}{38}$ | 0.5952 | ..... | $\frac{25}{42}$ | 0.6410 | $\frac{25}{39}$ | $\frac{25}{39}$ |
| 0.5532 | $\frac{26}{47}$ | $\frac{26}{47}$ | 0.5957 | $\frac{28}{47}$ | $\frac{28}{47}$ | 0.6415 | ..... | $\frac{34}{53}$ |
| 0.5556 | $\frac{10}{18}$ | ..... | 0.5965 | ..... | $\frac{34}{57}$ | 0.6429 | ..... | $\frac{18}{28}$ |
| 0.5556 | $\frac{15}{27}$ | $\frac{30}{54}$ | 0.5968 | ..... | $\frac{37}{62}$ | 0.6429 | ..... | $\frac{27}{42}$ |
| 0.5581 | $\frac{24}{43}$ | $\frac{24}{43}$ | 0.6000 | $\frac{9}{15}$ | $\frac{15}{25}$ | 0.6441 | ..... | $\frac{38}{59}$ |
| 0.5588 | ..... | $\frac{19}{34}$ | 0.6000 | $\frac{12}{20}$ | $\frac{18}{30}$ | 0.6452 | $\frac{20}{31}$ | $\frac{40}{62}$ |
| 0.5593 | ..... | $\frac{33}{59}$ | 0.6034 | ..... | $\frac{35}{58}$ | 0.6471 | $\frac{11}{17}$ | $\frac{22}{34}$ |
| 0.5600 | ..... | $\frac{14}{25}$ | 0.6038 | ..... | $\frac{32}{53}$ | 0.6471 | ..... | $\frac{33}{51}$ |
| 0.5606 | ..... | $\frac{37}{66}$ | 0.6047 | $\frac{26}{43}$ | $\frac{26}{43}$ | 0.6481 | ..... | $\frac{35}{54}$ |
| 0.5610 | $\frac{23}{41}$ | $\frac{23}{41}$ | 0.6053 | ..... | $\frac{23}{38}$ | 0.6486 | $\frac{24}{37}$ | $\frac{24}{37}$ |
| 0.5614 | ..... | $\frac{32}{57}$ | 0.6061 | ..... | $\frac{40}{66}$ | 0.6491 | ..... | $\frac{37}{57}$ |
| 0.5625 | $\frac{9}{16}$ | ..... | 0.6071 | ..... | $\frac{17}{28}$ | 0.6500 | $\frac{13}{20}$ | ..... |
| 0.5641 | $\frac{22}{39}$ | $\frac{22}{39}$ | 0.6078 | ..... | $\frac{31}{51}$ | 0.6512 | $\frac{28}{43}$ | $\frac{28}{43}$ |
| 0.5645 | ..... | $\frac{35}{62}$ | 0.6087 | $\frac{14}{23}$ | $\frac{28}{46}$ | 0.6515 | ..... | $\frac{43}{66}$ |
| 0.5652 | $\frac{13}{23}$ | $\frac{26}{46}$ | 0.6098 | $\frac{25}{41}$ | $\frac{25}{41}$ | 0.6522 | $\frac{15}{23}$ | $\frac{30}{46}$ |
| 0.5660 | ..... | $\frac{30}{53}$ | 0.6102 | ..... | $\frac{36}{59}$ | 0.6531 | $\frac{32}{49}$ | $\frac{32}{49}$ |
| 0.5667 | ..... | $\frac{17}{30}$ | 0.6111 | $\frac{11}{18}$ | $\frac{33}{54}$ | 0.6552 | $\frac{19}{29}$ | $\frac{38}{58}$ |
| 0.5676 | $\frac{21}{37}$ | $\frac{21}{37}$ | 0.6122 | $\frac{30}{49}$ | $\frac{30}{49}$ | 0.6579 | ..... | $\frac{25}{38}$ |
| 0.5686 | ..... | $\frac{29}{51}$ | 0.6129 | $\frac{19}{31}$ | $\frac{38}{62}$ | 0.6585 | $\frac{27}{41}$ | $\frac{27}{41}$ |
| 0.5690 | ..... | $\frac{33}{58}$ | 0.6140 | ..... | $\frac{35}{57}$ | 0.6596 | $\frac{31}{47}$ | $\frac{31}{47}$ |
| 0.5714 | ..... | $\frac{16}{28}$ | 0.6154 | $\frac{24}{39}$ | $\frac{24}{39}$ | 0.6604 | ..... | $\frac{35}{53}$ |
| 0.5714 | $\frac{12}{21}$ | $\frac{24}{42}$ | 0.6170 | $\frac{29}{47}$ | $\frac{29}{47}$ | 0.6610 | ..... | $\frac{39}{59}$ |
| 0.5714 | $\frac{28}{49}$ | $\frac{28}{49}$ | 0.6176 | ..... | $\frac{21}{34}$ | 0.6613 | ..... | $\frac{41}{62}$ |
| 0.5741 | ..... | $\frac{31}{54}$ | 0.6190 | $\frac{13}{21}$ | $\frac{26}{42}$ | 0.6667 | $\frac{12}{18}$ | $\frac{16}{24}$ |
| 0.5745 | $\frac{27}{47}$ | $\frac{27}{47}$ | 0.6207 | $\frac{18}{29}$ | $\frac{36}{58}$ | 0.6667 | $\frac{10}{15}$ | $\frac{20}{30}$ |
| 0.5758 | $\frac{19}{33}$ | $\frac{38}{66}$ | 0.6212 | ..... | $\frac{41}{66}$ | 0.6667 | $\frac{26}{39}$ | $\frac{26}{39}$ |
| 0.5763 | ..... | $\frac{34}{59}$ | 0.6216 | $\frac{23}{37}$ | $\frac{23}{37}$ | 0.6667 | $\frac{14}{21}$ | $\frac{28}{42}$ |
| 0.5789 | $\frac{11}{19}$ | $\frac{22}{38}$ | 0.6226 | ..... | $\frac{33}{53}$ | 0.6667 | ..... | $\frac{34}{51}$ |

* See explanatory note below Table 8.

### Accurate Angular Indexing Movements — 6 *

| Fractional Indexing Movement | B. & S., Becker, Hendey, K. & T. and Rockford | Cincinnati and LeBlond * | Fractional Indexing Movement | B. & S., Becker, Hendey, K. & T. and Rockford | Cincinnati and LeBlond * | Fractional Indexing Movement | B. & S., Becker, Hendey, K. & T. and Rockford | Cincinnati and LeBlond * |
|---|---|---|---|---|---|---|---|---|
| 0.6667 | $1\frac{8}{27}$ | $3\frac{6}{64}$ | 0.7119 | ..... | $4\frac{2}{69}$ | 0.7576 | $2\frac{5}{43}$ | $5\frac{9}{66}$ |
| 0.6667 | ..... | $3\frac{8}{67}$ | 0.7121 | ..... | $4\frac{7}{66}$ | 0.7581 | ..... | $4\frac{7}{62}$ |
| 0.6667 | $2\frac{4}{43}$ | $4\frac{4}{66}$ | 0.7143 | ..... | $2\frac{9}{28}$ | 0.7586 | $2\frac{2}{29}$ | $4\frac{4}{68}$ |
| 0.6724 | ..... | $3\frac{9}{68}$ | 0.7143 | $1\frac{5}{21}$ | $3\frac{9}{42}$ | 0.7593 | ..... | $4\frac{1}{64}$ |
| 0.6735 | $3\frac{3}{49}$ | $3\frac{3}{49}$ | 0.7143 | $3\frac{5}{49}$ | $3\frac{5}{49}$ | 0.7600 | ..... | $1\frac{9}{25}$ |
| 0.6739 | ..... | $3\frac{1}{46}$ | 0.7170 | ..... | $3\frac{8}{63}$ | 0.7609 | ..... | $3\frac{5}{46}$ |
| 0.6744 | $2\frac{9}{43}$ | $2\frac{9}{43}$ | 0.7174 | ..... | $3\frac{3}{46}$ | 0.7619 | $1\frac{6}{21}$ | $3\frac{3}{42}$ |
| 0.6757 | $2\frac{5}{37}$ | $2\frac{5}{37}$ | 0.7179 | $2\frac{8}{39}$ | $2\frac{8}{39}$ | 0.7627 | ..... | $4\frac{5}{69}$ |
| 0.6765 | ..... | $2\frac{3}{34}$ | 0.7193 | ..... | $4\frac{1}{57}$ | 0.7632 | ..... | $2\frac{9}{38}$ |
| 0.6774 | $2\frac{1}{31}$ | $4\frac{2}{62}$ | 0.7200 | ..... | $1\frac{8}{25}$ | 0.7647 | $1\frac{3}{17}$ | $2\frac{6}{34}$ |
| 0.6780 | ..... | $4\frac{9}{69}$ | 0.7209 | $3\frac{1}{43}$ | $3\frac{1}{43}$ | 0.7647 | ..... | $3\frac{9}{51}$ |
| 0.6786 | ..... | $1\frac{9}{58}$ | 0.7222 | $1\frac{3}{18}$ | $3\frac{9}{64}$ | 0.7660 | $3\frac{6}{47}$ | $3\frac{6}{47}$ |
| 0.6792 | ..... | $3\frac{9}{63}$ | 0.7234 | $3\frac{4}{47}$ | $3\frac{4}{47}$ | 0.7667 | ..... | $2\frac{3}{30}$ |
| 0.6800 | ..... | $1\frac{7}{25}$ | 0.7241 | $2\frac{1}{29}$ | $4\frac{2}{58}$ | 0.7674 | $3\frac{3}{43}$ | $3\frac{3}{43}$ |
| 0.6809 | $3\frac{2}{47}$ | $3\frac{2}{47}$ | 0.7255 | ..... | $3\frac{7}{51}$ | 0.7692 | $3\frac{0}{49}$ | $3\frac{0}{49}$ |
| 0.6818 | ..... | $4\frac{5}{66}$ | 0.7258 | ..... | $4\frac{5}{62}$ | 0.7719 | ..... | $4\frac{4}{57}$ |
| 0.6829 | $2\frac{8}{41}$ | $2\frac{8}{41}$ | 0.7273 | $2\frac{4}{33}$ | $4\frac{8}{66}$ | 0.7727 | ..... | $5\frac{1}{66}$ |
| 0.6842 | $1\frac{3}{19}$ | $2\frac{9}{58}$ | 0.7288 | ..... | $4\frac{3}{59}$ | 0.7736 | ..... | $4\frac{1}{53}$ |
| 0.6842 | ..... | $3\frac{9}{57}$ | 0.7297 | $2\frac{7}{37}$ | $2\frac{7}{37}$ | 0.7742 | $2\frac{4}{31}$ | $4\frac{8}{62}$ |
| 0.6852 | ..... | $3\frac{7}{54}$ | 0.7317 | $3\frac{9}{41}$ | $3\frac{9}{41}$ | 0.7755 | $3\frac{8}{49}$ | $3\frac{8}{49}$ |
| 0.6863 | ..... | $3\frac{5}{51}$ | 0.7333 | $1\frac{1}{15}$ | $2\frac{3}{30}$ | 0.7759 | ..... | $4\frac{5}{58}$ |
| 0.6875 | $1\frac{1}{16}$ | ..... | 0.7347 | $3\frac{9}{49}$ | $3\frac{9}{49}$ | 0.7778 | $1\frac{4}{18}$ | ..... |
| 0.6897 | $2\frac{9}{29}$ | $4\frac{9}{58}$ | 0.7353 | ..... | $2\frac{5}{34}$ | 0.7778 | $2\frac{1}{27}$ | $4\frac{2}{54}$ |
| 0.6905 | ..... | $2\frac{9}{42}$ | 0.7358 | ..... | $3\frac{9}{53}$ | 0.7797 | ..... | $4\frac{6}{59}$ |
| 0.6923 | $2\frac{7}{39}$ | $2\frac{7}{39}$ | 0.7368 | $1\frac{4}{19}$ | $2\frac{8}{38}$ | 0.7805 | $3\frac{2}{41}$ | $3\frac{2}{41}$ |
| 0.6935 | ..... | $4\frac{3}{62}$ | 0.7368 | ..... | $4\frac{2}{57}$ | 0.7826 | $1\frac{8}{23}$ | $3\frac{9}{46}$ |
| 0.6939 | $3\frac{4}{49}$ | $3\frac{4}{49}$ | 0.7381 | ..... | $3\frac{1}{42}$ | 0.7838 | $2\frac{9}{37}$ | $2\frac{9}{37}$ |
| 0.6949 | ..... | $4\frac{1}{59}$ | 0.7391 | $1\frac{7}{23}$ | $3\frac{4}{46}$ | 0.7843 | ..... | $4\frac{0}{51}$ |
| 0.6957 | $1\frac{6}{23}$ | $3\frac{2}{46}$ | 0.7407 | $2\frac{0}{27}$ | $4\frac{0}{54}$ | 0.7857 | ..... | $2\frac{2}{28}$ |
| 0.6970 | $2\frac{3}{33}$ | $4\frac{9}{66}$ | 0.7414 | ..... | $4\frac{3}{58}$ | 0.7857 | ..... | $3\frac{3}{42}$ |
| 0.6977 | $3\frac{9}{43}$ | $3\frac{9}{43}$ | 0.7419 | $2\frac{3}{31}$ | $4\frac{6}{62}$ | 0.7872 | $3\frac{7}{47}$ | $3\frac{7}{47}$ |
| 0.6981 | ..... | $3\frac{7}{53}$ | 0.7424 | ..... | $4\frac{9}{66}$ | 0.7879 | $2\frac{6}{33}$ | $5\frac{2}{66}$ |
| 0.7000 | $1\frac{4}{20}$ | $2\frac{1}{30}$ | 0.7436 | $2\frac{9}{39}$ | $2\frac{9}{39}$ | 0.7895 | $1\frac{5}{19}$ | $3\frac{0}{38}$ |
| 0.7018 | ..... | $4\frac{0}{57}$ | 0.7442 | $3\frac{2}{43}$ | $3\frac{2}{43}$ | 0.7895 | ..... | $4\frac{5}{57}$ |
| 0.7021 | $3\frac{3}{47}$ | $3\frac{3}{47}$ | 0.7447 | $3\frac{5}{47}$ | $3\frac{5}{47}$ | 0.7903 | ..... | $4\frac{9}{62}$ |
| 0.7027 | $2\frac{6}{37}$ | $2\frac{6}{37}$ | 0.7451 | ..... | $3\frac{8}{51}$ | 0.7907 | $3\frac{4}{43}$ | $3\frac{4}{43}$ |
| 0.7037 | $1\frac{9}{27}$ | $3\frac{8}{64}$ | 0.7458 | ..... | $4\frac{4}{59}$ | 0.7917 | ..... | $1\frac{9}{24}$ |
| 0.7059 | $1\frac{2}{17}$ | $2\frac{4}{34}$ | 0.7500 | $1\frac{2}{16}$ | $1\frac{8}{24}$ | 0.7925 | ..... | $4\frac{2}{53}$ |
| 0.7059 | ..... | $3\frac{6}{51}$ | 0.7500 | $1\frac{5}{20}$ | $2\frac{1}{28}$ | 0.7931 | $2\frac{3}{29}$ | $4\frac{6}{58}$ |
| 0.7069 | ..... | $4\frac{1}{58}$ | 0.7544 | ..... | $4\frac{3}{57}$ | 0.7941 | ..... | $2\frac{7}{34}$ |
| 0.7073 | $2\frac{9}{41}$ | $2\frac{9}{41}$ | 0.7547 | ..... | $4\frac{0}{53}$ | 0.7949 | $3\frac{1}{39}$ | $3\frac{1}{39}$ |
| 0.7083 | ..... | $1\frac{7}{24}$ | 0.7551 | $3\frac{7}{49}$ | $3\frac{7}{49}$ | 0.7959 | $3\frac{9}{49}$ | $3\frac{9}{49}$ |
| 0.7097 | $2\frac{2}{31}$ | $4\frac{4}{62}$ | 0.7561 | $3\frac{1}{41}$ | $3\frac{1}{41}$ | 0.7963 | ..... | $4\frac{3}{54}$ |
| 0.7105 | ..... | $2\frac{7}{38}$ | 0.7568 | $2\frac{8}{37}$ | $2\frac{8}{37}$ | 0.7966 | ..... | $4\frac{7}{59}$ |

* See explanatory note below Table 8.

## Accurate Angular Indexing Movements — 7 *

| Fractional Indexing Movement | B. & S., Becker, Hendey, K. & T. and Rockford | Cincinnati and LeBlond * | Fractional Indexing Movement | B. & S., Becker, Hendey, K. & T. and Rockford | Cincinnati and LeBlond * | Fractional Indexing Movement | B. & S., Becker, Hendey, K. & T. and Rockford | Cincinnati and LeBlond * |
|---|---|---|---|---|---|---|---|---|
| 0.8000 | $1\frac{9}{20}$ | $29\frac{5}{25}$ | 0.8431 | ..... | $4\frac{3}{61}$ | 0.8871 | ..... | $5\frac{5}{62}$ |
| 0.8000 | $1\frac{3}{15}$ | $24\frac{5}{30}$ | 0.8448 | ..... | $4\frac{9}{58}$ | 0.8889 | $1\frac{6}{18}$ | ..... |
| 0.8030 | ..... | $5\frac{3}{66}$ | 0.8462 | $3\frac{3}{39}$ | $3\frac{3}{39}$ | 0.8889 | $2\frac{4}{27}$ | $4\frac{8}{54}$ |
| 0.8039 | ..... | $4\frac{1}{61}$ | 0.8475 | ..... | $5\frac{9}{69}$ | 0.8913 | ..... | $4\frac{1}{46}$ |
| 0.8043 | ..... | $3\frac{7}{46}$ | 0.8478 | ..... | $3\frac{9}{46}$ | 0.8919 | $3\frac{8}{47}$ | $3\frac{3}{37}$ |
| 0.8049 | $3\frac{3}{41}$ | $3\frac{3}{41}$ | 0.8485 | $2\frac{8}{33}$ | $5\frac{9}{66}$ | 0.8929 | ..... | $2\frac{5}{28}$ |
| 0.8065 | $2\frac{5}{41}$ | $5\frac{9}{62}$ | 0.8491 | ..... | $4\frac{5}{53}$ | 0.8936 | $4\frac{2}{47}$ | $4\frac{2}{47}$ |
| 0.8070 | ..... | $4\frac{6}{67}$ | 0.8500 | $1\frac{7}{20}$ | ..... | 0.8939 | ..... | $5\frac{9}{66}$ |
| 0.8085 | $3\frac{8}{47}$ | $3\frac{8}{47}$ | 0.8511 | $4\frac{0}{47}$ | $4\frac{0}{47}$ | 0.8947 | $1\frac{7}{19}$ | $3\frac{4}{58}$ |
| 0.8095 | $1\frac{7}{21}$ | $3\frac{4}{42}$ | 0.8519 | $2\frac{3}{27}$ | $4\frac{6}{54}$ | 0.8947 | ..... | $5\frac{1}{57}$ |
| 0.8103 | ..... | $4\frac{7}{68}$ | 0.8529 | ..... | $2\frac{9}{34}$ | 0.8966 | $2\frac{6}{29}$ | $5\frac{2}{58}$ |
| 0.8108 | $3\frac{0}{37}$ | $3\frac{0}{37}$ | 0.8537 | $3\frac{5}{41}$ | $3\frac{5}{41}$ | 0.8974 | $3\frac{5}{39}$ | $3\frac{5}{39}$ |
| 0.8113 | ..... | $4\frac{3}{63}$ | 0.8548 | ..... | $5\frac{3}{62}$ | 0.8980 | $4\frac{4}{49}$ | $4\frac{4}{49}$ |
| 0.8125 | $1\frac{3}{16}$ | ..... | 0.8571 | ..... | $2\frac{4}{28}$ | 0.8983 | ..... | $5\frac{3}{59}$ |
| 0.8136 | ..... | $4\frac{8}{69}$ | 0.8571 | $1\frac{8}{21}$ | $3\frac{6}{42}$ | 0.9000 | $1\frac{8}{20}$ | $2\frac{7}{30}$ |
| 0.8140 | ..... | $3\frac{5}{43}$ | 0.8571 | $4\frac{2}{49}$ | $4\frac{2}{49}$ | 0.9020 | ..... | $4\frac{9}{51}$ |
| 0.8148 | ..... | $4\frac{4}{64}$ | 0.8596 | ..... | $4\frac{9}{57}$ | 0.9024 | $3\frac{7}{41}$ | $3\frac{7}{41}$ |
| 0.8158 | ..... | $3\frac{1}{38}$ | 0.8605 | $3\frac{7}{43}$ | $3\frac{7}{43}$ | 0.9032 | $2\frac{8}{31}$ | $5\frac{9}{62}$ |
| 0.8163 | $4\frac{0}{49}$ | $4\frac{0}{49}$ | 0.8621 | $2\frac{5}{29}$ | $5\frac{0}{58}$ | 0.9048 | $1\frac{9}{21}$ | $3\frac{8}{42}$ |
| 0.8182 | $2\frac{7}{33}$ | $5\frac{4}{66}$ | 0.8627 | ..... | $4\frac{4}{61}$ | 0.9057 | ..... | $4\frac{5}{53}$ |
| 0.8205 | $3\frac{2}{39}$ | $3\frac{2}{39}$ | 0.8636 | ..... | $5\frac{7}{66}$ | 0.9070 | $3\frac{9}{43}$ | $3\frac{9}{43}$ |
| 0.8214 | ..... | $2\frac{3}{28}$ | 0.8644 | ..... | $5\frac{1}{59}$ | 0.9074 | ..... | $4\frac{9}{54}$ |
| 0.8226 | ..... | $5\frac{1}{62}$ | 0.8649 | $3\frac{2}{37}$ | $3\frac{2}{37}$ | 0.9090 | $3\frac{9}{43}$ | $6\frac{9}{66}$ |
| 0.8235 | $1\frac{4}{17}$ | $2\frac{8}{34}$ | 0.8667 | $1\frac{3}{15}$ | $2\frac{6}{30}$ | 0.9118 | ..... | $3\frac{1}{34}$ |
| 0.8235 | ..... | $4\frac{2}{61}$ | 0.8679 | ..... | $4\frac{6}{53}$ | 0.9123 | ..... | $5\frac{2}{57}$ |
| 0.8246 | ..... | $4\frac{7}{57}$ | 0.8684 | ..... | $3\frac{3}{38}$ | 0.9130 | $2\frac{1}{23}$ | $4\frac{2}{46}$ |
| 0.8261 | $1\frac{9}{23}$ | $3\frac{8}{46}$ | 0.8696 | $2\frac{9}{23}$ | $4\frac{0}{46}$ | 0.9138 | ..... | $5\frac{3}{58}$ |
| 0.8276 | $2\frac{4}{29}$ | $4\frac{8}{58}$ | 0.8704 | ..... | $4\frac{7}{54}$ | 0.9149 | $4\frac{3}{47}$ | $4\frac{3}{47}$ |
| 0.8293 | $3\frac{4}{41}$ | $3\frac{4}{41}$ | 0.8710 | $2\frac{7}{31}$ | $5\frac{4}{62}$ | 0.9153 | ..... | $5\frac{4}{59}$ |
| 0.8298 | $3\frac{9}{47}$ | $3\frac{9}{47}$ | 0.8718 | $3\frac{4}{39}$ | $3\frac{4}{39}$ | 0.9167 | ..... | $2\frac{2}{24}$ |
| 0.8302 | ..... | $4\frac{4}{53}$ | 0.8723 | $4\frac{1}{47}$ | $4\frac{1}{47}$ | 0.9184 | $4\frac{5}{49}$ | $4\frac{5}{49}$ |
| 0.8305 | ..... | $4\frac{9}{59}$ | 0.8750 | $1\frac{4}{16}$ | $2\frac{1}{24}$ | 0.9189 | $3\frac{4}{37}$ | $3\frac{4}{37}$ |
| 0.8333 | $1\frac{5}{18}$ | $20\frac{4}{24}$ | 0.8772 | ..... | $5\frac{0}{57}$ | 0.9194 | ..... | $5\frac{7}{62}$ |
| 0.8333 | ..... | $25\frac{0}{30}$ | 0.8776 | $4\frac{3}{49}$ | $4\frac{3}{49}$ | 0.9200 | ..... | $2\frac{3}{25}$ |
| 0.8333 | ..... | $35\frac{4}{42}$ | 0.8780 | $3\frac{6}{41}$ | $3\frac{6}{41}$ | 0.9211 | ..... | $3\frac{5}{38}$ |
| 0.8333 | ..... | $45\frac{4}{54}$ | 0.8788 | $2\frac{9}{33}$ | $5\frac{8}{66}$ | 0.9216 | ..... | $4\frac{7}{51}$ |
| 0.8333 | ..... | $55\frac{6}{66}$ | 0.8793 | ..... | $5\frac{1}{58}$ | 0.9231 | $3\frac{6}{39}$ | $3\frac{6}{39}$ |
| 0.8367 | $4\frac{1}{49}$ | $4\frac{1}{49}$ | 0.8800 | ..... | $2\frac{2}{25}$ | 0.9242 | ..... | $6\frac{1}{66}$ |
| 0.8372 | $3\frac{6}{43}$ | $3\frac{6}{43}$ | 0.8810 | ..... | $3\frac{7}{42}$ | 0.9245 | ..... | $4\frac{9}{53}$ |
| 0.8378 | $3\frac{1}{37}$ | $3\frac{1}{37}$ | 0.8814 | ..... | $5\frac{2}{59}$ | 0.9259 | $2\frac{5}{27}$ | $5\frac{0}{54}$ |
| 0.8387 | $2\frac{6}{31}$ | $5\frac{2}{62}$ | 0.8824 | $1\frac{5}{17}$ | $3\frac{0}{34}$ | 0.9268 | $3\frac{8}{41}$ | $3\frac{8}{41}$ |
| 0.8400 | ..... | $2\frac{1}{25}$ | 0.8824 | ..... | $4\frac{5}{51}$ | 0.9286 | ..... | $2\frac{9}{28}$ |
| 0.8421 | ..... | $3\frac{2}{38}$ | 0.8837 | $3\frac{8}{43}$ | $3\frac{8}{43}$ | 0.9286 | ..... | $3\frac{9}{42}$ |
| 0.8421 | $1\frac{6}{19}$ | $4\frac{8}{57}$ | 0.8868 | ..... | $4\frac{7}{53}$ | 0.9298 | ..... | $5\frac{3}{57}$ |

* See explanatory note below Table 8.

## Accurate Angular Indexing Movements — 8 *

| Fractional Indexing Movement | B. & S., Becker, Hendey, K. & T. and Rockford | Cincinnati and LeBlond * | Fractional Indexing Movement | B. & S., Becker, Hendey, K. & T. and Rockford | Cincinnati and LeBlond * | Fractional Indexing Movement | B. & S., Becker, Hendey, K. & T. and Rockford | Cincinnati and LeBlond * |
|---|---|---|---|---|---|---|---|---|
| 0.9302 | $\frac{40}{43}$ | $\frac{40}{43}$ | 0.9500 | $\frac{19}{20}$ | .... | 0.9697 | $\frac{32}{33}$ | $\frac{64}{66}$ |
| 0.9310 | $\frac{27}{29}$ | $\frac{54}{58}$ | 0.9512 | $\frac{39}{41}$ | $\frac{39}{41}$ | 0.9706 | .... | $\frac{33}{34}$ |
| 0.9322 | .... | $\frac{55}{59}$ | 0.9516 | .... | $\frac{59}{62}$ | 0.9730 | $\frac{36}{37}$ | $\frac{36}{37}$ |
| 0.9333 | $\frac{14}{15}$ | $\frac{28}{30}$ | 0.9524 | $\frac{20}{21}$ | $\frac{40}{42}$ | 0.9737 | .... | $\frac{37}{38}$ |
| 0.9348 | .... | $\frac{43}{46}$ | 0.9535 | $\frac{41}{43}$ | $\frac{41}{43}$ | 0.9744 | $\frac{38}{39}$ | $\frac{38}{39}$ |
| 0.9355 | $\frac{29}{31}$ | $\frac{58}{62}$ | 0.9545 | .... | $\frac{63}{66}$ | 0.9756 | $\frac{40}{41}$ | $\frac{40}{41}$ |
| 0.9362 | $\frac{44}{47}$ | $\frac{44}{47}$ | 0.9565 | $\frac{22}{23}$ | $\frac{44}{46}$ | 0.9762 | .... | $\frac{41}{42}$ |
| 0.9375 | $\frac{15}{16}$ | .... | 0.9574 | $\frac{45}{47}$ | $\frac{45}{47}$ | 0.9767 | $\frac{42}{43}$ | $\frac{42}{43}$ |
| 0.9388 | $\frac{46}{49}$ | $\frac{46}{49}$ | 0.9583 | .... | $\frac{23}{24}$ | 0.9783 | .... | $\frac{45}{46}$ |
| 0.9394 | $\frac{31}{33}$ | $\frac{62}{66}$ | 0.9592 | $\frac{47}{49}$ | $\frac{47}{49}$ | 0.9787 | $\frac{46}{47}$ | $\frac{46}{47}$ |
| 0.9412 | $\frac{16}{17}$ | $\frac{32}{34}$ | 0.9600 | .... | $\frac{24}{25}$ | 0.9796 | $\frac{48}{49}$ | $\frac{48}{49}$ |
| 0.9412 | .... | $\frac{48}{51}$ | 0.9608 | .... | $\frac{49}{51}$ | 0.9804 | .... | $\frac{50}{51}$ |
| 0.9434 | .... | $\frac{50}{53}$ | 0.9623 | .... | $\frac{51}{53}$ | 0.9811 | .... | $\frac{52}{53}$ |
| 0.9444 | $\frac{17}{18}$ | $\frac{51}{54}$ | 0.9630 | $\frac{26}{27}$ | $\frac{52}{54}$ | 0.9815 | .... | $\frac{53}{54}$ |
| 0.9459 | $\frac{35}{37}$ | $\frac{35}{37}$ | 0.9643 | .... | $\frac{27}{28}$ | 0.9825 | .... | $\frac{56}{57}$ |
| 0.9474 | $\frac{18}{19}$ | $\frac{36}{38}$ | 0.9649 | .... | $\frac{55}{57}$ | 0.9828 | .... | $\frac{57}{58}$ |
| 0.9474 | .... | $\frac{54}{57}$ | 0.9655 | $\frac{28}{29}$ | $\frac{56}{58}$ | 0.9831 | .... | $\frac{58}{59}$ |
| 0.9483 | .... | $\frac{55}{58}$ | 0.9661 | .... | $\frac{57}{59}$ | 0.9839 | .... | $\frac{61}{62}$ |
| 0.9487 | $\frac{37}{39}$ | $\frac{37}{39}$ | 0.9667 | $\frac{29}{30}$ | $\frac{29}{30}$ | 0.9848 | .... | $\frac{65}{66}$ |
| 0.9492 | .... | $\frac{56}{59}$ | 0.9677 | $\frac{30}{31}$ | $\frac{60}{62}$ | ...... | .... | .... |

\* The foregoing tables may be used when indexing for angles in degrees, minutes or seconds. The tables are used as follows: Reduce the angle to seconds and divide the value thus obtained by 32,400. The quotient gives the number of complete turns and decimal fraction of a turn required. Then find the decimal (or nearest decimal) to this decimal fraction in the tables. Opposite this decimal will be found the fractional number indicating the indexing movement.

*Example:* — Assume that an angle of 10 degrees, 32 minutes, 12 seconds is to be indexed. Then, 10° 32′ 12″ = 37,932 seconds and 37,932 ÷ 32,400 = 1.1707; therefore this indexing can be made by one complete turn and 0.1707 part of a turn. The second table shows that 0.1707 part of a turn is obtained by moving 7 holes in the 41-hole circle.

*Example:* — Two slots are to be milled in the edge of a disk and the angle between their center-lines is 58 degrees 51 minutes and 53 seconds. Determine the indexing movement.

The angle 58° 51′ 53″ reduced to seconds = 211,913 seconds and 211,913 ÷ 32,400 = 6.5405; therefore this indexing movement requires six complete turns and 0.5405 part of a turn. The fifth table shows that 0.5405 part of a turn is obtained by moving 20 holes in the 37-hole circle.

The number of holes in the index circles of the indexing-heads made by the Brown & Sharpe Mfg. Co., Becker Milling Machine Co., Hendey Machine Co., Kearney & Trecker Co., and the Rockford Milling Machine Co. are the same. The index circles of the Cincinnati Milling Machine Co. differ from these; hence, a separate column is given in the table for the "Cincinnati" index-head. The R. K. LeBlond Machine Tool Co.'s dividing head has the same index circles as that of the Cincinnati Milling Machine Co., except that the former does not have the 24-, 25-, 28-, and 30-hole circles, but has, instead, 36-, 48-, and 56-hole circles. The movements in the 24- and 28-hole circles of the Cincinnati index-head may be made on the LeBlond index-head by taking double the number of holes in the 48-hole and 56-hole circles, respectively. In this way, the table can be used for practically all movements with LeBlond milling machines.

### Block or Multiple Indexing for Gear Cutting

| Teeth to be Cut | Number Indexed at Once | First Driver | First Follower | Second Driver | Second Follower | Turns of Locking Disk | Teeth to be Cut | Number Indexed at Once | First Driver | First Follower | Second Driver | Second Follower | Turns of Locking Disk |
|---|---|---|---|---|---|---|---|---|---|---|---|---|---|
| 25 | 4 | 100 | 50 | 72 | 30 | 4 | 77 | 4 | 100 | 70 | 96 | 44 | 2 |
| 26 | 3 | 100 | 50 | 90 | 52 | 4 | 78 | 5 | 100 | 30 | 90 | 78 | 2 |
| 27 | 2 | 100 | 50 | 60 | 54 | 4 | 80 | 3 | 100 | 50 | 90 | 80 | 2 |
| 28 | 3 | 100 | 50 | 90 | 56 | 4 | 81 | 7 | 100 | 30 | 84 | 52 | 2 |
| 29 | 3 | 100 | 50 | 90 | 58 | 4 | 82 | 5 | 100 | 30 | 90 | 82 | 2 |
| 30 | 7 | 100 | 30 | 84 | 40 | 4 | 84 | 5 | 100 | 30 | 90 | 84 | 2 |
| 31 | 3 | 100 | 50 | 90 | 62 | 4 | 85 | 4 | 100 | 50 | 96 | 68 | 2 |
| 32 | 3 | 100 | 50 | 90 | 64 | 4 | 86 | 5 | 100 | 30 | 90 | 86 | 2 |
| 33 | 4 | 100 | 50 | 80 | 44 | 4 | 87 | 7 | 100 | 30 | 84 | 58 | 2 |
| 34 | 3 | 100 | 50 | 90 | 68 | 4 | 88 | 5 | 100 | 30 | 90 | 88 | 2 |
| 35 | 4 | 100 | 50 | 96 | 56 | 4 | 90 | 7 | 100 | 30 | 70 | 50 | 2 |
| 36 | 5 | 100 | 48 | 80 | 40 | 4 | 91 | 3 | 100 | 70 | 72 | 52 | 2 |
| 37 | 5 | 100 | 30 | 90 | 74 | 4 | 92 | 5 | 100 | 30 | 90 | 92 | 2 |
| 38 | 5 | 100 | 30 | 90 | 76 | 4 | 93 | 7 | 100 | 30 | 84 | 62 | 2 |
| 39 | 5 | 100 | 30 | 90 | 78 | 4 | 94 | 5 | 100 | 30 | 90 | 94 | 2 |
| 40 | 3 | 100 | 50 | 90 | 80 | 4 | 95 | 4 | 100 | 50 | 96 | 76 | 2 |
| 41 | 5 | 100 | 30 | 90 | 82 | 4 | 96 | 5 | 100 | 30 | 90 | 96 | 2 |
| 42 | 5 | 100 | 30 | 90 | 84 | 4 | 98 | 5 | 100 | 30 | 90 | 98 | 2 |
| 43 | 5 | 100 | 30 | 90 | 86 | 4 | 99 | 10 | 100 | 30 | 80 | 44 | 2 |
| 44 | 5 | 100 | 30 | 90 | 88 | 4 | 100 | 7 | 100 | 50 | 84 | 40 | 2 |
| 45 | 7 | 100 | 50 | 70 | 30 | 4 | 102 | 5 | 100 | 30 | 60 | 68 | 2 |
| 46 | 5 | 100 | 30 | 90 | 92 | 4 | 104 | 5 | 100 | 60 | 90 | 52 | 2 |
| 47 | 5 | 100 | 30 | 90 | 94 | 4 | 105 | 4 | 100 | 70 | 96 | 60 | 2 |
| 48 | 5 | 100 | 30 | 90 | 96 | 4 | 108 | 7 | 100 | 30 | 70 | 60 | 2 |
| 49 | 5 | 100 | 30 | 90 | 98 | 4 | 110 | 7 | 100 | 50 | 84 | 44 | 2 |
| 50 | 7 | 100 | 50 | 84 | 40 | 4 | 111 | 5 | 100 | 74 | 80 | 40 | 2 |
| 51 | 4 | 100 | 30 | 96 | 68 | 2 | 112 | 5 | 100 | 60 | 90 | 56 | 2 |
| 52 | 5 | 100 | 30 | 90 | 52 | 2 | 114 | 7 | 100 | 30 | 84 | 76 | 2 |
| 54 | 5 | 100 | 30 | 90 | 54 | 2 | 115 | 8 | 100 | 50 | 96 | 46 | 2 |
| 55 | 4 | 100 | 50 | 96 | 44 | 2 | 116 | 5 | 100 | 60 | 90 | 58 | 2 |
| 56 | 5 | 100 | 30 | 90 | 56 | 2 | 117 | 8 | 100 | 30 | 96 | 78 | 2 |
| 57 | 4 | 100 | 30 | 96 | 76 | 2 | 119 | 3 | 100 | 70 | 72 | 68 | 2 |
| 58 | 5 | 100 | 30 | 90 | 58 | 2 | 120 | 7 | 100 | 50 | 70 | 40 | 2 |
| 60 | 7 | 100 | 30 | 84 | 40 | 2 | 121 | 4 | 60 | 66 | 96 | 44 | 2 |
| 62 | 5 | 100 | 30 | 90 | 62 | 2 | 123 | 7 | 100 | 30 | 84 | 82 | 2 |
| 63 | 5 | 100 | 30 | 80 | 56 | 2 | 124 | 5 | 100 | 60 | 90 | 62 | 2 |
| 64 | 5 | 100 | 30 | 90 | 64 | 2 | 125 | 7 | 100 | 50 | 84 | 50 | 2 |
| 65 | 4 | 100 | 50 | 96 | 52 | 2 | 126 | 5 | 100 | 50 | 50 | 42 | 2 |
| 66 | 5 | 100 | 44 | 80 | 40 | 2 | 128 | 5 | 100 | 60 | 90 | 64 | 2 |
| 67 | 5 | 100 | 30 | 90 | 67 | 2 | 129 | 7 | 100 | 30 | 84 | 86 | 2 |
| 68 | 5 | 100 | 30 | 90 | 68 | 2 | 130 | 7 | 100 | 50 | 84 | 52 | 2 |
| 69 | 5 | 100 | 46 | 80 | 40 | 2 | 132 | 5 | 100 | 88 | 80 | 40 | 2 |
| 70 | 3 | 100 | 50 | 90 | 70 | 2 | 133 | 4 | 100 | 70 | 96 | 76 | 2 |
| 72 | 5 | 100 | 30 | 90 | 72 | 2 | 134 | 5 | 100 | 60 | 90 | 67 | 2 |
| 74 | 5 | 100 | 30 | 90 | 74 | 2 | 135 | 7 | 100 | 50 | 84 | 54 | 2 |
| 75 | 7 | 100 | 30 | 84 | 50 | 2 | 136 | 5 | 100 | 60 | 90 | 68 | 2 |
| 76 | 5 | 100 | 30 | 90 | 76 | 2 | 138 | 5 | 100 | 92 | 80 | 40 | 2 |

### Block or Multiple Indexing for Gear Cutting

| Teeth to be Cut | Number Indexed at Once | First Driver | First Follower | Second Driver | Second Follower | Turns of Locking Disk | Teeth to be Cut | Number Indexed at Once | First Driver | First Follower | Second Driver | Second Follower | Turns of Locking Disk |
|---|---|---|---|---|---|---|---|---|---|---|---|---|---|
| 140 | 3 | 50 | 50 | 90 | 70 | 2 | 170 | 7 | 100 | 50 | 84 | 68 | 2 |
| 141 | 5 | 100 | 94 | 80 | 40 | 2 | 171 | 5 | 70 | 42 | 80 | 76 | 2 |
| 143 | 6 | 90 | 66 | 96 | 52 | 2 | 172 | 5 | 100 | 60 | 90 | 86 | 2 |
| 144 | 5 | 100 | 60 | 90 | 72 | 2 | 174 | 7 | 100 | 60 | 84 | 58 | 2 |
| 145 | 6 | 100 | 50 | 72 | 58 | 2 | 175 | 8 | 100 | 50 | 96 | 70 | 2 |
| 147 | 5 | 100 | 98 | 80 | 40 | 2 | 176 | 5 | 100 | 60 | 90 | 88 | 2 |
| 148 | 5 | 100 | 60 | 90 | 74 | 2 | 180 | 7 | 100 | 60 | 70 | 50 | 2 |
| 150 | 7 | 100 | 60 | 84 | 50 | 2 | 182 | 9 | 90 | 56 | 96 | 52 | 2 |
| 152 | 5 | 100 | 60 | 90 | 76 | 2 | 184 | 5 | 100 | 60 | 90 | 92 | 2 |
| 153 | 5 | 100 | 68 | 80 | 60 | 2 | 185 | 6 | 100 | 50 | 72 | 74 | 2 |
| 154 | 5 | 100 | 56 | 72 | 66 | 2 | 186 | 7 | 100 | 60 | 84 | 62 | 2 |
| 155 | 6 | 100 | 50 | 72 | 62 | 2 | 187 | 5 | 100 | 44 | 48 | 68 | 2 |
| 156 | 5 | 100 | 60 | 90 | 78 | 2 | 188 | 5 | 100 | 60 | 90 | 94 | 2 |
| 160 | 7 | 100 | 50 | 84 | 64 | 2 | 189 | 5 | 100 | 60 | 80 | 84 | 2 |
| 161 | 5 | 100 | 70 | 60 | 46 | 2 | 190 | 7 | 100 | 50 | 84 | 76 | 2 |
| 162 | 7 | 100 | 60 | 84 | 52 | 2 | 192 | 5 | 100 | 60 | 90 | 96 | 2 |
| 164 | 5 | 100 | 60 | 90 | 82 | 2 | 195 | 7 | 100 | 50 | 84 | 78 | 2 |
| 165 | 7 | 100 | 50 | 84 | 66 | 2 | 196 | 5 | 100 | 60 | 90 | 98 | 2 |
| 168 | 5 | 100 | 60 | 90 | 84 | 2 | 198 | 7 | 100 | 50 | 70 | 66 | 2 |
| 169 | 6 | 96 | 52 | 90 | 78 | 2 | 200 | 7 | 60 | 60 | 84 | 40 | 2 |

**Block or Multiple Indexing for Gear Cutting.** — With the block system of indexing, a number of teeth are indexed at one time, instead of cutting the teeth consecutively, and the gear is revolved several times before the teeth are all finished. For example, when cutting a gear having 25 teeth, the indexing mechanism is geared to index four teeth at once (see table) and the first time around, six widely separated tooth spaces are cut. The second time around, the cutter is one tooth behind the spaces previously milled. On the third indexing, the cutter has dropped back another tooth, thus finishing the gear (in this case) by indexing it around four times. The various combinations of change gears to use for block or multiple indexing are given in the accompanying table. The advantage claimed for block indexing is that the heat generated by the cutter (especially when cutting cast-iron gears of coarse pitch) is distributed more evenly about the rim and dissipated to a greater extent, thus avoiding distortion due to local heating and permitting higher speeds and feeds. The table given is intended for use with Brown & Sharpe automatic gear-cutting machines, but the gears for any other machine equipped with a similar indexing mechanism can be calculated. Assume, for example, that a gear cutter requires the following change gears for indexing a certain number of teeth: Driving gears having 20 and 30 teeth, respectively, and driven gears having 50 and 60 teeth. Then if it is desired to cut, say, every fifth tooth, multiply the fractions $\frac{20}{60}$ and $\frac{30}{50}$ by 5. Then, $\frac{20}{60} \times \frac{30}{50} \times \frac{5}{1} = \frac{1}{1}$. In this particular instance, then, the blank could be divided so that every fifth space would be cut, by using gears of equal size. The number of teeth in the gear and the number of teeth indexed in each block, must not have a common factor.

## Indexing Movements for 60-Tooth Worm-Wheel Dividing Head

| Divisions | Index Circle | No. of Turns | No. of Holes | Divisions | Index Circle | No. of Turns | No. of Holes | Divisions | Index Circle | No. of Turns | No. of Holes | Divisions | Index Circle | No. of Turns | No. of Holes |
|---|---|---|---|---|---|---|---|---|---|---|---|---|---|---|---|
| 2 | Any | 30 | .. | 50 | 60 | 1 | 12 | 98 | 49 | .. | 30 | 146 | 73 | .. | 30 |
| 3 | Any | 20 | .. | 51 | 17 | 1 | 3 | 99 | 33 | .. | 20 | 147 | 49 | .. | 20 |
| 4 | Any | 15 | .. | 52 | 26 | 1 | 4 | 100 | 60 | .. | 36 | 148 | 37 | .. | 15 |
| 5 | Any | 12 | .. | 53 | 53 | 1 | 7 | 101 | 101 | .. | 60 | 149 | 149 | .. | 60 |
| 6 | Any | 10 | .. | 54 | 27 | 1 | 3 | 102 | 17 | .. | 10 | 150 | 60 | .. | 24 |
| 7 | 21 | 8 | 12 | 55 | 33 | 1 | 3 | 103 | 103 | .. | 60 | 151 | 151 | .. | 60 |
| 8 | 26 | 7 | 13 | 56 | 28 | 1 | 2 | 104 | 26 | .. | 15 | 152 | 76 | .. | 30 |
| 9 | 21 | 6 | 14 | 57 | 19 | 1 | 1 | 105 | 21 | .. | 12 | 153 | 51 | .. | 20 |
| 10 | Any | 6 | .. | 58 | 29 | 1 | 1 | 106 | 53 | .. | 30 | 154 | 77 | .. | 30 |
| 11 | 33 | 5 | 15 | 59 | 59 | 1 | 1 | 107 | 107 | .. | 60 | 155 | 31 | .. | 12 |
| 12 | Any | 5 | .. | 60 | Any | 1 | .. | 108 | 27 | .. | 15 | 156 | 26 | .. | 10 |
| 13 | 26 | 4 | 16 | 61 | 61 | .. | 60 | 109 | 109 | .. | 60 | 157 | 157 | .. | 60 |
| 14 | 21 | 4 | 6 | 62 | 31 | .. | 30 | 110 | 33 | .. | 18 | 158 | 79 | .. | 30 |
| 15 | Any | 4 | .. | 63 | 21 | .. | 20 | 111 | 37 | .. | 20 | 159 | 53 | .. | 20 |
| 16 | 28 | 3 | 21 | 64 | 32 | .. | 30 | 112 | 28 | .. | 15 | 160 | 32 | .. | 12 |
| 17 | 17 | 3 | 9 | 65 | 26 | .. | 24 | 113 | 113 | .. | 60 | 161 | 161 | .. | 60 |
| 18 | 21 | 3 | 7 | 66 | 33 | .. | 30 | 114 | 19 | .. | 10 | 162 | 27 | .. | 10 |
| 19 | 19 | 3 | 3 | 67 | 67 | .. | 60 | 115 | 23 | .. | 12 | 163 | 163 | .. | 60 |
| 20 | Any | 3 | .. | 68 | 17 | .. | 15 | 116 | 29 | .. | 15 | 164 | 41 | .. | 15 |
| 21 | 21 | 2 | 18 | 69 | 23 | .. | 20 | 117 | 39 | .. | 20 | 165 | 33 | .. | 12 |
| 22 | 33 | 2 | 24 | 70 | 21 | .. | 18 | 118 | 59 | .. | 30 | 166 | 83 | .. | 30 |
| 23 | 23 | 2 | 14 | 71 | 71 | .. | 60 | 119 | 119 | .. | 60 | 167 | 167 | .. | 60 |
| 24 | 26 | 2 | 13 | 72 | 60 | .. | 50 | 120 | 26 | .. | 13 | 168 | 28 | .. | 10 |
| 25 | 60 | 2 | 24 | 73 | 73 | .. | 60 | 121 | 121 | .. | 60 | 169 | 169 | .. | 60 |
| 26 | 26 | 2 | 8 | 74 | 37 | .. | 30 | 122 | 61 | .. | 30 | 170 | 17 | .. | 6 |
| 27 | 27 | 2 | 6 | 75 | 60 | .. | 48 | 123 | 41 | .. | 20 | 171 | 57 | .. | 20 |
| 28 | 21 | 2 | 3 | 76 | 19 | .. | 15 | 124 | 31 | .. | 15 | 172 | 43 | .. | 15 |
| 29 | 29 | 2 | 2 | 77 | 77 | .. | 60 | 125 | 100 | .. | 48 | 173 | 173 | .. | 60 |
| 30 | Any | 2 | .. | 78 | 26 | .. | 20 | 126 | 21 | .. | 10 | 174 | 29 | .. | 10 |
| 31 | 31 | 1 | 29 | 79 | 79 | .. | 60 | 127 | 127 | .. | 60 | 175 | 35 | .. | 12 |
| 32 | 32 | 1 | 28 | 80 | 28 | .. | 21 | 128 | 32 | .. | 15 | 176 | 44 | .. | 15 |
| 33 | 33 | 1 | 27 | 81 | 27 | .. | 20 | 129 | 43 | .. | 20 | 177 | 59 | .. | 20 |
| 34 | 17 | 1 | 13 | 82 | 41 | .. | 30 | 130 | 26 | .. | 12 | 178 | 89 | .. | 30 |
| 35 | 21 | 1 | 15 | 83 | 83 | .. | 60 | 131 | 131 | .. | 60 | 179 | 179 | .. | 60 |
| 36 | 21 | 1 | 14 | 84 | 21 | .. | 15 | 132 | 33 | .. | 15 | 180 | 21 | .. | 7 |
| 37 | 37 | 1 | 23 | 85 | 17 | .. | 12 | 133 | 133 | .. | 60 | 181 | 181 | .. | 60 |
| 38 | 19 | 1 | 11 | 86 | 43 | .. | 30 | 134 | 67 | .. | 30 | 182 | 91 | .. | 30 |
| 39 | 26 | 1 | 14 | 87 | 29 | .. | 20 | 135 | 27 | .. | 12 | 183 | 61 | .. | 20 |
| 40 | 26 | 1 | 13 | 88 | 44 | .. | 30 | 136 | 68 | .. | 30 | 184 | 46 | .. | 15 |
| 41 | 41 | 1 | 19 | 89 | 89 | .. | 60 | 137 | 137 | .. | 60 | 185 | 37 | .. | 12 |
| 42 | 21 | 1 | 9 | 90 | 21 | .. | 14 | 138 | 23 | .. | 10 | 186 | 31 | .. | 10 |
| 43 | 43 | 1 | 17 | 91 | 91 | .. | 60 | 139 | 139 | .. | 60 | 187 | 187 | .. | 60 |
| 44 | 33 | 1 | 12 | 92 | 23 | .. | 15 | 140 | 21 | .. | 9 | 188 | 47 | .. | 15 |
| 45 | 21 | 1 | 7 | 93 | 31 | .. | 20 | 141 | 47 | .. | 20 | 189 | 63 | .. | 20 |
| 46 | 23 | 1 | 7 | 94 | 47 | .. | 30 | 142 | 71 | .. | 30 | 190 | 19 | .. | 6 |
| 47 | 47 | 1 | 13 | 95 | 19 | .. | 12 | 143 | 143 | .. | 60 | 191 | 191 | .. | 60 |
| 48 | 28 | 1 | 7 | 96 | 32 | .. | 20 | 144 | 60 | .. | 25 | 192 | 32 | .. | 10 |
| 49 | 49 | 1 | 11 | 97 | 97 | .. | 60 | 145 | 29 | .. | 12 | 193 | 193 | .. | 60 |

**Indexing for Rack Cutting.** — When racks are cut on a milling machine, there are two general methods of indexing. One is by using the graduated dial on the feed-screw and the other is by using an indexing attachment. The accompanying table shows the indexing movements when the first method is employed. This table applies to milling machines having feed-screws with the usual lead of ¼ inch and 250 dial graduations each equivalent to 0.001 inch of table movement.

$$\text{Actual rotation of feed-screw} = \frac{\text{Linear pitch of rack}}{\text{Lead of feed-screw}}$$

Multiply *decimal* part of turn (obtained by above formula) by 250, to obtain dial reading for fractional part of indexing movement, assuming that dial has 250 graduations.

### Indexing Movements for Cutting Rack Teeth on Milling Machine

These movements are for table feed-screws having the usual lead of ¼ inch

| Pitch of Rack Tooth | | Indexing, Movement | | Pitch of Rack Teeth | | Indexing, Movement | |
|---|---|---|---|---|---|---|---|
| Diametral Pitch | Linear or Circular | No. of Whole Turns | No. of .001 Inch Divisions | Diametral Pitch | Linear or Circular | No. of Whole Turns | No. of .001 Inch Divisions |
| 2 | 1.5708 | 6 | 70.8 | 12 | 0.2618 | 1 | 11.8 |
| 2¼ | 1.3963 | 5 | 146.3 | 13 | 0.2417 | 0 | 241.7 |
| 2½ | 1.2566 | 5 | 6.6 | 14 | 0.2244 | 0 | 224.4 |
| 2¾ | 1.1424 | 4 | 142.4 | 15 | 0.2094 | 0 | 208.4 |
| 3 | 1.0472 | 4 | 47.2 | 16 | 0.1963 | 0 | 196.3 |
| 3½ | 0.8976 | 3 | 147.6 | 17 | 0.1848 | 0 | 184.8 |
| 4 | 0.7854 | 3 | 35.4 | 18 | 0.1745 | 0 | 174.8 |
| 5 | 0.6283 | 2 | 128.3 | 19 | 0.1653 | 0 | 165.3 |
| 6 | 0.5236 | 2 | 23.6 | 20 | 0.1571 | 0 | 157.1 |
| 7 | 0.4488 | 1 | 198.8 | 22 | 0.1428 | 0 | 142.8 |
| 8 | 0.3927 | 1 | 142.7 | 24 | 0.1309 | 0 | 130.9 |
| 9 | 0.3491 | 1 | 99.1 | 26 | 0.1208 | 0 | 120.8 |
| 10 | 0.3142 | 1 | 64.2 | 28 | 0.1122 | 0 | 112.2 |
| 11 | 0.2856 | 1 | 35.6 | 30 | 0.1047 | 0 | 104.7 |

Note: The linear pitch of the rack equals the circular pitch of gear or pinion which is to mesh with the rack. The table gives both standard diametral pitches and their equivalent linear or circular pitches.

*Example:* Find indexing movement for cutting rack to mesh with a pinion of 10 diametral pitch.

Indexing movement equals 1 whole turn of feed-screw plus 64.2 thousandths or divisions on feed-screw dial. The feed-screw may be turned this fractional amount by setting dial back to its zero position for each indexing (without backward movement of feed-screw), or, if preferred, 64.2 (in this example) may be added to each successive dial position as shown below.

Dial reading for second position = 64.2 × 2 = 128.4 (complete movement = 1 turn + 64.2 additional divisions by turning feed-screw until dial reading is 128.4).

Third dial position = 64.2 × 3 = 192.6 (complete movement = 1 turn + 64.2 additional divisions by turning until dial reading is 192.6).

Fourth position = 64.2 × 4 − 250 = 6.8 (1 turn + 64.2 additional divisions by turning feed-screw until dial reading is 6.8 divisions past the zero mark); or, to simplify operation, set dial back to zero for fourth indexing (without moving feed-screw) and then repeat settings for the three previous indexings or whatever number can be made before making a complete turn of the dial.

# ALLOWANCES AND TOLERANCES FOR FITS

**Limits and Fits.** — Fits between cylindrical parts or, briefly cylindrical fits, govern the proper assembly and performance of countless mechanisms. Clearance fits permit relative freedom of motion between a shaft and a hole — axially, radially, or both. Interference fits secure a certain amount of tightness between parts, whether these are meant to remain permanently assembled or to be taken apart from time to time. Or again, two parts may be required to fit together snugly — without apparent tightness or looseness. The designer's problem is to specify these different types of fits in such a way that the shop can produce them. This involves the adoption of two manufacturing limits for the hole and two for the shaft, and, hence, the adoption of a manufacturing tolerance on each part.

In selecting and specifying limits and fits for various applications, it is essential in the interests of interchangeable manufacturing that (1) standard definitions of terms relating to limits and fits be used; (2) preferred basic sizes be selected wherever possible to reduce material and tooling costs; (3) limits be based upon a series of preferred tolerances and allowances; and (4) a uniform system of applying tolerances (preferably unilateral) be used. These principles have been incorporated in both the American and British standards for limits and fits. Information about these standards is given beginning on page 1537.

**Basic Dimensions.** — The basic size of a screw thread or machine part is the theoretical or nominal standard size from which variations are made. For example, a shaft may have a *basic* diameter of 2 inches, but a maximum variation of minus 0.010 inch may be permitted. The minimum hole should be of basic size in all cases where the use of standard tools represents the greatest economy. The maximum shaft should be of basic size in all cases where the use of standard purchased material, without further machining, represents the greatest economy, even though special tools are required to machine the mating part.

**Tolerances.** — Tolerance is the amount of variation permitted on dimensions or surfaces of machine parts. The tolerance is equal to the difference between the maximum and minimum limits of any specified dimension. For example, if the maximum limit for the diameter of a shaft is 2.000 inches and its minimum limit 1.990 inches, the tolerance for this diameter is 0.010 inch. By determining the maximum and minimum clearances required on operating surfaces, the extent of these tolerances is established. As applied to the fitting of machine parts, the word tolerance means the amount that duplicate parts are allowed to vary in size in connection with manufacturing operations, owing to unavoidable imperfections of workmanship. Tolerance may also be defined as the amount that duplicate parts are permitted to vary in size in order to secure sufficient accuracy without unnecessary refinement. The terms "tolerance" and "allowance" are often used interchangeably, but, according to common usage, *allowance* is a difference in dimensions prescribed in order to secure various classes of fits between different parts.

**Unilateral and Bilateral Tolerances.** — The term "unilateral tolerance" means that the total tolerance, as related to a basic dimension, is in *one* direction only. For example, if the basic dimension were 1 inch and the tolerance were expressed as 1.00 − 0.002, or as 1.00 + 0.002, these would be unilateral tolerances, since the total tolerance in each case is in one direction. On the contrary, if the tolerance were divided, so as to be partly plus and partly minus, it would be classed as "bilateral." Thus, $1.00 \begin{smallmatrix} +\ 0.001 \\ -\ 0.001 \end{smallmatrix}$ is an example of bilateral tolerance, because the total tolerance of 0.002 is given in two directions — plus and minus.

When unilateral tolerances are used, one of the three following methods should be used to express them:

(1) Specify limiting dimensions only as

Diameter of hole: 2.250, 2.252
Diameter of shaft: 2.249, 2.247

(2) One limiting size may be specified with its tolerances as

Diameter of hole: 2.250 + 0.002, − 0.000
Diameter of shaft: 2.249 + 0.000, − 0.002

(3) The nominal size may be specified for both parts, with a notation showing both allowance and tolerance, as

Diameter of hole: 2¼ + 0.002, − 0.000
Diameter of shaft: 2¼ − 0.001, − 0.003

Bilateral tolerances should be specified as such, usually with plus and minus tolerances of equal amount. Example of the expression of bilateral tolerances follow:

$$2 \pm 0.001 \quad \text{or} \quad 2 \begin{array}{c} + 0.001 \\ - 0.001 \end{array}$$

**Application of Tolerances.** — According to common practice, tolerances are applied in such a way as to show the permissible amount of dimensional variation in the direction that is less dangerous. When a variation in either direction is equally dangerous, a bilateral tolerance should be given. When a variation in one direction is more dangerous than a variation in another, a unilateral tolerance should be given in the less dangerous direction.

For non-mating surfaces, or atmospheric fits, the tolerances may be bilateral, or unilateral, depending entirely upon the nature of the variations that develop in manufacture. On mating surfaces, with but few exceptions, the tolerances should be unilateral.

Where tolerances are required on the distances between holes, usually they should be bilateral, as variation in either direction is usually equally dangerous. The variation in the distance between shafts carrying gears, however, should always be unilateral and plus; otherwise the gears might run too tight. A slight increase in the backlash between gears is seldom of much importance.

One exception to the use of unilateral tolerances on mating surfaces occurs when tapers are involved. In such cases either bilateral or unilateral tolerances may prove advisable, depending upon conditions. These should be determined in the same manner as the tolerances on the distances between holes. When a variation either in or out of the position of the mating taper surfaces is equally dangerous, the tolerances should be bilateral. When a variation in one direction is of less danger than a variation in the opposite direction, the tolerance should be unilateral and in the less dangerous direction.

**Locating Tolerance Dimensions** — Only one dimension in the same straight line can be controlled within fixed limits. That is the distance between the cutting surface of the tool and the locating or registering surface of the part being machined. Therefore, it is incorrect to locate any point or surface with tolerances from more than one point in the same straight line.

Every part of a mechanism must be located in each plane. Every operating part must be located with proper operating allowances. After such requirements of location are met, all other surfaces should have liberal clearances. Dimensions should be given between those points or surfaces that it is essential to hold in a specific relation to each other. This applies particularly to those surfaces in each plane which control the location of other component parts. Many dimensions are relatively unimportant in this respect. It is good practice in such cases to establish a common locating-point in each plane and give, so far as possible all such dimensions from these common locating-points. The locating points on the drawing, the locating or registering points used for machining the surfaces and the locating points for measuring should all be identical.

The initial dimensions placed on component drawings should be the exact dimensions that would be used if it were possible to work without tolerances. Tolerances should be given in that direction in which variations will cause the least harm or danger. When a variation in either direction is equally dangerous, the tolerances should be of equal amount in both directions, or bilateral. The initial clearance, or allowance, between operating parts should be as small as the operation of the mechanism will permit. The maximum clearance should be as great as the proper functioning of the mechanism will permit.

**Direction of Tolerances on Gages.** — The following fundamental principles have been adopted in connection with the American Standard Tolerances, Allowances and Gages for Metal Fits. The extreme sizes for all plain limit gages shall not exceed the extreme limits of the part to be gaged. All variations in the gages, whatever their cause or purpose, shall bring these gages within these extreme limits. Thus a gage which represents a minimum limit may be larger, but never smaller, than the minimum size specified for the part to be gaged, likewise the gage which represents a maximum limit may be smaller, but never larger, than the maximum size specified for the part to be gaged.

The final result sought by gaging is interchangeable manufacture in some degree. This means that the parts of a mechanism can be assembled without *fitting* one part to another and when assembled the parts will function properly.

Applied to manufactured material, the result sought is sufficient uniformity in size and contour to adapt the material without further fitting to the requirements of the industries. The fundamental principle involved in interchangeable manufacture requires that "a system of standardization and classification of fits shall establish a clearly defined line at which interference between mating parts begins." Hence, the standard or basic size, as physically represented by a correct standard master gage, represents the line at which this interference begins between mating parts. It is the minimum size of the external members of all mating parts of standardized practice, regardless of the kind of fit. It is the maximum size of internal members of all mating parts where interference begins or that fit metal to metal.

The limits of the component as physically represented by the limit master gages shall not be exceeded as a result either of tolerance or wear of the gages. "Go" gages, or the equivalent verification of all the factors involved in the fit, are necessary to prevent interference of mating parts. In the case of force fits, "go" gages are necessary to determine the maximum amount of interference between mating parts.

"Not go" gages, or the equivalent verification of the determining factor are necessary to prevent the maximum looseness of mating parts exceeding the limits specified. In the case of force fits, "not go" gages are necessary to determine the minimum amount of interference between mating parts.

**Gage Tolerance Based upon Work Tolerance.** — According to the plan to be described (which represents the practice of a prominent manufacturer of gages), a tolerance equal to 10 per cent of the tolerance on the work is generally allowed on ordinary working and inspection gages. Thus, if the work tolerance is 0.005 inch, the gage tolerance equals 0.0005 inch for both the working and inspection gages. There is a difference, however, between the maximum and minimum dimensions of the working and inspection gages. The minimum size of the working gage is made 10 per cent of the tolerance *larger* than the minimum size of the inspection gage, and the maximum size of the working gage is made 10 per cent of the tolerance *smaller* than the maximum size of the inspection gage.

Assume that the minimum and maximum diameters of a shaft are 1 and 1.005 inches respectively, the tolerance being 0.005 inch. The "not go" working gage will then measure 1.0005 inches, 10 per cent of the 0.005-inch tolerance being added to the minimum dimension of the shaft (1.000 + 10 per cent of 0.005 = 1.0005). The "go" working gage will measure 1.0045 inch, since 10 per cent of the tolerance, or 0.0005, is subtracted from the maximum size of the shaft (1.005 − 0.0005 = 1.0045).

When working gages are within the minimum and maximum limits allowed on the work and on the inspection gages, it is evident that all parts which pass the working gages will also pass the inspection gages. When the working gages are made the same size as the inspection gages, disputes are liable to arise and, moreover, working gages may wear faster and become larger than the inspection gages, provided they were both made the same size originally.

The tolerance of the gage itself should be properly applied to avoid any overlapping of dimensions between the working and inspection gages. This gage tolerance should be *minus* on "go" female gages and "not go" male gages, with the exception of the outside and root diameters of "not go" thread gages. The tolerance should be *plus* on "go" male gages and "not go" female gages, again excepting the outside and root diameters of "not go" thread gages. The dimensions on thread gages should be the same as corresponding diameters of "go" gages.

**Allowance for Forced Fits.** — The allowance per inch of diameter usually ranges from 0.001 inch to 0.0025 inch, 0.0015 being a fair average. Ordinarily the allowance per inch decreases as the diameter increases; thus the total allowance for a diameter of 2 inches might be 0.004 inch, whereas for a diameter of 8 inches the total allowance might not be over 0.009 or 0.010 inch. The parts to be assembled by forced fits are usually made cylindrical, although sometimes they are slightly tapered. The advantages of the taper form are that the possibility of abrasion of the fitted surfaces is reduced; that less pressure is required in assembling; and that the parts are more readily separated when renewal is required. On the other hand, the taper fit is less reliable, because if it loosens, the entire fit is free with but little axial movement. Some lubricant, such as white lead and lard oil mixed to the consistency of paint, should be applied to the pin and bore before assembling, to reduce the tendency of abrasion.

**Pressure for Forced Fits.** — The pressure required for assembling cylindrical parts depends not only upon the allowance for the fit, but also upon the area of the fitted surfaces, the pressure increasing in proportion to the distance that the inner member is forced in. The approximate ultimate pressure in tons can be determined by the use of the following formula in conjunction with the accompanying

**Pressure Factors**

| Diameter, Inches | Pressure Factor | Diameter, Inches | Pressure Factor | Diameter, Inches | Pressure Factor | Diameter, Inches | Pressure Factor | Diameter, Inches | Pressure Factor |
|---|---|---|---|---|---|---|---|---|---|
| 1 | 500 | 3½ | 132 | 6 | 75 | 9 | 48.7 | 14 | 30.5 |
| 1¼ | 395 | 3¾ | 123 | 6¼ | 72 | 9½ | 46.0 | 14½ | 29.4 |
| 1½ | 325 | 4 | 115 | 6½ | 69 | 10 | 43.5 | 15 | 28.3 |
| 1¾ | 276 | 4¼ | 108 | 6¾ | 66 | 10½ | 41.3 | 15½ | 27.4 |
| 2 | 240 | 4½ | 101 | 7 | 64 | 11 | 39.3 | 16 | 26.5 |
| 2¼ | 212 | 4¾ | 96 | 7¼ | 61 | 11½ | 37.5 | 16½ | 25.6 |
| 2½ | 189 | 5 | 91 | 7½ | 59 | 12 | 35.9 | 17 | 24.8 |
| 2¾ | 171 | 5¼ | 86 | 7¾ | 57 | 12½ | 34.4 | 17½ | 24.1 |
| 3 | 156 | 5½ | 82 | 8 | 55 | 13 | 33.0 | 18 | 23.4 |
| 3¼ | 143 | 5¾ | 78 | 8½ | 52 | 13½ | 31.7 | .... | .... |

table of "Pressure Factors." Assuming that $A$ = area of surface in contact in "fit"; $a$ = total allowance in inches; $P$ = ultimate pressure required, in tons; $F$ = pressure factor based upon assumption that the diameter of the hub is twice the diameter of the bore, that the shaft is of machine steel, and that the hub is of cast iron:

$$P = \frac{A \times a \times F}{2}$$

**Allowance for Given Pressure.** — By transposing the preceding formula, the approximate allowance for a required ultimate tonnage can be determined. Thus, $a = \frac{2P}{AF}$. The average ultimate pressure in tons commonly used ranges from 7 to 10 times the diameter in inches.

**Expansion Fits.** — In assembling certain classes of work requiring a very tight fit, the inner member is contracted by sub-zero cooling to permit insertion into the outer member and a tight fit is obtained as the temperature rises and the inner part expands. In order to obtain the sub-zero temperature, solid carbon dioxide or "dry ice" has been used but its temperature of about 109 degrees F. below zero will not contract some parts sufficiently to permit insertion in holes or recesses. Greater contraction may be obtained by using high purity liquid nitrogen which has a temperature of about 320 degrees F. below zero. During a temperature reduction from 75 degrees F. to —321 degrees F., the shrinkage per inch of diameter varies from about 0.002 to 0.003 inch for steel; 0.0042 inch for aluminum alloys; 0.0046 inch for magnesium alloys; 0.0033 inch for copper alloys; 0.0023 inch for monel metal; and 0.0017 inch for cast iron (not alloyed). The cooling equipment may vary from an insulated bucket to a special automatic unit, depending upon the kind and quantity of work. One type of unit is so arranged that parts are precooled by vapors from the liquid nitrogen before immersion. With another type, cooling is entirely by the vapor method.

**Shrinkage Fits.** — General practice seems to favor a smaller allowance for shrinkage fits than for forced fits, although in many shops the allowances are practically the same in each case, and for some classes of work, shrinkage allowances exceed those for forced fits. In any case, the shrinkage allowance varies to a great extent with the form and construction of the part which has to be shrunk into place. The thickness or amount of metal around the hole is the most important factor. The way in which the metal is distributed also has an influence on the results. Shrinkage allowances for locomotive driving wheel tires adopted by the American Railway Master Mechanics Association are as follows:

| Center diameter, inches | 38 | 44 | 50 | 56 | 62 | 66 |
|---|---|---|---|---|---|---|
| Allowance, inches | 0.040 | 0.047 | 0.053 | 0.060 | 0.066 | 0.070 |

Whether parts are to be assembled by forced or shrinkage fits depends upon conditions. For example, to press a tire over its wheel center, without heating, would ordinarily be a rather awkward and difficult job. On the other hand, pins, etc., are easily and quickly forced into place with a hydraulic press and there is the additional advantage of knowing the exact pressure required in assembling, whereas there is more or less uncertainty connected with a shrinkage fit, unless the stresses are calculated. Tests to determine the difference in the quality of shrinkage and forced fits showed that the resistance of a shrinkage fit to slippage was, for an axial pull, 3.66 times greater than that of a forced fit, and in rotation or torsion, 3.2 times greater. In each comparative test, the dimensions and allowances were the same.

**Allowances for Shrinkage Fits.** — The most important point to consider when calculating shrinkage fits is the stress in the hub at the bore, which depends chiefly upon the shrinkage allowance. If the allowance is excessive, the elastic limit of the material will be exceeded and permanent set will occur, or, in extreme cases, the ultimate strength of the metal will be exceeded and the hub will burst. The intensity of the grip of the fit and the resistance to slippage depends mainly upon the thickness of the hub; the greater the thickness, the stronger the grip, and *vice versa*. Assuming the modulus of elasticity for steel to be 30,000,000, and for cast iron, 15,000,000, the shrinkage allowance per inch of nominal diameter can be determined by the following formula, in which $A$ = allowance per inch of diameter; $T$ = true tangential tensile stress at inner surface of outer member; $C$ = factor taken from one of the accompanying tables, "Factors for Calculating Shrinkage Fit Allowances." For a cast-iron hub and steel shaft:

$$A = \frac{T(2+C)}{30,000,000} \qquad (1)$$

When both hub and shaft are of steel:

$$A = \frac{T(1+C)}{30,000,000} \qquad (2)$$

If the shaft is solid, the factor $C$ is taken from Table 1; if it is hollow and the hub is of steel, factor $C$ is taken from Table 2; if it is hollow and the hub is of cast iron, the factor is taken from Table 3.

**Table 1. Factors for Calculating Shrinkage Fit Allowances**

Values of Ratio $C$ for solid steel shafts of nominal diameter $D_1$, and hubs of steel or cast iron of nominal external and internal diameters $D_2$ and $D_1$, respectively.

| Ratio of Diameters $\frac{D_2}{D_1}$ | Steel Hub | Cast-iron Hub | Ratio of Diameters $\frac{D_2}{D_1}$ | Steel Hub | Cast-iron Hub |
|---|---|---|---|---|---|
| 1.5 | 0.227 | 0.234 | 2.8 | 0.410 | 0.432 |
| 1.6 | 0.255 | 0.263 | 3.0 | 0.421 | 0.444 |
| 1.8 | 0.299 | 0.311 | 3.2 | 0.430 | 0.455 |
| 2.0 | 0.333 | 0.348 | 3.4 | 0.438 | 0.463 |
| 2.2 | 0.359 | 0.377 | 3.6 | 0.444 | 0.471 |
| 2.4 | 0.380 | 0.399 | 3.8 | 0.450 | 0.477 |
| 2.6 | 0.397 | 0.417 | 4.0 | 0.455 | 0.482 |

*Example 1:* A steel crank web 15 inches outside diameter is to be shrunk on a 10-inch solid steel shaft. Required the allowance per inch of shaft diameter to produce a maximum tensile stress in the crank of 25,000 pounds per square inch, assuming the stresses in the crank to be equivalent to those in a ring of the diameter given.

The ratio of the external to the internal diameters equals $15 \div 10 = 1.5$; $T$ = 25,000 pounds; from Table 1, $C = 0.227$. Substituting in Formula (2):

$$A = \frac{25,000 \times (1 + 0.227)}{30,000,000} = 0.001 \text{ inch.}$$

*Example 2:* Find the allowance per inch of diameter for a 10-inch shaft having a 5-inch axial hole through it, other conditions being the same as in Example 1.

The ratio of external to internal diameters of the hub equals $15 \div 10 = 1.5$, before, and the ratio of external to internal diameters of the shaft equals $10 \div 5 = 2$. From Table 2, we find that factor $C = 0.455$; $T = 25,000$ pounds. Substituting these values in Formula (2):

$$A = \frac{25,000 \times (1 + 0.455)}{30,000,000} = 0.0012 \text{ inch.}$$

The increase in allowance, as compared with Example 1, is due to the fact that the hollow shaft is more compressible.

### Table 2. Factors for Calculating Shrinkage Fit Allowances

Values of Ratio $C$ for hollow steel shafts of external and internal diameters $D_1$ and $D_0$, respectively, and steel hubs of nominal external diameter $D_2$.

| $\frac{D_2}{D_1}$ | $\frac{D_1}{D_0}$ | $C$ | $\frac{D_2}{D_1}$ | $\frac{D_1}{D_0}$ | $C$ | $\frac{D_2}{D_1}$ | $\frac{D_1}{D_0}$ | $C$ |
|---|---|---|---|---|---|---|---|---|
| 1.5 | 2.0 | 0.455 | 2.4 | 2.0 | 0.760 | 3.4 | 2.0 | 0.876 |
| | 2.5 | 0.357 | | 2.5 | 0.597 | | 2.5 | 0.689 |
| | 3.0 | 0.313 | | 3.0 | 0.523 | | 3.0 | 0.602 |
| | 3.5 | 0.288 | | 3.5 | 0.481 | | 3.5 | 0.555 |
| 1.6 | 2.0 | 0.509 | 2.6 | 2.0 | 0.793 | 3.6 | 2.0 | 0.888 |
| | 2.5 | 0.400 | | 2.5 | 0.624 | | 2.5 | 0.698 |
| | 3.0 | 0.350 | | 3.0 | 0.546 | | 3.0 | 0.611 |
| | 3.5 | 0.322 | | 3.5 | 0.502 | | 3.5 | 0.562 |
| 1.8 | 2.0 | 0.599 | 2.8 | 2.0 | 0.820 | 3.8 | 2.0 | 0.900 |
| | 2.5 | 0.471 | | 2.5 | 0.645 | | 2.5 | 0.707 |
| | 3.0 | 0.412 | | 3.0 | 0.564 | | 3.0 | 0.619 |
| | 3.5 | 0.379 | | 3.5 | 0.519 | | 3.5 | 0.570 |
| 2.0 | 2.0 | 0.667 | 3.0 | 2.0 | 0.842 | 4.0 | 2.0 | 0.909 |
| | 2.5 | 0.524 | | 2.5 | 0.662 | | 2.5 | 0.715 |
| | 3.0 | 0.459 | | 3.0 | 0.580 | | 3.0 | 0.625 |
| | 3.5 | 0.422 | | 3.5 | 0.533 | | 3.5 | 0.576 |
| 2.2 | 2.0 | 0.718 | 3.2 | 2.0 | 0.860 | | ..... | ..... |
| | 2.5 | 0.565 | | 2.5 | 0.676 | | ..... | ..... |
| | 3.0 | 0.494 | | 3.0 | 0.591 | | ..... | ..... |
| | 3.5 | 0.455 | | 3.5 | 0.544 | | ..... | ..... |

*Example 3:* If the crank web in Example 1 is of cast iron and 4000 pounds per square inch is the maximum tensile stress in the hub, what is the allowance per inch of diameter?

$$\frac{D_2}{D_1} = 1.5; \quad T = 4000.$$

In Table 1, we find that $C = 0.234$. Substituting in Formula (1), for cast-iron hubs, $A = 0.0003$ inch, which, owing to the lower tensile strength of cast iron, is about one-third the shrinkage allowance in Example 1, although the stress is two-thirds of the elastic limit.

**Temperatures for Shrinkage Fits.** — The temperature to which the outer member in a shrinkage fit should be heated for clearance in assembling the parts depends on the total expansion required and on the coefficient $\alpha$ of linear expansion of the metal (that is, the increase in length of any section of the metal in any direction for an increase in temperature of 1 degree F.). The total expansion in diameter which is required consists of the total allowance for shrinkage and an added amount for clearance. The value of the coefficient $\alpha$ is, for nickel-steel, 0.000007; for steel

Table 3. Factors for Calculating Shrinkage Fit Allowances

| Values of Ratio $C$ for hollow steel shafts and cast-iron hubs. Notation as in Table 2. | | | | | | | | |
|---|---|---|---|---|---|---|---|---|
| $\dfrac{D_2}{D_1}$ | $\dfrac{D_1}{D_0}$ | $C$ | $\dfrac{D_2}{D_1}$ | $\dfrac{D_1}{D_0}$ | $C$ | $\dfrac{D_2}{D_1}$ | $\dfrac{D_1}{D_0}$ | $C$ |
| 1.5 | 2.0 | 0.468 | 2.4 | 2.0 | 0.798 | 3.4 | 2.0 | 0.926 |
|  | 2.5 | 0.368 |  | 2.5 | 0.628 |  | 2.5 | 0.728 |
|  | 3.0 | 0.322 |  | 3.0 | 0.549 |  | 3.0 | 0.637 |
|  | 3.5 | 0.296 |  | 3.5 | 0.506 |  | 3.5 | 0.587 |
| 1.6 | 2.0 | 0.527 | 2.6 | 2.0 | 0.834 | 3.6 | 2.0 | 0.941 |
|  | 2.5 | 0.414 |  | 2.5 | 0.656 |  | 2.5 | 0.740 |
|  | 3.0 | 0.362 |  | 3.0 | 0.574 |  | 3.0 | 0.647 |
|  | 3.5 | 0.333 |  | 3.5 | 0.528 |  | 3.5 | 0.596 |
| 1.8 | 2.0 | 0.621 | 2.8 | 2.0 | 0.864 | 3.8 | 2.0 | 0.953 |
|  | 2.5 | 0.488 |  | 2.5 | 0.679 |  | 2.5 | 0.749 |
|  | 3.0 | 0.427 |  | 3.0 | 0.594 |  | 3.0 | 0.656 |
|  | 3.5 | 0.393 |  | 3.5 | 0.547 |  | 3.5 | 0.603 |
| 2.0 | 2.0 | 0.696 | 3.0 | 2.0 | 0.888 | 4.0 | 2.0 | 0.964 |
|  | 2.5 | 0.547 |  | 2.5 | 0.698 |  | 2.5 | 0.758 |
|  | 3.0 | 0.479 |  | 3.0 | 0.611 |  | 3.0 | 0.663 |
|  | 3.5 | 0.441 |  | 3.5 | 0.562 |  | 3.5 | 0.610 |
| 2.2 | 2.0 | 0.753 | 3.2 | 2.0 | 0.909 | ..... | ..... | ...... |
|  | 2.5 | 0.592 |  | 2.5 | 0.715 |  | ..... | ...... |
|  | 3.0 | 0.518 |  | 3.0 | 0.625 | ..... | ..... | ...... |
|  | 3.5 | 0.477 |  | 3.5 | 0.576 |  | ..... | ...... |

in general, 0.0000065; for cast iron, 0.0000062. As an example, take an outer member of steel to be expanded 0.005 inch per inch of internal diameter, 0.001 being the shrinkage allowance and the remainder for clearance. Then:

$$\alpha \times t^\circ = 0.005$$

$$t = \frac{0.005}{0.0000065} = 769 \text{ degrees F.}$$

The value $t$ is the number of degrees F. which the temperature of the member must be raised above that of the room temperature.

## Johansson System of Tolerances with Diameter of Hole as Basic Size

| Diameter of Hole, Inches | Tolerances for Plug Gage used in Hole * | | | | Shaft Tolerances | |
|---|---|---|---|---|---|---|
| | Class A, Inch | | Class B, Inch | | Light Running Fit, Inch | |
| | Minimum | Maximum | Minimum | Maximum | Minimum | Maximum |
| 1/32 – 1/8 | −0.00016 | +0.00016 | −0.00008 | +0.00008 | −0.00083 | −0.00043 |
| 1/8 – 1/4 | −0.00024 | +0.00024 | −0.00012 | +0.00012 | −0.00122 | −0.00063 |
| 1/4 – 13/32 | −0.00031 | +0.00031 | −0.00016 | +0.00016 | −0.00165 | −0.00087 |
| 13/32 – 23/32 | −0.00043 | +0.00043 | −0.00024 | +0.00020 | −0.00217 | −0.00118 |
| 23/32 – 1⅛ | −0.00055 | +0.00055 | −0.00028 | +0.00028 | −0.00276 | −0.00157 |
| 1⅛ – 1⅞ | −0.00067 | +0.00067 | −0.00035 | +0.00031 | −0.00335 | −0.00197 |
| 1⅞ – 2 15/16 | −0.00083 | +0.00083 | −0.00043 | +0.00039 | −0.00402 | −0.00236 |
| 2 15/16 – 4 17/32 | −0.00098 | +0.00098 | −0.00051 | +0.00047 | −0.00473 | −0.00276 |
| 4 17/32 – 6⅞ | −0.00118 | +0.00118 | −0.00059 | +0.00059 | −0.00551 | −0.00315 |
| 6⅞ – 10 7/16 | −0.00138 | +0.00138 | −0.00071 | +0.00067 | −0.00630 | −0.00354 |
| 10 7/16 – 15¾ | −0.00157 | +0.00157 | −0.00079 | +0.00079 | −0.00709 | −0.00394 |

| Diameter of Hole, Inches | Shaft Tolerances | | | | | |
|---|---|---|---|---|---|---|
| | Running Fit, Inch | | Sliding Fit, Inch | | Push Fit, Inch | |
| | Minimum | Maximum | Minimum | Maximum | Minimum | Maximum |
| 1/32 – 1/8 | −0.00043 | −0.00020 | −0.00020 | −0.00008 | −0.00008 | +0.00012 |
| 1/8 – 1/4 | −0.00063 | −0.00031 | −0.00031 | −0.00012 | −0.00012 | +0.00020 |
| 1/4 – 13/32 | −0.00087 | −0.00043 | −0.00043 | −0.00016 | −0.00016 | +0.00028 |
| 13/32 – 23/32 | −0.00118 | −0.00059 | −0.00059 | −0.00020 | −0.00020 | +0.00031 |
| 23/32 – 1⅛ | −0.00157 | −0.00079 | −0.00079 | −0.00024 | −0.00024 | +0.00031 |
| 1⅛ – 1⅞ | −0.00197 | −0.00098 | −0.00098 | −0.00031 | −0.00031 | +0.00031 |
| 1⅞ – 2 15/16 | −0.00236 | −0.00118 | −0.00118 | −0.00039 | −0.00039 | +0.00028 |
| 2 15/16 – 4 17/32 | −0.00276 | −0.00138 | −0.00138 | −0.00047 | −0.00047 | +0.00024 |
| 4 17/32 – 6⅞ | −0.00315 | −0.00157 | −0.00157 | −0.00055 | −0.00055 | +0.00020 |
| 6⅞ – 10 7/16 | −0.00354 | −0.00177 | −0.00177 | −0.00067 | −0.00067 | +0.00020 |
| 10 7/16 – 15¾ | −0.00394 | −0.00197 | −0.00197 | −0.00075 | −0.00075 | +0.00020 |

| Diameter of Hole, Inches | Easy Driving Fit, Inch | | Close Driving Fit, Inch | | Forced Fit, Inch | |
|---|---|---|---|---|---|---|
| | Minimum | Maximum | Minimum | Maximum | Minimum | Maximum |
| 1/32 – 1/8 | +0.00012 | +0.00024 | +0.00024 | +0.00039 | +0.00039 | +0.00059 |
| 1/8 – 1/4 | +0.00020 | +0.00035 | +0.00035 | +0.00059 | +0.00059 | +0.00098 |
| 1/4 – 13/32 | +0.00028 | +0.00047 | +0.00047 | +0.00083 | +0.00083 | +0.00146 |
| 13/32 – 23/32 | +0.00031 | +0.00059 | +0.00059 | +0.00110 | +0.00110 | +0.00197 |
| 23/32 – 1⅛ | +0.00031 | +0.00071 | +0.00071 | +0.00142 | +0.00142 | +0.00252 |
| 1⅛ – 1⅞ | +0.00031 | +0.00087 | +0.00087 | +0.00177 | +0.00177 | +0.00319 |
| 1⅞ – 2 15/16 | +0.00028 | +0.00102 | +0.00102 | +0.00213 | +0.00213 | +0.00394 |
| 2 15/16 – 4 17/32 | +0.00024 | +0.00118 | +0.00118 | +0.00256 | +0.00256 | +0.00481 |
| 4 17/32 – 6⅞ | +0.00020 | +0.00138 | +0.00138 | +0.00303 | +0.00303 | +0.00579 |
| 6⅞ – 10 7/16 | +0.00020 | +0.00157 | +0.00157 | +0.00354 | +0.00354 | +0.00689 |
| 10 7/16 – 15¾ | +0.00020 | +0.00177 | +0.00177 | +0.00414 | +0.00414 | +0.00808 |

* Use column *A* for ordinary work, where greater tolerances are allowable. Use column *B* for more accurate work, where smaller tolerances are required. The values, in inches, in the table above have been given to five decimals to give the exact value of the dimensions in millimeters in the original tables.

**ANSI Standard Limits and Fits** (ANSI B4.1-1967, R1974). — This American National Standard for Preferred Limits and Fits for Cylindrical Parts presents definitions of terms applying to fits between plain (non-threaded) cylindrical parts and makes recommendations on preferred sizes, allowances, tolerances, and fits for use wherever they are applicable. This standard is in accord with the recommendations of American-British-Canadian (ABC) conferences (see page 1576) up to a diameter of 20 inches. Experimental work is being carried on with the objective of reaching agreement in the range above 20 inches. The recommendations in the Standard are presented for guidance and for use where they might serve to improve and simplify products, practices, and facilities. They should have application for a wide range of products.

As revised in 1967, and reaffirmed in 1974, the definitions in ANSI B4.1 have been expanded and some of the limits in certain classes have been changed.

**Factors Affecting Selection of Fits.** — Many factors, such as length of engagement, bearing load, speed, lubrication, temperature, humidity, and materials, must be taken into consideration in the selection of fits for a particular application, and modifications in the ANSI recommendations may be required to satisfy extreme conditions. Subsequent adjustments may also be found desirable as a result of experience in a particular application to suit critical functional requirements or to permit optimum manufacturing economy.

**Definitions.** — The following terms are defined in this standard:

*Nominal Size:* The nominal size is the designation which is used for the purpose of general identification.

*Dimension:* A dimension is a geometrical characteristic such as diameter, length, angle, or center distance.

*Size:* Size is a designation of magnitude. When a value is assigned to a dimension it is referred to as the size of that dimension. (It is recognized that the words "dimension" and "size" are both used at times to convey the meaning of magnitude.)

*Allowance:* An allowance is a prescribed difference between the maximum material limits of mating parts. (See definition of *Fit*). It is a minimum clearance (positive allowance) or maximum interference (negative allowance) between such parts.

*Tolerance:* A tolerance is the total permissible variation of a size. The tolerance is the difference between the limits of size.

*Basic Size:* The basic size is that size from which the limits of size are derived by the application of allowances and tolerances.

*Design Size:* The design size is the basic size with allowance applied, from which the limits of size are derived by the application of tolerances. Where there is no allowance, the design size is the same as the basic size.

*Actual Size:* An actual size is a measured size.

*Limits of Size:* The limits of size are the applicable maximum and minimum sizes.

*Maximum Material Limit:* A maximum material limit is that limit of size that provides the maximum amount of material for the part. Normally it is the maximum limit of size of an external dimension or the minimum limit of size of an internal dimension.*

*Minimum Material Limit:* A minimum material limit is that limit of size that provides the minimum amount of material for the part. Normally it is the minimum limit of size of an external dimension or the maximum limit of size of an internal dimension.*

* An example of exceptions: an exterior corner radius where the maximum radius is the minimum material limit and the minimum radius is the maximum material limit.

*Tolerance Limit:* A tolerance limit is the variation, positive or negative, by which a size is permitted to depart from the design size.

*Unilateral Tolerance:* A unilateral tolerance is a tolerance in which variation is permitted only in one direction from the design size.

*Bilateral Tolerance:* A bilateral tolerance is a tolerance in which variation is permitted in both directions from the design size.

*Unilateral Tolerance System:* A design plan which uses only unilateral tolerances is known as a Unilateral Tolerance System.

*Bilateral Tolerance System:* A design plan which uses only bilateral tolerances is known as a Bilateral Tolerance System.

*Fit:* Fit is the general term used to signify the range of tightness which may result from the application of a specific combination of allowances and tolerances in the design of mating parts.

*Actual Fit:* The actual fit between two mating parts is the relation existing between them with respect to the amount of clearance or interference which is present when they are assembled. (Fits are of three general types: clearance, transition, and interference.)

*Clearance Fit:* A clearance fit is one having limits of size so specified that a clearance always results when mating parts are assembled.

*Interference Fit:* An interference fit is one having limits of size so specified that an interference always results when mating parts are assembled.

*Transition Fit:* A transition fit is one having limits of size so specified that either a clearance or an interference may result when mating parts are assembled.

*Basic Hole System:* A basic hole system is a system of fits in which the design size of the hole is the basic size and the allowance, if any, is applied to the shaft.

*Basic Shaft System:* A basic shaft system is a system of fits in which the design size of the shaft is the basic size and the allowance, if any, is applied to the hole.

**Preferred Basic Sizes.** — In specifying fits, the basic size of mating parts shall be chosen from the decimal series or the fractional series in the following table. All dimensions are given in inches.

### Preferred Basic Sizes

| Decimal | | | Fractional | | | | |
|---|---|---|---|---|---|---|---|
| 0.010 | 2.00 | 8.50 | $\frac{1}{64}$ | 0.015625 | $2\frac{1}{4}$ | 2.2500 | $9\frac{1}{2}$ | 9.5000 |
| 0.012 | 2.20 | 9.00 | $\frac{1}{32}$ | 0.03125 | $2\frac{1}{2}$ | 2.5000 | 10 | 10.0000 |
| 0.016 | 2.40 | 9.50 | $\frac{1}{16}$ | 0.0625 | $2\frac{3}{4}$ | 2.7500 | $10\frac{1}{2}$ | 10.5000 |
| 0.020 | 2.60 | 10.00 | $\frac{3}{32}$ | 0.09375 | 3 | 3.0000 | 11 | 11.0000 |
| 0.025 | 2.80 | 10.50 | $\frac{1}{8}$ | 0.1250 | $3\frac{1}{4}$ | 3.2500 | $11\frac{1}{2}$ | 11.5000 |
| 0.032 | 3.00 | 11.00 | $\frac{5}{32}$ | 0.15625 | $3\frac{1}{2}$ | 3.5000 | 12 | 12.0000 |
| 0.040 | 3.20 | 11.50 | $\frac{3}{16}$ | 0.1875 | $3\frac{3}{4}$ | 3.7500 | $12\frac{1}{2}$ | 12.5000 |
| 0.05 | 3.40 | 12.00 | $\frac{1}{4}$ | 0.2500 | 4 | 4.0000 | 13 | 13.0000 |
| 0.06 | 3.60 | 12.50 | $\frac{5}{16}$ | 0.3125 | $4\frac{1}{4}$ | 4.2500 | $13\frac{1}{2}$ | 13.5000 |
| 0.08 | 3.80 | 13.00 | $\frac{3}{8}$ | 0.3750 | $4\frac{1}{2}$ | 4.5000 | 14 | 14.0000 |
| 0.10 | 4.00 | 13.50 | $\frac{7}{16}$ | 0.4375 | $4\frac{3}{4}$ | 4.7500 | $14\frac{1}{2}$ | 14.5000 |
| 0.12 | 4.20 | 14.00 | $\frac{1}{2}$ | 0.5000 | 5 | 5.0000 | 15 | 15.0000 |
| 0.16 | 4.40 | 14.50 | $\frac{9}{16}$ | 0.5625 | $5\frac{1}{4}$ | 5.2500 | $15\frac{1}{2}$ | 15.5000 |
| 0.20 | 4.60 | 15.00 | $\frac{5}{8}$ | 0.6250 | $5\frac{1}{2}$ | 5.5000 | 16 | 16.0000 |
| 0.24 | 4.80 | 15.50 | $\frac{11}{16}$ | 0.6875 | $5\frac{3}{4}$ | 5.7500 | $16\frac{1}{2}$ | 16.5000 |
| 0.30 | 5.00 | 16.00 | $\frac{3}{4}$ | 0.7500 | 6 | 6.0000 | 17 | 17.0000 |
| 0.40 | 5.20 | 16.50 | $\frac{7}{8}$ | 0.8750 | $6\frac{1}{2}$ | 6.5000 | $17\frac{1}{2}$ | 17.5000 |
| 0.50 | 5.40 | 17.00 | 1 | 1.0000 | 7 | 7.0000 | 18 | 18.0000 |
| 0.60 | 5.60 | 17.50 | $1\frac{1}{4}$ | 1.2500 | $7\frac{1}{2}$ | 7.5000 | $18\frac{1}{2}$ | 18.5000 |
| 0.80 | 5.80 | 18.00 | $1\frac{1}{2}$ | 1.5000 | 8 | 8.0000 | 19 | 19.0000 |
| 1.00 | 6.00 | 18.50 | $1\frac{3}{4}$ | 1.7500 | $8\frac{1}{2}$ | 8.5000 | $19\frac{1}{2}$ | 19.5000 |
| 1.20 | 6.50 | 19.00 | 2 | 2.0000 | 9 | 9.0000 | 20 | 20.0000 |
| 1.40 | 7.00 | 19.50 | ...... | ......... | ... | ....... | .... | ........ |
| 1.60 | 7.50 | 20.00 | | | | | | |
| 1.80 | 8.00 | ..... | | All dimensions are given in inches. | | | | |

Preferred Series of Tolerances and Allowances (In thousandths of an inch)

| | | | | | | | |
|---|---|---|---|---|---|---|---|
| 0.1 | 1 | 10 | 100 | 0.3 | 3 | 30 | ... |
| ... | 1.2 | 12 | 125 | ... | 3.5 | 35 | ... |
| 0.15 | 1.4 | 14 | ... | 0.4 | 4 | 40 | ... |
| ... | 1.6 | 16 | 160 | ... | 4.5 | 45 | ... |
| ... | 1.8 | 18 | ... | 0.5 | 5 | 50 | ... |
| 0.2 | 2 | 20 | 200 | 0.6 | 6 | 60 | ... |
| ... | 2.2 | 22 | ... | 0.7 | 7 | 70 | ... |
| 0.25 | 2.5 | 25 | 250 | 0.8 | 8 | 80 | ... |
| ... | 2.8 | 28 | ... | 0.9 | 9 | ... | ... |

**Standard Tolerances.** — The series of standard tolerances shown in Table 1 are so arranged that for any one grade they represent approximately similar production difficulties throughout the range of sizes. This table provides a suitable range from which appropriate tolerances for holes and shafts can be selected. This enables the use of standard gages. The tolerances shown in Table 1 have been used in the succeeding tables for different classes of fits.

Table 1.  ANSI Standard Tolerances (ANSI B4.1-1967, R1974)

| Nominal Size, Inches Over    To | Grade | | | | | | | | | |
|---|---|---|---|---|---|---|---|---|---|---|
| | 4 | 5 | 6 | 7 | 8 | 9 | 10 | 11 | 12 | 13 |
| | Tolerances in thousandths of an inch* | | | | | | | | | |
| 0–    0.12 | 0.12 | 0.15 | 0.25 | 0.4 | 0.6 | 1.0 | 1.6 | 2.5 | 4 | 6 |
| 0.12–  0.24 | 0.15 | 0.20 | 0.3 | 0.5 | 0.7 | 1.2 | 1.8 | 3.0 | 5 | 7 |
| 0.24–  0.40 | 0.15 | 0.25 | 0.4 | 0.6 | 0.9 | 1.4 | 2.2 | 3.5 | 6 | 9 |
| 0.40–  0.71 | 0.2 | 0.3 | 0.4 | 0.7 | 1.0 | 1.6 | 2.8 | 4.0 | 7 | 10 |
| 0.71–  1.19 | 0.25 | 0.4 | 0.5 | 0.8 | 1.2 | 2.0 | 3.5 | 5.0 | 8 | 12 |
| 1.19–  1.97 | 0.3 | 0.4 | 0.6 | 1.0 | 1.6 | 2.5 | 4.0 | 6 | 10 | 16 |
| 1.97–  3.15 | 0.3 | 0.5 | 0.7 | 1.2 | 1.8 | 3.0 | 4.5 | 7 | 12 | 18 |
| 3.15–  4.73 | 0.4 | 0.6 | 0.9 | 1.4 | 2.2 | 3.5 | 5 | 9 | 14 | 22 |
| 4.73–  7.09 | 0.5 | 0.7 | 1.0 | 1.6 | 2.5 | 4.0 | 6 | 10 | 16 | 25 |
| 7.09–  9.85 | 0.6 | 0.8 | 1.2 | 1.8 | 2.8 | 4.5 | 7 | 12 | 18 | 28 |
| 9.85– 12.41 | 0.6 | 0.9 | 1.2 | 2.0 | 3.0 | 5.0 | 8 | 12 | 20 | 30 |
| 12.41– 15.75 | 0.7 | 1.0 | 1.4 | 2.2 | 3.5 | 6 | 9 | 14 | 22 | 35 |
| 15.75– 19.69 | 0.8 | 1.0 | 1.6 | 2.5 | 4 | 6 | 10 | 16 | 25 | 40 |
| 19.69– 30.09 | 0.9 | 1.2 | 2.0 | 3 | 5 | 8 | 12 | 20 | 30 | 50 |
| 30.09– 41.49 | 1.0 | 1.6 | 2.5 | 4 | 6 | 10 | 16 | 25 | 40 | 60 |
| 41.49– 56.19 | 1.2 | 2.0 | 3 | 5 | 8 | 12 | 20 | 30 | 50 | 80 |
| 56.19– 76.39 | 1.6 | 2.5 | 4 | 6 | 10 | 16 | 25 | 40 | 60 | 100 |
| 76.39–100.9 | 2.0 | 3 | 5 | 8 | 12 | 20 | 30 | 50 | 80 | 125 |
| 100.9 –131.9 | 2.5 | 4 | 6 | 10 | 16 | 25 | 40 | 60 | 100 | 160 |
| 131.9 –171.9 | 3 | 5 | 8 | 12 | 20 | 30 | 50 | 80 | 125 | 200 |
| 171.9 –200 | 4 | 6 | 10 | 16 | 25 | 40 | 60 | 100 | 160 | 250 |

* All tolerances above heavy line are in accordance with American-British-Canadian (ABC) agreements.

### Relation of Machining Processes to Tolerance Grades

This chart may be used as a general guide to determine the machining processes that will, under normal conditions, produce work within the tolerance grades indicated.

(See also Relation of Surface Roughness to Dimensional Tolerances in the Surface Texture Section, page 2388.)

**ANSI Standard Fits.** — Tables 2 to 6 inclusive show a series of standard types and classes of fits on a unilateral hole basis, such that the fit produced by mating parts in any one class will produce approximately similar performance throughout the range of sizes. These tables prescribe the fit for any given size, or type of fit; they also prescribe the standard limits for the mating parts which will produce the fit. The fits listed in these tables contain all of those which appear in the approved American-British-Canadian proposal.

*Selection of Fits:* In selecting limits of size for any application, the type of fit is determined first, based on the use or service required from the equipment being designed; then the limits of size of the mating parts are established, to insure that the desired fit will be produced.

Theoretically, an infinite number of fits could be chosen, but the number of standard fits shown in the accompanying tables should cover most applications.

*Designation of Standard Fits:* Standard fits are designated by means of the following symbols which facilitate reference to classes of fit for educational purposes. The symbols are not intended to be shown on manufacturing drawings; instead, sizes should be specified on drawings.

The letter symbols used are as follows:

RC    Running or Sliding Clearance Fit
LC    Locational Clearance Fit
LT    Transition Clearance or Interference Fit
LN    Locational Interference Fit
FN    Force or Shrink Fit

These letter symbols are used in conjunction with numbers representing the class of fit; thus FN 4 represents a Class 4, force fit.

Each of these symbols (two letters and a number) represents a complete fit for which the minimum and maximum clearance or interference and the limits of size for the mating parts are given directly in the tables.

**Description of Fits.** — The classes of fits are arranged in three general groups: running and sliding fits, locational fits, and force fits.

*Running and Sliding Fits (RC):* Running and sliding fits, for which limits of clearance are given in Table 2, are intended to provide a similar running performance, with suitable lubrication allowance, throughout the range of sizes. The clearances for the first two classes, used chiefly as slide fits, increase more slowly with the diameter than for the other classes, so that accurate location is maintained even at the expense of free relative motion.

These fits may be described as follows:

RC 1 *Close sliding fits* are intended for the accurate location of parts which must assemble without perceptible play.

RC 2 *Sliding fits* are intended for accurate location, but with greater maximum clearance than class RC 1. Parts made to this fit move and turn easily but are not intended to run freely, and in the larger sizes may seize with small temperature changes.

RC 3 *Precision running fits* are about the closest fits which can be expected to run freely, and are intended for precision work at slow speeds and light journal pressures, but are not suitable where appreciable temperature differences are likely to be encountered.

RC 4 *Close running fits* are intended chiefly for running fits on accurate machinery with moderate surface speeds and journal pressures, where accurate location and minimum play is desired.

RC 5 and RC 6 *Medium running fits* are intended for higher running speeds, or heavy journal pressures, or both.

RC 7   *Free running fits* are intended for use where accuracy is not essential, or where large temperature variations are likely to be encountered, or under both these conditions.

RC 8 and RC 9   *Loose running fits* are intended for use where wide commercial tolerances may be necessary, together with an allowance, on the external member.

*Locational Fits* (LC, LT, and LN): Locational fits are fits intended to determine only the location of the mating parts; they may provide rigid or accurate location, as with interference fits, or provide some freedom of location, as with clearance fits. Accordingly, they are divided into three groups: clearance fits (LC), transition fits (LT), and interference fits (LN).

These are described as follows:

LC   *Locational clearance fits* are intended for parts which are normally stationary, but which can be freely assembled or disassembled. They range from snug fits for parts requiring accuracy of location, through the medium clearance fits for parts such as spigots, to the looser fastener fits where freedom of assembly is of prime importance.

LT   *Locational transition fits* are a compromise between clearance and interference fits, for application where accuracy of location is important, but either a small amount of clearance or interference is permissible.

LN   *Locational interference fits* are used where accuracy of location is of prime importance, and for parts requiring rigidity and alignment with no special requirements for bore pressure. Such fits are not intended for parts designed to transmit frictional loads from one part to another by virtue of the tightness of fit, as these conditions are covered by force fits.

*Force Fits* (FN): Force or shrink fits constitute a special type of interference fit, normally characterized by maintenance of constant bore pressures throughout the range of sizes. The interference therefore varies almost directly with diameter, and the difference between its minimum and maximum value is small, to maintain the resulting pressures within reasonable limits.

These fits are described as follows:

FN 1   *Light drive fits* are those requiring light assembly pressures, and produce more or less permanent assemblies. They are suitable for thin sections or long fits, or in cast-iron external members.

FN 2   *Medium drive fits* are suitable for ordinary steel parts, or for shrink fits on light sections. They are about the tightest fits that can be used with high-grade cast-iron external members.

FN 3   *Heavy drive fits* are suitable for heavier steel parts or for shrink fits in medium sections.

FN 4 and FN 5   *Force fits* are suitable for parts which can be highly stressed, or for shrink fits where the heavy pressing forces required are impractical.

**Graphical Representation of Limits and Fits.** — A visual comparison of the hole and shaft tolerances and the clearances or interferences provided by the various types and classes of fits can be obtained from the diagrams on page 1542. These diagrams have been drawn to scale for a nominal diameter of 1 inch.

**Use of Standard Fit Tables.** — *Example 1:* A Class RC 1 fit is to be used in assembling a mating hole and shaft of 2-inch nominal diameter. This class of fit was selected because the application required accurate location of the parts with no perceptible play (see description of RC 1 close sliding fits). From the data in Table 2, establish the limits of size and clearance of the hole and shaft.

Maximum hole = 2 + 0.0005 = 2.0005; minimum hole = 2 inches

(*Continued on page 1551*)

Graphical Representation of ANSI Standard Limits and Fits*

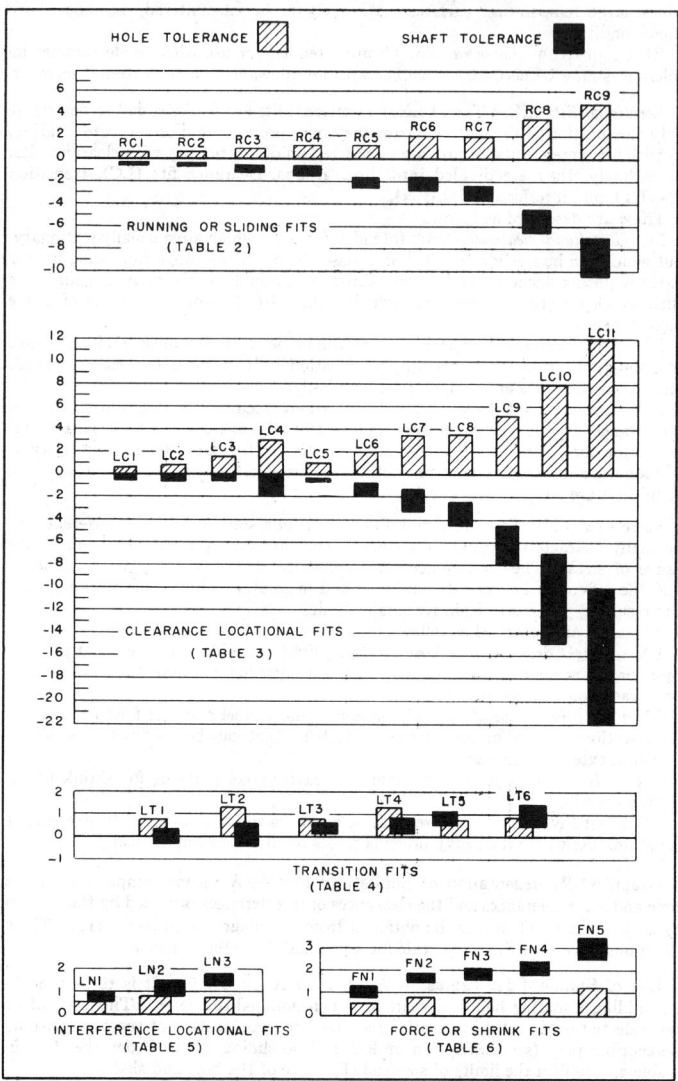

* Diagrams show disposition of hole and shaft tolerances (in thousandths of an inch) with respect to basic size (o) for a diameter of 1 inch.

**Table 2. American National Standard Running and Sliding Fits (ANSI B4.1-1967, R1974)**

Tolerance limits given in body of table are added or subtracted to basic size (as indicated by + or − sign) to obtain maximum and minimum sizes of mating parts.

Values shown below are in thousandths of an inch.

| Nominal Size Range, Inches Over — To | Class RC 1 Clearance | Hole H5 | Shaft g4 | Class RC 2 Clearance | Hole H6 | Shaft g5 | Class RC 3 Clearance | Hole H7 | Shaft f6 | Class RC 4 Clearance | Hole H8 | Shaft f7 |
|---|---|---|---|---|---|---|---|---|---|---|---|---|
| 0 – 0.12 | 0.1 / 0.45 | +0.2 / 0 | −0.1 / −0.25 | 0.1 / 0.55 | +0.25 / 0 | −0.1 / −0.3 | 0.3 / 0.95 | +0.4 / 0 | −0.3 / −0.55 | 0.3 / 1.3 | +0.6 / 0 | −0.3 / −0.7 |
| 0.12 – 0.24 | 0.15 / 0.5 | +0.2 / 0 | −0.15 / −0.3 | 0.15 / 0.65 | +0.3 / 0 | −0.15 / −0.35 | 0.4 / 1.12 | +0.5 / 0 | −0.4 / −0.7 | 0.4 / 1.6 | +0.7 / 0 | −0.4 / −0.9 |
| 0.24 – 0.40 | 0.2 / 0.6 | +0.25 / 0 | −0.2 / −0.35 | 0.2 / 0.85 | +0.4 / 0 | −0.2 / −0.45 | 0.5 / 1.5 | +0.6 / 0 | −0.5 / −0.9 | 0.5 / 2.0 | +0.9 / 0 | −0.5 / −1.1 |
| 0.40 – 0.71 | 0.25 / 0.75 | +0.3 / 0 | −0.25 / −0.45 | 0.25 / 0.95 | +0.4 / 0 | −0.25 / −0.55 | 0.6 / 1.7 | +0.7 / 0 | −0.6 / −1.0 | 0.6 / 2.3 | +1.0 / 0 | −0.6 / −1.3 |
| 0.71 – 1.19 | 0.3 / 0.95 | +0.4 / 0 | −0.3 / −0.55 | 0.3 / 1.2 | +0.5 / 0 | −0.3 / −0.7 | 0.8 / 2.1 | +0.8 / 0 | −0.8 / −1.3 | 0.8 / 2.8 | +1.2 / 0 | −0.8 / −1.6 |
| 1.19 – 1.97 | 0.4 / 1.1 | +0.4 / 0 | −0.4 / −0.7 | 0.4 / 1.4 | +0.6 / 0 | −0.4 / −0.8 | 1.0 / 2.6 | +1.0 / 0 | −1.0 / −1.6 | 1.0 / 3.6 | +1.6 / 0 | −1.0 / −2.0 |
| 1.97 – 3.15 | 0.4 / 1.2 | +0.5 / 0 | −0.4 / −0.7 | 0.4 / 1.6 | +0.7 / 0 | −0.4 / −0.9 | 1.2 / 3.1 | +1.2 / 0 | −1.2 / −1.9 | 1.2 / 4.2 | +1.8 / 0 | −1.2 / −2.4 |
| 3.15 – 4.73 | 0.5 / 1.5 | +0.6 / 0 | −0.5 / −0.9 | 0.5 / 2.0 | +0.9 / 0 | −0.5 / −1.1 | 1.4 / 3.7 | +1.4 / 0 | −1.4 / −2.3 | 1.4 / 5.0 | +2.2 / 0 | −1.4 / −2.8 |
| 4.73 – 7.09 | 0.6 / 1.8 | +0.7 / 0 | −0.6 / −1.1 | 0.6 / 2.3 | +1.0 / 0 | −0.6 / −1.3 | 1.6 / 4.2 | +1.6 / 0 | −1.6 / −2.6 | 1.6 / 5.7 | +2.5 / 0 | −1.6 / −3.2 |
| 7.09 – 9.85 | 0.6 / 2.0 | +0.8 / 0 | −0.6 / −1.2 | 0.6 / 2.6 | +1.2 / 0 | −0.6 / −1.4 | 2.0 / 5.0 | +1.8 / 0 | −2.0 / −3.2 | 2.0 / 6.6 | +2.8 / 0 | −2.0 / −3.8 |
| 9.85 – 12.41 | 0.8 / 2.3 | +0.9 / 0 | −0.8 / −1.4 | 0.8 / 2.9 | +1.2 / 0 | −0.8 / −1.7 | 2.5 / 5.7 | +2.0 / 0 | −2.5 / −3.7 | 2.5 / 7.5 | +3.0 / 0 | −2.5 / −4.5 |
| 12.41 – 15.75 | 1.0 / 2.7 | +1.0 / 0 | −1.0 / −1.7 | 1.0 / 3.4 | +1.4 / 0 | −1.0 / −2.0 | 3.0 / 6.6 | +2.2 / 0 | −3.0 / −4.4 | 3.0 / 8.7 | +3.5 / 0 | −3.0 / −5.2 |
| 15.75 – 19.69 | 1.2 / 3.0 | +1.0 / 0 | −1.2 / −2.0 | 1.2 / 3.8 | +1.6 / 0 | −1.2 / −2.2 | 4.0 / 8.1 | +2.5 / 0 | −4.0 / −5.6 | 4.0 / 10.5 | +4.0 / 0 | −4.0 / −6.5 |

See footnotes at end of table.

Table 2 (Continued). American National Standard Running and Sliding Fits (ANSI B4.1-1967, R1974)

Values shown below are in thousandths of an inch

| Nominal Size Range, Inches Over – To | Class RC 5 Clearance* | Class RC 5 Hole H8 | Class RC 5 Shaft e7 | Class RC 6 Clearance* | Class RC 6 Hole H9 | Class RC 6 Shaft e8 | Class RC 7 Clearance* | Class RC 7 Hole H9 | Class RC 7 Shaft d8 | Class RC 8 Clearance* | Class RC 8 Hole H10 | Class RC 8 Shaft c9 | Class RC 9 Clearance* | Class RC 9 Hole H11 | Class RC 9 Shaft |
|---|---|---|---|---|---|---|---|---|---|---|---|---|---|---|---|
| 0 – 0.12 | 0.6 / 1.6 | +0.6 / 0 | −0.6 / −1.0 | 0.6 / 2.2 | +1.0 / 0 | −0.6 / −1.2 | 1.0 / 2.6 | +1.0 / 0 | −1.0 / −1.6 | 2.5 / 5.1 | +1.6 / 0 | −2.5 / −3.5 | 4.0 / 8.1 | +2.5 / 0 | −4.0 / −5.6 |
| 0.12 – 0.24 | 0.8 / 2.0 | +0.7 / 0 | −0.8 / −1.3 | 0.8 / 2.7 | +1.2 / 0 | −0.8 / −1.5 | 1.2 / 3.1 | +1.2 / 0 | −1.2 / −1.9 | 2.8 / 5.8 | +1.8 / 0 | −2.8 / −4.0 | 4.5 / 9.0 | +3.0 / 0 | −4.5 / −6.0 |
| 0.24 – 0.40 | 1.0 / 2.5 | +0.9 / 0 | −1.0 / −1.6 | 1.0 / 3.3 | +1.4 / 0 | −1.0 / −1.9 | 1.6 / 3.9 | +1.4 / 0 | −1.6 / −2.5 | 3.0 / 6.6 | +2.2 / 0 | −3.0 / −4.4 | 5.0 / 10.7 | +3.5 / 0 | −5.0 / −7.2 |
| 0.40 – 0.71 | 1.2 / 2.9 | +1.0 / 0 | −1.2 / −1.9 | 1.2 / 3.8 | +1.6 / 0 | −1.2 / −2.2 | 2.0 / 4.6 | +1.6 / 0 | −2.0 / −3.0 | 3.5 / 7.9 | +2.8 / 0 | −3.5 / −5.1 | 6.0 / 12.8 | +4.0 / 0 | −6.0 / −8.8 |
| 0.71 – 1.19 | 1.6 / 3.6 | +1.2 / 0 | −1.6 / −2.4 | 1.6 / 4.8 | +2.0 / 0 | −1.6 / −2.8 | 2.5 / 5.7 | +2.0 / 0 | −2.5 / −3.7 | 4.5 / 10.0 | +3.5 / 0 | −4.5 / −6.5 | 7.0 / 15.5 | +5.0 / 0 | −7.0 / −10.5 |
| 1.19 – 1.97 | 2.0 / 4.6 | +1.6 / 0 | −2.0 / −3.0 | 2.0 / 6.1 | +2.5 / 0 | −2.0 / −3.6 | 3.0 / 7.1 | +2.5 / 0 | −3.0 / −4.6 | 5.0 / 11.5 | +4.0 / 0 | −5.0 / −7.5 | 8.0 / 18.0 | +6.0 / 0 | −8.0 / −12.0 |
| 1.97 – 3.15 | 2.5 / 5.5 | +1.8 / 0 | −2.5 / −3.7 | 2.5 / 7.3 | +3.0 / 0 | −2.5 / −4.3 | 4.0 / 8.8 | +3.0 / 0 | −4.0 / −5.8 | 6.0 / 13.5 | +4.5 / 0 | −6.0 / −9.0 | 9.0 / 20.5 | +7.0 / 0 | −9.0 / −13.5 |
| 3.15 – 4.73 | 3.0 / 6.6 | +2.2 / 0 | −3.0 / −4.4 | 3.0 / 8.7 | +3.5 / 0 | −3.0 / −5.2 | 5.0 / 10.7 | +3.5 / 0 | −5.0 / −7.2 | 7.0 / 15.5 | +5.0 / 0 | −7.0 / −10.5 | 10.0 / 24.0 | +9.0 / 0 | −10.0 / −15.0 |
| 4.73 – 7.09 | 3.5 / 7.6 | +2.5 / 0 | −3.5 / −5.1 | 3.5 / 10.0 | +4.0 / 0 | −3.5 / −6.0 | 6.0 / 12.5 | +4.0 / 0 | −6.0 / −8.5 | 8.0 / 18.0 | +6.0 / 0 | −8.0 / −12.0 | 12.0 / 28.0 | +10.0 / 0 | −12.0 / −18.0 |
| 7.09 – 9.85 | 4.0 / 8.6 | +2.8 / 0 | −4.0 / −5.8 | 4.0 / 11.3 | +4.5 / 0 | −4.0 / −6.8 | 7.0 / 14.3 | +4.5 / 0 | −7.0 / −9.8 | 10.0 / 21.5 | +7.0 / 0 | −10.0 / −14.5 | 15.0 / 34.0 | +12.0 / 0 | −15.0 / −22.0 |
| 9.85 – 12.41 | 5.0 / 10.0 | +3.0 / 0 | −5.0 / −7.0 | 5.0 / 13.0 | +5.0 / 0 | −5.0 / −8.0 | 8.0 / 16.0 | +5.0 / 0 | −8.0 / −11.0 | 12.0 / 25.0 | +8.0 / 0 | −12.0 / −17.0 | 18.0 / 38.0 | +12.0 / 0 | −18.0 / −26.0 |
| 12.41 – 15.75 | 6.0 / 11.7 | +3.5 / 0 | −6.0 / −8.2 | 6.0 / 15.5 | +6.0 / 0 | −6.0 / −9.5 | 10.0 / 19.5 | +6.0 / 0 | −10.0 / −13.5 | 14.0 / 29.0 | +9.0 / 0 | −14.0 / −20.0 | 22.0 / 45.0 | +14.0 / 0 | −22.0 / −31.0 |
| 15.75 – 19.69 | 8.0 / 14.5 | +4.0 / 0 | −8.0 / −10.5 | 8.0 / 18.0 | +6.0 / 0 | −8.0 / −12.0 | 12.0 / 22.0 | +6.0 / 0 | −12.0 / −16.0 | 16.0 / 32.0 | +10.0 / 0 | −16.0 / −22.0 | 25.0 / 51.0 | +16.0 / 0 | −25.0 / −35.0 |

All data above heavy lines are in accord with ABC agreements. Symbols H5, g4, etc. are hole and shaft designations in ABC system. Limits for sizes above 19.69 inches are also given in the ANSI Standard.

* Pairs of values shown represent minimum and maximum amounts of clearance resulting from application of standard tolerance limits.

Table 3.  American National Standard Clearance Locational Fits (ANSI B4.1-1967, R1974)

Tolerance limits given in body of table are added or subtracted to basic size (as indicated by + or − sign) to obtain maximum and minimum sizes of mating parts.

| Nominal Size Range, Inches | | Class LC 1 | | | Class LC 2 | | | Class LC 3 | | | Class LC 4 | | | Class LC 5 | | |
|---|---|---|---|---|---|---|---|---|---|---|---|---|---|---|---|---|
| | | | Standard Tolerance Limits | | | Standard Tolerance Limits | | | Standard Tolerance Limits | | | Standard Tolerance Limits | | | Standard Tolerance Limits |
| Over | To | Clearance | Hole H6 | Shaft h5 | Clearance | Hole H7 | Shaft h6 | Clearance | Hole H8 | Shaft h7 | Clearance | Hole H10 | Shaft h9 | Clearance | Hole H7 | Shaft g6 |
| | | | | | | | | Values shown below are in thousandths of an inch | | | | | | | | |
| 0 | 0.12 | 0 / 0.45 | +0.25 / 0 | 0 / −0.2 | 0 / 0.65 | +0.4 / 0 | 0 / −0.25 | 0 / 1 | +0.6 / 0 | 0 / −0.4 | 0 / 2.6 | +1.6 / 0 | 0 / −1.0 | 0.1 / 0.75 | +0.4 / 0 | −0.1 / −0.35 |
| 0.12 | 0.24 | 0 / 0.5 | +0.3 / 0 | 0 / −0.2 | 0 / 0.8 | +0.5 / 0 | 0 / −0.3 | 0 / 1.2 | +0.7 / 0 | 0 / −0.5 | 0 / 3.0 | +1.8 / 0 | 0 / −1.2 | 0.15 / 0.95 | +0.5 / 0 | −0.15 / −0.45 |
| 0.24 | 0.40 | 0 / 0.65 | +0.4 / 0 | 0 / −0.25 | 0 / 1.0 | +0.6 / 0 | 0 / −0.4 | 0 / 1.5 | +0.9 / 0 | 0 / −0.6 | 0 / 3.6 | +2.2 / 0 | 0 / −1.4 | 0.2 / 1.2 | +0.6 / 0 | −0.2 / −0.6 |
| 0.40 | 0.71 | 0 / 0.7 | +0.4 / 0 | 0 / −0.3 | 0 / 1.1 | +0.7 / 0 | 0 / −0.4 | 0 / 1.7 | +1.0 / 0 | 0 / −0.7 | 0 / 4.4 | +2.8 / 0 | 0 / −1.6 | 0.25 / 1.35 | +0.7 / 0 | −0.25 / −0.65 |
| 0.71 | 1.19 | 0 / 0.9 | +0.5 / 0 | 0 / −0.4 | 0 / 1.3 | +0.8 / 0 | 0 / −0.5 | 0 / 2 | +1.2 / 0 | 0 / −0.8 | 0 / 5.5 | +3.5 / 0 | 0 / −2.0 | 0.3 / 1.6 | +0.8 / 0 | −0.3 / −0.8 |
| 1.19 | 1.97 | 0 / 1.0 | +0.6 / 0 | 0 / −0.4 | 0 / 1.6 | +1.0 / 0 | 0 / −0.6 | 0 / 2.6 | +1.6 / 0 | 0 / −1 | 0 / 6.5 | +4.0 / 0 | 0 / −2.5 | 0.4 / 2.0 | +1.0 / 0 | −0.4 / −1.0 |
| 1.97 | 3.15 | 0 / 1.2 | +0.7 / 0 | 0 / −0.5 | 0 / 1.9 | +1.2 / 0 | 0 / −0.7 | 0 / 3 | +1.8 / 0 | 0 / −1.2 | 0 / 7.5 | +4.5 / 0 | 0 / −3 | 0.4 / 2.3 | +1.2 / 0 | −0.4 / −1.1 |
| 3.15 | 4.73 | 0 / 1.5 | +0.9 / 0 | 0 / −0.6 | 0 / 2.3 | +1.4 / 0 | 0 / −0.9 | 0 / 3.6 | +2.2 / 0 | 0 / −1.4 | 0 / 8.5 | +5.0 / 0 | 0 / −3.5 | 0.5 / 2.8 | +1.4 / 0 | −0.5 / −1.4 |
| 4.73 | 7.09 | 0 / 1.7 | +1.0 / 0 | 0 / −0.7 | 0 / 2.6 | +1.6 / 0 | 0 / −1.0 | 0 / 4.1 | +2.5 / 0 | 0 / −1.6 | 0 / 10.0 | +6.0 / 0 | 0 / −4 | 0.6 / 3.2 | +1.6 / 0 | −0.6 / −1.6 |
| 7.09 | 9.85 | 0 / 2.0 | +1.2 / 0 | 0 / −0.8 | 0 / 3.0 | +1.8 / 0 | 0 / −1.2 | 0 / 4.6 | +2.8 / 0 | 0 / −1.8 | 0 / 11.5 | +7.0 / 0 | 0 / −4.5 | 0.6 / 3.6 | +1.8 / 0 | −0.6 / −1.8 |
| 9.85 | 12.41 | 0 / 2.1 | +1.2 / 0 | 0 / −0.9 | 0 / 3.2 | +2.0 / 0 | 0 / −1.2 | 0 / 5 | +3.0 / 0 | 0 / −2.0 | 0 / 13.0 | +8.0 / 0 | 0 / −5 | 0.7 / 3.9 | +2.0 / 0 | −0.7 / −1.9 |
| 12.41 | 15.75 | 0 / 2.4 | +1.4 / 0 | 0 / −1.0 | 0 / 3.6 | +2.2 / 0 | 0 / −1.4 | 0 / 5.7 | +3.5 / 0 | 0 / −2.2 | 0 / 15.0 | +9.0 / 0 | 0 / −6 | 0.7 / 4.3 | +2.2 / 0 | −0.7 / −2.1 |
| 15.75 | 19.69 | 0 / 2.6 | +1.6 / 0 | 0 / −1.0 | 0 / 4.1 | +2.5 / 0 | 0 / −1.6 | 0 / 6.5 | +4 / 0 | 0 / −2.5 | 0 / 16.0 | +10.0 / 0 | 0 / −6 | 0.8 / 4.9 | +2.5 / 0 | −0.8 / −2.4 |

See footnotes at end of table.

Table 3 (Continued). ANSI Standard Clearance Locational Fits (ANSI B4.1-1967, R1974)

Values shown below are in thousandths of an inch

| Nominal Size Range, Inches, Over To | Class LC 6 Clearance* | LC 6 Hole H9 | LC 6 Shaft f8 | Class LC 7 Clearance* | LC 7 Hole H10 | LC 7 Shaft e9 | Class LC 8 Clearance* | LC 8 Hole H10 | LC 8 Shaft d9 | Class LC 9 Clearance* | LC 9 Hole H11 | LC 9 Shaft c10 | Class LC 10 Clearance* | LC 10 Hole H12 | LC 10 Shaft | Class LC 11 Clearance* | LC 11 Hole H13 | LC 11 Shaft |
|---|---|---|---|---|---|---|---|---|---|---|---|---|---|---|---|---|---|---|
| 0 - 0.12 | 0.3 / 1.9 | +1.0 / 0 | -0.3 / -0.9 | 0.6 / 3.2 | +1.6 / 0 | -0.6 / -1.6 | 1.0 / 2.0 | +1.6 / 0 | -1.0 / -2.0 | 2.5 / 6.6 | +2.5 / 0 | -2.5 / -4.1 | 4 / 12 | +4 / 0 | -4 / -8 | 5 / 17 | +6 / 0 | -5 / -11 |
| 0.12 - 0.24 | 0.4 / 2.3 | +1.2 / 0 | -0.4 / -1.1 | 0.8 / 3.8 | +1.8 / 0 | -0.8 / -2.0 | 1.2 / 4.2 | +1.8 / 0 | -1.2 / -2.4 | 2.8 / 7.6 | +3.0 / 0 | -2.8 / -4.6 | 4.5 / 14.5 | +5 / 0 | -4.5 / -9.5 | 6 / 20 | +7 / 0 | -6 / -13 |
| 0.24 - 0.40 | 0.5 / 2.8 | +1.4 / 0 | -0.5 / -1.4 | 1.0 / 4.6 | +2.2 / 0 | -1.0 / -2.4 | 1.6 / 5.2 | +2.2 / 0 | -1.6 / -3.0 | 3.0 / 8.7 | +3.5 / 0 | -3.0 / -5.2 | 5 / 17 | +6 / 0 | -5 / -11 | 7 / 25 | +9 / 0 | -7 / -16 |
| 0.40 - 0.71 | 0.6 / 3.2 | +1.6 / 0 | -0.6 / -1.6 | 1.2 / 5.6 | +2.8 / 0 | -1.2 / -2.8 | 2.0 / 6.4 | +2.8 / 0 | -2.0 / -3.6 | 3.5 / 10.3 | +4.0 / 0 | -3.5 / -6.3 | 6 / 20 | +7 / 0 | -6 / -13 | 8 / 28 | +10 / 0 | -8 / -18 |
| 0.71 - 1.19 | 0.8 / 4.0 | +2.0 / 0 | -0.8 / -2.0 | 1.6 / 7.1 | +3.5 / 0 | -1.6 / -3.6 | 2.5 / 8.0 | +3.5 / 0 | -2.5 / -4.5 | 4.5 / 13.0 | +5.0 / 0 | -4.5 / -8.0 | 7 / 23 | +8 / 0 | -7 / -15 | 10 / 34 | +12 / 0 | -10 / -22 |
| 1.19 - 1.97 | 1.0 / 5.1 | +2.5 / 0 | -1.0 / -2.6 | 2.0 / 8.5 | +4.0 / 0 | -2.0 / -4.5 | 3.6 / 9.5 | +4.0 / 0 | -3.0 / -5.5 | 5.0 / 15.0 | +6 / 0 | -5.0 / -9.0 | 8 / 28 | +10 / 0 | -8 / -18 | 12 / 44 | +16 / 0 | -12 / -28 |
| 1.97 - 3.15 | 1.2 / 6.0 | +3.0 / 0 | -1.0 / -3.0 | 2.5 / 10.0 | +4.5 / 0 | -2.5 / -5.5 | 4.0 / 11.5 | +4.5 / 0 | -4.0 / -7.0 | 6.0 / 17.5 | +7 / 0 | -6.0 / -10.5 | 10 / 34 | +12 / 0 | -10 / -22 | 14 / 50 | +18 / 0 | -14 / -32 |
| 3.15 - 4.73 | 1.4 / 7.1 | +3.5 / 0 | -1.4 / -3.6 | 3.0 / 11.5 | +5.0 / 0 | -3.0 / -6.5 | 5.0 / 13.5 | +5.0 / 0 | -5.0 / -8.5 | 7 / 21 | +9 / 0 | -7 / -12 | 11 / 39 | +14 / 0 | -11 / -25 | 16 / 60 | +22 / 0 | -16 / -38 |
| 4.73 - 7.09 | 1.6 / 8.1 | +4.0 / 0 | -1.6 / -4.1 | 3.5 / 13.5 | +6.0 / 0 | -3.5 / -7.5 | 6 / 16 | +6.0 / 0 | -6 / -10 | 8 / 24 | +10 / 0 | -8 / -14 | 12 / 44 | +16 / 0 | -12 / -28 | 18 / 68 | +25 / 0 | -18 / -43 |
| 7.09 - 9.85 | 2.0 / 9.3 | +4.5 / 0 | -2.0 / -4.8 | 4.0 / 15.5 | +7.0 / 0 | -4.0 / -8.5 | 7 / 18.5 | +7.0 / 0 | -7 / -11.5 | 10 / 29 | +12 / 0 | -10 / -17 | 16 / 52 | +18 / 0 | -16 / -34 | 22 / 78 | +28 / 0 | -22 / -50 |
| 9.85 - 12.41 | 2.2 / 10.2 | +5.0 / 0 | -2.2 / -5.2 | 4.5 / 17.5 | +8.0 / 0 | -4.5 / -9.5 | 8 / 20 | +8.0 / 0 | -8 / -12 | 12 / 32 | +12 / 0 | -12 / -20 | 20 / 60 | +20 / 0 | -20 / -40 | 28 / 88 | +30 / 0 | -28 / -58 |
| 12.41 - 15.75 | 2.5 / 12.0 | +6.0 / 0 | -2.5 / -6.0 | 5.0 / 20.0 | +9.0 / 0 | -5.0 / -11 | 8 / 23 | +9.0 / 0 | -8 / -14 | 14 / 37 | +14 / 0 | -14 / -23 | 22 / 66 | +22 / 0 | -22 / -44 | 30 / 100 | +35 / 0 | -30 / -65 |
| 15.75 - 19.69 | 2.8 / 12.8 | +6.0 / 0 | -2.8 / -6.8 | 5.0 / 21.0 | +10.0 / 0 | -5 / -11 | 9 / 25 | +10.0 / 0 | -9 / -15 | 16 / 42 | +16 / 0 | -16 / -26 | 25 / 75 | +25 / 0 | -25 / -50 | 35 / 115 | +40 / 0 | -35 / -75 |

All data above heavy lines are in accordance with American-British-Canadian (ABC) agreements. Symbols H6, H7, s6, etc., are hole and shaft designations in ABC system. Limits for sizes above 19.69 inches are not covered by ABC agreements but are given in the ANSI Standard.

\* Pairs of values shown represent minimum and maximum amounts of interference resulting from application of standard tolerance limits.

## Table 4. ANSI Standard Transition Locational Fits (ANSI B4.1-1967, R1974)

Values shown below are in thousandths of an inch

| Nominal Size Range, Inches (Over – To) | Class LT 1 Fit* | LT 1 Hole H7 | LT 1 Shaft js6 | Class LT 2 Fit* | LT 2 Hole H8 | LT 2 Shaft js7 | Class LT 3 Fit* | LT 3 Hole H7 | LT 3 Shaft k6 | Class LT 4 Fit* | LT 4 Hole H8 | LT 4 Shaft k7 | Class LT 5 Fit* | LT 5 Hole H7 | LT 5 Shaft n6 | Class LT 6 Fit* | LT 6 Hole H7 | LT 6 Shaft n7 |
|---|---|---|---|---|---|---|---|---|---|---|---|---|---|---|---|---|---|---|
| 0 – 0.12 | −0.12 / +0.52 | +0.4 / 0 | +0.12 / −0.12 | −0.2 / +0.8 | +0.6 / 0 | +0.2 / −0.2 | | | | | | | −0.5 / +0.15 | +0.4 / 0 | +0.5 / +0.25 | −0.65 / +0.15 | +0.4 / 0 | +0.65 / +0.25 |
| 0.12 – 0.24 | −0.15 / +0.65 | +0.5 / 0 | +0.15 / −0.15 | −0.25 / +0.95 | +0.7 / 0 | +0.25 / −0.25 | | | | | | | −0.6 / +0.2 | +0.5 / 0 | +0.6 / +0.3 | −0.8 / +0.2 | +0.5 / 0 | +0.8 / +0.3 |
| 0.24 – 0.40 | −0.2 / +0.8 | +0.6 / 0 | +0.2 / −0.2 | −0.3 / +1.2 | +0.9 / 0 | +0.3 / −0.3 | −0.5 / +0.5 | +0.6 / 0 | +0.5 / +0.1 | −0.7 / +0.8 | +0.9 / 0 | +0.7 / +0.1 | −0.8 / +0.2 | +0.6 / 0 | +0.8 / +0.4 | −1.0 / +0.2 | +0.6 / 0 | +1.0 / +0.4 |
| 0.40 – 0.71 | −0.2 / +0.9 | +0.7 / 0 | +0.2 / −0.2 | −0.35 / +1.35 | +1.0 / 0 | +0.35 / −0.35 | −0.5 / +0.6 | +0.7 / 0 | +0.5 / +0.1 | −0.8 / +0.9 | +1.0 / 0 | +0.8 / +0.1 | −0.9 / +0.2 | +0.7 / 0 | +0.9 / +0.5 | −1.2 / +0.2 | +0.7 / 0 | +1.2 / +0.5 |
| 0.71 – 1.19 | −0.25 / +1.05 | +0.8 / 0 | +0.25 / −0.25 | −0.4 / +1.6 | +1.2 / 0 | +0.4 / −0.4 | −0.6 / +0.7 | +0.8 / 0 | +0.6 / +0.1 | −0.9 / +1.1 | +1.2 / 0 | +0.9 / +0.1 | −1.1 / +0.2 | +0.8 / 0 | +1.1 / +0.6 | −1.4 / +0.2 | +0.8 / 0 | +1.4 / +0.6 |
| 1.19 – 1.97 | −0.3 / +1.3 | +1.0 / 0 | +0.3 / −0.3 | −0.5 / +2.1 | +1.6 / 0 | +0.5 / −0.5 | −0.7 / +0.9 | +1.0 / 0 | +0.7 / +0.1 | −1.1 / +1.5 | +1.6 / 0 | +1.1 / +0.1 | −1.3 / +0.3 | +1.0 / 0 | +1.3 / +0.7 | −1.7 / +0.3 | +1.0 / 0 | +1.7 / +0.7 |
| 1.97 – 3.15 | −0.3 / +1.5 | +1.2 / 0 | +0.3 / −0.3 | −0.6 / +2.4 | +1.8 / 0 | +0.6 / −0.6 | −0.8 / +1.1 | +1.2 / 0 | +0.8 / +0.1 | −1.3 / +1.7 | +1.8 / 0 | +1.3 / +0.1 | −1.5 / +0.4 | +1.2 / 0 | +1.5 / +0.8 | −2.0 / +0.4 | +1.2 / 0 | +2.0 / +0.8 |
| 3.15 – 4.73 | −0.4 / +1.8 | +1.4 / 0 | +0.4 / −0.4 | −0.7 / +2.9 | +2.2 / 0 | +0.7 / −0.7 | −1.0 / +1.3 | +1.4 / 0 | +1.0 / +0.1 | −1.5 / +2.1 | +2.2 / 0 | +1.5 / +0.1 | −1.9 / +0.4 | +1.4 / 0 | +1.9 / +1.0 | −2.4 / +0.4 | +1.4 / 0 | +2.4 / +1.0 |
| 4.73 – 7.09 | −0.5 / +2.1 | +1.6 / 0 | +0.5 / −0.5 | −0.8 / +3.3 | +2.5 / 0 | +0.8 / −0.8 | −1.1 / +1.5 | +1.6 / 0 | +1.1 / +0.1 | −1.7 / +2.4 | +2.5 / 0 | +1.7 / +0.1 | −2.2 / +0.4 | +1.6 / 0 | +2.2 / +1.2 | −2.8 / +0.4 | +1.6 / 0 | +2.8 / +1.2 |
| 7.09 – 9.85 | −0.6 / +2.4 | +1.8 / 0 | +0.6 / −0.6 | −0.9 / +3.7 | +2.8 / 0 | +0.9 / −0.9 | −1.4 / +1.6 | +1.8 / 0 | +1.4 / +0.2 | −2.0 / +2.6 | +2.8 / 0 | +2.0 / +0.2 | −2.6 / +0.4 | +1.8 / 0 | +2.6 / +1.4 | −3.4 / +0.4 | +1.8 / 0 | +3.4 / +1.4 |
| 9.85 – 12.41 | −0.6 / +2.6 | +2.0 / 0 | +0.6 / −0.6 | −1.0 / +4.0 | +3.0 / 0 | +1.0 / −1.0 | −1.4 / +1.8 | +2.0 / 0 | +1.4 / +0.2 | −2.2 / +2.8 | +3.0 / 0 | +2.2 / +0.2 | −2.6 / +0.6 | +2.0 / 0 | +2.6 / +1.4 | −3.4 / +0.6 | +2.0 / 0 | +3.4 / +1.6 |
| 12.41 – 15.75 | −0.7 / +2.9 | +2.2 / 0 | +0.7 / −0.7 | −1.0 / +4.5 | +3.5 / 0 | +1.0 / −1.0 | −1.6 / +2.0 | +2.2 / 0 | +1.6 / +0.2 | −2.4 / +3.3 | +3.5 / 0 | +2.4 / +0.2 | −3.0 / +0.6 | +2.2 / 0 | +3.0 / +1.6 | −3.8 / +0.6 | +2.2 / 0 | +3.8 / +1.6 |
| 15.75 – 19.69 | −0.8 / +3.3 | +2.5 / 0 | +0.8 / −0.8 | −1.2 / +5.2 | +4.0 / 0 | +1.2 / −1.2 | −1.8 / +2.3 | +2.5 / 0 | +1.8 / +0.2 | −2.7 / +3.8 | +4.0 / 0 | +2.7 / +0.2 | −3.4 / +0.7 | +2.5 / 0 | +3.4 / +1.8 | −4.3 / +0.7 | +2.5 / 0 | +4.3 / +1.8 |

All data above heavy lines are in accord with ABC agreements. Symbols H7, js6, etc. are hole and shaft designations in ABC system.
* Pairs of values shown represent maximum amount of interference (−) and maximum amount of clearance (+) resulting from application of standard tolerance limits.

**Table 5.   ANSI Standard Interference Locational Fits** (ANSI B4.1-1967, R1974)

Tolerance limits given in body of table are added or subtracted to basic size (as indicated by + or − sign) to obtain maximum and minimum sizes of mating parts.

| Nominal Size Range, Inches Over To | Class LN 1 Limits of Interference | Standard Limits Hole H6 | Shaft n5 | Class LN 2 Limits of Interference | Standard Limits Hole H7 | Shaft p6 | Class LN 3 Limits of Interference | Standard Limits Hole H7 | Shaft r6 |
|---|---|---|---|---|---|---|---|---|---|
| *Values shown below are given in thousandths of an inch* | | | | | | | | | |
| 0 − 0.12 | 0 / 0.45 | +0.25 / 0 | +0.45 / +0.25 | 0 / 0.65 | +0.4 / 0 | +0.65 / +0.4 | 0.1 / 0.75 | +0.4 / 0 | +0.75 / +0.5 |
| 0.12− 0.24 | 0 / 0.5 | +0.3 / 0 | +0.5 / +0.3 | 0 / 0.8 | +0.5 / 0 | +0.8 / +0.5 | 0.1 / 0.9 | +0.5 / 0 | +0.9 / +0.6 |
| 0.24− 0.40 | 0 / 0.65 | +0.4 / 0 | +0.65 / +0.4 | 0 / 1.0 | +0.6 / 0 | +1.0 / +0.6 | 0.2 / 1.2 | +0.6 / 0 | +1.2 / +0.8 |
| 0.40− 0.71 | 0 / 0.8 | +0.4 / 0 | +0.8 / +0.4 | 0 / 1.1 | +0.7 / 0 | +1.1 / +0.7 | 0.3 / 1.4 | +0.7 / 0 | +1.4 / +1.0 |
| 0.71− 1.19 | 0 / 1.0 | +0.5 / 0 | +1.0 / +0.5 | 0 / 1.3 | +0.8 / 0 | +1.3 / +0.8 | 0.4 / 1.7 | +0.8 / 0 | +1.7 / +1.2 |
| 1.19− 1.97 | 0 / 1.1 | +0.6 / 0 | +1.1 / +0.6 | 0 / 1.6 | +1.0 / 0 | +1.6 / +1.0 | 0.4 / 2.0 | +1.0 / 0 | +2.0 / +1.4 |
| 1.97− 3.15 | 0.1 / 1.3 | +0.7 / 0 | +1.3 / +0.8 | 0.2 / 2.1 | +1.2 / 0 | +2.1 / +1.4 | 0.4 / 2.3 | +1.2 / 0 | +2.3 / +1.6 |
| 3.15− 4.73 | 0.1 / 1.6 | +0.9 / 0 | +1.6 / +1.0 | 0.2 / 2.5 | +1.4 / 0 | +2.5 / +1.6 | 0.6 / 2.9 | +1.4 / 0 | +2.9 / +2.0 |
| 4.73− 7.09 | 0.2 / 1.9 | +1.0 / 0 | +1.9 / +1.2 | 0.2 / 2.8 | +1.6 / 0 | +2.8 / +1.8 | 0.9 / 3.5 | +1.6 / 0 | +3.5 / +2.5 |
| 7.09− 9.85 | 0.2 / 2.2 | +1.2 / 0 | +2.2 / +1.4 | 0.2 / 3.2 | +1.8 / 0 | +3.2 / +2.0 | 1.2 / 4.2 | +1.8 / 0 | +4.2 / +3.0 |
| 9.85−12.41 | 0.2 / 2.3 | +1.2 / 0 | +2.3 / +1.4 | 0.2 / 3.4 | +2.0 / 0 | +3.4 / +2.2 | 1.5 / 4.7 | +2.0 / 0 | +4.7 / +3.5 |
| 12.41−15.75 | 0.2 / 2.6 | +1.4 / 0 | +2.6 / +1.6 | 0.3 / 3.9 | +2.2 / 0 | +3.9 / +2.5 | 2.3 / 5.9 | +2.2 / 0 | +5.9 / +4.5 |
| 15.75−19.69 | 0.2 / 2.8 | +1.6 / 0 | +2.8 / +1.8 | 0.3 / 4.4 | +2.5 / 0 | +4.4 / +2.8 | 2.5 / 6.6 | +2.5 / 0 | +6.6 / +5.0 |

All data in this table are in accordance with American-British-Canadian (ABC) agreements.

Limits for sizes above 19.69 inches are not covered by ABC agreements but are given in the ANSI Standard.

Symbols H7, p6, etc. are hole and shaft designations in ABC system.

*Pairs of values shown represent minimum and maximum amounts of interference resulting from application of standard tolerance limits.

Table 6. ANSI Standard Force and Shrink Fits (ANSI B4.1-1967, R1974)

Values shown below are in thousandths of an inch

| Nominal Size Range, Inches Over | To | Class FN 1 Interference | Class FN 1 Standard Tolerance Limits Hole H6 | Class FN 1 Standard Tolerance Limits Shaft | Class FN 2 Interference | Class FN 2 Standard Tolerance Limits Hole H7 | Class FN 2 Standard Tolerance Limits Shaft s6 | Class FN 3 Interference | Class FN 3 Standard Tolerance Limits Hole H7 | Class FN 3 Standard Tolerance Limits Shaft t6 | Class FN 4 Interference | Class FN 4 Standard Tolerance Limits Hole H7 | Class FN 4 Standard Tolerance Limits Shaft u6 | Class FN 5 Interference | Class FN 5 Standard Tolerance Limits Hole H8 | Class FN 5 Standard Tolerance Limits Shaft x7 |
|---|---|---|---|---|---|---|---|---|---|---|---|---|---|---|---|---|
| 0 | 0.12 | 0.05 / 0.5 | +0.25 / 0 | +0.5 / +0.3 | 0.2 / 0.85 | +0.4 / 0 | +0.85 / +0.6 | | | | 0.3 / 0.95 | +0.4 / 0 | +0.95 / +0.7 | 0.3 / 1.3 | +0.6 / 0 | +1.3 / +0.9 |
| 0.12 | 0.24 | 0.1 / 0.6 | +0.3 / 0 | +0.6 / +0.4 | 0.2 / 1.0 | +0.5 / 0 | +1.0 / +0.7 | | | | 0.4 / 1.2 | +0.5 / 0 | +1.2 / +0.9 | 0.5 / 1.7 | +0.7 / 0 | +1.7 / +1.2 |
| 0.24 | 0.40 | 0.1 / 0.75 | +0.4 / 0 | +0.75 / +0.5 | 0.4 / 1.4 | +0.6 / 0 | +1.4 / +1.0 | | | | 0.6 / 1.6 | +0.6 / 0 | +1.6 / +1.2 | 0.5 / 2.0 | +0.9 / 0 | +2.0 / +1.4 |
| 0.40 | 0.56 | 0.1 / 0.8 | +0.4 / 0 | +0.8 / +0.5 | 0.5 / 1.6 | +0.7 / 0 | +1.6 / +1.2 | | | | 0.7 / 1.8 | +0.7 / 0 | +1.8 / +1.4 | 0.6 / 2.3 | +1.0 / 0 | +2.3 / +1.6 |
| 0.56 | 0.71 | 0.2 / 0.9 | +0.4 / 0 | +0.9 / +0.6 | 0.5 / 1.6 | +0.7 / 0 | +1.6 / +1.2 | | | | 0.7 / 1.8 | +0.7 / 0 | +1.8 / +1.4 | 0.8 / 2.5 | +1.0 / 0 | +2.5 / +1.8 |
| 0.71 | 0.95 | 0.2 / 1.1 | +0.5 / 0 | +1.1 / +0.7 | 0.6 / 1.9 | +0.8 / 0 | +1.9 / +1.4 | 0.8 / 2.1 | +0.8 / 0 | +2.1 / +1.6 | 0.8 / 2.1 | +0.8 / 0 | +2.1 / +1.6 | 1.0 / 3.0 | +1.2 / 0 | +3.0 / +2.2 |
| 0.95 | 1.19 | 0.3 / 1.2 | +0.5 / 0 | +1.2 / +0.8 | 0.6 / 1.9 | +0.8 / 0 | +1.9 / +1.4 | 1.0 / 2.6 | +1.0 / 0 | +2.6 / +2.0 | 1.0 / 2.3 | +0.8 / 0 | +2.3 / +1.8 | 1.3 / 3.3 | +1.2 / 0 | +3.3 / +2.5 |
| 1.19 | 1.58 | 0.3 / 1.3 | +0.6 / 0 | +1.3 / +0.9 | 0.8 / 2.4 | +1.0 / 0 | +2.4 / +1.8 | 1.2 / 2.8 | +1.0 / 0 | +2.8 / +2.2 | 1.5 / 3.1 | +1.0 / 0 | +3.1 / +2.5 | 1.4 / 4.0 | +1.6 / 0 | +4.0 / +3.0 |
| 1.58 | 1.97 | 0.4 / 1.4 | +0.6 / 0 | +1.4 / +1.0 | 0.8 / 2.4 | +1.0 / 0 | +2.4 / +1.8 | 1.3 / 3.2 | +1.2 / 0 | +3.2 / +2.5 | 1.8 / 3.4 | +1.0 / 0 | +3.4 / +2.8 | 2.4 / 5.0 | +1.6 / 0 | +5.0 / +4.0 |
| 1.97 | 2.56 | 0.6 / 1.8 | +0.7 / 0 | +1.8 / +1.3 | 0.8 / 2.7 | +1.2 / 0 | +2.7 / +2.0 | 1.8 / 3.7 | +1.2 / 0 | +3.7 / +3.0 | 2.3 / 4.2 | +1.2 / 0 | +4.2 / +3.5 | 3.2 / 6.2 | +1.8 / 0 | +6.2 / +5.0 |
| 2.56 | 3.15 | 0.7 / 1.9 | +0.7 / 0 | +1.9 / +1.4 | 1.0 / 2.9 | +1.2 / 0 | +2.9 / +2.2 | 2.1 / 4.4 | +1.4 / 0 | +4.4 / +3.5 | 2.8 / 4.7 | +1.2 / 0 | +4.7 / +4.0 | 4.2 / 7.2 | +1.8 / 0 | +7.2 / +6.0 |
| 3.15 | 3.94 | 0.9 / 2.4 | +0.9 / 0 | +2.4 / +1.8 | 1.4 / 3.7 | +1.4 / 0 | +3.7 / +2.8 | 2.6 / 4.9 | +1.4 / 0 | +4.9 / +4.0 | 3.6 / 5.9 | +1.4 / 0 | +5.9 / +5.0 | 4.8 / 8.4 | +2.2 / 0 | +8.4 / +7.0 |
| 3.94 | 4.73 | 1.1 / 2.6 | +0.9 / 0 | +2.6 / +2.0 | 1.6 / 3.9 | +1.4 / 0 | +3.9 / +3.0 | | | | 4.6 / 6.9 | +1.4 / 0 | +6.9 / +6.0 | 5.8 / 9.4 | +2.2 / 0 | +9.4 / +8.0 |

See footnotes at end of table.

Table 6 (Continued).   ANSI Standard Force and Shrink Fits (ANSI B4.1-1967, R1974)

| Nominal Size Range, Inches | | Class FN 1 | | | Class FN 2 | | | Class FN 3 | | | Class FN 4 | | | Class FN 5 | | |
|---|---|---|---|---|---|---|---|---|---|---|---|---|---|---|---|---|
| | | Interference* | Standard Tolerance Limits | | Interference* | Standard Tolerance Limits | | Interference* | Standard Tolerance Limits | | Interference* | Standard Tolerance Limits | | Interference* | Standard Tolerance Limits | |
| Over | To | | Hole H6 | Shaft | | Hole H7 | Shaft s6 | | Hole H7 | Shaft t6 | | Hole H7 | Shaft u6 | | Hole H8 | Shaft x7 |
| | | | | | | Values shown below are in thousandths of an inch | | | | | | | | | | |
| 4.73 | 5.52 | 1.2 / 2.9 | +1.0 / 0 | +2.9 / +2.2 | 1.9 / 4.5 | +1.6 / 0 | +4.5 / +3.5 | 3.4 / 6.0 | +1.6 / 0 | +6.0 / +5.0 | 5.4 / 8.0 | +1.6 / 0 | +8.0 / +7.0 | 7.5 / 11.6 | +2.5 / 0 | +11.6 / +10.0 |
| 5.52 | 6.30 | 1.5 / 3.2 | +1.0 / 0 | +3.2 / +2.5 | 2.4 / 5.0 | +1.6 / 0 | +5.0 / +4.0 | 3.4 / 6.0 | +1.6 / 0 | +6.0 / +5.0 | 5.4 / 8.0 | +1.6 / 0 | +8.0 / +7.0 | 9.5 / 13.6 | +2.5 / 0 | +13.6 / +12.0 |
| 6.30 | 7.09 | 1.8 / 3.5 | +1.0 / 0 | +3.5 / +2.8 | 2.9 / 5.5 | +1.6 / 0 | +5.5 / +4.5 | 4.4 / 7.0 | +1.6 / 0 | +7.0 / +6.0 | 6.4 / 9.0 | +1.6 / 0 | +9.0 / +8.0 | 9.5 / 13.6 | +2.5 / 0 | +13.6 / +12.0 |
| 7.09 | 7.88 | 1.8 / 3.8 | +1.2 / 0 | +3.8 / +3.0 | 3.2 / 6.2 | +1.8 / 0 | +6.2 / +5.0 | 5.2 / 8.2 | +1.8 / 0 | +8.2 / +7.0 | 7.2 / 10.2 | +1.8 / 0 | +10.2 / +9.0 | 11.2 / 15.8 | +2.8 / 0 | +15.8 / +14.0 |
| 7.88 | 8.86 | 2.3 / 4.3 | +1.2 / 0 | +4.3 / +3.5 | 3.2 / 6.2 | +1.8 / 0 | +6.2 / +5.0 | 5.2 / 8.2 | +1.8 / 0 | +8.2 / +7.0 | 8.2 / 11.2 | +1.8 / 0 | +11.2 / +10.0 | 13.2 / 17.8 | +2.8 / 0 | +17.8 / +16.0 |
| 8.86 | 9.85 | 2.3 / 4.3 | +1.2 / 0 | +4.3 / +3.5 | 4.2 / 7.2 | +1.8 / 0 | +7.2 / +6.0 | 6.2 / 9.2 | +1.8 / 0 | +9.2 / +8.0 | 10.2 / 13.2 | +1.8 / 0 | +13.2 / +12.0 | 13.2 / 17.8 | +2.8 / 0 | +17.8 / +16.0 |
| 9.85 | 11.03 | 2.8 / 4.9 | +1.2 / 0 | +4.9 / +4.0 | 4.0 / 7.2 | +2.0 / 0 | +7.0 / +6.0 | 7.0 / 10.2 | +2.0 / 0 | +10.2 / +9.0 | 10.0 / 13.2 | +2.0 / 0 | +13.2 / +12.0 | 15.0 / 20.0 | +3.0 / 0 | +20.0 / +18.0 |
| 11.03 | 12.41 | 2.8 / 4.9 | +1.2 / 0 | +4.9 / +4.0 | 5.0 / 8.2 | +2.0 / 0 | +8.2 / +7.0 | 7.0 / 10.2 | +2.0 / 0 | +10.2 / +9.0 | 12.0 / 15.2 | +2.0 / 0 | +15.2 / +14.0 | 17.0 / 22.0 | +3.0 / 0 | +22.0 / +20.0 |
| 12.41 | 13.98 | 3.1 / 5.5 | +1.4 / 0 | +5.5 / +4.5 | 5.8 / 9.4 | +2.2 / 0 | +9.4 / +8.0 | 7.8 / 11.4 | +2.2 / 0 | +11.4 / +10.0 | 13.8 / 17.4 | +2.2 / 0 | +17.4 / +16.0 | 18.5 / 24.2 | +3.5 / 0 | +24.2 / +22.0 |
| 13.98 | 15.75 | 3.6 / 6.1 | +1.4 / 0 | +6.1 / +5.0 | 5.8 / 9.4 | +2.2 / 0 | +9.4 / +8.0 | 9.8 / 13.4 | +2.2 / 0 | +13.4 / +12.0 | 15.8 / 19.4 | +2.2 / 0 | +19.4 / +18.0 | 21.5 / 27.2 | +3.5 / 0 | +27.2 / +25.0 |
| 15.75 | 17.72 | 4.4 / 7.0 | +1.6 / 0 | +7.0 / +6.0 | 6.5 / 10.6 | +2.5 / 0 | +10.6 / +9.0 | 9.5 / 13.6 | +2.5 / 0 | +13.6 / +12.0 | 17.5 / 21.6 | +2.5 / 0 | +21.6 / +20.0 | 24.0 / 30.5 | +4.0 / 0 | +30.5 / +28.0 |
| 17.72 | 19.69 | 4.4 / 7.0 | +1.6 / 0 | +7.0 / +6.0 | 7.5 / 11.6 | +2.5 / 0 | +11.6 / +10.0 | 11.5 / 15.6 | +2.5 / 0 | +15.6 / +14.0 | 19.5 / 23.6 | +2.5 / 0 | +23.6 / +22.0 | 26.0 / 32.5 | +4.0 / 0 | +32.5 / +30.0 |

All data above heavy lines are in accordance with American-British-Canadian (ABC) agreements.   Symbols H6, H7, s6, etc. are hole and shaft designations in ABC system.   Limits for sizes above 19.69 inches are not covered by ABC agreements but are given in the ANSI standard.
* Pairs of values shown represent minimum and maximum amounts of interference resulting from application of standard tolerance limits.

Maximum shaft = 2 − 0.0004 = 1.9996; minimum shaft = 2 − 0.0007 = 1.9993 inches

Minimum clearance = 0.0004; maximum clearance = 0.0012 inch

*Example 2:* Establish the limits for a Class LT 1 fit for a 2-inch diameter.

Maximum hole = 2 + 0.0012 = 2.0012; minimum hole = 2 inches

Maximum shaft = 2 + 0.0003 = 2.0003; minimum shaft = 2 −0.0003 = 1.9997 inches

Maximum resulting *interference* = 0.0003; maximum resulting *clearance* = 0.0015 inch

**Modified Standard Fits.** — Fits having the same limits of clearance or interference as those shown in Tables 2 to 6 may sometimes have to be produced by using holes or shafts having limits of size other than those shown in these tables. This may be accomplished by using either a *Bilateral Hole (System B)* or a *Basic Shaft System (Symbol S)*. Both methods will result in non-standard holes and shafts.

*Bilateral Hole Fits (Symbol B):* The common case is where holes are produced with fixed tools, such as drills or reamers; to provide a longer wear life for such tools a bilateral tolerance is desired.

The symbols used for these fits are identical with those used for standard fits except that they are followed by the letter B. Thus, LC 4B is a clearance locational fit, Class 4, except that it is produced with a bilateral hole.

The limits of clearance or interference are identical with those shown in Tables 2 to 6 for the corresponding fits.

The hole tolerance, however, is changed so that the plus limit is that for one grade finer than the value shown in the tables and the minus limit equals the amount by which the plus limit was lowered. The shaft limits are both lowered by the same amount as the lower limit of size of the hole. The finer grade of tolerance required to make these modifications may be obtained from Table 1. For example, an LC 4B fit for a 6-inch diameter hole would have tolerance limits of +4.0, −2.0 (+.0040 inch, −.0020 inch); the shaft would have tolerance limits of −2.0, −6.0 (−.0020 inch, −.0060 inch).

*Basic Shaft Fits (Symbol S):* For these fits the maximum size of the shaft is basic. The limits of clearance or interference are identical with those shown in Tables 2 to 6 for the corresponding fits and the symbols used for these fits are identical with those used for standard fits except that they are followed by the letter S. Thus, LC 4S is a clearance locational fit, Class 4, except that it is produced on a basic shaft basis.

The limits for hole and shaft as given in Tables 2 to 6 are increased for clearance fits (*decreased* for transition or interference fits) by the value of the upper shaft limit; that is, by the amount required to change the maximum shaft to the basic size.

**American National Standard Preferred Metric Limits and Fits.** — This standard (ANSI B4.2-1978) describes the ISO system of metric limits and fits for mating parts as approved for general engineering usage in the United States. It establishes: (1) the designation symbols used to define dimensional limits on drawings, material stock, related tools, gages, etc.; (2) the preferred basic sizes (first and second choices); (3) the preferred tolerance zones (first, second and third choices); (4) the preferred limits and fits for sizes (first choice only) up to and including 500 millimeters; and (5) the definitions of related terms.

The general terms "hole" and "shaft" can also be taken to refer to the space containing or contained by two parallel faces of any part, such as the width of a slot, the thickness of a key, etc.

**Definitions.** — The most important terms relating to limits and fits are shown in Fig. 1 and are defined as follows:

*Basic Size:* The size to which limits of deviation are assigned. The basic size is the same for both members of a fit. For example, it is designated by the numbers 40 in 40H7.

*Deviation:* The algebraic difference between a size and the corresponding basic size.

*Upper Deviation:* The algebraic difference between the maximum limit of size and the corresponding basic size.

*Lower Deviation:* The algebraic difference between the minimum limit of size and the corresponding basic size.

*Fundamental Deviation:* That one of the two deviations closest to the basic size. For example, it is designated by the letter H in 40H7.

*Tolerance:* The difference between the maximum and minimum size limits on a part.

*Tolerance Zone:* A zone representing the tolerance and its position in relation to the basic size.

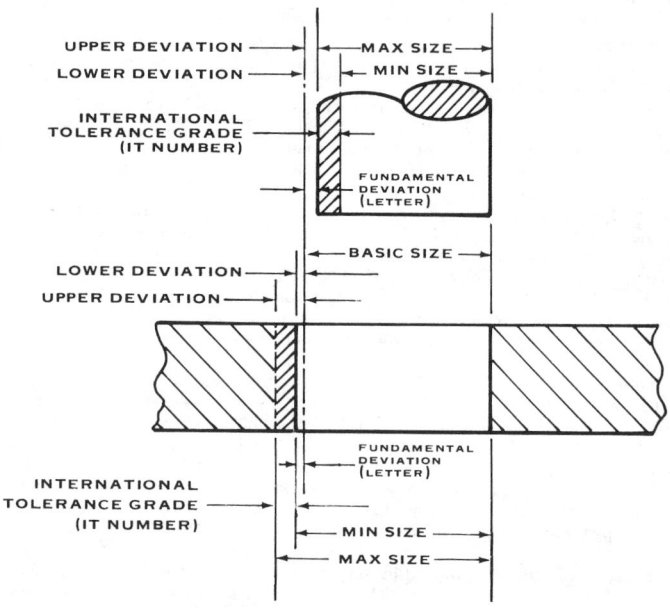

Fig. 1. Illustration of definitions.

*International Tolerance Grade (IT):* A group of tolerances which vary depending upon the basic size, but which provide the same relative level of accuracy within a given grade. For example, it is designated by the number 7 in 40H7 or as IT7.

*Hole Basis:* The system of fits where the minimum hole size is basic. The fundamental deviation for a hole basis system is H.

*Shaft Basis:* The system of fits where the maximum shaft size is basic. The fundamental deviation for a shaft basis system is h.

*Clearance Fit:* The relationship between assembled parts when clearance occurs under all tolerance conditions.

*Interference Fit:* The relationship between assembled parts when interference occurs under all tolerance conditions.

*Transition Fit:* The relationship between assembled parts when either a clearance or an interference fit can result, depending on the tolerance conditions of the mating parts.

**Tolerances Designation.** — An "International Tolerance grade" establishes the magnitude of the tolerance zone or the amount of part size variation allowed for external and internal dimensions alike (see Fig. 11). Tolerances are expressed in grade numbers which are consistent with International Tolerance grades identified by the prefix IT, such as IT6, IT11, etc. A smaller grade number provides a smaller tolerance zone.

A fundamental deviation establishes the position of the tolerance zone with respect to the basic size (see Fig. 1). Fundamental deviations are expressed by tolerance position letters. Capital letters are used for internal dimensions and lower case or small letters for external dimensions.

**Symbols.** — By combining the IT grade number and the tolerance position letter, the tolerance symbol is established which identifies the actual maximum and minimum limits of the part. The toleranced size is thus defined by the basic size of the part followed by a symbol composed of a letter and a number, such as 40H7, 40f7, etc.

A fit is indicated by the basic size common to both components, followed by a symbol corresponding to each component, the internal part symbol preceding the external part symbol, such as 40H8/f7.

Some methods of designating tolerances on drawings are:

$$\text{a. } 40\text{H8} \qquad \text{b. } 40\text{H8} \begin{pmatrix} 40.039 \\ 40.000 \end{pmatrix} \qquad \text{c. } \begin{pmatrix} 40.039 \\ 40.000 \end{pmatrix} 40\text{H8}$$

The values in parentheses indicate reference only.

**Table 1.　American National Standard Preferred Metric Sizes**
(ANSI B4.2-1978)

| Basic Size, mm | | Basic Size, mm | | Basic Size, mm | | Basic Size, mm | |
|---|---|---|---|---|---|---|---|
| 1st Choice | 2nd Choice | 1st Choice | 2nd Choice | 1st Choice | 2nd Choice | 1st Choice | 2nd Choice |
| 1 | ... | 6 | ... | 40 | ... | 250 | ... |
| ... | 1.1 | ... | 7 | ... | 45 | ... | 280 |
| 1.2 | ... | 8 | ... | 50 | ... | 300 | ... |
| ... | 1.4 | ... | 9 | ... | 55 | ... | 350 |
| 1.6 | ... | 10 | ... | 60 | ... | 400 | ... |
| ... | 1.8 | ... | 11 | ... | 70 | ... | 450 |
| 2 | ... | 12 | ... | 80 | ... | 500 | ... |
| ... | 2.2 | ... | 14 | ... | 90 | ... | 550 |
| 2.5 | ... | 16 | ... | 100 | ... | 600 | ... |
| ... | 2.8 | ... | 18 | ... | 110 | ... | 700 |
| 3 | ... | 20 | ... | 120 | ... | 800 | ... |
| ... | 3.5 | ... | 22 | ... | 140 | ... | 900 |
| 4 | ... | 25 | ... | 160 | ... | 1000 | ... |
| ... | 4.5 | ... | 28 | ... | 180 | ... | ... |
| 5 | ... | 30 | ... | 200 | ... | ... | ... |
| ... | 5.5 | ... | 35 | ... | 220 | ... | ... |

**Preferred Basic Sizes.** — The basic size of mating parts should, where possible, be chosen from the first choice sizes listed in Table 1. The sizes shown in this table have been selected from the preferred diameters of round metal products in the American National Standard for Preferred Metric Sizes for Round, Square, and Hexagonal Metal Products (ANSI B32.4-1974).

The preference rating has been based upon Renard's series of preferred numbers as given in American National Standard ANSI Z17.1-1973 (see Handbook page 116). The first choice sizes shown in Table 1 follow approximately the preferred number series R10 where succeeding numbers in the series increase by 25 per cent. The second choice series shown in Table 1 are rounded off from the R20 series of preferred numbers (12 per cent increments).

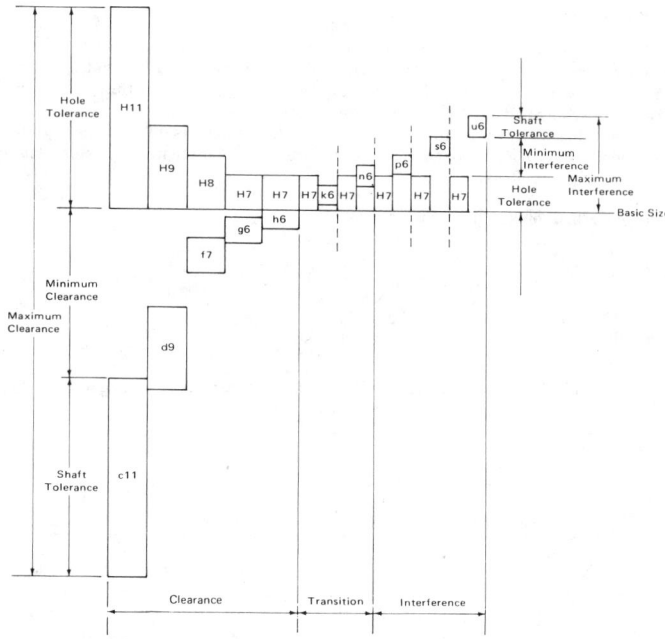

Fig. 2. Preferred hole basis fits.

The first choice sizes can be rationalized by selecting every second number in the series such as 1, 1.6, 2.5, 4, 6, 10, 16, etc.

Preferred sizes outside the range of 1 through 1000 are found by multiplying or dividing the sizes shown in this table by 1000 or multiples thereof.

**Preferred Fits.** — First choice tolerance zones are used to establish preferred fits in this Standard as shown in Figs. 2 and 3. A complete listing of first, second and third choice tolerance zones is given in the Standard.

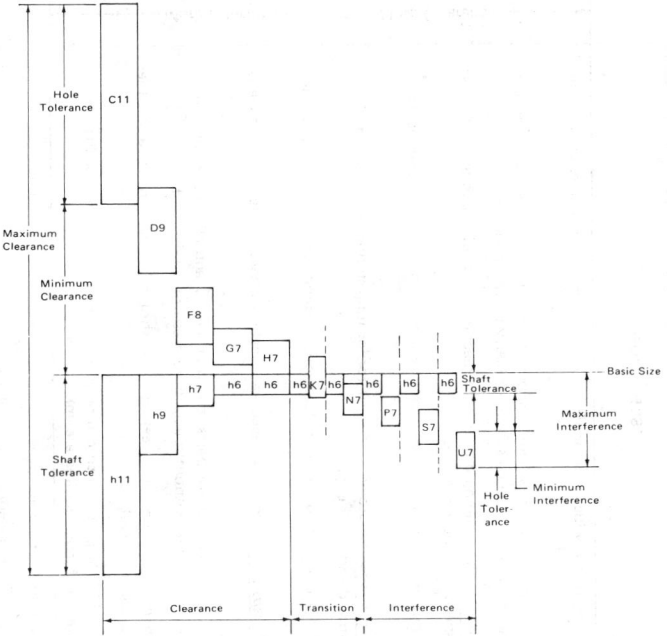

Fig. 3. Preferred shaft basis fits.

Hole basis fits have a fundamental deviation of H on the hole, and shaft basis fits have a fundamental deviation of h on the shaft and are shown in Fig. 2 for hole basis and Fig. 3 for shaft basis fits. A description of both types of fits which have the same relative fit condition is given in Fig. 4. Normally the hole basis system is preferred; however, when a common shaft mates with several holes, the shaft basis system should be used.

The hole basis and shaft basis fits shown in Fig. 4 are combined with the first choice sizes shown in Table 1 to form Tables 2, 3, 4, and 5 where specific limits as well as the resultant fits are tabulated.

If the required size is not tabulated in Tables 2 to 5, incl., then the preferred fit can be calculated from numerical values given in an appendix of ANSI B4.2-1978. It is anticipated that other fit conditions may be necessary to meet special requirements, and a preferred fit can be loosened or tightened simply by selecting a standard tolerance zone as given in the Standard. Information on how to calculate limit dimensions, clearances, and interferences, for non-preferred fits and sizes can also be found in an appendix of this Standard.

**Applications.** — Many factors such as length of engagement, bearing load, speed, lubrication, operating, temperatures, humidity, surface texture, and materials must be taken into account in fit selections for a particular application. Choice of other than

(Concluded on page 1565)

◄──── More Clearance    More Interference ────►

| ISO SYMBOL | | DESCRIPTION |
| Hole Basis | Shaft Basis | |
|---|---|---|
| H11/c11 | C11/h11 | Loose running fit for wide commercial tolerances or allowances on external members. |
| H9/d9 | D9/h9 | Free running fit not for use where accuracy is essential, but good for large temperature variations, high running speeds, or heavy journal pressures. |
| H8/f7 | F8/h7 | Close running fit for running on accurate machines and for accurate location at moderate speeds and journal pressures. |
| H7/g6 | G7/h6 | Sliding fit not intended to run freely, but to move and turn freely and locate accurately. |
| H7/h6 | H7/h6 | Locational clearance fit provides snug fit for locating stationary parts; but can be freely assembled and disassembled. |
| H7/k6 | K7/h6 | Locational transition fit for accurate location, a compromise between clearance and interference. |
| H7/n6 | N7/h6 | Locational transition fit for more accurate location where greater interference is permissible. |
| H7/p6[1] | P7/h6 | Locational interference fit for parts requiring rigidity and alignment with prime accuracy of location but without special bore pressure requirements. |
| H7/s6 | S7/h6 | Medium drive fit for ordinary steel parts or shrink fits on light sections, the tightest fit usable with cast iron. |
| H7/u6 | U7/h6 | Force fit suitable for parts which can be highly stressed or for shrink fits where the heavy pressing forces required are impractical. |

[1] Transition fit for basic sizes in range from 0 through 3 mm.

◄── Clearance Fits ──►◄── Transition Fits ──►◄── Interference Fits ──►

Fig. 4. Description of preferred fits.

Table 2. American National Standard Preferred Hole Basis Metric Clearance Fits (ANSI B4.2-1978)

| Basic Size* | | Loose Running | | | Free Running | | | Close Running | | | Sliding | | | Locational Clearance | | |
|---|---|---|---|---|---|---|---|---|---|---|---|---|---|---|---|---|
| | | Hole H11 | Shaft c11 | Fit† | Hole H9 | Shaft d9 | Fit† | Hole H8 | Shaft f7 | Fit† | Hole H7 | Shaft g6 | Fit† | Hole H7 | Shaft h6 | Fit† |
| 1 | Max | 1.060 | 0.940 | 0.180 | 1.025 | 0.980 | 0.070 | 1.014 | 0.994 | 0.030 | 1.010 | 0.998 | 0.018 | 1.010 | 1.000 | 0.016 |
| | Min | 1.000 | 0.880 | 0.060 | 1.000 | 0.955 | 0.020 | 1.000 | 0.984 | 0.006 | 1.000 | 0.992 | 0.002 | 1.000 | 0.994 | 0.000 |
| 1.2 | Max | 1.260 | 1.140 | 0.180 | 1.225 | 1.180 | 0.070 | 1.214 | 1.194 | 0.030 | 1.210 | 1.198 | 0.018 | 1.210 | 1.200 | 0.016 |
| | Min | 1.200 | 1.080 | 0.060 | 1.200 | 1.155 | 0.020 | 1.200 | 1.184 | 0.006 | 1.200 | 1.192 | 0.002 | 1.200 | 1.194 | 0.000 |
| 1.6 | Max | 1.660 | 1.540 | 0.180 | 1.625 | 1.580 | 0.070 | 1.614 | 1.594 | 0.030 | 1.610 | 1.598 | 0.018 | 1.610 | 1.600 | 0.016 |
| | Min | 1.600 | 1.480 | 0.060 | 1.600 | 1.555 | 0.020 | 1.600 | 1.584 | 0.006 | 1.600 | 1.592 | 0.002 | 1.600 | 1.594 | 0.000 |
| 2 | Max | 2.060 | 1.940 | 0.180 | 2.025 | 1.980 | 0.070 | 2.014 | 1.994 | 0.030 | 2.010 | 1.998 | 0.018 | 2.010 | 2.000 | 0.016 |
| | Min | 2.000 | 1.880 | .060 | 2.000 | 1.955 | 0.020 | 2.000 | 1.984 | 0.006 | 2.000 | 1.992 | 0.002 | 2.000 | 1.994 | 0.000 |
| 2.5 | Max | 2.560 | 2.440 | 0.180 | 2.525 | 2.480 | 0.070 | 2.514 | 2.494 | 0.030 | 2.510 | 2.498 | 0.018 | 2.510 | 2.500 | 0.016 |
| | Min | 2.500 | 2.380 | 0.060 | 2.500 | 2.455 | 0.020 | 2.500 | 2.484 | 0.006 | 2.500 | 2.492 | 0.002 | 2.500 | 2.494 | 0.000 |
| 3 | Max | 3.060 | 2.940 | 0.180 | 3.025 | 2.980 | 0.070 | 3.014 | 2.994 | 0.030 | 3.010 | 2.998 | 0.018 | 3.010 | 3.000 | 0.016 |
| | Min | 3.000 | 2.880 | 0.060 | 3.000 | 2.955 | 0.020 | 3.000 | 2.984 | 0.006 | 3.000 | 2.992 | 0.002 | 3.000 | 2.994 | 0.000 |
| 4 | Max | 4.075 | 3.930 | 0.220 | 4.030 | 3.970 | 0.090 | 4.018 | 3.990 | 0.040 | 4.012 | 3.996 | 0.024 | 4.012 | 4.000 | 0.020 |
| | Min | 4.000 | 3.855 | 0.070 | 4.000 | 3.940 | 0.030 | 4.000 | 3.978 | 0.010 | 4.000 | 3.988 | 0.004 | 4.000 | 3.992 | 0.000 |
| 5 | Max | 5.075 | 4.930 | 0.220 | 5.030 | 4.970 | 0.090 | 5.018 | 4.990 | 0.040 | 5.012 | 4.996 | 0.024 | 5.012 | 5.000 | 0.020 |
| | Min | 5.000 | 4.855 | 0.070 | 5.000 | 4.940 | 0.030 | 5.000 | 4.978 | 0.010 | 5.000 | 4.988 | 0.004 | 5.000 | 4.992 | 0.000 |
| 6 | Max | 6.075 | 5.930 | 0.220 | 6.030 | 5.970 | 0.090 | 6.018 | 5.990 | 0.040 | 6.012 | 5.996 | 0.024 | 6.012 | 6.000 | 0.020 |
| | Min | 6.000 | 5.855 | 0.070 | 6.000 | 5.940 | 0.030 | 6.000 | 5.978 | 0.010 | 6.000 | 5.988 | 0.004 | 6.000 | 5.992 | 0.000 |
| 8 | Max | 8.090 | 7.920 | 0.260 | 8.036 | 7.960 | 0.112 | 8.022 | 7.987 | 0.050 | 8.015 | 7.995 | 0.029 | 8.015 | 8.000 | 0.024 |
| | Min | 8.000 | 7.830 | 0.080 | 8.000 | 7.924 | 0.040 | 8.000 | 7.972 | 0.013 | 8.000 | 7.986 | 0.005 | 8.000 | 7.991 | 0.000 |
| 10 | Max | 10.090 | 9.920 | 0.260 | 10.036 | 9.960 | 0.112 | 10.022 | 9.987 | 0.050 | 10.015 | 9.995 | 0.029 | 10.015 | 10.000 | 0.024 |
| | Min | 10.000 | 9.830 | 0.080 | 10.000 | 9.924 | 0.040 | 10.000 | 9.972 | 0.013 | 10.000 | 9.986 | 0.005 | 10.000 | 9.991 | 0.000 |
| 12 | Max | 12.110 | 11.905 | 0.315 | 12.043 | 11.956 | 0.136 | 12.027 | 11.984 | 0.061 | 12.018 | 11.994 | 0.035 | 12.018 | 12.000 | 0.029 |
| | Min | 12.000 | 11.795 | 0.095 | 12.000 | 11.907 | 0.050 | 12.000 | 11.966 | 0.016 | 12.000 | 11.983 | 0.006 | 12.000 | 11.989 | 0.000 |
| 16 | Max | 16.110 | 15.905 | 0.315 | 16.043 | 15.950 | 0.136 | 16.027 | 15.984 | 0.061 | 16.018 | 15.994 | 0.035 | 16.018 | 16.000 | 0.029 |
| | Min | 16.000 | 15.795 | 0.095 | 16.000 | 15.907 | 0.050 | 16.000 | 15.966 | 0.016 | 16.000 | 15.983 | 0.006 | 16.000 | 15.989 | 0.000 |
| 20 | Max | 20.130 | 19.890 | 0.370 | 20.052 | 19.935 | 0.169 | 20.033 | 19.980 | 0.074 | 20.021 | 19.993 | 0.041 | 20.021 | 20.000 | 0.034 |
| | Min | 20.000 | 19.760 | 0.110 | 20.000 | 19.883 | 0.065 | 20.000 | 19.959 | 0.020 | 20.000 | 19.980 | 0.007 | 20.000 | 19.987 | 0.000 |

All dimensions are in millimeters.

*The sizes shown are first choice basic sizes (see Table 1). Preferred fits for other sizes can be calculated from data given in ANSI B4.2-1978.

†All fits shown in this table have clearance.

Table 2 (*Concluded*). American National Standard Preferred Hole Basis Metric Clearance Fits (ANSI B4.2-1978)

| Basic Size* | | Loose Running | | | Free Running | | | Close Running | | | Sliding | | | Locational Clearance | | |
|---|---|---|---|---|---|---|---|---|---|---|---|---|---|---|---|---|
| | | Hole H11 | Shaft c11 | Fit† | Hole H9 | Shaft d9 | Fit† | Hole H8 | Shaft f7 | Fit† | Hole H7 | Shaft g6 | Fit† | Hole H7 | Shaft h6 | Fit† |
| 25 | Max | 25.130 | 24.890 | 0.370 | 25.052 | 24.935 | 0.169 | 25.033 | 24.980 | 0.074 | 25.021 | 24.993 | 0.041 | 25.021 | 25.000 | 0.034 |
| | Min | 25.000 | 24.760 | 0.110 | 25.000 | 24.883 | 0.065 | 25.000 | 24.959 | 0.020 | 25.000 | 24.980 | 0.007 | 25.000 | 24.987 | 0.000 |
| 30 | Max | 30.130 | 29.890 | 0.370 | 30.052 | 29.935 | 0.169 | 30.033 | 29.980 | 0.074 | 30.021 | 29.993 | 0.041 | 30.021 | 30.000 | 0.034 |
| | Min | 30.000 | 29.760 | 0.110 | 30.000 | 29.883 | 0.065 | 30.000 | 29.959 | 0.020 | 30.000 | 29.980 | 0.007 | 30.000 | 29.987 | 0.000 |
| 40 | Max | 40.160 | 39.880 | 0.440 | 40.062 | 39.920 | 0.204 | 40.039 | 39.975 | 0.089 | 40.025 | 39.991 | 0.050 | 40.025 | 40.000 | 0.041 |
| | Min | 40.000 | 39.720 | 0.120 | 40.000 | 39.858 | 0.080 | 40.000 | 39.950 | 0.025 | 40.000 | 39.975 | 0.009 | 40.000 | 39.984 | 0.000 |
| 50 | Max | 50.160 | 49.870 | 0.450 | 50.062 | 49.920 | 0.204 | 50.039 | 49.975 | 0.089 | 50.025 | 49.991 | 0.050 | 50.025 | 50.000 | 0.041 |
| | Min | 50.000 | 49.710 | 0.130 | 50.000 | 49.858 | 0.080 | 50.000 | 49.950 | 0.025 | 50.000 | 49.975 | 0.009 | 50.000 | 49.984 | 0.000 |
| 60 | Max | 60.190 | 59.860 | 0.520 | 60.074 | 59.900 | 0.248 | 60.046 | 59.970 | 0.106 | 60.030 | 59.990 | 0.059 | 60.030 | 60.000 | 0.049 |
| | Min | 60.000 | 59.670 | 0.140 | 60.000 | 59.826 | 0.100 | 60.000 | 59.940 | 0.030 | 60.000 | 59.971 | 0.010 | 60.000 | 59.981 | 0.000 |
| 80 | Max | 80.190 | 79.850 | 0.530 | 80.074 | 79.900 | 0.248 | 80.046 | 79.970 | 0.106 | 80.030 | 79.990 | 0.059 | 80.030 | 80.000 | 0.049 |
| | Min | 80.000 | 79.660 | 0.150 | 80.000 | 79.826 | 0.100 | 80.000 | 79.940 | 0.030 | 80.000 | 79.971 | 0.010 | 80.000 | 79.981 | 0.000 |
| 100 | Max | 100.220 | 99.830 | 0.610 | 100.087 | 99.880 | 0.294 | 100.054 | 99.964 | 0.125 | 100.035 | 99.988 | 0.069 | 100.035 | 100.000 | 0.057 |
| | Min | 100.000 | 99.610 | 0.170 | 100.000 | 99.793 | 0.120 | 100.000 | 99.929 | 0.036 | 100.000 | 99.966 | 0.012 | 100.000 | 99.978 | 0.000 |
| 120 | Max | 120.220 | 119.820 | 0.620 | 120.087 | 119.880 | 0.294 | 120.054 | 119.964 | 0.125 | 120.035 | 119.988 | 0.069 | 120.035 | 120.000 | 0.057 |
| | Min | 120.000 | 119.600 | 0.180 | 120.000 | 119.793 | 0.120 | 120.000 | 119.929 | 0.036 | 120.000 | 119.966 | 0.012 | 120.000 | 119.978 | 0.000 |
| 160 | Max | 160.250 | 159.790 | 0.710 | 160.100 | 159.855 | 0.345 | 160.063 | 159.957 | 0.146 | 160.040 | 159.986 | 0.079 | 160.040 | 160.000 | 0.065 |
| | Min | 160.000 | 159.540 | 0.210 | 160.000 | 159.755 | 0.145 | 160.000 | 159.917 | 0.043 | 160.000 | 159.961 | 0.014 | 160.000 | 159.975 | 0.000 |
| 200 | Max | 200.290 | 199.760 | 0.820 | 200.115 | 199.830 | 0.400 | 200.072 | 199.950 | 0.168 | 200.046 | 199.985 | 0.090 | 200.046 | 200.000 | 0.075 |
| | Min | 200.000 | 199.470 | 0.240 | 200.000 | 199.715 | 0.170 | 200.000 | 199.904 | 0.050 | 200.000 | 199.956 | 0.015 | 200.000 | 199.971 | 0.000 |
| 250 | Max | 250.290 | 249.720 | 0.860 | 250.115 | 249.830 | 0.400 | 250.072 | 249.950 | 0.168 | 250.046 | 249.985 | 0.090 | 250.046 | 250.000 | 0.075 |
| | Min | 250.000 | 249.430 | 0.280 | 250.000 | 249.715 | 0.170 | 250.000 | 249.904 | 0.050 | 250.000 | 249.956 | 0.015 | 250.000 | 249.971 | 0.000 |
| 300 | Max | 300.320 | 299.670 | 0.970 | 300.130 | 299.810 | 0.450 | 300.081 | 299.944 | 0.189 | 300.052 | 299.983* | 0.101 | 300.052 | 300.000 | 0.084 |
| | Min | 300.000 | 299.350 | 0.330 | 300.000 | 299.680 | 0.190 | 300.000 | 299.892 | 0.056 | 300.000 | 299.951 | 0.017 | 300.000 | 299.968 | 0.000 |
| 400 | Max | 400.360 | 399.600 | 1.120 | 400.140 | 399.790 | 0.490 | 400.089 | 399.938 | 0.208 | 400.057 | 399.982 | 0.111 | 400.057 | 400.000 | 0.093 |
| | Min | 400.000 | 399.240 | 0.400 | 400.000 | 399.650 | 0.210 | 400.000 | 399.881 | 0.062 | 400.000 | 399.946 | 0.018 | 400.000 | 399.964 | 0.000 |
| 500 | Max | 500.400 | 499.520 | 1.280 | 500.155 | 499.770 | 0.540 | 500.097 | 499.932 | 0.228 | 500.063 | 499.980 | 0.123 | 500.063 | 500.000 | 0.103 |
| | Min | 500.000 | 499.120 | 0.480 | 500.000 | 499.615 | 0.230 | 500.000 | 499.869 | 0.068 | 500.000 | 499.940 | 0.020 | 500.000 | 499.960 | 0.000 |

All dimensions are in millimeters.
*The sizes shown are first choice basic sizes (see Table 1). Preferred fits for other sizes can be calculated from data given in ANSI B4.2-1978.
†All fits shown in this table have clearance.

Table 3. American National Standard Preferred Hole Basis Metric Transition and Interference Fits (ANSI B4.2-1978)

| Basic Size* | | Locational Transition | | | Locational Transition | | | Locational Interference | | | Medium Drive | | | Force | | |
|---|---|---|---|---|---|---|---|---|---|---|---|---|---|---|---|---|
| | | Hole H7 | Shaft k6 | Fit† | Hole H7 | Shaft n6 | Fit† | Hole H7 | Shaft p6 | Fit† | Hole H7 | Shaft s6 | Fit† | Hole H7 | Shaft u6 | Fit† |
| 1 | Max | 1.010 | 1.006 | +0.010 | 1.010 | 1.010 | +0.006 | 1.010 | 1.012 | +0.004 | 1.010 | 1.020 | -0.004 | 1.010 | 1.024 | -0.008 |
| | Min | 1.000 | 1.000 | -0.006 | 1.000 | 1.004 | -0.010 | 1.000 | 1.006 | -0.012 | 1.000 | 1.014 | -0.020 | 1.000 | 1.018 | -0.024 |
| 1.2 | Max | 1.210 | 1.206 | +0.010 | 1.210 | 1.210 | +0.006 | 1.210 | 1.212 | +0.004 | 1.210 | 1.220 | -0.004 | 1.210 | 1.224 | -0.008 |
| | Min | 1.200 | 1.200 | -0.006 | 1.200 | 1.204 | -0.010 | 1.200 | 1.206 | -0.012 | 1.200 | 1.214 | -0.020 | 1.200 | 1.218 | -0.024 |
| 1.6 | Max | 1.610 | 1.606 | +0.010 | 1.610 | 1.610 | +0.006 | 1.610 | 1.612 | +0.004 | 1.610 | 1.620 | -0.004 | 1.610 | 1.624 | -0.008 |
| | Min | 1.600 | 1.600 | -0.006 | 1.600 | 1.604 | -0.010 | 1.600 | 1.606 | -0.012 | 1.600 | 1.614 | -0.020 | 1.600 | 1.618 | -0.024 |
| 2 | Max | 2.010 | 2.006 | +0.010 | 2.010 | 2.010 | +0.006 | 2.010 | 2.012 | +0.004 | 2.010 | 2.020 | -0.004 | 2.010 | 2.024 | -0.008 |
| | Min | 2.000 | 2.000 | -0.006 | 2.000 | 2.004 | -0.010 | 2.000 | 2.006 | -0.012 | 2.000 | 2.014 | -0.020 | 2.000 | 2.018 | -0.024 |
| 2.5 | Max | 2.510 | 2.506 | +0.010 | 2.510 | 2.510 | +0.006 | 2.510 | 2.512 | +0.004 | 2.510 | 2.520 | -0.004 | 2.510 | 2.524 | -0.008 |
| | Min | 2.500 | 2.500 | -0.006 | 2.500 | 2.504 | -0.010 | 2.500 | 2.506 | -0.012 | 2.500 | 2.514 | -0.020 | 2.500 | 2.518 | -0.024 |
| 3 | Max | 3.010 | 3.006 | +0.010 | 3.010 | 3.010 | +0.006 | 3.010 | 3.012 | +0.004 | 3.010 | 3.020 | -0.004 | 3.010 | 3.024 | -0.008 |
| | Min | 3.000 | 3.000 | -0.006 | 3.000 | 3.004 | -0.010 | 3.000 | 3.006 | -0.012 | 3.000 | 3.014 | -0.020 | 3.000 | 3.018 | -0.024 |
| 4 | Max | 4.012 | 4.009 | +0.011 | 4.012 | 4.016 | +0.004 | 4.012 | 4.020 | 0.000 | 4.012 | 4.027 | -0.007 | 4.012 | 4.031 | -0.011 |
| | Min | 4.000 | 4.001 | -0.009 | 4.000 | 4.008 | -0.016 | 4.000 | 4.012 | -0.020 | 4.000 | 4.019 | -0.027 | 4.000 | 4.023 | -0.031 |
| 5 | Max | 5.012 | 5.009 | +0.011 | 5.012 | 5.016 | +0.004 | 5.012 | 5.020 | 0.000 | 5.012 | 5.027 | -0.007 | 5.012 | 5.031 | -0.011 |
| | Min | 5.000 | 5.001 | -0.009 | 5.000 | 5.008 | -0.016 | 5.000 | 5.012 | -0.020 | 5.000 | 5.019 | -0.027 | 5.000 | 5.023 | -0.031 |
| 6 | Max | 6.012 | 6.009 | +0.011 | 6.012 | 6.016 | +0.004 | 6.012 | 6.020 | 0.000 | 6.012 | 6.027 | -0.007 | 6.012 | 6.031 | -0.011 |
| | Min | 6.000 | 6.001 | -0.009 | 6.000 | 6.008 | -0.016 | 6.000 | 6.012 | -0.020 | 6.000 | 6.019 | -0.027 | 6.000 | 6.023 | -0.031 |
| 8 | Max | 8.015 | 8.010 | +0.014 | 8.015 | 8.019 | +0.005 | 8.015 | 8.024 | 0.000 | 8.015 | 8.032 | -0.008 | 8.015 | 8.037 | -0.013 |
| | Min | 8.000 | 8.001 | -0.010 | 8.000 | 8.010 | -0.019 | 8.000 | 8.015 | -0.024 | 8.000 | 8.023 | -0.032 | 8.000 | 8.028 | -0.037 |
| 10 | Max | 10.015 | 10.010 | +0.014 | 10.015 | 10.019 | +0.005 | 10.015 | 10.024 | 0.000 | 10.015 | 10.032 | -0.008 | 10.015 | 10.037 | -0.013 |
| | Min | 10.000 | 10.001 | -0.010 | 10.000 | 10.010 | -0.019 | 10.000 | 10.015 | -0.024 | 10.000 | 10.023 | -0.032 | 10.000 | 10.028 | -0.037 |
| 12 | Max | 12.018 | 12.012 | +0.017 | 12.018 | 12.023 | +0.006 | 12.018 | 12.029 | 0.000 | 12.018 | 12.039 | -0.010 | 12.018 | 12.044 | -0.015 |
| | Min | 12.000 | 12.001 | -0.012 | 12.000 | 12.012 | -0.023 | 12.000 | 12.018 | -0.029 | 12.000 | 12.028 | -0.039 | 12.000 | 12.033 | -0.044 |
| 16 | Max | 16.018 | 16.012 | +0.017 | 16.018 | 16.023 | +0.006 | 16.018 | 16.029 | 0.000 | 16.018 | 16.039 | -0.010 | 16.018 | 16.044 | -0.015 |
| | Min | 16.000 | 16.001 | -0.012 | 16.000 | 16.012 | -0.023 | 16.000 | 16.018 | -0.029 | 16.000 | 16.028 | -0.039 | 16.000 | 16.033 | -0.044 |
| 20 | Max | 20.021 | 20.015 | +0.019 | 20.021 | 20.028 | +0.006 | 20.021 | 20.035 | -0.001 | 20.021 | 20.048 | -0.014 | 20.021 | 20.054 | -0.020 |
| | Min | 20.000 | 20.002 | -0.015 | 20.000 | 20.015 | -0.028 | 20.000 | 20.022 | -0.035 | 20.000 | 20.035 | -0.048 | 20.000 | 20.041 | -0.054 |

All dimensions are in millimeters.
*The sizes shown are first choice basic sizes (see Table 1). Preferred fits for other sizes can be calculated from data given in ANSI B4.2-1978.
†A plus sign indicates clearance; a minus sign indicates interference.

Table 3 (Concluded). American National Standard Preferred Hole Basis Metric Transition and Interference Fits (ANSI B4.2-1978)

| Basic Size* | | Locational Transition | | | Locational Transition | | | Locational Interference | | | Medium Drive | | | Force | | |
|---|---|---|---|---|---|---|---|---|---|---|---|---|---|---|---|---|
| | | Hole H7 | Shaft k6 | Fit† | Hole H7 | Shaft n6 | Fit† | Hole H7 | Shaft p6 | Fit† | Hole H7 | Shaft s6 | Fit† | Hole H7 | Shaft u6 | Fit† |
| 25 | Max | 25.021 | 25.015 | +0.019 | 25.021 | 25.028 | +0.006 | 25.021 | 25.035 | -0.001 | 25.021 | 25.048 | -0.014 | 25.021 | 25.061 | -0.027 |
| | Min | 25.000 | 25.002 | -0.015 | 25.000 | 25.015 | -0.028 | 25.000 | 25.022 | -0.035 | 25.000 | 25.035 | -0.048 | 25.000 | 25.048 | -0.061 |
| 30 | Max | 30.021 | 30.015 | +0.019 | 30.021 | 30.028 | +0.006 | 30.021 | 30.035 | -0.001 | 30.021 | 30.048 | -0.014 | 30.021 | 30.061 | -0.027 |
| | Min | 30.000 | 30.002 | -0.015 | 30.000 | 30.015 | -0.028 | 30.000 | 30.022 | -0.035 | 30.000 | 30.035 | -0.048 | 30.000 | 30.048 | -0.061 |
| 40 | Max | 40.025 | 40.018 | +0.023 | 40.025 | 40.033 | +0.008 | 40.025 | 40.042 | -0.001 | 40.025 | 40.059 | -0.018 | 40.025 | 40.076 | -0.035 |
| | Min | 40.000 | 40.002 | -0.018 | 40.000 | 40.017 | -0.033 | 40.000 | 40.026 | -0.042 | 40.000 | 40.043 | -0.059 | 40.000 | 40.060 | -0.076 |
| 50 | Max | 50.025 | 50.018 | +0.023 | 50.025 | 50.033 | +0.008 | 50.025 | 50.042 | -0.001 | 50.025 | 50.059 | -0.018 | 50.025 | 50.086 | -0.045 |
| | Min | 50.000 | 50.002 | -0.018 | 50.000 | 50.017 | -0.033 | 50.000 | 50.026 | -0.042 | 50.000 | 50.043 | -0.059 | 50.000 | 50.070 | -0.086 |
| 60 | Max | 60.030 | 60.021 | +0.028 | 60.030 | 60.039 | +0.010 | 60.030 | 60.051 | -0.002 | 60.030 | 60.072 | -0.023 | 60.030 | 60.106 | -0.057 |
| | Min | 60.000 | 60.002 | -0.021 | 60.000 | 60.020 | -0.039 | 60.000 | 60.032 | -0.051 | 60.000 | 60.053 | -0.072 | 60.000 | 60.087 | -0.106 |
| 80 | Max | 80.030 | 80.021 | +0.028 | 80.030 | 80.039 | +0.010 | 80.030 | 80.051 | -0.002 | 80.030 | 80.078 | -0.029 | 80.030 | 80.121 | -0.072 |
| | Min | 80.000 | 80.002 | -0.021 | 80.000 | 80.020 | -0.039 | 80.000 | 80.032 | -0.051 | 80.000 | 80.059 | -0.078 | 80.000 | 80.102 | -0.121 |
| 100 | Max | 100.035 | 100.025 | +0.032 | 100.035 | 100.045 | +0.012 | 100.035 | 100.059 | -0.002 | 100.035 | 100.093 | -0.036 | 100.035 | 100.146 | -0.089 |
| | Min | 100.000 | 100.003 | -0.025 | 100.000 | 100.023 | -0.045 | 100.000 | 100.037 | -0.059 | 100.000 | 100.071 | -0.093 | 100.000 | 100.124 | -0.146 |
| 120 | Max | 120.035 | 120.025 | +0.032 | 120.035 | 120.045 | +0.012 | 120.035 | 120.059 | -0.002 | 120.035 | 120.101 | -0.044 | 120.035 | 120.166 | -0.109 |
| | Min | 120.000 | 120.003 | -0.025 | 120.000 | 120.023 | -0.045 | 120.000 | 120.037 | -0.059 | 120.000 | 120.079 | -0.101 | 120.000 | 120.144 | -0.166 |
| 160 | Max | 160.040 | 160.028 | +0.037 | 160.040 | 160.052 | +0.013 | 160.040 | 160.068 | -0.003 | 160.040 | 160.125 | -0.060 | 160.040 | 160.215 | -0.150 |
| | Min | 160.000 | 160.003 | -0.028 | 160.000 | 160.027 | -0.052 | 160.000 | 160.043 | -0.068 | 160.000 | 160.100 | -0.125 | 160.000 | 160.190 | -0.215 |
| 200 | Max | 200.046 | 200.033 | +0.042 | 200.046 | 200.060 | +0.015 | 200.046 | 200.079 | -0.004 | 200.046 | 200.151 | -0.076 | 200.046 | 200.265 | -0.190 |
| | Min | 200.000 | 200.004 | -0.033 | 200.000 | 200.031 | -0.060 | 200.000 | 200.050 | -0.079 | 200.000 | 200.122 | -0.151 | 200.000 | 200.236 | -0.265 |
| 250 | Max | 250.046 | 250.033 | +0.042 | 250.046 | 250.060 | +0.015 | 250.046 | 250.079 | -0.004 | 250.046 | 250.169 | -0.094 | 250.046 | 250.313 | -0.238 |
| | Min | 250.000 | 250.004 | -0.033 | 250.000 | 250.031 | -0.060 | 250.000 | 250.050 | -0.079 | 250.000 | 250.140 | -0.169 | 250.000 | 250.284 | -0.313 |
| 300 | Max | 300.052 | 300.036 | +0.048 | 300.052 | 300.066 | +0.018 | 300.052 | 300.088 | -0.004 | 300.052 | 300.202 | -0.118 | 300.052 | 300.382 | -0.298 |
| | Min | 300.000 | 300.004 | -0.036 | 300.000 | 300.034 | -0.066 | 300.000 | 300.056 | -0.088 | 300.000 | 300.170 | -0.202 | 300.000 | 300.350 | -0.382 |
| 400 | Max | 400.057 | 400.040 | +0.053 | 400.057 | 400.073 | +0.020 | 400.057 | 400.098 | -0.005 | 400.057 | 400.244 | -0.151 | 400.057 | 400.471 | -0.378 |
| | Min | 400.000 | 400.004 | -0.040 | 400.000 | 400.037 | -0.073 | 400.000 | 400.062 | -0.098 | 400.000 | 400.208 | -0.244 | 400.000 | 400.435 | -0.471 |
| 500 | Max | 500.063 | 500.045 | +0.058 | 500.063 | 500.080 | +0.023 | 500.063 | 500.108 | -0.005 | 500.063 | 500.292 | -0.189 | 500.063 | 500.580 | -0.477 |
| | Min | 500.000 | 500.005 | -0.045 | 500.000 | 500.040 | -0.080 | 500.000 | 500.068 | -0.108 | 500.000 | 500.252 | -0.292 | 500.000 | 500.540 | -0.580 |

All dimensions are in millimeters.
*The sizes shown are first choice basic sizes (see Table 1). Preferred fits for other sizes can be calculated from data given in ANSI B4.2-1978.
†A plus sign indicates clearance; a minus sign indicates interference.

Table 4. American National Standard Preferred Shaft Basis Metric Clearance Fits (ANSI B4.2-1978)

| Basic Size* | | Loose Running | | | Free Running | | | Close Running | | | Sliding | | | Locational Clearance | | |
|---|---|---|---|---|---|---|---|---|---|---|---|---|---|---|---|---|
| | | Hole $C_{11}$ | Shaft $h_{11}$ | Fit† | Hole $D_9$ | Shaft $h_9$ | Fit† | Hole $F_8$ | Shaft $h_7$ | Fit† | Hole $G_7$ | Shaft $h_6$ | Fit† | Hole $H_7$ | Shaft $h_6$ | Fit† |
| 1 | Max | 1.120 | 1.000 | 0.180 | 1.045 | 1.000 | 0.070 | 1.020 | 1.000 | 0.030 | 1.012 | 1.000 | 0.018 | 1.010 | 1.000 | 0.016 |
| | Min | 1.060 | 0.940 | 0.060 | 1.020 | 0.975 | 0.020 | 1.006 | 0.990 | 0.006 | 1.002 | 0.994 | 0.002 | 1.000 | 0.994 | 0.000 |
| 1.2 | Max | 1.320 | 1.200 | 0.180 | 1.245 | 1.200 | 0.070 | 1.220 | 1.200 | 0.030 | 1.212 | 1.200 | 0.018 | 1.210 | 1.200 | 0.016 |
| | Min | 1.260 | 1.140 | 0.060 | 1.220 | 1.175 | 0.020 | 1.206 | 1.190 | 0.006 | 1.202 | 1.194 | 0.002 | 1.200 | 1.194 | 0.000 |
| 1.6 | Max | 1.720 | 1.600 | 0.180 | 1.645 | 1.600 | 0.070 | 1.620 | 1.600 | 0.030 | 1.612 | 1.600 | 0.018 | 1.610 | 1.600 | 0.016 |
| | Min | 1.660 | 1.540 | 0.060 | 1.620 | 1.575 | 0.020 | 1.606 | 1.590 | 0.006 | 1.602 | 1.594 | 0.002 | 1.600 | 1.594 | 0.000 |
| 2 | Max | 2.120 | 2.000 | 0.180 | 2.045 | 2.000 | 0.070 | 2.020 | 2.000 | 0.030 | 2.012 | 2.000 | 0.018 | 2.010 | 2.000 | 0.016 |
| | Min | 2.060 | 1.940 | 0.060 | 2.020 | 1.975 | 0.020 | 2.006 | 1.990 | 0.006 | 2.002 | 1.994 | 0.002 | 2.000 | 1.994 | 0.000 |
| 2.5 | Max | 2.620 | 2.500 | 0.180 | 2.545 | 2.500 | 0.070 | 2.520 | 2.500 | 0.030 | 2.512 | 2.500 | 0.018 | 2.510 | 2.500 | 0.016 |
| | Min | 2.560 | 2.440 | 0.060 | 2.520 | 2.475 | 0.020 | 2.506 | 2.490 | 0.006 | 2.502 | 2.494 | 0.002 | 2.500 | 2.494 | 0.000 |
| 3 | Max | 3.120 | 3.000 | 0.180 | 3.045 | 3.000 | 0.070 | 3.020 | 3.000 | 0.030 | 3.012 | 3.000 | 0.018 | 3.010 | 3.000 | 0.016 |
| | Min | 3.060 | 2.940 | 0.060 | 3.020 | 2.975 | 0.020 | 3.006 | 2.990 | 0.006 | 3.002 | 2.994 | 0.002 | 3.000 | 2.994 | 0.000 |
| 4 | Max | 4.145 | 4.000 | 0.220 | 4.060 | 4.000 | 0.090 | 4.028 | 4.000 | 0.040 | 4.016 | 4.000 | 0.024 | 4.012 | 4.000 | 0.020 |
| | Min | 4.070 | 3.925 | 0.070 | 4.030 | 3.970 | 0.030 | 4.010 | 3.988 | 0.010 | 4.004 | 3.992 | 0.004 | 4.000 | 3.992 | 0.000 |
| 5 | Max | 5.145 | 5.000 | 0.220 | 5.060 | 5.000 | 0.090 | 5.028 | 5.000 | 0.040 | 5.016 | 5.000 | 0.024 | 5.012 | 5.000 | 0.020 |
| | Min | 5.070 | 4.925 | 0.070 | 5.030 | 4.970 | 0.030 | 5.010 | 4.988 | 0.010 | 5.004 | 4.992 | 0.004 | 5.000 | 4.992 | 0.000 |
| 6 | Max | 6.145 | 6.000 | 0.220 | 6.060 | 6.000 | 0.090 | 6.028 | 6.000 | 0.040 | 6.016 | 6.000 | 0.024 | 6.012 | 6.000 | 0.020 |
| | Min | 6.070 | 5.925 | 0.070 | 6.030 | 5.970 | 0.030 | 6.010 | 5.988 | 0.010 | 6.004 | 5.992 | 0.004 | 6.000 | 5.992 | 0.000 |
| 8 | Max | 8.170 | 8.000 | 0.260 | 8.076 | 8.000 | 0.112 | 8.035 | 8.000 | 0.050 | 8.020 | 8.000 | 0.029 | 8.015 | 8.000 | 0.024 |
| | Min | 8.080 | 7.910 | 0.080 | 8.040 | 7.964 | 0.040 | 8.013 | 7.985 | 0.013 | 8.005 | 7.991 | 0.005 | 8.000 | 7.991 | 0.000 |
| 10 | Max | 10.170 | 10.000 | 0.260 | 10.076 | 10.000 | 0.112 | 10.035 | 10.000 | 0.050 | 10.020 | 10.000 | 0.029 | 10.015 | 10.000 | 0.024 |
| | Min | 10.080 | 9.910 | 0.080 | 10.040 | 9.964 | 0.040 | 10.013 | 9.985 | 0.013 | 10.005 | 9.991 | 0.005 | 10.000 | 9.991 | 0.000 |
| 12 | Max | 12.205 | 12.000 | 0.315 | 12.093 | 12.000 | 0.136 | 12.043 | 12.000 | 0.061 | 12.024 | 12.000 | 0.035 | 12.018 | 12.000 | 0.029 |
| | Min | 12.095 | 11.890 | 0.095 | 12.050 | 11.957 | 0.050 | 12.016 | 11.982 | 0.016 | 12.006 | 11.989 | 0.006 | 12.000 | 11.989 | 0.000 |
| 16 | Max | 16.205 | 16.000 | 0.315 | 16.093 | 16.000 | 0.136 | 16.043 | 16.000 | 0.061 | 16.024 | 16.000 | 0.035 | 16.018 | 16.000 | 0.029 |
| | Min | 16.095 | 15.890 | 0.095 | 16.050 | 15.957 | 0.050 | 16.016 | 15.982 | 0.016 | 16.006 | 15.989 | 0.006 | 16.000 | 15.989 | 0.000 |
| 20 | Max | 20.240 | 20.000 | 0.370 | 20.117 | 20.000 | 0.169 | 20.053 | 20.000 | 0.074 | 20.028 | 20.000 | 0.041 | 20.021 | 20.000 | 0.034 |
| | Min | 20.110 | 19.870 | 0.110 | 20.065 | 19.948 | 0.065 | 20.020 | 19.979 | 0.020 | 20.007 | 19.987 | 0.007 | 20.000 | 19.987 | 0.000 |

All dimensions are in millimeters.

*The sizes shown are first choice basic sizes (see Table 1). Preferred fits for other sizes can be calculated from data given in ANSI B4.2-1978.

†All fits shown in this table have clearance.

**Table 4 (Concluded). American National Standard Preferred Shaft Basis Metric Clearance Fits** (ANSI B4.2-1978)

| Basic Size* | | Loose Running | | | Free Running | | | Close Running | | | Sliding | | | Locational Clearance | | |
|---|---|---|---|---|---|---|---|---|---|---|---|---|---|---|---|---|
| | | Hole C11 | Shaft h11 | Fit† | Hole D9 | Shaft h9 | Fit† | Hole F8 | Shaft h7 | Fit† | Hole G7 | Shaft h6 | Fit† | Hole H7 | Shaft h6 | Fit† |
| 25 | Max | 25.240 | 25.000 | 0.370 | 25.117 | 25.000 | 0.169 | 25.053 | 25.000 | 0.074 | 25.028 | 25.000 | 0.041 | 25.021 | 25.000 | 0.034 |
|  | Min | 25.110 | 24.870 | 0.110 | 25.065 | 24.948 | 0.065 | 25.020 | 24.979 | 0.020 | 25.007 | 24.987 | 0.007 | 25.000 | 24.987 | 0.000 |
| 30 | Max | 30.240 | 30.000 | 0.370 | 30.117 | 30.000 | 0.169 | 30.053 | 30.000 | 0.074 | 30.028 | 30.000 | 0.041 | 30.021 | 30.000 | 0.034 |
|  | Min | 30.110 | 29.870 | 0.110 | 30.065 | 29.948 | 0.065 | 30.020 | 29.979 | 0.020 | 30.007 | 29.987 | 0.007 | 30.000 | 29.987 | 0.000 |
| 40 | Max | 40.280 | 40.000 | 0.440 | 40.142 | 40.000 | 0.204 | 40.064 | 40.000 | 0.089 | 40.034 | 40.000 | 0.050 | 40.025 | 40.000 | 0.041 |
|  | Min | 40.120 | 39.840 | 0.120 | 40.080 | 39.938 | 0.080 | 40.025 | 39.975 | 0.025 | 40.009 | 39.984 | 0.009 | 40.000 | 39.984 | 0.000 |
| 50 | Max | 50.290 | 50.000 | 0.450 | 50.142 | 50.000 | 0.204 | 50.064 | 50.000 | 0.089 | 50.034 | 50.000 | 0.050 | 50.025 | 50.000 | 0.041 |
|  | Min | 50.130 | 49.840 | 0.130 | 50.080 | 49.938 | 0.080 | 50.025 | 49.975 | 0.025 | 50.009 | 49.984 | 0.009 | 50.000 | 49.984 | 0.000 |
| 60 | Max | 60.330 | 60.000 | 0.520 | 60.174 | 60.000 | 0.248 | 60.076 | 60.000 | 0.106 | 60.040 | 60.000 | 0.059 | 60.030 | 60.000 | 0.049 |
|  | Min | 60.140 | 59.810 | 0.140 | 60.100 | 59.926 | 0.100 | 60.030 | 59.970 | 0.030 | 60.010 | 59.981 | 0.010 | 60.000 | 59.981 | 0.000 |
| 80 | Max | 80.340 | 80.000 | 0.530 | 80.174 | 80.000 | 0.248 | 80.076 | 80.000 | 0.106 | 80.040 | 80.000 | 0.059 | 80.030 | 80.000 | 0.049 |
|  | Min | 80.150 | 79.810 | 0.150 | 80.100 | 79.926 | 0.100 | 80.030 | 79.970 | 0.030 | 80.010 | 79.981 | 0.010 | 80.000 | 79.981 | 0.000 |
| 100 | Max | 100.390 | 100.000 | 0.610 | 100.207 | 100.000 | 0.294 | 100.090 | 100.000 | 0.125 | 100.047 | 100.000 | 0.069 | 100.035 | 100.000 | 0.057 |
|  | Min | 100.170 | 99.780 | 0.170 | 100.120 | 99.913 | 0.120 | 100.036 | 99.965 | 0.036 | 100.012 | 99.978 | 0.012 | 100.000 | 99.978 | 0.000 |
| 120 | Max | 120.400 | 120.000 | 0.620 | 120.207 | 120.000 | 0.294 | 120.090 | 120.000 | 0.125 | 120.047 | 120.000 | 0.069 | 120.035 | 120.000 | 0.057 |
|  | Min | 120.180 | 119.780 | 0.180 | 120.120 | 119.913 | 0.120 | 120.036 | 119.965 | 0.036 | 120.012 | 119.978 | 0.012 | 120.000 | 119.978 | 0.000 |
| 160 | Max | 160.460 | 160.000 | 0.710 | 160.245 | 160.000 | 0.345 | 160.106 | 160.000 | 0.146 | 160.054 | 160.000 | 0.079 | 160.040 | 160.000 | 0.065 |
|  | Min | 160.210 | 159.750 | 0.210 | 160.145 | 159.900 | 0.145 | 160.043 | 159.960 | 0.043 | 160.014 | 159.975 | 0.014 | 160.000 | 159.975 | 0.000 |
| 200 | Max | 200.530 | 200.000 | 0.820 | 200.285 | 200.000 | 0.400 | 200.122 | 200.000 | 0.168 | 200.061 | 200.000 | 0.090 | 200.046 | 200.000 | 0.075 |
|  | Min | 200.240 | 199.710 | 0.240 | 200.170 | 199.885 | 0.170 | 200.050 | 199.954 | 0.050 | 200.015 | 199.971 | 0.015 | 200.000 | 199.971 | 0.000 |
| 250 | Max | 250.570 | 250.000 | 0.860 | 250.285 | 250.000 | 0.400 | 250.122 | 250.000 | 0.168 | 250.061 | 250.000 | 0.090 | 250.046 | 250.000 | 0.075 |
|  | Min | 250.280 | 249.710 | 0.280 | 250.170 | 249.885 | 0.170 | 250.050 | 249.954 | 0.050 | 250.015 | 249.971 | 0.015 | 250.000 | 249.971 | 0.000 |
| 300 | Max | 300.650 | 300.000 | 0.970 | 300.320 | 300.000 | 0.450 | 300.137 | 300.000 | 0.189 | 300.069 | 300.000 | 0.101 | 300.052 | 300.000 | 0.084 |
|  | Min | 300.330 | 299.680 | 0.330 | 300.190 | 299.870 | 0.190 | 300.056 | 299.948 | 0.056 | 300.017 | 299.968 | 0.017 | 300.000 | 299.968 | 0.000 |
| 400 | Max | 400.760 | 400.000 | 1.120 | 400.350 | 400.000 | 0.490 | 400.151 | 400.000 | 0.208 | 400.075 | 400.000 | 0.111 | 400.057 | 400.000 | 0.093 |
|  | Min | 400.400 | 399.640 | 0.400 | 400.210 | 399.860 | 0.210 | 400.062 | 399.943 | 0.062 | 400.018 | 399.964 | 0.018 | 400.000 | 399.964 | 0.000 |
| 500 | Max | 500.880 | 500.000 | 1.280 | 500.385 | 500.000 | 0.540 | 500.165 | 500.000 | 0.228 | 500.083 | 500.000 | 0.123 | 500.063 | 500.000 | 0.103 |
|  | Min | 500.480 | 499.600 | 0.480 | 500.230 | 499.845 | 0.230 | 500.068 | 499.937 | 0.068 | 500.020 | 499.960 | 0.020 | 500.000 | 499.960 | 0.000 |

All dimensions are in millimeters.
*The sizes shown are first choice basic sizes (see Table 1). Preferred fits for other sizes can be calculated from data given in ANSI B4.2-1978.
†All fits shown in this table have clearance.

**Table 5. American National Standard Preferred Shaft Basis Metric Transition and Interference Fits (ANSI B4.2-1978)**

| Basic Size* | | Locational Transition | | | Locational Transition | | | Locational Interference | | | Medium Drive | | | Force | | |
|---|---|---|---|---|---|---|---|---|---|---|---|---|---|---|---|---|
| | | Hole K7 | Shaft h6 | Fit† | Hole N7 | Shaft h6 | Fit† | Hole P7 | Shaft h6 | Fit† | Hole S7 | Shaft h6 | Fit† | Hole U7 | Shaft h6 | Fit† |
| 1 | Max | 1.000 | 1.000 | +0.006 | 0.996 | 1.000 | +0.002 | 0.994 | 1.000 | 0.000 | 0.986 | 1.000 | -0.008 | 0.982 | 1.000 | -0.012 |
| | Min | 0.990 | 0.994 | -0.010 | 0.986 | 0.994 | -0.014 | 0.984 | 0.994 | -0.016 | 0.976 | 0.994 | -0.024 | 0.972 | 0.994 | -0.028 |
| 1.2 | Max | 1.200 | 1.200 | +0.006 | 1.196 | 1.200 | +0.002 | 1.194 | 1.200 | 0.000 | 1.186 | 1.200 | -0.008 | 1.182 | 1.200 | -0.012 |
| | Min | 1.190 | 1.194 | -0.010 | 1.186 | 1.194 | -0.014 | 1.184 | 1.194 | -0.016 | 1.176 | 1.194 | -0.024 | 1.172 | 1.194 | -0.028 |
| 1.6 | Max | 1.600 | 1.600 | +0.006 | 1.596 | 1.600 | +0.002 | 1.594 | 1.600 | 0.000 | 1.586 | 1.600 | -0.008 | 1.582 | 1.600 | -0.012 |
| | Min | 1.590 | 1.594 | -0.010 | 1.586 | 1.594 | -0.014 | 1.584 | 1.594 | -0.016 | 1.576 | 1.594 | -0.024 | 1.572 | 1.594 | -0.028 |
| 2 | Max | 2.000 | 2.000 | +0.006 | 1.996 | 2.000 | +0.002 | 1.994 | 2.000 | 0.000 | 1.986 | 2.000 | -0.008 | 1.982 | 2.000 | -0.012 |
| | Min | 1.990 | 1.994 | -0.010 | 1.986 | 1.994 | -0.014 | 1.984 | 1.994 | -0.016 | 1.976 | 1.994 | -0.024 | 1.972 | 1.994 | -0.028 |
| 2.5 | Max | 2.500 | 2.500 | +0.006 | 2.496 | 2.500 | +0.002 | 2.494 | 2.500 | 0.000 | 2.486 | 2.500 | -0.008 | 2.482 | 2.500 | -0.012 |
| | Min | 2.490 | 2.494 | -0.010 | 2.486 | 2.494 | -0.014 | 2.484 | 2.494 | -0.016 | 2.476 | 2.494 | -0.024 | 2.472 | 2.494 | -0.028 |
| 3 | Max | 3.000 | 3.000 | +0.006 | 2.996 | 3.000 | +0.002 | 2.994 | 3.000 | 0.000 | 2.986 | 3.000 | -0.008 | 2.982 | 3.000 | -0.012 |
| | Min | 2.990 | 2.994 | -0.010 | 2.986 | 2.994 | -0.014 | 2.984 | 2.994 | -0.016 | 2.976 | 2.994 | -0.024 | 2.972 | 2.994 | -0.028 |
| 4 | Max | 4.003 | 4.000 | +0.011 | 3.996 | 4.000 | +0.004 | 3.992 | 4.000 | 0.000 | 3.985 | 4.000 | -0.007 | 3.981 | 4.000 | -0.011 |
| | Min | 3.991 | 3.992 | -0.009 | 3.984 | 3.992 | -0.016 | 3.980 | 3.992 | -0.020 | 3.973 | 3.992 | -0.027 | 3.969 | 3.992 | -0.031 |
| 5 | Max | 5.003 | 5.000 | +0.011 | 4.996 | 5.000 | +0.004 | 4.992 | 5.000 | 0.000 | 4.985 | 5.000 | -0.007 | 4.981 | 5.000 | -0.011 |
| | Min | 4.991 | 4.992 | -0.009 | 4.984 | 4.992 | -0.016 | 4.980 | 4.992 | -0.020 | 4.973 | 4.992 | -0.027 | 4.969 | 4.992 | -0.031 |
| 6 | Max | 6.003 | 6.000 | +0.011 | 5.996 | 6.000 | +0.004 | 5.992 | 6.000 | 0.000 | 5.985 | 6.000 | -0.007 | 5.981 | 6.000 | -0.011 |
| | Min | 5.991 | 5.992 | -0.009 | 5.984 | 5.992 | -0.016 | 5.980 | 5.992 | -0.020 | 5.973 | 5.992 | -0.027 | 5.969 | 5.992 | -0.031 |
| 8 | Max | 8.005 | 8.000 | +0.014 | 7.996 | 8.000 | +0.005 | 7.991 | 8.000 | 0.000 | 7.983 | 8.000 | -0.008 | 7.978 | 8.000 | -0.013 |
| | Min | 7.990 | 7.991 | -0.010 | 7.981 | 7.991 | -0.019 | 7.976 | 7.991 | -0.024 | 7.968 | 7.991 | -0.032 | 7.963 | 7.991 | -0.037 |
| 10 | Max | 10.005 | 10.000 | +0.014 | 9.996 | 10.000 | +0.005 | 9.991 | 10.000 | 0.000 | 9.983 | 10.000 | -0.008 | 9.978 | 10.000 | -0.013 |
| | Min | 9.990 | 9.991 | -0.010 | 9.981 | 9.991 | -0.019 | 9.976 | 9.991 | -0.024 | 9.968 | 9.991 | -0.032 | 9.963 | 9.991 | -0.037 |
| 12 | Max | 12.006 | 12.000 | +0.017 | 11.995 | 12.000 | +0.006 | 11.989 | 12.000 | 0.000 | 11.979 | 12.000 | -0.010 | 11.974 | 12.000 | -0.015 |
| | Min | 11.988 | 11.989 | -0.012 | 11.977 | 11.989 | -0.023 | 11.971 | 11.989 | -0.029 | 11.961 | 11.989 | -0.039 | 11.956 | 11.989 | -0.044 |
| 16 | Max | 16.006 | 16.000 | +0.017 | 15.995 | 16.000 | +0.006 | 15.989 | 16.000 | 0.000 | 15.979 | 16.000 | -0.010 | 15.974 | 16.000 | -0.015 |
| | Min | 15.988 | 15.989 | -0.012 | 15.977 | 15.989 | -0.023 | 15.971 | 15.989 | -0.029 | 15.961 | 15.989 | -0.039 | 15.956 | 15.989 | -0.044 |
| 20 | Max | 20.006 | 20.000 | +0.019 | 19.993 | 20.000 | +0.006 | 19.986 | 20.000 | -0.001 | 19.973 | 20.000 | -0.014 | 19.967 | 20.000 | -0.020 |
| | Min | 19.985 | 19.987 | -0.015 | 19.972 | 19.987 | -0.028 | 19.965 | 19.987 | -0.035 | 19.952 | 19.987 | -0.048 | 19.946 | 19.987 | -0.054 |

All dimensions are in millimeters.

*The sizes shown are first choice basic sizes (see Table 1). Preferred fits for other sizes can be calculated from data given in ANSI B4.2-1978.

†A plus sign indicates clearance and a minus sign indicates interference.

**Table 5 (Concluded). American National Standard Preferred Shaft Basis Metric Transition and Interference Fits (ANSI B4.2-1978)**

| Basic Size* | | Locational Transition | | | Locational Transition | | | Locational Interference | | | Medium Drive | | | Force | | |
|---|---|---|---|---|---|---|---|---|---|---|---|---|---|---|---|---|
| | | Hole K7 | Shaft h6 | Fit | Hole N7 | Shaft h6 | Fit | Hole P7 | Shaft h6 | Fit | Hole S7 | Shaft h6 | Fit | Hole U7 | Shaft h6 | Fit |
| 25 | Max | 25.006 | 25.000 | +0.019 | 24.993 | 25.000 | +0.006 | 24.986 | 25.000 | -0.001 | 24.973 | 25.000 | -0.014 | 24.960 | 25.000 | -0.027 |
| | Min | 24.985 | 24.987 | -0.015 | 24.972 | 24.987 | -0.028 | 24.965 | 24.987 | -0.035 | 24.952 | 24.987 | -0.048 | 24.939 | 24.987 | -0.061 |
| 30 | Max | 30.006 | 30.000 | +0.019 | 29.993 | 30.000 | +0.006 | 29.986 | 30.000 | -0.001 | 29.973 | 30.000 | -0.014 | 29.960 | 30.000 | -0.027 |
| | Min | 29.985 | 29.987 | -0.015 | 29.972 | 29.987 | -0.028 | 29.965 | 29.987 | -0.035 | 29.952 | 29.987 | -0.048 | 29.939 | 29.987 | -0.061 |
| 40 | Max | 40.007 | 40.000 | +0.023 | 39.992 | 40.000 | +0.008 | 39.983 | 40.000 | -0.001 | 39.966 | 40.000 | -0.018 | 39.949 | 40.000 | -0.035 |
| | Min | 39.982 | 39.984 | -0.018 | 39.967 | 39.984 | -0.033 | 39.958 | 39.984 | -0.042 | 39.941 | 39.984 | -0.059 | 39.924 | 39.984 | -0.076 |
| 50 | Max | 50.007 | 50.000 | +0.023 | 49.992 | 50.000 | +0.008 | 49.983 | 50.000 | -0.001 | 49.966 | 50.000 | -0.018 | 49.939 | 50.000 | -0.045 |
| | Min | 49.982 | 49.984 | -0.018 | 49.967 | 49.984 | -0.033 | 49.958 | 49.984 | -0.042 | 49.941 | 49.984 | -0.059 | 49.914 | 49.984 | -0.086 |
| 60 | Max | 60.009 | 60.000 | +0.028 | 59.991 | 60.000 | +0.010 | 59.979 | 60.000 | -0.002 | 59.958 | 60.000 | -0.023 | 59.924 | 60.000 | -0.057 |
| | Min | 59.979 | 59.981 | -0.021 | 59.961 | 59.981 | -0.039 | 59.949 | 59.981 | -0.051 | 59.928 | 59.981 | -0.072 | 59.894 | 59.981 | -0.106 |
| 80 | Max | 80.009 | 80.000 | +0.028 | 79.991 | 80.000 | +0.010 | 79.979 | 80.000 | -0.002 | 79.952 | 80.000 | -0.029 | 79.909 | 80.000 | -0.072 |
| | Min | 79.979 | 79.981 | -0.021 | 79.961 | 79.981 | -0.039 | 79.949 | 79.981 | -0.051 | 79.922 | 79.981 | -0.078 | 79.879 | 79.981 | -0.121 |
| 100 | Max | 100.010 | 100.000 | +0.032 | 99.990 | 100.000 | +0.012 | 99.976 | 100.000 | -0.002 | 99.942 | 100.000 | -0.036 | 99.889 | 100.000 | -0.089 |
| | Min | 99.975 | 99.978 | -0.025 | 99.955 | 99.978 | -0.045 | 99.941 | 99.978 | -0.059 | 99.907 | 99.978 | -0.093 | 99.854 | 99.978 | -0.146 |
| 120 | Max | 120.010 | 120.000 | +0.032 | 119.990 | 120.000 | +0.012 | 119.976 | 120.000 | -0.002 | 119.934 | 120.000 | -0.044 | 119.869 | 120.000 | -0.109 |
| | Min | 119.975 | 119.978 | -0.025 | 119.955 | 119.978 | -0.045 | 119.941 | 119.978 | -0.059 | 119.899 | 119.978 | -0.101 | 119.834 | 119.978 | -0.166 |
| 160 | Max | 160.012 | 160.000 | +0.037 | 159.988 | 160.000 | +0.013 | 159.972 | 160.000 | -0.003 | 159.915 | 160.000 | -0.060 | 159.825 | 160.000 | -0.150 |
| | Min | 159.972 | 159.975 | -0.028 | 159.948 | 159.975 | -0.052 | 159.932 | 159.975 | -0.068 | 159.875 | 159.975 | -0.125 | 159.785 | 159.975 | -0.215 |
| 200 | Max | 200.013 | 200.000 | +0.042 | 199.986 | 200.000 | +0.015 | 199.967 | 200.000 | -0.004 | 199.895 | 200.000 | -0.076 | 199.781 | 200.000 | -0.190 |
| | Min | 199.967 | 199.971 | -0.033 | 199.940 | 199.971 | -0.060 | 199.921 | 199.971 | -0.079 | 199.849 | 199.971 | -0.151 | 199.735 | 199.971 | -0.265 |
| 250 | Max | 250.013 | 250.000 | +0.042 | 249.986 | 250.000 | +0.015 | 249.967 | 250.000 | -0.004 | 249.877 | 250.000 | -0.094 | 249.733 | 250.000 | -0.238 |
| | Min | 249.967 | 249.971 | -0.033 | 249.940 | 249.971 | -0.060 | 249.921 | 249.971 | -0.079 | 249.831 | 249.971 | -0.169 | 249.687 | 249.971 | -0.313 |
| 300 | Max | 300.016 | 300.000 | +0.048 | 299.986 | 300.000 | +0.018 | 299.964 | 300.000 | -0.004 | 299.850 | 300.000 | -0.118 | 299.670 | 300.000 | -0.298 |
| | Min | 299.964 | 299.968 | -0.036 | 299.934 | 299.968 | -0.066 | 299.912 | 299.968 | -0.088 | 299.798 | 299.968 | -0.202 | 299.618 | 299.968 | -0.382 |
| 400 | Max | 400.017 | 400.000 | +0.053 | 399.984 | 400.000 | +0.020 | 399.959 | 400.000 | -0.005 | 399.813 | 400.000 | -0.151 | 399.586 | 400.000 | -0.378 |
| | Min | 399.960 | 399.964 | -0.040 | 399.927 | 399.964 | -0.073 | 399.902 | 399.964 | -0.098 | 399.756 | 399.964 | -0.244 | 399.529 | 399.964 | -0.471 |
| 500 | Max | 500.018 | 500.000 | +0.058 | 499.983 | 500.000 | +0.023 | 499.955 | 500.000 | -0.005 | 499.771 | 500.000 | -0.189 | 499.483 | 500.000 | -0.477 |
| | Min | 499.955 | 499.960 | -0.045 | 499.920 | 499.960 | -0.080 | 499.892 | 499.960 | -0.108 | 499.708 | 499.960 | -0.292 | 499.420 | 499.960 | -0.580 |

All dimensions are in millimeters.
*The sizes shown are first choice basic sizes (see Table 1). Preferred fits for other sizes can be calculated from data given in ANSI B4.2-1978.
†A plus sign indicates clearance and a minus sign indicates interference.

the preferred fits might be considered necessary to satisfy extreme conditions. Subsequent adjustments might also be desired as the result of experience in a particular application to suit critical functional requirements or to permit optimum manufacturing economy. Selection of a departure from these recommendations will depend upon consideration of the engineering and economic factors that might be involved; however, the benefits to be derived from the use of preferred fits should not be overlooked.

A general guide to machining processes which may normally be expected to produce work within the tolerances indicated by the IT grades given in ANSI B4.2-1978 is shown in the chart in Fig. 5.

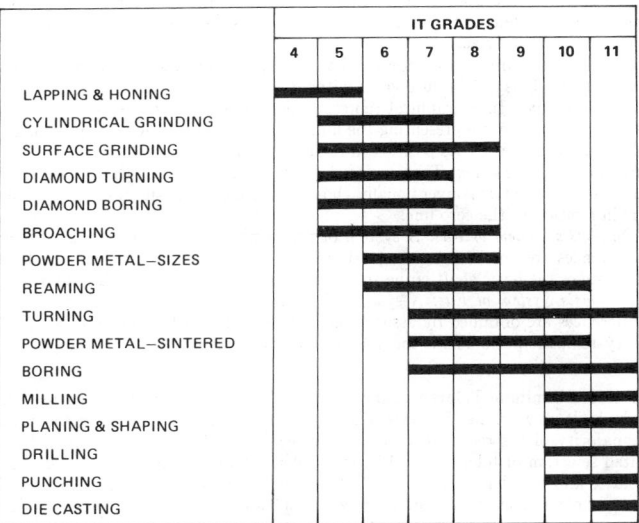

Fig. 5. Relation of machining processes to IT tolerance grades.

**British Standard for Metric ISO Limits and Fits.** — Based on ISO Recommendation R286, this British Standard (BS 4500:1969) is intended to provide a comprehensive range of metric limits and fits for engineering purposes, and meets the requirements of metrication in the United Kingdom. Sizes up to 3,150 mm are covered by the Standard, but the condensed information presented here embraces dimensions up to 500 mm only. The system is based on a series of tolerances graded to suit all classes of work from the finest to the most coarse, and the different types of fits that can be obtained range from coarse clearance to heavy interference. In the Standard, only cylindrical parts, designated holes and shafts are referred to explicitly, but it is emphasized that the recommendations apply equally well to other sections, and the general term *hole* or *shaft* can be taken to mean the space contained by or containing two parallel faces or tangent planes of any part, such as the width of a slot, or the thickness of a key. It is also strongly emphasized that the graded series of tolerances are intended for the most general application, and should be used wherever possible whether the features of the component involved are members of a fit or not.

**Definitions.** — The definitions given in the Standard include the following:

*Limits of Size:* The maximum and minimum sizes permitted for a feature.

*Basic Size:* The size by reference to which the limits of size are fixed. The basic size is the same for both members of a fit.

*Upper Deviation:* The algebraical difference between the maximum limit of size and the corresponding basic size. It is designated as ES for a hole, and as es for a shaft, which stands for the French term *écart supérieur.*

*Lower Deviation:* The algebraical difference between the minimum limit of size and the corresponding basic size. It is designated as EI for a hole, and as ei for a shaft, which stands for the French term *écart inférieur.*

*Zero Line:* In a graphical representation of limits and fits, the straight line to which the deviations are referred. The zero line is the line of zero deviation and represents the basic size.

*Tolerance:* The difference between the maximum limit of size and the minimum limit of size. It is an absolute value without sign.

*Tolerance Zone:* In a graphical representation of tolerances, the zone comprised between the two lines representing the limits of tolerance and defined by its magnitude (tolerance) and by its position in relation to the zero line.

*Fundamental Deviation:* That one of the two deviations, being the one nearest to the zero line, which is conventionally chosen to define the position of the tolerance zone in relation to the zero line.

*Shaft-basis System of Fits:* A system of fits in which the different clearances and interferences are obtained by associating various holes with a single shaft. In the ISO system, the basic shaft is the shaft the upper deviation of which is zero.

*Hole-basis System of Fits:* A system of fits in which the different clearances and interferences are obtained by associating various shafts with a single hole. In the ISO system, the basic hole is the hole the lower deviation of which is zero.

**Selected Limits of Tolerance, and Fits.** — The number of fit combinations that can be built up with the ISO system is very large. However, experience shows that the majority of fits required for usual engineering products can be provided by a limited selection of tolerances. Limits of tolerance for selected holes are shown in Table 1, and for shafts, in Table 2. Selected fits, based on combinations of the selected hole and shaft tolerances, are given in Table 3.

**Tolerances and Fundamental Deviations.** — There are 18 tolerance grades intended to meet the requirements of different classes of work, and they are designated IT 01, IT 02, and IT 1 to IT 16. (IT stands for ISO series of tolerances.) Table 4 shows the standardized numerical values for the 18 tolerance grades, which are known as standard tolerances. The system provides 27 fundamental deviations for sizes up to and including 500 mm, and Tables 5 and 6 contain the values for shafts and holes respectively. Upper case (capital) letters designate hole deviations, and the same letters in lower case designate shaft deviations. The deviation $j_s$ ($J_s$ for holes) is provided to meet the need for symmetrical bilateral tolerances. In this instance there is no fundamental deviation, and the tolerance zone, of whatever magnitude, is equally disposed about the zero line.

**Calculated Limits of Tolerance.** — The deviations and fundamental tolerances provided by the ISO system can be combined in any way that appears necessary to give a required fit. Thus, for example, the deviations H (basic hole) and f (clearance shaft) could be associated, and with each of these deviations any one of the tolerance grades IT 01 to IT 16 could be used. All the limits of tolerance of which the system is capable of providing for sizes up to and including 500 mm can be calculated from the standard tolerances given in Table 4, and the fundamental deviations given in Tables 5 and 6. The range includes limits of tolerance for shafts and holes used in small high-precision work and horology.

The system provides for the use of either hole-basis or shaft-basis fits, and the Standard includes details of procedures for converting from one type of fit to the other.

The limits of tolerance for a shaft or hole are designated by the appropriate letter indicating the fundamental deviation, followed by a suffix number denoting the tolerance grade. This suffix number is the numerical part of the tolerance grade designation. Thus, a hole tolerance with deviation H, and tolerance grade IT7 is designated H7. Likewise, a shaft with deviation p, and tolerance grade IT 6 is designated p6. The limits of size of a component feature are defined by the basic size, say 45 mm, followed by the appropriate tolerance designation, for example 45 H7 or 45 p6. A fit is indicated by combining the basic size common to both features with the designation appropriate to each of them, for example 45 H7-p6 or 45 H7/p6.

When calculating the limits of size for a shaft, the upper deviation es, or the lower deviation ei, is first obtained from Table 5, depending on the particular letter designation, and nominal dimension. If an upper deviation has been determined, the lower deviation ei = es − IT. The IT value is obtained from Table 4 for the particular tolerance grade being applied. If a lower deviation has been obtained from Table 5, the upper deviation es = ei + IT. When the upper deviation ES has been determined for a hole from Table 6, the lower deviation EI = ES − IT. If a lower deviation EI has been obtained from Table 6, then the upper deviation ES = EI + IT.

The upper deviations for holes K, M, and N with tolerance grades up to and including IT8, and for holes P to ZC with tolerance grades up to and including IT7, must be calculated by adding the delta (Δ) values given in Table 6 as indicated.

*Examples of Calculations:* The limits of size for a part of 133 mm basic size with a tolerance designation g9 are derived as follows:

From Table 5, the upper deviation (es) is − 0.014 mm. From Table 4, the tolerance grade (IT9) is 0.100 mm. The lower deviation (ei) = es − IT = −0.114 mm, and the limits of size are thus 132.986 and 132.886 mm.

The limits of size for a part 20 mm in size, with tolerance designation D3, are derived as follows: From Table 6, the lower deviation (EI) is + 0.065 mm. From Table 4, the tolerance grade (IT3) is 0.004 mm. The upper deviation (ES) = EI + IT = 0.069 mm, and thus the limits of size for the part are 20.069 and 20.065 mm.

The limits of size for a part 32 mm in size, with tolerance designation M5, which involves a delta value, are obtained as follows: From Table 6, the upper deviation ES is − 0.009 mm + Δ = −0.005 mm. (The delta value given at the end of this table for this size and grade IT 5, is 0.004 mm.) From Table 4, the tolerance grade (IT5) is 0.011 mm. The lower deviation (EI) = ES − IT = −0.016 mm, and thus the limits of size for the part are 31.995 and 31.984 mm.

Where the designations h and H or $j_s$ and $J_s$ are used, it is only necessary to refer to Table 4. For h and H, the fundamental deviation is always zero, and the disposition of the tolerance is always negative(−) for a shaft, and positive (+) for a hole. Thus, the limits for a part 40 mm in size, designated h8 are derived as follows: From Table 4, the tolerance grade (IT 8) is 0.039 mm, and the limits are therefore 40.000 and 39.961 mm.

The limits for a part 60 mm in size, designated $j_s7$ or $J_s7$ are derived as follows: From Table 1, the tolerance grade (IT 7) is 0.030 mm, and this value is divided equally about the basic size to give limits of 60.015 and 59.985 mm.

Table 1. British Standard Limits of Tolerance for Selected Holes
(Upper and Lower Deviations) (BS 4500:1969)

| Nominal Sizes, mm | | H7 | | H8 | | H9 | | H11 | |
|---|---|---|---|---|---|---|---|---|---|
| Over | Up to and Including | ES + | EI | ES + | EI | ES + | EI | ES + | EI |
| ... | 3 | 10 | 0 | 14 | 0 | 25 | 0 | 60 | 0 |
| 3 | 6 | 12 | 0 | 18 | 0 | 30 | 0 | 75 | 0 |
| 6 | 10 | 15 | 0 | 22 | 0 | 36 | 0 | 90 | 0 |
| 10 | 18 | 18 | 0 | 27 | 0 | 43 | 0 | 110 | 0 |
| 18 | 30 | 21 | 0 | 33 | 0 | 52 | 0 | 130 | 0 |
| 30 | 50 | 25 | 0 | 39 | 0 | 62 | 0 | 160 | 0 |
| 50 | 80 | 30 | 0 | 46 | 0 | 74 | 0 | 190 | 0 |
| 80 | 120 | 35 | 0 | 54 | 0 | 87 | 0 | 220 | 0 |
| 120 | 180 | 40 | 0 | 63 | 0 | 100 | 0 | 250 | 0 |
| 180 | 250 | 46 | 0 | 72 | 0 | 115 | 0 | 290 | 0 |
| 250 | 315 | 52 | 0 | 81 | 0 | 130 | 0 | 320 | 0 |
| 315 | 400 | 57 | 0 | 89 | 0 | 140 | 0 | 360 | 0 |
| 400 | 500 | 63 | 0 | 97 | 0 | 155 | 0 | 400 | 0 |

ES = Upper deviation.    EI = Lower deviation
The dimensions are given in 0.001 mm, except for the nominal sizes, which are in millimeters.

Table 2. British Standard Limits of Tolerance for Selected Shafts
(Upper and Lower Deviations) (BS 4500:1969)

| Nominal Sizes, mm | | c11 | | d10 | | e9 | | f7 | | g6 | | h6 | | k6 | | n6 | | p6 | | s6 | |
|---|---|---|---|---|---|---|---|---|---|---|---|---|---|---|---|---|---|---|---|---|---|
| Over | Up to and Incl. | es − | ei − | es − | ei − | es − | ei − | es − | ei − | es − | ei − | es − | ei − | es + | ei + | es + | ei + | es + | ei + | es + | ei + |
| ... | 3 | 60 | 120 | 20 | 60 | 14 | 39 | 6 | 16 | 2 | 8 | 0 | 6 | 6 | 0 | 10 | 4 | 12 | 6 | 20 | 14 |
| 3 | 6 | 70 | 145 | 30 | 78 | 20 | 50 | 10 | 22 | 4 | 12 | 0 | 8 | 9 | 1 | 16 | 8 | 20 | 12 | 27 | 19 |
| 6 | 10 | 80 | 170 | 40 | 98 | 25 | 61 | 13 | 28 | 5 | 14 | 0 | 9 | 10 | 1 | 19 | 10 | 24 | 15 | 32 | 23 |
| 10 | 18 | 95 | 205 | 50 | 120 | 32 | 75 | 16 | 34 | 6 | 17 | 0 | 11 | 12 | 1 | 23 | 12 | 29 | 18 | 39 | 28 |
| 18 | 30 | 110 | 240 | 65 | 149 | 40 | 92 | 20 | 41 | 7 | 20 | 0 | 13 | 15 | 2 | 28 | 15 | 35 | 22 | 48 | 35 |
| 30 | 40 | 120 | 280 | 80 | 180 | 50 | 112 | 25 | 50 | 9 | 25 | 0 | 16 | 18 | 2 | 33 | 17 | 42 | 26 | 59 | 43 |
| 40 | 50 | 130 | 290 | 80 | 180 | 50 | 112 | 25 | 50 | 9 | 25 | 0 | 16 | 18 | 2 | 33 | 17 | 42 | 26 | 59 | 43 |
| 50 | 65 | 140 | 330 | 100 | 220 | 60 | 134 | 30 | 60 | 10 | 29 | 0 | 19 | 21 | 2 | 39 | 20 | 51 | 32 | 72 | 53 |
| 65 | 80 | 150 | 340 | 100 | 220 | 60 | 134 | 30 | 60 | 10 | 29 | 0 | 19 | 21 | 2 | 39 | 20 | 51 | 32 | 78 | 59 |
| 80 | 100 | 170 | 390 | 120 | 260 | 72 | 159 | 36 | 71 | 12 | 34 | 0 | 22 | 25 | 3 | 45 | 23 | 59 | 37 | 93 | 71 |
| 100 | 120 | 180 | 400 | 120 | 260 | 72 | 159 | 36 | 71 | 12 | 34 | 0 | 22 | 25 | 3 | 45 | 23 | 59 | 37 | 101 | 79 |
| 120 | 140 | 200 | 450 | 145 | 305 | 85 | 185 | 43 | 83 | 14 | 39 | 0 | 25 | 28 | 3 | 52 | 27 | 68 | 43 | 117 | 92 |
| 140 | 160 | 210 | 460 | 145 | 305 | 85 | 185 | 43 | 83 | 14 | 39 | 0 | 25 | 28 | 3 | 52 | 27 | 68 | 43 | 125 | 100 |
| 160 | 180 | 230 | 480 | 145 | 305 | 85 | 185 | 43 | 83 | 14 | 39 | 0 | 25 | 28 | 3 | 52 | 27 | 68 | 43 | 133 | 108 |
| 180 | 200 | 240 | 530 | 170 | 355 | 100 | 215 | 50 | 96 | 15 | 44 | 0 | 29 | 33 | 4 | 60 | 31 | 79 | 50 | 151 | 122 |
| 200 | 225 | 260 | 550 | 170 | 355 | 100 | 215 | 50 | 96 | 15 | 44 | 0 | 29 | 33 | 4 | 60 | 31 | 79 | 50 | 159 | 130 |
| 225 | 250 | 280 | 570 | 170 | 355 | 100 | 215 | 50 | 96 | 15 | 44 | 0 | 29 | 33 | 4 | 60 | 31 | 79 | 50 | 169 | 140 |
| 250 | 280 | 300 | 620 | 190 | 400 | 110 | 240 | 56 | 108 | 17 | 49 | 0 | 32 | 36 | 4 | 66 | 34 | 88 | 56 | 190 | 158 |
| 280 | 315 | 330 | 650 | 190 | 400 | 110 | 240 | 56 | 108 | 17 | 49 | 0 | 32 | 36 | 4 | 66 | 34 | 88 | 56 | 202 | 170 |
| 315 | 355 | 360 | 720 | 210 | 440 | 125 | 265 | 62 | 119 | 18 | 54 | 0 | 36 | 40 | 4 | 73 | 37 | 98 | 62 | 226 | 190 |
| 355 | 400 | 400 | 760 | 210 | 440 | 125 | 265 | 62 | 119 | 18 | 54 | 0 | 36 | 40 | 4 | 73 | 37 | 98 | 62 | 244 | 208 |
| 400 | 450 | 440 | 840 | 230 | 480 | 135 | 290 | 68 | 131 | 20 | 60 | 0 | 40 | 45 | 5 | 80 | 40 | 108 | 68 | 272 | 232 |
| 450 | 500 | 480 | 880 | 230 | 480 | 135 | 290 | 68 | 131 | 20 | 60 | 0 | 40 | 45 | 5 | 80 | 40 | 108 | 68 | 292 | 252 |

es = upper deviation.    ei = lower deviation
The dimensions are given in 0.001 mm, except for the nominal sizes, which are in millimeters.

Table 3. British Standard Selected Fits. Minimum and Maximum Clearances (BS 4500:1969)

| Nominal Sizes, mm | | H11-c11 | | H9-d10 | | H9-e9 | | H8-f7 | | H7-g6 | | H7-h6 | | H7-k6 | | H7-n6 | | H7-p6 | | H7-s6 | |
|---|---|---|---|---|---|---|---|---|---|---|---|---|---|---|---|---|---|---|---|---|---|
| Over | Up to and Incl. | Min | Max | Min | Max | Min | Max | Min | Max | Min | Max | Min | Max | Min | Max | Min | Max | Min | Max | Min | Max |
| … | 3 | 60 | 180 | 20 | 85 | 14 | 64 | 6 | 30 | 2 | 18 | 0 | 16 | -6 | +10 | -10 | +6 | -12 | +4 | -20 | -4 |
| 3 | 6 | 70 | 220 | 30 | 108 | 20 | 80 | 10 | 40 | 4 | 24 | 0 | 20 | -9 | +11 | -16 | +4 | -20 | 0 | -27 | -7 |
| 6 | 10 | 80 | 260 | 40 | 134 | 25 | 97 | 13 | 50 | 5 | 29 | 0 | 24 | -10 | +14 | -19 | +5 | -24 | 0 | -32 | -8 |
| 10 | 18 | 95 | 315 | 50 | 163 | 32 | 118 | 16 | 61 | 6 | 35 | 0 | 29 | -12 | +17 | -23 | +6 | -29 | 0 | -39 | -10 |
| 18 | 30 | 110 | 370 | 65 | 201 | 40 | 144 | 20 | 74 | 7 | 41 | 0 | 34 | -15 | +19 | -28 | +6 | -35 | -1 | -48 | -14 |
| 30 | 40 | 120 | 440 | 80 | 242 | 50 | 174 | 25 | 89 | 9 | 50 | 0 | 41 | -18 | +23 | -33 | +8 | -42 | -1 | -59 | -18 |
| 40 | 50 | 130 | 450 | 80 | 242 | 50 | 174 | 25 | 89 | 9 | 50 | 0 | 41 | -18 | +23 | -33 | +8 | -42 | -1 | -59 | -18 |
| 50 | 65 | 140 | 520 | 100 | 294 | 60 | 208 | 30 | 106 | 10 | 59 | 0 | 49 | -21 | +28 | -39 | +10 | -51 | -2 | -72 | -23 |
| 65 | 80 | 150 | 530 | 100 | 294 | 60 | 208 | 30 | 106 | 10 | 59 | 0 | 49 | -21 | +28 | -39 | +10 | -51 | -2 | -78 | -29 |
| 80 | 100 | 170 | 610 | 120 | 347 | 72 | 246 | 36 | 125 | 12 | 69 | 0 | 57 | -25 | +32 | -45 | +12 | -59 | -2 | -93 | -36 |
| 100 | 120 | 180 | 620 | 120 | 347 | 72 | 246 | 36 | 125 | 12 | 69 | 0 | 57 | -25 | +32 | -45 | +12 | -59 | -2 | -101 | -44 |
| 120 | 140 | 200 | 700 | 145 | 405 | 85 | 285 | 43 | 146 | 14 | 79 | 0 | 65 | -28 | +37 | -52 | +13 | -68 | -3 | -117 | -52 |
| 140 | 160 | 210 | 710 | 145 | 405 | 85 | 285 | 43 | 146 | 14 | 79 | 0 | 65 | -28 | +37 | -52 | +13 | -68 | -3 | -125 | -60 |
| 160 | 180 | 230 | 730 | 145 | 405 | 85 | 285 | 43 | 146 | 14 | 79 | 0 | 65 | -28 | +37 | -52 | +13 | -68 | -3 | -133 | -68 |
| 180 | 200 | 240 | 820 | 170 | 470 | 100 | 330 | 50 | 168 | 15 | 90 | 0 | 75 | -33 | +42 | -60 | +15 | -79 | -4 | -151 | -76 |
| 200 | 225 | 260 | 840 | 170 | 470 | 100 | 330 | 50 | 168 | 15 | 90 | 0 | 75 | -33 | +42 | -60 | +15 | -79 | -4 | -159 | -84 |
| 225 | 250 | 280 | 860 | 170 | 470 | 100 | 330 | 50 | 168 | 15 | 90 | 0 | 75 | -33 | +42 | -60 | +15 | -79 | -4 | -169 | -94 |
| 250 | 280 | 300 | 940 | 190 | 530 | 110 | 370 | 56 | 189 | 17 | 101 | 0 | 84 | -36 | +48 | -66 | +18 | -88 | -4 | -190 | -106 |
| 280 | 315 | 330 | 970 | 190 | 530 | 110 | 370 | 56 | 189 | 17 | 101 | 0 | 84 | -36 | +48 | -66 | +18 | -88 | -4 | -202 | -118 |
| 315 | 355 | 360 | 1080 | 210 | 580 | 125 | 405 | 62 | 208 | 18 | 111 | 0 | 93 | -40 | +53 | -73 | +20 | -98 | -5 | -226 | -133 |
| 355 | 400 | 400 | 1120 | 210 | 580 | 125 | 405 | 62 | 208 | 18 | 111 | 0 | 93 | -40 | +53 | -73 | +20 | -98 | -5 | -244 | -151 |
| 400 | 450 | 440 | 1240 | 230 | 635 | 135 | 445 | 68 | 228 | 20 | 123 | 0 | 103 | -45 | +58 | -80 | +23 | -108 | -5 | -272 | -169 |
| 450 | 500 | 480 | 1280 | 230 | 635 | 135 | 445 | 68 | 228 | 20 | 123 | 0 | 103 | -45 | +58 | -80 | +23 | -108 | -5 | -292 | -189 |

The dimensions are given in 0.001 mm, except for the nominal sizes, which are in millimeters. Minus (−) sign indicates negative clearance, i.e., interference.

**Table 4. British Standard Limits and Fits** (BS 4500:1969)

| Nominal Sizes, mm | | Tolerance Grades | | | | | | | | | |
|---|---|---|---|---|---|---|---|---|---|---|---|
| Over | To | IT 01 | IT 0 | IT 1 | IT 2 | IT 3 | IT 4 | IT 5 | IT 6 | IT 7 | IT 8 |
| ... | 3 | 0.3 | 0.5 | 0.8 | 1.2 | 2 | 3 | 4 | 6 | 10 | 14 |
| 3 | 6 | 0.4 | 0.6 | 1 | 1.5 | 2.5 | 4 | 5 | 8 | 12 | 18 |
| 6 | 10 | 0.4 | 0.6 | 1 | 1.5 | 2.5 | 4 | 6 | 9 | 15 | 22 |
| 10 | 18 | 0.5 | 0.8 | 1.2 | 2 | 3 | 5 | 8 | 11 | 18 | 27 |
| 18 | 30 | 0.6 | 1 | 1.5 | 2.5 | 4 | 6 | 9 | 13 | 21 | 33 |
| 30 | 50 | 0.6 | 1 | 1.5 | 2.5 | 4 | 7 | 11 | 16 | 25 | 39 |
| 50 | 80 | 0.8 | 1.2 | 2 | 3 | 5 | 8 | 13 | 19 | 30 | 46 |
| 80 | 120 | 1 | 1.5 | 2.5 | 4 | 6 | 10 | 15 | 22 | 35 | 54 |
| 120 | 180 | 1.2 | 2 | 3.5 | 5 | 8 | 12 | 18 | 25 | 40 | 63 |
| 180 | 250 | 2 | 3 | 4.5 | 7 | 10 | 14 | 20 | 29 | 46 | 72 |
| 250 | 315 | 2.5 | 4 | 6 | 8 | 12 | 16 | 23 | 32 | 52 | 81 |
| 315 | 400 | 3 | 5 | 7 | 9 | 13 | 18 | 25 | 36 | 57 | 89 |
| 400 | 500 | 4 | 6 | 8 | 10 | 15 | 20 | 27 | 40 | 63 | 97 |

| Nominal Sizes, mm | | Tolerance Grades | | | | | | | |
|---|---|---|---|---|---|---|---|---|
| Over | To | IT 9 | IT 10 | IT 11 | IT 12 | IT 13 | IT 14† | IT 15† | IT 16† |
| ... | 3 | 25 | 40 | 60 | 100 | 140 | 250 | 400 | 600 |
| 3 | 6 | 30 | 48 | 75 | 120 | 180 | 300 | 480 | 750 |
| 6 | 10 | 36 | 58 | 90 | 150 | 220 | 360 | 580 | 900 |
| 10 | 18 | 43 | 70 | 110 | 180 | 270 | 430 | 700 | 1100 |
| 18 | 30 | 52 | 84 | 130 | 210 | 330 | 520 | 840 | 1300 |
| 30 | 50 | 62 | 100 | 160 | 250 | 390 | 620 | 1000 | 1600 |
| 50 | 80 | 74 | 120 | 190 | 300 | 460 | 740 | 1200 | 1900 |
| 80 | 120 | 87 | 140 | 220 | 350 | 540 | 870 | 1400 | 2200 |
| 120 | 180 | 100 | 160 | 250 | 400 | 630 | 1000 | 1600 | 2500 |
| 180 | 250 | 115 | 185 | 290 | 460 | 720 | 1150 | 1850 | 2900 |
| 250 | 315 | 130 | 210 | 320 | 520 | 810 | 1300 | 2100 | 3200 |
| 315 | 400 | 140 | 230 | 360 | 570 | 890 | 1400 | 2300 | 3600 |
| 400 | 500 | 155 | 250 | 400 | 630 | 970 | 1550 | 2500 | 4000 |

† Not applicable to sizes below 1 mm.

The dimensions are given in 0.001 mm, except for the nominal sizes which are in millimeters.

## Table 5. British Standard Fundamental Deviations for Shafts (BS 4500:1969)

| Nominal Size, mm Over | To | Grade — Fundamental (Upper) Deviation es (01 to 16) a* | b* | c | cd | d | e | ef | f | fg | g | h | js† | Fundamental (Lower) Dev'n ei — j 5-6 | j 7 | j 8 | k 4-7 | k ≤3 >7 |
|---|---|---|---|---|---|---|---|---|---|---|---|---|---|---|---|---|---|---|
| .. | 3 | -270 | -140 | -60 | -34 | -20 | -14 | -10 | -6 | -4 | -2 | 0 | | -2 | -4 | -6 | 0 | 0 |
| 3 | 6 | -270 | -140 | -70 | -46 | -30 | -20 | -14 | -10 | -6 | -4 | 0 | | -2 | -4 | : | +1 | 0 |
| 6 | 10 | -280 | -150 | -80 | -56 | -40 | -25 | -18 | -13 | -8 | -5 | 0 | | -2 | -5 | : | +1 | 0 |
| 10 | 14 | -290 | -150 | -95 | : | -50 | -32 | : | -16 | : | -6 | 0 | | -3 | -6 | : | +1 | 0 |
| 14 | 18 | -290 | -150 | -95 | : | -50 | -32 | : | -16 | : | -6 | 0 | | -3 | -6 | : | +1 | 0 |
| 18 | 24 | -300 | -160 | -110 | : | -65 | -40 | : | -20 | : | -7 | 0 | | -4 | -8 | : | +2 | 0 |
| 24 | 30 | -300 | -160 | -110 | : | -65 | -40 | : | -20 | : | -7 | 0 | | -4 | -8 | : | +2 | 0 |
| 30 | 40 | -310 | -170 | -120 | : | -80 | -50 | : | -25 | : | -9 | 0 | | -5 | -10 | : | +2 | 0 |
| 40 | 50 | -320 | -180 | -130 | : | -80 | -50 | : | -25 | : | -9 | 0 | | -5 | -10 | : | +2 | 0 |
| 50 | 65 | -340 | -190 | -140 | : | -100 | -60 | : | -30 | : | -10 | 0 | | -7 | -12 | : | +2 | 0 |
| 65 | 80 | -360 | -200 | -150 | : | -100 | -60 | : | -30 | : | -10 | 0 | | -7 | -12 | : | +2 | 0 |
| 80 | 100 | -380 | -220 | -170 | : | -120 | -72 | : | -36 | : | -12 | 0 | | -9 | -15 | : | +3 | 0 |
| 100 | 120 | -410 | -240 | -180 | : | -120 | -72 | : | -36 | : | -12 | 0 | | -9 | -15 | : | +3 | 0 |
| 120 | 140 | -460 | -260 | -200 | : | -145 | -85 | : | -43 | : | -14 | 0 | ±IT/2 | -11 | -18 | : | +3 | 0 |
| 140 | 160 | -520 | -280 | -210 | : | -145 | -85 | : | -43 | : | -14 | 0 | | -11 | -18 | : | +3 | 0 |
| 160 | 180 | -580 | -310 | -230 | : | -145 | -85 | : | -43 | : | -14 | 0 | | -11 | -18 | : | +3 | 0 |
| 180 | 200 | -660 | -340 | -240 | : | -170 | -100 | : | -50 | : | -15 | 0 | | -13 | -21 | : | +4 | 0 |
| 200 | 225 | -740 | -380 | -260 | : | -170 | -100 | : | -50 | : | -15 | 0 | | -13 | -21 | : | +4 | 0 |
| 225 | 250 | -820 | -420 | -280 | : | -170 | -100 | : | -50 | : | -15 | 0 | | -13 | -21 | : | +4 | 0 |
| 250 | 280 | -920 | -480 | -300 | : | -190 | -110 | : | -56 | : | -17 | 0 | | -16 | -26 | : | +4 | 0 |
| 280 | 315 | -1050 | -540 | -330 | : | -190 | -110 | : | -56 | : | -17 | 0 | | -16 | -26 | : | +4 | 0 |
| 315 | 355 | -1200 | -600 | -360 | : | -210 | -125 | : | -62 | : | -18 | 0 | | -18 | -28 | : | +4 | 0 |
| 355 | 400 | -1350 | -680 | -400 | : | -210 | -125 | : | -62 | : | -18 | 0 | | -18 | -28 | : | +4 | 0 |
| 400 | 450 | -1500 | -760 | -440 | : | -230 | -135 | : | -68 | : | -20 | 0 | | -20 | -32 | : | +5 | 0 |
| 450 | 500 | -1650 | -840 | -480 | : | -230 | -135 | : | -68 | : | -20 | 0 | | -20 | -32 | : | +5 | 0 |

The dimensions are in 0.001 mm, except the nominal sizes, which are in millimeters.

* Not applicable to sizes up to 1 mm. † In grades 7 to 11, the two symmetrical deviations ±IT/2 should be rounded if the IT value in micro-meters is an odd value by replacing it with the even value immediately below. For example, if IT = 175, replace it by 174.

**Table 5.** *(Continued).* British Standard Fundamental Deviations for Shafts (BS 4500:1969)

| Nominal Size, mm | | Grade 01 to 16 — Fundamental (Lower) Deviation ei | | | | | | | | | | | | | |
|---|---|---|---|---|---|---|---|---|---|---|---|---|---|---|---|
| Over | To | m | n | p | r | s | t | u | v | x | y | z | za | zb | zc |
| … | 3 | +2 | +4 | +6 | +10 | +14 | … | +18 | … | +20 | … | +26 | +32 | +40 | +60 |
| 3 | 6 | +4 | +8 | +12 | +15 | +19 | … | +23 | … | +28 | … | +35 | +42 | +50 | +80 |
| 6 | 10 | +6 | +10 | +15 | +19 | +23 | … | +28 | … | +34 | … | +42 | +52 | +67 | +97 |
| 10 | 14 | +7 | +12 | +18 | +23 | +28 | … | +33 | … | +40 | … | +50 | +64 | +90 | +130 |
| 14 | 18 | +7 | +12 | +18 | +23 | +28 | … | +33 | +39 | +45 | … | +60 | +77 | +108 | +150 |
| 18 | 24 | +8 | +15 | +22 | +28 | +35 | … | +41 | +47 | +54 | +63 | +73 | +98 | +136 | +188 |
| 24 | 30 | +8 | +15 | +22 | +28 | +35 | +41 | +48 | +55 | +64 | +75 | +88 | +118 | +160 | +218 |
| 30 | 40 | +9 | +17 | +26 | +34 | +43 | +48 | +60 | +68 | +80 | +94 | +112 | +148 | +200 | +274 |
| 40 | 50 | +9 | +17 | +26 | +34 | +43 | +54 | +70 | +81 | +97 | +114 | +136 | +180 | +242 | +325 |
| 50 | 65 | +11 | +20 | +32 | +41 | +53 | +66 | +87 | +102 | +122 | +144 | +172 | +226 | +300 | +405 |
| 65 | 80 | +11 | +20 | +32 | +43 | +59 | +75 | +102 | +120 | +146 | +174 | +210 | +274 | +360 | +480 |
| 80 | 100 | +13 | +23 | +37 | +51 | +71 | +91 | +124 | +146 | +178 | +214 | +258 | +335 | +445 | +585 |
| 100 | 120 | +13 | +23 | +37 | +54 | +79 | +104 | +144 | +172 | +210 | +254 | +310 | +400 | +525 | +690 |
| 120 | 140 | +15 | +27 | +43 | +63 | +92 | +122 | +170 | +202 | +248 | +300 | +365 | +470 | +620 | +800 |
| 140 | 160 | +15 | +27 | +43 | +65 | +100 | +134 | +190 | +228 | +280 | +340 | +415 | +535 | +700 | +900 |
| 160 | 180 | +15 | +27 | +43 | +68 | +108 | +146 | +210 | +252 | +310 | +380 | +465 | +600 | +780 | +1000 |
| 180 | 200 | +17 | +31 | +50 | +77 | +122 | +166 | +236 | +284 | +350 | +425 | +520 | +670 | +880 | +1150 |
| 200 | 225 | +17 | +31 | +50 | +80 | +130 | +180 | +258 | +310 | +385 | +470 | +575 | +740 | +960 | +1250 |
| 225 | 250 | +17 | +31 | +50 | +84 | +140 | +196 | +284 | +340 | +425 | +520 | +640 | +820 | +1050 | +1350 |
| 250 | 280 | +20 | +34 | +56 | +94 | +158 | +218 | +315 | +385 | +475 | +580 | +710 | +920 | +1200 | +1550 |
| 280 | 315 | +20 | +34 | +56 | +98 | +170 | +240 | +350 | +425 | +525 | +650 | +790 | +1000 | +1300 | +1700 |
| 315 | 355 | +21 | +37 | +62 | +108 | +190 | +268 | +390 | +475 | +590 | +730 | +900 | +1150 | +1500 | +1900 |
| 355 | 400 | +21 | +37 | +62 | +114 | +208 | +294 | +435 | +530 | +660 | +820 | +1000 | +1300 | +1650 | +2100 |
| 400 | 450 | +23 | +40 | +68 | +126 | +232 | +330 | +490 | +595 | +740 | +920 | +1100 | +1450 | +1850 | +2400 |
| 450 | 500 | +23 | +40 | +68 | +132 | +252 | +360 | +540 | +660 | +820 | +1000 | +1250 | +1600 | +2100 | +2600 |

The dimensions are in 0.001 mm, except the nominal sizes, which are in millimeters.

**Table 6. British Standard Fundamental Deviations for Holes** (BS 4500:1969)

| Nominal Size, mm | | Grade | | | | | | | | | | | | | | | | | | | | |
|---|---|---|---|---|---|---|---|---|---|---|---|---|---|---|---|---|---|---|---|---|---|---|
| | | 01 to 16 | | | | | | | | | | | | 6 | 7 | 8 | ≦8 | >8 | ≦8‡ | >8 | ≦8 | >8§ |
| | | Fundamental (Lower) Deviation EI | | | | | | | | | | | | Fundamental (Upper) Deviation ES | | | | | | | | |
| Over | To | A* | B* | C | CD | D | E | EF | F | FG | G | H | Js† | J | | | K†† | | M†† | | N†† | |
| ... | 3 | +270 | +140 | +60 | +34 | +20 | +14 | +10 | +6 | +4 | +2 | 0 | | +2 | +4 | +6 | 0 | 0 | −2 | −2 | −4 | −4 |
| 3 | 6 | +270 | +140 | +70 | +46 | +30 | +20 | +14 | +10 | +6 | +4 | 0 | | +5 | +6 | +10 | −1+△ | ... | −4+△ | −4 | −8+△ | 0 |
| 6 | 10 | +280 | +150 | +80 | +56 | +40 | +25 | +18 | +13 | +8 | +5 | 0 | | +5 | +8 | +12 | −1+△ | ... | −6+△ | −6 | −10+△ | 0 |
| 10 | 14 | +290 | +150 | +95 | ... | +50 | +32 | ... | +16 | ... | +6 | 0 | | +6 | +10 | +15 | −1+△ | ... | −7+△ | −7 | −12+△ | 0 |
| 14 | 18 | +290 | +150 | +95 | ... | +50 | +32 | ... | +16 | ... | +6 | 0 | | +6 | +10 | +15 | −1+△ | ... | −7+△ | −7 | −12+△ | 0 |
| 18 | 24 | +300 | +160 | +110 | ... | +65 | +40 | ... | +20 | ... | +7 | 0 | | +8 | +12 | +20 | −2+△ | ... | −8+△ | −8 | −15+△ | 0 |
| 24 | 30 | +300 | +160 | +110 | ... | +65 | +40 | ... | +20 | ... | +7 | 0 | | +8 | +12 | +20 | −2+△ | ... | −8+△ | −8 | −15+△ | 0 |
| 30 | 40 | +310 | +170 | +120 | ... | +80 | +50 | ... | +25 | ... | +9 | 0 | | +10 | +14 | +24 | −2+△ | ... | −9+△ | −9 | −17+△ | 0 |
| 40 | 50 | +320 | +180 | +130 | ... | +80 | +50 | ... | +25 | ... | +9 | 0 | | +10 | +14 | +24 | −2+△ | ... | −9+△ | −9 | −17+△ | 0 |
| 50 | 65 | +340 | +190 | +140 | ... | +100 | +60 | ... | +30 | ... | +10 | 0 | | +13 | +18 | +28 | −2+△ | ... | −11+△ | −11 | −20+△ | 0 |
| 65 | 80 | +360 | +200 | +150 | ... | +100 | +60 | ... | +30 | ... | +10 | 0 | | +13 | +18 | +28 | −2+△ | ... | −11+△ | −11 | −20+△ | 0 |
| 80 | 100 | +380 | +220 | +170 | ... | +120 | +72 | ... | +36 | ... | +12 | 0 | ±IT/2 | +16 | +22 | +34 | −3+△ | ... | −13+△ | −13 | −23+△ | 0 |
| 100 | 120 | +410 | +240 | +180 | ... | +120 | +72 | ... | +36 | ... | +12 | 0 | | +16 | +22 | +34 | −3+△ | ... | −13+△ | −13 | −23+△ | 0 |
| 120 | 140 | +460 | +260 | +200 | ... | +145 | +85 | ... | +43 | ... | +14 | 0 | | +18 | +26 | +41 | −3+△ | ... | −15+△ | −15 | −27+△ | 0 |
| 140 | 160 | +520 | +280 | +210 | ... | +145 | +85 | ... | +43 | ... | +14 | 0 | | +18 | +26 | +41 | −3+△ | ... | −15+△ | −15 | −27+△ | 0 |
| 160 | 180 | +580 | +310 | +230 | ... | +145 | +85 | ... | +43 | ... | +14 | 0 | | +18 | +26 | +41 | −3+△ | ... | −15+△ | −15 | −27+△ | 0 |
| 180 | 200 | +660 | +340 | +240 | ... | +170 | +100 | ... | +50 | ... | +15 | 0 | | +22 | +30 | +47 | −4+△ | ... | −17+△ | −17 | −31+△ | 0 |
| 200 | 225 | +740 | +380 | +260 | ... | +170 | +100 | ... | +50 | ... | +15 | 0 | | +22 | +30 | +47 | −4+△ | ... | −17−△ | −17 | −31+△ | 0 |
| 225 | 250 | +820 | +420 | +280 | ... | +170 | +100 | ... | +50 | ... | +15 | 0 | | +22 | +30 | +47 | −4+△ | ... | −17+△ | −17 | −31+△ | 0 |
| 250 | 280 | +920 | +480 | +300 | ... | +190 | +110 | ... | +56 | ... | +17 | 0 | | +25 | +36 | +55 | −4+△ | ... | −20+△ | −20 | −34+△ | 0 |
| 280 | 315 | +1050 | +540 | +330 | ... | +190 | +110 | ... | +56 | ... | +17 | 0 | | +25 | +36 | +55 | −4+△ | ... | −20+△ | −20 | −34+△ | 0 |
| 315 | 355 | +1200 | +600 | +360 | ... | +210 | +125 | ... | +62 | ... | +18 | 0 | | +29 | +39 | +60 | −4+△ | ... | −21+△ | −21 | −37+△ | 0 |
| 355 | 400 | +1350 | +680 | +400 | ... | +210 | +125 | ... | +62 | ... | +18 | 0 | | +29 | +39 | +60 | −4+△ | ... | −21+△ | −21 | −37+△ | 0 |
| 400 | 450 | +1500 | +760 | +440 | ... | +230 | +135 | ... | +68 | ... | +20 | 0 | | +33 | +43 | +66 | −5+△ | ... | −23+△ | −23 | −40+△ | 0 |
| 450 | 500 | +1650 | +840 | +480 | ... | +230 | +135 | ... | +68 | ... | +20 | 0 | | +33 | +43 | +66 | −5+△ | ... | −23+△ | −23 | −40+△ | 0 |

The dimensions are given in 0.001 mm, except the nominal sizes which are in millimeters. * Not applicable to sizes up to 1 mm. † In grades 7 to 11, the two symmetrical deviations ±IT/2 should be rounded if the IT value in micro-meters is an odd value, by replacing it with the even value below. For example, if IT = 175, replace it by 174.

†† When calculating deviations for holes K, M, and N with tolerance grades up to and including IT 8, and holes F to ZC with tolerance grades up to and including IT 7, the delta (△) values are added to the upper deviation ES. For example, for 25 P7, ES = −0.022 + 0.008 = −0.014 mm.

‡ Special case: for M6, ES = −9 for sizes from 250 to 315 mm, instead of −11. § Not applicable to sizes up to 1 mm.

**Table 6. (Continued). British Standard Fundamental Deviations for Holes** (BS 4500:1969)

For the Grade ≤7 section, column **P to ZC**: ▽ *Same deviation as for grades above 7 increased by Δ*

| Nominal Size, mm Over | To | Grade >7 — Fundamental (Upper) Deviation ES | | | | | | | | | | | | Values for delta (Δ)†† — Grade | | | | | |
|---|---|---|---|---|---|---|---|---|---|---|---|---|---|---|---|---|---|---|---|
| | | P | R | S | T | U | V | X | Y | Z | ZA | ZB | ZC | 3 | 4 | 5 | 6 | 7 | 8 |
| .. | 3 | −6 | −10 | −14 | | −18 | | −20 | | −26 | −32 | −40 | −60 | 0 | 0 | 0 | 0 | 0 | 0 |
| 3 | 6 | −12 | −15 | −19 | | −23 | | −28 | | −35 | −42 | −50 | −80 | 1 | 1.5 | 1 | 3 | 4 | 6 |
| 6 | 10 | −15 | −19 | −23 | | −28 | | −34 | | −42 | −52 | −67 | −97 | 1 | 1.5 | 1 | 3 | 6 | 7 |
| 10 | 14 | −18 | −23 | −28 | | −33 | | −40 | | −50 | −64 | −90 | −130 | 1 | 2 | 2 | 3 | 7 | 9 |
| 14 | 18 | −18 | −23 | −28 | | −33 | −39 | −45 | | −60 | −77 | −108 | −150 | 1 | 2 | 3 | 3 | 7 | 9 |
| 18 | 24 | −22 | −28 | −35 | | −41 | −47 | −54 | −63 | −73 | −98 | −136 | −188 | 1.5 | 2 | 3 | 4 | 8 | 12 |
| 24 | 30 | −22 | −28 | −35 | −41 | −48 | −55 | −64 | −75 | −88 | −118 | −160 | −218 | 1.5 | 2 | 3 | 4 | 8 | 12 |
| 30 | 40 | −26 | −34 | −43 | −48 | −60 | −68 | −80 | −94 | −112 | −148 | −200 | −274 | 1.5 | 3 | 4 | 5 | 9 | 14 |
| 40 | 50 | −26 | −34 | −43 | −54 | −70 | −81 | −97 | −114 | −136 | −180 | −242 | −325 | 1.5 | 3 | 4 | 5 | 9 | 14 |
| 50 | 65 | −32 | −41 | −53 | −66 | −87 | −102 | −122 | −144 | −172 | −226 | −300 | −405 | 2 | 3 | 5 | 6 | 11 | 16 |
| 65 | 80 | −32 | −43 | −59 | −75 | −102 | −120 | −146 | −174 | −210 | −274 | −360 | −480 | 2 | 3 | 5 | 6 | 11 | 16 |
| 80 | 100 | −37 | −51 | −71 | −91 | −124 | −146 | −178 | −214 | −258 | −335 | −445 | −585 | 2 | 4 | 5 | 7 | 13 | 19 |
| 100 | 120 | −37 | −54 | −79 | −104 | −144 | −172 | −210 | −254 | −310 | −400 | −525 | −690 | 2 | 4 | 5 | 7 | 13 | 19 |
| 120 | 140 | −43 | −63 | −92 | −122 | −170 | −202 | −248 | −300 | −365 | −470 | −620 | −800 | 3 | 4 | 6 | 7 | 15 | 23 |
| 140 | 160 | −43 | −65 | −100 | −134 | −190 | −228 | −280 | −340 | −415 | −535 | −700 | −900 | 3 | 4 | 6 | 7 | 15 | 23 |
| 160 | 180 | −43 | −68 | −108 | −146 | −210 | −252 | −310 | −380 | −465 | −600 | −780 | −1000 | 3 | 4 | 6 | 7 | 15 | 23 |
| 180 | 200 | −50 | −77 | −122 | −166 | −226 | −284 | −350 | −425 | −520 | −670 | −880 | −1150 | 3 | 4 | 6 | 7 | 17 | 26 |
| 200 | 225 | −50 | −80 | −130 | −180 | −258 | −310 | −385 | −470 | −575 | −740 | −960 | −1250 | 3 | 4 | 6 | 9 | 17 | 26 |
| 225 | 250 | −50 | −84 | −140 | −196 | −284 | −340 | −425 | −520 | −640 | −820 | −1050 | −1350 | 3 | 4 | 6 | 9 | 17 | 26 |
| 250 | 280 | −56 | −94 | −158 | −218 | −315 | −385 | −475 | −580 | −710 | −920 | −1200 | −1550 | 4 | 4 | 7 | 9 | 20 | 29 |
| 280 | 315 | −56 | −98 | −170 | −240 | −350 | −425 | −525 | −650 | −790 | −1000 | −1300 | −1700 | 4 | 4 | 7 | 9 | 20 | 29 |
| 315 | 355 | −62 | −108 | −190 | −268 | −390 | −475 | −590 | −730 | −900 | −1150 | −1500 | −1800 | 4 | 5 | 7 | 11 | 21 | 32 |
| 355 | 400 | −62 | −114 | −208 | −294 | −435 | −530 | −660 | −820 | −1000 | −1300 | −1650 | −2100 | 4 | 5 | 7 | 11 | 21 | 32 |
| 400 | 450 | −68 | −126 | −232 | −330 | −490 | −595 | −740 | −920 | −1100 | −1450 | −1850 | −2400 | 5 | 5 | 7 | 13 | 23 | 34 |
| 450 | 500 | −68 | −132 | −252 | −360 | −540 | −660 | −820 | −1000 | −1250 | −1600 | −2100 | −2600 | 5 | 5 | 7 | 13 | 23 | 34 |

The dimensions are given in 0.001 mm, except the nominal sizes which are in millimeters.

†† When calculating deviations for holes K, M, and N with tolerance grades up to and including IT 8, and holes P to ZC with tolerance grades up to and including IT 7, the delta (Δ) values are added to the upper deviation ES. For example, for 25 P7, ES = −0.022 + 0.008 = −0.014 mm.

**British Standard Preferred Numbers and Preferred Sizes.** — This British Standard (PD 6481:1977) gives recommendations for the use of preferred numbers and preferred sizes for functional characteristics and dimensions of various products.

The preferred number system is internationally standardized in ISO 3. It is also referred to as the Renard or R series (see Handbook page 116).

The series in the preferred number system are geometric series, i.e., in which there is a constant ratio between each figure and the succeeding one, within a decimal framework. Thus the R5 series has five steps between 1 and 10, the R10 series has 10 steps between 1 and 10, the R20 series, 20 steps and the R40 series, 40 steps, giving increases between steps of approximately 60, 25, 12 and 6 per cent, respectively.

The preferred size series have been developed from the preferred number series by rounding off the inconvenient numbers in the basic series and adjusting for linear measurement in millimeters. These series are shown in the table below.

Taking all normal considerations into account it is recommended that: (a) For ranges of values of the primary *functional* characteristics (outputs and capacities) of a series of products, the preferred number series R5 to R40 (see page 116) should be used and (b) whenever linear sizes are concerned, the preferred sizes as given in the table below should be used. The presentation of preferred sizes gives designers and users a logical selection and the benefits of rational variety reduction.

The second choice size given should only be used when it is not possible to employ the first choice, and the third choice should be applied only if a size from the second choice cannot be selected. With this procedure, common usage will tend to be concentrated on a limited range of sizes, and a contribution is thus made to variety reduction. However, the decision to use a particular size cannot be taken on the basis that one is first choice and the other not. Account must be taken of the effect on the design, the availability of tools, and other relevant factors.

### British Standard Preferred Sizes (PD 6481:1977)

| Choice | | | Choice | | | Choice | | | Choice | | | Choice | | | Choice | | |
|---|---|---|---|---|---|---|---|---|---|---|---|---|---|---|---|---|---|
| 1st | 2nd | 3rd | 1st | 2nd | 3rd | 1st | 2nd | 3rd | 1st | 2nd | 3rd | 1st | 2nd | 3rd | 1st | 2nd | 3rd |
| 1 | | | | | 5.2 | | | 23 | 65 | | | | | 122 | | | 188 |
| | 1.1 | | | 5.5 | | | | 24 | | | 66 | | 125 | | 190 | | |
| 1.2 | | | 6 | | | 25 | | | | 68 | | | | 128 | | | 192 |
| | | 1.3 | | | 6 | | | 26 | 70 | | | 130 | | | | 195 | |
| | 1.4 | | | | 6.2 | | 28 | | | 72 | | | | 132 | | | 198 |
| | | 1.5 | | 6.5 | | 30 | | | | | 74 | | 135 | | 200 | | |
| 1.6 | | | | | 6.8 | | 32 | | 75 | | | | | 138 | | | 205 |
| | | 1.7 | 7 | | | | | 34 | | | 76 | 140 | | | | 210 | |
| | 1.8 | | | | 7.5 | 35 | | | | 78 | | | | 142 | | | 215 |
| | | 1.9 | 8 | | | | | 36 | 80 | | | | 145 | | 220 | | |
| 2 | | | | | 8.5 | | 38 | | | | 82 | | | 148 | | | 225 |
| | | 2.1 | | 9 | | 40 | | | | 85 | | 150 | | | | 230 | |
| | 2.2 | | | | 9.5 | | 42 | | | | 88 | | | 152 | | | 235 |
| | | 2.4 | 10 | | | | | 44 | 90 | | | | 155 | | 240 | | |
| 2.5 | | | | 11 | | 45 | | | | | 92 | | | 158 | | | 245 |
| | | 2.6 | 12 | | | | | 46 | | 95 | | 160 | | | | 250 | |
| | 2.8 | | | | 13 | | 48 | | | | 98 | | | 162 | | | 255 |
| 3 | | | | 14 | | 50 | | | 100 | | | | 165 | | 260 | | |
| | | 3.2 | | | 15 | | 52 | | | | 102 | | | 168 | | | 265 |
| | 3.5 | | 16 | | | | | 54 | | 105 | | 170 | | | | 270 | |
| | | 3.8 | | | 17 | 55 | | | | | 108 | | | 172 | | | 275 |
| 4 | | | | 18 | | | | 56 | 110 | | | | 175 | | 280 | | |
| | | 4.2 | | | 19 | | 58 | | | | 112 | | | 178 | | | 285 |
| | 4.5 | | 20 | | | 60 | | | | 115 | | 180 | | | | 290 | |
| | | 4.8 | | | 21 | | 62 | | | | 118 | | | 182 | | | 295 |
| 5 | | | | 22 | | | | 64 | 120 | | | | 185 | | 300 | | |

For dimensions above 300, each series continues in a similar manner, i.e., the intervals between each series number are the same as between 200 and 300.

**ABC Proposed System of Limits and Fits.** — American-British-Canadian conferences held in 1952 and 1953 resulted in a draft proposal for an ABC system of limits and fits. The effect of this proposal was to provide a unified basis for each of the three countries' national standards on limits and fits. It was not intended that this document would be used in the form presented but rather that each country would republish it in its usual national form without, however, altering the technical contents. The ABC proposal comprises: (1) a graded system of fundamental tolerances for diameters 0.04 inch to approximately 20 inches, the tolerance values in each grade being related to the diameter by a given law or formula and each grade of tolerance being approximately related to others in the system in a Preferred Number ratio (R5 series, with 60 per cent increments); (2) a series of unilateral hole limits derived from the fundamental tolerances, for use with a hole basis system; (3) a series of shaft limits, one of these limits being determined by the allowance (according to some technical requirement), the other being derived from it by addition or subtraction of the fundamental tolerance; and (4) a selection of fits made by recommended association of certain of the holes and shafts, the selection being adequate to cover most engineering requirements.

The USA Standard for Preferred Limits and Fits for Cylindrical Parts (USAS B4.1-1967) is in accord with the recommendations contained in the ABC agreements and in addition has been extended to cover other areas considered important to American industry, mainly sizes above 20 inches.

**British Standard Limits and Fits for Engineering (B.S. 1916: Parts 1 and 2: 1953).** — Part 1 of this British Standard provides recommendations for fits between mating holes and shafts and is consistent with the American-British-Canadian Conference Agreement of February, 1953. Part 2 of B.S. 1916 is intended as a guide to the selection of fits specified in Part 1. The information contained in this guide is based upon a survey of present-day practice in industrial organizations throughout Britain, the United States and Canada and is a summary of practical experience rather than purely theoretical considerations.

*Primary and Secondary Selections of Fits:* In B.S. 1916C: 1954 Primary Selection of Fits (shown in Table 1) a simple selection of six fits is provided to meet the needs of a large proportion of requirements of normal engineering products. The secondary selection of fits in B.S. 1916D: 1954 (shown in Table 2) comprises five additional fits which, together with the primary selection, cover about 95 per cent of the fits used on the average good quality engineering product.

These fits are on the unilateral hole basis. The grade H8 hole tolerance corresponds to the "U" hole of the previous standard, B.S. 164: 1924, and is produced by boring or machine reaming. The grade H7 hole tolerance lies midway between the "B" and "U" holes of B.S. 164: 1924 and is produced by grinding, broaching, or careful reaming. The grade H6 hole tolerance corresponds with the "B" hole of B.S. 164: 1924, and can be produced by fine grinding or honing, and possibly hand reaming.

Using the unilateral "H" holes just described, the resulting fits in combination with the various shafts are:

*Clearance Fits:* Shaft d9 is generally suitable for loose running fits such as plummer block bearings and loose pulleys. Shaft e8 is recommended for general loose clearance fits and for properly lubricated bearings requiring appreciable clearance. Shafts f7 and f8 are recommended to provide a normal running fit. This fit is widely used as a normal grease-lubricated or oil-lubricated bearing where no substantial temperature differences are encountered. Typical applications are gear box shaft bearings and the bearings of small electric motors and pumps. Shafts g5 and g6 are expensive to manufacture since the clearances are small and they are not recommended for running fits except in precision equipment where shaft loadings are very light. Typical applications are the bearings for accurate link work and for piston and slide valves. In addition, they are often used for spigot or location fits. Shaft h7 has an upper limit of zero although in practice a slight clearance will usually be found. It is used for non-running parts and is useful for normal location and spigot fits and may be used as a precision sliding fit.

*Transition Fits:* Shaft j7 provides a transition fit averaging a slight clearance and is recommended for location fits requiring slightly less clearance than is given with the "h" shaft and where a slight interference is permissible. Typical applications are coupling spigots and recesses, gear rings clamped to steel hubs, etc. Shaft k6 is a true transition fit averaging virtually no clearance, and is recommended for location fits where a slight interference can be tolerated for the purpose, for example, of eliminating vibration.

*Interference Fits:* Shaft p6 gives a true interference fit with an H7 hole. It is the standard press fit for steel, cast iron, or brass-to-steel assemblies. The amount of interference is too small for satisfactory press fits to be obtained in materials of low modulus of elasticity such as light alloys. For these, shaft s6 is used and typical applications are collars pressed on to shafts, valve seatings, etc.

**Table 1. British Standard Primary Selection of Fits (B.S. 1916C: 1954)**

Tolerance limits given in body of table are added or subtracted to basic size (as indicated by + or − sign) to obtain maximum and minimum sizes of mating parts.

| Nominal Size Range, Inches | | Loose Clearance | | Average Running | | Precision Location | | Average Location | | Push | | Press (Ferrous parts) | |
|---|---|---|---|---|---|---|---|---|---|---|---|---|---|
| | | H8 | e8 | H8 | f8 | H7 | g6 | H8 | h7 | H7 | k6 | H7 | p6 |
| Over | To | | | | | | | | | | | | |
| in. 0.04 | in. 0.12 | +0.6 | −0.6 | +0.6 | −0.3 | +0.4 | −0.1 | +0.6 | −0 | +0.4 | ... | +0.4 | +0.65 |
| | | +0 | −1.2 | +0 | −0.9 | +0 | −0.35 | +0 | −0.4 | +0 | ... | +0 | +0.4 |
| 0.12 | 0.24 | +0.7 | −0.8 | +0.7 | −0.4 | +0.5 | −0.15 | +0.7 | −0 | +0.5 | ... | +0.5 | +0.8 |
| | | +0 | −1.5 | +0 | −1.1 | +0 | −0.45 | +0 | −0.5 | +0 | ... | +0 | +0.5 |
| 0.24 | 0.40 | +0.9 | −1.0 | +0.9 | −0.5 | +0.6 | −0.2 | +0.9 | −0 | +0.6 | +0.5 | +0.6 | +1.0 |
| | | +0 | −1.9 | +0 | −1.4 | +0 | −0.6 | 0 | −0.6 | +0 | +0.1 | +0 | +0.6 |
| 0.40 | 0.71 | +1.0 | −1.2 | +1.0 | −0.6 | +0.7 | −0.25 | +1.0 | −0 | +0.7 | +0.5 | +0.7 | +1.1 |
| | | +0 | −2.2 | +0 | −1.6 | +0 | −0.65 | +0 | −0.7 | +0 | +0.1 | +0 | +0.7 |
| 0.71 | 1.19 | +1.2 | −1.6 | +1.2 | −0.8 | +0.8 | −0.3 | +1.2 | −0 | +0.8 | +0.6 | +0.8 | +1.3 |
| | | +0 | −2.8 | +0 | −2.0 | +0 | −0.8 | +0 | −0.8 | +0 | +0.1 | +0 | +0.8 |
| 1.19 | 1.97 | +1.6 | −2.0 | +1.6 | −1.0 | +1.0 | −0.4 | +1.6 | −0 | +1.0 | +0.7 | +1.0 | +1.6 |
| | | +0 | −3.6 | +0 | −2.6 | +0 | −1.0 | +0 | −1.0 | +0 | +0.1 | +0 | +1.0 |
| 1.97 | 3.15 | +1.8 | −2.5 | +1.8 | −1.2 | +1.2 | −0.4 | +1.8 | −0 | +1.2 | +0.8 | +1.2 | +2.1 |
| | | +0 | −4.3 | +0 | −3.0 | +0 | −1.1 | +0 | −1.2 | +0 | +0.1 | +0 | +1.4 |
| 3.15 | 4.73 | +2.2 | −3.0 | +2.2 | −1.4 | +1.4 | −0.5 | +2.2 | −0 | +1.4 | +1.0 | +1.4 | +2.5 |
| | | +0 | −5.2 | +0 | −3.6 | +0 | −1.4 | +0 | −1.4 | +0 | +0.1 | +0 | +1.6 |
| 4.73 | 7.09 | +2.5 | −3.5 | +2.5 | −1.6 | +1.6 | −0.6 | +2.5 | −0 | +1.6 | +1.1 | +1.6 | +2.8 |
| | | +0 | −6.0 | +0 | −4.1 | +0 | −1.6 | +0 | −1.6 | +0 | +0.1 | +0 | +1.8 |
| 7.09 | 9.85 | +2.8 | −4.0 | +2.8 | −2.0 | +1.8 | −0.6 | +2.8 | −0 | +1.8 | +1.4 | +1.8 | +3.2 |
| | | +0 | −6.8 | +0 | −4.8 | +0 | −1.8 | +0 | −1.8 | +0 | +0.2 | +0 | +2.0 |
| 9.85 | 12.41 | +3.0 | −4.5 | +3.0 | −2.2 | +2.0 | −0.7 | +3.0 | −0 | +2.0 | +1.4 | +2.0 | +3.4 |
| | | +0 | −7.5 | +0 | −5.2 | +0 | −1.9 | +0 | −2.0 | +0 | +0.2 | +0 | +2.2 |
| 12.41 | 15.75 | +3.5 | −5.0 | +3.5 | −2.5 | +2.2 | −0.7 | +3.5 | −0 | +2.2 | +1.6 | +2.2 | +3.9 |
| | | +0 | −8.5 | +0 | −6.0 | +0 | −2.1 | +0 | −2.2 | +0 | +0.2 | +0 | +2.5 |
| 15.75 | 19.69 | +4.0 | −5.0 | +4.0 | −2.8 | +2.5 | −0.8 | +4.0 | −0 | +2.5 | +1.8 | +2.5 | +4.4 |
| | | +0 | −9.0 | +0 | −6.8 | +0 | −2.4 | +0 | −2.5 | +0 | +0.2 | +0 | +2.8 |

*Approximate Designation of Type of Fit* — *Tolerance Limits for Holes and Shafts* — Values shown below are in thousandths of an inch.

Symbols H8, e8, etc. are hole and shaft designations in ABC system.

Where fits other than those described are required British Standard 1916: **Parts 1 and 2: 1953** should be consulted.

### Table 2.    British Standard Secondary Selection of Fits (B.S. 1916D: 1954)

Tolerance limits given in body of table are added or subtracted to basic size (as indicated by + or − sign) to obtain maximum and minimum sizes of mating parts.

| Size Range, Inches | | Coarse or Loose | | Precision Running | | Precision Location | | Keying Fit | | Press (Non-ferrous parts) | |
|---|---|---|---|---|---|---|---|---|---|---|---|
| | | \<--- Approximate Designation of Type of Fit ---\> | | | | | | | | | |
| | | \<--- Tolerance Limits for Holes and Shafts ---\> | | | | | | | | | |
| Over | To | H8 | d9 | H7 | f7 | H6 | g5 | H8 | j7 | H7 | s6 |
| | | \<--- Values shown below are in thousandths of an inch ---\> | | | | | | | | | |
| 0.04 | 0.12 | +0.6 / +0 | −1.0 / −2.0 | +0.4 / +0 | −0.3 / −0.7 | +0.25 / +0 | −0.1 / −0.3 | +0.6 / +0 | +0.3 / −0.1 | +0.4 / +0 | +0.85 / +0.6 |
| 0.12 | 0.24 | +0.7 / +0 | −1.2 / −2.4 | +0.5 / +0 | −0.4 / −0.9 | +0 / +0 | −0.15 / −0.35 | +0.7 / +0 | +0.4 / −0.1 | +0.5 / +0 | +1.0 / +0.7 |
| 0.24 | 0.40 | +0.9 / +0 | −1.6 / −3.0 | +0.6 / +0 | −0.5 / −1.1 | +0.4 / +0 | −0.2 / −0.45 | +0.9 / +0 | +0.4 / −0.2 | +0.6 / +0 | +1.4 / +1.0 |
| 0.40 | 0.71 | +1.0 / +0 | −2.0 / −3.6 | +0.7 / +0 | −0.6 / −1.3 | +0.4 / +0 | −0.25 / −0.55 | +1.0 / +0 | +0.5 / −0.2 | +0.7 / +0 | +1.6 / +1.2 |
| 0.71 | 1.19 | +1.2 / +0 | −2.5 / −4.5 | +0.8 / +0 | −0.8 / −1.6 | +0.5 / +0 | −0.4 / −0.7 | +1.2 / +0 | +0.5 / −0.3 | +0.8 / +0 | +1.9 / +1.4 |
| 1.19 | 1.97 | +1.6 / +0 | −3.0 / −5.5 | +1.0 / +0 | −1.0 / −2.0 | +0.6 / +0 | −0.4 / −0.8 | +1.6 / +0 | +0.6 / −0.4 | +1.0 / +0 | +2.4 / +1.8 |
| 1.97 | 2.56 | +1.8 / +0 | −4.0 / −7.0 | +1.2 / +0 | −1.2 / −2.4 | +0.7 / +0 | −0.4 / −0.9 | +1.8 / +0 | +0.7 / −0.5 | +1.2 / +0 | +2.7 / +2.0 |
| 2.56 | 3.15 | | | | | | | | | +1.2 / +0 | +2.9 / +2.2 |
| 3.15 | 3.94 | +2.2 / +0 | −5.0 / −8.5 | +1.4 / +0 | −1.4 / −2.8 | +0.9 / +0 | −0.5 / −1.1 | +2.2 / +0 | +0.8 / −0.6 | +1.4 / +0 | +3.7 / +2.8 |
| 3.94 | 4.73 | | | | | | | | | +1.4 / +0 | +3.9 / +3.0 |
| 4.73 | 5.52 | +2.5 / +0 | −6.0 / −10.0 | +1.6 / +0 | −1.6 / −3.2 | +1.0 / +0 | −0.6 / −1.3 | +2.5 / +0 | +0.9 / −0.7 | +1.6 / +0 | +4.5 / +3.5 |
| 5.52 | 6.30 | | | | | | | | | +1.6 / +0 | +5.0 / +4.0 |
| 6.30 | 7.09 | | | | | | | | | +1.6 / +0 | +5.5 / +4.5 |
| 7.09 | 7.88 | +2.8 / +0 | −7.0 / −11.5 | +1.8 / +0 | −2.0 / −3.8 | +1.2 / +0 | −0.6 / −1.4 | +2.8 / +0 | +1.0 / −0.8 | +1.8 / +0 | +6.2 / +5.0 |
| 7.88 | 8.86 | | | | | | | | | +1.8 / +0 | +6.2 / +5.0 |
| 8.86 | 9.85 | | | | | | | | | +1.8 / +0 | +7.2 / +6.0 |
| 9.85 | 11.03 | +3.0 / +0 | −7.0 / −12.0 | +2.0 / +0 | −2.2 / −4.2 | +1.2 / +0 | −0.7 / −1.6 | +3.0 / +0 | +1.0 / −1.0 | +2.0 / +0 | +7.2 / +6.0 |
| 11.03 | 12.41 | | | | | | | | | +2.0 / +0 | +8.2 / +7.0 |
| 12.41 | 13.98 | +3.5 / +0 | −8.0 / −14.0 | +2.2 / +0 | −2.5 / −4.7 | +1.4 / +0 | −0.7 / −1.7 | +3.5 / +0 | +1.2 / −1.0 | +2.2 / +0 | +8.4 / +7.0 |
| 13.98 | 15.75 | | | | | | | | | +2.2 / +0 | +9.4 / +8.0 |
| 15.75 | 17.72 | +4.0 / +0 | −9.0 / −15.0 | +2.5 / +0 | −2.8 / −5.3 | +1.6 / +0 | −0.8 / −1.8 | +4.0 / +0 | +1.3 / −1.2 | +2.5 / +0 | +10.6 / +9.0 |
| 17.72 | 19.69 | | | | | | | | | +2.5 / +0 | +11.6 / +10.0 |

Symbols H8, d9, etc. are hole and shaft designations in ABC system.

**Length Differences per Inch for Change from Standard Temperature of 68 Degrees F**

| Temper-ature Deg. F. | Coefficient of thermal expansion of material per degree F. × 10⁶ | | | | | | | | | |
|---|---|---|---|---|---|---|---|---|---|---|
| | 1 | 2 | 3 | 4 | 5 | 10 | 15 | 20 | 25 | 30 |
| | Total change in length from standard, microinches per inch of length* | | | | | | | | | |
| 40 | −28 | −56 | −84 | −112 | −140 | −280 | −420 | −560 | −700 | −840 |
| 41 | −27 | −54 | −81 | −108 | −135 | −270 | −405 | −540 | −675 | −810 |
| 42 | −26 | −52 | −78 | −104 | −130 | −260 | −390 | −520 | −650 | −780 |
| 43 | −25 | −50 | −75 | −100 | −125 | −250 | −375 | −500 | −625 | −750 |
| 44 | −24 | −48 | −72 | − 96 | −120 | −240 | −360 | −480 | −600 | −720 |
| 45 | −23 | −46 | −69 | − 92 | −115 | −230 | −345 | −460 | −575 | −690 |
| 46 | −22 | −44 | −66 | − 88 | −110 | −220 | −330 | −440 | −550 | −660 |
| 47 | −21 | −42 | −63 | − 84 | −105 | −210 | −315 | −420 | −525 | −630 |
| 48 | −20 | −40 | −60 | − 80 | −100 | −200 | −300 | −400 | −500 | −600 |
| 49 | −19 | −38 | −57 | − 76 | − 95 | −190 | −285 | −380 | −475 | −570 |
| 50 | −18 | −36 | −54 | − 72 | − 90 | −180 | −270 | −360 | −450 | −540 |
| 51 | −17 | −34 | −51 | − 68 | − 85 | −170 | −255 | −340 | −425 | −510 |
| 52 | −16 | −32 | −48 | − 64 | − 80 | −160 | −240 | −320 | −400 | −480 |
| 53 | −15 | −30 | −45 | − 60 | − 75 | −150 | −225 | −300 | −375 | −450 |
| 54 | −14 | −28 | −42 | − 56 | − 70 | −140 | −210 | −280 | −350 | −420 |
| 55 | −13 | −26 | −39 | − 52 | − 65 | −130 | −195 | −260 | −325 | −390 |
| 56 | −12 | −24 | −36 | − 48 | − 60 | −120 | −180 | −240 | −300 | −360 |
| 57 | −11 | −22 | −33 | − 44 | − 55 | −110 | −165 | −220 | −275 | −330 |
| 58 | −10 | −20 | −30 | − 40 | − 50 | −100 | −150 | −200 | −250 | −300 |
| 59 | − 9 | −18 | −27 | − 36 | − 45 | − 90 | −135 | −180 | −225 | −270 |
| 60 | − 8 | −16 | −24 | − 32 | − 40 | − 80 | −120 | −160 | −200 | −240 |
| 61 | − 7 | −14 | −21 | − 28 | − 35 | − 70 | −105 | −140 | −175 | −210 |
| 62 | − 6 | −12 | −18 | − 24 | − 30 | − 60 | − 90 | −120 | −150 | −180 |
| 63 | − 5 | −10 | −15 | − 20 | − 25 | − 50 | − 75 | −100 | −125 | −150 |
| 64 | − 4 | − 8 | −12 | − 16 | − 20 | − 40 | − 60 | − 80 | −100 | −120 |
| 65 | − 3 | − 6 | − 9 | − 12 | − 15 | − 30 | − 45 | − 60 | − 75 | − 90 |
| 66 | − 2 | − 4 | − 6 | − 8 | − 10 | − 20 | − 30 | − 40 | − 50 | − 60 |
| 67 | − 1 | − 2 | − 3 | − 4 | − 5 | − 10 | − 15 | − 20 | − 25 | − 30 |
| 68 | 0 | 0 | 0 | 0 | 0 | 0 | 0 | 0 | 0 | 0 |
| 69 | 1 | 2 | 3 | 4 | 5 | 10 | 15 | 20 | 25 | 30 |
| 70 | 2 | 4 | 6 | 8 | 10 | 20 | 30 | 40 | 50 | 60 |
| 71 | 3 | 6 | 9 | 12 | 15 | 30 | 45 | 60 | 75 | 90 |
| 72 | 4 | 8 | 12 | 16 | 20 | 40 | 60 | 80 | 100 | 120 |
| 73 | 5 | 10 | 15 | 20 | 25 | 50 | 75 | 100 | 125 | 150 |
| 74 | 6 | 12 | 18 | 24 | 30 | 60 | 90 | 120 | 150 | 180 |
| 75 | 7 | 14 | 21 | 28 | 35 | 70 | 105 | 140 | 175 | 210 |
| 76 | 8 | 16 | 24 | 32 | 40 | 80 | 120 | 160 | 200 | 240 |
| 77 | 9 | 18 | 27 | 36 | 45 | 90 | 135 | 180 | 225 | 270 |
| 78 | 10 | 20 | 30 | 40 | 50 | 100 | 150 | 200 | 250 | 300 |
| 79 | 11 | 22 | 33 | 44 | 55 | 110 | 165 | 220 | 275 | 330 |
| 80 | 12 | 24 | 36 | 48 | 60 | 120 | 180 | 240 | 300 | 360 |
| 81 | 13 | 26 | 39 | 52 | 65 | 130 | 195 | 260 | 325 | 390 |
| 82 | 14 | 28 | 42 | 56 | 70 | 140 | 210 | 280 | 350 | 420 |
| 83 | 15 | 30 | 45 | 60 | 75 | 150 | 225 | 300 | 375 | 450 |
| 84 | 16 | 32 | 48 | 64 | 80 | 160 | 240 | 320 | 400 | 480 |
| 85 | 17 | 34 | 51 | 68 | 85 | 170 | 255 | 340 | 425 | 510 |
| 86 | 18 | 36 | 54 | 72 | 90 | 180 | 270 | 360 | 450 | 540 |
| 87 | 19 | 38 | 57 | 76 | 95 | 190 | 285 | 380 | 475 | 570 |
| 88 | 20 | 40 | 60 | 80 | 100 | 200 | 300 | 400 | 500 | 600 |
| 89 | 21 | 42 | 63 | 84 | 105 | 210 | 315 | 420 | 525 | 630 |
| 90 | 22 | 44 | 66 | 88 | 110 | 220 | 330 | 440 | 550 | 660 |
| 91 | 23 | 46 | 69 | 92 | 115 | 230 | 345 | 460 | 575 | 690 |
| 92 | 24 | 48 | 72 | 96 | 120 | 240 | 360 | 480 | 600 | 720 |
| 93 | 25 | 50 | 75 | 100 | 125 | 250 | 375 | 500 | 625 | 750 |
| 94 | 26 | 52 | 78 | 104 | 130 | 260 | 390 | 520 | 650 | 780 |
| 95 | 27 | 54 | 81 | 108 | 135 | 270 | 405 | 540 | 675 | 810 |

*See footnotes at end of table on following page.

**Finding Length Differences Due to Temperature Changes.** — The tables on this page and the preceding page give changes in length for changes in temperature from the standard reference temperature of 68 deg F (20 deg C). Thus, for example, in a steel bar with a coefficient of thermal expansion of .0000063 inch per inch per deg F, the increase in length at 73 deg F is 25 + 5 + 1.5 = 31.5 microinches per inch of length.

### Length Differences per Centimeter for Change from Standard Temperature of 20 Degrees Celsius (Centigrade)

| Temper- ature Deg. C. | Coefficient of thermal expansion of material per degree C. × 10⁶ | | | | | | | | | |
|---|---|---|---|---|---|---|---|---|---|---|
| | 1 | 2 | 3 | 4 | 5 | 10 | 15 | 20 | 25 | 30 |
| | Total change in length from standard, hundredths of microns (microns/100) per centimeter of length† | | | | | | | | | |
| 0 | −20 | −40 | −60 | −80 | −100 | −200 | −300 | −400 | −500 | −600 |
| 1 | −19 | −38 | −57 | −76 | − 95 | −190 | −285 | −380 | −475 | −570 |
| 2 | −18 | −36 | −54 | −72 | − 90 | −180 | −270 | −360 | −450 | −540 |
| 3 | −17 | −34 | −51 | −68 | − 85 | −170 | −255 | −340 | −425 | −510 |
| 4 | −16 | −32 | −48 | −64 | − 80 | −160 | −240 | −320 | −400 | −480 |
| 5 | −15 | −30 | −45 | −60 | − 75 | −150 | −225 | −300 | −375 | −450 |
| 6 | −14 | −28 | −42 | −56 | − 70 | −140 | −210 | −280 | −350 | −420 |
| 7 | −13 | −26 | −39 | −52 | − 65 | −130 | −195 | −260 | −325 | −390 |
| 8 | −12 | −24 | −36 | −48 | − 60 | −120 | −180 | −240 | −300 | −360 |
| 9 | −11 | −22 | −33 | −44 | − 55 | −110 | −165 | −220 | −275 | −330 |
| 10 | −10 | −20 | −30 | −40 | − 50 | −100 | −150 | −200 | −250 | −300 |
| 11 | − 9 | −18 | −27 | −36 | − 45 | − 90 | −135 | −180 | −225 | −270 |
| 12 | − 8 | −16 | −24 | −32 | − 40 | − 80 | −120 | −160 | −200 | −240 |
| 13 | − 7 | −14 | −21 | −28 | − 35 | − 70 | −105 | −140 | −175 | −210 |
| 14 | − 6 | −12 | −18 | −24 | − 30 | − 60 | − 90 | −120 | −150 | −180 |
| 15 | − 5 | −10 | −15 | −20 | − 25 | − 50 | − 75 | −100 | −125 | −150 |
| 16 | − 4 | − 8 | −12 | −16 | − 20 | − 40 | − 60 | − 80 | −100 | −120 |
| 17 | − 3 | − 6 | − 9 | −12 | − 15 | − 30 | − 45 | − 60 | − 75 | − 90 |
| 18 | − 2 | − 4 | − 6 | − 8 | − 10 | − 20 | − 30 | − 40 | − 50 | − 60 |
| 19 | − 1 | − 2 | − 3 | − 4 | − 5 | − 10 | − 15 | − 20 | − 25 | − 30 |
| 20 | 0 | 0 | 0 | 0 | 0 | 0 | 0 | 0 | 0 | 0 |
| 21 | 1 | 2 | 3 | 4 | 5 | 10 | 15 | 20 | 25 | 30 |
| 22 | 2 | 4 | 6 | 8 | 10 | 20 | 30 | 40 | 50 | 60 |
| 23 | 3 | 6 | 9 | 12 | 15 | 30 | 45 | 60 | 75 | 90 |
| 24 | 4 | 8 | 12 | 16 | 20 | 40 | 60 | 80 | 100 | 120 |
| 25 | 5 | 10 | 15 | 20 | 25 | 50 | 75 | 100 | 125 | 150 |
| 26 | 6 | 12 | 18 | 24 | 30 | 60 | 90 | 120 | 150 | 180 |
| 27 | 7 | 14 | 21 | 28 | 35 | 70 | 105 | 140 | 175 | 210 |
| 28 | 8 | 16 | 24 | 32 | 40 | 80 | 120 | 160 | 200 | 240 |
| 29 | 9 | 18 | 27 | 36 | 45 | 90 | 135 | 180 | 225 | 270 |
| 30 | 10 | 20 | 30 | 40 | 50 | 100 | 150 | 200 | 250 | 300 |
| 31 | 11 | 22 | 33 | 44 | 55 | 110 | 165 | 220 | 275 | 330 |
| 32 | 12 | 24 | 36 | 48 | 60 | 120 | 180 | 240 | 300 | 360 |
| 33 | 13 | 26 | 39 | 52 | 65 | 130 | 195 | 260 | 325 | 390 |
| 34 | 14 | 28 | 42 | 56 | 70 | 140 | 210 | 280 | 350 | 420 |
| 35 | 15 | 30 | 45 | 60 | 75 | 150 | 225 | 300 | 375 | 450 |
| 36 | 16 | 32 | 48 | 64 | 80 | 160 | 240 | 320 | 400 | 480 |
| 37 | 17 | 34 | 51 | 68 | 85 | 170 | 255 | 340 | 425 | 510 |
| 38 | 18 | 36 | 54 | 72 | 90 | 180 | 270 | 360 | 450 | 540 |
| 39 | 19 | 38 | 57 | 76 | 95 | 190 | 285 | 380 | 475 | 570 |
| 40 | 20 | 40 | 60 | 80 | 100 | 200 | 300 | 400 | 500 | 600 |

For intermediate coefficients add appropriate listed values. For example, a length change for a coefficient of 7 is the sum of values in the 5 and 2 columns. Fractional interpolation may be similarly calculated.

* Or hundredths of micron (microns/100) per centimeter (see table on preceding page).
† Or microinches per inch.

# MEASURING INSTRUMENTS AND GAGING METHODS

**Reading a Vernier.** — A general rule for taking readings with a vernier scale is as follows: Note the number of inches and sub-divisions of an inch that the zero mark of the vernier scale has moved along the true scale, and then add to this reading as many thousandths, or hundredths, or whatever fractional part of an inch the vernier reads to, as there are spaces between the vernier zero and that line on the vernier which coincides with one on the true scale. For example, if the zero line of a vernier which reads to thousandths is slightly beyond the 0.5 inch division on the main or true scale, as shown in Fig. 1, and graduation line 10 on the vernier exactly coincides with one on the true scale, the reading is 0.5 + 0.010 or 0.510 inch. In order to determine the reading or fractional part of an inch that can be obtained by a vernier, multiply the denominator of the finest sub-division given on the true scale by the total number of divisions on the vernier. For example, if one inch on the true scale is divided into 40 parts or fortieths (as in Fig. 1), and the vernier into twenty-five parts, the vernier will read to thousandths of an inch, as 25 × 40 = 1000. Similarly, if there are sixteen divisions to the inch on the true scale and a total of eight on the vernier, the latter will enable readings within one-hundred-twenty-eighths of an inch to be taken, as 8 × 16 = 128.

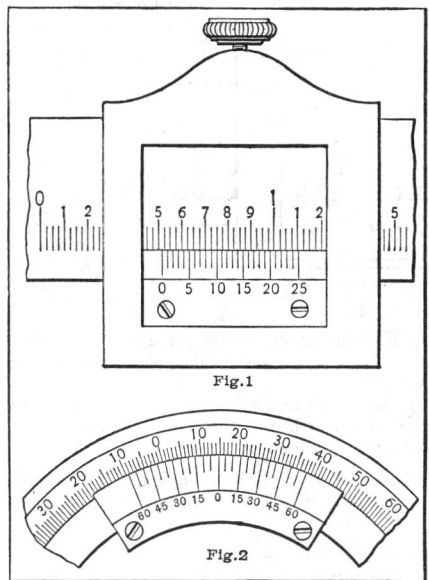

Fig.1

Fig.2

If the vernier is on a protractor, note the whole number of degrees passed by the vernier zero mark and then count the spaces between the vernier zero and that line which coincides with a graduation on the protractor scale. If the vernier indicates angles within five minutes or one-twelfth degree (as in Fig. 2), the number of spaces multiplied by 5 will, of course, give the number of minutes to be added to the whole number of degrees. The reading of the protractor set as illustrated would be 14 whole degrees (the number passed by the zero mark on the vernier) plus 30 minutes, as the graduation 30 on the vernier is the only one to the right of the vernier zero which exactly coincides with a line on the protractor scale. It will be noted that there are duplicate scales on the vernier, one being to the right and the other to the left of zero. The left-hand scale is used when the vernier zero is moved to the left of the zero of the protractor scale, whereas the right-hand graduations are used when the movement is to the right.

**Reading a Metric Vernier.** — The smallest graduation on the bar (true or main scale) of the metric vernier gage shown in Fig. 1, is 0.5 millimeter. The scale is numbered at each twentieth divison, and thus increments of 10, 20, 30, 40 millimeters, etc., are indicated. There are 25 divisions on the vernier scale, occupying the same length as 24 divisions on the bar, which is 12 millimeters. Therefore, one division on the vernier scale equals one twenty-fifth of 12 millimeters = 0.04 × 12 = 0.48 millimeter. Thus, the difference between one bar division (0.50 mm) and one vernier division (0.48 mm) is 0.50 − 0.48 = 0.02 millimeter, which is the minimum measuring increment that the gage provides. To permit direct readings, the vernier scale has graduations to represent tenths of a millimeter (0.1 mm) and fiftieths of a millimeter (0.02 mm).

To read a vernier gage, first note how many millimeters the zero line on the vernier is from the zero line on the bar. Next, find the graduation on the vernier

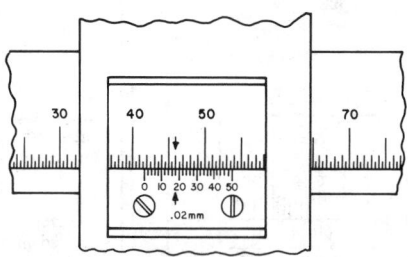

Fig. 1.

scale which exactly coincides with a graduation line on the bar, and note the value of the vernier scale graduation. This value is added to the value obtained from the bar, and the result is the total reading.

In the example shown in Fig. 1, the vernier zero is just past the 41.5 millimeters graduation on the bar. The 0.18 millimeter line on the vernier coincides with a line on the bar, and the total reading is therefore 41.5 + 0.18 = 41.68 mm.

**Dual Metric-Inch Vernier.** — The vernier gage shown in Fig. 1 has separate metric and inch 50-division vernier scales to permit measurements in either system.

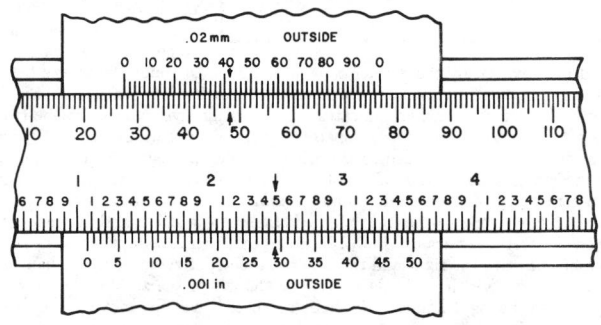

Fig. 1.

A 50-division vernier has more widely spaced graduations than the 25-division vernier shown on the previous pages, and is thus easier to read. On the bar, the smallest metric graduation is 1 millimeter, and the 50 divisions of the vernier occupy the same length as 49 divisions on the bar, which is 49 mm. Therefore, one division on the vernier scale equals one-fiftieth of 49 millimeters = 0.02 × 49 = 0.98 mm. Thus, the difference between one bar division (1.0 mm) and one vernier division (0.98 mm) is 0.02 mm, which is the minimum measuring increment the gage provides.

The vernier scale is graduated for direct reading to 0.02 mm. In the figure, the vernier zero is just past the 27 mm graduation on the bar, and the 0.42 mm graduation on the vernier coincides with a line on the bar. The total reading is therefore 27.42 mm.

The smallest inch graduation on the bar is 0.05 inch, and the 50 vernier divisions occupy the same length as 49 bar divisions, which is 2.45 inches. Therefore, one vernier division equals one-fiftieth of 2.45 inches = 0.02 × 2.45 = 0.049 inch. Thus, the difference between the length of a bar division and a vernier division is 0.050 − 0.049 = 0.001 inch. The vernier scale is graduated for direct reading to 0.001 inch. In the example, the vernier zero is past the 1.05 graduation on the bar, and the 0.029 graduation on the vernier coincides with a line on the bar. Thus, the total reading is 1.079 inches.

**Reading a Micrometer.** — To read a micrometer, count the number of whole divisions that are visible on the scale of the frame, multiply this number by 25 (the number of thousandths of an inch that each division represents) and add to the product the number of that division on the thimble which coincides with the axial zero line on the frame. The result will be the diameter expressed in thousandths of an inch. As the numbers 1, 2, 3, etc., opposite every fourth sub-division on the frame, indicate hundreds of thousandths, the reading can easily be taken mentally. Suppose the thimble were screwed out so that graduation 2, and three additional sub-divisions, were visible (as shown in Fig. 1), and that graduation 10 on the thimble coincided with the axial line on the frame. The reading then would be 0.200 + 0.075 + 0.010, or 0.285 inch.

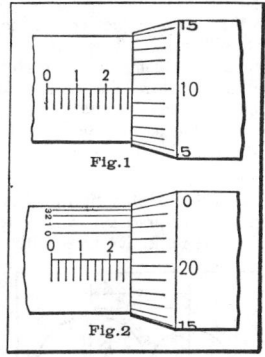

Fig. 1

Fig. 2

Some micrometers have a vernier scale on the frame in addition to the regular graduations, so that measurements within 0.0001 part of an inch can be taken. Micrometers of this type are read as follows: First determine the number of thousandths, as with an ordinary micrometer, and then find a line on the vernier scale that exactly coincides with one on the thimble; the number of this line represents the number of ten-thousandths to be added to the number of thousandths obtained by the regular graduations. The reading shown in the illustration, Fig. 2, is 0.270 + 0.0003 = 0.2703 inch.

Micrometers graduated according to the English system of measurement ordinarily have a table of decimal equivalents stamped on the sides of the frame, so that fractions such as sixty-fourths, thirty-seconds, etc., can readily be converted into decimals.

**Reading a Metric Micrometer.** — The spindle of an ordinary metric micrometer has 50 threads per 25 millimeters, and thus 50 complete revolutions from zero moves the spindle through a distance of 25 millimeters. One revolution therefore moves the spindle 0.5 millimeter. The longitudinal line on the sleeve is graduated with 1 millimeter divisions and 0.5 millimeter subdivisions. The thimble has 50 graduations, each being 0.01 millimeter (one-hundredth of a millimeter).

To read a metric micrometer, note the number of millimeter divisions visible on the scale of the sleeve, and add the total to the particular division on the thimble which coincides with the axial line on the sleeve. Suppose that the thimble were screwed out so that graduation 5, and one additional 0.5 subdivision were visible (as shown in Fig. 1), and that graduation 28 on the thimble coincided with the axial line on the sleeve. The reading then would be 5.00 + 0.5 + 0.28 = 5.78 mm.

Some micrometers are provided with a vernier scale on the sleeve in addition to the regular graduations to permit measurements within 0.002 millimeter to be made. Micrometers of this type are read as follows: First determine the number of whole millimeters (if any) and the number of hundredths of a millimeter, as with an ordinary micrometer, and then find a line on the sleeve vernier scale which exactly

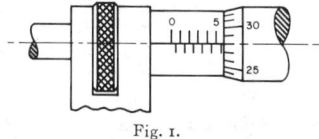

Fig. 1.

coincides with one on the thimble. The number of this coinciding vernier line represents the number of two-thousandths of a millimeter to be added to the reading already obtained. Thus, for example, a measurement of 2.958 millimeters would be obtained by reading 2.5 millimeters on the sleeve, adding 0.45 millimeter read from the thimble, and then adding 0.008 millimeter as determined by the vernier.

*Note:* 0.01 millimeter = 0.000393 inch, and 0.002 millimeter = 0.000078 inch (78 millionths). Therefore, metric micrometers provide smaller measuring increments than comparable inch unit micrometers — the smallest graduation of an ordinary inch reading micrometer is 0.001 inch; the vernier type has graduations down to 0.0001 inch. When using either a metric or inch micrometer, without a vernier, smaller readings than those graduated may of course be obtained by visual interpolation between graduations.

**Locating Holes by the Disk Method.**— When machining holes in comparatively small precision work, three carefully centered disks are sometimes used to align the respective holes with the lathe spindle. These disks are made to such

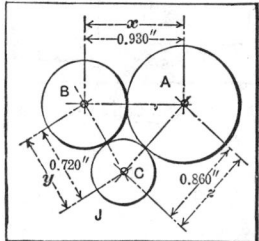

diameters that when their peripheries are in contact, each disk center will coincide with the position of the hole to be machined; the centers are then used for locating the work. The diameters of the disks can be found as follows: Subtract dimension $y$ from $x$, thus obtaining the difference between the radii of disks $C$ and $A$; add this difference to dimension $z$ to obtain the diameter of disk $A$; subtract the radius of disk $A$ from the center distance $x$ to obtain the radius of $B$; subtract the radius of $B$ from dimension $y$ to obtain the radius of $C$.

*Example:* 0.930 − 0.720 = 0.210, or the difference between the radii of disks $C$ and $A$. The diameter of $A$ equals 0.210 + 0.860 = 1.070 inch, and the radius equals 1.070 ÷ 2 = 0.535 inch. The radius of $B$ = 0.930 − 0.535 = 0.395 inch. The radius of $C$ = 0.720 − 0.395 = 0.325.

**Sine-bar.** — The sine-bar is used either for very accurate angular measurements or for locating work at a given angle as, for example, in surface grinding templets, gages, etc. The sine-bar is especially useful in measuring or checking angles when the limit of accuracy is 5 minutes or less. Some bevel protractors are equipped with verniers which read to 5 minutes but the setting depends upon the alignment of graduations whereas a sine-bar usually is located by positive contact with precision gage-blocks selected for whatever dimension is required for obtaining a given angle.

**Types of Sine-bars.** — A sine-bar consists of a hardened, ground and lapped steel bar which has very accurate cylindrical plugs of equal diameter attached to or near each end. The form illustrated by Fig. 1 has notched ends for receiving the cylindrical plugs which are held firmly against both faces of the notch. The standard center-to-center distance $C$ between the plugs is either 5 or 10 inches. The upper and lower sides of sine-bars are parallel to the center line of the plugs within very close limits. The body of the sine-bar ordinarily has several holes through it to reduce the weight. In the making of the sine-bar shown in Fig. 2, if too much material is removed from one locating notch, regrinding the shoulder at the opposite end would make it possible to obtain the correct center distance. That is the reason for this change in form. The type of sine-bar illustrated by Fig. 3 has the cylindrical disks or plugs attached to one side. These differences in form or arrangement do not, of course, affect the principle governing the use of the sine-bar. An accurate surface plate or master flat is always used in conjunction with a sine-bar in order to form the base from which the vertical measurements are made.

**Setting 5-inch Sine-bar to Given Angle.** — Since many sine-bars have a length of 5 inches, the accompanying table of constants is based upon that length. These constants represent the vertical distances $H$ for setting a 5-inch sine-bar to the required angle. Assume that the angle is 31° 20′; the table shows that height $H$

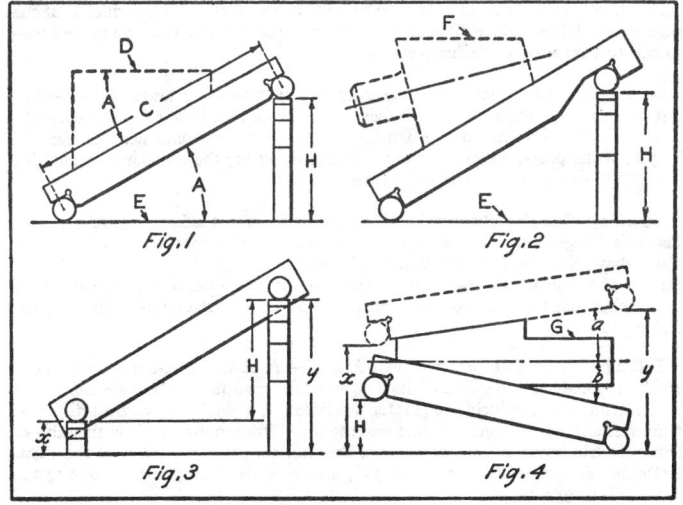

Fig. 1   Fig. 2   Fig. 3   Fig. 4

(Figs. 1, 2 and 3) should equal 2.6001 inches. Note: The constants in table equal five times sine of angle; thus the sine of 31° 20' (see table of trigonometric functions) is 0.52002, and 0.52002 × 5 = 2.6001 inches.

**Finding Angle when Height $H$ of Sine-bar is Known.** — In finding the angle equivalent to a given height $H$, the table of constants is used in reverse order. To illustrate, if the height $H$ is 1.4061 inches, the angle to which the sine-bar is set is 16° 20'. (Note: In using the regular table of sines, divide height $H$ by length of sine-bar, find sine equal to quotient and its angle; thus 1.4061 ÷ 5 = 0.2812. Table of sines shows that this is the sine of 16° 20'.)

**Checking Angle of Templet or Gage by Using Sine-bar.** — Place templet or gage on sine-bar as indicated by dotted lines, Fig. 1. Clamps may be used to hold work in place. Place upper end of sine-bar on gage-blocks having total height $H$ corresponding to the required angle. If upper edge $D$ of work is parallel with surface plate $E$, then angle $A$ of work equals angle $A$ to which sine-bar is set. Parallelism between edge $D$ and surface plate may be tested by checking the height at each end with a dial gage or some indicating type of comparator.

**Measuring Angle of Templet or Gage with Sine-bar.** — Adjust height of gage-blocks and sine-bar until edge $D$, Fig. 1, of gage or templet is parallel with surface plate $E$; then find angle corresponding to height $H$ of gage-blocks. For example, if height $H$ is 2.5938 inches when $D$ and $E$ are parallel, the table of sine-bar constants shows that angle $A$ of work is 31° 15'.

**Checking Taper per Foot with Sine-bar.** — As an example, assume that plug gage, Fig. 2, is supposed to have a taper of 6⅛ inches per foot and taper is to be checked by using a 5-inch sine-bar. The accompanying table, Tapers per Foot and Corresponding Angles, shows that the included angle for a taper of 6⅛ inches per foot is 28° 37' 59" or practically 28° 38'. The table of sine-bar constants shows that height $H$ of sine-bar should be 2.3960 inches; hence, if the upper surface $F$ of the gage is parallel to surface $E$ when sine-bar is at the height given, the angle corresponds to a taper of 6⅛ inches per foot.

**Setting Sine-bar which has Plugs Attached to Side.** — If lower plug does not rest directly upon the surface plate (see Fig. 3), note that height $H$ for setting the sine-bar is the difference between heights $x$ and $y$ or the difference between the heights of the plugs; otherwise the procedure in setting the sine-bar and checking angles is the same as previously described.

**Checking Templet Having Two Angles.** — Assume that angle $a$ of templet, Fig. 4, is 9 degrees, angle $b$ 12 degrees, and that edge $G$ is parallel to surface plate. Table shows that height $H$ equals 1.0395 inches for an angle $b$ of 12 degrees. For an angle $a$ of 9 degrees, the table shows that the difference between measurements $x$ and $y$ when sine-bar is in contact with the upper edge of the templet should equal 0.78215 inch.

**Setting 10-inch Sine-bar to Given Angle.** — A 10-inch sine-bar may be preferable in some cases because of its longer working surface or because the longer center distance is conducive to greater precision. To obtain the vertical distances $H$ for setting a 10-inch sine-bar, first find the sine of the required angle in the table of trigonometric functions and then move the decimal point one place to the right. Example: Sine of 39 degrees, is 0.62932; hence, vertical height $H$ for setting the sine-bar is 6.2932 inches.

## Constants for Setting a 5-inch Sine-bar

| Min. | 0° | 1° | 2° | 3° | 4° | 5° | 6° | 7° |
|---|---|---|---|---|---|---|---|---|
| 0 | 0.00000 | 0.08725 | 0.17450 | 0.26170 | 0.34880 | 0.43580 | 0.52265 | 0.60935 |
| 1 | .00145 | .08870 | .17595 | .26315 | .35025 | .43720 | .52410 | .61080 |
| 2 | .00290 | .09015 | .17740 | .26460 | .35170 | .43865 | .52555 | .61225 |
| 3 | .00435 | .09160 | .17885 | .26605 | .35315 | .44010 | .52700 | .61370 |
| 4 | .00580 | .09310 | .18030 | .26750 | .35460 | .44155 | .52845 | .61510 |
| 5 | 0.00725 | 0.09455 | 0.18175 | 0.26895 | 0.35605 | 0.44300 | 0.52985 | 0.61655 |
| 6 | .00875 | .09600 | .18320 | .27040 | .35750 | .44445 | .53130 | .61800 |
| 7 | .01020 | .09745 | .18465 | .27185 | .35895 | .44590 | .53275 | .61945 |
| 8 | .01165 | .09890 | .18615 | .27330 | .36040 | .44735 | .53420 | .62090 |
| 9 | .01310 | .10035 | .18760 | .27475 | .36185 | .44880 | .53565 | .62235 |
| 10 | 0.01455 | 0.10180 | 0.18905 | 0.27620 | 0.36330 | 0.45025 | 0.53710 | 0.62380 |
| 11 | .01600 | .10325 | .19050 | .27765 | .36475 | .45170 | .53855 | .62520 |
| 12 | .01745 | .10470 | .19195 | .27910 | .36620 | .45315 | .54000 | .62665 |
| 13 | .01890 | .10615 | .19340 | .28055 | .36765 | .45460 | .54145 | .62810 |
| 14 | .02035 | .10760 | .19485 | .28200 | .36910 | .45605 | .54290 | .62955 |
| 15 | 0.02180 | 0.10905 | 0.19630 | 0.28345 | 0.37055 | 0.45750 | 0.54435 | 0.63100 |
| 16 | .02325 | .11055 | .19775 | .28490 | .37200 | .45895 | .54580 | .63245 |
| 17 | .02475 | .11200 | .19920 | .28635 | .37345 | .46040 | .54725 | .63390 |
| 18 | .02620 | .11345 | .20065 | .28780 | .37490 | .46185 | .54865 | .63530 |
| 19 | .02765 | .11490 | .20210 | .28925 | .37635 | .46330 | .55010 | .63675 |
| 20 | 0.02910 | 0.11635 | 0.20355 | 0.29070 | 0.37780 | 0.46475 | 0.55155 | 0.63820 |
| 21 | .03055 | .11780 | .20500 | .29220 | .37925 | .46620 | .55300 | .63965 |
| 22 | .03200 | .11925 | .20645 | .29365 | .38070 | .46765 | .55445 | .64110 |
| 23 | .03345 | .12070 | .20795 | .29510 | .38215 | .46910 | .55590 | .64255 |
| 24 | .03490 | .12215 | .20940 | .29655 | .38360 | .47055 | .55735 | .64400 |
| 25 | 0.03635 | 0.12360 | 0.21085 | 0.29800 | 0.38505 | 0.47200 | 0.55880 | 0.64540 |
| 26 | .03780 | .12505 | .21230 | .29945 | .38650 | .47345 | .56025 | .64685 |
| 27 | .03925 | .12650 | .21375 | .30090 | .38795 | .47490 | .56170 | .64830 |
| 28 | .04070 | .12800 | .21520 | .30235 | .38940 | .47635 | .56315 | .64975 |
| 29 | .04220 | .12945 | .21665 | .30380 | .39085 | .47780 | .56455 | .65120 |
| 30 | 0.04365 | 0.13090 | 0.21810 | 0.30525 | 0.39230 | 0.47925 | 0.56600 | 0.65265 |
| 31 | .04510 | .13235 | .21955 | .30670 | .39375 | .48070 | .56745 | .65405 |
| 32 | .04655 | .13380 | .22100 | .30815 | .39520 | .48210 | .56890 | .65550 |
| 33 | .04800 | .13525 | .22245 | .30960 | .39665 | .48355 | .57035 | .65695 |
| 34 | .04945 | .13670 | .22390 | .31105 | .39810 | .48500 | .57180 | .65840 |
| 35 | 0.05090 | 0.13815 | 0.22535 | 0.31250 | 0.39955 | 0.48645 | 0.57325 | 0.65985 |
| 36 | .05235 | .13960 | .22680 | .31395 | .40100 | .48790 | .57470 | .66130 |
| 37 | .05380 | .14105 | .22825 | .31540 | .40245 | .48935 | .57615 | .66270 |
| 38 | .05525 | .14250 | .22970 | .31685 | .40390 | .49080 | .57760 | .66415 |
| 39 | .05670 | .14395 | .23115 | .31830 | .40535 | .49225 | .57900 | .66560 |
| 40 | 0.05820 | 0.14540 | 0.23265 | 0.31975 | 0.40680 | 0.49370 | 0.58045 | 0.66705 |
| 41 | .05965 | .14690 | .23410 | .32120 | .40825 | .49515 | .58190 | .66850 |
| 42 | .06110 | .14835 | .23555 | .32265 | .40970 | .49660 | .58335 | .66995 |
| 43 | .06255 | .14980 | .23700 | .32410 | .41115 | .49805 | .58480 | .67135 |
| 44 | .06400 | .15125 | .23845 | .32555 | .41260 | .49950 | .58625 | .67280 |
| 45 | 0.06545 | 0.15270 | 0.23990 | 0.32700 | 0.41405 | 0.50095 | 0.58770 | 0.67425 |
| 46 | .06690 | .15415 | .24135 | .32845 | .41550 | .50240 | .58915 | .67570 |
| 47 | .06835 | .15560 | .24280 | .32990 | .41695 | .50385 | .59060 | .67715 |
| 48 | .06980 | .15705 | .24425 | .33135 | .41840 | .50530 | .59200 | .67860 |
| 49 | .07125 | .15850 | .24570 | .33280 | .41985 | .50675 | .59345 | .68000 |
| 50 | 0.07270 | 0.15995 | 0.24715 | 0.33425 | 0.42130 | 0.50820 | 0.59490 | 0.68145 |
| 51 | .07415 | .16140 | .24860 | .33570 | .42275 | .50960 | .59635 | .68290 |
| 52 | .07565 | .16285 | .25005 | .33715 | .42420 | .51105 | .59780 | .68435 |
| 53 | .07710 | .16430 | .25150 | .33865 | .42565 | .51250 | .59925 | .68580 |
| 54 | .07855 | .16580 | .25295 | .34010 | .42710 | .51395 | .60070 | .68720 |
| 55 | 0.08000 | 0.16725 | 0.25440 | 0.34155 | 0.42855 | 0.51540 | 0.60215 | 0.68865 |
| 56 | .08145 | .16870 | .25585 | .34300 | .43000 | .51685 | .60355 | .69010 |
| 57 | .08290 | .17015 | .25730 | .34445 | .43145 | .51830 | .60500 | .69155 |
| 58 | .08435 | .17160 | .25875 | .34590 | .43290 | .51975 | .60645 | .69300 |
| 59 | .08580 | .17305 | .26028 | .34735 | .43435 | .52120 | .60790 | .69445 |
| 60 | 0.08725 | 0.17450 | 0.26170 | 0.34880 | 0.43580 | 0.52265 | 0.60935 | 0.69585 |

## Constants for Setting a 5-inch Sine-bar

| Min. | 8° | 9° | 10° | 11° | 12° | 13° | 14° | 15° |
|---|---|---|---|---|---|---|---|---|
| 0 | 0.69585 | 0.78215 | 0.86825 | 0.95405 | 1.0395 | 1.1247 | 1.2096 | 1.2941 |
| 1 | .69730 | .78360 | .86965 | .95545 | .0410 | .1261 | .2110 | .2955 |
| 2 | .69875 | .78505 | .87110 | .95690 | .0424 | .1276 | .2124 | .2969 |
| 3 | .70020 | .78650 | .87255 | .95835 | .0438 | .1290 | .2138 | .2983 |
| 4 | .70165 | .78790 | .87395 | .95975 | .0452 | .1304 | .2152 | .2997 |
| 5 | 0.70305 | 0.78935 | 0.87540 | 0.96120 | 1.0466 | 1.1318 | 1.2166 | 1.3011 |
| 6 | .70450 | .79080 | .87685 | .96260 | 0481 | .1332 | .2181 | .3025 |
| 7 | .70595 | .79225 | .87825 | .96405 | .0495 | .1346 | .2195 | .3039 |
| 8 | .70740 | .79365 | .87970 | .96545 | .0509 | .1361 | .2209 | .3053 |
| 9 | .70885 | .79510 | .88115 | .96690 | .0523 | .1375 | .2223 | .3067 |
| 10 | 0.71025 | 0.79655 | 0.88255 | 0.96830 | 1.0538 | 1.1389 | 1.2237 | 1.3081 |
| 11 | .71170 | .79795 | .88400 | .96975 | .0552 | .1403 | .2251 | .3095 |
| 12 | .71315 | .79940 | .88540 | .97115 | .0566 | .1417 | .2265 | .3109 |
| 13 | .71460 | .80085 | .88685 | .97260 | .0580 | .1431 | .2279 | .3123 |
| 14 | .71600 | .80230 | .88830 | .97405 | .0594 | .1446 | .2293 | .3137 |
| 15 | 0.71745 | 0.80370 | 0.88970 | 0.97545 | 1.0609 | 1.1460 | 1.2307 | 1.3151 |
| 16 | .71890 | .80515 | .89115 | .97690 | .0623 | .1474 | .2322 | .3165 |
| 17 | .72035 | .80660 | .89260 | .97830 | .0637 | .1488 | .2336 | .3179 |
| 18 | .72180 | .80800 | .89400 | .97975 | .0651 | .1502 | .2350 | .3193 |
| 19 | .72320 | .80945 | .89545 | .98115 | .0665 | .1516 | .2364 | .3207 |
| 20 | 0.72465 | 0.81090 | 0.89685 | 0.98260 | 1.0680 | 1.1531 | 1.2378 | 1.3221 |
| 21 | .72610 | .81230 | .89830 | .98400 | .0694 | .1545 | .2392 | .3235 |
| 22 | .72755 | .81375 | .89975 | .98545 | .0708 | .1559 | .2406 | .3250 |
| 23 | .72900 | .81520 | .90115 | .98685 | .0722 | .1573 | .2420 | .3264 |
| 24 | .73040 | .81665 | .90260 | .98830 | .0737 | .1587 | .2434 | .3278 |
| 25 | 0.73185 | 0.81805 | 0.90405 | 0.98970 | 1.0751 | 1.1601 | 1.2448 | 1.3292 |
| 26 | .73330 | .81950 | .90545 | .99115 | .0765 | .1615 | .2462 | .3306 |
| 27 | .73475 | .82095 | .90690 | .99255 | .0779 | .1630 | .2477 | .3320 |
| 28 | .73615 | .82235 | .90830 | .99400 | .0793 | .1644 | .2491 | .3334 |
| 29 | .73760 | .82380 | .90975 | .99540 | .0808 | .1658 | .2505 | .3348 |
| 30 | 0.73905 | 0.82525 | 0.91120 | 0.99685 | 1.0822 | 1.1672 | 1.2519 | 1.3362 |
| 31 | .74050 | .82665 | .91260 | .99825 | .0836 | .1686 | .2533 | .3376 |
| 32 | .74190 | .82810 | .91405 | .99970 | .0850 | .1700 | .2547 | .3390 |
| 33 | .74335 | .82955 | .91545 | 1.0011 | .0864 | .1714 | .2561 | .3404 |
| 34 | .74480 | .83100 | .91690 | .0026 | .0879 | .1729 | .2575 | .3418 |
| 35 | 0.74625 | 0.83240 | 0.91835 | 1.0039 | 1.0893 | 1.1743 | 1.2589 | 1.3432 |
| 36 | .74770 | .83385 | .91975 | .0054 | .0907 | .1757 | .2603 | .3446 |
| 37 | .74910 | .83530 | .92120 | .0068 | .0921 | .1771 | .2617 | .3460 |
| 38 | .75055 | .83670 | .92260 | .0082 | .0935 | .1785 | .2631 | .3474 |
| 39 | .75200 | .83815 | .92405 | .0096 | .0949 | .1799 | .2645 | .3488 |
| 40 | 0.75345 | 0.83960 | 0.92545 | 1.0110 | 1.0964 | 1.1813 | 1.2660 | 1.3502 |
| 41 | .75485 | .84100 | .92690 | .0125 | .0978 | .1828 | .2674 | .3516 |
| 42 | .75630 | .84245 | .92835 | .0139 | .0992 | .1842 | .2688 | .3530 |
| 43 | .75775 | .84390 | .92975 | .0153 | .1006 | .1856 | .2702 | .3544 |
| 44 | .75920 | .84530 | .93120 | .0168 | .1020 | .1870 | .2716 | .3558 |
| 45 | 0.76060 | 0.84675 | 0.93260 | 1.0182 | 1.1035 | 1.1884 | 1.2730 | 1.3572 |
| 46 | .76205 | .84820 | .93405 | .0196 | .1049 | .1898 | .2744 | .3586 |
| 47 | .76350 | .84960 | .93550 | .0210 | .1063 | .1912 | .2758 | .3600 |
| 48 | .76495 | .85105 | .93690 | .0225 | .1077 | .1926 | .2772 | .3614 |
| 49 | .76635 | .85250 | .93835 | .0239 | .1091 | .1941 | .2786 | .3628 |
| 50 | 0.76780 | 0.85390 | 0.93975 | 1.0253 | 1.1106 | 1.1955 | 1.2800 | 1.3642 |
| 51 | .76925 | .85535 | .94120 | .0267 | .1120 | .1969 | .2814 | .3656 |
| 52 | .77070 | .85680 | .94260 | .0281 | .1134 | .1983 | .2828 | .3670 |
| 53 | .77210 | .85820 | .94405 | .0296 | .1148 | .1997 | .2842 | .3684 |
| 54 | .77355 | .85965 | .94550 | .0310 | .1162 | .2011 | .2856 | .3698 |
| 55 | 0.77500 | 0.86110 | 0.94690 | 1.0324 | 1.1176 | 1.2025 | 1.2870 | 1.3712 |
| 56 | .77645 | .86250 | .94835 | .0338 | .1191 | .2039 | .2884 | .3726 |
| 57 | .77785 | .86395 | .94975 | .0353 | .1205 | .2054 | .2899 | .3740 |
| 58 | .77930 | .86540 | .95120 | .0367 | .1219 | .2068 | .2913 | .3754 |
| 59 | .78075 | .86680 | .95260 | .0381 | .1233 | .2082 | .2927 | .3768 |
| 60 | 0.78215 | 0.86825 | 0.95405 | 1.0395 | 1.1247 | 1.2096 | 1.2941 | 1.3782 |

### Constants for Setting a 5-inch Sine-bar

| Min. | 16° | 17° | 18° | 19° | 20° | 21° | 22° | 23° |
|---|---|---|---|---|---|---|---|---|
| 0 | 1.3782 | 1.4618 | 1.5451 | 1.6278 | 1.7101 | 1.7918 | 1.8730 | 1.9536 |
| 1 | .3796 | .4632 | .5464 | .6292 | .7114 | .7932 | .8744 | .9550 |
| 2 | .3810 | .4646 | .5478 | .6306 | .7128 | .7945 | .8757 | .9563 |
| 3 | .3824 | .4660 | .5492 | .6319 | .7142 | .7959 | .8771 | .9576 |
| 4 | .3838 | .4674 | .5506 | .6333 | .7155 | .7972 | .8784 | .9590 |
| 5 | 1.3852 | 1.4688 | 1.5520 | 1.6347 | 1.7169 | 1.7986 | 1.8797 | 1.9603 |
| 6 | .3865 | .4702 | .5534 | .6361 | .7183 | .8000 | .8811 | .9617 |
| 7 | .3879 | .4716 | .5547 | .6374 | .7196 | .8013 | .8824 | .9630 |
| 8 | .3893 | .4730 | .5561 | .6388 | .7210 | .8027 | .8838 | .9643 |
| 9 | .3907 | .4743 | .5575 | .6402 | .7224 | .8040 | .8851 | .9657 |
| 10 | 1.3921 | 1.4757 | 1.5589 | 1.6416 | 1.7237 | 1.8054 | 1.8865 | 1.9670 |
| 11 | .3935 | .4771 | .5603 | .6429 | .7251 | .8067 | .8878 | .9683 |
| 12 | .3949 | .4785 | .5616 | .6443 | .7265 | .8081 | .8892 | .9697 |
| 13 | .3963 | .4799 | .5630 | .6457 | .7278 | .8094 | .8905 | .9710 |
| 14 | .3977 | .4813 | .5644 | .6471 | .7292 | .8108 | .8919 | .9724 |
| 15 | 1.3991 | 1.4827 | 1.5658 | 1.6484 | 1.7306 | 1.8122 | 1.8932 | 1.9737 |
| 16 | .4005 | .4841 | .5672 | .6498 | .7319 | .8135 | .8946 | .9750 |
| 17 | .4019 | .4855 | .5686 | .6512 | .7333 | .8149 | .8959 | .9764 |
| 18 | .4033 | .4868 | .5699 | .6525 | .7347 | .8162 | .8973 | .9777 |
| 19 | .4047 | .4882 | .5713 | .6539 | .7360 | .8176 | .8986 | .9790 |
| 20 | 1.4061 | 1.4896 | 1.5727 | 1.6553 | 1.7374 | 1.8189 | 1.8999 | 1.9804 |
| 21 | .4075 | .4910 | .5741 | .6567 | .7387 | .8203 | .9013 | .9817 |
| 22 | .4089 | .4924 | .5755 | .6580 | .7401 | .8217 | .9026 | .9830 |
| 23 | .4103 | .4938 | .5768 | .6594 | .7415 | .8230 | .9040 | .9844 |
| 24 | .4117 | .4952 | .5782 | .6608 | .7428 | .8244 | .9053 | .9857 |
| 25 | 1.4131 | 1.4966 | 1.5796 | 1.6622 | 1.7442 | 1.8257 | 1.9067 | 1.9870 |
| 26 | .4145 | .4980 | .5810 | .6635 | .7456 | .8271 | .9080 | .9884 |
| 27 | .4159 | .4993 | .5824 | .6649 | .7469 | .8284 | .9094 | .9897 |
| 28 | .4173 | .5007 | .5837 | .6663 | .7483 | .8298 | .9107 | .9911 |
| 29 | .4187 | .5021 | .5851 | .6676 | .7496 | .8311 | .9120 | .9924 |
| 30 | 1.4201 | 1.5035 | 1.5865 | 1.6690 | 1.7510 | 1.8325 | 1.9134 | 1.9937 |
| 31 | .4214 | .5049 | .5879 | .6704 | .7524 | .8338 | .9147 | .9951 |
| 32 | .4228 | .5063 | .5893 | .6718 | .7537 | .8352 | .9161 | .9964 |
| 33 | .4242 | .5077 | .5906 | .6731 | .7551 | .8365 | .9174 | .9977 |
| 34 | .4256 | .5091 | .5920 | .6745 | .7565 | .8379 | .9188 | .9991 |
| 35 | 1.4270 | 1.5104 | 1.5934 | 1.6759 | 1.7578 | 1.8392 | 1.9201 | 2.0004 |
| 36 | .4284 | .5118 | .5948 | .6772 | .7592 | .8406 | .9215 | .0017 |
| 37 | .4298 | .5132 | .5961 | .6786 | .7605 | .8419 | .9228 | .0031 |
| 38 | .4312 | .5146 | .5975 | .6800 | .7619 | .8433 | .9241 | .0044 |
| 39 | .4326 | .5160 | .5989 | .6813 | .7633 | .8447 | .9255 | .0057 |
| 40 | 1.4340 | 1.5174 | 1.6003 | 1.6827 | 1.7646 | 1.8460 | 1.9268 | 2.0070 |
| 41 | .4354 | .5188 | .6017 | .6841 | .7660 | .8474 | .9282 | .0084 |
| 42 | .4368 | .5201 | .6030 | .6855 | .7673 | .8487 | .9295 | .0097 |
| 43 | .4382 | .5215 | .6044 | .6868 | .7687 | .8501 | .9308 | .0110 |
| 44 | .4396 | .5229 | .6058 | .6882 | .7701 | .8514 | .9322 | .0124 |
| 45 | 1.4410 | 1.5243 | 1.6072 | 1.6896 | 1.7714 | 1.8528 | 1.9335 | 2.0137 |
| 46 | .4423 | .5257 | .6085 | .6909 | .7728 | .8541 | .9349 | .0150 |
| 47 | .4437 | .5271 | .6099 | .6923 | .7742 | .8555 | .9362 | .0164 |
| 48 | .4451 | .5285 | .6113 | .6937 | .7755 | .8568 | .9376 | .0177 |
| 49 | .4465 | .5298 | .6127 | .6950 | .7769 | .8582 | .9389 | .0190 |
| 50 | 1.4479 | 1.5312 | 1.6141 | 1.6964 | 1.7782 | 1.8595 | 1.9402 | 2.0204 |
| 51 | .4493 | .5326 | .6154 | .6978 | .7796 | .8609 | .9416 | .0217 |
| 52 | .4507 | .5340 | .6168 | .6991 | .7809 | .8622 | .9429 | .0230 |
| 53 | .4521 | .5354 | .6182 | .7005 | .7823 | .8636 | .9443 | .0244 |
| 54 | .4535 | .5368 | .6196 | .7019 | .7837 | .8649 | .9456 | .0257 |
| 55 | 1.4549 | 1.5381 | 1.6209 | 1.7032 | 1.7850 | 1.8663 | 1.9469 | 2.0270 |
| 56 | .4563 | .5395 | .6223 | .7046 | .7864 | .8676 | .9483 | .0283 |
| 57 | .4577 | .5409 | .6237 | .7060 | .7877 | .8690 | .9496 | .0297 |
| 58 | .4591 | .5423 | .6251 | .7073 | .7891 | .8703 | .9510 | .0310 |
| 59 | .4604 | .5437 | .6264 | .7087 | .7905 | .8717 | .9523 | .0323 |
| 60 | 1.4618 | 1.5451 | 1.6278 | 1.7101 | 1.7918 | 1.8730 | 1.9536 | 2.0337 |

### Constants for Setting a 5-inch Sine-bar

| Min. | 24° | 25° | 26° | 27° | 28° | 29° | 30° | 31° |
|---|---|---|---|---|---|---|---|---|
| 0 | 2.0337 | 2.1131 | 2.1918 | 2.2699 | 2.3473 | 2.4240 | 2.5000 | 2.5752 |
| 1 | .0350 | .1144 | .1931 | .2712 | .3486 | .4253 | .5012 | .5764 |
| 2 | .0363 | .1157 | .1944 | .2725 | .3499 | .4266 | .5025 | .5777 |
| 3 | .0376 | .1170 | .1958 | .2738 | .3512 | .4278 | .5038 | .5789 |
| 4 | .0390 | .1183 | .1971 | .2751 | .3525 | .4291 | .5050 | .5802 |
| 5 | 2.0403 | 2.1197 | 2.1984 | 2.2764 | 2.3538 | 2.4304 | 2.5063 | 2.5814 |
| 6 | .0416 | .1210 | .1997 | .2777 | .3550 | .4317 | .5075 | .5826 |
| 7 | .0430 | .1223 | .2010 | .2790 | .3563 | .4329 | .5088 | .5839 |
| 8 | .0443 | .1236 | .2023 | .2803 | .3576 | .4342 | .5100 | .5851 |
| 9 | .0456 | .1249 | .2036 | .2816 | .3589 | .4355 | .5113 | .5864 |
| 10 | 2.0469 | 2.1262 | 2.2049 | 2.2829 | 2.3602 | 2.4367 | 2.5126 | 2.5876 |
| 11 | .0483 | .1276 | .2062 | .2842 | .3614 | .4380 | .5138 | .5889 |
| 12 | .0496 | .1289 | .2075 | .2855 | .3627 | .4393 | .5151 | .5901 |
| 13 | .0509 | .1302 | .2088 | .2868 | .3640 | .4405 | .5163 | .5914 |
| 14 | .0522 | .1315 | .2101 | .2881 | .3653 | .4418 | .5176 | .5926 |
| 15 | 2.0536 | 2.1328 | 2.2114 | 2.2893 | 2.3666 | 2.4431 | 2.5188 | 2.5938 |
| 16 | .0549 | .1341 | .2127 | .2906 | .3679 | .4444 | .5201 | .5951 |
| 17 | .0562 | .1354 | .2140 | .2919 | .3691 | .4456 | .5214 | .5963 |
| 18 | .0575 | .1368 | .2153 | .2932 | .3704 | .4469 | .5226 | .5976 |
| 19 | .0589 | .1381 | .2166 | .2945 | .3717 | .4482 | .5239 | .5988 |
| 20 | 2.0602 | 2.1394 | 2.2179 | 2.2958 | 2.3730 | 2.4494 | 2.5251 | 2.6001 |
| 21 | .0615 | .1407 | .2192 | .2971 | .3743 | .4507 | .5264 | .6013 |
| 22 | .0628 | .1420 | .2205 | .2984 | .3755 | .4520 | .5276 | .6025 |
| 23 | .0642 | .1433 | .2218 | .2997 | .3768 | .4532 | .5289 | .6038 |
| 24 | .0655 | .1447 | .2232 | .3010 | .3781 | .4545 | .5301 | .6050 |
| 25 | 2.0668 | 2.1460 | 2.2245 | 2.3023 | 2.3794 | 2.4558 | 2.5314 | 2.6063 |
| 26 | .0681 | .1473 | .2258 | .3036 | .3807 | .4570 | .5327 | .6075 |
| 27 | .0695 | .1486 | .2271 | .3048 | .3819 | .4583 | .5339 | .6087 |
| 28 | .0708 | .1499 | .2284 | .3061 | .3832 | .4596 | .5352 | .6100 |
| 29 | .0721 | .1512 | .2297 | .3074 | .3845 | .4608 | .5364 | .6112 |
| 30 | 2.0734 | 2.1525 | 2.2310 | 2.3087 | 2.3858 | 2.4621 | 2.5377 | 2.6125 |
| 31 | .0748 | .1538 | .2323 | .3100 | .3870 | .4634 | .5389 | .6137 |
| 32 | .0761 | .1552 | .2336 | .3113 | .3883 | .4646 | .5402 | .6149 |
| 33 | .0774 | .1565 | .2349 | .3126 | .3896 | .4659 | .5414 | .6162 |
| 34 | .0787 | .1578 | .2362 | .3139 | .3909 | .4672 | .5427 | .6174 |
| 35 | 2.0801 | 2.1591 | 2.2375 | 2.3152 | 2.3922 | 2.4684 | 2.5439 | 2.6187 |
| 36 | .0814 | .1604 | .2388 | .3165 | .3934 | .4697 | .5452 | .6199 |
| 37 | .0827 | .1617 | .2401 | .3177 | .3947 | .4709 | .5464 | .6211 |
| 38 | .0840 | .1630 | .2414 | .3190 | .3960 | .4722 | .5477 | .6224 |
| 39 | .0853 | .1643 | .2427 | .3203 | .3973 | .4735 | .5489 | .6236 |
| 40 | 2.0867 | 2.1656 | 2.2440 | 2.3216 | 2.3985 | 2.4747 | 2.5502 | 2.6249 |
| 41 | .0880 | .1670 | .2453 | .3229 | .3998 | .4760 | .5514 | .6261 |
| 42 | .0893 | .1683 | .2466 | .3242 | .4011 | .4773 | .5527 | .6273 |
| 43 | .0906 | .1696 | .2479 | .3255 | .4024 | .4785 | .5539 | .6286 |
| 44 | .0920 | .1709 | .2492 | .3268 | .4036 | .4798 | .5552 | .6298 |
| 45 | 2.0933 | 2.1722 | 2.2505 | 2.3280 | 2.4049 | 2.4811 | 2.5564 | 2.6310 |
| 46 | .0946 | .1735 | .2518 | .3293 | .4062 | .4823 | .5577 | .6323 |
| 47 | .0959 | .1748 | .2531 | .3306 | .4075 | .4836 | .5589 | .6335 |
| 48 | .0972 | .1761 | .2544 | .3319 | .4087 | .4848 | .5602 | .6348 |
| 49 | .0986 | .1774 | .2557 | .3332 | .4100 | .4861 | .5614 | .6360 |
| 50 | 2.0999 | 2.1787 | 2.2570 | 2.3345 | 2.4113 | 2.4874 | 2.5627 | 2.6372 |
| 51 | .1012 | .1801 | .2583 | .3358 | .4126 | .4886 | .5639 | .6385 |
| 52 | .1025 | .1814 | .2596 | .3371 | .4138 | .4899 | .5652 | .6397 |
| 53 | .1038 | .1827 | .2609 | .3383 | .4151 | .4912 | .5664 | .6409 |
| 54 | .1052 | .1840 | .2621 | .3396 | .4164 | .4924 | .5677 | .6422 |
| 55 | 2.1065 | 2.1853 | 2.2634 | 2.3409 | 2.4177 | 2.4937 | 2.5689 | 2.6434 |
| 56 | .1078 | .1866 | .2647 | .3422 | .4189 | .4949 | .5702 | .6446 |
| 57 | .1091 | .1879 | .2660 | .3435 | .4202 | .4962 | .5714 | .6459 |
| 58 | .1104 | .1892 | .2673 | .3448 | .4215 | .4975 | .5727 | .6471 |
| 59 | .1117 | .1905 | .2686 | .3460 | .4228 | .4987 | .5739 | .6483 |
| 60 | 2.1131 | 2.1918 | 2.2699 | 2.3473 | 2.4240 | 2.5000 | 2.5752 | 2.6496 |

Constants for Setting a 5-inch Sine-bar

| Min. | 32° | 33° | 34° | 35° | 36° | 37° | 38° | 39° |
|---|---|---|---|---|---|---|---|---|
| 0 | 2.6496 | 2.7232 | 2.7959 | 2.8679 | 2.9389 | 3.0091 | 3.0783 | 3.1466 |
| 1 | .6508 | .7244 | .7971 | .8690 | .9401 | .0102 | .0794 | .1477 |
| 2 | .6520 | .7256 | .7984 | .8702 | .9413 | .0114 | .0806 | .1488 |
| 3 | .6533 | .7268 | .7996 | .8714 | .9424 | .0125 | .0817 | .1500 |
| 4 | .6545 | .7280 | .8008 | .8726 | .9436 | .0137 | .0829 | .1511 |
| 5 | 2.6557 | 2.7293 | 2.8020 | 2.8738 | 2.9448 | 3.0149 | 3.0840 | 3.1522 |
| 6 | .6570 | .7305 | .8032 | .8750 | .9460 | .0160 | .0852 | .1534 |
| 7 | .6582 | .7317 | .8044 | .8762 | .9471 | .0172 | .0863 | .1545 |
| 8 | .6594 | .7329 | .8056 | .8774 | .9483 | .0183 | .0874 | .1556 |
| 9 | .6607 | .7341 | .8068 | .8786 | .9495 | .0195 | .0886 | .1567 |
| 10 | 2.6619 | 2.7354 | 2.8080 | 2.8798 | 2.9507 | 3.0207 | 3.0897 | 3.1579 |
| 11 | .6631 | .7366 | .8092 | .8809 | .9518 | .0218 | .0909 | .1590 |
| 12 | .6644 | .7378 | .8104 | .8821 | .9530 | .0230 | .0920 | .1601 |
| 13 | .6656 | .7390 | .8116 | .8833 | .9542 | .0241 | .0932 | .1612 |
| 14 | .6668 | .7402 | .8128 | .8845 | .9554 | .0253 | .0943 | .1624 |
| 15 | 2.6680 | 2.7414 | 2.8140 | 2.8857 | 2.9565 | 3.0264 | 3.0954 | 3.1635 |
| 16 | .6693 | .7427 | .8152 | .8869 | .9577 | .0276 | .0966 | .1646 |
| 17 | .6705 | .7439 | .8164 | .8881 | .9589 | .0288 | .0977 | .1658 |
| 18 | .6717 | .7451 | .8176 | .8893 | .9600 | .0299 | .0989 | .1669 |
| 19 | .6730 | .7463 | .8188 | .8905 | .9612 | .0311 | .1000 | .1680 |
| 20 | 2.6742 | 2.7475 | 2.8200 | 2.8916 | 2.9624 | 3.0322 | 3.1012 | 3.1691 |
| 21 | .6754 | .7487 | .8212 | .8928 | .9636 | .0334 | .1023 | .1703 |
| 22 | .6767 | .7499 | .8224 | .8940 | .9647 | .0345 | .1034 | .1714 |
| 23 | .6779 | .7512 | .8236 | .8952 | .9659 | .0357 | .1046 | .1725 |
| 24 | .6791 | .7524 | .8248 | .8964 | .9671 | .0369 | .1057 | .1736 |
| 25 | 2.6803 | 2.7536 | 2.8260 | 2.8976 | 2.9682 | 3.0380 | 3.1069 | 3.1748 |
| 26 | .6816 | .7548 | .8272 | .8988 | .9694 | .0392 | .1080 | .1759 |
| 27 | .6828 | .7560 | .8284 | .8999 | .9706 | .0403 | .1091 | .1770 |
| 28 | .6840 | .7572 | .8296 | .9011 | .9718 | .0415 | .1103 | .1781 |
| 29 | .6852 | .7584 | .8308 | .9023 | .9729 | .0426 | .1114 | .1792 |
| 30 | 2.6865 | 2.7597 | 2.8320 | 2.9035 | 2.9741 | 3.0438 | 3.1125 | 3.1804 |
| 31 | .6877 | .7609 | .8332 | .9047 | .9753 | .0449 | .1137 | .1815 |
| 32 | .6889 | .7621 | .8344 | .9059 | .9764 | .0461 | .1148 | .1826 |
| 33 | .6902 | .7633 | .8356 | .9070 | .9776 | .0472 | .1160 | .1837 |
| 34 | .6914 | .7645 | .8368 | .9082 | .9788 | .0484 | .1171 | .1849 |
| 35 | 2.6926 | 2.7657 | 2.8380 | 2.9094 | 2.9799 | 3.0495 | 3.1182 | 3.1860 |
| 36 | .6938 | .7669 | .8392 | .9106 | .9811 | .0507 | .1194 | .1871 |
| 37 | .6951 | .7681 | .8404 | .9118 | .9823 | .0519 | .1205 | .1882 |
| 38 | .6963 | .7694 | .8416 | .9130 | .9834 | .0530 | .1216 | .1893 |
| 39 | .6975 | .7706 | .8428 | .9141 | .9846 | .0542 | .1228 | .1905 |
| 40 | 2.6987 | 2.7718 | 2.8440 | 2.9153 | 2.9858 | 3.0553 | 3.1239 | 3.1916 |
| 41 | .7000 | .7730 | .8452 | .9165 | .9869 | .0565 | .1251 | .1927 |
| 42 | .7012 | .7742 | .8464 | .9177 | .9881 | .0576 | .1262 | .1938 |
| 43 | .7024 | .7754 | .8476 | .9189 | .9893 | .0588 | .1273 | .1949 |
| 44 | .7036 | .7766 | .8488 | .9200 | .9904 | .0599 | .1285 | .1961 |
| 45 | 2.7048 | 2.7778 | 2.8500 | 2.9212 | 2.9916 | 3.0611 | 3.1296 | 3.1972 |
| 46 | .7061 | .7790 | .8512 | .9224 | .9928 | .0622 | .1307 | .1983 |
| 47 | .7073 | .7802 | .8523 | .9236 | .9939 | .0634 | .1319 | .1994 |
| 48 | .7085 | .7815 | .8535 | .9248 | .9951 | .0645 | .1330 | .2005 |
| 49 | .7097 | .7827 | .8547 | .9259 | .9963 | .0657 | .1341 | .2016 |
| 50 | 2.7110 | 2.7839 | 2.8559 | 2.9271 | 2.9974 | 3.0668 | 3.1353 | 3.2028 |
| 51 | .7122 | .7851 | .8571 | .9283 | .9986 | .0680 | .1364 | .2039 |
| 52 | .7134 | .7863 | .8583 | .9295 | .9997 | .0691 | .1375 | .2050 |
| 53 | .7146 | .7875 | .8595 | .9307 | 3.0009 | .0703 | .1387 | .2061 |
| 54 | .7158 | .7887 | .8607 | .9318 | .0021 | .0714 | .1398 | .2072 |
| 55 | 2.7171 | 2.7899 | 2.8619 | 2.9330 | 3.0032 | 3.0725 | 3.1409 | 3.2083 |
| 56 | .7183 | .7911 | .8631 | .9342 | .0044 | .0737 | .1421 | .2095 |
| 57 | .7195 | .7923 | .8643 | .9354 | .0056 | .0748 | .1432 | .2106 |
| 58 | .7207 | .7935 | .8655 | .9365 | .0067 | .0760 | .1443 | .2117 |
| 59 | .7220 | .7947 | .8667 | .9377 | .0079 | .0771 | .1454 | .2128 |
| 60 | 2.7232 | 2.7959 | 2.8679 | 2.9389 | 3.0091 | 3.0783 | 3.1466 | 3.2139 |

### Constants for Setting a 5-inch Sine-bar

| Min. | 40° | 41° | 42° | 43° | 44° | 45° | 46° | 47° |
|---|---|---|---|---|---|---|---|---|
| 0 | 3.2139 | 3.2803 | 3.3456 | 3.4100 | 3.4733 | 3.5355 | 3.5967 | 3.6567 |
| 1 | .2150 | .2814 | .3467 | .4110 | .4743 | .5365 | .5977 | .6577 |
| 2 | .2161 | .2825 | .3478 | .4121 | .4754 | .5376 | .5987 | .6587 |
| 3 | .2173 | .2836 | .3489 | .4132 | .4764 | .5386 | .5997 | .6597 |
| 4 | .2184 | .2847 | .3499 | .4142 | .4774 | .5396 | .6007 | .6607 |
| 5 | 3.2195 | 3.2858 | 3.3510 | 3.4153 | 3.4785 | 3.5406 | 3.6017 | 3.6617 |
| 6 | .2206 | .2869 | .3521 | .4163 | .4795 | .5417 | .6027 | .6627 |
| 7 | .2217 | .2879 | .3532 | .4174 | .4806 | .5427 | .6037 | .6637 |
| 8 | .2228 | .2890 | .3543 | .4185 | .4816 | .5437 | .6047 | .6647 |
| 9 | .2239 | .2901 | .3553 | .4195 | .4827 | .5448 | .6058 | .6657 |
| 10 | 3.2250 | 3.2912 | 3.3564 | 3.4206 | 3.4837 | 3.5458 | 3.6068 | 3.6666 |
| 11 | .2262 | .2923 | .3575 | .4217 | .4848 | .5468 | .6078 | .6676 |
| 12 | .2273 | .2934 | .3586 | .4227 | .4858 | .5478 | .6088 | .6686 |
| 13 | .2284 | .2945 | .3597 | .4238 | .4868 | .5489 | .6098 | .6696 |
| 14 | .2295 | .2956 | .3607 | .4248 | .4879 | .5499 | .6108 | .6706 |
| 15 | 3.2306 | 3.2967 | 3.3618 | 3.4259 | 3.4889 | 3.5509 | 3.6118 | 3.6716 |
| 16 | .2317 | .2978 | .3629 | .4269 | .4900 | .5519 | .6128 | .6726 |
| 17 | .2328 | .2989 | .3640 | .4280 | .4910 | .5529 | .6138 | .6736 |
| 18 | .2339 | .3000 | .3650 | .4291 | .4921 | .5540 | .6148 | .6745 |
| 19 | .2350 | .3011 | .3661 | .4301 | .4931 | .5550 | .6158 | .6755 |
| 20 | 3.2361 | 3.3022 | 3.3672 | 3.4312 | 3.4941 | 3.5560 | 3.6168 | 3.6765 |
| 21 | .2373 | .3033 | .3683 | .4322 | .4952 | .5570 | .6178 | .6775 |
| 22 | .2384 | .3044 | .3693 | .4333 | .4962 | .5581 | .6188 | .6785 |
| 23 | .2395 | .3054 | .3704 | .4344 | .4973 | .5591 | .6198 | .6795 |
| 24 | .2406 | .3065 | .3715 | .4354 | .4983 | .5601 | .6208 | .6805 |
| 25 | 3.2417 | 3.3076 | 3.3726 | 3.4365 | 3.4993 | 3.5611 | 3.6218 | 3.6814 |
| 26 | .2428 | .3087 | .3736 | .4375 | .5004 | .5621 | .6228 | .6824 |
| 27 | .2439 | .3098 | .3747 | .4386 | .5014 | .5632 | .6238 | .6834 |
| 28 | .2450 | .3109 | .3758 | .4396 | .5024 | .5642 | .6248 | .6844 |
| 29 | .2461 | .3120 | .3769 | .4407 | .5035 | .5652 | .6258 | .6854 |
| 30 | 3.2472 | 3.3131 | 3.3779 | 3.4417 | 3.5045 | 3.5662 | 3.6268 | 3.6864 |
| 31 | .2483 | .3142 | .3790 | .4428 | .5056 | .5672 | .6278 | .6873 |
| 32 | .2494 | .3153 | .3801 | .4439 | .5066 | .5683 | .6288 | .6883 |
| 33 | .2505 | .3163 | .3811 | .4449 | .5076 | .5693 | .6298 | .6893 |
| 34 | .2516 | .3174 | .3822 | .4460 | .5087 | .5703 | .6308 | .6903 |
| 35 | 3.2527 | 3.3185 | 3.3833 | 3.4470 | 3.5097 | 3.5713 | 3.6318 | 3.6913 |
| 36 | .2538 | .3196 | .3844 | .4481 | .5107 | .5723 | .6328 | .6923 |
| 37 | .2550 | .3207 | .3854 | .4491 | .5118 | .5734 | .6338 | .6932 |
| 38 | .2561 | .3218 | .3865 | .4502 | .5128 | .5744 | .6348 | .6942 |
| 39 | .2572 | .3229 | .3876 | .4512 | .5138 | .5754 | .6358 | .6952 |
| 40 | 3.2583 | 3.3240 | 3.3886 | 3.4523 | 3.5149 | 3.5764 | 3.6368 | 3.6962 |
| 41 | .2594 | .3250 | .3897 | .4533 | .5159 | .5774 | .6378 | .6972 |
| 42 | .2605 | .3261 | .3908 | .4544 | .5169 | .5784 | .6388 | .6981 |
| 43 | .2616 | .3272 | .3918 | .4554 | .5180 | .5795 | .6398 | .6991 |
| 44 | .2627 | .3283 | .3929 | .4565 | .5190 | .5805 | .6408 | .7001 |
| 45 | 3.2638 | 3.3294 | 3.3940 | 3.4575 | 3.5200 | 3.5815 | 3.6418 | 3.7011 |
| 46 | .2649 | .3305 | .3950 | .4586 | .5211 | .5825 | .6428 | .7020 |
| 47 | .2660 | .3316 | .3961 | .4596 | .5221 | .5835 | .6438 | .7030 |
| 48 | .2671 | .3326 | .3972 | .4607 | .5231 | .5845 | .6448 | .7040 |
| 49 | .2682 | .3337 | .3982 | .4617 | .5242 | .5855 | .6458 | .7050 |
| 50 | 3.2693 | 3.3348 | 3.3993 | 3.4628 | 3.5252 | 3.5866 | 3.6468 | 3.7060 |
| 51 | .2704 | .3359 | .4004 | .4638 | .5262 | .5876 | .6478 | .7069 |
| 52 | .2715 | .3370 | .4014 | .4649 | .5273 | .5886 | .6488 | .7079 |
| 53 | .2726 | .3381 | .4025 | .4659 | .5283 | .5896 | .6498 | .7089 |
| 54 | .2737 | .3391 | .4036 | .4670 | .5293 | .5906 | .6508 | .7099 |
| 55 | 3.2748 | 3.3402 | 3.4046 | 3.4680 | 3.5304 | 3.5916 | 3.6518 | 3.7108 |
| 56 | .2759 | .3413 | .4057 | .4691 | .5314 | .5926 | .6528 | .7118 |
| 57 | .2770 | .3424 | .4068 | .4701 | .5324 | .5936 | .6538 | .7128 |
| 58 | .2781 | .3435 | .4078 | .4712 | .5335 | .5947 | .6548 | .7138 |
| 59 | .2792 | .3445 | .4089 | .4722 | .5345 | .5957 | .6558 | .7147 |
| 60 | 3.2803 | 3.3456 | 3.4100 | 3.4733 | 3.5355 | 3.5967 | 3.6567 | 3.7157 |

## Constants for Setting a 5-inch Sine-bar

| Min. | 48° | 49° | 50° | 51° | 52° | 53° | 54° | 55° |
|---|---|---|---|---|---|---|---|---|
| 0 | 3.7157 | 3.7735 | 3.8302 | 3.8857 | 3.9400 | 3.9932 | 4.0451 | 4.0957 |
| 1 | .7167 | .7745 | .8311 | .8866 | .9409 | .9940 | .0459 | .0966 |
| 2 | .7176 | .7754 | .8321 | .8875 | .9418 | .9949 | .0468 | .0974 |
| 3 | .7186 | .7764 | .8330 | .8884 | .9427 | .9958 | .0476 | .0982 |
| 4 | .7196 | .7773 | .8339 | .8894 | .9436 | .9967 | .0485 | .0991 |
| 5 | 3.7206 | 3.7783 | 3.8349 | 3.8903 | 3.9445 | 3.9975 | 4.0493 | 4.0999 |
| 6 | .7215 | .7792 | .8358 | .8912 | .9454 | .9984 | .0502 | .1007 |
| 7 | .7225 | .7802 | .8367 | .8921 | .9463 | .9993 | .0510 | .1016 |
| 8 | .7235 | .7811 | .8377 | .8930 | .9472 | 4.0001 | .0519 | .1024 |
| 9 | .7244 | .7821 | .8386 | .8939 | .9481 | .0010 | .0527 | .1032 |
| 10 | 3.7254 | 3.7830 | 3.8395 | 3.8948 | 3.9490 | 4.0019 | 4.0536 | 4.1041 |
| 11 | .7264 | .7840 | .8405 | .8958 | .9499 | .0028 | .0544 | .1049 |
| 12 | .7274 | .7850 | .8414 | .8967 | .9508 | .0036 | .0553 | .1057 |
| 13 | .7283 | .7859 | .8423 | .8976 | .9516 | .0045 | .0561 | .1066 |
| 14 | .7293 | .7869 | .8433 | .8985 | .9525 | .0054 | .0570 | .1074 |
| 15 | 3.7303 | 3.7878 | 3.8442 | 3.8994 | 3.9534 | 4.0062 | 4.0578 | 4.1082 |
| 16 | .7312 | .7887 | .8451 | .9003 | .9543 | .0071 | .0587 | .1090 |
| 17 | .7322 | .7897 | .8460 | .9012 | .9552 | .0080 | .0595 | .1099 |
| 18 | .7332 | .7906 | .8470 | .9021 | .9561 | .0089 | .0604 | .1107 |
| 19 | .7341 | .7916 | .8479 | .9030 | 9570 | .0097 | .0612 | .1115 |
| 20 | 3.7351 | 3.7925 | 3.8488 | 3.9039 | 3.9579 | 4.0106 | 4.0621 | 4.1124 |
| 21 | .7361 | .7935 | .8498 | .9049 | .9588 | .0115 | .0629 | .1132 |
| 22 | .7370 | .7944 | .8507 | .9058 | .9596 | .0123 | .0638 | .1140 |
| 23 | .7380 | .7954 | .8516 | .9067 | .9605 | .0132 | .0646 | .1148 |
| 24 | .7390 | .7963 | .8525 | .9076 | .9614 | .0141 | .0655 | .1157 |
| 25 | 3.7399 | 3.7973 | 3.8535 | 3.9085 | 3.9623 | 4.0149 | 4.0663 | 4.1165 |
| 26 | .7409 | .7982 | .8544 | .9094 | .9632 | .0158 | .0672 | .1173 |
| 27 | .7419 | .7992 | .8553 | .9103 | .9641 | .0167 | .0680 | .1181 |
| 28 | .7428 | .8001 | .8562 | .9112 | .9650 | .0175 | .0689 | .1190 |
| 29 | .7438 | .8011 | .8572 | .9121 | .9659 | .0184 | .0697 | .1198 |
| 30 | 3.7448 | 3.8020 | 3.8581 | 3.9130 | 3.9667 | 4.0193 | 4.0706 | 4.1206 |
| 31 | .7457 | .8029 | .8590 | .9139 | .9676 | .0201 | .0714 | .1214 |
| 32 | .7467 | .8039 | .8599 | .9148 | .9685 | .0210 | .0722 | .1223 |
| 33 | .7476 | .8048 | .8609 | .9157 | .9694 | .0219 | .0731 | .1231 |
| 34 | .7486 | .8058 | .8618 | .9166 | .9703 | .0227 | .0739 | .1239 |
| 35 | 3.7496 | 3.8067 | 3.8627 | 3.9175 | 3.9712 | 4.0236 | 4.0748 | 4.1247 |
| 36 | .7505 | .8077 | .8636 | .9184 | .9720 | .0244 | .0756 | .1255 |
| 37 | .7515 | .8086 | .8646 | .9193 | .9729 | .0253 | .0765 | .1264 |
| 38 | .7525 | .8096 | .8655 | .9202 | .9738 | .0262 | .0773 | .1272 |
| 39 | .7534 | .8105 | .8664 | .9212 | .9747 | .0270 | .0781 | .1280 |
| 40 | 3.7544 | 3.8114 | 3.8673 | 3.9221 | 3.9756 | 4.0279 | 4.0790 | 4.1288 |
| 41 | .7553 | .8124 | .8683 | .9230 | .9765 | .0288 | .0798 | .1296 |
| 42 | .7563 | .8133 | .8692 | .9239 | .9773 | .0296 | .0807 | .1305 |
| 43 | .7573 | .8143 | .8701 | .9248 | .9782 | .0305 | .0815 | .1313 |
| 44 | .7582 | .8152 | .8710 | .9257 | .9791 | .0313 | .0823 | .1321 |
| 45 | 3.7592 | 3.8161 | 3.8719 | 3.9266 | 3.9800 | 4.0322 | 4.0832 | 4.1329 |
| 46 | .7601 | .8171 | .8729 | .9275 | .9809 | .0331 | .0840 | .1337 |
| 47 | .7611 | .8180 | .8738 | .9284 | .9817 | .0339 | .0849 | .1346 |
| 48 | .7620 | .8190 | .8747 | .9293 | .9826 | .0348 | .0857 | .1354 |
| 49 | .7630 | .8199 | .8756 | .9302 | .9835 | .0356 | .0865 | .1362 |
| 50 | 3.7640 | 3.8208 | 3.8765 | 3.9311 | 3.9844 | 4.0365 | 4.0874 | 4.1370 |
| 51 | .7649 | .8218 | .8775 | .9320 | .9853 | .0374 | .0882 | .1378 |
| 52 | .7659 | .8227 | .8784 | .9329 | .9861 | .0382 | .0891 | .1386 |
| 53 | .7668 | .8236 | .8793 | .9338 | .9870 | .0391 | .0899 | .1395 |
| 54 | .7678 | .8246 | .8802 | .9347 | .9879 | .0399 | .0907 | .1403 |
| 55 | 3.7687 | 3.8255 | 3.8811 | 3.9355 | 3.9888 | 4.0408 | 4.0916 | 4.1411 |
| 56 | .7697 | .8265 | .8820 | .9364 | .9896 | .0416 | .0924 | .1419 |
| 57 | .7707 | .8274 | .8830 | .9373 | .9905 | .0425 | .0932 | .1427 |
| 58 | .7716 | .8283 | .8839 | .9382 | .9914 | .0433 | .0941 | .1435 |
| 59 | .7726 | .8293 | .8848 | .9391 | .9923 | .0442 | .0949 | .1443 |
| 60 | 3.7735 | 3.8302 | 3.8857 | 3.9400 | 3.9932 | 4.0451 | 4.0957 | 4.1452 |

**To Lay Out Angles Accurately.** — Angles can be laid out accurately without the use of a protractor, provided a table of natural tangents is available. *Example:* A line is to be drawn at an angle of 29 degrees 54 minutes with another line, as

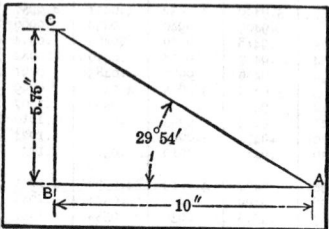

shown in the illustration. First draw a line *AB* of indefinite length, and then erect a perpendicular *BC* at a distance from *A* of, say 10 inches; then find in a table the tangent of 29 degrees 54 minutes, which equals 0.57503 for a radius of 1; hence, *BC* equals 0.57503 × 10 = 5.7503, or 5¾ inches. Measure 5¾ inches from *B* to *C*, at right angles to line *AB*, and draw a line from *A* to *C*, thus obtaining the required angle. Conversely, the angularity of two lines can be determined by measuring line *BC* and dividing by length *AB*; the quotient equals the tangent; then find the corresponding angle.

This method of laying out or measuring angles will give more accurate results than an ordinary protractor.

**Measuring Dovetail Slides.** — Dovetail slides which must be machined accurately to a given width are commonly gaged by using pieces of cylindrical rod or wire and measuring as indicated by the dimensions *x* and *y* of the accompanying

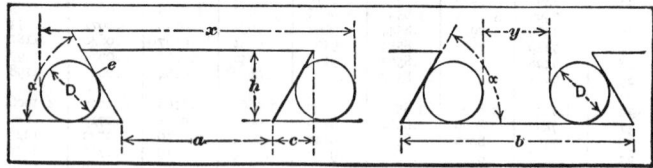

illustrations. To obtain dimension *x* for measuring male dovetails, add 1 to the cotangent of one-half the dovetail angle $\alpha$, multiply by diameter *D* of the rods used, and add the product to dimension *a*. To obtain dimension *y* for measuring a female dovetail, add 1 to the cotangent of one-half the dovetail angle $\alpha$, multiply by diameter *D* of the rod used, and subtract the result from dimension *b*. Expressing these rules as formulas:

$$x = D \left(1 + \cot \tfrac{1}{2}\,\alpha\right) + a.$$
$$y = b - D \left(1 + \cot \tfrac{1}{2}\,\alpha\right).$$

**Dimension *c*** equals $h \times \cot \alpha$.

The rod or wire used should be small enough so that the point of contact *e* is somewhat below the corner or edge of the dovetail.

**Taper Turning with Combined Feeds.** — When it is necessary to machine, on the boring mill, a conical surface which has such a large included angle that the tool-bar cannot be swiveled far enough to permit turning by the usual method, the combined vertical and horizontal feeds are sometimes used to obtain the required taper. Suppose a conical casting is to be turned to an angle $\alpha$ of 30 degrees (see illustration), and that the tool-head of the boring mill feeds horizontally ¼ inch per turn of the screw and has a vertical movement of ⁵⁄₁₆ inch per turn of the vertical feed shaft. If the two feeds are used simultaneously with the tool-bar at right angles to the table, the tool will move a distance $h$ of eight inches, while it moves downward a distance of six inches, thus turning the surface to an angle $\beta$. This angle $\beta$ is greater than the required angle $\alpha$, but if the tool-bar is swiveled to an angle $\gamma$, the tool, as it moves downward, will be advanced horizontally in addition to the regular horizontal feeding movement. Hence, if the tool-bar is set over to the proper angle $\gamma$, the surface can be turned to an angle $\alpha$. The problem, then,

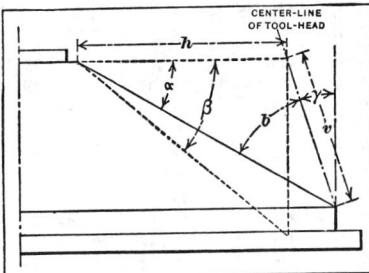

is to determine what the angle $\gamma$ should be for turning to a given angle $\alpha$.

Angle $\gamma$ can be calculated as follows: $\sin b = \dfrac{\sin \alpha \times h}{v}$ in which $h$ represents the rate of horizontal feed and $v$ the rate of vertical feed. Having angles $\alpha$ and $b$, the desired angle $\gamma$ is obtained by subtracting the sum of the former angles from 90 degrees. To illustrate (using the values given in the foregoing) the sine of 30 degrees is 0.5; then,

$$\sin b = \frac{0.5 \times \frac{1}{4}}{\frac{3}{16}} = 0.6666.$$ Hence, angle $b = 41$ degrees 48 minutes and $\gamma = 90°$ $- (30° + 41° 48') = 18$ degrees 12 minutes. If angle $\alpha$ were greater than angle $\beta$ obtained from the combined feeds with the tool-bar in the vertical position, it would be necessary to swing the lower end of the bar to the left rather than to the right of the vertical plane; that is, the lower end of the bar would be inclined to the left of the vertical an amount equal to the sum of angles $\alpha$ and $b$ subtracted from 90 degrees.

## Rules for Figuring Tapers

| Given | To Find | Rule |
|---|---|---|
| The taper per foot. | The taper per inch. | Divide the taper per foot by 12. |
| The taper per inch. | The taper per foot. | Multiply the taper per inch by 12. |
| End diameters and length of taper in inches. | The taper per foot. | Subtract small diameter from large; divide by length of taper, and multiply quotient by 12. |
| Large diameter and length of taper in inches, and taper per foot. | Diameter at small end in inches. | Divide taper per foot by 12; multiply by length of taper, and subtract result from large diameter. |
| Small diameter and length of taper in inches, and taper per foot. | Diameter at large end in inches. | Divide taper per foot by 12; multiply by length of taper, and add result to small diameter. |
| The taper per foot and two diameters in inches. | Distance between two given diameters in inches. | Subtract small diameter from large; divide remainder by taper per foot, and multiply quotient by 12. |
| The taper per foot. | Amount of taper in a certain length given in inches. | Divide taper per foot by 12; multiply by given length of tapered part. |

## Tapers per Foot and Corresponding Angles*

| Taper per Foot | Included Angle | | | Angle with Center Line | | | Taper per Foot | Included Angle | | | Angle with Center Line | | |
|---|---|---|---|---|---|---|---|---|---|---|---|---|---|
| 1/64 | 0° | 4' | 29" | 0° | 2' | 14" | 1 7/8 | 8° | 56' | 4" | 4° | 28' | 2" |
| 1/32 | 0 | 8 | 57 | 0 | 4 | 29 | 1 15/16 | 9 | 13 | 51 | 4 | 36 | 56 |
| 1/16 | 0 | 17 | 54 | 0 | 8 | 57 | 2 | 9 | 31 | 38 | 4 | 45 | 49 |
| 3/32 | 0 | 26 | 51 | 0 | 13 | 26 | 2 1/8 | 10 | 7 | 11 | 5 | 3 | 36 |
| 1/8 | 0 | 35 | 49 | 0 | 17 | 54 | 2 1/4 | 10 | 42 | 42 | 5 | 21 | 21 |
| 5/32 | 0 | 44 | 46 | 0 | 22 | 23 | 2 3/8 | 11 | 18 | 11 | 5 | 39 | 5 |
| 3/16 | 0 | 53 | 43 | 0 | 26 | 51 | 2 1/2 | 11 | 53 | 37 | 5 | 56 | 49 |
| 7/32 | 1 | 2 | 40 | 0 | 31 | 20 | 2 5/8 | 12 | 29 | 2 | 6 | 14 | 31 |
| 1/4 | 1 | 11 | 37 | 0 | 35 | 49 | 2 3/4 | 13 | 4 | 24 | 6 | 32 | 12 |
| 9/32 | 1 | 20 | 34 | 0 | 40 | 17 | 2 7/8 | 13 | 39 | 43 | 6 | 49 | 52 |
| 5/16 | 1 | 29 | 31 | 0 | 44 | 46 | 3 | 14 | 15 | 0 | 7 | 7 | 30 |
| 11/32 | 1 | 38 | 28 | 0 | 49 | 14 | 3 1/8 | 14 | 50 | 14 | 7 | 25 | 7 |
| 3/8 | 1 | 47 | 25 | 0 | 53 | 43 | 3 1/4 | 15 | 25 | 26 | 7 | 42 | 43 |
| 13/32 | 1 | 56 | 22 | 0 | 58 | 11 | 3 3/8 | 16 | 0 | 34 | 8 | 0 | 17 |
| 7/16 | 2 | 5 | 19 | 1 | 2 | 40 | 3 1/2 | 16 | 35 | 39 | 8 | 17 | 50 |
| 15/32 | 2 | 14 | 16 | 1 | 7 | 8 | 3 5/8 | 17 | 10 | 42 | 8 | 35 | 21 |
| 1/2 | 2 | 23 | 13 | 1 | 11 | 37 | 3 3/4 | 17 | 45 | 41 | 8 | 52 | 50 |
| 17/32 | 2 | 32 | 10 | 1 | 16 | 5 | 3 7/8 | 18 | 20 | 36 | 9 | 10 | 18 |
| 9/16 | 2 | 41 | 7 | 1 | 20 | 33 | 4 | 18 | 55 | 29 | 9 | 27 | 44 |
| 19/32 | 2 | 50 | 4 | 1 | 25 | 2 | 4 1/8 | 19 | 30 | 17 | 9 | 45 | 9 |
| 5/8 | 2 | 59 | 1 | 1 | 29 | 30 | 4 1/4 | 20 | 5 | 3 | 10 | 2 | 31 |
| 21/32 | 3 | 7 | 57 | 1 | 33 | 59 | 4 3/8 | 20 | 39 | 44 | 10 | 19 | 52 |
| 11/16 | 3 | 16 | 54 | 1 | 38 | 27 | 4 1/2 | 21 | 14 | 22 | 10 | 37 | 11 |
| 23/32 | 3 | 25 | 51 | 1 | 42 | 55 | 4 5/8 | 21 | 48 | 55 | 10 | 54 | 28 |
| 3/4 | 3 | 34 | 47 | 1 | 47 | 24 | 4 3/4 | 22 | 23 | 25 | 11 | 11 | 42 |
| 25/32 | 3 | 43 | 44 | 1 | 51 | 52 | 4 7/8 | 22 | 57 | 50 | 11 | 28 | 55 |
| 13/16 | 3 | 52 | 41 | 1 | 56 | 20 | 5 | 23 | 32 | 12 | 11 | 46 | 6 |
| 27/32 | 4 | 1 | 37 | 2 | 0 | 49 | 5 1/8 | 24 | 6 | 29 | 12 | 3 | 14 |
| 7/8 | 4 | 10 | 33 | 2 | 5 | 17 | 5 1/4 | 24 | 40 | 41 | 12 | 20 | 21 |
| 29/32 | 4 | 19 | 30 | 2 | 9 | 45 | 5 3/8 | 25 | 14 | 50 | 12 | 37 | 25 |
| 15/16 | 4 | 28 | 26 | 2 | 14 | 13 | 5 1/2 | 25 | 48 | 53 | 12 | 54 | 27 |
| 31/32 | 4 | 37 | 23 | 2 | 18 | 41 | 5 5/8 | 26 | 22 | 52 | 13 | 11 | 26 |
| 1 | 4 | 46 | 19 | 2 | 23 | 9 | 5 3/4 | 26 | 56 | 47 | 13 | 28 | 23 |
| 1 1/16 | 5 | 4 | 11 | 2 | 32 | 6 | 5 7/8 | 27 | 30 | 36 | 13 | 45 | 18 |
| 1 1/8 | 5 | 22 | 3 | 2 | 41 | 2 | 6 | 28 | 4 | 21 | 14 | 2 | 10 |
| 1 3/16 | 5 | 39 | 55 | 2 | 49 | 57 | 6 1/8 | 28 | 38 | 1 | 14 | 19 | 0 |
| 1 1/4 | 5 | 57 | 47 | 2 | 58 | 53 | 6 1/4 | 29 | 11 | 35 | 14 | 35 | 48 |
| 1 5/16 | 6 | 15 | 38 | 3 | 7 | 49 | 6 3/8 | 29 | 45 | 5 | 14 | 52 | 32 |
| 1 3/8 | 6 | 33 | 29 | 3 | 16 | 44 | 6 1/2 | 30 | 18 | 29 | 15 | 9 | 15 |
| 1 7/16 | 6 | 51 | 19 | 3 | 25 | 40 | 6 5/8 | 30 | 51 | 48 | 15 | 25 | 54 |
| 1 1/2 | 7 | 9 | 10 | 3 | 34 | 35 | 6 3/4 | 31 | 25 | 2 | 15 | 42 | 31 |
| 1 9/16 | 7 | 27 | 0 | 3 | 43 | 30 | 6 7/8 | 31 | 58 | 11 | 15 | 59 | 5 |
| 1 5/8 | 7 | 44 | 49 | 3 | 52 | 25 | 7 | 32 | 31 | 13 | 16 | 15 | 37 |
| 1 11/16 | 8 | 2 | 38 | 4 | 1 | 19 | 7 1/8 | 33 | 4 | 11 | 16 | 32 | 5 |
| 1 3/4 | 8 | 20 | 27 | 4 | 10 | 14 | 7 1/4 | 33 | 37 | 3 | 16 | 48 | 31 |
| 1 13/16 | 8 | 38 | 16 | 4 | 19 | 8 | 7 3/8 | 34 | 9 | 49 | 17 | 4 | 54 |

* For conversions into decimal degrees and radians see pages 277 and 278.

**Accurate Measurement of Angles and Tapers.** — When great accuracy is required in the measurement of angles, or when originating tapers, disks are commonly used. The principle of the disk method of taper measurement is that if two disks of unequal diameters are placed either in contact or a certain distance apart, lines tangent to their peripheries will represent an angle or taper, the degree of which depends upon the diameters of the two disks and the distance between them. The

gage shown in the accompanying illustration, which is a form commonly used for originating tapers or measuring angles accurately, is set by means of disks. This gage consists of two adjustable straight-edges $A$ and $A_1$, which are in contact with disks $B$ and $B_1$. The angle $\alpha$ or the taper between the straight-edges depends, of course, upon the diameters of the disks and the center distance $C$, and as

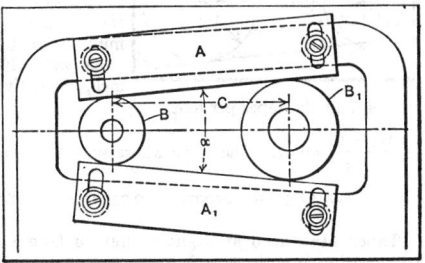

these three dimensions can be measured accurately, it is possible to set the gage to a given angle within very close limits. Moreover, if a record of the three dimensions is kept, the exact setting of the gage can be reproduced quickly at any time. The following rules may be used for adjusting a gage of this type, and cover all problems likely to arise in practice. Disks are also occasionally used for the setting of parts in angular positions for accurately machining them to a given angle; the rules will be found applicable to these conditions also.

**To Find Angle for Given Taper per Foot.** — When the taper in inches per foot is known, and the corresponding angle $\alpha$ is required. *Rule:* Divide the

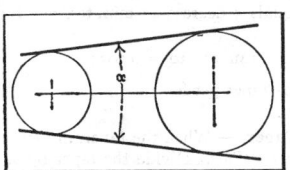

taper in inches per foot by 24; find the angle corresponding to the quotient, in a table of tangents, and double this angle.

*Example:* What angle $\alpha$ is equivalent to a taper of 1½ inch per foot?

$\frac{1.5}{24} = 0.0625$. The angle whose tangent is 0.0625 equals 3 degrees 35 minutes, nearly; then, 3 deg. 35 min. × 2 = 7 deg. 10 min.

**To Find Angle for Given Disk Dimensions.** — When the diameters $D$ and $d$ of the large and small disks and the center distance are given, to determine the

angle $\alpha$. *Rule:* Divide the difference between the disk diameters by twice the center distance; find the angle corresponding to the quotient, in a table of sines, and double the angle.

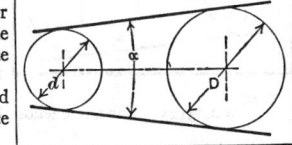

*Example:* If the disk diameters are 1 and 1.5 inch, respectively, and the center distance is 5 inches, find the included angle $\alpha$.

$\frac{1.5 - 1}{2 \times 5} = 0.05$. The angle whose sine is 0.05 equals 2 degrees 52 minutes; then, 2 deg. 52 min. × 2 = 5 deg. 44 min. = angle $\alpha$.

**To Find the Taper per Foot.** — When the diameters $D$ and $d$ of the large and small disks and the center distance $C$ are given, to determine the taper per

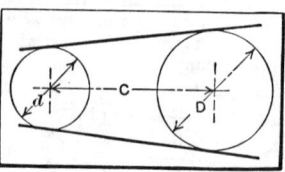

foot (measured at right angles to line through disk centers). *Rule:* Divide the difference between the disk diameters by twice the center distance; find the angle corresponding to the quotient, in a table of sines; then find the tangent corresponding to this angle, and multiply the tangent by 24.

*Example*: If disk diameters are 1 and 1.5 inch, respectively, and center distance is 5 inches, find the taper per foot.

$$\frac{1.5 - 1}{2 \times 5} = 0.05.$$ The angle whose sine is 0.05 equals 2 degrees 52 minutes;

$$\tan 2° \, 52' = 0.05007; \qquad 0.05007 \times 24 = 1.2017 \text{ inch taper per foot.}$$

**Taper Measured at Right Angles to One Side.** — When one side is taken as a base line, and the taper is measured at right angles to that side, use the following rule for determining the taper per foot. *Rule:* Divide the difference between the disk diameters $D$ and $d$ by twice the center distance $C$; find the angle corresponding to the quotient, in a table of sines; double this angle and find the corresponding tangent; then multiply the tangent by 12.

*Example:* If the disk diameters are 2 and 3 inches, respectively, and the center distance

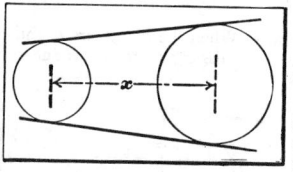

is 5 inches, what is the taper per foot measured at right angles to one side?

$$\frac{3 - 2}{2 \times 5} = 0.1.$$ The angle whose sine is 0.1 equals 5 degrees 45 minutes, nearly;

then, $\quad 2 \times 5$ deg. 45 min. $= 11$ deg. 30 min.; $\quad \tan 11° \, 30' = 0.20345;$

$$0.20345 \times 12 = 2.4414 \text{ inches taper per foot.}$$

**To Find Center Distance for a Given Taper.** — When the taper, in inches per foot, is given, to determine center distance $x$. *Rule:* Divide the taper by 24

and find the angle corresponding to the quotient in a table of tangents; then find the sine corresponding to this angle and divide the difference between the disk diameters by twice the sine.

*Example:* Gage is to be set to ¾ inch per foot, and disk diameters are 1.25 and 1.5 inch, respectively. Find the required center distance for the disks.

$$\frac{0.75}{24} = 0.03125.$$ The angle whose tangent is 0.03125 equals 1 degree 47.4 minutes;

$$\sin 1° \, 47.4' = 0.03123; \qquad 1.50 - 1.25 = 0.25 \text{ inch;}$$

$$\frac{0.25}{2 \times 0.03123} = 4.002 \text{ inches} = \text{center distance } x.$$

**To Find Center Distance for a Given Angle.** — When straight-edges must be set to a given angle $\alpha$, to determine center distance $x$ between disks of known diameter. *Rule:* Find the sine of half the angle $\alpha$ in a table of sines; divide the difference between the disk diameters by double this sine.

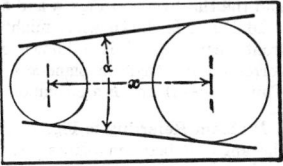

*Example:* If an angle $\alpha$ of 20 degrees is required, and the disks are 1 and 3 inches in diameter, respectively, find the required center distance $x$.

$$\frac{20}{2} = 10 \text{ degrees}; \qquad \sin 10° = 0.17365;$$

$$\frac{3-1}{2 \times 0.17365} = 5.759 \text{ inches} = \text{center distance } x.$$

**Center Distance when Taper is Measured from One Side.** — When taper is measured at right angles to one side, use the following rule for determining the

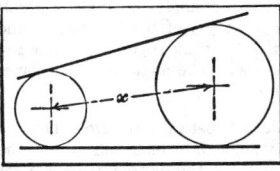

center distance $x$. *Rule:* Divide the taper in inches per foot by 12; find the angle corresponding to the quotient, in a table of tangents; find the sine of one-half this angle, and then divide the difference between the disk diameters by double the sine.

*Example:* If taper measured at right angles to one side is 6.9 inches per foot, and the disks are 2 and 5 inches in diameter, respectively, what is center distance $x$?

$\frac{6.9}{12} = 0.575$. The angle whose tangent is 0.575 equals 29 degrees 54 minutes;

then, $\qquad \frac{29 \text{ deg. } 54 \text{ min.}}{2} = 14 \text{ deg. } 57 \text{ min}; \qquad \sin 14° \, 57' = 0.25798;$

$$\frac{5-2}{2 \times 0.25798} = 5.814 \text{ inches, center distance.}$$

**Angular Measurements with Disks in Contact.** — When the two disks are to be in contact and the diameter of the small disk is known, the diameter $D$ of the large disk for a given angle $\alpha$ can be obtained

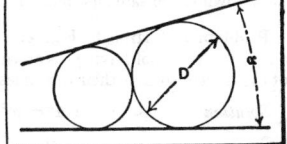

as follows. *Rule:* Multiply twice the diameter of the small disk by the sine of one-half the required angle; divide this product by 1 minus the sine of one-half the required angle; add the quotient to the diameter of the small disk to obtain the diameter of the large disk.

*Example:* The required angle $\alpha$ is 15 degrees. Find diameter of large disk, to be in contact with a standard 1-inch reference disk.

$$\sin 7° \, 30' = 0.13053. \qquad 2 \times 1 \times 0.13053 = 0.26106;$$

$$\frac{0.26106}{1 - 0.13053} = 0.3002. \qquad 1 + 0.3002 = 1.3002 = \text{diameter of large disk.}$$

**Compound Angles.** — Three types of compound angles are illustrated by Figs. 1–6. The first type is shown in Figs. 1, 2 and 3; the second type in Fig. 4; and the third type in Figs. 5 and 6.

In Fig. 1 is shown what might be considered as a thread-cutting tool without front clearance. $A$ is a known angle in plane $y$–$y$ of the top surface. $C$ is the corresponding angle in plane $x$–$x$ which is at some given angle $B$ with plane $y$–$y$. Thus, angles $A$ and $B$ are components of the compound angle $C$.

**Problem Referring to Fig. 1:** The angle $2\,A$ in plane $y$–$y$ is known, as is also the angle $B$ between planes $x$–$x$ and $y$–$y$. It is required to find the compound angle $2\,C$ in plane $x$–$x$.

*Solution:*         Let $2\,A = 60°$ and $B = 15°$

Then

$$\tan C = \tan A \cos B$$
$$\tan C = \tan 30° \cos 15°$$
$$\tan C = 0.57735 \times 0.96592$$
$$\tan C = 0.55767$$
$$C = 29° 8.8' \qquad 2\,C = 58° 17.6'$$

Fig. 2 shows a thread-cutting tool with front clearance angle $B$. Angle $A$ equals one-half the angle between the cutting edges in plane $y$–$y$ of the top surface and compound angle $C$ is one-half the angle between the cutting edges in a plane $x$–$x$ at right angles to the inclined front edge of the tool. The angle between planes $y$–$y$ and $x$–$x$ is, therefore, equal to clearance angle $B$.

**Problem Referring to Fig. 2:** Find the angle $2\,C$ between the front faces of a thread-cutting tool having a known clearance angle $B$, which will permit the grinding of these faces so that their top edges will form the desired angle $2\,A$ for cutting the thread.

*Solution:*         Let $2\,A = 60°$ and $B = 15°$

Then

$$\tan C = \frac{\tan A}{\cos B} = \frac{\tan 30°}{\cos 15°} = \frac{0.57735}{0.96592}$$
$$\tan C = 0.59772$$
$$C = 30° 52' \qquad 2\,C = 61° 44'$$

In Fig. 3 is shown a form-cutting tool in which $A$ is one-half the angle between the cutting edges in plane $y$–$y$ of the top surface; $B$ is the front clearance angle; and $C$ is one-half the angle between the cutting edges in plane $x$–$x$ at right angles to the front edges of the tool. The formula for finding angle $C$ when angles $A$ and $B$ are known is the same as that for Fig. 2

**Problem Referring to Fig. 3:** Find the angle $2\,C$ between the front faces of a form-cutting tool having a known clearance angle $B$ which will permit the grinding of these faces so that their top edges will form the desired angle $2\,A$ for form cutting.

*Solution:*         Let $2\,A = 46°$ and $B = 12°$

Then

$$\tan C = \frac{\tan A}{\cos B} = \frac{\tan 23°}{\cos 12°} = \frac{0.42447}{0.97815}$$
$$\tan C = 0.43395$$
$$C = 23° 27.5' \qquad 2\,C = 46° 55'$$

In Fig. 4 is shown a wedge-shaped block, the top surface of which is inclined at compound angle $C$ with the base in a plane at right angles with the base and at angle $R$ with the front edge. Angle $A$ in the vertical plane of the front of the plate

## Formulas for Compound Angles

$C$ = compound angle in plane $x$-$x$ and is the resultant of angles $A$ and $B$

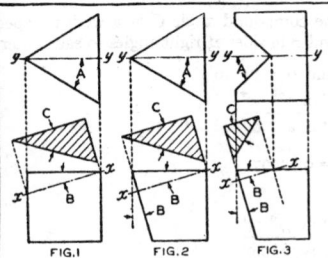

FIG.1 FIG.2 FIG.3

For given angles $A$ and $B$, find the resultant angle $C$ in plane $x$-$x$. Angle $B$ is measured in vertical plane $y$-$y$ of mid-section.

(Fig. 1) $\text{Tan } C = \tan A \times \cos B$

(Fig. 2) $\text{Tan } C = \dfrac{\tan A}{\cos B}$

(Fig. 3) (Same formula as for Fig. 2)

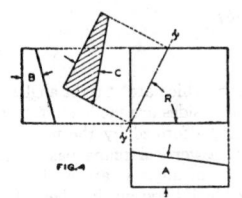

FIG.4

Fig. 4. In machining plate to angles $A$ and $B$, it is held at angle $C$ in plane $x$-$x$. Angle of rotation $R$ in plane parallel to base (or complement of $R$) is for locating plate so that plane $x$-$x$ is perpendicular to axis of pivot on angle-plate or work-holding vise.

$$\text{Tan } R = \frac{\tan B}{\tan A}; \quad \text{Tan } C = \frac{\tan A}{\cos R}$$

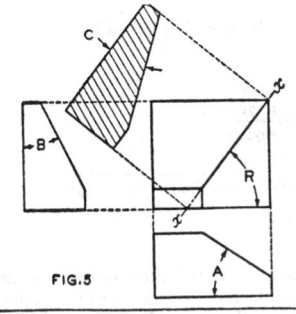

FIG.5

Fig. 5. Angle $R$ in plane parallel to base is angle from plane $x$-$x$ to side having angle $A$.

$$\text{Tan } R = \frac{\cot B}{\cot A}$$

$$\text{Cot } C = \sqrt{\cot^2 A + \cot^2 B}$$

Compound angle $C$ is angle in plane $x$-$x$ from base to corner formed by intersection of planes inclined to angles $A$ and $B$. This formula for $C$ may be used to find cot of complement of $C_1$, Fig. 6.

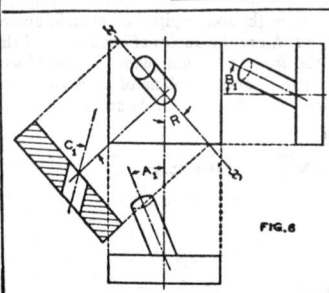

FIG.6

Fig. 6. Angles $A_1$ and $B_1$ are measured in vertical planes of front and side elevations. Plane $x$-$x$ is located by angle $R$ from center-line or from plane of angle $B_1$.

$$\text{Tan } C_1 = \sqrt{\tan^2 A_1 + \tan^2 B_1}$$

$$\text{Tan } R = \frac{\tan A_1}{\tan B_1}$$

The resultant angle $C_1$ would be required in drilling hole for pin.

and angle $B$ in the vertical plane of one side which is at right angles to the front are components of angle $C$.

**Problem Referring to Fig. 4:**  Find the compound angle $C$ of a wedge-shaped block having known component angles $A$ and $B$ in sides at right angles to each other.

*Solution:*     Let $A = 47°\ 14'$ and $B = 38°\ 10'$

$$\tan R = \frac{\tan B}{\tan A} = \frac{\tan 38°\ 10'}{\tan 47°\ 14'}$$

$$\tan R = \frac{0.78598}{1.0812} = 0.72695$$

$$R = 36°\ 0.9'$$

$$\tan C = \frac{\tan A}{\cos R} = \frac{\tan 47°\ 14'}{\cos 36°\ 0.9'}$$

$$\tan C = \frac{1.0812}{0.80887} = 1.3367$$

$$C = 53°\ 12'$$

In Fig. 5 is shown a four-sided block, two sides of which are at right angles to each other and to the base of the block. The other two sides are inclined at an oblique angle with the base. Angle $C$ is a compound angle formed by the intersection of these two inclined sides and the intersection of a vertical plane passing through $x$–$x$, and the base of the block. The components of angle $C$ are angles $A$ and $B$ and angle $R$ is the angle in the base plane of the block between the plane of angle $C$ and the plane of angle $A$.

**Problem Referring to Fig. 5:**  Find the angles $C$ and $R$ in the block shown in Fig. 5 when angles $A$ and $B$ are known.

*Solution:*  Let angle $A = 27°$ and $B = 36°$

$$\tan R = \frac{\cot B}{\cot A} = \frac{\cot 36°}{\cot 27°} = \frac{1.3764}{1.9626}$$

$$\tan R = 0.70131 \qquad R = 35°\ 2.5'$$

$$\cot C = \sqrt{\cot^2 A + \cot^2 B} = \sqrt{(1.9626)^2 + (1.3764)^2}$$

$$\cot C = \sqrt{5.74627572} = 2.3971$$

$$C = 22°\ 38.6'$$

**Problem Referring to Fig. 6:**  A rod or pipe is inserted into a rectangular block at an angle. Angle $C_1$ is the compound angle of inclination (measured from the vertical) in a plane passing through the center line of the rod or pipe and at right angles to the top surface of the block. Angles $A_1$ and $B_1$ are the angles of inclination of the rod or pipe when viewed respectively in the front and side planes of the block. Angle $R$ is the angle between the plane of angle $C_1$ and the plane of angle $B_1$. Find angles $C_1$ and $R$ when a rod or pipe is inclined at known angles $A_1$ and $B_1$.

*Solution:* Let $A_1 = 39°$ and $B_1 = 34°$

Then     $$\tan C_1 = \sqrt{\tan^2 A_1 + \tan^2 B_1} = \sqrt{(0.80978)^2 + (0.67451)^2}$$

$$\tan C_1 = \sqrt{1.1107074} = 1.0539$$

$$C_1 = 46°\ 30.2'$$

$$\tan R = \frac{\tan A_1}{\tan B_1} = \frac{0.80978}{0.67451}$$

$$\tan R = 1.2005 \qquad R = 50°\ 12.4'$$

# MILLING CUTTERS

**Selection of Milling Cutters.** — The most suitable type of milling cutter for a particular milling operation will depend upon, among other factors, the kind of cut to be made, the material to be cut, the number of parts to be machined, and the type of milling machine available. Solid cutters of small size will usually cost less, initially, than inserted blade types; for long-run production, inserted-blade cutters will probably have a lower over-all cost. Depending on either the material to be cut, or the amount of production involved, the use of carbide tipped cutters in preference to high-speed steel or other cutting tool materials may be justified.

Rake angles depend on both the cutter material and the work material. Carbide and cast alloy cutting tool materials generally have smaller rake angles than high-speed steel tool materials because of their lower edge strength and greater abrasion resistance. Soft work materials permit higher radial rake angles than hard materials; thin cutters permit zero or practically zero axial rake angles; and wide cutters operate smoother with high axial rake angles. (See page 1629.)

Cutting edge relief or clearance angles are usually from 3 to 6 degrees for hard or tough materials, 4 to 7 degrees for average materials, and 6 to 12 degrees for easily machined materials. (See page 1628.)

The number of teeth in the milling cutter is also a factor that should be given consideration as explained in the next paragraph.

**Number of Teeth in Milling Cutters.** — In determining the number of teeth a milling cutter should have for optimum performance, there is no universal rule that covers every case. There are, however, two factors which should be considered in making a choice: (1) The number of teeth should never be so great as to reduce the chip space between the teeth to a point where a free flow of chips is prevented; (2) The chip space should be smooth and without sharp corners that would cause clogging of the chips in the space.

For milling ductile materials which produce a continuous and curled chip, a cutter with large chip spaces is preferable. Such coarse tooth cutters permit an easier flow of the chips through the chip space than would be obtained with fine tooth cutters, and help to eliminate cutter "chatter." For cutting operations in thin materials, fine tooth cutters reduce cutter and workpiece vibration and the tendency for the cutter teeth to "straddle" the workpiece and dig in. For slitting copper and other soft non-ferrous materials, teeth that are either chamfered or alternately flat and V-shaped are best.

As a general rule, to give satisfactory performance the number of teeth in milling cutters should be such that *no more than two teeth at a time are engaged in the cut.* Based on this rule, The Cincinnati Milling Machine Co. recommends the use of the following formulas:

For face milling cutters,

$$T = \frac{6.3\,D}{W} \qquad\qquad (1)$$

For peripheral milling cutters,

$$T = \frac{12.6\,D\cos A}{D + 4d} \qquad\qquad (2)$$

where $T$ = number of teeth in cutter; $D$ = cutter diameter in inches; $W$ = width of cut in inches; $d$ = depth of cut in inches; and $A$ = helix angle of cutter.

To find the number of teeth that a cutter should have when other than two teeth in the cut at the same time is desired, Formulas (1) and (2) should be divided by two and the result multiplied by the number of teeth desired in the cut.

*Example:* Determine the required number of teeth in a face mill where $D = 6$ inches and $W = 4$ inches. Using Formula (1),

$$T = \frac{6.3 \times 6}{4} = 10 \text{ teeth, approximately}$$

*Example:* Determine the required number of teeth in a plain milling cutter where $D = 4$ inches and $d = \frac{1}{4}$ inch. Using Formula (2),

$$T = \frac{12.6 \times 4 \times \cos 0^\circ}{4 + 4 \times \frac{1}{4}} = 10 \text{ teeth, approximately}$$

In *high speed milling* with sintered carbide, high-speed steel, and cast non-ferrous cutting tool materials, a formula that permits full use of the power available at the motor driving the cutter but prevents overloading of the motor driving the milling machine is:

$$T = \frac{K \times H}{F \times N \times d \times W} \tag{3}$$

where $T$ = number of cutter teeth; $H$ = horsepower available at the cutter; $F$ = feed per tooth in inches; $N$ = revolutions per minute of cutter; $d$ = depth of cut in inches; $W$ = width of cut in inches; and $K$ = a constant which may be taken as 0.65 for average steel, 1.5 for cast iron, and 2.5 for aluminum. These values are conservative and take into account dulling of the cutter in service.

*Example:* Determine the required number of teeth in a sintered carbide tipped face mill for high speed milling of 200 Brinell hardness alloy steel if $H$ = 10 horsepower; $F$ = 0.008 inch; $N$ = 272 R.P.M.; $d$ = 0.125 inch; $W$ = 6 inches; and $K$ for alloy steel is 0.65. Using Formula (3),

$$T = \frac{0.65 \times 10}{0.008 \times 272 \times 0.125 \times 6} = 4 \text{ teeth, approximately}$$

**American National Standard Milling Cutters.** — According to American National Standard ANSI B94.19-1968, milling cutters may be classified in two general ways which are given as follows:

*By Type of Relief on Cutting Edges:* Milling cutters may be described on the basis of one of two methods of providing relief for the cutting edges. *Profile sharpened* cutters are those on which relief is obtained and which are resharpened by grinding a narrow land back of the cutting edges. Profile sharpened cutters may produce flat, curved, or irregular surfaces. *Form relieved* cutters are those which are so relieved that by grinding only the faces of the teeth the original form is maintained throughout the life of the cutters. Form relieved cutters may produce flat, curved or irregular surfaces.

*By Method of Mounting:* Milling cutters may be described by one of two methods used to mount the cutter. *Arbor type* cutters are those which have a hole for mounting on an arbor and usually have a keyway to receive a driving key. These are sometimes called *Shell type*. *Shank type* cutters are those which have a straight or tapered shank to fit the machine tool spindle or adapter.

**Explanation of the "Hand" of Milling Cutters.** — In the ANSI Standard the terms "right hand" and "left hand" are used to describe hand of rotation, hand of cutter and hand of flute helix.

*Hand of Rotation* or *Hand of Cut* is described as either "right hand" if the cutter revolves counterclockwise as it cuts when viewed from a position in front of a horizontal milling machine and facing the spindle or "left hand" if the cutter revolves clockwise as it cuts when viewed from the same position.

## American National Standard Plain Milling Cutters (ANSI B94.19-1968)

| Cutter Diameter | | | Range of Face Widths Nom.* | Hole Diameter | | |
|---|---|---|---|---|---|---|
| Nom. | Max. | Min. | | Nom. | Max. | Min. |
| Light-duty Cutters[1] | | | | | | |
| 2½ | 2.515 | 2.485 | ³⁄₁₆, ¼, ⁵⁄₁₆, ⅜, ½, ⅝, ¾, 1, 1½, 2, 2½ and 3 | 1 | 1.00075 | 1.0000 |
| 3 | 3.015 | 2.985 | ³⁄₁₆, ¼, ⁵⁄₁₆, ⅜ | 1 | 1.00075 | 1.0000 |
| 3 | 3.015 | 2.985 | ⅜, ½, ⅝, ¾, 1, 1¼, 1½, 2 and 3 | 1¼ | 1.2510 | 1.2500 |
| 4 | 4.015 | 3.985 | ¼, ⁵⁄₁₆ and ⅜ | 1 | 1.00075 | 1.0000 |
| 4 | 4.015 | 3.985 | ⅜, ½, ⅝, ¾, 1, 1½, 2, 3 and 4 | 1¼ | 1.2510 | 1.2500 |
| Heavy-duty Cutters[2] | | | | | | |
| 2½ | 2.515 | 2.485 | 2 | 1 | 1.00075 | 1.0000 |
| 2½ | 2.515 | 2.485 | 4 | 1 | 1.0010 | 1.0000 |
| 3 | 3.015 | 2.985 | 2, 2½, 3, 4 and 6 | 1¼ | 1.2510 | 1.2500 |
| 4 | 4.015 | 3.985 | 2, 3, 4 and 6 | 1½ | 1.5010 | 1.5000 |
| High-helix Cutters[3] | | | | | | |
| 3 | 3.015 | 2.985 | 4 and 6 | 1¼ | 1.2510 | 1.2500 |
| 4 | 4.015 | 3.985 | 4, 6 and 8 | 1½ | 1.5010 | 1.5000 |

All dimensions are in inches. All cutters are high-speed steel. Plain milling cutters are of cylindrical shape, having teeth on the peripheral surface only.

* *Tolerances on Face Widths:* Up to 1 inch, inclusive, ±0.001 inch; over 1 to 2 inches, inclusive, +0.010 inch; over 2 inches, 0.020 inch.

[1] Light-duty plain milling cutters with face widths under ¾ inch have straight teeth. Cutters with ¾-inch face and wider have helix angles of not less than 18 degrees nor greater than 25 degrees.

[2] Heavy-duty plain milling cutters have a helix angle of not less than 25 degrees nor greater than 45 degrees.

[3] High-helix plain milling cutters have a helix angle of not less than 45 degrees nor greater than 52 degrees.

*Hand of Cutter:* Some types of cutters require special consideration when referring to their hand. These are principally cutters with unsymmetrical forms or cutters with threaded holes. *Symmetrical* cutters may be reversed on the arbor in the same axial position and rotated in the cutting direction without altering the contour produced on the work-piece, and may be considered as either right or left hand. *Unsymmetrical* cutters reverse the contour produced on the work-piece when reversed on the arbor in the same axial position and rotated in the cutting direction. A *single-angle* cutter is considered to be a right-hand cutter if it revolves counterclockwise, or a left-hand cutter if it revolves clockwise, when cutting as viewed from the side of the larger diameter. The *hand of rotation* of a single angle milling cutter need not necessarily be the same as its *hand of cutter.* A *single corner rounding* cutter is considered to be a right-hand cutter if it revolves counterclockwise, or a left-hand cutter if it revolves clockwise, when cutting as viewed from the side of the smaller diameter.

## American National Standard Side Milling Cutters (ANSI B94.19-1968)

| Cutter Diameter | | | Range of Face Widths Nom.* | Hole Diameter | | |
|---|---|---|---|---|---|---|
| Nom. | Max. | Min. | | Nom. | Max. | Min. |
| | | | Side Cutters[1] | | | |
| 2 | 2.015 | 1.985 | 3/16, 1/4, 3/8 | 5/8 | 0.62575 | 0.6250 |
| 2½ | 2.515 | 2.485 | 1/4, 5/16, 3/8, 1/2 | 7/8 | 0.87575 | 0.8750 |
| 3 | 3.015 | 2.985 | 1/4, 5/16, 3/8, 7/16, 1/2 | 1 | 1.00075 | 1.0000 |
| 4 | 4.015 | 3.985 | 1/4, 3/8, 1/2, 5/8, 3/4, 7/8 | 1 | 1.00075 | 1.0000 |
| 4 | 4.015 | 3.985 | 1/2, 5/8, 3/4, 7/8 | 1¼ | 1.2510 | 1.2500 |
| 5 | 5.015 | 4.985 | 1/2, 5/8, 3/4 | 1 | 1.00075 | 1.0000 |
| 5 | 5.015 | 4.985 | 1/2, 5/8, 3/4, 1 | 1¼ | 1.2510 | 1.2500 |
| 6 | 6.015 | 5.985 | 1/2, 3/4 | 1 | 1.00075 | 1.0000 |
| 6 | 6.015 | 5.985 | 1/2, 5/8, 3/4, 1 | 1¼ | 1.2510 | 1.2500 |
| 7 | 7.015 | 6.985 | 3/4 | 1¼ | 1.2510 | 1.2500 |
| 8 | 8.015 | 7.985 | 3/4, 1 | 1¼ | 1.2510 | 1.2500 |
| | | | Staggered-tooth Side Cutters[2] | | | |
| 2½ | 2.515 | 2.485 | 1/4, 5/16, 3/8, 1/2 | 7/8 | 0.87575 | 0.8750 |
| 3 | 3.015 | 2.985 | 3/16, 1/4, 5/16, 3/8 | 1 | 1.00075 | 1.0000 |
| 3 | 3.015 | 2.985 | 1/2, 5/8, 3/4 | 1¼ | 1.2510 | 1.2500 |
| 4 | 4.015 | 3.985 | { 1/4, 5/16, 3/8, 7/16, 1/2, 5/8, 3/4 and 7/8 } | 1¼ | 1.2510 | 1.2500 |
| 5 | 5.015 | 4.985 | 1/2, 5/8, 3/4 | 1¼ | 1.2510 | 1.2500 |
| 6 | 6.015 | 5.985 | 3/8, 1/2, 5/8, 3/4, 7/8, 1 | 1¼ | 1.2510 | 1.2500 |
| 8 | 8.015 | 7.985 | 3/8, 1/2, 5/8, 3/4, 7/8, 1 | 1½ | 1.5010 | 1.5000 |
| | | | Half Side Cutters[3] | | | |
| 4 | 4.015 | 3.985 | 3/4 | 1¼ | 1.2510 | 1.2500 |
| 5 | 5.015 | 4.985 | 3/4 | 1¼ | 1.2510 | 1.2500 |
| 6 | 6.015 | 5.985 | 3/4 | 1¼ | 1.2510 | 1.2500 |
| 6 | 6.015 | 5.985 | 1 | 1½ | 1.5010 | 1.5000 |
| 7 | 7.015 | 6.985 | 3/4 | 1½ | 1.5010 | 1.5000 |
| 8 | 8.015 | 7.985 | 3/4, 1 | 1½ | 1.5010 | 1.5000 |

All dimensions are in inches. All cutters are high-speed steel. Side milling cutters are of cylindrical shape, having teeth on the periphery and on one or both sides.

* *Tolerances on Face Widths:* For side cutters, +0.002, −0.001 inch; for staggered-tooth side cutters up to 3/4 inch face width, inclusive, −0.0005 inch, and over 3/4 to 1 inch, inclusive, −0.0010 inch; and for half side cutters, +0.015 inch.

[1] Side milling cutters have straight peripheral teeth and side teeth on both sides.

[2] Staggered-tooth side milling cutters have peripheral teeth of alternate right- and left-hand helix and alternate side teeth.

[3] Half side milling cutters have side teeth on one side only. The peripheral teeth are helical of the same hand as the cut. Made either with right-hand or left-hand cut.

*Hand of Flute Helix:* Milling cutters may have *straight flutes* which means that their cutting edges are in planes parallel to the cutter axis. Milling cutters with flute helix in one direction only are described as having a right-hand helix if the flutes twist away from the observer in a clockwise direction when viewed from either end of the cutter or as having a left-hand helix if the flutes twist away from the observer in a counterclockwise direction when viewed from either end of the cutter. *Staggered tooth cutters* are milling cutters with every other flute of opposite (right and left hand) helix.

An illustration describing the various milling cutter elements of both a profile cutter and a form-relieved cutter is given on page 1609.

### American National Standard Staggered Teeth, T-Slot Milling Cutters with Brown & Sharpe Taper and Weldon Shanks (ANSI B94.19-1968)

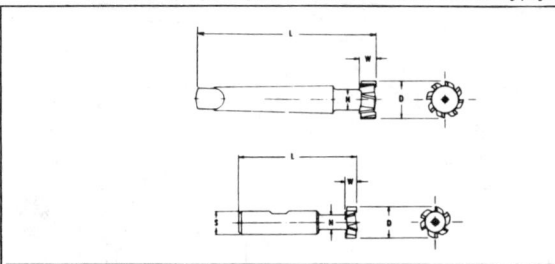

| Bolt Size | Cutter Diam., D | Face Width, W | Neck Diam., N | With B. & S. Taper* | | With Weldon Shank | |
|---|---|---|---|---|---|---|---|
| | | | | Length, L | Taper No. | Length, L | Diam., S |
| 1/4 | 9/16 | 15/64 | 17/64 | ... | ... | 2 19/32 | 1/2 |
| 5/16 | 21/32 | 17/64 | 21/64 | ... | ... | 2 11/16 | 1/2 |
| 3/8 | 25/32 | 21/64 | 13/32 | ... | ... | 3 1/4 | 3/4 |
| 1/2 | 31/32 | 25/64 | 17/32 | 5 | 7 | 3 7/16 | 3/4 |
| 5/8 | 1 1/4 | 31/64 | 21/32 | 5 1/4 | 7 | 3 15/16 | 1 |
| 3/4 | 1 15/32 | 5/8 | 25/32 | 6 7/8 | 9 | 4 7/16 | 1 |
| 1 | 1 27/32 | 53/64 | 1 1/32 | 7 1/4 | 9 | 4 13/16 | 1 1/4 |

All dimensions are in inches. All cutters are high-speed steel and only right-hand cutters are standard.

* For dimensions of Brown & Sharpe taper shanks, see information given on page 1731.

*Tolerances:* On $D$, −0.010 inch; on $W$, −0.005 inch; on $N$, −0.005 inch; on $L$, ±1/16 inch; on $S$, −0.0005 inch.

### American National Standard Form Relieved Corner Rounding Cutters with Weldon Shanks (ANSI B94.19-1968)

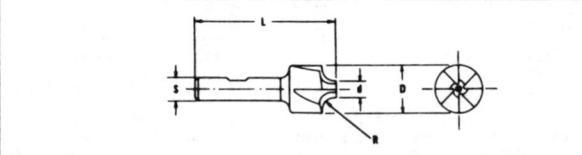

| Rad., R | Diam., D | Diam., d | S | L | Rad., R | Diam., D | Diam., d | S | L |
|---|---|---|---|---|---|---|---|---|---|
| 1/16 | 7/16 | 1/4 | 3/8 | 2 1/2 | 3/8 | 1 1/4 | 3/8 | 1/2 | 3 1/2 |
| 3/32 | 1/2 | 1/4 | 3/8 | 2 1/2 | 7/16 | 7/8 | 5/16 | 3/4 | 3 1/8 |
| 1/8 | 5/8 | 1/4 | 1/2 | 3 | 1/4 | 1 | 3/8 | 3/4 | 3 1/4 |
| 5/32 | 3/4 | 5/16 | 1/2 | 3 | 5/16 | 1 1/8 | 3/8 | 7/8 | 3 1/2 |
| 3/16 | 7/8 | 5/16 | 1/2 | 3 | 3/8 | 1 1/4 | 3/8 | 7/8 | 3 3/4 |
| 1/4 | 1 | 3/8 | 1/2 | 3 | 7/16 | 1 3/8 | 3/8 | 1 | 4 |
| 5/16 | 1 1/8 | 3/8 | 1/2 | 3 1/4 | 1/2 | 1 1/2 | 3/8 | 1 | 4 1/8 |

All dimensions are in inches. All cutters are high-speed steel. Right-hand cutters are standard.

*Tolerances:* On $D$, ±0.010 inch; on $d$, ±0.001 inch for cutters up to and including 1/8-inch radius, +0.002, −0.001 inch for cutters over 1/8-inch radius; on $S$, −0.0005 inch; and on $L$, ±1/16 inch.

**American National Standard Metal Slitting Saws (ANSI B94.19-1968)**

| Cutter Diameter | | | Range of Face Widths | Hole Diameter | | |
|---|---|---|---|---|---|---|
| Nom. | Max. | Min. | Nom.* | Nom. | Max. | Min. |
| **Plain Metal Slitting Saws[1]** | | | | | | |
| 2½ | 2.515 | 2.485 | 1/32, 3/64, 1/16, 3/32, 1/8 | 7/8 | 0.87575 | 0.8750 |
| 3 | 3.015 | 2.985 | 1/32, 3/64, 1/16, 3/32, 1/8 and 5/32 | 1 | 1.00075 | 1.0000 |
| 4 | 4.015 | 3.985 | 1/32, 3/64, 1/16, 3/32, 1/8, 5/32 and 3/16 | 1 | 1.00075 | 1.0000 |
| 5 | 5.015 | 4.985 | 1/16, 3/32, 1/8 | 1 | 1.00075 | 1.0000 |
| 5 | 5.015 | 4.985 | 1/8 | 1¼ | 1.2510 | 1.2500 |
| 6 | 6.015 | 5.985 | 1/16, 3/32, 1/8 | 1 | 1.00075 | 1.0000 |
| 6 | 6.015 | 5.985 | 1/8, 3/16 | 1¼ | 1.2510 | 1.2500 |
| 8 | 8.015 | 7.985 | 1/8 | 1 | 1.00075 | 1.0000 |
| 8 | 8.015 | 7.985 | 1/8 | 1¼ | 1.2510 | 1.2500 |
| **Metal Slitting Saws with Side Teeth[2]** | | | | | | |
| 2½ | 2.515 | 2.485 | 1/16, 3/32, 1/8 | 7/8 | 0.87575 | 0.8750 |
| 3 | 3.015 | 2.985 | 1/16, 3/32, 1/8, 5/32 | 1 | 1.00075 | 1.0000 |
| 4 | 4.015 | 3.985 | 1/16, 3/32, 1/8, 5/32, 3/16 | 1 | 1.00075 | 1.0000 |
| 5 | 5.015 | 4.985 | 1/16, 3/32, 1/8, 5/32, 3/16 | 1 | 1.00075 | 1.0000 |
| 5 | 5.015 | 4.985 | 1/8 | 1¼ | 1.2510 | 1.2500 |
| 6 | 6.015 | 5.985 | 1/16, 3/32, 1/8, 3/16 | 1 | 1.00075 | 1.0000 |
| 6 | 6.015 | 5.985 | 1/8, 3/16 | 1¼ | 1.2510 | 1.2500 |
| 8 | 8.015 | 7.985 | 1/8 | 1 | 1.00075 | 1.0000 |
| 8 | 8.015 | 7.985 | 1/8, 3/16 | 1¼ | 1.2510 | 1.2500 |
| **Metal Slitting Saws with Staggered Peripheral and Side Teeth[3]** | | | | | | |
| 3 | 3.015 | 2.985 | 3/16 | 1 | 1.00075 | 1.0000 |
| 4 | 4.015 | 3.985 | 3/16 | 1 | 1.00075 | 1.0000 |
| 5 | 5.015 | 4.985 | 3/16, 1/4 | 1 | 1.00075 | 1.0000 |
| 6 | 6.015 | 5.985 | 3/16, 1/4 | 1 | 1.00075 | 1.0000 |
| 6 | 6.015 | 5.985 | 3/16, 1/4 | 1¼ | 1.2510 | 1.2500 |
| 8 | 8.015 | 7.985 | 3/16, 1/4 | 1¼ | 1.2510 | 1.2500 |
| 10 | 10.015 | 9.985 | 3/16, 1/4 | 1¼ | 1.2510 | 1.2500 |
| 12 | 12.015 | 11.985 | 1/4, 5/16 | 1½ | 1.5010 | 1.5000 |

All dimensions are in inches. All saws are high-speed steel. Metal slitting saws are similar to plain or side milling cutters but are relatively thin.

* Tolerances on face widths are plus or minus 0.001 inch.

1 Plain metal slitting saws are relatively thin plain milling cutters having peripheral teeth only. They are furnished with or without hub and their sides are concaved to the arbor hole or hub.

2 Metal slitting saws with side teeth are relatively thin side milling cutters having both peripheral and side teeth.

3 Metal slitting saws with staggered peripheral and side teeth are relatively thin staggered tooth milling cutters having peripheral teeth of alternate right- and left-hand helix and alternate side teeth.

## Milling Cutter Terms

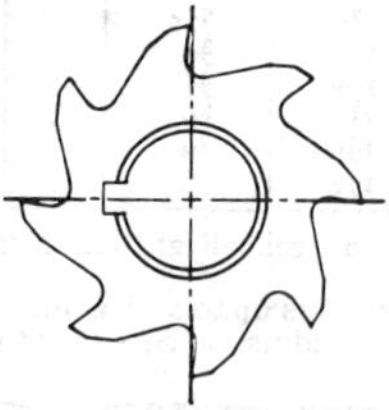

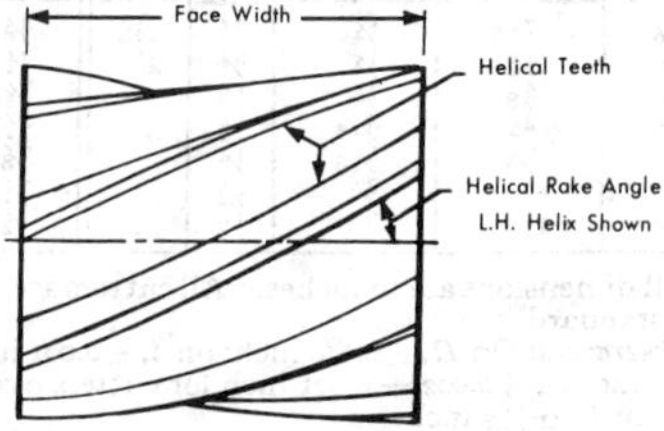

### American National Standard Single- and Double-angle Milling Cutters
(ANSI B94.19-1968)

| Cutter Diameter | | | Nominal Face Width* | Hole Diameter | | |
|---|---|---|---|---|---|---|
| Nom. | Max. | Min. | | Nom. | Max. | Min. |
| Single-angle Cutters[1] | | | | | | |
| †1¼ | 1.265 | 1.235 | ⁷⁄₁₆ | { ⅜-24 UNF-2B RH { ⅜-24 UNF-2B LH | | |
| †1⅝ | 1.640 | 1.610 | ⁹⁄₁₆ | ½-20 UNF-2B RH | | |
| 2¾ | 2.765 | 2.735 | ½ | 1 | 1.00075 | 1.0000 |
| 3 | 3.015 | 2.985 | ½ | 1¼ | 1.2510 | 1.2500 |
| Double-angle Cutters[2] | | | | | | |
| 2¾ | 2.765 | 2.735 | ½ | 1 | 1.00075 | 1.0000 |

All dimensions are in inches.   All cutters are high-speed steel.
* Face width tolerances are plus or minus 0.015 inch.
† These cutters have threaded holes, the sizes of which are given under "Hole Diameter."
[1] Single-angle milling cutters have peripheral teeth, one cutting edge of which lies in a conical surface and the other in the plane perpendicular to the cutter axis.   There are two types: one has a plain keywayed hole and has an included tooth angle of either 45 or 60 degrees plus or minus 10 minutes; the other has a threaded hole and has an included tooth angle of 60 degrees plus or minus 10 minutes.   Cutters with a right-hand threaded hole have a right-hand hand of rotation and a right-hand hand of cutter. Cutters with a left-hand threaded hole have a left-hand hand of rotation and a left-hand hand of cutter.   Cutters with plain keywayed holes are standard as either right-hand or left-hand cutters.
[2] Double-angle milling cutters have symmetrical peripheral teeth both sides of which lie in conical surfaces.   They are designated by the included angle, which may be 45, 60 or 90 degrees.   Tolerances are plus or minus 10 minutes for the half angle on each side of the center.

### Milling Cutter Terms (*Continued*)

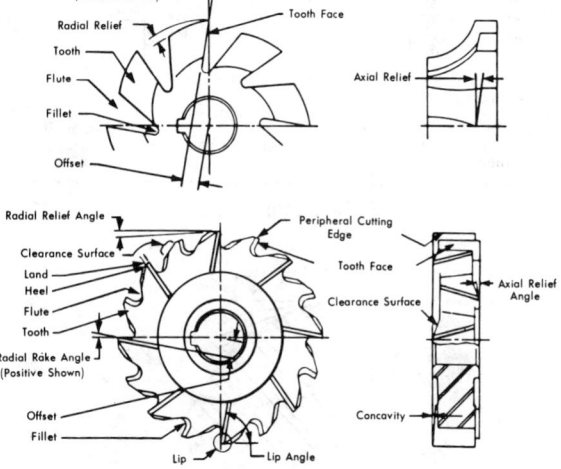

## American National Standard Shell Mills (ANSI B94.19-1968)

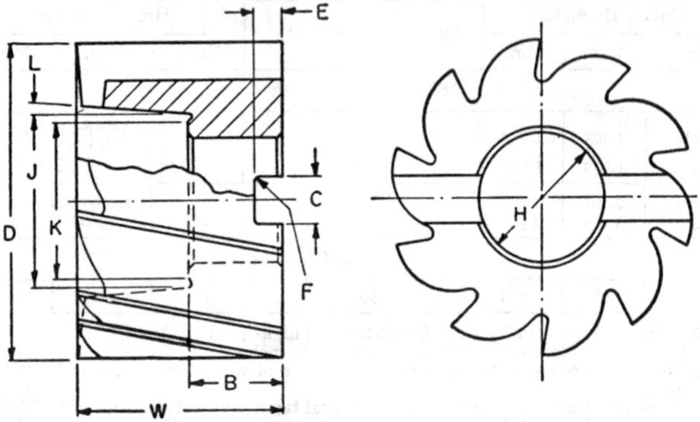
E
L
J
C
D
K
H
F
B
W

| Diam., *D* | Width, *W* | Diam., *H* | Length, *B* | Width, *C* | Depth, *E* | Radius, *F* | Diam., *J* | Diam., *K* | Angle, *L* |
|---|---|---|---|---|---|---|---|---|---|
| inches | inches | inches | inches | inches | inches | inches | inches | inches | degrees |
| 1¼ | 1 | ½ | ⅝ | ¼ | 5/32 | 1/64 | 1 1/16 | ⅝ | 0 |
| 1½ | 1⅛ | ½ | ⅝ | ¼ | 5/32 | 1/64 | 1 1/16 | ⅝ | 0 |
| 1¾ | 1¼ | ¾ | ¾ | 5/16 | 3/16 | 1/32 | 1 5/16 | ⅞ | 0 |
| 2 | 1⅜ | ¾ | ¾ | 5/16 | 3/16 | 1/32 | 1 5/16 | ⅞ | 0 |
| 2¼ | 1½ | 1 | ¾ | ⅜ | 7/32 | 1/32 | 1¼ | 1 3/16 | 0 |
| 2½* | 1⅝ | 1 | ¾ | ⅜ | 7/32 | 1/32 | 1⅜ | 1 3/16 | 0 |
| 2¾ | 1⅝ | 1 | ¾ | ⅜ | 7/32 | 1/32 | 1½ | 1 3/16 | 5 |
| 3* | 1¾ | 1¼ | ¾ | ½ | 9/32 | 1/32 | 1 21/32 | 1½ | 5 |
| 3½ | 1⅞ | 1¼ | ¾ | ½ | 9/32 | 1/32 | 1 11/16 | 1½ | 5 |
| 4* | 2¼ | 1½ | 1 | ⅝ | ⅜ | 1/16 | 2 1/32 | 1⅞ | 5 |
| 4½ | 2¼ | 1½ | 1 | ⅝ | ⅜ | 1/16 | 2 1/16 | 1⅞ | 10 |
| 5* | 2¼ | 1½ | 1 | ⅝ | ⅜ | 1/16 | 2 9/16 | 1⅞ | 10 |
| 6* | 2¼ | 2 | 1 | ¾ | 7/16 | 1/16 | 2 13/16 | 2½ | 15 |

All cutters are high-speed steel. Right-hand cutters with right-hand helix and square corners are standard.

* Left-hand cutters with left-hand helix and square corners are also standard.

*Tolerances:* On *D*, +1/64 inch; on *W*, ±1/64 inch; on *H*, +0.0005 inch; on *B*, +1/64 inch; on *C*, at least +0.008 but not more than +0.012 inch; on *E*, +1/64 inch; on *J*, ±1/64 inch; on *K*, ±1/64 inch.

## End Mill Terms

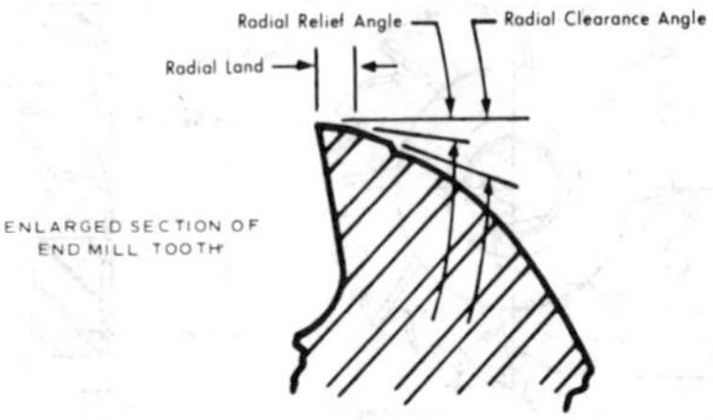
Radial Relief Angle
Radial Clearance Angle
Radial Land
ENLARGED SECTION OF END MILL TOOTH

### American National Standard Multiple- and Two-flute Single End Helical End Mills with Plain Straight and Weldon Shanks (ANSI B94.19-1968)

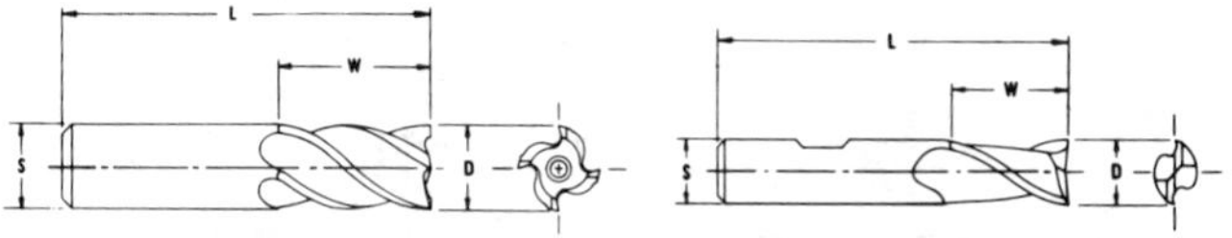

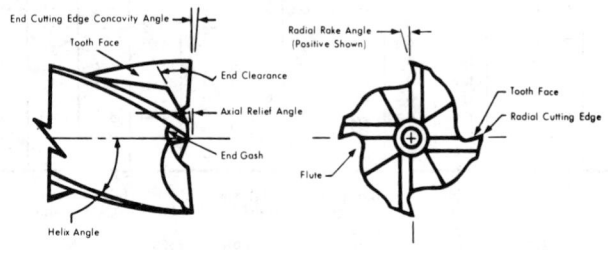

| Cutter Diameter, D | | | Shank Diameter, S | | Length of Cut, W | Length Overall, L |
|---|---|---|---|---|---|---|
| Nom. | Max. | Min. | Max. | Min. | | |
| *Multiple-flute with Plain Straight Shanks* | | | | | | |
| ⅛ | .130 | .125 | .125 | .1245 | ⁵⁄₁₆ | 1¼ |
| ³⁄₁₆ | .1925 | .1875 | .1875 | .1870 | ½ | 1⅜ |
| ¼ | .255 | .250 | .250 | .2495 | ⅝ | 1¹¹⁄₁₆ |
| ⅜ | .380 | .375 | .375 | .3745 | ¾ | 1¹³⁄₁₆ |
| ½ | .505 | .500 | .500 | .4995 | 1⁵⁄₁₆ | 2¼ |
| ¾ | .755 | .750 | .750 | .7495 | 1¼ | 2⅝ |
| *Two-flute for Keyway Cutting with Weldon Shanks* | | | | | | |
| ⅛ | .125 | .1235 | .125 | .1245 | ⅜ | 2⁵⁄₁₆ |
| ³⁄₁₆ | .1875 | .1860 | .1875 | .1870 | ⁷⁄₁₆ | 2⁵⁄₁₆ |
| ¼ | .250 | .2485 | .250 | .2495 | ½ | 2⁵⁄₁₆ |
| ⁵⁄₁₆ | .3125 | .3110 | .3125 | .3120 | ⁹⁄₁₆ | 2⁵⁄₁₆ |
| ⅜ | .375 | .3735 | .375 | .3745 | ⁹⁄₁₆ | 2⁵⁄₁₆ |
| ½ | .500 | .4985 | .500 | .4995 | 1 | 3 |
| ⅝ | .625 | .6235 | .625 | .6245 | 1⁵⁄₁₆ | 3⁷⁄₁₆ |
| ¾ | .750 | .7485 | .750 | .7495 | 1⁵⁄₁₆ | 3⁹⁄₁₆ |
| ⅞ | .875 | .8735 | .875 | .8745 | 1½ | 3¾ |
| 1 | 1.000 | .9985 | 1.000 | .9995 | 1⅝ | 4⅛ |
| 1¼ | 1.250 | 1.2485 | 1.250 | 1.2495 | 1⅝ | 4⅛ |
| 1½ | 1.500 | 1.4985 | 1.500 | 1.4995 | 1⅝ | 4⅛ |

All dimensions are in inches. All cutters are high-speed steel. Right-hand cutters with right-hand helix are standard.

The helix angle is not less than 10 degrees for multiple-flute cutters with plain straight shanks; the helix angle is optional with the manufacturer for two-flute cutters with Weldon shanks.

*Tolerances:* On *W*, ±¹⁄₃₂ inch; on *L*, ±¹⁄₁₆ inch.

### End Mill Terms (*Continued*)

End Cutting Edge Concavity Angle
Tooth Face
End Clearance
Axial Relief Angle
End Gash
Helix Angle

Radial Rake Angle
(Positive Shown)
Tooth Face
Radial Cutting Edge
Flute

ENLARGED SECTION
OF END MILL

**American National Standard Regular-, Long-, and Extra Long-length, Multiple-flute Medium Helix Single-end End Mills with Weldon Shanks** (ANSI B94.19-1968)

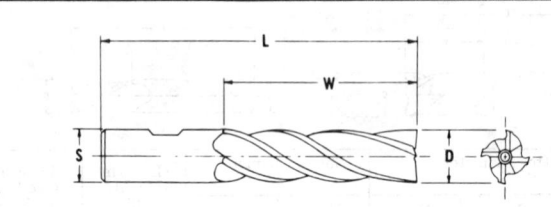

AS INDICATED BY THE DIMENSIONS GIVEN BELOW, SHANK DIAMETER S MAY BE LARGER, SMALLER, OR THE SAME AS THE CUTTER DIAMETER D.

| Cutter Diam., D | Regular Mills | | | | Long Mills | | | | Extra Long Mills | | | |
|---|---|---|---|---|---|---|---|---|---|---|---|---|
| | S | W | L | N* | S | W | L | N* | S | W | L | N* |
| 1/8† | 3/8 | 3/8 | 2 9/16 | 4 | ... | ... | ... | ... | ... | ... | ... | ... |
| 3/16† | 3/8 | 1/2 | 2 3/8 | 4 | ... | ... | ... | ... | ... | ... | ... | ... |
| 1/4† | 3/8 | 5/8 | 2 7/16 | 4 | 3/8 | 1 1/4 | 3 1/16 | 4 | 3/8 | 1 3/4 | 3 9/16 | 4 |
| 5/16† | 3/8 | 3/4 | 2 1/2 | 4 | 3/8 | 1 3/8 | 3 1/8 | 4 | 3/8 | 2 | 3 3/4 | 4 |
| 3/8† | 3/8 | 3/4 | 2 1/2 | 4 | 3/8 | 1 1/2 | 3 3/4 | 4 | 3/8 | 2 1/2 | 4 1/4 | 4 |
| 7/16 | 3/8 | 1 | 2 11/16 | 4 | 1/2 | 1 3/4 | 3 3/4 | 4 | ... | ... | ... | ... |
| 1/2 | 3/8 | 1 | 2 11/16 | 4 | 1/2 | 2 | 4 | 4 | 1/2 | 3 | 5 | 4 |
| 1/2† | 1/2 | 1 1/4 | 3 1/4 | 4 | ... | ... | ... | ... | ... | ... | ... | ... |
| 9/16 | 1/2 | 1 3/8 | 3 3/8 | 4 | ... | ... | ... | ... | ... | ... | ... | ... |
| 5/8 | 1/2 | 1 3/8 | 3 3/8 | 4 | 5/8 | 2 1/2 | 4 5/8 | 4 | 5/8 | 4 | 6 1/8 | 4 |
| 11/16 | 1/2 | 1 5/8 | 3 5/8 | 4 | ... | ... | ... | ... | ... | ... | ... | ... |
| 3/4 | 1/2 | 1 5/8 | 3 5/8 | 4 | 3/4 | 3 | 5 1/4 | 4 | 3/4 | 4 | 6 1/4 | 4 |
| 5/8† | 5/8 | 1 5/8 | 3 3/4 | 4 | ... | ... | ... | ... | ... | ... | ... | ... |
| 11/16 | 5/8 | 1 5/8 | 3 3/4 | 4 | ... | ... | ... | ... | ... | ... | ... | ... |
| 3/4† | 5/8 | 1 5/8 | 3 3/4 | 4 | ... | ... | ... | ... | ... | ... | ... | ... |
| 13/16 | 5/8 | 1 7/8 | 4 | 6 | ... | ... | ... | ... | ... | ... | ... | ... |
| 7/8 | 5/8 | 1 7/8 | 4 | 6 | 7/8 | 3 1/2 | 5 3/4 | 4 | 7/8 | 5 | 7 1/4 | 4 |
| 1 | 5/8 | 1 7/8 | 4 | 6 | 1 | 4 | 6 1/2 | 4 | 1 | 6 | 8 1/2 | 4 |
| 7/8 | 7/8 | 1 7/8 | 4 1/8 | 4 | ... | ... | ... | ... | ... | ... | ... | ... |
| 1 | 7/8 | 1 7/8 | 4 1/8 | 4 | ... | ... | ... | ... | ... | ... | ... | ... |
| 1 1/8 | 7/8 | 2 | 4 1/4 | 6 | 1 | 4 | 6 1/2 | 6 | ... | ... | ... | ... |
| 1 1/4 | 7/8 | 2 | 4 1/4 | 6 | 1 | 4 | 6 1/2 | 6 | 1 1/4 | 6 | 8 1/2 | 6 |
| 1 | 1 | 2 | 4 1/2 | 4 | ... | ... | ... | ... | ... | ... | ... | ... |
| 1 1/8 | 1 | 2 | 4 1/2 | 6 | ... | ... | ... | ... | ... | ... | ... | ... |
| 1 1/4 | 1 | 2 | 4 1/2 | 6 | ... | ... | ... | ... | ... | ... | ... | ... |
| 1 3/8 | 1 | 2 | 4 1/2 | 6 | ... | ... | ... | ... | ... | ... | ... | ... |
| 1 1/2 | 1 | 2 | 4 1/2 | 6 | 1 | 4 | 6 1/2 | 6 | ... | ... | ... | ... |
| 1 1/4 | 1 1/4 | 2 | 4 1/2 | 6 | 1 1/4 | 4 | 6 1/2 | 6 | ... | ... | ... | ... |
| 1 1/2 | 1 1/4 | 2 | 4 1/2 | 6 | 1 1/4 | 4 | 6 1/2 | 6 | 1 1/4 | 8 | 10 1/2 | 6 |
| 1 3/4 | 1 1/4 | 2 | 4 1/2 | 6 | 1 1/4 | 4 | 6 1/2 | 6 | ... | ... | ... | ... |
| 2 | 1 1/4 | 2 | 4 1/2 | 8 | 1 1/4 | 4 | 6 1/2 | 8 | ... | ... | ... | ... |

All dimensions are in inches. All cutters are high-speed steel. Helix angle is greater than 19 degrees but not more than 39 degrees. Right-hand cutters with right-hand helix are standard.

*Tolerances:* On D, +0.005 inch; on S, −0.0005 inch; on W, ±1/32 inch; on L, ±1/16 inch.
\* N = Number of flutes.
† In this size of regular mill a left-hand cutter with left-hand helix is also standard.

**American National Standard Two-flute, High Helix, Regular-, Long-, and Extra Long-length, Single-end End Mills with Weldon Shanks** (ANSI B94.19-1968)

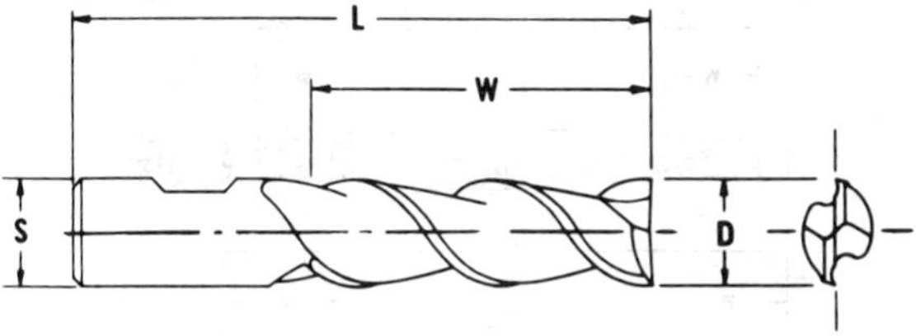


| Cutter Diam., *D* | Regular Mill | | | Long Mill | | | Extra Long Mill | | |
|---|---|---|---|---|---|---|---|---|---|
| | *S* | *W* | *L* | *S* | *W* | *L* | *S* | *W* | *L* |
| 1/4 | 3/8 | 5/8 | 2 7/16 | 3/8 | 1 1/4 | 3 1/16 | 3/8 | 1 3/4 | 3 9/16 |
| 5/16 | 3/8 | 3/4 | 2 1/2 | 3/8 | 1 3/8 | 3 1/8 | 3/8 | 2 | 3 3/4 |
| 3/8 | 3/8 | 3/4 | 2 1/2 | 3/8 | 1 1/2 | 3 1/4 | 3/8 | 2 1/2 | 4 1/4 |
| 7/16 | 3/8 | 1 | 2 11/16 | 1/2 | 1 3/4 | 3 3/4 | ... | ... | ... |
| 1/2 | 1/2 | 1 1/4 | 3 1/4 | 1/2 | 2 | 4 | 1/2 | 3 | 5 |
| 5/8 | 5/8 | 1 5/8 | 3 3/4 | 5/8 | 2 1/2 | 4 5/8 | 5/8 | 4 | 6 1/8 |
| 3/4 | 3/4 | 1 5/8 | 3 7/8 | 3/4 | 3 | 5 1/4 | 3/4 | 4 | 6 1/4 |
| 7/8 | 7/8 | 1 7/8 | 4 1/8 | ... | ... | ... | ... | ... | ... |
| 1 | 1 | 2 | 4 1/2 | 1 | 4 | 6 1/2 | 1 | 6 | 8 1/2 |
| 1 1/4 | 1 1/4 | 2 | 4 1/2 | 1 1/4 | 4 | 6 1/2 | 1 1/4 | 6 | 8 1/2 |
| 1 1/2 | 1 1/4 | 2 | 4 1/2 | 1 1/4 | 4 | 6 1/2 | 1 1/4 | 8 | 10 1/2 |
| 2 | 1 1/4 | 2 | 4 1/2 | 1 1/4 | 4 | 6 1/2 | ... | ... | ... |

All dimensions are in inches. All cutters are high-speed steel. Right-hand cutters with right-hand helix are standard. Helix angle is greater than 39 degrees.

*Tolerances:* On *D*, +0.003 inch; on *S*, −0.0005 inch; on *W*, ±1/32 inch; and on *L*, ±1/16 inch.

**American National Standard Combination Shanks* for End Mills** (ANSI B94.19-1968)

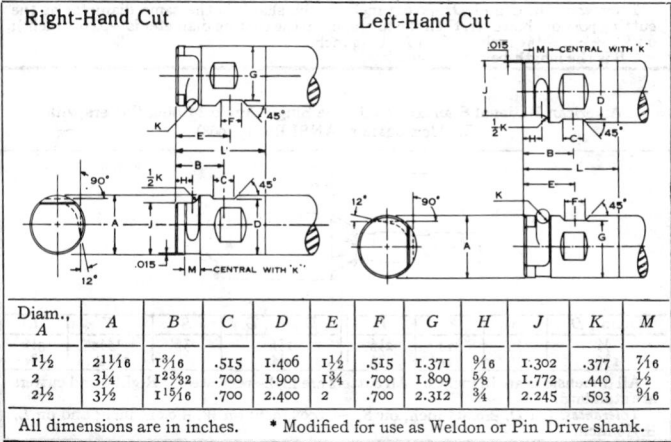

Right-Hand Cut

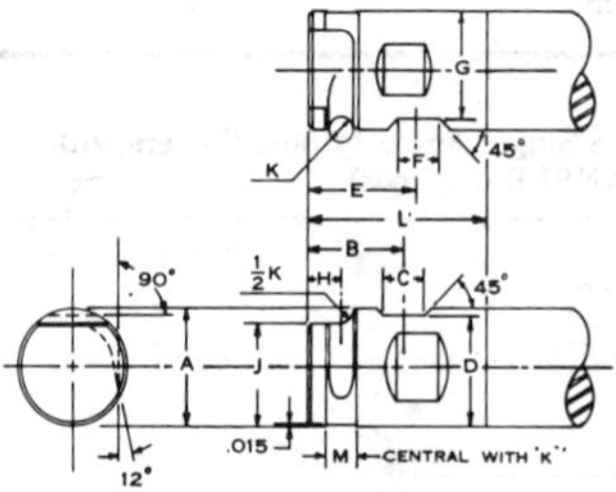


Left-Hand Cut

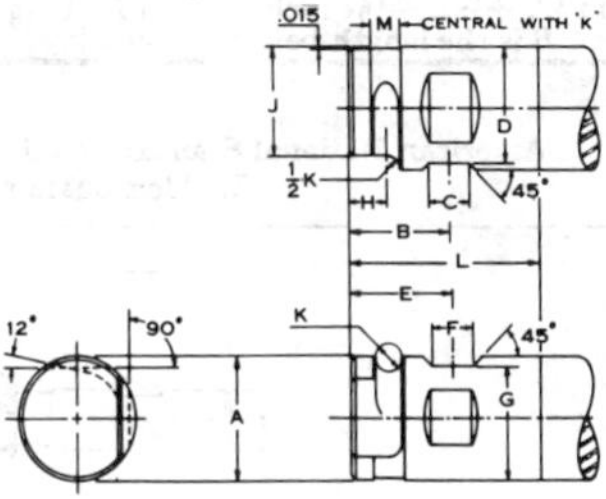


| Diam., *A* | *A* | *B* | *C* | *D* | *E* | *F* | *G* | *H* | *J* | *K* | *M* |
|---|---|---|---|---|---|---|---|---|---|---|---|
| 1 1/2 | 2 11/16 | 1 3/16 | .515 | 1.406 | 1 1/2 | .515 | 1.371 | 9/16 | 1.302 | .377 | 7/16 |
| 2 | 3 1/4 | 1 23/32 | .700 | 1.900 | 1 3/4 | .700 | 1.809 | 5/8 | 1.772 | .440 | 1/2 |
| 2 1/2 | 3 1/2 | 1 15/16 | .700 | 2.400 | 2 | .700 | 2.312 | 3/4 | 2.245 | .503 | 9/16 |

All dimensions are in inches. * Modified for use as Weldon or Pin Drive shank.

**American National Standard Stub-, Regular-, and Long-length, Four-flute, Medium Helix, Plain- and Ball-end, Double-end Miniature End Mills with $\frac{3}{16}$-inch Diameter Straight Shanks (ANSI B94.19-1968)**

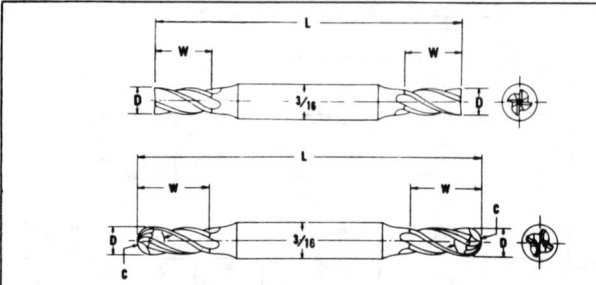

| Diam., C and D | Stub Length | | | | Regular Length | | | |
|---|---|---|---|---|---|---|---|---|
| | Plain End | | Ball End | | Plain End | | Ball End | |
| | W | L | W | L | W | L | W | L |
| $\frac{1}{16}$ | $\frac{3}{32}$ | 2 | ... | ... | $\frac{3}{16}$ | $2\frac{1}{4}$ | ... | ... |
| $\frac{3}{32}$ | $\frac{9}{64}$ | 2 | $\frac{9}{64}$ | 2 | $\frac{9}{32}$ | $2\frac{1}{4}$ | $\frac{9}{32}$ | $2\frac{1}{4}$ |
| $\frac{1}{8}$ | $\frac{3}{16}$ | 2 | $\frac{3}{16}$ | 2 | $\frac{3}{8}$ | $2\frac{1}{4}$ | $\frac{3}{8}$ | $2\frac{1}{4}$ |
| $\frac{5}{32}$ | $\frac{15}{64}$ | 2 | $\frac{15}{64}$ | 2 | $\frac{7}{16}$ | $2\frac{1}{4}$ | $\frac{7}{16}$ | $2\frac{1}{4}$ |
| $\frac{3}{16}$ | $\frac{9}{32}$ | 2 | $\frac{9}{32}$ | 2 | $\frac{1}{2}$ | $2\frac{1}{4}$ | $\frac{1}{2}$ | $2\frac{1}{4}$ |

| Diam., C and D | Long Length | | | | | |
|---|---|---|---|---|---|---|
| | Plain End | | | Ball End | | |
| | B* | W | L | B* | W | L |
| $\frac{1}{16}$ | $\frac{3}{8}$ | $\frac{7}{32}$ | $2\frac{1}{2}$ | ... | ... | ... |
| $\frac{3}{32}$ | $\frac{1}{2}$ | $\frac{9}{32}$ | $2\frac{5}{8}$ | $\frac{1}{2}$ | $\frac{9}{32}$ | $2\frac{5}{8}$ |
| $\frac{1}{8}$ | $\frac{3}{4}$ | $\frac{3}{4}$ | $3\frac{1}{8}$ | $\frac{3}{4}$ | $\frac{3}{4}$ | $3\frac{1}{8}$ |
| $\frac{5}{32}$ | $\frac{7}{8}$ | $\frac{7}{8}$ | $3\frac{1}{4}$ | $\frac{7}{8}$ | $\frac{7}{8}$ | $3\frac{1}{4}$ |
| $\frac{3}{16}$ | 1 | 1 | $3\frac{3}{8}$ | 1 | 1 | $3\frac{3}{8}$ |

All dimensions are in inches. All cutters are high-speed steel. Right-hand cutters with right-hand helix are standard. Helix angle is greater than 19 degrees but not more than 39 degrees.

*Tolerances:* On C and D, +0.003 inch (If the shank is the same diameter as the cutting portion, however, then the tolerance on the cutting diameter is −0.0025 inch.); on W, +$\frac{1}{32}$, −$\frac{1}{64}$ inch; and on L, ±$\frac{1}{16}$ inch.
* B is the length below the shank.

**American National Standard 60-degree Single-Angle Milling Cutters with Weldon Shanks (ANSI B94.19-1968)**

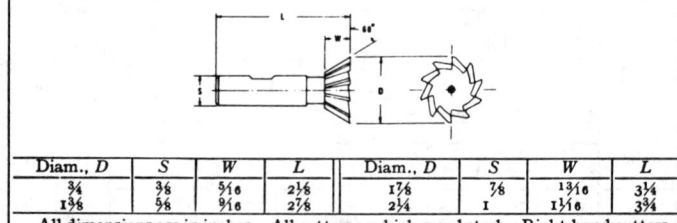

| Diam., D | S | W | L | Diam., D | S | W | L |
|---|---|---|---|---|---|---|---|
| $\frac{3}{4}$ | $\frac{3}{8}$ | $\frac{5}{16}$ | $2\frac{1}{8}$ | $1\frac{7}{8}$ | $\frac{7}{8}$ | $\frac{13}{16}$ | $3\frac{1}{4}$ |
| $1\frac{3}{8}$ | $\frac{5}{8}$ | $\frac{9}{16}$ | $2\frac{7}{8}$ | 1 | 1 | $1\frac{1}{16}$ | $3\frac{3}{4}$ |

All dimensions are in inches. All cutters are high-speed steel. Right-hand cutters are standard.

*Tolerances:* On D, ±0.015 inch; on S, −0.0005 inch; on W, ±0.015 inch; and on L, ±$\frac{1}{16}$ inch.

**American National Standard Stub-, Regular-, and Long-length, Two-flute, Medium Helix, Plain- and Ball-end, Double-end Miniature End Mills with 3/16-inch Diameter Straight Shanks (ANSI B94.19-1968)**

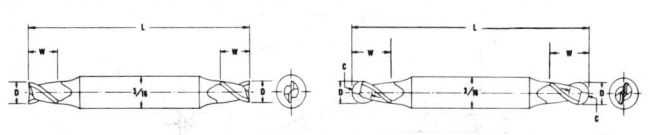

| Diam., C and D | Stub Length | | | | Regular Length | | | |
|---|---|---|---|---|---|---|---|---|
| | Plain End | | Ball End | | Plain End | | Ball End | |
| D | W | L | W | L | W | L | W | L |
| 1/32 | 3/64 | 2 | ... | ... | 3/32 | 2¼ | ... | ... |
| 3/64 | 1/16 | 2 | ... | ... | 9/64 | 2¼ | ... | ... |
| 1/16 | 3/32 | 2 | 3/32 | 2 | 3/16 | 2¼ | 3/16 | 2¼ |
| 5/64 | 1/8 | 2 | ... | ... | 15/64 | 2¼ | ... | ... |
| 3/32 | 9/64 | 2 | 9/64 | 2 | 9/32 | 2¼ | 9/32 | 2¼ |
| 7/64 | 5/32 | 2 | ... | ... | 21/64 | 2¼ | ... | ... |
| 1/8 | 3/16 | 2 | 3/16 | 2 | 3/8 | 2¼ | 3/8 | 2¼ |
| 9/64 | 7/32 | 2 | ... | ... | 13/32 | 2¼ | ... | ... |
| 5/32 | 15/64 | 2 | 15/64 | 2 | 7/16 | 2¼ | 7/16 | 2¼ |
| 11/64 | 1/4 | 2 | ... | ... | 1/2 | 2¼ | ... | ... |
| 3/16 | 9/32 | 2 | 9/32 | 2 | 1/2 | 2¼ | 1/2 | 2¼ |

| Diam., C and D | Long Length | | | | | |
|---|---|---|---|---|---|---|
| | Plain End | | | Ball End | | |
| D | B* | W | L | B* | W | L |
| 1/16 | 3/8 | 7/32 | 2½ | 3/8 | 7/32 | 2½ |
| 3/32 | 1/2 | 9/32 | 2⅝ | 1/2 | 9/32 | 2⅝ |
| 1/8 | 3/4 | 3/4 | 3⅛ | 3/4 | 3/4 | 3⅛ |
| 5/32 | 7/8 | 7/8 | 3¼ | 7/8 | 7/8 | 3¼ |
| 3/16 | 1 | 1 | 3⅜ | 1 | 1 | 3⅜ |

All dimensions are in inches. All cutters are high-speed steel. Right-hand cutters with right-hand helix are standard. Helix angle is greater than 19 degrees but not more than 39 degrees.

*Tolerances:* On C and D, −0.0015 inch for stub and regular length, +0.003 inch for long length (If the shank is the same diameter as the cutting portion, however, then the tolerance on the cutting diameter is −0.0025 inch.); on W, +1/32, −1/64 inch; and on L, ±1/16 inch.

\* B is the length below the shank.

**American National Standard Multiple Flute, Helical Series End Mills with Brown & Sharpe Taper Shanks\* (ANSI B94.19-1968)**

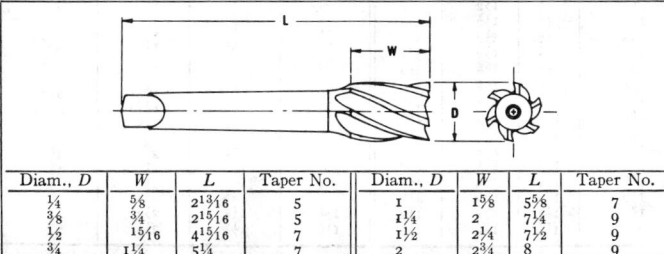

| Diam., D | W | L | Taper No. | Diam., D | W | L | Taper No. |
|---|---|---|---|---|---|---|---|
| 1/4 | 5/8 | 2¹³⁄₁₆ | 5 | 1 | 1⅝ | 5⅝ | 7 |
| 3/8 | 3/4 | 2¹⁵⁄₁₆ | 5 | 1¼ | 2 | 7¼ | 9 |
| 1/2 | 15/16 | 4¹⁵⁄₁₆ | 7 | 1½ | 2¼ | 7½ | 9 |
| 3/4 | 1¼ | 5¼ | 7 | 2 | 2¾ | 8 | 9 |

All dimensions are in inches. All cutters are high-speed steel. Right-hand cutters with right-hand helix are standard. Helix angle is not less than 10 degrees.

*Tolerances:* On D, +0.005 inch; on W, ±1/32 inch; and on L ±1/16 inch.

\* For dimensions of B & S taper shanks, see information given on page **1731**.

## American National Standard Stub- and Regular-length, Two-flute, Medium Helix, Plain- and Ball-end, Single-end End Mills with Weldon Shanks
(ANSI B94.19-1968)

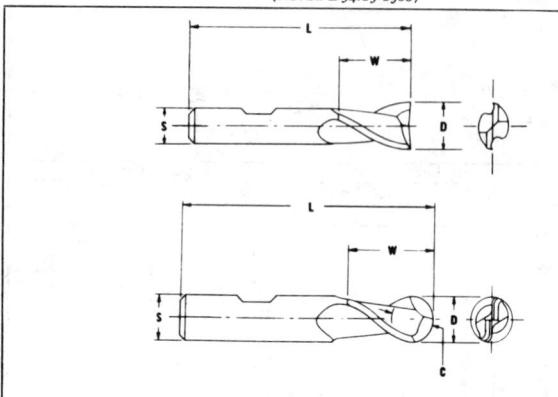

### Regular Length—Plain End

| Diam., D | S | W | L |
|---|---|---|---|
| 1/8 | 3/8 | 3/8 | 2 5/16 |
| 3/16 | 3/8 | 7/16 | 2 5/16 |
| 1/4 | 3/8 | 1/2 | 2 5/16 |
| 5/16 | 3/8 | 9/16 | 2 9/16 |
| 3/8 | 3/8 | 9/16 | 2 9/16 |
| 7/16 | 3/8 | 13/16 | 2 1/2 |
| 1/2 | 3/8 | 13/16 | 2 1/2 |
| 1/2 | 1/2 | 1 | 3 |
| 9/16 | 1/2 | 1 1/8 | 3 1/8 |
| 5/8 | 1/2 | 1 1/8 | 3 1/8 |
| 11/16 | 1/2 | 1 9/16 | 3 9/16 |
| 3/4 | 1/2 | 1 9/16 | 3 9/16 |
| 5/8 | 5/8 | 1 9/16 | 3 7/16 |
| 11/16 | 5/8 | 1 9/16 | 3 7/16 |
| 3/4 | 5/8 | 1 9/16 | 3 7/16 |
| 13/16 | 5/8 | 1 1/2 | 3 5/8 |
| 7/8 | 5/8 | 1 1/2 | 3 5/8 |
| 1 | 5/8 | 1 1/2 | 3 5/8 |
| 7/8 | 7/8 | 1 1/2 | 3 3/4 |
| 1 | 7/8 | 1 1/2 | 3 3/4 |
| 1 1/8 | 7/8 | 1 5/8 | 3 7/8 |
| 1 1/4 | 7/8 | 1 5/8 | 3 7/8 |
| 1 | 1 | 1 5/8 | 4 1/8 |
| 1 1/8 | 1 | 1 5/8 | 4 1/8 |
| 1 1/4 | 1 | 1 5/8 | 4 1/8 |
| 1 3/8 | 1 | 1 5/8 | 4 1/8 |
| 1 1/2 | 1 | 1 5/8 | 4 1/8 |
| 1 1/4 | 1 1/4 | 1 5/8 | 4 1/8 |
| 1 1/2 | 1 1/4 | 1 5/8 | 4 1/8 |
| 1 3/4 | 1 1/4 | 1 5/8 | 4 1/8 |
| 2 | 1 1/4 | 1 5/8 | 4 1/8 |

### Stub Length—Plain End

| Cutter Diam., D | Shank Diam., S | Length of Cut, W | Length Overall, L |
|---|---|---|---|
| 1/8 | 3/8 | 3/16 | 2 1/8 |
| 3/16 | 3/8 | 9/32 | 2 3/16 |
| 1/4 | 3/8 | 3/8 | 2 1/4 |

### Regular Length—Ball End

| Diam., C and D | Shank Diam., S | Length of Cut, W | Length Overall, L |
|---|---|---|---|
| 1/8 | 3/8 | 3/8 | 2 5/16 |
| 3/16 | 3/8 | 1/2 | 2 3/8 |
| 1/4 | 3/8 | 5/8 | 2 7/16 |
| 5/16 | 3/8 | 3/4 | 2 1/2 |
| 3/8 | 3/8 | 3/4 | 2 1/2 |
| 7/16 | 1/2 | 1 | 3 |
| 1/2 | 1/2 | 1 | 3 |
| 9/16 | 1/2 | 1 1/8 | 3 1/8 |
| 5/8 | 1/2 | 1 1/8 | 3 1/8 |
| 5/8 | 5/8 | 1 3/8 | 3 1/2 |
| 3/4 | 1/2 | 1 9/16 | 3 9/16 |
| 3/4 | 3/4 | 1 5/8 | 3 7/8 |
| 7/8 | 7/8 | 2 | 4 1/4 |
| 1 | 1 | 2 1/4 | 4 3/4 |
| 1 1/8 | 1 | 2 1/4 | 4 3/4 |
| 1 1/4 | 1 1/4 | 2 1/2 | 5 |
| 1 1/2 | 1 1/4 | 2 1/2 | 5 |

All dimensions are in inches. All cutters are high-speed steel. Right-hand cutters with right-hand helix are standard. Helix angle is greater than 19 degrees but not more than 39 degrees.

*Tolerances:* On C and D, −0.0015 inch for stub-length mills, +0.003 inch for regular-length mills; on S, −0.0005 inch; on W, ±1/32 inch; and on L, ±1/16 inch.

**American National Standard Long Length Single-end and Stub-, and Regular-length, Double-end, Plain- and Ball-end, Two-flute End Mills with Weldon Shanks** (ANSI B94.19-1968)

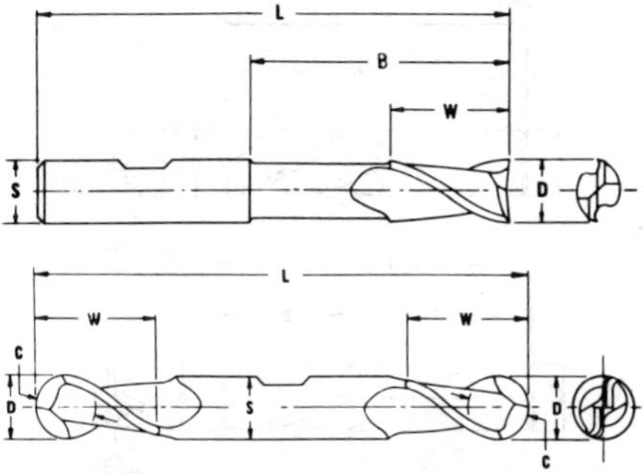


**Single End**

| Diam., C and D | Long Length—Plain End S | Long Length—Plain End B* | Long Length—Plain End W | Long Length—Plain End L | Long Length—Ball End S | Long Length—Ball End B* | Long Length—Ball End W | Long Length—Ball End L |
|---|---|---|---|---|---|---|---|---|
| 1/8 | ... | ... | ... | ... | 3/8 | 13/16 | 3/8 | 2 3/8 |
| 3/16 | ... | ... | ... | ... | 3/8 | 1 1/8 | 1/2 | 2 11/16 |
| 1/4 | 3/8 | 1 1/2 | 5/8 | 3 1/16 | 3/8 | 1 1/2 | 5/8 | 3 1/16 |
| 5/16 | 3/8 | 1 3/4 | 3/4 | 3 5/16 | 3/8 | 1 3/4 | 3/4 | 3 5/16 |
| 3/8 | 3/8 | 1 3/4 | 3/4 | 3 5/16 | 3/8 | 1 3/4 | 3/4 | 3 5/16 |
| 7/16 | ... | ... | ... | ... | 1/2 | 1 7/8 | 1 | 3 11/16 |
| 1/2 | 1/2 | 2 7/32 | 1 | 4 | 1/2 | 2 1/4 | 1 | 4 |
| 5/8 | 5/8 | 2 23/32 | 1 3/8 | 4 5/8 | 5/8 | 2 3/4 | 1 3/8 | 4 5/8 |
| 3/4 | 3/4 | 3 11/32 | 1 5/8 | 5 3/8 | 3/4 | 3 3/8 | 1 5/8 | 5 3/8 |
| 1 | 1 | 4 31/32 | 2 1/2 | 7 1/4 | 1 | 5 | 2 1/2 | 7 1/4 |
| 1 1/4 | 1 1/4 | 4 31/32 | 3 | 7 1/4 | ... | ... | ... | ... |

**Double End**

| Diam., C and D | Stub Length—Plain End S | Stub Length—Plain End W | Stub Length—Plain End L | Regular Length—Plain End S | Regular Length—Plain End W | Regular Length—Plain End L | Regular Length—Ball End S | Regular Length—Ball End W | Regular Length—Ball End L |
|---|---|---|---|---|---|---|---|---|---|
| 1/8 | 3/8 | 3/16 | 2 3/4 | 3/8 | 3/8 | 3 1/16 | 3/8 | 3/8 | 3 1/16 |
| 5/32 | 3/8 | 15/64 | 2 3/4 | 3/8 | 7/16 | 3 1/8 | ... | ... | ... |
| 3/16 | 3/8 | 9/32 | 2 3/4 | 3/8 | 7/16 | 3 1/8 | 3/8 | 7/16 | 3 1/8 |
| 7/32 | 3/8 | 21/64 | 2 7/8 | 3/8 | 1/2 | 3 1/8 | ... | ... | ... |
| 1/4 | 3/8 | 3/8 | 2 7/8 | 3/8 | 1/2 | 3 1/8 | 3/8 | 1/2 | 3 1/8 |
| 9/32 | ... | ... | ... | 3/8 | 9/16 | 3 1/8 | ... | ... | ... |
| 5/16 | ... | ... | ... | 3/8 | 9/16 | 3 1/8 | 3/8 | 9/16 | 3 1/8 |
| 11/32 | ... | ... | ... | 3/8 | 9/16 | 3 1/8 | ... | ... | ... |
| 3/8 | ... | ... | ... | 3/8 | 9/16 | 3 1/8 | 3/8 | 9/16 | 3 1/8 |
| 13/32 | ... | ... | ... | 1/2 | 13/16 | 3 3/4 | ... | ... | ... |
| 7/16 | ... | ... | ... | 1/2 | 13/16 | 3 3/4 | 1/2 | 13/16 | 3 3/4 |
| 15/32 | ... | ... | ... | 1/2 | 13/16 | 3 3/4 | ... | ... | ... |
| 1/2 | ... | ... | ... | 1/2 | 13/16 | 3 3/4 | 1/2 | 13/16 | 3 3/4 |
| 9/16 | ... | ... | ... | 5/8 | 1 1/8 | 4 1/2 | ... | ... | ... |
| 5/8 | ... | ... | ... | 5/8 | 1 1/8 | 4 1/2 | 5/8 | 1 1/8 | 4 1/2 |
| 11/16 | ... | ... | ... | 3/4 | 1 5/16 | 5 | ... | ... | ... |
| 3/4 | ... | ... | ... | 3/4 | 1 5/16 | 5 | 3/4 | 1 5/16 | 5 |
| 7/8 | ... | ... | ... | 7/8 | 1 9/16 | 5 1/2 | ... | ... | ... |
| 1 | ... | ... | ... | 1 | 1 5/8 | 5 7/8 | 1 | 1 5/8 | 5 7/8 |

All dimensions are in inches. All cutters are high-speed steel. Right-hand cutters with right-hand helix are standard. Helix angle is greater than 19 degrees but not more than 39 degrees.

*Tolerances:* On $C$ and $D$, $+0.003$ inch for single-end mills, $-0.0015$ inch for double-end mills; on $S$, $-0.0005$ inch; on $W$, $\pm 1/32$ inch; and on $L$, $\pm 1/16$ inch.

* $B$ is the length below the shank.

### American National Standard Regular-, Long-, and Extra Long-length, Three- and Four-flute, Medium Helix, Center Cutting, Single-end End Mills with Weldon Shanks (ANSI B94.19-1968)

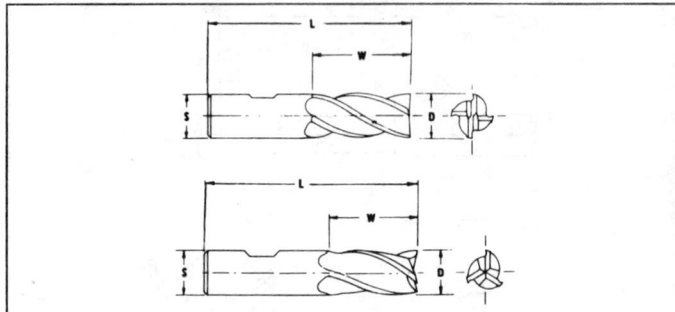

#### Four Flute

| Diam., D | Regular Length | | | Long Length | | | Extra Long Length | | |
|---|---|---|---|---|---|---|---|---|---|
| | S | W | L | S | W | L | S | W | L |
| 1/8 | 3/8 | 3/8 | 2 5/16 | ... | ... | ... | ... | ... | ... |
| 3/16 | 3/8 | 1/2 | 2 3/8 | ... | ... | ... | ... | ... | ... |
| 1/4 | 3/8 | 5/8 | 2 7/16 | 3/8 | 1 1/4 | 3 5/16 | 3/8 | 1 3/4 | 3 9/16 |
| 5/16 | 3/8 | 3/4 | 2 1/2 | 3/8 | 1 3/8 | 3 1/8 | 3/8 | 2 | 3 3/4 |
| 3/8 | 3/8 | 3/4 | 2 1/2 | 3/8 | 1 1/2 | 3 1/4 | 3/8 | 2 1/2 | 4 1/4 |
| 1/2 | 1/2 | 1 1/4 | 3 1/4 | 1/2 | 2 | 4 | 1/2 | 3 | 5 |
| 5/8 | 5/8 | 1 5/8 | 3 3/4 | 5/8 | 2 1/2 | 4 5/8 | 5/8 | 4 | 6 1/8 |
| 11/16 | 5/8 | 1 5/8 | 3 3/4 | ... | ... | ... | ... | ... | ... |
| 3/4 | 3/4 | 1 5/8 | 3 7/8 | 3/4 | 3 | 5 1/4 | 3/4 | 4 | 6 1/4 |
| 7/8 | 7/8 | 1 7/8 | 4 1/8 | 7/8 | 3 1/2 | 5 3/4 | 7/8 | 5 | 7 1/4 |
| 1 | 1 | 2 | 4 1/2 | 1 | 4 | 6 1/2 | 1 | 6 | 8 1/2 |
| 1 1/8 | 1 | 2 | 4 1/2 | ... | ... | ... | ... | ... | ... |
| 1 1/4 | 1 1/4 | 2 | 4 1/2 | 1 1/4 | 4 | 6 1/2 | 1 1/4 | 6 | 8 1/2 |
| 1 1/2 | 1 1/4 | 2 | 4 1/2 | ... | ... | ... | ... | ... | ... |

#### Three Flute

| Diam., D | S | W | L | Diam., D | S | W | L |
|---|---|---|---|---|---|---|---|
| Regular Length | | | | Regular Length (cont.) | | | |
| 1/8 | 3/8 | 3/8 | 2 5/16 | 1 1/8 | 1 | 2 | 4 1/2 |
| 3/16 | 3/8 | 1/2 | 2 3/8 | 1 1/4 | 1 | 2 | 4 1/2 |
| 1/4 | 3/8 | 5/8 | 2 7/16 | 1 1/2 | 1 | 2 | 4 1/2 |
| 5/16 | 3/8 | 3/4 | 2 1/2 | 1 1/4 | 1 1/4 | 2 | 4 1/2 |
| 3/8 | 3/8 | 3/4 | 2 1/2 | 1 1/2 | 1 1/4 | 2 | 4 1/2 |
| 7/16 | 3/8 | 1 | 2 11/16 | 1 3/4 | 1 1/4 | 2 | 4 1/2 |
| 1/2 | 3/8 | 1 | 2 11/16 | 2 | 1 1/4 | 2 | 4 1/2 |
| 1/2 | 1/2 | 1 1/4 | 3 1/4 | Long Length | | | |
| 9/16 | 1/2 | 1 3/8 | 3 3/8 | 1/4 | 3/8 | 1 1/4 | 3 1/16 |
| 5/8 | 1/2 | 1 3/8 | 3 3/8 | 5/16 | 3/8 | 1 3/8 | 3 1/8 |
| 3/4 | 1/2 | 1 5/8 | 3 5/8 | 3/8 | 3/8 | 1 1/2 | 3 3/4 |
| 5/8 | 5/8 | 1 5/8 | 3 3/4 | 7/16 | 1/2 | 1 3/4 | 3 3/4 |
| 3/4 | 5/8 | 1 5/8 | 3 3/4 | 1/2 | 1/2 | 2 | 4 |
| 7/8 | 5/8 | 1 7/8 | 4 | 5/8 | 5/8 | 2 1/2 | 4 5/8 |
| 1 | 5/8 | 1 7/8 | 4 | 3/4 | 3/4 | 3 | 5 1/4 |
| 3/4 | 3/4 | 1 5/8 | 3 7/8 | 1 | 1 | 4 | 6 1/2 |
| 7/8 | 3/4 | 1 7/8 | 4 1/8 | 1 1/4 | 1 1/4 | 4 | 6 1/2 |
| 1 | 3/4 | 1 7/8 | 4 1/8 | 1 1/2 | 1 1/4 | 4 | 6 1/2 |
| 1 | 7/8 | 1 7/8 | 4 1/8 | 1 3/4 | 1 1/4 | 4 | 6 1/2 |
| 1 | 1 | 2 | 4 1/2 | 2 | 1 1/4 | 4 | 6 1/2 |

All dimensions are in inches. All cutters are high-speed steel. Right-hand cutters with right-hand helix are standard. Helix angle is greater than 19 degrees but not more than 39 degrees.

*Tolerances:* On D, +0.003 inch; on S, −0.0005 inch; on W, ±1/32 inch; and on L, ±1/16 inch.

### American National Standard Stub- and Regular-length, Four-flute, Medium Helix, Double-end End Mills with Weldon Shanks (ANSI B94.19-1968)

| Diam., D | S | W | L | Diam., D | S | W | L | Diam., D | S | W | L |
|---|---|---|---|---|---|---|---|---|---|---|---|
| | | | | Stub Length | | | | | | | |
| 1/8 | 3/8 | 3/16 | 2 3/4 | 3/16 | 3/8 | 9/32 | 2 3/4 | 1/4 | 3/8 | 3/8 | 2 7/8 |
| 5/32 | 3/8 | 15/16 | 2 3/4 | 7/32 | 3/8 | 21/64 | 2 7/8 | ... | ... | ... | ... |
| | | | | Regular Length | | | | | | | |
| 1/8* | 3/8 | 3/8 | 3 1/16 | 11/32 | 3/8 | 3/4 | 3 1/2 | 5/8* | 5/8 | 1 3/8 | 5 |
| 5/32* | 3/8 | 7/16 | 3 1/8 | 3/8* | 3/8 | 3/4 | 3 1/2 | 11/16 | 3/4 | 1 5/8 | 5 5/8 |
| 3/16* | 3/8 | 1/2 | 3 1/4 | 13/32 | 1/2 | 7/8 | 4 1/8 | 3/4 | 3/4 | 1 5/8 | 5 5/8 |
| 7/32 | 3/8 | 9/16 | 3 1/4 | 7/16 | 1/2 | 1 | 4 1/8 | 13/16 | 7/8 | 1 7/8 | 6 1/8 |
| 1/4* | 3/8 | 5/8 | 3 3/8 | 15/32 | 1/2 | 1 | 4 1/8 | 7/8 | 7/8 | 1 7/8 | 6 1/8 |
| 9/32 | 3/8 | 11/16 | 3 3/8 | 1/2* | 1/2 | 1 | 4 1/8 | 1 | 1 | 1 7/8 | 6 3/8 |
| 5/16* | 3/8 | 3/4 | 3 1/2 | 9/16 | 5/8 | 1 | 5 | ... | ... | ... | ... |

All dimensions are in inches. All cutters are high-speed steel. Right-hand cutters with right-hand helix are standard. Helix angle is greater than 19 degrees but not more than 39 degrees.

*Tolerances:* On D, +0.003 inch (If the shank is the same diameter as the cutting portion, however, then the tolerance on the cutting diameter is −0.0025 inch.); on S, −0.0005 inch; on W, ±1/32 inch; and on L, ±1/16 inch.
* In this size of regular mill a left-hand cutter with a left-hand helix is also standard.

### American National Standard Three- and Four-flute, Regular Length, Medium Helix, Center Cutting, Double-end End Mills with Weldon Shanks (ANSI B94.19-1968)

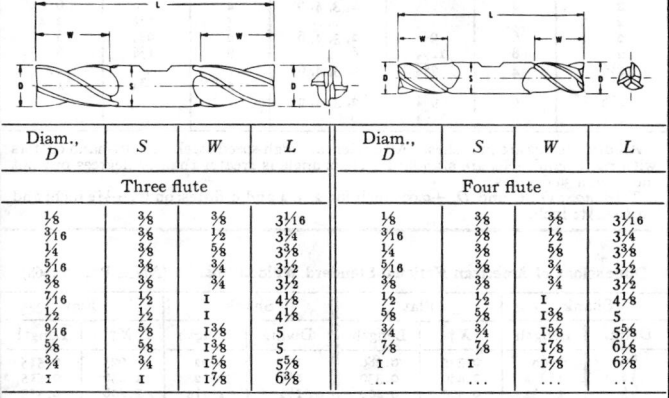

| Diam., D | S | W | L | Diam., D | S | W | L |
|---|---|---|---|---|---|---|---|
| | Three flute | | | | Four flute | | |
| 1/8 | 3/8 | 3/8 | 3 1/16 | 1/8 | 3/8 | 3/8 | 3 1/16 |
| 3/16 | 3/8 | 1/2 | 3 1/4 | 3/16 | 3/8 | 1/2 | 3 1/4 |
| 1/4 | 3/8 | 5/8 | 3 3/8 | 1/4 | 3/8 | 5/8 | 3 3/8 |
| 5/16 | 3/8 | 3/4 | 3 1/2 | 5/16 | 3/8 | 3/4 | 3 1/2 |
| 3/8 | 3/8 | 3/4 | 3 1/2 | 3/8 | 3/8 | 3/4 | 3 1/2 |
| 7/16 | 1/2 | 1 | 4 1/8 | 1/2 | 1/2 | 1 | 4 1/8 |
| 1/2 | 1/2 | 1 | 4 1/8 | 5/8 | 5/8 | 1 3/8 | 5 |
| 9/16 | 5/8 | 1 3/8 | 5 | 3/4 | 3/4 | 1 5/8 | 5 5/8 |
| 5/8 | 5/8 | 1 3/8 | 5 | 7/8 | 7/8 | 1 7/8 | 6 1/8 |
| 3/4 | 3/4 | 1 5/8 | 5 5/8 | 1 | 1 | 1 7/8 | 6 3/8 |
| 1 | 1 | 1 7/8 | 6 3/8 | ... | ... | ... | ... |

All dimensions are in inches. All cutters are high-speed steel. Right-hand cutters with right-hand helix are standard. Helix angle is greater than 19 degrees but not more than 39 degrees.

*Tolerances:* On D, −0.0015 inch; on S, −0.0005 inch; on W, ±1/32 inch; and on L, ±1/16 inch.

### American National Standard Plain- and Ball-end, Heavy Duty, Medium Helix, Single-end End Mills with 2-inch Diameter Shanks (ANSI B94.19-1968)

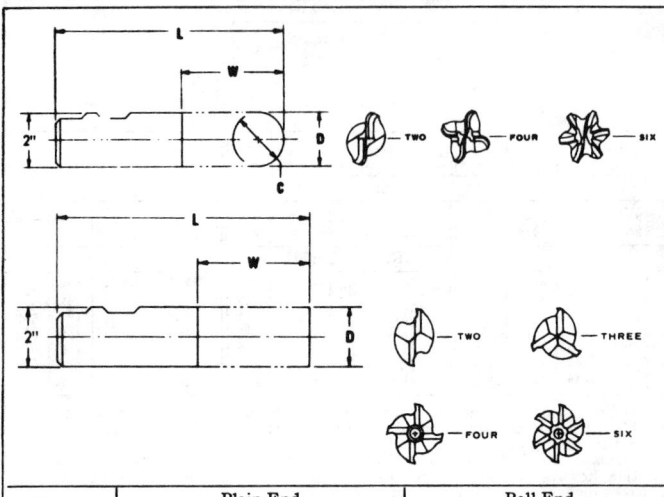

| Diam., C and D | Plain End | | | Ball End | | |
|---|---|---|---|---|---|---|
| | W | L | No. of Flutes | W | L | No. of Flutes |
| 2 | 2 | 5¾ | 2, 3, 4, 6 | ... | ... | ... |
| 2 | 3 | 6¾ | 2, 3 | ... | ... | ... |
| 2 | 4 | 7¾ | 2, 3, 4, 6 | 4 | 7¾ | 6 |
| 2 | ... | ... | ... | 5 | 8¾ | 2, 4 |
| 2 | 6 | 9¾ | 2, 3, 4, 6 | 6 | 9¾ | 6 |
| 2 | 8 | 11¾ | 6 | 8 | 11¾ | 6 |
| 2½ | 4 | 7¾ | 2, 3, 4, 6 | ... | ... | ... |
| 2½ | ... | ... | ... | 5 | 8¾ | 4 |
| 2½ | 6 | 9¾ | 2, 3, 4, 6 | ... | ... | ... |
| 2½ | 8 | 11¾ | 6 | ... | ... | ... |

All dimensions are in inches.  All cutters are high-speed steel.  Right-hand cutters with right-hand helix are standard.  Helix angle is greater than 19 degrees but not more than 39 degrees.

*Tolerances:* On C and D, ±0.005 inch for 2, 3, 4 and 6 flutes; on W, ±1/16 inch; and on L, ±1/16 inch.

### Dimensions of American National Standard Weldon Shanks (ANSI B94.19-1968)

| Shank | | Flat | | Shank | | Flat | |
|---|---|---|---|---|---|---|---|
| Diam. | Length | X† | Length | Diam. | Length | X† | Length |
| ⅜ | 1⁹⁄₁₆ | 0.325 | 0.280 | 1 | 2⁹⁄₃₂ | 0.925 | 0.515 |
| ½ | 1²⁵⁄₃₂ | 0.440 | 0.330 | 1¼ | 2⁹⁄₃₂ | 1.156 | 0.515 |
| ⅝ | 1²⁹⁄₃₂ | 0.560 | 0.400 | 1½ | 2¹¹⁄₁₆ | 1.406 | 0.515 |
| ¾ | 2¹⁄₃₂ | 0.675 | 0.455 | 2 | 3¼ | 1.900 | 0.700 |
| ⅞ | 2¹⁄₃₂ | 0.810 | 0.455 | 2½ | 3½ | 2.400 | 0.700 |

Centerline of flat is at half-length of shank except for 1½-, 2- and 2½-inch shanks where it is 1³⁄₁₆, 1²⁷⁄₃₂ and 1¹⁵⁄₁₆ from shank end, respectively.
† X is distance from bottom of flat to opposite side of shank.

**American National Standard Form Relieved, Concave, Convex, and Corner-rounding Arbor-type Cutters*** (ANSI B94.19-1968)

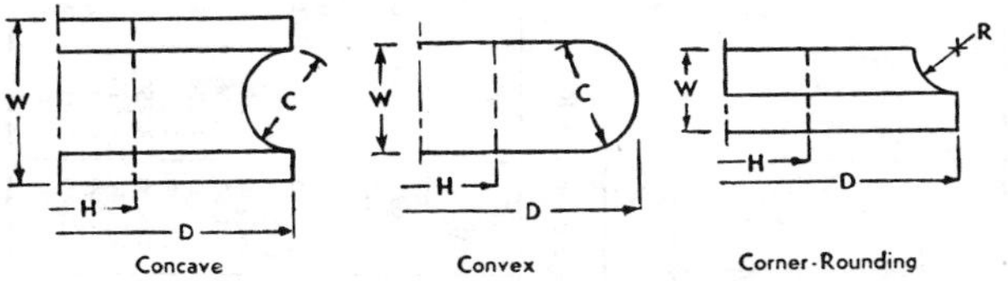


| Diameter C or Radius R | | | Cutter Diam. D[1] | Width W ±.010[2] | Diameter of Hole H | | |
|---|---|---|---|---|---|---|---|
| Nom. | Max. | Min. | | | Nom. | Max. | Min. |
| **Concave Cutters[3]** | | | | | | | |
| 1/8 | 0.1270 | 0.1240 | 2 1/4 | 1/4 | 1 | 1.00075 | 1.00000 |
| 3/16 | 0.1895 | 0.1865 | 2 1/4 | 3/8 | 1 | 1.00075 | 1.00000 |
| 1/4 | 0.2520 | 0.2490 | 2 1/2 | 7/16 | 1 | 1.00075 | 1.00000 |
| 5/16 | 0.3145 | 0.3115 | 2 3/4 | 9/16 | 1 | 1.00075 | 1.00000 |
| 3/8 | 0.3770 | 0.3740 | 2 3/4 | 5/8 | 1 | 1.00075 | 1.00000 |
| 7/16 | 0.4395 | 0.4365 | 3 | 3/4 | 1 | 1.00075 | 1.00000 |
| 1/2 | 0.5040 | 0.4980 | 3 | 1 3/16 | 1 | 1.00075 | 1.00000 |
| 5/8 | 0.6290 | 0.6230 | 3 1/2 | 1 | 1 1/4 | 1.251 | 1.250 |
| 3/4 | 0.7540 | 0.7480 | 3 3/4 | 1 3/16 | 1 1/4 | 1.251 | 1.250 |
| 7/8 | 0.8790 | 0.8730 | 4 | 1 3/8 | 1 1/4 | 1.251 | 1.250 |
| 1 | 1.0040 | 0.9980 | 4 1/4 | 1 9/16 | 1 1/4 | 1.251 | 1.250 |
| **Convex Cutters[3]** | | | | | | | |
| 1/8 | 0.1270 | 0.1230 | 2 1/4 | 1/8 | 1 | 1.00075 | 1.00000 |
| 3/16 | 0.1895 | 0.1855 | 2 1/4 | 3/16 | 1 | 1.00075 | 1.00000 |
| 1/4 | 0.2520 | 0.2480 | 2 1/2 | 1/4 | 1 | 1.00075 | 1.00000 |
| 5/16 | 0.3145 | 0.3105 | 2 3/4 | 5/16 | 1 | 1.00075 | 1.00000 |
| 3/8 | 0.3770 | 0.3730 | 2 3/4 | 3/8 | 1 | 1.00075 | 1.00000 |
| 7/16 | 0.4395 | 0.4355 | 3 | 7/16 | 1 | 1.00075 | 1.00000 |
| 1/2 | 0.5020 | 0.4980 | 3 | 1/2 | 1 | 1.00075 | 1.00000 |
| 5/8 | 0.6270 | 0.6230 | 3 1/2 | 5/8 | 1 1/4 | 1.251 | 1.250 |
| 3/4 | 0.7520 | 0.7480 | 3 3/4 | 3/4 | 1 1/4 | 1.251 | 1.250 |
| 7/8 | 0.8770 | 0.8730 | 4 | 7/8 | 1 1/4 | 1.251 | 1.250 |
| 1 | 1.0020 | 0.9980 | 4 1/4 | 1 | 1 1/4 | 1.251 | 1.250 |
| **Corner-rounding Cutters[4]** | | | | | | | |
| 1/8 | 0.1260 | 0.1240 | 2 1/2 | 1/4 | 1 | 1.00075 | 1.00000 |
| 1/4 | 0.2520 | 0.2490 | 3 | 13/32 | 1 | 1.00075 | 1.00000 |
| 3/8 | 0.3770 | 0.3740 | 3 3/4 | 9/16 | 1 1/4 | 1.251 | 1.250 |
| 1/2 | 0.5020 | 0.4990 | 4 1/4 | 3/4 | 1 1/4 | 1.251 | 1.250 |
| 5/8 | 0.6270 | 0.6240 | 4 1/4 | 15/16 | 1 1/4 | 1.251 | 1.250 |

All dimensions in inches. All cutters are high-speed steel and are form relieved.

* For key and keyway dimensions for these cutters, see page 1622.

[1] Tolerances on cutter diameter are +1/16, −1/16 inch for all sizes.

[2] Tolerance does not apply to convex cutters.

[3] Size of cutter is designated by specifying diameter $C$ of circular form.

[4] Size of cutter is designated by specifying radius $R$ of circular form.

American National Standard Keys and Keyways for Milling Cutters and Arbors (ANSI B94.19-1968)

ARBOR AND KEYSEAT     CUTTER HOLE AND KEYWAY     ARBOR AND KEY

| Nom. Arbor and Cutter Hole Diam. | Nom. Size Key (Square) | Arbor and Keyseat | | | | Hole and Keyway | | | | | Arbor and Key | | | |
|---|---|---|---|---|---|---|---|---|---|---|---|---|---|---|
| | | A Max. | A Min. | B Max. | B Min. | C Max. | C Min. | D† Min. | H Nom. | Corner Radius | E Max. | E Min. | F Max. | F Min. |
| 1/2 | 3/32 | 0.0947 | 0.0937 | 0.4531 | 0.4481 | 0.106 | 0.099 | 0.5578 | 3/64 | 0.020 | 0.0932 | 0.0927 | 0.5468 | 0.5408 |
| 5/8 | 1/8 | 0.126 | 0.125 | 0.5625 | 0.5575 | 0.137 | 0.130 | 0.6985 | 1/16 | 1/32 | 0.1245 | 0.1240 | 0.6875 | 0.6815 |
| 3/4 | 1/8 | 0.126 | 0.125 | 0.6875 | 0.6825 | 0.137 | 0.130 | 0.8225 | 1/16 | 1/32 | 0.1245 | 0.1240 | 0.8125 | 0.8065 |
| 7/8 | 1/8 | 0.126 | 0.125 | 0.8125 | 0.8075 | 0.137 | 0.130 | 0.9475 | 1/16 | 1/32 | 0.1245 | 0.1240 | 0.9375 | 0.9315 |
| 1 | 1/4 | 0.251 | 0.250 | 0.8438 | 0.8388 | 0.262 | 0.255 | 1.104 | 3/32 | 3/64 | 0.2495 | 0.2490 | 1.094 | 1.088 |
| 1 1/4 | 5/16 | 0.3135 | 0.3125 | 1.063 | 1.058 | 0.343 | 0.318 | 1.385 | 1/8 | 1/16 | 0.3120 | 0.3115 | 1.375 | 1.369 |
| 1 1/2 | 3/8 | 0.376 | 0.375 | 1.281 | 1.276 | 0.410 | 0.385 | 1.666 | 5/32 | 1/16 | 0.3745 | 0.3740 | 1.656 | 1.650 |
| 1 3/4 | 7/16 | 0.4385 | 0.4375 | 1.500 | 1.495 | 0.473 | 0.448 | 1.948 | 3/16 | 1/16 | 0.4370 | 0.4365 | 1.938 | 1.932 |
| 2 | 1/2 | 0.501 | 0.500 | 1.687 | 1.682 | 0.535 | 0.510 | 2.198 | 3/16 | 1/16 | 0.4995 | 0.4990 | 2.188 | 2.182 |
| 2 1/2 | 5/8 | 0.626 | 0.625 | 2.094 | 2.089 | 0.660 | 0.635 | 2.733 | 7/32 | 1/16 | 0.6245 | 0.6240 | 2.718 | 2.712 |
| 3 | 3/4 | 0.751 | 0.750 | 2.500 | 2.495 | 0.785 | 0.760 | 3.265 | 1/4 | 3/32 | 0.7495 | 0.7490 | 3.250 | 3.244 |
| 3 1/2 | 7/8 | 0.876 | 0.875 | 3.000 | 2.995 | 0.910 | 0.885 | 3.890 | 3/8 | 3/32 | 0.8745 | 0.8740 | 3.875 | 3.869 |
| 4 | 1 | 1.001 | 1.000 | 3.375 | 3.370 | 1.035 | 1.010 | 4.390 | 3/8 | 3/32 | 0.9995 | 0.9990 | 4.375 | 4.369 |
| 4 1/2 | 1 1/8 | 1.126 | 1.125 | 3.813 | 3.808 | 1.160 | 1.135 | 4.953 | 7/16 | 1/8 | 1.1245 | 1.1240 | 4.938 | 4.932 |
| 5 | 1 1/4 | 1.251 | 1.250 | 4.250 | 4.245 | 1.285 | 1.260 | 5.515 | 1/2 | 1/8 | 1.2495 | 1.2490 | 5.500 | 5.494 |

All dimensions given in inches.

† D max. is 0.010 inches larger than D min.

American National Standard Woodruff Keyseat Cutters† — Shank-type Straight-teeth and Arbor-type Staggered-teeth (ANSI B94.19-1968)

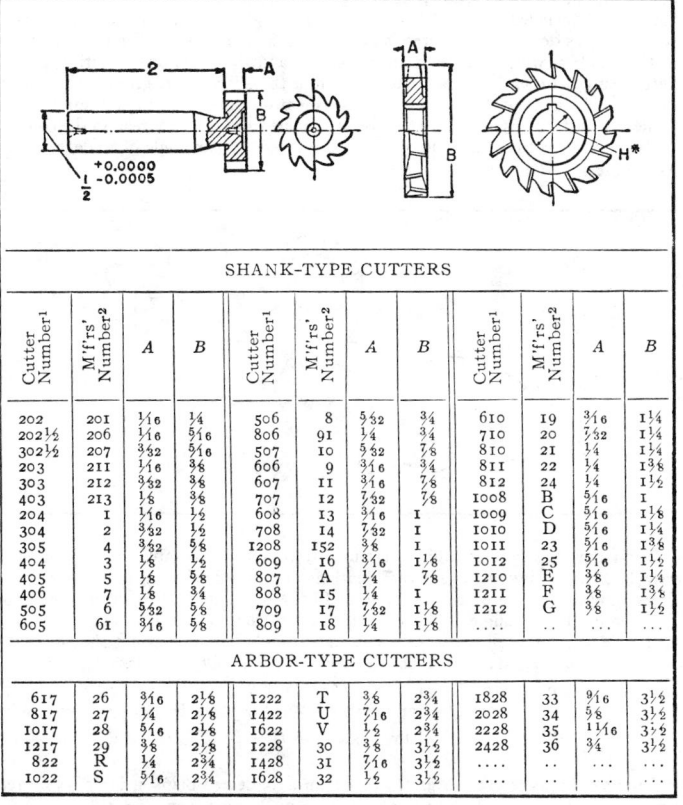

## SHANK-TYPE CUTTERS

| Cutter Number[1] | M'f'rs' Number[2] | A | B | Cutter Number[1] | M'f'rs' Number[2] | A | B | Cutter Number[1] | M'f'rs' Number[2] | A | B |
|---|---|---|---|---|---|---|---|---|---|---|---|
| 202 | 201 | 1/16 | 1/4 | 506 | 8 | 5/32 | 3/4 | 610 | 19 | 3/16 | 1 1/4 |
| 202½ | 206 | 1/16 | 5/16 | 806 | 91 | 1/4 | 3/4 | 710 | 20 | 7/32 | 1 1/4 |
| 302½ | 207 | 3/32 | 5/16 | 507 | 10 | 5/32 | 7/8 | 810 | 21 | 1/4 | 1 1/4 |
| 203 | 211 | 1/16 | 3/8 | 606 | 9 | 3/16 | 3/4 | 811 | 22 | 1/4 | 1 3/8 |
| 303 | 212 | 3/32 | 3/8 | 607 | 11 | 3/16 | 7/8 | 812 | 24 | 1/4 | 1 1/2 |
| 403 | 213 | 1/8 | 3/8 | 707 | 12 | 7/32 | 7/8 | 1008 | B | 5/16 | 1 |
| 204 | 1 | 1/16 | 1/2 | 608 | 13 | 3/16 | 1 | 1009 | C | 5/16 | 1 1/8 |
| 304 | 2 | 3/32 | 1/2 | 708 | 14 | 7/32 | 1 | 1010 | D | 5/16 | 1 1/4 |
| 305 | 4 | 3/32 | 5/8 | 1208 | 152 | 3/8 | 1 | 1011 | 23 | 5/16 | 1 3/8 |
| 404 | 3 | 1/8 | 1/2 | 609 | 16 | 3/16 | 1 1/8 | 1012 | 25 | 5/16 | 1 1/2 |
| 405 | 5 | 1/8 | 5/8 | 807 | A | 1/4 | 7/8 | 1210 | E | 3/8 | 1 1/4 |
| 406 | 7 | 1/8 | 3/4 | 808 | 15 | 1/4 | 1 | 1211 | F | 3/8 | 1 3/8 |
| 505 | 6 | 5/32 | 5/8 | 709 | 17 | 7/32 | 1 1/8 | 1212 | G | 3/8 | 1 1/2 |
| 605 | 61 | 3/16 | 5/8 | 809 | 18 | 1/4 | 1 1/8 | .... | .. | ... | ... |

## ARBOR-TYPE CUTTERS

| Cutter Number[1] | M'f'rs' Number[2] | A | B | Cutter Number[1] | M'f'rs' Number[2] | A | B | Cutter Number[1] | M'f'rs' Number[2] | A | B |
|---|---|---|---|---|---|---|---|---|---|---|---|
| 617 | 26 | 3/16 | 2 1/8 | 1222 | T | 3/8 | 2 3/4 | 1828 | 33 | 9/16 | 3 1/2 |
| 817 | 27 | 1/4 | 2 1/8 | 1422 | U | 7/16 | 2 3/4 | 2028 | 34 | 5/8 | 3 1/2 |
| 1017 | 28 | 5/16 | 2 1/8 | 1622 | V | 1/2 | 2 3/4 | 2228 | 35 | 11/16 | 3 1/2 |
| 1217 | 29 | 3/8 | 2 1/8 | 1228 | 30 | 3/8 | 3 1/2 | 2428 | 36 | 3/4 | 3 1/2 |
| 822 | R | 1/4 | 2 3/4 | 1428 | 31 | 7/16 | 3 1/2 | .... | .. | ... | ... |
| 1022 | S | 5/16 | 2 3/4 | 1628 | 32 | 1/2 | 3 1/2 | .... | .. | ... | ... |

All dimensions are given in inches. All cutters are high-speed steel.

[1] American Standard Number indicates the nominal key dimension or size cutter, that is, the last two digits give diameter B in eighths of an inch; the digits preceding the last two give width A in thirty-seconds. Thus, 204 indicates a cutter size 2/32 × 4/8 or 1/16 inch wide × 1/2 inch diameter.

[2] Manufacturers' Numbers formerly used to designate key and cutter sizes.

† For Woodruff key and key-slot dimensions, see pages 974 through 979.

* Diameter of hole H: for cutter numbers 617, 817, 1017, and 1217, H = 3/4 inch; for other cutters, H = 1 inch.

*Tolerances:* Face width A for shank-type cutters: 1/16- to 2/32-inch face, +0.0000, −0.0005; 3/16 to 7/32, −0.0002, −0.0007; 1/4, −0.0003, −0.0008; 5/16, −0.0004, −0.0009; 3/8, −0.0005, −0.0010 inch. Face width A for arbor-type cutters: 3/16 inch face, −0.0002, −0.0007; 1/4, −0.0003, −0.0008; 5/16, −0.0004, −0.0009; 3/8 and over, −0.0005, −0.0010 inch. Hole size H: +0.00075, −0.0000 inch. Diameter B furnished oversize to allow for sharpening; 1/32 inch in the case of arbor-type cutters.

**Setting-angles for Milling Straight Teeth of Uniform Land Width in End Mills, Angular Cutters, and Taper Reamers.** — The accompanying tables give setting-angles for the dividing head when straight teeth, having a land of uniform width throughout their length, are to be milled using single-angle fluting cutters. These setting-angles depend upon three factors: the number of teeth to be cut; the angle of the blank in which the teeth are to be cut; and the angle of the fluting cutter. Setting-angles for various combinations of these three factors are given in the tables. For example, assume that 12 teeth are to be cut on the end of an end mill using a 60-degree cutter. By following the horizontal line from 12 teeth, read in the column under 60 degrees that the dividing head should be set to an angle of 70 degrees and 32 minutes.

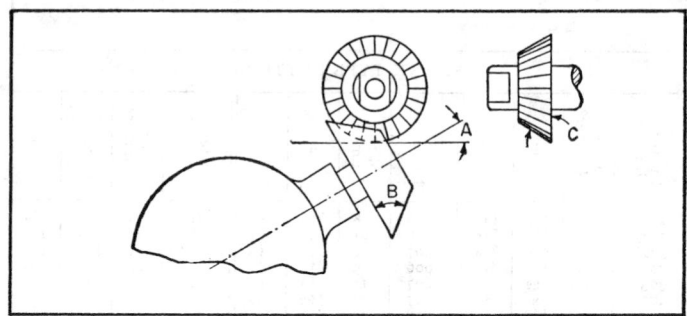

The following formulas, which were used to compile these tables, may be used to calculate the setting-angles for combinations of number of teeth, blank angle, and cutter angle not covered by the tables. In these formulas, $A$ = setting-angle for dividing head, $B$ = angle of blank in which teeth are to be cut, $C$ = angle of fluting cutter, $N$ = number of teeth to be cut, and $D$ and $E$ are angles not shown on the accompanying diagram and which are used only to simplify calculations.

$$\tan D = \cos (360°/N) \times \cot B \qquad (1)$$
$$\sin E = \tan (360°/N) \times \cot C \times \sin D \qquad (2)$$
$$\text{Setting-angle } A = D - E \qquad (3)$$

*Example:* Suppose 9 teeth are to be cut in a 35-degree blank using a 55-degree single-angle fluting cutter. In this case, $N = 9$, $B = 35°$, and $C = 55°$.

$\tan D = \cos (360°/9) \times \cot 35° = 0.76604 \times 1.4281 = 1.0940$; and $D = 47°34'$
$\sin E = \tan (360°/9) \times \cot 55° \times \sin 47°34' = 0.83910 \times 0.70021 \times 0.73806$
$\qquad = 0.43365$; and $E = 25°42'$
$$\text{Setting-angle } A = 47°34' - 25°42' = 21°52'$$

In the case of end mills and side mills the angle of the blank $B$ is 0 degrees and the following simplified formula may be used to find the setting angle $A$

$$\cos A = \tan (360°/N) \times \cot C \qquad (4)$$

*Example:* If in the previous example the blank angle was 0 degrees,
$\cos A = \tan (360°/9) \times \cot 55°$
$\qquad = 0.83910 \times 0.70021 = 0.58755$; and setting-angle $A = 54°1'$

## Angles of Elevation for Milling Straight Teeth in 0-, 5-, 10-, 15-, 20-, 25-, 30-, and 35-degree Blanks Using Single-angle Fluting Cutters

### 0° Blank (End Mill) | 5° Blank

| No. of Teeth | 90° | 80° | 70° | 60° | 50° | 90° | 80° | 70° | 60° | 50° |
|---|---|---|---|---|---|---|---|---|---|---|
| 6  | ... | 72° 13' | 50° 55' | ... | ... | 80° 4'  | 62° 34' | 41° 41' | ...     | ...     |
| 8  | ... | 79° 51' | 68° 39' | 54° 44' | 32° 57' | 82° 57' | 72° 52' | 61° 47' | 48° 0'  | 25° 40' |
| 10 | ... | 82° 38' | 74° 40' | 65° 12' | 52° 26' | 83° 50' | 76° 31' | 68° 35' | 59° 11' | 46° 4'  |
| 12 | ... | 84° 9'  | 77° 52' | 70° 32' | 61° 2'  | 84° 14' | 78° 25' | 72° 10' | 64° 52' | 55° 5'  |
| 14 | ... | 85° 8'  | 79° 54' | 73° 51' | 66° 10' | 84° 27' | 79° 36' | 74° 24' | 68° 23' | 60° 28' |
| 16 | ... | 85° 49' | 81° 20' | 76° 10' | 69° 40' | 84° 35' | 80° 25' | 75° 57' | 70° 49' | 64° 7'  |
| 18 | ... | 86° 19' | 82° 23' | 77° 52' | 72° 13' | 84° 41' | 81° 1'  | 77° 6'  | 72° 36' | 66° 47' |
| 20 | ... | 86° 43' | 83° 13' | 79° 11' | 74° 11' | 84° 45' | 81° 29' | 77° 59' | 73° 59' | 68° 50' |
| 22 | ... | 87° 2'  | 83° 52' | 80° 14' | 75° 44' | 84° 47' | 81° 50' | 78° 40' | 75° 4'  | 70° 26' |
| 24 | ... | 87° 18' | 84° 24' | 81° 6'  | 77° 0'  | 84° 49' | 82° 7'  | 79° 15' | 75° 57' | 71° 44' |

### 10° Blank | 15° Blank

| No. of Teeth | 90° | 80° | 70° | 60° | 50° | 90° | 80° | 70° | 60° | 50° |
|---|---|---|---|---|---|---|---|---|---|---|
| 6  | 70° 34' | 53° 50' | 34° 5'  | ...     | ...     | 61° 49' | 46° 12' | 28° 4'  | ...     | ...     |
| 8  | 76° 0'  | 66° 9'  | 55° 19' | 41° 56' | 20° 39' | 69° 15' | 59° 46' | 49° 21' | 36° 34' | 17° 34' |
| 10 | 77° 42' | 70° 31' | 62° 44' | 53° 30' | 40° 42' | 71° 40' | 64° 41' | 57° 8'  | 48° 12' | 36° 18' |
| 12 | 78° 30' | 72° 46' | 66° 37' | 59° 26' | 49° 50' | 72° 48' | 67° 13' | 61° 13' | 54° 14' | 45° 13' |
| 14 | 78° 56' | 74° 9'  | 69° 2'  | 63° 6'  | 55° 19' | 73° 26' | 68° 46' | 63° 46' | 57° 59' | 50° 38' |
| 16 | 79° 12' | 75° 5'  | 70° 41' | 65° 37' | 59° 1'  | 73° 50' | 69° 49' | 65° 30' | 60° 33' | 54° 20' |
| 18 | 79° 22' | 75° 45' | 71° 53' | 67° 27' | 61° 43' | 74° 5'  | 70° 33' | 66° 46' | 62° 26' | 57° 0'  |
| 20 | 79° 30' | 76° 16' | 72° 44' | 68° 52' | 63° 47' | 74° 16' | 71° 6'  | 67° 44' | 63° 52' | 59° 3'  |
| 22 | 79° 35' | 76° 40' | 73° 33' | 69° 59' | 65° 25' | 74° 24' | 71° 32' | 68° 29' | 65° 0'  | 60° 40' |
| 24 | 79° 39' | 76° 59' | 74° 9'  | 70° 54' | 66° 44' | 74° 30' | 71° 53' | 69° 6'  | 65° 56' | 61° 59' |

### 20° Blank | 25° Blank

| No. of Teeth | 90° | 80° | 70° | 60° | 50° | 90° | 80° | 70° | 60° | 50° |
|---|---|---|---|---|---|---|---|---|---|---|
| 6  | 53° 57' | 39° 39' | 23° 18' | ...     | ...     | 47° 0'  | 34° 6'  | 19° 33' | ...     | ...     |
| 8  | 62° 46' | 53° 45' | 43° 53' | 31° 53' | 14° 31' | 56° 36' | 48° 8'  | 38° 55' | 27° 47' | 11° 33' |
| 10 | 65° 47' | 59° 4'  | 51° 50' | 43° 18' | 32° 1'  | 60° 2'  | 53° 40' | 46° 47' | 38° 43' | 27° 47' |
| 12 | 67° 12' | 61° 49' | 56° 2'  | 49° 18' | 40° 40' | 61° 42' | 56° 33' | 51° 2'  | 44° 38' | 36° 10' |
| 14 | 68° 0'  | 63° 29' | 58° 39' | 53° 4'  | 46° 0'  | 62° 58' | 58° 19' | 53° 41' | 48° 20' | 41° 22' |
| 16 | 68° 30' | 64° 36' | 60° 26' | 55° 39' | 49° 38' | 63° 13' | 59° 29' | 55° 29' | 50° 53' | 44° 57' |
| 18 | 68° 50' | 65° 24' | 61° 44' | 57° 32' | 52° 17' | 63° 37' | 60° 19' | 56° 48' | 52° 46' | 47° 34' |
| 20 | 69° 3'  | 65° 59' | 62° 43' | 58° 58' | 54° 18' | 63° 53' | 60° 56' | 57° 47' | 54° 11' | 49° 33' |
| 22 | 69° 14' | 66° 28' | 63° 30' | 60° 7'  | 55° 55' | 64° 5'  | 61° 25' | 58° 34' | 55° 19' | 51° 9'  |
| 24 | 69° 21' | 66° 49' | 64° 7'  | 61° 2'  | 57° 12' | 64° 14' | 61° 47' | 59° 12' | 56° 13' | 52° 26' |

### 30° Blank | 35° Blank

| No. of Teeth | 90° | 80° | 70° | 60° | 50° | 90° | 80° | 70° | 60° | 50° |
|---|---|---|---|---|---|---|---|---|---|---|
| 6  | 40° 54' | 29° 22' | 16° 32' | ...     | ...     | 35° 32' | 25° 19' | 14° 3'  | ...     | ...     |
| 8  | 50° 46' | 42° 55' | 34° 24' | 24° 12' | 10° 14' | 45° 17' | 38° 5'  | 30° 18' | 21° 4'  | 8° 41'  |
| 10 | 54° 29' | 48° 30' | 42° 3'  | 34° 31' | 24° 44' | 49° 7'  | 43° 33' | 37° 35' | 30° 38' | 21° 40' |
| 12 | 56° 18' | 51° 26' | 46° 14' | 40° 12' | 32° 32' | 51° 3'  | 46° 30' | 41° 39' | 36° 2'  | 28° 55' |
| 14 | 57° 21' | 53° 15' | 48° 52' | 43° 49' | 37° 27' | 52° 8'  | 48° 19' | 44° 12' | 39° 28' | 33° 33' |
| 16 | 58° 0'  | 54° 27' | 50° 39' | 46° 19' | 40° 57' | 52° 50' | 49° 20' | 45° 56' | 41° 51' | 36° 45' |
| 18 | 58° 26' | 55° 18' | 51° 57' | 48° 7'  | 43° 20' | 53° 18' | 50° 21' | 47° 12' | 43° 36' | 39° 8'  |
| 20 | 58° 44' | 55° 55' | 52° 56' | 49° 30' | 45° 15' | 53° 38' | 50° 59' | 48° 14' | 44° 57' | 40° 57' |
| 22 | 58° 57' | 56° 24' | 53° 42' | 50° 36' | 46° 46' | 53° 53' | 51° 29' | 48° 56' | 46° 1'  | 42° 24' |
| 24 | 59° 8'  | 56° 48' | 54° 20' | 51° 30' | 48° 0'  | 54° 4'  | 51° 53' | 49° 32' | 46° 52' | 43° 35' |

# 1626 MILLING CUTTERS

## Angles of Elevation for Milling Straight Teeth in 40-, 45-, 50-, 55-, 60-, 65-, 70-, and 75-degree Blanks Using Single-angle Fluting Cutters

| No. of Teeth | 90° | 80° | 70° | 60° | 50° | 90° | 80° | 70° | 60° | 50° |
|---|---|---|---|---|---|---|---|---|---|---|
| | | **40° Blank** | | | | | **45° Blank** | | | |
| 6 | 30° 48' | 21° 48' | 11° 58' | ... | ... | 26° 34' | 18° 43' | 10° 11' | ... | ... |
| 8 | 40 7 | 33 36 | 26 33 | 18° 16' | 7° 23' | 35 16 | 29 25 | 23 8 | 15° 48' | 5° 58' |
| 10 | 43 57 | 38 51 | 33 32 | 27 3 | 18 55 | 38 58 | 34 21 | 29 24 | 23 40 | 16 10 |
| 12 | 45 54 | 41 43 | 37 14 | 32 3 | 25 33 | 40 54 | 37 5 | 33 0 | 28 18 | 22 13 |
| 14 | 47 3 | 43 29 | 39 41 | 35 19 | 29 51 | 42 1 | 38 46 | 35 17 | 31 18 | 26 9 |
| 16 | 47 45 | 44 39 | 41 21 | 37 33 | 32 50 | 42 44 | 39 54 | 36 52 | 33 24 | 28 57 |
| 18 | 48 14 | 45 29 | 42 34 | 39 13 | 35 5 | 43 13 | 40 42 | 38 1 | 34 56 | 30 1 |
| 20 | 48 35 | 46 7 | 43 30 | 40 30 | 36 47 | 43 34 | 41 18 | 38 53 | 36 8 | 32 37 |
| 22 | 48 50 | 46 36 | 44 13 | 41 30 | 38 8 | 43 49 | 41 46 | 39 34 | 37 5 | 34 53 |
| 24 | 49 1 | 46 58 | 44 48 | 42 19 | 39 15 | 44 0 | 42 7 | 40 7 | 37 50 | 35 55 |
| | | **50° Blank** | | | | | **55° Blank** | | | |
| 6 | 22° 45' | 15° 58' | 8° 38' | ... | ... | 19° 17' | 13° 30' | 7° 15' | ... | ... |
| 8 | 30 41 | 25 31 | 19 59 | 13° 33' | 5° 20' | 26 21 | 21 52 | 17 3 | 11° 30' | 4° 17' |
| 10 | 34 10 | 30 2 | 25 39 | 20 32 | 14 9 | 29 32 | 25 55 | 22 3 | 17 36 | 11 52 |
| 12 | 36 0 | 32 34 | 28 53 | 24 42 | 19 27 | 31 14 | 28 12 | 24 59 | 21 17 | 16 32 |
| 14 | 37 5 | 34 9 | 31 1 | 27 26 | 22 58 | 32 15 | 29 39 | 26 53 | 23 43 | 19 40 |
| 16 | 37 47 | 35 13 | 32 29 | 29 22 | 25 30 | 32 54 | 30 38 | 28 12 | 25 26 | 21 54 |
| 18 | 38 15 | 35 58 | 33 33 | 30 46 | 27 21 | 33 21 | 31 20 | 29 10 | 26 43 | 23 35 |
| 20 | 38 35 | 36 32 | 34 21 | 31 52 | 28 47 | 33 40 | 31 51 | 29 54 | 27 42 | 24 53 |
| 22 | 38 50 | 36 58 | 34 59 | 32 44 | 29 57 | 33 54 | 32 15 | 30 29 | 28 28 | 25 55 |
| 24 | 39 1 | 37 19 | 35 30 | 33 25 | 30 52 | 34 5 | 32 34 | 30 57 | 29 7 | 26 46 |
| | | **60° Blank** | | | | | **65° Blank** | | | |
| 6 | 16° 6' | 11° 12' | 6° 2' | ... | ... | 13° 7' | 9° 8' | 4° 53' | ... | ... |
| 8 | 22 13 | 18 24 | 14 19 | 9° 37' | 3° 44' | 18 15 | 15 6 | 11 42 | 7° 50' | 3° 1' |
| 10 | 25 2 | 21 56 | 18 37 | 14 49 | 10 5 | 20 40 | 18 4 | 15 19 | 12 9 | 8 15 |
| 12 | 26 34 | 23 57 | 21 10 | 17 59 | 14 13 | 21 59 | 19 48 | 17 28 | 14 49 | 11 32 |
| 14 | 27 29 | 25 14 | 22 51 | 20 6 | 16 44 | 22 48 | 20 55 | 18 54 | 16 37 | 13 48 |
| 16 | 28 5 | 26 7 | 24 1 | 21 37 | 18 40 | 23 18 | 21 39 | 19 53 | 17 53 | 15 24 |
| 18 | 28 29 | 26 44 | 24 52 | 22 44 | 20 6 | 23 40 | 22 11 | 20 37 | 18 50 | 16 37 |
| 20 | 28 46 | 27 11 | 25 30 | 23 35 | 21 14 | 23 55 | 22 35 | 21 10 | 19 33 | 17 34 |
| 22 | 29 0 | 27 34 | 26 2 | 24 17 | 22 8 | 24 6 | 22 53 | 21 36 | 20 8 | 18 20 |
| 24 | 29 9 | 27 50 | 26 26 | 24 50 | 22 52 | 24 15 | 23 8 | 21 57 | 20 36 | 18 57 |
| | | **70° Blank** | | | | | **75° Blank** | | | |
| 6 | 10° 18' | 7° 9' | 3° 48' | ... | ... | 7° 38' | 5° 19' | 2° 50' | ... | ... |
| 8 | 14 26 | 11 55 | 9 14 | 6° 9' | 2° 21' | 10 44 | 8 51 | 6 51 | 4° 34' | 1° 45' |
| 10 | 16 25 | 14 21 | 12 8 | 9 37 | 6 30 | 12 14 | 10 40 | 9 1 | 7 8 | 4 49 |
| 12 | 17 30 | 15 45 | 13 53 | 11 45 | 9 8 | 13 4 | 11 45 | 10 21 | 8 45 | 6 47 |
| 14 | 18 9 | 16 38 | 15 1 | 13 11 | 10 55 | 13 34 | 12 26 | 11 13 | 9 50 | 8 7 |
| 16 | 18 35 | 17 15 | 15 50 | 14 13 | 12 13 | 13 54 | 12 54 | 11 50 | 10 37 | 9 7 |
| 18 | 18 53 | 17 42 | 16 26 | 14 59 | 13 13 | 14 8 | 13 14 | 12 17 | 11 12 | 9 51 |
| 20 | 19 6 | 18 1 | 16 53 | 15 35 | 13 59 | 14 18 | 13 29 | 12 38 | 11 39 | 10 27 |
| 22 | 19 15 | 18 16 | 17 15 | 16 3 | 14 35 | 14 25 | 13 41 | 12 53 | 12 0 | 10 54 |
| 24 | 19 22 | 18 29 | 17 33 | 16 25 | 15 4 | 14 31 | 13 50 | 13 7 | 12 18 | 11 18 |

**Angles of Elevation for Milling Straight Teeth in 80- and 85-degree Blanks Using Single-angle Fluting Cutters**

| No. of Teeth | Angle of Fluting Cutter | | | | | | | | | |
|---|---|---|---|---|---|---|---|---|---|---|
| | 90° | 80° | 70° | 60° | 50° | 90° | 80° | 70° | 60° | 50° |
| | 80° Blank | | | | | 85° Blank | | | | |
| 6 | 5° 2′ | 3° 30′ | 1° 52′ | ... | ... | 2° 30′ | 1° 44′ | 0° 55′ | ... | ... |
| 8 | 7 6 | 5 51 | 4 31 | 3° 2′ | 1° 8′ | 3 32 | 2 55 | 2 15 | 1° 29′ | 0° 34′ |
| 10 | 8 7 | 7 5 | 5 59 | 4 44 | 3 11 | 4 3 | 3 32 | 2 59 | 2 21 | 1 35 |
| 12 | 8 41 | 7 48 | 6 52 | 5 48 | 4 29 | 4 20 | 3 53 | 3 25 | 2 53 | 2 15 |
| 14 | 9 2 | 8 16 | 7 28 | 6 32 | 5 24 | 4 30 | 4 7 | 3 43 | 3 15 | 2 42 |
| 16 | 9 15 | 8 35 | 7 51 | 7 3 | 6 3 | 4 37 | 4 17 | 3 56 | 3 30 | 3 1 |
| 18 | 9 24 | 8 48 | 8 10 | 7 26 | 6 33 | 4 42 | 4 24 | 4 5 | 3 43 | 3 16 |
| 20 | 9 31 | 8 58 | 8 24 | 7 44 | 6 56 | 4 46 | 4 29 | 4 12 | 3 52 | 3 28 |
| 22 | 9 36 | 9 6 | 8 35 | 7 59 | 7 15 | 4 48 | 4 33 | 4 18 | 3 59 | 3 37 |
| 24 | 9 40 | 9 13 | 8 43 | 8 11 | 7 30 | 4 50 | 4 36 | 4 22 | 4 5 | 3 45 |

**Spline-Shaft Milling Cutter.** — The most efficient method of forming splines on shafts is by hobbing, but special milling cutters may also be used. Since the cutter forms the space between adjacent splines, it must be made to suit the number of splines and the root diameter of the shaft. The cutter angle $B$ equals 360 degrees divided by the number of splines. The following formulas are for determining the chordal width $C$ at the root of the splines or the chordal width across the

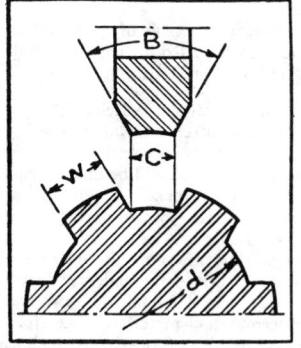

concave edge of the cutter. In these formulas, $A$ = angle between center line of spline and a radial line passing through the intersection of the root circle and one side of the spline; $W$ = width of spline; $d$ = root diameter of splined shaft; $C$ = chordal width at root circle between adjacent splines; $N$ = number of splines.

$$\text{Sin } A = \frac{W}{d} \; ; \; C = d \times \sin\left(\frac{180}{N} - A\right)$$

Splines of involute form are often used in preference to the straight-sided type. Dimensions of the American Standard involute splines and hobs are given in the section on splines.

## Cutter Grinding

**Wheels for Sharpening Milling Cutters.** — Milling cutters may be sharpened either by using the periphery of a disk wheel or the face of a cup wheel. The latter grinds the lands of the teeth flat, whereas the periphery of a disk wheel leaves the teeth slightly concave back of the cutting edges. The concavity produced by disk wheels reduces the effective clearance angle on the teeth, the effect being more pronounced for wheels of small diameter than for wheels of large diameter. For this reason, large diameter wheels are preferred when sharpening milling cutters with disk type wheels. Irrespective of what type of wheel is used to sharpen a milling cutter, any burrs resulting from grinding should be carefully removed by a hand stoning operation. Stoning also helps to reduce the roughness of grinding marks and improves the quality of the finish produced on the surface being machined, as

Specifications of Grinding Wheels for Sharpening Milling Cutters

| Cutter Material | Operation | Grinding Wheel | | | |
|---|---|---|---|---|---|
| | | Abrasive Material | Grain Size | Grade | Bond |
| Carbon Tool Steel | Roughing | Aluminum Oxide | 46–60 | K | Vitrified |
| | Finishing | Aluminum Oxide | 100 | H | Vitrified |
| High-speed Steel:<br>18–4–1 {<br>18–4–2 { | Roughing | Aluminum Oxide | 60 | K,H | Vitrified |
| | Finishing | Aluminum Oxide | 100 | H | Vitrified |
| | Roughing | Aluminum Oxide | 80 | F,G,H | Vitrified |
| | Finishing | Aluminum Oxide | 100 | H | Vitrified |
| Cast Non-Ferrous Tool Material | Roughing | Aluminum Oxide | 46 | H,K,L,N | Vitrified |
| | Finishing | Aluminum Oxide | 100–120 | H | Vitrified |
| Sintered Carbide | Roughing after Brazing | Silicon Carbide | 60 | G | Vitrified |
| | Roughing | Diamond | 100 | * | Resinoid |
| | Finishing | Diamond | Up to 500 | * | Resinoid |

* Not indicated in diamond wheel markings.

well as increasing the life of the cutter.  The accompanying table of specifications for grinding wheels used to sharpen milling cutters was compiled by The Cincinnati Milling Machine Co. in " A Treatise on Milling and Milling Machines."

**Wheel Speeds and Feeds for Sharpening Milling Cutters.** — Relatively low cutting speeds should be used when sharpening milling cutters to avoid tempering and heat checking.  Dry grinding is recommended in all cases except when diamond wheels are employed.  The surface speed of grinding wheels should be in the range of 4500 to 6500 feet per minute for grinding milling cutters of high-speed steel or cast non-ferrous tool material.  For sintered carbide cutters, 5000 to 5500 feet per minute should be used.

The maximum stock removed per pass of the grinding wheel should not exceed about 0.0004 inch for sintered carbide cutters; 0.003 inch for large high-speed steel and cast non-ferrous tool material cutters; and 0.0015 inch for narrow saws and slotting cutters of high-speed steel or cast non-ferrous tool material.  The stock removed per pass of the wheel may be increased for backing-off operations such as the grinding of secondary clearance behind the teeth since there is usually a sufficient body of metal to carry off the heat.

**Clearance Angles for Milling Cutter Teeth.** — The clearance angle provided on the cutting edges of milling cutters has an important bearing on cutter performance, cutting efficiency, and cutter life between sharpenings.  It is desirable in all cases to use a clearance angle as small as possible so as to leave more metal back of the cutting edges for better heat dissipation and to provide maximum support.  Excessive clearance angles not only weaken the cutting edges, but also increase the likelihood of " chatter " which will result in poor finish on the machined surface and reduce the life of the cutter.  According to The Cincinnati Milling Machine Co., milling cutters used for general purpose work and having diameters from ⅛ to 3 inches should have clearance angles from 13 to 5 degrees, respectively, decreasing proportionally as the diameter increases.  General purpose cutters over 3 inches in diameter should be provided with a clearance angle of 4 to 5 degrees. The land width is usually ⅟₆₄, ⅟₃₂, and ⅟₁₆ inch, respectively, for small, medium, and large cutters.

The primary clearance or relief angle for best results varies according to the material being milled about as follows: low carbon, high carbon, and alloy steels, 3 to 5 degrees; cast iron and medium and hard bronze, 4 to 7 degrees; brass, soft bronze, aluminum, magnesium, plastics, etc., 10 to 12 degrees. When milling cutters are resharpened, it is customary to grind a secondary clearance angle of 3 to 5 degrees behind the primary clearance angle to reduce the land width to its original value and thus avoid interference with the surface to be milled. A general formula for plain milling cutters, face mills, and form relieved cutters which gives the clearance angle $C$, in degrees, necessitated by the feed per revolution $F$, in inches, the width of land $L$, in inches, the depth of cut $d$, in inches, the cutter diameter $D$, in inches, and the Brinell hardness number $B$ of the work being cut is:

$$ C = \frac{45860}{DB}\left(1.5\,L + \frac{F}{\pi D}\sqrt{d(D-d)}\right) $$

**Rake Angles for Milling Cutters.** — In peripheral milling cutters, the rake angle is generally defined as the angle in degrees that the tooth face deviates from a radial line to the cutting edge. In face milling cutters, the teeth are inclined with respect to both the radial and axial lines. These angles are called *radial* and *axial* rake, respectively. The radial and axial rake angles may be positive, zero, or negative.

Positive rake angles should be used whenever possible for all types of high-speed steel milling cutters. For sintered carbide tipped cutters, zero and negative rake angles are frequently employed to provide more material back of the cutting edge to resist shock loads.

*Rake Angles for High-speed Steel Cutters:* Positive rake angles of 10 to 15 degrees are satisfactory for milling steels of various compositions with plain milling cutters. For softer materials such as magnesium and aluminum alloys, the rake angle may be 25 degrees or more. Metal slitting saws for cutting alloy steel usually have rake angles from 5 to 10 degrees, whereas zero and sometimes negative rake angles are used for saws to cut copper and other soft non-ferrous metals to reduce the tendency to "hog in." Form relieved cutters usually have rake angles of 0, 5, or 10 degrees. Commercial face milling cutters usually have 10 degrees positive radial and axial rake angles for general use in milling cast iron, forged and alloy steel, brass, and bronze; for milling castings and forgings of magnesium and free-cutting aluminum and their alloys, the rake angles may be increased to 25 degrees positive or more, depending on the operating conditions; a smaller rake angle is used for abrasive or difficult to machine aluminum alloys.

*Cast Non-ferrous Tool Material Milling Cutters:* Positive rake angles are generally provided on milling cutters using cast non-ferrous tool materials although negative rake angles may be used advantageously for some operations such as those where shock loads are encountered or where it is necessary to eliminate vibration when milling thin sections.

*Sintered Carbide Milling Cutters:* Peripheral milling cutters such as slab mills, slotting cutters, saws, etc., tipped with sintered carbide, generally have negative radial rake angles of 5 degrees for soft low carbon steel and 10 degrees or more for alloy steels. Positive axial rake angles of 5 and 10 degrees, respectively, may be provided, and for slotting saws and cutters, 0 degree axial rake may be used. On soft materials such as free-cutting aluminum alloys, positive rake angles of 10 to 20 degrees are used. For milling abrasive or difficult to machine aluminum alloys, small positive or even negative rake angles are used.

**Eccentric Type Radial Relief.** — When the radial relief angles on peripheral teeth of milling cutters are ground with a disc type grinding wheel in the conventional manner the ground surfaces on the lands are slightly concave, conforming approximately to the radius of the wheel. A flat land is produced when the radial relief angle is ground with a cup wheel. Another entirely different method of grinding the radial angle is by the eccentric method, which produces a slightly convex surface on the land. If the radial relief angle at the cutting edge is equal for all of the three types of land mentioned, it will be found that the land with the eccentric relief will drop away from the cutting edge a somewhat greater distance for a given distance around the land than will the others. This is evident from a study of Table 2 entitled, "Indicator Drops for Checking Radial Relief Angles on Peripheral Teeth." It is claimed that this feature is an advantage of the eccentric type relief.

The setup for grinding an eccentric relief is shown in Fig. 2. In this setup the point of contact between the cutter and the tooth rest must be in the same plane as the centers, or axes, of the grinding wheel and the cutter. A wide face is used on the grinding wheel, which is trued and dressed at an angle with respect to the axis of the cutter. An alternate method is to tilt the wheel at this angle. Then as the cutter is traversed and rotated past the grinding wheel while in contact with the tooth rest, an eccentric relief will be generated by the angular face of the wheel. This type of relief can only be ground on the peripheral teeth of milling cutters having helical flutes because the combination of the angular wheel face and the twisting motion of the cutter is required to generate the eccentric relief. Therefore, an eccentric relief cannot be ground on the peripheral teeth of straight fluted cutters.

Table 1 is a table of wheel angles for grinding an eccentric relief for different combinations of relief angles and helix angles. When angles are required that cannot be found in this table, the wheel angle, $W$, can be calculated by using the following formula, in which $R$ is the radial relief angle and $H$ is the helix angle of the flutes on the cutter.

$$\tan W = \tan R \times \tan H$$

**Indicator Drop Method of Checking Relief and Rake Angles.** — The most convenient and inexpensive method of checking the relief and rake angles on milling cutters is by the indicator drop method. Three tables, Tables 2, 3 and 4, of indicator drops

*(Continued on page 1633)*

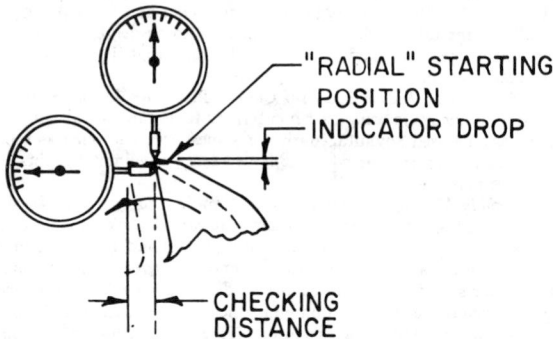

Fig. 1.   Setup for checking the radial relief angle by indicator drop method.

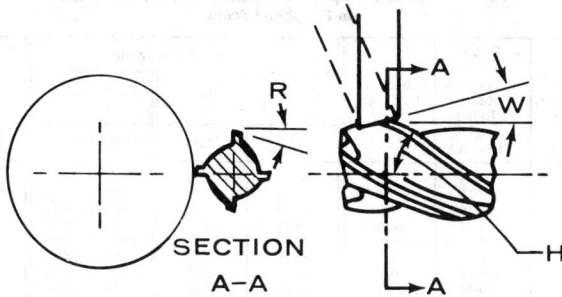

Fig. 2.   Setup for grinding eccentric type radial relief angle.

### Table 1.   Grinding Wheel Angles for Grinding Eccentric Type Radial Relief Angle

| Radial Relief Angle, R, Degrees | Helix Angle of Cutter Flutes, H, Degrees | | | | | | | |
|---|---|---|---|---|---|---|---|---|
| | 12 | 18 | 20 | 30 | 40 | 45 | 50 | 52 |
| | Wheel Angle, W, Degrees | | | | | | | |
| 1 | 0°13′ | 0°19′ | 0°22′ | 0°35′ | 0°50′ | 1°00′ | 1°12′ | 1°17′ |
| 2 | 0°26′ | 0°39′ | 0°44′ | 1°09′ | 1°41′ | 2°00′ | 2°23′ | 2°34′ |
| 3 | 0°38′ | 0°59′ | 1°06′ | 1°44′ | 2°31′ | 3°00′ | 3°34′ | 3°50′ |
| 4 | 0°51′ | 1°18′ | 1°27′ | 2°19′ | 3°21′ | 4°00′ | 4°46′ | 5°07′ |
| 5 | 1°04′ | 1°38′ | 1°49′ | 2°53′ | 4°12′ | 5°00′ | 5°57′ | 6°23′ |
| 6 | 1°17′ | 1°57′ | 2°11′ | 3°28′ | 5°02′ | 6°00′ | 7°08′ | 7°40′ |
| 7 | 1°30′ | 2°17′ | 2°34′ | 4°03′ | 5°53′ | 7°00′ | 8°19′ | 8°56′ |
| 8 | 1°43′ | 2°37′ | 2°56′ | 4°38′ | 6°44′ | 8°00′ | 9°30′ | 10°12′ |
| 9 | 1°56′ | 2°57′ | 3°18′ | 5°13′ | 7°34′ | 9°00′ | 10°41′ | 11°28′ |
| 10 | 2°09′ | 3°17′ | 3°40′ | 5°49′ | 8°25′ | 10°00′ | 11°52′ | 12°43′ |
| 11 | 2°22′ | 3°37′ | 4°03′ | 6°24′ | 9°16′ | 11°00′ | 13°03′ | 13°58′ |
| 12 | 2°35′ | 3°57′ | 4°25′ | 7°00′ | 10°07′ | 12°00′ | 14°13′ | 15°13′ |
| 13 | 2°49′ | 4°17′ | 4°48′ | 7°36′ | 10°58′ | 13°00′ | 15°23′ | 16°28′ |
| 14 | 3°02′ | 4°38′ | 5°11′ | 8°11′ | 11°49′ | 14°00′ | 16°33′ | 17°42′ |
| 15 | 3°16′ | 4°59′ | 5°34′ | 8°48′ | 12°40′ | 15°00′ | 17°43′ | 18°56′ |
| 16 | 3°29′ | 5°19′ | 5°57′ | 9°24′ | 13°32′ | 16°00′ | 18°52′ | 20°09′ |
| 17 | 3°43′ | 5°40′ | 6°21′ | 10°01′ | 14°23′ | 17°00′ | 20°01′ | 21°22′ |
| 18 | 3°57′ | 6°02′ | 6°45′ | 10°37′ | 15°15′ | 18°00′ | 21°10′ | 22°35′ |
| 19 | 4°11′ | 6°23′ | 7°09′ | 11°15′ | 16°07′ | 19°00′ | 22°19′ | 23°47′ |
| 20 | 4°25′ | 6°45′ | 7°33′ | 11°52′ | 16°59′ | 20°00′ | 23°27′ | 24°59′ |
| 21 | 4°40′ | 7°07′ | 7°57′ | 12°30′ | 17°51′ | 21°00′ | 24°35′ | 26°10′ |
| 22 | 4°55′ | 7°29′ | 8°22′ | 13°08′ | 18°44′ | 22°00′ | 25°43′ | 27°21′ |
| 23 | 5°09′ | 7°51′ | 8°47′ | 13°46′ | 19°36′ | 23°00′ | 26°50′ | 28°31′ |
| 24 | 5°24′ | 8°14′ | 9°12′ | 14°25′ | 20°29′ | 24°00′ | 27°57′ | 29°41′ |
| 25 | 5°40′ | 8°37′ | 9°38′ | 15°04′ | 21°22′ | 25°00′ | 29°04′ | 30°50′ |

### Table 2. Indicator Drops for Checking the Radial Relief Angle on Peripheral Teeth

| Cutter Diameter, Inch | Recommended Range of Radial Relief Angles, Degrees | Checking Distance, Inch | Indicator Drop, Inches | | | | Recommended Max. Primary Land Width, Inch |
|---|---|---|---|---|---|---|---|
| | | | For Flat and Concave Relief | | For Eccentric Relief | | |
| | | | Min. | Max. | Min. | Max. | |
| 1/16 | 20-25 | .005 | .0014 | .0019 | .0020 | .0026 | .007 |
| 3/32 | 16-20 | .005 | .0012 | .0015 | .0015 | .0019 | .007 |
| 1/8 | 15-19 | .010 | .0018 | .0026 | .0028 | .0037 | .015 |
| 5/32 | 13-17 | .010 | .0017 | .0024 | .0024 | .0032 | .015 |
| 3/16 | 12-16 | .010 | .0016 | .0023 | .0022 | .0030 | .015 |
| 7/32 | 11-15 | .010 | .0015 | .0022 | .0020 | .0028 | .015 |
| 1/4 | 10-14 | .015 | .0017 | .0028 | .0027 | .0039 | .020 |
| 9/32 | 10-14 | .015 | .0018 | .0029 | .0027 | .0039 | .020 |
| 5/16 | 10-13 | .015 | .0019 | .0027 | .0027 | .0035 | .020 |
| 11/32 | 10-13 | .015 | .0020 | .0028 | .0027 | .0035 | .020 |
| 3/8 | 10-13 | .015 | .0020 | .0029 | .0027 | .0035 | .020 |
| 13/32 | 9-12 | .020 | .0022 | .0032 | .0032 | .0044 | .025 |
| 7/16 | 9-12 | .020 | .0022 | .0033 | .0032 | .0043 | .025 |
| 15/32 | 9-12 | .020 | .0023 | .0034 | .0032 | .0043 | .025 |
| 1/2 | 9-12 | .020 | .0024 | .0034 | .0032 | .0043 | .025 |
| 9/16 | 9-12 | .020 | .0024 | .0035 | .0032 | .0043 | .025 |
| 5/8 | 8.11 | .020 | .0022 | .0032 | .0028 | .0039 | .025 |
| 11/16 | 8-11 | .030 | .0029 | .0045 | .0043 | .0059 | .035 |
| 3/4 | 8-11 | .030 | .0030 | .0046 | .0043 | .0059 | .035 |
| 13/16 | 8-11 | .030 | .0031 | .0047 | .0043 | .0059 | .035 |
| 7/8 | 8-11 | .030 | .0032 | .0048 | .0043 | .0059 | .035 |
| 15/16 | 7-10 | .030 | .0027 | .0043 | .0037 | .0054 | .035 |
| 1 | 7-10 | .030 | .0028 | .0044 | .0037 | .0054 | .035 |
| 1 1/8 | 7-10 | .030 | .0029 | .0045 | .0037 | .0053 | .035 |
| 1 1/4 | 6-9 | .030 | .0024 | .0040 | .0032 | .0048 | .035 |
| 1 3/8 | 6-9 | .030 | .0025 | .0041 | .0032 | .0048 | .035 |
| 1 1/2 | 6-9 | .030 | .0026 | .0041 | .0032 | .0048 | .035 |
| 1 5/8 | 6-9 | .030 | .0026 | .0042 | .0032 | .0048 | .035 |
| 1 3/4 | 6-9 | .030 | .0026 | .0042 | .0032 | .0048 | .035 |
| 1 7/8 | 6-9 | .030 | .0027 | .0043 | .0032 | .0048 | .035 |
| 2 | 6-9 | .030 | .0027 | .0043 | .0032 | .0048 | .035 |
| 2 1/4 | 5-8 | .030 | .0022 | .0038 | .0026 | .0042 | .040 |
| 2 1/2 | 5-8 | .030 | .0023 | .0039 | .0026 | .0042 | .040 |
| 2 3/4 | 5-8 | .030 | .0023 | .0039 | .0026 | .0042 | .040 |
| 3 | 5-8 | .030 | .0023 | .0039 | .0026 | .0042 | .040 |
| 3 1/2 | 5-8 | .030 | .0024 | .0040 | .0026 | .0042 | .047 |
| 4 | 5-8 | .030 | .0024 | .0040 | .0026 | .0042 | .047 |
| 5 | 4-7 | .030 | .0019 | .0035 | .0021 | .0037 | .047 |
| 6 | 4-7 | .030 | .0019 | .0035 | .0021 | .0037 | .047 |
| 7 | 4-7 | .030 | .0020 | .0036 | .0021 | .0037 | .060 |
| 8 | 4-7 | .030 | .0020 | .0036 | .0021 | .0037 | .060 |
| 10 | 4-7 | .030 | .0020 | .0036 | .0021 | .0037 | .060 |
| 12 | 4-7 | .030 | .0020 | .0036 | .0021 | .0037 | .060 |

### Table 3. Indicator Drops for Checking Relief Angles on Side Teeth and End Teeth

| Checking Distance, Inch | Given Relief Angle | | | | | | | | |
|---|---|---|---|---|---|---|---|---|---|
| | 1° | 2° | 3° | 4° | 5° | 6° | 7° | 8° | 9° |
| | Indicator Drop, inch | | | | | | | | |
| .005 | .00009 | .00017 | .00026 | .00035 | .0004 | .0005 | .0006 | .0007 | .0008 |
| .010 | .00017 | .00035 | .00052 | .0007 | .0009 | .0011 | .0012 | .0014 | .0016 |
| .015 | .00026 | .0005 | .00079 | .0010 | .0013 | .0016 | .0018 | .0021 | .0024 |
| .031 | .00054 | .0011 | .0016 | .0022 | .0027 | .0033 | .0038 | .0044 | .0049 |
| .047 | .00082 | .0016 | .0025 | .0033 | .0041 | .0049 | .0058 | .0066 | .0074 |
| .062 | .00108 | .0022 | .0032 | .0043 | .0054 | .0065 | .0076 | .0087 | .0098 |

Table 4.  Indicator Drops for Checking Rake Angles on Milling Cutter Face

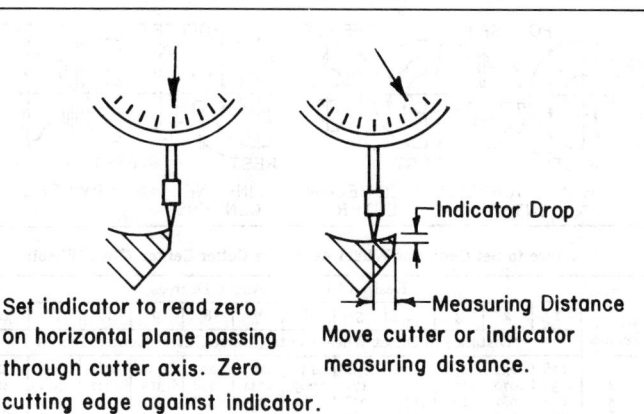

Set indicator to read zero on horizontal plane passing through cutter axis. Zero cutting edge against indicator.

Move cutter or indicator measuring distance.

Indicator Drop

Measuring Distance

| Rate Angle, Deg. | Measuring Distance, inch | | | | Rate Angle, Deg. | Measuring Distance, inch | | | |
|---|---|---|---|---|---|---|---|---|---|
| | .031 | .062 | .094 | .125 | | .031 | .062 | .094 | .125 |
| | Indicator Drop, inch | | | | | Indicator Drop, inch | | | |
| 1 | .0005 | .0011 | .0016 | .0022 | 11 | .0060 | .0121 | .0183 | .0243 |
| 2 | .0011 | .0022 | .0033 | .0044 | 12 | .0066 | .0132 | .0200 | .0266 |
| 3 | .0016 | .0032 | .0049 | .0066 | 13 | .0072 | .0143 | .0217 | .0289 |
| 4 | .0022 | .0043 | .0066 | .0087 | 14 | .0077 | .0155 | .0234 | .0312 |
| 5 | .0027 | .0054 | .0082 | .0109 | 15 | .0083 | .0166 | .0252 | .0335 |
| 6 | .0033 | .0065 | .0099 | .0131 | 16 | .0089 | .0178 | .0270 | .0358 |
| 7 | .0038 | .0076 | .0115 | .0153 | 17 | .0095 | .0190 | .0287 | .0382 |
| 8 | .0044 | .0087 | .0132 | .0176 | 18 | .0101 | .0201 | .0305 | .0406 |
| 9 | .0049 | .0098 | .0149 | .0198 | 19 | .0107 | .0213 | .0324 | .0430 |
| 10 | .0055 | .0109 | .0166 | .0220 | 20 | .0113 | .0226 | .0342 | .0455 |

are provided in this section, for checking radial relief angles on the peripheral teeth, relief angles on side and end teeth, and rake angles on the tooth faces.

The setup for checking the radial relief angle is illustrated in Fig. 1. Two dial test indicators are required, one of which should have a sharp pointed contact point. This indicator is positioned so that the axis of its spindle is vertical, passing through the axis of the cutter. The cutter may be held by its shank in the spindle of a tool and cutter grinder workhead, or between centers while mounted on a mandrel. The cutter is rotated to the position where the vertical indicator contacts a cutting edge. The second indicator is positioned with its spindle axis horizontal and with the contact point touching the tool face just below the cutting edge. With both indicators adjusted to read zero, the cutter is rotated a distance equal to the checking distance, as determined by the reading on the second indicator. Then the indicator drop is read on the vertical indicator and checked against the values in the tables. The indicator drops for radial relief angles ground by a disc type grinding wheel and those ground with a cup wheel are so nearly equal that the values are listed together; values for the eccentric type relief are listed separately, since they are larger. A similar procedure is used to check the relief angles on the side and end teeth of milling cutters; however, only one indicator is used. Also, instead of rotating the cutter, the indicator or the cutter must be moved a distance equal to the checking distance in a straight line.

**Various Set-ups Used in Grinding the Clearance Angle on Milling Cutter Teeth**

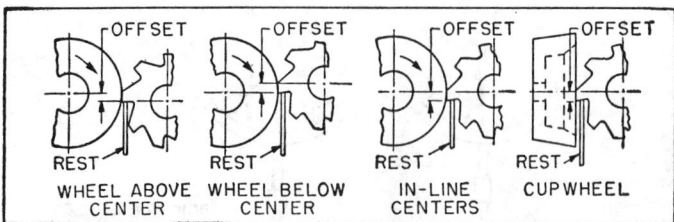

WHEEL ABOVE CENTER    WHEEL BELOW CENTER    IN-LINE CENTERS    CUP WHEEL

**Distance to Set Center of Wheel Above the Cutter Center (Disk Wheel)**

| Diam. of Wheel, Inches | Desired Clearance Angle, Degrees | | | | | | | | | | | |
|---|---|---|---|---|---|---|---|---|---|---|---|---|
| | 1 | 2 | 3 | 4 | 5 | 6 | 7 | 8 | 9 | 10 | 11 | 12 |
| | *Distance to Offset Wheel Center Above Cutter Center, Inches | | | | | | | | | | | |
| 3 | .026 | .052 | .079 | .105 | .131 | .157 | .183 | .209 | .235 | .260 | .286 | .312 |
| 4 | .035 | .070 | .105 | .140 | .174 | .209 | .244 | .278 | .313 | .347 | .382 | .416 |
| 5 | .044 | .087 | .131 | .174 | .218 | .261 | .305 | .348 | .391 | .434 | .477 | .520 |
| 6 | .052 | .105 | .157 | .209 | .261 | .314 | .366 | .417 | .469 | .521 | .572 | .624 |
| 7 | .061 | .122 | .183 | .244 | .305 | .366 | .427 | .487 | .547 | .608 | .668 | .728 |
| 8 | .070 | .140 | .209 | .279 | .349 | .418 | .488 | .557 | .626 | .695 | .763 | .832 |
| 9 | .079 | .157 | .236 | .314 | .392 | .470 | .548 | .626 | .704 | .781 | .859 | .936 |
| 10 | .087 | .175 | .262 | .349 | .436 | .523 | .609 | .696 | .782 | .868 | .954 | 1.040 |

* Calculated from the formula: Offset = Wheel Diameter × ½ × Sine of Clearance Angle.

**Distance to Set Center of Wheel Below the Cutter Center (Disk Wheel)**

| Diam. of Cutter, Inches | Desired Clearance Angle, Degrees | | | | | | | | | | | |
|---|---|---|---|---|---|---|---|---|---|---|---|---|
| | 1 | 2 | 3 | 4 | 5 | 6 | 7 | 8 | 9 | 10 | 11 | 12 |
| | *Distance to Offset Wheel Center Below Cutter Center, Inches | | | | | | | | | | | |
| 2 | .017 | .035 | .052 | .070 | .087 | .105 | .122 | .139 | .156 | .174 | .191 | .208 |
| 3 | .026 | .052 | .079 | .105 | .131 | .157 | .183 | .209 | .235 | .260 | .286 | .312 |
| 4 | .035 | .070 | .105 | .140 | .174 | .209 | .244 | .278 | .313 | .347 | .382 | .416 |
| 5 | .044 | .087 | .131 | .174 | .218 | .261 | .305 | .348 | .391 | .434 | .477 | .520 |
| 6 | .052 | .105 | .157 | .209 | .261 | .314 | .366 | .417 | .469 | .521 | .572 | .624 |
| 7 | .061 | .122 | .183 | .244 | .305 | .366 | .427 | .487 | .547 | .608 | .668 | .728 |
| 8 | .070 | .140 | .209 | .279 | .349 | .418 | .488 | .557 | .626 | .695 | .763 | .832 |
| 9 | .079 | .157 | .236 | .314 | .392 | .470 | .548 | .626 | .704 | .781 | .859 | .936 |
| 10 | .087 | .175 | .262 | .349 | .436 | .523 | .609 | .696 | .782 | .868 | .954 | 1.040 |

* Calculated from the formula: Offset = Cutter Diameter × ½ × Sine of Clearance Angle.

**Distance to Set Tooth Rest Below Center Line of Wheel and Cutter.** — When the clearance angle is ground with a disk type wheel by keeping the center line of the wheel in line with the center line of the cutter, the tooth rest should be lowered by an amount given by the following formula:

$$\text{Offset} = \frac{\text{Wheel Diam.} \times \text{Cutter Diam.} \times \text{Sine of One-half the Clearance Angle}}{\text{Wheel Diam.} + \text{Cutter Diam.}}$$

**Distance to Set Tooth Rest Below Cutter Center When Cup Wheel is Used.** — When the clearance is ground with a cup wheel, the tooth rest is set below the center of the cutter the same amount as given in the table for " Distance to Set Center of Wheel Below the Cutter Center (Disk Wheel)."

# REAMERS

**Hand Reamers.** — Hand reamers are made with both straight and helical flutes. The latter provide a shearing cut and are especially useful in reaming holes having keyways or grooves, as these are bridged over by the helical flutes, thus preventing binding or chattering. Hand reamers are made in both solid and expansion forms. The American standard dimensions for solid forms are given in the accompanying table. The expansion type is useful whenever, in connection with repair or other work, it is necessary to enlarge a reamed hole by a few thousandths of an inch. The expansion form is split through the fluted section and a slight amount of expansion is obtained by screwing in a tapering plug. The diameter increase may vary from 0.005 to 0.008 inch for reamers up to about 1 inch diameter and from 0.010 to 0.012 inch for diameters between 1 and 2 inches. Hand reamers are tapered slightly on the end to facilitate starting them properly. The actual diameter of the shanks of commercial reamers may be from 0.002 to 0.005 inch under the reamer size. That part of the shank which is squared should be turned smaller in diameter than the shank itself, so that, when applying a wrench, no burr may be raised which may mar the reamed hole if the reamer is passed clear through it.

When fluting reamers, the cutter is so set with relation to the center of the reamer blank that the tooth gets a slight negative rake; that is, the cutter should be set *ahead* of the center, as shown in the illustration accompanying the table giving the amount to set the cutter ahead of the radial line. The amount is so selected that a tangent to the circumference of the reamer at the cutting point makes an angle of approximately 95 degrees with the front face of the cutting edge.

**Amount to Set Cutter Ahead of Radial Line to Obtain Negative Front Rake**

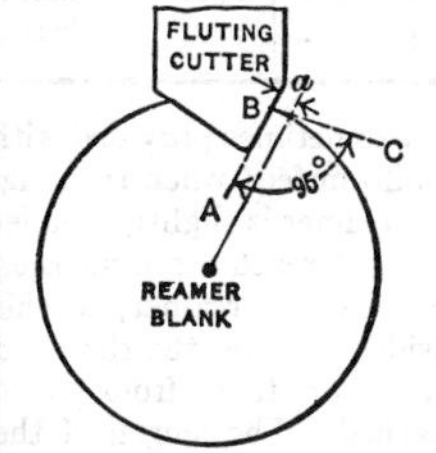

| | Size of Reamer | Dimension *a*, Inches | Size of Reamer | Dimension *a*, Inches | Size of Reamer | Dimension *a*, Inches |
|---|---|---|---|---|---|---|
| | ¼ | 0.011 | ⅞ | 0.038 | 2 | 0.087 |
| | ⅜ | 0.016 | 1 | 0.044 | 2¼ | 0.098 |
| | ½ | 0.022 | 1¼ | 0.055 | 2½ | 0.109 |
| | ⅝ | 0.027 | 1½ | 0.066 | 2¾ | 0.120 |
| | ¾ | 0.033 | 1¾ | 0.076 | 3 | 0.131 |

When fluting reamers, it is necessary to "break up the flutes"; that is, to space the cutting edges unevenly around the reamer. The difference in spacing should be very slight and need not exceed two degrees one way or the other. The manner in which the breaking up of the flutes is usually done is to move the index head to which the reamer is fixed a certain amount more or less than would be the case if the spacing were regular. A table is given showing the amount of this additional movement of the index crank for reamers with different numbers of flutes. When a reamer is provided with helical flutes, the angle of spiral should be such that the cutting edges make an angle of about 10 or at most 15 degrees with the axis of the reamer.

The relief of the cutting edges should be comparatively slight. An eccentric relief, that is, one where the land back of the cutting edge is convex, rather than flat, is used by one or two manufacturers, and is preferable for finishing reamers, as the reamer will hold its size longer. When hand reamers are used merely for removing stock, or simply for enlarging holes, the flat relief is better, because the reamer has a keener cutting edge. The width of the land of the cutting edges should be about 1/32 inch for a 1/4-inch, 1/16 inch for a 1-inch, and 3/32 inch for a 3-inch reamer.

**Irregular Spacing of Teeth in Reamers**

| Number of flutes in reamer | 4 | 6 | 8 | 10 | 12 | 14 | 16 |
|---|---|---|---|---|---|---|---|
| Index circle to use | 39 | 39 | 39 | 39 | 39 | 49 | 20 |
| Before cutting | Move Spindle the Number of Holes below More or Less than for Regular Spacing | | | | | | |
| 2d flute | 8 less | 4 less | 3 less | 2 less | 4 less | 3 less | 2 less |
| 3d flute | 4 more | 5 more | 5 more | 3 more | 4 more | 2 more | 2 more |
| 4th flute | 6 less | 7 less | 2 less | 5 less | 1 less | 2 less | 1 less |
| 5th flute | ..... | 6 more | 4 more | 2 more | 3 more | 4 more | 2 more |
| 6th flute | ..... | 5 less | 6 less | 2 less | 4 less | 1 less | 2 less |
| 7th flute | ..... | ..... | 2 more | 3 more | 4 more | 3 more | 1 more |
| 8th flute | ..... | ..... | 3 less | 2 less | 3 less | 2 less | 2 less |
| 9th flute | ..... | ..... | ..... | 5 more | 2 more | 1 more | 2 more |
| 10th flute | ..... | ..... | ..... | 1 less | 2 less | 3 less | 2 less |
| 11th flute | ..... | ..... | ..... | ..... | 3 more | 3 more | 1 more |
| 12th flute | ..... | ..... | ..... | ..... | 4 less | 2 less | 2 less |
| 13th flute | ..... | ..... | ..... | ..... | ..... | 2 more | 2 more |
| 14th flute | ..... | ..... | ..... | ..... | ..... | 3 less | 1 less |
| 15th flute | ..... | ..... | ..... | ..... | ..... | ..... | 2 more |
| 16th flute | ..... | ..... | ..... | ..... | ..... | ..... | 2 less |

**Threaded-end Hand Reamers.** — Hand reamers are sometimes provided with a thread at the extreme point in order to give them a uniform feed when reaming. The diameter on the top of this thread at the point of the reamer is slightly smaller than the reamer itself, and the thread tapers upward until it reaches a dimension of from 0.003 to 0.008 inch, according to size, below the size of the reamer; at this point the thread stops and a short neck about 1/16 inch wide separates the threaded portion from the actual reamer which is provided with a short taper from 5/16 to 7/16 inch long up to where the standard diameter is reached. The length of the threaded portion and the number of threads per inch for reamers of this kind are given in the accompanying table. The thread employed is a sharp V-thread.

**Dimensions for Threaded-end Hand Reamers**

| Sizes of Reamers | Length of Threaded Part | No. of Threads per Inch | Diam. of Thread at Point of Reamer | Sizes of Reamers | Length of Threaded Part | No. of Threads per Inch | Diam. of Thread at Point of Reamer |
|---|---|---|---|---|---|---|---|
| | | | Full diameter | | | | Full diameter |
| 1/8–5/16 | 3/8 | 32 | −0.006 | 1 1/32 –1 1/2 | 9/16 | 18 | −0.010 |
| 11/32–1/2 | 7/16 | 28 | −0.006 | 1 17/32–2 | 9/16 | 18 | −0.012 |
| 17/32–3/4 | 1/2 | 24 | −0.008 | 2 1/32 –2 1/2 | 9/16 | 18 | −0.015 |
| 25/32–1 | 9/16 | 18 | −0.008 | 2 17/32–3 | 9/16 | 18 | −0.020 |

**Fluted Chucking Reamers.** — Reamers of this type are used in turret lathes, screw machines, etc., for enlarging holes and finishing them smooth and to the required size. The best results are obtained with a floating type of holder which permits a reamer to align itself with the hole being reamed. These reamers are intended for removing a small amount of metal, 0.005 to 0.010 inch being common allowances. Fluted chucking reamers are provided either with a straight shank or a standard taper shank. (See table for standard dimensions.)

**Rose Chucking Reamers.** — The rose type of reamer is used for enlarging cored or other holes. The cutting edges at the end are ground to a 45-degree bevel. This type of reamer will remove considerable metal in one cut. The cylindrical part of the reamer has no cutting edges, but merely grooves cut for the full length of the reamer body, providing a way for the chips to escape and a channel for lubricant to reach the cutting edges. There is no relief on the cylindrical surface of the body part, but it is slightly back-tapered so that the diameter at the point with the beveled cutting edges is slightly larger than the diameter further back. The back-taper should not exceed 0.001 inch per inch. This form of reamer usually

### Fluting Cutters for Reamers

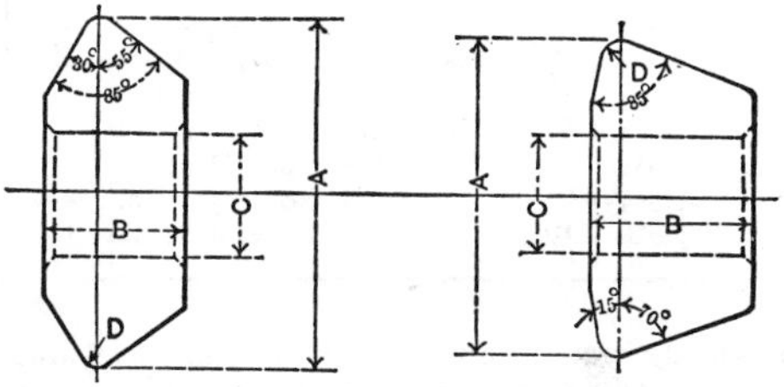

| Diameter of Reamer | Diameter of Fluting Cutter | Thickness of Fluting Cutter | Diameter of Hole in Cutter | Radius between Cutting Faces of Cutter | Diameter of Reamer | Diameter of Fluting Cutter | Thickness of Fluting Cutter | Diameter of Hole in Cutter | Radius between Cutting Faces of Cutter |
|---|---|---|---|---|---|---|---|---|---|
|  | A | B | C | D |  | A | B | C | D |
| 1/8 | 1¾ | 3/16 | ¾ | sharp corner, no radius. | 1 | 2¼ | ½ | 1 | 3/64 |
|  |  |  |  |  | 1¼ | 2¼ | 9/16 | 1 | 1/16 |
| 3/16 | 1¾ | 3/16 | ¾ | sharp corner, no radius. | 1½ | 2¼ | 5/8 | 1 | 1/16 |
|  |  |  |  |  | 1¾ | 2¼ | 5/8 | 1 | 5/64 |
| ¼ | 1¾ | 3/16 | ¾ | 1/64 | 2 | 2½ | ¾ | 1 | 5/64 |
| 3/8 | 2 | ¼ | ¾ | 1/64 | 2¼ | 2½ | ¾ | 1 | 5/64 |
| ½ | 2 | 5/16 | ¾ | 1/32 | 2½ | 2½ | 7/8 | 1 | 3/16 |
| 5/8 | 2 | 3/8 | ¾ | 1/32 | 2¾ | 2½ | 7/8 | 1 | 3/16 |
| ¾ | 2 | 7/16 | ¾ | 3/64 | 3 | 2½ | 1 | 1 | 3/16 |

**Dimensions of Formed Reamer Fluting Cutters**

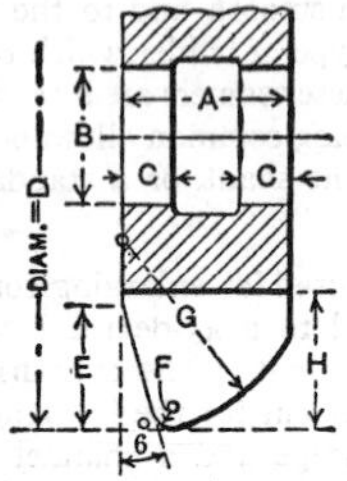


The making and maintenance of cutters of the formed type involves greater expense than the use of the angular cutters of which dimensions are given on the previous page; but the form of flute produced by the formed type of cutter is preferred by many reamer users. The claims made for the formed type of flute are that the chips can be more readily removed from the reamer, and that the reamer has greater strength and is less likely to crack or spring out of shape in hardening.

| Size of Reamers used for | Number of Teeth in Reamer | Diam. of Cutter | Width of Cutter | Diam. of Hole | Width of Bearing | Length of Bevel | Radius | Radius | Depth of Tooth | Number of Teeth in Cutter |
|---|---|---|---|---|---|---|---|---|---|---|
| | | *D* | *A* | *B* | *C* | *E* | *F* | *G* | *H* | |
| 1/8–3/16 | 6 | 1 3/4 | 3/16 | 7/8 | .... | 0.125 | 0.016 | 7/32 | 0.21 | 14 |
| 1/4–5/16 | 6 | 1 3/4 | 1/4 | 7/8 | .... | 0.152 | 0.022 | 9/32 | 0.25 | 13 |
| 3/8–7/16 | 6 | 1 7/8 | 3/8 | 7/8 | 1/8 | 0.178 | 0.029 | 1/2 | 0.28 | 12 |
| 1/2–11/16 | 6–8 | 2 | 7/16 | 7/8 | 1/8 | 0.205 | 0.036 | 9/16 | 0.30 | 12 |
| 3/4–1 | 8 | 2 1/8 | 1/2 | 7/8 | 5/32 | 0.232 | 0.042 | 11/16 | 0.32 | 12 |
| 1 1/16–1 1/2 | 10 | 2 1/4 | 9/16 | 7/8 | 5/32 | 0.258 | 0.049 | 3/4 | 0.38 | 11 |
| 1 9/16–2 1/8 | 12 | 2 3/8 | 5/8 | 7/8 | 3/16 | 0.285 | 0.056 | 27/32 | 0.40 | 11 |
| 2 1/4 –3 | 14 | 2 5/8 | 11/16 | 7/8 | 3/16 | 0.312 | 0.062 | 7/8 | 0.44 | 10 |

produces holes slightly larger than its size and it is, therefore, always made from 0.005 to 0.010 inch smaller than its nominal size, so that it may be followed by a fluted reamer for finishing. The grooves on the cylindrical portion are cut by a convex cutter having a width equal to from one-fifth to one-fourth the diameter of the rose reamer itself. The depth of the groove should be from one-eighth to one-sixth the diameter of the reamer. The teeth at the end of the reamer are milled with a 75-degree angular cutter; the width of the land of the cutting edge should be about one-fifth the distance from tooth to tooth. If an angular cutter is preferred to a convex cutter for milling the grooves on the cylindrical portion, because of the higher cutting speed possible when milling, an 80-degree angular cutter slightly rounded at the point may be used.

**Cutters for Fluting Rose Chucking Reamers.** — The cutters used for fluting rose chucking reamers on the end are 80-degree angular cutters for 1/4 and 5/16 inch diameter reamers; 75-degree angular cutters for 3/8 and 7/16 inch reamers; and 70-degree angular cutters for all larger sizes. The grooves on the cylindrical portion are milled with convex cutters of approximately the following sizes for given diameters of reamers: 5/32-inch convex cutter for 1/2-inch reamers; 5/16-inch cutter for 1-inch reamers; 3/8-inch cutter for 1 1/2-inch reamers; 13/32-inch cutters for 2-inch reamers; and 15/32-inch cutters for 2 1/2-inch reamers. The smaller sizes of reamers, from 1/4 to 3/8 inch in diameter, are often milled with regular double-angle reamer fluting cutters having a radius of 1/64 inch for 1/4-inch reamer, and 1/32 inch for 5/16- and 3/8-inch sizes.

## Vertical Adjustment of Tooth-rest for Grinding Clearance on Reamers

| Size of Reamer | Hand Reamer for Steel. Cutting Clearance Land 0.006 inch Wide | | Hand Reamer for Cast Iron and Bronze. Cutting Clearance Land 0.025 inch Wide | | Chucking Reamer for Cast Iron and Bronze. Cutting Clearance Land 0.025 inch Wide | | Rose Chucking Reamers for Steel |
|---|---|---|---|---|---|---|---|
| | For Cutting Clearance | For Second Clearance | For Cutting Clearance | For Second Clearance | For Cutting Clearance | For Second Clearance | For Cutting Clearance on Angular Edge at End |
| ½ | 0.012 | 0.052 | 0.032 | 0.072 | 0.040 | 0.080 | 0.080 |
| ⅝ | 0.012 | 0.062 | 0.032 | 0.072 | 0.040 | 0.090 | 0.090 |
| ¾ | 0.012 | 0.072 | 0.035 | 0.095 | 0.040 | 0.100 | 0.100 |
| ⅞ | 0.012 | 0.082 | 0.040 | 0.120 | 0.045 | 0.125 | 0.125 |
| 1 | 0.012 | 0.092 | 0.040 | 0.120 | 0.045 | 0.125 | 0.125 |
| 1⅛ | 0.012 | 0.102 | 0.040 | 0.120 | 0.045 | 0.125 | 0.125 |
| 1¼ | 0.012 | 0.112 | 0.045 | 0.145 | 0.050 | 0.160 | 0.160 |
| 1⅜ | 0.012 | 0.122 | 0.045 | 0.145 | 0.050 | 0.160 | 0.175 |
| 1½ | 0.012 | 0.132 | 0.048 | 0.168 | 0.055 | 0.175 | 0.175 |
| 1⅝ | 0.012 | 0.142 | 0.050 | 0.170 | 0.060 | 0.200 | 0.200 |
| 1¾ | 0.012 | 0.152 | 0.052 | 0.192 | 0.060 | 0.200 | 0.200 |
| 1⅞ | 0.012 | 0.162 | 0.056 | 0.196 | 0.060 | 0.200 | 0.200 |
| 2 | 0.012 | 0.172 | 0.056 | 0.216 | 0.064 | 0.224 | 0.225 |
| 2⅛ | 0.012 | 0.172 | 0.059 | 0.219 | 0.064 | 0.224 | 0.225 |
| 2¼ | 0.012 | 0.172 | 0.063 | 0.223 | 0.064 | 0.224 | 0.225 |
| 2⅜ | 0.012 | 0.172 | 0.063 | 0.223 | 0.068 | 0.228 | 0.230 |
| 2½ | 0.012 | 0.172 | 0.065 | 0.225 | 0.072 | 0.232 | 0.230 |
| 2⅝ | 0.012 | 0.172 | 0.065 | 0.225 | 0.075 | 0.235 | 0.235 |
| 2¾ | 0.012 | 0.172 | 0.065 | 0.225 | 0.077 | 0.237 | 0.240 |
| 2⅞ | 0.012 | 0.172 | 0.070 | 0.230 | 0.080 | 0.240 | 0.240 |
| 3 | 0.012 | 0.172 | 0.072 | 0.232 | 0.080 | 0.240 | 0.240 |
| 3⅛ | 0.012 | 0.172 | 0.075 | 0.235 | 0.083 | 0.240 | 0.240 |
| 3¼ | 0.012 | 0.172 | 0.078 | 0.238 | 0.083 | 0.243 | 0.245 |
| 3⅜ | 0.012 | 0.172 | 0.081 | 0.241 | 0.087 | 0.247 | 0.245 |
| 3½ | 0.012 | 0.172 | 0.084 | 0.244 | 0.090 | 0.250 | 0.250 |
| 3⅝ | 0.012 | 0.172 | 0.087 | 0.247 | 0.093 | 0.253 | 0.250 |
| 3¾ | 0.012 | 0.172 | 0.090 | 0.250 | 0.097 | 0.257 | 0.255 |
| 3⅞ | 0.012 | 0.172 | 0.093 | 0.253 | 0.100 | 0.260 | 0.255 |
| 4 | 0.012 | 0.172 | 0.096 | 0.256 | 0.104 | 0.264 | 0.260 |
| 4⅛ | 0.012 | 0.172 | 0.096 | 0.256 | 0.104 | 0.264 | 0.260 |
| 4¼ | 0.012 | 0.172 | 0.096 | 0.256 | 0.106 | 0.266 | 0.265 |
| 4⅜ | 0.012 | 0.172 | 0.096 | 0.256 | 0.108 | 0.268 | 0.265 |
| 4½ | 0.012 | 0.172 | 0.100 | 0.260 | 0.108 | 0.268 | 0.265 |
| 4⅝ | 0.012 | 0.172 | 0.100 | 0.260 | 0.110 | 0.270 | 0.270 |
| 4¾ | 0.012 | 0.172 | 0.104 | 0.264 | 0.114 | 0.274 | 0.275 |
| 4⅞ | 0.012 | 0.172 | 0.106 | 0.266 | 0.116 | 0.276 | 0.275 |
| 5 | 0.012 | 0.172 | 0.110 | 0.270 | 0.118 | 0.278 | 0.275 |

**Reamer Difficulties.** — There are certain frequently occurring problems in reaming for which it is necessary to apply remedial measures. These difficulties include the production of oversize holes, bellmouth holes and holes with a poor finish. The following is taken from suggestions for correction of these difficulties by the National Twist Drill and Tool Co. and Winter Brothers Co.*

*Oversize Holes:* The cutting of a hole oversize from the start of the reaming operations usually indicates a mechanical defect in the setup or reamer. Thus, the wrong reamer for the work-piece material may have been used or there may be inadequate work-piece support, inadequate or worn guide bushings, or misalignment of the spindles, bushings or work-piece or runout of the spindle or reamer holder. The reamer itself may be defective due to chamfer runout or runout of the cutting end due to a bent or non-concentric shank.

When reamers gradually start to cut oversize, it is due to pickup or galling, principally on the reamer margins. This condition is partly due to the work-piece material. Mild steels, certain cast irons and some aluminum alloys are particularly troublesome in this respect.

Corrective measures include reducing the reamer margin widths to about 0.005 to 0.010 inch, use of hard case surface treatments on high speed steel reamers, either alone or in combination with black oxide treatments, and the use of a high grade finish on the reamer faces, margins, and chamfer relief surfaces.

*Bellmouth Holes:* The cutting of a hole that becomes oversize at the entry end with the oversize decreasing gradually along its length always reflects misalignment of the cutting portion of the reamer with respect to the hole. The obvious solution is to provide improved guiding of the reamer by the use of accurate bushings and pilot surfaces. If this is not feasible, and the reamer is cutting in a vertical position, a flexible element may be employed to hold the reamer in such a way that it has both radial and axial float with the hope that the reamer will follow the original hole and prevent the bellmouth condition.

In horizontal setups where the reamer is held fixed and the work-piece rotated, any misalignment exerts a sideways force on the reamer as it is fed to depth, resulting in the formation of a tapered hole. This type of bellmouthing can frequently be reduced by shortening the bearing length of the cutting portion of the reamer. One way to do this is to reduce the reamer diameter by 0.010 to 0.030 inch, depending on size and length, behind a short full-diameter section, ⅛ to ½ inch long according to length and size, following the chamfer. The second method is to grind a high back taper, 0.008 to 0.015 inch per inch, behind the short full-diameter section. Either of these modifications reduces the length of the reamer tooth which can cause the bellmouth condition.

*Poor Finish:* The most obvious step towards producing a good finish is to reduce the reamer feed per revolution. Feeds as low as 0.0002 to 0.0005 inch per tooth have been used successfully. However, reamer life will be better if the maximum feasible feed is used.

The minimum practical amount of reaming stock allowance will often improve finish by reducing the volume of chips and the resulting heat generated on the cutting portion of the chamfer. Too little reamer stock, however, can be troublesome in that the reamer teeth may not cut freely but will actually deflect the work material out of the way. When this happens, excessive heat, poor finish and rapid reamer wear can occur.

Because of their superior abrasion resistance, carbide reamers are often used when fine finishes are required. When properly conditioned, carbide reamers can produce a large number of good quality holes. Careful honing of the carbide reamer edges is very important.

---

* *Metal Cuttings,* "Some Aspects of Reamer Design and Operation," April 1963.

## Dimensions of Centers for Reamers and Arbors

| Diameter of Arbor (A) | Largest Diameter of Center (B) | Number of Drill (C) | Depth of Hole (D) | Diameter of Arbor (A) | Largest Diameter of Center (B) | Letter of Drill (C) | Depth of Hole (D) |
|---|---|---|---|---|---|---|---|
| 3/4 | 3/8 | 25 | 7/16 | 2 1/2 | 11/16 | J | 27/32 |
| 13/16 | 13/32 | 20 | 1/2 | 2 5/8 | 45/64 | K | 7/8 |
| 7/8 | 7/16 | 17 | 17/32 | 2 3/4 | 23/32 | L | 29/32 |
| 15/16 | 15/32 | 12 | 9/16 | 2 7/8 | 47/64 | M | 29/32 |
| 1 | 1/2 | 8 | 19/32 | 3 | 3/4 | N | 15/16 |
| 1 1/8 | 33/64 | 5 | 5/8 | 3 1/8 | 49/64 | N | 31/32 |
| 1 1/4 | 17/32 | 3 | 21/32 | 3 1/4 | 25/32 | O | 31/32 |
| 1 3/8 | 35/64 | 2 | 21/32 | 3 3/8 | 51/64 | P | 1 |
| 1 1/2 | 9/16 | 1 | 11/16 | 3 1/2 | 13/16 | P | 1 |
| …. | …. | Letter | … | 3 5/8 | 53/64 | Q | 1 1/16 |
| 1 5/8 | 37/64 | A | | 3 3/4 | 27/32 | R | 1 1/16 |
| 1 3/4 | 19/32 | B | | 3 7/8 | 55/64 | R | 1 1/16 |
| 1 7/8 | 39/64 | C | | 4 | 7/8 | S | 1 1/8 |
| 2 | 5/8 | E | | 4 1/4 | 29/32 | T | 1 1/8 |
| 2 1/8 | 41/64 | F | | 4 1/2 | 15/16 | V | 1 3/16 |
| 2 1/4 | 21/32 | G | | 4 3/4 | 31/32 | W | 1 1/4 |
| 2 3/8 | 43/64 | H | | 5 | 1 | X | 1 1/4 |

| Diameter of Arbor (A) | Largest Diameter of Center (B) | Number of Drill (C) | Depth of Hole (D) |
|---|---|---|---|
| 1/4 | 1/8 | 55 | 5/32 |
| 5/16 | 5/32 | 52 | 3/16 |
| 3/8 | 3/16 | 48 | 7/32 |
| 7/16 | 7/32 | 43 | 1/4 |
| 1/2 | 1/4 | 39 | 5/16 |
| 9/16 | 9/32 | 33 | 11/32 |
| 5/8 | 5/16 | 30 | 3/8 |
| 11/16 | 11/32 | 29 | 13/32 |

### Straight Shank Center Reamers and Machine Countersinks (ANSI B94.2-1977)

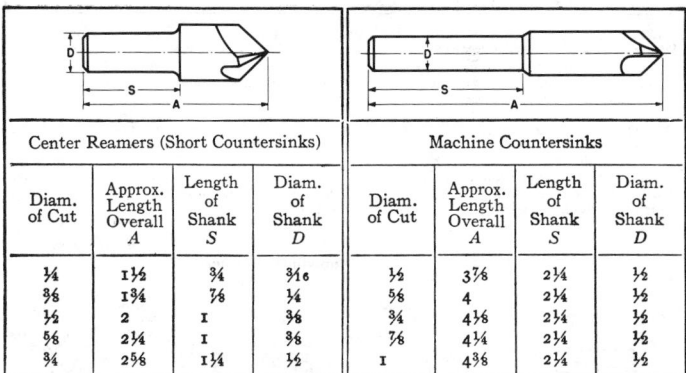

| Center Reamers (Short Countersinks) | | | | Machine Countersinks | | | |
|---|---|---|---|---|---|---|---|
| Diam. of Cut | Approx. Length Overall A | Length of Shank S | Diam. of Shank D | Diam. of Cut | Approx. Length Overall A | Length of Shank S | Diam. of Shank D |
| 1/4 | 1 1/2 | 3/4 | 3/16 | 1/2 | 3 7/8 | 2 1/4 | 1/2 |
| 3/8 | 1 3/4 | 7/8 | 1/4 | 5/8 | 4 | 2 1/4 | 1/2 |
| 1/2 | 2 | 1 | 3/8 | 3/4 | 4 1/8 | 2 1/4 | 1/2 |
| 5/8 | 2 1/4 | 1 | 3/8 | 7/8 | 4 1/4 | 2 1/4 | 1/2 |
| 3/4 | 2 5/8 | 1 1/4 | 1/2 | 1 | 4 3/8 | 2 1/4 | 1/2 |

All dimensions are given in inches. Material is high-speed steel. Reamers and countersinks have 3 or 4 flutes. Center reamers are standard with either 60, 82, 90 or 100 degrees included angle. Machine countersinks are standard with either 60 or 82 degrees included angle.

*Tolerances:* On length overall *A* the tolerance is ±1/8 inch for center reamers in a size range of from 1/4 to 3/8 inch, incl., and machine countersinks in a size range of from 1/2 to 5/8 inch, incl.; ±3/16 inch for center reamers, 1/2 to 3/4 inch, incl. and machine countersinks, 3/4 to 1 inch, incl. On shank diameter *D* the tolerance is −.0005 to −.002 inch. On shank length *S*, the tolerance is ±1/16 inch.

## Expansion Chucking Reamers — Straight and Taper Shanks* (ANSI B94.2-1977)

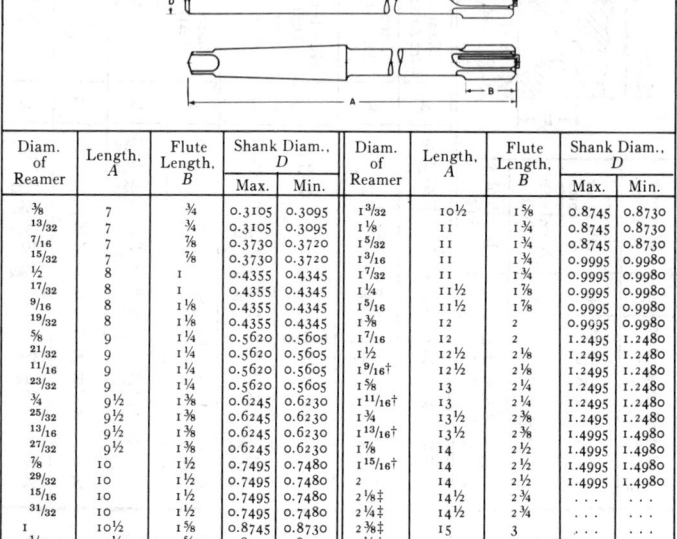

| Diam. of Reamer | Length, A | Flute Length, B | Shank Diam., D Max. | Shank Diam., D Min. | Diam. of Reamer | Length, A | Flute Length, B | Shank Diam., D Max. | Shank Diam., D Min. |
|---|---|---|---|---|---|---|---|---|---|
| 3/8 | 7 | 3/4 | 0.3105 | 0.3095 | 1 3/32 | 10 1/2 | 1 5/8 | 0.8745 | 0.8730 |
| 13/32 | 7 | 3/4 | 0.3105 | 0.3095 | 1 1/8 | 11 | 1 3/4 | 0.8745 | 0.8730 |
| 7/16 | 7 | 7/8 | 0.3730 | 0.3720 | 1 5/32 | 11 | 1 3/4 | 0.8745 | 0.8730 |
| 15/32 | 7 | 7/8 | 0.3730 | 0.3720 | 1 3/16 | 11 | 1 3/4 | 0.9995 | 0.9980 |
| 1/2 | 8 | 1 | 0.4355 | 0.4345 | 1 7/32 | 11 | 1 3/4 | 0.9995 | 0.9980 |
| 17/32 | 8 | 1 | 0.4355 | 0.4345 | 1 1/4 | 11 1/2 | 1 7/8 | 0.9995 | 0.9980 |
| 9/16 | 8 | 1 1/8 | 0.4355 | 0.4345 | 1 5/16 | 11 1/2 | 1 7/8 | 0.9995 | 0.9980 |
| 19/32 | 8 | 1 1/8 | 0.4355 | 0.4345 | 1 3/8 | 12 | 2 | 0.9995 | 0.9980 |
| 5/8 | 9 | 1 1/4 | 0.5620 | 0.5605 | 1 7/16 | 12 | 2 | 1.2495 | 1.2480 |
| 21/32 | 9 | 1 1/4 | 0.5620 | 0.5605 | 1 1/2 | 12 1/2 | 2 1/8 | 1.2495 | 1.2480 |
| 11/16 | 9 | 1 1/4 | 0.5620 | 0.5605 | 1 9/16† | 12 1/2 | 2 1/8 | 1.2495 | 1.2480 |
| 23/32 | 9 | 1 1/4 | 0.5620 | 0.5605 | 1 5/8 | 13 | 2 1/4 | 1.2495 | 1.2480 |
| 3/4 | 9 1/2 | 1 3/8 | 0.6245 | 0.6230 | 1 11/16† | 13 | 2 1/4 | 1.2495 | 1.2480 |
| 25/32 | 9 1/2 | 1 3/8 | 0.6245 | 0.6230 | 1 3/4 | 13 1/2 | 2 3/8 | 1.2495 | 1.2480 |
| 13/16 | 9 1/2 | 1 3/8 | 0.6245 | 0.6230 | 1 13/16† | 13 1/2 | 2 3/8 | 1.4995 | 1.4980 |
| 27/32 | 9 1/2 | 1 3/8 | 0.6245 | 0.6230 | 1 7/8 | 14 | 2 1/2 | 1.4995 | 1.4980 |
| 7/8 | 10 | 1 1/2 | 0.7495 | 0.7480 | 1 15/16† | 14 | 2 1/2 | 1.4995 | 1.4980 |
| 29/32 | 10 | 1 1/2 | 0.7495 | 0.7480 | 2 | 14 | 2 1/2 | 1.4995 | 1.4980 |
| 15/16 | 10 | 1 1/2 | 0.7495 | 0.7480 | 2 1/8‡ | 14 1/2 | 2 3/4 | ... | ... |
| 31/32 | 10 | 1 1/2 | 0.7495 | 0.7480 | 2 1/4‡ | 14 1/2 | 2 3/4 | ... | ... |
| 1 | 10 1/2 | 1 5/8 | 0.8745 | 0.8730 | 2 3/8‡ | 15 | 3 | ... | ... |
| 1 1/32 | 10 1/2 | 1 5/8 | 0.8745 | 0.8730 | 2 1/2‡ | 15 | 3 | ... | ... |
| 1 1/16 | 10 1/2 | 1 5/8 | 0.8745 | 0.8730 | ... | | | | |

All dimensions in inches. Material is high-speed steel. The number of flutes is as follows: 3/8- to 15/32-inch sizes, 4 to 6; 1/2- to 31/32-inch sizes, 6 to 8; 1- to 1 11/16-inch sizes, 8 to 10; 1 3/4- to 1 15/16-inch sizes, 8 to 12; 2- to 2 1/4-inch sizes, 10 to 12; 2 3/8- and 2 1/2-inch sizes, 10 to 14. The expansion feature of these reamers provides a means of adjustment which is important in reaming holes to close tolerances. When worn undersize, they may be expanded and reground to the original size.

*Taper is Morse taper: No. 1 for sizes 3/8 to 19/32 inch, incl.; No. 2 for sizes 5/8 to 29/32 incl.; No. 3 for sizes 15/16 to 1 7/32, incl.; No. 4 for sizes 1 1/4 to 1 5/8, incl.: and No. 5 for sizes 1 3/4 to 2 1/2, incl. For amount of taper see page 1728.

†Straight shank only.    ‡Taper shank only.

*Tolerances:* On reamer diameter, 3/8- to 1-inch sizes, incl., +.0001 to .0005 inch; over 1-inch size, +.0002 to .0006 inch. On length A and flute length B, 3/8- to 1-inch sizes, incl., ±1/16 inch; 1 1/32- to 2-inch sizes, incl., ±3/32 inch; over 2-inch sizes, ±1/8 inch.

## Illustration of Terms Applying to Reamers — 1

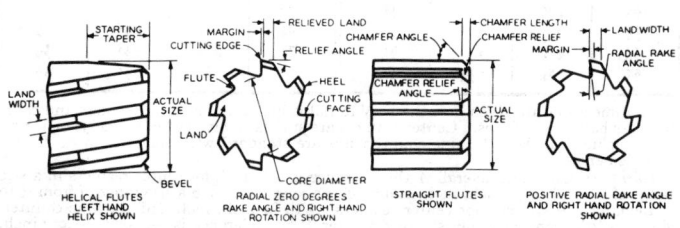

## Illustration of Terms Applying to Reamers — 2

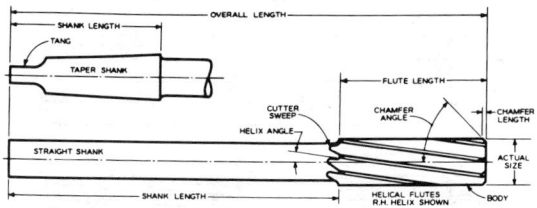

CHUCKING REAMER, STRAIGHT AND TAPER SHANK

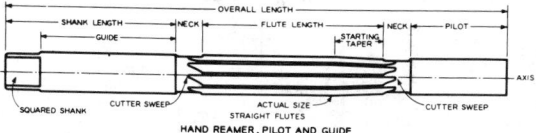

HAND REAMER, PILOT AND GUIDE

### American National Standard Fluted Taper Shank Chucking Reamers — Straight and Helical Flutes, Fractional Sizes (ANSI B94.2-1977)

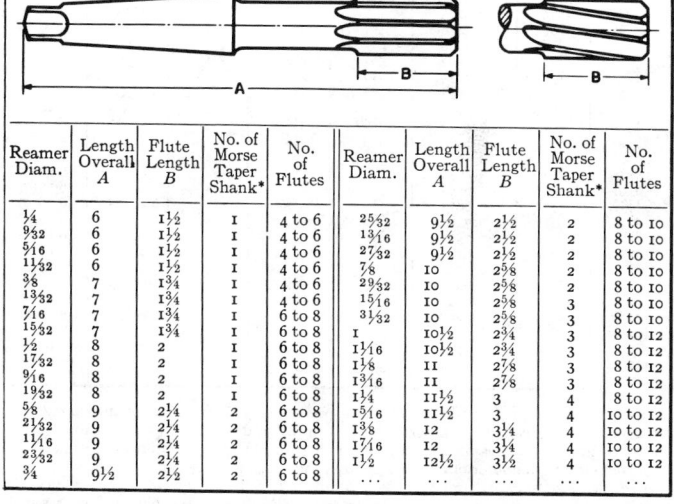

| Reamer Diam. | Length Overall A | Flute Length B | No. of Morse Taper Shank* | No. of Flutes | Reamer Diam. | Length Overall A | Flute Length B | No. of Morse Taper Shank* | No. of Flutes |
|---|---|---|---|---|---|---|---|---|---|
| $\frac{1}{4}$ | 6 | $1\frac{1}{2}$ | 1 | 4 to 6 | $\frac{25}{32}$ | $9\frac{1}{2}$ | $2\frac{1}{2}$ | 2 | 8 to 10 |
| $\frac{9}{32}$ | 6 | $1\frac{1}{2}$ | 1 | 4 to 6 | $\frac{13}{16}$ | $9\frac{1}{2}$ | $2\frac{1}{2}$ | 2 | 8 to 10 |
| $\frac{5}{16}$ | 6 | $1\frac{1}{2}$ | 1 | 4 to 6 | $\frac{27}{32}$ | $9\frac{1}{2}$ | $2\frac{1}{2}$ | 2 | 8 to 10 |
| $\frac{11}{32}$ | 6 | $1\frac{1}{2}$ | 1 | 4 to 6 | $\frac{7}{8}$ | 10 | $2\frac{5}{8}$ | 2 | 8 to 10 |
| $\frac{3}{8}$ | 7 | $1\frac{3}{4}$ | 1 | 4 to 6 | $\frac{29}{32}$ | 10 | $2\frac{5}{8}$ | 2 | 8 to 10 |
| $\frac{13}{32}$ | 7 | $1\frac{3}{4}$ | 1 | 4 to 6 | $\frac{15}{16}$ | 10 | $2\frac{5}{8}$ | 3 | 8 to 10 |
| $\frac{7}{16}$ | 7 | $1\frac{3}{4}$ | 1 | 6 to 8 | $\frac{31}{32}$ | 10 | $2\frac{5}{8}$ | 3 | 8 to 10 |
| $\frac{15}{32}$ | 7 | $1\frac{3}{4}$ | 1 | 6 to 8 | 1 | $10\frac{1}{2}$ | $2\frac{3}{4}$ | 3 | 8 to 12 |
| $\frac{1}{2}$ | 8 | 2 | 1 | 6 to 8 | $1\frac{1}{16}$ | $10\frac{1}{2}$ | $2\frac{3}{4}$ | 3 | 8 to 12 |
| $\frac{17}{32}$ | 8 | 2 | 1 | 6 to 8 | $1\frac{1}{8}$ | 11 | $2\frac{7}{8}$ | 3 | 8 to 12 |
| $\frac{9}{16}$ | 8 | 2 | 1 | 6 to 8 | $1\frac{3}{16}$ | 11 | $2\frac{7}{8}$ | 3 | 8 to 12 |
| $\frac{19}{32}$ | 8 | 2 | 1 | 6 to 8 | $1\frac{1}{4}$ | $11\frac{1}{2}$ | 3 | 4 | 8 to 12 |
| $\frac{5}{8}$ | 9 | $2\frac{1}{4}$ | 2 | 6 to 8 | $1\frac{5}{16}$ | $11\frac{1}{2}$ | 3 | 4 | 10 to 12 |
| $\frac{21}{32}$ | 9 | $2\frac{1}{4}$ | 2 | 6 to 8 | $1\frac{3}{8}$ | 12 | $3\frac{1}{4}$ | 4 | 10 to 12 |
| $\frac{11}{16}$ | 9 | $2\frac{1}{4}$ | 2 | 6 to 8 | $1\frac{7}{16}$ | 12 | $3\frac{1}{4}$ | 4 | 10 to 12 |
| $\frac{23}{32}$ | 9 | $2\frac{1}{4}$ | 2 | 6 to 8 | $1\frac{1}{2}$ | $12\frac{1}{2}$ | $3\frac{1}{2}$ | 4 | 10 to 12 |
| $\frac{3}{4}$ | $9\frac{1}{2}$ | $2\frac{1}{2}$ | 2 | 6 to 8 | ... | ... | ... | ... | ... |

All dimensions are given in inches.   Material is high-speed steel.
* American National Standard self-holding tapers (see Table 5, p. 1733).
*Tolerances:* On reamer diameter, $\frac{1}{4}$-inch size, plus .0001 to .0004 inch; $\frac{1}{4}$- to 1-inch size, incl., plus .0001 to .0005 inch; over 1-inch size, plus .0002 to .0006 inch.   On length overall A and flute length B, $\frac{1}{4}$- to 1-inch size, incl., plus or minus $\frac{1}{16}$ inch; $1\frac{1}{16}$- to $1\frac{1}{2}$-inch size, incl., plus or minus $\frac{3}{32}$ inch.

**Hand Reamers — Straight and Helical Flutes** (ANSI B94.2-1977)

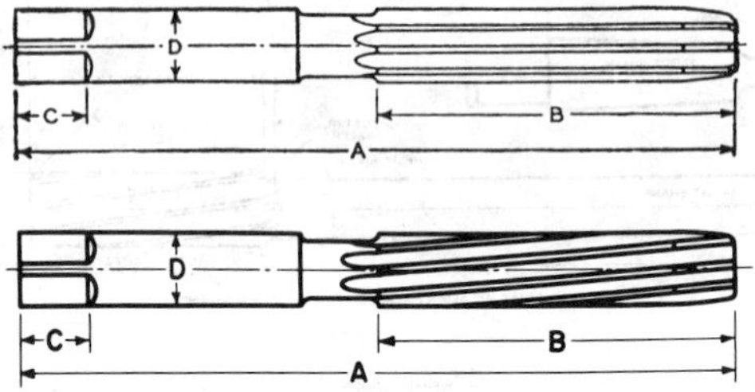


| Reamer Diameter | | | Length Overall *A* | Flute Length *B* | Square Length *C* | Size of Square | No. of Flutes |
|---|---|---|---|---|---|---|---|
| Straight Flutes | Helical Flutes | Decimal Equivalent | | | | | |
| 1/8 | ..... | 0.1250 | 3 | 1 1/2 | 5/32 | 0.095 | 4 to 6 |
| 5/32 | ..... | 0.1562 | 3 1/4 | 1 5/8 | 7/32 | 0.115 | 4 to 6 |
| 3/16 | ..... | 0.1875 | 3 1/2 | 1 3/4 | 7/32 | 0.140 | 4 to 6 |
| 7/32 | ..... | 0.2188 | 3 3/4 | 1 7/8 | 1/4 | 0.165 | 4 to 6 |
| 1/4 | 1/4 | 0.2500 | 4 | 2 | 1/4 | 0.185 | 4 to 6 |
| 9/32 | 9/32 | 0.2812 | 4 1/4 | 2 1/8 | 1/4 | 0.210 | 4 to 6 |
| 5/16 | 5/16 | 0.3125 | 4 1/2 | 2 1/4 | 5/16 | 0.235 | 4 to 6 |
| 11/32 | 11/32 | 0.3438 | 4 3/4 | 2 3/8 | 5/16 | 0.255 | 4 to 6 |
| 3/8 | 3/8 | 0.3750 | 5 | 2 1/2 | 3/8 | 0.280 | 4 to 6 |
| 13/32 | 13/32 | 0.4062 | 5 1/4 | 2 5/8 | 3/8 | 0.305 | 6 to 8 |
| 7/16 | 7/16 | 0.4375 | 5 1/2 | 2 3/4 | 7/16 | 0.330 | 6 to 8 |
| 15/32 | 15/32 | 0.4688 | 5 3/4 | 2 7/8 | 7/16 | 0.350 | 6 to 8 |
| 1/2 | 1/2 | 0.5000 | 6 | 3 | 1/2 | 0.375 | 6 to 8 |
| 17/32 | ..... | 0.5312 | 6 1/4 | 3 1/8 | 1/2 | 0.400 | 6 to 8 |
| 9/16 | 9/16 | 0.5625 | 6 1/2 | 3 1/4 | 9/16 | 0.420 | 6 to 8 |
| 19/32 | ..... | 0.5938 | 6 3/4 | 3 3/8 | 9/16 | 0.445 | 6 to 8 |
| 5/8 | 5/8 | 0.6250 | 7 | 3 1/2 | 5/8 | 0.470 | 6 to 8 |
| 21/32 | ..... | 0.6562 | 7 3/8 | 3 11/16 | 5/8 | 0.490 | 6 to 8 |
| 11/16 | 11/16 | 0.6875 | 7 3/4 | 3 7/8 | 11/16 | 0.515 | 6 to 8 |
| 23/32 | ..... | 0.7188 | 8 1/8 | 4 1/16 | 11/16 | 0.540 | 6 to 8 |
| 3/4 | 3/4 | 0.7500 | 8 3/8 | 4 3/16 | 3/4 | 0.560 | 6 to 8 |
| ..... | 13/16 | 0.8125 | 9 1/8 | 4 9/16 | 13/16 | 0.610 | 8 to 10 |
| 7/8 | 7/8 | 0.8750 | 9 3/4 | 4 7/8 | 7/8 | 0.655 | 8 to 10 |
| ..... | 15/16 | 0.9375 | 10 1/4 | 5 1/8 | 15/16 | 0.705 | 8 to 10 |
| 1 | 1 | 1.0000 | 10 7/8 | 5 7/16 | 1 | 0.750 | 8 to 10 |
| 1 1/8 | 1 1/8 | 1.1250 | 11 5/8 | 5 13/16 | 1 | 0.845 | 8 to 10 |
| 1 1/4 | 1 1/4 | 1.2500 | 12 1/4 | 6 1/8 | 1 | 0.935 | 8 to 12 |
| 1 3/8 | 1 3/8 | 1.3750 | 12 5/8 | 6 5/16 | 1 | 1.030 | 10 to 12 |
| 1 1/2 | 1 1/2 | 1.5000 | 13 | 6 1/2 | 1 1/8 | 1.125 | 10 to 14 |

All dimensions in inches. Material is high-speed steel. The nominal shank diameter *D* is the same as the reamer diameter. Helical-flute hand reamers with left-hand helical flutes are standard. Reamers are tapered slightly on the end to facilitate proper starting.

*Tolerances:* On diameter of reamer, up to 1/4-inch size, incl., +.0001 to .0004 inch; over 1/4- to 1-inch size, incl., +.0001 to .0005 inch; over 1-inch size, +.0002 to .0006 inch. On length overall *A* and flute length *B*, 1/8- to 1-inch size, incl., ±1/16 inch; 1 1/8- to 1 1/2-inch size, incl., ±3/32 inch. On length of square *C*, 1/8- to 1-inch size, incl., ±1/32 inch; 1 1/8- to 1 1/2-inch size, incl., ±1/16 inch. On shank diameter *D*, 1/8- to 1-inch size, incl., −.001 to −.005 inch; 1 1/8- to 1 1/2-inch size, incl., −.0015 to −.006 inch. On size of square, 1/8- to 1/2-inch size, incl., −.004 inch; 17/32- to 1-inch size, incl., −.006 inch; 1 1/8- to 1 1/2-inch size, incl., −.008 inch.

**American National Standard Expansion Hand Reamers — Straight and Helical Flutes, Squared Shank** (ANSI B94.2-1977)

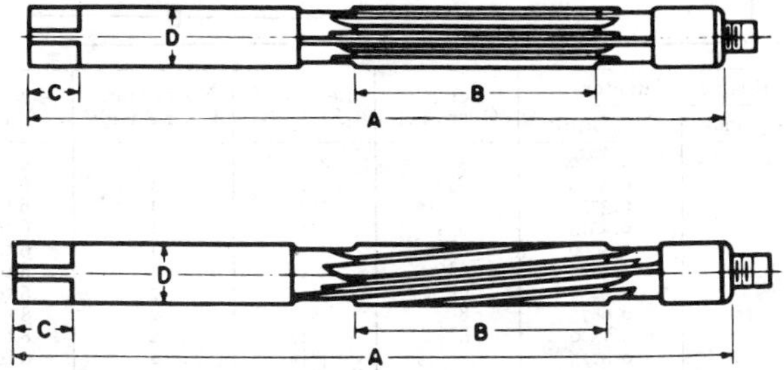


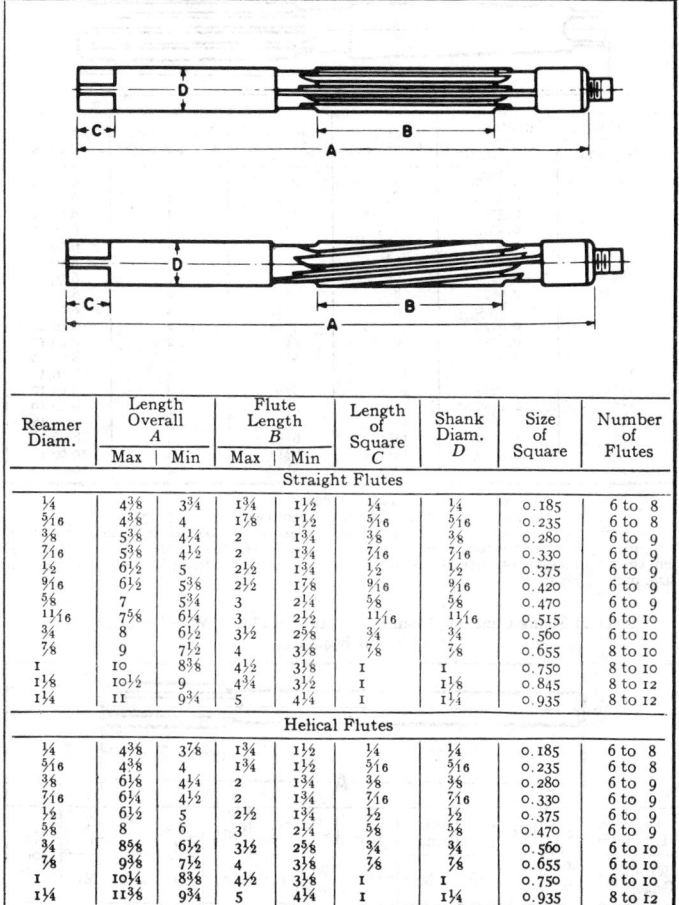

| Reamer Diam. | Length Overall A | | Flute Length B | | Length of Square C | Shank Diam. D | Size of Square | Number of Flutes |
|---|---|---|---|---|---|---|---|---|
| | Max | Min | Max | Min | | | | |
| **Straight Flutes** | | | | | | | | |
| 1/4 | 4 3/8 | 3 3/4 | 1 3/4 | 1 1/2 | 1/4 | 1/4 | 0.185 | 6 to 8 |
| 5/16 | 4 3/8 | 4 | 1 7/8 | 1 1/2 | 5/16 | 5/16 | 0.235 | 6 to 8 |
| 3/8 | 5 3/8 | 4 1/4 | 2 | 1 3/4 | 3/8 | 3/8 | 0.280 | 6 to 9 |
| 7/16 | 5 3/8 | 4 1/2 | 2 | 1 3/4 | 7/16 | 7/16 | 0.330 | 6 to 9 |
| 1/2 | 6 1/2 | 5 | 2 1/2 | 1 3/4 | 1/2 | 1/2 | 0.375 | 6 to 9 |
| 9/16 | 6 1/2 | 5 3/8 | 2 1/2 | 1 7/8 | 9/16 | 9/16 | 0.420 | 6 to 9 |
| 5/8 | 7 | 5 3/4 | 3 | 2 1/4 | 5/8 | 5/8 | 0.470 | 6 to 9 |
| 11/16 | 7 5/8 | 6 1/4 | 3 | 2 1/2 | 11/16 | 11/16 | 0.515 | 6 to 10 |
| 3/4 | 8 | 6 1/2 | 3 1/2 | 2 5/8 | 3/4 | 3/4 | 0.560 | 6 to 10 |
| 7/8 | 9 | 7 1/2 | 4 | 3 1/8 | 7/8 | 7/8 | 0.655 | 8 to 10 |
| 1 | 10 | 8 3/8 | 4 1/2 | 3 1/8 | 1 | 1 | 0.750 | 8 to 10 |
| 1 1/8 | 10 1/2 | 9 | 4 3/4 | 3 1/2 | 1 | 1 1/8 | 0.845 | 8 to 12 |
| 1 1/4 | 11 | 9 3/4 | 5 | 4 1/4 | 1 | 1 1/4 | 0.935 | 8 to 12 |
| **Helical Flutes** | | | | | | | | |
| 1/4 | 4 3/8 | 3 7/8 | 1 3/4 | 1 1/2 | 1/4 | 1/4 | 0.185 | 6 to 8 |
| 5/16 | 4 3/8 | 4 | 1 3/4 | 1 1/2 | 5/16 | 5/16 | 0.235 | 6 to 8 |
| 3/8 | 6 1/8 | 4 1/4 | 2 | 1 3/4 | 3/8 | 3/8 | 0.280 | 6 to 9 |
| 7/16 | 6 1/4 | 4 1/2 | 2 | 1 3/4 | 7/16 | 7/16 | 0.330 | 6 to 9 |
| 1/2 | 6 1/2 | 5 | 2 1/2 | 1 3/4 | 1/2 | 1/2 | 0.375 | 6 to 9 |
| 5/8 | 8 | 6 | 3 | 2 1/4 | 5/8 | 5/8 | 0.470 | 6 to 9 |
| 3/4 | 8 5/8 | 6 1/2 | 3 1/2 | 2 5/8 | 3/4 | 3/4 | 0.560 | 6 to 10 |
| 7/8 | 9 3/8 | 7 1/2 | 4 | 3 1/8 | 7/8 | 7/8 | 0.655 | 6 to 10 |
| 1 | 10 1/4 | 8 3/8 | 4 1/2 | 3 1/8 | 1 | 1 | 0.750 | 6 to 10 |
| 1 1/4 | 11 3/8 | 9 3/4 | 5 | 4 1/4 | 1 | 1 1/4 | 0.935 | 8 to 12 |

All dimensions are given in inches. Material is carbon steel. Reamers with helical flutes that are left hand are standard. Expansion hand reamers are primarily designed for work where it is necessary to enlarge reamed holes by a few thousandths. The pilots and guides on these reamers are ground undersize for clearance. The maximum expansion on these reamers is as follows: .006 inch for the 1/4- to 7/16-inch sizes, .010 inch for the 1/2- to 7/8-inch sizes and .012 inch for the 1- to 1 1/4-inch sizes.

*Tolerances:* On length overall $A$ and flute length $B$, ±1/16 inch for 1/4- to 1-inch sizes, ±3/32 inch for 1 1/8- to 1 1/4-inch sizes; on length of square $C$, ±1/32 inch for 1/4- to 1-inch sizes, ±1/16 inch for 1 1/8- to 1 1/4-inch sizes; on shank diameter $D$, −.001 to −.005 inch for 1/4- to 1-inch sizes, −.0015 to −.006 inch for 1 1/8- to 1 1/4-inch sizes; on size of square, −.004 inch for 1/4- to 1/2-inch sizes, −.006 inch for 9/16- to 1-inch sizes, and −.008 inch for 1 1/8- to 1 1/4-inch sizes.

### Taper Shank Jobbers Reamers — Straight Flutes (ANSI B94.2-1977)

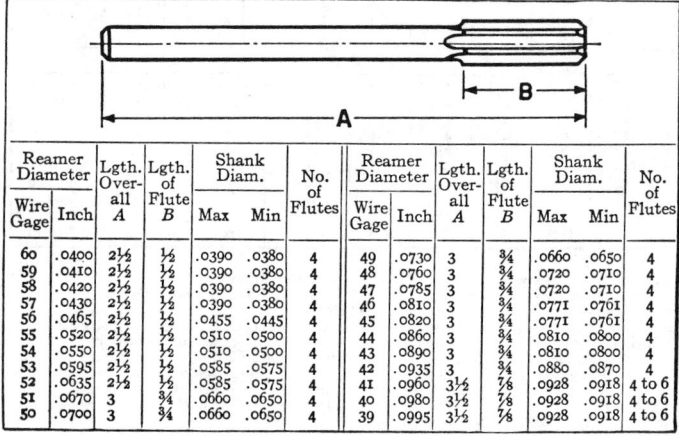

| Reamer Diameter | | Length Overall A | Length of Flute B | No. of Morse Taper Shank* | No. of Flutes |
|---|---|---|---|---|---|
| Fractional | Dec. Equiv. | | | | |
| $\frac{1}{4}$ | 0.2500 | $5\frac{3}{16}$ | 2 | 1 | 6 to 8 |
| $\frac{5}{16}$ | 0.3125 | $5\frac{1}{2}$ | $2\frac{1}{4}$ | 1 | 6 to 8 |
| $\frac{3}{8}$ | 0.3750 | $5\frac{13}{16}$ | $2\frac{1}{2}$ | 1 | 6 to 8 |
| $\frac{7}{16}$ | 0.4375 | $6\frac{1}{8}$ | $2\frac{3}{4}$ | 1 | 6 to 8 |
| $\frac{1}{2}$ | 0.5000 | $6\frac{7}{16}$ | 3 | 1 | 6 to 8 |
| $\frac{9}{16}$ | 0.5625 | $6\frac{3}{4}$ | $3\frac{1}{4}$ | 1 | 6 to 8 |
| $\frac{5}{8}$ | 0.6250 | $7\frac{9}{16}$ | $3\frac{1}{2}$ | 2 | 6 to 8 |
| $1\frac{1}{16}$ | 0.6875 | 8 | $3\frac{7}{8}$ | 2 | 8 to 10 |
| $\frac{3}{4}$ | 0.7500 | $8\frac{3}{8}$ | $4\frac{3}{16}$ | 2 | 8 to 10 |
| $\frac{13}{16}$ | 0.8125 | $8\frac{13}{16}$ | $4\frac{9}{16}$ | 2 | 8 to 10 |
| $\frac{7}{8}$ | 0.8750 | $9\frac{3}{16}$ | $4\frac{7}{8}$ | 2 | 8 to 10 |
| $\frac{15}{16}$ | 0.9375 | 10 | $5\frac{1}{8}$ | 3 | 8 to 10 |
| 1 | 1.0000 | $10\frac{3}{8}$ | $5\frac{7}{16}$ | 3 | 8 to 10 |
| $1\frac{1}{16}$ | 1.0625 | $10\frac{5}{8}$ | $5\frac{5}{8}$ | 3 | 8 to 10 |
| $1\frac{1}{8}$ | 1.1250 | $10\frac{7}{8}$ | $5\frac{13}{16}$ | 3 | 8 to 10 |
| $1\frac{3}{16}$ | 1.1875 | $11\frac{1}{8}$ | 6 | 3 | 8 to 12 |
| $1\frac{1}{4}$ | 1.2500 | $12\frac{9}{16}$ | $6\frac{1}{8}$ | 4 | 8 to 12 |
| $1\frac{3}{8}$ | 1.3750 | $12\frac{13}{16}$ | $6\frac{5}{16}$ | 4 | 10 to 12 |
| $1\frac{1}{2}$ | 1.5000 | $13\frac{1}{8}$ | $6\frac{1}{2}$ | 4 | 10 to 12 |

All dimensions in inches. Material is high-speed steel.
\* American National Standard self-holding tapers (see Table 5, p. 1733).
*Tolerances:* On reamer diameter, $\frac{1}{4}$-inch size, plus .0001 to .0004 inch; over $\frac{1}{4}$- to 1-inch size, incl., plus .0001 to .0005 inch; over 1-inch size, plus .0002 to .0006 inch. On overall length A and length of flute B, $\frac{1}{4}$- to 1-inch size, incl., $\pm\frac{1}{16}$ inch; and $1\frac{1}{16}$- to $1\frac{1}{2}$-inch size, incl., $\pm\frac{3}{32}$ inch.

### Straight Shank Chucking Reamers — Straight Flutes, Wire Gage Sizes — 1
(ANSI B94.2-1977)

| Reamer Diameter | | Lgth. Over-all A | Lgth. of Flute B | Shank Diam. | | No. of Flutes | Reamer Diameter | | Lgth. Over-all A | Lgth. of Flute B | Shank Diam. | | No. of Flutes |
|---|---|---|---|---|---|---|---|---|---|---|---|---|---|
| Wire Gage | Inch | | | Max | Min | | Wire Gage | Inch | | | Max | Min | |
| 60 | .0400 | $2\frac{1}{2}$ | $\frac{1}{2}$ | .0390 | .0380 | 4 | 49 | .0730 | 3 | $\frac{3}{4}$ | .0660 | .0650 | 4 |
| 59 | .0410 | $2\frac{1}{2}$ | $\frac{1}{2}$ | .0390 | .0380 | 4 | 48 | .0760 | 3 | $\frac{3}{4}$ | .0720 | .0710 | 4 |
| 58 | .0420 | $2\frac{1}{2}$ | $\frac{1}{2}$ | .0390 | .0380 | 4 | 47 | .0785 | 3 | $\frac{3}{4}$ | .0720 | .0710 | 4 |
| 57 | .0430 | $2\frac{1}{2}$ | $\frac{1}{2}$ | .0390 | .0380 | 4 | 46 | .0810 | 3 | $\frac{3}{4}$ | .0771 | .0761 | 4 |
| 56 | .0465 | $2\frac{1}{2}$ | $\frac{1}{2}$ | .0455 | .0445 | 4 | 45 | .0820 | 3 | $\frac{3}{4}$ | .0771 | .0761 | 4 |
| 55 | .0520 | $2\frac{1}{2}$ | $\frac{1}{2}$ | .0510 | .0500 | 4 | 44 | .0860 | 3 | $\frac{3}{4}$ | .0810 | .0800 | 4 |
| 54 | .0550 | $2\frac{1}{2}$ | $\frac{1}{2}$ | .0510 | .0500 | 4 | 43 | .0890 | 3 | $\frac{3}{4}$ | .0810 | .0800 | 4 |
| 53 | .0595 | $2\frac{1}{2}$ | $\frac{1}{2}$ | .0585 | .0575 | 4 | 42 | .0935 | 3 | $\frac{3}{4}$ | .0880 | .0870 | 4 |
| 52 | .0635 | $2\frac{1}{2}$ | $\frac{1}{2}$ | .0585 | .0575 | 4 | 41 | .0960 | $3\frac{1}{2}$ | $\frac{7}{8}$ | .0928 | .0918 | 4 to 6 |
| 51 | .0670 | 3 | $\frac{3}{4}$ | .0660 | .0650 | 4 | 40 | .0980 | $3\frac{1}{2}$ | $\frac{7}{8}$ | .0928 | .0918 | 4 to 6 |
| 50 | .0700 | 3 | $\frac{3}{4}$ | .0660 | .0650 | 4 | 39 | .0995 | $3\frac{1}{2}$ | $\frac{7}{8}$ | .0928 | .0918 | 4 to 6 |

## Straight Shank Chucking Reamers — Straight Flutes, Wire Gage Sizes — 2
### (ANSI B94.2-1977)

| Reamer Diameter | | Lgth. Over-all A | Lgth. of Flute B | Shank Diam. | | No. of Flutes | Reamer Diameter | | Lgth. Over-all A | Lgth. of Flute B | Shank Diam. | | No. of Flutes |
|---|---|---|---|---|---|---|---|---|---|---|---|---|---|
| Wire Gage | Inch | | | Max | Min | | Wire Gage | Inch | | | Max | Min | |
| 38 | .1015 | 3½ | ⅞ | .0950 | .0940 | 4 to 6 | 19 | .1660 | 4½ | 1⅛ | .1595 | .1585 | 4 to 6 |
| 37 | .1040 | 3½ | ⅞ | .0950 | .0940 | 4 to 6 | 18 | .1695 | 4½ | 1⅛ | .1595 | .1585 | 4 to 6 |
| 36 | .1065 | 3½ | ⅞ | .1030 | .1020 | 4 to 6 | 17 | .1730 | 4½ | 1⅛ | .1645 | .1635 | 4 to 6 |
| 35 | .1100 | 3½ | ⅞ | .1030 | .1020 | 4 to 6 | 16 | .1770 | 4½ | 1⅛ | .1704 | .1694 | 4 to 6 |
| 34 | .1110 | 3½ | ⅞ | .1055 | .1045 | 4 to 6 | 15 | .1800 | 4½ | 1⅛ | .1755 | .1745 | 4 to 6 |
| 33 | .1130 | 3½ | ⅞ | .1055 | .1045 | 4 to 6 | 14 | .1820 | 4½ | 1⅛ | .1755 | .1745 | 4 to 6 |
| 32 | .1160 | 3½ | ⅞ | .1120 | .1110 | 4 to 6 | 13 | .1850 | 4½ | 1⅛ | .1805 | .1795 | 4 to 6 |
| 31 | .1200 | 3½ | ⅞ | .1120 | .1110 | 4 to 6 | 12 | .1890 | 4½ | 1⅛ | .1805 | .1795 | 4 to 6 |
| 30 | .1285 | 3½ | ⅞ | .1190 | .1180 | 4 to 6 | 11 | .1910 | 5 | 1¼ | .1860 | .1850 | 4 to 6 |
| 29 | .1360 | 4 | 1 | .1275 | .1265 | 4 to 6 | 10 | .1935 | 5 | 1¼ | .1860 | .1850 | 4 to 6 |
| 28 | .1405 | 4 | 1 | .1350 | .1340 | 4 to 6 | 9 | .1960 | 5 | 1¼ | .1895 | .1885 | 4 to 6 |
| 27 | .1440 | 4 | 1 | .1350 | .1340 | 4 to 6 | 8 | .1990 | 5 | 1¼ | .1895 | .1885 | 4 to 6 |
| 26 | .1470 | 4 | 1 | .1430 | .1420 | 4 to 6 | 7 | .2010 | 5 | 1¼ | .1945 | .1935 | 4 to 6 |
| 25 | .1495 | 4 | 1 | .1430 | .1420 | 4 to 6 | 6 | .2040 | 5 | 1¼ | .1945 | .1935 | 4 to 6 |
| 24 | .1520 | 4 | 1 | .1460 | .1450 | 4 to 6 | 5 | .2055 | 5 | 1¼ | .2016 | .2006 | 4 to 6 |
| 23 | .1540 | 4 | 1 | .1460 | .1450 | 4 to 6 | 4 | .2090 | 5 | 1¼ | .2016 | .2006 | 4 to 6 |
| 22 | .1570 | 4 | 1 | .1510 | .1500 | 4 to 6 | 3 | .2130 | 5 | 1¼ | .2075 | .2065 | 4 to 6 |
| 21 | .1590 | 4½ | 1⅛ | .1530 | .1520 | 4 to 6 | 2 | .2210 | 6 | 1½ | .2173 | .2163 | 4 to 6 |
| 20 | .1610 | 4½ | 1⅛ | .1530 | .1520 | 4 to 6 | 1 | .2280 | 6 | 1½ | .2173 | .2163 | 4 to 6 |

All dimensions in inches. Material is high-speed steel.
*Tolerances:* On diameter of reamer, plus .0001 to plus .0004 inch. On overall length *A*, plus or minus ⅟₁₆ inch. On length of flute *B*, plus or minus ⅟₁₆ inch.

## Straight Shank Chucking Reamers — Straight Flutes, Letter Sizes (ANSI B94.2-1977)

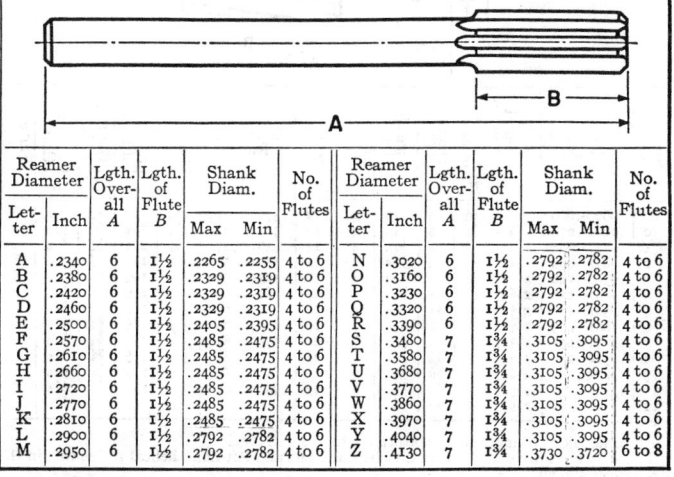

| Reamer Diameter | | Lgth. Over-all A | Lgth. of Flute B | Shank Diam. | | No. of Flutes | Reamer Diameter | | Lgth. Over-all A | Lgth. of Flute B | Shank Diam. | | No. of Flutes |
|---|---|---|---|---|---|---|---|---|---|---|---|---|---|
| Let-ter | Inch | | | Max | Min | | Let-ter | Inch | | | Max | Min | |
| A | .2340 | 6 | 1½ | .2265 | .2255 | 4 to 6 | N | .3020 | 6 | 1½ | .2792 | .2782 | 4 to 6 |
| B | .2380 | 6 | 1½ | .2329 | .2319 | 4 to 6 | O | .3160 | 6 | 1½ | .2792 | .2782 | 4 to 6 |
| C | .2420 | 6 | 1½ | .2329 | .2319 | 4 to 6 | P | .3230 | 6 | 1½ | .2792 | .2782 | 4 to 6 |
| D | .2460 | 6 | 1½ | .2329 | .2319 | 4 to 6 | Q | .3320 | 6 | 1½ | .2792 | .2782 | 4 to 6 |
| E | .2500 | 6 | 1½ | .2405 | .2395 | 4 to 6 | R | .3390 | 6 | 1½ | .2792 | .2782 | 4 to 6 |
| F | .2570 | 6 | 1½ | .2485 | .2475 | 4 to 6 | S | .3480 | 7 | 1¾ | .3105 | .3095 | 4 to 6 |
| G | .2610 | 6 | 1½ | .2485 | .2475 | 4 to 6 | T | .3580 | 7 | 1¾ | .3105 | .3095 | 4 to 6 |
| H | .2660 | 6 | 1½ | .2485 | .2475 | 4 to 6 | U | .3680 | 7 | 1¾ | .3105 | .3095 | 4 to 6 |
| I | .2720 | 6 | 1½ | .2485 | .2475 | 4 to 6 | V | .3770 | 7 | 1¾ | .3105 | .3095 | 4 to 6 |
| J | .2770 | 6 | 1½ | .2485 | .2475 | 4 to 6 | W | .3860 | 7 | 1¾ | .3105 | .3095 | 4 to 6 |
| K | .2810 | 6 | 1½ | .2485 | .2475 | 4 to 6 | X | .3970 | 7 | 1¾ | .3105 | .3095 | 4 to 6 |
| L | .2900 | 6 | 1½ | .2792 | .2782 | 4 to 6 | Y | .4040 | 7 | 1¾ | .3105 | .3095 | 4 to 6 |
| M | .2950 | 6 | 1½ | .2792 | .2782 | 4 to 6 | Z | .4130 | 7 | 1¾ | .3730 | .3720 | 6 to 8 |

All dimensions in inches. Material is high-speed steel.
*Tolerances:* On diameter of reamer, for sizes A to E, incl., plus .0001 to plus .0004 inch and for sizes F to Z, incl., plus .0001 to .0005 inch. On overall length *A*, plus or minus ⅟₁₆ inch. On length of flute *B*, plus or minus ⅟₁₆ inch.

**Straight Shank Chucking Reamers — Straight Flutes, Decimal Sizes (ANSI B94.2-1977)**

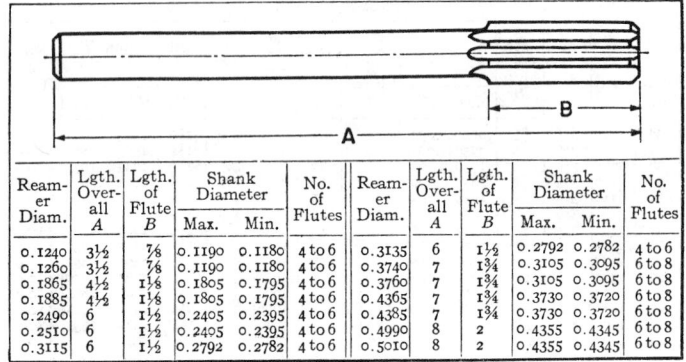

| Reamer Diam. | Lgth. Overall A | Lgth. of Flute B | Shank Diameter | | No. of Flutes | Reamer Diam. | Lgth. Overall A | Lgth. of Flute B | Shank Diameter | | No. of Flutes |
|---|---|---|---|---|---|---|---|---|---|---|---|
| | | | Max. | Min. | | | | | Max. | Min. | |
| 0.1240 | 3½ | ⅞ | 0.1190 | 0.1180 | 4 to 6 | 0.3135 | 6 | 1½ | 0.2792 | 0.2782 | 4 to 6 |
| 0.1260 | 3½ | ⅞ | 0.1190 | 0.1180 | 4 to 6 | 0.3740 | 7 | 1¾ | 0.3105 | 0.3095 | 6 to 8 |
| 0.1865 | 4½ | 1⅛ | 0.1805 | 0.1795 | 4 to 6 | 0.3760 | 7 | 1¾ | 0.3105 | 0.3095 | 6 to 8 |
| 0.1885 | 4½ | 1⅛ | 0.1805 | 0.1795 | 4 to 6 | 0.4365 | 7 | 1¾ | 0.3730 | 0.3720 | 6 to 8 |
| 0.2490 | 6 | 1½ | 0.2405 | 0.2395 | 4 to 6 | 0.4385 | 7 | 1¾ | 0.3730 | 0.3720 | 6 to 8 |
| 0.2510 | 6 | 1½ | 0.2405 | 0.2395 | 4 to 6 | 0.4990 | 8 | 2 | 0.4355 | 0.4345 | 6 to 8 |
| 0.3115 | 6 | 1½ | 0.2792 | 0.2782 | 4 to 6 | 0.5010 | 8 | 2 | 0.4355 | 0.4345 | 6 to 8 |

All dimensions in inches.    Material is high-speed steel.
*Tolerances:* On diameter of reamer, for 0.124 to 0.249-inch sizes, plus .0001 to plus .0004 inch and for 0.251 to 0.501-inch sizes, plus .0001 to plus .0005 inch.   On overall length $A$, plus or minus ¹⁄₁₆ inch.   On length of flute $B$, plus or minus ¹⁄₁₆ inch.

**Stub Screw Machine Reamers — Helical Flutes (ANSI B94.2-1977)**

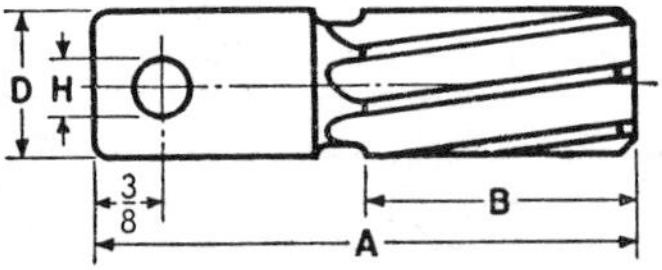

| Series No. | Diameter Range | Length Overall A | Length of Flute B | Diam. of Shank D | Size of Hole H | Flute No. | Series No. | Diameter Range | Length Overall A | Length of Flute B | Diam. of Shank D | Size of Hole H | Flute No. |
|---|---|---|---|---|---|---|---|---|---|---|---|---|---|
| 00 | .0600-.066 | 1¾ | ½ | ⅛ | ¹⁄₁₆ | 4 | 12 | .3761- .407 | 2½ | 1¼ | ½ | ³⁄₁₆ | 6 |
| 0 | .0661-.074 | 1¾ | ½ | ⅛ | ¹⁄₁₆ | 4 | 13 | .4071- .439 | 2½ | 1¼ | ½ | ³⁄₁₆ | 6 |
| 1 | .0741-.084 | 1¾ | ½ | ⅛ | ¹⁄₁₆ | 4 | 14 | .4391- .470 | 2½ | 1¼ | ½ | ³⁄₁₆ | 6 |
| 2 | .0841-.096 | 1¾ | ½ | ⅛ | ¹⁄₁₆ | 4 | 15 | .4701- .505 | 2½ | 1¼ | ½ | ³⁄₁₆ | 6 |
| 3 | .0961-.126 | 2 | ¾ | ⅛ | ¹⁄₁₆ | 4 | 16 | .5051- .567 | 3 | 1½ | ⅝ | ¼ | 6 |
| 4 | .1261-.158 | 2¼ | 1 | ¼ | ³⁄₃₂ | 4 | 17 | .5671- .630 | 3 | 1½ | ⅝ | ¼ | 6 |
| 5 | .1581-.188 | 2¼ | 1 | ¼ | ³⁄₃₂ | 4 | 18 | .6301- .692 | 3 | 1½ | ⅝ | ¼ | 6 |
| 6 | .1881-.219 | 2¼ | 1 | ¼ | ³⁄₃₂ | 6 | 19 | .6921- .755 | 3 | 1½ | ¾ | ⁵⁄₁₆ | 8 |
| 7 | .2191-.251 | 2¼ | 1 | ¼ | ³⁄₃₂ | 6 | 20 | .7551- .817 | 3 | 1½ | ¾ | ⁵⁄₁₆ | 8 |
| 8 | .2511-.282 | 2¼ | 1 | ⅜ | ⅛ | 6 | 21 | .8171- .880 | 3 | 1½ | ¾ | ⁵⁄₁₆ | 8 |
| 9 | .2821-.313 | 2¼ | 1 | ⅜ | ⅛ | 6 | 22 | .8801- .942 | 3 | 1½ | ¾ | ⁵⁄₁₆ | 8 |
| 10 | .3131-.344 | 2½ | 1¼ | ⅜ | ⅛ | 6 | 23 | .9421-1.010 | 3 | 1½ | ¾ | ⁵⁄₁₆ | 8 |
| 11 | .3441-.376 | 2½ | 1¼ | ⅜ | ⅛ | 6 | ... | ... | ... | ... | ... | ... | ... |

All dimensions in inches.    Material is high-speed steel.
These reamers are standard with right-hand cut and left-hand helical flutes within the size ranges shown.
*Tolerances:* On diameter of reamer, for sizes 00 to 7, incl., plus .0001 to plus .0004 inch and for sizes 8 to 23, incl., plus .0001 to plus .0005 inch.   On overall length $A$, plus or minus ¹⁄₁₆ inch.   On length of flute $B$, plus or minus ¹⁄₁₆ inch.   On diameter of shank $D$, minus .0005 to minus .002 inch.

American National Standard Straight Shank Rose Chucking and Chucking
Reamers — Straight and Helical Flutes: Fractional Sizes — 1 (ANSI B94.2-1977)

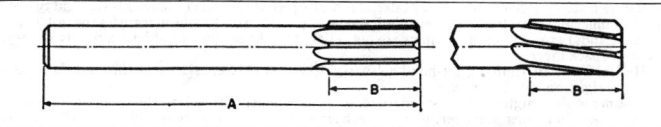

| Reamer Diameter | | Length Overall A | Flute Length B | Shank Diam. D | | No. of Flutes |
|---|---|---|---|---|---|---|
| Chucking | Rose Chucking | | | Max | Min | |
| 3⁄64‡ | ... | 2½ | ½ | 0.0455 | 0.0445 | 4 |
| 1⁄16 | ... | 2½ | ½ | 0.0585 | 0.0575 | 4 |
| 5⁄64 | ... | 3 | ¾ | 0.0720 | 0.0710 | 4 |
| 3⁄32 | ... | 3 | ¾ | 0.0880 | 0.0870 | 4 |
| 7⁄64 | ... | 3½ | ⅞ | 0.1030 | 0.1020 | 4 to 6 |
| ⅛ | ⅛‡ | 3½ | ⅞ | 0.1190 | 0.1180 | 4 to 6 |
| 9⁄64 | ... | 4 | 1 | 0.1350 | 0.1340 | 4 to 6 |
| 5⁄32 | ... | 4 | 1 | 0.1510 | 0.1500 | 4 to 6 |
| 11⁄64 | ... | 4½ | 1⅛ | 0.1645 | 0.1635 | 4 to 6 |
| 3⁄16 | 3⁄16‡ | 4½ | 1⅛ | 0.1805 | 0.1795 | 4 to 6 |
| 13⁄64 | ... | 5 | 1¼ | 0.1945 | 0.1935 | 4 to 6 |
| 7⁄32 | ... | 5 | 1¼ | 0.2075 | 0.2065 | 4 to 6 |
| 15⁄64 | ... | 6 | 1½ | 0.2265 | 0.2255 | 4 to 6 |
| ¼ | ¼‡ | 6 | 1½ | 0.2405 | 0.2395 | 4 to 6 |
| 17⁄64 | ... | 6 | 1½ | 0.2485 | 0.2475 | 4 to 6 |
| 9⁄32 | ... | 6 | 1½ | 0.2485 | 0.2475 | 4 to 6 |
| 19⁄64 | ... | 6 | 1½ | 0.2792 | 0.2782 | 4 to 6 |
| 5⁄16 | 5⁄16‡ | 6 | 1½ | 0.2792 | 0.2782 | 4 to 6 |
| 21⁄64 | ... | 6 | 1½ | 0.2792 | 0.2782 | 4 to 6 |
| 11⁄32 | ... | 6 | 1½ | 0.2792 | 0.2782 | 4 to 6 |
| 23⁄64 | ... | 7 | 1¾ | 0.3105 | 0.3095 | 4 to 6 |
| ⅜ | ⅜‡ | 7 | 1¾ | 0.3105 | 0.3095 | 4 to 6 |
| 25⁄64 | ... | 7 | 1¾ | 0.3105 | 0.3095 | 4 to 6 |
| 13⁄32 | ... | 7 | 1¾ | 0.3105 | 0.3095 | 4 to 6 |
| 27⁄64 | ... | 7 | 1¾ | 0.3730 | 0.3720 | 6 to 8 |
| 7⁄16 | 7⁄16‡ | 7 | 1¾ | 0.3730 | 0.3720 | 6 to 8 |
| 29⁄64 | ... | 7 | 1¾ | 0.3730 | 0.3720 | 6 to 8 |
| 15⁄32 | ... | 7 | 1¾ | 0.3730 | 0.3720 | 6 to 8 |
| 31⁄64 | ... | 8 | 2 | 0.4355 | 0.4345 | 6 to 8 |
| ½ | ½‡ | 8 | 2 | 0.4355 | 0.4345 | 6 to 8 |
| 17⁄32 | ... | 8 | 2 | 0.4355 | 0.4345 | 6 to 8 |
| 9⁄16 | ... | 8 | 2 | 0.4355 | 0.4345 | 6 to 8 |
| 19⁄32 | ... | 8 | 2 | 0.4355 | 0.4345 | 6 to 8 |
| ⅝ | ... | 9 | 2¼ | 0.5620 | 0.5605 | 6 to 8 |
| 21⁄32 | ... | 9 | 2¼ | 0.5620 | 0.5605 | 6 to 8 |
| 11⁄16 | ... | 9 | 2¼ | 0.5620 | 0.5605 | 6 to 8 |
| 23⁄32 | ... | 9 | 2¼ | 0.5620 | 0.5605 | 6 to 8 |
| ¾ | ... | 9½ | 2½ | 0.6245 | 0.6230 | 6 to 8 |
| 25⁄32 | ... | 9½ | 2½ | 0.6245 | 0.6230 | 8 to 10 |
| 13⁄16 | ... | 9½ | 2½ | 0.6245 | 0.6230 | 8 to 10 |
| 27⁄32 | ... | 9½ | 2½ | 0.6245 | 0.6230 | 8 to 10 |
| ⅞ | ... | 10 | 2⅝ | 0.7495 | 0.7480 | 8 to 10 |
| 29⁄32 | ... | 10 | 2⅝ | 0.7495 | 0.7480 | 8 to 10 |
| 15⁄16 | ... | 10 | 2⅝ | 0.7495 | 0.7480 | 8 to 10 |
| 31⁄32 | ... | 10 | 2⅝ | 0.7495 | 0.7480 | 8 to 10 |
| 1 | ... | 10½ | 2¾ | 0.8745 | 0.8730 | 8 to 12 |
| 1 1⁄16 | ... | 10½ | 2¾ | 0.8745 | 0.8730 | 8 to 12 |
| 1⅛ | ... | 11 | 2⅞ | 0.8745 | 0.8730 | 8 to 12 |
| 1 3⁄16 | ... | 11 | 2⅞ | 0.9995 | 0.9980 | 8 to 12 |
| 1¼ | ... | 11½ | 3 | 0.9995 | 0.9980 | 8 to 12 |
| 1 5⁄16† | ... | 11½ | 3 | 0.9995 | 0.9980 | 10 to 12 |
| 1⅜ | ... | 12 | 3¼ | 0.9995 | 0.9980 | 10 to 12 |
| 1 7⁄16† | ... | 12 | 3¼ | 1.2495 | 1.2480 | 10 to 12 |
| 1½ | ... | 12½ | 3½ | 1.2495 | 1.2480 | 10 to 12 |

For footnote see next page.

### American National Standard Straight Shank Rose Chucking and Chucking Reamers — Straight and Helical Flutes* Fractional Sizes — 2 (ANSI B94.2-1977)

All dimensions are given in inches. Material is high-speed steel. Chucking reamers are end cutting on the chamfer and the relief for the outside diameter is ground in back of the margin for the full length of land. Lands of rose chucking reamers are not relieved on the periphery but have a relatively large amount of back taper.

  * Helical flutes are right- or left-hand helix, right-hand cut, except sizes 1 1/16 through 1 1/2 inches, which are right-hand helix only.

  ‡ Reamer with straight flutes is standard only.      † Reamer with helical flutes is standard only.

*Tolerances:* On reamer diameter, up to 1/4-inch size, incl., +.0001 to .0004 inch; 1/4- to 1-inch size, incl., +.0001 to .0005 inch; over 1-inch size, +.0002 to .0006 inch. On length overall $A$ and flute length $B$, up to 1-inch size, incl., ±1/16 inch; 1 1/16- to 1 1/2-inch size, incl., ±3/32 inch.

### Shell Reamers — Straight and Helical Flutes (ANSI B94.2-1977)

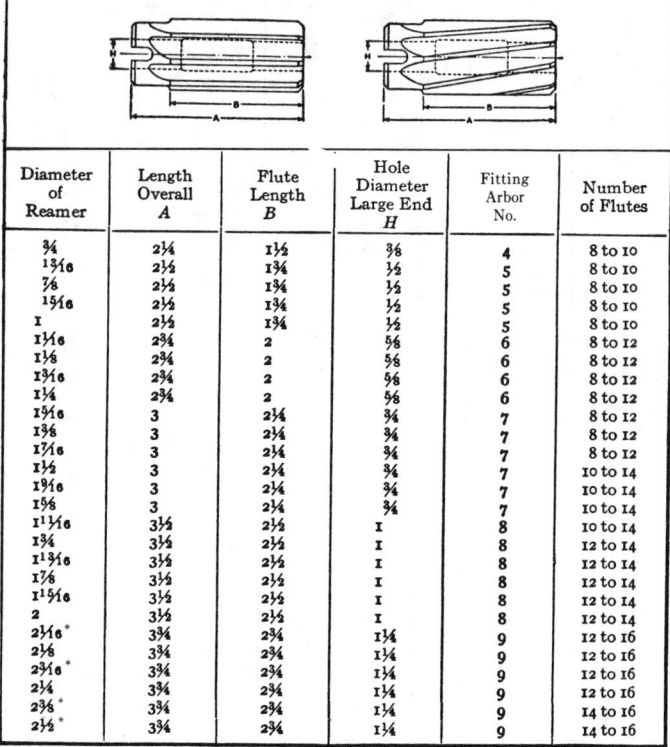

| Diameter of Reamer | Length Overall A | Flute Length B | Hole Diameter Large End H | Fitting Arbor No. | Number of Flutes |
|---|---|---|---|---|---|
| 3/4 | 2 1/4 | 1 1/2 | 3/8 | 4 | 8 to 10 |
| 13/16 | 2 1/2 | 1 3/4 | 1/2 | 5 | 8 to 10 |
| 7/8 | 2 1/2 | 1 3/4 | 1/2 | 5 | 8 to 10 |
| 15/16 | 2 1/2 | 1 3/4 | 1/2 | 5 | 8 to 10 |
| 1 | 2 1/2 | 1 3/4 | 1/2 | 5 | 8 to 10 |
| 1 1/16 | 2 3/4 | 2 | 5/8 | 6 | 8 to 12 |
| 1 1/8 | 2 3/4 | 2 | 5/8 | 6 | 8 to 12 |
| 1 3/16 | 2 3/4 | 2 | 5/8 | 6 | 8 to 12 |
| 1 1/4 | 2 3/4 | 2 | 5/8 | 6 | 8 to 12 |
| 1 5/16 | 3 | 2 1/4 | 3/4 | 7 | 8 to 12 |
| 1 3/8 | 3 | 2 1/4 | 3/4 | 7 | 8 to 12 |
| 1 7/16 | 3 | 2 1/4 | 3/4 | 7 | 8 to 12 |
| 1 1/2 | 3 | 2 1/4 | 3/4 | 7 | 10 to 14 |
| 1 9/16 | 3 | 2 1/4 | 3/4 | 7 | 10 to 14 |
| 1 5/8 | 3 | 2 1/4 | 3/4 | 7 | 10 to 14 |
| 1 11/16 | 3 1/2 | 2 1/2 | 1 | 8 | 10 to 14 |
| 1 3/4 | 3 1/2 | 2 1/2 | 1 | 8 | 12 to 14 |
| 1 13/16 | 3 1/2 | 2 1/2 | 1 | 8 | 12 to 14 |
| 1 7/8 | 3 1/2 | 2 1/2 | 1 | 8 | 12 to 14 |
| 1 15/16 | 3 1/2 | 2 1/2 | 1 | 8 | 12 to 14 |
| 2 | 3 1/2 | 2 1/2 | 1 | 8 | 12 to 14 |
| 2 1/16 * | 3 3/4 | 2 3/4 | 1 1/4 | 9 | 12 to 16 |
| 2 1/8 | 3 3/4 | 2 3/4 | 1 1/4 | 9 | 12 to 16 |
| 2 3/16 * | 3 3/4 | 2 3/4 | 1 1/4 | 9 | 12 to 16 |
| 2 1/4 | 3 3/4 | 2 3/4 | 1 1/4 | 9 | 12 to 16 |
| 2 3/8 * | 3 3/4 | 2 3/4 | 1 1/4 | 9 | 14 to 16 |
| 2 1/2 * | 3 3/4 | 2 3/4 | 1 1/4 | 9 | 14 to 16 |

All dimensions are given in inches. Material is high-speed steel. Helical flute shell reamers with left-hand helical flutes are standard. Shell reamers are designed as a sizing or finishing reamer and are held on an arbor provided with driving lugs. The holes in these reamers are ground with a taper of 1/8 inch per foot. * Helical flutes only.

*Tolerances:* On diameter of reamer, 3/4- to 1-inch size, incl., +.0001 to .0005 inch; over 1-inch size, +.0002 to .0006 inch. On length overall $A$ and flute length $B$, 3/4- to 1-inch size, incl., ±1/16 inch; 1 1/16- to 2-inch size, incl., ±3/32 inch; 2 1/16- to 2 1/2-inch size, incl., ±1/8 inch.

**American National Standard Arbors for Shell Reamers — Straight and Taper Shanks** (ANSI B94.2-1977)

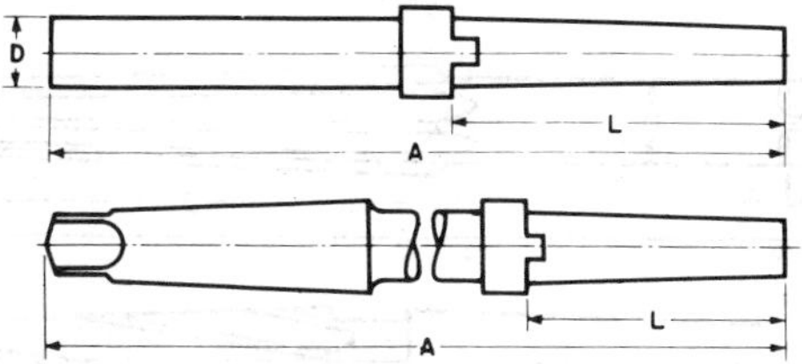


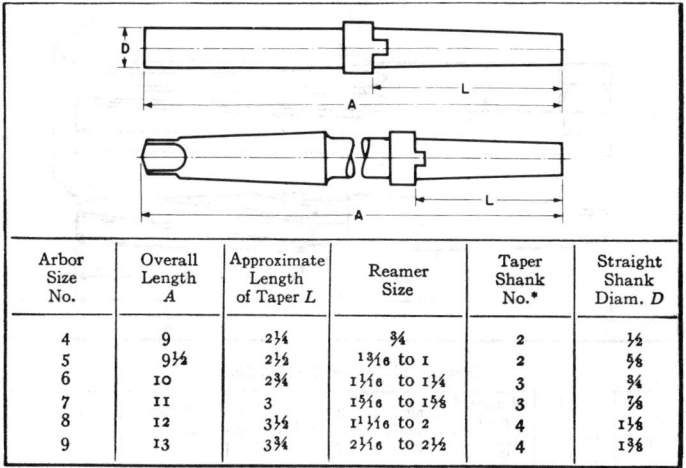

| Arbor Size No. | Overall Length *A* | Approximate Length of Taper *L* | Reamer Size | Taper Shank No.* | Straight Shank Diam. *D* |
|---|---|---|---|---|---|
| 4 | 9 | 2 1/4 | 3/4 | 2 | 1/2 |
| 5 | 9 1/2 | 2 1/2 | 13/16 to 1 | 2 | 5/8 |
| 6 | 10 | 2 3/4 | 1 1/16 to 1 1/4 | 3 | 3/4 |
| 7 | 11 | 3 | 1 5/16 to 1 5/8 | 3 | 7/8 |
| 8 | 12 | 3 1/2 | 1 11/16 to 2 | 4 | 1 1/8 |
| 9 | 13 | 3 3/4 | 2 1/16 to 2 1/2 | 4 | 1 3/8 |

All dimensions are given in inches. These arbors are designed to fit standard shell reamers (see table). End which fits reamer has taper of 1/8 inch per foot.

* American National Standard self-holding tapers (see Table 5, p. 1733).

*Tolerances:* On overall length *A*, in a range of arbor size numbers 4 to 5, incl., ±1/16 inch; arbor size nos. 6 to 8, incl., ±3/32 inch; arbor size no. 9, ±1/8 inch. On diameter of shank *D*, −.0005 to −.002 inch.

**American National Standard Driving Slots and Lugs for Shell Reamers or Shell Reamer Arbors** (ANSI B94.2-1977)

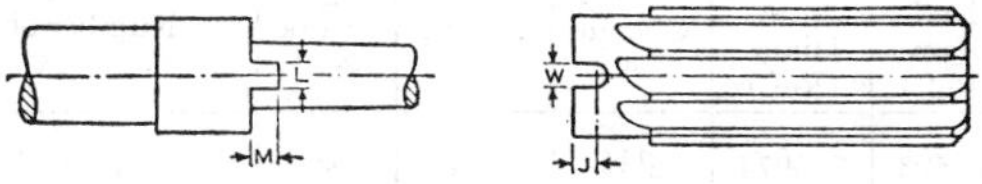


| Arbor Size No. | Fitting Reamer Sizes | Driving Slot | | Lug on Arbor | | Reamer Hole Diam. at Large End |
|---|---|---|---|---|---|---|
| | | Width *W* | Depth *J* | Width *L* | Depth *M* | |
| 4 | 3/4 | 5/32 | 3/16 | 9/64 | 5/32 | 0.375 |
| 5 | 13/16 to 1 | 3/16 | 1/4 | 11/64 | 7/32 | 0.500 |
| 6 | 1 1/16 to 1 1/4 | 3/16 | 1/4 | 11/64 | 7/32 | 0.625 |
| 7 | 1 5/16 to 1 5/8 | 1/4 | 5/16 | 15/64 | 9/32 | 0.750 |
| 8 | 1 11/16 to 2 | 1/4 | 5/16 | 15/64 | 9/32 | 1.000 |
| 9 | 2 1/16 to 2 1/2 | 5/16 | 3/8 | 19/64 | 11/32 | 1.250 |

All dimensions are given in inches. The hole in shell reamers has a taper of 1/8 inch per foot, with arbors tapered to correspond. Shell reamer arbor tapers are made to permit a driving fit with the reamer.

**American National Standard Morse Taper Finishing Reamers** (ANSI B94.2-1977)

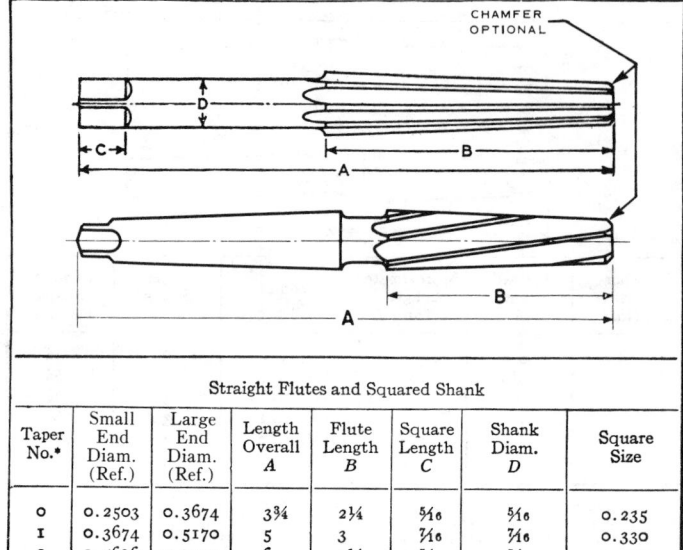

| Taper No.* | Small End Diam. (Ref.) | Large End Diam. (Ref.) | Length Overall A | Flute Length B | Square Length C | Shank Diam. D | Square Size |
|---|---|---|---|---|---|---|---|
| | | | **Straight Flutes and Squared Shank** | | | | |
| 0 | 0.2503 | 0.3674 | 3¾ | 2¼ | ⁵⁄₁₆ | ⁵⁄₁₆ | 0.235 |
| 1 | 0.3674 | 0.5170 | 5 | 3 | ⁷⁄₁₆ | ⁷⁄₁₆ | 0.330 |
| 2 | 0.5696 | 0.7444 | 6 | 3½ | ⅝ | ⅝ | 0.470 |
| 3 | 0.7748 | 0.9881 | 7¼ | 4¼ | ⅞ | ⅞ | 0.655 |
| 4 | 1.0167 | 1.2893 | 8½ | 5¼ | 1 | 1⅛ | 0.845 |
| 5 | 1.4717 | 1.8005 | 9¾ | 6¼ | 1⅛ | 1½ | 1.125 |

| Taper No.* | Small End Diam. (Ref.) | Large End Diam. (Ref.) | Length Overall A | Flute Length B | Taper Shank No.* | Number of Flutes |
|---|---|---|---|---|---|---|
| | | **Straight and Spiral Flutes and Taper Shank** | | | **Squared and Taper Shank** | |
| 0 | 0.2503 | 0.3674 | 5¹¹⁄₃₂ | 2¼ | 0 | 4 to 6 incl |
| 1 | 0.3674 | 0.5170 | 6⁵⁄₁₆ | 3 | 1 | 6 to 8 incl. |
| 2 | 0.5696 | 0.7444 | 7⅜ | 3½ | 2 | 6 to 8 incl. |
| 3 | 0.7748 | 0.9881 | 8⅞ | 4¼ | 3 | 8 to 10 incl. |
| 4 | 1.0167 | 1.2893 | 10⅞ | 5¼ | 4 | 8 to 10 incl. |
| 5 | 1.4717 | 1.8005 | 13⅛ | 6¼ | 5 | 10 to 12 incl. |

All dimensions are given in inches. Material is high-speed steel. The chamfer on the cutting end of the reamer is optional. Squared shank reamers are standard with straight flutes. Tapered shank reamers are standard with straight or spiral flutes. Spiral flute reamers are standard with left-hand spiral flutes. * Morse. For amount of taper see Table 1A, page 1728.

*Tolerances:* On overall length $A$ and flute length $B$, in taper numbers 0 to 3, incl., ±¹⁄₁₆ inch, in taper numbers 4 and 5, ±³⁄₃₂ inch. On length of square $C$, in taper numbers 0 to 3, incl., ±¹⁄₃₂ inch; in taper numbers 4 and 5, ±¹⁄₁₆ inch. On shank diameter $D$, −.0005 to −.002 inch. On size of square, in taper numbers 0 and 1, −.004 inch; in taper numbers 2 and 3, −.006 inch; in taper numbers 4 and 5, −.008 inch.

**Taper Pipe Reamers — Spiral Flutes** (ANSI B94.2-1977)

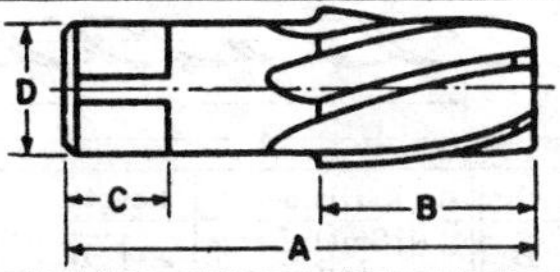


| Nom. Size | Diameter | | Length Overall *A* | Flute Length *B* | Square Length *C* | Shank Diameter *D* | Size of Square | No. of Flutes |
|---|---|---|---|---|---|---|---|---|
| | Large End | Small End | | | | | | |
| 1/8 | 0.362 | 0.316 | 2 1/8 | 3/4 | 3/8 | 0.4375 | 0.328 | 4 to 6 |
| 1/4 | 0.472 | 0.406 | 2 7/16 | 1 1/16 | 7/16 | 0.5625 | 0.421 | 4 to 6 |
| 3/8 | 0.606 | 0.540 | 2 9/16 | 1 1/16 | 1/2 | 0.7000 | 0.531 | 4 to 6 |
| 1/2 | 0.751 | 0.665 | 3 1/8 | 1 3/8 | 5/8 | 0.6875 | 0.515 | 4 to 6 |
| 3/4 | 0.962 | 0.876 | 3 1/4 | 1 3/8 | 11/16 | 0.9063 | 0.679 | 6 to 10 |
| 1 | 1.212 | 1.103 | 3 3/4 | 1 3/4 | 13/16 | 1.1250 | 0.843 | 6 to 10 |
| 1 1/4 | 1.553 | 1.444 | 4 | 1 3/4 | 15/16 | 1.3125 | 0.984 | 6 to 10 |
| 1 1/2 | 1.793 | 1.684 | 4 1/4 | 1 3/4 | 1 | 1.5000 | 1.125 | 6 to 10 |
| 2 | 2.268 | 2.159 | 4 1/2 | 1 3/4 | 1 1/8 | 1.8750 | 1.406 | 8 to 12 |

All dimensions are given in inches. These reamers are tapered 3/4 inch per foot and are intended for reaming holes to be tapped with American National Standard Taper Pipe Thread taps. Material is high-speed steel. Reamers are standard with left-hand spiral flutes.

*Tolerances:* On length overall *A* and flute length *B*, 1/8- to 3/4-inch size, incl., ±1/16 inch; 1- to 1 1/2-inch size, incl., ±3/32 inch; 2-inch size, ±1/8 inch. On length of square *C*, 1/8- to 3/4-inch size, incl., ±1/32 inch; 1- to 2-inch size, incl., ±1/16 inch. On shank diameter *D*, 1/8-inch size, −.0015 inch; 1/4- to 1-inch size, incl., −.002 inch; 1 1/4- to 2-inch size, incl., −.003 inch. On size of square, 1/8-inch size, −.004 inch; 1/4- to 3/4-inch size, incl., −.006 inch; 1- to 2-inch size, incl., −.008 inch.

**B & S Taper Reamers — Straight and Spiral Flutes, Squared Shank**

| Taper No.* | Diam., Small End | Diam., Large End | Overall Length | Square Length | Flute Length | Diam. of Shank | Size of Square | No. of Flutes |
|---|---|---|---|---|---|---|---|---|
| 1 | 0.1974 | 0.3176 | 4 3/4 | 1/4 | 2 7/8 | 9/32 | 0.210 | 4 to 6 |
| 2 | 0.2474 | 0.3781 | 5 1/8 | 5/16 | 3 1/8 | 11/32 | 0.255 | 4 to 6 |
| 3 | 0.3099 | 0.4510 | 5 1/2 | 3/8 | 3 3/8 | 13/32 | 0.305 | 4 to 6 |
| 4 | 0.3474 | 0.5017 | 5 7/8 | 7/16 | 3 11/16 | 7/16 | 0.330 | 4 to 6 |
| 5 | 0.4474 | 0.6145 | 6 3/8 | 1/2 | 4 | 9/16 | 0.420 | 4 to 6 |
| 6 | 0.4974 | 0.6808 | 6 7/8 | 5/8 | 4 3/8 | 5/8 | 0.470 | 4 to 6 |
| 7 | 0.5974 | 0.8011 | 7 1/2 | 3/4 | 4 7/8 | 3/4 | 0.560 | 6 to 8 |
| 8 | 0.7474 | 0.9770 | 8 1/8 | 13/16 | 5 1/2 | 13/16 | 0.610 | 6 to 8 |
| 9 | 0.8974 | 1.1530 | 8 7/8 | 7/8 | 6 1/8 | 1 | 0.750 | 6 to 8 |
| 10 | 1.0420 | 1.3376 | 9 3/4 | 1 | 6 7/8 | 1 1/8 | 0.845 | 6 to 8 |

These reamers are no longer ANSI Standard.

All dimensions are given in inches. Material is high-speed steel. The chamfer on the cutting end of the reamer is optional. All reamers are finishing reamers. Spiral flute reamers are standard with left-hand spiral flutes. B & S taper reamers are designed for use in reaming out Brown & Sharpe standard taper sockets.

* For taper per foot, see page 1731.

*Tolerances:* On length overall *A* and flute length *B*, taper nos. 1 to 7, incl., ±1/16 inch; taper nos. 8 to 10, incl., ±3/32 inch. On length of square *C*, taper nos. 1 to 9, incl., ±1/32 inch; taper no. 10, ±1/16 inch. On shank diameter *D*, −.0005 to −.002 inch. On size of square, taper nos. 1 to 3, incl., −.004 inch; taper nos. 4 to 9, incl., −.006 inch; taper no. 10, −.008 inch.

**American National Standard Die-Maker's Reamers** (ANSI B94.2-1977)

| Letter Size | Diameter Small End | Diameter Large End | Length A | Length B | Letter Size | Diameter Small End | Diameter Large End | Length A | Length B | Letter Size | Diameter Small End | Diameter Large End | Length A | Length B |
|---|---|---|---|---|---|---|---|---|---|---|---|---|---|---|
| AAA | 0.055 | 0.070 | 2 1/4 | 1 1/8 | G | 0.135 | 0.158 | 3 | 1 3/4 | O | 0.250 | 0.296 | 5 | 3 1/2 |
| AA | 0.065 | 0.080 | 2 1/4 | 1 1/8 | H | 0.145 | 0.169 | 3 1/4 | 1 7/8 | P | 0.275 | 0.327 | 5 1/2 | 4 |
| A | 0.075 | 0.090 | 2 1/4 | 1 1/8 | I | 0.160 | 0.184 | 3 1/4 | 1 7/8 | Q | 0.300 | 0.358 | 6 | 4 1/2 |
| B | 0.085 | 0.103 | 2 3/8 | 1 3/8 | J | 0.175 | 0.199 | 3 1/4 | 1 7/8 | R | 0.335 | 0.397 | 6 1/2 | 4 3/4 |
| C | 0.095 | 0.113 | 2 1/2 | 1 3/8 | K | 0.190 | 0.219 | 3 1/2 | 2 1/4 | S | 0.370 | 0.435 | 6 3/4 | 5 |
| D | 0.105 | 0.126 | 2 5/8 | 1 5/8 | L | 0.205 | 0.234 | 3 1/2 | 2 1/4 | T | 0.405 | 0.473 | 7 | 5 1/4 |
| E | 0.115 | 0.136 | 2 3/4 | 1 5/8 | M | 0.220 | 0.252 | 4 | 2 1/2 | U | 0.440 | 0.511 | 7 1/4 | 5 1/2 |
| F | 0.125 | 0.148 | 3 | 1 3/4 | N | 0.235 | 0.274 | 4 1/2 | 3 | ... | ..... | ..... | ... | ... |

All dimensions in inches. Material is high-speed steel. These reamers are designed for use in diemaking, have a taper of 3/4 degree included angle or 0.013 inch per inch, and have 2 or 3 flutes. Reamers are standard with left-hand spiral flutes. * May have conical end.

*Tolerances:* On length overall *A* and flute length *B*, ±1/16 inch.

**Taper Pin Reamers — Straight and Left-Hand Spiral Flutes, Squared Shank; and Left-Hand High-Spiral Flutes, Round Shank** (ANSI B94.2-1977)

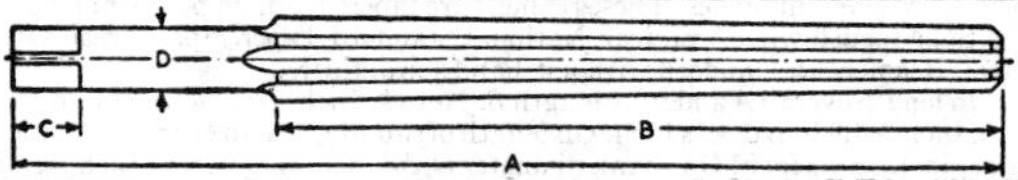


| No. of Taper Pin Reamer | Diameter at Large End of Reamer (Ref.) | Diameter at Small End of Reamer (Ref.) | Overall Length of Reamer *A* | Length of Flute *B* | Length of Square *C*† | Diameter of Shank *D* | Size of Square† |
|---|---|---|---|---|---|---|---|
| 8/0* | 0.0514 | 0.0351 | 1 5/8 | 25/32 | ... | 1/16 | ..... |
| 7/0 | 0.0666 | 0.0497 | 1 13/16 | 13/16 | 5/32 | 5/64 | 0.060 |
| 6/0 | 0.0806 | 0.0611 | 1 15/16 | 15/16 | 5/32 | 3/32 | 0.070 |
| 5/0 | 0.0966 | 0.0719 | 2 3/16 | 1 3/16 | 5/32 | 7/64 | 0.080 |
| 4/0 | 0.1142 | 0.0869 | 2 5/16 | 1 5/16 | 5/32 | 1/8 | 0.095 |
| 3/0 | 0.1302 | 0.1029 | 2 5/16 | 1 5/16 | 5/32 | 9/64 | 0.105 |
| 2/0 | 0.1462 | 0.1137 | 2 9/16 | 1 9/16 | 7/32 | 5/32 | 0.115 |
| 0 | 0.1638 | 0.1287 | 2 15/16 | 1 11/16 | 7/32 | 11/64 | 0.130 |
| 1 | 0.1798 | 0.1447 | 2 15/16 | 1 11/16 | 7/32 | 3/16 | 0.140 |
| 2 | 0.2008 | 0.1605 | 3 3/16 | 1 15/16 | 1/4 | 13/64 | 0.150 |
| 3 | 0.2294 | 0.1813 | 3 11/16 | 2 5/16 | 1/4 | 15/64 | 0.175 |
| 4 | 0.2604 | 0.2071 | 4 1/16 | 2 9/16 | 1/4 | 17/64 | 0.200 |
| 5 | 0.2994 | 0.2409 | 4 5/16 | 2 13/16 | 5/16 | 5/16 | 0.235 |
| 6 | 0.3540 | 0.2773 | 5 7/16 | 3 11/16 | 3/8 | 23/64 | 0.270 |
| 7 | 0.4220 | 0.3297 | 6 5/16 | 4 7/16 | 3/8 | 13/32 | 0.305 |
| 8 | 0.5050 | 0.3971 | 7 3/16 | 5 3/16 | 7/16 | 7/16 | 0.330 |
| 9 | 0.6066 | 0.4805 | 8 5/16 | 6 1/16 | 9/16 | 9/16 | 0.420 |
| 10 | 0.7216 | 0.5799 | 9 5/16 | 6 13/16 | 5/8 | 5/8 | 0.470 |

All dimensions in inches. Reamers have a taper ot 1/4 inch per foot, are made of high-speed steel. Straight flute reamers of carbon steel are also standard. The number of flutes is as follows: 3 or 4, for 7/0 to 4/0 sizes; 4 to 6, for 3/0 to 0 sizes; 5 or 6, for 1 to 5 sizes; 6 to 8, for 6 to 9 sizes; 7 or 8, for the 10 size in the case of straight- and spiral-flute reamers; and 2 or 3, for 8/0 to 8 sizes; 2 to 4, for the 9 and 10 sizes in the case of high-spiral flute reamers.

* Not applicable to straight and left-hand spiral fluted, squared shank reamers. † Not applicable to high-spiral flute reamers.

*Tolerances:* On length overall *A* and flute length *B*, ±1/16 inch. On length of square *C*, ±1/32 inch. On shank diameter *D*, −.001 to −.005 inch for straight- and spiral-flute reamers and −.0005 to −.002 inch for high-spiral flute reamers. On size of square, −.004 inch for 7/0 to 7 sizes and −.006 inch for 8 to 10 sizes.

### Chart to Facilitate Selection of Number and Sizes of Drills for Step-Drilling Prior to Taper Reaming

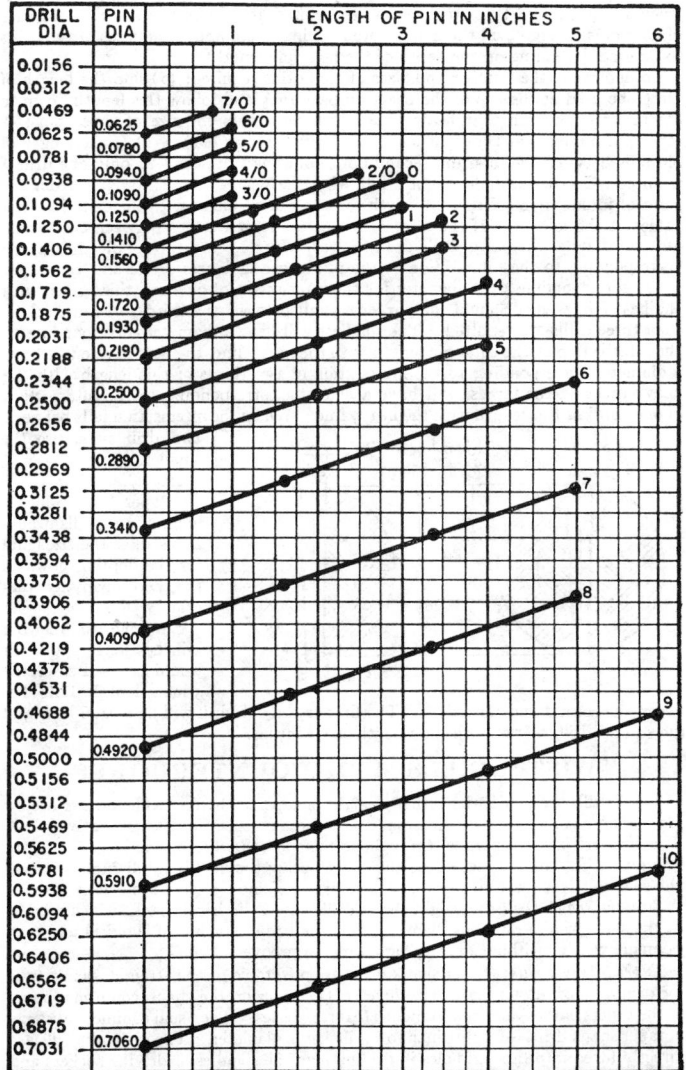

**Drilling Specifications for Taper Pins.**— When helically fluted taper pin reamers are used, the diameter of the through hole drilled prior to reaming is equal to the diameter at the small end of the taper pin. (See table, page 1119.) However, when straight fluted taper reamers are to be used, it may be necessary, in the case of long pins, to step drill the hole before reaming, the number and sizes of the drills to be used depending on the depth of the hole (pin length).

To determine the number and sizes of step drills required: (1) find the length of pin to be used at the top of the chart on page 1655 and follow this length down to the intersection with that heavy line which represents the size of taper pin (see taper pin numbers at the right-hand end of each heavy line). (2) If the length of pin falls between the first and second dots, counting from the left, only one drill is required. Its size is indicated by following the nearest horizontal line from the point of intersection (of the pin length) on the heavy line over to the drill diameter values at the left. (3) If the intersection of pin length comes between the second and third dots, then two drills are required. In this case the smaller has a size corresponding to the intersection of the pin length and heavy line and the larger is the corresponding drill diameter for the intersection of one-half this length with the heavy line. (4) Should the pin length fall between the third and fourth dots, then three drills are required. The smallest will have a diameter corresponding to the intersection of the total pin length with the heavy line, the next in size will have a diameter corresponding to the intersection of two-thirds of this length with the heavy line and the largest will have a diameter corresponding to the intersection of one-third of this length with the heavy line. Where the intersection falls between

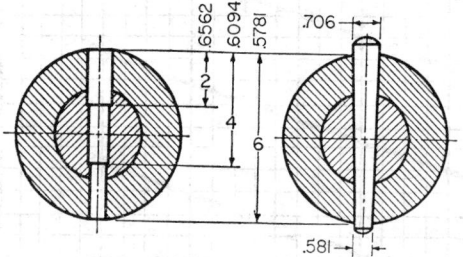

two drill sizes, use the smaller.

*Examples:* For a No. 10 taper pin 6-inches long, three drills would be used, of the sizes and for the depths shown in the accompanying diagram.

For a No. 10 taper pin 3-inches long, two drills would be used since the 3-inch length falls between the second and third dots. The first or through drill will be 0.6406 inch and the second drill, 0.6719 inch for a depth of 1½ inches.

# TWIST DRILLS AND COUNTERBORES

Twist drills are made with either straight or tapered shanks. The former are by far the more popular type in the smaller sizes because the price is comparatively low. The smaller sizes are parallel throughout their length whereas the larger sizes are ground with a "back taper" thus slightly reducing the drill body diameter nearest the shank. This feature is introduced to prevent binding when the drill is worn.

*Straight Shank Drills:* Straight shank drills have cylindrical shanks which may be of the same or of a different diameter than the body diameter of the drill and may be made with or without driving flats, tang, or grooves.

*Taper Shank Drills:* Taper shank drills are preferable to the straight shank type for drilling medium and large size holes. The taper on the shank conforms to one of the tapers in the American Standard (Morse) Series. The usual commercial range of taper shank drills is from ⅛ inch to 3½ inches in diameter. Taper shank drills are directly fitted into tapered holes in drilling machine spindles or driving sockets and generally have a driving tang.

## ANSI Standard Twist Drill Nomenclature

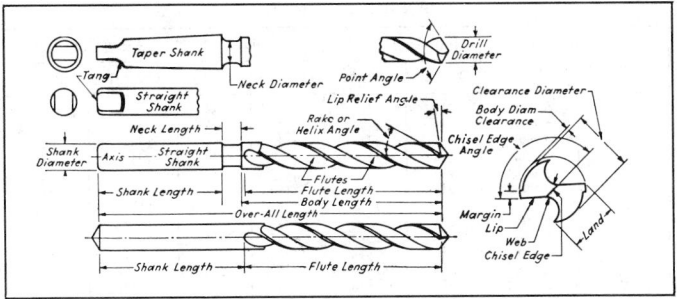

## ANSI Standard Straight Shank Twist Drills — From 0.0059 to 0.0785 Inch, Incl.
### (ANSI B94.11-1967, R1972)

| Drill Size | Drill Diam. | Jobbers Length Overall | Flute | Drill Size | Drill Diam. | Jobbers Length Overall | Flute |
|---|---|---|---|---|---|---|---|
| 97 | 0.0059 | $\tfrac{3}{4}$ | | 78 | 0.016 | $\tfrac{7}{8}$ | $\tfrac{3}{16}$ |
| 96 | 0.0063 | $\tfrac{3}{4}$ | | 77 | 0.018 | $\tfrac{7}{8}$ | $\tfrac{3}{16}$ |
| 95 | 0.0067 | $\tfrac{3}{4}$ | | 76 | 0.020 | $\tfrac{7}{8}$ | $\tfrac{3}{16}$ |
| 94 | 0.0071 | $\tfrac{3}{4}$ | | 75 | 0.021 | 1 | $\tfrac{1}{4}$ |
| 93 | 0.0075 | $\tfrac{3}{4}$ | | 74 | 0.0225 | 1 | $\tfrac{1}{4}$ |
| 92 | 0.0079 | $\tfrac{3}{4}$ | (Approximately 8 times diameter) | 73 | 0.024 | $1\tfrac{1}{8}$ | $\tfrac{5}{16}$ |
| 91 | 0.0083 | $\tfrac{3}{4}$ | | 72 | 0.025 | $1\tfrac{1}{8}$ | $\tfrac{5}{16}$ |
| 90 | 0.0087 | $\tfrac{3}{4}$ | | 71 | 0.026 | $1\tfrac{1}{4}$ | $\tfrac{3}{8}$ |
| 89 | 0.0091 | $\tfrac{3}{4}$ | | 70 | 0.028 | $1\tfrac{1}{4}$ | $\tfrac{3}{8}$ |
| 88 | 0.0095 | $\tfrac{3}{4}$ | | 69 | 0.0292 | $1\tfrac{3}{8}$ | $\tfrac{1}{2}$ |
| 87 | 0.010 | $\tfrac{3}{4}$ | | 68 | 0.031 | $1\tfrac{3}{8}$ | $\tfrac{1}{2}$ |
| 86 | 0.0105 | $\tfrac{3}{4}$ | | $\tfrac{1}{32}$ | 0.0312 | $1\tfrac{3}{8}$ | $\tfrac{1}{2}$ |
| 85 | 0.011 | $\tfrac{3}{4}$ | | 67 | 0.032 | $1\tfrac{3}{8}$ | $\tfrac{1}{2}$ |
| 84 | 0.0115 | $\tfrac{3}{4}$ | | 66 | 0.033 | $1\tfrac{3}{8}$ | $\tfrac{1}{2}$ |
| 83 | 0.012 | $\tfrac{3}{4}$ | | 65 | 0.035 | $1\tfrac{1}{2}$ | $\tfrac{5}{8}$ |
| 82 | 0.0125 | $\tfrac{3}{4}$ | | 64 | 0.036 | $1\tfrac{1}{2}$ | $\tfrac{5}{8}$ |
| 81 | 0.013 | $\tfrac{3}{4}$ | | 63 | 0.037 | $1\tfrac{1}{2}$ | $\tfrac{5}{8}$ |
| 80 | 0.0135 | $\tfrac{3}{4}$ | $\tfrac{1}{8}$ | 62 | 0.038 | $1\tfrac{1}{2}$ | $\tfrac{5}{8}$ |
| 79 | 0.0145 | $\tfrac{3}{4}$ | $\tfrac{1}{8}$ | 61 | 0.039 | $1\tfrac{5}{8}$ | $\tfrac{11}{16}$ |
| $\tfrac{1}{64}$ | 0.0156 | $\tfrac{3}{4}$ | $\tfrac{3}{16}$ | … | … | … | … |

| Drill Size | Drill Diam. | Jobbers Length Overall | Flute | Taper Length Overall | Flute | Screw Mach. Lgth. Overall | Flute |
|---|---|---|---|---|---|---|---|
| 60 | 0.0400 | $1\tfrac{5}{8}$ | $1\tfrac{1}{16}$ | $2\tfrac{1}{4}$ | $1\tfrac{1}{8}$ | $1\tfrac{3}{8}$ | $\tfrac{1}{2}$ |
| 59 | 0.041 | $1\tfrac{5}{8}$ | $1\tfrac{1}{16}$ | $2\tfrac{1}{4}$ | $1\tfrac{1}{8}$ | $1\tfrac{3}{8}$ | $\tfrac{1}{2}$ |
| 58 | 0.042 | $1\tfrac{5}{8}$ | $1\tfrac{1}{16}$ | $2\tfrac{1}{4}$ | $1\tfrac{1}{8}$ | $1\tfrac{3}{8}$ | $\tfrac{1}{2}$ |
| 57 | 0.043 | $1\tfrac{3}{4}$ | $\tfrac{3}{4}$ | $2\tfrac{1}{4}$ | $1\tfrac{1}{8}$ | $1\tfrac{3}{8}$ | $\tfrac{1}{2}$ |
| 56 | 0.0465 | $1\tfrac{3}{4}$ | $\tfrac{3}{4}$ | $2\tfrac{1}{4}$ | $1\tfrac{1}{8}$ | $1\tfrac{3}{8}$ | $\tfrac{1}{2}$ |
| $\tfrac{3}{64}$ | 0.0469 | $1\tfrac{3}{4}$ | $\tfrac{3}{4}$ | $2\tfrac{1}{4}$ | $1\tfrac{1}{8}$ | $1\tfrac{3}{8}$ | $\tfrac{1}{2}$ |
| 55 | 0.052 | $1\tfrac{7}{8}$ | $\tfrac{7}{8}$ | 3 | $1\tfrac{3}{4}$ | $1\tfrac{5}{8}$ | $\tfrac{5}{8}$ |
| 54 | 0.055 | $1\tfrac{7}{8}$ | $\tfrac{7}{8}$ | 3 | $1\tfrac{3}{4}$ | $1\tfrac{5}{8}$ | $\tfrac{5}{8}$ |
| 53 | 0.0595 | $1\tfrac{7}{8}$ | $\tfrac{7}{8}$ | 3 | $1\tfrac{3}{4}$ | $1\tfrac{5}{8}$ | $\tfrac{5}{8}$ |
| $\tfrac{1}{16}$ | 0.0625 | $1\tfrac{7}{8}$ | $\tfrac{7}{8}$ | 3 | $1\tfrac{3}{4}$ | $1\tfrac{5}{8}$ | $\tfrac{5}{8}$ |
| 52 | 0.0635 | $1\tfrac{7}{8}$ | $\tfrac{7}{8}$ | $3\tfrac{3}{4}$ | 2 | $1\tfrac{11}{16}$ | $1\tfrac{1}{16}$ |
| 51 | 0.067 | 2 | 1 | $3\tfrac{3}{4}$ | 2 | $1\tfrac{11}{16}$ | $1\tfrac{1}{16}$ |
| 50 | 0.070 | 2 | 1 | $3\tfrac{3}{4}$ | 2 | $1\tfrac{11}{16}$ | $1\tfrac{1}{16}$ |
| 49 | 0.073 | 2 | 1 | $3\tfrac{3}{4}$ | 2 | $1\tfrac{11}{16}$ | $1\tfrac{1}{16}$ |
| 48 | 0.076 | 2 | 1 | $3\tfrac{3}{4}$ | 2 | $1\tfrac{11}{16}$ | $1\tfrac{1}{16}$ |
| $\tfrac{5}{64}$ | 0.0781 | 2 | 1 | $3\tfrac{3}{4}$ | 2 | $1\tfrac{11}{16}$ | $1\tfrac{1}{16}$ |
| 47 | 0.0785 | 2 | 1 | $4\tfrac{1}{4}$ | $2\tfrac{1}{4}$ | $1\tfrac{11}{16}$ | $1\tfrac{1}{16}$ |

All diameters and lengths are given in inches.   Tolerances are given on page 166c.

## ANSI Standard Straight Shank Twist Drills — From 0.081 to 0.257 Inch, Incl.
(ANSI B94.11-1967, R1972)

| Drill Size | Drill Diam. | Jobbers Length Overall | Jobbers Length Flute | Taper Length Overall | Taper Length Flute | Screw Mach. Lgth. Overall | Screw Mach. Lgth. Flute |
|---|---|---|---|---|---|---|---|
| 46 | 0.081 | $2\frac{1}{8}$ | $1\frac{1}{8}$ | $4\frac{1}{4}$ | $2\frac{1}{4}$ | $1\frac{3}{4}$ | $\frac{3}{4}$ |
| 45 | 0.082 | $2\frac{1}{8}$ | $1\frac{1}{8}$ | $4\frac{1}{4}$ | $2\frac{1}{4}$ | $1\frac{3}{4}$ | $\frac{3}{4}$ |
| 44 | 0.086 | $2\frac{1}{8}$ | $1\frac{1}{8}$ | $4\frac{1}{4}$ | $2\frac{1}{4}$ | $1\frac{3}{4}$ | $\frac{3}{4}$ |
| 43 | 0.089 | $2\frac{1}{4}$ | $1\frac{1}{4}$ | $4\frac{1}{4}$ | $2\frac{1}{4}$ | $1\frac{3}{4}$ | $\frac{3}{4}$ |
| 42 | 0.0935 | $2\frac{1}{4}$ | $1\frac{1}{4}$ | $4\frac{1}{4}$ | $2\frac{1}{4}$ | $1\frac{3}{4}$ | $\frac{3}{4}$ |
| $\frac{3}{32}$ | 0.0938 | $2\frac{1}{4}$ | $1\frac{1}{4}$ | $4\frac{1}{4}$ | $2\frac{1}{4}$ | $1\frac{3}{4}$ | $\frac{3}{4}$ |
| 41 | 0.096 | $2\frac{3}{8}$ | $1\frac{3}{8}$ | $4\frac{5}{8}$ | $2\frac{1}{2}$ | $1\frac{13}{16}$ | $\frac{13}{16}$ |
| 40 | 0.098 | $2\frac{3}{8}$ | $1\frac{3}{8}$ | $4\frac{5}{8}$ | $2\frac{1}{2}$ | $1\frac{13}{16}$ | $\frac{13}{16}$ |
| 39 | 0.0995 | $2\frac{3}{8}$ | $1\frac{3}{8}$ | $4\frac{5}{8}$ | $2\frac{1}{2}$ | $1\frac{13}{16}$ | $\frac{13}{16}$ |
| 38 | 0.1015 | $2\frac{1}{2}$ | $1\frac{7}{16}$ | $4\frac{5}{8}$ | $2\frac{1}{2}$ | $1\frac{13}{16}$ | $\frac{13}{16}$ |
| 37 | 0.104 | $2\frac{1}{2}$ | $1\frac{7}{16}$ | $4\frac{5}{8}$ | $2\frac{1}{2}$ | $1\frac{13}{16}$ | $\frac{13}{16}$ |
| 36 | 0.1065 | $2\frac{1}{2}$ | $1\frac{7}{16}$ | $4\frac{5}{8}$ | $2\frac{1}{2}$ | $1\frac{13}{16}$ | $\frac{13}{16}$ |
| $\frac{7}{64}$ | 0.1094 | $2\frac{5}{8}$ | $1\frac{1}{2}$ | $4\frac{5}{8}$ | $2\frac{1}{2}$ | $1\frac{13}{16}$ | $\frac{13}{16}$ |
| 35 | 0.110 | $2\frac{5}{8}$ | $1\frac{1}{2}$ | $5\frac{1}{8}$ | $2\frac{3}{4}$ | $1\frac{7}{8}$ | $\frac{7}{8}$ |
| 34 | 0.111 | $2\frac{5}{8}$ | $1\frac{1}{2}$ | $5\frac{1}{8}$ | $2\frac{3}{4}$ | $1\frac{7}{8}$ | $\frac{7}{8}$ |
| 33 | 0.113 | $2\frac{5}{8}$ | $1\frac{1}{2}$ | $5\frac{1}{8}$ | $2\frac{3}{4}$ | $1\frac{7}{8}$ | $\frac{7}{8}$ |
| 32 | 0.116 | $2\frac{3}{4}$ | $1\frac{5}{8}$ | $5\frac{1}{8}$ | $2\frac{3}{4}$ | $1\frac{7}{8}$ | $\frac{7}{8}$ |
| 31 | 0.120 | $2\frac{3}{4}$ | $1\frac{5}{8}$ | $5\frac{1}{8}$ | $2\frac{3}{4}$ | $1\frac{7}{8}$ | $\frac{7}{8}$ |
| $\frac{1}{8}$ | 0.125 | $2\frac{3}{4}$ | $1\frac{5}{8}$ | $5\frac{1}{8}$ | $2\frac{3}{4}$ | $1\frac{7}{8}$ | $\frac{7}{8}$ |
| 30 | 0.1285 | $2\frac{3}{4}$ | $1\frac{5}{8}$ | $5\frac{3}{8}$ | 3 | $1\frac{15}{16}$ | $\frac{15}{16}$ |
| 29 | 0.136 | $2\frac{7}{8}$ | $1\frac{3}{4}$ | $5\frac{3}{8}$ | 3 | $1\frac{15}{16}$ | $\frac{15}{16}$ |
| 28 | 0.1405 | $2\frac{7}{8}$ | $1\frac{3}{4}$ | $5\frac{3}{8}$ | 3 | $1\frac{15}{16}$ | $\frac{15}{16}$ |
| $\frac{9}{64}$ | 0.1406 | $2\frac{7}{8}$ | $1\frac{3}{4}$ | $5\frac{3}{8}$ | 3 | $1\frac{15}{16}$ | $\frac{15}{16}$ |
| 27 | 0.144 | 3 | $1\frac{7}{8}$ | $5\frac{3}{8}$ | 3 | $2\frac{1}{16}$ | 1 |
| 26 | 0.147 | 3 | $1\frac{7}{8}$ | $5\frac{3}{8}$ | 3 | $2\frac{1}{16}$ | 1 |
| 25 | 0.1495 | 3 | $1\frac{7}{8}$ | $5\frac{3}{8}$ | 3 | $2\frac{1}{16}$ | 1 |
| 24 | 0.152 | $3\frac{1}{8}$ | 2 | $5\frac{3}{8}$ | 3 | $2\frac{1}{16}$ | 1 |
| 23 | 0.154 | $3\frac{1}{8}$ | 2 | $5\frac{3}{8}$ | 3 | $2\frac{1}{16}$ | 1 |
| $\frac{5}{32}$ | 0.1562 | $3\frac{1}{8}$ | 2 | $5\frac{3}{8}$ | 3 | $2\frac{1}{16}$ | 1 |
| 22 | 0.157 | $3\frac{1}{8}$ | 2 | $5\frac{3}{4}$ | $3\frac{3}{8}$ | $2\frac{1}{8}$ | $1\frac{1}{16}$ |
| 21 | 0.159 | $3\frac{1}{4}$ | $2\frac{1}{8}$ | $5\frac{3}{4}$ | $3\frac{3}{8}$ | $2\frac{1}{8}$ | $1\frac{1}{16}$ |
| 20 | 0.161 | $3\frac{1}{4}$ | $2\frac{1}{8}$ | $5\frac{3}{4}$ | $3\frac{3}{8}$ | $2\frac{1}{8}$ | $1\frac{1}{16}$ |
| 19 | 0.166 | $3\frac{1}{4}$ | $2\frac{1}{8}$ | $5\frac{3}{4}$ | $3\frac{3}{8}$ | $2\frac{1}{8}$ | $1\frac{1}{16}$ |
| 18 | 0.1695 | $3\frac{1}{4}$ | $2\frac{1}{8}$ | $5\frac{3}{4}$ | $3\frac{3}{8}$ | $2\frac{1}{8}$ | $1\frac{1}{16}$ |
| $\frac{11}{64}$ | 0.1719 | $3\frac{1}{4}$ | $2\frac{1}{8}$ | $5\frac{3}{4}$ | $3\frac{3}{8}$ | $2\frac{1}{8}$ | $1\frac{1}{16}$ |
| 17 | 0.173 | $3\frac{3}{8}$ | $2\frac{3}{16}$ | $5\frac{3}{4}$ | $3\frac{3}{8}$ | $2\frac{3}{16}$ | $1\frac{1}{8}$ |
| 16 | 0.177 | $3\frac{3}{8}$ | $2\frac{3}{16}$ | $5\frac{3}{4}$ | $3\frac{3}{8}$ | $2\frac{3}{16}$ | $1\frac{1}{8}$ |
| 15 | 0.180 | $3\frac{3}{8}$ | $2\frac{3}{16}$ | $5\frac{3}{4}$ | $3\frac{3}{8}$ | $2\frac{3}{16}$ | $1\frac{1}{8}$ |
| 14 | 0.182 | $3\frac{3}{8}$ | $2\frac{3}{16}$ | $5\frac{3}{4}$ | $3\frac{3}{8}$ | $2\frac{3}{16}$ | $1\frac{1}{8}$ |
| 13 | 0.185 | $3\frac{1}{2}$ | $2\frac{5}{16}$ | $5\frac{3}{4}$ | $3\frac{3}{8}$ | $2\frac{3}{16}$ | $1\frac{1}{8}$ |
| $\frac{3}{16}$ | 0.1875 | $3\frac{1}{2}$ | $2\frac{5}{16}$ | $5\frac{3}{4}$ | $3\frac{3}{8}$ | $2\frac{3}{16}$ | $1\frac{1}{8}$ |
| 12 | 0.189 | $3\frac{1}{2}$ | $2\frac{5}{16}$ | 6 | $3\frac{3}{8}$ | $2\frac{1}{4}$ | $1\frac{3}{16}$ |
| 11 | 0.191 | $3\frac{1}{2}$ | $2\frac{5}{16}$ | 6 | $3\frac{5}{8}$ | $2\frac{1}{4}$ | $1\frac{3}{16}$ |
| 10 | 0.1935 | $3\frac{5}{8}$ | $2\frac{7}{16}$ | 6 | $3\frac{5}{8}$ | $2\frac{1}{4}$ | $1\frac{3}{16}$ |
| 9 | 0.196 | $3\frac{5}{8}$ | $2\frac{7}{16}$ | 6 | $3\frac{5}{8}$ | $2\frac{1}{4}$ | $1\frac{3}{16}$ |
| 8 | 0.199 | $3\frac{5}{8}$ | $2\frac{7}{16}$ | 6 | $3\frac{5}{8}$ | $2\frac{1}{4}$ | $1\frac{3}{16}$ |
| 7 | 0.201 | $3\frac{5}{8}$ | $2\frac{7}{16}$ | 6 | $3\frac{5}{8}$ | $2\frac{1}{4}$ | $1\frac{3}{16}$ |
| $\frac{13}{64}$ | 0.2031 | $3\frac{5}{8}$ | $2\frac{7}{16}$ | 6 | $3\frac{5}{8}$ | $2\frac{1}{4}$ | $1\frac{3}{16}$ |
| 6 | 0.204 | $3\frac{3}{4}$ | $2\frac{1}{2}$ | 6 | $3\frac{5}{8}$ | $2\frac{3}{8}$ | $1\frac{1}{4}$ |
| 5 | 0.2055 | $3\frac{3}{4}$ | $2\frac{1}{2}$ | 6 | $3\frac{5}{8}$ | $2\frac{3}{8}$ | $1\frac{1}{4}$ |
| 4 | 0.209 | $3\frac{3}{4}$ | $2\frac{1}{2}$ | 6 | $3\frac{5}{8}$ | $2\frac{3}{8}$ | $1\frac{1}{4}$ |
| 3 | 0.213 | $3\frac{3}{4}$ | $2\frac{1}{2}$ | 6 | $3\frac{5}{8}$ | $2\frac{3}{8}$ | $1\frac{1}{4}$ |
| $\frac{7}{32}$ | 0.2188 | $3\frac{3}{4}$ | $2\frac{1}{2}$ | 6 | $3\frac{5}{8}$ | $2\frac{3}{8}$ | $1\frac{1}{4}$ |
| 2 | 0.221 | $3\frac{7}{8}$ | $2\frac{5}{8}$ | $6\frac{1}{8}$ | $3\frac{3}{4}$ | $2\frac{7}{16}$ | $1\frac{5}{16}$ |
| 1 | 0.228 | $3\frac{7}{8}$ | $2\frac{5}{8}$ | $6\frac{1}{8}$ | $3\frac{3}{4}$ | $2\frac{7}{16}$ | $1\frac{5}{16}$ |
| A | 0.234 | $3\frac{7}{8}$ | $2\frac{5}{8}$ | … | … | $2\frac{7}{16}$ | $1\frac{5}{16}$ |
| $\frac{15}{64}$ | 0.2344 | $3\frac{7}{8}$ | $2\frac{5}{8}$ | $6\frac{1}{8}$ | $3\frac{3}{4}$ | $2\frac{7}{16}$ | $1\frac{5}{16}$ |
| B | 0.238 | 4 | $2\frac{3}{4}$ | … | … | $2\frac{1}{2}$ | $1\frac{3}{8}$ |
| C | 0.242 | 4 | $2\frac{3}{4}$ | … | … | $2\frac{1}{2}$ | $1\frac{3}{8}$ |
| D | 0.246 | 4 | $2\frac{3}{4}$ | … | … | $2\frac{1}{2}$ | $1\frac{3}{8}$ |
| E & ¼ | 0.250 | 4 | $2\frac{3}{4}$ | $6\frac{1}{8}$ | $3\frac{3}{4}$ | $2\frac{1}{2}$ | $1\frac{3}{8}$ |
| F | 0.257 | $4\frac{1}{8}$ | $2\frac{7}{8}$ | … | … | $2\frac{5}{8}$ | $1\frac{7}{16}$ |

All diameters and lengths are given in inches.   Tolerances are given on page 1660.

**ANSI Standard Straight Shank Twist Drills — From 0.261 to 0.9062 Inch, Incl.**
(ANSI B94.11-1967, R1972)

| Drill Size | Drill Diam. | Jobbers Length | | Taper Length | | Screw Mach. Length | |
|---|---|---|---|---|---|---|---|
| | | Overall | Flute | Overall | Flute | Overall | Flute |
| G | 0.261 | $4\frac{1}{8}$ | $2\frac{7}{8}$ | ... | ... | $2\frac{5}{8}$ | $1\frac{7}{16}$ |
| $\frac{17}{64}$ | 0.2656 | $4\frac{1}{8}$ | $2\frac{7}{8}$ | $6\frac{1}{4}$ | $3\frac{7}{8}$ | $2\frac{5}{8}$ | $1\frac{7}{16}$ |
| H | 0.266 | $4\frac{1}{8}$ | $2\frac{7}{8}$ | ... | ... | $2\frac{11}{16}$ | $1\frac{1}{2}$ |
| I | 0.272 | $4\frac{1}{8}$ | $2\frac{7}{8}$ | ... | ... | $2\frac{11}{16}$ | $1\frac{1}{2}$ |
| J | 0.277 | $4\frac{1}{8}$ | $2\frac{7}{8}$ | ... | ... | $2\frac{11}{16}$ | $1\frac{1}{2}$ |
| K | 0.281 | $4\frac{1}{4}$ | $2\frac{15}{16}$ | ... | ... | $2\frac{11}{16}$ | $1\frac{1}{2}$ |
| $\frac{9}{32}$ | 0.2812 | $4\frac{1}{4}$ | $2\frac{15}{16}$ | $6\frac{1}{4}$ | $3\frac{7}{8}$ | $2\frac{11}{16}$ | $1\frac{1}{2}$ |
| L | 0.290 | $4\frac{1}{4}$ | $2\frac{15}{16}$ | ... | ... | $2\frac{3}{4}$ | $1\frac{9}{16}$ |
| M | 0.295 | $4\frac{3}{8}$ | $3\frac{1}{16}$ | ... | ... | $2\frac{3}{4}$ | $1\frac{9}{16}$ |
| $\frac{19}{64}$ | 0.2969 | $4\frac{3}{8}$ | $3\frac{1}{16}$ | $6\frac{3}{8}$ | 4 | $2\frac{3}{4}$ | $1\frac{9}{16}$ |
| N | 0.302 | $4\frac{3}{8}$ | $3\frac{1}{16}$ | ... | ... | $2\frac{13}{16}$ | $1\frac{5}{8}$ |
| $\frac{5}{16}$ | 0.3125 | $4\frac{1}{2}$ | $3\frac{3}{16}$ | $6\frac{3}{8}$ | 4 | $2\frac{13}{16}$ | $1\frac{5}{8}$ |
| O | 0.316 | $4\frac{1}{2}$ | $3\frac{3}{16}$ | ... | ... | $2\frac{15}{16}$ | $1\frac{11}{16}$ |
| P | 0.323 | $4\frac{5}{8}$ | $3\frac{5}{16}$ | ... | ... | $2\frac{15}{16}$ | $1\frac{11}{16}$ |
| $\frac{21}{64}$ | 0.3281 | $4\frac{5}{8}$ | $3\frac{5}{16}$ | $6\frac{1}{2}$ | $4\frac{1}{8}$ | $2\frac{15}{16}$ | $1\frac{11}{16}$ |
| Q | 0.332 | $4\frac{3}{4}$ | $3\frac{7}{16}$ | ... | ... | 3 | $1\frac{11}{16}$ |
| R | 0.339 | $4\frac{3}{4}$ | $3\frac{7}{16}$ | ... | ... | 3 | $1\frac{11}{16}$ |
| $\frac{11}{32}$ | 0.3438 | $4\frac{3}{4}$ | $3\frac{7}{16}$ | $6\frac{1}{2}$ | $4\frac{1}{8}$ | 3 | $1\frac{11}{16}$ |
| S | 0.348 | $4\frac{7}{8}$ | $3\frac{1}{2}$ | ... | ... | $3\frac{1}{16}$ | $1\frac{3}{4}$ |
| T | 0.358 | $4\frac{7}{8}$ | $3\frac{1}{2}$ | ... | ... | $3\frac{1}{16}$ | $1\frac{3}{4}$ |
| $\frac{23}{64}$ | 0.3594 | $4\frac{7}{8}$ | $3\frac{1}{2}$ | $6\frac{3}{4}$ | $4\frac{1}{4}$ | $3\frac{1}{16}$ | $1\frac{3}{4}$ |
| U | 0.368 | 5 | $3\frac{5}{8}$ | ... | ... | $3\frac{1}{8}$ | $1\frac{13}{16}$ |
| $\frac{3}{8}$ | 0.375 | 5 | $3\frac{5}{8}$ | $6\frac{3}{4}$ | $4\frac{1}{4}$ | $3\frac{1}{8}$ | $1\frac{13}{16}$ |
| V | 0.377 | 5 | $3\frac{5}{8}$ | ... | ... | $3\frac{1}{4}$ | $1\frac{7}{8}$ |
| W | 0.386 | $5\frac{1}{8}$ | $3\frac{3}{4}$ | ... | ... | $3\frac{1}{4}$ | $1\frac{7}{8}$ |
| $\frac{25}{64}$ | 0.3906 | $5\frac{1}{8}$ | $3\frac{3}{4}$ | 7 | $4\frac{3}{8}$ | $3\frac{1}{4}$ | $1\frac{7}{8}$ |
| X | 0.397 | $5\frac{1}{8}$ | $3\frac{3}{4}$ | ... | ... | $3\frac{9}{16}$ | $1\frac{15}{16}$ |
| Y | 0.404 | $5\frac{1}{4}$ | $3\frac{7}{8}$ | ... | ... | $3\frac{9}{16}$ | $1\frac{15}{16}$ |
| $\frac{13}{32}$ | 0.4062 | $5\frac{1}{4}$ | $3\frac{7}{8}$ | 7 | $4\frac{3}{8}$ | $3\frac{9}{16}$ | $1\frac{15}{16}$ |
| Z | 0.413 | $5\frac{1}{4}$ | $3\frac{7}{8}$ | ... | ... | $3\frac{3}{8}$ | 2 |
| $\frac{27}{64}$ | 0.4219 | $5\frac{3}{8}$ | $3\frac{15}{16}$ | $7\frac{1}{4}$ | $4\frac{5}{8}$ | $3\frac{3}{8}$ | 2 |
| $\frac{7}{16}$ | 0.4375 | $5\frac{1}{2}$ | $4\frac{1}{16}$ | $7\frac{1}{4}$ | $4\frac{5}{8}$ | $3\frac{7}{16}$ | $2\frac{1}{16}$ |
| $\frac{29}{64}$ | 0.4531 | $5\frac{5}{8}$ | $4\frac{3}{16}$ | $7\frac{1}{2}$ | $4\frac{3}{4}$ | $3\frac{9}{16}$ | $2\frac{1}{8}$ |
| $\frac{15}{32}$ | 0.4688 | $5\frac{3}{4}$ | $4\frac{5}{16}$ | $7\frac{1}{2}$ | $4\frac{3}{4}$ | $3\frac{5}{8}$ | $2\frac{1}{8}$ |
| $\frac{31}{64}$ | 0.4844 | $5\frac{7}{8}$ | $4\frac{3}{8}$ | $7\frac{3}{4}$ | $4\frac{3}{4}$ | $3\frac{11}{16}$ | $2\frac{3}{16}$ |
| $\frac{1}{2}$ | 0.5000 | 6 | $4\frac{1}{2}$ | $7\frac{3}{4}$ | $4\frac{3}{4}$ | $3\frac{3}{4}$ | $2\frac{1}{4}$ |
| $\frac{33}{64}$ | 0.5156 | $6\frac{3}{8}$ | $4\frac{13}{16}$ | 8 | $4\frac{3}{4}$ | $3\frac{7}{8}$ | $2\frac{3}{8}$ |
| $\frac{17}{32}$ | 0.5312 | $6\frac{3}{8}$ | $4\frac{13}{16}$ | 8 | $4\frac{3}{4}$ | $3\frac{7}{8}$ | $2\frac{3}{8}$ |
| $\frac{35}{64}$ | 0.5469 | $6\frac{5}{8}$ | $4\frac{13}{16}$ | $8\frac{1}{4}$ | $4\frac{7}{8}$ | 4 | $2\frac{1}{2}$ |
| $\frac{9}{16}$ | 0.5625 | $6\frac{5}{8}$ | $4\frac{13}{16}$ | $8\frac{1}{4}$ | $4\frac{7}{8}$ | 4 | $2\frac{1}{2}$ |
| $\frac{37}{64}$ | 0.5781 | $6\frac{5}{8}$ | $4\frac{13}{16}$ | $8\frac{3}{4}$ | $4\frac{7}{8}$ | $4\frac{1}{8}$ | $2\frac{5}{8}$ |
| $\frac{19}{32}$ | 0.5938 | $7\frac{1}{8}$ | $5\frac{3}{16}$ | $8\frac{3}{4}$ | $4\frac{7}{8}$ | $4\frac{1}{8}$ | $2\frac{5}{8}$ |
| $\frac{39}{64}$ | 0.6094 | $7\frac{1}{8}$ | $5\frac{3}{16}$ | $8\frac{3}{4}$ | $4\frac{7}{8}$ | $4\frac{1}{4}$ | $2\frac{3}{4}$ |
| $\frac{5}{8}$ | 0.625 | $7\frac{1}{8}$ | $5\frac{3}{16}$ | $8\frac{3}{4}$ | $4\frac{7}{8}$ | $4\frac{1}{4}$ | $2\frac{3}{4}$ |
| $\frac{41}{64}$ | 0.6406 | $7\frac{1}{8}$ | $5\frac{3}{16}$ | 9 | $5\frac{1}{8}$ | $4\frac{1}{2}$ | $2\frac{7}{8}$ |
| $\frac{21}{32}$ | 0.6562 | $7\frac{1}{8}$ | $5\frac{3}{16}$ | 9 | $5\frac{1}{8}$ | $4\frac{1}{2}$ | $2\frac{7}{8}$ |
| $\frac{43}{64}$ | 0.6719 | $7\frac{5}{8}$ | $5\frac{5}{8}$ | $9\frac{1}{4}$ | $5\frac{3}{8}$ | $4\frac{5}{8}$ | $2\frac{7}{8}$ |
| $\frac{11}{16}$ | 0.6875 | $7\frac{5}{8}$ | $5\frac{5}{8}$ | $9\frac{1}{4}$ | $5\frac{3}{8}$ | $4\frac{5}{8}$ | $2\frac{7}{8}$ |
| $\frac{45}{64}$ | 0.7031 | ... | ... | $9\frac{1}{2}$ | $5\frac{5}{8}$ | $4\frac{3}{4}$ | 3 |
| $\frac{23}{32}$ | 0.7188 | ... | ... | $9\frac{1}{2}$ | $5\frac{5}{8}$ | $4\frac{3}{4}$ | 3 |
| $\frac{47}{64}$ | 0.7344 | ... | ... | $9\frac{3}{4}$ | $5\frac{7}{8}$ | 5 | $3\frac{1}{8}$ |
| $\frac{3}{4}$ | 0.750 | ... | ... | $9\frac{3}{4}$ | $5\frac{7}{8}$ | 5 | $3\frac{1}{8}$ |
| $\frac{49}{64}$ | 0.7656 | ... | ... | $9\frac{7}{8}$ | 6 | $5\frac{1}{8}$ | $3\frac{1}{4}$ |
| $\frac{25}{32}$ | 0.7812 | ... | ... | $9\frac{7}{8}$ | 6 | $5\frac{1}{8}$ | $3\frac{1}{4}$ |
| $\frac{51}{64}$ | 0.7969 | ... | ... | 10 | $6\frac{1}{8}$ | $5\frac{1}{4}$ | $3\frac{3}{8}$ |
| $\frac{13}{16}$ | 0.8125 | ... | ... | 10 | $6\frac{1}{8}$ | $5\frac{1}{4}$ | $3\frac{3}{8}$ |
| $\frac{53}{64}$ | 0.8281 | ... | ... | 10 | $6\frac{1}{8}$ | $5\frac{3}{8}$ | $3\frac{1}{2}$ |
| $\frac{27}{32}$ | 0.8438 | ... | ... | 10 | $6\frac{1}{8}$ | $5\frac{3}{8}$ | $3\frac{1}{2}$ |
| $\frac{55}{64}$ | 0.8594 | ... | ... | 10 | $6\frac{1}{8}$ | $5\frac{1}{2}$ | $3\frac{1}{2}$ |
| $\frac{7}{8}$ | 0.875 | ... | ... | 10 | $6\frac{1}{8}$ | $5\frac{1}{2}$ | $3\frac{1}{2}$ |
| $\frac{57}{64}$ | 0.8906 | ... | ... | 10 | $6\frac{1}{8}$ | $5\frac{5}{8}$ | $3\frac{5}{8}$ |
| $\frac{29}{32}$ | 0.9062 | ... | ... | 10 | $6\frac{1}{8}$ | $5\frac{5}{8}$ | $3\frac{5}{8}$ |

All diameters and lengths are given in inches.    Tolerances are given on page 1660.

## ANSI Standard Straight Shank Twist Drills — From 0.9219 to 2.000 Inches, Incl.
(ANSI B94.11-1967, R1972)

| Drill Size | Drill Diam. | Jobbers Length | | Taper Length | | Screw Mach. Length | |
|---|---|---|---|---|---|---|---|
| | | Overall | Flute | Overall | Flute | Overall | Flute |
| 59/64 | 0.9219 | ... | ... | 10¾ | 6⅛ | 5¾ | 3¾ |
| 15/16 | 0.9375 | ... | ... | 10¾ | 6⅛ | 5¾ | 3¾ |
| 61/64 | 0.9531 | ... | ... | 11 | 6⅜ | 5⅞ | 3⅞ |
| 31/32 | 0.9688 | ... | ... | 11 | 6⅜ | 5⅞ | 3⅞ |
| 63/64 | 0.9844 | ... | ... | 11 | 6⅜ | 6 | 4 |
| 1 | 1.000 | ... | ... | 11 | 6⅜ | 6 | 4 |
| 1 1/64 | 1.0156 | ... | ... | 11⅛ | 6½ | ... | ... |
| 1 1/32 | 1.0312 | ... | ... | 11⅛ | 6½ | ... | ... |
| 1 3/64 | 1.0469 | ... | ... | 11¼ | 6⅝ | ... | ... |
| 1 1/16 | 1.0625 | ... | ... | 11¼ | 6⅝ | 6¼ | 4 |
| 1 5/64 | 1.0781 | ... | ... | 11½ | 6⅞ | ... | ... |
| 1 3/32 | 1.0938 | ... | ... | 11½ | 6⅞ | ... | ... |
| 1 7/64 | 1.1094 | ... | ... | 11¾ | 7⅛ | ... | ... |
| 1 1/8 | 1.125 | ... | ... | 11¾ | 7⅛ | 6⅜ | 4 |
| 1 9/64 | 1.1406 | ... | ... | 11⅞ | 7¼ | ... | ... |
| 1 5/32 | 1.1562 | ... | ... | 11⅞ | 7¼ | ... | ... |
| 1 11/64 | 1.1719 | ... | ... | 12 | 7⅜ | ... | ... |
| 1 3/16 | 1.1875 | ... | ... | 12 | 7⅜ | 6⅝ | 4¼ |
| 1 13/64 | 1.2031 | ... | ... | 12⅛ | 7½ | ... | ... |
| 1 7/32 | 1.2188 | ... | ... | 12⅛ | 7½ | ... | ... |
| 1 15/64 | 1.2344 | ... | ... | 12½ | 7⅞ | ... | ... |
| 1 1/4 | 1.250 | ... | ... | 12½ | 7⅞ | 6¾ | 4⅜ |
| 1 9/32 | 1.2812 | ... | ... | 14⅛ | 8½ | ... | ... |
| 1 5/16 | 1.3125 | ... | ... | 14¼ | 8⅝ | 7 | 4⅜ |
| 1 11/32 | 1.3438 | ... | ... | 14⅜ | 8¾ | ... | ... |
| 1 3/8 | 1.375 | ... | ... | 14½ | 8⅞ | 7⅛ | 4½ |
| 1 13/32 | 1.4062 | ... | ... | 14⅝ | 9 | ... | ... |
| 1 7/16 | 1.4375 | ... | ... | 14¾ | 9⅛ | 7⅜ | 4¾ |
| 1 15/32 | 1.4688 | ... | ... | 14⅞ | 9¼ | ... | ... |
| 1 1/2 | 1.500 | ... | ... | 15 | 9⅜ | 7½ | 4⅞ |
| 1 9/16 | 1.5625 | ... | ... | 15¼ | 9⅝ | 7¾ | 4⅞ |
| 1 5/8 | 1.625 | ... | ... | 15⅝ | 9⅞ | 7¾ | 4⅞ |
| 1 11/16 | 1.6875 | ... | ... | ... | ... | 8 | 5⅛ |
| 1 3/4 | 1.750 | ... | ... | 16¼ | 10½ | 8 | 5⅛ |
| 1 13/16 | 1.8125 | ... | ... | ... | ... | 8¼ | 5⅜ |
| 1 7/8 | 1.875 | ... | ... | ... | ... | 8¼ | 5⅜ |
| 1 15/16 | 1.9375 | ... | ... | ... | ... | 8½ | 5⅜ |
| 2 | 2.000 | ... | ... | ... | ... | 8½ | 5⅜ |

All diameters and lengths are given in inches. Tolerances are given in the table at the bottom of this page. Nominal shank diameters for drill sizes are nominal diameter of drill except that for screw machine length sizes 1 1/16 through 1 1/4 inches the shank diameter is 1 inch, sizes 1 5/16 through 1 1/2 inches the shank diameter is 1 1/4 inches, and sizes 1 9/16 through 2 inches the shank diameter is 1 1/2 inches.

### Diameter Tolerances for General Purpose Two, Three and Four Flute
ANSI Standard Twist Drills (ANSI B94.11-1967, R1972)

| Drill Size (Diameter measured at point) | Tolerance |
|---|---|
| No. 97 to No. 81, inclusive | Plus 0.0002 to minus 0.0002 |
| Over No. 81 to 1/8, inclusive | Plus 0.0000 to minus 0.0005 |
| Over 1/8 to 1/4, inclusive | Plus 0.0000 to minus 0.0007 |
| Over 1/4 to 1/2, inclusive | Plus 0.0000 to minus 0.0010 |
| Over 1/2 to 1, inclusive | Plus 0.0000 to minus 0.0012 |
| Over 1 to 2, inclusive | Plus 0.0000 to minus 0.0015 |
| Over 2 to 3 1/2, inclusive | Plus 0.0000 to minus 0.0020 |

All dimensions are given in inches except numbered drill sizes.

## ANSI Taper Shank Twist Drills — From 1/8 to 2 Inch (ANSI B94.11-1967, R1972)

| Drill Size (Inches) | Am. Std. Taper (Morse) | Overall Length (Inches) | Flute Length (Inches) |
|---|---|---|---|
| 1/8 | 1 | 5 1/8 | 1 7/8 |
| 9/64 | 1 | 5 3/8 | 2 1/8 |
| 5/32 | 1 | 5 3/8 | 2 1/8 |
| 11/64 | 1 | 5 3/8 | 2 1/2 |
| 3/16 | 1 | 5 3/4 | 2 1/2 |
| 13/64 | 1 | 6 | 2 3/4 |
| 7/32 | 1 | 6 | 2 3/4 |
| 15/64 | 1 | 6 1/8 | 2 7/8 |
| 1/4 | 1 | 6 1/8 | 2 7/8 |
| 17/64 | 1 | 6 1/4 | 3 |
| 9/32 | 1 | 6 1/4 | 3 |
| 19/64 | 1 | 6 3/8 | 3 1/8 |
| 5/16 | 1 | 6 3/8 | 3 1/8 |
| 21/64 | 1 | 6 1/2 | 3 1/4 |
| 11/32 | 1 | 6 1/2 | 3 1/4 |
| 23/64 | 1 | 6 3/4 | 3 1/2 |
| 3/8 | 1 | 6 3/4 | 3 1/2 |
| 25/64 | 1 | 7 | 3 5/8 |
| 13/32 | 1 | 7 | 3 5/8 |
| 27/64 | 1 | 7 1/4 | 3 7/8 |
| 7/16 | 1 | 7 1/4 | 3 7/8 |
| 29/64 | 1 | 7 1/2 | 4 1/8 |
| 15/32 | 1 | 7 1/2 | 4 1/8 |
| 31/64 | 2 | 8 1/4 | 4 3/8 |
| 1/2 | 2 | 8 1/4 | 4 3/8 |
| 33/64 | 2 | 8 1/2 | 4 5/8 |
| 17/32 | 2 | 8 1/2 | 4 5/8 |
| 35/64 | 2 | 8 3/4 | 4 7/8 |
| 9/16 | 2 | 8 3/4 | 4 7/8 |
| 37/64 | 2 | 8 3/4 | 4 7/8 |
| 19/32 | 2 | 8 3/4 | 4 7/8 |
| 39/64 | 2 | 8 3/4 | 4 7/8 |
| 5/8 | 2 | 8 3/4 | 4 7/8 |
| 41/64 | 2 | 9 | 5 1/8 |
| 21/32 | 2 | 9 | 5 1/8 |
| 43/64 | 2 | 9 1/4 | 5 3/8 |
| 11/16 | 2 | 9 1/4 | 5 3/8 |
| 45/64 | 2 | 9 1/2 | 5 5/8 |
| 23/32 | 2 | 9 1/2 | 5 5/8 |
| 47/64 | 2 | 9 3/4 | 5 7/8 |
| 3/4 | 2 | 9 3/4 | 5 7/8 |
| 49/64 | 2 | 9 7/8 | 6 |
| 25/32 | 2 | 9 7/8 | 6 |
| 51/64 | 3 | 10 3/4 | 6 1/8 |
| 13/16 | 3 | 10 3/4 | 6 1/8 |
| 53/64 | 3 | 10 3/4 | 6 1/8 |
| 27/32 | 3 | 10 3/4 | 6 1/8 |
| 55/64 | 3 | 10 3/4 | 6 1/8 |
| 7/8 | 3 | 10 3/4 | 6 1/8 |
| 57/64 | 3 | 10 3/4 | 6 1/8 |
| 29/32 | 3 | 10 3/4 | 6 1/8 |
| 59/64 | 3 | 10 3/4 | 6 1/8 |
| 15/16 | 3 | 10 3/4 | 6 1/8 |
| 61/64 | 3 | 11 | 6 3/8 |
| 31/32 | 3 | 11 | 6 3/8 |
| 63/64 | 3 | 11 | 6 3/8 |
| 1 | 3 | 11 | 6 3/8 |
| 1 1/64 | 3 | 11 1/8 | 6 1/2 |
| 1 1/32 | 3 | 11 1/8 | 6 1/2 |
| 1 3/64 | 3 | 11 1/4 | 6 5/8 |
| 1 1/16 | 3 | 11 1/4 | 6 5/8 |
| 1 5/64 | 4 | 12 1/2 | 6 7/8 |
| 1 3/32 | 4 | 12 1/2 | 6 7/8 |
| 1 7/64 | 4 | 12 3/4 | 7 1/8 |
| 1 1/8 | 4 | 12 3/4 | 7 1/8 |
| 1 9/64 | 4 | 12 7/8 | 7 1/4 |
| 1 5/32 | 4 | 12 7/8 | 7 1/4 |
| 1 11/64 | 4 | 13 | 7 3/8 |
| 1 3/16 | 4 | 13 | 7 3/8 |
| 1 13/64 | 4 | 13 1/8 | 7 1/2 |
| 1 7/32 | 4 | 13 1/8 | 7 1/2 |
| 1 15/64 | 4 | 13 1/2 | 7 7/8 |
| 1 1/4 | 4 | 13 1/2 | 7 7/8 |
| 1 17/64 | 4 | 14 1/8 | 8 1/2 |
| 1 9/32 | 4 | 14 1/8 | 8 1/2 |
| 1 19/64 | 4 | 14 1/4 | 8 5/8 |
| 1 5/16 | 4 | 14 1/4 | 8 5/8 |
| 1 21/64 | 4 | 14 3/8 | 8 3/4 |
| 1 11/32 | 4 | 14 3/8 | 8 3/4 |
| 1 23/64 | 4 | 14 1/2 | 8 7/8 |
| 1 3/8 | 4 | 14 1/2 | 8 7/8 |
| 1 25/64 | 4 | 14 5/8 | 9 |
| 1 13/32 | 4 | 14 5/8 | 9 |
| 1 27/64 | 4 | 14 3/4 | 9 1/8 |
| 1 7/16 | 4 | 14 3/4 | 9 1/8 |
| 1 29/64 | 4 | 14 7/8 | 9 1/4 |
| 1 15/32 | 4 | 14 7/8 | 9 1/4 |
| 1 31/64 | 4 | 15 | 9 3/8 |
| 1 1/2 | 4 | 15 | 9 3/8 |
| 1 33/64 | 5 | 16 3/8 | 9 3/8 |
| 1 17/32 | ... | ... | ... |
| 1 35/64 | 5 | 16 5/8 | 9 5/8 |
| 1 9/16 | ... | ... | ... |
| 1 37/64 | 5 | 16 7/8 | 9 7/8 |
| 1 19/32 | ... | ... | ... |
| 1 5/8 | 5 | 17 | 10 |
| 1 41/64 | ... | ... | ... |
| 1 21/32 | 5 | 17 1/8 | 10 1/8 |
| 1 43/64 | ... | ... | ... |
| 1 11/16 | 5 | 17 1/8 | 10 1/8 |
| 1 45/64 | ... | ... | ... |
| 1 23/32 | 5 | 17 1/8 | 10 1/8 |
| 1 47/64 | ... | ... | ... |
| 1 3/4 | 5 | 17 1/8 | 10 1/8 |
| 1 25/32 | 5 | 17 1/8 | 10 1/8 |
| 1 13/16 | 5 | 17 1/8 | 10 1/8 |
| 1 27/32 | 5 | 17 1/8 | 10 1/8 |
| 1 7/8 | 5 | 17 3/8 | 10 3/8 |
| 1 29/32 | 5 | 17 3/8 | 10 3/8 |
| 1 15/16 | 5 | 17 3/8 | 10 3/8 |
| 1 31/32 | 5 | 17 3/8 | 10 3/8 |
| 2 | 5 | 17 3/8 | 10 3/8 |

\* Some of the drill sizes listed are available with shanks smaller than regular, and some are available with shanks larger than regular.

Drill sizes from 2 to 3 1/2 inches diameter are not tabulated.

*Diameter Tolerances:* See table at bottom of page 1660.

**ANSI Standard Twist Drills.** — The standard ANSI B94.11-1967 (R1972) "Twist Drills, Straight Shank and Taper Shank Combined Drills and Countersinks" covers nomenclature, definitions, sizes and tolerances of straight and taper shank drills and combined drills and countersinks, plain and bell type. Straight shank drills are those having cylindrical shanks which may be the same or different diameter than the body of the drill. The shank may be with or without driving flats, tang, grooves or threads. Taper shank drills are those having conical shanks suitable for direct fitting into tapered holes in machine spindles, driving sleeves or sockets. Tapered shanks generally have a driving tang. Included in the Standard are core drills with three and four flutes which are drills commonly used for enlarging and finishing drilled, cast or punched holes. Conventional drills used for originating holes have two flutes. The combined drill and countersink is primarily used to produce center holes in work that will be held between machine centers. Dimensional tolerances for general purpose two, three and four flute twist drills are given which include tolerances on drill diameter at point, shank diameter (straight shank drills), back taper, flute length and overall length. Element tolerances for general purpose two flute twist drill elements such as included angle of point, lip height, centrality of web and flute spacing are also given.

In the accompanying Handbook tables will be found the drill diameter, American National Standard taper (when applicable), overall and flute lengths for jobbers length, taper length and screw machine length straight shank twist drills in a size range from 0.0059 to 2 inches and taper shank twist drills in a size range from 1/8 inch to 2 inches. Also in these tables will be found diameter tolerances for twist drills and various dimensions and tolerances for combined drills and countersinks.

**ANSI Standard Combined Drills and Countersinks, Plain and Bell Types**
(ANSI B94.11-1967, R1972)

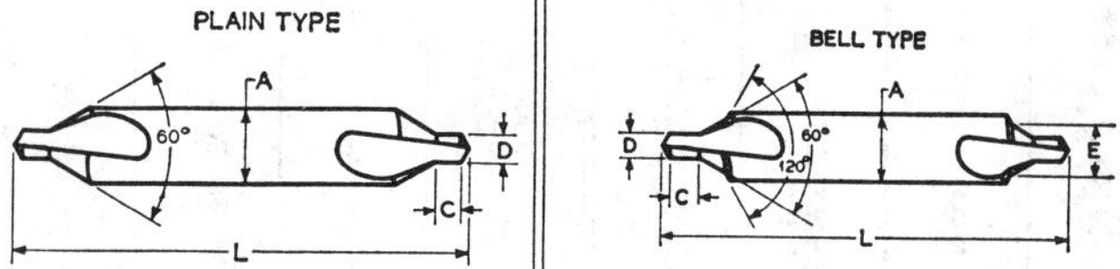


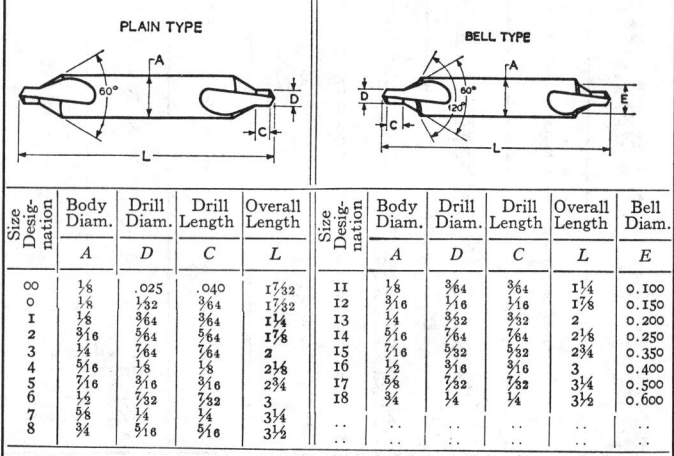

| Size Designation | Body Diam. A | Drill Diam. D | Drill Length C | Overall Length L | Size Designation | Body Diam. A | Drill Diam. D | Drill Length C | Overall Length L | Bell Diam. E |
|---|---|---|---|---|---|---|---|---|---|---|
| 00 | 1/8 | .025 | .040 | 1 7/32 | 11 | 1/8 | 3/64 | 3/64 | 1 1/4 | 0.100 |
| 0 | 1/8 | 1/32 | 3/64 | 1 7/32 | 12 | 3/16 | 1/16 | 1/16 | 1 7/8 | 0.150 |
| 1 | 1/8 | 3/64 | 3/64 | 1 1/4 | 13 | 1/4 | 3/32 | 3/32 | 2 | 0.200 |
| 2 | 3/16 | 5/64 | 5/64 | 1 7/8 | 14 | 5/16 | 7/64 | 7/64 | 2 1/8 | 0.250 |
| 3 | 1/4 | 7/64 | 7/64 | 2 | 15 | 7/16 | 5/32 | 5/32 | 2 3/4 | 0.350 |
| 4 | 5/16 | 1/8 | 1/8 | 2 1/8 | 16 | 1/2 | 3/16 | 3/16 | 3 | 0.400 |
| 5 | 7/16 | 3/16 | 3/16 | 2 3/4 | 17 | 5/8 | 7/32 | 7/32 | 3 1/4 | 0.500 |
| 6 | 1/2 | 7/32 | 7/32 | 3 | 18 | 3/4 | 1/4 | 1/4 | 3 1/2 | 0.600 |
| 7 | 5/8 | 1/4 | 1/4 | 3 1/4 | .. | .. | .. | .. | .. | .. |
| 8 | 3/4 | 5/16 | 5/16 | 3 1/2 | .. | .. | .. | .. | .. | .. |

All dimensions except size designation are in inches.

*Tolerance:* Plus 0.0000 minus 0.0020 on *A*; plus 0.0030 minus 0.0000 on *D*; plus 1/32 minus 1/16 on *L*; plus 0.0080 minus 0.0080 on *C* for sizes 00 through 2, 11 and 12; plus 1/64 minus 1/64 on *C* for sizes 3 through 8 and sizes 13 through 18.

**British Standard Metric Twist Drills.** — BS 328:Part 1:1959 (incorporating amendments issued March 1960 and March 1964) covers twist drills made to inch and metric dimensions which are intended for general engineering purposes. ISO recommendations are taken into account. The accompanying tables give the standard metric sizes of Morse taper shank twist drills and core drills, parallel shank jobbing and long series drills, and stub drills.

All drills are right-hand cutting unless otherwise specified, and normal, slow or quick helix angles may be provided. A 'back taper' is ground on the diameter from point to shank, which affords longitudinal clearance. Core drills may have three or four flutes, and are intended for opening up cast holes or enlarging machined holes, for example. The parallel shank jobber, and long series drills, and stub drills are made without driving tenons.

Morse taper shank drills with oversize dimensions are also listed, and Table 7 shows metric drill sizes superseding gage and letter size drills, which are now obsolete in the United Kingdom. To meet special requirements, the Standard lists non-standard sizes for the various types of drills.

The limits of tolerance on cutting diameters, as measured across the lands at the outer corners of a drill, shall be h8, in accordance with BS 1916, Limits and Fits for Engineering (Part 1, Limits and Tolerances), and Table 3 shows the values which are common to the different types of drills mentioned above.

The drills shall be permanently and legibly marked whenever possible, preferably by rolling, showing the size, and the manufacturer's name or trademark. If they are made from high speed steel, they shall be marked with the letters H.S. where practicable.

*Drill Elements:* The following definitions of drill elements are given.

*Axis:* The longitudinal centre-line.

*Body:* That portion of the drill extending from the extreme cutting end to the commencement of the shank.

*Shank:* That portion of the drill by which it is held and driven.

*Flutes:* The grooves in the body of the drill which provide lips and permit the removal of chips and allow cutting fluid to reach the lips.

*Web (Core):* The central portion of the drill situated between the roots of the flutes and extending from the point end towards the shank; the point end of the web or core forms the chisel edge.

*Lands:* The cylindrical-ground surfaces on the leading edges of the drill flutes. The width of the land is measured at right angles to the flute helix.

*Body Clearance:* That portion of the body surface reduced in diameter to provide diametral clearance.

*Heel:* The edge formed by the intersection of the flute surface and the body clearance.

*Point:* The sharpened end of the drill, consisting of all that part of the drill which is shaped to produce lips, faces, flanks and chisel edge.

*Face:* That portion of the flute surface adjacent to the lip on which the chip impinges as it is cut from the work.

*Flank:* That surface on a drill point which extends behind the lip to the following flute.

*Lip:* (cutting edge): The edge formed by the intersection of the flank and face.

*Relative Lip Height:* The relative position of the lips measured at the outer corners in a direction parallel to the drill axis.

*Outer Corner:* The corner formed by the intersection of the lip and the leading edge of the land.

*Chisel Edge:* The edge formed by the intersection of the flanks.

*Chisel Edge Corner:* The corner formed by the intersection of a lip and the chisel edge.

## Table 1. British Standard Morse Taper Shank Twist Drills and Core Drills — Standard Metric Sizes* (BS 328: Part 1: 1959)

| Diameter | Flute Length | Overall Length | Diameter | Flute Length | Overall Length | Diameter | Flute Length | Overall Length |
|---|---|---|---|---|---|---|---|---|
| 3.00 | 33 | 114 | 16.25 | 125 | 223 | 29.50 | 175 | 296 |
| 3.20 | 36 | 117 | 16.50 | | | 29.75 | | |
| 3.50 | 39 | 120 | 16.75 | | | 30.00 | | |
| 3.80 | 43 | 123 | 17.00 | | | 30.25 | 180 | 301 |
| 4.00 | | | 17.25 | 130 | 228 | 30.50 | | |
| 4.20 | | | 17.50 | | | 30.75 | | |
| 4.50 | 47 | 128 | 17.75 | | | 31.00 | | |
| 4.80 | 52 | 133 | 18.00 | | | 31.25 | | |
| 5.00 | | | 18.25 | 135 | 233 | 31.50 | | |
| 5.20 | | | 18.50 | | | 31.75 | 185 | 306 |
| 5.50 | 57 | 138 | 18.75 | | | 32.00 | 185 | 334 |
| 5.80 | | | 19.00 | | | 32.50 | | |
| 6.00 | | | 19.25 | 140 | 238 | 33.00 | | |
| 6.20 | 63 | 144 | 19.50 | | | 33.50 | | |
| 6.50 | | | 19.75 | | | 34.00 | 190 | 339 |
| 6.80 | 69 | 150 | 20.00 | | | 34.50 | | |
| 7.00 | | | 20.25 | 145 | 243 | 35.00 | | |
| 7.20 | | | 20.50 | | | 35.50 | | |
| 7.50 | | | 20.75 | | | 36.00 | 195 | 344 |
| 7.80 | 75 | 156 | 21.00 | | | 36.50 | | |
| 8.00 | | | 21.25 | 150 | 248 | 37.00 | | |
| 8.20 | | | 21.50 | | | 37.50 | | |
| 8.50 | | | 21.75 | | | 38.00 | 200 | 349 |
| 8.80 | 81 | 162 | 22.00 | | | 38.50 | | |
| 9.00 | | | 22.25 | | | 39.00 | | |
| 9.20 | | | 22.50 | 155 | 253 | 39.50 | | |
| 9.50 | | | 22.75 | | | 40.00 | | |
| 9.80 | 87 | 168 | 23.00 | | | 40.50 | 205 | 354 |
| 10.00 | | | 23.25 | 155 | 276 | 41.00 | | |
| 10.20 | | | 23.50 | | | 41.50 | | |
| 10.50 | | | 23.75 | 160 | 281 | 42.00 | | |
| 10.80 | 94 | 175 | 24.00 | | | 42.50 | | |
| 11.00 | | | 24.25 | | | 43.00 | 210 | 359 |
| 11.20 | | | 24.50 | | | 43.50 | | |
| 11.50 | | | 24.75 | | | 44.00 | | |
| 11.80 | | | 25.00 | | | 44.50 | | |
| 12.00 | 101 | 182 | 25.25 | 165 | 286 | 45.00 | | |
| 12.20 | | | 25.50 | | | 45.50 | 215 | 364 |
| 12.50 | | | 25.75 | | | 46.00 | | |
| 12.80 | | | 26.00 | | | 46.50 | | |
| 13.00 | | | 26.25 | | | 47.00 | | |
| 13.20 | | | 26.50 | | | 47.50 | | |
| 13.50 | 108 | 189 | 26.75 | 170 | 291 | 48.00 | 220 | 369 |
| 13.80 | | | 27.00 | | | 48.50 | | |
| 14.00 | | | 27.25 | | | 49.00 | | |
| 14.25 | 114 | 212 | 27.50 | | | 49.50 | | |
| 14.50 | | | 27.75 | | | 50.00 | | |
| 14.75 | | | 28.00 | | | 50.50 | 225 | 374 |
| 15.00 | | | 28.25 | 175 | 296 | 51.00 | 225 | 412 |
| 15.25 | 120 | 218 | 28.50 | | | 52.00 | | |
| 15.50 | | | 28.75 | | | 53.00 | | |
| 15.75 | | | 29.00 | | | 54.00 | 230 | 417 |
| 16.00 | | | 29.25 | | | 55.00 | | |

For footnotes see end of table.

**Table 1. (Continued). British Standard Morse Taper Shank Twist Drills and Core Drills — Standard Metric Sizes\* (BS 328: Part 1: 1959)**

| Diameter | Flute Length | Overall Length | Diameter | Flute Length | Overall Length | Diameter | Flute Length | Overall Length |
|---|---|---|---|---|---|---|---|---|
| 56.00 | 230 | 417 | 71.00 | 250 | 437 | 86.00 | | |
| 57.00 | | | 72.00 | | | 87.00 | | |
| 58.00 | 235 | 422 | 73.00 | 255 | 442 | 88.00 | 270 | 524 |
| 59.00 | | | 74.00 | | | 89.00 | | |
| 60.00 | | | 75.00 | | | 90.00 | | |
| 61.00 | | | 76.00 | 260 | 477 | 91.00 | | |
| 62.00 | 240 | 427 | | | | 92.00 | | |
| 63.00 | | | 77.00 | | | 93.00 | 275 | 529 |
| | | | 78.00 | | | 94.00 | | |
| 64.00 | | | 79.00 | 260 | 514 | 95.00 | | |
| 65.00 | 245 | 432 | 80.00 | | | | | |
| 66.00 | | | 81.00 | | | 96.00 | | |
| 67.00 | | | 82.00 | | | 97.00 | | |
| 68.00 | | | 83.00 | 265 | 519 | 98.00 | 280 | 534 |
| 69.00 | 250 | 437 | 84.00 | | | 99.00 | | |
| 70.00 | | | 85.00 | | | 100.00 | | |

All dimensions are in millimeters. Tolerances on diameters are given in table below.
\* The Morse taper shanks of these twist and core drills are as follows: 3.00 to 14.00 mm diameter, M.T. No. 1; 14.25 to 23.00 mm diameter, M.T. No. 2; 23.25 to 31.50 mm diameter, M.T. No. 3; 31.75 to 50.50 mm diameter, M.T. No. 4; 51.00 to 76.00 mm diameter, M.T. No. 5; 77.00 to 100.00 mm diameter, M.T. No. 6.
Table 2, below, shows twist drills that may be supplied with the shank and length over-size, but they should be regarded as non-preferred.

**Table 2. British Standard Morse Taper Shank Twist Drills — Metric Oversize Shank and Length Series† (BS 328: Part 1: 1959)**

| Diam. Range | Overall Length | M. T. No. | Diam. Range | Overall Length | M. T. No. | Diam. Range | Overall Length | M. T. No. |
|---|---|---|---|---|---|---|---|---|
| 12.00 to 13.20 | 199 | 2 | 22.50 to 23.00 | 276 | 3 | 45.50 to 47.50 | 402 | 5 |
| 13.50 to 14.00 | 206 | 2 | 26.75 to 28.00 | 319 | 4 | 48.00 to 50.00 | 407 | 5 |
| 18.25 to 19.00 | 256 | 3 | 29.00 to 30.00 | 324 | 4 | 50.50 | 412 | 5 |
| 19.25 to 20.00 | 251 | 3 | 30.25 to 31.50 | 329 | 4 | 64.00 to 67.00 | 499 | 6 |
| 20.25 to 21.00 | 266 | 3 | 40.50 to 42.50 | 392 | 5 | 68.00 to 71.00 | 504 | 6 |
| 21.25 to 22.25 | 271 | 3 | 43.00 to 45.00 | 397 | 5 | 72.00 to 75.00 | 509 | 6 |

Diameters and lengths are given in millimeters. For the individual sizes within the diameter ranges given, see Table 1.
† This series of drills should be regarded as non-preferred.

**Table 3. British Standard Limits of Tolerance on Diameter for Twist Drills and Core Drills — Metric Series (BS 328: Part 1: 1959)**

| Drill Size (Diameter measured across lands at outer corners) | Tolerance (h8) |
|---|---|
| 0 to 1 inclusive | Plus 0.000 to Minus 0.014 |
| Over 1 to 3 inclusive | Plus 0.000 to Minus 0.014 |
| Over 3 to 6 inclusive | Plus 0.000 to Minus 0.018 |
| Over 6 to 10 inclusive | Plus 0.000 to Minus 0.022 |
| Over 10 to 18 inclusive | Plus 0.000 to Minus 0.027 |
| Over 18 to 30 inclusive | Plus 0.000 to Minus 0.033 |
| Over 30 to 50 inclusive | Plus 0.000 to Minus 0.039 |
| Over 50 to 80 inclusive | Plus 0.000 to Minus 0.046 |
| Over 80 to 120 inclusive | Plus 0.000 to Minus 0.054 |

All dimensions are given in millimeters.

## Table 4. British Standard Parallel Shank Jobber Series Twist Drills — Standard Metric Sizes (BS 328: Part 1: 1959)

| Diameter | Flute Length | Overall Length | Diameter | Flute Length | Overall Length | Diameter | Flute Length | Overall Length | Diameter | Flute Length | Overall Length | Diameter | Flute Length | Overall Length |
|---|---|---|---|---|---|---|---|---|---|---|---|---|---|---|
| 0.20 | 2.5 | 19 | 1.75 | 22 | 46 | 5.40 | 57 | 93 | 10.20 | 87 | 133 | 10.20 | 87 | 133 |
| 0.22 |  |  | 1.80 |  |  | 5.50 |  |  | 10.30 |  |  |  |  |  |
| 0.25 | 3.0 | 19 | 1.85 |  |  | 5.60 |  |  | 10.40 |  |  |  |  |  |
| 0.28 |  |  | 1.90 |  |  | 5.70 |  |  | 10.50 |  |  |  |  |  |
| 0.30 | 4.0 | 19 | 1.95 | 24 | 49 | 5.80 |  |  | 10.60 |  |  |  |  |  |
| 0.32 | 4 | 19 | 2.00 |  |  | 5.90 |  |  | 10.70 | 94 | 142 |  |  |  |
| 0.35 |  |  | 2.05 |  |  | 6.00 |  |  | 10.80 |  |  |  |  |  |
| 0.38 |  |  | 2.10 |  |  | 6.10 | 63 | 101 | 10.90 |  |  |  |  |  |
| 0.40 | 5 | 20 | 2.15 | 27 | 53 | 6.20 |  |  | 11.00 |  |  |  |  |  |
| 0.42 |  |  | 2.20 |  |  | 6.30 |  |  | 11.10 |  |  |  |  |  |
| 0.45 |  |  | 2.25 |  |  | 6.40 |  |  | 11.20 |  |  |  |  |  |
| 0.48 |  |  | 2.30 |  |  | 6.50 |  |  | 11.30 |  |  |  |  |  |
| 0.50 | 6 | 22 | 2.35 |  |  | 6.60 |  |  | 11.40 |  |  |  |  |  |
| 0.52 |  |  | 2.40 | 30 | 57 | 6.70 |  |  | 11.50 |  |  |  |  |  |
| 0.55 | 7 | 24 | 2.45 |  |  | 6.80 | 69 | 109 | 11.60 |  |  |  |  |  |
| 0.58 |  |  | 2.50 |  |  | 6.90 |  |  | 11.70 |  |  |  |  |  |
| 0.60 |  |  | 2.55 |  |  | 7.00 |  |  | 11.80 |  |  |  |  |  |
| 0.62 | 8 | 26 | 2.60 |  |  | 7.10 |  |  | 11.90 | 101 | 151 |  |  |  |
| 0.65 |  |  | 2.65 | 33 | 61 | 7.20 |  |  | 12.00 |  |  |  |  |  |
| 0.68 | 9 | 28 | 2.70 |  |  | 7.30 |  |  | 12.10 |  |  |  |  |  |
| 0.70 |  |  | 2.75 |  |  | 7.40 |  |  | 12.20 |  |  |  |  |  |
| 0.72 |  |  | 2.80 |  |  | 7.50 |  |  | 12.30 |  |  |  |  |  |
| 0.75 |  |  | 2.85 |  |  | 7.60 | 75 | 117 | 12.40 |  |  |  |  |  |
| 0.78 | 10 | 30 | 2.90 |  |  | 7.70 |  |  | 12.50 |  |  |  |  |  |
| 0.80 |  |  | 2.95 |  |  | 7.80 |  |  | 12.60 |  |  |  |  |  |
| 0.82 |  |  | 3.00 |  |  | 7.90 |  |  | 12.70 |  |  |  |  |  |
| 0.85 |  |  | 3.10 | 36 | 65 | 8.00 |  |  | 12.80 |  |  |  |  |  |
| 0.88 | 11 | 32 | 3.20 |  |  | 8.10 |  |  | 12.90 |  |  |  |  |  |
| 0.90 |  |  | 3.30 |  |  | 8.20 |  |  | 13.00 |  |  |  |  |  |
| 0.92 |  |  | 3.40 | 39 | 70 | 8.30 |  |  | 13.10 |  |  |  |  |  |
| 0.95 |  |  | 3.50 |  |  | 8.40 |  |  | 13.20 |  |  |  |  |  |
| 0.98 | 12 | 34 | 3.60 | 39 | 70 | 8.50 |  |  | 13.30 | 108 | 160 |  |  |  |
| 1.00 |  |  | 3.70 |  |  | 8.60 | 81 | 125 | 13.40 |  |  |  |  |  |
| 1.05 |  |  | 3.80 | 43 | 75 | 8.70 |  |  | 13.50 |  |  |  |  |  |
| 1.10 | 14 | 36 | 3.90 |  |  | 8.80 |  |  | 13.60 |  |  |  |  |  |
| 1.15 |  |  | 4.00 |  |  | 8.90 |  |  | 13.70 |  |  |  |  |  |
| 1.20 | 16 | 38 | 4.10 |  |  | 9.00 |  |  | 13.80 |  |  |  |  |  |
| 1.25 |  |  | 4.20 |  |  | 9.10 |  |  | 13.90 |  |  |  |  |  |
| 1.30 |  |  | 4.30 | 47 | 80 | 9.20 |  |  | 14.00 |  |  |  |  |  |
| 1.35 | 18 | 40 | 4.40 |  |  | 9.30 |  |  | 14.25 | 114 | 169 |  |  |  |
| 1.40 |  |  | 4.50 |  |  | 9.40 |  |  | 14.50 |  |  |  |  |  |
| 1.45 |  |  | 4.60 |  |  | 9.50 |  |  | 14.75 |  |  |  |  |  |
| 1.50 |  |  | 4.70 |  |  | 9.60 | 87 | 133 | 15.00 |  |  |  |  |  |
| 1.55 | 20 | 43 | 4.80 | 52 | 86 | 9.70 |  |  | 15.25 | 120 | 178 |  |  |  |
| 1.60 |  |  | 4.90 |  |  | 9.80 |  |  | 15.50 |  |  |  |  |  |
| 1.65 |  |  | 5.00 |  |  | 9.90 |  |  | 15.75 |  |  |  |  |  |
| 1.70 |  |  | 5.10 |  |  | 10.00 |  |  | 16.00 |  |  |  |  |  |
|  |  |  | 5.20 |  |  | 10.10 |  |  |  |  |  |  |  |  |
|  |  |  | 5.30 |  |  |  |  |  |  |  |  |  |  |  |

All dimensions are in millimeters.   Tolerances on diameters are given in Table 3.

**Table 5.** British Standard Parallel Shank Long Series Twist Drills —
Standard Metric Sizes (BS 328: Part 1: 1959)

| Diameter | Flute Length | Overall Length | Diameter | Flute Length | Overall Length | Diameter | Flute Length | Overall Length |
|---|---|---|---|---|---|---|---|---|
| 2.00 | 56 | 85 | 6.80 | 102 | 156 | 12.70 | 134 | 205 |
| 2.05 | | | 6.90 | | | 12.80 | | |
| 2.10 | | | 7.00 | | | 12.90 | | |
| 2.15 | 59 | 90 | 7.10 | | | 13.00 | | |
| 2.20 | | | 7.20 | | | 13.10 | | |
| 2.25 | | | 7.30 | | | 13.20 | | |
| 2.30 | | | 7.40 | | | 13.30 | 140 | 214 |
| 2.35 | | | 7.50 | | | 13.40 | | |
| 2.40 | 62 | 95 | 7.60 | 109 | 165 | 13.50 | | |
| 2.45 | | | 7.70 | | | 13.60 | | |
| 2.50 | | | 7.80 | | | 13.70 | | |
| 2.55 | | | 7.90 | | | 13.80 | | |
| 2.60 | | | 8.00 | | | 13.90 | | |
| 2.65 | | | 8.10 | | | 14.00 | | |
| 2.70 | 66 | 100 | 8.20 | | | 14.25 | 144 | 220 |
| 2.75 | | | 8.30 | | | 14.50 | | |
| 2.80 | | | 8.40 | | | 14.75 | | |
| 2.85 | | | 8.50 | | | 15.00 | | |
| 2.90 | | | 8.60 | 115 | 175 | 15.25 | 149 | 227 |
| 2.95 | | | 8.70 | | | 15.50 | | |
| 3.00 | | | 8.80 | | | 15.75 | | |
| 3.10 | 69 | 106 | 8.90 | | | 16.00 | | |
| 3.20 | | | 9.00 | | | 16.25 | 154 | 235 |
| 3.30 | | | 9.10 | | | 16.50 | | |
| 3.40 | 73 | 112 | 9.20 | | | 16.75 | | |
| 3.50 | | | 9.30 | | | 17.00 | | |
| 3.60 | | | 9.40 | | | 17.25 | 158 | 241 |
| 3.70 | | | 9.50 | | | 17.50 | | |
| 3.80 | 78 | 119 | 9.60 | 121 | 184 | 17.75 | | |
| 3.90 | | | 9.70 | | | 18.00 | | |
| 4.00 | | | 9.80 | | | 18.25 | 162 | 247 |
| 4.10 | | | 9.90 | | | 18.50 | | |
| 4.20 | | | 10.00 | | | 18.75 | | |
| 4.30 | 82 | 126 | 10.10 | | | 19.00 | | |
| 4.40 | | | 10.20 | | | 19.25 | 166 | 254 |
| 4.50 | | | 10.30 | | | 19.50 | | |
| 4.60 | | | 10.40 | | | 19.75 | | |
| 4.70 | | | 10.50 | | | 20.00 | | |
| 4.80 | 87 | 132 | 10.60 | | | 20.25 | 171 | 261 |
| 4.90 | | | 10.70 | 128 | 195 | 20.50 | | |
| 5.00 | | | 10.80 | | | 20.75 | | |
| 5.10 | | | 10.90 | | | 21.00 | | |
| 5.20 | | | 11.00 | | | 21.25 | 176 | 268 |
| 5.30 | | | 11.10 | | | 21.50 | | |
| 5.40 | 91 | 139 | 11.20 | | | 21.75 | | |
| 5.50 | | | 11.30 | | | 22.00 | | |
| 5.60 | | | 11.40 | | | 22.25 | | |
| 5.70 | | | 11.50 | | | 22.50 | 180 | 275 |
| 5.80 | | | 11.60 | | | 22.75 | | |
| 5.90 | | | 11.70 | | | 23.00 | | |
| 6.00 | | | 11.80 | | | 23.25 | | |
| 6.10 | 97 | 148 | 11.90 | 134 | 205 | 23.50 | | |
| 6.20 | | | 12.00 | | | 23.75 | 185 | 282 |
| 6.30 | | | 12.10 | | | 24.00 | | |
| 6.40 | | | 12.20 | | | 24.25 | | |
| 6.50 | | | 12.30 | | | 24.50 | | |
| 6.60 | | | 12.40 | | | 24.75 | | |
| 6.70 | | | 12.50 | | | 25.00 | | |
| | | | 12.60 | | | | | |

All dimensions are in millimeters.   Tolerances on diameters are given in Table 3.

### Table 6. British Standard Stub Drills — Metric Sizes (BS 328: Part 1: 1959)

| Diameter | Flute Length | Overall Length | Diameter | Flute Length | Overall Length | Diameter | Flute Length | Overall Length | Diameter | Flute Length | Overall Length | Diameter | Flute Length | Overall Length |
|---|---|---|---|---|---|---|---|---|---|---|---|---|---|---|
| 0.50 | 3 | 20 | 5.00 | 26 | 62 | 9.50 | 40 | 84 | 14.00 | 54 | 107 |  |  |  |
| 0.80 | 5 | 24 | 5.20 |  |  |  |  |  | 14.50 | 56 | 111 |  |  |  |
| 1.00 | 6 | 26 |  |  |  | 9.80 | 43 | 89 | 15.00 |  |  |  |  |  |
| 1.20 | 8 | 30 | 5.50 | 28 | 66 | 10.00 |  |  | 15.50 | 58 | 115 |  |  |  |
| 1.50 | 9 | 32 | 5.80 |  |  | 10.20 |  |  | 16.00 |  |  |  |  |  |
| 1.80 | 11 | 36 | 6.00 |  |  | 10.50 |  |  | 16.50 | 60 | 119 |  |  |  |
| 2.00 | 12 | 38 | 6.20 | 31 | 70 | 10.80 | 47 | 95 | 17.00 |  |  |  |  |  |
| 2.20 | 13 | 40 | 6.50 |  |  | 11.00 |  |  | 17.50 | 62 | 123 |  |  |  |
| 2.50 | 14 | 43 | 6.80 | 34 | 74 | 11.20 |  |  | 18.00 |  |  |  |  |  |
| 2.80 | 16 | 46 | 7.00 |  |  | 11.50 |  |  | 18.50 | 64 | 127 |  |  |  |
| 3.00 |  |  | 7.20 |  |  | 11.80 |  |  | 19.00 |  |  |  |  |  |
| 3.20 | 18 | 49 | 7.50 |  |  | 12.00 | 51 | 102 | 19.50 | 66 | 131 |  |  |  |
| 3.50 | 20 | 52 | 7.80 | 37 | 79 | 12.20 |  |  | 20.00 |  |  |  |  |  |
| 3.80 | 22 | 55 | 8.00 |  |  | 12.50 |  |  | 21.00 | 68 | 136 |  |  |  |
| 4.00 |  |  | 8.20 |  |  | 12.80 |  |  | 22.00 | 70 | 141 |  |  |  |
| 4.20 |  |  | 8.50 |  |  | 13.00 |  |  | 23.00 | 72 | 146 |  |  |  |
| 4.50 | 24 | 58 | 8.80 | 40 | 84 | 13.20 |  |  | 24.00 | 75 | 151 |  |  |  |
| 4.80 | 26 | 62 | 9.00 |  |  | 13.50 | 54 | 107 | 25.00 |  |  |  |  |  |
|  |  |  | 9.20 |  |  | 13.80 |  |  |  |  |  |  |  |  |

All dimensions are given in millimeters. Tolerances on diameters are given in Table 3.

### Table 7. British Standard Drills — Metric Sizes Superseding Gauge and Letter Sizes* (BS 328: Part 1: 1959 Appendix B)

| Obsolete Drill Size | Recommended Metric Size (mm) | Obsolete Drill Size | Recommended Metric Size (mm) | Obsolete Drill Size | Recommended Metric Size (mm) | Obsolete Drill Size | Recommended Metric Size (mm) | Obsolete Drill Size | Recommended Metric Size (mm) |
|---|---|---|---|---|---|---|---|---|---|
| 80 | 0.35 | 58 | 1.05 | 36 | 2.70 | 14 | 4.60 | I | 6.90 |
| 79 | 0.38 | 57 | 1.10 | 35 | 2.80 | 13 | 4.70 | J | 7.00 |
| 78 | 0.40 | 56 | 3/64 in. | 34 | 2.80 | 12 | 4.80 |  |  |
| 77 | 0.45 |  |  | 33 | 2.85 | 11 | 4.90 | K | 9/32 in. |
| 76 | 0.50 | 55 | 1.30 | 32 | 2.95 |  |  | L | 7.40 |
|  |  | 54 | 1.40 | 31 | 3.00 | 10 | 4.90 | M | 7.50 |
| 75 | 0.52 | 53 | 1.50 |  |  | 9 | 5.00 | N | 7.70 |
| 74 | 0.58 | 52 | 1.60 | 30 | 3.30 | 8 | 5.10 | O | 8.00 |
| 73 | 0.60 | 51 | 1.70 | 29 | 3.50 | 7 | 5.10 |  |  |
| 72 | 0.65 |  |  | 28 | 9/64 in. | 6 | 5.20 |  |  |
| 71 | 0.65 | 50 | 1.80 | 27 | 3.70 |  |  | P | 8.20 |
|  |  | 49 | 1.85 | 26 | 3.70 | 5 | 5.20 | Q | 8.40 |
| 70 | 0.70 | 48 | 1.95 |  |  | 4 | 5.30 | R | 8.60 |
| 69 | 0.75 | 47 | 2.00 | 25 | 3.80 | 3 | 5.40 | S | 8.80 |
| 68 | 1/32 in. | 46 | 2.05 | 24 | 3.90 | 2 | 5.60 |  |  |
| 67 | 0.82 |  |  | 23 | 3.90 | 1 | 5.80 |  |  |
| 66 | 0.85 | 45 | 2.10 | 22 | 4.00 |  |  | U | 9.30 |
|  |  | 44 | 2.20 | 21 | 4.00 | A | 15/64 in. | V | 3/8 in. |
| 65 | 0.90 | 43 | 2.25 |  |  | B | 6.00 | W | 9.80 |
| 64 | 0.92 | 42 | 3/32 in. | 20 | 4.10 | C | 6.10 | X | 10.10 |
| 63 | 0.95 | 41 | 2.45 | 19 | 4.20 | D | 6.20 | Y | 10.30 |
| 62 | 0.98 |  |  | 18 | 4.30 | E | 1/4 in. | Z | 10.50 |
| 61 | 1.00 | 40 | 2.50 | 17 | 4.40 |  |  | ... | ... |
|  |  | 39 | 2.55 | 16 | 4.50 | F | 6.50 | ... | ... |
| 60 | 1.00 | 38 | 2.60 |  |  | G | 6.60 | ... | ... |
| 59 | 1.05 | 37 | 2.65 | 15 | 4.60 | H | 17/64 in. | ... | ... |

* Gauge and letter size drills are now obsolete in the United Kingdom and should not be used in the production of new designs. The table is given to assist users in changing over to the recommended standard sizes.

**Steels for Twist Drills.** — *Carbon Steel:* If the conditions are such that carbon steel drill speeds are sufficient for the purpose, high-speed-steel drills are not economical to use because the difference in performance as compared with the carbon steel tool does not compensate for the difference in price.

*High-Speed Steel:* For high surface speed drilling operations where carbon steels would fail by tempering, the properties of red hardness and abrasion resistance favor the use of high-speed steel.

*Cobalt High-Speed Steel:* These high-speed-steel drills are capable of withstanding cutting speeds beyond the range of conventional high-speed-steel drills and have superior resistance to abrasion but are not to be compared with tungsten-carbide tipped tools.

**Accuracy of Drilled Holes.** — Normally the diameter of drilled holes is not given a tolerance; the size of the hole is expected to be as close to the drill size as can be obtained. The accuracy of holes drilled with a two-fluted twist drill is influenced by many factors, which include: the accuracy of the drill point; the size of the drill; length and shape of the chisel edge; whether or not a bushing is used to guide the drill; the work material; length of the drill; runout of the spindle and the chuck; rigidity of the machine tool, workpiece, and the setup; also the cutting fluid used, if any. Usually when drilling most materials the diameter of the drilled holes will be oversize. The table below provides the results of tests reported by The Metal Cutting Tool Institute in which the diameter of over 2800 holes drilled in steel and cast iron were measured. The values in this table indicate what might be expected under average shop conditions; however, when the drill point is accurately ground and the other machining conditions are correct, the resulting hole size is more likely to be between the mean and average minimum values given in this table. If the drill is ground and used incorrectly, holes that are even larger than the average maximum values can result.

### Oversize Diameters in Drilling

| Drill Diam., Inch | Amount Oversize, Inch | | | Drill Diam., Inch | Amount Oversize, Inch | | |
|---|---|---|---|---|---|---|---|
| | Average Max. | Mean | Average Min. | | Average Max | Mean | Average Min. |
| 1/16 | .002 | .0015 | .001 | 1/2 | .008 | .005 | .003 |
| 1/8 | .0045 | .003 | .001 | 3/4 | .008 | .005 | .003 |
| 1/4 | .0065 | .004 | .0025 | 1 | .009 | .007 | .004 |

Courtesy of The Metal Cutting Institute

There are some conditions which will cause the drilled hole to be undersize. For example, holes drilled in light metals and in other materials having a high coefficient of thermal expansion may contract to a size that is smaller than the diameter of the drill as the metal surrounding the hole is cooled after having been heated by the drilling. The elastic action of the metal surrounding the hole may also cause the drilled hole to be undersize when drilling high strength materials with a drill that is dull at its outer corner.

The accuracy of the drill point has a great effect on the accuracy of the drilled hole. An inaccurately ground twist drill will produce holes that are excessively oversize. The drill point must be symmetrical; i.e., the point angles must be equal, as well as the lip lengths and the axial height of the lips. Any alteration to the lips or to the chisel edge, such as thinning the web, must be done with care to preserve the symmetry of the drill point. An adequate relief should be provided behind the chisel edge to prevent heel drag. On conventionally ground drill points this can be estimated by the chisel edge angle.

When drilling a hole, as the drill point starts to enter the workpiece, the drill will be unstable and will tend to wander. Then as the body of the drill enters the hole the drill will tend to stabilize. The result of this action is a tendency to drill a bellmouth shape in the hole at the entrance and perhaps beyond. Factors contributing to bellmouthing are: an unsymmetrically ground drill point; a large chisel edge length; inadequate relief behind the chisel edge; runout of the spindle and the chuck; using a slender drill that will bend easily; and lack of rigidity of the machine tool, workpiece, or the setup. Correcting these conditions as required will reduce the tendency for bellmouthing to occur and improve the accuracy of the hole diameter and its straightness. Starting the hole with a short stiff drill, such as a center drill, will quickly stabilize the drill that follows and reduce or eliminate bellmouthing; this procedure should always be used when drilling in a lathe, where the work is rotating. Bellmouthing can also be eliminated almost entirely and the accuracy of the hole improved by using a close fitting drill jig bushing placed close to the workpiece. Although specific recommendations cannot be made, many cutting fluids will help to increase the accuracy of the diameters of drilled holes. Double margin twist drills, available in the smaller sizes, will drill a more accurate hole than conventional twist drills having only a single margin at the leading edge of the land. The second land, located on the trailing edge of each land, provides greater stability in the drill bushing and in the hole. These drills are especially useful in drilling intersecting off-center holes. Single and double margin step drills, also available in the smaller sizes, will produce very accurate drilled holes, which are usually less than .002 inch larger than the drill size.

**Counterbores.** — Counterbores for screw holes are generally made in sets. Each set contains three counterbores: one with the body of the size of the screw head and the pilot the size of the hole to admit the body of the screw; one with the body the size of the head of the screw and the pilot the size of the tap drill; and

### Counterbores With Interchangeable Cutters and Guides

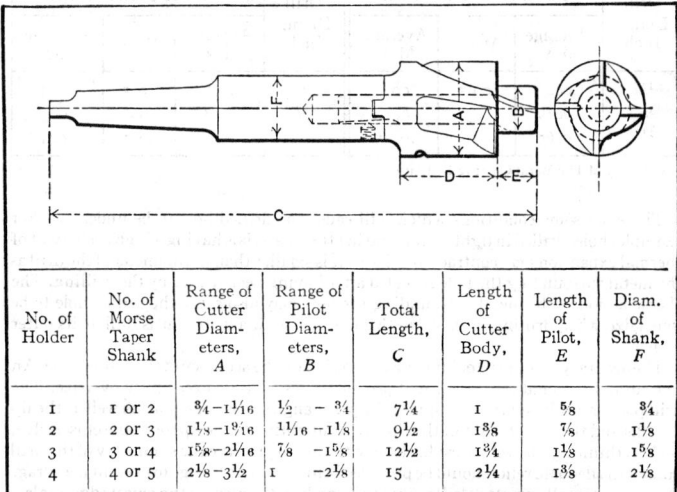

| No. of Holder | No. of Morse Taper Shank | Range of Cutter Diameters, A | Range of Pilot Diameters, B | Total Length, C | Length of Cutter Body, D | Length of Pilot, E | Diam. of Shank, F |
|---|---|---|---|---|---|---|---|
| 1 | 1 or 2 | $3/4 - 1\frac{1}{16}$ | $\frac{1}{2} - \frac{3}{4}$ | $7\frac{1}{4}$ | 1 | $\frac{5}{8}$ | $\frac{3}{4}$ |
| 2 | 2 or 3 | $1\frac{1}{8} - 1\frac{9}{16}$ | $1\frac{1}{16} - 1\frac{1}{8}$ | $9\frac{1}{2}$ | $1\frac{3}{8}$ | $\frac{7}{8}$ | $1\frac{1}{8}$ |
| 3 | 3 or 4 | $1\frac{5}{8} - 2\frac{1}{16}$ | $\frac{7}{8} - 1\frac{5}{8}$ | $12\frac{1}{2}$ | $1\frac{3}{4}$ | $1\frac{1}{8}$ | $1\frac{5}{8}$ |
| 4 | 4 or 5 | $2\frac{1}{8} - 3\frac{1}{2}$ | $1 - 2\frac{1}{8}$ | 15 | $2\frac{1}{4}$ | $1\frac{3}{8}$ | $2\frac{1}{8}$ |

**Dimensions of Solid Counterbores**

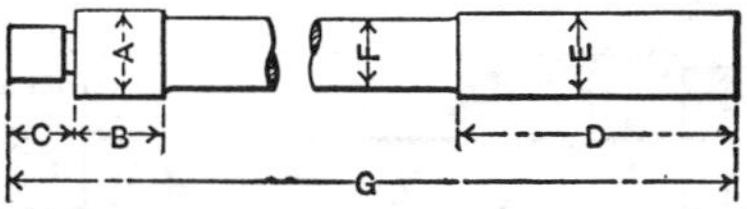


| A | B | C | D | E | F | G | A | B | C | D | E | F | G |
|---|---|---|---|---|---|---|---|---|---|---|---|---|---|
| 1/4 | 11/32 | 3/16 | 2 1/2 | 1/4 | 7/32 | 4 7/8 | 1 5/16 | 1 3/4 | 63/64 | 3 9/16 | 1 5/16 | 1 9/32 | 12 5/16 |
| 5/16 | 13/32 | 15/64 | 2 9/16 | 5/16 | 9/32 | 5 5/16 | 1 3/8 | 1 27/32 | 1 1/32 | 3 5/8 | 1 3/8 | 1 11/32 | 12 3/4 |
| 3/8 | 1/2 | 9/32 | 2 5/8 | 3/8 | 11/32 | 5 3/4 | 1 7/16 | 1 29/32 | 1 5/64 | 3 11/16 | 1 7/16 | 1 13/32 | 13 3/16 |
| 7/16 | 19/32 | 21/64 | 2 11/16 | 7/16 | 13/32 | 6 3/16 | 1 1/2 | 2 | 1 1/8 | 3 3/4 | 1 1/2 | 1 7/16 | 13 3/8 |
| 1/2 | 21/32 | 3/8 | 2 3/4 | 1/2 | 15/32 | 6 5/8 | 1 5/8 | 2 3/32 | 1 7/32 | 3 15/16 | 1 17/32 | 1 15/32 | 13 3/4 |
| 9/16 | 3/4 | 27/64 | 2 13/16 | 9/16 | 17/32 | 7 1/16 | 1 3/4 | 2 3/16 | 1 5/16 | 4 1/8 | 1 19/32 | 1 17/32 | 14 1/8 |
| 5/8 | 27/32 | 15/32 | 2 7/8 | 5/8 | 19/32 | 7 1/2 | 1 7/8 | 2 9/32 | 1 13/32 | 4 5/16 | 1 5/8 | 1 9/16 | 14 1/2 |
| 11/16 | 29/32 | 33/64 | 2 15/16 | 11/16 | 21/32 | 7 15/16 | 2 | 2 3/8 | 1 1/2 | 4 1/2 | 1 21/32 | 1 19/32 | 14 7/8 |
| 3/4 | 1 | 9/16 | 3 | 3/4 | 23/32 | 8 3/8 | 2 1/8 | 2 15/32 | 1 19/32 | 4 11/16 | 1 23/32 | 1 21/32 | 15 1/4 |
| 13/16 | 1 3/32 | 39/64 | 3 1/16 | 13/16 | 25/32 | 8 13/16 | 2 1/4 | 2 9/16 | 1 11/16 | 4 7/8 | 1 3/4 | 1 11/16 | 15 5/8 |
| 7/8 | 1 5/32 | 21/32 | 3 1/8 | 7/8 | 27/32 | 9 1/4 | 2 3/8 | 2 21/32 | 1 25/32 | 5 1/16 | 1 25/32 | 1 23/32 | 16 |
| 15/16 | 1 1/4 | 45/64 | 3 3/16 | 15/16 | 29/32 | 9 11/16 | 2 1/2 | 2 3/4 | 1 7/8 | 5 1/4 | 1 27/32 | 1 25/32 | 16 3/8 |
| 1 | 1 11/32 | 3/4 | 3 1/4 | 1 | 31/32 | 10 1/8 | 2 5/8 | 2 27/32 | 1 31/32 | 5 7/16 | 1 7/8 | 1 13/16 | 16 3/4 |
| 1 1/16 | 1 13/32 | 51/64 | 3 5/16 | 1 1/16 | 1 1/32 | 10 9/16 | 2 3/4 | 2 15/16 | 2 1/16 | 5 5/8 | 1 29/32 | 1 27/32 | 17 1/8 |
| 1 1/8 | 1 1/2 | 27/32 | 3 3/8 | 1 1/8 | 1 3/32 | 11 | 2 7/8 | 3 1/32 | 2 5/32 | 5 13/16 | 1 31/32 | 1 29/32 | 17 1/2 |
| 1 3/16 | 1 19/32 | 57/64 | 3 7/16 | 1 3/16 | 1 5/32 | 11 7/16 | 3 | 3 1/8 | 2 1/4 | 6 | 2 | 1 15/16 | 17 7/8 |
| 1 1/4 | 1 21/32 | 15/16 | 3 1/2 | 1 1/4 | 1 7/32 | 11 7/8 | ... | .... | .... | .... | .... | .... | .... |

the third with the body the size of the body of the screw and the pilot the size of the tap drill. Counterbores are usually provided with four flutes cut on a right-hand spiral. The angle of the spiral is 15 degrees with the center line of the counterbore, which corresponds to a lead of the flute equal to about twelve times the diameter of the body of the counterbore. Counterbores for brass are fluted straight.

Small counterbores are often made with three flutes, but should then have the size plainly stamped on them before fluting, as they cannot afterwards be conveniently measured. The flutes should be deep enough to come below the surface of the pilot. The counterbore should be relieved on the end of the body only, and not on the cylindrical surface. To facilitate the relieving process, a small neck is turned between the guide and the body for clearance. The amount of clearance on the cutting edges is, for general work, from 4 to 5 degrees. The accompanying table gives dimensions for straight shank counterbores. The same dimensions, except for the shank part, may be used for Morse taper shank counterbores. The number of shank used for counterbores with bodies of different diameters is usually as follows: Up to 1/2 inch diameter body, No. 1 Morse taper shank; from 9/16 to 7/8 inch, No. 2; from 15/16 to 1 3/8 inch, No. 3; from 1 7/16 to 2 inches, No. 4; from 2 1/16 to 3 inches, No. 5 shank.

**Lathe Arbors.** — Arbors are usually tapered about 0.006 inch per foot. The diameter or nominal size *D* in the table is at a distance *F* from the small end. The diameter *G* of the drills for the centers conforms to Stub's steel wire gage. The " width of flat," listed in the last column, is for the driving dog. The centers of arbors intended for very heavy duty may be made somewhat larger than those given in the table. As to hardening, the practice at the present time, among manufacturers, is to harden arbors all over, but for extremely accurate work, an arbor having hardened ends and a soft body is generally considered superior, as there is less tendency of distortion from internal stresses.

## Proportions of Solid Lathe Arbors

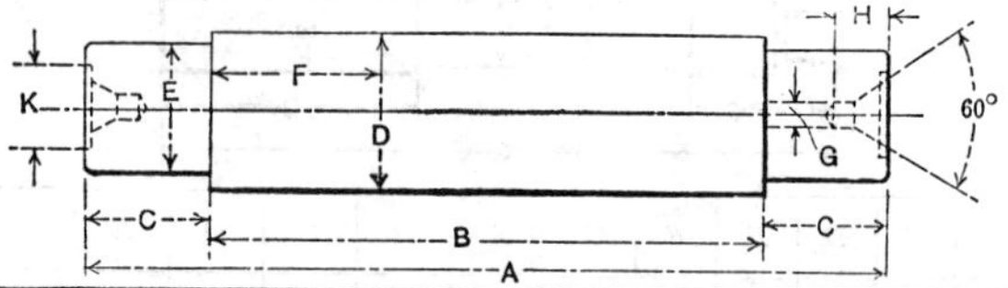


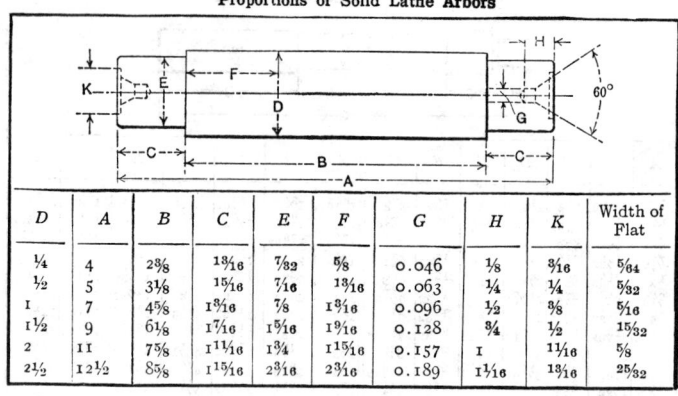

| *D* | *A* | *B* | *C* | *E* | *F* | *G* | *H* | *K* | Width of Flat |
|---|---|---|---|---|---|---|---|---|---|
| 1/4 | 4 | 2 3/8 | 13/16 | 7/32 | 5/8 | 0.046 | 1/8 | 3/16 | 5/64 |
| 1/2 | 5 | 3 1/8 | 15/16 | 7/16 | 13/16 | 0.063 | 1/4 | 1/4 | 5/32 |
| 1 | 7 | 4 5/8 | 1 3/16 | 7/8 | 1 3/16 | 0.096 | 1/2 | 3/8 | 5/16 |
| 1 1/2 | 9 | 6 1/8 | 1 7/16 | 1 5/16 | 1 9/16 | 0.128 | 3/4 | 1/2 | 15/32 |
| 2 | 11 | 7 5/8 | 1 11/16 | 1 3/4 | 1 15/16 | 0.157 | 1 | 11/16 | 5/8 |
| 2 1/2 | 12 1/2 | 8 5/8 | 1 15/16 | 2 3/16 | 2 3/16 | 0.189 | 1 1/16 | 13/16 | 25/32 |

## Boring-bar Couplings

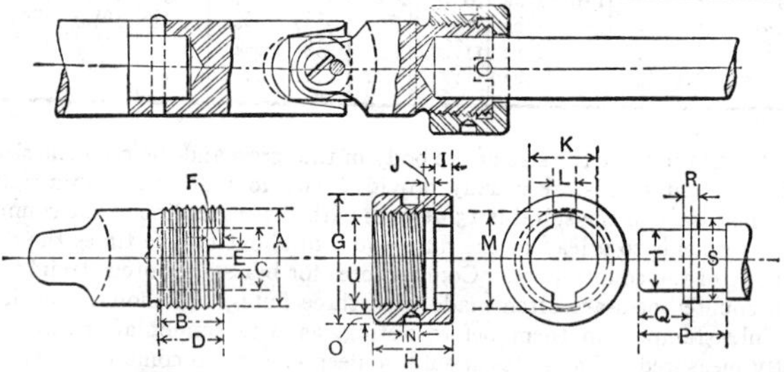


| *A* | *B* | *C* | *D* | *E* | *F* | *G* | *H* | *I* | *J* | *K* | *L* | *M* |
|---|---|---|---|---|---|---|---|---|---|---|---|---|
| 3/4 | 1/2 | 1/2 | 7/16 | 3/16 | 5/32 | 1 | 5/8 | 1/8 | 3/32 | 17/32 | 3/16 | 11/16 |
| 1 | 1/2 | 5/8 | 9/16 | 1/4 | 7/32 | 1 3/8 | 5/8 | 5/32 | 3/32 | 21/32 | 1/4 | 15/16 |
| 1 1/4 | 5/8 | 3/4 | 5/8 | 5/16 | 9/32 | 1 5/8 | 13/16 | 3/16 | 1/8 | 25/32 | 5/16 | 1 5/32 |
| 1 1/2 | 13/16 | 1 | 7/8 | 7/16 | 13/32 | 2 | 1 1/16 | 1/4 | 1/8 | 1 1/32 | 7/16 | 1 13/32 |
| 2 1/2 | 7/8 | 1 1/2 | 1 3/8 | 9/16 | 17/32 | 3 | 1 1/8 | 1/4 | 1/8 | 1 17/32 | 9/16 | 2 3/8 |
| 3 | 7/8 | 1 7/8 | 1 3/4 | 11/16 | 21/32 | 3 1/2 | 1 1/4 | 1/4 | 1/8 | 1 29/32 | 11/16 | 2 7/8 |

| *N* | *O* | *P* | *Q* | *R* | *S* | *T* | *U*, Bore | Threads per Inch | Used for Bars |
|---|---|---|---|---|---|---|---|---|---|
| 9/64 | 1/8 | 3/4 | 3/8 | 1/8 | 21/32 | 0.499 | 0.6610 | 16 | 3/8 to 1/2 |
| 11/64 | 5/32 | 7/8 | 15/32 | 3/16 | 7/8 | 0.624 | 0.9110 | 16 | 1/2 to 3/4 |
| 13/64 | 5/32 | 1 1/8 | 1/2 | 1/4 | 1 1/16 | 0.749 | 1.1615 | 12 | 3/4 to 1 |
| 15/64 | 7/32 | 1 3/8 | 11/16 | 3/8 | 1 5/16 | 0.999 | 1.4115 | 12 | 1 to 1 1/2 |
| 19/64 | 7/32 | 1 7/8 | 1 1/8 | 1/2 | 2 1/4 | 1.4985 | 2.4115 | 12 | 1 1/2 to 2 |
| 21/64 | 7/32 | 2 5/16 | 1 7/16 | 5/8 | 2 3/4 | 1.8735 | 2.9115 | 12 | 2 and over |

**Boring-bar Cutters — 1**

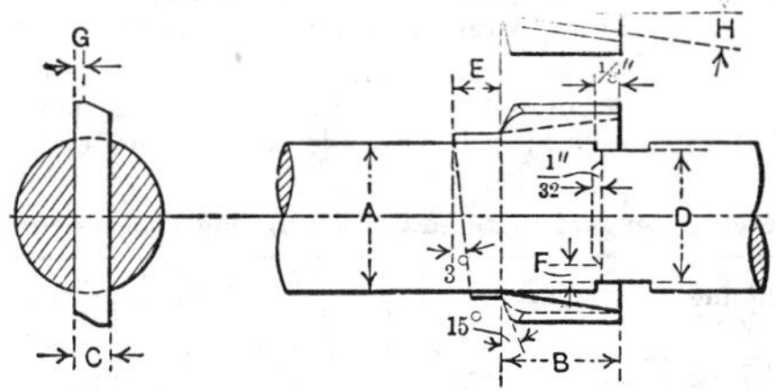


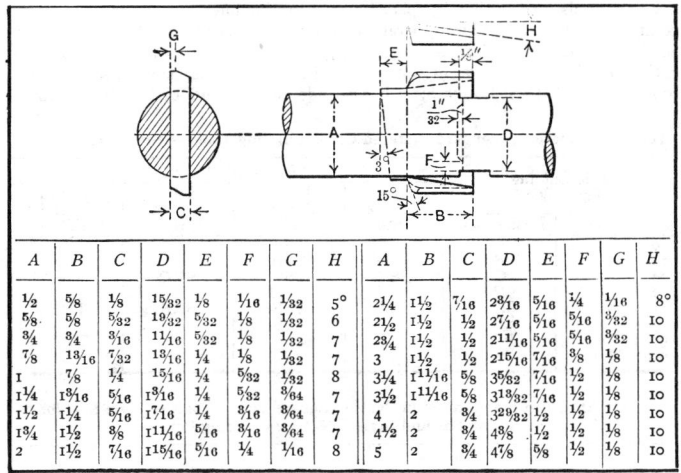

| A | B | C | D | E | F | G | H | A | B | C | D | E | F | G | H |
|---|---|---|---|---|---|---|---|---|---|---|---|---|---|---|---|
| 1/2 | 5/8 | 1/8 | 15/32 | 1/8 | 1/16 | 1/32 | 5° | 2 1/4 | 1 1/2 | 7/16 | 2 3/16 | 5/16 | 1/4 | 1/16 | 8° |
| 5/8 | 5/8 | 5/32 | 19/32 | 5/32 | 1/8 | 1/32 | 6 | 2 1/2 | 1 1/2 | 1/2 | 2 7/16 | 5/16 | 5/16 | 3/32 | 10 |
| 3/4 | 3/4 | 3/16 | 11/16 | 5/32 | 1/8 | 1/32 | 7 | 2 3/4 | 1 1/2 | 1/2 | 2 11/16 | 5/16 | 5/16 | 3/32 | 10 |
| 7/8 | 13/16 | 7/32 | 13/16 | 1/4 | 1/8 | 1/32 | 7 | 3 | 1 1/2 | 1/2 | 2 15/16 | 7/16 | 3/8 | 1/8 | 10 |
| 1 | 7/8 | 1/4 | 15/16 | 1/4 | 5/32 | 1/32 | 8 | 3 1/4 | 1 11/16 | 5/8 | 3 5/32 | 7/16 | 1/2 | 1/8 | 10 |
| 1 1/4 | 1 3/16 | 5/16 | 1 3/16 | 1/4 | 5/32 | 3/64 | 7 | 3 1/2 | 1 11/16 | 5/8 | 3 13/32 | 7/16 | 1/2 | 1/8 | 10 |
| 1 1/2 | 1 1/4 | 5/16 | 1 7/16 | 1/4 | 3/16 | 3/64 | 7 | 4 | 2 | 3/4 | 3 29/32 | 1/2 | 1/2 | 1/8 | 10 |
| 1 3/4 | 1 1/2 | 3/8 | 1 11/16 | 5/16 | 3/16 | 3/64 | 7 | 4 1/2 | 2 | 3/4 | 4 3/8 | 1/2 | 1/2 | 1/8 | 10 |
| 2 | 1 1/2 | 7/16 | 1 15/16 | 5/16 | 1/4 | 1/16 | 8 | 5 | 2 | 3/4 | 4 7/8 | 5/8 | 1/2 | 1/8 | 10 |

**Boring-bar Cutters — 2**

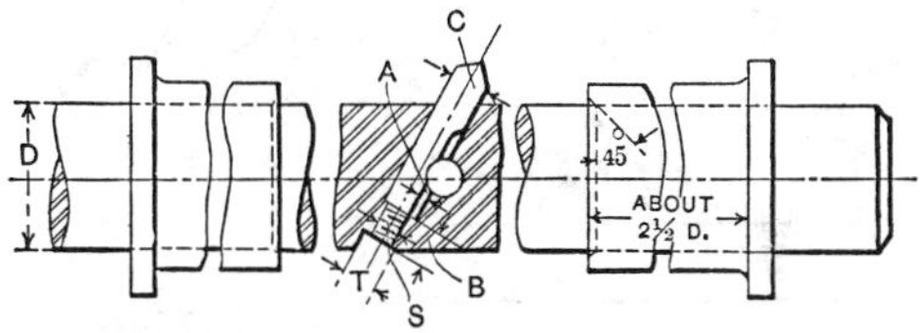


| Diameter of Bar, D | Diameter of Cutter, C | Diameter of Pin | Depth of Flat, A | Diameter of Screw, S | Diameter T of Counter-bore | Length B of Thread | Diameter of Bar, D | Diameter of Cutter, C | Diameter of Pin | Depth of Flat, A | Diameter of Screw, S | Diameter T of Counterbore | Length B of Thread |
|---|---|---|---|---|---|---|---|---|---|---|---|---|---|
| 3/8 | 1/8 | 1/16 | 1/64 | 1/16 | 1/8 | 5/64 | 1 1/2 | 3/8 | 9/32 | 3/64 | 5/16 | 3/8 | 3/8 |
| 7/16 | 1/8 | 3/32 | 1/64 | 1/16 | 1/8 | 5/64 | 1 5/8 | 7/16 | 5/16 | 3/64 | 5/16 | 3/8 | 3/8 |
| 1/2 | 1/8 | 3/32 | 1/64 | 1/16 | 1/8 | 5/64 | 1 3/4 | 7/16 | 5/16 | 3/64 | 3/8 | 7/16 | 1/2 |
| 9/16 | 1/8 | 1/8 | 1/32 | 1/8 | 5/32 | 1/8 | 1 7/8 | 1/2 | 11/32 | 1/16 | 3/8 | 7/16 | 1/2 |
| 5/8 | 3/16 | 1/8 | 1/32 | 1/8 | 5/32 | 1/8 | 2 | 1/2 | 3/8 | 1/16 | 7/16 | 1/2 | 9/16 |
| 11/16 | 3/16 | 5/32 | 1/32 | 1/8 | 3/16 | 5/32 | 2 1/8 | 9/16 | 13/32 | 1/16 | 7/16 | 1/2 | 9/16 |
| 3/4 | 3/16 | 5/32 | 1/32 | 1/8 | 3/16 | 5/32 | 2 1/4 | 9/16 | 13/32 | 1/16 | 1/2 | 9/16 | 5/8 |
| 13/16 | 1/4 | 5/32 | 1/32 | 3/16 | 1/4 | 3/16 | 2 3/8 | 5/8 | 7/16 | 5/64 | 1/2 | 9/16 | 5/8 |
| 7/8 | 1/4 | 5/32 | 1/32 | 3/16 | 1/4 | 3/16 | 2 1/2 | 5/8 | 15/32 | 5/64 | 9/16 | 5/8 | 11/16 |
| 15/16 | 1/4 | 3/16 | 1/32 | 3/16 | 1/4 | 1/4 | 2 5/8 | 11/16 | 1/2 | 5/64 | 9/16 | 5/8 | 11/16 |
| 1 | 1/4 | 3/16 | 1/32 | 3/16 | 1/4 | 1/4 | 2 3/4 | 11/16 | 1/2 | 5/64 | 5/8 | 11/16 | 3/4 |
| 1 1/8 | 5/16 | 7/32 | 1/32 | 1/4 | 5/16 | 5/16 | 2 7/8 | 3/4 | 17/32 | 3/32 | 5/8 | 3/4 | 7/8 |
| 1 1/4 | 5/16 | 1/4 | 3/64 | 1/4 | 5/16 | 5/16 | 3 | 3/4 | 9/16 | 3/32 | 5/8 | 3/4 | 7/8 |
| 1 3/8 | 3/8 | 1/4 | 3/64 | 1/4 | 5/16 | 5/16 | ... | ... | ... | ... | ... | ... | ... |

**Sintered Carbide Boring Tools.** — Industrial experience has shown that the shapes of tools used for boring operations need to be different from those of single-point tools ordinarily used for general applications such as lathe work. Accordingly, Section 3 of American Standard B5.36-1957 gives standard sizes, styles and designations for four basic types of sintered carbide boring tools, namely: solid carbide square; carbide-tipped square; solid carbide round; and carbide-tipped round

**Table 1. American Standard Sintered Carbide Boring Tools — Style Designations**

| Side Cutting Edge Angle E | | Boring Tool Styles | | | |
|---|---|---|---|---|---|
| Degrees | Designation | Solid Square (SS) | Tipped Square (TS) | Solid Round (SR) | Tipped Round (TR) |
| 0 | A | | TSA | | |
| 10 | B | | TSB | | |
| 30 | C | SSC | TSC | SRC | TRC |
| 40 | D | | TSD | | |
| 45 | E | SSE | TSE | SRE | TRE |
| 55 | F | | TSF | | |
| 90 (0° Rake) | G | | | | TRG |
| 90 (10° Rake) | H | | | | TRH |

**Table 2. American Standard Solid Carbide Square Boring Tools — Style SSC for 60°
Boring Bar and Style SSE for 45° Boring Bar**

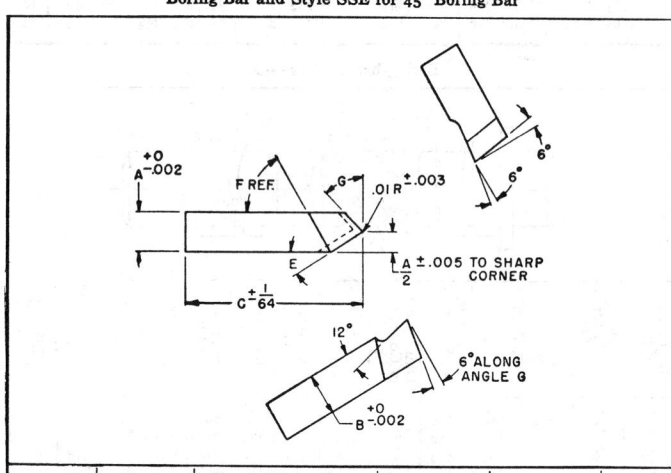

| Tool Desig-nation | Boring Bar Angle, Deg. From Axis | Shank Dimensions, Inches | | | Side Cutting Edge Angle E, Deg. | End Cutting Edge Angle G, Deg. | Shoulder Angle F, Deg. |
|---|---|---|---|---|---|---|---|
| | | Width A | Height B | Length C | | | |
| SSC-58 | 60 | ⁵⁄₃₂ | ⁵⁄₃₂ | 1 | 30 | 38 | 60 |
| SSE-58 | 45 | | | | 45 | 53 | 45 |
| SSC-610 | 60 | ³⁄₁₆ | ³⁄₁₆ | 1¼ | 30 | 38 | 60 |
| SSE-610 | 45 | | | | 45 | 53 | 45 |
| SSC-810 | 60 | ¼ | ¼ | 1¼ | 30 | 38 | 60 |
| SSE-810 | 45 | | | | 45 | 53 | 45 |
| SSC-1012 | 60 | ⁵⁄₁₆ | ⁵⁄₁₆ | 1½ | 30 | 38 | 60 |
| SSE-1012 | 45 | | | | 45 | 53 | 45 |

**Table 3.** American Standard Carbide-Tipped Square Boring Tools — Styles TSA and TSB for 90° Boring Bar, Styles TSC and TSD for 60° Boring Bar, and Styles TSE and TSF for 45° Boring Bar

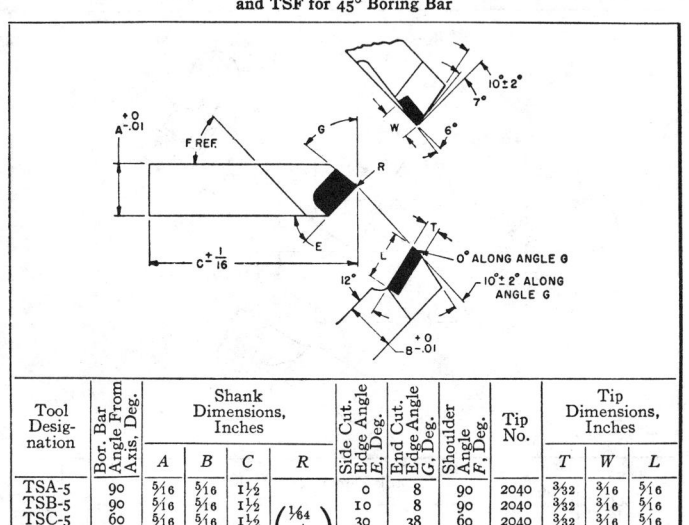

| Tool Designation | Bor. Bar Angle From Axis, Deg. | Shank Dimensions, Inches | | | | Side Cut. Edge Angle E, Deg. | End Cut. Edge Angle G, Deg. | Shoulder Angle F, Deg. | Tip No. | Tip Dimensions, Inches | | |
|---|---|---|---|---|---|---|---|---|---|---|---|---|
| | | A | B | C | R | | | | | T | W | L |
| TSA-5 | 90 | 5/16 | 5/16 | 1½ | | 0 | 8 | 90 | 2040 | 3/32 | 3/16 | 5/16 |
| TSB-5 | 90 | 5/16 | 5/16 | 1½ | | 10 | 8 | 90 | 2040 | 3/32 | 3/16 | 5/16 |
| TSC-5 | 60 | 5/16 | 5/16 | 1½ | 1/64 ± .005 | 30 | 38 | 60 | 2040 | 3/32 | 3/16 | 5/16 |
| TSD-5 | 60 | 5/16 | 5/16 | 1½ | | 40 | 38 | 60 | 2040 | 3/32 | 3/16 | 5/16 |
| TSE-5 | 45 | 5/16 | 5/16 | 1½ | | 45 | 53 | 45 | 2040 | 3/32 | 3/16 | 5/16 |
| TSF-5 | 45 | 5/16 | 5/16 | 1½ | | 55 | 53 | 45 | 2040 | 3/32 | 3/16 | 5/16 |
| TSA-6 | 90 | 3/8 | 3/8 | 1¾ | | 0 | 8 | 90 | 2040 | 3/32 | 3/16 | 5/16 |
| TSB-6 | 90 | 3/8 | 3/8 | 1¾ | | 10 | 8 | 90 | 2040 | 3/32 | 3/16 | 5/16 |
| TSC-6 | 60 | 3/8 | 3/8 | 1¾ | 1/64 ± .005 | 30 | 38 | 60 | 2040 | 3/32 | 3/16 | 5/16 |
| TSD-6 | 60 | 3/8 | 3/8 | 1¾ | | 40 | 38 | 60 | 2040 | 3/32 | 3/16 | 5/16 |
| TSE-6 | 45 | 3/8 | 3/8 | 1¾ | | 45 | 53 | 45 | 2040 | 3/32 | 3/16 | 5/16 |
| TSF-6 | 45 | 3/8 | 3/8 | 1¾ | | 55 | 53 | 45 | 2040 | 3/32 | 3/16 | 5/16 |
| TSA-7 | 90 | 7/16 | 7/16 | 2½ | | 0 | 8 | 90 | 2060 | 3/32 | ¼ | 3/8 |
| TSB-7 | 90 | 7/16 | 7/16 | 2½ | | 10 | 8 | 90 | 2060 | 3/32 | ¼ | 3/8 |
| TSC-7 | 60 | 7/16 | 7/16 | 2½ | 1/32 ± .010 | 30 | 38 | 60 | 2060 | 3/32 | ¼ | 3/8 |
| TSD-7 | 60 | 7/16 | 7/16 | 2½ | | 40 | 38 | 60 | 2060 | 3/32 | ¼ | 3/8 |
| TSE-7 | 45 | 7/16 | 7/16 | 2½ | | 45 | 53 | 45 | 2060 | 3/32 | ¼ | 3/8 |
| TSF-7 | 45 | 7/16 | 7/16 | 2½ | | 55 | 53 | 45 | 2060 | 3/32 | ¼ | 3/8 |
| TSA-8 | 90 | ½ | ½ | 2½ | | 0 | 8 | 90 | 2150 | 1/8 | 5/16 | 7/16 |
| TSB-8 | 90 | ½ | ½ | 2½ | | 10 | 8 | 90 | 2150 | 1/8 | 5/16 | 7/16 |
| TSC-8 | 60 | ½ | ½ | 2½ | 1/32 ± .010 | 30 | 38 | 60 | 2150 | 1/8 | 5/16 | 7/16 |
| TSD-8 | 60 | ½ | ½ | 2½ | | 40 | 38 | 60 | 2150 | 1/8 | 5/16 | 7/16 |
| TSE-8 | 45 | ½ | ½ | 2½ | | 45 | 53 | 45 | 2150 | 1/8 | 5/16 | 7/16 |
| TSF-8 | 45 | ½ | ½ | 2½ | | 55 | 53 | 45 | 2150 | 1/8 | 5/16 | 7/16 |
| TSA-10 | 90 | 5/8 | 5/8 | 3 | | 0 | 8 | 90 | 2220 | 5/32 | 3/8 | 9/16 |
| TSB-10 | 90 | 5/8 | 5/8 | 3 | | 10 | 8 | 90 | 2220 | 5/32 | 3/8 | 9/16 |
| TSC-10 | 60 | 5/8 | 5/8 | 3 | 1/32 ± .010 | 30 | 38 | 60 | 2220 | 5/32 | 3/8 | 9/16 |
| TSD-10 | 60 | 5/8 | 5/8 | 3 | | 40 | 38 | 60 | 2220 | 5/32 | 3/8 | 9/16 |
| TSE-10 | 45 | 5/8 | 5/8 | 3 | | 45 | 53 | 45 | 2220 | 5/32 | 3/8 | 9/16 |
| TSF-10 | 45 | 5/8 | 5/8 | 3 | | 55 | 53 | 45 | 2220 | 5/32 | 3/8 | 9/16 |
| TSA-12 | 90 | ¾ | ¾ | 3½ | | 0 | 8 | 90 | 2300 | 3/16 | 7/16 | 5/8 |
| TSB-12 | 90 | ¾ | ¾ | 3½ | | 10 | 8 | 90 | 2300 | 3/16 | 7/16 | 5/8 |
| TSC-12 | 60 | ¾ | ¾ | 3½ | 1/32 ± .010 | 30 | 38 | 60 | 2300 | 3/16 | 7/16 | 5/8 |
| TSD-12 | 60 | ¾ | ¾ | 3½ | | 40 | 38 | 60 | 2300 | 3/16 | 7/16 | 5/8 |
| TSE-12 | 45 | ¾ | ¾ | 3½ | | 45 | 53 | 45 | 2300 | 3/16 | 7/16 | 5/8 |
| TSF-12 | 45 | ¾ | ¾ | 3½ | | 55 | 53 | 45 | 2300 | 3/16 | 7/16 | 5/8 |

Table 4. American Standard Solid Carbide Round Boring Tools — Style SRC for 60° Boring Bar and Style SRE for 45° Boring Bar

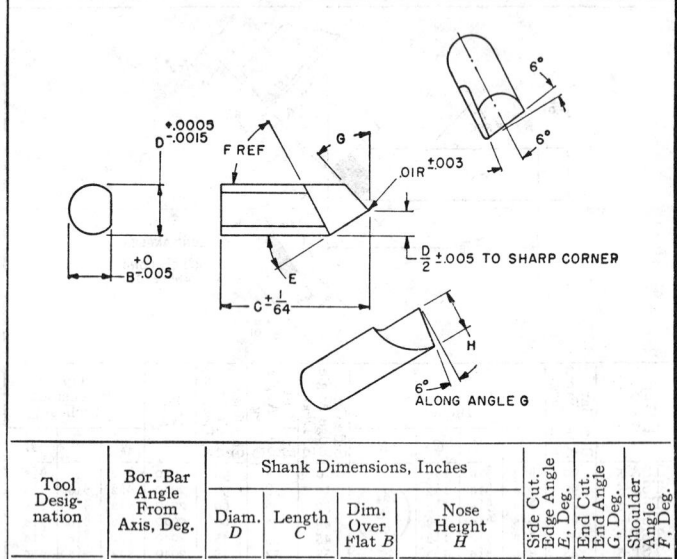

| Tool Desig-nation | Bor. Bar Angle From Axis, Deg. | Shank Dimensions, Inches | | | | Side Cut. Edge Angle E, Deg. | End Cut. End Angle G, Deg. | Shoulder Angle F, Deg. |
| | | Diam. D | Length C | Dim. Over Flat B | Nose Height H | | | |
|---|---|---|---|---|---|---|---|---|
| SRC-33 | 60 | 3/32 | 3/8 | 0.088 | 0.070 [+0.000] | 30 | 38 | 60 |
| SRE-33 | 45 | 3/32 | 3/8 | 0.088 | 0.070 [−0.005] | 45 | 53 | 45 |
| SRC-44 | 60 | 1/8 | 1/2 | 0.118 | 0.094 [+0.000] | 30 | 38 | 60 |
| SRE-44 | 45 | 1/8 | 1/2 | 0.118 | 0.094 [−0.005] | 45 | 53 | 45 |
| SRC-55 | 60 | 5/32 | 5/8 | 0.149 | 0.117 ±0.005 | 30 | 38 | 60 |
| SRE-55 | 45 | 5/32 | 5/8 | 0.149 | 0.117 ±0.005 | 45 | 53 | 45 |
| SRC-66 | 60 | 3/16 | 3/4 | 0.177 | 0.140 ±0.005 | 30 | 38 | 60 |
| SRE-66 | 45 | 3/16 | 3/4 | 0.177 | 0.140 ±0.005 | 45 | 53 | 45 |
| SRC-88 | 60 | 1/4 | 1 | 0.240 | 0.187 ±0.005 | 30 | 38 | 60 |
| SRE-88 | 45 | 1/4 | 1 | 0.240 | 0.187 ±0.005 | 45 | 53 | 45 |
| SRC-1010 | 60 | 5/16 | 1¼ | 0.300 | 0.235 ±0.005 | 30 | 38 | 60 |
| SRE-1010 | 45 | 5/16 | 1¼ | 0.300 | 0.235 ±0.005 | 45 | 53 | 45 |

boring tools. In addition to these ready-to-use standard boring tools, solid carbide round and square unsharpened boring tool bits are provided.

*Style Designations for Carbide Boring Tools:* Table 1 shows designations used to specify the styles of American Standard sintered carbide boring tools. The first letter denotes solid (S) or tipped (T). The second letter denotes square (S) or round (R). The side cutting edge angle is denoted by a third letter (A through H) to complete the style designation. Solid square and round bits with the mounting surfaces ground but the cutting edges unsharpened (Table 7) are designated using the same system except that the third letter indicating side cutting edge angle is omitted.

*Size Designation of Carbide Boring Tools:* Specific sizes of boring tools are identified by the addition of numbers after the style designation. The first number denotes

**Table 5.**  American Standard Carbide-Tipped Round Boring Tools — Style TRC for
60° Boring Bar and Style TRE for 45° Boring Bar

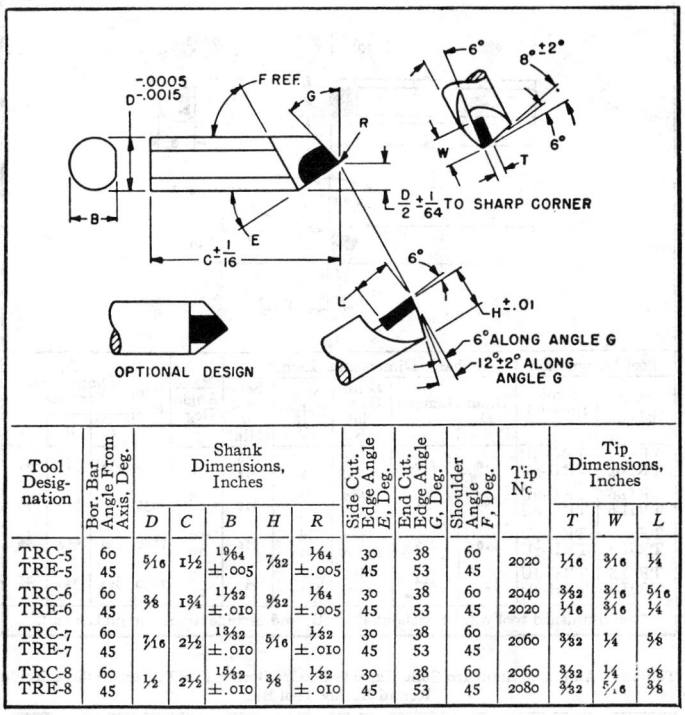

| Tool Designation | Bor. Bar Angle From Axis, Deg. | Shank Dimensions, Inches | | | | | Side Cut. Edge Angle E, Deg. | End Cut. Edge Angle G, Deg. | Shoulder Angle F, Deg. | Tip No | Tip Dimensions, Inches | | |
|---|---|---|---|---|---|---|---|---|---|---|---|---|---|
| | | D | C | B | H | R | | | | | T | W | L |
| TRC-5 | 60 | 5⁄16 | 1½ | 19⁄64 ±.005 | 7⁄32 | 1⁄64 ±.005 | 30 | 38 | 60 | 2020 | 1⁄16 | 3⁄16 | 1⁄4 |
| TRE-5 | 45 | | | | | | 45 | 53 | 45 | | | | |
| TRC-6 | 60 | 3⁄8 | 1¾ | 11⁄32 ±.010 | 9⁄32 | 1⁄64 ±.005 | 30 | 38 | 60 | 2040 | 3⁄32 | 3⁄16 | 5⁄16 |
| TRE-6 | 45 | | | | | | 45 | 53 | 45 | 2020 | 1⁄16 | 3⁄16 | 1⁄4 |
| TRC-7 | 60 | 7⁄16 | 2½ | 13⁄32 ±.010 | 5⁄16 | 1⁄32 ±.010 | 30 | 38 | 60 | 2060 | 3⁄32 | 1⁄4 | 3⁄8 |
| TRE-7 | 45 | | | | | | 45 | 53 | 45 | | | | |
| TRC-8 | 60 | 1⁄2 | 2½ | 15⁄32 ±.010 | 3⁄8 | 1⁄32 ±.010 | 30 | 38 | 60 | 2060 | 3⁄32 | 1⁄4 | 3⁄8 |
| TRE-8 | 45 | | | | | | 45 | 53 | 45 | 2080 | 3⁄32 | 5⁄16 | 3⁄8 |

the diameter or square size in number of $\frac{1}{32}$nds for types SS and SR and in number
of $\frac{1}{16}$ths for types TS and TR.  The second number denotes length in number of
$\frac{1}{8}$ths for types SS and SR.  For styles TRG and TRH, a letter "U" after the number
denotes a semi-finished tool (cutting edges unsharpened).  Complete designations
for the various standard sizes of carbide boring tools are given in Tables 2 through 7.
In the diagrams in the tables, angles shown without tolerance are $\pm 1°$.

*Examples of Tool Designation:* The designation TSC-8 indicates: a carbide-tipped
tool (T); square cross section (S); 30-degree side cutting edge angle (C); and $\frac{8}{16}$
or ½ inch square size.

The designation SRE-66 indicates: a solid carbide tool (S); round cross section
(R); 45 degree side cutting edge angle (E); $\frac{6}{32}$ or $\frac{3}{16}$ inch diameter (6); and $\frac{6}{8}$ or
¾ inch long (6).

The designation SS-610 indicates: a solid carbide tool (S); square cross section
(S); $\frac{6}{32}$ or $\frac{3}{16}$ inch square size (6); $\frac{10}{8}$ or 1¼ inches long (10).

It should be noted in this last example that the absence of a third letter (from
A to H) indicates that the tool has its mounting surfaces ground but that the cutting
edges are unsharpened.

**Table 6. American Standard Carbide-Tipped Round General-Purpose Square-End Boring Tools — Style TRG with 0° Rake and Style TRH with 10° Rake**

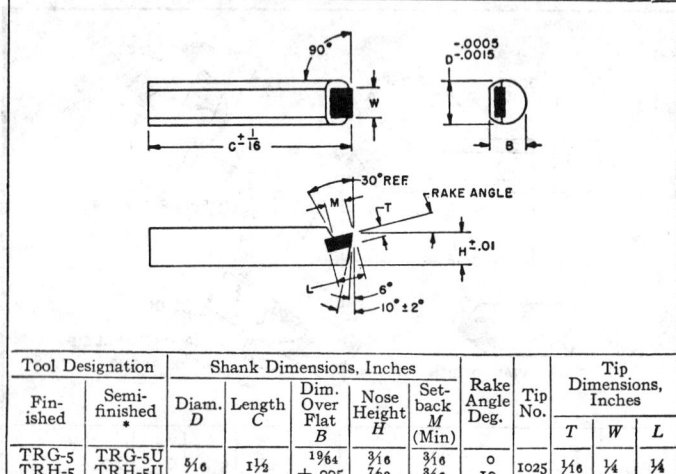

| Tool Designation | | Shank Dimensions, Inches | | | | | Rake Angle Deg. | Tip No. | Tip Dimensions, Inches | | |
|---|---|---|---|---|---|---|---|---|---|---|---|
| Finished | Semi-finished* | Diam. D | Length C | Dim. Over Flat B | Nose Height H | Set-back M (Min) | | | T | W | L |
| TRG-5<br>TRH-5 | TRG-5U<br>TRH-5U | 5/16 | 1½ | 19/64 ±.005 | 3/16<br>7/32 | 3/16<br>3/16 | 0<br>10 | 1025 | 1/16 | 1/4 | 1/4 |
| TRG-6<br>TRH-6 | TRG-6U<br>TRH-6U | 3/8 | 1¾ | 11/32 ±.010 | 7/32<br>1/4 | 3/16 | 0<br>10 | 1030 | 1/16 | 5/16 | 1/4 |
| TRG-7<br>TRH-7 | TRG-7U<br>TRH-7U | 7/16 | 2¼ | 13/32 ±.010 | 1/4<br>5/16 | 3/16 | 0<br>10 | 1080 | 3/32 | 5/16 | 3/8 |
| TRG-8<br>TRH-8 | TRG-8U<br>TRH-8U | 1/2 | 2½ | 15/32 ±.010 | 9/32<br>11/32 | 1/4 | 0<br>10 | 1090 | 3/32 | 3/8 | 3/8 |

* Semi-finished tool will be without Flat (B) and carbide unground on the end.

**Table 7. American Standard Solid Carbide Square Boring Tool Bits and Solid Carbide Round Boring Tool Bits**

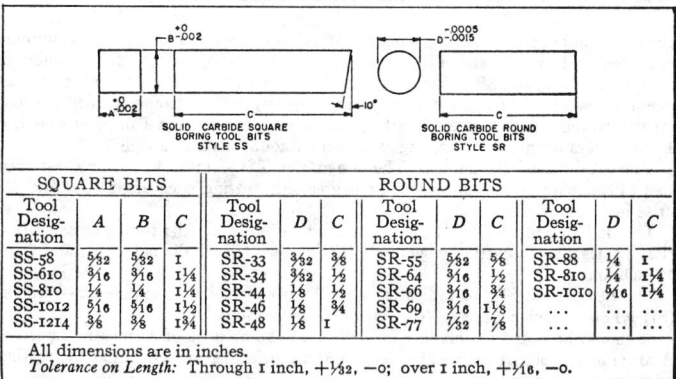

| SQUARE BITS | | | | ROUND BITS | | | | | | | | |
|---|---|---|---|---|---|---|---|---|---|---|---|---|
| Tool Designation | A | B | C | Tool Designation | D | C | Tool Designation | D | C | Tool Designation | D | C |
| SS-58 | 5/32 | 5/32 | 1 | SR-33 | 3/32 | 3/8 | SR-55 | 5/32 | 5/8 | SR-88 | 1/4 | 1 |
| SS-610 | 3/16 | 3/16 | 1¼ | SR-34 | 3/32 | 1/2 | SR-64 | 3/16 | 1/2 | SR-810 | 1/4 | 1¼ |
| SS-810 | 1/4 | 1/4 | 1¼ | SR-44 | 1/8 | 1/2 | SR-66 | 3/16 | 3/4 | SR-1010 | 5/16 | 1¼ |
| SS-1012 | 5/16 | 5/16 | 1½ | SR-46 | 1/8 | 3/4 | SR-69 | 3/16 | 1 1/8 | … | … | … |
| SS-1214 | 3/8 | 3/8 | 1¾ | SR-48 | 1/8 | 1 | SR-77 | 7/32 | 7/8 | … | … | … |

All dimensions are in inches.
*Tolerance on Length:* Through 1 inch, +1/32, −0; over 1 inch, +1/16, −0.

## Spade Drills and Drilling

Spade drills are used to produce holes ranging in size from about one inch to 6 inches diameter, and even larger. Very deep holes can be drilled and blades are available for core drilling, counterboring, and for bottoming to a flat or contoured shape. There are two principal parts to a spade drill, namely, the blade and the holder. The holder has a slot into which the blade fits; a wide slot at the back of the blade engages with a tongue in the holder slot to accurately locate the blade. A retaining screw holds the two parts together. The blade is usually made from high speed steel, although cast nonferrous metal and cemented carbide-tipped blades are also available. Spade drill holders are classified by a letter symbol designating the range of blade sizes that can be held and by their length. Standard stub, short, long, and extra long holders are available; for very deep holes, special holders having wear strips to support and guide the drill are often used. Long, extra long, and many short length holders have coolant holes to direct cutting fluid, under pressure, to the cutting edges. In addition to its function in cooling and lubricating the tool, the cutting fluid also flushes the chips out of the hole. The shank of the holder may be straight or tapered; automotive shanks as specials are also used. A holder and different shank designs are shown in Fig. 1; Fig. 2 shows some typical blades.

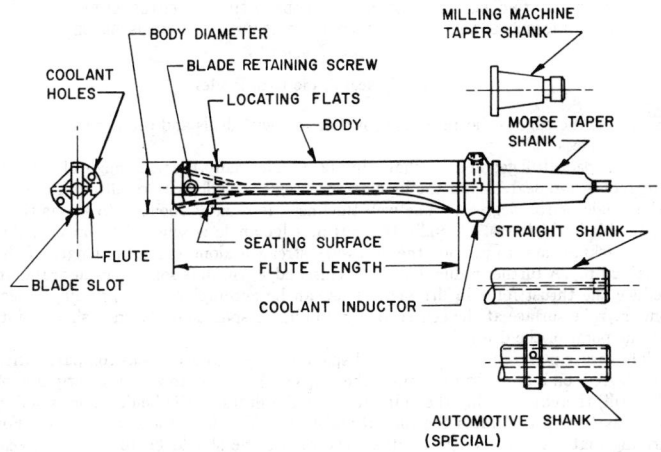

Fig. 1. Spade Drill Blade Holder

**Spade Drill Geometry.** — Metal separation from the work is accomplished in a like manner by both twist drills and spade drills, and the same mechanisms are involved in each case. The two cutting lips separate the metal by a shearing action that is identical to that of chip formation by a single-point cutting tool. At the chisel edge a much more complex condition exists. Here the metal is extruded sideways while at the same time it is sheared by the rotation of the blunt wedge-formed chisel edge. This accounts for the very high thrust force required to penetrate the work. The chisel edge of a twist drill is slightly rounded, while on spade drills it is a straight edge. For this reason it is likely that with spade drills it is more difficult for the extruded metal to escape from the region of the chisel edge.

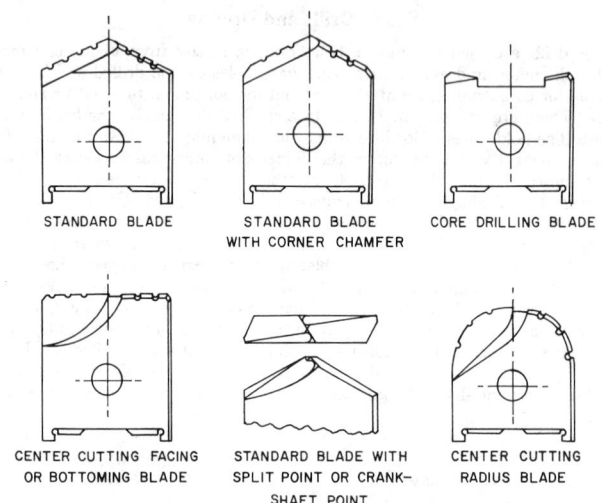

STANDARD BLADE

STANDARD BLADE
WITH CORNER CHAMFER

CORE DRILLING BLADE

CENTER CUTTING FACING
OR BOTTOMING BLADE

STANDARD BLADE WITH
SPLIT POINT OR CRANK-
SHAFT POINT

CENTER CUTTING
RADIUS BLADE

Fig. 2. Typical Spade Drill Blades

However, the edge is shorter in length than on twist drills and the thrust for spade drilling is less.

Basic spade drill geometry is shown in Fig. 3. Normally, the point angle of a standard tool is 130 degrees and the lip clearance angle is 18 degrees, resulting in a chisel edge angle of 108 degrees. The web thickness, is usually about ¼ to ⁵⁄₁₆ as thick as the blade thickness. Usually the cutting edge angle is selected to provide this web thickness and to provide the necessary strength along the entire length of the cutting lip. A further reduction of the chisel edge length is sometimes desirable to reduce the thrust force in drilling. This can be accomplished by grinding a secondary rake surface at the center, or by grinding a split point, or crankshaft point, on the point of the drill.

The larger point angle of a standard spade drill — 130 degrees as compared with 118 degrees on a twist drill — causes the chips to flow more toward the periphery of the drill, thereby allowing the chips to enter the flutes of the holder more readily. The rake angle facilitates the formation of the chip along the cutting lips. For drilling materials of average hardness the rake angle should be 10 to 12 degrees; for hard or tough steels it should be 5 to 7 degrees, and for soft and ductile materials it can be increased to 15 to 20 degrees. The rake surface may be flat or rounded, and the latter design is called radial rake. Radial rake is usually ground so that the rake angle is maximum at the periphery and decreases uniformly toward the center to provide greater cutting edge strength at the center. A flat rake surface is recommended for drilling hard and tough materials in order to reduce the tendency to chipping and to reduce heat damage.

A most important feature of the cutting edge is the chip splitters, which are also called chip breaker grooves. Functionally these grooves are chip dividers; instead of forming a single wide chip along the entire length of the cutting edge, these grooves cause several chips to form that can be readily disposed of through the flutes of the holder. Chip splitters must be carefully ground to prevent the chips from packing in

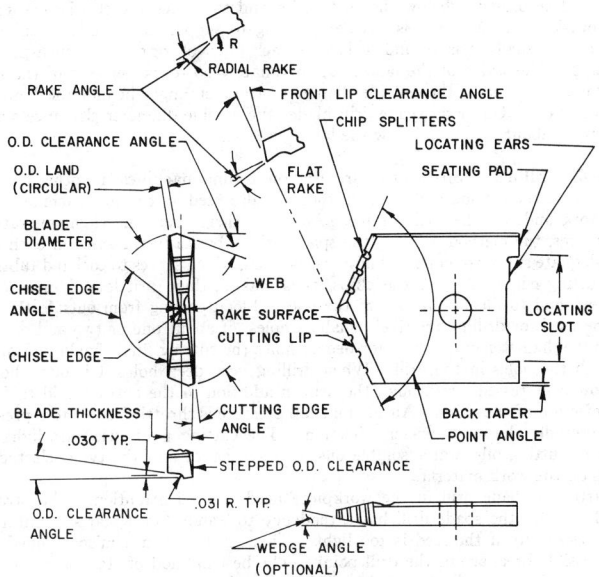

Fig. 3. Spade Drill Blade

the grooves which greatly reduces their effectiveness. They should be ground perpendicular to the cutting lip and parallel to the surface formed by the clearance angle. The grooves on the two cutting lips must not overlap when measured radially along the cutting lip. Figure 4 and the accompanying table show the groove form and dimensions.

On spade drills, the front lip clearance angle provides the relief. It may be ground on a drill grinding machine but usually it is ground flat. The normal front lip clearance angle is 8 degrees; in some instances a secondary relief angle of about 14 degrees is ground below the primary clearance. The wedge angle on the blade is optional. It is generally ground on thicker blades having a larger diameter to

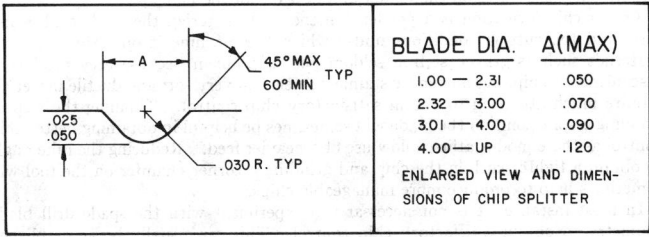

| BLADE DIA. | A(MAX) |
|---|---|
| 1.00 — 2.31 | .050 |
| 2.32 — 3.00 | .070 |
| 3.01 — 4.00 | .090 |
| 4.00 — UP | .120 |

ENLARGED VIEW AND DIMENSIONS OF CHIP SPLITTER

Fig. 4. Spade Drill Chip Splitter Dimensions

prevent heel dragging below the cutting lip and to reduce the chisel edge length. The outside diameter land is circular, serving to support and guide the blade in the hole. Usually it is ground to have a back taper of .001 to .002 inch per inch per side. The width of the land is approximately 20 to 25 per cent of the blade thickness. Normally the outside diameter clearance angle behind the land is 7 to 10 degrees. On many spade drill blades the outside diameter clearance surface is stepped about .030 inch below the land.

**Spade Drilling.** — Spade drills are used on drilling machines and other machine tools where the cutting tool rotates; they are also used on turning machines where the work and not the tool rotates. Although there are some slight operational differences, the method of using the spade drill is basically the same in each case. An adequate supply of cutting fluid must be used, which serves to cool and lubricate the cutting edges; to cool the chips, thus making them brittle and more easily broken; and to flush chips out of the hole. Flood cooling from outside the hole can be used for drilling relatively shallow holes, of about one or two and one-half times the diameter in depth. For deeper holes the cutting fluid should be injected through the holes in the drill. When drilling very deep holes it is often helpful to blow compressed air through the drill in addition to the cutting fluid to facilitate ejection of the chips. Air at full shop pressure is throttled down to a pressure that provides the most efficient ejection. The cutting fluids used are light and medium cutting oils, water soluble oils, and synthetics and the type selected depends on the work material.

Starting a spade drill in the workpiece needs special attention. The straight chisel edge on the spade drill has a tendency to wander as it starts to enter the work, especially if the feed is too light. This can result in a mispositioned hole and possible breakage of the drill point. The best method of starting the hole is to use a stub or short length spade drill holder and a blade of full size which should penetrate at least ⅛ inch at full diameter. The holder is then changed for a longer one as required to complete the hole to depth. Difficulties can be encountered if spotting with a center drill or starting drill is employed because the angles on these drills do not match the 130 degree point angle of the spade drill. Longer spade drills can be started without this starting procedure if the drill is guided by a jig bushing and if the holder is provided with wear strips.

Chip formation warrants the most careful attention as success in spade drilling is dependent on producing short, well-broken chips that can be easily ejected from the hole. Straight, stringy chips or chips that are wound like a clock spring cannot be ejected properly; they tend to pack around the blade, which may result in blade failure. The chip splitters must be functioning to produce a series of narrow chips along each cutting edge. Each chip must be broken and for drilling ductile materials they should be formed into a "C" or "figure 9" shape. Such chips will readily enter the flutes on the holder and flow out of the hole.

Proper chip formation is dependent on the work material, the spade drill geometry, and the cutting conditions under which the machine is operating. Brittle materials such as gray cast iron seldom pose a problem because they produce a discontinuous chip, but austenitic stainless steels and very soft and ductile materials require much attention to obtain satisfactory chip control. Thinning the web or grinding a split point on the blade will sometimes be helpful in obtaining better chip control as these modifications allow use of a heavier feed. Reducing the rake angle to obtain a tighter curl on the chip and grinding a corner chamfer on the tool will sometimes help to produce more manageable chips.

In most instances it is not necessary to experiment with the spade drill blade geometry to obtain satisfactory chip control. This can usually be accomplished by adjusting the cutting conditions; i.e., the cutting speed and the feed rate.

Normally the cutting speed for spade drilling should be 10 to 15 per cent lower than that for an equivalent twist drill, although the same speed can be used if a lower tool life is acceptable. The recommended cutting speeds for twist drills on pages 1774–1780 can be used as a starting point; however, they should be decreased by the percentage just given. It is essential to use a heavy feed rate when spade drilling to produce a thick chip and to force the chisel edge into the work. In ductile materials a light feed will produce a thin chip that is very difficult to break. The thick chip on the other hand, which often contains many rupture planes, will curl and break readily. Table 1 gives suggested feed rates for different drill sizes and materials. These should be used as a starting point and some adjustments may be necessary as experience is gained.

### Table 1. Feed Rates for Spade Drilling

| Material | Hardness, Bhn | Feed — Inches per Revolution | | | | | |
|---|---|---|---|---|---|---|---|
| | | Spade Drill Diameter — Inches | | | | | |
| | | 1–1¼ | 1¼–2 | 2–3 | 3–4 | 4–5 | 5–8 |
| Free Machining Steel | 100–240 | .014 | .016 | .018 | .022 | .025 | .030 |
| | 240–325 | .010 | .014 | .016 | .020 | .022 | .025 |
| Plain Carbon Steels | 100–225 | .012 | .015 | .018 | .022 | .025 | .030 |
| | 225–275 | .010 | .013 | .015 | .018 | .020 | .025 |
| | 275–325 | .008 | .010 | .013 | .015 | .018 | .020 |
| Free Machining Alloy Steels | 150–250 | .014 | .016 | .018 | .022 | .025 | .030 |
| | 250–325 | .012 | .014 | .016 | .018 | .020 | .025 |
| | 325–375 | .010 | .010 | .014 | .016 | .018 | .020 |
| Alloy Steels | 125–180 | .012 | .015 | .018 | .022 | .025 | .030 |
| | 180–225 | .010 | .012 | .016 | .018 | .022 | .025 |
| | 225–325 | .009 | .010 | .013 | .015 | .018 | .020 |
| | 325–400 | .006 | .008 | .010 | .012 | .014 | .016 |
| Tool Steels | | | | | | | |
| Water Hardening | 150–250 | .012 | .014 | .016 | .018 | .020 | .022 |
| Shock Resisting | 175–225 | .012 | .014 | .015 | .016 | .017 | .018 |
| Cold Work | 200–250 | .007 | .008 | .009 | .010 | .011 | .012 |
| Hot Work | 150–250 | .012 | .013 | .015 | .016 | .018 | .020 |
| Mold | 150–200 | .010 | .012 | .014 | .016 | .018 | .018 |
| Special Purpose | 150–225 | .010 | .012 | .014 | .016 | .016 | .018 |
| High Speed Steel | 200–240 | .010 | .012 | .013 | .015 | .017 | .018 |
| Gray Cast Iron | 110–160 | .020 | .022 | .026 | .028 | .030 | .034 |
| | 160–190 | .015 | .018 | .020 | .024 | .026 | .028 |
| | 190–240 | .012 | .014 | .016 | .018 | .020 | .022 |
| | 240–320 | .012 | .012 | .016 | .018 | .018 | .018 |
| Ductile or Nodular Iron | 140–190 | .014 | .016 | .018 | .020 | .022 | .024 |
| | 190–250 | .012 | .014 | .016 | .018 | .018 | .020 |
| | 250–300 | .010 | .012 | .016 | .018 | .018 | .018 |
| Malleable Iron | | | | | | | |
| Ferritic | 110–160 | .014 | .016 | .018 | .020 | .022 | .024 |
| Pearlitic | 160–220 | .012 | .014 | .016 | .018 | .020 | .020 |
| | 220–280 | .010 | .012 | .014 | .016 | .018 | .018 |
| Free Machining Stainless Steel | | | | | | | |
| Ferritic | ..... | .016 | .018 | .020 | .024 | .026 | .028 |
| Austenitic | ..... | .016 | .018 | .020 | .022 | .024 | .026 |
| Martensitic | ..... | .012 | .014 | .016 | .016 | .018 | .020 |
| Stainless Steel | | | | | | | |
| Ferritic | ..... | .012 | .014 | .018 | .020 | .020 | .022 |
| Austenitic | ..... | .012 | .014 | .016 | .018 | .020 | .020 |
| Martensitic | ..... | .010 | .012 | .012 | .014 | .016 | .018 |
| Aluminum Alloys | ..... | .020 | .022 | .024 | .028 | .030 | .040 |
| Copper Alloys | (Soft) | .016 | .018 | .020 | .026 | .028 | .030 |
| | (Hard) | .010 | .012 | .014 | .016 | .018 | .018 |
| Titanium Alloys | ..... | .008 | .010 | .012 | .014 | .014 | .016 |
| High Temperature Alloys | ..... | .008 | .010 | .012 | .012 | .014 | .014 |

**Power Consumption and Thrust for Spade Drilling** — In each individual setup there are factors and conditions influencing power consumption that cannot be accounted for in a simple equation; however, those given below will enable the user to estimate power consumption and thrust accurately enough for most practical purposes. They are based on experimentally derived values of unit horsepower, as given in Table 2. As a word of caution, these values are for sharp tools. In spade drilling, it is reasonable to estimate that a dull tool will increase the power consumption and the thrust by 25 to 50 per cent. The unit horsepower values in the table are for the power consumed at the cutting edge, to which must be added the power required to drive the machine tool itself, in order to obtain the horsepower required by the machine tool motor. This is calculated by dividing the horsepower at the cutter by a mechanical efficiency factor, $e_m$. This factor can be estimated to be .90 for a direct spindle drive with a belt, .75 for a back gear drive, and .70 to .80 for geared head drives. Thus, for spade drilling the formulas are:

$$hp_c = uhp \left( \frac{\pi D^2}{4} \right) fN$$

$$B_s = 148,500 \; uhp \, f \, D$$

$$hp_m = \frac{hp_c}{e_m}$$

$$f = \frac{f_m}{N}$$

Where: $hp_c$ = Horsepower at the cutter
$\quad\quad\;\; hp_m$ = Horsepower at the motor
$\quad\quad\;\;\; B_s$ = Thrust for spade drilling in pounds
$\quad\quad\; uhp$ = Unit horsepower
$\quad\quad\quad\; D$ = Drill diameter in inches
$\quad\quad\quad\;\; f$ = Feed in inches per revolution
$\quad\quad\quad f_m$ = Feed in inches per minute
$\quad\quad\quad\; N$ = Spindle speed in revolutions per minute
$\quad\quad\quad e_m$ = Mechanical efficiency factor

### Table 2.  Unit Horsepower for Spade Drilling

| Material | Hardness | Uhp | Material | Hardness | Uhp |
|---|---|---|---|---|---|
| Plain Carbon and Alloy Steel | 85–200 Bhn | .79 | Titanium Alloys | 250–375 Bhn | .72 |
| | 200–275 | .94 | High Temp Alloys | 200–360 Bhn | 1.44 |
| | 275–375 | 1.00 | Aluminum Alloys | ...... | .22 |
| | 375–425 | 1.15 | Magnesium Alloys | ...... | .16 |
| | 45–52 $R_c$ | 1.44 | Copper Alloys | 20–80 $R_B$ | .43 |
| Cast Irons | 110–200 Bhn | .5 | | 80–100 $R_B$ | .72 |
| | 200–300 | 1.08 | | | |
| Stainless Steels | 135–275 Bhn | .94 | | | |
| | 30–45 $R_c$ | 1.08 | | | |

*Example:* Estimate the horsepower and thrust required to drive a 2 inch diameter spade drill in AISI 1045 steel that is quenched and tempered to a hardness of 240 BHN. From the table (from Table 1.) on page 1683, the cutting speed, $V$, for drilling this material with a twist drill is 50 feet per minute. This value is reduced by 10 per cent for spade drilling and the speed selected is thus .9 × 50 = 45 feet per minute. The feed rate is .016 ipr and the unit horsepower is .94 (from Table 2.). The machine efficiency factor is estimated to be .80 and it will be assumed that a 50 per cent increase in the unit horsepower must be allowed for dull tools.

Step 1. Calculate the spindle speed from the following formula:

$$N = \frac{12V}{\pi D}$$

Where: $N$ = Spindle speed in revolutions per minute
$V$ = Cutting speed in feet per minute
$D$ = Drill diameter in inches

Thus: $N = \dfrac{12 \times 45}{\pi \times 2} = 86$ revolutions per minute

Step 2. Calculate the horsepower at the cutter:

$$hp_c = uhp \left(\frac{\pi D^2}{4}\right) fN = .94 \left(\frac{\pi \times 2^2}{4}\right) .016 \times 86 = 4$$

Step 3. Calculate the horsepower at the motor and provide for a 50 per cent power increase for the dull tool:

$$hp_m = \frac{hp_c}{e_m} = \frac{4}{.80} = 5 \text{ horsepower}$$

$$hp_m \text{ (with dull tool)} = 1.5 \times 5 = 7.5 \text{ horsepower}$$

Step 4. Estimate the spade drill thrust:

$$B_s = 148{,}500 \; uhp \, f \, D = 148{,}500 \times .94 \times .016 \times 2$$
$$= 4467 \text{ lbs (for sharp tool)}$$
$$B_s = 1.5 \times 4467$$
$$= 6700 \text{ lbs (for dull tool)}$$

**Trepanning.** — Cutting a groove in the form of a circle or boring or cutting a hole by removing the center or core in one piece is called trepanning. Shallow trepanning, also called face grooving, can be performed on a lathe using a single-point tool that is similar to a grooving tool but has a curved blade. Generally, the minimum outside diameter that can be cut by this method is about three inches and the maximum groove depth is about two inches. Trepanning is probably the most economical method of producing deep holes that are two inches, and larger, in diameter. Fast production rates can be achieved. The tool consists of a hollow bar, or stem, and a hollow cylindrical head to which a carbide or high-speed steel, single-point cutting tool is attached. Usually only one cutting tool is used although for some applications a multiple cutter head must be used; e.g., heads used to start the hole have multiple tools. In operation, the cutting tool produces a circular groove and a residue core which enters the hollow stem after passing through the head. On outside diameter exhaust trepanning tools the cutting fluid is applied through the stem and the chips are flushed around the outside of the tool; inside diameter exhaust tools flush the chips out through the stem with the cutting fluid applied from the outside. For starting the cut a tool that cuts a starting groove in the work must be used, or the trepanning tool must be guided by a bushing. For holes less than about five diameters deep, a machine that rotates the trepanning tool can be used. Often, an ordinary drill press is satisfactory; deeper holes should be machined on a lathe with the work rotating. A hole diameter tolerance of ± .010 inch can easily be obtained by trepanning and in some cases a tolerance of ± .001 inch has been held. Hole runout can be held to ± .003 inch per foot and, at times, to ± .001 inch per foot. On heat-treated metal a surface finish of 125 to 150 AA can be obtained and on annealed metals 100 to 250 AA is common.

# TAPS AND THREADING DIES

General dimensions and tap markings given in the ANSI Standard B94.9-1971 for hand taps, machine screw taps, spiral pointed taps, spiral fluted taps, nut taps, tapper taps, pulley taps, pipe taps, and STI (Screw Thread Insert) of Unified and American Standard Form are shown in the tables on the pages which follow. This Standard also gives the thread limits for taps with cut threads and ground threads. The thread limits for cut thread and ground thread taps for screw threads are given in Tables 13 through 16 and 19 and 20; thread limits for cut thread and ground thread taps for pipe threads are given in Tables 17a through 18c. Taps recommended for various classes of Unified and American Standard screw threads are given in Table 12 for numbered and fractional sizes.

**Hand Taps and Machine Screw Taps.** — Regular hand taps are taps with thread and shank approximately the same length, with a square to accommodate a driving mechanism. Hand taps were first identified as such because they were used by hand; however, for many years they have been generally machine driven.

Regular hand taps in 2, 3, 4 and 6 flutes are furnished in taper, plug or bottoming chamfers. Taper taps are chamfered 8 to 10 threads. Plug taps are chamfered 3 to 5 threads. Bottoming taps are chamfered approximately 1½ threads. For general dimensions see Tables 1a, 1b and 2.

*Regular Machine Screw Taps:* These are similar to regular hand taps, excepting that they are made in the numbered or machine screw sizes. See Tables 4a and 4b.

*Spiral-Pointed Taps:* Regular hand and machine screw taps having shallower flutes and wider lands. The cutting face of the first few threads is ground at an angle to force the chips ahead to prevent clogging in the flutes. See Tables 2 and 5.

*Spiral-Pointed Short-Flute Taps:* Regular hand and machine screw taps made with spiral point only. The balance of the threaded section is left unfluted. They are used for tapping thin material. See Tables 3 and 6.

*Regular Spiral-Fluted Taps:* Regular hand and machine screw taps having right-hand spiral flutes with a helix angle of from 25 to 35 degrees which are designed to help draw chips from the hole or bridge a keyway. See Tables 3 and 6.

*Fast Spiral-Fluted Taps:* These are similar to regular spiral-fluted taps except that the helix angle is from 45 to 60 degrees. See Tables 3 and 6.

**Nut Taps.** — Designed for nut tapping on a low-production basis. Approximately one-half to three quarters of the threaded portion is chamfered. This distributes the cutting load over a great number of teeth and facilitates the entering of the hole by the tap. See Table 7.

*Bent-Shank Tapper Taps:* These taps are designed for use in an automatic nut tapping machine. The tapped nuts pass over the bent shank and are ejected, so that continuous production is obtained without stopping or reversal. These taps are made in both fractional and machine screw sizes. See Table 10.

**Pulley Taps.** — These taps are exceptionally long hand taps; the extra length being in the shank. The diameter of the shank is the same as the full diameter of the thread. These taps are made available in various lengths. See Table 8.

**Pipe Taps.** — *Taper Pipe Taps:* For tapping standard taper pipe threads. See Table 11.

*Straight Pipe Taps:* For producing standard straight pipe threads. See Table 11.

**STI Screw Thread Insert Taps.** — These taps are oversize to the extent that the internal thread which they produce will accommodate a helical coil wire screw thread insert, which at final assembly will accept a screw thread of the nominal size and pitch. See Tables 9a and 9b.

**Definitions of Tap Terms.** — The definitions which follow are taken from ANSI B94.9 but include only the more important terms. Some tap terms are the same as screw thread terms; therefore, see also definitions on pages 1257 and 1261.

*Back Taper:* A gradual decrease in the diameter of the thread form on a tap from the chamfered end of the land towards the back which creates a slight radial relief in the threads.

*Base of Thread:* That which coincides with the cylindrical or conical surface from which the thread projects.

*Chamfer:* The tapering of the threads at the front end of each land or chaser of a tap by cutting away and relieving the crest of the first few teeth to distribute the cutting action over several teeth.

*Chamfer Angle:* The angle formed between the chamfer and the axis of the tap measured in an axial plane at the cutting edge.

*Chamfer Relief Angle:* The complement of the angle formed between a tangent to the relieved surface at the cutting edge and a radial line to the same point on the cutting edge.

*Class of Thread:* The designation of the class which determines the specification of the size, allowance, and tolerance to which a given threaded product is to be manufactured. It is not applicable to the tools used for threading.

## Tap Terms

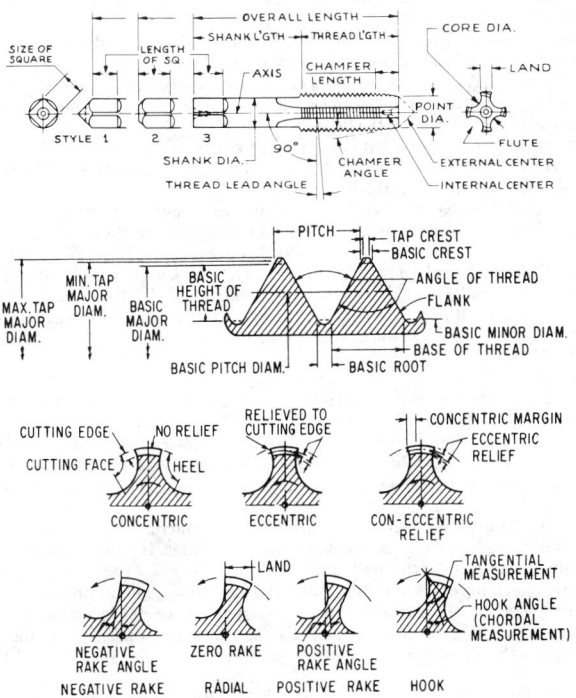

*Core Diameter:* The diameter of a circle which is tangent to the bottom of the flutes at a given point on the axis.

*Crest Clearance:* The radial distance between the root of the internal thread and the crest of the external thread of the coaxially assembled design forms of mating threads.

*First Full Thread:* The first full thread on the cutting edge back of the chamfer. It is at this point that rake, hook, and thread elements are measured.

*Flank Angle:* The flank angle is the angle between the individual flank and the perpendicular to the axis of the thread, measured in an axial plane. A flank angle of a symmetrical thread is commonly termed the "half-angle of thread."

*Flank-Leading:* (1) The flank of a thread facing toward the chamfered end of a threading tool. (2) The leading flank of a thread is the one which, when the thread is about to be assembled with a mating thread, faces the mating thread.

*Flank-Trailing:* The trailing flank of a thread is the one opposite the leading flank.

*Flutes:* The longitudinal channels formed in a tap to create cutting edges on the thread profile and to provide chip spaces and cutting fluid passages. On a parallel or straight thread tap they may be straight, angular or helical; on a taper thread tap they may be straight, angular or spiral.

*Flute–Angular:* A flute lying in a plane intersecting the tool axis at an angle.

*Flute–Helical:* A flute with uniform axial lead and constant helix in a helical path around the axis of a cylindrical tap.

*Flute–Spiral:* A flute with uniform axial lead in a spiral path around the axis of a conical tap.

*Flute Lead Angle:* The angle which a helical or spiral cutting edge at a given point makes with an axial plane through the same point.

*Flute–Straight:* A flute which forms a cutting edge lying in an axial plane.

*Front Taper:* A gradual increase in the diameter of the thread form on a tap from the leading end of the tool toward the back.

*Heel:* The edge of the land opposite the cutting edge.

*Hook Angle:* The inclination of a concave cutting face, usually specified either as Chordal Hook or Tangential Hook.

*Hook–Chordal Angle:* The angle between the chord passing through the root and crest of a thread form at the cutting face, and a radial line through the crest at the cutting edge.

*Hook–Tangential Angle:* The angle between a line tangent to a hook cutting face at the cutting edge and a radial line to the same point.

*Interrupted Thread Tap:* A tap having an odd number of lands with alternate teeth in the thread helix removed. In some cases alternate teeth are removed only for a portion of the thread length.

*Land:* One of the threaded sections between the flutes of a tap.

*Lead:* The distance a screw thread advances axially in one complete turn.

*Lead Error:* The deviation from prescribed limits.

*Lead Deviation:* The deviation from the basic nominal lead. *Progressive Lead Deviation:* (1) On a straight thread the deviation from a true helix where the thread helix advances uniformly. (2) On a taper thread the deviation from a true spiral where the thread spiral advances uniformly.

*Length of Thread:* The length of the thread of the tap includes the chamfered threads and the full threads but does not include an external center. It is indicated by the letter "B" in the illustrations at the heads of the tables.

*Limits:* The limits of size are the applicable maximum and minimum sizes.

*Major Diameter:* On a straight thread the major diameter is that of the major cylinder. On a taper thread the major diameter at a given position on the thread axis is that of the major cone at that position.

*Minor Diameter:* On a straight thread the minor diameter is that of the minor cylinder. On a taper thread the minor diameter at a given position on the thread axis is that of the minor cone at that position.

*Pitch Diameter (Simple Effective Diameter):* On a straight thread, the pitch diameter is the diameter of the imaginary coaxial cylinder, the surface of which would pass through the thread profiles at such points as to make the width of the groove equal to one-half the basic pitch. On a perfect thread this occurs at the point where the widths of the thread and groove are equal. On a taper thread, the pitch diameter at a given position on the thread axis is the diameter of the pitch cone at that position.

*Point Diameter:* The diameter at the cutting edge of the leading end of the chamfered section.

*Rake:* The angular relationship of the straight cutting face of a tooth with respect to a radial line through the crest of the tooth at the cutting edge. Positive rake means that the crest of the cutting face is angularly ahead of the balance of the cutting face of the tooth. Negative rake means that the crest of the cutting face is angularly behind the balance of the cutting face of the tooth. Zero rake means that the cutting face is directly on a radial line.

*Relief:* The removal of metal behind the cutting edge to provide clearance between the part being threaded and the threaded land.

*Relief–Center:* Clearance produced on a portion of the tap land by reducing the diameter of the entire thread form between cutting edge and heel.

*Relief–Chamfer:* The gradual decrease in land height from cutting edge to heel on the chamfered portion of the land on a tap to provide radial clearance for the cutting edge.

*Relief–Con-eccentric Thread:* Radial relief in the thread form starting back of a concentric margin.

*Relief–Double Eccentric Thread:* The combination of a slight radial relief in the thread form starting at the cutting edge and continuing for a portion of the land width, and a greater radial relief for the balance of the land.

*Relief–Eccentric Thread:* Radial relief in the thread form starting at the cutting edge and continuing to the heel.

*Relief–Flatted Land:* Clearance produced on a portion of the tap land by truncating the thread between cutting edge and heel.

*Relief–Grooved Land:* Clearance produced on a tap land by forming a longitudinal groove in the center of the land.

*Relief–Radial:* The clearance produced by removal of metal from behind the cutting edge. Taps should have the chamfer relieved and should have back taper, but may or may not have relief in the angle and on the major diameter of the threads. When the thread angle is relieved, starting at the cutting edge and continuing to the heel, the tap is said to have "eccentric" relief. If the thread angle is relieved back of a concentric margin (usually one-third of land width) the tap is said to have "con-eccentric" relief.

*Size–Actual:* The measured size of an element on an individual part.

*Size–Basic:* That size from which the limits of size are derived by the application of allowances and tolerances.

*Size–Functional:* The functional diameter of an external or internal thread is the pitch diameter of the enveloping thread of perfect pitch, lead and flank angles, having full depth of·engagement but clear at crests and roots, and of a specified length of engagement. It may be derived by adding to the pitch diameter in the case of an external thread, or subtracting from the pitch diameter in the case of an internal thread, the cumulative effects of deviations from specified profile, including variations in lead

*(Continued on page 1695)*

### Table 1a.  ANSI Standard Ground Thread Regular Hand Taps (ANSI B94.9-1971)

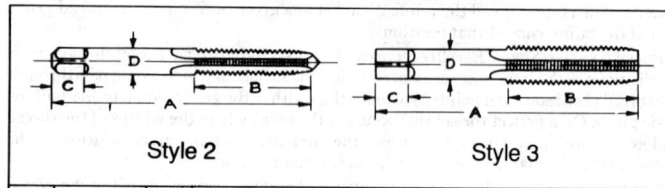

Style 2          Style 3

| Diam. of Tap | Threads per Inch | | No. of Flutes | Pitch Diameter Limits and Chamfers | | | | | Dimensions | | | | |
|---|---|---|---|---|---|---|---|---|---|---|---|---|---|
| | NC UNC | NF UNF | | H1 | H2 | H3 | H4 | H5 | Length Overall, A | Length of Thread, B | Length of Square, C | Diam. of Shank, D | Size of Square, E |
| 1/4 | 20 | .. | 4 | TPB | TPB | TPB | .. | PB | 2½ | 1 | 5/16 | 0.255 | 0.191 |
| 1/4 | .. | 28 | 4 | PB | PB | TPB | PB | .. | 2½ | 1 | 5/16 | 0.255 | 0.191 |
| 5/16 | 18 | .. | 4 | PB | TPB | TPB | .. | PB | 2 23/32 | 1 1/8 | 3/8 | 0.318 | 0.238 |
| 5/16 | .. | 24 | 4 | PB | PB | TPB | PB | .. | 2 23/32 | 1 1/8 | 3/8 | 0.318 | 0.238 |
| 3/8 | 16 | .. | 4 | PB | TPB | TPB | .. | PB | 2 15/16 | 1 1/4 | 7/16 | 0.381 | 0.286 |
| 3/8 | .. | 24 | 4 | PB | PB | TPB | PB | .. | 2 15/16 | 1 1/4 | 7/16 | 0.381 | 0.286 |
| 7/16 | 14 | 20 | 4 | PB | PB | TPB | .. | PB | 3 5/32 | 1 7/16 | 13/32 | 0.323 | 0.242 |
| 1/2 | 13 | 20 | 4 | PB | PB | TPB | .. | PB | 3 3/8 | 1 21/32 | 7/16 | 0.367 | 0.275 |
| 9/16 | 12 | .. | 4 | .. | P | TPB | .. | PB | 3 19/32 | 1 21/32 | 1/2 | 0.429 | 0.322 |
| 9/16 | .. | 18 | 4 | .. | P | TPB | .. | PB | 3 19/32 | 1 21/32 | 1/2 | 0.429 | 0.322 |
| 5/8 | 11 | .. | 4 | P | P | TPB | .. | PB | 3 13/16 | 1 13/16 | 9/16 | 0.480 | 0.360 |
| 5/8 | .. | 18 | 4 | P | P | TPB | .. | PB | 3 13/16 | 1 13/16 | 9/16 | 0.480 | 0.360 |
| 11/16† | .. | .. | 4 | .. | .. | TPB | .. | .. | 4 1/32 | 1 13/16 | 5/8 | 0.542 | 0.406 |
| 3/4 | 10 | .. | 4 | P | P | TPB | .. | PB | 4 1/4 | 2 | 11/16 | 0.590 | 0.442 |
| 3/4 | .. | 16 | 4 | P | P | TPB | .. | PB | 4 1/4 | 2 | 11/16 | 0.590 | 0.442 |
| 7/8§ | 9 | .. | 4 | .. | P | .. | TPB | .. | 4 11/16 | 2 7/32 | 3/4 | 0.697 | 0.523 |
| 7/8§ | .. | 14 | 4 | .. | P | .. | TPB | .. | 4 11/16 | 2 7/32 | 3/4 | 0.697 | 0.523 |
| 1§ | 8 | .. | 4 | .. | P | .. | TPB | .. | 5 1/8 | 2 1/2 | 13/16 | 0.800 | 0.600 |
| 1 | .. | 12 | 4 | .. | .. | .. | TPB | .. | 5 1/8 | 2 1/2 | 13/16 | 0.800 | 0.600 |
| 1‡ | .. | .. | 4 | .. | P | .. | TPB | .. | 5 1/8 | 2 1/2 | 13/16 | 0.800 | 0.600 |
| 1 1/8 | 7 | 12 | 4 | .. | .. | .. | TPB | .. | 5 7/16 | 2 9/16 | 7/8 | 0.896 | 0.672 |
| 1 1/4 | 7 | 12* | 4 | .. | .. | .. | TPB | .. | 5 3/4 | 2 9/16 | 1 | 1.021 | 0.766 |
| 1 3/8 | 6 | 12* | 4 | .. | .. | .. | TPB | .. | 6 1/16 | 3 | 1 1/16 | 1.108 | 0.831 |
| 1 1/2 | 6 | 12* | 4 | .. | .. | .. | TPB | .. | 6 3/8 | 3 | 1 1/8 | 1.233 | 0.925 |

Tolerances for General Dimensions

| Element | Diameter Range | Tolerance | Element | Diameter Range | Tolerance |
|---|---|---|---|---|---|
| Length Overall, A | 1/4 to 1 incl | ±1/32 | Diameter of Shank, D | 1/4 to 5/8 incl | −0.0015 |
| | 1 1/8 to 1 1/2 incl | ±1/16 | | 11/16 to 1 1/2 incl | −0.002 |
| Length of Thread, B | 1/4 to 1/2 incl | ±1/16 | Size of Square, E | 1/4 to 1 incl | −0.004 |
| | 9/16 to 1 1/2 incl | ±3/32 | | 9/16 to 1 incl | −0.006 |
| Length of Square, C | 1/4 to 1 incl | ±1/32 | | 1 1/8 to 1 1/2 incl | −0.008 |
| | 1 1/8 to 1 1/2 incl | ±1/16 | | | |

All dimensions are given in inches.

These taps are standard in high-speed steel.

* In these sizes NF-UNF thread taps have six flutes.

† This size has 11 or 16 threads per inch NS.

‡ This size has 14 threads per inch NS.

§ These sizes are also available with plug chamfer in H6 pitch diameter limits.

Chamfer designations are: T = taper, P = plug, and B = bottoming.

Style 2 taps, 3/8 inch and smaller, have external center on thread end (may be removed on bottoming taps) and external partial cone center on shank end with length of cone approximately one-quarter of diameter of shank.

Style 3 taps, larger than 3/8 inch, have internal center in thread and shank ends.

For standard thread limits see Table 14.

Table 1b.   ANSI Standard Cut Thread Regular Hand Taps (ANSI B94.9-1971)

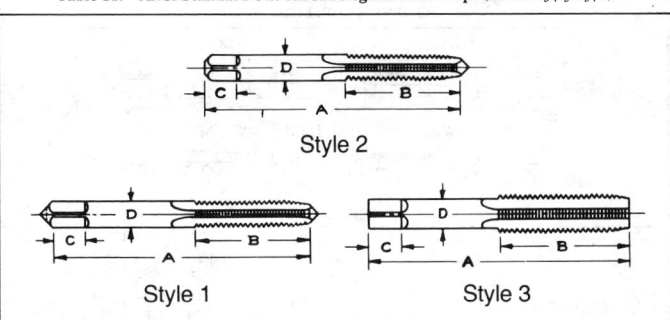

Style 2

Style 1

Style 3

| Diam. of Tap | Threads Per Inch | | | | | Dimensions | | | | | |
|---|---|---|---|---|---|---|---|---|---|---|---|
| | Carbon Steel | | | HS Steel | | No. of Flutes | Length Overall, A | Length of Thread, B | Length of Square, C | Diam. of Shank, D | Size of Square, E |
| | NC UNC | NF UNF | NS or UNS | NC UNC | NF UNF | | | | | | |
| 1/8 | .. | .. | 40 | .. | .. | 3 | 1 15/16 | 5/8 | 3/16 | 0.141 | 0.110 |
| 5/32 | .. | .. | 32 | .. | .. | 4 | 2 1/8 | 3/4 | 1/4 | 0.168 | 0.131 |
| 3/16 | .. | .. | 24, 32 | .. | .. | 4 | 2 3/8 | 7/8 | 1/4 | 0.194 | 0.152 |
| 7/32 | .. | .. | 24 | .. | .. | 4 | 2 3/8 | 15/16 | 9/32 | 0.220 | 0.165 |
| 1/4 | 20 | 28 | .. | 20 | 28 | 4 | 2 1/2 | 1 | 5/16 | 0.255 | 0.191 |
| 5/16 | 18 | 24 | .. | 18 | 24 | 4 | 2 23/32 | 1 1/8 | 3/8 | 0.318 | 0.238 |
| 3/8 | 16 | 24 | .. | 16 | 24 | 4 | 2 15/16 | 1 1/4 | 7/16 | 0.381 | 0.286 |
| 7/16 | 14 | 20 | .. | 14 | 20 | 4 | 3 5/32 | 1 7/16 | 13/32 | 0.323 | 0.242 |
| 1/2 | 13 | 20 | .. | 13 | 20 | 4 | 3 3/8 | 1 21/32 | 7/16 | 0.367 | 0.275 |
| 9/16 | 12 | 18 | .. | 12 | .. | 4 | 3 19/32 | 1 21/32 | 7/16 | 0.429 | 0.322 |
| 5/8 | 11 | 18 | .. | 11 | 18 | 4 | 3 13/16 | 1 13/16 | 9/16 | 0.480 | 0.360 |
| 3/4 | 10 | 16 | .. | 10 | 16 | 4 | 4 1/4 | 2 | 11/16 | 0.590 | 0.442 |
| 7/8 | 9 | 14 | .. | 9 | 14 | 4 | 4 11/16 | 2 7/32 | 3/4 | 0.697 | 0.523 |
| 1 | 8 | .. | 14* | 8 | .. | 4 | 5 1/8 | 2 1/2 | 13/16 | 0.800 | 0.600 |
| 1 1/8 | 7 | 12 | .. | .. | .. | 4 | 5 7/16 | 2 9/16 | 7/8 | 0.896 | 0.672 |
| 1 1/4 | 7 | 12† | .. | .. | .. | 4 | 5 3/4 | 2 9/16 | 1 | 1.021 | 0.766 |
| 1 3/8 | 6* | 12†* | .. | .. | .. | 4 | 6 1/16 | 3 | 1 1/16 | 1.108 | 0.831 |
| 1 1/2 | 6 | 12†* | .. | .. | .. | 4 | 6 3/8 | 3 | 1 1/8 | 1.233 | 0.925 |
| 1 3/4 | 5* | .. | .. | .. | .. | 6 | 7 | 3 3/16 | 1 1/4 | 1.430 | 1.072 |
| 2 | 4 1/2* | .. | .. | .. | .. | 6 | 7 5/8 | 3 9/16 | 1 3/8 | 1.644 | 1.233 |

Tolerances for General Dimensions

| Elements | Range | Tolerance | Elements | Range | Tolerance |
|---|---|---|---|---|---|
| Length Overall, A | 1/16 to 1 | ±1/32 | Diameter of Shank, D | 1/16 to 7/32 | −0.004 |
| | 1 1/8 to 2 | ±1/16 | | 1/4 to 1 | −0.005 |
| Length of Thread, B | 1/16 to 7/32 | ±3/64 | | 1 1/8 to 2 | −0.007 |
| | 1/4 to 1/2 | ±1/16 | Size of Square, E | 1/16 to 1/2 | −0.004 |
| | 9/16 to 1 1/2 | ±3/32 | | 9/16 to 1 | −0.006 |
| | 1 5/8 to 2 | ±1/8 | | 1 1/8 to 2 | −0.008 |
| Length of Square, C | 1/16 to 1 | ±1/32 | | | |
| | 1 1/8 to 2 | ±1/16 | | | |

All dimensions are given in inches.

\* Standard in plug chamfer only.

† In these sizes NF-UNF thread taps have six flutes.

Except where indicated, these taps are standard with taper, plug, or bottoming chamfer.

Cut thread taps, sizes 3/8 inch and smaller have optional style center on thread and shank ends; sizes larger than 3/8 inch have internal centers in thread and shank ends.

For standard thread limits see Table 13.

**Table 2. ANSI Standard Two- and Three-Fluted and Spiral-Pointed Hand Taps** (ANSI B94.9-1971)

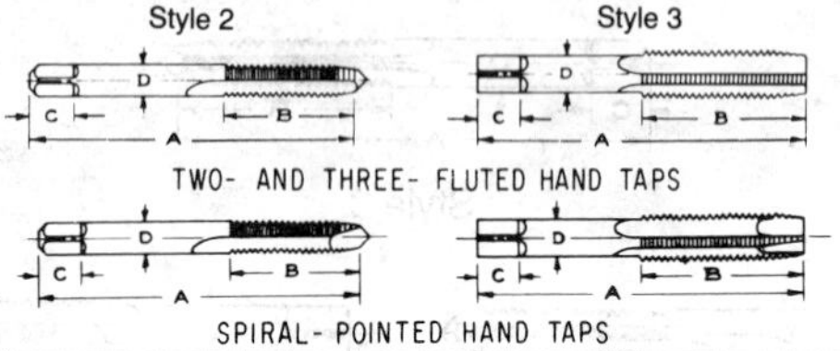


| Diam. of Tap | Threads per Inch NC UNC | Threads per Inch NF UNF | No. of Flutes | Pitch Diam. Limits and Chamfers*† H1 | H2 | H3 | H4 | H5 | Length Overall A | Length of Thread B | Length of Square C | Diam. of Shank D | Size of Square E |
|---|---|---|---|---|---|---|---|---|---|---|---|---|---|
| Ground Thread High Speed Steel Two- and Three-Fluted Hand Taps | | | | | | | | | | | | | |
| 1/4 | 20 | .. | 2 | .. | .. | PB | .. | .. | 2 1/2 | 1 | 5/16 | 0.255 | 0.191 |
| 1/4 | 20 | .. | 3 | P | P | PB | .. | PB | 2 1/2 | 1 | 5/16 | 0.255 | 0.191 |
| 1/4 | .. | 28 | 2, 3 | .. | .. | PB | .. | .. | 2 1/2 | 1 | 5/16 | 0.255 | 0.191 |
| 5/16 | 18 | .. | 2 | .. | .. | PB | .. | .. | 2 23/32 | 1 1/8 | 3/8 | 0.318 | 0.238 |
| 5/16 | 18 | .. | 3 | .. | P | PB | .. | PB | 2 23/32 | 1 1/8 | 3/8 | 0.318 | 0.238 |
| 5/16 | .. | 24 | 3 | .. | .. | PB | .. | .. | 2 23/32 | 1 1/8 | 3/8 | 0.318 | 0.238 |
| 3/8 | 16 | .. | 3 | P | .. | PB | .. | PB | 2 15/16 | 1 1/4 | 7/16 | 0.381 | 0.286 |
| 3/8 | .. | 24 | 3 | .. | .. | PB | .. | .. | 2 15/16 | 1 1/4 | 7/16 | 0.381 | 0.286 |
| 7/16 | 14 | .. | 3 | .. | .. | P | .. | .. | 3 5/32 | 1 7/16 | 13/32 | 0.323 | 0.242 |
| 7/16 | .. | 20 | 3 | .. | .. | P | .. | .. | 3 5/32 | 1 7/16 | 13/32 | 0.323 | 0.242 |
| 1/2 | 13 | .. | 3 | .. | .. | PB | .. | .. | 3 3/8 | 1 21/32 | 7/16 | 0.367 | 0.275 |
| 1/2 | .. | 20 | 3 | .. | .. | P | .. | .. | 3 3/8 | 1 21/32 | 7/16 | 0.367 | 0.275 |
| Ground and Cut Thread High-Speed Steel and Cut Thread Carbon Steel Spiral-Pointed Hand Taps | | | | | | | | | | | | | |
| 1/4 | 20 | .. | 2 | P | P | PB | .. | P | 2 1/2 | 1 | 5/16 | 0.255 | 0.191 |
| 1/4* | 20 | .. | 3 | .. | .. | P | .. | P | 2 1/2 | 1 | 5/16 | 0.255 | 0.191 |
| 1/4 | .. | 28‡ | 2 | P | P | PB | P | .. | 2 1/2 | 1 | 5/16 | 0.255 | 0.191 |
| 1/4* | .. | 28‡ | 3 | .. | P | .. | P | .. | 2 1/2 | 1 | 5/16 | 0.255 | 0.191 |
| 5/16 | 18 | .. | 2 | P | P | PB | .. | P | 2 23/32 | 1 1/8 | 3/8 | 0.318 | 0.238 |
| 5/16* | 18 | .. | 3 | .. | .. | P | .. | P | 2 23/32 | 1 1/8 | 3/8 | 0.318 | 0.238 |
| 5/16 | .. | 24‡ | 2 | P | P | PB | P | .. | 2 23/32 | 1 1/8 | 3/8 | 0.318 | 0.238 |
| 5/16* | .. | 24 | 3 | .. | P | .. | P | .. | 2 23/32 | 1 1/8 | 3/8 | 0.318 | 0.238 |
| 3/8 | 16 | .. | 3 | P | P | P | .. | P | 2 15/16 | 1 1/4 | 7/16 | 0.381 | 0.286 |
| 3/8 | .. | 24‡ | 3 | P | P | P | P | .. | 2 15/16 | 1 1/4 | 7/16 | 0.381 | 0.286 |
| 7/16* | 14 | 20 | 3 | .. | P | P | .. | P | 3 5/32 | 1 7/16 | 13/32 | 0.323 | 0.242 |
| 1/2 | 13‡ | 20* | 3 | P | P | P | .. | P | 3 3/8 | 1 21/32 | 7/16 | 0.367 | 0.275 |
| 5/8* | 11 | .. | 3 | .. | .. | P | .. | P | 3 13/16 | 1 13/16 | 9/16 | 0.480 | 0.360 |
| 3/4* | 10 | .. | 3 | .. | .. | P | .. | P | 4 1/4 | 2 | 11/16 | 0.590 | 0.442 |

Tolerances for General Dimensions

| Element | Diameter Range | Tolerance Ground Thread | Tolerance Cut Thread | Element | Diameter Range | Tolerance Ground Thread | Tolerance Cut Thread |
|---|---|---|---|---|---|---|---|
| Overall Length, *A* | 1/4 to 3/4 | ±1/32 | ±1/32 | Shank Diameter, *D* | 1/4 to 5/8<br>3/4 | −0.0015<br>−0.0020 | −0.005<br>....... |
| Thread Length, *B* | 1/4 to 1/2<br>5/8 to 3/4 | ±1/16<br>±3/32 | ±1/16<br>..... | Size of Square, *E* | 1/4 to 1/2<br>5/8 to 3/4 | −0.0040<br>−0.0060 | −0.004<br>....... |
| Square Length, *C* | 1/4 to 3/4 | ±1/32 | ±1/32 | | | | |

All dimensions are given in inches. Chamfer designations as follows: *P* = plug and *B* = bottoming. Ground thread taps—Style 2, 3/8 inch and smaller, have external center on thread end (may be removed on bottoming taps) and external partial cone center on shank end, with length of cone approximately 1/4 of shank diameter. Ground thread taps—Style 3, larger than 3/8 inch, have internal center in thread and shank ends. Cut thread taps, 3/8 inch and smaller have optional style center on thread and shank ends; sizes larger than 3/8 inch have internal centers in thread and shank ends. For standard thread limits see Tables 13 and 14.

*Applies only to ground thread high-speed steel taps. †Cut thread carbon and high-speed steel taps are standard with plug chamfer only. ‡Does not apply to carbon steel taps.

**Table 3.  Other Types of ANSI Standard Hand Taps** (ANSI B94.9-1971)

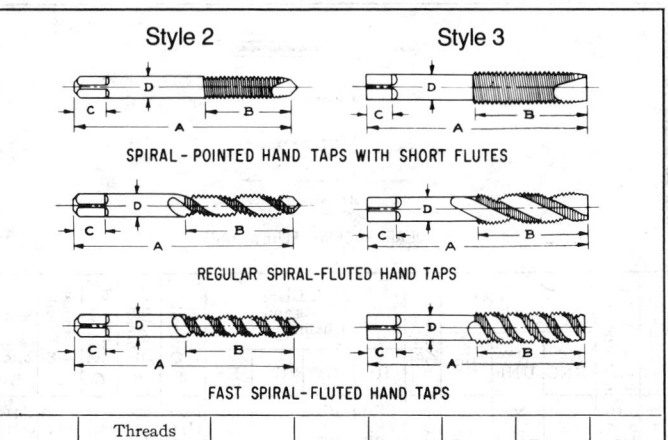

Style 2                    Style 3

SPIRAL-POINTED HAND TAPS WITH SHORT FLUTES

REGULAR SPIRAL-FLUTED HAND TAPS

FAST SPIRAL-FLUTED HAND TAPS

| Diam. of Tap | Threads per Inch | | Number of Flutes | Length Overall A | Length of Thread B | Length of Square C | Diam. of Shank D | Size of Square E |
|---|---|---|---|---|---|---|---|---|
| | NC UNC | NF UNF | | | | | | |
| ¼ | 20 | 28* | 2†, 3* | 2½ | 1 | 5/16 | 0.255 | 0.191 |
| 5/16 | 18 | 24* | 2‡, 3* | 2 23/32 | 1⅛ | 3/8 | 0.318 | 0.238 |
| ⅜ | 16 | 24* | 3 | 2 15/16 | 1¼ | 7/16 | 0.381 | 0.286 |
| 7/16* | 14 | 20*§ | 3 | 3 9/32 | 1 7/16 | 13/32 | 0.323 | 0.242 |
| ½ | 13 | 20* | 3 | 3⅜ | 1 21/32 | 7/16 | 0.367 | 0.275 |

| | | Tolerances for General Dimensions | | | | |
|---|---|---|---|---|---|---|
| Element | Diameter Range | Tolerance | Element | Diameter Range | Tolerance |
| Overall Length, A | ¼ to ½ | ±1/32 | Shank Diameter D | ¼ to ½ | −0.0015 |
| Thread Length, B | ¼ to ½ | ±1/16 | | | |
| Square Length, C | ¼ to ½ | ±1/32 | Size of Square, E | ¼ to ½ | −0.004 |

All dimensions are given in inches.  These standard taps are made of high-speed steel with ground threads.  For standard thread limits see Table 14.

*Spiral-Pointed Hand Taps with Short Flutes:* These taps are standard with plug chamfer only in H3 limit. They are provided with spiral point only. The balance of the threaded section is left unfluted.

*Regular Spiral-Fluted Hand Taps:* These taps are standard with plug or bottoming chamfer in H3 limit and have right-hand spiral flutes with a helix angle of from 25 to 35 degrees.

*Fast Spiral-Fluted Hand Taps:* These taps are standard with plug or bottoming chamfer in H3 limit and have right-hand spiral flutes with a helix angle of from 45 to 60 degrees.

Style 2 taps, ⅜ inch and smaller, have external center on thread end (may be removed on bottoming taps) and external partial cone center on shank end, with length of cone approximately ¼ of diameter of shank.

Style 3 taps, larger than ⅜ inch, have internal center in thread and shank ends.

*Does not apply to spiral-pointed hand taps with short flutes.

†Does not apply to fast-spiral fluted hand taps.

‡Applies only to spiral-pointed hand taps with short flutes.

§Does not apply to regular spiral-fluted hand taps.

**Table 4a. ANSI Standard Ground Thread Regular Machine Screw Taps** (ANSI-B94.9-1971)

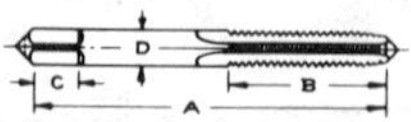


Standard Number of Flutes

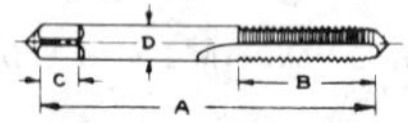


Optional Number of Flutes

REGULAR MACHINE SCREW TAPS

| Size | Basic Major Diameter | Threads per Inch | | | No. of Flutes | Pitch Diam. Limits and Chamfers† | | | | Length Overall | Length of Thread | Length of Square | Diameter of Shank | Size of Square |
|---|---|---|---|---|---|---|---|---|---|---|---|---|---|---|
| | | NC UNC | NF UNF | NS | | H1 | H2 | H3 | H7 | A | B | C | D | E |
| 0 | 0.060 | .. | 80 | .. | 2 | TPB | PB | .. | .. | 1 5/8 | 5/16 | 3/16 | 0.141 | 0.110 |
| 1 | 0.073 | 64 | .. | .. | 2 | TPB | P | .. | .. | 1 11/16 | 3/8 | 3/16 | 0.141 | 0.110 |
| 1 | 0.073 | .. | 72 | .. | 2 | TPB | PB | .. | .. | 1 11/16 | 3/8 | 3/16 | 0.141 | 0.110 |
| 2 | 0.086 | 56 | .. | .. | 2* | PB | PB | .. | .. | 1 3/4 | 7/16 | 3/16 | 0.141 | 0.110 |
| 2 | 0.086 | 56 | .. | .. | 3 | TPB | TPB | .. | .. | 1 3/4 | 7/16 | 3/16 | 0.141 | 0.110 |
| 2 | 0.086 | .. | 64 | .. | 3 | P | TPB | .. | .. | 1 3/4 | 7/16 | 3/16 | 0.141 | 0.110 |
| 3 | 0.099 | 48 | .. | .. | 2* | .. | PB | .. | .. | 1 13/16 | 1/2 | 3/16 | 0.141 | 0.110 |
| 3 | 0.099 | 48 | .. | .. | 3 | P | TPB | .. | .. | 1 13/16 | 1/2 | 3/16 | 0.141 | 0.110 |
| 3 | 0.099 | .. | 56 | .. | 3 | .. | TPB | .. | .. | 1 13/16 | 1/2 | 3/16 | 0.141 | 0.110 |
| 4 | 0.112 | .. | .. | 36 | 3 | .. | TPB | .. | .. | 1 7/8 | 9/16 | 3/16 | 0.141 | 0.110 |
| 4 | 0.112 | 40 | .. | .. | 2* | P | PB | .. | .. | 1 7/8 | 9/16 | 3/16 | 0.141 | 0.110 |
| 4 | 0.112 | 40 | .. | .. | 3 | TPB | TPB | .. | .. | 1 7/8 | 9/16 | 3/16 | 0.141 | 0.110 |
| 4 | 0.112 | .. | 48 | .. | 3 | P | TPB | .. | .. | 1 7/8 | 9/16 | 3/16 | 0.141 | 0.110 |
| 5 | 0.125 | 40 | .. | .. | 2* | .. | PB | .. | .. | 1 15/16 | 5/8 | 3/16 | 0.141 | 0.110 |
| 5 | 0.125 | 40 | .. | .. | 3 | PB | TPB | .. | .. | 1 15/16 | 5/8 | 3/16 | 0.141 | 0.110 |
| 5 | 0.125 | .. | 44 | .. | 2* | .. | P | .. | .. | 1 15/16 | 5/8 | 3/16 | 0.141 | 0.110 |
| 5 | 0.125 | .. | 44 | .. | 3 | .. | TPB | .. | .. | 1 15/16 | 5/8 | 3/16 | 0.141 | 0.110 |
| 6 | 0.138 | 32 | .. | .. | 2* | .. | PB | PB | .. | 2 | 11/16 | 3/16 | 0.141 | 0.110 |
| 6 | 0.138 | 32 | .. | .. | 3 | TPB | TPB | TPB | PB | 2 | 11/16 | 3/16 | 0.141 | 0.110 |
| 6 | 0.138 | .. | 40 | .. | 2* | .. | P | .. | .. | 2 | 11/16 | 3/16 | 0.141 | 0.110 |
| 6 | 0.138 | .. | 40 | .. | 3 | P | TPB | .. | .. | 2 | 11/16 | 3/16 | 0.141 | 0.110 |
| 8 | 0.164 | 32 | .. | .. | 2* | P | PB | PB | .. | 2 1/8 | 3/4 | 1/4 | 0.168 | 0.131 |
| 8 | 0.164 | 32 | .. | .. | 3* | PB | PB | PB | PB | 2 1/8 | 3/4 | 1/4 | 0.168 | 0.131 |
| 8 | 0.164 | 32 | .. | .. | 4 | TPB | TPB | TPB | PB | 2 1/8 | 3/4 | 1/4 | 0.168 | 0.131 |
| 8 | 0.164 | .. | 36 | .. | 4 | P | TPB | .. | .. | 2 1/8 | 3/4 | 1/4 | 0.168 | 0.131 |
| 10 | 0.190 | 24 | .. | .. | 2* | .. | PB | PB | .. | 2 3/8 | 7/8 | 1/4 | 0.194 | 0.152 |
| 10 | 0.190 | 24 | .. | .. | 3* | P | P | PB | PB | 2 3/8 | 7/8 | 1/4 | 0.194 | 0.152 |
| 10 | 0.190 | .. | 32 | .. | 2* | PB | PB | PB | .. | 2 3/8 | 7/8 | 1/4 | 0.194 | 0.152 |
| 10 | 0.190 | .. | 32 | .. | 3* | .. | PB | PB | PB | 2 3/8 | 7/8 | 1/4 | 0.194 | 0.152 |
| 10 | 0.190 | 24 | 32 | .. | 4 | TPB | TPB | TPB | PB | 2 3/8 | 7/8 | 1/4 | 0.194 | 0.152 |
| 12 | 0.216 | 24 | .. | .. | 4 | .. | .. | TPB | .. | 2 3/8 | 15/16 | 9/32 | 0.220 | 0.165 |
| 12 | 0.216 | .. | 28 | .. | 4 | P | .. | TPB | .. | 2 3/8 | 15/16 | 9/32 | 0.220 | 0.165 |

All dimensions are given in inches.

These taps are standard as high-speed steel taps with ground threads, with standard and optional number of flutes and pitch diameter limits and chamfers as given in the table.

*Optional number of flutes.

Chamfer designations are: $T$ = taper, $P$ = plug, and $B$ = bottoming.

Style 1 taps have external centers on thread and shank ends (may be removed on thread end of bottoming taps).

For standard thread limits see Table 15.

*Tolerances:* No. 0 to 12 size range—$A$, $\pm 1/32$; $B$, $\pm 3/64$; $C$, $\pm 1/32$; $D$, $-0.0015$ for ground threads, $-0.004$ for cut threads; $E$, $-0.004$. No. 14 size—$A$, $\pm 1/32$; $B$, $\pm 1/16$; $C$, $\pm 1/32$; $D$, $-0.005$; $E$, $-0.004$.

and flank angle over a specified length of engagement. The effects of taper, out-of-roundness, and surface defects may be positive or negative on either external or internal threads.

*Size–Nominal:* The designation which is used for the purpose of general identification.

*Spiral Flute:* See *Flutes.*

*Spiral Point:* The angular fluting in the cutting face of the land at the chamfered end. It is formed at an angle with respect to the tap axis of opposite hand to that of rotation. Its length is usually greater than the chamfer length and its angle with respect to the tap axis is usually made great enough to direct the chips ahead of the tap. The tap may or may not have longitudinal flutes.

*Thread Lead Angle:* On a straight thread, the lead angle is the angle made by the helix of the thread at the pitch line with a plane perpendicular to the axis. On a taper thread, the lead angle at a given axial position is the angle made by the conical spiral of the thread, with the plane perpendicular to the axis, at the pitch line.

**Table 4b. ANSI Standard Cut Thread Regular Machine Screw Taps** (ANSI B94.9-1971)

Style 1     Style 2

| Size | Basic Major Diameter | Threads per Inch | | | | | Number of Flutes | Dimensions | | | | |
|---|---|---|---|---|---|---|---|---|---|---|---|---|
| | | Carbon Steel | | | HS Steel | | | Length Overall, A | Length of Thread, B | Length of Square, C | Diameter of Shank, D | Size of Square, E |
| | | NC UNC | NF UNF | NS | NC UNC | NF UNF | | | | | | |
| 0 | 0.060 | ... | 80* | ... | ... | ... | 2 | 1 5/8 | 5/16 | 3/16 | 0.141 | 0.110 |
| 1 | 0.073 | 64* | 72* | ... | ... | ... | 2 | 1 11/16 | 3/8 | 3/16 | 0.141 | 0.110 |
| 2 | 0.086 | 56* | 64* | ... | ... | ... | 3 | 1 3/4 | 7/16 | 3/16 | 0.141 | 0.110 |
| 3 | 0.099 | 48* | 56* | ... | ... | ... | 3 | 1 13/16 | 1/2 | 3/16 | 0.141 | 0.110 |
| 4 | 0.112 | 40 | 48* | 36* | 40 | ... | 3 | 1 7/8 | 9/16 | 3/16 | 0.141 | 0.110 |
| 5 | 0.125 | 40 | ... | ... | 40 | ... | 3 | 1 15/16 | 5/8 | 3/16 | 0.141 | 0.110 |
| 6 | 0.138 | 32 | 40* | 36* | 32 | ... | 3 | 2 | 11/16 | 3/16 | 0.141 | 0.110 |
| 8 | 0.164 | 32 | 36* | 40* | 32 | ... | 4 | 2 1/8 | 3/4 | 1/4 | 0.168 | 0.131 |
| 10 | 0.190 | 24 | 32 | ... | 24 | 32 | 4 | 2 3/8 | 7/8 | 1/4 | 0.194 | 0.152 |
| 12 | 0.216 | 24 | 28* | ... | 24 | ... | 4 | 2 3/8 | 15/16 | 9/32 | 0.220 | 0.165 |
| 14 | 0.242 | ... | ... | 24* | ... | ... | 4 | 2 1/2 | 1 | 5/16 | 0.255 | 0.191 |

Tolerances for General Dimensions

| Element | Range | Tolerance | Element | Range | Tolerance |
|---|---|---|---|---|---|
| Length Overall, A | 0 to 14 incl | ±1/32 | Diameter of Shank, D | 0 to 12 incl | −0.004 |
| Length of Thread, B | 0 to 12 incl<br>14 | ±3/64<br>±1/16 | | 14 | −0.005 |
| Length of Square, C | 0 to 14 incl. | ±1/32 | Size of Square, E | 0 to 14 incl | −0.004 |

All dimensions are given in inches.

*These taps are standard with plug chamfer only. All others are standard with taper, plug or bottoming chamfer.

Styles 1 and 2 cut thread taps have optional style centers on thread and shank ends.

For standard thread limits see Table 16.

## Table 5. ANSI Standard Spiral-Pointed Machine Screw Taps (ANSI B94.9-1971)

Style 1

### High-Speed Steel Taps with Ground Threads

| Size | Basic Major Diameter | Threads per Inch NC UNC | NF UNF | NS | No. of Flutes | $H_1$ | $H_2$ | $H_3$ | $H_7$ | Length Overall A | Length of Thread B | Length of Square C | Diameter of Shank D | Size of Square E |
|---|---|---|---|---|---|---|---|---|---|---|---|---|---|---|
| 0 | 0.060 | .. | 80 | .. | 2 | PB | PB | .. | .. | $1\frac{7}{8}$ | $\frac{5}{16}$ | $\frac{3}{16}$ | 0.141 | 0.110 |
| 1 | 0.073 | 64 | 72 | .. | 2 | P | P | .. | .. | $1\frac{11}{16}$ | $\frac{3}{8}$ | $\frac{3}{16}$ | 0.141 | 0.110 |
| 2 | 0.086 | 56 | .. | .. | 2 | PB | PB | .. | .. | $1\frac{3}{4}$ | $\frac{7}{16}$ | $\frac{3}{16}$ | 0.141 | 0.110 |
| 2 | 0.086 | .. | 64 | .. | 2 | P | P | .. | .. | $1\frac{3}{4}$ | $\frac{7}{16}$ | $\frac{3}{16}$ | 0.141 | 0.110 |
| 3 | 0.099 | 48 | .. | .. | 2 | PB | PB | .. | .. | $1\frac{13}{16}$ | $\frac{1}{2}$ | $\frac{3}{16}$ | 0.141 | 0.110 |
| 3 | 0.099 | .. | 56 | .. | 2 | P | P | .. | .. | $1\frac{13}{16}$ | $\frac{1}{2}$ | $\frac{3}{16}$ | 0.141 | 0.110 |
| 4 | 0.112 | .. | .. | 36 | 2 | .. | P | .. | .. | $1\frac{7}{8}$ | $\frac{9}{16}$ | $\frac{3}{16}$ | 0.141 | 0.110 |
| 4 | 0.112 | 40 | .. | .. | 2 | PB | PB | .. | .. | $1\frac{7}{8}$ | $\frac{9}{16}$ | $\frac{3}{16}$ | 0.141 | 0.110 |
| 4 | 0.112 | .. | 48 | .. | 2 | PB | PB | .. | .. | $1\frac{7}{8}$ | $\frac{9}{16}$ | $\frac{3}{16}$ | 0.141 | 0.110 |
| 5 | 0.125 | 40 | .. | .. | 2 | P | PB | .. | .. | $1\frac{15}{16}$ | $\frac{5}{8}$ | $\frac{3}{16}$ | 0.141 | 0.110 |
| 5 | 0.125 | .. | 44 | .. | 2 | .. | P | .. | .. | $1\frac{15}{16}$ | $\frac{5}{8}$ | $\frac{3}{16}$ | 0.141 | 0.110 |
| 6 | 0.138 | 32 | .. | .. | 2 | PB | PB | PB | PB | 2 | $1\frac{1}{16}$ | $\frac{3}{16}$ | 0.141 | 0.110 |
| 6 | 0.138 | .. | 40 | .. | 2 | P | P | PB | PB | 2 | $1\frac{1}{16}$ | $\frac{3}{16}$ | 0.141 | 0.110 |
| 8 | 0.164 | 32 | .. | .. | 2 | PB | PB | PB | PB | $2\frac{1}{8}$ | $\frac{3}{4}$ | $\frac{1}{4}$ | 0.168 | 0.131 |
| 8 | 0.164 | .. | 36 | .. | 2 | P | P | .. | .. | $2\frac{1}{8}$ | $\frac{3}{4}$ | $\frac{1}{4}$ | 0.168 | 0.131 |
| 10 | 0.190 | 24 | .. | .. | 2 | P | PB | PB | P | $2\frac{3}{8}$ | $\frac{7}{8}$ | $\frac{1}{4}$ | 0.194 | 0.152 |
| 10 | 0.190 | .. | 32 | .. | 2 | PB | PB | PB | P | $2\frac{3}{8}$ | $\frac{7}{8}$ | $\frac{1}{4}$ | 0.194 | 0.152 |
| 12 | 0.216 | 24 | .. | .. | 2 | P | .. | PB | .. | $2\frac{3}{8}$ | $\frac{15}{16}$ | $\frac{9}{32}$ | 0.220 | 0.165 |
| 12 | 0.216 | .. | 28 | .. | 2 | .. | .. | .. | P | $2\frac{3}{8}$ | $\frac{15}{16}$ | $\frac{9}{32}$ | 0.220 | 0.165 |

### High-Speed and Carbon Steel Taps with Cut Threads

| Size | Basic Major Diameter | Carbon Steel NC UNC | NF UNF | HS Steel NC UNC | NF UNF | No. of Flutes | Length Overall A | Length of Thread B | Length of Square C | Diameter of Shank D | Size of Square E |
|---|---|---|---|---|---|---|---|---|---|---|---|
| 4 | 0.112 | .. | .. | 40 | .. | 2 | $1\frac{7}{8}$ | $\frac{9}{16}$ | $\frac{3}{16}$ | 0.141 | 0.110 |
| 5 | 0.125 | .. | .. | 40 | .. | 2 | $1\frac{15}{16}$ | $\frac{5}{8}$ | $\frac{3}{16}$ | 0.141 | 0.110 |
| 6 | 0.138 | 32 | .. | 32 | .. | 2 | 2 | $1\frac{1}{16}$ | $\frac{3}{16}$ | 0.141 | 0.110 |
| 8 | 0.164 | 32 | .. | 32 | .. | 2 | $2\frac{1}{8}$ | $\frac{3}{4}$ | $\frac{1}{4}$ | 0.168 | 0.131 |
| 10 | 0.190 | 24 | 32 | 24 | 32 | 2 | $2\frac{3}{8}$ | $\frac{7}{8}$ | $\frac{1}{4}$ | 0.194 | 0.152 |
| 12 | 0.216 | 24 | .. | 24 | .. | 2 | $2\frac{3}{8}$ | $\frac{15}{16}$ | $\frac{9}{32}$ | 0.220 | 0.165 |

### Tolerances for General Dimensions

| Element | Size Range | Tolerance Ground Thread | Cut Thread | Element | Size Range | Tolerance Ground Thread | Cut Thread |
|---|---|---|---|---|---|---|---|
| Overall Length, A | 0 to 12 | $\pm\frac{1}{32}$ | $\pm\frac{1}{32}$ | Shank Diameter, D | 0 to 12 | −0.0015 | −0.004 |
| Thread Length, B | 0 to 12 | $\pm\frac{3}{64}$ | $\pm\frac{3}{64}$ | Size of Square, E | 0 to 12 | −0.004 | −0.004 |
| Square Length, C | 0 to 12 | $\pm\frac{1}{32}$ | $\pm\frac{1}{32}$ | | | | |

All dimensions are given in inches. Chamfer designations are: P = plug and B = bottoming. The cut thread taps are standard with plug chamfer only. The Style 1 ground thread taps have external centers on thread and shank ends (may be removed on thread end of bottoming taps). The Style 1 cut thread taps have optional style centers on thread and shank ends. Standard thread limits for ground threads are given in Table 15 and those for cut threads in Table 16.

**Table 6. ANSI Standard Spiral-Pointed Machine Screw Taps with Short Flutes and Regular and Fast Spiral-Fluted Machine Screw Taps** (ANSI B94.9-1971)

Style 1

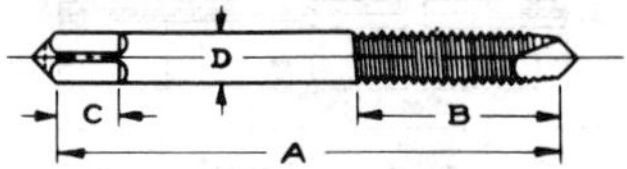


SPIRAL POINTED MACHINE SCREW TAP WITH SHORT FLUTES

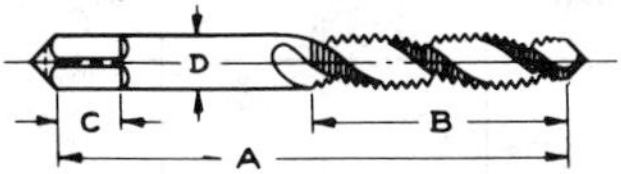


REGULAR SPIRAL-FLUTED MACHINE SCREW TAPS

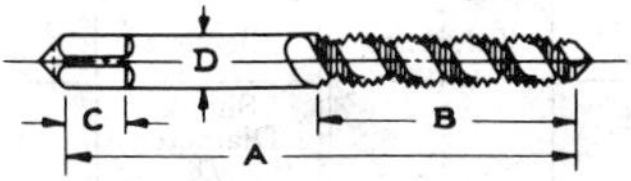


FAST SPIRAL-FLUTED MACHINE SCREW TAPS

| Size | Basic Major Diameter | Threads per Inch | | No. of Flutes | Pitch Dia. Limits & Chamfers* | | Length Overall *A* | Length of Thread *B* | Length of Square *C* | Diameter of Shank *D* | Size of Square *E* |
|---|---|---|---|---|---|---|---|---|---|---|---|
| | | NC UNC | NF UNF | | H2 | H3 | | | | | |
| 3† | 0.099 | 48 | .. | 2 | PB | .. | 1 13/16 | 1/2 | 3/16 | 0.141 | 0.110 |
| 4 | 0.112 | 40 | .. | 2 | PB | .. | 1 7/8 | 9/16 | 3/16 | 0.141 | 0.110 |
| 5 | 0.125 | 40 | .. | 2 | PB | .. | 1 15/16 | 5/8 | 3/16 | 0.141 | 0.110 |
| 6 | 0.138 | 32 | .. | 2 | .. | PB | 2 | 11/16 | 3/16 | 0.141 | 0.110 |
| 8 | 0.164 | 32 | .. | 2‡, 3† | .. | PB | 2 1/8 | 3/4 | 1/4 | 0.168 | 0.131 |
| 10 | 0.190 | 24 | 32 | 2‡, 3† | .. | PB | 2 3/8 | 7/8 | 1/4 | 0.194 | 0.152 |
| 12§ | 0.216 | 24 | .. | 2‡, 3† | .. | PB | 2 3/8 | 15/16 | 9/32 | 0.220 | 0.165 |

Tolerances for General Dimensions

| Element | Size Range | Tolerance | Element | Size Range | Tolerance |
|---|---|---|---|---|---|
| Overall Length, *A* | 3 to 12 | ±1/32 | Shank Diameter *D* | 3 to 12 | −0.0015 |
| Thread Length, *B* | 3 to 12 | ±3/64 | Size of Square, *E* | 3 to 12 | −0.004 |
| Square Length, *C* | 3 to 12 | ±1/32 | | | |

All dimensions are given in inches. These standard taps are made of high-speed steel with ground threads. For standard thread limits see Table 15.

*Spiral-Pointed Machine Screw Taps with Short Flutes:* These taps are standard with plug chamfer only. They are provided with a spiral point only; the balance of the threaded section is left unfluted. These Style 1 taps have external centers on thread and shank ends.

*Regular Spiral-Fluted Machine Screw Taps:* These taps have right-hand spiral flutes with a helix angle of from 25 to 35 degrees.

*Fast Spiral-Fluted Machine Screw Taps:* These taps have right-hand spiral flutes with a helix angle of from 45 to 60 degrees.

Both regular and fast spiral-fluted Style 1 taps have external centers on thread and shank ends (may be removed on thread end of bottoming taps).

Chamfer designations: P = plug and B = bottoming.

* Applies only to regular and fast spiral-fluted machine screw taps.

† Applies only to fast spiral-fluted machine screw taps.

‡ Does not apply to fast spiral-fluted machine screw taps.

§ Does not apply to regular spiral-fluted machine screw taps.

**Table 7. ANSI Standard Nut Taps** (ANSI B94.9-1971)

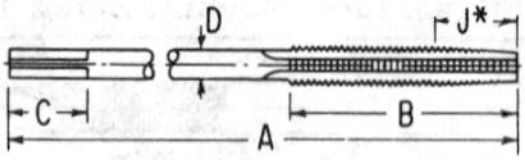


| Diam. of Tap | Threads per Inch NC UNC | Number of Flutes | Length Overall *A* | Length of Thread *B* | Length of Square *C* | Diameter of Shank *D* | Size of Square *E* |
|---|---|---|---|---|---|---|---|
| 1/4 | 20 | 4 | 5 | 1 5/8 | 9/16 | 0.185 | 0.139 |
| 5/16 | 18 | 4 | 5 1/2 | 1 13/16 | 5/8 | 0.240 | 0.180 |
| 3/8 | 16 | 4 | 6 | 2 | 11/16 | 0.294 | 0.220 |
| 1/2 | 13 | 4 | 7 | 2 1/2 | 7/8 | 0.400 | 0.300 |

Tolerances for General Dimensions

| Element | Diameter Range | Tolerance | Element | Diameter Range | Tolerance |
|---|---|---|---|---|---|
| Overall Length, *A* | 1/4 to 1/2 | ±1/16 | Shank Diameter *D* | 1/4 to 1/2 | −0.005 |
| Thread Length, *B* | 1/4 to 1/2 | ±1/16 | | | |
| Square Length, *C* | 1/4 to 1/2 | ±1/32 | Size of Square, *E* | 1/4 to 1/2 | −0.004 |

All dimensions are given in inches. These ground thread high-speed steel taps are standard in H3 limit only. All taps have an internal center in thread end. For standard limits see Table 14.

* Chamfer *J* is made 1/2 to 3/4 the thread length *B*.

**Table 8. ANSI Standard Pulley Taps** (ANSI B94.9-1971)

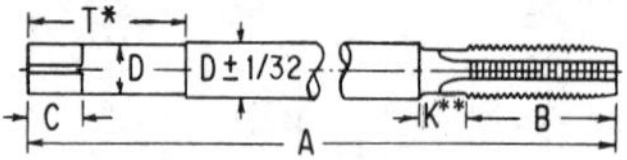


| Diam. of Tap | Threads per Inch NC UNC | Number of Flutes | Length Overall *A* | Length of Thread *B* | Length of Square *C* | Diam. of Shank *D* | Length of Close Tolerance *T** | Size of Square *E* | Length of Neck *K*** |
|---|---|---|---|---|---|---|---|---|---|
| 1/4 | 20 | 4 | 6,8 | 1 | 5/16 | 0.255 | 1 1/2 | 0.191 | 3/8 |
| 5/16 | 18 | 4 | 6,8 | 1 1/8 | 3/8 | 0.318 | 1 9/16 | 0.238 | 3/8 |
| 3/8 | 16 | 4 | 6,8,10 | 1 1/4 | 7/16 | 0.381 | 1 5/8 | 0.286 | 3/8 |
| 7/16 | 14 | 4 | 6,8 | 1 7/16 | 1/2 | 0.444 | 1 11/16 | 0.333 | 7/16 |
| 1/2 | 13 | 4 | 6,8,10,12 | 1 21/32 | 9/16 | 0.507 | 1 11/16 | 0.380 | 1/2 |
| 5/8 | 11 | 4 | 6,8,10.12 | 1 13/16 | 11/16 | 0.633 | 2 | 0.475 | 5/8 |
| 3/4 | 10 | 4 | 10,12 | 2 | 3/4 | 0.759 | 2 1/4 | 0.569 | 3/4 |

Tolerances for General Dimensions

| Element | Diameter Range | Tolerance | Element | Diameter Range | Tolerance |
|---|---|---|---|---|---|
| Overall Length, *A* | 1/4 to 3/4 | ±1/16 | Shank Diameter, *D* | 1/4 to 3/4 | −0.005 |
| Thread Length, *B* | 1/4 to 3/4 | ±1/16 | | | |
| Square Length, *C* | 1/4 to 3/4 | ±1/32 | Size of Square, *E* | 1/4 to 1/2<br>5/8 to 3/4 | −0.004<br>−0.006 |

All dimensions are given in inches. These ground thread high-speed taps are standard with plug chamfer in H3 limit only. All taps have an internal center in thread end. For standard limits see Table 14.

* *T* is minimum length of shank which is held to eccentricity tolerances.

** *K* neck optional with manufacturer.

**Table 9a. ANSI Standard Regular and Spiral-Pointed Hand Taps for Helical Coil Wire Screw Thread Inserts (STI) (ANSI B94.9-1971)**

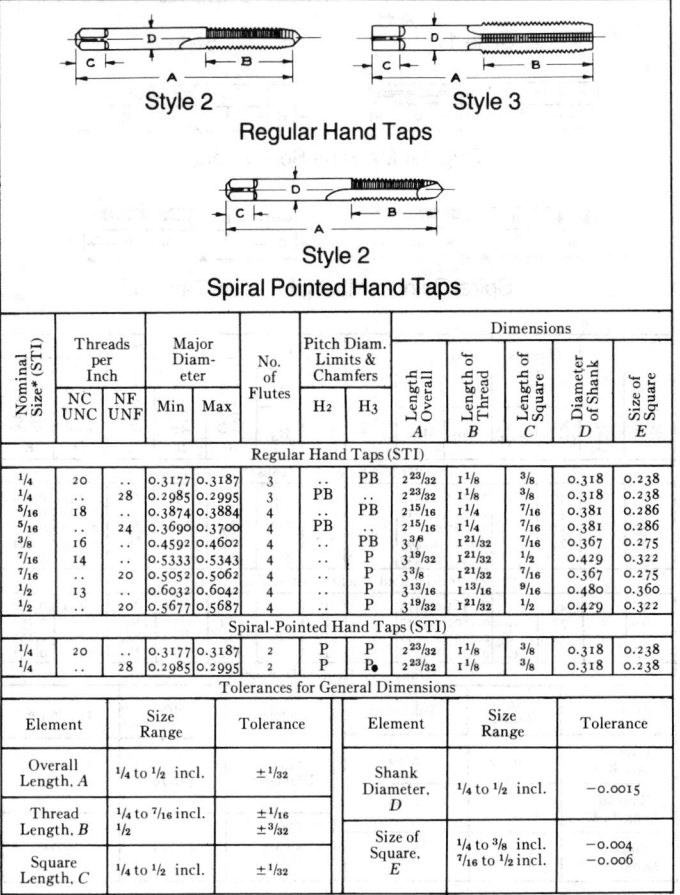

Style 2  Style 3

Regular Hand Taps

Style 2

Spiral Pointed Hand Taps

| Nominal Size* (STI) | Threads per Inch | | Major Diameter | | No. of Flutes | Pitch Diam. Limits & Chamfers | | Dimensions | | | | |
|---|---|---|---|---|---|---|---|---|---|---|---|---|
| | NC UNC | NF UNF | Min | Max | | H2 | H3 | Length Overall A | Length of Thread B | Length of Square C | Diameter of Shank D | Size of Square E |
| *Regular Hand Taps (STI)* | | | | | | | | | | | | |
| 1/4 | 20 | .. | 0.3177 | 0.3187 | 3 | .. | PB | 2 23/32 | 1 1/8 | 3/8 | 0.318 | 0.238 |
| 1/4 | .. | 28 | 0.2985 | 0.2995 | 3 | PB | .. | 2 23/32 | 1 1/8 | 3/8 | 0.318 | 0.238 |
| 5/16 | 18 | .. | 0.3874 | 0.3884 | 4 | .. | PB | 2 15/16 | 1 1/4 | 7/16 | 0.381 | 0.286 |
| 5/16 | .. | 24 | 0.3690 | 0.3700 | 4 | PB | .. | 2 15/16 | 1 1/4 | 7/16 | 0.381 | 0.286 |
| 3/8 | 16 | .. | 0.4592 | 0.4602 | 4 | .. | PB | 3 3/8 | 1 21/32 | 7/16 | 0.367 | 0.275 |
| 7/16 | 14 | .. | 0.5333 | 0.5343 | 4 | .. | P | 3 19/32 | 1 21/32 | 1/2 | 0.429 | 0.322 |
| 7/16 | .. | 20 | 0.5052 | 0.5062 | 4 | .. | P | 3 3/8 | 1 21/32 | 7/16 | 0.367 | 0.275 |
| 1/2 | 13 | .. | 0.6032 | 0.6042 | 4 | .. | P | 3 13/16 | 1 13/16 | 9/16 | 0.480 | 0.360 |
| 1/2 | .. | 20 | 0.5677 | 0.5687 | 4 | .. | P | 3 19/32 | 1 21/32 | 1/2 | 0.429 | 0.322 |
| *Spiral-Pointed Hand Taps (STI)* | | | | | | | | | | | | |
| 1/4 | 20 | .. | 0.3177 | 0.3187 | 2 | P | P | 2 23/32 | 1 1/8 | 3/8 | 0.318 | 0.238 |
| 1/4 | .. | 28 | 0.2985 | 0.2995 | 2 | P | P | 2 23/32 | 1 1/8 | 3/8 | 0.318 | 0.238 |

**Tolerances for General Dimensions**

| Element | Size Range | Tolerance | Element | Size Range | Tolerance |
|---|---|---|---|---|---|
| Overall Length, A | 1/4 to 1/2 incl. | ±1/32 | Shank Diameter, D | 1/4 to 1/2 incl. | −0.0015 |
| Thread Length, B | 1/4 to 7/16 incl. 1/2 | ±1/16 ±3/32 | Size of Square, E | 1/4 to 3/8 incl. 7/16 to 1/2 incl. | −0.004 −0.006 |
| Square Length, C | 1/4 to 1/2 incl. | ±1/32 | | | |

All dimensions are given in inches.

These taps are standard in high-speed steel with ground threads.

These taps are oversize to the extent that the internal thread which they produce will accommodate a helical coil wire screw thread insert, which at final assembly will accept a screw thread of the nominal size and pitch.

Chamfer designations are: P = plug and B = bottoming.

Style 2 taps, 5/16 inch and smaller, have external center on thread end (may be removed on bottoming taps) and external partial cone center on shank end, with length of cone approximately one-quarter of diameter of shank.

Style 3 taps, larger than 3/16 inch, have internal centers in thread and shank ends.

For standard thread limits see Table 19.

**Table 9b. ANSI Standard Regular and Spiral-Pointed Machine Screw Taps for Helical Coil Wire Screw Thread Inserts (STI)** (ANSI B94.9-1971)

Style 1

Style 2

Regular Machine Screw Taps

Spiral Pointed Machine Screw Taps

| Nominal Size* (STI) | Style No. | Threads per Inch NC UNC | Threads per Inch NF UNF | Major Diameter Min | Major Diameter Max | Number of Flutes | Pitch Diameter Limits H2 | Pitch Diameter Limits H3 | Dimensions Length Overall, A | Dimensions Length of Thread, B | Dimensions Length of Square, C | Dimensions Diam. of Shank, D | Dimensions Size of Square, E |
|---|---|---|---|---|---|---|---|---|---|---|---|---|---|
| Regular Machine Screw Taps (STI) |
| 4 | 1 | 40 | .. | 0.1463 | 0.1473 | 3 | PB | .. | 2 | 11/16 | 3/16 | 0.141 | 0.110 |
| 6 | 1 | 32 | .. | 0.1807 | 0.1817 | 3 | .. | PB | 2 3/8 | 7/8 | 1/4 | 0.194 | 0.152 |
| 8 | 1 | 32 | .. | 0.2067 | 0.2077 | 3 | .. | PB | 2 3/8 | 15/16 | 9/32 | 0.220 | 0.165 |
| 10 | 2 | 24 | .. | 0.2465 | 0.2475 | 3 | PB | .. | 2 1/2 | 1 | 5/16 | 0.255 | 0.191 |
| 10 | 2 | .. | 32 | 0.2327 | 0.2337 | 3 | PB | PB | 2 1/2 | 1 | 5/16 | 0.255 | 0.191 |
| Spiral-Pointed Machine Screw Taps (STI) |
| 4 | 1 | 40 | .. | 0.1463 | 0.1473 | 2 | P | .. | 2 | 11/16 | 3/16 | 0.141 | 0.110 |
| 6 | 1 | 32 | .. | 0.1807 | 0.1817 | 2 | P | P | 2 3/8 | 7/8 | 1/4 | 0.194 | 0.152 |
| 8 | 1 | 32 | .. | 0.2067 | 0.2077 | 2 | P | P | 2 3/8 | 15/16 | 9/32 | 0.220 | 0.165 |
| 10 | 2 | .. | 32 | 0.2327 | 0.2337 | 2 | P | .. | 2 1/2 | 1 | 5/16 | 0.255 | 0.191 |

**Tolerances for General Dimensions**

| Element | Size Range | Tolerance | Element | Size Range | Tolerance |
|---|---|---|---|---|---|
| Overall Length, A | 4 to 10 | ± 1/32 | Shank Diameter, D | 4 to 10 | −0.0015 |
| Thread Length, B | 4 to 8 / 10 | ± 3/64 / ± 1/16 | Size of Square, E | 4 to 10 | −0.004 |
| Square Length, C | 4 to 10 | ± 1/32 | | | |

All dimensions are given in inches.

These taps are standard in high-speed steel with ground threads.

These taps are oversize to the extent that the internal threads which they produce will accommodate a helical coil wire screw thread insert which at final assembly will accept a screw thread of the nominal size and pitch.

Chamfer designations are: P = plug and B = bottoming.

Style 1 regular and spiral pointed machine screw taps, sizes No. 8 and smaller, have external center on thread and shank ends (may be removed on thread end of bottoming taps).

Style 2 regular and spiral pointed machine screw taps, No. 10 size, have external center on thread end (may be removed on thread end of bottoming taps) and external partial cone center on shank end with length of cone approximately one-quarter of diameter of shank.

For standard thread limits see Table 19.

**Table 10. ANSI Standard Cut Thread Bent Shank Tapper Taps— Fractional and Numbered Sizes** (ANSI B94.9-1971)

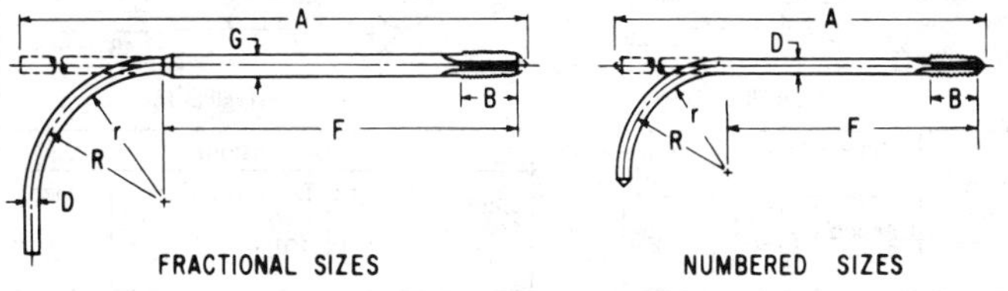


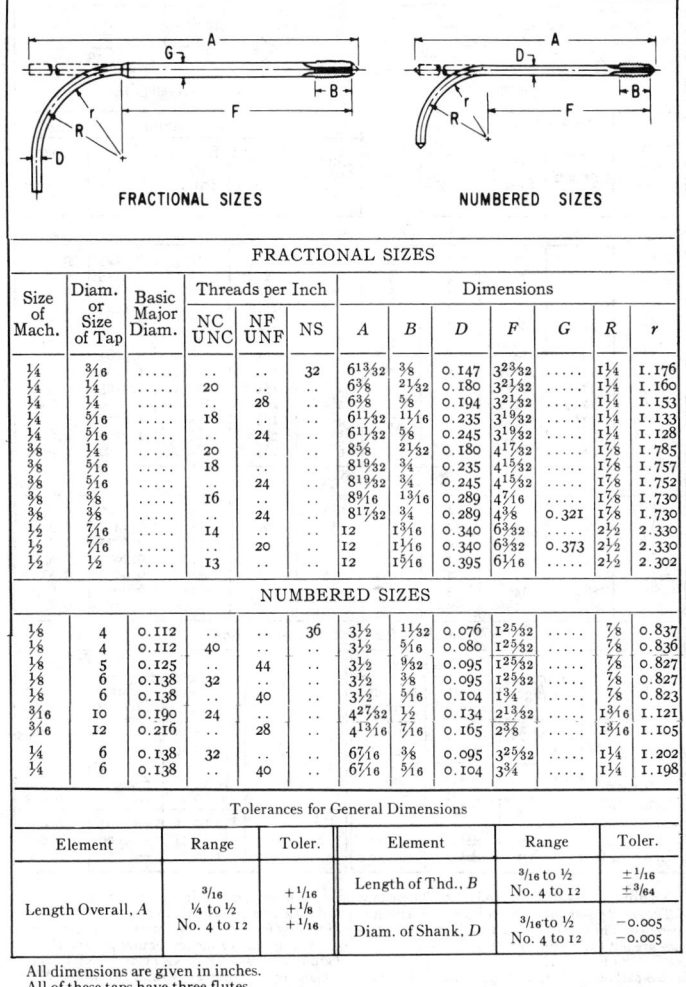

| Size of Mach. | Diam. or Size of Tap | Basic Major Diam. | Threads per Inch | | | Dimensions | | | | | | |
|---|---|---|---|---|---|---|---|---|---|---|---|---|
| | | | NC UNC | NF UNF | NS | *A* | *B* | *D* | *F* | *G* | *R* | *r* |
| FRACTIONAL SIZES | | | | | | | | | | | | |
| 1/4 | 3/16 | ..... | .. | .. | 32 | 6 13/32 | 3/8 | 0.147 | 3 23/32 | ..... | 1 1/4 | 1.176 |
| 1/4 | 1/4 | ..... | 20 | .. | .. | 6 3/8 | 21/32 | 0.180 | 3 21/32 | ..... | 1 1/4 | 1.160 |
| 1/4 | 1/4 | ..... | .. | 28 | .. | 6 3/8 | 5/8 | 0.194 | 3 21/32 | ..... | 1 1/4 | 1.153 |
| 1/4 | 5/16 | ..... | 18 | .. | .. | 6 11/32 | 11/16 | 0.235 | 3 19/32 | ..... | 1 1/4 | 1.133 |
| 1/4 | 5/16 | ..... | .. | 24 | .. | 6 11/32 | 5/8 | 0.245 | 3 19/32 | ..... | 1 1/4 | 1.128 |
| 3/8 | 1/4 | ..... | 20 | .. | .. | 8 5/8 | 21/32 | 0.180 | 4 17/32 | ..... | 1 7/8 | 1.785 |
| 3/8 | 5/16 | ..... | 18 | .. | .. | 8 19/32 | 3/4 | 0.235 | 4 15/32 | ..... | 1 7/8 | 1.757 |
| 3/8 | 5/16 | ..... | .. | 24 | .. | 8 19/32 | 3/4 | 0.245 | 4 15/32 | ..... | 1 7/8 | 1.752 |
| 3/8 | 3/8 | ..... | 16 | .. | .. | 8 9/16 | 13/16 | 0.289 | 4 7/16 | ..... | 1 7/8 | 1.730 |
| 3/8 | 3/8 | ..... | .. | 24 | .. | 8 17/32 | 3/4 | 0.289 | 4 3/8 | 0.321 | 1 7/8 | 1.730 |
| 1/2 | 7/16 | ..... | 14 | .. | .. | 12 | 1 3/16 | 0.340 | 6 3/32 | ..... | 2 1/2 | 2.330 |
| 1/2 | 7/16 | ..... | .. | 20 | .. | 12 | 1 1/16 | 0.340 | 6 3/32 | 0.373 | 2 1/2 | 2.330 |
| 1/2 | 1/2 | ..... | 13 | .. | .. | 12 | 1 5/16 | 0.395 | 6 1/16 | ..... | 2 1/2 | 2.302 |
| NUMBERED SIZES | | | | | | | | | | | | |
| 1/8 | 4 | 0.112 | .. | .. | 36 | 3 1/2 | 11/32 | 0.076 | 1 25/32 | ..... | 7/8 | 0.837 |
| 1/8 | 4 | 0.112 | 40 | .. | .. | 3 1/2 | 5/16 | 0.080 | 1 25/32 | ..... | 7/8 | 0.836 |
| 1/8 | 5 | 0.125 | .. | 44 | .. | 3 1/2 | 9/32 | 0.095 | 1 25/32 | ..... | 7/8 | 0.827 |
| 1/8 | 6 | 0.138 | 32 | .. | .. | 3 1/2 | 3/8 | 0.095 | 1 25/32 | ..... | 7/8 | 0.827 |
| 1/8 | 6 | 0.138 | .. | 40 | .. | 3 1/2 | 5/16 | 0.104 | 1 3/4 | ..... | 7/8 | 0.823 |
| 3/16 | 10 | 0.190 | 24 | .. | .. | 4 27/32 | 1/2 | 0.134 | 2 13/32 | ..... | 1 3/16 | 1.121 |
| 3/16 | 12 | 0.216 | .. | 28 | .. | 4 13/16 | 7/16 | 0.165 | 2 3/8 | ..... | 1 3/16 | 1.105 |
| 1/4 | 6 | 0.138 | 32 | .. | .. | 6 7/16 | 3/8 | 0.095 | 3 25/32 | ..... | 1 1/4 | 1.202 |
| 1/4 | 6 | 0.138 | .. | 40 | .. | 6 7/16 | 5/16 | 0.104 | 3 3/4 | ..... | 1 1/4 | 1.198 |

Tolerances for General Dimensions

| Element | Range | Toler. | Element | Range | Toler. |
|---|---|---|---|---|---|
| Length Overall, *A* | 3/16<br>1/4 to 1/2<br>No. 4 to 12 | +1/16<br>+1/8<br>+1/16 | Length of Thd., *B* | 3/16 to 1/2<br>No. 4 to 12 | ±1/16<br>±3/64 |
| | | | Diam. of Shank, *D* | 3/16 to 1/2<br>No. 4 to 12 | −0.005<br>−0.005 |

All dimensions are given in inches.

All of these taps have three flutes.

These bent shank tapper taps are standard in carbon steel with cut threads and are designed for use in automatic tapping machines.

Fractional size taps up to 3/8 inch, inclusive, have external center on thread end. Sizes 7/16 and larger have internal center in thread end.

All numbered sizes have external center on thread end.

For standard thread limits of fractional sizes, see Table 20.

For standard thread limits of numbered sizes, see Table 16.

**Table 11. ANSI Standard Taper and Straight Pipe Taps** (ANSI B94.9-1971)

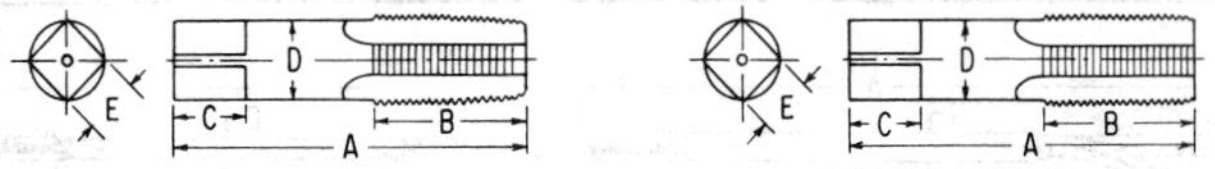


TAPER PIPE TAP STRAIGHT PIPE TAP

| Nominal Size | Threads per Inch | | Number of Flutes | | Dimensions | | | | |
|---|---|---|---|---|---|---|---|---|---|
| | Carbon Steel | High-Speed Steel | Regular | Interrupted | Length Overall *A* | Length of Thread *B* | Length of Square *C* | Diameter of Shank *D* | Size of Square *E* |
| **Taper Pipe Taps** | | | | | | | | | |
| 1/16 ‡ | .... | 27 | 4 | ... | 2 1/8 | 11/16 | 3/8 | 0.3125 | 0.234 |
| 1/8 | 27 | 27 | 4 | 5 | 2 1/8 | 3/4 | 3/8 | 0.3125 | 0.234 |
| 1/8 | 27 | 27 | 4 | 5 | 2 1/8 | 3/4 | 3/8 | 0.4375 | 0.328 |
| 1/4 | 18 | 18 | 4 | 5 | 2 7/16 | 1 1/16 | 7/16 | 0.5625 | 0.421 |
| 3/8 | 18 | 18 | 4 | 5 | 2 9/16 | 1 1/16 | 1/2 | 0.7000 | 0.531 |
| 1/2 | 14 | 14 | 4 | 5 | 3 1/8 | 1 3/8 | 5/8 | 0.6875 | 0.515 |
| 3/4 | 14 | 14 | 5 | 5 | 3 1/4 | 1 3/8 | 11/16 | 0.9063 | 0.679 |
| 1 | 11 1/2 | 11 1/2 | 5 | 5 | 3 3/4 | 1 3/4 | 13/16 | 1.1250 | 0.843 |
| 1 1/4 | 11 1/2 | 11 1/2 | 5 | 5 | 4 | 1 3/4 | 15/16 | 1.3125 | 0.984 |
| 1 1/2 | 11 1/2 | 11 1/2 | 7 | 7 * ‡ | 4 1/4 | 1 3/4 | 1 | 1.5000 | 1.125 |
| 2 | 11 1/2 | 11 1/2 | 7 | 7 * ‡ | 4 1/2 | 1 3/4 | 1 1/8 | 1.8750 | 1.406 |
| 2 1/2 † | 8 | ... | 8 | ... | 5 1/2 | 2 9/16 | 1 1/4 | 2.2500 | 1.687 |
| 3 † | 8 | ... | 8 | ... | 6 | 2 5/8 | 1 3/8 | 2.6250 | 1.968 |
| **Straight Pipe Taps** | | | | | | | | | |
| 1/8 ‡ | ... | 27 | 4 | ... | 2 1/8 | 3/4 | 3/8 | 0.3125 | 0.234 |
| 1/8 | ... | 27 | 4 | ... | 2 1/8 | 3/4 | 3/8 | 0.4375 | 0.328 |
| 1/4 | ... | 18 | 4 | ... | 2 7/16 | 1 1/16 | 7/16 | 0.5625 | 0.421 |
| 3/8 | ... | 18 | 4 | ... | 2 9/16 | 1 1/16 | 1/2 | 0.7000 | 0.531 |
| 1/2 | ... | 14 | 4 | ... | 3 1/8 | 1 3/8 | 5/8 | 0.6875 | 0.515 |
| 3/4 | ... | 14 | 5 | ... | 3 1/4 | 1 3/8 | 11/16 | 0.9063 | 0.679 |
| 1 | ... | 11 1/2 | 5 | ... | 3 3/4 | 1 3/4 | 13/16 | 1.1250 | 0.843 |

**Tolerances for General Dimensions**

| Element | Diameter Range | Tolerance | |
|---|---|---|---|
| | | Cut Thread | Ground Thread |
| Overall Length, *A* | 1/16 to 3/4 | ±1/32 | ±1/32 |
| | 1 to 3 | ±1/16 | ±1/16 |
| Thread Length, *B* | 1/16 to 3/4 | ±1/16 | ±1/16 |
| | 1 to 1 1/4 | ±3/32 | ±3/32 |
| | 1 1/2 to 3 | ±1/8 | ±1/8 |
| Square Length, *C* | 1/16 to 3/4 | ±1/32 | ±1/32 |
| | 1 to 3 | ±1/16 | ±1/16 |

| Element | Diameter Range | Tolerance | |
|---|---|---|---|
| | | Cut Thread | Ground Thread |
| Shank Diameter, *D* | 1/16 to 1/8 | ...... | −0.0015 |
| | 1/8 to 1/2 | −0.007 | ...... |
| | 1/4 to 1 | ....... | −0.002 |
| | 3/4 to 3 | −0.009 | ....... |
| | 1 1/4 to 2 | ....... | −0.003 |
| Size of Square, *E* | 1/16 to 1/8 | −0.004 | −0.004 |
| | 1/4 to 3/4 | −0.006 | −0.006 |
| | 1 to 3 | −0.008 | −0.008 |

All dimensions are given in inches. These taps have an internal center in the thread end.

*Taper Pipe Threads:* The 1/8-inch pipe tap is furnished with large size shank unless the small shank is specified. These taps have 2 to 3 1/2 threads chamfer. The first few threads on interrupted thread pipe taps are left full. The following styles and sizes are standard: 1/16 to 2 inches regular ground thread, NPT, NPTF, and ANPT: 1/8 to 2 inches interrupted ground thread, NPT, NPTF and ANPT: 1/8 to 3 inches carbon steel regular cut thread, NPT; 1/8 to 2 inches high-speed steel, regular cut thread, NPT; 1/8 to 1 1/4 inches high speed steel interrupted cut thread, NPT. For standard thread limits see Table 17.

*Straight Pipe Threads:* The 1/8-inch pipe tap is furnished with large size shank unless the small size is specified. These taps are standard with plug chamfer only. The following styles and sizes are standard: ground threads—1/8 to 1 inch, NPSC and NPSM; 1/8 to 3/4 inch, NPSF; cut threads—1/8 to 1 inch, NPSC and NPSM. For standard thread limits see Tables 18a, 18b, and 18c.

* Standard in NPT and NPTF form of thread only. † Cut thread taps only. ‡ Ground thread taps only.

Table 12.  Taps Recommended for Classes 1B*, 2B†, 3B, 2 and 3 Unified and American Screw Threads — Numbered and Fractional Sizes (ANSI B94.9-1971)

| Size | NC and UNC | NF and UNF | Class 2 | Class 3 | Class 2B† | Class 3B |
|---|---|---|---|---|---|---|
| | Threads Per Inch | | Recommended Tap‡ | | | |
| Numbered Size Taps | | | | | | |
| 0 | .. | 80 | G H1 | G H1 | G H2 | G H1 |
| 1 | 64 | .. | G H1 | G H1 | G H2 | G H1 |
| 1 | .. | 72 | G H1 | G H1 | G H2 | G H1 |
| 2 | 56 | .. | G H1 | G H1 | G H2 | G H1 |
| 2 | .. | 64 | G H1 | G H1 | G H2 | G H1 |
| 3 | 48 | .. | G H1 | G H1 | G H2 | G H1 |
| 3 | .. | 56 | G H1 | G H1 | G H2 | G H1 |
| 4 | 40 | .. | G H2 | G H1 | G H2 | G H2 |
| 4 | .. | 48 | G H1 | G H1 | G H2 | G H1 |
| 5 | 40 | .. | G H2 | G H1 | G H2 | G H2 |
| 5 | .. | 44 | G H1 | G H1 | G H2 | G H1 |
| 6 | 32 | .. | G H2 | G H1 | G H3 | G H2 |
| 6 | .. | 40 | G H2 | G H1 | G H3 | G H2 |
| 8 | 32 | .. | G H2 | G H1 | G H3 | G H2 |
| 8 | .. | 36 | G H2 | G H1 | G H3 | G H2 |
| 10 | 24 | .. | G H3 | G H1 | G H3 | G H3 |
| 10 | .. | 32 | G H2 | G H1 | G H3 | G H2 |
| 12 | 24 | .. | G H3 | G H1 | G H3 | G H3 |
| 12 | .. | 28 | G H3 | G H1 | G H3 | G H3 |
| Fractional Size Taps | | | | | | |
| 1/4 | 20 | .. | G H3 | G H2 | G H5 | G H3 |
| 1/4 | .. | 28 | G H2 | G H1 | G H4 | G H3 |
| 5/16 | 18 | .. | G H3 | G H2 | G H5 | G H3 |
| 5/16 | .. | 24 | G H3 | G H1 | G H4 | G H3 |
| 3/8 | 16 | .. | G H3 | G H2 | G H5 | G H3 |
| 3/8 | .. | 24 | G H3 | G H1 | G H4 | G H3 |
| 7/16 | 14 | .. | G H5 | G H3 | G H5 | G H3 |
| 7/16 | .. | 20 | G H3 | G H1 | G H5 | G H3 |
| 1/2 | 13 | .. | G H5 | G H3 | G H5 | G H3 |
| 1/2 | .. | 20 | G H3 | G H1 | G H5 | G H3 |
| 9/16 | 12 | .. | G H5 | G H3 | G H5 | G H3 |
| 9/16 | .. | 18 | G H3 | G H2 | G H5 | G H3 |
| 5/8 | 11 | .. | G H5 | G H3 | G H5 | G H3 |
| 5/8 | .. | 18 | G H3 | G H2 | G H5 | G H3 |
| 3/4 | 10 | .. | G H5 | G H3 | G H5 | G H5 |
| 3/4 | .. | 16 | G H3 | G H2 | G H5 | G H3 |
| 7/8 | 9 | .. | G H6 | G H4 | G H6 | G H4 |
| 7/8 | .. | 14 | G H4 | G H2 | G H6 | G H4 |
| 1 | 8 | .. | G H6 | G H4 | G H6 | G H4 |
| 1 | .. | 12 | G H4 | G H2 | G H6 | G H4 |
| 1 | | 14 NS | G H4 | G H2 | G H6 | G H4 |
| 1 1/8 | 7 | .. | G H8 | G H4 | G H8 | G H4 |
| 1 1/8 | .. | 12 | G H4 | G H4 | G H6 | G H4 |
| 1 1/4 | 7 | .. | G H8 | G H4 | G H8 | G H4 |
| 1 1/4 | .. | 12 | G H4 | G H4 | G H6 | G H4 |
| 1 3/8 | 6 | .. | G H8 | G H4 | G H8 | G H4 |
| 1 3/8 | .. | 12 | G H4 | G H4 | G H6 | G H4 |
| 1 1/2 | 6 | .. | G H8 | G H4 | G H8 | G H4 |
| 1 1/2 | .. | 12 | G H4 | G H4 | G H6 | G H4 |

* 1B tapped holes can be produced with cut thread taps. † Cut thread taps other than sizes 0, 1, and 2 may be used under normal conditions and in average materials for producing tapped holes to this classification.   ‡ See page 1712 for meaning of symbols G H2, etc.

The above recommended taps normally produce the Class of Thread indicated in average materials when used with reasonable care. However, if the tap specified does not give a satisfactory fit in the work, a choice of some other limit tap will be necessary.

Table 13. ANSI Standard Fractional-size Taps —
Cut Thread Limits (ANSI B94.9-1971)

| Tap Size | Threads per Inch | | | Major Diameter | | | Pitch Diameter | | |
|---|---|---|---|---|---|---|---|---|---|
| | NC UNC | NF UNF | NS or UN | Basic | Min. | Max. | Basic | Min. | Max. |
| 1/16 | .. | .. | 64 | 0.0625 | 0.0635 | 0.0650 | 0.0524 | 0.0526 | 0.0536 |
| 3/32 | .. | .. | 48 | 0.0938 | 0.0951 | 0.0966 | 0.0803 | 0.0805 | 0.0815 |
| 1/8 | .. | .. | 40 | 0.1250 | 0.1266 | 0.1286 | 0.1088 | 0.1090 | 0.1105 |
| 5/32 | .. | .. | 32 | 0.1563 | 0.1585 | 0.1605 | 0.1360 | 0.1365 | 0.1380 |
| 5/32 | .. | .. | 36 | 0.1563 | 0.1580 | 0.1600 | 0.1382 | 0.1384 | 0.1399 |
| 3/16 | .. | .. | 24 | 0.1875 | 0.1903 | 0.1923 | 0.1604 | 0.1609 | 0.1624 |
| 3/16 | .. | .. | 32 | 0.1875 | 0.1897 | 0.1917 | 0.1672 | 0.1677 | 0.1692 |
| 7/32 | .. | .. | 24 | 0.2188 | 0.2216 | 0.2236 | 0.1917 | 0.1922 | 0.1937 |
| 7/32 | .. | .. | 32 | 0.2188 | 0.2210 | 0.2230 | 0.1985 | 0.1990 | 0.2005 |
| 1/4 | 20 | .. | .. | 0.2500 | 0.2532 | 0.2557 | 0.2175 | 0.2180 | 0.2200 |
| 1/4 | .. | .. | 24 | 0.2500 | 0.2528 | 0.2553 | 0.2229 | 0.2234 | 0.2254 |
| 1/4 | .. | 28 | .. | 0.2500 | 0.2524 | 0.2549 | 0.2268 | 0.2273 | 0.2288 |
| 1/4 | .. | .. | 32* | 0.2500 | 0.2522 | 0.2547 | 0.2297 | 0.2302 | 0.2317 |
| 5/16 | 18 | .. | .. | 0.3125 | 0.3160 | 0.3185 | 0.2764 | 0.2769 | 0.2789 |
| 5/16 | .. | 24 | .. | 0.3125 | 0.3153 | 0.3178 | 0.2854 | 0.2859 | 0.2874 |
| 5/16 | .. | .. | 32* | 0.3125 | 0.3147 | 0.3172 | 0.2922 | 0.2927 | 0.2942 |
| 3/8 | 16 | .. | .. | 0.3750 | 0.3789 | 0.3814 | 0.3344 | 0.3349 | 0.3369 |
| 3/8 | .. | 24 | .. | 0.3750 | 0.3778 | 0.3803 | 0.3479 | 0.3484 | 0.3499 |
| 7/16 | 14 | .. | .. | 0.4375 | 0.4419 | 0.4449 | 0.3911 | 0.3916 | 0.3941 |
| 7/16 | .. | 20 | .. | 0.4375 | 0.4407 | 0.4437 | 0.4050 | 0.4055 | 0.4075 |
| 1/2 | 13 | .. | .. | 0.5000 | 0.5047 | 0.5077 | 0.4500 | 0.4505 | 0.4530 |
| 1/2 | .. | 20 | .. | 0.5000 | 0.5032 | 0.5062 | 0.4675 | 0.4680 | 0.4700 |
| 9/16 | 12 | .. | .. | 0.5625 | 0.5675 | 0.5705 | 0.5084 | 0.5089 | 0.5114 |
| 9/16 | .. | 18 | .. | 0.5625 | 0.5660 | 0.5690 | 0.5264 | 0.5269 | 0.5289 |
| 5/8 | 11 | .. | .. | 0.6250 | 0.6304 | 0.6334 | 0.5660 | 0.5665 | 0.5690 |
| 5/8 | .. | 18 | .. | 0.6250 | 0.6285 | 0.6315 | 0.5889 | 0.5894 | 0.5914 |
| 11/16 | .. | .. | 11 | 0.6875 | 0.6929 | 0.6969 | 0.6285 | 0.6290 | 0.6320 |
| 11/16 | .. | .. | 16 | 0.6875 | 0.6914 | 0.6954 | 0.6469 | 0.6474 | 0.6499 |
| 3/4 | 10 | .. | .. | 0.7500 | 0.7559 | 0.7599 | 0.6850 | 0.6855 | 0.6885 |
| 3/4 | .. | 16 | .. | 0.7500 | 0.7539 | 0.7579 | 0.7094 | 0.7099 | 0.7124 |
| 7/8 | 9 | .. | .. | 0.8750 | 0.8820 | 0.8860 | 0.8028 | 0.8038 | 0.8068 |
| 7/8 | .. | 14 | .. | 0.8750 | 0.8799 | 0.8839 | 0.8286 | 0.8296 | 0.8321 |
| 1 | 8 | .. | .. | 1.0000 | 1.0078 | 1.0118 | 0.9188 | 0.9198 | 0.9228 |
| 1 | .. | 12 | .. | 1.0000 | 1.0055 | 1.0095 | 0.9459 | 0.9469 | 0.9499 |
| 1 | .. | .. | 14 | 1.0000 | 1.0049 | 1.0089 | 0.9536 | 0.9546 | 0.9571 |
| 1 1/8 | 7 | .. | .. | 1.1250 | 1.1337 | 1.1382 | 1.0322 | 1.0332 | 1.0367 |
| 1 1/8 | .. | 12 | .. | 1.1250 | 1.1305 | 1.1350 | 1.0709 | 1.0719 | 1.0749 |
| 1 1/4 | 7 | .. | .. | 1.2500 | 1.2587 | 1.2632 | 1.1572 | 1.1582 | 1.1617 |
| 1 1/4 | .. | 12 | .. | 1.2500 | 1.2555 | 1.2600 | 1.1959 | 1.1969 | 1.1999 |
| 1 3/8 | 6 | .. | .. | 1.3750 | 1.3850 | 1.3895 | 1.2667 | 1.2677 | 1.2712 |
| 1 3/8 | .. | 12 | .. | 1.3750 | 1.3805 | 1.3850 | 1.3209 | 1.3219 | 1.3249 |
| 1 1/2 | 6 | .. | .. | 1.5000 | 1.5100 | 1.5145 | 1.3917 | 1.3927 | 1.3962 |
| 1 1/2 | .. | 12 | .. | 1.5000 | 1.5055 | 1.5100 | 1.4459 | 1.4469 | 1.4499 |
| 1 3/4 | 5 | .. | .. | 1.7500 | 1.7602 | 1.7657 | 1.6201 | 1.6216 | 1.6256 |
| 2 | 4 1/2 | .. | .. | 2.0000 | 2.0111 | 2.0166 | 1.8557 | 1.8572 | 1.8612 |

All dimensions are given in inches.
* NEF — UNEF.
*Lead Tolerance:* Plus or minus 0.003 inch max. per inch of thread.
*Angle Tolerance:* Plus or minus 35 min. in half angle or 53 min. in full angle for 4 1/2 to 5 1/2 thds. per in.; 40 min. half angle and 60 min. full angle for 6 to 9 thds.; 45 min. half angle and 68 min. full angle for 10 to 28 thds.; 60 min. half angle and 90 min. full angle for 30 to 64 thds. per in.

Table 14. ANSI Standard Fractional-size Taps — Ground Thread Limits (ANSI B94.9-1971)

| Size | Threads per Inch | | | Major Diameter | | | Basic Pitch Diam. | Pitch Diameter Limits | | | | | | | |
|---|---|---|---|---|---|---|---|---|---|---|---|---|---|---|---|
| | NC UNC | NF UNF | NS | Basic | Min. | Max. | | H1 Limit | | H2 Limit | | H3 & H4* Limits | | H4,* H5†& H6§ Limits | |
| | | | | | | | | Min. | Max. | Min. | Max. | Min. | Max. | Min. | Max. |
| 1/4 | 20 | .. | .. | 0.2500 | 0.2540 | 0.2550 | 0.2175 | 0.2175 | 0.2180 | 0.2180 | 0.2185 | 0.2185 | 0.2190 | 0.2195† | 0.2200† |
| 1/4 | .. | 28 | .. | 0.2500 | 0.2525 | 0.2535 | 0.2268 | 0.2268 | 0.2273 | 0.2273 | 0.2278 | 0.2278 | 0.2283 | 0.2283* | 0.2288* |
| 5/16 | 18 | .. | .. | 0.3125 | 0.3170 | 0.3180 | 0.2764 | 0.2764 | 0.2769 | 0.2769 | 0.2774 | 0.2774 | 0.2779 | 0.2784† | 0.2789† |
| 5/16 | .. | 24 | .. | 0.3125 | 0.3155 | 0.3165 | 0.2854 | 0.2854 | 0.2859 | 0.2859 | 0.2864 | 0.2864 | 0.2869 | 0.2869* | 0.2874* |
| 3/8 | 16 | .. | .. | 0.3750 | 0.3800 | 0.3810 | 0.3344 | 0.3344 | 0.3349 | 0.3349 | 0.3354 | 0.3354 | 0.3359 | 0.3364† | 0.3369† |
| 3/8 | .. | 24 | .. | 0.3750 | 0.3780 | 0.3790 | 0.3479 | 0.3479 | 0.3484 | 0.3484 | 0.3489 | 0.3489 | 0.3494 | 0.3494* | 0.3499* |
| 7/16 | 14 | .. | .. | 0.4375 | 0.4435 | 0.4445 | 0.3911 | 0.3911 | 0.3916 | 0.3916 | 0.3921 | 0.3921 | 0.3926 | 0.3931† | 0.3936† |
| 7/16 | .. | 20 | .. | 0.4375 | 0.4415 | 0.4425 | 0.4050 | 0.4050 | 0.4055 | 0.4055 | 0.4060 | 0.4060 | 0.4065 | 0.4070† | 0.4075† |
| 1/2 | 13 | .. | .. | 0.5000 | 0.5065 | 0.5075 | 0.4500 | 0.4500 | 0.4505 | 0.4505 | 0.4510 | 0.4510 | 0.4515 | 0.4520† | 0.4525† |
| 1/2 | .. | 20 | .. | 0.5000 | 0.5040 | 0.5050 | 0.4675 | 0.4675 | 0.4680 | 0.4680 | 0.4685 | 0.4685 | 0.4690 | 0.4695† | 0.4700† |
| 9/16 | 12 | .. | .. | 0.5625 | 0.5690 | 0.5700 | 0.5084 | | | 0.5089 | 0.5094 | 0.5094 | 0.5099 | 0.5104† | 0.5109† |
| 9/16 | .. | 18 | .. | 0.5625 | 0.5670 | 0.5680 | 0.5264 | | | 0.5269 | 0.5274 | 0.5274 | 0.5279 | 0.5284† | 0.5289† |
| 5/8 | 11 | .. | .. | 0.6250 | 0.6320 | 0.6330 | 0.5660 | 0.5660 | 0.5665 | 0.5665 | 0.5670 | 0.5670 | 0.5675 | 0.5680† | 0.5685† |
| 5/8 | .. | 18 | .. | 0.6250 | 0.6295 | 0.6305 | 0.5889 | 0.5889 | 0.5894 | 0.5894 | 0.5899 | 0.5899 | 0.5904 | 0.5909† | 0.5914† |
| 11/16 | .. | .. | 11 | 0.6875 | 0.6945 | 0.6955 | 0.6285 | | | | | 0.6295 | 0.6300 | | |
| 11/16 | .. | .. | 16 | 0.6875 | 0.6925 | 0.6935 | 0.6469 | | | | | 0.6479 | 0.6484 | | |
| 3/4 | 10 | .. | .. | 0.7500 | 0.7575 | 0.7590 | 0.6850 | 0.6850 | 0.6855 | 0.6855 | 0.6860 | 0.6860 | 0.6865 | 0.6870† | 0.6875† |
| 3/4 | .. | 16 | .. | 0.7500 | 0.7550 | 0.7560 | 0.7094 | 0.7094 | 0.7099 | 0.7099 | 0.7104 | 0.7104 | 0.7109 | 0.7114† | 0.7119† |
| 7/8 | 9 | .. | .. | 0.8750 | 0.8835 | 0.8850 | 0.8028 | | | 0.8033 | 0.8038 | 0.8043* | 0.8048* | 0.8053§ | 0.8058§ |
| 7/8 | .. | 14 | .. | 0.8750 | 0.8810 | 0.8820 | 0.8286 | | | 0.8291 | 0.8296 | 0.8301* | 0.8306* | 0.8311§ | 0.8316§ |
| 1 | 8 | .. | .. | 1.0000 | 1.0095 | 1.0110 | 0.9188 | | | 0.9193 | 0.9198 | 0.9203* | 0.9208* | 0.9213§ | 0.9218§ |
| 1 | .. | 12 | .. | 1.0000 | 1.0065 | 1.0075 | 0.9459 | | | | | 0.9474* | 0.9479* | | |
| 1 | .. | .. | 14 | 1.0000 | 1.0060 | 1.0070 | 0.9536 | | | 0.9541 | 0.9546 | 0.9551* | 0.9556* | | |

All dimensions are given in inches.
* H4 limit value.
† H5 limit value.
§ H6 limit value.

**Table 14** *(Continued)*.  **ANSI Standard Fractional-size Taps —**
**Ground Thread Limits** (ANSI B94.9-1971)

| Size | Threads per Inch | | | Major Diameter | | | Pitch Diameter Limits | | |
|---|---|---|---|---|---|---|---|---|---|
| | NC UNC | NF UNF | NS | Basic | Min. | Max. | Basic Pitch Diam. | H4 Limit | |
| | | | | | | | | Min. | Max. |
| 1⅛ | 7 | .. | .. | 1.1250 | 1.1350 | 1.1370 | 1.0322 | 1.0332 | 1.0342 |
| 1⅛ | .. | 12 | .. | 1.1250 | 1.1315 | 1.1325 | 1.0709 | 1.0719 | 1.0729 |
| 1¼ | 7 | .. | .. | 1.2500 | 1.2600 | 1.2620 | 1.1572 | 1.1582 | 1.1592 |
| 1¼ | .. | 12 | .. | 1.2500 | 1.2565 | 1.2575 | 1.1959 | 1.1969 | 1.1979 |
| 1⅜ | 6 | .. | .. | 1.3750 | 1.3870 | 1.3890 | 1.2667 | 1.2677 | 1.2687 |
| 1⅜ | .. | 12 | .. | 1.3750 | 1.3815 | 1.3825 | 1.3209 | 1.3219 | 1.3229 |
| 1½ | 6 | .. | .. | 1.5000 | 1.5120 | 1.5140 | 1.3917 | 1.3927 | 1.3937 |
| 1½ | .. | 12 | .. | 1.5000 | 1.5065 | 1.5075 | 1.4459 | 1.4469 | 1.4479 |

All dimensions are given in inches.

*Lead Tolerance:* Plus or minus 0.0005 inch per inch.

*Angle Tolerance:* Plus or minus 25 min. in half angle for 6 to 9 threads per inch; plus or minus 30 min. in half angle for 10 to 28 threads per inch.

For an explanation of the significance of the H4 limit value range see page 1712.

**Table 15.  ANSI Standard Machine Screw Taps —**
**Ground Thread Limits** (ANSI B94.9-1971)

| Size | Threads per Inch | | | Major Diameter | | | Pitch Diameter Limits* | | | | | | |
|---|---|---|---|---|---|---|---|---|---|---|---|---|---|
| | NC UNC | NF UNF | NS | Basic | Min. | Max. | Basic Pitch Diam. | H1 Limit | | H2 Limit | | H3 Limit | |
| | | | | | | | | Min. | Max. | Min. | Max. | Min. | Max. |
| 0 | .. | 80 | .. | 0.0600 | 0.0605 | 0.0615 | 0.0519 | 0.0519 | 0.0524 | 0.0524 | 0.0529 | ...... | ...... |
| 1 | 64 | .. | .. | 0.0730 | 0.0735 | 0.0745 | 0.0629 | 0.0629 | 0.0634 | 0.0634 | 0.0639 | ...... | ...... |
| 1 | .. | 72 | .. | 0.0730 | 0.0735 | 0.0745 | 0.0640 | 0.0640 | 0.0645 | 0.0645 | 0.0650 | ...... | ...... |
| 2 | 56 | .. | .. | 0.0860 | 0.0865 | 0.0875 | 0.0744 | 0.0744 | 0.0749 | 0.0749 | 0.0754 | ...... | ...... |
| 2 | .. | 64 | .. | 0.0860 | 0.0865 | 0.0875 | 0.0759 | 0.0759 | 0.0764 | 0.0764 | 0.0769 | ...... | ...... |
| 3 | 48 | .. | .. | 0.0990 | 0.1000 | 0.1010 | 0.0855 | 0.0855 | 0.0860 | 0.0860 | 0.0865 | ...... | ...... |
| 3 | .. | 56 | .. | 0.0990 | 0.0995 | 0.1005 | 0.0874 | 0.0874 | 0.0879 | 0.0879 | 0.0884 | ...... | ...... |
| 4 | .. | .. | 36 | 0.1120 | 0.1135 | 0.1145 | 0.0940 | ...... | ...... | 0.0945 | 0.0950 | ...... | ...... |
| 4 | 40 | .. | .. | 0.1120 | 0.1135 | 0.1145 | 0.0958 | 0.0958 | 0.0963 | 0.0963 | 0.0968 | ...... | ...... |
| 4 | .. | 48 | .. | 0.1120 | 0.1130 | 0.1140 | 0.0985 | 0.0985 | 0.0990 | 0.0990 | 0.0995 | ...... | ...... |
| 5 | 40 | .. | .. | 0.1250 | 0.1265 | 0.1275 | 0.1088 | 0.1088 | 0.1093 | 0.1093 | 0.1098 | ...... | ...... |
| 5 | .. | 44 | .. | 0.1250 | 0.1260 | 0.1270 | 0.1102 | ...... | ...... | 0.1107 | 0.1112 | ...... | ...... |
| 6 | 32 | .. | .. | 0.1380 | 0.1400 | 0.1410 | 0.1177 | 0.1177 | 0.1182 | 0.1182 | 0.1187 | 0.1187 | 0.1192 |
| 6 | .. | 40 | .. | 0.1380 | 0.1395 | 0.1405 | 0.1218 | 0.1218 | 0.1223 | 0.1223 | 0.1228 | ...... | ...... |
| 8 | 32 | .. | .. | 0.1640 | 0.1660 | 0.1670 | 0.1437 | 0.1437 | 0.1442 | 0.1442 | 0.1447 | 0.1447 | 0.1452 |
| 8 | .. | 36 | .. | 0.1640 | 0.1655 | 0.1665 | 0.1460 | 0.1460 | 0.1465 | 0.1465 | 0.1470 | ...... | ...... |
| 10 | 24 | .. | .. | 0.1900 | 0.1930 | 0.1940 | 0.1629 | 0.1629 | 0.1634 | 0.1634 | 0.1639 | 0.1639 | 0.1644 |
| 10 | .. | 32 | .. | 0.1900 | 0.1920 | 0.1930 | 0.1697 | 0.1697 | 0.1702 | 0.1702 | 0.1707 | 0.1707 | 0.1712 |
| 12 | 24 | .. | .. | 0.2160 | 0.2190 | 0.2200 | 0.1889 | 0.1889 | 0.1894 | ...... | ...... | 0.1899 | 0.1904 |
| 12 | .. | 28 | .. | 0.2160 | 0.2185 | 0.2195 | 0.1928 | 0.1928 | 0.1933 | ...... | ...... | 0.1938 | 0.1943 |

All dimensions are given in inches.

*Lead Tolerance:* Plus or minus 0.0005 inch per inch.

*Angle Tolerance:* Plus or minus 30 min. in half angle for 20 to 80 threads per inch.

For an explanation of the significance of the limit value ranges see page 1712.

*H7 limits (formerly designated as G) apply to same threads as H3 limits with the exception of the 12-24 and 12-28 threads. H7 limits have minimum and maximum major diameters 0.0020 inch larger than shown and minimum and maximum pitch diameters 0.0020 inch larger than shown for H3 limits.

Table 16.　ANSI Standard Machine Screw Taps —
Cut Thread Limits (ANSI B94.9-1971)

| Size | Threads per Inch | | | Major Diameter | | | Pitch Diameter | | |
|---|---|---|---|---|---|---|---|---|---|
| | NC UNC | NF UNF | NS or UN | Basic | Min. | Max. | Basic | Min. | Max. |
| 0 | .. | 80 | .. | 0.0600 | 0.0609 | 0.0624 | 0.0519 | 0.0521 | 0.0531 |
| 1 | 64 | .. | .. | 0.0730 | 0.0740 | 0.0755 | 0.0629 | 0.0631 | 0.0641 |
| 1 | .. | 72 | .. | 0.0730 | 0.0740 | 0.0755 | 0.0640 | 0.0642 | 0.0652 |
| 2 | 56 | .. | .. | 0.0860 | 0.0872 | 0.0887 | 0.0744 | 0.0746 | 0.0756 |
| 2 | .. | 64 | .. | 0.0860 | 0.0870 | 0.0885 | 0.0759 | 0.0761 | 0.0771 |
| 3 | 48 | .. | .. | 0.0990 | 0.1003 | 0.1018 | 0.0855 | 0.0857 | 0.0867 |
| 3 | .. | 56 | .. | 0.0990 | 0.1002 | 0.1017 | 0.0874 | 0.0876 | 0.0886 |
| 4 | .. | .. | 36 | 0.1120 | 0.1137 | 0.1157 | 0.0940 | 0.0942 | 0.0957 |
| 4 | 40 | .. | .. | 0.1120 | 0.1136 | 0.1156 | 0.0958 | 0.0960 | 0.0975 |
| 4 | .. | 48 | .. | 0.1120 | 0.1133 | 0.1153 | 0.0985 | 0.0987 | 0.1002 |
| 5 | 40 | .. | .. | 0.1250 | 0.1266 | 0.1286 | 0.1088 | 0.1090 | 0.1105 |
| 6 | 32 | .. | .. | 0.1380 | 0.1402 | 0.1422 | 0.1177 | 0.1182 | 0.1197 |
| 6 | .. | .. | 36 | 0.1380 | 0.1397 | 0.1417 | 0.1200 | 0.1202 | 0.1217 |
| 6 | .. | 40 | .. | 0.1380 | 0.1396 | 0.1416 | 0.1218 | 0.1220 | 0.1235 |
| 8 | 32 | .. | .. | 0.1640 | 0.1662 | 0.1682 | 0.1437 | 0.1442 | 0.1457 |
| 8 | .. | 36 | .. | 0.1640 | 0.1657 | 0.1677 | 0.1460 | 0.1462 | 0.1477 |
| 8 | .. | .. | 40 | 0.1640 | 0.1656 | 0.1676 | 0.1478 | 0.1480 | 0.1495 |
| 10 | 24 | .. | .. | 0.1900 | 0.1928 | 0.1948 | 0.1629 | 0.1634 | 0.1649 |
| 10 | .. | 32 | .. | 0.1900 | 0.1922 | 0.1942 | 0.1697 | 0.1702 | 0.1717 |
| 12 | 24 | .. | .. | 0.2160 | 0.2188 | 0.2208 | 0.1889 | 0.1894 | 0.1909 |
| 12 | .. | 28 | .. | 0.2160 | 0.2184 | 0.2204 | 0.1928 | 0.1933 | 0.1948 |
| 14 | .. | .. | 24 | 0.2420 | 0.2448 | 0.2473 | 0.2149 | 0.2154 | 0.2174 |

All dimensions are given in inches.
*Lead Tolerance:* Plus or minus 0.003 inch per inch of thread. *Angle Tolerance:* Plus or minus 45 min. in half angle and 68 min. in full angle for 20 to 28 threads per inch; plus or minus 60 min. in half angle and 90 min. in full angle for 30 or more threads per inch.

Table 17a.　ANSI Standard Taper Pipe Taps (NPT and NPTF) — Cut and
Ground Thread Tolerances (ANSI B94.9-1971)

| Nominal Size | Threads per Inch NPT or NPTF | Gage Measurement* | | | Taper per Foot, Inches | | | |
|---|---|---|---|---|---|---|---|---|
| | | Projection Inches | Tolerance Plus or Minus | | Cut Thread | | Ground Thread | |
| | | | Cut Thread | Ground Thread | Min. | Max. | Min. | Max. |
| 1/16 | 27 | 0.312 | 1/16 | 1/16 | 23/32 | 27/32 | 23/32 | 25/32 |
| 1/8 | 27 | 0.312 | 1/16 | 1/16 | 23/32 | 27/32 | 23/32 | 25/32 |
| 1/4 | 18 | 0.459 | 1/16 | 1/16 | 23/32 | 27/32 | 23/32 | 25/32 |
| 3/8 | 18 | 0.454 | 1/16 | 1/16 | 23/32 | 27/32 | 23/32 | 25/32 |
| 1/2 | 14 | 0.579 | 1/16 | 1/16 | 23/32 | 13/16 | 23/32 | 25/32 |
| 3/4 | 14 | 0.565 | 1/16 | 1/16 | 23/32 | 13/16 | 23/32 | 25/32 |
| 1 | 11½ | 0.678 | 3/32 | 3/32 | 23/32 | 13/16 | 23/32 | 25/32 |
| 1¼ | 11½ | 0.686 | 3/32 | 3/32 | 23/32 | 13/16 | 23/32 | 25/32 |
| 1½ | 11½ | 0.699 | 3/32 | 3/32 | 23/32 | 13/16 | 23/32 | 25/32 |
| 2 | 11½ | 0.667 | 3/32 | 3/32 | 23/32 | 13/16 | 23/32 | 25/32 |
| 2½ | 8 | 0.925 | 3/32 | 3/32 | 47/64 | 51/64 | 47/64 | 25/32 |
| 3 | 8 | 0.925 | 3/32 | 3/32 | 47/64 | 51/64 | 47/64 | 25/32 |

All dimensions are given in inches.
* Distance that small end of tap projects through L1 thread ring gage (see ANSI B2.2).
*Lead Tolerance:* Plus or minus 0.003 inch per inch of cut thread and plus or minus 0.0005 inch per inch of ground thread. *Angle Tolerance:* Plus or minus 40 min. in half angle and 60 min. in full angle for 8 cut threads per inch; plus or minus 45 min. in half angle and 68 min. in full angle for 11½ to 27 cut threads per inch; plus or minus 25 min. in half angle for 8 ground threads per inch; and plus and minus 30 min. in half angle for 11½ to 27 ground threads per inch.

Table 17b. ANSI Standard Taper Pipe Taps (NPT and NPTF) — Cut and
Ground Thread Limits (ANSI B94.9-1971)

| Threads per Inch, NPT or NPTF | Values to use in Formulas | | | | |
|---|---|---|---|---|---|
| | A | B | C | D | E |
| 27 | 0.0267 | 0.0296 | 0.0257 | 0.0234 | 0.0251 |
| 18 | 0.0408 | 0.0444 | 0.0401 | 0.0377 | 0.0395 |
| 14 | 0.0535 | 0.0571 | 0.0525 | 0.0515 | 0.0533 |
| 11½ | 0.0658 | 0.0696 | 0.0647 | 0.0614 | 0.0649 |
| 8 | 0.0966 | 0.1000 | 0.0946 | ...... | ...... |

If $M$ = measured pitch diameter in inches, then:
Major diam. min. = $M + A$      Minor diam. min. = $M - B$
Major diam. max. = $M + B$      Minor diam. max. = $M - C$
for cut and ground threads of the American Standard Pipe Form (NPT) and
Major diam. min. = $M + D$      Minor diam. min. = max. or smaller
Major diam. max. = $M + E$      Minor diam. max. = $M - E$
for ground threads of the American Standard Dryseal Pipe Form (NPTF)

Cut and Ground Thread American Standard Pipe Form Taps made to this table are to
be marked NPT. Ground Thread American Standard Dryseal Pipe Taps made to this
table are to be marked NPTF.

Table 18a. ANSI Standard Straight Pipe Taps (NPSF — Dryseal) —
Ground Thread Limits (ANSI B94.9-1971)

| Nominal Size, Inches | Threads per Inch | Major Diameter | | Pitch Diameter | | | Minor* Diam. Flat, Max. |
|---|---|---|---|---|---|---|---|
| | | Min. G | Max. H | Plug at Gaging Notch E | Min. K | Max. L | |
| ⅛ | 27 | 0.3932 | 0.3942 | 0.3736 | 0.3696 | 0.3701 | 0.004 |
| ¼ | 18 | 0.5239 | 0.5249 | 0.4916 | 0.4859 | 0.4864 | 0.005 |
| ⅜ | 18 | 0.6593 | 0.6603 | 0.6270 | 0.6213 | 0.6218 | 0.005 |
| ½ | 14 | 0.8230 | 0.8240 | 0.7784 | 0.7712 | 0.7717 | 0.005 |
| ¾ | 14 | 1.0335 | 1.0345 | 0.9889 | 0.9817 | 0.9822 | 0.005 |

Formulas For American Dryseal (NPSF) Ground Thread Taps

| Nominal Size, Inches | Major Diameter | | Pitch Diameter | | Max. Minor Diam. |
|---|---|---|---|---|---|
| | Min. G | Max. H | Min. K | Max. L | |
| ⅛ | $H - 0.0010$ | $K + Q - 0.0005$ | $L - 0.0005$ | $E - F$ | $M - Q$ |
| ¼ | $H - 0.0010$ | $K + Q - 0.0005$ | $L - 0.0005$ | $E - F$ | $M - Q$ |
| ⅜ | $H - 0.0010$ | $K + Q - 0.0005$ | $L - 0.0005$ | $E - F$ | $M - Q$ |
| ½ | $H - 0.0010$ | $K + Q - 0.0005$ | $L - 0.0005$ | $E - F$ | $M - Q$ |
| ¾ | $H - 0.0010$ | $K + Q - 0.0005$ | $L - 0.0005$ | $E - F$ | $M - Q$ |

Values to use in Formulas

| Threads per Inch | E | F | M | Q |
|---|---|---|---|---|
| 27 | Pitch diameter | 0.0035 | | 0.0251 |
| 18 | of plug at | 0.0052 | Actual measured | 0.0395 |
| 14 | gaging notch | 0.0067 | pitch diameter | 0.0533 |

All dimensions are given in inches.
* As specified or sharper.
*Note:* The major diameter of standard taper pipe plug gages and the minor diameter of
standard taper pipe ring gages used for gaging dryseal threads are truncated .20p minimum
to .25p maximum for all pitches.

*Lead Tolerance:* Plus or minus 0.0005 inch per inch of thread. *Angle Tolerance:* Plus or
minus 30 min. in half angle for 14 to 27 threads per inch.

**Table 18b. ANSI Standard Straight Pipe Taps (NPS) — Cut Thread Limits**
(ANSI B94.9-1971)

| Nominal Size | Threads per Inch, NPS, NPSC | Size at Gaging Notch | Pitch Diameter Min. | Pitch Diameter Max. | A | B | C |
|---|---|---|---|---|---|---|---|
| ⅛ | 27 | 0.3736 | 0.3721 | 0.3751 | 0.0267 | 0.0296 | 0.0257 |
| ¼ | 18 | 0.4916 | 0.4908 | 0.4938 | 0.0408 | 0.0444 | 0.0401 |
| ⅜ | 18 | 0.6270 | 0.6257 | 0.6292 | | | |
| ½ | 14 | 0.7784 | 0.7776 | 0.7811 | 0.0535 | 0.0571 | 0.0525 |
| ¾ | 14 | 0.9889 | 0.9876 | 0.9916 | | | |
| 1 | 11½ | 1.2386 | 1.2372 | 1.2412 | 0.0658 | 0.0696 | 0.0647 |

The following are approximate formulas, in which $M$ = measured pitch diameter in inches:

Major diam., min. = $M + A$

Major diam., max. = $M + B$  Minor diam., max. = $M - C$

All dimensions are given in inches.

*Lead Tolerance:* Plus or minus 0.003 inch per inch of thread. *Angle Tolerance:* All pitches, plus or minus 45 min. in half angle and 68 min. in full angle. Taps made to these specifications are to be marked NPS and used for NPSC thread form.

**Table 18c. ANSI Standard Straight Pipe Taps (NPS) — Ground Thread Limits**
(ANSI B94.9-1971)

| Nominal Size, Inches | Threads per Inch, NPS, NPSC, NPSM | Major Diameter Plug at Gaging Notch | Major Diameter Min. G | Major Diameter Max. H | Pitch Diameter Plug at Gaging Notch E | Pitch Diameter Min. K | Pitch Diameter Max. L |
|---|---|---|---|---|---|---|---|
| ⅛ | 27 | 0.3983 | 0.4022 | 0.4032 | 0.3736 | 0.3746 | 0.3751 |
| ¼ | 18 | 0.5286 | 0.5347 | 0.5357 | 0.4916 | 0.4933 | 0.4938 |
| ⅜ | 18 | 0.6640 | 0.6701 | 0.6711 | 0.6270 | 0.6287 | 0.6292 |
| ½ | 14 | 0.8260 | 0.8347 | 0.8357 | 0.7784 | 0.7086 | 0.7811 |
| ¾ | 14 | 1.0364 | 1.0447 | 1.0457 | 0.9889 | 0.9906 | 0.9916 |
| 1 | 11½ | 1.2966 | 1.3062 | 1.3077 | 1.2386 | 1.2402 | 1.2412 |

Formulas for NPS Ground Thread Taps

| Nominal Size | Major Diameter Min. G | Major Diameter Max. H | Minor Diam. Max. | Threads per Inch | A | B |
|---|---|---|---|---|---|---|
| | | | | 27 | 0.0296 | 0.0257 |
| ⅛ | $H - 0.0010$ | $(K + A) - 0.0010$ | $M - B$ | 18 | 0.0444 | 0.0401 |
| ¼ to ¾ | $H - 0.0010$ | $(K + A) - 0.0020$ | $M - B$ | 14 | 0.0571 | 0.0525 |
| 1 | $H - 0.0015$ | $(K + A) - 0.0021$ | $M - B$ | 11½ | 0.0696 | 0.0647 |

The maximum Pitch Diameter of tap is based upon an allowance deducted from the maximum product pitch diameter of NPSC or NPSM, whichever is smaller.

The minimum Pitch Diameter of tap is derived by subtracting the ground thread pitch diameter tolerance for actual equivalent size.

All dimensions are given in inches.

* In the formulas, $M$ equals the actual measured pitch diameter. *Lead tolerance:* Plus or minus 0.0005 inch per inch of thread. *Angle Tolerance:* All pitches, plus or minus 30 min. in half angle. Taps made to these specifications are to be marked NPS and used for NPSC and NPSM.

Table 19.　ANSI Standard Taps for Helical Coil Wire Screw Thread
Inserts (STI) — Ground Thread Limits (ANSI B94.9-1971)

| Nominal Size* (STI) | Threads per Inch | | Major Diameter | | Pitch Diameter Limits | | | |
|---|---|---|---|---|---|---|---|---|
| | | | | | H2 Limit | | H3 Limit | |
| | NC UNC | NF UNF | Minimum | Maximum | Minimum | Maximum | Minimum | Maximum |
| 4 | 40 | .. | 0.1463 | 0.1473 | 0.1288 | 0.1293 | .... | .... |
| 6 | 32 | .. | 0.1807 | 0.1817 | 0.1588 | 0.1593 | 0.1593 | 0.1598 |
| 8 | 32 | .. | 0.2067 | 0.2077 | 0.1848 | 0.1853 | 0.1853 | 0.1858 |
| 10 | 24 | .. | 0.2465 | 0.2475 | 0.2175 | 0.2180 | .... | .... |
| 10 | .. | 32 | 0.2327 | 0.2337 | 0.2108 | 0.2113 | 0.2113 | 0.2118 |
| 1/4 | 20 | .. | 0.3177 | 0.3187 | 0.2830 | 0.2835 | 0.2835 | 0.2840 |
| 1/4 | .. | 28 | 0.2985 | 0.2995 | 0.2737 | 0.2742 | 0.2742 | 0.2747 |
| 5/16 | 18 | .. | 0.3874 | 0.3884 | .... | .... | 0.3496 | 0.3501 |
| 3/8 | 16 | .. | 0.4592 | 0.4602 | .... | .... | 0.4166 | 0.4171 |
| 7/16 | 14 | .. | 0.5333 | 0.5343 | .... | .... | 0.4849 | 0.4854 |
| 7/16 | .. | 20 | 0.5052 | 0.5062 | .... | .... | 0.4710 | 0.4715 |
| 1/2 | 13 | .. | 0.6032 | 0.6042 | .... | .... | 0.5509 | 0.5514 |
| 1/2 | .. | 20 | 0.5677 | 0.5687 | .... | .... | 0.5335 | 0.5340 |

* These taps are oversize to the extent that the internal thread which they produce will accommodate a helical coil wire screw thread insert, which at final assembly will accept a screw thread of the nominal size and pitch.

STI basic thread dimensions are determined by adding twice the single thread height ($2 \times 0.649519$ p.) to the basic dimensions of the nominal screw size.

Table 20.　ANSI Standard Bent Shank Tapper Taps — Cut Thread Limits
(ANSI B94.9-1971)

| Size | Threads per Inch | | Major Diameter | | | Pitch Diameter | | |
|---|---|---|---|---|---|---|---|---|
| | NC UNC | NF UNF | Basic | Minimum | Maximum | Basic | Minimum | Maximum |
| Class 2 | | | | | | | | |
| 1/4 | 20 | .. | 0.2500 | 0.2527 | 0.2552 | 0.2175 | 0.2175 | 0.2195 |
| 1/4 | .. | 28 | 0.2500 | 0.2519 | 0.2544 | 0.2268 | 0.2263 | 0.2283 |
| 5/16 | 18 | .. | 0.3125 | 0.3155 | 0.3180 | 0.2764 | 0.2764 | 0.2784 |
| 5/16 | .. | 24 | 0.3125 | 0.3148 | 0.3173 | 0.2854 | 0.2849 | 0.2869 |
| 3/8 | 16 | .. | 0.3750 | 0.3784 | 0.3809 | 0.3344 | 0.3344 | 0.3364 |
| 3/8 | .. | 24 | 0.3750 | 0.3768 | 0.3793 | 0.3479 | 0.3469 | 0.3489 |
| 7/16 | 14 | .. | 0.4375 | 0.4414 | 0.4444 | 0.3911 | 0.3911 | 0.3936 |
| 7/16 | .. | 20 | 0.4375 | 0.4392 | 0.4422 | 0.4050 | 0.4035 | 0.4060 |
| 1/2 | 13 | .. | 0.5000 | 0.5042 | 0.5072 | 0.4500 | 0.4500 | 0.4525 |
| For Tapping Free Fit | | | | | | | | |
| 1/4 | 20 | .. | 0.2500 | 0.2542 | 0.2567 | 0.2175 | 0.2190 | 0.2210 |
| 5/16 | 18 | .. | 0.3125 | 0.3170 | 0.3195 | 0.2764 | 0.2779 | 0.2799 |
| 3/8 | 16 | .. | 0.3750 | 0.3799 | 0.3824 | 0.3344 | 0.3359 | 0.3379 |
| 7/16 | 14 | .. | 0.4375 | 0.4429 | 0.4459 | 0.3911 | 0.3926 | 0.3951 |
| 1/2 | 13 | .. | 0.5000 | 0.5057 | 0.5087 | 0.4500 | 0.4515 | 0.4540 |

All dimensions are given in inches.

*Lead Tolerance:* A maximum lead error of plus or minus .003 inch per inch of thread.

**Standard Marks or Symbols for Identifying Threads on Taps.** — All taps are marked with the nominal size, number of threads per inch and the proper symbol to identify the thread form. Taps having multiple threads are marked with diameter, number of threads per inch, form of thread and lead designated in fractions, also double, triple, etc. For example: A 1″-8 double thread special tap with National form of thread will be marked as follows: 1″-8NS Double ¼″ Lead.

Left-hand taps are marked "Left Hand" or "LH" as follows: 1″-8NS Double LH ¼″ Lead.

High speed steel taps are marked HS while carbon steel taps need not be marked with steel designation. Example of marking a ground thread tap:

¼-20 NC GH3 HS

Symbols to identify thread form are shown in the table below.

## Symbols Used for Standard Threads

| | |
|---|---|
| NC | American National Coarse Thread Series |
| *UNC | Unified Coarse Thread Series |
| NF | American National Fine Thread Series |
| *UNF | Unified Fine Thread Series |
| NEF | American National Extra-Fine Thread Series |
| *UNEF | Unified Extra-Fine Thread Series |
| N | American National 8, 12 and 16 Thread Series (8N, 12N, 16N) |
| *UN | Unified Constant Pitch Thread Series |
| NS | American National Thread—Special |
| *UNS | Unified Thread—Special |
| UNM | Unified Miniature Thread Series |
| NR | American National Thread with a 0.108p to 0.144p Controlled Root Radius |
| UNR | Unified Constant Pitch Thread Series with a 0.108p to 0.144p Controlled Root Radius |
| UNRC | Unified Coarse Thread Series with a 0.108p to 0.144p Controlled Root Radius |
| UNRF | Unified Fine Thread Series with a 0.108p to 0.144p Controlled Root Radius |
| UNJ | Unified Thread Series with a 0.15011p to 0.18042p Controlled Root Radius |
| UNJC | Unified Coarse Series with a 0.15011p to 0.18042p Controlled Root Radius |
| UNJF | Unified Fine Series with a 0.15011p to 0.18042p Controlled Root Radius |
| NH | American National Hose Coupling & Fire Hose Coupling Threads |
| NPS | For Tap Marking Only (See NPSC and NPSM) |
| NPSC | †American National Standard Straight Pipe Thread in Pipe Couplings (Tap Marked NPS) |
| NPSF | †Dryseal American National Std. Fuel Internal Straight Pipe Thread |
| NPSH | †American National Standard Straight Pipe Threads for Loose Fitting Mechanical Joints for Hose Couplings |
| NPSI | †Dryseal American National Standard Intermediate Internal Straight Pipe Thread |
| NPSL | †American National Standard Straight Pipe Thread for Loose-Fitting Mechanical Joints with Locknuts |
| NPSM | †American National Standard Straight Pipe Threads for Free-Fitting Mechanical Joints (Tap Marked NPS) |
| ANPT | Aeronautical National Form Taper Pipe Thread |
| NPT | †American National Standard Taper Pipe Thread |
| NPTF | †Dryseal American National Standard Taper Pipe Thread |
| NPTR | †American National Standard Taper Thread for Railing Joints (Tap Marked NPT) |
| NGO | National Gas Outlet Thread |
| NGS | National Gas Straight Thread |
| NGT | National Gas Taper Thread (See also "SGT") |
| PTF-SAE-SHORT | Dryseal SAE Short Taper Pipe Thread |
| ACME-C | Acme Thread Centralizing |
| ACME-G | Acme Thread-General Purpose |
| STUB ACME | Stub Acme Thread |
| AMO | American Standard Microscope Objective Thread |
| NBUTT | American Buttress Thread |
| V | A 60-degree "V" Thread with Truncated Crest and Root. The theoretical "V" form is usually flatted to the user's specifications. |
| SB | Manufacturers Stovebolt Standard Thread |
| STI | Special Thread for Helical Coil Wire Screw Thread Inserts |
| SGT | Special Gas Taper Thread |
| SPL-PTF | Dryseal special Taper Pipe Thread |

*Taps are not marked with "U" but with corresponding American Standard Form Symbol.
†Formerly designated USA (American).

**Designations for Ground Thread Taps.** — Designations for ground threads provide a wide selection of ground thread taps to secure the classes of thread tolerance desired with maximum wear life. They also allow the user to select a tap to limits which in general will suit the individual work and equipment conditions that he encounters.

While the changes from previous standards are major in character because they affect nomenclature, as well as pitch diameter limits in certain ranges, they permit an easy and economical transition from previous standards without obsolescence of existing inventory.

All standard ground thread taps were formerly designated as Commercial Ground, Commercial Ground High, or Precision Ground. Precision Ground thread taps were made to pitch diameter tolerances of .0005 inch, whereas Commercial Ground thread taps were made to pitch diameter tolerances varying from .001 to .0018 inch depending upon the diameter-pitch combination.

The ANSI Standard establishes only one classification of ground thread taps designated as "Ground Thread." The classification previously identified as Commercial Ground, Commercial Ground High, and Precision Ground no longer applies to ground taps.

The ANSI Standard for ground thread taps establishes pitch diameter tolerances for machine screw and hand taps, in size range No. 0 to 1 inch diameter in 0.0005 inch increments. The pitch diameter tolerances for taps over 1 inch to 1½ inch diameter, inclusive, are in 0.001 inch increments.

The pitch diameter limit designations for standard tolerance ranges, described in the following paragraphs, permit selection of the proper tap for the class of thread tolerance desired and maximum tap life.

Ground Thread taps made to pitch diameter tolerances above basic are designated "High" taps, and those made to tolerances below basic, "Low" taps and are identified by The letter "H" and "L" respectively. The tolerance ranges are indicated by a numeral after the "H" and "L" identification. This numeral discloses the number of half-thousandths (.0005) larger than basic of the maximum "H" tap or smaller than basic of the minimum "L" tap. The tolerance equals .0005 inch on taps 1-inch diameter and smaller and .001 inch on taps over 1-inch diameter to 1½-inch diameter, inclusive.

*Example 1:* A ¼"—20 NC Ground Thread Tap marked G H3 is identified as follows:

    G = the symbol for Ground Thread;
    H = the symbol for above basic; and
    3 = the symbol for pitch diameter limits which in this case are .001 to .0015" above basic.

*Example 2:* A ¼"—20 NC Ground Thread tap marked G L1 is identified as follows:

    G = symbol for Ground Thread;
    L = symbol for below basic; and
    1 = symbol for pitch diameter limits which in this case are .0000" to .0005" under basic.

*Example 3:* A 1¼"—7 NC Ground Thread tap marked G H4 is identified as follows:

    G = symbol for Ground Thread;
    H = symbol for above basic; and
    4 = symbol for pitch diameter limits which in this case are .0015" to .002" above basic.

It will be noted in these examples that the tolerance range numeral divided by 2 establishes, in thousandths of an inch, the amount that the maximum tap pitch diameter is above basic in the H series and the amount that the minimum tap pitch diameter is under basic in the L series.

The pitch diameter limit numbers and the corresponding limits for Ground Thread taps are as follows, H indicating above basic (High), and L indicating below basic (Low).

*Pitch Diameter Limit Numbers for Taps to 1" Diameter, Inclusive:*

$$L_1 = \text{Basic to basic minus } .0005''$$
$$H_1 = \text{Basic to basic plus } .0005''$$
$$H_2 = \text{Basic plus } .0005'' \text{ to basic plus } .001''$$
$$H_3 = \text{Basic plus } .001'' \text{ to basic plus } .0015''$$
$$H_4 = \text{Basic plus } .0015'' \text{ to basic plus } .002''$$
$$H_5 = \text{Basic plus } .002'' \text{ to basic plus } .0025''$$
$$H_6 = \text{Basic plus } .0025'' \text{ to basic plus } .003''$$
$$H_7 = \text{Basic plus } .003'' \text{ to basic plus } .0035''$$

*Pitch Diameter Limit Numbers for Taps Over 1" to 1 1/2" Diameter, Inclusive:*

$$H_4 = \text{Basic plus } .001'' \text{ to basic plus } .002''.$$

Tables 12 and 13 present a guide to assist in the selection of the proper tap to produce a desired class of thread tolerance under normal conditions.

**Acme and Square-threaded Taps.** — These taps are usually made in sets, three taps in a set undoubtedly being the most common. For very fine pitches, two taps in a set will be found sufficient, while as many as five taps in a set are used for coarse pitches. Tables are given herewith for proportioning both Acme and square-threaded taps when made in sets. One leading tap maker in cutting the threads of square-threaded taps makes them according to the following rules: The width of the groove between two threads is made equal to one-half the pitch of the thread, less 0.004 inch. This makes the width of the thread itself equal to one-half of the pitch, plus 0.004 inch. The depth of the thread is made equal to 0.45 times the pitch, plus 0.0025 inch. This latter rule produces a thread which for all the ordinarily used pitches for square-threaded taps has a depth less than the generally accepted standard depth, this latter depth being equal to one-half the pitch. The object of this shallow thread is to insure that if the hole to be threaded by the tap is not bored out so as to provide clearance at the bottom of the thread, the tap will cut its own clearance. The hole should, however, always be drilled out large enough so that the cutting of the clearance is not required of the tap.

Another maker follows under ordinary conditions the dimensions given in the accompanying tables, making the diameter at the end of the chamfer of the first tap equal to the root diameter of the thread, plus 0.010 inch. The diameter at the end of the chamfer of the second and third taps is made equal to the diameter of the straight portion of the next previous tap, minus 0.005 inch.

For Acme thread taps, this manufacturer makes the actual root diameter on the first tap 0.010 inch, and on the second tap 0.005 inch less than the standard root diameter. The finishing tap is made with standard root diameter, and a standard thread tool is used for all three taps in a set.

The table, "Dimensions of Acme Thread Taps in Sets of Three Taps" may be used for the length dimensions for Acme taps. The dimensions in this table apply to single-threaded taps. For multiple-threaded taps or taps with very coarse pitch, relative to the diameter, the length of the chamfered part of the thread may be increased. Square-threaded taps are made to the same table as Acme taps, with the exception of the figures in column $K$, which for square-threaded taps should be equal to the nominal diameter of the tap, no oversize allowance being customary in these taps. The first tap in a set of Acme taps (not square-threaded taps) should be turned taper in bottom of the thread for a distance of about one-quarter of the length of the threaded part. The taper should be so selected that the root diameter is about ⅟₃₂ inch smaller at the point than the proper root diameter of the tap. The first tap should preferably be provided with a short pilot at the point. For very coarse pitches, the first tap may be provided with spiral flutes at right angles to the angle of the thread. Acme and square-threaded taps should be relieved or backed off on the top of the thread of the chamfered portion on all of the taps in the set. When the taps are used as machine taps, rather than as hand taps, they should be relieved in the angle of the thread, as well as on the top, for the whole length of the chamfered portion. Acme taps should also always be relieved on the front side of the thread to within ⅟₃₂ inch of the cutting edge.

### Table for Making Acme Thread Taps in Sets of Three Taps

| No. of Threads per Inch | Amount in Inches to be Added to Root Diameter of Tap to Obtain Diameter of Straight Part of Thread of | | No. of Threads per Inch | Amount in Inches to be Added to Root Diameter of Tap to Obtain Diameter of Straight Part of Thread of | |
|---|---|---|---|---|---|
| | 1st Tap | 2d Tap | | 1st Tap | 2d Tap |
| 1 | 0.468 | 0.832 | 5 | 0.108 | 0.192 |
| 1½ | 0.318 | 0.566 | 5½ | 0.100 | 0.178 |
| 2 | 0.243 | 0.432 | 6 | 0.093 | 0.166 |
| 2½ | 0.198 | 0.352 | 7 | 0.082 | 0.146 |
| 3 | 0.168 | 0.298 | 8 | 0.074 | 0.132 |
| 3½ | 0.147 | 0.261 | 9 | 0.068 | 0.121 |
| 4 | 0.130 | 0.232 | 10 | 0.063 | 0.112 |
| 4½ | 0.118 | 0.210 | 12 | 0.055 | 0.098 |

### Table for Making Square-threaded Taps in Sets of Three Taps

| No. of Threads per Inch | Amount in Inches to be Added to Root Diameter of Tap to Obtain Diameter of Straight Part of Thread of | | No. of Threads per Inch | Amount in Inches to be Added to Root Diameter of Tap to Obtain Diameter of Straight Part of Thread of | |
|---|---|---|---|---|---|
| | 1st Tap | 2d Tap | | 1st Tap | 2d Tap |
| 1 | 0.410 | 0.800 | 5 | 0.082 | 0.160 |
| 1½ | 0.273 | 0.533 | 5½ | 0.075 | 0.146 |
| 2 | 0.205 | 0.400 | 6 | 0.068 | 0.133 |
| 2½ | 0.164 | 0.320 | 7 | 0.059 | 0.114 |
| 3 | 0.137 | 0.267 | 8 | 0.051 | 0.100 |
| 3½ | 0.117 | 0.229 | 9 | 0.046 | 0.089 |
| 4 | 0.102 | 0.200 | 10 | 0.041 | 0.080 |
| 4½ | 0.091 | 0.178 | 12 | 0.034 | 0.067 |

## Proportions of Acme and Square-threaded Taps Made in Sets

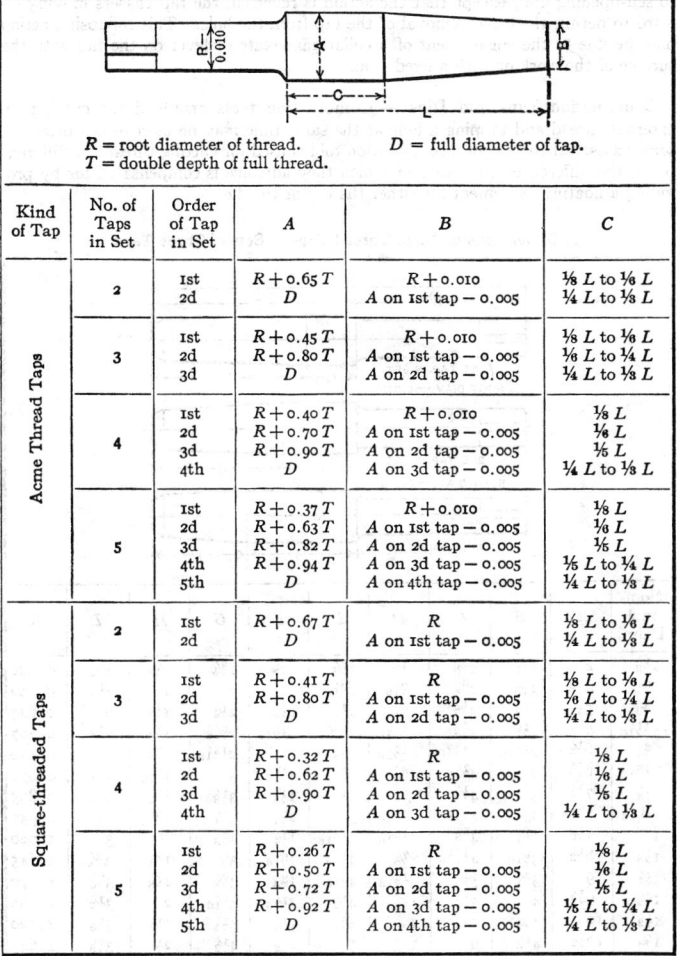

$R$ = root diameter of thread.    $D$ = full diameter of tap.
$T$ = double depth of full thread.

| Kind of Tap | No. of Taps in Set | Order of Tap in Set | $A$ | $B$ | $C$ |
|---|---|---|---|---|---|
| **Acme Thread Taps** | 2 | 1st | $R + 0.65\,T$ | $R + 0.010$ | ⅛ $L$ to ⅙ $L$ |
| | | 2d | $D$ | $A$ on 1st tap − 0.005 | ¼ $L$ to ⅓ $L$ |
| | 3 | 1st | $R + 0.45\,T$ | $R + 0.010$ | ⅛ $L$ to ⅙ $L$ |
| | | 2d | $R + 0.80\,T$ | $A$ on 1st tap − 0.005 | ⅙ $L$ to ¼ $L$ |
| | | 3d | $D$ | $A$ on 2d tap − 0.005 | ¼ $L$ to ⅓ $L$ |
| | 4 | 1st | $R + 0.40\,T$ | $R + 0.010$ | ⅛ $L$ |
| | | 2d | $R + 0.70\,T$ | $A$ on 1st tap − 0.005 | ⅙ $L$ |
| | | 3d | $R + 0.90\,T$ | $A$ on 2d tap − 0.005 | ⅕ $L$ |
| | | 4th | $D$ | $A$ on 3d tap − 0.005 | ¼ $L$ to ⅓ $L$ |
| | 5 | 1st | $R + 0.37\,T$ | $R + 0.010$ | ⅛ $L$ |
| | | 2d | $R + 0.63\,T$ | $A$ on 1st tap − 0.005 | ⅙ $L$ |
| | | 3d | $R + 0.82\,T$ | $A$ on 2d tap − 0.005 | ⅕ $L$ |
| | | 4th | $R + 0.94\,T$ | $A$ on 3d tap − 0.005 | ⅕ $L$ to ¼ $L$ |
| | | 5th | $D$ | $A$ on 4th tap − 0.005 | ¼ $L$ to ⅓ $L$ |
| **Square-threaded Taps** | 2 | 1st | $R + 0.67\,T$ | $R$ | ⅛ $L$ to ⅙ $L$ |
| | | 2d | $D$ | $A$ on 1st tap − 0.005 | ¼ $L$ to ⅓ $L$ |
| | 3 | 1st | $R + 0.41\,T$ | $R$ | ⅛ $L$ to ⅙ $L$ |
| | | 2d | $R + 0.80\,T$ | $A$ on 1st tap − 0.005 | ⅙ $L$ to ¼ $L$ |
| | | 3d | $D$ | $A$ on 2d tap − 0.005 | ¼ $L$ to ⅓ $L$ |
| | 4 | 1st | $R + 0.32\,T$ | $R$ | ⅛ $L$ |
| | | 2d | $R + 0.62\,T$ | $A$ on 1st tap − 0.005 | ⅙ $L$ |
| | | 3d | $R + 0.90\,T$ | $A$ on 2d tap − 0.005 | ⅕ $L$ |
| | | 4th | $D$ | $A$ on 3d tap − 0.005 | ¼ $L$ to ⅓ $L$ |
| | 5 | 1st | $R + 0.26\,T$ | $R$ | ⅛ $L$ |
| | | 2d | $R + 0.50\,T$ | $A$ on 1st tap − 0.005 | ⅙ $L$ |
| | | 3d | $R + 0.72\,T$ | $A$ on 2d tap − 0.005 | ⅕ $L$ |
| | | 4th | $R + 0.92\,T$ | $A$ on 3d tap − 0.005 | ⅕ $L$ to ¼ $L$ |
| | | 5th | $D$ | $A$ on 4th tap − 0.005 | ¼ $L$ to ⅓ $L$ |

**Adjustable Taps.** — Many adjustable taps are now used, especially for accurate work. Some taps of this class are made of a solid piece of steel which is split and provided with means of expanding sufficiently to compensate for wear. Most of the larger adjustable taps have inserted blades or chasers which are rigidly held, but capable of radial adjustment. The use of taps of this general class enables standard sizes to be maintained readily.

**Advantages of Collapsing Taps.** — Collapsing taps are similar in principle to self-opening dies, except that the action is reversed, the tap chasers moving inward to permit the rapid removal of the tap from the hole. This collapsing action may be due to the engagement of a collar gage-plate or lever on the tap with the surface of the work or with a fixed stop.

**Combination Taps and Dies.** — Combination tools arranged for cutting an external thread and tapping a hole at the same time may be used to advantage in some cases. If the tap of a combination tool is used for cutting threads of different pitch, the difference in the rate at which they advance is compensated for by providing a floating movement for either the tap or the die.

**Dimensions of Acme Thread Taps in Sets of Three Taps**

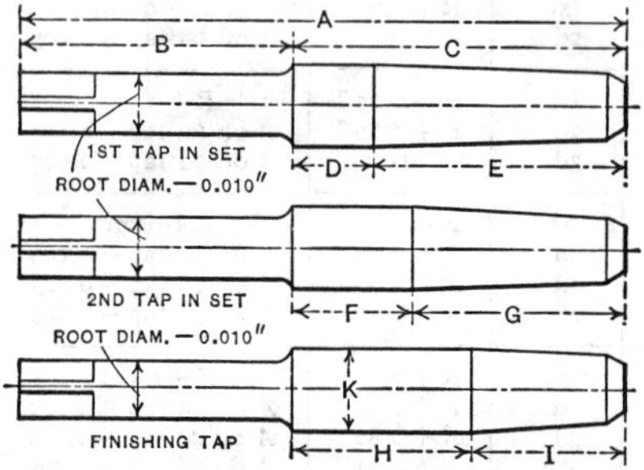


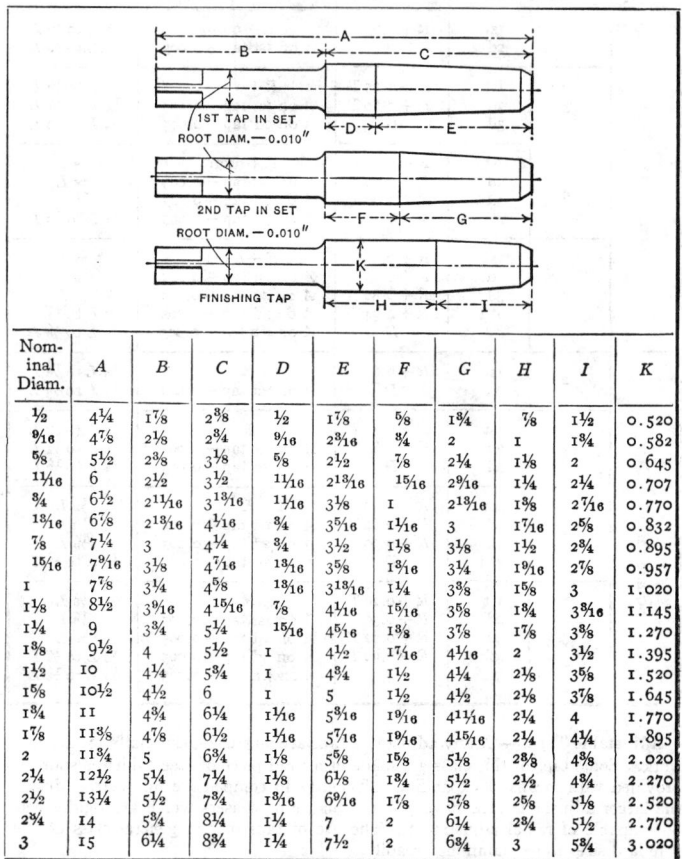

| Nominal Diam. | *A* | *B* | *C* | *D* | *E* | *F* | *G* | *H* | *I* | *K* |
|---|---|---|---|---|---|---|---|---|---|---|
| 1/2 | 4 1/4 | 1 7/8 | 2 3/8 | 1/2 | 1 7/8 | 5/8 | 1 3/4 | 7/8 | 1 1/2 | 0.520 |
| 9/16 | 4 7/8 | 2 1/8 | 2 3/4 | 9/16 | 2 3/16 | 3/4 | 2 | 1 | 1 3/4 | 0.582 |
| 5/8 | 5 1/2 | 2 3/8 | 3 1/8 | 5/8 | 2 1/2 | 7/8 | 2 1/4 | 1 1/8 | 2 | 0.645 |
| 11/16 | 6 | 2 1/2 | 3 1/2 | 11/16 | 2 13/16 | 15/16 | 2 9/16 | 1 1/4 | 2 1/4 | 0.707 |
| 3/4 | 6 1/2 | 2 11/16 | 3 13/16 | 11/16 | 3 1/8 | 1 | 2 13/16 | 1 3/8 | 2 7/16 | 0.770 |
| 13/16 | 6 7/8 | 2 13/16 | 4 1/16 | 3/4 | 3 5/16 | 1 1/16 | 3 | 1 7/16 | 2 5/8 | 0.832 |
| 7/8 | 7 1/4 | 3 | 4 1/4 | 3/4 | 3 1/2 | 1 1/8 | 3 1/8 | 1 1/2 | 2 3/4 | 0.895 |
| 15/16 | 7 9/16 | 3 1/8 | 4 7/16 | 13/16 | 3 5/8 | 1 3/16 | 3 1/4 | 1 9/16 | 2 7/8 | 0.957 |
| 1 | 7 7/8 | 3 1/4 | 4 5/8 | 13/16 | 3 13/16 | 1 1/4 | 3 3/8 | 1 5/8 | 3 | 1.020 |
| 1 1/8 | 8 1/2 | 3 9/16 | 4 15/16 | 7/8 | 4 1/16 | 1 5/16 | 3 5/8 | 1 3/4 | 3 3/16 | 1.145 |
| 1 1/4 | 9 | 3 3/4 | 5 1/4 | 15/16 | 4 5/16 | 1 3/8 | 3 7/8 | 1 7/8 | 3 3/8 | 1.270 |
| 1 3/8 | 9 1/2 | 4 | 5 1/2 | 1 | 4 1/2 | 1 7/16 | 4 1/16 | 2 | 3 1/2 | 1.395 |
| 1 1/2 | 10 | 4 1/4 | 5 3/4 | 1 | 4 3/4 | 1 1/2 | 4 1/4 | 2 1/8 | 3 5/8 | 1.520 |
| 1 5/8 | 10 1/2 | 4 1/2 | 6 | 1 | 5 | 1 1/2 | 4 1/2 | 2 1/8 | 3 7/8 | 1.645 |
| 1 3/4 | 11 | 4 3/4 | 6 1/4 | 1 1/16 | 5 3/16 | 1 9/16 | 4 11/16 | 2 1/4 | 4 | 1.770 |
| 1 7/8 | 11 3/8 | 4 7/8 | 6 1/2 | 1 1/16 | 5 7/16 | 1 9/16 | 4 15/16 | 2 1/4 | 4 1/4 | 1.895 |
| 2 | 11 3/4 | 5 | 6 3/4 | 1 1/8 | 5 5/8 | 1 5/8 | 5 1/8 | 2 3/8 | 4 3/8 | 2.020 |
| 2 1/4 | 12 1/2 | 5 1/4 | 7 1/4 | 1 1/8 | 6 1/8 | 1 3/4 | 5 1/2 | 2 1/2 | 4 3/4 | 2.270 |
| 2 1/2 | 13 1/4 | 5 1/2 | 7 3/4 | 1 3/16 | 6 9/16 | 1 7/8 | 5 7/8 | 2 5/8 | 5 1/8 | 2.520 |
| 2 3/4 | 14 | 5 3/4 | 8 1/4 | 1 1/4 | 7 | 2 | 6 1/4 | 2 3/4 | 5 1/2 | 2.770 |
| 3 | 15 | 6 1/4 | 8 3/4 | 1 1/4 | 7 1/2 | 2 | 6 3/4 | 3 | 5 3/4 | 3.020 |

**British Standard Screwing Taps — ISO Metric Series.** — This British Standard (BS 949:1969) provides dimensions and tolerances for taps for producing the following threads: ISO metric coarse, and fine pitch series (BS 3643:Part 2:1966 and Part 3:1967 respectively); ISO inch (Unified) pitch series; BS pipe threads (fastening, parallel, and taper); American National Pipe Threads (straight and taper), and British Standard Whitworth, British Standard Fine, and British Association threads.

ISO metric coarse pitch threads,are recommended as a first choice for use by industry in the United Kingdom, and ISO inch series threads are recommended as second choice. In the Standard, these threads are designated as preferred series, and the accompanying Handbook tables cover taps for those of ISO metric form only.

The Standard was first issued in 1941, and the latest revision is based on ISO Recommendation R529, Short Machine Taps and Hand Taps. Earlier editions of the Standard designated the commonly used screwing tap as a hand tap, but during the ISO work it was recognized that the majority of hand taps are used for machine tapping operations, and it was thus considered logical to include the term 'short

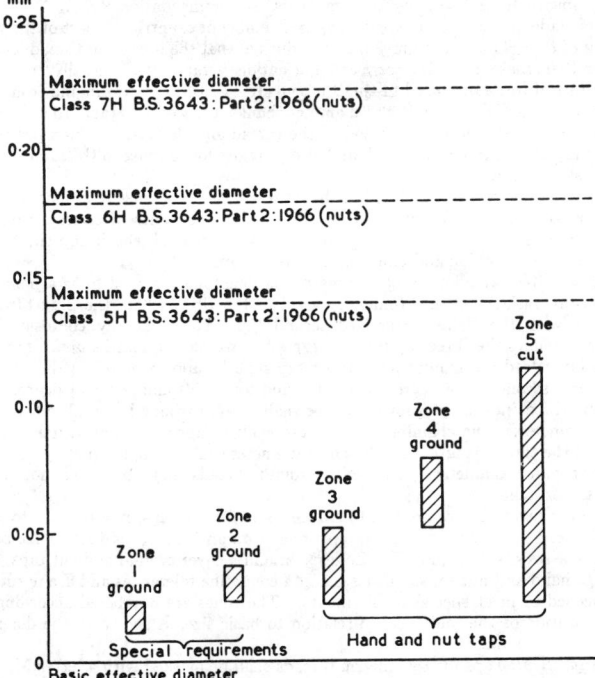

Fig. 1. Diagram Showing Thread Tolerance Zones for a Metric Tap in Accordance With BS949: 1969. The Example Relates to an M10 × 1.5 ISO Metric Coarse Pitch Thread.

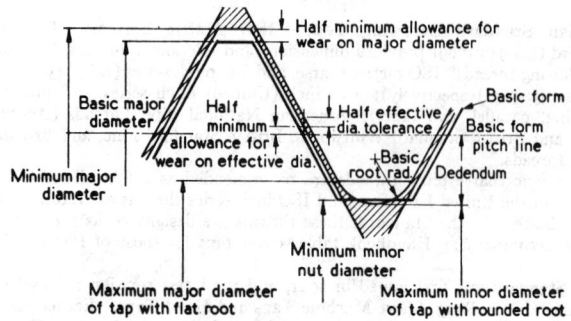

Fig. 2. Diagram showing Disposition of Tolerance Zones to Basic ISO Metric Thread Forms

machine taps,' which also distinguishes the tool from the longer nut tap. The shank dimensions and squares conform to ISO Recommendation R237.

Tests made under different conditions in a number of countries as part of the work leading to the ISO tap recommendation, indicated that the use of tap thread lengths shorter than those given in many existing national standards, generally enabled a better thread to be produced in the component. Thus, for taps ranging from 3 to 10 mm nominal diameter with full diameter shanks, the revised Standard specifies a recess or neck between the back end of the thread and the shank, which permits a shorter tap thread to be employed, and also provides for tapping to the same depths previously obtainable.

Thread tolerances are tabulated in the Standard, and ISO metric ground and cut taps are covered. Tables are also included giving the limits of tolerance on the shank diameters, and the dimensions of squares, the overall length and the thread length of short machine and hand taps, and nut taps.

In the United Kingdom, a bottoming tap is sometimes referred to as a plug tap, which is at variance with terminology used in the USA, where the plug tap corresponds with the British Standard second tap. To avoid any confusion, the Standard defines the three tap types as *taper tap*, *second tap*, and *bottoming tap*.

The taper lead or chamfer for a taper tap shall be approximately 4 degrees per side; for a second tap 8 degrees per side; and for a bottoming tap 23 degrees per side, and the point diameter for each type shall be approximately equal to the basic minor diameter. The chamfer of a nut tap shall be approximately equal to two-thirds of the thread length, and the point diameter shall be approximately equal to the basic minor diameter. Taps with ground threads may, but need not, have radial thread relief.

*Tolerance zones:* There are five tolerance zones for the taps, numbered 1 to 5 as shown in Fig. 1, which gives an example related to an M10 × 1.5 ISO coarse pitch thread, classes 5H, 6H, and 7H. Zones 3, 4, and 5 cover ground and cut taps, and apply to hand and nut taps. Zones 1 and 2 cover fine tolerances and are normally only needed to meet special requirements. The zones are numbered according to the disposition of the tolerances in relation to basic size, as shown in the diagram Fig. 2.

*Identification:* Taps for ISO metric threads shall be marked with a letter M, and show the particular diameter size, and the pitch — thus, for example, M6 × 1. The taps are right-hand cutting unless otherwise stated, and it will be assumed by the supplier that right-handed taps are required unless the purchaser specifies left-hand in an order.

## British Standard Short Machine Taps and Hand Taps — ISO Metric Coarse Pitch Threads (BS 949: 1969)

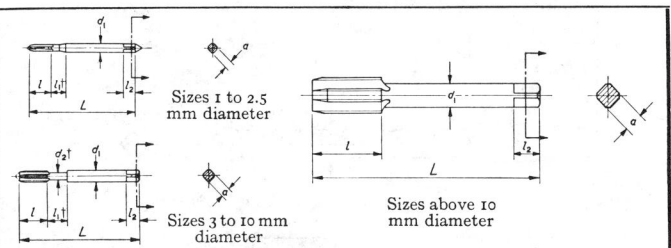

Sizes 1 to 2.5 mm diameter

Sizes above 10 mm diameter

Sizes 3 to 10 mm diameter

| Thread Designation | Basic Diameter | Pitch | Shank Diameter $d_1$ | Thread Length $l$ | Overall Length $L$ | Size of Square $a$ | Length of Square $l_2$ |
|---|---|---|---|---|---|---|---|
| M1 | 1.000 | 0.250 | 2.50 | 5.5 | 38.5 | 2.00 | 4 |
| M1.1 | 1.100 | 0.250 | 2.50 | 5.5 | 38.5 | 2.00 | 4 |
| M1.2 | 1.200 | 0.250 | 2.50 | 5.5 | 38.5 | 2.00 | 4 |
| M1.4 | 1.400 | 0.300 | 2.50 | 7 | 40 | 2.00 | 4 |
| M1.6 | 1.600 | 0.350 | 2.50 | 8 | 41 | 2.00 | 4 |
| M1.8 | 1.800 | 0.350 | 2.50 | 8 | 41 | 2.00 | 4 |
| M2 | 2.000 | 0.400 | 2.50 | 8 | 41 | 2.00 | 4 |
| M2.2 | 2.200 | 0.450 | 2.80 | 9.5 | 44.5 | 2.24 | 5 |
| M2.5 | 2.500 | 0.450 | 2.80 | 9.5 | 44.5 | 2.24 | 5 |
| M3 | 3.000 | 0.500 | 3.15 | 11 | 48 | 2.50 | 5 |
| M3.5 | 3.500 | 0.600 | 3.55 | 13 | 50 | 2.80 | 5 |
| M4 | 4.000 | 0.700 | 4.00 | 13 | 53 | 3.15 | 6 |
| M4.5 | 4.500 | 0.750 | 4.50 | 13 | 53 | 3.55 | 6 |
| M5 | 5.000 | 0.800 | 5.00 | 16 | 58 | 4.00 | 7 |
| M6 | 6.000 | 1.000 | 6.30 | 19 | 66 | 5.00 | 8 |
| (M7)* | 7.000 | 1.000 | 7.10 | 19 | 66 | 5.60 | 8 |
| M8 | 8.000 | 1.250 | 8.00 | 22 | 72 | 6.30 | 9 |
| (M9)* | 9.000 | 1.250 | 9.00 | 22 | 72 | 7.10 | 10 |
| M10 | 10.000 | 1.500 | 10.00 | 24 | 80 | 8.00 | 11 |
| (M11)* | 11.000 | 1.500 | 8.00 | 25 | 85 | 6.30 | 9 |
| M12 | 12.000 | 1.750 | 9.50 | 29 | 89 | 7.50 | 10 |
| M14 | 14.000 | 2.000 | 11.20 | 30 | 95 | 9.00 | 12 |
| M16 | 16.000 | 2.000 | 12.50 | 32 | 102 | 10.00 | 13 |
| M18 | 18.000 | 2.500 | 14.00 | 37 | 112 | 11.20 | 14 |
| M20 | 20.000 | 2.500 | 14.00 | 37 | 112 | 11.20 | 14 |
| M22 | 22.000 | 2.500 | 16.00 | 38 | 118 | 12.50 | 16 |
| M24 | 24.000 | 3.000 | 18.00 | 45 | 130 | 14.00 | 18 |
| M27 | 27.000 | 3.000 | 20.00 | 45 | 135 | 16.00 | 20 |
| M30 | 30.000 | 3.500 | 20.00 | 48 | 138 | 16.00 | 20 |
| M33 | 33.000 | 3.500 | 22.40 | 51 | 151 | 18.00 | 22 |
| M36 | 36.000 | 4.000 | 25.00 | 57 | 162 | 20.00 | 24 |
| M39 | 39.000 | 4.000 | 28.00 | 60 | 170 | 22.40 | 26 |
| M42 | 42.000 | 4.500 | 28.00 | 60 | 170 | 22.40 | 26 |
| M45 | 45.000 | 4.500 | 31.50 | 67 | 187 | 25.00 | 28 |
| M48 | 48.000 | 5.000 | 31.50 | 67 | 187 | 25.00 | 28 |
| M52 | 52.000 | 5.000 | 35.50 | 70 | 200 | 28.00 | 31 |
| M56 | 56.000 | 5.500 | 35.50 | 70 | 200 | 28.00 | 31 |
| M60 | 60.000 | 5.500 | 40.00 | 76 | 221 | 31.50 | 34 |
| M64 | 64.000 | 6.000 | 40.00 | 79 | 224 | 31.50 | 34 |
| M68 | 68.000 | 6.000 | 45.00 | 79 | 234 | 35.50 | 38 |

All dimensions are given in millimeters. The basic diameter and pitch dimensions are correct to three decimal places.

† Taps with M1, M1.1, and M1.2 threads have a neck length $l_1$ of 4.5 mm; those with M1.4, M1.6, and M1.8 threads 5 mm; those with an M2 thread 5.5 mm, and those with M2.2 and M2.5 threads 6 mm. The neck dimensions $d_2$ and $l_1$ of taps ranging from 3 to 10 mm diameter are (in millimeters): 2.12 dia. × 7 (M3 thread); 2.50 dia. × 7 (M3.5); 2.80 dia. × 8 (M4); 3.15 dia. × 8 (M4.5); 3.55 dia. × 9 (M5); 4.50 dia. × 11 (M6); 5.30 dia. × 11 (M7); 6.00 dia. × 13 (M8); 7.10 dia. × 14 (M9), and 7.5 dia. × 15 (M10).

* Designations shown in parentheses are non-preferred threads.

**British Standard Nut Taps — ISO Metric Coarse Pitch Threads (BS 949: 1969)**

| Thread Designation | Shank Diameter $d_1$ | Thread Length $l$† | Overall Length $L$‡ | Size of Square $a$ | Length of Square $l_2$ |
|---|---|---|---|---|---|
| M1.6 | 1.12 | 11 | 60 | ... | ... |
| M1.8 | 1.25 | 11 | 60 | ... | ... |
| M2 | 1.40 | 12 | 60 | ... | ... |
| M2.2 | 1.80 | 14 | 60 | ... | ... |
| M2.5 | 1.80 | 14 | 60 | ... | ... |
| M3 | 2.24 | 15 | 70 | 1.80 | 4 |
| M3.5 | 2.50 | 18 | 80 | 2.00 | 4 |
| M4 | 2.80 | 21 | 90 | 2.24 | 5 |
| M4.5 | 3.35 | 23 | 90 | 2.50 | 5 |
| M5 | 4.00 | 24 | 100 | 3.15 | 6 |
| M6 | 4.50 | 30 | 110 | 3.55 | 6 |
| (M7)* | 5.40 | 30 | 110 | 4.50 | 7 |
| M8 | 6.00 | 37.5 | 125 | 4.50 | 7 |
| (M9)* | 7.00 | 37.5 | 140 | 5.60 | 8 |
| M10 | 7.00 | 45 | 140 | 5.60 | 8 |
| (M11)* | 8.00 | 45 | 160 | 6.30 | 9 |
| M12 | 9.50 | 53 | 180 | 7.50 | 10 |
| M14 | 11.20 | 60 | 200 | 9.00 | 12 |
| M16 | 12.50 | 60 | 200 | 10.00 | 13 |
| M18 | 14.00 | 75 | 220 | 11.20 | 14 |
| M20 | 14.00 | 75 | 250 | 11.20 | 14 |
| M22 | 16.00 | 75 | 280 | 12.50 | 16 |
| M24 | 18.00 | 90 | 280 | 14.00 | 18 |
| M27 | 20.00 | 90 | 315 | 16.00 | 20 |
| M30 | 20.00 | 105 | 315 | 16.00 | 20 |
| M33 | 22.40 | 105 | 355 | 18.00 | 22 |
| M36 | 25.00 | 120 | 400 | 20.00 | 24 |
| M39 | 28.00 | 120 | 400 | 22.40 | 26 |
| M42 | 28.00 | 135 | 450 | 22.40 | 26 |
| M45 | 31.50 | 135 | 500 | 25.00 | 28 |
| M48 | 31.50 | 150 | 500 | 25.00 | 28 |
| M52 | 35.50 | 150 | 560 | 28.00 | 31 |
| M56 | 35.50 | 165 | 560 | 28.00 | 31 |
| M60 | 40.00 | 165 | 560 | 31.50 | 34 |
| M64 | 40.00 | 180 | | 31.50 | 34 |
| M68 | 45.00 | 180 | 630 | 35.50 | 38 |

All dimensions are given in millimeters. † Each thread length is equal to 30 pitches of the particular thread, rounded to the nearest 1.0 mm longer.
‡ Taps below 4 mm diameter may be provided with a male centre at either end to facilitate manufacture, and taps above 4 mm and below 8 mm diameter may have a male centre at one end only. There may be no square on taps below 3 mm diameter.
* Designations shown in parentheses are non-preferred threads.

**Tapping Square Threads.** — If it is necessary to tap square threads, this should be done by using a set of taps that will form the thread by a progressive cutting action, the taps varying in size in order to distribute the work, especially for threads of comparatively coarse pitch. From three to five taps may be required in a set, depending upon the pitch. Each tap should have a pilot to steady it. The pilot of the first tap has a smooth cylindrical end from 0.003 to 0.005 inch smaller than the hole, and the pilots of following taps should have teeth.

**Interrupted Thread Taps.** — On taps of this type each alternate tooth is omitted so that each of the teeth is followed by a space, and vice versa. This arrangement gives a freedom of action to each tooth that is impossible otherwise. In tapping out a given size hole, the interrupted thread tap removes the same amount of stock as the standard tap but actual practice shows that the resistance is from 30 to 50 per cent less. This would indicate that the greater resistance with the standard form of tap is caused by friction, which is more destructive to the tap than the wear due in the actual cutting operation. In tapping holes in material such as copper and boiler sheets, the tendency with the standard form of tap is to tear the threads that are being formed, owing to the wedging action of the cutting teeth and the lack of resistance offered by the metal to the pressure of the continuous row of cutting edges; the chips are carried forward in a mass in front of the cutting teeth, and unless the tap is frequently reversed and this mass of chips broken off, the thread may be mutilated or the tap itself broken.

**Steel for Taps.** — *Carbon steel* is still extensively used in the manufacture of taps both on production jobs and for general service tapping. A good steel to use should have a carbon content from about 1.15 to 1.25 per cent carbon. This steel, if uniform in its composition, can be hardened with a lead control well within a limit of plus or minus 0.003 inch in one inch of thread. Some carbon steels used in the manufacture of taps have a vanadium content of approximately 0.25 per cent. The vanadium has a tendency to add toughness to the steel as well as improving the cutting qualities and helps to prevent breakage particularly in small taps. The carbon content is the same as in regular carbon steels. This steel can be hardened also within the same lead range of plus or minus 0.003 inch in one inch of thread.

Expensive, non-shrinkable steels are obtainable in the market that show practically no change in either lead or the diameter of the tap when hardened but these are not used commercially on account of their high cost.

*High-speed steel* taps are used extensively for production tapping, with *cut threads* for a general line of commercial work, and *ground threads* where greater accuracy is essential. A high-speed steel suitable for taps should contain from 0.60 to 0.75 per cent carbon and from 17 to 19 per cent tungsten. This steel hardens at from 2200 to 2350 degrees F. The temper should be drawn at from 1000 to 1125 degrees F.

Due to the progress steel makers have made in the manufacture of high-speed steels and advances made by tap makers in the heat treatment, high-speed steel taps are especially good for production tapping and great savings have been accomplished by their use. The high-speed steel ground-thread tap is extensively used in production where accuracy is the controlling factor. The grinding of the blank and the threads corrects inaccuracies caused by distortion in heat treatment and insures accuracy for the pitch, outside, and shank diameters as well as in the lead of the thread. There are tap applications where so-called "surface treatments" can add appreciably to the life of a tap. These surface treatments may be roughly classified as nitrided, oxided, chrome plated, and vapor blasted.

**Hardness of Taps.** — Tests made over many years of tapping with high-speed steel ground-thread taps show that the ideal hardness of a tap should be about

63 Rockwell C. It should never be below 61 or above 65. If the hardness is below 61 Rockwell C., the corners of the teeth wear very rapidly; if the tap is harder than 65 Rockwell C., the teeth show a tendency of slightly chipping at the chamfer, especially when tapping hard or tough materials.

## Threading Dies

**Steel for Die Chasers.** — High-speed steels now are used extensively for die chasers. Such steels can now be hardened without injuring the finished surfaces; moreover, when the design of the chaser is such that grinding of the thread form after hardening is practicable, even keener cutting edges and smoother surfaces can be obtained. The National Acme Co. recommends high-speed steel for practically all threading chasers. In some cases, carbon steel has been recommended for threading copper and also for cutting Acme or worm threads.

**Types of Die Chasers and Methods of Forming Teeth.** — One common type of die chaser has the teeth formed on the end. The teeth of such chasers may be formed by three different methods. The first is by using a hob having helical teeth like the teeth of a tap. The second is by using a milling cutter which has annular rows of teeth that are perpendicular to the axis. This cutter is set to the helix angle of the screw thread when milling the chaser teeth, and it is fed across the end of the blank and forms a series of straight teeth. The third method is by using a milling cutter like the one just referred to. This cutter is set to the proper helix angle as in the previous case, and is then sunk into the end of the chaser blank, thus forming concave or circular teeth instead of straight teeth. Another form of milled chaser which has straight teeth is found in the Landis die-heads. In this case, the chasers are set tangentially to the work and the milled teeth, instead of being across the ends of the chasers, extend the full length of the chasers; the latter are sharpened by grinding on the ends.

The National Acme circular chasers are like circular forming tools. The circumferential thread grooves are annular and have no lead. This type of chaser is sharpened by grinding the cutting face back in a circumferential direction, the same as in grinding circular formed tools.

**Angle of Chamfer for Die Chasers.** — The leading side or "throat" of a die is chamfered to provide a more gradual cutting action, unless it is necessary to cut a full thread close to a shoulder. The throat angle should preferably be such that the work of cutting a thread to the full depth will be distributed over at least two or three teeth on the leading side of the die. The chamfer, according to common practice, extends from the root or base of the most advanced tooth in a set of chasers back to the top of the third tooth, which may be slightly beveled. Each chaser should be ground to the same angle so that each throat will be the same distance from the die axis. The angles recommended for Hartness dies vary according to the pitch of the thread, the angle being 15 degrees relative to the axis of the die for threads varying from 4 to 5½ per inch, 20 degrees for threads varying from 6 to 8 per inch, and 25 degrees for nine or more threads per inch.

**Relief of Die Chasers.** — The throat or chamfered edge of each chaser should have clearance back of the cutting edge or in a circumferential direction. This clearance should be just enough to insure free cutting. If there is not enough relief, the cutting action will be either prevented entirely or retarded, and the die will not advance as fast as it should. On the contrary, excessive relief tends to increase the rate of advance, assuming that the die is self-leading. It is of especial importance that die chasers for brass have as little clearance as possible, because the throats of the chasers steady the die when starting a thread. The sides of the chaser teeth are sometimes relieved instead of the leading sides or corners being chamfered, when a thread must be cut close to a shoulder in brass. This relieving is done by using a brass lap which has an angle of about 50 degrees for a 60-degree chaser tooth.

**Amount of Rake for Threading Dies.** — The front face of each die chaser should lie in a plane intersecting the axis of a die that is used for cutting threads on parts of cast brass, cast iron, or brittle materials of a granular structure. For the more tenacious or tougher materials which are not brittle, such as wrought iron, steel, copper and yellow brass, the chasers should have positive rake, the cutting faces lying in planes that are in advance of the die axis. Most aluminum castings, on account of the zinc in their composition, cut very much like cast brass, and should preferably be threaded with dies having little or no rake. Many of the dies used for cutting threads in machine steel have the front faces of the chasers located ahead of the die axis a distance equal to about one-fifth of the die radius. When grinding the front faces of die chasers, care should be taken to maintain the rake angles. According to experiments of the National Tube Co., the rake angles of dies for pipe threading should vary from 15 to 25 degrees, the latter angle being suitable for threading open-hearth steel pipe.

**Number of Chasers for Pipe Dies.** — To obtain the best results with pipe dies, the number of chasers should vary according to the size of the die, four chasers being used for diameters up to 1¼ inch; six chasers for diameters of from 1½ to 4 inches; eight chasers for diameters of from 4½ to 8 inches; twelve chasers for diameters of from 9 to 12 inches; fourteen chasers for diameters of from 13 to 16 inches; and sixteen chasers for diameters of from 17 to 20 inches. This information is based on experiments of the National Tube Co.

**Dies for Cutting Taper Threads.** — While short taper threads are commonly cut with solid dies, the type of die to use for accurate work, particularly when the length of thread exceeds ordinary die widths, is one having chasers which taper to correspond to the taper on the work and are arranged to move outward radially as the die moves along. Such dies are of the self-opening type. The radial outward movement of the chasers is controlled by a taper plate which allows the cam or scroll ring of the die-head to turn slowly as the die advances. When the thread is finished, the chasers spring out rapidly to clear the work. In dies of this class, the taper plate for controlling the movement of the chasers serves about the same purpose as the adjustable slide or bar of a lathe taper attachment.

**Dies for Cutting Square Threads.** — If dies are to be used for cutting square threads, the sides of the teeth should be relieved to prevent the teeth from binding and breaking. If the die is to be self-leading (the feeding movement not being controlled by a lead-screw), this side relief or clearance should be very slight because, if there is too much relief on the sides of the chaser teeth, the die will not be supported properly and is liable to cut a thread that is quite incorrect in lead. The use of a lead-screw is preferable when cutting square threads with a die. The Acme thread may be cut readily with dies.

**Positive Control of Die-feeding Movement.** — While most dies are self-leading, it is sometimes advisable to control positively the longitudinal motion of the die relative to the work. This control may be utilized merely to start the die, or the arrangement may be such that the longitudinal motion of the die is controlled positively throughout the entire screw-cutting operation. This positive action may be derived from a lead-screw or from a cam, depending upon the type of machine. A lead-screw is sometimes applied to a threading machine of the bolt-cutter type, especially when cutting square threads, or special forms. For screw-cutting operations of this kind, if the die follows its own lead, the accumulated error is often considerable. It is essential that the pitch of the die teeth correspond to the leading movement obtained from the lead-screw. The general method of cutting threads on the automatic screw machine is to use a cam that starts a die on the work and then allows the turret-slide to lag behind somewhat so that the die can lead itself on. The die-holder is designed to allow the die to follow its own lead or move independently of the turret-slide.

**Non-opening Dies.** — The non-opening dies are capable in some cases of hand adjustment, but the object of this adjustment is to vary the size of the die. There are four types of non-opening dies in common use, which may be designated as (1) solid dies, or those that are rigid and incapable of any adjustment for varying the diameter; (2) flexible dies, or those that are split in one or more places and may be adjusted to some extent by compressing or expanding; (3) sectional dies, or those formed of two adjustable sections; (4) rigid adjustable dies of the chaser type, having inserted chasers that may be adjusted radially within certain limits either for maintaining a standard size or for varying the size slightly. A non-opening die may be removed from the work in three different ways: (1) the work rotation may be reversed after the thread is cut; (2) the die itself may be reversed, thus unscrewing it from the threaded part; (3) the die may be revolved in the same direction as the work, but at a somewhat slower speed while cutting the thread and then at a faster rate so that the die backs off the threaded part.

**Automatic or Self-opening Dies.** — Dies which open automatically to permit their removal when the thread is finished, not only save time, but may prevent injury to the thread, which sometimes results when the chips wedge between the teeth of a non-opening die and the work. Self-opening dies differ both in regard to the mechanism for opening the die chasers automatically at the completion of a cut, and as to the method of resetting the chasers in the working or cutting position after the die has been removed. These dies, in general, are formed of two main sections, which have a certain relative motion for opening and closing the die. This motion may be parallel to the axis of the die, or it may be rotary. The radial movement of the chasers is derived from either cam surfaces or the conical surface of a slide in contact with the chasers. Dies of this class may be opened or tripped (1) by stopping the travel of the turret, (2) by the engagement of an outside tripping latch with a fixed stop, or (3) by the engagement of the end of the work with a tripping plate located in the center of the die back of the chasers. Most of the self-opening dies are of the non-revolving type, but some are designed for attachment to a revolving spindle.

**Spring Screw Threading Dies.** — These dies are usually tapped with a straight tap and hob, although this practice is somewhat objectionable on account of the fact that a slight inaccuracy is produced in the shape of the threads of the screw to be threaded when the prongs of the die are forced in by the adjustment of the clamp collar. It is, therefore, better to tap out this die from the back end with a tap that tapers an amount equal to the clearance required in the die when cutting. In that case, the die should be cut to the correct cutting size at the point, and not oversize, as is the case when it is cut with a straight tap. The amount of back taper may be made from about 0.005 to 0.010 inch per inch for iron and steel, and from 0.008 to 0.015 inch per inch for dies cutting brass, copper and metals of similar structure. Spring screw dies are generally made with four flutes, but for several reasons three flutes would be preferable. When four flutes are used, as a rule only two of the lands are cutting. If a die is made with three flutes, it should be fluted with a 60-degree angular cutter. If made with four flutes, the cutter should be a 48-, 45- or 40-degree angular cutter, according to the size of the die, the 48-degree cutter being used for the smallest dies and the 45-degree cutter for all ordinary sizes. Dies ½ inch in outside diameter or smaller are never made with more than three lands. If the die is not to cut close to a shoulder, about three threads should be chamfered off at the end. When dies are to cut close to a shoulder, not more than 1 or 1½ threads should be chamfered. The threads should be relieved on the chamfered part. It is common practice to make the length of the thread in a spring screw die about seven times the pitch. Tables are given herewith showing the general dimensions of spring screw threading dies as ordinarily manufactured, the length of the thread for various pitches, and the oversize required in taps for hobbing spring screw dies when these are cut with straight taps.

### Clamp Collars for Spring Screw Threading Dies

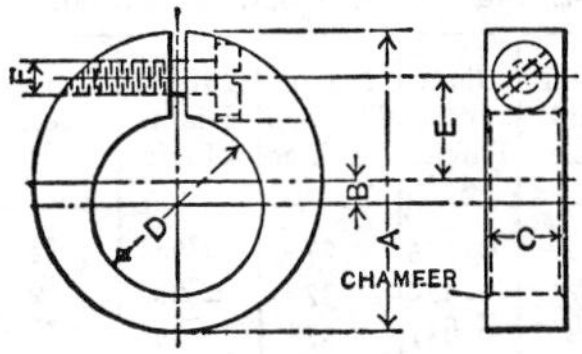

| D | A | B | C | E | F |
|---|---|---|---|---|---|
| ½ | 1 | 9/64 | 3/8 | 5/16 | 3/16 |
| ¾ | 1 5/16 | 9/64 | 7/16 | 13/32 | 7/32 |
| 1 | 1 11/16 | 11/64 | ½ | 9/16 | ¼ |
| 1 3/16 | 1 7/8 | 5/32 | 9/16 | 21/32 | ¼ |
| 1 ¼ | 1 7/8 | 1/8 | 9/16 | 23/32 | ¼ |
| 1 3/8 | 2 3/16 | 5/32 | 5/8 | ¾ | 5/16 |
| 1 5/8 | 2 5/8 | 7/32 | 11/16 | 7/8 | 3/8 |
| 2 | 3 1/8 | 9/32 | 13/16 | 1 1/16 | 3/8 |
| 2 ½ | 3 ¾ | 5/16 | 7/8 | 1 5/16 | 3/8 |
| 3 ¼ | 4 ½ | 5/16 | 15/16 | 1 11/16 | 3/8 |

### Spring Screw Threading Dies

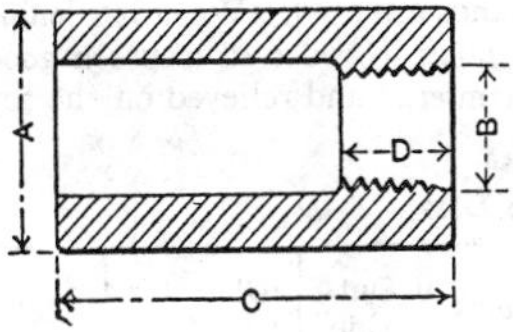

| Out-side Diam. | Diam. of Thread | Length of Die | Out-side Diam. | Diam. of Thread | Length of Die |
|---|---|---|---|---|---|
| A | B | C | A | B | C |
| ½ | 3/32–¼ | 1 ¼ | 1 3/8 | ½– ¾ | 2 ½ |
| ¾ | ¼–3/8 | 1 ¾ | 1 5/8 | 5/8–1 | 2 ½ |
| 1 | 5/16–½ | 2 | 2 | ¾–1 ¼ | 3 |
| 1 3/16 | 5/8–¾ | 2 ¼ | 2 ½ | 1 –1 ½ | 3 ½ |
| 1 ¼ | 3/8–¾ | 2 ½ | 3 ¼ | 1 5/8–2 1/8 | 4 |

### Length of Thread for Different Pitches

| No. of Threads per Inch | Length of Thread, D | No. of Threads per Inch | Length of Thread, D | No. of Threads per Inch | Length of Thread, D | No. of Threads per Inch | Length of Thread, D | No. of Threads per Inch | Length of Thread, D |
|---|---|---|---|---|---|---|---|---|---|
| 40 | 3/16 | 24 | 5/16 | 14 | ½ | 10 | ¾ | 6 | 1 3/16 |
| 36 | 7/32 | 20 | 3/8 | 13 | 9/16 | 9 | 13/16 | 5 ½ | 1 5/16 |
| 32 | ¼ | 18 | 13/32 | 12 | 5/8 | 8 | 7/8 | 5 | 1 7/16 |
| 28 | 9/32 | 16 | 7/16 | 11 | 11/16 | 7 | 1 | 4 ½ | 1 9/16 |

**Oversize of Taps for Hobbing Spring Screw Dies**

| No. of Threads per Inch | Over-size | No. of Threads per Inch | Over-size | No. of Threads per Inch | Over-size | No. of Threads per Inch | Over-size | No. of Threads per Inch | Over-size |
|---|---|---|---|---|---|---|---|---|---|
| 4½ | 0.015 | 8 | 0.007 | 13 | 0.006 | 22 | 0.005 | 40 | 0.003 |
| 5 | 0.013 | 9 | 0.007 | 14 | 0.005 | 24 | 0.004 | 48 | 0.003 |
| 5½ | 0.012 | 10 | 0.006 | 16 | 0.005 | 28 | 0.004 | 56 | 0.003 |
| 6 | 0.010 | 11 | 0.006 | 18 | 0.005 | 32 | 0.004 | 64 | 0.002 |
| 7 | 0.008 | 12 | 0.006 | 20 | 0.005 | 36 | 0.004 | 72 | 0.002 |

**Solid Round Gas Fixture Dies**

(For fixtures and thin brass tubing—60 degree V-thread with slight flat top and bottom)

| Nom-inal Size | Diam. of Thread | No. of Threads per Inch | Outside Diam. of Die | Thick-ness of Die | Nom-inal Size | Diam. of Thread | No. of Threads per Inch | Outside Diam. of Die | Thick-ness of Die |
|---|---|---|---|---|---|---|---|---|---|
| 0.148 | 0.148 | 32 | 5/8 | 1/4 | 1/2 | 0.515 | 27 | 1 7/16 | 3/8 |
| 0.196 | 0.196 | 32 | 5/8 | 1/4 | 9/16 | 0.578 | 27 | 1 7/16 | 3/8 |
| No. 4 | 0.246 | 27 | 5/8 | 1/4 | 5/8 | 0.637 | 27 | 1 7/16 | 3/8 |
| 1/4 | 0.260 | 27 | 1 | 5/16 | 3/4 | 0.770 | 27 | 2 | 1/2 |
| 5/16 | 0.342 | 27 | 1 | 5/16 | 7/8 | 0.885 | 27 | 2 | 1/2 |
| 3/8 | 0.390 | 27 | 1 7/16 | 3/8 | 1 | 1.006 | 27 | 2 | 1/2 |
| 7/16 | 0.459 | 27 | 1 7/16 | 3/8 | .... | .... | ..... | ..... | ..... |

**Round Split Adjustable Dies.** — These dies have three lands for sizes up to and including 3/16 inch. For all larger sizes, four lands are used. When hardening these dies, draw to a blue back of the clearance holes in order to insure a good spring temper. About three threads should be chamfered and relieved on the top of the chamfer on the leading side of the die.

**Round Split Adjustable Dies**

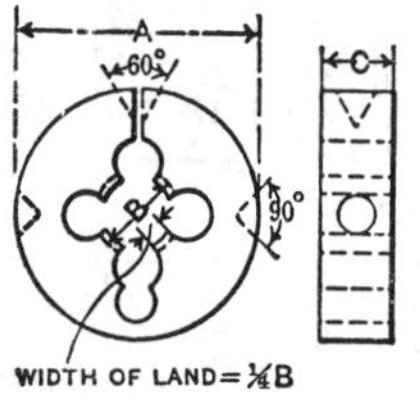
A
60°
90°
C
WIDTH OF LAND=¼B

| Out-side Diam. | Diameter of Thread | Thick-ness | Out-side Diam. | Diameter of Thread | Thick-ness |
|---|---|---|---|---|---|
| *A* | *B* | *C* | *A* | *B* | *C* |
| 5/8 | 1/16–17/64 | 1/4 | 1½ | 1/4–5/8 | 1/2 |
| 13/16 | 1/16–5/16 | 1/4 | 2 | 3/8–7/8 | 5/8 |
| 1 | 3/16–1/2 | 3/8 | 2½ | 1/2–1¼ | 11/16 |

**Hexagon Rethreading Dies.** — These dies are of hexagon form and resemble a nut. They are intended for repair work in the reconditioning of battered or rusty threads and are rotated either by a wrench, or by inserting in a bit brace socket. The commercial sizes cover the common range of bolt diameters ordinarily used in industrial applications.

# STANDARD TAPERS

Certain types of small tools and machine parts, such as twist drills, end mills, arbors, lathe centers, etc., are provided with taper shanks which fit into spindles or sockets of corresponding taper, thus providing not only accurate alignment between the tool or other part and its supporting member, but also more or less frictional resistance for driving the tool. There are several standards for " self-holding " tapers, but the Morse and the Brown & Sharpe are the standards most widely used by American manufacturers.

The name *self-holding* has been applied to the smaller tapers — like the Morse and the Brown & Sharpe — because, where the angle of the taper is only 2 or 3 degrees, the shank of a tool is so firmly seated in its socket that there is considerable frictional resistance to any force tending to turn or rotate the tool relative to the socket. The term " self-holding " is used to distinguish relatively small tapers from the larger or *self-releasing* type. A milling machine spindle having a taper of 3½ inches per foot is an example of a self-releasing taper. The included angle in this case is over 16 degrees and the tool or arbor requires a positive locking device to prevent slipping, but the shank may be released or removed more readily than one having a smaller taper of the self-holding type.

**Morse Taper.** — Dimensions relating to Morse standard taper shanks and sockets may be found in an accompanying table. The taper for different numbers of Morse tapers is slightly different, but it is approximately ⅝ inch per foot in most cases. The table gives the actual tapers, accurate to five decimal places. Morse taper shanks are used on a variety of tools, and exclusively on the shanks of twist drills. Dimensions for Morse Stub Taper Shanks are given in Table 1B.

**Brown & Sharpe Taper.** — This standard taper is used for taper shanks on tools such as end mills and reamers, the taper being approximately ½ inch per foot for all sizes except for taper No. 10, where the taper is 0.5161 inch per foot. Brown & Sharpe taper sockets are used for many arbors, collets, and machine tool spindles, especially milling machines and grinding machines. In many cases there are a number of different lengths of sockets corresponding to the same number of taper; all these tapers, however, are of the same diameter at the small end.

**Jarno Taper.** — The Jarno taper was originally proposed by Oscar J. Beale of the Brown & Sharpe Mfg. Co. This taper is based on such simple formulas that practically no calculations are required when the number of taper is known. The taper per foot of all Jarno taper sizes is 0.600 inch on the diameter. The diameter at the large end is as many eighths, the diameter at the small end is as many tenths, and the length as many half inches as are indicated by the number of the taper. For example, a No. 7 Jarno taper is ⅞ inch in diameter at the large end; ⁷⁄₁₀, or 0.700 inch at the small end; and ½, or 3½ inches long; hence, diameter at large end = No. of taper ÷ 8; diameter at small end = No. of taper ÷ 10; length of taper = No. of taper ÷ 2. The Jarno taper is used on various machine tools, especially profiling machines and die-sinking machines. It has also been used for the headstock and tailstock spindles of some lathes.

**American Standard Machine Tapers.** — This standard includes a self-holding series (Tables 5, 7, 8, 9 and 10) and a steep taper series, Table 6. The self-holding taper series consists of 22 sizes which are listed in Table 5. The reference gage for the self-holding tapers is a plug gage. Table 11 gives the dimensions and tolerances for both plug and ring gages applying to this series. Tables 7 to 10 inclusive give the dimensions for self-holding taper shanks and sockets which are classified as to (1) means of transmitting torque from spindle to the tool shank, and (2) means of retaining the shank in the socket. The steep machine tapers consist of a preferred

Table 1A.   Morse Standard Taper Shanks

| No. of Taper | Taper per Foot | Taper per Inch | Small End of Plug D | Diameter End of Socket A | Shank | | Depth of Hole H |
|---|---|---|---|---|---|---|---|
| | | | | | Length B | Depth S | |
| 0 | .62460 | .05205 | 0.252 | 0.3561 | 2 11/32 | 2 7/32 | 2 1/32 |
| 1 | .59858 | .04988 | 0.369 | 0.475 | 2 9/16 | 2 7/16 | 2 5/32 |
| 2 | .59941 | .04995 | 0.572 | 0.700 | 3 1/8 | 2 15/16 | 2 39/64 |
| 3 | .60235 | .05019 | 0.778 | 0.938 | 3 7/8 | 3 11/16 | 3 1/4 |
| 4 | .62326 | .05193 | 1.020 | 1.231 | 4 7/8 | 4 5/8 | 4 1/8 |
| 5 | .63151 | .05262 | 1.475 | 1.748 | 6 1/8 | 5 7/8 | 5 1/4 |
| 6 | .62565 | .05213 | 2.116 | 2.494 | 8 9/16 | 8 1/4 | 7 21/64 |
| 7 | .62400 | .05200 | 2.750 | 3.270 | 11 5/8 | 11 1/4 | 10 5/64 |

| Plug Depth P | Tang or Tongue | | | | Keyway | | Keyway to End K |
|---|---|---|---|---|---|---|---|
| | Thickness t | Length T | Radius R | Diam. | Width W | Length L | |
| 2 | .1562 | 1/4 | 5/32 | .235 | 1 1/64 | 9/16 | 1 15/16 |
| 2 1/8 | .2031 | 3/8 | 3/16 | .343 | 0.218 | 3/4 | 2 1/16 |
| 2 9/16 | .2500 | 7/16 | 1/4 | 17/32 | 0.266 | 7/8 | 2 1/2 |
| 3 3/16 | .3125 | 9/16 | 9/32 | 23/32 | 0.328 | 1 3/16 | 3 1/16 |
| 4 1/16 | .4687 | 5/8 | 5/16 | 31/32 | 0.484 | 1 1/4 | 3 7/8 |
| 5 3/16 | .6250 | 3/4 | 3/8 | 1 13/32 | 0.656 | 1 1/2 | 4 15/16 |
| 7 1/4 | .7500 | 1 1/8 | 1/2 | 2 | 0.781 | 1 3/4 | 7 |
| 10 | 1.1250 | 1 3/8 | 3/4 | 2 5/8 | 1.156 | 2 5/8 | 9 1/2 |

series (bold-face type, Table 6) and an intermediate series (light-face type). A self-holding taper is defined as "a taper with an angle small enough to hold a shank in place ordinarily by friction without holding means. (Sometimes referred to as slow taper.)" A steep taper is defined as "a taper having an angle sufficiently large to insure the easy or self-releasing feature." The term "gage line" indicates the basic diameter at or near the large end of the taper.

### Table 1B. Morse Stub Taper Shanks

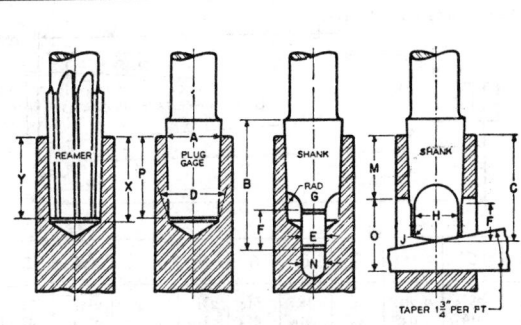

| No. of Taper | Taper per Foot* | Taper per Inch† | Small End of Plug,† D | Diam. End of Socket,* A | Shank Total Length, B | Shank Depth, C | Tang Thickness, E | Tang Length, F |
|---|---|---|---|---|---|---|---|---|
| 1 | .59858 | .049882 | .4314 | .475 | 1 9/16 | 1 1/8 | 13/64 | 5/16 |
| 2 | .59941 | .049951 | .6469 | .700 | 1 11/16 | 1 7/16 | 1/4 | 7/16 |
| 3 | .60235 | .050196 | .8753 | .938 | 2 | 1 3/4 | 25/64 | 9/16 |
| 4 | .62326 | .051938 | 1.1563 | 1.231 | 2 3/8 | 2 1/16 | 33/64 | 11/16 |
| 5 | .63151 | .052626 | 1.6526 | 1.748 | 3 | 2 11/16 | 3/4 | 15/16 |

| No. of Taper | Tang Radius of Mill, G | Tang Diameter, H | Socket Plug Depth, P | Socket Min. Depth of Tapered Hole Drilled, X | Socket Min. Depth of Tapered Hole Reamed, Y | Socket End to Tang Slot, M | Tang Slot Width, N | Tang Slot Length, O |
|---|---|---|---|---|---|---|---|---|
| 1 | 3/16 | 13/32 | 7/8 | 15/16 | 29/32 | 25/32 | 7/32 | 23/32 |
| 2 | 7/32 | 39/64 | 1 1/16 | 1 5/32 | 1 7/64 | 15/16 | 5/16 | 1 1/8 |
| 3 | 9/32 | 13/16 | 1 1/4 | 1 3/8 | 1 5/16 | 1 1/16 | 13/32 | 1 3/8 |
| 4 | 3/8 | 1 3/32 | 1 7/16 | 1 9/16 | 1 1/2 | 1 3/16 | 17/32 | 1 3/8 |
| 5 | 9/16 | 1 19/32 | 1 13/16 | 1 15/16 | 1 7/8 | 1 7/16 | 29/32 | 1 3/4 |

All dimensions in inches.
* These are basic dimensions.
† These dimensions are calculated for reference only.
Radius J is 3/64, 1/16, 5/64, 3/32, and 1/8 inch respectively for Nos. 1, 2, 3, 4, and 5 tapers.

**British Standard Tapers.** — British Standard 1660: Part I: 1950 "Self-holding Tapers" contains the basic dimensions and tolerances for the Nos. 2, 3, and 4, metric, the Nos. 1, 2, and 3, Brown & Sharpe and the Nos. 1, 2, 3, 4, 5, and 6 Morse tanged tapped end, and plain ended type self-holding tapers.

Part III of this same standard contains data for quick-release tapers.

Data for chuck tapers are shown in British Standard 1983:1953 "Accuracy of Chucks for Lathes and Drilling Machines" and include Jacobs Tapers Nos. 0, 1, 2, 2 short, and 3

**Table 2. Dimensions of Morse Taper Sleeves**

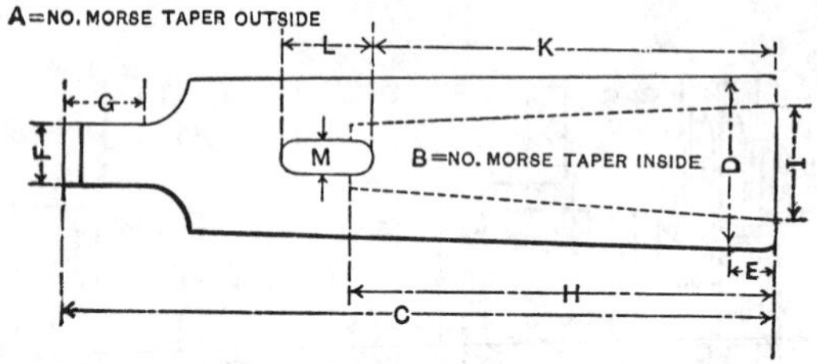


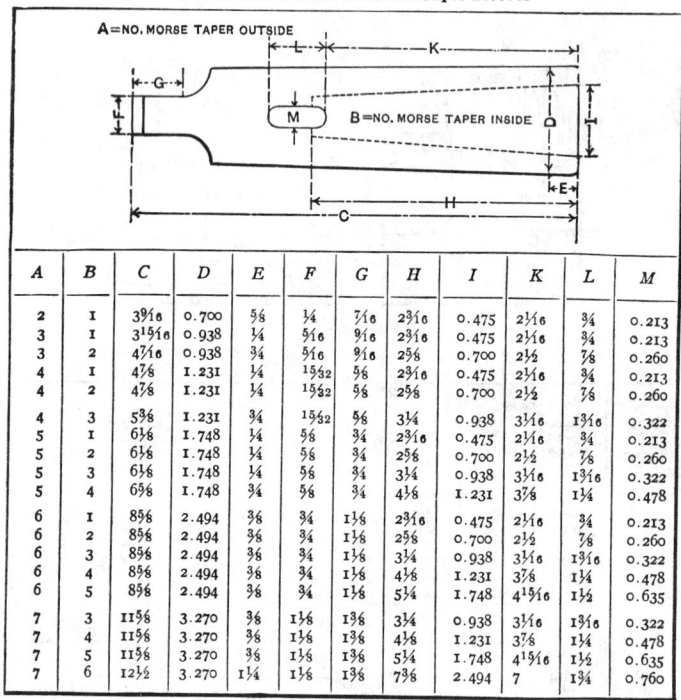

| *A* | *B* | *C* | *D* | *E* | *F* | *G* | *H* | *I* | *K* | *L* | *M* |
|---|---|---|---|---|---|---|---|---|---|---|---|
| 2 | 1 | 3 9/16 | 0.700 | 5/8 | 1/4 | 7/16 | 2 3/16 | 0.475 | 2 1/16 | 3/4 | 0.213 |
| 3 | 1 | 3 15/16 | 0.938 | 1/4 | 5/16 | 9/16 | 2 3/16 | 0.475 | 2 1/16 | 3/4 | 0.213 |
| 3 | 2 | 4 7/16 | 0.938 | 3/4 | 5/16 | 9/16 | 2 5/8 | 0.700 | 2 1/2 | 7/8 | 0.260 |
| 4 | 1 | 4 7/8 | 1.231 | 1/4 | 15/32 | 5/8 | 2 3/16 | 0.475 | 2 1/16 | 3/4 | 0.213 |
| 4 | 2 | 4 7/8 | 1.231 | 1/4 | 15/32 | 5/8 | 2 5/8 | 0.700 | 2 1/2 | 7/8 | 0.260 |
| 4 | 3 | 5 3/8 | 1.231 | 3/4 | 15/32 | 5/8 | 3 1/4 | 0.938 | 3 1/16 | 1 3/16 | 0.322 |
| 5 | 1 | 6 1/8 | 1.748 | 1/4 | 5/8 | 3/4 | 2 3/16 | 0.475 | 2 1/16 | 3/4 | 0.213 |
| 5 | 2 | 6 1/8 | 1.748 | 1/4 | 5/8 | 3/4 | 2 5/8 | 0.700 | 2 1/2 | 7/8 | 0.260 |
| 5 | 3 | 6 1/8 | 1.748 | 1/4 | 5/8 | 3/4 | 3 1/4 | 0.938 | 3 1/16 | 1 3/16 | 0.322 |
| 5 | 4 | 6 5/8 | 1.748 | 3/4 | 5/8 | 3/4 | 4 1/8 | 1.231 | 3 7/8 | 1 1/4 | 0.478 |
| 6 | 1 | 8 5/8 | 2.494 | 3/8 | 3/4 | 1 1/8 | 2 3/16 | 0.475 | 2 1/16 | 3/4 | 0.213 |
| 6 | 2 | 8 5/8 | 2.494 | 3/8 | 3/4 | 1 1/8 | 2 5/8 | 0.700 | 2 1/2 | 7/8 | 0.260 |
| 6 | 3 | 8 5/8 | 2.494 | 3/8 | 3/4 | 1 1/8 | 3 1/4 | 0.938 | 3 1/16 | 1 3/16 | 0.322 |
| 6 | 4 | 8 5/8 | 2.494 | 3/8 | 3/4 | 1 1/8 | 4 1/8 | 1.231 | 3 7/8 | 1 1/4 | 0.478 |
| 6 | 5 | 8 5/8 | 2.494 | 3/8 | 3/4 | 1 1/8 | 5 1/4 | 1.748 | 4 15/16 | 1 1/2 | 0.635 |
| 7 | 3 | 11 5/8 | 3.270 | 3/8 | 1 1/8 | 1 3/8 | 3 1/4 | 0.938 | 3 1/16 | 1 3/16 | 0.322 |
| 7 | 4 | 11 5/8 | 3.270 | 3/8 | 1 1/8 | 1 3/8 | 4 1/8 | 1.231 | 3 7/8 | 1 1/4 | 0.478 |
| 7 | 5 | 11 5/8 | 3.270 | 3/8 | 1 1/8 | 1 3/8 | 5 1/4 | 1.748 | 4 15/16 | 1 1/2 | 0.635 |
| 7 | 6 | 12 1/2 | 3.270 | 1 1/4 | 1 1/8 | 1 3/8 | 7 3/8 | 2.494 | 7 | 1 3/4 | 0.760 |

**Morse Taper Sockets — Hole and Shank Sizes**

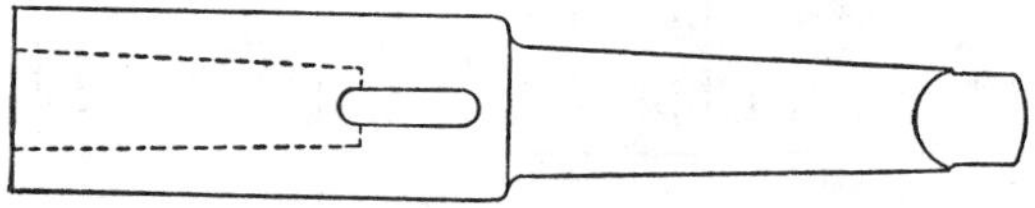

| Size | Morse Taper | | Size | Morse Taper | | Size | Morse Taper | |
|---|---|---|---|---|---|---|---|---|
| | Hole | Shank | | Hole | Shank | | Hole | Shank |
| 1 by 2 | No. 1 | No. 2 | 2 by 5 | No. 2 | No. 5 | 4 by 4 | No. 4 | No. 4 |
| 1 by 3 | No. 1 | No. 3 | 3 by 2 | No. 3 | No. 2 | 4 by 5 | No. 4 | No. 5 |
| 1 by 4 | No. 1 | No. 4 | 3 by 3 | No. 3 | No. 3 | 4 by 6 | No. 4 | No. 6 |
| 1 by 5 | No. 1 | No. 5 | 3 by 4 | No. 3 | No. 4 | 5 by 4 | No. 5 | No. 4 |
| 2 by 3 | No. 2 | No. 3 | 3 by 5 | No. 3 | No. 5 | 5 by 5 | No. 5 | No. 5 |
| 2 by 4 | No. 2 | No. 4 | 4 by 3 | No. 4 | No. 3 | 5 by 6 | No. 5 | No. 6 |

### Table 3. Brown & Sharpe Taper Shanks

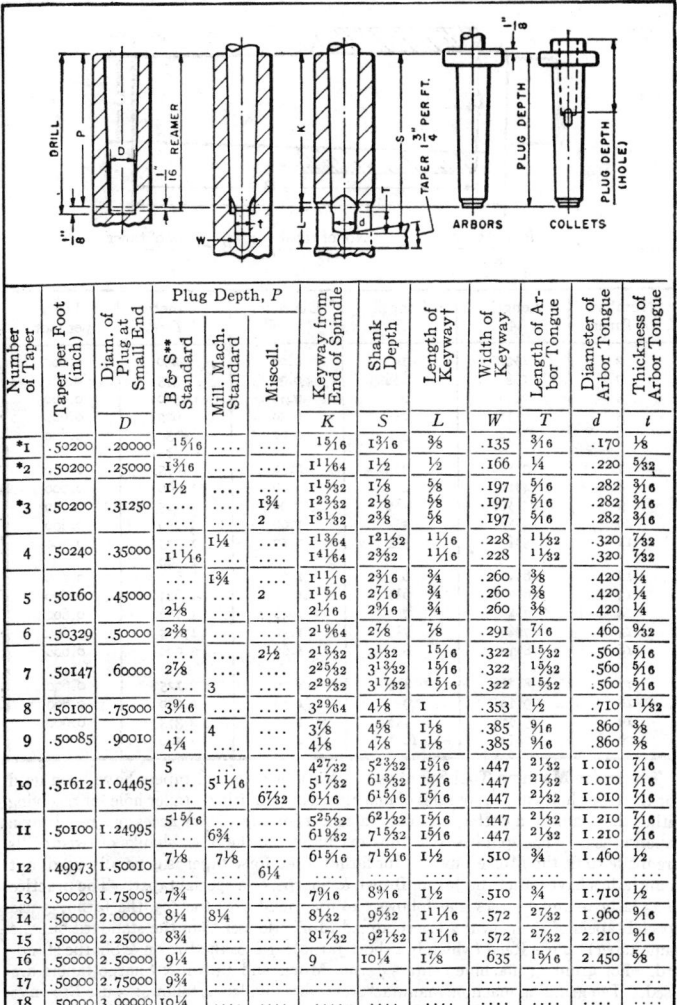

| Number of Taper | Taper per Foot (inch) | Diam. of Plug at Small End | Plug Depth, P | | | Keyway from End of Spindle | Shank Depth | Length of Keyway† | Width of Keyway | Length of Arbor Tongue | Diameter of Arbor Tongue | Thickness of Arbor Tongue |
| | | | B & S** Standard | Mill. Mach. Standard | Miscell. | | | | | | | |
| | | D | | | | K | S | L | W | T | d | t |
|---|---|---|---|---|---|---|---|---|---|---|---|---|
| *1 | .50200 | .20000 | 1 15/16 | .... | .... | 1 5/16 | 1 3/16 | 3/8 | .135 | 3/16 | .170 | 1/8 |
| *2 | .50200 | .25000 | 1 3/16 | .... | .... | 1 11/64 | 1 1/2 | 1/2 | .166 | 1/4 | .220 | 5/32 |
| *3 | .50200 | .31250 | 1 1/2 | .... | .... | 1 5/32 | 1 7/8 | 5/8 | .197 | 5/16 | .282 | 3/16 |
| | | | .... | .... | 1 3/4 | 1 23/32 | 2 1/8 | 5/8 | .197 | 5/16 | .282 | 3/16 |
| | | | .... | .... | 2 | 1 31/32 | 2 3/8 | 5/8 | .197 | 5/16 | .282 | 3/16 |
| 4 | .50240 | .35000 | 1 11/16 | 1 1/4 | .... | 1 13/64 | 1 21/32 | 1 1/16 | .228 | 1 1/32 | .320 | 7/32 |
| | | | .... | .... | .... | 1 41/64 | 2 3/32 | 1 1/16 | .228 | 1 1/32 | .320 | 7/32 |
| 5 | .50160 | .45000 | .... | 1 3/4 | .... | 1 11/16 | 2 3/16 | 3/4 | .260 | 3/8 | .420 | 1/4 |
| | | | .... | .... | 2 | 1 15/16 | 2 7/16 | 3/4 | .260 | 3/8 | .420 | 1/4 |
| | | | 2 1/8 | .... | .... | 2 1/16 | 2 9/16 | 3/4 | .260 | 3/8 | .420 | 1/4 |
| 6 | .50329 | .50000 | 2 3/8 | .... | .... | 2 19/64 | 2 7/8 | 7/8 | .291 | 7/16 | .460 | 9/32 |
| 7 | .50147 | .60000 | 2 7/8 | .... | 2 1/2 | 2 13/32 | 3 1/32 | 1 5/16 | .322 | 1 5/32 | .560 | 5/16 |
| | | | .... | .... | 3 | 2 25/32 | 3 13/32 | 1 5/16 | .322 | 1 5/32 | .560 | 5/16 |
| | | | .... | .... | | 2 29/32 | 3 17/32 | 1 5/16 | .322 | 1 5/32 | .560 | 5/16 |
| 8 | .50100 | .75000 | 3 9/16 | .... | .... | 3 29/64 | 4 1/8 | 1 | .353 | 1/2 | .710 | 1 1/32 |
| 9 | .50085 | .90010 | .... | 4 | .... | 3 7/8 | 4 5/8 | 1 1/8 | .385 | 9/16 | .860 | 3/8 |
| | | | 4 1/4 | .... | .... | 4 1/8 | 4 7/8 | 1 1/8 | .385 | 9/16 | .860 | 3/8 |
| 10 | .51612 | 1.04465 | 5 | .... | .... | 4 27/32 | 5 23/32 | 1 9/16 | .447 | 2 1/32 | 1.010 | 7/16 |
| | | | .... | 5 11/16 | .... | 5 17/32 | 6 13/32 | 1 9/16 | .447 | 2 1/32 | 1.010 | 7/16 |
| | | | .... | .... | 6 7/32 | 6 1/16 | 6 15/16 | 1 9/16 | .447 | 2 1/32 | 1.010 | 7/16 |
| 11 | .50100 | 1.24995 | 5 15/16 | .... | .... | 5 25/32 | 6 21/32 | 1 9/16 | .447 | 2 1/32 | 1.210 | 7/16 |
| | | | .... | 6 3/4 | .... | 6 19/32 | 7 15/32 | 1 9/16 | .447 | 2 1/32 | 1.210 | 7/16 |
| 12 | .49973 | 1.50010 | 7 1/8 | 7 1/8 | .... | 6 15/16 | 7 15/16 | 1 1/2 | .510 | 3/4 | 1.460 | 1/2 |
| | | | .... | .... | 6 1/4 | | | | | | | |
| 13 | .50020 | 1.75005 | 7 3/4 | .... | .... | 7 9/16 | 8 9/16 | 1 1/2 | .510 | 3/4 | 1.710 | 1/2 |
| 14 | .50000 | 2.00000 | 8 1/4 | 8 1/4 | .... | 8 1/32 | 9 5/32 | 1 11/16 | .572 | 2 7/32 | 1.960 | 9/16 |
| 15 | .50000 | 2.25000 | 8 3/4 | .... | .... | 8 17/32 | 9 21/32 | 1 11/16 | .572 | 2 7/32 | 2.210 | 9/16 |
| 16 | .50000 | 2.50000 | 9 1/4 | .... | .... | 9 | 10 1/4 | 1 7/8 | .635 | 1 5/16 | 2.450 | 5/8 |
| 17 | .50000 | 2.75000 | 9 3/4 | .... | .... | .... | .... | .... | .... | .... | .... | .... |
| 18 | .50000 | 3.00000 | 10 1/4 | .... | .... | .... | .... | .... | .... | .... | .... | .... |

\* Adopted by American Standards Association.
\*\* " B & S Standard " Plug Depths are not used in all cases.
† Special lengths of keyway are used instead of standard lengths in some places. Standard lengths need not be used when keyway is for driving only and not for admitting key to force out tool.

Table 4. Jarno Taper Shanks

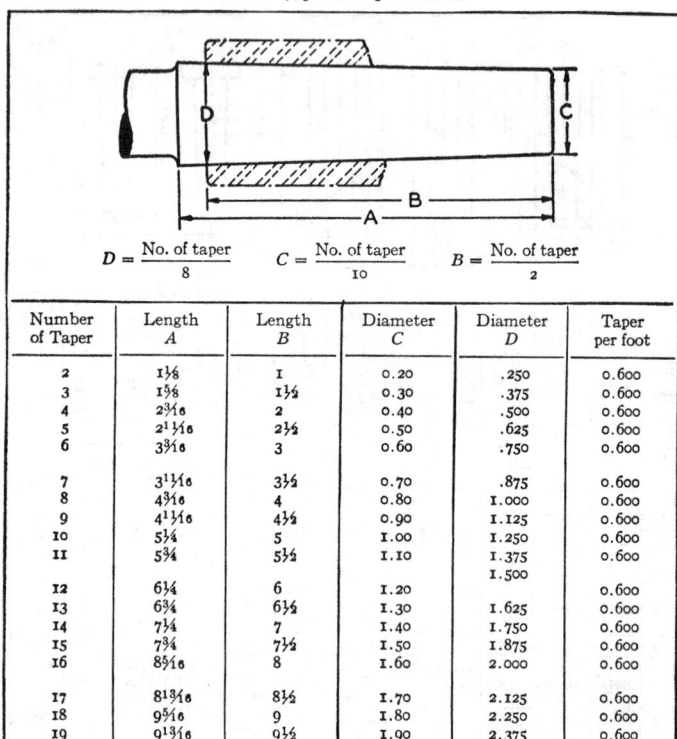

$$D = \frac{\text{No. of taper}}{8} \qquad C = \frac{\text{No. of taper}}{10} \qquad B = \frac{\text{No. of taper}}{2}$$

| Number of Taper | Length A | Length B | Diameter C | Diameter D | Taper per foot |
|---|---|---|---|---|---|
| 2 | 1⅛ | 1 | 0.20 | .250 | 0.600 |
| 3 | 1⅝ | 1½ | 0.30 | .375 | 0.600 |
| 4 | 2³⁄₁₆ | 2 | 0.40 | .500 | 0.600 |
| 5 | 2¹¹⁄₁₆ | 2½ | 0.50 | .625 | 0.600 |
| 6 | 3³⁄₁₆ | 3 | 0.60 | .750 | 0.600 |
| 7 | 3¹¹⁄₁₆ | 3½ | 0.70 | .875 | 0.600 |
| 8 | 4³⁄₁₆ | 4 | 0.80 | 1.000 | 0.600 |
| 9 | 4¹¹⁄₁₆ | 4½ | 0.90 | 1.125 | 0.600 |
| 10 | 5¼ | 5 | 1.00 | 1.250 | 0.600 |
| 11 | 5¾ | 5½ | 1.10 | 1.375 | 0.600 |
| 12 | 6¼ | 6 | 1.20 | 1.500 | 0.600 |
| 13 | 6¾ | 6½ | 1.30 | 1.625 | 0.600 |
| 14 | 7¼ | 7 | 1.40 | 1.750 | 0.600 |
| 15 | 7¾ | 7½ | 1.50 | 1.875 | 0.600 |
| 16 | 8⁵⁄₁₆ | 8 | 1.60 | 2.000 | 0.600 |
| 17 | 8¹³⁄₁₆ | 8½ | 1.70 | 2.125 | 0.600 |
| 18 | 9⁵⁄₁₆ | 9 | 1.80 | 2.250 | 0.600 |
| 19 | 9¹³⁄₁₆ | 9½ | 1.90 | 2.375 | 0.600 |
| 20 | 10⁵⁄₁₆ | 10 | 2.00 | 2.500 | 0.600 |

**Tapers for Machine Tool Spindles.** — Various standard tapers have been used for the taper holes in the spindles of machine tools requiring a taper hole for receiving either the shank of a cutter, an arbor, a center, or any tool or accessory requiring a tapering seat. The spindles of drilling machines and the taper shanks of twist drills are made to fit the Morse taper. For lathes, the Morse taper is generally used, but some lathes have either the Jarno, Brown & Sharpe, or a special taper. The practice of 33 lathe manufacturers is as follows: 20 use the Morse taper; 5, the Jarno; 3 use special tapers of their own; 2 use modified Morse (longer than the standard but the same taper); 2 use Reed (which is a short Jarno); 1 uses the Brown & Sharpe standard. For grinding machine centers Jarno, Morse and Brown & Sharpe tapers are used. Ten machine manufacturers were divided as follows: 3 use Brown & Sharpe; 3 use Morse, and 4 use Jarno. The Brown & Sharpe taper has been extensively used for milling machine and dividing head spindles. The standard milling machine spindle adopted in 1927 by the milling machine manufacturers of the National Machine Tool Builders' Association, has a taper of 3½ inches per foot. This comparatively steep taper was adopted to insure easy release of arbors.

Table 5.  American National Standard Self-holding Tapers — Basic Dimensions
(ANSI B5.10-1963, R1972)

| No. of Taper | Taper per Foot | Diam. at Gage Line (1) A | Means of Driving and Holding | | | | Origin of Series |
|---|---|---|---|---|---|---|---|
| .239 | 0.50200 | 0.23922 | | | | | Brown & Sharpe Taper Series |
| .299 | 0.50200 | 0.29968 | | | | | |
| .375 | 0.50200 | 0.37525 | | | | | |
| 1 | 0.59858 | 0.47500 | | | | | Morse Taper Series |
| 2 | 0.59941 | 0.70000 | | | | | |
| 3 | 0.60235 | 0.93800 | | | | | |
| 4 | 0.62326 | 1.23100 | Tang Drive With Shank Held in by Friction (See Table 7) | Tang Drive With Shank Held in by Key (See Table 8) | Key Drive With Shank Held in by Key (See Table 9) | Key Drive With Shank Held in by Draw-bolt (See Table 10) | |
| 4½ | 0.62400 | 1.50000 | | | | | |
| 5 | 0.63151 | 1.74800 | | | | | |
| 6 | 0.62565 | 2.49400 | | | | | |
| 7 (2) | 0.62400 | 3.27000 | | | | | |
| 200 | 0.750 | 2.000 | | | | | ¾ Inch per Foot Taper Series |
| 250 | 0.750 | 2.500 | | | | | |
| 300 | 0.750 | 3.000 | | | | | |
| 350 | 0.750 | 3.500 | | | | | |
| 400 | 0.750 | 4.000 | | | | | |
| 450 | 0.750 | 4.500 | | | | | |
| 500 | 0.750 | 5.000 | | | | | |
| 600 | 0.750 | 6.000 | | | | | |
| 800 | 0.750 | 8.000 | | | | | |
| 1000 | 0.750 | 10.000 | | | | | |
| 1200 | 0.750 | 12.000 | | | | | |

All dimensions given in inches.
(1) See illustrations above Tables 7, 8, 9 and 10.
(2) This size is continued in the Tang Drive series for the present to meet special needs.

Table 6.  ANSI Standard Steep Machine Tapers (ANSI B5.10-1963, R1972)

| No. of Taper | Taper per Foot (1) | Diam. at Gage Line (2) | Length Along Axis | No. of Taper | Taper per Foot (1) | Diam. at Gage Line (2) | Length Along Axis |
|---|---|---|---|---|---|---|---|
| 5 | 3.500 | 0.500 | 1¹⅟₁₆ | 35 | 3.500 | 1.500 | 2¼ |
| 10 | 3.500 | 0.625 | ⅞ | 40 | 3.500 | 1.750 | 2¹¹⁄₁₆ |
| 15 | 3.500 | 0.750 | 1¹⁄₁₆ | 45 | 3.500 | 2.250 | 3⁵⁄₁₆ |
| 20 | 3.500 | 0.875 | 1⁵⁄₁₆ | 50 | 3.500 | 2.750 | 4 |
| 25 | 3.500 | 1.000 | 1⁹⁄₁₆ | 55 | 3.500 | 3.500 | 5³⁄₁₆ |
| 30 | 3.500 | 1.250 | 1⅞ | 60 | 3.500 | 4.250 | 6⅜ |

All dimensions given in inches.
(1) This taper corresponds to an included angle of 16°, 35', 39.4".
The tapers numbered 10, 20, 30, 40, 50, and 60 that are printed in heavy-faced type are designated as the "Preferred Series." The tapers numbered 5, 15, 25, 35, 45, and 55 that are printed in light-faced type are designated as the "Intermediate Series."
(2) The basic diameter at gage line is at or near large end of taper.

### Table 7. American National Standard Taper Drive with Tang, Self-Holding Tapers
(ANSI B5.10-1963, R1972)

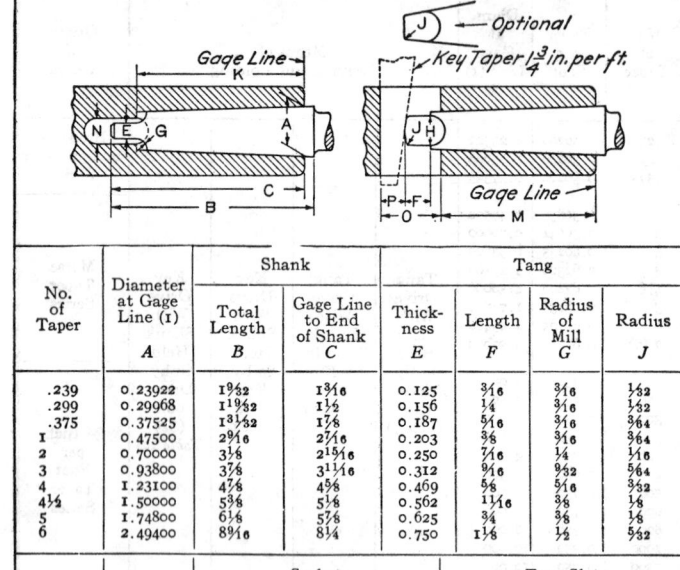

| No. of Taper | Diameter at Gage Line (1) | Shank | | Tang | | | |
|---|---|---|---|---|---|---|---|
| | | Total Length | Gage Line to End of Shank | Thickness | Length | Radius of Mill | Radius |
| | A | B | C | E | F | G | J |
| .239 | 0.23922 | 1 9/32 | 1 3/16 | 0.125 | 3/16 | 3/16 | 1/32 |
| .299 | 0.29968 | 1 19/32 | 1 1/2 | 0.156 | 1/4 | 3/16 | 1/32 |
| .375 | 0.37525 | 1 31/32 | 1 7/8 | 0.187 | 5/16 | 3/16 | 3/64 |
| 1 | 0.47500 | 2 9/16 | 2 7/16 | 0.203 | 3/8 | 3/16 | 3/64 |
| 2 | 0.70000 | 3 1/8 | 2 15/16 | 0.250 | 7/16 | 1/4 | 1/16 |
| 3 | 0.93800 | 3 7/8 | 3 11/16 | 0.312 | 9/16 | 9/32 | 5/64 |
| 4 | 1.23100 | 4 7/8 | 4 5/8 | 0.469 | 5/8 | 5/16 | 3/32 |
| 4 1/2 | 1.50000 | 5 3/8 | 5 1/4 | 0.562 | 11/16 | 3/8 | 1/8 |
| 5 | 1.74800 | 6 1/8 | 5 7/8 | 0.625 | 3/4 | 3/8 | 1/8 |
| 6 | 2.49400 | 8 9/16 | 8 1/4 | 0.750 | 1 1/8 | 1/2 | 5/32 |

| No. of Taper | Tang Diameter | Socket Minimum Depth K | | Gage Line to Tang Slot | Tang Slot | | Distance to Shank End |
|---|---|---|---|---|---|---|---|
| | | Drill | Ream | | Width | Length | |
| | H | | | M | N | O | P |
| .239 | 11/64 | 1 1/16 | 1 | 15/16 | 0.141 | 3/8 | 1/8 |
| .299 | 7/32 | 1 5/16 | 1 1/4 | 1 1/64 | 0.172 | 1/2 | 11/64 |
| .375 | 9/32 | 1 5/8 | 1 9/16 | 1 15/32 | 0.203 | 5/8 | 7/32 |
| 1 | 11/32 | 2 3/16 | 2 3/32 | 2 1/16 | 0.218 | 3/4 | 3/8 |
| 2 | 17/32 | 2 21/32 | 2 39/64 | 2 1/2 | 0.266 | 7/8 | 7/16 |
| 3 | 23/32 | 3 5/16 | 3 1/4 | 3 1/16 | 0.328 | 1 3/16 | 9/16 |
| 4 | 31/32 | 4 3/16 | 4 1/8 | 3 7/8 | 0.484 | 1 1/4 | 1/2 |
| 4 1/2 | 1 13/64 | 4 5/8 | 4 9/16 | 4 5/16 | 0.578 | 1 3/8 | 9/16 |
| 5 | 1 13/32 | 5 9/16 | 5 1/4 | 4 15/16 | 0.656 | 1 1/2 | 9/16 |
| 6 | 2 | 7 13/32 | 7 21/64 | 7 | 0.781 | 1 3/4 | 1/2 |

All dimensions are in inches. (1) See Table 11 for plug and ring gage dimensions.
*Tolerances:* For shank diameter A at gage line, +0.002 − 0.000; for hole diameter A, +0.000 − 0.002. For tang thickness E up to No. 5 inclusive, +0.000 − 0.006; No. 6, +0.000 − 0.008. For width N of tang slot up to No. 5 inclusive, +0.006 − 0.000; No. 6, +0.008 − 0.000. For centrality of tang E with center line of taper, .0025 (.005 total indicator variation). These centrality tolerances also apply to the tang slot N. On rate of taper, all sizes 0.002 per foot. This tolerance may be applied on *shanks* only in the direction which *increases* the rate of taper and on *sockets* only in the direction which *decreases* the rate of taper. Tolerances for fractional dimensions are plus or minus 0.010, unless otherwise specified.

**Table 8. ANSI Standard Taper Drive with Keeper Key, Self-Holding Tapers**
(ANSI B5.10-1963, R1972)

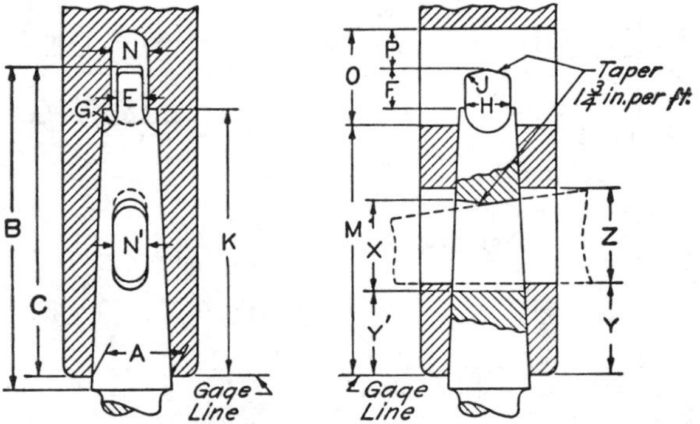


| No. of Taper | Diam. at Gage Line | Shank | | Tang | | | | | Socket | | |
|---|---|---|---|---|---|---|---|---|---|---|---|
| | | Total Length | Gage Line to End | Thickness | Length | Radius of Mill | Diameter | Radius | Min. Depth Hole, *K* | | Gage Line to Tang Slot |
| | *A* | *B* | *C* | *E* | *F* | *G* | *H* | *J* | Drill | Ream | *M* |
| 3 | 0.938 | 3 7/8 | 3 11/16 | 0.312 | 9/16 | 9/32 | 23/32 | 5/64 | 3 5/16 | 3 1/4 | 3 1/16 |
| 4 | 1.231 | 4 7/8 | 4 5/8 | 0.469 | 5/8 | 5/16 | 31/32 | 3/32 | 4 3/16 | 4 1/8 | 3 7/8 |
| 4 1/2 | 1.500 | 5 3/8 | 5 1/8 | 0.562 | 11/16 | 3/8 | 1 13/64 | 1/8 | 4 5/8 | 4 9/16 | 4 5/16 |
| 5 | 1.748 | 6 1/8 | 5 7/8 | 0.625 | 3/4 | 3/8 | 1 13/32 | 1/8 | 5 5/16 | 5 1/4 | 4 15/16 |
| 6 | 2.494 | 8 9/16 | 8 1/4 | 0.750 | 1 1/8 | 1/2 | 2 | 5/32 | 7 13/32 | 7 21/64 | 7 |
| 7 | 3.270 | 11 5/8 | 11 1/4 | 1.125 | 1 3/8 | 3/4 | 2 5/8 | 3/16 | 10 5/32 | 10 5/64 | 9 1/2 |

| No. of Taper | Tang Slot | | | Keeper Slot in Shank | | | Keeper Slot in Socket | | |
|---|---|---|---|---|---|---|---|---|---|
| | Width | Length | Shank End to Back of Slot | Gage Line to Bottom of Key Slot | Length | Width | Gage Line to Front of Key Slot | Length | Width |
| | *N* | *O* | *P* | *Y'* | *X* | *N'* | *Y* | *Z* | *N'* |
| 3 | 0.328 | 1 3/16 | 9/16 | 1 1/32 | 1 1/8 | 0.266 | 1 1/8 | 1 3/16 | 0.266 |
| 4 | 0.484 | 1 1/4 | 1/2 | 1 13/32 | 1 3/16 | 0.391 | 1 1/2 | 1 1/4 | 0.391 |
| 4 1/2 | 0.578 | 1 3/8 | 9/16 | 1 23/32 | 1 1/4 | 0.453 | 1 13/16 | 1 3/8 | 0.453 |
| 5 | 0.656 | 1 1/2 | 9/16 | 2 | 1 3/8 | 0.516 | 2 1/8 | 1 1/2 | 0.516 |
| 6 | 0.781 | 1 3/4 | 1/2 | 2 1/8 | 1 5/8 | 0.641 | 2 1/4 | 1 3/4 | 0.641 |
| 7 | 1.156 | 2 5/8 | 7/8 | 2 1/2 | 1 11/16 | 0.766 | 2 5/8 | 1 13/16 | 0.766 |

All dimensions are in inches. (1) See Table 11 for plug and ring gage dimensions. *Tolerances:* For shank diameter *A* at gage line, +0.002, −0; for hole diameter *A*, +0, −0.002. For tang thickness *E* up to No. 5 inclusive, +0, −0.006; larger than No. 5, +0, −0.008. For width of slots *N* and *N'* up to No. 5 inclusive, +0.006, −0; larger than No. 5, +0.008, −0. For centrality of tang *E* with center line of taper .0025 (.005 total indicator variation). These centrality tolerances also apply to slots *N* and *N'*. On rate of taper, see footnote in Table 7. Tolerances for fractional dimensions are ±0.010 unless otherwise specified.

**Table 9. ANSI Standard Nose Key Drive with Keeper Key, Self-Holding Tapers**
(ANSI B5.10-1963, R1972)

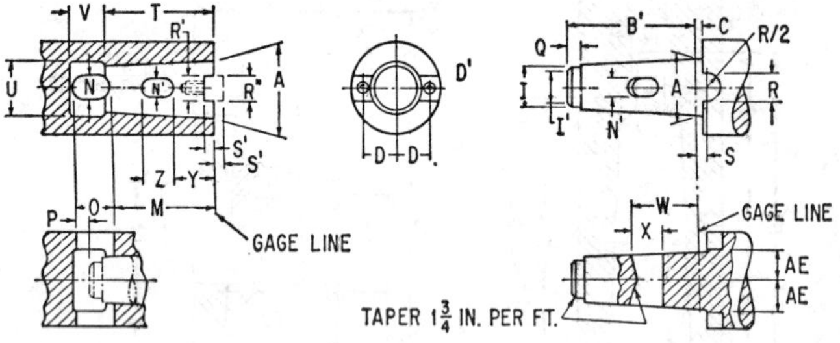


| Taper | *A* (1) | *B'* | *C* | *Q* | *I'* | *I* | *R* | *S* |
|---|---|---|---|---|---|---|---|---|
| 200 | 2.000 | 5⅛ | | ¼ | 1⅜ | 1⅝ | 1.010 | 9/16 |
| 250 | 2.500 | 5⅞ | | ¼ | 1⅜ | 2 1/16 | 1.010 | 9/16 |
| 300 | 3.000 | 6⅝ | Min. | ¼ | 1⅝ | 2½ | 2.010 | 9/16 |
| 350 | 3.500 | 7 7/16 | 0.003 | 5/16 | 2 | 2 15/16 | 2.010 | 9/16 |
| 400 | 4.000 | 8 3/16 | Max. | 5/16 | 2⅛ | 3 5/16 | 2.010 | 9/16 |
| 450 | 4.500 | 9 | 0.035 | ⅜ | 2⅜ | 3 13/16 | 3.010 | 13/16 |
| 500 | 5.000 | 9¾ | for | ⅜ | 2½ | 4¼ | 3.010 | 13/16 |
| 600 | 6.000 | 11 5/16 | all | 7/16 | 3 | 5 3/16 | 3.010 | 13/16 |
| 800 | 8.000 | 14⅜ | Sizes | ½ | 3½ | 7 | 4.010 | 1 1/16 |
| 1000 | 10.000 | 17 7/16 | | ⅝ | 4½ | 8¾ | 4.010 | 1 1/16 |
| 1200 | 12.000 | 20½ | | ¾ | 5⅜ | 10½ | 4.010 | 1 1/16 |
| **Taper** | ***D*** | **Screw*** | ***W*** | ***X*** | ***N'*** | ***R'*** | ***S'*** | ***T*** |
| 200 | 1 13/32 | ⅜ | 3 7/16 | 1 9/16 | 0.656 | 1.000 | ½ | 4¾ |
| 250 | 1 21/32 | ⅜ | 3 11/16 | 1 9/16 | 0.781 | 1.000 | ½ | 5½ |
| 300 | 2¼ | ⅜ | 4 1/16 | 1 9/16 | 1.031 | 2.000 | ½ | 6¼ |
| 350 | 2½ | ⅜ | 4⅞ | 2 | 1.031 | 2.000 | ½ | 6 15/16 |
| 400 | 2¾ | ⅜ | 5 5/16 | 2¼ | 1.031 | 2.000 | ½ | 7 11/16 |
| 450 | 3 | ½ | 5⅞ | 2 7/16 | 1.031 | 3.000 | ¾ | 8⅜ |
| 500 | 3¼ | ½ | 6 7/16 | 2⅝ | 1.031 | 3.000 | ¾ | 9⅛ |
| 600 | 3¾ | ½ | 7 7/16 | 3 | 1.281 | 3.000 | ¾ | 10 9/16 |
| 800 | 4¾ | ½ | 9 9/16 | 4 | 1.781 | 4.000 | 1 | 13½ |
| 1000 | ... | ... | 11½ | 4¾ | 2.031 | 4.000 | 1 | 16 5/16 |
| 1200 | ... | ... | 13¾ | 5¾ | 2.531 | 4.000 | 1 | 19 |
| **Taper** | ***U*** | ***V*** | ***M*** | ***N*** | ***O*** | ***P*** | ***Y*** | ***Z*** |
| 200 | 1 13/16 | 1 | 4½ | 0.656 | 1 9/16 | 15/16 | 2 | 1 11/16 |
| 250 | 2¼ | 1 | 5 3/16 | 0.781 | 1 15/16 | 1¼ | 2¼ | 1 11/16 |
| 300 | 2¾ | 1 | 5 15/16 | 1.031 | 2 3/16 | 1½ | 2⅝ | 1 11/16 |
| 350 | 3 3/16 | 1¼ | 6¾ | 1.031 | 2 3/16 | 1½ | 3 | 2⅛ |
| 400 | 3⅝ | 1¼ | 7½ | 1.031 | 2 3/16 | 1½ | 3¼ | 2⅜ |
| 450 | 4 3/16 | 1½ | 8 | 1.031 | 2¾ | 1¾ | 3⅝ | 2 9/16 |
| 500 | 4⅝ | 1½ | 8¾ | 1.031 | 2¾ | 1¾ | 4 | 2¾ |
| 600 | 5½ | 1¾ | 10⅛ | 1.281 | 3¼ | 2 1/16 | 4⅝ | 3¼ |
| 800 | 7⅜ | 2 | 12⅞ | 1.781 | 4¼ | 2¾ | 5¾ | 4¼ |
| 1000 | 9 3/16 | 2½ | 15¾ | 2.031 | 5 | 3 5/16 | 7 | 5 |
| 1200 | 11 | 3 | 18½ | 2.531 | 6 | 4 | 8¼ | 6 |

* Thread is UNF-2B for hole; UNF-2A for screw. (1) See Table 11 for plug and ring gage dimensions. All dimensions in inches.

*Tolerances:* For diameter *A* of hole at gage line, +0, −0.002; for diameter *A* of shank at gage line, +0.002, −0; for width of slots *N* and *N'*, +0.008, −0; for width of drive keyway *R'* in socket, +0, −0.001; for width of drive keyway *R* in shank, 0.010, −0; for centrality of slots *N* and *N'* with center line of spindle, 0.007; for centrality of keyway with spindle center line: for *R*, 0.004 and for *R'*, 0.002 T.I.V. On rate of taper, see footnote in Table 7. Fractional dimensions, ±0.010 unless otherwise specified.

Width of drive key is 0.001 less than width *R'* of keyway.

**Table 10. ANSI Standard Nose Key Drive with Drawbolt, Self-Holding Tapers**
(ANSI B5.10-1963, R1972)

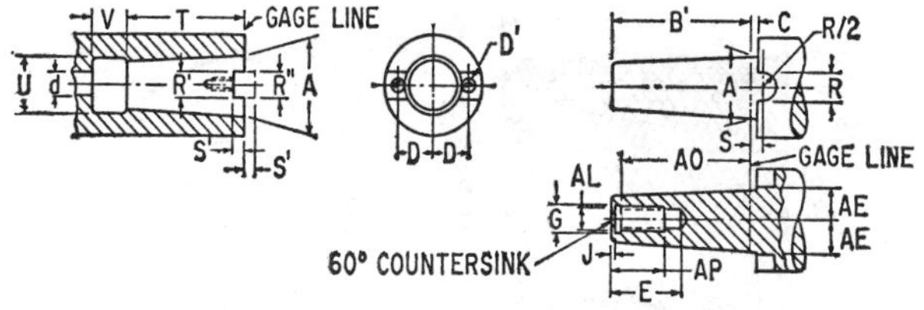


SOCKETS

| No. of Taper | Diam. at Gage Line* | Drive Key: Screw Holes: Center Line to Center of Screw | Drive Key: Screw Holes: UNF-2B Hole UNF-2A Screw | Drive Key: Width | Drive Keyway: Width | Drive Keyway: Depth | Gage Line to Front of Relief | Diam. of Relief | Depth of Relief | Diam. of Draw Bolt Hole |
|---|---|---|---|---|---|---|---|---|---|---|
| | *A* | *D* | *D'* | *R"* | *R'* | *S'* | *T* | *U* | *V* | *d* |
| 200 | 2.000 | 1 13/32 | 3/8 | 0.999 | 1.000 | 1/2 | 4 3/4 | 1 13/16 | 1 | 1 |
| 250 | 2.500 | 1 21/32 | 3/8 | 0.999 | 1.000 | 1/2 | 5 1/2 | 2 1/4 | 1 | 1 |
| 300 | 3.000 | 2 1/4 | 3/8 | 1.999 | 2.000 | 1/2 | 6 1/4 | 2 3/4 | 1 | 1 1/8 |
| 350 | 3.500 | 2 1/2 | 3/8 | 1.999 | 2.000 | 1/2 | 6 15/16 | 3 3/16 | 1 1/4 | 1 1/8 |
| 400 | 4.000 | 2 3/4 | 3/8 | 1.999 | 2.000 | 1/2 | 7 11/16 | 3 5/8 | 1 1/4 | 1 5/8 |
| 450 | 4.500 | 3 | 1/2 | 2.999 | 3.000 | 3/4 | 8 3/8 | 4 3/16 | 1 1/2 | 1 5/8 |
| 500 | 5.000 | 3 1/4 | 1/2 | 2.999 | 3.000 | 3/4 | 9 1/8 | 4 5/8 | 1 1/2 | 1 5/8 |
| 600 | 6.000 | 3 3/4 | 1/2 | 2.999 | 3.000 | 3/4 | 10 9/16 | 5 1/2 | 1 3/4 | 2 1/4 |
| 800 | 8.000 | 4 3/4 | 1/2 | 3.999 | 4.000 | 1 | 13 1/2 | 7 3/8 | 2 | 2 1/4 |
| 1000 | 10.000 | ... | ... | 3.999 | 4.000 | 1 | 16 5/16 | 9 3/16 | 2 1/2 | 2 1/4 |
| 1200 | 12.000 | ... | ... | 3.999 | 4.000 | 1 | 19 | 11 | 3 | 2 1/4 |

SHANKS

| No. of Taper | Length from Gage Line | Drawbar Hole: Diam. UNC-2B | Drawbar Hole: Depth of Drilled Hole | Drawbar Hole: Depth of Thread | Drawbar Hole: Diam. of Counter Bore | Drawbar Hole: Gage Line to First Thread | Drawbar Hole: Depth of 60° Chamfer | Drive Keyway: Width | Drive Keyway: Depth | Drive Keyway: Center Line to Bottom of Keyway |
|---|---|---|---|---|---|---|---|---|---|---|
| | *B'* | *AL* | *E* | *AP* | *G* | *AO* | *J* | *R* | *S* | *AE* |
| 200 | 5 1/8 | 7/8—9 | 2 7/16 | 1 3/4 | 29/32 | 4 25/32 | 1/8 | 1.010 | 9/16 | 1.005 |
| 250 | 5 7/8 | 7/8—9 | 2 7/16 | 1 3/4 | 29/32 | 5 17/32 | 1/8 | 1.010 | 9/16 | 1.255 |
| 300 | 6 5/8 | 1 —8 | 2 3/4 | 2 | 1 1/32 | 6 3/16 | 3/16 | 2.010 | 9/16 | 1.505 |
| 350 | 7 7/16 | 1 —8 | 2 3/4 | 2 | 1 1/32 | 7 | 3/16 | 2.010 | 9/16 | 1.755 |
| 400 | 8 3/16 | 1 1/2—6 | 4 | 3 | 1 17/32 | 7 1/2 | 5/16 | 2.010 | 9/16 | 2.005 |
| 450 | 9 | 1 1/2—6 | 4 | 3 | 1 17/32 | 8 5/16 | 5/16 | 3.010 | 13/16 | 2.255 |
| 500 | 9 3/4 | 1 1/2—6 | 4 | 3 | 1 17/32 | 9 1/16 | 5/16 | 3.010 | 13/16 | 2.505 |
| 600 | 11 5/16 | 2 —4 1/2 | 5 5/16 | 4 | 2 1/32 | 10 3/8 | 1/2 | 3.010 | 13/16 | 3.005 |
| 800 | 14 3/8 | 2 —4 1/2 | 5 5/16 | 4 | 2 1/32 | 13 7/16 | 1/2 | 4.010 | 1 1/16 | 4.005 |
| 1000 | 17 7/16 | 2 —4 1/2 | 5 5/16 | 4 | 2 1/32 | 16 1/2 | 1/2 | 4.010 | 1 1/16 | 5.005 |
| 1200 | 20 1/2 | 2 —4 1/2 | 5 5/16 | 4 | 2 1/32 | 19 9/16 | 1/2 | 4.010 | 1 1/16 | 6.005 |

All dimensions in inches. * See Table 11 for plug and ring gage dimensions.
Exposed length *C* is 0.003 minimum and 0.035 maximum for all sizes.
Drive key *D'* screw sizes are 3/8-24 UNF-2A up to taper No. 400 inclusive and 1/2-20 UNF-2A for larger tapers.

*Tolerances:* For diameter *A* of hole at gage line, +0.000, −0.002 for all sizes; for diameter *A* of shank at gage line, +0.002, −0.000 for all sizes; for width of drive keyway *R'* in socket, +0.000, −0.001; for width of drive keyway *R* in shank, +0.010, −0.000; for centrality of drive keyway *R* with center line of shank, 0.004 total indicator variation, and for drive keyway *R'*, with center line of spindle, 0.002. On rate of taper, see footnote in Table 7. Tolerances for fractional dimensions are ±0.010 unless otherwise specified.

Table 11.  ANSI Standard Plug and Ring Gages for the Self-Holding Taper Series
(ANSI B5.10-1963, R1972)

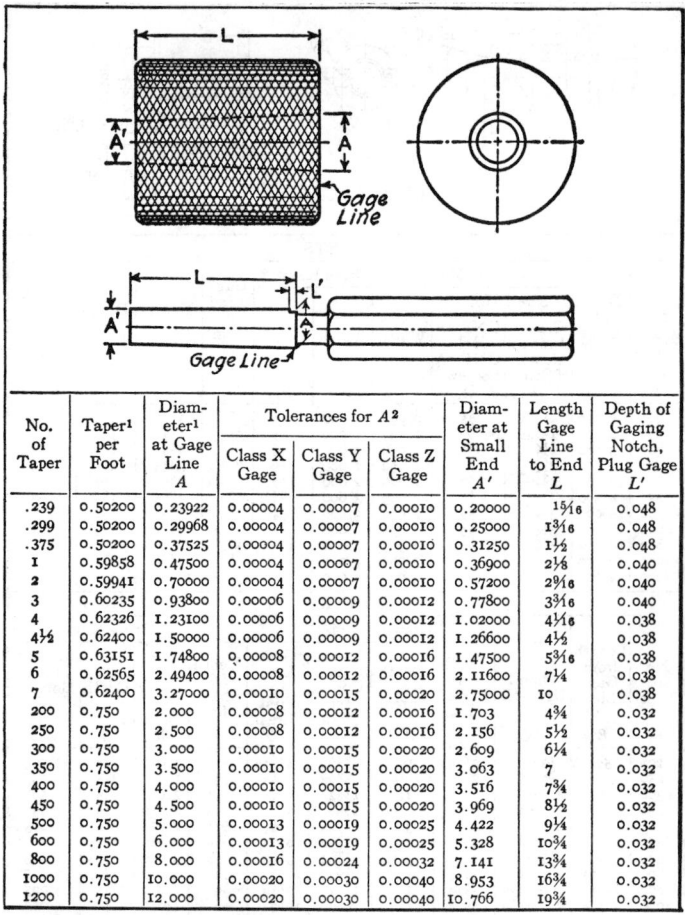

| No. of Taper | Taper[1] per Foot | Diameter[1] at Gage Line A | Tolerances for A[2] | | | Diameter at Small End A' | Length Gage Line to End L | Depth of Gaging Notch, Plug Gage L' |
|---|---|---|---|---|---|---|---|---|
| | | | Class X Gage | Class Y Gage | Class Z Gage | | | |
| .239 | 0.50200 | 0.23922 | 0.00004 | 0.00007 | 0.00010 | 0.20000 | 1 5/16 | 0.048 |
| .299 | 0.50200 | 0.29968 | 0.00004 | 0.00007 | 0.00010 | 0.25000 | 1 3/16 | 0.048 |
| .375 | 0.50200 | 0.37525 | 0.00004 | 0.00007 | 0.00010 | 0.31250 | 1 1/2 | 0.048 |
| 1 | 0.59858 | 0.47500 | 0.00004 | 0.00007 | 0.00010 | 0.36900 | 2 1/8 | 0.040 |
| 2 | 0.59941 | 0.70000 | 0.00004 | 0.00007 | 0.00010 | 0.57200 | 2 9/16 | 0.040 |
| 3 | 0.60235 | 0.93800 | 0.00006 | 0.00009 | 0.00012 | 0.77800 | 3 3/16 | 0.040 |
| 4 | 0.62326 | 1.23100 | 0.00006 | 0.00009 | 0.00012 | 1.02000 | 4 1/16 | 0.038 |
| 4½ | 0.62400 | 1.50000 | 0.00006 | 0.00009 | 0.00012 | 1.26600 | 4 1/2 | 0.038 |
| 5 | 0.63151 | 1.74800 | 0.00008 | 0.00012 | 0.00016 | 1.47500 | 5 3/16 | 0.038 |
| 6 | 0.62565 | 2.49400 | 0.00008 | 0.00012 | 0.00016 | 2.11600 | 7 1/4 | 0.038 |
| 7 | 0.62400 | 3.27000 | 0.00010 | 0.00015 | 0.00020 | 2.75000 | 10 | 0.038 |
| 200 | 0.750 | 2.000 | 0.00008 | 0.00012 | 0.00016 | 1.703 | 4 3/4 | 0.032 |
| 250 | 0.750 | 2.500 | 0.00008 | 0.00012 | 0.00016 | 2.156 | 5 1/2 | 0.032 |
| 300 | 0.750 | 3.000 | 0.00010 | 0.00015 | 0.00020 | 2.609 | 6 1/4 | 0.032 |
| 350 | 0.750 | 3.500 | 0.00010 | 0.00015 | 0.00020 | 3.063 | 7 | 0.032 |
| 400 | 0.750 | 4.000 | 0.00010 | 0.00015 | 0.00020 | 3.516 | 7 3/4 | 0.032 |
| 450 | 0.750 | 4.500 | 0.00010 | 0.00015 | 0.00020 | 3.969 | 8 1/2 | 0.032 |
| 500 | 0.750 | 5.000 | 0.00013 | 0.00019 | 0.00025 | 4.422 | 9 1/4 | 0.032 |
| 600 | 0.750 | 6.000 | 0.00013 | 0.00019 | 0.00025 | 5.328 | 10 3/4 | 0.032 |
| 800 | 0.750 | 8.000 | 0.00016 | 0.00024 | 0.00032 | 7.141 | 13 3/4 | 0.032 |
| 1000 | 0.750 | 10.000 | 0.00020 | 0.00030 | 0.00040 | 8.953 | 16 3/4 | 0.032 |
| 1200 | 0.750 | 12.000 | 0.00020 | 0.00030 | 0.00040 | 10.766 | 19 3/4 | 0.032 |

All dimensions in inches.

[1] The taper per foot and diameter $A$ at gage line are basic dimensions.  Dimensions in Column $A'$ are calculated for reference only.

[2] Tolerances for diameter $A$ are plus for plug gages and minus for ring gages.

The amounts of taper deviation for Class X, Class Y, and Class Z gages are the same, respectively, as the amounts shown for tolerances on diameter $A$.  Taper deviation is the permissible allowance from true taper at any point of diameter in the length of the gage. On taper *plug* gages, this deviation may be applied only in the direction which *decreases* the rate of taper.  On taper *ring* gages, this deviation may be applied only in the direction which *increases* the rate of taper.

**Jacobs Tapers and Threads for Drill Chucks and Spindles**

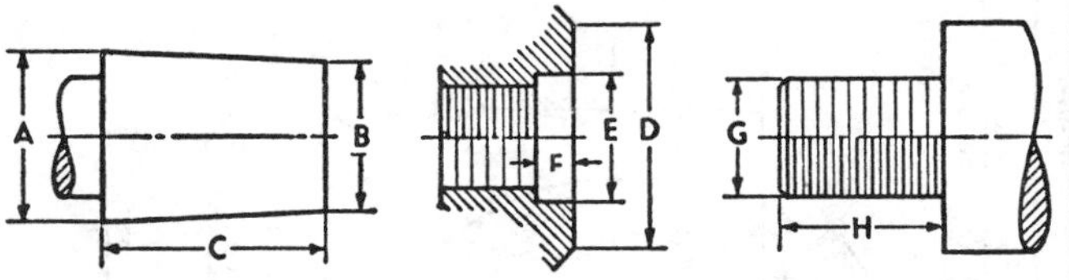


American Standard Thread Form

| Taper Series | *A* | *B* | *C* | Taper per Ft. | Taper Series | *A* | *B* | *C* | Taper per Ft. |
|---|---|---|---|---|---|---|---|---|---|
| No. 0 | .2500 | .22844 | .43750 | .59145 | No. 4 | 1.1240 | 1.0372 | 1.6563 | .62886 |
| No. 1 | .3840 | .33341 | .65625 | .92508 | No. 5 | 1.4130 | 1.3161 | 1.8750 | .62010 |
| No. 2 | .5590 | .48764 | .87500 | .97861 | No. 6 | 0.6760 | 0.6241 | 1.0000 | .62292 |
| No. 2[a] | .5488 | .48764 | .75000 | .97861 | No. 33 | 0.6240 | 0.5605 | 1.0000 | .76194 |
| No. 3 | .8110 | .74610 | 1.21875 | .63898 | ...... | ...... | ...... | ...... | ...... |

| Thread Size | Diameter *D* | | Diameter *E* | | Depth *F* | |
|---|---|---|---|---|---|---|
| | Max. | Min. | Max. | Min. | Max. | Min. |
| 5/16–24 | 0.531 | 0.516 | 0.3245 | 0.3195 | 0.135 | 0.115 |
| 5/16–24 | 0.633 | 0.618 | 0.3245 | 0.3195 | 0.135 | 0.115 |
| 3/8–24 | 0.633 | 0.618 | 0.385 | 0.380 | 0.135 | 0.115 |
| 1/2–20 | 0.860 | 0.845 | 0.510 | 0.505 | 0.135 | 0.115 |
| 5/8–11 | 1.125 | 1.110 | 0.635 | 0.630 | 0.166 | 0.146 |
| 5/8–16 | 1.125 | 1.110 | 0.635 | 0.630 | 0.166 | 0.146 |
| 45/64–16 | 1.250 | 1.235 | 0.713 | 0.708 | 0.166 | 0.146 |
| 3/4–16 | 1.250 | 1.235 | 0.760 | 0.755 | 0.166 | 0.146 |
| 1–8 | 1.437 | 1.422 | 1.036 | 1.026 | 0.281 | 0.250 |
| 1–10 | 1.437 | 1.422 | 1.036 | 1.026 | 0.281 | 0.250 |
| 1 1/2–8 | 1.871 | 1.851 | 1.536 | 1.526 | 0.343 | 0.312 |

| Thread Size | Threaded Spindle | | Plug Gage Pitch Diam. | | Ring Gage Pitch Diam. | |
|---|---|---|---|---|---|---|
| | *G*[b] | *H*[c] | Go | Not Go | Go | Not Go |
| 5/16–24 | 0.3120 | 0.3750 | 0.2854 | 0.2878 | 0.2840 | 0.2830 |
| 5/16–24 | 0.3120 | 0.4375 | 0.2854 | 0.2878 | 0.2840 | 0.2830 |
| 3/8–24 | 0.375 | 0.5625[d] | 0.3479 | 0.3503 | 0.3465 | 0.3455 |
| 1/2–20 | 0.500 | 0.5625 | 0.4675 | 0.4701 | 0.4660 | 0.4650 |
| 5/8–11 | 0.6250 | 0.6875 | 0.5600 | 0.5702 | 0.5640 | 0.5620 |
| 5/8–16 | 0.625 | 0.6875 | 0.5844 | 0.5876 | 0.5825 | 0.5815 |
| 45/64–16 | 0.703 | 0.6875 | 0.6625 | 0.6657 | 0.6605 | 0.6595 |
| 3/4–16 | 0.750 | 0.6875 | 0.7094 | 0.7126 | 0.7075 | 0.7065 |
| 1–8 | 1.000 | 1.000 | 0.9188 | 0.9242 | 0.9145 | 0.9135 |
| 1–10 | 1.000 | 1.000 | 0.9350 | 0.9395 | 0.9315 | 0.9305 |
| 1 1/2–8 | 1.500 | 1.000 | 1.4188 | 1.4242 | 1.4145 | 1.4135 |

*Usual Chuck Capacities for Different Taper Series Numbers:* No. 0 taper, drill diameters, 0–5/32 inch; No. 1, 0–1/4 inch; No. 2, 0–3/8 inch; No. 2 "Short," 0–1/4 or 0–5/16 inch; No. 3, 0–17/32, 1/8–5/8, 3/16–3/4 or 1/4–3/4 inch; No. 4, 1/8–3/4 inch; No. 5, 3/8–1 inch; No. 6, 0–1/2 inch; No. 33, 0–1/2 inch.

*Usual Chuck Capacities for Different Thread Sizes:* Size 5/16–24, drill diameters 0–5/32 or 0–3/16 inch; size 3/8–24, drill diameters 0–3/16, 0–1/4, 0–5/16 or 0–3/8 inch; size 1/2–20, drill diameters 0–1/4, 0–5/16, 0–3/8, 0–1/2 or 5/64–1/2 inch; sizes 5/8–11 and 5/8–16, drill diameters 1/8–5/8 inch; size 45/64–16, drill diameters 0–3/8, 0–1/2 or 1/8–5/8 inch; sizes 1–8 and 1–10, drill diameters 0–17/32 inch; and size 1 1/2–8, drill diameters 1/8–5/8 or 3/16–3/4 inch.

[a] These dimensions are for the No. 2 "short" taper.

[b] Tolerances for dimension *G* (screw thread major diameter) are as follows: −0.005 inch for screw sizes 5/16–24 to 3/4–16, inclusive and −0.015 inch for screw sizes 1–8 to 1 1/2–8, inclusive.

[c] Tolerances for dimension *H* are as follows: −0.030 inch for thread sizes 5/16–24 to 3/4–16, inclusive and −0.125 inch for thread sizes 1–8 to 1 1/2–8, inclusive.

[d] Length for Jacobs No. 1 B.S. chuck is 0.4375 inch.

Table 1.  Essential Dimensions of American National Standard Spindle Noses for Milling Machines (ANSI B5.18-1972)

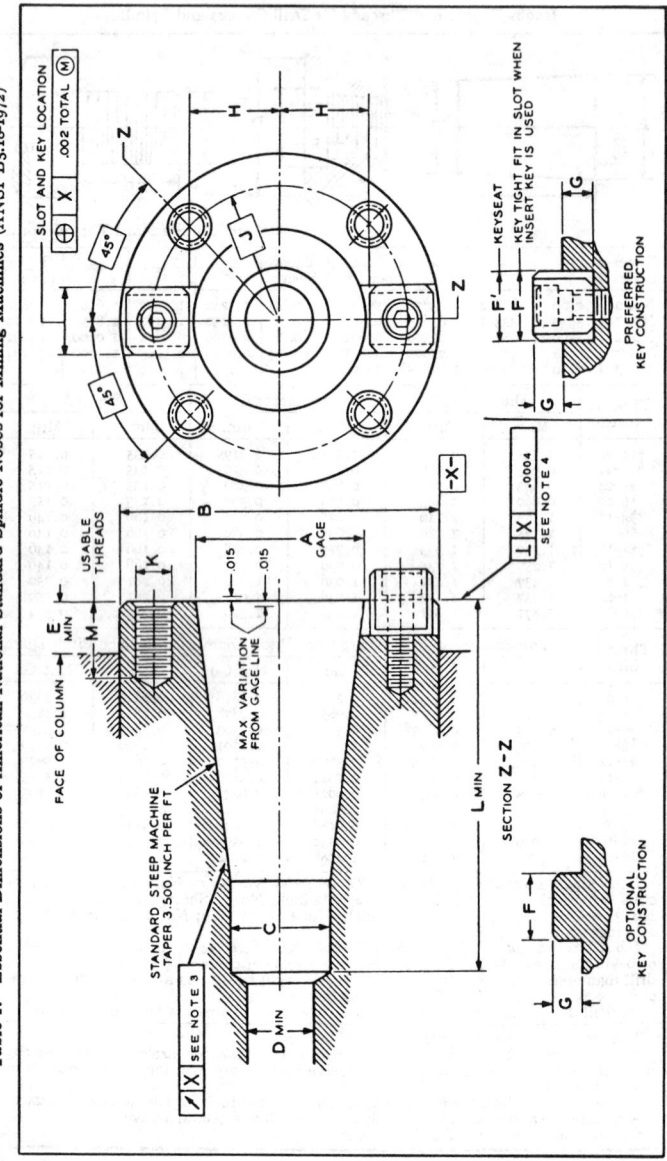

Table 1 (*Continued*). Essential Dimensions of American National Standard Spindle Noses for Milling Machines (ANSI B5.18-1972)

| Size No. | Gage Diam. of Taper A | Diam. of Spindle B | Pilot Diam. C | Clearance Hole for Draw-in Bolt Min. D | Minimum Dimension Spindle End to Column E | Width of Driving Key F | Width of Keyseat F' | Maximum Height of Driving Key G | Minimum Depth of Keyseat G' | Distance from Center to Driving Keys H | Radius of Bolt Hole Circle J | Size of Threads for Bolt Holes UNC-2B K | Full Depth of Arbor Hole in Spindle Min. L | Depth of Usable Thread for Bolt Hole M |
|---|---|---|---|---|---|---|---|---|---|---|---|---|---|---|
| 30 | 1.250 | 2.7493 2.7488 | 0.692 0.685 | 0.66 | 0.50 | 0.6255 0.6252 | 0.624 0.625 | 0.31 | 0.31 | 0.660 0.654 | 1.0625 (Note 1) | 0.375—16 | 2.88 | 0.62 |
| 40 | 1.750 | 3.4993 3.4988 | 1.005 0.997 | 0.66 | 0.62 | 0.6255 0.6252 | 0.624 0.625 | 0.31 | 0.31 | 0.910 0.904 | 1.3125 (Note 1) | 0.500—13 | 3.88 | 0.81 |
| 45 | 2.250 | 3.9993 3.9988 | 1.286 1.278 | 0.78 | 0.62 | 0.7505 0.7502 | 0.749 0.750 | 0.38 | 0.38 | 1.160 1.154 | 1.500 (Note 1) | 0.500—13 | 4.75 | 0.81 |
| 50 | 2.750 | 5.0618 5.0613 | 1.568 1.559 | 1.06 | 0.75 | 1.0006 1.0002 | 0.999 1.000 | 0.50 | 0.50 | 1.410 1.404 | 2.000 (Note 2) | 0.625—11 | 5.50 | 1.00 |
| 60 | 4.250 | 8.7180 8.7175 | 2.381 2.371 | 1.38 | 1.50 | 1.0006 1.0002 | 0.999 1.000 | 0.50 | 0.50 | 2.420 2.414 | 3.500 (Note 2) | 0.750—10 | 8.62 | 1.25 |

All dimensions are given in inches.

*Tolerances:*

Two-digit decimal dimensions ±0.010 unless otherwise specified.

A — Taper: Tolerance on rate of taper to be 0.001 inch per foot applied only in direction which decreases rate of taper.

F1 — Centrality of keyway with axis of taper 0.002 total at maximum material condition.   (0.002 Total indicator variation)

F — Centrality of solid key with axis of taper 0.002 total at maximum material condition.   (0.002 Total indicator variation)

Note 1: Holes spaced as shown and located within 0.006 inch diameter of true position.

Note 2: Holes spaced as shown and located within 0.010 inch diameter of true position.

Note 3: Maximum turnout on test plug:  0.0004 at 1 inch projection from gage line.
                                      : 0.0010 at 12 inch projection from gage line.

Note 4: Squareness of mounting face measured near mounting bolt hole circle.

Table 2.   Essential Dimensions of American National Standard Tool
Shanks for Milling Machines (ANSI B5.18-1972)

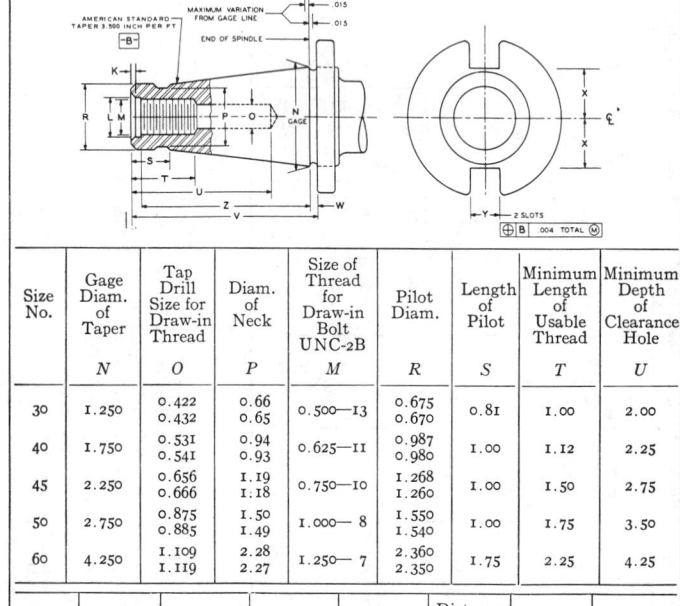

| Size No. | Gage Diam. of Taper | Tap Drill Size for Draw-in Thread | Diam. of Neck | Size of Thread for Draw-in Bolt UNC-2B | Pilot Diam. | Length of Pilot | Minimum Length of Usable Thread | Minimum Depth of Clearance Hole |
|---|---|---|---|---|---|---|---|---|
| | N | O | P | M | R | S | T | U |
| 30 | 1.250 | 0.422 0.432 | 0.66 0.65 | 0.500—13 | 0.675 0.670 | 0.81 | 1.00 | 2.00 |
| 40 | 1.750 | 0.531 0.541 | 0.94 0.93 | 0.625—11 | 0.987 0.980 | 1.00 | 1.12 | 2.25 |
| 45 | 2.250 | 0.656 0.666 | 1.19 1.18 | 0.750—10 | 1.268 1.260 | 1.00 | 1.50 | 2.75 |
| 50 | 2.750 | 0.875 0.885 | 1.50 1.49 | 1.000— 8 | 1.550 1.540 | 1.00 | 1.75 | 3.50 |
| 60 | 4.250 | 1.109 1.119 | 2.28 2.27 | 1.250— 7 | 2.360 2.350 | 1.75 | 2.25 | 4.25 |

| Size No. | Distance from Rear of Flange to End of Arbor | Clearance of Flange from Gage Diameter | Tool Shank Center-line to Driving Slot | Width of Driving Slot | Distance from Gage Line to Bottom of C'bore | Depth of 60° Center | Diameter of C'bore |
|---|---|---|---|---|---|---|---|
| | V | W | X | Y | Z | K | L |
| 30 | 2.75 | 0.045 0.075 | 0.640 0.625 | 0.635 0.645 | 2.50 | 0.05 0.07 | 0.525 0.530 |
| 40 | 3.75 | 0.045 0.075 | 0.890 0.875 | 0.635 0.645 | 3.50 | 0.05 0.07 | 0.650 0.655 |
| 45 | 4.38 | 0.105 0.135 | 1.140 1.125 | 0.760 0.770 | 4.06 | 0.05 0.07 | 0.775 0.780 |
| 50 | 5.12 | 0.105 0.135 | 1.390 1.375 | 1.010 1.020 | 4.75 | 0.05 0.12 | 1.025 1.030 |
| 60 | 8.25 | 0.105 0.135 | 2.400 2.385 | 1.010 1.020 | 7.81 | 0.05 0.12 | 1.307 1.312 |

All dimensions are given in inches.

*Tolerances:*
   Two digit decimal dimensions ±0.010 inch unless otherwise specified.
   *M* — Permissible for Class 2B "NoGo" gage to enter five threads before interference.
   *N* — Taper tolerance on rate of taper to be 0.001 inch per foot applied only in direction
   which increases rate of taper.
   *Y* — Centrality of drive slot with axis of taper shank 0.004 inch at maximum material
   condition.   (0.004 inch total indicator variation)

**Table 3.  American National Standard Draw-in Bolt Ends** (ANSI B5.18-1972)

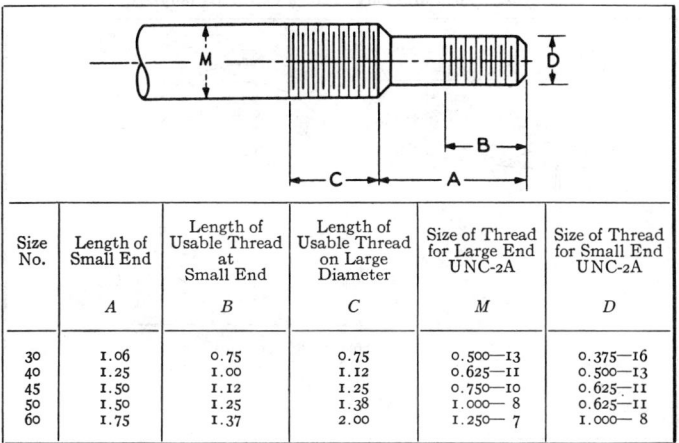

| Size No. | Length of Small End | Length of Usable Thread at Small End | Length of Usable Thread on Large Diameter | Size of Thread for Large End UNC-2A | Size of Thread for Small End UNC-2A |
|---|---|---|---|---|---|
| | A | B | C | M | D |
| 30 | 1.06 | 0.75 | 0.75 | 0.500—13 | 0.375—16 |
| 40 | 1.25 | 1.00 | 1.12 | 0.625—11 | 0.500—13 |
| 45 | 1.50 | 1.12 | 1.25 | 0.750—10 | 0.625—11 |
| 50 | 1.50 | 1.25 | 1.38 | 1.000— 8 | 0.625—11 |
| 60 | 1.75 | 1.37 | 2.00 | 1.250— 7 | 1.000— 8 |

All dimensions are given in inches.

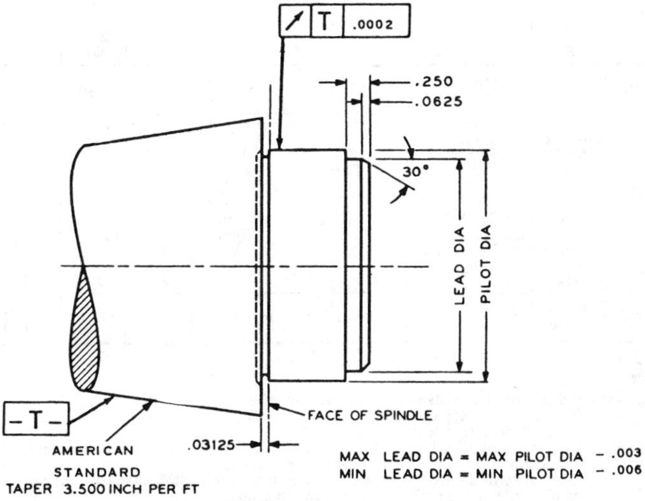

American National Standard Pilot Lead on Centering Plugs for
Flatback Milling Cutters (ANSI B5.18-1972)

## Table 4. Essential Dimensions for American National Standard Spindle Nose with Large Flange (ANSI B5.18-1972)

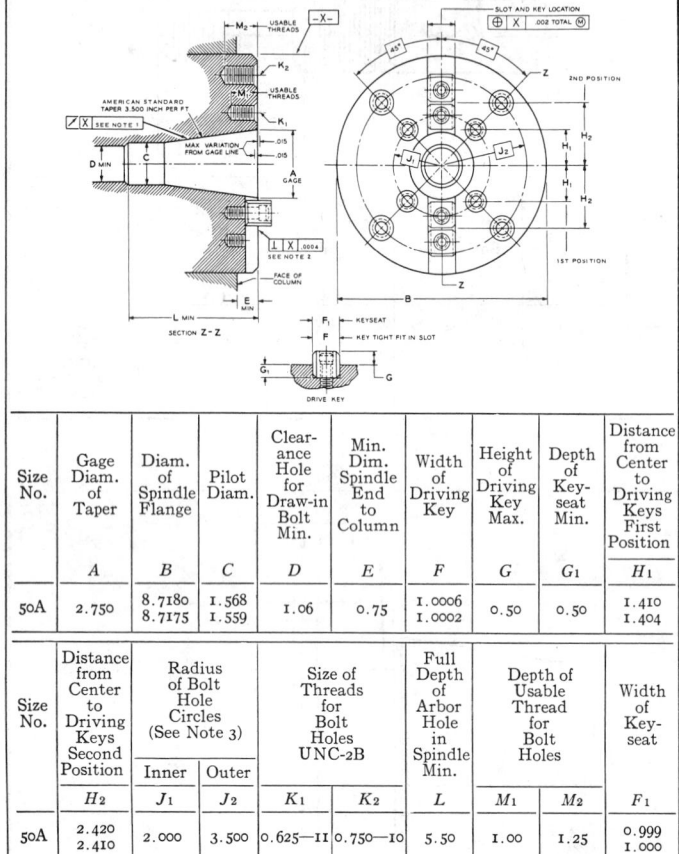

| Size No. | Gage Diam. of Taper | Diam. of Spindle Flange | Pilot Diam. | Clearance Hole for Draw-in Bolt Min. | Min. Dim. Spindle End to Column | Width of Driving Key | Height of Driving Key Max. | Depth of Keyseat Min. | Distance from Center to Driving Keys First Position |
|---|---|---|---|---|---|---|---|---|---|
| | A | B | C | D | E | F | G | G₁ | H₁ |
| 50A | 2.750 | 8.7180 8.7175 | 1.568 1.559 | 1.06 | 0.75 | 1.0006 1.0002 | 0.50 | 0.50 | 1.410 1.404 |

| Size No. | Distance from Center to Driving Keys Second Position | Radius of Bolt Hole Circles (See Note 3) | | Size of Threads for Bolt Holes UNC-2B | | Full Depth of Arbor Hole in Spindle Min. | Depth of Usable Thread for Bolt Holes | | Width of Keyseat |
|---|---|---|---|---|---|---|---|---|---|
| | H₂ | Inner J₁ | Outer J₂ | K₁ | K₂ | L | M₁ | M₂ | F₁ |
| 50A | 2.420 2.410 | 2.000 | 3.500 | 0.625—11 | 0.750—10 | 5.50 | 1.00 | 1.25 | 0.999 1.000 |

All dimensions are given in inches.

*Tolerances:*

Two-digit decimal dimensions ±0.010 unless otherwise specified.
  A — Tolerance on rate of taper to be 0.001 inch per foot applied only in direction which decreases rate of taper.
  F — Centrality of solid key with axis of taper 0.002 inch total at maximum material condition. (0.002 inch Total indicator variation)
  F₁ — Centrality of keyseat with axis of taper 0.002 inch total at maximum material condition. (0.002 inch Total indicator variation)

*Note 1:* Maximum runout on test plug: 0.0004 at 1 inch projection from gage line.
  : 0.0010 at 12 inch projection from gage line.
*Note 2:* Squareness of mounting face measured near mounting bolt hole circle.
*Note 3:* Holes located as shown and within 0.010 inch diameter of true position.

**American Standard Threaded and Tapered Spindles for Portable Air and Electric Tools** (ASA B5.38-1958)

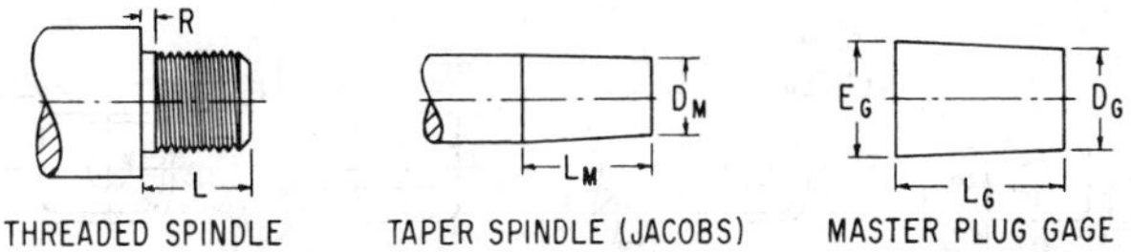


| Nom. Diam. and Thd. | Pitch Diam. Max. | Pitch Diam. Min. | R | L |
|---|---|---|---|---|
| 3/8—24 | 0.3479 | 0.3455 | 1/16 | 9/16* |
| 1/2—20 | 0.4675 | 0.4649 | 1/16 | 9/16 |
| 5/8—16 | 0.5844 | 0.5812 | 3/32 | 11/16 |
| 3/4—16 | 0.7094 | 0.7062 | 3/32 | 11/16 |

| No.† | $D_M$ | $L_M$ | $E_G$ | $D_G$ | $L_G$ | Taper per Foot†† |
|---|---|---|---|---|---|---|
| 1 | 0.335–0.333 | 0.656 | 0.38400 | 0.33341 | 0.65625 | 0.92508 |
| 2S** | 0.490–0.488 | 0.750 | 0.54880 | 0.48764 | 0.7500 | 0.97861 |
| 2 | 0.490–0.488 | 0.875 | 0.55900 | 0.48764 | 0.87500 | 0.97861 |
| 33 | 0.563–0.561 | 1.000 | 0.62401 | 0.56051 | 1.000 | 0.76194 |
| 6 | 0.626–0.624 | 1.000 | 0.67600 | 0.62409 | 1.000 | 0.62292 |
| 3 | 0.748–0.746 | 1.219 | 0.81100 | 0.74610 | 1.21875 | 0.63898 |

All dimensions in inches. Threads are per inch and right-hand.
* Also 7/16 inch. ** 2S stands for 2 Short.
† Jacobs taper number. †† Calculated from $E_G$, $D_G$, $L_G$ for the master plug gage.
*Tolerances:* On $R$, plus or minus 1/64 inch; on $L$, plus 0.000, minus 0.030 inch.

**American Standard Hexagonal Chucks for Portable Air and Electric Tools** (ASA B5.38-1958)

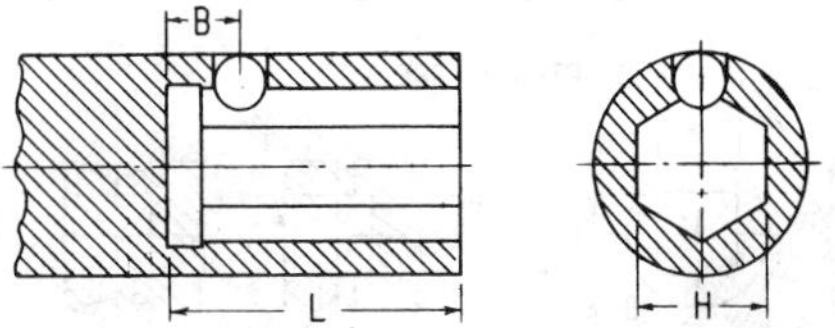


| Nominal Hexagon | H Min. | H Max. | B | L Max. | Nominal Hexagon | H Min. | H Max. | B | L Max. |
|---|---|---|---|---|---|---|---|---|---|
| 1/4 | 0.253 | 0.255 | 3/8 | 15/16 | 5/8 | 0.630 | 0.632 | 11/32 | 1 5/8 |
| 5/16 | 0.314 | 0.316 | 13/64 | 1 | 3/4 | 0.755 | 0.758 | 11/32 | 1 7/8 |
| 7/16 | 0.442 | 0.444 | 17/64 | 1 1/8 | ... | ... | ... | ... | ... |

All dimensions in inches.
Tolerance on $B$ is plus or minus 0.005 inch.

**American Standard Hexagonal Shanks for Portable Air and Electric Tools** (ASA B5.38-1958)

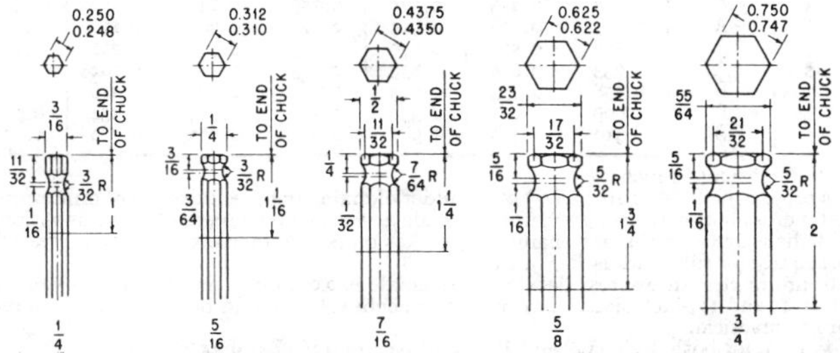

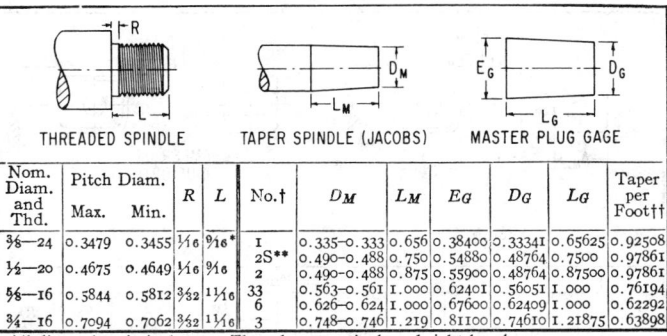

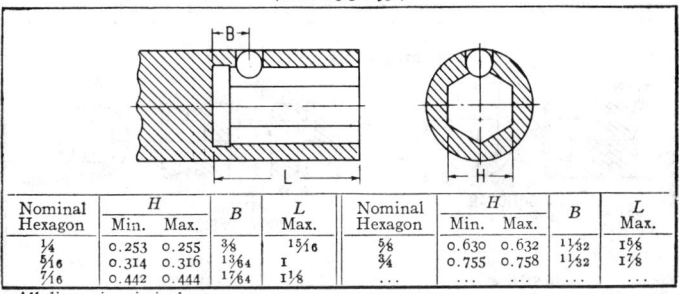

### American Standard Square Drives for Portable Air and Electric Tools
(ASA B5.38-1958)

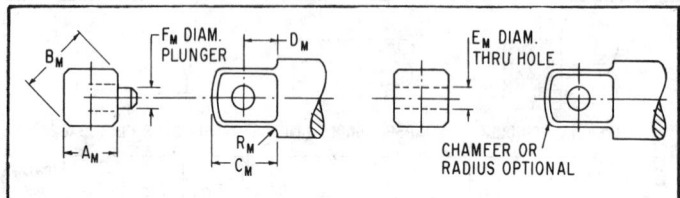

DESIGN A          DESIGN B

| Drive Size | Design | $A_M$ | | $B_M$ Max. | $C_M$ | | $D_M$ | | $E_M$ Min. | $F_M$ Max. | $R_M$ Max. |
|---|---|---|---|---|---|---|---|---|---|---|---|
| | | Max. | Min. | | Max. | Min. | Max. | Min. | | | |
| ¼ | A | 0.252 | 0.247 | 0.330 | 0.312 | 0.265 | 0.165 | 0.153 | ... | 0.078 | 0.015 |
| ⅜ | A | 0.377 | 0.372 | 0.500 | 0.438 | 0.406 | 0.227 | 0.215 | ... | 0.156 | 0.031 |
| ½ | A | 0.502 | 0.497 | 0.665 | 0.625 | 0.531 | 0.321 | 0.309 | ... | 0.187 | 0.031 |
| ⅝ | A | 0.627 | 0.622 | 0.834 | 0.656 | 0.594 | 0.321 | 0.309 | ... | 0.187 | 0.047 |
| ¾ | B | 0.752 | 0.747 | 1.000 | 0.938 | 0.750 | 0.415 | 0.403 | 0.216 | ... | 0.047 |
| 1 | B | 1.002 | 0.997 | 1.340 | 1.125 | 1.000 | 0.602 | 0.590 | 0.234 | ... | 0.063 |
| 1½ | B | 1.503 | 1.498 | 1.968 | 1.625 | 1.562 | 0.653 | 0.641 | 0.310 | ... | 0.094 |

*Male End* (column group heading)

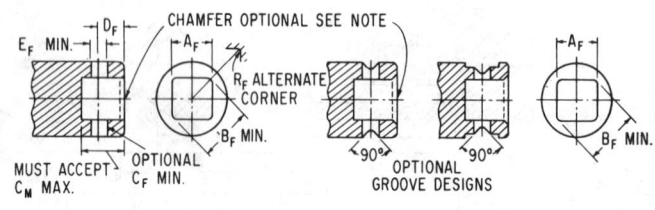

DESIGN A          DESIGN B

| Drive Size | Design | $A_F$ | | $B_F$ Min. | $D_F$ | | $E_F$ Min. | $R_F$ Max. |
|---|---|---|---|---|---|---|---|---|
| | | Max. | Min. | | Max. | Min. | | |
| ¼ | A | 0.258 | 0.253 | 0.335 | 0.159 | 0.147 | 0.090 | ... |
| ⅜ | A | 0.383 | 0.378 | 0.505 | 0.221 | 0.209 | 0.170 | ... |
| ½ | A | 0.508 | 0.503 | 0.670 | 0.315 | 0.303 | 0.201 | ... |
| ⅝ | A | 0.633 | 0.628 | 0.839 | 0.315 | 0.303 | 0.201 | ... |
| ¾ | B | 0.758 | 0.753 | 1.005 | 0.409 | 0.397 | 0.216 | 0.047 |
| 1 | B | 1.009 | 1.004 | 1.350 | 0.596 | 0.584 | 0.234 | 0.062 |
| 1½ | B | 1.510 | 1.505 | 1.983 | 0.647 | 0.635 | 0.310 | 0.125 |

*Female End* (column group heading)

All dimensions in inches.

Incorporating fillet radius ($R_M$) at shoulder of male tang precludes use of minimum diameter cross-hole in socket ($E_F$), unless female drive end is chamfered (shown as optional).

If female drive end is not chamfered, socket cross-hole diameter ($E_F$) is increased to compensate for fillet radius $R_M$, max.

Minimum clearance across flats male to female is 0.001 inch through ¾-inch size; 0.002 inch in 1- and 1½-inch sizes. For impact wrenches $A_M$ should be held as close to maximum as practical.

$C_F$, min. for both designs A and B should be equal to $C_M$, max.

**American Standard Abrasion Tool Spindles for Portable Air and Electric Tools**
(ASA B5.38-1958)

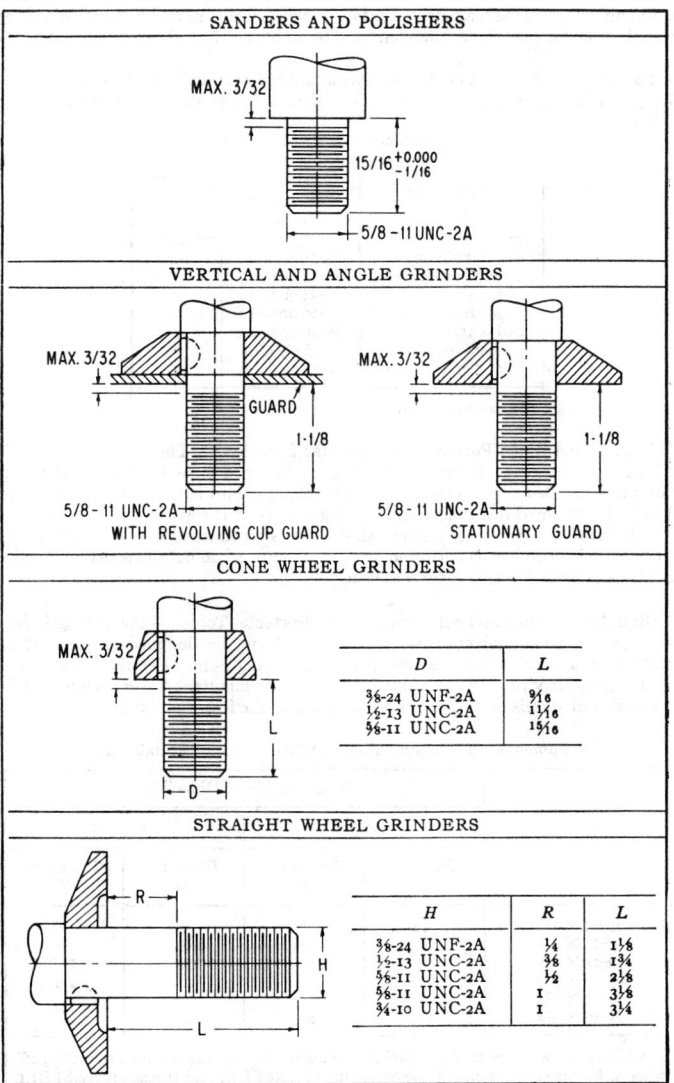

### SANDERS AND POLISHERS

MAX. 3/32

$15/16\ {}^{+0.000}_{-1/16}$

5/8 - 11 UNC-2A

### VERTICAL AND ANGLE GRINDERS

MAX. 3/32

GUARD

1-1/8

5/8 - 11 UNC-2A
WITH REVOLVING CUP GUARD

MAX. 3/32

1-1/8

5/8 - 11 UNC-2A
STATIONARY GUARD

### CONE WHEEL GRINDERS

MAX. 3/32

L

D

| D | L |
|---|---|
| 3/8-24 UNF-2A | 9/16 |
| 1/2-13 UNC-2A | 11/16 |
| 5/8-11 UNC-2A | 15/16 |

### STRAIGHT WHEEL GRINDERS

R

H

L

| H | R | L |
|---|---|---|
| 3/8-24 UNF-2A | 1/4 | 1 1/8 |
| 1/2-13 UNC-2A | 3/8 | 1 3/4 |
| 5/8-11 UNC-2A | 1/2 | 2 1/8 |
| 5/8-11 UNC-2A | 1 | 3 1/8 |
| 3/4-10 UNC-2A | 1 | 3 1/4 |

All dimensions in inches.      Threads are right-hand.

**Circular Saw Arbors.** — American Standard ASA B 5.38 "Driving and Spindle Ends for Portable and Air Electric Tools" calls for a round arbor of 5/8-inch diameter for nominal saw blade diameters of 6 to 8.5 inches, inclusive and a 3/4-inch diameter round arbor for saw blade diameters of 9 to 12 inches, inclusive.

**Spindles for Geared Chucks.** — Recommended threaded and tapered spindles for portable tool geared chucks of various sizes are as given in the following table:

**Recommended Spindle Sizes**

| Chuck Sizes, Inch | Recommended Spindles | |
|---|---|---|
| | Threaded | Taper* |
| 3/16 and 1/4 Light | 3/8–24 | 1 |
| 1/4 and 5/16 Medium | 3/8–24 or 1/2–20 | 2 Short |
| 3/8 Light | 3/8–24 or 1/2–20 | 2 |
| 3/8 Medium | 1/2–20 or 5/8–16 | 2 |
| 1/2 Light | 1/2–20 or 5/8–16 | 33 |
| 1/2 Medium | 5/8–16 or 3/4–16 | 6 |
| 5/8 and 3/4 Medium | 5/8–16 or 3/4–16 | 3 |

* Jacobs number.

**Vertical and Angle Portable Tool Grinder Spindles.** — The 5/8–11 spindle with a length of 1 1/8 inches shown on page 1747 is designed to permit the use of a jam nut with threaded cup wheels. When a revolving guard is used, the length of the spindle is measured from the wheel bearing surface of the guard. For unthreaded wheels with a 7/8-inch hole, a safety sleeve nut is recommended. The unthreaded wheel with 5/8-inch hole is not recommended because a jam nut alone may not resist the inertia effect when motor power is cut off.

**Straight Grinding Wheel Spindles for Portable Tools.** — Portable grinders with pneumatic or induction electric motors should be designed for the use of organic bond wheels rated 9500 feet per minute. Light-duty electric grinders may be designed for vitrified wheels rated 6500 feet per minute. Recommended maximum sizes of wheels of both types are as given in the following table:

**Recommended Maximum Grinding Wheel Sizes for Portable Tools**

| Spindle Size | Maximum Wheel Dimensions | | | |
|---|---|---|---|---|
| | 9500 fpm | | 6500 fpm | |
| | Diameter $D$ | Thickness $T$ | Diameter $D$ | Thickness $T$ |
| 3/8–24 × 1 1/8 | 2 1/2 | 1/2 | 4 | 1/2 |
| 1/2–13 × 1 3/4 | 4 | 3/4 | 5 | 3/4 |
| 5/8–11 × 2 1/8 | 8 | 1 | 8 | 1 |
| 5/8–11 × 3 1/8 | 6 | 2 | .. | ... |
| 5/8–11 × 3 1/8 | 8 | 1 1/2 | .. | ... |
| 3/4–10 × 3 1/4 | 8 | 2 | .. | ... |

Minimum $T$ with the first three spindles is about 1/8 inch to accommodate cutting off wheels. Flanges are assumed to be according to ANSI B7.1 and threads to ANSI B1.1.

# CUTTING SPEEDS AND FEEDS

**Work Materials.** — The large number of work materials that are commonly machined vary greatly in their basic structure and the ease with which they can be machined. Yet it is possible to group together certain materials having similar machining characteristics, for the purpose of recommending the cutting speed at which they can be cut. Most materials that are machined are metals and it has been found that the most important single factor influencing the ease with which a metal can be cut is its microstructure, followed by any cold work that may have been done to the metal, which increases its hardness. Metals that have a similar, but not necessarily the same microstructure, will tend to have similar machining characteristics. Thus, the grouping of the metals in the accompanying tables has been done on the basis of their microstructure.

With the exception of a few soft and gummy metals, experience has shown that harder metals are more difficult to cut than softer metals. Furthermore, any given metal is more difficult to cut when it is in a harder form than when it is softer. It is more difficult to penetrate the harder metal and more power is required to cut it. This in turn will generate a higher cutting temperature at any given cutting speed thereby making it necessary to use a slower speed, for the cutting temperature must always be kept within the limits that can be sustained by the cutting tool without failure. Hardness, then, is an important property that must be considered when machining a given metal. Hardness alone, however, cannot be used as a measure of cutting speed. For example, if pieces of AISI 11L17 and AISI 1117 steel both have a hardness of 150 Bhn, their recommended cutting speed for high speed steel tools will be 140 fpm and 130 fpm respectively. In some metals two entirely different microstructures can produce the same hardness. As an example, a fine pearlite microstructure and a tempered martensite microstructure can result in the same hardness in a steel. These microstructures will not machine alike. For practical purposes, however, information on hardness is usually easier to obtain than information on microstructure; thus, hardness alone is usually used to differentiate between different cutting speeds for machining a metal.

In some situations the hardness of a metal to be machined is not known. When this occurs, the material condition, which is listed separately in the tables, can be used as a guide. Lacking more specific information on the hardness of the metal to be machined, a knowledge of its material condition is helpful in selecting the cutting speed.

The surface of ferrous metal castings has a scale that is more difficult to machine than the metal below. Some scale is more difficult to machine than others, depending on the foundry sand used, the casting process, the method of cleaning the casting, and the type of metal cast. Special electrochemical treatments can be used in some cases that almost entirely eliminate the effect of the scale on machining, although castings so treated are not frequently encountered. Usually when casting scale is encountered, the cutting speed is reduced approximately 5 or 10 per cent. Difficult-to-machine surface scale can also be encountered when machining hot rolled or forged steel bars.

Metallurgical differences that affect machining characteristics are often found within a single piece of metal. The occurrence of hard spots in castings is an example. Different microstructures and hardness levels may occur within a casting as a result of variations in the cooling rate in different parts of the casting. Such variations are less severe in castings that have been heat treated. Steel bar stock is usually harder toward the outside than toward the center of the bar. Sometimes there are slight metallurgical differences along the length of a bar which can affect its cutting characteristics.

**Cutting Tool Materials.** — The recommended cutting speeds in the accompanying tables are given for high speed steel and for cemented carbides because these materials are the most commonly used and more data are available for them than for others. Other materials that are used to make cutting tools are cemented oxides or ceramics, cermets, cast non-ferrous alloys (stellite), single crystal diamonds, polycrystalline diamonds, and cubic boron nitride.

*High Speed Steel* designates a number of steels that have several properties which enhance their value as cutting tool material. They can be hardened to a high initial or room temperature hardness ranging from 63 $R_C$ to 65 $R_C$ for ordinary high speed steels and up to 70 $R_C$ for the so-called super-high speed steels. They can retain sufficient hardness at temperatures up to 1,000 to 1,100 F to enable them to cut at cutting speeds that will generate these tool temperatures, and they will return to their original hardness when cooled to room temperature. They harden very deeply enabling high speed steels to be ground to the tool shape from solid stock and to be reground many times without sacrificing hardness at the cutting edge. High speed steels can be made soft by annealing so that they can be machined into complex cutting tools such as drills, reamers, and milling cutters and then hardened.

The principal alloying elements of high speed steels are tungsten (W), molybdenum (Mo), chromium (Cr), vanadium (V), together with carbon (C). There are a number of grades of high speed steel which are divided into two types: tungsten high speed steels and molybdenum high speed steels. Tungsten high speed steels are designated by the prefix T before the number which designates the grade and molybdenum high speed steels are designated by the prefix letter M. Tungsten high speed steels were developed first and are still used. A common steel of this type is the T1 which has the following alloy composition: W − 18%, Cr − 4%, V − 1%, and C − .75%. Molybdenum high speed steels were developed because of a shortage in the supply of tungsten. They are less expensive than comparable grades of tungsten high speed steel and are, therefore, widely used at the present time. Molybdenum is used as a substitute for a part of the tungsten as shown by the following composition of the widely used M2 high speed steel: Mo − 5%, W − 6%, Cr − 4%, V − 2%, and C − .85%. Production experience and extensive laboratory tests have shown no significant superiority in either direction between comparable grades of tungsten or molybdenum high speed steel.

The addition of 5 to 12 per cent cobalt to high speed steel increases its hardness at the temperatures encountered in cutting thereby improving its wear resistance and cutting efficiency. Cobalt does slightly increase the brittleness of high speed steel making it susceptible to chipping at the cutting edge. For this reason, cobalt high speed steels are primarily made into single point cutting tools which are used to take heavy roughing cuts in abrasive materials and through rough abrasive surface scales.

The M40 series and T15 are a group of high hardness or so-called super-high-speed steels that can be hardened to 70 $R_C$; however, they tend to be brittle and difficult to grind. For cutting applications they are usually heat treated to 67-68 $R_C$ to reduce their brittleness and tendency to chip. The M40 series is appreciably easier to grind than T15. A typical composition is M42, having: Mo-9.6%, W-1.6%, Cr-3.75%, V-1.15%, Co-8.25%, C-1.08%. They are recommended for machining tough die steels and other difficult-to-cut materials; they are not recommended for applications where conventional high-speed steels perform well. High-speed steels made by the particle-metallurgy process are tougher and have an improved grindability when compared to similar grades made by the customary process. Conventional high-speed-steel alloys are made by this process as well as grades having a very high alloy content. The particle-metallurgy process enables high-speed steels to attain a higher alloy content than can be economically produced by the customary method. Tools made of these steels can be hardened about 1 $R_C$ higher without a sacrifice in toughness than comparable high-speed steels made by the customary process. They are particularly useful on applications involving intermittent cutting and where tool life is limited by chipping. All of these steels augment rather than replace the conventional high-speed steels.

*Cemented Carbides* are also called sintered carbides or simply carbides. They are harder than high-speed steels and have excellent wear resistance. They retain a very high degree of hardness at temperatures up to 1400 F and even higher; therefore, very fast cutting speeds can be used. When used at fast cutting speeds, they produce good surface finishes on the workpiece. Carbides are more brittle than high-speed steel and must, therefore, be used with more care.

Several hundred grades of carbides have been produced and attempts to classify these grades by area of application have not been entirely successful. There are four distinct types of carbides: 1) straight tungsten carbides; 2) crater resistant carbides; 3) titanium carbide; and 4) coated carbides. *Straight tungsten carbide* usually consists of approximately 94 to 97 per cent tungsten carbide with the remainder being cobalt. It is the most abrasion resistant cemented carbide and is used to machine gray cast iron, most non-ferrous metals, and non-metallic materials, where abrasion resistance is the primary criterion. When used to machine steel, straight tungsten carbide will rapidly form a crater on the face of the tool, which has a very adverse effect on the life of the tool. Titanium carbide is added to the tungsten carbide in order to counteract the rapid formation of the crater. In addition, tantalum carbide is also usually added to prevent the cutting edge from deforming when subject to the intense heat and pressure generated in taking a heavy cut. *Crater Resisting Carbides* containing titanium and tantalum carbide additions to the tungsten carbide are used to cut steels, alloy cast irons, and other materials that have a strong tendency to form a crater.

*Titanium Carbides*, available in several grades, are made entirely from titanium carbide plus small amounts of nickel and molybdenum. These carbides have an excellent resistance to cratering and to heat. Their high hot hardness enables them to operate at higher cutting speeds. They are, however, more brittle and less resistant to mechanical and thermal shocks; therefore, they are not recommended for taking heavy cuts and interrupted cuts. The fundamental abrasion resistance of titanium carbides is also lower and they are not recommended for cutting through scale or oxide films on steel. Although the resistance to cratering of titanium carbides is excellent, failure caused by crater formation can sometimes occur because the chip tends to curl very close to the cutting edge, thereby forming a small crater in this region which may break through.

*Coated Carbides* are available only as indexable inserts because the coating would be removed by grinding. The principal types of coating material are titanium carbide ($TiC$), titanium nitride ($TiN$), and aluminum oxide ($Al_2O_3$). A very thin layer (approximately .0002 in.) of coating material is deposited over a cemented carbide insert; the material below the coating is called the substrate. The overall performance of the coated carbide is limited by the substrate, which provides the required toughness and resistance to deformation and thermal shock. Although coated carbides are less resistant to mechanical shock than uncoated carbides, they have been used successfully on interrupted cuts and for milling. While both coated and uncoated carbides can fail as a result of thermal shock, certain grades of coated carbides have a better thermal shock resistance than many uncoated carbide grades. By having the right combination of coating material and substrate, and by having the proper cutting edge preparation, coated carbides will exhibit superior performance characteristics. Proper honing of the cutting edge by the carbide producer is very important in the application of coated carbide inserts. With an equal tool life, coated carbides can operate at a higher cutting speed than uncoated carbides. This may be 20 to 30 per cent and in some cases up to 50 per cent faster. Titanium carbide and titanium nitride coated carbides usually operate in the medium (200-800 fpm) cutting speed range, while aluminum oxide coated carbides are used in the higher (800-1600 fpm) cutting speed range.

*Carbide Grade Selection:* The selection of the best grade of carbide for a particular application is very important. An improper grade of carbide will result in a poor performance—in some cases it may even cause the cutting edge to fail before any significant amount of cutting has been done. Because of the many grades and the many variables that are involved, the carbide producers should be consulted to obtain recommendations for the application of their grades of carbide. A few general guidelines can be given which are useful to form an orientation. Metal cutting carbides usually range in hardness from about 89.5 HRA (Rockwell A Scale) to 93.0 HRA with the exception of titanium carbide which has a hardness range of 90.5 HRA to 93.5 HRA. Generally the harder carbides are more wear resistant and more brittle while the softer carbides are less wear resistant but tougher. A choice of hardness must be made to suit the given application. The very hard carbides are generally used for taking light finishing cuts. For other applications, select the carbide that has the highest hardness with sufficient strength to prevent chipping or breakage. A grade of straight tungsten carbide should always be used unless cratering is encountered. Straight tungsten carbides are used to machine gray cast iron, ferritic malleable iron, austenitic stainless steel, high temperature alloys, copper, brass, bronze, aluminum alloys, zinc alloy die castings, and plastics. Crater resistant carbides should be used to machine plain carbon steel, alloy steel, tool steel, pearlitic malleable iron, nodular iron, other highly alloyed cast irons, ferritic stainless steel, martensitic stainless steel, and certain high temperature alloys. Titanium carbides are recommended for taking high speed finishing and semi-finishing cuts on steel, especially the low carbon, low alloy steels, which are less abrasive and have a strong tendency to form a crater. They are also used to take light cuts on alloy cast iron and on some high nickel alloys. Non-ferrous materials which are essentially non-abrasive may also be machined with titanium carbides, such as some aluminum alloys and brass. Abrasive materials and others that should not be machined with titanium carbides include gray cast iron, titanium alloys, cobalt and nickel base superalloys, stainless steel, bronze, many aluminum alloys, fiberglass, plastics, and graphite. For materials that can successfully be machined with titanium carbide the cutting speed should be increased 30 to 60 per cent over the values given in the accompanying tables. The feed rate used should not exceed about .020 inch per revolution.

Coated carbides can be used to take cuts ranging from light finishing to heavy roughing on most materials that can be cut with these carbides. For titanium carbide and titanium nitride coatings, the cutting speed should be increased 20 to 30 per cent over the values in the tables; experience may show that in some cases an increase up to 50 per cent is possible. Aluminum oxide coated carbide should be operated at even higher cutting speeds. The coated carbides are recommended for machining all free machining steels, all plain carbon and alloy steels, tool steels, martensitic and ferritic stainless steels, precipitation hardening stainless steels, alloy cast iron, pearlitic and martensitic malleable iron, and nodular iron. They are also recommended for taking light finishing and roughing cuts on austenitic stainless steels. Coated carbides should not be used to machine nickel and cobalt base superalloys, titanium and titanium alloys, brass, bronze, aluminum alloys, pure metals, refractory metals, and non-metals such as fiberglass, graphite, and plastics.

*Ceramic Cutting Tool Materials* are made from finely powdered aluminum oxide particles sintered into a hard dense structure without a binder material. Aluminum oxide is also combined with titanium carbide to form a composite, which is called a cermet. These materials have a very high hot hardness enabling very high cutting speeds to be used. For example, ceramic cutting tools have been used to cut AISI 1040 steel at a cutting speed of 18,000 fpm with a satisfactory tool life. However, much lower cutting speeds, in the range of 1000 to 4000 fpm and lower are more common because of limitations placed by the machine tool, cutters, and chucks. Although most applications of ceramic and cermet cutting tool materials are for turning, they have also been used successfully for milling. They are relatively brittle and a special cutting edge preparation is required to prevent chipping or edge breakage. This consists of honing or grinding a narrow flat land, .002 to .006 inch wide, on the cutting edge which is made about 30 degrees with respect to the tool face. For some heavy duty applications a wider land is used. The setup should be as rigid as possible and the feed rate should not normally exceed .020 inch, although in some cases .030 inch has been used successfully. Ceramics and cermets are recommended for roughing and finishing operations on all cast irons, plain carbon and alloy steels, and the stainless steels. Materials up to a hardness of 60 HRC (Rockwell C Scale) can be cut with ceramic and cermet cutting tools. They should not be used to machine aluminum and aluminum alloys, magnesium alloys, titanium, and titanium alloys.

*Cast Nonferrous Alloys* are made from tungsten, tantalum, chromium, and cobalt plus carbon. Other alloying elements are also used to produce a high temperature resistant and wear resistant material. These alloys cannot be softened by heat treatment and must be cast and ground to shape. The room temperature hardness of cast nonferrous alloys is lower than for high-speed steel, but the hardness and wear resistance is retained to a higher temperature. They are generally marketed under trade names such as Stellite, Crobalt, and Tantung. As a starting point, the cutting speed for cast nonferrous tools can be 20 to 50 per cent greater than the recommended cutting speed for high-speed steel as given in the accompanying tables.

*Diamond* cutting tools are available in two forms: 1. single crystal natural diamonds shaped to a cutting edge and mounted on a tool holder on a boring bar; and, 2. polycrystalline diamond indexable inserts made from synthetic or natural diamond powders which have been compacted and sintered into a solid mass. Single crystal and polycrystalline diamond cutting tools are very wear resistant. They are, therefore, recommended for machining abrasive materials which cause other cutting tool materials to wear rapidly. Typical of the abrasive materials machined with single crystal and polycrystalline diamond tools are the following: fiberglass, 300 to 1000 fpm; fused silica, 900 to 950 fpm; reinforced melamine plastics, 350 to 1000 fpm; reinforced phenolic plastics, 350 to 1000 fpm; thermosetting plastics, 300 to 2000 fpm; teflon, 600 fpm; nylon, 200 to 300 fpm; mica, 300 to 1000 fpm; graphite, 200 to 2000 fpm; babbitt bearing metal, 700 fpm; and aluminum-silicon alloys, 1000 to 2000 fpm. Another important application of diamond cutting tools is to produce fine surface finishes on soft nonferrous metals that are difficult to finish by other methods. Surface finishes of 1 to 2 microinches can readily be obtained with single crystal diamond tools, and finishes down to 10 microinches can be obtained with polycrystalline diamond tools. In addition to babbitt and the aluminum-silicon alloys, other metals finished with diamond tools include: soft aluminum, 1000 to 2000 fpm; all wrought and cast aluminum alloys, 600 to 1500 fpm; copper, 1000 fpm; brass, 500 to 1000 fpm; bronze, 300 to 600 fpm; oilite bearing metal, 500 fpm; silver, gold, and platinum, 300 to 2500 fpm; and zinc, 1000 fpm. Ferrous alloys, such as cast iron and steel, should not be machined with diamond cutting tools because the high cutting temperatures generated will cause the diamond to transform into carbon.

*Cubic Boron Nitride* (CBN) is also known by its trade name, Borazon,® and when used as a cutting tool material it is designated by the symbol BZN.® Next to the diamond, it is the hardest known material. It will retain its hardness at a temperature of 1800 F and higher, making it an ideal cutting tool material for machining very hard and tough materials at cutting speeds beyond those possible with other cutting tool materials. Indexable inserts and cutting tool blanks made from this material consist of a layer, approximately .020 inch thick, of polycrystalline cubic boron nitride firmly bonded to the top of a cemented carbide substrate. Cubic boron nitride is recommended for rough and finish turning hardened plain carbon and alloy steels, hardened tool steels, hard cast irons, and particularly the superalloys. As a class, the superalloys are not as hard as hardened steel; however, their combination of high strength and tendency to deform plastically under the pressure of the cut, or gumminess, places them in the class of hard-to-machine materials. Conventional materials that can readily be machined with other cutting tool materials should not be machined with cubic boron nitride. Table 10 provides a representative sample of the materials machined with cubic boron nitride and gives the cutting conditions recommended for rough turning these materials. Finish turning operations on these materials with cubic boron nitride cutting tools produce excellent surface finishes. Round indexable inserts are recommended when taking severe cuts on these materials in order to provide maximum strength to the insert. When using square or triangular inserts, a large lead angle should be used. Normally the lead angle should be 15 degrees, and whenever possible, 45 degrees. A negative rake angle should always be used, which for most applications is negative 5 degrees. The relief angle should be 5 to 9 degrees. Although cubic boron nitride cutting tools can be used without a coolant, flooding the tool with a water soluble type coolant is recommended.

*Carbon Tool Steel* is used primarily to make the less expensive drills, taps, and reamers. It is seldom used to make single point cutting tools. Carbon steels are very shallow hardening although some have a small amount of vanadium and chromium added to improve their hardening quality. The cutting speed to use for plain carbon tool steel should be approximately one-half of the recommended speed for high-speed steel.

**Cutting Speed, Feed, Depth of Cut.** — The cutting conditions that determine the rate of metal removal are the cutting speed, the feed rate, and the depth of cut. These cutting conditions and the nature of the material to be cut determine the power required to take the cut. The cutting conditions must be adjusted to stay within the power available on the machine tool to be used.

The cutting conditions must also be considered in relation to the tool life. Tool life can be defined as the length of time that a cutting tool will cut before becoming dull or before it must be replaced. The end of tool life is defined as a given amount of wear on the flank of the tool by the ANSI Standard Specification For Tool Life Testing With Single-Point Tools-ANSI B94.34-1946(R1971) and B94.36-1956(R1971). These standards are followed when making scientific machinability tests with single-point cutting tools in order to achieve uniformity in testing procedure so that results from different machinability laboratories can readily be compared. It is not practicable or necessary to follow this standard in the shop; however, it should be understood that the cutting conditions and tool life are related. For example, the cutting speed and the feed may be increased if a shorter tool life is accepted. Furthermore, the decrease in the tool life will be proportionately greater than the increase in the cutting speed or the feed. Conversely, if the cutting speed or the feed is decreased, the increase in the tool life will be proportionately greater than the decrease in the cutting speed or the feed.

Tool life is influenced most by cutting speed, then by the feed rate, and least by the depth of cut. After the depth of cut is about 10 times greater than the feed rate, a further increase in the depth of cut will have no significant effect on the tool life. This characteristic of the performance of cutting tools is very important in determining the operating or cutting conditions for machining metals.

The first step in selecting the cutting conditions is to select the depth of cut. The depth of cut will be limited by the amount of metal that is to be machined from the workpiece, by the power available on the machine tool, by the rigidity of the workpiece and the cutting tool, and by the rigidity of the setup. Since the depth of cut has the least effect upon the tool life, always use the heaviest depth of cut that is possible.

The second step is to select the feed rate. In selecting the feed rate, consideration must be given to the power available on the machine tool, to the rigidity of the workpiece and the cutting tool, to the rigidity of the setup, and to the surface finish required on the finished workpiece. The available power must be considered in relation to the depth of cut previously selected in considering the feed. Select the maximum feed possible; however, it must not be greater than that which will produce an acceptable surface finish.

The third step is to select the cutting speed. The accompanying tables provide recommended cutting speeds. If previous experience has been had in machining a certain material, this may form the basis for selecting the cutting speed. In either case, however, the depth of cut should be selected first, followed by the feed, and the last to be selected should be the cutting speed.

**Use of Cutting Speed Tables.** — On the following pages tables of recommended cutting speeds are provided. The values in these tables are for average conditions and serve as a basis from which to start. In many cases they will be found to be the most satisfactory cutting speeds, while in others, modifications may offer advantages as experience is gained on a particular job. It is not possible to specify a single optimum cutting speed for a material that will fit every situation. Many factors unique to each job may make a modification desirable. These factors include: the size, type, model, and make of the machine tool; its power, rigidity, and the foundation on which it is standing; the workpiece configuration; the rigidity of the workpiece setup, or fixturing; safety aspects of the setup; the particular grade of cemented carbide used; the influence of a cutting fluid; and, the tool life desired. Except for certain difficult-to-machine materials, most materials can be cut successfully over a rather wide range of cutting speeds with a particular type of cutting tool material; however, not just any speed within this range should be used. A cutting speed that is too slow will result in a loss of production and in an increase in the cost of the part. Likewise, a cutting speed that is too fast can have the same result because the tool life will be too short and the production must frequently be interrupted to change tools. Moreover, the cost of sharpening or replacing the cutting tool must be considered. There is usually a narrow range of cutting speeds within which the most economical results will be obtained. When making a modification to the cutting speed the fundamental behaviour of cutting tools must be kept in mind; i.e., increasing the cutting speed will result in a proportionately larger reduction in the tool life; reducing the cutting speed will result in a proportionately larger increase in the tool life.

*Feed and Depth of Cut Factors:* Factors used to modify the cutting speeds in compensation for different feed rates and depths of cut are given in Table 5. These factors should be used only with the cutting speeds listed in Tables 1 through 4. Moreover, they do not apply when the hardness of a material exceeds approximately 350 to 400 HB. They should, however, be used in the other hardness ranges, at which most materials are machined. Experience in production and scientific investigations conducted in machinability laboratories have shown that a change in the feed or in the depth of cut will require a compensating change in the cutting speed if the tool life is to remain unchanged. For this reason a modification in the cutting speed should be made if the feed or depth of cut is changed significantly when machining the materials listed in the other tables.

*Cutting Tool Material:* The cutting speeds in the tables listed under cemented carbide are based on the assumption that a correct grade of uncoated straight tungsten carbide or an uncoated crater resisting carbide is used. Most carbide producers will be able to recommend a grade of carbide that can be used at the cutting speed given. An incorrect grade of carbide may, however, require a modification to the cutting speed, and in some instances it may result in an extremely short tool life. High speed steels are less sensitive in this respect; i.e., any type of high speed steel can cut most of the materials listed in the tables at the recommended cutting speed for high speed steel. It is true, however, that certain types of high speed steels do offer advantages in some applications, such as a longer tool life. Very hard and tough materials, such as the superalloys, should be cut with T15 high speed steel, or with one of the M-40 types, as for example M42. These premium high speed steels should not be used to cut most other materials, since they can be cut as well by other types of high speed steel.

*Coated Carbides:* A faster cutting speed can be used with coated carbide cutting tools than with uncoated tools when machining materials that can be cut by the coated grades. When using titanium carbide or titanium nitride coated cutting tools, the recommended cutting speed in the tables should be increased by 20 to 30 per cent;

### Summary of Principal Tables in the Speeds and Feeds Section

in some cases an increase up to 50 per cent is possible. The cutting speed should be increased even more when using aluminum oxide coated carbides.

*Titanium Carbides:* The cutting speed should be increased by 30 to 60 per cent over the values given in the tables when using titanium carbide cutting tools.

*Milling:* The recommended cutting speed for milling given in the tables can be used for all face milling operations, slab milling operations, slab milling type cuts taken with end milling cutters, and for shallow slotting cuts taken with either end milling cutters or with side milling cutters. When milling deep slots with end milling cutters or with side milling cutters, the cutting speed should be reduced approximately 10 per cent. Likewise, the cutting speed should be reduced about 10 per cent when taking wide side milling cuts with side milling cutters; no reduction in the cutting speed is required if the width of the side milling cut is small.

*Planing and Shaping:* The cutting speeds in Tables 1 through 4, and Tables 6 through 9 can be used for planing and shaping. The feed and depth of cut factors in Table 5 should also be used as explained previously. Very often other factors relating to the machine or the setup will require a reduction in the actual cutting speed used on the job.

*Surface Scale:* Certain heavy and abrasive surface scales encountered on castings and on some wrought metals require a reduction in the cutting speed, especially when drilling.

*Cutting Fluids:* Many cutting fluids permit a somewhat higher cutting speed to be used. It is not possible, however, to provide specific recommendations because each proprietary cutting fluid exhibits its own characteristics.

*Metric Units:* In metric practice the units used for the cutting speed is meters per minute, abbreviated m/min. or mpm. The cutting speeds in the tables are given in inch units, or feet per minute, which is abbreviated fpm. These units can be converted as follows: to obtain meters per minute, multiply feet per minute by 0.3; to obtain feet per minute, multiply meters per minute by 3.28.

*Cutting Speed Formulas:* Most machining operations are conducted on machine tools having a rotating spindle, and the cutting speed in feet or meters per minute must be converted to a spindle speed, or to revolutions per minute; this is accomplished by use of the following formulas:

For inch units only:

$$N = \frac{12V}{\pi D}$$

For metric units only:

$$N = \frac{1000V}{\pi D}$$

Where:   $N$ = Spindle speed; rpm

   $V$ = Cutting speed; fpm, or m/min

   $D$ = Diameter; in., or mm (For turning, $D$ is the outside diameter of the workpiece. For milling, drilling and reaming, $D$ is the diameter of the cutter.)

   $\pi$ = 3.14

The feed and depth-of-cut factors in Table 5 are only to be used together with Tables 1 through 4. While the values in Table 5 are for inch units, they can be used with metric units by converting the metric feed and depth of cut into inch units. First select the cutting speed from the appropriate table and then apply the factors from Table 5, using the formula given below:

$$V = V_o F_f F_d$$

Where:   $V$ = Cutting speed to be used; fpm, or m/min.

   $V_o$ = Cutting speed from tables; fpm, or m/min.

   $F_f$ = Feed Factor                (From Table 5)

   $F_d$ = Depth-of-cut factor        (From Table 5)

*Example:* Using both the inch and the metric formulas, calculate the spindle speed for turning a 1¼ inch (31.75 mm) bar of 200-220 HB AISI 1040 steel using depth of cut of .100 in. (2.54 mm) and a feed rate of .015 in. (0.38 mm/rev.). (A small insignificant difference in the answers may occur, which is caused by the conversion of the units and the rounding off of numbers.)

From Table 1: $V_o = 85$ fpm.
From Table 5: $F_f = .91$;
$F_d = 1.03$

$$V = V_o F_f F_d$$
$$= 85 \times .91 \times 1.03$$
$$= 80 \text{ fpm}$$
$$N = \frac{12V}{\pi D}$$
$$= \frac{12 \times 80}{\pi \times 1.250}$$
$$= 244 \text{ rpm}$$
$$V_o = 85 \times .3$$
$$= 25.5 \text{ m/min}$$
$$V = V_o F_f F_d$$
$$= 25.5 \times .91 \times 1.03$$
$$= 24 \text{ m/min}$$
$$N = \frac{1000V}{\pi D}$$
$$= \frac{1000 \times 24}{\pi \times 31.75}$$
$$= 241 \text{ rpm}$$

It is often necessary to calculate the cutting speed in feet per minute or in meters per minute, when the diameter of the workpiece or of the cutting tool and the spindle speed is known. In this event, the following formulas are used.

For inch units only:

$$V = \frac{\pi DN}{12}, \text{ where } D \text{ is in inches}$$

For metric units only:

$$V = \frac{\pi DN}{1000}, \text{ where } D \text{ is in millimeters}$$

*Example:* Calculate the cutting speed in feet per minute and in meters per minute when the spindle speed of a ¾ inch (19.05 mm) drill is 400 rpm.

$$V = \frac{\pi DN}{12} = \frac{\pi \times .75 \times 400}{12} = 78.5 \text{ fpm}$$

$$V = \frac{\pi DN}{1000} = \frac{\pi \times 19.05 \times 400}{1000} = 24 \text{ m/min}$$

**Table 1.  Recommended Cutting Speeds in Feet per Minute for Turning Plain Carbon and Alloy Steels**

| Material AISI and SAE Steels | Hardness, HB* | Material Condition* | Cutting Speed, fpm HSS | Cutting Speed, fpm Carbide |
|---|---|---|---|---|
| Free Machining Plain Carbon Steels (Resulphurized), 1212, 1213, 1215 | 100-150 | HR, A | 150 | 600 |
| | 150-200 | CD | 160 | 625 |
| 1108, 1109, 1115, 1117, 1118, 1120 1126, 1211 | 100-150 | HR, A | 130 | 500 |
| | 150-200 | CD | 120 | 525 |
| 1132, 1137, 1139, 1140, 1144, 1146, 1151 | 175-225 | HR, A, N, CD | 120 | 400 |
| | 275-325 | Q and T | 75 | 300 |
| | 325-375 | Q and T | 50 | 225 |
| | 375-425 | Q and T | 40 | 200 |
| Free Machining Plain Carbon Steels (Leaded), 11L17, 11L18, 12L13, 12L14 | 100-150 | HR, A, N, CD | 140 | 550 |
| | 150-200 | HR, A, N, CD | 145 | 560 |
| | 200-250 | N, CD | 110 | 400 |
| Plain Carbon Steels, 1006, 1008, 1009, 1010, 1012, 1015, 1016, 1017, 1018, 1019, 1020, 1021, 1022, 1023, 1024, 1025, 1026, 1513, 1514 | 100-125 | HR, A, N, CD | 120 | 450 |
| | 125-175 | HR, A, N, CD | 110 | 400 |
| | 175-225 | HR, N, CD | 90 | 350 |
| | 225-275 | CD | 70 | 300 |
| 1027, 1030, 1033, 1035, 1036, 1037, 1038, 1039, 1040, 1041, 1042, 1043, 1045, 1046, 1048, 1049, 1050, 1052, 1524, 1526, 1527, 1541 | 125-175 | HR, A, N, CD | 100 | 375 |
| | 175-225 | HR, A, N, CD | 85 | 325 |
| | 225-275 | N, CD, Q and T | 70 | 225 |
| | 275-325 | Q and T | 60 | 200 |
| | 325-375 | Q and T | 40 | 160 |
| | 375-425 | Q and T | 30 | 140 |
| 1055, 1060, 1064, 1065, 1070, 1074, 1078, 1080, 1084, 1086, 1090, 1095, 1548, 1551, 1552, 1561, 1566 | 125-175 | HR, A, N, CD | 100 | 370 |
| | 175-225 | HR, A, N, CD | 80 | 320 |
| | 225-275 | N, CD, Q and T | 65 | 220 |
| | 275-325 | Q and T | 50 | 180 |
| | 325-375 | Q and T | 35 | 150 |
| | 375-425 | Q and T | 30 | 130 |
| Free Machining Alloy Steels (Resulphurized), 4140, 4150 | 175-200 | HR, A, N, CD | 110 | 400 |
| | 200-250 | HR, N, CD | 90 | 350 |
| | 250-300 | Q and T | 65 | 300 |
| | 300-375 | Q and T | 50 | 225 |
| | 375-425 | Q and T | 40 | 165 |
| Free Machining Alloy Steels (Leaded), 41L30, 41L40, 41L47, 41L50, 43L47, 51L32, 52L100, 86L20, 86L40 | 150-200 | HR, A, N, CD | 120 | 430 |
| | 200-250 | HR, N, CD | 100 | 380 |
| | 250-300 | Q and T | 75 | 275 |
| | 300-375 | Q and T | 55 | 220 |
| | 375-425 | Q and T | 50 | 200 |
| Alloy Steels, 4012, 4023, 4024, 4028, 4118, 4320, 4419, 4422, 4427, 4615, 4620, 4621, 4626, 4718, 4720, 4815, 4817, 4820, 5015, 5117, 5120, 6118, 8115, 8615, 8617, 8620, 8622, 8625, 8627, 8720, 8822, 94B17 | 125-175 | HR, A, N, CD | 100 | 400 |
| | 175-225 | HR, N, CD | 90 | 350 |
| | 225-275 | CD, N, Q and T | 70 | 300 |
| | 275-325 | Q and T | 60 | 250 |
| | 325-375 | Q and T | 50 | 200 |
| | 375-425 | Q and T | 35 | 175 |

Based on a feed rate of .012 in. per rev. and a depth of cut of .125 in.
* Abbreviations designate: HR, hot rolled; CD, cold drawn; A, annealed; N, normalized; Q and T, quenched and tempered; and HB, Brinell hardness number.

# 1760 SPEEDS AND FEEDS

**Table 1** *(Concluded)*.  **Recommended Cutting Speeds in Feet per Minute for Turning Plain Carbon and Alloy Steels**

| Material AISI and SAE Steels | Hardness, HB* | Material Condition* | Cutting Speed, fpm | |
|---|---|---|---|---|
| | | | HSS | Carbide |
| Alloy Steels, 1330, 1335, 1340, 1345 4032, 4037, 4042, 4047, 4130, 4135, 4137, 4140, 4142, 4145, 4147, 4150, 4161, 4337, 4340, 50B44, 50B46, 50B50, 50B60, 5130, 5132, 5140, 5145, 5147, 5150, 5160, 51B60, 6150, 81B45, 8630, 8635, 8637, 8640, 8642, 8645, 8650, 8655, 8660, 8740, 9254, 9255, 9260, 9262, 94B30 | 175-225 225-275 275-325 325-375 375-425 | HR, A, N, CD N, CD, Q and T N, Q and T N, Q and T Q and T | 85 70 60 40 30 | 325 275 230 200 150 |
| Alloy Steels, E51100, E52100 | 175-225 225-275 275-325 325-375 375-425 | HR, A, CD N, CD, Q and T N, Q and T N, Q and T Q and T | 70 65 50 30 20 | 310 260 220 180 140 |
| Ultra High Strength Steels (Not AISI) AMS 6421 (98B37 Mod.), AMS 6422 (98BV40), AMS 6424, AMS 6427, AMS 6428, AMS 6430, AMS 6432, AMS 6433, AMS 6434, AMS 6436, AMS 6442, 300M, D6ac | 220-300 300-350 350-400 43-48 HRC 48-52 HRC | A N N Q and T Q and T | 65 50 35 25 10 | 270 200 150 120 80 |
| Maraging Steels (Not AISI) 18% Ni  Grade 200 18% Ni  Grade 250 18% Ni  Grade 300 18% Ni  Grade 350 | 250-325 50-52 HRC | A Maraged | 60 10 | 300 80 |
| Nitriding Steels (Not AISI) Nitralloy 125 Nitralloy 135 Nitralloy 135 Mod. Nitralloy 225 Nitralloy 230 Nitralloy N Nitralloy EZ Nitrex 1 | 200-250 300-350 | A N, Q and T | 70 30 | 300 225 |

Based on a feed rate of .012 in. per rev. and a depth of cut of .125 in.

\* Abbreviations designate: HR, hot rolled; CD, cold drawn; A, annealed; N, normalized; Q and T, quenched and tempered; HB, Brinell hardness number; and HRC, Rockwell C scale hardness number.

**Cutting Time for Turning, Boring, and Facing.** — The time required to turn a length of metal can be determined by the following formula in which $T$ = time in minutes, $L$ = length of cut in inches, $f$ = feed in inches per revolution, and $N$ = lathe spindle speed in revolutions per minute.

$$T = \frac{L}{fN}$$

When making job estimates, the time required to load and to unload the workpiece on the machine, and the machine handling time must be added to the cutting time for each length cut to obtain the floor-to-floor time.

**Table 2. Recommended Cutting Speeds in Feet per Minute for Turning Tool Steels**

| Material Tool Steels (AISI Types) | Hardness, HB* | Material Condition* | Cutting Speed, fpm | |
|---|---|---|---|---|
| | | | HSS | Carbide |
| Water Hardening W1, W2, W5 | 150-200 | A | 100 | 325 |
| Shock Resisting S1, S2, S5, S6, S7 | 175-225 | A | 70 | 300 |
| Cold Work, Oil Hardening O1, O2, O6, O7 | 175-225 | A | 70 | 250 |
| Cold Work, High Carbon High Chromium D2, D3, D4, D5, D7 | 200-250 | A | 45 | 175 |
| Cold Work, Air Hardening A2, A3, A8, A9, A10 | 200-250 | A | 70 | 250 |
| A4, A6 | 200-250 | A | 55 | 200 |
| A7 | 225-275 | A | 45 | 175 |
| Hot Work, Chromium Type H10, H11, H12, H13, H14, H19 | 150-200 | A | 80 | 300 |
| | 200-250 | A | 65 | 225 |
| | 325-375 | Q and T | 50 | 175 |
| | 48-50 HRC | Q and T | 20 | 95 |
| | 50-52 HRC | Q and T | 10 | 80 |
| | 52-54 HRC | Q and T | — | 60 |
| | 54-56 HRC | Q and T | — | 40 |
| Hot Work, Tungsten Type H21, H22, H23, H24, H25, H26 | 150-200 | A | 60 | 250 |
| | 200-250 | A | 50 | 200 |
| Hot Work, Molybdenum Type H41, H42, H43 | 150-200 | A | 55 | 225 |
| | 200-250 | A | 45 | 175 |
| Special Purpose, Low Alloy L2, L3, L6 | 150-200 | A | 75 | 325 |
| Mold P2, P3, P4, P5, P6 | 100-150 | A | 90 | 400 |
| P20, P21 | 150-200 | A | 80 | 350 |
| High Speed Steel M1, M2, M6, M10, T1, T2, T6 | 200-250 | A | 65 | 225 |
| M3-1, M4, M7, M30, M33, M34, M36, M41, M42, M43, M44, M46, M47, T5, T8 | 225-275 | A | 55 | 200 |
| T15, M3-2 | 225-275 | A | 45 | 170 |

Based on a feed rate of .012 in. per rev. and a depth of cut of .125 in.
* Abbreviations designate: A, annealed; Q and T, quenched and tempered; HB, Brinell hardness number; and HRC, Rockwell C scale hardness number.

**Cutting Speed for Tapping.** — A table of cutting speeds for tapping is not given. Several factors, singly or in combination, can cause very great differences in the permissible tapping speed. The principal factors affecting the tapping speed are the pitch of the thread, the chamfer length on the tap, the percentage of full thread to be cut, the length of the hole to be tapped, the cutting fluid used, whether the threads are straight or tapered, the machine tool used to perform the operation, and the material to be tapped.

The cutting speed for coarse pitch taps must be slower than for fine pitch taps

Table 3. Recommended Cutting Speeds in Feet per Minute for Turning
Stainless Steels

| Material Stainless Steels | Hardness, HB* | Material Condition* | Cutting Speed, fpm | |
|---|---|---|---|---|
| | | | HSS | Carbide |
| Free Machining Stainless Steels | | | | |
| (Ferritic), 430F, 430F Se | 135-185 | A | 110 | 400 |
| (Austenitic), 203EZ, 303, 303Se, 303MA, 303Pb, 303Cu, 303 Plus X | 135-185 | A | 100 | 350 |
| | 225-275 | CD | 80 | 325 |
| (Martensitic); 416, 416Se, 416 Plus X, 420F, 420F Se, 440F, 440F Se | 135-185 | A | 110 | 400 |
| | 185-240 | A, CD | 100 | 350 |
| | 275-325 | Q and T | 60 | 250 |
| | 375-425 | Q and T | 30 | 125 |
| Stainless Steels | | | | |
| (Ferritic), 405, 409, 429, 430, 434, 436, 442, 446, 502 | 135-185 | A | 90 | 300 |
| (Austenitic), 201, 202, 301, 302, 304, 304L, 305, 308, 321, 347, 348 | 135-185 | A | 75 | 225 |
| | 225-275 | CD | 65 | 200 |
| (Austenitic), 302B, 309, 309S, 310, 310S, 314, 316, 316L, 317, 330 | 135-185 | A | 70 | 225 |
| (Martensitic), 403, 410, 420, 501 | 135-175 | A | 95 | 350 |
| | 175-225 | A | 85 | 300 |
| | 275-325 | Q and T | 55 | 200 |
| | 375-425 | Q and T | 35 | 125 |
| (Martensitic), 414, 431, Greek Ascoloy | 225-275 | A | 60 | 250 |
| | 275-325 | Q and T | 50 | 200 |
| | 375-425 | Q and T | 30 | 125 |
| (Martensitic), 440A, 440B, 440C | 225-275 | A | 55 | 200 |
| | 275-325 | Q and T | 45 | 150 |
| | 375-425 | Q and T | 30 | 125 |
| (Precipitation Hardening) 15-5PH, 17-4PH, 17-7PH, AF-71, 17-14CuMo, AFC-77, AM-350, AM-355, AM-362, Custom 455, HNM, PH13-8, PH14-8Mo, PH15-7Mo, Stainless W | 150-200 | A | 60 | 225 |
| | 275-325 | H | 50 | 200 |
| | 325-375 | H | 40 | 130 |
| | 375-450 | H | 25 | 90 |

Based on a feed rate of .012 in. per rev. and a depth of cut of .125 in.
* Abbreviations designate: A, annealed; CD, cold drawn; Q and T, quenched and tempered; H, precipitation hardened; and HB, Brinell hardness number.

with the same diameter. Usually the difference in pitch becomes more pronounced as the diameter of the tap becomes larger and slight differences in the pitch of smaller diameter taps have little significant effect on the cutting speed. Unlike all other cutting tools, the feed per revolution of a tap cannot be independently adjusted—it is always equal to the lead of the thread and is always greater for coarse pitches than for fine pitches. Furthermore, the thread form of a coarse pitch thread is larger than that of a fine pitch thread; therefore, it is necessary to remove more metal when cutting a coarse pitch thread.

Taps with a long chamfer, such as starting or taper taps, can cut faster in a short hole than short chamfer taps, such as plug taps. In deep holes, however, short chamfer or plug taps can run faster than long chamfer taps. Bottoming taps must be run more slowly than either starting or plug taps. The chamfer helps to start the tap in the hole. It also functions to involve more threads, or thread form cutting edges, on the tap in cutting the thread in the hole. This reduces the cutting load on

**Table 8. Recommended Cutting Speeds in Feet per Minute for Turning, Milling, and Drilling Titanium and Titanium Alloys**

| Material Titanium and Titanium Alloys | Hardness, HB* | Material Condition* | Cutting Speed, fpm HSS | Carbide |
|---|---|---|---|---|
| **Commercially Pure** | | | | |
| 99.5 Ti | 110-150 | A | 110 | 400 |
| 99.1 Ti, 99.2 Ti | 180-240 | A | 90 | 300 |
| 99.0% Ti | 250-275 | A | 70 | 250 |
| **Low Alloyed** | | | | |
| 99.5 Ti-.15 Pd | 110-150 | A | 100 | 350 |
| 99.2 Ti-.15 Pd, 98.9 Ti-.8 Ni-.3 Mo | 180-250 | A | 85 | 280 |
| **Alpha Alloys and Alpha-Beta Alloys** | | | | |
| 5 Al-2.5 Sn, 8 Mn, 2 Al-11 Sn-5 Zr-1 Mo, 4 Al-3 Mo-1 V, 5 Al-6 Sn-2 Zr-1 Mo, 6 Al-2 Sn-4 Zr-2 Mo, 6 Al-2 Sn-4 Zr-6 Mo, 6 Al-2 Sn-4 Zr-2 Mo-.25 Si | 300-350 | A | 50 | 200 |
| 6 Al-4 V | 310-350 | A | 40 | 125 |
| 6 Al-6 V-2 Sn, 7 Al-4 Mo, 8 Al-1 Mo-1 V | 320-370 | A | 30 | 100 |
| 8 V-5 Fe-1 Al | 320-380 | A | 20 | 90 |
| 6 Al-4 V, 6 Al-2 Sn-4 Zr-2 Mo, 6 Al-2 Sn-4 Zr-6 Mo, 6 Al-2 Sn-4 Zr-2 Mo-.25 Sn | 320-380 | ST and A | 40 | 100 |
| 4 Al-3 Mo-1 V, 6 Al-6 V-2 Sn, 7 Al-4 Mo | 375-420 | ST and A | 20 | 80 |
| 1 Al-8 V-5 Fe | 375-440 | ST and A | 20 | 75 |
| **Beta Alloys** | | | | |
| 13 V-11 Cr-3 Al, 8 Mo-8 V-2 Fe-3 Al, 3 Al-8 V-6 Cr-4 Mo-4 Zr, 11.5 Mo-6 Zr-4.5 Sn | 275-350 | A, ST | 25 | 100 |
| | 350-440 | ST and A | 20 | 60 |

* Abbreviations designate: A, annealed; ST, solution treated; ST and A, solution treated and aged; and HB, Brinell hardness number.

cost and are expensive to sharpen. For these reasons a long tool life is desirable, and to obtain a long tool life relatively slow cutting speeds are used. In many instances slower cutting speeds are used because of the limitations of the machine in accelerating and stopping heavy broaching cutters. At other times the available power on the machine places a limit on the cutting speed that can be used; i.e., the cubic inches of metal removed per minute must be within the power capacity of the machine.

The cutting speeds for high speed steel broaches range from 3 to 50 feet per minute, although faster speeds have been used. In general, the harder and more difficult to machine materials are cut at a slower cutting speed and those that are easier to machine are cut at a faster speed. Some typical recommendations for high speed steel broaches are: AISI 1040, 10 to 30 fpm; AISI 1060, 10 to 25 fpm; AISI 4140, 10 to 25 fpm; AISI 41L40, 20 to 30 fpm; 201 austenitic stainless steel, 10 to 20 fpm; Class 20 gray cast iron, 20 to 30 fpm; Class 40 gray cast iron, 15 to 25 fpm; aluminum and magnesium alloys, 30 to 50 fpm; copper alloys, 20 to 30 fpm; commercially pure titanium, 20 to 25 fpm; alpha and beta titanium alloys, 5 fpm; and, the superalloys, 3 to 10 fpm. Surface broaching operations on gray iron castings have been conducted at a cutting speed of 150 fpm, using indexable insert cemented carbide broaching cutters. In selecting the speed for broaching, the cardinal principle of the performance of all metal cutting tools should be kept in mind;

**Table 9.   Recommended Cutting Speeds in Feet per Minute for Turning, Milling, and Drilling\* Superalloys**

| Material | Cutting Speed, fpm | | | |
|---|---|---|---|---|
| | Roughing | | Finishing | |
| | HSS | Carbide | HSS | Carbide |
| A-286 | 30-35 | 120-145 | 35-40 | 145-155 |
| AF2-1DA | 8-10 | 35-45 | 10-15 | 40-50 |
| Air Resist 213 | 15-20 | 55-65 | 20-25 | 70-85 |
| Air Resist 13, and 215 | 10-12 | 35-40 | 10-15 | 45-55 |
| Astroloy | 5-10 | 25-50 | 5-15 | 50-75 |
| B-1900 | 8-10 | 30-35 | 8-10 | 35-50 |
| CW-12M | 8-12 | 55-65 | 10-15 | 65-85 |
| Discalloy | 15-35 | 100-150 | 35-40 | 140-180 |
| FSX-H14 | 10-12 | 35-40 | 10-15 | 45-55 |
| GMR-235, and 235D | 8-10 | 30-35 | 8-10 | 40-50 |
| Hastelloy B, C, G, and X (wrought) | 15-20 | 60-90 | 20-25 | 80-100 |
| Hastelloy B, and C (cast) | 8-12 | 55-65 | 10-15 | 75-85 |
| Haynes 25, and 188 | 15-20 | 55-65 | 20-25 | 70-95 |
| Haynes 36, and 151 | 10-12 | 35-40 | 10-15 | 45-55 |
| HS 6, 21, 25, 31(X40), 36, and 151 | 10-12 | 35-40 | 10-15 | 45-55 |
| IN 100, and 738 | 8-10 | 30-35 | 8-10 | 35-50 |
| Incoloy 800, 801, and 802 | 30-35 | 120-160 | 35-40 | 145-180 |
| Incoloy 804, and 825 | 15-20 | 60-90 | 20-25 | 80-100 |
| Incoloy 901 | 10-20 | 30-60 | 20-35 | 40-80 |
| Inconel 625, 702, 706, 718 (wrought), 721, 722, X750, 751, 901, 600, and 604 | 15-20 | 35-60 | 20-25 | 60-90 |
| Inconel 700, and 702 | 10-12 | 40-65 | 12-15 | 65-70 |
| Inconel 713C, and 718 (cast) | 8-10 | 30-35 | 8-10 | 40-50 |
| J1300 | 15-25 | 80-100 | 20-30 | 100-125 |
| J1570 | 15-20 | 55-65 | 20-25 | 70-85 |
| M252 (wrought) | 15-20 | 65-75 | 20-25 | 75-85 |
| M252 (cast) | 8-10 | 30-35 | 8-10 | 40-50 |
| Mar-M200, M246, M421, and M432 | 8-10 | 30-35 | 10-12 | 35-50 |
| Mar-M905, and M918 | 15-20 | 55-65 | 20-25 | 70-85 |
| Mar-M302, M322, and M509 | 10-12 | 35-40 | 10-15 | 45-55 |
| N-12M | 8-12 | 55-65 | 10-15 | 75-85 |
| N-155 | 15-20 | 50-70 | 15-25 | 55-75 |
| Nasa Co-W-Re | 10-12 | 35-40 | 10-15 | 45-55 |
| Nimonic 75, and 80 | 15-20 | 65-75 | 20-25 | 75-85 |
| Nimonic 90 and 95 | 10-12 | 55-65 | 12-15 | 65-75 |
| Refractaloy 26 | 15-20 | 60-90 | 20-25 | 80-100 |
| René 41 | 10-15 | 35-60 | 12-20 | 55-80 |
| René 80, and 95 | 8-10 | 30-45 | 10-15 | 40-50 |
| S-590 | 10-20 | 60-90 | 15-30 | 80-100 |
| S-816 | 10-15 | 45-65 | 15-20 | 50-75 |
| TD-Nickel | 70-80 | 250-290 | 80-100 | 300-350 |
| Udimet 500, 700, and 710 | 10-15 | 30-50 | 12-20 | 40-60 |
| Udimet 630 | 10-20 | 30-80 | 20-25 | 80-100 |
| Unitemp 1753 | 8-10 | 35-45 | 10-15 | 40-50 |
| V-36 | 10-15 | 45-65 | 15-20 | 50-75 |
| V-57 | 30-35 | 120-160 | 35-40 | 145-180 |
| W-545 | 25-35 | 110-155 | 30-40 | 140-175 |
| WI-52 | 10-12 | 35-40 | 10-15 | 45-55 |
| Waspaloy | 10-30 | 30-60 | 25-35 | 50-95 |
| X-45 | 10-12 | 35-40 | 10-15 | 45-55 |
| 16-25-6 | 30-35 | 120-160 | 35-40 | 145-180 |
| 19-9DL | 25-35 | 110-150 | 30-40 | 140-180 |

\* For milling and drilling, use the cutting speeds recommended under roughing.

i.e., increasing the cutting speed results in a proportionately larger reduction in tool life, and conversely, reducing the cutting speed results in a proportionately larger increase in the tool life. When broaching most materials, a suitable cutting fluid should be used to obtain a good surface finish and a better tool life. Gray cast iron can be broached without using a cutting fluid although some shops prefer to use a soluble oil.

Table 10.   Representative Cutting Conditions for Rough Turning Hard-to-Machine
Materials with Single Point Cubic Boron Nitride (CBN) Cutting Tools.

| Material | Hardness, HRC | Cutting Speed, fpm | Feed, in./rev. | Depth of Cut, in. |
|---|---|---|---|---|
| Hardened Ferrous Alloys | | | | |
| AISI 8620 | 63 | 250 | .005 | .060 |
| AISI 52100 | 70 | 270 | .020 | .050 |
| A2, A6, Cold Work Tool Steel | 58 | 250 | .008 | .060 |
| D2, Cold Work Tool Steel | 54 | 250 | .008 | .060 |
| H10, Hot Work Tool Steel | 56 | 200 | .005 | .060 |
| S5, Shock Resisting Tool Steel | 60 | 400 | .008 | .060 |
| O1, Oil Hardening Tool Steel | 58 | 250 | .008 | .060 |
| M2, High Speed Steel | 62 | 250 | .008 | .060 |
| Chilled Gray Iron | 60 | 400 | .010 | .200 |
| Mechanite Cast Iron | 56 | 600 | .008 | .250 |
| Superalloys | | | | |
| Colmonoy | . . . | 600 | .006 | .125 |
| Incoloy 901 | . . . | 800 | .006 | .125 |
| Inconel 600 | . . . | 600 | .006 | .125 |
| Inconel 718 | . . . | 600 | .006 | .125 |
| K-Monel | . . . | 600 | .006 | .125 |
| René 41 | . . . | 600 | .006 | .125 |
| René 77 | . . . | 500 | .006 | .015 |
| René 95 (Hot Isostatic Pressed) | . . . | 900 | .005 | .125 |
| René 95 (Forged) | . . . | 450 | .005 | .125 |
| Stellite | . . . | 600 | .006 | .125 |
| Waspaloy | . . . | 600 | .003 | .060 |

**Thread Cutting with Single Point Cutting Tools.** — Whenever possible the cutting speed recommended for turning should be used to cut internal and external threads with single point thread cutting tools. This cutting speed can frequently be used on numerically controlled lathes, using either cemented carbide or high speed steel tools. There are occasions, however, when a slower cutting speed must be used as a result of the workpiece configuration, the setup, or when cutting certain difficult-to-machine threads, such as coarse pitch Acme threads. A slightly reduced cutting speed is sometimes used on numerically controlled lathes to obtain a longer tool life.

Thread cutting on an engine lathe is not necessarily a slow speed operation, although on these machines the operation is controlled manually. However, there must never be a compromise with safety; the operator must always be sure that he has control over the machine to the extent that he or others will not be injured, and that the machine or the workpiece will not be damaged. On some jobs a skilled operator can safely manipulate a lathe with such skill that a fast spindle speed can be used to cut the thread, in which case the cutting speed may be equal to that recommended for turning with high speed steel, or with cemented carbide in the case of the more difficult-to-machine materials. Other jobs require using a slower cutting speed, even when the thread cutting operation is performed by a highly skilled operator. Some of the reasons for cutting threads at a slower speed have been given in the previous paragraph. Other reasons involve the ability to safely manipulate the machine such as when cutting a thread close to a large shoulder, or when cutting an internal thread with the cutting tool feeding into the bore.

**Cutting Speed for Thread Chasing.** — Cutting threads with a self-opening die head is called thread chasing. The die head contains a set of thread chasers that cut the thread and feed the die head in a nut-and-screw-like action. The feed rate is determined entirely by the lead or pitch of the thread. Since the feed of the die head must not be

## Tool Trouble-Shooting Check List

| Problem | Tool Material | Remedy |
|---|---|---|
| Excessive flank wear — Tool life too short | Carbide | 1. Change to harder, more wear-resistant grade<br>2. Reduce the cutting speed<br>3. Reduce the cutting speed and increase the feed to maintain production<br>4. Reduce the feed<br>5. For work hardenable materials — increase the feed<br>6. Increase the lead angle<br>7. Increase the relief angles |
| | HSS | 1. Use a coolant<br>2. Reduce the cutting speed<br>3. Reduce the cutting speed and increase the feed to maintain production<br>4. Reduce the feed<br>5. For work hardenable materials — increase the feed<br>6. Increase the lead angle<br>7. Increase the relief angle |
| Excessive cratering | Carbide | 1. Use a crater-resistant grade<br>2. Use a harder, more wear-resistant grade<br>3. Reduce the cutting speed<br>4. Reduce the feed<br>5. Widen the chip breaker groove |
| | HSS | 1. Use a coolant<br>2. Reduce the cutting speed<br>3. Reduce the feed<br>4. Widen the chip breaker groove |
| Cutting edge chipping | Carbide | 1. Increase the cutting speed<br>2. Lightly hone the cutting edge<br>3. Change to a tougher grade<br>4. Use negative rake tools<br>5. Increase the lead angle<br>6. Reduce the feed<br>7. Reduce the depth of cut<br>8. Reduce the relief angles<br>9. If low cutting speed must be used — use a high additive EP cutting fluid |
| | HSS | 1. Use a high additive EP cutting fluid<br>2. Lightly hone the cutting edge before using<br>3. Increase the lead angle<br>4. Reduce the feed<br>5. Reduce the depth of cut<br>6. Use a negative rake angle<br>7. Reduce the relief angles |
| | Carbide and HSS | 1. Check the setup for cause if chatter occurs<br>2. Check the grinding procedure for tool over-heating<br>3. Reduce the tool overhang |
| Cutting edge deformation | Carbide | 1. Change to a grade containing more tantalum<br>2. Reduce the cutting speed<br>3. Reduce the feed |

SPEEDS AND FEEDS

**Tool Trouble-Shooting Check List** (Continued).

| Problem | Tool Material | Remedy |
|---------|---------------|--------|
| Poor surface finish | Carbide | 1. Increase the cutting speed<br>2. If low cutting speed must be used — use a high additive EP cutting fluid<br>3. For light cuts — use straight titanium carbide grade<br>4. Increase the nose radius<br>5. Reduce the feed<br>6. Increase the relief angles<br>7. Use positive rake tools |
| | HSS | 1. Use high additive EP cutting fluid<br>2. Increase the nose radius<br>3. Reduce the feed<br>4. Increase the relief angles<br>5. Increase the rake angles |
| | Diamond | 1. Use diamond tool for soft materials |
| Notching at the depth of cut line | Carbide and HSS | 1. Increase the lead angle<br>2. Reduce the feed |

too rapid, the thread lead and pitch place a limit on the spindle speed, and thereby on the cutting speed. Other factors affecting the cutting speed are the work material, the type and size of the thread, the thread tolerance, and the finish required. A cutting fluid should be used in most cases, which may also have an effect on the cutting speed. Much slower cutting speeds are recommended for thread chasing, as compared to turning. A cutting speed that is too fast will reduce the life of the thread chasers and may cause the threads cut to be rough or torn. Some typical cutting speeds recommended by one manufacturer of self-opening die heads are given below. These cutting speeds may have to be modified somewhat to suit existing conditions on each job.

| Material | Threads per Inch | | | |
|----------|-----|-----|-----|-----|
| | 3-7½ | 8-15 | 16-24 | 25-Up |
| | Cutting Speed (fpm) for Threads per Inch | | | |
| AISI 1010-1035 Steel | 20 | 30 | 40 | 50 |
| AISI 1112-1340 Steel | 20 | 30 | 40 | 50 |
| AISI 1040-1095 Steel | 15 | 20 | 25 | 30 |
| AISI 4130-4820 Steel | 8 | 10 | 15 | 20 |
| AISI 5120-52100 Steel | 8 | 10 | 15 | 20 |
| Stainless Steel | 8 | 10 | 15 | 20 |
| Gray Cast Iron | 25 | 40 | 50 | 80 |
| Aluminum Alloys | 50 | 100 | 150 | 200 |
| Brass Bar Stock | 50 | 100 | 150 | 200 |
| Phosphor Bronze | 40 | 80 | 100 | 150 |
| Zinc Die Castings | 50 | 100 | 150 | 200 |

**Feed Rate for Milling.** — Whenever the power feed is to be used to perform a milling operation, the table feed rate, in inches per minute, should always be calculated in order to achieve the best results. The table feed rate governs the production rate. Failure to calculate the table feed rate may result in overloading the milling cutter which can have serious consequences. Aside from the possible

### Table 11. Cutting Speeds in Feet per Minute for Milling Plain Carbon and Alloy Steels

| Material AISI and SAE Steels | Hardness, HB* | Material Condition* | Cutting Speed, fpm HSS | Cutting Speed, fpm Carbide |
|---|---|---|---|---|
| Free Machining Plain Carbon Steels (Resulphurized), 1212, 1213, 1215 | 100-150 | HR, A | 140 | 600 |
| | 150-200 | CD | 130 | 550 |
| 1108, 1109, 1115, 1117, 1118, 11120, 1126, 1211 | 100-150 | HR, A | 130 | 550 |
| | 150-200 | CD | 115 | 500 |
| 1132, 1137, 1139, 1140, 1144, 1146 1151 | 175-225 | HR, A, N, CD | 115 | 450 |
| | 275-325 | Q and T | 70 | 290 |
| | 325-375 | Q and T | 45 | 200 |
| | 375-425 | Q and T | 35 | 170 |
| Free Machining Plain Carbon Steels (Leaded), 11L17, 11L18, 12L13, 12L14 | 100-150 | HR, A, N, CD | 140 | 600 |
| | 150-200 | HR, A, N, CD | 130 | 625 |
| | 200-250 | N, CD | 110 | 400 |
| Plain Carbon Steels, 1006, 1008, 1009, 1010, 1012, 1015, 1016, 1017, 1018, 1019, 1020, 1021, 1022, 1023, 1024, 1025, 1026, 1513, 1514 | 100-125 | HR, A, N, CD | 110 | 425 |
| | 125-175 | HR, A, N, CD | 110 | 400 |
| | 175-225 | HR, N, CD | 90 | 350 |
| | 225-275 | CD | 65 | 250 |
| 1027, 1030, 1033, 1035, 1036, 1037, 1038, 1039, 1040, 1041, 1042, 1043, 1045, 1046, 1048, 1049, 1050, 1052, 1524, 1526, 1527, 1541 | 125-175 | HR, A, N, CD | 100 | 375 |
| | 175-225 | HR, A, N, CD | 85 | 325 |
| | 225-275 | N, CD, Q and T | 70 | 225 |
| | 275-325 | Q and T | 55 | 200 |
| | 325-375 | Q and T | 35 | 160 |
| | 375-425 | Q and T | 25 | 140 |
| 1055, 1060, 1064, 1065, 1070, 1074, 1078, 1080, 1084, 1086, 1090, 1095, 1548, 1551, 1552, 1561, 1566 | 125-175 | HR, A, N, CD | 90 | 350 |
| | 175-225 | HR, A, N, CD | 75 | 300 |
| | 225-275 | N, CD, Q and T | 60 | 200 |
| | 275-325 | Q and T | 45 | 160 |
| | 325-375 | Q and T | 30 | 145 |
| | 375-425 | Q and T | 15 | 125 |
| Free Machining Alloy Steels (Resulphurized), 4140, 4150 | 175-200 | HR, A, N, CD | 100 | 400 |
| | 200-250 | HR, N, CD | 90 | 350 |
| | 250-300 | Q and T | 60 | 280 |
| | 300-375 | Q and T | 45 | 220 |
| | 375-425 | Q and T | 35 | 160 |
| Free Machining Alloy Steels (Leaded), 41L30, 41L40, 41L47, 41L50, 43L47, 51L32, 52L100, 86L20, 86L40 | 150-200 | HR, A, N, CD | 115 | 425 |
| | 200-250 | HR, N, CD | 95 | 375 |
| | 250-300 | Q and T | 70 | 260 |
| | 300-375 | Q and T | 50 | 210 |
| | 375-425 | Q and T | 40 | 180 |
| Alloy Steels, 4012, 4023, 4024, 4028, 4118, 4320, 4419, 4422, 4427, 4615, 4620, 4621, 4626, 4718, 4720, 4815, 4817, 4820, 5015, 5117, 5120, 6118, 8115, 8615, 8617, 8620, 8622, 8625, 8627, 8720, 8822, 94B17 | 125-175 | HR, A, N, CD | 100 | 400 |
| | 175-225 | HR, N, CD | 90 | 350 |
| | 225-275 | CD, N, Q and T | 60 | 250 |
| | 275-325 | Q and T | 50 | 200 |
| | 325-375 | Q and T | 40 | 175 |
| | 375-425 | Q and T | 25 | 150 |

* Abbreviations designate: HR, hot rolled; CD, cold drawn; A, annealed; N, normalized; Q and T, quenched and tempered; HB, Brinell hardness number; and HRC, Rockwell C scale hardness number.

Table 11 (*Concluded*).  Cutting Speeds in Feet per Minute for Milling
Plain Carbon and Alloy Steels

| Material AISI and SAE Steels | Hardness, HB* | Material Condition* | Cutting Speed, fpm | |
|---|---|---|---|---|
| | | | HSS | Carbide |
| Alloy Steels, 1330, 1335, 1340, 1345, 4032, 4037, 4042, 4047, 4130, 4135, 4137, 4140, 4142, 4145, 4147, 4150, 4161, 4337, 4340, 50B44, 50B46, 50B50, 50B60, 5130, 5132, 5140, 5145, 5147, 5150, 5160, 51B60, 6150, 81B45, 8630, 8635, 8637, 8640, 8642, 8645, 8650, 8655, 8660, 8740, 9254, 9255, 9260, 9262, 94B30 | 175-225 225-275 275-325 325-375 375-425 | HR, A, N, CD N, CD, Q and T N, Q and T N, Q and T Q and T | 75 60 50 35 20 | 310 260 210 180 140 |
| Alloy Steels, E51100, E52100 | 175-225 225-275 275-325 325-375 375-425 | HR, A, CD N, CD, Q and T N, Q and T N, Q and T Q and T | 65 60 40 30 20 | 300 250 130 100 60 |
| Ultra High Strength Steels (Not AISI) AMS 6421 (98B37 Mod.), AMS 6422 (98BV40), AMS 6424, AMS 6427, AMS 6428, AMS 6430, AMS 6432, AMS 6433, AMS 6434, AMS 6436, AMS 6442, 300M, D6 ac | 220-300 300-350 350-400 43-48 HRC 48-52 HRC | A N N Q and T Q and T | 60 45 20 .. .. | 250 180 130 100 60 |
| Maraging Steels (Not AISI) 18% Ni Grade 200 18% Ni Grade 250 18% Ni Grade 300 18% Ni Grade 350 | 250-325 50-52 HRC | A Maraged | 50 .. | 250 60 |
| Nitriding Steels (Not AISI) Nitralloy 125 Nitralloy 135 Nitralloy 135 (Mod.) Nitralloy 225 Nitralloy 230 Nitralloy N Nitralloy EZ Nitrex 1 | 200-250 300-350 | A N, Q and T | 60 25 | 280 200 |

* Abbreviations designate: HR, hot rolled; CD, cold drawn; A, annealed; N, normalized; Q and T, quenched and tempered; HB, Brinell hardness number; and HRC, Rockwell C scale hardness number.

breakage of equipment, overloading the cutter will cause the tool life of the cutter to decrease. The tool life can also be decreased if the feed rate is too slow, which, in addition, certainly leads to a loss of production.

The basic feed rate for milling cutters is the feed per tooth ($f$) which is expressed in inches per tooth. There are many factors to consider in selecting the feed per tooth and no formula is available to resolve these factors. Among the factors to consider are: 1) the cutting tool material; 2) the work material and its hardness; 3) the width and the depth of the cut to be taken; 4) the type of milling cutter to be used and its size; 5) the surface finish to be produced; 6) the power available on the milling machine; 7) the rigidity of the milling machine, the workpiece, the setup of the workpiece, the milling cutter, and the cutter mounting.

As a guide to help in the selection of the feed rate, two tables are given; Table 15 is for high-speed steel cutters, and Table 16 is for cemented carbide cutters.

As a cardinal principle, always use the maximum feed rate that conditions will

Table 12. **Cutting Speed in Feet per Minute for Milling Tool Steels**

| Material Tool Steels (AISI Types) | Hardness, HB* | Material Condition* | Cutting Speed, fpm | |
|---|---|---|---|---|
| | | | HSS | Carbide |
| Water Hardening<br>W1, W2, W5 | 150-200 | A | 85 | 250 |
| Shock Resisting<br>S1, S2, S5, S6, S7 | 175-225 | A | 55 | 215 |
| Cold Work, Oil Hardening<br>O1, O2, O6, O7 | 175-225 | A | 50 | 200 |
| Cold Work, High Carbon High Chromium<br>D2, D3, D4, D5, D7 | 200-250 | A | 40 | 150 |
| Cold Work, Air Hardening | | | | |
| A2, A3, A8, A9, A10 | 200-250 | A | 50 | 200 |
| A4, A6 | 200-250 | A | 45 | 160 |
| A7 | 225-275 | A | 40 | 140 |
| Hot Work, Chromium Type<br>H10, H11, H12, H13, H14, H19 | 150-200 | A | 60 | 250 |
| | 200-250 | A | 50 | 200 |
| | 325-375 | Q and T | 30 | 150 |
| | 48-50 HRC | Q and T | — | 80 |
| | 50-52 HRC | Q and T | — | 60 |
| | 52-54 HRC | Q and T | — | 40 |
| | 54-56 HRC | Q and T | — | 20 |
| Hot Work, Tungsten Type<br>H21, H22, H23, H24, H25, H26 | 150-200 | A | 55 | 200 |
| | 200-250 | A | 45 | 170 |
| Hot Work, Molybdenum Type<br>H41, H42, H43 | 150-200 | A | 55 | 180 |
| | 200-250 | A | 45 | 140 |
| Special Purpose, Low Alloy<br>L2, L3, L6 | 150-200 | A | 65 | 300 |
| Mold | | | | |
| P2, P3, P4, P5, P6 | 100-150 | A | 75 | 350 |
| P20, P21 | 150-200 | A | 60 | 300 |
| High-Speed Steel | | | | |
| M1, M2, M6, M10, T1, T2, T6 | 200-250 | A | 50 | 175 |
| M3-1, M4, M7, M30, M33, M34,<br>M36, M41, M42, M43, M44, M46,<br>M47, T5, T8 | 225-275 | A | 40 | 150 |
| T15, M3-2 | 225-275 | A | 30 | 130 |

\* Abbreviations designate: A, annealed; Q and T, quenched and tempered; and HB, Brinell hardness number.

permit. Avoid, if possible, using a feed rate that is less than .001 inch per tooth because this will result in a decrease in the tool life of the cutter. When milling hard materials with small diameter end mills, such small feed rates may be necessary, but otherwise use as much feed as possible. Harder materials in general will require lower feed rates than softer materials. The width and the depth of cut also affect the feed rate; wider and deeper cuts must be fed somewhat more slowly than narrow and shallow cuts. A slower feed rate will result in a better surface finish; however, always use the heaviest feed rate that will produce the surface finish

**Table 13.  Recommended Cutting Speeds in Feet per Minute for Milling
Stainless Steels**

| Material Stainless Steels | Hardness, HB* | Material Condition* | Cutting Speed, fpm | |
|---|---|---|---|---|
| | | | HSS | Carbide |
| Free Machining Stainless Steels | | | | |
| (Ferritic), 430F, 430F Se | 135-185 | A | 95 | 375 |
| (Austenitic), 203EZ, 303, 303 Se, 303MA, 303Pb, 303Cu, 303 Plus X | 135-185 | A | 90 | 325 |
| | 225-275 | CD | 75 | 300 |
| (Martensitic), 416, 416 Se, 416 Plus X, 420F, 420F Se, 440F, 440F Se | 135-185 | A | 95 | 375 |
| | 185-240 | CD | 80 | 325 |
| | 275-325 | Q and T | 50 | 225 |
| | 375-425 | Q and T | 20 | 100 |
| Stainless Steels | | | | |
| (Ferritic), 405, 409, 429, 430, 434, 436, 442, 446, 502 | 135-185 | A | 75 | 275 |
| (Austenitic), 201, 202, 301, 302, 304, 304L, 305, 308, 321, 347, 348 | 135-185 | A | 60 | 200 |
| | 225-275 | CD | 50 | 180 |
| (Austenitic), 302B, 309, 309S, 310, 310S, 314, 316, 316L, 317, 330 | 135-185 | A | 50 | 200 |
| (Martensitic), 403, 410, 420, 501 | 135-175 | A | 75 | 325 |
| | 175-225 | A | 65 | 275 |
| | 275-325 | Q and T | 40 | 175 |
| | 375-425 | Q and T | 25 | 100 |
| (Martensitic), 414, 431, Greek Ascoloy | 225-275 | A | 55 | 225 |
| | 275-325 | Q and T | 45 | 180 |
| | 375-425 | Q and T | 25 | 100 |
| (Martensitic), 440A, 440B, 440C | 225-275 | A | 50 | 180 |
| | 275-325 | Q and T | 40 | 140 |
| | 375-425 | Q and T | 20 | 100 |
| (Precipitation Hardening) 15-5PH, 17-4PH, 17-7PH, AF-71, 17-14Cu Mo, AFC-77, AM-350, AM-355, AM-362, Custom 455, HNM, PH13-8, PH14-8Mo, PH15-7Mo, Stainless W | 150-200 | A | 60 | 200 |
| | 275-325 | H | 50 | 180 |
| | 325-375 | H | 40 | 110 |
| | 375-450 | H | 25 | 75 |

* Abbreviations designate: A, annealed; CD, cold drawn; Q and T, quenched and tempered; H, precipitation hardened; and HB, Brinell hardness number.

desired. Fine chips produced by fine feeds are dangerous when milling magnesium because spontaneous combustion can occur. Thus, when milling magnesium, a fast feed that will produce a relatively thick chip should be used. Cutting stainless steel produces a work hardened layer on the surface that has been cut. When milling this material, the feed should be large enough to allow each cutting edge on the cutter to penetrate below the work hardened layer produced by the previous cutting edge. The heavy feeds recommended for face milling cutters are to be used primarily by larger cutters on milling machines having an adequate amount of power. For smaller face milling cutters, start with the slower feeds and increase the feed as indicated by the performance of the cutter and the machine.

When planning a milling operation that is to entail the use of a high cutting speed and a fast feed rate, always check to determine if the power required to take the cut is within the capacity of the milling machine. Such cutting conditions are often encountered when milling with cemented carbide cutters. The large metal removal

(Continued on page 1780)

**Table 14.   Recommended Cutting Speeds in Feet per Minute for Milling Ferrous Cast Metals**

| Material Ferrous Cast Metals | Hardness, HB* | Material Condition* | Cutting Speed, fpm | |
|---|---|---|---|---|
| | | | HSS | Carbide |
| Gray Cast Iron | | | | |
| ASTM Class 20 | 120-150 | A | 100 | 425 |
| ASTM Class 25 | 160-200 | AC | 80 | 325 |
| ASTM Class 30, 35, and 40 | 190-220 | AC | 70 | 250 |
| ASTM Class 45 and 50 | 220-260 | AC | 50 | 190 |
| ASTM Class 55 and 60 | 250-260 | AC, HT | 30 | 110 |
| ASTM Type 1, 1b, 5 (Ni-Resist) | 100-215 | AC | 50 | 200 |
| ASTM Type 2, 3, 6 (Ni-Resist) | 120-175 | AC | 40 | 190 |
| ASTM Type 2b, 4 (Ni-Resist) | 150-250 | AC | 30 | 180 |
| Malleable Iron | | | | |
| (Ferritic), 32510, 35018 | 110-160 | MHT | 110 | 475 |
| (Pearlitic), 40010, 43010, 45006, 45008, 48005, 50005 | 160-200 | MHT | 80 | 375 |
| | 200-240 | MHT | 65 | 250 |
| (Martensitic), 53004, 60003, 60004 | 200-255 | MHT | 55 | 225 |
| (Martensitic), 70002, 70003 | 220-260 | MHT | 50 | 200 |
| (Martensitic), 80002 | 240-280 | MHT | 45 | 130 |
| (Martensitic), 90001 | 250-320 | MHT | 25 | 110 |
| Nodular (Ductile) Iron | | | | |
| (Ferritic), 60-40-18, 65-45-12 | 140-190 | A | 75 | 425 |
| (Ferritic-Pearlitic), 80-55-06 | 190-225 | AC | 60 | 325 |
| | 225-260 | AC | 50 | 200 |
| (Pearlitic-Martensitic), 100-70-03 | 240-300 | HT | 40 | 160 |
| (Martensitic), 120-90-02 | 270-330 | HT | 25 | 90 |
| | 330-400 | HT | — | 30 |
| Cast Steels | | | | |
| (Low Carbon), 1010, 1020 | 100-150 | AC, A, N | 100 | 375 |
| (Medium Carbon), 1030, 1040, 1050 | 125-175 | AC, A, N | 95 | 375 |
| | 175-225 | AC, A, N | 80 | 325 |
| | 225-300 | AC, HT | 60 | 250 |
| (Low Carbon Alloy), 1320, 2315, 2320, 4110, 4120, 4320, 8020, 8620 | 150-200 | AC, A, N | 85 | 325 |
| | 200-250 | AC, A, N | 75 | 300 |
| | 250-300 | AC, HT | 50 | 225 |
| (Medium Carbon Alloy), 1330, 1340, 2325, 2330, 4125, 4130, 4140, 4330, 4340, 8030, 80B30, 8040, 8430, 8440, 8630, 8640, 9525, 9530, 9535 | 175-225 | AC, A, N | 70 | 300 |
| | 225-250 | AC, A, N | 65 | 250 |
| | 250-300 | AC, HT | 50 | 200 |
| | 300-350 | AC, HT | 30 | 180 |
| | 350-400 | HT | .. | 125 |

\* Abbreviations designate: A, annealed; AC, as cast; N, normalized; HT, heat treated; MHT, malleablizing heat treatment; and HB, Brinell hardness number.

Table 15. Recommended Feed in Inches per Tooth ($f_t$) for Milling with High Speed Steel Cutters

| Material | Hardness, HB | End Mills Depth of Cut, .250 in. Cutter Diam., in. 1/2 | 3/4 | 1 and up | End Mills Depth of Cut, .050 in. Cutter Diam., in. 1/4 | 1/2 | 3/4 | 1 and up | Plain or Slab Mills | Form Relieved Cutters | Face Mills and Shell End Mills | Slotting and Side Mills |
|---|---|---|---|---|---|---|---|---|---|---|---|---|
| | | | | | | | | Feed per Tooth, inch | | | | |
| Free Machining Plain Carbon Steels | 100-185 | .001 | .003 | .004 | .001 | .002 | .003 | .004 | .003-.008 | .005 | .004-.012 | .002-.008 |
| Plain Carbon Steels, AISI 1006 to 1030; 1513 to 1522 | 100-150 | .001 | .003 | .003 | .001 | .002 | .003 | .004 | .003-.008 | .004 | .004-.012 | .002-.008 |
| | 150-200 | .001 | .002 | .003 | .001 | .002 | .002 | .003 | .003-.008 | .004 | .003-.012 | .002-.008 |
| AISI 1033 to 1095; 1524 to 1566 | 120-180 | .001 | .003 | .003 | .001 | .002 | .003 | .004 | .003-.008 | .004 | .004-.012 | .002-.008 |
| | 180-220 | .001 | .002 | .003 | .001 | .002 | .002 | .003 | .003-.008 | .004 | .003-.012 | .002-.008 |
| | 220-300 | .001 | .002 | .002 | .001 | .002 | .002 | .003 | .002-.006 | .003 | .002-.008 | .002-.006 |
| Alloy Steels having less than 3% Carbon. Typical examples: AISI 4012, 4023, 4027, 4118, 4320, 4422, 4427, 4615, 4620, 4626, 4720, 4820, 5015, 5120, 6118, 8115, 8620, 8627, 8720, 8822, 9310, 93B17 | 125-175 | .001 | .003 | .003 | .001 | .002 | .003 | .004 | .003-.008 | .004 | .004-.012 | .002-.008 |
| | 175-225 | .001 | .002 | .003 | .001 | .002 | .003 | .003 | .003-.008 | .004 | .003-.012 | .002-.008 |
| | 225-275 | .001 | .002 | .003 | .001 | .001 | .002 | .003 | .002-.006 | .003 | .003-.008 | .002-.006 |
| | 275-335 | .001 | .002 | .002 | .001 | .001 | .002 | .003 | .002-.005 | .003 | .002-.008 | .002-.005 |
| Alloy Steels have 3% Carbon or more. Typical examples: AISI 1330, 1340, 4032, 4037, 4130, 4140, 4150, 4340, 50B40, 50B60, 5130, 51B60, 6150, 81B45, 8630, 8640, 86B45, 8660, 8740, 94B30 | 175-225 | .001 | .002 | .003 | .001 | .002 | .003 | .004 | .003-.008 | .004 | .003-.012 | .002-.008 |
| | 225-275 | .001 | .002 | .003 | .001 | .001 | .002 | .003 | .002-.006 | .003 | .003-.010 | .002-.006 |
| | 275-325 | .001 | .002 | .002 | .001 | .001 | .002 | .003 | .002-.005 | .003 | .002-.008 | .002-.005 |
| | 325-375 | .001 | .002 | .002 | .001 | .001 | .002 | .002 | .002-.004 | .002 | .002-.008 | .002-.005 |
| Tool Steel | 150-200 | .001 | .002 | .002 | .001 | .002 | .003 | .003 | .003-.008 | .004 | .003-.010 | .002-.006 |
| | 200-250 | .001 | .002 | .002 | .001 | .002 | .002 | .003 | .002-.006 | .003 | .003-.008 | .002-.005 |
| Gray Cast Iron | 120-180 | .001 | .003 | .004 | .002 | .003 | .004 | .004 | .004-.012 | .005 | .005-.016 | .002-.010 |
| | 180-225 | .001 | .002 | .003 | .002 | .002 | .003 | .004 | .003-.010 | .004 | .004-.012 | .002-.008 |
| | 225-300 | .001 | .002 | .002 | .001 | .001 | .002 | .002 | .002-.006 | .003 | .002-.008 | .002-.005 |
| Ferritic Malleable Iron | 110-160 | .001 | .003 | .004 | .002 | .003 | .004 | .004 | .003-.010 | .005 | .005-.016 | .002-.010 |
| Pearlitic-Martensitic Malleable Iron | 160-200 | .001 | .003 | .004 | .001 | .002 | .003 | .004 | .003-.010 | .004 | .004-.012 | .002-.008 |
| | 200-240 | .001 | .002 | .003 | .001 | .002 | .003 | .003 | .003-.007 | .004 | .003-.010 | .002-.006 |
| | 240-300 | .001 | .002 | .002 | .001 | .001 | .002 | .002 | .002-.006 | .003 | .002-.008 | .002-.005 |

**Table 15** *(Concluded).* **Recommended Feed in Inches per Tooth ($f_t$) for Milling with High Speed Steel Cutters**

| Material | Hardness, HB | End Mills | | | | | | | Plain or Slab Mills | Form Relieved Cutters | Face Mills and Shell End Mills | Slotting and Side Mills |
|---|---|---|---|---|---|---|---|---|---|---|---|---|
| | | Depth of Cut, .250 in. (Cutter Diam., in.) | | | Depth of Cut, .050 in. (Cutter Diam., in.) | | | | | | | |
| | | ½ | ¾ | 1 and up | ¼ | ½ | ¾ | 1 and up | | | | |
| | | Feed per Tooth, inch | | | | | | | | | | |
| Cast Steel | 100-180 | .001 | .003 | .003 | .001 | .002 | .003 | .004 | .003-.008 | .004 | .003-.012 | .002-.008 |
| | 180-240 | .001 | .002 | .003 | .001 | .002 | .003 | .003 | .003-.008 | .004 | .003-.010 | .002-.006 |
| | 240-300 | .001 | .002 | .002 | .0005 | .002 | .002 | .002 | .002-.006 | .003 | .003-.008 | .002-.005 |
| Zinc Alloys (Die Castings) | ... | .002 | .003 | .004 | .001 | .003 | .004 | .006 | .003-.010 | .005 | .004-.015 | .002-.012 |
| Copper Alloys (Brasses & Bronzes) | 100-150 | .002 | .004 | .005 | .002 | .003 | .005 | .006 | .003-.015 | .004 | .004-.023 | .002-.010 |
| | 150-250 | .002 | .003 | .004 | .001 | .003 | .004 | .005 | .003-.015 | .004 | .003-.012 | .002-.008 |
| Free Cutting Brasses & Bronzes | 80-100 | .002 | .004 | .005 | .002 | .003 | .005 | .006 | .003-.015 | .004 | .004-.015 | .002-.010 |
| Cast Aluminum Alloys—As Cast | ... | .003 | .004 | .005 | .002 | .004 | .005 | .006 | .005-.016 | .006 | .005-.020 | .004-.012 |
| Cast Aluminum Alloys—Hardened | ... | .003 | .004 | .005 | .002 | .003 | .004 | .005 | .004-.012 | .005 | .005-.020 | .004-.012 |
| Wrought Aluminum Alloys—Cold Drawn | ... | .003 | .004 | .005 | .002 | .003 | .004 | .005 | .004-.014 | .004 | .005-.020 | .004-.012 |
| Wrought Aluminum Alloys—Hardened | ... | .002 | .003 | .004 | .001 | .002 | .003 | .004 | .003-.012 | .004 | .005-.020 | .004-.012 |
| Magnesium Alloys | ... | .003 | .004 | .005 | .003 | .004 | .005 | .007 | .005-.016 | .006 | .008-.020 | .005-.012 |
| Ferritic Stainless Steel | 135-185 | .001 | .002 | .003 | .001 | .002 | .003 | .003 | .002-.006 | .004 | .004-.008 | .002-.007 |
| Austenitic Stainless Steel | 135-185 | .001 | .002 | .003 | .001 | .002 | .003 | .003 | .003-.007 | .004 | .005-.008 | .002-.007 |
| | 185-275 | .001 | .002 | .003 | .001 | .002 | .002 | .002 | .003-.006 | .003 | .004-.006 | .002-.007 |
| Martensitic Stainless Steel | 135-185 | .001 | .002 | .002 | .001 | .002 | .003 | .003 | .003-.006 | .004 | .004-.010 | .002-.007 |
| | 185-225 | .001 | .002 | .002 | .001 | .002 | .002 | .003 | .003-.006 | .004 | .003-.008 | .002-.007 |
| | 225-300 | .0005 | .002 | .002 | .0005 | .001 | .002 | .002 | .002-.005 | .003 | .002-.006 | .002-.006 |
| Monel | 100-160 | .001 | .003 | .004 | .001 | .002 | .003 | .004 | .002-.006 | .004 | .002-.008 | .002-.006 |

Table 16. Recommended Feed in Inch per Tooth ($f_t$) for Milling with
Cemented Carbide Cutters

| Material | Hardness, HB | Face Mills | Slotting and Side Mills |
|---|---|---|---|
| | | Feed per Tooth, inch | |
| Free Machining Plain Carbon Steels | 100-185 | .008-.020 | .003-.010 |
| Plain Carbon Steels, AISI 1006 to 1030, 1513 to 1522 | 100-150 | .008-.020 | .003-.010 |
| | 150-200 | .008-.020 | .003-.010 |
| Plain Carbon Steels, AISI 1033 to 1095, 1524 to 1566 | 120-180 | .005-.020 | .003-.010 |
| | 180-220 | .005-.020 | .003-.010 |
| | 220-300 | .003-.012 | .003-.008 |
| Alloy Steels having less than .3% Carbon content. Typical examples: AISI 4012, 4023, 4027, 4118, 4320, 4422, 4427, 4615, 4620, 4626, 4720, 4820, 5015, 5120, 6118, 8115, 8620, 8627, 8720, 8822, 9310, 93B17 | 125-175 | .006-.020 | .003-.010 |
| | 175-225 | .006-.020 | .003-.010 |
| | 225-275 | .006-.016 | .003-.010 |
| | 275-325 | .004-.012 | .003-.008 |
| | 325-375 | .003-.008 | .003-.007 |
| Alloy Steels having .3% Carbon content, or more. Typical examples: AISI 1330, 1340, 4032, 4037, 4130, 4140, 4150, 4340, 50B40, 50B60, 5130, 51B60, 6150, 81B45, 8630, 8640, 86B45, 8660, 8740, 94B30 | 175-225 | .005-.020 | .003-.010 |
| | 225-275 | .004-.012 | .003-.008 |
| | 275-325 | .003-.010 | .003-.008 |
| | 325-375 | .003-.008 | .003-.007 |
| Tool Steels | 200-275 | .004-.012 | .003-.007 |
| | 275-325 | .003-.010 | .003-.006 |
| | 36-45 HRC | .003-.006 | .002-.005 |
| | 45-55 HRC | .003-.005 | .002-.003 |
| Ferritic Stainless Steels | 110-160 | .005-.015 | .003-.010 |
| Austenitic Stainless Steels | 135-185 | .005-.012 | .003-.010 |
| | 185-275 | .005-.010 | .003-.008 |
| Martensitic Stainless Steel | 135-185 | .005-.015 | .003-.010 |
| | 185-225 | .005-.010 | .003-.008 |
| | 225-300 | .004-.008 | .003-.007 |
| Precipitation Hardening Stainless Steels | Annealed | .004-.012 | .003-.010 |
| | 275-350 | .003-.008 | .002-.005 |
| | 350-450 | .002-.005 | .002-.004 |
| Cast Steel | 100-180 | .008-.020 | .003-.010 |
| | 180-240 | .005-.016 | .003-.010 |
| | 240-300 | .004-.012 | .003-.008 |
| Gray Cast Iron | 140-185 | .008-.020 | .005-.012 |
| | 185-225 | .008-.016 | .005-.010 |
| | 225-300 | .005-.012 | .004-.008 |
| Ferritic Malleable Iron | 110-160 | .005-.020 | .004-.012 |
| Pearlitic-Martensitic Malleable Iron | 160-200 | .005-.020 | .003-.010 |
| | 200-240 | .005-.016 | .003-.010 |
| | 240-300 | .004-.010 | .003-.008 |
| Nodular (Ductile) Iron | 140-200 | .008-.020 | .003-.010 |
| | 200-275 | .006-.014 | .003-.008 |
| | 275-325 | .005-.012 | .003-.007 |
| | 325-400 | .003-.008 | .002-.004 |
| Copper Alloys (Brasses and Bronzes) | 100-150 | .005-.020 | .003-.012 |
| | 150-250 | .004-.014 | .003-.010 |

**Table 16** (*Concluded*). **Recommended Feed in Inch per Tooth ($f_t$) for Milling with Cemented Carbide Cutters**

| Material | Hardness, HB | Face Mills | Slotting and Side Mills |
|---|---|---|---|
| | | Feed per Tooth, inch | |
| Wrought and Cast Aluminum Alloys | .... | .005-.020 | .005-.020 |
| Wrought and Cast Magnesium Alloys | .... | .005-.020 | .005-.020 |
| Superalloys | .... | .003-.010 | .002-.006 |
| Titanium Alloys | .... | .003-.010 | .002-.006 |
| Nickel Alloys | .... | .003-.010 | .002-.006 |
| Monel | .... | .003-.010 | .002-.006 |
| Plastics, Hard Rubber, etc. | .... | .003-.015 | .003-.012 |

rates that can be attained require a high horsepower output. An example of this type of calculation is given in the section on milling under "Horsepower for Machining." If the size of the cut must be reduced in order to stay within the power capacity of the machine, start by reducing the cutting speed rather than the feed rate in inches per tooth.

The formula for calculating the table feed rate, when the feed in inches per tooth is known, is given below:

$$f_m = f_t \, n_t \, N \qquad (4)$$

Where: $f_m$ = Milling machine table feed rate in inches per minute (ipm)
$f_t$ = Feed rate in inch per tooth (ipt)
$n_t$ = Number of teeth in the milling cutter
$N$ = Spindle speed of the milling machine in revolutions per minute (rpm)

*Example:* Calculate the feed rate for milling a piece of AISI 1040 steel having a hardness of 160 Bhn. The cutter is a 3-inch diameter high speed steel plain or slab milling cutter with 8 teeth. The width of the cut is 2 inches, the depth of cut is .062 inch and the cutting speed is 100 fpm. From the Table 16, the feed rate selected is .008 inch per tooth.

$$N = \frac{12 \, V}{\pi \, D} = \frac{12 \times 100}{3.14 \times 3} = 127 \text{ rpm} \qquad (1)$$
$$f_m = f_t \, n_t \, N = .008 \times 8 \times 127$$
$$= 8 \text{ ipm (approximately)}$$

**Feed Rates for Drilling.** — The feed rate for drilling is governed primarily by the size of the drill and by the material to be drilled. Other factors that also affect the feed rate that can be used are the workpiece configuration, the rigidity of the machine tool and the workpiece setup, and the length of the chisel edge. A chisel edge that is too long will result in a very significant increase in the thrust force, which may cause large deflections to occur on the machine tool and drill breakage. For ordinary twist drills the feed rate used is .001 to .003 in./rev. for drills smaller than ⅛ in.; .002 to .006 in./rev. for ⅛ to ¼ in. drills; .004 to .010 in./rev. for ¼ to ½ in. drills; .007 to .015 in./rev. for ½ to 1 in. drills; and, .010 to .025 in./rev. for drills larger than 1 inch. The lower values in the feed ranges should be used for hard materials such as tool steels, superalloys, and work hardening stainless steels; the higher values in the feed ranges should be used to drill soft materials such as aluminum and brass.

**Table 17.   Recommended Cutting Speeds in Feet per Minute for Drilling and Reaming Plain Carbon and Alloy Steels**

| Material AISI and SAE Steels | Hardness, HB* | Material Condition* | Cutting Speed, fpm | | |
|---|---|---|---|---|---|
| | | | Drilling | Reaming | |
| | | | HSS | HSS | Carbide |
| Free Machining Plain Carbon Steels (Resulphurized) 1212, 1213, 1214 | 100-150 | HR, A | 120 | 80 | 400 |
| | 150-200 | CD | 125 | 80 | 350 |
| 1108, 1109, 1115, 1117, 1118, 1120, 1126, 1211 | 100-150 | HR, A | 110 | 75 | 375 |
| | 150-200 | CD | 120 | 80 | 350 |
| 1132, 1137, 1139, 1140, 1144, 1146, 1151 | 175-225 | HR, A, N, CD | 100 | 65 | 350 |
| | 275-325 | Q and T | 70 | 45 | 250 |
| | 325-375 | Q and T | 45 | 30 | 175 |
| | 375-425 | Q and T | 35 | 20 | 100 |
| Free Machining Plain Carbon Steels (Leaded) 11L17, 11L18, 12L13, 12L14 | 100-150 | HR, A, N, CD | 130 | 85 | 400 |
| | 150-200 | HR, A, N, CD | 120 | 80 | 375 |
| | 200-250 | N, CD | 90 | 60 | 275 |
| Plain Carbon Steels, 1006, 1008, 1009, 1010, 1012, 1015, 1016, 1017, 1018, 1019, 1020, 1021, 1022, 1023, 1024, 1025, 1026, 1513, 1514 | 100-125 | HR, A, N, CD | 100 | 65 | 300 |
| | 125-175 | HR, A, N, CD | 90 | 60 | 275 |
| | 175-225 | HR, N, CD | 70 | 45 | 200 |
| | 225-275 | CD | 60 | 40 | 175 |
| 1027, 1030, 1033, 1035, 1036, 1037, 1038, 1039, 1040, 1041, 1042, 1043, 1045, 1046, 1048, 1049, 1050, 1052, 1524, 1526, 1527, 1541 | 125-175 | HR, A, N, CD | 90 | 60 | 250 |
| | 175-225 | HR, A, N, CD | 75 | 50 | 200 |
| | 225-275 | N, CD, Q and T | 60 | 40 | 150 |
| | 275-325 | Q and T | 50 | 30 | 120 |
| | 325-375 | Q and T | 35 | 20 | 100 |
| | 375-425 | Q and T | 25 | 15 | 80 |
| 1055, 1060, 1064, 1065, 1070, 1074, 1078, 1080, 1084, 1086, 1090, 1095, 1548, 1551, 1552, 1561, 1566 | 125-175 | HR, A, N, CD | 85 | 55 | 250 |
| | 175-225 | HR, A, N, CD | 70 | 45 | 200 |
| | 225-275 | N, CD, Q and T | 50 | 30 | 140 |
| | 275-325 | Q and T | 40 | 25 | 110 |
| | 325-375 | Q and T | 30 | 20 | 90 |
| | 375-425 | Q and T | 15 | 10 | 70 |
| Free Machining Alloy Steels (Resulphurized), 4140, 4150 | 175-200 | HR, A, N, CD | 90 | 60 | 250 |
| | 200-250 | HR, N, CD | 80 | 50 | 225 |
| | 250-300 | Q and T | 55 | 30 | 200 |
| | 300-375 | Q and T | 40 | 25 | 150 |
| | 375-425 | Q and T | 30 | 15 | 100 |
| Free Machining Alloy Steels (Leaded), 41L30, 41L40, 41L47, 41L50, 43L47, 51L32, 52L100, 86L20, 86L40 | 150-200 | HR, A, N, CD | 100 | 65 | 285 |
| | 200-250 | HR, N, CD | 90 | 60 | 250 |
| | 250-300 | Q and T | 65 | 40 | 200 |
| | 300-375 | Q and T | 45 | 30 | 150 |
| | 375-425 | Q and T | 30 | 15 | 110 |
| Alloy Steels, 4012, 4023, 4024, 4028, 4118, 4120, 4419, 4422, 4427, 4615, 4620, 4621, 4626, 4718, 4720, 4815, 4817, 4820, 5015, 5017, 5020, 6118, 8115, 8615, 8617, 8620, 8622, 8625, 8627, 8620, 8822, 94B17 | 125-175 | HR, A, N, CD | 85 | 55 | 250 |
| | 175-225 | HR, N, CD | 70 | 45 | 225 |
| | 225-275 | CD, N, Q and T | 55 | 35 | 200 |
| | 275-325 | Q and T | 50 | 30 | 150 |
| | 325-375 | Q and T | 35 | 25 | 125 |
| | 375-425 | Q and T | 25 | 15 | 90 |

* Abbreviations designate: A, annealed; HR, hot rolled; CD, cold drawn; N, normalized; Q and T, quenched and tempered; and HB, Brinell hardness number.

**Table 17** *(Concluded).* **Recommended Cutting Speeds in Feet per Minute for Drilling and Reaming Plain Carbon and Alloy Steels**

| Material AISI and SAE Steels | Hardness, HB* | Material Condition* | Cutting Speed, fpm | | |
|---|---|---|---|---|---|
| | | | Drilling | Reaming | |
| | | | HSS | HSS | Carbide |
| Alloy Steels, 1330, 1335, 1340, 1345, 4032, 4037, 4042, 4047, 4130, 4135, 4137, 4140, 4142, 4145, 4147 4150, 4160, 4337, 4340, 50B44, 50B46, 50B50, 50B60, 5130, 5132, 5140, 5145, 5147, 5150, 5160, 51B60, 6150, 81B45, 8630, 8635, 8637, 8640, 8642, 8645, 8650, 8655, 8660, 8740, 9254, 9255, 9260, 9262, 94B30 | 175-225 225-275 275-325 325-375 375-425 | HR, A, N, CD N, CD, Q and T N, Q and T N, Q and T Q and T | 75 60 45 30 20 | 50 40 30 15 15 | 200 175 150 100 80 |
| Alloy Steels, E51100, E52100 | 175-225 225-275 275-325 325-375 375-425 | HR, A, CD N, CD, Q and T N, Q and T N, Q and T Q and T | 60 50 35 30 20 | 40 30 25 20 10 | 200 125 100 80 50 |
| Ultra High Strength Steels (Not AISI), AMS6424, AMS6421 (98B37 Mod.), AMS6422 (98BV40), AMS6427, AMS6428, AMS6430, AMS6432, AMS6433, AMS6434, AMS6436, AMS6442, 300M, D6ac | 220-300 300-350 350-400 | A N N | 50 35 20 | 30 20 10 | 180 125 90 |
| Maraging Steels (Not AISI) 18% Nickel Grade 200 18% Nickel Grade 250 18% Nickel Grade 300 18% Nickel Grade 350 | 250-225 | A | 50 | 30 | 175 |
| Nitriding Steels Nitralloy 125 Nitralloy 135 Nitralloy 135 (Mod.) Nitralloy 225 Nitralloy 230 Nitralloy N Nitralloy EZ Nitrex 1 | 200-250 250-300 | A N, Q and T | 60 35 | 40 20 | 175 125 |

* Abbreviations designate: A, annealed; HR, hot rolled; CD, cold drawn; N, normalized; Q and T, quenched and tempered; and HB, Brinell hardness number.

**Drilling Difficulties.** — A drill split up the web is evidence of too much feed or insufficient lip clearance at the center due to improper grinding. The rapid wearing away of the extreme outer corners of the cutting edges indicates that the speed is too high. A drill chipping or breaking out at the cutting edges indicates that either the feed is too heavy or the drill has been ground with too much lip clearance. Nothing will "check" a high-speed drill quicker than to turn a stream of cold water on it after it has been heated while in use. It is equally bad to plunge it in cold water after the point has been heated in grinding. the small checks or cracks resulting from this practice will eventually chip out and cause rapid wear or breakage. Insufficient speed in drilling small holes with hand feed greatly increases the risk of breakage, especially at the moment the drill is breaking through the farther side of the work. This is due to the operator's inability to gage the feed when the drill is running too slowly.

Table 22. Recommended Cutting Speeds in Feet per Minute for
Drilling and Reaming Copper Alloys

| Material Copper Alloys (Copper Alloy Nos. as per the Copper Development Assn. Inc.) | Material Condition* | Cutting Speed, fpm | | |
|---|---|---|---|---|
| | | Drilling | Reaming | |
| | | HSS | HSS | Carbide |
| 314 Leaded Commercial Bronze<br>332 High Leaded Brass<br>340 Medium Leaded Brass<br>342 High Leaded Brass<br>353 High Leaded Brass<br>356 Extra High Leaded Brass<br>360 Free Cutting Brass<br>370 Free Cutting Muntz Metal<br>377 Forging Brass<br>385 Architectural Bronze<br>485 Leaded Naval Brass<br>544 Free Cutting Phosphor Bronze | A<br>CD | 160<br>175 | 160<br>175 | 320<br>360 |
| 226 Jewelry Bronze<br>230 Red Brass<br>240 Low Brass<br>260 Cartridge Brass 70%<br>268 Yellow Brass<br>280 Muntz Metal<br>335 Low Leaded Brass<br>365 Leaded Muntz Metal<br>368 Leaded Muntz Metal<br>443 Admiralty Brass (inhibited)<br>445 Admiralty Brass (inhibited)<br>651 Low Silicon Bronze<br>655 High Silicon Bronze<br>675 Manganese Bronze<br>687 Aluminum Brass<br>770 Nickel Silver<br>796 Leaded Nickel Silver | A<br>CD | 120<br>140 | 110<br>120 | 250<br>275 |
| 102 Oxygen Free Copper<br>110 Electrolytic Tough Pitch Copper<br>122 Phosphorus Deoxidized Copper<br>170 Beryllium Copper<br>172 Beryllium Copper<br>175 Beryllium Copper<br>210 Guilding, 95%<br>220 Commercial Bronze<br>502 Phosphor Bronze 1.25%<br>510 Phosphor Bronze 5%<br>521 Phosphor Bronze 8%<br>524 Phosphor Bronze 10%<br>614 Aluminum Bronze<br>706 Copper Nickel 10%<br>715 Copper Nickel 30%<br>745 Nickel Silver<br>752 Nickel Silver<br>754 Nickel Silver<br>757 Nickel Silver | A<br>CD | 60<br>65 | 50<br>60 | 180<br>200 |

* Abbreviations used in this column are as follows: A, annealed; CD, cold drawn

## Cutting Speeds and Equivalent R.P.M. for Drills of Number and Letter Sizes

| Size No. | Cutting Speed, Feet per Minute | | | | | | | | | | |
|---|---|---|---|---|---|---|---|---|---|---|---|
| | 30' | 40' | 50' | 60' | 70' | 80' | 90' | 100' | 110' | 130' | 150' |
| | Revolutions per Minute for Number Sizes | | | | | | | | | | |
| 1 | 503 | 670 | 838 | 1005 | 1173 | 1340 | 1508 | 1675 | 1843 | 2179 | 2513 |
| 2 | 518 | 691 | 864 | 1037 | 1210 | 1382 | 1555 | 1728 | 1901 | 2247 | 2593 |
| 4 | 548 | 731 | 914 | 1097 | 1280 | 1462 | 1645 | 1828 | 2010 | 2376 | 2741 |
| 6 | 562 | 749 | 936 | 1123 | 1310 | 1498 | 1685 | 1872 | 2060 | 2434 | 2809 |
| 8 | 576 | 768 | 960 | 1151 | 1343 | 1535 | 1727 | 1919 | 2111 | 2495 | 2879 |
| 10 | 592 | 790 | 987 | 1184 | 1382 | 1579 | 1777 | 1974 | 2171 | 2566 | 2961 |
| 12 | 606 | 808 | 1010 | 1213 | 1415 | 1617 | 1819 | 2021 | 2223 | 2627 | 3032 |
| 14 | 630 | 840 | 1050 | 1259 | 1469 | 1679 | 1889 | 2099 | 2309 | 2728 | 3148 |
| 16 | 647 | 863 | 1079 | 1295 | 1511 | 1726 | 1942 | 2158 | 2374 | 2806 | 3237 |
| 18 | 678 | 904 | 1130 | 1356 | 1582 | 1808 | 2034 | 2260 | 2479 | 2930 | 3380 |
| 20 | 712 | 949 | 1186 | 1423 | 1660 | 1898 | 2135 | 2372 | 2610 | 3084 | 3559 |
| 22 | 730 | 973 | 1217 | 1460 | 1703 | 1946 | 2190 | 2433 | 2676 | 3164 | 3649 |
| 24 | 754 | 1005 | 1257 | 1508 | 1759 | 2010 | 2262 | 2513 | 2764 | 3267 | 3769 |
| 26 | 779 | 1039 | 1299 | 1559 | 1819 | 2078 | 2338 | 2598 | 2858 | 3378 | 3898 |
| 28 | 816 | 1088 | 1360 | 1631 | 1903 | 2175 | 2447 | 2719 | 2990 | 3534 | 4078 |
| 30 | 892 | 1189 | 1487 | 1784 | 2081 | 2378 | 2676 | 2973 | 3270 | 3864 | 4459 |
| 32 | 988 | 1317 | 1647 | 1976 | 2305 | 2634 | 2964 | 3293 | 3622 | 4281 | 4939 |
| 34 | 1032 | 1376 | 1721 | 2065 | 2409 | 2753 | 3097 | 3442 | 3785 | 4474 | 5162 |
| 36 | 1076 | 1435 | 1794 | 2152 | 2511 | 2870 | 3228 | 3587 | 3945 | 4663 | 5380 |
| 38 | 1129 | 1505 | 1882 | 2258 | 2634 | 3010 | 3387 | 3763 | 4140 | 4892 | 5645 |
| 40 | 1169 | 1559 | 1949 | 2339 | 2729 | 3118 | 3508 | 3898 | 4287 | 5067 | 5846 |
| 42 | 1226 | 1634 | 2043 | 2451 | 2860 | 3268 | 3677 | 4085 | 4494 | 5311 | 6128 |
| 44 | 1333 | 1777 | 2221 | 2665 | 3109 | 3554 | 3999 | 4442 | 4886 | 5774 | 6662 |
| 46 | 1415 | 1886 | 2358 | 2830 | 3301 | 3773 | 4244 | 4716 | 5187 | 6130 | 7074 |
| 48 | 1508 | 2010 | 2513 | 3016 | 3518 | 4021 | 4523 | 5026 | 5528 | 6534 | 7539 |
| 50 | 1637 | 2183 | 2729 | 3274 | 3820 | 4366 | 4911 | 5457 | 6002 | 7094 | 8185 |
| 52 | 1805 | 2406 | 3008 | 3609 | 4211 | 4812 | 5414 | 6015 | 6619 | 7820 | 9023 |
| 54 | 2084 | 2778 | 3473 | 4167 | 4862 | 5556 | 6251 | 6945 | 7639 | 9028 | 10417 |
| Size | Revolutions per Minute for Letter Sizes | | | | | | | | | | |
| A | 491 | 654 | 818 | 982 | 1145 | 1309 | 1472 | 1636 | 1796 | 2122 | 2448 |
| B | 482 | 642 | 803 | 963 | 1124 | 1284 | 1445 | 1605 | 1765 | 2086 | 2407 |
| C | 473 | 631 | 789 | 947 | 1105 | 1262 | 1420 | 1578 | 1736 | 2052 | 2368 |
| D | 467 | 622 | 778 | 934 | 1089 | 1245 | 1400 | 1556 | 1708 | 2018 | 2329 |
| E | 458 | 611 | 764 | 917 | 1070 | 1222 | 1375 | 1528 | 1681 | 1968 | 2292 |
| F | 446 | 594 | 743 | 892 | 1040 | 1189 | 1337 | 1486 | 1635 | 1932 | 2229 |
| G | 440 | 585 | 732 | 878 | 1024 | 1170 | 1317 | 1463 | 1610 | 1903 | 2195 |
| H | 430 | 574 | 718 | 862 | 1005 | 1149 | 1292 | 1436 | 1580 | 1867 | 2154 |
| I | 421 | 562 | 702 | 842 | 983 | 1123 | 1264 | 1404 | 1545 | 1826 | 2106 |
| J | 414 | 552 | 690 | 827 | 965 | 1103 | 1241 | 1379 | 1517 | 1793 | 2068 |
| K | 408 | 544 | 680 | 815 | 951 | 1087 | 1223 | 1359 | 1495 | 1767 | 2039 |
| L | 395 | 527 | 659 | 790 | 922 | 1054 | 1185 | 1317 | 1449 | 1712 | 1976 |
| M | 389 | 518 | 648 | 777 | 907 | 1036 | 1166 | 1295 | 1424 | 1683 | 1942 |
| N | 380 | 506 | 633 | 759 | 886 | 1012 | 1139 | 1265 | 1391 | 1644 | 1897 |
| O | 363 | 484 | 605 | 725 | 846 | 967 | 1088 | 1209 | 1330 | 1571 | 1813 |
| P | 355 | 473 | 592 | 710 | 828 | 946 | 1065 | 1183 | 1301 | 1537 | 1774 |
| Q | 345 | 460 | 575 | 690 | 805 | 920 | 1035 | 1150 | 1266 | 1496 | 1726 |
| R | 338 | 451 | 564 | 676 | 789 | 902 | 1014 | 1127 | 1239 | 1465 | 1690 |
| S | 329 | 439 | 549 | 659 | 769 | 878 | 988 | 1098 | 1207 | 1427 | 1646 |
| T | 320 | 426 | 533 | 640 | 746 | 853 | 959 | 1066 | 1173 | 1387 | 1600 |
| U | 311 | 415 | 519 | 623 | 727 | 830 | 934 | 1038 | 1142 | 1349 | 1557 |
| V | 304 | 405 | 507 | 608 | 709 | 810 | 912 | 1013 | 1114 | 1317 | 1520 |
| W | 297 | 396 | 495 | 594 | 693 | 792 | 891 | 989 | 1088 | 1286 | 1484 |
| X | 289 | 385 | 481 | 576 | 672 | 769 | 865 | 962 | 1058 | 1251 | 1443 |
| Y | 284 | 378 | 473 | 567 | 662 | 756 | 851 | 945 | 1040 | 1229 | 1418 |
| Z | 277 | 370 | 462 | 555 | 647 | 740 | 832 | 925 | 1017 | 1202 | 1387 |

For fractional drill sizes, use table on page 1790.

## Revolutions per Minute for Various Cutting Speeds and Diameters
(Metric units)

| Diam., mm | Cutting Speed, Metres per Minute | | | | | | | | | | | |
|---|---|---|---|---|---|---|---|---|---|---|---|---|
| | 50 | 55 | 60 | 65 | 70 | 75 | 80 | 85 | 90 | 95 | 100 | 200 |
| | Revolutions per Minute | | | | | | | | | | | |
| 5 | 3183 | 3501 | 3820 | 4138 | 4456 | 4775 | 5093 | 5411 | 5730 | 6048 | 6366 | 12,732 |
| 6 | 2653 | 2918 | 3183 | 3448 | 3714 | 3979 | 4244 | 4509 | 4775 | 5039 | 5305 | 10,610 |
| 8 | 1989 | 2188 | 2387 | 2586 | 2785 | 2984 | 3183 | 3382 | 3581 | 3780 | 3979 | 7958 |
| 10 | 1592 | 1751 | 1910 | 2069 | 2228 | 2387 | 2546 | 2706 | 2865 | 3024 | 3183 | 6366 |
| 12 | 1326 | 1459 | 1592 | 1724 | 1857 | 1989 | 2122 | 2255 | 2387 | 2520 | 2653 | 5305 |
| 16 | 995 | 1094 | 1194 | 1293 | 1393 | 1492 | 1591 | 1691 | 1790 | 1890 | 1989 | 3979 |
| 20 | 796 | 875 | 955 | 1034 | 1114 | 1194 | 1273 | 1353 | 1432 | 1512 | 1592 | 3183 |
| 25 | 637 | 700 | 764 | 828 | 891 | 955 | 1019 | 1082 | 1146 | 1210 | 1273 | 2546 |
| 30 | 530 | 584 | 637 | 690 | 743 | 796 | 849 | 902 | 955 | 1008 | 1061 | 2122 |
| 35 | 455 | 500 | 546 | 591 | 637 | 682 | 728 | 773 | 819 | 864 | 909 | 1818 |
| 40 | 398 | 438 | 477 | 517 | 557 | 597 | 637 | 676 | 716 | 756 | 796 | 1592 |
| 45 | 354 | 389 | 424 | 460 | 495 | 531 | 566 | 601 | 637 | 672 | 707 | 1415 |
| 50 | 318 | 350 | 382 | 414 | 446 | 477 | 509 | 541 | 573 | 605 | 637 | 1273 |
| 55 | 289 | 318 | 347 | 376 | 405 | 434 | 463 | 492 | 521 | 550 | 579 | 1157 |
| 60 | 265 | 292 | 318 | 345 | 371 | 398 | 424 | 451 | 477 | 504 | 530 | 1061 |
| 65 | 245 | 269 | 294 | 318 | 343 | 367 | 392 | 416 | 441 | 465 | 490 | 979 |
| 70 | 227 | 250 | 273 | 296 | 318 | 341 | 364 | 387 | 409 | 432 | 455 | 909 |
| 75 | 212 | 233 | 255 | 276 | 297 | 318 | 340 | 361 | 382 | 403 | 424 | 849 |
| 80 | 199 | 219 | 239 | 259 | 279 | 298 | 318 | 338 | 358 | 378 | 398 | 796 |
| 90 | 177 | 195 | 212 | 230 | 248 | 265 | 283 | 301 | 318 | 336 | 354 | 707 |
| 100 | 159 | 175 | 191 | 207 | 223 | 239 | 255 | 271 | 286 | 302 | 318 | 637 |
| 110 | 145 | 159 | 174 | 188 | 203 | 217 | 231 | 246 | 260 | 275 | 289 | 579 |
| 120 | 133 | 146 | 159 | 172 | 186 | 199 | 212 | 225 | 239 | 252 | 265 | 530 |
| 130 | 122 | 135 | 147 | 159 | 171 | 184 | 196 | 208 | 220 | 233 | 245 | 490 |
| 140 | 114 | 125 | 136 | 148 | 159 | 171 | 182 | 193 | 205 | 216 | 227 | 455 |
| 150 | 106 | 117 | 127 | 138 | 149 | 159 | 170 | 180 | 191 | 202 | 212 | 424 |
| 160 | 99.5 | 109 | 119 | 129 | 139 | 149 | 159 | 169 | 179 | 189 | 199 | 398 |
| 170 | 93.6 | 103 | 112 | 122 | 131 | 140 | 150 | 159 | 169 | 178 | 187 | 374 |
| 180 | 88.4 | 97.3 | 106 | 115 | 124 | 133 | 141 | 150 | 159 | 168 | 177 | 354 |
| 190 | 83.8 | 92.1 | 101 | 109 | 117 | 126 | 134 | 142 | 151 | 159 | 167 | 335 |
| 200 | 79.6 | 87.5 | 95.5 | 103 | 111 | 119 | 127 | 135 | 143 | 151 | 159 | 318 |
| 220 | 72.3 | 79.6 | 86.8 | 94 | 101 | 109 | 116 | 123 | 130 | 137 | 145 | 289 |
| 240 | 66.3 | 72.9 | 79.6 | 86.2 | 92.8 | 99.5 | 106 | 113 | 119 | 126 | 132 | 265 |
| 260 | 61.2 | 67.3 | 73.4 | 79.6 | 85.7 | 91.8 | 97.9 | 104 | 110 | 116 | 122 | 245 |
| 280 | 56.8 | 62.5 | 68.2 | 73.9 | 79.6 | 85.3 | 90.9 | 96.6 | 102 | 108 | 114 | 227 |
| 300 | 53.1 | 58.3 | 63.7 | 69 | 74.3 | 79.6 | 84.9 | 90.2 | 95.5 | 101 | 106 | 212 |
| 350 | 45.5 | 50 | 54.6 | 59.1 | 63.7 | 68.2 | 72.8 | 77.3 | 81.8 | 99.1 | 91 | 182 |
| 400 | 39.8 | 43.8 | 47.7 | 51.7 | 55.7 | 59.7 | 63.7 | 67.6 | 71.6 | 75.6 | 79.6 | 159 |
| 450 | 35.4 | 38.9 | 42.4 | 46 | 49.5 | 53.1 | 56.6 | 60.1 | 63.6 | 67.2 | 70.7 | 141 |
| 500 | 31.8 | 35 | 38.2 | 41.4 | 44.6 | 47.7 | 50.9 | 54.1 | 57.3 | 60.5 | 63.6 | 127 |

**Planing Time.** — The approximate time required to plane a surface can be determined from the following formula in which $T$ = time in minutes, $L$ = length of stroke in feet, $V_c$ = cutting speed in feet per minute, $V_r$ = return speed in feet per minute, $W$ = width of surface to be planed in inches, $F$ = feed in inches, and 0.025 = approximate reversal time factor per stroke in minutes for most planers.

$$T = \frac{W}{F} \left[ L \times \left( \frac{1}{V_c} + \frac{1}{V_r} \right) + 0.025 \right]$$

**Speeds for Metal-cutting Saws.** — The following speeds and feeds for metal-cutting saws are recommended by Henry Disston & Sons, Inc. Speeds and feeds vary over a wide range depending upon the kind of machine as well as the size, shape, and kind of metal to be cut; hence, the following speeds are intended as a general guide only.

*Solid-tooth Circular Saws.* — Low-carbon steel: Speed, 60–90 feet per minute; feed, 3–6 inches per minute. Tool and alloy steel: Speed, 40–60 feet per minute; feed, ½₂–2½ inches per minute. Cast iron: Speed, 60–70 feet per minute; feed, 5 inches per minute. Brass: Speed, 500–800 feet per minute; feed, 80 inches per minute. Copper: Speed, 600–700 feet per minute; feed, 60 inches per minute. Aluminum: Speed, 800–1000 feet per minute; feed, 90 inches per minute.

"Hot saws" for sawing hot metal (structural shapes, rails, billets, etc., in rolling mills) should operate at speeds of 22,000 to 25,000 feet per minute.

*Inserted-tooth Metal-cutting Saws.* — Generally speaking, an inserted-tooth saw is not recommended smaller than 18 inches in diameter, although smaller sizes are made. Small saws usually are applied to work requiring fine or relatively fine teeth which are impossible when the teeth are of the inserted type. The following speeds are not suitable for all compositions but represent general practice. Steel: Speed, 40 feet per minute; feed, 2½–10 inches per minute. Cast iron: Speed, 40 feet per minute; feed, 5 inches per minute. Brass: Speed, 500 feet per minute; feed, 80 inches per minute. Copper: Speed, 750 feet per minute; feed, 60 inches per minute. Aluminum: Speed, 1200 feet per minute; feed, 90 inches per minute.

*Metal-cutting Band Saws.* — The speeds which follow apply to Disston "hard-edge" saws which have milled teeth. Aluminum sheets: Speed of blade, 1000–3000 feet per minute. Bakelite: Speed, 800–1000 feet per minute. Brass sheets and tubing: Speed, 700–1500 feet per minute. Carbon tool steel: Speed, 100–150 feet per minute. Cast iron: Speed, 100–150 feet per minute. Cold-rolled steel: Speed, 150–200 feet per minute. High-speed steel: Speed, 90–125 feet per minute. Malleable iron: Speed, 150–200 feet per minute. Hard rubber: Speed, 150–200 feet per minute. Slate: Speed, 100–150 feet per minute. The number of saw teeth per inch recommended for these different materials ranges from 8 to 14, and the saw user should be guided by the recommendations of the manufacturer.

**Hacksaw Speeds.** — The following hacksaw speeds are given on the authority of a leading manufacturer: The total amount of travel of the cutting blade in feet per minute, including forward and return strokes, should be as follows: For mild steel, 130; for annealed tool steel, 90; and for unannealed tool steel, 60 feet per minute. Thus in the case of a 6-inch stroke, for example, the revolutions per minute of the driving crank should be 130 for mild steel, 90 for annealed tool steel, and 60 for unannealed tool steel. All of these steels are cut with the use of a cutting compound. Bronze can ordinarily be cut at the same speed as mild steel when a suitable compound is used. Brass heats the blade very rapidly if cut dry, and must be cut with a cooling compound adapted to brass. It also fills up the teeth of the saw if not used with the right kind of compound, but with a suitable compound, it may be cut at the same speed as machine steel.

**Speeds for Turning Unusual Materials.** — *Slate*, on account of its peculiarly stratified formation, is rather difficult to turn, but if handled carefully, can be machined in an ordinary lathe. The cutting speed should be about the same as for cast iron. A sheet of fiber or pressed paper should be interposed between the chuck or steadyrest jaws and the slate, to protect the latter. Slate rolls must not be centered and run on the tailstock. A satisfactory method of supporting a slate roll having journals at the ends is to bore a piece of lignum vitæ to receive the turned end of the roll, and center it for the tailstock spindle.

*Rubber* can be turned at a peripheral speed of 200 feet per minute, although it is much easier to grind it with an abrasive wheel that is porous and soft. For cutting a rubber roll in two, the ordinary parting tool should not be used, but a tool shaped like a knife; such a tool severs the rubber without removing any material.

*Gutta percha* can be turned as easily as wood, but the tools must be sharp and a good soap-and-water lubricant used.

*Copper* can be turned easily at 200 feet per minute.

*Lime-stone* such as is used in the construction of pillars for balconies, etc., can be turned at 150 feet per minute, and the formation of ornamental contours is quite easy. *Marble* is a treacherous material to turn. It should be cut with a tool such as would be used for brass, but at a speed suitable for cast iron. It must be handled very carefully to prevent flaws in the surface.

The foregoing speeds are for high-speed steel tools. Tools tipped with tungsten carbide are adapted for cutting various non-metallic products which cannot be machined readily with steel tools, such as slate, marble, synthetic plastic materials, etc. In drilling slate and marble, use flat drills; and for plastic materials, tungsten-carbide-tipped twist drills. Cutting speeds ranging from 75 to 150 feet per minute have been used for drilling slate (without coolant) and a feed of 0.025 per revolution for drills ¾ and 1 inch in diameter.

**Estimating Machining Power.** — Knowledge of the power required to perform machining operations is useful when planning new machining operations, for optimizing existing machining operations, and to develop specifications for new machine tools that are to be acquired. The available power on any machine tool places a limit on the size of the cut that it can take. When much metal must be removed from the workpiece it is advisable to estimate the cutting conditions that will utilize the maximum power on the machine. Many machining operations require only light cuts to be taken for which the machine obviously has ample power; in this event, estimating the power required is a wasteful effort. Since conditions in different shops may vary and machine tools are not all designed alike, some variations between the estimated results and those obtained on the job are to be expected; however, by using the methods provided in this section a reasonable estimate of the power required can be made, which will suffice in most practical situations.

The measure of power in customary inch units is the horsepower; in SI metric units it is the kilowatt, which is used for both mechanical and electrical power. The power required to cut a material is dependent upon the rate at which the material is being cut and upon an experimentally determined power constant, $K_p$, which is also called the unit horsepower, unit power, or specific power consumption. The power constant is equal to the horsepower required to cut a material at a rate of one cubic inch per minute; in SI metric units the power constant is equal to the power in kilowatts required to cut a material at a rate of one cubic centimeter per second, or 1000 cubic millimeters per second ($1 \text{ cm}^3 = 1000 \text{ mm}^3$). Different values of the power constant are required for inch and for metric units, which are related as follows: to obtain the SI metric power constant, multiply the inch power constant by 2.73; to obtain the inch power constant, divide the SI metric power constant by 2.73. Values of the power constant appear in Tables 23, 24, and 25, which can be used for all machining operations except drilling and grinding. The values given are for sharp cutting tools.

**Table 23. Power Constants, $K_p$, for Wrought Steels, Using Sharp Cutting Tools**

| Material | Brinell Hardness Number | $K_p$ Inch Units | $K_p$ SI Metric Units |
|---|---|---|---|
| **Plain Carbon Steels** | | | |
| | 80-100 | .63 | 1.72 |
| | 100-120 | .66 | 1.80 |
| | 120-140 | .69 | 1.88 |
| | 140-160 | .74 | 2.02 |
| | 160-180 | .78 | 2.13 |
| | 180-200 | .82 | 2.24 |
| All Plain Carbon Steels | 200-220 | .85 | 2.32 |
| | 220-240 | .89 | 2.43 |
| | 240-260 | .92 | 2.51 |
| | 260-280 | .95 | 2.59 |
| | 280-300 | 1.00 | 2.73 |
| | 300-320 | 1.03 | 2.81 |
| | 320-340 | 1.06 | 2.89 |
| | 340-360 | 1.14 | 3.11 |
| **Free Machining Steels** | | | |
| | 100-120 | .41 | 1.12 |
| AISI 1108, 1109, 1110, 1115, 1116, 1117, 1118, 1119, | 120-140 | .42 | 1.15 |
| 1120, 1125, 1126, 1132 | 140-160 | .44 | 1.20 |
| | 160-180 | .48 | 1.31 |
| | 180-200 | .50 | 1.36 |
| | 180-200 | .51 | 1.39 |
| AISI 1137, 1138, 1139, 1140, 1141, 1144, 1145, 1146, | 200-220 | .55 | 1.50 |
| 1148, 1151 | 220-240 | .57 | 1.56 |
| | 240-260 | .62 | 1.69 |
| **Alloy Steels** | | | |
| | 140-160 | .62 | 1.69 |
| | 160-180 | .65 | 1.77 |
| | 180-200 | .69 | 1.88 |
| | 200-220 | .72 | 1.97 |
| AISI 4023, 4024, 4027, 4028, 4032, 4037, 4042, 4047, | 220-240 | .76 | 2.07 |
| 4137, 4140, 4142, 4145, 4147, 4150, 4340, 4640, 4815, | 240-260 | .80 | 2.18 |
| 4817, 4820, 5130, 5132, 5135, 5140, 5145, 5150, 6118, | 260-280 | .84 | 2.29 |
| 6150, 8637, 8640, 8642, 8645, 8650, 8740 | 280-300 | .87 | 2.38 |
| | 300-320 | .91 | 2.48 |
| | 320-340 | .96 | 2.62 |
| | 340-360 | 1.00 | 2.73 |
| | 140-160 | .56 | 1.53 |
| | 160-180 | .59 | 1.61 |
| | 180-200 | .62 | 1.69 |
| | 200-220 | .65 | 1.77 |
| AISI 4130, 4320, 4615, 4620, 4626, 5120, 8615, 8617, | 220-240 | .70 | 1.91 |
| 8620, 8622, 8625, 8630, 8720 | 240-260 | .74 | 2.02 |
| | 260-280 | .77 | 2.10 |
| | 280-300 | .80 | 2.18 |
| | 300-320 | .83 | 2.27 |
| | 320-340 | .89 | 2.43 |
| | 160-180 | .79 | 2.16 |
| | 180-200 | .83 | 2.27 |
| AISI 1330, 1335, 1340, E52100 | 200-220 | .87 | 2.38 |
| | 220-240 | .91 | 2.48 |
| | 240-260 | .95 | 2.59 |
| | 260-280 | 1.00 | 2.73 |

**Table 24. Power Constants, $K_p$, for Ferrous Cast Metals, Using Sharp Cutting Tools**

| Material | Brinell Hardness Number | $K_p$ Inch Units | $K_p$ SI Metric Units | Material | Brinell Hardness Number | $K_p$ Inch Units | $K_p$ SI Metric Units |
|---|---|---|---|---|---|---|---|
| | 100-120 | .28 | 0.76 | Malleable Iron | | | |
| | 120-140 | .35 | 0.96 | Ferritic | 150-175 | .42 | 1.15 |
| | 140-160 | .38 | 1.04 | | 175-200 | .57 | 1.56 |
| Gray Cast Iron | 160-180 | .52 | 1.42 | Pearlitic | 200-250 | .82 | 2.24 |
| | 180-200 | .60 | 1.64 | | 250-300 | 1.18 | 3.22 |
| | 200-220 | .71 | 1.94 | | | | |
| | 220-240 | .91 | 2.48 | Cast Steel | 150-175 | .62 | 1.69 |
| | | | | | 175-200 | .78 | 2.13 |
| | 150-175 | .30 | 0.82 | | 200-250 | .86 | 2.35 |
| Alloy Cast Iron | 175-200 | .63 | 1.72 | ... | ... | ... | ... |
| | 200-250 | .92 | 2.51 | ... | ... | ... | ... |

The value of the power constant is essentially unaffected by the cutting speed, the depth of cut, and by the cutting tool material. There are, however, factors that do affect the value of the power constant and thereby the power required to cut a material. These factors include the hardness and microstructure of the work material, the feed rate, the rake angle of the cutting tool, and the condition of the cutting edge of the tool, whether it is sharp or dull. Values in the power constant tables are given for different material hardness levels, whenever this information is available. Feed factors for the power constant are given in Table 26. All metal cutting tools wear as they are used and are expected to continue to cut as the cutting edge becomes worn. A worn cutting edge requires more power to cut than a sharp cutting edge. Factors to provide for tool wear are given in Table 27. In this table the extra-heavy duty category for milling and turning occurs only on operations where the tool is allowed to wear more than a normal amount before it is replaced, such as roll turning. In most instances the effect of the rake angle can be disregarded. The rake angle for which most of the data in the power constant tables are given is positive 14 degrees. Only when the deviation from this angle is large is it necessary to make an adjustment. Using a rake angle that is more positive reduces the power required approximately one per cent per degree; using a rake angle that is more negative increases the power required; again approximately one per cent per

**Table 25. Power Constant, $K_p$, for High Temperature Alloys, Tool Steel, Stainless Steel, and Nonferrous Metals, Using Sharp Cutting Tools**

| Material | Brinell Hardness Number | $K_p$ Inch Units | $K_p$ SI Metric Units | Material | Brinell Hardness Number | $K_p$ Inch Units | $K_p$ SI Metric Units |
|---|---|---|---|---|---|---|---|
| High Temperature Alloys | | | | | 150-175 | .60 | 1.64 |
| A286 | 165 | .82 | 2.24 | Stainless Steel | 175-200 | .72 | 1.97 |
| A286 | 285 | .93 | 2.54 | | 200-250 | .88 | 2.40 |
| Chromoloy | 200 | .78 | 3.22 | Zinc Die Cast Alloys | ... | .25 | 0.68 |
| Chromoloy | 310 | 1.18 | 3.00 | Copper (pure) | ... | .91 | 2.48 |
| Inco 700 | 330 | 1.12 | 3.06 | | | | |
| Inco 702 | 230 | 1.10 | 3.00 | Brass | | | |
| Hastelloy-B | 230 | 1.10 | 3.00 | Hard | ... | .83 | 2.27 |
| M-252 | 230 | 1.10 | 3.00 | Medium | ... | .50 | 1.36 |
| M-252 | 310 | 1.20 | 3.28 | Soft | ... | .25 | 0.68 |
| Ti-150A | 340 | .65 | 1.77 | Leaded | ... | .30 | 0.82 |
| U-500 | 375 | 1.10 | 3.00 | Bronze | | | |
| | | | | Hard | ... | .91 | 2.48 |
| Monel Metal | ... | 1.00 | 2.73 | Medium | ... | .50 | 1.36 |
| | | | | Soft | ... | .33 | 0.90 |
| | 175-200 | .75 | 2.05 | Aluminum | | | |
| | 200-250 | .88 | 2.40 | Cast | ... | .25 | 0.68 |
| Tool Steel | 250-300 | .98 | 2.68 | Rolled (hard) | ... | .33 | 0.90 |
| | 300-350 | 1.20 | 3.28 | | | | |
| | 350-400 | 1.30 | 3.55 | Magnesium Alloys | ... | .10 | 0.27 |

## Table 26.   Feed Factor, $C$, for Power Constants

| Inch Units | | | | SI Metric Units | | | |
|---|---|---|---|---|---|---|---|
| Feed in.* | $C$ | Feed in.* | $C$ | Feed mm† | $C$ | Feed mm† | $C$ |
| .001 | 1.60 | .014 | .97 | 0.02 | 1.70 | 0.35 | .97 |
| .002 | 1.40 | .015 | .96 | 0.05 | 1.40 | 0.38 | .95 |
| .003 | 1.30 | .016 | .94 | 0.07 | 1.30 | 0.40 | .94 |
| .004 | 1.25 | .018 | .92 | 0.10 | 1.25 | 0.45 | .92 |
| .005 | 1.19 | .020 | .90 | 0.12 | 1.20 | 0.50 | .90 |
| .006 | 1.15 | .022 | .88 | 0.15 | 1.15 | 0.55 | .88 |
| .007 | 1.11 | .025 | .86 | 0.18 | 1.11 | 0.60 | .87 |
| .008 | 1.08 | .028 | .84 | 0.20 | 1.08 | 0.70 | .84 |
| .009 | 1.06 | .030 | .83 | 0.22 | 1.06 | 0.75 | .83 |
| .010 | 1.04 | .032 | .82 | 0.25 | 1.04 | 0.80 | .82 |
| .011 | 1.02 | .035 | .80 | 0.28 | 1.01 | 0.90 | .80 |
| .012 | 1.00 | .040 | .78 | 0.30 | 1.00 | 1.00 | .78 |
| .013 | .98 | .060 | .72 | 0.33 | .98 | 1.50 | .72 |

\* Turning–in./rev; Milling–in./tooth; Planing and Shaping–in./stroke; Broaching–in./tooth.
†Turning–mm/rev; Milling–mm/tooth; Planing and Shaping–mm/stroke; Broaching–mm/tooth.

degree. Many indexable insert cutting tools have a pressed-in chip breaker or other cutting edge modifications, which have the effect of reducing the power required to cut a material. The extent of this effect cannot be predicted without a test of each design. Cutting fluids will also usually reduce the power required, when operating in the lower range of cutting speeds. Again, the extent of this effect cannot be predicted because each cutting fluid exhibits its own characteristics.

The machine tool transmits the power from the driving motor to the workpiece, where it is used to cut the material. How efficiently this is done is measured by the machine tool efficiency factor, $E$. Average values of this factor are given in Table 28. Formulas for calculating the metal removal rate, $Q$, for different machining operations are given in Table 29. These formulas are used together with those given below. The following formulas can be used with either customary inch or with SI metric units.

$$P_c = K_p C Q W \qquad\qquad\qquad (1)$$

$$P_m = \frac{P_c}{E} = \frac{K_p C Q W}{E} \qquad\qquad\qquad (2)$$

Where:   $P_c$ = Power at the cutting tool; hp, or kW
         $P_m$ = Power at the motor; hp, or kW
         $K_p$ = Power constant             (See Tables 23, 24, and 25)
         $Q$ = Metal removal rate; in.³/min, or cm³/s    (See Table 29)

## Table 27.   Tool Wear Factors, $W$

| Type of Operation | | $W$ |
|---|---|---|
| For all operations with sharp cutting tools | | 1.00 |
| Turning: | Finish turning (lightcuts) | 1.10 |
| | Normal rough and semi-finish turning | 1.30 |
| | Extra-heavy duty rough turning | 1.60-2.00 |
| Milling: | Slab milling | 1.10 |
| | End milling | 1.10 |
| | Light and medium face milling | 1.10-1.25 |
| | Extra-heavy duty face milling | 1.30-1.60 |
| Drilling: | Normal drilling | 1.30 |
| | Drilling hard-to-machine materials and drilling with a very dull drill | 1.50 |
| Broaching: | Normal broaching | 1.05-1.10 |
| | Heavy duty surface broaching | 1.20-1.30 |

For planing and shaping, use values given for turning.

**Table 28.    Machine Tool Efficiency Factors, E**

| Type of Drive | E | Type of Drive | E |
|---|---|---|---|
| Direct Belt Drive | .90 | Geared Head Drive | .70-.80 |
| Back Gear Drive | .75 | Oil-Hydraulic Drive | .60-.90 |

$C$ = Feed factor for power constant     (See Table 26)
$W$ = Tool wear factor     (See Table 27)
$E$ = Machine tool efficiency factor     (See Table 28)
$V$ = Cutting speed, fpm, or m/min
$N$ = Cutting speed, rpm
$f$ = Feed rate for turning; in./rev, or mm/rev
$f$ = Feed rate for planing and shaping; in./stroke, or mm/stroke
$f_t$ = Feed per tooth; in./tooth, or mm/tooth
$f_m$ = Feed rate; in./min, or mm/min
$d_t$ = Maximum depth of cut per tooth; in., or mm
$d$ = Depth of cut; in., or mm
$n_t$ = Number of teeth on milling cutter
$n_c$ = Number of teeth engaged in work
$w$ = width of cut; in. or mm

*Example:* A 180-200 Bhn AISI shaft is to be turned on a geared head lathe using a cutting speed of 350 fpm (107 m/min), a feed rate of .016 in./rev (0.40 mm/rev), and a depth of cut of .100 inch (2.54 mm). Estimate the power at the cutting tool and at the motor, using both the inch and metric data.

Inch units: $K_p$ = .62    (From Table 23)     $C$ = .94    (From Table 26)
            $W$ = 1.30   (From Table 27)     $E$ = .80    (From Table 28)

$$Q = 12\,Vfd = 12 \times 350 \times .016 \times .100 \quad \text{(From Table 29)}$$
$$= 6.72 \text{ in.}^3/\text{min}$$

$$P_c = K_p\,C\,Q\,W = .62 \times .94 \times 6.72 \times 1.30 = 5 \text{ hp}$$

$$P_m = \frac{P_c}{E} = \frac{5}{.80} = 6.25 \text{ hp}$$

SI metric units: $K_p$ = 1.69   (From Table 23)     $C$ = .94    (From Table 26)
                $W$ = 1.30   (From Table 27)     $E$ = .80    (From Table 28)

$$Q = \frac{V}{60}\,fd = \frac{107}{60} \times 0.40 \times 2.54 \quad \text{(From Table 29)}$$
$$= 1.81 \text{ cm}^3/\text{s}$$

$$P_c = K_p\,C\,Q\,W = 1.69 \times .94 \times 1.81 \times 1.30 = 3.74 \text{ kW}$$

$$P_m = \frac{P_c}{E} = \frac{3.74}{.80} = 4.675 \text{ kW}$$

**Table 29.    Formulas for Calculating the Metal Removal Rate, Q**

| Operation | Metal Removal Rate | |
|---|---|---|
| | For Inch Units Only $Q$ = in.$^3$/min | For SI Metric Units Only $Q$ = cm$^3$/s |
| Single Point Tools (Turning, Planing, and Shaping) | $12Vfd$ | $\dfrac{V}{60}\,fd$ |
| Milling | $f_m wd$ | $\dfrac{f_m wd}{60,000}$ |
| Surface Broaching | $12Vwn_c d_t$ | $\dfrac{V}{60}\,wn_c d_t$ |

Whenever possible the maximum power available on a machine tool should be used when heavy cuts must be taken. The cutting conditions for utilizing the maximum power should be selected in the following order: 1. select the maximum depth of cut that can be used; 2. select the maximum feed rate that can be used; and, 3. estimate the cutting speed that will utilize the maximum power available on the machine. This order is based on obtaining the longest tool life of the cutting tool while at the same time obtaining as much production as possible from the machine. *The life of a cutting tool is most affected by the cutting speed, then by the feed rate, and least of all by the depth of cut.* The maximum metal removal rate that a given machine is capable of machining from a given material is used as the basis for estimating the cutting speed that will utilize all of the power available on the machine.

*Example:* A .125 inch deep cut is to be taken on a 200-210 Bhn AISI 1050 steel part using a 10 hp geared head lathe. The feed rate selected for this job is .018 in./rev. Estimate the cutting speed that will utilize the maximum power available on the lathe.

$$K_p = .85 \quad \text{(From Table 23)} \qquad C = .92 \quad \text{(From Table 26)}$$
$$W = 1.30 \quad \text{(From Table 27)} \qquad E = .80 \quad \text{(From Table 28)}$$

$$Q_{max} = \frac{P_m E}{K_p\, C\, W} = \frac{10 \times .80}{.85 \times .92 \times 1.30} \qquad \left(P_m = \frac{K_p\, C\, Q\, W}{E}\right)$$
$$= 7.87 \text{ in.}^3/\text{min}$$

$$V = \frac{Q_{max}}{12\, f d} = \frac{7.87}{12 \times .018 \times .125} \qquad (Q = 12\, V f d)$$
$$= 290 \text{ fpm}$$

*Example:* A 160-180 Bhn gray iron casting that is 6 inches wide is to have ⅛ inch stock removed on a 10 hp milling machine, using an 8 inch diameter, 10 tooth, indexable insert cemented carbide face milling cutter. The feed rate selected for this cutter is .012 in./tooth, and all of the stock (.125 in.) will be removed in one cut. Estimate the cutting speed that will utilize the maximum power available on the machine.

$$K_p = .52 \quad \text{(From Table 24)} \qquad C = 1.00 \quad \text{(From Table 26)}$$
$$W = 1.20 \quad \text{(From Table 27)} \qquad E = .80 \quad \text{(From Table 28)}$$

$$Q_{max} = \frac{P_m E}{K_p C W} = \frac{10 \times .80}{.52 \times 1.00 \times 1.20} = 12.82 \text{ in.}^3/\text{min} \qquad \left(P_m = \frac{K_p C Q W}{E}\right)$$

$$f_m = \frac{Q_{max}}{wd} = \frac{12.82}{6 \times .125} = 17 \text{ in./min} \qquad (Q = f_m wd)$$

$$N = \frac{f_m}{f_t n_t} = \frac{17}{.012 \times 10} = 140 \text{ rpm} \qquad (f_m = f_t n_t N)$$

$$V = \frac{\pi D N}{12} = \frac{\pi \times 8 \times 140}{12} = 293 \text{ fpm} \qquad \left(N = \frac{12 V}{\pi D}\right)$$

**Estimating Drilling Thrust, Torque, and Power.** — Although the lips of a drill cut metal and produce a chip in the same manner as the cutting edges of other metal cutting tools, the chisel edge removes the metal by means of a very complex combination of extrusion and cutting. For this reason a separate method must be used to estimate the power required for drilling. Also, it is often desirable to know the magnitude of the thrust and the torque required to drill a hole. The formulas and tabular data provided in this section are based on information supplied by the National Twist Drill Division of Lear Siegler, Inc. The values in Tables 30 through 33 are for sharp drills and the tool

**Table 30.   Work Material Factor, $K_d$, for Drilling with a Sharp Drill**

| Work Material | Work Material Constant, $K_d$ |
|---|---|
| AISI 1117 (Resulfurized free machining mild steel) | 12,000 |
| Steel, 200 Bhn | 24,000 |
| Steel, 300 Bhn | 31,000 |
| Steel, 400 Bhn | 34,000 |
| Cast Iron, 150 Bhn | 14,000 |
| Most Aluminum Alloys | 7,000 |
| Most Magnesium Alloys | 4,000 |
| Most Brasses | 14,000 |
| Leaded Brass | 7,000 |
| Austenitic Stainless Steel (Type 316) | 24,000* for Torque 35,000* for Thrust |
| Titanium Alloy T16A   4V   40$R_c$ | 18,000* for Torque 29,000* for Thrust |
| Rene 41          40$R_c$ | 40,000*† min. |
| Hastelloy-C | 30,000* for Torque 37,000* for Thrust |

\* Values based upon a limited number of tests.       † Will increase with rapid wear.

wear factors are given in Table 27. For most ordinary drilling operations 1.30 can be used as the tool wear factor. When drilling most difficult-to-machine materials and when the drill is allowed to become very dull, 1.50 should be used as the value of this factor. It is usually more convenient to measure the web thickness at the drill point than the length of the chisel edge; for this reason, the approximate w/d ratio corresponding to each c/d ratio for a correctly ground drill is provided in Table 33. For most standard twist drills the c/d ratio is .18, unless the drill has been ground short or the web has been thinned. The c/d ratio of split point drills is .03. The formulas given below can be used for spade drills, as well as for twist drills. Separate formulas are required for use with customary inch units and for SI metric units.

For inch units only:

$$T = 2K_dF_fF_TBW + K_dd^2JW \tag{3}$$

$$M = K_dF_fF_MAW \tag{4}$$

$$P_c = \frac{MN}{63,025} \tag{5}$$

For SI metric units only:

$$T = 0.05\,K_dF_fF_TBW + 0.007\,K_dd^2JW \tag{6}$$

$$M = \frac{K_dF_fF_MAW}{40,000} = 0.000025\,K_dF_fF_MAW \tag{7}$$

$$P_c = \frac{MN}{9550} \tag{8}$$

Use with either inch or metric units:

$$P_m = \frac{P_c}{E} \tag{9}$$

Where:   $P_c$ = Power at the cutter; hp, or kW
$P_m$ = Power at the motor; hp, or kW
$M$ = Torque; in. lb, or N·m
$T$ = Thrust; lb, or N
$K_d$ = Work material factor                    (See Table 30)

## Table 31. Feed Factors, $F_f$, for Drilling

| Inch Units | | | | SI Metric Units | | | |
|---|---|---|---|---|---|---|---|
| Feed, in./rev | $F_f$ | Feed, in./rev | $F_f$ | Feed, mm/rev | $F_f$ | Feed, mm/rev | $F_f$ |
| .0005 | .0023 | .012 | .029 | 0.01 | .025 | 0.30 | .382 |
| .001 | .004 | .013 | .031 | 0.03 | .060 | 0.35 | .432 |
| .002 | .007 | .015 | .035 | 0.05 | .091 | 0.40 | .480 |
| .003 | .010 | .018 | .040 | 0.08 | .133 | 0.45 | .528 |
| .004 | .012 | .020 | .044 | 0.10 | .158 | 0.50 | .574 |
| .005 | .014 | .022 | .047 | 0.12 | .183 | 0.55 | .620 |
| .006 | .017 | .025 | .052 | 0.15 | .219 | 0.65 | .708 |
| .007 | .019 | .030 | .060 | 0.18 | .254 | 0.75 | .794 |
| .008 | .021 | .035 | .068 | 0.20 | .276 | 0.90 | .919 |
| .009 | .023 | .040 | .076 | 0.22 | .298 | 1.00 | 1.000 |
| .010 | .025 | .050 | .091 | 0.25 | .330 | 1.25 | 1.195 |

## Table 32. Drill Diameter Factors: $F_T$ for Thrust; $F_M$ for Torque

| Inch Units | | | | | | SI Metric Units | | | | | |
|---|---|---|---|---|---|---|---|---|---|---|---|
| Drill Diam., in. | $F_T$ | $F_M$ | Drill Diam., in. | $F_T$ | $F_M$ | Drill Diam., mm | $F_T$ | $F_M$ | Drill Diam., mm | $F_T$ | $F_M$ |
| .063 | .110 | .007 | .875 | .899 | .786 | 1.60 | 1.46 | 2.33 | 22.00 | 11.86 | 260.8 |
| .094 | .151 | .014 | .938 | .950 | .891 | 2.40 | 2.02 | 4.84 | 24.00 | 12.71 | 305.1 |
| .125 | .189 | .024 | 1.000 | 1.000 | 1.000 | 3.20 | 2.54 | 8.12 | 25.50 | 13.34 | 340.2 |
| .156 | .226 | .035 | 1.063 | 1.050 | 1.116 | 4.00 | 3.03 | 12.12 | 27.00 | 13.97 | 377.1 |
| .188 | .263 | .049 | 1.125 | 1.099 | 1.236 | 4.80 | 3.51 | 16.84 | 28.50 | 14.58 | 415.6 |
| .219 | .297 | .065 | 1.250 | 1.195 | 1.494 | 5.60 | 3.97 | 22.22 | 32.00 | 16.00 | 512.0 |
| .250 | .330 | .082 | 1.375 | 1.290 | 1.774 | 6.40 | 4.42 | 28.26 | 35.00 | 17.19 | 601.6 |
| .281 | .362 | .102 | 1.500 | 1.383 | 2.075 | 7.20 | 4.85 | 34.93 | 38.00 | 18.36 | 697.6 |
| .313 | .395 | .124 | 1.625 | 1.475 | 2.396 | 8.00 | 5.28 | 42.22 | 42.00 | 19.89 | 835.3 |
| .344 | .426 | .146 | 1.750 | 1.565 | 2.738 | 8.80 | 5.96 | 50.13 | 45.00 | 21.02 | 945.8 |
| .375 | .456 | .171 | 1.875 | 1.653 | 3.100 | 9.50 | 6.06 | 57.53 | 48.00 | 22.13 | 1062 |
| .438 | .517 | .226 | 2.000 | 1.741 | 3.482 | 11.00 | 6.81 | 74.90 | 50.00 | 22.86 | 1143 |
| .500 | .574 | .287 | 2.250 | 1.913 | 4.305 | 12.50 | 7.54 | 94.28 | 58.00 | 25.75 | 1493 |
| .563 | .632 | .355 | 2.500 | 2.081 | 5.203 | 14.50 | 8.49 | 123.1 | 64.00 | 27.86 | 1783 |
| .625 | .687 | .429 | 2.750 | 2.246 | 6.177 | 16.00 | 9.19 | 147.0 | 70.00 | 29.93 | 2095 |
| .688 | .741 | .510 | 3.000 | 2.408 | 7.225 | 17.50 | 9.87 | 172.8 | 76.00 | 31.96 | 2429 |
| .750 | .794 | .596 | 3.500 | 2.724 | 9.535 | 19.00 | 10.54 | 200.3 | 90.00 | 36.53 | 3293 |
| .813 | .847 | .689 | 4.000 | 3.031 | 12.13 | 20.00 | 10.98 | 219.7 | 100.00 | 39.81 | 3981 |

## Table 33. Chisel Edge Factors for Torque and Thrust

| c/d | Approx. w/d | Torque Factor A | Thrust Factor B | Thrust Factor J | c/d | Approx. w/d | Torque Factor A | Thrust Factor B | Thrust Factor J |
|---|---|---|---|---|---|---|---|---|---|
| .03 | .025 | 1.000 | 1.100 | .001 | .18 | .155 | 1.085 | 1.355 | .030 |
| .05 | .045 | 1.005 | 1.140 | .003 | .20 | .175 | 1.105 | 1.380 | .040 |
| .08 | .070 | 1.015 | 1.200 | .006 | .25 | .220 | 1.155 | 1.445 | .065 |
| .10 | .085 | 1.020 | 1.235 | .010 | .30 | .260 | 1.235 | 1.500 | .090 |
| .13 | .110 | 1.040 | 1.270 | .017 | .35 | .300 | 1.310 | 1.575 | .120 |
| .15 | .130 | 1.080 | 1.310 | .022 | .40 | .350 | 1.395 | 1.620 | .160 |

For drills of standard design, use $c/d = .18$.
For split point drills, use $c/d = .03$.
$c/d$ = Length of Chisel Edge ÷ Drill Diameter.
$w/d$ = Web Thickness at Drill Point ÷ Drill Diameter.

$F_f$ = Feed factor (See Table 31)
$F_T$ = Thrust factor for drill diameter (See Table 32)
$F_M$ = Torque factor for drill diameter (See Table 32)
$A$ = Chisel edge factor for torque (See Table 33)
$B$ = Chisel edge factor for thrust (See Table 33)
$J$ = Chisel edge factor for thrust (See Table 33)
$W$ = Tool wear factor (See Table 27)
$N$ = Spindle speed; rpm
$E$ = Machine tool efficiency factor (See Table 28)
$D$ = Drill diameter; in., or mm
$c$ = Chisel edge length; in., or mm (See Table 33)
$w$ = Web thickness at drill point; in. or mm (See Table 33)

*Example.* A standard ⅞ inch drill is to drill steel parts having a hardness of 200 Bhn on a drilling machine having an efficiency of .80. The spindle speed to be used is 350 rpm and the feed rate will be .008 in./rev. Calculate the thrust, torque, and power required to drill these holes:

$K_d$ = 24,000   (From Table 30)     $A$ = 1.085   (From Table 33)
$F_f$ = .021   (From Table 31)     $B$ = 1.355   (From Table 33)
$F_T$ = .899   (From Table 32)     $J$ = .030   (From Table 33)
$F_M$ = .786   (From Table 32)     $W$ = 1.30   (From Table 27)

$T = 2K_dF_fF_TBW + K_dd^2JW$
$\quad = 2 \times 24,000 \times .021 \times .899 \times 1.355 \times 1.30 + 24,000 \times .875^2 \times .030 \times 1.30$
$\quad = 2313$ lb

$M = K_dF_fF_mAW$
$\quad = 24,000 \times .021 \times .786 \times 1.085 \times 1.30 = 559$ in. lb

$$P_c = \frac{MN}{63,025} = \frac{559 \times 350}{63,025} = 3.1 \text{ hp} \qquad P_m = \frac{P_c}{E} = \frac{3.1}{.80} = 3.9 \text{ hp}$$

Twist drills are generally the most highly stressed of all metal cutting tools. They must not only resist the cutting forces on the lips, but also the drill torque resulting from these forces and the very large thrust force required to push the drill through the hole. Therefore, in many instances when drilling smaller holes, the twist drill places a limit on the power used while for very large holes, the machine may limit the power.

## Tool Wear

Metal cutting tools wear constantly when they are being used. A normal amount of wear should not be a cause for concern until the size of the worn region has reached the point where the tool should be replaced. Normal wear cannot be avoided and should be differentiated from abnormal tool breakage or excessively fast wear. Tool breakage and an excessive rate of wear indicate that the tool is not operating correctly and steps should be taken to correct this situation.

There are several basic mechanisms that cause tool wear. It is generally understood that tools wear as a result of abrasion which is caused by hard particles of work material plowing over the surface of the tool. Wear is also caused by diffusion or alloying between the work material and the tool material. In regions where the conditions of contact are favorable, the work material reacts with the tool material causing an attrition of the tool material. The rate of this attrition is dependent upon the temperature in the region of contact and the reactivity of the tool and the work materials with each other. Diffusion or alloying also occurs where particles of the work material are welded to the surface of the tool. These welded deposits are often quite visible in the form of a built-up edge, as particles or a layer of work material inside a crater or as small mounds attached to the face of the tool. The diffusion or alloying occurring between these deposits and the tool weakens the tool material below the weld. Frequently these deposits are again rejoined to the chip by welding or they are simply broken away by the force of collision with the passing chip. When this happens, a small amount of the tool material may remain attached to the deposit and be plucked from the surface of the tool, to be carried away with the chip. This mechanism can cause chips to be broken from the cutting edge and the formation of small craters on the tool face called pull-outs. It can also contribute to the enlargement of the larger crater that sometimes forms behind the cutting edge. Among the other mechanisms that can cause tool wear are severe thermal gradients and thermal shocks, which cause cracks to form near the cutting edge, ultimately leading to tool failure. This condition can be caused by improper tool grinding procedures, heavy interrupted cuts, or by the improper application of cutting fluids when machining at high cutting speeds. Chemical reactions between the active constituents in some cutting fluids sometimes accelerate the rate of tool wear. Oxidation of the heated metal near the cutting edge also contributes to tool wear, particularly when fast cutting speeds and high cutting temperatures are encountered. Breakage of the cutting edge caused by overloading, heavy shock loads, or improper tool design is not normal wear and should be corrected.

The wear mechanisms described bring about visible manifestations of wear on the tool which should be understood so that the proper corrective measures can be taken, when required. These visible signs of wear are described in the following paragraphs and the corrective measures that might be required are given in the accompanying Tool Trouble-Shooting Check List. The best procedure when trouble shooting is to try to correct only one condition at a time. When a correction has been made it should be checked. After one condition has been corrected, work can then start to correct the next condition.

*Flank Wear:* Tool wear occurring on the flank of the tool below the cutting edge is called flank wear. Flank wear always takes place and cannot be avoided. It should not give rise to concern unless the rate of flank wear is too fast or the flank wear land becomes too large in size. The size of the flank wear can be measured as the distance between the top of the cutting edge and the bottom of the flank wear land. In practice, a visual estimate is usually made instead of a precise measurement, although in many instances flank wear is ignored and the tool wear is "measured" by the loss of size on the part. The best measure of tool wear, however, is flank wear. When it becomes too large, the rubbing action of the wear land against the workpiece increases and the cutting edge must be replaced. Because conditions vary, it is not possible to give an exact amount of flank wear at which the tool should be replaced. Although there are many exceptions, as a rough estimate, high-speed steel tools should be replaced when the width of the flank wear land reaches .005 to .010 inch for finish turning and .030 to .060 inch for rough turning; and for cemented carbides .005 to .010 inch for finish turning and .020 to .040 inch for rough turning.

Under ideal conditions which, surprisingly, occur quite frequently, the width of the flank wear land will be very uniform along its entire length. When the depth of cut is uneven, such as when turning out-of-round stock, the bottom edge of the wear land may become somewhat slanted, the wear land being wider toward the nose. A jagged-appearing wear land usually is evidence of chipping at the cutting edge. Sometimes only one or two sharp depressions of the lower edge of the wear land will appear, to indicate that the cutting edge has chipped above these depressions. A deep notch will sometimes occur at the "depth of cut line", or that part of the cutting opposite the original surface of the work. This can be caused by a hard surface scale on the work, by a work-hardened surface layer on the work, or when machining high-temperature alloys. Often the size of the wear land is enlarged at the nose of the tool. This can be a sign of crater breakthrough near the nose or of chipping in this region. Under certain conditions, when machining with carbides, it can be an indication of deformation of the cutting edge in the region of the nose.

When a sharp tool is first used, the initial amount of flank wear is quite large in relation to the subsequent total amount. Under normal operating conditions, the width of the flank wear land will increase at a uniform rate until it reaches a critical size after which the cutting edge breaks down completely. This is called catastrophic failure and the cutting edge should be replaced before this occurs. When cutting at slow speeds with high-speed steel tools, there may be long periods when no increase in the flank wear can be observed. For a given work material and tool material, the rate of flank wear is primarily dependent on the cutting speed and then the feed rate.

*Cratering:* A deep crater will sometimes form on the face of the tool which is easily recognizable. The crater forms at a short distance behind the side cutting edge leaving a small shelf between the cutting edge and the edge of the crater. This shelf is sometimes covered with the built-up edge and at other times it is uncovered. Often the bottom of the crater is obscured with work material that is welded to the tool in this region. Under normal operating conditions, the crater will gradually enlarge until it breaks through a part of the cutting edge. Usually this occurs on the end cutting edge just behind the nose. When this takes place, the flank wear at the nose increases rapidly and complete tool failure follows shortly. Sometimes cratering cannot be avoided and a slow increase in the size of the crater is considered normal. However, if the rate of crater growth is rapid, leading to a short tool life, corrective measures must be taken.

*Cutting Edge Chipping:* Small chips are sometimes broken from the cutting edge which accelerates tool wear but does not necessarily cause immediate tool failure. Chipping can be recognized by the appearance of the cutting edge and the flank wear land. A sharp depression in the lower edge of the wear land is a sign of chipping and if this edge of the wear land has a jagged appearance it indicates that a large amount of chipping has taken place. Often the vacancy or cleft in the cutting edge that results from chipping is filled up with work material that is tightly welded in place. This occurs very rapidly when chipping is caused by a built-up edge on the face of the tool. In this manner the damage to the cutting edge is healed; however, the width of the wear land below the chip is usually increased and the tool life is shortened.

*Deformation:* Deformation occurs on carbide cutting tools when taking a very heavy cut using a slow cutting speed and a high feed rate. A large section of the cutting edge then becomes very hot and the heavy cutting pressure compresses the nose of the cutting edge, thereby lowering the face of the tool in the area of the nose. This reduces the relief under the nose, increases the width of the wear land in this region, and shortens the tool life.

*Surface Finish:* The finish on the machined surface does not necessarily indicate poor cutting tool performance unless there is a rapid deterioration. A good surface finish is, however, sometimes a requirement. The principal cause of a poor surface finish is the built-up edge which forms along the edge of the cutting tool. The elimination of the built-up edge will always result in an improvement of the surface finish. The most effective way to eliminate the built-up edge is to increase the cutting speed. When the cutting speed is increased beyond a certain critical cutting speed, there will be a rather sudden and large improvement in the surface finish. Cemented carbide tools can operate successfully at higher cutting speeds, where the built-up edge does not occur and where a good surface finish is obtained. Whenever possible, cemented carbide tools should be operated at cutting speeds where a good surface finish will result. There are times when this is not possible. Also, high-speed steel tools cannot be operated at the speed where the built-up edge does not form. In these cases the most effective method of obtaining a good surface finish is to employ a cutting fluid that has active sulphur or chlorine additives.

Cutting tool materials that do not alloy readily with the work material are also effective in obtaining an improved surface finish. Straight titanium carbide and diamond are the two principal tool materials that fall into this category.

The presence of feed marks can mar an otherwise good surface finish and attention must be paid to the feed rate and the nose radius of the tool if a good surface finish is desired. Changes in the tool geometry can also be helpful. A small "flat", or secondary cutting edge, ground on the end cutting edge behind the nose will some-times provide the desired surface finish. Finally, when the tool is in operation, the flank wear should not be allowed to become too large, particularly in the region of the nose where the finished surface of the tool is produced.

## Cutting Tools

**Tool Contour.** — Tools for turning, planing, etc., are made in straight, bent, offset, and other forms to locate the cutting edges in convenient positions for oper-ating on differently located surfaces. The contour or shape of the cutting edge may also be varied to suit different classes of work. Tool shapes, however, are not only related to the kind of operation, but, in the case of roughing tools particu-larly, the contour may have a decided effect upon the cutting efficiency of the tool. To illustrate, an increase in the side cutting-edge angle of a roughing tool, or in the nose radius, tends to permit higher cutting speeds because the chip will be thinner for a given feed rate. Such changes, however, may result in chattering or vibrations unless the work and the machine are rigid; hence, the most desirable contour may be a compromise between the ideal form and one that is needed to meet practical requirements.

**Terms and Definitions.** — The terms and definitions relating to single-point tools vary somewhat in different plants, but the following are in general use.

*Single-point Tool:* This term is applied to tools for turning, planing, boring, etc., which have a cutting edge at one end. This cutting edge may be formed on one end of a solid piece of steel, or the cutting part of the tool may consist of an insert or tip which is held to the body of the tool either by brazing, welding, or by mechanical means.

*Shank:* The shank is the main body of the tool. If the tool is an inserted cutter type, the shank supports the cutter or bit. (See diagram, Fig. 1.)

*Nose:* This is a general term sometimes used to designate the cutting end but usually it relates more particularly to the rounded tip of the cutting end.

*Face:* The surface against which the chips bear, as they are severed in turning or planing operations, is called the face.

*Flank:* The flank is that end surface that is adjacent to the cutting edge and below it when the tool is in a horizontal position as for turning.

*Base:* The base is that surface of the tool shank which bears again. the supporting tool-holder or block.

*Side Cutting Edge:* The side cutting edge is the cutting edge located on the side of the tool. Tools, such as shown in Fig. 1, do the bulk of the cutting with this cutting edge and are, therefore, sometimes called side cutting edge tools.

*End Cutting Edge:* The end cutting edge is the cutting edge located at the end of the tool. On side cutting edge tools, the end cutting edge can be used for light plunging and facing cuts. Cut-off tools and similar tools have only one cutting edge located on the end. These tools and other tools that are intended to cut primarily with the end cutting edge are sometimes called end cutting edge tools.

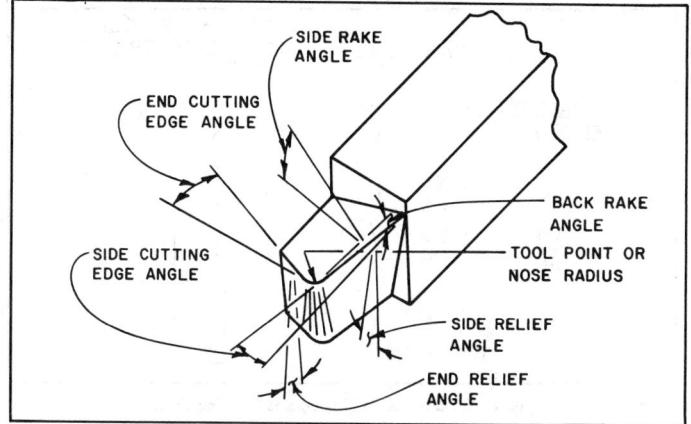

**Fig. 1. Terms Applied to Single-point Turning Tools.**

*Rake:* A metal cutting tool is said to have rake when the tool face or surface against which the chips bear as they are being severed, is inclined for the purpose of either increasing or diminishing the keenness or bluntness of the edge. The magnitude of the rake is most conveniently measured by two angles called the back rake angle and the side rake angle. The tool shown in Fig. 1 has rake. If the face of the tool did not incline but was parallel to the base, there would be no rake; the rake angles would be zero.

*Positive Rake:* If the inclination of the tool face is such as to make the cutting edge keener or more acute than when the rake angle is zero, the rake angle is defined as positive.

*Negative Rake:* If the inclination of the tool face makes the cutting edge less keen or more blunt than when the rake angle is zero, the rake is defined as negative.

*Back Rake:* The back rake is the inclination of the face toward or away from the end or the end cutting edge of the tool. When the inclination is away from the end cutting edge, as shown in Fig. 1, the back rake is positive. If the inclination is downward toward the end cutting edge it is negative.

*Side Rake:* The side rake is the inclination of the face toward or away from the side cutting edge. When the inclination is away from the side cutting edge, as shown in Fig. 1, the side rake is positive. If the inclination is toward the side cutting edge the side rake is negative.

*Relief:* The flanks below the side cutting edge and the end cutting edge must be relieved to allow these cutting edges to penetrate into the workpiece when taking a cut. If the flanks are not provided with relief, the cutting edges will rub against the workpiece and be unable to penetrate in order to form the chip. Relief is also provided below the nose of the tool to allow it to penetrate into the workpiece. The relief at the nose is usually a blend of the side relief and the end relief.

*End Relief Angle:* The end relief angle is a measure of the relief below the end cutting edge.

*Side Relief Angle:* The side relief angle is a measure of the relief below the side cutting edge.

*Back Rake Angle:* The back rake angle is a measure of the back rake. It is

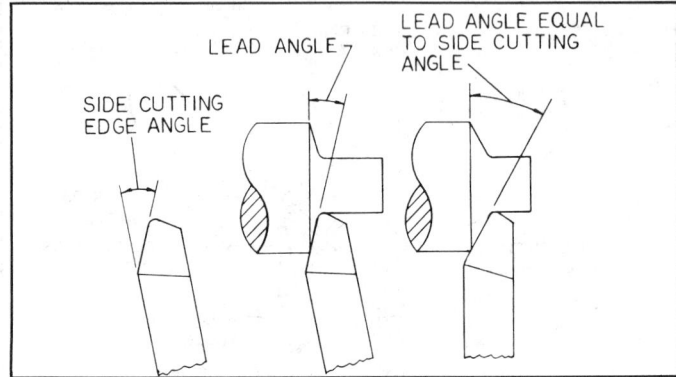

Fig. 2.   Lead Angle on Single-point Turning Tool.

measured in a plane that passes through the side cutting edge and is perpendicular to the base.   Thus, the back rake angle can be measured by measuring the inclination of the side cutting edge with respect to a line or plane that is parallel to the base. The back rake angle may be positive, negative, or zero depending upon the magnitude and direction of the back rake.

*Side Rake Angle:* The side rake angle is a measure of the side rake.   It is always measured in a plane that is perpendicular to the side cutting edge and perpendicular to the base.   Thus, the side rake angle is the angle of the inclination of the face perpendicular to the side cutting edge with reference to a line or a plane that is parallel to the base.

*End Cutting Edge Angle:* The end cutting edge angle is the angle made by the end cutting edge with respect to a plane perpendicular to the axis of the tool shank.   It is provided to allow the end cutting edge to clear the finish machined surface on the workpiece.

*Side Cutting Edge Angle:* The side cutting edge angle is the angle made by the side cutting edge and a plane that is parallel to the side of the shank.

*Nose Radius:* The nose radius is the radius of the nose of the tool.   The performance of the tool, in part, is influenced by nose radius and, for this reason, it must be carefully controlled.

*Lead Angle:* The lead angle, shown in Fig. 2, is not ground on the tool.   It is a tool setting angle which has a great influence on the performance of the tool.   The lead angle is bounded by the side cutting edge and a plane perpendicular to the workpiece surface when the tool is in position to cut; or, more correctly, the lead angle is the angle between the side cutting edge and a plane perpendicular to the direction of the feed travel.

*Solid Tool:* A solid tool is a cutting tool made from one piece of tool material.

*Brazed Tool:* A brazed tool is a cutting tool having a blank of cutting-tool material permanently brazed to a steel shank.

*Blank:* A blank is an unground piece of cutting-tool material from which a brazed tool is made.

*Tool Bit:* A tool bit is a relatively small cutting tool that is clamped in a holder in such a way that it can readily be removed and replaced.   It is intended primarily to be reground when dull and not indexed.

*Tool-bit Blank:* The tool-bit blank is an unground piece of cutting-tool material from which a tool bit can be made by grinding. It is available in standard sizes and shapes.

*Tool-bit Holder:* Made from forged steel, the tool-bit holder is used to hold the tool bit, to act as an extended shank for the tool bit, and to provide a means for clamping in the tool post.

*Straight-shank Tool-bit Holder:* A straight-shank tool-bit holder has a straight shank when viewed from the top. The axis of the tool bit is held parallel to the axis of the shank.

*Offset-shank Tool-bit Holder:* An offset-shank tool-bit holder has the shank bent to the right or left, as seen in Fig. 3. The axis of the tool bit is held at an angle with respect to the axis of the shank.

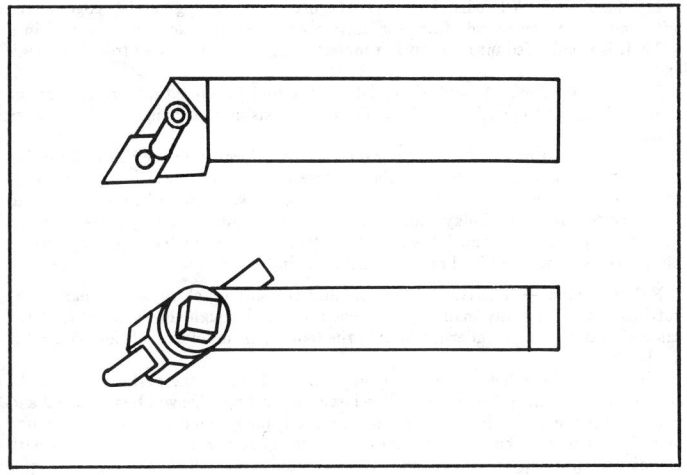

Fig. 3.   Top: Right-hand Offset-shank, Indexable Insert Holder.
Bottom: Right-hand Offset-shank Tool-bit Holder.

*Indexable Inserts:* An indexable insert is a relatively small piece of cutting-tool material that is geometrically shaped to have two or several cutting edges that are used until dull. It is then indexed on the holder to apply a sharp cutting edge. When all of the cutting edges have been dulled, the insert is discarded. The insert is held in a pocket or against other seating surfaces on an indexable insert holder by means of a mechanical clamping device that can be tightened or loosened easily.

*Indexable Insert Holder:* Made of steel, an indexable insert holder is used to hold indexable inserts. It is equipped with a mechanical clamping device that holds the inserts firmly in a pocket or against other seating surfaces.

*Straight-shank Indexable Insert Holder:* A straight-shank indexable insert tool-holder is essentially straight when viewed from the top, although the cutting edge of the insert may be oriented parallel, or at an angle, to the axis of the holder.

*Offset-shank Indexable Insert Holder:* An offset-shank indexable insert holder has the head end, or the end containing the insert pocket, offset to the right or left, as shown in Fig. 3.

*Side cutting Tool:* A side cutting tool has its major cutting edge on the side of the cutting part of the tool. The major cutting edge may be parallel or at an angle with respect to the axis of the tool.

*End cutting Tool:* An end cutting tool has its major cutting edge on the end of the cutting part of the tool. The major cutting edge may be perpendicular or at an angle, with respect to the axis of the tool.

*Curved Cutting-edge Tool:* A curved cutting-edge tool has a continuously variable side cutting edge angle. The cutting edge is usually in the form of a smooth, continuous curve along its entire length, or along a large portion of its length.

*Right-hand Tool:* A right-hand tool has the major, or working, cutting edge on the right-hand side when viewed from the cutting end with the face up. As used in a lathe, it is usually fed into the work from right to left, when viewed from the shank end.

*Left-hand Tool:* A left-hand tool has the major or working cutting edge on the left-hand side when viewed from the cutting end with the face up. As used in a lathe, it is usually fed into the work from left to right, when viewed from the shank end.

*Neutral-hand Tool:* A neutral-hand tool is a tool to cut either left to right or right to left; or the cut may be parallel to the axis of the shank as when plunge cutting.

*Chipbreaker:* A groove formed in or on a shoulder formed on the face of a turning tool back of the cutting edge, for the purpose of breaking up the chips and thus preventing the formation of long, continuous chips which would be dangerous to the operator and also bulky and cumbersome to handle. A chipbreaker of the shoulder type may be formed directly on the tool face or it may consist of a separate piece that is either held by brazing or by clamping.

**Relief Angles.** — The end relief angle and the side relief angle on single-point cutting tools are usually made equal to each other, although this is not always the case. The relief angle under the nose of the tool is a blend of the side and end relief angles.

The size of the relief angles has a pronounced effect on the performance of the cutting tool. If the relief angles are too large, the cutting edge will be weakened and in danger of breaking when a heavy cutting load is placed on it by a hard and tough material. On finish cuts, rapid wear of the cutting edge may cause problems with size control on the part. Relief angles that are too small will cause the rate of wear on the flank of the tool below the cutting edge to increase, thereby significantly reducing the tool life. In general, when cutting hard and tough materials, the relief angles should be 6 to 8 degrees for high-speed steel tools and 5 to 7 degrees for carbide tools. For medium steels, mild steels, cast iron, and other average work the recommended values of the relief angles are 8 to 12 degrees for high-speed steel tools and 5 to 10 degrees for carbides. Ductile materials having a relatively low modulus of elasticity should be cut using larger relief angles. For example, the relief angles recommended for turning copper, brass, bronze, aluminum, ferritic malleable iron, and similar metals are 12 to 16 degrees for high-speed steel tools and 8 to 14 degrees for carbides.

Larger relief angles generally tend to produce a better surface finish on the finish machined surface because less surface of the worn flank of the tool rubs against the workpiece. For this reason, single-point thread-cutting tools should be provided with relief angles that are as large as circumstances will permit. Problems encountered when machining stainless steel may be overcome by increasing the size of the relief angle. The relief angles used should never be smaller than necessary.

**Rake Angles.** — Machinability tests have confirmed that when the rake angle along which the chip slides, called the true rake angle, is made larger in the positive

both top and bottom surfaces. The positive-rake chipbreaker surface may be ground or precision molded on the insert.

Many materials, such as gray cast iron, form a discontinuous chip. For these materials an insert that has plain faces without chipbreaker grooves should always be used. Steels and other ductile materials form a continuous chip that must be broken into small segments when machined on lathes and planers having single-point, cemented-carbide and cemented-oxide cutting tools; otherwise, the chips can cause injury to the operator. In this case a chipbreaker must be used. Some inserts are made with chipbreaker grooves molded or ground directly on the insert. When inserts with plain faces are used, a cemented-carbide plate-type chipbreaker is clamped on top of the insert.

**Identification System for Indexable Inserts.** — The size of indexable inserts is determined by the diameter of an inscribed circle (I.C.), except for rectangular and parallelogram inserts where the length and width dimension is used. To describe an insert in its entirety a standard (ANSI B94.4-1965, R1972) identification system is used where each position number designates a feature of the insert. The ANSI standard system is given below, except in a few instances where a few nonstandard designations representing common usage are included:

$$\overset{\text{1}\quad\text{2}\quad\text{3}\quad\text{4}\quad\quad\text{5}\ \text{6}\ \text{7}\ \text{8}}{\textsf{S N M G} - \textsf{6 4 3 E}}$$

1. *Shape:* The shape of the insert is designated by a letter: **R** for round; **S**, square; **T**, triangle; **C**, 80° diamond; **D**, 55° diamond; **V**, 35° diamond; **M**, 86° diamond; **B**, 82° parallelogram; **F**, 70° parallelogram; **E**, 55° parallelogram; **L**, rectangle; **H**, hexagon; **O**, octagon; and **P**, pentagon.

2. *Relief Angle:* The relief in the insert is designated by a letter; when mounted on a holder, the actual relief angle may be different from that on the insert: **N**, for 0°; **A**, 3°; **B**, 5°; **C**, 7°; **P**, 10°; **E**, 20°; **F**, 25°; and **G**, 30°.

3. *Tolerances:* The third position is a letter to designate the tolerance of the inscribed circle diameter and the insert thickness, as given in the following table:

| Symbol | Tolerance | | Symbol | Tolerance | |
|---|---|---|---|---|---|
| | Inscribed Circle, Inch | Thickness, Inch | | Inscribed Circle, Inch | Thickness, Inch |
| **A** | ±.0002 | ±.001 | **E** | ±.001 | ±.001 |
| **B** | ±.0002 | ±.005 | **G** | ±.001 | ±.005 |
| **C** | ±.0005 | ±.001 | **M*** | ±.002, ±.004 | ±.005 |
| **D** | ±.0005 | ±.005 | **U*** | ±.005, ±.012 | ±.005 |

R = Blank with grind stock on all surfaces.
S = Blank with grind stock on top and bottom only.

\* Exact tolerance is determined by size of insert.

4. *Type:* The type of insert is designated by a letter. **A**, with hole; **B**, with hole and countersink; **C**, with hole and two countersinks; **D**, smaller than ¼ inch I.C. with hole; **E**, smaller than ¼ inch I.C. without hole; **F**, clamp-on type with chipbreaker; **G**, with hole and chipbreaker; **H**, with hole, one countersink and chipbreaker; **J**, with hole, two countersinks and chipbreaker; **K**, smaller than ¼ inch

I.C. with hole and chipbreaker; **L**, smaller than ¼ inch I.C. without hole and with chipbreaker; **P**†, 10° positive surface contour with hole and with chipbreaker; **S**†, 20° positive surface contour with hole and with chipbreaker; **Z**†, with hole and with chipbreaker on one side only.

5. *Size:* The size of the insert is designated by a number having the following meaning. For inserts less than ¼ inch I.C., the number represents the I.C. diameter in ½₂nds of an inch. For inserts ¼ inch I.C., and over, the number represents the I.C. diameter in ⅛ths of an inch. Rectangular and parallelogram inserts require two digits: 1st digit-number represents an ⅛th of an inch in width, and 2nd digit-number represents a ¼ of an inch in length.

6. *Thickness:* The thickness is designated by a number. For inserts less than ¼ inch I.C., the number represents ½₂nds inch of thickness. For inserts ¼ inch I.C., and over, the number represents the ¹⁄₁₆ths inch of thickness. For rectangular and parallelogram inserts, use width dimension in place of I.C. to designate thickness.

7. *Cutting Point Radius or Flat:* The cutting point, or nose radius, is designated by a number representing ¹⁄₆₄ths of an inch; a flat at the cutting point, or nose, is designated by a letter: **o** for sharp corner; **1**, ¹⁄₆₄-inch radius; **2**, ¹⁄₃₂-inch radius; **3**, ³⁄₆₄-inch radius; **4**, ¹⁄₁₆-inch radius; **6**, ³⁄₃₂-inch radius; **8**, ⅛-inch radius; and **A** for square insert with 45° chamfer; **B**, square insert with 45° chamfer and 4° sweep angle, right-handed or negative; **C**, square insert with 45° chamfer and 4° sweep angle, left-handed; **D**, square insert with 30° chamfer, right-hand or negative; **E**, square insert with 15° chamfer, right-hand or negative; **F**, square insert with 5° chamfer, right-hand or negative; **G**, square insert with 30° chamfer, left-hand; **H**, square insert with 15° chamfer, left-hand; **J**, square insert with 5° chamfer, left-hand; **K**, square insert with 30° double chamfer; **L**, square insert with 15° double chamfer; **M**, square insert with 5° double chamfer; **N**, truncated triangle insert; **P**, flat corner triangle insert, right-hand or negative; **R**, flatted corner triangle insert, left-hand; **S**, square negative insert with 10° double chamfer; **T**, square negative insert with 30° positive-rake chamfer; and **V**, octagon negative insert with 22½° corner chamfer.

8. *Other Conditions:* This position is intended primarily to specify the finish by means of a letter designation; it is not always used. **A** for ground all over, light-honed; **B**, ground all over, heavy-honed; **C**, ground top and bottom only, light-honed; **D**, ground top and bottom only, heavy-honed; **E**, unground insert, honed; **F**, unground insert, not honed; **T**, chamfer cutting edge; and **J**, polished.

**Indexable Insert Tool Holders.** — Indexable insert tool holders are made from a good grade of steel which is heat treated to a hardness of 44 to 48 $R_c$ for most normal applications. Accurate pockets are machined in the end of tool holders which serve to locate the insert in position and to provide surfaces against which the insert can be clamped. In almost all cases, a cemented carbide seat is provided which is held in the bottom of the pocket by a screw or by the clamping pin, if one is used. The seat is necessary to provide a flat bearing surface upon which the insert can rest and, in so doing, it adds materially to the ability of the insert to withstand the cutting load. The seating surface of the holder may provide a positive-, negative-, or a neutral-rake orientation to the insert when it is in position on the holder. Holders, therefore, are classified as positive, negative, or neutral rake.

There are four basic methods used to clamp the insert on the holder: 1. Top clamping; 2. Pin-lock clamping; 3. Multiple clamping using a top clamp and a pin lock; and, 4. Clamping the insert with a machine screw. All top clamps are actuated by a screw that forces the clamp directly against the insert. When required, a cemented-carbide, plate type chipbreaker is placed between the clamp and the insert. Pin-lock clamps require an insert having a hole: the pin acts against

† Common usage but not standard.

Table 3 (*Continued*). Indexable Insert Holder Application Guide

| Tool | Tool Holder Style | Insert Shape | Rake N-Negative P-Positive | Turn | Face | Turn and Face | Turn and Backface | Trace | Groove | Chamfer | Bore | Plane |
|---|---|---|---|---|---|---|---|---|---|---|---|---|
| | G | R | N | • | • | • | | | | | | |
| 10° 0° | G | C | N | • | • | • | | | | | | |
| | | | P | • | • | • | | | | | | |
| 38° | H | D | N | • | • | | | • | | | | |
| | | | | | | | | | | | | |
| -3° | J | T | N | | | | • | • | | | | |
| | | | P | | | | • | • | | | | |
| -3° | J | D | N | | | | • | • | | | | |
| | | | | | | | | | | | | |
| -3° | J | V | N | | | | • | • | | | | |
| | | | | | | | | | | | | |
| 15° | K | S | N | • | • | | | | | | • | |
| | | | P | • | • | | | | | | • | |
| 15° | K | C | N | • | • | | | | | | • | |
| | | | | | | | | | | | | |
| 5° -5° | L | C | N | | | • | • | | | | | |
| | | | | | | | | | | | | |
| 33° 27° | N | T | N | • | • | | | • | | | | |
| | | | P | • | • | | | • | | | | |
| 27° | N | D | N | • | • | | | • | | | | |
| | | | | | | | | | | | | |
| 45° | S | S | N | • | • | • | | • | | • | • | • |
| | | | P | • | • | • | | • | | • | • | • |
| 10° | W | S | N | • | • | | | | | | | |

strength and the maximum number of cutting edges available. Thus, a round insert is the strongest and has a maximum number of available cutting edges. It can be used with heavier feeds while producing a good surface finish. Round inserts are limited by their tendency to cause chatter, which may preclude using them. The square insert is the next most effective shape, providing good corner strength and more cutting edges than all other inserts except the round insert. The only limitation of this insert shape is that it must be used with a lead angle. Therefore, the square insert cannot be used for turning square shoulders or for back facing. Triangle inserts are the most versatile and can be used to perform more operations than any other insert shape. The 80-degree diamond insert is designed primarily for heavy turn and face operations, using the 100-degree corners, and for turning and back-facing square shoulders using the 80-degree corners. The 55- and 35-degree diamond inserts are intended primarily for tracing.

*Lead Angle:* Tool holders should be selected to provide the largest possible lead angle, although limitations are sometimes imposed by the nature of the job. For example, when turning and back-facing a shoulder, a negative lead angle must be used. Slender or small-diameter parts may deflect, causing difficulties in holding size or chatter when the lead angle is too large.

*End Cutting Edge Angle:* When tracing or contour turning, the plunge angle is determined by the end cutting edge angle. A two-degree minimum clearance angle should be provided between the workpiece surface and the end cutting edge of the insert. Table 4 provides the maximum plunge angle for holders commonly used to plunge when tracing. When severe cratering cannot be avoided, an insert having a small, end cutting edge angle is desirable to delay the crater breakthrough behind the nose. For very heavy cuts a small, end cutting edge angle will strengthen the corner of the tool.

Table 4.   Maximum Plunge Angle for Tracing or Contour Turning

| Tool Holder Style | Insert* Shape | Maximum Plunge Angle | Tool Holder Style | Insert* Shape | Maximum Plunge Angle |
|---|---|---|---|---|---|
| E | T | 58° | J | D | 30° |
| D and S | S | 43° | J | V | 50° |
| H | D | 71° | N | T | 55° |
| J | T | 25° | N | D | 58°–60° |

* Insert Shape: $S$ = square; $T$ = triangle; $D$ = 55° diamond; $V$ = 35° diamond.

**Indexable Insert Holders for NC.** — Indexable insert holders for numerical control lathes are usually made to more precise standards than ordinary holders. When made to these standards they are said to be qualified to prescribed tolerances, obtainable from the manufacturer. In some cases special shank modifications are required to clamp the holders in the machine, the modification being unique to the machine builder. When programming for NC turning, the programmed path is the path of the center of the tool tip, which is the center of the point, or nose radius, of the insert. Since the actual surfaces produced are the result of the path of the nose and the major cutting edge, it is necessary to compensate for the nose or point radius and the lead angle when writing the program. Table 5, developed by Kennametal Inc., provides the compensating dimensions for different holder styles. The reference point is determined by the intersection of extensions from the major and minor cutting edges, which would be the location of the point of a sharp pointed tool. The distances from this point to the nose radius are $L_1$ and $D_1$; $L_2$ and $D_2$ are the distances from the sharp point to the center of the nose radius. Threading tools have sharp corners and do not require a radius compensation. Other dimensions of importance in programming threading tools are also given.

The $C$ and $F$ dimensions are tool holder dimensions other than the shank size. In all cases the $C$ dimension is parallel to the length of the shank and the $F$ dimension is parallel to the side dimension; actual dimensions must be obtained from the manufacturer. For all K style holders the $C$ dimension is the distance from the end of the shank to the tangent point of the nose radius and the end cutting edge of the insert. For all other holders the $C$ dimension is from the end of the shank to a tangent to the nose radius of the insert. The $F$ dimension on all B, D, E, M, P, and V style holders is measured from the back side of the shank to the tangent point of the nose radius and the side cutting edge of the insert. For all A, F, G, J, K, and L style holders the $F$ dimension is the distance from the back side of the shank to the tangent of the nose radius of the insert. In all cases the nose radius is the standard radius corresponding to the inscribed circle as given in Table 2, on page 1818.

**Table 5. Insert Radius Compensation for NC Programming**

| Square Profile | | | | | | |
|---|---|---|---|---|---|---|
| **B Style\*** Also Applies to R Style | | Turning 15° Lead Angle | | | | |
| | | Rad. | $L$-$1$ | $L$-$2$ | $D$-$1$ | $D$-$2$ |
| | | $\frac{1}{64}$ | .0034 | .0190 | .0009 | .0110 |
| | | $\frac{1}{32}$ | .0067 | .0380 | .0019 | .0219 |
| | | $\frac{3}{64}$ | .0101 | .0570 | .0028 | .0329 |
| | | $\frac{1}{16}$ | .0135 | .0760 | .0037 | .0439 |
| **D Style\*** Also Applies to S Style | | Turning 45° Lead Angle | | | | |
| | | Rad. | $L$-$1$ | $L$-$2$ | $D$-$1$ | $D$-$2$ |
| | | $\frac{1}{64}$ | .0064 | .0219 | .0064 | 0 |
| | | $\frac{1}{32}$ | .0128 | .0439 | .0128 | 0 |
| | | $\frac{3}{64}$ | .0193 | .0658 | .0191 | 0 |
| | | $\frac{1}{16}$ | .0257 | .0877 | .0255 | 0 |
| **K Style\*** | | Facing 15° Lead Angle | | | | |
| | | Rad. | $L$-$1$ | $L$-$2$ | $D$-$1$ | $D$-$2$ |
| | | $\frac{1}{64}$ | .0009 | .0110 | .0034 | .0190 |
| | | $\frac{1}{32}$ | .0019 | .0219 | .0067 | .0380 |
| | | $\frac{3}{64}$ | .0028 | .0329 | .0101 | .0570 |
| | | $\frac{1}{16}$ | .0037 | .0439 | .0135 | .0760 |

All dimensions are given in inches.

\* $L$-$1$ and $D$-$1$ over sharp point to nose radius; and $L$-$2$ and $D$-$2$ over sharp point to center of nose radius. The $D$-$1$ dimension for the $B$, $E$, $D$, $M$, $P$, and $V$ style tools and the $L$-$1$ dimension on K style tools are over the sharp point of insert to a sharp point at the intersection of a line on the lead angle on the cutting edge of the insert and a line running perpendicular to the front of the tool that is tangent to the radius of the insert.

*Courtesy of Kennametal, Inc.*

Table 5 (*Continued*).  **Insert Radius Compensation for NC Programming**

| Triangle Profile | | |
|---|---|---|

**A Style\***
Also Applies to
G Style

Turning 0° Lead Angle

| Rad. | L-1 | L-2 | D-1 | D-2 |
|---|---|---|---|---|
| 1/64 | .0112 | .0269 | 0 | .0155 |
| 1/32 | .0225 | .0537 | 0 | .0310 |
| 3/64 | .0337 | .0806 | 0 | .0465 |
| 1/16 | .0450 | .1074 | 0 | .0620 |

**B Style\***
Also Applies to
R Style

Facing & Turning 15° Lead Angle

| Rad. | L-1 | L-2 | D-1 | D-2 |
|---|---|---|---|---|
| 1/64 | .0143 | .0300 | .0039 | .0080 |
| 1/32 | .0287 | .0600 | .0077 | .0161 |
| 3/64 | .0430 | .0899 | .0116 | .0241 |
| 1/16 | .0573 | .1198 | .0155 | .0321 |

**E Style\***

Turning 30° Lead Angle

| Rad. | L-1 | L-2 | D-1 | D-2 |
|---|---|---|---|---|
| 1/64 | .0152 | .0303 | .0088 | 0 |
| 1/32 | .0303 | .0606 | .0175 | 0 |
| 3/64 | .0455 | .0909 | .0262 | 0 |
| 1/16 | .0606 | .1212 | .0350 | 0 |

**F Style\***

Facing 90° Lead Angle

| Rad. | L-1 | L-2 | D-1 | D-2 |
|---|---|---|---|---|
| 1/64 | 0 | .0155 | .0112 | .0269 |
| 1/32 | 0 | .0310 | .0225 | .0537 |
| 3/64 | 0 | .0465 | .0337 | .0806 |
| 1/16 | 0 | .0620 | .0450 | .1074 |

**J Style\***

Turning & Facing 3° Lead Angle

| Rad. | L-1 | L-2 | D-1 | D-2 |
|---|---|---|---|---|
| 1/64 | .0104 | .0260 | .0013 | .0169 |
| 1/32 | .0208 | .0520 | .0025 | .0338 |
| 3/64 | .0312 | .0780 | .0038 | .0507 |
| 1/16 | .0416 | .1041 | .0051 | .0676 |

\* See footnote page 1823

**Table 5** *(Continued).* **Insert Radius Compensation for NC Programming**

| | 80° Diamond Profile | | | | | |
|---|---|---|---|---|---|---|
| G Style* | | Turning 0° Lead Angle | | | | |
| | | Rad. | *L-1* | *L-2* | *D-1* | *D-2* |
| | | $\frac{1}{64}$ | .0029 | .0185 | 0 | .0155 |
| | | $\frac{1}{32}$ | .0057 | .0370 | 0 | .0310 |
| | | $\frac{3}{64}$ | .0085 | .0554 | 0 | .0465 |
| | | $\frac{1}{16}$ | .0114 | .0739 | 0 | .0620 |
| L Style* | | Turning & Facing 5° Reverse Lead Angle | | | | |
| | | Rad. | *L-1* | *L-2* | *D-1* | *D-2* |
| | | $\frac{1}{64}$ | .0014 | .0171 | .0014 | .0171 |
| | | $\frac{1}{32}$ | .0029 | .0341 | .0029 | .0341 |
| | | $\frac{3}{64}$ | .0043 | .0512 | .0043 | .0512 |
| | | $\frac{1}{16}$ | .0057 | .0682 | .0057 | .0682 |
| F Style* | | Facing 0° Lead Angle | | | | |
| | | Rad. | *L-1* | *L-2* | *D-1* | *D-2* |
| | | $\frac{1}{64}$ | 0 | .0155 | .0029 | .0185 |
| | | $\frac{1}{32}$ | 0 | .0310 | .0057 | .0370 |
| | | $\frac{3}{64}$ | 0 | .0465 | .0085 | .0554 |
| | | $\frac{1}{16}$ | 0 | .0620 | .0114 | .0740 |
| B Style* Also Applies to R Style | | Turning 15° Lead Angle | | | | |
| | | Rad. | *L-1* | *L-2* | *D-1* | *D-2* |
| | | $\frac{1}{64}$ | .0009 | .0166 | .0003 | .0117 |
| | | $\frac{1}{32}$ | .0019 | .0333 | .0006 | .0232 |
| | | $\frac{3}{64}$ | .0029 | .0498 | .0009 | .0348 |
| | | $\frac{1}{16}$ | .0038 | .0663 | .0012 | .0464 |
| K Style* | | Facing 15° Lead Angle | | | | |
| | | Rad. | *L-1* | *L-2* | *D-1* | *D-2* |
| | | $\frac{1}{64}$ | .0003 | .0117 | .0009 | .0166 |
| | | $\frac{1}{32}$ | .0006 | .0232 | .0019 | .0333 |
| | | $\frac{3}{64}$ | .0009 | .0348 | .0029 | .0498 |
| | | $\frac{1}{16}$ | .0012 | .0464 | .0038 | .0663 |

* See footnote page 1823

**Table 5** (*Continued*).   **Insert Radius Compensation for NC Programming**

| 80° Diamond Profile (cont.) | | | | | | | |
|---|---|---|---|---|---|---|---|

### M Style*

| | Turning 40° Lead Angle | | | | |
|---|---|---|---|---|---|
| Rad. | *L-1* | *L-2* | *D-1* | *D-2* |
| 1/64 | .0086 | .0240 | .0072 | 0 |
| 1/32 | .0171 | .0479 | .0144 | 0 |
| 3/64 | .0257 | .0718 | .0215 | 0 |
| 1/16 | .0342 | .0958 | .0287 | 0 |

### 55° Profile

### J Style*

| | Profiling 3° Reverse Lead Angle | | | | |
|---|---|---|---|---|---|
| Rad. | *L-1* | *L-2* | *D-1* | *D-2* |
| 1/64 | .0133 | .0289 | .0014 | .0170 |
| 1/32 | .0266 | .0579 | .0028 | .0341 |
| 3/64 | .0399 | .0868 | .0043 | .0511 |
| 1/16 | .0533 | .1158 | .0057 | .0682 |

### P Style*

| | Profiling 27° 30′ Lead Angle | | | | |
|---|---|---|---|---|---|
| Rad. | *L-1* | *L-2* | *D-1* | *D-2* |
| 1/64 | .0177 | .0328 | .0092 | 0 |
| 1/32 | .0353 | .0656 | .0184 | 0 |
| 3/64 | .0530 | .0985 | .0276 | 0 |
| 1/16 | .0707 | .1313 | .0368 | 0 |

### 35° Profile

### J Style*

| | Profiling 3° Reverse Lead Angle | | | | |
|---|---|---|---|---|---|
| Rad. | *L-1* | *L-2* | *D-1* | *D-2* |
| 1/64 | .0327 | .0483 | .0024 | .0181 |
| 1/32 | .0654 | .0966 | .0049 | .0361 |
| 3/64 | .0980 | .1449 | .0073 | .0542 |
| 1/16 | .1307 | .1932 | .0097 | .0722 |

### V Style*

| | Profiling 17° 30′ Lead Angle | | | | |
|---|---|---|---|---|---|
| Rad. | *L-1* | *L-2* | *D-1* | *D-2* |
| 1/64 | .0342 | .0489 | .0108 | 0 |
| 1/32 | .0684 | .0978 | .0216 | 0 |
| 3/64 | .1026 | .1468 | .0324 | 0 |
| 1/16 | .1368 | .1957 | .0431 | 0 |

* See footnote page 1823

Table 5 (Continued).   Insert Radius Compensation for NC Programming

| Insert Size | Threading | | | | | |
|---|---|---|---|---|---|---|
| | T | R | U | Y | X | Z |
| 2 | ⁵⁄₃₂ Wide | .040 | .075 | .040 | .024 | .140 |
| 3 | ³⁄₁₆ Wide | .046 | .098 | .054 | .031 | .183 |
| 4 | ¼ Wide | .053 | .128 | .054 | .049 | .239 |
| 5 | ⅜ Wide | .099 | .190 | .... | .... | .... |

Buttress Threading   29° Acme   60° V-Threading

NTB-B   NTB-A   NA   NTF   NT

All dimensions are given in inches.

**Sharpening Of Twist Drills.** — Twist drills are cutting tools designed to perform concurrently several functions, such as penetrating directly into solid material, ejecting the removed chips outside the cutting area, maintaining the essentially straight direction of the advance movement and controlling the size of the drilled hole. The geometry needed for these multiple functions is incorporated into the design of the twist drill in such a manner that it can be retained even after repeated sharpening operations. Twist drills are actually resharpened many times during their service life, with the practically complete restitution of their original operational characteristics. However, in order to assure all the benefits which the design of the twist drill is capable of providing, the surface generated in the sharpening process must agree with the original form of the tool's operating surface, unless a change of shape is deliberately sought because the drill will be used for a different work material.

The principal elements of the tool geometry which are essential for the adequate cutting performance of twist drills are shown in Fig. 1.   The generally used values for these dimensions are the following:

*Point angle:* Commonly 118°, except for high strength steels, 118° to 135°; aluminum alloys, 90° to 140°; and magnesium alloys, 70° to 118°.

*Helix angle:* Commonly 24° to 32°, except for magnesium and copper alloys, 10° to 30°.

*Lip relief angle:* Commonly 10° to 15°, except for high strength or tough steels, 7° to 12°. The lower values of these angle ranges are used for drills of larger diameter, the higher values for the smaller diameters. For drills of diameters less than ¼ inch, the lip relief angles are increased beyond the listed max. values up to 24°. For soft and free machining materials, 12° to 18° except for diameters less than ¼ inch, 20° to 26°.

**Relief Grinding of the Tool Flanks.** — In the sharpening of the twist drill the tool flanks, containing the two cutting edges are ground.   Each of the flanks consists of a curved surface which provides the relief needed for the easy penetration and free cutting of the tool edges. In grinding the flanks, Fig. 2, the drill is swung around the axis *A* of an imaginary cone while resting in a support which holds the drill at one-half the point angle *B* with respect to the face of the grinding wheel. Feed *f* for stock removal is in the direction of the drill axis. That relief angle is

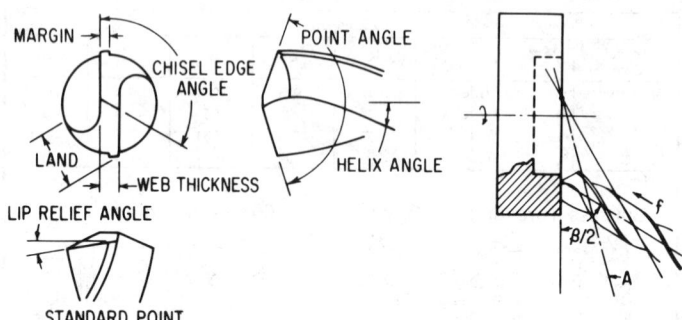

Fig. 1 (Left). The principal elements of tool geometry on twist drills. Fig. 2 (Right). In grinding the face of the twist drill the tool is swung around the axis $A$ of an imaginary cone, while resting in a support tilted by half of the point angle $\beta$ with respect to the face of the grinding wheel. Feed $f$ for stock removal is in the direction of the drill axis.

usually measured at the periphery of the twist drill and is also specified by that value. It is not a constant but should increase toward the center of the drill.

The relief grinding of the flank surfaces will generate the chisel angle on the web of the twist drill. The value of that angle, typically 55°, which can be measured, for example, with the protractor of an optical projector, is indicative of the correctness of the relief grinding.

**Drill Point Thinning.** — The chisel edge is the least efficient operating surface element of the twist drill because it does not cut, but actually squeezes or extrudes the work material. In order to improve the inefficient cutting conditions caused by the chisel edge, its width is often reduced in a drill-point thinning operation, resulting in a condition such as that shown in Fig. 3. Point thinning is particularly desirable on larger size drills and also on those which become shorter in usage, because the thickness of the web increases toward the shaft of the twist drill, thereby adding to the length of the chisel edge. The extent of point thinning is limited by the minimum strength of the web needed to avoid splitting of the drill point under the influence of cutting forces.

Both sharpening operations — the relieved face grinding and the point thinning — should be carried out in special drill grinding machines or with twist drill grinding fixtures mounted on general-purpose tool grinding machines, in order to assure the essential accuracy of the produced tool geometry. Off-hand grinding may be used

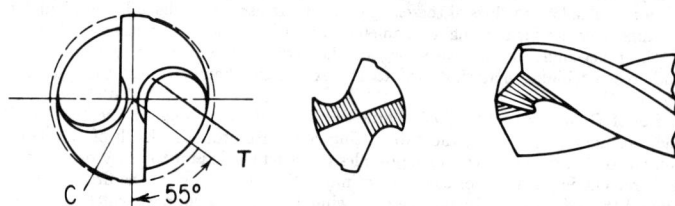

Fig. 3 (Left). The chisel edge $C$ after thinning the web by grinding off area $T$. Fig. 4 (Right). Split point or "crankshaft" type web thinning.

for the important web thinning when a special machine is not available; however, such operation requires skill and experience.

Improperly sharpened twist drills, e.g. those with unequal edge length or asymmetrical point angle, will tend to produce holes with poor diameter and directional control.

For deep holes and also drilling into stainless steel, titanium alloys, high temperature alloys, nickel alloys, very high strength materials and in some cases tool steels, split point grinding, resulting in a "crankshaft" type drill point, is recommended. In this type of pointing, see Fig. 4, the chisel edge is entirely eliminated, extending the positive rake cutting edges to the center of the drill, thereby greatly reducing the required thrust in drilling.

**Sharpening Carbide Tools.** — Cemented carbide indexable inserts are usually not resharpened but sometimes they require a special grind in order to form a contour on the cutting edge to suit a special purpose. Brazed type carbide cutting tools are resharpened after the cutting edge has become worn. On brazed carbide tools the cutting-edge wear should not be allowed to become excessive before the tool is resharpened. One method of determining when brazed carbide tools need resharpening is by periodic inspection of the flank wear and the condition of the face. Another method is to determine the amount of production which is normally obtained before excessive wear has taken place, or to determine the equivalent period of time. One disadvantage of this method is that slight variations in the work material will often cause the wear rate not to be uniform and the number of parts machined before regrinding will not be the same each time. Usually, sharpening should not require the removal of more than .005 to .010 inch of carbide.

*General Procedure in Carbide Tool Grinding:* The general procedure depends upon the kind of grinding operation required. If the operation is to resharpen a dull tool, a diamond wheel of 100 to 120 grain size is recommended although a finer wheel—up to 150 grain size—is sometimes used to obtain a better finish. If the tool is new or is a "standard" design and changes in shape are necessary, a 100-grit diamond wheel is recommended for roughing and a finer grit diamond wheel can be used for finishing. Some shops prefer to rough grind the carbide with a vitrified silicon carbide wheel, the finish grinding being done with a diamond wheel. A final operation commonly designated as lapping may or may not be employed for obtaining an extra-fine finish.

*Wheel Speeds:* — The speed of silicon carbide wheels usually is about 5000 feet per minute. The speeds of diamond wheels generally range from 5000 to 6000 feet per minute; yet lower speeds (550 to 3000 fpm) can be effective.

*Offhand Grinding.* — In grinding single-point tools (excepting chip breakers) the common practice is to hold the tool by hand, press it against the wheel face and traverse it continuously across the wheel face while the tool is supported on the machine rest or table which is adjusted to the required angle. This is known as "offhand grinding" to distinguish it from the machine grinding of cutters as in regular cutter grinding practice. The selection of wheels adapted to carbide tool grinding is very important.

**Silicon Carbide Wheels.** — The green colored silicon carbide wheels generally are preferred to the dark gray or gray-black variety, although the latter are sometimes used.

*Grain or Grit Sizes:* — For roughing, a grain size of 60 is very generally used. For finish grinding with silicon carbide wheels, a finer grain size of 100 or 120 is common. A silicon carbide wheel such as C60-I-7V may be used for grinding both the steel shank and carbide tip. However, for under-cutting steel shanks up to the carbide tip, it may be advantageous to use an aluminum oxide wheel suitable for high speed steel grinding.

*Grade:* — According to the standard system of marking, different grades from soft to hard are indicated by letters from A to Z. For carbide tool grinding fairly soft grades such as G, H, I and J are used. The usual grades for roughing are I or J and for finishing H, I and J. The grade should be such that a sharp free-cutting wheel will be maintained without excessive grinding pressure. Harder grades than those indicated tend to overheat and crack the carbide.

*Structure:* — The common structure numbers for carbide tool grinding are 7 and 8. The larger cup-wheels (10 to 14 inches) may be of the porous type and be designated as 12P. The standard structure numbers range from 1 to 15 with progressively higher numbers indicating less density and more open wheel structure.

**Diamond Wheels.** — Wheels with diamond-impregnated grinding faces, are fast and cool cutting and have a very low rate of wear. They are used extensively both for resharpening and for finish grinding of carbide tools when preliminary roughing is required. Diamond wheels are also adapted for sharpening multi-tooth cutters such as milling cutters, reamers, etc., which are ground in a cutter grinding machine.

*Resinoid bonded* wheels are commonly used for grinding chip breakers, milling cutters, reamers or other multi-tooth cutters. They are also applicable to precision grinding of carbide dies, gages, and various external, internal and surface grinding operations. Fast, cool cutting action is characteristic of these wheels.

*Metal bonded* wheels are often used for offhand grinding of single-point tools especially when durability or long life and resistance to grooving of the cutting face, are considered more important than the rate of cutting. *Vitrified bonded* wheels are used both for roughing of chipped or very dull tools and for ordinary resharpening and finishing. They provide rigidity for precision grinding, a porous structure for fast cool cutting, sharp cutting action and durability.

**Diamond Wheel Grit Sizes.** — For roughing with diamond wheels a grit size of 100 is the most common both for offhand and machine grinding. Grit sizes of 120 and 150 are frequently used in offhand grinding of single point tools (1) for resharpening, (2) for a combination roughing and finishing wheel and (3) for chip-breaker grinding. Grit sizes of 220 or 240 are used for ordinary finish grinding all types of tools (offhand and machine) and also for cylindrical, internal and surface finish grinding. Grits of 320 and 400 are used for "lapping" to obtain very fine finishes, and for hand hones. A grit of 500 is for lapping to a mirror finish on such work as carbide gages and boring or other tools for exceptionally fine finishes.

**Diamond Wheel Grades.** — Diamond wheels are made in several different grades to better adapt them to different classes of work. The grades vary for different types and shapes of wheels. Standard Norton grades are H, J and L for resinoid bonded wheels, grade N for metal bonded wheels and grades J, L, N and P for vitrified wheels. Harder and softer grades than standard may at times be used to advantage.

**Diamond Concentration.** — The relative amount (by carat weight) of diamond in the diamond section of the wheel is known as the "diamond concentration." Concentrations of 100 (high), 50 (medium) and 25 (low) ordinarily are supplied. A concentration of 50 represents one-half the diamond content of 100 (if the depth of the diamond is the same in each case) and 25 equals one-fourth the content of 100 or one-half the content of 50 concentration.

*100 Concentration:* — Recommended (especially in grit sizes up to about 220) for general machine grinding of carbides, and for grinding cutters and chip breakers. Vitrified and metal bonded wheels usually have 100 concentration.

*50 Concentration:* — In the finer grit sizes of 220, 240, 320, 400 and 500, a 50 concentration is recommended for offhand grinding with resinoid bonded cup-wheels.

*25 Concentration:* — A low concentration of 25 is recommended for offhand grinding with resinoid bonded cup-wheels with grit sizes of 100, 120 and 150.

*Depth of Diamond Section:* — The radial depth of the diamond section usually varies from 1/16 to 1/4 inch. The depth varies somewhat according to the wheel size and type of bond.

**Dry Versus Wet Grinding of Carbide Tools.** — In using silicon carbide wheels, grinding should be done either absolutely dry or with enough coolant to flood the wheel and tool. Satisfactory results may be obtained either by the wet or dry method. However, dry grinding is the most prevalent usually because, in wet grinding, operators tend to use an inadequate supply of coolant to obtain better visibility of the grinding operation and avoid getting wet; hence checking or cracking in many cases is more likely to occur in wet grinding than in dry grinding.

*Wet Grinding with Silicon Carbide Wheels:* — One advantage commonly cited in connection with wet grinding is that an ample supply of coolant permits using wheels about one grade harder than in dry grinding thus increasing the wheel life. Plenty of coolant also prevents thermal stresses and the resulting cracks, and there is less tendency for the wheel to load. A dust exhaust system also is unnecessary.

*Wet Grinding with Diamond Wheels:* — In grinding with diamond wheels the general practice is to use a coolant to keep the wheel face clean and promote free cutting. The amount of coolant may vary from a small stream to a coating applied to the wheel face by a felt pad.

**Coolants for Carbide Tool Grinding.** — In grinding either with silicon carbide or diamond wheels a coolant that is used extensively consists of water plus a small amount either of soluble oil, sal soda, or soda ash to prevent corrosion. One prominent manufacturer recommends for silicon carbide wheels about 1 ounce of soda ash per gallon of water and for diamond wheels kerosene. The use of kerosene is quite general for diamond wheels and usually it is applied to the wheel face by a felt pad. Another coolant recommended for diamond wheels consists of 80 per cent water and 20 per cent soluble oil.

**Peripheral Versus Flat Side Grinding.** — In grinding single point carbide tools with silicon carbide wheels, the roughing preparatory to finishing with diamond wheels may be done either by using the flat face of a cup-shaped wheel (side grinding) or the periphery of a "straight" or disk-shaped wheel. Even where side grinding is preferred, the periphery of a straight wheel may be used for heavy roughing

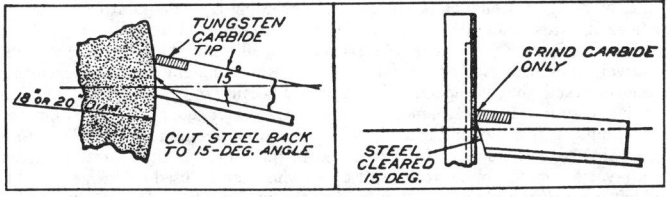

as in grinding back chipped or broken tools (see left-hand diagram). Reasons for preferring peripheral grinding include faster cutting with less danger of localized heating and checking especially in grinding broad surfaces. The advantages usually claimed for side grinding are that proper rake or relief angles are easier to obtain and the relief or land is ground flat.  The diamond wheels used for tool sharpening are designed for side grinding.  (See right-hand diagram.)

**Lapping Carbide Tools.** — Carbide tools may be finished by lapping, especially if an exceptionally fine finish is required on the work as, for example, tools used for precision boring or turning non-ferrous metals.  If the finishing is done by using a diamond wheel of very fine grit (such as 240, 320 or 400), the operation is often called "lapping."  A second lapping method is by means of a power-driven lapping disk charged with diamond dust, Norbide powder, or silicon carbide finishing compound.  A third method is by using a hand lap or hone usually of 320 or 400 grit.  In many plants the finishes obtained with carbide tools meet requirements without a special lapping operation.  In all cases any feather edge which may be left on tools should be removed and it is good practice to bevel the edges of roughing tools at 45 degrees to leave a chamfer 0.005 to 0.010 inch wide.  This is done by hand honing and the object is to prevent crumbling or flaking off at the edges when hard scale or heavy chip pressure is encountered.

*Hand Honing:* The cutting edge of carbide tools, and tools made from other tool materials, is sometimes hand honed before it is used in order to strengthen the cutting edge.  When interrupted cuts or heavy roughing cuts are to be taken, or when the grade of carbide is slightly too hard, hand honing is beneficial because it will prevent chipping, or even possibly, breakage of the cutting edge.  Whenever chipping is encountered, hand honing the cutting edge before use will be helpful.  It is important, however, to hone the edge lightly and only when necessary.  Heavy honing will always cause a reduction in tool life.  Normally, removing .002 to .004 inch from the cutting edge is sufficient.  When indexable inserts are used, the use of pre-honed inserts is preferred to hand honing although sometimes an additional amount of honing is required.  Hand honing of carbide tools in between cuts is sometimes done to defer grinding or to increase the life of a cutting edge on an indexable insert.  If correctly done, so as not to change the relief angle, this procedure is sometimes helpful.  If improperly done, it can result in a reduction in tool life.

**Chip Breaker Grinding.** — For this operation a straight diamond wheel is used on a universal tool and cutter grinder, a small surface grinder, or a special chip-breaker grinder.  A resinoid bonded wheel of the grade J or N commonly is used and the tool is held rigidly in an adjustable holder or vise.  The width of the diamond wheel usually varies from ⅛ to ¼ inch.  A vitrified bond may be used for wheels as thick as ¼ inch, and a resinoid bond for relatively narrow wheels.

**Summary of Miscellaneous Points.** — In grinding a single-point carbide tool, traverse it across the wheel face continuously to avoid localized heating.  This traverse movement should be quite rapid in using silicon carbide wheels and comparatively slow with diamond wheels.  A hand traversing and feeding movement, whenever practicable, is generally recommended because of greater sensitivity.  In grinding, maintain a constant, moderate pressure.  Never cool a hot tool by dipping it in a liquid, as this may crack the tip.  Wheel rotation should preferably be *against* the cutting edge or from the front face toward the back.  If the grinder is driven by a reversing motor, opposite sides of a cup wheel can be used for grinding right- and left-hand tools and with rotation against the cutting edge.  If it is necessary to grind the top face of a single-point tool, this should precede the grinding of the side and front relief, and top-face grinding should be minimized to maintain the tip thickness.  In machine grinding with a diamond wheel, limit the feed per traverse to 0.001 inch for 100 to 120 grit; 0.0005 inch for 150 to 240 grit; and 0.0002 inch for 320 grit and finer.

## Cutting Fluids for Machining

Cutting fluids can improve the performance of cutting tools by cooling the cutting tool and the workpiece, by lubricating the cutting and noncutting surfaces on the tool, by inhibiting seizure between the chip and the tool, and by flushing the chips from the work area. The results of the action of the cutting fluid can be a lowered tool force, improved tool life or a faster cutting speed, improved surface finish on the machined surface, and better control of workpiece accuracy. Cutting fluids are also used on grinding operations to cool the workpiece and to improve the surface finish.

Some cutting fluids are sometimes called coolants because of their cooling action. Cooling the cutting edge of the tool reduces the rate of tool wear and improves the tool life obtained at a given cutting speed. A faster cutting speed can be used if the tool life can be the same as for cutting dry. Sometimes difficulties can occur when applying a coolant to carbides. This is discussed further on. Cutting fluids also keep the workpiece cool and prevent dimensional changes that would affect precision measurements. On high-speed cutting operations the cooling action may be required to allow the operator to handle the workpiece. Friction always occurs between the sliding chip and the face of the tool, and between the worn flank of the tool and the workpiece. It also occurs on non-cutting tool surfaces which rub against the workpiece or over which the chip passes, when drilling, reaming, tapping, and broaching. Reducing the friction in any of these areas by using the lubricating qualities of the cutting fluid improves the surface finish on the work and lowers the tool force. In some applications the cutting fluid is selected to perform the usual function of a cutting fluid and, in addition, serve as a lubricant for the machine tool slides located near the cutting tools. Rust prevention on machined surfaces is another important function of the cutting fluid. The anti-seizure or anti-welding property of some cutting fluids is very important to attain a good finish on the machined surface. This property is generally only effective when cutting at the lower range (less than 200 fpm for steel) of cutting speed and, therefore, cutting fluids having anti-seizure properties usually are not used at higher speeds except, in some cases, for grinding. The anti-seizure properties inhibit the formation of the built-up edge and other adhesions to the cutting tool which are the principal cause of surface roughness on the machined surface. Other properties of cutting fluids that must also be considered are: the hygienic properties to prevent dermatitis on the hands of machine operators; the toxicity of the cutting fluid; the possibility of bacterial growth which causes unpleasant odors; the ease of handling, preparation, and concentration; the ability to resist foaming, smoking, and misting; and, the service life and cost.

There are four basic types of cutting fluids: 1) water-base soluble or emulsifying oils; 2) oil-base cutting fluids; 3) chemical or synthetic fluids; 4) gaseous fluids. Water-base soluble or emulsifying oils are sometimes classified as aqueous cutting fluids. Depending upon the application, they are mixed with water to a concentration of 1 part oil to 70 parts water up to as much as 1 part oil to 5 parts water. Soluble or emulsifying oils are very effective coolants and are used extensively as general-purpose cutting fluids for almost all machining operations and for grinding. Heavy-duty soluble oils and emulsions are available that contain certain additives which provide lubricating qualities and surface chemical effects in addition to their ability to act as a coolant. Improvements in tool life and surface finish are obtained through the use of heavy-duty soluble oils and emulsions. Chemical cutting fluids are also classified as synthetic cutting fluids. They are available to suit a wide range of applications. Some have excellent coolant properties and others combine cooling with very good properties as a lubricant. The most common gaseous cutting fluid is compressed air. Other gaseous fluids that have been used are carbon dioxide ($CO_2$), liquid argon, and liquid nitrogen. Oil base cutting fluids have been combined with compressed air to form a mist that is sprayed on the work and the tool and acts as a cutting fluid. In some cases this has been very effective. Care must be exercised with this type of cooling to prevent injurious effects to the machine operator. Sprayed or brushed on cutting fluids used to perform difficult operations such as tapping have also proven to be effective.

There are no fixed rules or formulas for selecting cutting fluids, except the rule of an actual test with a given tool or operation and material. The following paragraphs treat the more widely used types of cutting fluids used in machine shop practice.

**Emulsifying or Soluble Oils.** — Soluble oils are extensively used for machining operations, particularly when an inexpensive cooling medium meets practical requirements. When an emulsifying oil is mixed with water, an emulsion is formed rather than a solution, but the name "soluble oil" has been generally accepted because the oil apparently dissolves or goes into solution with the water. To obtain a mixture of oil and water, an emulsifying agent is required and soap has proved to be very effective. The emulsion formed contains an infinite number of minute and invisible oil particles which give the mixture a milky or creamy white color. The proportions of the mixture should be determined by test. If the mixture is too weak, it may cause corrosion of both work and machine. The mixture may range from 1 part soluble oil to 5 parts water up to 1 part soluble oil and 50 or even 100 parts water. Soluble oils are easily mixed; but since the emulsion is of the oil-in-water type, it should always be prepared by pouring the *oil* into the water instead of pouring the water into the oil. The emulsification is aided at temperatures somewhat above 40 degrees F., but the use of boiling water may have a reverse tendency. Agitation or stirring increases the emulsifying tendency unless too violent when there may be a reverse tendency. Even though a soluble oil may serve readily with hard water, it may be desirable to soften the water for obtaining one of the leaner mixtures or to eliminate scum formation. To prevent corrosion when a soluble oil is used, there should be sufficient air circulation to evaporate the water from the machined surface, thus leaving the oil to form a protective coating. In other words, if rapid drying is prevented either by lack of air circulation or by high humidity, rusting is likely to occur.

*Regular Soluble Oils:* These are primarily considered to be coolants; however, heavy-duty soluble oils are available which contain additives that provide lubrication and other surface effects in addition. Because of their superior cooling properties when compared to oil-base cutting fluids, they drastically reduce or entirely eliminate the problem of smoking and fogging when heavy cuts are taken and, they have replaced oil-base fluids on some applications. True solution type cutting fluids and temperature-soluble dispersions have been developed which are superior to emulsions for machining and grinding titanium. These fluids have a strong detergent action and the coolant system of the machine must be thoroughly cleaned before they are used.

*Paste Compounds:* Emulsions may be formed by mixing paste compounds with water. A paste made of saponified mineral oil with a high soap content has the consistency of grease. Paste compounds are not used as extensively as soluble oils but they are often applied particularly in connection with grinding operations. These paste compounds do not emulsify with water as readily as the soluble oils and this is particularly true with hard water. Hot water and steam may be required in some cases to obtain the best mixing results.

**Straight Cutting Oils.** — Straight cutting oils may be either in the mineral class or in the fatty or animal class such as the lard oil.

*Straight Mineral Oils:* The chief use of mineral oils in connection with machining processes is for blending with base cutting oils for obtaining whatever properties are required for different machining operations. Straight mineral oils are applicable to light machining operations especially on non-ferrous metals or in machining free-cutting steel, particularly if both cooling and lubricating effects are required. The low viscosity grades are preferable because of their cooling and penetrating properties. Straight mineral oils are sometimes used on automatics in preference to an emulsion which might interfere with proper lubrication of adjacent machine slides or other parts.

*Straight Fatty Oils:* Straight fatty oils have in the past been used quite extensively, especially for the heavier machining operations, but certain objectionable features have resulted in the substitution of other cutting fluids. The cost of a fatty oil such as lard oil is not only high but a straight animal oil tends to become rancid and produce an objectionable odor. Bacteria breed in such oils and cause skin troubles among machine operators. Sulphurized cutting oils have supplanted the fatty oils to a considerable degree, for machining operations requiring chip lubrication. Fatty oils, however, may be used in cases where a sulphurized mineral oil would tarnish the machined parts unless some preventive treatment is applied.

**Mineral Lard Oil.** — Mixtures of mineral oil and lard oil may be used to obtain a cutting fluid having greater lubricating value than a straight mineral oil and much lower cost than a straight lard oil. They may be used when the chip pressures are moderate. The proportion of lard oil and mineral oil depends upon the character of the machining operation and the need for both cooling and lubricating effects. For light machining, straight mineral oil may be satisfactory; for heavier duty or where the metal is removed at comparatively high rate, the mineral oil may contain from 10 to 40 per cent of lard oil. As a general rule, the percentage of lard oil is increased with the hardness of the stock to be machined. The No. 1 or prime lard oil is the most commonly used animal fatty oil. There are, however, synthetic fatty oils which are used in conjunction with mineral oils to obtain the so-called mineral lard oils. Mineral lard oils are frequently used in connection with automatic screw machine practice. The Saybolt viscosity (which is controlled by the mineral oil) generally varies from 150 to 225 seconds at 100 degrees F. for the blended oil.

**Sulphurized and Chlorinated Cutting Oils.** — The oils in these two general classes are especially useful when there is high chip-bearing pressure as in machining alloy or other tough steels. The sulphurized and chlorinated base oils are blended with mineral oils and produce metallic oxides or a *metallic*-film lubrication instead of fluid-film lubrication, as a result of the heat generated by the cutting tool. While mineral oils may contain natural sulphur, it has little or no value in cutting oils. There are two methods of adding sulphur. The sulphur may be cooked in and bonded with a fatty oil such as lard or sperm oil, thus obtaining a compound or "sulphur base" which is added to the mineral oil. The transparent oils are produced by this method, and the required viscosity is obtained by blending the base with a mineral oil. A second method is to add the sulphur direct to the mineral oil by the application of heat. The result is a dark or opaque oil which is suitable where work visibility is not required as in pipe threading and miscellaneous roughing operations. Chlorine, like sulphur, is added to mineral oils to obtain cutting fluids suitable for high chip-bearing pressures or to assist in obtaining extremely high pressure (EP) lubrication.

*Chemical or Synthetic Cutting Fluids.* — These cutting and grinding fluids are true solutions of organic and inorganic materials in water. They contain little or no petroleum products. They may be classified into two types: 1) those with added lubricants and wetting agents, and 2) those without.

Chemical cutting fluids containing lubricants and wetting agents are recommended for use on tough machining operations such as tapping, threading, sawing, broaching, gear shaving, and gear cutting. They are also recommended for turning,

boring, drilling, and milling.· They are excellent rust inhibitors, have a high heat
conductivity, and their excellent lubricity keeps the machine tool slides moving
smoothly. They tend to produce good finishes on the machined surfaces and they
do not require degreasing of the parts which is an advantage in production. They
do not combine readily with hard water and have a tendency to foam which makes
them unsuitable for use on disc-wheel type surface grinders.

Chemical cutting fluids that do not contain lubricants or wetting agents do not
foam readily and are excellent coolants and rust inhibitors. They are clear solutions
and are recommended for surface grinding operations. They are not recommended
for the severe machining operations where their lack of lubricity is a disadvantage.
Evaporation of the water will leave behind deposits that can accumulate and interfere with the operation of machine slides. They do resist the growth of undesirable
bacteria and do not form a scum in hard water.

**Aqueous Solutions.** — When the function of a cutting fluid is merely to cool
the work and possibly wash away chips, water containing some alkali has often been
used, although these aqueous solutions have been replaced quite largely by modern
cutting fluids. They do not, of course, provide the lubricating film that is important for many classes of work and they also cause corrosion of both work and the
machine. These aqueous solutions may contain carbonate of soda, borax, caustic
soda, etc. Lard oil and soft soap may also be used to improve the properties. An
inexpensive mixture for turning, milling, etc., is made in the following proportions:
1 pound of sal-soda (carbonate of soda), 1 quart of lard oil, 1 quart of soft soap, and
enough water to make 10 or 12 gallons. This mixture is boiled for one-half hour,
preferably by passing a steam coil through it. If the solution should have an
objectionable smell, this can be eliminated by adding about 2 pounds of unslaked
lime.

**Cutting Fluids for Different Materials and Operations.** — The selection of the
cutting fluid depends upon the machinability of the metal, the severity of the
operation, the cutting tool material, and the overall cost. Other factors, too, must
be considered. Some shops standardize on a few cutting fluids which are to serve all
purposes. In other cases, one cutting fluid must be used for all of the operations
performed on a machine. Sometimes, in such cases, a very severe operating condition is alleviated by applying the "right" cutting fluid manually while the machine
supplies the cutting fluid for the other operations through its coolant system. There
are many excellent proprietary cutting fluids to select from and it is not possible to
specify from among these. While the following recommendations represent good
practice, they are to serve as a guide only, and it is not intended to say that other
cutting fluids will not, in certain specific cases, also be effective.

*Steels:* Caution should be used when using a cutting fluid on steel that is being
turned at a high cutting speed with cemented carbide cutting tools. See "The
Application of Cutting Fluids to Carbides" later. Frequently this operation is
performed dry. If a cutting fluid is used, it should be a soluble oil mixed to a
consistency of about 1 part oil to 20 to 30 parts water. A sulphurized mineral oil is
recommended for reaming with carbide tipped reamers although a heavy-duty
soluble oil has also been used successfully.

The cutting fluid recommended for machining steel with high speed cutting tools
depends largely on the severity of the operation. For ordinary turning, boring,
drilling, and milling on medium and low strength steels, use a soluble oil having a
consistency of 1 part oil to 10 to 20 parts water. For tool steels and tough alloy
steels, a heavy-duty soluble oil having a consistency of 1 part oil to 10 parts water is
recommended for turning and milling, while for drilling and reaming a light sul-

## Cutting Fluids Recommended for Machining Operations — 1

*Soluble Oils.* — Types of oils or paste compounds which form emulsions when mixed with water: Soluble oils are used extensively in machining both ferrous and non-ferrous metals when the cooling quality is paramount and the chip-bearing pressure is not excessive. Care should be taken in selecting the proper soluble oil for precision grinding operations. Grinding coolants should be free from fatty materials that tend to load the wheel, thus affecting the finish on the machined part. Soluble coolants should contain rust preventive constituents to prevent corrosion.

*Base Oils.* — Various types of highly sulphurized and chlorinated oils containing inorganic, animal, or fatty materials. This "base stock" usually is "cut back" or blended with a lighter oil, unless the chip-bearing pressures are high, as in cutting alloy steel, in which case the base stock may be used straight. Base oils usually have a viscosity range of from 300 to 900 seconds at 100 degrees F.

*Mineral Oils.* — This group includes all types of oils extracted from petroleum such as paraffin oil, mineral seal oil, and kerosene. Mineral oils are often blended with base stocks, but they are generally used in the original form for light machining operations on both free-machining steels and non-ferrous metals. The coolants in this class should be of a type that has a relatively high flash point. Care should be taken to see that they are nontoxic, so that they will not be injurious to the operator. The heavier mineral oils (paraffin oils) usually have a viscosity of about 100 seconds at 100 degrees F. Mineral seal oil and kerosene have a viscosity of 35 to 60 seconds at 100 degrees F.

| Material to be Cut | Turning | Milling |
|---|---|---|
| Aluminum (Note 1–See notes on next page) | Mineral Oil with 10 Per Cent Fat (or) Soluble Oil | Soluble Oil (96 Per Cent Water) (or) Mineral Seal Oil (or) Mineral Oil |
| Alloy Steels (Note 2) | 25 Per Cent Sulphur base Oil* with 75 Per Cent Mineral Oil | 10 Per Cent Lard Oil with 90 Per Cent Mineral Oil |
| Brass | Mineral Oil with 10 Per Cent Fat | Soluble Oil (96 Per Cent Water) |
| Tool Steels and Low-carbon Steels | 25 Per Cent Lard Oil with 75 Per Cent Mineral Oil | Soluble Oil |
| Copper | Soluble Oil | Soluble Oil |
| Monel Metal | Soluble Oil | Soluble Oil |
| Cast Iron (Note 3) | Dry | Dry |
| Malleable Iron | Soluble Oil | Soluble Oil |
| Bronze | Soluble Oil | Soluble Oil |
| Magnesium (Note 5) | 10 Per Cent Lard Oil with 90 Per Cent Mineral Oil | Mineral Seal Oil |

**Cutting Fluids Recommended for Machining Operations — 2**

| Material to be Cut | Drilling | Tapping |
|---|---|---|
| Aluminum<br>(Note 4) | Soluble Oil<br>(75 to 90 Per Cent Water)<br>(or) 10 Per Cent Lard Oil with<br>90 Per Cent Mineral Oil | Lard Oil<br>(or) Sperm Oil<br>(or) Wool Grease<br>(or) 25 Per Cent Sulphur-base<br>Oil* Mixed with Mineral Oil |
| Alloy Steels<br>(Note 2) | Soluble Oil | 30 Per Cent Lard Oil with<br>70 Per Cent Mineral Oil |
| Brass | Soluble Oil<br>(75 to 90 Per Cent Water)<br>(or) 30 Per Cent Lard Oil with<br>70 Per Cent Mineral Oil | 10 to 20 Per Cent Lard Oil<br>with Mineral Oil |
| Tool Steels and<br>Low-carbon Steels | Soluble Oil | 25 to 40 Per Cent Lard Oil<br>with Mineral Oil<br>(or) 25 Per Cent Sulphur-base<br>Oil* with 75 Per Cent<br>Mineral Oil |
| Copper | Soluble Oil | Soluble Oil |
| Monel Metal | Soluble Oil | 25 to 40 Per Cent Lard Oil<br>Mixed with Mineral Oil<br>(or) Sulphur-base Oil* Mixed<br>with Mineral Oil |
| Cast Iron<br>(Note 3) | Dry | Dry<br>(or) 25 Per Cent Lard Oil with<br>75 Per Cent Mineral Oil |
| Malleable Iron | Soluble Oil | Soluble Oil |
| Bronze | Soluble Oil | 20 Per Cent Lard Oil with<br>80 Per Cent Mineral Oil |
| Magnesium<br>(Note 5) | 60-second Mineral Oil | 20 Per Cent Lard Oil with<br>80 Per Cent Mineral Oil |

*Note* 1. In machining aluminum, several varieties of coolants may be used. For rough machining, where the stock removal is sufficient to produce heat, water soluble mixtures can be used with good results to dissipate the heat. Other oils that may be recommended are straight mineral seal oil; a 50–50 mixture of mineral seal oil and kerosene; a mixture of 10 per cent lard oil with 90 per cent kerosene; and a 100-second mineral oil cut back with mineral seal oil or kerosene.

*\*Note* 2. The sulphur-base oil referred to contains 4½ per cent sulphur compound. Base oils are usually dark in color. As a rule, they contain sulphur compounds resulting from a thermal or catalytic refinery process. When so processed, they are more suitable for industrial coolants than when they have had such compounds as flowers of sulphur added by hand. The adding of sulphur compounds by hand to the coolant reservoir is of temporary value only, and the non-uniformity of the solution may affect the machining operation.

*Note* 3. A soluble oil or low-viscosity mineral oil may be used in machining cast iron to prevent excessive metal dust.

*Note* 4. Sulphurized oils ordinarily are not recommended for tapping aluminum; however, for some tapping operations they have proved very satisfactory, although the work should be slushed in a solvent right after machining to prevent discoloration.

*Note* 5. When machining magnesium, use an anhydrous non-acid oil. Water solubles or emulsions should never be used because water would intensify an accidental chip fire.

phurized mineral-fatty oil is used.  For tough operations such as tapping, threading, and broaching, a sulphochlorinated mineral-fatty oil is recommended for tool steels and high-strength steels, and a heavy sulphurized mineral-fatty oil or a sulpho-chlorinated mineral oil can be used for medium- and low-strength steels.  Straight sulphurized mineral oils are often recommended for machining tough, stringy low carbon steels in order to reduce tearing and rough surface finishes.

*Stainless Steel:* For ordinary turning and milling a heavy-duty soluble oil mixed to a consistency of 1 part oil to 5 parts water is recommended.  Broaching, threading, drilling, and reaming can be done with best results by using a sulphochlorinated mineral-fatty oil.

*Copper Alloys:* Most brasses, bronzes, and copper are stained when exposed to cutting oils containing active sulphur and chlorine; thus, sulphurized and sulpho-chlorinated oils should not be used.  For most operations a straight soluble oil, mixed to 1 part oil and 20 to 25 parts water is satisfactory.  For very severe operations and for automatic screw machine work a mineral-fatty oil is used.  A typical mineral-fatty oil might contain 5 to 10 per cent lard oil with the remainder mineral oil.

*Monel Metal:* When turning this material, an emulsion gives a slightly longer tool life than a sulpherized mineral oil, but the latter aids in chip breakage, which is frequently desirable.

*Aluminum Alloys:* Aluminum and aluminum alloys are frequently machined dry. When a cutting fluid is used it should be selected for its ability to act as a coolant. Soluble oils mixed to a consistency of 1 part oil to 20 to 30 parts water can be used. Mineral oil-base cutting fluids, when used to machine aluminum alloys, are frequently cut back to increase their viscosity in order to obtain good cooling characteristics and to flow easily to cover the tool and the work.  For example, a mineral-fatty oil or a mineral plus a sulphurized fatty oil can be cut back by the addition of as much as 50 per cent kerosene.

*Cast Iron:* Ordinarily, cast iron is machined dry.  Some increase in tool life can be obtained or a faster cutting speed can be used with a chemical cutting fluid or a soluble oil mixed to consistency of 1 part oil and 20 to 40 parts water.  A soluble oil is sometimes used to reduce the amount of dust around the machine.

*Magnesium:* A water soluble oil must never be used to machine magnesium.  Thin magnesium chips are subject to spontaneous combustion and, in the event of a fire, the water will release hydrogen which will intensify the fire.  Usually, magnesium is machined dry.  Light mineral oil or mineral-fatty oils are sometimes used to cool the work and to reduce the fire hazard.

*Grinding:* Soluble oil emulsions or emulsions made from paste compounds are used extensively in precision grinding operations.  For cylindrical grinding use 1 part oil to 40 to 50 parts water.  Solution type fluids and translucent grinding emulsions are particularly suited for many fine-finish grinding applications.  Mineral oil-base grinding fluids are recommended for many applications where a fine surface finish is required on the ground surface.  Mineral oils are used with vitrified wheels but are not recommended for wheels with rubber or shellac bonds.  Under certain conditions the oil vapor mist caused by the action of the grinding wheel can be ignited by the grinding sparks and explode.  To quench the grinding spark a secondary coolant line to direct a flow of grinding oil below the grinding wheel is recommended.

*Broaching:* For steel, use a heavy mineral oil such as sulphurized oil of 300 to 500 Saybolt viscosity at 100 degrees F. to provide both adequate lubricating effect and a dampening of the shock loads.  Soluble oil emulsions may be used for the lighter broaching operations.

**The Application of Cutting Fluids to Carbides.** — Carbide turning, boring, and similar operations on lathes are performed dry or with the help of soluble oil or chemical cutting fluids. The effectiveness of cutting fluids in improving tool life or by permitting higher cutting speeds to be used, is less with carbides than with high-speed steel tools. Furthermore, the effectiveness of the cutting fluid is lessened as the cutting speed is increased. Cemented carbides are very sensitive to sudden changes in temperature and to temperature gradients within the carbide. Thermal shocks to the carbide will cause thermal cracks to form near the cutting edge which are a prelude to tool failure. An unsteady or interrupted flow of the coolant reaching the cutting edge will generally cause thermal cracks to form near the cutting edge which results in a rapid failure of the tool. The flow of the chip over the face of the tool can cause an interruption to the flow of the coolant reaching the cutting edge even though a steady stream of coolant is directed at the tool. When a cutting fluid is used and frequent tool breakage is encountered, it is often best to cut dry. When a cutting fluid must be used to keep the workpiece cool for size control or to allow it to be handled by the operator, special precautions must be used. Sometimes applying the coolant from the front and the side of the tool simultaneously is helpful. On lathes equipped with overhead shields, it is very effective to apply the coolant from below the tool into the space between the shoulder of the work and the tool flank, in addition to applying the coolant from the top. Another method is not to direct the coolant stream at the cutting tool at all but to direct it at the workpiece above or behind the cutting tool.

The danger of thermal cracking is great when milling with carbide cutters. The nature of the milling operation itself tends to promote thermal cracking because the cutting edge is constantly heated to a high temperature and rapidly cooled as it enters and leaves the workpiece. For this reason, carbide milling operations should be performed dry.

Lower cutting-edge temperatures diminish the danger of thermal cracking. The cutting-edge temperatures usually encountered when reaming with solid carbide or carbide tipped reamers are generally such that thermal cracking is not apt to occur except when reaming certain difficult-to-machine metals. Therefore, cutting fluids are very effective when used on carbide reamers. Practically every kind of cutting fluid has been used, depending on the job material encountered. For difficult surface-finish problems in holes, heavy duty soluble oils, sulphurized mineral-fatty oils, and sulphochlorinated mineral-fatty oils have been used successfully. In some cases, the grade and the hardness of the carbide also have an effect on the surface finish of the hole.

**Application of Cutting Fluids.** — Cutting fluids should be applied where the cutting action is taking place and at the highest possible velocity without causing splashing. As a general rule, it is preferable to supply from 3 to 5 gallons per minute for each single-point tool on a machine such as a turret lathe or automatic. The temperature of the cutting fluid should be kept below 110 degrees F. If the volume of fluid used is not sufficient to maintain the proper temperature, means of cooling it should be provided.

**Cutting Fluids for Machining Magnesium.** — In machining magnesium, it is the general but not invariable practice in the United States to use a cutting fluid, whereas, in England, magnesium usually is machined dry except in cases where heat generated by high cutting speeds would not be dissipated rapidly enough without a cutting fluid. This condition may exist when, for example, small tools without much heat-conducting capacity are employed on automatics.

The cutting fluid for magnesium should be an anhydrous oil having, at most, a very low acid content. Various mineral-oil cutting fluids are used for magnesium.

To secure adequate cooling, the supply of fluid should be large (4 to 5 gallons per minute) and the viscosity low; however, to avoid too low a flash point, a compromise between cooling capacity and flash point is necessary. *Soluble oils or emulsions should never be used for machining magnesium.* Compressed air may be preferable to a fluid because it leaves the chips or swarf clean and dry.

A cutting fluid serves primarily to cool the work and also eliminate a possible fire hazard, especially when dull tools are operated at high speeds with fine feeds. Even when using sharp tools, the cut should not be less than 0.001 inch because fine chips are more likely to become ignited at high speeds. While a variety of mineral oils may be used, the following properties are recommended: Specific gravity 0.79 to 0.86; viscosity (Saybolt) at 100 degrees F., up to 55 seconds; flash point, minimum value (closed cup), 160 degrees F.; saponification No., 16 (max.); free acid (max.) 0.2 per cent. Oil-water emulsions, while good coolants, are objectionable because water will greatly intensify any accidental chip fire.

**Machining Magnesium.** — Magnesium alloys are readily machined and with relatively low power consumption per cubic inch of metal removed. The usual practice is to employ high cutting speeds with relatively coarse feeds and deep cuts. Exceptionally fine finishes can be obtained so that grinding to improve the finish usually is unnecessary. The horsepower normally required in machining magnesium varies from 0.15 to 0.30 per cubic inch per minute. While this value is low, especially in comparison with power required for cast iron and steel, the total amount of power for machining magnesium usually is high because of the exceptionally rapid rate at which metal is removed.

Carbide tools are recommended for maximum efficiency, although high-speed steel frequently is employed. Tools should be designed so as to dispose of chips readily or without excessive friction, by employing polished chip-bearing surfaces, ample chip spaces, large clearances, and small contact areas. *Keen-edged tools should always be used.*

*Feeds and Speeds for Magnesium:* Speeds ordinarily range up to 5000 feet per minute for rough- and finish-turning, up to 3000 feet per minute for rough-milling, and up to 9000 feet per minute for finish-milling. For rough-turning, the following combinations of speed in feet per minute, feed per revolution, and depth of cut are recommended: Speed 300 to 600 feet per minute — feed 0.030 to 0.100 inch, depth of cut 0.5 inch; speed 600 to 1000 — feed 0.020 to 0.080, depth of cut 0.4; speed 1000 to 1500 — feed 0.010 to 0.060, depth of cut 0.3; speed 1500 to 2000 — feed 0.010 to 0.040, depth of cut 0.2; speed 2000 to 5000 — feed 0.010 to 0.030, depth of cut 0.15.

*Lathe Tool Angles for Magnesium:* The true or actual rake angle resulting from back and side rakes usually varies from 10 to 15 degrees. Back rake varies from 10 to 20, and side rake from 0 to 10 degrees. Reduced back rake may be employed to obtain better chip breakage. The back rake may also be reduced to from 2 to 8 degrees on form tools or other broad tools to prevent chatter.

*Parting Tools:* For parting tools, the back rake varies from 15 to 20 degrees, the front end relief 8 to 10 degrees, the side relief measured perpendicular to the top face 8 degrees, the side relief measured in the plane of the top face from 3 to 5 degrees.

*Milling Magnesium:* In general, the coarse-tooth type of cutter is recommended. The number of teeth or cutting blades may be one-third to one-half the number normally used; however, the two-blade fly-cutter has proved to be very satisfactory. As a rule, the land relief or primary peripheral clearance is 10 degrees followed by secondary clearance of 20 degrees. The lands should be narrow, the width being about 3/64 to 1/16 inch. The rake, which is positive, is about 15 degrees.

For rough-milling and speeds in feet per minute up to 900 — feed, inch per tooth, 0.005 to 0.025, depth of cut up to 0.5; for speeds 900 to 1500 — feed 0.005 to 0.020, depth of cut up to 0.375; for speeds 1500 to 3000 — feed 0.005 to 0.010, depth of cut up to 0.2.

*Drilling Magnesium:* If the depth of a hole is less than five times the drill diameter, an ordinary twist drill with highly polished flutes may be used. The included angle of the point may vary from 70 degrees to the usual angle of 118 degrees. The relief angle is about 12 degrees. The drill should be kept sharp and the outer corners rounded to produce a smooth finish and prevent burr formation. For deep hole drilling, use a drill having a helix angle of 40 to 45 degrees with large polished flutes of uniform cross-section throughout the drill length to facilitate the flow of chips. A pyramid-shaped "spur" or "pilot point" at the tip of the drill will reduce the "spiraling or run-off."

Drilling speeds vary from 300 to 2000 feet per minute with feeds per revolution ranging from 0.015 to 0.050.

*Reaming Magnesium:* Reamers up to 1 inch in diameter should have four flutes; larger sizes six flutes. These flutes may be either parallel with the axis or have a negative helix angle of 10 degrees. The positive rake angle varies from 5 to 8 degrees, the relief angle from 4 to 7 degrees, and the clearance angle from 15 to 20 degrees.

*Tapping Magnesium:* Standard taps may be used unless Class 3 or Class 4 tolerances are required, in which case the tap should be designed for magnesium. A high-speed steel concentric type with a ground thread is recommended. The concentric form, which eliminates the radial thread relief, prevents jamming of chips while the tap is being backed out of the hole. The positive rake angle at the front may vary from 10 to 25 degrees and the "heel rake angle" at the back of the tooth from 3 to 5 degrees. The chamfer extends over two to three threads. For holes up to ¼ inch in diameter, two-fluted taps are recommended; for sizes from ½ to ¾ inch, three flutes; and for larger holes, four flutes. Tapping speeds ordinarily range from 75 to 200 feet per minute, and mineral oil cutting fluid should be used.

*Threading Dies for Magnesium:* Threading dies for use on magnesium should have about the same cutting angles as taps. Narrow lands should be used to provide ample chip space. Either solid or self-opening dies may be used. The latter type is recommended when maximum smoothness is required. Threads may be cut at speeds up to 1000 feet per minute.

*Grinding Magnesium:* As a general rule, magnesium is ground dry. The highly inflammable dust should be formed into a sludge by means of a spray of water or low-viscosity mineral oil. Accumulations of dust or sludge should be avoided. For surface grinding, when a fine finish is desirable, a low-viscosity mineral oil may be used.

**Machining Aluminum.** — Some of the alloys of aluminum have been machined successfully without any lubricant or cutting compound, but in order to obtain the best results, some form of lubricant is desirable. For many purposes, a soluble cutting oil is good.

Tools for aluminum and aluminum alloys should have larger relief and rake angles than tools for cutting steel. For high-speed steel turning tools the following angles are recommended: relief angles, 14 to 16 degrees; back rake angle, 5 to 20 degrees; side rake angle, 15 to 35 degrees. For very soft alloys even larger side rake angles are sometimes used. High silicon aluminum alloys and some others have a very abrasive effect on the cutting tool. While these alloys can be cut successfully with high-speed-steel tools, cemented carbides are recommended because of their superior abrasion resistance. The tool angles recommended for cemented carbide turning tools are: relief angles, 12 to 14 degrees; back rake angle, 0 to 15 degrees; side rake angle, 8 to 30 degrees.

Cut-off tools and necking tools for machining aluminum and its alloys should have from 12 to 20 degrees back rake angle and the end relief angle should be from 8 to 12 degrees. Excellent threads can be cut with single-point tools in even the softest aluminum. Experience seems to vary somewhat regarding the rake angle for single-point thread cutting tools. Some prefer to use a rather large back and side rake angle although this requires a modification in the included angle of the tool to produce the correct thread contour. When both rake angles are zero, the included angle of the tool is ground equal to the included angle of the thread. Excellent threads have been cut in aluminum with zero rake angle thread-cutting tools using large relief angles, which are 16 to 18 degrees opposite the front side of the thread and 12 to 14 degrees opposite the back side of the thread. In either case, the cutting edges should be ground and honed to a keen edge. It is sometimes advisable to give the face of the tool a few strokes with a hone between cuts when chasing the thread in order to remove any built-up edge on the cutting edge.

Fine surface finishes are often difficult to obtain on aluminum and aluminum alloys, particularly the softer metals. When a fine finish is required, the cutting tool should be honed to a keen edge and the surfaces of the face and the flank will also benefit by being honed smooth. Tool wear is inevitable; however, it should not be allowed to progress too far before the tool is changed or sharpened. A sulphurized mineral oil or a heavy-duty soluble oil will sometimes be helpful in obtaining a satisfactory surface finish. For best results, however, a diamond cutting tool is recommended. Excellent surface finishes can be obtained on even the softest aluminum and aluminum alloys with these tools.

Although ordinary milling cutters can be used successfully in shops where aluminum parts are only machined occasionally, the best results are obtained with coarse-tooth, large helix-angle cutters having large rake and clearance angles. Clearance angles up to 10 to 12 degrees are recommended. When slab milling and end milling a profile, using the peripheral teeth on the end mill, climb milling (also called down milling) will generally produce a better finish on the machined surface than conventional (or up) milling. Face milling cutters should have a large axial rake angle. Standard twist drills can be used without difficulty in drilling aluminum and aluminum alloys although high helix-angle drills are preferred. The wide flutes and high helix-angle in these drills helps to clear the chips. In some cases the use of split-point drills is preferred. Carbide tipped twist drills can be used for drilling aluminum and its alloys which may afford advantages in some production applications. Ordinary hand and machine taps can be used to tap aluminum and its alloys although spiral-fluted ground thread taps give superior results. Experience has shown that such taps should have a right-hand ground flute when intended to cut right-hand threads and the helix angle should be similar to that used in an ordinary twist drill.

**Machining Plastics.** — Molded plastic parts do not require machining, as a general rule, unless the tolerances are exceptionally small or there are undercuts, angular holes, or other openings difficult or impracticable to reproduce in a mold. It is common practice, however, to machine laminated phenolic plastics and also cast phenolic plastics, as well as sheet, bar, and tube stock such as is commonly used for making parts such as pinions or gears. The machining characteristics of different plastics vary somewhat so that the general recommendations given may require some modification to obtain the best results for a given class of work. Although plastics are poor conductors of heat, they usually are machined dry or without a cutting fluid. In some cases, either an air jet, water, or a soap solution is used. The maximum speed at which a cutting tool can operate without excessive heating should be determined by actual test.

*Turning and Boring Plastics:* Tools of high-speed steel, Stellite and carbide are commonly used in machining plastics. The general practice in turning and boring is similar to that for brass so far as the feed and speed are concerned. Speeds usually vary from 250 to 500 feet per minute for high-speed steel tools and from 500 to 1500 for Stellite and carbide tools. According to the Haynes Stellite Co., hot-set molded parts require tools having less clearance and more rake than those used for steel or other metals. The cold-set acetate and polystyrene molded parts should be turned with tools having no rake and plenty of clearance. Parts molded of "Vinylite" plastic can usually be machined satisfactorily with ordinary metal-cutting tools, provided the front and side clearances are about double those required for machine steel.

*Cutting Off Plastics:* Tools for cutting off should have greater front and side clearances than for steel. The cutting speed should be about half that employed for a turning operation.

*Drilling Plastics:* Standard twist drills may be used for small holes up to ⅛ or 3⁄16 inch, but for larger sizes particularly, it is preferable to use the commercial high-speed steel drills designed especially for plastics. These drills are made both in wire gage and fractional sizes. They have relatively large flutes to provide greater space for chips, and polished flutes are preferable. For large-quantity production, carbide-tipped drills are recommended. Extra clearance back of the edges of the flutes tends to reduce friction and heating. Frequent removal of the drill may be necessary, especially in drilling comparatively deep holes. A feed of 0.007 to 0.015 inch per revolution is a common range. Drilling speeds usually vary from 150 to 300 feet per minute. With carbide-tipped drills, the speeds may be as high as 12,000 to 15,000 R.P.M. The drill point should have an included angle of 55 to 60 degrees for thin sections and 90 degrees for the thicker sections. The clearance angle usually is 15 degrees. To avoid excessive heating and aid in chip removal, an air jet may be directed into the hole. In some cases, a soap solution is effective as a cutting fluid. If the drill-spindle movement is cam-operated, design the cam to advance the drill tip slowly at the beginning or for about 0.010 inch, then continue at the full rate of speed and allow a dwell of about 2 to 5 revolutions at the bottom of the hole; then withdraw the drill rapidly.

*Tapping and Threading Plastics:* For tapping small holes in hot- or cold-set molded parts, a high-speed steel nitrided chromium-plated tap is recommended. The rake may vary from 0 to 5 degrees negative. The size of the hole should allow for about three-fourths of the standard thread depth. Small holes generally are tapped dry. If a coolant is used, water is preferable to oil. Tapping speeds usually vary from 40 to 55 per minute. Threaded brass inserts or bushings are often used, in which case they may be tapped either before or after insertion. In cutting a 60-degree thread, such as the American standard, use a tool that is ground to cut on one side only and feed it in at an angle of 30 degrees by setting the compound rest at this angle.

*Drill Jigs for Plastics:* In the design of jigs for drilling, close-fitting drill bushing should be avoided. They may increase not only the friction on the drill but also the tendency of the chips to plug up the drill flutes. If the operation is such that a drill bushing is absolutely essential, a floating leaf or templet should be employed. When using a templet, the hole should be spotted with the templet in place, using the drill size corresponding to the final hole size; then the templet should be removed and the hole completed. Pilot holes should be avoided, except in special instances when the hole is to be reamed or counterbored.

*Sawing Hot-set Molded Parts:* The sawing of molded hot-set parts is done chiefly on circular and band saws. Band saws are to be recommended at times for straight cutting because they run cooler than circular saws. Band saw manufacturers advocate saw teeth set to clear, some advocating one-half the thickness of the blade on each side so that saws give a width of cut double their thickness. Narrower saws and more set are needed for cutting curves than for straight cuts. Band saws just soft enough to permit filing are recommended, but saws must be kept sharp. Dull saws cause chipping and might result in saw breakage. Sawing is usually done dry but some recommend water for cooling. Saw teeth should have little set and eight to nine teeth per inch. The speed is 1800 to 2500 feet per minute.

*Sawing Cold-set Molded Parts:* Sawing of cold-set polystyrene or cellulose acetate can be done with circular saws having nine to twelve teeth per inch for thin sheets and six teeth per inch for thickness over ¼ inch. Saws six to nine inches in diameter are run at speeds of 3000 to 3600 R.P.M. and should be hollow ground. They usually are ½₂ to ¹⁄₁₆ inch thick. The use of a water spray gives a cleaner cut. One large saw manufacturer recommends that pieces be cut with a stream of water running in the kerf while the saw is cutting. This applies to both circular and band saws; otherwise, the cold-set type of material will fuse. The circular saws recommended are 14 inches, 12 and 9 gage, 130 teeth, 10 degrees rake to be operated at 3000 R.P.M. and made of a special alloy steel stock.

For band sawing, manufacturers suggest a band saw which is 19 to 20 gage, having twenty points to the inch and hardened and tempered. The saw should be operated at 4000 to 4500 feet per minute.

*Milling Hot-set and Cold-set Molded Parts:* Milling of molded parts is not as a rule feasible, but where it is required, milling speeds and feeds of the range used for brass are recommended. A speed of 400 feet with carbon steel cutters and 1200-1600 feet is recommended with carbide cutters. Single and double bladed fly cutters are sometimes used at high speed with fine cuts. Where little material has to be removed, a high-speed woodworking shaper with a carbide tipped tool can be used to advantage. It is desirable and may be necessary to use an air blast to assure proper chip removal from the milling cutter. Wherever possible it is recommended that spiral milling cutters be used, and that the number of teeth in the cutter head be such that at least two of them are in contact with the work at all times.

The same general rules that apply to turning, facing, and boring operations also hold for milling Vinyl molded parts. Standard cutters can be used, but higher speeds are feasible if extra clearances are ground on the cutter blades.

**Machining Non-metallic Gear Blanks.** — Laminated phenolic plastics are extensively used for non-metallic gears and pinions. A non-metallic pinion should preferably be used in conjunction with either a hardened steel or a cast-iron gear, thus providing a durable and comparatively noiseless drive. Small- or medium-sized gears may be formed by molding, provided the quantity is large enough to warrant the cost of a mold. Most non-metallic gears, however, are machined from blanks cut from laminated phenolic plastics which have physical properties superior to the molded gears. These blanks may be cut either by punching in a die, by shearing, or by sawing. An efficient method of producing small blanks is to punch them from thin sheet stock. If necessary, the face width of the gear or pinion may be increased by riveting together two or more of these blanks. Gear blanks may also be cut either from a laminated plastic bar or thick tube. Before cutting the gear teeth, the blanks are machined, as by grinding, to obtain a concentric blank of the required diameter. The larger gears are sawed from sheet stock and then turned to size. Gear blanks which are ready for the gear-cutting operation are supplied by manufacturers of laminated plastic materials. The gear teeth are cut with standard gear-cutting equipment. Metal reinforcing end plates and bushings are commonly employed in connection with laminated gears.

*Punching Operations:* Most grades of laminated phenolics can be punched either hot or cold. The die must be kept sharp, however, in order to produce good results. The minimum clearance between individual punchings, and also between punchings and the edge of the material strips, should be about three times the thickness of the material. For hot punching or shearing, the material is heated in a steam or electric oven, designed to give a uniform temperature throughout the heating chamber. The material is left in the oven just long enough to be uniformly heated to oven temperature. Further heating causes brittleness. Temperatures of 100 to 120 degrees C. (212 to 248 degrees F.) are recommended. The heating time ranges from five minutes for 1/16-inch material to thirty minutes for 1/4-inch material.

Dies for punching laminated plastics are designed the same as for punching metal, except that smaller clearances are allowed between punch and die. In cold punching, this clearance is small, approaching a "sliding fit." The strippers are close fitting and backed with strong springs.

Because these plastic materials expand after being compressed in a die, blanks will be larger than the die diameter and holes will be smaller. On hot punchings, allowance should be made for shrinkage of the material after punching. This shrinkage varies with the grade of material, thickness of piece, and the temperature of the material during the punching operation. For very small holes and blanks, allowances for shrinkage are often neglected, while for large pieces and accurate work, they must be carefully considered. As an example, suppose a 1-inch diameter hole is to be punched hot in 3/32-inch thick stock; this would require a die of 1.009 inches and a punch of 1.007 inches in diameter. If this piece is punched cold, however, the die should be 1.005 inches in diameter and the punch 1.003 inches in diameter.

*Shearing Laminated Plastics:* Shears suitable for thin metal sheet are used for cutting laminated plastics. The knife must be kept sharp. In trimming paper-base grades cold, clean-cut edges are obtained with thicknesses of 1/32 inch and under, and when trimmed hot, up to 1/8 inch. Fabric-base grades are trimmed cold up to 1/16 inch, and hot up to 1/8 inch. Greater thicknesses can be sheared if the condition of the edge is not important.

*Sawing Plastics:* Material up to 1 inch thick may be cut with a 12- to 16-inch circular saw at about 3000 R.P.M., and material 1 inch thick and over, with a 16-inch saw at about 2400 R.P.M. A saw used for roughing cuts has bevel teeth, seven to the inch, while a smooth saw, with no set — similar to that for metal — is used for finishing cuts. For use on all thick material, the saws should be hollow-ground to prevent binding. The smaller the projection of the saw above the material on the sawing table, the better will be the sawed edges. A thin sheet of plastic or other material placed under the piece to be sawed, is of advantage when extreme smoothness of cut is desired.

A band saw is used for sawing round blanks from plate stock. The usual band saw is of the bevel-tooth type, with some set, and has three to seven teeth per inch. It is run at 3000 feet per minute.

**Machining Zinc Alloy Die-Castings.** — Machining of zinc alloy die-castings is mostly done without a lubricant. For particular work, especially deep drilling and tapping, a lubricant such as lard oil and kerosene (about half and half) or a 50–50 mixture of kerosene and machine oil may be used to advantage. A mixture of turpentine and kerosene has been found effective on certain difficult jobs.

In drilling, standard carbon steel drills are used for shallow holes and high-speed drills of the high-spiral type are recommended for deep holes. The standard 118-degree angle of point between cutting edges is recommended. A lip clearance of 12 degrees is satisfactory for most drilling, but in some cases may be increased up to 15 degrees. Flutes that are larger than normal offer the advantage of providing plenty of chip clearance. Straight flute drills have been found useful in enlarging existing holes. Peripheral speeds of from 200 to 300 feet per minute are generally found satisfactory for high-speed steel drills and about half this speed for carbon steel drills.

*Threading:* Button or acorn dies are satisfactory for threading small diameters of work. Either radial or tangent type chasers may be employed for the larger diameters. For radial type chasers, one manufacturer recommends a 10-degree radial hook for straight threads, a 7-degree radial hook for tapered threads; and a surface speed of 50 feet per minute for 3½ to 7½ threads per inch, 100 feet per minute for 8 to 11 threads per inch, and 200 feet per minute for 12 to 32 threads per inch. In using tangent type chasers with zinc alloys, the cutting edge must be on or very near the center to avoid rapid wearing of the chasers just behind the cutting edge. A 5-degree positive rake is recommended.

*Reaming:* In reaming, tools with six straight flutes are commonly used, although tools with eight flutes irregularly spaced have been found to yield better results by one manufacturer. Many standard reamers have a land that is too wide for best results. A land about 0.015 inch wide is recommended but this may often be ground down to around 0.007 or even 0.005 inch to obtain freer cutting, less tendency to loading, and reduced heating.

*Turning:* Tools of high-speed steel are commonly employed although the application of Stellite and carbide tools, even on short runs, is feasible. For steel or Stellite, a positive top rake of from 0 to 20 degrees and an end clearance of about 15 degrees is commonly recommended. Where side cutting is involved, a side clearance of about 4 degrees minimum is recommended. With carbide tools, the end clearance should not exceed 6 to 8 degrees and the top rake should be from 5 to 10 degrees positive. For boring, facing, and other lathe operations, rake and clearance angles are about the same as for tools used in turning.

## Machining Monel and Nickel Alloys. —

These alloys are machined with high-speed steel and with cemented carbide cutting tools. High-speed steel lathe tools usually have a back rake of 6 to 8 degrees, a side rake of 10 to 15 degrees, and relief angles of 8 to 12 degrees. Broad-nose finishing tools have a back rake of 20 to 25 degrees and an end relief angle of 12 to 15 degrees. In most instances standard commercial cemented-carbide tool holders and tool shanks can be used which provide an acceptable tool geometry. Honing the cutting edge lightly will help if chipping is encountered.

The most satisfactory tool materials for machining Monel and the softer nickel alloys, such as Nickel 200 and Nickel 230, are M2 and T5 for high-speed steel, and crater resistant grades of cemented carbides. For the harder nickel alloys such as K Monel, Permanickel, Duranickel, and Nitinol alloys, the recommended tool materials are T15, M41, M42, M43, and for high-speed steel M42. For carbides a grade of crater resistant carbide is recommended when the hardness is less than 300 Bhn and when the hardness is more than 300 Bhn a grade of straight tungsten carbide will often work best although some crater resistant grades will also work well.

A sulphurized oil or a water-soluble oil is recommended for rough and finish turning. A sulphurized oil is also recommended for milling, threading, tapping, reaming, and broaching. Recommended cutting speeds for Monel and the softer nickel alloys are 70 to 100 fpm for high-speed steel tools and 200 to 300 fpm for cemented carbide tools. For the harder nickel alloys, the recommended speed for high-speed steel is 40 to 70 fpm for a hardness up to 300 Bhn and for a higher hardness, 10 to 20 fpm; for cemented carbides, 175 to 225 fpm when the hardness is less than 300 Bhn and for a higher hardness, 30 to 70 fpm.

Nickel alloys have a high tendency to work harden. To minimize work hardening caused by machining, the cutting tools should be provided with adequate relief angles and positive rake angles. Furthermore, the cutting edges should be kept sharp and replaced when dull to prevent burnishing of the work surface. The depth of cut and feed should be sufficiently large to ensure that the tool penetrates the work without rubbing.

## KNURLING AND KNURLING TOOLS

Knurling tools may be of the cylindrical type, or the flat type. The cylindrical-type of knurling tool comprises a tool holder and one or more knurls having a centrally-located mounting hole and either straight or diagonal teeth on its periphery. The pattern on the periphery of the knurl is reproduced on the work as the work blank and the knurl rotate together with sufficient pressure to displace the metal on the work.

The flat-type of knurling tool is a knurling die, commonly used in reciprocating types of rolling machines. Dies may be made with either single or duplex faces having either straight or diagonal teeth.

**Knurls for Knurling in the Lathe.** — The knurls commonly used for lathe work have helical teeth and ordinarily are divided into three classes known as coarse, medium, and fine. The pitch of the teeth of coarse knurls (measured parallel to the axis of the work) is about 14 teeth to the inch; for medium knurls, about 21 teeth to the inch; and for fine teeth, 33 teeth to the inch. The medium pitch of knurl is the most commonly used. In some cases, where the knurled surface is to be used in a press fit assembly, knurls of special pitch are required so that the knurled surface will be raised the precise amount needed for proper assembly.

To prevent forming a double set of projections when knurling, the knurling tool should be fed in with considerable pressure at the start and the pressure relieved somewhat before the power feed is engaged. A slow surface speed and a liberal supply of oil should be used.

**Straight Knurls.** — It is important to select a suitable angle for the teeth for knurling different materials. A knurl with a blunt angle tooth will work better on soft materials than one with a more acute angle. The following angles are satisfactory: Brass and hard copper, 90 degrees; machine steel, 70 degrees; and drill rod and tool steel, 60 degrees. The depth of the knurl teeth $d$ may be computed from the formula $d = 0.5 \times p \times \cot \alpha/2$, where $p$ is the circular pitch of the knurl and $\alpha$ is the included angle of the teeth.

**Concave Knurls.** — The radius of a concave knurl should not be the same as the radius of the piece to be knurled. If the knurl and the work are of the same radius, the material compressed by the knurl will be forced down on the shoulder $D$ and spoil the appearance of the work. A design of concave knurl is shown in the accompanying illustration, and all the important dimensions are designated by

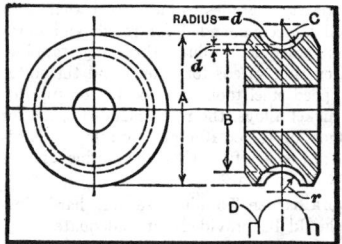

letters. To find these dimensions, the pitch of the knurl required must be known, and also, approximately, the throat diameter $B$. This diameter must suit the knurl holder used, and be such that the circumference contains an even number of teeth with the required pitch. When these dimensions have been decided upon, all the other unknown factors can be found by the following formula: Let $R$ = radius of piece to be knurled; $r$ = radius of concave part of knurl; $C$ = radius of cutter or hob for cutting the teeth in the knurl; $B$ = diameter over concave part of knurl (throat diameter); $A$ = outside diameter of knurl; $d$ = depth of tooth in knurl; $P$ = pitch of knurl (number of teeth per inch circumference); $p$ = circular pitch of knurl; then, $r = R + \frac{1}{2}d$; $C = r + d$; $A = B + 2r - (3d + 0.010$ inch); and $d = 0.5 \times p \times \cot \alpha/2$, where $\alpha$ is the included angle of the teeth.

As the depth of the tooth is usually very slight, the throat diameter $B$ will be accurate enough for all practical purposes for calculating the pitch, and it is not necessary to take into consideration the pitch circle. For example, assume that the pitch of a knurl is 32, that the throat diameter $B$ is 0.5561 inch, that the radius $R$ of the piece to be knurled is $\frac{1}{16}$ inch, and that the angle of the teeth is 90 degrees; find the dimensions of the knurl. Using the notation given:

$$p = \frac{1}{P} = \frac{1}{32} = 0.03125 \text{ inch;}$$

$$d = 0.5 \times 0.03125 \times \cot 45° = 0.0156 \text{ inch;}$$

$$r = \frac{1}{16} + \frac{0.0156}{2} = 0.0703 \text{ inch;}$$

$$C = 0.0703 + 0.0156 = 0.0859 \text{ inch;}$$

$$A = 0.5561 + 0.1406 - (0.0468 + 0.010) = 0.6399 \text{ inch.}$$

**ANSI Standard Knurls and Knurling.** — ANSI Standard B94.6-1966 (R1972) covers knurling tools with standardized diametral pitches and their dimensional relations with respect to the work in the production of straight, diagonal, and diamond knurling on cylindrical surfaces having teeth of uniform pitch parallel to the cylinder axis or at a helix angle not exceeding 45 degrees with the work axis.

These knurling tools and the recommendations for their use are equally applicable to general purpose and precision knurling. The advantage of this ANSI Standard system is the provision by which good tracking (the ability of teeth to mesh as the tool penetrates the work blank in successive revolutions) is obtained by tools designed on the basis of diametral pitch instead of TPI (teeth per inch) when used with work blank diameters that are multiples of $\frac{1}{64}$ inch for 64 and 128 diametral pitch or $\frac{1}{32}$ inch for 96 and 160 diametral pitch. The use of knurls and work blank diameters which will permit good tracing should improve the uniformity and appearance of knurling, eliminate the costly trial and error methods, reduce the failure of knurling tools and production of defective work, and decrease the number of tools required. Preferred sizes for cylindrical knurls are given in Table 1 and detailed specifications appear in Table 2.

The term *Diametral Pitch* applies to the quotient obtained by dividing the total number of teeth in the circumference of the work by the basic blank diameter; in the case of the knurling tool it would be the total number of teeth in the circumference divided by the *nominal* diameter. In the Standard the diametral pitch and number of teeth are always measured in a transverse plane which is perpendicular to the axis of rotation for diagonal as well as straight knurls and knurling.

**Formulas Applicable to Knurls and Knurled Work.** — Formulas (1) through (5) apply to knurled work produced by cylindrical knurls having either straight or diagonal teeth, when the axis of the knurl is parallel to the axis of the work.

$$P = \text{diametral pitch} = N_w \div D_w \qquad (1)$$

$$D_w = \text{work blank diameter} = N_w \div P \qquad (2)$$

$$N_w = \text{no. of teeth on work} = P \times D_w \qquad (3)$$

$$a = \text{"addendum" of tooth on work} = (D_{ow} - D_w) \div 2 \qquad (4)$$

$$h = \text{tooth depth (see Table 2)}$$

$$D_{ow} = \text{knurled diameter (outside diameter after knurling)} = D_w + 2a \qquad (5)$$

*Formulas for Diagonal and Diamond Knurling with Straight Tooth Knurling Tools Set at an Angle to the Work Axis:*

**If,**    $\psi$ = angle between tool axis and work axis;

     $P$ = diametral pitch on tool;

     $P_\psi$ = diametral pitch produced on work blank (as measured in the transverse plane) by setting tool axis at an angle $\psi$ with respect to work blank axis;

     $D_w$ = diameter of work blank; and

     $N_w$ = number of teeth produced on work blank (as measured in the transverse plane);

**then,**
$$P_\psi = P \cos \psi \tag{6}$$

**and,**
$$N_w = D_w P \cos \psi \tag{7}$$

For example, if 30 degree diagonal knurling were to be produced on 1-inch diameter stock with a 160 pitch straight knurl:

$$N_w = D_w P \cos 30° = 1.000 \times 160 \times 0.86603 = 138.56 \text{ teeth}$$

Good tracking is theoretically possible by changing the helix angle as follows to correspond to a whole number of teeth (138):

$$\cos \psi = N_w \div D_w P = 138 \div 1 \times 160 = 0.8625$$

$$\psi = 30\tfrac{1}{2} \text{ degrees, approximately}$$

Whenever it is more practical to machine the stock, good tracking can be obtained by reducing the work blank diameter as follows to correspond to a whole number of teeth (138):

$$D_w = \frac{N_w}{P \cos \psi} = \frac{138}{160 \times 0.866} = 0.996 \text{ inch}$$

*Formulas for Cylindrical Knurls:*

$$P = \text{diametral pitch of knurl} = N_t \div D_{nt} \tag{8}$$

$$D_{nt} = \text{nominal diameter of knurl} = N_t \div P \tag{9}$$

$$N_t = \text{no. of teeth on knurl} = P \times D_{nt} \tag{10}$$

$$*P_{nt} = \text{circular pitch on nominal diameter} = \pi \div P \tag{11}$$

$$*P_{ot} = \text{circular pitch on major diameter} = \pi D_{ot} \div N_t \tag{12}$$

* *Note:* For diagonal knurls, $P_{nt}$ and $P_{ot}$ are the transverse circular pitches which are measured in the plane perpendicular to the axis of rotation.

$$D_{ot} = \text{major diameter of knurl} = D_{nt} - (N_t Q \div \pi) \tag{13}$$

$$Q = P_{nt} - P_{ot} = \text{tracking correction factor in Formula (13)} \tag{14}$$

*Tracking Correction Factor Q:* Use of the preferred pitches for cylindrical knurls, Table 1, results in good tracking on all fractional work-blank diameters which are multiples of $\frac{1}{64}$ inch for 64 and 128 diametral pitch, and $\frac{1}{32}$ inch for 96 and 160 diametral pitch; an indication of good tracking is evenness of marking on the work surface during the first revolution of the work.

The many variables involved in knurling practice, such as hardness of the work material, elasticity of tool holders and machines, penetration of the work by the knurl during the first revolution of the work, etc., require that an empirical correction method be used to determine what actual circular pitch is needed at the major diameter of the knurl to produce good tracking and the required circular pitch on the workpiece. The tracking correction factor, $Q$, in Table 2 is based on experimental work and experience and is used in the calculation of the major diameter of the knurl, Formula (13).

**Table 1. ANSI Standard Preferred Sizes for Cylindrical Type Knurls\***
(ANSI B94.6-1966, R1972)

| Nominal Outside Diameter $D_{nt}$ | Width of Face F | Diameter of Hole A | Standard Pitches, P | | | |
|---|---|---|---|---|---|---|
| | | | 64 | 96 | 128 | 160 |
| | | | Number of Teeth, $N_t$, for Standard Pitches | | | |
| ½ | 3/16 | 3/16 | 32 | 48 | 64 | 80 |
| 5/8 | ¼ | ¼ | 40 | 60 | 80 | 100 |
| ¾ | 3/8 | ¼ | 48 | 72 | 96 | 120 |
| 7/8 | 3/8 | ¼ | 56 | 84 | 112 | 140 |
| Additional Sizes for Bench and Engine Lathe Tool Holders | | | | | | |
| 5/8 | 5/16 | 7/32 | 40 | 60 | 80 | 100 |
| ¾ | 5/8 | ¼ | 48 | 72 | 96 | 120 |
| 1 | 3/8 | 9/16 | 64 | 96 | 128 | 160 |

\* The 96 diametral pitch knurl should be given preference in the interest of tool simplification; the 64 pitch knurl should be avoided as much as possible.

**Table 2. ANSI Standard Specifications for Cylindrical Knurls with Straight or Diagonal Teeth\*** (ANSI B94.6-1966, R1972)

| Diametral Pitch P | Nominal Diameter, $D_{nt}$, | | | | | Tracking Correction Factor Q | Tooth Depth, h, +0.0015, −0.0000 | | Radius at Root R |
|---|---|---|---|---|---|---|---|---|---|
| | ½ | 5/8 | ¾ | 7/8 | 1 | | Straight | Diagonal | |
| | Major Diameter of Knurl, $D_{ot}$, +0.0000, −0.0015 | | | | | | | | |
| 64 | 0.4932 | 0.6165 | 0.7398 | 0.8631 | 0.9864 | 0.0006676 | 0.024 | 0.021 | 0.0070 0.0050 |
| 96 | 0.4960 | 0.6200 | 0.7440 | 0.8680 | 0.9920 | 0.0002618 | 0.016 | 0.014 | 0.0060 0.0040 |
| 128 | 0.4972 | 0.6215 | 0.7458 | 0.8701 | 0.9944 | 0.0001374 | 0.012 | 0.010 | 0.0045 0.0030 |
| 160 | 0.4976 | 0.6220 | 0.7464 | 0.8708 | 0.9952 | 0.00009425 | 0.009 | 0.008 | 0.0040 0.0025 |

All dimensions except diametral pitch are in inches.
Approximate angle of space between sides of adjacent teeth for both straight and diagonal teeth is 80 degrees. The permissible eccentricity of teeth for all knurls is 0.002 inch maximum (total indicator reading).
Number of teeth in a knurl equals diametral pitch multiplied by nominal diameter.
\* Diagonal teeth have 30-degree helix angle.

**Table 3. ANSI Standard Recommended Tolerances on Knurled Diameters**
(ANSI B94.6-1966, R1972)

| Tolerance Class | Diametral Pitch | | | | | | | |
|---|---|---|---|---|---|---|---|---|
| | 64 | 96 | 128 | 160 | 64 | 96 | 128 | 160 |
| | Tolerance on Knurled Outside Diameter | | | | Tolerance on Work-Blank Diameter Before Knurling | | | |
| I | +.005 −.012 | +.004 −.010 | +.003 −.008 | +.002 −.006 | ±.0015 | ±.0010 | ±.0007 | ±.0005 |
| II | +.000 −.010 | +.000 −.009 | +.000 −.008 | +.000 −.006 | ±.0015 | ±.0010 | ±.0007 | ±.0005 |
| III | +.000 −.006 | +.000 −.005 | +.000 −.004 | +.000 −.003 | +.000 −.0015 | +.0000 −.0010 | +.000 −.0007 | +.0000 −.0005 |

**Recommended Tolerance on Knurled Outside Diameters.** — The three classes of American National Standard tolerances for the outside diameter of knurled work are given in Table 3. These classes and recommended applications are:

*Class I:* Tolerances in this classification may be applied to straight, diagonal and raised diamond knurling where the knurled outside diameter of the work need not be held to close dimensional tolerances. Such applications include knurling for decorative effect, grip on thumb screws, and inserts for moldings and castings

*Class II:* Tolerances in this classification may be applied to straight knurling only and are recommended for applications requiring closer dimensional control of the knurled outside diameter than provided for by Class I tolerances.

*Class III:* Tolerances in this classification may be applied to straight knurling only and are recommended for applications requiring closest possible dimensional control of the knurled outside diameter. Such applications include knurling for close fits.

*Note:* The width of the knurling should not exceed the diameter of the blank, and knurling wider than the knurling tool cannot be produced unless the knurl starts at the end of the work.

## SCREW MACHINE FEEDS AND SPEEDS

**Feeds and Speeds for Automatic Screw Machine Tools.** — Approximate feeds and speeds for standard screw machine tools are given in the accompanying table.

**Knurling in Automatic Screw Machines.** — When knurling is done from the cross slide, it is good practice to feed the knurl gradually to the center of the work, starting to feed when the knurl touches the work and then passing off the center of the work with a quick rise of the cam. The knurl should also dwell for a certain number of revolutions, depending on the pitch of the knurl and the kind of material being knurled.

When two knurls are employed for spiral and diamond knurling from the turret, the knurls can be operated at a higher rate of feed for producing a spiral than they can for producing a diamond pattern. The reason for this is that in the first case the knurls work in the same groove, whereas in the latter case they work independently of each other.

**Revolutions Required for Top Knurling.** — The depth of the teeth and the feed per revolution govern the number of revolutions required for top knurling from the cross slide. If $R$ is the radius of the stock, $d$ is the depth of the teeth, $c$ is the distance the knurl travels from the point of contact to the center of the work at the feed required for knurling, and $r$ is the radius of the knurl; then

$$c = \sqrt{(R + r)^2 - (R + r - d)^2}$$

For example, if the stock radius $R$ is $\frac{5}{32}$ inch, depth of teeth $d$ is 0.0156 inch, and radius of knurl $r$ is 0.3125 inch, then

$$c = \sqrt{(0.1562 + 0.3125)^2 - (0.1562 + 0.3125 - 0.0156)^2}$$
$$= 0.120 \text{ inch} = \text{cam rise required.}$$

Assume that it is required to find the number of revolutions to knurl a piece of brass $\frac{5}{16}$ inch in diameter using a 32 pitch knurl. The included angle of the teeth for brass is 90 degrees, the circular pitch is 0.03125 inch, and the calculated tooth depth is 0.0156 inch. The distance $c$ (as determined in the previous example) is 0.120 inch. Referring to the accompanying table of feeds and speeds, the feed for top knurling brass is 0.005 inch per revolution. The number of revolutions required for knurling is, therefore, 0.120 ÷ 0.005 = 24 revolutions. If conditions permit, the higher feed of 0.008 inch per revolution given in the table may be used, in which case 15 revolutions are required for knurling.

**Cams for Threading.** — The tables "Spindle Revolutions and Cam Rise for Threading" beginning on page 1855 give the revolutions required for threading various lengths and pitches and the corresponding rise for the cam lobe. To illustrate the use of these tables, suppose a set of cams is required for threading a screw to the length of $\frac{3}{8}$ inch in a B. & S. machine. Assume that the spindle speed is

## Approximate Cutting Speeds and Feeds for Standard Automatic Screw Machine Tools — Brown and Sharpe

| Tool | Cut | | Material to be Machined | | | | | | |
| | Width or Depth, Inches | Diam. of Hole, Inches | Brass | Mild or Soft Steel | | | Tool Steel, .80-1.00% C | | |
| | | | Feed, Inches per Rev. | Feed, Inches per Rev. | Surface Speed, Feet per Min. Carbon Tools | Surface Speed, Feet per Min. H.S.S. Tools | Feed, Inches per Rev. | Surface Speed, Feet per Min. Carbon Tools | Surface Speed, Feet per Min. H.S.S. Tools |
|---|---|---|---|---|---|---|---|---|---|
| Boring tools | 0.005 | … | … | .008 | 50 | 110 | .004 | 30 | 60 |
| Box tools, roller rest / Single chip finishing | 1/32 | … | .012 | .010 | 70 | 150 | .005 | 40 | 75 |
|  | 1/16 | … | .010 | .008 | 70 | 150 | .004 | 40 | 75 |
|  | 1/8 | … | .008 | .007 | 70 | 150 | .003 | 40 | 75 |
|  | 3/16 | … | .006 | .006 | 70 | 150 | .002 | 40 | 75 |
|  | 1/4 | … | .006 | .005 | 70 | 150 | .0015 | 40 | 75 |
| Finishing | 0.005 | … | .010 | .010 | 70 | 150 | .006 | 40 | 75 |
| Center drills | … | Under 1/8 | .003 | .0015 | 50 | 110 | .001 | 30 | 75 |
|  | … | Over 1/8 | .006 | .0035 | 50 | 110 | .002 | 30 | 75 |
| Cutoff tools — Angular | … | … | .0015 | .0006 | 80 | 150 | .0004 | 50 | 85 |
| Cutoff tools — Circular | 3/64-1/16 | … | .0035 | .0015 | 80 | 150 | .001 | 50 | 85 |
| Cutoff tools — Straight | 1/16-1/8 | … | .0035 | .0015 | 80 | 150 | .001 | 50 | 85 |
| Stock diameter under 1/8 in. | … | … | .002 | .0008 | 80 | 150 | .0005 | 50 | 85 |
| Dies — Button | … | 0.02 | .0014 | .001 | 30 | 40 | … | 14 | 20 |
| Dies — Chaser | … | 0.04 | .002 | .0014 | 30 | 40 | … | 16 | … |
| Drills, twist cut | … | 1/16 | .004 | .002 | 40 | 60 | .0006 | 30 | 45 |
|  | … | 3/32 | .006 | .0025 | 40 | 60 | .0008 | 30 | 45 |
|  | … | 1/8 | .009 | .0035 | 40 | 60 | .0012 | 30 | 45 |
|  | … | 3/16 | .012 | .004 | 40 | 60 | .0016 | 30 | 45 |
|  | … | 1/4 | .014 | .005 | 40 | 75 | .002 | 30 | 60 |
|  | … | 5/16 | .016 | .005 | 40 | 75 | .003 | 30 | 60 |
|  | … | 3/8-5/8 | .016 | .006 | 40 | 75 | .0035 | 30 | 60 |
|  | … | | | | | 85 | .004 | | 60 |
| Form tools, circular | 1/8 | … | .002 | .0009 | 80 | 150 | .0006 | 50 | 85 |
|  | 1/4 | … | .002 | .0008 | 80 | 150 | .0005 | 50 | 85 |
|  | 3/8 | … | .0015 | .0007 | 80 | 150 | .0004 | 50 | 85 |
|  | 1/2 | … | .0012 | .0006 | 80 | 150 | .0004 | 50 | 85 |
|  | 5/8 | … | .001 | .0005 | 80 | 150 | .0003 | 50 | 85 |
|  | 3/4 | … | .001 | .0005 | 80 | 150 | .0003 | 50 | 85 |
|  | 1 | … | .001 | .0004 | 80 | 150 | … | 50 | 85 |

Compiled by Brown and Sharpe.

## Approximate Cutting Speeds and Feeds for Standard Automatic Screw Machine Tools — Brown and Sharpe (Continued)

| Tool | Cut: Width or Depth, Inches | Cut: Diam. of Hole, Inches | Brass† Feed, Inches per Rev. | Mild or Soft Steel Feed, Inches per Rev. | Mild or Soft Steel Surface Speed, Carbon Tools | Mild or Soft Steel Surface Speed, H.S.S. Tools | Tool Steel, .80–1.00% C Feed, Inches per Rev. | Tool Steel, .80–1.00% C Surface Speed, Carbon Tools | Tool Steel, .80–1.00% C Surface Speed, H.S.S. Tools |
|---|---|---|---|---|---|---|---|---|---|
| Hollow mills and balance turning tools — Turned diam. under 5⁄32 in. | 1⁄32 | … | .012 | .010 | 70 | 150 | .008 | 40 | 85 |
| | 1⁄16 | … | .010 | .009 | 70 | 150 | .006 | 40 | 85 |
| Hollow mills and balance turning tools — Turned diam. over 5⁄32 in. | 1⁄32 | … | .017 | .014 | 70 | 150 | .010 | 40 | 85 |
| | 1⁄16 | … | .015 | .012 | 70 | 150 | .008 | 40 | 85 |
| | 1⁄8 | … | .012 | .010 | 70 | 150 | .008 | 40 | 85 |
| | 3⁄16 | … | .010 | .008 | 70 | 150 | .006 | 40 | 85 |
| | 1⁄4 | … | .009 | .007 | 70 | 150 | .0045 | 40 | 85 |
| Knee tools | 1⁄32 | … | … | .010 | 70 | 150 | .008 | 40 | 85 |
| Knurling tools — Turret | On | … | .020 | .015 | 150 | … | .010 | 105 | … |
| | Off | … | .040 | .030 | 150 | … | .025 | 105 | … |
| Knurling tools — Side or swing | … | … | .004 | .004 | 150 | … | .002 | 105 | … |
| | … | … | .006 | .003 | 150 | … | .003 | 105 | … |
| Knurling tools — Top | … | … | .005 | .006 | 150 | … | .002 | 105 | … |
| | … | … | .008 | … | 150 | … | .004 | 105 | … |
| Pointing and facing tools | … | … | .001 | .0008 | 70 | 150 | .0005 | 40 | 80 |
| | … | … | .0025 | .002 | 70 | 150 | .0008 | 40 | 80 |
| Reamers and bits | .003–.004 | 1⁄8 or less | .010–.007 | .008–.006 | 70 | 105 | .006–.004 | 40 | 60 |
| | .004–.008 | 1⁄8 or over | .010 | .010 | 70 | 105 | .006–.008 | 40 | 60 |
| Recessing tools — End cut | … | … | .001 | .0006 | 70 | 150 | .0004 | 40 | 75 |
| | … | … | .005 | .003 | 70 | 150 | .002 | 40 | 75 |
| Recessing tools — Inside cut | 1⁄16–1⁄8 | … | .0025 | .002 | 70 | 105 | .0015 | 40 | 60 |
| | … | … | .008 | .0006 | 70 | 105 | .0004 | 40 | 60 |
| Swing tools, forming | 1⁄8 | … | .002 | .0007 | 70 | 150 | .0005 | 40 | 85 |
| | 1⁄4 | … | .0012 | .0005 | 70 | 150 | .0003 | 40 | 85 |
| | 3⁄8 | … | .001 | .0004 | 70 | 150 | .0002 | 40 | 85 |
| | 1⁄2 | … | .0008 | .0003 | 70 | 150 | .0002 | 40 | 85 |
| Turning, straight and taper* | 1⁄32 | … | .008 | .006 | 70 | 150 | .0035 | 40 | 85 |
| | 1⁄16 | … | .006 | .004 | 70 | 150 | .003 | 40 | 85 |
| | 1⁄8 | … | .005 | .003 | 70 | 150 | .002 | 40 | 85 |
| | 3⁄16 | … | .004 | .0025 | 70 | 150 | .0015 | 40 | 85 |
| Taps | … | … | … | … | 25 | 30 | … | 12 | 15 |

† Use maximum spindle speed on machine.    * For taper turning use feed slow enough for greatest depth of cut.

## Spindle Revolutions and Cam Rise for Threading — 1

**Number of Threads per Inch**

First Line: Revolutions of Spindle for Threading.   Second Line: Rise on Cam for Threading

Each cell shows *revolutions of spindle / rise on cam*.

| Length of Threaded Portion | 14 | 16 | 18 | 20 | 24 | 28 | 30 | 32 | 36 | 40 | 48 | 56 | 64 | 72 | 80 |
|---|---|---|---|---|---|---|---|---|---|---|---|---|---|---|---|
| 1/16 | … | … | … | … | 3.00 / 0.106 | 5.00 / 0.157 | 5.00 / 0.147 | 5.00 / 0.138 | 5.50 / 0.134 | 5.50 / 0.121 | 6.00 / 0.110 | 8.00 / 0.129 | 8.50 / 0.120 | 9.00 / 0.113 | 9.50 / 0.107 |
| 1/8 | … | 3.50 / 0.186 | 3.50 / 0.165 | 4.00 / 0.170 | 4.50 / 0.159 | 6.50 / 0.204 | 7.00 / 0.205 | 7.00 / 0.193 | 7.00 / 0.171 | 8.00 / 0.176 | 9.00 / 0.165 | 11.50 / 0.185 | 12.50 / 0.176 | 13.50 / 0.169 | 14.50 / 0.163 |
| 3/16 | 4.00 / 0.243 | 4.50 / 0.239 | 5.00 / 0.236 | 5.50 / 0.234 | 6.00 / 0.213 | 8.50 / 0.267 | 8.50 / 0.249 | 9.00 / 0.248 | 10.00 / 0.244 | 10.50 / 0.231 | 12.00 / 0.220 | 15.00 / 0.241 | 16.50 / 0.232 | 18.00 / 0.225 | 19.50 / 0.219 |
| 1/4 | 5.00 / 0.304 | 5.50 / 0.292 | 6.00 / 0.283 | 6.50 / 0.276 | 7.50 / 0.266 | 10.00 / 0.314 | 10.50 / 0.308 | 11.00 / 0.303 | 12.00 / 0.293 | 13.00 / 0.286 | 15.00 / 0.275 | 18.50 / 0.297 | 20.50 / 0.288 | 22.50 / 0.281 | 24.50 / 0.276 |
| 5/16 | 6.00 / 0.364 | 6.50 / 0.345 | 7.00 / 0.330 | 8.00 / 0.340 | 9.00 / 0.319 | 12.00 / 0.377 | 12.50 / 0.367 | 13.00 / 0.358 | 14.50 / 0.354 | 15.50 / 0.341 | 18.00 / 0.330 | 22.00 / 0.354 | 24.50 / 0.345 | 27.00 / 0.338 | 29.50 / 0.332 |
| 3/8 | 7.00 / 0.425 | 7.50 / 0.398 | 8.50 / 0.401 | 9.00 / 0.383 | 10.50 / 0.372 | 13.50 / 0.424 | 14.50 / 0.425 | 15.00 / 0.413 | 16.50 / 0.403 | 18.00 / 0.396 | 21.00 / 0.385 | 25.50 / 0.410 | 28.50 / 0.401 | 31.50 / 0.394 | 34.50 / 0.388 |
| 7/16 | 7.50 / 0.455 | 8.50 / 0.451 | 9.50 / 0.446 | 10.50 / 0.446 | 12.00 / 0.425 | 15.50 / 0.487 | 16.00 / 0.469 | 17.00 / 0.468 | 19.00 / 0.464 | 20.50 / 0.451 | 24.00 / 0.440 | 29.00 / 0.466 | 32.50 / 0.457 | 36.00 / 0.450 | 39.50 / 0.444 |
| 1/2 | 8.50 / 0.516 | 9.50 / 0.504 | 10.50 / 0.496 | 11.50 / 0.489 | 13.50 / 0.478 | 17.00 / 0.534 | 18.00 / 0.528 | 19.00 / 0.523 | 21.00 / 0.513 | 23.00 / 0.506 | 27.00 / 0.495 | 32.50 / 0.522 | 36.50 / 0.513 | 40.50 / 0.506 | 44.50 / 0.501 |
| 9/16 | 9.50 / 0.577 | 10.50 / 0.558 | 11.50 / 0.543 | 13.00 / 0.553 | 15.00 / 0.531 | 19.00 / 0.597 | 20.00 / 0.587 | 21.00 / 0.578 | 23.50 / 0.574 | 25.50 / 0.561 | 30.00 / 0.550 | 36.00 / 0.579 | 40.50 / 0.570 | 45.00 / 0.563 | 49.50 / 0.559 |
| 5/8 | 10.50 / 0.637 | 11.50 / 0.611 | 13.00 / 0.614 | 14.00 / 0.595 | 16.50 / 0.584 | 20.50 / 0.644 | 22.00 / 0.645 | 23.00 / 0.633 | 25.50 / 0.623 | 28.00 / 0.616 | 33.00 / 0.605 | 39.50 / 0.635 | 44.50 / 0.626 | 49.50 / 0.619 | 54.50 / 0.613 |
| 11/16 | 11.00 / 0.668 | 12.50 / 0.664 | 14.00 / 0.661 | 15.50 / 0.659 | 18.00 / 0.638 | 22.50 / 0.707 | 23.50 / 0.689 | 25.00 / 0.688 | 28.00 / 0.684 | 30.50 / 0.671 | 36.00 / 0.665 | 43.00 / 0.691 | 48.50 / 0.682 | 54.00 / 0.675 | 59.50 / 0.679 |
| 3/4 | 12.00 / 0.728 | 13.50 / 0.717 | 15.00 / 0.708 | 16.50 / 0.701 | 19.50 / 0.691 | 24.00 / 0.754 | 25.50 / 0.748 | 27.00 / 0.743 | 30.00 / 0.733 | 33.00 / 0.726 | 39.00 / 0.715 | 46.50 / 0.747 | 52.50 / 0.738 | 58.50 / 0.731 | 64.50 / 0.726 |

## Spindle Revolutions and Cam Rise for Threading — 2

First Line: Revolutions of Spindle for Threading. Second Line: Rise on Cam for Threading.

| Length of Threaded Portion | Number of Threads per Inch | | | | | | | | | | | | | | |
|---|---|---|---|---|---|---|---|---|---|---|---|---|---|---|---|
| | 14 | 16 | 18 | 20 | 24 | 28 | 30 | 32 | 36 | 40 | 48 | 56 | 64 | 72 | 80 |
| 13⁄16 | 13.00 / 0.789 | 14.50 / 0.770 | 16.00 / 0.755 | 18.00 / 0.765 | 21.00 / 0.744 | 26.00 / 0.817 | 27.50 / 0.807 | 29.00 / 0.798 | 32.50 / 0.794 | 35.50 / 0.781 | 42.00 / 0.770 | 50.00 / 0.804 | 56.50 / 0.795 | 63.00 / 0.788 | 69.50 / 0.782 |
| ⅞ | 14.00 / 0.850 | 15.50 / 0.823 | 17.50 / 0.826 | 19.00 / 0.808 | 22.50 / 0.797 | 27.50 / 0.864 | 29.50 / 0.865 | 31.00 / 0.853 | 34.50 / 0.843 | 38.00 / 0.836 | 45.00 / 0.825 | 53.50 / 0.860 | 60.50 / 0.851 | 67.50 / 0.844 | 74.50 / 0.838 |
| 15⁄16 | 14.50 / 0.880 | 16.50 / 0.876 | 18.50 / 0.873 | 20.50 / 0.871 | 24.00 / 0.850 | 29.50 / 0.927 | 31.00 / 0.909 | 33.00 / 0.908 | 37.00 / 0.904 | 40.50 / 0.891 | 48.00 / 0.880 | 57.00 / 0.916 | 64.50 / 0.907 | 72.00 / 0.900 | 79.50 / 0.894 |
| 1 | 15.50 / 0.941 | 17.50 / 0.929 | 19.50 / 0.920 | 21.50 / 0.914 | 25.50 / 0.903 | 31.00 / 0.974 | 33.00 / 0.968 | 35.00 / 0.963 | 39.00 / 0.953 | 43.00 / 0.946 | 51.00 / 0.918 | 60.50 / 0.972 | 68.50 / 0.963 | 76.50 / 0.956 | 84.50 / 0.951 |
| 1⅛ | 17.50 / 1.062 | 19.50 / 1.035 | 22.00 / 1.038 | 24.00 / 1.020 | 28.50 / 1.009 | 34.50 / 1.083 | 37.00 / 1.084 | 39.00 / 1.073 | 43.50 / 1.061 | 48.00 / 1.056 | 57.00 / 1.045 | 67.50 / 1.084 | 76.50 / 1.076 | 85.50 / 1.069 | 94.50 / 1.063 |
| 1¼ | 19.00 / 1.153 | 21.50 / 1.142 | 24.00 / 1.133 | 26.50 / 1.126 | 31.50 / 1.115 | 38.00 / 1.193 | 40.50 / 1.187 | 43.00 / 1.183 | 48.00 / 1.171 | 53.00 / 1.166 | 63.00 / 1.155 | 74.50 / 1.197 | 84.50 / 1.188 | 94.50 / 1.181 | 104.5 / 1.176 |
| 1⅜ | 21.00 / 1.275 | 23.50 / 1.248 | 26.50 / 1.251 | 29.00 / 1.233 | 34.50 / 1.211 | 41.50 / 1.303 | 44.50 / 1.304 | 47.00 / 1.293 | 52.50 / 1.281 | 58.00 / 1.276 | 69.00 / 1.265 | 81.50 / 1.310 | 92.50 / 1.301 | 103.5 / 1.294 | ...... |
| 1½ | 22.50 / 1.366 | 25.50 / 1.354 | 28.50 / 1.345 | 31.50 / 1.339 | 37.50 / 1.328 | 45.00 / 1.413 | 48.00 / 1.406 | 51.00 / 1.403 | 57.00 / 1.391 | 63.00 / 1.386 | 75.00 / 1.375 | 88.50 / 1.422 | 100.5 / 1.413 | ...... | ...... |
| 1⅝ | 24.50 / 1.487 | 27.50 / 1.460 | 31.00 / 1.463 | 34.00 / 1.445 | 40.50 / 1.434 | 48.50 / 1.523 | 52.00 / 1.524 | 55.00 / 1.513 | 61.50 / 1.501 | 68.00 / 1.496 | 81.00 / 1.485 | 95.50 / 1.535 | ...... | ...... | ...... |
| 1¾ | 26.00 / 1.578 | 29.50 / 1.566 | 33.00 / 1.558 | 36.50 / 1.551 | 43.50 / 1.540 | 52.00 / 1.633 | 55.50 / 1.626 | 59.00 / 1.623 | 66.00 / 1.610 | 73.00 / 1.606 | 87.00 / 1.595 | 102.5 / 1.647 | ...... | ...... | ...... |
| 1⅞ | 28.00 / 1.700 | 31.50 / 1.673 | 35.50 / 1.676 | 39.00 / 1.658 | 46.50 / 1.646 | 55.50 / 1.743 | 59.50 / 1.743 | 63.00 / 1.733 | 70.50 / 1.720 | 78.00 / 1.716 | 93.00 / 1.705 | ...... | ...... | ...... | ...... |
| 2 | 29.50 / 1.791 | 33.50 / 1.779 | 37.50 / 1.770 | 41.50 / 1.764 | 49.50 / 1.752 | 59.00 / 1.853 | 63.00 / 1.846 | 67.00 / 1.843 | 75.00 / 1.830 | 83.00 / 1.826 | 99.00 / 1.815 | ...... | ...... | ...... | ...... |

2400 revolutions per minute; the number of revolutions to complete one piece, 400; time required to make one piece, 10 seconds; pitch of the thread, $\frac{1}{32}$ inch or 32 threads per inch. By referring to the table, under 32 threads per inch, and opposite $\frac{3}{8}$ inch (length of threaded part), the number of revolutions required is found to be 15 and the rise required for the cam, 0.413 inch.

Cams of this type are often cut on a circular milling attachment. When this method is employed, the number of minutes the attachment should be revolved for each 0.001 inch rise, is first determined. As 15 revolutions are required for threading and 400 for completing one piece, that part of the cam surface required for the actual threading operation equals $15 \div 400 = 0.0375$, which is equivalent to 810 minutes of the circumference. As the total rise, in this case, through an arc of 810 minutes is 0.413 inch, the number of minutes for each 0.001 inch rise equals $810 \div 413 = 1.96$ or, approximately, two minutes. If the attachment is graduated to read to five minutes, the cam will be fed laterally 0.0025 inch each time it is turned five minutes.

**Practical Points on Cam and Tool Design.** — The following general rules are given to aid in designing cams and special tools for automatic screw machines, and apply particularly to Brown and Sharpe machines:

1. Use the highest speeds recommended for the material used that the various tools will stand.

2. Use the arrangement of circular tools best suited for the class of work.

3. Decide on the quickest and best method of arranging the operations before designing the cams.

4. Do not use turret tools for forming when the cross-slide tools can be used to better advantage.

5. Make the shoulder on the circular cutoff tool large enough so that the clamping screw will grip firmly.

6. Do not use too narrow a cutoff blade.

7. Allow 0.005 to 0.010 inch for the circular tools to approach the work and 0.003 to 0.005 inch for the cutoff tool to pass the center.

8. When cutting off work, the feed of the cutoff tool should be decreased near the end of the cut where the piece breaks off.

9. When a thread is cut up to a shoulder, the piece should be grooved or necked to make allowance for the lead on the die. This requires an extra projection on the forming tool and also an extra amount of rise on the cam.

10. Allow sufficient clearance for tools to pass one another.

11. Always make a diagram of the cross-slide tools in position on the work when difficult operations are to be performed; it is also necessary to make a diagram of the tools held in the turret.

12. Do not drill a hole the depth of which is more than 3 times the diameter of the drill, but rather use two or more drills as required. If there are not sufficient holes in the turret for the extra drills needed, make provision for withdrawing the drill clear of the hole and then advancing it into the hole again.

13. Do not run drills at low speeds. Feeds and speeds recommended in the table of feeds and speeds on pages 1853 and 1854 should be followed as far as is practicable.

14. When the turret tools operate farther in than the face of the chuck, see that they will clear the chuck when the turret is revolved.

15. See that the body of all turret tools will clear the side of the chute when the turret is revolved.

16. Use a balance turning tool or a hollow mill for roughing cuts.

17. The rise on the thread lobe should be reduced so that the spindle will reverse when the tap or die holder is drawn out.

18. When bringing another tool into position after a threading operation, allow clearance before revolving the turret.

19. Make provision to revolve the turret rapidly, especially when pieces are being made in from three to five seconds and when only a few tools are used in the turret. It is sometimes convenient to use two sets of tools.

20. When using a belt-shifting attachment for threading, clearance should be allowed, as it requires extra time to shift the belt.

21. When laying out a set of cams for operating on a piece which requires to be slotted, cross-drilled or burred, allowance should be made on the lead cam so that the transferring arm can descend and ascend to and from the work without coming in contact with any of the turret tools.

22. Always allow a vacant hole in the turret when it is necessary to use the transferring arm.

23. When designing special tools allow as much clearance as possible. Do not make them so that they will just clear each other, as a slight inaccuracy in the dimensions will then often cause trouble.

24. When designing special tools having intricate movements, avoid springs as much as possible, and use positive actions.

**Stock for Screw Machine Products.** — The amount of stock required for the production of 1000 pieces on the automatic screw machine can be obtained directly from the table "Stock required for Screw Machine Products." To use this table, add to the length of the work the width of the cut-off tool blade; then the number of feet of material required for 1000 pieces can be found opposite the figure thus obtained, in the column headed "Feet per 1000 Parts." Screw machine stock usually comes in bars 10 feet long, and in compiling this table an allowance was made for chucking on each bar.

The table can be extended by using the following formula, in which $F$ = number of feet required for 1000 pieces; $L$ = length of piece in inches; $W$ = width of cut-off tool blade in inches.

$$F = (L + W) \times 84.$$

The amount to add to the length of the work, or the width of the cut-off tool, is given in the following, which is standard in a number of machine shops:

| Diameter of Stock, Inches | Width of Cut-off Tool Blade, Inches |
|---|---|
| 0.000–0.250 | 0.045 |
| 0.251–0.375 | 0.062 |
| 0.376–0.625 | 0.093 |
| 0.626–1.000 | 0.125 |
| 1.000–1.500 | 0.156 |

It is sometimes convenient to know the weight of a certain number of pieces, when estimating the price. The weight of round bar stock can be found by means of the following formulas, in which $W$ = weight in pounds; $D$ = diameter of stock in inches; $F$ = length in feet:

For brass stock:  $W = D^2 \times 2.86 \times F.$
For steel stock:  $W = D^2 \times 2.675 \times F.$
For iron stock:  $W = D^2 \times 2.65 \times F.$

## Stock Required for Screw Machine Products

The table gives the amount of stock, in feet, required for 1000 pieces, when the length of the finished part plus the thickness of the cut-off tool blade is known. Allowance has been made for chucking. To illustrate, if length of cut-off tool and work equals 0.140 inch, 11.8 feet of stock is required for the production of 1000 parts.

| Length of Piece and Cut-off Tool | Feet per 1000 Parts | Length of Piece and Cut-off Tool | Feet per 1000 Parts | Length of Piece and Cut-off Tool | Feet per 1000 Parts | Length of Piece and Cut-off Tool | Feet per 1000 Parts |
|---|---|---|---|---|---|---|---|
| 0.050 | 4.2 | 0.430 | 36.1 | 0.810 | 68.1 | 1.380 | 116.0 |
| 0.060 | 5.0 | 0.440 | 37.0 | 0.820 | 68.9 | 1.400 | 117.6 |
| 0.070 | 5.9 | 0.450 | 37.8 | 0.830 | 69.7 | 1.420 | 119.3 |
| 0.080 | 6.7 | 0.460 | 38.7 | 0.840 | 70.6 | 1.440 | 121.0 |
| 0.090 | 7.6 | 0.470 | 39.5 | 0.850 | 71.4 | 1.460 | 122.7 |
| 0.100 | 8.4 | 0.480 | 40.3 | 0.860 | 72.3 | 1.480 | 124.4 |
| 0.110 | 9.2 | 0.490 | 41.2 | 0.870 | 73.1 | 1.500 | 126.1 |
| 0.120 | 10.1 | 0.500 | 42.0 | 0.880 | 73.9 | 1.520 | 127.7 |
| 0.130 | 10.9 | 0.510 | 42.9 | 0.890 | 74.8 | 1.540 | 129.4 |
| 0.140 | 11.8 | 0.520 | 43.7 | 0.900 | 75.6 | 1.560 | 131.1 |
| 0.150 | 12.6 | 0.530 | 44.5 | 0.910 | 76.5 | 1.580 | 132.8 |
| 0.160 | 13.4 | 0.540 | 45.4 | 0.920 | 77.3 | 1.600 | 134.5 |
| 0.170 | 14.3 | 0.550 | 46.2 | 0.930 | 78.2 | 1.620 | 136.1 |
| 0.180 | 15.1 | 0.560 | 47.1 | 0.940 | 79.0 | 1.640 | 137.8 |
| 0.190 | 16.0 | 0.570 | 47.9 | 0.950 | 79.8 | 1.660 | 139.5 |
| 0.200 | 16.8 | 0.580 | 48.7 | 0.960 | 80.7 | 1.680 | 141.2 |
| 0.210 | 17.6 | 0.590 | 49.6 | 0.970 | 81.5 | 1.700 | 142.9 |
| 0.220 | 18.5 | 0.600 | 50.4 | 0.980 | 82.4 | 1.720 | 144.5 |
| 0.230 | 19.3 | 0.610 | 51.3 | 0.990 | 83.2 | 1.740 | 146.2 |
| 0.240 | 20.2 | 0.620 | 52.1 | 1.000 | 84.0 | 1.760 | 147.9 |
| 0.250 | 21.0 | 0.630 | 52.9 | 1.020 | 85.7 | 1.780 | 149.6 |
| 0.260 | 21.8 | 0.640 | 53.8 | 1.040 | 87.4 | 1.800 | 151.3 |
| 0.270 | 22.7 | 0.650 | 54.6 | 1.060 | 89.1 | 1.820 | 152.9 |
| 0.280 | 23.5 | 0.660 | 55.5 | 1.080 | 90.8 | 1.840 | 154.6 |
| 0.290 | 24.4 | 0.670 | 56.3 | 1.100 | 92.4 | 1.860 | 156.3 |
| 0.300 | 25.2 | 0.680 | 57.1 | 1.120 | 94.1 | 1.880 | 158.0 |
| 0.310 | 26.1 | 0.690 | 58.0 | 1.140 | 95.8 | 1.900 | 159.7 |
| 0.320 | 26.9 | 0.700 | 58.8 | 1.160 | 97.5 | 1.920 | 161.3 |
| 0.330 | 27.7 | 0.710 | 59.7 | 1.180 | 99.2 | 1.940 | 163.0 |
| 0.340 | 28.6 | 0.720 | 60.5 | 1.200 | 100.8 | 1.960 | 164.7 |
| 0.350 | 29.4 | 0.730 | 61.3 | 1.220 | 102.5 | 1.980 | 166.4 |
| 0.360 | 30.3 | 0.740 | 62.2 | 1.240 | 104.2 | 2.000 | 168.1 |
| 0.370 | 31.1 | 0.750 | 63.0 | 1.260 | 105.9 | 2.100 | 176.5 |
| 0.380 | 31.9 | 0.760 | 63.9 | 1.280 | 107.6 | 2.200 | 184.9 |
| 0.390 | 32.8 | 0.770 | 64.7 | 1.300 | 109.2 | 2.300 | 193.3 |
| 0.400 | 33.6 | 0.780 | 65.5 | 1.320 | 110.9 | 2.400 | 201.7 |
| 0.410 | 34.5 | 0.790 | 66.4 | 1.340 | 112.6 | 2.500 | 210.1 |
| 0.420 | 35.3 | 0.800 | 67.2 | 1.360 | 114.3 | 2.600 | 218.5 |

**American Standard Single-point Tools and Tool Posts.** — American Standard ASA B5.22-1950 gives the preferred sizes of tool bits, tool shanks, and tool holders. Included in the standard are standard types of single-point tools. (For sintered carbide blanks and cutting tools see page 1863.) Fig. 1 shows the single-screw tool post used on most small lathes, and Fig. 2 illustrates a design for shapers and some light planers. A *tool-post ring collar* (Fig. 1) has a spherical seat and *rocker base* for supporting the tool and providing simple means of adjusting the height of the tool point. Shims may also be used for this purpose; or the ring collar may have diametrically spaced steps on the upper face to provide for adjustment.

*Serrated wedges* are sometimes used with square tool posts on screw machines or with open-side tool posts for obtaining vertical adjustment. An *adjustable collar and nut* fitted about the base of a round tool post is sometimes used on screw machines. The *open-side tool post* for lathes (Fig. 3) may have either a rocker base or a flat tool baseplate. The *four-way turret tool post* (Fig. 4) used on manufacturing lathes, turret lathes and vertical boring mills is an indexing open-side type with either fixed base or rocker. The *strap-and-stud clamp type* of tool holder (Fig. 5) is generally used on large lathes, boring mills, shapers, slotters, and planers. The studs in the case of a planer are attached (threaded or recessed) to the clapper (Fig. 6) or in sliding T-nuts (Fig. 5) and pass through the serrated tool baseplate against which the tool bears.

**Table 1. American Standard Dimensions of Tool Shanks, Tool-post Openings, and Lathe Center Height** (ASA B5.22-1950)

| Shank Section Figs. 1, 2, 4, and 5 | | | Lathe Tool-Post Opening Figs. 1, 3, 4, and 5 | | |
|---|---|---|---|---|---|
| Max. $w$ | Nominal $w \times h$ | Max. $h$ | Min. $B$ | Nominal $B \times D$ | Min. $D$ |
| 0.40 | * 3/8 × 3/4 | 0.85 | 0.49 | 1/2 × 1 3/8 | 1.27 |
| 0.48 | 7/16 × 7/8 | 0.99 | 0.57 | 9/16 × 1 1/2 | 1.48 |
| 0.56 | * 1/2 × 1 | 1.15 | 0.68 | 11/16 × 1 3/4 | 1.72 |
| 0.67 | * 5/8 × 1 1/4 | 1.34 | 0.81 | 13/16 × 2 | 2.00 |
| 0.80 | * 3/4 × 1 1/2 | 1.56 | 0.96 | 1 × 2 3/8 | 2.34 |
| 0.95 | † 7/8 × 1 3/4 | 1.81 | 1.14 | 1 3/16 × 2 3/4 | 2.71 |
| 1.13 | * 1 × 2 | 2.11 | 1.35 | 1 3/8 × 3 1/8 | 3.16 |
| 1.34 | † 1 1/4 × 2 1/4 | 2.43 | 1.61 | 1 5/8 × 3 11/16 | 3.65 |
| 1.60 | † 1 1/2 × 2 3/4 | 2.86 | 1.91 | 1 7/8 × 4 1/4 | 4.29 |

| Planer and Shaper Tool-Post Opening Figs. 1, 2, and 5 | | | Lathe Center Height Figs. 1 and 3 | | |
|---|---|---|---|---|---|
| Min. $B$ | Nominal $B \times E$ | Min. $E$ | Max. $C$ | Nominal $C$ | Min. $C$ |
| 0.49 | 1/2 × 1 1/16 | 1.04 | 0.93 | 7/8 | 0.85 |
| 0.57 | 9/16 × 1 1/4 | 1.19 | 1.09 | 1 | 0.99 |
| 0.68 | 11/16 × 1 7/16 | 1.38 | 1.26 | 1 3/16 | 1.15 |
| 0.81 | 13/16 × 1 5/8 | 1.61 | 1.47 | 1 3/8 | 1.34 |
| 0.96 | 1 × 1 7/8 | 1.87 | 1.72 | 1 9/16 | 1.56 |
| 1.14 | 1 3/16 × 2 3/16 | 2.17 | 1.99 | 1 13/16 | 1.81 |
| 1.35 | 1 3/8 × 2 9/16 | 2.53 | 2.32 | 2 1/8 | 2.11 |
| 1.61 | 1 5/8 × 3 | 2.92 | 2.67 | 2 7/16 | 2.43 |
| 1.91 | 1 7/8 × 3 7/16 | 3.43 | 3.15 | 2 7/8 | 2.86 |

All dimensions are given in inches.
* Size listed in Table 2. † Size listed in Table 3.

**American Standard Tool Posts** (ASA B5.22-1950)

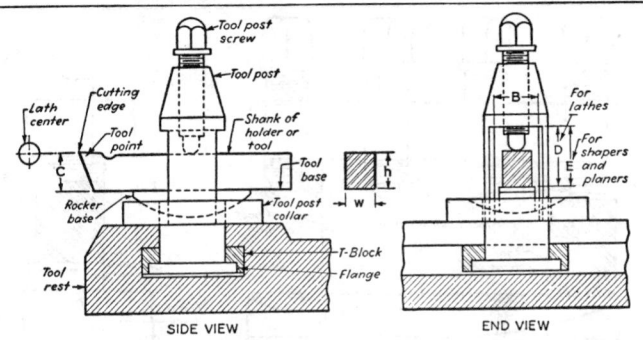

Fig. 1.   Single-screw Tool-post with Rocker Base

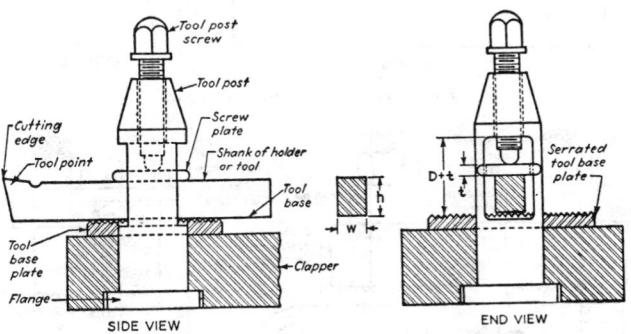

Fig. 2.   Single-screw Tool-post with Screw Plate and Serrated Base

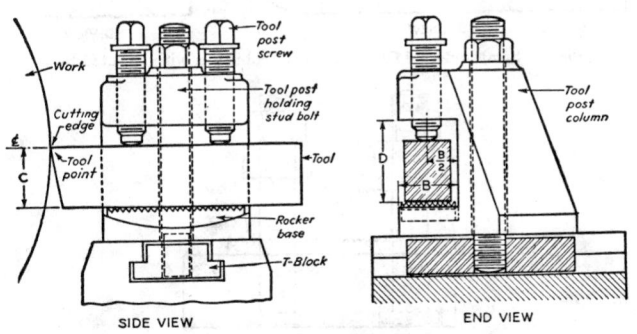

Fig. 3.   Open-side Tool-post with Serrated Rocker Base

American Standard Tool Posts (ASA B5.22-1950)

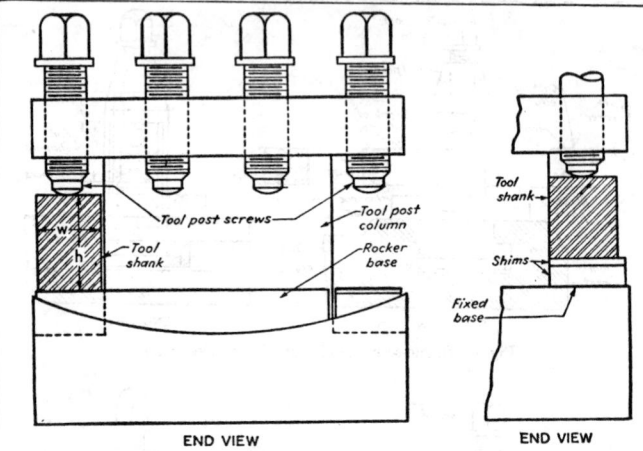

END VIEW                          END VIEW

Fig. 4.   Four-way Open-side Tool-post

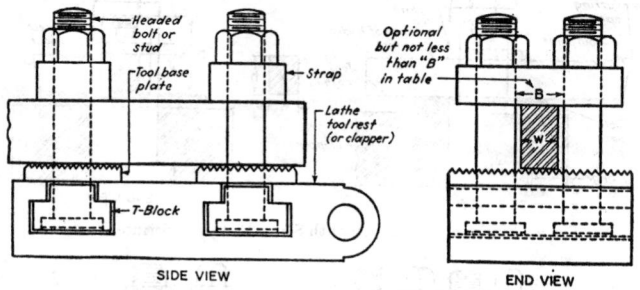

SIDE VIEW                         END VIEW

Fig. 5.   Strap-and-stud Clamp Type of Tool-holder with Serrated Base

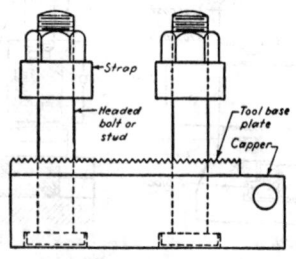

Fig. 6.   Planer Studs

Table 2. American Standard Lengths for Solid Square and Rectangular Tools*
(ASA B5.22-1950)

| Square Shanks | | | | Rectangular Shanks | | | |
|---|---|---|---|---|---|---|---|
| Square Section | Length | Square Section | Length | Rectangular Section | Length | Rectangular Section | Length |
| 3/16 | 2 | 5/8 | 4½ | ¼ × ½ | 4 | 5/8 × 1¼ | †6½ or 8 |
| ¼ | 2½ | ¾ | 6 | 5/16 × 5/8 | 4½ | ¾ × 1¼ | †6½ or 9 |
| 5/16 | 2½ | 1 | 7 | 3/8 × ¾ | 5 | 1 × 1½ | 10 |
| 3/8 | 3 | 1¼ | 8 | ½ × ¾ | 5¼ | 1 × 2 | 12 |
| ½ | 4 | 1½ | 10 | ½ × 1 | 5½ or 7 | 1½ × 2 | ........ |

All dimensions in inches.
* Solid tools include those having the full section of the cutting end only of metal-cutting material.
† The shorter lengths are normally for turret lathes, boring mills, etc.

Table 3. American Standard Sizes for Tool-holder Shanks (ASA B5.22-1950)

| Shank Cross Section | Length of Shank | | | | | | |
|---|---|---|---|---|---|---|---|
| | Turning Tools | Cut-off, Side-cut | Boring Tools | Threading Tools | Knurling Tools | Carbide Tipped | Planer Tools |
| 5/16 × ½ | 4 | .... | | .... | .... | .... | .... |
| 5/16 × ¾ | 4½ | 4½ | * | 5 | 5 | .... | .... |
| 3/8 × 7/8 | 5 | 5 | * | 5 | 5† | .... | .... |
| 3/8 × 15/16 | .... | .... | .... | .... | .... | 6 | .... |
| ½ × 1 | .... | .... | .... | .... | .... | .... | 6 |
| ½ × 1 1/8 | 5 | 6 | * | 6 | 6† | .... | .... |
| ½ × 1¼ | .... | .... | .... | .... | .... | 7 | .... |
| 5/8 × 1¼ | .... | .... | .... | .... | .... | .... | 8½ |
| 5/8 × 1 3/8 | 7 | 7 | * | 7 | 7† | .... | .... |
| 5/8 × 1½ | .... | .... | .... | .... | .... | 8 | .... |
| ¾ × 1½ | .... | .... | .... | .... | .... | .... | 10 |
| ¾ × 1 5/8 | 8 | 8 | * | .... | .... | .... | .... |
| ¾ × 1¾ | .... | .... | .... | .... | .... | 9 | .... |
| 7/8 × 1¾ | 9 | 9 | * | .... | 9 | .... | .... |
| 7/8 × 1 7/8 | .... | .... | .... | .... | .... | 10 | .... |
| 1 × 2 | 11 | .... | * | .... | .... | .... | .... |
| 1 × 2 1/8 | .... | .... | .... | .... | .... | 12 | .... |
| 1 1/8 × 1¾ | .... | .... | .... | .... | .... | .... | 13 |
| 1¼ × 2¼ | 13 | .... | .... | .... | .... | .... | .... |
| 1 3/8 × 2 | .... | .... | .... | .... | .... | .... | 16 |
| 1 7/8 × 2¼ | .... | .... | .... | .... | .... | .... | 19 |
| 2 1/8 × 2¾ | .... | .... | .... | .... | .... | .... | 22 |

* Asterisks (*) indicate boring tool cross sections available. Lengths not specified.
† Tolerance of plus ½ inch may apply.

**American Standard Sintered Carbide Blanks and Cutting Tools.** — Section 1 of American Standard B5.36-1957 provides standard sizes and designations for eight styles of sintered carbide blanks. These blanks are the unground solid carbide from which either solid or tipped cutting tools are made. Tipped cutting tools are made by brazing a blank onto a shank to produce the cutting tool; these differ from carbide *insert* cutting tools which consist of a carbide insert mechanically held in a tool holder (see page 1871). A typical single-point carbide-tipped cutting tool is shown in the accompanying diagram.

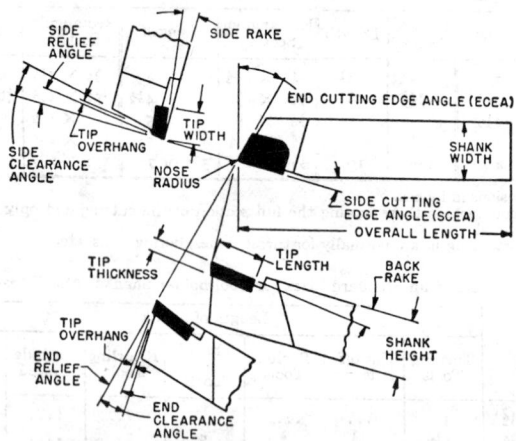

A typical single-point carbide tipped cutting tool. The side rake, side relief, and the clearance angles are *normal* to the side-cutting edge, rather than the shank, to facilitate its being ground on a tilting-table grinder. The end-relief and clearance angles are *normal* to the end-cutting edge. The end-rake angle is parallel to the side-cutting edge

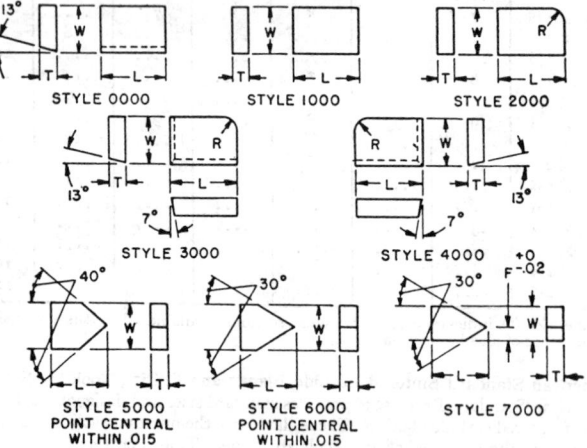

Eight styles of sintered carbide blanks. Standard dimensions for these blanks are given in Table 4

Table 4.  **American Standard Sizes and Designations for Carbide Blanks**
(ASA B5.36-1957)

(See diagram on page 1864)

| T | W | L | Style 1000 | Style 2000 | T | W | L | Style 0000 | Style 1000 | Style 3000 | Style 4000 |
|---|---|---|---|---|---|---|---|---|---|---|---|
|   |   |   | Blank Designation | | | | | Blank Designation | | | |
| 1/16 | 1/8 | 5/8 | 1010 | 2010 | 1/4 | 3/8 | 9/16 | 0350 | 1350 | 3350 | 4350 |
| 1/16 | 5/32 | 1/4 | 1015 | 2015 | 1/4 | 3/8 | 3/4 | 0360 | 1360 | 3360 | 4360 |
| 1/16 | 3/16 | 1/4 | 1020 | 2020 | 1/4 | 7/16 | 5/8 | 0370 | 1370 | 3370 | 4370 |
| 1/16 | 1/4 | 1/4 | 1025 | 2025 | 1/4 | 1/2 | 3/4 | 0380 | 1380 | 3380 | 4380 |
| 1/16 | 1/4 | 5/16 | 1030 | 2030 | 1/4 | 9/16 | 1 | 0390 | 1390 | 3390 | 4390 |
|   |   |   |   |   | 1/4 | 5/8 | 5/8 | 0400 | 1400 | 3400 | 4400 |
| 3/32 | 1/8 | 3/4 | 1035 | 2035 | 1/4 | 3/4 | 3/4 | 0405 | 1405 | 3405 | 4405 |
| 3/32 | 3/16 | 5/16 | 1040 | 2040 | 1/4 | 3/4 | 1 | 0410 | 1410 | 3410 | 4410 |
| 3/32 | 3/16 | 1/2 | 1050 | 2050 | 1/4 | 1 | 1 | 0415 | 1415 | 3415 | 4415 |
| 3/32 | 1/4 | 3/8 | 1060 | 2060 |   |   |   |   |   |   |   |
| 3/32 | 1/4 | 1/2 | 1070 | 2070 | 5/16 | 7/16 | 5/8 | 0420 | 1420 | 3420 | 4420 |
|   |   |   |   |   | 5/16 | 7/16 | 15/16 | 0430 | 1430 | 3430 | 4430 |
| 3/32 | 5/16 | 3/8 | 1080 | 2080 | 5/16 | 1/2 | 3/4 | 0440 | 1440 | 3440 | 4440 |
| 3/32 | 3/8 | 3/8 | 1090 | 2090 | 5/16 | 1/2 | 1 | 0450 | 1450 | 3450 | 4450 |
| 3/32 | 3/8 | 1/2 | 1100 | 2100 | 5/16 | 5/8 | 1 | 0460 | 1460 | 3460 | 4460 |
| 3/32 | 7/16 | 1/2 | 1105 | 2105 | 5/16 | 3/4 | 3/4 | 0470 | 1470 | 3470 | 4470 |
|   |   |   |   |   | 5/16 | 3/4 | 1 | 0475 | 1475 | 3475 | 4475 |
| 1/8 | 3/16 | 3/4 | 1110 | 2110 | 5/16 | 3/4 | 1 1/4 | 0480 | 1480 | 3480 | 4480 |
| 1/8 | 1/4 | 1/2 | 1120 | 2120 |   |   |   |   |   |   |   |
| 1/8 | 1/4 | 5/8 | 1130 | 2130 | 3/8 | 1/2 | 3/4 | 0490 | 1490 | 3490 | 4490 |
| 1/8 | 1/4 | 3/4 | 1140 | 2140 | 3/8 | 1/2 | 1 | 0500 | 1500 | 3500 | 4500 |
| 1/8 | 5/16 | 7/16 | 1150 | 2150 | 3/8 | 5/8 | 1 | 0510 | 1510 | 3510 | 4510 |
| 1/8 | 5/16 | 1/2 | 1160 | 2160 | 3/8 | 5/8 | 1 1/4 | 0515 | 1515 | 3515 | 4515 |
|   |   |   |   |   | 3/8 | 3/4 | 1 1/4 | 0520 | 1520 | 3520 | 4520 |
| 1/8 | 5/16 | 5/8 | 1170 | 2170 | 3/8 | 3/4 | 1 1/2 | 0525 | 1525 | 3525 | 4525 |
| 1/8 | 3/8 | 1/2 | 1180 | 2180 |   |   |   |   |   |   |   |
| 1/8 | 3/8 | 3/4 | 1190 | 2190 | 1/2 | 3/4 | 1 | 0530 | 1530 | 3530 | 4530 |
| 1/8 | 1/2 | 1/2 | 1200 | 2200 | 1/2 | 3/4 | 1 1/4 | 0540 | 1540 | 3540 | 4540 |
| 1/8 | 1/2 | 3/4 | 1210 | 2210 | 1/2 | 3/4 | 1 1/2 | 0550 | 1550 | 3550 | 4550 |
| 1/8 | 3/4 | 3/4 | 1215 | 2215 |   |   |   |   |   |   |   |

| T | W | L | F | Style 5000 | Style 6000 | Style 7000 |
|---|---|---|---|---|---|---|
| 5/32 | 3/8 | 9/16 | 1220 | 2220 | | |
| 5/32 | 3/8 | 3/4 | 1230 | 2230 | | |
| 5/32 | 5/8 | 5/8 | 1240 | 2240 | | |
| 1/16 | 1/4 | 5/16 | ... | 5030 | ... | ... |
| 3/16 | 5/16 | 7/16 | 1250 | 2250 | | |
| 3/16 | 5/16 | 5/8 | 1260 | 2260 | | |
| 3/32 | 1/4 | 3/8 | 1/16 | ... | ... | 7060 |
| 3/16 | 3/8 | 1/2 | 1270 | 2270 | 3/32 | 1/4 | 5/16 | 3/8 | ... | 5080 | 6080 | ... |
| 3/16 | 3/8 | 5/8 | 1280 | 2280 | 3/32 | 3/8 | 1/2 | ... | 5100 | 6100 | ... |
| 3/16 | 3/8 | 3/4 | 1290 | 2290 | 3/32 | 7/16 | 1/2 | ... | 5105 | ... | ... |
| 3/16 | 7/16 | 5/8 | 1300 | 2300 | 1/8 | 5/16 | 5/8 | 3/32 | ... | ... | 7170 |
| 3/16 | 7/16 | 13/16 | 1310 | 2310 | 1/8 | 1/2 | 1/2 | ... | 5200 | 6200 | ... |
| 3/16 | 1/2 | 1/2 | 1320 | 2320 | 5/32 | 3/8 | 3/4 | 1/8 | ... | ... | 7230 |
| 3/16 | 1/2 | 3/4 | 1330 | 2330 | 5/32 | 5/8 | 5/8 | ... | 5240 | 6240 | ... |
| 3/16 | 3/4 | 3/4 | 1340 | 2340 | 3/16 | 3/4 | 3/4 | ... | 5340 | 6340 | ... |
|   |   |   |   | 1/4 | 1 | 3/4 | ... | 5410 | ... | ... |

All dimensions are in inches.

*Maximum Variation from Flatness (All Styles):* For dimensions through 1/2 inch, 0.003 inch; dimensions over 1/2 through 1 inch, 0.005 inch; and dimensions over 1 through 2 inches, 0.006 inch.

*Tolerances for Length, Width, and Thickness (Except Width of Styles 5000 and 6000):* Dimensions through 3/8 inch, +0.015, −0 inch; over 3/8 through 1 inch, +0.020, −0 inch; and over 1 through 2 inches, +0.040, −0 inch.

*Tolerance on Width of Styles 5000 and 6000:* Widths through 3/8 inch, −0.010 to −0.025 inch; and over 3/8 through 1 inch, −0.010 to −0.030 inch.

*Radii Furnished with Styles 2000, 3000, and 4000:* For widths of 1/8 through 1/4 inch, $R$ = 1/8 inch; over 1/4 through 3/8 inch, $R$ = 3/16 inch; and for widths over 3/8 inch, $R$ = 1/4 inch.

**Table 5. Standard Style A Tipped Tools (ASA B5.36-1957)**

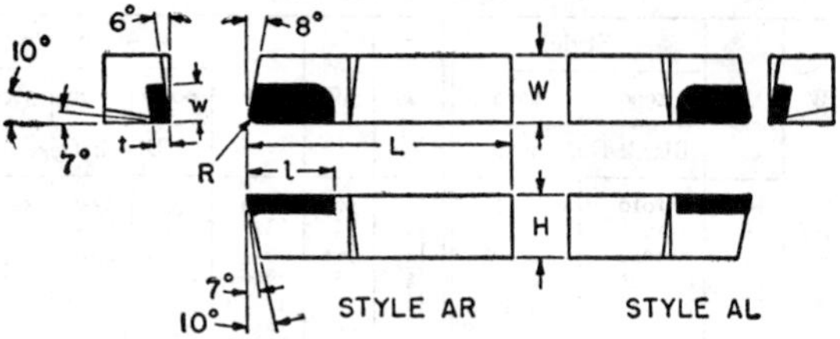


| Catalog Number | | Shank Dimensions | | | Tip Catalog Number | Tip Dimensions | | |
|---|---|---|---|---|---|---|---|---|
| Style AR* | Style AL* | Width *W* | Height *H* | Length *L* | | Thickness *t* | Width *w* | Length *l* |
| SQUARE SHANK | | | | | | | | |
| AR 4 | AL 4 | 1/4 | 1/4 | 1 1/2 | 2040 | 3/32 | 3/16 | 5/16 |
| AR 5 | AL 5 | 5/16 | 5/16 | 2 1/4 | 2070 | 3/32 | 1/4 | 1/2 |
| AR 6 | AL 6 | 3/8 | 3/8 | 2 1/2 | 2070 | 3/32 | 1/4 | 1/2 |
| AR 7 | AL 7 | 7/16 | 7/16 | 3 | 2070 | 3/32 | 1/4 | 1/2 |
| AR 8 | AL 8 | 1/2 | 1/2 | 3 1/2 | 2170 | 1/8 | 5/16 | 5/8 |
| AR 10 | AL 10 | 5/8 | 5/8 | 4 | 2230 | 5/32 | 3/8 | 3/4 |
| AR 12 | AL 12 | 3/4 | 3/4 | 4 1/2 | 2310 | 3/16 | 7/16 | 1 3/16 |
| AR 16 | AL 16 | 1 | 1 | 7 | 3390<br>4390 | 1/4 | 9/16 | 1 |
| AR 20 | AL 20 | 1 1/4 | 1 1/4 | 8 | 3460<br>4460 | 5/16 | 5/8 | 1 |
| AR 24 | AL 24 | 1 1/2 | 1 1/2 | 8 | 3510<br>4510 | 3/8 | 5/8 | 1 |
| RECTANGULAR SHANK | | | | | | | | |
| AR 44 | AL 44 | 1/2 | 1 | 7 | 2260 | 3/16 | 5/16 | 5/8 |
| AR 54 | AL 54 | 5/8 | 1 | 6 | 3360<br>4360 | 1/4 | 3/8 | 3/4 |
| AR 55 | AL 55 | 5/8 | 1 1/4 | 8 | 3360<br>4360 | 1/4 | 3/8 | 3/4 |
| AR 64 | AL 64 | 3/4 | 1 | 6 | 3380<br>4380 | 1/4 | 1/2 | 3/4 |
| AR 66 | AL 66 | 3/4 | 1 1/2 | 8 | 3430<br>4430 | 5/16 | 7/16 | 1 5/16 |
| AR 85 | AL 85 | 1 | 1 1/4 | 8 | 3460<br>4460 | 5/16 | 5/8 | 1 |
| AR 86 | AL 86 | 1 | 1 1/2 | 8 | 3510<br>4510 | 3/8 | 5/8 | 1 |
| AR 88 | AL 88 | 1 | 2 | 10 | 3510<br>4510 | 3/8 | 5/8 | 1 |
| AR 90 | AL 90 | 1 1/2 | 2 | 8 | 3540<br>4540 | 1/2 | 3/4 | 1 1/4 |

* "A" is straight shank, 0 deg., SCEA (side-cutting-edge angle). "R" is right-cut. "L" is left-cut. Where a pair of tip numbers is shown, the upper number applies to AR tools, the lower to AL tools.

**Styles and Sizes of Single-point, Sintered-carbide-tipped Tools.** — American Standard ASA B5.36-1957 distinguishes eight different styles of single-point, carbide-tipped tools. These styles are designated by letters A to G inclusive; each style is made for both right-hand and left-hand cutting as indicated by the letters R or L following the style letter. Dimensions of tips and shanks for the various styles of tools are given in Tables 5 to 12.

A number follows the letters of the tool style and hand designation, and for square

**Table 6. Standard Style B Tipped Tools with 15-degree Side-cutting-edge Angle** (ASA B5.36-1957)

| Catalog Number | | Shank Dimensions | | | Tip Catalog Number | Tip Dimensions | | |
|---|---|---|---|---|---|---|---|---|
| Style BR | Style BL | Width W | Height H | Length L | | Thickness t | Width w | Length l |
| SQUARE SHANK | | | | | | | | |
| BR 6 | BL 6 | 3/8 | 3/8 | 2½ | 1070 | 3/32 | 1/4 | 1/2 |
| BR 8 | BL 8 | 1/2 | 1/2 | 3½ | 2170 | 1/8 | 5/16 | 5/8 |
| BR 10 | BL 10 | 5/8 | 5/8 | 4 | 2230 | 5/32 | 3/8 | 3/4 |
| BR 12 | BL 12 | 3/4 | 3/4 | 4½ | 2310 | 3/16 | 7/16 | 13/16 |
| BR 16 | BL 16 | 1 | 1 | 7 | 3390 / 4390 | 1/4 | 9/16 | 1 |
| BR 20 | BL 20 | 1¼ | 1¼ | 8 | 3460 / 4460 | 5/16 | 5/8 | 1 |
| BR 24 | BL 24 | 1½ | 1½ | 8 | 3510 / 4510 | 3/8 | 5/8 | 1 |
| RECTANGULAR SHANK | | | | | | | | |
| BR 44 | BL 44 | 1/2 | 1 | 7 | 2260 | 3/16 | 5/16 | 5/8 |
| BR 54 | BL 54 | 5/8 | 1 | 6 | 3360 / 4360 | 1/4 | 3/8 | 3/4 |
| BR 55 | BL 55 | 5/8 | 1¼ | 8 | 3360 / 4360 | 1/4 | 3/8 | 3/4 |
| BR 64 | BL 64 | 3/4 | 1 | 6 | 3380 / 4380 | 1/4 | 1/2 | 3/4 |
| BR 66 | BL 66 | 3/4 | 1½ | 8 | 3430 / 4430 | 5/16 | 7/16 | 15/16 |
| BR 85 | BL 85 | 1 | 1¼ | 8 | 3460 / 4460 | 5/16 | 5/8 | 1 |
| BR 86 | BL 86 | 1 | 1½ | 8 | 3510 / 4510 | 3/8 | 5/8 | 1 |
| BR 88 | BL 88 | 1 | 2 | 10 | 3510 / 4510 | 3/8 | 5/8 | 1 |
| BR 90 | BL 90 | 1½ | 2 | 8 | 3540 / 4540 | 1/2 | 3/4 | 1¼ |

Tools BR and BL 4, 5, 7, and 55 are also in the Standard but are not included here. Where a pair of tip numbers is shown, the upper number applies to BR tools, the lower to BL tools.

shank tools, represents the number of sixteenths of an inch of width, $W$, and height, $H$. In the case of rectangular shanks, the first digit of the number indicates the number of eighths of an inch in the shank width, $W$, and the second digit the number of quarters of an inch in the shank height, $H$. To limit the number of characters to be stamped on tool shanks the 1½ by 2 inch tools have been arbitrarily given the number 90.

Table 7. Standard Style C, Square-end Tipped Tools (ASA B5.36-1957)

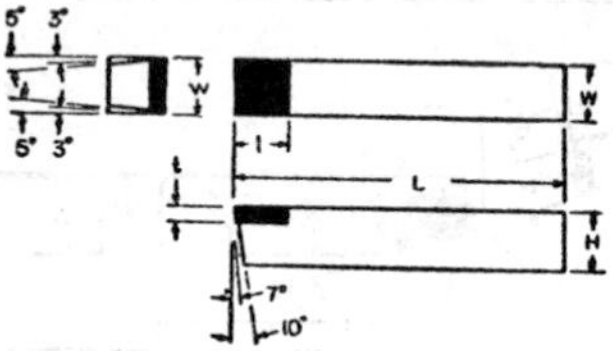

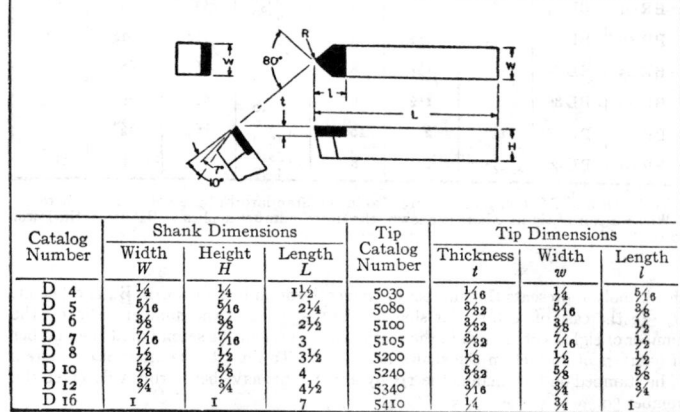

| Catalog Number | Shank Dimensions | | | Tip Catalog Number | Tip Dimensions | | |
|---|---|---|---|---|---|---|---|
| | Width W | Height H | Length L | | Thickness t | Width w | Length l |
| SQUARE SHANK | | | | | | | |
| C 4 | ¼ | ¼ | 1½ | 1030 | ¹⁄₁₆ | ¼ | ⁵⁄₁₆ |
| C 5 | ⁵⁄₁₆ | ⁵⁄₁₆ | 2¼ | 1080 | ³⁄₃₂ | ⁵⁄₁₆ | ⅜ |
| C 6 | ⅜ | ⅜ | 2½ | 1090 | ³⁄₃₂ | ⅜ | ⅜ |
| C 7 | ⁷⁄₁₆ | ⁷⁄₁₆ | 3 | 1105 | ³⁄₃₂ | ⁷⁄₁₆ | ½ |
| C 8 | ½ | ½ | 3½ | 1200 | ⅛ | ½ | ½ |
| C 10 | ⅝ | ⅝ | 4 | 1240 | ⁵⁄₃₂ | ⅝ | ⅝ |
| C 12 | ¾ | ¾ | 4½ | 1340 | ³⁄₁₆ | ¾ | ¾ |
| C 16 | 1 | 1 | 7 | 1410 | ¼ | 1 | ¾ |
| C 20 | 1¼ | 1¼ | 8 | 1480 | ⁵⁄₁₆ | 1¼ | ¾ |
| RECTANGULAR SHANK | | | | | | | |
| C 44 | ½ | 1 | 7 | 1320 | ³⁄₁₆ | ½ | ½ |
| C 54 | ⅝ | 1 | 6 | 1400 | ¼ | ⅝ | ⅝ |
| C 55 | ⅝ | 1¼ | 8 | 1400 | ¼ | ⅝ | ⅝ |
| C 64 | ¾ | 1 | 6 | 1405 | ¼ | ¾ | ¾ |
| C 66 | ¾ | 1½ | 8 | 1470 | ⁵⁄₁₆ | ¾ | ¾ |
| C 86 | 1 | 1½ | 8 | 1475 | ⁵⁄₁₆ | 1 | ¾ |

A chamfer of 0.015 inch at 45 degrees on each corner is permissible.

Table 8. Standard Style D, 80-degree Nose-angle
Tipped Tools (ASA B5.36-1957)

| Catalog Number | Shank Dimensions | | | Tip Catalog Number | Tip Dimensions | | |
|---|---|---|---|---|---|---|---|
| | Width W | Height H | Length L | | Thickness t | Width w | Length l |
| D 4 | ¼ | ¼ | 1½ | 5030 | ¹⁄₁₆ | ¼ | ⁵⁄₁₆ |
| D 5 | ⁵⁄₁₆ | ⁵⁄₁₆ | 2¼ | 5080 | ³⁄₃₂ | ⁵⁄₁₆ | ⅜ |
| D 6 | ⅜ | ⅜ | 2½ | 5100 | ³⁄₃₂ | ⅜ | ½ |
| D 7 | ⁷⁄₁₆ | ⁷⁄₁₆ | 3 | 5105 | ³⁄₃₂ | ⁷⁄₁₆ | ½ |
| D 8 | ½ | ½ | 3½ | 5200 | ⅛ | ½ | ½ |
| D 10 | ⅝ | ⅝ | 4 | 5240 | ⁵⁄₃₂ | ⅝ | ⅝ |
| D 12 | ¾ | ¾ | 4½ | 5340 | ³⁄₁₆ | ¾ | ¾ |
| D 16 | 1 | 1 | 7 | 5410 | ¼ | ¾ | 1 |

**Table 9.  Standard Style E, 60-degree Nose-angle, Tipped Tools (ASA B5.36-1957)**

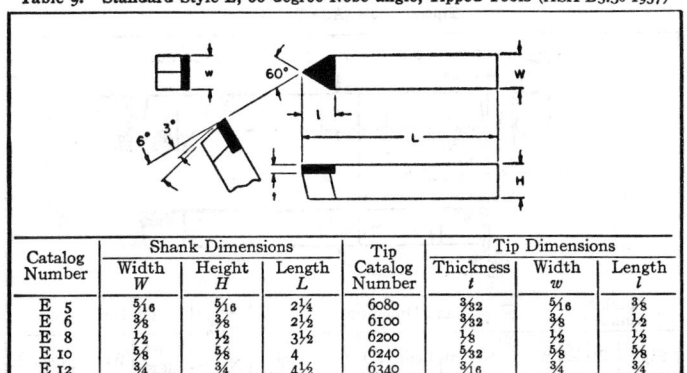

| Catalog Number | Shank Dimensions | | | Tip Catalog Number | Tip Dimensions | | |
|---|---|---|---|---|---|---|---|
| | Width W | Height H | Length L | | Thickness t | Width w | Length l |
| E 5 | 5/16 | 5/16 | 2 1/4 | 6080 | 3/32 | 5/16 | 3/8 |
| E 6 | 3/8 | 3/8 | 2 1/2 | 6100 | 3/32 | 3/8 | 1/2 |
| E 8 | 1/2 | 1/2 | 3 1/2 | 6200 | 1/8 | 1/2 | 1/2 |
| E 10 | 5/8 | 5/8 | 4 | 6240 | 5/32 | 5/8 | 5/8 |
| E 12 | 3/4 | 3/4 | 4 1/2 | 6340 | 3/16 | 3/4 | 3/4 |

**Table 10.  Standard Style F, Offset, End-cutting, Tipped Tools (ASA B5.36-1957)**

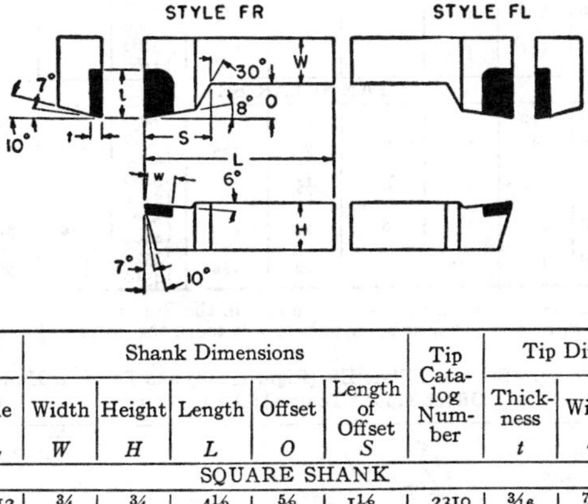

STYLE FR          STYLE FL

| Catalog Number | | Shank Dimensions | | | | | Tip Catalog Number | Tip Dimensions | | |
|---|---|---|---|---|---|---|---|---|---|---|
| Style FR | Style FL | Width W | Height H | Length L | Offset O | Length of Offset S | | Thickness t | Width w | Length l |
| SQUARE SHANK | | | | | | | | | | |
| FR 12 | FL 12 | 3/4 | 3/4 | 4 1/2 | 5/8 | 1 1/8 | 2310 | 3/16 | 7/16 | 13/16 |
| FR 16 | FL 16 | 1 | 1 | 7 | 3/4 | 1 7/16 | 3390 / 4390 | 1/4 | 9/16 | 1 |
| FR 20 | FL 20 | 1 1/4 | 1 1/4 | 8 | 3/4 | 1 3/8 | 3460 / 4460 | 5/16 | 5/8 | 1 |
| RECTANGULAR SHANK | | | | | | | | | | |
| FR 44 | FL 44 | 1/2 | 1 | 6 | 1/2 | 1 | 2260 | 3/16 | 5/16 | 5/8 |
| FR 55 | FL 55 | 5/8 | 1 1/4 | 7 | 5/8 | 1 1/4 | 3360 / 4360 | 1/4 | 3/8 | 3/4 |
| FR 64 | FL 64 | 3/4 | 1 | 6 | 5/8 | 1 5/16 | 3380 / 4380 | 1/4 | 1/2 | 3/4 |
| FR 85 | FL 85 | 1 | 1 1/4 | 8 | 3/4 | 1 1/2 | 3460 / 4460 | 5/16 | 5/8 | 1 |
| FR 86 | FL 86 | 1 | 1 1/2 | 8 | 3/4 | 1 9/16 | 3510 / 4510 | 3/8 | 5/8 | 1 |

Tools FR and FL 8, 10, 24, 66, and 90 are also in the Standard.  Where a pair of tip numbers is shown, the upper number applies to FR tools, the lower to FL tools.

**Table 11. American Standard Style G, Offset, Side-cutting, Tipped Tools (ASA B5.36-1957)**

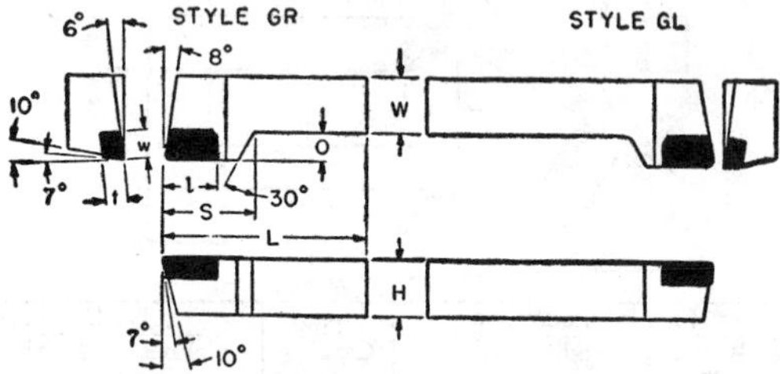

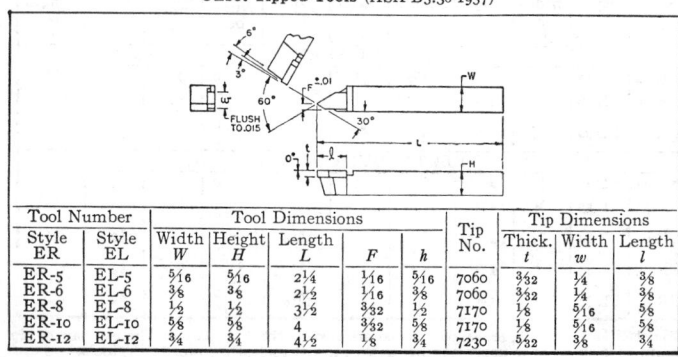

| Catalog Number | | Shank Dimensions | | | | | Tip Catalog Number | Tip Dimensions | | |
|---|---|---|---|---|---|---|---|---|---|---|
| Style GR | Style GL | Width W | Height H | Length L | Offset O | Length of Offset S | | Thickness t | Width w | Length l |
| SQUARE SHANK | | | | | | | | | | |
| GR 12 | GL 12 | ¾ | ¾ | 4½ | ⅜ | 1⅜ | 2310 | 3/16 | 7/16 | 1 3/16 |
| GR 16 | GL 16 | 1 | 1 | 7 | ½ | 1⅝ | {3390 / 4390} | ¼ | 9/16 | 1 |
| GR 20 | GL 20 | 1¼ | 1¼ | 8 | ¾ | 1¾ | {3460 / 4460} | 5/16 | ⅝ | 1 |
| RECTANGULAR SHANK | | | | | | | | | | |
| GR 44 | GL 44 | ½ | 1 | 6 | ¼ | 1⅛ | 2260 | 3/16 | 5/16 | ⅝ |
| GR 55 | GL 55 | ⅝ | 1¼ | 7 | ⅜ | 1⅛ | {3360 / 4360} | ¼ | ⅜ | ¾ |
| GR 64 | GL 64 | ¾ | 1 | 6 | ½ | 1⅜ | {3380 / 4380} | ¼ | ½ | ¾ |
| GR 85 | GL 85 | 1 | 1¼ | 8 | ½ | 1⅝ | {3460 / 4460} | 5/16 | ⅝ | 1 |
| GR 86 | GL 86 | 1 | 1½ | 8 | ½ | 1⅝ | {3510 / 4510} | ⅜ | ⅝ | 1 |

Tools GR and GL 8, 10, 24, 66, and 90 are also in the Standard. Where a pair of tip numbers is shown, the upper number applies to GR tools, the lower to GL tools.

**Table 12. Standard Style ER (Right Hand) and EL (Left Hand) Offset Tipped Tools (ASA B5.36-1957)**

| Tool Number | | Tool Dimensions | | | | | Tip No. | Tip Dimensions | | |
|---|---|---|---|---|---|---|---|---|---|---|
| Style ER | Style EL | Width W | Height H | Length L | F | h | | Thick. t | Width w | Length l |
| ER-5 | EL-5 | 5/16 | 5/16 | 2¼ | 1/16 | 5/16 | 7060 | 3/32 | ¼ | ⅜ |
| ER-6 | EL-6 | ⅜ | ⅜ | 2½ | 1/16 | ⅜ | 7060 | 3/32 | ¼ | ⅜ |
| ER-8 | EL-8 | ½ | ½ | 3½ | 3/32 | ½ | 7170 | ⅛ | 5/16 | ⅝ |
| ER-10 | EL-10 | ⅝ | ⅝ | 4 | 3/32 | ⅝ | 7170 | ⅛ | 5/16 | ⅝ |
| ER-12 | EL-12 | ¾ | ¾ | 4½ | ⅛ | ¾ | 7230 | 5/32 | ⅜ | ¾ |

**Single-point Tool Nose Radii.** — The tool nose radii recommended in the American Standard are as follows: For square-shank tools up to and including ⅜-inch square tools, ¹⁄₆₄ inch; for those over ⅜-inch square through 1¼-inches square, ¹⁄₃₂ inch; and for those above 1¼-inches square, ¹⁄₁₆ inch. For rectangular-shank tools with shank section of ½ × 1 inch through 1 × 1½ inches, the nose radii are ¹⁄₃₂ inch, and for 1 × 2 inch-shanks, the nose radius is ¹⁄₁₆ inch.

**Single-point Tool Angle Tolerances.** — The tool angles shown on the diagrams in the Tables 5 through 12 are general recommendations. Tolerances applicable to these angles are ±1 degree on all angles except end and side clearance angles; for these the tolerance is ±2 degrees.

**Single-point Tool Tip Overhang.** — As shown on the diagram, page 1864, the tip of the brazed carbide blank overhangs the shank of the tool by a small amount. The amount of overhang is either ¹⁄₃₂ or ¹⁄₁₆ inch depending on the size of the tool. These overhangs are not shown in the diagrams accompanying Tables 5 through 12.

For tools shown in Tables 5, 6, 7, 8, 10, and 11, the maximum overhang is ¹⁄₃₂ inch for shank size numbers 4, 5, 6, 7, 8, 10, 12, and 14; for other shank sizes in these tables, the maximum overhang is ¹⁄₁₆ inch. In Tables 9 and 12 all tools have a maximum overhang of ¹⁄₃₂ inch.

**Solid Sintered Carbide Inserts and Their Holders.** — Section 4 of American Standard B5.36-1957 gives standard sizes, styles, and designations of solid sintered carbide inserts and their holders. These inserts are described in the following paragraphs; their holders are described in the following section.

Solid carbide inserts are clamped in holders for machining operations; they are available in four standard cross sections, circular, triangular, square, and diamond.

*Circular Inserts:* These are designated by the letters SC followed by a number indicating the number of ¹⁄₃₂nds in the nominal cross-section dimension. For example, SC-16 indicates a circular-section insert of ¹⁶⁄₃₂ or ½-inch diameter; SC-12 indicates a circular-section insert of ¹²⁄₃₂ or ⅜-inch diameter; SC-8 indicates a circular-section insert of ⁸⁄₃₂ or ¼-inch diameter.

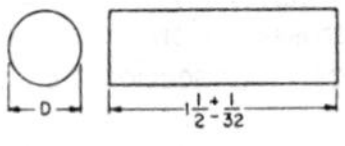

| | Insert Designation | Diameter D | Insert Designation | Diameter D |
|---|---|---|---|---|
| | SC-8 | ¼ | SC-16 | ½ |
| | SC-12 | ⅜ | ... | ... |

*Triangular Inserts:* These are designated by the letters TB followed by (1) a one- or two-digit number indicating the number of ¹⁄₃₂nds in the circle that can be inscribed in the triangular cross section; and (2) followed by a number indicating the ¹⁄₆₄ths in the radius at each of the corners. Thus, TB-164 indicates a triangular-section insert of ¹⁶⁄₃₂ or ½-inch diameter inscribed circle with ⁴⁄₆₄ or ¹⁄₁₆-inch corner radii.

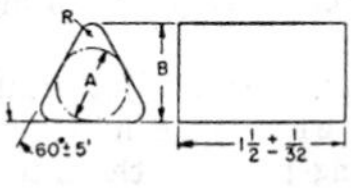

| | Insert Designation | A ±.001 | B ±.001 | R | Insert Designation | A ±.001 | B ±.001 | R |
|---|---|---|---|---|---|---|---|---|
| | TB-82 | ¼ | .344 | ¹⁄₃₂ | TB-168 | ½ | .625 | ⅛ |
| | TB-122 | ⅜ | .531 | ¹⁄₃₂ | TB-204 | ⅝ | .875 | ¹⁄₁₆ |
| | TB-124 | ⅜ | .500 | ¹⁄₁₆ | TB-208 | ⅝ | .812 | ⅛ |
| | TB-164 | ½ | .688 | ¹⁄₁₆ | ... | ... | ... | ... |

*Square Inserts:* These are designated by the letters SQ followed by (1) a one- or two-digit number indicating the size of the square in $\frac{1}{32}$nds; and (2) by a number indicating the $\frac{1}{64}$ths in the radius on each corner. Thus, SQ-164 indicates a square-section insert of $\frac{16}{32}$ or $\frac{1}{2}$-inch side with $\frac{4}{64}$ or $\frac{1}{16}$-inch corner radii.

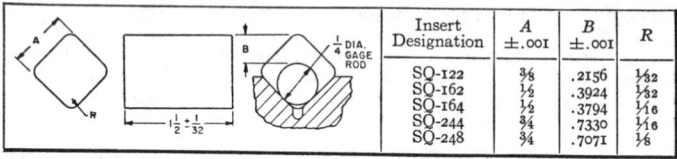

| Insert Designation | $A$ ±.001 | $B$ ±.001 | $R$ |
|---|---|---|---|
| SQ-122 | $\frac{3}{8}$ | .2156 | $\frac{1}{32}$ |
| SQ-162 | $\frac{1}{2}$ | .3924 | $\frac{1}{32}$ |
| SQ-164 | $\frac{1}{2}$ | .3794 | $\frac{1}{16}$ |
| SQ-244 | $\frac{3}{4}$ | .7330 | $\frac{1}{16}$ |
| SQ-248 | $\frac{3}{4}$ | .7071 | $\frac{1}{8}$ |

*Diamond Inserts:* These are designated by the letters DB through DH in which the D indicates diamond section and the second letter indicates the number of degrees in the included angle in each of two opposite corners (the cutting corners of the insert); the letter B indicates an included angle of 85°; C = 80°; D = 75°; E = 70°; F = 65°; G = 60°; and H = 55°. These letter designations are followed by (1) a number designating the $\frac{1}{32}$nds in the diameter of the inscribed circle to the nearest $\frac{1}{32}$-inch nominal size; and (2) followed by the number of $\frac{1}{64}$ths in the corner radius. Thus, DC-152 indicates: a diamond-section insert (D), 80 degrees included angle (C), an inscribed circle of $\frac{15}{32}$-inch nominal diameter which is nearest the actual diameter of 0.464 (15), and a corner radius of $\frac{2}{64}$ or $\frac{1}{32}$ inch (2).

| Insert Designation | $A$ ±.001 | $Z$ ±5' | $B$ ±.001 |
|---|---|---|---|
| DC-152 | .464 | 80° | .385 |
| DH-152 | .464 | 55° | .573 |

**Holders for Solid Carbide Inserts.** — The dimensions of holders for solid carbide inserts of circular, triangular, square, and diamond cross-section are given in Tables 1 to 8. These American Standard holders (ASA B5.36-1957) are designated by letters for the style and by numbers for the shank cross-section.

The first letter in the style designation indicates the insert cross-section (T for triangular, S for square, R for round, or D for diamond).

The second letter in the style designation indicates the insert position in the holder: A = 10° end cutting edge angle on the round and 0° side cutting edge angle on the triangle as shown in Tables 2 and 4; B = 37° end cutting edge angle on the round and 15° side cutting edge angle on the triangle and square, as shown in Tables 3, 5, and 7; F = end cutting tools with 0° end cutting edge angle on the triangle and 15° end cutting edge angle on the square, as shown in Tables 6 and 8

The third letter indicates the holder as either right (R) or left (L) hand.

*Shank Size Designation:* The style designation is followed by a number to indicate the shank cross-section. For square shank sizes the number represents the number of $\frac{1}{16}$ths of width and height; for rectangular shank sizes, the first digit represents the number of $\frac{1}{8}$ths of shank width and is followed by a second digit representing the number of $\frac{1}{4}$ths of shank height.

*Clamping and Adjusting Inserts:* The provisions for clamping, vertical adjustment, and angular clearances for the insert holders in Tables 2 to 8 are left to the individual manufacturers. In the adjusted position, the lowest cutting point of the insert should be in the plane of the top of the shank.

**Table 1.  A, B, C Dimensions of Insert Holders in Tables 2 to 8***

| Holder Shank-Size Number | Holder Dimensions | | |
|---|---|---|---|
| | $A$ $+0$ $-.010$ | $B$ | $C$ $\pm\frac{1}{8}$ |
| −12 | $\frac{3}{4}$ | $\frac{3}{4}$ $+0.000$ $-0.010$ | 6 |
| −16 | 1 | 1 $+0.000$ $-0.010$ | 6 |
| −64 | $\frac{3}{4}$ | 1 $+0.000$ $-0.010$ | 6 |
| −65 | $\frac{3}{4}$ | $1\frac{1}{4}$ $+0.000$ $-0.015$ | 6 |
| −85 | 1 | $1\frac{1}{4}$ $+0.000$ $-0.015$ | 6 |
| −86 | 1 | $1\frac{1}{2}$ $+0.000$ $-0.015$ | 7 |

All dimensions are in inches.
* Other dimensions are given in Tables 2 to 8, and a description of the style and size designation system appears on page 1872.

Dimensions and designations for inserts for these holders are given on pages 1871 and 1872.

**Table 2.  Holders for Round Inserts**

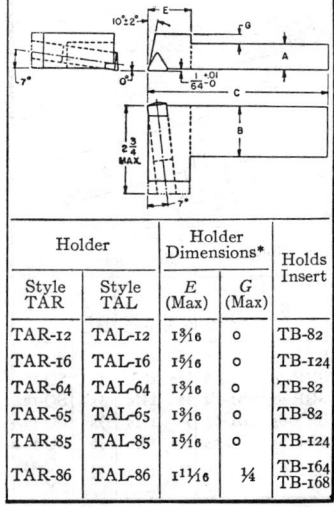

| Holder | | Holder Dimensions* | | Holds Insert |
|---|---|---|---|---|
| Style RAR | Style RAL | $E$ (Max) | $G$ (Max) | |
| RAR-12 | RAL-12 | $1\frac{3}{16}$ | 0 | SC-12 |
| RAR-16 | RAL-16 | $1\frac{5}{16}$ | $\frac{1}{16}$ | SC-16 |
| RAR-64 | RAL-64 | $1\frac{3}{16}$ | 0 | SC-12 |
| RAR-65 | RAL-65 | $1\frac{3}{16}$ | 0 | SC-12 |
| RAR-85 | RAL-85 | $1\frac{5}{16}$ | $\frac{1}{16}$ | SC-16 |
| RAR-86 | RAL-86 | $1\frac{5}{16}$ | 0 | SC-16 |

*See, also, Table 1.

**Table 3.  Holders for Round Inserts**

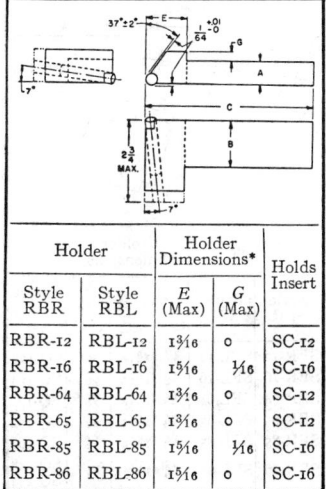

| Holder | | Holder Dimensions* | | Holds Insert |
|---|---|---|---|---|
| Style RBR | Style RBL | $E$ (Max) | $G$ (Max) | |
| RBR-12 | RBL-12 | $1\frac{3}{16}$ | 0 | SC-12 |
| RBR-16 | RBL-16 | $1\frac{5}{16}$ | $\frac{1}{16}$ | SC-16 |
| RBR-64 | RBL-64 | $1\frac{3}{16}$ | 0 | SC-12 |
| RBR-65 | RBL-65 | $1\frac{3}{16}$ | 0 | SC-12 |
| RBR-85 | RBL-85 | $1\frac{5}{16}$ | $\frac{1}{16}$ | SC-16 |
| RBR-86 | RBL-86 | $1\frac{5}{16}$ | 0 | SC-16 |

* See, also, Table 1.

**Table 4.  Holders for Triangular Inserts**

| Holder | | Holder Dimensions* | | Holds Insert |
|---|---|---|---|---|
| Style TAR | Style TAL | $E$ (Max) | $G$ (Max) | |
| TAR-12 | TAL-12 | $1\frac{3}{16}$ | 0 | TB-82 |
| TAR-16 | TAL-16 | $1\frac{5}{16}$ | 0 | TB-124 |
| TAR-64 | TAL-64 | $1\frac{3}{16}$ | 0 | TB-82 |
| TAR-65 | TAL-65 | $1\frac{3}{16}$ | 0 | TB-82 |
| TAR-85 | TAL-85 | $1\frac{5}{16}$ | 0 | TB-124 |
| TAR-86 | TAL-86 | $1\frac{11}{16}$ | $\frac{1}{4}$ | TB-164 TB-168 |

* See, also, Table 1.

**Table 5. Holders for Triangular Inserts**

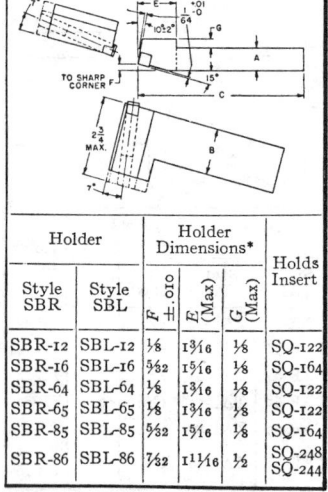

| Holder | | Holder Dimensions* | | | Holds Insert |
|---|---|---|---|---|---|
| Style TBR | Style TBL | $F \pm .010$ | $E$ (Max) | $G$ (Max) | |
| TBR-12 | TBL-12 | $\frac{9}{64}$ | $1\frac{3}{16}$ | $\frac{1}{8}$ | TB-82 |
| TBR-16 | TBL-16 | $\frac{3}{16}$ | $1\frac{5}{16}$ | $\frac{1}{8}$ | TB-124 |
| TBR-64 | TBL-64 | $\frac{9}{64}$ | $1\frac{3}{16}$ | $\frac{1}{8}$ | TB-82 |
| TBR-65 | TBL-65 | $\frac{9}{64}$ | $1\frac{3}{16}$ | $1\frac{1}{8}$ | TB-82 |
| TBR-85 | TBL-85 | $\frac{3}{16}$ | $1\frac{5}{16}$ | $\frac{1}{8}$ | TB-124 |
| TBR-86 | TBL-86 | $\frac{1}{4}$ | $1\frac{11}{16}$ | $\frac{1}{4}$ | TB-164 TB-168 |

* See, also, Table 1.

**Table 6. Holders for Triangular Inserts**

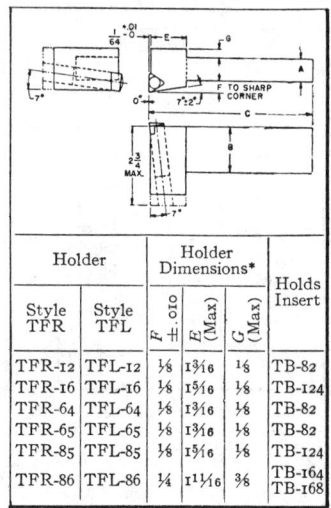

| Holder | | Holder Dimensions* | | | Holds Insert |
|---|---|---|---|---|---|
| Style TFR | Style TFL | $F \pm .010$ | $E$ (Max) | $G$ (Max) | |
| TFR-12 | TFL-12 | $\frac{1}{8}$ | $1\frac{3}{16}$ | $1\frac{1}{8}$ | TB-82 |
| TFR-16 | TFL-16 | $\frac{1}{8}$ | $1\frac{5}{16}$ | $\frac{1}{8}$ | TB-124 |
| TFR-64 | TFL-64 | $\frac{1}{8}$ | $1\frac{3}{16}$ | $\frac{1}{8}$ | TB-82 |
| TFR-65 | TFL-65 | $\frac{1}{8}$ | $1\frac{3}{16}$ | $\frac{1}{8}$ | TB-82 |
| TFR-85 | TFL-85 | $\frac{1}{8}$ | $1\frac{5}{16}$ | $\frac{1}{8}$ | TB-124 |
| TFR-86 | TFL-86 | $\frac{1}{4}$ | $1\frac{11}{16}$ | $\frac{3}{8}$ | TB-164 TB-168 |

* See, also, Table 1.

**Table 7. Holders for Square Inserts**

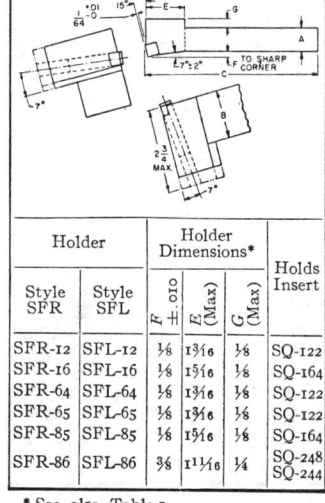

| Holder | | Holder Dimensions* | | | Holds Insert |
|---|---|---|---|---|---|
| Style SBR | Style SBL | $F \pm .010$ | $E$ (Max) | $G$ (Max) | |
| SBR-12 | SBL-12 | $\frac{1}{8}$ | $1\frac{3}{16}$ | $\frac{1}{8}$ | SQ-122 |
| SBR-16 | SBL-16 | $\frac{5}{32}$ | $1\frac{5}{16}$ | $\frac{1}{8}$ | SQ-164 |
| SBR-64 | SBL-64 | $\frac{1}{8}$ | $1\frac{3}{16}$ | $\frac{1}{8}$ | SQ-122 |
| SBR-65 | SBL-65 | $\frac{1}{8}$ | $1\frac{3}{16}$ | $\frac{1}{8}$ | SQ-122 |
| SBR-85 | SBL-85 | $\frac{5}{32}$ | $1\frac{5}{16}$ | $\frac{1}{8}$ | SQ-164 |
| SBR-86 | SBL-86 | $\frac{7}{32}$ | $1\frac{11}{16}$ | $\frac{1}{2}$ | SQ-248 SQ-244 |

* See, also, Table 1.

**Table 8. Holders for Square Inserts**

| Holder | | Holder Dimensions* | | | Holds Insert |
|---|---|---|---|---|---|
| Style SFR | Style SFL | $F \pm .010$ | $E$ (Max) | $G$ (Max) | |
| SFR-12 | SFL-12 | $\frac{1}{8}$ | $1\frac{3}{16}$ | $\frac{1}{8}$ | SQ-122 |
| SFR-16 | SFL-16 | $\frac{1}{8}$ | $1\frac{5}{16}$ | $\frac{1}{8}$ | SQ-164 |
| SFR-64 | SFL-64 | $\frac{1}{8}$ | $1\frac{3}{16}$ | $\frac{1}{8}$ | SQ-122 |
| SFR-65 | SFL-65 | $\frac{1}{8}$ | $1\frac{3}{16}$ | $\frac{1}{8}$ | SQ-122 |
| SFR-85 | SFL-85 | $\frac{1}{8}$ | $1\frac{5}{16}$ | $\frac{1}{8}$ | SQ-164 |
| SFR-86 | SFL-86 | $\frac{3}{8}$ | $1\frac{11}{16}$ | $\frac{1}{4}$ | SQ-248 SQ-244 |

* See, also, Table 1.

**Table Giving Step Dimensions and Angles on Straight or Dovetailed Forming Tools.** — The accompanying table "Dimensions of Steps and Angles on Straight Forming Tools" gives the required dimensions and angles within its range, direct or without calculation.

*To Find Dimension of Step:* The upper section of the table is used in determining the dimensions of steps. The radial depth of the step or the actual cutting depth $D$ (see left-hand diagram) is given in the first column of the table. The columns which follow give the corresponding depths $d$ for a front clearance angle of either

10, 15 or 20 degrees, as the case may be. To illustrate the use of the table, suppose a tool is required for turning the part shown in Fig. 1 which has diameters of 0.75, 1.25 and 1.75 inch, respectively. The difference between the largest and the smallest radius is 0.5 inch, which is the depth of one step. Assume that the clearance angle is 15 degrees. First, locate 0.5 in the column headed "Radial Depth of Step $D$"; then find depth $d$ in the column headed "when $C = 15°$." As will be seen, this depth is 0.48296 inch. Practically the same procedure is followed in de-

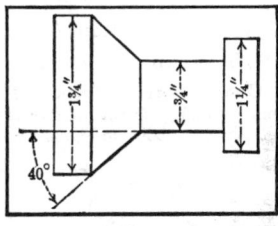

Fig. 1

termining the depth of the second step on the tool. The difference in the radii in this case equals 0.25. Since this value is not given directly in the table, we first find the depth equivalent to 0.200 and then add to it the depth equivalent to 0.050. Thus, we have in this case 0.19138 + 0.04829 = 0.24147. In using this table, it is assumed that the top face of the tool is set at the height of the axis of the work.

*To Find Angle:* The lower section of the table applies to angles when they are measured relative to the axis of the work. The application of the table will again be illustrated by using the part shown in Fig. 1. The angle in this case is 40 degrees (which is also the angle in the plane of the cutting face of the tool). If the clearance angle is 15 degrees, then the angle measured in plane $x$-$x$ square to the face of the tool is shown by the table to be 39° 1' — a reduction of practically one degree.

**Straight Forming Tools With Rake.** — If a straight forming tool has rake, the depth $x$ of each step (see Fig. 2), measured perpendicular to the front or clearance face, is affected not only by the clearance angle but by the rake angle $F$ and the radii $R$ and $r$ of the steps on the work. First it is

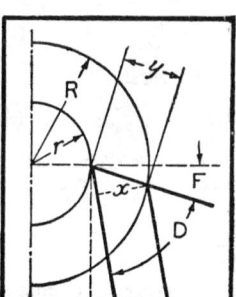

necessary to find three angles. These angles, which will be designated as $A$, $B$ and $C$, are not shown on the drawing.

Angle $A = 180°$ − rake angle $F$; $\sin B = \dfrac{r \sin A}{R}$ ;

angle $C = 180° − (A + B)$

$y = \dfrac{R \sin C}{\sin A}$ ; angle $D$ of tool $= 90° − (E + F)$;

depth $x = y \sin D$

If the work has two or more shoulders, the depth $x$ for other steps on the tool may be determined for each radius $r$. If the work has curved or angular forms, it is more practical to use a tool without rake because its profile, in the plane of the cutting face, duplicates that of the work.

Fig. 2

*Example:* Assume that radius $R$ equals 0.625 and radius $r$ equals 0.375 inch so that the step on the work has a radial depth of 0.25 inch. The tool has a rake angle $F$ of 10 degrees and a clearance angle $E$ of 15 degrees. Then angle $A = 180 - 10 = 170$ degrees.

$$\sin B = \frac{0.375 \times 0.17365}{0.625} = 0.10419$$

Angle $B = 5° \, 59'$ nearly. Angle $C = 180 - (170° + 5° \, 59') = 4° \, 1'$

$$\text{Dimension } y = \frac{0.625 \times 0.07005}{0.17365} = 0.25212$$

Angle $D = 90° - (15 + 10) = 65$ degrees.

Depth $x$ of step $= 0.25212 \times 0.90631 = 0.2285$ inch.

**Circular Forming Tools.** — To provide sufficient periphery clearance on circular forming tools, the cutting face is off-set with relation to the center of the tool a distance $C$ as shown in Fig. 3. Whenever a circular tool has two or more diameters,

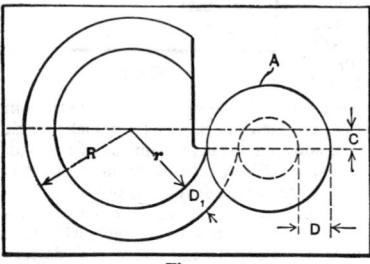

Fig. 3

the difference in the radii of the steps on the tool will, therefore, not correspond exactly to the difference in the steps on the work. The form produced with the tool also changes, although the change is very slight, unless the amount of off-set $C$ is considerable. Assume that a circular tool is required to produce the piece $A$ having two diameters as shown. If the difference $D_1$ between the large and small radii of the tool were made equal to dimension $D$ required on the work, $D$ would be a certain amount over-size,

depending upon the off-set $C$ of the cutting edge. The following formulas can be used to determine the radii of circular forming tools for turning parts to different diameters:

Let $R$ = largest radius of tool in inches; $D$ = difference in radii of steps on work; $C$ = amount cutting edge is off-set from center of tool; $r$ = required radius in inches; then:

$$r = \sqrt{\left(\sqrt{R^2 - C^2} - D\right)^2 + C^2}. \tag{1}$$

If the small radius $r$ is given and the large radius $R$ is required, then:

$$R = \sqrt{\left(\sqrt{r^2 - C^2} + D\right)^2 + C^2}. \tag{2}$$

To illustrate, if $D$ (Fig. 3) is to be ⅛ inch, the large radius $R$ is 1⅛ inch, and $C$ is ⁵⁄₃₂ inch, what radius $r$ would be required to compensate for the off-set $C$ of the cutting edge? Inserting these values in Formula (1):

$$r = \sqrt{\left(\sqrt{(1\tfrac{1}{8})^2 - (\tfrac{5}{32})^2} - \tfrac{1}{8}\right)^2 + (\tfrac{5}{32})^2} = 1.0014 \text{ inch.}$$

The value of $r$ is thus found to be 1.0014 inch; hence the diameter $= 2 \times 1.0014 = 2.0028$ inches instead of 2 inches, as would have been the case if the cutting edge had been exactly on the center-line. Formulas for circular tools used on different makes of screw machines can be simplified when the values $R$ and $C$ are constant for each size of machine. The accompanying table "Formulas for Circular Form-

**Formulas for Circular Forming Tools**

(For notation, see Fig. 3)

| Make of Machine | Size of Machine | Radius $R$, Inches | Offset $C$, Inches | Radius $r$, Inches |
|---|---|---|---|---|
| Brown & Sharpe | No. 00 | 0.875 | 0.125 | $r = \sqrt{(0.8660 - D)^2 + 0.0156}$ |
| | No. 0 | 1.125 | 0.15625 | $r = \sqrt{(1.1141 - D)^2 + 0.0244}$ |
| | No. 2 | 1.50 | 0.250 | $r = \sqrt{(1.4790 - D)^2 + 0.0625}$ |
| | No. 6 | 2.00 | 0.3125 | $r = \sqrt{(1.975 - D)^2 + 0.0976}$ |
| Acme | No. 51 | 0.75 | 0.09375 | $r = \sqrt{(0.7441 - D)^2 + 0.0088}$ |
| | No. 515 | 0.75 | 0.09375 | $r = \sqrt{(0.7441 - D)^2 + 0.0088}$ |
| | No. 52 | 1.0 | 0.09375 | $r = \sqrt{(0.9956 - D)^2 + 0.0088}$ |
| | No. 53 | 1.1875 | 0.125 | $r = \sqrt{(1.1809 - D)^2 + 0.0156}$ |
| | No. 54 | 1.250 | 0.15625 | $r = \sqrt{(1.2402 - D)^2 + 0.0244}$ |
| | No. 55 | 1.250 | 0.15625 | $r = \sqrt{(1.2402 - D)^2 + 0.0244}$ |
| | No. 56 | 1.50 | 0.1875 | $r = \sqrt{(1.4882 - D)^2 + 0.0352}$ |
| Cleveland | ¼″ | 0.625 | 0.03125 | $r = \sqrt{(0.6242 - D)^2 + 0.0010}$ |
| | ⅜″ | 0.84375 | 0.0625 | $r = \sqrt{(0.8414 - D)^2 + 0.0039}$ |
| | ⅝″ | 1.15625 | 0.0625 | $r = \sqrt{(1.1546 - D)^2 + 0.0039}$ |
| | ⅞″ | 1.1875 | 0.0625 | $r = \sqrt{(1.1859 - D)^2 + 0.0039}$ |
| | 1¼″ | 1.375 | 0.0625 | $r = \sqrt{(1.3736 - D)^2 + 0.0039}$ |
| | 2″ | 1.375 | 0.0625 | $r = \sqrt{(1.3736 - D)^2 + 0.0039}$ |
| | 2¼″ | 1.625 | 0.125 | $r = \sqrt{(1.6202 - D)^2 + 0.0156}$ |
| | 2¾″ | 1.875 | 0.15625 | $r = \sqrt{(1.8685 - D)^2 + 0.0244}$ |
| | 3¼″ | 1.875 | 0.15625 | $r = \sqrt{(1.8685 - D)^2 + 0.0244}$ |
| | 4¼″ | 2.50 | 0.250 | $r = \sqrt{(2.4875 - D)^2 + 0.0625}$ |
| | 6″ | 2.625 | 0.250 | $r = \sqrt{(2.6131 - D)^2 + 0.0625}$ |

ing Tools" gives the standard values of $R$ and $C$ for circular tools used on different automatics. The formulas for determining the radius $r$ (see column at right-hand side of table) contain a constant which represents the value of the expression $\sqrt{R^2 - C^2}$ in formula (1).

The table "Constants for Determining Diameters of Circular Forming Tools" has been compiled to facilitate proportioning tools of this type. It gives constants for computing the various diameters of forming tools, when the cutting face of the tool is either ⅛, ³⁄₁₆, ¼ or ⁵⁄₁₆ inch below the horizontal center-line. As there is no standard distance for the location of the cutting face, the table has been prepared to correspond with distances commonly used. As an example, suppose the tool is required for a part having three diameters of 1.75, 0.75 and 1.25 inch, respectively, as shown in Fig. 1, and that the largest diameter of the tool is 3 inches and its cutting face is ¼ inch below the horizontal center-line. The first step would be to determine approximately the respective diameters of the forming tool and then correct these diameters by the use of the table. To produce the three diameters shown in Fig. 1, with a 3-inch forming tool, the tool diameters would be approximately 2, 3 and 2.5 inches, respectively. The first dimension (2 inches) is 1 inch less in diameter than that of the tool, and the necessary correction should be given in the column "Correction for Difference in Diameter"; but as the table is only extended to half-inch differences, it will be necessary to obtain this particular correction in two steps. On the line for 3-inch diameter and under corrections for ½ inch, we find 0.0085; then in line with 2½ and under the same heading, we find 0.0129; hence the total correction would be 0.0085 + 0.0129 = 0.0214 inch. This correction is added to the approximate diameter, making the exact diameter of the first step 2 + 0.0214 = 2.0214 inches. The next step would be computed in the same way, by noting on the 3-inch line the correction for ½ inch and adding it to the approximate diameter of the second step, giving an exact diameter of 2.5 + 0.0085 = 2.5085 inches. Therefore, to produce the part shown in Fig. 1, the tool should have three steps of 3, 2.0214 and 2.5085 inches, respectively, provided the cutting face is ¼ inch below the center. All diameters are computed in this way, from the largest diameter of the tool.

The tables "Corrected Diameters of Circular Forming Tools" are especially applicable to tools used on Brown & Sharpe automatic screw machines. Directions for using these tables are given at the end of Table 4.

**Circular Tools Having Top Rake.** — Circular forming tools without top rake are satisfactory for brass, but tools for steel or other tough metals cut better when there is a rake angle of 10 or 12 degrees. For such tools the small radius $r$ (see Fig. 3) for an outside radius $R$ may be found by the formula:

$$r = \sqrt{P^2 + R^2 - 2\,PR\cos\theta}$$

To find the value of $P$ proceed as follows: $\sin\phi$ = small rad. on work × sin rake angle ÷ large rad. on work. Angle $\beta$ = rake angle − $\phi$. $P$ = large rad. on work × sin $\beta$ ÷ sin rake angle. Angle $\theta$ = rake angle + $\delta$. Sin $\delta$ = vertical height $C$ from center of tool to center of work ÷ $R$. It is assumed that the point of tool is to be set at same height as the work center.

### Dimensions for Circular Cut-off Tools

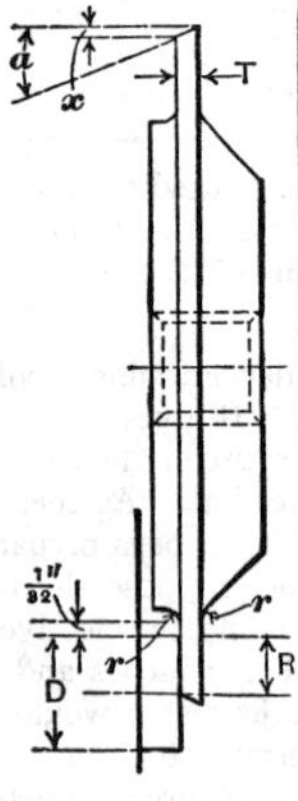

| Diam. of Stock | Soft Brass, Copper $a = 23$ Deg. | | Norway Iron, Machine Steel $a = 15$ Deg. | | Drill Rod, Tool Steel $a = 12$ Deg. | |
|---|---|---|---|---|---|---|
| | $T$ | $x$ | $T$ | $x$ | $T$ | $x$ |
| ¹⁄₁₆ | 0.031 | 0.013 | 0.039 | 0.010 | 0.043 | 0.009 |
| ⅛ | 0.044 | 0.019 | 0.055 | 0.015 | 0.062 | 0.013 |
| ³⁄₁₆ | 0.052 | 0.022 | 0.068 | 0.018 | 0.076 | 0.016 |
| ¼ | 0.062 | 0.026 | 0.078 | 0.021 | 0.088 | 0.019 |
| ⁵⁄₁₆ | 0.069 | 0.029 | 0.087 | 0.023 | 0.098 | 0.021 |
| ⅜ | 0.076 | 0.032 | 0.095 | 0.025 | 0.107 | 0.023 |
| ⁷⁄₁₆ | 0.082 | 0.035 | 0.103 | 0.028 | 0.116 | 0.025 |
| ½ | 0.088 | 0.037 | 0.110 | 0.029 | 0.124 | 0.026 |
| ⁹⁄₁₆ | 0.093 | 0.039 | 0.117 | 0.031 | 0.131 | 0.028 |
| ⅝ | 0.098 | 0.042 | 0.123 | 0.033 | 0.137 | 0.029 |
| ¹¹⁄₁₆ | 0.103 | 0.044 | 0.129 | 0.035 | 0.145 | 0.031 |
| ¾ | 0.107 | 0.045 | 0.134 | 0.036 | 0.152 | 0.032 |
| ¹³⁄₁₆ | 0.112 | 0.047 | 0.141 | 0.038 | 0.158 | 0.033 |
| ⅞ | 0.116 | 0.049 | 0.146 | 0.039 | 0.164 | 0.035 |
| ¹⁵⁄₁₆ | 0.120 | 0.051 | 0.151 | 0.040 | 0.170 | 0.036 |
| 1 | 0.124 | 0.053 | 0.156 | 0.042 | 0.175 | 0.037 |

The length of the blade equals radius of stock $R + x + r + \frac{1}{32}$ inch (for notation see illustration above); $r = \frac{1}{16}$ inch for ⅜- to ¾-inch stock, and ³⁄₃₂ inch for ¾- to 1-inch stock.

## Constants for Determining Diameters of Circular Forming Tools

| Diam. of Tool | Radius of Tool | Cutting Face 1/8 Inch Below Center — Correction for 1/8 Inch Difference in Diam. | Correction for 1/4 Inch Difference in Diam. | Correction for 1/2 Inch Difference in Diam. | Cutting Face 3/16 Inch Below Center — Correction for 1/8 Inch Difference in Diam. | Correction for 1/4 Inch Difference in Diam. | Correction for 1/2 Inch Difference in Diam. | Cutting Face 1/4 Inch Below Center — Correction for 1/8 Inch Difference in Diam. | Correction for 1/4 Inch Difference in Diam. | Correction for 1/2 Inch Difference in Diam. | Cutting Face 5/16 Inch Below Center — Correction for 1/8 Inch Difference in Diam. | Correction for 1/4 Inch Difference in Diam. | Correction for 1/2 Inch Difference in Diam. |
|---|---|---|---|---|---|---|---|---|---|---|---|---|---|
| 1 | 0.500 | ..... | ..... | ..... | ..... | ..... | ..... | ..... | ..... | ..... | ..... | ..... | ..... |
| 1⅛ | 0.5625 | 0.0036 | ..... | ..... | 0.0086 | ..... | ..... | 0.0167 | ..... | ..... | 0.0298 | ..... | ..... |
| 1¼ | 0.625 | 0.0028 | 0.0065 | ..... | 0.0067 | 0.0154 | ..... | 0.0128 | 0.0296 | ..... | 0.0221 | 0.0519 | ..... |
| 1⅜ | 0.6875 | 0.0023 | ..... | ..... | 0.0054 | ..... | ..... | 0.0102 | ..... | ..... | 0.0172 | ..... | ..... |
| 1½ | 0.750 | 0.0019 | 0.0042 | 0.0107 | 0.0045 | 0.0099 | 0.0253 | 0.0083 | 0.0185 | 0.0481 | 0.0138 | 0.0310 | 0.0829 |
| 1⅝ | 0.8125 | 0.0016 | ..... | ..... | 0.0037 | ..... | ..... | 0.0069 | ..... | ..... | 0.0114 | ..... | ..... |
| 1¾ | 0.875 | 0.0014 | 0.0030 | ..... | 0.0032 | 0.0069 | ..... | 0.0058 | 0.0128 | ..... | 0.0095 | 0.0210 | ..... |
| 1⅞ | 0.9375 | 0.0012 | ..... | ..... | 0.0027 | ..... | ..... | 0.0050 | ..... | ..... | 0.0081 | ..... | ..... |
| 2 | 1.000 | 0.0010 | 0.0022 | 0.0052 | 0.0024 | 0.0051 | 0.0121 | 0.0044 | 0.0094 | 0.0223 | 0.0070 | 0.0152 | 0.0362 |
| 2⅛ | 1.0625 | 0.0009 | ..... | ..... | 0.0021 | ..... | ..... | 0.0038 | ..... | ..... | 0.0061 | ..... | ..... |
| 2¼ | 1.125 | 0.0008 | 0.0017 | ..... | 0.0018 | 0.0040 | ..... | 0.0034 | 0.0072 | ..... | 0.0054 | 0.0116 | ..... |
| 2⅜ | 1.1875 | 0.0007 | ..... | ..... | 0.0016 | ..... | ..... | 0.0029 | ..... | ..... | 0.0048 | ..... | ..... |
| 2½ | 1.250 | 0.0006 | 0.0014 | 0.0031 | 0.0015 | 0.0031 | 0.0071 | 0.0027 | 0.0057 | 0.0129 | 0.0043 | 0.0092 | 0.0208 |
| 2⅝ | 1.3125 | 0.0006 | ..... | ..... | 0.0013 | ..... | ..... | 0.0024 | ..... | ..... | 0.0038 | ..... | ..... |
| 2¾ | 1.375 | 0.0005 | 0.0011 | ..... | 0.0012 | 0.0026 | ..... | 0.0022 | 0.0046 | ..... | 0.0035 | 0.0073 | ..... |
| 2⅞ | 1.4375 | 0.0005 | ..... | ..... | 0.0011 | ..... | ..... | 0.0020 | ..... | ..... | 0.0032 | ..... | ..... |
| 3 | 1.500 | 0.0004 | 0.0009 | 0.0021 | 0.0010 | 0.0021 | 0.0047 | 0.0018 | 0.0038 | 0.0085 | 0.0029 | 0.0061 | 0.0135 |
| 3⅛ | 1.5625 | 0.0004 | ..... | ..... | 0.0009 | ..... | ..... | 0.0017 | ..... | ..... | 0.0027 | ..... | ..... |
| 3¼ | 1.625 | 0.0003 | 0.0008 | ..... | 0.0008 | 0.0018 | ..... | 0.0015 | 0.0032 | ..... | 0.0024 | 0.0051 | ..... |
| 3⅜ | 1.6875 | 0.0003 | ..... | ..... | 0.0008 | ..... | ..... | 0.0014 | ..... | ..... | 0.0023 | ..... | ..... |
| 3½ | 1.750 | 0.0003 | 0.0007 | 0.0015 | 0.0007 | 0.0015 | 0.0033 | 0.0013 | 0.0028 | 0.0060 | 0.0021 | 0.0044 | 0.0095 |
| 3⅝ | 1.8125 | 0.0003 | ..... | ..... | 0.0007 | ..... | ..... | 0.0012 | ..... | ..... | 0.0019 | ..... | ..... |
| 3¾ | 1.875 | 0.0002 | 0.0006 | ..... | 0.0006 | 0.0013 | ..... | 0.0011 | 0.0024 | ..... | 0.0018 | 0.0038 | ..... |

Corrected Diameters of Circular Forming Tools — 1

| Length c on Tool | Number of B. & S. Automatic Screw Machine | | | Length c on Tool | Number of B. & S. Automatic Screw Machine | | |
|---|---|---|---|---|---|---|---|
| | No. 00 | No. 0 | No. 2 | | No. 00 | No. 0 | No. 2 |
| 0.001 | 1.7480 | 2.2480 | 2.9980 | 0.058 | 1.6353 | 2.1352 | 2.8857 |
| 0.002 | 1.7460 | 2.2460 | 2.9961 | 0.059 | 1.6333 | 2.1332 | 2.8837 |
| 0.003 | 1.7441 | 2.2441 | 2.9941 | 0.060 | 1.6313 | 2.1312 | 2.8818 |
| 0.004 | 1.7421 | 2.2421 | 2.9921 | 0.061 | 1.6294 | 2.1293 | 2.8798 |
| 0.005 | 1.7401 | 2.2401 | 2.9901 | 0.062 | 1.6274 | 2.1273 | 2.8778 |
| 0.006 | 1.7381 | 2.2381 | 2.9882 | 1/16 | 1.6264 | 2.1263 | 2.8768 |
| 0.007 | 1.7362 | 2.2361 | 2.9862 | 0.063 | 1.6254 | 2.1253 | 2.8759 |
| 0.008 | 1.7342 | 2.2341 | 2.9842 | 0.064 | 1.6234 | 2.1233 | 2.8739 |
| 0.009 | 1.7322 | 2.2321 | 2.9823 | 0.065 | 1.6215 | 2.1213 | 2.8719 |
| 0.010 | 1.7302 | 2.2302 | 2.9803 | 0.066 | 1.6195 | 2.1194 | 2.8699 |
| 0.011 | 1.7282 | 2.2282 | 2.9783 | 0.067 | 1.6175 | 2.1174 | 2.8680 |
| 0.012 | 1.7263 | 2.2262 | 2.9763 | 0.068 | 1.6155 | 2.1154 | 2.8660 |
| 0.013 | 1.7243 | 2.2243 | 2.9744 | 0.069 | 1.6136 | 2.1134 | 2.8640 |
| 0.014 | 1.7223 | 2.2222 | 2.9724 | 0.070 | 1.6116 | 2.1115 | 2.8621 |
| 0.015 | 1.7203 | 2.2203 | 2.9704 | 0.071 | 1.6096 | 2.1095 | 2.8601 |
| 1/64 | 1.7191 | 2.2191 | 2.9692 | 0.072 | 1.6076 | 2.1075 | 2.8581 |
| 0.016 | 1.7184 | 2.2183 | 2.9685 | 0.073 | 1.6057 | 2.1055 | 2.8561 |
| 0.017 | 1.7164 | 2.2163 | 2.9665 | 0.074 | 1.6037 | 2.1035 | 2.8542 |
| 0.018 | 1.7144 | 2.2143 | 2.9645 | 0.075 | 1.6017 | 2.1016 | 2.8522 |
| 0.019 | 1.7124 | 2.2123 | 2.9625 | 0.076 | 1.5997 | 2.0996 | 2.8503 |
| 0.020 | 1.7104 | 2.2104 | 2.9606 | 0.077 | 1.5978 | 2.0976 | 2.8483 |
| 0.021 | 1.7085 | 2.2084 | 2.9586 | 0.078 | 1.5958 | 2.0956 | 2.8463 |
| 0.022 | 1.7065 | 2.2064 | 2.9566 | 5/64 | 1.5955 | 2.0954 | 2.8461 |
| 0.023 | 1.7045 | 2.2045 | 2.9547 | 0.079 | 1.5938 | 2.0937 | 2.8443 |
| 0.024 | 1.7025 | 2.2025 | 2.9527 | 0.080 | 1.5918 | 2.0917 | 2.8424 |
| 0.025 | 1.7005 | 2.2005 | 2.9507 | 0.081 | 1.5899 | 2.0897 | 2.8404 |
| 0.026 | 1.6986 | 2.1985 | 2.9488 | 0.082 | 1.5879 | 2.0877 | 2.8384 |
| 0.027 | 1.6966 | 2.1965 | 2.9468 | 0.083 | 1.5859 | 2.0857 | 2.8365 |
| 0.028 | 1.6946 | 2.1945 | 2.9448 | 0.084 | 1.5839 | 2.0838 | 2.8345 |
| 0.029 | 1.6926 | 2.1925 | 2.9428 | 0.085 | 1.5820 | 2.0818 | 2.8325 |
| 0.030 | 1.6907 | 2.1906 | 2.9409 | 0.086 | 1.5800 | 2.0798 | 2.8306 |
| 0.031 | 1.6887 | 2.1886 | 2.9389 | 0.087 | 1.5780 | 2.0778 | 2.8286 |
| 1/32 | 1.6882 | 2.1881 | 2.9384 | 0.088 | 1.5760 | 2.0759 | 2.8266 |
| 0.032 | 1.6867 | 2.1866 | 2.9369 | 0.089 | 1.5740 | 2.0739 | 2.8247 |
| 0.033 | 1.6847 | 2.1847 | 2.9350 | 0.090 | 1.5721 | 2.0719 | 2.8227 |
| 0.034 | 1.6827 | 2.1827 | 2.9330 | 0.091 | 1.5701 | 2.0699 | 2.8207 |
| 0.035 | 1.6808 | 2.1807 | 2.9310 | 0.092 | 1.5681 | 2.0679 | 2.8187 |
| 0.036 | 1.6788 | 2.1787 | 2.9290 | 0.093 | 1.5661 | 2.0660 | 2.8168 |
| 0.037 | 1.6768 | 2.1767 | 2.9271 | 3/32 | 1.5647 | 2.0645 | 2.8153 |
| 0.038 | 1.6748 | 2.1747 | 2.9251 | 0.094 | 1.5642 | 2.0640 | 2.8148 |
| 0.039 | 1.6729 | 2.1727 | 2.9231 | 0.095 | 1.5622 | 2.0620 | 2.8128 |
| 0.040 | 1.6709 | 2.1708 | 2.9211 | 0.096 | 1.5602 | 2.0600 | 2.8109 |
| 0.041 | 1.6689 | 2.1688 | 2.9192 | 0.097 | 1.5582 | 2.0581 | 2.8089 |
| 0.042 | 1.6669 | 2.1668 | 2.9172 | 0.098 | 1.5563 | 2.0561 | 2.8069 |
| 0.043 | 1.6649 | 2.1649 | 2.9152 | 0.099 | 1.5543 | 2.0541 | 2.8050 |
| 0.044 | 1.6630 | 2.1629 | 2.9133 | 0.100 | 1.5523 | 2.0521 | 2.8030 |
| 0.045 | 1.6610 | 2.1609 | 2.9113 | 0.101 | 1.5503 | 2.0502 | 2.8010 |
| 0.046 | 1.6590 | 2.1589 | 2.9093 | 0.102 | 1.5484 | 2.0482 | 2.7991 |
| 3/64 | 1.6573 | 2.1572 | 2.9076 | 0.103 | 1.5464 | 2.0462 | 2.7971 |
| 0.047 | 1.6570 | 2.1569 | 2.9073 | 0.104 | 1.5444 | 2.0442 | 2.7951 |
| 0.048 | 1.6550 | 2.1549 | 2.9054 | 0.105 | 1.5425 | 2.0422 | 2.7932 |
| 0.049 | 1.6531 | 2.1529 | 2.9034 | 0.106 | 1.5405 | 2.0403 | 2.7912 |
| 0.050 | 1.6511 | 2.1510 | 2.9014 | 0.107 | 1.5385 | 2.0383 | 2.7892 |
| 0.051 | 1.6491 | 2.1490 | 2.8995 | 0.108 | 1.5365 | 2.0363 | 2.7873 |
| 0.052 | 1.6471 | 2.1470 | 2.8975 | 0.109 | 1.5346 | 2.0343 | 2.7853 |
| 0.053 | 1.6452 | 2.1451 | 2.8955 | 7/64 | 1.5338 | 2.0336 | 2.7846 |
| 0.054 | 1.6432 | 2.1431 | 2.8936 | 0.110 | 1.5326 | 2.0324 | 2.7833 |
| 0.055 | 1.6412 | 2.1411 | 2.8916 | 0.111 | 1.5306 | 2.0304 | 2.7814 |
| 0.056 | 1.6392 | 2.1391 | 2.8896 | 0.112 | 1.5287 | 2.0284 | 2.7794 |
| 0.057 | 1.6373 | 2.1372 | 2.8877 | 0.113 | 1.5267 | 2.0264 | 2.7774 |

## Corrected Diameters of Circular Forming Tools — 2

| Length $c$ on Tool | Number of B. & S. Automatic Screw Machine | | | Length $c$ on Tool | Number of B. & S. Automatic Screw Machine | | |
|---|---|---|---|---|---|---|---|
| | No. 00 | No. 0 | No. 2 | | No. 00 | No. 0 | No. 2 |
| 0.113 | 1.5267 | 2.0264 | 2.7774 | 0.171 | 1.4124 | 1.9119 | 2.6634 |
| 0.114 | 1.5247 | 2.0245 | 2.7755 | 11/64 | 1.4107 | 1.9103 | 2.6617 |
| 0.115 | 1.5227 | 2.0225 | 2.7735 | 0.172 | 1.4104 | 1.9099 | 2.6614 |
| 0.116 | 1.5208 | 2.0205 | 2.7715 | 0.173 | 1.4084 | 1.9080 | 2.6595 |
| 0.117 | 1.5188 | 2.0185 | 2.7696 | 0.174 | 1.4065 | 1.9060 | 2.6575 |
| 0.118 | 1.5168 | 2.0166 | 2.7676 | 0.175 | 1.4045 | 1.9040 | 2.6556 |
| 0.119 | 1.5148 | 2.0146 | 2.7656 | 0.176 | 1.4025 | 1.9021 | 2.6536 |
| 0.120 | 1.5129 | 2.0126 | 2.7637 | 0.177 | 1.4006 | 1.9001 | 2.6516 |
| 0.121 | 1.5109 | 2.0106 | 2.7617 | 0.178 | 1.3986 | 1.8981 | 2.6497 |
| 0.122 | 1.5089 | 2.0087 | 2.7597 | 0.179 | 1.3966 | 1.8961 | 2.6477 |
| 0.123 | 1.5070 | 2.0067 | 2.7578 | 0.180 | 1.3947 | 1.8942 | 2.6457 |
| 0.124 | 1.5050 | 2.0047 | 2.7558 | 0.181 | 1.3927 | 1.8922 | 2.6438 |
| 0.125 | 1.5030 | 2.0027 | 2.7538 | 0.182 | 1.3907 | 1.8902 | 2.6418 |
| 0.126 | 1.5010 | 2.0008 | 2.7519 | 0.183 | 1.3888 | 1.8882 | 2.6398 |
| 0.127 | 1.4991 | 1.9988 | 2.7499 | 0.184 | 1.3868 | 1.8863 | 2.6379 |
| 0.128 | 1.4971 | 1.9968 | 2.7479 | 0.185 | 1.3848 | 1.8843 | 2.6359 |
| 0.129 | 1.4951 | 1.9948 | 2.7460 | 0.186 | 1.3829 | 1.8823 | 2.6339 |
| 0.130 | 1.4932 | 1.9929 | 2.7440 | 0.187 | 1.3809 | 1.8804 | 2.6320 |
| 0.131 | 1.4912 | 1.9909 | 2.7420 | 3/16 | 1.3799 | 1.8794 | 2.6310 |
| 0.132 | 1.4892 | 1.9889 | 2.7401 | 0.188 | 1.3789 | 1.8784 | 2.6300 |
| 0.133 | 1.4872 | 1.9869 | 2.7381 | 0.189 | 1.3770 | 1.8764 | 2.6281 |
| 0.134 | 1.4853 | 1.9850 | 2.7361 | 0.190 | 1.3750 | 1.8744 | 2.6261 |
| 0.135 | 1.4833 | 1.9830 | 2.7342 | 0.191 | 1.3730 | 1.8725 | 2.6241 |
| 0.136 | 1.4813 | 1.9810 | 2.7322 | 0.192 | 1.3711 | 1.8705 | 2.6222 |
| 0.137 | 1.4794 | 1.9790 | 2.7302 | 0.193 | 1.3691 | 1.8685 | 2.6202 |
| 0.138 | 1.4774 | 1.9771 | 2.7282 | 0.194 | 1.3671 | 1.8665 | 2.6182 |
| 0.139 | 1.4754 | 1.9751 | 2.7263 | 0.195 | 1.3652 | 1.8646 | 2.6163 |
| 0.140 | 1.4734 | 1.9731 | 2.7243 | 0.196 | 1.3632 | 1.8626 | 2.6143 |
| 9/64 | 1.4722 | 1.9719 | 2.7231 | 0.197 | 1.3612 | 1.8606 | 2.6123 |
| 0.141 | 1.4715 | 1.9711 | 2.7224 | 0.198 | 1.3592 | 1.8587 | 2.6104 |
| 0.142 | 1.4695 | 1.9692 | 2.7204 | 0.199 | 1.3573 | 1.8567 | 2.6084 |
| 0.143 | 1.4675 | 1.9672 | 2.7184 | 0.200 | 1.3553 | 1.8547 | 2.6064 |
| 0.144 | 1.4655 | 1.9652 | 2.7165 | 0.201 | ........ | 1.8527 | 2.6045 |
| 0.145 | 1.4636 | 1.9632 | 2.7145 | 0.202 | ........ | 1.8508 | 2.6025 |
| 0.146 | 1.4616 | 1.9613 | 2.7125 | 0.203 | ........ | 1.8488 | 2.6006 |
| 0.147 | 1.4596 | 1.9593 | 2.7106 | 13/64 | ........ | 1.8486 | 2.6003 |
| 0.148 | 1.4577 | 1.9573 | 2.7086 | 0.204 | ........ | 1.8468 | 2.5986 |
| 0.149 | 1.4557 | 1.9553 | 2.7066 | 0.205 | ........ | 1.8449 | 2.5966 |
| 0.150 | 1.4537 | 1.9534 | 2.7047 | 0.206 | ........ | 1.8429 | 2.5947 |
| 0.151 | 1.4517 | 1.9514 | 2.7027 | 0.207 | ........ | 1.8409 | 2.5927 |
| 0.152 | 1.4498 | 1.9494 | 2.7007 | 0.208 | ........ | 1.8390 | 2.5908 |
| 0.153 | 1.4478 | 1.9474 | 2.6988 | 0.209 | ........ | 1.8370 | 2.5888 |
| 0.154 | 1.4458 | 1.9455 | 2.6968 | 0.210 | ........ | 1.8350 | 2.5868 |
| 0.155 | 1.4439 | 1.9435 | 2.6948 | 0.211 | ........ | 1.8330 | 2.5849 |
| 0.156 | 1.4419 | 1.9415 | 2.6929 | 0.212 | ........ | 1.8311 | 2.5829 |
| 5/32 | 1.4414 | 1.9410 | 2.6924 | 0.213 | ........ | 1.8291 | 2.5809 |
| 0.157 | 1.4399 | 1.9395 | 2.6909 | 0.214 | ........ | 1.8271 | 2.5790 |
| 0.158 | 1.4380 | 1.9376 | 2.6889 | 0.215 | ........ | 1.8252 | 2.5770 |
| 0.159 | 1.4360 | 1.9356 | 2.6870 | 0.216 | ........ | 1.8232 | 2.5751 |
| 0.160 | 1.4340 | 1.9336 | 2.6850 | 0.217 | ........ | 1.8212 | 2.5731 |
| 0.161 | 1.4321 | 1.9317 | 2.6830 | 0.218 | ........ | 1.8193 | 2.5711 |
| 0.162 | 1.4301 | 1.9297 | 2.6811 | 7/32 | ........ | 1.8178 | 2.5697 |
| 0.163 | 1.4281 | 1.9277 | 2.6791 | 0.219 | ........ | 1.8173 | 2.5692 |
| 0.164 | 1.4262 | 1.9238 | 2.6772 | 0.220 | ........ | 1.8153 | 2.5672 |
| 0.165 | 1.4242 | 1.9238 | 2.6752 | 0.221 | ........ | 1.8133 | 2.5653 |
| 0.166 | 1.4222 | 1.9218 | 2.6732 | 0.222 | ........ | 1.8114 | 2.5633 |
| 0.167 | 1.4203 | 1.9198 | 2.6713 | 0.223 | ........ | 1.8094 | 2.5613 |
| 0.168 | 1.4183 | 1.9178 | 2.6693 | 0.224 | ........ | 1.8074 | 2.5594 |
| 0.169 | 1.4163 | 1.9159 | 2.6673 | 0.225 | ........ | 1.8055 | 2.5574 |
| 0.170 | 1.4144 | 1.9139 | 2.6654 | 0.226 | ........ | 1.8035 | 2.5555 |

### Corrected Diameters of Circular Forming Tools — 3

| Length c on Tool | No. of B. & S. Machine | | Length c on Tool | No. of B. & S. Machine | | Length c on Tool | No. 2 B. & S Machine |
|---|---|---|---|---|---|---|---|
| | No. 0 | No. 2 | | No. 0 | No. 2 | | |
| 0.227 | 1.8015 | 2.5535 | 0.284 | 1.6894 | 2.4418 | 0.341 | 2.3303 |
| 0.228 | 1.7996 | 2.5515 | 0.285 | 1.6874 | 2.4398 | 0.342 | 2.3284 |
| 0.229 | 1.7976 | 2.5496 | 0.286 | 1.6854 | 2.4378 | 0.343 | 2.3264 |
| 0.230 | 1.7956 | 2.5476 | 0.287 | 1.6835 | 2.4359 | 11/32 | 2.3250 |
| 0.231 | 1.7936 | 2.5456 | 0.288 | 1.6815 | 2.4340 | 0.344 | 2.3245 |
| 0.232 | 1.7917 | 2.5437 | 0.289 | 1.6795 | 2.4320 | 0.345 | 2.3225 |
| 0.233 | 1.7897 | 2.5417 | 0.290 | 1.6776 | 2.4300 | 0.346 | 2.3206 |
| 0.234 | 1.7877 | 2.5398 | 0.291 | 1.6756 | 2.4281 | 0.347 | 2.3186 |
| 15/64 | 1.7870 | 2.5390 | 0.292 | 1.6736 | 2.4261 | 0.348 | 2.3166 |
| 0.235 | 1.7858 | 2.5378 | 0.293 | 1.6717 | 2.4242 | 0.349 | 2.3147 |
| 0.236 | 1.7838 | 2.5358 | 0.294 | 1.6697 | 2.4222 | 0.350 | 2.3127 |
| 0.237 | 1.7818 | 2.5339 | 0.295 | 1.6677 | 2.4203 | 0.351 | 2.3108 |
| 0.238 | 1.7799 | 2.5319 | 0.296 | 1.6658 | 2.4183 | 0.352 | 2.3088 |
| 0.239 | 1.7779 | 2.5300 | 19/64 | 1.6641 | 2.4166 | 0.353 | 2.3069 |
| 0.240 | 1.7759 | 2.5280 | 0.297 | 1.6638 | 2.4163 | 0.354 | 2.3049 |
| 0 241 | 1.7739 | 2.5260 | 0.298 | 1.6618 | 2.4144 | 0.355 | 2.3030 |
| 0.242 | 1.7720 | 2.5241 | 0.299 | 1.6599 | 2.4124 | 0.356 | 2.3010 |
| 0.243 | 1.7700 | 2.5221 | 0.300 | 1.6579 | 2.4105 | 0.357 | 2.2991 |
| 0.244 | 1.7680 | 2.5201 | 0.301 | ........ | 2.4085 | 0.358 | 2.2971 |
| 0.245 | 1.7661 | 2.5182 | 0.302 | ........ | 2.4066 | 0.359 | 2.2952 |
| 0.246 | 1.7641 | 2.5162 | 0.303 | ........ | 2.4046 | 23/64 | 2.2945 |
| 0.247 | 1.7621 | 2.5143 | 0.304 | ........ | 2.4026 | 0.360 | 2.2932 |
| 0.248 | 1.7602 | 2.5123 | 0.305 | ........ | 2.4007 | 0.361 | 2.2913 |
| 0.249 | 1.7582 | 2.5104 | 0.306 | ........ | 2.3987 | 0.362 | 2.2893 |
| 0.250 | 1.7562 | 2.5084 | 0.307 | ........ | 2.3968 | 0.363 | 2.2874 |
| 0.251 | 1.7543 | 2.5064 | 0.308 | ........ | 2.3948 | 0.364 | 2.2854 |
| 0.252 | 1.7523 | 2.5045 | 0.309 | ........ | 2.3929 | 0.365 | 2.2835 |
| 0.253 | 1.7503 | 2.5025 | 0.310 | ........ | 2.3909 | 0.366 | 2.2815 |
| 0.254 | 1.7484 | 2.5005 | 0.311 | ........ | 2.3890 | 0.367 | 2.2796 |
| 0.255 | 1.7464 | 2.4986 | 0.312 | ........ | 2.3870 | 0.368 | 2.2776 |
| 0.256 | 1.7444 | 2.4966 | 5/16 | ........ | 2.3860 | 0.369 | 2.2757 |
| 0.257 | 1.7425 | 2.4947 | 0.313 | ........ | 2.3851 | 0.370 | 2.2737 |
| 0.258 | 1.7405 | 2.4927 | 0.314 | ........ | 2.3831 | 0.371 | 2.2718 |
| 0.259 | 1.7385 | 2.4908 | 0.315 | ........ | 2.3811 | 0.372 | 2.2698 |
| 0.260 | 1.7366 | 2.4888 | 0.316 | ........ | 2.3792 | 0.373 | 2.2679 |
| 0.261 | 1.7346 | 2.4868 | 0.317 | ........ | 2.3772 | 0.374 | 2.2659 |
| 0.262 | 1.7326 | 2.4849 | 0.318 | ........ | 2.3753 | 0.375 | 2.2640 |
| 0.263 | 1.7306 | 2.4829 | 0.319 | ........ | 2.3733 | 0.376 | 2.2620 |
| 0.264 | 1.7287 | 2.4810 | 0.320 | ........ | 2.3714 | 0.377 | 2.2601 |
| 0.265 | 1.7267 | 2.4790 | 0.321 | ........ | 2.3694 | 0.378 | 2.2581 |
| 17/64 | 1.7255 | 2.4778 | 0.322 | ........ | 2.3675 | 0.379 | 2.2562 |
| 0.266 | 1.7248 | 2.4770 | 0.323 | ........ | 2.3655 | 0.380 | 2.2542 |
| 0.267 | 1.7228 | 2.4751 | 0.324 | ........ | 2.3636 | 0.381 | 2.2523 |
| 0.268 | 1.7208 | 2.4731 | 0.325 | ........ | 2.3616 | 0.382 | 2.2503 |
| 0.269 | 1.7189 | 2.4712 | 0.326 | ........ | 2.3596 | 0.383 | 2.2484 |
| 0.270 | 1.7169 | 2.4692 | 0.327 | ........ | 2.3577 | 0.384 | 2.2464 |
| 0.271 | 1.7149 | 2.4673 | 0.328 | ........ | 2.3557 | 0.385 | 2.2445 |
| 0.272 | 1.7130 | 2.4653 | 21/64 | ........ | 2.3555 | 0.386 | 2.2425 |
| 0.273 | 1.7110 | 2.4633 | 0.329 | ........ | 2.3538 | 0.387 | 2.2406 |
| 0.274 | 1.7090 | 2.4614 | 0.330 | ........ | 2.3518 | 0.388 | 2.2386 |
| 0.275 | 1.7071 | 2.4594 | 0.331 | ........ | 2.3499 | 0.389 | 2.2367 |
| 0.276 | 1.7051 | 2.4575 | 0.332 | ........ | 2.3479 | 0.390 | 2.2347 |
| 0.277 | 1.7031 | 2.4555 | 0.333 | ........ | 2.3460 | 25/64 | 2.2335 |
| 0.278 | 1.7012 | 2.4535 | 0.334 | ........ | 2.3440 | 0.391 | 2.2328 |
| 0.279 | 1.6992 | 2.4516 | 0.335 | ........ | 2.3421 | 0.392 | 2.2308 |
| 0.280 | 1.6972 | 2.4496 | 0.336 | ........ | 2.3401 | 0.393 | 2.2289 |
| 0.281 | 1.6953 | 2.4477 | 0.337 | ........ | 2.3381 | 0.394 | 2.2269 |
| 9/32 | 1.6948 | 2.4472 | 0.338 | ........ | 2.3362 | 0.395 | 2.2250 |
| 0.282 | 1.6933 | 2.4457 | 0.339 | ........ | 2.3342 | 0.396 | 2.2230 |
| 0.283 | 1.6913 | 2.4438 | 0.340 | ........ | 2.3323 | 0.397 | 2.2211 |

### Corrected Diameters of Circular Forming Tools — 4

| Length $c$ on Tool | No. 2 B. & S. Machine | Length $c$ on Tool | No. 2 B. & S. Machine | Length $c$ on Tool | No. 2 B. & S. Machine | Length $c$ on Tool | No. 2 B. & S. Machine |
|---|---|---|---|---|---|---|---|
| 0.398 | 2.2191 | 0.423 | 2.1704 | 0.449 | 2.1199 | 0.474 | 2.0713 |
| 0.399 | 2.2172 | 0.424 | 2.1685 | 0.450 | 2.1179 | 0.475 | 2.0694 |
| 0.400 | 2.2152 | 0.425 | 2.1666 | 0.451 | 2.1160 | 0.476 | 2.0674 |
| 0.401 | 2.2133 | 0.426 | 2.1646 | 0.452 | 2.1140 | 0.477 | 2.0655 |
| 0.402 | 2.2113 | 0.427 | 2.1627 | 0.453 | 2.1121 | 0.478 | 2.0636 |
| 0.403 | 2.2094 | 0.428 | 2.1607 | $^{29}/_{64}$ | 2.1118 | 0.479 | 2.0616 |
| 0.404 | 2.2074 | 0.429 | 2.1588 | 0.454 | 2.1101 | 0.480 | 2.0597 |
| 0.405 | 2.2055 | 0.430 | 2.1568 | 0.455 | 2.1082 | 0.481 | 2.0577 |
| 0.406 | 2.2035 | 0.431 | 2.1549 | 0.456 | 2.1063 | 0.482 | 2.0558 |
| $^{13}/_{32}$ | 2.2030 | 0.432 | 2.1529 | 0.457 | 2.1043 | 0.483 | 2.0538 |
| 0.407 | 2.2016 | 0.433 | 2.1510 | 0.458 | 2.1024 | 0.484 | 2.0519 |
| 0.408 | 2.1996 | 0.434 | 2.1490 | 0.459 | 2.1004 | 0.485 | 2.0500 |
| 0.409 | 2.1977 | 0.435 | 2.1471 | 0.460 | 2.0985 | 0.486 | 2.0480 |
| 0.410 | 2.1957 | 0.436 | 2.1452 | 0.461 | 2.0966 | 0.487 | 2.0461 |
| 0.411 | 2.1938 | 0.437 | 2.1432 | 0.462 | 2.0946 | 0.488 | 2.0441 |
| 0.412 | 2.1919 | $^{7}/_{16}$ | 2.1422 | 0.463 | 2.0927 | 0.489 | 2.0422 |
| 0.413 | 2.1899 | 0.438 | 2.1413 | 0.464 | 2.0907 | 0.490 | 2.0403 |
| 0.414 | 2.1880 | 0.439 | 2.1393 | 0.465 | 2.0888 | 0.491 | 2.0383 |
| 0.415 | 2.1860 | 0.440 | 2.1374 | 0.466 | 2.0868 | 0.492 | 2.0364 |
| 0.416 | 2.1841 | 0.441 | 2.1354 | 0.467 | 2.0849 | 0.493 | 2.0344 |
| 0.417 | 2.1821 | 0.442 | 2.1335 | 0.468 | 2.0830 | 0.494 | 2.0325 |
| 0.418 | 2.1802 | 0.443 | 2.1315 | $^{15}/_{32}$ | 2.0815 | 0.495 | 2.0306 |
| 0.419 | 2.1782 | 0.444 | 2.1296 | 0.469 | 2.0810 | 0.496 | 2.0286 |
| 0.420 | 2.1763 | 0.445 | 2.1276 | 0.470 | 2.0791 | 0.497 | 2.0267 |
| 0.421 | 2.1743 | 0.446 | 2.1257 | 0.471 | 2.0771 | 0.498 | 2.0247 |
| $^{27}/_{64}$ | 2.1726 | 0.447 | 2.1237 | 0.472 | 2.0752 | 0.499 | 2.0228 |
| 0.422 | 2.1724 | 0.448 | 2.1218 | 0.473 | 2.0733 | 0.500 | 2.0209 |

**Method of Using Tables for "Corrected Diameters of Circular Forming Tools."** — These tables are especially applicable to the Brown & Sharpe automatic screw machines. The maximum diameter $D$ of forming tools for these machines should be as follows: For No. 00 machine, 1¾ inch; for No. 0 machine, 2¼ inches; for No. 2 machine, 3 inches. To find the other diameters of the tool for any piece

### Dimensions of Forming Tools for B. & S. Automatic Screw Machines

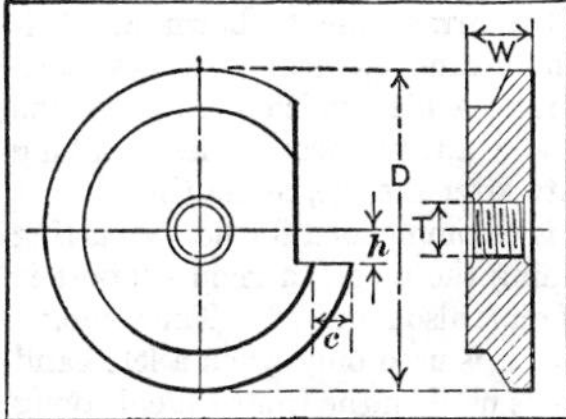

| No. of Machine | Max. Diam., $D$ | $h$ | $T$ | $W$ |
|---|---|---|---|---|
| 00 | 1¾ | ⅛ | ⅜–16 | ¼ |
| 0 | 2¼ | ⁵⁄₃₂ | ½–14 | ⁵⁄₁₆ |
| 2 | 3 | ¼ | ⅝–12 | ⅜ |
| 6 | 4 | ⁵⁄₁₆ | ¾–12 | ⅜ |

to be formed, proceed as follows: Subtract the smallest diameter of the work from that diameter of the work which is to be formed by the required tool diameter; divide the remainder by 2; locate the quotient obtained in the column headed "Length $c$ on Tool," and opposite the figure thus located and in the column headed by the number of the machine used, read off directly the diameter to which the tool is to be made. The quotient obtained, which is located in the column headed "Length $c$ on Tool," is the length $c$ as shown in the illustration above.

*Example:* — A piece of work is to be formed on a No. o machine to two diameters, one being ¼ inch and one 0.550 inch; find the diameters of the tool. The maximum tool diameter is 2¼ inches. This will be the diameter which will cut the ¼ inch diameter of the work. To find the other diameter, proceed according to the rule given: 0.550 − ¼ = 0.300; 0.300 ÷ 2 = 0.150. In Table 2, opposite 0.150, we find that the required tool diameter is 1.9534 inch. These tables are for tools without rake

**Arrangement of Circular Tools.** — When applying circular tools to automatic screw machines, their arrangement has an important bearing on the results obtained. The various ways of arranging the circular tools, with relation to the rotation of the spindle, are shown at *A, B, C* and *D*, in the illustration. These diagrams represent the view obtained when looking towards the chuck. The arrangement shown at *A* gives good results on long forming operations on brass and steel for the reason that the pressure of the cut on the front tool is downward; the support is more rigid than when the forming tool is turned upside down on the front slide as shown at *B*; here the stock, turning up towards the tool, has a tendency to lift

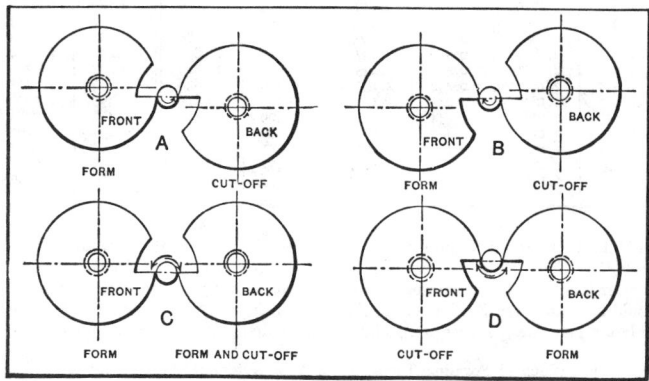

the cross-slide, causing chattering; therefore, the arrangement shown at *A* is recommended when a high finish is desired. The arrangement at *B* works satisfactorily for short steel pieces which do not require a high finish; it allows the chips to drop clear of the work, and is especially advantageous when making screws, when the forming and cut-off tools operate after the die, as no time is lost in reversing the spindle. The arrangement at *C* is recommended for heavy cutting on large work, when both tools are used for forming the piece; a rigid support is then necessary for both tools and a good supply of oil is also required. The arrangement at *D* is objectionable and should be avoided; it is used only when a left-hand thread is cut on the piece and when the cut-off tool is used on the front slide, leaving the heavy cutting to be performed from the rear slide. In all "cross-forming" work, it is essential that the spindle be kept in good condition, and that the collet or chuck have a parallel contact upon the bar which is being formed.

**Feeds and Speeds for Forming Tools.** — Approximate feeds and speeds for forming tools are given in the table beginning on page 1853. The feeds and speeds are average values, and if the job at hand has any features out of the ordinary the figures given should be altered accordingly.

Former ASA Standard Forming Tool Blanks (ASA B5.7-1954). — The standard covers both circular and dovetail forming tool blanks such as are commonly used on hand screw machines and on automatic screw machines. It also includes the mounting and clamping elements of the tool-holders used on these machines. The purpose of the standard is to provide interchangeability and also to permit a reduction in the number of forming tool blanks required. In order to establish a minimum number of blank sizes, the machines have been classified for reference purposes into six different groups of comparable stock capacities as shown by the accompanying Table 1. Each group of machines takes a definite size tool and the holders are provided with suitable mounting or clamping devices. The machine group numbers have been assigned arbitrarily to identify the size of the tool with the machines on which that size may be used. This standard is not intended to provide interchangeability of holders for various makes of machines except in connection with the mounting or clamp details.

Table 1. Machine Classifications for Former ASA Standard Forming Tool Blanks

| Group Number | Type of Machine | Maximum Capacity* | Group Number | Type of Machine | Maximum Capacity* |
|---|---|---|---|---|---|
| 1 | No. 00 Brown & Sharpe | 3/8 | | 2 5/8 New Britain | 2 5/8 |
| | No. 19 Brown & Sharpe | 3/8 | | 1 3/4 Gridley | 1 3/4 |
| | Index "0" | 7/16 | | 1 3/4 Greenlee | 1 3/4 |
| | 3/8 Cleveland | 5/8 | 4 | No. 4 Brown & Sharpe | 1 7/8 |
| | 3/8 Gridley | 3/8 | | 2 Greenlee | 2 |
| | 1/2 Davenport | 7/8 | | 2 Gridley | 2 |
| | 9/16 Acme Gridley | 9/16 | | 2 Cleveland | 2 1/2 |
| | No. 0 Brown & Sharpe | 5/8 | | 2 × 2 3/4 Cleveland | 3 1/4 |
| 2 | 5/8 Cleveland | 3/4 | | 2 1/4 Cleveland | 2 1/2 |
| | 5/8 × 7/8 Cleveland | 1 1/16 | | 2 1/4 × 2 3/4 Cleveland | 3 1/4 |
| | 1 New Britain | 1 | | 2 1/4 Gridley | 2 1/4 |
| | 1 3/8 New Britain | 1 3/8 | | 2 1/4 Greenlee | 2 1/4 |
| | 7/8 Greenlee | 1 | | No. 6 Brown & Sharpe | 2 3/8 |
| | 7/8 × 1 1/4 Cleveland | 7/8 | | 2 5/8 Gridley | 2 5/8 |
| | 7/8 Gridley | 7/8 | | 2 3/4 × 3 3/4 Cleveland | 3 1/4 |
| | 1 Acme Gridley | 1 | 5 | 2 3/4 × 4 Cleveland | 2 3/4 |
| | No. 2 Brown & Sharpe | 1 1/8 | | 3 Gridley | 3 |
| | 1 1/4 Gridley | 1 1/4 | | 3 5/16 Gridley | 3 5/16 |
| | 1 1/4 Cleveland | 1 1/4 | | 3 1/4 Gridley | 3 1/4 |
| 3 | 1 1/4 Cleveland | 1 3/8 | | 3 1/2 Gridley | 3 1/2 |
| | 1 1/4 × 1 1/2 Cleveland | 1 3/4 | | 4 Gridley | 4 |
| | 1 1/4 × 1 1/2 Cleveland | 1 1/2 | | 4 Cleveland | 4 |
| | 1 1/4 Greenlee | 1 1/4 | | 4 1/4 Cleveland | 4 1/4 |
| | 1 3/8 Gridley | 1 3/8 | | 4 1/2 Cleveland | 4 1/2 |
| | 1 1/2 Greenlee | 1 1/2 | 6 | 4 1/4 Gridley | 4 1/4 |
| | 1 5/8 Gridley | 1 5/8 | | 4 1/2 Gridley | 4 1/2 |
| | 1 5/8 Gridley | 1 5/8 | | 5 Gridley | 5 |
| | 1 5/8 Acme Gridley | 1 5/8 | | 5 1/2 Cleveland | 5 1/2 |
| 4 | 1 5/8 New Britain | 1 5/8 | | 6 3/4 Cleveland | 6 3/4 |
| | 2 1/4 New Britain | 2 1/4 | | 7 3/4 Cleveland | 7 3/4 |

* The group classification numbers apply to all machine models of the respective makes listed having the maximum capacities indicated. Dimensions in inches.

Table 2. Former ASA Standard Dimensions of Finished Blanks for Circular Forming Tools With Threaded Mounting Hole

The circular tool blanks in Table 2, which are provided with threaded mounting holes, are used for tools having nominal outside diameters of 1¾, 2¼, and 3 inches. For the larger blank diameters listed in Table 3, the mounting holes are unthreaded and counterbored. This method of mounting is for tool blanks having nominal outside diameters of 3½, 4, and 5 inches.

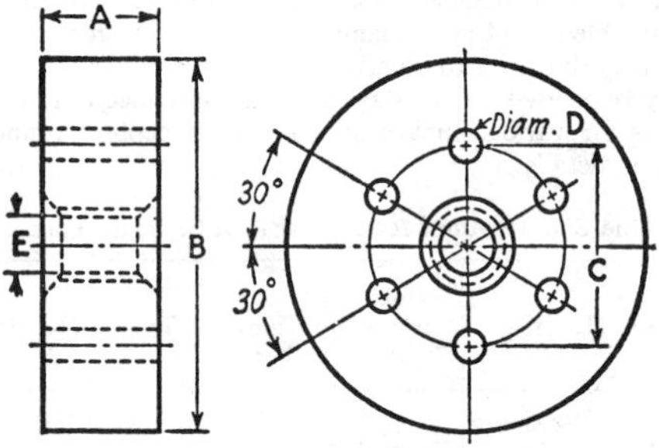


| Group Number | Nominal Blank Size Notes 1, 2 | Max. Diam. *B* Note 3 | Max. Width *A* Note 3 | Diam. Adjusting Hole Circle *C* | Diam. Adjusting Holes *D* Notes 4, 5, 6 | Threaded Hole *E* | |
|---|---|---|---|---|---|---|---|
| | | | | | | Diam. *E* Notes 7, 8 | No. Threads per Inch |
| | 1¾ × ¼ | 1 25/32 | 9/32 | No pin hole | | ⅜ | 16 |
| | 1¾ × ⅜ | 1 25/32 | 13/32 | | | ⅜ | 16 |
| 1 | 1¾ × ½ | 1 25/32 | 17/32 | | | ⅜ | 16 |
| | 1¾ × ¾ | 1 25/32 | 25/32 | | | ⅜ | 16 |
| | 1¾ × 1 | 1 25/32 | 1 1/32 | | | ⅜ | 16 |
| | 2¼ × ⅜ | 2 9/32 | 13/32 | 1⅜ | 3/16 | ½ | 13 |
| | 2¼ × ½ | 2 9/32 | 17/32 | 1⅜ | 3/16 | ½ | 13 |
| 2 | 2¼ × ¾ | 2 9/32 | 25/32 | 1⅜ | 3/16 | ½ | 13 |
| | 2¼ × 1 | 2 9/32 | 1 1/32 | 1⅜ | 3/16 | ½ | 13 |
| | 2¼ × 1¼ | 2 9/32 | 1 9/32 | 1⅜ | 3/16 | ½ | 13 |
| | 3 × ½ | 3 1/32 | 17/32 | 1½ | 3/16 | ⅝ | 11 |
| | 3 × ¾ | 3 1/32 | 25/32 | 1½ | 3/16 | ⅝ | 11 |
| 3 | 3 × 1 | 3 1/32 | 1 1/32 | 1½ | 3/16 | ⅝ | 11 |
| | 3 × 1¼ | 3 1/32 | 1 9/32 | 1½ | 3/16 | ⅝ | 11 |
| | 3 × 1½ | 3 1/32 | 1 17/32 | 1½ | 3/16 | ⅝ | 11 |

All dimensions are given in inches.

(1) Blanks are designated by giving the nominal outside diameter and width.

(2) Blanks made of high-speed steel shall be stamped H.S.

(3) The tolerance on diameter *B* and width *A* is −1/64 inch.

(4) Six adjusting holes shall have a minimum depth of ¼ inch, greater depth or through holes being optional. Adjusting holes shall be equally spaced on the circle *C* and be slightly chamfered.

(5) Diameter of adjusting holes *D* shall be ± 0.002.

(6) Diameter *C* of location circle shall be ± 0.002.

(7) The threaded hole *E* shall be made to Unified and American Standard, Class 2B. Both ends of threaded hole *E* shall be chamfered 35 degrees.

(8) Commercial blanks may not, in all cases, have the standard number of threads per inch listed in this table.

**Table 7. Typical Circular Forming Tool-holder for Machines in Groups Nos. 4, 5, and 6**

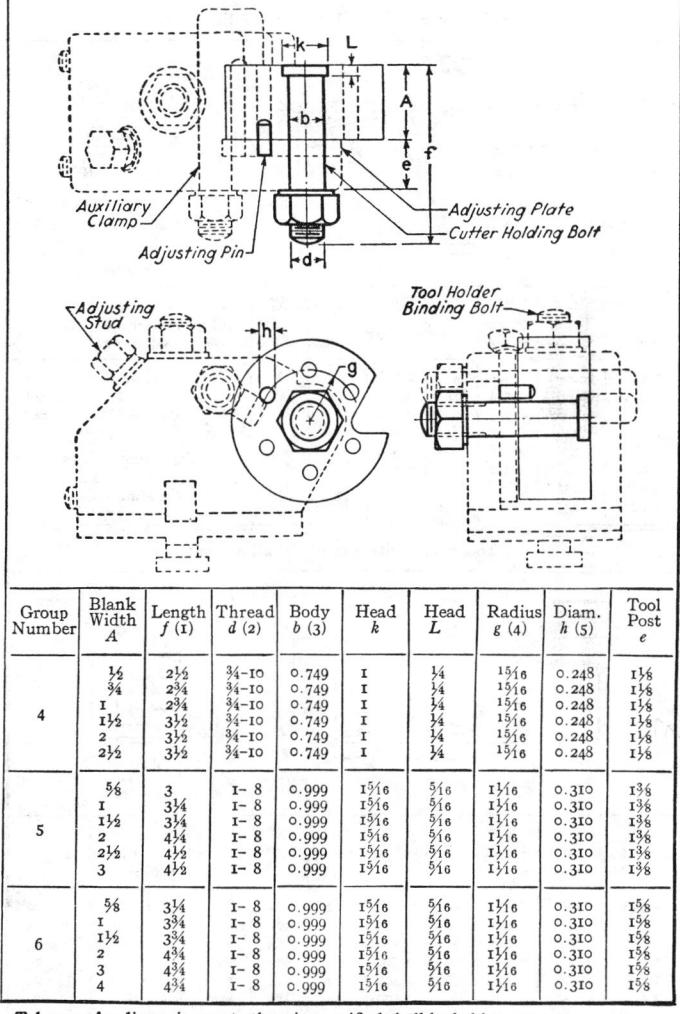

| Group Number | Blank Width A | Length f (1) | Thread d (2) | Body b (3) | Head k | Head L | Radius g (4) | Diam. h (5) | Tool Post e |
|---|---|---|---|---|---|---|---|---|---|
| 4 | 1/2 | 2 1/2 | 3/4-10 | 0.749 | 1 | 1/4 | 1 5/16 | 0.248 | 1 1/8 |
| | 3/4 | 2 3/4 | 3/4-10 | 0.749 | 1 | 1/4 | 1 5/16 | 0.248 | 1 1/8 |
| | 1 | 2 3/4 | 3/4-10 | 0.749 | 1 | 1/4 | 1 5/16 | 0.248 | 1 1/8 |
| | 1 1/2 | 3 1/2 | 3/4-10 | 0.749 | 1 | 1/4 | 1 5/16 | 0.248 | 1 1/8 |
| | 2 | 3 1/2 | 3/4-10 | 0.749 | 1 | 1/4 | 1 5/16 | 0.248 | 1 1/8 |
| | 2 1/2 | 3 1/2 | 3/4-10 | 0.749 | 1 | 1/4 | 1 5/16 | 0.248 | 1 1/8 |
| 5 | 5/8 | 3 | 1-8 | 0.999 | 1 5/16 | 5/16 | 1 1/16 | 0.310 | 1 3/8 |
| | 1 | 3 1/4 | 1-8 | 0.999 | 1 5/16 | 5/16 | 1 1/16 | 0.310 | 1 3/8 |
| | 1 1/2 | 3 3/4 | 1-8 | 0.999 | 1 5/16 | 5/16 | 1 1/16 | 0.310 | 1 3/8 |
| | 2 | 4 1/4 | 1-8 | 0.999 | 1 5/16 | 5/16 | 1 1/16 | 0.310 | 1 3/8 |
| | 2 1/2 | 4 1/2 | 1-8 | 0.999 | 1 5/16 | 5/16 | 1 1/16 | 0.310 | 1 3/8 |
| | 3 | 4 1/2 | 1-8 | 0.999 | 1 5/16 | 5/16 | 1 1/16 | 0.310 | 1 3/8 |
| 6 | 5/8 | 3 1/4 | 1-8 | 0.999 | 1 5/16 | 5/16 | 1 1/16 | 0.310 | 1 5/8 |
| | 1 | 3 3/4 | 1-8 | 0.999 | 1 5/16 | 5/16 | 1 1/16 | 0.310 | 1 5/8 |
| | 1 1/2 | 3 3/4 | 1-8 | 0.999 | 1 5/16 | 5/16 | 1 1/16 | 0.310 | 1 5/8 |
| | 2 | 4 3/4 | 1-8 | 0.999 | 1 5/16 | 5/16 | 1 1/16 | 0.310 | 1 5/8 |
| | 3 | 4 3/4 | 1-8 | 0.999 | 1 5/16 | 5/16 | 1 1/16 | 0.310 | 1 5/8 |
| | 4 | 4 3/4 | 1-8 | 0.999 | 1 5/16 | 5/16 | 1 1/16 | 0.310 | 1 5/8 |

Tolerance for dimensions not otherwise specified shall be held to ±0.010.
(1) Tolerance for length of bolt $f$ is ±1/32.
(2) The screw thread $d$ shall be made to Unified and American Standard, Class 3A.
(3) The body of the screw $b$ shall have tolerance of +0.000, −0.001.
(4) Tolerance for the radius of the adjusting pin hole location $g$ is ±0.001.
(5) Tolerance for the diameter of the adjusting pin $h$ is +0.000, −0.001.

Table 8. Former ASA Standard Typical Dovetail Forming Tool-holders

### Tool-holder with Collar-head Screw Adjustment

| M' +.010 -.000 | C +.010 -.000 | N' +.002 -.000 | O' ±.003 | B ±.0003 | R +.010 -.000 | R' +.010 -.000 | n +.010 -.000 | m +.010 -.000 | q +.000 -.002 | J UNF-3A |
|---|---|---|---|---|---|---|---|---|---|---|
| 0.732 | 0.489 | 0.305 | 1/4 | 0.156 | 1/32 | 1/64 | 11/64 | 9/16 | 0.122 | 1/4 -28 |
| 0.951 | 0.690 | 0.524 | 17/64 | 0.156 | 1/32 | 1/64 | 15/64 | 11/16 | 0.122 | 5/16-24 |
| 1.250 | 0.9085 | 0.567 | 3/8 | 0.250 | 1/16 | 1/32 | 15/64 | 11/16 | 0.122 | 5/16-24 |
| 1.614 | 1.110 | 0.248 | 33/64 | 0.500 | 1/16 | 1/32 | | | | |
| 1.882 | 1.378 | 0.516 | 33/64 | 0.500 | 1/16 | 1/32 | | | | |
| 2.000 | 1.532 | 0.634 | 31/64 | 0.500 | 1/16 | 1/32 | These dimensions are not given in the standard. | | | |
| 2.238 | 1.734 | 0.872 | 33/64 | 0.500 | 1/16 | 1/32 | | | | |
| 2.883 | 2.235 | 1.175 | 41/64 | 0.625 | 1/16 | 1/32 | | | | |

### Tool-holder with Hook-bolt Adjustment

| M' +.010 -.000 | C +.010 -.000 | N' +.002 -.000 | O' ±.003 | B ±.0003 | R +.010 -.000 | R' +.010 -.000 | n +.010 -.000 | x +.010 -.000 | y +.010 -.000 | J • |
|---|---|---|---|---|---|---|---|---|---|---|
| 0.732 | 0.489 | 0.305 | 1/4 | 0.156 | 1/32 | 1/64 | 7/32 | 1/8 | 1/8 | 1/4 -28 |
| 0.951 | 0.690 | 0.524 | 17/64 | 0.156 | 1/32 | 1/64 | 1/4 | 3/16 | 3/16 | 5/16-24 |
| 1.250 | 0.9085 | 0.567 | 3/8 | 0.250· | 1/16 | 1/32 | 1/4 | 1/4 | 1/4 | 5/16-24 |
| 1.614 | 1.110 | 0.248 | 33/64 | 0.500 | 1/16 | 1/32 | 9/32 | 1/4 | 1/4 | 3/8 -24 |
| 1.882 | 1.378 | 0.516 | 33/64 | 0.500 | 1/16 | 1/32 | 9/32 | 5/16 | 5/16 | 3/8 -24 |
| 2.000 | 1.532 | 0.634 | 31/64 | 0.500 | 1/16 | 1/32 | 5/16 | 5/16 | 5/16 | 7/16-20 |
| 2.238 | 1.734 | 0.872 | 33/64 | 0.500 | 1/16 | 1/32 | 5/16 | 5/16 | 5/16 | 7/16-20 |
| 2.883 | 2.235 | 1.175 | 41/64 | 0.625 | 1/16 | 1/32 | 5/16 | 5/16 | 5/16 | 7/16-20 |

All dimensions are given in inches.
• The adjusting screw thread is UNF-3A; adjusting nut thread is UNF-2B.

**Precision Gage Blocks.** — Precision gage blocks are usually purchased in sets comprising a specific number of blocks of different sizes. The nominal gage lengths of individual blocks in a set are determined mathematically so that particular desired lengths can be obtained by combining selected blocks. They are made to several different tolerance grades which categorize them as master blocks, calibration blocks, inspection blocks, and workshop blocks. *Master blocks* are employed as basic reference standards; *calibration blocks* are used for high precision gaging work and calibrating inspection blocks; *inspection blocks* are used as toolroom standards and for checking and setting limit and comparator gages, for example. The *workshop blocks* are working gages used as shop standards for a variety of direct precision measurements and gaging applications, including sine bar settings.

Federal Specification GGG-G-15B, Gage Blocks (see below), lists typical sets, and gives details of materials, design, and manufacturing requirements, and tolerance grades. When there is in a set no single block of the exact size that is wanted, two or more blocks are combined by "wringing" them together. This is achieved by first placing one block crosswise on the other and applying some pressure. Then a swiveling motion is applied and the blocks are twisted to a parallel position. This causes them to adhere firmly to one another.

When combining blocks for a given dimension, the object is to use as few blocks as possible to obtain the dimension. The procedure for selecting blocks is based on successively eliminating the right-hand figure of the desired dimension.

*Example.* Referring to gage block set number 1 in Table 1, determine the blocks required to obtain 3.6742 inches. *Step 1:* Eliminate .0002 by selecting a .1002 block. Subtract .1002 from 3.6742 = 3.5740. *Step 2:* Eliminate .004 by selecting a .124 block. Subtract .124 from 3.5740 = 3.450. *Step 3:* Eliminate .450 with a block this size. Subtract .450 from 3.450 = 3.000. *Step 4:* Select a 3.000 inch block. The combined blocks are .1002 + .124 + .450 + 3.000 = 3.6742 inches.

**Federal Specification for Gage Blocks, Inch and Metric Sizes.** — This Specification, GGG-G-15B, November 6, 1970, which supersedes GGG-G-15a, September 22, 1964, covers design, manufacturing, and purchasing details for precision gage blocks in inch and metric sizes up to and including 20 inches and 500 millimeters gage lengths. The shapes of blocks are designated Style 1, which is rectangular; Style 2, which is square with a center accessory hole, and Style 3, which defines other shapes as may be specified by the purchaser. Blocks may be made from steel, chromium-plated steel, chromium carbide, tungsten carbide, and other materials to specification. There are four tolerance grades, which are designated Grade 0.5 (formerly Grade AAA in the GGG-G-15a issue of the Specification); Grade 1 (formerly Grade AA); Grade 2 (formerly Grade A+); and Grade 3 (a compromise between former Grades A and B). Grade 0.5 blocks are special reference gages used for extremely high precision gaging work, and are not recommended for general use. Grade 1 blocks are laboratory reference standards used for calibrating inspection gage blocks and high precision gaging work. Grade 2 blocks are used as inspection and toolroom standards, and Grade 3 blocks are used as shop standards.

Inch and metric sizes of blocks in specific sets are given in Tables 1 and 2. It is not a complete list of available sizes, and it should be noted that some gage blocks must be ordered as specials, some may not be available in all materials, and some may not be available from all manufacturers. Gage block set number 4 (88 blocks), listed in the Specification, is not given in Table 1. It is the same as set number 1 (81 blocks) but contains seven additional blocks measuring 0.0625, 0.078125, 0.093750, 0.100025, 0.100050, 0.100075, and 0.109375 inch. In Table 2, gage block set number 3M (112 blocks) is not given. It is similar to set number 2M (88 blocks), and the chief difference is the inclusion of a larger number of blocks in the 0.5 millimeter increment series up to 24.5 mm.

**Table 1. Gage Block Sets\* — Inch Sizes** (Federal Specification GGG-G-15B†)

### SET NUMBER 1 (81 BLOCKS)

First Series: 0.0001 Inch Increments (9 Blocks)

| | | | | | | | | |
|---|---|---|---|---|---|---|---|---|
| .1001 | .1002 | .1003 | .1004 | .1005 | .1006 | .1007 | .1008 | .1009 |

Second Series: 0.001 Inch Increments (49 Blocks)

| | | | | | | | | | |
|---|---|---|---|---|---|---|---|---|---|
| .101 | .102 | .103 | .104 | .105 | .106 | .107 | .108 | .109 | .110 |
| .111 | .112 | .113 | .114 | .115 | .116 | .117 | .118 | .119 | .120 |
| .121 | .122 | .123 | .124 | .125 | .126 | .127 | .128 | .129 | .130 |
| .131 | .132 | .133 | .134 | .135 | .136 | .137 | .138 | .139 | .140 |
| .141 | .142 | .143 | .144 | .145 | .146 | .147 | .148 | .149 | |

Third Series: 0.50 Inch Increments (19 Blocks)

| | | | | | | | | | |
|---|---|---|---|---|---|---|---|---|---|
| .050 | .100 | .150 | .200 | .250 | .300 | .350 | .400 | .450 | .500 |
| .550 | .600 | .650 | .700 | .750 | .800 | .850 | .900 | .950 | |

Fourth Series: 1.000 Inch Increments (4 Blocks)

| | | | |
|---|---|---|---|
| 1.000 | 2.000 | 3.000 | 4.000 |

### SET NUMBER 5 (21 BLOCKS)

First Series: 0.0001 Inch Increments (9 Blocks)

| | | | | | | | | |
|---|---|---|---|---|---|---|---|---|
| .0101 | .0102 | .0103 | .0104 | .0105 | .0106 | .0107 | .0108 | .0109 |

Second Series: 0.001 Inch Increments (11 Blocks)

| | | | | | | | | | | |
|---|---|---|---|---|---|---|---|---|---|---|
| .010 | .011 | .012 | .013 | .014 | .015 | .016 | .017 | .018 | .019 | .020 |

One Block 0.01005 Inch

### SET NUMBER 6 (28 BLOCKS)

First Series: 0.0001 Inch Increments (9 Blocks)

| | | | | | | | | |
|---|---|---|---|---|---|---|---|---|
| .0201 | .0202 | .0203 | .0204 | .0205 | .0206 | .0207 | .0208 | .0209 |

Second Series: 0.001 Inch Increments (9 Blocks)

| | | | | | | | | |
|---|---|---|---|---|---|---|---|---|
| .021 | .022 | .023 | .024 | .025 | .026 | .027 | .028 | .029 |

Third Series: 0.010 Inch Increments (9 Blocks)

| | | | | | | | | |
|---|---|---|---|---|---|---|---|---|
| .010 | .020 | .030 | .040 | .050 | .060 | .070 | .080 | .090 |

One Block 0.02005 Inch

### LONG GAGE BLOCK SET NUMBER 7 (8 BLOCKS)

Whole Inch Series (8 Blocks)

| | | | | | | | |
|---|---|---|---|---|---|---|---|
| 5 | 6 | 7 | 8 | 10 | 12 | 16 | 20 |

### SET NUMBER 8 (36 BLOCKS)

First Series: 0.0001 Inch Increments (9 Blocks)

| | | | | | | | | |
|---|---|---|---|---|---|---|---|---|
| .1001 | .1002 | .1003 | .1004 | .1005 | .1006 | .1007 | .1008 | .1009 |

Second Series: 0.001 Inch Increments (11 Blocks)

| | | | | | | | | | | |
|---|---|---|---|---|---|---|---|---|---|---|
| .100 | .101 | .102 | .103 | .104 | .105 | .106 | .107 | .108 | .109 | .110 |

Third Series: 0.010 Inch Increments (8 Blocks)

| | | | | | | | |
|---|---|---|---|---|---|---|---|
| .120 | .130 | .140 | .150 | .160 | .170 | .180 | .190 |

Fourth Series: 0.100 Inch Increments (4 Blocks)

| | | | |
|---|---|---|---|
| .200 | .300 | .400 | .500 |

Whole Inch Series (3 Blocks)

| | | |
|---|---|---|
| 1 | 2 | 4 |

One Block 0.050 Inch

### SET NUMBER 9 (20 BLOCKS)

First Series: 0.0001 Inch Increments (9 Blocks)

| | | | | | | | | |
|---|---|---|---|---|---|---|---|---|
| .0501 | .0502 | .0503 | .0504 | .0505 | .0506 | .0507 | .0508 | .0509 |

Second Series: 0.001 Inch Increments (10 Blocks)

| | | | | | | | | | |
|---|---|---|---|---|---|---|---|---|---|
| .050 | .051 | .052 | .053 | .054 | .055 | .056 | .057 | .058 | .059 |

One Block 0.05005 Inch

\* Set number 4 is not shown, and the Specification does not list a set 2 or 3.
† Arranged here in incremental series for convenience of use.

**Table 2. Gage Block Sets\* — Metric Sizes** (Federal Specification GGG-G-15B†)

| SET NUMBER 1M (47 BLOCKS) | | | | | | | | |
|---|---|---|---|---|---|---|---|---|
| First Series: 0.01 Millimeter Increments (9 Blocks) | | | | | | | | |
| 1.01 | 1.02 | 1.03 | 1.04 | 1.05 | 1.06 | 1.07 | 1.08 | 1.09 |
| Second Series: 0.10 Millimeter Increments (9 Blocks) | | | | | | | | |
| 1.10 | 1.20 | 1.30 | 1.40 | 1.50 | 1.60 | 1.70 | 1.80 | 1.90 |
| Third Series: 1.0 Millimeter Increments (24 Blocks) | | | | | | | | |
| 1.0 2.0 3.0 4.0 5.0 6.0 7.0 8.0 9.0 10.0 | | | | | | | | |
| 11 12 13 14 15 16 17 18 19 20 21 22 23 24 | | | | | | | | |
| Fourth Series: 20 Millimeter Increments (3 Blocks) | | | | | | | | |
| 40 60 80 | | | | | | | | |
| One Block 1.005 mm    One Block 100 mm | | | | | | | | |

| SET NUMBER 2M (88 BLOCKS) | | | | | | | | | |
|---|---|---|---|---|---|---|---|---|---|
| First Series: 0.001 Millimeter Increments (9 Blocks) | | | | | | | | | |
| 1.001 | 1.002 | 1.003 | 1.004 | 1.005 | 1.006 | 1.007 | 1.008 | 1.009 | |
| Second Series: 0.01 Millimeter Increments (49 Blocks) | | | | | | | | | |
| 1.01 | 1.02 | 1.03 | 1.04 | 1.05 | 1.06 | 1.07 | 1.08 | 1.09 | 1.10 |
| 1.11 | 1.12 | 1.13 | 1.14 | 1.15 | 1.16 | 1.17 | 1.18 | 1.19 | 1.20 |
| 1.21 | 1.22 | 1.23 | 1.24 | 1.25 | 1.26 | 1.27 | 1.28 | 1.29 | 1.30 |
| 1.31 | 1.32 | 1.33 | 1.34 | 1.35 | 1.36 | 1.37 | 1.38 | 1.39 | 1.40 |
| 1.41 | 1.42 | 1.43 | 1.44 | 1.45 | 1.46 | 1.47 | 1.48 | 1.49 | |
| Third Series: 0.50 Millimeter Increments (19 Blocks) | | | | | | | | | |
| 0.5 | 1.0 | 1.5 | 2.0 | 2.5 | 3.0 | 3.5 | 4.0 | 4.5 | 5.0 |
| 5.5 | 6.0 | 6.5 | 7.0 | 7.5 | 8.0 | 8.5 | 9.0 | 9.5 | |
| Fourth Series: 10 Millimeter Increments (10 Blocks) | | | | | | | | | |
| 10 20 30 40 50 60 70 80 90 100 | | | | | | | | | |
| One Block 1.0005 mm | | | | | | | | | |

| SET NUMBER 4M (64 BLOCKS) | | | | | | | | | |
|---|---|---|---|---|---|---|---|---|---|
| First Series: 0.001 Millimeter Increments (9 Blocks) | | | | | | | | | |
| 2.001 | 2.002 | 2.003 | 2.004 | 2.005 | 2.006 | 2.007 | 2.008 | 2.009 | |
| Second Series: 0.01 Millimeter Increments (49 Blocks) | | | | | | | | | |
| 2.01 | 2.02 | 2.03 | 2.04 | 2.05 | 2.06 | 2.07 | 2.08 | 2.09 | 2.10 |
| 2.11 | 2.12 | 2.13 | 2.14 | 2.15 | 2.16 | 2.17 | 2.18 | 2.19 | 2.20 |
| 2.21 | 2.22 | 2.23 | 2.24 | 2.25 | 2.26 | 2.27 | 2.28 | 2.29 | 2.30 |
| 2.31 | 2.32 | 2.33 | 2.34 | 2.35 | 2.36 | 2.37 | 2.38 | 2.39 | 2.40 |
| 2.41 | 2.42 | 2.43 | 2.44 | 2.45 | 2.46 | 2.47 | 2.48 | 2.49 | |
| Third Series: 0.10 Millimeter Increments (5 Blocks) | | | | | | | | | |
| 2.50 2.60 2.70 2.80 2.90 | | | | | | | | | |
| One Block 2.00 mm | | | | | | | | | |

| SET NUMBER 5M (24 BLOCKS) | | | | | | | | |
|---|---|---|---|---|---|---|---|---|
| First Series: 0.01 Millimeter Increments (9 Blocks) | | | | | | | | |
| .41 | .42 | .43 | .44 | .45 | .46 | .47 | .48 | .49 |
| Second Series: 0.05 Millimeter Increments (5 Blocks) | | | | | | | | |
| .20 .25 .30 .35 .40 | | | | | | | | |
| Third Series: 0.05 Millimeter Increments (10 Blocks) | | | | | | | | |
| .50 | .55 | .60 | .65 | .70 | .75 | .80 | .85 | .90 .95 |

| LONG GAGE BLOCK SET NUMBER 6M (8 BLOCKS) |
|---|
| Whole Millimeter Series (8 Blocks) |
| 125 150 175 200 250 300 400 500 |

*Note:* Gage blocks measuring 1.09 millimeters and under in set number 1M, and blocks measuring 1.5 millimeters and under in set number 2M, are not available in tolerance grade 0.5.

\* Set number 3M is not listed.

† Arranged here in incremental series for convenience of use.

## Woodruff Key-slot Gages — Former S.A.E. Production Standard

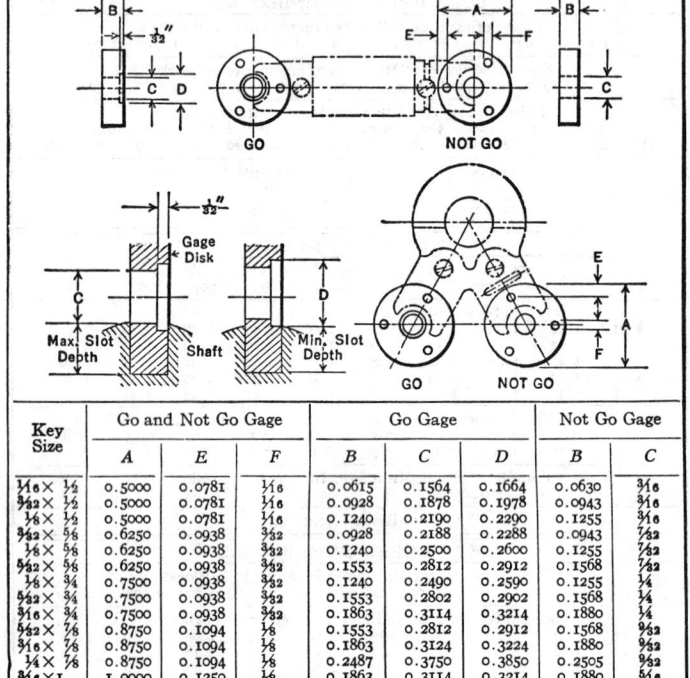

| Key Size | Go and Not Go Gage | | | Go Gage | | | Not Go Gage | |
|---|---|---|---|---|---|---|---|---|
| | A | E | F | B | C | D | B | C |
| 1/16 × 1/2 | 0.5000 | 0.0781 | 1/16 | 0.0615 | 0.1564 | 0.1664 | 0.0630 | 3/16 |
| 3/32 × 1/2 | 0.5000 | 0.0781 | 1/16 | 0.0928 | 0.1878 | 0.1978 | 0.0943 | 3/16 |
| 1/8 × 1/2 | 0.5000 | 0.0781 | 1/16 | 0.1240 | 0.2190 | 0.2290 | 0.1255 | 3/16 |
| 3/32 × 5/8 | 0.6250 | 0.0938 | 3/32 | 0.0928 | 0.2188 | 0.2288 | 0.0943 | 7/32 |
| 1/8 × 5/8 | 0.6250 | 0.0938 | 3/32 | 0.1240 | 0.2500 | 0.2600 | 0.1255 | 7/32 |
| 5/32 × 5/8 | 0.6250 | 0.0938 | 3/32 | 0.1553 | 0.2812 | 0.2912 | 0.1568 | 7/32 |
| 1/8 × 3/4 | 0.7500 | 0.0938 | 3/32 | 0.1240 | 0.2490 | 0.2590 | 0.1255 | 1/4 |
| 5/32 × 3/4 | 0.7500 | 0.0938 | 3/32 | 0.1553 | 0.2802 | 0.2902 | 0.1568 | 1/4 |
| 3/16 × 3/4 | 0.7500 | 0.0938 | 3/32 | 0.1863 | 0.3114 | 0.3214 | 0.1880 | 1/4 |
| 5/32 × 7/8 | 0.8750 | 0.1094 | 1/8 | 0.1553 | 0.2812 | 0.2912 | 0.1568 | 9/32 |
| 3/16 × 7/8 | 0.8750 | 0.1094 | 1/8 | 0.1863 | 0.3124 | 0.3224 | 0.1880 | 9/32 |
| 1/4 × 7/8 | 0.8750 | 0.1094 | 1/8 | 0.2487 | 0.3750 | 0.3850 | 0.2505 | 9/32 |
| 3/16 × 1 | 1.0000 | 0.1250 | 1/8 | 0.1863 | 0.3114 | 0.3214 | 0.1880 | 5/16 |
| 1/4 × 1 | 1.0000 | 0.1250 | 1/8 | 0.2487 | 0.3740 | 0.3840 | 0.2505 | 5/16 |
| 5/16 × 1 | 1.0000 | 0.1250 | 1/8 | 0.3111 | 0.4364 | 0.4464 | 0.3130 | 5/16 |
| 3/16 × 1 1/8 | 1.1250 | 0.1562 | 1/8 | 0.1863 | 0.3444 | 0.3544 | 0.1880 | 11/32 |
| 1/4 × 1 1/8 | 1.1250 | 0.1562 | 1/8 | 0.2487 | 0.4070 | 0.4170 | 0.2505 | 11/32 |
| 5/16 × 1 1/8 | 1.1250 | 0.1562 | 1/8 | 0.3111 | 0.4694 | 0.4794 | 0.3130 | 11/32 |
| 1/4 × 1 1/4 | 1.2500 | 0.1719 | 1/8 | 0.2487 | 0.4060 | 0.4160 | 0.2505 | 3/8 |
| 5/16 × 1 1/4 | 1.2500 | 0.1719 | 1/8 | 0.3111 | 0.4684 | 0.4784 | 0.3130 | 3/8 |
| 3/8 × 1 1/4 | 1.2500 | 0.1719 | 1/8 | 0.3735 | 0.5310 | 0.5410 | 0.3755 | 3/8 |
| 1/4 × 1 3/8 | 1.3750 | 0.1875 | 1/8 | 0.2487 | 0.4370 | 0.4470 | 0.2505 | 7/16 |
| 5/16 × 1 3/8 | 1.3750 | 0.1875 | 1/8 | 0.3111 | 0.4994 | 0.5094 | 0.3130 | 7/16 |
| 3/8 × 1 3/8 | 1.3750 | 0.1875 | 1/8 | 0.3735 | 0.5620 | 0.5720 | 0.3755 | 7/16 |
| 1/4 × 1 1/2 | 1.5000 | 0.2187 | 1/8 | 0.2487 | 0.4680 | 0.4780 | 0.2505 | 1/2 |
| 5/16 × 1 1/2 | 1.5000 | 0.2187 | 1/8 | 0.3111 | 0.5304 | 0.5404 | 0.3130 | 1/2 |
| 3/8 × 1 1/2 | 1.5000 | 0.2187 | 1/8 | 0.3735 | 0.5930 | 0.6030 | 0.3755 | 1/2 |

All dimensions in inches. Tolerances: All diameters A plus 0.0002 minus 0.000 for Go gage; plus or minus 0.001 for Not Go. All widths B plus 0.0002 minus 0.000 for both Go and Not-Go gages. All diameters C plus 0.0004 minus 0.000 for Go gage and plus or minus 0.005 for drilled holes in Not-Go gage. All diameters D plus 0.0004 minus 0.000.

The detail sectional views above table show use of hole C and counterbore D for checking the maximum and minimum slot depths.

The three equally spaced holes F provide for shifting the disk to three positions relative to the handle, thus increasing the disk life three times as compared with a single-position disk. Use oil-hardening steel, or its equivalent.

The straight handle is intended for gaging key sizes up to 5/16 by 1 1/8 and the V-shaped handle for larger sizes.

Table 1. American National Standard Plain Cylindrical Plug Gaging
Members — Taper Lock Design (ANSI B47.1-1974)

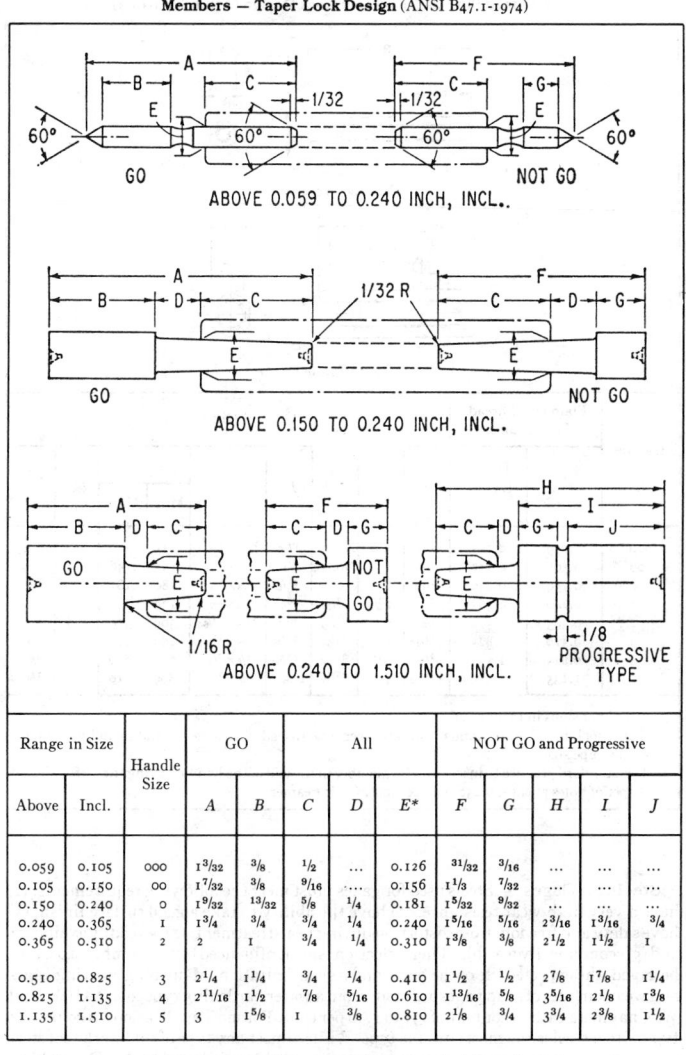

ABOVE 0.059 TO 0.240 INCH, INCL.

ABOVE 0.150 TO 0.240 INCH, INCL.

ABOVE 0.240 TO 1.510 INCH, INCL.

PROGRESSIVE TYPE

| Range in Size | | Handle Size | GO | | All | | | NOT GO and Progressive | | | | |
|---|---|---|---|---|---|---|---|---|---|---|---|---|
| Above | Incl. | | A | B | C | D | E* | F | G | H | I | J |
| 0.059 | 0.105 | 000 | $1^{3}/_{32}$ | $3/_{8}$ | $1/_{2}$ | ... | 0.126 | $31/_{32}$ | $3/_{16}$ | ... | ... | ... |
| 0.105 | 0.150 | 00 | $1^{7}/_{32}$ | $3/_{8}$ | $9/_{16}$ | ... | 0.156 | $1^{1}/_{3}$ | $7/_{32}$ | ... | ... | ... |
| 0.150 | 0.240 | 0 | $1^{9}/_{32}$ | $13/_{32}$ | $5/_{8}$ | $1/_{4}$ | 0.181 | $1^{5}/_{32}$ | $9/_{32}$ | ... | ... | ... |
| 0.240 | 0.365 | 1 | $1^{3}/_{4}$ | $3/_{4}$ | $3/_{4}$ | $1/_{4}$ | 0.240 | $1^{5}/_{16}$ | $5/_{16}$ | $2^{3}/_{16}$ | $1^{3}/_{16}$ | $3/_{4}$ |
| 0.365 | 0.510 | 2 | 2 | 1 | $3/_{4}$ | $1/_{4}$ | 0.310 | $1^{3}/_{8}$ | $3/_{8}$ | $2^{1}/_{2}$ | $1^{1}/_{2}$ | 1 |
| 0.510 | 0.825 | 3 | $2^{1}/_{4}$ | $1^{1}/_{4}$ | $3/_{4}$ | $1/_{4}$ | 0.410 | $1^{1}/_{2}$ | $1/_{2}$ | $2^{7}/_{8}$ | $1^{7}/_{8}$ | $1^{1}/_{4}$ |
| 0.825 | 1.135 | 4 | $2^{11}/_{16}$ | $1^{1}/_{2}$ | $7/_{8}$ | $5/_{16}$ | 0.610 | $1^{13}/_{16}$ | $5/_{8}$ | $3^{5}/_{16}$ | $2^{1}/_{8}$ | $1^{3}/_{8}$ |
| 1.135 | 1.510 | 5 | 3 | $1^{5}/_{8}$ | 1 | $3/_{8}$ | 0.810 | $2^{1}/_{8}$ | $3/_{4}$ | $3^{3}/_{4}$ | $2^{3}/_{8}$ | $1^{1}/_{2}$ |

All dimensions in inches. Tapers of all plug-gage shanks 0.25 inch per foot.

*Maximum diameters E are given (minimum diameters are 0.001 inch less up to 0.365-inch plug size, inclusive, and 0.002 inch less for larger sizes).

See Table 2 for handles.

### Table 2. American National Standard Handles for Plain Cylindrical and Thread Plug Gages—Taper Lock Design (ANSI B47.1-1974)

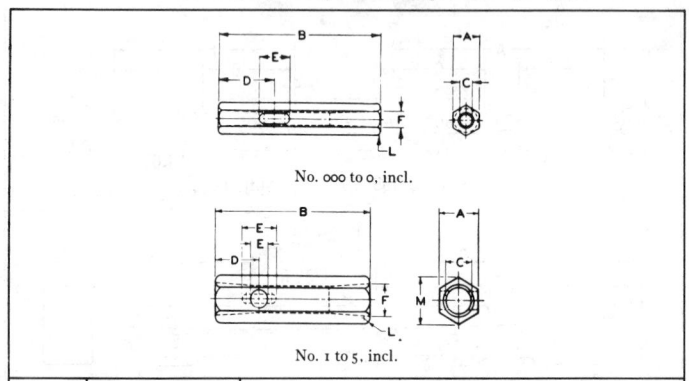

No. 000 to 0, incl.

No. 1 to 5, incl.

| Handle Size No. | Plain and Thread Plug Diams. | | General Dimensions | | | | | | | | |
|---|---|---|---|---|---|---|---|---|---|---|---|
| | Above | To and Including | A | B | C, Drill Size | D | E | F Min. | F Max. | G, No. | L |
| 000 | 0.059 | 0.105 | 3/16 | 1 1/2 | No. 34 | 9/16 | 5/64 × 1/4 | .125 | .126 | 000 | 1/32 |
| 00 | 0.105 | 0.150 | 1/4 | 1 3/4 | No. 29 | 5/8 | 3/32 × 5/16 | .155 | .156 | 0 | 1/32 |
| 0 | 0.150 | 0.240 | 5/16 | 2 | No. 20 | 11/16 | 1/8 × 3/8 | .180 | .181 | 2 | 1/32 |
| 1 | 0.240 | 0.365 | 3/8 | 2 3/4 | 7/32 | 25/32 | 1/8 × 1/2 | .239 | .240 | 4 | 1/16 |
| 2 | 0.365 | 0.510 | 1/2 | 3 | 0.29 | 25/32 | 1/8 × 15/64 | .309 | .310 | 6 | 1/16 |
| 3 | 0.510 | 0.825 | 11/16 | 3 1/4 | 25/64 | 27/32 | 1/8 × 11/32 | .409 | .410 | 7 | 3/32 |
| 4 | 0.825 | 1.135 | 7/8 | 3 3/8 | 37/64 | 63/64 | 1/8 × 3/8 | .609 | .610 | 10 | 3/32 |
| 5 | 1.135 | 1.510 | 1 11/16 | 4 | 25/32 | 1 1/8 | 1/8 × 7/16 | .809 | .810 | 11 | 1/8 |

All dimensions in inches.

Taper lock handles are standard for all taper pipe thread plug gages to and including 2-inch nominal pipe size.

It is standard practice to insert the GO member in the end of the handle having the drift hole.

Taper of holes at ends, 0.250 inch per foot, G-pin reamer.

**Steels for Gages.** — Steels used for gages must meet the following requirements: (1) have a very high wear resistance; (2) have the ability to take a good surface finish; (3) have safety and freedom from distortion during heat-treatment; (4) be readily available; and (5) economically feasible. Their selection is also influenced by the number of parts to be gaged, the expected life of the gage, and its configuration. Many plug and ring gages and wear surfaces of snap gages are made from a water hardening type tool steel (W1 and W2), having a carbon content of .95 to 1.30 per cent. For most applications, a cold work type of tool steel is the most suitable type. While almost any type of cold work tool steel can be used, the most wear resistant types are A7 and D7, followed by A3, D3, and D4. For thread gages, where distortion must be kept to a minimum, type D4 is recommended. Other cold work tool steels include oil hardening o1, o2, and o6 types, air hardening A1 and A2 types, and high-carbon, high-chromium D2, D3, and D5 types.

**Table 3. American National Standard Plain Cylindrical Plug Gaging Members — Trilock Design** (ANSI B47.1-1974)

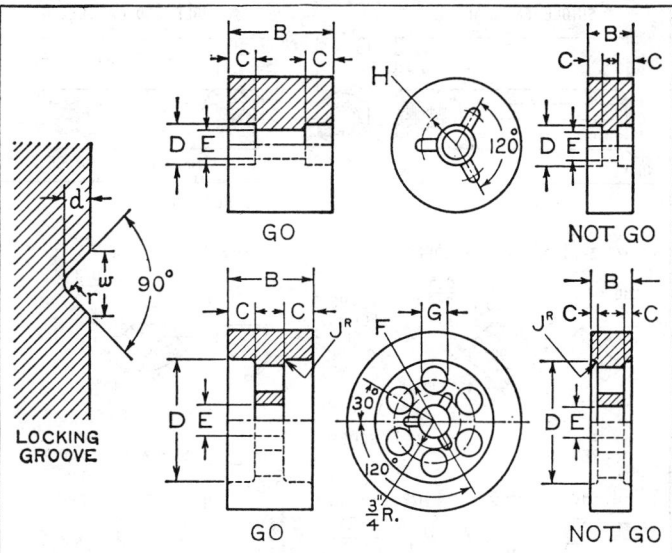

GO    NOT GO    LOCKING GROOVE    GO    NOT GO

Locking groove width for Nos. 2½, 3½ and 4½: *w* = .120–.130, *d* = .050 min., *r* = .020–.030; for Nos. 5½, 6 and 7: *w* = .188–.198, *d* = .073 min., *r* = .030–.050.

| Handle Size No. | Range of Diameters | | GO Gages | | NOT GO | | GO and NOT GO | | | |
|---|---|---|---|---|---|---|---|---|---|---|
| | Above | To and Including | B | C | B | C | D | E | F | H |
| 2½ | .760 | .947 | 1¼ | ¼ | ¾ | ¼ | 25⁄64 | 17⁄64 | .... | 15⁄64 |
| 3½ | .947 | 1.135 | 1⅜ | ¼ | ¾ | ¼ | 25⁄64 | 17⁄64 | .... | 21⁄64 |
| 4½ | 1.135 | 1.510 | 1½ | ⅜ | ¾ | ¼ | 37⁄64 | 25⁄64 | .... | 27⁄64 |
| 5½ | 1.510 | 2.010 | 1⅞ | ½ | ⅞ | 5⁄16 | 25⁄32 | 17⁄32 | .... | 9⁄16 |
| 6 | 2.010 | 2.510 | 2 | ½ | ⅞ | 5⁄16 | 25⁄32 | 17⁄32 | .... | ⅝ |
| | 2.510 | 3.010 | 2 | ⅝ | 1 | ⅛ | 1⅞ | 29⁄32 | .... | |
| | 3.010 | 3.510 | 2 | ⅝ | 1 | ⅛ | 2¼ | 29⁄32 | .... | G |
| | 3.510 | 4.010 | 2⅛ | 11⁄16 | 1 | ⅛ | 2⅝ | 29⁄32 | .... | |
| | 4.010 | 4.510 | 2⅛ | 11⁄16 | 1 | ⅛ | 3 | 29⁄32 | 1 1⁄16 | ¾ |
| | 4.510 | 5.010 | 2⅛ | 11⁄16 | 1 | ⅛ | 3 7⁄16 | 29⁄32 | 1 3⁄16 | 13⁄16 |
| 7 | 5.010 | 5.510 | 2⅛ | 11⁄16 | 1 | ⅛ | 3⅞ | 29⁄32 | 1¼ | ⅞ |
| | 5.510 | 6.010 | 2⅛ | 11⁄16 | 1 | ⅛ | 4 5⁄16 | 29⁄32 | 1⅜ | 1 |
| | 6.010 | 6.510 | 2⅛ | 11⁄16 | 1 | ⅛ | 4¾ | 29⁄32 | 1½ | 1⅛ |
| | 6.510 | 7.010 | 2⅛ | 11⁄16 | 1 | ⅛ | 5¼ | 29⁄32 | 1⅝ | 1¼ |
| | 7.010 | 7.510 | 2⅛ | 11⁄16 | 1 | ⅛ | 5¾ | 29⁄32 | 1¾ | 1⅜ |
| | 7.510 | 8.010 | 2⅛ | 11⁄16 | 1 | ⅛ | 6¼ | 29⁄32 | 1⅞ | 1½ |

All dimensions in inches. Radius *J* is ³⁄₁₆ inch for NOT GO gages above 2.510 to 8.010 inch diameters and for GO gages above 2.010 to 3.010 inch diameters. For GO gages above 3.010 to 8.010 inch diameters, *J* is ⁵⁄₁₆ inch. Three equally spaced locking grooves (see enlarged section) are required on both ends of all gages. Not shown are progressive plug gage members, sizes 1.510 to 2.010, incl. and 2.010 to 2.510. See Table 4 for handle dimensions.

**Table 4. American National Standard Handles for Plain Cylindrical Thread and Spline Plug Gages — Trilock Design (ANSI B47.1-1974)**

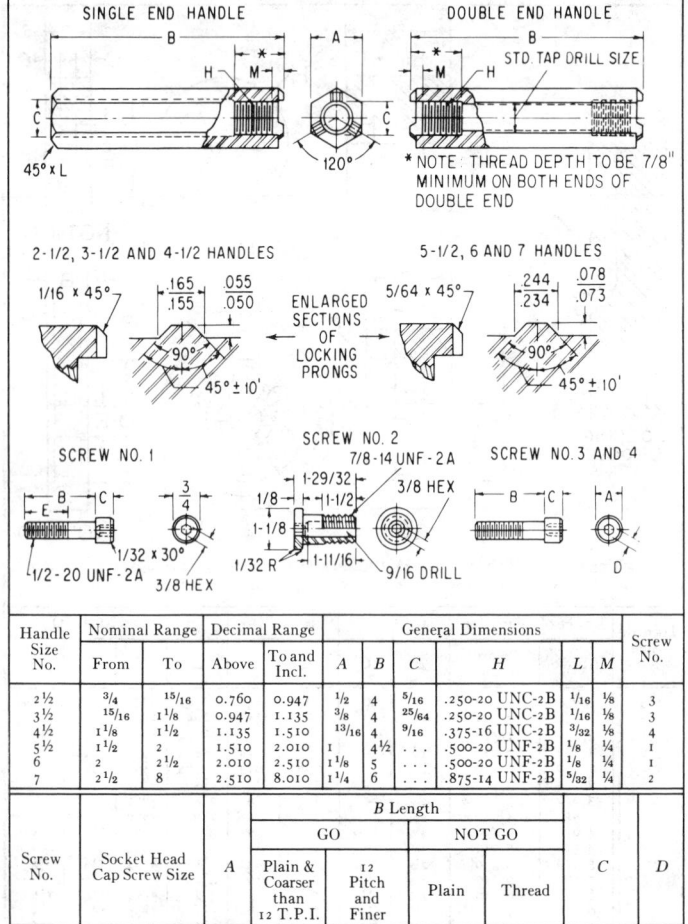

| Handle Size No. | Nominal Range From | Nominal Range To | Decimal Range Above | Decimal Range To and Incl. | A | B | C | H | L | M | Screw No. |
|---|---|---|---|---|---|---|---|---|---|---|---|
| 2½ | ¾ | 15/16 | 0.760 | 0.947 | ½ | 4 | 5/16 | .250-20 UNC-2B | 1/16 | 1/8 | 3 |
| 3½ | 15/16 | 1⅛ | 0.947 | 1.135 | ⅜ | 4 | 25/64 | .250-20 UNC-2B | 1/16 | 1/8 | 3 |
| 4½ | 1⅛ | 1½ | 1.135 | 1.510 | 13/16 | 4 | 9/16 | .375-16 UNC-2B | 3/32 | 1/8 | 4 |
| 5½ | 1½ | 2 | 1.510 | 2.010 | I | 4½ | ... | .500-20 UNF-2B | 1/8 | ¼ | I |
| 6 | 2 | 2½ | 2.010 | 2.510 | 1⅛ | 5 | ... | .500-20 UNF-2B | 1/8 | ¼ | I |
| 7 | 2½ | 8 | 2.510 | 8.010 | 1¼ | 6 | ... | .875-14 UNF-2B | 5/32 | ¼ | 2 |

| Screw No. | Socket Head Cap Screw Size | A | B Length GO Plain & Coarser than 12 T.P.I. | B Length GO 12 Pitch and Finer | B Length NOT GO Plain | B Length NOT GO Thread | C | D |
|---|---|---|---|---|---|---|---|---|
| 3 | .250-20 UNC-2A | ⅜ | 1½ | 1½ | I | I | ¼ | 3/16 |
| 4 | .375-16 UNC-2A | 9/16 | 1½ | 1¼ | 1¼ | 1¼ | *⅜ | 5/16 |

| Plug | B | C | E |
|---|---|---|---|
| GO-For Fine Thds. 16 T.P.I. and Finer | 1½ | 5/16 | 1¼ |
| NOT GO | 1½ | 5/16 | 1¼ |
| GO | 2¼ | ½ | 1¼ |
| Progressive | 3 | ½ | 1¼ |

All dimensions in inches.    *C for NOT GO screw is ¼ inch.

**American National Standard ANSI B47.1-1974.** — This standard covers standard designs for plain and thread plug gage blanks to 12.010 inches maximum gaging diameter; plain and thread ring gage blanks to 12.260 inches maximum gaging diameter; involute and serrated spline plug and ring gage blanks to 8.000 inches major diameter, and straight-sided spline plug and ring gage blanks to major diameters of 8.000 inches for plugs and 6.000 inches for rings; machine taper plug and ring gage blanks to 5.000 inches gaging diameter; adjustable snap gages to 12 inches; adjustable length gages to any desired length; dial indicators up to 3¾ inches nominal bezel diameter; and master disks up to 8.010 inches in diameter. Recommended general designs covering taper plug and ring gages for special applications, flush-pin gages, and flat plug gages are also included.

**Definitions of Gage Terms.** — The definitions which follow apply to certain terms used in connection with the American Gage Design Standards.

*American Gage Design Standard:* The caption "American Gage Design Standard" has been adopted to designate gages made to the design specifications promulgated by the American Gage Design Committee. (This Committee was formed in 1926 to simplify gaging practice through the adoption of standard designs for gage blanks and component parts. The designs developed by the Committee are availabe to everyone and will minimize the necessity for the manufacture of special gages of the simpler types.)

*Anvil:* The gaging member of a gage when constructed as a fixed nonadjustable block, or as the integral jaw of the gage.

*Flange:* That external portion of a large ring gage which is reduced in section for the purpose of lightening the gage.

*Gaging Button:* An adjustable gaging member of an adjustable snap or length gage consisting of a shank and a flanged portion, the latter constituting the gaging section.

*Gaging Member:* That integral unit of a gage which is accurately finished to size and is employed for size control of the work. In taper lock plug gages, the gaging member consists of a shank and a gaging section.

*Snap Gage Pin:* A straight, unflanged adjustable gaging member of an adjustable snap gage.

*Plain Adjustable Snap Gage:* A complete external caliper gage employed for the size control of plain external dimensions, comprising an open frame, in both jaws of which gaging members are provided, one or more pairs of which can be set and locked to any predetermined size within the range of the gage.

*Plain Solid Snap Gage:* A complete external caliper gage employed for the size control of plain external dimensions, comprising an open frame and jaws, the latter carrying gaging members in the form of fixed, parallel, nonadjustable anvils.

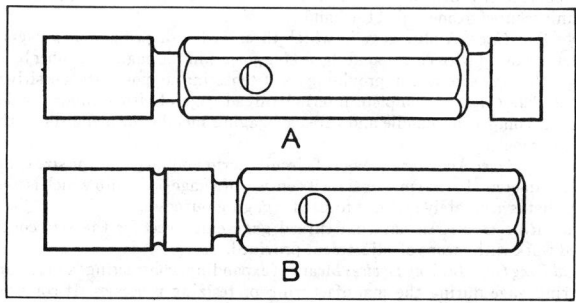

A

B

Fig. 1

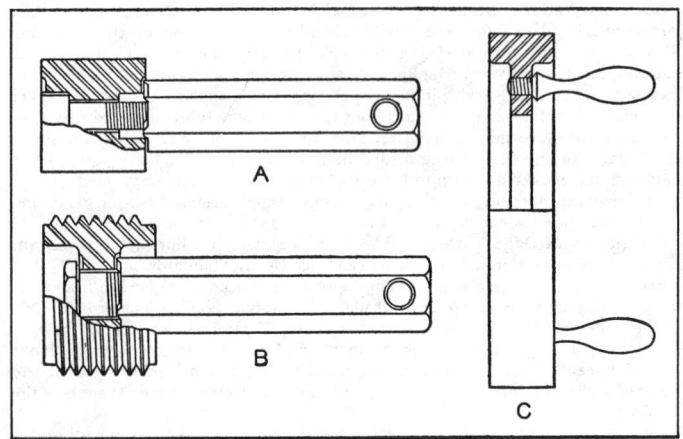

Fig. 2

*Taper Lock:* Term designating that construction in which the gaging member has a taper shank, which is forced into a taper hole in the handle.

*Lightening Holes:* Drilled holes provided in the heavier sizes of gaging members for the purpose of reducing the weight of the gage.

*Marking Disk:* A plate which can be attached to a gage frame to provide, when suitably marked, a means of identification for the gage.

*Annular Plug Gage:* A shell type plug gage in which the gaging member is in the form of a ring, the external surface of which is the gaging section, the central portion of the web being machined away for the purpose of reducing weight, ball handles being provided for convenience in handling. This construction is employed for plain and thread plug gages in the ranges above 8.010 inches.

*Plain Cylindrical Plug Gage:* A complete internal gage of single- or double-ended type for the size control of holes and other applications. It consists of handle and gaging member or members, with suitable locking means.

*Progressive Cylindrical Plug Gage:* A complete internal gage consisting of handle and gaging member in which the "GO" and "NOT GO" gaging sections are combined in a single unit secured to one end of the handle.

*Trilock Plug Gage:* A plug gage in which three wedge-shaped *locking prongs* on the handle are engaged with corresponding *locking grooves* in the gaging member by means of a single through screw, thus providing a self-centering support with a positive lock.

*Thread Plug Gage:* A complete internal thread gage of either single- or double-ended type, comprising handle and threaded gaging member or members, with suitable locking means.

*Plain Ring Gage:* An external gage of circular form employed for the size control of external diameters. In the smaller sizes it consists of a gage body into which is pressed a *bushing*, that is accurately finished to size for gaging purposes.

*Thread Ring Gage:* An external thread gage employed for the size control of threaded work with means of adjustment provided.

*Thread Ring Gage Locking Device:* Means of expanding, contracting, and locking the thread ring gage during the manufacturing or resizing processes. It comprises an adjusting screw, a locking screw and a sleeve.

**Table 5. American National Standard Plain Cylindrical Plug Gaging
Members — Annular Design** (ANSI B47.1-1974)

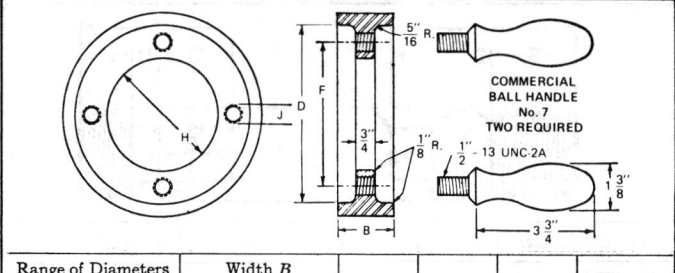

| Range of Diameters | | Width B | | Diam. D | Diam. F | Diam. H | Tapped Hole J* |
|---|---|---|---|---|---|---|---|
| Above | To and Including | GO | NOT GO | | | | |
| 8.010 | 8.510 | 2¼ | 1 | 6½ | 5¼ | 4 | ½-13 |
| 8.510 | 9.010 | 2¼ | 1 | 6¹⁵⁄₁₆ | 5⅝ | 4⅜ | ½-13 |
| 9.010 | 9.510 | 2¼ | 1 | 7⅜ | 6 | 4¾ | ½-13 |
| 9.510 | 10.010 | 2¼ | 1 | 7¹³⁄₁₆ | 6½ | 5⅛ | ½-13 |
| 10.010 | 10.510 | 2¼ | 1 | 8¼ | 7 | 5½ | ½-13 |
| 10.510 | 11.010 | 2¼ | 1 | 8¹¹⁄₁₆ | 7½ | 5⅞ | ½-13 |
| 11.010 | 11.510 | 2¼ | 1 | 9⅛ | 8 | 6¼ | ½-13 |
| 11.510 | 12.010 | 2¼ | 1 | 9⁹⁄₁₆ | 8½ | 6⅝ | ½-13 |

All dimensions in inches.
*A Class 2B fit is specified for the tapped holes (½-13 UNC-2B). The web of the gage is provided with four tapped holes for convenience in bolting to faceplate during manufacture. Two of these holes are employed for attaching No. 7 ball handles to the gaging member. (For handle dimensions, see illustration above.)

*Shank:* That portion of the gaging member which is employed for fixing the gaging member in the handle or frame.

**Taper Lock Design of Gage.** — This type of gage is simple and is economical to produce and maintain. The gaging member has a taper shank which is forced into a taper hole in the handle. When properly assembled, the taper lock gage possesses the rigidity of a solid gage and is entirely free of shake or "wink." Drift slots or drift holes are provided near the GO end of the handle, permitting gaging members to be removed when replacement is necessary. In the case of double-end gages, the other end is removed by running a rod through the hollow handle.

Complete dimensional tolerances have been established for the mating parts of gaging members and handles, insuring absolute interchangeability of gages and handles wherever manufactured.

**Trilock Design of Gage.** — Considerations of rigidity of construction and long life have dictated the trilock design for the size range above 0.760 to and including 8.010 inches. With this construction there is no chance for shake or "wink" to interfere with the sensitive feel so necessary in gages of this type. Three wedge-shaped locking prongs on the handles are engaged with corresponding grooves in the gaging member by a single through screw, thus providing a self-centering support with a positive lock. This results in a degree of rigidity equivalent to that of a solid gage. The useful life of the plug, furthermore, is materially increased, for when one end is worn the plug can be reversed; it is then, for most purposes, as good as new.

Table 6. American National Standard Thread Plug Gaging
Members — Taper Lock Design (ANSI B47.1-1974)

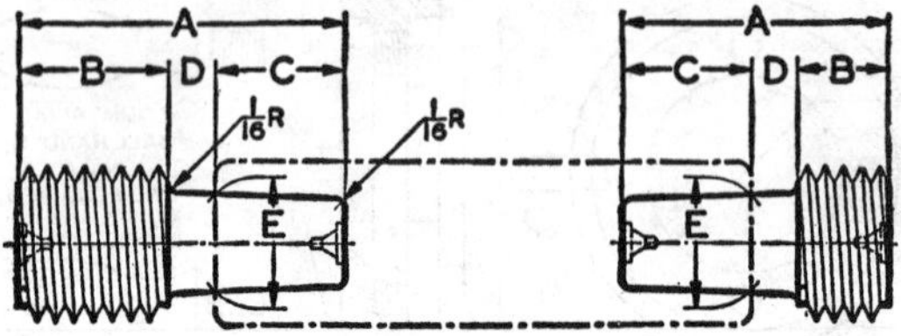

| Range of Diameters | | GO Gages | | | | GO and NOT GO | |
| Above | To and Including | A | B | C | D | Min. E | Max. E |
|---|---|---|---|---|---|---|---|
| 0.240 | 0.365 | 1½ | ½ | ¾ | ¼ | 0.239 | 0.240 |
| 0.365 | 0.510 | 1¾ | ¾ | ¾ | ¼ | 0.309 | 0.310 |
| 0.510 | 0.825 | 1⅞ | ⅞ | ¾ | ¼ | 0.408 | 0.410 |
| 0.825 | 1.135 | 2³⁄₁₆ | 1 | ⅞ | ⁵⁄₁₆ | 0.608 | 0.610 |
| *1.135 | 1.510 | 2⅝ | 1¼ | 1 | ⅜ | 0.808 | 0.810 |
| †1.135 | 1.510 | 2⅜ | 1 | 1 | ⅜ | 0.808 | 0.810 |

| Range of Diameters | | HI Gages‡ | | | | Handle Size No. | Taper of Gage Shank |
| Above | To and Including | A | B | C | D | | |
|---|---|---|---|---|---|---|---|
| 0.240 | 0.365 | 1⁵⁄₁₆ | ⁵⁄₁₆ | ¾ | ¼ | 1 | 0.25 inch per foot for all gage shanks |
| 0.365 | 0.510 | 1⅜ | ⅜ | ¾ | ¼ | 2 | |
| 0.510 | 0.825 | 1½ | ½ | ¾ | ¼ | 3 | |
| 0.825 | 1.135 | 1¹³⁄₁₆ | ⅝ | ⅞ | ⁵⁄₁₆ | 4 | |
| *1.135 | 1.510 | 2⅛ | ¾ | 1 | ⅜ | 5 | |
| †1.135 | 1.510 | 2⅛ | ¾ | 1 | ⅜ | 5 | |

*Less than 12 threads per inch.    †12 threads per inch and over.    ‡Formerly known as NOT GO gages.
All dimensions in inches. For dimensions of handles see Table 2.
The "HI" plug must have not less than 3 full threads.
Taper lock gaging members are standard for all taper pipe thread plug gages up to and including 2 inch nominal pipe size.

The construction is protected by carefully worked out dimensional limits, and interchangeability is insured between gaging members and handles, wherever manufactured.

**Annular Design of Gages above 8.010 Inches.** — Because of the fact that large plug gages are heavy and difficult to handle, it was necessary to adopt a design for the range above 8.010 inches which would have the lightest possible section consistent with strength and permanence. The annular design having a rim and web of properly proportioned section, the center being bored out for purposes of weight reduction, has therefore been adopted as standard. The web is provided with four tapped holes for convenience in bolting to a face plate during manufacturing. Two of these are further employed for fixing ball handles in the gaging member. Details of construction are shown in Table 5.

**Table 7. American National Standard Thread Plug Gaging Members — Reversible or Trilock Design** (ANSI B47.1-1974)

Thread plug gaging members are of the same general form as the plain cylindrical gaging members of the reversible type. See illustrations accompanying Table 3 for dimensions indicated by the letters *B*, *C*, *D*, *E*, *F*, *G* in the table below.

| Handle Size No. | Range of Diameters | | GO Gages | | | | | |
|---|---|---|---|---|---|---|---|---|
| | Above | To and Including | *B* | *C* | *B* | *C* | *B* | *C* |
| | | | 7 Threads per Inch and Under | | 8 to 12 Threads per Inch | | 13 Threads per Inch and Over | |
| 2½ | 0.760 | 0.947 | 1 1/4 | 1/4 | 1 | 1/4 | 1 | 1/4 |
| 3½ | 0.947 | 1.135 | 1 3/8 | 1/4 | 1 1/8 | 1/4 | 1 | 1/4 |
| 4½ | 1.135 | 1.510 | 1 1/2 | 3/8 | 1 1/4 | 3/8 | 1 | 3/8 |
| | | | 7 Threads per Inch and Under | | 8 to 14 Threads per Inch | | 16 Threads per Inch and Over | |
| 5½ | 1.510 | 2.010 | 1 7/8 | 1/2 | 1 1/4 | 3/8 | 7/8 | 5/16 |
| 6 | 2.010 | 2.510 | 2 | 1/2 | 1 3/8 | 3/8 | 7/8 | 5/16 |
| | 2.510 | 3.010 | 2 | 5/8 | 1 1/2 | 3/8 | 1 | 1/8 |
| | 3.010 | 3.510 | 2 | 5/8 | 1 1/2 | 3/8 | 1 | 1/8 |
| | 3.510 | 4.010 | 2 1/8 | 11/16 | 1 1/2 | 3/8 | 1 | 1/8 |
| | 4.010 | 4.510 | 2 1/8 | 11/16 | 1 1/2 | 3/8 | 1 | 1/8 |
| | 4.510 | 5.010 | 2 1/8 | 11/16 | 1 1/2 | 3/8 | 1 | 1/8 |
| 7 | 5.010 | 5.510 | 2 1/8 | 11/16 | 1 1/2 | 3/8 | 1 | 1/8 |
| | 5.510 | 6.010 | 2 1/8 | 11/16 | 1 1/2 | 3/8 | 1 | 1/8 |
| | 6.010 | 6.510 | 2 1/8 | 11/16 | 1 1/2 | 3/8 | 1 | 1/8 |
| | 6.510 | 7.010 | 2 1/8 | 11/16 | 1 1/2 | 3/8 | 1 | 1/8 |
| | 7.010 | 7.510 | 2 1/8 | 11/16 | 1 1/2 | 3/8 | 1 | 1/8 |
| | 7.510 | 8.010 | 2 1/8 | 11/16 | 1 1/2 | 3/8 | 1 | 1/8 |

| Handle Size No. | Range of Diameters | | HI Gages* All Pitches | | GO and HI Gages All Pitches | | | |
|---|---|---|---|---|---|---|---|---|
| | Above | To and Including | *B* | *C* | *D* | *E* | *F* | *H*† |
| 2½ | 0.760 | 0.947 | 3/4 | 1/4 | 25/64 | 17/64 | ...... | 15/64 |
| 3½ | 0.947 | 1.135 | 3/4 | 1/4 | 25/64 | 17/64 | ...... | 21/64 |
| 4½ | 1.135 | 1.510 | 3/4 | 1/4 | 37/64 | 25/64 | ...... | 27/64 |
| 5½ | 1.510 | 2.010 | 7/8 | 5/16 | 25/32 | 17/32 | ...... | 9/16 |
| 6 | 2.010 | 2.510 | 7/8 | 5/16 | 25/32 | 17/32 | ...... | 5/8 |
| | 2.510 | 3.010 | 1 | 1/8 | 1 7/8 | 29/32 | ...... | |
| | 3.010 | 3.510 | 1 | 1/8 | 2 1/4 | 29/32 | ...... | *G* |
| | 3.510 | 4.010 | 1 | 1/8 | 2 5/8 | 29/32 | ...... | |
| | 4.010 | 4.510 | 1 | 1/8 | 3 | 29/32 | 1 1/16 | 3/4 |
| | 4.510 | 5.010 | 1 | 1/8 | 3 7/16 | 29/32 | 1 3/16 | 13/16 |
| 7 | 5.010 | 5.510 | 1 | 1/8 | 3 7/8 | 29/32 | 1 1/4 | 7/8 |
| | 5.510 | 6.010 | 1 | 1/8 | 4 5/16 | 29/32 | 1 3/8 | 1 |
| | 6.010 | 6.510 | 1 | 1/8 | 4 3/4 | 29/32 | 1 1/2 | 1 1/8 |
| | 6.510 | 7.010 | 1 | 1/8 | 5 1/4 | 29/32 | 1 5/8 | 1 1/4 |
| | 7.010 | 7.510 | 1 | 1/8 | 5 3/4 | 29/32 | 1 3/4 | 1 3/8 |
| | 7.510 | 8.010 | 1 | 1/8 | 6 1/4 | 29/32 | 1 7/8 | 1 1/2 |

All dimensions in inches. *Formerly known as NOT GO gages.

†*H* for GO gages of 2.510 to 8.010 inch diameters is 5/16 inch for 7 tpi and coarser, 1/4 inch for 8 to 14 tpi, except 3/16 inch for 2.510 to 3.010 inch diameters, 3/16 inch for 16 tpi and finer; and 3/16 inch for "HI" gages of 2.510 to 8.010 inch diameters, all pitches.

Three equally spaced locking grooves (see enlarged sections in Table 3) are required on both ends of all gages. See Table 4 for handle dimensions.

"HI" gages having 16 tpi or more are relieved on both ends 1/32 inch below sharp roots of the threads, and 1/8 inch from each end.

## Table 8. American National Standard Thread Plug Gaging Members—Annular Design (ANSI B47.1-1974)

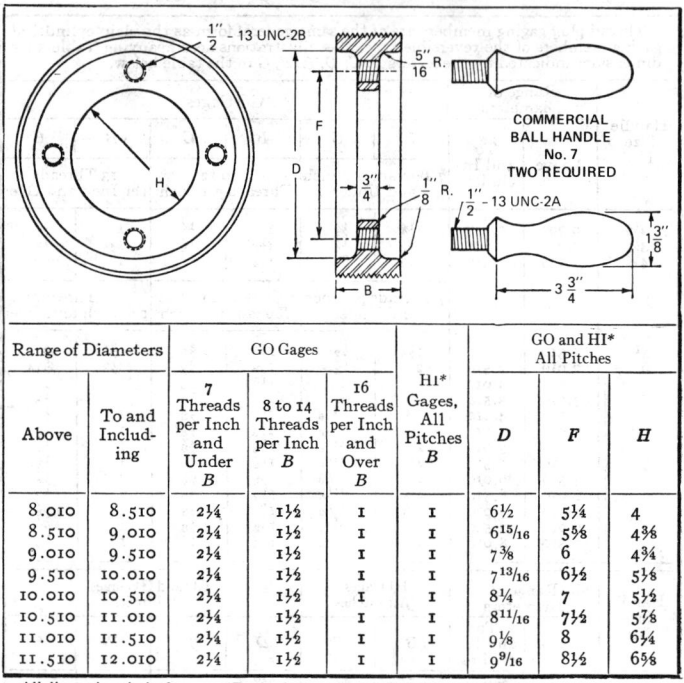

| Range of Diameters | | GO Gages | | | H1* | GO and HI* All Pitches | | |
|---|---|---|---|---|---|---|---|---|
| Above | To and Including | 7 Threads per Inch and Under B | 8 to 14 Threads per Inch B | 16 Threads per Inch and Over B | Gages, All Pitches B | D | F | H |
| 8.010 | 8.510 | 2¼ | 1½ | 1 | 1 | 6½ | 5¼ | 4 |
| 8.510 | 9.010 | 2¼ | 1½ | 1 | 1 | 6¹⁵⁄₁₆ | 5⅝ | 4⅜ |
| 9.010 | 9.510 | 2¼ | 1½ | 1 | 1 | 7⅜ | 6 | 4¾ |
| 9.510 | 10.010 | 2¼ | 1½ | 1 | 1 | 7¹³⁄₁₆ | 6½ | 5⅛ |
| 10.010 | 10.510 | 2¼ | 1½ | 1 | 1 | 8¼ | 7 | 5½ |
| 10.510 | 11.010 | 2¼ | 1½ | 1 | 1 | 8¹¹⁄₁₆ | 7½ | 5⅞ |
| 11.010 | 11.510 | 2¼ | 1½ | 1 | 1 | 9⅛ | 8 | 6¼ |
| 11.510 | 12.010 | 2¼ | 1½ | 1 | 1 | 9⁹⁄₁₆ | 8½ | 6⅝ |

All dimensions in inches.     *Formerly known as NOT GO gages.

**Handles for Plain Cylindrical and Thread Plug Gages.**—Handles for wire type gages are hexagonal with hexagonal collet nuts, and are provided in both single- and double-end types. Handles for taper lock gages are provided with tapered holes in both ends as shown in Table 2. Handles for trilock gages are provided in both single- and double-end types as shown in Table 4. Commercial ball handles are employed for the annular plug gage and for certain of the large ring gages. See Fig. 2 and Table 5.

Handles as designed for all gages offer a feature of economy in that they may be disassembled from gaging members when the latter are worn out or discarded for any other reason, and then reassembled with new gaging members, thus giving them, with reasonable care, practically indefinite life.

**Thread Plug Gages.**—The wire type, taper lock, trilock, and annular designs have been adopted for thread plug gage blanks and handles. The designs are patterned after the plain cylindrical plug gage blanks with the exception that the length of thread gaging members is slightly different in some instances. The use of taper lock blanks and handles for pipe thread plug gages is standard to and including 2 inches nominal pipe size.

**Plain Ring Gages.**—The use of the solid ring gage design for size control being well established, the work of the Standards Committee on plain ring gages was concerned chiefly with matters of proportion.

**Table 9. American National Standard Plain Ring Gages** (ANSI B47.1-1974)

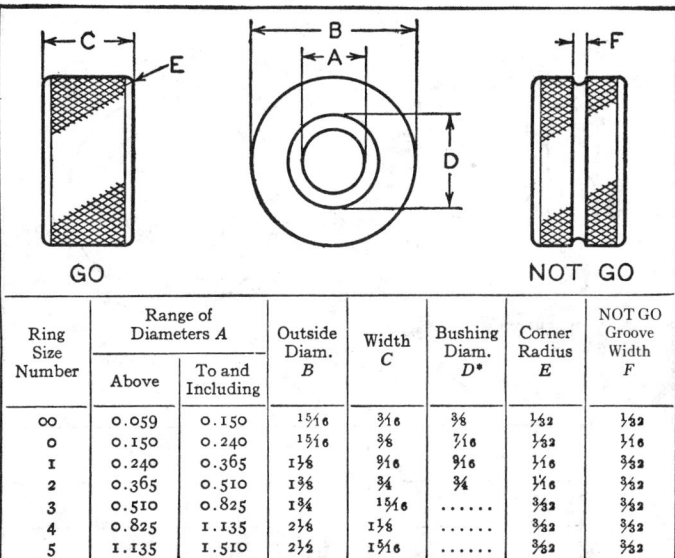

GO                                    NOT GO

| Ring Size Number | Range of Diameters A | | Outside Diam. B | Width C | Bushing Diam. D* | Corner Radius E | NOT GO Groove Width F |
|---|---|---|---|---|---|---|---|
| | Above | To and Including | | | | | |
| ∞ | 0.059 | 0.150 | 15⁄16 | 3⁄16 | 3⁄8 | 1⁄32 | 1⁄32 |
| 0 | 0.150 | 0.240 | 15⁄16 | 3⁄8 | 7⁄16 | 1⁄32 | 1⁄16 |
| 1 | 0.240 | 0.365 | 1⅛ | 9⁄16 | 9⁄16 | 1⁄16 | 3⁄32 |
| 2 | 0.365 | 0.510 | 1⅜ | 3⁄4 | 3⁄4 | 1⁄16 | 3⁄32 |
| 3 | 0.510 | 0.825 | 1¾ | 15⁄16 | ...... | 3⁄32 | 3⁄32 |
| 4 | 0.825 | 1.135 | 2⅛ | 1⅛ | ...... | 3⁄32 | 3⁄32 |
| 5 | 1.135 | 1.510 | 2½ | 15⁄16 | ...... | 3⁄32 | 3⁄32 |

* The bushings may be 1⁄16 inch longer than the ring thickness but are ground flush after hole is finished. Ring sizes 0, 00, 1, and 2 are solid or bushed; 3, 4 and 5 are solid.

In the smaller sizes of plain ring gages a hardened bushing may be pressed into a soft gage body in place of the one-piece ring gage. This design is optional in the range above 0.059 to and including 0.510 inch. However, the single-piece gage may be employed in this range, and it is standard in all cases above 0.501 inch. Gages in sizes above 1.510 inch are flanged in order to eliminate unnecessary weight and to facilitate handling.

No dimensional difference exists between GO and NOT GO blanks of identical size range and service class, but an annular groove is provided in the periphery of NOT GO blanks as a means of identification.

Gages in sizes above 5.510 inches may be provided with two .500-13 UNC-2B tapped holes in the web, 180 degrees apart, to accommodate No. 7 ball handles.

**Thread Ring Gages.**—All American Gage Design Standard thread ring gage blanks are equipped with an effective device for adjusting and locking the gage in the manufacturing and resizing processes. Of the many locking devices considered, the single-unit locking device was finally adopted as standard, as it provides a minimum diameter of blank for a given size range, and provides a simple adjustment and positive lock without introducing any mechanical stresses into the gage body which might tend to create distortion after setting.

As shown in Fig. 3, the adjusting screw is threaded externally and internally and split longitudinally. Turning this screw to the right exerts pressure on the sleeve against the shoulder in the right-hand side of the gage, thus spreading the ring. Once the ring has been properly adjusted by means of the adjusting screw, the adjustment is tightened by

**Table 10. American National Standard Plain Ring Gages** (ANSI B47.1-1974)

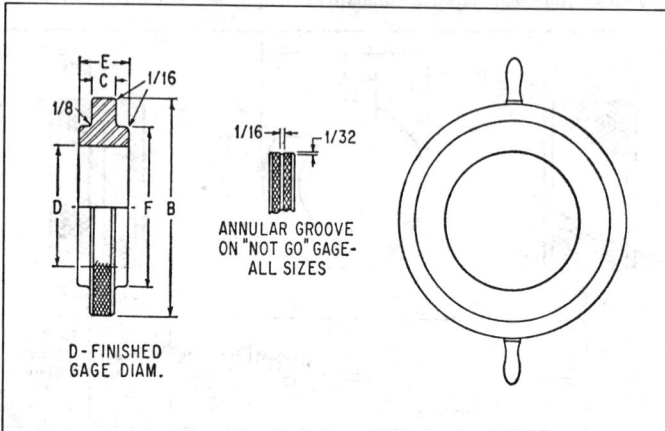

D - FINISHED GAGE DIAM.

ANNULAR GROOVE ON "NOT GO" GAGE-ALL SIZES

| Ring Size No. | Diam., D | | Inspection or Working Ring Gages | | | |
|---|---|---|---|---|---|---|
| | Above | To and Incl. | B | C | E | F |
| 6 | 1.510 | 2.010 | 4 | 1/2 | 1 1/2 | 2 7/8 |
| 7 | 2.010 | 2.510 | 4 1/2 | 9/16 | 1 1/2 | 3 3/8 |
| 8 | 2.510 | 3.010 | 5 | 5/8 | 1 1/2 | 4 |
| 9 | 3.010 | 3.510 | 5 1/2 | 11/16 | 1 1/2 | 4 1/2 |
| 10 | 3.510 | 4.010 | 6 3/8 | 3/4 | 1 1/2 | 5 1/8 |
| 11 | 4.010 | 4.760 | 7 1/4 | 7/8 | 1 1/2 | 5 7/8 |
| 12 | 4.760 | 5.510 | 8 1/4 | 1 | 1 1/2 | 6 5/8 |
| 13 | 5.510 | 6.260 | 9 1/4 | 1 | 1 1/2 | 7 3/8 |
| 14 | 6.260 | 7.010 | 10 1/4 | 1 | 1 1/2 | 8 1/8 |
| 15 | 7.010 | 7.760 | 11 1/4 | 1 | 1 1/2 | 8 7/8 |
| 16 | 7.760 | 8.510 | 12 1/4 | 1 | 1 1/2 | 9 5/8 |
| 17 | 8.510 | 9.260 | 13 1/4 | 1 | 1 1/2 | 10 3/8 |
| 18 | 9.260 | 10.010 | 14 1/4 | 1 | 1 1/2 | 11 1/8 |
| 19 | 10.010 | 10.760 | 15 1/4 | 1 | 1 1/2 | 11 7/8 |
| 20 | 10.760 | 11.510 | 16 1/4 | 1 | 1 1/2 | 12 5/8 |
| 21 | 11.510 | 12.260 | 17 1/4 | 1 | 1 1/2 | 13 3/8 |

All dimensions in inches.

Diameters above 5.510 inches are provided with commercial ball handles as shown by the illustration at the right.

The NOT GO gage has an annular groove cut around the knurled flange, as shown, and the GO gage flange is without a groove.

means of the locking screw. The tightening of the locking screw exerts a pull between the shoulder immediately under its head and the internal threads of the adjusting screw which causes the adjusting screw to expand into the threads in the wall of the gage, the thrust of this action being taken up longitudinally by the sleeve. Therefore the clamping is accomplished by expansion of the adjusting screw equally in all directions and not by the application of any eccentric forces that tend to distort the gage or upset the adjustment. The locking pressure is taken up centrally by the locking screw itself as the reacting support is directly under the head of the locking screw in the form of a shoulder in the gage. The sleeve, being accurately fitted, serves as a large dowel to maintain the alignment of the gage.

**Table 10a.  American National Standard Thread Ring Gage
Adjusting Screws and Sleeves** (ANSI B47.1-1974)

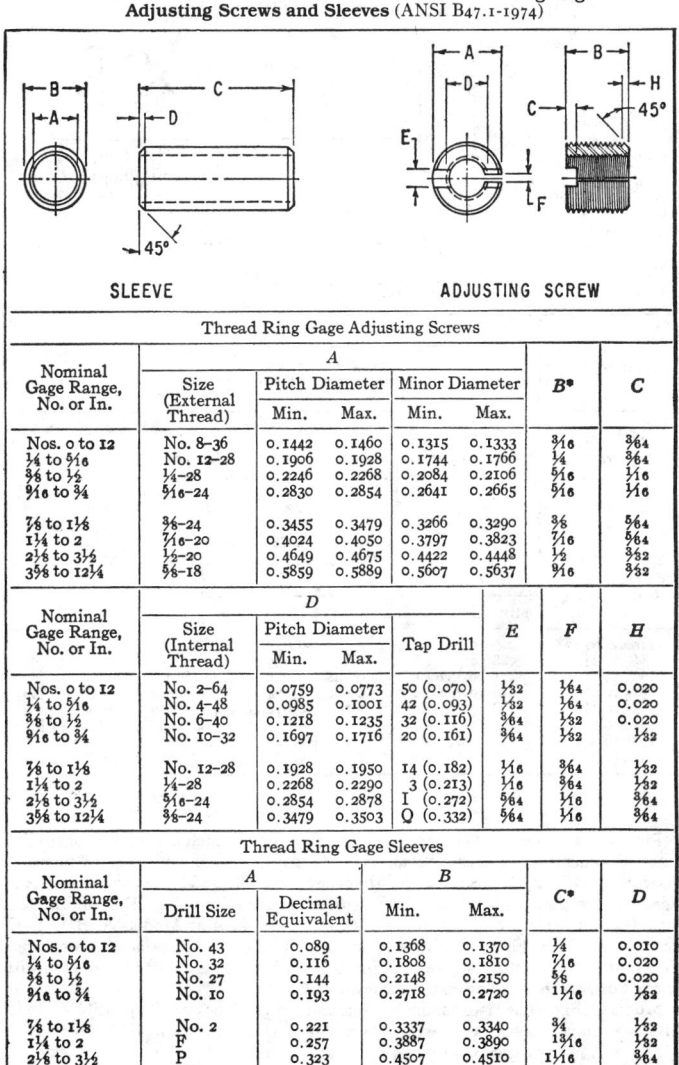

SLEEVE                ADJUSTING SCREW

### Thread Ring Gage Adjusting Screws

| Nominal Gage Range, No. or In. | Size (External Thread) | A Pitch Diameter Min. | A Pitch Diameter Max. | A Minor Diameter Min. | A Minor Diameter Max. | B* | C |
|---|---|---|---|---|---|---|---|
| Nos. 0 to 12 | No. 8-36 | 0.1442 | 0.1460 | 0.1315 | 0.1333 | 3⁄16 | 3⁄64 |
| 1⁄4 to 5⁄16 | No. 12-28 | 0.1906 | 0.1928 | 0.1744 | 0.1766 | 1⁄4 | 3⁄64 |
| 3⁄8 to 1⁄2 | 1⁄4-28 | 0.2246 | 0.2268 | 0.2084 | 0.2106 | 5⁄16 | 1⁄16 |
| 9⁄16 to 3⁄4 | 5⁄16-24 | 0.2830 | 0.2854 | 0.2641 | 0.2665 | 5⁄16 | 1⁄16 |
| 7⁄8 to 1 1⁄8 | 3⁄8-24 | 0.3455 | 0.3479 | 0.3266 | 0.3290 | 3⁄8 | 5⁄64 |
| 1 1⁄4 to 2 | 7⁄16-20 | 0.4024 | 0.4050 | 0.3797 | 0.3823 | 7⁄16 | 5⁄64 |
| 2 1⁄8 to 3 1⁄2 | 1⁄2-20 | 0.4649 | 0.4675 | 0.4422 | 0.4448 | 1⁄2 | 3⁄32 |
| 3 5⁄8 to 12 1⁄4 | 5⁄8-18 | 0.5859 | 0.5889 | 0.5607 | 0.5637 | 9⁄16 | 3⁄32 |

| Nominal Gage Range, No. or In. | Size (Internal Thread) | D Pitch Diameter Min. | D Pitch Diameter Max. | Tap Drill | E | F | H |
|---|---|---|---|---|---|---|---|
| Nos. 0 to 12 | No. 2-64 | 0.0759 | 0.0773 | 50 (0.070) | 1⁄32 | 1⁄64 | 0.020 |
| 1⁄4 to 5⁄16 | No. 4-48 | 0.0985 | 0.1001 | 42 (0.093) | 1⁄32 | 1⁄64 | 0.020 |
| 3⁄8 to 1⁄2 | No. 6-40 | 0.1218 | 0.1235 | 32 (0.116) | 3⁄64 | 1⁄32 | 0.020 |
| 9⁄16 to 3⁄4 | No. 10-32 | 0.1697 | 0.1716 | 20 (0.161) | 3⁄64 | 1⁄32 | 1⁄32 |
| 7⁄8 to 1 1⁄8 | No. 12-28 | 0.1928 | 0.1950 | 14 (0.182) | 1⁄16 | 3⁄64 | 1⁄32 |
| 1 1⁄4 to 2 | 1⁄4-28 | 0.2268 | 0.2290 | 3 (0.213) | 1⁄16 | 3⁄64 | 1⁄32 |
| 2 1⁄8 to 3 1⁄2 | 5⁄16-24 | 0.2854 | 0.2878 | I (0.272) | 5⁄64 | 1⁄16 | 3⁄64 |
| 3 5⁄8 to 12 1⁄4 | 3⁄8-24 | 0.3479 | 0.3503 | Q (0.332) | 5⁄64 | 1⁄16 | 3⁄64 |

### Thread Ring Gage Sleeves

| Nominal Gage Range, No. or In. | A Drill Size | A Decimal Equivalent | B Min. | B Max. | C* | D |
|---|---|---|---|---|---|---|
| Nos. 0 to 12 | No. 43 | 0.089 | 0.1368 | 0.1370 | 1⁄4 | 0.010 |
| 1⁄4 to 5⁄16 | No. 32 | 0.116 | 0.1808 | 0.1810 | 7⁄16 | 0.020 |
| 3⁄8 to 1⁄2 | No. 27 | 0.144 | 0.2148 | 0.2150 | 5⁄8 | 0.020 |
| 9⁄16 to 3⁄4 | No. 10 | 0.193 | 0.2718 | 0.2720 | 11⁄16 | 1⁄32 |
| 7⁄8 to 1 1⁄8 | No. 2 | 0.221 | 0.3337 | 0.3340 | 3⁄4 | 1⁄32 |
| 1 1⁄4 to 2 | F | 0.257 | 0.3887 | 0.3890 | 13⁄16 | 1⁄32 |
| 2 1⁄8 to 3 1⁄2 | P | 0.323 | 0.4507 | 0.4510 | 1 1⁄16 | 3⁄64 |
| 3 5⁄8 to 12 1⁄4 | 25⁄64 | 0.391 | 0.5707 | 0.5710 | 1 1⁄2 | 3⁄64 |

All dimensions in inches.
* Tolerance = ±1⁄64 in.

**Table 10b. American National Standard Thread Ring Gage Locking Screws** (ANSI B47.1-1974)

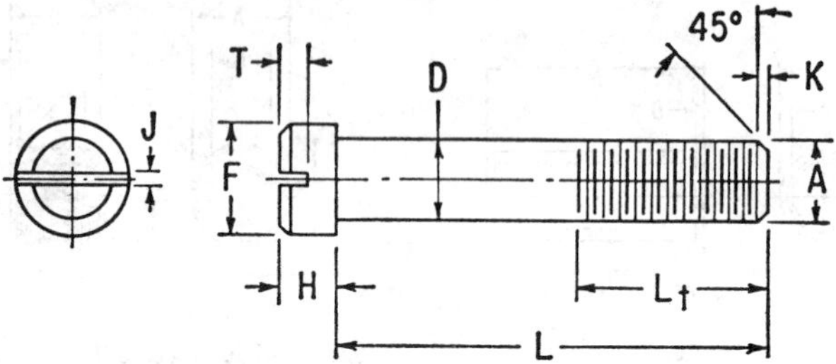


| Nominal Gage Range, No. or In. | A | | | D | | F | |
|---|---|---|---|---|---|---|---|
| | Size | Pitch Diameter | | | | | |
| | | Min. | Max. | Min. | Max. | Min. | Max. |
| Nos. 0 to 12 | .086-64 UNF-3A | 0.0744 | 0.0759 | 0.0840 | 0.0860 | 0.136 | 0.140 |
| 1/4 to 5/16 | .112-48 UNF-3A | 0.0967 | 0.0985 | 0.1096 | 0.1120 | 0.178 | 0.183 |
| 3/8 to 1/2 | .138-40 UNF-3A | 0.1198 | 0.1218 | 0.1353 | 0.1380 | 0.221 | 0.226 |
| 9/16 to 3/4 | .190-32 UNF-3A | 0.1674 | 0.1697 | 0.1867 | 0.1900 | 0.306 | 11/32 |
| 7/8 to 1 1/8 | .216-28 UNF-3A | 0.1904 | 0.1928 | 0.2127 | 0.2160 | 0.337 | 11/32 |
| 1 1/4 to 2 | .250-28 UNF-3A | 0.2243 | 0.2268 | 0.2464 | 0.2500 | 0.367 | 3/8 |
| 2 1/8 to 3 1/2 | .3125-24 UNF-3A | 0.2827 | 0.2854 | 0.3084 | 0.3125 | 0.429 | 7/16 |
| 3 5/8 to 12 1/4 | .375-24 UNF-3A | 0.3450 | 0.3479 | 0.3705 | 0.3750 | 0.553 | 9/16 |

| Nominal Gage Range, No. or In. | H | | J | | T | | K | L ±1/32 | $L_t$ |
|---|---|---|---|---|---|---|---|---|---|
| | Min. | Max. | Min. | Max. | Min. | Max. | | | |
| Nos. 0 to 12 | 0.066 | 0.083 | 0.023 | 0.031 | 0.025 | 0.037 | 0.010 | 29/64 | 3/16 |
| 1/4 to 5/16 | 0.088 | 0.107 | 0.031 | 0.039 | 0.035 | 0.048 | 0.020 | 23/32 | 5/16 |
| 3/8 to 1/2 | 0.111 | 0.132 | 0.039 | 0.048 | 0.045 | 0.060 | 0.020 | 1 | 7/16 |
| 9/16 to 3/4 | 0.156 | 0.180 | 0.050 | 0.060 | 0.064 | 0.083 | 1/32 | 1 1/16 | 7/16 |
| 7/8 to 1 1/8 | 0.178 | 0.205 | 0.056 | 0.067 | 0.074 | 0.094 | 1/32 | 1 3/16 | 1/2 |
| 1 1/4 to 2 | 0.207 | 0.237 | 0.064 | 0.075 | 0.087 | 0.109 | 1/32 | 1 23/64 | 9/16 |
| 2 1/8 to 3 1/2 | 0.262 | 0.295 | 0.072 | 0.084 | 0.110 | 0.137 | 3/64 | 1 23/32 | 5/8 |
| 3 5/8 to 12 1/4 | 0.315 | 0.355 | 0.081 | 0.094 | 0.133 | 0.164 | 3/64 | 2 3/16 | 3/4 |

All dimensions in inches.

**Plain Adjustable Snap Gages.** — This type of gage which is used for controlling external dimensions consists of an open C-shaped frame with gaging members inserted in both jaws and so arranged that they can be set and locked to any predetermined size within the range of adjustment. The American Gage Design Standard includes four designs of adjustable snap gages designated as Models A, B, C, MC, and E. Model A, Fig. 4, has four gaging pins. Model B is the same as A, excepting that four gaging "buttons" are used instead of pins. These buttons are pins with flanged or enlarged gaging ends (as shown at C).

Model C (Fig. 4) has two gaging buttons and a single block anvil opposite. Model MC is a miniature snap gage with two gaging buttons and a single block anvil.

Model E has two gaging buttons, either square or round, and a single block anvil extending beyond the gaging buttons.

Models A, B and C are made in sixteen different sizes as indicated by frame Nos. 1 to 16. Model A is for all diameters up to 12 inches inclusive; Model B is for

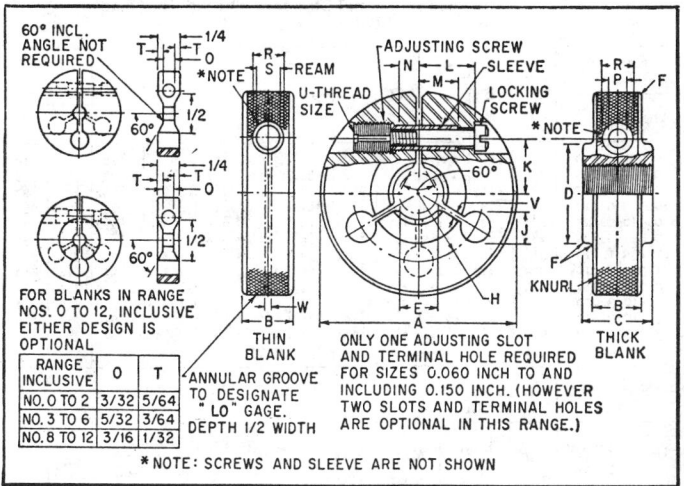

Fig. 3.

diameters ranging from ½ to 11¼ inches inclusive; Model C is for all diameters up to 11⅝ inches inclusive; Model MC is for diameters up to 0.760 inch inclusive; and Model E is for all diameters up to 5¹¹⁄₁₆ inches, inclusive.

The gaging pins or buttons may be adjusted by screws which bear against the ends. The locking device for the pins or buttons consists of a locking screw, a locking bushing, and a locking nut. The locking bushing and nut are beveled on one side to an angle of 30 degrees, and, as the screw is tightened, these beveled surfaces are drawn against corresponding flats on the shank of the gaging button or pin. This locking device was adopted because it has stood the test of time. The front edge or corner of the pins or buttons are beveled where they first engage the work. In the development of these gages exceptional care was taken to insure embodying all of the best features of snap-gage design including:

1. A design of frame which has proved to be exceptionally rigid under severe tests. 2. Reduction of weight to as low a point as strength of materials permits. 3. Distribution of metal to assure a nice balance and feel. 4. An effective and proved locking device. 5. Suitable construction of gaging pins, buttons, and anvils to give ample rigidity and maintain accuracy. 6. Ease and simplicity of adjustment. 7. Provision for sealing. 8. Careful selection of limits and tolerances to preserve accuracy and permit interchangeability.

The frames of models A, B, C have been so designed that common patterns can be used for all three. Frames are of the conventional C or semi-circular type, of cast iron with solid web. Particular attention has been given to weight, which approximates the average of former proprietary designs.

The straight gaging pins are of circular cross-section, an arcuate bevel being provided at the front edge where they first engage the work. The flanged gaging buttons are provided with either square or circular heads, the former being chamfered on their forward edges, and the latter being provided with an arcuate bevel where they first engage the work. The gap between GO and NOT GO has been kept to a minimum.

**Table 11. American National Standard Thread Ring Gages**
(ANSI B47-.1-1974)

| Range of Diams. Above–To and Incl. | A | B | C | D | E | F | H | J | K | L | M ±1/64 | N ±1/64 |
|---|---|---|---|---|---|---|---|---|---|---|---|---|
| 0.059–0.150 | 1 | 1/4 | ... | ... | ... | 1/32 | 5/16 | 5/32 | 5/16 | 7/32 | 5/32 | 1/16 |
| 0.150–0.240 | 1 | 1/4 | ... | ... | ... | 1/32 | 5/16 | 5/32 | 5/16 | 7/32 | 5/32 | 1/16 |
| 0.240–0.365 | 1 3/8 | 11/32 | ... | ... | 5/32 | 1/32 | 7/16 | 3/16 | 3/8 | 11/32 | 1/4 | 1/8 |
| 0.365–0.510 | 1 3/4 | 7/16 | ... | ... | 3/16 | 3/64 | 19/32 | 1/4 | 15/32 | 1/2 | 3/8 | 3/16 |
| 0.510–0.825 | 2 3/16 | 9/16 | 3/4 | 1 1/16 | 11/32 | 1/16 | 3/4 | 5/16 | 11/16 | 17/32 | 13/32 | 7/32 |
| 0.825–1.135 | 2 5/8 | 11/16 | 15/16 | 1 1/2 | 9/16 | 1/16 | 31/32 | 5/16 | 7/8 | 17/32 | 13/32 | 9/32 |
| 1.135–1.510 | 3 1/4 | 3/4 | 1 1/8 | 1 7/8 | 27/32 | 1/16 | 1 3/16 | 3/8 | 1 1/8 | 5/8 | 7/16 | 5/16 |
| 1.510–2.010 | 3 3/4 | 13/16 | 1 1/4 | 2 3/8 | 1 3/16 | 3/32 | 1 7/16 | 3/8 | 1 3/8 | 5/8 | 7/16 | 5/16 |
| 2.010–2.510 | 4 1/2 | 7/8 | 1 5/16 | 2 7/8 | 1 19/32 | 3/32 | 1 3/4 | 7/16 | 1 11/16 | 13/16 | 9/16 | 7/16 |
| 2.510–3.010 | 5 | 7/8 | 1 3/8 | 3 3/8 | 2 | 3/32 | 2 | 7/16 | 1 15/16 | 13/16 | 9/16 | 7/16 |
| 3.010–3.510 | 5 1/2 | 15/16 | 1 7/16 | 3 7/8 | 2 7/16 | 3/32 | 2 7/32 | 7/16 | 2 3/16 | 13/16 | 9/16 | 7/16 |
| 3.510–4.010 | 6 3/8 | 15/16 | 1 1/2 | 4 5/8 | 2 15/16 | 3/32 | 2 5/8 | 1/2 | 2 9/16 | 1 | 3/4 | 5/8 |
| 4.010–4.760 | 7 1/4 | 1 | 1 1/2 | 5 3/8 | 3 3/8 | 3/32 | 3 1/32 | 1/2 | 3 | 1 | 3/4 | 5/8 |

| Range of Diams. | Drill Size No. or In. | | Ream | | U | | | V | W |
|---|---|---|---|---|---|---|---|---|---|
| | | | | | | Pitch Diam. | | | |
| Above–To and Incl. | P | R | Min. | Max. | Size | Min. | Max. | | |
| 0.059–0.150 | 41 | 11/64 | 0.1370 | 0.1373 | No. 8–36 | 0.1460 | 0.1478 | 0.010* | 1/32 |
| 0.150–0.240 | 41 | 11/64 | 0.1370 | 0.1373 | No. 8–36 | 0.1460 | 0.1478 | 1/64 | 1/32 |
| 0.240–0.365 | 31 | 7/32 | 0.1810 | 0.1813 | No. 12–28 | 0.1928 | 0.1950 | 1/32 | 1/16 |
| 0.365–0.510 | 25 | 17/64 | 0.2150 | 0.2153 | 1/4–28 | 0.2268 | 0.2290 | 1/32 | 3/32 |
| 0.510–0.825 | 7 | 21/64 | 0.2720 | 0.2723 | 5/16–24 | 0.2854 | 0.2878 | 1/16 | 3/32 |
| 0.825–1.135 | 1 | 25/64 | 0.3340 | 0.3344 | 3/8–24 | 0.3479 | 0.3503 | 1/16 | 3/32 |
| 1.135–1.510 | 17/64 | 29/64 | 0.3890 | 0.3894 | 7/16–20 | 0.4050 | 0.4076 | 1/16 | 3/32 |
| 1.510–2.010 | 17/64 | 29/64 | 0.3890 | 0.3894 | 7/16–20 | 0.4050 | 0.4076 | 1/16 | 1/8 |
| 2.010–2.510 | 21/64 | 33/64 | 0.4510 | 0.4515 | 1/2–20 | 0.4675 | 0.4701 | 3/32 | 1/8 |
| 2.510–3.010 | 21/64 | 33/64 | 0.4510 | 0.4515 | 1/2–20 | 0.4675 | 0.4701 | 3/32 | 1/8 |
| 3.010–3.510 | 21/64 | 33/64 | 0.4510 | 0.4515 | 1/2–20 | 0.4675 | 0.4701 | 3/32 | 1/8 |
| 3.510–4.010 | 13/32 | 41/64 | 0.5710 | 0.5715 | 5/8–18 | 0.5889 | 0.5919 | 3/32 | 1/8 |
| 4.010–4.760 | 13/32 | 41/64 | 0.5710 | 0.5715 | 5/8–18 | 0.5889 | 0.5919 | 3/32 | 1/8 |

All dimensions in inches. See Fig. 3 on opposite page.
* Approximate.
Blanks for the range 0.059 to 0.240, incl., may be either counterbored or milled as shown in figure on page 1915.
Use thin gage blanks for all LO (formerly NOT GO) gages. For GO gages use thin blanks for 0.059 to 0.510-inch sizes, all pitches; above 0.510 to 1.135, pitches 12 t. p. i. and finer except for 9/16-12; and above 1.135 to 4.760, pitches 10 t. p. i. and finer. Use thick blanks for above 0.510 to 1.135 inch sizes, coarser than 12 t. p. i.; and above 1.135 to 4.760, coarser than 10 t. p. i.
See Tables 10a and 10b for adjusting screw, locking screw and sleeve dimensions.

**Other American National Standard Gages.** — In addition to the types of gages previously mentioned there are covered by American National Standard ANSI B47.1-1974 the following types of gages (this Standard should be consulted for detailed specifications):

*Taper plug and ring gages for checking taper lock handles and gaging members:* The taper limits established by the American Gage Design Committee for taper lock handles and shanks may be readily maintained by the use of taper plug and ring gages specified for this purpose. A complete set consists of a taper plug, a taper ring, and a taper check plug for each size range.

*Involute, serrated, and straight-sided spline plug and ring gage blanks:* A series of blanks for producing gages used in the gaging of internal splines comprises the

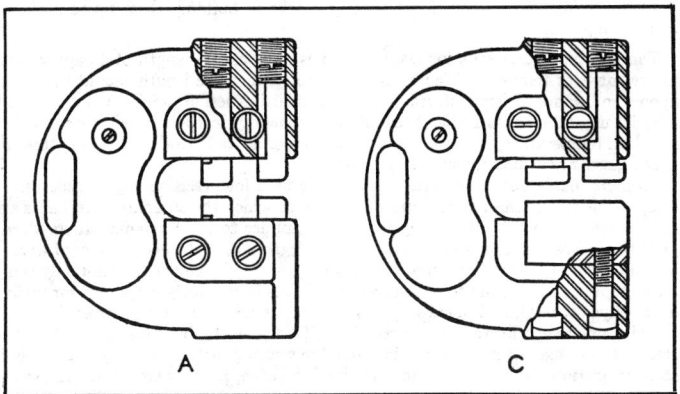

Fig. 4.

following: Spline plug gages and relieved-type spline plug gages, with integral handles, for sizes to and including 2 inches; spline gaging members, pilot-type spline gaging members, and pilot relieved-type spline gaging members, with handles for same, for sizes above 1.5 to and including 8 inches; spline tapered tooth plug gages, with either integral or detachable handles, for sizes to and including 8 inches; plug gage stands; and spline tapered tooth plug gages for sizes which are up to and including 8 inches.

The series of blanks for producing gages used in the gaging of external splines comprises the following: Spline ring gages and pilot-type spline ring gages for sizes to and including 8 inches; prong-type spline relation ring gages in two models, one for space widths greater than 0.070 inch and one for widths of 0.070 inch or less; double-end prong-type spline alignment ring gages for sizes above 1 inch to and including 6 inches; built-up double-end snap gages for tooth thicknesses to and including 0.750 inch.

*Machine taper plug and ring gage blanks:* Gage blanks for Morse, Brown & Sharpe, Jarno, ASA, and Series Nos. 10 to 50 milling machine tapers are also provided in this Standard.

**Glass Gages.** — Glass gages have proved satisfactory for gaging parts of steel and brass, but they are not recommended for aluminum. The cost of glass gages is relatively low compared with steel. They also resist corrosion and have low thermal conductivity. If a glass gage is dropped, breakage may occur, but this is preferable to distortion which, in the case of a steel gage, is not always apparent.

**Sapphire Gages.** — The sapphire is next to the diamond in the mineral hardness scale; hence, plug and ring gages of sapphire offer exceptional resistance to abrasion and wear; in fact they may be made without wear allowance. Sapphire gages resist ordinary shocks and are not subject to burring. Exceptional gaging records have been made with sapphire gages.

**Carbide Gages.** — Cemented carbide is another wear and corrosion-resistant material that has proven successful for gage blocks, plug gages and ring gages. Carbide gage blocks are practically non-magnetic and their adhesive factor facilitates the wringing together of the thinner blocks in building them up to a given dimension.

**Temperature Standards for Gages.** — Inasmuch as the length of a gage varies somewhat with temperature changes, it is evident that the length should be based upon some standard temperature. In the standardization of precision gages for in-dustrial use, 68 degrees F. has been adopted generally in the United States as the standard temperature, because it is the common or average working temperature to which gages are ordinarily subjected in practice.

Formerly 62 degrees F. was the temperature used for precision gage standardization, as this is the temperature, approximately, at which the standard yard bar is at the correct length. While 62 degrees F. still applies to the fundamental standard yard bar in Washington, a temperature of 68 degrees F. is the generally used work-ing standard for the calibration of industrial gages. This temperature not only con-forms to average working temperatures, but it has been widely employed for many other physical tests, and moreover, it is the exact equivalent of 20 degrees C.

This same temperature of 20 degrees C., or 68 degrees F., has been adopted as the standard for gage work and other industrial measuring instruments, by engineering standardization bodies in Germany, Holland, Sweden, and Switzerland. In Great Britain 62 degrees F., applies to the fundamental standard yard bar, but 68 degrees F. is the temperature for industrial gage and instrument calibration.

*Standard for Metric Instruments:* Two temperatures — 0 degrees C. (32 degrees F.) and 20 degrees C. (68 degrees F.) — are employed for the industrial standardization of metric measuring instruments. The 0 degrees C. temperature is the standard at which the fundamental standard meter bar is of correct length, but as this temper-ature is far below ordinary working temperatures, materials having different co-efficients of expansion would show measurable differences in length when the tem-peratures were increased from 0 degrees C. to ordinary working temperatures. For this reason the director of the International Bureau of Weights and Measures recom-mended the following practice, which, incidentally, has been very generally adopted in France.

Gages and other measuring instruments used in the manufacture of metal parts should be so made that when calibrated at a temperature of 20 degrees C., they will have an assumed coefficient of expansion of eleven millionths per unit of length per degree centigrade. In other words, at 20 degrees C. the actual length of such stand-ards will be 220 millionths per unit of length longer than the corresponding subdi-vision of the fundamental standard of the meter at 0 degrees C. This assumed co-efficient of eleven millionths is approximately correct for steel and cast iron, and the error due to the difference between this arbitrary coefficient and the actual coef-ficient of ordinary gage materials is so small that it may safely be ignored in industrial gage standardization.

**Inspection of Tapped Holes.** — One method of inspecting tapped holes is to inspect the tap first and then test the tapped holes periodically with " Go " and " Not Go " gages. The tap can be checked for wear by testing the tapped holes with a " Go " gage. A generally accepted plan consists of using a " Go " thread plug gage and " Not Go " thread plug gage for the minor diameter. Taps may be inspected by measuring the various elements, such as the pitch diameter, angle, and lead. Another method consists in tapping a hole with each tap before it is placed in use; the tapped hole is then checked with a " Go " and " Not Go " plug gage. " Go " and " Not Go " plain cylindrical plug gages are used for inspecting the minor diameter of the tapped hole. Plain ring or snap gages are used for inspecting the major diameter of the screw. When used, it is recommended that the " Go " in-spection gage be a ring gage and the " Not Go " inspection gage be a snap gage The working gages may be combined as a " Go " and " Not Go " snap gage.

**Specifications for Thread Gages.** — The following specifications of the National Screw Thread Commission will be helpful in the design and construction of gages used for producing threaded work. Specific information regarding the tolerances for working gages, inspection gages, and "setting" gages will be found on pages 1402 to 1408. The table on 1405 includes two grades of standard tolerances.

*Gage Steel.* — Gages may be made of a good grade of machine steel pack-hardened, or of straight carbon steel of not less than 1 per cent carbon; or preferably of an oil-hardening steel of approximately 1.10 per cent carbon and 1.40 per cent chromium. The handles should be made of a good grade of machine steel plainly marked to identify the gage.

*Plug Gages.* — All plug gages, whether plain or threaded, should be single-ended. Plug gages of 2 inches and less in diameter should be made with a plug inserted in the handle and fastened thereto by means of a pin. Plug gages of more than 2 inches in diameter should have the gaging blank so made as to be reversible. This can be accomplished by having a finished hole in the gage blank fitting a shouldered projection on the end of the handle, the gage blank being held on with a nut and keyed in the case of a threaded plug gage. The "go" plug gage should be noticeably longer than the "not go" gage, or some distinguishing feature in the design of the handle should be used to serve as a ready means of identification, such as a chamfer on the handle of the "go" gage.

*Plain Ring Gages.* — Both the "go" and "not go" gages should have their outside diameters knurled if made circular. The "go" gage should have a decided chamfer in order to provide a ready means of identification for distinguishing the "go" from the "not go" gage.

*Snap Gages.* — Snap gages may be either adjustable or nonadjustable. It is recommended that all snap gages up to and including ⅛ inch be of the built-up type. For larger snap gages, forged blanks, flat plate stock or other suitable construction may be used. Sufficient clearance beyond the mouth of the gage should be provided to permit the gaging of cylindrical work. Snap gages for measuring lengths and diameters may have one gaging dimension only, or may have a maximum and minimum gaging dimension, both on one end, or maximum and minimum gaging dimension on opposite ends of the gage. When the maximum and minimum gaging dimensions are placed on opposite ends of the gage, the maximum or "go" end of the snap gage should be distinguished from the minimum or "not go" end by having the corners of the gage on the "not go" end decidedly chamfered.

*Plug Thread Gages.* — End threads on plug thread gages should not be chamfered, but the first half turn of the end thread should be flattened to avoid a feather edge.

*Dirt Grooves.* — Inspection and working thread plug gages should be provided with dirt grooves which extend into the gage for a depth of from one to four threads.

*Length of Thread.* — The length of thread parallel to the axis of the gage should, for all standard "go" thread plug and ring gages, be at least as much as the quantity expressed in the following formula, in which $L$ = length of thread and $D$ = basic major diameter of thread: $L = 1.5\,D$.

For threaded work of shorter length of engagement than $1.5\,D$, the length of thread on the "go" gage may be correspondingly shorter.

*"Not Go" Gage for Pitch Diameter only.* — All "not go" thread plug gages should be made to check the pitch diameter only. This necessitates removal of the crest of the thread so that the dimension of the major diameter is never greater than that specified for the "go" gage, and also removing the portion of the thread at the root of the standard thread form.

*Ring Thread Gages.* — All ring thread gages should be made adjustable. The "go" gage should be distinguished from the "not go" gage by having a decided chamfer, and both gages should have their outside diameter knurled if made circular. The end threads on ring thread gages should not be chamfered but the first half turn of the end thread should be flattened to avoid a feather edge.

"Not go" ring thread gages should be made to check the pitch diameter only. This necessitates removal of the crest of the thread (so that the dimension of the minor diameter is never less than that specified for the maximum or "go" gage) and also removing the portion of the thread at the root of the standard form.

**The Marking of Gages.** — The maximum and minimum limits or sizes of gages may be marked in different ways. In the case of a plug gage, for example, the larger end may be marked either "max." (maximum), "high," " + " (plus), or "not go," while the small end would be marked "min." (minimum), "low," " − " (minus), or "go." In the case of a snap gage, the maximum dimension would be marked "max.," "high," " +," or "go," and the minimum, "min.," "low," " −," or "not go." The markings "max.," "min.," "high," and "low" refer to the dimension, while the markings "go" and "not go" refer to the use of the gage. When plug gages are marked "max." and "min.," it is evident that the "min." size is intended to pass into the hole while the "max." size is not supposed to enter. With a snap gage, however, the conditions are reversed: the "max." size should pass over the shaft, while the "min." size should not. Were the gages marked "go" and "not go," the meaning of these words would, in both cases, be the same, which is an advantage. That part of a gage marked "go" would pass over or into the work, while the part marked "not go" would not pass over or into the work.

Working and inspection double-ended plug gages should have the "go" end longer than the "not go" end. Working and inspection double-ended snap gages should have the "go" end rounded to a radius of about ⅛ inch, while the "not go" end should be beveled for a distance of about ⅛ inch. This makes it possible to see at a glance which is the "go" and which the "not go" end.

In marking the sizes on gages, the marking, when expressed in inches, should always be carried to at least three decimals, whether the last decimal is a 0 or not; for example, 0.370 and 0.200, etc. When the exact size requires more than three decimals, as, for example, 0.5798, the required number of decimals should, of course, be stamped on the gage. When the size is expressed in millimeters, the marking should be carried to at least two decimals. For example, 6.00 and 7.40; and if more than two decimals are required to express the exact size, the required number of decimals will, of course, be given.

**Allowance for Lapping Thread Gages.** — The allowance for lapping usually varies from 0.0002 to 0.0005 inch, although, in some cases, the allowance may be as high as 0.001 inch or more, the amount left for lapping increasing with the size of the gage.

As to the material for laps, some gage-makers prefer cast iron and others, soft steel. It is essential to use laps which are accurate as to lead and thread form, although some laps intended for correcting errors have thread angles which are slightly greater or less than the standard, the object being to change the angle of the gage thread more readily.

**Abrasives for Lapping Gages.** — Flour of emery is extensively used for lapping gages, and artificial abrasives are also used. The abrasive is mixed with some oil such as lard oil, sperm oil, or possibly kerosene oil. When a very slow cutting abrasive is required and the amount to be removed by lapping is small, rouge and lard oil may be used. Information regarding the different kinds of artificial abrasives adapted for lapping may be obtained from the manufacturers. After lapping a gage, it should be washed in gasoline before measuring it. If the gage has been heated appreciably as the result of lapping, it should be cooled in water down to the room temperature before measuring.

**American National Standard Dial Indicators** (ANSI B47.1-1974)

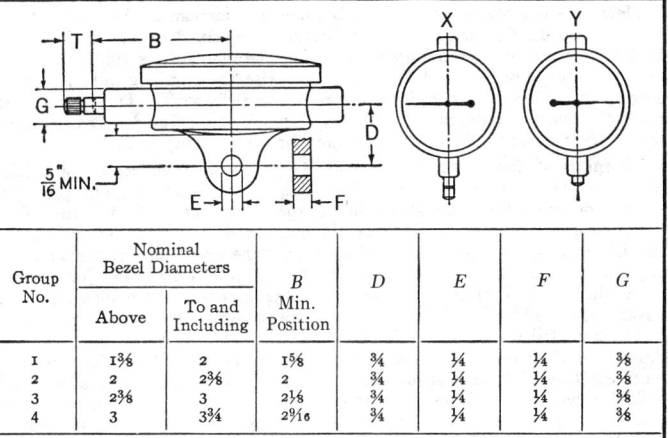

| Group No. | Nominal Bezel Diameters | | B Min. Position | D | E | F | G |
|---|---|---|---|---|---|---|---|
| | Above | To and Including | | | | | |
| 1 | 1⅜ | 2 | 1⅝ | ¾ | ¼ | ¼ | ⅜ |
| 2 | 2 | 2⅜ | 2 | ¾ | ¼ | ¼ | ⅜ |
| 3 | 2⅜ | 3 | 2⅛ | ¾ | ¼ | ¼ | ⅜ |
| 4 | 3 | 3¾ | 2⁹⁄₁₆ | ¾ | ¼ | ¼ | ⅜ |

All dimensions in inches.     C is ¼ inch.

These standard basic dimensions are to permit mounting, interchangeably, dial gages of various makes and models. To secure uniformity between total spindle movement $T$ and the magnification, total travel $T$ should be equivalent to 2½ revolutions of indicator hand, except for special applications requiring greater movement. The indicator hand should swing to the 9 o'clock position or ¼ revolution to left of zero (as shown at $X$) when spindle is in its outer or extended position, to permit measuring on both the plus and minus sides of zero without making a full revolution of the indicating hand. Diagram $Y$ shows the position of the hand when the spindle has moved its full range inward.

**Spirit Levels.** — The accuracy of a spirit level depends upon the curvature of the glass tube. This tube is ground on the inside to a barrel shape, except in cheap levels which have a tube bent to the approximate curve. The bent tube type is only for work which does not require great accuracy. The tube is nearly filled with spirits of wine, ether, or some similar fluid and is hermetically sealed at each end. The larger radius of curvature the glass has, the more sensitive will be the level. The following table gives the curvature for various degrees of sensitiveness, the graduations being in tenths of an inch:

| | Seconds | | | | | Min. | Deg. |
|---|---|---|---|---|---|---|---|
| Angular Value of Each Graduation | 2 | 5 | 10 | 20 | 30 | 1 | 1 |
| Corresponding Diameter of Curvature in Feet | 1718 | 687 | 343 | 171 | 114 | 57 | 0.95 |

The air space in a ground glass is much longer than in a bent one, being ordinarily from ¼ to ⅓ the length of the tube. Modern levels are graduated to tenths and twentieths of an inch, except when they are divided according to the metric system. The leveling glass or "bubble" is generally fixed in a brass tube with plaster-of-paris. This method is satisfactory for all levels having an accuracy of about five seconds angular measurement to each one-tenth inch graduation. For finer levels, it is better to fix one end only with plaster-of-paris and the other end with cork.

# PUNCHES, DIES, AND PRESS WORK

**Clearance between Punches and Dies.** — The amount of clearance between a punch and die for blanking and perforating is governed by the thickness and kind of stock to be operated upon. For thin material such as tin, for example, the punch should be a close sliding fit, as, otherwise, the punching will have ragged edges, but for heavier stock there should be some clearance. The clearance between the punch and die in cutting heavy material, lessens the danger of breaking the punch and reduces the pressure required for the punching operation.

**Meaning of the Term "Clearance."** — There is a difference of opinion among diemakers as to the method of designating clearance. The prevailing practice of fifteen firms specializing in die work is as follows: Ten of these firms define clearance as the space between the punch and die on *one side*, or one-half the difference between the punch and die sizes. The remaining five firms consider clearance as the total difference between the punch and die sizes; for example, if the die is round, clearance equals die diameter minus punch diameter. The advantage of designating clearance as the space on each side is particularly evident in the case of dies of irregular form or of angular shape. While the practice of designating clearance as the difference between the punch and die diameters, may be satisfactory in the case of round dies, it leads to confusion when the dies are of special unsymmetrical forms. The term "clearance" should not be used in specifications without indicating clearly just what it means. According to the practice of one manufacturer of dies, the term "cutting clearance" is used to indicate the space between the punch and die on each side, and the term "die clearance" refers to the angular clearance provided below the cutting edge so that the parts will clear as they fall through the die. The term "clearance" as here used means the space on one side only; hence, for round dies, clearance equals die radius minus punch radius.

**Clearances Generally Allowed.** — For brass and soft steel, most dies are given a clearance on one side equal to the stock thickness multiplied by 0.05 or 0.06; but one-half of this clearance is preferred for some classes of work, and a clearance equal to the stock thickness multiplied by 0.10 may give the cleanest fracture for certain other operations such, for example, as punching holes in ductile steel boiler plate.

**Where Clearance is Applied.** — Whether clearance is deducted from the diameter of the punch or added to the diameter of the die depends upon the nature of the work. If a blank of given size is required, the die is made to that size and the punch is made smaller. Inversely, when holes of a given size are required, the punch is made to the diameter wanted and the die is made larger. Therefore, for blanking to a given size, the clearance is deducted from the size of the punch, and for perforating, the clearance is added to the size of the die.

**Effect of Clearance on Working Pressure.** — Clearance not only affects the smoothness of the fracture, but also the pressure required for punching or blanking. This pressure is greatest when the punch diameter is small compared to the thickness of the stock. In one test, for example, a punching pressure of about 32,000 pounds was required to punch ¾-inch holes into ⁵⁄₁₆-inch mild steel plate when the clearance was about 10 per cent. With a clearance of about 4½ per cent, the pressure increased to 33,000 pounds and a clearance of 2¾ per cent resulted in a pressure of 34,500 pounds.

Soft ductile metal requires more clearance than hard metal, although it has been common practice to increase the clearance for the harder metals. In punching holes in fairly hard steel, a clean fracture was obtained with a clearance of only 0.03 times stock thickness.

**Angular Clearance for Dies.** — The amount of angular clearance ordinarily given a blanking die varies from one to two degrees, although dies that are to be used for producing a comparatively small number of blanks are sometimes given a clearance angle of four or five degrees to facilitate making the die quickly. When a large number of blanks are required, a clearance of about one degree is used. There are two methods of giving clearance to dies: In one case the clearance extends to the top face of the die; in the other, there is a space about ⅛ inch below the cutting edge which is left practically straight, or having a very small amount of clearance. For very soft metal, such as soft, thin brass, the first method is employed, but for harder material, such as hard brass, steel, etc., it is better to have a very shallow clearance for a short distance below the cutting edge. When a die is made in this way, thousands of blanks can be cut with little variation in their size, as grinding the die face will not enlarge the hole to any appreciable extent.

**Lubricants for Press Work.** — Dies are often run without lubrication, but they will last longer if oiled slightly. The oil is applied to the stock either from a saturated felt-roller, brush or pad, or by coating one sheet thickly and then feeding it through the rolls. By the latter method, the rolls are coated with sufficient lubricant for a number of sheets, and a very thin coat is applied to the material so that the work does not have to be cleaned, as is sometimes necessary when a felt-roller or pad is used. Lard or sperm oil is used when punching iron, steel or copper. For drawing steel, the following mixture is recommended: 25 per cent flaked graphite; 25 per cent beef tallow; and 50 per cent lard oil. This mixture should be heated and the work dipped into it. Oildag mixed with heavy grease is also used for steel, and a thin mixture of grease (preferably tallow) and white lead has proved satisfactory. The following compound is also used for drawing sheet steel of a mild grade: Mix one pound of white lead, one quart of fish oil, three ounces of black lead, and one pint of water. These ingredients should be boiled until thoroughly mixed. For drawing brass and copper, a solution obtained by dissolving soap in hot water is often used. (Ivory soap has given good results.) The quantity of soap to use depends upon the thickness of the metal, a thin solution being preferable for thin stock. For cutting aluminum, use kerosene, and for drawing aluminum, use kerosene or vaseline of a cheap grade. Lard oil is also applied to aluminum when drawing deep shells. Aluminum should never be worked without a lubricant. For many classes of die work, no lubricant is required, especially when the metal is of a "greasy" nature, like tin plate, for instance.

**Annealing Drawn Shells.** — When drawing steel, iron, brass or copper, annealing is necessary after two or three draws have been made, as the metal is hardened by the drawing process. For steel and brass, anneal between every other reduction, at least. Tin plate or stock that cannot be annealed without spoiling the finish must ordinarily be drawn to size in one or two operations. Aluminum can be drawn deeper and with less annealing than the other commercial metals, provided the proper grade is used. In case it is necessary to anneal aluminum, this can be done by heating it in a muffle furnace, care being taken to see that the temperature does not exceed 700 degrees F.

**Drawing Brass.** — When drawing brass shells or cup-shaped articles, it is usually possible to make the depth of the first draw equal to the diameter of the shell. By heating brass to a temperature just below what would show a dull red in a dark room, it is possible to draw difficult shapes, otherwise almost impossible, and to get shapes with square corners.

**Drawing Rectangular Shapes.** — When square or rectangular shapes are to be drawn, the radius of the corners should be as large as possible, because it is in the corners that defects occur when drawing. Moreover, the smaller the radius, the less the depth which can be obtained in the first draw. The maximum depths which can be drawn with corners of a given radii are approximately as follows: With a radius of ³⁄₃₂ to ³⁄₁₆ inch, depth of draw, 1 inch; radius ³⁄₁₆ to ³⁄₈ inch, depth 1½ inch; radius ³⁄₈ to ½ inch, depth, 2 inches; radius ½ to ¾ inch, depth, 3 inches. These figures are taken from actual practice and can doubtless be exceeded slightly when using extra good metal. If the box needs to be quite deep and the radius is quite small, two or more drawing operations will be necessary.

**Speeds and Pressures for Presses.** — The speeds for presses equipped with cutting dies depend largely upon the kind of material being worked, and its thickness. For punching and shearing ordinary metals not over ¼ inch thick, the speeds usually range between 50 and 200 strokes per minute, 100 strokes per minute being a fair average. For punching metal over ¼ inch thick, geared presses with speeds ranging from 25 to 75 strokes per minute are commonly employed.

The cutting pressures required depend upon the shearing strength of the material, and the actual area of the surface being severed. For round holes the pressure required equals the circumference of the hole × the thickness of the stock × the shearing strength. To allow for some excess pressure, the tensile strength may be substituted for the shearing strength; the tensile strength for these calculations may be roughly assumed as follows: Mild steel, 60,000 pounds per square inch; wrought iron, 50,000 pounds; bronze, 40,000 pounds; copper, 30,000 pounds; aluminum, 20,000 pounds; zinc, 10,000 pounds; tin and lead, 5,000 pounds.

**Pressure required for Punching.** — The following approximate rule may be used for rapidly finding the pressure in tons required for punching circular holes in sheet steel: Multiply the diameter of the hole in inches by the thickness of the sheet steel and multiply this product by 80. The result is the pressure in tons required. To find the pressure required for punching holes in brass, multiply the diameter of the hole by the thickness, and multiply this product by 65.

*Example:* — What pressure is required for punching a hole 2 inches in diameter through ¼-inch steel stock? According to the rule, 2 × ¼ × 80 = 40 tons.

If a hole is not circular, use as a factor, instead of the diameter of the hole, one-third of the total length of the outline of the hole to be punched.

*Example:* — What pressure is required for punching a 1 inch square hole in ¼ inch thick steel? According to the rule, the total length of the outline of the square is 4 inches. One-third of this is 1⅓, and the pressure required is equal to 1⅓ × ¼ × 80 = 26⅔ tons.

*Example:* — What pressure is required for punching a 1 by 2 inch rectangular hole in ¼ inch thick brass? According to the rule, the total length of the outline of the rectangle is 6 inches. One-third of this is 2, and the pressure required is equal to 2 × ¼ × 65 = 32½ tons.

**Shut Height of Press.** — The term "shut height" as applied to power presses, indicates the die space when the slide is at the bottom of its stroke and the slide connection has been adjusted upward as far as possible. The "shut height" is the distance from the lower face of the slide, either to the top of the bed or to the top of the bolster plate, there being two methods of determining it; hence, this term should always be accompanied by a definition explaining its meaning. According to one press manufacturer, the safest plan is to define "shut height" as the distance from the top of the bolster to the bottom of the slide, with the stroke down and the adjustment up, because most dies are mounted on bolster plates of standard thickness, and a misunderstanding which results in providing too much die space is less serious than having insufficient die space. It is believed that the expression "shut height" was applied first to dies rather than to presses, the shut height of a die being the distance from the bottom of the lower section to the top of the upper section or punch, excluding the shank, and measured when the punch is in the lowest working position.

**Diameters of Shell Blanks.** — The diameters of blanks for drawing plain cylindrical shells can be obtained from the accompanying table, which gives a very close approximation for thin stock. The blank diameters given in this table are for sharp-cornered shells and are found by the following formula:

$$D = \sqrt{d^2 + 4\,dh}, \tag{1}$$

in which $D$ = diameter of flat blank; $d$ = diameter of finished shell; $h$ = height of finished shell.

*Example:* — If the diameter of the finished shell is to be 1.5 inch, and the height, 2 inches, the trial diameter of the blank would be found as follows:

$$D = \sqrt{1.5^2 + 4 \times 1.5 \times 2} = \sqrt{14.25} = 3.78 \text{ inches.}$$

For a round-cornered cup, the following formula, in which $r$ equals the radius of the corner, will give fairly accurate diameters, provided the radius does not exceed, say, ¼ the height of the shell:

$$D = \sqrt{d^2 + 4\,dh} - r. \tag{2}$$

These formulas are based on the assumption that the thickness of the drawn shell is the same as the original thickness of the stock, and that the blank is so proportioned that its area will equal the area of the drawn shell. This method of calculating the blank diameter is quite accurate for thin material, when there is only a slight reduction in the thickness of the metal incident to drawing; but when heavy stock is drawn and the thickness of the finished shell is much less than the original thickness of the stock, the blank diameter obtained from Formulas (1) or (2) will be too large, because when the stock is drawn thinner, there is an increase in area. When an appreciable reduction in thickness is to be made, the blank diameter can be obtained by first determining the "mean height" of the drawn shell by the following formula. This formula is only approximately correct, but will give results sufficiently accurate for most work:

$$M = \frac{ht}{T} \tag{3}$$

in which $M$ = approximate mean height of drawn shell; $h$ = height of drawn shell; $t$ = thickness of shell; $T$ = thickness of metal before drawing.

After determining the mean height, the blank diameter for the required shell diameter is obtained from the table previously referred to, the mean height being used instead of the actual height.

*Example:* — Suppose a shell 2 inches in diameter and 3¾ inches high is to be drawn, and that the original thickness of the stock is 0.050 inch, and thickness of drawn shell, 0.040 inch. To what diameter should the blank be cut? Using Formula (3) to obtain the mean height:

$$M = \frac{ht}{T} = \frac{3.75 \times 0.040}{0.050} = 3 \text{ inches.}$$

According to the table, the blank diameter for a shell 2 inches in diameter and 3 inches high is 5.29 inches. This formula is accurate enough for all practical purposes, unless the reduction in the thickness of the metal is greater than about one-fifth the original thickness. When there is considerable reduction, a blank calculated by this formula produces a shell that is too long. This, however, is an error in the right direction, as the edges of drawn shells are ordinarily trimmed.

If the shell has a rounded corner, the radius of the corner should be deducted from the figures given in the table. For example, if the shell referred to in the foregoing example had a corner of ¼-inch radius, the blank diameter would equal 5.29 — 0.25 = 5.04 inches.

Another formula which is sometimes used for obtaining blank diameters for shells, when there is a reduction in the thickness of the stock, is as follows:

$$D = \sqrt{a^2 + \left(a^2 - b^2\right)\frac{h}{t}} \qquad (4)$$

In this formula $D$ = blank diameter; $a$ = outside diameter; $b$ = inside diameter; $t$ = thickness of shell at bottom; $h$ = depth of shell. This formula is based on the cubic contents of the drawn shell. It is assumed that the shells are cylindrical, and no allowance is made for a rounded corner at the bottom, or for trimming the shell after drawing. To allow for trimming, add the required amount to depth $h$. When a shell is of irregular cross-section, if its weight is known, the blank diameter can be determined by the following formula:

$$D = 1.1284 \sqrt{\frac{W}{wt}} \qquad (5)$$

in which $D$ = blank diameter in inches; $W$ = weight of shell; $w$ = weight of metal per cubic inch; $t$ = thickness of the shell.

In the construction of dies for producing shells, especially of irregular form, a common method of procedure is to make the drawing parts first. The actual blank diameter can then be determined by trial. One method is to cut a trial blank as near to size as can be estimated. The outline of this blank is then scribed on a flat sheet, after which the blank is drawn. If the finished shell shows that the blank is not of the right diameter, a new trial blank is cut either larger or smaller than the size indicated by the line previously scribed, this line acting as a guide. If a model shell is available, the blank diameter can also be determined as follows: First cut a blank somewhat large, and from the same material used for making the model; then, reduce the size of the blank until its weight equals the weight of the model.

**Depth and Diameter Reductions of Drawn Shells.** —The depth to which metal can be drawn in one operation depends upon the quality and kind of material, its thickness, the slant or angle of the dies, and the amount that the stock is thinned or "ironed" in drawing. A general rule for determining the depth to which cylindrical shells can be drawn in one operation is as follows: The depth or length of the first draw should never be greater than the diameter of the shell. If the shell is to have a flange at the top, it may not be practicable to draw as deeply as is indicated by this rule, unless the metal is extra good, because the stock is subjected to a higher tensile stress, owing to the larger blank which is necessary for forming the flange. According to another rule, the depth given the shell on the first draw should equal one-third the diameter of the blank. Ordinarily, it is possible to draw sheet steel of any thickness up to ¼ inch, so that the diameter of the first shell equals about six-tenths of the blank diameter. When drawing plain shells, the amount that the diameter is reduced for each draw must be governed by the quality of the metal and its susceptibility to drawing. The reduction for various thicknesses of metal is about as follows:

| Approximate thickness of sheet steel | | | | | | ¹⁄₁₆ | ⅛ | ³⁄₁₆ | ¼ | ⁵⁄₁₆ |
|---|---|---|---|---|---|---|---|---|---|---|
| Possible reduction in diameter for each succeeding step, per cent | | | | | | 20 | 15 | 12 | 10 | 8 |

## Diameters of Blanks for Drawn Shells

Height of Shell

| Diam. of Shell | ¼ | ½ | ¾ | 1 | 1¼ | 1½ | 1¾ | 2 | 2¼ | 2½ | 2¾ | 3 | 3¼ | 3½ | 3¾ | 4 | 4½ | 5 | 5½ | 6 |
|---|---|---|---|---|---|---|---|---|---|---|---|---|---|---|---|---|---|---|---|---|
| ¼ | 0.56 | 0.75 | 0.90 | 1.03 | 1.14 | 1.25 | 1.35 | 1.44 | 1.52 | 1.60 | 1.68 | 1.75 | 1.82 | 1.89 | 1.95 | 2.01 | 2.14 | 2.25 | 2.36 | 2.46 |
| ½ | 0.87 | 1.12 | 1.32 | 1.50 | 1.66 | 1.80 | 1.94 | 2.06 | 2.18 | 2.29 | 2.40 | 2.50 | 2.60 | 2.69 | 2.78 | 2.87 | 3.04 | 3.21 | 3.36 | 3.50 |
| ¾ | 1.14 | 1.44 | 1.68 | 1.89 | 2.08 | 2.25 | 2.41 | 2.56 | 2.70 | 2.84 | 2.97 | 3.09 | 3.21 | 3.33 | 3.44 | 3.54 | 3.75 | 3.95 | 4.13 | 4.31 |
| 1 | 1.41 | 1.73 | 2.00 | 2.24 | 2.45 | 2.65 | 2.83 | 3.00 | 3.16 | 3.32 | 3.46 | 3.61 | 3.74 | 3.87 | 4.00 | 4.12 | 4.36 | 4.58 | 4.80 | 5.00 |
| 1¼ | 1.68 | 2.01 | 2.30 | 2.56 | 2.79 | 3.01 | 3.21 | 3.40 | 3.58 | 3.75 | 3.91 | 4.07 | 4.22 | 4.37 | 4.51 | 4.64 | 4.91 | 5.15 | 5.39 | 5.62 |
| 1½ | 1.94 | 2.29 | 2.60 | 2.87 | 3.12 | 3.36 | 3.57 | 3.78 | 3.97 | 4.15 | 4.33 | 4.50 | 4.66 | 4.82 | 4.98 | 5.12 | 5.41 | 5.68 | 5.94 | 6.18 |
| 1¾ | 2.19 | 2.56 | 2.88 | 3.17 | 3.44 | 3.68 | 3.91 | 4.13 | 4.34 | 4.53 | 4.72 | 4.91 | 5.08 | 5.26 | 5.41 | 5.58 | 5.88 | 6.17 | 6.45 | 6.71 |
| 2 | 2.45 | 2.83 | 3.16 | 3.46 | 3.74 | 4.00 | 4.24 | 4.47 | 4.69 | 4.90 | 5.10 | 5.29 | 5.48 | 5.66 | 5.83 | 6.00 | 6.32 | 6.63 | 6.93 | 7.21 |
| 2¼ | 2.70 | 3.09 | 3.44 | 3.75 | 4.04 | 4.31 | 4.56 | 4.80 | 5.03 | 5.25 | 5.46 | 5.66 | 5.86 | 6.05 | 6.23 | 6.41 | 6.75 | 7.07 | 7.39 | 7.69 |
| 2½ | 2.96 | 3.36 | 3.71 | 4.03 | 4.33 | 4.61 | 4.87 | 5.12 | 5.36 | 5.59 | 5.81 | 6.02 | 6.22 | 6.42 | 6.61 | 6.80 | 7.16 | 7.50 | 7.82 | 8.14 |
| 2¾ | 3.21 | 3.61 | 3.98 | 4.31 | 4.62 | 4.91 | 5.18 | 5.44 | 5.68 | 5.92 | 6.15 | 6.37 | 6.58 | 6.79 | 6.99 | 7.18 | 7.55 | 7.91 | 8.25 | 8.58 |
| 3 | 3.46 | 3.87 | 4.24 | 4.58 | 4.90 | 5.20 | 5.48 | 5.74 | 6.00 | 6.25 | 6.48 | 6.71 | 6.93 | 7.14 | 7.35 | 7.55 | 7.94 | 8.31 | 8.66 | 9.00 |
| 3¼ | 3.71 | 4.13 | 4.51 | 4.85 | 5.18 | 5.48 | 5.77 | 6.04 | 6.31 | 6.56 | 6.80 | 7.04 | 7.27 | 7.49 | 7.70 | 7.91 | 8.31 | 8.69 | 9.06 | 9.41 |
| 3½ | 3.97 | 4.39 | 4.77 | 5.12 | 5.45 | 5.77 | 6.06 | 6.34 | 6.61 | 6.87 | 7.12 | 7.36 | 7.60 | 7.83 | 8.05 | 8.26 | 8.67 | 9.07 | 9.45 | 9.81 |
| 3¾ | 4.22 | 4.64 | 5.03 | 5.39 | 5.73 | 6.05 | 6.35 | 6.64 | 6.91 | 7.18 | 7.44 | 7.69 | 7.92 | 8.16 | 8.38 | 8.61 | 9.03 | 9.44 | 9.83 | 10.20 |
| 4 | 4.47 | 4.90 | 5.29 | 5.66 | 6.00 | 6.32 | 6.63 | 6.93 | 7.21 | 7.48 | 7.75 | 8.00 | 8.25 | 8.49 | 8.72 | 8.94 | 9.38 | 9.80 | 10.20 | 10.58 |
| 4¼ | 4.72 | 5.15 | 5.55 | 5.92 | 6.27 | 6.60 | 6.91 | 7.22 | 7.50 | 7.78 | 8.05 | 8.31 | 8.56 | 8.81 | 9.04 | 9.28 | 9.72 | 10.15 | 10.56 | 10.96 |
| 4½ | 4.98 | 5.41 | 5.81 | 6.19 | 6.54 | 6.87 | 7.19 | 7.50 | 7.79 | 8.08 | 8.35 | 8.62 | 8.87 | 9.12 | 9.37 | 9.60 | 10.06 | 10.50 | 10.92 | 11.32 |
| 4¾ | 5.22 | 5.66 | 6.07 | 6.45 | 6.80 | 7.15 | 7.47 | 7.78 | 8.08 | 8.37 | 8.65 | 8.92 | 9.18 | 9.44 | 9.69 | 9.93 | 10.40 | 10.84 | 11.27 | 11.69 |
| 5 | 5.48 | 5.92 | 6.32 | 6.71 | 7.07 | 7.42 | 7.75 | 8.06 | 8.37 | 8.66 | 8.94 | 9.22 | 9.49 | 9.75 | 10.00 | 10.25 | 10.72 | 11.18 | 11.62 | 12.04 |
| 5¼ | 5.73 | 6.17 | 6.58 | 6.97 | 7.33 | 7.68 | 8.02 | 8.34 | 8.65 | 8.95 | 9.24 | 9.52 | 9.79 | 10.05 | 10.31 | 10.56 | 11.05 | 11.51 | 11.96 | 12.39 |
| 5½ | 5.98 | 6.42 | 6.84 | 7.23 | 7.60 | 7.95 | 8.29 | 8.62 | 8.93 | 9.23 | 9.53 | 9.81 | 10.08 | 10.36 | 10.62 | 10.87 | 11.37 | 11.84 | 12.30 | 12.74 |
| 5¾ | 6.23 | 6.68 | 7.09 | 7.49 | 7.86 | 8.22 | 8.56 | 8.89 | 9.21 | 9.52 | 9.81 | 10.10 | 10.38 | 10.66 | 10.92 | 11.18 | 11.69 | 12.17 | 12.63 | 13.08 |
| 6 | 6.48 | 6.93 | 7.35 | 7.75 | 8.12 | 8.49 | 8.83 | 9.17 | 9.49 | 9.80 | 10.10 | 10.39 | 10.68 | 10.95 | 11.23 | 11.49 | 12.00 | 12.49 | 12.96 | 13.42 |

For example, if a shell made of $\frac{1}{16}$ inch stock is 3 inches in diameter after the first draw, it can be reduced 20 per cent on the next draw, and so on until the required diameter is obtained. These figures are based upon the assumption that the shell is annealed after the first drawing operation, and at least between every two of the following operations. Necking operations — that is, the drawing out of a short portion of the lower part of the cup into a long neck — may be done without such frequent annealings. In double-action presses, where the inside of the cup is supported by a bushing during drawing, the reductions possible may be increased to 30, 24, 18, 15 and 12 per cent, respectively. (The latter figures may also be used for brass in single-action presses.)

When a hole is to be pierced at the bottom of a cup and the remaining metal is to be drawn after the hole has been pierced or punched, always pierce from the opposite direction to that in which the stock is to be drawn after piercing. In extreme cases, it is necessary to machine the metal around the pierced hole in order to prevent the starting of cracks or flaws in the subsequent drawing operations.

The foregoing figures represent conservative practice and it is often possible to make greater reductions than are indicated by these figures, especially when using a good drawing metal. Taper shells require smaller reductions than cylindrical shells, because the metal tends to wrinkle if the shell to be drawn is much larger than the punch. The amount that the stock is "ironed" or thinned out while being drawn must also be considered, because a reduction in gage or thickness means greater pressure of the punch against the bottom of the shell; hence the amount that the shell diameter is reduced for each drawing operation must be lessened when much ironing is necessary. The extent to which a shell can be ironed in one drawing operation ranges between 0.002 and 0.004 inch per side, and should not exceed 0.001 inch on the final draw, if a good finish is required.

**Allowances for Bending Sheet Metal.** — In bending steel, brass, bronze or other metals, the problem is to find the length of straight stock required for each bend; then these lengths are added to the lengths of the straight sections to obtain the total length of the material before bending.

If $L$ = length, in inches, of straight stock required before bending; $T$ = thickness in inches; $R$ = inside radius of bend in inches.

For 90-degree bends in soft brass and soft copper

$$\text{Table 1 or } L = (0.55 \times T) + (1.57 \times R) \tag{1}$$

For 90-degree bends in half-hard copper and brass, soft steel, and aluminum

$$\text{Table 2 or } L = (0.64 \times T) + (1.57 \times R) \tag{2}$$

For 90-degree bends in bronze, hard copper, cold-rolled steel and spring steel

$$\text{Table 3 or } L = (0.71 \times T) + (1.57 \times R) \tag{3}$$

*Angle of Bend Other Than 90 Degrees:* For angles other than 90 degrees, find length $L$, using tables or formulas, and multiply $L$ by angle of bend, in degrees, divided by 90, to find length of stock before bending. In using this rule, note that *angle of bend* is the angle through which the material has actually been bent; hence, it is not in all cases the angle as given on a drawing. To illustrate, in Fig. 1 (see diagram), the angle on the drawing is 60 degrees, but the angle of bend $A$ is 120 degrees ($180 - 60 = 120$); in Fig. 2, the angle of bend $A$ is 60 degrees; in Fig. 3, angle $A$ is $90 - 30 = 60$ degrees. The Formulas (1), (2) and (3) are based upon extensive experiments of the Westinghouse Electric & Mfg. Co. They apply to parts bent with simple tools or on the bench, where limits of plus or minus $\frac{1}{64}$ inch are specified. If a part has two or more bends of the same radius, it is, of course, only necessary to obtain the length required for one of the bends and then multiply by the number of bends, thus obtaining the total allowance for the bent sections.

## Table 1. Lengths of Straight Stock Required for 90-Degree Bends in Soft Copper and Soft Brass

| Radius R of Bend, Inches | Thickness T of Material, Inch | | | | | | | | | | | | |
|---|---|---|---|---|---|---|---|---|---|---|---|---|---|
| | 1/64 | 1/32 | 3/64 | 1/16 | 5/64 | 3/32 | 1/8 | 5/32 | 3/16 | 7/32 | 1/4 | 9/32 | 5/16 |
| 1/32 | 0.058 | 0.066 | 0.075 | 0.083 | 0.092 | 0.101 | 0.118 | 0.135 | 0.152 | 0.169 | 0.187 | 0.204 | 0.221 |
| 3/64 | 0.083 | 0.091 | 0.100 | 0.108 | 0.117 | 0.126 | 0.143 | 0.160 | 0.177 | 0.194 | 0.212 | 0.229 | 0.246 |
| 1/16 | 0.107 | 0.115 | 0.124 | 0.132 | 0.141 | 0.150 | 0.167 | 0.184 | 0.201 | 0.218 | 0.236 | 0.253 | 0.270 |
| 3/32 | 0.156 | 0.164 | 0.173 | 0.181 | 0.190 | 0.199 | 0.216 | 0.233 | 0.250 | 0.267 | 0.285 | 0.302 | 0.319 |
| 1/8 | 0.205 | 0.213 | 0.222 | 0.230 | 0.239 | 0.248 | 0.265 | 0.282 | 0.299 | 0.316 | 0.334 | 0.351 | 0.368 |
| 5/32 | 0.254 | 0.262 | 0.271 | 0.279 | 0.288 | 0.297 | 0.314 | 0.331 | 0.348 | 0.365 | 0.383 | 0.400 | 0.417 |
| 3/16 | 0.303 | 0.311 | 0.320 | 0.328 | 0.337 | 0.346 | 0.363 | 0.380 | 0.397 | 0.414 | 0.432 | 0.449 | 0.466 |
| 7/32 | 0.353 | 0.361 | 0.370 | 0.378 | 0.387 | 0.396 | 0.413 | 0.430 | 0.447 | 0.464 | 0.482 | 0.499 | 0.516 |
| 1/4 | 0.401 | 0.409 | 0.418 | 0.426 | 0.435 | 0.444 | 0.461 | 0.478 | 0.495 | 0.512 | 0.530 | 0.547 | 0.564 |
| 9/32 | 0.450 | 0.458 | 0.467 | 0.475 | 0.484 | 0.493 | 0.510 | 0.527 | 0.544 | 0.561 | 0.579 | 0.596 | 0.613 |
| 5/16 | 0.499 | 0.507 | 0.516 | 0.524 | 0.533 | 0.542 | 0.559 | 0.576 | 0.593 | 0.610 | 0.628 | 0.645 | 0.662 |
| 11/32 | 0.549 | 0.557 | 0.566 | 0.574 | 0.583 | 0.592 | 0.609 | 0.626 | 0.643 | 0.660 | 0.678 | 0.695 | 0.712 |
| 3/8 | 0.598 | 0.606 | 0.615 | 0.623 | 0.632 | 0.641 | 0.658 | 0.675 | 0.692 | 0.709 | 0.727 | 0.744 | 0.761 |
| 13/32 | 0.646 | 0.654 | 0.663 | 0.671 | 0.680 | 0.689 | 0.706 | 0.723 | 0.740 | 0.757 | 0.775 | 0.792 | 0.809 |
| 7/16 | 0.695 | 0.703 | 0.712 | 0.720 | 0.729 | 0.738 | 0.755 | 0.772 | 0.789 | 0.806 | 0.824 | 0.841 | 0.858 |
| 15/32 | 0.734 | 0.742 | 0.751 | 0.759 | 0.768 | 0.777 | 0.794 | 0.811 | 0.828 | 0.845 | 0.863 | 0.880 | 0.897 |
| 1/2 | 0.794 | 0.802 | 0.811 | 0.819 | 0.828 | 0.837 | 0.854 | 0.871 | 0.888 | 0.905 | 0.923 | 0.940 | 0.957 |
| 9/16 | 0.892 | 0.900 | 0.909 | 0.917 | 0.926 | 0.935 | 0.952 | 0.969 | 0.986 | 1.003 | 1.021 | 1.038 | 1.055 |
| 5/8 | 0.990 | 0.998 | 1.007 | 1.015 | 1.024 | 1.033 | 1.050 | 1.067 | 1.084 | 1.101 | 1.119 | 1.136 | 1.153 |
| 11/16 | 1.089 | 1.097 | 1.106 | 1.114 | 1.123 | 1.132 | 1.149 | 1.166 | 1.183 | 1.200 | 1.218 | 1.235 | 1.252 |
| 3/4 | 1.187 | 1.195 | 1.204 | 1.212 | 1.221 | 1.230 | 1.247 | 1.264 | 1.281 | 1.298 | 1.316 | 1.333 | 1.350 |
| 13/16 | 1.286 | 1.294 | 1.303 | 1.311 | 1.320 | 1.329 | 1.346 | 1.363 | 1.380 | 1.397 | 1.415 | 1.432 | 1.449 |
| 7/8 | 1.384 | 1.392 | 1.401 | 1.409 | 1.418 | 1.427 | 1.444 | 1.461 | 1.478 | 1.495 | 1.513 | 1.530 | 1.547 |
| 15/16 | 1.481 | 1.489 | 1.498 | 1.506 | 1.515 | 1.524 | 1.541 | 1.558 | 1.575 | 1.592 | 1.610 | 1.627 | 1.644 |
| 1 | 1.580 | 1.588 | 1.597 | 1.605 | 1.614 | 1.623 | 1.640 | 1.657 | 1.674 | 1.691 | 1.709 | 1.726 | 1.743 |
| 1 1/16 | 1.678 | 1.686 | 1.695 | 1.703 | 1.712 | 1.721 | 1.738 | 1.755 | 1.772 | 1.789 | 1.807 | 1.824 | 1.841 |
| 1 1/8 | 1.777 | 1.785 | 1.794 | 1.802 | 1.811 | 1.820 | 1.837 | 1.854 | 1.871 | 1.888 | 1.906 | 1.923 | 1.940 |
| 1 3/16 | 1.875 | 1.883 | 1.892 | 1.900 | 1.909 | 1.918 | 1.935 | 1.952 | 1.969 | 1.986 | 2.004 | 2.021 | 2.038 |
| 1 1/4 | 1.972 | 1.980 | 1.989 | 1.997 | 2.006 | 2.015 | 2.032 | 2.049 | 2.066 | 2.083 | 2.101 | 2.118 | 2.135 |

Table 2.  Lengths of Straight Stock Required for 90-Degree Bends in Half-Hard Brass and Sheet Copper, Soft Steel and Aluminum

| Radius R of Bend, Inches | Thickness T of Material, Inch | | | | | | | | | | | | |
|---|---|---|---|---|---|---|---|---|---|---|---|---|---|
| | 1/64 | 1/32 | 3/64 | 1/16 | 5/64 | 3/32 | 1/8 | 5/32 | 3/16 | 7/32 | 1/4 | 9/32 | 5/16 |
| 1/32 | 0.059 | 0.069 | 0.079 | 0.089 | 0.099 | 0.109 | 0.129 | 0.149 | 0.169 | 0.189 | 0.209 | 0.229 | 0.249 |
| 3/64 | 0.084 | 0.094 | 0.104 | 0.114 | 0.124 | 0.134 | 0.154 | 0.174 | 0.194 | 0.214 | 0.234 | 0.254 | 0.274 |
| 1/16 | 0.108 | 0.118 | 0.128 | 0.138 | 0.148 | 0.158 | 0.178 | 0.198 | 0.218 | 0.238 | 0.258 | 0.278 | 0.298 |
| 3/32 | 0.157 | 0.167 | 0.177 | 0.187 | 0.197 | 0.207 | 0.227 | 0.247 | 0.267 | 0.287 | 0.307 | 0.327 | 0.347 |
| 1/8 | 0.206 | 0.216 | 0.226 | 0.236 | 0.246 | 0.256 | 0.276 | 0.296 | 0.316 | 0.336 | 0.356 | 0.376 | 0.396 |
| 5/32 | 0.255 | 0.265 | 0.275 | 0.285 | 0.295 | 0.305 | 0.325 | 0.345 | 0.365 | 0.385 | 0.405 | 0.425 | 0.445 |
| 3/16 | 0.304 | 0.314 | 0.324 | 0.334 | 0.344 | 0.354 | 0.374 | 0.394 | 0.414 | 0.434 | 0.454 | 0.474 | 0.494 |
| 7/32 | 0.354 | 0.364 | 0.374 | 0.384 | 0.394 | 0.404 | 0.424 | 0.444 | 0.464 | 0.484 | 0.504 | 0.524 | 0.544 |
| 1/4 | 0.402 | 0.412 | 0.422 | 0.432 | 0.442 | 0.452 | 0.472 | 0.492 | 0.512 | 0.532 | 0.552 | 0.572 | 0.592 |
| 9/32 | 0.451 | 0.461 | 0.471 | 0.481 | 0.491 | 0.501 | 0.521 | 0.541 | 0.561 | 0.581 | 0.601 | 0.621 | 0.641 |
| 5/16 | 0.500 | 0.510 | 0.520 | 0.530 | 0.540 | 0.550 | 0.570 | 0.590 | 0.610 | 0.630 | 0.650 | 0.670 | 0.690 |
| 11/32 | 0.550 | 0.560 | 0.570 | 0.580 | 0.590 | 0.600 | 0.620 | 0.640 | 0.660 | 0.680 | 0.700 | 0.720 | 0.740 |
| 3/8 | 0.599 | 0.609 | 0.619 | 0.629 | 0.639 | 0.649 | 0.669 | 0.689 | 0.709 | 0.729 | 0.749 | 0.769 | 0.789 |
| 13/32 | 0.647 | 0.657 | 0.667 | 0.677 | 0.687 | 0.697 | 0.717 | 0.737 | 0.757 | 0.777 | 0.797 | 0.817 | 0.837 |
| 7/16 | 0.696 | 0.706 | 0.716 | 0.726 | 0.736 | 0.746 | 0.766 | 0.786 | 0.806 | 0.826 | 0.846 | 0.866 | 0.886 |
| 15/32 | 0.746 | 0.756 | 0.766 | 0.776 | 0.786 | 0.796 | 0.816 | 0.836 | 0.856 | 0.876 | 0.896 | 0.916 | 0.936 |
| 1/2 | 0.795 | 0.805 | 0.815 | 0.825 | 0.835 | 0.845 | 0.865 | 0.885 | 0.905 | 0.925 | 0.945 | 0.965 | 0.985 |
| 9/16 | 0.893 | 0.903 | 0.913 | 0.923 | 0.933 | 0.943 | 0.963 | 0.983 | 1.003 | 1.023 | 1.043 | 1.063 | 1.083 |
| 5/8 | 0.991 | 1.001 | 1.011 | 1.021 | 1.031 | 1.041 | 1.061 | 1.081 | 1.101 | 1.121 | 1.141 | 1.161 | 1.181 |
| 11/16 | 1.090 | 1.100 | 1.110 | 1.120 | 1.130 | 1.140 | 1.160 | 1.180 | 1.200 | 1.220 | 1.240 | 1.260 | 1.280 |
| 3/4 | 1.188 | 1.198 | 1.208 | 1.218 | 1.228 | 1.238 | 1.258 | 1.278 | 1.298 | 1.318 | 1.338 | 1.358 | 1.378 |
| 13/16 | 1.287 | 1.297 | 1.307 | 1.317 | 1.327 | 1.337 | 1.357 | 1.377 | 1.397 | 1.417 | 1.437 | 1.457 | 1.477 |
| 7/8 | 1.385 | 1.395 | 1.405 | 1.415 | 1.425 | 1.435 | 1.455 | 1.475 | 1.495 | 1.515 | 1.535 | 1.555 | 1.575 |
| 15/16 | 1.482 | 1.492 | 1.502 | 1.512 | 1.522 | 1.532 | 1.552 | 1.572 | 1.592 | 1.612 | 1.632 | 1.652 | 1.672 |
| 1 | 1.581 | 1.591 | 1.601 | 1.611 | 1.621 | 1.631 | 1.651 | 1.671 | 1.691 | 1.711 | 1.731 | 1.751 | 1.771 |
| 1 1/16 | 1.679 | 1.689 | 1.699 | 1.709 | 1.719 | 1.729 | 1.749 | 1.769 | 1.789 | 1.809 | 1.829 | 1.849 | 1.869 |
| 1 1/8 | 1.778 | 1.788 | 1.798 | 1.808 | 1.818 | 1.828 | 1.848 | 1.868 | 1.888 | 1.908 | 1.928 | 1.948 | 1.968 |
| 1 3/16 | 1.876 | 1.886 | 1.896 | 1.906 | 1.916 | 1.926 | 1.946 | 1.966 | 1.986 | 2.006 | 2.026 | 2.046 | 2.066 |
| 1 1/4 | 1.973 | 1.983 | 1.993 | 2.003 | 2.013 | 2.023 | 2.043 | 2.063 | 2.083 | 2.103 | 2.123 | 2.143 | 2.163 |

**Table 3.  Lengths of Straight Stock Required for 90-Degree Bends in Hard Copper, Bronze, Cold-Rolled Steel, and Spring Steel**

| Radius R of Bend, Inches | Thickness T of Material, Inch | | | | | | | | | | | | |
|---|---|---|---|---|---|---|---|---|---|---|---|---|---|
| | 1/64 | 1/32 | 3/64 | 1/16 | 5/64 | 3/32 | 1/8 | 5/32 | 3/16 | 7/32 | 1/4 | 9/32 | 5/16 |
| 1/32 | 0.060 | 0.071 | 0.082 | 0.093 | 0.104 | 0.116 | 0.138 | 0.160 | 0.182 | 0.204 | 0.227 | 0.249 | 0.271 |
| 3/64 | 0.085 | 0.096 | 0.107 | 0.118 | 0.129 | 0.141 | 0.163 | 0.185 | 0.207 | 0.229 | 0.252 | 0.274 | 0.296 |
| 1/16 | 0.109 | 0.120 | 0.131 | 0.142 | 0.153 | 0.165 | 0.187 | 0.209 | 0.231 | 0.253 | 0.276 | 0.298 | 0.320 |
| 3/32 | 0.158 | 0.169 | 0.180 | 0.191 | 0.202 | 0.214 | 0.236 | 0.258 | 0.280 | 0.302 | 0.325 | 0.347 | 0.369 |
| 1/8 | 0.207 | 0.218 | 0.229 | 0.240 | 0.251 | 0.263 | 0.285 | 0.307 | 0.329 | 0.351 | 0.374 | 0.396 | 0.418 |
| 5/32 | 0.256 | 0.267 | 0.278 | 0.289 | 0.300 | 0.312 | 0.334 | 0.356 | 0.378 | 0.400 | 0.423 | 0.445 | 0.467 |
| 3/16 | 0.305 | 0.316 | 0.327 | 0.338 | 0.349 | 0.361 | 0.383 | 0.405 | 0.427 | 0.449 | 0.472 | 0.494 | 0.516 |
| 7/32 | 0.355 | 0.366 | 0.377 | 0.388 | 0.399 | 0.411 | 0.433 | 0.455 | 0.477 | 0.499 | 0.522 | 0.544 | 0.566 |
| 1/4 | 0.403 | 0.414 | 0.425 | 0.436 | 0.447 | 0.459 | 0.481 | 0.503 | 0.525 | 0.547 | 0.570 | 0.592 | 0.614 |
| 9/32 | 0.452 | 0.463 | 0.474 | 0.485 | 0.496 | 0.508 | 0.530 | 0.552 | 0.574 | 0.596 | 0.619 | 0.641 | 0.663 |
| 5/16 | 0.501 | 0.512 | 0.523 | 0.534 | 0.545 | 0.557 | 0.579 | 0.601 | 0.623 | 0.645 | 0.668 | 0.690 | 0.712 |
| 11/32 | 0.551 | 0.562 | 0.573 | 0.584 | 0.595 | 0.607 | 0.629 | 0.651 | 0.673 | 0.695 | 0.718 | 0.740 | 0.762 |
| 3/8 | 0.600 | 0.611 | 0.622 | 0.633 | 0.644 | 0.656 | 0.678 | 0.700 | 0.722 | 0.744 | 0.767 | 0.789 | 0.811 |
| 13/32 | 0.648 | 0.659 | 0.670 | 0.681 | 0.692 | 0.704 | 0.726 | 0.748 | 0.770 | 0.792 | 0.815 | 0.837 | 0.859 |
| 7/16 | 0.697 | 0.708 | 0.719 | 0.730 | 0.741 | 0.753 | 0.775 | 0.797 | 0.819 | 0.841 | 0.864 | 0.886 | 0.908 |
| 15/32 | 0.736 | 0.747 | 0.758 | 0.769 | 0.780 | 0.792 | 0.814 | 0.836 | 0.858 | 0.880 | 0.903 | 0.925 | 0.947 |
| 1/2 | 0.796 | 0.807 | 0.818 | 0.829 | 0.840 | 0.852 | 0.874 | 0.896 | 0.918 | 0.940 | 0.963 | 0.985 | 1.007 |
| 9/16 | 0.894 | 0.905 | 0.916 | 0.927 | 0.938 | 0.950 | 0.972 | 0.994 | 1.016 | 1.038 | 1.061 | 1.083 | 1.105 |
| 5/8 | 0.992 | 1.003 | 1.014 | 1.025 | 1.036 | 1.048 | 1.070 | 1.092 | 1.114 | 1.136 | 1.159 | 1.181 | 1.203 |
| 11/16 | 1.091 | 1.102 | 1.113 | 1.124 | 1.135 | 1.147 | 1.169 | 1.191 | 1.213 | 1.235 | 1.258 | 1.280 | 1.302 |
| 3/4 | 1.189 | 1.200 | 1.211 | 1.222 | 1.233 | 1.245 | 1.267 | 1.289 | 1.311 | 1.333 | 1.356 | 1.378 | 1.400 |
| 13/16 | 1.288 | 1.299 | 1.310 | 1.321 | 1.332 | 1.344 | 1.366 | 1.388 | 1.410 | 1.432 | 1.455 | 1.477 | 1.499 |
| 7/8 | 1.386 | 1.397 | 1.408 | 1.419 | 1.430 | 1.442 | 1.464 | 1.486 | 1.508 | 1.530 | 1.553 | 1.575 | 1.597 |
| 15/16 | 1.483 | 1.494 | 1.505 | 1.516 | 1.527 | 1.539 | 1.561 | 1.583 | 1.605 | 1.627 | 1.650 | 1.672 | 1.694 |
| 1 | 1.582 | 1.593 | 1.604 | 1.615 | 1.626 | 1.638 | 1.660 | 1.682 | 1.704 | 1.726 | 1.749 | 1.771 | 1.793 |
| 1 1/16 | 1.680 | 1.691 | 1.702 | 1.713 | 1.724 | 1.736 | 1.758 | 1.780 | 1.802 | 1.824 | 1.847 | 1.869 | 1.891 |
| 1 1/8 | 1.779 | 1.790 | 1.801 | 1.812 | 1.823 | 1.835 | 1.857 | 1.879 | 1.901 | 1.923 | 1.946 | 1.968 | 1.990 |
| 1 3/16 | 1.877 | 1.888 | 1.899 | 1.910 | 1.921 | 1.933 | 1.955 | 1.977 | 1.999 | 2.021 | 2.044 | 2.066 | 2.088 |
| 1 1/4 | 1.974 | 1.985 | 1.996 | 2.007 | 2.018 | 2.030 | 2.052 | 2.074 | 2.096 | 2.118 | 2.141 | 2.163 | 2.185 |

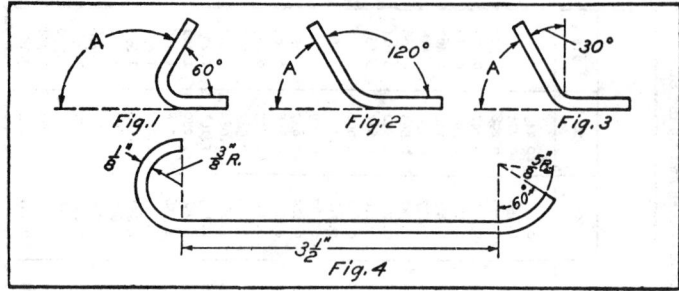

*Example Showing Application of Formulas:* Find the length before bending of
the part illustrated by Fig. 4. Soft steel is to be used.

For bend at left-hand end (180-degree bend)

$$L = [(0.64 \times 0.125) + (1.57 \times 0.375)] \times \frac{180}{90} = 1.338$$

For bend at right-hand end (60-degree bend)

$$L = [(0.64 \times 0.125) + (1.57 \times 0.625)] \times \frac{60}{90} = 0.707$$

Total length before bending = 3.5 + 1.338 + 0.707 = 5.545 inches

**Other Bending Allowance Formulas.** — When bending sheet steel or brass,
add from 1/8 to 1/2 of the thickness of the stock, for *each bend*, to the sum of the
inside dimensions of the finished piece, to get the length of the straight blank. The
harder the material the greater the allowance (1/8 of the thickness is added for soft

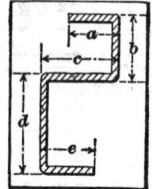

stock and 1/2 of the thickness for hard material). The data
given in the table, "Allowances for Bends in Sheet Metal,"
refer more particularly to the bending of sheet metal for
counters, bank fittings and general office fixtures, for which
purpose it is not absolutely essential to have the sections
of the bends within very close limits. Absolutely accurate
data for this work cannot be deduced, as the stock varies
considerably as to hardness, etc. The figures given apply
to sheet steel, aluminum, brass and bronze. Experience has
demonstrated that for the semisquare corners, such as are
formed in a V-die, the amount to be deducted from the sum of the outside
bend dimensions, as shown in the accompanying illustration by the sum of the
letters from *a* to *e*, is as follows: $X = 1.67\ BG$, where $X$ = the amount to be
deducted; $B$ = the number of bends; and $G$ = the decimal equivalent of the
gage. The values of $X$ for different gages and numbers of bends are given in the
table. Its application may be illustrated by an example: A strip having two
bends is to have outside dimensions of 2, 1½ and 2 inches, and is made of stock
0.125 inch thick. The sum of the outside dimensions is thus 5½ inches, and from
the table the amount to be deducted is found to be 0.416; hence the blank will be
5.5 − 0.416 = 5.084 inches long.

The lower part of the table applies to square bends which are either drawn through
a block of steel made to the required shape, or else drawn through rollers in a draw-
bench. The pressure applied not only gives a much sharper corner, but it also
elongates the material more than in the V-die process. In this case, the deduction is
$X = 1.33\ BG$.

### Allowances for Bends in Sheet Metal

| Square Bends | Gage | Thickness, Inches | Amount to be Deducted from the Sum of the Outside Bend Dimensions, Inches | | | | | | |
|---|---|---|---|---|---|---|---|---|---|
| | | | 1 Bend | 2 Bends | 3 Bends | 4 Bends | 5 Bends | 6 Bends | 7 Bends |
| Formed in a Press by a V-die | 18 | 0.0500 | 0.083 | 0.166 | 0.250 | 0.333 | 0.416 | 0.500 | 0.583 |
| | 16 | 0.0625 | 0.104 | 0.208 | 0.312 | 0.416 | 0.520 | 0.625 | 0.729 |
| | 14 | 0.0781 | 0.130 | 0.260 | 0.390 | 0.520 | 0.651 | 0.781 | 0.911 |
| | 13 | 0.0937 | 0.156 | 0.312 | 0.468 | 0.625 | 0.781 | 0.937 | 1.093 |
| | 12 | 0.1093 | 0.182 | 0.364 | 0.546 | 0.729 | 0.911 | 1.093 | 1.276 |
| | 11 | 0.1250 | 0.208 | 0.416 | 0.625 | 0.833 | 1.041 | 1.250 | 1.458 |
| | 10 | 0.1406 | 0.234 | 0.468 | 0.703 | 0.937 | 1.171 | 1.406 | 1.643 |
| Rolled or Drawn in a Draw-bench | 18 | 0.0500 | 0.066 | 0.133 | 0.200 | 0.266 | 0.333 | 0.400 | 0.466 |
| | 16 | 0.0625 | 0.083 | 0.166 | 0.250 | 0.333 | 0.416 | 0.500 | 0.583 |
| | 14 | 0.0781 | 0.104 | 0.208 | 0.312 | 0.416 | 0.521 | 0.625 | 0.729 |
| | 13 | 0.0937 | 0.125 | 0.250 | 0.375 | 0.500 | 0.625 | 0.750 | 0.875 |
| | 12 | 0.1093 | 0.145 | 0.291 | 0.437 | 0.583 | 0.729 | 0.875 | 1.020 |
| | 11 | 0.1250 | 0.166 | 0.333 | 0.500 | 0.666 | 0.833 | 1.000 | 1.166 |
| | 10 | 0.1406 | 0.187 | 0.375 | 0.562 | 0.750 | 0.937 | 1.125 | 1.312 |

## Drop-Forging Dies

**Steel for Drop-forging Dies.** — Practically all drop-forging dies are made of high-grade open-hearth steel. A 60-point carbon steel is mostly used, although steel as low as 40-point and as high as 85-point carbon is employed in some cases. A special hardening treatment is required for the low-carbon steel, which more than offsets the saving in price, and, except in special cases, there is no advantage in using high-carbon steels, owing to the expense. The average 60-point carbon steel die, if properly hardened, should last for from 15,000 to 40,000 forgings, and sometimes as many as 70,000 forgings can be made from one set of dies. When making dies for large forgings, it is often thought advisable to use 80-point carbon steel, and not harden the dies. This obviates the danger from "checking" or cracking in hardening, and the un-hardened steel is hard enough to resist the tendency to stretch. A steel that is quite high in carbon should always be used for dies that are intended for making forgings from tool steel or any other hard steel.

**Allowance for Shrinkage.** — When making dies for small cold-trimmed steel forgings, the proper allowance for shrinkage is 3/16 inch to the foot, or 0.015 inch to the inch. Such forgings are finished at a bright red heat and the rate of shrinkage is considerable. When making dies for hot-trimmed steel forgings of medium and large sizes, the shrinkage allowance is 1/8 inch to the foot, or 0.010 inch to the inch. Hot-trimmed forgings receive the finishing blow while comparatively cold, and shrink a smaller amount than the cold-trimmed forgings. The foregoing allowances are used for all dimensions of the die impression, such as depth, width or length. The shrinkage allowance for dies to be used in forging bronze or copper is practically the same as that for steel.

**Draft Allowance.** — The amount of draft in a drop-forging die varies from 3 to 10 degrees. If the die is for a thin forging of uniform section, 3 degrees is ample.

but if the forging is deep and has narrow ribs which are apt to stick, at least 7 degrees is necessary. If a die is used for forging a piece that is ring-shaped or has an annular part, the central plug that forms the interior of the ring should have a draft of 10 degrees, because, as the forging cools while being worked, it tends to shrink around the plug and if the draft is insufficient, it will stick in the die. With the foregoing exception, most drop-forging dies have a 7-degree draft. For convenience in laying out, it is well to remember that a 7-degree taper is approximately equal to a ⅛-inch taper to the inch, and a 10-degree taper, ³⁄₁₆ inch to the inch.

**Locating Impression in the Die.** — When laying out a drop-forging die, the impression should be located so that the heaviest end of the forging will be at the front of the die-block. This makes the forging easier to handle and also permits the use of a fairly large sprue. There should be at least 1½ inch left all around between the impression and the outside edge of the block. This also holds true for any part of the die, such as the edger, anvil or forming impression. If the forging has a hub or other projection that extends some distance from the main part on one side, the upper or top die should contain this deeper impression.

**Obtaining Weight of Forging from Lead Proof.** — After the upper and lower dies have been completed, shrinkage allowances and the general finish of the impressions are ordinarily tested by taking a "lead proof," and by weighing the lead, an approximate idea of the weight of the finished forging can be obtained. Roughly speaking, the finished forging will weigh two-thirds as much as the lead proof. The shrinkage of lead is practically the same as that of steel, so that the finished forging will also measure about the same as the one made of lead. In case of dies for eye-bolts and similar work, this rule must be disregarded, because the plugs that form the central opening will prevent the lead from shrinking naturally. When taking the lead proof, the die impressions are dusted with powdered chalk, and after the dies are clamped together, the molten lead is poured.

**Amount of Flash for Drop-forging Dies.** — Theoretically, there should be just enough forging metal in a die to fill the impression, and no more, but this is, of course, not practicable, as there is always some stock that must be disposed of after the impression is filled. To take care of this excess metal, dies are relieved all around the impression by milling a flat shallow recess about ¹⁄₆₄ inch deep and ⅝ inch wide. These dimensions are for dies of average size; in comparatively large dies this recess or "flash" would be a little deeper and wider. Both the upper and lower dies are flashed in this way. In addition, the upper die is "back-flashed," which means that there is a deeper recess, sometimes called the "gutter," milled around the impression at a distance of ¼ inch from the impression at every point. This back-flash is ³⁄₆₄ inch deep and acts as a relief for the excess metal after it has been squeezed from the flash proper. Only the finishing impression is provided with a flash and back-flash.

**The Break-down of Drop-forging Dies.** — The width of section used as a break-down (also known as the edger or side cut) should be enough wider than the forging to give plenty of room for the work of forging. A forging 1 inch thick should have a break-down 1½ inch wide, and about the same proportions should be followed for forgings of other widths. The break-down should have a section corresponding with the gate and sprue of the die impression, but it should be made slightly longer, so that the forging will not be stretched when struck in the impression.

**Hardening Drop-forging Dies.** — Dies to be carburized should always be packed for hardening in cast-iron or sheet-iron boxes containing a mixture of fresh bone and charcoal. The ordinary mixture is half bone and half charcoal. More bone gives greater hardness and more charcoal, less hardness, for a given heat; hence, the proportions should be varied according to requirements. The die should be packed face down on a one- or two-inch layer of this mixture and be settled so that the impression is filled. Sometimes the face is coated, before packing, with a thick paste of linseed oil and powdered bone-black, to protect the delicate edges from oxidation when in contact with the air. Fill the space between the sides of the die and the box with the bone and charcoal mixture, and cover over with a thick layer of wet clay paste to prevent the charcoal from burning out. Dies made of steel having less than 60-point carbon content should always be carburized. Open-hearth steel dies containing 60-point carbon or over can be hardened without carburizing.

**Heating the Die.** — An oil or gas furnace is recommended for heating, although a coal or coke-fired muffle furnace, capable of maintaining a temperature of at least 1600 degrees F., may be used, provided the temperature can be held constant. A temperature indicating device is necessary. The die should be put into the furnace as soon as the latter is lighted. If the correct quenching temperature for the steel is, say, 1500 degrees F., the furnace should be checked when the pyrometer indicates 1400 degrees, the die being allowed to "soak" at that heat for three or four hours. Then the heat should be slowly raised to 1500 degrees and held at that point one or two hours longer, according to the size of the die. Five hours is the minimum total time for heating, and seven or eight hours is much safer. A 60-point carbon die should be quenched between 1425 and 1450 degrees F.

**Cooling the Heated Die.** — When cooling, the face of the die should receive a sufficient flow of cold water to cause it to harden to the greatest possible depth. The back of the die should, at the same time, be cooled to make the shrinkage of the face and back equal, and to prevent warping. A good form of cooling tank is one having a large supply pipe extending up through the bottom for cooling the die face, and a smaller pipe above the tank to cool the back. Unless a jet of water under pressure is applied to the face of the die, the sunken parts of the impression will not harden equally with the face. Dies should not be cooled in a tank of still water, because steam forms in the die cavity which prevents the water from entering, thus causing the formation of soft spots. To overcome this, the water must be forced into the impression by pressure sufficient to overcome the resistance of the steam thus formed. Oil should not be used for hardening hammer dies, as its cooling action is not great enough to produce a sufficient depth of hardening. Hammer dies which are simply surface hardened will not withstand the heavy blows received in service. To secure a greater hardening effect, brine of about 40 per cent solution is used by some die-makers.

**Tempering Dies.** — Dies should be tempered and drawn as soon as they are cool enough to remove from the tank. The dies should be heated in an oil bath, and quenched in water or cool oil. Any high-grade cylinder oil of high flash-point is suitable. Low-grade oils smoke unpleasantly and will not stand high temperatures. The drawing temperature of die steels is about 450 degrees F., for average conditions. The corners of the die and the cut-off should be drawn to a purple color with the aid of a blow torch.

**Dies for Bronze and Copper Forgings.** — Dies for producing drop-forgings from bronze or copper differ from those used for steel or iron forgings principally in the matter of finish. Owing to the softness of copper and bronze, the metal is driven into very minute impressions in the surface of the dies; hence, these surfaces must be perfectly free from scratches, in order to insure a smooth finish on the work. Even though these metals are soft, the hammering necessary when forging is very hard on the dies, and to prevent them from dishing or spreading, tool steel is ordinarily used, unless the forgings are extra large and heavy. The shrinkage, draft and finish allowances on this class of drop-forging dies are practically the same as on dies for steel and iron.

**Trimming Dies for Drop-forgings.** — Hot-trimming dies are made of a special grade of steel known as hot-trimming die-stock. The objection to using ordinary tool steel for hot-trimming dies is that the edges of a hardened die check badly after the die has been used for a short time, and this checking is followed by a breaking away of the steel around the edges, thus rendering the die unfit for use. This special steel requires no hardening, and after the die is in use, the edges toughen and give better service than the best hardened tool steel. The usual form of punch for hot-trimming dies merely supports the forging while it is being pushed through. If the forging has a broad, flat top face, the punch need only be a little more than a flat piece that covers the forging and acts as a pusher. Such punches are commonly made of cast iron. Cold-trimming dies are made from good tool steel of from 1.00 to 1.25 per cent carbon, and hardened and drawn to a dark straw color. The punches for cold trimmers are also made of tool steel and are hardened and drawn to a very dark straw color. These punches are hardened to prevent them from upsetting at the edges. As with hot-trimming punches, the punch should fit the die loosely, but it should support the forging at every point while it is being pushed through the die. There are two instances in which trimming punches should fit the dies as closely as the average punching die for sheet metal work; first, when trimming forgings on which the fin comes at the corner of the forging; second, forgings that are formed all in one die, the other die being flat. In these two cases, unless the dies fit very well, there will be burrs at the trimmed edges.

**Standard Tolerances for Forgings.** — The tolerances adopted by the Drop Forging Association in 1937 (see accompanying Tables 1 to 5) apply to forgings under 100 pounds each. Forging tolerances may either be "special" or "regular." *Special tolerances* are those which are particularly noted in the specifications and may state any or all tolerances in any way as required. Special tolerances apply only to the particular dimensions noted. In all cases where special tolerances are not specified, regular tolerances apply.

*Regular tolerances* are divided into two divisions — "Commercial Standard" and "Close Standard." "Commercial Standard" tolerances are for general forging practice, but when extra close work is desired involving additional expense and care in the production of forgings, "Close Standard" may be specified. When no standard is specified, "Commercial Standard" shall apply.

Regular tolerances are applicable to (1) thickness; (2) width, including shrinkage and die wear, mismatching, and trimmed size; (3) draft angle; (4) quantity in shipment; (5) fillets and corners.

*Thickness Tolerances:* Thickness tolerances shall apply to the overall thickness of a forging. (See Table 1.)

*Width and Length Tolerances:* Width and length tolerances shall be alike and shall apply to the width or length of a forging. When applied to drop hammer forgings, they shall apply to the width or length in a direction parallel to the main or fundamental parting plane of the die, but only to such dimensions as are enclosed by and actually formed by the die. When applied to upset forgings, they shall apply to the width or length in a direction perpendicular to the direction of travel of the ramp.

Width and length tolerances consist of the three subdivisions following: (a) Shrinkage and die wear tolerance; (b) mismatching tolerance; (c) trimmed size tolerance. The latter must not be greater nor less than the limiting sizes at the parting

**Table 1. Standard Tolerances for Forgings**

Adopted, 1937, by Drop Forging Association for forgings under 100 pounds each

| Thickness Tolerances, Inch* | | | | | | | | |
|---|---|---|---|---|---|---|---|---|
| Net Weights, Pounds, up to — | Commercial | | Close | | Net Weights, Pounds, up to — | Commercial | | Close | |
| | − | + | − | + | | − | + | − | + |
| .2 | .008 | .024 | .004 | .012 | 20 | .026 | .078 | .013 | .039 |
| .4 | .009 | .027 | .005 | .015 | 30 | .030 | .090 | .015 | .045 |
| .6 | .010 | .030 | .005 | .015 | 40 | .034 | .102 | .017 | .051 |
| .8 | .011 | .033 | .006 | .018 | 50 | .038 | .114 | .019 | .057 |
| 1 | .012 | .036 | .006 | .018 | 60 | .042 | .126 | .021 | .063 |
| 2 | .015 | .045 | .008 | .024 | 70 | .046 | .138 | .023 | .069 |
| 3 | .017 | .051 | .009 | .027 | 80 | .050 | .150 | .025 | .075 |
| 4 | .018 | .054 | .009 | .027 | 90 | .054 | .162 | .027 | .081 |
| 5 | .019 | .057 | .010 | .030 | 100 | .058 | .174 | .029 | .087 |
| 10 | .022 | .066 | .011 | .033 | | | | | |

* Thickness tolerances apply to the over-all thickness. For drop-hammer forgings, they apply to the thickness in a direction perpendicular to the main or fundamental parting plane of the die. For upset forgings, they apply to the thickness in the direction parallel to the travel of the ram, but only to such dimensions as are enclosed by and actually formed by the die.

**Table 2. Standard Tolerances for Forgings**

Adopted, 1937, by Drop Forging Association for forgings under 100 pounds each

| Shrinkage | | Plus | Die Wear | | Mismatching | | |
|---|---|---|---|---|---|---|---|
| Lengths or widths up to — in. | Commercial + or − | Close + or − | Net wt. up to — lbs. | Commercial + or − | Close + or − | Net Weights, Pounds, up to — | Commercial | Close |
| 1 | .003 | .002 | 1 | .032 | .016 | 1 | .015 | .010 |
| 2 | .006 | .003 | 3 | .035 | .018 | 7 | .018 | .012 |
| 3 | .009 | .005 | 5 | .038 | .019 | 13 | .021 | .014 |
| 4 | .012 | .006 | 7 | .041 | .021 | 19 | .024 | .016 |
| 5 | .015 | .008 | 9 | .044 | .022 | 25 | .027 | .018 |
| 6 | .018 | .009 | 11 | .047 | .024 | 31 | .030 | .020 |

For each additional inch under shrinkage, add 0.003 to the commercial tolerance and 0.0015 to the close tolerance. For example, if length or width is 12 inches, the commercial tolerance is plus or minus 0.036 and the close tolerance plus or minus 0.018.

For each additional 2 pounds under die wear, add 0.003 to the commercial tolerance and 0.0015 to the close tolerance. Thus, if the net weight is 21 pounds, the die wear commercial tolerance is 0.062 plus or minus, and the close tolerance 0.031 plus or minus.

For each additional 6 pounds under mismatching, add 0.003 to the commercial tolerance and 0.002 to the close tolerance. Thus, if the net weight is 37 pounds, the mismatching commercial tolerance is 0.033 and the close tolerance 0.022.

### Table 3. Standard Tolerances for Forgings

Adopted, 1937, by Drop Forging Association for forgings under 100 pounds each

| Draft angle tolerances — the permissible variations from the standard or nominal draft angle | | | | | | | |
|---|---|---|---|---|---|---|---|
| Drop-Hammer Forgings | | | | Upset Forgings | | | |
| Location of Surface | Nominal Angle Degrees | Commercial Limits | Close Limits | Location of Surface | Nominal Angle Degrees | Commercial Limits | Close Limits |
| Outside | 7 | 0–10 | 0–8 | Outside | 3 | 0–5 | 0–4 |
| Holes and Depressions } | 10 | 0–13 | ... | Holes and Depressions } | 5 | 0–8 | 0–7 |
|  | 7 | ... | 0–8 |  |  |  |  |

### Table 4. Standard Tolerances for Forgings

Adopted, 1937, by Drop Forging Association for forgings under 100 pounds each

| Quantity Tolerances | | | | | |
|---|---|---|---|---|---|
| Number of Pieces on Order | Permissible Variation | | Number of Pieces on Order | Permissible Variation | |
|  | Over-run, Pieces | Under-run, Pieces |  | Over-run, Per cent | Under-run, Per cent |
| 1- 2 | 1 | 0 | 100- 199 | 10 | 5.0 |
| 3- 5 | 2 | 1 | 200- 299 | 9 | 4.5 |
| 6-19 | 3 | 1 | 300- 599 | 8 | 4.0 |
| 20-29 | 4 | 2 | 600- 1,249 | 7 | 3.5 |
| 30-39 | 5 | 2 | 1,250- 2,999 | 6 | 3.0 |
| 40-49 | 6 | 3 | 3,000- 9,999 | 5 | 2.5 |
| 50-59 | 7 | 3 | 10,000- 39,999 | 4 | 2.0 |
| 60-69 | 8 | 4 | 40,000-299,999 | 3 | 1.5 |
| 70-79 | 9 | 4 | 300,000 up | 2 | 1.0 |
| 80-99 | 10 | 5 |  |  |  |

These quantity tolerances represent the permissible over-run or under-run allowed for each release or part shipment of an order. Any shipping quantity within the limits of over-run or under-run shall be considered as completing the order.

### Table 5. Standard Tolerances for Forgings

Adopted, 1937, by Drop Forging Association for forgings under 100 pounds each

| Maximum Radii of Fillets and Corners, Inch | | | | | |
|---|---|---|---|---|---|
| Net Weights, Pounds, up to — | Commercial | Close | Net Weights, Pounds, up to — | Commercial | Close |
| .3 | 3/32 | 3/64 | 10 | 3/16 | 3/32 |
| 1 | 1/8 | 1/16 | 30 | 7/32 | 7/64 |
| 3 | 5/32 | 5/64 | 100 | 1/4 | 1/8 |

plane, imposed by the sum of the draft angle tolerances and the shrinkage and die wear tolerances.

*Shrinkage and Die Wear:* Shrinkage and die wear tolerances shall apply to that part of the forging formed by a single die block only. They shall not apply to any dimension crossing the parting plane. They shall be the sum of the shrinkage tolerances and the die wear tolerances as given in Table 2 (left-hand section). The shrinkage tolerances and die wear tolerances shall not be applied separately, but shall only be used as the sum of the two. These tolerances shall not be so applied as to include draft.

*Mismatching Tolerance:* Mismatching is the displacement of a point in that part of a forging formed by one die block of a pair, from its desired position when located from the part of the forging formed in the other die block of the pair. Mismatching does not include any displacement caused by variation in thickness of the forging, but is only the displacement in a plane parallel to the main or fundamental parting plane of the dies. Mismatching tolerances are independent of, and in addition to, any other tolerances. See Table 2 (right-hand section).

*Fillet and Corner Tolerances:* Fillet and corner tolerances apply to all meeting surfaces even though drawings or models indicate sharp corners, unless such drawings or models have or indicate (even though actual dimensions are not specified) fillet or corner dimensions of larger radii than the standards in Table 5, in which case such actual or indicated larger dimensions shall be considered specified and the tolerances shall be "special tolerances."

Where a corner tolerance applies on the meeting of two drafted surfaces, the tolerance shall apply to the narrow end of such meeting and the radius will increase toward the wide end. The total increase in the radius will equal the length of the drafted surface in inches, multiplied by the tangent of the nominal draft angle.

# BROACHES AND BROACHING

**The Broaching Process.** — The broaching process may be applied in machining holes or other internal surfaces and also to many flat or other external surfaces. Internal broaching is applied in forming either symmetrical or irregular holes, grooves, or slots in machine parts, especially when the size or shape of the opening, or its length in proportion to diameter or width, make other machining processes impracticable. Broaching originally was utilized for such work as cutting keyways, machining round holes into square, hexagonal, or other shapes, forming splined holes, and for a large variety of other internal operations. The development of broaching machines and broaches finally resulted in extensive application of the process to external, flat, and other surfaces. Most external or surface broaching is done on machines of vertical design, but horizontal machines are also used for some classes of work. The broaching process is very rapid, accurate, and it leaves a finish of good quality. It is employed extensively in automotive and other plants where duplicate parts must be produced in large quantities and frequently to given dimensions within small tolerances.

**Types of Broaches.** — A number of typical broaches and the operations for which they are intended are shown by the diagrams, Fig. 1. Broach $A$ produces a round-cornered, square hole. Prior to broaching square holes, it is usually the practice to drill a round hole having a diameter $d$ somewhat larger than the width of the square. Hence, the sides are not completely finished, but this unfinished part is not objectionable in most cases. In fact, this clearance space is an advantage during the broaching operation in that it serves as a channel for the broaching lubricant; moreover, the broach has less metal to remove. Broach $B$ is for finishing

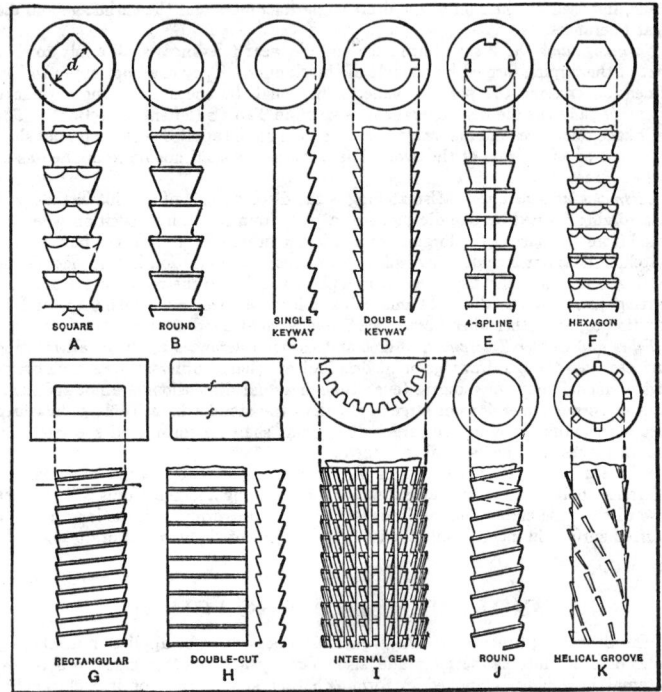

Fig. 1. Types of Broaches

round holes. Broaching is superior to reaming for some classes of work, because the broach will hold its size for a much longer period, thus insuring greater accuracy. Broaches *C* and *D* are for cutting single and double keyways, respectively. Broach *C* is of rectangular section and, when in use, slides through a guiding bushing which is inserted in the hole. Broach *E* is for forming four integral splines in a hub. The broach at *F* is for producing hexagonal holes. Rectangular holes are finished by broach *G*. The teeth on the sides of this broach are inclined in opposite directions, which has the following advantages: The broach is stronger than it would be if the teeth were opposite and parallel to each other; thin work cannot drop between the inclined teeth, as it tends to do when the teeth are at right angles, because at least two teeth are always cutting; the inclination in opposite directions neutralizes the lateral thrust. The teeth on the edges are staggered, the teeth on one side being midway between the teeth on the other edge, as shown by the dotted line. A double cut broach is shown at *H*. This type is for finishing, simultaneously, both sides *f* of a slot, and for similar work. Broach *I* is the style used for forming the teeth in internal gears. It is practically a series of gear-shaped cutters, the outside diameters of which gradually increase toward the finishing end of the broach. Broach *J* is for round holes but differs from style *B* in that it has a continuous helical cutting edge. Some prefer this form because it gives a shearing cut. Broach

$K$ is for cutting a series of helical grooves in a hub or bushing. In helical broaching, either the work or the broach is rotated to form the helical grooves as the broach is pulled through.

In addition to the typical broaches shown in Fig. 1, many special designs are now in use for performing more complex operations. Two surfaces on opposite sides of a casting or forging are sometimes machined simultaneously by twin broaches and, in other cases, three or four broaches are drawn through a part at the same time, for finishing as many duplicate holes or surfaces. Notable developments have been made in the design of broaches for external or " surface " broaching.

**Pitch of Broach Teeth.** — The pitch of broach teeth depends upon the depth of cut or chip thickness, length of cut, the cutting force required and power of the broaching machine. In the pitch formulas which follow

$L$ = length, in inches, of layer to be removed by broaching

$d$ = depth of cut per tooth as shown by Table 1 (For internal broaches, $d$ = depth of cut as measured on one side of broach or one-half difference in diameters of successive teeth in case of a round broach)

$F$ = a factor. (For brittle types of material, $F$ = 3 or 4 for roughing teeth, and 6 for finishing teeth. For ductile types of material, $F$ = 4 to 7 for roughing teeth and 8 for finishing teeth.)

$b$ = width in inches, of layer to be removed by broaching

$P$ = pressure required in tons per square inch, of an area equal to depth of cut times width of cut, in inches (Table 2)

$T$ = usable capacity, in tons, of broaching machine = 70 per cent of maximum tonnage

The minimum pitch shown by Formula (1) is based upon the receiving capacity of the chip space. The minimum, however, should not be less than 0.2 inch unless a smaller pitch is required for exceptionally short cuts to provide at least two teeth in contact simultaneously, with the part being broached. A reduction below 0.2 inch is seldom required in surface broaching but it may be necessary in connection with internal broaching.

$$\text{Minimum pitch} = 3\sqrt{LdF} \tag{1}$$

Whether the minimum pitch may be used or not depends upon the power of the available machine. The factor $F$ in the formula provides for the increase in volume as the material is broached into chips. If a broach has adjustable inserts for the finishing teeth, the pitch of the finishing teeth may be smaller than the pitch of the roughing teeth because of the smaller depth $d$ of the cut. The higher value of $F$ for finishing teeth prevents the pitch from becoming too small, so that the spirally curled chips will not be crowded into too small a space. The pitch of the roughing and finishing teeth should be equal for broaches without separate inserts (notwithstanding the different values of $d$ and $F$) so that some of the finishing teeth may be ground into roughing teeth after wear makes this necessary.

$$\text{Allowable pitch} = \frac{dLbP}{T} \tag{2}$$

If the pitch obtained by Formula (2) is larger than the minimum obtained by Formula (1), this larger value should be used because it is based upon the usable power of the machine. As the notation indicates, 70 per cent of the maximum tonnage $T$ is taken as the usable capacity. The 30 per cent reduction is to provide a margin for the increase in broaching load resulting from the gradual dulling of the cutting edges. The procedure in calculating both minimum and allowable pitches will be illustrated by an example.

## Table 1. Designing Data for Surface Broaches

| Material to be Broached | Depth of Cut per Tooth, Inch | | Face Angle or Rake, Degrees | Clearance Angle, Degrees | |
|---|---|---|---|---|---|
| | Roughing* | Finishing | | Roughing | Finishing |
| Steel, High Tensile Strength.............. | 0.0015-0.002 | 0.0005 | 10-12 | 1.5-3 | 0.5-1 |
| Steel, Medium Tensile Strength............. | 0.0025-0.005 | 0.0005 | 14-18 | 1.5-3 | 0.5-1 |
| Cast Steel................ | 0.0025-0.005 | 0.0005 | 10 | 1.5-3 | 0.5 |
| Malleable Iron............ | 0.0025-0.005 | 0.0005 | 7 | 1.5-3 | 0.5 |
| Cast Iron, Soft........... | 0.006 -0.010 | 0.0005 | 10-15 | 1.5-3 | 0.5 |
| Cast Iron, Hard.......... | 0.003 -0.005 | 0.0005 | 5 | 1.5-3 | 0.5 |
| Zinc Die Castings........ | 0.005 -0.010 | 0.0010 | 12 ** | 5 | 2 |
| Cast Bronze.............. | 0.010 -0.025 | 0.0005 | 8 | 0 | 0 |
| Wrought Aluminum Alloys | 0.005 -0.010 | 0.0010 | 15 ** | 3 | 1 |
| Cast Aluminum Alloys.... | 0.005 -0.010 | 0.0010 | 12 ** | 3 | 1 |
| Magnesium Die Castings.. | 0.010 -0.015 | 0.0010 | 20 ** | 3 | 1 |

* The lower depth-of-cut values for roughing are recommended when work is not **very** rigid, the tolerance is small, a good finish is required, or length of cut is comparatively **short**.

** In broaching these materials, smooth surfaces for tooth and chip spaces are **especially** recommended.

## Table 2. Broaching Pressure P for Use in Pitch Formula (2)

| Material to be Broached | Depth $d$ of Cut per Tooth, Inch | | | | | Pressure $P$, Side-cutting Broaches |
|---|---|---|---|---|---|---|
| | 0.024 | 0.010 | 0.004 | 0.002 | 0.001 | |
| | Pressure $P$ in Tons per Square Inch | | | | | |
| Steel, High Ten. Strength...... | .. | ... | ... | 250 | 312 | 200- .004″ cut |
| Steel, Med. Ten. Strength...... | .. | ... | 158 | 185 | 243 | 143- .006″ cut |
| Cast Steel.................... | .. | ... | 128 | 158 | ... | 115- .006″ cut |
| Malleable Iron................. | .. | ... | 108 | 128 | ... | 100- .006″ cut |
| Cast Iron..................... | .. | 115 | 115 | 143 | ... | 115- .020″ cut |
| Cast Brass.................... | .. | 50 | 50 | ... | ... | .............. |
| Brass, Hot Pressed............ | .. | 85 | 85 | ... | ... | .............. |
| Zinc Die Castings............. | .. | 70 | 70 | ... | ... | .............. |
| Cast Bronze.................. | 35 | 35 | ... | ... | ... | .............. |
| Wrought Aluminum............ | .. | 70 | 70 | ... | ... | .............. |
| Cast Aluminum................ | .. | 85 | 85 | ... | ... | .............. |
| Magnesium Alloy.............. | 35 | 35 | ... | ... | ... | .............. |

*Example:* Determine pitch of broach for cast iron when $L$ = 9 inches; $d$ = 0.004; and $F$ = 4.

$$\text{Minimum pitch} = 3\sqrt{9 \times 0.004 \times 4} = 1.14$$

Next, apply Formula (2). Assume that $b$ = 3 and $T$ = 10; for cast iron **and** depth $d$ of 0.004, $P$ = 115 (Table 2). Then,

$$\text{Allowable pitch} = \frac{0.004 \times 9 \times 3 \times 115}{10} = 1.24$$

This pitch is safely above the minimum. If in this case the usable tonnage of an available machine were, say, 8 tons instead of 10 tons, the pitch as shown by Formula (2) might be increased to about 1.5 inches, thus reducing the number of teeth cutting simultaneously and, consequently, the load on the machine; or the cut per tooth might be reduced instead of increasing the pitch, especially if only a few teeth are in cutting contact, as might be the case with a short length of cut. If the usable tonnage in the preceding example were, say, 15, then a pitch of 0.84 would be obtained by Formula (2); hence the pitch in this case should not be less than the minimum of approximately 1.14 inches.

**Depth of Cut per Tooth.** — The term " depth of cut " as applied to surface or external broaches means the difference in the heights of successive teeth. This term, as applied to internal broaches for round, hexagonal or other holes, may indicate the total increase in the diameter of successive teeth; however, to avoid confusion, the term as here used means in all cases and regardless of the type of broach, the depth of cut as measured on one side.

In broaching free cutting steel, the Broaching Tool Institute recommends 0.003 to 0.006 inch depth of cut for surface broaching; 0.002 to 0.003 inch for multi-spline broaching; and 0.0007 to 0.0015 inch for round hole broaching. The accompanying

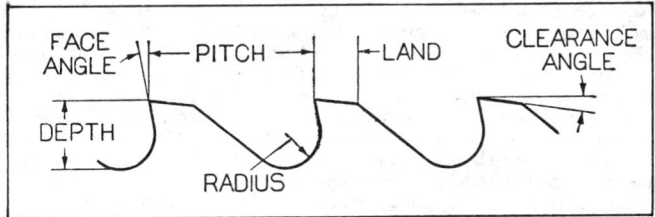

Terms Commonly Used in Broach Design

table contains data from a German source and applies specifically to surface broaches. All data relating to depth of cut are intended as a general guide only. While depth of cut is based primarily upon the machinability of the material, some reduction from the depth thus established may be required particularly when the work supporting fixture in surface broaching is not sufficiently rigid to resist the thrust from the broaching operation. In some cases, the pitch and cutting length may be increased to reduce the thrust force. Another possible remedy in surface broaching certain classes of work is to use a side-cutting broach instead of the ordinary depth cutting type. A broach designed for side cutting takes relatively deep narrow cuts which extend nearly to the full depth required. The side cutting section is followed by teeth arranged for depth cutting to obtain the required size and surface finish on the work. In general, small tolerances in surface broaching require a reduced cut per tooth to minimize work deflection resulting from the pressure of the cut.

**Face Angle or Rake.** — The face angle (see diagram) of broach teeth affects the chip flow and varies considerably for different materials. While there are some variations in practice, even for the same material, the angles given in the accompanying table are believed to represent commonly used values. Some broach designers increase the rake angle for finishing teeth in order to improve the finish on the work.

**Clearance Angle.** — The clearance angle (see illustration) for roughing steel varies from 1.5 to 3 degrees and for finishing steel from 0.5 to 1 degree. Some recommend the same clearance angles for cast iron and others, larger clearance angles varying from 2 to 4 or 5 degrees. Additional data will be found in Table 1.

**Land Width.** — The width of the land usually is about 0.25 × pitch. It varies, however, from about one-fourth to one-third of the pitch. The land width is selected so as to obtain the proper balance between tooth strength and chip space.

**Depth of Broach Teeth.** — The tooth depth as established experimentally and on the basis of experience, usually varies from about 0.37 to 0.40 of the pitch. This depth is measured radially from the cutting edge to the bottom of the tooth fillet.

**Radius of Tooth Fillet.** — The " gullet " or bottom of the chip space between the teeth should have a rounded fillet to strengthen the broach, facilitate curling of the chips, and safeguard against cracking in connection with the hardening operation. One rule is to make the radius equal to one-fourth the pitch. Another is to make it equal 0.4 to 0.6 the tooth depth. A third method preferred by some broach designers is to make the radius equal one-third of the sum obtained by adding together the land width, one-half the tooth depth, and one-fourth of the pitch.

**Total Length of Broach.** — After the depth of cut per tooth has been determined, the total amount of material to be removed by a broach is divided by this decimal to ascertain the number of cutting teeth required. This number of teeth multiplied by the pitch gives the length of the active portion of the broach. By adding to this dimension the distance over three or four straight teeth, the length of a pilot to be provided at the finishing end of the broach, and the length of a shank which must project through the work and the faceplate of the machine to the draw-head, the over-all length of the broach is found. This calculated length is often greater than the stroke of the machine, or greater than is practical for a broach of the diameter required. In such cases, a set of broaches must be used.

**Chip Breakers.** — The teeth of broaches frequently have rounded chip-breaking grooves located at intervals along the cutting edges. These grooves break up wide curling chips and prevent them from clogging the chip spaces, thus reducing the cutting pressure and strain on the broach. These chip-breaking grooves are on the roughing teeth only. They are staggered and applied to both round and flat or surface broaches. The grooves are formed by a round edged grinding wheel and usually vary in width from about 1/32 to 3/32 inch depending upon the size of broach. The more ductile the material, the wider the chip breaker grooves should be and the smaller the distance between them. Narrow slotting broaches may have the right- and left-hand corners of alternate teeth beveled to obtain chip-breaking action.

**Shear Angle.** — The teeth of surface broaches ordinarily are inclined so they are not at right angles to the broaching movement. The object of this inclination is to obtain a shearing cut which results in smoother cutting action and an improvement in surface finish. The shearing cut also tends to eliminate troublesome vibration. Shear angles for surface broaches are not suitable for broaching slots or any profiles that resist the outward movement of the chips. When the teeth are inclined, the fixture should be designed to resist the resulting thrusts unless it is practicable to incline the teeth of right- and left-hand sections in opposite directions to neutralize the thrust. The shear angle usually varies from 10 to 25 degrees.

**Steels for Broaches.** — High-speed steels are commonly used at the present time. The 18-4-2 or 18-4-3 types are more resistant to abrasion and wear than steels with a lower vanadium content although steels with one per cent vanadium or less are also used. A broach steel said to have very high resistance to wear and suitable for broaching steels having tensile strengths above 130,000 pounds per square inch, contains 10 per cent tungsten, 4 per cent chromium, 1.7 per cent vanadium, and 0.8 per cent carbon. The hardening heat recommended is 2250–2300 degrees F. and the tempering heat 1000–1040 degrees F. A steel of low tungsten content that is also recommended contains 2.5 per cent tungsten, 4 per cent chromium, 3 per cent vanadium, 2.5 per cent molybdenum, and 1.2–1.3 per cent carbon. The hardening heat is 1975–2050 degrees F. and the tempering heat 1060–1075 degrees F.

**Cutting Oils for Broaching.—** For broaching steel, a good grade of sulphur base oil is preferable, as a general rule. Cast iron may be broached dry but good results have also been obtained with a mixture of kerosene and soluble oil. Another mixture consists of one part soluble oil and twenty parts water.

**Types of Broaching Machines. —** Broaching machines may be divided into horizontal and vertical designs, and they may be classified further according to the method of operation, as, for example, whether a broach in a vertical machine is pulled up or pulled down in forcing it through the work. Horizontal machines usually pull the broach through the work in internal broaching but short rigid broaches may be pushed through. External surface broaching is also done on some machines of horizontal design, but usually vertical machines are employed for flat or other external broaching. Although parts usually are broached by traversing the broach itself, some machines are designed to hold the broach or broaches stationary during the actual broaching operation. This principle has been applied both to internal and surface broaching.

*Vertical Duplex Type:* The vertical duplex type of surface broaching machine has two slides or rams which move in opposite directions and operate alternately. While the broach connected to one slide is moving downward on the cutting stroke, the other broach and slide is returning to the starting position, and this returning time is utilized for reloading the fixture on that side; consequently, the broaching operation is practically continuous. Each ram or slide may be equipped to perform a separate operation on the same part when two operations are required.

*Pull-up Type:* Vertical hydraulically operated machines which pull the broach or broaches up through the work are used for internal broaching of holes of various shapes, for broaching bushings, splined holes, small internal gears, etc. A typical machine of this kind is so designed that all broach handling is done automatically.

*Pull-down Type:* The various movements in the operating cycle of a hydraulic pull-down type of machine equipped with an automatic broach-handling slide, are the reverse of the pull-up type. The broaches for a pull-down type of machine have shanks on each end, there being an upper one for the broach-handling slide and a lower one for pulling the broaches through the work.

*Hydraulic Operation:* Modern broaching machines, as a general rule, are operated hydraulically rather than by mechanical means. Hydraulic operation is efficient, flexible in the matter of speed adjustments, low in maintenance cost, and the "smooth" action required for fine precision finishing may be obtained. The hydraulic pressures required, which frequently are 800 to 1000 pounds per square inch, are obtained from a motor-driven pump forming part of the machine. Broaching machines for general use are so designed that the length of the stroke can be adjusted to suit the length of the broach. The cutting speeds of broaching machines may be varied for different materials and operations. These speeds frequently are between 20 and 30 feet per minute, and the return speeds often are double the cutting speed, or higher, to reduce the idle period.

### Causes of Broaching Difficulties

| Broaching Difficulty | Possible Causes |
|---|---|
| Stuck broach | Insufficient machine capacity; dulled teeth; clogged chip gullets; failure of power during cutting stroke.<br>To remove a stuck broach, workpiece and broach are removed from the machine as a unit; never try to back out broach by reversing machine. If broach does not loosen by tapping workpiece lightly and trying to slide it off its starting end, mount workpiece and broach in a lathe and turn down workpiece to the tool surface. Workpiece may be sawed longitudinally into several sections in order to free the broach.<br>Check broach design, perhaps tooth relief (back off) angle is too small or depth of cut per tooth is too great. |
| Galling and pickup | Lack of homogeneity of material being broached — uneven hardness, porosity; improper or insufficient coolant; poor broach design, mutilated broach; dull broach; improperly sharpened broach; improperly designed or outworn fixtures.<br>Good broach design will do away with possible chip build-up on tooth faces and excessive heating. Grinding of teeth should be accurate so that the correct gullet contour is maintained. Contour should be fair and smooth. |
| Broach breakage | Overloading; broach dullness; improper sharpening; interrupted cutting stroke; backing up broach with workpiece in fixture; allowing broach to pass entirely through guide hole; ill fitting and/or sharp edged key; crooked holes; untrue locating surface; excessive hardness of workpiece; insufficient clearance angle; sharp corners on pull end of broach.<br>When grinding bevels on pull end of broach use wheel that is not too pointed. |
| Chatter | Too few teeth in cutting contact simultaneously; excessive hardness of material being broached; loose or poorly constructed tooling; surging of ram due to load variations.<br>Chatter can be alleviated by changing the broaching speed, by using shear cutting teeth instead of right angle teeth, and by changing the coolant and the face and relief angles of the teeth. |
| Drifting or misalignment of tool during cutting stroke | Lack of proper alignment when broach is sharpened in grinding machine, which may be caused by dirt in the female center of the broach; inadequate support of broach during the cutting stroke, on a horizontal machine especially; body diameter too small; cutting resistance variable around I.D. of hole due to lack of symmetry of surfaces to be cut; variations in hardness around I.D. of hole; too few teeth in cutting contact. |
| Streaks in broached surface | Lands too wide; presence of forging, casting or annealing scale; metal pickup; presence of grinding burrs and grinding and cleaning abrasives. |
| Rings in the broached hole | Due to surging resulting from uniform pitch of teeth; presence of sharpening burrs on broach; tooth clearance angle too large; locating face not smooth or square; broach not supported for all cutting teeth passing through the work. The use of differential tooth spacing or shear cutting teeth helps in preventing surging. Sharpening burrs on a broach may be removed with a wood block. |

**Broaching Difficulties.** — The accompanying table has been compiled from information supplied by the National Broach and Machine Co. and presents some of the common broaching difficulties, their causes and means of correction.

# FILES

**Definitions of File Terms.** — The following file terms apply to hand files but not to rotary files and burs.

*Axis:* Imaginary line extending the entire length of a file equidistant from faces and edges.

*Back:* The convex side of a file having the same or similar cross-section as a half-round file.

*Bastard Cut:* A grade of file coarseness between coarse and second cut of American pattern files and rasps.

*Blank:* A file in any process of manufacture before being cut.

*Blunt:* A file whose cross-sectional dimensions from point to tang remain unchanged.

*Coarse Cut:* The coarsest of all American pattern file and rasp cuts.

*Coarseness:* Term describing the relative number of teeth per unit length, the coarsest having the least number of file teeth per unit length; the smoothest, the most. American pattern files and rasps have four degrees of coarseness: coarse, bastard, second and smooth. Swiss pattern files usually have seven degrees of coarseness: 00, 0, 1, 2, 3, 4, 6 (from coarsest to smoothest). Curved tooth files have three degrees of coarseness: standard, fine and smooth.

*Curved Cut:* File teeth which are made in curved contour across the file blank.

*Cut:* Term used to describe file teeth with respect to their coarseness or their character (single, double, rasp, curved, special).

*Double Cut:* A file tooth arrangement formed by two series of cuts, namely the overcut followed, at an angle, by the upcut.

*Edge:* Surface joining faces of a file. May have teeth or be smooth.

*Face:* Widest cutting surface or surfaces that are used for filing.

*Heel or Shoulder:* That portion of a file that abuts the tang.

*Hopped:* A term used among file makers to represent a very wide skip or spacing between file teeth.

*Length:* The distance from the heel to the point.

*Overcut:* The first series of teeth put on a double-cut file.

*Point:* The front end of a file; the end opposite the tang.

*Rasp Cut:* A file tooth arrangement of round-topped teeth, usually not connected, that are formed individually by means of a narrow, punch-like tool.

*Re-cut:* A worn-out file which has been re-cut and re-hardened after annealing and grinding off the old teeth.

*Safe Edge:* An edge of a file that is made smooth or uncut, so that it will not injure that portion or surface of the workpiece with which it may come in contact during filing.

*Second Cut:* A grade of file coarseness between bastard and smooth of American pattern files and rasps.

*Set:* To blunt the sharp edges or corners of file blanks before and after the overcut is made, in order to prevent weakness and breakage of the teeth along such edges or corners when the file is put to use.

*Shoulder or Heel:* See *Heel or Shoulder*.

*Single Cut:* A file tooth arrangement where the file teeth are composed of single unbroken rows of parallel teeth formed by a single series of cuts.

*Smooth Cut:* An American pattern file and rasp cut that is smoother than second cut.

*Tang:* The narrowed portion of a file which engages the handle.

*Upcut:* The series of teeth superimposed on the overcut, and at an angle to it, on a double-cut file.

**File Characteristics.** — Files are classified according to their shape or cross-section and according to the pitch or spacing of their teeth and the nature of the cut.

*Cross-section and Outline:* The cross-section may be quadrangular, circular, triangular, or some special shape. The outline or contour may be tapered or blunt. In the former, the point is more or less reduced in width and thickness by a gradually narrowing section that extends for one-half to two-thirds of the length. In the latter, the cross-section remains uniform from tang to point.

*Cut:* The character of the teeth is designated as single, double, rasp or curved. The *single cut file* (or *float* as the coarser cuts are sometimes called) has a single series of parallel teeth extending across the face of the file at an angle of from 45 to 85 degrees with the axis of the file. This angle depends upon the form of the file and the nature of the work for which it is intended. The single cut file is customarily used with a light pressure to produce a smooth finish. The *double cut file* has a multiplicity of small pointed teeth inclining toward the point of the file arranged in two series of diagonal rows that cross each other. For general work, the angle of the first series of rows is from 40 to 45 degrees and of the second from 70 to 80 degrees. For *double cut finishing files* the first series has an angle of about 30 degrees and the second, from 80 to 87 degrees. The second, or upcut, is almost always deeper than the first or overcut. Double cut files are usually employed, under heavier pressure, for fast metal removal and where a rougher finish is permissible. The *rasp* is formed by raising a series of individual rounded teeth from the surface of the file blank with a sharp, narrow, punch-like cutting tool and is used with a relatively heavy pressure on soft substances for fast removal of material. The curved tooth file has teeth that are in the form of parallel arcs extending across the face of the file, the middle portion of each arc being closest to the point of the file. The teeth are usually single cut and are relatively coarse. They may be formed by steel displacement but are more commonly formed by milling.

With reference to coarseness of cut the terms *coarse, bastard, second* and *smooth cuts* are used, the coarse or bastard files being used on the heavier classes of work and the second or smooth cut files for the finishing or more exacting work. These degrees of coarseness are only comparable when files of the same length are compared, as the number of teeth per inch of length decreases as the length of the file increases. The number of teeth per inch varies considerably for different sizes and shapes and for files of different makes. The coarseness range for the curved tooth files is given as standard, fine and smooth. In the case of Swiss pattern files, a series of numbers is used to designate coarseness instead of names; Nos. oo, o, 1, 2, 3, 4 and 6 being the most common with No. oo the coarsest and No. 6 the finest.

**Classes of Files.** — There are five main classes of files: mill or saw files; machinists' files; curved tooth files; Swiss pattern files; and rasps. The first two classes are commonly referred to as American pattern files.

*Mill or Saw Files:* These are used for sharpening mill or circular saws, large cross-cut saws; for lathe work; for draw filing; for filing brass and bronze; and for smooth filing generally. *Cantsaw files* (1) have an obtuse isosceles triangular section, a blunt outline, are single cut and are used for sharpening saws having "M"-shaped teeth and teeth of less than 60-degree angle. *Crosscut files* (2) have a narrow triangular section with short side rounded, a blunt outline, are single cut and are used to sharpen crosscut saws. The rounded portion is used to deepen the gullets of saw teeth and the sides are used to sharpen the teeth themselves. *Double ender files* (3) have a triangular section, are tapered from the middle to both ends, are tangless, are single cut and are used reversibly for sharpening saws. The *mill file* (4), itself, is usually single cut, tapered in width, and often has two square cutting edges in addition to the cutting sides. Either or both edges may be rounded, however, for filing the gullets between saw teeth. The *blunt mill file* has a uniform rectangular cross-section from tip to tang. The *triangular saw files* or *taper saw files* (5) have an equilateral triangular section, are tapered, are single cut and are used for filing saws with 60-degree angle teeth. They come in taper, slim taper, extra slim taper and double extra slim taper thicknesses. *Blunt triangular* and *blunt hand saw files* are without taper. *Web saw files* (6) have a diamond-shaped section, a blunt outline, are single cut and are used for sharpening pulpwood or web saws.

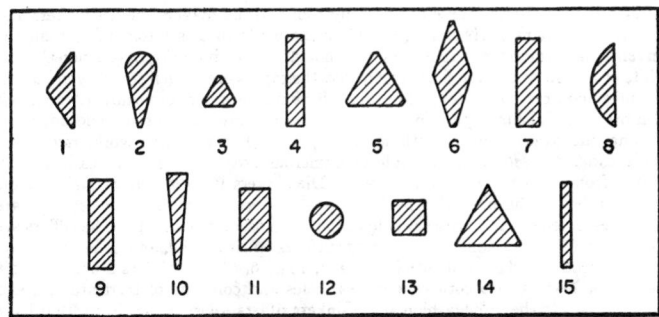

*Machinists' Files:* These files are used throughout industry where metal must be removed rapidly and finish is of secondary importance. Except for certain exceptions in the round and half-round shapes, all are double cut. *Flat files* (7) have a rectangular section, are tapered in width and thickness, are cut on both sides and edges and are used for general utility work. *Half round files* (8) have a circular segmental section, are tapered in width and thickness, have their flat side double cut, their rounded side mostly double but sometimes single cut, and are used to file rounded holes, concave corners, etc. in general filing work. *Hand files* (9) are similar to flat files but taper in thickness only. One edge is uncut or "safe." *Knife files* (10) have a "knife-blade" section, are tapered in width only, are double cut, and are used by tool and die makers on work having acute angles. *Machinist's General Purpose files* have a rectangular section, are tapered and have single cut teeth divided by angular serrations which produce short cutting edges These edges help stock removal but still leave a smooth finish and are suitable for use on various materials including aluminum, bronze, cast iron, malleable iron, mild steels and annealed tool steels. *Pillar files* (11) are similar to hand files but are thicker and not as wide. *Round files* (12) have a circular section, are tapered, single cut, and are generally used to file circular openings or curved surfaces. *Square files* (13) have a square section, are tapered, and are used for filing slots, keyways and for general surface filing where a heavier section is preferred. *Three square files* (14) have an equilateral triangular section and are tapered on all sides. They are double cut and have sharp corners as contrasted with taper triangular files which are single cut and have somewhat rounded corners. They are used for filing accurate internal angles, for clearing out square corners, and for filing taps and cutters. *Warding files* (15) have a rectangular section, and taper in width to a narrow point. They are used for general narrow space filing. *Wood files* are made in the same sections as flat and half round files but with coarser teeth especially suited for working on wood.

*Curved Tooth Files:* Regular curved tooth files are made in both rigid and flexible forms. The rigid type has either a tang for a conventional handle or is made plain with a hole at each end for mounting in a special holder. The flexible type is furnished plain for use in special holders only. The curved tooth files come in standard, fine and smooth cuts and in parallel flat, square, pillar, pillar narrow, half round and shell types. A special curved tooth file is available with teeth divided by long angular serrations. The teeth are cut in an " off center " arc. When moved across the work toward one edge of the file a fast cutting action is provided; when moved toward the other edge, a smoothing action; thus the file is made to serve a dual purpose.

*Swiss Pattern Files:* These are used by tool and die makers, model makers and delicate instrument parts finishers. They are made to closer tolerances than the conventional American pattern files although with similar cross-sections. The points of the Swiss pattern files are smaller, the tapers are longer and they are available in much finer cuts. They are primarily finishing tools for removing burrs left from previous finishing operations, truing up narrow grooves, notches and keyways, cleaning out corners and smoothing small parts. For very fine work, *round* and *square handled needle files*, available in numerous cross-sectional shapes in overall lengths from 4 to 7¾ inches, are used. Die sinkers use *die sinkers files* and *die sinkers rifflers*. The files, also made in many different cross-sectional shapes, are 3½ inches in length and are available in the cut Nos. o, 1, 2 and 4. The rifflers are from 5½ to 6¾ inches long, have cutting surfaces on either end, and come in numerous cross-sectional shapes in cut Nos. o, 2, 3, 4 and 6. These rifflers are used by die makers for getting into corners, crevices, holes and contours of intricate dies and molds. Used in the same fashion as die sinkers rifflers, *silversmiths rifflers*, that have a much heavier cross-section, are available in lengths from 6⅞ to 8 inches and in cut Nos. o, 1, 2 and 3. *Blunt machine files* in cut Nos. oo, o, and 2 for use in ordinary and bench filing machines are available in many different cross-sectional shapes, in lengths from 3 to 8 inches.

*Rasps:* Rasps are employed for work on relatively soft substances such as wood, leather, and lead where fast removal of material is required. They come in rectangular and half round cross-sections, the latter with and without a sharp edge.

*Special Purpose Files:* Falling under one of the preceding five classes of files, but modified to meet the requirements of some particular function, are a number of special purpose files. The *long angle lathe file* is used for filing work that is rotating in a lathe. The long tooth angle provides a clean shear, eliminates drag or tear and is self clearing. This file has safe or uncut edges to protect shoulders of the work which are not to be filed. The *foundry file* has especially sturdy teeth with heavy set edges for the snagging of castings — the removing of fins, sprues and other projections. The *die casting file* has extra strong teeth on corners and edges as well as sides for working on die castings of magnesium, zinc, or aluminum alloys. A special file for stainless steel is designed to stand up under the abrasive action of stainless steel alloys. *Aluminum rasps* and *files* are designed to eliminate clogging. A special tooth construction is used in one type of aluminum file which breaks up the filings, allows the file to clear itself and overcomes chatter. A *brass file* is designed so that with a little pressure the sharp, high-cut teeth bite deep while with less pressure, their short upcut angle produces a smoothing effect. The *lead float* has coarse, single cut teeth at almost right angles to the file axis. These shear away the metal under ordinary pressure and produce a smoothing effect under light pressure. The *shear tooth file* has a coarse single cut with a long angle for soft metals or alloys, plastics, hard rubber and wood. *Chain saw files* are designed to sharpen all types of chain saw teeth. These files come in round, rectangular, square and diamond-shaped sections. The round and square sectioned files have either double or single cut teeth, the rectangular files have single cut teeth and the diamond-shaped files have double cut teeth.

**Filing Methods.** — In *straight filing*, the file is pushed lengthwise, straight ahead or slightly diagonally across the work. The amount of pressure to apply should be sufficient to enable the file to cut at all times. Too much pressure tends to clog the teeth and break them off. Too little pressure, especially on harder materials tends to dull the teeth quickly. In *drawfiling*, the axis of the file is held at right angles to the direction of motion of the stroke This method produces a somewhat finer finish than straight filing. In *lathe filing* the file is stroked constantly across the revolving work and not held rigid or stationary. A slight gliding or lateral motion assists the file to clear itself of chips, and also avoids producing ridges or scores.

**Rotary Files and Burs.** — Rotary files and burs are used with power-operated tools, such as flexible- or stationary-shaft machines, drilling machines, lathes, and portable electric or pneumatic tools, for abrading or smoothing metals and other materials. Corners can be broken and chamfered, burrs and fins removed, holes and slots enlarged or elongated, and scale removed in die-sinking, metal pattern-making, mold finishing, toolmaking and casting operations.

The difference between rotary files and rotary burs, as defined by most companies, is that the former have teeth cut by hand with hammer and chisel whereas the latter, have teeth or flutes ground from the solid blank after hardening, or milled from the solid blank before hardening. (At least one company, however, prefers to differentiate the two by use and size: The larger-sized general purpose tools with $\frac{1}{4}$-inch shanks, whether hand cut or ground, are referred to as rotary files; the smaller-shanked — $\frac{1}{8}$-inch — and correspondingly smaller-headed tools used by diesinkers and jewelers are referred to as burs.) Rotary files are made from high-speed steel and rotary burs from high-speed steel or cemented carbide in various cuts such as double extra coarse, extra coarse or rough, coarse or standard, medium, fine, and smooth. Standard shanks are $\frac{1}{4}$ inch in diameter.

**Use of Rotary Files and Burs.** — The choice between a rotary file and a bur depends on the type of job and the preference of the user. The rotary files, with their hand-cut interrupted teeth or flutes, are better suited for work on tough, dense metals, such as die steels, steel forgings, and electric and gas welds. Generally the burs are more efficient on non-ferrous metals.

In using rotary files or burs, the tool should be moved at an even rate and pressure to avoid producing an uneven surface. The machine chuck should grip the rotary file or bur shank near the head or cut section for accurate control. For efficient operation, the tools should be kept sharp by regrinding.

The speed at which the tool should be operated varies with the skill and technique of the operator, the type of power used, the material being removed, the type of operation, and the size of the bur. Approximate speeds of medium-cut rotary files and burs for general applications are given in the accompanying table. There are many instances, however, where speeds much higher than those given in the table have been used successfully. The best method is to determine from experience what speeds will give the most satisfactory results.

It is advisable to start using a bur at a fairly low speed and increase the speed as the bur becomes broken in until the best speed is found for the particular operation being performed. For finishing operations, higher than normal speeds with decreased pressure provide a better finish, although possibly with some sacrifice in rate of stock removal. Excessive speeds sometimes make it difficult to control the bur, and should be avoided. Speeds slower than normal usually result in a poorer finish and tend to decrease the rate of stock removal, although the latter is dependent on such factors as the pressure applied, characteristics of the material being operated on, and the manner in which the bur is used.

**Effectiveness of Rotary Files and Burs.** — There is very little difference in the efficiency of rotary files or burs when used in electric tools and when used in air tools, provided the speeds have been reasonably well selected. Flexible-shaft and other machines used as a source of power for these tools have a limited number of speeds which govern the revolutions per minute at which the tools can be operated.

The carbide bur may be used on hard or soft materials with equally good results. The principal difference in construction of the carbide bur is that its teeth or flutes are provided with a negative rather than a radial rake. Carbide burs are relatively brittle, and must be treated more carefully than ordinary burs. They should be kept cutting freely, in order to prevent too much pressure, which might result in crumbling of the cutting edges.

At the same speeds, both high-speed steel and carbide burs remove approximately the same amount of metal. However, when carbide burs are used at their most efficient speeds, the rate of stock removal may be as much as four times that of ordinary burs. In certain cases, speeds much higher than those shown in the table can be used. It has been demonstrated that a carbide bur will last up to 100 times as long as a high-speed steel bur of corresponding size and shape.

### Approximate Speeds of Rotary Files and Burs*

| Tool Diam., Inches | Medium Cut High-Speed Steel Bur or File | | | | | Carbide Bur | |
|---|---|---|---|---|---|---|---|
| | | | | | | Medium Cut | Fine Cut |
| | Mild Steel | Cast Iron | Bronze | Alumi-num | Magne-sium | Any Material | |
| | Speed, Revolutions per Minute | | | | | | |
| ⅛ | 4600 | 7000 | 15,000 | 20,000 | 30,000 | 45,000 | 30,000 |
| ¼ | 3450 | 5250 | 11,250 | 15,000 | 22,500 | 30,000 | 20,000 |
| ⅜ | 2750 | 4200 | 9000 | 12,000 | 18,000 | 24,000 | 16,000 |
| ½ | 2300 | 3500 | 7500 | 10,000 | 15,000 | 20,000 | 13,350 |
| ⅝ | 2000 | 3100 | 6650 | 8900 | 13,350 | 18,000 | 12,000 |
| ¾ | 1900 | 2900 | 6200 | 8300 | 12,400 | 16,000 | 10,650 |
| ⅞ | 1700 | 2600 | 5600 | 7500 | 11,250 | 14,500 | 9650 |
| 1 | 1600 | 2400 | 5150 | 6850 | 10,300 | 13,000 | 8650 |
| 1⅛ | 1500 | 2300 | 4850 | 6500 | 9750 | ..... | ..... |
| 1¼ | 1400 | 2100 | 4500 | 6000 | 9000 | ..... | ..... |

*As recommended by the Nicholson File Company.

# HACK SAW BLADES

Hack saw blades recommended for regular stock production are given in the accompanying table. According to American National Standard Specifications for Hack Saw Blades ANSI B121.1-1970, tooth set is wavy for blades with 24 or 32 teeth. Blades having other than 24 or 32 teeth are alternate-set (each tooth set alternately) or raker-set (every third tooth unset).

**Types of Hack Saw Blades.** — According to this American National Standard a *welded composite blade* is made up of two or more pieces of steel joined together, the cutting edge or edges of which consist of high-speed steel. A *standard steel blade* contains not more than 1.25 per cent of tungsten or an equivalent alloying element. A *high-speed steel blade* has an alloy content such that the steel retains its red hardness at not less than 1000 degrees F. within limits of 60 to 66 on the Rockwell "C" hardness scale.

**Rules for Use of Hand Hack Saw Blades.** — When using hack saw blades in hand frames, at least three consecutive teeth should be in contact with the stock being cut at all times, particularly when tubing or other thin-walled shapes are being cut. For general work where blades are not changed for each individual job a blade with 18 teeth per inch is recommended. When inserting the blade the teeth are always pointed away from the user which means that the blade cuts on the forward stroke. Blades are best operated at from 40 to 60 cutting strokes per minute. For thin sections and soft metals a light pressure is used; thick stock is best cut with coarse-tooth blades and a heavy pressure.

**Rules for Use of Power Hack Saw Blades.** — At least three consecutive teeth should always be in contact with the stock being cut. Blades with finer teeth should be used on hard stock and thin sections. The teeth of the blade should be pointed correctly; that is, toward the cut at the start of the power or cutting stroke. The blade should be kept tight at all times. A cutting lubricant should be used on all metals being cut with the exception of cast iron and copper. Except on soft or very thin-walled sections, apply all the pressure the teeth will stand without breaking off. As the teeth show signs of dulling, increase the pressure; when dull, replace. When replacing a blade after half finishing a cut, the workpiece should be turned around and a new cut started that will meet the old. Recommended cutting speeds for power driven hack saws are given on page 1794.

**American National Standard Hack Saw Blades** (ANSI B121.1-1970)

| Nominal Length | Number of Teeth per Inch | Width | Thickness | Length Overall | Center to Center of Pinholes | Pinhole Diameter |
|---|---|---|---|---|---|---|
| | | | Hand Hack Saw Blades — All Steel Types | | | |
| 10 | 18, 24, 32 | ½ | 0.025 | 10⅜ | 9⅞ | 0.156 |
| 12 | 14, 18, 24, 32 | ½ | 0.025 | 12⅜ | 11⅞ | 0.156 |
| | | | Power Hack Saw Blades — Types with High-Speed Steel Cutting Sections | | | |
| 12 | 14, 18 | ⅝ | 0.032 | 12½ | 11⅞ | 0.188 |
| 12 | 10, 14 | 1 | 0.050 | 12¾ | 11⅞ | 0.281 |
| 14 | 10, 14 | 1 | 0.050 | 14⅜ | 13½ | 0.281 |
| 14 | 6, 10 | 1¼ | 0.062 | 14½ | 13½ | 0.281 |
| 14 | 4, 6 | 1½ | 0.075 | 14½ | 13½ | 0.281 |
| 17 | 10, 14 | 1 | 0.050 | 17⅜ | 16½ | 0.281 |
| 17 | 4, 6, 10 | 1¼ | 0.062 | 17½ | 16½ | 0.281 |
| 18 | 6, 10 | 1¼ | 0.062 | 18½ | 17½ | 0.281 |
| 18 | 4, 6 | 1½ | 0.075 | 18½ | 17½ | 0.281 |
| 18 | 4, 6 | 1¾ | 0.088 | 18¾ | 17½ | 0.281 |
| 21 | 4, 6 | 1¾ | 0.088 | 22¼ | 21 | 0.281 |
| 24 | 4, 6 | 1¾ | 0.088 | 25¼ | 24 | 0.391 |
| 24 | 3, 4 | 2 | 0.100 | 25¼ | 24 | 0.391 |
| 30 | 4 | 2½ | 0.100 | 32 | 30 | 0.391 |
| 36 | 2½ | 4½ | 0.125 | 38 | 36 | 0.500 |

All dimensions are given in inches including tolerances.

*Tolerances.* — Hand hack saw blades: width, +0, −1⁄64; thickness (high-speed steel), ±0.003; thickness (standard steel), +0.001, −0.003; overall length, +0, −1⁄16; pinholes (center to center), ±1⁄64; and pinhole diameter, +0.010, −0. Power hack saw blades: width (double-edge blades), +¼, −0; width (single-edge blades, ⅝ in.), +1⁄16, −0; width (single-edge blades, 1 in.), +3⁄32, −0; width (single-edge blades, over 1 in.), +⅛, −0; thickness, ±0.003; overall length, ±1⁄16; pinholes (center to center), ±1⁄32; and pinhole diameter, +0.010, −0.

## JIGS AND FIXTURES

**Material for Jig Bushings.** — Bushings are generally made of a good grade of tool steel to insure hardening at a fairly low temperature and to lessen the danger of fire cracking. They can also be made from machine steel, which will answer all practical purposes, provided the bushings are properly casehardened to a depth of about ⅟₁₆ inch. Sometimes bushings for guiding tools may be made of cast iron, but only when the cutting tool is of such a design that no cutting edges come within the bushing itself. For example, bushings used simply to support the smooth surface of a boring-bar or the shank of a reamer might, in some instances, be made of cast iron, but hardened steel bushings should always be used for guiding drills, reamers, taps, etc., when the cutting edges come in direct contact with the guiding surfaces. If the outside diameter of the bushing is very large, as compared with the diameter of the cutting tool, the cost of the bushing can sometimes be reduced by using an outer cast-iron body and inserting a hardened tool steel bushing.

**Hardening Jig Bushings.** — When hardening bushings made of tool steel they should be brought to an even red heat in a clean fire; the heating should never be hurried. Gas furnaces are excellent for heating, but a clean charcoal fire will answer the purpose. As soon as the bushing has been brought to an even red heat, it should be dipped in water just warm enough to take off the chill. Heat bushing to a "sizzling" heat, and leave it in the air to cool.

**American National Standard Jig Bushings.** — Specifications for the following types of jig bushings are given in American National Standard B94.33-1962 (R1971): Head Type Press Fit Wearing Bushings, Type H (See Tables 1 and 3); Headless Type Press Fit Wearing Bushings, Type P (See Tables 2 and 3); Slip Type Renewable Wearing Bushings, Type S (See Tables 4 and 5); Fixed Type Renewable Wearing Bushings, Type F (See Tables 5 and 6); Headless Type Liner Bushings, Type L (See Table 7); and Head Type Liner Bushings, Type HL (See Table 8). Specifications for locking mechanisms are also given (See Table 9).

**Jig Bushing Definitions.** — *Renewable bushings:* Renewable wearing bushings to guide the tool are for use in liners which in turn are installed in the jig. They are used where the bushing will wear out or become obsolete before the jig or where several bushings are to be interchangeable in one hole. Renewable wearing bushings are divided into two classes, "Fixed" and "Slip." Fixed renewable bushings are installed in the liner with the intention of leaving them in place until worn out. Slip renewable bushings are interchangeable in a given size of liner and, to facilitate removal, they are usually made with a knurled head. They are most frequently used where two or more operations requiring different inside diameters are performed in a single jig, such as where drilling is followed by reaming, tapping, spot facing, counterboring or some other secondary operation.

*Press fit bushings:* Press fit wearing bushings to guide the tool are for installation directly in the jig without the use of a liner and are employed principally where the bushings are used for short production runs and will not require replacement. They are intended also for short center distances.

*Liner bushings:* Liner bushings are provided with and without heads and are permanently installed in a jig to receive the renewable wearing bushings. They are sometimes called master bushings.

**Jig Plate Thickness.** — The standard lengths of the press fit portion of jig bushings as established are based on standardized uniform jig plate thicknesses of ⁵⁄₁₆, ⅜, ½, ¾, 1, 1⅜, 1¾, 2⅛, 2½, and 3 inches.

**Table 3** (*Concluded*). **Specifications for Head Type H and Headless Type P Press Fit Wearing Bushings** (ANSI B94.33-1962, R1971)

Hole sizes are in accordance with American National Standard Twist Drill Sizes. The maximum and minimum values of the hole size, *A*, shall be as follows:

| Nominal Size of Hole | Maximum | Minimum |
|---|---|---|
| Above 0.0000 to 1/4 in. incl. | Nominal + 0.0004 in. | Nominal + 0.0001 in. |
| Above 1/4 to 3/4 in. incl. | Nominal + 0.0005 in. | Nominal + 0.0001 in. |
| Above 3/4 to 1 1/2 in. incl. | Nominal + 0.0006 in. | Nominal + 0.0002 in. |
| Above 1 1/2 | Nominal + 0.0007 in. | Nominal + 0.0003 in. |

Bushings in the size range from 0.0135 through 0.3125 will be counterbored to provide for lubrication and chip clearance.

Bushings without counterbore are optional and will be furnished upon request.

The size of the counterbore shall be inside diameter of the bushing + 1/32 inch.

The included angle at the bottom of the counterbore shall be 118 deg, ± 2 deg.

The depth of the counterbore shall be in accordance with the chart below to provide adequate drill bearing.

| Body Length *C*, In. | Drill Bushing Hole Size | | | | | | | | | | | |
|---|---|---|---|---|---|---|---|---|---|---|---|---|
| | 0.0135 to 0.0625 in. | | 0.0630 to 0.0995 in. | | 0.1015 to 0.1405 in. | | 0.1406 to 0.1875 in. | | 0.1890 to 0.2500 in. | | 0.257 to 0.3125 in. | |
| | P | H | P | H | P | H | P | H | P | H | P | H |
| | Minimum Drill Bearing Length — Inch | | | | | | | | | | | |
| 1/4 | X | 1/4 | X | X | X | X | X | X | X | X | X | X |
| 5/16 | X | 1/4 | X | X | X | X | X | X | X | X | X | X |
| 3/8 | 1/4 | 1/4 | X | X | X | X | X | X | X | X | X | X |
| 1/2 | 1/4 | 1/4 | X | 5/16 | X | 5/16 | X | 3/8 | X | X | X | X |
| 3/4 | + | + | 3/8 | 3/8 | 3/8 | 3/8 | X | 3/8 | X | X | X | X |
| 1 | + | + | + | + | + | + | 5/8 | 5/8 | 5/8 | 5/8 | 5/8 | 5/8 |
| 1 3/8 | + | + | + | + | + | + | + | + | 5/8 | 5/8 | 5/8 | 5/8 |
| 1 3/4 | + | + | + | + | + | + | + | + | 5/8 | 5/8 | 5/8 | 5/8 |

X — Indicates no counterbore. + — Indicates not American National Standard length.

**Table 4. American National Standard Slip Type Renewable Wearing Bushings — Type S** (ANSI B94.33-1962, R1971)

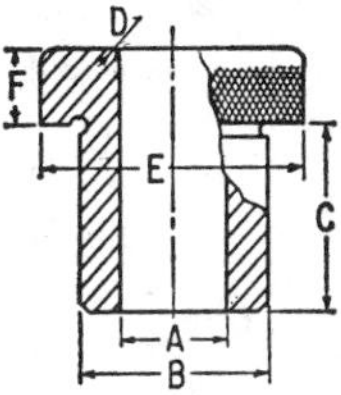


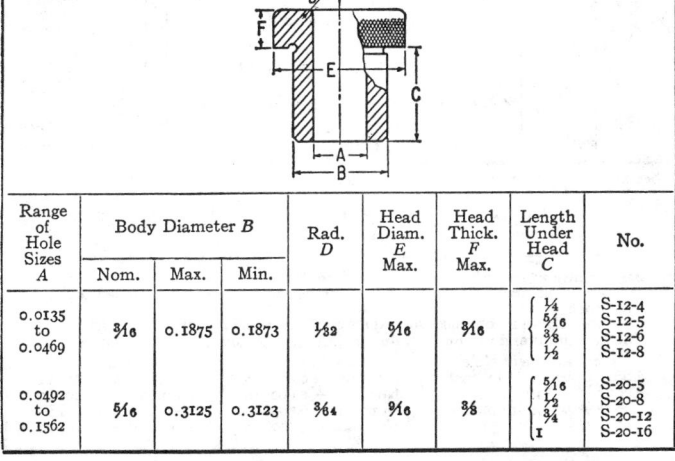

| Range of Hole Sizes *A* | Body Diameter *B* | | | Rad. *D* | Head Diam. *E* Max. | Head Thick. *F* Max. | Length Under Head *C* | No. |
|---|---|---|---|---|---|---|---|---|
| | Nom. | Max. | Min. | | | | | |
| 0.0135 to 0.0469 | 3/16 | 0.1875 | 0.1873 | 1/32 | 5/16 | 3/16 | 1/4<br>5/16<br>3/8<br>1/2 | S-12-4<br>S-12-5<br>S-12-6<br>S-12-8 |
| 0.0492 to 0.1562 | 5/16 | 0.3125 | 0.3123 | 3/64 | 9/16 | 3/8 | 5/16<br>1/2<br>3/4<br>1 | S-20-5<br>S-20-8<br>S-20-12<br>S-20-16 |

**Table 4** (*Concluded*). **American National Standard Slip Type Renewable Wearing Bushings — Type S** (ANSI B94.33-1962, R1971)

| Range of Hole Sizes *A* | Body Diameter *B* Nom. | Body Diameter *B* Max. | Body Diameter *B* Min. | Rad. *D* | Head Diam. *E* Max. | Head Thick. *F* Max. | Length Under Head *C* | No. |
|---|---|---|---|---|---|---|---|---|
| 0.1570 to 0.3125 | ½ | 0.5000 | 0.4998 | 3/64 | 13/16 | 7/16 | 5/16<br>½<br>¾<br>1<br>1⅜<br>1¾ | S-32-5<br>S-32-8<br>S-32-12<br>S-32-16<br>S-32-22<br>S-32-28 |
| 0.3160 to 0.5000 | ¾ | 0.7500 | 0.7498 | 3/32 | 1 1/16 | 7/16 | ½<br>¾<br>1<br>1⅜<br>1¾<br>2⅛ | S-48-8<br>S-48-12<br>S-48-16<br>S-48-22<br>S-48-28<br>S-48-34 |
| 0.5156 to 0.7500 | 1 | 1.0000 | 0.9998 | 3/32 | 1 7/16 | 7/16 | ½<br>¾<br>1<br>1⅜<br>1¾<br>2⅛<br>2½ | S-64-8<br>S-64-12<br>S-64-16<br>S-64-22<br>S-64-28<br>S-64-34<br>S-64-40 |
| 0.7656 to 1.0000 | 1⅜ | 1.3750 | 1.3747 | 3/32 | 1 13/16 | 7/16 | ¾<br>1<br>1⅜<br>1¾<br>2⅛<br>2½ | S-88-12<br>S-88-16<br>S-88-22<br>S-88-28<br>S-88-34<br>S-88-40 |
| 1.0156 to 1.3750 | 1¾ | 1.7500 | 1.7497 | ⅛ | 2 5/16 | ⅝ | 1<br>1⅜<br>1¾<br>2⅛<br>2½<br>3 | S-112-16<br>S-112-22<br>S-112-28<br>S-112-34<br>S-112-40<br>S-112-48 |
| 1.3906 to 1.7500 | 2¼ | 2.2500 | 2.2496 | ⅛ | 2 13/16 | ⅝ | 1<br>1⅜<br>1¾<br>2⅛<br>2½<br>3 | S-144-16<br>S-144-22<br>S-144-28<br>S-144-34<br>S-144-40<br>S-144-48 |

See also Table 5 for additional specifications.

**Table 5. Specifications for Slip Type S and Fixed Type F Renewable Wearing Bushings** (ANSI B94.33-1962, R1971)

All dimensions given in inches.

Tolerance on fractional dimensions where not otherwise specified shall be ± 0.010 inch.

Hole sizes are in accordance with American National Standard Twist Drill Sizes.

The maximum and minimum values of hole size, *A*, shall be as follows:

| Nominal Size of Hole | Maximum | Minimum |
|---|---|---|
| Above 0.0000 to ¼ in. incl. | Nominal + 0.0004 in. | Nominal + 0.0001 in. |
| Above ¼ to ¾ in. incl. | Nominal + 0.0005 in. | Nominal + 0.0001 in. |
| Above ¾ to 1½ in. incl. | Nominal + 0.0006 in. | Nominal + 0.0002 in. |
| Above 1½ | Nominal + 0.0007 in. | Nominal + 0.0003 in. |

**Table 5** *(Concluded)*. **Specifications for Slip Type S and Fixed Type F Renewable Wearing Bushings** (ANSI B94.33-1962, R1971)

The head design shall be in accordance with the manufacturer's practice.
Head of slip type is usually knurled.
When renewable wearing bushings are used with liner bushings of the head type, the length under the head will still be equal to the thickness of the jig plate, because the head of the liner bushing will be countersunk into the jig plate.
Diameter $A$ must be concentric to diameter $B$ within 0.0005 T.I.V. on finish ground bushings.
Size and type of chamfer on lead end to be manufacturer's option.
Bushings in the size range from 0.0135 through 0.3125 will be counterbored to provide for lubrication and chip clearance.
Bushings without counterbore are optional and will be furnished upon request.
The size of the counterbore shall be inside diameter of the bushings + $\frac{1}{32}$ inch.
The included angle at the bottom of the counterbore shall be 118 deg, ± 2 deg.
The depth of the counterbore shall be in accordance with the chart below to provide adequate drill bearing.

| Body Length $C$, In. | Drill Bushing Hole Size | | | | | | | | | | | |
|---|---|---|---|---|---|---|---|---|---|---|---|---|
| | 0.0135 to 0.0625 in. | | 0.0630 to 0.0995 in. | | 0.1015 to 0.1405 in. | | 0.1406 to 0.1875 in. | | 0.1890 to 0.2500 in. | | 0.2500 to 0.3125 in. | |
| | S | F | S | F | S | F | S | F | S | F | S | F |
| | Minimum Drill Bearing Length — Inch | | | | | | | | | | | |
| ¼ | ¼ | ¼ | ⅜ | ⅜ | X | X | X | X | X | X | X | X |
| 5⁄16 | ¼ | ¼ | ⅜ | ⅜ | ⅜ | ⅜ | ⅜ | ⅜ | ⅜ | ⅜ | X | X |
| ⅜ | ¼ | ¼ | ⅜ | ⅜ | ⅜ | ⅜ | ⅜ | ⅜ | ⅜ | ⅜ | X | X |
| ½ | ¼ | ¼ | ⅜ | ⅜ | ⅜ | ⅜ | ⅜ | ⅜ | ⅜ | ⅜ | X | X |
| ¾ | ¼ | ¼ | ⅜ | ⅜ | ⅜ | ⅜ | ⅜ | ⅜ | ⅝ | ⅝ | ⅝ | ⅝ |
| 1 | 5⁄16 | 5⁄16 | ⅜ | ⅜ | ⅜ | ⅜ | ⅝ | ⅝ | ⅝ | ⅝ | ⅝ | ⅝ |
| 1⅜ | + | + | + | + | + | + | ⅝ | ⅝ | ⅝ | ⅝ | ⅝ | ⅝ |
| 1¾ | + | + | + | + | + | + | ⅞ | ⅞ | ⅞ | ⅞ | ⅞ | ⅞ |

X — Indicates no counterbore.  + — Indicates not American National Standard length.

**Table 6. American National Standard Fixed Type Renewable Wearing Bushings — Type F** (ANSI B94.33-1962, R1971)

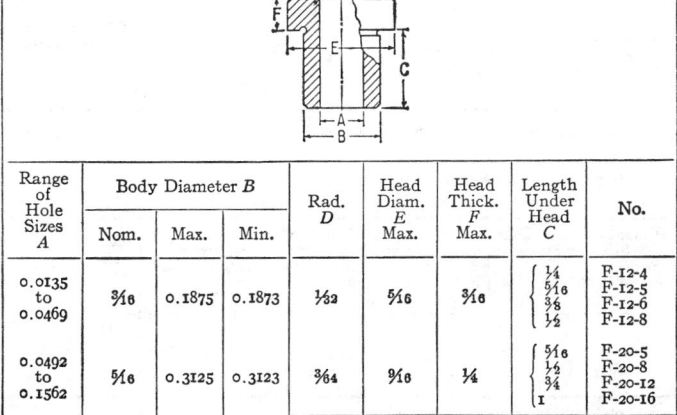

| Range of Hole Sizes $A$ | Body Diameter $B$ | | | Rad. $D$ | Head Diam. $E$ Max. | Head Thick. $F$ Max. | Length Under Head $C$ | No. |
|---|---|---|---|---|---|---|---|---|
| | Nom. | Max. | Min. | | | | | |
| 0.0135 to 0.0469 | 3⁄16 | 0.1875 | 0.1873 | 1⁄32 | 5⁄16 | 3⁄16 | ¼<br>5⁄16<br>⅜<br>½ | F-12-4<br>F-12-5<br>F-12-6<br>F-12-8 |
| 0.0492 to 0.1562 | 5⁄16 | 0.3125 | 0.3123 | 3⁄64 | 9⁄16 | ¼ | 5⁄16<br>½<br>¾<br>1 | F-20-5<br>F-20-8<br>F-20-12<br>F-20-16 |

Table 6 *(Concluded)*. **American National Standard Fixed Type Renewable Wearing Bushings — Type F** (ANSI B94.33-1962, R1971)

| Range of Hole Sizes A | Body Diameter B Nom. | Max. | Min. | Rad. D | Head Diam. E Max. | Head Thick. F Max. | Length Under Head C | No. |
|---|---|---|---|---|---|---|---|---|
| 0.1570 to 0.3125 | 1/2 | 0.5000 | 0.4998 | 3/64 | 13/16 | 1/4 | 5/16 | F-32-5 |
|  |  |  |  |  |  |  | 1/2 | F-32-8 |
|  |  |  |  |  |  |  | 3/4 | F-32-12 |
|  |  |  |  |  |  |  | 1 | F-32-16 |
|  |  |  |  |  |  |  | 1 3/8 | F-32-22 |
|  |  |  |  |  |  |  | 1 3/4 | F-32-28 |
| 0.3160 to 0.5000 | 3/4 | 0.7500 | 0.7498 | 3/32 | 1 1/16 | 1/4 | 1/2 | F-48-8 |
|  |  |  |  |  |  |  | 3/4 | F-48-12 |
|  |  |  |  |  |  |  | 1 | F-48-16 |
|  |  |  |  |  |  |  | 1 3/8 | F-48-22 |
|  |  |  |  |  |  |  | 1 3/4 | F-48-28 |
|  |  |  |  |  |  |  | 2 1/8 | F-48-34 |
| 0.5156 to 0.7500 | 1 | 1.0000 | 0.9998 | 3/32 | 1 7/16 | 3/8 | 1/2 | F-64-8 |
|  |  |  |  |  |  |  | 3/4 | F-64-12 |
|  |  |  |  |  |  |  | 1 | F-64-16 |
|  |  |  |  |  |  |  | 1 3/8 | F-64-22 |
|  |  |  |  |  |  |  | 1 3/4 | F-64-28 |
|  |  |  |  |  |  |  | 2 1/8 | F-64-34 |
|  |  |  |  |  |  |  | 2 1/2 | F-64-40 |
| 0.7656 to 1.0000 | 1 3/8 | 1.3750 | 1.3747 | 3/32 | 1 13/16 | 3/8 | 3/4 | F-88-12 |
|  |  |  |  |  |  |  | 1 | F-88-16 |
|  |  |  |  |  |  |  | 1 3/8 | F-88-22 |
|  |  |  |  |  |  |  | 1 3/4 | F-88-28 |
|  |  |  |  |  |  |  | 2 1/8 | F-88-34 |
|  |  |  |  |  |  |  | 2 1/2 | F-88-40 |
| 1.0156 to 1.3750 | 1 3/4 | 1.7500 | 1.7497 | 1/8 | 2 5/16 | 3/8 | 1 | F-112-16 |
|  |  |  |  |  |  |  | 1 3/8 | F-112-22 |
|  |  |  |  |  |  |  | 1 3/4 | F-112-28 |
|  |  |  |  |  |  |  | 2 1/8 | F-112-34 |
|  |  |  |  |  |  |  | 2 1/2 | F-112-40 |
|  |  |  |  |  |  |  | 3 | F-112-48 |
| 1.3906 to 1.7500 | 2 1/4 | 2.2500 | 2.2496 | 1/8 | 2 13/16 | 3/8 | 1 | F-144-16 |
|  |  |  |  |  |  |  | 1 3/8 | F-144-22 |
|  |  |  |  |  |  |  | 1 3/4 | F-144-28 |
|  |  |  |  |  |  |  | 2 1/8 | F-144-34 |
|  |  |  |  |  |  |  | 2 1/2 | F-144-40 |
|  |  |  |  |  |  |  | 3 | F-144-48 |

See also Table 5 for additional specifications.

Table 7. **American National Standard Headless Type Liner Bushings — Type L** (ANSI B94.33-1962, R1971)

| Range of Hole Sizes in Renewable Bushings | Inside Diameter A Nom. | Max. | Min. | Body Diameter B Nom. | Unfinished Max. | Min. | Finished Max. | Min. | Rad. D | Over-all Length C | No. |
|---|---|---|---|---|---|---|---|---|---|---|---|
| 0.0135 to 0.0469 | 3/16 | 0.1879 | 0.1876 | 5/16 | 0.3341 | 0.3288 | 0.3141 | 0.3138 | 1/32 | 1/4 | L-20-4 |
|  |  |  |  |  |  |  |  |  |  | 5/16 | L-20-5 |
|  |  |  |  |  |  |  |  |  |  | 3/8 | L-20-6 |
|  |  |  |  |  |  |  |  |  |  | 1/2 | L-20-8 |

For illustration of bushing, see following page.

Table 7 (*Concluded*).  American National Standard Headless Type Liner Bushings — Type L (ANSI B94.33-1962, R1971)

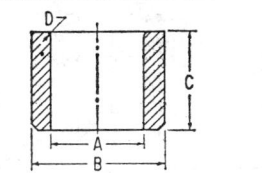

| Range of Hole Sizes in Renewable Bushings | Inside Diameter A | | | Body Diameter B | | | | | Rad. D | Over-all Length C | No. |
|---|---|---|---|---|---|---|---|---|---|---|---|
| | Nom. | Max. | Min. | Nom. | Unfinished | | Finished | | | | |
| | | | | | Max. | Min. | Max. | Min. | | | |
| 0.0492 to 0.1562 | 5/16 | 0.3129 | 0.3126 | 1/2 | 0.520 | 0.515 | 0.5017 | 0.5014 | 3/64 | 5/16<br>1/2<br>3/4<br>1 | L-32-5<br>L-32-8<br>L-32-12<br>L-32-16 |
| 0.1570 to 0.3125 | 1/2 | 0.5005 | 0.5002 | 3/4 | 0.770 | 0.765 | 0.7518 | 0.7515 | 1/16 | 5/16<br>1/2<br>3/4<br>1<br>1 3/8<br>1 3/4 | L-48-5<br>L-48-8<br>L-48-12<br>L-48-16<br>L-48-22<br>L-48-28 |
| 0.3160 to 0.5000 | 3/4 | 0.7506 | 0.7503 | 1 | 1.020 | 1.015 | 1.0018 | 1.0015 | 1/16 | 1/2<br>3/4<br>1<br>1 3/8<br>1 3/4<br>2 1/8 | L-64-8<br>L-64-12<br>L-64-16<br>L-64-22<br>L-64-28<br>L-64-34 |
| 0.5156 to 0.7500 | 1 | 1.0007 | 1.0004 | 1 3/8 | 1.395 | 1.390 | 1.3772 | 1.3768 | 3/32 | 1/2<br>3/4<br>1<br>1 3/8<br>1 3/4<br>2 1/8<br>2 1/2 | L-88-8<br>L-88-12<br>L-88-16<br>L-88-22<br>L-88-28<br>L-88-34<br>L-88-40 |
| 0.7656 to 1.0000 | 1 3/8 | 1.3760 | 1.3756 | 1 3/4 | 1.770 | 1.765 | 1.7523 | 1.7519 | 3/32 | 3/4<br>1<br>1 3/8<br>1 3/4<br>2 1/8<br>2 1/2 | L-112-12<br>L-112-16<br>L-112-22<br>L-112-28<br>L-112-34<br>L-112-40 |
| 1.0156 to 1.3750 | 1 3/4 | 1.7512 | 1.7508 | 2 1/4 | 2.270 | 2.265 | 2.2525 | 2.2521 | 3/32 | 1<br>1 3/8<br>1 3/4<br>2 1/8<br>2 1/2<br>3 | L-144-16<br>L-144-22<br>L-144-28<br>L-144-34<br>L-144-40<br>L-144-48 |
| 1.3906 to 1.7500 | 2 1/4 | 2.2515 | 2.2510 | 2 3/4 | 2.770 | 2.765 | 2.7526 | 2.7522 | 1/8 | 1<br>1 3/8<br>1 3/4<br>2 1/8<br>2 1/2<br>3 | L-176-16<br>L-176-22<br>L-176-28<br>L-176-34<br>L-176-40<br>L-176-48 |

All dimensions given in inches.

Tolerance on fractional dimensions where not otherwise specified shall be ± 0.010 inch.

The body diameter, B for unfinished bushings is 0.015 to 0.020 in. larger than the nominal diameter in order to provide grinding stock for fitting to jig plate holes.

Diameter A must be concentric to diameter B within 0.0005 T.I.V. on finish ground bushings.

**Table 8.  American National Standard Head Type Liner Bushings — Type HL**
(ANSI B94.33-1962, R1971)

| Range of Hole Sizes in Renewable Bushings | Inside Diameter A | | | Body Diameter B | | | | | Rad. D | Head Diam. E | Head Thick. F — Max. | Over-all Length C | No. |
|---|---|---|---|---|---|---|---|---|---|---|---|---|---|
| | Nom. | Max. | Min. | Nom. | Un-finished | | Finished | | | | | | |
| | | | | | Max. | Min. | Max. | Min. | | | | | |
| 0.0135 to 0.1562 | 5/16 | 0.3129 | 0.3126 | 1/2 | 0.520 | 0.515 | 0.5017 | 0.5014 | 3/64 | 5/8 | 3/32 | 5/16<br>1/2<br>3/4<br>1 | HL-32-5<br>HL-32-8<br>HL-32-12<br>HL-32-16 |
| 0.1570 to 0.3125 | 1/2 | 0.5005 | 0.5002 | 3/4 | 0.770 | 0.765 | 0.7518 | 0.7515 | 1/16 | 7/8 | 3/32 | 5/16<br>1/2<br>3/4<br>1<br>1 3/8<br>1 3/4 | HL-48-5<br>HL-48-8<br>HL-48-12<br>HL-48-16<br>HL-48-22<br>HL-48-28 |
| 0.3160 to 0.5000 | 3/4 | 0.7506 | 0.7503 | 1 | 1.020 | 1.015 | 1.0018 | 1.0015 | 1/16 | 1 1/8 | 1/8 | 1/2<br>3/4<br>1<br>1 3/8<br>1 3/4<br>2 1/8 | HL-64-8<br>HL-64-12<br>HL-64-16<br>HL-64-22<br>HL-64-28<br>HL-64-34 |
| 0.5156 to 0.7500 | 1 | 1.0007 | 1.0004 | 1 3/8 | 1.395 | 1.390 | 1.3772 | 1.3768 | 3/32 | 1 1/2 | 1/8 | 1/2<br>3/4<br>1<br>1 3/8<br>1 3/4<br>2 1/8<br>2 1/2 | HL-88-8<br>HL-88-12<br>HL-88-16<br>HL-88-22<br>HL-88-28<br>HL-88-34<br>HL-88-40 |
| 0.7656 to 1.0000 | 1 3/8 | 1.3760 | 1.3756 | 1 3/4 | 1.770 | 1.765 | 1.7523 | 1.7519 | 3/32 | 1 7/8 | 3/16 | 3/4<br>1<br>1 3/8<br>1 3/4<br>2 1/8<br>2 1/2 | HL-112-12<br>HL-112-16<br>HL-112-22<br>HL-112-28<br>HL-112-34<br>HL-112-40 |
| 1.0156 to 1.3750 | 1 3/4 | 1.7512 | 1.7508 | 2 1/4 | 2.270 | 2.265 | 2.2525 | 2.2521 | 3/32 | 2 3/8 | 3/16 | 1<br>1 3/8<br>1 3/4<br>2 1/8<br>2 1/2<br>3 | HL-144-16<br>HL-144-22<br>HL-144-28<br>HL-144-34<br>HL-144-40<br>HL-144-48 |
| 1.3906 to 1.7500 | 2 1/4 | 2.2515 | 2.2510 | 2 3/4 | 2.770 | 2.765 | 2.7526 | 2.7522 | 1/8 | 2 7/8 | 3/16 | 1<br>1 3/8<br>1 3/4<br>2 1/8<br>2 1/2<br>3 | HL-176-16<br>HL-176-22<br>HL-176-28<br>HL-176-34<br>HL-176-40<br>HL-176-48 |

All dimensions in inches.  See also footnotes to Table 7.

Table 9.  American National Standard Locking Mechanisms for Jig Bushings
(ANSI B94.33-1962, R1971)

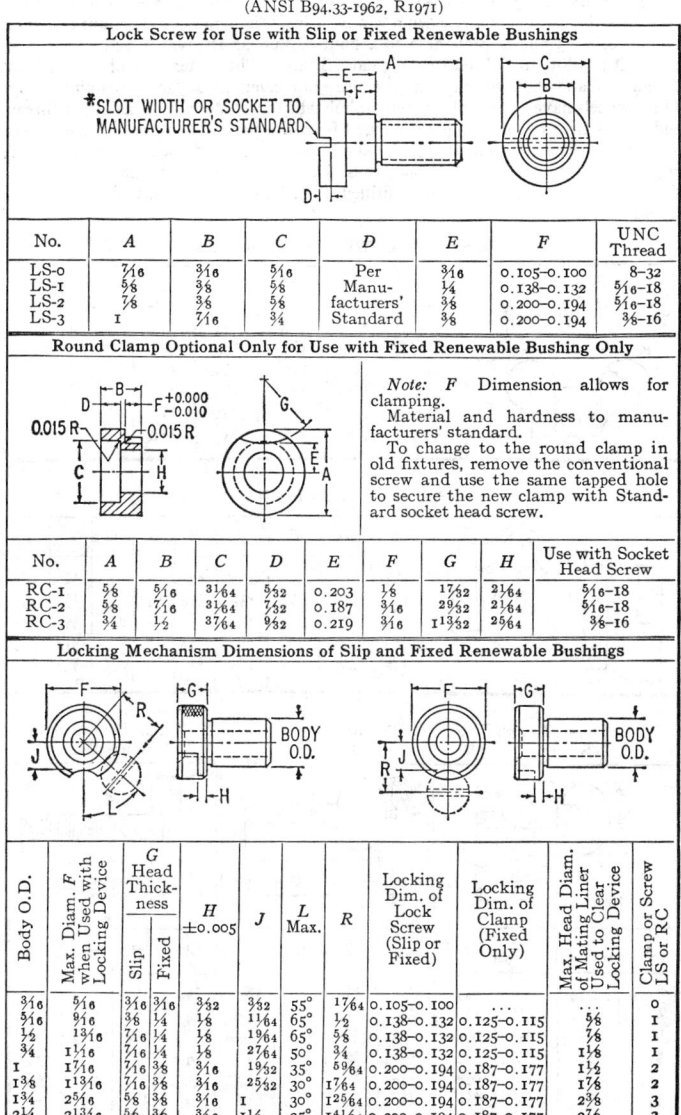

### Lock Screw for Use with Slip or Fixed Renewable Bushings

*SLOT WIDTH OR SOCKET TO MANUFACTURER'S STANDARD

| No. | A | B | C | D | E | F | UNC Thread |
|---|---|---|---|---|---|---|---|
| LS-0 | $\frac{7}{16}$ | $\frac{3}{16}$ | $\frac{5}{16}$ | Per Manu-facturers' Standard | $\frac{3}{16}$ | 0.105–0.100 | 8–32 |
| LS-1 | $\frac{5}{8}$ | $\frac{3}{8}$ | $\frac{5}{8}$ | | $\frac{1}{4}$ | 0.138–0.132 | $\frac{5}{16}$–18 |
| LS-2 | $\frac{7}{8}$ | $\frac{3}{8}$ | $\frac{5}{8}$ | | $\frac{3}{8}$ | 0.200–0.194 | $\frac{5}{16}$–18 |
| LS-3 | 1 | $\frac{7}{16}$ | $\frac{3}{4}$ | | $\frac{3}{8}$ | 0.200–0.194 | $\frac{3}{8}$–16 |

### Round Clamp Optional Only for Use with Fixed Renewable Bushing Only

*Note:* F Dimension allows for clamping.
Material and hardness to manufacturers' standard.
To change to the round clamp in old fixtures, remove the conventional screw and use the same tapped hole to secure the new clamp with Standard socket head screw.

| No. | A | B | C | D | E | F | G | H | Use with Socket Head Screw |
|---|---|---|---|---|---|---|---|---|---|
| RC-1 | $\frac{5}{8}$ | $\frac{5}{16}$ | $\frac{31}{64}$ | $\frac{5}{32}$ | 0.203 | $\frac{1}{8}$ | $\frac{17}{32}$ | $\frac{21}{64}$ | $\frac{5}{16}$–18 |
| RC-2 | $\frac{5}{8}$ | $\frac{7}{16}$ | $\frac{31}{64}$ | $\frac{7}{32}$ | 0.187 | $\frac{3}{16}$ | $\frac{29}{32}$ | $\frac{21}{64}$ | $\frac{5}{16}$–18 |
| RC-3 | $\frac{3}{4}$ | $\frac{1}{2}$ | $\frac{37}{64}$ | $\frac{9}{32}$ | 0.219 | $\frac{3}{16}$ | $1\frac{13}{32}$ | $\frac{25}{64}$ | $\frac{3}{8}$–16 |

### Locking Mechanism Dimensions of Slip and Fixed Renewable Bushings

| Body O.D. | Max. Diam. F when Used with Locking Device | G Head Thickness Slip | G Head Thickness Fixed | H ±0.005 | J | L Max. | R | Locking Dim. of Lock Screw (Slip or Fixed) | Locking Dim. of Clamp (Fixed Only) | Max. Head Diam. of Mating Liner Used to Clear Locking Device | Clamp or Screw LS or RC |
|---|---|---|---|---|---|---|---|---|---|---|---|
| $\frac{3}{16}$ | $\frac{5}{16}$ | $\frac{3}{16}$ | $\frac{3}{16}$ | $\frac{3}{32}$ | $\frac{3}{32}$ | 55° | $\frac{17}{64}$ | 0.105–0.100 | ... | ... | 0 |
| $\frac{5}{16}$ | $\frac{9}{16}$ | $\frac{3}{8}$ | $\frac{1}{4}$ | $\frac{1}{8}$ | $\frac{11}{64}$ | 65° | $\frac{1}{2}$ | 0.138–0.132 | 0.125–0.115 | $\frac{5}{8}$ | 1 |
| $\frac{1}{2}$ | $1\frac{3}{16}$ | $\frac{7}{16}$ | $\frac{1}{4}$ | $\frac{1}{8}$ | $\frac{19}{64}$ | 65° | $\frac{5}{8}$ | 0.138–0.132 | 0.125–0.115 | $\frac{7}{8}$ | 1 |
| $\frac{3}{4}$ | $1\frac{1}{16}$ | $\frac{7}{16}$ | $\frac{1}{4}$ | $\frac{1}{8}$ | $\frac{27}{64}$ | 50° | $\frac{3}{4}$ | 0.138–0.132 | 0.125–0.115 | $1\frac{1}{8}$ | 1 |
| 1 | $1\frac{5}{16}$ | $\frac{7}{16}$ | $\frac{3}{8}$ | $\frac{3}{16}$ | $\frac{19}{32}$ | 35° | $\frac{59}{64}$ | 0.200–0.194 | 0.187–0.177 | $1\frac{1}{2}$ | 2 |
| $1\frac{3}{8}$ | $1\frac{13}{16}$ | $\frac{7}{16}$ | $\frac{3}{8}$ | $\frac{3}{16}$ | $\frac{25}{32}$ | 30° | $1\frac{7}{64}$ | 0.200–0.194 | 0.187–0.177 | $1\frac{7}{8}$ | 2 |
| $1\frac{3}{4}$ | $2\frac{5}{16}$ | $\frac{5}{8}$ | $\frac{3}{8}$ | $\frac{3}{16}$ | 1 | 30° | $1\frac{25}{64}$ | 0.200–0.194 | 0.187–0.177 | $2\frac{3}{8}$ | 3 |
| $2\frac{1}{4}$ | $2\frac{13}{16}$ | $\frac{5}{8}$ | $\frac{3}{8}$ | $\frac{3}{16}$ | $1\frac{1}{4}$ | 25° | $1\frac{41}{64}$ | 0.200–0.194 | 0.187–0.177 | $2\frac{7}{8}$ | 3 |

**Wing-nuts for Jigs. — Star Handwheels.** — Wing-nuts are used on hook bolts or swiveling eye-bolts, when a comparatively light pressure is required. The thumb- or wing-nut is preferable to a knurled nut, as it gives a better grip and makes it possible to tighten the bolt more firmly. The dimensions of an excellent design of handwheel for use on jigs, etc., are given in an accompanying table. These wheels have a rather long stem or hub which provides a good length of thread and brings the grip or handle far enough from the jig body to prevent the fingers or knuckles from striking it. The "star" design of handle also permits a good grip. By having the casting solid, these handwheels can be tapped out for any size thread, or a plain hole can be drilled when it is desired to attach the handles to round stock.

### Dimensions of Wing or Thumb Nuts

| | A | B | C | D | E | F | G |
|---|---|---|---|---|---|---|---|
| | 3/16 | 5/8 | 1 3/16 | 5/16 | 3/8 | 7/16 | 1/8 |
| | 1/4 | 3/4 | 1 1/2 | 15/32 | 1/2 | 17/32 | 5/32 |
| | 5/16 | 3/4 | 1 1/2 | 15/32 | 1/2 | 17/32 | 5/32 |
| | 3/8 | 13/16 | 1 3/4 | 17/32 | 9/16 | 5/8 | 5/32 |
| | 7/16 | 7/8 | 2 | 21/32 | 5/8 | 11/16 | 3/16 |
| | 1/2 | 1 1/16 | 2 1/2 | 3/4 | 13/16 | 7/8 | 3/16 |

### Star Handwheels for Jigs

| | A | B | C | D | E | F | G | H | I |
|---|---|---|---|---|---|---|---|---|---|
| | 3/4 | 1 3/4 | 1 | 1 | 3/8 | 3/16 | 5/16 | 1/8 | 1/8 |
| | 1 | 1 7/8 | 1 1/4 | 1 1/8 | 7/16 | 5/16 | 7/16 | 1/8 | 1/8 |
| | 1 1/8 | 2 | 1 1/2 | 1 3/8 | 1/2 | 3/8 | 9/16 | 3/16 | 3/16 |
| | 1 1/2 | 2 1/8 | 2 | 1 5/8 | 9/16 | 1/2 | 11/16 | 3/16 | 3/16 |
| | 1 5/8 | 2 1/4 | 2 1/2 | 1 3/4 | 5/8 | 7/8 | 7/8 | 3/16 | 1/4 |

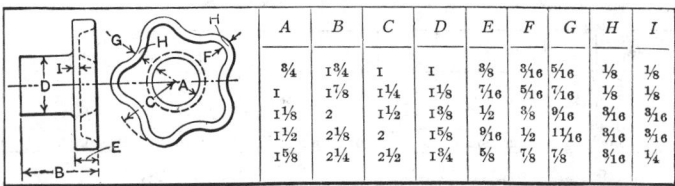

| Shoulder Thumb-Screws | | | | Regular Thumb-Screws | | | | Thumb-Screws with Wide Grip | | | |
|---|---|---|---|---|---|---|---|---|---|---|---|
| A | B | C | D | A | B | C | D | A | B | C | D |
| 3/16 | 9/16 | 1/2 | 3/32 | 3/16 | 3/4 | 5/8 | 3/32 | 3/16 | 1/4 | 7/8 | 3/32 |
| 1/4 | 1 | 11/16 | 1/8 | 1/4 | 15/16 | 3/4 | 1/8 | 1/4 | 5/16 | 1 | 1/8 |
| 5/16 | 1 1/8 | 3/4 | 5/32 | 5/16 | 1 1/8 | 7/8 | 5/32 | 3/8 | 7/16 | 1 1/2 | 5/32 |
| 3/8 | 1 1/4 | 13/16 | 5/32 | 3/8 | 1 1/4 | 15/16 | 5/32 | 7/16 | 1/2 | 1 3/4 | 3/16 |
| 7/16 | 1 1/2 | 15/16 | 3/16 | 7/16 | 1 1/2 | 1 1/16 | 3/16 | 1/2 | 9/16 | 2 | 3/16 |
| 1/2 | 1 5/8 | 1 1/8 | 3/16 | 1/2 | 1 5/8 | 1 3/16 | 3/16 | 5/8 | 11/16 | 2 1/2 | 7/32 |

## Wing Nuts, Wing Screws and Thumb Screws

**Wing Nuts.** — A wing nut is a nut having wings designed for manual turning without driver or wrench. As covered by ANSI Standard B18.17-1968 (R1975) wing nuts are classified first, by type on the basis of the method of manufacture; and second, by style on the basis of design characteristics. They consist of:

*Type A:* Type A wing nuts are cold forged or cold formed solid nuts having wings of moderate height. In some sizes they are produced in regular, light, and heavy series to best suit the requirements of specific applications. Dimensions are given in Table 1.

*Type B:* Type B wing nuts are hot forged solid nuts available in two wing styles: Style 1, having wings of moderate height; and Style 2, having high wings. Dimensions are given in Table 2.

*Type C:* Type C wing nuts are die cast solid nuts and are available in three wing styles: Style 1, having wings of moderate height; Style 2, having low wings; and Style 3, having high wings. In some sizes, the Style 1 nuts are produced in regular, light, and heavy series to best suit the requirements of specific applications. Dimensions are given in Table 3.

*Type D:* Type D wing nuts are stamped sheet metal nuts and are available in three styles: Style 1, having wings of moderate height; Style 2, having low wings; and Style 3, having wings of moderate height and a larger bearing surface. In some sizes, Styles 2 and 3 are produced in regular, light, and heavy series to best suit the requirements of specific applications. Dimensions are given in Table 4.

**Specification of Wing Nuts.** — When specifying wing nuts, the following data should be included in the designation and should appear in the following sequence: nominal size (number, fraction or decimal equivalent), threads per inch, type, style and/or series, material, and finish.

*Examples:*

10–32 Type A Wing Nut, Regular Series, Steel, Zinc Plated.

.250–20 Type C Wing Nut, Style 1, Zinc Alloy, Plain.

**Threads for Wing Nuts.** — Threads are in conformance with the ANSI Standard Unified Thread, Class 2B for all types of wing nuts except type D which have a modified Class 2B thread. Because of the method of manufacture, the minor diameter of the thread in type D nuts may be somewhat larger than the Unified Thread Class 2B maximum but shall in no case exceed the minimum pitch diameter.

**Materials and Finish for Wing Nuts.** — Types A, B, and D wing nuts are normally supplied as specified by the user in carbon steel, brass or corrosion resistant steel of good quality and adaptable to the manufacturing process. Type C wing nuts are made from die cast zinc alloy. Unless otherwise specified, wing nuts are supplied with a plain (unplated or uncoated) finish.

**Wing Screws.** — A wing screw is a screw having a wing-shaped head designed for manual turning without a driver or wrench. As covered by ANSI Standard B18.17-1968 (R1975) wing screws are classified first, by type on the basis of the method of manufacture, and second, by style on the basis of design characteristics. They consist of the following:

*Type A:* Type A wing screws are of two-piece construction having cold formed or cold forged wing portions of moderate height. In some sizes they are produced in regular, light, and heavy series to best suit the requirements of specific applications. Dimensions are given in Table 5.

*Type B:* Type B wing screws are of hot forged one-piece construction available

**Table 1.   American National Standard Type A Wing Nuts (ANSI B18.17-1968, R1975)**

| Nominal Size or Basic Major Diameter of Thread* | Thds. per Inch | Series† | Nut Blank Size (Ref) | A Wing Spread Max | A Min | B Wing Height Max | B Min | C Wing Thick. Max | C Min | D Between Wings Max | D Min | E Boss Diam. Max | E Min | G Boss Height Max | G Min |
|---|---|---|---|---|---|---|---|---|---|---|---|---|---|---|---|
| 3 (0.0990) | 48, 56 | Hvy. | AA | 0.72 | 0.59 | 0.41 | 0.28 | 0.11 | 0.07 | 0.21 | 0.17 | 0.33 | 0.29 | 0.14 | 0.10 |
| 4 (0.1120) | 40, 38 | Hvy. | AA | 0.72 | 0.59 | 0.41 | 0.28 | 0.11 | 0.07 | 0.21 | 0.17 | 0.33 | 0.29 | 0.14 | 0.10 |
| 5 (0.1250) | 40, 44 | Lgt. | AA | 0.72 | 0.59 | 0.41 | 0.28 | 0.11 | 0.07 | 0.21 | 0.17 | 0.33 | 0.29 | 0.14 | 0.10 |
|  |  | Hvy. | A | 0.91 | 0.78 | 0.47 | 0.34 | 0.14 | 0.10 | 0.27 | 0.22 | 0.43 | 0.39 | 0.18 | 0.14 |
| 6 (0.1380) | 32, 40 | Lgt. | AA | 0.72 | 0.59 | 0.41 | 0.28 | 0.11 | 0.07 | 0.21 | 0.17 | 0.33 | 0.29 | 0.14 | 0.10 |
|  |  | Hvy. | A | 0.91 | 0.78 | 0.47 | 0.34 | 0.14 | 0.10 | 0.27 | 0.22 | 0.43 | 0.39 | 0.18 | 0.14 |
| 8 (0.1640) | 32, 36 | Lgt. | A | 0.91 | 0.78 | 0.47 | 0.34 | 0.14 | 0.10 | 0.27 | 0.22 | 0.43 | 0.39 | 0.18 | 0.14 |
|  |  | Hvy. | B | 1.10 | 0.97 | 0.57 | 0.43 | 0.18 | 0.14 | 0.33 | 0.26 | 0.50 | 0.45 | 0.22 | 0.17 |
| 10 (0.1900) | 24, 32 | Lgt. | A | 0.91 | 0.78 | 0.47 | 0.34 | 0.14 | 0.10 | 0.27 | 0.22 | 0.43 | 0.39 | 0.18 | 0.14 |
|  |  | Hvy. | B | 1.10 | 0.97 | 0.57 | 0.43 | 0.18 | 0.14 | 0.33 | 0.26 | 0.50 | 0.45 | 0.22 | 0.17 |
| 12 (0.2160) | 24, 28 | Lgt. | B | 1.10 | 0.97 | 0.57 | 0.43 | 0.18 | 0.14 | 0.33 | 0.26 | 0.50 | 0.45 | 0.22 | 0.17 |
|  |  | Hvy. | C | 1.25 | 1.12 | 0.66 | 0.53 | 0.21 | 0.17 | 0.39 | 0.32 | 0.58 | 0.51 | 0.25 | 0.20 |
| ¼ (0.2500) | 20, 28 | Lgt. | B | 1.10 | 0.97 | 0.57 | 0.43 | 0.18 | 0.14 | 0.33 | 0.26 | 0.50 | 0.45 | 0.22 | 0.17 |
|  |  | Reg. | C | 1.25 | 1.12 | 0.66 | 0.53 | 0.21 | 0.17 | 0.39 | 0.32 | 0.58 | 0.51 | 0.25 | 0.20 |
|  |  | Hvy. | D | 1.44 | 1.31 | 0.79 | 0.65 | 0.24 | 0.20 | 0.48 | 0.42 | 0.70 | 0.64 | 0.30 | 0.26 |
| 5⁄16 (0.3125) | 18, 24 | Lgt. | C | 1.25 | 1.12 | 0.66 | 0.53 | 0.21 | 0.17 | 0.39 | 0.32 | 0.58 | 0.51 | 0.25 | 0.20 |
|  |  | Reg. | D | 1.44 | 1.31 | 0.79 | 0.65 | 0.24 | 0.20 | 0.48 | 0.42 | 0.70 | 0.64 | 0.30 | 0.26 |
|  |  | Hvy. | E | 1.94 | 1.81 | 1.00 | 0.87 | 0.33 | 0.26 | 0.65 | 0.54 | 0.93 | 0.86 | 0.39 | 0.35 |
| 3⁄8 (0.3750) | 16, 24 | Lgt. | D | 1.44 | 1.31 | 0.79 | 0.65 | 0.24 | 0.20 | 0.48 | 0.42 | 0.70 | 0.64 | 0.30 | 0.26 |
|  |  | Reg. | E | 1.94 | 1.81 | 1.00 | 0.87 | 0.33 | 0.26 | 0.65 | 0.54 | 0.93 | 0.86 | 0.39 | 0.35 |
| 7⁄16 (0.4375) | 14, 20 | Lgt. | E | 1.94 | 1.81 | 1.00 | 0.87 | 0.33 | 0.26 | 0.65 | 0.54 | 0.93 | 0.86 | 0.39 | 0.35 |
|  |  | Hvy. | F | 2.76 | 2.62 | 1.44 | 1.31 | 0.40 | 0.34 | 0.90 | 0.80 | 1.19 | 1.13 | 0.55 | 0.51 |
| ½ (0.5000) | 13, 20 | Lgt. | E | 1.94 | 1.81 | 1.00 | 0.87 | 0.33 | 0.26 | 0.65 | 0.54 | 0.93 | 0.86 | 0.39 | 0.35 |
|  |  | Hvy. | F | 2.76 | 2.62 | 1.44 | 1.31 | 0.40 | 0.34 | 0.90 | 0.80 | 1.19 | 1.13 | 0.55 | 0.51 |
| 9⁄16 (0.5825) | 12, 18 | Hvy. | F | 2.76 | 2.62 | 1.44 | 1.31 | 0.40 | 0.34 | 0.90 | 0.80 | 1.19 | 1.13 | 0.55 | 0.51 |
| 5⁄8 (0.6250) | 11, 18 | Hvy. | F | 2.76 | 2.62 | 1.44 | 1.31 | 0.40 | 0.34 | 0.90 | 0.80 | 1.19 | 1.13 | 0.55 | 0.51 |
| ¾ (0.7500) | 10, 16 | Hvy. | F | 2.76 | 2.62 | 1.44 | 1.31 | 0.40 | 0.34 | 0.90 | 0.80 | 1.19 | 1.13 | 0.55 | 0.51 |

All dimensions in inches.
* Where specifying nominal size in decimals, zeros in the fourth decimal place are omitted.
Note: Lgt. = Light; Hvy. = Heavy; Reg. = Regular. Sizes shown in **bold face** are preferred.

in two wing styles: Style 1, having wings of moderate height; and Style 2, having high wings.   Dimensions are given in Table 5.

*Type C:* Type C wing screws are available in two styles: Style 1, of a one-piece die cast construction having wings of moderate height; and Style 2, of a two-piece construction having a die cast wing portion of moderate height.   Dimensions are given in Table 6.

*Type D:* Type D wing screws are of two-piece welded construction having stamped

Table 5.   American National Standard Types A and B Wing Screws
(ANSI B18.17-1968, R1975)

Type A     STYLE 1     STYLE 2     Type B

SHANK PERMANENTLY INSERTED INTO WING PORTION — See Footnote 1

| Nominal Size or Basic Screw Diameter* | Thds. per Inch | Series† | Head Blank Size (Ref) | A Wing Spread Max | Min | B Wing Height Max | Min | C Wing Thick. Max | Min | E Boss Diam. Max | Min | G Boss Height Max | Min | L Practical Screw Lengths Max | Min |
|---|---|---|---|---|---|---|---|---|---|---|---|---|---|---|---|
| **TYPE A** | | | | | | | | | | | | | | | |
| 4 (0.1120) | 40 | Hvy. | AA | 0.72 | 0.59 | 0.41 | 0.28 | 0.11 | 0.07 | 0.33 | 0.29 | 0.14 | 0.10 | 0.75 | 0.25 |
| 6 (0.1380) | 32 | Lgt. | AA | 0.72 | 0.59 | 0.41 | 0.28 | 0.11 | 0.07 | 0.33 | 0.29 | 0.14 | 0.10 | }0.75 | 0.25 |
|  |  | **Hvy.** | **A** | **0.91** | **0.78** | **0.47** | **0.34** | **0.14** | **0.10** | **0.43** | **0.39** | **0.18** | **0.14** |  |  |
| 8 (0.1640) | 32 | **Lgt.** | **A** | **0.91** | **0.78** | **0.47** | **0.34** | **0.14** | **0.10** | **0.43** | **0.39** | **0.18** | **0.14** | }0.75 | 0.38 |
|  |  | Hvy. | B | 1.10 | 0.97 | 0.57 | 0.43 | 0.18 | 0.14 | 0.50 | 0.45 | 0.22 | 0.17 |  |  |
| 10 (0.1900) | 24, 32 | **Lgt.** | **A** | **0.91** | **0.78** | **0.47** | **0.34** | **0.14** | **0.10** | **0.43** | **0.39** | **0.18** | **0.14** | 1.00 | 0.38 |
|  |  | Hvy. | B | 1.10 | 0.97 | 0.57 | 0.43 | 0.18 | 0.14 | 0.50 | 0.45 | 0.22 | 0.17 |  |  |
| 12 (0.2160) | 24 | **Lgt.** | **B** | **1.10** | **0.97** | **0.57** | **0.43** | **0.18** | **0.14** | **0.50** | **0.45** | **0.22** | **0.17** | 1.00 | 0.38 |
|  |  | Hvy. | C | 1.25 | 1.12 | 0.66 | 0.53 | 0.21 | 0.17 | 0.58 | 0.51 | 0.25 | 0.20 |  |  |
| ¼ (0.2500) | 20 | **Lgt.** | **B** | **1.10** | **0.97** | **0.57** | **0.43** | **0.18** | **0.14** | **0.50** | **0.45** | **0.22** | **0.17** | }1.50 | 0.50 |
|  |  | Reg. | C | 1.25 | 1.12 | 0.66 | 0.53 | 0.21 | 0.17 | 0.58 | 0.51 | 0.25 | 0.20 |  |  |
|  |  | Hvy. | D | 1.44 | 1.31 | 0.79 | 0.65 | 0.24 | 0.20 | 0.70 | 0.64 | 0.30 | 0.26 |  |  |
| 5⁄16 (0.3125) | 18 | **Lgt.** | **C** | **1.25** | **1.12** | **0.66** | **0.53** | **0.21** | **0.17** | **0.58** | **0.51** | **0.25** | **0.20** | }1.50 | 0.50 |
|  |  | Reg. | D | 1.44 | 1.31 | 0.79 | 0.65 | 0.24 | 0.20 | 0.70 | 0.64 | 0.30 | 0.26 |  |  |
|  |  | Hvy. | E | 1.94 | 1.81 | 1.00 | 0.87 | 0.33 | 0.26 | 0.93 | 0.86 | 0.39 | 0.35 |  |  |
| 3⁄8 (0.3750) | 16 | **Lgt.** | **D** | **1.44** | **1.31** | **0.79** | **0.65** | **0.24** | **0.20** | **0.70** | **0.64** | **0.30** | **0.26** | }2.00 | 0.75 |
|  |  | Reg. | E | 1.94 | 1.81 | 1.00 | 0.87 | 0.33 | 0.26 | 0.93 | 0.86 | 0.39 | 0.35 |  |  |
|  |  | Hvy. | F | 2.76 | 2.62 | 1.44 | 1.31 | 0.40 | 0.34 | 1.19 | 1.13 | 0.55 | 0.51 |  |  |
| 7⁄16 (0.4375) | 14 | **Lgt.** | **E** | **1.94** | **1.81** | **1.00** | **0.87** | **0.33** | **0.26** | **0.93** | **0.86** | **0.39** | **0.35** | }4.00 | 1.00 |
|  |  | Hvy. | F | 2.76 | 2.62 | 1.44 | 1.31 | 0.40 | 0.34 | 1.19 | 1.13 | 0.55 | 0.51 |  |  |
| ½ (0.5000) | 13 | **Lgt.** | **E** | **1.94** | **1.81** | **1.00** | **0.87** | **0.33** | **0.26** | **0.93** | **0.86** | **0.39** | **0.35** | }4.00 | 1.00 |
|  |  | Hvy. | F | 2.76 | 2.62 | 1.44 | 1.31 | 0.40 | 0.34 | 1.19 | 1.13 | 0.55 | 0.51 |  |  |
| 5⁄8 (0.6250) | 11 | Hvy. | F | 2.76 | 2.62 | 1.44 | 1.31 | 0.40 | 0.34 | 1.19 | 1.13 | 0.55 | 0.51 | 4.00 | 1.25 |
| **TYPE B, STYLE 1** | | | | | | | | | | | | | | | |
| 10 (0.1900) | 24 | .... | ... | 0.97 | 0.91 | 0.45 | 0.39 | 0.15 | 0.12 | 0.39 | 0.36 | 0.28 | 0.22 | 2.00 | 0.50 |
| ¼ (0.2500) | 20 | .... | ... | 1.16 | 1.09 | 0.56 | 0.50 | 0.17 | 0.14 | 0.47 | 0.44 | 0.34 | 0.28 | 3.00 | 0.50 |
| 5⁄16 (0.3125) | 18 | .... | ... | 1.44 | 1.38 | 0.67 | 0.61 | 0.18 | 0.15 | 0.55 | 0.52 | 0.41 | 0.34 | 3.00 | 0.50 |
| 3⁄8 (0.3750) | 16 | .... | ... | 1.72 | 1.66 | 0.80 | 0.73 | 0.20 | 0.17 | 0.63 | 0.60 | 0.47 | 0.41 | 4.00 | 0.50 |
| 7⁄16 (0.4375) | 14 | .... | ... | 2.00 | 1.94 | 0.91 | 0.84 | 0.21 | 0.18 | 0.71 | 0.68 | 0.53 | 0.47 | 3.00 | 1.00 |
| ½ (0.5000) | 13 | .... | ... | 2.31 | 2.22 | 1.06 | 0.94 | 0.23 | 0.20 | 0.79 | 0.76 | 0.62 | 0.50 | 3.00 | 1.00 |
| 5⁄8 (0.6250) | 11 | .... | ... | 2.84 | 2.72 | 1.31 | 1.19 | 0.27 | 0.23 | 0.96 | 0.92 | 0.75 | 0.62 | 2.50 | 1.00 |
| **TYPE B, STYLE 2** | | | | | | | | | | | | | | | |
| 10 (0.1900) | 24 | .... | ... | 1.01 | 0.95 | 0.78 | 0.72 | 0.14 | 0.11 | 0.39 | 0.36 | 0.28 | 0.22 | 1.25 | 0.50 |
| ¼ (0.2500) | 20 | .... | ... | 1.22 | 1.16 | 0.94 | 0.88 | 0.16 | 0.13 | 0.47 | 0.44 | 0.34 | 0.28 | 2.00 | 0.50 |
| 5⁄16 (0.3125) | 18 | .... | ... | 1.43 | 1.37 | 1.09 | 1.03 | 0.17 | 0.14 | 0.55 | 0.52 | 0.41 | 0.34 | 2.00 | 0.50 |
| 3⁄8 (0.3750) | 16 | .... | ... | 1.63 | 1.57 | 1.25 | 1.19 | 0.18 | 0.15 | 0.63 | 0.60 | 0.47 | 0.41 | 2.00 | 0.50 |

All dimensions in inches.  Sizes shown in **bold face** are preferred.
[1] Plain point, unless alternate point from styles shown in Table 8 is specified by user.
* Where specifying nominal size in decimals, zeros in the fourth decimal place are omitted.
† Hvy. = Heavy; Lgt. = Light; Reg. = Regular.

## Table 6.  American National Standard Types C and D Wing Screws
(ANSI B18.17-1968, R1975)

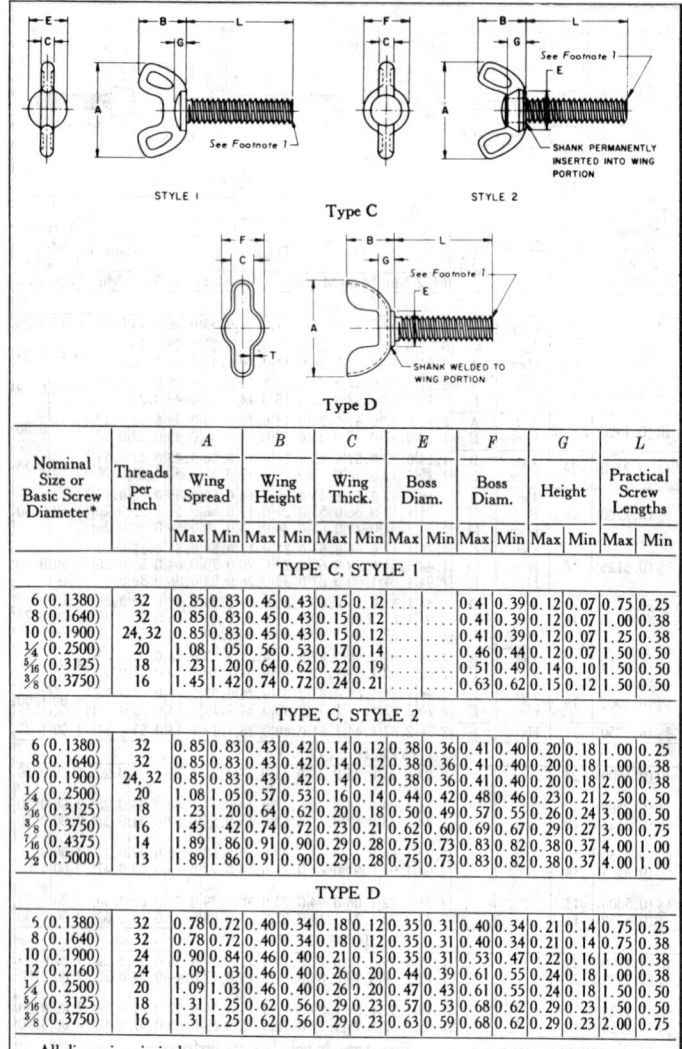

Type C

Type D

| Nominal Size or Basic Screw Diameter* | Threads per Inch | A Wing Spread | | B Wing Height | | C Wing Thick. | | E Boss Diam. | | F Boss Diam. | | G Height | | L Practical Screw Lengths | |
|---|---|---|---|---|---|---|---|---|---|---|---|---|---|---|---|
| | | Max | Min | Max | Min | Max | Min | Max | Min | Max | Min | Max | Min | Max | Min |
| TYPE C, STYLE 1 | | | | | | | | | | | | | | | |
| 6 (0.1380) | 32 | 0.85 | 0.83 | 0.45 | 0.43 | 0.15 | 0.12 | .... | .... | 0.41 | 0.39 | 0.12 | 0.07 | 0.75 | 0.25 |
| 8 (0.1640) | 32 | 0.85 | 0.83 | 0.45 | 0.43 | 0.15 | 0.12 | .... | .... | 0.41 | 0.39 | 0.12 | 0.07 | 1.00 | 0.38 |
| 10 (0.1900) | 24, 32 | 0.85 | 0.83 | 0.45 | 0.43 | 0.15 | 0.12 | .... | .... | 0.41 | 0.39 | 0.12 | 0.07 | 1.25 | 0.38 |
| ¼ (0.2500) | 20 | 1.08 | 1.05 | 0.56 | 0.53 | 0.17 | 0.14 | .... | .... | 0.46 | 0.44 | 0.12 | 0.07 | 1.50 | 0.50 |
| ⁵⁄₁₆ (0.3125) | 18 | 1.23 | 1.20 | 0.64 | 0.62 | 0.22 | 0.19 | .... | .... | 0.51 | 0.49 | 0.14 | 0.10 | 1.50 | 0.50 |
| ⅜ (0.3750) | 16 | 1.45 | 1.42 | 0.74 | 0.72 | 0.24 | 0.21 | .... | .... | 0.63 | 0.62 | 0.15 | 0.12 | 1.50 | 0.50 |
| TYPE C, STYLE 2 | | | | | | | | | | | | | | | |
| 6 (0.1380) | 32 | 0.85 | 0.83 | 0.43 | 0.42 | 0.14 | 0.12 | 0.38 | 0.36 | 0.41 | 0.40 | 0.20 | 0.18 | 1.00 | 0.25 |
| 8 (0.1640) | 32 | 0.85 | 0.83 | 0.43 | 0.42 | 0.14 | 0.12 | 0.38 | 0.36 | 0.41 | 0.40 | 0.20 | 0.18 | 1.00 | 0.38 |
| 10 (0.1900) | 24, 32 | 0.85 | 0.83 | 0.43 | 0.42 | 0.14 | 0.12 | 0.38 | 0.36 | 0.41 | 0.40 | 0.20 | 0.18 | 2.00 | 0.38 |
| ¼ (0.2500) | 20 | 1.08 | 1.05 | 0.57 | 0.53 | 0.16 | 0.14 | 0.44 | 0.42 | 0.48 | 0.46 | 0.23 | 0.21 | 2.50 | 0.50 |
| ⁵⁄₁₆ (0.3125) | 18 | 1.23 | 1.20 | 0.64 | 0.62 | 0.20 | 0.18 | 0.50 | 0.49 | 0.57 | 0.55 | 0.26 | 0.24 | 3.00 | 0.50 |
| ⅜ (0.3750) | 16 | 1.45 | 1.42 | 0.74 | 0.72 | 0.23 | 0.21 | 0.62 | 0.60 | 0.69 | 0.67 | 0.29 | 0.27 | 3.00 | 0.75 |
| ⁷⁄₁₆ (0.4375) | 14 | 1.89 | 1.86 | 0.91 | 0.90 | 0.29 | 0.28 | 0.75 | 0.73 | 0.83 | 0.82 | 0.38 | 0.37 | 4.00 | 1.00 |
| ½ (0.5000) | 13 | 1.89 | 1.86 | 0.91 | 0.90 | 0.29 | 0.28 | 0.75 | 0.73 | 0.83 | 0.82 | 0.38 | 0.37 | 4.00 | 1.00 |
| TYPE D | | | | | | | | | | | | | | | |
| 6 (0.1380) | 32 | 0.78 | 0.72 | 0.40 | 0.34 | 0.18 | 0.12 | 0.35 | 0.31 | 0.40 | 0.34 | 0.21 | 0.14 | 0.75 | 0.25 |
| 8 (0.1640) | 32 | 0.78 | 0.72 | 0.40 | 0.34 | 0.18 | 0.12 | 0.35 | 0.31 | 0.40 | 0.34 | 0.21 | 0.14 | 0.75 | 0.38 |
| 10 (0.1900) | 24 | 0.90 | 0.84 | 0.46 | 0.40 | 0.21 | 0.15 | 0.35 | 0.31 | 0.53 | 0.47 | 0.22 | 0.16 | 1.00 | 0.38 |
| 12 (0.2160) | 24 | 1.09 | 1.03 | 0.46 | 0.40 | 0.26 | 0.20 | 0.44 | 0.39 | 0.61 | 0.55 | 0.24 | 0.18 | 1.00 | 0.38 |
| ¼ (0.2500) | 20 | 1.09 | 1.03 | 0.46 | 0.40 | 0.26 | 0.20 | 0.47 | 0.43 | 0.61 | 0.55 | 0.24 | 0.18 | 1.50 | 0.50 |
| ⁵⁄₁₆ (0.3125) | 18 | 1.31 | 1.25 | 0.62 | 0.56 | 0.29 | 0.23 | 0.57 | 0.53 | 0.68 | 0.62 | 0.29 | 0.23 | 1.50 | 0.50 |
| ⅜ (0.3750) | 16 | 1.31 | 1.25 | 0.62 | 0.56 | 0.29 | 0.23 | 0.63 | 0.59 | 0.68 | 0.62 | 0.29 | 0.23 | 2.00 | 0.75 |

All dimensions in inches.
[1] Plain point, unless alternate point from styles shown in Table 8 is specified by user.
* Where specifying nominal size in decimals, zeros in the fourth decimal place are omitted.

Table 7.   American National Standard Types A and B Thumb Screws
(ANSI B18.17-1968, R1975)

Type A   Type B

| Nominal Size or Basic Screw Diameter* | Threads per Inch | A Head Width | | B Head Height | | C Head Thick. | | C' Head Thick. | | E Shoulder Diameter | | L Practical Screw Lengths | |
|---|---|---|---|---|---|---|---|---|---|---|---|---|---|
| | | Max | Min | Max | Min | Max | Min | Max | Min | Max | Min | Max | Min |
| **TYPE A, REGULAR** | | | | | | | | | | | | | |
| 6 (0.1380) | 32 | 0.31 | 0.29 | 0.33 | 0.31 | 0.05 | 0.04 | .... | .... | 0.25 | 0.23 | 0.75 | 0.25 |
| 8 (0.1640) | 32 | 0.36 | 0.34 | 0.38 | 0.36 | 0.06 | 0.05 | .... | .... | 0.31 | 0.29 | 0.75 | 0.38 |
| 10 (0.1900) | 24, 32 | 0.42 | 0.40 | 0.48 | 0.46 | 0.06 | 0.05 | .... | .... | 0.35 | 0.32 | 1.00 | 0.38 |
| 12 (0.2160) | 24 | 0.48 | 0.46 | 0.54 | 0.52 | 0.06 | 0.05 | .... | .... | 0.40 | 0.38 | 1.00 | 0.38 |
| 1/4 (0.2500) | 20 | 0.55 | 0.52 | 0.64 | 0.61 | 0.07 | 0.05 | .... | .... | 0.47 | 0.44 | 1.50 | 0.50 |
| 5/16 (0.3125) | 18 | 0.70 | 0.67 | 0.78 | 0.75 | 0.09 | 0.07 | .... | .... | 0.59 | 0.56 | 1.50 | 0.50 |
| 3/8 (0.3750) | 16 | 0.83 | 0.80 | 0.95 | 0.92 | 0.11 | 0.09 | .... | .... | 0.76 | 0.71 | 2.00 | 0.75 |
| **TYPE A, HEAVY** | | | | | | | | | | | | | |
| 10 (0.1900) | 24 | 0.89 | 0.83 | 0.84 | 0.72 | 0.18 | 0.16 | 0.10 | 0.08 | 0.33 | 0.31 | 2.00 | 0.50 |
| 1/4 (0.2500) | 20 | 1.05 | 0.99 | 0.94 | 0.81 | 0.24 | 0.22 | 0.10 | 0.08 | 0.40 | 0.38 | 3.00 | 0.50 |
| 5/16 (0.3125) | 18 | 1.21 | 1.15 | 1.00 | 0.88 | 0.27 | 0.25 | 0.11 | 0.09 | 0.46 | 0.44 | 4.00 | 0.50 |
| 3/8 (0.3750) | 16 | 1.41 | 1.34 | 1.16 | 1.03 | 0.30 | 0.28 | 0.11 | 0.09 | 0.55 | 0.53 | 4.00 | 0.50 |
| 7/16 (0.4375) | 14 | 1.59 | 1.53 | 1.22 | 1.09 | 0.36 | 0.34 | 0.13 | 0.11 | 0.71 | 0.69 | 2.50 | 1.00 |
| 1/2 (0.5000) | 13 | 1.81 | 1.72 | 1.28 | 1.16 | 0.40 | 0.38 | 0.14 | 0.12 | 0.83 | 0.81 | 3.00 | 1.00 |
| **TYPE B, REGULAR** | | | | | | | | | | | | | |
| 6 (0.1380) | 32 | 0.45 | 0.43 | 0.28 | 0.26 | 0.08 | 0.06 | 0.03 | 0.02 | .... | .... | 1.00 | 0.25 |
| 8 (0.1640) | 32 | 0.51 | 0.49 | 0.32 | 0.30 | 0.09 | 0.07 | 0.04 | 0.02 | .... | .... | 1.00 | 0.38 |
| 10 (0.1900) | 24, 32 | 0.58 | 0.54 | 0.39 | 0.36 | 0.10 | 0.08 | 0.05 | 0.03 | .... | .... | 2.00 | 0.38 |
| 12 (0.2160) | 24 | 0.71 | 0.67 | 0.45 | 0.43 | 0.11 | 0.09 | 0.05 | 0.03 | .... | .... | 2.00 | 0.38 |
| 1/4 (0.2500) | 20 | 0.83 | 0.80 | 0.52 | 0.48 | 0.16 | 0.14 | 0.06 | 0.03 | .... | .... | 2.50 | 0.50 |
| 5/16 (0.3125) | 18 | 0.96 | 0.91 | 0.64 | 0.60 | 0.17 | 0.14 | 0.09 | 0.06 | .... | .... | 3.00 | 0.50 |
| 3/8 (0.3750) | 16 | 1.09 | 1.03 | 0.71 | 0.67 | 0.22 | 0.18 | 0.11 | 0.08 | .... | .... | 3.00 | 0.75 |
| 7/16 (0.4375) | 14 | 1.40 | 1.35 | 0.96 | 0.91 | 0.27 | 0.24 | 0.14 | 0.11 | .... | .... | 4.00 | 1.00 |
| 1/2 (0.5000) | 13 | 1.54 | 1.46 | 1.09 | 1.03 | 0.33 | 0.29 | 0.15 | 0.11 | .... | .... | 4.00 | 1.00 |
| **TYPE B, HEAVY** | | | | | | | | | | | | | |
| 10 (0.1900) | 24 | 0.89 | 0.83 | 0.78 | 0.66 | 0.18 | 0.16 | 0.08 | 0.06 | .... | .... | 2.00 | 0.50 |
| 1/4 (0.2500) | 20 | 1.05 | 0.99 | 0.81 | 0.72 | 0.24 | 0.22 | 0.11 | 0.09 | .... | .... | 3.00 | 0.50 |
| 5/16 (0.3125) | 18 | 1.21 | 1.15 | 0.88 | 0.78 | 0.27 | 0.25 | 0.11 | 0.09 | .... | .... | 4.00 | 0.50 |
| 3/8 (0.3750) | 16 | 1.41 | 1.34 | 0.94 | 0.84 | 0.30 | 0.28 | 0.14 | 0.12 | .... | .... | 4.00 | 0.50 |
| 7/16 (0.4375) | 14 | 1.59 | 1.53 | 1.00 | 0.91 | 0.36 | 0.34 | 0.14 | 0.12 | .... | .... | 3.00 | 1.00 |
| 1/2 (0.5000) | 13 | 1.81 | 1.72 | 1.09 | 0.97 | 0.40 | 0.38 | 0.18 | 0.16 | .... | .... | 3.00 | 1.00 |

All dimensions in inches.
[1] Plain point, unless alternate point from styles shown in Table 8 is specified by user.
* Where specifying nominal size in decimals, zeros in the fourth decimal place are omitted.

### Table 8.   American National Standard Alternate Points for Wing and Thumb Screws
(ANSI B18.17-1968, R1975)

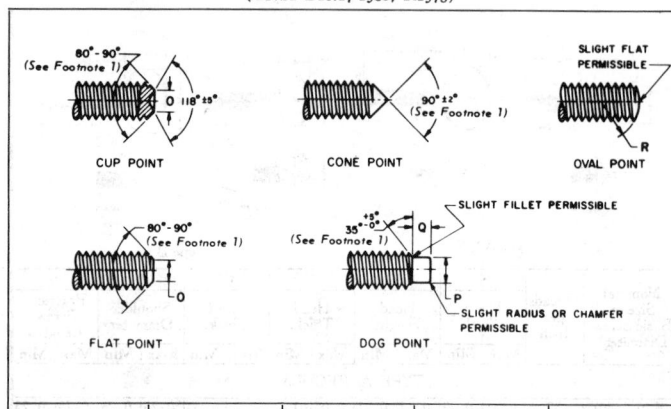

| Nominal Size or Basic Screw Diameter* | O Cup and Flat Point Diameter | | P Dog Point† Diameter | | Q Dog Point† Length | | R Oval Point Radius | |
|---|---|---|---|---|---|---|---|---|
| | Max | Min | Max | Min | Max | Min | Max | Min |
| 4 (0.1120) | 0.061 | 0.051 | 0.075 | 0.070 | 0.061 | 0.051 | 0.099 | 0.084 |
| 6 (0.1380) | 0.074 | 0.064 | 0.092 | 0.087 | 0.075 | 0.065 | 0.140 | 0.109 |
| 8 (0.1640) | 0.087 | 0.076 | 0.109 | 0.103 | 0.085 | 0.075 | 0.156 | 0.125 |
| 10 (0.1900) | 0.102 | 0.088 | 0.127 | 0.120 | 0.095 | 0.085 | 0.172 | 0.141 |
| 12 (0.2160) | 0.115 | 0.101 | 0.144 | 0.137 | 0.115 | 0.105 | 0.188 | 0.156 |
| ¼ (0.2500) | 0.132 | 0.118 | 0.156 | 0.149 | 0.130 | 0.120 | 0.219 | 0.188 |
| ⁵⁄₁₆ (0.3125) | 0.172 | 0.156 | 0.203 | 0.195 | 0.161 | 0.151 | 0.256 | 0.234 |
| ⅜ (0.3750) | 0.212 | 0.194 | 0.250 | 0.241 | 0.193 | 0.183 | 0.312 | 0.281 |
| ⁷⁄₁₆ (0.4375) | 0.252 | 0.232 | 0.297 | 0.287 | 0.224 | 0.214 | 0.359 | 0.328 |
| ½ (0.5000) | 0.291 | 0.270 | 0.344 | 0.334 | 0.255 | 0.245 | 0.406 | 0.375 |
| ⅝ (0.6250) | 0.371 | 0.347 | 0.469 | 0.456 | 0.321 | 0.305 | 0.500 | 0.469 |

All dimensions in inches.
[1] The external point angles specified shall apply to those portions of the angles which lie below the thread root diameter, it being recognized the angle within the thread profile may be varied due to the manufacturing processes.
* Where specifying nominal size in decimals, zeros in the fourth decimal place are omitted.
† The axis of dog points shall not be eccentric with the axis of the screw by more than 3 percent of the basic screw diameter or 0.005 in., whichever is the smaller.

Type D wing screws are normally supplied in carbon steel but also may be made from corrosion resistant steel, brass or other materials.

Thumb screws of all types are normally made from a good commercial quality of carbon steel having a maximum ultimate tensile strength of 48,000 psi. Where so specified, carbon steel thumb screws are case hardened. They are also made from corrosion resistant steel, brass, and other materials as agreed upon by the manufacturer and user.

Unless otherwise specified, wing screws and thumb screws are supplied with a plain (unplated or uncoated) finish.

Collar-head Screws

| A | B | C | D | E | F | Diameter G and Threads per Inch |
|---|---|---|---|---|---|---|
| 5/8 | 1/16 | 3/16 | 5/16 | 3/16 | 1/2 | No. 10 — 32 |
| 7/8 | 1/16 | 3/16 | 5/16 | 3/16 | 5/8 | No. 10 — 32 |
| 1 1/8 | 1/16 | 3/16 | 5/16 | 3/16 | 7/8 | No. 10 — 32 |
| 1 3/8 | 1/16 | 3/16 | 5/16 | 3/16 | 1 | No. 10 — 32 |
| 1 5/8 | 1/16 | 3/16 | 5/16 | 3/16 | 1 1/4 | No. 10 — 32 |
| 7/8 | 3/32 | 1/4 | 7/16 | 1/4 | 5/8 | No. 14 — 24 |
| 1 1/8 | 3/32 | 1/4 | 7/16 | 1/4 | 7/8 | No. 14 — 24 |
| 1 3/8 | 3/32 | 1/4 | 7/16 | 1/4 | 1 | No. 14 — 24 |
| 1 5/8 | 3/32 | 1/4 | 7/16 | 1/4 | 1 1/4 | No. 14 — 24 |
| 1 7/8 | 3/32 | 1/4 | 7/16 | 1/4 | 1 7/16 | No. 14 — 24 |
| 7/8 | 1/8 | 5/16 | 9/16 | 5/16 | 5/8 | 5/16 — 18 |
| 1 1/4 | 1/8 | 5/16 | 9/16 | 5/16 | 1 | 5/16 — 18 |
| 1 5/8 | 1/8 | 5/16 | 9/16 | 5/16 | 1 1/4 | 5/16 — 18 |
| 2 | 1/8 | 5/16 | 9/16 | 5/16 | 1 1/2 | 5/16 — 18 |
| 1 | 1/8 | 3/8 | 11/16 | 3/8 | 3/4 | 3/8 — 16 |
| 1 3/4 | 1/8 | 3/8 | 11/16 | 3/8 | 1 5/16 | 3/8 — 16 |
| 2 1/2 | 1/8 | 3/8 | 11/16 | 3/8 | 1 7/8 | 3/8 — 16 |
| 1 3/8 | 3/16 | 7/16 | 3/4 | 7/16 | 1 | 7/16 — 14 |
| 2 1/8 | 3/16 | 7/16 | 3/4 | 7/16 | 1 5/8 | 7/16 — 14 |
| 2 1/2 | 3/16 | 7/16 | 3/4 | 7/16 | 1 7/8 | 7/16 — 14 |
| 1 3/4 | 3/16 | 1/2 | 7/8 | 1/2 | 1 5/16 | 1/2 — 13 |
| 2 1/2 | 3/16 | 1/2 | 7/8 | 1/2 | 1 7/8 | 1/2 — 13 |
| 3 1/4 | 3/16 | 1/2 | 7/8 | 1/2 | 2 3/8 | 1/2 — 13 |

**Clamping Screws, Screw Bushings and Studs.** — Collar-head screws are used on jigs and fixtures in conjunction with clamps, straps and latches for clamping purposes (see table for dimensions). Rocking collar-screws are used with clamps for rough work, since they adapt themselves to any irregularities of the work and give a full bearing on the clamps in any position the work may assume.

Shoulder-screws are used for fastening clamping blocks, latches, or any parts that must move through a limited distance while still remaining permanently fastened to the tool.

Quarter-turn thumb-screws may be rapidly manipulated and are especially of use in box jigs. An objection to this type of screw is that the wear takes place on the boss on which it acts. Half-turn thumb-screws are also used in box jigs when the quarter-turn thumb-screw cannot be used on account of the work or bushing protruding through the end of the jig. These screws are used in pairs, one on each side of the jig cover.

Screw bushings are generally avoided when accurate work is required, as a threaded bushing is likely to be out of true. Sometimes, however, no other type of bushing is adapted for the work in hand. (See table of "Aligning Screw Bushings.") Studs are used in jig design for locating work with holes in it. The accompanying table shows a recommended form of collar stud.

## Rocking Collar-screws

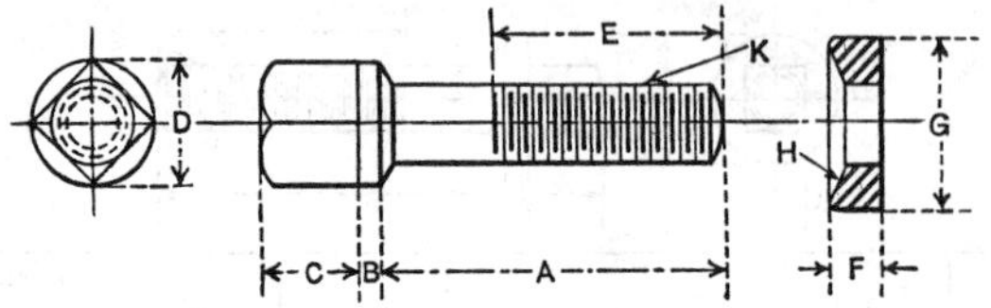


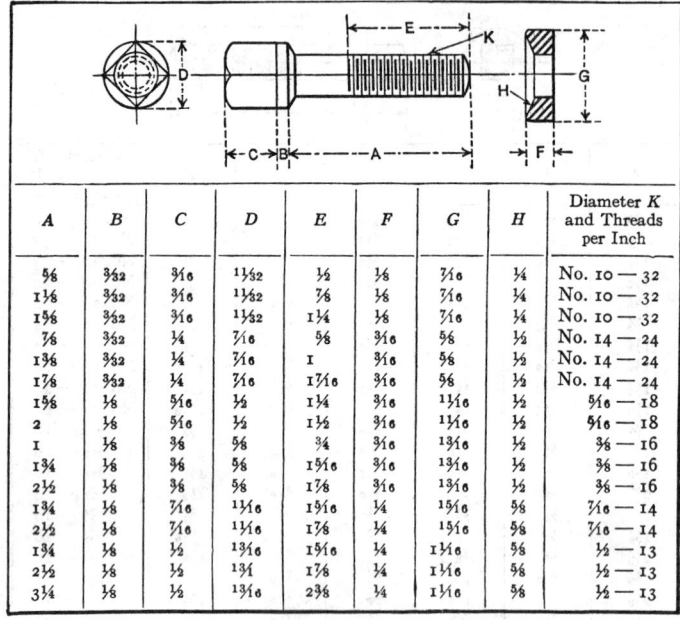

| *A* | *B* | *C* | *D* | *E* | *F* | *G* | *H* | Diameter *K* and Threads per Inch |
|---|---|---|---|---|---|---|---|---|
| 5/8 | 3/32 | 3/16 | 11/32 | 1/2 | 1/8 | 7/16 | 1/4 | No. 10 — 32 |
| 1 1/8 | 3/32 | 3/16 | 11/32 | 7/8 | 1/8 | 7/16 | 1/4 | No. 10 — 32 |
| 1 5/8 | 3/32 | 3/16 | 11/32 | 1 1/4 | 1/8 | 7/16 | 1/4 | No. 10 — 32 |
| 7/8 | 3/32 | 1/4 | 7/16 | 5/8 | 3/16 | 5/8 | 1/2 | No. 14 — 24 |
| 1 3/8 | 3/32 | 1/4 | 7/16 | 1 | 3/16 | 5/8 | 1/2 | No. 14 — 24 |
| 1 7/8 | 3/32 | 1/4 | 7/16 | 1 7/16 | 3/16 | 5/8 | 1/2 | No. 14 — 24 |
| 1 5/8 | 1/8 | 5/16 | 1/2 | 1 1/4 | 3/16 | 11/16 | 1/2 | 5/16 — 18 |
| 2 | 1/8 | 5/16 | 1/2 | 1 1/2 | 3/16 | 11/16 | 1/2 | 5/16 — 18 |
| 1 | 1/8 | 3/8 | 5/8 | 3/4 | 3/16 | 13/16 | 1/2 | 3/8 — 16 |
| 1 3/4 | 1/8 | 3/8 | 5/8 | 1 5/16 | 3/16 | 13/16 | 1/2 | 3/8 — 16 |
| 2 1/2 | 1/8 | 3/8 | 5/8 | 1 7/8 | 3/16 | 13/16 | 1/2 | 3/8 — 16 |
| 1 3/4 | 1/8 | 7/16 | 11/16 | 1 5/16 | 1/4 | 15/16 | 5/8 | 7/16 — 14 |
| 2 1/2 | 1/8 | 7/16 | 11/16 | 1 7/8 | 1/4 | 15/16 | 5/8 | 7/16 — 14 |
| 1 3/4 | 1/8 | 1/2 | 13/16 | 1 5/16 | 1/4 | 1 1/16 | 5/8 | 1/2 — 13 |
| 2 1/2 | 1/8 | 1/2 | 13/1 | 1 7/8 | 1/4 | 1 1/16 | 5/8 | 1/2 — 13 |
| 3 1/4 | 1/8 | 1/2 | 13/16 | 2 3/8 | 1/4 | 1 1/16 | 5/8 | 1/2 — 13 |

## Shoulder-screws

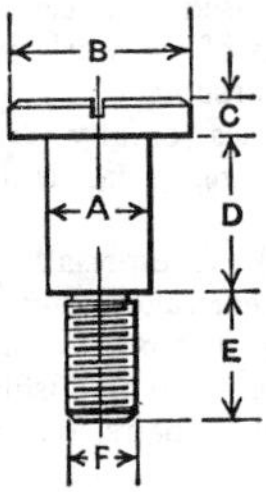


| *A* | *B* | *C* | *D* | *E* | Diameter *F* and Threads per inch |
|---|---|---|---|---|---|
| 0.249 | 1/2 | 1/8 | 1/2 to 3/4 | 7/16 | No. 10 — 32 |
| 0.249 | 1/2 | 1/8 | 3/16 to 7/16 | 5/16 | No. 10 — 32 |
| 0.3115 | 5/8 | 5/32 | 3/16 to 1/2 | 3/8 | No. 14 — 24 |
| 0.3115 | 5/8 | 5/32 | 9/16 to 7/8 | 1/2 | No. 14 — 24 |
| 0.374 | 3/4 | 5/32 | 1/4 to 7/16 | 3/8 | No. 14 — 24 |
| 0.374 | 3/4 | 5/32 | 1/2 to 11/16 | 1/2 | No. 14 — 24 |
| 0.4365 | 13/16 | 5/32 | 3/8 to 9/16 | 1/2 | 5/16 — 18 |
| 0.4365 | 13/16 | 5/32 | 7/8 to 1 1/8 | 3/4 | 5/16 — 18 |
| 0.499 | 7/8 | 3/16 | 3/8 to 9/16 | 1/2 | 3/8 — 16 |
| 0.499 | 7/8 | 3/16 | 15/16 to 1 1/4 | 3/4 | 3/8 — 16 |
| 0.5615 | 1 | 3/16 | 1/2 to 3/4 | 5/8 | 7/16 — 14 |
| 0.5615 | 1 | 3/16 | 1 1/4 to 1 1/2 | 7/8 | 7/16 — 14 |
| 0.6235 | 1 1/8 | 1/4 | 1/2 to 5/8 | 5/8 | 1/2 — 13 |
| 0.6235 | 1 1/8 | 1/4 | 1 1/8 to 1 5/8 | 7/8 | 1/2 — 13 |
| 0.686 | 1 1/4 | 1/4 | 1 1/8 to 1 3/4 | 1 | 1/2 — 13 |
| 0.7485 | 1 3/8 | 5/16 | 3/4 to 1 1/8 | 7/8 | 5/8 — 11 |
| 0.7485 | 1 3/8 | 5/16 | 1 1/4 to 2 | 1 1/8 | 5/8 — 11 |

### Quarter-turn Thumb-screws

| | A | B | C | D | E | Diameter F and Threads per Inch |
|---|---|---|---|---|---|---|
| | 7/16 | 1/2 | 1 1/16 | 3/16 | 5/16 | No. 10 — 32 |
| | 5/8 | 1/2 | 1 1/16 | 3/16 | 1/2 | No. 10 — 32 |
| | 7/8 | 1/2 | 1 1/16 | 3/16 | 5/8 | No. 10 — 32 |
| | 1 3/8 | 1/2 | 1 1/16 | 3/16 | 1 | No. 10 — 32 |
| | 5/8 | 9/16 | 1 3/16 | 1/4 | 1/2 | No. 14 — 24 |
| | 1 1/8 | 9/16 | 1 3/16 | 1/4 | 7/8 | No. 14 — 24 |
| | 1 5/8 | 9/16 | 1 3/16 | 1/4 | 1 1/4 | No. 14 — 24 |
| | 5/8 | 5/8 | 1 | 5/16 | 1/2 | 5/16 — 18 |
| | 1 1/4 | 5/8 | 1 | 5/16 | 1 | 5/16 — 18 |
| | 2 | 5/8 | 1 | 5/16 | 1 1/2 | 5/16 — 18 |
| | 3/4 | 1 1/16 | 1 3/16 | 3/8 | 9/16 | 3/8 — 16 |
| | 1 | 1 1/16 | 1 3/16 | 3/8 | 3/4 | 3/8 — 16 |
| | 1 3/4 | 1 1/16 | 1 3/16 | 3/8 | 1 5/16 | 3/8 — 16 |
| | 2 1/2 | 1 1/16 | 1 3/16 | 3/8 | 1 7/8 | 3/8 — 16 |
| | 1 | 3/4 | 1 5/16 | 7/16 | 3/4 | 7/16 — 14 |
| | 1 3/4 | 3/4 | 1 5/16 | 7/16 | 1 5/16 | 7/16 — 14 |
| | 2 1/2 | 3/4 | 1 5/16 | 7/16 | 1 1/4 | 7/16 — 14 |
| | 1 3/4 | 3/4 | 1 5/16 | 1/2 | 1 9/16 | 1/2 — 13 |
| | 2 1/2 | 3/4 | 1 5/16 | 1/2 | 1 1/4 | 1/2 — 13 |
| | 3 1/4 | 3/4 | 1 5/16 | 1/2 | 1 5/8 | 1/2 — 13 |

### Half-turn Thumb-screws

| | A | B | C | D | E | F | Diameter G and Threads per Inch |
|---|---|---|---|---|---|---|---|
| | 7/16 | 1/2 | 3/16 | 1 1/16 | 5/16 | 3/4 | No. 10 — 32 |
| | 5/8 | 1/2 | 3/16 | 1 1/16 | 1/2 | 3/4 | No. 10 — 32 |
| | 7/8 | 1/2 | 3/16 | 1 1/16 | 5/8 | 3/4 | No. 10 — 32 |
| | 1 3/8 | 1/2 | 3/16 | 1 1/16 | 1 | 3/4 | No. 10 — 32 |
| | 5/8 | 9/16 | 3/16 | 1 3/16 | 1/2 | 7/8 | No. 14 — 24 |
| | 1 1/8 | 9/16 | 3/16 | 1 3/16 | 7/8 | 7/8 | No. 14 — 24 |
| | 1 5/8 | 9/16 | 3/16 | 1 3/16 | 1 1/4 | 7/8 | No. 14 — 24 |
| | 5/8 | 5/8 | 3/16 | 1 | 1/2 | 1 | 5/16 — 18 |
| | 1 1/4 | 5/8 | 3/16 | 1 | 1 | 1 | 5/16 — 18 |
| | 2 | 5/8 | 3/16 | 1 | 1 1/2 | 1 | 5/16 — 18 |
| | 3/4 | 1 1/16 | 1/4 | 1 3/16 | 9/16 | 1 1/4 | 3/8 — 16 |
| | 1 | 1 1/16 | 1/4 | 1 3/16 | 3/4 | 1 1/4 | 3/8 — 16 |
| | 1 3/4 | 1 1/16 | 1/4 | 1 3/16 | 1 5/16 | 1 1/4 | 3/8 — 16 |
| | 2 1/2 | 1 1/16 | 1/4 | 1 3/16 | 1 7/8 | 1 1/4 | 3/8 — 16 |
| | 1 | 3/4 | 1/4 | 1 5/16 | 3/4 | 1 3/8 | 7/16 — 14 |
| | 1 3/4 | 3/4 | 1/4 | 1 5/16 | 1 5/16 | 1 3/8 | 7/16 — 14 |
| | 2 1/2 | 3/4 | 1/4 | 1 5/16 | 1 1/4 | 1 3/8 | 7/16 — 14 |
| | 1 3/4 | 3/4 | 1/4 | 1 5/16 | 1 5/16 | 1 1/2 | 1/2 — 13 |
| | 2 1/2 | 3/4 | 1/4 | 1 5/16 | 1 1/4 | 1 1/2 | 1/2 — 13 |
| | 3 1/4 | 3/4 | 1/4 | 1 5/16 | 1 5/8 | 1 1/2 | 1/2 — 13 |

### Aligning Screw Bushings

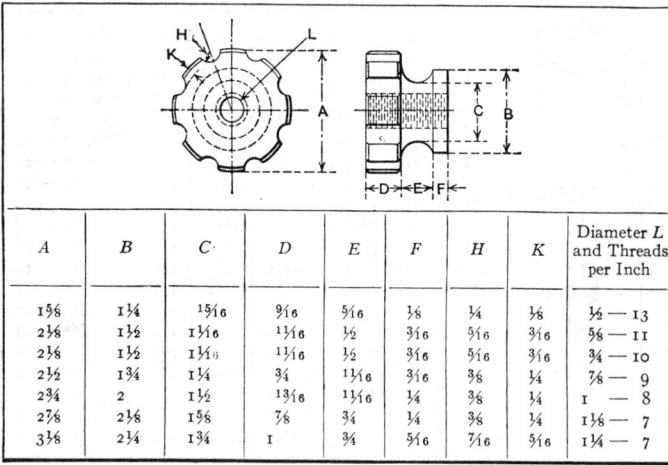

| | B | D | E | F | Diameter G and Threads per Inch |
|---|---|---|---|---|---|
| | 7/8 | 1/8 | 5/16 | 11/16 | 1/2 — 13 |
| | 1 | 9/32 | 5/16 | 7/8 | 5/8 — 11 |
| | 1 1/4 | 3/16 | 5/16 | 1 | 3/4 — 10 |
| | 1 1/2 | 3/16 | 7/16 | 1 1/4 | 1 — 14 |
| | 1 7/8 | 1/4 | 1/2 | 1 5/8 | 1 1/4 — 12 |
| | 2 1/4 | 1/4 | 5/8 | 2 | 1 1/2 — 12 |
| A and C, according | 2 3/4 | 1/4 | 3/4 | 2 3/8 | 1 3/4 — 8 |
| to requirements | 3 1/4 | 1/4 | 7/8 | 2 3/4 | 2 — 8 |

### Collar Studs — Hardened and Ground

| | A | B | C | D | E | F |
|---|---|---|---|---|---|---|
| | 0.251 | 0.249 | 1/2 | 1/2 | 3/16 | 1/4 — 5/8 |
| | 0.3135 | 0.3115 | 9/16 | 1/2 | 3/16 | 5/16 — 11/16 |
| | 0.376 | 0.374 | 5/8 | 5/8 | 3/16 | 3/8 — 3/4 |
| | 0.4385 | 0.4365 | 11/16 | 5/8 | 3/16 | 7/16 — 13/16 |
| | 0.501 | 0.499 | 3/4 | 5/8 | 1/4 | 1/2 — 7/8 |
| | 0.5635 | 0.5615 | 7/8 | 5/8 | 1/4 | 1/2 — 7/8 |
| | 0.626 | 0.624 | 1 | 3/4 | 1/4 | 1/2 — 1 1/16 |
| | 0.6885 | 0.6865 | 1 1/8 | 3/4 | 1/4 | 5/8 — 1 3/16 |
| | 0.751 | 0.749 | 1 1/4 | 7/8 | 1/4 | 5/8 — 1 3/8 |

### Hand Nuts

| A | B | C | D | E | F | H | K | Diameter L and Threads per Inch |
|---|---|---|---|---|---|---|---|---|
| 1 5/8 | 1 1/4 | 15/16 | 9/16 | 5/16 | 1/8 | 1/4 | 1/8 | 1/2 — 13 |
| 2 1/8 | 1 1/2 | 1 1/16 | 11/16 | 1/2 | 3/16 | 5/16 | 3/16 | 5/8 — 11 |
| 2 3/8 | 1 1/2 | 1 11/16 | 11/16 | 1/2 | 3/16 | 5/16 | 3/16 | 3/4 — 10 |
| 2 1/2 | 1 3/4 | 1 1/4 | 3/4 | 11/16 | 3/16 | 3/8 | 1/4 | 7/8 — 9 |
| 2 3/4 | 2 | 1 1/2 | 13/16 | 11/16 | 1/4 | 3/8 | 1/4 | 1 — 8 |
| 2 7/8 | 2 1/8 | 1 5/8 | 7/8 | 3/4 | 1/4 | 3/8 | 1/4 | 1 1/8 — 7 |
| 3 1/8 | 2 1/4 | 1 3/4 | 1 | 3/4 | 5/16 | 7/16 | 5/16 | 1 1/4 — 7 |

**Dimensions of Jig-Screw Latches**

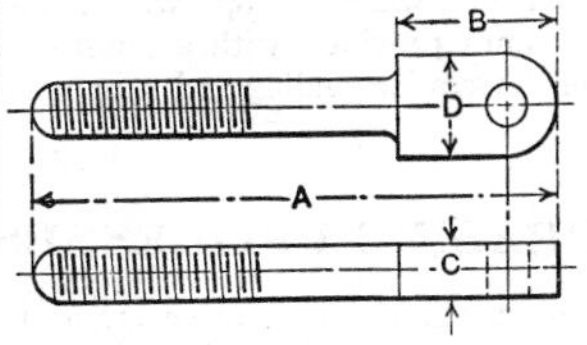


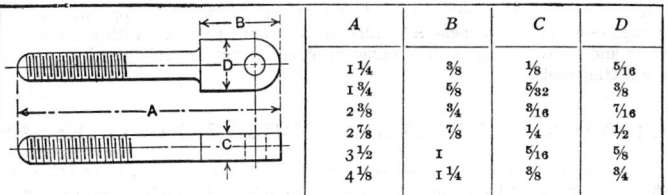

| *A* | *B* | *C* | *D* |
|---|---|---|---|
| 1 1/4 | 3/8 | 1/8 | 5/16 |
| 1 3/4 | 5/8 | 5/32 | 3/8 |
| 2 3/8 | 3/4 | 3/16 | 7/16 |
| 2 7/8 | 7/8 | 1/4 | 1/2 |
| 3 1/2 | 1 | 5/16 | 5/8 |
| 4 1/8 | 1 1/4 | 3/8 | 3/4 |

**Dimensions of Latch Nuts**

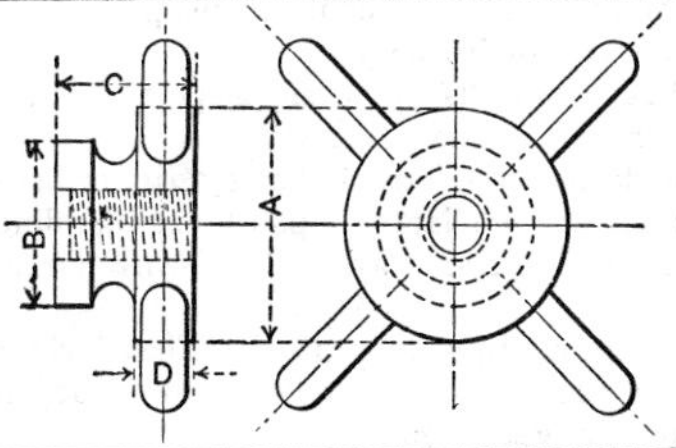


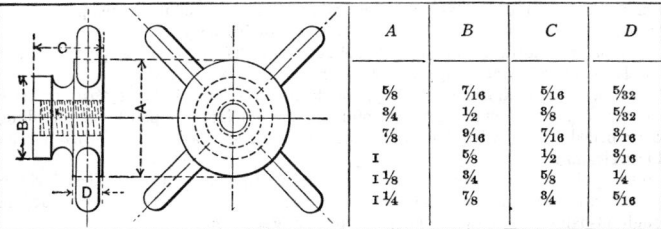

| *A* | *B* | *C* | *D* |
|---|---|---|---|
| 5/8 | 7/16 | 5/16 | 5/32 |
| 3/4 | 1/2 | 3/8 | 5/32 |
| 7/8 | 9/16 | 7/16 | 3/16 |
| 1 | 5/8 | 1/2 | 3/16 |
| 1 1/8 | 3/4 | 5/8 | 1/4 |
| 1 1/4 | 7/8 | 3/4 | 5/16 |

**Standard Jig Feet**

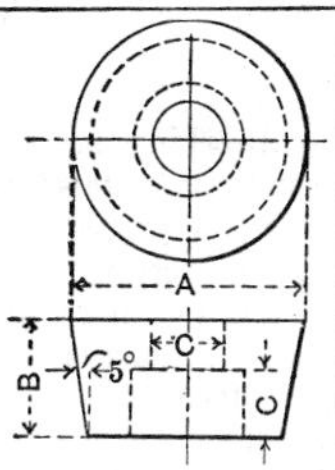


| *A* | *B* | *C* | *A* | *B* | *C* |
|---|---|---|---|---|---|
| 3/8 | 3/16 | 1/8 | 13/16 | 13/32 | 7/32 |
| 7/16 | 7/32 | 9/64 | 3/4 | 3/8 | 1/4 |
| 1/2 | 1/4 | 5/32 | 7/8 | 7/16 | 9/32 |
| 9/16 | 9/32 | 11/64 | 1 | 1/2 | 5/16 |
| 5/8 | 5/16 | 3/16 | ...... | ...... | ...... |

**Screws for Jig Feet**

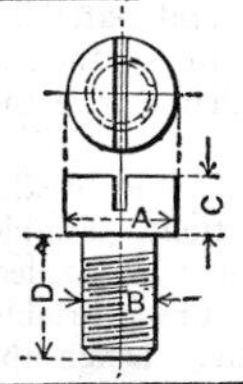


| *A* | *B* | *C* | *D* | *A* | *B* | *C* | *D* |
|---|---|---|---|---|---|---|---|
| 0.160 | 1/8 | 0.110 | 9/32 | 0.299 | 7/32 | 0.192 | 7/16 |
| 0.191 | 9/64 | 0.123 | 5/16 | 0.343 | 1/4 | 0.219 | 15/32 |
| 0.213 | 5/32 | 0.137 | 11/32 | 0.386 | 9/32 | 0.246 | 1/2 |
| 0.233 | 11/64 | 0.150 | 3/8 | 0.426 | 5/16 | 0.273 | 17/32 |
| 0.256 | 3/16 | 0.164 | 13/32 | ..... | .... | ..... | .... |

**Definition of Jig and Fixture.** — The distinction between a jig and fixture is not easy to define, but, as a general rule, it is as follows: A jig either holds or is held on the work, and, at the same time, contains guides for the various cutting tools, whereas a fixture holds the work while the cutting tools are in operation, but does not contain any special arrangements for guiding the tools. A fixture, therefore, must be securely held or fixed to the machine on which the operation is performed — hence the name. A fixture is sometimes provided with a number of gages and stops, but not with bushings or other devices for guiding and supporting the cutting tools.

# GRINDING AND OTHER ABRASIVE PROCESSES

Processes and equipment discussed under this heading use abrasive grains for shaping workpieces by means of machining or related methods. Abrasive grains are hard crystals either found in nature or manufactured. The most commonly used materials are aluminum oxide, silicon carbide and diamond. Other materials such as garnet, zirconia, glass and even walnut shells are used for some applications. Abrasive products are used in three basic forms by industry:

a. *Bonded* to form a solid shaped tool such as disks (the basic shape of grinding wheels), cylinders, rings, cups, segments or sticks to name a few.

b. *Coated* on backings made of paper or cloth, in the form of sheets, strips or belts.

c. *Loose*, held in some liquid or solid carrier, (for lapping, polishing, tumbling) or propelled by centrifugal force, air or water pressure against the work surface (blast cleaning).

The applications for abrasive processes are multiple and varied. They include:

a. *Cleaning* of surfaces, also the coarse removal of excess material — such as rough off-hand grinding in foundries to remove gates and risers.

b. *Shaping*, such as in form grinding and tool sharpening.

c. *Sizing*, a general objective, but of primary importance in precision grinding.

d. *Surface finish improvement*, either primarily as in lapping, honing and polishing or as a secondary objective in other types of abrasive processes.

e. *Separating*, as in cut-off or slicing operations.

The main field of application of abrasive processes is in metalworking, because of the capacity of abrasive grains to penetrate into even the hardest metals and alloys. However, the great hardness of the abrasive grains makes the process also preferred for working other hard materials, such as stones, glass and certain types of plastics. Abrasive processes are also chosen for working relatively soft materials, such as wood, rubber, etc., for such reasons as high stock removal rates, long-lasting cutting ability, good form control, and fine finish of the worked surface.

## Grinding Wheels

**Abrasive Materials.** — In earlier times, only natural abrasives were available. From about the beginning of this century, however, manufactured abrasives, primarily silicon carbide and aluminum oxide, have replaced the natural materials; even natural diamonds have been almost completely supplanted by synthetics. Superior and controllable properties, and dependable uniformity characterize the manufactured abrasives.

Both silicon carbide and aluminum oxide abrasives are very hard and brittle. This brittleness, called friability, is controllable for different applications. Friable abrasives break easily, thus forming sharp edges. This decreases the force needed to penetrate into the work material and the heat generated during cutting. Friable abrasives are most commonly used for precision and finish grinding. Tough abrasives resist fracture and last longer. They are used for rough grinding, snagging and off-hand grinding.

As a general rule, although subject to variation:

1. Aluminum oxide abrasives are used for grinding plain and alloyed steel in a soft or hardened condition.

2. Silicon carbide abrasives are selected for cast iron, nonferrous metals and nonmetallic materials.

3. Diamond is the best type of abrasive for grinding cemented carbides. It is also used for grinding glass, ceramics, and in some cases, hardened tool steel.

4. Borazon, the trade name for cubic boron nitride, is a man-made abrasive particularly effective for grinding hardened tool steels. It produces a superior surface integrity on the ground surface, close tolerances are easier to maintain, and the grinding wheel lasts longer.

**Bond Properties and Grinding Wheel Grades.** — The four main types of bonds used for grinding wheels are the vitrified, resinoid, rubber and metal.

*Vitrified bonds* are used for more than half of all grinding wheels made, and are preferred because of their strength and other desirable qualities. Being inert, glass-like materials, vitrified bonds are not affected by water or by the chemical composition of different grinding fluids. Vitrified bonds also withstand the high temperatures generated during normal grinding operations. The structure of vitrified wheels can be controlled over a wide range of strength and porosity. Vitrified wheels, however, are more sensitive to impact than those made with organic bonds.

*Resinoid bonds* are selected for wheels subjected to impact, or sudden loads or very high operating speeds. They are preferred for snagging, portable grinder uses or roughing operations. The higher flexibility of this type of bond — essentially a filled thermosetting plastic — helps it withstand rough treatment.

*Rubber bonds* are even more flexible than the resinoid type, and for that reason are used for producing a high finish and for resisting sudden rises in load. Rubber bonded wheels are commonly used for wet cut-off wheels because of the nearly burr-free cuts they produce, and for centerless grinder regulating wheels to provide a stronger grip and more reliable workpiece control.

*Metal bonds* are used in diamond wheels for grinding carbides. Their very high strength and toughness help to retain the costly diamond abrasive firmly bonded throughout its entire useful life. Metal bonded wheels — either with diamond or aluminum oxide abrasive — are also used in electrical discharge grinding and electrochemical grinding where an electrically conductive wheel is needed.

In addition to the basic properties of the various bond materials, each can also be applied in different proportions, thereby controlling the grade of the grinding wheel.

*Grinding wheel grades* commonly associated with hardness, express the amount of bond material in a grinding wheel, and hence the strength by which the bond retains the individual grains.

During grinding, the forces generated when cutting the work material tend to dislodge the abrasive grains. As the grains get dull and if they don't fracture to resharpen themselves, the cutting forces will eventually tear the grains from their supporting bond. For a "soft" wheel the cutting forces will dislodge the abrasive grains before they have an opportunity to fracture. When a "hard" wheel is used, the situation is reversed. Because of the extra bond in the wheel the grains are so firmly held that they never break loose and the wheel becomes glazed. During most grinding operations it is desirable to have an intermediate wheel where there is a continual slow wearing process composed of both grain fracture and dislodgement.

The grades of the grinding wheels are designated by capital letters used in alphabetical order to express increasing "hardness" from A to Z.

**Grinding Wheel Structure.** — The individual grains, which are encased and held together by the bond material, do not fill the entire volume of the grinding wheel; the intermediate open space is needed for several functional purposes such as heat dissipation, coolant application and particularly for the temporary storage of chips. It follows that the spacing of the grains must be greater for coarse grains which cut thicker chips and for large contact areas within which the chips have to be retained on the surface of the wheel before being disposed of. On the other hand, a wide spacing reduces the number of grains which contact the work surface within a given advance distance, thereby producing a coarser finish.

In general, denser structures are specified for grinding hard materials, for high speed grinding operations, when the contact area is narrow and also for producing fine finishes and/or accurate forms. Wheels with open structure are used for tough materials, high stock removal rates and extended contact areas, such as grinding with the face of the wheel. There are, however, several exceptions to these basic rules, an important one being the grinding of parts made by powder metallurgy, such as cemented carbides; although they represent one of the hardest industrial materials, they require wheels with an open structure.

Most kinds of general grinding operations when carried out with the periphery of the wheel call for medium spacing of the grains. The structure of the grinding wheels is expressed by numerals from 1 to 16, ranging from dense to open. In some cases "induced porosity" is used with open structure wheels. This term means that the grinding wheel manufacturer has placed filler material (which later burns out when the wheel is fired to vitrify the bond) in the grinding wheel mix. These fillers create large "pores" between grain clusters without changing the total volume of the "pores" in the grinding wheel. Thus an A46-H12V wheel and an A46-H12VP wheel will contain the same amounts of bond, abrasive and air space. In the former a large number of relatively small pores will be distributed throughout the wheel. The latter will have a smaller number of larger pores.

**American National Standard Grinding Wheel Markings.** — ANSI Standard B74.13-1970, "Markings for Identifying Grinding Wheels and Other Bonded Abrasives," applies to grinding wheels and other bonded abrasives, segments, bricks, sticks, hones, rubs, and other shapes which are for removing material, or producing a desired surface or dimension. It does not apply to diamond wheels or to specialities such as sharpening stones and provides only a standard system of markings. Wheels having the same standard markings but made by different wheel manufacturers may not — and probably will not — produce exactly the same grinding action. This desirable result cannot be accomplished because of the impossibility of closely correlating any measurable physical properties of bonded abrasive products in terms of their grinding action.

American National Standards Institute has also issued an Identification Code for Diamond Wheel Shapes which is discussed on page 2004.

**Sequence of Markings.** — The accompanying illustration taken from ANSI B74.13-1970 shows the makeup of a typical wheel or bonded abrasive marking.

| | 1 | 2 | 3 | 4 | 5 | 6 |
|---|---|---|---|---|---|---|
| Prefix | Abrasive Type | Grain Size | Grade | Structure | Bond Type | Manufacturer's Record |

$$51 - A - 36 - L - 5 - V - 23$$

The meaning of each letter and number in this or other markings is indicated by the following complete list.

1. *Abrasive Letters:* The letter (A) is used for aluminum oxide and (C) for silicon carbide. The manufacturer may designate some particular type in either of these broad classes, by using his own symbol as a prefix (Example, 51A).

2. *Grain Size:* The grain sizes commonly used and varying from coarse to fine are indicated by the following numbers: 8, 10, 12, 14, 16, 20, 24, 30, 36, 46, 54, 60, 70, 80, 90, 100, 120, 150, 180, 220. The following additional sizes are used occasionally: 240, 280, 320, 400, 500, 600. The wheel manufacturer may add to the regular grain number an additional symbol to indicate a special grain combination.

3. *Grade:* Grades are indicated by letters of the alphabet from A to Z in all bonds or processes. Wheel grades from A to Z range from soft to hard.

4. *Structure:* The use of a structure symbol is optional. The structure is indicated by Nos. 1 to 16 (or higher, if necessary) with progressively higher numbers.

5. *Bond or Process:* Bonds are indicated by the following letters: V, vitrified; S, silicate; E, shellac or elastic; R, rubber; RF, rubber reinforced; B, resinoid (synthetic resins); BF, resinoid reinforced; O, oxychloride.

6. *Manufacturer's Record:* The sixth position may be used for manufacturer's private factory records; this is optional.

**American National Standard Shapes and Sizes of Grinding Wheels.** — The ANSI Standard B74.2-1974, which includes shapes and sizes of grinding wheels, gives a wide variety of grinding wheel shape and size combinations. These are suitable for the majority of applications. While grinding wheels can be manufactured to shapes and dimensions different from those listed, it is advisable, for reasons of cost and inventory control, to avoid using special shapes and sizes, unless technically warranted.

Standard shapes and size ranges as given in this Standard together with typical applications are shown in Table 1.

The operating surface of the grinding wheel is often referred to as the wheel face. In the majority of cases it is the periphery of the grinding wheel which, when not specified otherwise, has a straight profile. However, other face shapes can also be supplied by the grinding wheel manufacturers, and also reproduced during usage by appropriate truing. ANSI B74.2-1974 offers thirteen different shapes for grinding wheel faces, which are shown in Table 2.

**The Selection of Grinding Wheels.** — In selecting a grinding wheel, the determining factors are the composition of the work material, the type of grinding machine, the size range of the wheels used and the expected grinding results, in this approximate order.

The Norton Company has developed, as the result of extensive test series, a method of grinding wheel recommendation that is more flexible and also better adapted to taking into consideration pertinent factors of the job, than are listings based solely on workpiece categories. This approach is the basis for Tables 3–6, inclusive. Tool steels and constructional steels are considered in the detailed recommendations presented in these tables.

Table 3 assigns most of the standardized tool steels to five different grindability groups. The AISI-SAE tool steel designations are used.

After having defined the grindability group of the tool steel to be ground, the operation to be carried out is found in the first column of Table 4. The second column in this table distinguishes between different grinding wheel size ranges, because wheel size is a factor in determining the contact area between wheel and work, thus affecting the apparent hardness of the grinding wheel. Distinction is also made between wet and dry grinding.

Finally, the last two columns define the essential characteristics of the recommended types of grinding wheels under the headings of first and second choice, respectively. Where letters are used *preceding* A, the standard designation for aluminum oxide, they indicate a degree of friability different from the regular, thus:

(Continued on page 1991)

Table 1.  Standard Shapes and Size Ranges of Grinding Wheels
(Based on ANSI B74.2-1974)

Type 1. STRAIGHT WHEEL
For peripheral grinding.

| Applications | Size Ranges of Principal Dimensions, Inches | | |
| --- | --- | --- | --- |
| | $D$ = Diam. | $T$ = Thick. | $H$ = Hole |
| CUTTING OFF. (Organic bonds only) | 6 to 48 | 1/32 to 3/8 | 5/8 to 6 |
| CYLINDRICAL GRINDING. Between centers | 12 to 48 | 1/2 to 6 | 5 to 20 |
| CYLINDRICAL GRINDING. Centerless grinding wheels | 14 to 30 | 1 to 8 | 5 to 12 |
| CYLINDRICAL GRINDING. Centerless regulating wheels | 8 to 14 | 1 to 12 | 3 to 6 |
| INTERNAL GRINDING. | 1/4 to 4 | 1/4 to 2 | 3/32 to 7/8 |
| OFFHAND GRINDING. Grinding on the periphery — General purpose — For wet tool grinding only | 6 to 36 / 30 or 36 | 1/2 to 4 / 3 or 4 | 1/2 to 3 / 20 |
| SAW GUMMING. (F-type face) | 6 to 12 | 1/4 to 1 1/2 | 1/2 to 1 1/4 |
| SNAGGING. Floor stand machines | 12 to 24 | 1 to 3 | 1 1/4 to 2 1/2 |
| SNAGGING. Floor stand machines (Organic bond, wheel speed over 6500 sfpm) | 20 to 36 | 2 to 4 | 6 to 12 |
| SNAGGING. Mechanical grinders (Organic bond; wheel speed up to 16,500 SFPM) | 24 | 2 to 3 | 12 |
| SNAGGING. Portable machines | 3 to 8 | 1/4 to 1 | 3/8 to 5/8 |
| SNAGGING. Portable machines (Reinforced organic bond, 17,000 SFPM) | 6 or 8 | 3/4 or 1 | 1 |
| SNAGGING. Swing frame machines | 12 to 24 | 2 to 3 | 3 1/2 to 12 |
| SURFACE GRINDING. Horizontal spindle machines | 6 to 24 | 1/2 to 6 | 1 1/4 to 12 |
| TOOL GRINDING. Broaches, cutters, mills, reamers, taps, etc. | 6 to 10 | 1/4 to 1/2 | 5/8 to 5 |

Type 2. CYLINDRICAL WHEEL
Side grinding wheel — mounted on the diameter; may also be mounted in a chuck or on a plate.

| | | $W$ = Wall |
| --- | --- | --- |
| SURFACE GRINDING. Vertical spindle machines | 8 to 20 | 4 or 5 | 1 to 4 |

Throughout table heavy arrows indicate grinding surfaces.

**Table 1** (*Continued*).  **Standard Shapes and Size Ranges of Grinding Wheels**
(Based on ANSI B74.2-1974)

| Applications | Size Ranges of Principal Dimensions, Inches | | |
|---|---|---|---|
| | $D$ = Diam. | $T$ = Thick. | $H$ = Hole |

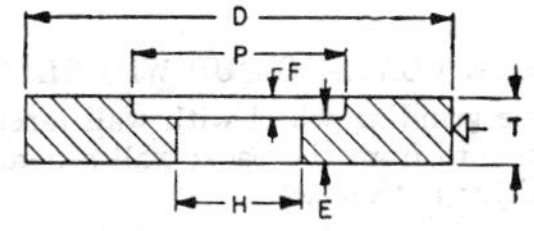

Type 5. WHEEL, recessed one side

For peripheral grinding. Allows wider faced wheels than the available mounting thickness, and also grinding clearance for the nut and flange.

| Applications | $D$ = Diam. | $T$ = Thick. | $H$ = Hole |
|---|---|---|---|
| CYLINDRICAL GRINDING.<br>Between centers | 12 to 36 | 1½ to 4 | 5 to 12 |
| CYLINDRICAL GRINDING.<br>Centerless regulating wheel | 8 to 14 | 3 to 6 | 3 to 5 |
| INTERNAL GRINDING. | ⅜ to 4 | ⅜ to 2 | ⅛ to ⅞ |
| SURFACE GRINDING.  Horizontal<br>spindle machines | 7 to 24 | ¾ to 6 | 1¼ to 12 |

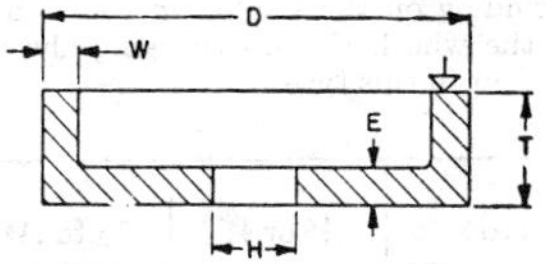

Type 6. STRAIGHT-CUP WHEEL

Side grinding wheel, in whose dimensioning the wall thickness ($W$) takes precedence over the diameter of the recess.  Hole is either ⅞″ diam. or ⅝-11UNC-2B threaded for the snagging wheels and ½ or 1¼″ for the tool grinding wheels.

| Applications | $D$ = Diam. | $T$ = Thick. | $W$ = Wall |
|---|---|---|---|
| SNAGGING.  Portable machines, organic bond only (With straight hole or with threaded hole insert) | 4 to 6 | 2 | ¾ to 1½ |
| TOOL GRINDING.  Broaches, cutters,<br>mills, reamers, taps, etc. | 2 to 6 | 1¼ to 2 | 5⁄16 to ⅜ |

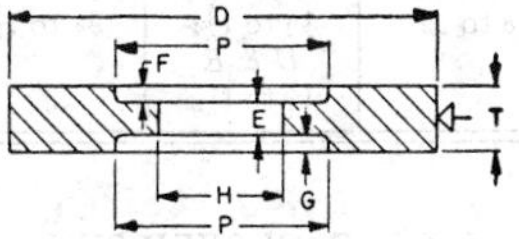

Type 7. WHEEL, recessed two sides

Peripheral grinding.  Recesses allow grinding clearance for both flanges and also narrower mounting thickness than overall thickness.

| Applications | $D$ = Diam. | $T$ = Thick. | $H$ = Hole |
|---|---|---|---|
| CYLINDRICAL GRINDING.<br>Between centers | 12 to 36 | 1½ to 4 | 5 or 12 |
| CYLINDRICAL GRINDING.<br>Centerless regulating wheel | 8 to 14 | 4 to 20 | 3 to 6 |
| SURFACE GRINDING.  Horizontal<br>spindle machines | 12 to 24 | 2 to 6 | 5 to 12 |

Table 1 *(Continued)*. **Standard Shapes and Size Ranges of Grinding Wheels**
(Based on ANSI B74.2-1974)

| Applications | Size Ranges of Principal Dimensions, Inches | | |
|---|---|---|---|
| | $D$ = Diam. | $T$ = Thick. | $H$ = Hole |
| 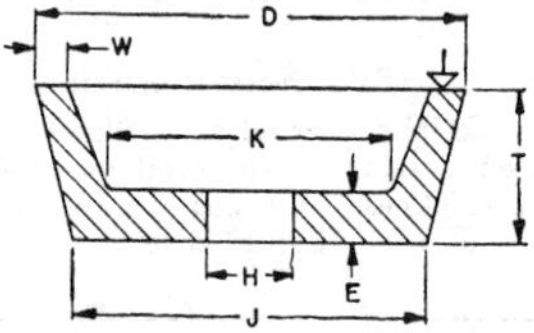 Type 11. FLARING-CUP WHEEL — Side grinding wheel with wall tapered outward from the back; wall generally thicker in the back. | | | |
| SNAGGING. Portable machines Organic bonds only. Also available with ⅝-11UNC-2B threaded hole insert | 4 to 6 | 2 | ⅞ |
| TOOL GRINDING. Broaches, cutters, mills, reamers, taps, etc. | 2 to 5 | 1¼ to 2 | ½ to 1¼ |
| 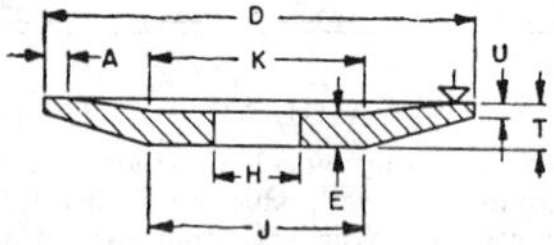 Type 12. DISH WHEEL — Grinding on the side or on the U-face of the wheel, the U-face being always present in this type. | | | |
| TOOL GRINDING. Broaches, cutters, mills, reamers, taps, etc. | 3 to 8 | ½ or ¾ | ½ to 1¼ |
| 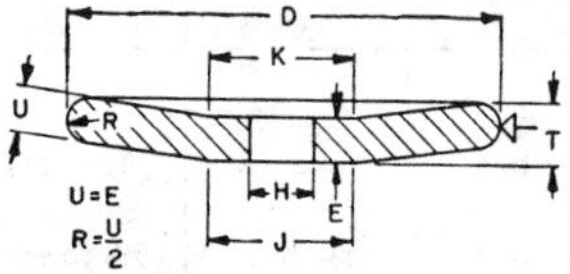 Type 13. SAUCER WHEEL — Peripheral grinding wheel, resembling the shape of a saucer, with cross section equal throughout. $U=E$, $R=\frac{U}{2}$ | | | |
| SAW GUMMING. Saw tooth shaping and sharpening | 8 to 12 | ½ to 1¾ U & E ¼ to 1½ | ¾ to 1¼ |
| 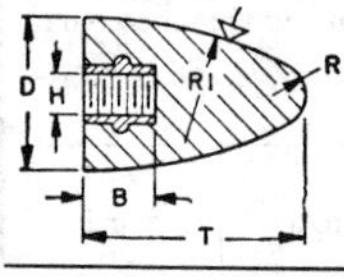 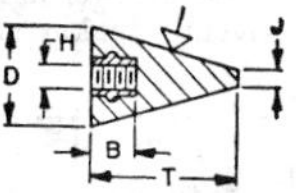 Type 16. CONE, CURVED SIDE. Type 17. CONE, STRAIGHT SIDE SQUARE TIP | | | |
| SNAGGING. Portable machine Threaded holes | 1¼ to 3 | 2 to 3½ | ⅜-24UNF-2B to ⅝-11UNC-2B |

**Table 1** (*Continued*). **Standard Shapes and Size Ranges of Grinding Wheels**
(Based on ANSI B74.2-1974)

| Applications | Size Ranges of Principal Dimensions, Inches | | |
|---|---|---|---|
| | D = Diam. | T = Thick. | H = Hole |
| Type 18. PLUG, SQUARE END  Type 18R. PLUG, ROUND END  $R = D/2$ | | | |
| Type 19. PLUGS, CONICAL END, SQUARE TIP | | | |
| SNAGGING.  Portable machine  Threaded holes | 1¼ to 3 | 2 to 3½ | ⅜-24UNF-2B to ⅝-11UNC-2B |
| Type 20. WHEEL, RELIEVED ONE SIDE  Peripheral grinding wheel, one side flat, the other side relieved to a flat. | | | |
| CYLINDRICAL GRINDING.  Between centers | 12 to 36 | ¾ to 4 | 5 to 20 |
| Type 21. WHEEL, RELIEVED TWO SIDES  Both sides relieved to a flat. | | | |
| Type 22. WHEEL, RELIEVED ONE SIDE, RECESSED OTHER SIDE  One side relieved to a flat. | | | |
| Type 23. WHEEL, RELIEVED AND RECESSED SAME SIDE  The other side is straight. | | | |
| CYLINDRICAL GRINDING.  Between centers, with wheel periphery | 20 to 36 | 2 to 4 | 12 or 20 |

Table I (*Concluded*). **Standard Shapes and Size Ranges of Grinding Wheels**
(Based on ANSI B74.2-1974

| Applications | Size Ranges of Principal Dimensions, Inches | | |
|---|---|---|---|
| | $D$ = Diam. | $T$ = Thick. | $H$ = Hole |
| Type 24. WHEEL, RELIEVED AND RECESSED ONE SIDE, RECESSED OTHER SIDE <br> One side recessed, the other side is relieved to a recess. | | | |
| Type 25. WHEEL, RELIEVED AND RECESSED ONE SIDE, RELIEVED OTHER SIDE <br> One side relieved to a flat, the other side relieved to a recess. | | | |
| Type 26. WHEEL, RELIEVED AND RECESSED BOTH SIDES | | | |
| CYLINDRICAL GRINDING. <br> Between centers, with the periphery of the wheel | 20 to 36 | 2 to 4 | 12 or 20 |
| Types 27 & 27A. WHEEL, DEPRESSED CENTER <br> 27. *Portable Grinding:* Grinding normally done by contact with work at approx. a 15° angle with face of the wheel. <br> 27A. *Cutting-off:* Using the periphery as grinding face. | | | |
| CUTTING OFF. Reinforced organic bonds only | 20 to 30 | $U = E = \frac{3}{16}$ or $\frac{1}{4}$ | 1 or $1\frac{1}{2}$ |
| SNAGGING. Portable machine | 3 to 9 | $U$ = Uniform thick. $\frac{1}{8}$ to $\frac{3}{8}$ | $\frac{3}{8}$ or $\frac{7}{8}$ |
| Type 28. WHEEL, DEPRESSED CENTER (SAUCER SHAPED GRINDING FACE) <br> Grinding at approx. 15° angle with wheel face. | | | |
| SNAGGING. Portable machine | 7 and 9 | $U$ = Uniform thickness $\frac{1}{4}$ | $\frac{7}{8}$ |

Table 2.   Standard Shapes of Grinding Wheel Faces (ANSI B74.2-1974)

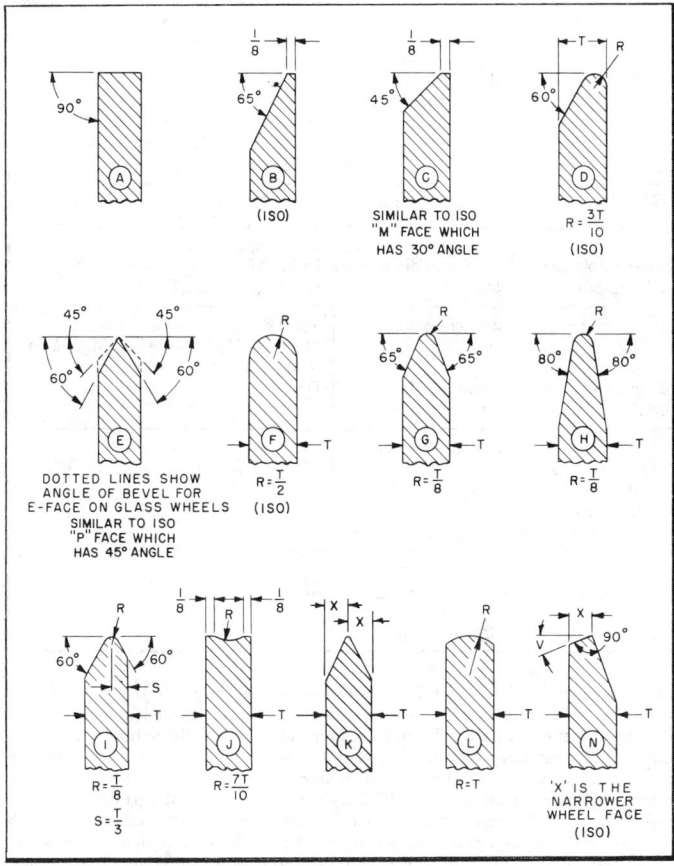

SF = semi friable (Norton equivalent 16A) and F = friable (Norton equivalent 32A and 38A).   The suffix P, where applied, expresses a degree of porosity that is more open than the regular.

Recommendations, similar in principle, yet somewhat less discriminating have been developed by the Norton Company for *constructional steels*.   These materials can be ground either in their original state (soft) or in their after-hardened state (directly or following carburization).   Constructional steels must be distinguished from structural steels which are used primarily by the building industry in mill shapes, without or with a minimum of machining.

Table 3· Classification of Tool Steels by their Relative Grindability

| Relative Grindability Group | AISI-SAE Designation of Tool Steels |
|---|---|
| GROUP 1 — Any area of work surface<br><br>High grindability tool and die steels<br>(Grindability index greater than 12) | W1, W2, W4, W5<br>S1, S2, S4, S5<br>O1, O2, O6, O7<br>A2, A4, A5, A6<br>H11, H12, H13, H14, H20, H21,<br>   H22, H23, H24, H25, H26<br>L1, L2, L3, L6, L7 |
| GROUP 2 — Small area of work surface (as found in tools)<br><br>Medium grindability tool and die steels<br>(Grindability index 3 to 12) | H41, H42, H43<br>T1, T2, T4, T7, T8<br>M1, M2, M8, M10, M30<br>F1, F2, F3<br>D1, D2, D3, D4, D5, D6 |
| GROUP 3 — Small area of work surface (as found in tools)<br><br>Low grindability tool and die steels<br>(Grindability index between 1.0 and 3) | T5, T6<br>M3, M6, M7, M34, M35, M36, M41,<br>   M43, M44<br>D7 |
| GROUP 4 — Large area of work surface (as found in dies)<br>Medium and low grindability tool and die steels<br>(Grindability index between 1.0 and 12) | All steels found in Groups 2 and 3 |
| GROUP 5 — Any area of work surface<br>Very low grindability tool and die steels<br>(Grindability index less than 1.0) | T9, T15<br>M4, M15 |

Constructional steels are either plain carbon or alloy type steels assigned in the AISI-SAE specifications to different groups, according to the predominant types of alloying elements. In the following recommendations no distinction is made because of different compositions since that factor generally, has a minor effect on grinding wheel choice in the case of constructional steels. However, separate recommendations are made for soft (Table 5) and hardened (Table 6) constructional steels. For the relatively rare case where the use of a single type of wheel for both soft and hardened steel materials is considered more important than the selection of the best suited types for each condition of the work materials, Table 5 lists "All-Around" wheels in its last column.

For applications where cool cutting properties of the wheel are particularly important, Table 6 lists, as a second alternative, porous-type wheels. The sequence of choices as presented in these tables does not necessarily represent a second, or third best; it can also apply to conditions where the first choice did not provide optimum results and by varying slightly the composition of the grinding wheel, as indicated in the subsequent choices, the performance experience by the first choice might actually be improved.

(Continued on page 1998)

**Table 4.  Grinding Wheel Recommendations for Hardened Tool Steels
According to their Grindability**

| Operation | Wheel or Rim Diameter, Inches | First Choice Specifications | Second Choice Specifications |
|---|---|---|---|
| | GROUP 1 STEELS | | |
| Surfacing | | | |
|   Surfacing wheels | 14 and smaller | Wet FA46-I8V | SFA46-G12VP |
| | 14 and smaller | Dry FA46-H8V | FA46-F12VP |
| | Over 14 | Wet FA36-I8V | SFA36-I8V |
|   Segments or Cylinders | 1½ rim or less | Wet FA30-H8V | FA30-F12VP |
|   Cups | ¾ rim or less | Wet FA36-H8V | FA46-F12VP |
| | (for rims wider than 1½ inches, go one grade softer in available specifications) | | |
| Cutter sharpening | | | |
|   Straight wheel | .............. | Wet  FA46-K8V | FA60-K8V |
| | .............. | Dry  FA46-J8V | FA46-H12VP |
|   Dish shape | .............. | Dry  FA60-J8V | FA60-H12VP |
|   Cup shape | .............. | Dry  FA46-L8V | FA60-H12VP |
| | .............. | Wet SFA46-L5V | SFA60-L5V |
| Form tool grinding | 8 and smaller | Wet FA60-L8V  to  FA100-M7V | |
| | 8 and smaller | Dry FA60-K8V  to  FA100-L8V | |
| | 10 and larger | Wet FA60-L8V  to  FA80-M6V | |
| Cylindrical | 14 and smaller | Wet SFA60-L5V | .............. |
| | 16 and larger | Wet SFA60-M5V | .............. |
| Centerless | .............. | Wet SFA60-M5V | .............. |
| Internal | | | |
|   Production grinding | Under ½ | Wet SFA80-N6V | SFA80-N7V |
| | ½ to 1 | Wet SFA60-M5V | SFA60-M6V |
| | Over 1 to 3 | Wet SFA54-L5V | SFA54-L6V |
| | Over 3 | Wet SFA46-L5V | SFA46-K5V |
|   Tool room grinding | Under ½ | Dry  FA80-L6V | SFA80-L7V |
| | ½ to 1 | Dry  FA70-K7V | SFA70-K7V |
| | Over 1 to 3 | Dry  FA60-J8V | FA60-H12VP |
| | Over 3 | Dry  FA46-J8V | FA54-H12VP |
| | GROUP 2 STEELS | | |
| Surfacing | | | |
|   Straight wheels | 14 and smaller | Wet FA46-I8V | FA46-G12VP |
| | 14 and smaller | Dry FA46-H8V | FA46-F12VP |
| | Over 14 | Wet FA46-H8V | SFA46-I8V |
|   Segments or Cylinders | 1½ rim or less | Wet FA30-G8V | FA36-E12VP |
|   Cups | ¾ rim or less | Wet FA36-H8V | FA46-F12VP |
| | (for rims wider than 1½ inches, go one grade softer in available specifications) | | |

**Table 4.** *(Continued)* **Grinding Wheel Recommendations for Hardened Tool Steels According to their Grindability**

| Operation | Wheel or Rim Diameter, Inches | First Choice Specifications | Second Choice Specifications |
|---|---|---|---|
| GROUP 2 STEELS (Continued) | | | |
| Cutter sharpening Straight wheel | ............. | Wet FA46-L5V | FA60-K8V |
| | ............. | Dry FA46-J8V | FA60-H12VP |
| Dish shape | ............. | Dry FA60-J5V | FA60-G12VP |
| Cup shape | ............. | Dry FA46-K5V | FA60-G12VP |
| | ............. | Wet FA46-L5V | FA60-J8V |
| Form tool grinding | 8 and smaller | Wet FA60-K8V to | FA120-L8V |
| | 8 and smaller | Dry FA80-K8V to | FA150-K8V |
| | 10 and larger | Wet FA60-K8V to | FA120-L8V |
| Cylindrical | 14 and less | Wet FA60-L5V | SFA60-L5V |
| | 16 and larger | Wet FA60-K5V | SFA60-K5V |
| Centerless | ............. | Wet FA60-M5V | SFA60-M5V |
| Internal Production grinding | Under ½ | Wet FA80-L6V | SFA80-L6V |
| | ½ to 1 | Wet FA70-K5V | SFA70-K5V |
| | Over 1 to 3 | Wet FA60-J8V | SFA60-J7V |
| | Over 3 | Wet FA54-J8V | SFA54-J8V |
| | | | SFA80-K7V |
| Tool room grinding | Under ½ | Dry FA80-I8V | |
| | ½ to 1 | Dry FA70-J8V | SFA70-J7V |
| | Over 1 to 3 | Dry FA60-I8V | FA60-G12VP |
| | Over 3 | Dry FA54-I8V | FA54-G12VP |
| GROUP 3 STEELS | | | |
| Surfacing Straight wheels | 14 and smaller | Wet FA60-I8V | FA60-G12VP |
| | 14 and smaller | Dry FA60-H8V | FA60-F12VP |
| | Over 14 | Wet FA60-H8V | SFA60-I8V |
| Segments or Cylinders Cups | 1½ rim or less | Wet FA46-G8V | FA46-E12VP |
| | ¾ rim or less | Wet FA46-G8V | FA46-E12VP |
| (for rims wider than 1½ inches, go one grade softer in available specifications) | | | |
| Cutter sharpening Straight wheel | ............. | Wet FA46-J8V | FA60-J8V |
| | ............. | Dry FA46-I8V | FA46-G12VP |
| Dish shape | ............. | Dry FA60-H8V | FA60-F12VP |
| Cup shape | ............. | Dry FA46-I8V | FA60-F12VP |
| | ............. | Wet FA46-J8V | FA60-J8V |
| Form tool grinding | 8 and smaller | Wet FA80-K8V to | FA150-L9V |
| | 8 and smaller | Dry FA100-J8V to | FA150-K8V |
| | 10 and larger | Wet FA80-J8V to | FA150-J8V |

**Table 4.** *(Continued)* **Grinding Wheel Recommendations for Hardened Tool Steels According to their Grindability**

| Operation | Wheel or Rim Diameter, Inches | First Choice Specifications | Second Choice Specifications |
|---|---|---|---|
| GROUP 3 STEELS (Continued) | | | |
| Cylindrical | 14 and less | Wet FA80-L5V | SFA80-L6V |
| | 16 and larger | Wet FA60-L6V | SFA60-K5V |
| Centerless | .............. | Wet FA60-L5V | SFA60-L5V |
| Internal | | | |
| Production grinding | Under ½ | Wet FA90-L6V | SFA90-L6V |
| | ½ to 1 | Wet FA80-L6V | SFA80-L6V |
| | Over 1 to 3 | Wet FA70-K5V | SFA70-K5V |
| | Over 3 | Wet FA60-J5V | SFA60-J5V |
| Tool room grinding | Under ½ | Dry FA90-K8V | SFA90-K7V |
| | ½ to 1 | Dry FA80-J8V | SFA80-J7V |
| | Over 1 to 3 | Dry FA70-I8V | FA70-G12VP |
| | Over 3 | Dry FA60-I8V | FA60-G12VP |
| GROUP 4 STEELS | | | |
| Surfacing | | | |
| Straight wheels | 14 and smaller | Wet FA60-I8V | C60-JV |
| | 14 and smaller | Wet FA60-H8V | C60-IV |
| | Over 14 | Wet FA46-H8V | C60-HV |
| Segments | 1½ rim or less | Wet FA46-G8V | C46-HV |
| Cylinders | 1½ rim or less | Wet FA46-G8V | C60-HV |
| Cups | ¾ rim or less | Wet FA46-G8V | C60-IV |
| (for rims wider than 1½ inches, go one grade softer in available specifications) | | | |
| Form tool grinding | 8 and smaller | Wet FA60-J8V to | FA150-K8V |
| | 8 and smaller | Dry FA80-I8V to | FA180-J8V |
| | 10 and larger | Wet FA60-J8V to | FA150-K8V |
| Cylindrical | 14 and less | Wet FA80-K8V | C60-KV |
| | 16 and larger | Wet FA60-J8V | C60-KV |
| Internal | | | |
| Production grinding | Under ½ | Wet FA90-L6V | C90-LV |
| | ½ to 1 | Wet FA80-K5V | C80-KV |
| | Over 1 to 3 | Wet FA70-J8V | C70-JV |
| | Over 3 | Wet FA60-I8V | C60-IV |
| Tool room grinding | Under ½ | Dry FA90-K8V | C90-KV |
| | ½ to 1 | Dry FA80-J8V | C80-JV |
| | Over 1 to 3 | Dry FA70-I8V | C70-IV |
| | Over 3 | Dry FA60-H8V | C60-HV |

**Table 4.** (*Concluded*) Grinding Wheel Recommendations for Hardened Tool Steels According to their Grindability

| Operation | Wheel (or Rim) Diameter, Inches | First Choice Specifications | Second Choice Specifications | Third Choice Specifications |
|---|---|---|---|---|
| | | GROUP 5 STEELS | | |
| **Surfacing** | | | | |
| Straight wheels | 14 and smaller | Wet SFA60-H8V | FA60-E12VP | C60-IV |
| | 14 and smaller | Dry SFA80-H8V | FA80-E12VP | C80-HV |
| | Over 14 | Wet SFA60-H8V | FA60-E12VP | C60-HV |
| Segments or Cylinders | 1½ rim or less | Wet SFA46-G8V | FA46-E12VP | C46-GV |
| Cups | ¾ rim or less | Wet SFA60-G8V | FA60-E12VP | C60-GV |
| | (for rims wider than 1½ inches, go one grade softer in available specifications) | | | |
| **Cutter sharpening** | | | | |
| Straight wheels | .......... | Wet SFA60-I8V | SFA60-G12VP | .......... |
| | .......... | Dry SFA60-H8V | SFA80-F12VP | .......... |
| Dish shape | .......... | Dry SFA80-H8V | SFA80-F12VP | .......... |
| | .......... | Dry SFA60-I8V | SFA60-G12VP | .......... |
| Cup shape | .......... | Wet SFA60-J8V | SFA60-H12VP | .......... |
| **Form tool grinding** | 8 and smaller | Wet FA80-J8V | FA180-J9V | FA80-H12VP |
| | 8 and smaller | Dry FA100-I8V | FA220-J9V | FA80-G12VP |
| | 10 and larger | Wet FA80-J8V | FA180-J9V | .......... |
| **Cylindrical** | 14 and less | Wet FA80-I8V | C80-KV | .......... |
| | 16 and larger | Wet FA80-I8V | C80-KV | .......... |
| **Centerless** | .......... | Wet FA80-J5V | C80-LV | .......... |
| **Internal** | | | | |
| Production grinding | Under ½ | Wet FA100-L8V | C90-MV | .......... |
| | ½ to 1 | Wet FA90-K8V | C80-LV | .......... |
| | Over 1 to 3 | Wet FA80-J8V | C70-KV | FA80-H12VP |
| | Over 3 | Wet FA70-I8V | C60-IV | FA70-G12VP |
| Tool room grinding | Under ½ | Dry FA100-K8V | C90-KV | .......... |
| | ½ to 1 | Dry FA90-J8V | C80-JV | .......... |
| | Over 1 to 3 | Dry FA80-I8V | C70-IV | FA80-G12VP |
| | Over 3 | Dry FA70-I8V | C60-IV | FA70-G12VP |

**Table 5. Grinding Wheel Recommendations for Constructional Steels (Soft)**

| Grinding Operation | Wheel or Rim Diameter, Inches | First Choice | | Alternate Choice (Porous type) | All-Around Wheel |
|---|---|---|---|---|---|
| **Surfacing** | | | | | |
| Straight wheels | 14 and smaller | Wet | FA46-J8V | FA46-H12VP | FA46-J8V |
| | 14 and smaller | Dry | FA46-I8V | FA46-H12VP | FA46-I8V |
| | Over 14 | Wet | FA36-J8V | FA36-H12VP | FA36-J8V |
| Segments | 1½ rim or less | Wet | FA24-H8V | FA30-F12VP | FA24-H8V |
| Cylinders | 1½ rim or less | Wet | FA24-I8V | FA30-G12VP | FA24-H8V |
| Cups | ¾ rim or less | Wet | FA24-H8V | FA30-F12VP | FA30-H8V |
| | (For wider rims, go one grade softer) | | | | |
| Cylindrical | 14 and smaller | Wet | SFA60-M5V | .......... | SFA60-L5V |
| | 16 and larger | Wet | SFA54-M5V | .......... | SFA54-L5V |
| Centerless | .............. | Wet | SFA54-N5V | .......... | SFA60-M5V |
| Internal | Under ½ | Wet | SFA60-M5V. | .......... | SFA80-L6V |
| | ½ to 1 | Wet | SFA60-L5V | .......... | SFA60-K5V |
| | Over 1 to 3 | Wet | SFA54-K5V | ......... | SFA54-J5V |
| | Over 3 | Wet | SFA46-K5V | .......... | SFA46-J5V |

**Table 6. Grinding Wheel Recommendations for Constructional Steels (Hardened or Carburized)**

| Grinding Operation | Wheel or Rim Diameter, Inches | First Choice | | Alternate Choice (Porous Type) |
|---|---|---|---|---|
| **Surfacing** | | | | |
| Straight wheels | 14 and smaller | Wet | FA46-I8V | FA46-G12VP |
| | 14 and smaller | Dry | FA46-H8V | FA46-F12VP |
| | Over 14 | Wet | FA36-I8V | FA36-G12VP |
| Segments or Cylinders | 1½ rim or less | Wet | FA30-H8V | FA36-F12VP |
| Cups | ¾ rim or less | Wet | FA36-H8V | FA46-F12VP |
| | (For wider rims, go one grade softer) | | | |
| Forms and Radius Grinding | 8 and smaller | Wet | FA60-L7V to | FA100-M8V |
| | 8 and smaller | Dry | FA60-K8V to | FA100-L8V |
| | 10 and larger | Wet | FA60-L7V to | FA 80-M7V |
| **Cylindrical** | | | | |
| Work diameter | | | | |
| 1" and smaller | 14 and smaller | Wet | SFA80-L6V | .............. |
| Over 1" | 14 and smaller | Wet | SFA80-K5V | .............. |
| 1" and smaller | 16 and larger | Wet | SFA60-L5V | .............. |
| Over 1" | 16 and larger | Wet | SFA60-L5V | .............. |
| Centerless | .............. | Wet | SFA80-M6V | .............. |
| **Internal** | Under ½ | Wet | SFA80-N6V | .............. |
| | ½ to 1 | Wet | SFA60-M5V | .............. |
| | Over 1 to 3 | Wet | SFA54-L5V | .............. |
| | Over 3 | Wet | SFA46-K5V | .............. |
| | Under ½ | Dry | FA80-L6V | .............. |
| | ½ to 1 | Dry | FA70-K8V | .............. |
| | Over 1 to 3 | Dry | FA60-J8V | FA60-H12VP |
| | Over 3 | Dry | FA46-J8V | FA54-H12VP |

**Variations from General Grinding Wheel Recommendations.** — Recommendations for the selection of grinding wheels are usually based on average values with regard to both operational conditions and process objectives. In case of variations from such average values the composition of the grinding wheels must be adjusted for obtaining optimum results. While it is impossible to list and to appraise all possible variations and to define their effects on the selection of the best suited grinding wheels, some guidance can be obtained from experience. The following tabulation indicates the general directions in which the characteristics of the initially selected grinding wheel may have to be altered in order to approach optimum performance. Variations in a sense opposite to those mentioned will call for wheel characteristic changes in reverse.

| Conditions or Objectives | Direction of Change |
|---|---|
| To increase cutting rate | Coarser grain, softer bond, higher porosity |
| To retain wheel size and/or form | Finer grain, harder bond |
| For small or narrow work surface | Finer grain, harder bond |
| For larger wheel diameter | Coarser grain |
| To improve finish or work | Finer grain, harder bond, or resilient bond |
| For increased work speed or feed rate | Harder bond |
| For increased wheel speed | Generally softer bond, except for high speed grinding which requires harder bond for added wheel strength |
| For interrupted or coarse work surface | Harder bond |
| For thin walled parts | Softer bond |
| To reduce load on the machine drive motor | Softer bond |

**Dressing and Truing Grinding Wheels.** — The perfect grinding wheel operating under ideal conditions will be self sharpening; i.e., as the abrasive grains become dull, they will tend to fracture and be dislodged from the wheel by the grinding forces, thereby exposing new, sharp abrasive grains. While in precision machine grinding this ideal may be partially attained in some instances, it is almost never attained completely. Usually, the grinding wheel must be dressed and trued after mounting on the precision grinding machine spindle and periodically thereafter.

Dressing may be defined as any operation performed on the face of a grinding wheel that improves its cutting action. Truing is a dressing operation but is more precise, i.e., the face of the wheel may be made parallel to the spindle or made into a radius or special shape. Regularly applied truing is also needed for the accurate size control of the work, particularly in automatic grinding. The tools and processes generally used in grinding wheel dressing and truing are listed and described in Table 1.

Table 1.  Tools and Methods for Grinding Wheel Dressing and Truing

| Designation | Description | Application |
|---|---|---|
| Rotating Hand Dressers | Freely rotating discs, either star shaped with protruding points, or discs with corrugated or twisted perimeter, supported in a fork-type handle, the lugs of which can lean on the tool rest of the grinding machine. | Preferred for bench or floor type grinding machines, also for use on heavy portable grinders (snagging grinders) where free cutting properties of the grinding wheel are primarily sought and the accuracy of the trued profile is not critical. |
| Abrasive Sticks | Made of silicon carbide grains with a hard bond. Applied directly or supported in a handle. Less frequently abrasive sticks are also made of boron carbide. | Usually hand held and use limited to smaller size wheels. Because it also shears the grains of the grinding wheel, it is often used for roughing or pre-shaping, prior to final dressing with, e.g., a diamond. |
| Abrasive Wheels (Rolls) | Silicon carbide grains in a hard vitrified bond are cemented on ball bearing mounted spindles. Used either as hand tools with handles or rigidly held in a supporting member of the grinding machine. Generally freely rotating; also available with adjustable brake for diamond wheel dressing. | Preferred for large grinding wheels as a diamond saver, but also for the improved control of the dressed surface characteristics. When skewing the abrasive dresser wheel by a few degrees out of parallel with the grinding wheel axis, the basic crushing action is supplemented with wiping and shearing, thus producing the desired degree of wheel surface smoothness. |
| Single-point Diamonds | A diamond stone of selected size is mounted in a steel nib of cylindrical shape with or without head, dimensioned to fit the truing spindle of specific grinding machines. Proper orientation and retainment of the diamond point in the setting is an important requirement. | The most widely used tool for dressing and truing grinding wheels in precision grinding. Permits precisely controlled dressing action by regulating infeed and cross feed rate of the truing device. Most dependable in duplicating the movements of the truing spindle when latter is guided by cams or templates for accurate form truing. |
| Single-point Form Truing Diamonds | Selected diamonds having symmetrically located natural edges with precisely lapped diamond points, controlled cone angles and vertex radius, and the axis coinciding with that of the nib. | Truing operations requiring very accurately controlled, and often, also steeply inclined wheel profiles, such as in thread and gear grinding, where one or more diamond points participate in generating the resulting wheel periphery form, are dependent on specially designed and made truing diamonds and nibs. |

Table 1 (*Continued*). Tools and Methods for Grinding Wheel Dressing and Truing

| Designation | Description | Application |
|---|---|---|
| Cluster Type Diamond Dresser | Several, usually seven, smaller diamond stones are mounted in spaced relationship across the working surface of the nib. In some types of tools more than a single layer of such clusters is set at parallel levels in the matrix, the deeper positioned layer becoming active after the preceding layer has worn away. | Intended for straight face dressing and permits the utilization of smaller, less expensive diamond stones. In use, the holder is canted at a 3° to 10° angle, bringing two to five points into contact with the wheel. The multiple point contact permits faster cross feed rates during truing than used with single point diamonds for generating a specific degree of wheel face finish. |
| Impregnated Matrix Type Diamond Dressers | The operating surface consists of a layer of small, randomly distributed, yet rather uniformly spaced diamonds which are retained in a bond holding the points in an essentially common plane. Supplied either with straight or canted shaft, the latter being used to cancel the tilt of angular truing posts. | For the truing of wheel surfaces consisting of a single or several flat elements. The nib face should be held tangent to the grinding wheel periphery or parallel with a flat working surface. Offers economic advantages where technically applicable because of using less expensive diamond splinters presented in a manner permitting efficient utilization. |
| Form Generating Truing Devices | Swivelling diamond holder post with adjustable pivot location, arm length and swivel arc, mounted on angularly adjustable cross slides with controlled traverse movement, permit the generation of various straight and circular profile elements, kept in specific mutual locations. | Such devices are made in various degrees of complexity for the positionally controlled interrelation of several different profile elements. Limited to regular straight and circular sections, yet offers great flexibility of setup, very accurate adjustment and unique versatility for handling a large variety of frequently changing profiles. |
| Contour Duplicating Truing Devices | The form of a master, called cam or template, having the actual profile which is to be produced on the wheel, or its magnified version, is translated into the path of the diamond point by means of mechanical linkage, a fluid actuator or a pantograph device. | Preferred single-point truing method for profiles to be produced in quantities warranting the making of special profile bars or templates. Used also in small and medium volume production when the complexity of the profile to be produced, excludes alternate methods of form generation. |

**Table 1** *(Concluded)*. **Tools and Methods for Grinding Wheel Dressing and Truing**

| Designation | Description | Application |
|---|---|---|
| Grinding Wheel Contouring by Crush Truing | A hardened steel or carbide roll which is free to rotate and has the desired form of the workpiece, is fed gradually into the grinding wheel which runs at slow speed. The roll will, by crushing action produce its reverse form in the wheel. Crushing produces a free-cutting wheel face with sharp grains. | Requires grinding machines designed for crush truing, having stiff spindle bearings, rigid construction, slow wheel speed for truing, etc. Due to the cost of crush rolls and equipment the process is used for repetitive work only. It is one of the most efficient methods for precisely duplicating complex wheel profiles which are capable of grinding in the 8 mu in. AA range. Applicable for both surface and cylindrical grinding. |
| Rotating Diamond Roll Type Grinding Wheel Truing | Special rolls made to agree with specific profile specifications have their periphery coated with a large number of uniformly distributed diamonds, held in a matrix into which the individual stones are set by hand (for larger diamonds) or bonded by a plating process (for smaller elements). | The diamond rolls must be rotated by an air, hydraulic or electric motor at about one fourth of the grinding wheel surface speed and in opposite direction to the wheel rotation. While the initial costs are substantially higher than for single-point diamond truing, the savings in truing time warrants the method's application in large volume production of profile-ground components. |
| Diamond Dressing Blocks | Made as flat blocks for straight wheel surfaces, are also available for radius dressing and profile truing. The working surface consists of a layer of electroplated diamond grains, uniformly distributed and capable of truing even closely toleranced profiles. | For straight wheels it can reduce dressing time and offers easy installation on surface grinders, where the blocks mount on the magnetic plate. Recommended for small and medium volume production for truing intricate profiles on regular surface grinders, because the higher pressure developed in crush dressing is avoided. |

**Guidelines for Truing and Dressing With Single Point Diamonds.** — The diamond nib should be canted at an angle of 10 to 15 degrees in the direction of the wheel rotation and also, if possible, by the same amount in the direction of the cross feed traverse during the truing (see diagram). The dragging effect resulting from this "angling", combined with the occasional rotation of the diamond nib in its holder, will prolong the diamond life by limiting the extent of wear facets and will also tend to produce a pyramid shape of the diamond tip. The diamond may also be set to contact the wheel by about ⅛ to ¼ inch below its centerline.

*Depth of cut:* This should not exceed 0.001 inch per pass for general work, and will have to be reduced to 0.0002 to 0.0004 inch per pass for wheels with fine grains used for precise finishing work.

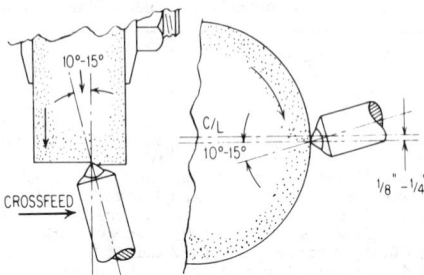

*Diamond crossfeed rate:* This may be varied to some extent depending on the required wheel surface: faster crossfeed for free cutting, and slower crossfeed for fine-finish producing characteristics. Such variations, however, must always stay within the limits set by the grain size of the wheel. Thus the advance rate of the truing diamond per wheel revolution should not exceed the diameter of a grain or be less than half of that rate. Consequently, the diamond crossfeed must be slower for a large wheel than for a smaller wheel having the same grain size number.

Typical crossfeed values for frequently used grain sizes are given in Table 2.

**Table 2. Typical Diamond Truing Diamond Cross Feeds**

| Grain Size................................. | 30 | 36 | 46 | 50 |
|---|---|---|---|---|
| Cross Feed per Wheel Rev., inch.... | .014–.024 | .012–.019 | .008–.014 | .007–.012 |
| Grain Size................................. | 60 | 80 | 120 | ....... |
| Cross Feed per Wheel Rev., inch.... | .006–.010 | .004–.007 | .0025–.004 | ....... |

These values can then easily be converted into the more conveniently used inch-per-minute units, simply by multiplying them by the actual rpm of the grinding wheel. Example: For a 20-inch diameter wheel, Grain No. 46, running at 1200 rpm:

Crossfeed rate for roughing-cut truing — about 17 ipm

for finishing-cut truing — about 10 ipm

*Coolant* should be applied before the diamond is making contact with the wheel and must be continued in generous supply while truing.

*The speed of the grinding wheel* should be at the regular grinding rate, or not much lower. For that reason the feed wheels of centerless grinding machines have usually an additional speed rate higher than functionally needed, that speed being provided for wheel truing only.

*The initial approach of the diamond* to the wheel surface must be carried out carefully to prevent sudden contact with the diamond, resulting in penetration in excess of the selected depth of cut. It should be noted that the highest point of a worn wheel is often in its center portion and not at the edge from where the crossfeed of the diamond starts.

*The general conditions* of the truing device are important for best truing results and also for assuring extended diamond life. Rigid truing spindle, well-seated diamond nib and firmly set diamond point are mandatory. Sensitive infeed and smooth traverse movement at uniform speed must also be maintained.

*Resetting of the diamond point.* Never let the diamond point wear to a degree where the grinding wheel is contacting the steel nib. This can damage the setting of the diamond point and result in its loss. Expert resetting of a worn diamond can repeatedly add to its useful life, even when applied to lighter work because of reduced size.

## Size Selection Guide for Single-Point Truing Diamonds. — There are no rigid rules for determining the proper size of diamond for any particular truing application because of the very large number of factors affecting that choice. Several of these factors are related to the condition, particularly the rigidity, of the grinding machine and truing device, as well as such characteristics of the diamond itself as purity, crystalline structure, etc. While these factors are difficult to evaluate in a generally applicable manner, the expected effects of several other conditions can be appraised and also considered in the selection of the proper diamond size.

The recommended sizes in Table 3 must be considered as informative only and as representing minimum values for generally favorable conditions. Factors calling for larger diamond sizes than listed are the following:

— Silicon carbide wheels (Table 3 refers to aluminum oxide wheels)
— Dry truing
— Grains sizes coarser than No. 46
— Bonds harder than M
— Wheel speed substantially higher than 6500 sfm.

It is advisable to consider any single or pair of these factors as justifying the selection of one size larger diamond. As an example: for using an SiC wheel, with grain size No. 36 and hardness P, select a diamond that is two sizes larger than that shown in Table 3 for the actually used wheel size.

**Table 3.  Recommended Minimum Sizes for Single-point Truing Diamonds**

| Diamond Size in Carats* | Index Number (Wheel Diam. × Width in Inches) | Examples of Max. Grinding Wheel Dimensions | |
|---|---|---|---|
| | | Diameter | Width |
| 0.25 | 3 | 4 | 0.75 |
| 0.35 | 6 | 6 | 1 |
| 0.50 | 10 | 8 | 1.25 |
| 0.60 | 15 | 10 | 1.50 |
| 0.75 | 21 | 12 | 1.75 |
| 1.00 | 30 | 12 | 2.50 |
| 1.25 | 48 | 14 | 3.50 |
| 1.50 | 65 | 16 | 4.00 |
| 1.75 | 80 | 20 | 4.00 |
| 2.00 | 100 | 20 | 5.00 |
| 2.50 | 150 | 24 | 6.00 |
| 3.00 | 200 | 24 | 8.00 |
| 3.50 | 260 | 30 | 8.00 |
| 4.00 | 350 | 36 | 10.00 |

* One carat equals 0.2 gram.

Single-point diamonds are available as loose stones, but are preferably procured from specialized manufacturers supplying the diamonds set into steel nibs. Expert setting, comprising both the optimum orientation of the stone and its firm retainment, are mandatory for assuring adequate diamond life and also satisfactory truing. Because the holding devices for truing diamonds are not yet standardized, the required nib dimensions vary depending on the make and type of different grinding machines. Some nibs are made with angular heads, usually hexagonal, to permit occasional rotation of the nib either manually, with a wrench, or automatically.

## Diamond Wheels

**Diamond Wheels.** — A diamond wheel is a special type of grinding wheel in which the abrasive elements are diamond grains held in a bond and applied to form a layer on the operating face of a non-abrasive core. Diamond wheels are used for grinding very hard or highly abrasive materials. Primary applications are the grinding of cemented carbides, such as the sharpening of carbide cutting tools; the grinding of glass, ceramics, asbestos and cement products; and the cutting and slicing of germanium and silicon.

**Shapes of Diamond Wheels.** — The industry-wide accepted standard (ANSI B74.1-1966) specifies ten basic diamond wheel core shapes which are shown in Table 1 with the applicable designation symbols. The applied diamond abrasive layer may have different cross-sectional shapes. Those standardized are shown in Table 2. The third aspect which is standardized is the location of the diamond section on the wheel as shown by the diagrams in Table 3. Finally, modifications of the general core shape together with pertinent designation letters are given in Table 4.

The characteristics of the wheel shape listed in these four tables make up the components of the standard designation symbol for diamond wheel shapes. An example of that symbol with arbitrarily selected components is shown in Figure 1.

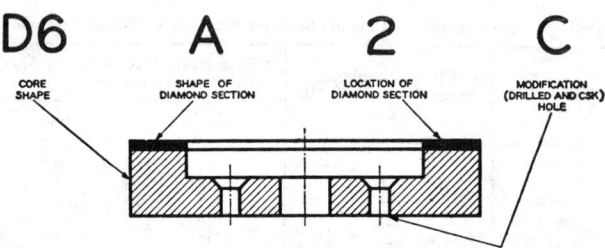

Fig. 1. — A Typical Diamond Wheel Shape Designation Symbol

An explanation of these components is as follows:

*Basic Core Shape:* This portion of the symbol indicates the basic shape of the core on which the diamond abrasive section is mounted. The shape is actually designated by a number preceded by the letter "D." The various core shapes and their designations are given in Table 1.

*Diamond Cross-Section Shape:* This, the second component, consisting of one or two letters, denotes the cross-sectional shape of the diamond abrasive section. The various shapes and their corresponding letter designations are given in Table 2.

*Diamond Section Location:* The third component of the symbol consists of a number which gives the location of the diamond section, i.e., periphery, side, corner, etc. An explanation of these numbers is shown in Table 3.

*Modification:* The fourth component of the symbol is a letter designating some modification, such as drilled and counterbored holes for mounting or special relieving of diamond section or core. This modification position of the symbol is used only when required. The modifications and their designations are given in Table 4.

Table 1. Diamond Wheel Core Shapes and Designations

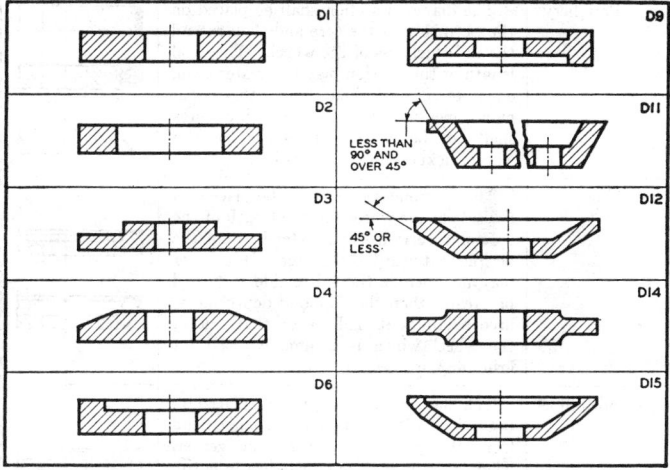

Table 2. Diamond Cross-sections and Designations

**Table 3.  Designations for Location of Diamond Section on Diamond Wheel**
(ANSI B74.1-1966)

| Designation No. and Location | Description | Illustration |
|---|---|---|
| 1 — Periphery | The diamond section shall be placed on the periphery of the core and shall extend the full thickness of the wheel.  The axial length of this section may be greater than, equal to, or less than the depth of diamond, measured radially.  A hub or hubs shall not be considered as part of the wheel thickness for this definition. | |
| 2 — Side | The diamond section shall be placed on the side of the wheel and the length of the diamond section shall extend from the periphery toward the center.  It may or may not include the entire side and shall be greater than the diamond depth measured axially.  It shall be on that side of the wheel which is commonly used for grinding purposes. | |
| 3 — Both Sides | The diamond sections shall be placed on both sides of the wheel and shall extend from the periphery toward the center.  They may or may not include the entire sides, and the radial length of the diamond section shall exceed the axial diamond depth. | |
| 4 — Inside Bevel or Arc | This designation shall apply to the general wheel types 2, 6, 11, 12, and 15 and shall locate the diamond section on the side wall.  This wall shall have an angle or arc extending from a higher point at the wheel periphery to a lower point toward the wheel center. | |
| 5 — Outside Bevel or Arc | This designation shall apply to the general wheel types, 2, 6, 11, and 15 and shall locate the diamond section on the side wall.  This wall shall have an angle or arc extending from a lower point at the wheel periphery to a higher point toward the wheel center. | |
| 6 — Part of Periphery | The diamond section shall be placed on the periphery of the core but shall not extend the full thickness of the wheel and shall not reach to either side. | |

Table 3 *(Continued)*.  **Designations for Location of Diamond Section on Diamond Wheel** (ANSI B74.1-1966)

| Designation No. and Location | Description | Illustration |
|---|---|---|
| 7 — Part of Side | The diamond section shall be placed on the side of the core and shall not extend to the wheel periphery. It may or may not extend to the center. | 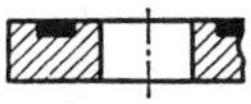 |
| 8 — Through-out | This shall designate wheels of solid diamond abrasive section without cores. | 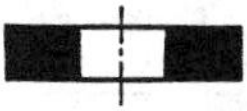 |
| 9 — Corner | This shall designate a location which would commonly be considered to be on the periphery except that the diamond section shall be on one corner but shall not extend to the other corner. | 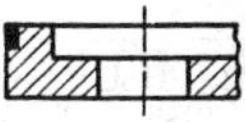 |
| 10 — Annular | This shall designate a location of the diamond abrasive section on the inner annular surface of the wheel. | 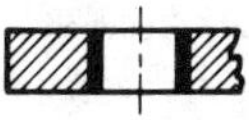 |

**Composition of Diamond Wheels.** — For designating the composition of diamond wheels a second symbol is used, an example of which is shown in Figure 2.

| Abrasive | Grit Size | Grade | Concentration | Bond Type | Bond Modification | Depth of Diamond Section | Manufacturer's Identification Symbol |
|---|---|---|---|---|---|---|---|
| ASD | 100 — R | | 100 | B | 56 | 1/8 | * |

Fig. 2.  Designation symbols for composition of diamond wheels.

The meaning of each component in this symbol is indicated by the following list:

1. *Abrasive Type:* The letter (D) is used for natural diamond; (SD) or (MD) for manufactured diamond; (CD) for coated diamond; and (ASD) for armored diamond.

2. *Grit Size:* The diamond grains are of different sizes, referred to as grit and are expressed in numbers indicating screen mesh sizes. Commonly used grit sizes range from 24 to 500; for special applications, even finer.

3. *Grade:* The grade designates the relative hardness, indicating the grain retaining strength of the bond. Letter symbols are used in alphabetical order from "soft" to "hard." Commonly used grades for various bond materials are as follows: for resinoid bond, from H to R; for metallic bond, from L to R; and for vitrified bond from J to T.

4. *Concentration:* The individual diamond grains are held in a bond material and the proportion of diamond grains in terms of the unit volume of the abrasive conglomerate is termed concentration. The highest concentration, designated as 100 per cent contains 72 carats of diamond grains for each cubic inch of conglomerate. Other commonly used concentrations are 75, 50 and 25 per cent (low) indicating a proportionately reduced diamond content.

5. *Bond:* Commonly, resinoid (B) bonds are used. For applications where

**Table 4.  Designation Letters for Modifications of Diamond Wheels**
(ANSI B74.1-1966)

| Designation Letter* | Description | Illustration |
|---|---|---|
| B—Drilled and Counterbored | Holes drilled and counterbored in core. | |
| C—Drilled and Countersunk | Holes drilled and countersunk in core. | |
| H—Plain Hole | Straight hole drilled in core. | |
| M—Holes Plain and Threaded | Mixed holes, some plain, some threaded, are in core. | |
| P—Relieved One Side | Core relieved on one side of wheel. Thickness of core is less than wheel thickness. | |
| R—Relieved Two Sides | Core relieved on both sides of wheel. Thickness of core is less than wheel thickness. | |
| S—Segmented Diamond Section | Wheel has segmental diamond section mounted on core. (Clearance between segments has no bearing on definition.) | |
| SS—Segmental and Slotted | Wheel has separated segments mounted on a slotted core. | |
| T—Threaded Holes | Threaded holes are in core. | |
| Q—Diamond Inserted | Three surfaces of the diamond section are partially or completely enclosed by the core. | |
| V—Diamond Inverted | Any diamond cross section, which is mounted on the core so that the interior point of any angle, or the concave side of any arc, is exposed shall be considered inverted. *Exception:* Diamond cross section AH shall be placed on the core with the concave side of the arc exposed. | |

* Y—Diamond Inserted and Inverted.  See definitions for Q and V.

stronger retention of the grains, i.e., greater hardness is needed, metallic (M) or vitrified (V) bonds may provide better service. (The letters in parenthesis are the respective identification symbols).

6. *Bond Modification:* Occasionally numerals selected by individual manufacturers are used to designate a special bond modification. Examples are: resinoid — 56 and 69.

7. *Abrasive Layer:* The thickness of the abrasive layer applied to the core of the diamond wheel, also called the depth of impregnation or depth of diamond section, may vary for different wheel types. Thicknesses of commonly used sections are $\frac{1}{32}$, $\frac{1}{16}$, $\frac{1}{8}$, and $\frac{1}{4}$ inch. The thicker sections are specified for heavier stock removal rates and for diamond wheels which are used extensively.

8. *Manufacturer's Identification Symbol:* This symbol, when used, appears in the eighth position.

**The Selection of Diamond Wheels.** — Two general aspects must be defined: (a) The shape of the wheel, also referred to as the basic wheel type and (b) The specification of the abrasive portion.

**Table 5. General Diamond Wheel Recommendations for Wheel Type and Abrasive Specification**

| Typical Application or Operation | Basic Wheel Type | Abrasive Specification |
|---|---|---|
| Single Point Tools (offhand grinding) | D6A2C | *Rough:* MD100-N100-B$\frac{1}{8}$<br>*Finish:* MD220-P75-B$\frac{1}{8}$ |
| Single Point Tools (machine ground) | D6A2H | *Rough:* MD180-J100-B$\frac{1}{8}$<br>*Finish:* MD320-L75-B$\frac{1}{8}$ |
| Chip Breakers | D1A1 | MD150-R100-B$\frac{1}{8}$ |
| Multitooth Tools and Cutters (face mills, end mills, reamers, broaches, etc.)<br>Sharpening and Backing off | D11V9 | *Rough:* MD100-R100-B$\frac{1}{8}$<br>*Combination:* MD150-R100-B$\frac{1}{8}$<br>*Finish:* MD220-R100-B$\frac{1}{8}$ |
| Fluting | D12A2 | MD180-N100-B$\frac{1}{8}$ |
| Saw Sharpening | D12A2 | MD180-R100-B$\frac{1}{8}$ |
| Surface Grinding (horizontal spindle) | D1A1 | *Rough:* MD120-N100-B$\frac{1}{8}$<br>*Finish:* MD240-P100-B$\frac{1}{8}$ |
| Surface Grinding (vertical spindle) | D2A2T | MD80-R75-B$\frac{1}{8}$ |
| Cylindrical or Centertype Grinding | D1A1 | MD120-P100-B$\frac{1}{8}$ |
| Internal Grinding | D1A1 | MD150-N100-B$\frac{1}{8}$ |
| Slotting and Cutoff | D1A1R | MD150-R100-B$\frac{1}{4}$ |
| Lapping | Disc | MD400-L50-B$\frac{1}{16}$ |
| Hand Honing | DH1, DH2 | *Rough:* MD220-B$\frac{1}{16}$<br>*Finish:* MD320-B$\frac{1}{16}$ |

General recommendations for the dry grinding, with resin bond diamond wheels, of most grades of cemented carbides of average surface to ordinary finishes at normal removal with average size wheels, as published by Cincinnati Milacron, are listed in Table 5.

A further set of variables are *the dimensions of the wheel*, which must be adapted to the available grinding machine and, in some cases, to the configuration of the work.

For modifying the general abrasive specifications, given in Table 5, in order to better adapt the wheel to the actual operational conditions, the following suggestions may be considered:

— Use softer wheel grades for harder grades of carbides, for grinding larger areas or larger or wider wheel faces.

— Use harder wheel grades for softer grades of carbides, for grinding smaller areas, for using smaller and narrower face wheels and for light cuts.

— Use fine grit sizes for harder grades of carbides and to obtain better finishes.

— Use coarser grit sizes for softer grades of carbides and for roughing cuts.

— Use higher diamond concentration for harder grades of carbides, for larger diameter or wider face wheels, for heavier cuts and for obtaining better finish.

**Guidelines for the Handling and Operation of Diamond Wheels.** — Grinding machines used for grinding with diamond wheels should be of the precision type, in good service condition, with true running spindles and smooth slide movements.

*Mounting of Diamond Wheels:* Wheel mounts should be used which permit the precise centering of the wheel, resulting in a runout of less than 0.001 inch axially and 0.0005 inch radially. These conditions should be checked with a 0.0001-inch type dial indicator. Once mounted and centered, the diamond wheel should be retained on its mount and stored in that condition when temporarily removed from the machine.

*Truing and Dressing:* Resinoid bonded diamond wheels seldom require dressing, but when necessary a soft silicon carbide stick may be hand-held against the wheel. Peripheral and cup type wheels may be sharpened by grinding the cutting face with a 60 to 80 grit silicon carbide wheel. This can be done with the diamond wheel mounted on the spindle of the machine, and with the silicon carbide wheel driven at a relatively slow speed by a specially designed table-mounted grinder or by a small table-mounted tool post grinder. The diamond wheel can be mounted on a special arbor and ground on a lathe with a tool post grinder; peripheral wheels can be ground on a cylindrical grinder or with a special brake-controlled truing device with the wheel mounted on the machine on which it is used. Cup and face type wheels are often lapped on a cast iron or glass plate using a 100 grit silicon carbide abrasive. Care must be used to lap the face parallel to the back, otherwise they must be ground to restore parallelism. Peripheral diamond wheels can be trued and dressed by grinding a silicon carbide block or a special diamond impregnated bronze block in a manner similar to surface grinding. Conventional diamonds must not be used for truing and dressing diamond wheels.

**Speeds and Feeds in Diamond Grinding.** — General recommendations are as follows:

*Wheel Speeds:* The generally recommended wheel speeds for diamond grinding are in the range of 5000 to 6000 surface feet per minute, with this upper limit as a maximum to avoid harmful "overspeeding". Exceptions from that general rule are diamond wheels with coarse grains and high concentration (100 per cent) where the wheel wear in dry surface grinding can be reduced by lowering the speed to

2500–3000 sfpm. However, this lower speed range can cause rapid wheel breakdown in finer grit wheels or in those with reduced diamond concentration.

*Work Speeds:* In diamond grinding, work rotation and table traverse are usually established by experience, adjusting these values to the selected infeed so as to avoid excessive wheel wear.

*Infeed Per Pass:* This is often referred to as downfeed and is generally a function of the grit size of the wheel. The following are general values which may be increased for raising the productivity, or lowered to improve finish or to reduce wheel wear.

| Wheel Grit Size Range | Infeed per Pass |
|---|---|
| 100 to 120 | 0.001 inch |
| 150 to 220 | 0.0005 inch |
| 250 and finer | 0.00025 inch |

## Grinding Wheel Safety

**Safety in Operating Grinding Wheels.** — Grinding wheels, although capable of exceptional cutting performance due to hardness and wear resistance, are prone to damage caused by improper handling and operation. Vitrified wheels, comprising the major part of grinding wheels used in industry, are held together by an inorganic bond which is actually a type of pottery product and therefore brittle and breakable. Although most of the organic bond types are somewhat more resistant to shocks, it must be realized that all grinding wheels are conglomerates of individual grains joined by a bond material whose strength is limited by the need of releasing the dull abrasive grains during use.

It must also be understood that during the grinding process very substantial forces act on the grinding wheel, including the centrifugal force due to rotation, the grinding forces resulting from the resistance of the work material, and shocks caused by sudden contact with the work. To be able to resist these forces, the grinding wheel must have a substantial minimum strength throughout that is well beyond that needed to hold the wheel together under static conditions.

Finally, a damaged grinding wheel can disintegrate during grinding, liberating dormant forces which are constrained by the resistance of the bond, thus presenting great hazards to both operator and equipment.

In order to avoid the breakage of the operating wheel and, should such a mishap occur, to prevent damage or injury, specific precautions have to be applied. These have been formulated into rules and regulations and are set forth in the American National Standard ANSI B7.1-1970, entitled the American Standard Safety Code for the Use, Care and Protection of Abrasive Wheels.

**Handling, Storage and Inspection.** — Grinding wheels shall be hand carried, or transported, with proper support, by truck or conveyor. A grinding wheel must not be rolled around on its periphery.

The storage area, positioned not far from the location of the grinding machines, shall be free from excessive temperature variations and humidity. Specially built racks are recommended on which the smaller or thin wheels are stacked lying on their sides and the larger wheels in an upright position on two-point cradle supports consisting of appropriately spaced wooden bars. Partitions should separate either the individual wheels, or a small group of identical wheels. Good accessibility to the stored wheels reduces the need of undesirable handling.

Inspection will primarily be directed at detecting visible damage, mostly originating from handling and shipping. Cracks which are not obvious can usually be detected by "ring testing"; this consists of suspending the wheel from its hole and tapping it with a non-metallic implement. A clear metallic tone, a "ring" should be heard; a dead sound being indicative of a possible crack or cracks in the wheel.

**Machine Conditions.** — The general design of the grinding machines must assure safe operation under normal conditions. The bearings and grinding wheel spindle must be dimensioned to withstand the expected forces and ample driving power should be provided to assure maintenance of the rated spindle speed. For the protection of the operator, stationary machines used for dry grinding should have provision made for connection to an exhaust system and when used for off-hand grinding, a work support must be available.

Wheel guards are particularly important protection elements and their material specifications, wall thicknesses and construction principles should agree with the Standard's specifications. The exposure of the wheel should be just enough to avoid interference with the grinding operation. The need for access of the work to the grinding wheel will define the boundary of guard opening, particularly in the direction of the operator.

**Grinding Wheel Mounting.** — The mass and speed of the operating grinding wheel makes it particularly sensitive to unbalance. Vibrations which result from such conditions are harmful to the machine, particularly the spindle bearings and also affect the ground surface, i.e., wheel unbalance causes chatter marks and interferes with size control. While grinding wheels are shipped from the manufacturer's plant in a balanced condition, the retainment of the balanced state after the mounting of the wheel is quite uncertain. For that reason a balancing of the mounted wheel is indicated, and is particularly important for medium and large size wheels, as well as for producing accurate and smooth surfaces. The most common way of balancing mounted wheels is by using balancing flanges with adjustable weights. The wheel and balancing flanges are mounted on a short balancing arbor, the two concentric and round stub ends of which are supported in a balancing stand. Such stands are of two types: (a) the parallel straight-edged which must be set up precisely level; (b) the disc type having two pairs of ball bearing mounted overlapping discs, which form a V for containing the arbor ends without hindering the free rotation of the wheel mounted on that arbor. The wheel will then only rotate when out of balance and its heavy spot is not in the lowest position. Rotating the wheel by hand to different positions will move the heavy spot, should such exist, from the bottom to a higher location where it can reveal its presence by causing the wheel to turn. Having detected the presence and location of the heavy spot, its effect can be cancelled by displacing the weights in the circular groove of the flange until a balanced condition has been accomplished.

Flanges are commonly used means for holding grinding wheels on the machine spindle. For that purpose, the wheel can either be mounted directly through its hole or by means of a sleeve which slips over a tapered section of the machine spindle. In either case, the flanges must be of equal diameter, usually not less than one-third of the new wheel's diameter. The purpose is to securely hold the wheel between the flanges, yet without interfering with the grinding operation even when the wheel becomes worn down to the point where it is ready to be discarded. Blotters or flange facings of compressible material should cover the entire contact area of the flanges.

One of the flanges is usually fixed while the other is loose and can be removed and adjusted along the machine spindle. The movable flange is held against the mounted grinding wheel by means of a nut engaging a threaded section of the machine spindle. The sense of that thread should be opposed to the direction of the wheel rotation, which is the direction in which the nut must be turned for removing it.

**Safe Operating Speeds.** — Safe grinding processes, while predicated on the proper use of the previously discussed equipment and procedures, are greatly dependent on the application of adequate operating speeds.

The Standard establishes maximum speeds at which grinding wheels can be operated, assigning the various types of wheels to several classification groups. Different values are listed according to bond type and also to wheel strength, distinguishing between low, medium and high strength wheels.

There is presented in Table 1, for the purpose of general information, an abbreviated version of the Standard's specification. However, for the actually permissible limits, the authoritative source is the manufacturer's tag on the wheel which, particularly for wheels of lower strength, might specify speeds below those of the table. All grinding wheels of 6 inches or greater diameter must be test run in the wheel manufacturer's plant at a speed which for all wheels having operating speeds in excess of 5000 sfpm is 1.5 times the maximum speed marked on the tag of the wheel.

The table shows the permissible wheel speeds in surface feet per minute (sfpm) units, whereas the tags on the grinding wheels state, for the convenience of the user, the maximum operating speed in revolutions per minute (rpm). The sfpm unit has the advantage of remaining valid for worn wheels too, whose rotational speed may be increased to the applicable sfpm value. The conversion from either to the other of these two kinds of units is a matter of simple calculation using the formulas:

$$\text{sfpm} = \text{rpm} \times \frac{D}{12} \times \pi \quad \text{or}$$

$$\text{rpm} = \frac{\text{sfpm} \times 12}{D \times \pi}$$

Where D = maximum diameter of the grinding wheel, in inches. Table 2 showing the conversion values from surface speed into rotational speed can be used for the direct reading of the rpm values corresponding to several different wheel diameters and surface speeds.

*Special Speeds:* Continuing progress in grinding methods has led to the recognition of certain advantages which can result from operating grinding wheels above, sometimes even higher than twice, the speeds considered earlier as the safe limits of grinding wheel operations. While advantages from the application of high speed grinding are limited to specific processes, the Standard admits, and offers code regulations for the use of wheels at special high speeds. These regulations define the structural requirements of the grinding machine and the responsibilities of the grinding wheel manufacturers, as well as of the users. High speed grinding should not be applied unless the machines, particularly guards, spindle assemblies and drive motors, are suitable for such methods. Also, appropriate grinding wheels, expressly made for special high speeds must be used and, of course, the maximum operating speeds indicated on the wheel's tag must never be exceeded.

**Portable Grinders.** — The above discussed rules and regulations, devised primarily for stationery grinding machines are, in general, applicable for portable grinders too. In addition, the details of various other regulations, specially applicable to different types of portable grinders are discussed in the Standard, which should be consulted, particularly in the case of safety-wise critical uses of portable grinding machines.

**Maximum Peripheral Speeds for Grinding Wheels** (Based on ANSI B7.1-1970)

| Classification No. | Type of Wheels | Maximum Operating Speeds, sfpm, Depending on Strength of Bond | |
|---|---|---|---|
| | | Inorganic Bonds | Organic Bonds |
| 1 | Basically straight wheels — except classifications 6 and 7<br>Types 1, 4, 5, 7, 20, 21, 22, 23, 24, 25, 26<br>Dish wheels — Type 12<br>Saucer wheels — Type 13<br>Cones and plugs — Types 16, 17, 18, 19 | 5,500 to 6,500 | 6,500 to 9,500 |
| 2 | Cylinder wheels — Type 2<br>Segments | 5,000 to 6,000 | 5,000 to 7,000 |
| 3 | Cup shape tool grinding wheels — Types 6 and 11 (for fixed base machines) | 4,500 to 6,000 | 6,000 to 8,500 |
| 4 | Cup shape snagging wheels — Types 6 and 11 (for portable machines) | 4,500 to 6,500 | 6,000 to 9,500 |
| 5 | Abrasive discs | 5,500 to 6,500 | 5,500 to 8,500 |
| 6 | Reinforced wheels — except cutting-off wheels (depending on diameter and thickness) | ............ | 9,500 to 16,000 |
| 7 | Type 1 wheels for bench and pedestal grinders, also in certain sizes for surface grinders | 5,500 to 7,550 | 6,500 to 9,500 |
| 8 | Diamond wheels<br>Cutting-off (specific shapes only)<br>Metal bond<br>Resin bond<br>Vitrified bond | to 16,000<br>to 12,000<br>............<br>to 6,500 | to 16,000<br>............<br>to 9,500<br>............ |
| 9 | Cutting-off wheels — Larger than 16-inch diameter (incl. reinforced organic) | ............ | 9,500 to 14,200 |
| 10 | Cutting-off wheels — 16-inch diameter and smaller (incl. reinforced organic) | ............ | 9,500 to 16,000 |
| 11 | Thread and flute grinding wheels | 8,000 to 12,000 | 8,000 to 12,000 |
| 12 | Crankshaft and camshaft grinding wheels | 5,500 to 8,500 | 6,500 to 9,500 |

* Values in this table are for general information only.
Under no conditions should a wheel be operated faster than the maximum operating speed established by the manufacturer.

**Revolution per Minute for Various Grinding Speeds and Wheel Diameters**

| Wheel Diameter, Inch | Peripheral (Surface) Speed, Feet per Minute | | | | | | | | | | | | | | | | Wheel Diameter, Inch |
|---|---|---|---|---|---|---|---|---|---|---|---|---|---|---|---|---|---|
| | 16,000 | 14,000 | 12,000 | 10,000 | 9,500 | 9,000 | 8,500 | 8,000 | 7,500 | 7,000 | 6,500 | 6,000 | 5,500 | 5,000 | 4,500 | 4,000 | |
| | Revolutions per Minute | | | | | | | | | | | | | | | | |
| 1 | 61,116 | 53,474 | 45,836 | 38,196 | 36,287 | 34,377 | 32,467 | 30,558 | 28,647 | 26,737 | 24,828 | 22,918 | 21,008 | 19,098 | 17,189 | 15,279 | 1 |
| 2 | 30,558 | 26,737 | 22,918 | 19,098 | 18,143 | 17,188 | 16,238 | 15,278 | 14,328 | 13,368 | 12,414 | 11,459 | 10,504 | 9,549 | 8,594 | 7,639 | 2 |
| 3 | 20,372 | 17,826 | 15,278 | 12,732 | 12,115 | 11,459 | 10,822 | 10,186 | 9,549 | 8,913 | 8,276 | 7,639 | 7,003 | 6,366 | 5,729 | 5,093 | 3 |
| 4 | 15,278 | 13,368 | 11,459 | 9,549 | 9,072 | 8,595 | 8,116 | 7,640 | 7,162 | 6,685 | 6,207 | 5,729 | 5,252 | 4,775 | 4,297 | 3,820 | 4 |
| 5 | 12,224 | 10,696 | 9,168 | 7,640 | 7,258 | 6,876 | 6,494 | 6,112 | 5,730 | 5,348 | 4,966 | 4,584 | 4,202 | 3,820 | 3,438 | 3,056 | 5 |
| 6 | 10,186 | 8,913 | 7,639 | 6,366 | 6,048 | 5,729 | 5,411 | 5,092 | 4,775 | 4,456 | 4,138 | 3,820 | 3,501 | 3,183 | 2,865 | 2,546 | 6 |
| 7 | 8,732 | 7,640 | 6,548 | 5,456 | 5,183 | 4,911 | 4,638 | 4,366 | 4,092 | 3,820 | 3,547 | 3,274 | 3,001 | 2,728 | 2,455 | 2,183 | 7 |
| 8 | 7,640 | 6,685 | 5,729 | 4,775 | 4,535 | 4,297 | 4,058 | 3,820 | 3,580 | 3,342 | 3,103 | 2,865 | 2,626 | 2,387 | 2,148 | 1,910 | 8 |
| 9 | 6,792 | 5,940 | 5,092 | 4,244 | 4,032 | 3,820 | 3,606 | 3,396 | 3,182 | 2,970 | 2,758 | 2,546 | 2,334 | 2,122 | 1,910 | 1,698 | 9 |
| 10 | 6,112 | 5,348 | 4,584 | 3,820 | 3,629 | 3,438 | 3,247 | 3,056 | 2,865 | 2,674 | 2,483 | 2,292 | 2,101 | 1,910 | 1,719 | 1,528 | 10 |
| 12 | 5,092 | 4,456 | 3,820 | 3,183 | 3,023 | 2,864 | 2,705 | 2,546 | 2,386 | 2,228 | 2,069 | 1,910 | 1,751 | 1,591 | 1,432 | 1,273 | 12 |
| 14 | 4,366 | 3,820 | 3,274 | 2,728 | 2,592 | 2,455 | 2,319 | 2,182 | 2,046 | 1,910 | 1,773 | 1,637 | 1,500 | 1,364 | 1,228 | 1,091 | 14 |
| 16 | 3,820 | 3,342 | 2,865 | 2,387 | 2,268 | 2,149 | 2,029 | 1,910 | 1,791 | 1,672 | 1,552 | 1,432 | 1,313 | 1,194 | 1,074 | 955 | 16 |
| 18 | 3,396 | 2,970 | 2,546 | 2,122 | 2,016 | 1,910 | 1,803 | 1,698 | 1,591 | 1,485 | 1,379 | 1,273 | 1,167 | 1,061 | 955 | 849 | 18 |
| 20 | 3,056 | 2,674 | 2,292 | 1,910 | 1,814 | 1,719 | 1,623 | 1,528 | 1,432 | 1,337 | 1,241 | 1,146 | 1,050 | 955 | 859 | 764 | 20 |
| 22 | 2,776 | 2,430 | 2,084 | 1,736 | 1,649 | 1,562 | 1,476 | 1,388 | 1,302 | 1,215 | 1,128 | 1,042 | 955 | 868 | 781 | 694 | 22 |
| 24 | 2,546 | 2,228 | 1,910 | 1,591 | 1,512 | 1,433 | 1,353 | 1,274 | 1,194 | 1,115 | 1,034 | 955 | 875 | 796 | 716 | 637 | 24 |
| 26 | 2,352 | 2,056 | 1,762 | 1,468 | 1,395 | 1,322 | 1,248 | 1,176 | 1,101 | 1,028 | 955 | 881 | 808 | 734 | 661 | 588 | 26 |
| 28 | 2,182 | 1,910 | 1,637 | 1,364 | 1,296 | 1,228 | 1,159 | 1,092 | 1,023 | 955 | 887 | 818 | 750 | 682 | 614 | 546 | 28 |
| 30 | 2,036 | 1,782 | 1,528 | 1,274 | 1,210 | 1,146 | 1,082 | 1,018 | 955 | 891 | 828 | 764 | 700 | 637 | 573 | 509 | 30 |
| 32 | 1,910 | 1,672 | 1,432 | 1,194 | 1,134 | 1,074 | 1,014 | 954 | 895 | 836 | 776 | 716 | 656 | 597 | 537 | 477 | 32 |
| 34 | 1,796 | 1,572 | 1,348 | 1,124 | 1,067 | 1,011 | 955 | 898 | 843 | 786 | 730 | 674 | 618 | 562 | 505 | 449 | 34 |
| 36 | 1,698 | 1,484 | 1,273 | 1,061 | 1,007 | 954 | 902 | 848 | 795 | 742 | 690 | 637 | 583 | 530 | 477 | 424 | 36 |
| 38 | 1,608 | 1,408 | 1,206 | 1,006 | 955 | 904 | 854 | 804 | 754 | 704 | 653 | 603 | 553 | 503 | 452 | 402 | 38 |
| 40 | 1,528 | 1,338 | 1,146 | 956 | 908 | 860 | 812 | 764 | 716 | 669 | 620 | 573 | 525 | 478 | 430 | 382 | 40 |
| 42 | 1,464 | 1,272 | 1,090 | 908 | 863 | 818 | 775 | 732 | 682 | 636 | 591 | 545 | 500 | 454 | 409 | 366 | 42 |
| 44 | 1,388 | 1,216 | 1,042 | 868 | 824 | 780 | 737 | 694 | 651 | 608 | 564 | 521 | 478 | 434 | 390 | 347 | 44 |
| 46 | 1,332 | 1,164 | 1,000 | 832 | 791 | 750 | 708 | 666 | 624 | 582 | 541 | 500 | 458 | 416 | 375 | 333 | 46 |
| 48 | 1,272 | 1,116 | 956 | 796 | 756 | 716 | 676 | 636 | 597 | 558 | 517 | 478 | 438 | 398 | 358 | 318 | 48 |
| 53 | 1,152 | 1,006 | 864 | 720 | 683 | 648 | 612 | 576 | 539 | 503 | 468 | 432 | 395 | 360 | 324 | 288 | 53 |
| 60 | 1,020 | 892 | 774 | 638 | 606 | 574 | 542 | 510 | 478 | 446 | 414 | 387 | 350 | 319 | 287 | 255 | 60 |
| 72 | 849 | 742 | 637 | 530 | 504 | 477 | 451 | 424 | 398 | 371 | 345 | 318 | 291 | 265 | 239 | 212 | 72 |

## Cylindrical Grinding

Cylindrical grinding designates a general category of various grinding methods which have the common characteristic of rotating the workpiece around a fixed axis while grinding outside surface sections in controlled relation to that axis of rotation.

The form of the part or section being ground in this process is frequently cylindrical, hence the designation of the general category. However, the shape of the part may be tapered or of curvilinear profile; the position of the ground surface may also be perpendicular to the axis; and it is possible to grind concurrently several surface sections, adjacent or separated, of equal or different diameters, located in parallel or mutually inclined planes, etc., as long as the condition of a common axis of rotation is satisfied.

*Size Range of Work Pieces and Machines:* Cylindrical grinding is applied in the manufacture of miniature parts, such as instrument components and, at the opposite extreme, for grinding rolling mill rolls weighing several tons. Accordingly, there are cylindrical grinding machines of many different types, each adapted to a specific work size range. Machine capacities are usually expressed by such factors as maximum work diameter, work length and weight, complemented, of course, by many other significant data.

**Plain, Universal and Limited-Purpose Cylindrical Grinding Machines.** — The plain cylindrical grinding machine is considered the basic type of this general category, and is used for grinding parts with cylindrical or slightly tapered form.

The universal cylindrical grinder can be used, in addition to grinding the basic cylindrical forms, for the grinding of parts with steep tapers, of surfaces normal to the part axis, including the entire face of the workpiece, and also for internal grinding independently or in conjunction with the grinding of the part's outer surfaces. Such variety of part configurations requiring grinding is typical of work in the tool room, which constitutes the major area of application for universal cylindrical grinding machines.

Limited-purpose cylindrical grinders are needed for special work configurations and for high volume production, where productivity is more important than flexibility of adaptation. Examples of limited-purpose cylindrical grinding machines are crankshaft and camshaft grinders, polygonal grinding machines, roll grinders, etc.

**Traverse or Plunge Grinding.** — In traverse grinding the machine table carrying the work performs a reciprocating movement of specific travel length for transporting the rotating workpiece along the face of the grinding wheel. At each, or at alternate stroke ends, the wheel slide advances for the gradual feeding of the wheel into the work. The length of the surface which can be ground by this method is generally limited only by the stroke length of the machine table. In large roll grinders the relative movement between work and wheel is accomplished by the traverse of the wheel slide along a stationary machine table.

In plunge grinding the machine table, after having been set, is locked and, while the part is rotating, the wheel slide continually advances at a preset rate, until the finish size of the part is reached. The width of the grinding wheel is a limiting factor of the section length which can be ground in this process. Plunge grinding is required for profiled surfaces and for the simultaneous grinding of multiple surfaces of different diameters or located in different planes.

When the configuration of the part does not make the use of either method mandatory, the choice may be made on the basis of the following general considerations: traverse grinding usually produces a better finish, while the productivity of plunge grinding is generally higher.

Table 1. Work Holding Methods and Devices for Cylindrical Grinding

| Designation | Description | Discussion |
|---|---|---|
| Centers, non-rotating ("dead"), with drive plate | Headstock with non-rotating spindle holds the center. Around the spindle an independently supported sleeve carries the drive plate for rotating the work. Tailstock for opposite center. | The simplest method of holding the work between two opposite centers is also the potentially most accurate, as long as correctly prepared and located centerholes are used in the work. |
| Centers, driving type | Work held between two centers obtains its rotation from the concurrently applied drive by the live headstock spindle and live tailstock spindle. | Eliminates the drawback of the common center-type grinding with driver plate which requires a dog attached to the workpiece. Driven spindles permit the grinding of the work up to both ends. |
| Chuck, geared or cam actuated. | Two, three or four jaws moved radially through mechanical elements, hand or power operated, exert concentrically acting clamping force. | Adaptable to workpieces of different configurations and within a generally wide capacity of the chuck. Flexible in uses which, however, do not include high precision work. |
| Chuck, diaphragm | Force applied by hand or power of a flexible diaphragm causes the attached jaws to deflect temporarily for accepting the work which is held when force is released. | Rapid action and flexible adaptation to different work configurations by means of special jaws offer varied uses for the grinding of disk shaped and similar parts. |
| Collets | Holding devices with externally or internally acting clamping force, easily adaptable to power actuation, assuring high centering accuracy. | Limited to parts with previously machined or ground holding surfaces, because of the small range of clamping movement of the collet jaws. |
| Face plate | Has four independently actuated jaws, any or several of which may be used, or entirely removed, using the base plate for supporting special clamps. | Used for holding bulky parts, or those of awkward shape, which are ground in small quantities not warranting special fixtures. |
| Magnetic plate | Flat plates, with pole distribution adapted to the work, are mounted on the spindle like chucks and may be used for work with locating face normal to the axis. | Applicable for light cuts such as are frequent in tool making, where the rapid clamping action, and easy access to both the O.D. and the exposed face, are sometimes of advantage. |
| Steady rests | Two basic types are used: (a) The two-jaw type supporting the work from the back (back rest), leaving access by the wheel; (b) The three-jaw type (center-rest). | A complementary work-holding device, used in conjunction with primary work holders, to provide additional support, particularly to long and/or slender parts. |
| Special fixtures | Single-purpose devices, designed for a particular workpiece, primarily for providing special locating elements. | Typical workpieces requiring special fixturing are, as examples: crankshafts where the holding is combined with balancing functions; internal gears located on the pitch circle of the teeth for O.D. grinding. |

**Work Holding on Cylindrical Grinding Machines.** — The manner in which the work is located and held in the machine during the grinding process determines the configuration of the part which can be adapted for cylindrical grinding and affects the resulting accuracy of the ground surface. The method of work holding also affects the attainable production rate, because the mounting and dismounting of the part can represent a substantial portion of the total operational time.

Whatever method is used for holding the part on cylindrical types of grinding machines, two basic conditions must be satisfied: (1) The part should be located with respect to its correct axis of rotation, and (2) The work drive must cause the part to rotate, at a specific speed, around the established axis. The lengthwise location of the part, although controlled, is not too critical in traverse grinding; however, in plunge grinding, particularly when shoulder sections are also involved, it must be assured with great accuracy.

Table 1 presents a listing, with brief discussions, of work holding methods and devices which are the most frequently used in cylindrical grinding.

**Table 2.   Wheel Recommendations for Cylindrical Grinding**

| Material | Wheel Marking | Material | Wheel Marking |
|---|---|---|---|
| Aluminum | SFA46-18V | Forgings | A46-M5V |
| Armatures (laminated) | SFA100-18V | Gages (plug) | SFA80-K8V |
| Axles (auto & railway) | A54-M5V | General purpose grinding | SFA54-L5V |
| Brass | C36-KV | Glass | BFA220-011V |
| Bronze | | Gun barrels | |
| Soft | C36-KV | Spotting and O.D. | BFA60-M5V |
| Hard | A46-M5V | Nitralloy | |
| Bushings (hardened steel) | BFA60-L5V | Before nitriding | A60-K5V |
| Bushings (cast iron) | C36-JV | After nitriding | |
| Cam lobes (cast alloy) | | Commercial finish | SFA60-18V |
| Roughing | BFA54-N5V | High finish | C100-1V |
| Finishing | A70-P6B | Reflective finish | C500-19E |
| Cam lobes (hardened steel) | | Pistons (aluminum) | SFA46-18V |
| Roughing | BFA54-L5V | (cast iron) | C36-KV |
| Finishing | BFA80-T8B | Plastics | C46-JV |
| Cast iron | C36-JV | Rubber | |
| Chromium plating | | Soft | SFA20-K5B |
| Commercial finish | SFA60-J8V | Hard | C36-KB |
| High finish | A150-K5E | Spline shafts | SFA60-N5V |
| Reflective finish | C500-19E | Sprayed metal | C60-JV |
| Commutators (copper) | C60-M4E | Steel | |
| Crankshafts (airplane) | | Soft | |
| Pins | BFA46-K5V | 1″ dia. and smaller | SFA60-M5V |
| Bearings | A46-L5V | over 1″ dia. | SFA46-L5V |
| Crankshafts (automotive | | Hardened | |
| pins and bearings) | | 1″ dia. and smaller | SFA80-L8V |
| Finishing | A54-N5V | over 1″ dia. | SFA60-K5V |
| Roughing & finishing | A54-O5V | 300 series stainless | SFA46-K8V |
| Regrinding | A54-M5V | Stellite | BFA46-M5V |
| Regrinding, sprayed | | Titanium | C60-JV |
| metal | C60-JV | Valve stems (automotive) | BFA54-N5V |
| Drills | BFA54-N5V | Valve tappets | BFA54-M5V |

*Note:* Prefixes to the standard designation "A" of aluminum oxide indicate modified abrasives:
  BFA = Blended Friable (a blend of regular and friable)
  SFA = Semi-Friable

**Selection of Grinding Wheels for Cylindrical Grinding.** — For cylindrical grinding, as for grinding in general, the primary factor to be considered in wheel selection is the work material. Other factors are the amount of excess stock and its rate of removal (speeds and feeds), the desired accuracy and surface finish, the ratio of wheel and work diameter, wet or dry grinding, etc. In view of these many variables, it is not practical to set up a complete list of grinding wheel recommendations with general validity. Instead, examples of recommendations embracing a wide range of typical applications and assuming common practices are presented in Table 2. This is intended as a guide for the starting selection of grinding-wheel specifications which, in case of a not entirely satisfactory performance, can be refined subsequently. The content of the table is a version of the grinding-wheel recommendations for cylindrical grinding by the Norton Company using, however, non-proprietary designations for the abrasive types and bonds.

**Operational Data for Cylindrical Grinding.** — In cylindrical grinding, similarly to other metalcutting processes, the applied speed and feed rates must be adjusted to the operational conditions as well as to the objectives of the process. Grinding differs, however, from other types of metalcutting methods in regard to the cutting speed of the tool which, in grinding, is generally not a variable; it should be maintained at, or close to the optimum rate, commonly 6500 feet per minute peripheral speed.

In establishing the proper process values for grinding, of prime consideration are the work material, its condition (hardened or soft), and the type of operation (roughing or finishing). Other influencing factors are the characteristics of the grinding machine (stability, power), the specifications of the grinding wheel, the material allowance, the rigidity and balance of the workpiece, as well as several grinding process conditions, such as wet or dry grinding, the manner of wheel truing, etc.

Variables of the cylindrical grinding process, often referred to as *grinding data*, comprise the speed of work rotation (measured as the surface speed of the work); the infeed (in inches per pass for traverse grinding, or in inches per minute for plunge grinding); and, in the case of traverse grinding, the speed of the reciprocating table movement (expressed either in feet per minute, or as a fraction of the wheel width for each revolution of the work).

For the purpose of starting values in setting up a cylindrical grinding process, a brief listing of basic data for common cylindrical grinding conditions and involving frequently used materials, is presented in Table 3.

These data, which are, in general, considered conservative, are based on average operating conditions and may be modified subsequently,

— reducing the values in case of unsatisfactory quality of the grinding or the occurrence of failures;
— increasing the rates for raising the productivity of the process, particularly for rigid workpieces, substantial stock allowance, etc.

**High Speed Cylindrical Grinding.** — The maximum peripheral speed of the wheels in regular cylindrical grinding is generally 6500 feet per minute; the commonly used grinding wheels and machines are designed to operate efficiently at this speed. Recently, efforts were made to raise the productivity of different grinding methods, including cylindrical grinding, by increasing the peripheral speed of the grinding wheel to a substantially higher than traditional level, such as 12,000 feet per minute or more. Such methods are designated by the distinguishing term of high speed grinding.

Table 3.   Basic Process Data For Cylindrical Grinding

| TRAVERSE GRINDING | | | | | | |
|---|---|---|---|---|---|---|
| Work Material | Material Condition | Work Surface Speed, fpm | Infeed, Inch/Pass | | Traverse for Each Work Revolution, In Fractions of The Wheel Width | |
| | | | Roughing | Finishing | Roughing | Finishing |
| Plain Carbon Steel | Annealed | 100 | .002 | .0005 | 1/2 | 1/6 |
| | Hardened | 70 | .002 | .0003 to .0005 | 1/4 | 1/8 |
| Alloy Steel | Annealed | 100 | .002 | .0005 | 1/2 | 1/6 |
| | Hardened | 70 | .002 | .0002 to .0005 | 1/4 | 1/8 |
| Tool Steel | Annealed | 60 | .002 | .0005 max. | 1/2 | 1/6 |
| | Hardened | 50 | .002 | .0001 to .0005 | 1/4 | 1/8 |
| Copper Alloys | Annealed or Cold Drawn | 100 | .002 | .0005 max. | 1/3 | 1/6 |
| Aluminum Alloys | Cold Drawn or Solution Treated | 150 | .002 | .0005 max. | 1/3 | 1/6 |

| PLUNGE GRINDING | | |
|---|---|---|
| Work Material | Infeed Per Revolution of The Work, Inch | |
| | Roughing | Finishing |
| Steel, soft | 0.0005 | 0.0002 |
| Plain carbon steel, hardened | 0.0002 | 0.000050 |
| Alloy and tool steel, hardened | 0.0001 | 0.000025 |

For high speed grinding, special grinding machines have been built with high dynamic stiffness and static rigidity, equipped with powerful drive motors, extra-strong spindles and bearings, reinforced wheel guards, etc., and using grinding wheels expressly made and tested for operating at high peripheral speeds. The higher stock-removal rate accomplished by high speed grinding represents an advantage when the work configuration and material permit, and the removable stock allowance warrants its application.

*CAUTION:* High speed grinding must *not* be applied on standard types of equipment, such as general models of grinding machines and regular grinding wheels. Operating grinding wheels, even temporarily, at higher than approved speed constitutes a grave safety hazard.

**Areas and Degrees of Automation in Cylindrical Grinding.** — Power drive for the work rotation and for the reciprocating table traverse are fundamental machine movements which, once set for a certain rate, will function without requiring additional attention. Loading and removing the work, starting and stopping the main movements, and applying infeed by hand wheel, are carried out by the operator on cylindrical grinding machines in their basic degree of mechanization. Such equipment is still frequently used in tool room and jobbing type work.

More advanced levels of automation can be developed for cylindrical grinders and are being applied in different degrees, particularly in the following principal respects:

a. *Infeed*, in which different rates are provided for rapid approach, roughing and finishing, followed by a spark-out period, with presetting of the advance rates, the cutoff points, and the duration of time-related functions.

b. *Automatic cycling* actuated by a single lever to start work rotation, table reciprocation, grinding-fluid supply and infeed, followed at the end of the operation by wheel slide retraction, the successive stopping of the table movement, the work rotation and the fluid supply.

c. *Table traverse dwells* (tarry) in the extreme positions of the travel, over preset periods, to assure uniform exposure to the wheel contact of the entire work section.

d. *Mechanized work loading*, clamping and, after termination of the operation, unloading, combined with appropriate work-feeding devices such as indexing type drums.

e. *Size control* by in-process or post-process measurements. Signals originated by the gage will control the advance movement or cause automatic compensation of size variations by adjusting the cutoff points of the infeed.

f. *Automatic wheel dress-off* at preset frequency, combined with appropriate compensation in the infeed movement.

g. *Numerical control*, programmed on punched cards or tape, obviates the time-consuming setups for repetitive work which is carried out in small- or medium-size lots. As an application example: shafts with several sections of different lengths and diameters can be ground automatically in a single operation, grinding the sections in consecutive order to close dimensional limits, controlled by an in-process gage, which is also automatically set by means of the program card.

The choice of the grinding machine functions to be automated and the extent of automation will generally be guided by economic considerations, after a thorough review of the available standard and optional equipment.

**Cylindrical Grinding Troubles and Their Correction.** — Troubles that may be encountered in cylindrical grinding may be classified as work defects (chatter, checking, burning, scratching, and inaccuracies), improperly operating machines (jumpy in-feed or traverse), and wheel defects (too hard or soft action, loading, glazing, and breakage). The Landis Tool Company lists some of these troubles, their causes, and corrections as follows:

**Chatter.** — Sources of chatter include: (1) Faulty coolant, (2) wheel out of balance, (3) wheel out of round, (4) wheel too hard, (5) improper dressing, (6) faulty work support or rotation, (7) improper operation, (8) faulty traverse, (9) work vibration, (10) outside vibration transmitted to machine, (11) interference, (12) wheel base, and (13) headstock. Suggested procedures for correction of these troubles are:

(*1*) *Faulty coolant:* Clean tanks and lines. Replace dirty or heavy coolant with correct mixture.

(*2*) *Out-of-balance wheel:* Rebalance on mounting before and after dressing. Run wheel without coolant to remove excess water. Store a removed wheel on its side to keep retained water from causing a false heavy side. Tighten wheel mounting flanges. Make sure wheel center fits spindle.

(*3*) *Wheel out of round:* True before and after balancing. True sides to face.

(*4*) *Wheel too hard:* Use coarser grit, softer grade, more open bond. See "Wheel Defects."

(*5*) *Improper dressing:* Use sharp diamond and hold rigidly close to wheel. It must not overhang too far. Check diamond in mounting.

(*6*) *Faulty work support or rotation:* Use sufficient number of work rests and adjust them more carefully. Use proper angles in centers of work. Clean dirt from footstock spindle and be sure spindle is tight. Make certain that work centers fit properly in spindles.

(*7*) *Improper operation:* Reduce rate of wheel feed.

(*8*) *Faulty traverse:* See "Uneven Traverse or In-feed of Wheel Head."

(*9*) *Work vibration:* Reduce work speed. Check workpiece for balance.

(*10*) *Outside vibration transmitted to machine:* Check and make sure that machine is level and sitting solidly on foundation. Isolate machine or foundation.

(*11*) *Interference:* Check all guards for clearance.

(*12*) *Wheel base:* Check spindle bearing clearance. Use belts of equal lengths or uniform cross-section on motor drive. Check drive motor for unbalance. Check balance and fit of pulleys. Check wheel feed mechanism to see that all parts are tight.

(*13*) *Headstock:* Put belts of same length and cross-section on motor drive. Incorrect work speeds. Check drive motor for unbalance. Make certain that headstock spindle is not loose. Check work center fit in spindle. Check wear of face plate and jackshaft bearings.

**Spirals on Work (traverse lines with same lead on work as rate of traverse).** — Sources of spirals include: (1) Machine parts out of line, and (2) truing. Suggested procedures for correction of these troubles are:

(*1*) *Machine parts out of line:* Check wheel base, headstock, and footstock for proper alignment.

(*2*) *Truing:* Point truing tool down 3 degrees at the workwheel contact line. Make edges of wheel face round.

**Check Marks on Work.** — Sources of check marks include: (1) Improper operation, (2) improper heat treatment, (3) improper size control, (4) improper wheel, and (5) improper dressing. Suggested procedures for correction of these troubles are:

(*1*) *Improper operation:* Make wheel act softer. See "Wheel Defects." Do not force wheel into work. Use greater volume of coolant and a more even flow. Affirm the correct positioning of coolant nozzles to direct a copious flow of clean coolant at the proper location.

(*2*) *Improper heat treatment:* Take corrective measures in heat treating operations.

(*3*) *Improper size control:* Make sure that engineering establishes reasonable size limits. See that they are maintained.

(*4*) *Improper wheel:* Make wheel act softer. Use softer grade wheel. Review the grain size and type of abrasive. A finer grit or more friable abrasive or both may be called for.

(*5*) *Improper dressing:* Make sure you have a sharp, good quality diamond that is well set. Increase speed of the dressing cycle. Make sure diamond is not cracked.

**Burning and Discoloration of Work.** — Sources of burning and discoloration are: (1) Improper operation, and (2) improper wheel. Suggested procedures for correction of these troubles are:

(*1*) *Improper operation:* Decrease rate of in-feed. Don't stop work while in contact with wheel.

(*2*) *Improper wheel:* Use softer wheel or obtain softer effect. See "Wheel Defects." Use greater volume of coolant.

**Isolated Deep Marks on Work.** — Source of trouble is improper wheel. Use finer wheel and consider a change in abrasive type.

**Fine Spiral or Thread on Work.** — Sources of this trouble are: (1) Improper operation and (2) faulty wheel dressing. Suggested procedures for corrections of these troubles are:

(1) *Improper operation:* Reduce wheel pressure. Use more work rests. Reduce traverse with respect to work rotation. Use different traverse rates to break up pattern when making numerous passes. Keep edge of wheel from penetrating by dressing wheel face parallel to work.

(2) *Faulty wheel dressing:* Use slower or more even dressing traverse. Set dressing tool at least 3 degrees down and 30 degrees to the side from time to time. Tighten holder. Don't take too deep a cut. Round off wheel edges. Start dressing cut from wheel edge.

**Narrow and Deep Regular Marks on Work.** — Source of trouble is wheel too coarse. Use finer grain size.

**Wide, Irregular Marks of Varying Depth on Work.** — Source of trouble is wheel too soft. Use harder grade wheel. See "Wheel Defects."

**Widely Spaced Spots on Work.** — Source of trouble is oil spots or glazed areas on wheel face. Balance and true wheel. Keep oil from wheel face.

**Irregular "Fish-tail" Marks of Various Lengths and Widths on Work.** — Source of trouble is dirty coolant. Clean tank frequently. Use filter for fine finish grinding. Flush wheel guards after dressing or when changing to finer wheel.

**Wavy Traverse Lines on Work.** — Source of trouble is wheel edges. Round off. Check for loose thrust on spindle and correct if necessary.

**Irregular Marks on Work.** — Cause is loose dirt. Keep machine clean.

**Deep, Irregular Marks on Work.** — Source of trouble is loose wheel flanges. Tighten and make sure blotters are used.

**Isolated Deep Marks on Work.** — Source of trouble is: (1) Grains pull out, coolant too strong; (2) coarse grains or foreign matter in wheel face; and (3) improper dressing. Respective suggested procedures for corrections of these troubles are: (1) Decrease soda content in coolant mixture, (2) dress out, and (3) use sharper dressing tool. Brush wheel after dressing with stiff bristle brush.

**Grain Marks on Work.** — Source of trouble is: (1) Improper finishing cut; (2) grain sizes of roughing and finishing wheels differ too much; (3) dressing too coarse; and (4) wheel too coarse or too soft. Respective suggested procedures for corrections of these troubles are: (1) Start with high work and traverse speeds; finish with high work speed and slow traverse, letting wheel "spark-out" completely; (2) finish out better with roughing wheel or use finer roughing wheel; (3) use shallower and slower cut; and (4) use finer grain size or harder grade wheel.

**Inaccuracies in Work.** — Work out-of-round, out-of-parallel, or tapered. Source of trouble is: (1) Misalignment of machine parts, (2) work centers, (3) improper operation, (4) coolant, (5) wheel, (6) improper dressing, (7) spindle bearings, and (8) work. Suggested procedures for corrections of these troubles are:

(1) *Misalignment of machine parts:* Check headstock and tailstock for alignment and proper clamping.

(2) *Work centers:* Centers in work must be deep enough to clear center point. Keep work centers clean and lubricated. Check play of footstock spindle and see that footstock spindle is clean and tightly seated. Regrind work center if worn. Work centers must fit taper of work center holes. Footstock must be checked for proper tension.

(3) *Improper operation:* Don't let wheel traverse beyond end of work. Decrease wheel pressure so work won't spring. Use harder wheel or change feeds and speeds to make wheel act harder. Allow work to "spark-out." Decrease feed rate. Use proper number of work rests. Allow proper amount of tarry. Workpiece must be balanced if odd shape.

(4) *Coolant:* Use greater volume of coolant.

(5) *Wheel:* Rebalance wheel on mounting before and after truing.

(6) *Improper dressing:* Use same positions and machine conditions for dressing as in grinding.

(7) *Spindle bearings:* Check clearance.

(8) *Work:* Work must come to machine in reasonably accurate form.

**Inaccurate Work Sizing (when wheel is fed to same position, it grinds one piece to correct size, another oversize, and still another undersize).** — Sources of trouble are: (1) Improper work support or rotation, (2) wheel out of balance, (3) loaded wheel, (4) improper infeed, (5) improper traverse, (6) coolant, (7) misalignment, and (8) work. Suggested procedures for corrections of these troubles are:

(1) *Improper work support or rotation:* Keep work centers clean and lubricated. Regrind work center tips to proper angle. Be sure footstock spindle is tight. Use sufficient work rests properly spaced.

(2) *Wheel out of balance:* Balance wheel on mounting before and after truing.

(3) *Loaded wheel:* See "Wheel Defects."

(4) *Improper infeed:* Check forward stops of rapid feed and slow feed. When readjusting position of wheel base by means of the fine feed, move the wheel base back after making the adjustment and then bring it forward again to take up backlash and relieve strain in feed-up parts. Check wheel spindle bearings. Don't let excessive lubrication of wheel base slide cause "floating." Check and tighten wheel feed mechanism. Check parts for wear. Check pressure in hydraulic system. Set in-feed cushion properly. Check pistons to see that they are not sticking.

(5) *Improper traverse:* Check traverse hydraulic system and the operating pressure. Prevent excessive lubrication of carriage ways with resultant "floating" condition. Check to see if carriage traverse piston rods are binding. Carriage rack and driving gear must not bind. Change length of tarry period.

(6) *Coolant:* Use greater volume of clean coolant.

(7) *Misalignment:* Check level and alignment of machine.

(8) *Work:* Work pieces may vary too much in length permitting uneven center pressure.

**Uneven Traverse or In-feed of Wheel Head.** — Sources of uneven traverse or in-feed of wheel head are: (1) Carriage and wheel head, (2) hydraulic system, (3) interference, (4) unbalanced conditions, and (5) wheel out of balance. Suggested procedures for correction of these troubles are:

(1) *Carriage and wheel head:* Ways may be scored. Be sure to use recommended oil for both lubrication and hydraulic system. Make sure ways are not too smooth that they press out oil film. Check lubrication of ways. Check wheel feed mechanism, traverse gear and carriage rack clearance. Prevent binding of carriage traverse cylinder rods.

(2) *Hydraulic systems:* Remove air and check pressure of hydraulic oil. Check pistons and valves for oil leakage and for gumminess caused by incorrect oil. Check worn valves or pistons that permit leakage.

(3) *Interference:* Make sure guard strips do not interfere.

(4) *Unbalanced conditions:* Eliminate loose pulleys, unbalanced wheel drive motor, uneven belts, or high spindle keys.

(5) *Wheel out of balance:* Balance wheel on mounting before and after truing.

**Wheel Defects.** — When *wheel is acting too hard*, such defects as glazing, some loading, lack of cut, chatter, and burning of work result. Suggested procedures for correction of these faults are: (1) Increase work and traverse speeds as well as rate of in-feed; (2) decrease wheel, speed, diameter, or width; (3) dress more sharply; (4) use thinner coolant; (5) don't tarry at end of traverse; (6) select softer wheel grade and coarser grain size; (7) avoid gummy coolant; and (8) on hardened work select finer grit, more fragile abrasive or both to get penetration. Use softer grade.

When *wheel is acting too soft*, such defects as wheel marks, tapered work, short wheel life, and not-holding-cut result. Suggested procedures for correction of these faults are: (1) Decrease work and traverse speeds as well as rate of in-feed; (2) increase wheel speed, diameter, or width; (3) dress with little in-feed and slow traverse; (4) use heavier coolants; (5) don't let wheel run off work at end of traverse; and (6) select harder wheel or less fragile grain or both.

**Wheel Loading and Glazing.** — Sources of the trouble of wheel loading or glazing are: (1) Incorrect wheel, (2) improper dress, (3) faulty operation, (4) faulty coolant, and (5) gummy coolant. Suggested procedures for correction of these faults are:

(*1*) *Incorrect wheel:* Use coarser grain size, more open bond, or softer grade.

(*2*) *Improper dressing:* Keep wheel sharp with sharp dresser, clean wheel after dressing, use faster dressing traverse, and deeper dressing cut.

(*3*) *Faulty operation:* Control speeds and feeds to soften action of wheel. Use less in-feed to prevent loading; more in-feed to stop glazing.

(*4*) *Faulty coolant:* Use more, cleaner and thinner coolant, and less oily coolant.

(*5*) *Gummy coolant:* To stop wheel glazing, increase soda content and avoid the use of soluble oils if water is hard. In using soluble oil coolant with hard water a suitable conditioner or "softener" should be used.

**Wheel Breakage.** — Suggested procedures for the correction of a radial break with three or more pieces are: (1) Reduce wheel speed to or below rated speed; (2) mount wheel properly, use blotters, tight arbors, even flange pressure and be sure to keep out dirt between flange and wheel; (3) use plenty of coolant to prevent over-heating; (4) use less in-feed; and (5) don't allow wheel to become jammed on work.

A radial break with two pieces may be caused by excessive side strain. To prevent an irregular wheel break, don't let wheel become jammed on work; allow striking of wheel; and never use wheels that have been damaged in handling. In general, do not use a wheel that is too tight on the arbor since the wheel is apt to break when started. Prevent excessive hammering action of wheel. Follow rules of the American National Standard Safety Code for the Use, Care, and Protection of Abrasive Wheels (ANSI B7.1).

## Centerless Grinding

In centerless grinding the work is supported on a work rest blade and is between the grinding wheel and a regulating wheel. The regulating wheel generally is a rubber bonded abrasive wheel. In the normal grinding position the grinding wheel forces the work downward against the work rest blade and also against the regulating wheel. The latter imparts a uniform rotation to the work giving it its same peripheral speed which is adjustable.

The higher the work center is placed above the line joining the centers of the grinding and regulating wheels the quicker the rounding action. Rounding action is also increased by a high work speed and a slow rate of traverse (if a through-feed operation). It is possible to have a higher work center when using softer wheels,

CENTERLESS GRINDING

as their use gives decreased contact pressures and the tendency of the workpiece to lift off the work rest blade is lessened.

Long rods or bars are sometimes ground with their centers below the line-of-centers of the wheels to eliminate the whipping and chattering due to slight bends or kinks in the rods or bars, as they are held more firmly down on the blade by the wheels.

There are three general methods of centerless grinding which may be described as through-feed, in-feed, and end-feed methods.

**Through-feed Method of Grinding.** — The through-feed method is applied to straight cylindrical parts. The work is given an axial movement by the regulating wheel and passes between the grinding and regulating wheels from one side to the other. The rate of feed depends upon the diameter and speed of the regulating wheel and its inclination which is adjustable. It may be necessary to pass the work between the wheels more than once, the number of passes depending upon such factors as the amount of stock to be removed, the roundness and straightness of the unground work, and the limits of accuracy required.

The work rest fixture also contains adjustable guides on either side of the wheels that directs the work to and from the wheels in a straight line.

**In-feed Method of Centerless Grinding.** — When parts have shoulders, heads or some part larger than the ground diameter, the in-feed method usually is employed. This method is similar to the "plungecut" form grinding on a center type of grinder. The length or sections to be ground in any one operation is limited by the width of the wheel. As there is no axial feeding movement, the regulating wheel is set with its axis approximately parallel to that of the grinding wheel, there being a slight inclination to keep the work tight against the end stop.

**End-feed Method of Grinding.** — The end-feed method is applied only to taper work. The grinding wheel, regulating wheel, and the work rest blade are set in a fixed relation to each other and the work is fed in from the front mechanically or manually to a fixed end stop. Either the grinding or regulating wheel, or both, are dressed to the proper taper.

**Automatic Centerless Grinding.** — The grinding of relatively small parts may be done automatically by equipping the machine with a magazine, gravity chute, or hopper feed, provided the shape of the part will permit using these feed mechanisms.

**Internal Centerless Grinding.** — Internal grinding machines based upon the centerless principle utilize the outside diameter of the work as a guide for grinding the bore which is concentric with the outer surface. In addition to straight and tapered bores, interrupted and "blind" holes can be ground by the centerless method. When two or more grinding operations must be performed on the same part, such as roughing and finishing, the work can be rechucked in the same location as often as required.

**Centerless Grinding Troubles.** — A number of troubles and some corrective measures compiled by Cincinnati Grinders Inc. are listed here for the through-feed and in-feed methods of centerless grinding.

*Chattermarks* are caused by having the work center too high above the line joining the line joining the centers of the grinding and regulating wheels; using too hard or too fine a grinding wheel; using too steep an angle on the work support blade; using too thin a work support blade; "play" in the set-up due to loosely clamped members; having the grinding wheel fit loosely on the spindle; having vibration either transmitted to the machine or caused by a defective drive in the machine; having the grinding wheel out-of-balance; using too heavy a stock removal; and having the grinding wheel or the regulating wheel spindles not properly adjusted.

*Feed lines or spiral marks* in through-feed grinding are caused by too sharp a corner on the exit side of the grinding wheel which may be alleviated by dressing the grinding wheel to a slight taper about ½ inch from the edge, dressing the edge to a slight radius, or swiveling the regulating wheel a bit.

*Scored work* is caused by burrs, abrasive grains, or removed material being imbedded in or fused to the work support blade. This condition may be alleviated by using a coolant with increased lubricating properties and if this does not help a softer grade wheel should be used.

*Work not ground round* may be due to the work center not being high enough above the line joining the centers of the grinding and regulating wheels. Placing the work center higher and using a softer grade wheel should help to alleviate this condition.

*Work not ground straight* in through-feed grinding may be due to an incorrect setting of the guides used in introducing and removing the work from the wheels, and the existence of convex or concave faces on the regulating wheel. For example if the work is tapered on the front end, the work guide on the entering side is deflected toward the regulating wheel. If tapered on the back end then the work guide on the exit side is deflected towards the regulating wheel. If both ends are tapered then both work guides are deflected towards the regulating wheel. The same barrel-shaped pieces are also obtained if the face of the regulating wheel is convex at the line of contact with the work. Conversely the work would be ground with hollow shapes if the work guides were deflected toward the grinding wheel or if the face of the regulating wheel were concave at the line of contact with the work. The use of a warped work rest blade may also result in the work not being ground straight and the blade should be removed and checked with a straight edge.

In in-feed grinding, in order to keep the wheel faces straight which will insure straightness of the cylindrical pieces being ground, the first item to be checked is the straightness and the angle of inclination of the work rest blade. If this is satisfactory then one of three corrective measures may be taken: the first might be to swivel the regulating wheel to compensate for the taper, the second might be to true the grinding wheel to that angle that will give a perfectly straight workpiece and the third might be to change the inclination of the regulating wheel (this is true only for correcting very slight tapers up to 0.0005 inch).

*Difficulties in sizing* the work in in-feed grinding are generally due to a worn in-feed mechanism and may be overcome by adjusting the in-feed nut.

*Flat spots* on the workpiece in in-feed grinding usually occur when grinding heavy work and generally when the stock removal is light. This condition is due to insufficient driving power between the work and the regulating wheel which may be alleviated by equipping the work rest with a roller that exerts a force against the workpiece; and by feeding the workpiece to the end stop using the upper slide.

## Surface Grinding

The term surface grinding implies, in current technical usage, the grinding of surfaces which are essentially flat. Several methods of surface grinding, however, are adapted and used to produce surfaces characterized by parallel straight line elements in one direction, while normal to that direction the contour of the surface may consist of several straight line sections at different angles to each other (e.g., the guideways of a lathe bed); in other cases the contour may be curved or profiled (e.g., a thread cutting chaser).

**Advantages of Surface Grinding.** — Alternate methods for machining work surfaces similar to those produced by surface grinding and milling and, to a much more limited degree, planing. Surface grinding, however, has several advantages over alternate methods which are carried out with metal-cutting tools. Examples of such potential advantages are:

(1) Grinding is applicable to very hard and/or abrasive work materials, without significant effect on the efficiency of the stock removal.

(2) The desired form and dimensional accuracy of the work surface can be obtained to a much higher degree and in a more consistent manner.

(3) Surface textures of very high finish and—when the appropriate system is utilized—also with the required lay, are generally produced.

(4) Tooling for surface grinding is, as a rule, substantially less expensive, particularly for producing profiled surfaces, the shapes of which may be dressed into the wheel, often with simple devices, in processes which are much more economical than the making and the maintenance of form cutters.

(5) Fixturing for work holding is generally very simple in surface grinding, particularly when magnetic chucks are applicable, although the mechanical holding fixture can also be simpler, because of the lesser clamping force required than in milling or planing.

(6) Parallel surfaces on opposite sides of the work are produced accurately, either in consecutive operations using the first ground surface as a dependable reference plane, or, simultaneously, in double face grinding which usually operates without the need for holding the parts by clamping.

(7) Surface grinding is well adapted to process automation, particularly for size control, but also for mechanized work handling in the large volume production of a wide range of component parts.

**Principal Systems of Surface Grinding.** — Flat surfaces can be ground with different surface portions of the wheel, by different arrangements of the work and wheel, as well as by different interrelated movements. The various systems of surface grinding, with their respective capabilities, can best be reviewed by considering two major distinguishing characteristics:

*(1) The operating surface of the grinding wheel,* which may be the periphery or the face (the side);

*(2) The movement of the work during the process,* which may be traverse (generally reciprocating) or rotary (continuous), depending on the design of a particular category of surface grinders.

The accompanying table provides a concise review of the principal surface grinding systems, defined by the preceding characteristics. It should be noted that there are surface grinders built for specific applications, which do not fit exactly into any one of these major categories.

**Selection of Grinding Wheels for Surface Grinding.** — The most practical way to select a grinding wheel for surface grinding is to base the selection on the work material. Table 1a gives the grinding wheel recommendations for Types 1, 5, and 7 straight wheels used on reciprocating and rotary table surface grinders with horizontal spindles. Table 1b gives the grinding wheel recommendations for Type 2 cylinder wheels, Type 6 cup wheels, and segments used on vertical spindle surface grinders.

The wheel markings of the tables are those used by the Norton Company, complementing the basic standard markings with Norton symbols. The complementary symbols used in these tables, that is, those preceding the letter designating A (aluminum oxide) or C (silicon carbide) indicate the special type of basic abrasive which has the friability best suited for particular work materials. Those preceding A (aluminum oxide) are:

**Table 1a. Grinding Wheel Recommendations for Surface Grinding—Using Straight Wheel Types 1, 5, and 7**

| Horizontal-spindle, reciprocating-table surface grinders | | |
|---|---|---|
| Material | Wheels less than 16 inches in diameter | Wheels 16 inches in diameter and over |
| Cast iron | 37C36-K8V or 23A46-18VBE | 23A36-I8VBE |
| Non-ferrous metals | 37C36-K8V | 37C36-K8V |
| Soft steel | 23A46-J8VBE | 23A36-J8VBE |
| Hardened steel— broad contact | 32A46-H8VBE or 32A60-F12VBEP | 32A36-H8VBE or 32A36-F12VBEP |
| Hardened steel— narrow contact or interrupted cut | 32A46-I8VBE | 32A36-J8VBE |
| General purpose wheel | 23A46-H8VBE | 23A36-I8VBE |
| Cemented carbides | Diamond wheels* | Diamond wheels* |
| Horizontal-spindle, rotary-table surface grinders | | |
| Material | | Wheels of any diameter |
| Cast iron | | 37C36-K8V or 23A46-I8VBE |
| Non-ferrous metals | | 37C36-K8V |
| Soft steel | | 23A46-J8VBE |
| Hardened steel—broad contact | | 32A46-I8VBE |
| Hardened steel—narrow contact or interrupted cut | | 32A46-J8VBE |
| General purpose wheel | | 23A46-I8VBE |
| Cemented carbides—roughing | | Diamond wheels* |

*Courtesy of Norton Company*

\* General diamond wheel recommendations are listed in Table 5, page 2009.

57—a versatile abrasive suitable for grinding steel in either a hard or soft state,
38—the most friable abrasive,
32—the abrasive suited for tool steel grinding,
23—an abrasive with intermediate grinding action, and
19—the abrasive produced for less heat-sensitive steels.
Those preceding C (silicon carbide) are:
37—a general application abrasive, and
39—an abrasive for grinding hard cemented carbide.

**Table 1b. Grinding Wheel Recommendations for Surface Grinding—Using Type 2 Cylinder Wheels, Type 6 Cup Wheels, and Segments**

| Material | Type 2 Cylinder Wheels | Type 6 Cup Wheels | Segments |
|---|---|---|---|
| High tensile cast iron and non-ferrous metals | 37C24-HKV | 37C24-HVK | 37C24-HVK |
| Soft steel, malleable cast iron, steel castings, boiler plate | 23A24-18VBE or 23A30-G12VBEP | 23A24-I8VBE | 23A24-I8VSM or 23A30-H12VSM |
| Hardened steel— broad contact | 32A46-G8VBE or 32A36-E12VBEP | 32A46-G8VBE or 32A60-E12VBEP | 32A36-G8VBE or 32A46-E12VBEP |
| Hardened steel— narrow contact or interrupted cut | 32A46-H8VBE | 32A60-H8VBE | 32A46-G8VBE or 32A60-G12VBEP |
| General purpose use | 23A30-H8VBE or 23A30-E12VBEP | .... | 23A30-H8VSM or 23A30-G12VSM |

*Courtesy of Norton Company*

**Principal Systems of Surface Grinding—Diagrams**

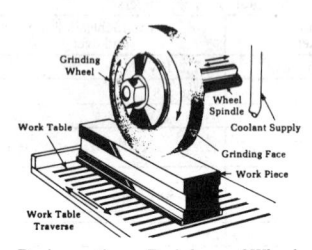

Reciprocating—Periphery of Wheel

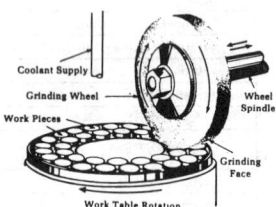

Rotary—Periphery of Wheel

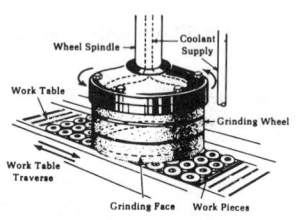

Reciprocating—Face (Side) of Wheel

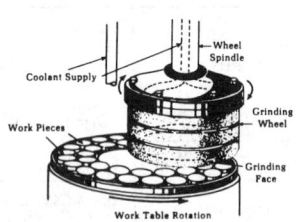

Rotary—Face (Side) of Wheel

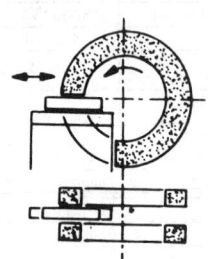

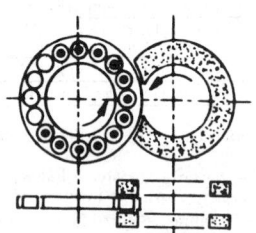

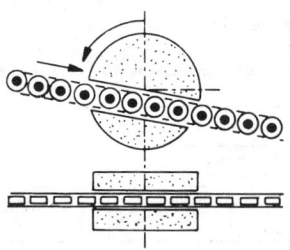

Traverse Along Straight Line or
Arcuate Path—Face (Side) of Wheel

**Principal Systems of Surface Grinding—Principles of Operation**

Effective Grinding Surface—Periphery of Wheel
Movement of Work—Reciprocating

Work is mounted on the horizontal machine table which then traverses in a reciprocating movement at a speed generally selected from a steplessly variable range. The transverse movement, called cross feed of the table or of the wheel slide, operates at the end of the reciprocating stroke and assures the gradual exposure of the entire work surface which commonly exceeds the width of the wheel. The depth of the cut is controlled by the downfeed of the wheel, applied in increments at the reversal of the transverse movement.

Effective Grinding Surface—Periphery of Wheel
Movement of Work—Rotary

Work is mounted, usually on the full diameter magnetic chuck of the circular machine table which rotates at a pre-set constant or automatically varying speed, the latter maintaining an approximately equal peripheral speed of the work surface area being ground. The wheelhead, installed on a cross slide, traverses over the table along a radial path, moving in alternating directions, toward and away from the center of the table. Infeed is by vertical movement of the saddle along the guideways of the vertical column, at the end of the radial wheelhead stroke. The saddle contains the guideways along which the wheelhead slide reciprocates.

Effective Grinding Surface—Face (Side) of Wheel
Movement of Work—Reciprocating

Operation is similar to the reciprocating table type peripheral surface grinder, but grinding is with the face, usually with the rim of a cup shaped wheel, or a segmental wheel for large model machines. Capable of covering a much wider area of the work surface than the peripheral grinder, thus frequently no need for cross feed. Provides efficient stock removal, but is less adaptable than the reciprocating table type peripheral grinder.

Effective Grinding Surface—Face (Side) of Wheel
Movement of Work—Rotary

The grinding wheel, usually of segmental type, is set in a position to cover either an annular area near the periphery of the table or, more commonly, to reach beyond the table center. A large circular magnetic chuck generally covers the entire table surface and facilitates the mounting of workpieces, even of fixtures, when needed. The uninterrupted passage of the work in contact with the large wheel face permits a very high rate of stock removal and the machine, with single or double wheelhead, can be adapted also to automatic operation with continuous part feed by mechanized work handling.

Effective Grinding Surface—Face (Side) of Wheel
Movement of Work—Traverse Along Straight or Arcuate Path

Operates with practically the entire face of the wheel which is designated as an abrasive disc (hence "disc grinding") because of its narrow width in relation to the large diameter. Built either for one or, more frequently, for two discs operating with opposed faces for the simultaneous grinding of both sides of the workpiece. The parts pass between the operating faces of the wheel (a) pushed-in and retracted by the drawer-like movement of a feeding slide; (b) in an arcuate movement carried in the nests of a rotating feed wheel; (c) nearly diagonally advancing along a rail. Very well adapted to fully mechanized work handling.

The last letters (two or three) which may follow the bond designation V (vitrified) or B (resinoid) refer to: (1) bond modification, "BE" being especially suitable for surface grinding: (2) special structure, "P" type being distinctively porous; and (3) for segments made of 23 A type abrasives, the term 12 VSM implies porous structure, and the letter "P" is not needed.

**Process Data for Surface Grinding.** — In surface grinding, similarly to other metalcutting processes, the speed and feed rates that are applied must be adjusted to

**Table 2.   Basic Process Data for Peripheral Surface Grinding on Reciprocating Table Surface Grinders**

| Work Material | Hardness | Material Condition | Wheel Speed, fpm | Table Speed, fpm | Downfeed, in. per pass | | Cross Feed per Pass, fraction of wheel width |
|---|---|---|---|---|---|---|---|
| | | | | | Rough | Finish | |
| Plain carbon steel | 52 R_c max. | Annealed, Cold drawn | 5500 to 6500 | 50 to 100 | 0.003 | 0.0005 max. | 1/4 |
| | 52 to 65 R_c | Carburized and/or quenched and tempered | 5500 to 6500 | 50 to 100 | 0.003 | 0.0005 max. | 1/10 |
| Alloy steels | 52 R_c max. | Annealed or quenched and tempered | 5500 to 6500 | 50 to 100 | 0.003 | 0.001 max. | 1/4 |
| | 52 to 65 R_c | Carburized and/or quenched and tempered | 5500 to 6500 | 50 to 100 | 0.003 | 0.0005 max. | 1/10 |
| Tool steels | 150 to 275 BHN | Annealed | 5500 to 6500 | 50 to 100 | 0.002 | 0.0005 max. | 1/5 |
| | 56 to 65 R_c | Quenched and tempered | 5500 to 6500 | 50 to 100 | 0.002 | 0.0005 max. | 1/10 |
| Nitriding steels | 200 to 350 BHN | Normalized, annealed | 5500 to 6500 | 50 to 100 | 0.003 | 0.001 max. | 1/4 |
| | 60 to 65 R_c | Nitrided | 5500 to 6500 | 50 to 100 | 0.003 | 0.0005 max. | 1/10 |
| Cast steels | 52 R_c max. | Normalized, annealed | 5500 to 6500 | 50 to 100 | 0.003 | 0.001 max. | 1/4 |
| | Over 52 R_c | Carburized and/or quenched and tempered | 5500 to 6500 | 50 to 100 | 0.003 | 0.0005 max. | 1/10 |
| Gray irons | 52 R_c max. | As cast, anneal-ed, and/or quenched and tempered | 5000 to 6500 | 50 to 100 | 0.003 | 0.001 max. | 1/3 |
| Ductile irons | 52 R_c max. | As cast, an-nealed or quenched and tempered | 5500 to 6500 | 50 to 100 | 0.003 | 0.001 max. | 1/5 |
| Stainless steels, martensitic | 135 to 235 BHN | Annealed or cold drawn | 5500 to 6500 | 50 to 100 | 0.002 | 0.0005 max. | 1/4 |
| | Over 275 BHN | Quenched and tempered | 5500 to 6500 | 50 to 100 | 0.001 | 0.0005 max. | 1/8 |
| Aluminum alloys | 30 to 150 BHN | As cast, cold drawn or treated | 5500 to 6500 | 50 to 100 | 0.003 | 0.001 max. | 1/3 |

**Table 3. Common Faults and Possible Causes in Surface Grinding**

| CAUSES \ FAULTS | Work not flat | Work not parallel | Poor size holding | Burnishing of work | Burning or checking | Feed lines | Chatter marks | Scratches on surface | Poor finish | Wheel loading | Wheel glazing | Rapid wheel wear | Not firmly seated | Work sliding on chuck |
|---|---|---|---|---|---|---|---|---|---|---|---|---|---|---|
| **WORK CONDITION** | | | | | | | | | | | | | | |
| Heat treat stresses | ✓ | | | | | | | | | | | | | |
| Work too thin | ✓ | ✓ | | | | | | | | | | | ✓ | |
| Work warped | ✓ | ✓ | | | | | | | | | | | | |
| Abrupt section changes | ✓ | | | | | | | | | | | | | |
| **GRINDING WHEEL** | | | | | | | | | | | | | | |
| Grit too fine | | | | | | | | | | ✓ | ✓ | | | |
| Grit too coarse | | | | | | | | ✓ | ✓ | | | | | |
| Grade too hard | | | | ✓ | ✓ | | | | ✓ | ✓ | ✓ | | | |
| Grade too soft | ✓ | ✓ | | | | | | | | | | ✓ | | |
| Wheel not balanced | | | | | | | ✓ | | | | | | | |
| Dense structure | | | | ✓ | ✓ | | | | | ✓ | ✓ | | | |
| **TOOLING AND COOLANT** | | | | | | | | | | | | | | |
| Improper coolant | | | | | | | | | | ✓ | | | | |
| Insufficient coolant | | | | ✓ | ✓ | | | | | ✓ | | | | |
| Dirty coolant | | | | | | | | ✓ | | | | | | |
| Diamond loose or chipped | | | | | | | ✓ | ✓ | | | | | | |
| Diamond dull | | | | ✓ | ✓ | | | ✓ | ✓ | ✓ | ✓ | | | |
| No or poor magnetic force | ✓ | ✓ | ✓ | | | | | | | | | | | |
| Chuck surface worn or burred | ✓ | ✓ | ✓ | | | | | | | | | | | ✓ |
| **MACHINE AND SETUP** | | | | | | | | | | | | | | |
| Chuck not aligned | ✓ | ✓ | ✓ | | | | | | | | | | | |
| Vibrations in machine | | | | | | | ✓ | ✓ | | | | | | |
| Plane of movement out of parallel | ✓ | ✓ | | | | | | | | | | | | |
| **OPERATIONAL CONDITIONS** | | | | | | | | | | | | | | |
| Too low work speed | | | | | | | | | | ✓ | ✓ | | | |
| Too light feed | | | | ✓ | ✓ | | | | | | ✓ | | | |
| Too heavy cut | | | | | | | | | | | | ✓ | | ✓ |
| Chuck retained swarf | ✓ | ✓ | | | | | | | | | | | ✓ | ✓ |
| Chuck loading improper | ✓ | ✓ | | | | | | | | | | | ✓ | |
| Insufficient blocking of parts | ✓ | ✓ | | | | | | | | | | | | |
| Wheel runs off the work | | ✓ | | | | | | | | | | | | |
| Wheel dressing too fine | | | | | | | | ✓ | ✓ | ✓ | ✓ | | | |
| Wheel edge not chamfered | ✓ | ✓ | | | | | | | | | | | | |
| Loose dirt under guard | | | | | | ✓ | | ✓ | | | | ✓ | | |

the operational conditions as well as to the objectives of the process. Grinding differs, however, from other types of metal cutting methods in regard to the cutting speed of the tool; the peripheral speed of the grinding wheel is maintained within a narrow range, generally 5500 to 6500 surface feet per minute. Speed ranges different from the common one are used in particular processes which require special wheels and equipment.

In establishing the proper process values for grinding, of prime consideration are the work material, its condition, and the type of operation (roughing or finishing). Table 2 gives basic process data for peripheral surface grinding on reciprocating table surface grinders. For different work materials and hardness ranges data are given regarding table speeds, downfeed (infeed) rates and cross feed, the latter as a function of the wheel width.

**Common Faults and Possible Causes in Surface Grinding.** — Approaching the ideal performance with regard to both the quality of the ground surface and the efficiency of surface grinding, requires the monitoring of the process and the correction of conditions adverse to the attainment of that goal.

Defective, or just not entirely satisfactory surface grinding may have any one or more of several causes. Exploring and determining the cause for eliminating its harmful effects is facilitated by knowing the possible sources of the experienced undesirable performance. Table 3, associating the common faults with their possible causes, is intended to aid in determining the actual cause, the correction of which should restore the desired performance level.

While the table lists the more common faults in surface grinding, and points out their frequent causes, other types of improper performance and/or other causes, in addition to those indicated, are not excluded.

# Offhand Grinding

Offhand grinding consists of holding the wheel to the work or the work to the wheel and grinding to broad tolerances and includes such operations as certain types of tool sharpening, weld grinding, snagging castings and other rough grinding. Types of machines that are used for rough grinding in foundries are floor- and bench-stand machines. Wheels for these machines vary from 6 to 30 inches in diameter. Portable grinding machines (electric, flexible shaft, or air-driven) are used for cleaning and smoothing castings.

Many rough grinding operations on castings can be best done with shaped wheels, such as cup wheels (including plate mounted) or cone wheels, and it is advisable to have a good assortment of such wheels on hand to do the odd jobs the best way.

**Floor- and Bench-Stand Grinding.** — The most common method of rough grinding is on double-end floor and bench stands. In machine shops, welding shops, and automotive repair shops, these grinders are usually provided with a fairly coarse grit wheel on one end for miscellaneous rough grinding and a finer grit wheel on the other end for sharpening tools. The pressure exerted is a very important factor in selecting the proper grinding wheel. If grinding is to be done mostly on hard sharp fins, then durable, coarse and hard wheels are required, but if grinding is mostly on large gate and riser pads, then finer and softer wheels should be used for best cutting action.

**Portable Grinding.** — Portable grinding machines are usually classified as air grinders, flexible shaft grinders, and electric grinders. The electric grinders are of two types; namely, those driven by standard 60 cycle current and so-called high-cycle grinders. Portable grinders are used for grinding down and smoothing weld seams; cleaning metal before welding; grinding out imperfections, fins and parting lines in castings and smoothing castings; grinding punch press dies and patterns to proper size and shape; and grinding manganese steel castings.

Wheels used on portable grinders are of three bond types; namely, resinoid, rubber, and vitrified. By far the largest percentage is resinoid. Rubber bond is used for relatively thin wheels and where a good finish is required. Some of the smaller wheels such as cone and plug wheels are vitrified bonded.

Grit sizes most generally used in wheels from 4 to 8 inches in diameter are 16, 20, and 24. In the still smaller diameters, finer sizes are used, such as 30, 36, and 46.

The particular grit size to use depends chiefly on the kind of grinding to be done. If the work consists of sharp fins and the machine has ample power, a coarse grain size combined with a fairly hard grade should be used. If the job is more in the nature of smoothing or surfacing and a fairly good finish is required, then finer and softer wheels are called for.

**Swing-Frame Grinding.** — This type of grinding is employed where a considerable amount of material is to be removed as on snagging large castings. It may be possible to remove 10 times as much material from steel castings using swing-frame grinders as with portable grinders; and 3 times as much material as with high-speed floor-stand grinders.

The largest field of application for swing-frame machines is on castings which are too heavy to handle on a floor stand; but often it is found that comparatively large gates and risers on smaller castings can be ground more quickly with swing-frame grinders, even if fins and parting lines have to be ground on floor stands as a second operation.

In foundries, the swing-frame machines are usually suspended from a trolley on a jib that can be swung out of the way when placing the work on the floor with the help of an overhead crane. In steel mills when grinding billets, a number of swing-frame machines are usually suspended from trolleys on a line of beams which facilitate their use as required.

The grinding wheels used on swing-frame machines are made with coarser grit sizes and harder grades than wheels used on floor stands for the same work. The reason is that greater grinding pressures can be obtained on the swing-frame machines.

**Mounted Wheels and Mounted Points.** — These wheels and points are used in hard-to-get-at places and are available with a vitrified bond. The wheels are available with aluminum oxide or silicon carbide abrasive grains. The aluminum oxide wheels are used to grind tough and tempered die steels and the silicon carbide wheels, cast iron, chilled iron, bronze, and other non-ferrous metals.

The illustrations on pages 2036 and 2037 give the standard shapes of mounted wheels and points as published by the Grinding Wheel Institute. A note about the maximum operating speed for these wheels is given at the bottom of the first page of illustrations.

## Abrasive Belt Grinding

Abrasive belts are used in the metalworking industry for removing stock, light cleaning up of metal surfaces, grinding welds, deburring, breaking and polishing hole edges, and finish grinding of sheet steel. The types of belts that are used may be coated with aluminum oxide (the most common coating) for stock removal and finishing of all alloy steels, high-carbon steel, and tough bronzes; and silicon carbide for use on hard, brittle, and low-tensile strength metals which would include aluminum and cast irons.

Table 1 is a guide to the selection of the proper abrasive belt, lubricant, and contact wheel. This table is entered on the basis of the material used and type of operation to be done and gives the abrasive belt specifications (type of bonding and

## Standard Shapes of Mounted Wheels and Points — 1

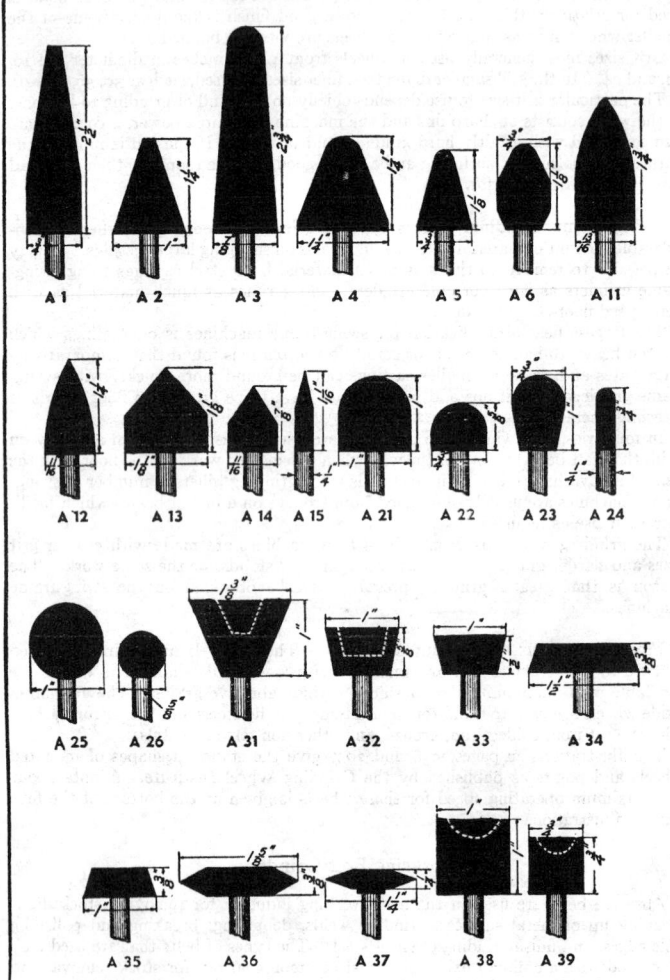

The maximum speeds of mounted vitrified wheels and points of average grade range from about 38,000 to 152,000 R.P.M. for diameters of 1 inch down to ¼ inch. However, the safe operating speed usually is limited by the critical speed (speed at which vibration or whip tends to become excessive) which varies according to wheel or point dimensions, spindle diameter, and overhang.

## Standard Shapes of Mounted Wheels and Points — 2

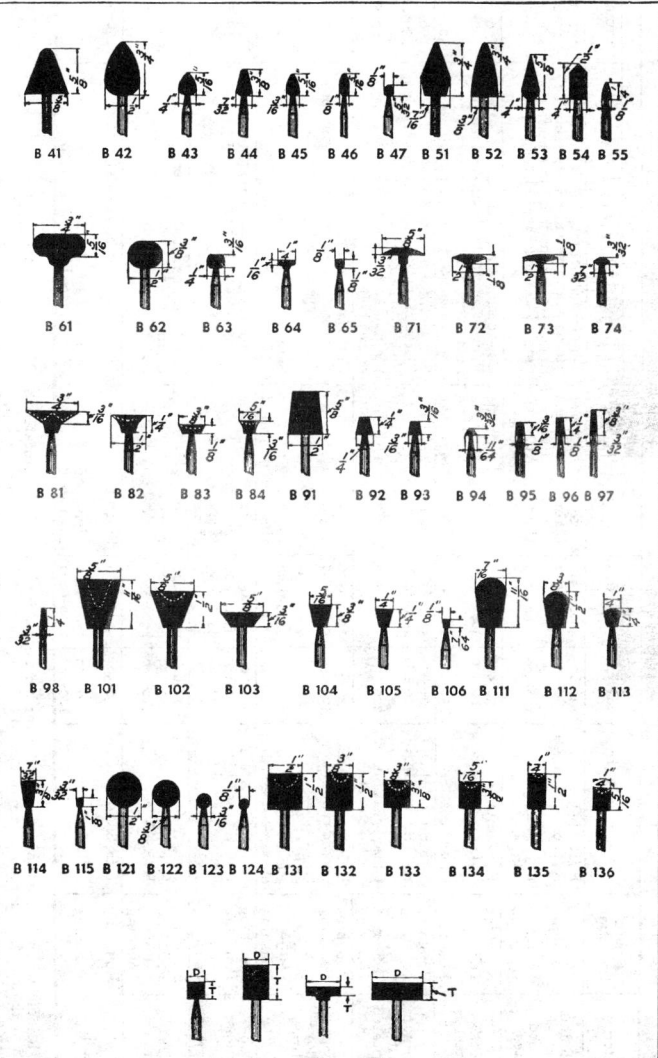

B 41   B 42   B 43   B 44   B 45   B 46   B 47   B 51   B 52   B 53   B 54   B 55

B 61   B 62   B 63   B 64   B 65   B 71   B 72   B 73   B 74

B 81   B 82   B 83   B 84   B 91   B 92   B 93   B 94   B 95   B 96   B 97

B 98   B 101   B 102   B 103   B 104   B 105   B 106   B 111   B 112   B 113

B 114   B 115   B 121   B 122   B 123   B 124   B 131   B 132   B 133   B 134   B 135   B 136

Group W

Table 1. Guide to the Selection and Application of Abrasive Belts

| Material | Type of Operation | Abrasive Belt* | Grit | Belt Speed, fpm | Type of Grease Lubricant | Contact Wheel Type | Durometer Hardness |
|---|---|---|---|---|---|---|---|
| Hot- and Cold-Rolled Steel | Roughing | R/R Al₂O₃ | 24-60 | 4000-6500 | Light-body or none | Cog-tooth, serrated rubber | 70-90 |
| | Polishing | R/G or R/R Al₂O₃ | 80-150 | 4500-7000 | Light-body or none | Plain or serrated rubber, sectional or finger-type cloth wheel, free belt | 20-60 |
| | Fine Polishing | R/G or electro-coated Al₂O₃ cloth | 180-500 | 4500-7000 | Heavy or with abrasive compound | Smooth-faced rubber or cloth | 20-40 |
| Stainless Steel | Roughing | R/R Al₂O₃ | 50-80 | 3500-5000 | Light-body or none | Cog-tooth, serrated rubber | 70-90 |
| | Polishing | R/G or R/R Al₂O₃ | 80-120 | 4000-5500 | Light-body or none | Plain or serrated rubber, sectional or finger-type cloth wheel, free belt | 30-60 |
| | Fine Pol. | Closed-coat SiC | 150-280 | 4500-5500 | Heavy or oil mist | Smooth-faced rubber or cloth | 20-40 |
| Aluminum, Cast or Fabricated | Roughing | R/R SiC or Al₂O₃ | 24-80 | 5000-6500 | Light | Cog-tooth, serrated rubber | 70-90 |
| | Polishing | R/G SiC or Al₂O₃ | 100-180 | 4500-6500 | Light | Plain or serrated rubber, sectional or finger-type cloth wheel, free belt | 30-50 |
| | Fine Polishing | Closed-coat SiC or electro-coated Al₂O₃ | 220-320 | 4500-6500 | Heavy or with abrasive compound | Plain faced rubber, finger-type cloth or free belt | 20-50 |
| Copper Alloys or Brass | Roughing | R/R SiC or Al₂O₃ | 36-80 | 2200-4500 | Light-body | Cog-tooth, serrated rubber | 70-90 |
| | Polishing | Closed-coat SiC or electro-coated Al₂O₃ or R/G SiC or Al₂O₃ | 100-150 | 4000-6500 | Light-body | Plain or serrated rubber, sectional or finger-type cloth wheel, free belt | 30-50 |
| | Fine Polishing | Closed-coat SiC or electro-coated Al₂O₃ | 180-320 | 4000-6500 | Light or with abrasive compound | Same as for polishing | 20-30 |
| Non-ferrous Die-castings | Roughing | R/R SiC or Al₂O₃ | 24-80 | 4500-6500 | Light-body | Hard wheel depending on application | 50-70 |
| | Polishing | R/G SiC or Al₂O₃ | 100-180 | 4500-6500 | Light-body | Plain rubber, cloth or free belt | 30-50 |
| | Fine Polishing | Electro-coated Al₂O₃ or closed-coat SiC | 220-320 | 4500-6500 | Heavy or with abrasive compound | Plain or finger-type cloth wheel, or free belt | 20-30 |
| Cast Iron | Roughing | R/R Al₂O₃ | 24-60 | 2000-4000 | None | Cog-tooth, serrated rubber | 70-90 |
| | Polishing | R/R Al₂O₃ | 80-150 | 4000-5500 | None | Serrated rubber | 30-70 |
| | Fine Polishing | R/R Al₂O₃ | 120-240 | 4500-5500 | Light-body | Smooth-faced rubber | 30-40 |
| Titanium | Roughing | R/R SiC or Al₂O₃ | 36-50 | 700-1500 | Sulphur-chlorinated | Small-diameter, cog-tooth serrated rubber | 70-80 |
| | Polishing | R/R SiC | 60-120 | 1200-2000 | Light-body | Standard serrated rubber | 50 |
| | Fine Pol. | R/R SiC | 120-240 | 1200-2000 | Light-body | Smooth-faced rubber or cloth | 20-40 |

* R/R indicates that both the making and sizing bond coats are resin. R/G indicates that the making coat is glue and the sizing coat is resin. The abbreviations Al₂O₃ for aluminum oxide and SiC for silicon carbide are used. Almost all R/R and R/G Al₂O₃ and SiC belts have a heavy-drill weight cloth backing. Most electro-coated Al₂O₃ and closed-coat SiC belts have a jeans weight cloth backing.

**Table 2. Guide to the Selection and Application of Contact Wheels**

| Surface | Material | Hardness and Density | Purposes | Wheel Action | Comments |
|---|---|---|---|---|---|
| Cog-tooth | Rubber | 70 to 90 durometer | Roughing | Fast cutting, allows long belt life. | For cutting down projections on castings and weld beads. |
| Standard serrated | Rubber | 40 to 50 durometer, medium density | Roughing | Leaves rough- to medium-ground surface. | For smoothing projections and face defects. |
| X-shaped serrations | Rubber | 20 to 50 durometer | Roughing and polishing | Flexibility of rubber allows entry into contours. Medium polishing, light removal. | Same as for standard serrated wheels but preferred for soft non-ferrous metals. |
| Plain face | Rubber | 20 to 70 durometer | Roughing and polishing | Plain wheel face allows controlled penetration of abrasive grain. Softer wheels give better finishes. | For large or small flat faces. |
| Flat flexible | Compressed canvas | About nine densities from very hard to very soft | Roughing and polishing | Hard wheels can remove metal, but not as quickly as cog-tooth rubber wheels. Softer wheels polish well. | Good for medium-range grinding and polishing. |
| Flat flexible | Solid sectional canvas | Soft, medium, and hard | Polishing | Uniform polishing. Avoids abrasive pattern on work. Adjusts to contours. Can be preformed for contours. | A low-cost wheel with uniform density at the face. Handles all types of polishing. |
| Flat flexible | Buff section canvas | Soft | Contour polishing | For fine polishing and finishing. | Can be widened or narrowed by adding or removing sections. Low cost. |
| Flat flexible | Sponge rubber inserts | 5 to durometer, soft | Polishing | Uniform polishing and finishing. Polishes and blends contours. | Has replaceable segments. Polishes and blends contours. Segments allow density changes. |
| Flexible | Fingers of canvas attached to hub | Soft | Polishing | Uniform polishing and finishing. | For polishing and finishing. |
| Flat flexible | Rubber segments | Varies in hardness | Roughing and polishing | Grinds or polishes depending on density and hardness of inserts. | For portable machines. Uses replaceable segments that save on wheel costs and allow density changes. |
| Flat flexible | Inflated rubber | Air pressure controls hardness | Roughing and polishing | Uniform finishing. | Adjusts to contours. |

abrasive grain size and material), the range of speeds at which the belt may best be operated, the type of lubricant to use, and the type and hardness of the contact wheel to use. Table 2 serves as a guide in the selection of contact wheels. This table is entered on the basis of the type of contact wheel surface and the contact wheel material. The table gives the hardness and/or density, the type of abrasive belt grinding for which the contact wheel is intended, the character of the wheel action and such comments as the uses, and hints for best use. Both tables are intended only as guides for general shop practice; selections may be altered to suit individual requirements.

There are three types of abrasive belt grinding machines. One type employs a contact wheel behind the belt at the point of contact of the workpiece to the belt and facilitates a high rate of stock removal. Another type uses an accurate parallel ground platen over which the abrasive belt passes and facilitates the finishing of precision parts. A third type which has no platens or contact wheel is used for finishing parts having uneven surfaces or contours. In this type there is no support behind the belt at the point of contact of the belt with the workpiece. Some machines are so constructed that besides grinding against a platen or a contact wheel the workpiece may be moved and ground against an unsupported portion of the belt, thereby in effect making it a dual machine.

Although abrasive belts at the time of their introduction were used dry, since the advent of the improved waterproof abrasive belts, they have been used with coolants, oil-mists, and greases to aid the cutting action. The application of a coolant to the area of contact retards loading, resulting in a cool, free cutting action, a good finish and a long belt life.

## Abrasive Cutting

Abrasive cut-off wheels are used for cutting steel, brass and aluminum bars and tubes of all shapes and hardnesses, ceramics, plastics, insulating materials, glass and cemented carbides. Originally a tool or stock room procedure, this method has developed into a high speed production operation. While the abrasive cut-off machine and cut-off wheel can be said to have revolutionized the practice of cutting-off materials, the metal saw continues to be the more economical method for cutting-off large cross-sections of certain materials. However, there are innumerable materials and shapes that can be cut with much greater speed and economy by the abrasive wheel method. On conventional chop-stroke abrasive cutting machines using 16-inch diameter wheels, 2-inch diameter bar stock is the maximum size that can be cut with satisfactory wheel efficiency, but bar stock up to 6 inches in diameter can be cut efficiently on oscillating-stroke machines. Tubing up to 3½ inches in diameter can also be cut efficiently.

Abrasive wheels are commonly available in four types of bonds: Resinoid, rubber, shellac and fiber or fabric reinforced. In general, resinoid bonded cut-off wheels are used for dry cutting where burrs and some burn are not objectionable and rubber bonded wheels are used for wet cutting where cuts are to be smooth, clean and free from burrs. Shellac bonded wheels have a soft, free cutting quality which makes them particularly useful in the tool room where tool steels are to be cut without discoloration. Fiber reinforced bonded wheels are able to withstand severe flexing and side pressures and fabric reinforced bonded wheels which are highly resistant to breakage caused by extreme side pressures, are fast cutting and have a low rate of wear.

The types of abrasives available in cut-off wheels are: Aluminum oxide, for cutting steel and most other metals; silicon carbide, for cutting non-metallic materials such as carbon, tile, slate, ceramics, etc.; and diamond, for cutting cemented carbides. The method of denoting abrasive type, grain size, grade, structure and bond type by using a system of markings is the same as for grinding wheels (see page 1984). Maximum wheel speeds given in the American National Standard Safety Code for The Use, Care, and Protection of Abrasive Wheels (ANSI B7.1-1970) range from 9500 to 14,200 surface feet per minute for organic bonded cut-off wheels larger than 16 inches in diameter and from 9500 to 16,000 surface feet per minute for organic bonded cut-off wheels 16 inches in diameter and smaller. Maximum wheel speeds specified by the manufacturer should never be exceeded even though they may be lower than those given in the ANSI Code.

There are four basic types of abrasive cutting machines: Chop-stroke, oscillating stroke, horizontal stroke and work rotating. Each of these four types may be designed for dry cutting or for wet cutting (includes submerged cutting).

The accompanying table based upon information made available by The Carborundum Co. gives some of the probable causes of cutting off difficulties that might be experienced when using abrasive cut-off wheels.

**Probable Causes of Cutting-Off Difficulties**

| Difficulty | Probable Cause |
|---|---|
| Angular Cuts and Wheel Breakage | (1) Inadequate clamping which allows movement of work while the wheel is in the cut. The work should be clamped on both sides of the cut.<br>(2) Work vise higher on one side than the other causing wheel to be pinched.<br>(3) Wheel vibration resulting from worn spindle bearings.<br>(4) Too fast feeding into the cut when cutting wet. |
| Burning of Stock | (1) Insufficient power or drive allowing wheel to stall.<br>(2) Cuts too heavy for grade of wheel being used.<br>(3) Wheel fed through the work too slowly. This causes a heating up of the material being cut. This difficulty encountered chiefly in dry cutting. |
| Excessive Wheel Wear | (1) Too rapid cutting when cutting wet.<br>(2) Grade of wheel too hard for work, resulting in excessive heating and burning out of bond.<br>(3) Inadequate coolant supply in wet cutting.<br>(4) Grade of wheel too soft for work.<br>(5) Worn spindle bearings allowing wheel vibration. |
| Excessive Burring | (1) Feeding too slowly when cutting dry.<br>(2) Grit size in wheel too coarse.<br>(3) Grade of wheel too hard.<br>(4) Wheel too thick for job. |

## Honing Process

The hone-abrading process for obtaining cylindrical forms with precise dimensions and surfaces can be applied to internal cylindrical surfaces with a wide range of diameters such as engine cylinders, bearing bores, pin holes, etc. and also to some external cylindrical surfaces. The process is used to: (1) eliminate inaccuracies resulting from previous operations by generating a true cylindrical form with respect to roundness and straightness within minimum dimensional limits; (2) generate final dimensional size accuracy within low tolerance, as may be required for interchangeability of parts; (3) provide rapid and economical stock removal consistent with accomplishment of the other results; (4) generate surface finishes of a specified degree of surface smoothness, with high surface quality.

**Amount and Rate of Stock Removal.** — Honing may be employed to increase bore diameters by as much as 0.100 inch or as little as 0.001 inch. The amount of stock removed by the honing process is entirely a question of processing economy. If other operations are performed before honing then the bulk of the stock should be taken off by the operation that can do it most economically. In large diameter bores that have been distorted in heat treating, it may be necessary to remove as much as 0.030 to 0.040 inches from the diameter to make the bore round and straight. For out-of-round or tapered bores, a good " rule of thumb " is to leave twice as much stock (on the diameter) for honing as there is error in the bore. Another general rule is: For bores over one inch in diameter, leave 0.001 to 0.0015 inch stock per inch of diameter. For example, 0.002 to 0.003 inch of stock is left in two-inch bores and 0.010 to 0.015 inch in ten-inch bores. Where parts are to be honed for finish only, the amount of metal to be left for removing tool marks may be as little as 0.0002 to 0.001 inch on the diameter.

In general, the honing process can be employed to remove stock from bore diameters at the rate of 0.009 to 0.012 inch per minute on cast-iron parts and from 0.005 to 0.008 inch per minute on steel parts having a hardness of 60 to 65 Rockwell C. These rates are based on parts having a length equal to three or four times the diameter. Stock has been removed from long parts such as gun barrels, at the rate of 65 cubic inches per hour. Recommended honing speeds for cast iron range from 110 to 200 surface feet per minute of rotation and from 50 to 110 lineal feet per minute of reciprocation. For steel, rotating surface speeds range from 50 to 110 feet per minute and reciprocation speeds from 20 to 90 lineal feet per minute. The exact rotation and reciprocation speeds to be used depend upon the size of the work, the amount and characteristics of the material to be removed and the quality of the finish desired. In general, the harder the material to be honed, the lower the speed. Interrupted bores are usually honed at faster speeds than plain bores.

**Formula for Rotative Speeds.** — Empirical formulas for determining rotative speeds for honing have been developed by the Micromatic Hone Corp. These formulas take into consideration the type of material being honed, its hardness and its surface characteristics; the abrasive area; and the type of surface pattern and degree of surface roughness desired. Because of the wide variations in material characteristics, abrasives available, and types of finishes specified, these formulas should be considered as a guide only in determining which of the available speeds (pulley or gear combinations) should be used for any particular application.

The formula for rotative speed, $S$, in surface feet per minute is:

$$S = \frac{K \times D}{W \times N}$$

The formula for rotative speed in revolutions per minute is:

$$\text{R.P.M.} = \frac{R}{W \times N}$$

where, $K$ and $R$ are factors taken from the table on the following page, $D$ is the diameter of the bore in inches, $W$ is the width of the abrasive stone or stick in inches, and $N$ is the number of stones.

Although the actual speed of the abrasive is the resultant of both the rotative speed and the reciprocation speed, this latter quantity is seldom solved for or used. The reciprocation speed is not determined empirically but by testing under operating conditions. Changing the reciprocation speed affects the dressing action of the abrasive stones, therefore, the reciprocation speed is adjusted to provide for a desired surface finish which is usually a well lubricated bearing surface that will not scuff.

### Table of Factors for Use in Rotative Speed Formulas

| Character of Surface[1] | Material | Hardness[2] | | | | | |
|---|---|---|---|---|---|---|---|
| | | Soft | | Medium | | Hard | |
| | | Factors | | | | | |
| | | K | R | K | R | K | R |
| Base Metal | Cast Iron | 110 | 420 | 80 | 300 | 60 | 230 |
| | Steel | 80 | 300 | 60 | 230 | 50 | 190 |
| Dressing Surface | Cast Iron | 150 | 570 | 110 | 420 | 80 | 300 |
| | Steel | 110 | 420 | 80 | 300 | 60 | 230 |
| Severe Dressing | Cast Iron | 200 | 760 | 150 | 570 | 110 | 420 |
| | Steel | 150 | 570 | 110 | 420 | 80 | 300 |

[1] The character of the surface is classified according to its effect on the abrasive; *Base Metal* being a honed, ground or fine bored section that has little dressing action on the grit; *Dressing Surface* being a rough bored, reamed or broached surface or any surface broken by cross holes or ports; *Severe Dressing* being a surface interrupted by keyways, undercuts or burrs that dress the stones severely. If over half of the stock is to be removed after the surface is cleaned up, the speed should be computed using the *Base Metal* factors for $K$ and $R$.

[2] Hardness designations of soft, medium and hard cover the following ranges on the Rockwell "C" hardness scale, respectively: 15 to 45, 45 to 60 and 60 to 70.

### Possible Adjustments for Eliminating Undesirable Honing Conditions

| Undesirable Condition | Adjustment Required to Correct Condition[1] | | | | | | | | |
|---|---|---|---|---|---|---|---|---|---|
| | Abrasive[2] | | | | Other | | | | |
| | Friability | Grain Size | Hardness | Structure | Feed Pressure | Reciprocation | R.P.M. | Runout Time | Stroke Length |
| Abrasive Glazing | + | − − | − − | + | + + | + + | − − | − | o |
| Abrasive Loading | o | − − | − | − | + + | + | − − | o | o |
| Too Rough Surface Finish | o | + + | + + | − | − | − | + + | + | o |
| Too Smooth Surface Finish | o | − − | − − | + | + | + | − − | − | o |
| Poor Stone Life | − | + | + + | − | − | − | + | o | o |
| Slow Stock Removal | + | − − | − | + | + + | + + | − − | o | o |
| Taper — Large at Ends | o | o | o | o | o | o | o | o | − |
| Taper — Small at Ends | o | o | o | o | o | o | o | o | + |

Compiled by Micromatic Hone Corp.

[1] The + and + + symbols generally indicate that there should be an increase or addition while the − and − − symbols indicate that there should be a reduction or elimination. In each case, the double symbol indicates that the contemplated change would have the greatest effect. The o symbol means that a change would have no effect.

[2] For the abrasive adjustments the + and + + symbols indicate a more friable grain, a finer grain, a harder grade or a more open structure and the − and − − symbols just the reverse.

**Abrasive Stones for Honing.** — Honing stones consist of aluminum oxide, silicon carbide or diamond abrasive grits, held together in stick form by a vitrified clay, resinoid or metal bond. The grain and grade of abrasive to be used in any particular honing operation depend upon the quality of finish desired, the amount of stock to be removed, the material being honed and other factors. The following general rules may be followed in the application of abrasive for honing: (1) Silicon-carbide abrasive is commonly used for honing cast iron, while aluminum-oxide abrasive is generally used on steel; (2) The harder the material being honed, the softer the abrasive stick used; (3) A rapid reciprocating speed will tend to make the abrasive cut fast because the dressing action on the grits will be severe; and (4) To improve the finish, use a finer abrasive grit, incorporate more multi-direction action, allow more "run-out" time after honing to size, or increase the speed of rotation. Surface roughnesses ranging from less than 1 micro-inch r.m.s. to a relatively coarse roughness can be obtained by judicious choice of abrasive and honing time but the most common range is from 3 to 50 micro-inches r.m.s.

**Adjustments for Eliminating Honing Conditions.** — The accompanying table indicates adjustments that may be made to correct certain undesirable conditions encountered in honing. Only one change should be made at a time and its effect noted before making other adjustments.

**Tolerances.** — For bore diameters above 4 inches the tolerance of honed surfaces with respect to roundness and straightness ranges from 0.0005 to 0.001 inch; for bore diameters from 1 to 4 inches, 0.0003 to 0.0005 inch; and for bore diameters below 1 inch, 0.00005 to 0.0003 inch.

# Laps and Lapping

**Material for Laps.** — Laps are usually made of soft cast iron, copper, brass or lead. In general, the best material for laps to be used on very accurate work is soft, close-grained cast iron. If the grinding, prior to lapping, is of inferior quality, or an excessive allowance has been left for lapping, copper laps may be preferable. They can be charged more easily and cut more rapidly than cast iron, but do not produce as good a finish. Whatever material is used, the lap should be softer than the work, as, otherwise, the latter will become charged with the abrasive and cut the lap, the order of the operation being reversed. A common and inexpensive form of lap for holes is made of lead which is cast around a tapering steel arbor. The arbor usually has a groove or keyway extending lengthwise, into which the lead flows, thus forming a key that prevents the lap from turning. When the lap has worn slightly smaller than the hole and ceases to cut, the lead is expanded or stretched a little by the driving in of the arbor. When this expanding operation has been repeated two or three times, the lap usually must be trued or replaced with a new one, owing to distortion.

The tendency of lead laps to lose their form is an objectionable feature. They are, however, easily molded, inexpensive, and quickly charged with the cutting abrasive. A more elaborate form for holes is composed of a steel arbor and a split cast-iron or copper shell which is sometimes prevented from turning by a small dowel pin. The lap is split so that it can be expanded to accurately fit the hole being operated upon. For hardened work, some toolmakers prefer copper to either cast iron or lead. For holes varying from ¼ to ½ inch in diameter, copper or brass is sometimes used; cast iron is used for holes larger than ½ inch in diameter. The arbors for these laps should have a taper of about ¼ or ⅜ inch per foot. The length of the lap should be somewhat greater than the length of the hole, and the thickness of the shell or lap proper should be from ⅛ to ⅙ its diameter.

External laps are commonly made in the form of a ring, there being an outer ring or holder and an inner shell which forms the lap proper. This inner shell is made of cast iron, copper, brass or lead. Ordinarily the lap is split and screws are provided in the holder for adjustment. The length of an external lap should at least equal the diameter of the work, and might well be longer. Large ring laps usually have a handle for moving them across the work.

**Laps for Flat Surfaces.** — Laps for producing plane surfaces are made of cast iron. In order to secure accurate results, the lapping surface must be a true plane. A flat lap that is used for roughing or "blocking down" will cut better if the surface is scored by narrow grooves. These are usually located about ½ inch apart and extend both lengthwise and crosswise, thus forming a series of squares similar to those on a checker-board. An abrasive of No. 100 or 120 emery and lard oil can be used for charging the roughing lap. For finer work, a lap having an unscored surface is used, and the lap is charged with a finer abrasive. After a lap is charged, all loose abrasive should be washed off with gasoline, for fine work, and when lapping, the surface should be kept moist, preferably with kerosene. Gasoline will cause the lap to cut a little faster, but it evaporates so rapidly that the lap soon becomes dry and the surface caked and glossy in spots. Loose emery should not be applied while lapping, for if the lap is well charged with abrasive in the beginning, is kept well moistened and not crowded too hard, it will cut for a considerable time. The pressure upon the work should be just enough to insure constant contact. The lap can be made to cut only so fast, and if excessive pressure is applied it will become "stripped" in places. The causes of scratches are: Loose abrasive on the lap; too much pressure on the work, and poorly graded abrasive. To produce a perfectly smooth surface free from scratches, the lap should be charged with a very fine abrasive.

**Grading Abrasives for Lapping.** — For high-grade lapping, abrasives can be evenly graded as follows: A quantity of flour-emery or other abrasive is placed in a heavy cloth bag, which is gently tapped, causing very fine particles to be sifted through. When a sufficient quantity has been obtained in this way, it is placed in a dish of lard or sperm oil. The largest particles will then sink to the bottom and in about one hour the oil should be poured into another dish, care being taken not to disturb the sediment at the bottom. The oil is then allowed to stand for several hours, after which it is poured again, and so on, until the desired grade is obtained.

**Charging Laps.** — To charge a flat cast-iron lap, spread a very thin coating of the prepared abrasive over the surface and press the small cutting particles into the lap with a hard steel block. There should be as little rubbing as possible. When the entire surface is apparently charged, clean and examine for bright spots; if any are visible, continue charging until the entire surface has a uniform gray appearance. When the lap is once charged, it should be used without applying more abrasive until it ceases to cut. If a lap is over-charged and an excessive amount of abrasive is used, there is a rolling action between the work and lap which results in inaccuracy. The surface of a flat lap is usually finished true, prior to charging, by scraping and testing with a standard surface-plate, or by the well-known method of scraping-in three plates together, in order to secure a plane surface. In any case, the bearing marks or spots should be uniform and close together. These spots can be blended by covering the plates evenly with a fine abrasive and rubbing them together. While the plates are being ground in, they should be carefully tested and any high spots which may form should be reduced by rubbing them down with a smaller block.

To charge cylindrical laps for internal work, spread a thin coating of prepared abrasive over the surface of a hard steel block, preferably by rubbing lightly with a cast-iron or copper block; then insert an arbor through the lap and roll the latter over the steel block, pressing it down firmly to imbed the abrasive into the surface of the lap. For external cylindrical laps, the inner surface can be charged by rolling-in the abrasive with a hard steel roller that is somewhat smaller in diameter than the lap. The taper cast-iron blocks which are sometimes used for lapping taper holes can also be charged by rolling-in the abrasive, as previously described; there is usually one roughing and one finishing lap, and when charging the former, it may be necessary to vary the charge in accordance with any error which might exist in the taper.

**Rotary Diamond Lap.** — This style of lap is used for accurately finishing very small holes, which, because of their size, cannot be ground. While the operation is referred to as lapping, it is, in reality, a grinding process, the lap being used the same as a grinding wheel. Laps employed for this work are made of mild steel, soft material being desirable because it can be charged readily. Charging is usually done by rolling the lap between two hardened steel plates. The diamond dust and a little oil is placed on the lower plate, and as the lap revolves, the diamond is forced into its surface. After charging, the lap should be washed in benzine. The rolling plates should also be cleaned before charging with dust of a finer grade. It is very important not to force the lap when in use, especially if it is a small size. The lap should just make contact with the high spots and gradually grind them off. If a diamond lap is lubricated with kerosene, it will cut freer and faster. These small laps are run at very high speeds, the rate depending upon the lap diameter. Soft work should never be ground with diamond dust because the dust will leave the lap and charge the work.

When using a diamond lap, it should be remembered that such a lap will not produce sparks like a regular grinding wheel; hence, it is easy to crowd the lap and "strip" some of the diamond dust. To prevent this, a sound intensifier or "harker" should be used. This is placed against some stationary part of the grinder spindle, and indicates when the lap touches the work, the sound produced by the slightest contact being intensified.

**Grading Diamond Dust.** — The grades of diamond dust used for charging laps are designated by numbers, the fineness of the dust increasing as the numbers increase. The diamond, after being crushed to powder in a mortar, is thoroughly mixed with high-grade olive oil. This mixture is allowed to stand five minutes and then the oil is poured into another receptacle. The coarse sediment which is left is removed and labeled No. o, according to one system. The oil poured from No. o is again stirred and allowed to stand ten minutes, after which it is poured into another receptacle and the sediment remaining is labeled No. 1. This operation is repeated until practically all of the dust has been recovered from the oil, the time that the oil is allowed to stand being increased as shown by the following table. This is done in order to obtain the smaller particles that require a longer time for precipitation:

| | |
|---|---|
| To obtain No. 1 — 10 minutes. | To obtain No. 4 — 2 hours. |
| To obtain No. 2 — 30 minutes. | To obtain No. 5 — 10 hours. |
| To obtain No. 3 — 1 hour. | To obtain No. 6 — until oil is clear. |

The No. o or coarse diamond which is obtained from the first settling is usually washed in benzine, and re-crushed unless very coarse dust is required. This No. o grade is sometimes known as "ungraded" dust. In some places the time for settling, in order to obtain the various numbers, is greater than that given in the table.

**Cutting Properties of Laps and Abrasives.** — In order to determine the cutting properties of abrasives when used with different lapping materials and lubricants, a series of tests was conducted, the results of which were given in a paper by W. A. Knight and A. A. Case, presented before the American Society of Mechanical Engineers. In connection with these tests, a special machine was used, the construction being such that quantitative results could be obtained with various combinations of abrasive, lubricant, and lap material. These tests were confined to surface lapping.

It was not the intention to test a large variety of abrasives, three being selected as representative; namely, Naxos emery, carborundum, and alundum. Abrasive No. 150 was used in each case, and seven different lubricants, five different pressures, and three different lap materials were employed. The lubricants were lard oil, machine oil, kerosene, gasoline, turpentine, alcohol, and soda water.

These tests indicated throughout that there is, for each different combination of lap and lubricant, a definite size of grain that will give the maximum amount of cutting. With all the tests, except when using the two heavier lubricants, some reduction in the size of the grain below that used in the tests (No. 150) seemed necessary before the maximum rate of cutting was reached. This reduction, however, was continuous and soon passed below that which gave the maximum cutting rate.

**Cutting Qualities with Different Laps.** — The surfaces of the steel and cast-iron laps were finished by grinding. The hardness of the different laps, as determined by the scleroscope was, for cast-iron, 28; steel, 18; copper, 5. The total amount ground from the test-pieces with each of the three laps showed that, taking the whole number of tests as a standard, there is scarcely any difference between the steel and cast iron, but that copper has somewhat better cutting qualities, although, when comparing the laps on the basis of the highest and lowest values obtained with each lap, steel and cast iron are as good for all practical purposes as copper, when the proper abrasive and lubricant are used.

**Wear of Laps.** — The wear of laps depends upon the material from which they are made and the abrasive used. The wear on all laps was about twice as fast with carborundum as with emery, while with alundum the wear was about one and one-fourth times that with emery. On an average, the wear of the copper lap was about three times that of the cast-iron lap. This is not absolute wear, but wear in proportion to the amount ground from the test-pieces.

**Lapping Abrasives.** — As to the qualities of the three abrasives tested, it was found that carborundum usually began at a lower rate than the other abrasives, but, when once started, its rate was better maintained. The performance gave a curve that was more nearly a straight line. The charge or residue as the grinding proceeded remained cleaner and sharper and did not tend to become pasty or mucklike, as is so frequently the case with emery. When using a copper lap, carborundum shows but little gain over the cast-iron and steel laps, whereas, with emery and alundum, the gain is considerable.

**Effect of Different Lapping Lubricants.** — The action of the different lubricants, when tested, was found to depend upon the kind of abrasive and the lap material.

*Lard and Machine Oil.* — The test showed that lard oil, without exception, gave the higher rate of cutting, and that, in general, the initial rate of cutting is higher with the lighter lubricants, but falls off more rapidly as the test continues. The lowest results were obtained with machine oil, when using an emery-charged, cast-iron lap. When using lard oil and a carborundum-charged steel lap, the highest results were obtained.

*Gasoline and Kerosene.* — On the cast-iron lap, gasoline was superior to any of the lubricants tested. Considering all three abrasives, the relative value of gasoline, when applied to the different laps, is as follows: Cast iron, 127; copper, 115; steel, 106. Kerosene, like gasoline, gives the best results on cast iron and the poorest on steel. The values obtained by carborundum were invariably higher than those obtained with emery, except when using gasoline and kerosene on a copper lap.

*Turpentine and Alcohol.* — Turpentine was found to do good work with carborundum on any lap. With emery, turpentine did fair work on the copper lap, but, with the emery on cast-iron and steel laps, it was distinctly inferior. Alcohol gives the lowest results with emery on the cast-iron and steel laps.

*Soda Water.* — Soda water gives medium results with almost any combination of lap and abrasives, the best work being on the copper lap and the poorest, on the steel lap. On the cast-iron lap, soda water is better than machine or lard oil, but not so good as gasoline or kerosene. Soda water when used with alundum on the copper lap, gave the highest results of any of the lubricants used with that particular combination.

**Lapping Pressures.** — Within the limits of the pressures used, that is, up to 25 pounds per square inch, the rate of cutting was found to be practically proportional to the pressure. The higher pressures of 20 and 25 pounds per square inch are not so effective on the copper lap as on the other materials.

**Wet and Dry Lapping.** — With the "wet method" of using a surface lap, there is a surplus of oil and abrasive on the surface of the lap. As the specimen being lapped is moved over it, there is more or less movement or shifting of the abrasive particles. With the "dry method," the lap is first charged by rubbing or rolling the abrasive into its surface. All surplus oil and abrasive are then washed off, leaving a clean surface, but one that has embedded uniformly over it small particles of the abrasive. It is then like the surface of a very fine oilstone and will cut away hardened steel that is rubbed over it. While this has been termed the dry method, in practice, the lap surface is kept moistened with kerosene or gasoline.

Experiments on dry lapping were carried out on the cast-iron, steel, and copper laps used in the previous tests, and also on one of tin made expressly for the purpose. Carborundum alone was used as the abrasive and a uniform pressure of 15 pounds per square inch was applied to the specimen throughout the tests. In dry lapping, much depends upon the manner of charging the lap. The rate of cutting decreased much more rapidly after the first 100 revolutions than with the wet method. Considering the amounts ground off during the first 100 revolutions, and the best result obtained with each lap taken as the basis of comparison, it was found that with a tin lap, charged by rolling No. 150 carborundum into the surface, the rate of cutting, when dry, approached that obtained with the wet method. With the other lap materials, the rate with the dry method was about one-half that of the wet method.

**Summary of Lapping Tests.** — The initial rate of cutting does not greatly differ for different abrasives. There is no advantage in using an abrasive coarser than No. 150. The rate of cutting is practically proportional to the pressure. The wear of the laps is in the following proportions: cast iron, 1.00; steel, 1.27; copper, 2.62. In general, copper and steel cut faster than cast iron, but, where permanence of form is a consideration, cast iron is the superior metal. Gasoline and kerosene are the best lubricants to use with a cast-iron lap. Machine and lard oil are the best lubricants to use with copper or steel laps. They are, however, least effective on a cast-iron lap. In general, wet lapping is from 1.2 to 6 times as fast as dry lapping, depending upon the material of the lap and the manner of charging.

## Power Brush Finishing

Power brush finishing is a production method of metal finishing that employs wire, elastomer bonded wire, or non-metallic (cord, natural fiber or synthetic) brushing wheels in automatic machines, semi-automatic machines and portable air tools to smooth or roughen surfaces, remove surface oxidation and weld scale or remove burrs.

**Description of Brushes.** — Brushes work in the following ways: the wire points of a brush can be considered to act as individual cutting tools so that the brush, in effect, is a multiple-tipped cutting tool. The fill material, as it is rotated, contacts the surface of the work and imparts an impact action which produces a coldworking effect. The type of finish produced depends upon the wheel material, wheel speed, and how the wheel is applied. Brushes differ in the following ways: (1) fill material (wire — carbon steel, stainless steel; synthetic; Tampico; and cord), (2) length of fill material (or trim), and (3) the density of the fill material.

To aid in wheel selection and use, the accompanying table made up from information supplied by The Osborn Manufacturing Company lists the characteristics and major uses of brushing wheels.

**Use of Brushes.** — The brushes should be located so as to bring the full face of the brush in contact with the work. Full face contact is necessary to avoid grooving the brush. Operations that are set up with the brush face not in full contact with the work require some provision for dressing the brush face. When the tips of a brush, used with full face contact, become dull during use with subsequent loss of working clearance, reconditioning and resharpening is necessary. This is accomplished simply and efficiently by alternately reversing the direction of rotation during use.

**Deburring and Producing a Radius on the Tooth Profile of Gears.** — The brush employed for deburring and producing a radius on the tooth profile of gears is a short trim, dense, wire-fill radial brush. The brush should be set up so as to brush across the edge as shown in Fig. 1A. Line contact brushing, as shown in Fig. 1B should be avoided because the brush face will wear non-uniformly; and the wire points, being flexible, tend to flare to the side, thus minimizing the effectiveness of the brushing operation. When brushing gears, the brushes are spaced and contact the tooth profile on the center line of the gear as shown in Fig. 2. This facilitates using brush reversal to maintain the wire brushing points at their maximum cutting efficiency.

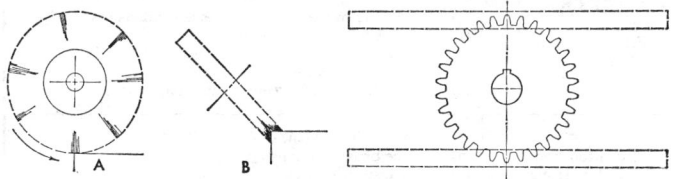

Fig. 1. (Left) — Methods of Brushing an Edge; (A) Correct, (B) Incorrect
Fig. 2. (Right) — Setup for Deburring Gears

The setup for brushing spline bores differs from brushing gears in that the brushes are located off-center, as illustrated in Fig. 3. When helical gears are brushed, it is sometimes necessary to favor the acute side of the gear tooth to develop a generous radius prior to shaving.    This can be accomplished by locating the brushes as shown in Fig. 4.    Elastomer bonded wire-filled brushes are used for deburring fine pitch gears.    These brushes remove the burrs without leaving any secondary roll.    The use of bonded brushes is necessary when the gears are not shaved after hobbing or gear shaping.

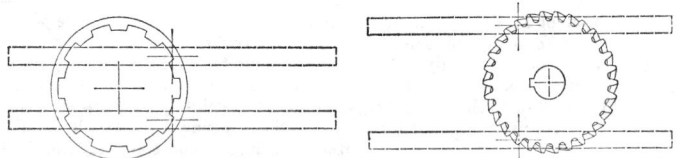

Fig. 3. (Left) — Setup for Brushing Broached Splines
Fig. 4. (Right) — Setup for Finishing Helical Gears

**Adjustments for Eliminating Undesirable Conditions in Power Brush Finishing**

| Undesirable Condition | Possible Adjustments for Eliminating Condition |
|---|---|
| Brush works too slowly | (1) Decrease trim length and increase fill density.<br>(2) Increase filament diameter.<br>(3) Increase surface speed by increasing R.P.M. or outside diameter. |
| Brush works too fast | (1) Reduce filament diameter.<br>(2) Reduce surface speed by reducing R.P.M. or outside diameter.<br>(3) Reduce fill density.<br>(4) Increase trim length. |
| Action of brush peens burr to adjacent surface | (1) Decrease trim length and increase fill density.<br>(2) If wire brush tests indicate metal too ductile (burr is peened rather than removed), change to nonmetallic brush such as a treated Tampico brush used with a burring compound. |
| Finer or smoother finish required | (1) Decrease trim length and increase fill density.<br>(2) Decrease filament diameter.<br>(3) Try treated Tampico or cord brushes with suitable compounds at recommended speeds.<br>(4) Use auxiliary buffing compound with brush. |
| Finish too smooth and lustrous | (1) Increase trim length.<br>(2) Reduce brush fill density.<br>(3) Reduce surface speed.<br>(4) Increase filament diameter. |
| Brushing action not sufficiently uniform | (1) Devise hand-held or mechanical fixture or machine which will avoid irregular off-hand manipulation.<br>(2) Increase trim length and decrease fill density. |

Characteristics and Applications of Brushes Used in Power Brush Finishing

| Brush Type | Description | Operating Speed Range, sfpm | Uses | Remarks |
|---|---|---|---|---|
| Radial, short trim dense wire fill | Develops very little impact action but maximum cutting action. | 6500 | Removal of burrs from gear teeth and sprockets. Produces blends and radii at juncture of intersecting surfaces. | Brush should be set up so as to brush across any edge. Reversal of rotation needed to maintain maximum cutting efficiency of brush points. |
| Radial, medium to long trim twisted knot wire fill | Normally used singly and on portable tools. Brush is versatile and provides high impact action. | 7500-9500 for high speeds. 1200 for slow speeds. | For cleaning welds in the automotive and pipeline industries. Also for cleaning surfaces prior to painting, stripping rubber flash from molded products and cleaning mesh-wire conveyor belts. | Surface speed plays an important role since at low speeds the brush is very flexible and at high speeds it is extremely hard and fast cutting. |
| Radial, medium to long trim crimped wire fill | With the 4- to 8-inch diameter brush, part is hand held. With the 10- to 15-inch diameter brush, part is held by machine. | 4500-6000 | Serves as utility tool on bench grinder for removing feather grinding burrs, machining burrs, and for cleaning and producing a satin or matte finish. | Good for hand held parts as brush is soft enough to conform to irregular surfaces and hard-to-reach areas. Smaller diameter brushes are not recommended for high-production operations. |
| Radial, sectional; non-metallic fill (treated and untreated Tampico or cord) | Provides means for improving finish or improving surface for plating. Works best with grease base deburring or buffing compound. | 5500-6500 7500 for polishing | For producing radii and improving surface finish. Removes the sharp peaks that fixed abrasives leave on a surface so that surface will accept a uniform plating. Polishing marks and draw marks can be successfully blended. | Brush is selective to an edge which means that it removes metal from an edge but not from adjoining surfaces. It will produce a very uniform radius without peening or rolling any secondary metal. |
| Radial, wide-face, nonmetallic fill (natural fibers or synthetics) | Can be used with flow-through mounting which facilitates feeding of cold water and hot alkaline solution's through brush face to prevent buildup. | 750-1200 for cleaning steel  600 when used with slurries | For cleaning steel. Used in electrolytic tinplate lines, continuous galvanizing and annealing lines, and cold reduction lines. Used to produce dull or matte-type finishes on stainless steel and synthetics. | Speeds above 3600 sfpm will not appreciably improve operation as brush wear will be excessive. Avoid excessive pressures. Ammeters should be installed in drive-motor circuit to indicate brushing pressure. |

Characteristics and Applications of Brushes Used in Power Brush Finishing (*Concluded*)

| Brush Type | Description | Operating Speed Range, sfpm | Uses | Remarks |
|---|---|---|---|---|
| Radial, wide face, metallic fill | This brush is made to customer's specifications. It is dynamically balanced at the speed at which it will operate. | 2000–4000 | Removes buildup of aluminum oxide from work rolls in aluminum mill. Removes lime or magnesium coatings from certain types of steel. Burnishes hot-dipped galvanized steel to produce a minimum spangled surface. | Each brush should have its own drive. An ammeter should be present in drive-motor circuit to measure brushing pressure. If strip is being brushed, a steel backup roll should be opposite the brush roll. |
| Radial, wide face, strip (interrupted brush face) | Performs cleaning operations that would cause a solid face brush to become loaded and unusable. | When cleaning conveyor belts, brush speed is 2 to 3 times that of conveyor belt. | Need for cleaning conveyor belts, brush back material which would normally foul snubber pulley and return idlers. | Designed for medium- to light-duty work. Brush face does not load. |
| Radial, Cup Flared End, and Straight End, wire fill elastomer bonded | Extremely fast cutting with maximum operator safety. No loss of wire through fatigue. Always has uniform face. | 3600–9000 | For removing oxide weld scale, burrs, and insulation from wire. | Periodic reversing of brush direction will result in a brush life ten times greater than non-bonded wheels. Fast cutting action necessitates precise holding of part with respect to brush. |
| Cup, twisted knot wire fill | Fast cutting wheel used on portable tools to clean welds, scale, rust, and other oxides. | 8000–10,000 4500–6500 for deburring and producing a radius around periphery of holes. | Used in shipyards and in structural steel industry. For cleaning outside diameter of pipe and removing burrs and producing radii on heat exchanger tube sheets and laminations for stator cores. | Fast acting brush cleans large areas economically. Setup time is short. |
| Radial, wire or treated Tampico or cord | For use with standard centerless grinders. Brush will not remove metal from a cylindrical surface. Parts must be ground to size before brushing. | .... | For removing feather grinding burrs and improving surface finish. Parts of 25 microinches can be finished down to 15 to 10 microinches. Parts of 10 to 12 microinches can be finished down to 7 to 4 microinches. | Follows centerless grinding principles, except that accuracy in pressure and adjustment is not critical. A machine no longer acceptable for grinding can be used for brushing. |

## Polishing and Buffing

The terms "polishing" and "buffing" are sometimes applied to similar classes of work in different plants, but according to approved usage of the terms, there is the following distinction: Polishing is any operation performed with wheels having abrasive glued to the working surfaces, whereas, buffing is done with wheels having the abrasive applied loosely instead of imbedding it into glue; moreover, buffing is not so harsh an operation as ordinary polishing, and it is commonly utilized for obtaining very fine surfaces having a "grainless finish."

**Polishing Wheels.** — The principal materials from which polishing wheels are made are wood, leather, canvas, cotton cloth, felt, paper, walrus or sea-horse hide, sheepskin, impregnated rubber, canvas composition, and wool. Leather and canvas are the materials most commonly used in polishing wheel construction. Bullneck leather wheels are made of oak tanned bullneck leather cut into disks of uniform thickness and cemented together. Wooden wheels covered with leather to which emery or some other abrasive is glued, are employed extensively for polishing flat surfaces, especially when good edges must be maintained. Canvas wheels are made in various ways; wheels having disks that are cemented together are very hard and used for rough, coarse work, whereas those having sewed disks are made of varying densities by sewing together a larger or smaller number of disks into sections and gluing them. Wheels in which the disks are held together by sewing and which are not stiffened by the use of glue, usually require metal side plates to support the canvas disks. Muslin wheels are made from sewed buffs glued together, but the outer edges of a wheel frequently are left open or free from glue to provide an open face of any desired depth. Wool felt wheels are flexible and resilient, and the density may be varied by sewing two or more disks together and then cementing these to form a wheel. Solid wheels made of Spanish or Mexican felt are quite popular for fine finishing but have little value as general utility wheels. Paper wheels are made from strawboard paper disks and are cemented together under pressure to form a very hard wheel for rough work. Softer wheels are similarly made from felt paper. Walrus leather or sea-horse hide may be used for fine polishing, but these wheels are expensive. The "compress" canvas wheel is commonly used in place of walrus wheels. This compress type of wheel has a cushion of polishing material formed by pieces of leather, canvas, felt, or whatever material is used, which are held in a crosswise radial position by two side plates attached to the wheel hub. This cushion of polishing material may be varied in density to suit the requirements; it may readily be shaped to conform to the curvature of the work and this shape can be maintained. Sheepskin polishing wheels and also paper wheels are used very little at the present time.

**Polishing Operations and Abrasives.** — Polishing operations on such parts as chisels, hammers, screwdrivers, wrenches, and other parts which are given a fine finish but are not plated, usually require four operations which are "roughing," "dry fining," "greasing" and "coloring." The roughing is frequently regarded as a solid grinding wheel job. Sometimes there are two steps to the greasing operation — rough and fine greasing. For some hardware, such as the cheaper screwdrivers, wrenches, etc., the operations of roughing and dry fining are considered sufficient. For knife blades and cutlery the roughing operation is performed with solid grinding wheels and the polishing is known as fine or blue glazing, but these terms are never used when referring to the polishing of hardware parts, plumbers' supplies, etc. A term used in finishing German silver, white metal, and similar materials is "sand-buffing," which, in distinction from the ordinary buffing operation that is used only to produce a very high finish, actually removes considerable metal, as in rough polishing or flexible grinding. For sand-buffing, rotten-stone and pumice are loosely applied.

Aluminum oxide abrasives are widely used for polishing high tensile strength metals such as carbon and alloy steels, tough iron, and non-ferrous alloys. Silicon carbide abrasives are recommended for hard, brittle substances such as grey iron, cemented carbide tools, and also materials of low tensile strength such as brass, aluminum and copper.

**Buffing Wheels.** — Buffing wheels, as defined by the Metal Finishers' Equipment Association, are wheels manufactured from disks (either whole or pieced) of bleached or unbleached cotton or woolen cloth, and they are used as the agent for carrying abrasive powders, such as tripoli, crocus, rouge, lime, etc., which are mixed with waxes or greases as a bond. There are two main classes of buffs known as the "pieced-sewed" buffs, which are made from various weaves and weights of cloths, and the "full disk" buffs which are made from the best sheeting and shirting. Bleached cloth is harder and stiffer than unbleached cloth, and is used for the faster cutting buffs. Coarsely woven unbleached cloth is recommended for highly colored work on soft metals, while the finer woven unbleached cloths are better adapted for the harder metals. A stiff buff when working at the usual speed is not suitable for "cutting down" soft metal or for use on light plated ware, but is used on the harder metals and for heavy nickel-plated articles.

**Speed of Polishing Wheels.** — The proper speed for polishing is governed to some extent by the nature of the work, but for ordinary operations the polishing wheel, according to one manufacturer, should have a peripheral speed of about 7500 feet per minute. If run at a lower rate of speed, the work tends to tear the polishing material from the wheel too readily, and the work is not as good in quality. Another manufacturer recommends the following speeds: Muslin, felt or sea-horse polishing wheels having wood or iron centers should be run at peripheral speeds varying from 3000 to 7000 feet per minute. It is rarely necessary to exceed 6000 feet per minute, and for most purposes 4000 feet per minute is sufficient. If the wheels are kept in good condition, in perfect balance, and are suitably mounted on substantial buffing lathes, they can safely be used for speeds within the limits given.

**Grain Numbers of Emery.** — The numbers commonly used in designating the different grains of emery, corundum and other abrasives are 10, 12, 14, 16, 18, 20, 24, 30, 36, 40, 46, 54, 60, 70, 80, 90, 100, 120, 150, 180 and 200, ranging from coarse to fine. These numbers represent the number of meshes per linear inch in the grading sieve. An abrasive finer than No. 200 is known as "flour" and the degree of fineness is designated by the letters CF, F, FF, FFF, FFFF and PCF or SF, ranging from coarse to fine. The methods of grading flour-emery adopted by different manufacturers do not exactly agree, the letters differing somewhat for the finer grades.

**Grades of Emery Cloth.** — The coarseness of emery cloth is indicated by letters and numbers corresponding to the grain number of loose emery. The letters and numbers for grits ranging from fine to coarse are as follows: FF, F, 120, 100, 90, 80, 70, 60, 54, 46, 40. For large work roughly filed, use coarse cloth such as Nos. 46 or 54, and then finer grades to obtain the required polish. If the work has been carefully filed, a good polish can be obtained with Nos. 60 and 90 cloth, and a brilliant polish by finishing with No. 120 and flour-emery.

**Mixture for Cementing Emery Cloth to a Lapping Wheel.** — Use 4½ pounds of rosin; 3 pounds of paraffine; 9 ounces of vaseline; melt the ingredients and mix them thoroughly. Heat the surface of the lapping wheel and spread on the mixture; then rub the emery cloth down so as to exclude all air from between the surface of the wheel and cloth. The surface of the lapping wheel should be clean before the cement is applied.

## Oilstones

Two most commonly used natural oilstones are the Washita and the Arkansas. The Washita is a coarser and more rapidly cutting stone than the Arkansas and is generally considered the most satisfactory for sharpening woodworkers' tools. Arkansas stones are harder and denser and have an exceedingly sharp fine grain which will cut and polish very hard metals and produce keen, smooth edges.

Artificial stones are commonly furnished in a variety of shapes in three grits: fine, medium and coarse. Coarse stones are used where fast cutting is required without regard to fine finish, such as in the case of sharpening very large and dull or nicked tools. Medium stones are used for sharpening mechanics' tools and fine stones for sharpening tools requiring a very fine, keen edge.

**Truing Surfaces of Oilstones.** — Oilstones which have uneven surfaces can be trued by the following method: Secure a cast-iron block having a true surface, and cover the surface with loose emery mixed with water; then place the oilstone upon the cast-iron block and grind it true. This method is applicable to either coarse oilstones or fine razor hones. Stones of special shape may be formed by planing a groove of corresponding shape in the cast-iron block and drawing the stone through the groove, using emery and water as an abrasive.

**Care of Oilstones.** — Oilstones should be properly cared for: first, in order to retain the original life and sharpness of the grit; second, to keep the surface flat and even; third, to prevent glazing. The following instructions are given by the Pike Mfg. Co.: An oilstone should be kept clean and moist; allowing it to remain dry a long time, or exposing it to the air, tends to harden it. A new stone should be soaked in oil for several days before using (with the exception of Pike India and Pike Crystolon). If the stone is kept in a dry place, it should be placed in a box having a closed cover, and a few drops of fresh, clean oil should be placed on it. To restore an even flat surface on an oilstone, grind it on the side of a grindstone, or rub it down with sand-stone or an emery brick.

An oilstone can be prevented from glazing by the proper use of oil or water. Either oil or water will prevent the particles of steel that are cut away from the tool being sharpened from filling the surface of the stone. Plenty of water should be used on all coarse-grained natural stones; on medium- or fine-grained natural stones, such as Arkansas or Washita, as well as on all artificial stones, oil should be used invariably, as water is not thick enough to keep the steel particles out of the pores. To further prevent glazing, dirty oil should always be wiped off the stone as soon as possible after using. This is very important, for if the oil is left on the stone, it dries in, carrying steel dust with it. Cotton waste is one of the best things for cleaning a stone. If a stone does become glazed or gummed up, cleaning with gasoline or ammonia will usually restore its cutting qualities; but if this treatment is not effective, scour the stone with loose emery or a piece of sand-paper fastened to a flat board.

## Balancing Rotating Parts

**Static Balancing.** — There are several methods of testing the standing or static balance of a rotating part. A simple method that is sometimes used for flywheels, etc., is illustrated by the diagram, Fig. 1. An accurate shaft is inserted through the bore of the finished wheel, which is then mounted on carefully leveled "parallels" $A$. If the wheel is in an unbalanced state, it will turn until the heavy side is downward. When it will stand in any position as the result of counter-balancing and reducing the heavy portions, it is said to be in standing or static balance. Another test which is used for disk-shaped parts is shown in Fig. 2. The disk $D$

is mounted on a vertical arbor attached to an adjustable cross-slide $B$. The latter is carried by a table $C$, which is supported by a knife-edged bearing. A pendulum having an adjustable screw-weight $W$ at the lower end, is suspended from cross-slide $B$. To test the static balance of disk $D$, slide $B$ is adjusted until pointer $E$ of the pendulum coincides with the center of a stationary scale $F$. Disk $D$ is then turned halfway around without moving the slide, and if the indicator remains stationary, it shows that the disk is in balance for this particular position. The test is then repeated for ten or twelve other positions, and the heavy sides are reduced, usually by drilling out the required amount of metal. There are several other devices for testing the static balance which are designed on this same principle.

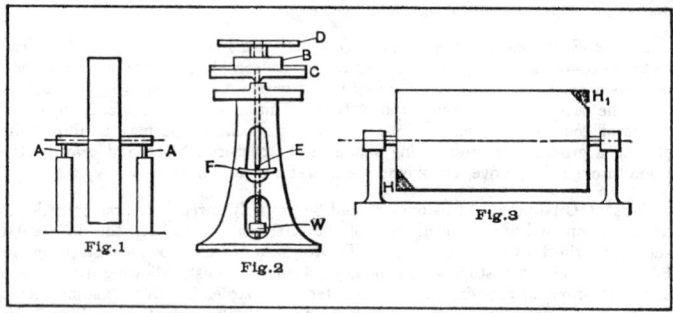

Fig. 1    Fig. 2    Fig. 3

**Running or Dynamic Balance.**—A cylindrical body may be in perfect static balance and not be in a balanced state when rotating at high speed. If the part is in the form of a thin disk, static balancing, if carefully done, may be accurate enough for high speeds, but if the rotating part is long in proportion to its diameter, and the unbalanced portions are at opposite ends or in different planes, the balancing must be done so as to counteract the centrifugal force of these heavy parts when they are rotating rapidly. This is known as a running balance or dynamic balancing. To illustrate, if a heavy section is located at $H$ (Fig. 3), and another correspondingly heavy section at $H_1$, one may exactly counter-balance the other when the cylinder is stationary, and this static balance may be sufficient for a part rigidly mounted and rotating at a comparatively slow speed; but when the speed is very high, as in the case of turbine rotors, etc., the heavy masses $H$ and $H_1$, being in different planes, are in an unbalanced state owing to the effect of centrifugal force, which results in excessive strains and injurious vibrations. Theoretically, to obtain a perfect running balance, the exact position of the heavy sections should be located and the balancing effected either by reducing their weight or by adding counterweights opposite each section and in the same plane at the proper radius; but if the rotating part is rigidly mounted on a stiff shaft, a running balance that is sufficiently accurate for practical purposes can be obtained by means of comparatively few counter-balancing weights located with reference to the unbalanced parts.

**Balancing Machines.** — Several types of machines have been developed for testing the running or dynamic balance of machine parts. Some balancing machines are designed primarily for wheels, disks and comparatively narrow face parts, whereas others are arranged to test various classes of work, such as crankshafts, rotors of generators and motors, pulleys, spindles, etc. Balancing machines are widely used, particularly when rotative speeds are high and the requirements are exacting in regard to vibration.

**Balancing Calculations.** — As indicated previously, centrifugal forces caused by an unbalanced mass or masses in a rotating machine member cause additional loads on the bearings which are transmitted to the housing or frame and to other machine members. Such dynamically unbalanced conditions can occur even though static balance (balance at zero speed) exists. Dynamic balance can be achieved by the addition of one or two masses rotating about the same axis and at the same speed as the unbalanced masses. A single unbalanced mass can be balanced by one counterbalancing mass located 180 degrees opposite and in the same plane of rotation as the unbalanced mass, if the product of their respective radii and masses are equal; i.e., $M_1r_1 = M_2r_2$. Two or more unbalanced masses rotating in the same plane can be balanced by a single mass rotating in the same plane, or by two masses rotating about the same axis in two separate planes. Likewise, two or more unbalanced masses rotating in different planes about a common axis can be balanced by two masses rotating about the same axis in separate planes. When the unbalanced masses are in separate planes they may be in static balance but not in dynamic balance; ie., they may be balanced when not rotating but unbalanced when rotating. If a system is in dynamic balance, it will remain in balance at all speeds, although this is not strictly true at the critical speed of the system. (See Critical Speeds.)

In all of the equations that follow the symbol $M$ denotes either mass in kilograms or in slugs, or weight in pounds. Either mass or weight units may be used and the equations may be used with metric or with customary English units without change; however, in a given problem the units must be all metric or all customary English.

**Counterbalancing Several Masses Located in a Single Plane.** — In all balancing problems, it is the product of counterbalancing mass (or weight) and its radius that is calculated; it is thus necessary to select either the mass or the radius and then calculate the other value from the product of the two quantities. Design considerations usually make this decision self-evident. The angular position of the counterbalancing mass must also be calculated. Referring to Fig. 1:

$$M_B r_B = \sqrt{(\Sigma Mr \cos \theta)^2 + (\Sigma Mr \sin \theta)^2} \qquad (1)$$

$$\tan \theta_B = \frac{-(\Sigma Mr \sin \theta)}{-(\Sigma Mr \cos \theta)} = \frac{y}{x} \qquad (2)$$

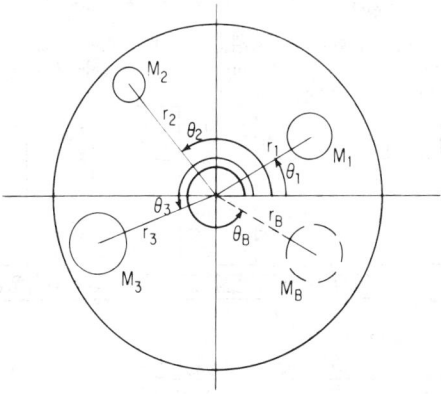

Fig. 1

Table 1. Relationship of the Signs of the Functions of the Angle
with Respect to the Quadrant in Which They Occur

| II \| I |  | Angle $\theta$ |  |  |  |
|---|---|---|---|---|---|
| | | 0° to 90° | 90° to 180° | 180° to 270° | 270° to 360° |
| | | Signs of the Functions |  |  |  |
| tan | | $\dfrac{+y}{+x}$ | $\dfrac{+y}{-x}$ | $\dfrac{-y}{-x}$ | $\dfrac{-y}{+x}$ |
| sine | | $\dfrac{+y}{+r}$ | $\dfrac{+y}{+r}$ | $\dfrac{-y}{+r}$ | $\dfrac{-y}{+r}$ |
| cosine | | $\dfrac{+x}{+r}$ | $\dfrac{-x}{+r}$ | $\dfrac{-x}{+r}$ | $\dfrac{+x}{+r}$ |

where:

$M_1, M_2, M_3, \ldots M_n$ = any unbalanced mass or weight, kg or lb

$M_B$ = counterbalancing mass or weight, kg or lb

$r$ = radius to center of gravity of any unbalanced mass or weight, mm or in.

$r_B$ = radius to center of gravity of counterbalancing mass or weight, mm or in.

$\theta$ = angular position of $r$ of any unbalanced mass or weight, degrees

$\theta_B$ = angular position of $r_B$ of counterbalancing mass or weight, degrees

$x$ and $y$ = see Table 1.

Table 1 is helpful in finding the angular position of the counterbalancing mass or weight. It indicates the range of the angles within which this angular position occurs by noting the plus and minus signs of the numerator and the denominator of the terms in Equation 2. In a like manner, Table 1 is helpful in determining the *sign* of the sine or cosine functions for angles ranging from 0 to 360 degrees. Balancing problems are usually solved most conveniently by arranging the arithmetical calculations in a tabular form.

*Example:* Referring to Fig. 1, the particular values of the unbalanced weights have been entered in the table below. Calculate the magnitude of the counterbalancing weight if its radius is to be 10 inches.

| $M$ | | $r$ | $\theta$ | $\cos \theta$ | $\sin \theta$ | $Mr \cos \theta$ | $Mr \sin \theta$ |
|---|---|---|---|---|---|---|---|
| No. | lb. | in. | deg. | | | | |
| 1 | 10 | 10 | 30 | 0.8660 | 0.5000 | 86.6 | 50.0 |
| 2 | 5 | 20 | 120 | −0.5000 | 0.8660 | −50.0 | 86.6 |
| 3 | 15 | 15 | 200 | −0.9397 | −0.3420 | −211.4 | −77.0 |
| | | | | | | −174.8 = | 59.6 = |
| | | | | | | $\Sigma Mr \cos \theta$ | $\Sigma Mr \sin \theta$ |

$$M_B = \frac{\sqrt{(\Sigma Mr \cos \theta)^2 + (\Sigma Mr \sin \theta)^2}}{r_B} = \frac{\sqrt{(-174.8)^2 + (59.6)^2}}{10}$$

$$M_B = 18.5 \text{ lb}$$

$$\tan \theta_B = \frac{-(\Sigma Mr \sin \theta)}{-(\Sigma Mr \cos \theta)} = \frac{-(59.6)}{-(-174.8)} = \frac{-y}{+x}$$

$$\theta_B = 341° \, 10'$$

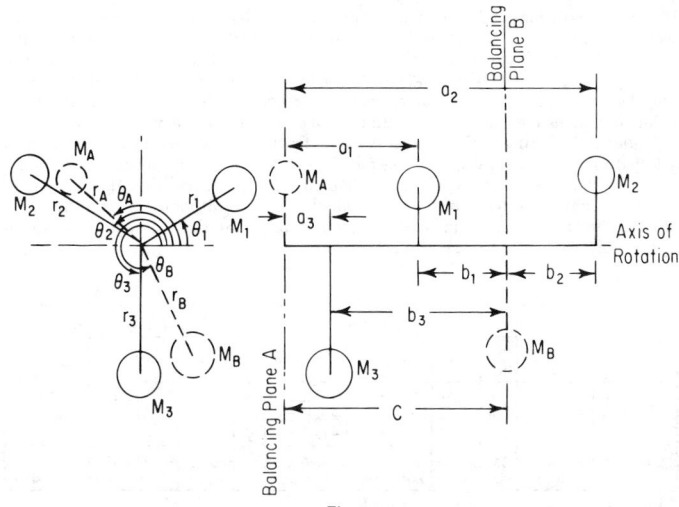

Fig. 2

**Counterbalancing Masses Located in Two or More Planes.**—Unbalanced masses or weights rotating about a common axis in two separate planes of rotation form a couple, which must be counterbalanced by masses or weights, also located in two separate planes, call them planes $A$ and $B$, and rotating about the same common axis (see *Couples*, page 288). In addition, they must be balanced in the direction perpendicular to the axis, as before. Since two counterbalancing masses are required, two separate equations are required to calculate the product of each mass or weight and its radius, and two additional equations are required to calculate the angular positions. The planes $A$ and $B$ selected as balancing planes may be any two planes separated by any convenient distance $c$, along the axis of rotation. In Fig. 2:

For balancing plane $A$:

$$M_A r_A = \frac{\sqrt{(\Sigma Mrb \cos \theta)^2 + (\Sigma Mrb \sin \theta)^2}}{c} \qquad (3)$$

$$\tan \theta_A = \frac{-(\Sigma Mrb \sin \theta)}{-(\Sigma Mrb \cos \theta)} = \frac{y}{x} \qquad (4)$$

For balancing plane $B$:

$$M_B r_B = \frac{\sqrt{(\Sigma Mra \cos \theta)^2 + (\Sigma Mra \sin \theta)^2}}{c} \qquad (5)$$

$$\tan \theta_B = \frac{-(\Sigma Mra \sin \theta)}{-(\Sigma Mra \cos \theta)} = \frac{y}{x} \qquad (6)$$

Where: $M_A$ and $M_B$ are the mass or weight of the counterbalancing masses in the balancing planes $A$ and $B$, respectively; $r_A$ and $r_B$ are the radii and $\theta_A$ and $\theta_B$ are the angular positions of the balancing masses in these planes. $M$, $r$, and $\theta$ are the mass or weight, radius, and angular positions of the unbalanced masses, with the subscripts defining the particular mass to which the values are assigned. The length $c$, the distance between the balancing planes, is always a positive value. The axial dimensions, $a$ and $b$, may be either positive or negative, depending upon their position relative to the balancing plane; e.g., in Fig. 2, the dimension $b_2$ would be negative

*Example*: Referring to Fig. 2, a set of values for the masses and dimensions has been selected and put into convenient table form below. The separation of balancing planes, $c$, is assumed as being 15 inches. If in balancing plane $A$, the radius of the counterbalancing weight is selected to be 10 inches; calculate the magnitude of the counterbalancing mass and its position. If in balancing plane $B$, the counterbalancing mass is selected to be 10 lb; calculate its radius and position.

| Plane | $M$ lb | $r$ in. | $\theta$ deg. | Balancing Plane A | | | | Balancing Plane B | | | |
|---|---|---|---|---|---|---|---|---|---|---|---|
| | | | | $b$ in. | $Mrb$ | $Mrb\cos\theta$ | $Mrb\sin\theta$ | $a$ in. | $Mra$ | $Mra\cos\theta$ | $Mra\sin\theta$ |
| 1 | 10 | 8 | 30 | 6 | 480 | 415.7 | 240.0 | 9 | 720 | 623.5 | 360.0 |
| 2 | 8 | 10 | 135 | −6 | −480 | 339.4 | − 339.4 | 21 | 1680 | −1187.9 | 1187.9 |
| 3 | 12 | 9 | 270 | 12 | 1296 | 0.0 | −1296.0 | 3 | 324 | 0.0 | − 324.0 |
| A | ? | 10 | ? | 15* | ... | 755.1 = $\Sigma Mrb\cos\theta$ | −1395.4 = $\Sigma Mrb\sin\theta$ | 0 | ... | − 564.4 = $\Sigma Mra\cos\theta$ | 1223.9 = $\Sigma Mra\sin\theta$ |
| B | 10 | ? | ? | 0 | ... | | | 15* | ... | | |

*15 inches = distance $c$ between planes $A$ and $B$.

For balancing plane $A$:

$$M_A = \frac{\sqrt{(\Sigma Mrb\cos\theta)^2 + (\Sigma Mrb\sin\theta)^2}}{r_A c} = \frac{\sqrt{(755.1)^2 + (-1395.4)^2}}{10(15)}$$

$$M_A = 10.6 \text{ lb}$$

$$\tan\theta_A = \frac{-(\Sigma Mrb\sin\theta)}{-(\Sigma Mrb\cos\theta)} = \frac{-(-1395.4)}{-(755.1)} = \frac{+y}{-x}$$

$$\theta_A = 118° 25'$$

For balancing plane $B$:

$$r_B = \frac{\sqrt{(\Sigma Mra\cos\theta)^2 + (\Sigma Mra\sin\theta)^2}}{M_B c} = \frac{\sqrt{(-564.4)^2 + (1223.9)^2}}{10(15)}$$

$$= 8.985 \text{ in.}$$

$$\tan\theta_B = \frac{-(\Sigma Mra\sin\theta)}{-(\Sigma Mra\cos\theta)} = \frac{-(1223.9)}{-(-564.4)} = \frac{-y}{+x}$$

$$\theta_B = 294° 45'$$

**Balancing Lathe Fixtures.** — Lathe fixtures rotating at a high speed require balancing. Often it is assumed that the center of gravity of the workpiece and fixture, and of the counterbalancing masses are in the same plane; however, this is not usually the case. Counterbalancing masses are required in two separate planes to prevent excessive vibration or bearing loads at high speeds. Usually a single counterbalancing

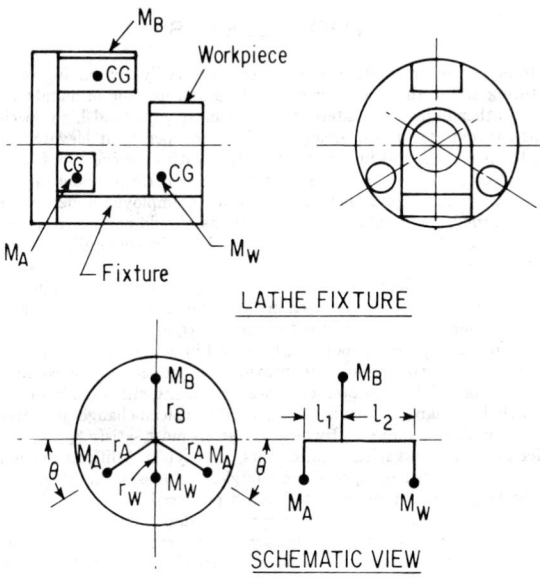

LATHE FIXTURE

SCHEMATIC VIEW

Fig. 3

mass is placed in one plane selected to be 180 degrees directly opposite the combined center of gravity of the workpiece and the fixture. Two equal counterbalancing masses are then placed in the second counterbalancing plane, equally spaced on each side of the fixture. Referring to Fig. 3, the two counterbalancing masses $M_A$ and the two angles $\theta$ are equal. For the design in this illustration, the following formulas can be used to calculate the magnitude of the counterbalancing masses. Since their angular positions are fixed by the design, they are not calculated.

$$M_B = \frac{M_w r_w (l_1 + l_2)}{r_B l_1} \tag{7}$$

$$M_A = \frac{M_B r_B - M_w r_w}{2 r_A \sin \theta} \tag{8}$$

In these formulas $M_w$ and $r_w$ denote the mass or weight and the radius of the combined center of gravity of the workpiece and the fixture.

*Example:* In Fig. 3 the combined weight of the workpiece and the fixture is 18.5 lb. The following dimensions were determined from the layout of the fixture and by calculating the centers of gravity: $r_w = 2$ in.; $r_A = 6.25$ in.; $r_B = 6$ in.; $l_1 = 3$ in.; $l_2 = 5$ in.; and, $\theta = 30°$. Calculate the weights of the counterbalancing masses.

$$M_B = \frac{M_w r_w (l_1 + l_2)}{r_B l_1} = \frac{18.5 \times 2 \times 8}{6 \times 3} = 16.44 \text{ lb}$$

$$M_A = \frac{M_B r_B - M_w r_w}{2 r_A \sin \theta} = \frac{(16.25 \times 6) - (18.5 \times 2)}{(2 \times 6.25) \sin 30} = 9.86 \text{ lb (each weight).}$$

# TOOL STEELS

Tool steels, as the designation implies, serve primarily for making tools used in manufacturing and in the trades for the working and forming of metals, also wood, plastics and other industrial materials. Tools must withstand high specific loads, often concentrated at exposed areas, may have to operate at elevated or rapidly changing temperatures and in continual contact with abrasive types of work materials, are often subjected to shocks or may have to perform under other varieties of adverse conditions. Nevertheless, the tool when employed under circumstances which are regarded as normal operating conditions, should not suffer major damage, untimely wear resulting in the dulling of the edges, or be susceptible to detrimental metallurgical changes.

Tools for less demanding uses, such as ordinary handtools, including hammers, chisels, files, mining bits, etc., are often made of standard AISI steels which are not considered as belonging to any of the tool steel categories.

The steel for most types of tools must be used in a heat treated state, generally hardened and tempered, in order to provide the properties needed for the particular application. The adaptability to heat treatment with a minimum of harmful effects, which dependably results in the intended beneficial changes in material properties, is still another requirement which tool steels must satisfy.

In order to meet such varied requirements, steel types of different chemical composition, often produced by special metallurgical processes, have been developed. Due to the large number of tool steel types produced by the steel mills, which generally are made available with proprietary designations, it became rather difficult for the user to select those types which are most suitable for any specific application, unless the recommendations of a particular steel producer or producers are obtained.

Substantial clarification has resulted from the development of a classification system which is now widely accepted throughout the industry, on the part of both the producers and the users of tool steels. That system is used in the following as a base for providing concise information on tool steel types, their properties and methods of tool steel selection.

The tool steel classification system establishes seven basic categories of tool and die steels. These categories are associated with the predominant applicational characteristics of the tool steel types they comprise. A few of these categories are composed of several groups to distinguish between families of steel types which, while serving the same general purpose, differ in regard to one or more dominant characteristics.

In order to offer an easily applicable guide for the selection of tool steel types best suited for a particular application, the subsequent discussions and tables are based on the above mentioned application-related categories. As an introduction to the detailed surveys, a concise discussion is presented of the principal tool steel characteristics which govern the suitability for varying service purposes and operational conditions. A brief review of the major steel alloying elements and of the effect of these constituents on the significant characteristics of tool steels is also given in the sections which follow.

**The Properties of Tool Steels.** — Tool steels must possess certain properties to a higher than ordinary degree, in order to make them adaptable for uses which require the ability of sustaining heavy loads and performing dependably even under adverse conditions.

The extent and the type of loads, the characteristics of the operating conditions and the expected performance with regard to both the duration and the level of consistency, are the principal considerations, in combination with the aspects of cost, which govern the selection of tool steels for specific applications.

While it is not possible to define and to apply exact parameters for measuring the significant tool steel characteristics, certain properties can be determined which may greatly assist in appraising the suitability of various types of tool steels for specific uses.

Because tool steels are generally heat treated to make them adaptable to the intended use by enhancing the desirable properties, *the behavior of the steel during heat treatment* is of prime importance. The behavior of the steel comprises, in this respect, both the resistance to harmful effects and the attainment of the desirable properties. The following are considered the major properties related to heat treatment:

*Safety in Hardening:* This designation expresses the ability of the steel to withstand the harmful effects of exposure to very high heat and particularly to the sudden temperature changes during quenching without harmful effects. One way of obtaining this property is by means of alloying elements which reduce the critical speed at which the quenching must be carried out, thus permitting the use of less harsh acting quenching media such as oil, salt or just still air.

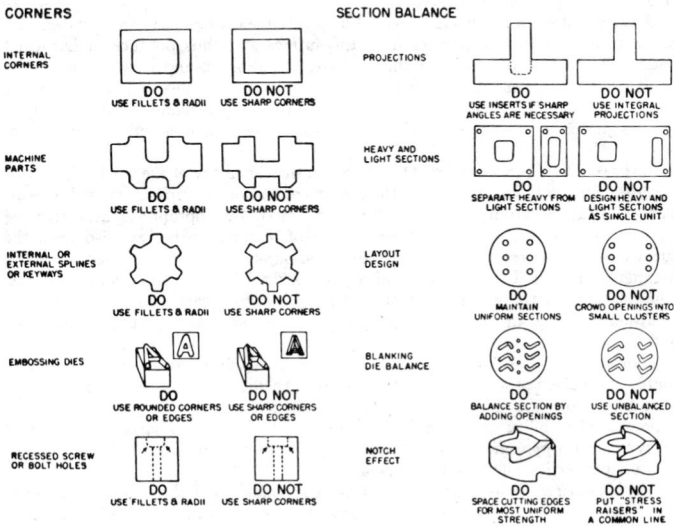

Tool and die design tips to reduce breakage in heat treatment.
*Courtesy of Society of Automotive Engineers, Inc.*

The most common harm which parts made of tool steel can suffer during heat treat is the development of cracks. In addition to the composition of the steel and the applied heat treating process, the configuration of the part can also affect the sensitivity to cracking. The accompanying figure illustrates a few design characteristics related to cracking and warpage in heat treatment; the observation of these design tips, which call for generous filleting, avoidance of sharp angles and major changes without transition in the cross section, is particularly advisable when using tool steel types with a low index value for safety in hardening.

In current practice this property of tool steels is rated in the order of decreasing safety (that is increasing sensitivity) as Highest, Very High, High, Medium and Low safety, expressed in the Tables 4–9 by the letters A, B, C, D, and E.

*Distortions in Heat Treating:* In parts made of tool steels such distortions are often a consequence of inadequate design (see previous illustration) or improper heat treatment (e.g. lack of stress relieving). However, certain types of tool steels display different degrees of sensitivity to distortion. Those which are less stable require safer design of the parts for which they are used, more careful heat treatment including the proper support of long and slender parts, or thin sections, and possibly greater grinding allowance for permitting subsequent correction of the distorted shape. In some cases parts made of a type of steel generally sensitive to distortions can be heat treated with very little damage when the requirements of the part call for a relatively shallow hardened layer over a soft core. However, for intricate shapes and large tools such steel types should be selected which possess superior non-deforming properties. The ratings used in Tables 4–9 express the non-deforming properties (stability of shape in heat treatment) of the steel types and start with the lowest distortion (the best stability) designated as A; the greatest susceptibility to distortion being designated as E.

*Depth of Hardening:* This is indicated by a relative rating based on how deep the phase transformation penetrates from the surface and thus produces a hardened layer. Because of the effect of the heat treating process, and particularly of the applied quenching medium on the depth of hardness, reference is made in Tables 4–9 to the quench which results in the listed relative hardenability values. These are designated by letters A, B and C, expressing deep, medium and shallow depth, respectively.

*Resistance to Decarburization:* Depending on the chemistry of the steel, a higher or lower sensitivity to losing a part of the carbon content of the surface exposed to heat, exists. That sensitivity can partially be balanced by appropriate heat treating equipment and processes. Also the amount of material to be removed from the surface after heat treatment, usually by grinding, should be determined in such a manner as to avoid the retention of a decarburized layer on functional surfaces. The relative resistance of individual tool steel types to decarburization during heat treatment is rated in Tables 4–9 from High to Low, expressed by the letters A, B, and C.

Tool steels must be workable with generally available means, without requiring highly specialized processes. The tools made of these steels must, of course, perform adequately, often under adverse environmental and burdensome operational conditions. The ability of the individual types of tool steels to satisfy, to different degrees, such *applicational requirements* can also be appraised on the basis of significant properties, such as the following:

*Machinability:* Tools are precision products, whose final shape and dimensions must be produced by machining, a process to which not all tool steel types lend themselves equally well. That difference in machinability is particularly evident in tool steels which, depending on their chemical composition, may contain substantial amounts of metallic carbides, beneficial to increased wear resistance, yet detrimental to the service life of tools with which the steel has to be worked. The microstructure of the steel type can also affect the ease of machining and, in some types, certain phase conditions such as those due to low carbon content, may cause difficulties in achieving a fine surface finish. Certain types of tool steels have their machinability improved by the addition of slight amounts of sulfur or lead. In the selection of tool steels machinability may be considered as a factor in the cost of making the tool, and also for reasons of feasibility, particularly in the case of intricate tool shapes. The ratings in Tables 4–9, starting with A for the greatest ease of machining, to E for the lowest machinability, refer to working of the steel in an unhardened condition. That characteristic is not necessarily identical with the grindability which expresses how well the steel is adapted to grinding after heat treating. The ease of grinding, however, may become an important consideration in tool steel selection, particularly in the case of tools like cutting tools and dies, which require regular sharpening, involving extensive grinding. AVCO Bay State Abrasives Company has compiled information on the relative grindability of frequently used types of tool steels. A simplified version of that information is presented in Table 1 which assigns the listed tool steel types to one of the following grindability grades: High (A), Medium (B), Low (C) and Very Low (D), expressing decreasinging ratios of volume of metal removed to wheel wear.

*Hot Hardness:* This property designates the steel's resistance to the softening effect of elevated temperature. This characteristic is related to the tempering temperature of the type of steel which is controlled by various alloying elements, such as tungsten, molybdenum, vanadium, cobalt and chromium.

**Table 1. Relative Grindability of Selected Types of Frequently Used Tool Steels**

| AISI Tool Steel Type | H41 | H42 | H43 | Other H | D2 | D3 | D5 | D7 | A Types | O Types | L Types | F Types |
|---|---|---|---|---|---|---|---|---|---|---|---|---|
| Relative Grindability Index | B | B | B | A | B | B | B | C | A | A | A | B |

| AISI High Speed Tool Steel Type | M1 | M2 | M3 (1) | M3 (2) | M4 | M7 | M8 | M10 | M15 | M36 | M43 | T1 | T2 | T3 | T5 | T6 | T15 |
|---|---|---|---|---|---|---|---|---|---|---|---|---|---|---|---|---|---|
| Relative Grindability Index | A | B | C | C | D | B | A | B | D | B | B | A | B | C | B | B | D |

Hot hardness is a necessary property of tools used for hot work, like forging, casting and hot extrusion. Hot hardness is also important in cutting tools operated at high speed which generates sufficient heat to raise their temperature well above the level where ordinary steels lose their hardness; hence the designation *high speed steels* which refers to a family of tool steels developed for such uses. Frequently it is the degree of the tool steel's resistance to softening at elevated temperature which governs important process data, such as the applicable cutting speed. In the ratings of Tables 4–9, tool steel types having the highest hot hardness are marked with A, subsequent letters expressing gradually decreasing capacity to endure elevated temperature without losing hardness.

*Wear Resistance:* The gradual erosion of the tool's operating surface, most conspicuously occurring at the exposed edges, is known as wear. Resistance to wear prolongs the useful life of the tool, by delaying the degradation of its surface through abrasive contact with the work at regular operating temperatures; these vary according to the type of the process. Wear resistance is observable experimentally and measurable by comparison. Certain types of metallic carbides embedded into the steel matrix are considered to be the prime contributing factors to wear resistance, besides the hardness of the heat treated steel material. The ratings of Tables 4–9, starting with A for the best to E for poor, are based on conditions considered as normal in operations for which various types of tool materials are primarily used.

*Toughness:* In tool steels this property expresses their ability to sustain shocks, suddenly applied and relieved loads, or major impacts without breaking. Steels used for making tools must also be able to absorb such forces with only a minimum of elastic deformation and without permanent deformation to any extent which would interfere with the proper functioning of the tool. Certain types of tool steels, particularly those with high carbon content and without the presence of beneficial alloying constituents, tend to be the most sensitive to shocks, although they can also be made to act tougher when used for tools which permit a hardened case to be supported by a soft core. Tempering improves toughness, while generally reducing hardness. The rating indexes in Tables 4–9, A for the highest toughness, through E for the types most sensitive to shocks, apply to tools heat treated to hardness values normally used for the particular type of tool steel.

**Common Tool Faults and Failures.** — While the proper selection of the steel grade used for any particular type of tool is of great importance, it should be recognized that many of the failures experienced in common practice originate from other causes than those related to the tool material.

To permit a better appraisal of the actual causes of failure and possible corrective action, a general, although not complete, list of common tool faults, resulting failures and corrective actions is shown in Table 2. In this list the potential failure causes are grouped into four categories. The possibility of more than a single cause being responsible for the experienced failure should not be excluded.

Finally, it must be remembered that the proper usage of tools is indispensable for obtaining satisfactory performance and tool life. Using the tools properly involves, for example, the avoidance of damage to the tool, overload, excessive speeds and feeds; the application of adequate coolant when called for; assurance of a rigid setup; proper alignment; and firm tool and work holding.

**The Effect of Alloying Elements on Tool Steel Properties.** — *Carbon (C):* The presence of carbon, usually in excess of 0.60 per cent for non-alloyed types, is essential for assuring the hardenability of steels to the levels needed for tools. Raising the carbon content by different amounts up to a maximum of about 1.3 per cent increases the hardness slightly and the wear resistance considerably. The amount of carbon in tool steels is specified for attaining certain properties (such as in the water-hardening category where higher carbon content may be chosen to improve wear resistance, although to the detriment of toughness) or, in the alloyed types of tool steels, in conformance with the other constituents for producing well balanced metallurgical and performance properties.

*Manganese (Mn):* In smaller amounts, to about 0.60 per cent, manganese is added for reducing brittleness and to improve forgeability. Larger amounts of manganese improve hardenability, permitting oil quenching for non-alloyed carbon steels, thus reducing deformation, although with regard to several other properties, manganese is not an equivalent replacement for the regular alloying elements.

*Silicon (Si):* In itself silicon may not be considered an alloying element of tool steels, but it is needed as a deoxidizer and improves the hot-forming properties of the steel. In combination with certain alloying elements the silicon content is sometimes raised up to about 2 per cent for increasing the strength and toughness of steels used for tools which have to sustain shock loads.

*Tungsten (W):* This is one of the important alloying elements of tool steels, particularly because of two valuable properties: it improves "hot hardness", that is, the resistance of the steel to the softening effect of elevated temperature, and it also forms hard, abrasion resistant carbides, thus improving the wear properties of tool steels.

*Vanadium (V):* This element contributes to the refinement of the carbide structure and thus improves the forgeability of alloy tool steels. Vanadium has a very strong tendency to form a hard carbide, which improves both the hardness and the wear properties of tool steels, however, a large amount of vanadium carbide makes the grinding of the tool very difficult (causes low grindability).

*Molybdenum (Mo):* In small amounts molybdenum improves certain metallurgical properties of alloy steels, such as deep hardening and toughness. It is used often in

Table 2. Common Tool Faults, Failures and Cures — 1

| Fault Description | Probable Failure | Possible Cure |
|---|---|---|
| IMPROPER TOOL DESIGN | | |
| Drastic section changes — widely different thicknesses of adjacent wall sections or protruding elements | In liquid quenching the thin section will cool and then harden more rapidly than the adjacent thicker section, setting up stresses which may exceed the strength of the steel. | Make such parts of two pieces or use an air hardening tool steel which avoids the harsh action of a liquid quench. |
| Sharp corners on shoulders or in square holes | Cracking can occur, particularly in liquid quenching, due to stress concentrations. | Apply fillets to the corners and/or use an air hardening tool steel. |
| Sharp cornered keyways | Failure may arise during service, usually considered as caused by fatigue. | The use of round keyways should be preferred when the general configuration of the part makes it prone to failure due to square keyways. |
| Abrupt section changes in battering tools | Pneumatic tools, due to impact in service, are particularly sensitive to stress concentrations which lead to fatigue failures. | Use taper transitions, which are better than even generous fillets. |
| Functional inadequacy of tool design — e.g., insufficient guidance for a punch | Excessive wear or breakage in service may occur. | Assure solid support, avoid unnecessary play, adapt travel length to operational conditions (e.g. punch to penetrate to four-fifths of thickness in hard work material.) |
| Improper tool clearance, such as in blanking and punching tools | Deformed and burred parts may be produced, excessive tool wear or breakage can result. | Adapt clearances to material conditions and dimensions for reducing tool load and also for obtaining clean sheared surfaces. |
| FAULTY CONDITION OR INADEQUATE GRADE OF TOOL STEEL | | |
| Improper tool steel grade selection | Typical failures: Chipping — insufficient toughness. Wear — poor abrasion resistance. Softening—inadequate "red hardness". | Choose the tool steel grade by following recommendations and improve selection when needed, guided by property ratings. |
| Material defects — voids, streaks, tears, flakes, surface cooling cracks, etc | When not recognized during material inspection, tools made of defective steel often prove to be useless. | Obtain tool steels from reliable sources and inspect tool material for detectable defects. |
| Decarburized surface layer ("bark") in rolled tool steel bars | Cracking may originate from the decarburized layer, or it will not harden ("soft skin"). | Provide allowance for stock to be removed from all surfaces of hot rolled tool steel. Recommended amounts are listed in tool steel catalogs and vary according to section size, generally about 10 percent for smaller and 5 percent for larger diameters. |
| Brittleness caused by poor carbide distribution in high alloy tool steels | Excessive brittleness can cause chipping or breakage during the service of the tool. | Bars with large diameter (above about four inches) tend to be prone to nonuniform carbide distribution. Choose upset forged discs instead of large diameter bars. |

**Table 2.   Common Tool Faults, Failures and Cures — 2**

| Fault Description | Probable Failure | Possible Cure |
|---|---|---|
| **FAULTY CONDITION OR INADEQUATE GRADE OF TOOL STEEL** | | |
| Unfavorable grain flow | Improper grain flow of the steel used for milling cutters and similar tools can cause the breaking out of teeth. | Upset forged discs made with an upset ratio of about 2 to 1 (starting to upset thickness) display radial grain flow. Highly stressed tools, e.g., gear shaper cutters, may require the cross forging of blanks. |
| **HEAT TREATMENT FAULTS** | | |
| Improper preparation for heat treatment. Certain tools may require stress relieving or annealing, and often pre-heating too | Tools highly stressed during machining or forming, unless stress relieved, may aggravate the thermal stresses of heat treatment, thus causing cracks. Excessive temperature gradients developed in non-preheated tools with different section thicknesses can cause warpage. | Stress relieve, when needed before hardening. Anneal prior to heavy machining or cold forming (e.g., hobbing). Pre-heat tools (a) having substantial section thickness variations, or (b) requiring high quenching temperatures, as those made of high speed tool steels. |
| Overheating during hardening; quenching from too high a temperature | Causes grain coarsening and a sensitivity to cracking which is more pronounced in tools with drastic section changes. | Overheated tools have a characteristic microstructure which aids recognition of the cause of failure and indicates the need for improved temperature control. |
| Low hardening temperature | The tool may not harden at all, or in its outer portion only, thereby setting up stresses which can lead to cracks. | Controlling both the temperature of the furnace and the time of holding the tool on quenching temperature, will prevent this not too frequent deficiency. |
| Inadequate composition or condition of the quenching media | Water hardening tool steels are particularly sensitive to inadequate quenching media, which can cause soft spots or even violent cracking. | For water hardening tool steels use water free of dissolved air and contaminants, also assure sufficient quantity and proper agitation of the quench. |
| Improper handling during and after quenching | Cracking of the tools, particularly of those with sharp corners, during the heat treatment can result from holding the part too long in the quench or incorrectly applied tempering. | Following the steel producer's specifications is a safe way for assuring proper heat treat handling. In general, the tool should be left in the quench until it reaches a temperature of 150 to 200°F, and is then transferred promptly into a warm tempering furnace. |
| Insufficient tempering | The omission of double tempering for steel types which require it may cause early failure by heat checking in hot work or make the tool abnormally sensitive to grinding checks. | Apply double tempering for highly alloyed tool steel of the high speed, hot work and high chromium categories, for removing stresses caused by martensite formed during the first tempering phase. Second temper also increases hardness of most high speed steels. |

Table 2.   Common Tool Faults, Failures and Cures — 3

| Fault Description | Probable Failure | Possible Cure |
|---|---|---|
| **HEAT TREATMENT FAULTS** | | |
| Decarburization and carburization | Unless hardened in a nuetral atmosphere the original carbon content of the tool surface may be changed: Reduced (decarburization) causing a soft layer which wears rapidly. Increased (carburization) which, when excessive, may cause brittleness. | Heating in neutral atmosphere or well maintained salt bath and controlling the furnace temperature and the time during which the tool is subjected to heating can usually keep the carbon unbalance within acceptable limits. |
| **GRINDING DAMAGES** | | |
| Excessive stock removal rate causing the heating of the part surface beyond the applied tempering temperature | Scorched tool surface displaying temper colors varying from yellow to purple, depending on the degree of heat. This causes softening of the ground surface. When coolant is used, a local rehardening can take place, often resulting in cracks. | Prevention: by reducing speed and feed, or using coarser, softer, more open structured grinding wheel, with ample coolant. Correction: by subsequent light stock removal eliminate the discolored layer. Not always a cure, because the effects of abusive grinding may not be corrected. |
| Improper grinding wheel specifications; grain too fine or bond too hard | Intense localized heating during the grinding may set up surface stresses causing grinding cracks. These are either parallel but at right angles to the direction of grinding or, when more advanced, form a network. May need cold etch or magnetic particle testing to become recognizable. | Prevention: by correcting the grinding wheel specifications. Correction: in the case of shallow (.002 to .004 inch) cracks, by removing the damaged layer, when permitted by the design of the tool, using very light grinding passes. |
| Incorrectly dressed or loaded grinding wheel | Heating of the work surface can cause scorching or cracking. Incorrect dressing can also cause a poor finish of the ground work surface. | Dress wheel with sharper diamond and faster diamond advance to produce coarser wheel surface. Alternate dressing methods, like crush-dressing, can improve wheel surface conditions. Dress wheel regularly to avoid loading or glazing of the wheel surface. |
| Inadequate coolant, with regard to composition, amount, distribution and cleanliness | Introducing into the tool surface heat which is not adequately dissipated or absorbed by the coolant, can cause softening, or even the development of cracks. | Improve coolant supply and quality, or reduce stock removal rate in order to generate less heat in the grinding. |
| Damage caused by abusive abrasive cutoff | The intensive heat developed during this process can cause a hardening of the steel surface, or may even result in cracks. | Reduce rate of advance, adopt wheel specifications better suited for the job. Use ample coolant or, when harmful effect not eliminated, replace abrasive cutoff by some cooler acting stock separation method (e.g., sawing or lathe cutoff) unless damaged surface is being removed by subsequent machining. |

*Note:* Illustrated examples of tool failures from causes such as listed above may be found in "The Tool Steel Trouble Shooter" handbook, published by Bethlehem Steel Corporation.

larger amounts in certain high speed tool steels to replace tungsten, primarily for economic reasons, often with nearly equivalent results.

*Cobalt (Co):* As an alloying element of tool steels, cobalt increases hot hardness and is used in applications where that property is needed. Substantial addition of cobalt, however, raises the critical quenching temperature of the steel with a tendency to increase the decarburization of the surface, and also reduces toughness.

*Chromium (Cr):* This element is added in amounts of several per cent to high alloy tool steels, and up to 12 per cent to types in which chromium is the major alloying element. It improves hardenability and, together with high carbon, provides both wear resistance and toughness, a combination valuable in certain tool applications. However, high chromium raises the hardening temperature of the tool steel, and thus can cause proneness to hardening deformations. A high percentage of chromium also affects the grindability of the tool steel.

*Nickel (Ni):* Generally in combination with other alloying elements, particularly chromium, nickel is used to improve the toughness and, to some extent, the wear resistance of tool steels.

The addition of more than one alloy to a steel often produces what is called a synergistic effect. Thus, the combined effects of two or more alloy elements may be greater than the sum of the individual effects of each alloy.

**Classification of Tool Steels.** — Steels for tools must satisfy a number of different, often conflicting requirements. The need for specific steel properties arising from widely varying applications has led to the development of many compositions of tool steels, each intended to meet a particular combination of applicational requirements. The resultant diversity of tool steels, their number being continually expanded by the addition of new developments, made it extremely difficult for the user to select the type best suited to his needs, or to find equivalent alternatives for specific types available from particular sources.

As a cooperative industrial effort under the sponsorship of AISI and SAE, a tool classification system has been developed in which the commonly used tool steels are grouped into seven major categories. These categories, several of which contain more than a single group, are listed in the following with the letter symbols used for

| Category Designation | Letter Symbol | Group Designation |
|---|---|---|
| High Speed Tool Steels | M | Molybdenum types |
| | T | Tungsten types |
| Hot Work Tool Steels | H1–H19 | Chromium types |
| | H20–H39 | Tungsten types |
| | H40–H59 | Molybdenum types |
| Cold Work Tool Steels | D | High carbon, high chromium types |
| | A | Medium alloy, air hardening types |
| | O | Oil hardening types |
| Shock Resisting Tool Steels | S | . . . . . |
| Mold Steels | P | . . . . . |
| Special Purpose Tool Steels | L | Low alloy types |
| | F | Carbon tungsten types |
| Water Hardening Tool Steels | W | . . . . . |

Table 3. Quick Reference Guide for Tool Steel Selection — 1

| Application Areas | High Speed Tool Steels, M and T | Hot Work Tool Steels H | Cold Work Tool Steels, D, A and O | Shock Resisting Tool Steels, S | Mold Steels, P | Special Purpose Tool Steels, L and F | Water Hardening Tool Steels, W |
|---|---|---|---|---|---|---|---|
| | | | | Tool Steel Categories and AISI Letter Symbols | | | |
| | | | Examples of Typical Application | | | | |
| Cutting Tools<br>Single point types (lathe, planer, boring)<br>Milling cutters<br>Drills<br>Reamers<br>Taps<br>Threading dies<br>Form cutters | General purpose production tools: M2, T1<br>For increased abrasion resistance: M3, M4, and M10<br>Heavy-duty work calling for high hot hardness: T5, T15<br>Heavy-duty work calling for high abrasion resistance: M42, M44 | | Tools with keen edges (knives, razors)<br>Tools for operations where no high speed is involved, yet stability in heat treatment and substantial abrasion resistance are needed | Pipe cutter wheels | | | Uses which do not require hot hardness or high abrasion resistance.<br>Examples with carbon content of applicable group:<br>Taps (1.05/1.10%C)<br>Reamers (1.10/1.15%C)<br>Twist drills (1.20/1.25%C)<br>Files (1.35/1.40%C) |
| Hot Forging Tools and Dies<br>Dies and inserts<br>Forging machine plungers and pierces | For combining hot hardness with high abrasion resistance: M2, T1 | Dies for presses and hammers: H20, H21<br>For severe conditions over extended service periods: H22 to H26, also H43 | Hot trimming dies: D2 | Hot trimming dies<br>Blacksmith tools<br>Hot swaging dies | | | Smith's tools (.65/.70%C)<br>Hot chisels (.70/.75%C)<br>Drop forging dies (.90/1.00%C)<br>Applications limited to short-run production |
| Hot Extrusion Tools and Dies<br>Extrusion dies and mandrels,<br>Dummy blocks<br>Valve extrusion tools | Brass extrusion dies: T1 | Extrusion dies and dummy blocks: H20 to H26<br>For tools which are exposed to less heat: H10 to H19 | | Compression molding: S1 | | | |

Table 3. Quick Reference Guide for Tool Steel Selection — 2

| Application Areas | High Speed Tool Steels, M and T | Hot Work Tool Steels, H | Cold Work Tool Steels, D, A and O | Shock Resisting Tool Steels, S | Mold Steels, P | Special Purpose Tool Steels, L and F | Water Hardening Tool Steels, W |
|---|---|---|---|---|---|---|---|
| | | | Examples of Typical Application | | | | |
| Cold Forming Dies | Burnishing tools: M1, T1 | Cold heading die casings: H13 | Drawing dies: or Coining tools: O1, D2 | Hobbing and short-run applications: S1, S7 | | Blanking, forming, and trimmer dies when toughness has precedence over abrasion resistance: L6 | Cold heading dies: W1 or W2 (C ≈ 1.00%) Bending dies: W1 (C ≈ 1.00%) |
| Bending, forming, drawing, and deep drawing dies and punches | | | Forming and bending dies: A2 Thread rolling dies: D2 | Rivet sets and rivet busters | | | |
| Shearing Tools | Special dies for cold and hot work: T1 For work requiring high abrasion resistance: M2, M3 | For shearing knives: H11, H12 For severe hot shearing applications: M21, M25 | Dies for medium runs: A2, A6 also O1 and O4 | Cold and hot shear blades | | Knives for work requiring high toughness: L6 | Trimming dies (.90/.95%C) Cold blanking and punching dies (1.00%C) |
| Dies for piercing, punching, and trimming | | | Dies for long runs: D2, D3 Trimming dies (also for hot trimming): A2 | Hot punching and piercing tools | | | |
| Shear blades | | For aluminum and lead: H11 and H13 For brass: H21 | A2 and A6 O1 | Boilermaker's tools | | | |
| Die Casting and Molding Dies | | | | | Plastic molds: P2 to P4, and P20 | | |
| Structural Parts for Severe Service Conditions | Roller bearings for high temperature environment: T1 Lathe centers: M2 and T1 | For aircraft components (landing gears, arrester hooks, rocket cases): H11 | Lathe centers: D2, D3 Arbors: O1 Bushings: A4 Gages: D2 | Pawls Clutch parts | | Spindles, clutch parts (where high toughness is needed): L6 | Spring steel (1.10/1.15%C) |
| Battering Tools for Hand and Power Tool Use | | | | Pneumatic chisels for cold work: S5 For higher performance: S7 | | | For intermittent use: W1 (.80%C) |

their identification. The individual types of tool steels within each category are identified by suffix numbers following the letter symbols.

The subsequent detailed discussion of tool steels will be in agreement with these categories, showing for each type the percentages of the major alloying elements. However, these values are for identification only; tool steels of different producers may, in the mean analysis of the individual types, deviate from the listed percentages.

**The Selection of Tool Steels for Particular Applications.** While the advice of the specialized steel producer is often sought as a reliable source of information, in many cases the engineer is faced with the task of making his own tool steel selection. It must be realized that frequently the designation of the tool or of the process will not define the particular tool steel type which is best suited for the job. For that reason tool steel selection tables naming a single type for each listed applicational category cannot take into consideration such, often conflicting work factors, as ease of tool fabrication and maintenance (resharpening), productivity, product quality and tooling cost.

When data related to past experience with tool steels for identical or similar applications are not available, a tool steel selection procedure may be followed which is based on information supplied in this Handbook section and comprises these steps:

1. For identifying the AISI category which contains the sought type of steel the Quick Reference Table on pages 2071 and 2072 should be consulted.
2. Within the defined category
   (a) find from the listed applications of the most frequently used types of tool steels that particular type which corresponds to the job on hand; or
   (b) evaluate from the table of property ratings the best compromise between any conflicting properties (e.g., compromising on wear resistance for obtaining better toughness).

For those willing to refine even further the first choice or to improve on it in the case of not entirely satisfactory experience in one or more meaningful respects, the identifying analyses of the different types of tool steels within each general category may provide additional guidance. In this procedure the general discussion of the effects of different alloying elements on the properties of tools steels, in a previous section, will probably be found useful.

The following two examples illustrate the procedure for refining an original choice with the purpose of adopting a tool steel grade best suited to a particular set of conditions:

*Example 1. Workpiece — Trimming Dies:* For the manufacture of a type of trimming die the first choice was grade A2, because for the planned medium rate of production the lower material cost was considered an advantage.

A subsequent rise in the production rate indicated the use of a higher alloy tool steel, such as D2, whose increased abrasion resistance would permit longer runs between regrinds.

Then a still further increase in the abrasion resistant properties was sought, which led to the use of D7, the high carbon and high chromium content of which provided an excellent edge retainment, although at the cost of greatly reduced grindability. Finally, it became a matter of economic appraisal, whether the somewhat shorter tool regrind intervals (for D2), or the more expensive tool sharpening (for D7) constituted the lesser burden.

*Example 2. Workpiece — Circular form cutter made of high speed tool steel for use on multiple-spindle automatic turning machines:* The first choice from the Quick Reference Guide may be the classical tungsten base high speed tool steel T1, because of its good performance and ease of heat treatment, or its alternate in the molybdenum high speed tool steel category, the type M2.

In practice, neither of these grades provided a tool which could hold its edge and profile over the economical tool change time, because of the abrasive properties of the work material and the high cutting speed applied in the cycle. An overrating of the problem resulted in reaching for the top of the scale, making the tool from T15, a high alloy high speed tool steel (high vanadium and high cobalt).

While the performance of the tools made of T15 was excellent, the cost of this steel type was rather high, and the grinding of the tool, both for making it and in the regularly needed re-sharpening, proved to be very time consuming and expensive. Therefore an intermediate tool steel type was tried, the M3 which, while providing added abrasion resistance (due to increased carbon and vanadium content) was less expensive and much easier to grind than the T15.

**High Speed Tool Steels.** — Their primary application is for tools used for the working of metals at high cutting speeds. Cutting metal at high speed generates heat, the penetration of the cutting tool edge into the work material requires great hardness and strength, and the continued frictional contact of the tool with both the parent material and the detached chips can only be sustained by an abrasion resistant tool edge. Accordingly, the dominant properties of high speed steel are (a) resistance to the softening effect of elevated temperature, (b) great hardness penetrating to substantial depth from the surface, and (c) excellent abrasion resistance.

High speed tool steels are listed in the AISI specifications in two groups: Molybdenum types and Tungsten types, these designations expressing the dominant alloying element of the respective group.

**Molybdenum Type High Speed Tool Steels.** — In distinction to the traditional tungsten base high speed steels, the tool steels listed in this category are considered to have molybdenum as the principal alloying constituent, this element being also used in the designation of the group. Actually, in several types listed in this category other significant elements like tungsten and cobalt might be present in equal, or even greater amount. The available range of types comprises also high speed tool steels with higher than usual carbon and vanadium content; these alloying elements have been increased to obtain better abrasion resistance although such a change in composition may adversely affect the machinability and the grindability of the steel. The series in whose AISI identification numbers 4 is the first digit were developed for attaining exceptionally high hardness in heat treatment which, for these types, usually comprises triple tempering rather than the double tempering generally applied for high speed tool steels.

**Properties and Applications of Frequently Used Molybdenum Types.** — *AISI M1:* This was developed as a substitute for the classical T1 to save on the alloying element tungsten by replacing most of it with molybdenum. In most uses this steel is an acceptable substitute, although it requires greater care or more advanced equipment for its heat treatment than the tungsten alloyed type it replaces. Selected for cutting tools like drills, taps, milling cutters, reamers, lathe tools used for lighter cuts, and also for shearing dies.

*AISI M2:* Similar to M1, yet with substantial tungsten content replacing a part of the molybdenum. It is one of the general-purpose high speed tool steels, combining the economic advantages of the molybdenum type steels with greater ease of hardening, excellent wear resistance and improved toughness. It is a preferred steel type for the manufacture of general-purpose lathe tools; of most categories of multiple-edge cutting tools, like milling cutters, taps, dies, reamers and also for form tools in lathe operations.

*AISI M3:* A high speed tool steel with increased vanadium content for improved wear resistance, yet still below the level where vanadium would interfere with the

## Table 4. Molybdenum High Speed Steels

Identifying Chemical Composition and Typical Heat Treatment Data

| AISI Type | M1 | M2 | M3 Cl.1 | M3 Cl.2 | M4 | M6 | M7 | M10 | M30 | M33 | M34 | M36 | M41 | M42 | M43 | M44 | M46 | M47 |
|---|---|---|---|---|---|---|---|---|---|---|---|---|---|---|---|---|---|---|
| **Identifying Chemical Elements in Per Cent** | | | | | | | | | | | | | | | | | | |
| C | .80 | .85; 1.00 | 1.05 | 1.20 | 1.30 | .80 | 1.00 | .85; 1.00 | .80 | .90 | .90 | .80 | 1.10 | 1.10 | 1.20 | 1.15 | 1.25 | 1.10 |
| W | 1.50 | 6.00 | 6.00 | 6.00 | 5.50 | 4.00 | 1.75 | ... | 2.00 | 1.50 | 2.00 | 6.00 | 6.75 | 1.50 | 2.75 | 5.25 | 2.00 | 1.50 |
| Mo | 8.00 | 5.00 | 5.00 | 5.00 | 4.50 | 4.00 | 8.75 | 8.00 | 8.00 | 9.50 | 8.00 | 5.00 | 3.75 | 9.50 | 8.00 | 6.25 | 8.25 | 9.50 |
| Cr | 4.00 | 4.00 | 4.00 | 4.00 | 4.00 | 4.00 | 4.00 | 4.00 | 4.00 | 4.00 | 4.00 | 4.00 | 4.25 | 3.75 | 3.75 | 4.25 | 4.00 | 3.75 |
| V | 1.00 | 2.00 | 2.40 | 3.00 | 4.00 | 1.50 | 2.00 | 2.00 | 1.25 | 1.15 | 2.00 | 2.00 | 2.00 | 1.15 | 1.60 | 2.25 | 3.20 | 1.25 |
| Co | ... | ... | ... | ... | ... | 12.00 | ... | ... | 5.00 | 8.00 | 8.00 | 8.00 | 5.00 | 8.00 | 8.25 | 12.00 | 8.25 | 5.00 |
| **Heat Treat Data** | | | | | | | | | | | | | | | | | | |
| Hardening Temperature Range, °F | 2150-2225 | 2175-2225 | 2200-2250 | 2200-2250 | 2200-2250 | 2150-2200 | 2150-2225 | 2150-2225 | 2200-2250 | 2200-2250 | 2200-2250 | 2225-2275 | 2175-2220 | 2175-2210 | 2175-2220 | 2190-2240 | 2175-2225 | 2150-2200 |
| Tempering Temperature Range, °F | 1000-1100 | 1000-1166 | 1000-1100 | 1000-1100 | 1000-1100 | 1000-1100 | 1000-1100 | 1000-1100 | 1000-1100 | 1000-1100 | 1000-1100 | 1000-1100 | 1000-1100 | 950-1100 | 950-1100 | 1000-1166 | 975-1050 | 975-1100 |
| Approx. Tempered Hardness, Rc | 65-66 | 65-66 | 66-61 | 66-61 | 66-61 | 66-61 | 66-61 | 65-66 | 65-66 | 65-66 | 65-66 | 65-60 | 70-65 | 70-65 | 70-65 | 70-62 | 69-67 | 70-65 |

Relative Ratings of Properties (A = greatest to E = least)

| | M1 | M2 | M3 Cl.1 | M3 Cl.2 | M4 | M6 | M7 | M10 | M30 | M33 | M34 | M36 | M41 | M42 | M43 | M44 | M46 | M47 |
|---|---|---|---|---|---|---|---|---|---|---|---|---|---|---|---|---|---|---|
| **Characteristics in Heat Treatment** | | | | | | | | | | | | | | | | | | |
| Safety in Hardening | D | D | D | D | D | D | D | D | D | D | D | D | D | D | D | D | D | D |
| Depth of Hardening | A | A | A | A | A | A | A | A | A | A | A | A | A | A | A | A | A | A |
| Resistance to Decarburization | C | B | B | B | B | C | C | C | C | C | C | C | C | C | C | C | C | C |
| Stability of Shape in Heat Treatment — Quenching Medium Air or Salt | C | C | C | C | C | C | C | C | C | C | C | C | C | C | C | C | C | C |
| Stability of Shape in Heat Treatment — Quenching Medium Oil | D | D | D | D | D | D | D | D | D | D | D | D | D | D | D | D | D | D |
| **Service Properties** | | | | | | | | | | | | | | | | | | |
| Machinability | D | D | D | D/E | D | D | D | D | D | D | D | D | D | D | D | D | D | D |
| Hot Hardness | B | B | B | B | B | A | B | B | A | A | A | A | A | A | A | A | A | A |
| Wear Resistance | B | B | B | B | A | B | B | B | B | B | B | B | B | B | B | A | B | B |
| Toughness | E | E | E | E | E | E | E | E | E | E | E | E | E | E | E | E | E | E |

ease of grinding. Preferred for cutting tools requiring the improved wear resistance, like broaches, form tools, milling cutters, chasers, reamers, etc.

*AISI M7:* The chemical composition of this type is similar to that of M1, except for the higher carbon and vanadium content which raises the cutting efficiency without materially reducing the toughness. Because of its sensitivity to decarburization heat treatment in a salt bath or a controlled atmosphere is advisable. Used for blanking and trimming dies, shear blades, lathe tools and thread rolling dies.

*AISI M10:* While the relatively high vanadium content assures excellent wear and cutting properties, the only slightly increased carbon does not cause brittleness to an extent which is harmful in many applications. Form cutters and single point lathe tools, broaches, planer tools, punches, blanking dies, shear blades, etc., are examples of typical uses.

*AISI M42:* In applications where high hardness both at regular and at elevated temperatures is needed, this type of high speed steel with high cobalt content can provide excellent service. Typical applications are tool bits, form tools, shaving tools, fly cutters, roll turning tools, thread rolling dies, etc. Important uses are found for M42, and also for other types of the "M40" group in the working of "difficult-to-machine" type alloys.

**Tungsten Type High Speed Tool Steels.** — For several decades following their introduction the tungsten base high speed steels were the only types available for cutting operations involving the generation of substantial heat, and are still preferred by users who do not have that kind of advanced heat treating equipment which the efficient hardening of the molybdenum type high speed tool steels requires. Most types of tungsten high speed steel display excellent resistance to decarburization and can be brought to good hardness by means of simple heat treating equipment. However, even in the case of the tungsten type high speed steels, heat treatment by using modern methods and furnaces can appreciably improve the metallurgical qualities of the hardened material and the performance of the cutting tools made of these steels.

**Properties and Applications of Frequently Used Tungsten Types.** — *AISI T1:* Also mentioned as the 18-4-1 type with reference to the nominal percentage of its principal alloying elements (W-Cr-V), it is considered as the classical type of high speed tool steel. The chemical composition of T1 was developed around the turn of this century, and has since changed very little. T1 is still considered as perhaps the best general-purpose high speed tool steel because of the comparative ease of its machining and heat treatment. It combines a high degree of cutting ability with relative toughness. T1 steel is used for all types of multiple-edge cutting tools like drills, reamers, milling cutters, threading taps and dies, light- and medium-duty lathe tools, also for punches, dies, machine knives, as well as structural parts which are subjected to elevated temperatures, like lathe centers, certain types of antifriction bearings, etc.

*AISI T2:* Similar to T1 except for somewhat higher carbon content and twice the vanadium contained in the former grade. Its handling ease, both in machining and heat treating, is comparable to that of T1, although it should be held at the quenching temperature slightly longer, particularly when the heating is carried out in a controlled atmosphere furnace. The applications are similar to that of T1, however, because of its increased wear resistance T2 is preferred for tools required for finer cuts, and where the form or size retention of the tool is particularly important, such as for form and finishing tools.

*AISI T5:* The essential characteristic of this type of high speed steel, its superior red hardness, stems from its substantial cobalt content which, combined with the

## Table 5. Tungsten High Speed Tool Steels

Identifying Chemical Composition Heat and Treatment Data

| | | T1 | T2 | T4 | T5 | T6 | T8 | T15 |
|---|---|---|---|---|---|---|---|---|
| **Identifying Chemical Elements in Per Cent** | C | .75 | .80 | .75 | .80 | .80 | .75 | 1.50 |
| | W | 18.00 | 18.00 | 18.00 | 18.00 | 20.00 | 14.00 | 12.00 |
| | Cr | 4.00 | 4.00 | 4.00 | 4.00 | 4.50 | 4.00 | 4.00 |
| | V | 1.00 | 2.00 | 1.00 | 2.00 | 1.50 | 2.00 | 5.00 |
| | Co | ... | ... | 5.00 | ... | ... | 5.00 | 5.00 |
| **Heat Treat Data** | Hardening Temperature Range, °F | 2300-2375 | 2300-2375 | 2300-2375 | 2325-2375 | 2325-2375 | 2300-2375 | 2200-2300 |
| | Tempering Temperature Range, °F | 1000-1100 | 1000-1100 | 1000-1100 | 1000-1100 | 1000-1100 | 1000-1100 | 1000-1200 |
| | Approx. Tempered Hardness, $R_c$ | 65-66 | 66-61 | 66-62 | 65-66 | 65-66 | 65-60 | 68-63 |
| | Relative Ratings of Properties (A = greatest to E = least) | | | | | | | |
| **Characteristics in Heat Treatment** | Safety in Hardening | C | C | D | D | D | D | D |
| | Depth of Hardening | A | A | A | A | A | A | A |
| | Resistance to Decarburization | A | A | B | C | C | B | B |
| | Stability of Shape in Heat Treatment — Quenching Medium, Air or Salt | C | C | C | C | C | C | C |
| | Stability of Shape in Heat Treatment — Quenching Medium, Oil | D | D | D | D | D | D | D |
| **Service Properties** | Machinability | D | D | D | D | D/E | D | D/E |
| | Hot Hardness | B | B | A | A | A | A | A |
| | Wear Resistance | B | B | B | B | B | B | A |
| | Toughness | E | E | E | E | E | E | E |

relatively high amount of vanadium, also produces excellent wear resistance. In heat treatment the tendency for decarburization must be considered and heating in controlled, slightly reducing atmosphere is recommended. This type of high speed tool steel is mainly used for single point tools and inserts, it is well adapted for working at high speeds and feeds, for cutting hard materials and those which produce discontinuous chips, also for non-ferrous metals and, in general, for all kinds of tools needed for hogging (removing great bulks of material).

*AISI T15:* The performance qualities of this high alloy tool steel surpass most of those found in other grades of high speed tool steels. The high vanadium content, supported by uncommonly high carbon assures superior cutting ability and wear resistance. The addition of high cobalt increases the "hot hardness," and therefore tools made of T15 can sustain cutting speed in excess of those commonly applicable to tools made of steel. The machining and heat treatment of T15 does not cause extraordinary problems, although for best results, heating to high temperature is often applied in its heat treatment, and double, or even triple tempering is recommended. On the other hand, T15 is rather difficult to grind because of the presence of large amounts of very hard metallic carbides, therefore it is considered to have a very low "grindability" index. The main uses are in the field of high speed cutting and the working of hard metallic materials, T15 being often considered to represent in its application a transition from the regular high speed tool steels to cemented carbides. Lathe tool bits, form cutters, solid and inserted blade milling cutters are examples of uses for cutting tools, however the application of this steel type is not confined to this primary area of utilization; excellent results may also be attained for such tools as cold work dies, punches, blanking and forming dies, etc. The low toughness rating of the T15 steel excludes its application for operations which involve shock or sudden variations in load.

**Hot Work Tool Steels.** — For tools which in their regular service are in contact with hot metals over a shorter or longer period of time, with or without cooling being applied, a family of special tool steels has been developed and these are comprised under the general designation of hot work steels. The essential property of these steels is the capability of sustaining elevated temperature without seriously affecting the usefulness of the tools made of them. Depending on the purpose of the tools for which they were developed, the particular types of hot work tool steels have different dominant properties and are assigned to either of three groups, based primarily on their principal alloying elements.

**Hot Work Tool Steels — Chromium Types.** — As referred to in the group designation, the chromium content is considered the characteristic element of these tool steels. Their predominant properties are high hardenability, excellent toughness and great ductility, even at the cost of wear resistance. Some members of this family are made with the addition of tungsten, and in one type, cobalt as well. These alloying elements improve the resistance to the softening effect of elevated temperatures, but reduce ductility.

**Properties and Applications of Frequently Used Chromium Types.** — *AISI H11:* This hot work tool steel of the chromium-molybdenum-vanadium type has excellent ductility, can be machined easily and retains its strength at temperatures up to 1000 degrees F. These properties, combined with relatively good abrasion and shock resistance, account for the varied fields of application of H11, which include the following typical uses: (a) structural applications where high strength is needed at elevated operating temperatures, as for jet engine components and (b) hot work tools, particularly of the kind whose service involves shocks and drastic cooling of the tool, such as in extrusion tools, pierce and draw punches, bolt header dies, etc.

**Table 6. Hot Work Tool Steels**

Identifying Chemical Composition and Typical Heat Treatment Data

| | | Chromium Types | | | | | | Tungsten Types | | | | | | Molybdenum Types | | |
|---|---|---|---|---|---|---|---|---|---|---|---|---|---|---|---|---|
| **AISI** | **Type** | H10 | H11 | H12 | H13 | H14 | H19 | H21 | H22 | H23 | H24 | H25 | H26 | H41 | H42 | H43 |
| **Identifying Elements in %** | C | .40 | .35 | .35 | .35 | .40 | .40 | .35 | .35 | .35 | .45 | .25 | .50 | .65 | .60 | .55 |
| | W | ... | ... | 1.50 | ... | 5.00 | 4.25 | 9.00 | 11.00 | 12.00 | 15.00 | 15.00 | 18.00 | 1.50 | 6.00 | ... |
| | Mo | 2.50 | 1.50 | 1.50 | 1.50 | ... | ... | ... | ... | ... | ... | ... | ... | 8.00 | 5.00 | 8.00 |
| | Cr | 3.25 | 5.00 | 5.00 | 5.00 | 5.00 | 4.25 | 3.50 | 2.00 | 12.00 | 3.00 | 4.00 | 4.00 | 4.00 | 4.00 | 4.00 |
| | V | .40 | .40 | .40 | 1.00 | ... | 2.00 | ... | ... | ... | ... | ... | 1.00 | 1.00 | 2.00 | 2.00 |
| | Co | ... | ... | ... | ... | ... | 4.25 | ... | ... | ... | ... | ... | ... | ... | ... | ... |
| **Heat Treat Data** | Hardening Temperature Range °F | 1850-1900 | 1825-1875 | 1825-1875 | 1825-1900 | 1850-1950 | 2000-2200 | 2000-2200 | 2000-2200 | 2000-2300 | 2000-2250 | 2100-2300 | 2150-2300 | 2000-2175 | 2050-2225 | 2000-2175 |
| | Tempering Temperature Range °F | 1000-1200 | 1000-1200 | 1000-1200 | 1000-1200 | 1100-1200 | 1000-1300 | 1100-1250 | 1100-1250 | 1200-1500 | 1050-1200 | 1050-1250 | 1050-1250 | 1050-1200 | 1050-1200 | 1050-1200 |
| | Approx. Tempered Hardness, $R_c$ | 56-39 | 54-38 | 55-38 | 53-38 | 47-40 | 59-40 | 54-36 | 52-39 | 47-30 | 55-45 | 44-35 | 58-43 | 60-50 | 60-50 | 58-45 |

Relative Ratings of Properties (A = greatest to D = least)

| | | H10 | H11 | H12 | H13 | H14 | H19 | H21 | H22 | H23 | H24 | H25 | H26 | H41 | H42 | H43 |
|---|---|---|---|---|---|---|---|---|---|---|---|---|---|---|---|---|
| **Characteristics in Heat Treatment** | Safety in Hardening | A | A | A | A | A | B | B | B | B | B | B | B | C | C | C |
| | Depth of Hardening | A | A | A | A | A | A | A | A | A | A | A | A | A | A | A |
| | Resistance to Decarburization | B | B | B | B | B | B | B | B | B | B | B | B | C | B | C |
| | Stability of shape in heat treatment — Quenching medium — Air or Salt | B | B | B | B | C | C | C | C | ... | C | C | C | C | C | C |
| | Oil | ... | ... | ... | ... | ... | ... | ... | ... | ... | ... | ... | ... | ... | ... | ... |
| | Machinability | C/D | C/D | C/D | C/D | C | C | C | C | B | C | B | C | C | C | D |
| **Service Properties** | Hot Hardness | C | C | C | C | D | D | C | C | D | B | B | B | B | B | B |
| | Wear Resistance | D | D | D | D | C | C | C | C/D | C/D | C | D | C | C | C | C |
| | Toughness | C | B | B | B | C | C | C | C | D | C | C | C | D | C | D |

*AISI H12:* The properties of this type are comparable to those of H11, with increased abrasion resistance and hot hardness, resulting from the addition of tungsten, yet in an amount which does not affect the good toughness of this steel type. The applications, based on these properties, are hot work tools which often have to withstand severe impact, such as various punches, bolt header dies, trimmer dies, hot shear blades, also H12 is used to make aluminum extrusion dies and die-casting dies.

*AISI H13:* This type differs from the preceding ones particularly in properties related to the addition of about 1 per cent vanadium, which contributes to increased hot hardness, abrasion resistance and reduced sensitivity for heat checking. Such properties are needed in die casting, particularly of aluminum, where the tools are subjected to drastic cooling at high operating temperatures. Besides die-casting dies, H13 is also widely used for extrusion dies, trimmer dies, hot gripper and header dies, hot shear blades, etc.

*AISI H19:* This high-alloyed hot work tool steel, containing chromium, tungsten, cobalt and vanadium, has excellent resistance to abrasion and shocks at elevated temperature. It is particularly well adapted to severe hot work uses where the tool, in order to retain its size and shape, must withstand wear and the washing out effect of hot work material. Typical applications include brass extrusion dies and dummy blocks, inserts for forging and valve extrusion dies, press forging dies, hot punches, etc.

**Hot Work Tool Steels — Tungsten Types.** — Substantial amounts of tungsten, yet very low carbon content characterizes the hot work tool steels of this group. These tool steels have been developed for applications where the tool is in contact with the hot work material over extended periods of time, therefore the resistance of the steel to the softening effect of elevated temperatures is of prime importance, even to the extent of accepting a lower degree of toughness.

**Properties and Applications of Frequently used Tungsten Types.** — *AISI H21:* This is a medium tungsten alloyed hot work tool steel with substantially increased abrasion resistance over the chromium alloyed types, yet possessing a degree of toughness which represents a transition between the chromium and the higher alloyed tungsten steel types. The principal applications are for tools subjected to continued abrasion, yet to only a limited amount of shock loads, like tools for the extrusion of brass, both dies and dummy blocks, piercers for forging machines, inserts for forging tools, hot nut tools, etc. Another typical application is dies for the hot extrusion of automobile valves.

*AISI H24:* The comparatively high tungsten content (about 14 per cent) results in good hardness, great compression strength and excellent abrasion resistance, but makes it sensitive to shock loads. Taking these properties into account, the principal applications include extrusion dies for brass in long run operations, hot forming and gripper dies with shallow impressions, punches which are subjected to great wear yet only to moderate shocks, hot shear blades, etc.

*AISI H26:* This high alloyed tungsten type hot work steel resembles in its composition the tungsten type high speed steel AISI T1, except for the somewhat lower carbon content for improved toughness. The high amount of tungsten provides the maximum resistance to the softening effect of elevated temperature and assures excellent wear-resistant properties, including the withstanding of the washing-out effect of certain processes. However, this type is less resistant to thermal shocks than the chromium hot work steels. Typical applications comprise extrusion dies for long production runs, extrusion mandrels operated without cooling, hot piercing punches, hot forging dies and inserts. It is also used as special structural steel for springs operating at elevated temperatures.

**Hot Work Tool Steels — Molybdenum Types.** — These are closely related to certain types of molybdenum high speed steels and possess excellent resistance to the softening effect of elevated temperature but their ductility is rather low. These steel types are generally available on special orders only.

**Properties and Applications of Frequently Used Molybdenum Types.** — *AISI H43:* The principal constituents of this hot work steel, chromium, molybdenum and vanadium, provide excellent abrasion and wear-resistant properties at elevated temperatures. H43 has a good resistance against the development of heat checks and a toughness adequate for many different applications. These include tools and operations which tend to cause surface wear in high temperature work, like hot headers, punch and die inserts, hot heading and hot nut dies, as well as different kinds of punches operating at high temperature in service involving considerable wear.

**Cold Work Tool Steels.** — Tool steels of this category are primarily intended for die work, although their use is by no means restricted to that general field. Cold work tool steels are extensively used for tools whose regular service does not involve elevated temperatures. They are available in chemical compositions adjusted to the varying requirements of a wide range of different applications. According to their predominant properties, characterized either by the chemical composition or by the quenching medium in heat treatment, the cold work tool steels are assigned to three different groups.

**Cold Work Tool Steels — High Carbon, High Chromium Types.** — The chemical composition of tool steels of this family is characterized by the very high chromium content, in the order of 12 to 13 per cent, and the uncommonly high carbon content in the range of about 1.50 to 2.30 per cent. Additional alloying elements which are present in different amounts in some of the steel types of this group are vanadium, molybdenum and cobalt, each of which contributes desirable properties. The predominant properties of the whole group are: (a) excellent dimensional stability in heat treatment where, with one exception, air quench is used; (b) great wear resistance particularly in the types with the highest carbon content; and (c) rather good machinability.

**Properties and Applications of Frequently Used High Carbon-High Chromium Types.** — *AISI D2:* An air hardening die steel with high carbon, high chromium content with several desirable tool steel properties, such as abrasion resistance, high hardness as well as non-deforming characteristics. The carbon content of this type, although relatively high, is not particularly detrimental to its machining. The ease of working can further be improved by selecting the same basic type with the addition of sulfur. Several steel producers supply the sulfurized version of D2, in which the uniformly distributed sulfide particles substantially improve the machinability and the resulting surface finish. The applications comprise primarily cold working press tools for shearing (blanking and stamping dies, punches, shear blades), for forming (bending, seaming), also for thread rolling dies, solid gages and wear-resistant structural parts. Dies for hot trimming of forgings are also made of D2 when heat treated to a lower hardness for the purpose of increasing toughness.

*AISI D3:* The high carbon content of this high chromium tool steel type results in excellent resistance to wear and abrasion and also provides superior compressive strength as long as the pressure is gradually applied, without exerting sudden shocks. In hardening, an oil quench is used, without affecting the excellent non-deforming properties of this type. Its deep hardening properties make it particularly suitable for tools which require repeated regrinding during their service life, such as different types of dies and punches. The more important applications comprise blanking, stamping and trimming dies and punches for long production runs; forming, bending and drawing tools; also structural elements like plug and ring gages, lathe centers, etc., in applications where high wear resistance is important.

**Cold Work Tool Steels — Oil Hardening Types.** — With a relatively low percentage of alloying elements, yet with a substantial amount of manganese, these less expensive types of tool steels attain good depth of hardness in an oil quench, although at the cost of reduced resistance to deformation. Their good machinability supports general purpose applications, yet because of relatively low wear resistance they are mostly selected for comparatively short-run work.

**Properties and Applications of Frequently Used Oil Hardening Types.** — *AISI O1:* A low alloy tool steel which is hardened in oil and exhibits only a low tendency to shrinking or warping. It is used for cutting tools, the operation of which does not generate high heat, such as taps and threading dies, reamers, broaches, and for press tools like blanking, trimming, and forming dies in short- or medium-run operations.

*AISI O2:* Manganese is the dominant alloying element in this type of oil hardening tool steel which has good non-deforming properties, can be machined easily and performs satisfactorily in low volume production. The low hardening temperature results in good safety in hardening, both with regard to form stability and freedom from cracking. The combination of handling ease including free-machining properties, with good wear resistance, makes this type of tool steel adaptable to a wide range of common applications such as cutting tools for low and medium speed operations; forming tools including thread rolling dies; structural parts such as bushings, fixed gages, and also for plastic molding dies.

*AISI O6:* This oil hardening type of tool steel belongs to a group often designated as graphitic because of the presence of small particles of graphitic carbon which are uniformly dispersed throughout the steel. Usually about one-third of the total carbon is present as free graphite in nodular form, which contributes to the uncommon ease of machining. In the service of parts made of this type of steel the free graphite acts like a lubricant, reducing wear and galling. The ease of hardening is also excellent, requiring only comparatively low quenching temperature. Deep hardness penetration is produced and the oil quench causes very little dimensional change. The principal applications of the O6 tool steel are in the field of structural parts, like arbors, bushings, bodies for inserted tool cutters and shanks for cutting tools, jigs and machine parts and fixed gages like plugs, rings, snap gages, etc. It is also used for blanking, forming, and trimming dies and punches, in applications where the stability of the tool material is more important than high wear resistance.

**Cold Work Tool Steels — Medium Alloy Air Hardening Types.** — The desirable non-deforming properties of the high chromium types are approached by the members of this family, with substantially lower alloy content which, however, is sufficient to permit hardening by air quenching. The machinability is good, and the comparatively low wear resistance is balanced by relatively high toughness, a property which in certain applications may be considered of prime importance.

**Properties and Applications of Frequently Used Medium Alloy Air Hardening Types.** — *AISI A2:* The lower chromium content, about 5 per cent, makes this air hardening tool steel less expensive than the high chromium types, without affecting its nondeforming properties. The somewhat reduced wear resistance is balanced by greater toughness, making this type suitable for press work where the process calls for tough tool materials. The machinability is improved by the addition of about 0.12 percent sulfur, offered as a variety of the basic composition by several steel producers. The prime uses of this tool steel type are punches for

## Table 7.  Cold Work Tool Steels

### Identifying Chemical Composition and Typical Heat Treatment Data

| | High Carbon High Chromium Types | | | | | Medium Alloy Air Hardening Types | | | | | | | | Oil Hardening Types | | | |
|---|---|---|---|---|---|---|---|---|---|---|---|---|---|---|---|---|---|
| **Types** | D2 | D3 | D4 | D5 | D7 | A2 | A3 | A4 | A6 | A7 | A8 | A9 | A10 | O1 | O2 | O6 | O7 |
| **Identifying Elements in %** | | | | | | | | | | | | | | | | | |
| C | 1.50 | 2.25 | 2.25 | 1.50 | 2.35 | 1.00 | 1.25 | 1.00 | .70 | 2.25 | .55 | .50 | 1.35 | .90 | .90 | 1.45 | 1.20 |
| Mn | | | | | | | | 2.00 | 2.00 | | | | 1.80 | 1.00 | 1.60 | 1.00 | |
| Si | | | | | | | | | | | | | 1.25 | | | 1.00 | |
| W | | | | | | | | | | 1.00 | 1.25 | | | .50 | | | 1.75 |
| Mo | 1.00 | | 1.00 | 1.00 | 1.00 | 1.00 | 1.00 | 1.00 | 1.25 | 1.00 | 1.25 | 1.40 | 1.50 | | | .25 | |
| Cr | 12.00 | 12.00 | 12.00 | 12.00 | 12.00 | 5.00 | 5.00 | 1.00 | 1.00 | 5.25 | 5.00 | 5.00 | | .50 | | | .75 |
| V | 1.00 | | | | 4.00 | | 1.00 | | | 4.75 | | 1.00 | | | | | |
| Co | | | | 3.00 | | | | | | | | | | | | | |
| Ni | | | | | | | | | | | | 1.50 | 1.80 | | | | |
| **Heat Treatment Data** | | | | | | | | | | | | | | | | | |
| Hardening Temperature Range °F | 1800-1875 | 1700-1800 | 1775-1850 | 1800-1875 | 1850-1950 | 1700-1800 | 1750-1850 | 1500-1600 | 1525-1600 | 1750-1800 | 1800-1850 | 1800-1875 | 1450-1500 | 1450-1500 | 1400-1475 | 1450-1500 | 1550-1525 |
| Quenching Medium | Air | Oil | Air | Air | Air | Air | Air | Air | Air | Air | Air | Air | Air | Oil | Oil | Oil | Oil |
| Tempering Temperature Range °F | 400-1000 | 400-1000 | 400-1000 | 400-1000 | 300-1000 | 350-1000 | 350-1000 | 350-800 | 300-800 | 300-1000 | 350-1100 | 950-1150 | 350-800 | 350-500 | 350-500 | 350-600 | 350-550 |
| Approx. Tempered Hardness, $R_c$ | 61-54 | 61-54 | 61-54 | 61-54 | 65-58 | 62-57 | 65-57 | 62-54 | 60-54 | 67-57 | 60-50 | 56-35 | 62-55 | 62-57 | 62-57 | 63-58 | 64-58 |

### Relative Ratings of Properties (A = greatest to E = least)

| | D2 | D3 | D4 | D5 | D7 | A2 | A3 | A4 | A6 | A7 | A8 | A9 | A10 | O1 | O2 | O6 | O7 |
|---|---|---|---|---|---|---|---|---|---|---|---|---|---|---|---|---|---|
| **Service Characteristics in Ht. T.** | | | | | | | | | | | | | | | | | |
| Safety in Hardening | A | C | A | A | A | A | A | A | A | A | A | A | A | B | B | B | B |
| Depth of Hardening | A | A | A | A | A | A | A | A | A | A | A | A | A | B | B | B | B |
| Resistance to Decarburization | B | B | B | B | B | B | B | A/B | A/B | B | B | B | A/B | A | A | A | A |
| Stability of Shape in Heat Treatment | A | B | A | A | A | A | A | A | A | A | A | A | A | B | B | B | B |
| **Service Properties** | | | | | | | | | | | | | | | | | |
| Machinability | E | E | E | E | E | D | D | D/E | D/E | E | D | D | C/D | C | C | B | C |
| Hot Hardness | C | C | C | C | C | C | C | D | D | C | C/D | C/D | D | E | E | E | E |
| Wear Resistance | B/C | B | B | B/C | A | C | B | C/D | C/D | A | C/D | C/D | C | D | D | D | D |
| Toughness | E | E | E | E | E | E | D | D | D | E | C | C | D | C | C | E | C |

blanking and forming, cold and hot trimming dies (the latter heat treated to a lower hardness), thread rolling dies and also plastic molding dies.

*AISI A6:* The composition of this type of tool steel makes it adaptable to air hardening from a relatively low temperature, which is comparable to that of oil hardening types, yet offering improved stability in the heat treating. Its reduced tendency to heat treat distortions makes this tool steel type well adapted for die work, forming tools, gages, etc., which do not require the highest degree of wear-resistance.

### Shock Resisting, Mold and Special-purpose Tool Steels.

There are fields of tool application in which specific properties of the tool steels have dominant significance, determining to a great extent the performance and the service life of the tools which are made of these materials. To meet these requirements special types of tool steels have been developed. These individual types grew into families with members which, while similar in their major characteristics, provide related properties to different degrees. Originally developed for a specific use, the resulting particular properties of some of these tool steels made them desirable for other uses as well. In the tool steel classification system they are shown in three groups.

### Shock Resisting Tool Steels.

These are made with low carbon content for increased toughness, even at the expense of wear resistance which is generally low. Each member of this group also contains alloying elements, different in composition and amount, selected to provide properties particularly adjusted to specific applications. Such varying properties are the degree of toughness (generally high in all members), hot hardness, abrasion resistance and machinability.

### Properties and Applications of Frequently Used Shock Resisting Types.

*AISI S1:* This chromium-tungsten alloyed tool steel combines, in its hardened state, great toughness with high hardness and strength. Although it has a low carbon content for reasons of good toughness, the carbon forming alloys contribute to deep hardenability and abrasion resistance. When high wear resistance is also required, this property can be improved by carburizing the surface of the tool while still retaining its shock resistant characteristics. Primary uses are for battering tools, including hand and pneumatic chisels. The chemical composition, particularly the silicon and tungsten content, provide good hot hardness too, up to operating temperatures of about 1050 °F, making this tool steel type also adaptable for such hot work applications involving shock loads, as headers, piercers, forming tools, drop forge die inserts and heavy shear blades.

*AISI S2:* This steel type serves primarily for hand chisels and pneumatic tools, although it also has limited applications for hot work. While its wear resistance properties are only moderate, S2 is sometimes used for forming and thread rolling applications, when the resistance to rupturing is more important than extended service life. For hot work applications this steel requires heat treatment in a neutral atmosphere in order to safely avoid either carburization or decarburization of the surface. Such conditions make this tool steel type particularly susceptible to failure in hot work uses.

*AISI S5:* This is essentially a silicon-manganese type tool steel with small additions of chromium, molybdenum and vanadium for the purpose of improved deep hardening and refinement of the grain structure. The most important properties of this steel are its high elastic limit and good ductility resulting in excellent shock-resisting characteristics, when used at atmospheric temperatures. Its recommended quenching medium is oil, although a water quench may also be applied as long as the design of the tools avoids sharp corners or drastic sectional changes. Typical applications include pneumatic tools in severe service, like chipping chisels,

## Table 8. Shock Resisting, Mold, and Special Purpose Tool Steels

Identifying Chemical Compositions and Typical Heat Treatment Data

| Category | Shock Resisting Tool Steels | | | | Mold Steels | | | | | | | Special Purpose Tool Steels | | | | |
|---|---|---|---|---|---|---|---|---|---|---|---|---|---|---|---|---|
| **Types** | S1 | S2 | S5 | S7 | P2 | P3 | P4 | P5 | P6 | P20 | P21[a] | L2[b] | L3[b] | L6 | F1 | F2 |
| **Identifying Elements in Per Cent** | | | | | | | | | | | | | | | | |
| C | .50 | .50 | .55 | .50 | .07 | .10 | .07 | .10 | .10 | .35 | .20 | .50/1.10 | 1.00 | .70 | 1.00 | 1.25 |
| Mn | | | .80 | | | | | | | | | | | | | |
| Si | | 1.00 | 2.00 | | | | | | | | | | | | | |
| W | 2.50 | | | | | | | | | | | | | | 1.25 | 3.50 |
| Mo | | .50 | .40 | 1.40 | .20 | | .75 | | | .40 | | | | .25 | | |
| Cr | 1.50 | | | 3.25 | 2.00 | .60 | 5.00 | 2.25 | 1.50 | 1.25 | | 1.00 | 1.50 | .75 | | |
| V | | | | | | | | | | | | .20 | .20 | | | |
| Ni | | | | | .50 | 1.25 | | | 3.50 | | 4.00 | | | 1.50 | | |
| **Heat Treat Data** | | | | | | | | | | | | | | | | |
| Hardening Temperature | 1650-1750 | 1550-1650 | 1600-1700 | 1700-1750 | 1525-1550* | 1475-1525* | 1775-1825* | 1550-1600* | 1450-1500* | 1500-1600* | Sol'n treat.[c] | 1550-1700 | 1500-1600 | 1450-1550 | 1450-1600 | 1450-1600 |
| Tempering Temp. Range, °F | 400-1200 | 350-800 | 350-800 | 400-1150 | 350-500 | 350-500 | 350-900 | 350-500 | 350-450 | 900-1100 | Aged | 350-1000 | 350-600 | 350-1000 | 350-500 | 350-500 |
| Approx. Tempered Hardness Rc | 58-40 | 60-50 | 60-50 | 57-45 | 64-58† | 64-58† | 64-58† | 64-58† | 61-58† | 37-28† | 40-30 | 63-45 | 63-56 | 62-45 | 63-60 | 65-62 |
| **Heat Treatment Characteristics** — Relative Ratings of Properties (A = greatest to E = least) | | | | | | | | | | | | | | | | |
| Safety in Hardening | C | E | C | B/C | C | C | B | B | A | C | A | D | D | C | E | E |
| Depth of Hardening | B | B | B | A | B‡ | B‡ | B‡ | B‡ | A‡ | B | A | B | B | A | C | C |
| Resist. to Decarb. | B | C | C | B | A | A | A | A | B | A | A | A | A | A | A | A |
| Quench Med. — Air | | | | A | | | B | | | C | A | | | | | |
| Quench Med. — Oil | D | D | D | | C/D | C/D | | C | C | C/D | | D | E | E | C | |
| Quench Med. — Water | D/E | D/E | D/E | | | | | | | | | D/E | | | | E |
| Stability of Shape in Heat Treatment | D | D | D | B | C | C | B | C | C | C | A | D | D | C | E | E |
| **Service Properties** | | | | | | | | | | | | | | | | |
| Machinability | C/D | C/D | C/D | D | C/D | D | D/E | D | D | C/D | D | C | E | E | C | D |
| Hot Hardness | D | D/E | E | C | D | D | D | D | E | D/E | D | E | E | E | C | E |
| Wear Resistance | D/E | D/E | D/E | D/E | D | D | C | D | D | D/E | D | D | D | D | D | B/C |
| Toughness | B | A | A | A | C | C | C | C | C | C | D | B | D | D | D | E |

*After carburizing  †Carburized case  ‡Core Hardenability

bQuenched in oil  eSometimes brine is used.

aContains also about 1.20 per cent Al.  cSolution treated in hardening.

also shear blades, heavy duty punches, bending rolls, etc. Occasionally this steel is also used for structural applications, like shanks for carbide tools and machine parts subject to shocks.

**Mold Steels.** — These differ from all other types of tool steels by their very low carbon content, generally requiring carburizing for obtaining a hard operating surface. A special property of most steel types in this group is the adaptability to shaping by impression (hobbing) instead of by conventional machining. They also have high resistance to decarburization in heat treatment and dimensional stability, characteristics which obviate the need for grinding following heat treatment. The molding dies for plastic materials require an excellent surface finish, even to the degree of high luster; the generally high chromium content of these types of tool steels greatly aids in meeting this requirement.

**Properties and Applications of Frequently Used Mold Steel Types.** — *AISI P2 and P4:* Essentially, both types of tool steels were developed for the same special purpose, that is, the making of plastic molding dies. The application conditions of plastic molding dies require high core strength, good wear resistance at elevated temperature and excellent surface finish. Both types are carburizing steels which possess good dimensional stability. Because hobbing, that is, sinking the cavity by pressing into the tool material a punch representing the inverse replica of the cavity, is the process by which many of the plastic molding die cavities are produced, the "hobbability" of the tool steels used for this purpose is an important requirement. The different chemistry of these two types of mold steels is responsible for the high core hardness of the P4, which makes this latter type better suited for applications requiring high strength at elevated temperature.

*AISI P6:* This nickel-chromium type plastic mold steel has exceptional core strength and develops a deep carburized case. Due to the high nickel-chromium content the cavities of dies made of this steel type are produced by machining rather than by hobbing. An outstanding characteristic of this steel type is the high luster which is produced by the polishing of the hard case surface.

*AISI P20:* This is a general type mold steel which is adaptable to both through hardening and carburized case hardening. In through hardening an oil quench is used and a relatively lower, yet deeply penetrating hardness is obtained, such as is needed for zinc die-casting dies and injection plastic molds. Carburizing produces, after the direct quenching and tempering, a very hard case and comparatively high core hardness. When thus heat treated this steel is particularly well adapted for making compression, transfer and plunger type plastic molds.

**Special-purpose Tool Steels.** — These consist of several low alloy types of tool steels which were developed to provide transitional types between the more commonly used basic types of tool steels, and thereby contribute to the balancing of certain conflicting properties such as wear resistance and toughness; to offer intermediate depth of hardening; and to be less expensive than the higher alloyed types of tool steels.

**Properties and Applications of Frequently Used Special-purpose Types.** — *AISI D6:* This is a low-alloy type special-purpose tool steel. The comparatively safe hardening and the fair non-deforming properties, combined with the service advantage of good toughness in comparison to most other oil hardening types, explains the acceptance of this steel with a rather special chemical composition. The uses of L6 are limited to tools whose toughness requirements prevail over abrasion-resistant properties, such as forming rolls and forming and trimmer dies in applications where a combination of moderate shock and wear-resistant properties are sought. The areas of use also include structural parts, like clutch members, pawls, knuckle pins, etc., which must withstand shock loads and still display good wear properties.

*AISI F2:* This carbon-tungsten type is one of the most abrasion resistant of all water hardening tool steels. However, it is sensitive to thermal changes, such as are involved in heat treatment and is also susceptible to distortions. Consequently, its use is limited to tools of simple shape in order to avoid cracking in hardening. The shallow hardening characteristics of F2, which result in a tough core, are desirable properties for certain tool types which, at the same time, require the excellent wear-resistant properties of this tool steel type.

**Water Hardening Tool Steels.** — Steel types in this category are made without, or with only a minimum amount of alloying elements and need in their heat treatment the harsh quenching action of water or brine, hence the general designation of the category.

Water hardening steels are usually available with different percentages of carbon, to provide properties required for different applications; the classification system lists a carbon range of 0.60 to 1.40 per cent. In practice, however, the steel mills produce these steels in a few varieties of differing carbon content, often giving proprietary designations to each particular group. Typical carbon content limits of frequently used water hardening tool steels are 0.70–0.90, 0.90–1.10, 1.05–1.20 and 1.20–1.30 per cent. The appropriate group should be chosen according to the intended use, as indicated in the steel selection guide for this category, keeping in mind that while higher carbon content results in deeper hardness penetration, it also reduces toughness.

The general system distinguishes the following four grades: (1) special, (2) extra, (3) standard and (4) commercial, listed in the order of decreasing quality. The differences between these grades, which are not offered by all steel mills, are defined in principle only. The distinguishing characteristics are purity and consistency, resulting from different degrees of process refinement and inspection steps applied in making the steel. Higher qualities are selected for assuring dependable uniformity and performance of the tools made of the steel.

Since the groups with higher carbon content are more sensitive to heat treatment defects and are generally used for the more demanding applications, the better grades are usually chosen for the high carbon types and the lower grades for applications where steels with lower carbon content only are needed.

Water hardening tool steels, while being the least expensive, have several drawbacks which, however, are quite acceptable in many types of applications. Such limiting properties are the tendency to deformation in heat treatment due to harsh effects of the applied quenching medium, the sensitivity to heat during the use of the tools made of these steels, the only fair degree of toughness and the shallow penetration of hardness. However, this last mentioned property may prove a desirable characteristic in certain applications, such as, e.g., cold heading dies, because the relatively shallow hard case is supported by tough, although softer core.

The AISI designation for water hardening tool steels is W, followed by a numeral indicating the type, primarily defined by the steel's chemical composition, as shown in the following table.

**Recommended Applications of Water Hardening Type W-1 (Plain carbon) Tool Steels. —**

*Group I (C-0.70 to 0.90%):* Relatively tough and therefore preferred for tools which are subjected to shocks or abusive treatment. For such applications as: hand tools — chisels, screwdriver blades, cold punches, nail sets, etc., and fixture elements — vise jaws, anvil faces, chuck jaws, etc.

**Table 9.   Water Hardening Tool Steels — Identifying Chemical Composition and Heat Treatment Data**

| | AISI Types | | W1 | W2 | W5 |
|---|---|---|---|---|---|
| **Identifying Elements in Per Cent** | C | | 0.60 to 1.40 | 0.60 to 1.40 | 1.10 |
| | | | Varying carbon content may be available | | |
| | V | | ... | 0.25 | ... |
| | Cr | | These elements are adjusted | | 0.50 |
| | Mn | | to satisfy the hardening requirements | | |
| | Si | | | | |
| **Heat Treatment Data** | Hardening Temperature Ranges °F Varying with Carbon Content | 0.60 – 0.80% | 1450 to 1500 | | |
| | | 0.85 – 1.05 % | 1425 to 1550 | | |
| | | 1.10 – 1.40% | 1400 to 1525 | | |
| | Quenching Medium | | Brine or Water | | |
| | Tempering Temperature Range, °F | | 350 to 650 | | |
| | Approx. Tempered Hardness, $R_c$ | | 64 to 50 | | |

Relative Ratings of Properties (A = greatest to E = least)

| Characteristics in Heat Treatment | | | | Service Properties | | | |
|---|---|---|---|---|---|---|---|
| Safety in Hardening | Depth of Hardening | Resistance to Decarburization | Stability of Shape in Heat Treatment | Machinability | Hot Hardness | Wear Resistance | Toughness |
| D | C | A | E | A | E | D/E | C/D |

*Group II (C-0.90 to 1.10%):* Combines greater hardness with fair toughness, resulting in improved cutting capacity and moderate ability to sustain shock loads. For such applications as: hand tools — knives, center punches, pneumatic chisels; cutting tools — reamers, hand taps and threading dies, wood augers; die parts — drawing and heading dies, shear knives, cutting and forming dies; and fixture elements — drill bushings, lathe centers, collets, fixed gages.

*Group III (C-1.05 to 1.20%):* The higher carbon content increases the depth of hardness penetrations, yet reduces toughness, thus the resistance to shock loads. Preferred for applications where wear resistance and cutting ability are the prime considerations.  For such applications as: hand tools — woodworking chisels, paper knives; cutting tools (for low speed applications) — milling cutters, reamers;

planer tools, thread chasers, center drills; and die parts — cold blanking, coining, bending dies.

*Group IV (C-1.20 to 1.30%):* The high carbon content produces a hard case of considerable depth, with improved wear resistance, yet sensitive to shock and concentrated stresses. Selected for applications where the capacity to withstand abrasive wear is needed, and also where the retention of a keen edge or the original shape of the tool is important. For such applications as: cutting tools — (a) for finishing work, like cutters, reamers, (b) for cutting chilled cast iron and forming tools — for ferrous and nonferrous metals, burnishing tools.

By adding small amounts of alloying elements to W-steel types 2, and 5, certain characteristics which are desirable for specific applications, are improved. The vanadium in type 2 contributes to retaining a greater degree of fine grain structure after heat treating. Chromium in type 5 improves the deep hardening characteristics of the steel, a property being needed for large sections, and also assists in maintaining a keen cutting edge, which is desirable in cutting tools, like broaches, reamers, threading taps and dies.

## Mill Production Forms of Tool Steels. — Tool steels are produced in many different forms, although not all those listed in the following are always readily available; certain forms and shapes being made on special orders only.

*Hot Finished Bars and Cold Finished Bars:* These are the most commonly produced forms of tool steels. Bars can be furnished in many different cross sections, the round shape being the most common. Sizes can vary over a wide range, with a more limited number of standard stock sizes. Various conditions may also be available, however, technological limitations prevent all conditions applying to every size, shape or type of steel. Tool steel bars may be supplied in one of the following conditions and surface finishes:

Conditions: Hot rolled or forged (natural); Hot rolled or forged and annealed; Hot rolled or forged and heat treated; Cold or hot drawn (as drawn); and Cold or Hot drawn and annealed.

Finishes: Hot rolled finish (scale not removed); Pickled or blast cleaned; Cold drawn; Turned or machined; Rough ground; Centerless ground or precision flat ground; and Polished (rounds only).

Other forms in which tool steels are supplied are the following:

*Rolled or Forged Special Shapes:* These are usually produced on special orders only, for the purpose of reducing material loss and machining time in the large-volume manufacture of certain frequently used types of tools.

*Forgings:* All types of tool steels may be supplied in the form of forgings, which are usually specified for special shapes and also for dimensions which are beyond the range covered by bars.

*Wires:* Tool steel wires are produced either by hot or cold drawing and are specified for the purpose of obtaining special shapes, controlled dimensional accuracy, improved surface finish or special mechanical properties. Round wire is commonly produced within an approximate size range of 0.015 to 0.500 inch, which also indicates the limits within which other shapes of tool steel wires, like oval, square, rectangular, etc., may be produced.

*Drill Rods:* These are produced in round, rectangular, square, hexagonal and octagonal shapes, usually with tight dimensional tolerances in order to eliminate subsequent machining, thereby offering manufacturing economies for the users.

*Hot Rolled Plates and Sheets, and Cold Rolled Strips:* Such forms of tool steel are generally specified for the high volume production of specific tool types.

*Tool Bits:* These are semifinished tools and are used by clamping in a tool holder or shank in a manner permitting ready replacement. Tool bits are commonly made of high speed types of tool steels, mostly in square, but also in round, rectangular and other shapes. Tool bits are made of hot rolled bars and are commonly, yet not exclusively, supplied in hardened and ground form, ready for use after the appropriate cutting edges are ground, usually in the user's plant.

*Hollow Bars:* These are generally produced by trepanning, boring or drilling of solid round rods and are used for making tools or structural parts of annular shapes, like rolls, ring gages, bushings, etc.

**Tolerances of Dimensions.** — Such tolerances have been developed and published by the Iron and Steel Institute (AISI) as a compilation of available industry experience which, however, does not exclude the establishment of closer tolerances, particularly for hot rolled products manufactured in large quantities. The tolerances differ for various categories of production processes (e.g., forged, hot rolled, cold drawn, centerless ground) and of general shapes. See Handbook pages 471 and 472.

**Allowances for Machining.** — These serve to assure freedom from soft spots and defects of the tool surface, thereby preventing failures in the heat treatment or in service. After removing from its surface a layer of specific thickness, known as the allowance, the bar or other form of tool steel material should have a surface without decarburization and other surface defects, such as scale marks or seams. The industry-wide accepted machining allowance values for tool steels in different conditions, shapes and size ranges are spelled out in AISI specifications and are generally also listed in the tool steel catalogs of the producer companies.

**Decarburization Limits.** — The heating of the steel for its production operation causes the oxidation of the exposed surfaces resulting in the loss of carbon. That condition, termed decarburization, penetrates to a certain depth from the surface, depending on the applied process, the shape and the dimensions of the product. The toleranced values of decarburization must be considered as one of the factors for defining the machining allowances, which must also compensate for expected variations of size and shape, the dimensional effects of heat treatment, and so forth. Decarburization can be present not only in hot rolled and forged, but also in rough turned and cold drawn conditions.

**Advances in Tool Steel Making Technology.** — In recent years significant advances in processes for tool steel production have been made which offer more homogeneous materials of greater density and higher purity for applications where such extremely high quality is required. Two of these newer methods of tool steel production are of particular interest.

*Vacuum melted tool steels:* These are produced by the consumable electrode method involving a remelting of the steel originally produced by conventional processes. Inside a vacuum-tight shell which has been evacuated, the electrode cast of tool steel of the desired chemical analysis is lowered into a water cooled copper mold where it strikes a low voltage-high amperage arc causing the electrode to be consumed by gradual melting. The undesirable gases and volatiles are drawn off by the vacuum, and the inclusions float on the surface of the pool, accumulating on the top of the produced ingot, to be removed later by cropping. In the field of tool steels the consumable electrode vacuum melting (CVM) process is applied primarily to the production of special grades of hot work and high speed tool steels.

*High speed tool steels produced by powder metallurgy:* The steel produced by conventional methods is reduced to a fine powder by a gas atomization process. The powder is compacted by a hot isostatic method with pressures in the range of 15,000 to 17,000 psi. The compacted billets are hot rolled to the final bar size, yielding a tool-steel material which has 100 per cent theoretical density. High speed tool steels produced by the P/M method offer a tool material providing increased tool wear life and high impact strength, of particular advantage in interrupted cuts.

## Trade Names of AISI Classified Tool Steels*—1

| AISI Type | Al-Tech | Atlas Steels | Bethlehem | Braeburn | Carpenter | Columbia | Crucible | Jessop | Latrobe | Simonds | Teledyne-Vasco | Universal-Cyclops |
|---|---|---|---|---|---|---|---|---|---|---|---|---|
| HIGH SPEED TOOL STEELS — TUNGSTEN TYPES |||||||||||||
| T1 | LXX | Spartan-7 | Bethlehem T-1 | Vinco | Star Zenith | Clarite | Rex AA | Supremus | Electrite No. I XL | Red Streak | Red-Cut Superior | … |
| T2 | ML | … | … | Twinvan | … | Vanite | … | Supremus Extra | Electrite No. 19 | Lock Port Special | E.V.M. | … |
| T4 | Panther Special | … | … | Cobalt | … | Acmite | Rex AAA | Purple Label | Electrite Cobalt | Tunco | Red Cut Cobalt | … |
| T5 | Super Panther | Nipigon | … | Bonded Carbide JR | … | Cobide | … | Purple Label Extra | Electrite Super Cobalt | Super Cobalt | Circle C | … |
| T6 | … | … | … | … | … | Cobite II | … | King Cobalt | Electrite Ultra Cobalt | … | … | … |
| T8 | … | … | … | … | … | Maxite | Rex 95 | Jessop T8 | … | … | … | … |
| T15 | Panther 5 | Sabre | … | Braeburn T15 | … | Maxite 15 | CPM Rex T-15 | Jessop T15 | Electrite Dynavan | … | Vasco Supreme | … |
| HIGH SPEED TOOL STEELS — MOLYBDENUM TYPES |||||||||||||
| M1 | LMW | Mohican-8 | Bethlehem M-1 | Mocut | Starmax, Starmax FM | Molite 1 | Rex TMO | Mogul | Electrite Tatmo | STM | 8-N-2 | Motung |
| M2 | DBL-2 | Sixix | Bethlehem M-2 | Braemow, Mocarb | Speed Star, Speed Star FM | Molite 2 | Rex M-2, CPM Rex M-2 | Mustang | Electrite Double Six M2 XL | Molva T | Vasco M-2 | Motung 652 |
| M3 Class I | DBL-2½ | Atlas M-3 | … | Braevan | … | Molite 3 Class I | Rex M-3-I | Jessop M3 Class I | Electrite Corsair XL | Molva TC1 | Van Cut | Unicut |

* Source: Committee of Tool Steel Producers, American Iron and Steel Institute, 1000 16th St., N.W., Washington, D.C. 20036

Trade Names of AISI Classified Tool Steels — 2

HIGH SPEED TOOL STEELS — MOLYBDENUM TYPES

| AISI Type | Al-Tech | Atlas Steels | Bethlehem | Braeburn | Carpenter | Columbia | Crucible | Jessop | Latrobe | Simonds | Teledyne Vasco | Universal-Cyclops |
|---|---|---|---|---|---|---|---|---|---|---|---|---|
| M3 Class 2 | DBL-3 | …… | …… | Braevan 2 | …… | Molite 3 Class 2 | Rex M-3-2, CPM Rex M-3-2 | Jessop M3 Class 2 | Electrite Crusader XL | Molva TC 2 | Van Cut Type 2 | Unicut 2 |
| M4 | DBL-4 | Atlas M-4 | …… | Braefour | Four Star | Molite 4 | CPM Rex M-4 | Jessop M4 | Electrite Stark | Molva HC | Neatro | Cyclops M4 |
| M6 | …… | …… | …… | Congo | …… | …… | …… | …… | …… | …… | …… | …… |
| M7 | LMW-V | Atlas M-7 | Bethlehem M-7 | Motuf | Seven Star | Molite 7 | Rex M-7 | Jessop M7 | Electrite Tatmo V | Molva C | Vasco M-7 | Motung CV |
| M10 | VLM | Atlas M-10 | Bethlehem M-10 | Motemp | Ten Star | Molite 10 | Rex VM | Jessop M10 | Electrite TNW | Molva | Van Lom | …… |
| M30 | Super LMW | …… | …… | Como | …… | Molite 30 | …… | …… | Electrite Lacomo | …… | 8-N-2 Cobalt | Super Motung |
| M33 | Super LMW Extra | …… | …… | Braeburn M33 | …… | Molite 33 | Rex M-33 | …… | Electrite Kelvan | STMCO | 8-N-2 Cobalt 8 | Super Motung 33 |
| M34 | Super LMW Special | Atlas M-34 | …… | …… | …… | Molite 34 | …… | …… | Electrite Tatmo Cobalt | …… | …… | …… |
| M36 | Super DBL | …… | …… | Moco | …… | Molite 36 | …… | …… | Electrite CO-6 | …… | Victory Cobalt | …… |
| M41 | …… | …… | …… | …… | …… | Molite 41 | Rex 49 | Jessop RC 70 | …… | …… | …… | …… |
| M42 | Exocut | Atlas M-42 | …… | Braemax | Super Star | Molite 42 | Rex M-42, CPM Rex M-42 | …… | Electrite Dynamax | …… | Hypercut | Cyclops M-42 |
| M43 | …… | …… | …… | …… | …… | …… | …… | …… | Electrite Dynacut | …… | …… | …… |
| M44 | …… | …… | Braecut | …… | …… | …… | …… | …… | …… | …… | …… | …… |

Producer

## Trade Names of AISI Classified Tool Steels — 3

| AISI Type | Al-Tech | Atlas Steels | Bethlehem | Braeburn | Carpenter | Columbia | Crucible | Jessop | Latrobe | Simonds | Teledyne Vasco | Universal-Cyclops |
|---|---|---|---|---|---|---|---|---|---|---|---|---|
| HIGH SPEED TOOL STEELS — MOLYBDENUM TYPES | | | | | | | | | | | | |
| M46 | AL-46 | …… | …… | …… | …… | …… | Rex M-46 | …… | …… | …… | …… | …… |
| M47 | Exohard | …… | …… | …… | …… | …… | …… | …… | …… | …… | …… | …… |
| HOT WORK TOOL STEELS — CHROMIUM TYPES | | | | | | | | | | | | |
| H10 | AL-173 | …… | …… | Pressurdie 6 | …… | Columbia H10 | Peerless 56 | …… | Dart | …… | …… | …… |
| H11 | Potomac A | Atlas H-11 | Cromo-V | Pressurdie 3L | 882, 882 FM | Firedie | Nu-Die, Halcomb 218 | Dica B (Modified) | Dycast No. 1 | Howard A | Hot Form No. 2 | Thermold H11 |
| H12 | Potomac | Crodi | Cromo-W | Pressurdie 2 | 345, 345 FM | Alcodie | Chro-Mow | Dica B | LPD | Howard B | Hot Form No. 1 | Thermold H12 |
| H13 | Potomac M | Dievac | Cromo-High-V | Pressurdie 3 | 883, 883 FM | Firedie 13 | Nu-Die V | Dica B Vanadium | VDC, Viscount 20, Viscount 44 | Howard C | Hot Form V | Thermold H13 |
| H14 | …… | Red Indian | …… | Pressurdie 1 | …… | Firedie 14 | …… | …… | Lumdie | …… | …… | …… |
| H19 | B-47 | Atlas H-19 | …… | Pressurdie C | …… | Firedie 19 | Halcomb 425 | …… | Lesco 19 | …… | W.C.C. | …… |
| HOT WORK TOOL STEELS — TUNGSTEN TYPES | | | | | | | | | | | | |
| H21 | Atlas A | Seneca | 57 HW | T-Alloy A | TK | Formite 21 | Peerless A | 2-BLC | CLW | …… | Marvel | Thermold H21 |
| H22 | Atlas B | …… | …… | T-Alloy | TK (Modified) | …… | …… | 2-BMC | …… | …… | …… | …… |
| H23 | …… | …… | …… | HCA | …… | Formite 23 | …… | …… | Kalkos | …… | W.W. Hot Work | …… |
| H24 | …… | …… | …… | T-Alloy B | …… | Formite 24 | …… | 2-BHC | CHW | …… | SC Special | …… |
| H25 | Mohawk | …… | …… | T-Alloy C | …… | …… | …… | …… | EHW No. I | …… | Forge-Die | …… |
| H26 | …… | Spartan-5 | Special HS-55 | Vinco Hot Work | Carpenter H26 | Clarite HW 26 | Rex AA PX | …… | Electrite No. 5 | …… | Red Cut Superior J | …… |

## Trade Names of AISI Classified Tool Steels — 4

| AISI Type | Al-Tech | Atlas Steels | Bethlehem | Braeburn | Carpenter | Columbia | Crucible | Jessop | Latrobe | Simonds | Teledyne Vasco | Universal-Cyclops |
|---|---|---|---|---|---|---|---|---|---|---|---|---|
| **HOT WORK TOOL STEELS — MOLYBDENUM TYPE** | | | | | | | | | | | | |
| H42 | …… | …… | …… | …… | …… | …… | …… | Mustang L.C. | Electrite No. 7 | …… | …… | …… |
| **COLD WORK TOOL STEELS — HIGH CARBON HIGH CHROMIUM TYPES** | | | | | | | | | | | | |
| D2 | Ontario | FNS | Lehigh H | Superior 3 | 610,610FM | Atmodie | Airdi 150 | CNS-1 | Olympic FM | CCM | Ohio Die | Ultradie 3 |
| D3 | Huron | …… | Lehigh S | Superior 1 | Hampden | Superdie | …… | CNS-2 | GSN | HCCM | …… | …… |
| D4 | …… | NN | …… | AT 2 | …… | Atmodie 4 | HYCC | CNS-3 | GSN + MO | …… | Crocar | …… |
| D5 | AL-D-5 | …… | …… | Superior 2 | Carpenter D5 | Atmodie 5 | …… | 3 C Special | Cobalt Chrome FM | …… | …… | …… |
| D7 | Huron V | …… | …… | …… | Carpenter D7 | …… | HYCV | Truwear | BR-4 FM | ARS | …… | …… |
| **SPECIAL PURPOSE TOOL STEELS — LOW ALLOY TYPES** | | | | | | | | | | | | |
| L2 | Albany Caroga | …… | Tough M | …… | …… | Columbia L2 | Halvan | ET-6 | Crown, Superb | …… | Vanadium Type H | Cyclops L2 |
| L6 | Tioga | Atlas L-6 | Bethalloy | …… | R.D.S. | Nicrodie | Champaloy | ET-4 | NDS | …… | Nikro M | Cyclops L6 |
| **COLD WORK TOOL STEELS — MEDIUM ALLOY AIR HARDENING TYPES** | | | | | | | | | | | | |
| A2 | Sagamore | Cromoloy | A-H5 | Airque | 484, 484 FM | E-Z-Die | Airkool | Windsor | Select B FM | Airtrue | Air Hard | Sparta |
| A3 | …… | …… | …… | Airque V | …… | …… | …… | …… | …… | …… | …… | …… |
| A4 | …… | …… | Air-4 | …… | …… | …… | …… | …… | …… | …… | …… | …… |
| A6 | Apache | Nutherm | A-6 | …… | Vega | Uni-Die | CSM 6 | Jess-Air | …… | …… | …… | Lo-Air |

## Trade Names of AISI Classified Tool Steels — 5

| AISI Type | Al-Tech | Atlas Steels | Bethlehem | Braeburn | Carpenter | Columbia | Crucible | Jessop | Latrobe | Simonds | Teledyne Vasco | Timken | Universal-Cyclops |
|---|---|---|---|---|---|---|---|---|---|---|---|---|---|
| COLD WORK TOOL STEELS — MEDIUM ALLOY AIR HARDENING TYPES | | | | | | | | | | | | | |
| A7 | Sagamore V | … | … | … | … | E-Z-Die V | Airkool V | BX 3 | BR-3 | A7W | Chrome-wear | … | … |
| A8 | AL-158 | … | Cromo-W55 | Pressurdie 16 | Carpenter A8 | Columbi A-8 | … | … | MGR | Airtrue LC | Hot Form No. 3 | … | … |
| A9 | … | … | … | … | Carpenter A9 | Formdie | … | … | … | … | … | … | … |
| A10 | … | … | … | … | … | … | … | … | … | … | … | Graph-Air | Thermold J |
| COLD WORK TOOL STEELS — OIL HARDENING TYPES | | | | | | | | | | | | | |
| O1 | Saratoga | Keewatin | BTR | Kiski | Carpenter O-1 | EXL-Die | Ketos | Truform | Badger | Teenax | Colonial No. 6 | … | Wando |
| O2 | Deward | … | … | … | Stentor | … | … | Special Oil Hardening | … | … | … | … | … |
| O6 | Oilgraph | … | O-6 Graphitic | … | … | Col-Graph | Halgraph | Truglide | … | … | … | Graph-Mo | … |
| O7 | Utica | … | … | … | … | Tapdie | … | … | W Tap | BFD | Red Star Tungsten | … | … |
| SHOCK RESISTING TOOL STEELS | | | | | | | | | | | | | |
| S1 | Seminole | Falcon-6 | 67 Chisel | Vibro | Excelo | Buster Alloy | Atha Pneu | Top Notch | XL Chisel | Com-mando | Par-Exc | … | … |
| S2 | … | … | Imperial | … | Solar | … | … | RTS | … | Havoc | … | … | … |
| S5 | AL-602, AL-609 | Monark-2 | Omega | Alloy 10 | 481 Collet | Silico Alloy | La Belle Silicon #2 | 259 Grade | Lanark | Orleans | Mosil | … | Venango Special |
| S6 | … | … | … | … | … | Columbia S6 | La Belle HT | … | … | … | … | … | Cyclops S5 |
| S7 | AL-7 | … | Bearcat | … | Carpenter S-7 | Shock-Die | Crucible S-7 | Super Shock 7 | … | … | Simoch | … | … |

## Trade Names of AISI Classified Tool Steels — 6

| AISI Type | Al-Tech | Atlas Steels | Bethlehem | Braeburn | Carpenter | Columbia | Crucible | Jessop | Latrobe | Simonds | Teledyne Vasco |
|---|---|---|---|---|---|---|---|---|---|---|---|
| | | | | | | Producer | | | | | |
| | | | | | MOLD STEELS | | | | | | |
| P2 | … | … | Duramold B | … | … | … | … | … | … | … | … |
| P4 | … | … | Duramold A | … | Super Samson | … | … | … | … | … | … |
| P5 | … | … | … | … | Samson Extra | … | … | … | … | … | Vasco Chro-mold VM |
| P6 | … | Super Impacto PQ | Duramold N | … | No. 158 | … | … | … | … | … | … |
| P20 | Almold-20 | Mold Special | Bethlehem P-20 | … | … | … | CSM #2 | P20 | … | … | … |
| P21 | … | … | … | … | … | … | … | … | Cascade | … | … |
| | | | | WATER HARDENING TOOL STEELS | | | | | | | |
| W1 | Pompton | X-10, X-12, Alpha, XX-95 | X, XCL, XX, Cold Header Die | Extra | Comet Green Label, H-9 Double Header, No. 11 Special, Titan, Reading Tap | Columbia Special, Extra, Extra, Headerdie, Standard | Sanderson Extra, Labelle Cold Header, Black Diamond | Washington Lion, Lion Extra, New Process Cold Header | Carbon Types | Red Label, Blue Label, Diamond S, Green Label | Colonial No. 14 |
| W2 | Python | Asa-10 | Best, Superior | Coldie | Nitro Special Vanadium | Vanadium Extra, Standard | Alva Extra | Lion Van., Lion Extra | Carbon Vanadium Types | Red Label Extra | Colonial No. 7 |
| W5 | Crow | Atlas "Q" | Bethlehem W-5 | Braeburn W-5 | U.D.R. | Waterdie Extra, Standard | … | W-5 | CFS | … | … |

**Numbering Systems for Metals and Alloys.** — Several different numbering systems have been developed for metals and alloys by various trade associations, professional engineering societies, standards organizations, and by private industries for their own use. The numerical code used to identify the metal or alloy may or may not be related to a specification, which is a statement of the technical and commercial requirements that the product must meet. Numbering systems in use include those developed by the American Iron and Steel Institute (AISI), Society of Automotive Engineers (SAE), American Society for Testing and Materials (ASTM), American National Standards Institute (ANSI), Steel Founders Society of America, American Society of Mechanical Engineers (ASME), American Welding Society (AWS), Aluminum Association, Copper Development Association, U.S. Department of Defense (Military Specifications), and the General Accounting Office (Federal Specifications).

The Unified Numbering System (UNS) was developed through a joint effort of the ASTM and the SAE to provide a means of correlating the different numbering systems for metals and alloys that have a commercial standing. This system avoids the confusion caused when more than one identification number is used to specify the same material, or when the same number is assigned to two entirely different materials. It is important to understand that a UNS number is not a specification; it is an identification number for metals and alloys for which detailed specifications are provided elsewhere. There are seventeen series of UNS numbers, which are shown in Table 1. Each UNS number consists of a letter prefix followed by five digits. In some cases the letter is suggestive of the family of metals identified by the series, such as A for aluminum and C for copper. Whenever possible, the numbers in the UNS number groups contain numbering sequences taken directly from other systems in order to facilitate the identification of the material; e.g., the corresponding UNS number for AISI 1020 steel is G10200. The UNS numbers corresponding to the commonly used AISI-SAE numbers that are used to identify plain carbon alloy and tool steels are given in Table 2.

**Table 1. Unified Numbering System (UNS) for Metals and Alloys**

| UNS Series | Metal |
|---|---|
| Nonferrous Metals and Alloys | |
| A00001 to A99999 | Aluminum and aluminum alloys |
| C00001 to C99999 | Copper and copper alloys |
| E00001 to E99999 | Rare earth and rare earth-like metals and alloys |
| L00001 to L99999 | Low melting metals and alloys |
| M00001 to M99999 | Miscellaneous nonferrous metals and alloys |
| P00001 to P99999 | Precious metals and alloys |
| R00001 to R99999 | Reactive and refractory metals and alloys |
| Z00001 to Z99999 | Zinc and zinc alloys |
| | |
| Ferrous Metals and Alloys | |
| D00001 to D99999 | Specified mechanical property steels |
| F00001 to F99999 | Cast irons |
| G00001 to G99999 | AISI and SAE carbon and alloy steels (except tool steels) |
| H00001 to H99999 | AISI H-steels |
| J00001 to J99999 | Cast steels (except tool steels) |
| K00001 to K99999 | Miscellaneous steels and ferrous alloys |
| S00001 to S99999 | Heat and corrosion resistant (stainless) steels |
| T00001 to T99999 | Tool steels |

## Table 2. AISI and SAE Numbers and Their Corresponding UNS Numbers for Plain Carbon, Alloy, and Tool Steels

| AISI-SAE Numbers | UNS Numbers | AISI-SAE Numbers | UNS Numbers | AISI-SAE Numbers | UNS Numbers | AISI-SAE Numbers | UNS Numbers |
|---|---|---|---|---|---|---|---|
| Plain Carbon Steels | | | | | | | |
| 1005 | G10050 | 1030 | G10300 | 1070 | G10700 | 1566 | G15660 |
| 1006 | G10060 | 1035 | G10350 | 1078 | G10780 | 1110 | G11100 |
| 1008 | G10080 | 1037 | G10370 | 1080 | G10800 | 1117 | G11170 |
| 1010 | G10100 | 1038 | G10380 | 1084 | G10840 | 1118 | G11180 |
| 1012 | G10120 | 1039 | G10390 | 1086 | G10860 | 1137 | G11370 |
| 1015 | G10150 | 1040 | G10400 | 1090 | G10900 | 1139 | G11390 |
| 1016 | G10160 | 1042 | G10420 | 1095 | G10950 | 1140 | G11400 |
| 1017 | G10170 | 1043 | G10430 | 1513 | G15130 | 1141 | G11410 |
| 1018 | G10180 | 1044 | G10440 | 1522 | G15220 | 1144 | G11440 |
| 1019 | G10190 | 1045 | G10450 | 1524 | G15240 | 1146 | G11460 |
| 1020 | G10200 | 1046 | G10460 | 1526 | G15260 | 1151 | G11510 |
| 1021 | G10210 | 1049 | G10490 | 1527 | G15270 | 1211 | G12110 |
| 1022 | G10220 | 1050 | G10500 | 1541 | G15410 | 1212 | G12120 |
| 1023 | G10230 | 1053 | G10530 | 1548 | G15480 | 1213 | G12130 |
| 1025 | G10250 | 1055 | G10550 | 1551 | G15510 | 1215 | G12150 |
| 1026 | G10260 | 1059 | G10590 | 1552 | G15520 | 12L14 | G12144 |
| 1029 | G10290 | 1060 | G10600 | 1561 | G15610 | ... | ... |
| Alloy Steels | | | | | | | |
| 1330 | G13300 | 4150 | G41500 | 5140 | G51400 | 8642 | G86420 |
| 1335 | G13350 | 4161 | G41610 | 5150 | G51500 | 8645 | G86450 |
| 1340 | G13400 | 4320 | G43200 | 5155 | G51550 | 8655 | G86550 |
| 1345 | G13450 | 4340 | G43400 | 5160 | G51600 | 8720 | G87200 |
| 4023 | G40230 | E4340 | E43406 | E51100 | G51986 | 8740 | G87400 |
| 4024 | G40240 | 4615 | G46150 | E52100 | G52986 | 8822 | G88220 |
| 4027 | G40270 | 4620 | G46200 | 6118 | G61180 | 9260 | G92600 |
| 4028 | G40280 | 4626 | G46260 | 6150 | G61500 | 50B44 | G50441 |
| 4037 | G40370 | 4720 | G47200 | 8615 | G86150 | 50B46 | G50461 |
| 4047 | G40470 | 4815 | G48150 | 8617 | G86170 | 50B50 | G50501 |
| 4118 | G41180 | 4817 | G48170 | 8620 | G86200 | 50B60 | G50601 |
| 4130 | G41300 | 4820 | G48200 | 8622 | G86220 | 51B60 | G51601 |
| 4137 | G41370 | 5117 | G51170 | 8625 | G86250 | 81B45 | G81451 |
| 4140 | G41400 | 5120 | G51200 | 8627 | G86270 | 94B17 | G94171 |
| 4142 | G41420 | 5130 | G51300 | 8630 | G86300 | 94B30 | G94301 |
| 4145 | G41450 | 5132 | G51320 | 8637 | G86370 | ... | ... |
| 4147 | G41470 | 5135 | G51350 | 8640 | G86400 | ... | ... |
| Tool Steels (AISI and UNS Only) | | | | | | | |
| M1 | T11301 | T6 | T12006 | A6 | T30106 | P4 | T51604 |
| M2 | T11302 | T8 | T12008 | A7 | T30107 | P5 | T51605 |
| M4 | T11304 | T15 | T12015 | A8 | T30108 | P6 | T51606 |
| M6 | T11306 | H10 | T20810 | A9 | T30109 | P20 | T51620 |
| M7 | T11307 | H11 | T20811 | A10 | T30110 | P21 | T51621 |
| M10 | T11310 | H12 | T20812 | D2 | T30402 | F1 | T60601 |
| M3-1 | T11313 | H13 | T20813 | D3 | T30403 | F2 | T60602 |
| M3-2 | T11323 | H14 | T20814 | D4 | T30404 | L2 | T61202 |
| M30 | T11330 | H19 | T20819 | D5 | T30405 | L3 | T61203 |
| M33 | T11333 | H21 | T20821 | D7 | T30407 | L6 | T61206 |
| M34 | T11334 | H22 | T20822 | O1 | T31501 | W1 | T72301 |
| M36 | T11336 | H23 | T20823 | O2 | T31502 | W2 | T72302 |
| M41 | T11341 | H24 | T20824 | O6 | T31506 | W5 | T72305 |
| M42 | T11342 | H25 | T20825 | O7 | T31507 | CA2 | T90102 |
| M43 | T11343 | H26 | T20826 | S1 | T41901 | CD2 | T90402 |
| M44 | T11344 | H41 | T20841 | S2 | T41902 | CD5 | T90405 |
| M46 | T11346 | H42 | T20842 | S4 | T41904 | CH12 | T90812 |
| M47 | T11347 | H43 | T20843 | S5 | T41905 | CH13 | T90813 |
| T1 | T12001 | A2 | T30102 | S6 | T41906 | CO1 | T91501 |
| T2 | T12002 | A3 | T30103 | S7 | T41907 | CS5 | T91905 |
| T4 | T12004 | A4 | T30104 | P2 | T51602 | ... | ... |
| T5 | T12005 | A5 | T30105 | P3 | T51603 | ... | ... |

## Standard Steels — Compositions, Applications, and Heat-treatments

The standard steel compositions of the Society of Automotive Engineers (SAE), Inc., given in the accompanying table, are considered adequate for practically all parts made of ferrous materials that are necessary for the production of automotive apparatus, and include grades that have been found commercially available and technically adequate for the service required of such parts. Definite applications of SAE steels are not specified as the selection of a proper steel for a given part must depend upon an intimate knowledge of a number of important factors, such as the availability and price of the material, the detailed design of the part and the severity of the service to be imposed, whether the part is to be forged or machined and its machineability; hence only general applications are indicated. (See following text and tables.)

**Numbering Systems for Steel.** — The primary numbering systems for identifying steels are the American Iron and Steel Institute (AISI) and the Society of Automotive Engineers (SAE) systems. Although they are entirely separate systems, they are closely coordinated and are nearly identical. The basic AISI-SAE numbering system for plain carbon and alloy steels is shown in the table on page 2106. All steels are identified by four numbers, except certain chromium steels which have five numbers. The first two numbers identify the type of steel and the last two numbers indicate the approximate amount of carbon in hundredths of a per cent. Placed between the first and second pair of numbers, the letter L indicates that the steel contains lead to improve its machinability; when placed in this position, the letter B indicates a boron steel. The prefix E, as in E52100, indicates a steel made by the basic electric furnace method; the suffix H, as in 4150H, indicates a steel that is produced to specific hardenability limits. The Unified Numbering System (UNS) for metals and alloys is also used to designate steels (see pages 2097 and 2098 ).

**Applications and Heat-treatments.** — In applying the detailed heat-treatments (see tables, " Typical Heat-treatments for SAE Steels "), it is recommended, in order to obtain uniform physical properties, that the final quench be made from the lowest temperature that will develop the maximum physical properties, bearing in mind that with thinner sections lower temperatures are required than with thicker sections. It is important to bear in mind when using the information on heat-treatments, that it is based on the tests and experience of steel manufacturers and consumers and is intended only as a guide in selecting the proper steels and their heat-treatment. The notes on heat-treatments and physical properties are not to be considered in any way a part of the standard specifications for SAE steels. They are added solely for the information of users of the steels and the guidance of purchasers in the selection of proper materials for different purposes. They should not be incorporated in the customer's specifications when ordering steel.

Variations in the effect of the usual forms of heat-treatment may be due to personnel, variations in the steel composition or manufacture, and to changes in local conditions such as control and precision of heat-treating equipment. In order to minimize the effects of such variations, steel users should keep each heat of steel separate in the stock-room and during processing so that the necessary adjustments of treatment can be made.

*Water and Brine Quenching:* When selecting a steel, the user should always keep in mind the importance of obtaining the desired strength and hardness without the necessity of resorting to drastic forms of quenching. Water and brine quenching may be considered as drastic treatments when applied to carbon and simple alloy steels containing more than 0.35 per cent carbon. Oil quenching minimizes distortion, whereas water and brine accentuate it. Improperly applied drastic quenching may lead to serious cracking or spalling. Frequently the necessity for drastic heat-treatment can be avoided by proper control of hardenability. Such procedure is especially important in the case of parts of intricate shape and with sudden changes in section. This caution is not intended to condemn the practice of water or brine quenching in all cases, as there are frequent instances where the production of suitable parts can be obtained in no other manner. It is intended as a warning to the uninitiated to go slowly until the necessary experience and skill have been acquired.

**Carburizing.** — The process of carburizing as considered in these notes refers to the various dry or pack hardening methods as well as to the newer processes utilizing gases and molten baths as the carburizing medium. The procedure after carburizing is usually divided into two methods: (1) Quench direct; (2) Cool slowly or in box.

The first method refers to removal of the work from the furnace or from the carburizing box and quenching the parts while they are at or slightly below the carburizing temperature or by quenching from gas carburizing furnaces. The second method is to allow the work to cool slowly without any quenching, in the box or container or in a cooling chamber provided in the furnace. The relative value of these two methods is dependent upon the type of steel treated, the method of carburizing, the kind of furnace installation and the physical results desired. Tempering of parts after carburizing, cyaniding and activated bath treatments is sometimes omitted in commercial practice but is included in the accompanying recommendations as being in accord with good heat treating practice. Parts carburized in activated baths should be treated similarly to other carburized work and may be given any of the hardening treatments shown under the specific steels.

**Jominy Hardenability Test.** — The Jominy end-quench test is a standard method of determining and designating *hardenability* of steel. This test may be used in comparing the hardenability either of successive heats of steel, or of steels of different compositions. The probable hardness of steel for new parts, when production experience is not available, may also be predicted, provided cooling rates occurring during quenching are known.

The Jominy test consists in water-quenching one end of a test bar which is held vertically with the lower end ½ inch above a ½-inch round quenching orifice. The quenching water has a temperature of 40 to 85 degrees F. When the test bar is not in position, this column of water must rise to a free height of 2½ inches above the opening. The test bar (except for shallow-hardening steels) has a diameter of 1 inch and a length of either 3 or 4 inches, depending upon the method of suspending it vertically in the quenching fixture. All decarburization must be removed from the test bar in machining it to the standard diameter and the bar must be normalized. Preparatory to quenching, the test bar is heated to the specified hardening temperature and held at this temperature for 30 minutes. The heated specimen remains in the fixture for at least 10 minutes, and the surrounding air must be free from currents during cooling. The fixture must be dry at the beginning of each test, and the time between removal of the specimen from the furnace and the beginning of the quench shall not be more than 5 seconds.

Two flats 180 degrees apart are ground not less than 0.015 inch deep along the entire length of the test bar. Then Rockwell C hardness measurements are made at ⅟16-inch intervals along these flat surfaces. To illustrate the method of recording the result, assume that a test bar hardened above Rockwell C 50 to a distance of 1½ inches (2⁴⁄16) from the quenched end. This result would be indicated by the marking J₅₀ = 24. Similarly, a hardness of Rockwell C 30 out ⁵⁄16 inch from the end would be designated by J₃₀ = 5. The last figure in each case equals the distance from the end of the test bar in sixteenths of an inch.

### Table 1.  Composition of Standard Steels

These compositions are applicable either to open-hearth or electric furnace steels. For the latter, the maximum phosphorus and sulphur content is 0.025 per cent. The carbon and manganese contents shown differ slightly from those for structural shapes.

| SAE Number | AISI Number* | Carbon C | Manganese Mn | Phosphorus P (Max.) | Sulphur S (Max.) |
|---|---|---|---|---|---|
| | | CARBON STEELS | | | |
| 1006 | C1006 | 0.08 max. | 0.25-0.40 | 0.040 | 0.050 |
| 1008 | C1008 | 0.10 max. | 0.25-0.50 | 0.040 | 0.050 |
| 1010 | C1010 | 0.08-0.13 | 0.30-0.60 | 0.040 | 0.050 |
| 1015 | C1015 | 0.13-0.18 | 0.30-0.60 | 0.040 | 0.050 |
| 1016 | C1016 | 0.13-0.18 | 0.60-0.90 | 0.040 | 0.050 |
| 1017 | C1017 | 0.15-0.20 | 0.30-0.60 | 0.040 | 0.050 |
| 1018 | C1018 | 0.15-0.20 | 0.60-0.90 | 0.040 | 0.050 |
| 1019 | C1019 | 0.15-0.20 | 0.70-1.00 | 0.040 | 0.050 |
| 1020 | C1020 | 0.18-0.23 | 0.30-0.60 | 0.040 | 0.050 |
| 1021 | C1021 | 0.18-0.23 | 0.60-0.90 | 0.040 | 0.050 |
| 1022 | C1022 | 0.18-0.23 | 0.70-1.00 | 0.040 | 0.050 |
| 1024 | C1024 | 0.19-0.25 | 1.35-1.65 | 0.040 | 0.050 |
| 1025 | C1025 | 0.22-0.28 | 0.30-0.60 | 0.040 | 0.050 |
| 1026 | C1026 | 0.22-0.28 | 0.60-0.90 | 0.040 | 0.050 |
| 1027 | C1027 | 0.22-0.29 | 1.20-1.50 | 0.040 | 0.050 |
| 1030 | C1030 | 0.28-0.34 | 0.60-0.90 | 0.040 | 0.050 |
| 1033 | C1033 | 0.30-0.36 | 0.70-1.00 | 0.040 | 0.050 |
| 1034 | C1034 | 0.32-0.38 | 0.50-0.80 | 0.040 | 0.050 |
| 1035 | C1035 | 0.32-0.38 | 0.60-0.90 | 0.040 | 0.050 |
| 1036 | C1036 | 0.30-0.37 | 1.20-1.50 | 0.040 | 0.050 |
| 1038 | C1038 | 0.35-0.42 | 0.60-0.90 | 0.040 | 0.050 |
| 1039 | C1039 | 0.37-0.44 | 0.70-1.00 | 0.040 | 0.050 |
| 1040 | C1040 | 0.37-0.44 | 0.60-0.90 | 0.040 | 0.050 |
| 1041 | C1041 | 0.36-0.44 | 1.35-1.65 | 0.040 | 0.050 |
| 1042 | C1042 | 0.40-0.47 | 0.60-0.90 | 0.040 | 0.050 |
| 1043 | C1043 | 0.40-0.47 | 0.70-1.00 | 0.040 | 0.050 |
| 1045 | C1045 | 0.43-0.50 | 0.60-0.90 | 0.040 | 0.050 |
| 1046 | C1046 | 0.43-0.50 | 0.70-1.00 | 0.040 | 0.050 |
| 1049 | C1049 | 0.46-0.53 | 0.60-0.90 | 0.040 | 0.050 |
| 1050 | C1050 | 0.48-0.55 | 0.60-0.90 | 0.040 | 0.050 |
| 1052 | C1052 | 0.47-0.55 | 1.20-1.50 | 0.040 | 0.050 |
| 1055 | C1055 | 0.50-0.60 | 0.60-0.90 | 0.040 | 0.050 |
| 1060 | C1060 | 0.55-0.65 | 0.60-0.90 | 0.040 | 0.050 |
| 1062 | C1062 | 0.54-0.65 | 0.85-1.15 | 0.040 | 0.050 |
| 1064 | C1064 | 0.60-0.70 | 0.50-0.80 | 0.040 | 0.050 |
| 1065 | C1065 | 0.60-0.70 | 0.60-0.90 | 0.040 | 0.050 |
| 1066 | C1066 | 0.60-0.71 | 0.85-1.15 | 0.040 | 0.050 |
| 1070 | C1070 | 0.65-0.75 | 0.60-0.90 | 0.040 | 0.050 |
| 1074 | C1074 | 0.70-0.80 | 0.50-0.80 | 0.040 | 0.050 |
| 1078 | C1078 | 0.72-0.85 | 0.30-0.60 | 0.040 | 0.050 |
| 1080 | C1080 | 0.75-0.88 | 0.60-0.90 | 0.040 | 0.050 |
| 1085 | C1085 | 0.80-0.93 | 0.70-1.00 | 0.040 | 0.050 |
| 1086 | C1086 | 0.82-0.95 | 0.30-0.50 | 0.040 | 0.050 |
| 1090 | C1090 | 0.85-0.98 | 0.60-0.90 | 0.040 | 0.050 |
| 1095 | C1095 | 0.90-1.03 | 0.30-0.50 | 0.040 | 0.050 |

* American Iron and Steel Institute.

**Quality Variations of Carbon and Alloy Steels.** — Carbon steels may be produced with chemical composition (carbon, manganese, phosphorus, sulfur, and silicon) within the specified limits of a given grade and still have characteristics that are dissimilar. Each grade and quality variation thereof has a proper and useful place, depending upon the end products to be made and the methods of fabrication.

(Continued on page 2107.)

Table 2. Compositions of Free-Cutting Steels[a]

| SAE Number | AISI Number | C | Mn | P | S |
|---|---|---|---|---|---|
| 1111[b] | B1111[c] C1211 | 0.13 max | 0.60-0.90 | 0.07-0.12 | 0.08-0.15 |
| 1112[b] | B1112[c] C1212 | 0.13 max | 0.70-1.00 | 0.07-0.12 | 0.16-0.23 |
| 1113[b] | B1113[c] C1213 | 0.13 max | 0.70-1.00 | 0.07-0.12 | 0.24-0.33 |
| 12L14[d] | — | 0.15 max | 0 80-1.20 | 0.04-0.09 | 0.28-0.35 |
| Open Hearth | | | | Max | |
| 1108 | C1108 | 0.08-0.13 | 0.50-0.80 | 0.040 | 0.08-0.13 |
| 1109 | C1109 | 0.08-0.13 | 0.60-0.90 | 0.040 | 0.08-0.13 |
| 1114 | C1114 | 0.10-0.16 | 1.00-1.30 | 0.040 | 0.08-0.13 |
| 1115 | C1115 | 0.13-0.18 | 0.60-0.90 | 0.040 | 0.08-0.13 |
| 1117 | C1117 | 0.14-0.20 | 1.00-1.30 | 0.040 | 0.08-0.13 |
| 1118 | C1118 | 0.14-0.20 | 1.30-1.60 | 0.040 | 0.08-0.13 |
| 1119 | C1119 | 0.14-0.20 | 1.00-1.30 | 0.040 | 0.24-0.33 |
| 1120 | C1120 | 0.18-0.23 | 0.70-1.00 | 0.040 | 0.08-0.13 |
| 1126 | C1126 | 0.23-0.29 | 0.70-1.00 | 0.040 | 0.08-0.13 |
| 1132 | C1132 | 0.27-0.34 | 1.35-1.65 | 0.040 | 0.08-0.13 |
| 1137 | C1137 | 0.32-0.39 | 1.35-1.65 | 0.040 | 0.08-0.13 |
| 1138 | C1138 | 0.34-0.40 | 0.70-1.00 | 0.040 | 0.08-0.13 |
| 1139 | C1139 | 0.35-0.43 | 1.35-1.65 | 0.040 | 0.12-0.20 |
| 1140 | C1140 | 0.37-0.44 | 0.70-1.00 | 0.040 | 0.08-0.13 |
| 1141 | C1141 | 0.37-0.45 | 1.35-1.65 | 0.040 | 0.08-0.13 |
| 1144 | C1144 | 0.40-0.48 | 1.35-1.65 | 0.040 | 0.24-0.33 |
| 1145 | C1145 | 0.42-0.49 | 0.70-1.00 | 0.040 | 0.04-0.07 |
| 1146 | C1146 | 0.42-0.49 | 0.70-1.00 | 0.040 | 0.08-0.13 |
| 1151 | C1151 | 0.48-0.55 | 0.70-1.00 | 0.040 | 0.08-0.13 |

[a] When silicon is required, the following limits and ranges are commonly used for basic open hearth steel grades: for steel designations up to but excluding SAE 1114, 0.10% max; for SAE 1114 and over, 0.10% max or the ranges of 0.10 to 0.20% or 0.15 to 0.30%; except that limits for SAE 1139 are 0.15 to 0.30%.

[b] These steels may be produced by the Bessemer basic open hearth or basic electric steel-making practices.

[c] Because of the technological nature of the process, acid Bessemer steels are not furnished with specified silicon content.

[d] Lead 0.15 to 0.35%.

Table 3. Compositions of Alloy Steels[*]

| SAE No. | AISI No. | C | Mn | P | S | Si | Ni | Cr |
|---|---|---|---|---|---|---|---|---|
| 1330 | 1330 | 0.28-0.33 | 1.60-1.90 | (a) | (a) | (d) | — | — |
| 1335 | 1335 | 0.33-0.38 | 1.60-1.90 | (a) | (a) | (d) | — | — |
| 1340 | 1340 | 0.38-0.43 | 1.60-1.90 | (a) | (a) | (d) | — | — |
| 1345 | 1345 | 0.43-0.48 | 1.60-1.90 | (a) | (a) | (d) | — | — |
| 2517[1] | E2517 | 0.15-0.20 | 0.45-0.60 | (b) | (b) | (d) | 4.75-5.25 | — |
| 3135 | 3135 | 0.33-0.38 | 0.60-0.80 | (a) | (a) | (d) | 1.10-1.40 | 0.55-0.75 |
| 3140 | 3140 | 0.38-0.43 | 0.70-0.90 | (a) | (a) | (d) | 1.10-1.40 | 0.55-0.75 |
| 3310[1] | E3310 | 0.08-0.13 | 0.45-0.60 | (b) | (b) | (d) | 3.25-3.75 | 1.40-1.75 |

[*] See footnotes at end of table.

Table 3 *(Continued)*. Compositions of Alloy Steels*

| SAE No. | AISI No. | C | Mn | P | S | Si | Ni | Cr | Mo |
|---|---|---|---|---|---|---|---|---|---|
| 4012 | 4012 | 0.09–0.14 | 0.75–1.00 | (a) | (a) | (d) | — | —· | 0.15–0.25 |
| 4023 | 4023 | 0.20–0.25 | 0.70–0.90 | (a) | (a) | (d) | — | — | 0.20–0.30 |
| 4024 | 4024 | 0.20–0.25 | 0.70–0.90 | (a) | (c) | (d) | — | — | 0.20–0.30 |
| 4027 | 4027 | 0.25–0.30 | 0.70–0.90 | (a) | (a) | (d) | — | — | 0.20–0.30 |
| 4028 | 4028 | 0.25–0.30 | 0.70–0.90 | (a) | (c) | (d) | — | — | 0.20–0.30 |
| 4032 | 4032 | 0.30–0.35 | 0.70–0.90 | (a) | (a) | (d) | — | — | 0.20–0.30 |
| 4037 | 4037 | 0.35–0.40 | 0.70–0.90 | (a) | (a) | (d) | — | — | 0.20–0.30 |
| 4042 | 4042 | 0.40–0.45 | 0.70–0.90 | (a) | (a) | (d) | — | — | 0.20–0.30 |
| 4047 | 4047 | 0.45–0.50 | 0.70–0.90 | (a) | (a) | (d) | — | — | 0.20–0.30 |
| 4063 | 4063 | 0.60–0.67 | 0.75–1.00 | (a) | (a) | (d) | — | — | 0.20–0.30 |
| 4118 | 4118 | 0.18–0.23 | 0.70–0.90 | (a) | (a) | (d) | — | 0.40–0.60 | 0.08–0.15 |
| 4130 | 4130 | 0.28–0.33 | 0.40–0.60 | (a) | (a) | (d) | — | 0.80–1.10 | 0.15–0.25 |
| 4135 | 4135 | 0.33–0.38 | 0.70–0.90 | (a) | (a) | (d) | — | 0.80–1.10 | 0.15–0.25 |
| 4137 | 4137 | 0.35–0.40 | 0.70–0.90 | (a) | (a) | (d) | — | 0.80–1.10 | 0.15–0.25 |
| 4140 | 4140 | 0.38–0.43 | 0.75–1.00 | (a) | (a) | (d) | — | 0.80–1.10 | 0.15–0.25 |
| 4142 | 4142 | 0.40–0.45 | 0.75–1.00 | (a) | (a) | (d) | — | 0.80–1.10 | 0.15–0.25 |
| 4145 | 4145 | 0.43–0.48 | 0.75–1.00 | (a) | (a) | (d) | — | 0.80–1.10 | 0.15–0.25 |
| 4147 | 4147 | 0.45–0.50 | 0.75–1.00 | (a) | (a) | (d) | — | 0.80–1.10 | 0.15–0.25 |
| 4150 | 4150 | 0.48–0.53 | 0.75–1.00 | (a) | (a) | (d) | — | 0.80–1.10 | 0.15–0.25 |
| 4320 | 4320 | 0.17–0.22 | 0.45–0.65 | (a) | (a) | (d) | 1.65–2.00 | 0.40–0.60 | 0.20–0.30 |
| 4337 | 4337 | 0.35–0.40 | 0.60–0.80 | (a) | (a) | (d) | 1.65–2.00 | 0.70–0.90 | 0.20–0.30 |
| 4340 | 4340 | 0.38–0.43 | 0.60–0.80 | (a) | (a) | (d) | 1.65–2.00 | 0.70–0.90 | 0.20–0.30 |
| E4340[2] | E4340 | 0.38–0.43 | 0.65–0.85 | (b) | (b) | (d) | 1.65–2.00 | 0.70–0.90 | 0.20–0.30 |
| 4422 | — | 0.20–0.25 | 0.70–0.90 | (a) | (a) | (d) | — | — | 0.35–0.45 |
| 4427 | — | 0.24–0.29 | 0.70–0.90 | (a) | (a) | (d) | — | — | 0.35–0.45 |
| 4520 | — | 0.18–0.23 | 0.45–0.65 | (a) | (a) | (d) | — | — | 0.45–0.60 |
| 4615 | 4615 | 0.13–0.18 | 0.45–0.65 | (a) | (a) | (d) | 1.65–2.00 | — | 0.20–0.30 |
| 4617 | 4617 | 0.15–0.20 | 0.45–0.65 | (a) | (a) | (d) | 1.65–2.00 | — | 0.20–0.30 |
| 4620 | 4620 | 0.17–0.22 | 0.45–0.65 | (a) | (a) | (d) | 1.65–2.00 | — | 0.20–0.30 |
| 4621 | 4621 | 0.18–0.23 | 0.70–0.90 | (a) | (a) | (d) | 1.65–2.00 | — | 0.20–0.30 |
| 4640 | 4640 | 0.38–0.43 | 0.60–0.80 | (a) | (a) | (d) | 1.65–2.00 | — | 0.20–0.30 |
| 4718 | — | 0.16–0.21 | 0.70–0.90 | — | — | — | 0.90–1.20 | 0.35–0.55 | 0.30–0.40 |
| 4720 | 4720 | 0.17–0.22 | 0.50–0.70 | (a) | (a) | (d) | 0.90–1.20 | 0.35–0.55 | 0.15–0.25 |
| 4815 | 4815 | 0.13–0.18 | 0.40–0.60 | (a) | (a) | (d) | 3.25–3.75 | — | 0.20–0.30 |
| 4817 | 4817 | 0.15–0.20 | 0.40–0.60 | (a) | (a) | (d) | 3.25–3.75 | — | 0.20–0.30 |
| 4820 | 4820 | 0.18–0.23 | 0.50–0.70 | (a) | (a) | (d) | 3.25–3.75 | — | 0.20–0.30 |
| 5015 | 5015 | 0.12–0.17 | 0.30–0.50 | (a) | (a) | (d) | — | 0.30–0.50 | — |
| 50B40 | 50B40 | 0.38–0.43 | 0.75–1.00 | (a) | (a) | (d) | — | 0.40–0.60 | — |
| 50B44 | 50B44 | 0.43–0.48 | 0.75–1.00 | (a) | (a) | (d) | — | 0.40–0.60 | — |
| 5046 | 5046 | 0.43–0.50 | 0.75–1.00 | (a) | (a) | (d) | — | 0.20–0.35 | — |
| 50B46[2] | 50B46 | 0.43–0.50 | 0.75–1.00 | (a) | (a) | (d) | — | 0.20–0.35 | — |
| 50B50[3] | 50B50 | 0.48–0.53 | 0.75–1.00 | (a) | (a) | (d) | — | 0.40–0.60 | — |
| 50B60[3] | 50B60 | 0.55–0.65 | 0.75–1.00 | (a) | (a) | (d) | — | 0.40–0.60 | — |
| 5115 | — | 0.13–0.18 | 0.70–0.90 | (a) | (a) | (d) | — | 0.70–0.90 | — |
| 5120 | 5120 | 0.17–0.22 | 0.70–0.90 | (a) | (a) | (d) | — | 0.70–0.90 | — |
| 5130 | 5130 | 0.28–0.33 | 0.70–0.90 | (a) | (a) | (d) | — | 0.80–1.10 | — |
| 5132 | 5132 | 0.30–0.35 | 0.60–0.80 | (a) | (a) | (d) | — | 0.75–1.00 | — |
| 5135 | 5135 | 0.33–0.38 | 0.60–0.80 | (a) | (a) | (d) | — | 0.80–1.05 | — |

* See footnotes at end of table.

Table 3 *(Continued)*. Compositions of Alloy Steels

| SAE No. | AISI No. | C | Mn | P | S | Si | Ni | Cr | Mo | V |
|---|---|---|---|---|---|---|---|---|---|---|
| 5140 | 5140 | 0.38–0.43 | 0.70–0.90 | (a) | (a) | (d) | — | 0.70–0.90 | — | — |
| 5145 | 5145 | 0.43–0.48 | 0.70–0.90 | (a) | (a) | (d) | — | 0.70–0.90 | — | — |
| 5147 | 5147 | 0.45–0.52 | 0.70–0.95 | (a) | (a) | (d) | — | 0.85–1.15 | — | — |
| 5150 | 5150 | 0.48–0.53 | 0.70–0.90 | (a) | (a) | (d) | — | 0.70–0.90 | — | — |
| 5155 | 5155 | 0.50–0.60 | 0.70–0.90 | (a) | (a) | (d) | — | 0.70–0.90 | — | — |
| 5160 | 5160 | 0.55–0.65 | 0.75–1.00 | (a) | (a) | (d) | — | 0.70–0.90 | — | — |
| 51B60[3] | 51B60 | 0.55–0.65 | 0.75–1.00 | (a) | (a) | (d) | — | 0.70–0.90 | — | — |
| 50100[1] | E50100 | 0.95–1.10 | 0.25–0.45 | (b) | (b) | (d) | — | 0.40–0.60 | — | — |
| 51100[1] | E51100 | 0.95–1.10 | 0.25–0.45 | (b) | (b) | (d) | — | 0.90–1.15 | — | — |
| 52100[1] | E52100 | 0.95–1.10 | 0.25–0.45 | (b) | (b) | (d) | — | 1.30–1.60 | — | — |
| 6118 | — | 0.16–0.21 | 0.50–0.70 | (a) | — | (a) | — | 0.50–0.70 | — | (e) |
| 6120 | 6120 | 0.17–0.22 | 0.70–0.90 | (a) | (a) | (d) | — | 0.70–0.90 | — | .10 |
| 6150 | 6150 | 0.48–0.53 | 0.70–0.90 | (a) | (a) | (d) | — | 0.80–1.10 | — | .15 |
| 8115 | 8115 | 0.13–0.18 | 0.70–0.90 | (a) | (a) | (d) | 0.20–0.40 | 0.30–0.50 | 0.08–0.15 | — |
| 81B45[3] | 81B45 | 0.43–0.48 | 0.75–1.00 | (a) | (a) | (d) | 0.20–0.40 | 0.35–0.55 | 0.08–0.15 | — |
| 8615 | 8615 | 0.13–0.18 | 0.70–0.90 | (a) | (a) | (d) | 0.40–0.70 | 0.40–0.60 | 0.15–0.25 | — |
| 8617 | 8617 | 0.15–0.20 | 0.70–0.90 | (a) | (a) | (d) | 0.40–0.70 | 0.40–0.60 | 0.15–0.25 | — |
| 8620 | 8620 | 0.18–0.23 | 0.70–0.90 | (a) | (a) | (d) | 0.40–0.70 | 0.40–0.60 | 0.15–0.25 | — |
| 8622 | 8622 | 0.20–0.25 | 0.70–0.90 | (a) | (a) | (d) | 0.40–0.70 | 0.40–0.60 | 0.15–0.25 | — |
| 8625 | 8625 | 0.23–0.28 | 0.70–0.90 | (a) | (a) | (d) | 0.40–0.70 | 0.40–0.60 | 0.15–0.25 | — |
| 8627 | 8627 | 0.25–0.30 | 0.70–0.90 | (a) | (a) | (d) | 0.40–0.70 | 0.40–0.60 | 0.15–0.25 | — |
| 8630 | 8630 | 0.28–0.33 | 0.70–0.90 | (a) | (a) | (d) | 0.40–0.70 | 0.40–0.60 | 0.15–0.25 | — |
| 8637 | 8637 | 0.35–0.40 | 0.75–1.00 | (a) | (a) | (d) | 0.40–0.70 | 0.40–0.60 | 0.15–0.25 | — |
| 8640 | 8640 | 0.38–0.43 | 0.75–1.00 | (a) | (a) | (d) | 0.40–0.70 | 0.40–0.60 | 0.15–0.25 | — |
| 8642 | 8642 | 0.40–0.45 | 0.75–1.00 | (a) | (a) | (d) | 0.40–0.70 | 0.40–0.60 | 0.15–0.25 | — |
| 8645 | 8645 | 0.43–0.48 | 0.75–1.00 | (a) | (a) | (d) | 0.40–0.70 | 0.40–0.60 | 0.15–0.25 | — |
| 86B45[3] | 86B45 | 0.43–0.48 | 0.75–1.00 | (a) | (a) | (d) | 0.40–0.70 | 0.40–0.60 | 0.15–0.25 | — |
| 8650 | 8650 | 0.48–0.53 | 0.75–1.00 | (a) | (a) | (d) | 0.40–0.70 | 0.40–0.60 | 0.15–0.25 | — |
| 8655 | 8655 | 0.50–0.60 | 0.75–1.00 | (a) | (a) | (d) | 0.40–0.70 | 0.40–0.60 | 0.15–0.25 | — |
| 8660 | 8660 | 0.55–0.65 | 0.75–1.00 | (a) | (a) | (d) | 0.40–0.70 | 0.40–0.60 | 0.15–0.25 | — |
| 8720 | 8720 | 0.18–0.23 | 0.70–0.90 | (a) | (a) | (d) | 0.40–0.70 | 0.40–0.60 | 0.20–0.30 | — |
| 8740 | 8740 | 0.38–0.43 | 0.75–1.00 | (a) | (a) | (d) | 0.40–0.70 | 0.40–0.60 | 0.20–0.30 | — |
| 8742 | 8742 | 0.40–0.45 | 0.75–1.00 | (a) | (a) | (d) | 0.40–0.70 | 0.40–0.60 | 0.20–0.30 | |
| 8822 | 8822 | 0.20–0.25 | 0.75–1.00 | (a) | (a) | (d) | 0.40–0.70 | 0.40–0.60 | 0.30–0.40 | — |
| 9254 | — | 0.50–0.60 | 0.50–0.80 | (a) | (a) | (f) | — | 0.50–0.80 | — | — |
| 9255 | 9255 | 0.50–0.60 | 0.70–0.95 | (a) | (a) | (g) | — | — | — | — |
| 9260 | 9260 | 0.55–0.65 | 0.70–1.00 | (a) | (a) | (g) | — | — | — | — |
| 9262 | 9262 | 0.55–0.65 | 0.75–1.00 | (a) | (a) | (g) | — | 0.25–0.40 | — | — |
| 9310[1] | E9310 | 0.08–0.13 | 0.45–0.65 | (b) | (b) | (d) | 3.00–3.50 | 1.00–1.40 | 0.08–0.15 | — |
| 9315[1] | — | 0.13–0.18 | 0.45–0.65 | (b) | (b) | (d) | 3.00–3.50 | 1.00–1.40 | 0.08–0.15 | — |
| 9317[1] | — | 0.15–0.20 | 0.45–0.65 | (b) | (b) | (d) | 3.00–3.50 | 1.00–1.40 | 0.08–0.15 | — |
| 94B15[2] | 94B15 | 0.13–0.18 | 0.75–1.00 | (a) | (a) | (d) | 0.30–0.60 | 0.30–0.50 | 0.08–0.15 | — |
| 94B17[2] | 94B17 | 0.15–0.20 | 0.75–1.00 | (a) | (a) | (d) | 0.30–0.60 | 0.30–0.50 | 0.08–0.15 | — |
| 94B30 | 94B30 | 0.28–0.33 | 0.75–1.00 | (a) | (a) | (d) | 0.30–0.60 | 0.30–0.50 | 0.08–0.15 | — |
| 94B40 | 94B40 | 0.38–0.43 | 0.75–1.00 | (a) | (a) | (d) | 0.30–0.60 | 0.30–0.50 | 0.08–0.15 | — |
| 9840 | 9840 | 0.38–0.43 | 0.70–0.90 | (a) | (a) | (d) | 0.85–1.15 | 0.70–0.90 | 0.20–0.30 | — |
| 9850 | 9850 | 0.48–0.53 | 0.70–0.90 | (a) | (a) | (d) | 0.85–1.15 | 0.70–0.90 | 0.20–0.30 | — |

Where no letter prefix to the AISI number is shown, the steel is predominately open hearth. [1] Electric furnace steel. [2] Electric furnace steel for aircraft use only. [3] Boron content is 0.0005% min.

(a) 0.040%. (b) 0.025%. (c) 0.035 to 0.050%. (d) 0.20 to 0.35%. (e) 0.10 to 0.15%. (f) 1.20 to 1.60%. (g) 1.80 to 2.20%.

Table 4.  Compositions of Wrought Chromium-Nickel Austenitic Steels
(Not Hardenable by Heat Treatment)

| SAE No.[1] | AISI Type | C Max | Mn Max | Si Max | P Max | S Max | Cr Range | Ni Range | Other Elements |
|---|---|---|---|---|---|---|---|---|---|
| 30201 | 201 | .15 | 5.5–7.5 | 1.0 | .060 | .03 | 16–18 | 3.5–5.5 | N, .25 max |
| 30202 | 202 | .15 | 7.5–10 | 1.0 | .060 | .03 | 17–19 | 4.0–6.0 | N, .25 max |
| 30301 | 301 | .15 | 2.0 | 1.0 | .045 | .03 | 16–18 | 6.0–8.0 | — |
| 30302 | 302 | .15 | 2.0 | 1.0 | .045 | .03 | 17–19 | 8.0–10.0 | — |
| 30302B | 302B | .15 | 2.0 | 2.0–3.0 | .045 | .03 | 17–19 | 8.0–10.0 | — |
| 30303 | 303 | .15 | 2.0 | 1.0 | .20 | .15 min | 17–19 | 8.0–10.0 | Zr or Mo, .60 max[2] |
| 30303 Se | 303 Se | .15 | 2.0 | 1.0 | .20 | .06 | 17–19 | 8.0–10.0 | Se, .15 min |
| 30304 | 304 | .08 | 2.0 | 1.0 | .045 | .03 | 18–20 | 8.0–12.0 | — |
| 30304L | 304L | .03 | 2.0 | 1.0 | .045 | .03 | 18–20 | 8.0–12.0 | — |
| 30305 | 305 | .12 | 2.0 | 1.0 | .045 | .03 | 17–19 | 10.0–13.0 | — |
| 30308 | 308 | .08 | 2.0 | 1.0 | .045 | .03 | 19–21 | 10.0–12.0 | — |
| 30309 | 309 | .20 | 2.0 | 1.0 | .045 | .03 | 22–24 | 12.0–15.0 | — |
| 30309S | 309S | .08 | 2.0 | 1.0 | .045 | .03 | 22–24 | 12.0–15.0 | — |
| 30310 | 310 | .25 | 2.0 | 1.5 | .045 | .03 | 24–26 | 19.0–22.0 | — |
| 30310S | 310S | .08 | 2.0 | 1.5 | .045 | .03 | 24–26 | 19.0–22.0 | — |
| 30314 | 314 | .25 | 2.0 | 1.5–3.0 | .045 | .03 | 23–26 | 19.0–22.0 | — |
| 30316 | 316 | .08 | 2.0 | 1.0 | .045 | .03 | 16–18 | 10.0–14.0 | Mo, 2.0–3.0 |
| 30316L[3] | 316L | .03 | 2.0 | 1.0 | .045 | .03 | 16–18 | 10.0–14.0 | Mo, 2.0–3.0 |
| 30317 | 317 | .08 | 2.0 | 1.0 | .045 | .03 | 18–20 | 11.0–15.0 | Mo, 3.0–4.0 |
| 30321[4] | 321 | .08 | 2.0 | 1.0 | .045 | .03 | 17–19 | 9.0–12.0 | Ti, 5 × C min |
| 30325 | — | .25 | .60–.90 | 1.0–2.0 | .045 | .03 | 7–10 | 19.0–23.0 | Cu, 1.0–1.5 |
| 30330 | — | .25 | 2.0 | 1.5[1] | .045 | .04 | 14–17 | 33.0–37.0 | — |
| 30330A | — | .40–.50 | 2.0 | 1.5[1] | .045 | .04 | 14–17 | 33.0–37.0 | — |
| 30347 | 347 | .08 | 2.0 | 1.0 | .045 | .03 | 17–19 | 9.0–13.0 | Cb-Ta, 10 × C min |
| 30348 | 348 | .08 | 2.0 | 1.0 | .045 | .03 | 17–19 | 9.0–13.0 | Cb-Ta, 10 × C min; Ta, .1 max |

[1] To minimize carbon or nitrogen pick-up, 0.75–1.50 Si is recommended for high temperature applications involving carbon or nitrogen atmospheres.
[2] At producer's option; reported only when intentionally added.
[3] 10.0–15.0 Ni permitted for tubular products.
[4] 9.0–13.0 Ni permitted for tubular products.

Table 5.  Compositions of Wrought Stainless Martensitic
Chromium Steels (Hardenable)

| SAE No.[1] | AISI Type[1] | C Max | Mn Max | Si Max | P Max | S Max | Cr Range | Ni Range | Other Elements |
|---|---|---|---|---|---|---|---|---|---|
| 51403 | 403 | .15 | 1.00 | 0.5 | .04 | .03 | 11.5–13.0 | — | — |
| 51410 | 410 | .15 | 1.00 | 1.0 | .04 | .03 | 11.5–13.5 | — | — |
| 51414 | 414 | .15 | 1.00 | 1.0 | .04 | .03 | 11.5–13.5 | 1.25–2.5 | — |
| 51416 | 416 | .15 | 1.25 | 1.0 | .06 | .15 min | 12.0–14.0 | — | Zr or Mo, .60 max[2] |
| 51416 Se | 416 Se | .15 | 1.25 | 1.0 | .06 | .06 | 12.0–14.0 | — | Se, .15 min |
| 51420 | 420 | Over.15 | 1.00 | 1.0 | .04 | .03 | 12.0–14.0 | — | — |
| 51420F | — | .30–.40 | 1.25 | 1.0 | .06 | .15 min | 12.0–14.0 | — | Zr or Mo, .60 max[2] |
| 51420F Se | — | .30–.40 | 1.25 | 1.0 | .06 | .06 | 12.0–14.0 | — | Se, .15 min |
| 51431 | 431 | .20 | 1.00 | 1.0 | .04 | .03 | 15.0–17.0 | 1.25–2.5 | — |
| 51440A | 440A | .60–.75 | 1.00 | 1.0 | .04 | .03 | 16.0–18.0 | — | Mo, .75 max |
| 51440B | 440B | .75–.95 | 1.00 | 1.0 | .04 | .03 | 16.0–18.0 | — | Mo, .75 max |
| 51440C | 440C | .95–1.20 | 1.00 | 1.0 | .04 | .03 | 16.0–18.0 | — | Mo, .75 max |
| 51440F | — | .95–1.20 | 1.25 | 1.0 | .06 | .15 min | 16.0–18.0 | — | Zr or Mo, .75 max[2] |
| 51440F Se | — | .95–1.20 | 1.25 | 1.0 | .06 | .06 | 16.0–18.0 | — | Se, .15 min |
| 51501 | 501 | Over .10 | 1.00 | 1.0 | .04 | .03 | 4.0–6.0 | — | Mo, .40–.65 |

[1] Suffix F is used to denote a free machining steel.  Suffixes A, B, and C are used to denote three types of steel differing only in carbon content.
[2] At producer's option; reported only when intentionally added.

Table 6. Compositions of Wrought Stainless Ferritic Chromium Steels
(Not Hardenable by Heat Treatment)

| SAE No.[1] | AISI Type[1] | C Max | Mn Max | Si Max | P Max | S Max | Cr Range | Other Elements |
|---|---|---|---|---|---|---|---|---|
| 51405[2] | 405 | .08 | 1.00 | 1.00 | .04 | .03 | 11.5–14.5 | Al, .10–.30 |
| 51430 | 430 | .12 | 1.00 | 1.00 | .04 | .03 | 14.0–18.0 | — |
| 51430F | 430F | .12 | 1.25 | 1.00 | .06 | .15 min | 14.0–18.0 | Zr or Mo, .60 max[3] |
| 51430F Se | 430F Se | .12 | 1.25 | 1.00 | .06 | .06 | 14.0–18.0 | Se, .15 min |
| 51442 | — | .20 | 1.00 | 1.00 | .04 | .035 | 18.0–23.0 | — |
| 51446 | 446 | .20 | 1.50 | 1.00 | .04 | .030 | 23.0–27.0 | N, .25 max |
| 51502 | 502 | .10 | 1.00 | 1.00 | .04 | .030 | 4.0– 6.0 | Mo, .40–.65 |

[1] Suffix F is used to denote a free machining steel.
[2] Essentially nonhardenable by heat treatment.
[3] At producer's option; reported only when intentionally added.

### AISI-SAE System of Designating Carbon and Alloy Steels

| AISI-SAE Designation | Type of Steel and Nominal Alloy Content |
|---|---|
| | Carbon Steels |
| 10XX | Plain Carbon (Mn 1.00% max.) |
| 11XX | Resulfurized |
| 12XX | Resulfurized and Rephosphorized |
| 15XX | Plain Carbon (Max. Mn range 1.00 to 1.65%) |
| | Manganese Steels |
| 13XX | Mn 1.75 |
| | Nickel Steels |
| 23XX | Ni 3.50 |
| 25XX | Ni 5.00 |
| | Nickel-Chromium Steels |
| 31XX | Ni 1.25; Cr 0.65 and 0.80 |
| 32XX | Ni 1.75; Cr 1.07 |
| 33XX | Ni 3.50; Cr 1.50 and 1.57 |
| 34XX | Ni 3.00; Cr 0.77 |
| | Molybdenum Steels |
| 40XX | Mo 0.20 and 0.25 |
| 44XX | Mo 0.40 and 0.52 |
| | Chromium-Molybdenum Steels |
| 41XX | Cr 0.50, 0.80, and 0.95; Mo 0.12, 0.20, 0.25, and 0.30 |
| | Nickel-Chromium-Molybdenum Steels |
| 43XX | Ni 1.82; Cr 0.50 and 0.80; Mo 0.25 |
| 43BVxx | Ni 1.82; Cr 0.50; Mo 0.12 and 0.35; V 0.03 min. |
| 47XX | Ni 1.05; Cr 0.45; Mo 0.20 and 0.35 |
| 81XX | Ni 0.30; Cr 0.40; Mo 0.12 |
| 86XX | Ni 0.55; Cr 0.50; Mo 0.20 |
| 87XX | Ni 0.55; Cr 0.50; Mo 0.25 |
| 88XX | Ni 0.55; Cr 0.50; Mo 0.35 |
| 93XX | Ni 3.25; Cr 1.20; Mo 0.12 |
| 94XX | Ni 0.45; Cr 0.40; Mo 0.12 |
| 97XX | Ni 0.55; Cr 0.20; Mo 0.20 |
| 98XX | Ni 1.00; Cr 0.80; Mo 0.25 |
| | Nickel-Molybdenum Steels |
| 46XX | Ni 0.85 and 1.82; Mo 0.20 and 0.25 |
| 48XX | Ni 3.50; Mo 0.25 |
| | Chromium Steels |
| 50XX | Cr 0.27, 0.40, 0.50, and 0.65 |
| 51XX | Cr 0.80, 0.87, 0.92, 0.95, 1.00, and 1.05 |
| 50XXX | Cr 0.50; C 1.00 min. |
| 51XXX | Cr 1.02; C 1.00 min. |
| 52XXX | Cr 1.45; C 1.00 min. |
| | Chromium-Vanadium Steels |
| 61XX | Cr 0.60, 0.80, and 0.95; V 0.10 and 0.15 min |
| | Tungsten-Chromium Steels |
| 72XX | W 1.75; Cr 0.75 |
| | Silicon-Manganese Steels |
| 92XX | Si 1.40 and 2.00; Mn 0.65, 0.82, and 0.85; Cr 0.00 and 0.65 |
| | High-Strength Low-Alloy Steels |
| 9XX | Various SAE grades |

In all phases of steel production, various practices are employed which determine the quality and types of the finished material.

*Quality Classifications:* The term "quality" as it technically relates to steel products may be indicative of many conditions such as the degree of internal soundness, relative uniformity of composition, relative freedom from injurious surface imperfections, and finish. Steel quality also relates to general suitability for particular applications. Sheet steel surface requirements may be broadly identified as to the end use by the suffix E for exposed parts requiring a good painted surface and suffix U for unexposed parts for which surface finish is unimportant.

Carbon Steel may be obtained in a number of fundamental qualities which reflect various degrees of the quality conditions mentioned above. Some of those qualities may be modified by such requirements as limited Austenitic Grain Size, Specified Discard, Macroetch Test, Special Heat Treating, Maximum Incidental Alloy Elements, Restricted Chemical Composition and Nonmetallic Inclusions. In addition, several of the products have special qualities which are intended for specific end uses or fabricating practices.

Alloy Steels also may be obtained in special qualities with such requirements as Extensometer Test, Fracture Test, Impact Test, Macroetch Test, Nonmetallic Inclusion Test, Special Hardenability Test, and Grain Size Test.

For complete descriptions of the qualities and supplementary requirements for carbon and alloy steels, reference should be made to the latest applicable American Iron and Steel Institute (AISI) Steel Products Manual Section.

## High Strength, Low Alloy Steel, SAE 950. — High strength, low alloy steel represents a specific type of steel in which enhanced mechanical properties and, in most cases, good resistance to atmospheric corrosion are obtained by the addition of moderate amounts of one or more alloying elements other than carbon.

Steels of this type are normally furnished in the hot rolled or annealed condition to minimum mechanical properties. They are not intended for quenching and tempering. The user should not subject them to such treatment without assuming responsibility for the ensuing mechanical properties. Where these steels are used for fabrication by welding, no preheat or postheat is required. In certain complex structures, stress relieving may be desirable. These steels may be obtained in the standard shapes or forms normally available in carbon steel.

*Application:* These steels, because of their enhanced strength, corrosion and erosion resistance, and their high strength-to-weight ratio and service life, are adapted particularly for use in mobile equipment and other structures where substantial weight savings are generally desirable. Typical applications are automotive bumper face bars, truck bodies, frames and structural members, scrapers, dump wagons, cranes, shovels, booms, chutes, conveyors, railroad and industrial cars.

### Certain Minimum Properties of SAE 950 Steel as Furnished by the Mill

| Property * | Thickness or Diameter, inches | | | | |
|---|---|---|---|---|---|
| | Up to 0.0709, inclusive | 0.0710 to 0.2299, inclusive | 0.2300 to ½, inclusive | Over ½ to 1, inclusive | Over 1 to 2, inclusive |
| Minimum yield point, psi...... | 50,000 | 50,000 | 50,000 | 47,000 | 45,000 |
| Minimum tensile strength, psi.. | 70,000 | 70,000 | 70,000 | 67,000 | 65,000 |
| Elongation in 2 in., %......... | 20 | 22 | 22 | 22 | 22 |

* For severe cold forming operations requiring greater ductility, relaxation of the yield point and tensile strength requirements is commonly negotiated between producer and consumer.

**Carbon Steels.** — *SAE steels 1006, 1008, 1010, 1015:* These steels are the lowest carbon steels of the plain carbon type, and are selected where cold formability is the primary requisite of the user. They are produced both as rimmed and killed steels. Rimmed steel is used for sheet, strip, rod, and wire where excellent surface finish or good drawing qualities are required, such as body and fender stock, hoods, lamps, oil pans, and other deep drawn and formed products. It is also used for cold heading wire for tacks, and rivets and low carbon wire products. Killed steel (usually aluminum killed or special killed) is used for difficult stampings or where non-aging properties are needed. Killed steels (usually silicon killed) should be used in preference to rimmed steel for forging or heat treating applications.

These steels have relatively low tensile values and should not be selected where much strength is desired. Within the carbon range of the group, strength and hardness will increase with increase in carbon and/or with cold work, but such increases in strength are at the sacrifice of ductility or the ability to withstand cold deformation. Where cold rolled strip is used the proper temper designation should be specified to obtain the desired properties.

When under 0.15 carbon, the steels are susceptible to serious grain growth, causing brittleness, which may occur as the result of a combination of critical strain (from cold work) followed by heating to certain elevated temperatures. If cold worked parts formed from these steels are to be later heated to temperatures in excess of 1100 degrees F., the user should exercise care to avoid trouble from this cause. When this condition develops it can be overcome by heating the parts to a temperature well in excess of the upper critical point, or at least 1750 degrees F.

Steels in this group, being nearly pure iron or ferritic in structure, do not machine freely and should be avoided for cut screws and operations requiring broaching or smooth finish on turning. The machinability of bar, rod and wire products is improved by cold drawing. Steels in this group are readily welded.

*SAE 1016, 1017, 1018, 1019, 1020, 1021, 1022, 1023, 1024, 1025, 1026, 1027, 1030:* Steels in this group, due to the carbon range covered, have increased strength and hardness, and reduced cold formability compared to the lowest carbon group. For heat treating purposes they are known as carburizing or case hardening grades. When uniform response to heat treatment is required, or for forgings, killed steel is preferred; for other uses, semi-killed or rimmed steel may be indicated, depending on the combination of properties desired. Rimmed steels can ordinarily be supplied up to 0.25 carbon.

Selection of one of these steels for carburizing applications depends on the nature of the part, the properties desired, and the processing practice preferred. Increase in carbon gives greater core hardness with a given quench, or permits the use of thicker sections. Increase in manganese improves the hardenability of both the core and case; in carbon steels this is the only change in composition that will increase case hardenability. The higher manganese variants also machine much better. For carburizing applications SAE 1016, 1018, and 1019 are widely used for thin sections or water quenched parts. SAE 1022 and 1024 are used for heavier sections or where oil quenching is desired, and SAE 1024 is sometimes used for such parts as transmission and rear axle gears. SAE 1027 is used for parts given a light case to obtain satisfactory core properties without drastic quenching. SAE 1025 and 1030, while not usually regarded as carburizing types, are sometimes used in this manner for larger sections or where greater core hardness is needed.

For cold formed or headed parts the lowest manganese grades (SAE 1017, 1020, and 1025) offer the best formability at their carbon level. SAE 1020 is used for fan blades and some frame members, and SAE 1020 and 1025 are widely used for low strength bolts. The next higher manganese types (SAE 1018, 1021, and 1026) provide increased strength.

All of these steels may be readily welded or brazed by the common commercial methods. SAE 1020 is frequently used for welded tubing. These steels are used for numerous forged parts, the lower carbon grades where high strength is not essential. Forgings from the lower carbon steels usually machine better in the as forged condition without annealing, or after normalizing.

*SAE 1030, 1033, 1034, 1035, 1036, 1038, 1039, 1040, 1041, 1042, 1043, 1045, 1046, 1049, 1050, 1052:* These steels, of the medium carbon type, are selected for uses where higher mechanical properties are needed and are frequently further hardened and strengthened by heat treatment or by cold work. These grades are ordinarily produced as killed steels.

Steels in this group are suitable for a wide variety of automotive type applications. The particular carbon and manganese level selected is affected by a number of factors. Increase in the mechanical properties required in section thickness, or in depth of hardening, ordinarily indicates either higher carbon or manganese or both. The heat treating practice preferred, particularly the quenching medium, has a great effect on the steel selected. In general, any of the grades over 0.30 carbon may be selectively hardened by induction or flame methods.

The lower carbon and manganese steels in this group find usage for certain types of cold formed parts. SAE 1030 is used for shift and brake levers, SAE 1034 and 1035 are used in the form of wire and rod for cold upsetting such as bolts, and SAE 1038 for bolts and studs. In practically all cases the parts cold formed from these steels are heat treated prior to use. Stampings are usually limited to flat parts or simple bends. The higher carbon SAE 1038, 1040, and 1042 are frequently cold drawn to specified physical properties for use without heat treatment for some applications, such as cylinder head studs.

All of this group of steels are used for forgings, the selection being governed by the section size and the physical properties desired after heat treatment. Thus SAE 1030 and 1035 are used for shifter forks and many small forgings where moderate properties are desired, but the deeper hardening SAE 1036 is used for more critical parts where a higher strength level and more uniformity is essential, such as some front suspension parts. Forgings such as connecting rods, steering arms, truck front axles, axle shafts, and tractor wheels are commonly made from the SAE 1038 to 1045 group. Larger forgings at similar strength levels need more carbon and perhaps more manganese. Examples are crankshafts from SAE 1046 and 1052. These steels are also used for small forgings where high hardness after oil quenching is desired. Suitable heat treatment is necessary on forgings from this group to provide machinability. These steels are also widely used for parts machined from bar stock, the selection following an identical pattern to that described for forgings. They are used both with and without heat treatment, depending on the application and the level of properties needed. As a class they are considered good for normal machining operations. It is also possible to weld these steels by most commercial methods, but precautions should be taken to avoid cracking from too rapid cooling.

*SAE 1055, 1060, 1062, 1064, 1065, 1066, 1070, 1074, 1078, 1080, 1085, 1086, 1090, 1095:* Steels in this group are of the high carbon type, having more carbon than is required to achieve maximum as quenched hardness. They are used for applications where the higher carbon is needed to improve wear characteristics for cutting edges, to make springs, and for special purposes. Selection of a particular grade is affected by the nature of the part, its end use, and the manufacturing methods available.

In general, cold forming methods are not practical on this group of steels, being limited to flat stampings and springs coiled from small diameter wire. Practically all parts from these steels are heat treated before use, with some variations in heat treating methods to obtain optimum properties for the particular use to which the steel is to be put.

Uses in the spring industry include SAE 1065 for pretempered wire and SAE 1066 for cushion springs of hard drawn wire, SAE 1064 may be used for small washers and thin stamped parts, SAE 1074 for light flat springs formed from annealed stock, and SAE 1080 and 1085 for thicker flat springs. SAE 1085 is also used for heavier coil springs. Valve spring wire and music wire are special products.

Due to good wear properties when properly heat treated, the high carbon steels find wide usage in the farm implement industry. SAE 1070 has been used for plow beams, SAE 1074 for plow shares, and SAE 1078 for such parts as rake teeth, scrapers, cultivator shovels and plow shares. SAE 1085 has been used for scraper blades, disks, and for spring tooth harrows. SAE 1086 and 1090 find use as mower and binder sections, twine holders, and knotter disks.

**Free Cutting Steels.** — *SAE 1111, 1112, 1113:* This class of steels is intended for those uses where easy machining is the primary requirement. They are characterized by a higher sulphur content than comparable carbon steels. This results in some sacrifice of cold forming properties, weldability, and forging characteristics. In general the uses are similar to those for carbon steels of similar carbon and manganese content.

These steels are commonly known as Bessemer screw stock, and are considered the best machining steels available, machinability improving within the group as sulphur increases. They are used for a wide variety of machined parts. While of excellent strength in the cold drawn condition, they have an unfavorable property of cold shortness and are not commonly used for vital parts. These steels may be cyanided or carburized but when uniform response to heat treating is necessary, open hearth steels are recommended.

*SAE 1109, 1114, 1115, 1116, 1117, 1118, 1119, 1120, 1126:* Steels in this group are used where a combination of good machinability and more uniform response to heat treatment is needed. The lower carbon varieties are used for small parts which are to be cyanided or carbonitrided. SAE 1116, 1117, 1118, and 1119 carry more manganese for better hardenability, permitting oil quenching after case hardening heat treatments in many instances. The higher carbon SAE 1120 and 1126 provide more core hardness when this is needed.

*SAE 1132, 1137, 1138, 1140, 1141, 1144, 1145, 1146, 1151:* This group of steels has characteristics comparable to carbon steels of the same carbon level, except for changes due to higher sulphur as noted previously.

They are widely used for parts where a large amount of machining is necessary, or where threads, splines or other operations offer special tooling problems. SAE 1137, for example, is widely used for nuts and bolts and studs with machined threads. The higher manganese SAE 1132, 1137, 1141, and 1144 offer greater hardenability, the higher carbon types being suitable for oil quenching for many parts. All of these steels may be selectively hardened by induction or flame heating if desired.

**Carburizing Grades of Alloy Steels.** *Properties of the Case:* The properties of carburized and hardened cases depend upon the carbon and alloy content, the structure of the case, and the degree and distribution of residual stresses. The carbon content of the case depends upon the details of the carburizing process, and the response of iron and the alloying elements present, to carburization. The original carbon content of the steel has little or no effect upon the carbon content produced in the case. The hardenability of the case therefore depends upon the alloy content of the steel and the final carbon content produced by carburizing, but not upon the initial carbon content of the steel.

With complete carbide solution the effect of alloying elements upon the hardenability of the case, will in general be the same as the effect of these elements upon the hardenability of the core. As an exception to this, any element which inhibits carburizing may reduce the hardenability of the case. It is also true that some elements which raise the hardenability of the core may tend to produce more retained austenite and consequently somewhat lower hardness in the case.

Alloy steels are frequently used for case hardening because the required surface hardness can be obtained by moderate speeds of quenching. This may mean less distortion than would be encountered with water quenching. It is usually desirable to select a steel which will attain a minimum surface hardness of 58 or 60 Rockwell C after carburizing and oil quenching. Where section sizes are large, a high hardenability alloy steel may be necessary, while for medium and light sections, low hardenability steels will suffice.

In general, the case hardening alloy steels may be divided into two classes so far as the hardenability of the case is concerned. Only the general type of steel (SAE 3300–4100, etc.) is given. As the original carbon content of the steel has no effect upon the carbon content of the case, the last two digits in the specification numbers are not meaningful so far as the case is concerned.

(a) — *High Hardenability Case.*— *SAE 2500, 3300, 4300, 4800, 9300*

As these are high alloy steels, both the case and the core have high hardenability. These types of steel are used particularly for carburized parts having thick sections, such as bevel drive pinions and heavy gears. Good case properties can be obtained by oil quenching. These steels are likely to have retained austenite in the case after carburizing and quenching, consequently special precautions or treatments, such as refrigeration, may be required.

(b) — *Medium Hardenability Case.*— *SAE 1300, 2300, 4000, 4100, 4600, 5100, 8600, 8700*

Carburized cases of these steels have medium hardenability which means that their hardenability is intermediate between that of plain carbon steel and the higher alloy carburizing steels just described. In general, these steels can be used for average size case hardened automotive parts such as gears, pinions, piston pins, ball studs, universal crosses, crankshafts, etc. Satisfactory case hardness should be produced in most cases by oil quenching.

*Core Properties:* The core properties of case hardened steels depend upon both carbon and alloy content of the steel. Each of the general types of alloy case hardening steel is usually made with two or more carbon contents so as to produce different hardenability in the core.

The most desirable hardness for the core depends upon the design and functioning of the individual part. In general, where high compressive loads are encountered, relatively high core hardness is beneficial in supporting the case. Low core hardnesses may be desirable where great toughness is essential.

The case hardening steels may be divided into three general classes depending upon hardenability of the core.

(a) — *Low Hardenability Core.* — *SAE 4017, 4023, 4024, 4027[1], 4028[1], 4608, 4615, 4617[1], 8615[1], 8617[1]*

(b) — *Medium Hardenability Core.* — *SAE 1320, 2317, 2512, 2515[1], 3115, 3120, 4032, 4119, 4317, 4620, 4621, 4812, 4815[1], 5115, 5120, 8620, 8622, 8720, 9420*

(c) — *High Hardenability Core.* — *SAE 2517, 3310, 3316, 4320, 4817, 4820, 9310, 9315, 9317*

*Heat Treatments:* In general, all of the alloy carburizing steels are made fine grain and most are suitable for direct quenching from the carburizing temperature. Several other types of heat treatment involving single and double quenching are also used for most of these steels. (See tables of Typical Heat Treatments for SAE Steels.)

---

[1] Borderline classifications might be considered in the next higher hardenability group.

**Directly Hardenable Grades of Alloy Steels.** — These steels may be considered in five groups on the basis of approximate mean carbon content of the SAE specification. In general, the last two figures of the specification agree with the mean carbon content. Consequently the heading " .30–.37 Mean Carbon Content of SAE Specification " includes steels such as SAE 1330, 3135, and 4137.

| Mean Carbon Content of SAE Specification | Common Applications |
|---|---|
| (a) — .30–.37 per cent | Heat treated parts requiring moderate strength and great toughness. |
| (b) — .40–.42 per cent | Heat treated parts requiring higher strength and good toughness. |
| (c) — .45–.50 per cent | Heat treated parts requiring fairly high hardness and strength with moderate toughness. |
| (d) — .50–.62 per cent | Springs and hand tools. |
| (e) — 1.02 per cent | Ball and roller bearings. |

It is necessary to deviate from the above plan in the classification of the carbon molybdenum steels. When carbon molybdenum steels are used, it is customary to specify higher carbon content for any given application than would be specified for other alloy steels, due to the low alloy content of these steels. For example, SAE 4063 is used for the same applications as SAE 4140, 4145 and 5150. Consequently in the following discussion, the carbon molybdenum steels have been shown in the groups where they belong on the basis of applications rather than carbon content.

For the present discussion, steels of each carbon content are divided into two or three groups on the basis of hardenability. Transformation ranges and consequently heat treating practices vary somewhat with different alloying elements even though the hardenability is not changed.

*.30–.37 Mean Carbon Content of SAE Specification.* — These steels are frequently used for water quenched parts of moderate section size and for oil quenched parts of small section size. Typical applications of these steels are connecting rods, steering arms and steering knuckles, axle shafts, bolts, studs, screws, and other parts requiring strength and toughness where section size is small enough to permit obtaining the desired physical properties with the customary heat treatment.

Steels falling in this classification may be subdivided into two groups on the basis of hardenability:

(a) — Low Hardenability: SAE 1330, 1335, 4037, 4042, 4130, 5130, 5132, 8630

(b) — Medium Hardenability: SAE 2330, 3130, 3135, 4137, 5135, 8632, 8635, 8637, 8735, 9437

*.40–.42 Mean Carbon Content of SAE Specification.* — In general, these steels are used for medium and large size parts requiring high degree of strength and toughness. The choice of the proper steel depends upon the section size and the mechanical properties which must be produced. The low and medium hardenability steels are used for average size automotive parts such as steering knuckles, axle shafts, propeller shafts, etc. The high hardenability steels are used particularly for large axles and shafts for large aircraft parts.

These steels are usually considered as oil quenching steels, although some large parts made of the low and medium hardenability classifications may be quenched in water under properly controlled conditions.

These steels may be divided into three groups on the basis of hardenability:

(a) — Low Hardenability: SAE 1340, 4047, 5140, 9440

(b) — Medium Hardenability: SAE 2340, 3140, 3141, 4053, 4063, 4140, 4640, 8640, 8641, 8642, 8740, 8742, 9442

(c) — High Hardenability: SAE 4340, 9840

*.45-.50 Mean Carbon Content of SAE Specification.* — These steels are used primarily for gears and other parts requiring fairly high hardness as well as strength and toughness. Such parts are usually oil quenched and a minimum of 90 per cent martensite in the as quenched condition is desirable.

(a) — Low Hardenability: SAE 5045, 5046, 5145, 9747, 9763

(b) — Medium Hardenability: SAE 2345, 3145, 3150, 4145, 5147, 5150, 8645, 8647, 8650, 8745, 8747, 8750, 9445, 9845

(c) — High Hardenability: SAE 4150, 9850

*.50-.62 Mean Carbon Content of SAE Specification.* — These steels are used primarily for springs and hand tools. The hardenability necessary depends upon the thickness of the material and the quenching practice.

(a) — Medium Hardenability: SAE 4068, 5150, 5152, 6150, 8650, 9254, 9255, 9260, 9261

(b) — High Hardenability: SAE 8653, 8655, 8660, 9262

*1.02 Mean Carbon Content of SAE Specification.* — *SAE 50100, 51100, 52100*

These are straight chromium electric furnace steels used primarily for the races and balls or rollers of anti-friction bearings. They are also used for other parts requiring high hardness and wear resistance. The compositions of the three steels are identical, except for a variation in chromium, with a corresponding variation in hardenability.

(a) — Low Hardenability: SAE 50100

(b) — Medium Hardenability: SAE 51100, 52100

*Resulphurized Steel.* — Some of the alloy steels, SAE 4024, 4028 and 8641, are made resulphurized so as to give better machinability at a relatively high hardness. In general, increased sulphur results in decreased transverse ductility, notched impact toughness, and weldability.

## Chromium Nickel Austenitic Steels (*Not capable of heat treatment*). —

*SAE 30301:* This steel is capable of attaining high tensile strength and ductility by moderate or severe cold working. It is used largely in the cold rolled or cold drawn condition in the form of sheet, strip and wire. Its corrosion resistance is good but not equal to SAE 30302.

*SAE 30302:* This is the most widely used of the general purpose austenitic chromium nickel stainless steels. It is used for deep drawing largely in the annealed condition. It can be worked to high tensile strengths but with slightly lower ductility than SAE 30301.

*SAE 30303F:* This is a free machining type recommended for the manufacture of parts produced on automatic machines. Caution must be used in forging this steel.

*SAE 30304:* This is similar to SAE 30302 but somewhat superior in corrosion resistance and having superior welding properties for certain types of equipment.

*SAE 30305:* Similar to SAE 30304 but capable of lower hardness. Has greater ductility with slower work hardening tendency.

*SAE 30309:* This steel has high heat resisting qualities and is resistant to oxidation at temperatures up to about 1800 deg. F.

*SAE 30310:* This steel has the highest heat resisting properties of any of the chromium nickel steels listed herewith and is used to resist oxidation at temperatures up to about 1900 deg. F.

*SAE 30316:* This steel is recommended for use in parts where unusual resistance to chemical or salt water corrosion is necessary. It has superior creep strength at elevated temperatures.

*SAE 30317:* This steel is similar to SAE 30316 but has the highest corrosion resistance of all these alloys in many environments.

*SAE 30321:* This steel is recommended for use in the manufacture of welded structures where heat treatment after welding is not feasible. It is also recommended for use where temperatures up to 1600 deg. F. are encountered in service.

*SAE 30325:* Used for such parts as heat control shafts.

*SAE 30347:* This steel is similar to SAE 30321 with the following additional statement. This columbium alloy is sometimes preferred to titanium because less columbium is lost in the welding operation.

**Stainless Chromium Irons and Steels.** — *SAE 51410:* This is a general purpose stainless steel capable of heat treatment to show good physical properties. It is used for general stainless applications, both in the heat treated and annealed condition but it is not as resistant to corrosion as SAE 51430 in either the annealed or heat treated condition.

*SAE 51414:* This is a corrosion and heat resisting nickel-bearing chromium steel with somewhat better corrosion resistance than SAE 51410. It will attain slightly higher mechanical properties when heat treated than SAE 51410. It is used in the form of tempered strip or wire, and in bars and forgings for heat treated parts.

*SAE 51416F:* This is a free machining grade for the manufacture of parts produced in automatic screw machines.

*SAE 51420:* This steel is capable of heat treating to a relatively high hardness. It will harden to a maximum of approximately 500 Brinell. It has its maximum corrosion resisting qualities only in the fully hardened condition. It is used for cutlery, hardened pump shafts, etc.

*SAE 51420F:* This is similar to SAE 51420 except for its free machining properties.

*SAE 51430:* This is a steel of a high chromium type not capable of heat treatment and is recommended for use in parts of moderate draw. Corrosion and heat resistance are superior to SAE 51410.

*SAE 51430F:* This is similar to SAE 51430 except for its free machining properties.

*SAE 51431:* This is a nickel bearing chromium steel designed for heat treatment to high mechanical properties. Its corrosion resistance is superior to other hardenable steels.

*SAE 51440A:* A hardenable chromium steel with greater quenched hardness than SAE 51420 and greater toughness than SAE 51440B and 51440C. Maximum corrosion resistance is obtained in the fully hardened and polished condition.

*SAE 51440B:* A hardenable chromium steel with greater quenched hardness than SAE 51440A. Maximum corrosion resistance is obtained in the fully hardened and polished condition. Capable of hardening to 50–60 Rockwell C depending upon carbon content.

*SAE 51440C:* This steel has the greatest quenched hardness and wear resistance upon heat treatment of any corrosion or heat resistant steel.

*SAE 51440F:* The same as SAE 51440C, except for its free machining characteristics.

*SAE 51442:* A corrosion and heat resisting chromium steel with corrosion resisting properties slightly better than SAE 51430 and with good scale resistance up to 1600 deg. F.

*SAE 51446:* A corrosion and heat resisting steel with maximum amount of chromium consistent with commercial malleability. Used principally for parts which must resist high temperatures in service without scaling. Resists oxidation up to 2000 deg. F.

*SAE 51501:* Used for its heat and corrosion resistance and good mechanical properties at temperatures up to approximately 1000 deg. F.

## General Applications of SAE Steels

These applications are intended as a general guide only since the selection may depend upon the exact character of the service, cost of material, machinability when machining is required, or other factors. When more than one steel is recommended for a given application, information on the characteristics of each steel listed will be found in the section beginning on page 2108.

| Application | SAE No. | Application | SAE No. |
|---|---|---|---|
| Adapters................ | 1145 | Chain pins, transmission .. | 4320 |
| Agricultural steel........ | 1070 | "      "      "      .. | 4815 |
| "          " ........ | 1080 | "      "      " | 4820 |
| Aircraft forgings........ | 4140 | Chains, transmission..... | 3135 |
| Axles, front or rear....... | 1040 | "          " ..... | 3140 |
| "    "    " ........ | 4140 | Clutch disks............ | 1060 |
| Axle shafts............. | 1045 | "      " ............ | 1070 |
| "      " ............. | 2340 | "      " ............ | 1085 |
| "      " ............. | 2345 | Clutch springs........... | 1060 |
| "      " ............. | 3135 | Coil springs............. | 4063 |
| "      " ............. | 3140 | Cold-headed bolts........ | 4042 |
| "      " ............. | 3141 | Cold-heading steel........ | 30905 |
| "      " ............. | 4063 | Cold-heading wire or rod.. | rimmed* |
| "      " ............. | 4340 | "      "      " .. | 1035 |
| Ball-bearing races........ | 52100 | Cold-rolled steel........ | 1070 |
| Balls for ball bearings..... | 52100 | Connecting-rods......... | 1040 |
| Body stock for cars....... | rimmed* | "          " ........ | 3141 |
| Bolts, anchor........... | 1040 | Connecting-rod bolts..... | 3130 |
| Bolts and screws........ | 1035 | Corrosion resisting...... | 51710 |
| Bolts, cold-headed....... | 4042 | "          " ...... | 30805 |
| Bolts, connecting-rod..... | 3130 | Covers, transmission...... | rimmed* |
| Bolts, heat-treated........ | 2330 | Crankshafts............. | 1045 |
| Bolts, heavy-duty........ | 4815 | "          " | 1145 |
| "      "      " ........ | 4820 | "          " | 3135 |
| Bolts, steering-arm....... | 3130 | "          " | 3140 |
| Brake levers............ | 1030 | "          " | 3141 |
| "      " ............ | 1040 | Crankshafts, Diesel engine. | 4340 |
| Bumper bars............ | 1085 | Cushion springs......... | 1060 |
| Cams, free-wheeling...... | 4615 | Cutlery, stainless........ | 51335 |
| "      "      ...... | 4620 | Cylinder studs........... | 3130 |
| Camshafts............. | 1020 | Deep-drawing steel....... | rimmed* |
| "      " ............. | 1040 | "      "      " ...... | 30905 |
| Carburized parts........ | 1020 | Differential gears........ | 4023 |
| "          " ........ | 1022 | Disks, clutch............ | 1070 |
| "          " ........ | 1024 | "      " ............ | 1060 |
| "          " ........ | 1320 | Ductile steel............ | 30905 |
| "          " ........ | 2317 | Fan blades............. | 1020 |
| "          " ........ | 2515 | Fatigue resisting......... | 4340 |
| "          " ........ | 3310 | "          " ........ | 4640 |
| "          " ........ | 3115 | Fender stock for cars..... | rimmed* |
| "          " ........ | 3120 | Forgings, aircraft........ | 4140 |
| "          " ........ | 4023 | Forgings, carbon steel..... | 1040 |
| "          " ........ | 4032 | "      "      " ..... | 1045 |
| "          " ........ | 1117 | Forgings, heat-treated..... | 3240 |
| "          " ........ | 1118 | "      "      " ..... | 5140 |

* The "rimmed" and "killed" steels listed are in the SAE 1008, 1010 and 1015 group. See general description of these steels.

## General Applications of SAE Steels

These applications are intended as a general guide only since the selection may depend upon the exact character of the service, cost of material, machinability when machining is required, or other factors. When more than one steel is recommended for a given application, information on the characteristics of each steel listed will be found in the section beginning on page 2108.

| Application | SAE No. | Application | SAE No. |
|---|---|---|---|
| Forgings, heat-treated.... | 6150 | Key stock................ | 1030 |
| Forgings, high-duty....... | 6150 | "     "  ................ | 2330 |
| Forgings, small or medium. | 1035 | "     "  ................ | 3130 |
| Forgings, large.......... | 1036 | Leaf springs............. | 1085 |
| Free-cutting carbon steel.. | 1111 | "     "  ............. | 9260 |
| "     "     "     " | 1113 | Levers, brake............ | 1030 |
| Free-cutting chro.-ni. steel. | 30615 | "     "  ............ | 1040 |
| Free-cutting mang. steel... | 1132 | Levers, gear shift........ | 1030 |
| "     "     "     " ... | 1137 | Levers, heat-treated...... | 2330 |
| Gears, carburized........ | 1320 | Lock-washers............ | 1060 |
| "     "  ........ | 2317 | Mower knives............ | 1085 |
| "     "  ........ | 3115 | Mower sections.......... | 1070 |
| "     "  ........ | 3120 | Music wire.............. | 1085 |
| "     "  ........ | 3310 | Nuts.................... | 3130 |
| "     "  ........ | 4119 | Nuts, heat-treated....... | 2330 |
| "     "  ........ | 4125 | Oil-pans, automobile...... | rimmed* |
| "     "  ........ | 4320 | Pinions, carburized...... | 3115 |
| "     "  ........ | 4615 | "     "  ..... | 3120 |
| "     "  ........ | 4620 | "     "  ...... | 4320 |
| "     "  ........ | 4815 | Piston-pins.............. | 3115 |
| "     "  ........ | 4820 | "     "  .............. | 3120 |
| Gears, heat-treated....... | 2345 | Plow beams............. | 1070 |
| Gears, car and truck..... | 4027 | Plow disks.............. | 1080 |
| "     "     "     "  .... | 4032 | Plow shares............. | 1080 |
| Gears, cyanide-hardening.. | 5140 | Propeller shafts......... | 2340 |
| Gears, differential........ | 4023 | "     "  ......... | 2345 |
| Gears, high duty......... | 4640 | "     "  ......... | 4140 |
| "     "     "  ........ | 6150 | Races, ball-bearing....... | 52100 |
| Gears, oil-hardening...... | 3145 | Ring gears.............. | 3115 |
| "     "     "  ...... | 3150 | "     "  .............. | 3120 |
| "     "     "  ...... | 4340 | "     "  .............. | 4119 |
| "     "     "  ...... | 5150 | Rings, snap............. | 1060 |
| Gears, ring............. | 1045 | Rivets.................. | rimmed* |
| "     "  ............. | 3115 | Rod and wire........... | killed* |
| "     "  ............. | 3120 | Rod, cold-heading....... | 1035 |
| "     "  ............. | 4119 | Roller bearings.......... | 4815 |
| Gears, transmission....... | 3115 | Rollers for bearings...... | 52100 |
| "     "  ....... | 3120 | Screws and bolts........ | 1035 |
| "     "  ....... | 4119 | Screw stock, Bessemer.... | 1111 |
| Gears, truck and bus..... | 3310 | "     "     "  .... | 1112 |
| "     "     "     "  ..... | 4320 | "     "     "  ..... | 1113 |
| Gear shift levers........ | 1030 | Screw stock, open hearth... | 1115 |
| Harrow disks............ | 1080 | Screws, heat-treated...... | 2330 |
| "     "  ........ | 1095 | Seat springs............. | 1095 |
| Hay-rake teeth.......... | 1095 | Shafts, axle............. | 1045 |

* The "rimmed" and "killed" steels listed are in the SAE 1008, 1010 and 1015 group. See general description of these steels.

## General Applications of SAE Steels

These applications are intended as a general guide only since the selection may depend upon the exact character of the service, cost of material, machinability when machining is required, or other factors. When more than one steel is recommended for a given application, information on the characteristics of each steel listed will be found in the section beginning on page 2108.

| Application | SAE No. | Application | SAE No. |
|---|---|---|---|
| Shafts, cyanide-hardening.. | 5140 | Steel, cold-heading....... | 30905 |
| Shafts, heavy-duty....... | 4340 | Steel, free-cutting carbon.. | 11111 |
| "    "    " ....... | 6150 | "    "    "    " .. | 1113 |
| "    "    " ....... | 4615 | Steel, free-cutting chro.-ni.. | 30615 |
| "    "    " ....... | 4620 | Steel, free-cutting mang... | 1132 |
| Shafts, oil-hardening...... | 5150 | "    "    "    " .. | 0000 |
| Shafts, propeller......... | 2340 | Steel, minimum distortion. | 4615 |
| "    " ......... | 2345 | "    "    " .. | 4620 |
| "    " ......... | 4140 | "    "    " . | 4640 |
| Shafts, transmission...... | 4140 | Steel, soft ductile ....... | 30905 |
| Sheets and strips......... | rimmed* | Steering arms............ | 4042 |
| Snap rings.............. | 1060 | Steering-arm bolts........ | 3130 |
| Spline shafts............ | 1045 | Steering knuckles........ | 3141 |
| "    " ............ | 1320 | Steering-knuckle pins..... | 4815 |
| "    " ............ | 2340 | "    "    " ..... | 4820 |
| "    " ............ | 2345 | Studs................... | 1040 |
| "    " ............ | 3115 | "    ................... | 1111 |
| "    " ............ | 3120 | Studs, cold-headed....... | 4042 |
| "    " ............ | 3135 | Studs, cylinder.......... | 3130 |
| "    " ............ | 3140 | Studs, heat-treated....... | 2330 |
| "    " ............ | 4023 | Studs, heavy-duty........ | 4815 |
| Spring clips............. | 1060 | "    "    " ....... | 4820 |
| Springs, coil............. | 1095 | Tacks................... | rimmed* |
| "    " ............. | 4063 | Thrust washers.......... | 1060 |
| "    " ............. | 6150 | Thrust washers, oil-harden. | 5150 |
| Springs, clutch.......... | 1060 | Transmission shafts...... | 4140 |
| Springs, cushion......... | 1060 | Tubing.................. | 1040 |
| Springs, leaf............. | 1085 | Tubing, front axle....... | 4140 |
| "    " ............. | 1095 | Tubing, seamless......... | 1030 |
| "    " ............. | 4063 | Tubing, welded.......... | 1020 |
| "    " ............. | 4068 | Universal joints.......... | 1145 |
| "    " ............. | 9260 | Valve springs............ | 1060 |
| "    " ............. | 6150 | Washers, lock............ | 1060 |
| Springs, hard-drawn coiled. | 1066 | Welded structures........ | 30705 |
| Springs, oil-hardening..... | 5150 | Wire and rod............ | killed* |
| Springs, oil-tempered wire. | 1066 | Wire, cold-heading....... | rimmed* |
| Springs, seat............ | 1095 | "    "    " ....... | 1035 |
| Springs, valve........... | 1060 | Wire, hard-drawn spring .. | 1045 |
| Spring wire.............. | 1045 | "    "    "    " ... | 1055 |
| Spring wire, hard-drawn... | 1055 | Wire, music............. | 1085 |
| Spring wire, oil-tempered.. | 1055 | Wire, oil-tempered spring.. | 1055 |
| Stainless irons........... | 51210 | Wrist-pins, automobile.... | 1020 |
| "    ............ | 51710 | Yokes................... | 1145 |
| Steel, cold-rolled........ | 1070 | | |

* The "rimmed" and "killed" steels listed are in the SAE 1008, 1010 and 1015 group. See general description of these steels.

Expected Minimum Mechanical Properties, Conventional Practice, of Cold Drawn Carbon Steel Rounds, Squares, and Hexagons — 1

| Size, in. | As Cold Drawn | | | | | Cold Drawn Followed by Low-Temperature Stress Relief | | | | | Cold Drawn Followed by High-Temperature Stress Relief | | | | |
|---|---|---|---|---|---|---|---|---|---|---|---|---|---|---|---|
| | Strength — Tensile, 1000 lb/in.² | Strength — Yield, 1000 lb/in.² | Elongation in 2 in., Per cent | Reduction in Area, Per cent | Hardness, Bhn | Strength — Tensile, 1000 lb/in.² | Strength — Yield, 1000 lb/in.² | Elongation in 2 in., Per cent | Reduction in Area, Per cent | Hardness, Bhn | Strength — Tensile, 1000 lb/in.² | Strength — Yield, 1000 lb/in.² | Elongation in 2 in., Per cent | Reduction in Area, Per cent | Hardness, Bhn |
| **AISI 1018 and 1025 Steels** | | | | | | | | | | | | | | | |
| 5/8–7/8 | 70 | 60 | 18 | 40 | 143 | ... | ... | ... | ... | ... | 65 | 45 | 20 | 45 | 131 |
| Over 7/8–1¼ | 65 | 55 | 16 | 40 | 131 | ... | ... | ... | ... | ... | 60 | 45 | 20 | 45 | 121 |
| Over 1¼–2 | 60 | 50 | 15 | 35 | 121 | ... | ... | ... | ... | ... | 55 | 45 | 16 | 40 | 111 |
| Over 2–3 | 55 | 45 | 15 | 35 | 111 | ... | ... | ... | ... | ... | 50 | 40 | 15 | 40 | 101 |
| **AISI 1117 and 1118 Steels** | | | | | | | | | | | | | | | |
| 5/8–7/8 | 75 | 65 | 15 | 40 | 149 | 80 | 70 | 15 | 40 | 163 | 70 | 50 | 18 | 45 | 143 |
| Over 7/8–1¼ | 70 | 60 | 15 | 40 | 143 | 75 | 65 | 15 | 40 | 149 | 65 | 50 | 16 | 45 | 131 |
| Over 1¼–2 | 65 | 55 | 13 | 35 | 131 | 70 | 60 | 13 | 35 | 143 | 60 | 50 | 15 | 40 | 121 |
| Over 2–3 | 60 | 50 | 12 | 30 | 121 | 65 | 55 | 12 | 35 | 131 | 55 | 45 | 15 | 40 | 111 |
| **AISI 1035 Steel** | | | | | | | | | | | | | | | |
| 5/8–7/8 | 85 | 75 | 13 | 35 | 170 | 90 | 80 | 13 | 35 | 179 | 80 | 60 | 16 | 45 | 163 |
| Over 7/8–1¼ | 80 | 70 | 12 | 35 | 163 | 85 | 75 | 12 | 35 | 170 | 75 | 60 | 15 | 45 | 149 |
| Over 1¼–2 | 75 | 65 | 12 | 35 | 149 | 80 | 70 | 12 | 35 | 163 | 70 | 60 | 15 | 40 | 143 |
| Over 2–3 | 70 | 60 | 10 | 30 | 143 | 75 | 65 | 10 | 30 | 149 | 65 | 55 | 12 | 35 | 131 |
| **AISI 1040 and 1140 Steels** | | | | | | | | | | | | | | | |
| 5/8–7/8 | 90 | 80 | 12 | 35 | 179 | 95 | 85 | 12 | 35 | 187 | 85 | 65 | 15 | 45 | 170 |
| Over 7/8–1¼ | 85 | 75 | 12 | 35 | 170 | 90 | 80 | 12 | 35 | 179 | 80 | 65 | 15 | 45 | 163 |
| Over 1¼–2 | 80 | 70 | 10 | 30 | 163 | 85 | 75 | 10 | 30 | 170 | 75 | 60 | 15 | 40 | 149 |
| Over 2–3 | 75 | 65 | 10 | 30 | 149 | 80 | 70 | 10 | 30 | 163 | 70 | 55 | 12 | 35 | 143 |

*Source:* AISI Committee of Hot Rolled and Cold Finished Bar Producers as published in 1974 DATABOOK issue of the American Society for Metals' METAL PROGRESS magazine and used with its permission.

Expected Minimum Mechanical Properties, Conventional Practice, of Cold Drawn Carbon Steel Rounds, Squares, and Hexagons — 2

| Size, in. | As Cold Drawn | | | | | Cold Drawn Followed by Low-Temperature Stress Relief | | | | | Cold Drawn Followed by High-Temperature Stress Relief | | | | |
|---|---|---|---|---|---|---|---|---|---|---|---|---|---|---|---|
| | Strength 1000 lb/in.² | | Elonga-tion in 2 in., Per cent | Reduc-tion in Area, Per cent | Hard-ness, Bhn | Strength 1000 lb/in.² | | Elonga-tion in 2 in., Per cent | Reduc-tion in Area, Per cent | Hard-ness, Bhn | Strength 1000 lb/in.² | | Elonga-tion in 2 in., Per cent | Reduc-tion in Area, Per cent | Hard-ness, Bhn |
| | Tensile | Yield | | | | Tensile | Yield | | | | Tensile | Yield | | | |
| **AISI 1045, 1145, and 1146 Steels** | | | | | | | | | | | | | | | |
| 5⁄8–7⁄8 | 95 | 85 | 12 | 35 | 187 | 100 | 90 | 12 | 35 | 197 | 90 | 70 | 15 | 45 | 179 |
| Over 7⁄8–1¼ | 90 | 80 | 11 | 30 | 179 | 95 | 85 | 11 | 30 | 187 | 85 | 70 | 15 | 45 | 170 |
| Over 1¼–2 | 85 | 75 | 10 | 30 | 170 | 90 | 80 | 10 | 30 | 179 | 80 | 65 | 15 | 40 | 163 |
| Over 2–3 | 80 | 70 | 10 | 30 | 163 | 85 | 75 | 10 | 25 | 170 | 75 | 60 | 12 | 35 | 149 |
| **AISI 1050, 1137, and 1151 Steels** | | | | | | | | | | | | | | | |
| 5⁄8–7⁄8 | 100 | 90 | 11 | 35 | 197 | 105 | 95 | 11 | 35 | 212 | 95 | 75 | 15 | 45 | 187 |
| Over 7⁄8–1¼ | 95 | 85 | 11 | 30 | 187 | 100 | 90 | 11 | 30 | 197 | 90 | 75 | 15 | 40 | 179 |
| Over 1¼–2 | 90 | 80 | 10 | 30 | 179 | 95 | 85 | 10 | 30 | 187 | 85 | 70 | 15 | 40 | 170 |
| Over 2–3 | 85 | 75 | 10 | 30 | 170 | 90 | 80 | 10 | 25 | 179 | 80 | 65 | 12 | 35 | 163 |
| **AISI 1141 Steel** | | | | | | | | | | | | | | | |
| 5⁄8–7⁄8 | 105 | 95 | 11 | 30 | 212 | 110 | 100 | 11 | 30 | 223 | 100 | 80 | 15 | 45 | 197 |
| Over 7⁄8–1¼ | 100 | 90 | 10 | 30 | 197 | 105 | 95 | 10 | 30 | 212 | 95 | 80 | 15 | 40 | 187 |
| Over 1¼–2 | 95 | 85 | 10 | 30 | 187 | 100 | 90 | 10 | 25 | 197 | 90 | 75 | 15 | 40 | 179 |
| Over 2–3 | 90 | 80 | 10 | 20 | 179 | 95 | 85 | 10 | 20 | 187 | 85 | 70 | 12 | 30 | 170 |
| **AISI 1144 Steel** | | | | | | | | | | | | | | | |
| 5⁄8–7⁄8 | 110 | 100 | 10 | 30 | 223 | 115 | 105 | 10 | 30 | 229 | 105 | 85 | 15 | 40 | 212 |
| Over 7⁄8–1¼ | 105 | 95 | 10 | 30 | 212 | 110 | 105 | 10 | 30 | 223 | 100 | 85 | 15 | 40 | 197 |
| Over 1¼–2 | 100 | 90 | 10 | 25 | 197 | 105 | 95 | 10 | 25 | 212 | 95 | 80 | 15 | 35 | 187 |
| Over 2–3 | 95 | 85 | 10 | 20 | 187 | 100 | 90 | 10 | 20 | 197 | 90 | 75 | 12 | 30 | 179 |

*Source:* AISI Committee of Hot Rolled and Cold Finished Bar Producers as published in 1974 DATABOOK issue of the American Society for Metals' METAL PROGRESS magazine and used with its permission.

## Typical Mechanical Properties of Selected Carbon and Alloy Steels — 1
(Hot Rolled, Normalized, and Annealed)

| AISI No.* | Treatment | Strength | | Elonga-tion, Per cent | Reduc-tion in Area, Per cent | Hard-ness, Bhn | Impact Strength (Izod), ft-lb |
|---|---|---|---|---|---|---|---|
| | | Tensile | Yield | | | | |
| | | lb/in.² | | | | | |
| 1015 | As-rolled | 61,000 | 45,500 | 39.0 | 61.0 | 126 | 81.5 |
| | Normalized (1700 F) | 61,500 | 47,000 | 37.0 | 69.6 | 121 | 85.2 |
| | Annealed (1600 F) | 56,000 | 41,250 | 37.0 | 69.7 | 111 | 84.8 |
| 1020 | As-rolled | 65,000 | 48,000 | 36.0 | 59.0 | 143 | 64.0 |
| | Normalized (1600 F) | 64,000 | 50,250 | 35.8 | 67.9 | 131 | 86.8 |
| | Annealed (1600 F) | 57,250 | 42,750 | 36.5 | 66.0 | 111 | 91.0 |
| 1022 | As-rolled | 73,000 | 52,000 | 35.0 | 67.0 | 149 | 60.0 |
| | Normalized (1700 F) | 70,000 | 52,000 | 34.0 | 67.5 | 143 | 86.5 |
| | Annealed (1600 F) | 65,250 | 46,000 | 35.0 | 63.6 | 137 | 89.0 |
| 1030 | As-rolled | 80,000 | 50,000 | 32.0 | 57.0 | 179 | 55.0 |
| | Normalized (1700 F) | 75,000 | 50,000 | 32.0 | 60.8 | 149 | 69.0 |
| | Annealed (1550 F) | 67,250 | 49,500 | 31.2 | 57.9 | 126 | 51.2 |
| 1040 | As-rolled | 90,000 | 60,000 | 25.0 | 50.0 | 201 | 36.0 |
| | Normalized (1650 F) | 85,500 | 54,250 | 28.0 | 54.9 | 170 | 48.0 |
| | Annealed (1450 F) | 75,250 | 51,250 | 30.2 | 57.2 | 149 | 32.7 |
| 1050 | As-rolled | 105,000 | 60,000 | 20.0 | 40.0 | 229 | 23.0 |
| | Normalized (1650 F) | 108,500 | 62,000 | 20.0 | 39.4 | 217 | 20.0 |
| | Annealed (1450 F) | 92,250 | 53,000 | 23.7 | 39.9 | 187 | 12.5 |
| 1060 | As-rolled | 118,000 | 70,000 | 17.0 | 34.0 | 241 | 13.0 |
| | Normalized (1650 F) | 112,500 | 61,000 | 18.0 | 37.2 | 229 | 9.7 |
| | Annealed (1450 F) | 90,750 | 54,000 | 22.5 | 38.2 | 179 | 8.3 |
| 1080 | As-rolled | 140,000 | 85,000 | 12.0 | 17.0 | 293 | 5.0 |
| | Normalized (1650 F) | 146,500 | 76,000 | 11.0 | 20.6 | 293 | 5.0 |
| | Annealed (1450 F) | 89,250 | 54,500 | 24.7 | 45.0 | 174 | 4.5 |
| 1095 | As-rolled | 140,000 | 83,000 | 9.0 | 18.0 | 293 | 3.0 |
| | Normalized (1650 F) | 147,000 | 72,500 | 9.5 | 13.5 | 293 | 4.0 |
| | Annealed (1450 F) | 95,250 | 55,000 | 13.0 | 20.6 | 192 | 2.0 |
| 1117 | As-rolled | 70,600 | 44,300 | 33.0 | 63.0 | 143 | 60.0 |
| | Normalized (1650 F) | 67,750 | 44,000 | 33.5 | 63.8 | 137 | 62.8 |
| | Annealed (1575 F) | 62,250 | 40,500 | 32.8 | 58.0 | 121 | 69.0 |
| 1118 | As-rolled | 75,600 | 45,900 | 32.0 | 70.0 | 149 | 80.0 |
| | Normalized (1700 F) | 69,250 | 46,250 | 33.5 | 65.9 | 143 | 76.3 |
| | Annealed (1450 F) | 65,250 | 41,250 | 34.5 | 66.8 | 131 | 78.5 |
| 1137 | As-rolled | 91,000 | 55,000 | 28.0 | 61.0 | 192 | 61.0 |
| | Normalized (1650 F) | 97,000 | 57,500 | 22.5 | 48.5 | 197 | 47.0 |
| | Annealed (1450 F) | 84,750 | 50,000 | 26.8 | 53.9 | 174 | 36.8 |
| 1141 | As-rolled | 98,000 | 52,000 | 22.0 | 38.0 | 192 | 8.2 |
| | Normalized (1650 F) | 102,500 | 58,750 | 22.7 | 55.5 | 201 | 38.8 |
| | Annealed (1500 F) | 86,800 | 51,200 | 25.5 | 49.3 | 163 | 25.3 |
| 1144 | As-rolled | 102,000 | 61,000 | 21.0 | 41.0 | 212 | 39.0 |
| | Normalized (1650 F) | 96,750 | 58,000 | 21.0 | 40.4 | 197 | 32.0 |
| | Annealed (1450 F) | 84,750 | 50,250 | 24.8 | 41.3 | 167 | 48.0 |

* All grades are fine-grained except those in the 1100 series which are coarse-grained. Austenitizing temperatures are given in parentheses. Heat-treated specimens were oil quenched.

*Source:* Bethlehem Steel Corp. and Republic Steel Corp. as published in 1974 DATA-BOOK issue of the American Society for Metals' METAL PROGRESS magazine and used with its permission.

## Typical Mechanical Properties of Selected Carbon and Alloy Steels — 2
(Hot Rolled, Normalized, and Annealed)

| AISI No.* | Treatment | Strength | | Elongation, Per cent | Reduction in Area, Per cent | Hardness, Bhn | Impact Strength (Izod), ft-lb |
|---|---|---|---|---|---|---|---|
| | | Tensile | Yield | | | | |
| | | lb/in.² | | | | | |
| 1340 | Normalized (1600 F) | 121,250 | 81,000 | 22.0 | 62.9 | 248 | 68.2 |
| | Annealed (1475 F) | 102,000 | 63,250 | 25.5 | 57.3 | 207 | 52.0 |
| 3140 | Normalized (1600 F) | 129,250 | 87,000 | 19.7 | 57.3 | 262 | 39.5 |
| | Annealed (1500 F) | 100,000 | 61,250 | 24.5 | 50.8 | 197 | 34.2 |
| 4130 | Normalized (1600 F) | 97,000 | 63,250 | 25.5 | 59.5 | 197 | 63.7 |
| | Annealed (1585 F) | 81,250 | 52,250 | 28.2 | 55.6 | 156 | 45.5 |
| 4140 | Normalized (1600 F) | 148,000 | 95,000 | 17.7 | 46.8 | 302 | 16.7 |
| | Annealed (1500 F) | 95,000 | 60,500 | 25.7 | 56.9 | 197 | 40.2 |
| 4150 | Normalized (1600 F) | 167,500 | 106,500 | 11.7 | 30.8 | 321 | 8.5 |
| | Annealed (1500 F) | 105,750 | 55,000 | 20.2 | 40.2 | 197 | 18.2 |
| 4320 | Normalized (1640 F) | 115,000 | 67,250 | 20.8 | 50.7 | 235 | 53.8 |
| | Annealed (1560 F) | 84,000 | 61,625 | 29.0 | 58.4 | 163 | 81.0 |
| 4340 | Normalized (1600 F) | 185,500 | 125,000 | 12.2 | 36.3 | 363 | 11.7 |
| | Annealed (1490 F) | 108,000 | 68,500 | 22.0 | 49.9 | 217 | 37.7 |
| 4620 | Normalized (1650 F) | 83,250 | 53,125 | 29.0 | 66.7 | 174 | 98.0 |
| | Annealed (1575 F) | 74,250 | 54,000 | 31.3 | 60.3 | 149 | 69.0 |
| 4820 | Normalized (1580 F) | 109,500 | 70,250 | 24.0 | 59.2 | 229 | 81.0 |
| | Annealed (1500 F) | 98,750 | 67,250 | 22.3 | 58.8 | 197 | 68.5 |
| 5140 | Normalized (1600 F) | 115,000 | 68,500 | 22.7 | 59.2 | 229 | 28.0 |
| | Annealed (1525 F) | 83,000 | 42,500 | 28.6 | 57.3 | 167 | 30.0 |
| 5150 | Normalized (1600 F) | 126,250 | 76,750 | 20.7 | 58.7 | 255 | 23.2 |
| | Annealed (1520 F) | 98,000 | 51,750 | 22.0 | 43.7 | 197 | 18.5 |
| 5160 | Normalized (1575 F) | 138,750 | 77,000 | 17.5 | 44.8 | 269 | 8.0 |
| | Annealed (1495 F) | 104,750 | 40,000 | 17.2 | 30.6 | 197 | 7.4 |
| 6150 | Normalized (1600 F) | 136,250 | 89,250 | 21.8 | 61.0 | 269 | 26.2 |
| | Annealed (1500 F) | 96,750 | 59,750 | 23.0 | 48.4 | 197 | 20.2 |
| 8620 | Normalized (1675 F) | 91,750 | 51,750 | 26.3 | 59.7 | 183 | 73.5 |
| | Annealed (1600 F) | 77,750 | 55,875 | 31.3 | 62.1 | 149 | 82.8 |
| 8630 | Normalized (1600 F) | 94,250 | 62,250 | 23.5 | 53.5 | 187 | 69.8 |
| | Annealed (1550 F) | 81,750 | 54,000 | 29.0 | 58.9 | 156 | 70.2 |
| 8650 | Normalized (1600 F) | 148,500 | 99,750 | 14.0 | 40.4 | 302 | 10.0 |
| | Annealed (1465 F) | 103,750 | 56,000 | 22.5 | 46.4 | 212 | 21.7 |
| 8740 | Normalized (1600 F) | 134,750 | 88,000 | 16.0 | 47.9 | 269 | 13.0 |
| | Annealed (1500 F) | 100,750 | 60,250 | 22.2 | 46.4 | 201 | 29.5 |
| 9255 | Normalized (1650 F) | 135,250 | 84,000 | 19.7 | 43.4 | 269 | 10.0 |
| | Annealed (1550 F) | 112,250 | 70,500 | 21.7 | 41.1 | 229 | 6.5 |
| 9310 | Normalized (1630 F) | 131,500 | 82,750 | 18.8 | 58.1 | 269 | 88.0 |
| | Annealed (1550 F) | 119,000 | 63,750 | 17.3 | 42.1 | 241 | 58.0 |

* All grades are fine-grained except those in the 1100 series which are coarse-grained. Austenitizing temperatures are given in parentheses. Heat-treated specimens were oil quenched.
  *Source:* Bethlehem Steel Corp. and Republic Steel Corp. as published in 1974 DATA-BOOK issue of the American Society for Metals' METAL PROGRESS magazine and used with its permission.

## Typical Mechanical Properties of Selected Carbon and Alloy Steels — 3
(Quenched and Tempered)

| AISI No.* | Tempering Temperature, deg F | Strength | | Elonga- tion, Per cent | Reduction in Area, Per cent | Hard- ness, Bhn |
|---|---|---|---|---|---|---|
| | | Tensile | Yield | | | |
| | | 1000 lb/in.² | | | | |
| 1030† | 400 | 123 | 94 | 17 | 47 | 495 |
| | 600 | 116 | 90 | 19 | 53 | 401 |
| | 800 | 106 | 84 | 23 | 60 | 302 |
| | 1000 | 97 | 75 | 28 | 65 | 255 |
| | 1200 | 85 | 64 | 32 | 70 | 207 |
| 1040† | 400 | 130 | 96 | 16 | 45 | 514 |
| | 600 | 129 | 94 | 18 | 52 | 444 |
| | 800 | 122 | 92 | 21 | 57 | 352 |
| | 1000 | 113 | 86 | 23 | 61 | 269 |
| | 1200 | 97 | 72 | 28 | 68 | 201 |
| 1040 | 400 | 113 | 86 | 19 | 48 | 262 |
| | 600 | 113 | 86 | 20 | 53 | 255 |
| | 800 | 110 | 80 | 21 | 54 | 241 |
| | 1000 | 104 | 71 | 26 | 57 | 212 |
| | 1200 | 92 | 63 | 29 | 65 | 192 |
| 1050† | 400 | 163 | 117 | 9 | 27 | 514 |
| | 600 | 158 | 115 | 13 | 36 | 444 |
| | 800 | 145 | 110 | 19 | 48 | 375 |
| | 1000 | 125 | 95 | 23 | 58 | 293 |
| | 1200 | 104 | 78 | 28 | 65 | 235 |
| 1050 | 400 | . . . | . . . | . . . | . . . | . . . |
| | 600 | 142 | 105 | 14 | 47 | 321 |
| | 800 | 136 | 95 | 20 | 50 | 277 |
| | 1000 | 127 | 84 | 23 | 53 | 262 |
| | 1200 | 107 | 68 | 29 | 60 | 223 |
| 1060 | 400 | 160 | 113 | 13 | 40 | 321 |
| | 600 | 160 | 113 | 13 | 40 | 321 |
| | 800 | 156 | 111 | 14 | 41 | 311 |
| | 1000 | 140 | 97 | 17 | 45 | 277 |
| | 1200 | 116 | 76 | 23 | 54 | 229 |
| 1080 | 400 | 190 | 142 | 12 | 35 | 388 |
| | 600 | 189 | 142 | 12 | 35 | 388 |
| | 800 | 187 | 138 | 13 | 36 | 375 |
| | 1000 | 164 | 117 | 16 | 40 | 321 |
| | 1200 | 129 | 87 | 21 | 50 | 255 |
| 1095† | 400 | 216 | 152 | 10 | 31 | 601 |
| | 600 | 212 | 150 | 11 | 33 | 534 |
| | 800 | 199 | 139 | 13 | 35 | 388 |
| | 1000 | 165 | 110 | 15 | 40 | 293 |
| | 1200 | 122 | 85 | 20 | 47 | 235 |
| 1095 | 400 | 187 | 120 | 10 | 30 | 401 |
| | 600 | 183 | 118 | 10 | 30 | 375 |
| | 800 | 176 | 112 | 12 | 32 | 363 |
| | 1000 | 158 | 98 | 15 | 37 | 321 |
| | 1200 | 130 | 80 | 21 | 47 | 269 |
| 1137 | 400 | 157 | 136 | 5 | 22 | 352 |
| | 600 | 143 | 122 | 10 | 33 | 285 |
| | 800 | 127 | 106 | 15 | 48 | 262 |
| | 1000 | 110 | 88 | 24 | 62 | 229 |
| | 1200 | 95 | 70 | 28 | 69 | 197 |

* All grades are fine-grained except those in the 1100 series which are coarse-grained. Austenitizing temperatures are given in parentheses. Heat-treated specimens were oil quenched unless otherwise indicated.    † Water quenched.

*Source:* Bethlehem Steel Corp. and Republic Steel Corp. as published in 1974 DATA-BOOK issue of the American Society for Metals' METAL PROGRESS magazine and used with its permission.

**Typical Mechanical Properties of Selected Carbon and Alloy Steels — 4**
(Quenched and Tempered)

| AISI No.* | Tempering Temperature, deg F | Strength Tensile | Strength Yield | Elongation, Per cent | Reduction in Area, Per cent | Hardness, Bhn |
|---|---|---|---|---|---|---|
| | | 1000 lb/in.² | 1000 lb/in.² | | | |
| 1137† | 400 | 217 | 169 | 5 | 17 | 415 |
| | 600 | 199 | 163 | 9 | 25 | 375 |
| | 800 | 160 | 143 | 14 | 40 | 311 |
| | 1000 | 120 | 105 | 19 | 60 | 262 |
| | 1200 | 94 | 77 | 25 | 69 | 187 |
| 1141 | 400 | 237 | 176 | 6 | 17 | 461 |
| | 600 | 212 | 186 | 9 | 32 | 415 |
| | 800 | 169 | 150 | 12 | 47 | 331 |
| | 1000 | 130 | 111 | 18 | 57 | 262 |
| | 1200 | 103 | 86 | 23 | 62 | 217 |
| 1144 | 400 | 127 | 91 | 17 | 36 | 277 |
| | 600 | 126 | 90 | 17 | 40 | 262 |
| | 800 | 123 | 88 | 18 | 42 | 248 |
| | 1000 | 117 | 83 | 20 | 46 | 235 |
| | 1200 | 105 | 73 | 23 | 55 | 217 |
| 1330† | 400 | 232 | 211 | 9 | 39 | 459 |
| | 600 | 207 | 186 | 9 | 44 | 402 |
| | 800 | 168 | 150 | 15 | 53 | 335 |
| | 1000 | 127 | 112 | 18 | 60 | 263 |
| | 1200 | 106 | 83 | 23 | 63 | 216 |
| 1340 | 400 | 262 | 231 | 11 | 35 | 505 |
| | 600 | 230 | 206 | 12 | 43 | 453 |
| | 800 | 183 | 167 | 14 | 51 | 375 |
| | 1000 | 140 | 120 | 17 | 58 | 295 |
| | 1200 | 116 | 90 | 22 | 66 | 252 |
| 4037 | 400 | 149 | 110 | 6 | 38 | 310 |
| | 600 | 138 | 111 | 14 | 53 | 295 |
| | 800 | 127 | 106 | 20 | 60 | 270 |
| | 1000 | 115 | 95 | 23 | 63 | 247 |
| | 1200 | 101 | 61 | 29 | 60 | 220 |
| 4042 | 400 | 261 | 241 | 12 | 37 | 516 |
| | 600 | 234 | 211 | 13 | 42 | 455 |
| | 800 | 187 | 170 | 15 | 51 | 380 |
| | 1000 | 143 | 128 | 20 | 59 | 300 |
| | 1200 | 115 | 100 | 28 | 66 | 238 |
| 4130† | 400 | 236 | 212 | 10 | 41 | 467 |
| | 600 | 217 | 200 | 11 | 43 | 435 |
| | 800 | 186 | 173 | 13 | 49 | 380 |
| | 1000 | 150 | 132 | 17 | 57 | 315 |
| | 1200 | 118 | 102 | 22 | 64 | 245 |
| 4140 | 400 | 257 | 238 | 8 | 38 | 510 |
| | 600 | 225 | 208 | 9 | 43 | 445 |
| | 800 | 181 | 165 | 13 | 49 | 370 |
| | 1000 | 138 | 121 | 18 | 58 | 285 |
| | 1200 | 110 | 95 | 22 | 63 | 230 |
| 4150 | 400 | 280 | 250 | 10 | 39 | 530 |
| | 600 | 256 | 231 | 10 | 40 | 495 |
| | 800 | 220 | 200 | 12 | 45 | 440 |
| | 1000 | 175 | 160 | 15 | 52 | 370 |
| | 1200 | 139 | 122 | 19 | 60 | 290 |

* All grades are fine-grained except those in the 1100 series which are coarse-grained. Austenitizing temperatures are given in parentheses. Heat-treated specimens were oil quenched unless otherwise indicated. † Water quenched.

*Source:* Bethlehem Steel Corp. and Republic Steel Corp. as published in 1974 DATA-BOOK issue of the American Society for Metals' METAL PROGRESS magazine and used with its permission.

## Typical Mechanical Properties of Selected Carbon and Alloy Steels — 5
### (Quenched and Tempered)

| AISI No.* | Tempering Temperature, deg F | Strength Tensile | Strength Yield | Elonga-tion, Per cent | Reduction in Area, Per cent | Hard-ness, Bhn |
|---|---|---|---|---|---|---|
| | | 1000 lb/in.² | | | | |
| 4340 | 400 | 272 | 243 | 10 | 38 | 520 |
| | 600 | 250 | 230 | 10 | 40 | 486 |
| | 800 | 213 | 198 | 10 | 44 | 430 |
| | 1000 | 170 | 156 | 13 | 51 | 360 |
| | 1200 | 140 | 124 | 19 | 60 | 280 |
| 5046 | 400 | 253 | 204 | 9 | 25 | 482 |
| | 600 | 205 | 168 | 10 | 37 | 401 |
| | 800 | 165 | 135 | 13 | 50 | 336 |
| | 1000 | 136 | 111 | 18 | 61 | 282 |
| | 1200 | 114 | 95 | 24 | 66 | 235 |
| 50B46 | 400 | ... | ... | ... | ... | 560 |
| | 600 | 258 | 235 | 10 | 37 | 505 |
| | 800 | 202 | 181 | 13 | 47 | 405 |
| | 1000 | 157 | 142 | 17 | 51 | 322 |
| | 1200 | 128 | 115 | 22 | 60 | 273 |
| 50B60 | 400 | ... | ... | ... | ... | 600 |
| | 600 | 273 | 257 | 8 | 32 | 525 |
| | 800 | 219 | 201 | 11 | 34 | 435 |
| | 1000 | 163 | 145 | 15 | 38 | 350 |
| | 1200 | 130 | 113 | 19 | 50 | 290 |
| 5130 | 400 | 234 | 220 | 10 | 40 | 475 |
| | 600 | 217 | 204 | 10 | 46 | 440 |
| | 800 | 185 | 175 | 12 | 51 | 379 |
| | 1000 | 150 | 136 | 15 | 56 | 305 |
| | 1200 | 115 | 100 | 20 | 63 | 245 |
| 5140 | 400 | 260 | 238 | 9 | 38 | 490 |
| | 600 | 229 | 210 | 10 | 43 | 450 |
| | 800 | 190 | 170 | 13 | 50 | 365 |
| | 1000 | 145 | 125 | 17 | 58 | 280 |
| | 1200 | 110 | 96 | 25 | 66 | 235 |
| 5150 | 400 | 282 | 251 | 5 | 37 | 525 |
| | 600 | 252 | 230 | 6 | 40 | 475 |
| | 800 | 210 | 190 | 9 | 47 | 410 |
| | 1000 | 163 | 150 | 15 | 54 | 340 |
| | 1200 | 117 | 118 | 20 | 60 | 270 |
| 5160 | 400 | 322 | 260 | 4 | 10 | 627 |
| | 600 | 290 | 257 | 9 | 30 | 555 |
| | 800 | 233 | 212 | 10 | 37 | 461 |
| | 1000 | 169 | 151 | 12 | 47 | 341 |
| | 1200 | 130 | 116 | 20 | 56 | 269 |
| 51B60 | 400 | ... | ... | ... | ... | 600 |
| | 600 | ... | ... | ... | ... | 540 |
| | 800 | 237 | 216 | 11 | 36 | 460 |
| | 1000 | 175 | 160 | 15 | 44 | 355 |
| | 1200 | 140 | 126 | 20 | 47 | 290 |
| 6150 | 400 | 280 | 245 | 8 | 38 | 538 |
| | 600 | 250 | 228 | 8 | 39 | 483 |
| | 800 | 208 | 193 | 10 | 43 | 420 |
| | 1000 | 168 | 155 | 13 | 50 | 345 |
| | 1200 | 137 | 122 | 17 | 58 | 282 |

* All grades are fine-grained except those in the 1100 series which are coarse-grained. Austenitizing temperatures are given in parentheses. Heat-treated specimens were oil quenched.

*Source:* Bethlehem Steel Corp. and Republic Steel Corp. as published in 1974 DATA-BOOK issue of the American Society for Metals' METAL PROGRESS magazine and used with its permission.

## Typical Mechanical Properties of Selected Carbon and Alloy Steels — 6
### (Quenched and Tempered)

| AISI No.* | Tempering Temperature, deg F | Strength Tensile | Strength Yield | Elongation, Per cent | Reduction in Area, Per cent | Hardness, Bhn |
|---|---|---|---|---|---|---|
| | | 1000 lb/in.² | | | | |
| 81B45 | 400 | 295 | 250 | 10 | 33 | 550 |
| | 600 | 256 | 228 | 8 | 42 | 475 |
| | 800 | 204 | 190 | 11 | 48 | 405 |
| | 1000 | 160 | 149 | 16 | 53 | 338 |
| | 1200 | 130 | 115 | 20 | 55 | 280 |
| 8630 | 400 | 238 | 218 | 9 | 38 | 465 |
| | 600 | 215 | 202 | 10 | 42 | 430 |
| | 800 | 185 | 170 | 13 | 47 | 375 |
| | 1000 | 150 | 130 | 17 | 54 | 310 |
| | 1200 | 112 | 100 | 23 | 63 | 240 |
| 8640 | 400 | 270 | 242 | 10 | 40 | 505 |
| | 600 | 240 | 220 | 10 | 41 | 460 |
| | 800 | 200 | 188 | 12 | 45 | 400 |
| | 1000 | 160 | 150 | 16 | 54 | 340 |
| | 1200 | 130 | 116 | 20 | 62 | 280 |
| 86B45 | 400 | 287 | 238 | 9 | 31 | 525 |
| | 600 | 246 | 225 | 9 | 40 | 475 |
| | 800 | 200 | 191 | 11 | 41 | 395 |
| | 1000 | 160 | 150 | 15 | 49 | 335 |
| | 1200 | 131 | 127 | 19 | 58 | 280 |
| 8650 | 400 | 281 | 243 | 10 | 38 | 525 |
| | 600 | 250 | 225 | 10 | 40 | 490 |
| | 800 | 210 | 192 | 12 | 45 | 420 |
| | 1000 | 170 | 153 | 15 | 51 | 340 |
| | 1200 | 140 | 120 | 20 | 58 | 280 |
| 8660 | 400 | ... | ... | ... | ... | 580 |
| | 600 | ... | ... | ... | ... | 535 |
| | 800 | 237 | 225 | 13 | 37 | 460 |
| | 1000 | 190 | 176 | 17 | 46 | 370 |
| | 1200 | 155 | 138 | 20 | 53 | 315 |
| 8740 | 400 | 290 | 240 | 10 | 41 | 578 |
| | 600 | 249 | 225 | 11 | 46 | 495 |
| | 800 | 208 | 197 | 13 | 50 | 415 |
| | 1000 | 175 | 165 | 15 | 55 | 363 |
| | 1200 | 143 | 131 | 20 | 60 | 302 |
| 9255 | 400 | 305 | 297 | 1 | 3 | 601 |
| | 600 | 281 | 260 | 4 | 10 | 578 |
| | 800 | 233 | 216 | 8 | 22 | 477 |
| | 1000 | 182 | 160 | 15 | 32 | 352 |
| | 1200 | 144 | 118 | 20 | 42 | 285 |
| 9260 | 400 | ... | ... | ... | ... | 600 |
| | 600 | ... | ... | ... | ... | 540 |
| | 800 | 255 | 218 | 8 | 24 | 470 |
| | 1000 | 192 | 164 | 12 | 30 | 390 |
| | 1200 | 142 | 118 | 20 | 43 | 295 |
| 94B30 | 400 | 250 | 225 | 12 | 46 | 475 |
| | 600 | 232 | 206 | 12 | 49 | 445 |
| | 800 | 195 | 175 | 13 | 57 | 382 |
| | 1000 | 145 | 135 | 16 | 65 | 307 |
| | 1200 | 120 | 105 | 21 | 69 | 250 |

* All grades are fine-grained except those in the 1100 series which are coarse-grained. Austenitizing temperatures are given in parentheses. Heat-treated specimens were oil quenched.

*Source:* Bethlehem Steel Corp. and Republic Steel Corp. as published in 1974 DATA-BOOK issue of the American Society for Metals' METAL PROGRESS magazine and used with its permission.

**Corrosion-resistant Steels.** — Many different terms and trade names have been applied to corrosion-resistant steels. "Stainless Steel" is a term commonly used to indicate any or all rustless steels or iron alloys designed to resist atmospheric corrosion, the attack of hot or cold acids, and scaling at elevated temperatures. However, "Stainless Steel" is strictly a trade name, originally applied to cutlery steels containing no more than 0.70 per cent carbon and from 9 to 16 per cent chromium which were patented in 1916 by the English metallurgist Brearley, and the genuine "Stainless Steel" produced in this country is a straight chrome-iron alloy made under patents owned by the American Stainless Steel Co., Pittsburgh, Pa.

*Applications.* — The applications of stainless steels may be divided broadly into two groups: (1) Where corrosion resistance is required, including resistance to high-temperature oxidation; (2) where unusual mechanical properties of hardness, strength, toughness or ductility are required, including resistance to wear and abrasion. Corrosion-resistant steels cover a wide range of compositions and physical properties. The common applications include cutlery; surgical and dental instruments; poppet valves for internal-combustion engines; turbine blades; pump shafts; architectural trim; polished parts of automobiles; chemical, dairy, laundry, and oil equipment, etc. The chromium content commonly ranges from 10 or 12 to 18 or 20 per cent, some steels having less and some more than these minimum or maximum values. The "18-8" stainless steel often referred to is a steel having about 18 per cent chromium and 8 per cent nickel.

*Stainless Steel with Free Machining Qualities.* — The high-chromium stainless steel alloys first produced were extremely difficult to machine, and grinding and polishing operations were also difficult and expensive. By producing this steel with a high sulphur content or by the addition of selenium, free machining qualities can be obtained. Such stainless steels contain approximately 0.10 per cent carbon, 18 per cent chromium, 8 per cent nickel, and 0.30 per cent sulphur (or 0.25 per cent selenium instead of sulphur). They can be machined in automatic screw machines with regular tools at speeds equal to, or closely approximating, those used for ordinary Bessemer screw stock. These materials can also be easily drilled, tapped, and threaded with dies. Wire and tubing can be cold-drawn by simply using the lime coat and lubricants regularly employed for drawing ordinary steel.

*Characteristics of 18-8 Stainless Steel.* — The chrome-nickel stainless steel known as 18-8 is made to have a tensile strength of from 90,000 to 100,000 pounds per square inch in the annealed state. The elongation varies from 60 to 70 per cent. Cold-working will increase the tensile strength to from 120,000 to 125,000 pounds per square inch. Because of the high feeding pressure required for drilling 18-8 stainless steels, a specially heat-treated high-speed drill with a heavy web section has been introduced. The web should be thinned at the point and a sulphur-base oil used as a cutting fluid.

**General Properties of Alloy Steels.** — Alloy or "special" steels are combinations of iron and carbon with some other element, such as nickel, chromium, tungsten, vanadium, manganese and molybdenum. All of these metals give certain distinct properties to the steel, but in all cases the principal quality is the increase in hardness and toughness.

*Nickel steel* usually contains from 3 to 3.5 per cent nickel (ordinarily not over 5 per cent), and from 0.20 to 0.40 per cent carbon. This steel is used for armor plate, ammunition, bridge construction, rails, etc. One of the reasons why nickel steel is adapted for armor plate is that it does not crack when perforated by a projectile. The Krupp steel used for armor plate contains approximately 3.5 per cent nickel, 1.5 per cent chromium and 0.25 per cent carbon. The advantages claimed for nickel steel for railroad rails are its increased resistance to abrasion and high elastic limit. On sharp curves, it has been estimated that a nickel steel rail will outlast four ordinary rails.

*Chromium steel* is well adapted for armor-piercing projectiles, owing to its hardness, toughness and stiffness, and is extensively used for this purpose. Chromium steel is also used in the construction of safes and for castings subjected to unusually severe stresses, such as those used in rock-crushing machinery, etc. The percentage of chromium used in chromium steels varies over quite a wide range in the low-chromium and high-chromium steels.

*Tungsten steel* is largely employed for high-speed metal cutting tools and magnet steels. It has also been used in the manufacture of armor plate and armor-piercing projectiles, in which case it is combined either with nickel or chromium or with both of these metals. The property that tungsten imparts to steel is that of hardening in the air, after heating to the required temperature. This steel usually contains from 5 to 15 per cent tungsten (although the percentage is sometimes as high as 24 per cent) and from 0.4 to 2 per cent carbon.

*Vanadium steels* ordinarily contain from 0.16 to 0.25 per cent vanadium. The effect of vanadium is to increase the tensile strength and elastic limit, and it gives the steel the valuable property of resisting, to an unusual degree, repeated stresses. Vanadium steel is especially adapted for springs, car axles, gears subjected to severe service, and for all parts which must withstand constant vibration and varying stresses.

*Manganese steel* (also known as Hadfield manganese steel) contains about 12 per cent manganese and from 0.8 to 1.25 per cent carbon. If there is only 1.5 per cent manganese, the steel is very brittle, and additional manganese increases this brittleness until the quantity has reached 4 to 5.5 per cent, when the steel can be pulverized under the hammer. With a further increase of manganese, the steel becomes ductile and very hard, these qualities being at their highest degree when the manganese content is 12 per cent. The ductility of the steel is brought out by sudden cooling, the process being opposite that employed for carbon steel.

*Molybdenum steels* have properties similar to tungsten steels, except that a smaller quantity of molybdenum than of tungsten is required to secure similar results.

## TESTING THE HARDNESS OF METALS

**Brinell Hardness Test.** — The Brinell test for determining the hardness of metallic materials consists in applying a known load to the surface of the material to be tested through a hardened steel ball of known diameter. The diameter of the resulting permanent impression in the metal is measured and the Brinell Hardness Number (BHN) is then calculated from the following formula in which $D$ = diameter of ball in millimeters, $d$ = measured diameter at the rim of the impression in millimeters, and $P$ = applied load in kilograms.

$$ BHN = \frac{\text{load on indenting tool in kilograms}}{\text{surface area of indentation in sq. mm.}} = \frac{P}{\dfrac{\pi D}{2}(D - \sqrt{D^2 - d^2})} $$

If the steel ball were not deformed under the applied load and if the impression were truly spherical, then the above formula would be a general one, and any combination of applied load and size of ball could be used. The impression, however, is not quite a spherical surface since there must always be some deformation of the steel ball and some recovery of form of the metal in the impression; hence for a standard Brinell test, the size and characteristics of the ball and the magnitude of the applied load must be standardized. In the standard Brinell test, a ball 10 millimeters in diameter and a load of 3000, 1500, or 500 kilograms is used. It is desirable, although not mandatory, that the test load be of such magnitude that the diameter of the impression be in the range of 2.50 to 4.75 millimeters. The following test loads and approximate Brinell numbers for this range of impression diameters are: 3000 kg., 160 to 600 BHN; 1500 kg., 80 to 300 BHN; 500 kg., 26 to 100 BHN. In making a Brinell test the load should be applied steadily and without a jerk for at least 15 seconds in the case of iron and steel, and at least 30 seconds in testing other metals. A minimum period of two minutes, for example, has been recommended for magnesium and magnesium alloys. (For the softer metals, loads of 250 kg., 125 kg., or 100 kg., are sometimes used.)

According to the American Society for Testing and Materials Standard E10-66, a steel ball may be used on material having a BHN not over 450, a Hultgren ball on material not over 500, or a carbide ball on material not over 630. The Brinell hardness test is not recommended for material having a BHN over 630.

**Rockwell Hardness Test.** — The Rockwell hardness tester is essentially a machine that measures hardness by determining the depth of penetration of a penetrator into the specimen under certain fixed conditions of test. The penetrator may be either a steel ball or a diamond sphero-conical penetrator. The hardness number is related to the depth of indentation and the number is higher the harder the material. A minor load of 10 kg. is first applied which causes an initial penetration; the dial is set at zero on the black-figure scale, and the major load is applied. This major load is customarily 60 kg. or 100 kg. when a steel ball is used as a penetrator, but other loads may be used when found necessary. The ball penetrator is $\frac{1}{16}$ inch in diameter normally; but other penetrators of larger diameter, such as $\frac{1}{8}$ inch, may be employed for soft metals. When a diamond sphero-conical penetrator is employed the load usually is 150 kg. Experience decides the best combination of load and penetrator for use. After the major load is applied and removed, according to standard procedure, the reading is taken while the minor load is still applied.

**The Rockwell Hardness Scales.** — The various Rockwell scales and their applications are shown in the table below. The type of penetrator and load used with each are shown in Tables 1 and 2 which give comparative hardness values for different hardness scales.

| Scale | Testing Application |
|---|---|
| A | For tungsten carbide and other extremely hard materials. Also for thin, hard sheets. |
| B | For materials of medium hardness such as low and medium carbon steels in the annealed condition. |
| C | For materials harder than Rockwell B-100. |
| D | Where somewhat lighter load is desired than on C scale, as on case hardened pieces. |
| E | For very soft materials such as bearing metals. |
| F | Same as E scale but using $\frac{1}{16}$-inch ball. |
| G | For metals harder than tested on B scale. |
| H & K | For softer metals. |
| 15–N; 30–N; 45–N | Where shallow impression or small area is desired. For hardened steel and hard alloys. |
| 15–T; 30–T; 45–T | Where shallow impression or small area is desired for materials softer than hardened steel. |

**Shore's Scleroscope.** — The scleroscope is an instrument which measures the hardness of the work in terms of elasticity. A diamond-tipped hammer is allowed to drop from a known height on the metal to be tested. As this hammer strikes the metal, it rebounds, and the harder the metal, the greater the rebound. The extreme height of the rebound is recorded, and an average of a number of readings taken on a single piece will give a good indication of the hardness of the work. The surface smoothness of the work affects the reading of the instrument. The readings are also affected by the contour and mass of the work and the depth of the case, in carburized work, the soft core of light-depth carburizing, pack-hardening, or cyanide hardening, absorbing the force of the hammer fall and decreasing the rebound. The hammer weighs about 40 grains, the height of the rebound of hardened steel is in the neighborhood of 100 on the scale, or about 6¼ inches, while the total fall is about 10 inches or 255 millimeters.

**Vickers Hardness Test.** — The Vickers test is similar in principle to the Brinell test. The standard Vickers penetrator is a square-based diamond pyramid having an included point angle of 136 degrees. The numerical value of the hardness number equals the applied load in kilograms divided by the area of the pyramidal impression. A smooth, firmly supported, flat surface is required. The load, which usually is applied for 30 seconds, may either be 5, 10, 20, 30, 50 or 120 kilograms. The 50-kilogram load is usually employed. The hardness number is based upon the diagonal length of the square impression. The Vickers test, which is considered very accurate, may, with proper load regulation, be applied to thin sheets as well as to larger sections.

**Knoop Hardness Numbers.** — The Knoop hardness test is applicable to extremely thin metal, plated surfaces, exceptionally hard and brittle materials, very shallow carburized or nitrided surfaces, or whenever the applied load must be kept below 3600 grams. The Knoop indentor is a diamond ground to an elongated pyramidal form and it produces an indentation having long and short diagonals with a ratio of approximately 7 to 1. The longitudinal angle of the indentor is 172 degrees 30 minutes and the transverse angle 130 degrees. The Tukon Tester in which the Knoop indentor is used is fully automatic under electronic control. The Knoop hardness number equals load in kilograms divided by the projected area of indentation in square millimeters. The indentation number corresponding to the long diagonal and for a given load, may be determined from a table computed for a theoretically perfect indentor. The load, which may be varied from 25 to 3600 grams, is applied for a definite period and always normal to the surface tested. Lapped plane surfaces free from scratches are required.

**Monotron Hardness Indicator.** — With this instrument, a diamond-ball impressor point ¾ mm. in diameter is forced into the material to a depth of 9/5000 inch and the pressure required to produce this constant impression indicates the hardness. One of two dials shows the pressure in kilograms and pounds, and the other shows the depth of the impression in millimeters and inches. Readings in Brinell numbers may be obtained by means of a scale designated as $M - 1$.

**Keep's Test.** — With this apparatus a standard steel drill is caused to make a definite number of revolutions, while it is pressed with standard force against the specimen to be tested. The hardness is automatically recorded on a diagram on which a dead soft material gives a horizontal line, while a material as hard as the drill itself gives a vertical line, intermediate hardness being represented by the corresponding angle between 0 and 90 degrees.

### Table 1. Comparative Hardness Scales for Steel — 1

| Rockwell C-Scale Hardness Number | Diamond Pyramid Hardness Number Vickers | Brinell Hardness Number 10-mm. Ball, 3000-kgf Load | | | Rockwell Hardness Number | | Rockwell Superficial Hardness Number Superficial Diam. Penetrator | | | Shore Scleroscope Hardness Number |
|---|---|---|---|---|---|---|---|---|---|---|
| | | Standard Ball | Hultgren Ball | Tungsten Carbide Ball | A-Scale 60-kgf Load Diam. Penetrator | D-Scale 100-kgf Load Diam. Penetrator | 15-N Scale 15-kgf Load | 30-N Scale 30-kgf Load | 45-N Scale 45-kgf Load | |
| 68 | 940 | .... | .... | .... | 85.6 | 76.9 | 93.2 | 84.4 | 75.4 | 97 |
| 67 | 900 | .... | .... | .... | 85.0 | 76.1 | 92.9 | 83.6 | 74.2 | 95 |
| 66 | 865 | .... | .... | .... | 84.5 | 75.4 | 92.5 | 82.8 | 73.3 | 92 |
| 65 | 832 | .... | .... | 739 | 83.9 | 74.5 | 92.2 | 81.9 | 72.0 | 91 |
| 64 | 800 | .... | .... | 722 | 83.4 | 73.8 | 91.8 | 81.1 | 71.0 | 88 |
| 63 | 772 | .... | .... | 705 | 82.8 | 73.0 | 91.4 | 80.1 | 69.9 | 87 |
| 62 | 746 | .... | .... | 688 | 82.3 | 72.2 | 91.1 | 79.3 | 68.8 | 85 |
| 61 | 720 | .... | .... | 670 | 81.8 | 71.5 | 90.7 | 78.4 | 67.7 | 83 |
| 60 | 697 | .... | 613 | 654 | 81.2 | 70.7 | 90.2 | 77.5 | 66.6 | 81 |
| 59 | 674 | .... | 599 | 634 | 80.7 | 69.9 | 89.8 | 76.6 | 65.5 | 80 |
| 58 | 653 | .... | 587 | 615 | 80.1 | 69.2 | 89.3 | 75.7 | 64.3 | 78 |
| 57 | 633 | .... | 575 | 595 | 79.6 | 68.5 | 88.9 | 74.8 | 63.2 | 76 |
| 56 | 613 | .... | 561 | 577 | 79.0 | 67.7 | 88.3 | 73.9 | 62.0 | 75 |
| 55 | 595 | .... | 546 | 560 | 78.5 | 66.9 | 87.9 | 73.0 | 60.9 | 74 |
| 54 | 577 | .... | 534 | 543 | 78.0 | 66.1 | 87.4 | 72.0 | 59.8 | 72 |
| 53 | 560 | .... | 519 | 525 | 77.4 | 65.4 | 86.9 | 71.2 | 58.6 | 71 |
| 52 | 544 | 500 | 508 | 512 | 76.8 | 64.6 | 86.4 | 70.2 | 57.4 | 69 |
| 51 | 528 | 487 | 494 | 496 | 76.3 | 63.8 | 85.9 | 69.4 | 56.1 | 68 |
| 50 | 513 | 475 | 481 | 481 | 75.9 | 63.1 | 85.5 | 68.5 | 55.0 | 67 |
| 49 | 498 | 464 | 469 | 469 | 75.2 | 62.1 | 85.0 | 67.6 | 53.8 | 66 |
| 48 | 484 | 451 | 455 | 455 | 74.7 | 61.4 | 84.5 | 66.7 | 52.5 | 64 |
| 47 | 471 | 442 | 443 | 443 | 74.1 | 60.8 | 83.9 | 65.8 | 51.4 | 63 |
| 46 | 458 | 432 | 432 | 432 | 73.6 | 60.0 | 83.5 | 64.8 | 50.3 | 62 |
| 45 | 446 | 421 | 421 | 421 | 73.1 | 59.2 | 83.0 | 64.0 | 49.0 | 60 |
| 44 | 434 | 409 | 409 | 409 | 72.5 | 58.5 | 82.5 | 63.1 | 47.8 | 58 |
| 43 | 423 | 400 | 400 | 400 | 72.0 | 57.7 | 82.0 | 62.2 | 46.7 | 57 |
| 42 | 412 | 390 | 390 | 390 | 71.5 | 56.9 | 81.5 | 61.3 | 45.5 | 56 |
| 41 | 402 | 381 | 381 | 381 | 70.9 | 56.2 | 80.9 | 60.4 | 44.3 | 55 |
| 40 | 392 | 371 | 371 | 371 | 70.4 | 55.4 | 80.4 | 59.5 | 43.1 | 54 |
| 39 | 382 | 362 | 362 | 362 | 69.9 | 54.6 | 79.9 | 58.6 | 41.9 | 52 |
| 38 | 372 | 353 | 353 | 353 | 69.4 | 53.8 | 79.4 | 57.7 | 40.8 | 51 |
| 37 | 363 | 344 | 344 | 344 | 68.9 | 53.1 | 78.8 | 56.8 | 39.6 | 50 |
| 36 | 354 | 336 | 336 | 336 | 68.4 | 52.3 | 78.3 | 55.9 | 38.4 | 49 |
| 35 | 345 | 327 | 327 | 327 | 67.9 | 51.5 | 77.7 | 55.0 | 37.2 | 48 |
| 34 | 336 | 319 | 319 | 319 | 67.4 | 50.8 | 77.2 | 54.2 | 36.1 | 47 |
| 33 | 327 | 311 | 311 | 311 | 66.8 | 50.0 | 76.6 | 53.3 | 34.9 | 46 |
| 32 | 318 | 301 | 301 | 301 | 66.3 | 49.2 | 76.1 | 52.1 | 33.7 | 44 |
| 31 | 310 | 294 | 294 | 294 | 65.8 | 48.4 | 75.6 | 51.3 | 32.5 | 43 |
| 30 | 302 | 286 | 286 | 286 | 65.3 | 47.7 | 75.0 | 50.4 | 31.3 | 42 |
| 29 | 294 | 279 | 279 | 279 | 64.7 | 47.0 | 74.5 | 49.5 | 30.1 | 41 |
| 28 | 286 | 271 | 271 | 271 | 64.3 | 46.1 | 73.9 | 48.6 | 28.9 | 41 |
| 27 | 279 | 264 | 264 | 264 | 63.8 | 45.2 | 73.3 | 47.7 | 27.8 | 40 |

*Note:* The values in this table shown in bold faced type correspond to those shown in American Society for Testing and Materials Specification E140-67.

### Table 1.  Comparative Hardness Scales for Steel — 2

| Rockwell C-Scale Hardness Number | Diamond Pyramid Hardness Number Vickers | Brinell Hardness Number 10-mm. Ball, 3000-kgf Load | | | Rockwell Hardness Number | | Rockwell Superficial Hardness Number Superficial Brale Penetrator | | | Shore Scleroscope Hardness Number |
|---|---|---|---|---|---|---|---|---|---|---|
| | | Standard Ball | Hultgren Ball | Tungsten Carbide Ball | A-Scale 60-kgf Load Diam. Penetrator | D-Scale 100-kgf Load Diam. Penetrator | 15-N Scale 15-kgf Load | 30-N Scale 30-kgf Load | 45-N Scale 45-kgf Load | |
| 26 | 272 | 258 | 258 | 258 | 63.3 | 44.6 | 72.8 | 46.8 | 26.7 | 38 |
| 25 | 266 | 253 | 253 | 253 | 62.8 | 43.8 | 72.2 | 45.9 | 25.5 | 38 |
| 24 | 260 | 247 | 247 | 247 | 62.4 | 43.1 | 71.6 | 45.0 | 24.3 | 37 |
| 23 | 254 | 243 | 243 | 243 | 62.0 | 42.1 | 71.0 | 44.0 | 23.1 | 36 |
| 22 | 248 | 237 | 237 | 237 | 61.5 | 41.6 | 70.5 | 43.2 | 22.0 | 35 |
| 21 | 243 | 231 | 231 | 231 | 61.0 | 40.9 | 69.9 | 42.3 | 20.7 | 35 |
| 20 | 238 | 226 | 226 | 226 | 60.5 | 40.1 | 69.4 | 41.5 | 19.6 | 34 |
| (18) | 230 | 219 | 219 | 219 | .... | .... | .... | .... | .... | 33 |
| (16) | 222 | 212 | 212 | 212 | .... | .... | .... | .... | .... | 32 |
| (14) | 213 | 203 | 203 | 203 | .... | .... | .... | .... | .... | 31 |
| (12) | 204 | 194 | 194 | 194 | .... | .... | .... | .... | .... | 29 |
| (10) | 196 | 187 | 187 | 187 | .... | .... | .... | .... | .... | 28 |
| (8) | 188 | 179 | 179 | 179 | .... | .... | .... | .... | .... | 27 |
| (6) | 180 | 171 | 171 | 171 | .... | .... | .... | .... | .... | 26 |
| (4) | 173 | 165 | 165 | 165 | .... | .... | .... | .... | .... | 25 |
| (2) | 166 | 158 | 158 | 158 | .... | .... | .... | .... | .... | 24 |
| (o) | 160 | 152 | 152 | 152 | .... | .... | .... | .... | .... | 24 |

*Note:* The values in this table shown in bold faced type correspond to those shown in American Society for Testing and Materials Specification E140–67.

Values in ( ) are beyond the normal range and are given for information only.

**Comparison of Hardness Scales.** — Tables 1 and 2 show comparisons of various hardness scales. All such tables are based on the assumption that the metal tested is homogeneous to a depth several times that of the indentation. To the extent that the metal being tested is not homogeneous, errors are introduced because different loads and different shapes of penetrators meet the resistance of metal of varying hardness, depending on the depth of indentation. Another source of error is introduced in comparing the hardness of different materials as measured on different hardness scales. This arises from the fact that in any hardness test, metal that is severely cold-worked actually supports the penetrator and different metals, different alloys, and different analyses of the same type of alloy have different cold-working properties. In spite of the possible inaccuracies introduced by such factors, it is of considerable value to be able to compare hardness values in a general way.

The data shown in Table 1 are based upon extensive tests on carbon and alloy steels mostly in the heat-treated condition, but have been found to be reliable on constructional alloy steels and tool steels in the as-forged, annealed, normalized, quenched and tempered conditions, providing they are homogeneous. These hardness comparisons are not as accurate for special cases such as high manganese steel, 18–8 stainless steel and other austenitic steels, nickel base alloys, as well as constructional alloy steels and nickel base alloys in the cold-worked condition.

The data shown in Table 2 are for hardness measurements of unhardened steel, steel of soft temper, grey and malleable cast iron, and most non-ferrous metals. Again these hardness comparisons are not as accurate for annealed metals of high Rockwell B hardness such as austenitic stainless steel, nickel and high nickel alloys and cold-worked metals of low B-scale hardness such as aluminum and the softer alloys.

Table 2. Comparative Hardness Scales for Unhardened Steel, Soft-temper Steel, Grey and Malleable Cast Iron, and Non-ferrous Alloys* — 1

| Rockwell Hardness Number | | | Rockwell Superficial Hardness Number | | | Rockwell Hardness Number | | | Brinell Hardness Number | |
|---|---|---|---|---|---|---|---|---|---|---|
| Rockwell B scale 1/16" Ball Penetrator 100 kg. Load | Rockwell F scale 1/16" Ball Penetrator 60 kg. Load | Rockwell G scale 1/16" Ball Penetrator 150 kg. Load | Rockwell Superficial 15-T scale 1/16" Ball Penetrator 15 kg. Load | Rockwell Superficial 30-T scale 1/16" Ball Penetrator 30 kg. Load | Rockwell Superficial 45-T scale 1/16" Ball Penetrator 45 kg. Load | Rockwell E scale 1/8" Ball Penetrator 100 kg. Load | Rockwell K scale 1/8" Ball Penetrator 150 kg. Load | Rockwell A scale "Brale" Penetrator 60 kg. Load | Brinell Scale 10 mm Standard Ball — 500 kg. Load | Brinell Scale 10 mm. Standard Ball — 3000 kg. Load |
| 100 | .... | 82.5 | 93.0 | 82.0 | 72.0 | .... | .... | 61.5 | 201 | 240 |
| 99 | .... | 81.0 | 92.5 | 81.5 | 71.0 | .... | .... | 61.0 | 195 | 234 |
| 98 | .... | 79.0 | .... | 81.0 | 70.0 | .... | .... | 60.0 | 189 | 228 |
| 97 | .... | 77.5 | 92.0 | 80.5 | 69.0 | .... | .... | 59.5 | 184 | 222 |
| 96 | .... | 76.0 | .... | 80.0 | 68.0 | .... | .... | 59.0 | 179 | 216 |
| 95 | .... | 74.0 | 91.5 | 79.0 | 67.0 | .... | .... | 58.0 | 175 | 210 |
| 94 | .... | 72.5 | .... | 78.5 | 66.0 | .... | .... | 57.5 | 171 | 205 |
| 93 | .... | 71.0 | 91.0 | 78.0 | 65.5 | .... | .... | 57.0 | 167 | 200 |
| 92 | .... | 69.0 | 90.5 | 77.5 | 64.5 | .... | 100 | 56.5 | 163 | 195 |
| 91 | .... | 67.5 | .... | 77.0 | 63.5 | .... | 99.5 | 56.0 | 160 | 190 |
| 90 | .... | 66.0 | 90.0 | 76.0 | 62.5 | .... | 98.5 | 55.5 | 157 | 185 |
| 89 | .... | 64.0 | 89.5 | 75.5 | 61.5 | .... | 98.0 | 55.0 | 154 | 180 |
| 88 | .... | 62.5 | .... | 75.0 | 60.5 | .... | 97.0 | 54.0 | 151 | 176 |
| 87 | .... | 61.0 | 89.0 | 74.5 | 59.5 | .... | 96.5 | 53.5 | 148 | 172 |
| 86 | .... | 59.0 | 88.5 | 74.0 | 58.5 | .... | 95.5 | 53.0 | 145 | 169 |
| 85 | .... | 57.5 | .... | 73.5 | 58.0 | .... | 94.5 | 52.5 | 142 | 165 |
| 84 | .... | 56.0 | 88.0 | 73.0 | 57.0 | .... | 94.0 | 52.0 | 140 | 162 |
| 83 | .... | 54.0 | 87.5 | 72.0 | 56.0 | .... | 93.0 | 51.0 | 137 | 159 |
| 82 | .... | 52.5 | .... | 71.5 | 55.0 | .... | 92.0 | 50.5 | 135 | 156 |
| 81 | .... | 51.0 | 87.0 | 71.0 | 54.0 | .... | 91.0 | 50.0 | 133 | 153 |
| 80 | .... | 49.0 | 86.5 | 70.0 | 53.0 | .... | 90.5 | 49.5 | 130 | 150 |
| 79 | .... | 47.5 | .... | 69.5 | 52.0 | .... | 89.5 | 49.0 | 128 | 147 |
| 78 | .... | 46.0 | 86.0 | 69.0 | 51.0 | .... | 88.5 | 48.5 | 126 | 144 |
| 77 | .... | 44.0 | 85.5 | 68.0 | 50.0 | .... | 88.0 | 48.0 | 124 | 141 |
| 76 | .... | 42.5 | .... | 67.5 | 49.0 | .... | 87.0 | 47.0 | 122 | 139 |
| 75 | 99.5 | 41.0 | 85.0 | 67.0 | 48.5 | .... | 86.0 | 46.5 | 120 | 137 |
| 74 | 99.0 | 39.0 | .... | 66.0 | 47.5 | .... | 85.0 | 46.0 | 118 | 135 |
| 73 | 98.5 | 37.5 | 84.5 | 65.5 | 46.5 | .... | 84.5 | 45.5 | 116 | 132 |
| 72 | 98.0 | 36.0 | 84.0 | 65.0 | 45.5 | .... | 83.5 | 45.0 | 114 | 130 |
| 71 | 97.5 | 34.5 | .... | 64.0 | 44.5 | 100 | 82.5 | 44.5 | 112 | 127 |
| 70 | 97.0 | 32.5 | 83.5 | 63.5 | 43.5 | 99.5 | 81.5 | 44.0 | 110 | 125 |
| 69 | 96.0 | 31.0 | 83.0 | 62.5 | 42.5 | 99.0 | 81.0 | 43.5 | 109 | 123 |
| 68 | 95.5 | 29.5 | .... | 62.0 | 41.5 | 98.0 | 80.0 | 43.0 | 107 | 121 |
| 67 | 95.0 | 28.0 | 82.5 | 61.5 | 40.5 | 97.5 | 79.0 | 42.5 | 106 | 119 |
| 66 | 94.5 | 26.5 | 82.0 | 61.0 | 39.5 | 97.0 | 78.0 | 42.0 | 104 | 117 |
| 65 | 94.0 | 25.0 | .... | 60.0 | 38.5 | 96.0 | 77.5 | .... | 102 | 116 |
| 64 | 93.5 | 23.5 | 81.5 | 59.5 | 37.5 | 95.5 | 76.5 | 41.5 | 101 | 114 |
| 63 | 93.0 | 22.0 | 81.0 | 58.5 | 36.5 | 95.0 | 75.5 | 41.0 | 99 | 112 |
| 62 | 92.0 | 20.5 | .... | 58.0 | 35.5 | 94.5 | 74.5 | 40.5 | 98 | 110 |
| 61 | 91.5 | 19.0 | 80.5 | 57.0 | 34.5 | 93.5 | 74.0 | 40.0 | 96 | 108 |
| 60 | 91.0 | 17.5 | .... | 56.5 | 33.5 | 93.0 | 73.0 | 39.5 | 95 | 107 |
| 59 | 90.5 | 16.0 | 80.0 | 56.0 | 32.0 | 92.5 | 72.0 | 39.0 | 94 | 106 |

* See note at end of table.

**Table 2. Comparative Hardness Scales for Unhardened Steel, Soft-temper Steel, Grey and Malleable Cast-iron, and Non-ferrous Alloys\* — 2**

| Rockwell Hardness Number | | | Rockwell Superficial Hardness Number | | | Rockwell Hardness Number | | | | Brinell |
|---|---|---|---|---|---|---|---|---|---|---|
| Rockwell B scale 1/16" Ball Penetrator 100 kg. Load | Rockwell F scale 1/16" Ball Penetrator 60 kg. Load | Rockwell G scale 1/16" Ball Penetrator 150 kg. Load | Rockwell Superficial 15-T scale 1/16" Ball Penetrator 15 kg. Load | Rockwell Superficial 30-T scale 1/16" Ball Penetrator 30 kg. Load | Rockwell Superficial 45-T scale 1/16" Ball Penetrator 45 kg. Load | Rockwell E scale 1/8" Ball Penetrator 100 kg. Load | Rockwell H scale 1/8" Ball Penetrator 60 kg. Load | Rockwell K scale 1/8" Ball Penetrator 150 kg. Load | Rockwell A scale "Brale" Penetrator 60 kg. Load | Brinell Scale 10-mm. Standard Ball — 500 kg. Load |
| 58 | 90.0 | 14.5 | 79.5 | 55.0 | 31.0 | 92.0 | | 71.0 | 38.5 | 92 |
| 57 | 89.5 | 13.0 | | 54.5 | 30.0 | 91.0 | | 70.5 | 38.0 | 91 |
| 56 | 89.0 | 11.5 | 79.0 | 54.0 | 29.0 | 90.5 | | 69.5 | | 90 |
| 55 | 88.0 | 10.0 | 78.5 | 53.0 | 28.0 | 90.0 | | 68.5 | 37.5 | 89 |
| 54 | 87.5 | 8.5 | | 52.5 | 27.0 | 89.5 | | 68.0 | 37.0 | 87 |
| 53 | 87.0 | 7.0 | 78.0 | 51.5 | 26.0 | 89.0 | | 67.0 | 36.5 | 86 |
| 52 | 86.5 | 5.5 | 77.5 | 51.0 | 25.0 | 88.0 | | 66.0 | 36.0 | 85 |
| 51 | 86.0 | 4.0 | | 50.5 | 24.0 | 87.5 | | 65.0 | 35.5 | 84 |
| 50 | 85.5 | 2.5 | 77.0 | 49.5 | 23.0 | 87.0 | | 64.5 | 35.0 | 83 |
| 50 | 85.5 | 2.5 | 77.0 | 49.5 | 23.0 | 87.0 | | 64.5 | 35.0 | 83 |
| 49 | 85.0 | 1.0 | 76.5 | 49.0 | 22.0 | 86.5 | | 63.5 | | 82 |
| 48 | 84.5 | | | 48.5 | 20.5 | 85.5 | | 62.5 | 34.5 | 81 |
| 47 | 84.0 | | 76.0 | 47.5 | 19.5 | 85.0 | | 61.5 | 34.0 | 80 |
| 46 | 83.0 | | 75.5 | 47.0 | 18.5 | 84.5 | | 61.0 | 33.5 | .. |
| 45 | 82.5 | | | 46.0 | 17.5 | 84.0 | | 60.0 | 33.0 | 79 |
| 44 | 82.0 | | 75.0 | 45.5 | 16.5 | 83.5 | | 59.0 | 32.5 | 78 |
| 43 | 81.5 | | 74.5 | 45.0 | 15.5 | 82.5 | | 58.0 | 32.0 | 77 |
| 42 | 81.0 | | | 44.0 | 14.5 | 82.0 | | 57.5 | 31.5 | 76 |
| 41 | 80.5 | | 74.0 | 43.5 | 13.5 | 81.5 | | 56.5 | 31.0 | 75 |
| 40 | 79.5 | | 73.5 | 43.0 | 12.5 | 81.0 | | 55.5 | | .. |
| 39 | 79.0 | | | 42.0 | 11.0 | 80.0 | | 54.5 | 30.5 | 74 |
| 38 | 78.5 | | 73.0 | 41.5 | 10.0 | 79.5 | | 54.0 | 30.0 | 73 |
| 37 | 78.0 | | 72.5 | 40.5 | 9.0 | 79.0 | | 53.0 | 29.5 | 72 |
| 36 | 77.5 | | | 40.0 | 8.0 | 78.5 | 100 | 52.0 | 29.0 | .. |
| 35 | 77.0 | | 72.0 | 39.5 | 7.0 | 78.0 | 99.5 | 51.5 | 28.5 | 71 |
| 34 | 76.5 | | 71.5 | 38.5 | 6.0 | 77.0 | 99.0 | 50.5 | 28.0 | 70 |
| 33 | 75.5 | | | 38.0 | 5.0 | 76.5 | | 49.5 | | 69 |
| 32 | 75.0 | | 71.0 | 37.5 | 4.0 | 76.0 | 98.5 | 48.5 | 27.5 | .. |
| 31 | 74.5 | | | 36.5 | 3.0 | 75.5 | 98.0 | 48.0 | 27.0 | 68 |
| 30 | 74.0 | | 70.5 | 36.0 | 2.0 | 75.0 | | 47.0 | 26.5 | 67 |
| 29 | 73.5 | | 70.0 | 35.5 | 1.0 | 74.0 | 97.5 | 46.0 | 26.0 | .. |
| 28 | 73.0 | | | 34.5 | | 73.5 | 97.0 | 45.0 | 25.5 | 66 |
| 27 | 72.5 | | 69.5 | 34.0 | | 73.0 | 96.5 | 44.5 | 25.0 | .. |
| 26 | 72.0 | | 69.0 | 33.0 | | 72.5 | | 43.5 | 24.5 | 65 |
| 25 | 71.0 | | | 32.5 | | 72.0 | 96.0 | 42.5 | | 64 |
| 24 | 70.5 | | 68.5 | 32.0 | | 71.0 | 95.5 | 41.5 | 24.0 | .. |
| 23 | 70.0 | | 68.0 | 31.0 | | 70.5 | | 41.0 | 23.5 | 63 |
| 22 | 69.5 | | | 30.5 | | 70.0 | 95.0 | 40.0 | 23.0 | .. |
| 21 | 69.0 | | 67.5 | 29.5 | | 69.5 | 94.5 | 39.0 | 22.5 | 62 |
| 20 | 68.5 | | | 29.0 | | 68.5 | | 38.0 | 22.0 | .. |
| 19 | 68.0 | | 67.0 | 28.5 | | 68.0 | 94.0 | 37.5 | 21.5 | 61 |
| 18 | 67.0 | | 66.5 | 27.5 | | 67.5 | 93.5 | 36.5 | | .. |

\* See note at end of table.

Table 2.  Comparative Hardness Scales for Unhardened Steel, Soft-temper Steel, Grey and Malleable Cast Iron, and Non-ferrous Alloys* — 3

(Compiled by Wilson Mechanical Instrument Co.)

| Rockwell Hardness Number | | | Rockwell Superficial Hardness Number | | | Rockwell Hardness Number | | | | Brinell |
|---|---|---|---|---|---|---|---|---|---|---|
| Rockwell B scale 1/16" Ball Penetrator 100 kg. Load | Rockwell F scale 1/16" Ball Penetrator 60 kg. Load | Rockwell G scale 1/16" Ball Penetrator 150 kg. Load | Rockwell Superficial 15-T scale 1/16" Ball Penetrator 15 kg. Load | Rockwell Superficial 30-T scale 1/16" Ball Penetrator 30 kg. Load | Rockwell Superficial 45-T scale 1/16" Ball Penetrator 45 kg. Load | Rockwell E scale 1/8" Ball Penetrator 100 kg. Load | Rockwell H scale 1/8" Ball Penetrator 60 kg. Load | Rockwell K scale 1/8" Ball Penetrator 150 kg. Load | Rockwell A scale "Brale" Penetrator 60 kg. Load | Brinell Scale 10-mm. Standard Ball — 500 kg. Load |
| 17 | 66.5 | .... | .... | .... | 27.0 | .... | 67.0 | 93.0 | 35.5 | 21.0 | 60 |
| 16 | 66.0 | .... | .... | 66.0 | 26.0 | .... | 66.5 | .... | 35.0 | 20.5 | .. |
| 15 | 65.5 | .... | .... | 65.5 | 25.5 | .... | 65.5 | 92.5 | 34.0 | 20.0 | 59 |
| 14 | 65.0 | .... | .... | .... | 25.0 | .... | 65.0 | 92.0 | 33.0 | .... | .. |
| 13 | 64.5 | .... | .... | 65.0 | 24.0 | .... | 64.5 | .... | 32.0 | .... | 58 |
| 12 | 64.0 | .... | .... | 64.5 | 23.5 | .... | 64.0 | 91.5 | 31.5 | .... | .. |
| 11 | 63.5 | .... | .... | .... | 23.0 | .... | 63.5 | 91.0 | 30.5 | .... | .. |
| 10 | 63.0 | .... | .... | 64.0 | 22.0 | .... | 62.5 | 90.5 | 29.5 | .... | 57 |
| 9 | 62.0 | .... | .... | .... | 21.5 | .... | 62.0 | .... | 29.0 | .... | .. |
| 8 | 61.5 | .... | .... | 63.5 | 20.5 | .... | 61.5 | 90.0 | 28.0 | .... | .. |
| 7 | 61.0 | .... | .... | 63.0 | 20.0 | .... | 61.0 | 89.5 | 27.0 | .... | 56 |
| 6 | 60.5 | .... | .... | .... | 19.5 | .... | 60.5 | .... | 26.0 | .... | .. |
| 5 | 60.0 | .... | .... | 62.5 | 18.5 | .... | 60.0 | 89.0 | 25.5 | .... | 55 |
| 4 | 59.5 | .... | .... | 62.0 | 18.0 | .... | 59.0 | 88.5 | 24.5 | .... | .. |
| 3 | 59.0 | .... | .... | .... | 17.0 | .... | 58.5 | 88.0 | 23.5 | .... | .. |
| 2 | 58.0 | .... | .... | 61.5 | 16.5 | .... | 58.0 | .... | 23.0 | .... | 54 |
| 1 | 57.5 | .... | .... | 61.0 | 16.0 | .... | 57.5 | 87.5 | 22.0 | .... | .. |
| 0 | 57.0 | .... | .... | .... | 15.0 | .... | 57.0 | 87.0 | 21.0 | .... | 53 |

* Not applicable to annealed metals of high B-scale hardness such as austenitic stainless steels, nickel and high-nickel alloys nor to cold-worked metals of low B-scale hardness such as aluminum and the softer alloys.

**Turner's Sclerometer.** — In making this test a weighted diamond point is drawn, once forward and once backward, over the smooth surface of the material to be tested. The hardness number is the weight in grams required to produce a standard scratch.

**Mohs's Hardness Scale.** — Hardness, in general, is determined by what is known as Mohs's scale, a standard for hardness which is mainly applied to non-metallic elements and minerals. In this hardness scale there are ten degrees or steps, each designated by a mineral, the difference in hardness of the different steps being determined by the fact that any member in the series will scratch any of the preceding members. This scale is as follows:

1. Talc; 2. gypsum; 3. calcite; 4. fluor spar; 5. apatite; 6. orthoclase; 7. quartz; 8. topaz; 9. sapphire or corundum; 10. diamond.

These minerals, arbitrarily selected as standards, are successively harder, from talc, the softest of all minerals, to diamond, the hardest. This scale, which is now universally used for non-metallic minerals, is, however, not applied to metals.

**Relation Between Hardness and Tensile Strength.** — The approximate relationship between the hardness and tensile strength is shown by the following formula in which $B$ = Brinell hardness number.

Tensile strength = $B \times 515$ (for Brinell numbers up to 175).

Tensile strength = $490 \times B$ (for Brinell numbers larger than 175).

These formulas give the tensile strength in pounds per square inch and apply to steels. This definite relationship between hardness and tensile strength does not apply to non-ferrous metals with the possible exception of certain aluminum alloys.

# HEAT-TREATMENT—STANDARD STEELS

**Heat-Treating Definitions.** — This glossary of heat-treating terms has been adopted by the American Foundrymen's Association, the American Society for Metals, the American Society for Testing Materials and the Society of Automotive Engineers. Since it is not intended to be a specification but is strictly a set of definitions, temperatures have purposely been omitted.

*Aging:* A change in a metal by which its structure recovers from an unstable condition produced by quenching (quench aging) or by cold working (strain aging). The change in structure consists in precipitation, often submicroscopic, and is marked by a change in physical properties. Aging which takes place slowly at room temperature may be accelerated by a slight increase in temperature. See also *Stress Relieving.*

*Annealing:* A process involving heating and cooling applied usually to induce softening. The term is also used to cover treatments intended to remove stresses; alter mechanical or physical properties; produce a definite microstructure; remove gases. Certain specific heat treatments of iron-base alloys covered by the term annealing are black annealing, blue annealing, box annealing, bright annealing, full annealing, graphitizing, malleablizing, process annealing.

*Annealing, Black:* A process of box annealing iron-base alloy sheets after hot rolling, shearing and pickling. The process does not impart a black color to the product if properly done. The name originated in the appearance of the hot-rolled material before pickling and annealing.

*Annealing, Blue:* A process of softening iron-base alloys in the form of hot-rolled sheet, in which the sheet is heated in the open furnace to a temperature within the transformation range and cooled in air; the formation of a bluish oxide on the surface is incidental.

*Annealing, Box:* A process of annealing which, to prevent oxidation, is carried out in a suitable closed metal container with or without packing material. The charge is usually heated slowly to a temperature below, but sometimes above or within, the transformation temperature range and cooled slowly. It is also called *Close Annealing* or *Pot Annealing.*

*Annealing, Bright:* A process of annealing which is usually carried out in a controlled furnace atmosphere so that surface oxidation is reduced to a minimum and the surface remains relatively bright.

*Annealing, Flame:* A process in which the surface of an iron-base alloy is softened by localized heat applied by a high-temperature flame.

*Annealing, Full:* A softening process in which an iron-base alloy is heated to a temperature above the transformation range and, after being held for a proper time at this temperature, is cooled slowly to a temperature below the transformation range. The objects are ordinarily allowed to cool slowly in the furnace, although they may be removed from the furnace and cooled in some medium which assures a slow rate of cooling.

*Annealing, Inverse:* A heat treatment, analogous to *Precipitation Hardening,* applied to cast iron usually to increase its hardness and strength.

*Annealing, Process:* — A process commonly applied in the sheet and wire industries, in which an iron-base alloy is heated to a temperature close to, but below, the lower limit of the transformation range and subsequently cooled. This process is applied for the purpose of softening for further cold working.

*Austempering:* A trade name for a patented heat treating process consisting in quenching an iron-base alloy from a temperature above the transformation range in a medium having a suitably high rate of heat abstraction, and maintaining the alloy, until transformation is complete, at a temperature which is below that of pearlite formation and above that of martensite formation. The temperature for austenite transformation is chosen on the basis of the properties desired.

*Bluing:* A treatment of the surface of iron-base alloys, usually in the form of sheet or strip, on which, by the action of air or steam at a suitable temperature, a thin blue oxide film is formed on the initially scale-free surface, as a means of improving appearance and resistance to corrosion. This term is also used to denote a heat treatment of springs after fabrication, to reduce the internal stress created by coiling and forming.

*Brunorizing:* The trade name for a special treatment applied to steel rails which, after cooling to a temperature below the transformation range, are reheated to a temperature slightly above that range, and then are allowed to cool in the air, the ends of the rails being partially quenched by jets of compressed air.

*Burnt:* A term applied to a metal permanently damaged by being heated to a temperature close to the melting point. The damage may involve melting of some constituent or penetration by, and reaction of the metal with, a gas such as oxygen, or by segregation of component elements of the metal.

*Carburizing:* A process in which carbon is introduced into a solid iron-base alloy by heating above the transformation temperature range while in contact with a carbonaceous material which may be a solid, liquid or gas. Carburizing is frequently followed by quenching to produce a hardened case. The term carbonizing is sometimes used erroneously in place of carburizing.

*Case:* (1) The surface layer of an iron-base alloy which has been suitably altered in composition and can be made substantially harder than the interior or core by a process of case hardening. (2) The term case is also used to designate the hardened surface layer of a piece of steel that is large enough to have a distinctly softer core or center.

*Cementation:* The process of introducing elements into the outer layer of metal objects by means of high-temperature diffusion.

*Controlled Cooling:* A term used to describe a process by which a steel object is cooled from an elevated temperature, usually from the final hot forming operation in a predetermined manner of cooling to avoid hardening, cracking, or internal damage.

*Core:* (1) The interior portion of an iron-base alloy which after case hardening is substantially softer than the surface layer or case. (2) The term core is also used to designate the relatively soft central portion of certain hardened tool steels.

*Critical Range* or *Critical Temperature Range:* Synonymous with *Transformation Range*, which is preferred.

*Cyaniding:* A process of case hardening an iron-base alloy by the simultaneous absorption of carbon and nitrogen by heating in a cyanide salt. Cyaniding is usually followed by quenching to produce a hard case.

*Decarburization:* The loss of carbon from the surface of an iron-base alloy as the result of heating in a medium which reacts with the carbon.

*Drawing:* Drawing, or drawing the temper, is synonymous with *Tempering*, which is preferable.

*Graphitizing:* An annealing process applied to certain iron-base alloys, such as cast iron or some steels with high carbon and silicon contents, by which the combined carbon is wholly or in part transformed to graphitic or free carbon. See *Temper Carbon.*

*Hardening:* Any process of increasing hardness of metal by suitable treatment, usually involving heating and cooling.

*Hardening, Age:* See *Aging*

*Hardening, Case:* A process of surface hardening involving a change in the composition of the outer layer of an iron-base alloy followed by appropriate thermal treatment. Typical case-hardening processes are *Carburizing, Cyaniding, Carbo-Nitriding* and *Nitriding.*

*Hardening, Flame:* A process of heating the surface layer of an iron-base alloy above the transformation temperature range by means of a high-temperature flame, followed by quenching.

*Hardening, Precipitation:* A process of hardening an alloy in which a constituent precipitates from a supersaturated solid solution. See also *Aging.*

*Hardening, Secondary:* An increase in hardness following the normal softening that occurs during the tempering of certain alloy steels.

*Heating, Differential:* A heating process by which the temperature is made to vary throughout the object being heated so that on cooling different portions may have such different physical properties as may be desired.

*Heating, Induction:* A process of local heating by electrical induction.

*Heat Treatment:* A combination of heating and cooling operations applied to a metal or alloy in the solid state to obtain desired conditions or properties. Heating for the sole purpose of hot working is excluded from the meaning of this definition.

*Heat Treatment, Solution:* A treatment in which an alloy is heated to a suitable temperature and held at this temperature for a sufficient length of time to allow a desired constituent to enter into solid solution, followed by rapid cooling to hold the constituent in solution. The material is then in a supersaturated, unstable state, and may subsequently exhibit *Age Hardening.*

*Homogenizing:* A high-temperature heat-treatment process intended to eliminate or to decrease chemical segregation by diffusion.

*Malleablizing:* A process of annealing white cast iron in which the combined carbon is wholly or in part transformed to graphitic or free carbon, and, in some cases, part of the carbon is removed completely. See *Temper Carbon.*

*Nitriding:* A process of case hardening in which an iron-base alloy of special composition is heated in an atmosphere of ammonia or in contact with nitrogenous material. Surface hardening is produced by the absorption of nitrogen without quenching.

*Nitriding, Carbo:* A process of case hardening an iron-base alloy by the simultaneous absorption of carbon and nitrogen by heating in a gaseous atmosphere of suitable composition, followed by either quenching or cooling slowly as required.

*Normalizing:* A process in which an iron-base alloy is heated to a temperature above the transformation range and subsequently cooled in still air at room temperature.

*Overheated:* A metal is said to have been overheated if, after exposure to an unduly high temperature, it develops an undesirably coarse grain structure but is not permanently damaged. The structure damaged by overheating can be corrected by suitable heat treatment or by mechanical work or by a combination of the two. In this respect it differs from a *Burnt* structure.

*Patenting:* A process of heat treatment applied to medium or high carbon steel in wire making prior to the wire drawing or between drafts. It consists in heating to a temperature above the transformation range, followed by cooling to a temperature below that range in air or in a bath of molten lead or salt maintained at a temperature appropriate to the carbon content of the steel and the properties required of the finished product.

*Preheating:* (1) A general term used to describe a heating applied preliminary to some further thermal or mechanical treatment. (2) A term specifically applied to tool steel to describe a process in which the steel is heated slowly and uniformly to a temperature below the hardening temperature and is then transferred to a furnace in which the temperature is substantially above the preheating temperature.

*Quenching:* A process of rapid cooling from an elevated temperature, by contact with liquids, gases or solids.

*Quenching, Differential:* A quenching process by which only certain desired portions of the object are quenched and hardened.

*Quenching, Hot:* A process of quenching iron-base alloys in a medium, the temperature of which is substantially higher than atmospheric temperature.

*Quenching, Pot:* A process of quenching carburized parts directly from the carburizing box or pot.

*Sandberg Sorbitic Treatment:* A treatment in which carbon steel objects are moderately hardened, either wholly or in part. It consists in cooling the parts to be hardened through the transformation range at a moderately rapid rate by the application of jets of air, steam, or atomized water and then allowing the residual heat in the object to effect a tempering operation.

*Soaking:* Prolonged heating of a metal at a selected temperature.

*Spheroidizing:* Any process of heating and cooling steel that produces a rounded or globular form of carbide in the structure.

*Stress Relieving:* A process to reduce internal residual stresses in a metal object by heating the object to a suitable temperature and holding for a proper time at that temperature. This treatment may be applied to relieve stresses induced by casting, quenching, normalizing, machining, cold working or welding. Stress relieving is sometimes termed *Aging*.

*Temper Carbon:* The free or graphitic carbon which comes out of solution usually in the form of rounded nodules in the structure during *Graphitizing* or *Malleablizing*.

*Tempering:* A process of reheating hardened or normalized steel to a temperature below the transformation temperature range, followed by any desired rate of cooling.

*Transformation Range:* In ferrous alloys the transformation range on heating is the temperature interval within which austenite forms. The transformation range on cooling is the temperature interval in which austenite disappears. Distinction must be made between the two ranges. They may overlap but never coincide. The limiting temperatures of the ranges depend on the composition of the alloy and, particularly for the cooling, on the rate of change of temperature.

**Structure of Fully Annealed Carbon Steel.**—In carbon steel that has been fully annealed, there are normally present, apart from such impurities as phosphorus and sulphur, two constituents: the element iron in a form metallurgically known as *ferrite* and the chemical compound iron carbide in the form metallurgically known as *cementite*. This latter constituent consists of 6.67 per cent carbon and 93.33 per cent iron. A certain proportion of these two constituents will be present as a mechanical mixture. This mechanical mixture, the amount of which depends upon the carbon content of the steel, consists of alternate bands or layers of ferrite and cementite. Under the microscope it frequently has the appearance of mother-of-pearl and hence has been named *pearlite*. Pearlite contains about 0.85 per cent carbon and 99.15 per cent iron, neglecting impurities. A fully annealed steel containing 0.85 per cent carbon would consist entirely of pearlite. Such a steel is known as *eutectoid* steel and has a laminated structure characteristic of a eutectic alloy. Steel which has less than 0.85 per cent carbon (*hypo-eutectoid* steel) has an excess of ferrite above that required to mix with the cementite present to form pearlite, hence both ferrite and pearlite are present in the fully annealed state. Steel having a carbon content greater than 0.85 per cent (*hyper-eutectoid* steel) has an excess of cementite over that required to mix with the ferrite to form pearlite, hence both cementite and pearlite are present in the fully annealed state. The structural constitution of carbon steel in terms of ferrite, cementite, pearlite and austenite for different carbon contents and at different temperatures is shown by the accompanying diagram.

**Effect of Heating Fully Annealed Carbon Steel.**—When carbon steel in the fully annealed state is heated above the lower critical point, which is some tempera-

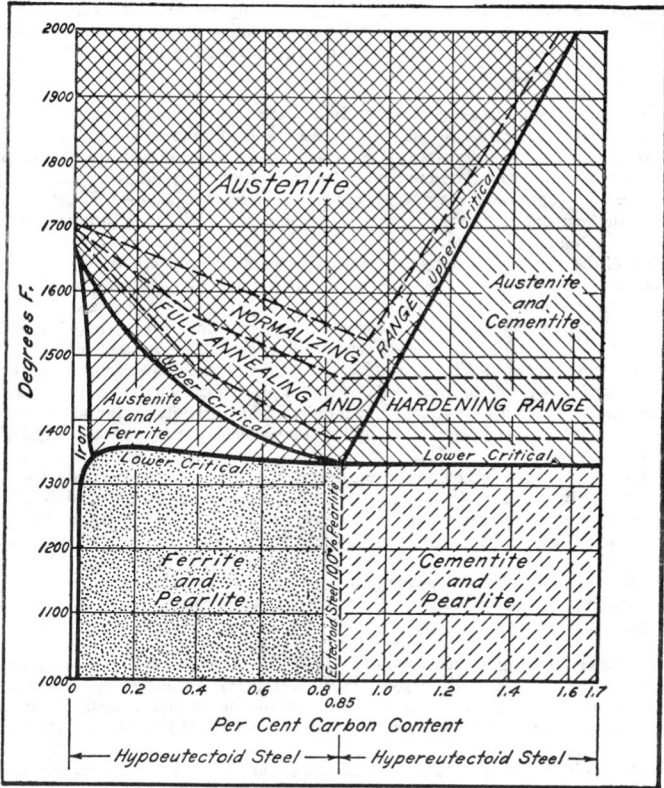

ture in the range of 1335 to 1355 degrees F. (depending upon the carbon content), the alternate bands or layers of ferrite and cementite which make up the pearlite begin to merge into each other. This process continues until the pearlite is thoroughly "dissolved," forming what is known as *austenite*. If the temperature of the steel continues to rise and there is present, in addition to the pearlite, any excess ferrite or cementite, this also will begin to dissolve into the austenite until finally only austenite will be present. The temperature at which the excess ferrite or cementite is completely dissolved in the austenite is called the *upper critical point*. This temperature varies with the carbon content of the steel much more widely than the lower critical point (see diagram).

**Effect of Slow Cooling Carbon Steel.** — If carbon steel which has been heated to the point where it consists entirely of austenite is slowly cooled, the process of transformation which took place during the heating will be reversed but the upper and lower critical points will occur at somewhat lower temperatures than they do on heating. Assuming that the steel was originally fully annealed, its structure upon returning to atmospheric temperature after slow cooling will be the same as before in terms of the proportions of ferrite or cementite and pearlite present. The austenite will have entirely disappeared.

**Effect of Rapid Cooling or Quenching Carbon Steel.** — Observations have shown that as the rate at which carbon steel is cooled from an austenitic state is increased, the temperature at which the austenite begins to change into pearlite drops more and more below the slow cooling transformation temperature of about 1300 degrees F. (For example, a 0.80 per cent carbon steel that is cooled at such a rate that the temperature drops 500 degrees in one second will show transformation of austenite beginning at 930 degrees F.) As the cooling rate is increased, the laminations of the pearlite formed by the transformation of the austenite become finer and finer up to the point where they cannot be detected under a high power microscope, while the steel itself increases in hardness and tensile strength. As the rate of cooling is still further increased, this transformation temperature suddenly drops down to around 500 degrees F. or lower, depending upon the carbon content of the steel. The cooling rate at which this sudden drop in transformation temperature takes place is called the *critical cooling rate*. When a piece of carbon steel is quenched at this rate or faster, a new structure is formed. The austenite is transformed into *martensite* which is characterized by an angular needlelike structure and a very high hardness.

If carbon steel is subjected to a severe quench or to extremely rapid cooling, a small percentage of the austenite, instead of being transformed into martensite during the quenching operation, may be retained. Over a period of time, however, this remaining austenite tends to be gradually transformed into martensite even though the steel is not subjected to further heating or cooling. Since martensite has a lower density than austenite, such a change or " ageing " as it is called, often results in an appreciable increase in volume or " growth " and the setting up of new internal stresses in the steel.

**Steel Heat-Treating Furnaces.** — Various types of furnaces heated by gas, oil or electricity, are used for the heat-treatment of steel. These include the oven or box type in various modifications for " in-and-out " or for continuous loading and unloading; the retort type; the pit type; the pot type; and the salt-bath electrode type.

*Oven or Box Furnaces:* This type of furnace has a box or oven-shaped heating chamber. The " in-and-out " oven furnaces are loaded by hand or by a track-mounted car which, when rolled into the furnace, forms the bottom of the heating chamber. The car type is used where heavy or bulky pieces must be handled. Some oven type furnaces are provided with a full muffle or a semi-muffle which is an enclosed refractory chamber into which the parts to be heated are placed. The full-muffle, being fully enclosed, prevents any flames or burning gases from coming in contact with the work and permits a special atmosphere to be used to protect or condition the work. The semi-muffle, which is open at the top, protects the work from direct impingement of the flame although it does not shut the work off from the hot gases. In the direct-heat type oven furnace, the work is open to the flame. In the electric oven furnace, a retort is provided if gas atmospheres are to be employed to confine the gas and prevent it from attacking the heating elements. Where muffles are used, they must be replaced periodically and a greater amount of fuel is required than in a direct-heat type of oven furnace.

For continuous loading and unloading, there are several types such as rotary hearth car; roller-, furnace belt-, walking-beam or pusher-conveyor; and a continuous-kiln type through which track-mounted cars are run. In the continuous type of furnace, the work may pass through several zones maintained at different temperatures for preheating, heating, soaking, and cooling.

*Retort Furnace:* This is a vertical type of furnace provided with a cylindrical metal retort into which the parts to be heat-treated are suspended either individually, if large enough, or in a container of some sort. The use of a retort permits special gas atmospheres to be employed for carburizing, nitriding, etc.

*Pit Type Furnace:* This is a vertical furnace arranged for the loading of parts in a metal basket. The parts are heated by convection, the basket, when lowered into place, fitting into the furnace chamber in such a way as to provide a dead-air space to prevent direct heating.

*Pot Type Furnace:* This furnace is used for the immersion method of heat-treating small parts. A cast-alloy pot is employed to hold a bath of molten lead or salt in which the parts are placed for heating.

*Salt Bath Electrode Furnace:* In this type of electric furnace, heating is accomplished by means of electrodes suspended directly in the salt bath. The patented grouping and design of electrodes provide an electromagnetic action which results in an automatic stirring action throughout the bath. This tends to produce an even temperature throughout.

## Hardening

**Basic Steps in Hardening.** — The operation of hardening steel consists fundamentally of two steps. The first step is to heat the steel to some temperature above (usually at least 100 degrees F. above) its transformation point so that it becomes entirely austenitic in structure. The second step is to quench the steel at some rate faster than the critical rate (which depends on the carbon content, the amounts of alloying elements present other than carbon, and the grain size of the austenite) to produce a martensitic structure. The hardness of a martensitic steel depends upon its carbon content and ranges from about 460 Brinell at 0.20 per cent carbon to about 710 Brinell above 0.50 carbon. In comparison, ferrite has a hardness of about 90 Brinell, pearlite about 240 Brinell, and cementite around 550 Brinell.

**Critical Points of Decalescence and Recalescence.** — The critical or transformation point at which pearlite is transformed into austenite as it is being heated is also called the *decalescence point.* If the temperature of the steel was observed as it passed through the decalescence point, it would be noted that it would continue to absorb heat without appreciably rising in temperature, although the immediate surroundings were hotter than the steel. Similarly, the critical or transformation point at which austenite is transformed back into pearlite upon cooling is called the *recalescence point.* When this point is reached, the steel will give out heat so that its temperature instead of continuing to fall, will momentarily increase.

The recalescent point is lower than the decalescence point by anywhere from 85 to 215 degrees F., and the lower of these points does not manifest itself unless the higher one has first been fully passed. These critical points have a direct relation to the hardening of steel. Unless a temperature sufficient to reach the decalescence point is obtained, so that the pearlite is changed into austenite, no hardening action can take place; and unless the steel is cooled suddenly before it reaches the recalescence point, thus preventing the changing back again from austenite to pearlite, no hardening can take place. The critical points vary for different kinds of steel and must be determined by tests in each case. It is the variation in the critical points that makes it necessary to heat different steels to different temperatures when hardening.

**Hardening Temperatures.** — The maximum temperature to which a steel is heated before quenching to harden it is called the hardening temperature. Hardening temperatures vary for different steels and different classes of service, although, in general, it may be said that the hardening temperature for any given steel is above the lower critical point of that steel. Just how far above this point the hardening temperature lies for any particular steel depends on three factors: (1) The chemical composition of the steel; (2) the amount of excess ferrite (if the steel has less than 0.85 per cent carbon content) or the amount of excess cementite (if the steel has more than 0.85 per cent carbon content) that is to be dissolved in the austenite; and (3) the maximum grain size permitted, if desired.

The general range of full hardening temperatures for carbon steels is shown by the diagram. This range is merely indicative of general practice and is not intended to represent absolute hardening temperature limits. It can be seen that for steels of less than 0.85 per cent carbon content, the hardening range is above the upper critical point — that is, above the temperature at which all of the excess ferrite has been dissolved in the austenite. On the other hand, for steels of more than 0.85 per cent carbon content, the hardening range lies somewhat below the upper critical point. This indicates that in this hardening range some of the excess cementite still remains undissolved in the austenite. If steel of more than 0.85 per cent carbon content were heated above the upper critical point and then quenched, the resulting grain size would be excessively large.

At one time it was considered desirable to heat steel only to the minimum temperature at which it would fully harden, one of the reasons being to avoid grain growth that takes place at higher temperature. It is now realized that no such rule as this can be applied generally since there are factors other than hardness which must be taken into consideration. For example, in many cases toughness can be impaired by too low a temperature just as much as by too high a temperature. It is true, however, that too high hardening temperatures result in warpage, distortion, increased scale, and decarburization.

**Hardening Temperatures for Carbon Tool Steels.** — The best hardening temperatures for any given tool steel are dependent upon the type of tool and the intended class of service. Wherever possible, the specific recommendations of the tool steel manufacturer should be followed. General recommendations for hardening temperatures of carbon tool steels based on carbon content are as follows: For steel of 0.65 to 0.80 per cent carbon content, 1450 to 1550 degrees F.; for steel of 0.80 to 0.95 per cent carbon content, 1410 to 1460 degrees F.; for steel of 0.95 to 1.10 per cent content, 1390 to 1430 degrees F.; and for steels of 1.10 per cent and over carbon content, 1380 to 1420 degrees F. For a given hardening temperature range, the higher temperatures tend to produce deeper hardness penetration and increased compressional strength while the lower temperatures tend to result in shallower hardness penetration but increased resistance to splitting or bursting stresses.

**Determining Hardening Temperatures.** — A hardening temperature can be specified directly or it may be specified indirectly as a certain temperature rise above the lower critical point of the steel. Where the temperature is specified directly, a pyrometer of the type which indicates the furnace temperature or a pyrometer of the type which indicates the work temperature may be employed. If the pyrometer shows furnace temperature, care must be taken to allow sufficient time for the work to reach the furnace temperature after the pyrometer indicates that the required hardening temperature has been attained. If the pyrometer indicates work temperature, then, where the work-piece is large, time must be allowed for the interior of the work to reach the temperature of the surface which is the temperature indicated by the pyrometer.

Where the hardening temperature is specified as a given temperature rise above the critical point of the steel, a pyrometer which indicates the temperature of the work should be used. The critical point, as well as the given temperature rise, can be more accurately determined with this type of pyrometer. As the work is heated, its temperature, as indicated by the pyrometer, rises steadily until the lower critical or decalescence point of the steel is reached. At this point, the temperature of the work ceases to rise and the pyrometer indicating or recording pointer remains stationary or fluctuates slightly. After a certain elapsed period, depending upon the rate at which heat is furnished to the work, the internal changes in structure of the steel which take place at the lower critical point are completed and the temperature of the work again begins to rise. Since a small fluctuation in temperature may occur in the interval during which structural changes are taking place, for uniform practice the critical point may be considered as the temperature at which the pointer first becomes stationary.

**Heating Steel in Liquid Baths.** — The liquid baths commonly used for heating steel tools preparatory to hardening are molten lead, sodium cyanide, barium chloride, a mixture of barium and potassium chloride and other metallic salts. The molten substance is retained in a crucible or pot and the heat required may be obtained from gas, oil, or electricity. The principal advantages of heating baths are as follows: No part of the work can be heated to a temperature above that of the bath; the temperature can be easily maintained at whatever degree has proved, in practice, to give the best results; the submerged steel can be heated uniformly, and the finished surfaces are protected against oxidation.

**Salt Baths.** — Molten baths of various salt mixtures or compounds are used extensively for heat-treating operations such as hardening and tempering; they are also utilized for annealing ferrous and non-ferrous metals. Commercial salt-bath mixtures are available which meet a wide range of temperature and other metallurgical requirements. For example, there are neutral baths for heating tool and die steels without carburizing the surfaces; baths for carburizing the surfaces of low-carbon steel parts; baths adapted for the usual tempering temperatures of, say, 300 to 1100 degrees F.; and baths which may be heated to temperatures up to approximately 2400 degrees F. for hardening high-speed steels. Salt baths are also adapted for local or selective hardening, the type of bath being selected to suit the requirements. For example, a neutral bath may be used for annealing the ends of tubing or other parts, or an activated cyanide bath for carburizing the ends of shafts or other parts. Surfaces which are not to be carburized are protected by copper plating. When the work is immersed, the unplated parts are subjected to the carburizing action.

Baths may consist of a mixture of sodium, potassium, barium, and calcium chlorides or nitrates of sodium, potassium, barium, and calcium in varying proportions, to which sodium carbonate and sodium cyanide are sometimes added to prevent decarburization. Various proportions of these salts provide baths of different properties. Potassium cyanide is seldom used as sodium cyanide costs less. The specific gravity of a salt bath is not as high as that of a lead bath; consequently the work may be suspended in a salt bath and does not have to be held below the surface as in a lead bath.

**The Lead Bath.** — The lead bath is extensively used, but is not adapted to the high temperatures required for hardening high-speed steel, as it begins to vaporize at about 1190 degrees F. As the temperature increases, the lead volatilizes and gives off poisonous vapors; hence, lead furnaces should be equipped with hoods to carry away the fumes. Lead baths are generally used for temperatures below 1500 or 1600 degrees F. They are often employed for heating small pieces which must be hardened in quantities. It is important to use pure lead that is free from sulphur. The work should be pre-heated before plunging it into the molten lead.

**Defects in Hardening.** — Uneven heating is the cause of most of the defects in hardening. Cracks of a circular form, from the corners or edges of a tool, indicate uneven heating in hardening. Cracks of a vertical nature and dark-colored fissures indicate that the steel has been burned and should be put on the scrap heap. Tools which have hard and soft places have been either unevenly heated, unevenly cooled, or " soaked," a term used to indicate prolonged heating. A tool not thoroughly moved about in the hardening fluid will show hard and soft places, and have a tendency to crack. Tools which are hardened by dropping them to the bottom of the tank, sometimes have soft places, owing to contact with the floor or sides.

**Scale on Hardened Steel.** — The formation of scale on the surface of hardened steel is due to the contact of oxygen with the heated steel; hence, to prevent scale, the heated steel must not be exposed to the action of the air. When using an oven heating furnace, the flame should be so regulated that it is not visible in the heating chamber. The heated steel should be exposed to the air as little as possible, when transferring it from the furnace to the quenching bath. An old method of preventing scale and retaining a fine finish on dies used in jewelry manufacture, small taps, etc. is as follows: Fill the die impression with powdered boracic acid and place near the fire until the acid melts; then add a little more acid to insure covering all the surfaces. The die is then hardened in the usual way. If the boracic acid does not come entirely off in the quenching bath, immerse the work in boiling water. Dies hardened by this method are said to be as durable as those heated without the acid.

**Hardening or Quenching Baths.** — The purpose of a quenching bath is to remove heat from the steel being hardened at a rate that is faster than the critical cooling rate. Generally speaking, the more rapid the rate of heat extraction above the cooling rate, the higher will be the resulting hardness. To obtain the different rates of cooling required by different classes of work, baths of various kinds are used. These include plain or fresh water, brine, caustic soda solutions, oils of various classes, oil-water emulsions, baths of molten salt or lead for high-speed steels and air cooling for some high-speed steel tools when a slow rate of cooling is required. To minimize distortion and cracking where such tendencies are present, without sacrificing depth of hardness penetration, a quenching medium should be selected that will cool rapidly at the higher temperatures and more slowly at the lower temperatures, i.e., below 750 degrees F. Oil quenches in general meet this requirement.

*Oil Quenching Baths:* Oil is used very extensively as a quenching medium as it results in a good proportion of hardness, toughness, and freedom from warpage when used with standard steels. Oil baths are used extensively for alloy steels. Various kinds of oils are employed such as prepared mineral oils and vegetable, animal and fish oils, either singly or in combination. Prepared mineral quenching oils are widely used because they have good quenching characteristics, are chemically stable, do not have an objectionable odor, and are relatively inexpensive. Special compounded oils of the soluble type are used in many plants instead of such oils as fish oil, linseed oil, cottonseed oil, etc. The soluble properties enable the oil to form an emulsion with water.

Oil cools steel at a slower rate than water, but the rate is fast enough for alloy steel. Oils have different cooling rates, however, and this rate may vary through the initial and final stages of the quenching operation. Faster cooling in the initial stage and slower cooling at lower temperatures is preferable because there is less danger of cracking the steel. The temperature of quenching oil baths should range ordinarily between 90 and 130 degrees F. A fairly constant temperature may be maintained either by circulating the oil through cooling coils or by using a tank provided with a cold water jacket.

A good quenching oil should possess a flash and fire point sufficiently high to be safe under the conditions used and 350 degrees F. should be about the minimum point. The specific heat of the oil regulates the hardness and toughness of the quenched steel; and the greater the specific heat, the higher will be the hardness produced. Specific heats of quenching oils vary from 0.20 to 0.75, the specific heats of fish, animal and vegetable oils usually being from 0.2 to 0.4, and of soluble and mineral oils from 0.5 to 0.7. The efficient temperature range for quenching oil is from 90 to 140 degrees F.

**Quenching in Water.** — Many carbon tool steels are hardened by immersing them in a bath of fresh water, but water is not an ideal quenching medium. Contact between the water and work and the cooling of the hot steel is impaired by the formation of gas bubbles or an insulating vapor film especially in holes, cavities or pockets. The result is uneven cooling and in some cases excessive strains which may cause the tool to crack; in fact, there is greater danger of cracking in a fresh water bath than in one containing salt water or brine.

In order to secure more even cooling and reduce danger of cracking, either rock salt (8 or 9 per cent) or caustic soda (3 to 5 per cent) may be added to the bath in order to eliminate or prevent the formation of a vapor film or gas pockets, thus promoting rapid early cooling. Brine is commonly used and ¾ pound of rock salt per gallon of water is equivalent to about 8 per cent of salt. Brine is not inherently a more severe or drastic quenching medium than plain water, although it may seem to be because the brine makes better contact with the heated steel and, consequently, cooling is more effective. In still bath quenching, a slow up-and-down movement of the tool is preferable to a violent swishing around.

The temperature of water-base quenching baths should preferably be kept around 70 degrees F., but 70 to 90 or 100 degrees F. is a safe range. The temperature of the hardening bath has a great deal to do with the hardness obtained. The higher the temperature of the quenching water, the more nearly does its effect approach that of oil; and if boiling water is used for quenching, it will have an effect even more gentle than that of oil — in fact, it would leave the steel nearly soft. Parts of irregular shape are sometimes quenched in a water bath that has been warmed somewhat to prevent sudden cooling and cracking.

When water is used, it should be " soft " as unsatisfactory results will be obtained with " hard " water. Any contamination of water-base quenching liquids by soap tends to decrease their rate of cooling. A water bath having 1 or 2 inches of oil on the top is sometimes employed to advantage for quenching tools made of high-carbon steel as the oil through which the work first passes reduces the sudden quenching action of the water.

The bath should be amply large to dissipate the heat rapidly and the temperature should be kept about constant so that successive pieces will be cooled at the same rate. Irregularly shaped parts should be immersed so that the heaviest or thickest section enters the bath first. After immersion, the part to be hardened should be agitated in the bath; the agitation reduces the tendency of the formation of a vapor coating on certain surfaces, and a more uniform rate of cooling is obtained. The work should never be dropped to the bottom of the bath until quite cool.

*Flush or Local Quenching by Pressure-spraying:* When dies for cold heading, drawing, extruding, etc., or other tools, require a hard working surface and a relatively soft but tough body, the quenching may be done by spraying water under pressure against the interior or other surfaces to be hardened. Special spraying fixtures are used to hold the tool and apply the spray where the hardening is required. The pressure-spray prevents the formation of gas pockets previously referred to in connection with the fresh water quenching bath; hence fresh water is effective for flush quenching and there is no advantage in using brine.

**Quenching in Molten Salt Bath.**—A molten salt bath may be used in preference to oil for quenching high-speed steel. The object in using a liquid salt bath for quenching (instead of an oil bath) is to obtain maximum hardness with minimum cooling stresses and distortion which might result in cracking expensive tools, especially if there are irregular sections. The temperature of the quenching bath may be around 1100 or 1200 degrees F. Quenching is followed by cooling to room temperature and then the tool is tempered or drawn in a bath having a temperature range of 950 to 1100 degrees F. In many cases, the tempering temperature is about 1050 degrees F.

**Tanks for Quenching Baths.** — The main point to be considered in a quenching bath is to keep it at a uniform temperature, so that successive pieces quenched will be subjected to the same heat. The next consideration is to keep the bath agitated, so that it will not be of different temperatures in different places; if thoroughly agitated and kept in motion, as is the case with the bath shown in Fig. 1, it is not even necessary to keep the pieces in motion in the bath, as steam will not be likely to form around the pieces quenched. Experience has proved that if a piece is held still in a thoroughly agitated bath, it will come out much straighter than if it has been moved around in an unagitated bath. This is an important consideration, especially when hardening long pieces. It is, besides, no easy matter to keep heavy and long pieces in motion unless it be done by mechanical means.

In Fig. 1 is shown a water or brine tank for quenching baths. Water is forced by a pump or other means through the supply pipe into the intermediate space between the outer and inner tank. From the intermediate space it is forced into the inner tank through holes as indicated. The water returns to the storage tank by overflowing from the inner tank into the outer one and then through the overflow pipe as indicated. In Fig. 3 is shown another water or brine tank of a more common type. In this case the water or brine is pumped from the storage tank and continuously returned to it. If the storage tank contains a large volume of water, there is no need of a special means for cooling. Otherwise, arrangements must be made for cooling the water after it has passed through the tank. The bath is agitated by the force with which the water is pumped into it. The holes at *A* are drilled at an angle, so as to throw the water toward the center of the tank. In Fig. 2 is shown an oil quenching tank in which water is circulated in an outer surrounding tank for keeping the oil bath cool. Air is forced into the oil bath to keep it agitated. Fig. 4 shows the ordinary type of quenching tank cooled by water forced through a coil of pipe. This can be used for either oil, water or brine. Fig. 5 shows a similar type of quenching tank, but with two coils of pipe. Water flows through one of these and steam through the other. By this means it is possible to keep the bath at a constant temperature.

**Interrupted Quenching.** — *Austempering, martempering,* and *isothermal quenching* are three methods of interrupted quenching that have been developed to obtain greater toughness and ductility for given hardnesses and to avoid the difficulties of quench cracks, internal stresses, and warpage, frequently experienced when the conventional method of quenching steel directly and rapidly from above the transformation point to atmospheric temperature is employed. In each of these three methods, quenching is begun when the work has reached some temperature above the transformation point and is conducted at a rate faster than the critical rate. The rapid cooling of the steel is interrupted, however, at some temperature above that at which martensite begins to form. The three methods differ in the temperature range at which interruption of the rapid quench takes place, the length of time that the steel is held at this temperature, and whether the subsequent cooling to atmospheric temperature is rapid or slow, and is or is not preceded by a tempering operation.

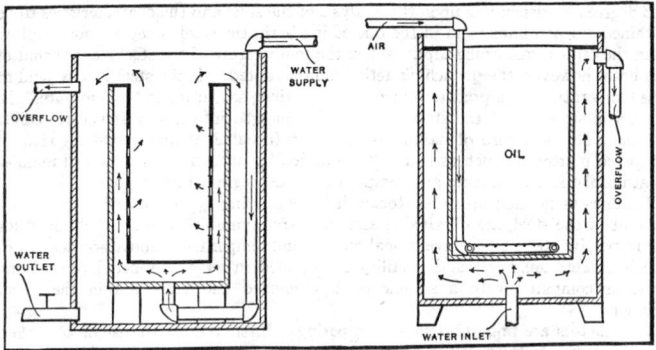

Fig. 1.　　　　　　　　　　Fig. 2

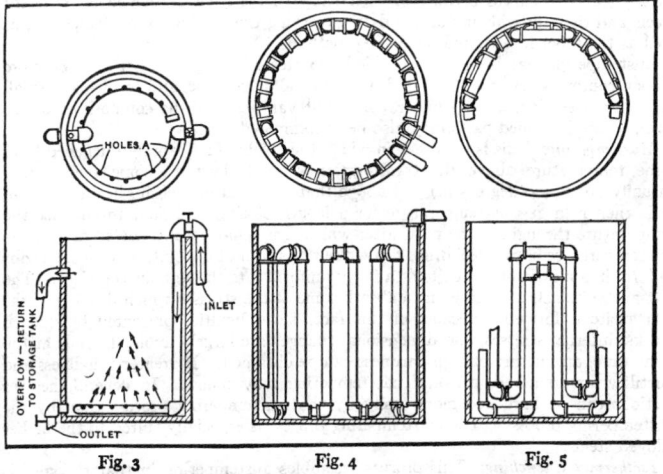

Fig. 3　　　　　Fig. 4　　　　　Fig. 5

One of the reasons for maintaining the steel at a constant temperature for a definite period of time is to permit the inside sections of the piece to reach the same temperature as the outer sections so that when transformation of the structure does take place, it will occur at about the same rate and period of time throughout the piece. In order to maintain the constant temperature required in interrupted quenching, a quenching arrangement for absorbing and dissipating a large quantity of heat without increase in temperature is needed. Molten salt baths equipped for water spray or air cooling around the exterior of the bath container have been used for this purpose.

*Austempering:* This is a patented heat-treating process in which steels are quenched in a bath maintained at some constant temperature in the range of 350 to

800 degrees F., depending upon the analysis of the steel and the characteristics to be obtained. Upon immersion in the quenching bath, the steel is cooled more rapidly than the critical quenching rate. When the temperature of the steel reaches that of the bath, however, the quenching action is interrupted. If the steel is now held at this temperature for a predetermined length of time, say, from 10 to 60 minutes, the austenitic structure of the steel is gradually changed into a new structure, called *bainite*. The structure of bainite is accicular (needlelike) and resembles that of tempered martensite such as is usually obtained by quenching in the usual manner to atmospheric temperature and tempering at 400 degrees F. or higher.

Hardnesses ranging up to 60 Rockwell C, depending upon the carbon and alloy content of the steel, are obtainable and compare favorably with those obtained for the respective steels by a conventional quench and tempering to above 400 degrees F. Much greater toughness and ductility is obtained in an austempered piece, however, as compared with a similar piece quenched and tempered in the usual manner.

Two factors are important in austempering. First, the steel must be quenched rapidly enough to the specified sub-transformation temperature to avoid any formation of pearlite, and second, it must be held at this temperature until the transformation from austenite to bainite is completed. Time and temperature transformation curves (called S-curves because of their shape) have been developed for different steels and these provide important data governing the conduct of austempering, as well as the other interrupted quenching methods.

Austempering has been applied chiefly to steels having 0.60 per cent or more carbon content with or without additional low alloy content, and to pieces of small diameter or section, usually under 1 inch but varying with the composition of the steel. Case hardened parts may also be austempered.

*Martempering:* This is a process in which the steel is first rapidly quenched from some temperature above the transformation point down to some temperature (usually about 400 degrees F.) just above that at which martensite begins to form. It is then held at this temperature for a length of time sufficient to equalize the temperature throughout the part, after which it is removed and cooled in air. As the temperature of the steel drops below the transformation point, martensite begins to form in a matrix of austenite at a fairly uniform rate throughout the piece. The soft austenite acts as a cushion to absorb some of the stresses which develop as the martensite is formed. Because of this fact, the difficulties presented by quench cracks, internal stresses, and dimensional changes are largely avoided, while at the same time, a structure of high hardness can be obtained. If greater toughness and ductility are required, conventional tempering may follow. In general, heavier sections can be hardened more easily by the martempering process than by the austempering process. The martempering process is especially suited to the higher alloyed steels.

*Isothermal Quenching:* This process resembles austempering in that the steel is first rapidly quenched from above the transformation point down to a temperaure which is above that at which martensite begins to form and is held at this temperature until the austenite is completely transformed into bainite. The constant temperature to which the piece is quenched and then maintained is usually 450 degrees F. or above. The process differs from austempering in that after transformation to a bainite structure has been completed, the steel is immersed in another bath and is brought up to some higher temperature, depending upon the characteristics desired, and is maintained at this temperature for a definite period of time, followed by cooling in air. Thus, tempering to obtain the desired toughness or ductility takes place immediately after the structure of the steel has changed to bainite and before it is cooled to atmospheric temperature.

## Tempering

The object of *tempering* or *drawing* is to reduce the brittleness in hardened steel and to remove the internal strains caused by the sudden cooling in the quenching bath. The tempering process consists in heating the steel by various means to a certain temperature and then cooling it. When steel is in a fully hardened condition, its structure consists largely of *martensite*. On reheating to a temperature of from about 300 to 750 degrees F., a softer and tougher structure known as *troostite* is formed. If the steel is reheated to a temperature of from 750 to 1290 degrees F., a structure known as *sorbite* is formed which has somewhat less strength than troostite but much greater ductility.

**Tempering Temperatures.** — If steel is heated in an oxidizing atmosphere a film of oxide forms on the surface which changes color as the temperature increases. These oxide colors (see table) have been used extensively in the past as a means of gaging the correct amount of temper; but since these colors are affected to some extent by the composition of the metal, the method is not dependable.

The availability of reliable pyrometers in combination with tempering baths of oil, salt or lead make it possible to heat the work uniformly and to a given temperature within close limits.

Suggested temperatures for tempering various tools are given in the accompanying table.

**Tempering in Oil.** — Oil baths are extensively used for tempering tools (especially in quantity), the work being immersed in oil heated to the required temperature, which is indicated by a thermometer. It is important that the oil have a uniform temperature throughout and that the work be immersed long enough to acquire this temperature. Cold steel should not be plunged into a bath heated for tempering, owing to the danger of cracking it. The steel should either be preheated to about 300 degrees F., before placing it in the bath, or the latter should be at a comparatively low temperature before immersing the steel, and then be heated to the required degree. A temperature of from 650 to 700 degrees F. can be obtained with heavy tempering oils; for higher temperatures, either a bath of nitrate salts or a lead bath is generally used.

In tempering, the best method is to immerse the pieces to be tempered in the oil before starting to heat the latter. They are then heated with the oil. After the pieces tempered are taken out of the oil bath, they should immediately be dipped in a tank of caustic soda, and after that in a tank of hot water. This will remove all oil which might adhere to the tools. The following tempering oil has given satisfactory results: mineral oil, 94 per cent; saponifiable oil, 6 per cent; specific gravity, 0.920; flash point, 550 degrees F.; fire test, 625 degrees F.

**Tempering in Salt Baths.** — Molten salt baths may be used for tempering or drawing operations. Nitrate baths are particularly adapted for the usual drawing temperature range of, say, 300 to 1100 degrees F. Tempering in an oil bath usually is limited to temperatures of 500 to 600 degrees, and some heat-treating specialists recommend the use of a salt bath for temperatures above 350 or 400 degrees, as it is considered more efficient and economical. Tempering in a bath (salt or oil) has several advantages, such as ease in controlling the temperature range and maintenance of a uniform temperature. The work is also heated much more rapidly in a molten bath. While a gas- or oil-fired muffle or semi-muffle furnace may be used for tempering, a salt bath or oil bath is preferable. A salt bath is recommended for tempering high-speed steel, although furnaces may also be used. The bath or furnace temperature should be increased gradually, say, from 300 to 400 degrees up to the tempering temperature which may range from 1050 to 1150 degrees F. for high-speed steel.

**Tempering in a Lead Bath.** — The lead bath is commonly used for heating steel in connection with tempering, as well as for hardening. The bath is first heated to the temperature at which the steel should be tempered; the pre-heated work is then placed in the bath long enough to acquire this temperature, after which it is removed and cooled. As the melting temperature of pure lead is about 620 degrees F., tin is commonly added to it to lower the temperature sufficiently for tempering. Reductions in temperature can be obtained by varying the proportions of lead and tin, as shown by the table, " Temperatures of Lead Bath Alloys."

Temperatures of Lead Bath Alloys

| Parts Lead | Parts Tin | Melting Temp., Deg. F. | Parts Lead | Parts Tin | Melting Temp., Deg. F. | Parts Lead | Parts Tin | Melting Temp., Deg. F. |
|---|---|---|---|---|---|---|---|---|
| 200 | 8 | 560 | 39 | 8 | 510 | 19 | 8 | 460 |
| 100 | 8 | 550 | 33 | 8 | 500 | 17 | 8 | 450 |
| 75 | 8 | 540 | 28 | 8 | 490 | 16 | 8 | 440 |
| 60 | 8 | 530 | 24 | 8 | 480 | 15 | 8 | 430 |
| 48 | 8 | 520 | 21 | 8 | 470 | 14 | 8 | 420 |

**To Prevent Lead from Sticking to Steel.** — To prevent hot lead from sticking to parts heated in it, mix common whiting with wood alcohol, and paint the part that is to be heated. Water can be used instead of alcohol, but in that case the paint must be thoroughly dry, as otherwise the moisture will cause the lead to " fly." Another method is to make a thick paste according to the following formula: Pulverized charred leather, 1 pound; fine wheat flour, 1½ pounds; fine table salt, 2 pounds. Coat the tool with this paste and heat slowly until dry, then proceed to harden. Still another method is to heat the work to a blue color, or about 600 degrees F., and then dip it in a strong solution of salt water, prior to heating in the lead bath. The lead is sometimes removed from parts having fine projections or teeth, by using a stiff brush just before immersing in the cooling bath. This is necessary to prevent the formation of soft spots.

**Tempering in Sand.** — The sand bath is used for tempering certain classes of work. One method is to deposit the sand on an iron plate or in a shallow box which has burners beneath it. With this method of tempering, tools such as boiler punches, etc., can be given a varying temper by placing them endwise in the sand. As the temperature of the sand bath is higher toward the bottom, a tool can be so placed that the color of the lower end will be a deep dark blue when the middle portion is a very dark straw, and the working end or top a light straw color, the hardness gradually increasing from the bottom up.

**Double Tempering.** — In tempering high-speed steel tools, it is common practice to repeat the tempering operation or " double temper " the steel. This is done by heating the steel to the tempering temperature (say 1050 degrees F.) and holding it at that temperature for two hours. It is then cooled to room temperature, reheated to 1050 degrees F. for another two-hour period, and again cooled to room temperature. After the first tempering operation, some untempered martensite remains in the steel. This martensite is not only tempered by a second tempering operation but is relieved of internal stresses, thus improving the steel for service conditions. The hardening temperature for the higher alloy steels may affect the hardness after tempering. For example, molybdenum high-speed steel when heated to 2100 degrees F. had a hardness of 61 Rockwell C after tempering whereas a temperature of 2250 degrees F. resulted in a hardness of 64.5 Rockwell C after tempering.

Temperatures as Indicated by the Color of Plain Carbon Steel

| Degrees Centigrade | Degrees Fahrenheit | Color of Steel | Degrees Centigrade | Degrees Fahrenheit | Color of Steel |
|---|---|---|---|---|---|
| 221.1 | 430 | Very pale yellow | 265.6 | 510 | Spotted red-brown |
| 226.7 | 440 | Light yellow | 271.1 | 520 | Brown-purple |
| 232.2 | 450 | Pale straw-yellow | 276.7 | 530 | Light purple |
| 237.8 | 460 | Straw-yellow | 282.2 | 540 | Full purple |
| 243.3 | 470 | Deep straw-yellow | 287.8 | 550 | Dark purple |
| 248.9 | 480 | Dark yellow | 293.3 | 560 | Full blue |
| 254.4 | 490 | Yellow-brown | 298.9 | 570 | Dark blue |
| 260.0 | 500 | Brown-yellow | 337.8 | 640 | Light blue |

Tempering Temperatures for Various Plain Carbon Steel Tools

| Degrees F. | Class of Tool |
|---|---|
| 495 to 500 | Taps ½ inch or over, for use on automatic screw machines. |
| 495 to 500 | Nut taps ½ inch and under. |
| 515 to 520 | Taps ¼ inch and under, for use on automatic screw machines. |
| 525 to 530 | Thread dies to cut thread close to shoulder. |
| 500 to 510 | Thread dies for general work. |
| 495 | Thread dies for tool steel or steel tube. |
| 525 to 540 | Dies for bolt threader threading to shoulder. |
| 460 to 470 | Thread rolling dies. |
| 430 to 435 | Hollow mills (solid type) for roughing on automatic screw machine. |
| 485 | Knurls. |
| 450 | Twist drills for hard service. |
| 450 | Centering tools for automatic screw machine. |
| 430 | Forming tools for automatic screw machine. |
| 430 to 435 | Cut-off tools for automatic screw machine. |
| 440 to 450 | Profile cutters for milling machine. |
| 430 | Formed milling cutters. |
| 435 to 440 | Milling cutters. |
| 430 to 440 | Reamers. |
| 460 | Counterbores and countersinks. |
| 480 | Cutters for tube or pipe-cutting machine. |
| 460 and 520 | Snaps for pneumatic hammers — harden full length, temper to 460 degrees, then bring point to 520 degrees. |

## Annealing, Spheroidizing, and Normalizing

Annealing of steel is a heat-treating process in which the steel is heated to some elevated temperature, usually in or near the critical range, is held at this temperature for some period of time, and is then cooled, usually at a slow rate. Spheroidizing and normalizing may be considered as special cases of annealing.

The *full annealing* of carbon steel consists in heating it slightly above the *upper* critical point for hypo-eutectoid steels (steels of less than 0.85 per cent carbon content) and slightly above the *lower* critical point for hyper-eutectoid steels (steels of more than 0.85 per cent carbon content), holding it at this temperature until it is uniformly heated and then slowly cooling it to 1000 degrees F. or below. The resulting structure is layer-like or lamellar in character due to the pearlite which is formed during the slow cooling.  Annealing is employed (1) to soften steel for machining, cutting, stamping, etc., or for some particular service; (2) to alter ductility, toughness, electrical or magnetic characteristics or other physical properties; (3) to refine the crystal structure; (4) to produce grain reorientation; or (5) to relieve stresses and hardness resulting from cold working.

The *spheroidizing* of steel, according to the American Society of Metals, is " any process of heating and cooling that produces a rounded or globular form of carbide." High-carbon steels are spheroidized to improve their machinability especially in continuous cutting operations such as are performed by lathes and screw machines. In low-carbon steels, spheroidizing may be employed to meet certain strength requirements before subsequent heat-treatment.  Spheroidizing also tends to increase resistance to abrasion.

The *normalizing* of steel consists in heating it to some temperature above that used for annealing, usually about 100 degrees F. above the upper critical range, and then cooling it in still air at room temperature.  Normalizing is intended to put the steel into a uniform, unstressed condition of proper grain size and refinement so that it will properly respond to further heat-treatments.  It is particularly important in the case of forgings which are to be later heat-treated.  Normalizing may or may not (depending upon the composition) leave steel in a sufficiently soft state for machining with available tools.  In some cases, annealing for machinability is preceded by normalizing and the combined treatment — frequently called a *double anneal* — produces a better result than a simple anneal.

**Annealing Practice.** — For carbon steels, the following annealing temperatures are recommended by the American Society of Testing Materials: Steels of less than 0.12 per cent carbon content, 1600 to 1700 degrees F.; steels of 0.12 to 0.29 per cent carbon content, 1550 to 1600 degrees F.; steels of 0.30 to 0.49 per cent carbon content, 1500 to 1550 degrees F.; and for 0.50 to 1.00 per cent carbon steels, from 1450 to 1500 degrees F.  Slightly lower temperatures are satisfactory for steels having more than 0.75 per cent manganese content.  Heating should be uniform to avoid the formation of additional stresses.  In the case of large work-pieces, the heating should be slow enough so that the temperature of the interior does not lag too far behind that of the surface.

It has been found that in annealing steel, the higher the temperature to which it is heated to produce an austenitic structure, the greater the tendency of the structure to become lamellar (pearlitic) in cooling.  On the other hand, the closer the austenitizing temperature to the critical temperature, the greater is the tendency of the annealed steel to become spheroidal.

*Rate of Cooling:*  After heating the steel to some temperature within the annealing range, it should be cooled slowly enough to permit the development of the desired softness and ductility.  In general, the slower the cooling rate, the greater the resulting softness and ductility.  Steel of a high carbon content should be cooled more slowly than steel of a low carbon content; also the higher the alloy content, the slower is the cooling rate usually required.  Where extreme softness and ductility are not required, the steel may be cooled in the annealing furnace to some temperature well below the critical point, say, to about 1000 degrees F. and then removed and cooled in air.

**Annealing by Constant Temperature Transformation.** — It has been found that steel which has been heated above the critical point so that it has an austenitic structure can be transformed into a lamellar (pearlitic) or a spheroidal structure by holding it for a definite period of time at some constant sub-critical temperature.  In other words, it is feasible to anneal steel by means of a constant-temperature transformation as well as by the conventional continuous cooling method.  When the constant-temperature transformation method is employed, the steel, after being heated to some temperature above the critical and held at this temperature until it is austenitized, is cooled as rapidly as feasible to some relatively high sub-critical transformation temperature. The selection of this temperature is governed by the desired microstructure and hardness required and is taken from a transformation time and temperature curve (often called a TTT curve.) As drawn for a particular steel, such a curve shows the length of time required to transform that steel from an austenitic state at various sub-critical temperatures. After being held at the selected sub-critical temperature for the required length of time, the steel is cooled to room temperature,— again, as rapidly as feasible. This rapid cooling down to the selected transformation temperature and then down to room temperature has a negligible effect on the structure of the steel and makes possible in many instances a considerable saving in time over the conventional slow cooling method of annealing.

The softest condition in steel can be developed by heating it to a temperature usually less than 100 degrees F. above the lower critical point and then cooling it to some temperature, usually less than 100 degrees below the critical point, where it is held until the transformation is completed. Certain steels require a very lengthy period of time for transformation of the austenite when held at a constant temperature within this range. For such steels a practical procedure is to allow most of the transformation to take place in this temperature range where a soft product is formed and then to finish the transformation at a lower temperature where the time for the completion of the transformation is short.

**Spheroidizing Practice.** — A common method of spheroidizing steel consists in heating it to or slightly below the lower critical point, holding it at this temperature for a period of time, and then cooling it slowly to about 1000 degrees F. or below. The length of time at which the steel is held at the spheroidizing temperature largely governs the degree of spheroidization. High-carbon steel may be spheroidized by subjecting it to a temperature that alternately rises and falls between a point within and a point without the critical range. Tool steel may be spheroidized by heating to a temperature slightly above the critical range and then, after being held at this temperature for a period of time, cooling without removal from the furnace.

**Normalizing Practice.** — When using the lower carbon steels, simple normalizing is often sufficient to place the steel in its best condition for machining and will lessen distortion in carburizing or hardening. In the medium and higher carbon steels, combined normalizing and annealing constitutes the best practice. For unimportant parts, the normalizing may be omitted entirely or annealing practiced only when the steel is otherwise difficult to machine. Both processes are recommended in the following heat-treatments (for S. A. E. steels) as representing the best metallurgical practice. The temperatures recommended for normalizing and annealing have been made indefinite in many instances because of the many different types of furnace used in various plants and the difference in results desired.

## Casehardening

In order to harden low-carbon steel it is necessary to increase the carbon content of the surface of the steel so that a thin outer " case " can be hardened by heating the steel to the hardening temperature and then quenching it. The process, therefore, involves two separate operations. The first is the *carburizing* operation for impregnating the outer surface with sufficient carbon, and the second operation is that of heat-treating the carburized parts so as to obtain a hard outer case and, at the same time, give the " core " the required physical properties. The term " casehardening " is ordinarily used to indicate the complete process of carburizing and hardening.

**Carburization.** — Carburization is the result of heating iron or steel to a temperature below its melting point in the presence of a solid, liquid, or gaseous material which decomposes so as to liberate carbon when heated to the temperature used. In this way, it is possible to obtain by the gradual penetration, diffusion, or absorption of the carbon by the steel, a " zone " or " case " of higher carbon content at the outer surfaces than that of the original object. When a carburized object is rapidly cooled or quenched in water, oil, brine, etc., from the proper temperature, this case becomes hard, leaving the inside of the piece soft, but of great toughness.

**Use of Carbonaceous Mixtures.** — When carburizing materials of the solid class are used, the casehardening process consists in packing steel articles in metal boxes or pots, with a carbonaceous compound surrounding the steel objects. The boxes or pots are sealed and placed in a carburizing oven or furnace maintained usually at a heat of from about 1650 to 1700 degrees F. for a length of time depending upon the extent of the carburizing action desired. The carbon from the carburizing compound will then be absorbed by the steel on the surfaces desired, and the low-carbon steel is converted into high-carbon steel at these portions, while the internal sections and the insulated parts of the object retain practically their original low-carbon content. The result is a steel of a dual structure, a high-carbon and a low-carbon steel in the same piece. The carburized steel may now be heat-treated by heating and quenching, in much the same way as high-carbon steel is hardened, in order to develop the properties of hardness and toughness; but as the steel is, in reality, two steels in one, one high-carbon and one low-carbon, the correct heat-treatment after carburizing includes two distinct processes, one suitable for the high-carbon portion or the " case," as it is generally termed, and one suitable for the low-carbon portion or core. The method of heat-treatment varies according to the kind of steel used. Usually an initial heating and slow cooling is followed by reheating to 1400–1450 degrees F., quenching in oil or water, and a final tempering. More definite information is given in the following section on S.A.E. steels.

*Carburizers:* There are many commercial carburizers on the market in which the materials used as the generator may be hard and soft wood charcoal, animal charcoal, coke, coal, beans and nuts, bone and leather, or various combinations of these. The energizers may be barium, cyanogen, and ammonium compounds, various salts, soda ash, or lime and oil hydrocarbons.

**Pack-Hardening.** — When cutting tools, gages, and other parts made from high-carbon steels, are heated for hardening while packed in some carbonaceous material in order to protect delicate edges, corners or finished surfaces, the process usually is known as pack-hardening. Thus, the purpose is to protect the work, prevent scale formation, insure uniform heating, and minimize the danger of cracking and warpage. The work is packed, as in carburizing, and in the same type of receptacle. Common hardwood charcoal often is used, especially if it has had an initial heating to eliminate shrinkage and discharge its more impure gases. The lowest temperature required for hardening should be employed for pack-hardening — usually 1400 to 1450 degrees F. for carbon steels. Pack-hardening has also been applied to high-speed steels, but modern developments in heat-treating salts have made it possible to harden high-speed steel without decarburization, injury to sharp edges, or marring the finished surfaces. See paragraph on Salt Baths.

**Cyanide Hardening.** — When low-carbon steel requires a very hard outer surface but does not need high shock-resisting qualities, the cyanide hardening process may be employed to produce what is known as superficial hardness. This superficial hardening is the result of carburizing a very thin outer skin (which may be only a few thousandths inch thick) by immersing the steel in a bath containing sodium cyanide. The temperatures usually vary from 1450 to 1650 degrees F. and the percentage of sodium cyanide in the bath extends over a wide range, depending upon the steel used and properties required.

**Nitriding Process.** — Nitriding is a process for surface hardening certain alloy steels by heating the steel in an atmosphere of nitrogen (ammonia gas) at approximately 950 degrees F. The steel is then cooled slowly. Finish machined surfaces hardened by nitriding are subject to minimum distortion. The physical properties, such as toughness, high impact strength, etc., can be imparted to the core by previous heat-treatments and are unaffected by drawing temperatures up to 950 degrees F. The "Nitralloy" steels suitable for this process may readily be machined in the heat-treated as well as in the annealed state, and they forge as easily as alloy steels of the same carbon content. Certain heat-treatments must be applied prior to nitriding, the first being annealing to relieve rolling, forging, or machining strains. Parts or sections not requiring heat-treating should be machined or ground to the exact dimensions required. Close tolerances must be maintained in finish machining, but allowances for growth due to adsorption of nitrogen should be made, and this usually amounts to about 0.0005 inch for a case depth of 0.02 inch. Parts requiring heat-treatment for definite physical properties are forged or cut from annealed stock; heat-treated for the desired physical properties, rough machined, normalized, and finish machined. If quenched and drawn parts are normalized afterwards, the drawing and normalizing temperatures should be alike. The normalizing temperature may be below but should never be above the drawing temperature.

**Liquid Carburizing.** — Activated liquid salt baths are now used extensively for carburizing. Sodium cyanide and other salt baths are used. The salt bath is heated by electrodes immersed in it, the bath itself acting as the conductor and resistor. One or more groups of electrodes, with two or more electrodes per group, may be used. The heating is accompanied by a stirring action to insure uniform temperature and carburizing activity throughout the bath. The temperature may be controlled by a thermocouple immersed in the bath and connecting with a pyrometer designed to provide automatic regulation. The advantages of liquid baths include rapid action; uniform carburization; minimum distortion; and elimination of the packing and unpacking required when carbonaceous mixtures are used. In selective carburizing, the portions of the work which are not to be carburized are copper plated and the entire piece is then immersed in an activated cyanide bath. The copper inhibits any carburizing action on the plated parts, and this method offers a practical solution for selectively carburizing any portion of a steel part.

**Gas Carburizing.** — When carburizing gases are used, the mixture varies with the type of case and quality of product desired. The gaseous hydrocarbons most widely used are methane (natural gas), propane, and butane. These carbon bearing gases are mixed with air, with manufactured gases of several types, with flue gas, or with other specially prepared " diluent " gases. It is necessary to maintain a continuous fresh stream of carburizing gases to the carburizing retort or muffle, as well as to remove the spent gases from the muffle continuously, in order to obtain the correct mixture of gases inside the muffle. A slight pressure is maintained on the muffle to exclude foreign gases.

The horizontal rotary type of gas carburizing furnace has a retort or muffle which revolves slowly. This type is adapted to small parts such as ball and roller bearings, chain links, small axles, bolts, etc. With this type of furnace very large pieces such as gears, for example, may be injured by successive shocks due to tumbling within the rotor.

The vertical pit type of gas carburizer has a stationary rotor which is placed vertically in a pit. The work, instead of circulating in the gases as with the rotary type, is stationary and the gases circulate around it. This type is applicable to long large shafts or other parts or shapes that cannot be rolled in a rotary type of furnace. There are three types of continuous gas furnaces which may be designated as (1) direct quench and manually operated, (2) direct quench and mechanically operated, and (3) cooling-zone type. Where production does not warrant using a large continuous type furnace, a horizontal muffle furnace of the batch type may be used, especially if the quantities of work are varied and the production not continuous.

**Casehardening Steels.** — A low-carbon steel containing, say, from 0.15 to 0.20 per cent of carbon is suitable for casehardening. In addition to straight-carbon steels, the low-carbon alloy steels are employed. The alloys add to casehardened parts the same advantageous properties which they give to other classes of steel. Various steels for casehardening will be found in the section on S.A.E. steels.

**To Clean Work after Casehardening.** — To clean work, especially if knurled where dirt is likely to stick into crevices after casehardening, wash it in caustic soda (1 part soda to 10 parts water). In making this solution, the soda should be put into hot water gradually, and the mixture stirred until the soda is thoroughly dissolved. A still more effective method of cleaning is to dip the work into a mixture of 1 part sulphuric acid and 2 parts water. Leave the pieces in this mixture about three minutes; then wash them off immediately in a soda solution.

**Flame Hardening.** — This method of hardening is especially applicable to the selective hardening of large steel forgings or castings which must be finish-machined prior to heat-treatment, or which because of size or shape cannot be heat-treated by using a furnace or bath. An oxy-acetylene torch is used to heat quickly the surface to be hardened; this surface is then quenched to secure a hardened layer which may vary in depth from a mere skin to ¼ inch and with hardness ranging from 400 to 700 Brinell. A multi-flame torch-head may be equipped with quenching holes or a spray nozzle back of the flame. This is not a carburizing or a case-hardening process as the torch is only a heating medium. Most authorities recommend tempering or drawing of the hardened surface at temperatures between 200 and 350 degrees F. This treatment may be done in a standard furnace, an oil bath, or with a gas flame. It should follow the hardening process as close as possible. Medium-carbon and many low-alloy steels are suitable for hardening. Plain carbon steels ranging from 0.35 to 0.60 per cent carbon will give hardnesses of from 400 to 700 Brinell. Steels in the 0.40 to 0.45 per cent carbon range are preferred, as they have excellent core properties and produce hardnesses of from 400 to 500 Brinell without checking or cracking. Higher carbon steels will give greater hardnesses, but extreme care must be taken to prevent cracking. This requires careful control of the quenching operation.

*Spinning Method of Flame Hardening:* This method is employed on circular objects that can be rotated or spun past a stationary flame. It may be subdivided according to the speed of rotation, as, first, where the part is rotated slowly in front of a stationary flame and the quench is applied immediately after the flame. This method is used on large circular pieces such as track wheels and bearing surfaces. There will be a narrow band of material with lower hardness between adjacent torches if more than one path of the flame is required to harden the surface. There will also be an area of lower hardness where the flame is extinguished. A second method is applicable to small rollers or pinions. The work is spun at a speed of 50 to 150 R.P.M. in front of the flame until the entire piece has reached the proper temperature; then it is quenched as a unit by a cooling spray or ejecting into a cooling bath.

*The Progressive Method:* With this method the torch travels along the face of the work while the work remains stationary. It is used to harden lathe ways, gear teeth, and track rails.

*The Stationary or Spot-hardening Method:* When this method is employed, the work and torch are both stationary. When the spot to be hardened reaches the quenching temperature, the flame is removed and the quench applied.

*The Combination Method:* This is a combination of the spinning and progressive methods. It is used for long bearing surfaces. The work rotates slowly past the torch as the torch travels longitudinally across the face of the work at the rate of the torch width per revolution of the work.

The equipment for the stationary method of flame-hardening consists merely of an acetylene torch, an oxy-acetylene supply, and a suitable means of quenching; but when the other methods are employed, work-handling tools are essential and specially designed torches are desirable. A lathe is ideally suited for the spinning or combination hardening method, while a planer is adapted for progressive hardening. Production jobs, such as the hardening of gears, require specially designed machines. These machines reduce handling and hardening time, as well as assuring consistent results.

**Induction Hardening.** — The hardening of steel by means of induction heating and subsequent quenching in either liquid or air, is particularly applicable to parts which require localized hardening or controlled depth of hardening and to irregularly shaped parts, such as cams which require uniform surface hardening around their contour. Advantages offered by induction hardening are: (1) a short heating cycle which may range from a fraction of a second to several seconds (heat energy can be induced in a piece of steel at the rate of 100 to 250 B.T.U. per square inch per minute by induction heating, as compared with a rate of 3 B.T.U. per square inch per minute for the same material at room temperature when placed in a furnace with a wall temperature of 2000 degrees F.); (2) absence of tendency to produce oxidation or decarburization; (3) exact control of depth and area of hardening; (4) close regulation of degree of hardness obtained by automatic timing of heating and quenching cycles; (5) minimum amount of warpage or distortion; and (6) possibility of substituting carbon steels for higher cost alloy steels.

The principal advantage of induction hardening to the designer lies in its application to localized zones. Thus, specific areas in a given part can be heat treated separately to the respective hardnesses required. Parts can be designed so that the stresses at any given point in the finished piece can be relieved by local heating. Parts can be designed in which welded or brazed assemblies are built up prior to heat treating with only internal surfaces or projections requiring hardening.

**Types of Induction Heating Equipment.** — Induction heating is secured by placing the metal part inside or close to an " applicator " coil of one or more turns, through which alternating current is passed. The coil, formed to suit the general class of work to be heated, is usually made of copper tubing through which water is passed to prevent overheating of the coil itself. In most cases, the work piece is held either in a fixed position or is rotated slowly within or close to the applicator coil. Where the length of work is too great to permit heating in a fixed position, progressive heating may be employed. Thus, a rod or tube of steel may be fed through an applicator coil of one or more turns so that the heating zone travels progressively along the entire length of the work piece.

The frequency of the alternating current used and the type of generator employed to supply this current to the applicator coil depend upon the character of the work to be done. There are three types of equipment used commercially to produce high-frequency current for induction heating: (1) motor generator sets which deliver current at frequencies of approximately 1000, 2000, 3000, and 10,000 cycles;

(2) spark gap oscillator units which produce frequencies ranging from 80,000 to 300,000 cycles; and (3) vacuum tube oscillator sets, which produce currents at frequencies ranging from 350,000 to 15,000,000 cycles or more.

**Depth of Heat Penetration.** — Generally speaking, the higher the frequency used, the shallower the depth of heat penetration. For heating clear through, for deep hardening and for large work pieces, low power concentrations and low frequencies are usually employed. For very shallow and closely controlled depths of heating, as in surface hardening, and in localized heat treating of small work pieces, currents at high frequencies are employed.

For example, a ½-inch round bar of hardenable steel will be heated through its entire structure quite rapidly by an induced current of 2000 cycles. After quenching, the bar would show through hardness with a decrease in hardness from surface to center. The same piece of steel could be readily heated and surface hardened to a depth of 0.100 inch with current at 9600 cycles, and to an even shallower depth with current at 100,000 cycles. A ¼-inch bar, however, would not reach a sufficiently high temperature at 2000 cycles to permit hardening but at 9600 cycles through hardening would be accomplished, while current at over 100,000 cycles would be needed for surface hardening.

**Types of Steel for Induction Hardening.** — Most of the standard types of steels can be hardened by induction heating, providing the carbon content is sufficient to produce the desired degree of hardness by quenching. Thus, low carbon steels with a carburized case, medium and high carbon steels (both plain and alloy), and cast iron with a portion of the carbon in combined form, may be used for this purpose. In the case of alloy steels, induction heating should be limited primarily to the shallow hardening type, i.e., those of low alloy content, otherwise the severe quench usually required may result in a highly-stressed surface with consequent reduced load-carrying capacity and danger of cracking.

**Through Hardening, Annealing and Normalizing by Induction.** — For through hardening, annealing and normalizing by induction, low power concentrations are desirable to prevent too great a temperature differential between surface and interior of the work. A satisfactory rate of heating is obtained when the total power input to the work is slightly greater than the radiation losses at the desired temperature. If possible, as low a frequency should be used as is consistent with good electrical coupling. A number of applicator coils may be connected in a series so that several work pieces can be heated simultaneously, thus reducing the power input to each. Widening the spacing between work and applicator coil also will reduce the amount of power delivered to the work.

**Induction Surface Hardening.** — As indicated in " Depth of Heat Penetration," currents at much higher frequencies are required in induction surface hardening than in through hardening by induction. In general, the smaller the work piece, the thinner the section, or the shallower the depth to be hardened, the higher will be the frequency required. High power concentrations are also needed to make possible a short heating period so that an undue amount of heat will not be conducted to adjacent or interior areas, where a change in hardness is not desired. Generators of large capacity and applicator coils of but a few turns, or even a single turn, provide the necessary concentration of power in the localized area to be hardened.

Induction heating of internal surfaces, such as the interior of a hollow cylindrical part or the inside of a hole, can be accomplished readily with applicator coils shaped to match the cross-section of the opening which may be round, square, eliptical, etc. If the internal surface is of short length, a multiturn applicator coil extending along its entire length may be employed. Where the power available is insufficient to heat the entire internal surface at once, progressive heating is used. For this purpose an

Table 2. Typical Heat Treatments for SAE Alloy Steels

| | | | Carburizing Grades | | | | |
|---|---|---|---|---|---|---|---|
| SAE No. | Normalize[1] | Cycle Anneal[3] | Carburized Deg. F. | Cool * | Reheat Deg. F. | Cool * | Temper[3] Deg. F |
| 1320 | yes | ... | 1650–1700 | E | 1400–1450[6] | E | 250–350 |
| | yes | ... | 1650–1700 | E | 1475–1525[7] | E | 250–350 |
| | yes | ... | 1650–1700 | C | 1400–1450[6] | E | 250–350 |
| | yes | ... | 1650–1700 | C | 1500–1550[7] | E | 250–350 |
| | yes | ... | 1650–1700 | E[5] | .......... | .. | 250–350 |
| | yes | ... | 1500–1650[4] | E | .......... | .. | 250–350 |
| 2317 | yes | yes | 1650–1700 | E | 1375–1425[6] | E | 250–350 |
| | yes | yes | 1650–1700 | E | 1450–1500[7] | E | 250–350 |
| | yes | yes | 1650–1700 | C | 1375–1425[6] | E | 250–350 |
| | yes | yes | 1650–1700 | C | 1475–1525[7] | E | 250–350 |
| | yes | yes | 1650–1700 | E[5] | .......... | .. | 250–350 |
| | yes | yes | 1450–1650[4] | E | .......... | .. | 250–350 |
| 2512 to 2517 | yes[2] | ... | 1650–1700 | C | 1325–1375[6] | E | 250–350 |
| | yes[2] | ... | 1650–1700 | C | 1425–1475[7] | E | 250–350 |
| | yes | ... | 1650–1700 | E | 1400–1450[6] | E | 250–350 |
| | yes | ... | 1650–1700 | E | 1475–1525[7] | E | 250–350 |
| 3115 & 3120 | yes | ... | 1650–1700 | C | 1400–1450[6] | E | 250–350 |
| | yes | ... | 1650–1700 | C | 1500–1550[7] | E | 250–350 |
| | yes | ... | 1650–1700 | E[5] | .......... | .. | 250–350 |
| | yes | ... | 1500–1650[4] | E | .......... | .. | 250–350 |
| 3310 & 3316 | yes[2] | ... | 1650–1700 | E | 1400–1450[6] | E | 250–350 |
| | yes[2] | ... | 1650–1700 | C | 1475–1500[7] | E | 250–350 |
| 4017 to 4032<br>4119 & 4125 | yes | yes | 1650–1700 | E[5] | .......... | .. | 250–350 |
| | yes | ... | 1650–1700 | E[5] | .......... | .. | 250–350 |
| 4317 & 4320<br>4608 to 4621 | yes | yes | 1650–1700 | E | 1425–1475[6] | E | 250–350 |
| | yes | yes | 1650–1700 | E | 1475–1525[7] | E | 250–350 |
| | yes | yes | 1650–1700 | C | 1425–1475[6] | E | 250–350 |
| | yes | yes | 1650–1700 | C | 1475–1525[7] | E | 250–350 |
| | yes | yes | 1650–1700 | E[5] | .......... | .. | 250–350 |
| | yes | yes | 1650–1700 | E[5] | .......... | .. | 250–350 |
| | yes | ... | 1500–1650[4] | E | .......... | .. | 250–350 |
| 4812 to 4820 | yes[2] | yes | 1650–1700 | E | 1375–1425[6] | E | 250–350 |
| | yes[2] | yes | 1650–1700 | E | 1450–1500[7] | E | 250–35c |
| | yes[2] | yes | 1650–1700 | C | 1375–1425[6] | E | 250–350 |
| | yes[2] | yes | 1650–1700 | C | 1450–1500[7] | E | 250–350 |
| | ... | ... | 1650–1700 | E[5] | .......... | .. | 250–350 |
| 5115 & 5120 | yes | ... | 1650–1700 | E | 1425–1475[6] | E | 250–350 |
| | yes | ... | 1650–1700 | E | 1500–1550[7] | E | 250–350 |
| | yes | ... | 1650–1700 | C | 1425–1475[6] | E | 250–350 |
| | yes | ... | 1650–1700 | C | 1500–1550[7] | E | 250–350 |
| | yes | ... | 1500–1650[4] | E | .......... | .. | 250–350 |
| 8615 to 8625<br>8720 | yes | yes | 1650–1700 | E | 1475–1525[6] | E | 250–350 |
| | yes | yes | 1650–1700 | E | 1525–1575[7] | E | 250–350 |
| | yes | yes | 1650–1700 | C | 1475–1525[6] | E | 250–350 |
| | yes | yes | 1650–1700 | C | 1525–1575[7] | E | 250–350 |
| | yes | yes | 1650–1700 | E[5] | .......... | .. | 250–350 |
| | yes | yes | 1500–1650[4] | E | .......... | .. | 250–350 |
| 9310 to 9317 | yes[2] | ... | 1650–1700 | E | 1400–1450[6] | E | 250–350 |
| | yes[2] | ... | 1650–1700 | C | 1500–1525 | E | 250–350 |

* Symbols: C = Cool slowly; E = Oil.

[1] Normalizing temperatures should not be less than 50 degrees F. higher than the carburizing temperature. Follow by air cooling.

[2] After normalizing, reheat to temperatures of 1000–1200 degrees F. and hold approximately 4 hours.

[3] Where cycle annealing is desired, heat to normalizing temperature — hold for uniformity — cool rapidly to 1000–1250 degrees F.; hold 1 to 3 hours, then air or furnace cool to obtain a structure suitable for machining and finish.

[4] This treatment is for activated or cyanide baths, and parts may be given refining heats as indicated for other heat treating processes.

[5] This treatment applicable to fine-grained steels only. When fine-grained steels are employed, a second reheat is often unnecessary.

[6] This treatment when case hardness only is paramount.

[7] This treatment when higher core hardness is desired.

[8] Tempering treatment is optional. Tempering is generally employed for partial stress relief and improved resistance to cracking from grinding operations.

Table 2 (*Continued*).    Typical Heat Treatments for SAE Alloy Steels

| | Directly Hardenable Grades | | | | |
|---|---|---|---|---|---|
| SAE No. | Normalize Deg. F. | Anneal Deg. F. | Harden Deg. F. | Quench * | Temper Deg. F. |
| 1330 | ......... | ......... | 1525–1575 | B | To desired hardness |
| | 1600–1700 and/or | 1500–1600 | 1525–1575 | B | To desired hardness |
| 1335 & 1340 | ......... | ......... | 1500–1550 | E | To desired hardness |
| | 1600–1700 and/or | 1500–1600 | 1525–1575 | E | To desired hardness |
| 2330 | ......... | ......... | 1450–1500 | E | To desired hardness |
| | 1600–1700 and/or | 1400–1500 | 1450–1500 | E | To desired hardness |
| 2340 & 2345 | ......... | ......... | 1425–1475 | E | To desired hardness |
| | 1600–1700 and/or | 1400–1500 | 1425–1475 | E | To desired hardness |
| 3130 | 1600–1700 | ......... | 1500–1550 | B | To desired hardness |
| 3135 to 3141 | ......... | ......... | 1500–1550 | E | To desired hardness |
| | 1600–1700 and/or | 1450–1550 | 1500–1550 | E | To desired hardness |
| 3145 & 3150 | ......... | ......... | 1500–1550 | E | To desired hardness |
| | 1600–1700 and/or | 1400–1500 | 1500–1550 | E | To desired hardness |
| 4037 & 4042 | ......... | 1525–1575 | 1500–1575 | E | Gears, 350–450 / To desired hardness |
| 4047 & 4053 | ......... | 1450–1550 | 1500–1575 | E | To desired hardness |
| 4063 & 4068 | ......... | 1450–1550 | 1475–1550 | E | To desired hardness |
| 4130 | 1600–1700 and/or | 1450–1550 | 1600–1650 | B | To desired hardness |
| 4137 & 4140 | 1600–1700 and/or | 1450–1550 | 1550–1600 | E | To desired hardness |
| 4145 & 4150 | 1600–1700 and/or | 1450–1550 | 1500–1600 | E | To desired hardness |
| 4340 | 1600–1700 and draw 1100–1225 | | 1475–1525 | E | To desired hardness |
| 4640 | 1600–1700 and/or | 1450–1550 | 1450–1500 | E | To desired hardness |
| | 1600–1700 and/or | 1450–1550 | 1450–1500 | E | Gears, 350–450 |
| 5045 & 5046 | 1600–1700 and/or | 1450–1550 | 1475–1500 | E | 250–300 |
| 5130 & 5132 | 1650–1750 and/or | 1450–1550 | 1500–1550 | G | To desired hardness |
| 5135 to 5145 | 1650–1750 and/or | 1450–1550 | 1500–1550 | E | To desired hardness / Gears, 350–400 |
| 5147 to 5152 | 1650–1750 and/or | 1450–1550 | 1475–1550 | E | To desired hardness / Gears, 350–400 |
| 50100 | ......... | 1350–1450 | 1425–1475 | H | To desired hardness |
| 51100 52100 | ......... | 1350–1450 | 1500–1600 | E | To desired hardness |
| 6150 | 1650–1750 and/or | 1550–1650 | 1600–1650 | E | To desired hardness |
| 9254 to 9262 | ......... | ......... | 1500–1650 | E | To desired hardness |
| 8627 to 8632 | 1600–1700 and/or | 1450–1550 | 1550–1650 | B | To desired hardness |
| 8635 to 8641 | 1600–1700 and/or | 1450–1550 | 1525–1575 | E | To desired hardness |
| 8642 to 8653 | 1600–1700 and/or | 1450–1550 | 1500–1550 | E | To desired hardness |
| 8655 & 8660 | 1650–1750 and/or | 1450–1550 | 1475–1550 | E | To desired hardness |
| 8735 & 8740 | 1600–1700 and/or | 1450–1550 | 1525–1575 | E | To desired hardness |
| 8745 & 8750 | 1600–1700 and/or | 1450–1550 | 1500–1550 | E | To desired hardness |
| 9437 & 9440 | 1600–1700 and/or | 1450–1550 | 1550–1600 | E | To desired hardness |
| 9442 to 9747 | 1600–1700 and/or | 1450–1550 | 1500–1600 | E | To desired hardness |
| 9840 | 1600–1700 and/or | 1450–1550 | 1500–1550 | E | To desired hardness |
| 9845 & 9850 | 1600–1700 and/or | 1450–1550 | 1500–1550 | E | To desired hardness |

| | Heat Treating Grades — Chromium-Nickel Austenitic Steels | | | | |
|---|---|---|---|---|---|
| SAE No. | Normalize | Anneal[9] | Harden Deg. F. | Quenching Medium | Temper |
| 30301 to 30347 | .... | 1800–2100 | .... | Water or Air | .... |

* Symbols: B = Water or oil; E = Oil; G = Water, caustic solution or oil; H = Water.
[9] Quench to produce full austenitic structure using water or air in accordance with thickness of section. Annealing temperatures given cover process and full annealing as now used by industry, the lower end of the range being used for process annealing.

Table 2 (*Continued*).  Typical Heat Treatments for SAE Alloy Steels

| SAE No.[*] | Normalize | Sub-critical Anneal, Deg. F. | Full Anneal Deg. F. | Harden Deg. F. | Quenching Medium | Temper Deg. F. |
|---|---|---|---|---|---|---|
| Heat Treating Grades — Stainless Chromium Irons and Steels | | | | | | |
| 51410 | ..... / ..... | 1300–1350[10] / .......... | 1550–1650[11] / .......... | } 1750–1850 | Oil or air | To desired hardness |
| 51414 | ..... / ..... | 1200–1250[10] / .......... | .......... / .......... | } 1750–1850 | Oil or air | To desired hardness |
| 51416 | ..... / ..... | 1300–1350[10] / .......... | 1550–1650[11] / .......... | } 1750–1850 | Oil or air | To desired hardness |
| 51420 / 51420F | ..... / ..... | 1350–1450[10] / .......... | 1550–1650[11] / .......... | } 1800–1850 | Oil or air | To desired hardness |
| 51430 | ..... | 1400–1500[12] | .......... | .......... | .......... | .................... |
| 51430F | ..... | 1250–1500[12] | .......... | .......... | .......... | .................... |
| 51431 | ..... | 1150–1225[10] | .......... | 1800–1900 | Oil or air | To desired hardness |
| 51440A / 51440B / 51440C / 51440F | ..... | 1350–1440[10] | 1550–1650[11] | 1850–1950 | Oil or air | To desired hardness |
| 51442 | ..... | 1400–1500[12] | .......... | .......... | .......... | .................... |
| 51446 | ..... | 1500–1650[12] | .......... | .......... | .......... | .................... |
| 51501 | ..... | 1325–1375[10] | 1525–1600[11] | 1600–1700 | Oil or air | To desired hardness |

[*] Suffixes A, B and C denote three types of steel differing in carbon content only. Suffix F denotes a free machining steel.
[10] Usually air cooled, but may be furnace cooled.
[11] Cool slowly in furnace.
[12] Cool rapidly in air.

## Heat-Treating High-Speed Steels

**Cobaltcrom Steel.** — This is a tungstenless alloy steel or high-speed steel which contains approximately 1.5 per cent carbon, 12.5 per cent chromium, and 3.5 per cent cobalt. Tools such as dies, milling cutters, etc., made from cobaltcrom steel can be cast to shape in suitable molds, the teeth of cutters being formed so that it is necessary only to grind them.

Before the blanks can be machined, they must be annealed; this operation is performed by pack-annealing at the temperature of 1800 degrees F., for a period of from three to six hours, according to the size of the castings being annealed. The following directions are given for the hardening of blanking and trimming dies, milling cutters, and similar tools made from cobaltcrom steel: Heat slowly in a hardening furnace to about 1830 degrees F., and hold the temperature at this point until the tools are thoroughly soaked. Then reduce the temperature about 50 degrees, withdraw the tools from the furnace, and allow them to cool in the atmosphere. As soon as the red color disappears from the cooling tool, place it in quenching oil until cold. The slight drop of 50 degrees in temperature while the tool is still in the hardening furnace is highly important in order to obtain proper results. The steel will be injured if the tool is heated above 1860 degrees F. In cooling milling cutters or other rotary tools, it is suggested that they be suspended on a wire to insure a uniform rate of cooling.

Tools that are subjected to shocks or vibration, such as pneumatic rivet sets, shear blades, etc., should be heated slowly to 1650 degrees F., after which the temperature should be reduced to about 1610 degrees F., at which point the tool should be removed from the furnace and permitted to cool in the atmosphere. There is no appreciable scaling present in the hardening of cobaltcrom steel tools.

**Preheating Tungsten High-Speed Steel.** — Tungsten high-speed steel must be hardened at a very high temperature; consequently, tools made from such steel are seldom hardened without at least one preheating stage to avoid internal strain. This applies especially to milling cutters, taps, and other tools having thin teeth and thick bodies and to forming tools of irregular shape and section. The tools should be heated slowly and carefully to a temperature somewhat below the critical point of the steel, usually in the range of 1500 to 1600 degrees F. By so limiting the preheating temperature, the operation is not unduly sensitive and the tool may be safely left in the furnace until it reaches a uniform temperature throughout its length and cross section.

A single stage of preheating is customary for tools that are simple in form and are not more than from 1 to 1½ inches in thickness. For large, intricate tools, two stages of preheating are frequently used. The first brings the tool up to a temperature of about 1100 to 1200 degrees F., and the second raises its temperature to 1550 to 1600 degrees F. A preheating time of 5 minutes for each ¼ inch in tool thickness has been recommended for a furnace temperature of 1600 degrees F. This is where a single stage of preheating is used and the furnace capacity is sufficient for maintaining practically constant temperature when the tools are changed. To prevent undue chilling, it is common practice to insert a single tool or a small lot in the hardening furnace as often as a tool or lot is removed, rather than to insert a full charge of cold metal at one time.

Preheating is usually done in a simple type of oven furnace heated by gas, electricity, or oil. Atmospheric control is seldom used, although in the case of 18–4–1 steel a slightly reducing atmosphere (2 to 6 per cent carbon monoxide) has been found to produce the least amount of scale and will result in a better surface after final hardening.

**Hardening of Tungsten High-Speed Steel.** — All tungsten high-speed steels must be heated to a temperature close to their fusion point to develop their maximum efficiency as metal-cutting tools. This requires a hardening temperature ranging from 2200 to 2500 degrees F. The effect of changes in the hardening temperature on the cutting efficiency of several of the more common high-speed steels are shown in Table 1. The figures given are ratios, the value 1.00 for each steel being assigned to the highest observed cutting speed for that steel. The figures for different steels therefore cannot be directly compared with each other, except to note changes in the point of maximum cutting efficiency.

The figures in the table refer to tools heated in an oven type furnace in which a neutral atmosphere is maintained. The available data indicate that a steel reaches its best cutting qualities at a temperature approximately 50 degrees F. lower than the figures in the table if it is hardened in a bath type furnace. It is, however, desirable, in any case, to use a hardening temperature approximately 50 degrees lower than that giving maximum cutting qualities, in order to avoid possibility of overheating the tool.

*Length of Time for Heating:* The cutting efficiency of a tool is affected by the time that it is kept at the hardening temperature, almost as much as by the hardening temperature itself. It has been common practice to heat a tool for hardening until a " sweat " appeared on its surface. This sweat is presumably a melting of the oxide film on the surface of a tool heated in an oxidizing atmosphere. It does not show when the tool is heated in an inert atmosphere. This method of determining the proper heating time is at best an approximation and indicates only the temperature on the outside of the tool rather than the condition of the interior. As such, it cannot be relied upon to give consistent results.

The only safe method is to heat the tool for a definite predetermined time, based on the size and the thickness of metal which the heat must penetrate to reach the

Table 1.  Relation of Hardening Temperature to Cutting Efficiency

| Hardening Temperature, Degrees F. | Typical Analyses of High-Speed Steels | | | |
|---|---|---|---|---|
| | 18-4-1 | 14-4-2 | 18-4-1 Cobalt | 14-4-2 Cobalt |
| 2200 | 0.86 | 0.83 | 0.84 | 0.85 |
| 2250 | 0.88 | 0.88 | 0.86 | 0.88 |
| 2300 | 0.90 | 0.93 | 0.90 | 0.91 |
| 2350 | 0.95 | 0.98 | 0.94 | 0.94 |
| 2400 | 0.99 | 0.98 | 0.98 | 0.98 |
| 2450 | 1.00 | . . . | 0.99 | 1.00 |
| 2500 | 0.98 | . . . | 1.00 | 0.97 |

Table 2.  Length of Heating Time for Through Hardening

| High-Speed Steel Tool Thickness, in Inches | Time in Furnace at High Heat, in Minutes | High-Speed Steel Tool Thickness, in Inches | Time in Furnace at High Heat, in Minutes | High-Speed Steel Tool Thickness, in Inches | Time in Furnace at High Heat, in Minutes |
|---|---|---|---|---|---|
| ¼ | 2 | 1½ | 7 | 5 | 18 |
| ½ | 3 | 2 | 8 | 6 | 20 |
| ¾ | 4 | 3 | 12 | 8 | 25 |
| 1 | 5 | 4 | 15 | 10 | 30 |

interior.  The values given in Table 2 are based on a series of experiments to determine the relative cutting efficiency of a group of tools hardened in an identical manner, except for variations in the time the tools were kept at the hardening temperature.  The time given is based on that required to harden throughout a tool resting on a conducting hearth; the tool receives heat freely from three sides, on its large top surface and its smaller side surfaces.  (The table does not apply to a disk lying flat on the hearth.)  In the case of a tool having a projecting cutting edge, such as a tap, the thickness or depth of the projection portion on which the cutting edge is formed should be used in referring to the table.

The time periods given in Table 2 are based on complete penetration of the hardening.  For very thick tools, the practical procedure is to harden to a depth sufficient to produce an adequate cutting edge, leaving the interior of the tool relatively soft.

Where atmosphere control is not provided, it will be found impracticable in many cases to use both the temperature for maximum cutting efficiency, given in Table 1, and the heating time, given in Table 2, because abnormal scaling, grain growth, and surface decarburization of the tool will result.  The principal value of an accurate control of the furnace atmosphere appears to lie in the fact that its use makes possible the particular heat-treatment that produces the best structure in the tool without destruction of the tool surface or grain.

**Quenching Tungsten High-Speed Steel.** — High-speed steel is usually quenched in oil.  The oil bath offers a convenient quench; it calls for no unusual care in handling and brings about a uniform and satisfactory rate of cooling, which does not vary appreciably with the temperature of the oil.  Some authorities believe it desirable to withdraw the tool from the oil bath for a few seconds after it has reached a dull red.  It is also believed desirable to move the tool around in the quenching oil, particularly immediately after it has been placed in it, to prevent the formation of a gas film on the tool. Such a film is usually a poor conductor of heat and slows up the rate of cooling.

*Salt Bath:* Quenching in a lead or salt bath at from 1000 to 1200 degrees F. has the advantage that cooling of the tool from hardening to room temperature is accomplished in two stages, thus reducing the possibility of setting up internal strains which may tend to crack the tool. The quenching temperature is sufficiently below the lower critical point for a tool so quenched to be allowed to cool to room temperature in still air. This type of quench is particularly advantageous for tools of complicated section which would easily develop hardening cracks. The salt quench has the advantage that the tool sinks and requires only a support, while the same tool will float in the lead bath and must be held under the surface. It is believed that the lead quench gives a somewhat higher matrix hardness, and is of advantage for tools that tend to fail by nose abrasion. Tools treated as described are brittle unless given a regular tempering treatment, as the 1100-degree F. quenching temperature is not a substitute for later tempering at the same temperature, after the tool has cooled to room temperature.

*Air Cooling:* Many high-speed steel tools are quenched in air, either in a stream of dry compressed air or in still air. Small sections harden satisfactorily in still air, but heavier sections should be subjected to air under pressure. One advantage of air cooling is that the tool can be kept straight and free from distortion, although it is likely that there will be more scale on a tool thus quenched than when oil, lead, or salt is used. Thin flat tools, which must be kept straight and flat, can be cooled advantageously between steel plates.

## Straightening High-Speed Tools when Quenching.

The final straightness required in a tool must be considered when it is quenched. When a number of similar tools are to be hardened, a jig can be used to advantage for holding the tools while quenching. When long slender tools are quenched without holders, they frequently warp and must be straightened later. The best time for this straightening is during the first few minutes after the tools have been quenched, as the steel is then quite pliable and may be bent without difficulty. The straightening must be done at once, as the tools become hard in a few minutes.

## Anneal Before Rehardening.

Tools that are too soft after hardening must be annealed before rehardening. A quick anneal, such as previously described, is all that is required to put such a tool into the proper condition for rehardening. This treatment is absolutely essential. In the case of milling cutters and forming tools of irregular section, a full anneal should be used.

## Tempering or Drawing Tungsten High-Speed Steel.

The tempering or drawing temperature for high-speed steel tools usually varies from 900 to 1200 degrees F. This temperature is higher for turning and planing tools than for such tools as milling cutters, forming tools, etc. If the temperature is below 800 degrees F., the tool is likely to be too brittle. The general idea is to temper tools at the highest temperature likely to occur in actual service. Since this temperature ordinarily would not be known, the general practice is to temper at whatever temperature experience with that particular steel and tool has proved to be the best. The furnace used for tempering usually is kept at a temperature of from 1000 to 1100 degrees F. for ordinary high-speed steels and from 1200 to 1300 degrees F. for steels of the cobalt type. These furnace temperatures apply to tools of the class used on lathes and planers. Such tools, in service, frequently heat to the point of visible redness. Milling cutters, forming tools, or any other tools for lighter duty, may be tempered as low as 850 or 900 degrees F. When the tool has reached the temperature of the furnace it should be held at this temperature for from one to several hours until it has been heated evenly throughout. It should then be allowed to cool gradually in the air and in a place that is dry and free from air drafts. In tempering, the tool should not be quenched, as this tends to produce strains which may result later in cracks.

**Annealing Tungsten High-Speed Steel.** — The following method of annealing high-speed steel has been used extensively. Use an iron box or pipe of sufficient size to allow at least one-half inch of packing between the pieces of steel to be annealed and the sides of the box or pipe. It is not necessary that each piece of steel be kept separate from every other piece, but only that the steel be prevented from touching the sides of the annealing pipe or box. Pack carefully with powdered charcoal, fine dry lime or mica (preferably charcoal), and cover with an air-tight cap, or lute with fire clay; heat slowly to 1600 to 1650 degrees F. and keep at this heat from 2 to 8 hours, depending upon the size of the pieces to be annealed. A piece 2 by 1 by 8 inches requires about three hours. Cool as slowly as possible, and do not expose to the air until cold as cooling in air is likely to cause partial hardening. A good way is to allow the box or pipe to remain in the furnace until cold.

**Hardening Molybdenum High-Speed Steels.** — In Table 3 are given the compositions of several molybdenum high-speed steels which are widely used for general

Table 3.  Composition of Molybdenum High-Speed Steels

| Element | Molybdenum-Tungsten | | Molybdenum-Vanadium | Tungsten-Molybdenum |
|---|---|---|---|---|
| | Type Ia (Per Cent) | Type Ib* (Per Cent) | Type II (Per Cent) | Type III (Per Cent) |
| Caron | 0.70–0.85 | 0.76–0.82 | 0.70–0.90 | 0.75–0.90 |
| Tungsten | 1.25–2.00 | 1.60–2.30 | ........ | 5.00–6.00 |
| Chromium | 3.00–5.00 | 3.70–4.20 | 3.00–5.00 | 3.50–5.00 |
| Vanadium | 0.90–1.50 | 1.05–1.35 | 1.50–2.25 | 1.25–1.75 |
| Molybdenum | 8.00–9.50 | 8.00–9.00 | 7.50–9.50 | 3.50–5.50 |
| Cobalt | See footnote | 4.50–5.50 | See footnote | See footnote |

*Cobalt may be used in any of these steels in varying amounts up to 9 per cent, and the vanadium content may be as high as 2.25 per cent. When cobalt is used in Type III the vanadium content may be as high as 2.25 per cent. When cobalt is used in Type III steel, this steel becomes susceptible to decarburization. As an illustration of the use of cobalt, Type Ib steel is included. This is steel T10 in the U. S. Navy Specification 46S37, dated November 1, 1939.

commercial tool applications. The general method of hardening molybdenum high-speed steels resembles that used for 18–4–1 tungsten high-speed steel except that the hardening temperatures are lower and more precautions must be taken to avoid decarburization, especially on tools made from Type I or Type II steels, when the surface is not ground after hardening. Either salt baths or atmosphere controlled furnaces are recommended for hardening molybdenum high-speed steel.

The usual method is to preheat uniformly in a separate furnace to 1250 to 1550 degrees F. and then transfer to a high-heat furnace maintained at some temperature in the hardening temperature range given in Table 4. Single point cutting tools, in general, should be hardened at the upper end of the temperature range indicated by Table 4. Slight grain coarsening is not objectionable in such tools when they are properly supported in service and are not subjected to chattering; however, when

Table 4.  Heat-Treatment of Molybdenum High-Speed Steels

| Heat-Treating Operation | Molybdenum-Tungsten | Molybdenum-Vanadium | Tungsten-Molybdenum |
|---|---|---|---|
| | Types Ia and Ib* (Temp., in Deg. F.) | Type II (Temp., in Deg. F.) | Type III (Temp., in Deg. F.) |
| Forging | 1850–2000 | 1850–2000 | 1900–2050 |
| Not below | 1600 | 1600 | 1600 |
| Annealing | 1450–1550 | 1450–1550 | 1450–1550 |
| Strain Relief | 1150–1350 | 1150–1350 | 1150–1350 |
| Preheating | 1250–1500 | 1250–1500 | 1250–1550 |
| Hardening† | 2150–2250* | 2150–2250 | 2175–2275 |
| Salt | 2150–2225 | 2150–2225 | 2150–2250 |
| Tempering | 950–1100 | 950–1100 | 950–1100 |

*Under similar conditions Type Ib steel requires a slightly higher hardening heat than Type Ia.

†The higher side of the hardening range should be used for large sections, and the lower side for small sections.

these tools are used for intermittent cuts, it is better to use the middle of the temperature range.  All other cutting tools, such as drills, countersinks, taps, milling cutters, reamers, broaches, form tools, etc., should be hardened in the middle of the range shown.  For certain tools, such as slender taps, cold punches, blanking and trimming dies, etc., where greater toughness to resist shocks is required, the lower end of the hardening temperature range should be used.

Molybdenum high-speed steels can be pack-hardened following the same practice as is used for tungsten high-speed steels, but keeping on the lower side of the hardening range (approximately 1850 degrees F.).  Special surface treatments such as nitriding by immersion in molten cyanide that are used for tungsten high-speed steels are also applicable to molybdenum high-speed tools.

When heated in an open fire or in furnaces without atmosphere control, these steels do not sweat like 18–4–1 steels; consequently, determining the proper time in the high-heat chamber is a matter of experience.  This time approximates that used with 18–4–1 steels, although it may be slightly longer when the lower part of the hardening range is used.  Much can be learned by preliminary hardening of test pieces and checking up on the hardness fracture and structure.  It is difficult to give the exact heating time, as this is affected by temperature, type of furnace, size and shape, and furnace atmosphere.  Rate of heat transfer is most rapid in salt baths, and slowest in controlled-atmosphere furnaces with high carbon-monoxide content.

**Quenching and Tempering of Molybdenum High-speed Tools.** — Quenching may be done in oil, air, or molten bath.  To reduce the possibility of breakage and undue distortion of intricately shaped tools, it is advisable to quench in a molten bath at approximately 1100 degrees F.  The tool also may be quenched in oil and removed while still red, or at approximately 1100 degrees F.  The tool is then cooled in air to room temperature, and tempered immediately to avoid cracking.

When straightening is necessary, it should be done after quenching and before cooling to room temperature prior to tempering.

To temper, the tools should be reheated slowly and uniformly to 950 to 1100 degrees F.  For general work, 1050 degrees F. is most common.  The tools should be held at this temperature at least one hour. Two hours is a safer minimum, and four hours is maximum. The time and temperature depend on the hardness and toughness required. Where tools are subjected to more or less shock, multiple temperings are suggested.

**Protective Coatings for Molybdenum Steels.** — Borax may be applied by sprinkling it lightly over the steel when the latter is heated in a furnace to a slow temperature (1200 to 1400 degrees F.). Small tools may be rolled in a box of borax before heating. Another method more suitable for finished tools is to apply the borax or boric acid in the form of a supersaturated water solution. In such cases, the tools are immersed in the solution at 180 to 212 degrees F., or it may be applied with a brush or spray. Pieces so treated are heated as usual, taking care in handling to insure good adherence. Special protective coatings or paints, when properly applied, have been found extremely useful. They do not fuse or run at the temperatures used, and therefore do not affect the furnace hearth. When applying these coatings, it is necessary to have a surface free from scale or grease to insure good adherence. They may be sprayed or brushed on, and usually one thin coat is sufficient. Heavy coats tend to pit the surface of the tool and also are difficult to remove. Tools covered with these coatings should be allowed to dry before they are charged into the preheat furnace. After hardening and tempering, the coating can be easily removed by light blasting with sand or steel shot. When tools are lightly ground, these coatings come off immediately.

**Nitriding High-speed Steel Tools.** — Nitriding as applied to high-speed steel is for the purpose of increasing tool life by producing a very hard skin or case, the thickness of which ordinarily is from 0.001 to 0.002 inch. Nitriding is done after the tool has been fully heat-treated and finish ground. (The process differs entirely from that which is applied to certain alloy steels in order to surface harden them by heating in an atmosphere of nitrogen or ammonia gas.) The temperature of the high-speed steel nitriding bath, which is a mixture of sodium and potassium cyanides, is equal to or slightly lower than the tempering temperature. For ordinary tools, this temperature usually varies from about 1025 to 1050 degrees F.; but if the tools are exceptionally fragile, the range may be reduced to 950 or 1000 degrees F. Accurate temperature control is essential to prevent exceeding the final tempering temperature. The nitriding time may vary from 10 or 15 minutes to 30 minutes or longer, as determined by experiment. The shorter periods are applied to tools for iron or steel, or any shock-resisting tools, and the longer periods are for tools used in machining non-ferrous metals and plastics. This nitriding process is applied to tools such as hobs, reamers, taps, box-tools, form tools, milling cutters, etc. Nitriding may increase tool life 50 to 200 per cent, or more, but it should always be preceded by correct heat-treatment.

*Nitriding Bath Mixtures and Temperatures:* A mixture of 60 per cent sodium cyanide and 40 per cent potassium cyanide is commonly used. This mixture has a melting point of 925 degrees F. which is gradually reduced to 800 degrees F. as the cyanate content of the bath increases. A more economical mixture of 70 per cent sodium cyanide and 30 per cent potassium cyanide may be used if the operating temperature of the bath is only 1050 degrees F. Nitriding bath temperature should not exceed 1100 degrees F. because higher temperatures accelerate the formation of carbonate at the expense of the essential cyanide. A third mixture suitable for nitriding consists of 55 per cent sodium cyanide, 25 per cent potassium chloride and 20 per cent sodium carbonate. This mixture melts at 930 degrees F.

**Equipment for Hardening High-speed Steel.** — Equipment for hardening high-speed steel consists of a hardening furnace capable of maintaining a temperature of 2350 to 2450 degrees F.; a preheating furnace capable of maintaining a temperature of 1700 to 1800 degrees F., and of sufficient size to hold a number of pieces of the work; a tempering (drawing) furnace capable of maintaining a temperature of 1000 to 1200 degrees F. as a general rule; and a water-cooled tank of quenching oil.

High-speed steels usually are heated for hardening either in some type of electric furnace or in a gas-fired furnace of the muffle type. The small furnaces used for high-speed steel seldom are oil-fired. It is desirable to use automatic temperature control and, where an oven type of furnace is employed, a controlled atmosphere is advisable because of the variations in cutting qualities caused by hardening under uncontrolled conditions. Some furnaces in both electric and fuel-fired types are equipped with a salt bath suitable for high-speed steel hardening temperatures. Salt baths have the advantage of providing protection against the atmosphere during the heating period. A type of salt developed for commercial use is water-soluble, so that all deposits from the hardening bath may be removed by immersion in water after quenching in oil or salt, or after air-cooling. One type of electric furnace heats the salt bath internally by electrodes immersed in it. The same type is also applied to various heat-treating operations, such as cyanide hardening, liquid carburizing, tempering, and annealing.

An open-forge fire has many disadvantages, especially in hardening cutters or other tools that cannot be ground all over after hardening. The air blast decarburizes the steel and lack of temperature control makes it impossible to obtain uniform results. Electric and gas furnaces provide continuous uniform heat, and the temperature may be regulated accurately, especially when pyrometers are used. In shops equipped with only one furnace for carbon steel and one for high-speed steel, the tempering can be done in the furnace used for hardening carbon steel after the preheating is finished and the steel has been removed for hardening.

**Heating High-speed Steel for Forging.** — Care should be taken not to heat high-speed steel for forging too abruptly. In winter, the steel may be extremely cold when brought into the forge shop. If the steel is put directly into the hot forge fire, it is likely to develop cracks which will show up later in the finished tool. It should, therefore, be warmed gradually before heating for forging.

## Sub-zero Treatment of Steel

The sub-zero treatment consists in subjecting the steel, after hardening and either before or after tempering, to a sub-zero temperature (which usually ranges from −100 degrees F. to −120 degrees F.) and for a period of time varying with the size or volume of the tool, gage or other part. Commercial equipment is available for obtaining these low temperatures.

The sub-zero treatment is employed by most gage manufacturers in order to stabilize precision gages and prevent subsequent changes in size or form. Sub-zero treatment is also applied to some high-speed steel cutting tools. The object in this case is to increase the durability or life of the tools; however, up to the present time the results of tests by metallurgists and tool engineers often differ considerably and in some instances are contradictory. Methods of procedure also vary, especially in regard to the order and number of operations in the complete heat-treating and cooling cycle.

**Changes Resulting From Sub-zero Treatment.** — When steel is at the hardening temperature, it contains a solid solution of carbon and iron known as *austenite*. When the steel is hardened by sudden cooling, most of the austenite which is relatively soft, tough and ductile even at room temperatures, is transformed to marten-

such parts as automotive cylinders, pistons, piston rings, crankcases, brake drums; for certain machine tool castings, for certain types of dies, for parts of crushing and grinding machinery, and for applications where the casting must resist scaling at high temperatures. Machinable alloy cast irons having tensile strengths up to 70,000 pounds per square inch or even higher may be produced.

**Malleable-iron Castings.** — Malleable iron is produced by the annealing or graphitization of white iron castings. The graphitization in this case produces temper carbon which is graphite in the form of compact rounded aggregates. Malleable castings are used for many industrial applications where strength, ductility, machinability, and resistance to shock are important factors. In manufacturing these castings, the usual procedure is to first produce a hard, brittle white iron from a charge of pig iron and scrap. These hard white-iron castings are then placed in stationary batch-type furnaces or car-bottom furnaces and the graphitization (malleablizing) of the castings is accomplished by means of a suitable annealing heat treatment. During this annealing period the temperature is slowly (50 hours) increased to as much as 1650 or 1700 degrees F. after which time it is slowly (60 hours) cooled. The American National Standard G48.1-1969 — Specifications for Malleable Iron Castings (ASTM A 47-68) specifies the following grades and their properties: No. 32520, having a minimum tensile strength of 50,000 pounds per square inch, a minimum yield strength of 32,500 psi., and a minimum elongation in 2 inches of 10 per cent; and No. 35018, having a minimum tensile strength of 53,000 psi., a minimum yield strength of 35,000 psi., and a minimum elongation in 2 inches of 18 per cent.

*Cupola Malleable Iron.* — Another method of producing malleable iron involves initially the use of a cupola or a cupola in conjunction with an air furnace. This type of malleable iron, called cupola malleable iron, exhibits good fluidity and will produce sound castings. It is used in the making of pipe fittings, valves, and similar parts and possesses the useful property of being well suited to galvanizing. The American National Standard G49.1-1948 (R1969) — Specifications for Cupola Malleable Iron [ASTM A 197-47(1971)] calls for a minimum tensile strength of 40,000 pounds per square inch; a minimum yield strength of 30,000 psi.; and a minimum elongation in 2 inches of 5 per cent.

*Pearlitic Malleable Iron.* — This type of malleable iron contains some combined carbon in various forms. It may be produced either by stopping the heat treatment of regular malleable iron during production before the combined carbon contained therein has all been transformed to graphite or by reheating regular malleable iron above the transformation range. Pearlitic malleable irons exhibit a wide range of properties and are used in place of steel castings or forgings or to replace malleable iron when a greater strength or wear resistance is required. Some forms are made rigid to resist deformation while others will undergo considerable deformation before breaking. This material has been used in axle housings, differential housings, camshafts, and crankshafts for automobiles; machine parts; ordnance equipment; and tools. Tension test requirements of pearlitic malleable iron castings called for in ASTM Specification A 220-71 are given in the accompanying table.

**Tension Test Requirements of Pearlitic Malleable Iron Castings** (ASTM A 220-71)

| Casting Grade Numbers | | 40010 | 45008 | 45006 | 50005 | 60004 | 70003 | 80002 | 90001 |
|---|---|---|---|---|---|---|---|---|---|
| Min. Tensile Strength | 1000's Lbs. per Sq In. | 60 | 65 | 65 | 70 | 80 | 85 | 95 | 105 |
| Min. Yield Strength | | 40 | 45 | 45 | 50 | 60 | 70 | 80 | 90 |
| Min. Elong. in 2 In., Per Cent | | 10 | 8 | 6 | 5 | 4 | 3 | 2 | 1 |

**Nodular Cast Iron (Ductile Iron).** — A distinguishing feature of this relatively new type of cast iron, also known as spheroidal graphite iron, is that the graphite is present in ball-like form instead of in flakes as in ordinary gray cast iron. The addition of small amounts of magnesium- or cerium-bearing alloys together with special processing produces this spheroidal graphite structure and results in a casting of high strength and appreciable ductility. Its toughness is intermediate between that of cast iron and steel and its shock resistance is comparable to ordinary grades of mild carbon steel. Melting point and fluidity are similar to those of the high carbon cast irons. It exhibits good pressure tightness under high stress and can be welded and brazed. It can be softened by annealing or hardened by normalizing and air cooling or oil quenching and drawing.

Two grades of this iron are specified in ASTM A 339-55 — Standard Specification for Nodular Iron Castings. Their nomenclature and properties are as follows: Grade 80-60-03 having a minimum tensile strength of 80,000 pounds per square inch, a minimum yield strength of 60,000 psi., and a minimum elongation in 2 inches of 3 per cent; and grade 60-45-10 having a minimum tensile strength of 60,000 psi., a minimum yield strength of 45,000 psi. and a minimum elongation in 2 inches of 10 per cent. The minimum properties for grade 80-60-03 can usually be obtained in the as-cast condition but to obtain minimum properties in the 60-45-10 grade, a heat treatment (graphitizing anneal) is generally required. Tensile strengths of 100,000 psi. and higher can be obtained by oil quenching and tempering but as the hardness and strength are increased the ductility is decreased. Properties similar to those obtainable with pearlitic malleable iron can be obtained. Several other types are commercially available to meet specific needs. A heat-resistant type, alloyed with silicon, is particularly resistant to growth and oxidation but is restricted to low-shock applications. An austenitic type with high nickel content has high corrosion resistance as well as good strength at elevated temperatures and can be used at temperatures up to 1400 degrees F.

Nodular cast iron can be cast in molds containing metal chills if wear-resisting surfaces are desired. Hard carbide areas will form in a manner similar to the forming of areas of chilled cast iron in gray iron castings. Surface hardening by flame or induction methods is also feasible. Nodular cast iron can be machined with the same ease as gray cast iron. It finds use as crankshafts, pistons, and cylinder heads in the automotive industry; forging hammer anvils, cylinders, guides, and control levers in the heavy machinery field; and wrenches, clamp frames, face-plates, chuck bodies, and dies for forming metals in the tool and die field.

**Steel Castings.** — Steel castings are especially adapted for machine parts that must withstand shocks or heavy loads. They are stronger than either wrought iron, cast iron, or malleable iron and are very tough. The steel used for making steel castings may be produced either by the open-hearth, electric arc, side-blow converter, or electric induction methods. The raw materials used are steel scrap, pig iron, and iron ore, the materials and their proportions varying according to the process and the type of furnace used. The open-hearth method is used when large tonnages are continually required while a small electric furnace might be used for steels of widely differing analyses, which are required in small lot production. The high frequency induction furnace is used for small quantity production of expensive steels of special composition such as high-alloy steels. Foundries having a limited floor space and desiring nearly continuous pouring, use the converter method. Steel castings are used for such parts as hydroelectric turbine wheels, forging presses, gears, railroad car frames, valve bodies, pump casings, bridge components, mining machinery, marine equipment, engine casings, etc. They are of special importance in the construction, mining, and railroad equipment industries.

### Table 1. Physical Properties of Steel Castings

For general information only.  Not for use as design or specification limit values.

| Tensile Strength, Lbs. per Sq. In. | Yield Point, Lbs. per Sq. In. | Elongation in 2 In., Per Cent | Brinell Hardness Number | Type of Heat Treatment | Application Indicating Properties |
|---|---|---|---|---|---|
| | | | Structural Grades of Carbon Steel Castings | | |
| 60,000 | 30,000 | 32 | 120 | Annealed | Low electric resistivity. Desirable magnetic properties.  Carburizing and case hardening grades. Weldability. |
| 65,000 | 35,000 | 30 | 130 | Normalized | Good weldability. Medium strength with good machinability and high ductility. |
| 70,000 | 38,000 | 28 | 140 | Normalized | |
| 80,000 | 45,000 | 26 | 160 | Normalized and tempered | High strength carbon steels with good machinability, toughness and good fatigue resistance. |
| 85,000 | 50,000 | 24 | 175 | Normalized and tempered | |
| 100,000 | 70,000 | 20 | 200 | Quenched and tempered | Wear resistance. Hardness. |
| | | | Engineering Grades of Low Alloy Steel Castings | | |
| 70,000 | 45,000 | 26 | 150 | Normalized and tempered | Good weldability. Medium strength with high toughness and good machinability.  For high temperature service. |
| 80,000 | 50,000 | 24 | 170 | Normalized and tempered | |
| 90,000 | 60,000 | 22 | 190 | Normalized and tempered* | Certain steels of these classes have good high temperature properties and deep hardening properties.  Toughness. |
| 100,000 | 68,000 | 20 | 209 | Normalized and tempered* | |
| 110,000 | 85,000 | 20 | 235 | Quenched and tempered | Impact resistance.  Good low temperature properties for certain steels. Deep hardening.  Good combination of strength and toughness. |
| 120,000 | 95,000 | 16 | 245 | Quenched and tempered | |
| 150,000 | 125,000 | 12 | 300 | Quenched and tempered | Deep hardening.  High strength.  Wear and fatigue resistance. |
| 175,000 | 148,000 | 8 | 340 | Quenched and tempered | High strength and hardness.  Wear resistance. High fatigue resistance. |
| 200,000 | 170,000 | 5 | 400 | Quenched and tempered | |

* Quench and temper heat treatments may also be employed for these classes.
The values listed above have been compiled by the Steel Founders' Society of America as those normally expected in the production of steel castings.  The castings are classified according to tensile strength values which are given in the first column. Specifications covering steel castings are prepared by the American Society for Testing Materials, the Association of American Railroads, the Society of Automotive Engineers, the United States Government (Federal and Military Specifications), etc. These specifications appear in publications issued by these organizations.

CASTINGS

**Carbon Steel Castings.** — Two general classes of steel castings are the carbon steel castings and the alloy steel castings. Carbon steel castings may be designated as low-carbon, medium-carbon, and high-carbon. Low-carbon steel castings have a carbon content of less than 0.20 per cent (most are produced in the 0.16 to 0.19 per cent range). Other elements present are: manganese, 0.50 to 0.85 per cent; silicon, 0.25 to 0.70 per cent; phosphorus, 0.05 per cent max.; and sulphur, 0.06 per cent max. Their tensile strengths (annealed condition) range from 40,000 to 70,000 pounds per square inch. Medium-carbon steel castings have a carbon content of from 0.20 to 0.50 per cent. Other elements present are: manganese, 0.50 to 1.00 per cent; silicon, 0.20 to 0.80 per cent; phosphorus, 0.05 per cent max.; and sulphur, 0.06 per cent max. Their tensile strengths range from 65,000 to 105,000 pounds per square inch depending, in part, upon heat treatment. High-carbon steel castings have a carbon content of more than 0.50 per cent and also contain: manganese, 0.50 to 1.00 per cent; silicon, 0.20 to 0.70 per cent; and phosphorus and sulphur, 0.05 per cent max. each. Fully annealed high-carbon steel castings exhibit tensile strengths of from 95,000 to 125,000 pounds per square inch. See Table 1 for grades and properties of carbon steel castings.

**Table 2. Chemical Composition and Physical Properties of Heat-Resistant Steel Castings (ASTM A 297-55)**

| ACI Type and ASTM Grade | Chemical Composition (Per Cent) | | | | | | | | Minimum Physical Properties | | |
|---|---|---|---|---|---|---|---|---|---|---|---|
| | Carbon | Manganese, max. | Silicon, max. | Phosphorus, max. | Sulphur, max. | Chromium | Nickel | Molybdenum, max.* | Tensile Strength, Lbs per Sq In. | Yield Point, Lbs per Sq In. | Elongation in 2 In., Per Cent |
| HC | 0.50 max. | 1.00 | 2.00 | 0.05 | 0.05 | 26.0 to 30.0 | 4.00 max. | 0.50 | 55,000 | ... | ... |
| HE | 0.20 to 0.50 | 2.00 | 2.00 | 0.05 | 0.05 | 26.0 to 30.0 | 8.00 to 11.0 | 0.50 | 85,000 | 40,000 | 9 |
| HF | 0.20 to 0.40 | 2.00 | 2.00 | 0.05 | 0.05 | 18.0 to 23.0 | 8.0 to 12.0 | 0.50 | 70,000 | 35,000 | 25 |
| HH | 0.20 to 0.50 | 2.00 | 2.00 | 0.05 | 0.05 | 24.0 to 28.0 | 11.0 to 14.0 | 0.50 | 75,000 | 35,000 | 15 |
| HI | 0.20 to 0.50 | 2.00 | 2.00 | 0.05 | 0.05 | 26.0 to 30.0 | 14.0 to 18.0 | 0.50 | 75,000 | 35,000 | 15 |
| HK | 0.20 to 0.60 | 2.00 | 2.00 | 0.05 | 0.05 | 24.0 to 28.0 | 18.0 to 22.0 | 0.50 | 75,000 | 35,000 | 15 |
| HT | 0.35 to 0.75 | 2.00 | 2.50 | 0.05 | 0.05 | 13.0 to 17.0 | 33.0 to 37.0 | 0.50 | 65,000 | ... | 4 |
| HU | 0.35 to 0.75 | 2.00 | 2.50 | 0.05 | 0.05 | 17.0 to 21.0 | 37.0 to 41.0 | 0.50 | 65,000 | ... | 4 |
| HW | 0.35 to 0.75 | 2.00 | 2.50 | 0.05 | 0.05 | 10.0 to 14.0 | 58.0 to 62.0 | 0.50 | 60,000 | ... | ... |
| HX | 0.35 to 0.75 | 2.00 | 2.50 | 0.05 | 0.05 | 15.0 to 19.0 | 64.0 to 68.0 | 0.50 | 60,000 | ... | ... |

* Castings having a specified molybdenum range agreed upon by the manufacturer and the purchaser may also be furnished under these specifications.

Table 3. Chemical Composition and Physical Properties of Corrosion-Resistant Steel Castings (ASTM A 296-55)

| ACI Type and ASTM Grade | Chemical Composition (Per Cent) | | | | | | | | Minimum Physical Properties | | |
|---|---|---|---|---|---|---|---|---|---|---|---|
| | Carbon, max. | Manganese, max. | Silicon, max. | Phosphorus, max. | Sulphur, max. | Chromium | Nickel | Molybdenum | Tensile Strength, Lbs per Sq In. | Yield Point, Lbs per Sq In. | Elongation in 2 In., Per Cent |
| CA-15 | 0.15 | 1.00 | 1.50 | 0.05 | 0.05 | 11.5 to 14.0 | 1.00 max. | 0.50 max. | 90,000 | 65,000 | 18 |
| CB-30[4] | 0.30 | 1.00 | 1.00 | 0.05 | 0.05 | 18.0 to 21.0 | 2.00 max. | ... | 65,000 | 30,000 | ... |
| CC-50 | 0.50 | 1.00 | 1.00 | 0.05 | 0.05 | 26.0 to 30.0 | 4.00 max. | ... | 55,000 | ... | ... |
| CE-30 | 0.30 | 1.50 | 2.00 | 0.05 | 0.05 | 26.0 to 30.0 | 8.0 to 11.0 | ... | 80,000 | 40,000 | 10 |
| CF-8 | 0.08 | 1.50 | 2.00 | 0.05 | 0.05 | 18.0 to 21.0 | 8.0 to 11.0 | | 70,000 | 28,000 | 35 |
| CF-8C[3] | 0.08 | 1.50 | 2.00 | 0.05 | 0.05 | 18.0 to 21.0 | 9.0 to 12.0 | ... | 70,000 | 30,000 | 30 |
| CF-8M | 0.08 | 1.50 | 2.00 | 0.05 | 0.05 | 18.0 to 21.0 | 9.0 to 12.0 | 2.0 to 3.0 | 70,000 | 30,000 | 30 |
| CF-16F[1] | 0.16 | 1.50 | 2.00 | 0.05[1] | 0.05[1] | 18.0 to 21.0 | 9.0 to 12.0 | ...[1] | 70,000 | 30,000 | 25 |
| CF-20 | 0.20 | 1.50 | 2.00 | 0.05 | 0.05 | 18.0 to 21.0 | 8.0 to 11.0 | ... | 70,000 | 30,000 | 30 |
| CG-12 | 0.12 | 1.50 | 2.00 | 0.05 | 0.05 | 20.0 to 23.0 | 10.0 to 13.0 | ... | 70,000 | 28,000 | 35 |
| CH-20[2] | 0.20[2] | 1.50 | 2.00 | 0.05 | 0.05 | 22.0 to 26.0 | 12.0 to 15.0 | ... | 70,000[5] | 30,000 | 30 |
| CK-20 | 0.20 | 2.00 | 2.00 | 0.05 | 0.05 | 23.0 to 27.0 | 19.0 to 22.0 | ... | 65,000 | 28,000 | 30 |

[1] For free machining properties the composition of grade CF-16F may contain suitable combinations of selenium, phosphorus, and molybdenum (grade CF-16F) or of sulfur and molybdenum (grade CF-16Fa) as follows: Grade CF-16F — selenium, 0.20 to 0.35 per cent; phosphorus, 0.17 per cent max.; and molybdenum, 1.50 per cent max. Grade CF-16Fa — sulfur, 0.20 to 0.40 per cent; and molybdenum, 0.40 to 0.80 per cent.

[2] For the more severe general corrosive conditions, and when so specified, the carbon content shall not exceed 0.10 per cent. This low-carbon grade shall be designated as Grade CH-10.

[3] This grade shall have a columbium content of not less than eight times the carbon content and not more than 1.00 per cent.

[4] For this grade a copper content of 0.90 to 1.20 per cent is optional.

[5] A tensile strength of 65,000 pounds per square inch, min., is permitted when the carbon content is 0.06 per cent max. or the silicon content is 1.00 per cent max., or both.

**Alloy Steel Castings.** — Although it is difficult to clearly distinguish alloy cast steels from carbon cast steels, it might be said that alloy cast steels are those in which special alloying elements such as manganese, chromium, nickel, molybdenum, vanadium have been added in sufficient quantities to obtain or increase certain desirable properties. Alloy cast steels are comprised of two groups — the low-alloy steels with their alloy content totaling less than 8 per cent and the high-alloy steels with their alloy content totaling 8 per cent or more. The addition of these various alloying elements, in conjunction with suitable heat-treatments, makes it possible to secure steel castings having a wide range of physical properties. The three accompanying tables give information on these steels. The lower portion of Table 1 gives the engineering grades of low-alloy cast steels grouped according to tensile strengths and gives physical properties normally expected in the production of steel castings. Tables 2 and 3 give the standard designations and chemical composition ranges of high-alloy castings which may be classified according to heat or corrosion resistance. The grades given in these tables are recognized in whole or in part by the Alloy Casting Institute (ACI), the American Society for Testing Materials (ASTM), and the Society of Automotive Engineers (SAE).

*Austenitic Manganese Cast Steel.* — Austenitic manganese cast steel is an important high-alloy cast steel which provides a high degree of shock and wear resistance. Its composition normally falls within the following ranges: carbon, 1.00 to 1.40 per cent; manganese, 10.00 to 14.00 per cent; silicon, 0.30 to 1.00 per cent; sulphur, 0.06 per cent max.; phosphorus, 0.10 per cent, max. In the as-cast condition, austenitic manganese steel is quite brittle. In order to strengthen and toughen the steel, it is heated to between 1830 and 1940 degrees F. and quenched in cold water. Physical properties of quenched austenitic manganese steel that has been cast to size are as follows: tensile strength, 80,000 to 100,000 pounds per square inch; shear strength (single shear), 84,000 pounds per square inch; elongation in 2 inches, 15 to 35 per cent; reduction in area, 15 to 35 per cent; and Brinell hardness number, 180 to 220. When cold worked, the surface of such a casting increases to a Brinell hardness of from 450 to 550. In many cases the surfaces are cold worked to maximum hardness to assure immediate hardness in use. Heat-treated austenitic manganese steel is machined only with great difficulty since it hardens at and slightly ahead of the point of contact of the cutting tool. Grinding wheels mounted on specially adapted machines are used for boring, planing, keyway cutting, and similar operations on this steel. Where grinding cannot be employed and machining must be resorted to, high-speed tool steel or cemented carbide tools are used with heavy, rigid equipment and slow, steady operation. In any event, this procedure tends to be both tedious and expensive. Austenitic manganese cast steel can be arc-welded with manganese-nickel steel welding rods containing from 3 to 5 per cent nickel, 10 to 15 per cent manganese, and, usually, 0.60 to 0.80 per cent carbon.

## Finishing Operations for Castings

**Removal of Gates and Risers from Castings.** — After the molten iron or steel has solidified and cooled, the castings are removed from their molds, either manually or by placing them on vibratory machines and shaking the sand loose from the castings. The gates and risers which are not broken off in the shake-out are removed by impact, sawing, shearing, or burning-off methods. In the impact method, a hammer is used to knock off the gates and risers. Where the possibility exists that the fracture would extend into the casting itself, the gates or risers are first notched to assure fracture in the proper place. Some risers have a necked down section at which the riser breaks off when struck. Sprue-cutter machines are also used to shear off gates. These machines facilitate the removal of a number of small castings from a central runner. Band saws, power saws, and abrasive cut-off wheel machines are also used to remove gates and risers. The use of band saws permits following the contour of the casting when removing unwanted appendages. Abrasive cut-off wheels are used when the castings are too hard or difficult to saw. Oxy-acetylene cutting torches are used to cut off gates and risers and also to gouge out or remove surface defects on castings. These torches are used on steel castings where the gates and risers are of a relatively large size. Surface defects are subsequently repaired by conventional welding methods.

Any unwanted material in the form of fins, gate and riser pads which come above the casting surface, chaplets, parting-line flash, etc. is removed by chipping with pneumatic hammers, or by grinding with such equipment as floor or bench-stand grinders, portable grinders, and swing-frame grinders.

**Blast Cleaning of Castings.** — Blast cleaning of castings is performed to remove adhering sand, remove cores, to improve the casting appearance, or to prepare the castings for their final finishing operation which includes painting, machining, or assembling. Scale produced as a result of heat-treating can also be removed. A variety of machines are used to handle all sizes of castings. The methods employed include blasting with sand, metal shot, or grit; and hydraulic cleaning or tumbling. In blasting, sharp sand, shot or grit is carried by a stream of compressed air or water or by centrifugal force (gained as a result of whirling in a rapidly rotating machine) and directed against the casting surface by means of nozzles. The operation is usually performed in cabinets or enclosed booths. In some set-ups the castings are placed on a revolving table and the nozzles which are either mechanically- or hand-held are played over all the casting surfaces. Tumbling machines are also employed for cleaning, the castings being placed in large revolving drums together with slugs, balls, pins, metal punchings or some abrasive, such as sandstone or granite chips, slag, silica, sand, or pumice. Quite frequently the tumbling and blasting methods are used together, the parts in this instance being tumbled and blasted simultaneously. Castings may also be cleaned by hydro-blasting. In this method the castings are placed in a water-tight room and a water and sand mixture under high pressure is played over the castings by means of nozzles. The action of the water and sand mixture cleans the castings very effectively.

**Heat-treatment of Steel Castings.** — Steel castings can be heat-treated to bring about diffusion of carbon or alloying elements, softening, hardening, stress-relieving, toughening, improved machinability, increased wear resistance, and removal of hydrogen entrapped at the surface of the casting. Heat treatment of steel castings of a given composition follows closely that of wrought steel of similar composition. For discussion of types of heat treatment refer to the Heat-treatment of Steel section of this Handbook.

## Pattern Materials — Shrinkage, Draft and Finish Allowances

**Woods for Patterns.** — Woods commonly used for patterns are white pine, mahogany, cherry, maple, birch, white wood and fir. For most patterns, white pine is considered superior because it is easily worked, readily takes glue and varnish, and is fairly durable. For medium- and small-sized patterns, especially if they are to be extensively used, a harder wood is preferable. Mahogany is often used for patterns of this class, although many prefer cherry. As mahogany has a close grain, it is not as susceptible to atmospheric changes as a wood of coarser grain. Mahogany is superior in this respect to cherry, but is more expensive. In selecting cherry, never use young timber. Maple and birch are employed quite extensively, especially for turned parts, as they take a good finish. White wood is sometimes substituted for pine, but it is inferior to the latter in being more susceptible to atmospheric changes.

**Selection of Wood.** — It is very important to select wood for patterns that is well seasoned; that is, it should either be kiln dried or kept one or two years before using, the time depending upon the size of the lumber. During the seasoning or drying process the moisture leaves the wood cells and the wood shrinks, the shrinkage being almost entirely across the grain rather than in a lengthwise direction. Naturally, after this change takes place, the wood is less liable to warp, although it will absorb moisture in damp weather. Patterns also tend to absorb moisture from the damp sand of molds, and to minimize troubles from this source they are covered with varnish. Green or water-soaked lumber should not be put in a drying room, because the ends will dry out faster than the rest of the log, thus causing cracks. In a log there is what is called the "sap wood" and the "heart wood." The outer layers form the sap wood which is not as firm as the heart wood and is more likely to warp; hence, it should be avoided, if possible.

**Pattern Varnish.** — Patterns intended for repeated use are varnished to protect them against moisture, especially when in the damp molding sand. The varnish used should dry quickly to give a smooth surface that readily draws from the sand. Yellow shellac varnish is generally used. It is made by dissolving gum shellac in grain alcohol. Wood alcohol is sometimes substituted, but is inferior. The color of the varnish is commonly changed for covering core prints, in order to readily distinguish the prints from the body of the pattern. Black shellac varnish (which is the color generally used) is made by the addition of lamp black. This should be of good quality and free from grit. Red varnish can be made by adding Chinese vermilion. All coloring powders should be well pulverized. At least three coats of varnish should be applied to patterns, the surfaces being rubbed down with sand paper after applying the preliminary coats, in order to obtain a smooth surface.

**Glue for Patterns.** — There are many qualities of glue both in the liquid, sheet and pulverized form. Animal glue in the sheet or flake form is generally used for pattern work. As a rule, the best quality is of amber color and the flakes are rather thin. Where glue is used in small quantities, the pulverized form has the advantage of being quickly prepared. Freshly made glue is the strongest and, if of good quality, can be drawn out into thin threads. Whenever practicable, glued joints should be reinforced by nails or screws. A joint to be glued should be accurately fitted because glue does not get a grip unless the parts are in close contact.

Before applying glue, clean the surfaces of sand paper dust or other foreign material, so that the glue can enter the pores of the wood. This is very important. If the end grain must be glued, first apply a sizing coat to fill the openings among the fibers. When the sizing coat is dry, apply a second coat to the surface and unite. If the preliminary coat is not applied, the open end grain is liable to absorb the glue so rapidly as to weaken the joint. The hot glue should be thin enough to spread easily. It can be thicker, however, for pine than for wood of closer grain, like mahogany, because (aside from the quality of the glue) the holding or binding property depends upon the extent to which the glue enters the pores of the wood. All glued joints should be firmly pressed together with clamps immediately after applying the glue. The latter should be given plenty of time in which to set; ten or twelve hours in a dry place should be sufficient.

**Shrinkage Allowance.** — The shrinkage allowances ordinarily specified for patterns to compensate for the contraction of castings in cooling are as follows: cast iron, $\frac{3}{32}$ to $\frac{1}{8}$ inch per foot; common brass, $\frac{3}{16}$ inch per foot; yellow brass, $\frac{7}{32}$ inch per foot; bronze, $\frac{5}{32}$ inch per foot; aluminum, $\frac{1}{8}$ to $\frac{5}{32}$ inch per foot; magnesium, $\frac{1}{8}$ to $1\frac{1}{64}$ inch per foot; steel casting, $\frac{3}{16}$ inch per foot. These shrinkage allowances are approximate values only since the exact allowance in any case depends upon the size and shape of the casting and the resistance of the mold to the normal contraction of the casting during cooling. It is, therefore, possible that more than one shrinkage allowance will be required for different parts of the same pattern. Another factor that affects shrinkage allowance is the molding method, which may vary to such an extent from one foundry to another, that different shrinkage allowances for each would have to be used for the same pattern. For these reasons it is recommended that patterns be made at the foundry where the castings are to be produced to eliminate difficulties due to lack of accurate knowledge of shrinkage requirements.

An example of how casting shape can affect shrinkage allowance is given in the Steel Castings Handbook. In this example a straight round steel bar required a shrinkage allowance of approximately $\frac{9}{32}$ inch per foot. The same bar but with a large knob on each end required a shrinkage allowance of only $\frac{3}{16}$ inch per foot. A third steel bar with large flanges at each end required a shrinkage allowance of only $\frac{7}{64}$ inch per foot. This example would seem to indicate that the best practice in designing castings and making patterns is to obtain shrinkage values from the foundry that is to make the casting since there can be no fixed allowances.

**Draft for Patterns.** — The draft or the amount of taper given to patterns to facilitate withdrawing them from the mold, depends somewhat upon the size and shape of the pattern. A general rule is to taper each side $\frac{1}{8}$ inch for each foot of surface to be drawn. The average amount for small patterns is about $\frac{1}{16}$ inch per foot, although in some cases it can be less, but, as a rule, there should be at least $\frac{1}{32}$ inch draft. The draft slopes away from the pattern "face" which is usually uppermost in the mold when the pattern is drawn. Some patterns do not require draft because none of the surfaces are at right angles to the face. In some cases, very small patterns are made without draft.

**Finish Allowance.** — The amount added to a pattern to allow for machining the casting varies widely. It depends upon the method of machining, the size of the casting, and the importance of having a clean surface, free from flaws or defective spots. If castings are to be finished from the rough upon disk grinders, very little allowance is necessary; in fact, the molder may rap a pattern enough to allow for the finish. On small castings to be finished by milling or planing about $\frac{1}{8}$ inch is usually allowed for the machining operation, whereas large castings for engine beds, flywheels, pump cylinders, etc., often have an allowance of $\frac{3}{4}$ to 1 inch.

**Standard Pattern Colors.** — The color markings described in the following are recommended as standard for foundry patterns and core-boxes of wood construction. These standard colors have been accepted by the Bureau of Standards, the American Foundrymen's Association, the Steel Founders' Society of America, the Malleable Iron Research Institute, and by numerous other associations as well as prominent manufacturers.

(1) Surfaces to be left *unfinished* are to be painted *black*.

(2) Surfaces to be *machined* are to be painted *red*.

(3) *Seats of and for loose pieces* are to be marked by *red stripes on a yellow background*.

(4) *Core prints* and seats for loose core prints are to be painted *yellow*.

(5) *Stop-offs* are to be indicated by *diagonal black stripes on a yellow base*.

The colors may be obtained by mixing suitable inexpensive pigments with varnish or shellac to produce the type of coating desired.

**Metal Patterns.** — Metal patterns are especially adapted to molding machine practice, owing to their durability and superiority in retaining the required shape. The original master pattern is generally made of wood, the casting obtained from the wood pattern being finished to make the metal pattern. The materials commonly used are brass, cast iron, aluminum and steel. Brass patterns should have a rather large percentage of tin, as this gives a good surface for the casting. Cast iron is generally used for patterns of large size, as it is cheaper than brass and more durable. Cast-iron patterns are largely used on molding machines. Aluminum patterns are light but they shrink considerably. White metal is sometimes used when it is necessary to avoid shrinkage. The gates for the mold may be cast or made of sheet brass. Some patterns are made of vulcanized rubber, especially for light match-board work.

**Obtaining Weight of Casting from Pattern Weight.** — To obtain the approximate weight of a casting, multiply the weight of the pattern by the factor given in the accompanying table. For example, if the weight of a white-pine pattern is 4 pounds, what is the weight of a solid cast-iron casting obtained from that pattern? Casting weight = 4 × 16 = 64 pounds. If the casting is cored, fill the core-boxes with dry sand, and multiply the weight of the sand by one of the following factors: For cast iron, 4; for brass, 4.65; for aluminum, 1.4. Then subtract the product of the sand weight and the factor just given from the weight of the solid casting, to obtain the weight of the cored casting. As the weight of

Factors for Obtaining Weight of Casting from Pattern Weight

| Pattern Material | Factors | | | | |
|---|---|---|---|---|---|
| | Cast Iron | Aluminum | Copper | Zinc | Brass, 70% Copper, 30% Zinc |
| White pine............... | 16.00 | 5.70 | 19.60 | 15.00 | 19.00 |
| Mahogany, Honduras ..... | 12.00 | 4.50 | 14.70 | 11.50 | 14.00 |
| Cherry.................. | 10.50 | 3.80 | 13.00 | 10.00 | 12.50 |
| Cast iron............... | 1.00 | 0.35 | 1.22 | 0.95 | 1.17 |
| Aluminum .............. | 2.85 | 1.00 | 3.44 | 2.70 | 3.30 |

wood varies considerably, the results obtained by the use of the table are only approximate, the factors being based on the average weight of the woods listed. For metal patterns, the results are more accurate.

**Branch Pipes for Exhausting Shavings from Wood Working Machines.** — The sizes of branch pipes, given in the accompanying table, are correct for pipes not exceeding twenty feet in length. Where branch pipes contain a number of elbows, or exceed twenty feet in length, the area should be proportionately increased. Where the work is light and the branch pipes short, smaller connections than those given can sometimes be used. The area of the main duct should be equal to, or slightly larger than, the sum of the areas of the connecting branches. This proportion should be carefully maintained. If the main pipe is too small, the suction will be impaired, and if it is too large, the velocity of the air may be reduced to such an extent that the material being exhausted will settle in the bottom of the pipe, thereby reducing the area. If the main pipe is unusually long (exceeding 100 feet) the area should be increased from 10 to 20 per cent. Avoid abrupt turns in the piping, and never enter branch pipes at right angles to the main pipe, but always connect them at an angle of from 30 to 45 degrees. Branch pipes should never enter the main at the bottom but always at the side, and two pipes should not enter directly opposite each other.

**Branch Pipes for Exhausting Shavings from Wood Working Machines**

| Type of Machine | Diam. of Branch Pipe, Inches | Type of Machine | Diam. of Branch Pipe, Inches |
|---|---|---|---|
| Saws: | | Planer knives: | |
| Rip ⎤ | | Length of knife, 24 inches.. | 6 |
| Cut-off ⎱ 18 inches diam. | | Length of knife, 30 inches. | 7 |
| Split ⎰ or less....... | 4 | Matcher heads, each........ | 5 |
| Swing ⎰18 to 24 inches | | Door tenoner.............. | 5 |
| Bracket ⎱ diam........ | 5 | Sash tenoner.............. | 5 |
| Groove ⎦ | | Sticker machines, each head. | 4 |
| Heavy cut-off, 24 to 42 | | Sand drum, 24 inches long.. | 4 |
| inches diameter........ | 6 | Sand drum, 30 inches long... | 5 |
| Band................... | 4 | Floor sweep-up............. | 6 |
| Band resaw, ¾ to 1 inch.. | 5 | Heavy timber planer, each | |
| Band resaw, 1½ to 2½ ins.. | 6 | head.................... | 7 |
| Planer knives: | | Diagonal planer for doors... | 7 |
| Length of knife, 5 inches.. | 4 | Diagonal planer for doors, | |
| Length of knife, 10 inches. | 5 | with sand drum.......... | 8 |
| Length of knife, 14 inches. | 6 | ......................... | ...... |

**Speed of Circular Saws for Wood**

| Size of Saw, Ins. | Rev. per Min. | Size of Saw, Ins. | Rev. per Min. | Size of Saw, Ins. | Rev. per Min. | Size of Saw, Ins. | Rev. per Min. | Size of Saw, Ins. | Rev. per Min. |
|---|---|---|---|---|---|---|---|---|---|
| 8 | 4500 | 20 | 1800 | 32 | 1125 | 44 | 840 | 56 | 650 |
| 10 | 3600 | 22 | 1650 | 34 | 1050 | 46 | 800 | 58 | 625 |
| 12 | 3000 | 24 | 1500 | 36 | 1000 | 48 | 750 | 60 | 600 |
| 14 | 2600 | 26 | 1400 | 38 | 950 | 50 | 725 | 64 | 550 |
| 16 | 2200 | 28 | 1300 | 40 | 900 | 52 | 700 | 68 | 525 |
| 18 | 2000 | 30 | 1200 | 42 | 870 | 54 | 675 | 72 | 500 |

## Extrusion of Metals

**The Extrusion Process.** — By means of the extrusion process, certain fairly plastic metals are formed into various shapes by forcing the metal, which is usually heated, under high pressure through an aperture of the shape to be produced. In this manner, a continuous bar or pipe of the cross-section of the aperture or die is produced. Many different non-ferrous alloys are used in connection with the extrusion process. These include various compositions of bronze and brass, aluminum alloys, copper, tin, etc., the alloy depending, of course, upon the application. The S.A.E. standard aluminum compositions Nos. 260, 24, 282, and 281 are used for extruded shapes, as well as in the form of sheets, bars, rods, and wire. The advantages of the extrusion process are that it permits parts of unusual cross-section to be produced cheaply. On account of the high pressure under which the metal is extruded, its structure becomes more compact and its strength is increased. The surfaces are smooth and free from flaws and other defects. The dimensions of the extruded shapes can be gaged with accuracy, so that they can be used directly with no or very little additional finishing.   The extruded bar may be passed through a drawing die if a special degree of accuracy is required.

**Extruded Shapes.** — Extruded bars, rods and sections are produced in a large variety of standard and special shapes.   These include, in addition to the common shapes, many more or less special cross-sections for producing small pinions, ratchet wheels, segment gears, and many other parts, with machining in many cases reduced to a cutting-off operation.   The extrusion process is not restricted to solid bars and rods as it is applied in producing various hollow forms in bronze, brass and other compositions.   The structural shapes for aluminum alloys include S.A.E. standards for equal, unequal and bulb angles; shallow, deep and bulb channels; tees and zees. Many extruded sections are employed for art metal work.   The extrusion process is often used instead of rolling because it is much simpler and less expensive to provide extrusion tools for a new section or special shape than to provide rolls.   These special shapes may be odd sizes of so-called standard sections or they may be sections that are nearly standard but are changed slightly.

**Extruding Blooms for Aluminum Tubing.** — An important class of extruded products includes those sections on which further work is to be done, such as tube blooms and sheet slabs, as well as rectangular billets.   Practically all drawn aluminum tubing is made from extruded blooms.   The principal reason for this is that the bloom that goes on to the draw- or push-bench must be as nearly perfect in finish and dimensions as it is possible to make it.   Extrusion seems to be the only practical method of obtaining these results.   It is possible to pierce aluminum-alloy billets, but the extruded blooms are much more nearly perfect, resulting in lower scrap losses and a better finished product.

**Extruding Alloys Difficult to Roll.** — Another class of extruded products consists of those made from alloys that cannot be rolled.   Certain aluminum-magnesium alloys are so brittle in the hot cast condition that they break up badly on rolling, but can be extruded without great difficulty.   On account of its very desirable physical properties, one of these alloys is used to a considerable extent for drawing fine wire.   It has been found to be almost impossible to roll cast ingots of this alloy; hence billets about $1\frac{1}{2}$ inches square are extruded, and then put on the rod mill, where they are rolled without great difficulty.

While it is true that certain alloys can be extruded with less difficulty than they can be rolled, this is not true of all alloys.   Pure aluminum can be rolled and extruded more easily than any of its alloys.   Those alloys of the duralumin class, containing about 4 per cent copper, require about three times as much power to roll as pure aluminum.   In color and surface finish, extruded products are quite superior to rolled shapes.   This is very important when aluminum is used for decorative purposes, particularly for architectural work.

**Extruding Collapsible Tubes.** — The collapsible tubes used for tooth paste, artists' paints, etc., are extruded from disks or slugs of pure tin containing about one-half of one per cent copper to harden the metal slightly.   These tubes are formed by forcing the slightly lubricated metal through a small annular space between a punch and die, using a pressure of 50 to 100 tons, depending upon the size of the tube.

# Die-Casting

Die-casting is a method of producing finished castings by forcing molten metal into a suitable mold, which is arranged to open after the metal has solidified so that the casting can be removed.   The die-casting process makes it possible to secure accuracy and uniformity in castings, and machining costs are either eliminated altogether or are greatly reduced. The greatest advantage of the die-casting process is due to the fact that parts are accurately and, usually, completely finished when taken from the dies. When the dies are properly made, castings may be accurate within 0.001 inch or even less and a limit of 0.002 and 0.003 inch can be maintained easily on many classes of work.

Die-castings are extensively used in the manufacture of such products as cash registers, meters, time-controlling devices, small housings, washing machines, and parts for a great variety of mechanisms. Lugs and gear teeth are cast in place and both external and internal screw threads can be cast. Holes can be formed within about 0.001 inch of size and the most accurate bearings require only a finish-reaming operation. Figures and letters may be cast sunken or in relief on wheels for counting or printing devices, and owing to numerous developments, many shapes which formerly were believed too intricate for die-casting are now produced successfully by this process.

As to the limitations of the die-casting process it may be mentioned that the cost of dies is high, and, therefore, die-casting is applicable only when a large number of duplicate parts are required. The stronger and harder metals cannot be die-cast so that the process is not applicable for casting parts which must necessarily be made of iron or steel, although special alloys have been developed for die-casting which have considerable tensile and compressive strength.

**Alloys Used for Die-Casting.** — The alloys used in modern die-casting practice may be divided into six main classifications as follows: (1) Zinc-base alloys; (2) tin-base alloys; (3) lead-base alloys; (4) aluminum-base alloys; (5) copper-base or brass and bronze alloys; and (6) magnesium-base alloys.

**Zinc-base Die-casting Alloys.** — The alloys in this group are produced by alloying zinc with aluminum and copper. S.A.E. alloy No. 903 contains aluminum 3.5 to 4.3; copper, maximum, 0.10; magnesium, 0.03 to 0.08 per cent; iron, maximum, 0.10; cadmium, maximum, 0.005; tin, maximum, 0.005; lead, maximum, 0.007 and the remainder zinc. S.A.E. alloy No. 921 contains aluminum, 3.5 to 4.5; copper, 2.5 to 3.5; magnesium, 0.02 to 0.10; iron, lead, cadmium and tin, the same as No. 903 and the remainder, zinc. S.A.E. No. 925 contains copper 0.75 to 1.25; magnesium, 0.02 to 0.08; tin, 0.002; aluminum, iron, lead and cadmium, the same as No. 903 and the remainder, zinc. Alloy No. 921 has been largely replaced by alloys 903 and 925 because the latter possess greater permanence of properties and dimensions. Alloy No. 925 is much stronger and harder than alloy No. 903 but at elevated temperatures is subject to growth in dimensions and loss of impact strength and in this respect is inferior to No. 903.

**Tin-base Alloys.** — In this group tin is alloyed with copper, antimony, and lead. S.A.E. Alloy No. 10 contains, as the principal ingredients, in percentages, tin, 90; copper, 4 to 5; antimony, 4 to 5; lead, maximum, 0.35. This high-quality babbitt mixture is used for main-shaft and connecting-rod bearings or bronze-backed bearings in the automotive and aircraft industries. S.A.E. No. 110 contains tin, 87.75; antimony, 7.0 to 8.5; copper, maximum, 2.25 to 3.75 and other constituents the same as No. 10. S.A.E. No. 11, which contains a little more copper and antimony and about 4 per cent less tin than No. 10, is also used for bearings or other applications requiring a high-class tin-base alloy. These tin-base compositions are used chiefly for automotive bearings but they are also used for different classes of die-castings, especially for milking machines, soda fountains, syrup pumps, and similar apparatus requiring resistance against the action of acids, alkalies, and moisture.

**Lead-base Alloys.** — These alloys are employed usually where a cheap non-corrosive metal is needed and strength is relatively unimportant. Such alloys are used for parts which must withstand the action of strong mineral acids and for parts of X-ray apparatus. S.A.E. Composition No. 13 contains (in percentages) lead, 86; antimony, 9.25 to 10.75; tin, 4.5 to 5.5 per cent. S.A.E. Specification No. 14 contains less lead and more antimony and copper. The lead content is 76; antimony, 14 to 16; and tin, 9.25 to 10.75 per cent. These alloys, Nos. 13 and 14, are inexpensive owing to the high lead content and may be used for bearings which are large and subjected to light service. They are also suitable for some die-castings, but should not be substituted for an alloy with a high tin content.

**Aluminum-base Alloys.** — Aluminum die-castings are used for many parts requiring lightness, strength, and resistance to corrosion. These alloys will take and hold a high polish, and are used for vacuum cleaners and other household utensils, camera parts, motor and instrument housings, etc. There are many compositions limited practically to two general groups, namely, the aluminum-copper and the aluminum-silicon alloys. S.A.E. Alloy No. 312 (generally known as No. 12) is an inexpensive general-purpose alloy, and has been used in the United States more than any other aluminum casting alloy. The main elements, in addition to aluminum, are: Copper, 7 to 9; iron, maximum, 2.5; and silicon, 1 to 2 per cent. The tensile strength should be about 33,000 pounds per square inch. A typical aluminum-silicon alloy contains, in addition to aluminum: Silicon, 12; iron, 2, maximum; copper, 0.60 maximum. The tensile strength is about 30,000 to 33,000 pounds per square inch. S.A.E. Alloy No. 305 contains 11 to 13 per cent silicon, and is especially resistant to salt-water corrosion. These alloys because of their fluidity are adapted for thin-walled castings or for complicated castings consisting of both thin and heavy sections. The tensile strength of ¼-inch round test bars cast to size should be about 33,000 pounds per square inch, with an elongation of 1.5 per cent.

**Copper-base Alloys.** — In producing die-castings, the use of alloys having relatively high melting temperatures naturally presents difficulties not occurring with lower melting points, especially in regard to the life of the die-casting dies. Thus, in casting copper-base alloys, it has been necessary to develop special alloy steels and casting methods. The well-known plunger type and "gooseneck" types of die-casting machines are not suitable for brass or bronze alloys. The latter, when cast, are handled in small charges and forced into the die at an unusual speed and pressure. Zinc, for instance, is cast at pressures of 800 to 1000 pounds per square inch; aluminum, at from 400 to 500; and brass, at about 20,000 pounds per square inch. A typical copper-base alloy contains (in percentages) about 57 to 59 copper, 40 to 42 zinc, and 0.5 to 1.5 tin, and the tensile strength is around 65,000 to 75,000 pounds per square inch.

**Magnesium-base Alloys.** — Magnesium alloy die-castings are used where extreme lightness or a high strength-weight ratio is required. S.A.E. alloy No. 501 is a widely used composition and combines good mechanical properties with good casting characteristics. It contains: Aluminum, 8.3 to 9.7; manganese, min., 0.13; zinc, 0.4 to 1.0; silicon, max., 0.5; copper, max., 0.05; nickel, max., 0.03; other impurities, max., 0.3 per cent and the remainder, magnesium. Typical physical properties are a tensile strength of 31,000 pounds per square inch and a yield strength of 20,000 pounds per square inch.

**Dies for Die-casting Machines.** — The dies or molds of die-casting machines are generally made of steel although cast-iron and also non-metallic materials of a refractory nature have been used, the latter being intended especially for bronze or brass castings, which, owing to their comparatively high melting temperatures, would injure ordinary steel dies. The steel most generally used is a low-carbon steel. Chromium-vanadium and tungsten steels have been employed when dies must withstand relatively high temperatures.

The making of these dies requires considerable skill and experience. They must be so designed that the metal will rapidly flow to all parts of the impression and at the same time allow the air to excape through very small grooves cut into the parting of the die, as otherwise blow-holes or air-pockets will result. In order to secure solid castings, the gates and vents must be located with reference to the particular shape to be cast. Shrinkage is another important feature, especially on accurate work. The amount usually varies from 0.002 to 0.007 inch per inch, but to determine the exact shrinkage allowance for an alloy containing three or four metals is difficult except by experiment.

**Die-casting Bearing Metals in Place.** — Practically all the metals that are suitable for bearings can be conveniently die-cast in place. Automobile connecting-rods are an example of work to which this process has been applied successfully. In this case, after the bearings are cast in place, they are finished by reaming. The best metals for the bearings, and those that also can be die-cast most readily, are the babbitts containing about 85 per cent tin with the remainder copper and antimony. These metals should not contain over 9 per cent copper. The copper constitutes the hardening element in the bearing. A recommended composition for a high-class bearing metal is 85 per cent tin, 10 per cent antimony, and 5 per cent copper. The antimony may vary from 7 to 10 per cent and the copper from 5 to 8 per cent. As bearing metals with so high a percentage of tin are expensive, a number of bearing metals have been developed containing a high percentage of lead. One of these metals contains from 95 to 98 per cent lead which has been treated in the electric furnace. After die-casting, the metal becomes harder upon seasoning a few days. In die-casting bearings, the work is located from the bolt holes which are drilled previous to die-casting. It is important that the bolt holes be accurately drilled with relation to the remainder of the machined surfaces.

## Precision Investment Casting

This highly developed process is capable of great casting accuracy and also the formation of intricate contours. The process may be utilized when metals are too hard to machine or otherwise fabricate; when it is the only practical method of producing a part; or when it is more economical than any other method of obtaining work of the quality required. Precision investment casting is especially applicable in producing either exterior or interior contours of intricate form with surfaces so located that they could not be machined readily if at all. This process provides efficient, accurate means of producing such parts as turbine blades, airplane or other parts which have high melting points and must withstand exceptionally high temperatures, and many other products. The accuracy and finish of precision investment castings may either eliminate machining entirely or reduce it to a minimum. The quantity that may be produced economically, may range from a few to thousands of duplicate parts.

**Materials Which May Be Cast.** — The precision investment process may be applied to a wide range of both ferrous and non-ferrous alloys. In industrial applications these include aluminum, and bronze alloys, stellite, Hastelloys, stainless and other alloy steels, and iron castings especially where thick and thin sections are encountered. In producing investment castings, it is possible to control the process in various ways so as to change the porosity or density of castings, obtain hardness variations in different sections, and vary the corrosion resistance and strength by special alloying.

**General Procedure in Making Investment Castings.** — Precision investment casting is similar in principle to the " lost wax " process which has long been used in manufacturing jewelry, ornamental pieces, dentures, inlays, and other items required in dentistry. When this process is employed, both the pattern and mold used in producing the casting are destroyed after each casting operation, but they may both be replaced readily. The " dispensable pattern " (or cluster of duplicate patterns) is first formed in a permanent mold or die and then is used to form the cavity in the mold or " investment " in which the casting (or castings) is made. The investment or casting mold consists of a refractory material contained within a reinforcing steel flask. The pattern is made of wax, plastic, or a mixture of the two. The material used is evacuated from the investment to form a cavity (without parting lines) for receiving the metal to be cast. Evacuation of the pattern (by the application of sufficient heat to melt and vaporize it) and the use of a master mold or die for reproducing it quickly and accurately in making duplicate castings, are distinguishing features of this casting process. Modern applications of the process include many developments such as variations in the preparation of molds, patterns, investments, etc., as well as in the casting procedure. Application of the process requires specialized knowledge and experience.

**Master Mold for Making Dispensable Patterns.** — Duplicate patterns for each casting operation are made by injecting the wax or other pattern material into the master mold or die which usually is made either of carbon steel or of a soft metal alloy. Rubber, alloy steels, and other materials are used in some cases. The mold cavity commonly is designed to form a cluster of patterns for multiple casting. The mold cavity is not, as a rule, an exact duplicate of the part to be cast because it is necessary to allow for shrinkage and perhaps to compensate for distortion which might affect the accuracy of the cast product. In producing master molds, there is considerable variation in practice. One general method is to form the cavity by machining; another is by pouring a molten alloy around a master pattern which usually is made of monel metal or of a high alloy stainless steel. Unless the cavity is machined, a master pattern is required. Sometimes, a sample of the product itself may be used as a master pattern, when, for example, a slight reduction in size due to shrinkage is not objectionable. The dispensable pattern material, which may consist of waxes, plastics, or a combination of these materials, is injected into the mold either by pressure, by gravity, or by the centrifugal method. The mold is made in sections to permit removal of the dispensable pattern. The mold while in use may be kept at the correct temperature either by electrical means, by using steam, or a water jacket.

**Shrinkage Allowances for Patterns.** — The shrinkage allowance varies considerably for different materials. In casting accurate parts, experimental preliminary casting operations may be necessary to determine the required shrinkage allowance and possible effects of distortion. Shrinkage allowances, in inches per inch, usually average about 0.022 for steel, 0.012 for gray iron, 0.016 for brass, 0.012 to 0.022 for bronze, 0.014 for aluminum and magnesium alloys. (See also page 2184.)

**Investment Materials.** — The investment materials which surround the dispensable pattern are made according to various formulas which may be simple or complex. The investment is in liquid form and the degree of liquidity depends chiefly upon the intricacy of the pattern forms. The investment must be fluid enough to enter such places as small holes, form sharp contours, threads, etc., and fill completely every part of the investment cavity. The accuracy of the product and its surface finish may be affected by the liquidity of the investment material. Porosity is another factor. If it is excessive, the finish will be marred but there should be enough porosity to permit escape of air when the molten metal enters the cavity. The procedure in " investing " or applying the investment materials, varies according to these materials, the kind of metal to be cast and properties required in the casting. The hardening or setting up of the material may merely involve standing for a given time or hardening may be at a controlled temperature.

**Evacuation of the Pattern.** — In cases where heat is applied to control investment hardening or setting up, this heat may also result in evacuating most of the pattern material; however, complete evacuation is essential and methods are employed to remove all traces of the pattern. Evacuation may require closely regulated temperature and the use of boiling water, low pressure steam, or the application of a vacuum to the sprue bottom.

**Casting Operations.** — The temperature of the flask for casting may range all the way from a chilled condition up to 2000 degrees F. or higher, depending upon the metal to be cast, the size and shape of the casting or cluster, and the desired metallurgical conditions. Metals while being cast are nearly always subjected to centrifugal force or other pressure. The procedure is governed by the kind of alloy, the size of investment cavity, and its contours or shape.

**Investment Removal.** — The investments surrounding the casting or cluster is removed by destroying it. The investment may be soluble in water but those used for ferrous castings are broken by using pneumatic tools, hammers, or by shot or abrasive blasting and tumbling to remove all of the particles. Gates, sprues and runners may be removed from the castings by an abrasive cutting wheel or a band saw. The shape of the cluster and machinability of the material are factors governing the selection of the method.

**Accuracy of Investment Castings.** — The accuracy of precision investment castings may, in general, compare favorably with many machined parts. The over-all tolerance varies with the size of the work, the kind of metal, and the skill and experience back of the casting operations. Under normal conditions, the tolerances may vary from plus or minus 0.005 or 0.006 inch per inch, down to 0.0015 to 0.002 inch per inch, and even smaller tolerances are possible on very small dimensions. Where tolerances applying to a lengthwise dimension must be smaller than would be normal for the casting process, the casting gate may be at one end to permit controlling the length by a grinding operation when the gate is removed.

**Casting Milling Cutters by Investment Method.** — Possible applications of precision investment casting in tool manufacture and in other industrial applications, are indicated by its use in producing high-speed steel milling cutters of various forms and sizes. Thousands of these cutters have been precision cast in the Ford plant. Removal of the risers, sand-blasting to improve the appearance, and grinding the cutting edges are the only machining operations required. The bore is used as cast. Numerous tests have shown that the life of these cutters compares favorably with high-speed steel cutters made in the usual way.

**Casting Weights and Sizes.** — Investment castings may vary in weight from a fractional part of an ounce up to 25 pounds or more. Although the range of weights representing the practice of different firms specializing in investment casting may vary from about ½ pound up to 10 or 20 pounds, a practical limit of 3 or 4 pounds is common. The length of investment castings ordinarily does not exceed 5 or 6 inches but much longer parts may be cast. While it is possible to cast sections having a thickness of only a few thousandths of an inch, the preferable minimum thickness, as a general rule, is about 0.020 inch for alloys of high castability and 0.040 inch for alloys of low castability.

## Flame Spraying Process

In this process—the forerunner of which was called the metal spraying process—metals, alloys, ceramics, and cermets are deposited on metallic or other surfaces. The object may be to build up worn or undersize parts, provide wear-resisting or corrosion-resisting surfaces, correct defective castings, etc.

Different types of equipment are available that provide the means of depositing the coatings on the surfaces. In one, wire is fed automatically through the nozzle of the spray gun; then a combustible gas, oxygen, and compressed air serve to melt and blow the atomized metal against the surface to be coated. The gas usually used is acetylene but other gases may be used. Any desired thickness of metal may be deposited and the metals include steels, ranging from low to high carbon content, various brass and bronze compositions, babbitt metal, tin, zinc, lead, nickel, copper and aluminum. The movement of the spray gun, in covering a given surface, is controlled either mechanically or by hand. In enlarging worn or undersize shafts, spindles, etc., it is common practice to clamp the gun in a lathe toolholder and use the feed mechanism to traverse the gun at a uniform rate while the metal is being deposited upon the rotating workpiece. The spraying operation may be followed by machining or grinding to obtain a more precise dimension.

Some typical production applications using the wire process are the coating of automotive exhaust valves, refinishing of transfer ink rollers for the printing industry, and the rebuilding of worn truck clutch plates. Other production applications include the metallizing of glass meter box windows, the spraying of aluminum onto cloth gauze to produce electrolytic condenser plates, and the spraying of zinc or copper for coating ceramic insulators.

With another type of equipment, metal, refractory, and ceramic powders are used instead of wire. Ordinarily this equipment employs the use of two gases, oxygen and a fuel gas. The fuel gas is usually acetylene but in some instances hydrogen may be used. When hand held, a small reservoir supplies the powder to the equipment but a larger reservoir is used for lathe-mounted equipment or for large-scale production work. The four basic types of coating powders used with this equipment are ceramics, oxidation-resistant metals and alloys, self-bonding alloys, and alloys for fused coatings. These powders are used to produce wear-resistant, corrosion-resistant, heat-resistant, and electrically conductive coatings.

Still other equipment employs the use of plasma flame with which vapors of materials are raised to a higher energy level than the ordinary gaseous state. Its use raises the temperature ceiling and provides a controlled atmosphere by permitting employment of an inert or chemically inactive gas so that chemical action, such as oxidation, during the heating and application of the spray material can be controlled. The temperatures that can be obtained with commercially available plasma equipment often exceed 30,000 degrees F. but for most plasma flame spray processes the temperature range of from 12,000 to 20,000 degrees F. is optimum. Plasma flame spray materials include alumina, zirconia, tungsten, molybdenum, tantalum, copper, aluminum, carbides, and nickel-base alloys.

Regardless of the equipment used, what is important is the proper preparation of the surface that will receive the sprayed coating. Preparation activities include the degreasing or solvent cleaning of the surface, undercutting of the surface to provide room for the proper coating thickness, abrasive or grit blasting the substrate to provide a roughened surface, grooving (in the case of flat surfaces) or rough threading (in the case of cylindrical work) the surface to be coated, preheating the base metal. Methods of obtaining a bond between the sprayed material and the substrate are: heating the base, roughening the base, or spraying a "self-bonding" material onto a smooth surface; however, heating alone is seldom used in machine element work as the elevated temperatures required to obtain the proper bond causes problems of warpage and surface corrosion.

# METAL JOINING, CUTTING, AND SURFACING

Metals may be joined without using fasteners by employing soldering, brazing, and welding. Soldering involves the use of a non-ferrous metal whose melting point is below that of the base metal and in all cases below 800 degrees F. Brazing entails the use of a non-ferrous filler metal with a melting point below that of the base metal but above 800 degrees F. In fusion welding, abutting metal surfaces are made molten, are joined in the molten state and then allowed to cool. The use of a filler metal and the application of pressure are considered to be optional in the practice of fusion welding.

## Soldering

Soldering employs lead- or tin-base alloys with melting points below 800 degrees F. and is commonly referred to as soft soldering. Use of hard solders, silver solders and spelter solders which have silver, copper, or nickel bases and have melting points above 800 degrees F. is known as brazing. Soldering is used to provide a convenient joint that does not require any great mechanical strength. It is used in a great many instances in combination with mechanical staking, crimping or folding; the solder being used only to seal against leakage, or to assure electrical contact. The accompanying table gives some of the properties and uses of various solders that are available.

**Forms Available.** — Soft solders can be obtained in bar, cake, wire, pig, slab, ingot, ribbon, segment, powder and foil-form for various uses to which they are put. In bar form they are commonly used for hand soldering. The pigs, ingots, and slabs are used in operations that employ melting kettles. The ribbon, segment, powder and foil forms are used for special applications and the cake form is used for wiping. Wire forms are either solid or they contain acid or rosin cores for fluxing. These wire forms, both solid and core-containing, are used in hand and automatic machine applications. Prealloyed powders, suspended in a fluxing medium, are frequently applied by brush and upon heating, consistently wet the solderable surfaces to produce a satisfactory joint.

**Fluxes for Soldering.** — The surfaces of the metals being joined in the soldering operation must be clean in order to obtain an efficient joint. Fluxes clean the surfaces of the metal in the joint area by removing the oxide coating present, keep the area clean by preventing formation of oxide films, and lower the surface tension of the solder thereby increasing its wetting properties. Rosin, tallow, and stearin are mild fluxes which prevent oxidation but are not too effective in removing oxides present. Rosin is used for electrical applications since the residue is non-corrosive and non-conductive. Zinc chloride and ammonium chloride (sal ammoniac), used separately or in combination, are common fluxes that remove oxide films readily. The residue from these fluxes may in time cause trouble, due to their corrosive effects, if they are not removed or neutralized. Washing with water containing about 5 ounces of sodium citrate (for non-ferrous soldering) or 1 ounce of trisodium phosphate (for ferrous and non-ferrous soldering) per gallon followed by a clear water rinse or washing with commercial water-soluble detergents are methods of inactivating and removing this residue.

**Methods of Application.** — Solder is applied using a soldering iron, a torch, a solder bath, electric induction or resistance or heating, a stream of hot neutral gas or by wiping. Clean surfaces which are hot enough to melt the solder being applied or accept molten solder are necessary to obtain a good clean bond. Parts being soldered should be free of oxides, dirt, oil, and scale. Scraping and the use of abrasives as well as fluxes are resorted to for preparing surfaces for soldering. The

Properties of Soft Solder Alloys (Appendix, ASTM:B 32-70)

| Sn | Pb | Sb | Ag | Specific Gravity† | Solidus | Liquidus | Uses |
|----|----|----|----|----|----|----|----|
| 70 | 30 | ... | ... | 8.32 | 361 | 378 | For coating metals. |
| 63 | 37 | ... | ... | 8.40 | 361 | 361 | As lowest melting solder for dip and hand soldering methods. |
| 60 | 40 | ... | ... | 8.65 | 361 | 374 | "Fine Solder." For general purposes, but particularly where the temperature requirements are critical. |
| 50 | 50 | ... | ... | 8.85 | 361 | 421 | For general purposes. Most popular of all. |
| 45 | 55 | ... | ... | 8.97 | 361 | 441 | For automobile radiator cores and roofing seams. |
| 40 | 60 | ... | ... | 9.30 | 361 | 460 | Wiping solder for joining lead pipes and cable sheaths. For automobile radiator cores and heating units. |
| 35 | 65 | ... | ... | 9.50 | 361 | 477 | General purpose and wiping solder. |
| 30 | 70 | ... | ... | 9.70 | 361 | 491 | For machine and torch soldering. |
| 25 | 75 | ... | ... | 10.00 | 361 | 511 | For machine and torch soldering. |
| 20 | 80 | ... | ... | 10.20 | 361 | 531 | For coating and joining metals. For filling dents or seams in automobile bodies. |
| 15 | 85 | ... | ... | 10.50 | 440§ | 550 | For coating and joining metals. |
| 10 | 90 | ... | ... | 10.80 | 514§ | 570 | For coating and joining metals. |
| 5 | 95 | ... | ... | 11.30 | 518 | 594 | For coating and joining metals. |
| 40 | 58 | 2 | ... | 9.23 | 365 | 448 | Same uses as (50-50) tin-lead but not recommended for use on galvanized iron. |
| 35 | 63.2 | 1.8 | ... | 9.44 | 365 | 470 | For wiping and all uses except on galvanized iron. |
| 30 | 68.4 | 1.6 | ... | 9.65 | 364 | 482 | For torch soldering or machine soldering, except on galvanized iron. |
| 25 | 73.7 | 1.3 | ... | 9.96 | 364 | 504 | For torch and machine soldering, except on galvanized iron. |
| 20 | 79 | 1 | ... | 10.17 | 363 | 517 | For machine soldering and coating of metals, tipping, and like uses, but not recommended for use on galvanized iron. |
| 95 | .... | 5 | ... | 7.25 | 452 | 464 | For joints on copper in electrical, plumbing and heating work. |
| ... | 97.5 | ... | 2.5 | 11.35 | 579 | 579 | For use on copper, brass, and similar metals with torch heating. Not recommended in humid environments due to its known susceptibility to corrosion. |
| 1 | 97.5 | ... | 1.5 | 11.28 | 588 | 588 | For use on copper, brass, and similar metals with torch heating. |

* Abbreviations of alloying elements are as follows: Sn, tin; Pb, lead; Sb, antimony; and Ag, silver.

† The specific gravity multiplied by 0.0361 equals the density in pounds per cubic inch.

‡ The alloys are completely solid below the lower point given, designated "solidus," and completely liquid above the higher point given, designated "liquidus." In the range of temperatures between these two points the alloys are partly solid and partly liquid.

§ For some engineering design purposes, it is well to consider these alloys as having practically no mechanical strength above 360 degrees F.

procedures followed in soldering aluminum, magnesium and stainless steel differ somewhat from conventional soldering techniques and are indicated in the material which follows:

*Soldering Aluminum:* Two properties of aluminum which tend to make it more difficult to solder are its high thermal conductivity and the tenacity of its ever-present oxide film. Aluminum soldering is performed in a temperature range of from 550 to 770 degrees F., compared to 375 to 400 degrees F. temperature range for ordinary metals, because of the metal's high thermal conductivity. Two methods can be used, one using flux and one using abrasion. The method employing flux is most widely used and is known as flow soldering. In this method flux dissolves the aluminum oxide and keeps it from re-forming. The flux should be fluid at soldering temperatures so that the solder can displace it in the joint. In the friction method the oxide film is mechanically abraded with a soldering iron, wire brush, or multi-toothed tool while being covered with molten solder. The molten solder keeps the oxygen in the atmosphere from reacting with the newly-exposed aluminum surface; thus wetting of the surface can take place.

The alloys that are used in soldering aluminum generally contain from 50 to 75 per cent tin with the remainder, zinc. The following aluminum alloys are listed in order of ease of soldering: commercial and high-purity aluminum, wrought alloys containing not more than 1 per cent manganese or magnesium, and finally the heat-treatable alloys which are the most difficult. Cast and forged aluminum parts are not generally soldered.

*Soldering Magnesium:* Magnesium is not ordinarily soldered to itself or other metals. Soldering is generally used for filling small surface defects, voids or dents in castings or sheets where the soldered area is not to be subjected to any load. Two solders can be used: one with a composition of 60 per cent cadmium, 30 per cent zinc, and 10 per cent tin has a melting point of 315 degrees F.; the other has a melting point of 500 degrees F. and has a nominal composition of 90 per cent cadmium and 10 per cent zinc.

The surfaces to be soldered are cleaned to a bright metallic luster by abrasive methods before soldering. The parts are preheated with a torch to the approximate melting temperature of the solder being used. The solder is applied and the surface under the molten solder is rubbed vigorously with a sharp pointed tool or wire brush. This action results in the wetting of the magnesium surface. To completely wet the surface, the solder is kept molten and the rubbing action continued. The use of flux is not recommended.

*Soldering Stainless Steel:* Stainless steel is somewhat more difficult to solder than other common metals. This is true because of a tightly adhering oxide film on the surface of the metal and because of its low thermal conductivity. The surface of the stainless steel must be thoroughly cleaned. This can be done by abrasion or by clean white pickling with acid. Muriatic (hydrochloric) acid saturated with zinc or combinations of this mixture and 25 per cent additional muriatic acid, or 10 per cent additional acetic acid, or 10 to 20 per cent additional water solution of ortho-phosphoric acid may all be used as fluxes for soldering stainless steel. Tin-lead solder can be used successfully. Because of the low thermal conductivity of stainless steel a large soldering iron is needed to bring the surfaces to the proper temperature. The proper temperature is reached when the solder flows freely into the area of the joint. Removal of the corrosive flux is important in order to prevent joint failure. Soap and water or a suitable commercial detergent may be used to remove the flux residue.

**Ultrasonic Fluxless Soldering.** — This more recently introduced method of soldering makes use of ultrasonic vibrations which facilitates the penetration of surface films by the molten solder thus eliminating the need for flux. The equipment offered by one manufacturer consists of an ultrasonic generator, ultrasonic soldering head which includes a transducer coupling, soldering tip, tip heater, and heating platen. Metals that can be soldered by this method include aluminum, copper, brass, silver, magnesium, germanium, and silicon.

# Brazing

Brazing is a metal joining process which uses a non-ferrous filler metal with a melting point below that of the base metals but above 800 degrees F. The filler metal wets the base metal when molten in a manner similar to that of a solder and its base metal. There is a slight diffusion of the filler metal into the hot, solid base metal or a surface alloying of the base and filler metal. The molten metal flows between the close-fitting metals because of capillary forces.

**Filler Metals for Brazing Applications.** — Brazing filler metals have melting points that are lower than those of the base metals being joined and have the ability when molten to flow readily into closely fitted surfaces by capillary action. The commonly used brazing metals may be considered as grouped into the seven standard classifications shown in Table 1. These are aluminum-silicon; copper-phosphorus; silver; nickel; copper and copper-zinc; magnesium; and precious metals. The accompanying Table 1 gives the designations, nominal compositions, physical properties, and uses of these classifications.

The solidus and liquidus are given in the table instead of the melting and flow points in order to avoid confusion. The solidus is the highest temperature at which the metal is completely solid or in other words the temperature above which the melting starts. The liquidus is the lowest temperature at which the metal is completely liquid or putting it another way the temperature below which the solidification starts.

**Fluxes for Brazing.** — In order to obtain a sound joint the surfaces in and adjacent to the joint must be free from dirt, oil, and oxides or other foreign matter at the time of brazing. Cleaning may be achieved by chemical or mechanical means. Some of the mechanical means employed are filing, grinding, scratch brushing and machining. The chemical means include the use of trisodium phosphate, carbon tetrachloride, and trichloroethylene for removing oils and greases. If the surface has been machined and a coolant or cutting oil used, then these materials must also be removed prior to the brazing operation.

Fluxes are used mainly to prevent the formation of oxides and to remove any oxides on the base and filler metals. They also promote free flow of the filler metal during the course of the brazing operation. They are made available in the following forms: powders, pastes or solutions, gases or vapors, and as coatings on the brazing rods. In the powder form a flux can be sprinkled along the joint, provided that the joint has been preheated sufficiently to permit the sprinkled flux to adhere and not be blown away by the torch flame during brazing. A thin paste or solution is easily applied and when spread on evenly, with no bare spots, gives a very satisfactory flux coating. Gases or vapors are used in controlled atmosphere furnace brazing where large amounts of assemblies are mass-brazed. Coatings on the brazing rods protect the filler metal from becoming oxidized and eliminate the need for dipping rods into the flux but it is recommended that flux be applied to the base metal since it may become oxidized in the heating operation. No matter which flux is used, however, it performs its task only if it is chemically active at the brazing temperature.

Chemical compounds incorporated into brazing fluxes include borates (sodium, potassium, lithium, etc.), fused borax, fluoborates (potassium, sodium, etc.), fluorides (sodium, potassium, lithium, etc.), chlorides (sodium, potassium, lithium), acids (boric, calcined boric acid), alkalies (potassium hydroxide, sodium hydroxide), wetting agents, and water (either as water of crystallization or as an addition for paste fluxes).

The accompanying Table 2 provides a guide which will aid in the selection of brazing fluxes that are available commercially.

**Methods of Steadying Work for Brazing.** — Pieces to be joined by brazing after being properly jointed may be held in a stable position by means of clamping devices, spot welds, or mechanical means such as crimping, staking or spinning. When using clamping devices care must be taken to avoid the use of devices containing springs for applying pressure because springs tend to lose their properties under the influence of heat. Care must also be taken to be sure that the clamping devices are no larger than is necessary for strength considerations, since a large metal mass in contact with the base metal near the brazing area would tend to conduct heat away from the area too quickly and result in an inefficient braze. Thin sections that are to be brazed are frequently held together by spot welds. It must be remembered that these spot welds may interfere with the flow of the molten brazing alloy and appropriate steps taken to be sure that the alloy is placed where it can flow into all portions of the joint.

**Methods of Supplying Heat for Brazing.** — The methods of supplying heat for brazing form the basis of the classification of the different brazing methods and are given as follows:

*Torch or Blowpipe Brazing:* Air-gas, oxy-acetylene, air-acetylene and oxy-other fuel gas blowpipes are used to bring the areas of the joint and the filler material to the proper heat of brazing. The flames should generally be neutral or slightly reducing but in some instances some types of bronze welding require a slightly oxidizing flame.

*Dip Brazing:* Baths of molten alloy covered with flux or baths of molten salts are used. The parts, properly jigged, are dipped into the baths for brazing. In the case of the salt bath the filler metal has previously been inserted between the parts being joined or if in the form of wire has been wrapped around the area of the joint; flowing into the joint after immersion in the bath because of capillary action.

*Furnace Brazing:* Furnaces that are heated electrically or by gas or oil with auxiliary equipment that maintains a reducing or protective atmosphere and controlled temperatures therein are used for brazing a large number of units.

*Resistance Brazing:* Heat is supplied by means of hot or incandescent electrodes. The heat is produced by the resistance of the electrodes to the flow of electricity and the filler metal is frequently used as an insert between the parts being joined.

*Induction Brazing:* Parts to be joined are heated by being placed near a coil carrying an electric current. Eddy current losses of the induced electric current are dissipated in the form of heat. This method is both quick and clean.

**Brazing High-speed Steel Tips to Carbon Steel Shanks.** — A method which is used in a large manufacturing plant is as follows: A seat is formed in the tool shank to receive the tip. A welding compound or flux is used in welding the tip to the shank. The flux is placed on the seat of the shank and the tip is then put on top of the flux in the desired position. The tool is placed in the preheating chamber of the furnace and heated to 1550–1600 degrees F., allowing sufficient time for complete penetration of the heat. The tool is then removed from the preheating chamber and the tip is pressed firmly to the seat of the shank to insure a close contact between the two pieces. Then the tool is placed in the main furnace chamber and heated rapidly to a temperature of approximately 2250 to 2400 de-

Table I. Brazing Filler Metals [Based on Specification and Appendix of AWS (American Welding Society) A5.8-69]

| AWS Classification[1] | Nominal Composition,[2] Per Cent | | | | | | Temperature, Degrees F. | | | Standard Form[3] | Uses |
|---|---|---|---|---|---|---|---|---|---|---|---|
| | Ag | Cu | Zn | Al | Ni | Ot | Solidus | Liquidus | Brazing Range | | |
| BAlSi-2 | … | … | … | 92.5 | … | Si; 7.5 | 1070 | 1135 | 1110–1150 | 5 | For joining the following aluminum alloys: 1060, EC, 1100, 3003, 3004, 5005, 5050, 6053, 6061, 6062, 6063, 6951 and cast alloys A612 and C612. All of these filler metals are suitable for furnace and dip brazing. BAlSi-3, -4 and -5 are suitable for torch brazing. Used with lap and tee joints rather than butt joints. Joint clearances run from .006 to .025 inch. |
| BAlSi-3 | … | 4 | … | 86 | … | Si; 10 | 970 | 1085 | 1066–1120 | 2, 3, 5 | |
| BAlSi-4 | … | … | … | 88 | … | Si; 12 | 1070 | 1080 | 1080–1120 | 2, 3, 4, 5 | |
| BAlSi-5 | … | … | … | 90 | … | Si; 10 | 1070 | 1095 | 1090–1120 | 5 | |
| BCuP-1 | … | 95 | … | … | … | P; 5 | 1310 | 1695 | 1450–1700 | 1 | For joining copper and its alloys with some limited use on silver, tungsten and molybdenum. Not for use on ferrous or nickel-base alloys. Are used for cupro-nickels but caution should be exercised when nickel content is greater than 30 per cent. Suitable for all brazing processes. Lap joints recommended but butt joints may be used. Clearances used range from .001 to .005 inch. |
| BCuP-2 | … | 93 | … | … | … | P; 7 | 1310 | 1460 | 1350–1550 | 2, 3, 4 | |
| BCuP-3 | 5 | 89 | … | … | … | P; 6 | 1190 | 1485 | 1300–1500 | 2, 3, 4 | |
| BCuP-4 | 6 | 87 | … | … | … | P; 7 | 1190 | 1335 | 1300–1450 | 2, 3, 4 | |
| BCuP-5 | 15 | 80 | … | … | … | P; 5 | 1190 | 1475 | 1300–1500 | 1, 2, 3, 4 | |
| BAg-1 | 45 | 15 | 16 | … | … | Cd; 24 | 1125 | 1145 | 1145–1400 | 1, 2, 4 | For joining most ferrous and nonferrous metals except aluminum and magnesium. These filler metals have good brazing properties and are suitable for preplacement in the joint or for manual feeding into the joint. All methods of heating may be used. Lap joints are generally used; however, butt joints may be used. Joint clearances of .002 to .005 inch are recommended. Flux is generally required. |
| BAg-1a | 50 | 15.5 | 16.5 | … | … | Cd; 18 | 1160 | 1175 | 1175–1400 | 1, 2, 4 | |
| BAg-2 | 35 | 26 | 21 | … | … | Cd; 18 | 1125 | 1295 | 1295–1550 | 1, 2, 4 | |
| BAg-2a | 30 | 27 | 23 | … | … | Cd; 20 | 1125 | 1310 | 1310–1550 | 1, 2, 4 | |
| BAg-3 | 50 | 15.5 | 15.5 | … | 3 | Cd; 16 | 1170 | 1270 | 1270–1500 | 1, 2, 4 | |
| BAg-4 | 40 | 30 | 28 | … | 2 | … | 1240 | 1435 | 1435–1650 | 1, 2 | |
| BAg-5 | 45 | 30 | 25 | … | … | … | 1250 | 1370 | 1370–1550 | 1, 2 | |
| BAg-6 | 50 | 34 | 16 | … | … | … | 1270 | 1425 | 1425–1600 | 1, 2 | |
| BAg-7 | 56 | 22 | 17 | … | … | Sn; 5 | 1145 | 1205 | 1205–1400 | 1, 2 | |
| BAg-8 | 72 | 28 | … | … | … | … | 1435 | 1435 | 1435–1650 | 1, 2, 4 | |
| BAg-8a | 72 | 27.8 | … | … | … | Li; .2 | 1410 | 1410 | 1410–1600 | 1, 2 | |
| BAg-13 | 54 | 40 | 5 | … | 1 | … | 1325 | 1575 | 1575–1775 | 1, 2 | |
| BAg-13a | 56 | 42 | … | … | 2 | … | 1420 | 1640 | 1600–1800 | 1, 2 | |
| BAg-18 | 60 | 30 | … | … | … | Sn; 10 | 1115 | 1325 | 1325–1550 | 1, 2 | |
| BAg-19 | 92.5 | 7.3 | … | … | … | Li; .2 | 1435 | 1635 | 1610–1800 | 1, 2 | |

For footnotes see bottom of continued table on next page.

Table 1 (Continued). **Brazing Filler Metals** (Based on Specification and Appendix of AWS A5.8-69)

| AWS Classification[1] | Nominal Composition[2] Per Cent | | | | | | Temperature, Degrees F. | | | Standard Form[3] | Uses |
|---|---|---|---|---|---|---|---|---|---|---|---|
| | Ni | Cu | Cr | B | Si | Ot | Solidus | Liquidus | Brazing Range | | |
| BNi-1 | 74 | ... | 14 | 3.5 | 4 | Fe, 4.5 | 1790 | 1900 | 1950–2200 | 1, 2, 3, 4 | For brazing AISI 300 and 400 series stainless steels, and nickel- and cobalt-base alloys. Particularly suited to vacuum systems and vacuum tube applications because of their very low vapor pressure. The limiting element is chromium in those alloys in which it is employed. |
| BNi-2 | 82.5 | ... | 7 | 3 | 4.5 | Fe, 3 | 1780 | 1830 | 1850–2150 | 1, 2, 3, 4 | |
| BNi-3 | 91 | ... | ... | 3 | 4.5 | Fe, 1.5 | 1800 | 1900 | 1850–2150 | 1, 2, 3, 4 | |
| BNi-4 | 93.5 | ... | ... | 1.5 | 3.5 | Fe, 1.5 | 1800 | 1950 | 1850–2150 | 1, 2, 3, 4 | |
| BNi-5 | 71 | ... | 19 | ... | 10 | ... | 1975 | 2075 | 2100–2200 | 1, 2, 3, 4 | |
| BNi-6 | 89 | ... | ... | ... | ... | P, 11 | 1610 | 1610 | 1700–1875 | 1, 2, 3, 4 | |
| BNi-7 | 77 | ... | 13 | ... | ... | P, 10 | 1630 | 1630 | 1700–1900 | 1, 2, 3, 4 | |
| BCu-1 | ... | 100 | ... | ... | ... | ... | 1980 | 1980 | 2000–2100 | 1, 2, 3 | For joining various ferrous and nonferrous metals. They can also be used with various brazing processes. Avoid overheating the Cu-Zn alloys. Lap and butt joints are commonly used. |
| BCu-1a | ... | 99 | ... | ... | ... | Ot, 1 | 1980 | 1980 | 2000–2100 | 4 | |
| BCu-2 | ... | 86.5 | ... | ... | ... | O, 13.5 | 1980 | 1980 | 2000–2100 | 6 | |
| RBCuZn-A | ... | 59 | ... | ... | ... | Zn, 41 | 1630 | 1650 | 1670–1750 | 1, 2, 3 | |
| RBCuZn-D | 10 | 48 | ... | ... | ... | Zn, 42 | 1690 | 1715 | 1720–1800 | 1, 2, 3 | |
| BMg-1 | ... | ... | ... | ... | ... | [4] | 830 | 1100 | 1120–1160 | 2, 3 | BMg-1 and BMg-2a are used for joining AZ10A, K1A, and M1A magnesium-base metals. BMg-2a is also used for joining AZ31B and ZE10A metals. Joint clearances used run .004 to .010 inch. |
| BMg-2a | ... | ... | ... | ... | ... | [5] | 770 | 1050 | 1080–1130 | 1, 2, 3, 4 | |
| BAu-1 | ... | 63 | ... | ... | ... | Au, 37 | 1815 | 1860 | 1860–2000 | 2, 4, 5 | For brazing of iron, nickel, and cobalt-base metals where resistance to oxidation or corrosion is required. Low rate of interaction with base metal facilitates use on thin base metals. Used with induction, furnace, or resistance heating in a reducing atmosphere or in a vacuum and with no flux. For other applications, a borax-boric acid flux is used. |
| BAu-2 | ... | 20.5 | ... | ... | ... | Au, 79.5 | 1635 | 1635 | 1635–1850 | 2, 4, 5 | |
| BAu-3 | 3 | 62.5 | ... | ... | ... | Au, 34.5 | 1785 | 1885 | 1885–1995 | 2, 4, 5 | |
| BAu-4 | 18.5 | ... | ... | ... | ... | Au, 81.5 | 1740 | 1740 | 1740–1840 | 2, 4, 5 | |

[1] These classifications contain chemical symbols preceded by "B" which stands for brazing filler metal.   [2] These are nominal compositions. Trace elements may be present in small amounts and are not shown. Abbreviations used are: Ag, silver; Cu, copper; Zn, zinc; Al, aluminum; Ni, nickel; Ot, other; Si, silicon; P, phosphorus; Cd, cadmium; Sn, tin; Li, lithium; Cr, chromium; B, boron; Fe, iron; O, oxygen; Mg, magnesium; and Au, gold.   [3] Numbers specify standard forms as follows: 1, strip; 2, wire; 3, rod; 4, powder; 5, sheet; and 6, paste.   [4] Al, 9; Zn, 2; Mg, 89.   [5] Al, 12; Zn, 5; Mg, 83.

Table 2.   Guide to Selection of Brazing Filler Metals and Fluxes

| Base Metals Being Brazed | Filler Metals Recommended* | Flux | | | | |
|---|---|---|---|---|---|---|
| | | American Welding Society Brazing Flux Type No. | Effective Temperature Range, Degrees F. | Ingredients Contained Therein | Form Supplied In | Method of Use† |
| All brazeable aluminum alloys | BAlSi | 1 | 700 to 1190 | Chlorides Fluorides | Powder | 1, 2 3, 4 |
| All brazeable magnesium alloys | BMg | 2 | 900 to 1200 | Chlorides Fluorides | Powder | 3, 4 |
| Alloys such as aluminum-bronze; aluminum-brass containing additions of aluminum of 0.5 per cent or more | BCuZn BCuP | 4§ | 1050 to 1800 | Chlorides Fluorides Borates Wetting agent | Paste or Powder | 1, 2, 3 |
| Titanium and zirconium in base alloys | BAg | 6 | 700 to 1600 | Chlorides Fluorides Wetting agent | Paste or Powder | 1, 2, 3 |
| Any other brazeable alloys not listed above | All brazing filler metals except BAlSi and BMg | 3 | 700 to 2000 | Boric acid Borates Fluorides Fluoborates (must contain fluorine compound) Wetting agent | Paste, Powder, or Liquid | 1, 2, 3 |
| | All brazing filler metals except BAlSi, BMg, and BAg 1 through BAg 7 | 5 | 1000 to 2200 | Borax Boric acid Borates Wetting agent No fluorine in any form | Paste, Powder, or Liquid | 1, 2, 3 |

* Abbreviations used in this column are as follows: B, brazing filler metal; Al, aluminum; Si, silicon; Mg, magnesium; Cu, copper; Zn, zinc; P, phosphorus; and Ag, silver.

† Explanation of numbering system used is as follows: 1 — dry powder is sprinkled in joint region; 2 — heated metal filler rod is dipped into powder or paste; 3 — flux is mixed with alcohol, water, monochlorobenzene, etc. to form a paste or slurry; 4 — flux is used molten in a bath.

§ Types 1 and 3 fluxes, alone or in combination, may be used with some of these base metals also.

grees F., depending upon the kind of material used for the tip and its hardening requirements. The tool is next removed from the furnace and sufficient pressure applied to the tip to insure perfect cohesion. The press used for this purpose should be equipped with a pivoted pressure shoe and this shoe must be preheated to prevent cracking of the tip.

The hardening is accomplished at the same time as the tipping operation when tools are tipped with low-cobalt high-speed and high-cobalt high-speed steel. Tools that cannot be ground after hardening are often heated in barium chloride or some similar salt bath. After the pressing and welding operation, the tool is cooled to room temperature under an air blast or quenched in oil. It is advisable to maintain the oil quenching bath at a temperature of 150 to 200 degrees F. After the tipping operation, the tool should be reheated uniformly in an open furnace to a temperature of 1050 to 1150 degrees F. and allowed to cool in air. The hardened tools should have a minimum Rockwell C hardness of 63. Tips are cut from bar stock according to dimensions on standard detail drawings. The material generally used for the shanks of tools tipped with the cutting materials regularly employed contains 0.50 to 0.63 per cent carbon; 0.60 to 0.90 per cent manganese; 0.04 per cent phosphorus; and 0.15 per cent silicon.

**Brazing Carbide Tips to Steel Shanks.** — Sintered carbide tips or blanks are attached to steel shanks by brazing. Shanks usually are made of low-alloy steels having carbon contents ranging from 0.40 to 0.60 per cent. One prominent manufacturer of carbide tools recommends the following shank steels in the order listed: (1) High-silicon steel containing approximately 0.55 carbon, 0.85 manganese, 0.30 vanadium, 2.10 silicon, 0.25 chromium, and maximum sulphur and phosphorus contents of 0.025; (2) S.A.E. 2340 steel; (3) any low-alloy steel having 0.40 to 0.60 carbon.

*Shank Preparation:* The carbide tip usually is inserted into a milled recess or seat, but some prefer to omit the recess and braze the tip to the top of the shank. When a recess is used, the bottom should be flat to provide a firm even support for the tip. The corner radius of the seat should preferably be somewhat smaller than the radius on the tip to avoid contact and insure support along each side of the recess.

*Cleaning:* All surfaces to be brazed must be absolutely clean. Surfaces of the tip may be cleaned by grinding lightly or by sand-blasting. Cleaning with carbon tetrachloride is also recommended.

*Brazing Materials and Equipment:* The brazing metal may be copper, naval brass such as Tobin bronze, or silver solder. A flux such as borax is used to protect the clean surfaces and prevent oxidation. Heating may be done either in a furnace or by means of an oxy-acetylene torch or an oxy-hydrogen torch. Copper brazing usually is done in a furnace, although an oxy-hydrogen torch with excess hydrogen is sometimes used. An oxy-acetylene torch usually is employed for silver brazing or soldering.

*Brazing Procedure:* One method of brazing with a torch is to first place a thin sheet material, such as copper foil, around and beneath the carbide tip, the top of which is covered with flux. The flame is applied to the under side of the tool shank, and, when the materials melt, the tip is pressed firmly into its seat with tongs or with the end of a rod. If the brazing material is in the form of wire or rod, this may be used to coat or tin the surfaces of the recess after the flux melts and runs freely. The tip is then inserted, flux is applied to the top, and the heating is continued until the coatings melt and run freely. A firm which supplies carbide tips with nickel-coated surfaces ready for silver soldering, suggests the following procedure: The tip, after coating with flux, is placed in the recess and the shank end is heated. Then a small piece of silver solder, having a melting point of 1325 degrees F., is placed on top of the tip. When this solder melts, it runs over the nickel-coated surfaces while the tip is held firmly in its seat. In all carbide tip brazing, the brazed tool should be cooled slowly to avoid cracking due to unequal contraction between the steel and carbide. To insure slow cooling, the tool may be buried in powdered charcoal, graphite, asbestos, mica, or lime.

## Welding

Welding or more specifically fusion welding may be defined as a group of processes in which metals are joined by bringing abutting surfaces to a molten state. Welding may be performed with or without the application of pressure and with or without the use of a filler metal. Heat may be provided by an electric arc, a gas flame, a chemical reaction, or the electrical resistance of the metals being joined to current passed through the joint.

The use of welding offers many advantages. It lends flexibility to machine designs. It facilitates lightweight construction and permits the use of standard rolled shapes. Standard shapes may be rolled or formed and joined with welding to cut costs by reducing material, machining, and finishing. As a joining process it is used not only for fabrication but also for the repair and maintenance of broken and worn parts. It requires relatively low capital investment costs.

**Welding Processes.** — Welding processes may be classified into the following broad categories: arc welding, gas welding, resistance welding, thermit welding, induction welding, and forge welding. Each broad category may be further modified into other classifications, i.e., shielded metal-arc, unshielded carbon-arc, spot, seam, oxy-acetylene, etc. Descriptions and uses according to these modifications are given in the accompanying Table 1.

**Materials Used in Welds.** — Materials used in welds are available in the form of electrodes or welding rods. Electrodes are used with arc-welding equipment and welding rods with gas welding equipment. Coiled wire for use in automatic welding machinery is also employed as a weld material. Electrodes and welding rods are available in a great many different chemical compositions which satisfy practically all the welding situations encountered. Although the electrodes are available as proprietary alloys a number of them have been classified jointly by the American Welding Society and American Society for Testing Materials and assigned AWS-ASTM classification numbers or letters.

The AWS-ASTM classification numbers or letters indicate one or more of the following: the type of covering or coating on the electrode; the chemical composition of the filler metal, either in rod or electrode form, or as deposited; the minimum tensile and yield strengths and ductility of the filler metal as deposited; and a range of standard sizes (diameter of core) of a standard length or a range of standard lengths. Different coatings applied to electrodes provide shielding to protect the arc from the atmosphere, and determine operating characteristics of the electrode. Some coatings contain iron powder that deposits as filler metal with the core wire in the weld. The AWS-ASTM classification numbers, compositions, properties and related information are given in specifications published by the American Welding Society and American Society for Testing Materials.

Electrodes contain fluxes in their coatings. Submerged arc fluxes are externally applied granular materials which completely cover and hide the arc. In gas welding, fluxes are used in the form of liquids, slurries or powders. Liquids or slurries may be used as a dip for welding rods or they may be used to paint the surface of the base metal. When powders are used the rod is heated and then dipped into the powder. The powder may also be sprinkled onto the base metal. Inert gases

Table 1.    Welding Processes — Designation, Description, and Use

| Designation | | Description | Uses |
|---|---|---|---|
| Induction Welding | | Heat result of electric current induced in pieces to be joined. Workpieces are placed in or very close to inductor coil. | For welding steel pieces in thicknesses of from 0.010 inch to ⅜ inch and in diameters up to 3 inches by an economical mass-production method. |
| Arc Welding — Shielded Metal Electrode | Shielded Stud | Heat result of an electric arc struck between a metal stud and work. At proper temperature stud is brought against work under nominal pressure of stud welding gun. Shielding provided by ceramic ferrule surrounding the stud which reduces chance of oxidation. | For welding studs or fasteners onto various items such as decking, machine bases, and engine blocks. Used in the shipbuilding, railroad, building construction, equipment, and automotive industries. Both manual and machine welding performed. |
| | Submerged Arc | Heat result of an electric arc or arcs maintained between bare or lightly coated electrodes and work. Shielding provided by decomposition of fusible granular material. Pressure not used. Filler metal obtained from electrode or supplementary welding rod. | Most commonly used for welding low-carbon and high-strength, low-alloy steels. Straight chromium and austenitic chromium-nickel steels are weldable in all thicknesses. Higher alloy air-hardening steels weldable using special techniques. Welding done automatically. |
| | Inert Gas Metal-Arc | Heat result of an electric arc struck between an electrode (usually tungsten) and work. Shielding provided by inert gases (helium or argon). Use of pressure and filler metal optional. | For welding sheet steel, aluminum, and magnesium. Also for welding deoxidized copper, dissimilar metals, and for surfacing. Both manual and machine welding are performed. |
| | Atomic Hydrogen | Heat result of an electric arc maintained between two electrodes in an atmosphere of hydrogen which provides shielding. Use of pressure and filler metal optional. | For welding practically all non-ferrous and ferrous metals and alloys. Flux required for aluminum, copper, copper alloys and some stainless steels. For repairing steel molds and dies. Production welding of tubing and fabrication of thin stock. Both manual and automatic welding are performed. |
| | Shielded Metal-Arc | Heat result of electric arc struck between flux-coated electrode and base metal. Shielding provided by gases of decomposed flux coating. No pressure used. Filler metal supplied by consumable electrode. | For welding all ferrous metals particularly cast-iron, low-carbon, medium-carbon, high-carbon, low-alloy high-tensile, stainless, corrosion-resistant and copper bearing steels; and also the following non-ferrous metals: copper alloys, aluminum, nickel, nickel alloys, and bronze. These welding processes are used very widely, the main fields being structural, transportation, piping, appliance and machinery fabrication. Both manual and machine welding is performed employing these processes. |

**Table 1 (Continued). Welding Processes — Designation, Description, and Use**

| Designation | | Description | Uses |
|---|---|---|---|
| Arc Welding—Unshielded Carbon Electrode | Carbon-Arc | Heat result of an electric arc struck between an electrode and work. Use of pressure and filler metal optional. | For welding steel, galvanized steel, copper, brass, bronze and aluminum. Both manual and automatic set-ups are used in these welding processes. All of these processes are also used for preheating small areas which are to be welded later, as well as for cutting metals. The twin-arc process may be used for brazing and soldering copper, galvanized and tinned parts both in light and heavy gages. |
| | Twin Carbon-Arc | Heat result of an electric arc maintained between two electrodes. Use of pressure and filler metal optional. | |
| Gas Welding | Air-Acetylene | Heat result of burning of mixture of air and acetylene. Lower flame temperatures achieved than with other gas welding processes. Pressure not used. Use of filler metal optional. Fluxes may be used. | For welding lead in thicknesses up to ¼ inch. Also used for torch brazing employing silver brazing alloys and for soft-soldering electrical connections and copper pipe joints. Manual equipment is used. |
| | Oxy-Acetylene | Heat result of burning of mixture of oxygen and acetylene. The use of pressure and filler metal optional. Fluxes may be used. | For welding many ferrous and non-ferrous metals. Heat obtained as high as original casting temperature. Metals of considerable thickness may be welded. Used in maintenance and repair. Both manual and machine equipment used. |
| | Oxy-Hydrogen | Heat result of burning of mixture of oxygen and hydrogen. The use of filler metal is optional. No pressure is used. Lower flame temperature than oxy-acetylene flame. | For low-temperature applications such as welding low-melting point metals, brazing and braze-welding. Aluminum, magnesium and lead are welded using this process. Manual equipment used. |
| | Gas Pressure | Heat result of burning of mixture of acetylene and oxygen. Pressure used. Filler metal not used. Pressure may be applied while heating (closed joint method) or upsetting may be performed after heating (open-joint method). | For welding many steels of varying carbon contents including also low- and high-alloy steels. Non-ferrous alloys such as Monel, nickel-chromium, and copper-silicon alloys are welded by this process. Machine equipment used. |
| Thermit Welding | Pressure | Heat result of chemical reaction between a metal oxide (usually iron oxide) and aluminum both of which are in a fine-powder form. In the reaction, the aluminum displaces the metal in the oxide to form aluminum oxide. The metal liquefies and is used as a filler metal when a filler metal is used. Use of pressure optional. More non-pressure welding done. | For fabrication and repair of heavy sections of ferrous metals such as heavy machinery; stern frames of ships, sheet steel mill equipment; rail welding for railroads and mines; and locomotive repairs. |
| | Non-pressure | | |

**Table I** (*Continued*). Welding Processes — Designation, Description, and Use

| Designation | | Description | Uses |
|---|---|---|---|
| Resistance Welding | Spot | Heat result of electrical resistance offered by workpieces held under pressure between two electrodes. Weld formed immediately between workpieces. | For welding a great many different metals in a great range of different sizes and shapes. For welding landing gears, wing assemblies in the aircraft industry; bodies, cabs, seats in the automotive industry; trusses and partitions in the building construction industry; telephone equipment instruments in the communications industry; etc. All of these welding processes are performed by machine. |
| | Seam | Heat result of electrical resistance offered by workpieces held under pressure between two circular electrodes. Resulting weld is a continuous series of overlapping spot welds. | |
| | Projection | Heat result of electrical resistance offered by workpieces held under pressure between two electrodes. Welds are formed at projections, or other raised surfaces on workpieces specified by design. | |
| | Flash | Heat result of electrical resistance between two abutting surfaces. Pressure applied after heating which causes flashing, upsetting and an expulsion of metal from joint. | For welding sheets and other extended sections end-to-end. Used in high-production fabrication set-ups. Automotive and aircraft products, appliances and refrigerators are flash welded. Machine equipment used. |
| | Upset | Heat result of electrical resistance between two abutting surfaces. Pressure applied before and maintained during heating period. | For welding small ferrous and non-ferrous strips as well as the welding of longitudinal butt joints in tubing and pipe and transverse butt joints in heavy steel rings. Machine equipment used. |
| | Percussion | Heat result of an electric arc produced by a rapid discharge of stored electrical energy with pressure percussively applied during or immediately following the discharge. | For welding separate pieces of rod, tube or pipe to one another or to flat surfaces. For joining dissimilar metals not economically welded by flash welding method. |
| Forge Welding | Hammer | Heat result of external heating by furnace. Workpieces when at proper temperature are joined by blows of a hand or machine hammer. | Largely replaced by other welding methods but now used for some metal manufacturing and hardware applications. |
| | Die | Heat result of external heating by furnace. When proper temperature is reached, metals being joined are brought together under pressure with a bell or a mandrel and tube rolls. | Used in the fabrication of large diameter water transmission pipes and penstocks. |
| | Roll | Heat result of external heating by furnace. When proper temperature is reached, metals being joined are brought together under pressure between plate rolls. | For welding cladding metal to steel plate in the making of clad steels. |

**Table 2. Recommended Processes for Welding[1]**

| Metals and Alloys | | Shielded metal-arc (coated electrode) | Submerged-arc | Atomic hydrogen | Inert-gas tungsten-arc | Inert-gas metal-arc | Flash welding | Seam welding | Spot welding | Gas welding (oxy-acetylene) | Brazing-furnace | Brazing-torch | Thermit |
|---|---|---|---|---|---|---|---|---|---|---|---|---|---|
| | | Recommendation Based on Metal Used | | | | | | | | | | | |
| Steels | Low carbon; SAE 1010, 1020 | R | R | S | S | S | R | R | R | R | R | S | S |
| | Medium carbon; SAE 1030, 1050 | R | R | S | S | S | R | S | R | R | R | S | S |
| | Wrought alloy; SAE 4130, 4340 | R | R | S | S | S | R | N | R | S | S | N | S |
| | Austenitic stainless; AISI 301, 309, 316 | R | R | R | R | R | R | R | R | S | R,S | S | N |
| | Other stainless; AISI 405, 430 | R | S | S | S | S | S | S | S | S | S | S | N |
| High temperature alloys; 19-9DL, 16-25-6 | | R | S | S | S | S | S | R | S | S | N | N | N |
| Cast iron, gray iron | | S | N | N | S | N | N | D | D | R | N | R | S |
| Non-ferrous Alloys | Aluminum and its alloys | S | N | S | R | R | S | S | R | S | R | R | D |
| | Nickel and its alloys | R | S | S | R | R | S | S | R | S | S | R | N |
| | Copper and its alloys | N | N | N | R | R | S | N | S | S | S | R | N |
| | Magnesium and its alloys | D | D | N | R | S | N | N | S | N | N | N | D |
| | Silver | N | N | R | R | S | S | N | N | R | S | R | N |
| | Gold, platinum, iridium | N | N | R | R | S | S | N | S | R | S | R | N |
| | Titanium and its alloys | D | D | D | R | R | S | S | S | D | S[2] | S[2] | D |
| | Uranium, molybdenum, vanadium, zirconium, tungsten | D | D | N | R | S[2] | S | S | S | N | N | N | D |
| Joint Design[3] | | Recommendation Based on Joint Used | | | | | | | | | | | |
| Butt Joint | Light section | S | S | R | R | N | N | D | D | R | N | S | D |
| | Heavy section | R | R | S | S | R | R | D | D | S | N | S | R |
| Lap Joint | Light section | R | S | S | R | N | D | R | R | R | R | R | D |
| | Heavy section | R | R | S | S | R | D | R | R | S | R | R | D |
| Fillet Joint | Light section | R | S | S | R | N | D | D | D | R | R | R | D |
| | Heavy section | R | R | S | S | R | N | D | D | S | R | R | D |
| Edge Joint | Light section | N | N | R | R | N | D | R | D | R | D | S | D |
| | Heavy section | R | S | S | S | S | D | R | D | S | D | S | D |
| Overlay Welding | | R | R | R | R | R | D | D | D | R | N | S | N |

[1] From data supplied by John J. Chyle, Director of Welding Research, A. O. Smith Corporation, Milwaukee 1, Wis. Symbols used in the table have the following meanings: R — recommended, S — satisfactory, N — not recommended, D — does not apply.

[2] Questionable as to whether it can be satisfactorily welded.

[3] Light sections include sizes from 0.005 inch up to 0.078 inch; heavy sections include sizes 0.078 inch and larger.

**Table 3.** Resistance Welding — Ease of Weldability of Some Common Metals and Alloys[1]

| Metals and Alloys | Aluminum | Brass | Copper | Galvanized Iron | Inconel | Lead | Magnesium | Molybdenum | Monel | Nichrome | Nickel | Nickel Silver | Phosphor Bronze | Stainless Steel | Steel | Tantalum | Terneplate | Tin Plate | Titanium | Tungsten | Zinc |
|---|---|---|---|---|---|---|---|---|---|---|---|---|---|---|---|---|---|---|---|---|---|
| Aluminum | 1 | 4 | 5 | 5 | 6 | 5 | 3 | 6 | 6 | 6 | 6 | 6 | 6 | 5 | 5 | 5 | 6 | 5 | 5 | 6 | 5 |
| Brass | 4 | 1 | 3 | 5 | 5 | 5 | 5 | 5 | 5 | 5 | 3 | 3 | 2 | 5 | 5 | 4 | 4 | 4 | 5 | 5 | 4 |
| Copper | 5 | 3 | 4 | 5 | 5 | 5 | 5 | 5 | 4 | 4 | 4 | 3 | 3 | 5 | 5 | 4 | 5 | 5 | 4 | 5 | 5 |
| Galvanized Iron | 5 | 5 | 5 | 2 | 5 | 4 | 6 | 6 | 4 | 4 | 4 | 5 | 5 | 3 | 3 | 4 | 3 | 3 | 5 | 5 | 3 |
| Inconel | 6 | 5 | 5 | 5 | 1 | 5 | 5 | 3 | 2 | 3 | 2 | 2 | 5 | 2 | 2 | 3 | 4 | 4 | 4 | 3 | 5 |
| Lead | 5 | 5 | 5 | 4 | 5 | 3 | 6 | 6 | 5 | 5 | 5 | 5 | 5 | 5 | 5 | 5 | 5 | 5 | 5 | 5 | 3 |
| Magnesium | 3 | 5 | 5 | 6 | 5 | 6 | 1 | 3 | 6 | 6 | 6 | 6 | 6 | 5 | 5 | 5 | 6 | 5 | 5 | 6 | 5 |
| Molybdenum | 6 | 5 | 5 | 6 | 3 | 6 | 3 | 3 | 3 | 3 | 3 | 4 | 4 | 3 | 3 | 3 | 4 | 4 | 4 | 4 | 5 |
| Monel | 6 | 5 | 4 | 4 | 2 | 5 | 6 | 3 | 1 | 3 | 3 | 3 | 4 | 2 | 2 | 3 | 4 | 4 | 4 | 3 | 5 |
| Nichrome | 6 | 5 | 4 | 4 | 3 | 5 | 6 | 3 | 3 | 1 | 2 | 2 | 4 | 3 | 3 | 4 | 5 | 5 | 4 | 3 | 6 |
| Nickel | 6 | 3 | 4 | 4 | 2 | 5 | 6 | 3 | 3 | 2 | 1 | 2 | 3 | 3 | 3 | 4 | 5 | 4 | 4 | 4 | 6 |
| Nickel Silver | 6 | 3 | 3 | 5 | 2 | 5 | 6 | 4 | 3 | 2 | 2 | 1 | 2 | 4 | 4 | 4 | 4 | 4 | 4 | 5 | 6 |
| Phosphor Bronze | 6 | 2 | 3 | 5 | 5 | 5 | 6 | 4 | 4 | 4 | 3 | 2 | 2 | 5 | 5 | 4 | 4 | 4 | 5 | 5 | 4 |
| Stainless Steel | 5 | 5 | 5 | 3 | 2 | 5 | 5 | 3 | 2 | 3 | 3 | 4 | 5 | 1 | 1 | 2 | 3 | 3 | 3 | 3 | 5 |
| Steel | 5 | 5 | 5 | 3 | 2 | 5 | 5 | 3 | 2 | 3 | 3 | 4 | 5 | 1 | 1 | 1 | 3 | 3 | 3 | 3 | 5 |
| Tantalum | 5 | 4 | 4 | 4 | 3 | 5 | 5 | 3 | 3 | 4 | 4 | 4 | 4 | 2 | 1 | 2 | 5 | 5 | 4 | 4 | 5 |
| Terneplate | 6 | 4 | 5 | 3 | 4 | 5 | 6 | 4 | 4 | 5 | 5 | 4 | 4 | 3 | 3 | 5 | 2 | 3 | 5 | 5 | 3 |
| Tin Plate | 5 | 4 | 5 | 3 | 4 | 5 | 5 | 4 | 4 | 5 | 4 | 4 | 4 | 3 | 3 | 5 | 3 | 3 | 5 | 5 | 5 |
| Titanium | 5 | 5 | 4 | 5 | 4 | 5 | 5 | 4 | 4 | 4 | 4 | 5 | 5 | 3 | 3 | 4 | 5 | 5 | 1 | 5 | 5 |
| Tungsten | 6 | 5 | 5 | 5 | 3 | 5 | 6 | 4 | 3 | 3 | 4 | 4 | 5 | 3 | 3 | 4 | 5 | 5 | 5 | 2 | 5 |
| Zinc | 5 | 4 | 5 | 3 | 5 | 3 | 5 | 5 | 5 | 6 | 6 | 6 | 4 | 5 | 5 | 5 | 3 | 5 | 5 | 5 | 3 |

[1] *Key:* 1 — Welds excellently, 2 — Welds well, 3 — Welds fairly well, 4 — Welds poorly, 5 — Welds very poorly, 6 — Welding not practical. This data was made available by Sciaky Bros., Inc., Chicago 38, Illinois.

such as helium and argon are used to provide a protective shield for the deposition of the filler metal from the electrode.

**Materials That Can Be Welded.** — Almost all metallic materials can be welded, some more easily and by more processes than others. The accompanying Table 2 gives a list of common metals and alloys and their relative ease of weldability with respect to many of the available welding processes. It also lists different types of joints and gives the recommended and satisfactory welding methods that may be used in making these joints. In using the table as a guide to the selection of a possible welding process, it is suggested that the recommended welding processes prescribed by a particular type joint be determined and then compared with the welding processes prescribed as recommended or satisfactory for the metals or alloys to be used.

The accompanying Table 3 gives the ease of weldability of many similar and dissimilar metals and alloys employing the resistance welding processes. It must be remembered that these ratings are nominal and inclusive of the whole range of resistance welding processes i.e., spot, seam, projection, etc. and are therefore just a general indication of weldability. Other factors which have to be considered and which may change the ease of weldability are thickness, plating, treatment (hot or cold), and the shape of the piece itself.

**Typical Weldments.** — Welded design is used for nearly every type of machine, ranging from steel mill equipment to jigs and fixtures. One advantage in designing for welding is that individual sections of a part can be considered as separate components which are then joined into one piece to form the entire machine part. Design is thereby simplified. Some of the elements which can be considered separately are shown in the illustrations which follow giving examples of welded construction and rolled shapes available for use in weldments.

## Checklist of Good Arc-Welding Practices

The checklist of arc-welding practices which follows has been prepared by The Lincoln Electric Company and has been designed to call attention to considerations that affect the quality and cost of arc welding. Anyone using the information in the checklist is less likely to run into difficulties when arc welding and especially those who are novices in the field.

**Work.** — *Material:* For easiest welding, use steel within the following preferred analysis range: carbon, 0.13 to 0.20 per cent; manganese, 0.40 to 0.60 per cent; silicon, 0.10 per cent max.; sulphur, 0.035 per cent max.; and phosphorus, 0.035 per cent max. Steels with a higher alloy content may require special treatment, such as a preheat, in order to produce sound welds.

*Thickness:* Exceptionally thin material (20 gage or thinner) and very heavy material (1 inch or thicker) may require special procedures or fixtures.

*Accessibility:* Joints must be accessible to the operator, if he is to weld them satisfactorily in a reasonable length of time.

*Position:* Material $3/16$ inch and thicker is welded fastest and best in the flat position, while lighter material is best welded in a 45-degree downhill position.

*Fixtures:* Proper fixtures will reduce handling time, improve position, and maintain alignment of the work.

*Cleanliness:* The joint must be clean for best results. Dirty joints are frequently the cause of porous and cracked welds and also increase welding time.

*Rigidity:* Heavy, rigid parts should be welded from the center out toward the edges to avoid locked stresses in the work that may cause weld cracks.

**Examples of Welded Construction — 1**

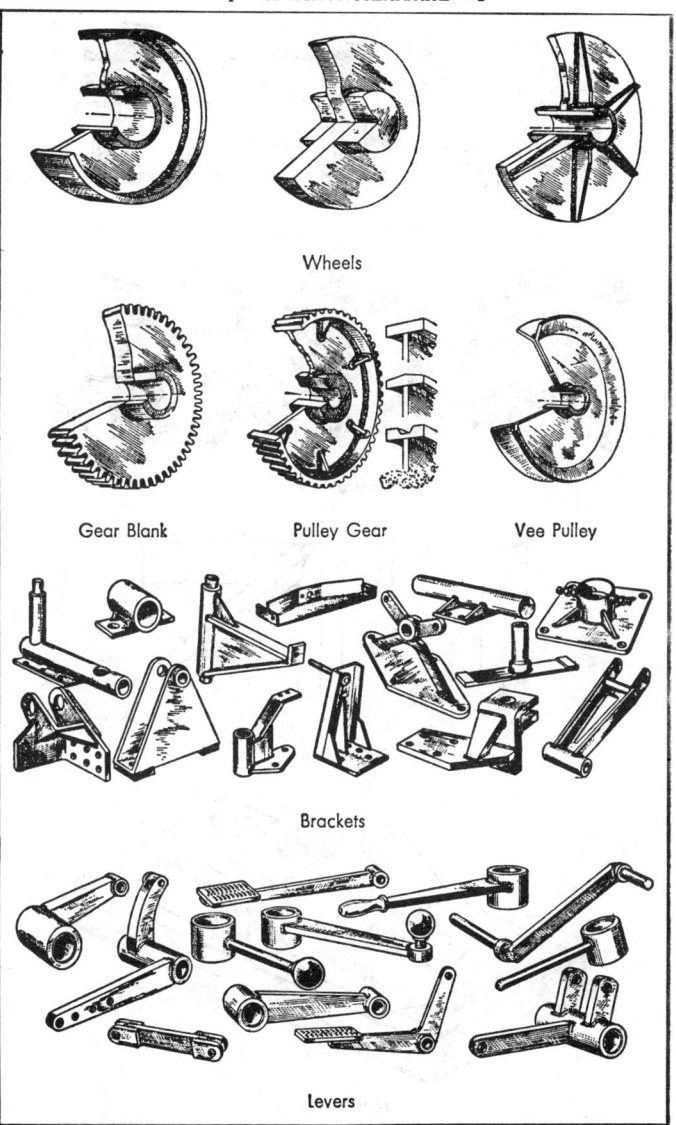

Wheels

Gear Blank      Pulley Gear      Vee Pulley

Brackets

Levers

Examples of Welded Construction — 2

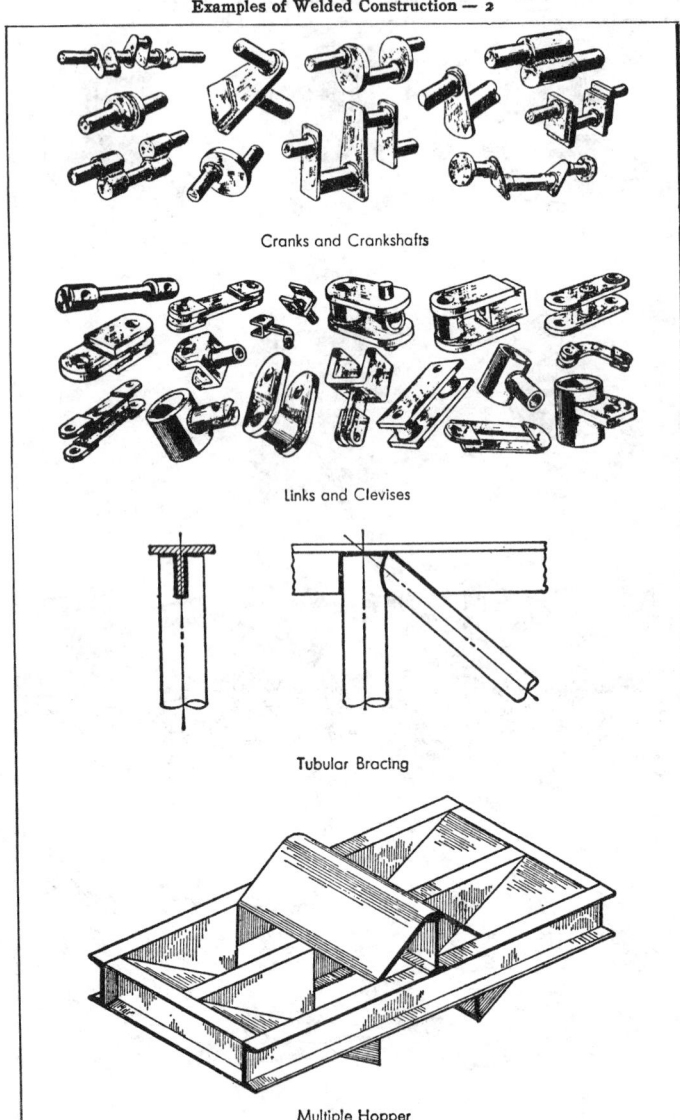

Cranks and Crankshafts

Links and Clevises

Tubular Bracing

Multiple Hopper

## Examples of Welded Construction — 3

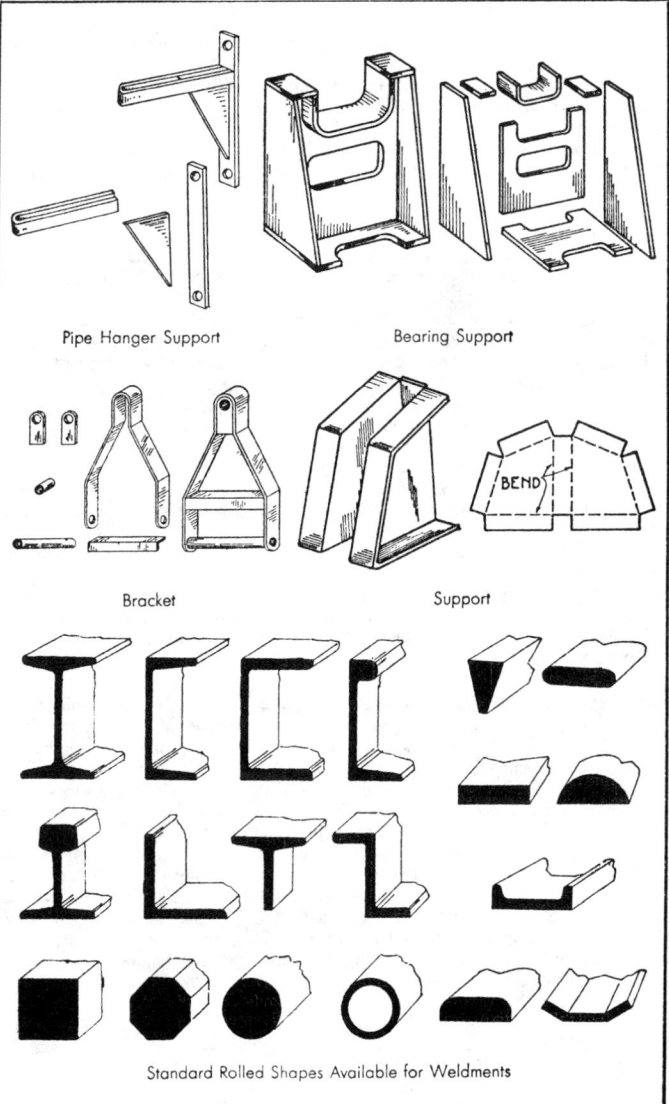

Pipe Hanger Support

Bearing Support

Bracket

Support

BEND

Standard Rolled Shapes Available for Weldments

*Joint Design:* Proper joint design provides a minimum amount of weld metal, yet provides adequate penetration.

**Process.** — *Manual Welding:* Welding with manual electrodes is the most versatile method of welding.

*Automatic Submerged Arc Welding:* Automatic welding is substantially faster than manual welding, but is more limited in application. Quality of deposits is consistently excellent.

*Semi-Automatic Submerged Arc Welding:* Semi-automatic welding combines the advantages of versatility of manual welding and the high speed and consistent quality of automatic welding.

**Welding Materials.** — *Manual Electrode:* For most efficient welding with best quality, select the type and size of the electrode, as well as the amount of current used on it, to meet the requirements of the material, joint, and position of the particular job on which it is being used.

*Electrode and Flux for Automatic and Semi-Automatic Welding:* Selection of electrode and flux for these welding operations affects the quality of the weld, as well as the welding speeds attainable on any particular job that is being performed in the welding shop.

**Welding Procedures.** — *Current:* Use the highest current possible consistent with good bead shape and quality. Alternating current reduces arc blow, while direct current gives best operation on critical applications.

*Travel Speed:* The speed with which the electrode is moved across the work will determine the size and shape of the weld. Too fast a speed produces a rough bead with undercut.

*Grounding:* The location of the ground may affect arc blow, particularly on lighter material.

*Sequence:* The sequence may affect the amount of distortion created by the welding operation. Where distortion is critical, a specific sequence must be established.

*Equipment:* Equipment must be of the proper type, with adequate capacity and controls, that is best suited to the application.

**American National Standard Welding Symbols.** — Graphical symbols for welding provide a means of conveying complete welding information from the designer to the welder by means of drawings. The symbols and their method of use (examples of which are given in the table following this section) are part of the American National Standards Institute's Standard ANSI Y32.3-1969 sponsored by the American Society of Mechanical Engineers and the American Institute of Electrical Engineers and collaborated by the American Welding Society. In the standard a distinction is made between the terms *weld symbol* and *welding symbol.* *Weld symbol* is the ideograph used to indicate the type of weld desired, whereas *welding symbol* connotes a symbol made up of as many as eight parts conveying explicit welding instructions. The eight elements which may appear in a welding symbol are: reference line, arrow, basic weld symbols, dimensions and other data, supplementary symbols, finish symbols, tail, and specification, process or other reference. The standard location of elements of a welding symbol are shown in Fig. 1.

*Reference Line:* This is the basis of the welding symbol. All other elements are oriented with respect to this line. The arrow is affixed to one end and a tail, when necessary, is affixed to the other.

*Arrow:* This connects the reference line to one side of the joint in the case of groove, fillet, flange, and flash or upset welding symbols. This side of the joint is

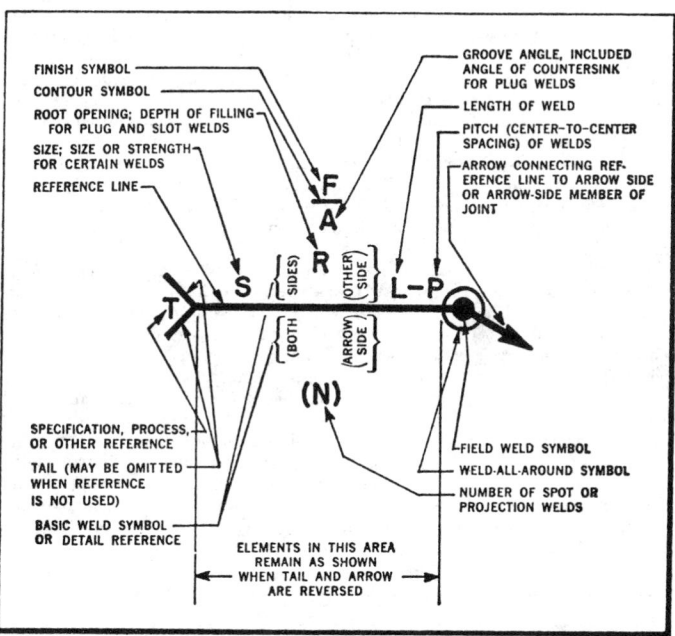

Fig. 1

| BASIC WELD SYMBOLS | | | | | | | |
|---|---|---|---|---|---|---|---|
| FILLET | PLUG OR SLOT | SPOT OR PROJECTION | SEAM | BACK OR BACKING | SURFACING | FLANGE | |
| | | | | | | EDGE | CORNER |
| △ | ▭ | ○ | ⊕ | ⌣ | ⌣⌣ | ⎰⎱ | ⎰⎱ |

| {FLASH OR UPSET} | GROVE | | | | | | |
|---|---|---|---|---|---|---|---|
| SQUARE | V | BEVEL | U | J | | FLARE-V | FLARE-BEVEL |
| ‖ | ∨ | ∨ | ∪ | ∪ | | ⟩⟨ | ⟩⟨ |

| NONPREFERRED SYMBOLS: USE PREFERRED SYMBOL WITH PROCESS REFERENCE IN THE TAIL | | | | |
|---|---|---|---|---|
| ARC·SEAM OR ARC·SPOT | RESISTANCE SPOT | PROJECTION | RESISTANCE SEAM | FLASH OR UPSET |
| ▰ | ✕ | ✕ | ✕✕✕ | ❘ |

Fig. 2

known as the *arrow side* of the joint. The opposite side is known as the *other side* of the joint. In the case of plug, slot, spot, projection, and seam welding symbols the arrow connects the reference line to the outer surface of one of the members of the joint at the center line of the weld. In this case the member to which the arrow points is the *arrow-side* member; the other member is the *other-side* member. In the case of bevel and J-groove welding symbols a two-directional arrow pointing toward a member indicates that the member is to be chamfered.

*Basic weld symbols:* These designate the type of welding to be performed. The basic symbols which are shown in Fig. 2 are placed approximately in the center of the reference line, either above or below it or on both sides of it. Welds on the *arrow side* of the joint are shown by placing the weld symbols on the lower side of the reference line. Welds on the *other side* of the joint are shown by placing the weld symbol on the upper side of the reference line. Where the reference line is vertical, the side to the reader's right is considered to be the lower side of the joint. Nonpreferred symbols are also shown in Fig. 2. Their use is to be avoided. Use of preferred symbols with process reference in the tail is recommended. When encountering resistance and flash or upset welding symbols employing nonpreferred weld symbols it will be noted that the symbols are similarly placed except that they are superimposed directly over the reference line and extend equally above and below it.

*Supplementary symbols:* These convey additional information relative to the extent of welding, where the welding is to be performed, and the contour of the weld bead. The "weld all around" and the "field" symbols are placed at the end of the reference line at the base of the arrow as shown in Fig. 3 which gives the supplementary weld symbols. The contour symbols are placed above (on *other side*) or below (on *arrow side*) the weld symbol.

| SUPPLEMENTARY SYMBOLS | | | | | |
|---|---|---|---|---|---|
| WELD ALL AROUND | FIELD WELD | MELT-THRU | CONTOUR | | |
| | | | FLUSH | CONVEX | CONCAVE |
| | | | — | | |

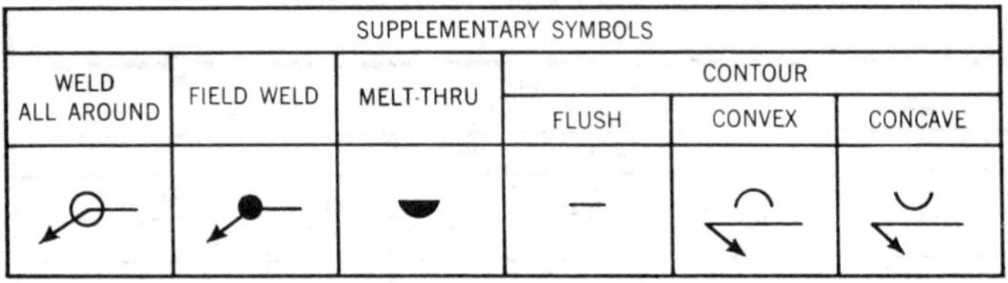

Fig. 3

*Dimensions:* These include the size, length, spacing, etc. of the weld. The size of the weld is given to the left of the basic weld symbol and the length to the right. If the length is followed by a dash and another number, this number indicates the center-to-center spacing of intermittent welds. Other pertinent information such as groove angles, included angle of countersink for plug welds and the designation of the number of spot or projection welds are also located above (on *other side*) or below (on *arrow side*) the weld symbol. The number designating the number of spot or projection welds is always enclosed in parentheses.

*Finish symbols:* These symbols always appear above (on *other side*) or below (on *arrow side*) the supplementary contour symbol. Finish symbols should conform to the user's standard, at least until they are standardized by the American National Standards Institute.

*Tail:* The tail, which appears on the end of the reference line opposite the arrow end, is used when a specification, process or other reference is made in the welding symbol.

*Specification, process or other references:* These are placed in the tail of the welding symbol. They are in accordance with the American standard where they are listed and do not have to be used if a note is placed on the drawing indicating that the welding is done to some specific specification or that instructions are given elsewhere as to the welding procedure to be used. They are used to indicate the various types of welding processes which are not otherwise shown by symbols.

**Welding Codes, Rules Regulations, and Specifications.** — Codes recommending procedures for obtaining specified results in the welding of various structures have been established by societies, institutes, bureaus, and associations, as well as state and federal departments. The latest codes, rules, etc. may be obtained from these agencies, whose names and addresses are listed according to the fields in which their codes, rules, etc., are pertinent.

*Pressure Vessels:*
American Society of Mechanical Engineers, 345 E. 47th St., N. Y., N. Y. 10017.
American Petroleum Institute, 1271 Ave. of the Americas, N. Y., N. Y. 10020.
U. S. Government Printing Office, Washington, D. C. 20402.
American National Standards Institute, 1430 Broadway, N. Y., N. Y. 10018.
Insurance Services Offices, 160 Water St., N. Y., N. Y. 10038.
Naval Ship Engineering Center, Dept. of the Navy, Hyattsville, Md. 20782

*Tanks:*
American Welding Society, 2501 N.W. 7th St., Miami, Florida 33125.

*Piping:*
American National Standards Institute, 1430 Broadway, N. Y., N. Y. 10018.
American Welding Society, 2501 N.W. 7th St., Miami, Florida 33125.
Mechanical Contractors Association of America, 666 Third Ave., N. Y., N. Y. 10017.
Superintendent, State Division of Safety and Hygiene, Columbus, Ohio.

*Structural and Bridges:*
American Welding Society, 2501 N.W. 7th St., Miami, Florida 33125.
Naval Facilities Engineering Command, Dept. of the Navy, Washington, D. C. 20390.
American Institute of Steel Construction, Inc., 101 Park Ave., N.Y., N.Y. 10017.

*Ships:*
American Welding Society, 2501 N.W. 7th St., Miami, Florida 33125.
Lloyd's Register of Shipping, 17 Battery Pl., N. Y., N. Y. 10004.
Naval Ship Engineering Center, Dept. of the Navy, Hyattsville, Md. 20782.
American Bureau of Shipping, 45 Broad St., N. Y., N. Y. 10004.

*Aircraft Construction:*
American Welding Society, 2501 N.W. 7th St., Miami, Florida 33125.
Dept. of Transportation, Federal Aviation Administration, Washington, D. C. 20590.
AF/LGM, Dept. of the Air Force, Washington, D. C. 20330.

*Electric Welding Machinery:*
American National Standards Institute, 1430 Broadway, N. Y., N. Y. 10018.
National Electrical Manufacturers Assn., 155 E. 44th St., N. Y., N. Y. 10017.

*General Rules:*
American Welding Society, 2501 N.W. 7th St., Miami, Florida 33125.
American National Standards Institute, 1430 Broadway, N. Y., N. Y. 10018.

*Welding Electrodes and Rods:*
American Welding Society, 2501 N.W. 7th St., Miami, Florida 33125.
American Society for Testing and Materials, 1916 Race St., Philadelphia, Pa. 19103.
National Electrical Manufacturers Assn., 155 E. 44th St., N. Y., N. Y. 10017.

Application of American National Standard Welding Symbols — 1

| Desired Weld | Symbol | Symbol Meaning |
|---|---|---|
| | | Symbol indicates fillet weld on *arrow side* of the joint. |
| | | Symbol indicates square-groove weld on *other side* of the joint. |
| | | Symbol indicates bevel-groove weld on both sides of joint. Break in arrow indicates bevels on upper member of joint. Breaks in arrows are used on symbols designating bevel and J-groove welds. |
| | | Symbol indicates plug weld on *arrow side* of joint. |
| | | Symbol indicates resistance-seam weld. Weld symbol appears on both sides of reference line pointing up the fact that *arrow* and *other side* of joint references have no significance. |
| | | Symbol indicates electron beam seam weld on *other side* of joint. |

Application of American National Standard Welding Symbols — 2

| Desired Weld | Symbol | Symbol Meaning |
|---|---|---|
| <br>GROOVE WELD MADE BEFORE WELDING OTHER SIDE<br><br>BACK WELD | | Symbol indicates single-pass back weld. |
| | | Symbol indicates a built-up surface $\frac{1}{8}$ inch thick. |
| | | Symbol indicates a bead-type back weld on the *other side* of joint, and a J-groove grooved horizontal member (shown by break in arrow) and fillet weld on *arrow side* of the joint. |
| | | Symbol indicates two fillet welds, both with $\frac{1}{2}$-inch leg dimensions. |
| | | Symbol indicates a $\frac{1}{2}$-inch fillet weld on *arrow side* of the joint and a $\frac{1}{4}$-inch fillet weld on *far side* of the joint. |
| | <br>ORIENTATION SHOWN ON DRAWING | Symbol indicates a fillet weld on *arrow side* of joint with $\frac{1}{4}$- and $\frac{1}{2}$-inch legs. Orientation of legs must be shown on drawing. |

Application of American National Standard Welding Symbols — 3

| Desired Weld | Symbol | Symbol Meaning |
|---|---|---|
| | | Symbol indicates a 24-inch long fillet weld on the *arrow side* of the joint. |
| LOCATE WELDS AT ENDS OF JOINT | | Symbol indicates a series of intermittent fillet welds each 2 inches long and spaced 5 inches apart on centers directly opposite each other on both sides of the joint. |
| LOCATE WELDS AT ENDS OF JOINT | | Symbol indicates a series of intermittent fillet welds each 3 inches long and spaced 10 inches apart on centers. The centers of the welds on one side of the joint are displaced from those on the other. |
| | | Symbol indicates a fillet weld around the perimeter of the member. |
| | | Symbol indicates a ¼-inch V-groove weld with a ⅛-inch root penetration. |
| | | Symbol indicates a ¼-inch bevel weld with a 5⁄16-inch root penetration plus a subsequent ⅜-inch fillet weld. |

Application of American National Standard Welding Symbols — 4

| Desired Weld | Symbol | Symbol Meaning |
|---|---|---|
| | | Symbol indicates a bevel weld with a root opening of ³⁄₁₆ inch. |
| | | Symbol indicates a V-groove weld with a groove angle of 65 degrees on the *arrow side* and 90 degrees on the *other side.* |
| | | Symbol indicates a flush surface with the reinforcement removed by chipping on the *other side* of the joint and a smooth grind on the *arrow side.* The symbols C and G should be the user's standard finish symbols. |
| | | Symbol indicates a 2-inch U-groove weld with a 25-degree groove angle and no root opening for both sides of the joint. |
| | | Symbol indicates plug welds of 1-inch diameter, a depth of filling of ½ inch and a 60-degree angle of countersink spaced 6 inches apart on centers. |
| | | Symbol indicates all-around bevel and square-groove weld of these studs. |

Application of American National Standard Welding Symbols — 5

| Desired Weld | Symbol | Symbol Meaning |
|---|---|---|
| | | Symbol indicates an electron beam seam weld with a minimum acceptable joint strength of 200 pounds per lineal inch. |
| | | Symbol indicates four .10-inch diameter electron beam spot welds located at random. |
| | | Symbol indicates a fillet weld on the *other side* of joint and a flare-bevel-groove weld and a fillet weld on the *arrow side* of the joint. |
| | | Symbol indicates gas tungsten-arc seam weld on *arrow side* of joint. |
| | | Symbol indicates edge-flange weld on *arrow side* of joint and flare-V-groove weld on *other side* of joint. |
| | | Symbol indicates melt-thru weld. By convention this symbol is placed on the opposite side of the reference line from the corner-flange symbol. |

## Hard-Facing

Hard-facing is the process of welding onto parts or surfaces a coating, edge, or point of a metal highly capable of resisting one or more of the following: abrasion, corrosion, heat, and impact. The process can be applied equally well to new parts or old worn parts. The most common welding methods used to apply hard-facing materials include the oxy-acetylene gas, shielded-metal-arc, submerged arc, atomic hydrogen shielded arc, and inert-gas-shielded arc (consuming and nonconsuming electrode).

**Hard-Facing Materials.** — The first thing to be considered in the selection of a hard-facing material is to know what type of service the part in question is to undergo. Besides this, other considerations such as machinability, cost of hard-facing material, porosity of the deposit, appearance in use, and the ease of application may play a large part in the selection of a hard-facing material. Only generalized information can be given that can be used to guide in the selection of a material as the choice is dependent upon experience with a particular type of service. Generally the greater the hardness of the facing material the greater is its resistance to abrasion and shock or impact wear. Many hardenable materials may be used for hard-facing such as carbon steels, low-alloy steels, medium-alloy steels, and medium-high alloys but these are not outstanding. Some of the materials that might be considered more outstanding are high speed steel, austenitic manganese steel, austenitic high-chromium iron, cobalt-chromium alloy, copper-base alloy, and nickel-chromium-boron alloy.

**High-Speed Steels.** — These metals are available in the form of welding rods (RFe5) and electrodes (EFe5) for hard-facing where hardness is required at service temperatures up to 1100 degrees F. and where wear resistance and toughness are also required. Typical surfacing operations are done on cutting tools, shear blades, reamers, forming dies, shearing dies, guides, ingot tongs and broaches using these metals.

*Hardness:* They have a hardness of 55 to 60 on the Rockwell C scale in the as-welded condition and a hardness of 30 Rockwell C in the annealed condition. At a temperature of 1100 degrees F. the as-deposited hardness of 60 Rockwell C falls off very slowly to 47 Rockwell C. At about 1200 degrees F. the maximum Rockwell C hardness is 30.

*Resistance Properties:* As deposited the alloys can withstand only medium impact but when tempered the impact resistance is increased appreciably. Deposits of these alloys will oxidize readily because of their high molybdenum content but can withstand atmospheric corrosion. They do not withstand liquid corrosives.

*Other Properties or Characteristics:* The metals are well suited for metal-to-metal wear especially at elevated temperatures. They retain their hardness at elevated temperatures and can take a high polish. For machining, these alloys must first be annealed. Full hardness may be regained by a subsequent heat treatment of the metal.

**Austenitic Manganese Steels.** — These metals are available in the form of electrodes (EFeMn) for hard-facing when dealing with metal-to-metal wear and impact. Uses include facing rock-crushing equipment and railway frogs and crossings.

*Hardness:* Hardness of the as-deposited metals are 170 to 230 BHN but they can be work-hardened to 450 to 550 BHN very readily. For all practical purposes these metals have no hot hardness as they become brittle when reheated above 500 to 600 degrees F.

*Resistance Properties:* These metals are considered to be the outstanding engineering materials with respect to impact resistance. Their corrosion and oxidation resistance are similar to ordinary carbon steels. Their resistance to abrasion is only mediocre against hard abrasives like quartz.

*Other Properties or Characteristics:* The yield strength of the deposited metal in compression is low, but any compressive deformation rapidly raises it until plastic flow ceases. This property is an asset in impact wear situations. Machining is difficult with ordinary tools and equipment; finished surfaces for the most part are usually ground.

**Austenitic High-Chromium Irons.** — These metals are available in rod (RFeCr-A) and electrode (EFeCr-A) form for facing agricultural machinery parts, coke chutes, steel mill guides, sand blasting equipment and brick-making machinery.

*Hardness:* The as-welded deposit ranges in hardness from 51 to 62 Rockwell C. Under impact the deposit work hardens somewhat but the resulting deformation also leads to cracking and impact service is therefore avoided. Hot hardness decreases slowly at temperatures up to 800 and 900 degrees F. At 900 degrees F. the instantaneous hardness is 43 Rockwell C. In three minutes under load the hardness drops to 37 Rockwell C. At 1200 degrees F. the instantaneous hardness is 5 Rockwell C. The decrease in hardness during hot testing is practically recovered on cooling to ordinary temperatures.

*Resistance Properties:* Deposits will withstand only light impact without cracking. Dynamic compression stresses above 60,000 pounds per square inch should be avoided. These metals exhibit good oxidation resistance up to 1800 degrees F. and can be considered for hot wear applications where hot plasticity is not objectionable. They are not very resistant to corrosion from liquids and will rust in moist air but are more stable than ordinary iron and steel. Resistance to low stress scratching is outstanding and is related to the amount of hard carbides present. However, under high stress grinding abrasion, performance is only mediocre and they are not deemed advantageous for such service.

*Other Properties or Characteristics:* The deposited metals have a yield strength (0.1 per cent offset) of between 80,000 to 140,000 pounds per square inch in compression and an ultimate strength of from 150,000 to 280,000 pounds per square inch. Their tensile strength is low and therefore tension uses are avoided in design. These deposits are considered to be commercially unmachinable and are also very difficult to grind. When ground, a grinding wheel of aluminum oxide abrasive with a 24-grit size and a hard (Q) and medium spaced resinoid bond is recommended for off-hand high-speed work and a slightly softer (P) vitrified bond for off-hand low-speed work.

**Cobalt-Base Alloys.** — These metals are available in both rod (RCoCr) and electrode (ECoCr) form and are frequently used to surface the contact surfaces of exhaust valves in aircraft, truck, and bus engines. Other uses include parts such as valve trim in steam engines, and on pump shafts where conditions of corrosion and erosion are encountered. Several metals with a greater carbon content are available (CoCr-B, CoCr-C) which are used in applications requiring greater hardness and abrasion resistance but where impact resistance is not mandatory or expected to be a factor.

*Hardness:* Hardness ranges on the Rockwell C scale for gas welded deposits are as follows: CoCr-A, 38 to 47; CoCr-B, 45 to 49; and CoCr-C, 48 to 58. For arc-welded deposits hardness ranges (Rockwell C) as follows: CoCr-A, 23 to 47; CoCr-B, 34 to 47; and CoCr-C, 43 to 58. The values for the arc weld deposits depend for the most part on the base metal dilution. The greater the dilution the lower the hardness. Many surfacing alloys are softened permanently by heating to elevated temperatures, however, these metals are an exception. They do exhibit lower hardness values when hot but return to their approximate original hardness values upon cooling. Elevated temperature strength and hardness are outstanding properties of this group. Their use at 1200 degrees F. and above is considered advantageous between 1000 and 1200 degrees F. their advantages are not definitely established and at temperatures below 1000 degrees F. other surfacing metals may prove better.

*Resistance Properties:* In the temperature range from 1000 to 1200 degrees F. weld deposits of these metals have a great resistance to creep. Tough martensitic steel deposits are considered superior to cobalt-base deposits in both flow resistance and toughness. The chromium in the deposited metal promotes the formation of a thin, tightly adherent scale which provides a scaling resistance to combustion products of internal combustion engines including deposits from leaded fuels. These metals are corrosion resistant in such media as air, food and certain acids. It would be well to conduct field tests to determine specific corrosion resistance for the instance in question.

*Other Properties or Characteristics:* They are able to take a high polish and have a low coefficient of friction and are therefore well suited for metal-to-metal wear. Machining of these deposits is difficult; the difficulty increases in proportion to the increase in carbon content. CoCr-A metals are preferably machined with sintered carbide tools. CoCr-C deposits are finished by grinding.

**Copper-Base Alloys.** — These metals are available in the rod (RCuAl-A2, RCuAl-B, RCuAl-C, RCuAl-D, RCuAl-E, RCuSi-A, RCuSn, RCuSn-D, RCuSn-E, and RCuZn-E) and electrode (ECuAl-A2, ECuAl-B, ECuAl-C, ECuAl-D, ECuAl-E, ECuSi, ECuSn-A, ECuSn-C, ECuSn-E, and ECuZn-E) form and are used in depositing overlays and inlays for bearing, corrosion-resistant and wear-resistant surfaces. The CuAl-A2 rods and electrodes are used for surfacing bearing surfaces between the hardness ranges of 130 to 190 BHN as well as corrosion-resistant surfaces. The CuAl-B and CuAl-C rods and electrodes are used for surfacing bearing surfaces of hardness ranges 140 to 290 BHN. The CuAl-D and CuAl-E rods and electrodes are used to surface bearing and wear-resistant surfaces requiring the higher hardnesses of 230 to 390 BHN such as are on gears, cams, wear plates, and dies. The copper-tin (CuSn) metals are used where a lower hardness is required for surfacing, corrosion-resistant surfaces and sometimes for wear-resistant applications.

*Hardness:* Hardness of a deposit depends upon the welding process employed and the manner of depositing the metal. Deposits made by the inert-gas metal-arc process (both consumable and non-consumable electrode) will be higher in hardness than deposits made with the gas, metal-arc and carbon-arc processes since lower losses of aluminum, tin, silicon and zinc are achieved due to the better shielding from oxidation. Copper-base alloy metals are not recommended for use at elevated temperatures since their hardness and mechanical properties decrease consistently as the temperature goes above 400 degrees F.

*Resistance Properties:* The highest impact resistance of the copper-base alloy metals is exhibited by the CuAl-A2 deposit. As the aluminum content increases the impact resistance decreases markedly. CuSi weld deposits have good impact properties. CuSn metals as deposited have low impact resistance and CuZn-E deposits have a very low impact resistance. Deposits of the CuAl filler metals form a protective oxide coating upon exposure to the atmosphere. Oxidation resistance of CuSi deposits is fair and that of CuSn deposits are comparable to pure copper. These metals with the exception of the CuSn-E and CuZn-E alloys are widely used to resist many acids, mild alkalies and salt water. None of the copper-base alloy deposits are recommended for use where severe abrasion is encountered in service.

CuAl filler metals are used to overlay surfaces subjected to excessive wear from metal-to-metal contact such as gears, cams, sheaves, wear plates, and dies.

*Other Properties or Characteristics:* All copper-base alloy metals are used for overlays and inlays for bearing surfaces with the exception of the CuSi metals. Metals selected for bearing surfaces should have a Brinell hardness of 50 to 75 units below that of the mating metal surface. Slight porosity is generally acceptable in bearing service as a porous deposit is able to retain oil for lubricating purposes. CuAl deposits in compression have elastic limits ranging from 25,000 to 65,000 pounds per square inch and ultimate strengths of 120,000 to 171,000 pounds per square inch. The elastic limit and ultimate strength of CuSi deposits in compression are 22,000 pounds per square inch and 60,000 pounds per square inch, respectively. CuZn-E deposits in compression only have an elastic limit of about 5000 pounds per square inch and an ultimate strength of 20,000 pounds per square inch. All copper-base alloy deposits can be machined.

**Nickel-Chromium-Boron Alloys.** — These metals are available in both rod (RNiCr) and electrode (ENiCr) form and their deposits have good metal-to-metal wear resistance, good low-stress scratch abrasion resistance, corrosion resistance, and retention of hardness at elevated temperatures. These properties make their use suitable for seal rings, cement pump screws, valves, screw conveyors and cams. Three different formulations of these metals are recognized (NiCr-A, NiCr-B, and NiCr-C).

*Hardness:* Hardness of the deposited NiCr-A from rods range from 35 to 40 Rockwell C; of NiCr-B rods, 45 to 50 Rockwell C; of NiCr-C rods, 56 to 62 Rockwell C. Hardness of the deposited NiCr-A from electrodes ranges from 24 to 35 Rockwell C; of NiCr-B from electrodes, 30 to 45 Rockwell C; and of NiCr-C electrodes, 35 to 56. The lower hardness values and greater ranges of hardness values of the electrode deposits are attributed to the dilution of deposit and base metals. Hot Rockwell C hardness values of NiCr-A electrode deposits range from 30 to 19 in the temperature range from 600 to 1000 degrees F. from instantaneous loading to a 3-minute loading interval. NiCr-A rod deposits range from 34 to 24 in the same temperature range and under the same load conditions. Hot Rockwell C hardness values of NiCr-B electrode deposits range from 41 to 26 in the temperature range from 600 to 1000 degrees F. from instantaneous loading to a 3-minute loading interval. NiCr-B rod deposits range from 46 to 37 in the same temperature range and under the same load conditions. Hot Rockwell C hardness values of NiCr-C electrode deposits range from 49 to 31 in the temperature range from 600 to 1000 degrees F. from instantaneous loading to a 3-minute loading interval. NiCr-C rod deposits range from 55 to 40 in the same temperature range and under the same load conditions.

*Resistance Properties:* Deposits of these metal alloys will withstand light impact fairly well. When plastic deformation occurs cracks are more likely to appear in the NiCr-C deposit than in the NiCr-A and NiCr-B deposits. NiCr deposits are oxidation resistant up to 1800 degrees F. Their use above 1750 degrees F. is not recommended since fusion may begin near this temperature. NiCr deposits are completely resistant to atmospheric, steam, salt water, and salt spray corrosion and also resistant to the milder acids and many common corrosive chemicals. It would be well to conduct field tests when a corrosion application is contemplated. These metals are not recommended for high stress grinding abrasion. NiCr deposits have good metal-to-metal wear resistance, take a high polish under wearing conditions and are particularly resistant to galling. These properties are especially evident in the NiCr-C alloy.

*Other Properties or Characteristics:* In compression these alloys have an elastic limit of 42,000 pounds per square inch. Their yield strength in compression is

92,000 pounds per square inch (0.01 per cent offset), 150,000 pounds per square inch (0.10 per cent offset) and 210,000 pounds per square inch (0.20 per cent offset). Deposits of NiCr filler metals may be machined with tungsten carbide tools using slow speeds, light feeds and heavy tool shanks. They are also finished by grinding using a soft-to-medium vitrified silicon carbide wheel.

## Cutting Metals with an Oxidizing Flame

The oxy-hydrogen and oxy-acetylene flames are especially adapted to cutting metals. When iron or steel is heated to a high temperature, it has a great affinity for oxygen and readily combines with it to form different oxides, which causes the metal to be disintegrated and burned with great rapidity. The metal cutting or burning torch operates on this principle. A torch tip is designed to pre-heat the metal, which is then burned or oxidized by a jet of pure oxygen. The kerf or path left by the flame is suggestive of a saw cut when the cutting torch has been properly adjusted and used. The traversing motion of the torch along the work may be controlled either by hand or mechanically.

**The Cutting Torch.** — The ordinary cutting torch consists of a heating jet using oxygen and acetylene, oxygen and hydrogen, or, in fact, any other gas which, when combined with oxygen, will produce sufficient heat. By the use of this heating jet, the metal is first brought to a sufficiently high temperature, and an auxiliary jet of pure oxygen is then turned onto the red-hot metal, when the action just referred to takes place. Some cutting torches have a number of pre-heating flame ports surrounding the central oxygen port, so that a pre-heating flame will precede the oxygen regardless of the way in which the torch is moved. This arrangement has been used to advantage in mechanically guided torches. The rate of cutting varies with the thickness of the steel, the size of the tip and the oxygen pressure.

**Adjustment and Use of Cutting Torch.** — When using the cutting torch for the cutting of steel plate, the pre-heating flame first comes into contact with the edge of the plate and quickly raises it to a white hot temperature, and then the oxygen valve is opened by pulling a trigger on the torch and, as the pure oxygen comes into contact with the heated metal, the latter is burned or oxidized.

**Metals that can be Cut.** — Metals such as wrought iron and steels of comparatively low carbon content can be cut readily with the cutting torch. High carbon steels may be cut successfully if pre-heated to a temperature that depends somewhat on the carbon content. The higher the carbon content, the greater the degree of pre-heating. A black heat is sufficient for ordinary tool steel, but a low red may be required for some of the alloy tool steels. Brass and bronze plates have been cut by interposing them between steel plates.

**Cutting Stainless Steel.** — Stainless steel can be cut readily by the flux-injection method. The elements which give stainless steel their desirable properties produce oxides which reduce the operation to a slow melting away process when the conventional oxy-acetylene cutting equipment is used. By injecting a suitable flux directly into the stream of cutting oxygen before it enters the torch, the obstructing oxides are removed. A portable flux feeding unit is designed to inject a predetermined amount of the flux powder. The rate of flux flow is accurately regulated by a vibrator type of dispenser with rheostat control. The flux-injection method is applicable either to machine cutting or to a hand controlled torch. The operating procedure and speed of cutting are practically the same as in cutting mild steel.

*Cutting Cast Iron.* — The cutting of cast iron with the oxy-acetylene torch is practicable although it cannot be cut as readily as steel. The ease of cutting seems to depend largely on the physical character of the cast iron, very soft cast iron being more difficult to cut than harder varieties. The cost is much higher than that for cutting the same thickness of steel, because of the larger pre-heating flame necessary and the larger oxygen consumption. In spite of this, however, this method is economical in many cases. The slag from a cast iron cut contains considerable melted cast iron, while in the case of steel, the slag is practically free from particles of the metal. This indicates that cast iron cutting is partly a melting operation. Increased speed and decreased cost can often be obtained by feeding a steel rod, about ¼ inch in diameter, into the top of the cut, just beneath the torch tip. This furnishes a large amount of slag which flows over the cut and increases the temperature of the cast iron. Special tips are used owing to the amount of heat and oxygen required.

**Mechanically Guided Torches.** — Cutting torches used for cutting openings in plates or blocks or for cutting parts to some definite outline, often are guided mechanically. Torches guided by pantograph mechanisms are especially adapted for tracing the outline to be cut from a pattern or drawing. Other designs are preferable for straight-line cutting and one type is designed for circular cutting.

**Cutting Steel Castings.** — When cutting steel castings, care should be taken to prevent burning pockets in the metal when the flame strikes a blow-hole. If a blow-hole is penetrated, the molten oxide will splash into the cavity and the flame will be diverted. The presence of the blow-hole is generally indicated by excessive sparks. The operator should immediately move the torch back along the cut and direct it at an angle so as to strike the metal beneath the blow-hole and burn it away if possible beyond the cavity, when cutting in the normal position may be resumed.

**Thickness of Metal that can be Cut.** — The maximum thickness of metal that can be cut by these high-temperature flames depends largely upon the gases used and the pressure of the oxygen, which may be as high as 150 pounds per square inch; the thicker the metal the higher the pressure required. When using the oxy-acetylene flame, it might be practicable to cut iron or steel up to 12 or 14 inches in thickness, whereas, the oxy-hydrogen flame has been used to cut steel plates 24 inches thick. The oxy-hydrogen flame will cut thicker material principally because it is longer than the oxy-acetylene flame and can penetrate to the full depth of the cut, thus keeping all the oxide in a molten condition so that it can easily be blown out by the oxygen cutting jet. A mechanically guided torch will cut thick material more satisfactorily than a hand-guided torch, because the flame is directed straight into the cut and does not wobble, as it tends to do when the torch is held by hand. With any flame, the cut is less accurate and the kerf wider, as the thickness of the metal increases. When cutting light material, the kerf might be ⅟₁₆ inch wide, whereas, for heavy stock it might be ¼ or ⅜ inch wide.

## Arc Cutting of Metals

**Arc Cutting.** — According to the "Procedure Handbook of Arc-Welding Design & Practice" published by The Lincoln Electric Co. a steel may be readily cut with great accuracy by means of the oxy-acetylene torch. All metals, however, do not cut as easily as steel. Cast iron, stainless steels, manganese steels and non-ferrous materials are not as readily cut and shaped with the oxy-acetylene cutting process because of their reluctance to oxidize. In these cases, arc cutting is often used to good advantage.

The cutting of steel is a chemical action. The oxygen combines readily with the iron to form iron oxide. In cast iron, this action is hindered by the presence of carbon in graphite form. Thus, cast iron cannot be cut as readily as steel; higher temperatures are necessary and cutting is slower. In steel, the action starts at bright red heat, whereas, in cast iron, the temperature must be more near the melting point in order to obtain a sufficient reaction.

Due to its very high temperature, the rate of cutting is usually fairly high. However, as the process is essentially one of melting without any great action tending to force the molten metal out of cut, some provision must be made for permitting the metal to flow readily away from the cut. This is usually done by starting at some point from which the molten metal may readily flow. This method is followed until the desired amount of metal has been melted away.

As an example, the general method is to apply the electric arc on the under side of the work, starting at a lower corner, working toward the center on the lower surface, and then up the side, repeating this action as many times as necessary. This will allow the molten metal to flow out of the cut.

A carbon electrode is generally used. Graphite electrodes are used to some extent because they permit use of higher currents. Shielded-arc type electrodes are also effective. In starting a cut, the arc is held at the point selected for the initial cut as, for example, a lower corner. When the metal begins to flow and run off, the arc is moved along at a rate to permit the metal to continuously flow out of the cut.

The width of the cut is dependent upon the ability of the operator to follow a straight line, the size electrode used, and the thickness of material. The width of the cut is greater on thick sections than on thin.

A relatively new development, the arc-air process, is also widely used for cutting and gouging. The process is essentially the melting of metal, any metal, with an electric arc and simultaneously mechanically removing the molten metal by means of a high velocity air jet, external and parallel to the electrode. This process is not dependent on oxidation and, for this reason, works on metals which do not readily oxidize, as well as those that do.

The equipment consists of a torch with a concentric cable which carries both air and current. The electrode is usually carbon graphite, though coated metal may also be used. An air line from an ordinary air compressor and the cable from a DC welding machine are both attached to the end of a concentric cable, which carries both the air and the current to the torch. The lever at the bottom of the torch controls air flow. The electrode is held in a rotating head which allows it to be set at any angle, but maintains the air stream always directed at the proper location.

An ordinary DC welding machine with reverse polarity is used. The current depends on the size of the electrode and varies from 70 amperes on $\frac{5}{32}$ inch electrodes to 600 amperes on $\frac{1}{2}$ inch electrodes. Although the higher the current density, the more efficient the process becomes.

The necessary air is obtained from an ordinary compressor. The torch is designed to operate at 90 to 100 psi pressure, which is the usual line pressure in most shops, and, since this is not a critical value, no regulator is needed.

The torch is used by holding the electrode at a leading angle and striking an arc between the electrode and the material to be cut. The air blast is directed immediately behind the point of arcing, and the electrode is pushed forward at a rapid rate with the air jet on continuously. The depth of the groove is determined by the angle of the electrode and the speed of travel.

Because of the small area, the metal being cut is instantly brought to the molten stage and speed of travel is very rapid; the surrounding metal does not reach a very high temperature; and there is little distortion or crack propagation during the operation.

The speed varies somewhat with individual shop conditions, such as operator technique and the current used. To give a general idea, a groove ⅜ inch wide and ¼ inch deep is usually cut at about 3 fpm, or just about as fast as the operator can move.

**Tool and Die Welding.** — Arc welding is extensively used in the building and repairing of tools and dies. Inexpensive and truly functional tools and dies can be readily made using mild steel or medium carbon steel as the base material and surfacing the cutting or wearing surfaces with weld metal. Tool steel or high-speed steel can be quickly applied either in the fabrication or repair of dies and cutting tools with standard arc welding electrodes.

**Chromium Plating.** — Chromium plating is an electrolytic process of depositing chromium on metals either as a protection against corrosion or to increase the surface wearing qualities. The value of chromium-plating plug and ring gages has probably been more thoroughly demonstrated than any other one application of this treatment. Chromium-plated gages not only wear longer, but when worn, the chromium may be removed and the gage replated and reground to size.

In general, chromium-plated tools have operated well, giving greatly improved performance on nearly all classes of materials such as brass, bronze, copper, nickel, aluminum, cast iron, steel, plastics, asbestos compositions, and similar materials. Increased cutting life has been obtained with chromium-plated drills, taps, reamers, files, broaches, tool tips, saws, thread chasers, and the like. Dies for stamping, drawing, hot-forging, die-casting, and for molding plastic materials have shown greatly increased life after being plated with hard chromium.

Special care is essential in grinding and lapping tools preparatory to plating the cutting edges, because the chromium deposit is influenced materially by the grain structure and hardness of the base metal. The thickness of the plating may vary from 0.0001 to 0.001 or 0.002 inch, the thicker platings being used to build up undersize tools such as taps and reamers. Procedure followed by Westinghouse in the hard chromium-plating of tools, as well as parts salvaged by depositing chromium to increase diameters, is as follows: (1) Degrease with solvent; (2) mount the tools on racks; (3) clean in an anodic alkali bath held at a temperature of 82 degrees C. for from three to five minutes; (4) rinse in boiling water; (5) immerse in a 20 per cent hydro-chloric acid solution for two to three seconds; (6) rinse in cold water; (7) rinse in hot water; (8) etch in a reverse-current chromic acid bath for two to five minutes; (9) place work immediately in the chromium-plating bath; and (10) remove hydrogen embrittlement, if necessary, by immersing the plated tools for two hours in an oil bath maintained at 177 degrees C.

Chromium has a very low coefficient of friction. The static coefficient of friction for steel on chromium-plated steel is 0.17, and the sliding coefficient of friction is 0.16. This compares with static coefficient of friction for steel on steel of 0.30 and a sliding coefficient of friction of 0.20. The static coefficient of friction for steel on babbitt is 0.25, and the sliding coefficient of friction 0.20, whereas for chromium-plated steel on babbitt, the static coefficient of friction is 0.15, and the sliding coefficient of friction 0.13. These figures apply to highly polished bearing surfaces. Articles that are to be chromium-plated in order to resist frictional wear should be highly polished before plating so that full advantage can be taken of the low coefficient of friction that is characteristic of chromium. Chromium resists attack by almost all organic and inorganic compounds, except muriatic and sulphuric acids. The melting point of chromium is 2930 degrees F., and it remains bright up to 1200 degrees F. Above this temperature, it forms a light adherent oxide, which does not readily become detached. For this reason, chromium has been used successfully for protecting articles that must resist high temperatures, even above 2000 degrees F.

## Heat-treatment of Non-Ferrous Alloys

The solution and precipitation methods of heat-treatment may be applied to certain non-ferrous alloys such as wrought aluminum and also to some of the magnesium or Dowmetal alloys (see page 2255).

**Wrought Aluminum Alloys.** — The wrought alloys of aluminum may be divided into two classes depending upon the manner in which their harder tempers are produced. One class comprises the alloys in which strain-hardening, by definite amounts of cold work following the last annealing operation, produces the varying degrees of strength and hardness. The alloys in the other class depend primarily upon heat-treatment processes to develop their higher mechanical properties. While there is a wide range of tensile properties in both classes of alloys, the highest combinations of strength and ductility available in the widest range of products are to be found in the heat-treated alloys. In the aluminum alloys which respond to heat-treatment, the alloying constituents which give the increased strength and hardness are substances which are more soluble in solid aluminum at high temperatures than at low temperatures.

**Solution Heat-treatment.** — The first step in heat-treatment, frequently called the "solution heat-treatment," consists in heating the alloy to some temperature below the melting point, usually in the range of 900 to 1000 degrees F. for aluminum alloys, to put as much as possible of the alloying constituent into solid solution. The alloy is held at this temperature for some period of time, usually from 20 to 60 minutes for aluminum alloys, according to the thickness of the piece. This permits the entire piece to reach a uniform temperature and the dissolving of the alloying elements in the solid solution to take place throughout. In effect, the alloying constituent has been dissolved in the aluminum and dispersed as completely as when sugar is dissolved in water. The alloy is then quenched and — in contrast to steel after quenching — is in a relatively soft condition.

**Precipitation Heat-treatment.** — After quenching, the alloy undergoes an aging process which, if carried out at elevated temperatures, is called a "precipitation heat-treatment," because during this stage some of the alloying constituent which is held in solid solution precipitates from the solid solution in the form of extremely fine particles. This precipitation may occur spontaneously at room temperature, as is the case in the so-called "natural aging" of certain alloys, or it may require a "precipitation heat-treatment" or "artificial aging" at about 300 degrees F., in the case of certain other alloys to produce increased hardness and tensile strength.

**Heat Treatment of Copper Alloys.** — Precipitation hardening of copper alloys is useful in producing materials which have high strength and high electrical conductivity. In the case of beryllium copper alloys the most favorable properties are obtained through the use of both the solution and precipitation methods of heat-treatment. The solution heat-treatment is generally accomplished by heating and keeping the alloy at temperatures ranging from 1450 to 1650 degrees F. in a circulating air furnace for a certain length of time and then water quenching it. The temperature and length of time depend upon the composition of the alloy and physical dimensions of the part being treated. The precipitation heat-treatment may then be accomplished by heating and holding the alloy at a temperature from 600 to 900 degrees F. in a circulating air furnace (salt baths are not recommended) for 3 hours followed by an uncontrolled air cooling. The temperature to be used depends upon the composition of the alloy. Where special physical properties are desired, the times and temperatures may be varied.

# BRASS, BRONZE, ALUMINUM AND OTHER ALLOYS

**Cast Brass and Bronze.** — The following information on S.A.E. Standard Brass and Bronze Castings includes typical applications of the different alloys in the automotive industry, the composition in percentage, and physical properties based upon standard test bars cast to size with only a minimum amount of machining to remove the fin gate. Standard specimens of wrought material are taken parallel to the direction of rolling and all rods, bars and shapes are tested in full size when practicable.

**Red Brass Castings. — S.A.E. Standard No. 40.** — Red brass is used for water-pump impellers, fittings for gasoline and oil lines, small bushings, small miscellaneous castings. This is a free-cutting brass with good casting and finished properties.

*Composition of No. 40:* Copper, 84 to 86; tin, 4 to 6; lead, 4 to 6; zinc, 4 to 6; iron, max., 0.25; nickel, max., 0.75; phosphorus, max., 0.05; aluminum, 0.00; sulphur, max., 0.05; antimony, max., 0.25; other impurities, max., 0.15 per cent.

*Physical Properties:* Tensile strength, 26,000 pounds per square inch; yield point, 12,000 pounds per square inch; elongation in 2 inches (or proportionate gage length), 15 per cent.

**Yellow Brass Castings — S.A.E. Standard No. 41.** — Yellow brass is used for radiator parts, fittings for water-cooling systems, battery terminals, miscellaneous castings. This alloy is intended for commercial castings when cheapness and good machining properties are essential.

*Composition of No. 41:* Copper, 62 to 67; lead, 1.50 to 3.50; tin, max., 1; iron, max., 0.75; nickel, max., 0.25; phosphorus, max., 0.03; aluminum, max., 0.30; sulphur, max., 0.05; antimony, max., 0.15; other impurities, max., 0.15 per cent; zinc, remainder.

*Physical Properties:* Tensile strength, 20,000 pounds per square inch; elongation in 2 inches (or proportionate gage length), 15 per cent.

**Manganese Bronze Castings — S.A.E. Standard No. 43.** — This alloy is intended for castings requiring strength and toughness. It is used for such automotive parts as gear-shifter forks; counters, spiders; brackets and similar fittings; parts for starting motors; landing-gear and tail-skid castings for airplanes.

*Composition of No. 43:* Copper, 55 to 60; zinc, 38 to 42; tin, max., 1.50; manganese, max., 3.50; aluminum, max., 1.50; iron, max., 2; lead, max., 0.40 per cent.

*Physical Properties:* Tensile strength, 65,000 pounds per square inch; elongation in 2 inches (or proportionate gage length), 25 per cent.

**High Tensile Manganese Bronze Castings — S.A.E. Standard No. 430.** — This alloy is intended for use in castings where high strength and toughness are required such as marine propellers, shafts and gears.

*Composition of No. 430:* Copper, 60 to 68; iron, 2 to 4; aluminum, 3 to 6; manganese, 2.5 to 5; tin, max., 0.50; lead, max., 0.20, and nickel, max., 0.50 per cent; zinc, remainder.

*Physical Properties:* This alloy is manufactured in two grades, distinguished by chemical composition: Grade A being in the lower, and Grade B in the higher range of manganese, aluminum and iron content. Tensile strength, Grade A, 90,000, and Grade B, 110,000 pounds per square inch; elongation in 2 inches, Grade A, 20, and Grade B, 12 per cent.

**Cast Brass to be Brazed — S.A.E. Standard No. 44.** — This brass is used for water-pipe fittings which are to be brazed. It begins to melt at about 1830 degrees F. and is entirely melted at approximately 1870 degrees F. The alloy or spelter used for brazing must have a lower melting temperature. Silver solder may be used.

*Composition of No. 44:* Copper, 83 to 86; zinc, 14 to 17; lead, max., 0.50; iron, max., 0.15 per cent.

**Brazing Solder — S.A.E. Standard No. 45.** — This solder begins to melt at approximately 1560 degrees F. and is entirely melted at about 1600 degrees F. It may be used by melting in a crucible under a flux of borax, with or without the addition of boric acid. The part to be brazed is dipped into the melted solder. When used in powdered form, this solder, mixed with a flux, is applied to the material and then melted either by means of a brazing torch or by using a furnace.

*Composition of No. 45:* Copper, 48 to 52; lead, max., 0.50; iron, max., 0.10 per cent; zinc, remainder.

**Hard Bronze Castings — S.A.E. Standard No. 62.** — This is a strong general utility bronze suitable for severe working conditions and heavy pressures. Typical applications include gears; bearings; bushings for severe service; valve guides; valve-tappet guides; camshaft bearings; fuel pump, timer and distributor parts; connecting-rod bushings; piston-pins; rocker lever; steering sector and hinge bushings; starting-motor parts.

*Composition of No. 62:* Copper, 86 to 89; tin, 9 to 11; lead, max., 0.20; iron, max., 0.06; zinc, 1 to 3 per cent.

*Physical Properties:* Tensile strength, 30,000 pounds per square inch; yield point, 15,000 pounds per square inch; elongation in 2 inches (or proportionate gage length), 14 per cent.

**Leaded Gun Metal Castings — S.A.E. Standard No. 63.** — This general-utility bronze combines strength with fair machining qualities. It is especially good for bushings subjected to heavy loads and severe working conditions. It is also used for fittings subjected to moderately high water or oil pressures.

*Composition of No. 63:* Copper, 86 to 89; tin, 9 to 11; phosphorus, max., 0.25; zinc and other impurities, max., 0.50; lead, 1 to 2.50 per cent.

*Physical Properties:* Tensile strength, 30,000 pounds per square inch; yield point, 12,000 pounds per square inch; elongation in 2 inches (or proportionate gage length), 10 per cent.

**Phosphor Bronze Castings — S.A.E. Standard No. 64.** — This alloy is excellent when anti-friction qualities are important and where resistance to wear and scuffing are desired. It is used for such parts as wrist-pins, piston-pins, valve rocker-arm bushings, fuel and water-pump bushings, steering-knuckle bushings, aircraft control bushings.

*Properties of No. 64:* Copper, 78.50 to 81.50; tin, 9 to 11; lead, 9 to 11; phosphorus, 0.05 to 0.25; zinc, max., 0.75; other impurities, max., 0.25 per cent.

*Physical Properties:* Tensile strength, 25,000 pounds per square inch; yield point, 12,000 pounds per square inch; elongation in 2 inches (or proportionate gage length), 8 per cent.

**Phosphor Gear Bronze Castings — S.A.E. Standard No. 65.** — This bronze is not used regularly but it may be employed for gears and worm wheels where the requirements are severe and a very hard bronze is necessary.

*Properties of No. 65:* Copper, 88 to 90; tin, 10 to 12; phosphorus, 0.10 to 0.30; nickel, max., 0.05; lead, zinc, and other impurities, max., 0.50 per cent.

*Physical Properties:* Tensile strength, 35,000 pounds per square inch; yield point, 20,000 pounds per square inch; elongation in 2 inches (or proportionate gage length), 10 per cent.

**Bronze Backing for Lined Bearings — S.A.E. Standard No. 66.** — This is an inexpensive but suitable alloy for bronze-backed bearings of connecting-rods or main engine bearings.

*Composition:* Copper, 83 to 86; tin, 4.50 to 6; lead, 8 to 10; zinc, max., 2; other impurities, max., 0.25 per cent.

*Physical Properties:* Tensile strength, 25,000 pounds per square inch; yield point, 12,000 pounds per square inch; elongation in 2 inches, 8 per cent.

**Bronze Bearing Castings — S.A.E. Standard No. 660.** — This composition is widely used for bronze bearings. Typical applications in the automotive industry include such parts as spring bushings, torque tube bushings, steering-knuckle bushings, piston-pin bushings, thrust washers, etc.

*Composition of No. 660:* Copper, 81 to 85; tin, 6.50 to 7.50; lead, 6 to 8; zinc, 2 to 4; iron, max., 0.20; antimony, max., 0.20; other impurities, max., 0.50 per cent.

*Physical Properties:* Tensile strength, 30,000 pounds per square inch; yield point, 14,000 pounds per square inch; elongation in 2 inches, 18 per cent.

**Cast Aluminum Bronze — S.A.E. Standard No. 68.** — This alloy has considerable strength, resistance to corrosion, hardness equal to manganese bronze, and good bearing qualities under certain conditions. It is used for worm-wheels, gears, valve guides, valve seats, and forgings.

*Composition of No. 68:* Copper, (Grade A) 87 to 89, (Grade B) 89.50 to 90.50; aluminum, (Grade A) 7 to 9, (Grade B) 9.50 to 10.50; iron, (Grade A) 2.50 to 4, (Grade B) not over 1; tin, max., (Grade A) 0.5, (Grade B) 0.2; total other impurities, (Grade A), 1, (Grade B) 0.5 per cent.

*Physical Properties:* Tensile strength, (Grades A and B) as cast, 65,000 pounds per square inch; tensile strength, (Grade B) as heat-treated, quenched and drawn, 80,000 pounds per square inch; yield point, (Grades A and B) as cast, 25,000 pounds per square inch; yield point, (Grade B) as heat-treated, 50,000 pounds per square inch. Elongation in 2 inches, (Grade A) as cast, 20 per cent; (Grade B) 15 per cent; (Grade B) as heat-treated, 4 per cent.

## Wrought Copper and Copper Alloys

**Brass Sheet and Strip — S.A.E. Standard No. 70.** — There are two grades designated as 70A (Cartridge Brass) and 70C (Yellow Brass). Tempers range from quarter hard through extra spring. These are given in the accompanying table. The numbers following each temper designation in the table represent the amount of reduction in B. & S. gage numbers when the brass sheets are rolled. The greater the reduction, the harder the brass.

This alloy is used to make radiator cores and tanks in the automotive industry; bead chain, flashlight shells, socket and screw shells in the electrical industry; and eyelets, fasteners, springs and stampings in the hardware industry.

*Composition of No. 70A:* Copper, 68.5 to 71.5; lead, max., 0.07; iron, max., 0.05; zinc, remainder.

*Composition of No. 70C:* Copper, 64.0 to 68.5; lead, max., 0.15; iron, max., 0.05; zinc, remainder.

*Mechanical Properties:* Tensile strengths and Rockwell hardness numbers are given in the accompanying table.

**Aluminum Bronze Rods, Bars, and Shapes — S.A.E. Standard No. 701.** — This alloy is commonly used for bushings, gears, valve parts, bearings, sleeves, screws, pins, and fabricated sections. It is also used where strength at elevated temperatures, a low coefficient of friction against steel, or a combination of strength and corrosion resistance is required. Alloy grades are: 701B, 701C and 701D.

**Hardness and Ultimate Strength of No. 70 Sheet Brass by Tempers**

| Temper of Brass Sheet (S.A.E. No. 70) and Equivalent Reduction in B. & S. Gage Numbers | | Rockwell Hardness Numbers | | | | Ultimate Strength Pounds per Square Inch | |
|---|---|---|---|---|---|---|---|
| | | B Scale 1/16" Ball- 100 kg. Load | | Superficial 30-T Scale 1/16" Ball- 30 kg. Load | | | |
| Temper | Gage Nos. | Min. | Max. | Min. | Max. | Min. | Max. |
| Grade A | | | | | | | |
| Quarter Hard | 1 | 40 | 65 | 43 | 60 | 49,000 | 59,000 |
| Half Hard | 2 | 60 | 77 | 56 | 68 | 57,000 | 67,000 |
| Three-Quarter Hard | 3 | 72 | 82 | 65 | 72 | 64,000 | 74,000 |
| Hard | 4 | 79 | 86 | 70 | 74 | 71,000 | 81,000 |
| Extra Hard | 6 | 85 | 91 | 74 | 77 | 83,000 | 92,000 |
| Spring | 8 | 89 | 93 | 76 | 78 | 91,000 | 100,000 |
| Extra Spring | 10 | 91 | 95 | 77 | 79 | 95,000 | 104,000 |
| Grade C | | | | | | | |
| Quarter Hard | 1 | 40 | 65 | 43 | 60 | 49,000 | 59,000 |
| Half Hard | 2 | 57 | 74 | 54 | 66 | 55,000 | 65,000 |
| Three-Quarter Hard | 3 | 70 | 80 | 65 | 71 | 62,000 | 72,000 |
| Hard | 4 | 76 | 84 | 68 | 73 | 68,000 | 78,000 |
| Extra Hard | 6 | 83 | 89 | 73 | 76 | 79,000 | 89,000 |
| Spring | 8 | 87 | 92 | 75 | 78 | 86,000 | 95,000 |
| Extra Spring | 10 | 88 | 93 | 76 | 79 | 90,000 | 99,000 |

The hardness numbers equivalent to such temper designations as "quarter hard," "half hard," etc., vary over a wide range as shown by the table above. The hardness number represented by a given temper designation depends not only upon the kind of annealing and thickness of a given material, but may be affected decidedly by the composition or type of alloy. "Quarter hard" red brass sheet (S.A.E. No. 79), for example, may have a Rockwell hardness varying from 50 to 95 which differs considerably from the minimum and maximum numbers given in the table above opposite "quarter hard."

Hardness tests of the indentation type, such as Rockwell or Brinell, are generally used for thin materials; however, if the sheet is very thin, the test may be for comparison only with other sheets of the same composition and thickness. When the penetration is deep relative to the thickness, there may be an apparent decrease of hardness due to the flow or punching-through of the material because of lack of lateral support; however, when the penetration is even greater relative to thickness, there may be an apparent *increase* in hardness due to the pressure of the penetrator on the anvil of the instrument.

*Composition of No. 701B:* Copper, 80.0 to 93.0; aluminum, 6.5 to 11.0; iron, max., 4.00; nickel, max., 1.00; manganese, max., 1.50; silicon, max., 2.25; tin, max., 0.60; zinc, max., 1.0; tellurium, max., 0.6; other elements, max., 0.50.

*Composition of No. 701C:* Copper, 78.0; aluminum, 9.0 to 11.0; iron, 2.0 to 4.0; nickel, 4.0 to 5.5; manganese, max., 1.50; silicon, max., 0.25; tin, max., 0.20; other elements, max. 0.50.

*Composition of No. 701D:* Copper, 88.0 to 92.5; aluminum, 6.0 to 8.0; iron, 1.5 to 3.5; other elements, max., 0.50.

*Mechanical Properties:* Minimum tensile strengths of the No. 701B alloy grade range from 70,000 to 80,000 psi, minimum yield strengths from 30,000 to 40,000 psi and minimum elongations in 2 inches from 12 to 9 per cent depending on the shape or size of rod or bar. Minimum tensile strengths of the No. 701C alloy grade range from 85,000 to 100,000 psi, minimum yield strengths from 42,500 to 50,000 psi and minimum elongations in 2 inches from 10 to 5 per cent depending on size.

This alloy must withstand cold bending without fracture through an angle of 120 degrees around a pin, the diameter of which is equal to twice the diameter of round rod or four times the thickness of bar or other shapes.

**Copper Sheet and Strip — S.A.E. Standard No. 71.** — This alloy is used for building fronts, roofing, radiators, chemical process equipment, rotating bands, and vats.

*Composition of No. 71:* Copper, min., 99.90 (plus silver). In one type of sheet used in the automotive industry 6 to 10 troy ounces of silver may be added to one ton (avoirdupois) of copper. This is sufficient to raise the recrystallization temperature appreciably.

*Mechanical Properties:* Minimum tensile strengths range from 30,000 to 52,000 psi depending on temper. Generally the higher the strength the harder the temper.

**Free Cutting Brass Rod — S.A.E. Standard No. 72.** — This alloy is used for small screw machine parts, pins, nuts, plugs, screws, valve discs and caps.

*Composition of No. 72:* Copper, 60.0 to 63.0; lead, 2.5 to 3.7; iron, max., 0.35; other elements, max., 0.50; zinc, remainder.

*Mechanical Properties:* In the soft temper the minimum tensile strength ranges from 40,000 to 48,000 psi, the minimum yield strength from 15,000 to 20,000 psi and the minimum elongation in 2 inches from 25 to 15 per cent as the thickness decreases down from over 2 inches. In the half hard temper the minimum tensile strength ranges from 45,000 to 57,000 psi, the minimum yield strength ranges from 15,000 to 25,000 psi and the minimum elongation in 2 inches from 20 to 7 per cent as the size decreases down from over 2 inches. In the hard temper the minimum tensile strength is 80,000 psi and the minimum yield strength 45,000 psi for thicknesses of $\frac{1}{8}$ to $\frac{3}{16}$ inch. The minimum tensile strength is 70,000 psi, the minimum yield strength is 35,000 psi and the minimum elongation in 2 inches is 4 per cent for thicknesses over $\frac{3}{16}$ to $\frac{5}{16}$ inch.

**Naval Brass Rods, Bars, Forgings, and Shapes — S.A.E. No. 73.** — This material is intended for use where brass rod that is stronger, tougher, and more corrosion resistant than commercial brass rod is required. Uses include forgings, water pump and propeller shafts, studs and nuts, bushings, turnbuckle barrels, adjusting stud ends, and screw machine parts.

*Composition of No. 73:* Copper, 59.0 to 62.0; tin, 0.50 to 1.00; lead, max., 0.20; iron, max., 0.10; other elements, max., 0.10; zinc, remainder.

*Mechanical Properties:* Rods and bars in the soft temper have a minimum tensile strength ranging from 50,000 to 54,000 psi, a minimum yield strength of 20,000 psi and a minimum elongation in 4 times the diameter or thickness of 30 per cent as the size decreases. Rods and bars in the half hard temper have a minimum tensile strength ranging from 54,000 to 60,000 psi, a minimum yield strength ranging from 22,000 to 27,000 psi and a minimum elongation in 4 times the diameter or thickness ranging from 30 to 22 per cent as the size decreases. Rods and bars in the hard temper have a minimum tensile strength ranging from 54,000 to 67,000 psi, a minimum yield strength ranging from 22,000 to 45,000 psi and a minimum elongation in 4 times the diameter or thickness ranging from 30 to 13 per cent as the size decreases.

**Seamless Brass Tubes — S.A.E. Standard No. 74.** — The alloys comprising these tubes are identified by the letters *A, B, C,* and *D.* Nos. 74A and 74D are used for condenser and heat exchanger tubes and flexible hose. Nos. 74B and 74C are general purpose materials used for water pipe radiator and ornamental work. The tubes may be formed, bent, upset, squeezed, swaged, flared, roll threaded and knurled.

*Composition of No. 74A (Muntz Metal):* Copper, 59.0 to 63.0; lead, max., 0.30; iron, max., 0.07; zinc, remainder.

*Composition of No. 74B (Yellow Brass):* Copper, 65.0 to 68.0; lead, 0.20 to 0.80; iron, max., 0.07; zinc, remainder.

*Composition of No. 74C (Cartridge Brass):* Copper, 68.5 to 71.5; lead, max., 0.07; iron, max., 0.05; zinc, remainder.

*Composition of No. 74D (Red Brass, 85%):* Copper, 84.0 to 86.0; lead, max., 0.06; iron, max., 0.05; zinc, remainder.

*Mechanical Properties of No. 74A:* This tube in drawn temper exhibits a minimum tensile strength of 54,000 psi. Common tempers of this tube include light annealed, drawn general purpose, and hard drawn.

*Mechanical Properties of Nos. 74B and 74C:* These tubes in drawn temper exhibit a minimum tensile strength of 54,000 psi. In hard temper they exhibit a minimum tensile strength of 66,000 psi. Common tempers of these tubes include drawn general purpose and hard drawn.

*Mechanical Properties of No. 74D:* This tube in light, drawn, and hard tempers exhibits minimum tensile strengths of 44,000, 44,000, and 57,000 psi, respectively. Common tempers of this tube include light, drawn general purpose, and hard.

**Copper Tubes — S.A.E. Standard No. 75.** — These tubes which contain a minimum of 99.90 per cent deoxidized copper are used for general engineering purposes, including gasoline, hydraulic and oil lines.

*Mechanical Properties:* In the light drawn temper the minimum tensile strength is 36,000 psi and the maximum tensile strength is 47,000 psi. In the drawn general purpose temper the minimum tensile strength is 36,000 psi and in the hard drawn temper (applying to tubes up to 1 inch outside diameter, inclusive, with wall thicknesses from 0.020 to 0.120 inch; tubes over 1 to 2 inches outside diameter, inclusive with wall thicknesses from 0.035 to 0.180 inch; and tubes over 2 to 4 inches outside diameter with wall thicknesses from 0.060 to 0.250 inch) the minimum tensile strength is 45,000 psi.

**Phosphor Bronze Sheet and Strip — S.A.E. Standard No. 77.** — Typical uses for this sheet and strip include springs, switch parts, sleeve bushings, clutch discs, diaphragms, fuse clips, and fasteners. There are two grades of this alloy, 77A and 77C. Six tempers are applied to this alloy, namely, soft, half hard, hard, extra hard, spring and extra spring.

*Composition of No. 77A:* Tin, 3.5 to 5.8; phosphorus, 0.03 to 0.35; lead, max., 0.05; iron, max., 0.10; zinc, max., 0.30; antimony, max., 0.01; copper, tin, and phosphorus, min., 99.50.

*Composition of No. 77C:* Tin, 7.0 to 9.0; phosphorus, 0.03 to 0.35; lead, max., 0.05; iron, max., 0.10; zinc, max., 0.20; antimony, max., 0.01; copper, tin, and phosphorus, min., 99.50.

*Mechanical Properties:* The minimum tensile strength of the No. 77A alloy ranges from 40,000 to 96,000 psi as the temper ranges from soft to extra spring. The minimum tensile strength of the No. 77C alloy ranges from 53,000 to 110,000 psi as the temper ranges from soft to extra spring.

**Red Brass and Low Brass Sheet and Strip — S.A.E. Standard No. 79.** — There are two grades designated as 79A (Red Brass, 85 per cent) and 79B (Low Brass, 80 per cent). Common tempers of No. 79A strip are quarter hard, half hard, extra hard, and spring. Common temper of No. 79A sheet is half hard. Common tempers of No. 79B strip are quarter hard, half hard, hard, and spring. Typical uses include weather strip, trim, conduit, sockets, fasteners, radiator cores and costume jewelry.

*Composition of No. 79A:* Copper, 84.0 to 86.0; lead, 0.05; iron, 0.05; zinc, remainder.

*Composition of No. 79B:* Copper, 78.5 to 81.5; lead, 0.05; iron, 0.05; zinc, remainder.

*Mechanical Properties:* Minimum tensile strengths of the 79A alloy range from 44,000 to 82,000 psi as the temper ranges from quarter hard to extra spring and minimum tensile strengths of the 79B alloy range from 48,000 to 89,000 psi as the temper ranges from quarter hard to extra spring.

**Brass Wire — S.A.E. Standard No. 80.** — This wire is used for making springs, locking wire, rivets, screws, and for wrapping turnbuckles. There are two grades, 80A and 80B.

*Composition of No. 80A:* Copper, 68.5 to 71.5; lead, max., 0.07; iron, max., 0.05; zinc, remainder.

*Composition of No. 80B:* Copper, 63.0 to 68.5; lead, max., 0.10; iron, max., 0.05; zinc, remainder.

*Mechanical Properties:* Minimum tensile strengths of the 80A and 80B alloys range from 50,000 to 120,000 psi as tempers range from eighth hard to spring.

**Phosphor Bronze Wire and Rod — S.A.E. Standard No. 81.** — This alloy is used for springs, switch parts, fasteners, and cotter pins. It should withstand being bent cold through an angle of 120 degrees without fracture, around a pin with a diameter twice the diameter of the wire.

*Composition of No. 81:* Tin, 3.50 to 5.80; phosphorus, 0.03 to 0.35; lead, max., 0.05; iron, max., 0.10; zinc, max., 0.30; copper, tin, and phosphorus, min., 99.50.

*Mechanical Properties:* Minimum tensile strengths of hard drawn wire in coils range from 145,000 to 105,000 as the wire diameter ranges from 0.025 to 0.500 inch. Minimum tensile strengths of spring temper rods range from 125,000 to 90,000 psi as the rod diameter ranges from 0.025 to 0.500 inch.

**Annealed Copper Wire — S.A.E. Standard No. 83.** — This wire is used primarily for electrical purposes but it is also used for metal spraying and copper brazing. No composition limits are specified for this wire but the copper should be of such quality and purity that when drawn and annealed should exhibit the mechanical properties (maximum tensile strength and minimum elongation) and electrical characteristics called for in the standard. Its electrical resistivity should not exceed 875.20 ohms per mile-lb (100 per cent electrical conductivity IACS, International Annealed Copper Standard) at a temperature of 20 degrees C.

*Mechanical Properties:* Maximum tensile strengths of annealed wire range from 36,000 to 38,000 psi for wire diameters ranging from 0.4600 down to over 0.0201 inch. Minimum elongations in 10 inches of annealed wire range from 15 to 35 per cent as the wire diameter ranges from over 0.0030 to 0.4600 inch.

**Brass Forgings — S.A.E. Standard No. 88.** — Typical uses for this alloy are forgings and pressings of all kinds.

*Composition of No. 88:* Copper, 58.0 to 61.0; lead, 1.50 to 2.50; iron, max., 0.30; other elements, max., 0.50; zinc, remainder.

*Mechanical Properties:* Hot-pressed forgings made of this alloy should have tensile strengths ranging from 45,000 to 60,000 psi and elongations in 2 inches ranging from 25 to 60 per cent.

## Aluminum and Aluminum Alloys

Pure aluminum is a silver-white metal characterized by a slightly bluish cast. It has a specific gravity of 2.70, resists the corrosive effects of many chemicals and has a malleability approaching that of gold. When alloyed with other metals numerous properties are obtained which make these alloys useful over a wide range of applications.

Aluminum alloys are light in weight compared to steel, brass, nickel or copper; can be fabricated by all common processes; are available in a wide range of sizes, shapes and forms; resist corrosion; readily accept a wide range of surface finishes; have good electrical and thermal conductivities; and are highly reflective to both heat and light.

**Characteristics of Aluminum and Aluminum Alloys.** — Aluminum and its alloys lose part of their strength at elevated temperatures, although some alloys retain good strength at temperatures from 400 to 500 degrees F. At subzero temperatures, however, their strength increases without loss of ductility so that aluminum is a particularly useful metal for low-temperature applications.

When aluminum surfaces are exposed to the atmosphere, a thin invisible oxide skin forms immediately which protects the metal from further oxidation. This self-protecting characteristic gives aluminum its high resistance to corrosion. Unless exposed to some substance or condition which destroys this protective oxide coating, the metal remains protected against corrosion. Aluminum is highly resistant to weathering, even in industrial atmospheres. It is also corrosion resistant to many acids. Alkalis are among the few substances that attack the oxide skin and therefore are corrosive to aluminum. Although the metal can safely be used in the presence of certain mild alkalis with the aid of inhibitors, in general, direct contact with alkaline substances should be avoided. Direct contact with certain other metals should be avoided in the presence of an electrolyte; otherwise galvanic corrosion of the aluminum may take place in the vicinity of the contact area. Where other metals must be fastened to aluminum, the use of a bituminous paint coating or insulating tape is recommended.

Aluminum is one of the two common metals having an electrical conductivity high enough for use as an electric conductor. The conductivity of electric-conductor (EC) grade is about 62 per cent that of the International Annealed Copper Standard. Because aluminum has less than one-third the specific gravity of copper, however, a pound of aluminum will go almost twice as far as a pound of copper when used for this purpose. Alloying lowers the conductivity somewhat so that wherever possible the EC grade is used in electric conductor applications.

Aluminum has nonsparking and nonmagnetic characteristics which make the metal useful for electrical shielding purposes such as in bus bar housings or enclosures for other electrical equipment and for use around inflammable or explosive substances.

Aluminum can be cast by any method known to foundrymen. It can be rolled to any desired thickness down to foil thinner than paper and in sheet form can be stamped, drawn, spun or rolled-formed. The metal also may be hammered or forged. Aluminum wire, drawn from rolled rod, may be stranded into cable of any desired size and type. The metal may be extruded into a variety of shapes. It may be turned, milled, bored, or machined in machines often operating at their maximum speeds. Aluminum rod and bar may readily be employed in the high-speed manufacture of automatic screw-machine parts.

Almost any method of joining is applicable to aluminum — riveting, welding or brazing. A wide variety of mechanical aluminum fasteners simplifies the assembly of many products. Resin bonding of aluminum parts has been successfully employed, particularly in aircraft components.

For the majority of applications, aluminum needs no protective coating. Mechanical finishes such as polishing, sand blasting or wire brushing meet the majority of needs. When additional protection is desired, chemical, electrochemical and paint finishes are all used. Vitreous enamels have recently been developed for aluminum, and the metal may also be electroplated.

**Temper Designations for Aluminum Alloys.** — The temper designation system adopted by The Aluminum Association and used in industry pertains to all forms of wrought and cast aluminum and aluminum alloys except ingot. It is based on the sequences of basic treatments used to produce the various tempers. The temper designation follows the alloy designation, being separated by a dash.

Basic temper designations consist of letters. Subdivisions of the basic tempers, where required, are indicated by one or more digits following the letter. These designate specific sequences of basic treatments, but only operations recognized as significantly influencing the characteristics of the product are indicated. Should some other variation of the same sequence of basic operations be applied to the same alloy, resulting in different characteristics, then additional digits are added.

The basic temper designations and subdivisions are as follows:

-F  *as fabricated:* Applies to products which acquire some temper from shaping processes not having special control over the amount of strain-hardening or thermal treatment. For wrought products, there are no mechanical property limits.

-O  *annealed, recrystallized (wrought products only):* Applies to the softest temper of wrought products.

-H  *strain-hardened (wrought products only):* Applies to products which have their strength increased by strain-hardening with or without supplementary thermal treatments to produce partial softening.

The -H is always followed by two or more digits.
The first digit indicates the specific combination of basic operations, as follows:

-H1  *strain-hardened only:* Applies to products which are strain-hardened to obtain the desired mechanical properties without supplementary thermal treatment.

   The number following this designation indicates the degree of strain-hardening.

-H2  *strain-hardened and then partially annealed:* Applies to products which are strain-hardened more than the desired final amount and then reduced in strength to the desired level by partial annealing. For alloys that age-soften at room temperature, the -H2 tempers have approximately the same ultimate strength as the corresponding -H3 tempers. For other alloys, the -H2 tempers have approximately the same ultimate strengths as the corresponding -H1 tempers and slightly higher elongations.

   The number following this designation indicates the degree of strain-hardening remaining after the product has been partially annealed.

-H3  *strain-hardened and then stabilized:* Applies to products which are strain-hardened and then stabilized by a low temperature heating to slightly lower their strength and increase ductility. This designation applies only to the magnesium-containing alloys which, unless stabilized, gradually age-soften at room temperature.

   The number following this designation indicates the degree of strain-hardening remaining after the product has been strain-hardened a specific amount and then stabilized.

The second digit following the designations $-H_1$, $-H_2$, and $-H_3$ indicates the final degree of strain-hardening. Numeral 8 has been assigned to indicate tempers having a final degree of strain-hardening equivalent to that resulting from approximately 75 per cent reduction of area. Tempers between $-O$ (annealed) and 8 (full hard) are designated by numerals 1 through 7. Material having an ultimate strength about midway between that of the $-O$ temper and that of the 8 temper is designated by the numeral 4 (half hard); between $-O$ and 4 by the numeral 2 (quarter hard); between 4 and 8 by the numeral 6 (three-quarter hard); etc. (NOTE. For two-digit $-H$ tempers whose second figure is odd, the standard limits for ultimate strength are exactly midway between those for the adjacent two-digit $-H$ tempers whose second figures are even). Numeral 9 designates extra hard tempers.

The third digit, when used, indicates a variation of a two-digit $-H$ temper. It is used when the degree of control of temper or the mechanical properties are different from but close to those for the two-digit $-H$ temper designation to which it is added. (NOTE. The minimum ultimate strength of a three-digit $-H$ temper is at least as close to that of the corresponding two-digit $-H$ temper as it is to the adjacent two-digit $-H$ tempers.) Numerals 1 through 9 may be arbitrarily assigned and registered with The Aluminum Association for an alloy and product to indicate a specific degree of control of temper or specific mechanical property limits. Zero has been assigned to indicate degrees of control of temper or mechanical property limits negotiated between the manufacturer and purchaser which are not used widely enough to justify registration with The Aluminum Association.

The following three-digit $-H$ temper designations have been assigned for wrought products in all alloys:

$-H_{111}$  Applies to products which are strain-hardened less than the amount required for a controlled $H_{11}$ temper.

$-H_{112}$  Applies to products which acquire some temper from shaping processes not having special control over the amount of strain-hardening or thermal treatment, but for which there are mechanical property limits or mechanical property testing is required.

$-H_{311}$  Applies to products which are strain-hardened less than the amount required for a controlled $H_{31}$ temper.

The following three-digit $-H$ temper designations have been assigned for

| Patterned or Embossed Sheet | Fabricated From |
|---|---|
| $-H_{114}$ | $-O$ temper |
| $-H_{124}, -H_{224}, -H_{324}$ | $-H_{11}, -H_{21}, -H_{31}$ temper, respectively |
| $-H_{134}, -H_{234}, -H_{334}$ | $-H_{12}, -H_{22}, -H_{32}$ temper, respectively |
| $-H_{144}, -H_{244}, -H_{344}$ | $-H_{13}, -H_{23}, -H_{33}$ temper, respectively |
| $-H_{154}, -H_{254}, -H_{354}$ | $-H_{14}, -H_{24}, -H_{34}$ temper, respectively |
| $-H_{164}, -H_{264}, -H_{364}$ | $-H_{15}, -H_{25}, -H_{35}$ temper, respectively |
| $-H_{174}, -H_{274}, -H_{374}$ | $-H_{16}, -H_{26}, -H_{36}$ temper, respectively |
| $-H_{184}, -H_{284}, -H_{384}$ | $-H_{17}, -H_{27}, -H_{37}$ temper, respectively |
| $-H_{194}, -H_{294}, -H_{394}$ | $-H_{18}, -H_{28}, -H_{38}$ temper, respectively |
| $-H_{195}, -H_{395}$ | $-H_{19}, -H_{39}$, temper, respectively |

$-W$  *solution heat-treated:* An unstable temper applicable only to alloys which spontaneously age at room temperature after solution heat-treatment. This designation is specific only when the period of natural aging is indicated: for example, $-W_{1/2}$ hour.

**–T** *thermally treated to produce stable tempers other than –F, –O, or –H:* Applies to products which are thermally treated, with or without supplementary strain-hardening, to produce stable tempers.

The –T is always followed by one or more digits. Numerals 2 through 10 have been assigned to indicate specific sequences of basic treatments, as follows:

**–T2** *annealed (cast products only):* Designates a type of annealing treatment used to improve ductility and increase dimensional stability of castings.

**–T3** *solution heat-treated and then cold worked:* Applies to products which are cold worked to improve strength, or in which the effect of cold work in flattening or straightening is recognized in applicable specifications.

**–T4** *solution heat-treated and naturally aged to a substantially stable condition:* Applies to products which are not cold worked after solution heat-treatment, or in which the effect of cold work in flattening or straightening may not be recognized in applicable specifications.

**–T5** *artificially aged only:* Applies to products which are artificially aged after an elevated-temperature rapid-cool fabrication process, such as casting or extrusion, to improve mechanical properties and/or dimensional stability.

**–T6** *solution heat-treated and then artificially aged:* Applies to products which are not cold worked after solution heat-treatment, or in which the effect of cold work in flattening or straightening may not be recognized in applicable specifications.

**–T7** *solution heat-treated and then stabilized:* Applies to products which are stabilized to carry them beyond the point of maximum hardness, providing control of growth and/or residual stress.

**–T8** *solution heat-treated, cold worked, and then artificially aged:* Applies to products which are cold worked to improve strength, or in which the effect of cold work in flattening or straightening is recognized in applicable specifications.

**–T9** *solution heat-treated, artificially aged, and then cold worked:* Applies to products which are cold worked to improve strength.

**–T10** *artificially aged and then cold worked:* Applies to products which are artificially aged after an elevated-temperature rapid-cool fabrication process, such as casting or extrusion, and then cold worked to improve strength.

A period of natural aging at room temperature may occur between or after the operations listed for tempers –T3 through –T10. Control of this period is exercised when it is metallurgically important.

Additional digits may be added to designations –T2 through –T10 to indicate a variation in treatment which significantly alters the characteristics of the product. These may be arbitrarily assigned and registered with The Aluminum Association for an alloy and product to indicate a specific treatment or specific mechanical property limits.

These additional digits have been assigned for wrought products in all alloys:

**–TX51** *stress-relieved by stretching:* Applies to products which are stress-relieved by stretching the following amounts after solution heat-treatment:

|  |  |
|---|---|
| Plate | 1½ to 3 per cent permanent set |
| Rod, Bar and Shapes | 1 to 3 per cent permanent set |

Applies directly to plate and rolled or cold-finished rod and bar.

These products receive no further straightening after stretching.

Applies to extruded rod, bar and shapes when designated as follows:

–TX510 Applies to extruded rod, bar and shapes which receive no further straightening after stretching.

–TX511 Applies to extruded rod, bar and shapes which receive minor straightening after stretching to comply with standard tolerances.

–TX52 *stress-relieved by compressing:* Applies to products which are stress-relieved by compressing after solution heat-treatment, to produce a nominal permanent set of 2½ per cent.

–TX53 *stress-relieved by thermal treatment*

The following two-digit –T temper designations have been assigned for wrought products in all alloys:

–T42 Applies to products solution heat-treated by the user which attain mechanical properties different from those of the –T4 temper.

–T62 Applies to products solution heat-treated and artificially aged by the user which attain mechanical properties different from those of the –T6 temper. (NOTE. Exceptions not conforming to the definitions given for –T42 and –T62 are 4032–T62, 6101–T62, 6061–T62, 6062–T62, 6063–T42 and 6463 –T42.)

**Aluminum Alloy Designation Systems.** — Aluminum casting alloys are listed in many specifications of various standardizing agencies. These include Federal Specifications, Military Specifications, ASTM Specifications and SAE Specifications, to mention some. The numbering systems used by each differ and are not always correlatable. Casting alloys are available from producers who use a commercial numbering system and this numbering system is the one used in the tables of aluminum casting alloys which are given further along in this section.

A system of four-digit numerical designations for wrought aluminum and wrought aluminum alloys was adopted by The Aluminum Association in 1954. This system is used by the commercial producers and is similar to the one used by the SAE; the difference being the addition of two prefix letters.

The first digit of the designation identifies the alloy type, 1, indicating an aluminum of 99.00 per cent or greater purity; 2, copper; 3, manganese; 4, silicon; 5, magnesium; 6, magnesium and silicon; 7, zinc; 8, some element other than those aforementioned; 9, unused (not assigned at present). If the second digit in the designation is zero, it indicates that there is no special control on individual impurities; while integers 1 through 9, indicate special control on one or more individual impurities.

In the 1000 series group for aluminum of 99.00 per cent or greater purity, the last two of the four digits indicate to the nearest hundredth the amount of aluminum above 99.00 per cent. Thus designation 1030 indicates 99.30 per cent minimum aluminum. In the 2000 to 8000 series groups the last two of the four digits have no significance but are used to identify different alloys in the group. At the time of adoption of this designation system most of the existing commercial designation numbers were used as these last two digits, as for example, 14S became 2014, 3S became 3003, and 75S became 7075. When new alloys are developed and are commercially used these last two digits are assigned consecutively beginning with –01, skipping any numbers previously assigned at the time of initial adoption.

Experimental alloys are also designated in accordance with this system but they are indicated by the prefix X. The prefix is dropped upon standardization.

**Heat-treatability of Wrought Aluminum Alloys.** — In high-purity form, aluminum is soft and ductile. Most commercial uses, however, require greater strength than pure aluminum affords. This is achieved in aluminum first by the addition of other elements to produce various alloys, which singly or in combination impart strength to the metal. Further strengthening is possible by means which classify the alloys roughly into two categories, non-heat-treatable and heat-treatable.

*Non-heat-treatable alloys:* The initial strength of alloys in this group depends upon the hardening effect of elements such as manganese, silicon, iron and magnesium, singly or in various combinations. The non-heat-treatable alloys are usually designated, therefore, in the 1000, 3000, 4000, or 5000 series. Since these alloys are work-hardenable, further strengthening is made possible by various degrees of cold working, denoted by the "H" series of tempers. Alloys containing appreciable amounts of magnesium when supplied in strain-hardened tempers are usually given a final elevated-temperature treatment called *stabilizing* for property stability.

*Heat-treatable alloys:* The initial strength of alloys in this group is enhanced by the addition of alloying elements such as copper, magnesium, zinc, and silicon. Since these elements singly or in various combinations show increasing solid solubility in aluminum with increasing temperature, it is possible to subject them to thermal treatments which will impart pronounced strengthening.

The first step, called *heat-treatment* or *solution heat-treatment*, is an elevated-temperature process designed to put the soluble element in solid solution. This is followed by rapid quenching, usually in water, which momentarily "freezes" the structure and for a short time renders the alloy very workable. It is at this stage that some fabricators retain this more workable structure by storing the alloys at below freezing temperatures until they are ready to form them. At room or elevated temperatures the alloys are not stable after quenching, however, and precipitation of the constituents from the supersaturated solution begins. After a period of several days at room temperature, termed *aging* or *room-temperature precipitation*, the alloy is considerably stronger. Many alloys approach a stable condition at room temperature, but some alloys, particularly those containing magnesium and silicon or magnesium and zinc, continue to age-harden for long periods of time at room temperature.

By heating for a controlled time at slightly elevated temperatures, even further strengthening is possible and properties are stabilized. This process is called *artificial aging* or *precipitation hardening*. By the proper combination of solution heat-treatment, quenching, cold working and artificial aging, the highest strengths are obtained.

**Clad Aluminum Alloys.** — The heat-treatable alloys in which copper or zinc are major alloying constituents, are less resistant to corrosive attack than the majority of non-heat-treatable alloys. To increase the corrosion resistance of these alloys in sheet and plate form they are often clad with high-purity aluminum, a low magnesium-silicon alloy, or an alloy containing 1 per cent zinc. The cladding, usually from 2½ to 5 per cent of the total thickness on each side, not only protects the composite due to its own inherently excellent corrosion resistance but also exerts a galvanic effect which further protects the core material.

Special composites may be obtained such as clad non-heat-treatable alloys for extra corrosion protection, for brazing purposes, or for special surface finishes. Some alloys in wire and tubular form are clad for similar reasons and on an experimental basis extrusions also have been clad.

**Characteristics of Principal Aluminum Alloy Series Groups.** — 1000 series: These alloys are characterized by high corrosion resistance, high thermal and electrical conductivity, low mechanical properties and good workability. Moderate increases in strength may be obtained by strain-hardening. Iron and silicon are the major impurities.

2000 series: Copper is the principal alloying element in this group. These alloys require solution heat-treatment to obtain optimum properties; in the heat-treated condition mechanical properties are similar to, and sometimes exceed, those of mild steel. In some instances artificial aging is employed to further increase the mechanical properties. This treatment materially increases yield strength, with attendant loss in elongation; its effect on tensile (ultimate) strength is not as great. The alloys in the 2000 series do not have as good corrosion resistance as most other aluminum alloys and under certain conditions they may be subject to intergranular corrosion. Therefore, these alloys in the form of sheet are usually clad with a high-purity alloy or a magnesium-silicon alloy of the 6000 series which provides galvanic protection to the core material and thus greatly increases resistance to corrosion. Alloy 2024 is perhaps the best known and most widely used aircraft alloy.

3000 series: Manganese is the major alloying element of alloys in this group, which are generally non-heat-treatable. Because only a limited percentage of manganese, up to about 1.5 per cent, can be effectively added to aluminum, it is used as a major element in only a few instances. One of these, however, is the popular 3003, used for moderate-strength applications requiring good workability.

4000 series: The major alloying element of this group is silicon, which can be added in sufficient quantities to cause substantial lowering of the melting point without producing brittleness in the resulting alloys. For these reasons aluminum-silicon alloys are used in welding wire and as brazing alloys where a lower melting point than that of the parent metal is required. Most alloys in this series are non-heat-treatable, but when used in welding heat-treatable alloys they will pick up some of the alloying constituents of the latter and so respond to heat-treatment to a limited extent. The alloys containing appreciable amounts of silicon become dark gray when anodic oxide finishes are applied, and hence are in demand for architectural applications.

5000 series: Magnesium is one of the most effective and widely used alloying elements for aluminum. When it is used as the major alloying element or with manganese, the result is a moderate to high strength non-heat-treatable alloy. Magnesium is considerably more effective than manganese as a hardener, about 0.8 per cent magnesium being equal to 1.25 per cent manganese, and it can be added in considerably higher quantities. Alloys in this series possess good welding characteristics and good resistance to corrosion in marine atmospheres. However, certain limitations should be placed on the amount of cold work and the safe operating temperatures permissible for the higher magnesium content alloys (over about 3½ per cent for operating temperatures over about 150 deg. F.) to avoid susceptibility to stress corrosion.

6000 series: Alloys in this group contain silicon and magnesium in approximate proportions to form magnesium silicide, thus making them capable of being heat-treated. The major alloy in this series is 6061, one of the most versatile of the heat-treatable alloys. Though less strong than most of the 2000 or 7000 alloys, the magnesium-silicon (or magnesium-silicide) alloys possess good formability and corrosion resistance, with medium strength. Alloys in this heat-treatable group may be formed in the −T4 temper (solution heat-treated but not artificially aged) and then reach full −T6 properties by artificial aging.

7000 series: Zinc is the major alloying element in this group, and when coupled with a smaller percentage of magnesium results in heat-treatable alloys of very high strength. Usually other elements such as copper and chromium are also added in small quantities. Notable member of this group is 7075, which is among the highest strength aluminum alloys available and is used in air-frame structures and for highly stressed parts.

### Table 1. Aluminum Casting Alloys — Typical Mechanical Properties[1]

| Alloy[2] and Temper | Strength, ksi[8] | | | Elongation in 2 In., % | Alloy[2] and Temper | Strength, ksi[8] | | | Elongation in 2 In., % |
|---|---|---|---|---|---|---|---|---|---|
| | Ultimate Tensile | Yield[3] (0.2% offset) | Ultimate Shear | | | Ultimate Tensile | Yield[3] (0.2% offset) | Ultimate Shear | |
| Sand Casting Alloys | | | | | | | | | |
| 43-F | 19 | 8 | 14 | 8.0 | 220-T4 | 48 | 26 | 34 | 16.0 |
| 108-F | 21 | 14 | 17 | 2.5 | 319-F | 27 | 18 | 22 | 2.0 |
| 112-F | 24 | 15 | 20 | 1.5 | 319-T5 | 30 | 26 | 24 | 1.5 |
| 113-F | 24 | 15 | 20 | 1.5 | 319-T6 | 36 | 24 | 29 | 2.0 |
| 122-T61 | 41 | 40 | 32 | ....[4] | 355-T51 | 28 | 23 | 22 | 1.5 |
| A140-F | 33 | 28 | .. | 1.0 | 355-T6 | 35 | 25 | 28 | 3.0 |
| 142-T21 | 27 | 18 | 21 | 1.0 | 355-T7 | 38 | 36 | 28 | 0.5 |
| 142-T571 | 32 | 30 | 26 | 0.5 | 355-T71 | 35 | 29 | 26 | 1.5 |
| 142-T77 | 30 | 23 | 24 | 2.0 | A355-T51 | 28 | 24 | 22 | 1.5 |
| A142-T75 | 31 | .. | .. | 2.0 | 356-T51 | 25 | 20 | 20 | 2.0 |
| 195-T4[5] | 32 | 16 | 26 | 8.5 | 356-T6 | 33 | 24 | 26 | 3.5 |
| 195-T6 | 36 | 24 | 30 | 5.0 | 356-T7 | 34 | 30 | 24 | 2.0 |
| 195-T62 | 41 | 32 | 33 | 2.0 | 356-T71 | 28 | 21 | 20 | 3.5 |
| 212-F | 23 | 14 | 20 | 2.0 | A612-F | 35[6] | 25[6] | 26 | 5.0[6] |
| 214-F | 25 | 12 | 20 | 9.0 | 750-T5 | 20 | 11 | 14 | 8.0 |
| B214-F | 20 | 13 | 17 | 2.0 | A750-T5 | 20 | 11 | 14 | 5.0 |
| F214-F | 21 | 12 | 17 | 3.0 | B750-T5 | 27 | 22 | 18 | 2.0 |
| Permanent-Mold Casting Alloys | | | | | | | | | |
| 43-F | 23 | 9 | 16 | 10.0 | 333-F | 34 | 19 | 27 | 2.0 |
| A108-F | 28 | 16 | 22 | 2.0 | 333-T5 | 34 | 25 | 27 | 1.0 |
| 113-F | 28 | 19 | 22 | 2.0 | 333-T6 | 42 | 30 | 33 | 1.5 |
| C113-F | 30 | 24 | 24 | 1.5 | 333-T7 | 37 | 28 | 28 | 2.0 |
| 122-T551 | 37 | 35 | 30 | ....[4] | 344-F | 28 | 10 | .. | 10.0 |
| 122-T65 | 48 | 36 | 36 | ....[4] | 355-T51 | 30 | 24 | 24 | 2.0 |
| A132-T551 | 36 | 28 | 28 | 0.5 | 355-T6 | 42 | 27 | 34 | 4.0 |
| A132-T65 | 47 | 43 | 36 | 0.5 | 355-T62 | 45 | 40 | 36 | 1.5 |
| F132-T5 | 36 | 28 | .. | 1.0 | 355-T71 | 36 | 31 | 27 | 3.0 |
| 138-F | 30 | 24 | 24 | 1.5 | C355-T61 | 46 | 34 | 32 | 6.0 |
| 142-T571 | 40 | 34 | 30 | 1.0 | 356-T6 | 38 | 27 | 30 | 5.0 |
| 142-T61 | 47 | 42 | 35 | 0.5 | 356-T7 | 32 | 24 | 25 | 6.0 |
| B195-T4[5] | 37 | 19 | 30 | 9.0 | A356-T61 | 41 | 30 | 28 | 10.0 |
| B195-T6 | 40 | 26 | 32 | 5.0 | C612-F | 35[6] | 18[6] | .. | 8.0[6] |
| B195-T7 | 39 | 20 | 30 | 4.5 | 750-T5 | 23 | 11 | 15 | 12.0 |
| A214-F | 27 | 16 | 22 | 7.0 | A750-T5 | 20 | 11 | 14 | 5.0 |
| ........ | .. | .. | .. | .. | B750-T5 | 32 | 23 | 21 | 5.0 |
| Die-Casting Alloys | | | | | | | | | |
| 13 | 43 | 21 | 28 | 2.5 | A360[7] | 46 | 24 | 29 | 5.0 |
| 43 | 33 | 16 | 21 | 9.0 | 364 | 43 | 23 | 26 | 7.5 |
| A214 | 40 | 22 | 26 | 10.0 | 380 | 48 | 24 | 31 | 3.0 |
| 218 | 45 | 27 | 29 | 8.0 | A380[7] | 47 | 23 | 30 | 4.0 |
| 360 | 47 | 25 | 30 | 3.0 | 384 | 47 | 25 | 30 | 1.0 |

[1] The properties given in the table are typical or average and should not be used as specifications. They have been determined by tests conducted by the Aluminum Company of America. The specimens used in the tensile tests conducted for the sand and permanent-mold casting alloys were standard half-inch tensile test specimens individually cast in green-sand molds and were tested without machining off the surface, while those for the die-casting alloys were ASTM standard round die-cast test specimens, ¼ inch in diameter, produced on a cold chamber (high pressure) die-casting machine.

The modulus of elasticity varies somewhat with each alloy, but an average value of 10,300,000 pounds per square inch can be used for most calculations.

[2] Alloy designation numbers are those of the Aluminum Co. of America.

[3] Values shown are tensile yield strength values. Compressive yield strength values (0.2 per cent offset, specimens having an $l/r$ ratio of 12) of sand and permanent-mold casting alloys are on the average 3.3 per cent greater than the values shown.

[4] Less than 0.5 per cent.

[5] After several weeks at room temperature the alloy's properties approach those of the -T6 condition.

[6] From tests made approximately 30 days after casting.

[7] Prefix "A," indicates lower control limits of impurities, notably iron, as compared to those alloys of the same number designation without the prefix.

[8] Strength, given in thousands of pounds per square inch; ksi being kilopounds per square inch.

Table 2. Aluminum Casting Alloys — Nominal Composition, Characteristics, and Typical Uses

| Alloy | Usual Commercial Tempers | Types of Castings | | | Nominal Composition,[1] Per Cent | | | | | Approximate Relative Ratings of Various Characteristics[2] | | | | | | | | Typical Uses |
|---|---|---|---|---|---|---|---|---|---|---|---|---|---|---|---|---|---|---|
| | | Sand | Permanent-Mold | Die | Cu | Si | Mg | Zn | Ni | Strength | Castability | Resistance to Corrosion | Weldability (Torch and Arc) | Machinability | Electrical Conductivity | Pressure Tightness | Hardness | |
| 43 | F | √ | √ | | … | 5.0 | … | … | … | D | A | B | A | D | C | A | D | General-purposes and pipe fittings. |
| 108 | F | √ | | | 4.0 | 3.0 | … | … | … | C | B | B | B | B | C | B | C | General-purposes, manifolds and valve bodies. |
| A108 | F | | √ | √ | 4.5 | 5.5 | … | … | … | C | B | B | B | B | C | B | C | General-purposes and ornamental grills. |
| 113 | F | | √ | | 7.0 | 2.0 | … | … | … | C | B | D | C | B | C | B | C | Housings, covers, and hand wheels. |
| C113 | F | √ | √ | | 7.0 | 3.5 | … | … | … | B | B | D | C | B | C | B | B | Automotive cylinder heads and timing gears. |
| 122 | T551, T61, T65 | √ | √ | | 10.0 | … | 0.2 | … | 2.5 | B,A | C | D | C | A | D | C | A | Bushings, bearing caps and meter parts. |
| A132 | T551, T65 | | √ | | 0.8 | 12.0 | 1.2 | … | … | B | C | D | C | C | C | B | B,A | Heavy-duty diesel pistons. |
| 138 | F | | √ | | 10.0 | 4.0 | 0.3 | … | … | B | C | D | B | B | D | C | B | Sole plates for electric hand irons. |
| 142 | T21, T571, T61, T77 | √ | √ | | 4.0 | … | 1.5 | … | 2.0 | C,A | C | C | C | B | D | C | C,B | Heavy-duty pistons and air-cooled cylinder heads. |
| 195 | T4, T6, T62 | √ | | | 4.5 | 0.8 | … | … | … | B | C | D | C | B | B,C | C | C,B | Machinery and aircraft structural members. |
| B195 | T4, T6, T7 | | √ | | 4.5 | 2.5 | … | … | … | C | C | D | C | B | C | C | C,B | Aircraft fittings and gear housings. |
| 212 | F | √ | | √ | 8.0 | 1.2 | … | … | … | C | C | C | C | A | C | … | C | General purposes. |
| 214 | F | √ | | | … | … | 3.8 | … | … | C | C | A | B | A | C | D | C | Chemical, marine and architectural applications. |
| A214 | F | | √ | √ | … | … | 3.8 | 1.8 | … | C | C | A | B | A | C | D | C | Cooking utensils and ornamental hardware. |
| B214 | F | √ | | | … | 1.8 | 3.8 | 1.8 | … | C | C | A | B | B | C | D | C | Cooking utensils and pipe fittings. |
| F214 | F | | √ | | … | 0.5 | 3.8 | … | … | C | C | A | B | A | C | D | C | Alumilite-coated architectural parts. |

The information presented in this table has been compiled from data prepared by the Aluminum Company of America whose alloy designation numbers are used.

[1] Shown are the amounts of principal alloying elements. Aluminum and normal impurities constitute the remainder. Abbreviations used for column headings are as follows: Cu, copper; Si, silicon; Mg, magnesium; Zn, zinc; and Ni, nickel. [2] A, B, C, and D are arbitrary relative ratings in decreasing order of merit. Where two letters are shown, that on the left is for the softest temper listed and that on the right for the hardest temper listed.

**Table 2 (Concluded).** Aluminum Casting Alloys — Nominal Composition, Characteristics, and Typical Uses

| Alloy | Usual Commercial Tempers | Types of Castings | | | Nominal Composition, Per Cent [1] | | | | Approximate Relative Ratings of Various Characteristics [2] | | | | | | | | Typical Uses |
|---|---|---|---|---|---|---|---|---|---|---|---|---|---|---|---|---|---|
| | | Sand | Permanent-Mold | Die | Cu | Ni | Mg | Other | Strength | Castability | Resistance to Corrosion | Weldability (Torch and Arc) | Machinability | Electrical Conductivity | Pressure Tightness | Hardness | |
| 218 | F | | | ✓ | … | … | 8.0 | … | A | D | A | … | A | D | D | … | Marine fittings and hardware. |
| 220 | T4 | ✓ | | | … | … | 10.0 | … | A | D | A | D | A | D | B | C | Aircraft structural members. |
| 319 | F, T6 | ✓ | ✓ | | 3.5 | … | … | Si, 6.3 | C,B | B | C | B,C | B | C,B | B | C,B | Engine parts and cylinder heads. |
| 333 | F, T5, T6, T7 | | ✓ | | 3.8 | … | … | Si, 9.0 | B,A | B | C | C | B | D,C | B | C,B | Engine and gas meter parts. |
| 355 | T51, T6, T61, T62, T7, T71 | ✓ | ✓ | | 1.3 | … | 0.50 | Si, 5.0 | C,B | A | B | B | B | C,B | A | C,B | Crankcases, housings and aircraft fittings. |
| C355 | T61 | ✓ | ✓ | | 1.3 | … | 0.50 | Si, 5.0 | A | A | B | B | B | B | A | B | Aircraft, missile and other structural uses. |
| 356 | T51, T6, T7, T71, T61 | ✓ | ✓ | | … | … | 0.30 | Si, 7.0 | C,B | A | B | B | B | C,B | A | C,B | Transmission cases, truck and bridge railing parts. |
| A356 | T61 | ✓ | ✓ | | … | … | 0.30 | Si, 7.0 | A | A | B | B | B | B | A | B | Aircraft, missile and other structural uses. |
| 360 | F | | ✓ | ✓ | … | … | 0.50 | Si, 9.5 | A | A | B | B | C | D | A | B | Cover plates and instrument cases. |
| 364 | F, T5 | | ✓ | ✓ | … | … | 0.30 | Si, 8.5 [3] | A | A | B | … | C | D | A | … | Ductile, impact-resistant castings. |
| 380 | F | | | ✓ | … | … | … | Si, 9.0 | B | A | C | … | C | D | A | … | General-purposes. |
| A612 | F | ✓ | ✓ | | 0.50 | … | 0.7 | Zn, 6.5 | B | B | B | C | A | C | D | C | General-purposes and for brazing. |
| C612 | F | ✓ | ✓ | | 0.50 | … | 0.35 | Zn, 6.5 | C | C | C | C | A | B | D | C | Torque converter blades and brazed parts. |
| 750 | T5 | ✓ | ✓ | | 1.0 | 1.0 | … | Sn, 6.5 | D | D | C | D | A | A | D | D | Bearings. |
| A750 | T5 | ✓ | ✓ | | 1.0 | 0.50 | … | Sn, 6.5 [4] | D | D | C | D | A | A | D | D | Bearings. |
| B750 | T5 | ✓ | ✓ | | 2.0 | 1.2 | 0.75 | Sn, 6.5 | B,C | D | C | D | A | A | D | B,C | Bearings. |

The information presented in this table has been compiled from data prepared by the Aluminum Company of America whose alloy designation numbers are used.

[1] Shown are the amounts of principal alloying elements. Aluminum and normal impurities constitute the remainder. Abbreviations used for column headings are as follows: Cu, copper; Si, silicon; Mg, magnesium; Zn, zinc; Ni, nickel; and Sn, tin. [2] A, B, C, and D are arbitrary relative ratings in decreasing order of merit. Where two letters are shown, that on the left is for the softest temper listed and that on the right for the hardest temper listed. [3] Alloy also contains chromium, 0.35 and beryllium, 0.03. [4] Alloy also contains Si, 2.5.

## Table 1. Wrought Aluminum Alloys — Typical Mechanical Properties[1]

| Alloy and Temper | Strength, ksi[2] | | | Elongation in 2 In.,[3] % | | Alloy and Temper | Strength, ksi[2] | | | Elongation in 2 In.,[3] % | | Alloy and Temper | Strength, ksi[2] | | | Elongation in 2 In.,[3] % | |
|---|---|---|---|---|---|---|---|---|---|---|---|---|---|---|---|---|---|
| | UT | Y | US | 1/16 | 1/2 | | UT | Y | US | 1/16 | 1/2 | | UT | Y | US | 1/16 | 1/2 |
| EC-O[4] | 10 | 4 | 8 | .. | 3 | 2014*-T451 | 61 | 37 | 37 | 22 | .. | 3003-H18 | 29 | 27 | 16 | 4 | 10 |
| -H12 | 14 | 12 | 9 | .. | .. | *-T6 | 68 | 60 | 41 | 10 | .. | *-O | 16 | 6 | 11 | 30 | 40 |
| -H14 | 16 | 14 | 10 | .. | .. | *-T651 | 68 | 60 | 41 | 10 | .. | *-H12 | 19 | 18 | 12 | 10 | 20 |
| -H16 | 18 | 16 | 11 | .. | .. | 2017-O | 26 | 10 | 18 | .. | 22 | *-H14 | 22 | 21 | 14 | 8 | 16 |
| -H18 | 21 | 19 | .. | .. | .. | -T4 | 62 | 40 | 38 | .. | 22 | *-H16 | 26 | 25 | 15 | 5 | 14 |
| -H19 | 27 | 24 | 15 | .. | 3 | -T451 | 62 | 40 | 38 | .. | 22 | *-H18 | 29 | 27 | 16 | 4 | 10 |
| 2EC-T6 | 32 | 28 | 22 | .. | 19 | 2018-T61 | 61 | 46 | 39 | .. | 12 | 3004-O | 26 | 10 | 16 | 20 | 25 |
| -T61 | 25 | 20 | 17 | .. | 22 | 2024-O | 27 | 11 | 18 | 20 | 22 | -H32 | 31 | 25 | 17 | 10 | 17 |
| -T62 | 30 | 25 | 21 | .. | 20 | -T3 | 70 | 50 | 41 | 18 | .. | -H34 | 35 | 29 | 18 | 9 | 12 |
| -T64 | 17 | 9 | 13 | .. | 24 | -T36 | 72 | 57 | 42 | 13 | .. | -H36 | 38 | 33 | 20 | 5 | 9 |
| 1060-O | 10 | 4 | 7 | 43 | .. | -T4 | 68[6] | 47[6] | 41 | 20 | 19 | -H38 | 41 | 36 | 21 | 5 | 6 |
| -H12 | 12 | 11 | 8 | 16 | .. | -T351 | 68[6] | 47[6] | 41 | 20 | 19 | *-O | 26 | 10 | 16 | 20 | 25 |
| -H14 | 14 | 13 | 9 | 12 | .. | -T6 | 69 | 57 | 41 | .. | 10 | *-H32 | 31 | 25 | 17 | 10 | 17 |
| -H16 | 16 | 15 | 10 | 8 | .. | -T81 | 70 | 65 | 43 | 6 | .. | *-H34 | 35 | 29 | 18 | 9 | 12 |
| -H18 | 19 | 18 | 11 | 6 | .. | -T851 | 70 | 65 | 43 | 6 | .. | *-H36 | 38 | 33 | 20 | 5 | 9 |
| 1100-O | 13 | 5 | 9 | 35 | 40 | -T86 | 75 | 71 | 45 | 6 | .. | *-H38 | 41 | 36 | 21 | 5 | 6 |
| -H12 | 16 | 15 | 10 | 12 | 25 | *-O | 26 | 11 | 18 | 20 | .. | 3105-H25 | 26 | 24 | 16 | 8 | 20 |
| -H14 | 18 | 17 | 11 | 9 | 20 | *-T3 | 65 | 45 | 40 | 18 | .. | 4032-T6 | 55 | 46 | 38 | .. | 9 |
| -H16 | 21 | 20 | 12 | 6 | 17 | *-T36 | 67 | 53 | 41 | 11 | .. | 5005-O | 18 | 6 | 11 | .. | .. |
| -H18 | 24 | 22 | 13 | 5 | 15 | *-T4 | 64 | 42 | 40 | 19 | .. | -H12 | 20 | 19 | 14 | 10 | .. |
| 1345-O | 12 | 4 | 8 | 40 | .. | *-T351 | 64 | 42 | 40 | 19 | .. | -H14 | 23 | 22 | 14 | 6 | .. |
| -H12 | 14 | 12 | 9 | 14 | .. | *-T81 | 65 | 60 | 40 | 6 | .. | -H16 | 26 | 25 | 15 | 5 | .. |
| -H14 | 16 | 14 | 10 | 11 | .. | *-T851 | 65 | 60 | 40 | 6 | .. | -H18 | 29 | 28 | 16 | 4 | .. |
| -H16 | 18 | 16 | 11 | 7 | .. | *-T86 | 70 | 66 | 42 | 6 | .. | -H32 | 20 | 17 | 14 | 11 | .. |
| -H18 | 21 | 19 | 12 | 6 | .. | 2025-T6 | 58 | 37 | 35 | .. | 19 | -H34 | 23 | 20 | 14 | 8 | .. |
| -H19 | 28 | 25 | 16 | 4 | .. | 2117-T4 | 43 | 24 | 28 | .. | 27 | -H36 | 26 | 24 | 15 | 6 | .. |
| 2011-T3 | 55 | 43 | 32 | .. | 15 | 2218-T72 | 48 | 37 | 30 | .. | 11 | -H38 | 29 | 27 | 16 | 5 | .. |
| -T6 | 57 | 39 | 34 | .. | 17 | 2219-O | 25 | 10 | .. | 20 | .. | 5050-O | 21 | 8 | 15 | 24 | .. |
| -T8 | 59 | 45 | 35 | .. | 12 | -T31 | 54 | 37 | 33 | 17 | .. | -H32 | 25 | 21 | 17 | 9 | .. |
| 2014-O | 27 | 14 | 18 | .. | 18 | -T37 | 60 | 49 | 37 | 11 | .. | -H34 | 28 | 24 | 18 | 8 | .. |
| -T4 | 62 | 42[5] | 38 | .. | 20 | -T62 | 61 | 42 | 37 | 11 | .. | -H36 | 30 | 26 | 19 | 7 | .. |
| -T451 | 62 | 42[5] | 38 | .. | 20 | -T81 | 70 | 53 | 41 | 11 | .. | -H38 | 32 | 29 | 20 | 6 | .. |
| -T6 | 70[6] | 60[6] | 42 | .. | 13 | -T87 | 70 | 58 | 41 | 10 | .. | 5052-O | 28 | 13 | 18 | 25 | 30 |
| -T651 | 70[6] | 60[6] | 42 | .. | 13 | 3003-O | 16 | 6 | 11 | 30 | 40 | -H32 | 33 | 28 | 20 | 12 | 18 |
| *-O | 25 | 10 | 18 | 21 | .. | -H12 | 19 | 18 | 12 | 10 | 16 | -H34 | 38 | 31 | 21 | 10 | 16 |
| *-T3 | 63 | 40 | 37 | 20 | .. | -H14 | 22 | 21 | 14 | 8 | 16 | -H36 | 40 | 35 | 23 | 8 | 14 |
| *-T4 | 61 | 37 | 37 | 22 | .. | -H16 | 26 | 25 | 15 | 5 | 14 | -H38 | 42 | 37 | 24 | 7 | 14 |

* Alclad alloy is designated.

[1] These typical properties compiled from a listing prepared by the Aluminum Company of America are average for various forms, sizes and methods of manufacture, and may not describe any one particular product. The moduli of elasticity (average of tension and compression) in pounds per square inch for these wrought alloys range from $10.0 \times 10^6$ to $10.8 \times 10^6$ except for the 4032-T6 alloy whose modulus is $11.4 \times 10^6$. The compression modulus for any particular alloy is about 2 per cent greater than the tension modulus.

[2] Strengths are given in units of thousands of pounds per square inch. Symbols heading columns are as follows: UT, ultimate tensile strength; Y, yield strength (0.2 per cent offset); and US, ultimate shearing strength.

[3] The elongations are listed under two column headings, 1/16 and 1/2. The values under 1/16 refer to 1/16-inch *thick* specimens and those under 1/2 to 1/2-inch *diameter* specimens except that EC-O and EC-H19 with values of 23 and 1.5, respectively, refer to 10-inch wire specimens.

[4] Electrical conductor grade, 99.60 per cent aluminum, minimum.

[5] Die forgings will have a yield strength approximately 20 per cent lower than this value.

[6] Extruded products more than 3/4 inch thick will have strengths 15 to 20 per cent higher than this value.

Table I (Concluded). Wrought Aluminum Alloys — Typical Mechanical Properties[1]

| Alloy and Temper | Strength, ksi[2] | | | Elongat'n in 2 In.,[3] % | | Alloy and Temper | Strength, ksi[2] | | | Elongat'n in 2 In.,[3] % | | Alloy and Temper | Strength, ksi[2] | | | Elongat'n in 2 In.,[3] % | |
|---|---|---|---|---|---|---|---|---|---|---|---|---|---|---|---|---|---|
| | UT | Y | US | 1/16 | 1/2 | | UT | Y | US | 1/16 | 1/2 | | UT | Y | US | 1/16 | 1/2 |
| 5056-O | 42 | 22 | 26 | .. | 35 | 5456-H311 | 47 | 33 | 27 | 18 | .. | 6063-O | 13 | 7 | 10 | .. | .. |
| -H18 | 63 | 59 | 34 | .. | 10 | -H321 | 51 | 37 | 30 | 16 | 16 | -T4 | 25 | 13 | 16 | 22 | .. |
| -H38 | 60 | 50 | 32 | .. | 15 | -H323 | 51 | 38 | 30 | 10 | .. | -T42 | 22 | 13 | 14 | 20 | 33 |
| 5083-O | 42 | 21 | 25 | 22 | 25 | -H343 | 56 | 43 | 33 | 8 | .. | -T5 | 27 | 21 | 17 | 12 | 22 |
| -H112[4] | 44 | 28 | 26 | 16 | .. | 5457-O | 19 | 7 | 12 | 25 | .. | -T6 | 35 | 31 | 22 | 12 | 18 |
| -H113 | 46 | 33 | 28 | 16 | 16 | -H25 | 27 | 23 | 16 | 10 | .. | -T83 | 37 | 35 | 22 | 9 | .. |
| -H323 | 47 | 36 | 27 | 10 | .. | -H26 | 29 | 24 | 17 | 10 | .. | -T831 | 30 | 27 | 18 | 10 | .. |
| -H343 | 52 | 41 | 30 | 8 | .. | -H38 | 32 | 30 | 18 | 7 | .. | -T832 | 42 | 39 | 27 | 12 | .. |
| 5086-O | 38 | 17 | 23 | 22 | 30 | 5557-O | 16 | 6 | 10 | 25 | .. | -T835 | 48 | 43 | 30 | 8 | .. |
| -H32 | 42 | 30 | 25 | 12 | 16 | -H25 | 25 | 21 | 15 | 10 | .. | 6151-T6 | 48 | 43 | 32 | .. | 17 |
| -H34 | 47 | 37 | 28 | 10 | 14 | -H26 | 27 | 23 | 16 | 10 | .. | 6262-T9 | 58 | 55 | 35 | .. | 10 |
| -H36 | 50 | 41 | 29 | 8 | .. | -H38 | 30 | 28 | 18 | 8 | .. | 6463-O | 13 | 7 | 10 | .. | .. |
| -H112 | 39 | 19 | 23 | 14 | .. | 5652-O | 28 | 13 | 18 | 25 | 30 | -T4 | 25 | 13 | 16 | 22 | .. |
| 5154-O | 35 | 17 | 22 | 27 | 30 | -H32 | 33 | 28 | 20 | 12 | 18 | -T42 | 22 | 13 | 14 | 20 | .. |
| -H32 | 39 | 30 | 23 | 15 | 18 | -H34 | 38 | 31 | 21 | 10 | 14 | -T5 | 27 | 21 | 17 | 12 | .. |
| -H34 | 42 | 33 | 24 | 13 | 16 | -H36 | 40 | 35 | 23 | 8 | 10 | -T6 | 35 | 31 | 22 | 12 | .. |
| -H36 | 45 | 36 | 26 | 12 | 14 | -H38 | 42 | 37 | 24 | 7 | 8 | 6563-T4 | 20 | 11 | .. | 20 | .. |
| -H38 | 48 | 39 | 28 | 10 | .. | 6053-O | 16 | 8 | 11 | .. | 35 | -T6 | 28 | 23 | .. | 12 | .. |
| -H112 | 35 | 17 | 22 | 25 | .. | -T6 | 37 | 32 | 23 | .. | 13 | 6951-O | 16 | 6 | 11 | 30 | .. |
| 5254-O | 35 | 17 | 22 | 27 | 30 | 6061-O | 18 | 8 | 12 | 25 | 30 | -T6 | 39 | 33 | 26 | 13 | .. |
| -H32 | 39 | 30 | 23 | 15 | 18 | -T4 | 35 | 21 | 24 | 22 | 25 | 7075-O | 33 | 15 | 22 | 17 | 16 |
| -H34 | 42 | 33 | 24 | 13 | 16 | -T451 | 35 | 21 | 24 | 22 | 25 | -T6 | 83[6] | 73[6] | 48 | 11 | 11 |
| -H36 | 45 | 36 | 26 | 12 | 14 | -T6 | 45[5] | 40[5] | 30 | 12 | 17 | -T651 | 83[6] | 73[6] | 48 | 11 | 11 |
| -H38 | 48 | 39 | 28 | 10 | .. | -T651 | 45[5] | 40[5] | 30 | 12 | 17 | -T73 | 73 | 63 | .. | .. | 13 |
| -H112 | 35 | 17 | 22 | 25 | .. | -T81 | 55 | 52 | 32 | .. | 15 | *-O | 32 | 14 | 22 | 17 | .. |
| 5357-O | 19 | 7 | 12 | 25 | .. | -T91 | 59 | 57 | 33 | .. | 12 | *-T6 | 76 | 67 | 46 | 11 | .. |
| -H25 | 27 | 23 | 16 | 10 | .. | -T913 | 67 | 66 | 35 | .. | 10 | *-T651 | 76 | 67 | 46 | 11 | .. |
| -H26 | 29 | 24 | 17 | 10 | .. | *-O | 17 | 7 | 11 | 25 | .. | 7079-T6 | 78 | 68 | 45 | .. | 14 |
| -H38 | 32 | 30 | 18 | 6 | .. | *-T451 | 33 | 19 | 22 | 22 | .. | -T651 | 78 | 68 | 45 | .. | 14 |
| 5454-O | 36 | 17 | 23 | 22 | 25 | *-T6 | 42 | 37 | 27 | 12 | .. | 7178-O | 33 | 15 | 22 | 15 | 16 |
| -H32 | 40 | 30 | 24 | 10 | 18 | *-T651 | 42 | 37 | 27 | 12 | .. | -T6 | 88[7] | 78[7] | 52 | 10 | 11 |
| -H34 | 44 | 35 | 26 | 10 | 16 | 6062-O | 18 | 8 | 12 | .. | 30 | -T651 | 88[7] | 78[7] | 52 | 10 | 11 |
| -H112 | 36 | 18 | 23 | 20 | .. | -T4 | 35 | 21 | 24 | .. | 25 | *-O | 32 | 14 | 22 | 16 | .. |
| -H311 | 38 | 26 | 23 | 18 | .. | -T451 | 35 | 21 | 24 | .. | 25 | *-T6 | 81 | 71 | 49 | 10 | .. |
| 5456-O | 45 | 23 | 27 | 24 | 20 | -T6 | 45 | 40 | 30 | .. | 17 | *-T651 | 81 | 71 | 49 | 10 | .. |
| -H24 | 54 | 41 | 31 | 12 | .. | -T651 | 45 | 40 | 30 | .. | 17 | .... | .. | .. | .. | .. | .. |
| -H112 | 45 | 24 | 27 | 22 | .. | | | | | | | | | | | | |

* Alclad alloy is designated.

[1] These typical properties compiled from a listing prepared by the Aluminum Company of America are average for various forms, sizes and methods of manufacture, and may not describe any one particular product. The moduli of elasticity (average of tension and compression) in pounds per square inch for these wrought alloys range from 10.0 × 10⁶ to 10.4 × 10⁶. The compression modulus for any particular alloy is about 2 per cent greater than the tension modulus.

[2] Strengths are given in units of thousands of pounds per square inch. Symbols heading columns are as follows: UT, ultimate tensile strength; Y, yield strength (0.2 per cent offset); and US, ultimate shearing strength.

[3] The elongations are listed under two column headings, 1/16 and 1/2. The values under 1/16 refer to 1/16-inch *thick* specimens and those under 1/2 to 1/2-inch *diameter* specimens.

[4] Applies to extrusions only.

[5] Die forgings will have strengths approximately 5 per cent higher than this value.

[6] Extruded products will have strengths approximately 10 per cent higher than this value and die forgings have strengths approximately 4 per cent lower than this value.

[7] Extruded products will have strengths approximately 10 per cent higher than this value.

## Table 2. Wrought Aluminum Alloys — Composition, Forms, Tempers and Properties[1]

| Alloy | Cu | Si | Mn | Mg | Cr | Ot | Commercial Forms Available In[3] | Usual Tempers | Corrosion Resistance | Workability (Cold) | Machinability | Brazeability | Gas | Arc | Resistance[22] | Forgeability |
|---|---|---|---|---|---|---|---|---|---|---|---|---|---|---|---|---|
| EC | .... | .... | .... | .... | .... | 5 | W, R, B, RS | -O | A | A | D | A | A | A | B | ... |
| | | | | | | | | -H12 | A | A | D | A | A | A | A | ... |
| | | | | | | | | -H14 | A | A | C | A | A | A | A | ... |
| | | | | | | | | -H16 | A | A | C | A | A | A | A | ... |
| | | | | | | | | -H19 | A | C | C | A | A | A | A | ... |
| 1060 | .... | .... | .... | .... | .... | 5 | ..... | -O | A | ... | ... | ... | A | A | A | ... |
| | | | | | | | | -H14 | A | ... | ... | ... | A | A | A | ... |
| 1100 | .... | .... | .... | .... | .... | 6 | S, P, W, R, B, RI, F | -O | A | A | D | A | A | A | B | A |
| | | | | | | | | -H12 | A | A | D | A | A | A | A | A |
| | | | | | | | | -H14 | A | A | C | A | A | A | A | A |
| | | | | | | | | -H16 | A | B | C | A | A | A | A | A |
| | | | | | | | | -H18 | A | C | C | A | A | A | A | A |
| 2011 | 5.5 | .... | .... | .... | .... | 7 | W, R, B, D8 | -T3 | D | C | A | D | D | D | D | ... |
| | | | | | | | | -T8 | C | D | A | D | D | D | D | ... |
| 2014 | 4.4 | 0.8 | 0.8 | .40 | .... | .... | P, R, B, E, F | -T4 | C | C | B | D | D | B | B | C |
| | | | | | | | | -T6 | C | D | B | D | D | B | B | C |
| 2017 | 4.0 | .... | .50 | .50 | .... | .... | W, R, B, RI | -T4 | C | C | B | D | D | B | B | ... |
| 2018 | 4.0 | .... | .... | 0.6 | .... | 9 | F | -T61 | C | ... | B | D | D | B | B | C |
| 2024 | 4.5 | .... | 0.6 | 1.5 | .... | .... | S, P, W, R, B, E, D, RI | -T3 | C | C | B | D | D | B | B | ... |
| | | | | | | | | -T4 | C | C | B | D | D | B | B | ... |
| | | | | | | | | -T36 | C | D | B | D | D | B | B | ... |
| | | | | | | | | -T81 | C | D | B | D | D | B | B | ... |
| 2117 | 2.5 | .... | .... | .30 | .... | .... | W10, R11, RI | -T4 | C | B | C | D | B | B | B | ... |
| 2218 | 4.0 | .... | .... | 1.5 | .... | 9 | F | -T72 | C | ... | B | D | ... | B | ... | D |
| 2219 | 6.3 | .... | .30 | .... | .... | 12 | S, P, F | -T31 | C | C | B | D | A | A | A | C |
| | | | | | | | | -T37 | C | D | B | D | A | A | A | C |
| | | | | | | | | -T81 | C | D | B | D | A | A | A | C |
| | | | | | | | | -T87 | C | D | B | D | A | A | A | C |
| 3003 | .... | .... | 1.2 | .... | .... | .... | S, P, W, R, B, E, D, F | -O | A | A | D | A | A | A | B | A |
| | | | | | | | | -H12 | A | A | D | A | A | A | A | A |
| | | | | | | | | -H14 | A | B | C | A | A | A | A | A |
| | | | | | | | | -H16 | A | C | C | A | A | A | A | A |
| | | | | | | | | -H18 | A | C | C | A | A | A | A | A |
| 3004 | .... | .... | 1.2 | 1.0 | .... | .... | S, P | -O | A | A | D | B | B | A | B | ... |
| | | | | | | | | -H32 | A | B | D | B | B | B | A | ... |
| | | | | | | | | -H34 | A | B | C | B | B | B | A | ... |
| | | | | | | | | -H36 | A | C | C | B | B | B | A | ... |
| | | | | | | | | -H38 | A | C | C | B | B | B | A | ... |
| 4032 | 0.9 | 12.2 | .... | 1.1 | .... | 13 | F | -T6 | C | ... | ... | D | D | B | C | ... |

See footnotes at end of table.

Table 2 (*Continued*).  Wrought Aluminum Alloys — Composition, Forms, Tempers and Properties[1]

| Alloy | Nominal Composition,[2] Per Cent | | | | | | Commercial Forms Available In[3] | Usual Tempers | Corrosion Resistance | Workability (Cold) | Machinability | Brazeability | Weldability | | | Forgeability |
|---|---|---|---|---|---|---|---|---|---|---|---|---|---|---|---|---|
| | Cu | Si | Mn | Mg | Cr | Ot | | | | | | | Gas | Arc | Resistance[22] | |
| 5005 | .... | .... | .... | 0.8 | .... | .... | S, P, W, R[14] | -O<br>-H12<br>-H14<br>-H16<br>-H18<br>-H32<br>-H34<br>-H36<br>-H38 | A<br>A<br>A<br>A<br>A<br>A<br>A<br>A<br>A | A<br>A<br>B<br>C<br>C<br>A<br>B<br>C<br>C | D<br>D<br>C<br>C<br>C<br>D<br>C<br>C<br>C | B<br>B<br>B<br>B<br>B<br>B<br>B<br>B<br>B | A<br>A<br>A<br>A<br>A<br>A<br>A<br>A<br>A | A<br>A<br>A<br>A<br>A<br>A<br>A<br>A<br>A | B<br>A<br>A<br>A<br>A<br>A<br>A<br>A<br>A | ...<br>...<br>...<br>...<br>...<br>...<br>...<br>...<br>... |
| 5050 | .... | .... | .... | 1.4 | .... | .... | S, P, D[8] | -O<br>-H32<br>-H34<br>-H36<br>-H38 | A<br>A<br>A<br>A<br>A | A<br>B<br>B<br>C<br>C | D<br>D<br>C<br>C<br>C | B<br>B<br>B<br>B<br>B | A<br>A<br>A<br>A<br>A | A<br>A<br>A<br>A<br>A | B<br>A<br>A<br>A<br>A | ...<br>...<br>...<br>...<br>... |
| 5052 | .... | .... | .... | 2.5 | .25 | .... | S, P, W, R, B, D[8] | -O<br>-H32<br>-H34<br>-H36<br>-H38 | A<br>A<br>A<br>A<br>A | A<br>B<br>B<br>C<br>C | D<br>D<br>C<br>C<br>C | C<br>C<br>C<br>C<br>C | A<br>A<br>A<br>A<br>A | A<br>A<br>A<br>A<br>A | B<br>A<br>A<br>A<br>A | ...<br>...<br>...<br>...<br>... |
| 5056 | .... | .... | .10 | 5.2 | .10 | .... | W, R[11], RI | -O<br>-H38 | A<br>A | A<br>C | D<br>C | D<br>D | C<br>C | A<br>A | B<br>A | ...<br>... |
| 5083 | .... | .... | 0.8 | 4.45 | .10 | .... | P, F | -O<br>-H113<br>-H323<br>-H343 | A<br>A<br>A<br>B | B<br>C<br>C<br>C | D<br>D<br>D<br>C | D<br>D<br>D<br>D | C<br>C<br>C<br>C | A<br>A<br>A<br>A | B<br>A<br>A<br>A | ...<br>...<br>...<br>... |
| 5086 | .... | .... | .45 | 4.0 | .10 | .... | S, P, E | -O<br>-H32<br>-H34<br>-H36<br>-H38 | A<br>A<br>A<br>B<br>B | A<br>B<br>B<br>C<br>C | D<br>D<br>C<br>C<br>C | D<br>D<br>D<br>D<br>D | C<br>C<br>C<br>C<br>C | A<br>A<br>A<br>A<br>A | A<br>A<br>A<br>A<br>A | ...<br>...<br>...<br>...<br>... |
| 5154 | .... | .... | .... | 3.5 | .25 | .... | W[15], E[16] | -O<br>-H32<br>-H34<br>-H36<br>-H112 | A<br>A<br>A<br>A<br>A | A<br>B<br>B<br>C<br>B | D<br>D<br>C<br>C<br>D | D<br>D<br>D<br>D<br>D | C<br>C<br>C<br>C<br>C | A<br>A<br>A<br>A<br>A | B<br>A<br>A<br>A<br>A | ...<br>...<br>...<br>...<br>... |
| 5357 | .... | .... | .30 | 1.0 | .... | .... | S, P | -O<br>-H25<br>-H26<br>-H38 | A<br>A<br>A<br>A | A<br>B<br>B<br>C | D<br>C<br>C<br>C | B<br>B<br>B<br>B | A<br>A<br>A<br>A | A<br>A<br>A<br>A | B<br>A<br>A<br>A | ...<br>...<br>...<br>... |
| 5454 | .... | .... | 0.8 | 2.75 | .10 | .... | S, P, E | -O<br>-H32<br>-H34<br>-H36<br>-H38 | A<br>A<br>B<br>B<br>B | A<br>B<br>B<br>C<br>C | D<br>D<br>C<br>C<br>C | D<br>D<br>D<br>D<br>D | C<br>C<br>C<br>C<br>C | A<br>A<br>A<br>A<br>A | B<br>A<br>A<br>A<br>A | ...<br>...<br>...<br>...<br>... |
| 5456 | .... | .... | 0.8 | 5.25 | .10 | .... | S, P, E | -O<br>-H321<br>-H323<br>-H343 | A<br>A<br>A<br>B | B<br>B<br>C<br>C | D<br>C<br>C<br>C | D<br>D<br>D<br>D | C<br>C<br>C<br>C | A<br>A<br>A<br>A | B<br>A<br>A<br>A | ...<br>...<br>...<br>... |

See footnotes at end of table.

**Table 2** (*Concluded*). **Wrought Aluminum Alloys — Composition, Forms, Tempers and Properties[1]**

| Alloy | Nominal Composition,[2] Per Cent | | | | | | Commercial Forms Available In[3] | Temper and Temper Properties[4] | | | | | | | | |
|---|---|---|---|---|---|---|---|---|---|---|---|---|---|---|---|---|
| | | | | | | | | Usual Tempers | Corrosion Resistance | Workability (Cold) | Machinability | Brazeability | Weldability | | | Forgeability |
| | Cu | Si | Mn | Mg | Cr | Ot | | | | | | | Gas | Arc | Resistance[22] | |
| 5457 | .... | .... | .30 | 1.0 | .... | .... | S | -O | A | A | D | B | A | A | B | ... |
| | | | | | | | | -H25 | A | B | C | B | A | A | A | ... |
| | | | | | | | | -H26 | A | B | C | B | A | A | A | ... |
| | | | | | | | | -H38 | A | C | C | B | A | A | A | ... |
| 5557 | .... | .... | .25 | 0.6 | .... | .... | S | -O | A | A | D | B | A | A | B | ... |
| | | | | | | | | -H25 | A | B | C | B | A | A | A | ... |
| | | | | | | | | -H26 | A | B | C | B | A | A | A | ... |
| | | | | | | | | -H38 | A | C | C | B | A | A | A | ... |
| 6061 | 0.25 | 0.6 | .... | 1.0 | .25 | .... | S, P, W, R, B, RS, E, D, RI, F | -O | A | A | D | A | A | A | B | ... |
| | | | | | | | | -T4 | A | B | C | A | A | A | A | ... |
| | | | | | | | | -T6 | A | C | C | A | A | A | A | ... |
| 6062 | 0.25 | 0.6 | .... | 1.0 | .06 | .... | E[17], D[8] | -O | A | A | D | A | A | A | B | ... |
| | | | | | | | | -T4 | A | B | C | A | A | A | A | ... |
| | | | | | | | | -T6 | A | C | C | A | A | A | A | ... |
| 6063 | .... | .40 | .... | 0.7 | .... | .... | E, D | -O | A | A | D | A | A | A | B | ... |
| | | | | | | | | -T4 | A | B | C | A | A | A | A | ... |
| | | | | | | | | -T5 | A | B | C | A | A | A | A | ... |
| | | | | | | | | -T6 | A | C | C | A | A | A | A | ... |
| | | | | | | | | -T42 | A | A | C | A | A | A | A | ... |
| | | | | | | | | -T83 | A | C | C | A | A | A | A | ... |
| | | | | | | | | -T831 | A | C | C | A | A | A | A | ... |
| | | | | | | | | -T832 | A | C | C | A | A | A | A | ... |
| 6066 | 0.9 | 1.3 | 0.9 | 1.1 | .... | .... | ..... | -O | B | B | D | A | A | A | B | ... |
| | | | | | | | | -T4 | B | C | B | A | A | A | A | ... |
| | | | | | | | | -T6 | B | C | B | A | A | A | A | ... |
| 7075 | 1.6 | .... | .... | 2.5 | .30 | 18 | S,P,W,R,B,E,F | -T6 | C | D | B | D | D | D | B | D |
| 7079 | 0.6 | .... | .20 | 3.3 | .20 | 19 | P, E[20], F | -T6 | C | D | D | B | B | D | B | D |
| 7178 | 2.0 | .... | .... | 2.7 | .30 | 21 | S, P, E | -T6 | C | D | B | B | D | C | B | D |

[1] This selected listing of wrought aluminum alloys has been compiled from listings prepared by the Aluminum Company of America and does not pretend to include all available alloys and tempers. More detailed information can be obtained from one or more of the commercial aluminum producers. [2] Values shown are per cent of alloying elements. Aluminum and normal impurities constitute the remainder. The following abbreviations are used: Cu, copper; Si, silicon; Mn, manganese; Mg, magnesium; Cr, chromium; and Ot, other. Only footnote designation numbers are used in the Ot headed column and the other alloying element or elements are given in the footnotes. [3] Abbreviations used are: S, sheet; P, plate; W, wire; R, rod; B, bar; RS, rolled shapes; E, extruded shapes, tube and pipe; D, drawn tube and pipe; RI, rivets; and F, forgings and forging stock. [4] Corrosion Resistance, etc. ratings A, B, C and D are relative ratings in decreasing order of merit. Weldability and Brazeability are defined as follows: A. Generally weldable by all commercial procedures and methods. B. Weldable with special technique or on specific applications which justify preliminary trials or testing to develop welding procedure and weld performance. C. Limited weldability because of crack sensitivity or loss in resistance to corrosion, and all mechanical properties. D. No commonly used welding methods have so far been developed. [5] 99.60 per cent aluminum, min. [6] 99.00 per cent aluminum, min. [7] Also contains lead, 0.50 and bismuth, 0.50. [8] Tube only. [9] Also contains nickel, 2.0. [10] Rivet wire only. [11] Rivet rod and redraw rod only. [12] Also contains titanium, 0.06; vanadium, 0.10; and zirconium, 0.18. [13] Also contains nickel, 0.9. [14] Redraw rod only. [15] Wire, straight length welding rod spooled I.G. electrode. [16] "Special" for pipe. [17] Extruded shapes and tube only. [18] Also contains zinc, 5.6. [19] Also contains zinc, 4.3. [20] Extruded shapes only. [21] Also contains zinc, 6.8. [22] Spot and seam.

## Magnesium Alloys

Pure magnesium is a relatively soft, silver-white metal. In its pure state it does not possess sufficient strength for many commercial uses. When alloyed with certain other metals, chiefly aluminum, manganese and zinc, a wide range of useful properties is obtained. Some of these alloys are characterized by their strength, others by their toughness, and still others by their thermal conductivity. The chief characteristic is extreme lightness, the average specific gravity being only 1.80.

**Compositions of Dowmetal Alloys.** — Some magnesium alloys are designated by the trade name "Dowmetal." Dowmetal alloys are available for sand and permanent mold castings, die-castings, press and hammer forgings, extruded shapes, plates, sheets and strips. Their compositions vary more or less for different applications. The nominal compositions of these alloys are given in the accompanying table.

**Nominal Compositions of Dowmetal Alloys \***

| Element | Dowmetal Alloy | | | | | | | |
|---|---|---|---|---|---|---|---|---|
| | C | FS-1 | G | H | J-1 | M | O-1 | R |
| | Composition, Per Cent | | | | | | | |
| Aluminum........ | 9.0 | 3.0 | 10.0 | 6.0 | 6.5 | .... | 8.5 | 9.0 |
| Manganese....... | 0.1 | 0.3 | 0.1 | 0.2 | 0.2 | 1.5 | 0.2 | 0.2 |
| Zinc... .......... | 2.0 | 1.0 | .... | 3.0 | 1.0 | .... | 0.5 | 0.6 |
| Magnesium ...... | 88.9 | 95.7 | 89.9 | 90.8 | 92.3 | 98.5 | 90.8 | 90.2 |

\* Note: These compositions vary slightly depending on the application of the alloy. Other elements, such as silicon, copper, nickel and iron, present in small amounts as impurities, not shown.

**Applications of Dowmetal Alloys.** — Recommendations for the use of various Dowmetal alloys in the form of castings, forgings, extruded shapes and rolled forms are given below.

*Sand Castings:* Dowmetal C and Dowmetal H are most commonly used for sand castings. Where pressure tightness is the governing factor especially in thin walled castings, Dowmetal C is preferred. Where high strength is the important factor, Dowmetal H is indicated. Both alloys can be heat-treated for improvement of various properties. Dowmetal M has casting characteristics inferior to those of Dowmetals C and H but it does have good welding characteristics and is used for such applications as tank fittings where castings must be welded into place.

*Permanent Mold Castings:* Dowmetal C is most widely used for permanent mold castings because of its combination of good casting characteristics, mechanical properties and corrosion resistance. Dowmetal G has better casting characteristics than Dowmetal C but its corrosion resistance and mechanical properties are slightly inferior. Dowmetal H does not have as good foundry characteristics as C and G, hence it is used only for special applications.

*Die-castings:* Dowmetal R is the magnesium alloy most widely used for die-castings. It possesses a desirable combination of good casting characteristics and good mechanical properties. Special alloys are available for die-castings if specific properties are required.

*Forgings:* The most widely used Dowmetal alloys for forging are J-1 and O-1, due to their superior mechanical properties. Dowmetal O-1 is used where maximum strength is required, whereas Dowmetal J-1 is indicated where greater formability and weldability are desired. Dowmetal O-1 can be heat-treated for improved mechanical properties. Dowmetal *M* is used where maximum formability and weldability are desired and strength is not of paramount importance.

*Extruded Shapes:* Dowmetals FS-1, J-1, *M* and O-1 are available as extruded materials. The alloys are made in all forms of extrusions with the exception that Dowmetal O-1 is not available as tubing. Dowmetal J-1 is a general-purpose extrusion alloy of improved strength. Dowmetal *M* is a moderate strength alloy with the best weldability and hot formability. Dowmetal O-1 has the highest strength in the extruded condition and is heat treatable.

*Sheet, Plate, and Strip:* Dowmetals FS-1, J-1 and *M* are available as rolled magnesium products. Each alloy is available in the annealed condition, denoted by the letter "*a*" following the alloy designation, or in the hard rolled condition denoted by the letter "*h*". Dowmetal J-1 is also available as plate in the as rolled condition designated as J-1r. Dowmetal J-1 has the best mechanical properties of the rolled alloys and is used in applications where strength is most important. Where better formability is desired along with good shear and tensile strengths, Dowmetal FS-1 is used. Dowmetal *M* is used where maximum weldability and formability, low cost, and moderate strength are desired.

**Physical Properties of Dowmetal Alloys.** — The physical properties of the various Dowmetal alloys varies with their composition and condition, i.e., whether as cast, heat-treated, aged, stabilized, etc. The range in tensile strengths of these alloys is given below.

*Casting Alloys:* In the cast condition ultimate tensile strengths range from 14,000 to 29,000 pounds per square inch and yield strengths from 4500 to 14,000 pounds per square inch. In the heat-treated condition ultimate tensile strengths range from 33,000 to 40,000 pounds per square inch and ultimate strengths from 12,000 to 23,000 pounds per square inch.

*Forging Alloys:* The ultimate tensile strengths range from 36,000 to 50,000 pounds per square inch and the yield strengths from 23,000 to 34,000 pounds per square inch.

*Extruded Alloys:* For bars and rods the ultimate tensile strengths range from 38,000 to 50,000 pounds per square inch and the yield strengths from 26,000 to 34,000 pounds per square inch.

For shapes, the ultimate tensile strength ranges from 34,000 to 49,000 pounds per square inch and the yield strength from 20,000 to 32,000 pounds per square inch. For tubing the ultimate tensile strength ranges from 33,000 to 40,000 pounds per square inch and the yield strength is 21,000 pounds per square inch.

*Rolled Alloys:* In the annealed condition the ultimate tensile strength ranges from 33,000 to 43,000 pounds per square inch and the yield strength from 15,000 to 26,000 pounds per square inch. In the hard rolled condition the ultimate tensile strength ranges from 37,000 to 47,000 pounds per square inch and the yield strength from 29,000 to 34,000 pounds per square inch. In the as rolled condition the ultimate tensile strength of the J-1 alloy is 43,000 pounds per square inch and the yield strength 28,000 pounds per square inch.

**Heat-treatment of Dowmetal Alloys.** — Dowmetal castings may be used as cast or in a heat-treated condition. Heat-treatment is not required for general use. However, when increased tensile strength, ductility and toughness are required, without change of yield strength or hardness, castings are "solution heat-treated." This solution heat-treatment is performed in specially designed ovens at temperatures varying from 630 to 785 degrees F., depending upon the alloy, and is followed by air-cooling. Castings so treated are in the best condition for shock resistance. If castings require high yield strength but are not subject to shock, they are solution heat-treated and aged. This aging or "precipitation" is done at about 350 degrees F.

## S.A.E. Cast Magnesium Alloys

**S.A.E. Standard No. 50 Alloy.** — This alloy is used for most commercial applications. It is used in the "as cast," "heat treated," or "heat treated and aged" condition as may be required.

*Composition of No. 50:* Aluminum, 5.3 to 6.7; manganese, min., 0.15; zinc, 2.5 to 3.5; silicon, max., 0.5; copper, max., 0.05; nickel, max., 0.03; other impurities, max., 0.3 per cent and the remainder, magnesium.

*Physical Properties:* For sand castings as in the "as cast," "heat treated" and "heat treated and aged" conditions the minimum tensile strengths are respectively: 24,000, 30,000 and 32,000 pounds per square inch; the minimum yield strengths are respectively: 10,000, 10,000 and 16,000 pounds per square inch and the elongations in 2 inches are respectively: 4, 6 and 2 per cent.

**S.A.E. Standard No. 500 Alloy.** — This is a sand casting alloy to be used particularly where maximum pressure tightness is required. It may be used in the "as cast," "heat treated" or "heat treated and aged" condition as may be required.

*Composition of No. 500:* Aluminum 8.3 to 9.7; manganese, min., 0.10; zinc, 1.7 to 2.3; silicon, max., 0.5; copper, max., 0.05; nickel, max., 0.03; other impurities, max., 0.3 per cent and the remainder, magnesium.

*Physical Properties:* For sand castings in the "as cast," "heat treated" and "heat treated and aged" conditions, the minimum tensile strengths are respectively: 20,000, 30,000 and 32,000 pounds per square inch; the yield strengths are respectively: 10,000, 10,000 and 17,000 pounds per square inch and the elongations in 2 inches are respectively: 1, 6 and 1 per cent.

## S.A.E. Wrought Magnesium Alloys

**S.A.E. Standard No. 51 Alloy.** — This alloy is used where maximum salt water resistance and weldability are desired. It is used in the annealed temper for applications requiring maximum formability, such as aircraft tanks and wheel fairings.

*Composition of No. 51:* Manganese, min., 1.20; silicon, max., 0.3; copper, max., 0.05; nickel, max., 0.03; other impurities, max., 0.3 per cent and the remainder, magnesium.

*Physical Properties:* Standard tensile test specimens machined from plate or sheet stock in thicknesses between 0.016 inch and 0.025 inch have a minimum tensile strength of 32,000 pounds per square inch in the hard rolled temper, a maximum tensile strength of 35,000 pounds per square inch in the annealed temper and an elongation in 2 inches of 4 per cent in the hard rolled temper and 12 per cent in the annealed temper.

**S.A.E. Standard No. 510 Alloy.** — This alloy is generally used where moderate formability and mechanical properties are required.

*Composition of No. 510:* Aluminum, 3.3 to 4.7; manganese, min., 0.20; zinc, max., 0.3; silicon, max., 0.5; copper, max., 0.05; nickel, max., 0.03; other impurities, max., 0.3 per cent and the remainder, magnesium.

*Physical Properties:* Standard tensile test specimens machined from plate or sheet stock in thicknesses between 0.16 and 0.125 inch have a tensile strength of 36,000 pounds per square inch, minimum in the hard rolled temper and 38,000 pounds per square inch, maximum in the annealed temper; a yield strength of 25,000 pounds per square inch in the hard rolled temper and an elongation in 2 inches of 4 per cent in the hard rolled temper and 10 per cent in the annealed temper.

**S.A.E. Standard No. 511 Alloy.** — This alloy is used where high mechanical properties are required. It is available in the hard rolled and annealed tempers.

*Composition of No. 511:* Aluminum 5.8 to 7.2; manganese, min., 0.15; zinc, max., 0.3; silicon, max., 0.5; copper, max., 0.05; nickel, 0.03; other impurities, 0.3 per cent and the remainder, magnesium.

*Physical Properties:* Standard tension test specimens machined from plate or sheet stock in thicknesses between 0.016 inch and 0.125 inch have tensile strength of 39,000 pounds per square inch, minimum in the hard rolled temper and 42,000 pounds per square inch, maximum in the annealed temper; a yield strength of 28,000 pounds per square inch in the hard rolled temper and an elongation in 2 inches of 3 per cent in the hard rolled temper and 10 per cent in the annealed temper.

**S.A.E. Standard No. 52 Alloy.** — This is a general purpose alloy with moderate strength and fair weldability. It is especially suited for the production of thin wall tubing and other sections requiring good extrusion characteristics.

*Composition of No. 52:* Aluminum, 2.4 to 3.0; manganese, min., 0.20; zinc, 0.7 to 1.3; silicon, max., 0.5; copper, max., 0.05; nickel, max., 0.03; other impurities, max., 0.3 per cent and the remainder magnesium.

*Physical Properties:* Standard test specimens machined from solid bar stock and structural shapes have a minimum tensile strength of 37,000 pounds per square inch in extruded bars up to 1½ inches and 34,000 pounds per square inch in structural shapes; a yield strength of 25,000 pounds per square inch in the former and 17,000 pounds per square inch in the latter and an elongation in 2 inches of 12 per cent in the former and 10 per cent in the latter.

**S.A.E. Standard No. 520 Alloy.** — This alloy is used for extruded bars, rods and shapes with good strength and fair weldability.

*Composition of No. 520:* Aluminum, 5.8 to 7.2; manganese, min., 0.15; zinc, 0.4 to 1.0; silicon, max., 0.5; iron, max., 0.05; nickel, max., 0.03; other impurities, max., 0.3 per cent and the remainder, magnesium.

*Physical Properties:* Standard tension test specimens machined from solid bar stock and structural shapes have a minimum tensile strength of 40,000 pounds per square inch in extruded bars up to 1½ inches and 38,000 pounds per square inch in structural shapes; a yield strength of 26,000 pounds per square inch in the former and 23,000 pounds per square inch in the latter and an elongation in 2 inches of 12 per cent in the former and 10 per cent in the latter.

**S.A.E. Standard No. 522 Alloy.** — This is an extrusion alloy used for applications requiring maximum weldability.

*Composition of No. 522:* Manganese, min. 1.2; silicon, max. 0.3; copper, max. 0.05; nickel, max. 0.03; and calcium, 0.3 per cent; remainder, magnesium.

*Physical Properties:* Tensile strength is 30,000 pounds per square inch for extruded bars, ¼ inch to 1½ inches; 29,000 pounds per square inch for structural shapes and 28,000 pounds per square inch for hollow shapes. Elongation in 2 inches is 3 per cent for extruded bars, ¼ inch to 1½ inches and 2 per cent for structural and hollow shapes.

**S.A.E. Standard Nos. 53, 531, 532 and 533 Alloys.** — These are forging alloys. Nos. 53 and 533 are suitable for hammer forging. The former has somewhat better physical properties but the latter may be readily welded and contains no tin. No.

533 may also be press forged. Hammer forgings are normally more economical than press forgings but can only be used for applications involving moderate stresses. Press forging alloys Nos. 531 and 532 are used in applications involving higher stresses. No. 532 is stronger than No. 531 but more difficult to forge and is usually employed only for comparatively simple forgings requiring highest physical properties.

*Composition of No. 53:* Aluminum, 3.0 to 4.0; manganese, min. 0.2; zinc, max. 0.3; silicon, max. 0.3; copper, max. 0.05; nickel, max. 0.005; iron, max. 0.005 and tin, 4.0 to 6.0 per cent; remainder, magnesium.

*Composition of No. 531:* Aluminum, 5.8 to 7.2; manganese, minimum 0.15; zinc, 0.4 to 1.5; silicon, maximum 0.3; copper, max. 0.05; nickel, maximum 0.005 and iron, maximum 0.005 per cent; remainder, magnesium.

*Composition of No. 532:* Aluminum, 7.8 to 9.2; manganese, minimum 0.12; zinc, 0.2 to 0.8; silicon, maximum 0.3; copper, maximum 0.05; nickel, maximum 0.005 and iron, maximum 0.005 per cent; remainder, magnesium.

*Composition of No. 533:* Manganese, minimum 1.2; silicon, maximum 0.3; copper, maximum 0.05 and nickel, maximum 0.03 per cent; remainder, magnesium.

*Physical Properties:* In the as forged condition, No. 53 has a minimum tensile strength of 36,000; No. 531, 38,000; No. 532, 42,000; and No. 533, 30,000 pounds per square inch. Nos. 53 and 531 have a yield strength of 22,000; No. 532, 26,000; and No. 533, 18,000 pounds per square inch. No. 53 has a minimum elongation in 2 inches of 7; No. 531, 6; No. 532, 5; and No. 533, 3 per cent.

## Nickel and Nickel Alloys

**Nickel.** — Nickel is noted for its corrosion resistance, good electrical conductivity and high heat-transfer properties. It is used to fabricate process equipment for handling pure foods and drugs, electrical contact parts, and radio and X-ray tube elements.

*Approximate Composition:* (Commercially pure wrought nickel:) Nickel (including cobalt), 99.4; copper, 0.1; iron, 0.15; manganese, 0.25; silicon, 0.05; carbon, 0.05; and sulphur, 0.005. (Cast nickel:) Nickel, 97.0; copper, 0.3; iron, 0.25; manganese, 0.5; silicon, 1.6; and carbon, 0.5.

*Average Physical Properties:* Wrought nickel in the annealed, hot-rolled, cold-drawn, and hard temper cold-rolled conditions exhibits yield strengths (0.2 per cent offset) of 20,000, 25,000, 70,000, and 95,000 pounds per square inch, respectively; tensile strengths of 70,000, 75,000, 95,000, and 105,000 pounds per square inch, respectively; elongations in 2 inches of 40, 40, 25, and 5 per cent, respectively; and Brinell hardnesses of 100, 110, 170, and 210, respectively.

**Low-Carbon Nickel.** — A special type of nickel that is corrosion resistant and has a high ductility and heat resistance. It lends itself well to spinning and cold coining or forging and is used in the manufacture of tubing and molds for the beverage and food industries.

*Approximate Composition:* Nickel, 99.4; copper, 0.05; iron, 0.1; silicon, 0.15; manganese, 0.2; carbon, 0.01; and sulphur, 0.005.

*Average Physical Properties:* Annealed low-carbon nickel exhibits a yield strength (0.2 per cent offset) of 15,000 pounds per square inch, a tensile strength of 60,000 pounds per square inch, an elongation in 2 inches of 50 per cent and a Brinell hardness of 90.

**Duranickel.** — This age-hardenable alloy has good spring and low-sparking properties and is slightly magnetic after heat treatment. Items such as corrosion-resistant paper machine shaker springs, diaphragms, and extrusion dies for plastics are made from it.

*Approximate Composition:* Nickel, 93.7; copper, 0.05; iron, 0.35; aluminum, 4.4; silicon, 0.5; manganese, 0.3; carbon, 0.17; and sulphur, 0.005.

*Average Physical Properties:* In the hot-rolled, hot-rolled and age-hardened, cold-drawn, and cold-drawn and age-hardened conditions this alloy exhibits yield strengths (0.2 per cent offset) of 50,000, 130,000, 90,000, and 135,000 pounds per square inch, respectively; tensile strengths of 105,000, 170,000, 120,000, and 175,000 pounds per square inch, respectively; elongations in 2 inches of 35, 15, 25, and 15 per cent, respectively; and Brinell hardnesses of 180, 320, 220, and 340, respectively.

**Monel.** — This general purpose alloy is corrosion-resistant, strong, tough and has a silvery-white color. It is used for making abrasion- and heat-resistant valves and pump parts, propeller shafts, laundry machines, chemical processing equipment, etc.

*Approximate Composition:* Nickel, 67; copper, 30; iron, 1.4; silicon, 0.1; manganese, 1; carbon, 0.15; and sulphur 0.01.

*Average Physical Properties:* Wrought Monel in the annealed, hot-rolled, cold-drawn, and hard temper cold-rolled conditions exhibits yield strengths (0.2 per cent offset) of 35,000, 50,000, 80,000, and 100,000 pounds per square inch, respectively; tensile strengths of 75,000, 90,000, 100,000, and 110,000 pounds per square inch, respectively; elongations in 2 inches of 40, 35, 25, and 5 per cent, respectively; and Brinell hardnesses of 125, 150, 190, and 240, respectively.

**"R" Monel.** — This free-cutting, corrosion resistant alloy is used for automatic screw machine products such as bolts, screws and precision parts.

*Approximate Composition:* Nickel, 67; copper, 30; iron, 1.4; silicon, 0.05; manganese, 1; carbon, 0.15; and sulphur, 0.035.

*Average Physical Properties:* In the hot-rolled and cold-drawn conditions this alloy exhibits yield strengths (0.2 per cent offset) of 45,000 and 75,000 pounds per square inch, respectively; tensile strengths of 85,000 and 90,000 pounds per square inch, respectively; elongations in 2 inches of 35, and 25 per cent, respectively; and Brinell hardnesses of 145 and 180, respectively.

**"K" Monel.** — This strong and hard alloy, comparable to heat-treated alloy steel, is age-hardenable, non-magnetic and has low-sparking properties. It is used for corrosive applications where the material is to be machined or formed, then age hardened. Pump and valve parts, scrapers, and instrument parts are made from this alloy.

*Approximate Composition:* Nickel, 66; copper, 29; iron, 0.9; aluminum, 2.75; silicon, 0.5; manganese, 0.75; carbon, 0.15; and sulphur, 0.005.

*Average Physical Properties:* In the hot-rolled, hot-rolled and age-hardened, cold-drawn, and cold-drawn and age-hardened conditions the alloy exhibits yield strengths (0.2 per cent offset) of 45,000, 110,000, 85,000, and 115,000 pounds per square inch, respectively; tensile strengths of 100,000, 150,000, 115,000, and 155,000 pounds per square inch, respectively; elongations in 2 inches of 40, 25, 25, and 20 per cent, respectively; and Brinell hardnesses of 160, 280, 210, and 290, respectively.

**"KR" Monel.** — This strong, hard, age-hardenable and non-magnetic alloy is more readily machinable than "K" Monel. It is used for making valve stems, small parts for pumps, and screw machine products requiring an age-hardening material that is corrosion-resistant.

*Approximate Composition:* Nickel, 66; copper, 29; iron, 0.9; aluminum, 2.75; silicon, 0.5; manganese, 0.75; carbon, 0.28; and sulphur, 0.005.

*Average Physical Properties:* Essentially the same as "K" Monel.

**"S" Monel.** — This extra hard casting alloy is non-galling, corrosion-resisting, non-magnetic, age-hardenable and has low-sparking properties. It is used for gall-resistant pump and valve parts which have to withstand high temperatures, corrosive chemicals and severe abrasion.

*Approximate Composition:* Nickel, 63; copper, 30; iron, 2; silicon, 4; manganese, 0.75; carbon, 0.1; and sulphur, 0.015.

*Average Physical Properties:* In the annealed sand-cast, as-cast sand-cast, and age-hardened sand-cast conditions it exhibits yield strengths (0.2 per cent offset) of 70,000, 100,000, and 100,000 pounds per square inch, respectively; tensile strengths of 90,000, 130,000, and 130,000 pounds per square inch, respectively; elongations in 2 inches of 3, 2, and 2 per cent, respectively; and Brinell hardnesses of 275, 320, and 350, respectively.

**"H" Monel.** — An extra hard casting alloy with good ductility, intermediate strength and hardness that is used for pumps, impellers and steam nozzles.

*Approximate Composition:* Nickel, 63; copper, 31; iron, 2; silicon, 3; manganese, 0.75; carbon, 0.1; and sulphur, 0.015.

*Average Physical Properties:* In the as-cast sand-cast condition this alloy exhibits a yield strength (0.2 per cent offset) of 60,000 pounds per square inch, a tensile strength of 100,000 pounds per square inch, an elongation in 2 inches of 15 per cent and a Brinell hardness of 210.

**Inconel.** — This heat resistant alloy retains its strength at high heats, resists oxidation and corrosion, has a high creep strength and is non-magnetic. It is used for high temperature applications (up to 2000 degrees F.) such as engine exhaust manifolds and furnace and heat treating equipment. Springs operating at temperatures up to 700 degrees F. are also made from it.

*Approximate Composition:* Nickel, 76; copper, 0.20; iron, 7.5; chromium, 15.5; silicon, 0.25; manganese, 0.25; carbon, 0.08; and sulphur, 0.007.

*Physical Properties:* Wrought Inconel in the annealed, hot-rolled, cold-drawn, and hard temper cold-rolled conditions exhibits yield strengths (0.2 per cent offset) of 35,000, 60,000, 90,000, and 110,000 pounds per square inch, respectively; tensile strengths of 85,000, 100,000, 115,000, and 135,000 pounds per square inch, respectively; elongations in 2 inches of 45, 35, 20, and 5 per cent, respectively; and Brinell hardnesses of 150, 180, 200, and 260, respectively.

**Inconel "X".** — This alloy has a low creep rate, is age-hardenable and non-magnetic, resists oxidation and exhibits a high strength at elevated temperatures. Uses include the making of bolts and turbine rotors used at temperatures up to 1500 degrees F., aviation brake drum springs and relief valve and turbine springs with low load-loss or relaxation for temperatures up to 1000 degrees F.

*Approximate Composition:* Nickel, 73; copper, 0.2 maximum; iron, 7; chromium, 15; aluminum, 0.7; silicon, 0.4; manganese, 0.5; carbon, 0.04; sulphur, 0.007; columbium, 1; and titanium, 2.5.

*Average Physical Properties:* Wrought Inconel " X " in the annealed and age-hardened hot-rolled conditions exhibits yield strengths (0.2 per cent offset) of 50,000 and 120,000 pounds per square inch, respectively; tensile strengths of 115,000 and 180,000 pounds per square inch, respectively; elongations in 2 inches of 50 and 25 per cent, respectively; and Brinell hardnesses of 200 and 360, respectively.

## Titanium and Titanium Alloys

**Titanium.** — This metal is used in its commercially pure state and in alloy form (being alloyed with manganese or ferrochromium) for applications requiring a metal with properties of light weight, high strength, and good temperature- and corrosion-resistance. Titanium and its alloys weigh approximately 44 per cent less

## Properties of Titanium and Titanium Alloys

| Type* | Yield Strength, Pounds per Sq. In. | Ult. Tensile Strength, Pounds per Sq. In. | Per Cent Elongation in 2 In. |
|---|---|---|---|
| Titanium, pure | 40,000 to 85,000 | 60,000 to 110,000 | 30 to 20 |
| 3% Ferrochromium or 4% Manganese | 75,000 to 110,000 | 100,000 to 125,000 | 20 to 15 |
| 6% Ferrochromium | 110,000 to 125,000 | 120,000 to 155,000 | 18 to 10 |
| 7% Manganese | 120,000 to 160,000 | 130,000 to 170,000 | 18 to 7 |

* The percentage of chief alloying elements is given, except for the first entry which is commercially pure titanium.

than stainless or alloy steels, are equal or greater in yield and ultimate tensile strength than structural alloys in common use, withstand temperatures up to 800 degrees F. and higher temperatures up to 2000 degrees F. for short periods and are resistant to the corrosive effects of salt water and many acids, alkalis and other chemicals. It is available in the form of plates, sheets, strip, forgings, ingots, bars, rods, and wire.

*Composition and Properties:* The accompanying table gives the nominal compositions, yield strengths, tensile strengths and elongations of titanium and some of its alloys.

## Copper-Silicon and Beryllium Copper Alloys

**Everdur.** — This copper-silicon alloy is available in five slightly different nominal compositions for applications which require high strength, good fabricating and fusing qualities, immunity to rust, free-machining and a corrosion resistance equivalent to copper. The following table gives the nominal compositions and tensile strengths, yield strengths, and per cent elongations for various tempers and forms.

## Nominal Composition and Properties of Everdur

| Desig. No. | Nominal Composition[1] | | | | | Temper[2] | Strength, Thousands of Pounds per Square Inch | | Per Cent Elongation |
|---|---|---|---|---|---|---|---|---|---|
| | Cu | Si | Mn | Pb | Al | | Tensile | Yield | |
| 1010 | 95.80 | 3.10 | 1.10 | ... | ... | A | 52 | 15 | 35* |
| | | | | | | HRA | 50 | 18 | 40 |
| | | | | | | CRA | 52 | 18 | 35 |
| | | | | | | CRHH | 71 | 40 | 10 |
| | | | | | | CRH | 87 | 60 | 3 |
| | | | | | | H | 70 to 85 | 38 to 50 | 17 to 8* |
| 1015 | 98.25 | 1.50 | 0.25 | ... | ... | AP | 38 | 10 | 35 |
| | | | | | | HP | 50 | 40 | 7 |
| | | | | | | XHB | 75 to 85 | 45 to 55 | 8 to 6* |
| 1012 | 95.60 | 3.00 | 1.00 | 0.40 | ... | A | 52 | 15 | 35* |
| | | | | | | H | 85 | 50 | 13 to 8* |
| 1000 | 94.90 | 4.00 | 1.10 | ... | ... | AC | 45 | .... | 15 |
| 1014 | 90.75 | 2.00 | ... | ... | 7.25 | A | 75 to 90 | 37.5 to 45 | 12 to 9* |

Designation numbers are those of The American Brass Co.

The values given for the tensile strength, yield strength and elongation are all minimum values. Where ranges are shown, the first values given are for the largest diameter or largest size specimens. Yield strength values were determined at 0.50 per cent elongation under load.

* Per cent elongation in 4 times the diameter or thickness of the specimen. All other values are per cent elongation in 2 inches.

[1] The following chemical symbols are used: Cu for copper, Si for silicon, Mn for manganese, Pb for lead, and Al for aluminum.

[2] Symbols used are: HRA for hot-rolled and annealed tank plates; CRA for cold-rolled sheets and strips; CRHH for cold-rolled half hard strips; and CRH for cold-rolled hard strips. For round, square, hexagonal, and octagonal rods: A for annealed; H for hard; and XHB for extra-hard bolt temper (in coils for cold-heading). For pipe and tube: AP for annealed; and HP for hard. For castings: AC for as cast.

*Uses:* (1010) Hot-rolled-and-annealed plates for unfired pressure vessels, and rods for hot forging, hot upsetting, and machining. (1015) Cold-headed-and-roll-threaded bolts and cold-drawn seamless tubes for electrical metallic tubing and rigid conduit. (1012) Screw machine products. (1000) Castings. (1014) Hot forgings and for free machining applications; not for cold working or welding.

**Beryllium Copper.** — These alloys which contain copper, beryllium, cobalt and in the case of one alloy, silver, fall into two groups. One group whose beryllium content is greater than one per cent is characterized by its high strength and hardness and the other, whose beryllium content is less than 1 per cent, by its high electrical and thermal conductivity. The alloys have many applications in the electrical and aircraft industries or wherever strength, corrosion resistance, conductivity, non-magnetic and non-sparking properties are essential. Beryllium copper is obtainable in the form of strips, rods and bars, wire, platers bars, billets, tubes, and casting ingots.

*Composition and Physical Properties:* The accompanying table lists some of the more common wrought and casting alloys and gives some of their physical properties.

Nominal Composition and Properties of Beryllium Copper Alloys

| Alloy¹ | Composition² | | | | Form | Temper³ | Tensile Strength, Thousands of Lbs. per Sq. In. | % Elong. in 2 In. | Rockwell Hardness, Scale-Range |
|---|---|---|---|---|---|---|---|---|---|
| | Be | Co | Ag | Cu | | | | | |
| | Per Cent | | | | | | | | |
| 25 | 1.80 to 2.05 | 0.18 to 0.30 | ... | Bal. | Strip | A<br>H<br>HT | 60 to 78<br>100 to 120<br>190 to 215 | 35 to 60<br>2 to 7<br>1 to 3 | B-45 to 78<br>B-96 to 102<br>C-40 to 45 |
| | | | | | Rod | A<br>AT | 60 to 85<br>165 to 190 | 35 to 60<br>4 to 10 | B-45 to 85<br>C-36 to 41 |
| | | | | | Wire | A<br>AT | 58 to 78<br>165 to 190 | 35 to 55<br>3 to 8 | .....<br>..... |
| 165 | 1.60 to 1.80 | 0.18 to 0.30 | ... | Bal. | Strip | A<br>H<br>HT | 60 to 78<br>100 to 120<br>180 to 200 | 35 to 60<br>2 to 7<br>1 to 3 | B-45 to 78<br>B-96 to 102<br>C-39 to 41 |
| 10 | 0.40 to 0.70 | 2.35 to 2.70 | ... | Bal. | Strip | A<br>H<br>HT | 38 to 55<br>70 to 85<br>110 to 130 | 20 to 35<br>5 to 8<br>5 to 12 | B-20 to 45<br>B-70 to 80<br>B-95 to 102 |
| | | | | | Rod | A<br>AT | 38 to 55<br>100 to 120 | 20 to 35<br>10 to 25 | B-20 to 45<br>B-92 to 100 |
| 50 | 0.25 to 0.50 | 1.40 to 1.70 | 0.90 to 1.10 | Bal. | Rod | A<br>AT | 38 to 55<br>100 to 120 | 20 to 35<br>10 to 25 | B-20 to 45<br>B-92 to 100 |
| 20C | 1.90 to 2.15 | 0.35 to 0.65 | ... | Bal. | ... | AT | 150 to 175 | 1 to 3 | C-38 to 45 |
| 275C | 2.50 to 2.75 | 0.35 to 0.65 | ... | Bal. | ... | AT | 140 to 165 | 1 to 2 | C-42 to 48 |
| 10C | 0.45 to 0.75 | 2.35 to 2.70 | ... | Bal. | ... | AT | 100 to 120 | 5 to 12 | B-92 to 103 |

¹ Alloys with number designations are wrought alloys and those with number and letter designations are casting alloys. Designations are those of The Beryllium Corp.

² Chemical symbols are used to designate the constituent metals: Be, beryllium; Co, cobalt; Ag, silver; Cu, copper.

³ Temper and condition symbol designations: A, solution annealed; H, hard; HT, heat treated from hard; AT, heat treated from solution annealed.

## Powdered Metal Process

This is a process by means of which metal parts in large quantities can be made by the compressing and sintering of various powdered metals such as brass, bronze and iron. The compressing of the metal powder into the shape of the part to be made is done by accurately formed dies and punches in special types of presses known as briquetting machines. The "green" compressed pieces are then sintered in an atmosphere controlled furnace at high temperatures, causing the metal powder to be bonded together into a solid mass. A subsequent sizing or pressing operation and supplementary heat treatments may also be employed in some cases. The physical properties of the final product are usually comparable to those of cast or wrought products of the same composition. Using closely controlled conditions, steel of high hardness and tensile strength has also been made by this process.

Any desired porosity from 5 to 50 per cent can be obtained in the final product. Large quantities of porous bronze and iron bearings which are impregnated with oil for self-lubrication, have been made by this process. Other porous powder metal products are being used for the filtering of liquids and gases. Where continuous porosity is desired in the final product, the voids between particles are kept connected or open by mixing one per cent of zinc stearate or other finely powdered metallic soap throughout the metal powder before briquetting and then boiling this out in a low temperature baking before the piece is sintered.

The dense type of powdered metal products include refractory metal wire and sheet, cemented carbide tools, and electrical contact materials (products which could not be made as satisfactorily by other processes) and gears or other complex shapes which might also have been made by die-casting or the precise machining of wrought or cast metal.

**Advantages of Powdered Metal Process.** — This process is advantageous when irregular curves, eccentrics, radial projections, or recesses are required. Where a part has irregular holes, keyways, flat sides, splines or square holes that are not easily machined, powdered metal parts may solve the problem. Tapered holes and counterbores are easily produced. Axial projections can be formed but the permissible size depends on the extent to which the powder will flow into the die recesses. Projections not more than one-quarter the length of the part are practicable. Slots, grooves, blind holes, and recesses of varied depths are also obtainable.

**Limiting Factors in Powdered Metal Process.** — The number and variety of shapes which may be obtained are limited by the lack of plastic flow of powders, i.e., the difficulty with which they can be made to flow around corners. Tolerances in diameter usually cannot be held closer than 0.001 inch and tolerances in length are limited to 0.005 inch. This difference in diameter and length tolerances may be due to the elasticity of the powder and spring of the press.

**Factors Affecting Design of Briquetting Tools.** — High-speed steel is recommended for dies and punches and oil-hardening steel for strippers and knock-outs. One manufacturer specifies dimensional tolerances of 0.0002 inch and super-finished surfaces for these tools. Because of the high pressures employed and the abrasive character of certain refractory materials used in some powdered metal compositions, there is frequently a tendency toward severe wear of dies and punches. In such cases, carbide inserts, chrome plating, or highly resistant die steels are employed. With regard to the shape of the die, corner radii, fillets, and bevels should be used to avoid sharp corners. Feather edges, threads, and reentrant angles are usually impracticable. The making of punches and dies is particularly exacting because allowances must be made for change in dimensions due to growth after briquetting and shrinkage or growth during sintering.

## Etching and Etching Fluids

**Etching Fluids for Different Metals.** — A common method of etching names or simple designs upon steel is to apply a thin, even coating of beeswax or some similar substance which will resist acid; then mark the required lines or letters in the wax with a sharp-pointed scriber, thus exposing the steel (where the wax has been removed by the scriber point) to the action of an acid, which is finally applied. To apply a very thin coating of beeswax, place the latter in a silk cloth, warm the piece to be etched, and rub the pad over it. Regular coach varnish is also used instead of wax, as a "resist."

An etching fluid ordinarily used for carbon steel consists of nitric acid, 1 part; water, 4 parts. It may be necessary to vary the amount of water, as the exact proportion depends upon the carbon in the steel and whether it is hard or soft. For hard steel, use nitric acid, 2 parts; acetic acid, 1 part. For high-speed steel, nickel or brass, use nitro-hydrochloric acid (nitric, 1 part; hydrochloric, 4 parts). For high-speed steel it is sometimes better to add a little more nitric acid. For etching bronze, use nitric acid, 100 parts; muriatic acid, 5 parts. For brass, nitric acid, 16 parts; water, 160 parts; dissolve 6 parts potassium chlorate in 100 parts of water; then mix the two solutions and apply.

A fluid which may be used either for producing a frosted effect or for deep etching (depending upon the time it is allowed to act) is composed of 1 ounce sulphate of copper (blue vitriol); ¼ ounce alum; ½ teaspoonful of salt; 1 gill of vinegar, and 20 drops of nitric acid. For aluminum, use a solution composed of alcohol, 4 ounces; acetic acid, 6 ounces; antimony chloride, 4 ounces; water, 40 ounces.

Various acid-resisting materials are used for covering the surfaces of steel rules, etc., prior to marking off the lines on a graduating machine. When the graduation lines are fine and very closely spaced, as on machinists' scales which are divided into hundredths or sixty-fourths, it is very important to use a thin resist that will cling to the metal and prevent any under-cutting of the acid; the resist should also enable fine lines to be drawn without tearing or crumbling as the tool passes through it. One resist that has been extensively used is composed of about 50 per cent of asphaltum, 25 per cent of beeswax, and, in addition, a small percentage of Burgundy pitch, black pitch, and turpentine. A thin covering of this resisting material is applied to the clean polished surface to be graduated and, after it is dry, the work is ready for the graduating machine. For some classes of work, paraffin is used for protecting the surface surrounding the graduation lines which are to be etched. The method of application consists in melting the paraffin and raising its temperature high enough so that it will flow freely; then the work is held at a slight angle and the paraffin is poured on its upper surface. The melted paraffin forms a thin protective coating.

## Coloring Metals

**General Requirements in the Coloring of Metal Surfaces.** — Copper is more susceptible to coloring processes and materials than any of the other metals, and hence the alloys containing large percentages of copper are readily given various shades of yellow, brown, red, blue, purple, and black. Alloys with smaller percentages of copper (or none at all) can be given various colors, but not as easily as if copper were the principal ingredient, and the higher the copper content, the more readily can the alloy be colored. The shades, and even the colors, can be altered by varying the density of the solution, its temperature and the length of time the object is immersed. They can also be altered by finishing the work in different ways. If a cotton buff is used, one shade will be produced; a scratch brush will produce another, etc. Thus to color work the same shade as that of a former lot, all the data in connection with these operations must be preserved so they can be repeated with exactness.

**Cleaning Metals for Coloring.** — Metal surfaces to be colored chemically must first be thoroughly cleaned. To remove grease from small parts, dip in benzine, ether or some other solvent for the grease. Boil large pieces in a solution of one part caustic soda and ten parts water. For zinc, tin or britannia metal, do not use caustic soda, but a bath composed of one part carbonate of soda or potash and ten parts water. After boiling, wash in clean water. Do not touch the clean surfaces with the fingers, but handle the objects by the use of tongs or wires.

**Pickling Solutions or Dips for Coloring.** — The grease removal should be followed by chemical cleansing, which principally serves the purpose of removing the greenish or brownish films which form on copper, brass, bronze, etc. The composition of the bath or mixture for pickling varies for different metals. For copper and its alloys, a mixture of 100 parts concentrated sulphuric acid (66 degrees Baumé) and 75 parts nitric acid (40 degrees Baumé) is sometimes used. If the metal is to be given a luster instead of a mat or dull finish, add about 1 part common salt to 100 parts of the pickling solution, by weight. A better dip for a mat surface consists of 90 parts nitric acid (36 degrees Baumé), 45 parts concentrated sulphuric acid, 1 part salt, and from 1 to 5 parts of sulphate of zinc, by weight. The composition of copper-zinc alloys will produce different color tones in the same dip and will affect the results of chemical coloring. After pickling, washing in water is necessary.

Another good method of removing these films is to soak the work in a pickle composed of spent aquafortis until a black scale is formed, and then dip it for a few minutes into a solution of 64 parts water, 64 parts commercial sulphuric acid, 32 parts aquafortis, and 1 part hydrochloric acid. After that the work should be thoroughly rinsed several times with distilled water.

**Coloring Brass.** — Polished brass pieces can be given various shades from golden yellow to orange by immersing them for a certain length of time in a solution composed of 5 parts, by weight, of caustic soda, 50 parts water and 10 parts copper carbonate. When the desired shade is reached, the work must be well washed with water and dried in sawdust. Golden yellow may be produced as follows: Dissolve 100 grains lead acetate in 1 pint of water and add a solution of sodium hydrate until the precipitate which first forms is re-dissolved; then add 300 grains red potassium ferro-cyanide. With the solution at ordinary temperatures, the work will assume a golden yellow, but heating the solution darkens the color, until at 125 degrees F. it has changed to a brown.

**To Produce a Rich Gold Color.** — Brass can be given a rich gold color by boiling it in a solution composed of 2 parts, by weight, of saltpeter, 1 part common salt, 1 part alum, 24 parts water and 1 part hydrochloric acid. Another method is to apply a mixture of 3 parts alum, 6 parts saltpeter, 3 parts sulphate of zinc, and 3 parts common salt. After applying this mixture the work is heated over a hot plate until it becomes black, after which it is washed with water, rubbed with vinegar, and again washed and dried.

**White Colors or Coatings.** — The white color or coating that is given to such brass articles as pins, hooks and eyes, buttons, etc., can be produced by dipping them in a solution made as follows: Dissolve 2 ounces fine-grain silver in nitric acid, then add 1 gallon distilled water, and put this into a strong solution of sodium chloride. The silver will precipitate in the form of chloride, and must be washed until all traces of the acid are removed. Testing the last rinse water with litmus paper will show when the acid has disappeared; then mix this chloride of silver with an equal amount of potassium bitartrate (cream of tartar), and add enough water to give it the consistency of cream. The work is then immersed in this solution and stirred around until properly coated, after which it is rinsed in hot water and dried in sawdust.

**Silvering.** — A solution for silvering, that is applicable to such work as gage or clock dials, etc., can be made by grinding together in a mortar 1 ounce of very dry chloride of silver, 2 ounces cream of tartar, and 3 ounces common salt, then add enough water to obtain the desired consistency and rub it onto the work with a soft cloth. This will give brass or bronze surfaces a dead-white thin silver coating, but it will tarnish and wear if not given a coat of lacquer. The ordinary silver lacquers that can be applied cold are the best. Before adding the water, the mixture, as it leaves the mortar, can be kept a long time if put in very dark colored bottles, but if left in the light it will decompose.

**To Give Brass a Green Tint.** — One solution that will produce the verde antique, or rust green, is composed of 3 ounces crystallized chloride of iron, 1 pound ammonium chloride, 8 ounces verdigris, 10 ounces common salt, 4 ounces potassium bitartrate and 1 gallon of water. If the objects to be colored are large, the solution can be put on with a brush. Several applications may be required to give the desired depth of color. Small work should be immersed and the length of time it remains in the solution will govern the intensity of the color.

**Blackening Brass.** — There are many different processes and solutions for blackening brass. Trioxide of arsenic, white arsenic or arsenious acid are different names for the chemical that is most commonly used. It is the cheapest chemical for producing black on brass, copper, nickel, German silver, etc., but has a tendency to fade, especially if not properly applied, although a coat of lacquer will preserve it a long time. A good black can be produced by immersing the work in a solution composed of 2 ounces white arsenic, 5 ounces cyanide of potassium, and 1 gallon of water. This should be boiled in an enamel or agate vessel, and used hot. Another cheap solution is composed of 8 ounces of sugar of lead, 8 ounces hyposulphite of soda and 1 gallon of water. This must also be used hot and the work afterwards lacquered to prevent fading. When immersed, the brass first turns yellow, then blue and then black, the latter being a deposit of sulphide of lead.

**Preservation of Color.** — After a part has been given the desired color, it is usually washed in water and then dried with clean sawdust. The colored surfaces of alloys are commonly protected and preserved by coating with a colorless lacquer, such as japan lacquer. Small parts are coated by dipping, and large ones by rubbing the lacquer on. The lacquer is hard after drying, and insoluble in most fluids; hence, it can be washed without injury.

**Niter Process of Bluing Steel.** — The niter process of bluing iron and steel is as follows: The niter or nitrate of potash (often called saltpeter) is melted in an iron pot and heated to about 600 degrees F. The parts to be blued are cleaned and polished and then immersed in the molten niter until a uniform color of the desired shade has been obtained. This requires only a few seconds. The articles are then removed and allowed to cool, after which the adhering niter is washed off in water. Parts which will not warp may be immersed immediately after removing from the niter bath. After cleaning, dry in sawdust, and then apply some suitable oil, such as linseed, to prevent rusting. To secure uniform coloring, a pyrometer should be used to gage the temperature of the niter, because a higher heat than 600 degrees F. will produce a dark color, whereas a lower heat will give a lighter shade.

**Bluing Steel by Heat-treatment.** — Polished steel parts can be given a blue color by heating in hot sand. wood ashes, or pulverized charcoal. Place the substance in an iron receptacle and stir constantly, while heating, in order to heat uniformly. Heat just hot enough to char a pine stick. The parts to be blued must be absolutely free from grease. They are placed in the heated substance until the desired color is obtained. Further coloring is then checked by immersing in oil. Small parts are sometimes heated by a Bunsen burner or by laying upon a heated plate. For a light blue color, heat in sand or wood ashes, and for a dark blue, use pulverized charcoal. The quality of the color depends largely upon the fineness of the finish. Still another method of coloring by heat is to immerse the parts in a molten bath of potassium nitrate and sodium nitrate. The coloring is then checked by plunging the work into boiling water.

**Blue-black Finish.** — To obtain a blue-black finish on small steel parts, use a mixture of 16 parts, by weight, of saltpeter and 2 parts of black oxide of manganese. This mixture is heated to a temperature of 750 degrees F. and the objects are immersed in it. The oxide of manganese is deposited on the work and must, therefore, be frequently replenished in the mixture.

**Black Finish.** — To obtain a black rust-protecting finish on hardened parts, temper, after hardening, in "heavy" cylinder oil; then immediately place the part with the oil on it in an oven having a temperature of from 300 to 350 degrees F. Remove the work in from 5 to 8 minutes, when the black finish is baked onto it.

**Gun Metal Finish.** — Several different chemical solutions have been used successfully for giving steel a gun metal finish or black color. Among these are the following: 1. Bismuth chloride, one part; copper chloride, one part; mercury chloride, two parts; hydrochloric acid, six parts; and water, fifty parts. 2. Ferric chloride, one part; alcohol, eight parts; and water, eight parts. 3. Copper sulphate, two parts; hydrochloric acid, three parts; nitric acid, seven parts; and perchloride of iron, eighty-eight parts. Other solutions have been prepared from nitric ether, nitric acid, copper sulphate, iron chloride, alcohol and water and from nitric acid, copper sulphate, iron chloride and water. The method of applying these and finishing the work is practically the same in all cases.

The surface is given a very thin coating with a soft brush or sponge that has been well squeezed, and is then allowed to dry. The work is then put in a closed retort to which steam is admitted and maintained at a temperature of about 100 degrees F., until the parts are covered with a slight rust. They are then boiled in clean water for about fifteen minutes and allowed to dry. A coating of black oxide will cover the surface, and this is scratch brushed. After brushing, the surface will show a grayish black. By repeating the sponging, steaming and brushing operations several times, a shiny black lasting surface will be obtained. For the best finishes, these operations are repeated as many as eight times.

Another process employs a solution of mercury chloride and ammonium chloride which is applied to the work three times and dried each time. A solution of copper sulphate, ferric chloride, nitric acid, alcohol and water is then applied three times and dried as before. A third solution of ferrous chloride, nitric acid and water is applied three times, and the work is boiled in clean water and dried each time. Finally, a solution of potassium chloride is applied and the work boiled and dried three times. The work is then scratch brushed and given a thin coating of oil. Ordnance for the French Government is treated in this way. The above methods are applicable to hardened and tempered steels, as a temperature of 100 degrees F. does not affect the hardness of the steel. For steels that will stand 600 degrees temperature without losing the desired hardness, better and much cheaper methods have been devised.

The American Gas Furnace Co. has developed a process employing a furnace with a revolving retort. The work is charged in this, together with well-burnt bone.

A chemical solution that gasifies when it enters the furnace is then injected into this retort while the work is heated to the proper temperature. This solution has been named "Carbonia." The color does not form a coating on the outside, as with the other processes, but a thin layer of the metal itself is turned to the proper color. By varying the temperature of the furnace, the time the work is in it, and the chemical, different colors can be produced from light straw to brown, blue, purple and black, or gun metal finish. Rough or sand-blasted surfaces will have a frosted appearance, while smooth polished surfaces will have a shiny brilliant appearance.

**Browning Iron and Steel.** — A good brown color can be obtained as follows: Coat the steel with ammonia and dry it in a warm place; then coat it with muriatic or nitric acid and dry it in a warm place; then place the steel in a solution of tannin or gallic acid and again dry it. The color can be deepened by placing the work near the fire, but it should be withdrawn the minute the desired shade is reached or it will turn black.

**To Produce a Bronze Color.** — A bronze-like color can be produced by exposing iron or steel parts to the vapors of heated *aqua regia*, dipping them in melted petroleum jelly, and then heating them until it begins to decompose, when it is wiped off with a soft cloth. Another method of producing this bronze-brown color is to slightly heat the work, evenly cover the surfaces with a paste of antimony chloride (known as "bronzing salt"), and let the object stand until the desired color is obtained. The paste can be made more active by adding a little nitric acid.

**To Produce a Gray Color.** — A gray color on steel can be obtained by immersing the work in a heated solution of ten grains of antimony chloride, ten grains of gallic acid, 400 grains of ferric chloride and five fluid ounces of water. The first color to appear is pale blue, and this passes through the darker blues to the purple, and, finally, to the gray. If immersed long enough, the metal will assume the gray color, but any of the intermediate colors may be produced. When used cold, this is also one of the bronzing solutions.

**Mottled Coloring.** — Mottled colors on steel can be produced by heating the objects to a good cherry-red for several minutes in cyanide of potassium, then pouring the cyanide off, and placing the receptacle containing the work back on the fire for five minutes. The contents are then quickly dumped into clean water. To heighten the colors, boil afterward in water and oil.

**Coppering Solution.** — A coppering solution for coating finished surfaces in order that lay-out lines may be more easily seen, is composed of the following ingredients: To 4 ounces of distilled water (or rain water) add all the copper sulphate (blue vitriol) it will dissolve; then add 10 drops of sulphuric acid. Test by applying to a piece of steel, and, if necessary, add four or five drops of acid. The surface to be coppered should be polished and free from grease. Apply the solution with clean waste, and, if a bright copper coating is not obtained, add a few more drops of the solution; then scour the surface with fine emery cloth, and apply rapidly a small quantity of fresh solution.

**White Coatings for Laying Out Lines.** — Powdered chalk or whiting mixed with alcohol is commonly used for coating finished metal surfaces preparatory to laying out lines for machining operations. Alcohol is preferable to water, because it will dry quicker and does not tend to rust the surface. This mixture can be applied with a brush and is more convenient than a coppering solution for general work. For many purposes, the surface can be coated satisfactorily by simply rubbing dry chalk over it.

# MATERIALS

## The Elements — Symbols, Atomic Numbers and Weights, Melting Points

| Name of Element | Symbol | Atomic Number | Atomic Weight[1] | Melting Point Deg. C.[2] | Name of Element | Symbol | Atomic Number | Atomic Weight[1] | Melting Point Deg. C.[2] |
|---|---|---|---|---|---|---|---|---|---|
| Actinium.... | Ac | 89 | 227 | ........ | Molybdenum.. | Mo | 42 | 95.95 | 2625 |
| Aluminum.. | Al | 13 | 26.97 | 660.0 | Neodymium.. | Nd | 60 | 144.27 | 840 |
| Americium.. | Am | 95 | (241) | ........ | Neon......... | Ne | 10 | 20.183 | −248.6 |
| Antimony... | Sb | 51 | 121.76 | 630.5 | Neptunium... | Np | 93 | (237) | ........ |
| Argon....... | A | 18 | 39.944 | −189.4 | Nickel........ | Ni | 28 | 58.69 | 1455 |
| Arsenic...... | As | 33 | 74.91 | 814 | Niobium...... | Nb | 41 | 92.91 | 2500 |
| Astatine..... | At | 85 | (210) | ........ | Nitrogen...... | N | 7 | 14.008 | −210.0 |
| Barium..... | Ba | 56 | 137.36 | 704 | Osmium...... | Os | 76 | 190.2 | 2700 |
| Berkelium... | Bk | 97 | ........ | ........ | Oxygen...... | O | 8 | 16.0000 | −218.8 |
| Beryllium... | Be | 4 | 9.013 | 1280 | Palladium..... | Pd | 46 | 106.7 | 1554 |
| Bismuth.... | Bi | 83 | 209.00 | 271.3 | Phosphorus... | P | 15 | 30.98 | ........ |
| Boron...... | B | 5 | 10.82 | 2040 | Platinum..... | Pt | 78 | 195.23 | 1773.5 |
| Bromine.... | Br | 35 | 79.916 | −7.2 | Plutonium.... | Pu | 94 | (239) | ........ |
| Cadmium.... | Cd | 48 | 112.41 | 320.9 | Polonium..... | Po | 84 | 210 | ........ |
| Calcium..... | Ca | 20 | 40.08 | 850 | Potassium..... | K | 19 | 39.096 | 63 |
| Californium.. | Cf | 98 | (248) | ........ | Praseodymium | Pr | 59 | 140.92 | 940 |
| Carbon...... | C | 6 | 12.011 | 3700 | Promethium.. | Pm | 61 | (147) | ........ |
| Cerium...... | Ce | 58 | 140.13 | 600 | Proactinium.. | Pa | 91 | 231 | 3000 |
| Cesium..... | Cs | 55 | 132.91 | 28 | Radium....... | Ra | 88 | 226.05 | 700 |
| Chlorine.... | Cl | 17 | 35.457 | −101 | Radon........ | Rn | 86 | 222 | −71 |
| Chromium... | Cr | 24 | 52.01 | 1800 | Rhenium..... | Re | 75 | 186.31 | 3170 |
| Cobalt...... | Co | 27 | 58.94 | 1495 | Rhodium..... | Rh | 45 | 102.91 | 1966 |
| Columbium[3] | | | ........ | ........ | Rubidium.... | Rb | 37 | 85.48 | 39 |
| Copper...... | Cu | 29 | 63.54 | 1083.2 | Ruthenium.... | Ru | 44 | 101.1 | 2500 |
| Curium..... | Cm | 96 | (245) | ........ | Samarium..... | Sm | 62 | 150.43 | >1300 |
| Dysprosium.. | Dy | 66 | 162.46 | ........ | Scandium..... | Sc | 21 | 45.10 | 1200 |
| Erbium...... | Er | 68 | 167.2 | ........ | Selenium..... | Se | 34 | 78.96 | ........ |
| Europium... | Eu | 63 | 152.0 | ........ | Silicon........ | Si | 14 | 28.06 | 1415 |
| Fluorine.... | F | 9 | 19.00 | −223 | Silver......... | Ag | 47 | 107.880 | 960.5 |
| Francium.... | Fr | 87 | (223) | ........ | Sodium....... | Na | 11 | 22.991 | 97.7 |
| Gadolinium. | Gd | 64 | 156.9 | ........ | Strontium..... | Sr | 38 | 87.63 | 770 |
| Gallium..... | Ga | 31 | 69.72 | 29.78 | Sulphur....... | S | 16 | 32.066 | ........ |
| Germanium.. | Ge | 32 | 72.60 | 958 | Tantalum..... | Ta | 73 | 180.95 | 3000 |
| Gold........ | Au | 79 | 197.0 | 1063.0 | Technetium... | Tc | 43 | (99) | ........ |
| Hafnium.... | Hf | 72 | 178.6 | 1700 | Tellurium..... | Te | 52 | 127.61 | 450 |
| Helium..... | He | 2 | 4.003 | −271.4 | Terbium...... | Tb | 65 | 158.93 | 327 |
| Holmium... | Ho | 67 | 164.94 | ........ | Thallium..... | Tl | 81 | 204.39 | 300 |
| Hydrogen... | H | 1 | 1.0080 | −259.2 | Thorium...... | Th | 90 | 232.05 | 1800 |
| Indium..... | In | 49 | 114.76 | 156.4 | Thulium...... | Tm | 69 | 168.94 | ........ |
| Iodine....... | I | 53 | 126.92 | 114 | Tin.......... | Sn | 50 | 118.70 | 231.9 |
| Iridium..... | Ir | 77 | 192.2 | 2454 | Titanium..... | Ti | 22 | 47.90 | 1820 |
| Iron........ | Fe | 26 | 55.85 | 1539 | Tungsten[4]... | W | 74 | ........ | ........ |
| Krypton.... | Kr | 36 | 83.7 | −157.0 | Uranium...... | U | 92 | 238.07 | 1133 |
| Lanthanum. | La | 57 | 138.92 | 826 | Vanadium.... | V | 23 | 50.95 | 1735 |
| Lead........ | Pb | 82 | 207.21 | 327.4 | Wolfram..... | W | 74 | 183.92 | 3410 |
| Lithium..... | Li | 3 | 6.940 | 186 | Xenon....... | Xe | 54 | 131.3 | −112 |
| Lutetium.... | Lu | 71 | 174.99 | ........ | Ytterbium.... | Yb | 70 | 173.04 | 1800 |
| Magnesium.. | Mg | 12 | 24.32 | 650 | Yttrium...... | Y | 39 | 88.92 | 1490 |
| Manganese.. | Mn | 25 | 54.94 | 1260 | Zinc.......... | Zn | 30 | 65.38 | 419.5 |
| Mercury..... | Hg | 80 | 200.61 | −38.87 | Zirconium.... | Zr | 40 | 91.22 | 1750 |

[1] International Atomic Weight (1953.) Value in parenthesis is for the mass number of the most stable known isotope. [2] Melting point data from U. S. National Bureau of Standards, U. S. Government Printing Office publication, 1947. [3] See Niobium. [4] See Wolfram.

## Specific Gravity and Properties of Metals

| Metal or Composition | Chemical Symbol | Specific Gravity | Weight per Cubic Inch, Pound† | Weight per Cubic Foot, Pounds† | Melting Point, Deg. F. | Linear Expansion per Unit Length per Deg. F. | Temp, Deg. F.* | Electric Conductivity Silver = 100 |
|---|---|---|---|---|---|---|---|---|
| Aluminum | Al | 2.70 | 0.0975 | 168.5 | 1220 | 0.0001244 | 68 | 63.0 |
| Antimony | Sb | 6.618 | 0.2390 | 413.0 | 1167 | 0.0000755 | 68 | 3.59 |
| Barium | Ba | 3.78 | 0.1365 | 235.9 | 1562 | ... | ... | 30.61 |
| Bismuth | Bi | 9.781 | 0.3532 | 610.3 | 520 | 0.0000077 | 68 | 1.40 |
| Boron | B | 2.535 | 0.0916 | 158.2 | 4172 | ... | ... | ... |
| Brass: 8oC., 2oZ. | ... | 8.60 | 0.3105 | 536.6 | 1823 | ... | ... | ... |
| 7oC., 3oZ. | ... | 8.44 | 0.3048 | 526.7 | 1706 | 0.00001 | 76 to 212 | ... |
| 6oC., 4oZ. | ... | 8.36 | 0.3018 | 521.7 | 1652 | ... | ... | ... |
| 5oC., 5oZ. | ... | 8.20 | 0.2961 | 511.7 | 1616 | ... | ... | ... |
| Bronze: 9oC., 1oT. | ... | 8.78 | 0.3171 | 547.9 | 1841 | 0.0001 | ... | ... |
| Cadmium | Cd | 8.648 | 0.3123 | 539.6 | 610 | ... | ... | 24.38 |
| Calcium | Ca | 1.54 | 0.0556 | 96.1 | 1490 | ... | ... | 21.77 |
| Chromium | Cr | 6.93 | 0.2502 | 432.4 | 2939 | 0.0000683 | 68 | 16.00 |
| Cobalt | Co | 8.71 | 0.3145 | 543.5 | 2696 | 0.0000900 | 68 | 16.93 |
| Copper | Cu | 8.89 | 0.3210 | 554.7 | 1981 | 0.0000778 | 68 | 97.61 |
| Gold | Au | 19.3 | 0.6969 | 1204.3 | 1945 | 0.0000361 | 68 | 76.61 |
| Iridium | Ir | 22.42 | 0.8096 | 1399.0 | 4262 | 0.0000361 | 68 | 13.52 |
| Iron, cast | Fe | 7.03-7.73 | 0.254-0.279 | 438.7-482.4 | 1990-2300 | 0.0000655 | 68 | ... |
| Iron, wrought | Fe | 7.80-7.90 | 0.282-0.285 | 486.7-493.0 | 2750 | 0.0000661 | 212 | 14.57 |
| Lead | Pb | 11.342 | 0.4096 | 707.7 | 621 | 0.0000163 | 68-212 | 8.42 |
| Magnesium | Mg | 1.741 | 0.0628 | 108.6 | 1204 | 0.0001444 | 68 | 39.44 |
| Manganese | Mn | 7.3 | 0.2636 | 455.5 | 2300 | 0.0001294 | 68 | 15.75 |
| Mercury (68° F.) | Hg | 13.546 | 0.4892 | 845.3 | -38 | ... | ... | 1.75 |
| Molybdenum | Mo | 10.2 | 0.3683 | 636.5 | 4748 | 0.0000294 | 68 | 17.60 |
| Nickel | Ni | 8.8 | 0.3178 | 549.1 | 2651 | 0.0000700 | 68 | 12.89 |
| Platinum | Pt | 21.37 | 0.7717 | 1333.5 | 3224 | 0.0000496 | 68 | 14.43 |
| Potassium | K | 0.870 | 0.0314 | 54.3 | 144 | ... | ... | 19.62 |
| Silver | Ag | 10.42-10.53 | 0.376-0.380 | 650.2-657.1 | 1761 | 0.0001025 | 68 | 100.00 |
| Sodium | Na | 0.9712 | 0.0351 | 60.6 | 207 | ... | ... | 31.98 |
| Steel, Carbon | ... | ...... | 0.283-0.284 | 489.0-490.8 | 2500 | 0.0000633 | 68 | 12.00 |
| Tantalum | Ta | 16.6 | 0.5998 | 1035.8 | 5162 | 0.0000361 | 68 | 54.63 |
| Tellurium | Te | 6.25 | 0.2257 | 390.0 | 846 | 0.0000093 | 104 | 0.001 |
| Tin | Sn | 7.29 | 0.2633 | 454.9 | 449 | 0.0001496 | 64-212 | 14.39 |
| Titanium | Ti | 4.5 | 0.1621 | 280.1 | 3272 | 0.0000049 | 68-392 | 13.73 |
| Tungsten | W | 18.6-19.1 | 0.672-0.690 | 1161-1192 | 6098 | 0.0000239 | 32-212 | 14.00 |
| Uranium | U | 18.7 | 0.6753 | 1166.9 | <3362 | ... | ... | 16.47 |
| Vanadium | V | 5.6 | 0.2022 | 349.4 | 3110 | 0.0000043 | 32-104 | 4.95 |
| Zinc | Zn | 7.04-7.16 | 0.254-0.259 | 439.3-446.8 | 788 | 0.000017 | 68 | 29.57 |

*Temperature given for each metal is that at which expansion coefficient shown in previous column was determined     †Avoirdupois

## Melting Points of Alloys of Low Fusing Point

| Composition in Per Cent | | | | Melting Point, Degrees F. | Composition in Per Cent | | | Melting Point, Degrees F. |
|---|---|---|---|---|---|---|---|---|
| Bismuth | Lead | Tin | Cadmium | | Bismuth | Lead | Tin | |
| 50.0 | 25.0 | 12.5 | 12.5 | 149 | 20.0 | 40.0 | 40.0 | 293 |
| 50.1 | 26.6 | 13.3 | 10.0 | 158 | 19.0 | 38.0 | 43.0 | 298 |
| 38.4 | 30.8 | 15.4 | 15.4 | 160 | 18.1 | 36.2 | 45.7 | 304 |
| 27.5 | 27.5 | 10.5 | 34.5 | 167 | 17.3 | 34.6 | 48.1 | 311 |
| 50.0 | 34.5 | 9.3 | 6.2 | 171 | 16.6 | 33.2 | 50.2 | 316 |
| 50.0 | 25.0 | 25.0 | .... | 187 | 16.0 | 36.0 | 48.0 | 311 |
| 50.0 | 31.2 | 18.8 | .... | 201 | 15.3 | 38.8 | 45.9 | 309 |
| 55.6 | .... | 33.3 | 11.1 | 203 | 14.8 | 40.2 | 45.0 | 307 |
| 50.0 | .... | 25.0 | 25.0 | 203 | 14.0 | 43.0 | 43.0 | 309 |
| 47.0 | 35.5 | 17.5 | .... | 208 | 13.7 | 44.8 | 41.5 | 320 |
| 42.1 | 42.1 | 15.8 | .... | 226 | 13.3 | 46.6 | 40.1 | 329 |
| 40.0 | 40.0 | 20.0 | .... | 235 | 12.8 | 49.0 | 38.2 | 342 |
| 36.5 | 36.5 | 27.0 | .... | 243 | 12.5 | 50.0 | 37.5 | 352 |
| 33.3 | 33.4 | 33.3 | .... | 253 | 11.7 | 46.8 | 41.5 | 333 |
| 30.8 | 38.4 | 30.8 | .... | 266 | 11.4 | 45.6 | 43.0 | 329 |
| 28.5 | 43.0 | 28.5 | .... | 270 | 11.2 | 44.4 | 44.4 | 320 |
| 25.0 | 50.0 | 25.0 | .... | 300 | 10.8 | 43.2 | 46.0 | 318 |
| 23.5 | 47.0 | 29.5 | .... | 304 | 10.5 | 42.0 | 47.5 | 320 |
| 22.2 | 44.4 | 33.4 | .... | 289 | 10.2 | 41.0 | 48.8 | 322 |
| 21.0 | 42.0 | 37.0 | .... | 289 | 10.0 | 40.0 | 50.0 | 324 |

## Weights of American Woods, in Pounds per Cubic Foot
(United States Department of Agriculture)

| Species | Green | Airdry | Species | Green | Airdry |
|---|---|---|---|---|---|
| Alder, red | 46 | 28 | Hickory, pecan | 62 | 45 |
| Ash, black | 52 | 34 | Hickory, true | 63 | 51 |
| Ash, commercial white | 48 | 41 | Honeylocust | 61 | .. |
| Ash, Oregon | 46 | 38 | Larch, western | 48 | 36 |
| Aspen | 43 | 26 | Locust, black | 58 | 48 |
| Basswood | 42 | 26 | Maple, bigleaf | 47 | 34 |
| Beech | 54 | 45 | Maple, black | 54 | 40 |
| Birch | 57 | 44 | Maple, red | 50 | 38 |
| Birch, paper | 50 | 38 | Maple, silver | 45 | 33 |
| Cedar, Alaska | 36 | 31 | Maple, sugar | 56 | 44 |
| Cedar, eastern red | 37 | 33 | Oak, red | 64 | 44 |
| Cedar, northern white | 28 | 22 | Oak, white | 63 | 47 |
| Cedar, southern white | 26 | 23 | Pine, lodgepole | 39 | 29 |
| Cedar, western red | 27 | 23 | Pine, northern white | 36 | 25 |
| Cherry, black | 45 | 35 | Pine, Norway | 42 | 34 |
| Chestnut | 55 | 30 | Pine, ponderosa | 45 | 28 |
| Cottonwood, eastern | 49 | 28 | Pines, southern yellow: | | |
| Cottonw'd, northern black | 46 | 24 | Pine, loblolly | 53 | 36 |
| Cypress, southern | 51 | 32 | Pine, longleaf | 55 | 41 |
| Douglas fir, coast region | 38 | 34 | Pine, shortleaf | 52 | 36 |
| Douglas fir, Rocky Mt. reg. | 35 | 30 | Pine, sugar | 52 | 25 |
| Elm, American | 54 | 35 | Pine, western white | 35 | 27 |
| Elm, rock | 53 | 44 | Poplar, yellow | 38 | 28 |
| Elm, slippery | 56 | 37 | Redwood | 50 | 28 |
| Fir, balsam | 45 | 25 | Spruce, eastern | 34 | 28 |
| Fir, commercial white | 46 | 27 | Spruce, Engelmann | 39 | 23 |
| Gum, black | 45 | 35 | Spruce, Sitka | 33 | 28 |
| Gum, red | 50 | 34 | Sycamore | 52 | 34 |
| Hemlock, eastern | 50 | 28 | Tamarack | 47 | 37 |
| Hemlock, western | 41 | 29 | Walnut, black | 58 | 38 |

**Specific gravity** is a number indicating how many times a certain volume of a material is heavier than an equal volume of water. As the density of water differs slightly at different temperatures, it is the usual custom to make comparisons on the basis that the water has a temperature of 62 degrees F. The weight of one cubic inch of pure water at 62 degrees F. is 0.0361 pound. If the specific gravity of any material is known, the weight of a cubic inch of the material can, therefore, be found by multiplying its specific gravity by 0.0361.

*Example:* — The specific gravity of cast iron is 7.2. Find the weight of 5 cubic inches of cast iron.

$$7.2 \times 0.0361 \times 5 = 1.2996 \text{ pound.}$$

To find the weight per cubic foot of a material, the specific gravity of which is known, multiply the specific gravity by 62.355.

If the weight of a cubic inch of a material is known, the specific gravity is found by dividing the weight per cubic inch by 0.0361.

*Example:* — The weight of a cubic inch of gold is 0.697 pound. Find the specific gravity.

$$0.697 \div 0.0361 = 19.31.$$

If the weight per cubic foot of a material is known, the specific gravity is found by multiplying this weight by 0.01604.

**Average Specific Gravity of Miscellaneous Substances**

| Substance | Sp. Gr. | Weight per Cubic Foot, Lbs. | Substance | Sp. Gr. | Weight per Cubic Foot, Lbs. |
|---|---|---|---|---|---|
| Asbestos.............. | 2.4 | 150 | Gypsum.............. | 2.4 | 150 |
| Asphaltum........... | 1.4 | 87 | Ice................. | 0.9 | 56 |
| Borax................ | 1.8 | 112 | Iron slag............ | 2.7 | 168 |
| Brick, common....... | 1.8 | 112 | Limestone........... | 2.6 | 162 |
| Brick, fire........... | 2.3 | 143 | Marble.............. | 2.7 | 168 |
| Brick, hard.......... | 2.0 | 125 | Masonry............. | 2.4 | 150 |
| Brick, pressed........ | 2.2 | 137 | Mica............... | 2.8 | 175 |
| Brickwork, in mortar... | 1.6 | 100 | Mortar.............. | 1.5 | 94 |
| Brickwork, in cement.. | 1.8 | 112 | Phosphorus.......... | 1.8 | 112 |
| Cement, Portland (set). | 3.1 | 193 | Plaster of Paris....... | 1.8 | 112 |
| Chalk................ | 2.3 | 143 | Quartz.............. | 2.6 | 162 |
| Charcoal............. | 0.4 | 25 | Salt, common........ | ... | 48 |
| Coal, anthracite....... | 1.5 | 94 | Sand, dry............ | ... | 100 |
| Coal, bituminous...... | 1.3 | 81 | Sand, wet............ | ... | 125 |
| Concrete............. | 2.2 | 137 | Sandstone........... | 2.3 | 143 |
| Earth, loose.......... | ... | 75 | Slate................ | 2.8 | 175 |
| Earth, rammed....... | ... | 100 | Soapstone........... | 2.7 | 168 |
| Emery............... | 4.0 | 249 | Sulphur............. | 2.0 | 125 |
| Glass................ | 2.6 | 162 | Tar, bituminous...... | 1.2 | 75 |
| Granite.............. | 2.7 | 168 | Tile................. | 1.8 | 112 |
| Gravel............... | ... | 109 | Trap rock........... | 3.0 | 187 |

The weight per cubic foot is calcuated on the basis of the specific gravity except for those substances that occur in bulk, heaped or loose form. In these instances, only the weights per cubic foot are given since the voids present in representative samples make the values of the specific gravities insignificant.

**Specific Gravity of Liquids.** — The specific gravity of liquids is the number which indicates how much a certain volume of the liquid weighs compared with an equal volume of water, the same as in the case of solid bodies. The density of liquids is also often expressed in degrees on the hydrometer, an instrument for determining the density of liquids, provided with graduations made to an arbitrary scale. The hydrometer consists of a glass tube with a bulb at one end containing air, and arranged with a weight at the bottom so as to float in an upright position in the liquid, the density of which is to be measured. The depth to which the hydrometer sinks in the liquid is read off on the graduated scale. The most commonly used hydrometer is the Baumé. The value of the degrees on the Baumé scale differs according to whether the liquid is heavier or lighter than water. The specific gravity for liquids heavier than water equals 145 ÷ (145 − degrees Baumé). For liquids lighter than water, the specific gravity equals 140 ÷ (130 + degrees Baumé).

**Specific Gravity of Gases.** — The specific gravity of gases is the number which indicates their weight in comparison with that of an equal volume of air. The specific gravity of air is 1, and the comparison is made at 32 degrees F.

### Specific Gravity of Gases
(At 32 degrees F.)

| Gas | Sp. Gr. | Gas | Sp. Gr. | Gas | Sp. Gr. |
|---|---|---|---|---|---|
| Air................. | 1.000 | Ether vapor..... | 2.586 | Marsh gas...... | 0.555 |
| Acetylene.......... | 0.920 | Ethylene........ | 0.967 | Nitrogen........ | 0.971 |
| Alcohol vapor....... | 1.601 | Hydrofluoric acid. | 2.370 | Nitric oxide..... | 1.039 |
| Ammonia.......... | 0.592 | Hydrochloric acid. | 1.261 | Nitrous oxide... | 1.527 |
| Carbon dioxide...... | 1.520 | Hydrogen....... | 0.069 | Oxygen......... | 1.106 |
| Carbon monoxide.... | 0.967 | Illuminating gas.. | 0.400 | Sulphur dioxide. | 2.250 |
| Chlorine........... | 2.423 | Mercury vapor... | 6.940 | Water vapor.... | 0.623 |

1 cubic foot of air at 32 degrees F. and atmospheric pressure weighs 0.0807 pound.

### Average Weights and Volumes of Fuels

Anthracite coal, 1 cubic foot = 55 to 65 pounds.

        1 ton (2240 pounds) = 34 to 41 cubic feet.

Bituminous coal, 1 cubic foot = 50 to 55 pounds.

        1 ton (2240 pounds) = 41 to 45 cubic feet.

Charcoal, 1 cubic foot = 18 to 18.5 pounds.

        1 ton (2240 pounds) = 120 to 124 cubic feet.

Coke, 1 cubic foot = 28 pounds.

        1 ton (2240 pounds) = 80 cubic feet.

The average weight of a bushel of charcoal is 20 pounds; of a bushel of coke, 40 pounds; of a bushel of anthracite coal, 67 pounds; and of a bushel of bituminous coal, 60 pounds.

**Weight of Wood.** — The weight of seasoned wood per cord is approximately as follows, assuming about 70 cubic feet of *solid wood* per cord: Beech, 3300 pounds; chestnut, 2600 pounds; elm, 2900 pounds; maple, 3100 pounds; poplar, 2200 pounds; white pine, 2200 pounds; red oak, 3300 pounds; white oak, 3500 pounds.

**Weight per Foot of Wood, Board Measure.** — The following is the weight in pounds of various kinds of woods, commercially known as dry timber, per foot board measure: White oak, 4.16; white pine, 1.98; Douglas fir, 2.65; short-leaf yellow pine, 2.65; red pine, 2.60; hemlock, 2.08; spruce, 2.08; cypress, 2.39; cedar, 1.93; chestnut, 3.43; Georgia yellow pine, 3.17; California spruce, 2.08.

## Specific Gravity of Liquids

| Liquid | Sp. Gr. | Liquid | Sp. Gr. | Liquid | Sp. Gr. |
|---|---|---|---|---|---|
| Acetic acid........... | 1.06 | Fluoric acid.... | 1.50 | Petroleum oil... | 0.82 |
| Alcohol, commerical... | 0.83 | Gasoline...... | 0.70 | Phosphoric acid. | 1.78 |
| Alcohol, pure......... | 0.79 | Kerosene...... | 0.80 | Rape oil....... | 0.92 |
| Ammonia............. | 0.89 | Linseed oil.... | 0.94 | Sulphuric acid... | 1.84 |
| Benzine............. | 0.69 | Mineral oil.... | 0.92 | Tar............ | 1.00 |
| Bromine............. | 2.97 | Muriatic acid.. | 1.20 | Turpentine oil... | 0.87 |
| Carbolic acid........ | 0.96 | Naphtha...... | 0.76 | Vinegar........ | 1.08 |
| Carbon disulphide..... | 1.26 | Nitric acid..... | 1.50 | Water.......... | 1.00 |
| Cotton-seed oil........ | 0.93 | Olive oil....... | 0.92 | Water, sea...... | 1.03 |
| Ether, sulphuric....... | 0.72 | Palm oil....... | 0.97 | Whale oil....... | 0.92 |

## Degrees on Baumé's Hydrometer Converted into Specific Gravity

| Deg. Baumé | Specific Gravity | | Deg. Baumé | Specific Gravity | | Deg. Baumé | Specific Gravity | |
|---|---|---|---|---|---|---|---|---|
| | Liquids Heavier than Water | Liquids Lighter than Water | | Liquids Heavier than Water | Liquids Lighter than Water | | Liquids Heavier than Water | Liquids Lighter than Water |
| 0 | 1.000 | ...... | 27 | 1.229 | 0.892 | 54 | 1.593 | 0.761 |
| 1 | 1.007 | ...... | 28 | 1.239 | 0.886 | 55 | 1.611 | 0.757 |
| 2 | 1.014 | ...... | 29 | 1.250 | 0.881 | 56 | 1.629 | 0.753 |
| 3 | 1.021 | ...... | 30 | 1.261 | 0.875 | 57 | 1.648 | 0.749 |
| 4 | 1.028 | ...... | 31 | 1.272 | 0.870 | 58 | 1.667 | 0.745 |
| 5 | 1.036 | ...... | 32 | 1.283 | 0.864 | 59 | 1.686 | 0.741 |
| 6 | 1.043 | ...... | 33 | 1.295 | 0.859 | 60 | 1.706 | 0.737 |
| 7 | 1.051 | ...... | 34 | 1.306 | 0.854 | 61 | 1.726 | 0.733 |
| 8 | 1.058 | ...... | 35 | 1.318 | 0.849 | 62 | 1.747 | 0.729 |
| 9 | 1.066 | ...... | 36 | 1.330 | 0.843 | 63 | 1.768 | 0.725 |
| 10 | 1.074 | 1.000 | 37 | 1.343 | 0.838 | 64 | 1.790 | 0.721 |
| 11 | 1.082 | 0.993 | 38 | 1.355 | 0.833 | 65 | 1.813 | 0.718 |
| 12 | 1.090 | 0.986 | 39 | 1.368 | 0.828 | 66 | 1.836 | 0.714 |
| 13 | 1.099 | 0.979 | 40 | 1.381 | 0.824 | 67 | 1.859 | 0.710 |
| 14 | 1.107 | 0.972 | 41 | 1.394 | 0.819 | 68 | 1.883 | 0.707 |
| 15 | 1.115 | 0.966 | 42 | 1.408 | 0.814 | 69 | 1.908 | 0.704 |
| 16 | 1.124 | 0.959 | 43 | 1.422 | 0.809 | 70 | 1.933 | 0.700 |
| 17 | 1.133 | 0.952 | 44 | 1.436 | 0.805 | 71 | 1.959 | 0.696 |
| 18 | 1.142 | 0.946 | 45 | 1.450 | 0.800 | 72 | 1.986 | 0.693 |
| 19 | 1.151 | 0.940 | 46 | 1.465 | 0.796 | 73 | 2.014 | 0.689 |
| 20 | 1.160 | 0.933 | 47 | 1.480 | 0.791 | 74 | 2.042 | 0.686 |
| 21 | 1.169 | 0.927 | 48 | 1.495 | 0.787 | 75 | 2.071 | 0.683 |
| 22 | 1.179 | 0.921 | 49 | 1.510 | 0.782 | 76 | 2.101 | 0.679 |
| 23 | 1.189 | 0.915 | 50 | 1.526 | 0.778 | 77 | 2.132 | 0.676 |
| 24 | 1.198 | 0.909 | 51 | 1.542 | 0.773 | 78 | 2.164 | 0.673 |
| 25 | 1.208 | 0.903 | 52 | 1.559 | 0.769 | 79 | 2.197 | 0.669 |
| 26 | 1.219 | 0.897 | 53 | 1.576 | 0.765 | 80 | 2.230 | 0.666 |

## Weights of Non-Ferrous Metal Sheets — 1

| American Wire or Brown & Sharpe Gage No. | Thickness, Inch | Approximate Weight, Pounds per Square Foot | | | | |
|---|---|---|---|---|---|---|
| | | Copper* | Yellow Brass | Tobin Bronze | 5 Per Cent Phosphor-Bronze | Everdur 1010 |
| 0000 | 0.4600 | 21.33 | 20.27 | 20.14 | 21.20 | 20.40 |
| 000 | 0.4096 | 18.99 | 18.05 | 17.93 | 18.88 | 18.17 |
| 00 | 0.3648 | 16.92 | 16.07 | 16.41 | 16.81 | 16.18 |
| 0 | 0.3249 | 15.06 | 14.32 | 14.23 | 14.98 | 14.41 |
| 1 | 0.2893 | 13.41 | 12.75 | 12.67 | 13.33 | 12.83 |
| 2 | 0.2576 | 11.94 | 11.35 | 11.28 | 11.87 | 11.43 |
| 3 | 0.2294 | 10.64 | 10.11 | 10.04 | 10.57 | 10.17 |
| 4 | 0.2043 | 9.473 | 9.002 | 8.943 | 9.414 | 9.061 |
| 5 | 0.1819 | 8.434 | 8.015 | 7.963 | 8.382 | 8.068 |
| 6 | 0.1620 | 7.512 | 7.138 | 7.092 | 7.465 | 7.185 |
| 7 | 0.1443 | 6.691 | 6.358 | 6.317 | 6.649 | 6.400 |
| 8 | 0.1285 | 5.958 | 5.662 | 5.625 | 5.921 | 5.699 |
| 9 | 0.1144 | 5.304 | 5.041 | 5.008 | 5.272 | 5.074 |
| 10 | 0.1019 | 4.725 | 4.490 | 4.461 | 4.696 | 4.519 |
| 11 | 0.0907 | 4.206 | 3.997 | 3.971 | 4.180 | 4.023 |
| 12 | 0.0808 | 3.747 | 3.560 | 3.537 | 3.723 | 3.584 |
| 13 | 0.0720 | 3.338 | 3.173 | 3.152 | 3.318 | 3.193 |
| 14 | 0.0641 | 2.972 | 2.825 | 2.807 | 2.954 | 2.843 |
| 15 | 0.0571 | 2.648 | 2.516 | 2.500 | 2.631 | 2.532 |
| 16 | 0.0508 | 2.355 | 2.238 | 2.223 | 2.341 | 2.253 |
| 17 | 0.0453 | 2.100 | 1.996 | 1.983 | 2.087 | 2.009 |
| 18 | 0.0403 | 1.869 | 1.776 | 1.764 | 1.857 | 1.787 |
| 19 | 0.0359 | 1.665 | 1.582 | 1.572 | 1.654 | 1.592 |
| 20 | 0.0320 | 1.484 | 1.410 | 1.401 | 1.475 | 1.419 |
| 21 | 0.0285 | 1.321 | 1.256 | 1.248 | 1.314 | 1.264 |
| 22 | 0.0253 | 1.178 | 1.119 | 1.112 | 1.170 | 1.127 |
| 23 | 0.0226 | 1.048 | 0.9958 | 0.9893 | 1.041 | 1.002 |
| 24 | 0.0201 | 0.9320 | 0.8857 | 0.8799 | 0.9263 | 0.8915 |
| 25 | 0.0179 | 0.8300 | 0.7887 | 0.7836 | 0.8248 | 0.7939 |
| 26 | 0.0159 | 0.7373 | 0.7006 | 0.6960 | 0.7327 | 0.7052 |
| 27 | 0.0142 | 0.6584 | 0.6257 | 0.6216 | 0.6544 | 0.6298 |
| 28 | 0.0126 | 0.5842 | 0.5552 | 0.5516 | 0.5806 | 0.5588 |
| 29 | 0.0113 | 0.5240 | 0.4979 | 0.4947 | 0.5207 | 0.5012 |
| 30 | 0.0100 | 0.4637 | 0.4406 | 0.4377 | 0.4608 | 0.4435 |
| 31 | 0.0089 | 0.4127 | 0.3922 | 0.3897 | 0.4102 | 0.3947 |
| 32 | 0.0080 | 0.3709 | 0.3525 | 0.3502 | 0.3686 | 0.3548 |
| 33 | 0.0071 | 0.3292 | 0.3129 | 0.3109 | 0.3272 | 0.3149 |
| 34 | 0.0063 | 0.2921 | 0.2776 | 0.2758 | 0.2903 | 0.2794 |
| 35 | 0.0056 | 0.2597 | 0.2468 | 0.2452 | 0.2581 | 0.2484 |
| 36 | 0.0050 | 0.2318 | 0.2203 | 0.2189 | 0.2304 | 0.2218 |
| 37 | 0.0045 | 0.2087 | 0.1983 | 0.1970 | 0.2074 | 0.1996 |
| 38 | 0.0040 | 0.1855 | 0.1763 | 0.1752 | 0.1844 | 0.1774 |
| 39 | 0.0035 | 0.1623 | 0.1542 | 0.1532 | 0.1613 | 0.1552 |
| 40 | 0.0031 | 0.1437 | 0.1366 | 0.1357 | 0.1429 | 0.1375 |

* Copper sheets can also be obtained in fractional-inch thicknesses varying by sixteenths of an inch from 1/16 to 2 inches.

## Weights of Non-Ferrous Metal Sheets — 2

| American Wire or Brown & Sharpe Gage No. | Thickness, Inch | Approximate Weight, Pounds per Square Foot | | | | |
|---|---|---|---|---|---|---|
| | | S.A.E. Aluminum Alloys Nos. 26 and 27 | S.A.E. Aluminum Alloy No. 28 | Aluminum Commercially Pure (99 to 99.4 Per Cent) | Nickel Silver 18%* | Nickel Silver 20%–30% |
| 0000 | 0.4600 | 6.680 | 6.410 | 6.490 | 20.93 | 21.20 |
| 000 | 0.4096 | 5.950 | 5.710 | 5.780 | 18.64 | 18.88 |
| 00 | 0.3648 | 5.290 | 5.090 | 5.140 | 16.60 | 16.81 |
| 0 | 0.3249 | 4.720 | 4.530 | 4.580 | 14.78 | 14.97 |
| 1 | 0.2893 | 4.200 | 4.030 | 4.080 | 13.16 | 13.33 |
| 2 | 0.2576 | 3.738 | 3.591 | 3.632 | 11.72 | 11.87 |
| 3 | 0.2294 | 3.329 | 3.198 | 3.234 | 10.44 | 10.57 |
| 4 | 0.2043 | 2.964 | 2.848 | 2.880 | 9.296 | 9.414 |
| 5 | 0.1819 | 2.640 | 2.536 | 2.565 | 8.277 | 8.382 |
| 6 | 0.1620 | 2.351 | 2.258 | 2.284 | 7.372 | 7.466 |
| 7 | 0.1443 | 2.094 | 2.012 | 2.034 | 6.566 | 6.649 |
| 8 | 0.1285 | 1.865 | 1.792 | 1.812 | 5.847 | 5.921 |
| 9 | 0.1144 | 1.660 | 1.595 | 1.613 | 5.206 | 5.272 |
| 10 | 0.1019 | 1.479 | 1.420 | 1.437 | 4.637 | 4.696 |
| 11 | 0.0907 | 1.316 | 1.264 | 1.279 | 4.127 | 4.179 |
| 12 | 0.0808 | 1.172 | 1.126 | 1.139 | 3.677 | 3.724 |
| 13 | 0.0720 | 1.045 | 1.004 | 1.015 | 3.276 | 3.318 |
| 14 | 0.0641 | 0.930 | 0.894 | 0.904 | 2.917 | 2.954 |
| 15 | 0.0571 | 0.829 | 0.796 | 0.805 | 2.598 | 2.631 |
| 16 | 0.0508 | 0.737 | 0.708 | 0.716 | 2.312 | 2.341 |
| 17 | 0.0453 | 0.657 | 0.631 | 0.639 | 2.061 | 2.087 |
| 18 | 0.0403 | 0.585 | 0.562 | 0.568 | 1.834 | 1.857 |
| 19 | 0.0359 | 0.5210 | 0.5010 | 0.5060 | 1 634 | 1.655 |
| 20 | 0.0320 | 0.4640 | 0.4460 | 0.4510 | 1.456 | 1.474 |
| 21 | 0.0285 | 0.4140 | 0.3970 | 0.4020 | 1.297 | 1.313 |
| 22 | 0.0253 | 0.3671 | 0.3527 | 0.3567 | 1.156 | 1.171 |
| 23 | 0.0226 | 0.3280 | 0.3150 | 0.3186 | 1.028 | 1.041 |
| 24 | 0.0201 | 0.2917 | 0.2802 | 0.2834 | 0.9146 | 0.9262 |
| 25 | 0.0179 | 0.2597 | 0.2495 | 0.2524 | 0.8145 | 0.8248 |
| 26 | 0.0159 | 0.2307 | 0.2216 | 0.2242 | 0.7235 | 0.7327 |
| 27 | 0.0142 | 0.2060 | 0.1980 | 0.2002 | 0.6462 | 0.6544 |
| 28 | 0.0126 | 0.1828 | 0.1756 | 0.1776 | 0.5734 | 0.5807 |
| 29 | 0.0113 | 0.1640 | 0.1575 | 0.1593 | 0.5142 | 0.5207 |
| 30 | 0.0100 | 0.1451 | 0.1394 | 0.1410 | 0.4550 | 0.4608 |
| 31 | 0.0089 | 0.1296 | 0.1245 | 0.1259 | 0.4050 | 0.4101 |
| 32 | 0.0080 | 0.1154 | 0.1108 | 0.1121 | 0.3640 | 0.3686 |
| 33 | 0.0071 | 0.1027 | 0.0987 | 0.0998 | 0.3231 | 0.3272 |
| 34 | 0.0063 | 0.0914 | 0.0878 | 0.0888 | 0.2867 | 0.2903 |
| 35 | 0.0056 | 0.0814 | 0.0782 | 0.0791 | 0.2548 | 0.2580 |
| 36 | 0.0050 | 0.0726 | 0.0697 | 0.0705 | 0.2275 | 0.2304 |
| 37 | 0.0045 | 0.0646 | 0.0620 | 0.0627 | 0.2048 | 0.2074 |
| 38 | 0.0040 | 0.0576 | 0.0553 | 0.0560 | 0.1820 | 0.1843 |
| 39 | 0.0035 | 0.0512 | 0.0492 | 0.0498 | 0.1593 | 0.1613 |
| 40 | 0.0031 | 0.0456 | 0.0438 | 0.0443 | 0.1411 | 0.1429 |

* Multiply weights in this column by 0.9905 for 10 per cent nickel-silver and by 0.9937 for 15 per cent nickel-silver.

### Weights of Magnesium Alloy Sheets and Plates — 1

Weights are for Dowmetal M (S.A.E. No. 51). See also *Note 2* at end of table.

| Thickness, Inch | American or B. & S. Gage No. | Weight, Pounds per Square Foot | Standard Widths, Inches | Standard Lengths, Inches | | | | |
|---|---|---|---|---|---|---|---|---|
| | | | | 60 | 72 | 96 | 120 | 144 |
| | | | | Weights in Pounds | | | | |
| 0.016 | 26 | 0.147 | 24 | 1.47 | 1.76 | .... | .... | .... |
| 0.018 | 25 | 0.166 | 24 | 1.66 | 1.99 | .... | .... | .... |
| 0.020 | 24 | 0.184 | 24 | 1.84 | 2.21 | 2.94 | .... | .... |
| | | | 30 | 2.30 | 2.76 | .... | .... | .... |
| 0.023 | 23 | 0.212 | 24 | 2.12 | 2.54 | 3.39 | .... | .... |
| | | | 30 | 2.65 | 3.18 | .... | .... | .... |
| 0.025 | 22 | 0.230 | 24 | 2.3 | 2.76 | 3.68 | .... | .... |
| | | | 30 | 2.87 | 3.45 | .... | .... | .... |
| 0.028 | 21 | 0.258 | 24 | 2.58 | 3.10 | 4.13 | .... | .... |
| | | | 30 | 3.22 | 3.87 | .... | .... | .... |
| 0.032 | 20 | 0.295 | 24 | 2.95 | 3.54 | 4.72 | 5.9 | .... |
| | | | 30 | 3.69 | 4.42 | 5.90 | 7.38 | .... |
| | | | 36 | 4.42 | 5.31 | 7.08 | .... | .... |
| 0.036 | 19 | 0.332 | 24 | 3.32 | 3.98 | 5.31 | 6.64 | .... |
| | | | 30 | 4.15 | 4.98 | 6.64 | 8.30 | .... |
| | | | 36 | 4.98 | 5.98 | 7.97 | .... | .... |
| 0.040 | 18 | 0.369 | 24 | 3.69 | 4.43 | 5.90 | 7.38 | .... |
| | | | 30 | 4.61 | 5.53 | 7.38 | 9.22 | .... |
| | | | 36 | 5.53 | 6.64 | 8.85 | 11.1 | .... |
| 0.045 | 17 | 0.415 | 24 | 4.15 | 4.98 | 6.64 | 8.30 | .... |
| | | | 30 | 5.19 | 6.22 | 8.30 | 10.4 | .... |
| | | | 36 | 6.22 | 7.47 | 9.96 | 12.5 | .... |
| 0.051 | 16 | 0.470 | 24 | 4.70 | 5.64 | 7.52 | 9.40 | 11.3 |
| | | | 30 | 5.87 | 7.05 | 9.40 | 11.7 | 14.1 |
| | | | 36 | 7.05 | 8.46 | 11.3 | 14.1 | 16.9 |
| | | | 42 | 8.22 | 9.87 | 13.2 | 16.4 | 19.7 |
| 0.057 | 15 | 0.525 | 24 | 5.25 | 6.30 | 8.4 | 10.5 | 12.6 |
| | | | 30 | 6.57 | 7.88 | 10.5 | 13.1 | 15.8 |
| | | | 36 | 7.88 | 9.45 | 12.6 | 15.8 | 18.9 |
| | | | 42 | 9.19 | 10.1 | 14.7 | 18.4 | 22.1 |
| 0.064 | 14 | 0.590 | 24 | 5.9 | 7.07 | 9.45 | 11.8 | 14.2 |
| | | | 30 | 7.38 | 8.85 | 11.8 | 14.7 | 17.7 |
| | | | 36 | 8.85 | 10.6 | 14.2 | 17.7 | 21.2 |
| | | | 42 | 10.3 | 12.4 | 16.5 | 20.6 | 24.8 |
| 0.072 | 13 | 0.662 | 24 | 6.54 | 7.85 | 10.5 | 13.1 | 15.7 |
| | | | 30 | 8.17 | 9.80 | 13.1 | 16.3 | 19.6 |
| | | | 36 | 9.80 | 11.8 | 15.7 | 19.6 | 23.5 |
| | | | 42 | 11.4 | 13.7 | 18.3 | 22.9 | 27.4 |
| 0.081 | 12 | 0.746 | 24 | 7.46 | 8.96 | 11.9 | 14.9 | 17.9 |
| | | | 30 | 9.34 | 11.2 | 14.9 | 18.6 | 22.4 |
| | | | 36 | 11.2 | 13.4 | 17.9 | 22.4 | 26.8 |
| | | | 42 | 13.0 | 15.7 | 20.9 | 26.1 | 31.3 |
| | | | 48 | 14.9 | 17.9 | 23.9 | 29.8 | 35.8 |
| 0.091 | 11 | 0.838 | 24 | 8.38 | 10.1 | 13.4 | 16.8 | 20.1 |
| | | | 30 | 10.5 | 12.6 | 16.8 | 20.9 | 25.1 |
| | | | 36 | 12.6 | 15.1 | 20.1 | 25.1 | 30.2 |
| | | | 42 | 14.7 | 17.6 | 23.5 | 29.3 | 35.2 |
| | | | 48 | 16.8 | 20.1 | 26.8 | 33.5 | 40.2 |
| 0.102 | 10 | 0.940 | 24 | 9.4 | 11.3 | 15.0 | 18.8 | 22.6 |
| | | | 30 | 11.7 | 14.1 | 18.8 | 23.5 | 28.2 |
| | | | 36 | 14.1 | 16.9 | 22.6 | 28.2 | 33.8 |
| | | | 42 | 16.4 | 19.7 | 26.3 | 32.9 | 39.4 |
| | | | 48 | 18.8 | 22.6 | 30.0 | 37.6 | 45.1 |

### Weights of Magnesium Alloy Sheets and Plates — 2

| Thickness, Inch | American or B. & S. Gage No. | Weight, Pounds per Square Foot | Standard Widths, Inches | Standard Lengths, Inches | | | | |
|---|---|---|---|---|---|---|---|---|
| | | | | 60 | 72 | 96 | 120 | 144 |
| | | | | Weights in Pounds | | | | |
| 0.114 | 9 | 1.05 | 24 | 10.5 | 12.6 | 16.8 | 21.0 | 25.2 |
| | | | 30 | 13.1 | 15.7 | 21.0 | 26.2 | 31.5 |
| | | | 36 | 15.7 | 18.9 | 25.2 | 31.5 | 37.8 |
| | | | 42 | 18.4 | 22.1 | 29.4 | 36.7 | 44.1 |
| | | | 48 | 21.0 | 25.2 | 33.6 | 42.0 | 50.4 |
| 0.128 | 8 | 1.18 | 24 | 11.8 | 14.2 | 18.9 | 23.6 | 28.3 |
| | | | 30 | 14.7 | 17.7 | 23.6 | 29.4 | 35.4 |
| | | | 36 | 17.7 | 21.2 | 28.3 | 35.4 | 42.5 |
| | | | 42 | 20.6 | 24.8 | 33.0 | 41.3 | 49.5 |
| | | | 48 | 23.6 | 28.3 | 37.8 | 47.2 | 56.6 |
| 0.156 | 5⁄32 | 1.44 | 24 | 14.4 | 17.3 | 23.1 | 28.8 | 34.6 |
| | | | 30 | 18.0 | 21.6 | 28.8 | 36.0 | 43.2 |
| | | | 36 | 21.6 | 25.9 | 34.6 | 43.2 | 51.9 |
| | | | 42 | 25.2 | 30.2 | 40.3 | 50.4 | 60.5 |
| | | | 48 | 28.8 | 34.6 | 46.1 | 57.6 | 69.1 |
| 0.188 | 3⁄16 | 1.73 | 24 | 17.3 | 20.8 | 27.7 | 34.6 | 41.5 |
| | | | 30 | 21.6 | 25.9 | 34.6 | 43.3 | 51.9 |
| | | | 36 | 25.9 | 31.2 | 41.5 | 51.9 | 62.3 |
| | | | 42 | 30.3 | 36.4 | 48.4 | 60.5 | 72.6 |
| | | | 48 | 34.6 | 41.5 | 55.4 | 69.2 | 82.5 |
| 0.219 | 7⁄32 | 2.02 | 24 | 20.2 | 24.2 | 32.3 | 40.4 | 48.5 |
| | | | 30 | 25.2 | 30.3 | 40.4 | 50.5 | 60.6 |
| | | | 36 | 30.3 | 36.4 | 48.5 | 60.6 | 72.7 |
| | | | 42 | 35.4 | 42.4 | 56.6 | 70.7 | 84.9 |
| | | | 48 | 40.4 | 48.5 | 64.7 | 80.8 | .... |
| 0.250 | 1⁄4 | 2.30 | 24 | 23.0 | 27.6 | 36.8 | 46.0 | 55.2 |
| | | | 30 | 28.8 | 34.5 | 46.0 | 57.5 | 69.0 |
| | | | 36 | 34.5 | 41.4 | 55.2 | 69.0 | 82.8 |
| | | | 42 | 40.3 | 48.3 | 54.4 | 80.5 | .... |
| | | | 48 | 46.0 | 55.2 | 73.6 | 92.0 | .... |
| 0.313 | 5⁄16 | 2.88 | 24 | 28.8 | 34.6 | 46.1 | 57.6 | 69.1 |
| | | | 30 | 36.0 | 43.2 | 57.6 | 72.0 | 86.4 |
| | | | 36 | 43.2 | 51.8 | 69.1 | 86.4 | .... |
| | | | 42 | 50.4 | 60.5 | 80.6 | .... | .... |
| | | | 48 | 57.6 | 69.1 | 92.2 | .... | .... |
| 0.375 | 3⁄8 | 3.46 | 24 | 34.6 | 41.5 | 55.4 | 69.2 | 83.0 |
| | | | 30 | 43.3 | 51.9 | 69.2 | 86.5 | .... |
| | | | 36 | 51.9 | 62.3 | 83.0 | .... | .... |
| | | | 42 | 60.6 | 72.7 | .... | .... | .... |
| | | | 48 | 69.2 | 83.0 | .... | .... | .... |
| 0.437 | 7⁄16 | 4.03 | 24 | 40.3 | 48.4 | 64.5 | .... | .... |
| | | | 30 | 50.4 | 60.5 | 80.6 | .... | .... |
| | | | 36 | 60.5 | 72.5 | .... | .... | .... |
| | | | 42 | 70.5 | 84.6 | .... | .... | .... |
| | | | 48 | 80.6 | .... | .... | .... | .... |
| 0.500 | 1⁄2 | 4.61 | 24 | 46.1 | 55.3 | 73.8 | 92.2 | .... |
| | | | 30 | 57.6 | 69.2 | 92.2 | .... | .... |
| | | | 36 | 69.2 | 83.0 | .... | .... | .... |
| | | | 42 | 80.7 | .... | .... | .... | .... |
| | | | 48 | 92.2 | .... | .... | .... | .... |

*Note 1:* Sheet is defined as rolled metal up to 0.250 inch thickness. Material 0.250 inch thick and over is designated as plate.

*Note 2:* The weights in this table are for Dowmetal M(S.A.E. No. 51). To obtain comparable weights for Dowmetal FS-1 (S.A.E. No. 510), multiply the weights shown by 1.005. For Dowmetal J-1 (S.A.E. No. 511), multiply the weights shown by 1.022.

### Weights per Thousand of Square Brass Blanks — 1

In estimating costs for press work, the material cost is the most important item. To illustrate the use of the tables, assume that a 6⅝ inch round blank is required and that the estimate is to be based upon a blank 6⅞ inches square, there being a ¼-inch margin allowance. If brass of No. 22 B & S gage is to be used, the weight in pounds per thousand square blanks is found in the table under No. 22 gage

and opposite the square blank size of 6⅞. This table is based upon the following formula:

Weight per thousand = blank area × thickness × 0.306 × 1000.

To obtain weight of sheet steel blanks multiply figure in table by 0.928. For sheet aluminum, multiply by 0.32. For sheet copper, multiply by 1.051. For circular blanks, multiply by 0.7854.

| Square Blanks | | B. & S. Gage Numbers and Decimal Equivalents | | | | | | |
|---|---|---|---|---|---|---|---|---|
| Sizes, Inches | Areas, Sq. In. | No. 16 0.05082 | No. 17 0.04526 | No. 18 0.04030 | No. 19 0.03589 | No. 20 0.03196 | No. 21 0.02846 | No. 22 0.02535 |
| ½ | 0.25 | 3.89 | 3.46 | 3.08 | 2.75 | 2.44 | 2.18 | 1.94 |
| ⅝ | 0.39 | 6.07 | 5.41 | 4.82 | 4.29 | 3.82 | 3.40 | 3.03 |
| ¾ | 0.56 | 8.71 | 7.79 | 6.94 | 6.18 | 5.50 | 4.90 | 4.36 |
| ⅞ | 0.77 | 11.97 | 10.60 | 9.44 | 8.41 | 7.49 | 6.67 | 5.94 |
| 1 | 1.00 | 15.55 | 13.85 | 12.33 | 10.98 | 9.78 | 8.71 | 7.76 |
| 1⅛ | 1.27 | 19.69 | 17.53 | 15.61 | 13.90 | 12.38 | 11.03 | 9.82 |
| 1¼ | 1.56 | 24.29 | 21.63 | 19.26 | 17.15 | 15.28 | 13.60 | 12.12 |
| 1⅜ | 1.89 | 29.41 | 26.19 | 23.32 | 20.77 | 18.49 | 16.47 | 14.67 |
| 1½ | 2.25 | 34.99 | 31.16 | 27.75 | 24.71 | 22.00 | 19.59 | 17.45 |
| 1⅝ | 2.64 | 41.07 | 36.58 | 32.57 | 29.00 | 25.83 | 23.00 | 20.49 |
| 1¾ | 3.06 | 47.62 | 42.41 | 37.76 | 33.63 | 29.95 | 26.67 | 23.75 |
| 1⅞ | 3.52 | 54.68 | 48.70 | 43.36 | 38.61 | 34.39 | 30.62 | 27.27 |
| 2 | 4.00 | 62.20 | 55.40 | 49.33 | 43.93 | 39.12 | 34.84 | 31.03 |
| 2⅛ | 4.52 | 70.23 | 62.54 | 55.69 | 49.60 | 44.17 | 39.33 | 35.03 |
| 2¼ | 5.06 | 78.72 | 70.11 | 62.42 | 55.59 | 49.51 | 44.08 | 39.27 |
| 2⅜ | 5.64 | 87.72 | 78.13 | 69.56 | 61.95 | 55.17 | 49.13 | 43.76 |
| 2½ | 6.25 | 97.19 | 86.56 | 77.07 | 68.64 | 61.12 | 54.43 | 48.48 |
| 2⅝ | 6.89 | 107.15 | 95.44 | 84.98 | 75.68 | 67.39 | 60.01 | 53.45 |
| 2¾ | 7.56 | 117.60 | 104.73 | 93.25 | 83.05 | 73.95 | 65.86 | 58.62 |
| 2⅞ | 8.27 | 128.54 | 114.48 | 101.93 | 90.78 | 80.84 | 71.99 | 64.12 |
| 3 | 9.00 | 139.96 | 124.65 | 110.99 | 98.84 | 88.02 | 78.38 | 69.81 |
| 3⅛ | 9.77 | 151.87 | 135.25 | 120.43 | 107.25 | 95.51 | 85.05 | 75.76 |
| 3¼ | 10.56 | 164.22 | 146.25 | 130.22 | 115.97 | 103.27 | 91.96 | 81.91 |
| 3⅜ | 11.39 | 177.13 | 157.75 | 140.46 | 125.09 | 111.39 | 99.19 | 88.35 |
| 3½ | 12.25 | 190.50 | 169.66 | 151.06 | 134.53 | 119.80 | 106.68 | 95.02 |
| 3⅝ | 13.14 | 204.34 | 181.98 | 162.04 | 144.31 | 128.51 | 114.43 | 101.93 |
| 3¾ | 14.06 | 218.65 | 194.72 | 173.39 | 154.41 | 137.50 | 122.44 | 109.06 |
| 3⅞ | 15.02 | 233.57 | 208.02 | 185.22 | 164.95 | 146.89 | 130.81 | 116.51 |
| 4 | 16.00 | 248.81 | 221.59 | 197.31 | 175.72 | 156.48 | 139.34 | 124.11 |
| 4⅛ | 17.02 | 264.68 | 235.72 | 209.89 | 186.92 | 166.45 | 148.22 | 132.03 |
| 4¼ | 18.06 | 280.85 | 250.12 | 222.71 | 198.34 | 176.62 | 157.28 | 140.09 |
| 4⅜ | 19.14 | 297.64 | 265.08 | 236.03 | 210.20 | 187.18 | 166.69 | 148.47 |
| 4½ | 20.25 | 314.91 | 280.45 | 249.72 | 222.39 | 198.04 | 176.35 | 157.08 |
| 4⅝ | 21.39 | 332.63 | 296.24 | 263.78 | 234.91 | 209.19 | 186.28 | 165.92 |

## Weights per Thousand of Square Brass Blanks — 2

| Square Blanks | | B. & S. Gage Numbers and Decimal Equivalents | | | | | | |
|---|---|---|---|---|---|---|---|---|
| Sizes, Inches | Areas, Sq. In. | No. 16 0.05082 | No. 17 0.04526 | No. 18 0.04030 | No. 19 0.03589 | No. 20 0.03196 | No. 21 0.02846 | No. 22 0.02535 |
| 4¾ | 22.56 | 350.83 | 312.45 | 278.20 | 247.76 | 220.63 | 196.47 | 175.00 |
| 4⅞ | 23.77 | 369.65 | 329.20 | 293.13 | 261.05 | 232.46 | 207.01 | 184.39 |
| 5 | 25.00 | 388.77 | 346.24 | 308.30 | 274.56 | 244.49 | 217.72 | 193.93 |
| 5⅛ | 26.26 | 408.37 | 363.69 | 323.83 | 288.40 | 256.82 | 228.69 | 203.70 |
| 5¼ | 27.56 | 428.58 | 381.69 | 339.86 | 302.67 | 269.53 | 240.01 | 213.79 |
| 5⅜ | 28.89 | 449.27 | 400.11 | 356.27 | 317.28 | 282.54 | 251.60 | 224.10 |
| 5½ | 30.25 | 470.42 | 418.95 | 373.04 | 332.22 | 295.84 | 263.44 | 234.65 |
| 5⅝ | 31.64 | 492.03 | 438.20 | 390.18 | 347.48 | 309.43 | 275.55 | 245.43 |
| 5¾ | 33.06 | 514.11 | 457.87 | 407.69 | 363.08 | 323.32 | 287.91 | 256.45 |
| 5⅞ | 34.52 | 536.82 | 478.09 | 425.69 | 379.11 | 337.60 | 300.62 | 267.78 |
| 6 | 36.00 | 559.83 | 498.58 | 443.94 | 395.36 | 352.07 | 313.52 | 279.26 |
| 6⅛ | 37.52 | 583.47 | 519.64 | 462.70 | 412.06 | 366.94 | 326.75 | 291.05 |
| 6¼ | 39.06 | 607.42 | 540.96 | 481.68 | 428.97 | 382.00 | 340.16 | 302.99 |
| 6⅜ | 40.64 | 631.99 | 562.85 | 501.16 | 446.32 | 397.45 | 353.92 | 315.25 |
| 6½ | 42.25 | 657.03 | 585.14 | 521.02 | 464.00 | 413.19 | 367.95 | 327.74 |
| 6⅝ | 43.89 | 682.53 | 607.86 | 541.24 | 482.01 | 429.23 | 382.23 | 340.46 |
| 6¾ | 45.56 | 708.50 | 630.99 | 561.84 | 500.36 | 445.57 | 396.77 | 353.41 |
| 6⅞ | 47.27 | 735.09 | 654.67 | 582.92 | 519.14 | 462.29 | 411.66 | 366.68 |
| 7 | 49.00 | 762.00 | 678.63 | 604.26 | 538.13 | 479.21 | 426.73 | 380.10 |
| 7⅛ | 50.76 | 789.36 | 703.00 | 625.96 | 557.46 | 496.42 | 442.06 | 393.75 |
| 7¼ | 52.56 | 817.36 | 727.93 | 648.16 | 577.23 | 514.02 | 457.73 | 407.71 |
| 7⅜ | 54.39 | 845.81 | 753.28 | 670.73 | 597.33 | 531.92 | 473.67 | 421.91 |
| 7½ | 56.25 | 874.74 | 779.04 | 693.66 | 617.76 | 550.11 | 489.87 | 436.34 |
| 7⅝ | 58.14 | 904.13 | 805.21 | 716.97 | 638.51 | 568.60 | 506.33 | 451.00 |
| 7¾ | 60.06 | 933.99 | 831.80 | 740.65 | 659.60 | 587.37 | 523.05 | 465.89 |
| 7⅞ | 62.02 | 964.47 | 858.95 | 764.82 | 681.12 | 606.54 | 540.12 | 481.10 |
| 8 | 64.00 | 995.26 | 886.37 | 789.24 | 702.87 | 625.90 | 557.36 | 496.45 |

| Square Blank Sizes, Inches | B. & S. Gage Numbers and Decimal Equivalents | | | | | | |
|---|---|---|---|---|---|---|---|
| | No. 23 0.02257 | No. 24 0.02010 | No. 25 0.01790 | No. 26 0.01594 | No. 27 0.01420 | No. 28 0.01264 | No. 29 0.01126 | No. 30 0.01003 |
| ½ | 1.73 | 1.54 | 1.37 | 1.22 | 1.09 | 0.97 | 0.86 | 0.77 |
| ⅝ | 2.70 | 2.40 | 2.14 | 1.91 | 1.70 | 1.51 | 1.35 | 1.20 |
| ¾ | 3.88 | 3.46 | 3.08 | 2.74 | 2.44 | 2.18 | 1.94 | 1.73 |
| ⅞ | 5.29 | 4.71 | 4.19 | 3.73 | 3.33 | 2.96 | 2.64 | 2.35 |
| 1 | 6.91 | 6.15 | 5.48 | 4.88 | 4.35 | 3.87 | 3.45 | 3.07 |
| 1⅛ | 8.74 | 7.79 | 6.93 | 6.18 | 5.50 | 4.90 | 4.36 | 3.89 |
| 1¼ | 10.79 | 9.61 | 8.56 | 7.62 | 6.79 | 6.04 | 5.38 | 4.79 |
| 1⅜ | 13.06 | 11.63 | 10.36 | 9.22 | 8.22 | 7.31 | 6.52 | 5.80 |
| 1½ | 15.54 | 13.84 | 12.32 | 10.97 | 9.78 | 8.70 | 7.75 | 6.91 |
| 1⅝ | 18.24 | 16.24 | 14.47 | 12.88 | 11.48 | 10.21 | 9.10 | 8.11 |
| 1¾ | 21.15 | 18.83 | 16.77 | 14.94 | 13.31 | 11.84 | 10.55 | 9.40 |
| 1⅞ | 24.28 | 21.63 | 19.26 | 17.15 | 15.28 | 13.60 | 12.11 | 10.79 |
| 2 | 27.63 | 24.60 | 21.91 | 19.51 | 17.38 | 15.47 | 13.78 | 12.28 |
| 2⅛ | 31.19 | 27.78 | 24.74 | 22.03 | 19.62 | 17.47 | 15.56 | 13.86 |
| 2¼ | 34.96 | 31.13 | 27.73 | 24.69 | 22.00 | 19.58 | 17.44 | 15.54 |
| 2⅜ | 38.96 | 34.70 | 30.90 | 27.51 | 24.51 | 21.82 | 19.44 | 17.31 |

### Weights per Thousand of Square Brass Blanks — 3

| Square Blank Sizes, Inches | B. & S. Gage Numbers and Decimal Equivalents | | | | | | | |
|---|---|---|---|---|---|---|---|---|
| | No. 23 0.02257 | No. 24 0.02010 | No. 25 0.01790 | No. 26 0.01594 | No. 27 0.01420 | No. 28 0.01264 | No. 29 0.01126 | No. 30 0.01003 |
| 2½ | 43.17 | 38.44 | 34.23 | 30.49 | 27.16 | 24.17 | 21.53 | 19.18 |
| 2⅝ | 47.59 | 42.38 | 37.71 | 33.61 | 29.94 | 26.65 | 23.74 | 21.15 |
| 2¾ | 52.23 | 46.51 | 41.42 | 36.88 | 32.86 | 29.25 | 26.06 | 23.21 |
| 2⅞ | 57.09 | 50.84 | 45.28 | 40.32 | 35.92 | 31.97 | 28.48 | 25.37 |
| 3 | 62.16 | 55.36 | 49.00 | 43.90 | 39.11 | 34.81 | 31.01 | 27.62 |
| 3⅛ | 67.45 | 60.07 | 53.49 | 47.64 | 42.44 | 37.77 | 33.65 | 29.97 |
| 3¼ | 72.93 | 64.95 | 57.84 | 51.51 | 45.89 | 40.84 | 36.39 | 32.51 |
| 3⅜ | 78.67 | 70.06 | 62.39 | 55.56 | 49.49 | 44.06 | 39.25 | 34.96 |
| 3½ | 84.60 | 75.34 | 67.10 | 59.75 | 53.23 | 47.38 | 42.21 | 37.60 |
| 3⅝ | 90.75 | 80.82 | 71.97 | 64.09 | 57.10 | 50.82 | 45.27 | 40.33 |
| 3¾ | 97.10 | 86.48 | 77.01 | 68.58 | 61.09 | 54.38 | 48.44 | 43.15 |
| 3⅞ | 103.73 | 92.38 | 82.27 | 73.26 | 65.26 | 58.09 | 51.75 | 46.10 |
| 4 | 110.50 | 98.41 | 87.64 | 78.04 | 69.52 | 61.89 | 55.13 | 49.11 |
| 4⅛ | 117.55 | 104.68 | 93.23 | 83.02 | 73.96 | 65.83 | 58.64 | 52.24 |
| 4¼ | 124.73 | 111.08 | 98.92 | 88.09 | 78.47 | 69.85 | 62.23 | 55.43 |
| 4⅜ | 132.19 | 117.72 | 104.84 | 93.36 | 83.17 | 74.03 | 65.95 | 58.74 |
| 4½ | 139.86 | 124.55 | 110.92 | 98.77 | 87.99 | 78.32 | 69.77 | 62.15 |
| 4⅝ | 147.73 | 131.56 | 117.16 | 104.33 | 92.94 | 82.73 | 73.70 | 65.65 |
| 4¾ | 155.81 | 138.76 | 123.57 | 110.04 | 98.03 | 87.26 | 77.73 | 69.24 |
| 4⅞ | 164.17 | 146.20 | 130.20 | 115.94 | 103.29 | 91.94 | 81.90 | 72.95 |
| 5 | 172.66 | 153.77 | 136.94 | 121.94 | 108.63 | 96.70 | 86.14 | 76.73 |
| 5⅛ | 181.36 | 161.51 | 143.84 | 128.09 | 114.10 | 101.57 | 90.48 | 80.60 |
| 5¼ | 190.34 | 169.51 | 150.96 | 134.43 | 119.75 | 106.60 | 94.96 | 84.59 |
| 5⅜ | 199.53 | 177.69 | 158.24 | 140.92 | 125.53 | 111.74 | 99.54 | 88.67 |
| 5½ | 208.92 | 186.06 | 165.69 | 147.55 | 131.44 | 117.00 | 104.23 | 92.84 |
| 5⅝ | 218.52 | 194.60 | 173.30 | 154.33 | 137.48 | 122.38 | 109.02 | 97.11 |
| 5¾ | 228.33 | 203.34 | 181.08 | 161.25 | 143.65 | 127.87 | 113.91 | 101.47 |
| 5⅞ | 238.41 | 212.32 | 189.08 | 168.38 | 150.00 | 133.52 | 118.94 | 105.95 |
| 6 | 248.63 | 221.42 | 197.19 | 175.60 | 156.43 | 139.24 | 124.04 | 110.49 |
| 6⅛ | 259.13 | 230.77 | 205.51 | 183.01 | 163.03 | 145.12 | 129.28 | 115.16 |
| 6¼ | 269.76 | 240.24 | 213.95 | 190.52 | 169.72 | 151.08 | 134.58 | 119.88 |
| 6⅜ | 280.68 | 249.96 | 222.60 | 198.23 | 176.59 | 157.19 | 140.03 | 124.73 |
| 6½ | 291.80 | 259.86 | 231.42 | 206.08 | 183.58 | 163.42 | 145.57 | 129.67 |
| 6⅝ | 303.12 | 269.95 | 240.40 | 214.08 | 190.71 | 169.76 | 151.23 | 134.71 |
| 6¾ | 314.66 | 280.22 | 249.55 | 222.23 | 197.97 | 176.22 | 156.98 | 139.83 |
| 6⅞ | 326.47 | 290.74 | 258.92 | 230.57 | 205.40 | 182.83 | 162.87 | 145.08 |
| 7 | 338.41 | 301.38 | 268.39 | 239.00 | 212.91 | 189.52 | 168.83 | 150.39 |
| 7⅛ | 350.57 | 312.20 | 278.03 | 247.59 | 220.56 | 196.33 | 174.90 | 155.79 |
| 7¼ | 363.00 | 323.28 | 287.89 | 256.37 | 228.38 | 203.29 | 181.10 | 161.32 |
| 7⅜ | 375.64 | 334.53 | 297.92 | 265.39 | 236.34 | 210.37 | 187.40 | 166.93 |
| 7½ | 388.49 | 345.97 | 308.10 | 274.37 | 244.42 | 217.57 | 193.81 | 172.64 |
| 7⅝ | 401.54 | 357.60 | 318.46 | 283.59 | 252.63 | 224.88 | 200.32 | 178.44 |
| 7¾ | 414.80 | 369.41 | 328.97 | 292.95 | 260.97 | 232.30 | 206.94 | 184.33 |
| 7⅞ | 428.34 | 381.46 | 339.71 | 302.51 | 269.49 | 239.88 | 213.69 | 190.35 |
| 8 | 442.01 | 393.64 | 350.55 | 312.17 | 278.09 | 247.54 | 220.52 | 196.43 |

WEIGHTS OF MATERIALS

Pounds of Round Brass Rod per Thousand Pieces — 1*

| Length, Inches† | Diameter of Stock, Inches | | | | | | | | | | | | | | | |
|---|---|---|---|---|---|---|---|---|---|---|---|---|---|---|---|---|
| | 1/16 | 3/32 | 1/8 | 5/32 | 3/16 | 7/32 | 1/4 | 9/32 | 5/16 | 11/32 | 3/8 | 13/32 | 7/16 | 15/32 | 1/2 | 17/32 |
| 1/32 | 0.03 | 0.07 | 0.12 | 0.19 | 0.3 | 0.4 | 0.5 | 0.6 | 0.7 | 0.9 | 1.1 | 1.3 | 1.5 | 1.7 | 1.9 | 2.2 |
| 1/16 | 0.06 | 0.13 | 0.24 | 0.36 | 0.6 | 0.8 | 1.0 | 1.1 | 1.4 | 1.7 | 2.1 | 2.4 | 2.8 | 3.3 | 3.7 | 4.2 |
| 3/32 | 0.09 | 0.19 | 0.36 | 0.55 | 0.8 | 1.0 | 1.4 | 1.7 | 2.2 | 2.6 | 3.1 | 3.7 | 4.3 | 4.9 | 5.6 | 6.3 |
| 1/8 | 0.12 | 0.26 | 0.48 | 0.73 | 1.0 | 1.4 | 1.8 | 2.3 | 2.9 | 3.5 | 4.2 | 4.9 | 5.7 | 6.6 | 7.5 | 8.5 |
| 5/32 | 0.15 | 0.33 | 0.59 | 0.92 | 1.3 | 1.8 | 2.3 | 2.9 | 3.6 | 4.4 | 5.3 | 6.2 | 7.2 | 8.2 | 9.4 | 10.6 |
| 3/16 | 0.18 | 0.39 | 0.71 | 1.1 | 1.5 | 2.1 | 2.8 | 3.5 | 4.4 | 5.3 | 6.3 | 7.4 | 8.6 | 9.9 | 11.2 | 12.7 |
| 7/32 | 0.21 | 0.46 | 0.83 | 1.2 | 1.8 | 2.5 | 3.2 | 4.1 | 5.1 | 6.2 | 7.4 | 8.7 | 10.1 | 11.5 | 13.1 | 14.8 |
| 1/4 | 0.24 | 0.52 | 0.95 | 1.4 | 2.1 | 2.8 | 3.7 | 4.7 | 5.8 | 7.1 | 8.4 | 9.9 | 11.5 | 13.2 | 15.0 | 17.0 |
| 9/32 | 0.27 | 0.59 | 1.1 | 1.6 | 2.3 | 3.2 | 4.2 | 5.3 | 6.6 | 8.0 | 9.5 | 11.2 | 13.0 | 14.8 | 16.9 | 19.1 |
| 5/16 | 0.30 | 0.66 | 1.2 | 1.8 | 2.6 | 3.6 | 4.7 | 5.9 | 7.3 | 8.9 | 10.6 | 12.4 | 14.4 | 16.5 | 18.8 | 21.2 |
| 11/32 | 0.33 | 0.72 | 1.3 | 2.0 | 2.9 | 3.9 | 5.1 | 6.5 | 8.1 | 9.7 | 11.6 | 13.6 | 15.8 | 18.1 | 20.7 | 23.3 |
| 3/8 | 0.36 | 0.79 | 1.4 | 2.2 | 3.1 | 4.3 | 5.6 | 7.1 | 8.8 | 10.6 | 12.7 | 14.9 | 17.3 | 19.8 | 22.5 | 25.5 |
| 13/32 | 0.39 | 0.85 | 1.6 | 2.3 | 3.4 | 4.7 | 6.1 | 7.7 | 9.5 | 11.5 | 13.7 | 16.1 | 18.7 | 21.5 | 24.4 | 27.6 |
| 7/16 | 0.42 | 0.92 | 1.7 | 2.5 | 3.7 | 5.0 | 6.5 | 8.3 | 10.3 | 12.4 | 14.8 | 17.4 | 20.2 | 23.1 | 26.3 | 29.7 |
| 15/32 | 0.45 | 0.99 | 1.8 | 2.7 | 3.9 | 5.4 | 7.0 | 8.9 | 11.0 | 13.3 | 15.9 | 18.6 | 21.6 | 24.8 | 28.2 | 31.8 |
| 1/2 | 0.48 | 1.0 | 1.9 | 2.8 | 4.2 | 5.7 | 7.5 | 9.5 | 11.7 | 14.2 | 16.9 | 19.9 | 23.1 | 26.4 | 30.1 | 34.0 |
| 17/32 | 0.51 | 1.1 | 2.1 | 3.1 | 4.5 | 6.1 | 8.0 | 10.1 | 12.5 | 15.1 | 18.0 | 21.1 | 24.5 | 28.1 | 31.9 | 36.1 |
| 9/16 | 0.54 | 1.1 | 2.2 | 3.3 | 4.7 | 6.5 | 8.4 | 10.7 | 13.2 | 16.0 | 19.0 | 22.4 | 26.0 | 29.7 | 33.8 | 38.2 |
| 19/32 | 0.57 | 1.2 | 2.3 | 3.4 | 5.0 | 6.8 | 8.9 | 11.3 | 14.0 | 16.9 | 20.1 | 23.6 | 27.4 | 31.4 | 35.7 | 40.3 |
| 5/8 | 0.60 | 1.3 | 2.4 | 3.6 | 5.3 | 7.2 | 9.4 | 11.9 | 14.7 | 17.8 | 21.2 | 24.9 | 28.9 | 33.1 | 37.6 | 42.5 |
| 21/32 | 0.63 | 1.3 | 2.5 | 3.8 | 5.5 | 7.6 | 9.8 | 12.5 | 15.4 | 18.6 | 22.2 | 26.1 | 30.3 | 34.7 | 39.5 | 44.6 |
| 11/16 | 0.66 | 1.4 | 2.6 | 4.0 | 5.8 | 7.9 | 10.3 | 13.1 | 16.2 | 19.5 | 23.3 | 27.4 | 31.7 | 36.4 | 41.4 | 46.7 |
| 23/32 | 0.69 | 1.5 | 2.8 | 4.2 | 6.0 | 8.3 | 10.8 | 13.7 | 16.9 | 20.4 | 24.3 | 28.6 | 33.2 | 38.0 | 43.2 | 48.8 |
| 3/4 | 0.72 | 1.5 | 2.9 | 4.4 | 6.3 | 8.6 | 11.3 | 14.3 | 17.6 | 21.3 | 25.4 | 29.8 | 34.6 | 39.7 | 45.1 | 51.0 |
| 25/32 | 0.75 | 1.6 | 3.0 | 4.6 | 6.6 | 9.0 | 11.7 | 14.9 | 18.4 | 22.2 | 26.5 | 31.1 | 36.1 | 41.3 | 47.0 | 53.1 |
| 13/16 | 0.78 | 1.7 | 3.1 | 4.7 | 6.8 | 9.4 | 12.2 | 15.5 | 19.1 | 23.1 | 27.5 | 32.3 | 37.5 | 43.0 | 48.9 | 55.2 |
| 27/32 | 0.81 | 1.7 | 3.2 | 4.9 | 7.1 | 9.7 | 12.7 | 16.1 | 19.8 | 24.0 | 28.6 | 33.6 | 39.0 | 44.6 | 50.8 | 57.3 |
| 7/8 | 0.84 | 1.8 | 3.4 | 5.1 | 7.4 | 10.1 | 13.1 | 16.7 | 20.6 | 24.9 | 29.6 | 34.8 | 40.4 | 46.3 | 52.6 | 59.5 |

* Multiply given weight by 1.273 for square stock, 1.103 for hexagonal stock, and 1.055 for octagonal stock.

† Length selected should allow for waste due to cutting off, etc.

## Pounds of Round Brass Rod per Thousand Pieces —2*

| Length, Inches† | Diameter of Stock, Inches | | | | | | | | | | | | | | | |
|---|---|---|---|---|---|---|---|---|---|---|---|---|---|---|---|---|
|  | 1/16 | 3/32 | 1/8 | 5/32 | 3/16 | 7/32 | 1/4 | 9/32 | 5/16 | 11/32 | 3/8 | 13/32 | 7/16 | 15/32 | 1/2 | 17/32 |
| 29/32 | 0.87 | 1.9 | 3.5 | 5.3 | 7.6 | 10.4 | 13.6 | 17.3 | 21.3 | 25.8 | 30.7 | 36.1 | 41.9 | 47.9 | 54.5 | 61.6 |
| 15/16 | 0.90 | 1.9 | 3.6 | 5.5 | 7.9 | 10.8 | 14.1 | 17.9 | 22.1 | 26.7 | 31.8 | 37.3 | 43.3 | 49.6 | 56.4 | 63.7 |
| 31/32 | 0.93 | 2.0 | 3.7 | 5.7 | 8.2 | 11.2 | 14.6 | 18.5 | 22.8 | 27.5 | 32.8 | 38.6 | 44.7 | 51.3 | 58.3 | 65.8 |
| 1 | 0.96 | 2.1 | 3.7 | 5.8 | 8.4 | 11.5 | 15.0 | 19.1 | 23.5 | 28.4 | 33.9 | 39.8 | 46.1 | 52.9 | 60.2 | 68.0 |
| 1 1/16 | 1.0 | 2.2 | 4.0 | 6.2 | 9.0 | 12.3 | 16.0 | 20.3 | 25.0 | 30.2 | 36.0 | 42.3 | 49.1 | 56.2 | 63.9 | 72.2 |
| 1 1/8 | 1.1 | 2.3 | 4.2 | 6.6 | 9.5 | 13.0 | 16.9 | 21.5 | 26.5 | 32.0 | 38.1 | 44.8 | 52.0 | 59.5 | 67.7 | 76.5 |
| 1 3/16 | 1.1 | 2.5 | 4.4 | 6.9 | 10.0 | 13.7 | 17.8 | 22.7 | 28.0 | 33.8 | 40.2 | 47.3 | 54.8 | 62.8 | 71.5 | 80.7 |
| 1 1/4 | 1.2 | 2.6 | 4.7 | 7.3 | 10.6 | 14.4 | 18.8 | 23.9 | 29.4 | 35.6 | 42.4 | 49.8 | 57.8 | 66.2 | 75.2 | 85.0 |
| 1 5/16 | 1.2 | 2.7 | 4.9 | 7.7 | 11.1 | 15.2 | 19.7 | 25.1 | 30.9 | 37.3 | 44.5 | 52.2 | 60.6 | 69.5 | 79.0 | 89.2 |
| 1 3/8 | 1.3 | 2.9 | 5.1 | 8.0 | 11.6 | 15.9 | 20.7 | 26.3 | 32.4 | 39.1 | 46.6 | 54.7 | 63.5 | 72.8 | 82.8 | 93.5 |
| 1 7/16 | 1.3 | 3.0 | 5.4 | 8.4 | 12.1 | 16.6 | 21.6 | 27.5 | 33.9 | 40.9 | 48.7 | 57.2 | 66.4 | 76.1 | 86.5 | 97.7 |
| 1 1/2 | 1.4 | 3.1 | 5.6 | 8.8 | 12.7 | 17.3 | 22.6 | 28.7 | 35.3 | 42.7 | 50.8 | 59.7 | 69.3 | 79.4 | 90.3 | 102.0 |
| 1 9/16 | 1.5 | 3.3 | 5.9 | 9.2 | 13.2 | 18.1 | 23.5 | 29.9 | 36.8 | 44.5 | 53.0 | 62.2 | 72.2 | 82.7 | 94.1 | 106.0 |
| 1 5/8 | 1.5 | 3.4 | 6.1 | 9.5 | 13.7 | 18.8 | 24.4 | 31.0 | 38.3 | 46.2 | 55.1 | 64.7 | 75.1 | 86.0 | 97.8 | 110.0 |
| 1 11/16 | 1.6 | 3.5 | 6.3 | 9.9 | 14.3 | 19.5 | 25.4 | 32.2 | 39.7 | 48.0 | 57.2 | 67.2 | 78.0 | 89.3 | 101.0 | 114.0 |
| 1 3/4 | 1.6 | 3.6 | 6.6 | 10.3 | 14.8 | 20.2 | 26.3 | 33.4 | 41.2 | 49.8 | 59.3 | 69.7 | 80.9 | 92.6 | 105.0 | 119.0 |
| 1 13/16 | 1.7 | 3.8 | 6.8 | 10.6 | 15.3 | 20.9 | 27.3 | 34.6 | 42.7 | 51.6 | 61.4 | 72.2 | 83.8 | 95.9 | 109.0 | 123.0 |
| 1 7/8 | 1.8 | 3.9 | 7.0 | 11.0 | 15.9 | 21.7 | 28.2 | 35.8 | 44.2 | 53.4 | 63.6 | 74.7 | 86.7 | 99.3 | 112.0 | 127.0 |
| 1 15/16 | 1.8 | 4.0 | 7.3 | 11.4 | 16.4 | 22.4 | 29.2 | 37.0 | 45.6 | 55.1 | 65.7 | 77.1 | 89.5 | 102.0 | 116.0 | 131.0 |
| 2 | 1.9 | 4.2 | 7.5 | 11.7 | 16.9 | 23.1 | 30.1 | 38.2 | 47.1 | 56.9 | 67.8 | 79.6 | 92.4 | 105.0 | 120.0 | 136.0 |
| 2 1/8 | 2.0 | 4.4 | 8.0 | 12.5 | 18.0 | 24.6 | 32.0 | 40.6 | 50.1 | 60.5 | 72.0 | 84.6 | 98.2 | 112.0 | 127.0 | 144.0 |
| 2 1/4 | 2.1 | 4.7 | 8.4 | 13.2 | 19.0 | 26.0 | 33.9 | 43.0 | 53.0 | 64.0 | 76.3 | 89.6 | 104.0 | 119.0 | 135.0 | 153.0 |
| 2 3/8 | 2.2 | 5.0 | 8.9 | 13.9 | 20.1 | 27.5 | 35.7 | 45.4 | 56.0 | 67.6 | 80.5 | 94.6 | 109.0 | 125.0 | 143.0 | 161.0 |
| 2 1/2 | 2.4 | 5.2 | 9.4 | 14.7 | 21.2 | 28.9 | 37.6 | 47.8 | 58.9 | 71.2 | 84.8 | 99.6 | 115.0 | 132.0 | 150.0 | 170.0 |
| 2 5/8 | 2.5 | 5.5 | 9.9 | 15.4 | 22.2 | 30.4 | 39.5 | 50.2 | 61.9 | 74.7 | 89.0 | 104.0 | 121.0 | 139.0 | 158.0 | 178.0 |
| 2 3/4 | 2.6 | 5.8 | 10.3 | 16.1 | 23.3 | 31.8 | 41.4 | 52.6 | 64.8 | 78.3 | 93.2 | 109.0 | 127.0 | 145.0 | 165.0 | 187.0 |
| 2 7/8 | 2.7 | 6.0 | 10.8 | 16.9 | 24.3 | 33.3 | 43.3 | 55.0 | 67.8 | 81.8 | 97.5 | 114.0 | 132.0 | 152.0 | 173.0 | 195.0 |
| 3 | 2.8 | 6.3 | 11.3 | 17.6 | 25.4 | 34.7 | 45.2 | 57.4 | 70.7 | 85.4 | 101.0 | 119.0 | 138.0 | 158.0 | 180.0 | 204.0 |

* Multiply given weight by 1.273 for square stock, 1.103 for hexagonal stock, and 1.055 for octagonal stock.
† Length selected should allow for waste due to cutting off, etc.

## Pounds of Round Brass Rod per Thousand Pieces — 3*

| Length, Inches† | Diameter of Stock, Inches | | | | | | | | | | | | | | | |
|---|---|---|---|---|---|---|---|---|---|---|---|---|---|---|---|---|
| | 9/16 | 19/32 | 5/8 | 11/16 | 3/4 | 13/16 | 7/8 | 15/16 | 1 | 1 1/16 | 1 1/8 | 1 3/16 | 1 1/4 | 1 5/16 | 1 3/8 | 1 1/2 |
| 1/32 | 2.4 | 2.7 | 3.0 | 3.6 | 4.3 | 5.0 | 5.8 | 6.7 | 7.6 | 8.5 | 9.6 | 10.7 | 11.8 | 13.0 | 14.0 | 17.0 |
| 1/16 | 4.7 | 5.3 | 5.9 | 7.1 | 8.5 | 9.9 | 11.5 | 13.2 | 15.0 | 17.0 | 19.0 | 21.2 | 23.6 | 26.0 | 28.0 | 34.0 |
| 3/32 | 7.1 | 7.9 | 8.8 | 10.7 | 12.7 | 14.9 | 17.3 | 19.8 | 22.6 | 25.5 | 28.6 | 31.8 | 35.4 | 39.0 | 43.0 | 51.0 |
| 1/8 | 9.5 | 10.6 | 11.8 | 14.2 | 17.0 | 19.9 | 23.1 | 26.4 | 30.1 | 34.0 | 38.1 | 42.5 | 47.2 | 52.0 | 57.0 | 68.0 |
| 5/32 | 11.9 | 13.3 | 14.7 | 17.8 | 21.2 | 24.9 | 28.9 | 33.1 | 37.7 | 42.5 | 47.7 | 53.1 | 59.0 | 65.0 | 71.0 | 85.0 |
| 3/16 | 14.2 | 15.9 | 17.7 | 21.4 | 25.5 | 29.8 | 34.6 | 39.7 | 45.2 | 51.0 | 57.2 | 63.7 | 70.8 | 78.0 | 85.0 | 102.0 |
| 7/32 | 16.6 | 18.6 | 20.6 | 24.9 | 29.7 | 34.8 | 40.4 | 46.3 | 52.7 | 59.5 | 66.7 | 74.4 | 82.6 | 91.0 | 100.0 | 119.0 |
| 1/4 | 19.0 | 21.2 | 23.6 | 28.5 | 34.0 | 39.8 | 46.2 | 52.9 | 60.3 | 68.0 | 76.3 | 85.0 | 94.4 | 104.0 | 113.0 | 136.0 |
| 9/32 | 21.4 | 23.9 | 26.5 | 32.1 | 38.2 | 44.8 | 52.0 | 59.5 | 67.8 | 76.5 | 85.8 | 95.6 | 106.0 | 117.0 | 128.0 | 153.0 |
| 5/16 | 23.8 | 26.6 | 29.5 | 35.7 | 42.5 | 49.8 | 57.8 | 66.2 | 75.4 | 85.0 | 95.4 | 106.0 | 118.0 | 130.0 | 142.0 | 170.0 |
| 11/32 | 26.1 | 29.2 | 32.4 | 39.2 | 46.7 | 54.7 | 63.5 | 72.8 | 82.9 | 93.5 | 104.0 | 116.0 | 129.0 | 143.0 | 156.0 | 187.0 |
| 3/8 | 28.5 | 31.9 | 35.4 | 42.8 | 51.0 | 59.7 | 69.3 | 79.4 | 90.4 | 102.0 | 114.0 | 127.0 | 141.0 | 156.0 | 170.0 | 204.0 |
| 13/32 | 30.9 | 34.5 | 38.3 | 46.4 | 55.2 | 64.7 | 75.1 | 86.0 | 98.0 | 110.0 | 124.0 | 138.0 | 153.0 | 169.0 | 184.0 | 221.0 |
| 7/16 | 33.3 | 37.2 | 41.3 | 49.9 | 59.5 | 69.7 | 80.9 | 92.6 | 105.0 | 119.0 | 133.0 | 148.0 | 165.0 | 182.0 | 199.0 | 238.0 |
| 15/32 | 35.7 | 39.9 | 44.2 | 53.5 | 63.7 | 74.7 | 86.7 | 99.3 | 113.0 | 127.0 | 143.0 | 159.0 | 177.0 | 195.0 | 213.0 | 255.0 |
| 1/2 | 38.0 | 42.5 | 47.2 | 57.1 | 68.0 | 79.6 | 92.4 | 105.0 | 120.0 | 136.0 | 152.0 | 170.0 | 188.0 | 208.0 | 227.0 | 272.0 |
| 17/32 | 40.4 | 45.2 | 50.1 | 60.6 | 72.2 | 84.6 | 98.2 | 112.0 | 128.0 | 144.0 | 162.0 | 180.0 | 200.0 | 221.0 | 241.0 | 289.0 |
| 9/16 | 42.8 | 47.8 | 53.1 | 64.2 | 76.5 | 89.6 | 104.0 | 119.0 | 135.0 | 153.0 | 171.0 | 191.0 | 212.0 | 234.0 | 256.0 | 306.0 |
| 19/32 | 45.2 | 50.5 | 56.0 | 67.8 | 80.7 | 94.6 | 109.0 | 125.0 | 143.0 | 161.0 | 181.0 | 201.0 | 224.0 | 247.0 | 270.0 | 323.0 |
| 5/8 | 47.6 | 53.2 | 59.0 | 71.4 | 85.0 | 99.6 | 115.0 | 132.0 | 150.0 | 170.0 | 190.0 | 212.0 | 236.0 | 260.0 | 284.0 | 340.0 |
| 21/32 | 49.9 | 55.8 | 61.9 | 74.9 | 89.2 | 104.0 | 121.0 | 139.0 | 158.0 | 178.0 | 200.0 | 223.0 | 247.0 | 273.0 | 298.0 | 357.0 |
| 11/16 | 52.3 | 58.5 | 64.9 | 78.5 | 93.5 | 109.0 | 127.0 | 145.0 | 165.0 | 187.0 | 209.0 | 233.0 | 259.0 | 286.0 | 313.0 | 374.0 |
| 23/32 | 54.7 | 61.1 | 67.8 | 82.1 | 97.7 | 114.0 | 132.0 | 152.0 | 173.0 | 195.0 | 219.0 | 244.0 | 271.0 | 299.0 | 327.0 | 391.0 |
| 3/4 | 57.1 | 63.8 | 70.8 | 85.6 | 102.0 | 119.0 | 138.0 | 158.0 | 180.0 | 204.0 | 228.0 | 255.0 | 283.0 | 312.0 | 341.0 | 408.0 |
| 25/32 | 59.5 | 66.5 | 73.7 | 89.2 | 106.0 | 124.0 | 144.0 | 165.0 | 188.0 | 212.0 | 238.0 | 265.0 | 295.0 | 325.0 | 355.0 | 425.0 |
| 13/16 | 61.8 | 69.1 | 76.7 | 92.8 | 110.0 | 129.0 | 150.0 | 172.0 | 196.0 | 221.0 | 248.0 | 276.0 | 306.0 | 338.0 | 369.0 | 442.0 |
| 27/32 | 64.2 | 71.8 | 79.6 | 96.4 | 114.0 | 134.0 | 156.0 | 178.0 | 203.0 | 229.0 | 257.0 | 287.0 | 318.0 | 351.0 | 384.0 | 459.0 |
| 7/8 | 66.6 | 74.4 | 82.6 | 100.0 | 119.0 | 139.0 | 161.0 | 185.0 | 211.0 | 238.0 | 267.0 | 297.0 | 330.0 | 364.0 | 398.0 | 476.0 |

* Multiply given weight by 1.273 for square stock, 1.103 for hexagonal stock, and 1.055 for octagonal stock.

† Length selected should allow for waste due to cutting off, etc.

Pounds of Round Brass Rod per Thousand Pieces — 4*

| Length, Inches† | Diameter of Stock, Inches | | | | | | | | | | | | | | | |
|---|---|---|---|---|---|---|---|---|---|---|---|---|---|---|---|---|
| | 9/16 | 19/32 | 5/8 | 11/16 | 3/4 | 13/16 | 7/8 | 15/16 | 1 | 1 1/16 | 1 1/8 | 1 3/16 | 1 1/4 | 1 5/16 | 1 3/8 | 1 1/2 |
| 29/32 | 69.0 | 77.1 | 85.5 | 103 | 123 | 144 | 167 | 191 | 218 | 246 | 276 | 308 | 342 | 377 | 412 | 493 |
| 15/16 | 71.4 | 79.8 | 88.5 | 107 | 127 | 149 | 173 | 198 | 226 | 255 | 286 | 318 | 354 | 390 | 426 | 510 |
| 31/32 | 73.7 | 82.4 | 91.4 | 110 | 131 | 154 | 179 | 205 | 233 | 263 | 295 | 329 | 365 | 403 | 441 | 527 |
| 1 | 76.1 | 85.1 | 94.4 | 114 | 136 | 159 | 184 | 211 | 241 | 272 | 305 | 340 | 377 | 416 | 455 | 544 |
| 1 1/16 | 80.9 | 90.4 | 100.0 | 121 | 144 | 169 | 196 | 225 | 256 | 289 | 324 | 361 | 401 | 442 | 483 | 578 |
| 1 1/8 | 85.5 | 95.7 | 106.0 | 128 | 153 | 179 | 208 | 238 | 271 | 306 | 343 | 382 | 424 | 468 | 512 | 612 |
| 1 3/16 | 90.4 | 101.0 | 112.0 | 135 | 161 | 189 | 219 | 251 | 286 | 323 | 362 | 403 | 448 | 494 | 540 | 646 |
| 1 1/4 | 95.2 | 106.0 | 118.0 | 142 | 170 | 199 | 231 | 264 | 301 | 340 | 381 | 425 | 472 | 520 | 569 | 680 |
| 1 5/16 | 99.9 | 111.0 | 123.0 | 149 | 178 | 209 | 242 | 278 | 316 | 357 | 400 | 446 | 495 | 546 | 597 | 714 |
| 1 3/8 | 104.0 | 117.0 | 129.0 | 157 | 187 | 219 | 254 | 291 | 331 | 374 | 419 | 467 | 519 | 572 | 626 | 748 |
| 1 7/16 | 109.0 | 122.0 | 135.0 | 164 | 195 | 229 | 265 | 304 | 346 | 391 | 438 | 488 | 542 | 598 | 654 | 782 |
| 1 1/2 | 114.0 | 127.0 | 141.0 | 171 | 204 | 239 | 277 | 317 | 361 | 408 | 457 | 510 | 566 | 624 | 683 | 816 |
| 1 9/16 | 119.0 | 133.0 | 147.0 | 178 | 212 | 249 | 289 | 331 | 377 | 425 | 477 | 531 | 590 | 650 | 711 | 850 |
| 1 5/8 | 123.0 | 138.0 | 153.0 | 185 | 221 | 259 | 300 | 344 | 392 | 442 | 496 | 552 | 613 | 676 | 739 | 884 |
| 1 11/16 | 128.0 | 143.0 | 159.0 | 192 | 229 | 268 | 312 | 357 | 407 | 459 | 515 | 574 | 637 | 702 | 768 | 918 |
| 1 3/4 | 133.0 | 148.0 | 165.0 | 199 | 238 | 278 | 323 | 370 | 422 | 476 | 534 | 595 | 660 | 728 | 796 | 952 |
| 1 13/16 | 138.0 | 154.0 | 171.0 | 207 | 246 | 288 | 335 | 383 | 437 | 493 | 553 | 616 | 684 | 754 | 825 | 986 |
| 1 7/8 | 142.0 | 159.0 | 177.0 | 214 | 255 | 298 | 346 | 397 | 452 | 510 | 572 | 637 | 708 | 780 | 853 | 1020 |
| 1 15/16 | 147.0 | 164.0 | 182.0 | 221 | 263 | 308 | 358 | 410 | 467 | 527 | 591 | 659 | 731 | 806 | 882 | 1054 |
| 2 | 152.0 | 170.0 | 188.0 | 228 | 272 | 318 | 369 | 423 | 482 | 544 | 610 | 680 | 755 | 832 | 910 | 1088 |
| 2 1/8 | 161.0 | 180.0 | 200.0 | 242 | 289 | 338 | 393 | 450 | 512 | 578 | 648 | 722 | 802 | 884 | 967 | 1156 |
| 2 1/4 | 171.0 | 191.0 | 212.0 | 257 | 306 | 358 | 416 | 476 | 542 | 612 | 686 | 765 | 849 | 936 | 1024 | 1224 |
| 2 3/8 | 180.0 | 202.0 | 224.0 | 271 | 323 | 378 | 439 | 503 | 573 | 646 | 725 | 807 | 896 | 988 | 1081 | 1292 |
| 2 1/2 | 190.0 | 212.0 | 236.0 | 285 | 340 | 398 | 462 | 529 | 603 | 680 | 763 | 850 | 944 | 1040 | 1138 | 1360 |
| 2 5/8 | 199.0 | 223.0 | 247.0 | 299 | 357 | 418 | 485 | 556 | 633 | 714 | 801 | 892 | 991 | 1092 | 1195 | 1428 |
| 2 3/4 | 209.0 | 234.0 | 259.0 | 314 | 374 | 438 | 508 | 582 | 663 | 748 | 839 | 935 | 1038 | 1144 | 1252 | 1496 |
| 2 7/8 | 218.0 | 244.0 | 271.0 | 328 | 391 | 458 | 531 | 609 | 693 | 782 | 877 | 977 | 1085 | 1196 | 1309 | 1564 |
| 3 | 228.0 | 255.0 | 283.0 | 342 | 408 | 478 | 554 | 635 | 723 | 816 | 915 | 1020 | 1132 | 1248 | 1366 | 1632 |

\* Multiply given weight by 1.273 for square stock, 1.103 for hexagonal stock, and 1.055 for octagonal stock.

† Length selected should allow for waste due to cutting off, etc.

**Weights of Copper and Brass Rods and Bars, Pounds per Foot**

| Diam. or Distance across Flats | COPPER | | | FREE-CUTTING BRASS | | | ROMAN BRONZE NAVAL BRASS | | |
|---|---|---|---|---|---|---|---|---|---|
| | Round | Square | Hexagon | Round | Square | Hexagon | Round | Square | Hexagon |
| | Pounds per Lineal Foot | | | | | | | | |
| 1/32 | ... | ... | ... | .00283 | .00360 | .00312 | ... | ... | ... |
| 1/16 | .0119 | .0151 | .0131 | .0113 | .0144 | .0125 | .0112 | .0142 | .0123 |
| 3/32 | ... | ... | ... | .0254 | .0324 | .0280 | ... | ... | ... |
| 1/8 | .0476 | .0606 | .0525 | .0452 | .0576 | .0499 | .0448 | .0570 | .0494 |
| 5/32 | ... | ... | ... | .0706 | .0899 | .0779 | ... | ... | ... |
| 3/16 | .107 | .136 | .118 | .102 | .130 | .112 | .101 | .128 | .111 |
| 7/32 | ... | ... | ... | .138 | .176 | .153 | ... | ... | ... |
| 1/4 | .190 | .242 | .210 | .181 | .230 | .199 | .179 | .228 | .198 |
| 9/32 | ... | ... | ... | .229 | .291 | .252 | ... | ... | ... |
| 5/16 | .297 | .379 | .328 | .283 | .360 | .312 | .280 | .356 | .309 |
| 11/32 | ... | ... | ... | .342 | .435 | .377 | ... | ... | ... |
| 3/8 | .428 | .545 | .472 | .407 | .518 | .449 | .403 | .513 | .444 |
| 13/32 | ... | ... | ... | .478 | .608 | .527 | ... | ... | ... |
| 7/16 | .583 | .742 | .643 | .554 | .705 | .611 | .548 | .698 | .605 |
| 15/32 | ... | ... | ... | .636 | .809 | .701 | ... | ... | ... |
| 1/2 | .761 | .969 | .839 | .723 | .921 | .798 | .716 | .912 | .790 |
| 17/32 | ... | ... | ... | .817 | 1.04 | .901 | ... | ... | ... |
| 9/16 | .963 | 1.23 | 1.06 | .915 | 1.17 | 1.01 | .907 | 1.15 | 1.00 |
| 19/32 | ... | ... | ... | 1.02 | 1.30 | 1.13 | ... | ... | ... |
| 5/8 | 1.19 | 1.51 | 1.31 | 1.13 | 1.44 | 1.25 | 1.12 | 1.43 | 1.23 |
| 21/32 | ... | ... | ... | 1.25 | 1.59 | 1.37 | ... | ... | ... |
| 11/16 | 1.44 | 1.83 | 1.59 | 1.37 | 1.74 | 1.51 | 1.35 | 1.72 | 1.49 |
| 23/32 | ... | ... | ... | 1.49 | 1.90 | 1.65 | ... | ... | ... |
| 3/4 | 1.71 | 2.18 | 1.89 | 1.63 | 2.07 | 1.80 | 1.61 | 2.05 | 1.78 |
| 25/32 | ... | ... | ... | 1.77 | 2.25 | 1.95 | ... | ... | ... |
| 13/16 | 2.01 | 2.56 | 2.22 | 1.91 | 2.43 | 2.11 | 1.89 | 2.41 | 2.09 |
| 27/32 | ... | ... | ... | 2.06 | 2.62 | 2.27 | ... | ... | ... |
| 7/8 | 2.33 | 2.97 | 2.57 | 2.22 | 2.82 | 2.44 | 2.19 | 2.79 | 2.42 |
| 29/32 | ... | ... | ... | 2.38 | 3.03 | 2.62 | ... | ... | ... |
| 15/16 | 2.68 | 3.41 | 2.95 | 2.54 | 3.24 | 2.80 | 2.52 | 3.21 | 2.78 |
| 31/32 | ... | ... | ... | 2.72 | 3.46 | 2.99 | ... | ... | ... |
| 1 | 3.04 | 3.88 | 3.36 | 2.89 | 3.68 | 3.19 | 2.87 | 3.65 | 3.16 |
| 1 1/16 | 3.44 | 4.38 | 3.79 | 3.27 | 4.16 | 3.60 | 3.23 | 4.12 | 3.57 |
| 1 1/8 | 3.85 | 4.91 | 4.25 | 3.66 | 4.66 | 4.04 | 3.63 | 4.62 | 4.00 |
| 1 3/16 | 4.29 | 5.47 | 4.73 | 4.08 | 5.20 | 4.50 | 4.04 | 5.14 | 4.46 |
| 1 1/4 | 4.76 | 6.06 | 5.25 | 4.52 | 5.76 | 4.99 | 4.48 | 5.70 | 4.94 |
| 1 5/16 | 5.24 | 6.68 | 5.78 | 4.98 | 6.35 | 5.50 | 4.94 | 6.28 | 5.44 |
| 1 3/8 | 5.76 | 7.33 | 6.35 | 5.47 | 6.97 | 6.03 | 5.42 | 6.90 | 5.97 |
| 1 7/16 | 6.29 | 8.01 | 6.94 | 5.98 | 7.61 | 6.59 | 5.92 | 7.54 | 6.53 |
| 1 1/2 | 6.85 | 8.72 | 7.55 | 6.51 | 8.29 | 7.18 | 6.45 | 8.21 | 7.11 |
| 1 9/16 | 7.43 | 9.46 | 8.20 | 7.06 | 8.99 | 7.79 | 6.99 | 8.91 | 7.72 |
| 1 5/8 | 8.04 | 10.2 | 8.86 | 7.64 | 9.73 | 8.43 | 7.57 | 9.63 | 8.34 |
| 1 11/16 | 8.67 | 11.0 | 9.56 | 8.24 | 10.5 | 9.09 | 8.16 | 10.4 | 9.00 |
| 1 3/4 | 9.32 | 11.9 | 10.3 | 8.86 | 11.3 | 9.77 | 8.77 | 11.2 | 9.68 |
| 1 13/16 | 10.0 | 12.7 | 11.0 | 9.51 | 12.1 | 10.5 | 9.41 | 12.0 | 10.4 |
| 1 7/8 | 10.7 | 13.6 | 11.8 | 10.2 | 13.0 | 11.2 | 10.1 | 12.8 | 11.1 |
| 1 15/16 | 11.4 | 14.6 | 12.6 | 10.9 | 13.8 | 12.0 | 10.8 | 13.7 | 11.9 |
| 2 | 12.2 | 15.5 | 13.4 | 11.6 | 14.7 | 12.8 | 11.5 | 14.6 | 12.6 |
| 2 1/8 | 13.7 | 17.5 | 15.2 | 13.1 | 16.6 | 14.4 | 12.9 | 16.5 | 14.3 |
| 2 1/4 | 15.4 | 19.6 | 17.0 | 14.6 | 18.7 | 16.2 | 14.5 | 18.5 | 16.0 |
| 2 3/8 | 17.2 | 21.9 | 18.9 | 16.3 | 20.8 | 18.0 | 16.2 | 20.6 | 17.8 |
| 2 1/2 | 19.0 | 24.2 | 21.0 | 18.1 | 23.0 | 19.9 | 17.9 | 22.8 | 19.8 |
| 2 5/8 | 21.0 | 26.7 | 23.1 | 19.9 | 25.4 | 22.0 | 19.7 | 25.1 | 21.8 |
| 2 3/4 | 23.0 | 29.3 | 25.4 | 21.9 | 27.9 | 24.1 | 21.7 | 27.6 | 23.9 |
| 2 7/8 | 25.2 | 32.0 | 27.7 | 23.9 | 30.5 | 26.4 | 23.7 | 30.2 | 26.1 |
| 3 | 27.4 | 34.9 | 30.2 | 26.0 | 33.2 | 28.7 | 25.8 | 32.8 | 28.4 |

These weights are based on the following densities in pounds per cubic inch: Copper, 0.323; Free-cutting Brass, 0.307; Roman Bronze, 0.304; and Naval Brass, 0.304. Variations from these weights must be expected in practice.

*Source: Weights and Data, 1956 Edition, Revere Copper and Brass, Inc., with permission.*

## Areas and Weights of Aluminum Alloy Wire, Rods and Bars

| B&S Gage No. | Wire Diam., Inch | Wire Weight Pounds per Foot | Diam. or Distance Across Flats, Inches | Round Rods Area, Square Inches | Round Rods Weight, Pounds per Foot | Square Rods Area, Square Inches | Square Rods Weight, Pounds per Foot | Hexagon Rods Area, Square Inches | Hexagon Rods Weight, Pounds per Foot |
|---|---|---|---|---|---|---|---|---|---|
| 00 | .3648 | .1228 | 1/32 | .0008 | .0009 | .0010 | .0011 | .0008 | .0009 |
| 0 | .3249 | .0974 | 1/16 | .0031 | .0036 | .0039 | .0046 | .0034 | .0040 |
| 1 | .2893 | .0772 | 3/32 | .0069 | .0081 | .0087 | .0103 | .0088 | .0134 |
| 2 | .2576 | .0612 | 1/8 | .0123 | .0145 | .0156 | .0183 | .0135 | .0158 |
| 3 | .2294 | .0485 | 5/32 | .0192 | .0225 | .0244 | .0286 | .0211 | .0248 |
| 4 | .2043 | .0385 | 3/16 | .0276 | .0324 | .0352 | .0413 | .0305 | .0358 |
| 5 | .1819 | .0305 | 7/32 | .0376 | .0442 | .0476 | .0559 | .0414 | .0486 |
| 6 | .1620 | .0242 | 1/4 | .0491 | .0577 | .0625 | .0734 | .0542 | .0631 |
| 7 | .1443 | .0193 | 9/32 | .0621 | .0730 | .0707 | .0967 | .0685 | .0804 |
| 8 | .1285 | .0152 | 5/16 | .0767 | .0901 | .0977 | .1148 | .0845 | .0990 |
| 9 | .1144 | .0121 | 11/32 | .0928 | .1090 | .1181 | .1387 | .1022 | .1200 |
| 10 | .1019 | .0096 | 3/8 | .1104 | .1297 | .1406 | .1652 | .1218 | .1431 |
| 11 | .0907 | .0076 | 13/32 | .1296 | .1522 | .1650 | .1938 | .1429 | .1679 |
| 12 | .0808 | .0060 | 7/16 | .1503 | .1766 | .1914 | .2249 | .1658 | .1948 |
| 13 | .0720 | .0048 | 15/32 | .1726 | .2028 | .2192 | .2574 | .1902 | .2235 |
| 14 | .0641 | .0038 | 1/2 | .1963 | .2306 | .2500 | .2937 | .2165 | .2543 |
| 15 | .0571 | .0031 | 17/32 | .2216 | .2603 | .2822 | .3309 | .2444 | .2872 |
| 16 | .0508 | .0023 | 9/16 | .2485 | .2919 | .3164 | .3717 | .2740 | .3219 |
| 17 | .0453 | .0019 | 19/32 | .2769 | .3253 | .3525 | .4141 | .3053 | .3587 |
| 18 | .0403 | .0015 | 5/8 | .3068 | .3604 | .3906 | .4589 | .3381 | .3974 |
| 19 | .0359 | .0012 | 21/32 | .3382 | .3973 | .4306 | .5058 | .3730 | .4382 |
| 20 | .0320 | .00094 | 11/16 | .3713 | .4361 | .4727 | .5553 | .4091 | .4810 |
| 21 | .0285 | .00075 | 23/32 | .4057 | .4767 | .5165 | .6068 | .4474 | .5256 |
| 22 | .0253 | .00059 | 3/4 | .4418 | .5190 | .5625 | .6608 | .4870 | .5722 |
| 23 | .0226 | .00049 | 25/32 | .4793 | .5631 | .6103 | .7169 | .5286 | .6210 |
| 24 | .0201 | .00038 | 13/16 | .5185 | .6091 | .6602 | .7756 | .5715 | .6716 |
| 25 | .0179 | .00029 | 27/32 | .5590 | .6569 | .7119 | .8363 | .6162 | .7239 |
| 26 | .0159 | .00023 | 7/8 | .6013 | .7064 | .7656 | .8994 | .6626 | .7789 |
| 27 | .0142 | .00019 | 29/32 | .6450 | .7577 | .8213 | .9649 | .7108 | .8350 |
| 28 | .0126 | .00015 | 15/16 | .6903 | .8110 | .8789 | 1.032 | .7601 | .8931 |
| 29 | .0113 | .00012 | 31/32 | .7370 | .8660 | .9385 | 1.102 | .8123 | .9543 |
| 30 | .0100 | .000093 | 1 | .7854 | .9227 | 1.000 | 1.174 | .8650 | 1.106 |
| 31 | .0089 | .000074 | 1 1/16 | .8866 | 1.041 | 1.129 | 1.326 | .9766 | 1.147 |
| 32 | .0080 | .000058 | 1 1/8 | .9940 | 1.167 | 1.266 | 1.486 | 1.095 | 1.286 |
| 33 | .0071 | .000046 | 1 3/16 | 1.107 | 1.301 | 1.410 | 1.657 | 1.220 | 1.433 |
| 34 | .0063 | .000036 | 1 1/4 | 1.227 | 1.441 | 1.562 | 1.836 | 1.351 | 1.587 |
| 35 | .0056 | .000029 | 1 5/16 | 1.353 | 1.589 | 1.723 | 2.024 | 1.490 | 1.750 |
| 36 | .0050 | .000023 | 1 3/8 | 1.484 | 1.744 | 1.891 | 2.221 | 1.635 | 1.921 |
| | | | 1 7/16 | 1.622 | 1.906 | 2.066 | 2.428 | 1.787 | 2.100 |
| | | | 1 1/2 | 1.767 | 2.076 | 2.250 | 2.643 | 1.946 | 2.286 |
| | | | 1 9/16 | 1.917 | 2.252 | 2.441 | 2.868 | 2.114 | 2.484 |
| | | | 1 5/8 | 2.073 | 2.436 | 2.641 | 3.102 | 2.286 | 2.686 |
| | | | 1 11/16 | 2.236 | 2.627 | 2.848 | 3.346 | 2.466 | 2.897 |
| | | | 1 3/4 | 2.405 | 2.825 | 3.062 | 3.598 | 2.652 | 3.115 |
| | | | 1 13/16 | 2.580 | 3.031 | 3.285 | 3.860 | 2.844 | 3.342 |
| | | | 1 7/8 | 2.761 | 3.243 | 3.516 | 4.130 | 3.044 | 3.576 |
| | | | 1 15/16 | 2.948 | 3.463 | 3.844 | 4.516 | 3.250 | 3.819 |
| | | | 2 | 3.141 | 3.690 | 4.000 | 4.699 | 3.464 | 4.069 |
| | | | 2 1/8 | 3.546 | 4.166 | 4.516 | 5.305 | 3.910 | 4.594 |
| | | | 2 1/4 | 3.976 | 4.671 | 5.062 | 5.947 | 4.384 | 5.150 |
| | | | 2 3/8 | 4.430 | 5.204 | 5.641 | 6.627 | 4.884 | 5.738 |
| | | | 2 1/2 | 4.908 | 5.766 | 6.250 | 7.342 | 5.412 | 6.358 |
| | | | 2 5/8 | 5.411 | 6.357 | 6.891 | 8.095 | 5.967 | 7.010 |
| | | | 2 3/4 | 5.939 | 6.977 | 7.562 | 8.884 | 6.549 | 7.693 |
| | | | 2 7/8 | 6.491 | 7.626 | 8.266 | 9.710 | 7.158 | 8.409 |
| | | | 3 | 7.068 | 8.304 | 9.000 | 10.57 | 7.794 | 9.156 |
| | | | 3 1/8 | 7.669 | 9.010 | 9.766 | 11.47 | ... | ... |
| | | | 3 1/4 | 8.295 | 9.745 | 10.56 | 12.41 | ... | ... |
| | | | 3 3/8 | 8.946 | 10.51 | 11.39 | 13.38 | ... | ... |

Round Rods (Cont.)

| Diam., Inches | Area, Square Inches | Weight, Pounds per Foot |
|---|---|---|
| 3 1/2 | 9.621 | 11.30 |
| 3 3/4 | 11.04 | 12.97 |
| 4 | 12.56 | 14.76 |
| 4 1/4 | 14.18 | 16.66 |
| 4 1/2 | 15.90 | 18.68 |
| 4 3/4 | 17.72 | 20.81 |
| 5 | 19.63 | 23.06 |
| 5 1/4 | 21.64 | 25.43 |
| 5 1/2 | 23.75 | 27.99 |
| 5 3/4 | 25.96 | 30.50 |
| 6 | 28.27 | 33.21 |
| 6 1/2 | 33.18 | 38.98 |
| 7 | 38.48 | 45.21 |

Weights are based on a density of 0.0979 pounds per cubic inch.
Source: *Aluminum Data Book*, Reynolds Metals Co., with permission.

### Areas and Weights of Magnesium Alloy Rods and Bars

| Round Bars | | | Square Bars | | | Hexagonal Bars | | |
|---|---|---|---|---|---|---|---|---|
| Diameter, Inches | Area, Square Inches | Weight, Pounds per Foot | Width, Inches | Area, Square Inches | Weight, Pounds per Foot | Width Across Flats, Inches | Area, Square Inches | Weight, Pounds per Foot |
| 1/8 | 0.012 | 0.009 | 1/8 | 0.016 | 0.012 | 1/4 | 0.054 | 0.041 |
| 3/16 | 0.028 | 0.022 | 3/16 | 0.035 | 0.027 | 5/16 | 0.084 | 0.064 |
| 1/4 | 0.049 | 0.038 | 1/4 | 0.063 | 0.048 | 3/8 | 0.122 | 0.094 |
| 5/16 | 0.077 | 0.059 | 5/16 | 0.098 | 0.075 | 7/16 | 0.166 | 0.127 |
| 3/8 | 0.110 | 0.084 | 3/8 | 0.141 | 0.108 | 1/2 | 0.216 | 0.166 |
| 7/16 | 0.150 | 0.115 | 7/16 | 0.191 | 0.147 | 9/16 | 0.274 | 0.210 |
| 1/2 | 0.196 | 0.151 | 1/2 | 0.250 | 0.192 | 5/8 | 0.338 | 0.259 |
| 9/16 | 0.248 | 0.190 | 9/16 | 0.316 | 0.243 | 11/16 | 0.409 | 0.314 |
| 5/8 | 0.307 | 0.236 | 5/8 | 0.391 | 0.300 | 3/4 | 0.487 | 0.374 |
| 11/16 | 0.371 | 0.285 | 11/16 | 0.473 | 0.363 | 13/16 | 0.571 | 0.438 |
| 3/4 | 0.442 | 0.339 | 3/4 | 0.563 | 0.432 | 7/8 | 0.663 | 0.509 |
| 13/16 | 0.518 | 0.398 | 13/16 | 0.660 | 0.507 | 15/16 | 0.761 | 0.584 |
| 7/8 | 0.601 | 0.462 | 7/8 | 0.766 | 0.588 | 1 | 0.865 | 0.665 |
| 15/16 | 0.690 | 0.530 | 15/16 | 0.879 | 0.675 | 1 1/16 | 0.977 | 0.750 |
| 1 | 0.785 | 0.603 | 1 | 1.00 | 0.768 | 1 1/8 | 1.09 | 0.837 |
| 1 1/16 | 0.887 | 0.681 | 1 1/16 | 1.13 | 0.868 | 1 3/16 | 1.22 | 0.937 |
| 1 1/8 | 0.994 | 0.763 | 1 1/8 | 1.27 | 0.975 | 1 1/4 | 1.35 | 1.04 |
| 1 3/16 | 1.11 | 0.852 | 1 3/16 | 1.41 | 1.08 | 1 5/16 | 1.49 | 1.14 |
| 1 1/4 | 1.23 | 0.945 | 1 1/4 | 1.56 | 1.20 | 1 3/8 | 1.64 | 1.26 |
| 1 5/16 | 1.35 | 1.04 | 1 5/16 | 1.72 | 1.32 | 1 7/16 | 1.79 | 1.37 |
| 1 3/8 | 1.48 | 1.14 | 1 3/8 | 1.89 | 1.45 | 1 1/2 | 1.95 | 1.50 |
| 1 7/16 | 1.62 | 1.24 | 1 7/16 | 2.07 | 1.59 | 1 9/16 | 2.11 | 1.62 |
| 1 1/2 | 1.77 | 1.36 | 1 1/2 | 2.25 | 1.73 | 1 5/8 | 2.28 | 1.75 |
| 1 9/16 | 1.92 | 1.47 | 1 9/16 | 2.44 | 1.87 | 1 11/16 | 2.46 | 1.89 |
| 1 5/8 | 2.07 | 1.59 | 1 5/8 | 2.64 | 2.03 | 1 3/4 | 2.65 | 2.03 |
| 1 11/16 | 2.24 | 1.72 | 1 11/16 | 2.85 | 2.19 | 1 13/16 | 2.84 | 2.18 |
| 1 3/4 | 2.40 | 1.84 | 1 3/4 | 3.06 | 2.38 | 1 7/8 | 3.04 | 2.33 |
| 1 13/16 | 2.58 | 1.98 | 1 13/16 | 3.29 | 2.53 | 1 15/16 | 3.25 | 2.50 |
| 1 7/8 | 2.76 | 2.12 | 1 7/8 | 3.52 | 2.70 | 2 | 3.46 | 2.66 |
| 1 15/16 | 2.95 | 2.26 | 1 15/16 | 3.75 | 2.88 | 2 1/8 | 3.91 | 3.00 |
| 2 | 3.14 | 2.41 | 2 | 4.00 | 3.07 | 2 1/4 | 4.38 | 3.36 |
| 2 1/8 | 3.55 | 2.73 | 2 1/8 | 4.52 | 3.47 | 2 3/8 | 4.88 | 3.75 |
| 2 1/4 | 3.98 | 3.06 | 2 1/4 | 5.06 | 3.89 | 2 1/2 | 5.41 | 4.15 |
| 2 3/8 | 4.43 | 3.40 | 2 3/8 | 5.64 | 4.33 | 2 5/8 | 5.96 | 4.58 |
| 2 1/2 | 4.91 | 3.77 | 2 1/2 | 6.25 | 4.80 | 2 3/4 | 6.54 | 5.02 |
| 2 5/8 | 5.41 | 4.15 | 2 5/8 | 6.89 | 5.29 | 2 7/8 | 7.14 | 5.48 |
| 2 3/4 | 5.94 | 4.56 | 2 3/4 | 7.56 | 5.81 | 3 | 7.79 | 5.98 |
| 2 7/8 | 6.48 | 4.98 | 2 7/8 | 8.25 | 6.34 | 3 1/4 | 9.14 | 7.02 |
| 3 | 7.07 | 5.43 | 3 | 9.00 | 6.91 | 3 1/2 | 10.6 | 8.14 |
| 3 1/8 | 7.67 | 5.89 | 3 1/8 | 9.77 | 7.50 | 3 3/4 | 12.2 | 9.35 |
| 3 1/4 | 8.29 | 6.37 | 3 1/4 | 10.6 | 8.14 | 4 | 13.8 | 10.6 |
| 3 3/8 | 8.95 | 6.87 | 3 3/8 | 11.4 | 8.75 | 4 1/2 | 17.5 | 13.4 |
| 3 1/2 | 9.62 | 7.39 | 3 1/2 | 12.2 | 9.37 | 5 | 21.6 | 16.6 |
| 3 5/8 | 10.3 | 7.91 | 3 5/8 | 13.1 | 10.1 | 5 1/2 | 26.2 | 20.1 |
| 3 3/4 | 11.0 | 8.45 | 3 3/4 | 14.0 | 10.8 | 6 | 31.1 | 23.9 |
| 3 7/8 | 11.8 | 9.06 | 3 7/8 | 15.0 | 11.5 | 6 1/2 | 36.6 | 28.1 |
| 4 | 12.6 | 9.68 | 4 | 16.0 | 12.3 | 7 | 42.4 | 32.6 |
| 4 1/4 | 14.2 | 10.9 | 4 1/4 | 18.1 | 13.9 | .... | .... | .... |
| 4 1/2 | 15.9 | 12.2 | 4 1/2 | 20.2 | 15.5 | .... | .... | .... |
| 4 3/4 | 17.7 | 13.6 | 4 3/4 | 22.6 | 17.3 | .... | .... | .... |
| 5 | 19.6 | 15.1 | 5 | 25.0 | 19.5 | .... | .... | .... |
| 5 1/2 | 23.8 | 18.3 | 5 1/2 | 30.2 | 23.2 | .... | .... | .... |
| 6 | 28.3 | 21.7 | 6 | 36.0 | 27.6 | .... | .... | .... |
| .... | .... | .... | .... | .... | .. | ... | .... | .... |

*Note:* Weights in table are for Dowmetal M (S.A.E. No. 522). For weights of Dowmetal Fs-1 (S.A.E. No. 52) multiply weights shown by 1.005; for Dowmetal J-1 (S.A.E. No. 520) and Dowmetal O-1, multiply by 1.022; for pure magnesium, multiply by 0.989.

### Theoretical Weights of Round, Square and Hexagon Steel Bars

| Thickness or Diameter, Inches | Round Bars | | Square Bars | | Hexagon Bars | |
|---|---|---|---|---|---|---|
| | Weight, Pounds Per Inch | Weight, Pounds Per Foot | Weight, Pounds Per Inch | Weight, Pounds Per Foot | Weight, Pounds Per Inch | Weight, Pounds Per Foot |
| $\frac{1}{32}$ | 0.0002 | 0.0026 | 0.0003 | 0.0033 | 0.0002 | 0.0028 |
| $\frac{1}{16}$ | 0.0009 | 0.0104 | 0.0011 | 0.0133 | 0.0010 | 0.0115 |
| $\frac{3}{32}$ | 0.0020 | 0.0235 | 0.0025 | 0.0299 | 0.0022 | 0.0259 |
| $\frac{1}{8}$ | 0.0035 | 0.0417 | 0.0044 | 0.0531 | 0.0038 | 0.0460 |
| $\frac{5}{32}$ | 0.0054 | 0.0652 | 0.0069 | 0.0830 | 0.0060 | 0.0719 |
| $\frac{3}{16}$ | 0.0078 | 0.0939 | 0.0100 | 0.1195 | 0.0086 | 0.1035 |
| $\frac{7}{32}$ | 0.0106 | 0.1278 | 0.0136 | 0.1627 | 0.0117 | 0.1409 |
| $\frac{1}{4}$ | 0.0139 | 0.1669 | 0.0177 | 0.2125 | 0.0153 | 0.1840 |
| $\frac{9}{32}$ | 0.0176 | 0.2112 | 0.0224 | 0.2689 | 0.0194 | 0.2329 |
| $\frac{5}{16}$ | 0.0217 | 0.2608 | 0.0277 | 0.3320 | 0.0240 | 0.2875 |
| $\frac{11}{32}$ | 0.0263 | 0.3155 | 0.0335 | 0.4018 | 0.0290 | 0.3479 |
| $\frac{3}{8}$ | 0.0313 | 0.3755 | 0.0398 | 0.4781 | 0.0345 | 0.4141 |
| $\frac{13}{32}$ | 0.0367 | 0.4407 | 0.0468 | 0.5611 | 0.0405 | 0.4860 |
| $\frac{7}{16}$ | 0.0426 | 0.5111 | 0.0542 | 0.6508 | 0.0470 | 0.5636 |
| $\frac{15}{32}$ | 0.0489 | 0.5867 | 0.0623 | 0.7471 | 0.0538 | 0.6470 |
| $\frac{1}{2}$ | 0.0556 | 0.6676 | 0.0708 | 0.8500 | 0.0613 | 0.7361 |
| $\frac{17}{32}$ | 0.0628 | 0.7536 | 0.0800 | 0.9596 | 0.0693 | 0.8310 |
| $\frac{9}{16}$ | 0.0704 | 0.8449 | 0.0896 | 1.076 | 0.0776 | 0.9317 |
| $\frac{19}{32}$ | 0.0785 | 0.9414 | 0.0999 | 1.199 | 0.0865 | 1.038 |
| $\frac{5}{8}$ | 0.0869 | 1.043 | 0.1107 | 1.328 | 0.0958 | 1.150 |
| $\frac{21}{32}$ | 0.0958 | 1.150 | 0.1220 | 1.464 | 0.1057 | 1.268 |
| $1\frac{1}{16}$ | 0.1052 | 1.262 | 0.1339 | 1.607 | 0.1160 | 1.392 |
| $\frac{23}{32}$ | 0.1150 | 1.380 | 0.1464 | 1.756 | 0.1268 | 1.521 |
| $\frac{3}{4}$ | 0.1252 | 1.502 | 0.1594 | 1.913 | 0.1380 | 1.656 |
| $\frac{25}{32}$ | 0.1358 | 1.630 | 0.1729 | 2.075 | 0.1498 | 1.797 |
| $\frac{13}{16}$ | 0.1469 | 1.763 | 0.1870 | 2.245 | 0.1620 | 1.944 |
| $\frac{27}{32}$ | 0.1584 | 1.901 | 0.2017 | 2.421 | 0.1747 | 2.096 |
| $\frac{7}{8}$ | 0.1704 | 2.044 | 0.2169 | 2.603 | 0.1879 | 2.254 |
| $\frac{29}{32}$ | 0.1828 | 2.193 | 0.2327 | 2.792 | 0.2015 | 2.418 |
| $\frac{15}{16}$ | 0.1956 | 2.347 | 0.2490 | 2.988 | 0.2157 | 2.588 |
| $\frac{31}{32}$ | 0.2088 | 2.506 | 0.2659 | 3.191 | 0.2303 | 2.763 |
| 1 | 0.2225 | 2.670 | 0.2833 | 3.400 | 0.2454 | 2.944 |
| $1\frac{1}{16}$ | 0.2512 | 3.015 | 0.3199 | 3.838 | 0.2770 | 3.324 |
| $1\frac{1}{8}$ | 0.2816 | 3.380 | 0.3586 | 4.303 | 0.3106 | 3.727 |
| $1\frac{3}{16}$ | 0.3138 | 3.766 | 0.3995 | 4.795 | 0.3460 | 4.152 |
| $1\frac{1}{4}$ | 0.3477 | 4.172 | 0.4427 | 5.313 | 0.3834 | 4.601 |
| $1\frac{5}{16}$ | 0.3833 | 4.600 | 0.4881 | 5.857 | 0.4227 | 5.072 |
| $1\frac{3}{8}$ | 0.4207 | 5.049 | 0.5357 | 6.428 | 0.4639 | 5.567 |
| $1\frac{7}{16}$ | 0.4598 | 5.518 | 0.5855 | 7.026 | 0.5070 | 6.085 |
| $1\frac{1}{2}$ | 0.5007 | 6.008 | 0.6375 | 7.650 | 0.5521 | 6.625 |
| $1\frac{9}{16}$ | 0.5433 | 6.519 | 0.6917 | 8.301 | 0.5991 | 7.189 |
| $1\frac{5}{8}$ | 0.5876 | 7.051 | 0.7482 | 8.978 | 0.6479 | 7.775 |

Based on a weight of 0.2833 lbs. per cubic in. (489.6 lbs. per cubic ft.).

### Theoretical Weights of Round, Square and Hexagon Steel Bars (*Continued*)

| Thickness or Diameter, Inches | Round Bars | | Square Bars | | Hexagon Bars | |
|---|---|---|---|---|---|---|
| | Weight, Pounds Per Inch | Weight, Pounds Per Foot | Weight, Pounds Per Inch | Weight, Pounds Per Foot | Weight, Pounds Per Inch | Weight, Pounds Per Foot |
| 1 11/16 | 0.6337 | 7.604 | 0.8068 | 9.682 | 0.6988 | 8.385 |
| 1 3/4 | 0.6815 | 8.178 | 0.8677 | 10.41 | 0.7515 | 9.018 |
| 1 13/16 | 0.7310 | 8.773 | 0.9308 | 11.17 | 0.8060 | 9.67 |
| 1 7/8 | 0.7823 | 9.388 | 0.9961 | 11.95 | 0.8626 | 10.35 |
| 1 15/16 | 0.8354 | 10.02 | 1.064 | 12.76 | 0.9211 | 11.05 |
| 2 | 0.8901 | 10.68 | 1.133 | 13.60 | 0.9815 | 11.78 |
| 2 1/16 | 0.9466 | 11.36 | 1.205 | 14.46 | 1.044 | 12.53 |
| 2 1/8 | 1.005 | 12.06 | 1.279 | 15.35 | 1.108 | 13.30 |
| 2 3/16 | 1.065 | 12.78 | 1.356 | 16.27 | 1.174 | 14.09 |
| 2 1/4 | 1.127 | 13.52 | 1.434 | 17.21 | 1.242 | 14.91 |
| 2 5/16 | 1.190 | 14.28 | 1.515 | 18.18 | 1.312 | 15.75 |
| 2 3/8 | 1.255 | 15.06 | 1.598 | 19.18 | 1.384 | 16.61 |
| 2 7/16 | 1.322 | 15.87 | 1.683 | 20.20 | 1.458 | 17.49 |
| 2 1/2 | 1.391 | 16.69 | 1.771 | 21.25 | 1.534 | 18.40 |
| 2 5/8 | 1.533 | 18.40 | 1.952 | 23.43 | 1.691 | 20.29 |
| 2 3/4 | 1.683 | 20.19 | 2.143 | 25.71 | 1.856 | 22.27 |
| 2 7/8 | 1.839 | 22.07 | 2.342 | 28.10 | 2.028 | 24.34 |
| 3 | 2.003 | 24.03 | 2.550 | 30.60 | 2.208 | 26.50 |
| 3 1/8 | 2.173 | 26.08 | 2.767 | 33.20 | 2.396 | 28.75 |
| 3 1/4 | 2.350 | 28.21 | 2.993 | 35.91 | 2.592 | 31.10 |
| 3 3/8 | 2.535 | 30.42 | 3.227 | 38.73 | 2.795 | 33.54 |
| 3 1/2 | 2.726 | 32.71 | 3.471 | 41.65 | 3.006 | 36.07 |
| 3 5/8 | 2.924 | 35.09 | 3.723 | 44.68 | 3.224 | 38.69 |
| 3 3/4 | 3.129 | 37.55 | 3.984 | 47.81 | 3.451 | 41.41 |
| 3 7/8 | 3.341 | 40.10 | 4.254 | 51.05 | 3.684 | 44.21 |
| 4 | 3.560 | 42.73 | 4.533 | 54.40 | 3.926 | 47.11 |
| 4 1/8 | 3.786 | 45.44 | 4.821 | 57.85 | 4.175 | 50.10 |
| 4 1/4 | 4.019 | 48.23 | 5.118 | 61.41 | 4.432 | 53.18 |
| 4 3/8 | 4.259 | 51.11 | 5.423 | 65.08 | 4.700 | 56.36 |
| 4 1/2 | 4.506 | 54.07 | 5.738 | 68.85 | 4.970 | 59.63 |
| 4 5/8 | 4.760 | 57.12 | 6.061 | 72.73 | 5.248 | 62.98 |
| 4 3/4 | 5.021 | 60.25 | 6.393 | 76.71 | 5.536 | 66.44 |
| 4 7/8 | 5.289 | 63.46 | 6.734 | 80.80 | 5.831 | 69.98 |
| 5 | 5.563 | 66.76 | 7.083 | 85.00 | 6.134 | 73.61 |
| 5 1/8 | 5.845 | 70.14 | 7.442 | 89.30 | 6.445 | 77.34 |
| 5 1/4 | 6.133 | 73.60 | 7.809 | 93.71 | 6.763 | 81.16 |
| 5 3/8 | 6.429 | 77.15 | 8.186 | 98.23 | 7.089 | 85.07 |
| 5 1/2 | 6.732 | 80.78 | 8.571 | 102.85 | 7.422 | 89.07 |
| 5 5/8 | 7.041 | 84.49 | 8.965 | 107.58 | 7.763 | 93.16 |
| 5 3/4 | 7.357 | 88.29 | 9.368 | 112.41 | 8.112 | 97.35 |
| 5 7/8 | 7.681 | 92.17 | 9.779 | 117.35 | 8.470 | 101.63 |
| 6 | 8.011 | 96.13 | 10.200 | 122.40 | 8.833 | 106.00 |

Based on a weight of 0.2833 lbs. per cubic in. (489.6 lbs. per cubic ft.).
Source: *Steel Products Manual*, American Iron & Steel Institute, with permission.

Weights of Flat Rolled Steel per Lineal Foot in Pounds

| U. S. St'd Gage for Plate | Width of Flat Steel, Inches | | | | | | |
|---|---|---|---|---|---|---|---|
| | ⅛ | ³⁄₁₆ | ¼ | ⁵⁄₁₆ | ⅜ | ⁷⁄₁₆ | ½ |
| 0000000 | 0.2126 | 0.3189 | 0.4252 | 0.5315 | 0.6378 | 0.7441 | 0.8504 |
| 000000 | 0.1992 | 0.2998 | 0.3984 | 0.4980 | 0.5976 | 0.6972 | 0.7968 |
| 00000 | 0.1860 | 0.2790 | 0.3720 | 0.4650 | 0.5580 | 0.6510 | 0.7440 |
| 0000 | 0.1728 | 0.2592 | 0.3456 | 0.4320 | 0.5184 | 0.6048 | 0.6912 |
| 000 | 0.1594 | 0.2391 | 0.3188 | 0.3985 | 0.4782 | 0.5579 | 0.6376 |
| 00 | 0.1462 | 0.2193 | 0.2924 | 0.3655 | 0.4386 | 0.5117 | 0.5848 |
| 0 | 0.1328 | 0.1992 | 0.2656 | 0.3320 | 0.3984 | 0.4648 | 0.5312 |
| 1 | 0.1196 | 0.1794 | 0.2392 | 0.2990 | 0.3588 | 0.4186 | 0.4784 |
| 2 | 0.1130 | 0.1695 | 0.2260 | 0.2825 | 0.3390 | 0.3955 | 0.4520 |
| 3 | 0.1062 | 0.1593 | 0.2124 | 0.2655 | 0.3186 | 0.3717 | 0.4248 |
| 4 | 0.0996 | 0.1494 | 0.1992 | 0.2490 | 0.2988 | 0.3486 | 0.3984 |
| 5 | 0.0930 | 0.1395 | 0.1860 | 0.2325 | 0.2790 | 0.3255 | 0.3720 |
| 6 | 0.0864 | 0.1296 | 0.1728 | 0.2160 | 0.2592 | 0.3024 | 0.3456 |
| 7 | 0.0798 | 0.1197 | 0.1596 | 0.1995 | 0.2394 | 0.2793 | 0.3192 |
| 8 | 0.0730 | 0.1095 | 0.1460 | 0.1825 | 0.2190 | 0.2555 | 0.2920 |
| 9 | 0.0664 | 0.0996 | 0.1328 | 0.1660 | 0.1992 | 0.2324 | 0.2656 |
| 10 | 0.0598 | 0.0897 | 0.1196 | 0.1495 | 0.1794 | 0.2093 | 0.2392 |
| 11 | 0.0532 | 0.0798 | 0.1064 | 0.1330 | 0.1596 | 0.1862 | 0.2128 |
| 12 | 0.0466 | 0.0699 | 0.0932 | 0.1165 | 0.1398 | 0.1631 | 0.1864 |
| 13 | 0.0398 | 0.0597 | 0.0796 | 0.0995 | 0.1194 | 0.1393 | 0.1592 |
| 14 | 0.0332 | 0.0498 | 0.0664 | 0.0830 | 0.0996 | 0.1162 | 0.1328 |
| 15 | 0.0298 | 0.0447 | 0.0596 | 0.0745 | 0.0894 | 0.1043 | 0.1192 |
| 16 | 0.0266 | 0.0399 | 0.0532 | 0.0665 | 0.0798 | 0.0931 | 0.1064 |
| 17 | 0.0240 | 0.0360 | 0.0480 | 0.0600 | 0.0720 | 0.0840 | 0.0960 |
| 18 | 0.0212 | 0.0318 | 0.0424 | 0.0530 | 0.0636 | 0.0742 | 0.0848 |
| 19 | 0.0186 | 0.0279 | 0.0372 | 0.0465 | 0.0558 | 0.0651 | 0.0744 |
| 20 | 0.0160 | 0.0240 | 0.0320 | 0.0400 | 0.0480 | 0.0560 | 0.0640 |
| 21 | 0.0146 | 0.0219 | 0.0292 | 0.0365 | 0.0438 | 0.0511 | 0.0584 |
| 22 | 0.0133 | 0.0201 | 0.0268 | 0.0335 | 0.0402 | 0.0469 | 0.0536 |
| 23 | 0.0120 | 0.0180 | 0.0240 | 0.0300 | 0.0360 | 0.0420 | 0.0480 |
| 24 | 0.0106 | 0.0159 | 0.0212 | 0.0265 | 0.0318 | 0.0371 | 0.0424 |
| 25 | 0.0094 | 0.0141 | 0.0188 | 0.0235 | 0.0282 | 0.0329 | 0.0376 |
| 26 | 0.0080 | 0.0120 | 0.0160 | 0.0200 | 0.0240 | 0.0280 | 0.0320 |
| 27 | 0.0074 | 0.0111 | 0.0148 | 0.0185 | 0.0222 | 0.0259 | 0.0296 |
| 28 | 0.0066 | 0.0099 | 0.0132 | 0.0165 | 0.0198 | 0.0231 | 0.0264 |
| 29 | 0.0060 | 0.0090 | 0.0120 | 0.0150 | 0.0180 | 0.0210 | 0.0240 |
| 30 | 0.0054 | 0.0081 | 0.0108 | 0.0135 | 0.0162 | 0.0189 | 0.0212 |
| 31 | 0.0046 | 0.0069 | 0.0092 | 0.0115 | 0.0138 | 0.0161 | 0.0184 |
| 32 | 0.0044 | 0.0066 | 0.0088 | 0.0110 | 0.0132 | 0.0154 | 0.0176 |
| 33 | 0.0040 | 0.0060 | 0.0080 | 0.0100 | 0.0120 | 0.0140 | 0.0160 |
| 34 | 0.0036 | 0.0054 | 0.0072 | 0.0090 | 0.0108 | 0.0126 | 0.0144 |
| 35 | 0.0034 | 0.0051 | 0.0068 | 0.0085 | 0.0102 | 0.0119 | 0.0136 |
| 36 | 0.0030 | 0.0045 | 0.0060 | 0.0075 | 0.0090 | 0.0105 | 0.0120 |
| 37 | 0.0028 | 0.0042 | 0.0056 | 0.0070 | 0.0084 | 0.0098 | 0.0112 |
| 38 | 0.0026 | 0.0039 | 0.0052 | 0.0065 | 0.0078 | 0.0091 | 0.0104 |

**Weights of Flat Rolled Steel Per Lineal Foot in Pounds** (*Continued*)

| U. S. St'd Gage for Plate | Width of Flat Steel, Inches | | | | | | | |
|---|---|---|---|---|---|---|---|---|
| | 9⁄16 | 5⁄8 | 11⁄16 | 3⁄4 | 13⁄16 | 7⁄8 | 1 | 2 |
| 0000000 | 0.9567 | 1.0630 | 1.1693 | 1.2756 | 1.3819 | 1.4882 | 1.7008 | 3.4016 |
| 000000 | 0.8964 | 0.9960 | 1.0956 | 1.1952 | 1.2948 | 1.3944 | 1.5936 | 3.1872 |
| 00000 | 0.8370 | 0.9300 | 1.0230 | 1.1160 | 1.2090 | 1.3020 | 1.4880 | 2.9760 |
| 0000 | 0.7776 | 0.8640 | 0.9500 | 1.0368 | 1.1232 | 1.2096 | 1.3824 | 2.7648 |
| 000 | 0.7173 | 0.7970 | 0.8767 | 0.9564 | 1.0361 | 1.1158 | 1.2752 | 2.5504 |
| 00 | 0.6579 | 0.7310 | 0.8041 | 0.8772 | 0.9505 | 1.0234 | 1.1696 | 2.3392 |
| 0 | 0.5976 | 0.6640 | 0.7300 | 0.7968 | 0.8632 | 0.9296 | 1.0624 | 2.1248 |
| 1 | 0.5382 | 0.5980 | 0.6578 | 0.7176 | 0.7774 | 0.8372 | 0.9568 | 1.9136 |
| 2 | 0.5085 | 0.5650 | 0.6215 | 0.6780 | 0.7345 | 0.7910 | 0.9040 | 1.8080 |
| 3 | 0.4779 | 0.5310 | 0.5841 | 0.6372 | 0.6903 | 0.7434 | 0.8496 | 1.6992 |
| 4 | 0.4482 | 0.4980 | 0.5478 | 0.5976 | 0.6474 | 0.6972 | 0.7968 | 1.5936 |
| 5 | 0.4185 | 0.4650 | 0.5115 | 0.5580 | 0.6045 | 0.6510 | 0.7440 | 1.4880 |
| 6 | 0.3888 | 0.4320 | 0.4752 | 0.5184 | 0.5616 | 0.6048 | 0.6912 | 1.3824 |
| 7 | 0.3591 | 0.3990 | 0.4389 | 0.4788 | 0.5187 | 0.5586 | 0.6384 | 1.2768 |
| 8 | 0.3285 | 0.3650 | 0.4015 | 0.4380 | 0.4745 | 0.5110 | 0.5840 | 1.1680 |
| 9 | 0.2988 | 0.3320 | 0.3652 | 0.3984 | 0.4316 | 0.4648 | 0.5312 | 1.0624 |
| 10 | 0.2691 | 0.2990 | 0.3289 | 0.3588 | 0.3887 | 0.4186 | 0.4784 | 0.9568 |
| 11 | 0.2394 | 0.2660 | 0.2926 | 0.3192 | 0.3458 | 0.3724 | 0.4256 | 0.8512 |
| 12 | 0.2097 | 0.2330 | 0.2563 | 0.2796 | 0.3029 | 0.3262 | 0.3728 | 0.7456 |
| 13 | 0.1791 | 0.1990 | 0.2189 | 0.2388 | 0.2587 | 0.2786 | 0.3184 | 0.6368 |
| 14 | 0.1494 | 0.1660 | 0.1826 | 0.1992 | 0.2158 | 0.2324 | 0.2656 | 0.5312 |
| 15 | 0.1341 | 0.1490 | 0.1639 | 0.1788 | 0.1937 | 0.2086 | 0.2384 | 0.4768 |
| 16 | 0.1197 | 0.1330 | 0.1463 | 0.1596 | 0.1729 | 0.1862 | 0.2128 | 0.4256 |
| 17 | 0.1080 | 0.1200 | 0.1320 | 0.1440 | 0.1560 | 0.1680 | 0.1920 | 0.3840 |
| 18 | 0.0954 | 0.1060 | 0.1166 | 0.1272 | 0.1378 | 0.1484 | 0.1696 | 0.3392 |
| 19 | 0.0837 | 0.0930 | 0.1023 | 0.1116 | 0.1209 | 0.1302 | 0.1488 | 0.2976 |
| 20 | 0.0720 | 0.0800 | 0.0880 | 0.0960 | 0.1040 | 0.1120 | 0.1280 | 0.2560 |
| 21 | 0.0657 | 0.0730 | 0.0803 | 0.0876 | 0.0949 | 0.1022 | 0.1168 | 0.2336 |
| 22 | 0.0603 | 0.0670 | 0.0737 | 0.0804 | 0.0871 | 0.0938 | 0.1072 | 0.2144 |
| 23 | 0.0540 | 0.0600 | 0.0660 | 0.0720 | 0.0780 | 0.0840 | 0.0960 | 0.1920 |
| 24 | 0.0477 | 0.0530 | 0.0583 | 0.0636 | 0.0689 | 0.0742 | 0.0848 | 0.1696 |
| 25 | 0.0423 | 0.0470 | 0.0517 | 0.0564 | 0.0611 | 0.0658 | 0.0752 | 0.1504 |
| 26 | 0.0360 | 0.0400 | 0.0440 | 0.0480 | 0.0520 | 0.0560 | 0.0640 | 0.1280 |
| 27 | 0.0333 | 0.0370 | 0.0407 | 0.0444 | 0.0481 | 0.0518 | 0.0592 | 0.1184 |
| 28 | 0.0297 | 0.0330 | 0.0363 | 0.0396 | 0.0429 | 0.0462 | 0.0528 | 0.1056 |
| 29 | 0.0270 | 0.0300 | 0.0330 | 0.0360 | 0.0390 | 0.0420 | 0.0480 | 0.0960 |
| 30 | 0.0243 | 0.0270 | 0.0297 | 0.0324 | 0.0351 | 0.0378 | 0.0432 | 0.0864 |
| 31 | 0.0207 | 0.0230 | 0.0253 | 0.0276 | 0.0299 | 0.0322 | 0.0368 | 0.0736 |
| 32 | 0.0198 | 0.0220 | 0.0242 | 0.0264 | 0.0286 | 0.0308 | 0.0352 | 0.0704 |
| 33 | 0.0180 | 0.0200 | 0.0220 | 0.0240 | 0.0260 | 0.0280 | 0.0320 | 0.0640 |
| 34 | 0.0162 | 0.0180 | 0.0198 | 0.0216 | 0.0234 | 0.0252 | 0.0288 | 0.0576 |
| 35 | 0.0153 | 0.0170 | 0.0187 | 0.0204 | 0.0221 | 0.0238 | 0.0272 | 0.0544 |
| 36 | 0.0135 | 0.0150 | 0.0165 | 0.0180 | 0.0195 | 0.0210 | 0.0240 | 0.0480 |
| 37 | 0.0126 | 0.0140 | 0.0154 | 0.0168 | 0.0182 | 0.0196 | 0.0224 | 0.0448 |
| 38 | 0.0117 | 0.0130 | 0.0143 | 0.0156 | 0.0169 | 0.0182 | 0.0208 | 0.0416 |

**Weights of Flat Rolled Steel per Lineal Foot in Pounds** (*Continued*)

| U.S. St'd Gage for Plate | Width of Flat Steel, Inches | | | | | | | |
|---|---|---|---|---|---|---|---|---|
| | 3 | 4 | 5 | 6 | 7 | 8 | 9 | 10 |
| 0000000 | 5.1024 | 6.8032 | 8.504 | 10.204 | 11.905 | 13.606 | 15.307 | 17.008 |
| 000000 | 4.7808 | 6.3744 | 7.968 | 9.561 | 11.155 | 12.748 | 14.342 | 15.936 |
| 00000 | 4.4640 | 5.9520 | 7.440 | 8.928 | 10.416 | 11.904 | 13.392 | 14.880 |
| 0000 | 4.1472 | 5.5296 | 6.912 | 8.294 | 9.676 | 11.059 | 12.441 | 13.824 |
| 000 | 3.8256 | 5.1008 | 6.376 | 7.651 | 8.926 | 10.201 | 11.476 | 12.752 |
| 00 | 3.5088 | 4.6784 | 5.848 | 7.017 | 8.187 | 9.356 | 10.526 | 11.696 |
| 0 | 3.1872 | 4.2496 | 5.312 | 6.374 | 7.436 | 8.499 | 9.561 | 10.624 |
| 1 | 2.8704 | 3.8272 | 4.784 | 5.740 | 6.697 | 7.654 | 8.611 | 9.568 |
| 2 | 2.7120 | 3.6160 | 4.520 | 5.424 | 6.328 | 7.232 | 8.136 | 9.040 |
| 3 | 2.5488 | 3.3984 | 4.248 | 5.097 | 5.947 | 6.796 | 7.646 | 8.496 |
| 4 | 2.3904 | 3.1872 | 3.984 | 4.780 | 5.577 | 6.374 | 7.171 | 7.968 |
| 5 | 2.2320 | 2.9760 | 3.720 | 4.464 | 5.208 | 5.952 | 6.696 | 7.440 |
| 6 | 2.0736 | 2.7648 | 3.456 | 4.147 | 4.838 | 5.529 | 6.220 | 6.912 |
| 7 | 1.9152 | 2.5536 | 3.192 | 3.830 | 4.468 | 5.107 | 5.745 | 6.384 |
| 8 | 1.7520 | 2.3360 | 2.920 | 3.504 | 4.088 | 4.672 | 5.256 | 5.840 |
| 9 | 1.5936 | 2.1248 | 2.656 | 3.187 | 3.718 | 4.249 | 4.780 | 5.312 |
| 10 | 1.4352 | 1.9136 | 2.392 | 2.870 | 3.348 | 3.827 | 4.305 | 4.784 |
| 11 | 1.2768 | 1.7024 | 2.128 | 2.553 | 2.979 | 3.404 | 3.830 | 4.256 |
| 12 | 1.1184 | 1.4912 | 1.864 | 2.236 | 2.609 | 2.982 | 3.355 | 3.728 |
| 13 | 0.9552 | 1.2736 | 1.592 | 1.910 | 2.228 | 2.547 | 2.865 | 3.184 |
| 14 | 0.7968 | 1.0624 | 1.328 | 1.593 | 1.859 | 2.124 | 2.390 | 2.656 |
| 15 | 0.7152 | 0.9536 | 1.192 | 1.430 | 1.668 | 1.907 | 2.145 | 2.384 |
| 16 | 0.6384 | 0.8512 | 1.064 | 1.276 | 1.489 | 1.702 | 1.915 | 2.128 |
| 17 | 0.5760 | 0.7680 | 0.960 | 1.152 | 1.344 | 1.536 | 1.728 | 1.920 |
| 18 | 0.5088 | 0.6786 | 0.848 | 1.027 | 1.187 | 1.357 | 1.526 | 1.696 |
| 19 | 0.4464 | 0.5952 | 0.744 | 0.892 | 1.041 | 1.190 | 1.319 | 1.488 |
| 20 | 0.3840 | 0.5120 | 0.640 | 0.768 | 0.896 | 1.024 | 1.152 | 1.280 |
| 21 | 0.3504 | 0.4672 | 0.584 | 0.700 | 0.817 | 0.934 | 1.051 | 1.168 |
| 22 | 0.3216 | 0.4288 | 0.536 | 0.643 | 0.750 | 0.857 | 0.964 | 1.072 |
| 23 | 0.2880 | 0.3840 | 0.480 | 0.576 | 0.672 | 0.768 | 0.864 | 0.960 |
| 24 | 0.2544 | 0.3392 | 0.424 | 0.508 | 0.593 | 0.678 | 0.763 | 0.848 |
| 25 | 0.2256 | 0.3008 | 0.376 | 0.451 | 0.526 | 0.601 | 0.676 | 0.752 |
| 26 | 0.1920 | 0.2560 | 0.320 | 0.384 | 0.448 | 0.512 | 0.596 | 0.640 |
| 27 | 0.1776 | 0.2368 | 0.296 | 0.355 | 0.414 | 0.473 | 0.532 | 0.592 |
| 28 | 0.1584 | 0.2112 | 0.264 | 0.316 | 0.369 | 0.422 | 0.475 | 0.528 |
| 29 | 0.1440 | 0.1920 | 0.240 | 0.288 | 0.336 | 0.384 | 0.432 | 0.480 |
| 30 | 0.1272 | 0.1768 | 0.216 | 0.259 | 0.302 | 0.353 | 0.388 | 0.432 |
| 31 | 0.1104 | 0.1472 | 0.184 | 0.220 | 0.257 | 0.297 | 0.331 | 0.368 |
| 32 | 0.1056 | 0.1408 | 0.176 | 0.211 | 0.246 | 0.281 | 0.316 | 0.352 |
| 33 | 0.0960 | 0.1280 | 0.160 | 0.192 | 0.224 | 0.256 | 0.288 | 0.320 |
| 34 | 0.0864 | 0.1152 | 0.144 | 0.172 | 0.201 | 0.230 | 0.259 | 0.288 |
| 35 | 0.0816 | 0.1088 | 0.136 | 0.163 | 0.190 | 0.217 | 0.244 | 0.272 |
| 36 | 0.0720 | 0.0960 | 0.120 | 0.148 | 0.168 | 0.192 | 0.216 | 0.240 |
| 37 | 0.0674 | 0.0896 | 0.112 | 0.134 | 0.156 | 0.179 | 0.201 | 0.224 |
| 38 | 0.0624 | 0.0832 | 0.104 | 0.124 | 0.145 | 0.166 | 0.187 | 0.208 |

## Weights of Flat Rolled Steel per Lineal Foot in Pounds (*Continued*)

| U.S. St'd Gage for Plate | Width of Flat Steel, Inches | | | | | | | |
|---|---|---|---|---|---|---|---|---|
| | 11 | 12 | 13 | 14 | 15 | 16 | 18 | 20 |
| 0000000 | 18.708 | 20.409 | 22.110 | 23.811 | 25.512 | 27.212 | 30.614 | 34.016 |
| 000000 | 17.529 | 19.123 | 20.716 | 22.310 | 23.904 | 25.497 | 28.684 | 31.872 |
| 00000 | 16.368 | 17.856 | 19.344 | 20.832 | 22.320 | 23.808 | 26.784 | 29.760 |
| 0000 | 15.206 | 16.588 | 17.971 | 19.353 | 20.736 | 22.118 | 24.883 | 27.648 |
| 000 | 14.027 | 15.302 | 16.577 | 17.852 | 19.128 | 20.403 | 22.953 | 25.504 |
| 00 | 12.865 | 14.035 | 15.204 | 16.374 | 17.544 | 18.713 | 21.052 | 23.392 |
| 0 | 11.686 | 12.748 | 13.811 | 14.873 | 15.936 | 16.998 | 19.123 | 21.248 |
| 1 | 10.524 | 11.481 | 12.438 | 13.395 | 14.352 | 15.308 | 17.222 | 19.136 |
| 2 | 9.944 | 10.848 | 11.752 | 12.656 | 13.560 | 14.464 | 16.272 | 18.080 |
| 3 | 9.345 | 10.195 | 11.044 | 11.894 | 12.744 | 13.593 | 15.292 | 16.992 |
| 4 | 8.764 | 9.561 | 10.358 | 11.155 | 11.952 | 12.748 | 14.342 | 15.936 |
| 5 | 8.184 | 8.928 | 9.672 | 10.416 | 11.160 | 11.804 | 13.392 | 14.880 |
| 6 | 7.603 | 8.294 | 8.985 | 9.676 | 10.368 | 11.059 | 12.441 | 13.824 |
| 7 | 7.022 | 7.660 | 8.299 | 8.937 | 9.576 | 10.214 | 11.491 | 12.768 |
| 8 | 6.424 | 7.008 | 7.592 | 8.176 | 8.760 | 9.344 | 10.512 | 11.680 |
| 9 | 5.843 | 6.374 | 6.905 | 7.436 | 7.968 | 8.499 | 9.561 | 10.624 |
| 10 | 5.262 | 5.740 | 6.219 | 6.697 | 7.176 | 7.654 | 8.611 | 9.568 |
| 11 | 4.681 | 5.107 | 5.532 | 5.958 | 6.384 | 6.809 | 7.660 | 8.512 |
| 12 | 4.100 | 4.473 | 4.846 | 5.219 | 5.592 | 5.964 | 6.710 | 7.456 |
| 13 | 3.502 | 3.820 | 4.139 | 4.457 | 4.776 | 5.094 | 5.731 | 6.368 |
| 14 | 2.921 | 3.187 | 3.452 | 3.718 | 3.984 | 4.249 | 4.780 | 5.312 |
| 15 | 2.622 | 2.860 | 3.099 | 3.337 | 3.576 | 3.814 | 4.291 | 4.768 |
| 16 | 2.340 | 2.553 | 2.766 | 2.979 | 3.192 | 3.404 | 3.830 | 4.256 |
| 17 | 2.112 | 2.304 | 2.496 | 2.688 | 2.880 | 3.072 | 3.456 | 3.840 |
| 18 | 1.865 | 2.055 | 2.204 | 2.374 | 2.544 | 2.714 | 3.056 | 3.392 |
| 19 | 1.636 | 1.785 | 1.934 | 2.083 | 2.232 | 2.380 | 2.638 | 2.976 |
| 20 | 1.408 | 1.536 | 1.664 | 1.792 | 1.920 | 2.048 | 2.304 | 2.560 |
| 21 | 1.284 | 1.401 | 1.518 | 1.635 | 1.752 | 1.868 | 2.102 | 2.336 |
| 22 | 1.179 | 1.286 | 1.393 | 1.500 | 1.608 | 1.715 | 1.929 | 2.144 |
| 23 | 1.056 | 1.152 | 1.248 | 1.344 | 1.440 | 1.536 | 1.728 | 1.920 |
| 24 | 0.932 | 1.017 | 1.102 | 1.187 | 1.272 | 1.356 | 1.526 | 1.696 |
| 25 | 0.827 | 0.902 | 0.977 | 1.052 | 1.128 | 1.203 | 1.353 | 1.504 |
| 26 | 0.704 | 0.768 | 0.832 | 0.896 | 0.960 | 1.024 | 1.192 | 1.380 |
| 27 | 0.651 | 0.710 | 0.769 | 0.828 | 0.888 | 0.947 | 1.065 | 1.184 |
| 28 | 0.580 | 0.633 | 0.686 | 0.739 | 0.792 | 0.844 | 0.950 | 1.056 |
| 29 | 0.528 | 0.576 | 0.624 | 0.672 | 0.720 | 0.768 | 0.864 | 0.960 |
| 30 | 0.475 | 0.518 | 0.561 | 0.604 | 0.648 | 0.707 | 0.777 | 0.864 |
| 31 | 0.404 | 0.441 | 0.478 | 0.515 | 0.552 | 0.594 | 0.662 | 0.736 |
| 32 | 0.387 | 0.422 | 0.457 | 0.492 | 0.528 | 0.563 | 0.633 | 0.714 |
| 33 | 0.352 | 0.384 | 0.416 | 0.448 | 0.480 | 0.512 | 0.576 | 0.640 |
| 34 | 0.316 | 0.345 | 0.374 | 0.403 | 0.432 | 0.460 | 0.518 | 0.576 |
| 35 | 0.299 | 0.326 | 0.359 | 0.380 | 0.408 | 0.435 | 0.489 | 0.544 |
| 36 | 0.264 | 0.296 | 0.312 | 0.336 | 0.360 | 0.384 | 0.432 | 0.480 |
| 37 | 0.246 | 0.279 | 0.291 | 0.313 | 0.336 | 0.358 | 0.403 | 0.448 |
| 38 | 0.228 | 0.249 | 0.270 | 0.291 | 0.312 | 0.332 | 0.374 | 0.416 |

**Weight of Flat Rolled Steel Bars in Pounds per Lineal Foot**

(One cubic foot of rolled steel weighs 489.6 pounds.)

| Thickness of Bar, Ins. | Width of Bar, Inches | | | | | | | | | | | | | | | | | |
|---|---|---|---|---|---|---|---|---|---|---|---|---|---|---|---|---|---|---|
| | 1/4 | 1/2 | 3/4 | 1 | 1 1/4 | 1 1/2 | 1 3/4 | 2 | 2 1/4 | 2 1/2 | 2 3/4 | 3 | 3 1/4 | 3 1/2 | 3 3/4 | 4 | 4 1/4 | 4 1/2 |
| 1/16 | 0.053 | 0.106 | 0.159 | 0.21 | 0.27 | 0.32 | 0.37 | 0.42 | 0.48 | 0.53 | 0.58 | 0.64 | 0.69 | 0.74 | 0.80 | 0.85 | 0.90 | 0.96 |
| 1/8 | 0.106 | 0.212 | 0.319 | 0.42 | 0.53 | 0.64 | 0.74 | 0.85 | 0.96 | 1.06 | 1.17 | 1.27 | 1.38 | 1.49 | 1.59 | 1.70 | 1.81 | 1.91 |
| 3/16 | 0.159 | 0.319 | 0.478 | 0.64 | 0.80 | 0.96 | 1.12 | 1.28 | 1.43 | 1.59 | 1.75 | 1.91 | 2.07 | 2.23 | 2.39 | 2.55 | 2.71 | 2 87 |
| 1/4 | 0.213 | 0.425 | 0.638 | 0.85 | 1.06 | 1.28 | 1.49 | 1.70 | 1.91 | 2.13 | 2.34 | 2.55 | 2.76 | 2.98 | 3.19 | 3.40 | 3.61 | 3.83 |
| 5/16 | 0.266 | 0.531 | 0.797 | 1.06 | 1.33 | 1.59 | 1.86 | 2.13 | 2.39 | 2.66 | 2.92 | 3.19 | 3.45 | 3.72 | 3.98 | 4.25 | 4.52 | 4.78 |
| 3/8 | 0.319 | 0.638 | 0.956 | 1.28 | 1.59 | 1.91 | 2.23 | 2.55 | 2.87 | 3.19 | 3.51 | 3.83 | 4.14 | 4.46 | 4.78 | 5.10 | 5.42 | 5.74 |
| 7/16 | 0.372 | 0.744 | 1.12 | 1.49 | 1.86 | 2.23 | 2.60 | 2.98 | 3.35 | 3.72 | 4.09 | 4.46 | 4.83 | 5.21 | 5.58 | 5.95 | 6.32 | 6.69 |
| 1/2 | 0.425 | 0.850 | 1.28 | 1.70 | 2.13 | 2.55 | 2.98 | 3.40 | 3.83 | 4.25 | 4.68 | 5.10 | 5.53 | 5.95 | 6.38 | 6.80 | 7.22 | 7.65 |
| 9/16 | 0.478 | 0.956 | 1.43 | 1.91 | 2.39 | 2.87 | 3.35 | 3.83 | 4.30 | 4.78 | 5.26 | 5.74 | 6.22 | 6.69 | 7.17 | 7.65 | 8.13 | 8.61 |
| 5/8 | 0.531 | 1.06 | 1.59 | 2.13 | 2.66 | 3.19 | 3.72 | 4.25 | 4.78 | 5.31 | 5.84 | 6.38 | 6.91 | 7.44 | 7.97 | 8.50 | 9.03 | 9.56 |
| 11/16 | 0.584 | 1.17 | 1.75 | 2.34 | 2.92 | 3.51 | 4.09 | 4.68 | 5.26 | 5.84 | 6.43 | 7.01 | 7.60 | 8.18 | 8.77 | 9.35 | 9.93 | 10.52 |
| 3/4 | 0.638 | 1.28 | 1.91 | 2.55 | 3.19 | 3.83 | 4.46 | 5.10 | 5.74 | 6.38 | 7.01 | 7.65 | 8.29 | 8.93 | 9.56 | 10.20 | 10.84 | 11.48 |
| 13/16 | 0.691 | 1.38 | 2.07 | 2.76 | 3.45 | 4.14 | 4.83 | 5.53 | 6.22 | 6.91 | 7.60 | 8.29 | 8.98 | 9.67 | 10.36 | 11.05 | 11.74 | 12.43 |
| 7/8 | 0.744 | 1.49 | 2.23 | 2.98 | 3.72 | 4.46 | 5.21 | 5.95 | 6.69 | 7.44 | 8.18 | 8.93 | 9.67 | 10.41 | 11.16 | 11.90 | 12.64 | 13.39 |
| 15/16 | 0.797 | 1.59 | 2.39 | 3.19 | 3.98 | 4.78 | 5.58 | 6.38 | 7.17 | 7.97 | 8.77 | 9.56 | 10.36 | 11.16 | 11.95 | 12.75 | 13.55 | 14.34 |
| 1 | 0.850 | 1.70 | 2.55 | 3.40 | 4.25 | 5.10 | 5.95 | 6.80 | 7.65 | 8.50 | 9.35 | 10.20 | 11.05 | 11.90 | 12.75 | 13.60 | 14.45 | 15.30 |
| 1 1/16 | 0.903 | 1.81 | 2.71 | 3.61 | 4.52 | 5.42 | 6.32 | 7.23 | 8.13 | 9.03 | 9.93 | 10.84 | 11.74 | 12.64 | 13.55 | 14.45 | 15.35 | 16.26 |
| 1 1/8 | 0.956 | 1.91 | 2.87 | 3.83 | 4.78 | 5.74 | 6.69 | 7.65 | 8.61 | 9.56 | 10.52 | 11.48 | 12.43 | 13.39 | 14.34 | 15.30 | 16.26 | 17.21 |
| 1 3/16 | 1.01 | 2.02 | 3.03 | 4.04 | 5.05 | 6.06 | 7.07 | 8.08 | 9.08 | 10.09 | 11.10 | 12.11 | 13.12 | 14.13 | 15.14 | 16.15 | 17.16 | 18.17 |
| 1 1/4 | 1.06 | 2.12 | 3.19 | 4.25 | 5.31 | 6.38 | 7.44 | 8.50 | 9.56 | 10.63 | 11.69 | 12.75 | 13.81 | 14.88 | 15.94 | 17.00 | 18.06 | 19.13 |
| 1 5/16 | 1.12 | 2.23 | 3.34 | 4.46 | 5.58 | 6.69 | 7.81 | 8.93 | 10.04 | 11.16 | 12.27 | 13.39 | 14.50 | 15.62 | 16.73 | 17.85 | 18.97 | 20.08 |
| 1 3/8 | 1.17 | 2.34 | 3.50 | 4.68 | 5.84 | 7.01 | 8.18 | 9.35 | 10.52 | 11.69 | 12.86 | 14.03 | 15.19 | 16.36 | 17.53 | 18.70 | 19.87 | 21.04 |
| 1 7/16 | 1.22 | 2.44 | 3.66 | 4.89 | 6.11 | 7.33 | 8.55 | 9.78 | 11.00 | 12.22 | 13.44 | 14.66 | 15.88 | 17.11 | 18.33 | 19.55 | 20.77 | 21.99 |
| 1 1/2 | 1.27 | 2.55 | 3.82 | 5.10 | 6.38 | 7.65 | 8.93 | 10.20 | 11.48 | 12.75 | 14.03 | 15.30 | 16.58 | 17.85 | 19.13 | 20.40 | 21.68 | 22.95 |
| 1 9/16 | 1.33 | 2.66 | 3.98 | 5.31 | 6.64 | 7.97 | 9.30 | 10.63 | 11.95 | 13.28 | 14.61 | 15.92 | 17.27 | 18.59 | 19.92 | 21.25 | 22.58 | 23.91 |
| 1 5/8 | 1.38 | 2.76 | 4.14 | 5.53 | 6.91 | 8.29 | 9.67 | 11.05 | 12.43 | 13.81 | 15.19 | 16.58 | 17.96 | 19.34 | 20.72 | 22.10 | 23.48 | 24 86 |
| 1 11/16 | 1.43 | 2.87 | 4.30 | 5.74 | 7.17 | 8.61 | 10.04 | 11.48 | 12.91 | 14.34 | 15.78 | 17.21 | 18.65 | 20.08 | 21.52 | 22.95 | 24.38 | 25.82 |
| 1 3/4 | 1.49 | 2.97 | 4.46 | 5.95 | 7.44 | 8.93 | 10.41 | 11.90 | 13.39 | 14.88 | 16.36 | 17.85 | 19.34 | 20.83 | 22.31 | 23.80 | 25.29 | 26.78 |
| 1 13/16 | 1.54 | 3.08 | 4.62 | 6.16 | 7.70 | 9.24 | 10.78 | 12.33 | 13.87 | 15.41 | 16.95 | 18.49 | 20.03 | 21.57 | 23.11 | 24.65 | 26.19 | 27.73 |
| 1 7/8 | 1.59 | 3.19 | 4.78 | 6.38 | 7.97 | 9.56 | 11.16 | 12.75 | 14.34 | 15.94 | 17.53 | 19.13 | 20.72 | 22.31 | 23.91 | 25.50 | 27.09 | 28.69 |
| 1 15/16 | 1.65 | 3.29 | 4.94 | 6.59 | 8.23 | 9.88 | 11.53 | 13.18 | 14.82 | 16.47 | 18.12 | 19.76 | 21.41 | 23.06 | 24.70 | 26.35 | 28.00 | 29.64 |
| 2 | 1.70 | 3.40 | 5.10 | 6.80 | 8.50 | 10.20 | 11.90 | 13.60 | 15.30 | 17.00 | 18.70 | 20.40 | 22.10 | 23.80 | 25.50 | 27.20 | 28.90 | 30.60 |

## Weight of Flat Rolled Steel Bars in Pounds per Lineal Foot

(One cubic foot of rolled steel weighs 489.6 pounds.)

| Thickness of Bar, Ins. | Width of Bar, Inches | | | | | | | | | | | | | | | | | |
|---|---|---|---|---|---|---|---|---|---|---|---|---|---|---|---|---|---|---|
| | 4¾ | 5 | 5¼ | 5½ | 5¾ | 6 | 6½ | 7 | 7½ | 8 | 8½ | 9 | 9½ | 10 | 10½ | 11 | 11½ | 12 |
| 1/16 | 1.01 | 1.06 | 1.11 | 1.17 | 1.22 | 1.27 | 1.38 | 1.49 | 1.59 | 1.70 | 1.81 | 1.91 | 2.02 | 2.12 | 2.23 | 2.34 | 2.44 | 2.55 |
| 1/8 | 2.02 | 2.12 | 2.23 | 2.34 | 2.44 | 2.55 | 2.76 | 2.97 | 3.18 | 3.40 | 3.61 | 3.82 | 4.04 | 4.25 | 4.46 | 4.67 | 4.89 | 5.10 |
| 3/16 | 3.03 | 3.19 | 3.35 | 3.51 | 3.67 | 3.83 | 4.14 | 4.46 | 4.78 | 5.10 | 5.42 | 5.74 | 6.06 | 6.38 | 6.69 | 7.01 | 7.33 | 7.65 |
| 1/4 | 4.04 | 4.25 | 4.46 | 4.68 | 4.89 | 5.10 | 5.53 | 5.95 | 6.38 | 6.80 | 7.23 | 7.65 | 8.08 | 8.50 | 8.93 | 9.35 | 9.78 | 10.20 |
| 5/16 | 5.05 | 5.31 | 5.58 | 5.84 | 6.11 | 6.38 | 6.91 | 7.44 | 7.97 | 8.50 | 9.03 | 9.56 | 10.09 | 10.63 | 11.16 | 11.69 | 12.22 | 12.75 |
| 3/8 | 6.06 | 6.38 | 6.69 | 7.01 | 7.33 | 7.65 | 8.29 | 8.93 | 9.56 | 10.20 | 10.84 | 11.48 | 12.11 | 12.75 | 13.39 | 14.03 | 14.66 | 15.30 |
| 7/16 | 7.07 | 7.44 | 7.81 | 8.18 | 8.55 | 8.93 | 9.67 | 10.41 | 11.16 | 11.90 | 12.64 | 13.39 | 14.13 | 14.88 | 15.62 | 16.36 | 17.11 | 17.85 |
| 1/2 | 8.08 | 8.50 | 8.93 | 9.35 | 9.78 | 10.20 | 11.05 | 11.90 | 12.75 | 13.60 | 14.45 | 15.31 | 16.15 | 17.00 | 17.85 | 18.70 | 19.55 | 20.40 |
| 9/16 | 9.08 | 9.56 | 10.04 | 10.52 | 11.00 | 11.48 | 12.43 | 13.39 | 14.34 | 15.30 | 16.26 | 17.21 | 18.17 | 19.13 | 20.08 | 21.04 | 21.99 | 22.95 |
| 5/8 | 10.09 | 10.63 | 11.16 | 11.69 | 12.22 | 12.75 | 13.81 | 14.88 | 15.94 | 17.00 | 18.06 | 19.13 | 20.19 | 21.25 | 22.31 | 23.38 | 24.44 | 25.50 |
| 11/16 | 11.10 | 11.69 | 12.27 | 12.86 | 13.44 | 14.03 | 15.19 | 16.36 | 17.53 | 18.70 | 19.87 | 21.04 | 22.21 | 23.38 | 24.54 | 25.71 | 26.88 | 28.05 |
| 3/4 | 12.11 | 12.75 | 13.39 | 14.03 | 14.67 | 15.30 | 16.58 | 17.85 | 19.13 | 20.40 | 21.68 | 22.95 | 24.23 | 25.50 | 26.78 | 28.05 | 29.33 | 30.66 |
| 13/16 | 13.12 | 13.81 | 14.50 | 15.19 | 15.88 | 16.58 | 17.96 | 19.34 | 20.72 | 22.10 | 23.48 | 24.86 | 26.26 | 27.63 | 29.01 | 30.39 | 31.77 | 33.15 |
| 7/8 | 14.13 | 14.88 | 15.62 | 16.36 | 17.11 | 17.85 | 19.34 | 20.83 | 22.31 | 23.80 | 25.29 | 26.78 | 28.26 | 29.75 | 31.24 | 32.73 | 34.21 | 35.70 |
| 15/16 | 15.14 | 15.94 | 16.73 | 17.53 | 18.33 | 19.13 | 20.72 | 22.31 | 23.91 | 25.50 | 27.09 | 28.69 | 30.28 | 31.88 | 33.47 | 35.06 | 36.66 | 38.25 |
| 1 | 16.15 | 17.00 | 17.85 | 18.70 | 19.55 | 20.40 | 22.10 | 23.80 | 25.50 | 27.20 | 28.90 | 30.60 | 32.30 | 34.00 | 35.70 | 37.40 | 39.10 | 40.80 |
| 1 1/16 | 17.16 | 18.06 | 18.97 | 19.87 | 20.77 | 21.68 | 23.48 | 25.29 | 27.09 | 28.90 | 30.71 | 32.51 | 34.32 | 36.13 | 37.93 | 39.74 | 41.54 | 43.35 |
| 1 1/8 | 18.17 | 19.13 | 20.08 | 21.04 | 21.99 | 22.95 | 24.86 | 26.78 | 28.69 | 30.60 | 32.51 | 34.43 | 36.34 | 38.25 | 40.16 | 42.08 | 43.99 | 45.90 |
| 1 3/16 | 19.18 | 20.19 | 21.20 | 22.21 | 23.22 | 24.23 | 26.24 | 28.26 | 30.28 | 32.30 | 34.32 | 36.34 | 38.36 | 40.38 | 42.39 | 44.41 | 46.43 | 48.45 |
| 1 1/4 | 20.19 | 21.25 | 22.31 | 23.38 | 24.44 | 25.50 | 27.63 | 29.75 | 31.88 | 34.00 | 36.13 | 38.25 | 40.38 | 42.50 | 44.63 | 46.75 | 48.88 | 51.00 |
| 1 5/16 | 21.20 | 22.31 | 23.43 | 24.54 | 25.66 | 26.78 | 29.01 | 31.24 | 33.47 | 35.70 | 37.93 | 40.16 | 42.39 | 44.63 | 46.86 | 49.09 | 51.32 | 53.55 |
| 1 3/8 | 22.21 | 23.38 | 24.54 | 25.71 | 26.88 | 28.05 | 30.39 | 32.73 | 35.06 | 37.40 | 39.74 | 42.08 | 44.41 | 46.75 | 49.09 | 51.43 | 53.76 | 56.10 |
| 1 7/16 | 23.22 | 24.44 | 25.66 | 26.88 | 28.10 | 29.33 | 31.77 | 34.21 | 36.66 | 39.10 | 41.54 | 43.99 | 46.43 | 48.88 | 51.32 | 53.76 | 56.21 | 58.65 |
| 1 1/2 | 24.23 | 25.50 | 26.78 | 28.05 | 29.33 | 30.60 | 33.15 | 35.70 | 38.25 | 40.80 | 43.35 | 45.90 | 48.45 | 51.00 | 53.55 | 56.10 | 58.65 | 61.20 |
| 1 9/16 | 25.24 | 26.56 | 27.89 | 29.22 | 30.55 | 31.88 | 34.53 | 37.19 | 39.84 | 42.50 | 45.16 | 47.81 | 50.47 | 53.13 | 55.78 | 58.44 | 61.09 | 63.75 |
| 1 5/8 | 26.25 | 27.63 | 29.01 | 30.39 | 31.77 | 33.15 | 35.91 | 38.68 | 41.44 | 44.20 | 46.96 | 49.73 | 52.49 | 55.25 | 58.01 | 60.78 | 63.54 | 66.30 |
| 1 11/16 | 27.25 | 28.69 | 30.12 | 31.56 | 32.99 | 34.43 | 37.29 | 40.16 | 43.03 | 45.90 | 48.77 | 51.64 | 54.51 | 57.38 | 60.24 | 63.11 | 65.98 | 68.85 |
| 1 3/4 | 28.26 | 29.75 | 31.24 | 32.73 | 34.21 | 35.70 | 38.68 | 41.65 | 44.63 | 47.60 | 50.58 | 53.55 | 56.53 | 59.50 | 62.48 | 65.45 | 68.43 | 71.40 |
| 1 13/16 | 29.27 | 30.81 | 32.35 | 33.89 | 35.43 | 36.98 | 40.06 | 43.14 | 46.22 | 49.30 | 52.38 | 55.46 | 58.54 | 61.63 | 64.71 | 67.79 | 70.87 | 73.95 |
| 1 7/8 | 30.28 | 31.88 | 33.47 | 35.06 | 36.66 | 38.25 | 41.44 | 44.63 | 47.81 | 51.00 | 54.19 | 57.38 | 60.56 | 63.75 | 66.94 | 70.13 | 73.31 | 76.50 |
| 1 15/16 | 31.29 | 32.94 | 34.58 | 36.23 | 37.88 | 39.53 | 42.82 | 46.11 | 49.41 | 52.70 | 55.99 | 59.29 | 62.58 | 65.88 | 69.17 | 72.46 | 75.76 | 79.05 |
| 2 | 32.30 | 34.00 | 35.70 | 37.40 | 39.10 | 40.80 | 44.20 | 47.60 | 51.00 | 54.40 | 57.80 | 61.20 | 64.60 | 68.00 | 71.40 | 74.80 | 78.20 | 81.66 |

## Areas and Weights of Fillets of Steel, Cast Iron and Brass

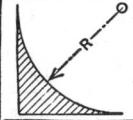

Calculations are based on the following weights:

Steel.........489.6 pounds per cubic foot.
Cast iron.....450 pounds per cubic foot.
Cast brass....504 pounds per cubic foot.

| Radius R, Inches | Area, Square Inches | Weight of Steel | | Weight of Cast Iron | | Weight of Cast Brass | |
|---|---|---|---|---|---|---|---|
| | | Per Foot | Per Inch | Per Foot | Per Inch | Per Foot | Per Inch |
| ¼ | 0.0134 | 0.0455 | 0.0038 | 0.0418 | 0.0035 | 0.0469 | 0.0040 |
| 5/16 | 0.0209 | 0.0712 | 0.0059 | 0.0655 | 0.0054 | 0.0733 | 0.0061 |
| ⅜ | 0.0302 | 0.1027 | 0.0085 | 0.0945 | 0.0078 | 0.1058 | 0.0088 |
| 7/16 | 0.0411 | 0.1397 | 0.0116 | 0.1285 | 0.0107 | 0.1439 | 0.0120 |
| ½ | 0.0536 | 0.1825 | 0.0152 | 0.1679 | 0.0140 | 0.1880 | 0.0157 |
| 9/16 | 0.0679 | 0.2310 | 0.0192 | 0.2125 | 0.0177 | 0.2380 | 0.0200 |
| ⅝ | 0.0834 | 0.2847 | 0.0237 | 0.2619 | 0.0218 | 0.2932 | 0.0244 |
| 11/16 | 0.1014 | 0.3447 | 0.0287 | 0.3171 | 0.0264 | 0.3550 | 0.0300 |
| ¾ | 0.1207 | 0.4105 | 0.0342 | 0.3777 | 0.0315 | 0.4228 | 0.0352 |
| 13/16 | 0.1416 | 0.4817 | 0.0401 | 0.4432 | 0.0369 | 0.4962 | 0.0414 |
| ⅞ | 0.1643 | 0.5580 | 0.0465 | 0.5134 | 0.0428 | 0.5747 | 0.0479 |
| 15/16 | 0.1886 | 0.6405 | 0.0534 | 0.5893 | 0.0491 | 0.6597 | 0.0550 |
| 1 | 0.2146 | 0.7300 | 0.0608 | 0.6716 | 0.0559 | 0.7519 | 0.0626 |
| 1⅛ | 0.2716 | 0.9250 | 0.0771 | 0.8510 | 0.0709 | 0.9527 | 0.0794 |
| 1¼ | 0.3353 | 1.140 | 0.0950 | 1.049 | 0.0874 | 1.174 | 0.0979 |
| 1⅜ | 0.4057 | 1.379 | 0.1150 | 1.268 | 0.1057 | 1.420 | 0.1183 |
| 1½ | 0.4828 | 1.642 | 0.1368 | 1.511 | 0.1259 | 1.691 | 0.1410 |
| 1⅝ | 0.5668 | 1.930 | 0.1608 | 1.776 | 0.1479 | 1.988 | 0.1657 |
| 1¾ | 0.6572 | 2.235 | 0.1862 | 2.056 | 0.1713 | 2.302 | 0.1920 |
| 1⅞ | 0.7545 | 2.565 | 0.2137 | 2.360 | 0.1970 | 2.642 | 0.2202 |
| 2 | 0.8585 | 2.917 | 0.2431 | 2.684 | 0.2237 | 3.005 | 0.2504 |
| 2⅛ | 0.9692 | 3.292 | 0.2743 | 3.029 | 0.2502 | 3.391 | 0.2826 |
| 2¼ | 1.086 | 3.695 | 0.3079 | 3.399 | 0.2832 | 3.806 | 0.3172 |
| 2⅜ | 1.210 | 4.115 | 0.3429 | 3.786 | 0.3155 | 4.238 | 0.3532 |
| 2½ | 1.341 | 4.560 | 0.3800 | 4.195 | 0.3496 | 4.697 | 0.3914 |
| 2⅝ | 1.478 | 5.030 | 0.4192 | 4.628 | 0.3857 | 5.181 | 0.4317 |
| 2¾ | 1.623 | 5.507 | 0.4589 | 5.066 | 0.4222 | 5.672 | 0.4727 |
| 2⅞ | 1.774 | 6.027 | 0.5022 | 5.545 | 0.4621 | 6.208 | 0.5017 |
| 3 | 1.931 | 6.565 | 0.5471 | 5.940 | 0.4950 | 6.762 | 0.5635 |
| 3⅛ | 2.096 | 7.125 | 0.5937 | 6.555 | 0.5462 | 7.339 | 0.6116 |
| 3¼ | 2.267 | 7.700 | 0.6417 | 7.084 | 0.5903 | 7.931 | 0.6609 |
| 3⅜ | 2.444 | 8.300 | 0.6917 | 7.636 | 0.6363 | 8.549 | 0.7124 |
| 3½ | 2.629 | 8.925 | 0.7438 | 8.211 | 0.6926 | 9.193 | 0.7661 |
| 3⅝ | 2.820 | 9.575 | 0.7979 | 8.809 | 0.7341 | 9.862 | 0.8220 |
| 3¾ | 3.018 | 10.27 | 0.8523 | 9.448 | 0.7873 | 10.58 | 0.8817 |
| 3⅞ | 3.222 | 10.97 | 0.9142 | 10.09 | 0.8408 | 11.30 | 0.9417 |
| 4 | 3.434 | 11.65 | 0.9709 | 10.72 | 0.8933 | 12.00 | 1.000 |
| 4¼ | 3.876 | 13.15 | 1.096 | 12.10 | 1.008 | 13.54 | 1.130 |
| 4½ | 4.346 | 14.77 | 1.231 | 13.59 | 1.132 | 15.21 | 1.270 |
| 4¾ | 4.842 | 16.45 | 1.371 | 15.13 | 1.261 | 16.94 | 1.412 |
| 5 | 5.365 | 18.25 | 1.521 | 16.79 | 1.400 | 18.80 | 1.570 |

### Tin Plate Base Weight and Thickness

| Base * Weight, Lbs., and Symbols | Weight per Sq. Ft. | Approx. Thickness, Inch | Base Weight, Lbs., and Symbols | Weight per Sq. Ft. | Approx. Thickness, Inch | Base Weight, Lbs., and Symbols | Weight per Sq. Ft. | Approx. Thickness, Inch |
|---|---|---|---|---|---|---|---|---|
| 55 | 0.253 | 0.006 | 128–IXL | 0.588 | 0.015 | 255–7 X | 1.125 | 0.028 |
| 60 | 0.276 | 0.007 | 135–IX | 0.620 | 0.015 | 268–D 4 X | 1.230 | 0.031 |
| 65 | 0.298 | 0.007 | 139–DC | 0.638 | 0.016 | 275–8 X | 1.263 | 0.032 |
| 70 | 0.321 | 0.008 | 155–2 X | 0.712 | 0.018 | 295–9 X | 1.355 | 0.034 |
| 75 | 0.344 | 0.009 | 175–3 X | 0.804 | 0.020 | 315–10 X | 1.447 | 0.036 |
| 80 | 0.367 | 0.009 | 180–DX | 0.827 | 0.021 | 335–11 X | 1.539 | 0.038 |
| 85 | 0.390 | 0.010 | 195–4 X | 0.895 | 0.022 | 355–12 X | 1.631 | 0.041 |
| 90 | 0.413 | 0.010 | 210–D 2 X | 0.964 | 0.024 | 375–13 X | 1.722 | 0.043 |
| 95 | 0.436 | 0.011 | 215–5 X | 0.988 | 0.025 | 395–14 X | 1.814 | 0.045 |
| 100–ICL | 0.459 | 0.011 | 235–6 X | 1.08 | 0.027 | 415–15 X | 1.906 | 0.048 |
| 107–IC | 0.491 | 0.012 | 240–D 3 X | 1.10 | 0.027 | 435–16 X | 1.998 | 0.050 |

* Weight of standard " base box " containing 112 sheets, 14 X 20 inches.

### Sheet Zinc Gage
(Matthiessen & Hegeler Zinc Co.)

| Gage No. | Thickness, Inches | Gage No. | Thickness, Inches | Gage No. | Thickness, Inches | Gage No. | Thickness, Inches |
|---|---|---|---|---|---|---|---|
| 1 | 0.002 | 8 | 0.016 | 15 | 0.040 | 22 | 0.090 |
| 2 | 0.004 | 9 | 0.018 | 16 | 0.045 | 23 | 0.100 |
| 3 | 0.006 | 10 | 0.020 | 17 | 0.050 | 24 | 0.125 |
| 4 | 0.008 | 11 | 0.024 | 18 | 0.055 | 25 | 0.250 |
| 5 | 0.010 | 12 | 0.028 | 19 | 0.060 | 26 | 0.375 |
| 6 | 0.012 | 13 | 0.032 | 20 | 0.070 | 27 | 0.500 |
| 7 | 0.014 | 14 | 0.036 | 21 | 0.080 | 28 | 1.000 |

### American "Russia-Iron" Gage

| Gage No. | Thickness, Ins. | Gage No. | Thickness, Ins. | Gage No. | Thickness, Ins. | Gage No. | Thickness, Ins. | Gage No. | Thickness, Ins. |
|---|---|---|---|---|---|---|---|---|---|
| 7 | 0.015 | 9 | 0.017 | 11 | 0.020 | 13 | 0.024 | 15 | 0.027 |
| 8 | 0.016 | 10 | 0.018 | 12 | 0.021 | 14 | 0.025 | 16 | 0.030 |

### Weight in Pounds per Square Foot of Zinc Plate

| Gage No. | Weight in Pounds per Sq. Foot | Gage No. | Weight in Pounds per Sq. Foot | Gage No. | Weight in Pounds per Sq. Foot | Gage No. | Weight in Pounds per Sq. Foot |
|---|---|---|---|---|---|---|---|
| 1 | 0.07 | 8 | 0.60 | 15 | 1.50 | 22 | 3.37 |
| 2 | 0.15 | 9 | 0.67 | 16 | 1.68 | 23 | 3.75 |
| 3 | 0.22 | 10 | 0.75 | 17 | 1.87 | 24 | 4.70 |
| 4 | 0.30 | 11 | 0.90 | 18 | 2.06 | 25 | 9.40 |
| 5 | 0.37 | 12 | 1.05 | 19 | 2.25 | 26 | 14.00 |
| 6 | 0.45 | 13 | 1.20 | 20 | 2.62 | 27 | 18.75 |
| 7 | 0.52 | 14 | 1.35 | 21 | 3.00 | 28 | 37.50 |

**Table giving Number of Pieces in One Pound, when Weight of One Hundred Pieces is Known**

*Example:* — 100 pieces weigh 11 pounds 5 ounces. From the table, there are then 8.84 pieces in one pound.

| Ounces | Pounds | | | | | | | | |
|---|---|---|---|---|---|---|---|---|---|
| | 0 | 1 | 2 | 3 | 4 | 5 | 6 | 7 | 8 |
| 0 | ...... | 100.00 | 50.00 | 33.33 | 25.00 | 20.00 | 16.67 | 14.29 | 12.50 |
| 1 | 1600.00 | 94.12 | 48.48 | 32.65 | 24.61 | 19.75 | 16.49 | 14.16 | 12.40 |
| 2 | 800.00 | 88.88 | 47.06 | 32.00 | 24.24 | 19.51 | 16.33 | 14.03 | 12.31 |
| 3 | 533.33 | 84.21 | 45.71 | 31.37 | 23.88 | 19.27 | 16.16 | 13.91 | 12.21 |
| 4 | 400.00 | 80.00 | 44.44 | 30.77 | 23.52 | 19.05 | 16.00 | 13.79 | 12.12 |
| 5 | 320.00 | 76.19 | 43.24 | 30.19 | 23.19 | 18.82 | 15.84 | 13.67 | 12.03 |
| 6 | 266.66 | 72.73 | 42.11 | 29.63 | 22.86 | 18.60 | 15.69 | 13.56 | 11.94 |
| 7 | 228.57 | 69.57 | 41.03 | 29.09 | 22.53 | 18.39 | 15.53 | 13.44 | 11.85 |
| 8 | 200.00 | 66.67 | 40.00 | 28.57 | 22.22 | 18.18 | 15.38 | 13.33 | 11.76 |
| 9 | 177.78 | 64.00 | 39.02 | 28.07 | 21.92 | 17.98 | 15.24 | 13.22 | 11.68 |
| 10 | 160.00 | 61.54 | 38.09 | 27.58 | 21.62 | 17.78 | 15.09 | 13.11 | 11.59 |
| 11 | 145.45 | 59.26 | 37.21 | 27.12 | 21.33 | 17.58 | 14.95 | 13.01 | 11.51 |
| 12 | 133.33 | 57.14 | 36.36 | 26.67 | 21.05 | 17.39 | 14.81 | 12.90 | 11.43 |
| 13 | 123.08 | 55.17 | 35.56 | 26.23 | 20.78 | 17.20 | 14.68 | 12.80 | 11.35 |
| 14 | 114.29 | 53.33 | 34.78 | 25.81 | 20.51 | 17.02 | 14.54 | 12.70 | 11.27 |
| 15 | 106.67 | 51.61 | 34.04 | 25.40 | 20.25 | 16.84 | 14.41 | 12.60 | 11.19 |

| Ounces | Pounds | | | | | | | | |
|---|---|---|---|---|---|---|---|---|---|
| | 9 | 10 | 11 | 12 | 13 | 14 | 15 | 16 | 17 |
| 0 | 11.11 | 10.00 | 9.09 | 8.33 | 7.69 | 7.14 | 6.66 | 6.25 | 5.88 |
| 1 | 11.03 | 9.94 | 9.04 | 8.29 | 7.65 | 7.11 | 6.64 | 6.23 | 5.86 |
| 2 | 10.96 | 9.88 | 8.99 | 8.25 | 7.62 | 7.08 | 6.61 | 6.20 | 5.84 |
| 3 | 10.89 | 9.82 | 8.94 | 8.20 | 7.58 | 7.04 | 6.59 | 6.17 | 5.82 |
| 4 | 10.81 | 9.76 | 8.89 | 8.16 | 7.54 | 7.01 | 6.56 | 6.15 | 5.80 |
| 5 | 10.74 | 9.69 | 8.84 | 8.12 | 7.51 | 6.98 | 6.53 | 6.13 | 5.78 |
| 6 | 10.67 | 9.64 | 8.79 | 8.08 | 7.47 | 6.95 | 6.50 | 6.11 | 5.76 |
| 7 | 10.59 | 9.58 | 8.74 | 8.04 | 7.44 | 6.92 | 6.47 | 6.08 | 5.74 |
| 8 | 10.53 | 9.52 | 8.69 | 8.00 | 7.41 | 6.89 | 6.45 | 6.06 | 5.72 |
| 9 | 10.46 | 9.47 | 8.65 | 7.96 | 7.37 | 6.86 | 6.43 | 6.04 | 5.70 |
| 10 | 10.39 | 9.41 | 8.60 | 7.92 | 7.34 | 6.84 | 6.40 | 6.01 | 5.68 |
| 11 | 10.32 | 9.36 | 8.56 | 7.88 | 7.31 | 6.81 | 6.37 | 5.98 | 5.66 |
| 12 | 10.25 | 9.30 | 8.51 | 7.84 | 7.27 | 6.78 | 6.35 | 5.96 | 5.64 |
| 13 | 10.19 | 9.25 | 8.46 | 7.80 | 7.24 | 6.75 | 6.32 | 5.94 | 5.62 |
| 14 | 10.13 | 9.19 | 8.42 | 7.76 | 7.21 | 6.72 | 6.30 | 5.92 | 5.60 |
| 15 | 10.06 | 9.14 | 8.38 | 7.73 | 7.17 | 6.69 | 6.27 | 5.90 | 5.58 |

| Ounces | Pounds | | | | | | | | |
|---|---|---|---|---|---|---|---|---|---|
| | 18 | 19 | 20 | 21 | 22 | 23 | 24 | 25 | 26 |
| 0 | 5.56 | 5.26 | 5.00 | 4.76 | 4.54 | 4.35 | 4.16 | 4.00 | 3.84 |
| 1 | 5.54 | 5.24 | 4.98 | 4.74 | 4.53 | 4.34 | 4.15 | 3.99 | 3.83 |
| 2 | 5.52 | 5.23 | 4.96 | 4.73 | 4.52 | 4.33 | 4.14 | 3.98 | 3.82 |
| 3 | 5.50 | 5.21 | 4.95 | 4.71 | 4.51 | 4.32 | 4.13 | 3.97 | 3.81 |
| 4 | 5.48 | 5.19 | 4.94 | 4.70 | 4.50 | 4.30 | 4.12 | 3.96 | 3.81 |
| 5 | 5.46 | 5.18 | 4.92 | 4.69 | 4.49 | 4.29 | 4.11 | 3.95 | 3.80 |
| 6 | 5.44 | 5.16 | 4.90 | 4.68 | 4.48 | 4.28 | 4.10 | 3.94 | 3.79 |
| 7 | 5.42 | 5.14 | 4.89 | 4.66 | 4.46 | 4.27 | 4.09 | 3.93 | 3.78 |
| 8 | 5.40 | 5.12 | 4.88 | 4.65 | 4.44 | 4.25 | 4.08 | 3.92 | 3.77 |
| 9 | 5.38 | 5.11 | 4.86 | 4.64 | 4.43 | 4.24 | 4.07 | 3.91 | 3.76 |
| 10 | 5.36 | 5.09 | 4.84 | 4.63 | 4.42 | 4.23 | 4.06 | 3.90 | 3.75 |
| 11 | 5.34 | 5.08 | 4.83 | 4.61 | 4.41 | 4.22 | 4.05 | 3.89 | 3.74 |
| 12 | 5.32 | 5.07 | 4.82 | 4.60 | 4.39 | 4.21 | 4.04 | 3.88 | 3.73 |
| 13 | 5.31 | 5.05 | 4.81 | 4.59 | 4.38 | 4.19 | 4.03 | 3.87 | 3.73 |
| 14 | 5.30 | 5.04 | 4.79 | 4.57 | 4.37 | 4.18 | 4.02 | 3.86 | 3.72 |
| 15 | 5.28 | 5.02 | 4.78 | 4.56 | 4.36 | 4.17 | 4.01 | 3.85 | 3.71 |

# THERMAL ENERGY OR HEAT

**Thermometer Scales.** — There are two thermometer scales in general use: the Fahrenheit (F), which is used in the United States and in other countries still using the English system of units, and the Celsius (C) or Centrigrade used throughout the rest of the world.

In the Fahrenheit thermometer, the freezing point of water is marked at 32 degrees on the scale and the boiling point, at atmospheric pressure, at 212 degrees. The distance between these two points is divided into 180 degrees. On the Celsius scale, the freezing point of water is at 0 degrees and the boiling point at 100 degrees. The following formulas may be used for converting temperatures given on any one of the scales to the other scale:

$$\text{Degrees Fahrenheit} = \frac{9 \times \text{degrees C}}{5} + 32$$

$$\text{Degrees Celsius} = \frac{5 \times (\text{degrees F} - 32)}{9}$$

Tables appear on the pages which follow that can be used to convert degrees Celsius into degrees Fahrenheit or vice versa. In the event that the conversions are not covered in the tables use those applicable portions of the formulas given above for converting.

**Absolute Temperature and Absolute Zero.** — A point has been determined on the thermometer scale, by theoretical considerations, which is called the absolute zero and beyond which a further decrease in temperature is inconceivable. This point is located at − 273.2 degrees Celsius or − 459.7 degrees F. A temperature reckoned from this point, instead of from the zero on the ordinary thermometers, is called absolute temperature. Absolute temperature in degrees C is known as "degrees Kelvin" or the "Kelvin scale" (K) and absolute temperature in degrees F is known as "degrees Rankine" or the "Rankine scale" (R).

$$\text{Degrees Kelvin} = \text{degrees C} + 273.2$$

$$\text{Degrees Rankine} = \text{degrees F} + 459.7$$

**Measures of the Quantity of Thermal Energy.** — The unit of quantity of thermal energy used in the United States is the British thermal unit, which is the quantity of heat or thermal energy required to raise the temperature of one pound of pure water one degree F. (American National Standard abbreviation, Btu; conventional British symbol, B.Th.U.) The French thermal unit or *kilogram calorie*, is the quantity of heat or thermal energy required to raise the temperature of one kilogram of pure water one degree C. One kilogram calorie = 3.968 British thermal units = 1000 gram calories. The number of foot-pounds of mechanical energy equivalent to one British thermal unit is called the *mechanical equivalent of heat*, and equals 778 foot-pounds.

In the modern metric or SI system of units, the unit for thermal energy is the *joule* (J); a commonly used multiple being the kilojoule (kJ) or 1000 joules. See page 2356 for an explanation of the SI System. One kilojoule = 0.9478 Btu. Also in the SI System, the *watt* (W), equal to joule per second (J/s), is used for power, where one watt = 3.412 Btu per hour.

**Fahrenheit — Celsius (Centrigrade) Conversion.** — A simple way to convert a Fahrenheit temperature reading into a Celsius temperature reading or vice versa is to enter the accompanying table in the center or boldface column of figures. These figures refer to the temperature in either Fahrenheit of Celsius degrees. If it is desired to convert from Fahrenheit to Celsius degrees, consider the center column as a table of Fahrenheit temperatures and read the corresponding Celsius temperature in the column at the left. If it is desired to convert from Celsius to Fahrenheit degrees, consider the center column as a table of Celsius values, and read the corresponding Fahrenheit temperature on the right.

**Interpolation Factors**

| deg C | | deg F | deg C | | deg F |
|---|---|---|---|---|---|
| 0.56 | 1 | 1.8 | 3.33 | 6 | 10.8 |
| 1.11 | 2 | 3.6 | 3.89 | 7 | 12.6 |
| 1.67 | 3 | 5.4 | 4.44 | 8 | 14.4 |
| 2.22 | 4 | 7.2 | 5.00 | 9 | 16.2 |
| 2.78 | 5 | 9.0 | 5.56 | 10 | 18.0 |

Interpolation factors are given for use with that portion of the table in which the center column advances in increments of 10. To illustrate, suppose it is desired to find the Fahrenheit equivalent of 314 degrees C. The equivalent of 310 degrees C, found in the body of the main table, is seen to be 590.0 degrees F. The Fahrenheit equivalent of a 4-degree C difference is seen to be 7.2, as read in the table of interpolating factors. The answer is the sum or 597.2 degrees F.

**Fahrenheit — Celsius (Centigrade) Conversion Table**

| deg C | | deg F | deg C | | deg F | deg C | | deg F | deg C | | deg.F |
|---|---|---|---|---|---|---|---|---|---|---|---|
| −273 | −459.4 | ... | −101 | −150 | −238 | − 8.3 | 17 | 62.6 | 9.4 | 49 | 120.2 |
| −268 | −450 | ... | − 96 | −140 | −220 | − 7.8 | 18 | 64.4 | 10.0 | 50 | 122.0 |
| −262 | −440 | ... | − 90 | −130 | −202 | − 7.2 | 19 | 66.2 | 10.6 | 51 | 123.8 |
| −257 | −430 | ... | − 84 | −120 | −184 | − 6.7 | 20 | 68.0 | 11.1 | 52 | 125.6 |
| −251 | −420 | ... | − 79 | −110 | −166 | − 6.1 | 21 | 69.8 | 11.7 | 53 | 127.4 |
| −246 | −410 | ... | − 73 | −100 | −148 | − 5.6 | 22 | 71.6 | 12.2 | 54 | 129.2 |
| −240 | −400 | ... | − 68 | − 90 | −130 | − 5.0 | 23 | 73.4 | 12.8 | 55 | 131.0 |
| −234 | −390 | ... | − 62 | − 80 | −112 | − 4.4 | 24 | 75.2 | 13.3 | 56 | 132.8 |
| −229 | −380 | ... | − 57 | − 70 | − 94 | − 3.9 | 25 | 77.0 | 13.9 | 57 | 134.6 |
| −223 | −370 | ... | − 51 | − 60 | − 76 | − 3.3 | 26 | 78.8 | 14.4 | 58 | 136.4 |
| −218 | −360 | ... | −46 | −50 | −58 | − 2.8 | 27 | 80.6 | 15.0 | 59 | 138.2 |
| −212 | −350 | ... | −40 | −40 | −40 | − 2.2 | 28 | 82.4 | 15.6 | 60 | 140.0 |
| −207 | −340 | ... | −34 | −30 | −22 | − 1.7 | 29 | 84.2 | 16.1 | 61 | 141.8 |
| −201 | −330 | ... | −29 | −20 | − 4 | − 1.1 | 30 | 86.0 | 16.7 | 62 | 143.6 |
| −196 | −320 | ... | −23 | −10 | 14 | − 0.6 | 31 | 87.8 | 17.2 | 63 | 145.4 |
| −190 | −310 | ... | −17.8 | 0 | 32 | 0— | 32 | 89.6 | 17.8 | 64 | 147.2 |
| −184 | −300 | ... | −17.2 | 1 | 33.8 | 0.6 | 33 | 91.4 | 18.3 | 65 | 149.0 |
| −179 | −290 | ... | −16.7 | 2 | 35.6 | 1.1 | 34 | 93.2 | 18.9 | 66 | 150.8 |
| −173 | −280 | ... | −16.1 | 3 | 37.4 | 1.7 | 35 | 95.0 | 19.4 | 67 | 152.6 |
| −169 | −273 | −459.4 | −15.6 | 4 | 39.2 | 2.2 | 36 | 96.8 | 20.0 | 68 | 154.4 |
| −168 | −270 | −454 | −15.0 | 5 | 41.0 | 2.7 | 37 | 98.6 | 20.6 | 69 | 156.2 |
| −162 | −260 | −436 | −14.4 | 6 | 42.8 | 3.3 | 38 | 100.4 | 21.1 | 70 | 158.0 |
| −157 | −250 | −418 | −13.9 | 7 | 44.6 | 3.9 | 39 | 102.2 | 21.7 | 71 | 159.8 |
| −151 | −240 | −400 | −13.3 | 8 | 46.4 | 4.4 | 40 | 104.0 | 22.2 | 72 | 161.6 |
| −146 | −230 | −382 | −12.8 | 9 | 48.2 | 5.0 | 41 | 105.8 | 22.8 | 73 | 163.4 |
| −140 | −220 | −364 | −12.2 | 10 | 50.0 | 5.6 | 42 | 107.6 | 23.3 | 74 | 165.2 |
| −134 | −210 | −346 | −11.7 | 11 | 51.8 | 6.1 | 43 | 109.4 | 23.9 | 75 | 167.0 |
| −129 | −200 | −328 | −11.1 | 12 | 53.6 | 6.7 | 44 | 111.2 | 24.4 | 76 | 168.8 |
| −123 | −190 | −310 | −10.6 | 13 | 55.4 | 7.2 | 45 | 113.0 | 25.0 | 77 | 170.6 |
| −118 | −180 | −292 | −10.0 | 14 | 57.2 | 7.8 | 46 | 114.8 | 25.6 | 78 | 172.4 |
| −112 | −170 | −274 | − 9.4 | 15 | 59.0 | 8.3 | 47 | 116.6 | 26.1 | 79 | 174.2 |
| −107 | −160 | −256 | − 8.9 | 16 | 60.8 | 8.9 | 48 | 118.4 | 26.7 | 80 | 176.0 |

## Fahrenheit — Celsius (Centigrade) Conversion Table (*Continued*)

| deg C | | deg F | deg C | | deg F | deg C | | deg F | deg C | | deg F |
|---|---|---|---|---|---|---|---|---|---|---|---|
| 27.2 | 81 | 177.8 | 58.3 | 137 | 278.6 | 89.4 | 193 | 379.4 | 304.4 | 580 | 1076 |
| 27.8 | 82 | 179.6 | 58.9 | 138 | 280.4 | 90.0 | 194 | 381.2 | 310.0 | 590 | 1094 |
| 28.3 | 83 | 181.4 | 59.4 | 139 | 282.2 | 90.6 | 195 | 383.0 | 315.6 | 600 | 1112 |
| 28.9 | 84 | 183.2 | 60.0 | 140 | 284.0 | 91.1 | 196 | 384.8 | 321.1 | 610 | 1130 |
| 29.4 | 85 | 185.0 | 60.6 | 141 | 285.8 | 91.7 | 197 | 386.6 | 326.7 | 620 | 1148 |
| 30.0 | 86 | 186.8 | 61.1 | 142 | 287.6 | 92.2 | 198 | 388.4 | 332.2 | 630 | 1166 |
| 30.6 | 87 | 188.6 | 61.7 | 143 | 289.4 | 92.8 | 199 | 390.2 | 337.8 | 640 | 1184 |
| 31.1 | 88 | 190.4 | 62.2 | 144 | 291.2 | 93.3 | 200 | 392.0 | 343.3 | 650 | 1202 |
| 31.7 | 89 | 192.2 | 62.8 | 145 | 293.0 | 93.9 | 201 | 393.8 | 348.9 | 660 | 1220 |
| 32.2 | 90 | 194.0 | 63.3 | 146 | 294.8 | 94.4 | 202 | 395.6 | 354.4 | 670 | 1238 |
| 32.8 | 91 | 195.8 | 63.9 | 147 | 296.6 | 95.0 | 203 | 397.4 | 360.0 | 680 | 1256 |
| 33.3 | 92 | 197.6 | 64.4 | 148 | 298.4 | 95.6 | 204 | 399.2 | 365.6 | 690 | 1274 |
| 33.9 | 93 | 199.4 | 65.0 | 149 | 300.2 | 96.1 | 205 | 401.0 | 371.1 | 700 | 1292 |
| 34.4 | 94 | 201.2 | 65.6 | 150 | 302.0 | 96.7 | 206 | 402.8 | 376.7 | 710 | 1310 |
| 35.0 | 95 | 203.0 | 66.1 | 151 | 303.8 | 97.2 | 207 | 404.6 | 382.2 | 720 | 1328 |
| 35.6 | 96 | 204.8 | 66.7 | 152 | 305.6 | 97.8 | 208 | 406.4 | 387.8 | 730 | 1346 |
| 36.1 | 97 | 206.6 | 67.2 | 153 | 307.4 | 98.3 | 209 | 408.2 | 393.3 | 740 | 1364 |
| 36.7 | 98 | 208.4 | 67.8 | 154 | 309.2 | 98.9 | 210 | 410.0 | 398.9 | 750 | 1382 |
| 37.2 | 99 | 210.2 | 68.3 | 155 | 311.0 | 99.4 | 211 | 411.8 | 404.4 | 760 | 1400 |
| 37.8 | 100 | 212.0 | 68.9 | 156 | 312.8 | 100.0 | 212 | 413.6 | 410.0 | 770 | 1418 |
| 38.3 | 101 | 213.8 | 69.4 | 157 | 314.6 | 104.4 | 220 | 428.0 | 415.6 | 780 | 1436 |
| 38.9 | 102 | 215.6 | 70.0 | 158 | 316.4 | 110.0 | 230 | 446.0 | 421.1 | 790 | 1454 |
| 39.4 | 103 | 217.4 | 70.6 | 159 | 318.2 | 115.6 | 240 | 464.0 | 426.7 | 800 | 1472 |
| 40.0 | 104 | 219.2 | 71.1 | 160 | 320.0 | 121.1 | 250 | 482.0 | 432.2 | 810 | 1490 |
| 40.6 | 105 | 221.0 | 71.7 | 161 | 321.8 | 126.7 | 260 | 500.0 | 437.8 | 820 | 1508 |
| 41.1 | 106 | 222.8 | 72.2 | 162 | 323.6 | 132.2 | 270 | 518.0 | 443.3 | 830 | 1526 |
| 41.7 | 107 | 224.6 | 72.8 | 163 | 325.4 | 137.8 | 280 | 536.0 | 448.9 | 840 | 1544 |
| 42.2 | 108 | 226.4 | 73.3 | 164 | 327.2 | 143.3 | 290 | 554.0 | 454.4 | 850 | 1562 |
| 42.8 | 109 | 228.2 | 73.9 | 165 | 329.0 | 148.9 | 300 | 572.0 | 460.0 | 860 | 1580 |
| 43.3 | 110 | 230.0 | 74.4 | 166 | 330.8 | 154.4 | 310 | 590.0 | 465.6 | 870 | 1598 |
| 43.9 | 111 | 231.8 | 75.0 | 167 | 332.6 | 160.0 | 320 | 608.0 | 471.1 | 880 | 1616 |
| 44.4 | 112 | 233.6 | 75.6 | 168 | 334.4 | 165.6 | 330 | 626.0 | 476.7 | 890 | 1634 |
| 45.0 | 113 | 235.4 | 76.1 | 169 | 336.2 | 171.1 | 340 | 644.0 | 482.2 | 900 | 1652 |
| 45.6 | 114 | 237.2 | 76.7 | 170 | 338.0 | 176.7 | 350 | 662.0 | 487.8 | 910 | 1670 |
| 46.1 | 115 | 239.0 | 77.2 | 171 | 339.8 | 182.2 | 360 | 680.0 | 493.3 | 920 | 1688 |
| 46.7 | 116 | 240.8 | 77.8 | 172 | 341.6 | 187.8 | 370 | 698.0 | 498.9 | 930 | 1706 |
| 47.2 | 117 | 242.6 | 78.3 | 173 | 343.4 | 193.3 | 380 | 716.0 | 504.4 | 940 | 1724 |
| 47.8 | 118 | 244.4 | 78.9 | 174 | 345.2 | 198.9 | 390 | 734.0 | 510.0 | 950 | 1742 |
| 48.3 | 119 | 246.2 | 79.4 | 175 | 347.0 | 204.4 | 400 | 752.0 | 515.6 | 960 | 1760 |
| 48.9 | 120 | 248.0 | 80.0 | 176 | 348.8 | 210 | 410 | 770.0 | 521.1 | 970 | 1778 |
| 49.4 | 121 | 249.8 | 80.6 | 177 | 350.6 | 215.6 | 420 | 788 | 526.7 | 980 | 1796 |
| 50.0 | 122 | 251.6 | 81.1 | 178 | 352.4 | 221.1 | 430 | 806 | 532.2 | 990 | 1814 |
| 50.6 | 123 | 253.4 | 81.7 | 179 | 354.2 | 226.7 | 440 | 824 | 537.8 | 1000 | 1832 |
| 51.1 | 124 | 255.2 | 82.2 | 180 | 356.0 | 232.2 | 450 | 842 | 565.6 | 1050 | 1922 |
| 51.7 | 125 | 257.0 | 82.8 | 181 | 357.8 | 237.8 | 460 | 860 | 593.3 | 1100 | 2012 |
| 52.2 | 126 | 258.8 | 83.3 | 182 | 359.6 | 243.3 | 470 | 878 | 621.1 | 1150 | 2102 |
| 52.8 | 127 | 260.6 | 83.9 | 183 | 361.4 | 248.9 | 480 | 896 | 648.9 | 1200 | 2192 |
| 53.3 | 128 | 262.4 | 84.4 | 184 | 363.2 | 254.4 | 490 | 914 | 676.7 | 1250 | 2282 |
| 53.9 | 129 | 264.2 | 85.0 | 185 | 365.0 | 260.0 | 500 | 932 | 704.4 | 1300 | 2372 |
| 54.4 | 130 | 266.0 | 85.6 | 186 | 366.8 | 265.6 | 510 | 950 | 732.2 | 1350 | 2462 |
| 55.0 | 131 | 267.8 | 86.1 | 187 | 368.6 | 271.1 | 520 | 968 | 760.0 | 1400 | 2552 |
| 55.6 | 132 | 269.6 | 86.7 | 188 | 370.4 | 276.7 | 530 | 986 | 787.8 | 1450 | 2642 |
| 56.1 | 133 | 271.4 | 87.2 | 189 | 372.2 | 282.2 | 540 | 1004 | 815.6 | 1500 | 2732 |
| 56.7 | 134 | 273.2 | 87.8 | 190 | 374.0 | 287.8 | 550 | 1022 | 1093.9 | 2000 | 3632 |
| 57.2 | 135 | 275.0 | 88.3 | 191 | 375.8 | 293.3 | 560 | 1040 | 1648.9 | 3000 | 5432 |
| 57.8 | 136 | 276.8 | 88.9 | 192 | 377.6 | 298.9 | 570 | 1058 | 2760.0 | 5000 | 9032 |

Above 1000 in the center column, the table increases in increments of 50. To convert 1462 degrees F to Celsius, for instance, add to the Celsius equivalent of 1400 degrees F ten times the interpolation factor for 6 and the interpolation factor for 2 or 760.0 + 33.3 + 1.11, which equals 794.4.

## Coefficients of Heat Transmission

Heat transmitted, in British thermal units, per second, through metal 1 inch thick, per square inch of surface, for a temperature difference of 1° F.

| Metal | Btu per Second | Metal | Btu per Second | Metal | Btu per Second |
|---|---|---|---|---|---|
| Aluminum | 0.00203 | German silver | 0.00050 | Steel, soft | 0.00062 |
| Antimony | 0.00022 | Iron | 0.00089 | Silver | 0.00610 |
| Brass, yellow | 0.00142 | Lead | 0.00045 | Tin | 0.00084 |
| Brass, red | 0.00157 | Mercury | 0.00011 | Zinc | 0.00170 |
| Copper | 0.00404 | Steel, hard | 0.00034 | | |

## Coefficients of Heat Radiation

Heat radiated, in British thermal units, per square foot of surface per hour, for a temperature difference of 1° F.

| Surface | Btu per Hour | Surface | Btu per Hour |
|---|---|---|---|
| Cast-iron, new | 0.6480 | Sawdust | 0.7215 |
| Cast-iron, rusted | 0.6868 | Sand, fine | 0.7400 |
| Copper, polished | 0.0327 | Silver, polished | 0.0266 |
| Glass | 0.5948 | Tin, polished | 0.0439 |
| Iron, ordinary | 0.5662 | Tinned iron, polished | 0.0858 |
| Iron, sheet-, polished | 0.0920 | Water | 1.0853 |
| Oil | 1.4800 | | |

## Freezing Mixtures

| Mixture | Temperature Change, Degrees F. | |
|---|---|---|
| | From | To |
| Common salt (NaCl), 1 part; snow, 3 parts | 32 | ±0 |
| Common salt (NaCl), 1 part; snow, 1 part | 32 | −0.4 |
| Calcium chloride (CaCl₂), 3 parts; snow, 2 parts | 32 | −27 |
| Calcium chloride (CaCl₂), 2 parts; snow, 1 part | 32 | −44 |
| Sal ammoniac (NH₄Cl), 5 parts; saltpeter (KNO₃), 5 parts; water, 16 parts | 50 | +10 |
| Sal ammoniac (NH₄Cl), 1 part; saltpeter (KNO₃), 1 part; water, 1 part | 46 | −11 |
| Ammonium nitrate (NH₄NO₃), 1 part; water, 1 part | 50 | +3 |
| Potassium hydrate (KOH), 4 parts; snow, 3 parts | 32 | −35 |

**Ignition Temperatures.** — The following temperatures are required to ignite the different substances specified: Phosphorus, transparent, 120 degrees F.; bisulphide of carbon, 300 degrees F.; gun cotton, 430 degrees F.; nitro-glycerine, 490 degrees F.; phosphorus, amorphous, 500 degrees F.; rifle powder, 550 degrees F.; charcoal, 660 degrees F.; dry pine wood, 800 degrees F.; dry oak wood, 900 degrees F.

**Latent Heat.** — When a body changes from the solid to the liquid state or from the liquid to the gaseous state, a certain amount of heat is used to accomplish this change. This heat does not raise the temperature of the body and is called latent heat. When the body changes again from the gaseous to the liquid, or from the liquid to the solid state, this quantity of heat is given out by it. The *latent heat of fusion* is the heat supplied to a solid body at the melting point; this heat is absorbed by the body although its temperature remains nearly stationary during the whole operation of melting. The *latent heat of evaporation* is the heat that must be supplied to a liquid at the boiling point to transform the liquid into a vapor. The latent heat is generally given in British thermal units per pound. When it is said that the latent heat of evaporation of water is 966.6, this means that it takes 966.6 heat units to evaporate one pound of water after it has been raised to the boiling point, 212 degrees F.

Latent Heat of Fusion

| Substance | Btu per Pound | Substance | Btu per Pound | Substance | Btu per Pound |
|---|---|---|---|---|---|
| Bismuth | 22.75 | Paraffine | 63.27 | Sulphur | 16.86 |
| Beeswax | 76.14 | Phosphorus | 9.06 | Tin | 25.65 |
| Cast iron, gray | 41.40 | Lead | 10.00 | Zinc | 50.63 |
| Cast iron, white | 59.40 | Silver | 37.92 | Ice | 144.00 |

Latent Heat of Evaporation

| Liquid | Btu per Pound | Liquid | Btu per Pound | Liquid | Btu per Pound |
|---|---|---|---|---|---|
| Alcohol, ethyl | 371.0 | Bisulphide of carbon | 160.0 | Sulphur dioxide | 164.0 |
| Alcohol, methyl | 481.0 | Ether | 162.8 | Turpentine | 133.0 |
| Ammonia | 529.0 | | | Water | 966.6 |

Boiling Points of Various Substances at Atmospheric Pressure

| Substance | Boiling Point, Degrees F. | Substance | Boiling Point, Degrees F. | Substance | Boiling Point, Degrees F. |
|---|---|---|---|---|---|
| Aniline | 363 | Chloroform | 140 | Saturated brine | 226 |
| Alcohol | 173 | Ether | 100 | Sulphur | 833 |
| Ammonia | −28 | Linseed oil | 597 | Sulphuric acid | 590 |
| Benzine | 176 | Mercury | 676 | Water, pure | 212 |
| Bromine | 145 | Napthaline | 428 | Water, sea | 213.2 |
| Carbon bisulphide | 118 | Nitric acid | 248 | Wood alcohol | 150 |
| | | Oil of turpentine | 315 | | |

**Specific Heat.** — The specific heat of a substance is the ratio of the heat required to raise the temperature of a certain weight of the given substance one degree F. to that required to raise the temperature of the same weight of water one degree. As the specific heat is not constant at all temperatures, it is generally assumed that it is determined by raising the temperature from 62 to 63 degrees F. For most substances, however, it is practically constant for temperatures up to 212 degrees F.

### Average Specific Heats (Btu/lb-F) of Various Substances

| Substance | Specific Heat | Substance | Specific Heat |
|---|---|---|---|
| Alcohol (absolute) | 0.700 | Kerosene | 0.500 |
| Alcohol (density 0.8) | 0.622 | Lead | 0.031 |
| Aluminum | 0.214 | Limestone | 0.217 |
| Antimony | 0.051 | Magnesia | 0.222 |
| Benzine | 0.450 | Marble | 0.210 |
| Brass | 0.094 | Masonry, brick | 0.200 |
| Brickwork | 0.200 | Mercury | 0.033 |
| Cadmium | 0.057 | Naphtha | 0.310 |
| Charcoal | 0.200 | Nickel | 0.109 |
| Chalk | 0.215 | Oil, machine | 0.400 |
| Coal | 0.240 | Oil, olive | 0.350 |
| Coke | 0.203 | Phosphorus | 0.189 |
| Copper, 32° to 212° F. | 0.094 | Platinum | 0.032 |
| Copper, 32° to 572° F. | 0.101 | Quartz | 0.188 |
| Corundum | 0.198 | Sand | 0.195 |
| Ether | 0.503 | Silica | 0.191 |
| Fusel oil | 0.564 | Silver | 0.056 |
| Glass | 0.194 | Soda | 0.231 |
| Gold | 0.031 | Steel, mild | 0.116 |
| Graphite | 0.201 | Steel, high carbon | 0.117 |
| Ice | 0.504 | Stone (generally) | 0.200 |
| Iron, cast | 0.130 | Sulphur | 0.178 |
| Iron, wrought, 32° to 212° F. | 0.110 | Sulphuric acid | 0.330 |
| 32° to 392° F. | 0.115 | Tin | 0.056 |
| 32° to 572° F. | 0.122 | Turpentine | 0.472 |
| 32° to 662° F. | 0.126 | Water | 1.000 |
| Iron, at high temperatures: | | Wood, fir | 0.650 |
| 1382° to 1832° F. | 0.213 | Wood, oak | 0.570 |
| 1750° to 1840° F. | 0.218 | Wood, pine | 0.467 |
| 1920° to 2190° F. | 0.199 | Zinc | 0.095 |

### Specific Heat of Gases (Btu/lb-F)

| Gas | Constant Pressure | Constant Volume | Gas | Constant Pressure | Constant Volume |
|---|---|---|---|---|---|
| Acetic acid | 0.412 | | Chloroform | 0.157 | |
| Air | 0.238 | 0.168 | Hydrogen | 3.409 | 2.412 |
| Alcohol | 0.453 | 0.399 | Nitrogen | 0.244 | 0.173 |
| Ammonia | 0.508 | 0.399 | Oxygen | 0.217 | 0.155 |
| Carbonic acid | 0.217 | 0.171 | Ethylene | 0.404 | 0.332 |
| Carbonic oxide | 0.245 | 0.176 | Steam | 0.480 | 0.346 |
| Chlorine | 0.121 | | | | |

**Heat Loss from Uncovered Steam Pipes.** — The loss of heat from a bare steam or hot water pipe varies with the difference between the temperature inside the pipe and that of the surrounding air. The loss is 2.15 Btu per hour, per square foot of pipe surface, per degree F. of temperature difference when the latter is 100 degrees; for a difference of 200 degrees, the loss is 2.66 Btu; for 300 degrees, 3.26 Btu; for 400 degrees, 4.03 Btu; for 500 degrees, 5.18 Btu. Thus, if the pipe area is 1.18 square feet per foot of length, and the temperature difference 300 degrees F., the loss per hour per foot of length = 1.18 × 300 × 3.26 = 1154 Btu.

### Values of Thermal Conductivity (k) and of Conductance (C) of Common Building and Insulating Materials

| Type of Material | Thickness, in. | k or C* |
|---|---|---|
| **BUILDING** | | |
| Batt: | ... | ... |
| Mineral Fiber | 2 to 2¾ | 0.14 |
| Mineral Fiber | 3 to 3½ | 0.09 |
| Mineral Fiber | 3½ to 6½ | 0.05 |
| Mineral Fiber | 6 to 7 | 0.04 |
| Mineral Fiber | 8½ | 0.03 |
| Block: | ... | ... |
| Cinder | 4 | 0.90 |
| Cinder | 8 | 0.58 |
| Cinder | 12 | 0.53 |
| Block: | ... | ... |
| Concrete | 4 | 1.40 |
| Concrete | 8 | 0.90 |
| Concrete | 12 | 0.78 |
| Board: | ... | ... |
| Asbestos Cement | ¼ | 16.5 |
| Plaster | ½ | 2.22 |
| Plywood | ¾ | 1.07 |
| Brick: | ... | ... |
| Common | 1 | 5.0 |
| Face | 1 | 9.0 |
| Concrete (poured) | 1 | 12.0 |
| Floor: | ... | ... |
| Wood Subfloor | ¾ | 1.06 |
| Hardwood Finish | ¾ | 1.47 |
| Tile | Avg. | 20.0 |
| Glass: | ... | ... |
| Architectural | ... | 10.00 |
| Mortar: | ... | ... |
| Cement | 1 | 5.0 |
| Plaster: | ... | ... |
| Sand | ⅜ | 13.30 |
| Sand and Gypsum | ½ | 11.10 |
| Stucco | 1 | 5.0 |
| Roofing: | ... | ... |
| Asphalt Roll | Avg. | 6.50 |
| Shingle, asb. cem. | Avg. | 4.76 |
| Shingle, asphalt | Avg. | 2.27 |
| Shingle, wood | Avg. | 1.06 |

| Type of Material | Thickness, in. | k or C* | Max. Temp., °F | Density, Lb per cu. ft. | k* |
|---|---|---|---|---|---|
| **BUILDING** (*Continued*) | | | | | |
| Siding: | ... | ... | ... | ... | ... |
| Metal‡ | Avg. | 1.61 | ... | ... | ... |
| Wood, Med. Density | 7/16 | 1.49 | ... | ... | ... |
| Stone: | ... | ... | ... | ... | ... |
| Lime or Sand | 1 | 12.50 | ... | ... | ... |
| Wall Tile: | ... | ... | ... | ... | ... |
| Hollow Clay, 1-Cell | 4 | 0.9 | ... | ... | ... |
| Hollow Clay, 2-Cell | 8 | 0.54 | ... | ... | ... |
| Hollow Clay, 3-Cell | 12 | 0.40 | ... | ... | ... |
| Hollow Gypsum | Avg. | 0.7 | ... | ... | ... |
| **INSULATING** | | | | | |
| Blanket, Mineral Fiber: | ... | ... | ... | ... | ... |
| Felt | ... | ... | 400 | 3 to 8 | 0.26 |
| Rock or Slag | ... | ... | 1200 | 6 to 12 | 0.26† |
| Glass | ... | ... | 350 | 0.65 | 0.33 |
| Blanket, Hairfelt | ... | ... | 350 | 0.65 | 0.31 |
| | ... | ... | 180 | 10 | 0.29 |
| Board, Block and Pipe | ... | ... | ... | ... | ... |
| Insulation: | ... | ... | ... | ... | ... |
| Amosite | ... | ... | 1500 | 15 to 18 | 0.32† |
| Asbestos Paper | ... | ... | 700 | 30 | 0.40† |
| Glass or Slag (for Pipe) | ... | ... | 350 | 3 to 4 | 0.23 |
| Glass or Slag (for Pipe) | ... | ... | 1000 | 10 to 15 | 0.33† |
| Glass, Cellular | ... | ... | 800 | 9 | 0.40 |
| Magnesia (85%) | ... | ... | 600 | 11 to 12 | 0.35† |
| Mineral Fiber | ... | ... | 100 | 15 | 0.29 |
| Polystyrene, Beaded | ... | ... | 170 | 1 | 0.28 |
| Polystyrene, Rigid | ... | ... | 170 | 1.8 | 0.25 |
| Rubber, Rigid Foam | ... | ... | 150 | 4.5 | 0.22 |
| Wood Felt | ... | ... | 180 | 20 | 0.31 |
| Loose Fill: | ... | ... | ... | ... | ... |
| Cellulose | ... | ... | ... | 2.5 to 3 | 0.27 |
| Mineral Fiber | ... | ... | ... | 2 to 5 | 0.28 |
| Perlite | ... | ... | ... | 5 to 8 | 0.37 |
| Silica Aerogel | ... | ... | ... | 7.6 | 0.17 |
| Vermiculite | ... | ... | ... | 7 to 8.2 | 0.47 |
| Mineral Fiber Cement | ... | ... | ... | ... | ... |
| Clay Binder | ... | ... | 1800 | 24 to 30 | 0.49† |
| Hydraulic Binder | ... | ... | 1200 | 30 to 40 | 0.75† |

* Units are in Btu/hr-ft²-°F. Where thickness is given as 1 inch, the value given is thermal conductivity ($k$); for other thicknesses the value given is thermal conductance ($C$). All values are for a test mean temperature of 75°F, except those designated with (†), which are for 100°F.    † See * footnote.    ‡ Over hollowback sheathing.    *Source:* American Society of Heating, Refrigerating and Air-Conditioning Engineers, Inc.: HANDBOOK OF FUNDAMENTALS.

### Linear Expansion of Various Substances Between 32 and 212 Deg. Fahr.

(For linear expansion of metals see "Specific Gravity and Properties of Metals")
Expansion of volume = 3 × linear expansion.

| Substance | Linear Expansion for 1 Deg. Fahr. | Substance | Linear Expansion for 1 Deg. Fahr. |
|---|---|---|---|
| Brick..................... | 0.0000030 | Masonry, brick, from.... | 0.0000026 |
| Cement, Portland....... | 0.0000060 | to...... | 0.0000050 |
| Concrete................ | 0.0000080 | Plaster.................. | 0.0000092 |
| Ebonite................. | 0.0000428 | Porcelain................ | 0.0000020 |
| Glass, thermometer..... | 0.0000050 | Quartz, from............ | 0.0000043 |
| Glass, hard.............. | 0.0000040 | to................ | 0.0000079 |
| Granite.................. | 0.0000044 | Slate.................... | 0.0000058 |
| Marble, from............ | 0.0000031 | Sandstone............... | 0.0000065 |
| to.............. | 0.0000079 | Wood (pine)............. | 0.0000028 |

# PROPERTIES, COMPRESSION AND FLOW OF AIR

**Properties of Air.** — Air is a mechanical mixture composed of 78 per cent, by volume, of nitrogen, 21 per cent of oxygen and 1 per cent of argon. The density of pure air at 32 degrees F and atmospheric pressure (29.92 inches of mercury or 14.70 pounds per square inch) is 0.08073 pound per cubic foot. The volume of a pound of air at the same temperature and pressure is 12.387 cubic feet. The density of air at any other temperature or pressure is:

$$W = \frac{1.325 \times B}{T}$$

in which $W$ = density in pounds per cubic foot; $B$ = height of barometric pressure in inches of mercury; $T$ = absolute temperature Rankine.

**Volume and Weight of Air at Different Temperatures, at Atmospheric Pressure**

| Temperature, Degrees Fahr. | Volume of 1 Pound of Air in Cubic Feet | Density, Pounds per Cubic Foot | Temperature, Degrees Fahr. | Volume of 1 Pound of Air in Cubic Feet | Density, Pounds per Cubic Foot | Temperature, Degrees Fahr. | Volume of 1 Pound of Air in Cubic Feet | Density, Pounds per Cubic Foot |
|---|---|---|---|---|---|---|---|---|
| 0   | 11.57 | 0.0864 | 172 | 15.92 | 0.0628 | 800  | 31.75 | 0.0315 |
| 12  | 11.88 | 0.0842 | 182 | 16.18 | 0.0618 | 900  | 34.25 | 0.0292 |
| 22  | 12.14 | 0.0824 | 192 | 16.42 | 0.0609 | 1000 | 37.31 | 0.0268 |
| 32  | 12.39 | 0.0807 | 202 | 16.67 | 0.0600 | 1100 | 39.37 | 0.0254 |
| 42  | 12.64 | 0.0791 | 212 | 16.92 | 0.0591 | 1200 | 41.84 | 0.0239 |
| 52  | 12.89 | 0.0776 | 230 | 17.39 | 0.0575 | 1300 | 44.44 | 0.0225 |
| 62  | 13.14 | 0.0761 | 250 | 17.89 | 0.0559 | 1400 | 46.95 | 0.0213 |
| 72  | 13.39 | 0.0747 | 275 | 18.52 | 0.0540 | 1500 | 49.51 | 0.0202 |
| 82  | 13.64 | 0.0733 | 300 | 19.16 | 0.0522 | 1600 | 52.08 | 0.0192 |
| 92  | 13.89 | 0.0720 | 325 | 19.76 | 0.0506 | 1700 | 54.64 | 0.0183 |
| 102 | 14.14 | 0.0707 | 350 | 20.41 | 0.0490 | 1800 | 57.14 | 0.0175 |
| 112 | 14.41 | 0.0694 | 375 | 20.96 | 0.0477 | 2000 | 62.11 | 0.0161 |
| 122 | 14.66 | 0.0682 | 400 | 21.69 | 0.0461 | 2200 | 67.11 | 0.0149 |
| 132 | 14.90 | 0.0671 | 450 | 22.94 | 0.0436 | 2400 | 72.46 | 0.0138 |
| 142 | 15.17 | 0.0659 | 500 | 24.21 | 0.0413 | 2600 | 76.92 | 0.0130 |
| 152 | 15.41 | 0.0649 | 600 | 26.60 | 0.0376 | 2800 | 82.64 | 0.0121 |
| 162 | 15.67 | 0.0638 | 700 | 29.59 | 0.0338 | 3000 | 87.72 | 0.0114 |

The absolute zero from which all temperatures must be counted when dealing with the weight and volume of gases is assumed to be −459.7 degrees F. Hence, to obtain the absolute temperature $T$ used in the formula above, add to the temperature observed on a regular Fahrenheit thermometer the value 459.7.

In obtaining the value of $B$, 1 inch of mercury at 32 degrees F may be taken as equal to a pressure of 0.491 pound per square inch.

*Example.* — What would be the weight of a cubic foot of air at atmospheric pressure (29.92 inches of mercury) at 100 degrees F?

$$W = \frac{1.325 \times 29.92}{100 + 459.7} = 0.0708 \text{ pound.}$$

Density of Air at Different Pressures and Temperatures

Gage Pressure, Pounds

Density in Pounds per Cubic Foot

| Temp. of Air, Degrees Fahr. | 0 | 5 | 10 | 20 | 30 | 40 | 50 | 60 | 80 | 100 | 120 | 150 | 200 | 250 | 300 |
|---|---|---|---|---|---|---|---|---|---|---|---|---|---|---|---|
| -20 | 0.0900 | 0.1205 | 0.1515 | 0.2125 | 0.274 | 0.336 | 0.397 | 0.458 | 0.580 | 0.702 | 0.825 | 1.010 | 1.318 | 1.625 | 1.930 |
| -10 | 0.0882 | 0.1184 | 0.1485 | 0.2090 | 0.268 | 0.328 | 0.388 | 0.448 | 0.567 | 0.687 | 0.807 | 0.989 | 1.288 | 1.588 | 1.890 |
| 0 | 0.0864 | 0.1160 | 0.1455 | 0.2040 | 0.263 | 0.321 | 0.380 | 0.438 | 0.555 | 0.672 | 0.790 | 0.968 | 1.260 | 1.553 | 1.850 |
| 10 | 0.0846 | 0.1136 | 0.1425 | 0.1995 | 0.257 | 0.314 | 0.372 | 0.429 | 0.543 | 0.658 | 0.774 | 0.947 | 1.233 | 1.520 | 1.810 |
| 20 | 0.0828 | 0.1112 | 0.1395 | 0.1955 | 0.252 | 0.307 | 0.364 | 0.420 | 0.533 | 0.645 | 0.757 | 0.927 | 1.208 | 1.489 | 1.770 |
| 30 | 0.0811 | 0.1088 | 0.1366 | 0.1916 | 0.246 | 0.301 | 0.357 | 0.412 | 0.522 | 0.632 | 0.742 | 0.908 | 1.184 | 1.460 | 1.735 |
| 40 | 0.0795 | 0.1067 | 0.1338 | 0.1876 | 0.241 | 0.295 | 0.350 | 0.404 | 0.511 | 0.619 | 0.727 | 0.890 | 1.161 | 1.431 | 1.701 |
| 50 | 0.0780 | 0.1045 | 0.1310 | 0.1839 | 0.237 | 0.290 | 0.343 | 0.396 | 0.501 | 0.607 | 0.713 | 0.873 | 1.139 | 1.403 | 1.668 |
| 60 | 0.0764 | 0.1025 | 0.1283 | 0.1803 | 0.232 | 0.284 | 0.336 | 0.388 | 0.493 | 0.596 | 0.700 | 0.856 | 1.116 | 1.376 | 1.636 |
| 80 | 0.0736 | 0.0988 | 0.1239 | 0.1738 | 0.224 | 0.274 | 0.324 | 0.374 | 0.473 | 0.572 | 0.673 | 0.824 | 1.074 | 1.325 | 1.573 |
| 100 | 0.0710 | 0.0954 | 0.1197 | 0.1676 | 0.215 | 0.264 | 0.312 | 0.360 | 0.455 | 0.551 | 0.648 | 0.794 | 1.035 | 1.276 | 1.517 |
| 120 | 0.0686 | 0.0921 | 0.1155 | 0.1618 | 0.208 | 0.255 | 0.302 | 0.348 | 0.440 | 0.533 | 0.626 | 0.767 | 1.001 | 1.234 | 1.465 |
| 140 | 0.0663 | 0.0889 | 0.1115 | 0.1565 | 0.201 | 0.246 | 0.291 | 0.336 | 0.426 | 0.516 | 0.606 | 0.742 | 0.968 | 1.194 | 1.416 |
| 150 | 0.0652 | 0.0874 | 0.1096 | 0.1541 | 0.198 | 0.242 | 0.286 | 0.331 | 0.419 | 0.508 | 0.596 | 0.730 | 0.953 | 1.175 | 1.392 |
| 175 | 0.0626 | 0.0840 | 0.1054 | 0.1482 | 0.191 | 0.233 | 0.275 | 0.318 | 0.403 | 0.488 | 0.573 | 0.701 | 0.914 | 1.128 | 1.337 |
| 200 | 0.0603 | 0.0809 | 0.1014 | 0.1427 | 0.184 | 0.225 | 0.265 | 0.305 | 0.388 | 0.470 | 0.552 | 0.674 | 0.879 | 1.084 | 1.287 |
| 225 | 0.0581 | 0.0779 | 0.0976 | 0.1373 | 0.177 | 0.216 | 0.255 | 0.295 | 0.374 | 0.452 | 0.531 | 0.649 | 0.846 | 1.043 | 1.240 |
| 250 | 0.0560 | 0.0751 | 0.0941 | 0.1323 | 0.170 | 0.208 | 0.247 | 0.284 | 0.360 | 0.436 | 0.513 | 0.627 | 0.817 | 1.007 | 1.197 |
| 275 | 0.0541 | 0.0726 | 0.0910 | 0.1278 | 0.164 | 0.201 | 0.238 | 0.274 | 0.348 | 0.421 | 0.494 | 0.605 | 0.789 | 0.972 | 1.155 |
| 300 | 0.0523 | 0.0707 | 0.0881 | 0.1237 | 0.159 | 0.194 | 0.230 | 0.265 | 0.336 | 0.407 | 0.478 | 0.585 | 0.762 | 0.940 | 1.118 |
| 350 | 0.0491 | 0.0658 | 0.0825 | 0.1160 | 0.149 | 0.183 | 0.216 | 0.249 | 0.316 | 0.382 | 0.449 | 0.549 | 0.715 | 0.883 | 1.048 |
| 400 | 0.0463 | 0.0621 | 0.0779 | 0.1090 | 0.140 | 0.172 | 0.203 | 0.235 | 0.297 | 0.360 | 0.423 | 0.517 | 0.674 | 0.831 | 0.987 |
| 450 | 0.0437 | 0.0586 | 0.0735 | 0.1033 | 0.133 | 0.163 | 0.192 | 0.222 | 0.281 | 0.340 | 0.399 | 0.488 | 0.637 | 0.786 | 0.934 |
| 500 | 0.0414 | 0.0555 | 0.0696 | 0.0978 | 0.126 | 0.154 | 0.182 | 0.210 | 0.266 | 0.322 | 0.379 | 0.463 | 0.604 | 0.746 | 0.885 |
| 550 | 0.0394 | 0.0528 | 0.0661 | 0.0930 | 0.120 | 0.146 | 0.173 | 0.200 | 0.253 | 0.306 | 0.359 | 0.440 | 0.573 | 0.749 | 0.841 |
| 600 | 0.0376 | 0.0504 | 0.0631 | 0.0885 | 0.114 | 0.139 | 0.165 | 0.190 | 0.241 | 0.292 | 0.343 | 0.419 | 0.547 | 0.675 | 0.801 |

**Relation between Pressure, Temperature and Volume of Air.** — This relationship is expressed by the formula:

$$\frac{P \times V}{T} = 53.3,$$

in which $P$ = absolute pressure in pounds per square foot; $V$ = volume in cubic feet of one pound of air at the given pressure and temperature; $T$ = absolute temperature in degrees R.

*Example.* — What is the volume of one pound of air at a pressure of 24.7 pounds per square inch and at a temperature of 210 degrees F?

$$\frac{24.7 \times 144 \times V}{210 + 459.7} = 53.3, \text{ or } V = \frac{53.3 \times 669.7}{24.7 \times 144} = 10.04 \text{ cubic feet.}$$

**Relation Between Barometric Pressure, and Pressures in Pounds per Square Inch and Square Foot**

| Barometer, Inches | Pressure in Pounds per Square Inch | Pressure in Pounds per Square Foot | Barometer, Inches | Pressure in Pounds per Square Inch | Pressure in Pounds per Square Foot | Barometer, Inches | Pressure in Pounds per Square Inch | Pressure in Pounds per Square Foot |
|---|---|---|---|---|---|---|---|---|
| 28.00 | 13.75 | 1980 | 29.25 | 14.36 | 2068 | 30.50 | 14.98 | 2156 |
| 28.25 | 13.87 | 1997 | 29.50 | 14.48 | 2086 | 30.75 | 15.10 | 2174 |
| 28.50 | 13.99 | 2015 | 29.75 | 14.61 | 2103 | 31.00 | 15.22 | 2192 |
| 28.75 | 14.12 | 2033 | 30.00 | 14.73 | 2121 | 31.25 | 15.34 | 2210 |
| 29.00 | 14.24 | 2050 | 30.25 | 14.85 | 2139 | ..... | ..... | ...... |

**Expansion and Compression of Air.** — The formula for the relationship of pressure, temperature and volume of air just given indicates that when the pressure remains constant the volume is directly proportional to the absolute temperature. If the temperature remains constant, the volume is inversely proportional to the absolute pressure. Theoretically, air (as well as other gases) can be expanded or compressed according to different laws. *Adiabatic* expansion or compression takes place when the air is expanded or compressed without transmission of heat to or from it; as for example, if the air could be expanded or compressed in a cylinder of an absolutely non-conducting material. Let:

$P_1$ = initial absolute pressure in pounds per square foot;
$V_1$ = initial volume in cubic feet;
$T_1$ = initial absolute temperature in degrees R;
$P_2$ = absolute pressure in pounds per square foot, after compression;
$V_2$ = volume in cubic feet, after compression;
$T_2$ = absolute temperature in degrees R, after compression.

Then:

$$\frac{V_2}{V_1} = \left(\frac{P_1}{P_2}\right)^{0.71} \qquad \frac{P_2}{P_1} = \left(\frac{V_1}{V_2}\right)^{1.41} \qquad \frac{T_2}{T_1} = \left(\frac{V_1}{V_2}\right)^{0.41}$$

$$\frac{V_2}{V_1} = \left(\frac{T_1}{T_2}\right)^{2.46} \qquad \frac{P_2}{P_1} = \left(\frac{T_2}{T_1}\right)^{3.46} \qquad \frac{T_2}{T_1} = \left(\frac{P_2}{P_1}\right)^{0.29}$$

These formulas are also applicable if all pressures are in pounds per square inch; if all volumes are in cubic inches; or if any other consistent set of units is used for pressure or volume.

*Isothermal* expansion or compression takes place when the gas is expanded or compressed with an addition or transmission of sufficient heat to maintain a constant temperature. Let:

$P_1$ = initial absolute pressure in pounds per square foot;
$V_1$ = initial volume in cubic feet;
$P_2$ = absolute pressure in pounds per square foot, after compression;
$V_2$ = volume in cubic feet, after compression;
$R$ = 53.3
$T$ = temperature in degrees Rankine maintained during isothermal expansion or contraction.

Then:

$$P_1 \times V_1 = P_2 \times V_2 = RT.$$

*Example.* — A volume of 165 cubic feet of air, at a pressure of 15 pounds per square inch, is compressed adiabatically to a pressure of 80 pounds per square inch. What will be the volume at this pressure?

$$V_2 = V_1 \left(\frac{P_1}{P_2}\right)^{0.71} = 165 \left(\frac{15}{80}\right)^{0.71} = 50 \text{ cubic feet, approx.}$$

*Example.* — The same volume of air is compressed isothermally from 15 to 80 pounds per square inch. What will be the volume after compression?

$$V_2 = \frac{P_1 \times V_1}{P_2} = \frac{15 \times 165}{80} = 31 \text{ cubic feet.}$$

**Foot-pounds of Work Required in Compression of Air**
Initial Pressure = 1 atmosphere = 14.7 pounds per square inch

| Gage Pressure in Pounds per Square Inch | Isothermal Compression | Adiabatic Compression | Actual Compression | Gage Pressure in Pounds per Square Inch | Isothermal Compression | Adiabatic Compression | Actual Compression |
|---|---|---|---|---|---|---|---|
| | Foot-pounds Required per Cubic Foot of Air at Initial Pressure | | | | Foot-pounds Required per Cubic Foot of Air at Initial Pressure | | |
| 5 | 619.6 | 649.5 | 637.5 | 55 | 3393.7 | 4188.9 | 3870.8 |
| 10 | 1098.2 | 1192.0 | 1154.6 | 60 | 3440.4 | 4422.8 | 4029.8 |
| 15 | 1488.3 | 1661.2 | 1592.0 | 65 | 3577.6 | 4645.4 | 4218.2 |
| 20 | 1817.7 | 2074.0 | 1971.4 | 70 | 3706.3 | 4859.6 | 4398.1 |
| 25 | 2102.6 | 2451.6 | 2312.0 | 75 | 3828.0 | 5063.9 | 4569.5 |
| 30 | 2353.6 | 2794.0 | 2617.8 | 80 | 3942.9 | 5259.7 | 4732.9 |
| 35 | 2578.0 | 3111.0 | 2897.8 | 85 | 4051.5 | 5450.0 | 4890.1 |
| 40 | 2780.8 | 3405.5 | 3155.6 | 90 | 4155.7 | 5633.1 | 5042.1 |
| 45 | 2966.0 | 3681.7 | 3395.4 | 95 | 4254.3 | 5819.3 | 5187.3 |
| 50 | 3136.2 | 3942.3 | 3619.8 | 100 | 4348.1 | 5981.2 | 5327.9 |

**Work Required in Compression of Air.** — The total work required for compression and expulsion of air, adiabatically compressed, is:

$$\text{Total work in foot-pounds} = 3.46 \, P_1 V_1 \left[\left(\frac{P_2}{P_1}\right)^{0.29} - 1\right]$$

in which $P_1$ = initial absolute pressure in pounds per square foot;
$\quad\quad\quad P_2$ = absolute pressure in pounds per square foot, after compression;
$\quad\quad\quad V_1$ = initial volume in cubic feet.

The total work required for isothermal compression is:

$$\text{Total work in foot-pounds} = P_1 V_1 \log_e \frac{V_1}{V_2}$$

in which $P_1$, $P_2$ and $V_1$ denote the same quantities as in the previous equation, and $V_2$ = volume of air in cubic feet, after compression.

The work required to compress air isothermally, that is, when the heat of compression is removed as rapidly as produced, is considerably less than the work required for compressing air adiabatically, or when all the heat is retained. In actual practice, neither of these two theoretical extremes are obtainable, but the power required for air compression is about the medium between the powers that would be required for each. The accompanying table gives the average number of foot-pounds of work required to compress air.

**Horsepower Required to Compress Air.** — In the accompanying tables is given the horsepower required for compressing one cubic foot of free air per minute (isothermally and adiabatically) from atmospheric pressure (14.7 pounds per square inch) to various gage pressures, for one-, two- and three-stage compression. The formula for calculating the horsepower required to compress, adiabatically, a given volume of free air to a given pressure is:

$$\text{H. P.} = \frac{144\, NPVn}{33000\,(n-1)} \left[ \left(\frac{P_2}{P}\right)^{\frac{n-1}{Nn}} - 1 \right]$$

in which $N$ = number of stages in which compression is accomplished;
$\quad\quad\quad P$ = atmospheric pressure in pounds per square inch;
$\quad\quad\quad P_2$ = absolute terminal pressure in pounds per square inch;
$\quad\quad\quad V$ = volume of air, in cubic feet, compressed per minute, at atmospheric pressure;
$\quad\quad\quad n$ = exponent of the compression curve = 1.41 for adiabatic compression.

For different methods of compression and for one cubic foot of air per minute, this formula may be simplified as follows:

For one-stage compression: H. P. = 0.015 $P$ $(R^{0.29} - 1)$
For two-stage compression: H. P. = 0.030 $P$ $(R^{0.145} - 1)$
For three-stage compression: H. P. = 0.045 $P$ $(R^{0.0975} - 1)$
For four-stage compression: H. P. = 0.060 $P$ $(R^{0.0725} - 1)$

In these latter formulas $R = \dfrac{P_2}{P}$ = number of atmospheres to be compressed.

The formula for calculating the horsepower required to compress isothermally a given volume of free air to a given pressure is:

$$\text{H. P.} = \frac{144\, PV}{33000} \left( \log_e \frac{P_2}{P} \right)$$

Natural logarithms are obtained by multiplying common logarithms by 2.30259. See also the tables of natural logarithms beginning on page 143.

## Horsepower Required to Compress Air

Horsepower Required for Compressing One Cubic Foot of Free Air per Minute (Isothermally and Adiabatically) from Atmospheric Pressure (14.7 pounds per square inch) to Various Gage Pressures. — Single-stage Compression

(Initial Temperature of Air, 60° F. — Jacket-cooling not considered)

| Gage Pressure, Pounds | Absolute Pressure, Pounds | Number of Atmospheres | Isothermal Compression | | Adiabatic Compression | | | |
|---|---|---|---|---|---|---|---|---|
| | | | Mean Effective Pressure * | Horsepower | Mean Effective Pressure,* Theoretical | Mean Eff. Pressure plus 15 per cent Friction | Horsepower, Theoretical | Horsepower plus 15 per cent Friction |
| 5 | 19.7 | 1.34 | 4.13 | 0.018 | 4.46 | 5.12 | 0.019 | 0.022 |
| 10 | 24.7 | 1.68 | 7.57 | 0.033 | 8.21 | 9.44 | 0.036 | 0.041 |
| 15 | 29.7 | 2.02 | 11.02 | 0.048 | 11.46 | 13.17 | 0.050 | 0.057 |
| 20 | 34.7 | 2.36 | 12.62 | 0.055 | 14.30 | 16.44 | 0.062 | 0.071 |
| 25 | 39.7 | 2.70 | 14.68 | 0.064 | 16.94 | 19.47 | 0.074 | 0.085 |
| 30 | 44.7 | 3.04 | 16.30 | 0.071 | 19.32 | 22.21 | 0.084 | 0.096 |
| 35 | 49.7 | 3.38 | 17.90 | 0.078 | 21.50 | 24.72 | 0.094 | 0.108 |
| 40 | 54.7 | 3.72 | 19.28 | 0.084 | 23.53 | 27.05 | 0.103 | 0.118 |
| 45 | 59.7 | 4.06 | 20.65 | 0.090 | 25.40 | 29.21 | 0.111 | 0.127 |
| 50 | 64.7 | 4.40 | 21.80 | 0.095 | 27.23 | 31.31 | 0.119 | 0.136 |
| 55 | 69.7 | 4.74 | 22.95 | 0.100 | 28.90 | 33.23 | 0.126 | 0.145 |
| 60 | 74.7 | 5.08 | 23.90 | 0.104 | 30.53 | 35.10 | 0.133 | 0.153 |
| 65 | 79.7 | 5.42 | 24.80 | 0.108 | 32.10 | 36.91 | 0.140 | 0.161 |
| 70 | 84.7 | 5.76 | 25.70 | 0.112 | 33.57 | 38.59 | 0.146 | 0.168 |
| 75 | 89.7 | 6.10 | 26.62 | 0.116 | 35.00 | 40.25 | 0.153 | 0.175 |
| 80 | 94.7 | 6.44 | 27.52 | 0.120 | 36.36 | 41.80 | 0.159 | 0.182 |
| 85 | 99.7 | 6.78 | 28.21 | 0.123 | 37.63 | 43.27 | 0.164 | 0.189 |
| 90 | 104.7 | 7.12 | 28.93 | 0.126 | 38.89 | 44.71 | 0.169 | 0.195 |
| 95 | 109.7 | 7.46 | 29.60 | 0.129 | 40.11 | 46.12 | 0.175 | 0.201 |
| 100 | 114.7 | 7.80 | 30.30 | 0.132 | 41.28 | 47.46 | 0.180 | 0.207 |
| 110 | 124.7 | 8.48 | 31.42 | 0.137 | 43.56 | 50.09 | 0.190 | 0.218 |
| 120 | 134.7 | 9.16 | 32.60 | 0.142 | 45.69 | 52.53 | 0.199 | 0.229 |
| 130 | 144.7 | 9.84 | 33.75 | 0.147 | 47.72 | 54.87 | 0.208 | 0.239 |
| 140 | 154.7 | 10.52 | 34.67 | 0.151 | 49.64 | 57.08 | 0.216 | 0.249 |
| 150 | 164.7 | 11.20 | 35.59 | 0.155 | 51.47 | 59.18 | 0.224 | 0.258 |
| 160 | 174.7 | 11.88 | 36.30 | 0.158 | 53.70 | 61.80 | 0.234 | 0.269 |
| 170 | 184.7 | 12.56 | 37.20 | 0.162 | 55.60 | 64.00 | 0.242 | 0.278 |
| 180 | 194.7 | 13.24 | 38.10 | 0.166 | 57.20 | 65.80 | 0.249 | 0.286 |
| 190 | 204.7 | 13.92 | 38.80 | 0.169 | 58.80 | 67.70 | 0.256 | 0.294 |
| 200 | 214.7 | 14.60 | 39.50 | 0.172 | 60.40 | 69.50 | 0.263 | 0.303 |

* Mean Effective Pressure (MEP) is defined as that single pressure rise, above atmospheric, which would require the same horsepower as the actual varying pressures during compression.

## Horsepower Required to Compress Air

Horsepower Required for Compressing One Cubic Foot of Free Air per Minute (Isothermally and Adiabatically) from Atmospheric Pressure (14.7 pounds per square inch) to Various Gage Pressures. — Two-stage Compression
(Initial Temperature of Air, 60° F. — Jacket-cooling not considered)

| Gage Pressure, Pounds | Absolute Pressure, Pounds | Number of Atmospheres | Correct Ratio of Cylinder Volumes | Intercooler Gage Pressure | Isothermal Compression | | Adiabatic Compression | | | | Percentage of Saving over One-stage Compression |
|---|---|---|---|---|---|---|---|---|---|---|---|
| | | | | | Mean Effective Pressure * | Horsepower | Mean Eff. Pressure,* Theoretical | Mean Eff. Pressure plus 15 per cent Friction | Horsepower, Theoretical | H.P. plus 15 per cent Friction | |
| 50 | 64.7 | 4.40 | 2.10 | 16.2 | 21.80 | 0.095 | 24.30 | 27.90 | 0.106 | 0.123 | 10 9 |
| 60 | 74.7 | 5.08 | 2.25 | 18.4 | 23.90 | 0.104 | 27.20 | 31.30 | 0.118 | 0.136 | 11.3 |
| 70 | 84.7 | 5.76 | 2.40 | 20.6 | 25.70 | 0.112 | 29.31 | 33.71 | 0.128 | 0.147 | 12.3 |
| 80 | 94.7 | 6.44 | 2.54 | 22.7 | 27.52 | 0.120 | 31.44 | 36.15 | 0.137 | 0.158 | 13.8 |
| 90 | 104.7 | 7.12 | 2.67 | 24.5 | 28.93 | 0.126 | 33.37 | 38.36 | 0.145 | 0.167 | 14.2 |
| 100 | 114.7 | 7.80 | 2.79 | 26.3 | 30.30 | 0.132 | 35.20 | 40.48 | 0.153 | 0.176 | 15.0 |
| 110 | 124.7 | 8.48 | 2.91 | 28.1 | 31.42 | 0.137 | 36.82 | 42.34 | 0.161 | 0.185 | 15.2 |
| 120 | 134.7 | 9.16 | 3.03 | 29.8 | 32.60 | 0.142 | 38.44 | 44.20 | 0.168 | 0.193 | 15.6 |
| 130 | 144.7 | 9.84 | 3.14 | 31.5 | 33.75 | 0.147 | 39.86 | 45.83 | 0.174 | 0.200 | 16.3 |
| 140 | 154.7 | 10.52 | 3.24 | 32.9 | 34.67 | 0.151 | 41.28 | 47.47 | 0.180 | 0.207 | 16.7 |
| 150 | 164.7 | 11.20 | 3.35 | 34.5 | 35.59 | 0.155 | 42.60 | 48.99 | 0.186 | 0.214 | 16.9 |
| 160 | 174.7 | 11.88 | 3.45 | 36.1 | 36.30 | 0.158 | 43.82 | 50.39 | 0.191 | 0.219 | 18.4 |
| 170 | 184.7 | 12.56 | 3.54 | 37.3 | 37.20 | 0.162 | 44.93 | 51.66 | 0.196 | 0.225 | 19.0 |
| 180 | 194.7 | 13.24 | 3.64 | 38.8 | 38.10 | 0.166 | 46.05 | 52.95 | 0.201 | 0.231 | 19.3 |
| 190 | 204.7 | 13.92 | 3.73 | 40.1 | 38.80 | 0.169 | 47.16 | 54.22 | 0.206 | 0.236 | 19.5 |
| 200 | 214.7 | 14.60 | 3.82 | 41.4 | 39.50 | 0.172 | 48.18 | 55.39 | 0.210 | 0.241 | 20.1 |
| 210 | 224.7 | 15.28 | 3.91 | 42.8 | 40.10 | 0.174 | 49.35 | 56.70 | 0.216 | 0.247 | ..... |
| 220 | 234.7 | 15.96 | 3.99 | 44.0 | 40.70 | 0.177 | 50.30 | 57.70 | 0.220 | 0.252 | ..... |
| 230 | 244.7 | 16.64 | 4.08 | 45.3 | 41.30 | 0.180 | 51.30 | 59.10 | 0.224 | 0.257 | ..... |
| 240 | 254.7 | 17.32 | 4.17 | 46.6 | 41.90 | 0.183 | 52.25 | 60.10 | 0.228 | 0.262 | ..... |
| 250 | 264.7 | 18.00 | 4.24 | 47.6 | 42.70 | 0.186 | 52.84 | 60.76 | 0.230 | 0.264 | ..... |
| 260 | 274.7 | 18.68 | 4.32 | 48.8 | 43.00 | 0.188 | 53.85 | 62.05 | 0.235 | 0.270 | ..... |
| 270 | 284.7 | 19.36 | 4.40 | 50.0 | 43.50 | 0.190 | 54.60 | 62.90 | 0.238 | 0.274 | ..... |
| 280 | 294.7 | 20.04 | 4.48 | 51.1 | 44.00 | 0.192 | 55.50 | 63.85 | 0.242 | 0.278 | ..... |
| 290 | 304.7 | 20.72 | 4.55 | 52.2 | 44.50 | 0.194 | 56.20 | 64.75 | 0.246 | 0.282 | ..... |
| 300 | 314.7 | 21.40 | 4.63 | 53.4 | 45.80 | 0.197 | 56.70 | 65.20 | 0.247 | 0.283 | ..... |
| 350 | 364.7 | 24.80 | 4.98 | 58.5 | 49.20 | 0.206 | 60.15 | 69.16 | 0.262 | 0.301 | ..... |
| 400 | 414.7 | 28.20 | 5.31 | 63.3 | 49.20 | 0.214 | 63.19 | 72.65 | 0.276 | 0.317 | ..... |
| 450 | 464.7 | 31.60 | 5.61 | 67.8 | 51.20 | 0.223 | 65.93 | 75.81 | 0.287 | 0.329 | ..... |
| 500 | 514.7 | 35.01 | 5.91 | 72.1 | 52.70 | 0.229 | 68.46 | 78.72 | 0.298 | 0.342 | ..... |

* See footnote on page 2312.

## Horsepower Required to Compress Air

Horsepower Required for Compressing One Cubic Foot of Free Air per Minute (Isothermally and Adiabatically) from Atmospheric Pressure (14.7 pounds per square inch) to Various Gage Pressures. — Three-stage Compression

(Initial Temperature of Air, 60° F. — Jacket-cooling not considered)

| Gage Pressure, Pounds | Absolute Pressure, Pounds | Number of Atmospheres | Correct Ratio of Cylinder Volumes | Intercooler Gage Pressure, First and Second Stages | Isothermal Compression | | Adiabatic Compression | | | | Precentage of Saving over Two-stage Compression |
|---|---|---|---|---|---|---|---|---|---|---|---|
| | | | | | Mean Effective Pressure * | Horsepower | Mean Eff. Pressure,* Theoretical | Mean Eff. Pressure plus 15 per cent Friction | Horsepower, Theoretical | H.P. plus 15 per cent Friction | |
| 100 | 114.7 | 7.8 | 1.98 | 14.4– 42.9 | 30.30 | 0.132 | 33.30 | 38.30 | 0.145 | 0.167 | 5.23 |
| 150 | 164.7 | 11.2 | 2.24 | 18.2– 59.0 | 35.59 | 0.155 | 40.30 | 46.50 | 0.175 | 0.202 | 5.92 |
| 200 | 214.7 | 14.6 | 2.44 | 21.2– 73.0 | 39.50 | 0.172 | 45.20 | 52.00 | 0.196 | 0.226 | 6.67 |
| 250 | 264.7 | 18.0 | 2.62 | 23.8– 86.1 | 42.70 | 0.186 | 49.20 | 56.60 | 0.214 | 0.246 | 6.96 |
| 300 | 314.7 | 21.4 | 2.78 | 26.1– 98.7 | 45.30 | 0.197 | 52.70 | 60.70 | 0.229 | 0.264 | 7.28 |
| 350 | 364.7 | 24.8 | 2.92 | 28.2–110.5 | 47.30 | 0.206 | 55.45 | 63.80 | 0.242 | 0.277 | 7.64 |
| 400 | 414.7 | 28.2 | 3.04 | 30.0–121.0 | 49.20 | 0.214 | 58.25 | 66.90 | 0.253 | 0.292 | 8.33 |
| 450 | 464.7 | 31.6 | 3.16 | 31.8–132.3 | 51.20 | 0.223 | 60.40 | 69.40 | 0.263 | 0.302 | 8.36 |
| 500 | 514.7 | 35.0 | 3.27 | 33.4–142.4 | 52.70 | 0.229 | 62.30 | 71.70 | 0.273 | 0.314 | 8.38 |
| 550 | 564.7 | 38.4 | 3.38 | 35.0–153.1 | 53.75 | 0.234 | 65.00 | 74.75 | 0.283 | 0.326 | 8.80 |
| 600 | 614.7 | 41.8 | 3.47 | 36.3–162.3 | 54.85 | 0.239 | 66.85 | 76.90 | 0.291 | 0.334 | 8.86 |
| 650 | 664.7 | 45.2 | 3.56 | 37.6–171.5 | 56.00 | 0.244 | 67.90 | 78.15 | 0.296 | 0.340 | 9.02 |
| 700 | 714.7 | 48.6 | 3.65 | 38.9–180.8 | 57.15 | 0.249 | 69.40 | 79.85 | 0.303 | 0.348 | 9.18 |
| 750 | 764.7 | 52.0 | 3.73 | 40.1–189.8 | 58.10 | 0.253 | 70.75 | 81.40 | 0.309 | 0.355 | .... |
| 800 | 814.7 | 55.4 | 3.82 | 41.4–199.5 | 59.00 | 0.257 | 72.45 | 83.25 | 0.315 | 0.362 | .... |
| 850 | 864.7 | 58.8 | 3.89 | 42.5–207.8 | 60.20 | 0.262 | 73.75 | 84.90 | 0.321 | 0.369 | .... |
| 900 | 914.7 | 62.2 | 3.95 | 43.4–214.6 | 60.80 | 0.265 | 74.80 | 86.00 | 0.326 | 0.375 | .... |
| 950 | 964.7 | 65.6 | 4.03 | 44.6–224.5 | 61.72 | 0.269 | 76.10 | 87.50 | 0.331 | 0.381 | .... |
| 1000 | 1014.7 | 69.0 | 4.11 | 45.7–233.5 | 62.40 | 0.272 | 77.20 | 88.80 | 0.336 | 0.383 | .... |
| 1050 | 1064.7 | 72.4 | 4.15 | 46.3–238.3 | 63.10 | 0.275 | 78.10 | 90.10 | 0.340 | 0.391 | .... |
| 1100 | 1114.7 | 75.8 | 4.23 | 47.5–248.3 | 63.80 | 0.278 | 79.10 | 91.10 | 0.344 | 0.396 | .... |
| 1150 | 1164.7 | 79.2 | 4.30 | 48.5–256.8 | 64.40 | 0.281 | 80.15 | 92.20 | 0.349 | 0.401 | .... |
| 1200 | 1214.7 | 82.6 | 4.33 | 49.0–261.3 | 65.00 | 0.283 | 81.00 | 93.15 | 0.353 | 0.405 | .... |
| 1250 | 1264.7 | 86.0 | 4.42 | 50.3–272.3 | 65.60 | 0.286 | 82.00 | 94.30 | 0.357 | 0.411 | .... |
| 1300 | 1314.7 | 89.4 | 4.48 | 51.3–280.8 | 66.30 | 0.289 | 82.90 | 95.30 | 0.362 | 0.416 | .... |
| 1350 | 1364.7 | 92.8 | 4.53 | 52.0–287.3 | 66.70 | 0.291 | 84.00 | 96.60 | 0.366 | 0.421 | .... |
| 1400 | 1414.7 | 96.2 | 4.58 | 52.6–293.5 | 67.00 | 0.292 | 84.60 | 97.30 | 0.368 | 0.423 | .... |
| 1450 | 1464.7 | 99.6 | 4.64 | 53.5–301.5 | 67.70 | 0.295 | 85.30 | 98.20 | 0.371 | 0.426 | .... |
| 1500 | 1514.7 | 103.0 | 4.69 | 54.3–309.3 | 68.30 | 0.298 | 85.80 | 98.80 | 0.374 | 0.430 | .... |
| 1550 | 1564.7 | 106.4 | 4.74 | 55.0–317.3 | 68.80 | 0.300 | 86.80 | 99.85 | 0.378 | 0.434 | .... |
| 1600 | 1614.7 | 109.8 | 4.79 | 55.8–323.3 | 69.10 | 0.302 | 87.60 | 100.80 | 0.382 | 0.438 | .... |

* See footnote on page 2312.

**Flow of Air in Pipes.** — The following formulas are used:

$$v = \sqrt{\frac{25,000\, dp}{L}} \qquad p = \frac{Lv^2}{25,000\, d}$$

in which   $v$ = velocity of air in feet per second;
   $p$ = loss of pressure due to flow through the pipes in ounces per square inch;
   $d$ = inside diameter of pipe in inches;
   $L$ = length of pipe in feet.

**The** quantity of air discharged in cubic feet per second is the product of the **velocity** as obtained from the formula above and the area of the pipe in square **feet.** The horsepower required to drive air through a pipe equals the volume of air in cubic feet per second multiplied by the pressure in pounds per square foot, and this product divided by 550.

**Volume of Air Transmitted, in Cubic Feet per Minute, Through Pipes**

| Velocity of Air in Feet per Second | Actual Inside Diameter of Pipe, Inches | | | | | | | | | |
|---|---|---|---|---|---|---|---|---|---|---|
| | 1 | 2 | 3 | 4 | 6 | 8 | 10 | 12 | 16 | 24 |
| 1 | 0.33 | 1.31 | 2.95 | 5.2 | 11.8 | 20.9 | 32.7 | 47.1 | 83.8 | 188 |
| 2 | 0.65 | 2.62 | 5.89 | 10.5 | 23.6 | 41.9 | 65.4 | 94.2 | 167.5 | 377 |
| 3 | 0.98 | 3.93 | 8.84 | 15.7 | 35.3 | 62.8 | 98.2 | 141.4 | 251.3 | 565 |
| 4 | 1.31 | 5.24 | 11.78 | 20.9 | 47.1 | 83.8 | 131.0 | 188.0 | 335.0 | 754 |
| 5 | 1.64 | 6.55 | 14.7 | 26.2 | 59.0 | 104.0 | 163.0 | 235.0 | 419.0 | 942 |
| 6 | 1.96 | 7.85 | 17.7 | 31.4 | 70.7 | 125.0 | 196.0 | 283.0 | 502.0 | 1131 |
| 7 | 2.29 | 9.16 | 20.6 | 36.6 | 82.4 | 146.0 | 229.0 | 330.0 | 586.0 | 1319 |
| 8 | 2.62 | 10.50 | 23.5 | 41.9 | 94.0 | 167.0 | 262.0 | 377.0 | 670.0 | 1508 |
| 9 | 2.95 | 11.78 | 26.5 | 47.0 | 106.0 | 188.0 | 294.0 | 424.0 | 754.0 | 1696 |
| 10 | 3.27 | 13.1 | 29.4 | 52.0 | 118.0 | 209.0 | 327.0 | 471.0 | 838.0 | 1885 |
| 12 | 3.93 | 15.7 | 35.3 | 63.0 | 141.0 | 251.0 | 393.0 | 565.0 | 1005.0 | 2262 |
| 15 | 4.91 | 19.6 | 44.2 | 78.0 | 177.0 | 314.0 | 491.0 | 707.0 | 1256.0 | 2827 |
| 18 | 5.89 | 23.5 | 53.0 | 94.0 | 212.0 | 377.0 | 589.0 | 848.0 | 1508.0 | 3393 |
| 20 | 6.55 | 26.2 | 59.0 | 105.0 | 235.0 | 419.0 | 654.0 | 942.0 | 1675.0 | 3770 |
| 24 | 7.86 | 31.4 | 71.0 | 125.0 | 283.0 | 502.0 | 785.0 | 1131.0 | 2010.0 | 4524 |
| 25 | 8.18 | 32.7 | 73.0 | 131.0 | 294.0 | 523.0 | 818.0 | 1178.0 | 2094.0 | 4712 |
| 28 | 9.16 | 36.6 | 82.0 | 146.0 | 330.0 | 586.0 | 916.0 | 1319.0 | 2346.0 | 5278 |
| 30 | 9.80 | 39.3 | 88.0 | 157.0 | 353.0 | 628.0 | 982.0 | 1414.0 | 2513.0 | 5655 |

**Flow of Compressed Air in Pipes.** — When there is a comparatively small difference of pressure at the two ends of the pipe, the volume of flow in cubic feet per minute is found by the formula:

$$V = 58\sqrt{\frac{pd^5}{WL}}$$

in which   $V$ = volume of air in cubic feet per minute;
   $p$ = difference in pressure at the two ends of the pipe in pounds per square inch;
   $d$ = inside diameter of pipe in inches;
   $W$ = weight in pounds of one cubic foot of entering air;
   $L$ = length of pipe in feet.

## Velocity of Escaping Compressed Air

| Pressure Above Atmospheric Pressure | | | Theoretical Velocity, Feet per Second | Pressure Above Atmospheric Pressure | | | Theoretical Velocity, Feet per Second |
|---|---|---|---|---|---|---|---|
| In Atmospheres | In Inches Mercury | In Lbs. per Sq. In. | | In Atmospheres | In Inches Mercury | In Lbs. per Sq. In. | |
| 0.010 | 0.30 | 0.147 | 94.4 | 0.680 | 20.4 | 10.0 | 780 |
| 0.066 | 2.10 | 1.00 | 246.0 | 0.809 | 24.3 | 12.0 | 855 |
| 0.100 | 3.00 | 1.47 | 299.0 | 1.0 | 30.0 | 14.7 | 946 |
| 0.136 | 4.08 | 2.00 | 348.0 | 2.0 | 60.0 | 29.4 | 1094 |
| 0.204 | 6.12 | 3.00 | 427.0 | 5.0 | 150.0 | 73.5 | 1219 |
| 0.272 | 8.16 | 4.00 | 493.0 | 10.0 | 300.0 | 147.0 | 1275 |
| 0.340 | 10.20 | 5.00 | 552.0 | 20.0 | 600.0 | 294.0 | 1304 |
| 0.408 | 12.24 | 6.00 | 604.0 | 40.0 | 1200.0 | 588.0 | 1323 |
| 0.500 | 15.00 | 7.35 | 673.0 | 100.0 | 3000.0 | 1470.0 | 1331 |
| 0.544 | 16.32 | 8.00 | 697.0 | 200.0 | 6000.0 | 2940.0 | 1334 |
| 0.611 | 18.34 | 9.00 | 741.0 | ...... | ...... | ...... | ....... |

The theoretical velocities in the table above must be reduced by multiplying by a "coefficient of discharge," which varies with the orifice and the pressure. The following coefficients are used for orifices in thin plate and short tubes.

| Type of Orifice | Pressures in Atmospheres Above Atmospheric Pressure | | | | | | |
|---|---|---|---|---|---|---|---|
| | 0.01 | 0.1 | 0.5 | 1 | 5 | 10 | 100 |
| Orifice in thin plate........ | 0.65 | 0.64 | 0.57 | 0.54 | 0.45 | 0.44 | 0.42 |
| Orifice in short tube...... | 0.83 | 0.82 | 0.71 | 0.67 | 0.53 | 0.51 | 0.49 |

**Velocity of Air under Low Pressures.** — The table "Velocity of Air under Low Pressures" gives the theoretical velocity for the discharge of air into the atmosphere. These theoretical velocities are modified by multiplying them by a coefficient varying with the form of the orifice. For an orifice with sharp edges in a thin plate, this coefficient equals 0.65. For a plate with the inside of the orifice rounded, the coefficient equals 0.70 to 0.75, and for a well-shaped nozzle, 0.93.

## Velocity of Air Under Low Pressures
(Temperature 62° F.)

| Gage Pressure, Ounces per Square Inch | Theoretical Velocity, Feet per Second | Gage Pressure, Ounces per Square Inch | Theoretical Velocity, Feet per Second | Gage Pressure, Ounces per Square Inch | Theoretical Velocity, Feet per Second | Gage Pressure, Ounces per Square Inch | Theoretical Velocity, Feet per Second | Gage Pressure, Ounces per Square Inch | Theoretical Velocity, Feet per Second |
|---|---|---|---|---|---|---|---|---|---|
| 0.006 | 6.61 | 0.115 | 29.5 | 0.346 | 51.2 | 0.866 | 80.9 | 2.308 | 132.0 |
| 0.012 | 9.35 | 0.173 | 36.2 | 0.404 | 55.3 | 1.153 | 93.5 | 2.597 | 140.0 |
| 0.023 | 13.20 | 0.231 | 41.8 | 0.461 | 59.1 | 1.442 | 104.0 | 2.885 | 148.0 |
| 0.040 | 17.40 | 0.260 | 44.3 | 0.519 | 62.7 | 1.731 | 114.0 | 3.462 | 162.0 |
| 0.058 | 20.90 | 0.289 | 46.7 | 0.577 | 66.1 | 2.020 | 124.0 | .... | .... |

# PRESSURES AND FLOW OF WATER

**Water Pressures.** — Water is composed of two gases, hydrogen and oxygen, in the ratio of two volumes of the former to one of the latter. In the English System of measure, water boils under atmospheric pressure at 212 degrees F. and freezes at 32 degrees F. Its greatest density is 62.425 pounds per cubic foot, at 39.1 degrees F. In metric SI measure, water boils under atmospheric pressure at 100°C (Celsius) and freezes at 0°C. Its density is equal to one kilogram per liter, where one liter is one cubic decimeter. Also in metric SI, pressure is given in pascals (Pa) or the equivalent Newton per square meter. See page 2407 for additional information on metric SI.

For higher temperatures, the pressure slightly decreases in the proportion indicated by the table "Weight of Water per Cubic Foot at Different Temperatures." The pressure per square inch is equal in all directions, downwards, upwards and sideways. Water can be compressed only in a very slight degree, the compressibility being so slight that even at the depth of a mile, a cubic foot of water weighs only about one-half pound more than at the surface.

### Pressure in Pounds per Square Inch for Different Heads of Water

| Head, Feet | 0 | 1 | 2 | 3 | 4 | 5 | 6 | 7 | 8 | 9 |
|---|---|---|---|---|---|---|---|---|---|---|
| 0  | .... | 0.43 | 0.87 | 1.30 | 1.73 | 2.16 | 2.60 | 3.03 | 3.46 | 3.90 |
| 10 | 4.33 | 4.76 | 5.20 | 5.63 | 6.06 | 6.49 | 6.93 | 7.36 | 7.79 | 8.23 |
| 20 | 8.66 | 9.09 | 9.53 | 9.96 | 10.39 | 10.82 | 11.26 | 11.69 | 12.12 | 12.56 |
| 30 | 12.99 | 13.42 | 13.86 | 14.29 | 14.72 | 15.15 | 15.59 | 16.02 | 16.45 | 16.89 |
| 40 | 17.32 | 17.75 | 18.19 | 18.62 | 19.05 | 19.48 | 19.92 | 20.35 | 20.78 | 21.22 |
| 50 | 21.65 | 22.08 | 22.52 | 22.95 | 23.38 | 23.81 | 24.25 | 24.68 | 25.11 | 25.55 |
| 60 | 25.98 | 26.41 | 26.85 | 27.28 | 27.71 | 28.14 | 28.58 | 29.01 | 29.44 | 29.88 |
| 70 | 30.31 | 30.74 | 31.18 | 31.61 | 32.04 | 32.47 | 32.91 | 33.34 | 33.77 | 34.21 |
| 80 | 34.64 | 35.07 | 35.51 | 35.94 | 36.37 | 36.80 | 37.24 | 37.67 | 38.10 | 38.54 |
| 90 | 38.97 | 39.40 | 39.84 | 40.27 | 40.70 | 41.13 | 41.57 | 42.00 | 42.43 | 42.87 |

### Heads of Water in Feet Corresponding to Certain Pressures in Pounds per Square Inch

| Pressure, Lbs. | 0 | 1 | 2 | 3 | 4 | 5 | 6 | 7 | 8 | 9 |
|---|---|---|---|---|---|---|---|---|---|---|
| 0  | .... | 2.3 | 4.6 | 6.9 | 9.2 | 11.5 | 13.9 | 16.2 | 18.5 | 20.8 |
| 10 | 23.1 | 25.4 | 27.7 | 30.0 | 32.3 | 34.6 | 36.9 | 39.3 | 41.6 | 43.9 |
| 20 | 46.2 | 48.5 | 50.8 | 53.1 | 55.4 | 57.7 | 60.0 | 62.4 | 64.7 | 67.0 |
| 30 | 69.3 | 71.6 | 73.9 | 76.2 | 78.5 | 80.8 | 83.1 | 85.4 | 87.8 | 90.1 |
| 40 | 92.4 | 94.7 | 97.0 | 99.3 | 101.6 | 103.9 | 106.2 | 108.5 | 110.8 | 113.2 |
| 50 | 115.5 | 117.8 | 120.1 | 122.4 | 124.7 | 127.0 | 129.3 | 131.6 | 133.9 | 136.3 |
| 60 | 138.6 | 140.9 | 143.2 | 145.5 | 147.8 | 150.1 | 152.4 | 154.7 | 157.0 | 159.3 |
| 70 | 161.7 | 164.0 | 166.3 | 168.6 | 170.9 | 173.2 | 175.5 | 177.8 | 180.1 | 182.4 |
| 80 | 184.8 | 187.1 | 189.4 | 191.7 | 194.0 | 196.3 | 198.6 | 200.9 | 203.2 | 205.5 |
| 90 | 207.9 | 210.2 | 212.5 | 214.8 | 217.1 | 219.4 | 221.7 | 224.0 | 226.3 | 228.6 |

## Comparison of Different Methods of Measuring Pressures

| Ounces and Pounds per Square Inch, and Inches of Water and Mercury | | | | | | | |
|---|---|---|---|---|---|---|---|
| Ounces per Square Inch | Pounds per Square Inch | Inches of Water | Inches of Mercury | Ounces per Square Inch | Pounds per Square Inch | Inches of Water | Inches of Mercury |
| 0.25 | 0.016 | 0.433 | 0.0319 | 8 | 0.500 | 13.856 | 1.020 |
| 0.50 | 0.031 | 0.866 | 0.0638 | 9 | 0.562 | 15.588 | 1.148 |
| 1 | 0.062 | 1.732 | 0.1275 | 10 | 0.625 | 17.320 | 1.275 |
| 2 | 0.125 | 3.464 | 0.2551 | 11 | 0.687 | 19.052 | 1.403 |
| 3 | 0.187 | 5.196 | 0.3826 | 12 | 0.750 | 20.784 | 1.531 |
| 4 | 0.250 | 6.928 | 0.5102 | 13 | 0.812 | 22.516 | 1.658 |
| 5 | 0.312 | 8.660 | 0.6377 | 14 | 0.875 | 24.248 | 1.786 |
| 6 | 0.375 | 10.392 | 0.7653 | 15 | 0.937 | 25.980 | 1.913 |
| 7 | 0.437 | 12.124 | 0.8928 | 16 | 1.000 | 27.712 | 2.041 |

| Pounds per Square Inch, Inches and Feet of Water and Inches of Mercury | | | | | | | |
|---|---|---|---|---|---|---|---|
| Pounds per Square Inch | Inches of Water | Feet of Water | Inches of Mercury | Pounds per Square Inch | Inches of Water | Feet of Water | Inches of Mercury |
| 1 | 27.71 | 2.31 | 2.041 | 14 | 387.97 | 32.33 | 28.57 |
| 2 | 55.42 | 4.62 | 4.081 | 14.7 | 407.37 | 33.95 | 30.00 |
| 3 | 83.14 | 6.93 | 6.122 | 15 | 415.68 | 34.64 | 30.61 |
| 4 | 110.85 | 9.24 | 8.163 | 16 | 443.40 | 36.95 | 32.65 |
| 5 | 138.56 | 11.55 | 10.20 | 17 | 471.11 | 39.26 | 34.69 |
| 6 | 166.27 | 13.86 | 12.24 | 18 | 498.82 | 41.57 | 36.73 |
| 7 | 193.99 | 16.17 | 14.28 | 19 | 526.53 | 43.88 | 38.77 |
| 8 | 221.70 | 18.47 | 16.33 | 20 | 554.25 | 46.19 | 40.81 |
| 9 | 249.41 | 20.78 | 18.37 | 21 | 581.96 | 48.50 | 42.85 |
| 10 | 277.12 | 23.09 | 20.41 | 22 | 609.67 | 50.81 | 44.89 |
| 11 | 304.84 | 25.40 | 22.45 | 23 | 637.38 | 53.12 | 46.94 |
| 12 | 332.55 | 27.71 | 24.49 | 24 | 665.10 | 55.42 | 48.98 |
| 13 | 360.26 | 30.02 | 26.53 | 25 | 692.81 | 57.73 | 51.02 |

## Volume of Water at Different Temperatures

| Degrees Fahr. | Volume | Degrees Fahr. | Volume | Degrees Fahr. | Volume | Degrees Fahr. | Volume |
|---|---|---|---|---|---|---|---|
| 39.1 | 1.00000 | 86 | 1.00425 | 131 | 1.01423 | 176 | 1.02872 |
| 50 | 1.00025 | 95 | 1.00586 | 140 | 1.01678 | 185 | 1.03213 |
| 59 | 1.00083 | 104 | 1.00767 | 149 | 1.01951 | 194 | 1.03570 |
| 68 | 1.00171 | 113 | 1.00967 | 158 | 1.02241 | 203 | 1.03943 |
| 77 | 1.00286 | 122 | 1.01186 | 167 | 1.02548 | 212 | 1.04332 |

## Weight of Water per Cubic Foot at Different Temperatures

| Temperature, Degrees F. | Weight per Cubic Foot, Pounds | Temperature, Degrees F. | Weight per Cubic Foot, Pounds | Temperature, Degrees F. | Weight per Cubic Foot, Pounds | Temperature, Degrees F. | Weight per Cubic Foot, Pounds | Temperature, Degrees F. | Weight per Cubic Foot, Pounds | Temperature, Degrees F. | Weight per Cubic Foot, Pounds |
|---|---|---|---|---|---|---|---|---|---|---|---|
| 32 | 62.42 | 130 | 61.56 | 220 | 59.63 | 320 | 56.66 | 420 | 52.6 | 520 | 47.6 |
| 40 | 62.42 | 140 | 61.37 | 230 | 59.37 | 330 | 56.30 | 430 | 52.2 | 530 | 47.0 |
| 50 | 62.41 | 150 | 61.18 | 240 | 59.11 | 340 | 55.94 | 440 | 51.7 | 540 | 46.3 |
| 60 | 62.37 | 160 | 60.98 | 250 | 58.83 | 350 | 55.57 | 450 | 51.2 | 550 | 45.6 |
| 70 | 62.31 | 170 | 60.77 | 260 | 58.55 | 360 | 55.18 | 460 | 50.7 | 560 | 44.9 |
| 80 | 62.23 | 180 | 60.55 | 270 | 58.26 | 370 | 54.78 | 470 | 50.2 | 570 | 44.1 |
| 90 | 62.13 | 190 | 60.32 | 280 | 57.96 | 380 | 54.36 | 480 | 49.7 | 580 | 43.3 |
| 100 | 62.02 | 200 | 60.12 | 290 | 57.65 | 390 | 53.94 | 490 | 49.2 | 590 | 42.6 |
| 110 | 61.89 | 210 | 59.88 | 300 | 57.33 | 400 | 53.50 | 500 | 48.7 | 600 | 41.8 |
| 120 | 61.74 | 212 | 59.83 | 310 | 57.00 | 410 | 53.00 | 510 | 48.1 | ... | ... |

## Table of Horsepower Due to Certain Head of Water

The table gives the horsepower of 1 cubic foot of water per minute, and is based on an efficiency of 85 per cent.

| Heads in Feet | Horse-power | Heads in Feet | Horse-power | Heads in Feet | Horse-power | Heads in Feet | Horse-power | Heads in Feet | Horse-power |
|---|---|---|---|---|---|---|---|---|---|
| 1 | 0.0016 | 170 | 0.274 | 340 | 0.547 | 520 | 0.837 | 1250 | 2.012 |
| 10 | 0.0161 | 180 | 0.290 | 350 | 0.563 | 540 | 0.869 | 1300 | 2.093 |
| 20 | 0.0322 | 190 | 0.306 | 360 | 0.580 | 560 | 0.901 | 1350 | 2.173 |
| 30 | 0.0483 | 200 | 0.322 | 370 | 0.596 | 580 | 0.934 | 1400 | 2.254 |
| 40 | 0.0644 | 210 | 0.338 | 380 | 0.612 | 600 | 0.966 | 1450 | 2.334 |
| 50 | 0.0805 | 220 | 0.354 | 390 | 0.628 | 650 | 1.046 | 1500 | 2.415 |
| 60 | 0.0966 | 230 | 0.370 | 400 | 0.644 | 700 | 1.127 | 1550 | 2.495 |
| 70 | 0.1127 | 240 | 0.386 | 410 | 0.660 | 750 | 1.207 | 1600 | 2.576 |
| 80 | 0.1288 | 250 | 0.402 | 420 | 0.676 | 800 | 1.288 | 1650 | 2.656 |
| 90 | 0.1449 | 260 | 0.418 | 430 | 0.692 | 850 | 1.368 | 1700 | 2.737 |
| 100 | 0.1610 | 270 | 0.435 | 440 | 0.708 | 900 | 1.449 | 1750 | 2.818 |
| 110 | 0.1771 | 280 | 0.451 | 450 | 0.724 | 950 | 1.529 | 1800 | 2.898 |
| 120 | 0.1932 | 290 | 0.467 | 460 | 0.740 | 1000 | 1.610 | 1850 | 2.978 |
| 130 | 0.2093 | 300 | 0.483 | 470 | 0.757 | 1050 | 1.690 | 1900 | 3.059 |
| 140 | 0.2254 | 310 | 0.499 | 480 | 0.773 | 1100 | 1.771 | 1950 | 3.139 |
| 150 | 0.2415 | 320 | 0.515 | 490 | 0.789 | 1150 | 1.851 | 2000 | 3.220 |
| 160 | 0.2576 | 330 | 0.531 | 500 | 0.805 | 1200 | 1.932 | 2100 | 3.381 |

**Flow of Water in Pipes.** — The quantity of water that will be discharged through a pipe depends primarily on the head and also upon the diameter of the pipe, the character of the interior surface, and the number and shape of the bends. The head may be either the actual distance between the levels of the surface of water in a reservoir and the point of discharge, or it may be caused by mechanically applied pressure, as by pumping, in which case the head is calculated as the vertical distance corresponding to the pressure. One pound per square inch is equal to 2.309 feet head, or 1 foot head is equal to a pressure of 0.433 pound per square inch.

All formulas for finding the amount of water that will flow through a pipe in a given time are approximate. The formula below will give results within 5 or 10 per cent of actual results, if applied to pipe lines carefully laid and in a fair condition.

$$V = C\sqrt{\frac{hD}{L + 54\,D}}$$

in which  $V$ = approximate mean velocity in feet per second;
$C$ = coefficient from the accompanying table;
$D$ = diameter of pipe in feet;
$h$ = total head in feet;
$L$ = total length of pipe line in feet.

### Values of Coefficient C

| Diam. of Pipe | | C | Diam. of Pipe | | C | Diam. of Pipe | | C |
|---|---|---|---|---|---|---|---|---|
| Feet | Inches | | Feet | Inches | | Feet | Inches | |
| 0.1 | 1.2 | 23 | 0.8 | 9.6 | 46 | 3.5 | 42 | 64 |
| 0.2 | 2.4 | 30 | 0.9 | 10.8 | 47 | 4.0 | 48 | 66 |
| 0.3 | 3.6 | 34 | 1.0 | 12.0 | 48 | 5.0 | 60 | 68 |
| 0.4 | 4.8 | 37 | 1.5 | 18.0 | 53 | 6.0 | 72 | 70 |
| 0.5 | 6.0 | 39 | 2.0 | 24.0 | 57 | 7.0 | 84 | 72 |
| 0.6 | 7.2 | 42 | 2.5 | 30.0 | 60 | 8.0 | 96 | 74 |
| 0.7 | 8.4 | 44 | 3.0 | 36.0 | 62 | 10.0 | 120 | 77 |

*Example.* — A pipe line, 1 mile long, 12 inches in diameter, discharges water under a head of 100 feet. Find the velocity and quantity of discharge.

From the table, the coefficient $C$ is found to be 48 for a pipe 1 foot in diameter, hence:

$$V = 48\sqrt{\frac{100 \times 1}{5280 + 54 \times 1}} = 6.57 \text{ feet per second.}$$

To find the discharge in cubic feet per second, multiply the velocity found by the area of cross-section of the pipe in square feet:

$$6.57 \times 0.7854 = 5.16 \text{ cubic feet per second.}$$

The loss of head due to a bend in the pipe is most frequently given in the equivalent length of straight pipe, which would cause the same loss in head as the bend. Experiments show that a right-angle bend should have a radius of about three times the diameter of the pipe. Assuming this curvature, then, if $D$ is the diameter of the pipe in inches and $L$ is the length of straight pipe in feet, which causes the same loss of head as the bend in the pipe, the following formula gives the equivalent length of straight pipe that should be added to compensate for a right-angle bend:

$$L = 4\,D \div 3.$$

Thus the loss of head due to a right-angle bend in a six-inch pipe would be equal to that in 8 feet of straight pipe. Experiments undertaken to determine the losses due to valves in pipe lines indicate that a fully open gate valve in a pipe causes a loss of head corresponding to that in a length of pipe equal to six diameters.

## Flow of Water Through Nozzles in Cubic Feet per Second

| Head in Feet, at Nozzle | Pressure, Pounds per Square Inch | Theoretical Velocity, Feet per Second | Diameter of Nozzle, Inches | | | | | | | |
|---|---|---|---|---|---|---|---|---|---|---|
| | | | 1 | 1½ | 2 | 2½ | 3 | 3½ | 4 | 4½ |
| 5 | 2.17 | 17.93 | 0.10 | 0.22 | 0.39 | 0.61 | 0.88 | 1.20 | 1.56 | 2.04 |
| 10 | 4.33 | 25.36 | 0.14 | 0.31 | 0.55 | 0.86 | 1.24 | 1.69 | 2.21 | 2.87 |
| 20 | 8.66 | 35.86 | 0.19 | 0.44 | 0.78 | 1.22 | 1.76 | 2.39 | 3.13 | 4.07 |
| 30 | 12.99 | 43.92 | 0.24 | 0.54 | 0.96 | 1.50 | 2.16 | 2.93 | 3.83 | 4.98 |
| 40 | 17.32 | 50.72 | 0.28 | 0.62 | 1.10 | 1.73 | 2.49 | 3.39 | 4.43 | 5.75 |
| 50 | 21.65 | 56.71 | 0.31 | 0.70 | 1.24 | 1.93 | 2.78 | 3.79 | 4.95 | 6.43 |
| 60 | 25.99 | 62.12 | 0.34 | 0.76 | 1.35 | 2.12 | 3.05 | 4.15 | 5.42 | 7.04 |
| 70 | 30.32 | 67.10 | 0.37 | 0.82 | 1.46 | 2.29 | 3.29 | 4.48 | 5.86 | 7.61 |
| 80 | 34.65 | 71.73 | 0.39 | 0.88 | 1.56 | 2.44 | 3.52 | 4.79 | 6.26 | 8.13 |
| 90 | 38.98 | 76.08 | 0.42 | 0.94 | 1.66 | 2.59 | 3.73 | 5.08 | 6.64 | 8.63 |
| 100 | 43.31 | 80.20 | 0.44 | 0.99 | 1.75 | 2.73 | 3.94 | 5.38 | 7.00 | 9.09 |
| 120 | 51.97 | 87.88 | 0.49 | 1.08 | 1.87 | 3.00 | 4.31 | 5.87 | 7.67 | 9.96 |
| 140 | 60.63 | 94.89 | 0.52 | 1.17 | 2.07 | 3.23 | 4.66 | 6.35 | 8.28 | 10.76 |
| 160 | 69.29 | 101.45 | 0.56 | 1.25 | 2.21 | 3.46 | 4.98 | 6.78 | 8.86 | 11.50 |
| 180 | 77.96 | 107.59 | 0.59 | 1.32 | 2.34 | 3.67 | 5.28 | 7.19 | 9.39 | 12.20 |
| 200 | 86.62 | 113.41 | 0.62 | 1.39 | 2.47 | 3.87 | 5.57 | 7.57 | 9.90 | 12.86 |
| 250 | 108.50 | 126.80 | 0.70 | 1.56 | 2.76 | 4.32 | 6.22 | 8.47 | 11.07 | 14.38 |
| 300 | 130.20 | 138.91 | 0.76 | 1.71 | 3.03 | 4.74 | 6.82 | 9.27 | 12.13 | 15.75 |
| 350 | 151.90 | 150.04 | 0.82 | 1.84 | 3.27 | 5.12 | 7.37 | 10.02 | 13.10 | 17.01 |
| 400 | 173.60 | 160.40 | 0.88 | 1.97 | 3.50 | 5.47 | 7.87 | 10.71 | 14.00 | 18.19 |
| 450 | 195.30 | 170.12 | 0.93 | 2.09 | 3.71 | 5.80 | 8.35 | 11.36 | 14.85 | 19.39 |
| 500 | 216.00 | 179.33 | 0.99 | 2.21 | 3.91 | 6.11 | 8.80 | 11.98 | 15.65 | 20.34 |

| Head in Feet, at Nozzle | Pressure, Pounds per Square Inch | Theoretical Velocity, Feet per Second | Diameter of Nozzle, Inches | | | | | | | |
|---|---|---|---|---|---|---|---|---|---|---|
| | | | 5 | 6 | 7 | 8 | 9 | 10 | 11 | 12 |
| 5 | 2.17 | 17.93 | 2.44 | 3.52 | 4.81 | 6.3 | 7.9 | 9.8 | 12.8 | 14.1 |
| 10 | 4.33 | 25.36 | 3.46 | 4.98 | 6.78 | 8.8 | 11.2 | 13.8 | 16.7 | 19.9 |
| 20 | 8.66 | 35.86 | 4.88 | 7.04 | 9.58 | 12.5 | 15.8 | 19.6 | 23.7 | 28.2 |
| 30 | 12.99 | 43.92 | 5.99 | 8.62 | 11.74 | 15.3 | 19.4 | 23.9 | 29.0 | 34.5 |
| 40 | 17.32 | 50.72 | 6.92 | 9.96 | 13.56 | 17.7 | 22.4 | 27.7 | 33.5 | 39.8 |
| 50 | 21.65 | 56.71 | 7.73 | 11.13 | 15.16 | 19.8 | 25.0 | 30.9 | 37.4 | 44.5 |
| 60 | 25.99 | 62.12 | 8.44 | 12.19 | 16.60 | 21.7 | 27.4 | 33.9 | 41.0 | 48.8 |
| 70 | 30.32 | 67.10 | 9.15 | 13.17 | 17.93 | 23.4 | 29.6 | 36.6 | 44.3 | 52.7 |
| 80 | 34.65 | 71.73 | 9.78 | 14.08 | 19.17 | 25.0 | 31.7 | 39.1 | 47.3 | 56.4 |
| 90 | 38.98 | 76.08 | 10.38 | 14.93 | 20.35 | 26.6 | 33.6 | 41.5 | 50.2 | 59.7 |
| 100 | 43.31 | 80.20 | 10.94 | 15.74 | 21.44 | 28.0 | 35.4 | 43.7 | 52.9 | 63.0 |
| 120 | 51.97 | 87.88 | 11.99 | 17.25 | 23.49 | 30.7 | 38.8 | 47.9 | 58.0 | 69.0 |
| 140 | 60.63 | 94.89 | 12.94 | 18.63 | 25.36 | 33.1 | 41.9 | 51.7 | 62.6 | 74.5 |
| 160 | 69.29 | 101.45 | 13.84 | 19.91 | 27.12 | 35.4 | 44.8 | 55.3 | 67.0 | 79.7 |
| 180 | 77.96 | 107.59 | 14.67 | 21.12 | 28.76 | 37.6 | 47.5 | 58.7 | 71.0 | 84.5 |
| 200 | 86.62 | 113.41 | 15.47 | 22.26 | 30.31 | 39.6 | 50.1 | 61.8 | 74.8 | 89.1 |
| 250 | 108.50 | 126.80 | 17.29 | 24.86 | 33.89 | 44.3 | 56.0 | 69.2 | 83.7 | 99.6 |
| 300 | 130.20 | 138.91 | 18.90 | 27.27 | 37.13 | 48.5 | 61.4 | 75.8 | 91.7 | 109.1 |
| 350 | 151.90 | 150.04 | 20.46 | 29.45 | 40.10 | 52.4 | 66.3 | 81.8 | 99.0 | 117.8 |
| 400 | 173.60 | 160.40 | 21.88 | 31.49 | 42.87 | 56.0 | 70.9 | 87.5 | 105.9 | 126.0 |
| 450 | 195.30 | 170.12 | 23.20 | 33.39 | 45.26 | 59.4 | 75.2 | 92.8 | 112.2 | 133.6 |
| 500 | 216.00 | 179.33 | 24.46 | 35.20 | 47.93 | 62.6 | 79.2 | 97.8 | 118.4 | 140.8 |

## Theoretical Velocity of Water Due to Head in Feet

| Head in Feet | Theoretical Velocity, Feet per Second | Theoretical Velocity, Feet per Minute | Head in Feet | Theoretical Velocity, Feet per Second | Theoretical Velocity, Feet per Minute | Head in Feet | Theoretical Velocity, Feet per Second | Theoretical Velocity, Feet per Minute |
|---|---|---|---|---|---|---|---|---|
| 1 | 8.02 | 481 | 48 | 55.60 | 3336 | 95 | 78.22 | 4693 |
| 2 | 11.34 | 682 | 49 | 56.17 | 3370 | 96 | 78.63 | 4718 |
| 3 | 13.90 | 834 | 50 | 56.74 | 3405 | 97 | 79.04 | 4742 |
| 4 | 16.05 | 963 | 51 | 57.31 | 3438 | 98 | 79.44 | 4767 |
| 5 | 17.94 | 1077 | 52 | 57.87 | 3472 | 99 | 79.85 | 4791 |
| 6 | 19.66 | 1179 | 53 | 58.42 | 3505 | 100 | 80.25 | 4815 |
| 7 | 21.23 | 1274 | 54 | 58.97 | 3538 | 105 | 82.23 | 4934 |
| 8 | 22.70 | 1362 | 55 | 59.51 | 3571 | 110 | 84.17 | 5050 |
| 9 | 24.07 | 1445 | 56 | 60.05 | 3603 | 115 | 86.06 | 5163 |
| 10 | 25.38 | 1523 | 57 | 60.59 | 3635 | 120 | 87.91 | 5274 |
| 11 | 26.61 | 1597 | 58 | 61.12 | 3667 | 125 | 89.72 | 5383 |
| 12 | 27.80 | 1668 | 59 | 61.64 | 3698 | 130 | 91.50 | 5490 |
| 13 | 28.93 | 1736 | 60 | 62.16 | 3730 | 135 | 93.24 | 5594 |
| 14 | 30.03 | 1802 | 61 | 62.68 | 3761 | 140 | 94.95 | 5697 |
| 15 | 31.08 | 1865 | 62 | 63.19 | 3791 | 145 | 96.63 | 5798 |
| 16 | 32.10 | 1926 | 63 | 63.70 | 3822 | 150 | 98.28 | 5897 |
| 17 | 33.09 | 1985 | 64 | 64.20 | 3852 | 155 | 99.91 | 5994 |
| 18 | 34.05 | 2043 | 65 | 64.70 | 3882 | 160 | 101.50 | 6090 |
| 19 | 34.98 | 2099 | 66 | 65.19 | 3912 | 165 | 103.08 | 6185 |
| 20 | 35.89 | 2153 | 67 | 65.69 | 3941 | 170 | 104.63 | 6278 |
| 21 | 36.77 | 2206 | 68 | 66.17 | 3970 | 175 | 106.16 | 6370 |
| 22 | 37.64 | 2258 | 69 | 66.66 | 4000 | 180 | 107.66 | 6460 |
| 23 | 38.49 | 2309 | 70 | 67.14 | 4028 | 185 | 109.15 | 6549 |
| 24 | 39.31 | 2359 | 71 | 67.62 | 4057 | 190 | 110.61 | 6637 |
| 25 | 40.12 | 2407 | 72 | 68.09 | 4086 | 195 | 112.06 | 6724 |
| 26 | 40.92 | 2455 | 73 | 68.56 | 4114 | 200 | 113.49 | 6809 |
| 27 | 41.70 | 2502 | 74 | 69.03 | 4142 | 205 | 114.90 | 6894 |
| 28 | 42.46 | 2548 | 75 | 69.50 | 4170 | 210 | 116.29 | 6978 |
| 29 | 43.21 | 2593 | 76 | 69.96 | 4198 | 215 | 117.66 | 7060 |
| 30 | 43.95 | 2637 | 77 | 70.42 | 4225 | 220 | 119.03 | 7142 |
| 31 | 44.68 | 2681 | 78 | 70.87 | 4252 | 225 | 120.38 | 7222 |
| 32 | 45.40 | 2724 | 79 | 71.33 | 4280 | 230 | 121.70 | 7302 |
| 33 | 46.10 | 2766 | 80 | 71.78 | 4307 | 235 | 123.02 | 7381 |
| 34 | 46.79 | 2783 | 81 | 72.22 | 4333 | 240 | 124.32 | 7459 |
| 35 | 47.48 | 2848 | 82 | 72.67 | 4360 | 245 | 125.60 | 7537 |
| 36 | 48.15 | 2889 | 83 | 73.11 | 4387 | 250 | 126.88 | 7613 |
| 37 | 48.81 | 2929 | 84 | 73.55 | 4413 | 255 | 128.15 | 7649 |
| 38 | 49.47 | 2968 | 85 | 73.99 | 4439 | 260 | 129.39 | 7764 |
| 39 | 50.12 | 3007 | 86 | 74.42 | 4465 | 270 | 131.86 | 7912 |
| 40 | 50.75 | 3045 | 87 | 74.85 | 4491 | 280 | 134.28 | 8057 |
| 41 | 51.38 | 3083 | 88 | 75.28 | 4517 | 290 | 136.66 | 8200 |
| 42 | 52.01 | 3120 | 89 | 75.71 | 4542 | 300 | 138.99 | 8340 |
| 43 | 52.62 | 3157 | 90 | 76.13 | 4568 | 310 | 141.29 | 8478 |
| 44 | 53.23 | 3194 | 91 | 76.55 | 4593 | 320 | 143.55 | 8613 |
| 45 | 53.83 | 3230 | 92 | 76.97 | 4618 | 330 | 145.78 | 8761 |
| 46 | 54.43 | 3266 | 93 | 77.39 | 4643 | 340 | 147.97 | 8878 |
| 47 | 55.02 | 3301 | 94 | 77.80 | 4668 | 350 | 150.13 | 9008 |

# PIPE AND PIPE FITTINGS

**Selected List of Pipe Standards with their ANSI, ASTM, and API Designations**

| ASTM or API | ANSI | Title |
|---|---|---|
| ASTM A53 | B125.1 | Welded and Seamless Steel Pipe |
| ASTM A106 | B125.30 | Seamless Carbon Steel Pipe for High-Temperature Service |
| ASTM A120 | B125.2 | Black and Hot-Dipped Zinc-Coated (Galvanized) Welded and Seamless Steel Pipe for Ordinary Uses |
| ASTM A134 | B125.55 | Electric-Fusion (Arc)-Welded Steel Plate Pipe (Sizes 16 in. and Over) |
| ASTM A135 | B125.3 | Electric-Resistance-Welded Steel Pipe |
| ASTM A139 | B125.31 | Electric-Fusion (Arc)-Welded Steel Plate Pipe (Sizes 4 in. and Over) |
| ASTM A155 | B125.4 | Electric-Fusion-Welded Steel Pipe for High-Pressure Service |
| ASTM A211 | B125.56 | Spiral-Welded Steel or Iron Pipe |
| ASTM A312 | B125.16 | Seamless and Welded Austenitic Stainless Steel Pipe |
| ASTM A333 | B125.17 | Seamless and Welded Steel Pipe for Low-Temperature Service |
| ASTM A335 | B125.24 | Seamless Ferritic Alloy Steel Pipe for High-Temperature Service |
| ASTM A358 | B125.57 | Electric-Fusion-Welded Austenitic Chromium-Nickel Alloy Steel Pipe for High-Temperature Service |
| ASTM A369 | B125.27 | Carbon and Ferritic Alloy Steel Forged and Bored Pipe for High-Temperature Service |
| ASTM A376 | B125.25 | Seamless Austenitic Steel Pipe for High-Temperature Central-Station Service |
| ASTM A381 | B125.35 | Metal-Arc-Welded Steel Pipe for High-Pressure Transmission Systems |
| ASTM A405 | B125.26 | Seamless Ferritic Alloy Steel Pipe Specially Heat Treated for High-Temperature Service |
| ASTM A523 | G62.5 | Plain End Seamless and Electric-Resistance-Welded Steel Pipe for High Pressure Pipe-Type Cable Circuits |
| ASTM A524 | B125.37 | Seamless Carbon Steel Pipe for Process Piping |
| ASTM A530 | B125.20 | General Requirements for Specialized Carbon and Alloy Steel Pipe |
| API 5L | . . . | Line Pipe |
| API 5LX | . . . | High-Test Line Pipe |
| API 5LS | . . . | Spiral Weld Line Pipe |

**Pipe Standards.** — A listing of pipe standards is given in the table above with the American National Standards Institute (ANSI), American Society for Testing and Materials (ASTM), and the American Petroleum Institute (API) designations. Copies of these standards may be purchased from these respective organizations.

**Wrought Steel Pipe.** — ANSI B36.10-1975 covers dimensions of welded and seamless wrought steel pipe, for high or low temperatures or pressures.

The word *pipe* as distinguished from *tube* is used to apply to tubular products of dimensions commonly used for pipelines and piping systems. Pipe dimensions of sizes 12 inches and smaller have outside diameters numerically larger than the corresponding nominal sizes whereas outside diameters of tubes are identical to nominal sizes.

*Size:* The size of all pipe is identified by the nominal pipe size. The manufacture of pipe in the nominal sizes of ⅛ inch to 12 inches, inclusive, is based on a standardized outside diameter (OD). This OD was originally selected so that pipe with a standard OD and having a wall thickness which was typical of the period would have an inside diameter (ID) approximately equal to the nominal size. Although there is now no such

### Table 1. American National Standard Weights and Dimensions of Welded and Seamless Wrought Steel Pipe (ANSI B36.10-1975)

| Nom. Size and (O.D.), in. | Wall Thick., in. | Plain End Wgt., lb/ft | Sch. No. | Other* |
|---|---|---|---|---|
| ⅛ (0.405) | 0.068 | 0.24 | 40 | 5L STD |
| | 0.095 | 0.31 | 80 | 5L XS |
| ¼ (0.540) | 0.088 | 0.42 | 40 | 5L STD |
| | 0.119 | 0.54 | 80 | 5L XS |
| ⅜ (0.675) | 0.091 | 0.57 | 40 | 5L STD |
| | 0.126 | 0.74 | 80 | 5L XS |
| ½ (0.840) | 0.109 | 0.85 | 40 | 5L STD |
| | 0.147 | 1.09 | 80 | 5L XS |
| | 0.188 | 1.31 | 160 | ... |
| | 0.294 | 1.71 | ... | 5L XXS |
| ¾ (1.050) | 0.113 | 1.13 | 40 | 5L STD |
| | 0.154 | 1.47 | 80 | 5L XS |
| | 0.219 | 1.94 | 160 | ... |
| | 0.308 | 2.44 | ... | 5L XXS |
| 1 (1.315) | 0.133 | 1.68 | 40 | 5L STD |
| | 0.179 | 2.17 | 80 | 5L XS |
| | 0.250 | 2.84 | 160 | ... |
| | 0.358 | 3.66 | ... | 5L XXS |
| 1¼ (1.660) | 0.140 | 2.27 | 40 | 5L STD |
| | 0.191 | 3.00 | 80 | 5L XS |
| | 0.250 | 3.76 | 160 | ... |
| | 0.382 | 5.21 | ... | 5L XXS |
| 1½ (1.900) | 0.145 | 2.72 | 40 | 5L STD |
| | 0.200 | 3.63 | 80 | 5L XS |
| | 0.281 | 4.86 | 160 | ... |
| | 0.400 | 6.41 | ... | 5L XXS |
| 2 (2.375) | 0.083 | 2.03 | ... | 5L, 5LX ... |
| | 0.109 | 2.64 | ... | 5L, 5LX ... |
| | 0.125 | 3.00 | ... | 5L, 5LX ... |
| | 0.141 | 3.36 | ... | 5L, 5LX ... |
| | 0.154 | 3.65 | 40 | 5L, 5LX STD |
| | 0.172 | 4.05 | ... | 5L, 5LX ... |
| | 0.188 | 4.39 | ... | 5L, 5LX ... |
| | 0.218 | 5.02 | 80 | 5L, 5LX XS |
| | 0.250 | 5.67 | ... | 5L, 5LX ... |
| | 0.281 | 6.28 | ... | 5L, 5LX ... |
| | 0.344 | 7.46 | 160 | ... |
| | 0.436 | 9.03 | ... | 5L, 5LX XXS |
| 2½ (2.875) | 0.083 | 2.47 | ... | 5L, 5LX ... |
| | 0.109 | 3.22 | ... | 5L, 5LX ... |
| | 0.125 | 3.67 | ... | 5L, 5LX ... |
| | 0.141 | 4.12 | ... | 5L, 5LX ... |
| | 0.156 | 4.53 | ... | 5L, 5LX ... |
| | 0.172 | 4.97 | ... | 5L, 5LX ... |
| | 0.188 | 5.40 | ... | 5L, 5LX ... |
| | 0.203 | 5.79 | 40 | 5L, 5LX STD |
| | 0.216 | 6.13 | ... | 5L, 5LX ... |
| | 0.250 | 7.01 | ... | 5L, 5LX ... |
| | 0.276 | 7.66 | 80 | 5L, 5LX XS |
| | 0.375 | 10.01 | 160 | ... |
| | 0.552 | 13.69 | ... | 5L, 5LX XXS |
| 3 (3.500) | 0.083 | 3.03 | ... | 5L, 5LX ... |
| | 0.109 | 3.95 | ... | 5L, 5LX ... |
| | 0.125 | 4.51 | ... | 5L, 5LX ... |
| | 0.141 | 5.06 | ... | 5L, 5LX ... |
| | 0.156 | 5.57 | ... | 5L, 5LX ... |
| | 0.172 | 6.11 | ... | 5L, 5LX ... |
| | 0.188 | 6.65 | ... | 5L, 5LX ... |
| | 0.216 | 7.58 | 40 | 5L, 5LX STD |
| | 0.250 | 8.68 | ... | 5L, 5LX ... |
| | 0.281 | 9.66 | ... | 5L, 5LX ... |
| | 0.300 | 10.25 | 80 | 5L, 5LX XS |
| | 0.438 | 14.32 | 160 | ... |
| | 0.600 | 18.58 | ... | 5L, 5LX XXS |
| 3½ (4.000) | 0.083 | 3.47 | ... | 5L, 5LX ... |
| | 0.109 | 4.53 | ... | 5L, 5LX ... |
| | 0.125 | 5.17 | ... | 5L, 5LX ... |
| | 0.141 | 5.81 | ... | 5L, 5LX ... |
| | 0.156 | 6.40 | ... | 5L, 5LX ... |
| | 0.172 | 7.03 | ... | 5L, 5LX ... |
| | 0.188 | 7.65 | ... | 5L, 5LX ... |
| | 0.226 | 9.11 | 40 | 5L, 5LX STD |
| | 0.250 | 10.01 | ... | 5L, 5LX ... |
| | 0.281 | 11.16 | ... | 5L, 5LX ... |
| | 0.318 | 12.50 | 80 | 5L, 5LX XS |
| 4 (4.500) | 0.083 | 3.92 | ... | 5L, 5LX |
| | 0.109 | 5.11 | ... | 5L |
| | 0.125 | 5.84 | ... | 5L, 5LX ... |
| | 0.141 | 6.56 | ... | 5L, 5LX ... |
| | 0.156 | 7.24 | ... | 5L, 5LX ... |
| | 0.172 | 7.95 | ... | 5L, 5LX ... |
| | 0.188 | 8.66 | ... | 5L, 5LX ... |
| | 0.203 | 9.32 | ... | 5L, 5LX ... |
| | 0.219 | 10.01 | ... | 5L, 5LX ... |
| | 0.237 | 10.79 | 40 | 5L, 5LX STD |
| | 0.250 | 11.35 | ... | 5L, 5LX ... |
| | 0.281 | 12.66 | ... | 5L, 5LX ... |
| | 0.312 | 13.96 | ... | 5L, 5LX ... |
| | 0.337 | 14.98 | 80 | 5L, 5LX XS |
| | 0.438 | 19.00 | 120 | 5L, 5LX ... |
| | 0.531 | 22.51 | 160 | 5L, 5LX ... |
| | 0.674 | 27.54 | ... | 5L, 5LX XXS |
| 5 (5.563) | 0.083 | 4.86 | ... | 5L ... |
| | 0.125 | 7.26 | ... | 5L ... |
| | 0.156 | 9.01 | ... | 5L ... |
| | 0.188 | 10.79 | ... | 5L ... |
| | 0.219 | 12.50 | ... | 5L ... |
| | 0.258 | 14.62 | 40 | 5L STD |
| | 0.281 | 15.85 | ... | 5L ... |
| | 0.312 | 17.50 | ... | 5L ... |
| | 0.344 | 19.17 | ... | 5L ... |
| | 0.375 | 20.78 | 80 | 5L XS |
| | 0.500 | 27.04 | 120 | 5L ... |
| | 0.625 | 32.96 | 160 | 5L ... |
| | 0.750 | 38.55 | ... | 5L XXS |

* Wall thicknesses listed in American Petroleum Institute (API) Standards 5L and 5LX are indicated but wall thicknesses listed in API Standard 5LS are not indicated. For these see ANSI B36.10-1975 or API 5LS Standard. Commercial designations are: STD = Standard; XS = Extra Strong; and XXS = Double Extra Strong.

Table 1 (Continued).　**American National Standard Weights and Dimensions of Welded and Seamless Wrought Steel Pipe** (ANSI B36.10-1975)

| Nom. Size and (O.D.), in.† | Wall Thick., in. | Plain End Wgt., lb/ft | Sch. No. | Other* |
|---|---|---|---|---|
| | 0.083 | 5.80 | ... | 5L, 5LX ... |
| | 0.109 | 7.59 | ... | 5L, 5LX ... |
| | 0.125 | 8.68 | ... | 5L, 5LX ... |
| | 0.141 | 9.76 | ... | 5L, 5LX ... |
| | 0.156 | 10.78 | ... | 5L, 5LX ... |
| | 0.172 | 11.85 | ... | 5L, 5LX ... |
| | 0.188 | 12.92 | ... | 5L, 5LX ... |
| | 0.203 | 13.92 | ... | 5L, 5LX ... |
| | 0.219 | 14.98 | ... | 5L, 5LX ... |
| | 0.250 | 17.02 | ... | 5L, 5LX ... |
| 6 (6.625) | 0.280 | 18.97 | 40 | 5L, 5LX STD |
| | 0.312 | 21.04 | ... | 5L, 5LX ... |
| | 0.344 | 23.08 | ... | 5L, 5LX ... |
| | 0.375 | 25.03 | ... | 5L, 5LX ... |
| | 0.432 | 28.57 | 80 | 5L, 5LX XS |
| | 0.500 | 32.71 | ... | 5L, 5LX ... |
| | 0.562 | 36.39 | 120 | 5L, 5LX ... |
| | 0.625 | 40.05 | ... | 5L, 5LX ... |
| | 0.719 | 45.35 | 160 | 5L, 5LX ... |
| | 0.864 | 53.16 | ... | 5L XXS |
| | 0.125 | 11.35 | ... | 5L, 5LX ... |
| | 0.156 | 14.11 | ... | 5L, 5LX ... |
| | 0.188 | 16.94 | ... | 5L, 5LX ... |
| | 0.203 | 18.26 | ... | 5LX |
| | 0.219 | 19.66 | ... | 5L, 5LX ... |
| | 0.250 | 22.36 | 20 | 5L, 5LX ... |
| | 0.277 | 24.70 | 30 | 5L, 5LX ... |
| | 0.312 | 27.70 | ... | 5L, 5LX ... |
| | 0.322 | 28.55 | 40 | 5L, 5LX STD |
| | 0.344 | 30.42 | ... | 5L, 5LX ... |
| 8 (8.625) | 0.375 | 33.04 | ... | 5L, 5LX ... |
| | 0.406 | 35.64 | 60 | ... |
| | 0.438 | 38.30 | ... | 5L, 5LX ... |
| | 0.500 | 43.39 | 80 | 5L, 5LX XS |
| | 0.562 | 48.40 | ... | 5L, 5LX ... |
| | 0.594 | 50.95 | 100 | ... |
| | 0.625 | 53.40 | ... | 5L, 5LX ... |
| | 0.719 | 60.71 | 120 | 5L, 5LX ... |
| | 0.812 | 67.76 | 140 | ... |
| | 0.875 | 72.42 | ... | 5L XXS |
| | 0.906 | 74.69 | 160 | ... |
| | 0.156 | 17.65 | ... | 5L, 5LX ... |
| | 0.188 | 21.21 | ... | 5L, 5LX ... |
| | 0.203 | 22.87 | ... | 5LX |
| | 0.219 | 24.63 | ... | 5L, 5LX ... |
| | 0.250 | 28.04 | 20 | 5L, 5LX ... |
| | 0.279 | 31.20 | ... | 5L, 5LX ... |
| 10 (10.750) | 0.307 | 34.24 | 30 | 5L, 5LX ... |
| | 0.344 | 38.23 | ... | 5L, 5LX ... |
| | 0.365 | 40.48 | 40 | 5L, 5LX STD |
| | 0.438 | 48.24 | ... | 5L, 5LX ... |
| | 0.500 | 54.74 | 60 | 5L, 5LX XS |
| | 0.562 | 61.15 | ... | 5L, 5LX ... |
| | 0.594 | 64.43 | 80 | ... |
| | 0.625 | 67.58 | ... | 5L, 5LX ... |

| Nom. Size and (O.D.), in.† | Wall Thick., in. | Plain End Wgt., lb/ft | Sch. No. | Other* |
|---|---|---|---|---|
| 10 (10.750) | 0.719 | 77.03 | 100 | 5L, 5LX ... |
| | 0.812 | 86.18 | ... | 5L |
| | 0.844 | 89.29 | 120 | ... |
| | 1.000 | 104.13 | 140 | XXS |
| | 1.125 | 115.64 | 160 | ... |
| | 0.172 | 23.11 | ... | 5L, 5LX ... |
| | 0.188 | 25.22 | ... | 5L, 5LX ... |
| | 0.203 | 27.20 | ... | 5LX |
| | 0.219 | 29.31 | ... | 5L, 5LX ... |
| | 0.250 | 33.38 | 20 | 5L, 5LX ... |
| | 0.281 | 37.42 | ... | 5L, 5LX .. |
| | 0.312 | 41.45 | ... | 5L, 5LX ... |
| | 0.330 | 43.77 | 30 | 5L, 5LX ... |
| | 0.344 | 45.58 | ... | 5L, 5LX ... |
| | 0.375 | 49.56 | ... | 5L, 5LX STD |
| | 0.406 | 53.52 | 40 | 5LX |
| 12 (12.750) | 0.438 | 57.59 | ... | 5L, 5LX ... |
| | 0.500 | 65.42 | ... | 5L, 5LX XS |
| | 0.562 | 73.15 | 60 | 5L, 5LX ... |
| | 0.625 | 80.93 | ... | 5L, 5LX ... |
| | 0.688 | 88.63 | 80 | 5L, 5LX ... |
| | 0.750 | 96.12 | ... | 5L, 5LX ... |
| | 0.812 | 103.53 | ... | 5L, 5LX ... |
| | 0.844 | 107.32 | 100 | ... |
| | 0.875 | 110.97 | ... | 5L, 5LX ... |
| | 1.000 | 125.49 | 120 | XXS |
| | 1.125 | 139.67 | 140 | ... |
| | 1.312 | 160.27 | 160 | ... |
| | 0.188 | 27.73 | ... | 5L, 5LX ... |
| | 0.203 | 29.91 | ... | 5L |
| | 0.210 | 30.93 | ... | 5LX |
| | 0.219 | 32.23 | ... | 5LX |
| | 0.250 | 36.71 | 10 | 5L, 5LX ... |
| | 0.281 | 41.17 | ... | 5L, 5LX ... |
| | 0.312 | 45.61 | 20 | 5L, 5LX ... |
| | 0.344 | 50.17 | ... | 5L, 5LX ... |
| | 0.375 | 54.57 | 30 | 5L, 5LX STD |
| | 0.406 | 58.94 | ... | 5LX |
| | 0.438 | 63.44 | 40 | 5L, 5LX ... |
| | 0.469 | 67.78 | ... | 5LX |
| 14 (14.000) | 0.500 | 72.09 | ... | 5L, 5LX XS |
| | 0.562 | 80.66 | ... | 5L, 5LX ... |
| | 0.594 | 85.05 | 60 | ... |
| | 0.625 | 89.28 | ... | 5L, 5LX ... |
| | 0.688 | 97.81 | ... | 5L, 5LX ... |
| | 0.750 | 106.13 | 80 | 5L, 5LX ... |
| | 0.812 | 114.37 | ... | 5L, 5LX ... |
| | 0.875 | 122.65 | ... | 5L, 5LX ... |
| | 0.938 | 130.85 | 100 | 5L, 5LX ... |
| | 1.094 | 150.79 | 120 | ... |
| | 1.250 | 170.21 | 140 | ... |
| | 1.406 | 189.11 | 160 | ... |
| | 2.000 | 256.32 | ... | ... |
| | 2.125 | 269.50 | ... | ... |
| | 2.200 | 277.25 | ... | ... |
| | 2.500 | 307.05 | ... | ... |

* Wall thicknesses listed in American Petroleum Institute (API) Standards 5L and 5LX are indicated but wall thicknesses listed in API Standard 5LS are not indicated. For these see ANSI B36.10-1975 or API 5LS Standard. Commercial Designations are: STD = Standard; XS = Extra Strong; and XXS = Double Extra Strong.

† For sizes larger than 14 inches see ANSI B36.10-1975 Standard.

relation between the existing standard thicknesses, ODs and nominal sizes, these nominal sizes and standard ODs continue in use as "standard."

The manufacture of pipe in nominal sizes of 14-inch OD and larger proceeds on the basis of an OD corresponding to the nominal size.

*Weight:* The nominal weights of steel pipe are calculated values and are tabulated in Table 1. They are based on the following formula:

$$W_{pe} = 10.68(D - t)t$$

where $W_{pe}$ = nominal plain end weight rounded to the nearest 0.01 lb/ft.
$D$ = outside diameter to the nearest 0.001 in.
$t$ = specified wall thickness rounded to the nearest 0.001 in.

*Wall thickness:* The nominal wall thicknesses are given in Table 1 which also indicates the wall thicknesses in API Standards 5L and 5LX. Thicknesses listed in API Standard 5LS are not indicated but may be found in that Standard or in ANSI B36.10-1975.

The wall thickness designations "Standard," "Extra-Strong," and "Double Extra-Strong" have been commercially used designations for many years. The Schedule Numbers were subsequently added as a convenient designation for use in ordering pipe. "Standard" and Schedule 40 are identical for nominal pipe sizes up to 10 inch, inclusive. All larger sizes of "Standard" have ⅜-inch wall thickness. "Extra-Strong" and Schedule 80 are identical for nominal pipe sizes up to 8 inch, inclusive. All larger sizes of "Extra-Strong" have ½-inch wall thickness.

Pipe of sizes and wall thickness other than those of "Standard," "Extra-Strong," "Double Extra-Strong" and Schedule Number were adopted from API Standards 5L, 5LX, and 5LS. It was not considered practical to establish Schedule Numbers or new designations for them.

*Wall Thickness Selection:* When the selection of wall thickness depends primarily on capacity to resist internal pressure under given conditions, the designer shall compute the exact value of wall thickness suitable for conditions for which the pipe is required as prescribed in the "ASME Boiler and Pressure Vessel Code," "ANSI B31 Code for Pressure Piping," or other similar codes, whichever governs the construction. A thickness can then be selected from Table 1 to suit the value computed to fulfill the conditions for which the pipe is desired.

**Bursting Pressure of Pipes.** — The bursting pressure of pipes can be determined approximately by the following formula (Barlow's): $P = [2T \times S] \div O$ in which $P$ = bursting pressure in pounds per square inch; $T$ = thickness of wall, in inches; $O$ = outside diameter of pipe, in inches; $S$ = tensile strength of material, in pounds per square inch. The value of $S$ as determined by actual bursting tests is 40,000 pounds for butt-welded steel pipe, and 50,000 pounds for lap-welded steel pipe. The accuracy of the foregoing formula has been verified by an exhaustive series of tests conducted by the National Tube Co. In these tests, all types of pipe and tubing were burst, and a number of different methods of plugging the ends were employed to obtain results for different strains. These results were carefully checked with all available formulas and the Barlow formula came closer to the experimental results than any of the others. The *working* pressure for a pipe is usually taken as ⅙, ⅛, or ¹⁄₁₀ of the bursting pressure.

**Making Screwed Joints Tight.** — When making up screwed joints, the threads should be clean, and red or white lead, or some standard pipe joint cement or lubricant, should be applied to the threads in order to decrease the friction of the bearing surfaces of the threads; the joint should not be screwed up fast enough to produce excessive friction. Friction of the threads produces heat, thus causing the metal of the pipe to expand before the joint is properly made, with the result that, when the pipe cools again and contracts in the flange or fitting, the joint may be loose and cause leakage when the pressure is turned on in the piping system.

**Table 2. Properties of American National Standard Schedule 40 Welded and Seamless Wrought Steel Pipe**

| Diameter, Inches | | | Wall Thickness, Inches | Cross-Sectional Area of Metal | Weight per Foot, Pounds | | Capacity per Foot of Length | | Length of Pipe in Feet to Contain | | Properties of Sections | | |
|---|---|---|---|---|---|---|---|---|---|---|---|---|---|
| Nominal | Inside Actual | Outside Actual | | | Of Pipe | Of Water in Pipe | In Cubic Inches | In Gallons | One Cubic Foot | One Gallon | Moment of Inertia | Radius of Gyration | Section Modulus |
| ⅛ | 0.269 | 0.405 | 0.068 | 0.072 | 0.24 | 0.025 | 0.682 | 0.003 | 2532. | 338.7 | 0.00106 | 0.122 | 0.00525 |
| ¼ | 0.364 | 0.540 | 0.088 | 0.125 | 0.42 | 0.045 | 1.249 | 0.005 | 1384. | 185.0 | 0.00331 | 0.163 | 0.01227 |
| ⅜ | 0.493 | 0.675 | 0.091 | 0.167 | 0.57 | 0.083 | 2.291 | 0.010 | 754.4 | 100.8 | 0.00729 | 0.209 | 0.02160 |
| ½ | 0.622 | 0.840 | 0.109 | 0.250 | 0.85 | 0.132 | 3.646 | 0.016 | 473.9 | 63.35 | 0.01709 | 0.261 | 0.04070 |
| ¾ | 0.824 | 1.050 | 0.113 | 0.333 | 1.13 | 0.231 | 6.399 | 0.028 | 270.0 | 36.10 | 0.0374 | 0.334 | 0.07055 |
| 1 | 1.049 | 1.315 | 0.133 | 0.494 | 1.68 | 0.374 | 10.37 | 0.045 | 166.6 | 22.27 | 0.08734 | 0.421 | 0.1328 |
| 1¼ | 1.380 | 1.660 | 0.140 | 0.669 | 2.27 | 0.648 | 17.95 | 0.078 | 96.28 | 12.87 | 0.1947 | 0.539 | 0.2346 |
| 1½ | 1.610 | 1.900 | 0.145 | 0.799 | 2.72 | 0.882 | 24.43 | 0.106 | 70.73 | 9.456 | 0.3099 | 0.623 | 0.3262 |
| 2 | 2.067 | 2.375 | 0.154 | 1.075 | 3.65 | 1.454 | 40.27 | 0.174 | 42.91 | 5.737 | 0.6658 | 0.787 | 0.5607 |
| 2½ | 2.469 | 2.875 | 0.203 | 1.704 | 5.79 | 2.074 | 57.45 | 0.249 | 30.08 | 4.021 | 1.530 | 0.947 | 1.064 |
| 3 | 3.068 | 3.500 | 0.216 | 2.228 | 7.58 | 3.202 | 88.71 | 0.384 | 19.48 | 2.604 | 3.017 | 1.163 | 1.724 |
| 3½ | 3.548 | 4.000 | 0.226 | 2.680 | 9.11 | 4.283 | 118.6 | 0.514 | 14.56 | 1.947 | 4.788 | 1.337 | 2.394 |
| 4 | 4.026 | 4.500 | 0.237 | 3.174 | 10.79 | 5.515 | 152.8 | 0.661 | 11.31 | 1.512 | 7.233 | 1.510 | 3.215 |
| 5 | 5.047 | 5.563 | 0.258 | 4.300 | 14.62 | 8.666 | 240.1 | 1.04 | 7.198 | 0.9622 | 15.16 | 1.878 | 5.451 |
| 6 | 6.065 | 6.625 | 0.280 | 5.581 | 18.97 | 12.52 | 346.7 | 1.50 | 4.984 | 0.6663 | 28.14 | 2.245 | 8.496 |
| 8 | 7.981 | 8.625 | 0.322 | 8.399 | 28.55 | 21.67 | 600.3 | 2.60 | 2.878 | 0.3848 | 72.49 | 2.938 | 16.81 |
| 10 | 10.020 | 10.750 | 0.365 | 11.91 | 40.48 | 34.16 | 946.3 | 4.10 | 1.826 | 0.2441 | 160.7 | 3.674 | 29.91 |
| 12 | 11.938 | 12.750 | 0.406 | 15.74 | 53.52 | 48.49 | 1343. | 5.81 | 1.286 | 0.1720 | 300.2 | 4.364 | 47.09 |
| 16 | 15.000 | 16.000 | 0.500 | 24.35 | 82.77 | 76.55 | 2121. | 9.18 | 0.8149 | 0.1089 | 732.0 | 5.484 | 91.50 |
| 18 | 16.876 | 18.000 | 0.562 | 30.79 | 104.7 | 96.90 | 2684. | 11.62 | 0.6438 | 0.0861 | 1172. | 6.168 | 130.2 |
| 20 | 18.812 | 20.000 | 0.594 | 36.21 | 123.1 | 120.4 | 3335. | 14.44 | 0.5181 | 0.0693 | 1706. | 6.864 | 170.6 |
| 24 | 22.624 | 24.000 | 0.688 | 50.39 | 171.3 | 174.1 | 4844. | 20.88 | 0.3582 | 0.0479 | 3426. | 8.246 | 285.5 |
| 32 | 30.624 | 32.000 | 0.688 | 67.68 | 230.1 | 319.1 | 8839. | 38.26 | 0.1955 | 0.0261 | 8299. | 11.07 | 518.7 |

*Note:* Torsional section modulus equals twice section modulus.

**Table 3. Properties of American National Standard Schedule 80 Welded and Seamless Wrought Steel Pipe**

| Diameter, Inches | | | Wall Thickness, Inches | Cross-Sectional Area of Metal | Weight per Foot, Pounds | | Capacity per Foot of Length | | Length of Pipe in Feet to Contain | | Properties of Sections | | |
|---|---|---|---|---|---|---|---|---|---|---|---|---|---|
| Nominal | Inside Actual | Outside Actual | | | Of Pipe | Of Water in Pipe | In Cubic Inches | In Gallons | One Cubic Foot | One Gallon | Moment of Inertia | Radius of Gyration | Section Modulus |
| 1/8 | 0.215 | 0.405 | 0.095 | 0.093 | 0.315 | 0.016 | 0.436 | 0.0019 | 3966. | 530.2 | 0.00122 | 0.115 | 0.00600 |
| 1/4 | 0.302 | 0.540 | 0.119 | 0.157 | 0.537 | 0.031 | 0.860 | 0.0037 | 2010. | 268.7 | 0.00377 | 0.155 | 0.01395 |
| 3/8 | 0.423 | 0.675 | 0.126 | 0.217 | 0.739 | 0.061 | 1.686 | 0.0073 | 1025. | 137.0 | 0.00862 | 0.199 | 0.02554 |
| 1/2 | 0.546 | 0.840 | 0.147 | 0.320 | 1.088 | 0.101 | 2.810 | 0.0122 | 615.0 | 82.22 | 0.02008 | 0.250 | 0.04780 |
| 3/4 | 0.742 | 1.050 | 0.154 | 0.433 | 1.474 | 0.187 | 5.189 | 0.0225 | 333.0 | 44.52 | 0.04479 | 0.321 | 0.08531 |
| 1 | 0.957 | 1.315 | 0.179 | 0.639 | 2.172 | 0.312 | 8.632 | 0.0374 | 200.2 | 26.76 | 0.1056 | 0.407 | 0.1606 |
| 1 1/4 | 1.278 | 1.660 | 0.191 | 0.881 | 2.997 | 0.556 | 15.39 | 0.0667 | 112.3 | 15.01 | 0.2418 | 0.534 | 0.2913 |
| 1 1/2 | 1.500 | 1.900 | 0.200 | 1.068 | 3.631 | 0.766 | 21.21 | 0.0918 | 81.49 | 10.89 | 0.3912 | 0.605 | 0.4118 |
| 2 | 1.939 | 2.375 | 0.218 | 1.477 | 5.022 | 1.279 | 35.43 | 0.1534 | 48.77 | 6.519 | 0.8680 | 0.766 | 0.7309 |
| 2 1/2 | 2.323 | 2.875 | 0.276 | 2.254 | 7.661 | 1.836 | 50.86 | 0.2202 | 33.98 | 4.542 | 1.924 | 0.924 | 1.339 |
| 3 | 2.900 | 3.500 | 0.300 | 3.016 | 10.25 | 2.861 | 79.26 | 0.3431 | 21.80 | 2.914 | 3.895 | 1.136 | 2.225 |
| 3 1/2 | 3.364 | 4.000 | 0.318 | 3.678 | 12.50 | 3.850 | 106.7 | 0.4617 | 16.20 | 2.166 | 6.280 | 1.307 | 3.140 |
| 4 | 3.826 | 4.500 | 0.337 | 4.407 | 14.98 | 4.980 | 138.0 | 0.5972 | 12.53 | 1.674 | 9.611 | 1.477 | 4.272 |
| 5 | 4.813 | 5.563 | 0.375 | 6.112 | 20.78 | 7.882 | 218.3 | 0.9451 | 7.915 | 1.058 | 20.67 | 1.839 | 7.432 |
| 6 | 5.761 | 6.625 | 0.432 | 8.405 | 28.57 | 11.29 | 312.8 | 1.354 | 5.524 | 0.738 | 40.49 | 2.195 | 12.22 |
| 8 | 7.625 | 8.625 | 0.500 | 12.76 | 43.39 | 19.78 | 548.0 | 2.372 | 3.153 | 0.422 | 105.7 | 2.878 | 24.52 |
| 10 | 9.562 | 10.750 | 0.594 | 18.95 | 64.42 | 31.11 | 861.7 | 3.730 | 2.005 | 0.268 | 245.2 | 3.597 | 45.62 |
| 12 | 11.374 | 12.750 | 0.688 | 26.07 | 88.63 | 44.02 | 1219. | 5.278 | 1.417 | 0.189 | 475.7 | 4.271 | 74.62 |
| 14 | 12.500 | 14.000 | 0.750 | 31.22 | 106.1 | 53.16 | 1473. | 6.375 | 1.173 | 0.157 | 687.4 | 4.692 | 98.19 |
| 16 | 14.312 | 16.000 | 0.844 | 40.19 | 136.6 | 69.69 | 1931. | 8.357 | 0.895 | 0.120 | 1158. | 5.366 | 144.7 |
| 18 | 16.124 | 18.000 | 0.938 | 50.28 | 170.9 | 88.46 | 2450. | 10.61 | 0.705 | 0.094 | 1835. | 6.041 | 203.9 |
| 20 | 17.938 | 20.000 | 1.031 | 61.44 | 208.9 | 109.5 | 3033. | 13.13 | 0.570 | 0.076 | 2772. | 6.716 | 277.2 |
| 22 | 19.750 | 22.000 | 1.125 | 73.78 | 250.8 | 132.7 | 3676. | 15.91 | 0.470 | 0.063 | 4031. | 7.391 | 366.4 |

*Note:* Torsional section modulus equals twice section modulus.

## Volume of Flow at 1 Foot Per-Minute Velocity in Pipe and Tube*

| Nominal Diam, Inches | Schedule 40 Pipe | | | Schedule 80 Pipe | | | Type K Copper Tube | | | Type L Copper Tube | | |
|---|---|---|---|---|---|---|---|---|---|---|---|---|
| | Cu. Ft. per Minute | Gallons per Minute | Pounds 60 F Water per Min. | Cu. Ft. per Minute | Gallons per Minute | Pounds 60 F Water per Min. | Cu. Ft. per Minute | Gallons per Minute | Pounds 60 F Water per Min. | Cu. Ft. per Minute | Gallons per Minute | Pounds 60 F Water per Min. |
| ⅛ | 0.0004 | 0.003 | 0.025 | 0.0003 | 0.002 | 0.016 | 0.0002 | 0.0014 | 0.012 | 0.0002 | 0.002 | 0.014 |
| ¼ | 0.0007 | 0.005 | 0.044 | 0.0005 | 0.004 | 0.031 | 0.0005 | 0.0039 | 0.033 | 0.0005 | 0.004 | 0.034 |
| ⅜ | 0.0013 | 0.010 | 0.081 | 0.0010 | 0.007 | 0.061 | 0.0009 | 0.0066 | 0.055 | 0.0010 | 0.008 | 0.063 |
| ½ | 0.0021 | 0.016 | 0.132 | 0.0016 | 0.012 | 0.102 | 0.0015 | 0.0113 | 0.094 | 0.0016 | 0.012 | 0.101 |
| ¾ | 0.0037 | 0.028 | 0.232 | 0.0030 | 0.025 | 0.213 | 0.0030 | 0.0267 | 0.189 | 0.0034 | 0.025 | 0.210 |
| 1 | 0.0062 | 0.046 | 0.387 | 0.0050 | 0.037 | 0.312 | 0.0054 | 0.0404 | 0.338 | 0.0057 | 0.043 | 0.358 |
| 1¼ | 0.0104 | 0.078 | 0.649 | 0.0088 | 0.067 | 0.555 | 0.0085 | 0.0632 | 0.53 | 0.0087 | 0.065 | 0.545 |
| 1½ | 0.0141 | 0.106 | 0.882 | 0.0123 | 0.092 | 0.765 | 0.0196 | 0.1465 | 1.22 | 0.0124 | 0.093 | 0.770 |
| 2 | 0.0233 | 0.174 | 1.454 | 0.0206 | 0.154 | 1.280 | 0.0209 | 0.1565 | 1.31 | 0.0215 | 0.161 | 1.34 |
| 2½ | 0.0332 | 0.248 | 2.073 | 0.0294 | 0.220 | 1.830 | 0.0323 | 0.2418 | 2.02 | 0.0331 | 0.248 | 2.07 |
| 3 | 0.0514 | 0.383 | 3.201 | 0.0460 | 0.344 | 2.870 | 0.0461 | 0.3446 | 2.88 | 0.0473 | 0.354 | 2.96 |
| 3½ | 0.0682 | 0.513 | 4.287 | 0.0617 | 0.458 | 3.720 | 0.0625 | 0.4675 | 3.91 | 0.0640 | 0.479 | 4.00 |
| 4 | 0.0884 | 0.660 | 5.516 | 0.0800 | 0.597 | 4.970 | 0.0811 | 0.6068 | 5.07 | 0.0841 | 0.622 | 5.20 |
| 5 | 0.1390 | 1.040 | 8.674 | 0.1260 | 0.947 | 7.940 | 0.1159 | 0.9415 | 7.87 | 0.1296 | 0.969 | 8.10 |
| 6 | 0.2010 | 1.500 | 12.52 | 0.1820 | 1.355 | 11.300 | 0.1797 | 1.3440 | 11.2 | 0.1862 | 1.393 | 11.6 |
| 8 | 0.3480 | 2.600 | 21.68 | 0.3180 | 2.380 | 19.800 | 0.3135 | 2.3446 | 19.6 | 0.3253 | 2.434 | 20.3 |
| 10 | 0.5476 | 4.10 | 34.18 | 0.5560 | 4.165 | 31.130 | 0.4867 | 3.4405 | 30.4 | 0.5050 | 3.777 | 21.6 |
| 12 | 0.7773 | 5.81 | 48.52 | 0.7060 | 5.280 | 44.040 | 0.6978 | 5.2194 | 43.6 | 0.7291 | 5.454 | 45.6 |
| 14 | 0.9396 | 7.03 | 58.65 | 0.8520 | 6.380 | 53.180 | — | — | — | — | — | — |
| 16 | 1.227 | 9.18 | 76.60 | 1.1170 | 8.360 | 69.730 | — | — | — | — | — | — |
| 18 | 1.553 | 11.62 | 96.95 | 1.4180 | 10.610 | 88.500 | — | — | — | — | — | — |
| 20 | 1.931 | 14.44 | 120.5 | 1.7550 | 13.130 | 109.510 | — | — | — | — | — | — |

* To obtain volume of flow at any other velocity, multiply values in table by velocity in feet per minute.

### Seamless Drawn Brass and Copper Pipe

Made to correspond with iron pipe and to fit iron pipe fittings (American Tube Works).

| Diameter | | | Approximate Weight per Foot, Pounds | | Diameter | | | Approximate Weight per Foot, Pounds | |
|---|---|---|---|---|---|---|---|---|---|
| Iron Pipe Size | Approx. Outside Diam. | Exact Outside Diam. | Brass | Copper | Iron Pipe Size | Approx. Outside Diam. | Exact Outside Diam. | Brass | Copper |
| ⅛ | ⅜ | 0.405 | 0.25 | 0.26 | 2½ | 2⅞ | 2.875 | 5.75 | 6.05 |
| ¼ | ⁹⁄₁₆ | 0.540 | 0.43 | 0.45 | 3 | 3½ | 3.500 | 8.30 | 8.74 |
| ⅜ | 11⁄16 | 0.675 | 0.62 | 0.65 | 3½ | 4 | 4.000 | 10.90 | 11.47 |
| ½ | 13⁄16 | 0.840 | 0.90 | 0.95 | 4 | 4½ | 4.500 | 12.70 | 13.37 |
| ¾ | 1¹⁄₁₆ | 1.050 | 1.25 | 1.32 | 4½ | 5 | 5.000 | 13.90 | 14.63 |
| 1 | 1⁵⁄₁₆ | 1.315 | 1.70 | 1.79 | 5 | 5⁹⁄₁₆ | 5.563 | 15.75 | 16.58 |
| 1¼ | 1⅝ | 1.660 | 2.50 | 2.63 | 6 | 6⅝ | 6.625 | 18.31 | 19.27 |
| 1½ | 1⅞ | 1.900 | 3.00 | 3.16 | 7 | 7⅝ | 7.625 | 23.73 | 24.98 |
| 2 | 2⅜ | 2.375 | 4.00 | 4.21 | ..... | ..... | ..... | ..... | .... |

**Threading Pipe.** — Clean, smooth pipe threads are essential to a good joint and depend largely upon the rake or lip angle and lead of the chasers, and the clearance, chip space and number of chasers in the die-head. The lip angle should vary from 15 to 25 degrees, depending upon the style and condition of the chasers and chaser holders. The chip space in front of the chasers should be large enough to allow room for accumulation of chips and at the same time provide means of

### Length of Thread on Pipe Required to Make a Tight Joint
(Crane Co.)

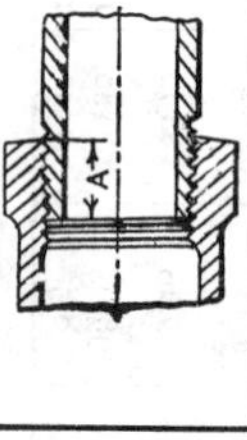

| | Size of Pipe, Inches | Dimension A, Inches | Size of Pipe, Inches | Dimension A, Inches | Size of Pipe, Inches | Dimension A, Inches |
|---|---|---|---|---|---|---|
| | ⅛ | ¼ | 1½ | 11⁄16 | 5 | 1¼ |
| | ¼ | ⅜ | 2 | ¾ | 6 | 1⁵⁄₁₆ |
| | ⅜ | ⅜ | 2½ | 15⁄16 | 7 | 1⅜ |
| | ½ | ½ | 3 | 1 | 8 | 1⁷⁄₁₆ |
| | ¾ | ⁹⁄₁₆ | 3½ | 1¹⁄₁₆ | 9 | 1½ |
| | 1 | 11⁄16 | 4 | 1⅛ | 10 | 1⅝ |
| | 1¼ | 11⁄16 | 4½ | 1³⁄₁₆ | 12 | 1¾ |

Dimensions do not allow for variation in tapping or threading.

lubricating the chasers. This is an important point, as insufficient chip space will cause the chips to clog and tear the threads. The lead of the chaser is the angle which is machined or ground on the leading or front side, to enable the die to start readily on the pipe, and also to distribute the work of cutting over a number of threads. To secure a good thread, the lead should cover the first three threads. As the heaviest cutting is done by this beveled part, it should have a slightly greater clearance angle than the rest of the threads on the chaser. When re-grinding chasers which have become dull on the lead, care should be taken to give each chaser the same length of lead, as otherwise the work will be unevenly distributed.

(Continued on page 2335)

**Sizes and Weights in Pounds per Foot of Seamless Brass Tubes** *

| Outside Diam. of Tube, Inches | Thickness — Stub's or Birmingham Gage | | | | | | | | | | | |
|---|---|---|---|---|---|---|---|---|---|---|---|---|
| | 3 | 4 | 5 | 6 | 7 | 8 | 9 | 10 | 11 | 12 | 13 | 14 |
| | Decimal Equivalent of Gage Number, Inch | | | | | | | | | | | |
| | 0.259 | 0.238 | 0.220 | 0.203 | 0.180 | 0.165 | 0.148 | 0.134 | 0.120 | 0.109 | 0.095 | 0.083 |
| 1/8 | .... | .... | .... | .... | .... | .... | .... | .... | .... | .... | .... | .... |
| 3/16 | .... | .... | .... | .... | .... | .... | .... | .... | .... | .... | .... | .... |
| 1/4 | .... | .... | .... | .... | .... | .... | .... | .... | 0.18 | 0.177 | 0.170 | 0.160 |
| 5/16 | .... | .... | .... | .... | .... | .... | .... | .... | 0.27 | 0.256 | 0.238 | 0.220 |
| 3/8 | .... | .... | .... | .... | .... | 0.40 | 0.39 | 0.37 | 0.35 | 0.335 | 0.307 | 0.280 |
| 7/16 | .... | .... | .... | .... | .... | 0.52 | 0.49 | 0.47 | 0.44 | 0.413 | 0.376 | 0.340 |
| 1/2 | .... | .... | .... | 0.70 | 0.66 | 0.64 | 0.60 | 0.57 | 0.53 | 0.492 | 0.444 | 0.400 |
| 9/16 | .... | .... | .... | 0.84 | 0.79 | 0.76 | 0.71 | 0.66 | 0.61 | 0.571 | 0.513 | 0.460 |
| 5/8 | 1.09 | 1.06 | 1.03 | 0.99 | 0.92 | 0.88 | 0.81 | 0.76 | 0.70 | 0.649 | 0.581 | 0.520 |
| 11/16 | 1.28 | 1.23 | 1.19 | 1.13 | 1.05 | 0.99 | 0.92 | 0.86 | 0.79 | 0.728 | 0.650 | 0.580 |
| 3/4 | 1.47 | 1.41 | 1.35 | 1.28 | 1.18 | 1.11 | 1.03 | 0.95 | 0.87 | 0.807 | 0.718 | 0.640 |
| 13/16 | 1.65 | 1.58 | 1.50 | 1.43 | 1.31 | 1.23 | 1.13 | 1.05 | 0.96 | 0.885 | 0.787 | 0.700 |
| 7/8 | 1.84 | 1.75 | 1.66 | 1.57 | 1.44 | 1.35 | 1.24 | 1.15 | 1.04 | 0.964 | 0.855 | 0.759 |
| 15/16 | 2.03 | 1.92 | 1.82 | 1.72 | 1.57 | 1.47 | 1.35 | 1.24 | 1.13 | 1.042 | 0.924 | 0.819 |
| 1 | 2.22 | 2.09 | 1.98 | 1.87 | 1.70 | 1.59 | 1.45 | 1.34 | 1.22 | 1.12 | 0.99 | 0.88 |
| 1 1/8 | 2.60 | 2.44 | 2.30 | 2.16 | 1.96 | 1.83 | 1.67 | 1.53 | 1.39 | 1.28 | 1.13 | 1.00 |
| 1 1/4 | 2.97 | 2.78 | 2.61 | 2.45 | 2.22 | 2.07 | 1.88 | 1.73 | 1.56 | 1.44 | 1.27 | 1.12 |
| 1 3/8 | 3.35 | 3.12 | 2.93 | 2.75 | 2.48 | 2.30 | 2.10 | 1.92 | 1.74 | 1.59 | 1.40 | 1.24 |
| 1 1/2 | 3.72 | 3.47 | 3.25 | 3.04 | 2.74 | 2.54 | 2.31 | 2.11 | 1.91 | 1.75 | 1.54 | 1.36 |
| 1 5/8 | 4.09 | 3.81 | 3.57 | 3.33 | 3.00 | 2.78 | 2.52 | 2.31 | 2.08 | 1.91 | 1.68 | 1.48 |
| 1 3/4 | 4.47 | 4.15 | 3.88 | 3.62 | 3.26 | 3.02 | 2.74 | 2.50 | 2.26 | 2.06 | 1.82 | 1.60 |
| 1 7/8 | 4.84 | 4.50 | 4.20 | 3.92 | 3.52 | 3.26 | 2.95 | 2.69 | 2.43 | 2.22 | 1.95 | 1.72 |
| 2 | 5.21 | 4.84 | 4.52 | 4.21 | 3.78 | 3.50 | 3.16 | 2.89 | 2.60 | 2.38 | 2.09 | 1.84 |
| 2 1/8 | 5.59 | 5.18 | 4.84 | 4.50 | 4.04 | 3.73 | 3.38 | 3.08 | 2.78 | 2.54 | 2.23 | 1.96 |
| 2 1/4 | 5.96 | 5.53 | 5.15 | 4.80 | 4.30 | 3.97 | 3.59 | 3.27 | 2.95 | 2.69 | 2.36 | 2.08 |
| 2 3/8 | 6.34 | 5.87 | 5.47 | 5.09 | 4.56 | 4.21 | 3.80 | 3.47 | 3.12 | 2.85 | 2.50 | 2.20 |
| 2 1/2 | 6.71 | 6.21 | 5.79 | 5.38 | 4.82 | 4.45 | 4.02 | 3.66 | 3.30 | 3.01 | 2.64 | 2.32 |
| 2 5/8 | 7.08 | 6.56 | 6.11 | 5.67 | 5.08 | 4.69 | 4.23 | 3.85 | 3.47 | 3.17 | 2.77 | 2.44 |
| 2 3/4 | 7.46 | 6.90 | 6.42 | 5.97 | 5.34 | 4.92 | 4.44 | 4.05 | 3.64 | 3.32 | 2.91 | 2.56 |
| 2 7/8 | 7.83 | 7.24 | 6.74 | 6.26 | 5.60 | 5.16 | 4.66 | 4.24 | 3.81 | 3.48 | 3.05 | 2.68 |
| 3 | 8.20 | 7.59 | 7.06 | 6.55 | 5.86 | 5.40 | 4.87 | 4.43 | 3.99 | 3.64 | 3.19 | 2.79 |
| 3 1/8 | 8.58 | 7.93 | 7.38 | 6.85 | 6.12 | 5.64 | 5.08 | 4.63 | 4.16 | 3.79 | 3.32 | 2.91 |
| 3 1/4 | 8.95 | 8.27 | 7.69 | 7.14 | 6.38 | 5.88 | 5.30 | 4.82 | 4.33 | 3.95 | 3.46 | 3.03 |
| 3 3/8 | 9.33 | 8.62 | 8.01 | 7.43 | 6.64 | 6.11 | 5.51 | 5.01 | 4.51 | 4.11 | 3.60 | 3.15 |
| 3 1/2 | 9.70 | 8.96 | 8.33 | 7.72 | 6.90 | 6.35 | 5.72 | 5.21 | 4.68 | 4.27 | 3.73 | 3.27 |
| 3 5/8 | 10.07 | 9.30 | 8.65 | 8.02 | 7.16 | 6.59 | 5.94 | 5.40 | 4.85 | 4.42 | 3.87 | 3.39 |
| 3 3/4 | 10.45 | 9.65 | 8.96 | 8.31 | 7.42 | 6.83 | 6.15 | 5.59 | 5.03 | 4.58 | 4.01 | 3.51 |
| 3 7/8 | 10.82 | 9.99 | 9.28 | 8.60 | 7.68 | 7.07 | 6.37 | 5.79 | 5.20 | 4.74 | 4.15 | 3.63 |

To determine weight per foot of a tube of a given *inside diameter*, add to weights in above list the weights in pounds per foot given below under corresponding gage numbers.

| Gage No. | 3 | 4 | 5 | 6 | 7 | 8 | 9 | 10 | 11 | 12 | 13 | 14 |
|---|---|---|---|---|---|---|---|---|---|---|---|---|
| Weight Added | 1.549 | 1.308 | 1.117 | 0.951 | 0.748 | 0.628 | 0.506 | 0.414 | 0.332 | 0.274 | 0.208 | 0.159 |

* Bridgeport Brass Co.

**Sizes and Weights in Pounds per Foot of Seamless Brass Tubes**

| Outside Diam. of Tube, Inches | Thickness — Stub's or Birmingham Gage | | | | | | | | | | |
|---|---|---|---|---|---|---|---|---|---|---|---|
| | 15 | 16 | 17 | 18 | 19 | 20 | 21 | 22 | 23 | 24 | 25 |
| | Decimal Equivalent of Gage Number, Inch | | | | | | | | | | |
| | 0.072 | 0.065 | 0.058 | 0.049 | 0.042 | 0.035 | 0.032 | 0.028 | 0.025 | 0.022 | 0.020 |
| 1/8 | .... | 0.045 | 0.045 | 0.043 | 0.040 | 0.036 | 0.034 | 0.031 | 0.029 | 0.026 | 0.024 |
| 3/16 | 0.096 | 0.092 | 0.087 | 0.078 | 0.070 | 0.062 | 0.057 | 0.051 | 0.047 | 0.042 | 0.039 |
| 1/4 | 0.148 | 0.139 | 0.129 | 0.114 | 0.101 | 0.087 | 0.080 | 0.072 | 0.065 | 0.058 | 0.053 |
| 5/16 | 0.200 | 0.186 | 0.170 | 0.149 | 0.131 | 0.112 | 0.104 | 0.092 | 0.083 | 0.074 | 0.067 |
| 3/8 | 0.252 | 0.233 | 0.212 | 0.184 | 0.161 | 0.137 | 0.127 | 0.112 | 0.101 | 0.090 | 0.082 |
| 7/16 | 0.304 | 0.279 | 0.254 | 0.220 | 0.192 | 0.163 | 0.150 | 0.132 | 0.119 | 0.106 | 0.096 |
| 1/2 | 0.356 | 0.326 | 0.296 | 0.255 | 0.222 | 0.188 | 0.173 | 0.152 | 0.137 | 0.121 | 0.111 |
| 9/16 | 0.408 | 0.373 | 0.338 | 0.290 | 0.252 | 0.213 | 0.196 | 0.173 | 0.155 | 0.137 | 0.125 |
| 5/8 | 0.460 | 0.420 | 0.380 | 0.326 | 0.283 | 0.238 | 0.219 | 0.193 | 0.173 | 0.153 | 0.140 |
| 11/16 | 0.511 | 0.467 | 0.421 | 0.361 | 0.313 | 0.264 | 0.242 | 0.213 | 0.191 | 0.169 | 0.154 |
| 3/4 | 0.563 | 0.514 | 0.463 | 0.396 | 0.343 | 0.289 | 0.265 | 0.233 | 0.209 | 0.185 | 0.169 |
| 13/16 | 0.615 | 0.561 | 0.505 | 0.432 | 0.373 | 0.314 | 0.288 | 0.253 | 0.227 | 0.201 | 0.183 |
| 7/8 | 0.667 | 0.608 | 0.547 | 0.467 | 0.404 | 0.339 | 0.311 | 0.274 | 0.245 | 0.217 | 0.197 |
| 15/16 | 0.719 | 0.655 | 0.589 | 0.502 | 0.434 | 0.365 | 0.334 | 0.294 | 0.263 | 0.232 | 0.211 |
| 1 | 0.77 | 0.70 | 0.63 | 0.54 | 0.46 | 0.389 | 0.358 | 0.314 | 0.281 | 0.248 | 0.226 |
| 1 1/8 | 0.87 | 0.79 | 0.71 | 0.61 | 0.52 | 0.439 | 0.404 | 0.354 | 0.317 | 0.280 | 0.255 |
| 1 1/4 | 0.98 | 0.89 | 0.80 | 0.68 | 0.59 | 0.490 | 0.450 | 0.395 | 0.354 | 0.312 | 0.284 |
| 1 3/8 | 1.08 | 0.98 | 0.88 | 0.75 | 0.65 | 0.540 | 0.496 | 0.435 | 0.390 | 0.343 | 0.313 |
| 1 1/2 | 1.19 | 1.08 | 0.96 | 0.82 | 0.71 | 0.591 | 0.542 | 0.476 | 0.426 | 0.375 | 0.342 |
| 1 5/8 | 1.29 | 1.17 | 1.05 | 0.89 | 0.77 | 0.641 | 0.588 | 0.516 | 0.462 | 0.407 | 0.371 |
| 1 3/4 | 1.39 | 1.26 | 1.13 | 0.96 | 0.83 | 0.692 | 0.635 | 0.556 | 0.498 | 0.439 | 0.399 |
| 1 7/8 | 1.50 | 1.36 | 1.22 | 1.03 | 0.89 | 0.742 | 0.681 | 0.597 | 0.534 | 0.470 | 0.428 |
| 2 | 1.60 | 1.45 | 1.30 | 1.10 | 0.95 | 0.793 | 0.727 | 0.637 | 0.570 | 0.502 | 0.457 |
| 2 1/8 | 1.71 | 1.55 | 1.38 | 1.17 | 1.01 | 0.843 | 0.773 | 0.678 | 0.606 | 0.534 | 0.486 |
| 2 1/4 | 1.81 | 1.64 | 1.47 | 1.24 | 1.07 | 0.894 | 0.819 | 0.718 | 0.642 | 0.566 | 0.515 |
| 2 3/8 | 1.91 | 1.73 | 1.55 | 1.32 | 1.13 | 0.944 | 0.866 | 0.758 | 0.678 | 0.597 | 0.544 |
| 2 1/2 | 2.02 | 1.83 | 1.63 | 1.39 | 1.19 | 0.995 | 0.912 | 0.799 | 0.714 | 0.629 | 0.573 |
| 2 5/8 | 2.12 | 1.92 | 1.72 | 1.46 | 1.25 | 1.045 | 0.958 | 0.839 | 0.750 | 0.661 | ..... |
| 2 3/4 | 2.23 | 2.01 | 1.80 | 1.53 | 1.31 | 1.096 | 1.004 | 0.880 | 0.786 | 0.693 | ..... |
| 2 7/8 | 2.33 | 2.11 | 1.89 | 1.60 | 1.37 | 1.146 | 1.050 | 0.920 | 0.822 | 0.724 | ..... |
| 3 | 2.43 | 2.20 | 1.97 | 1.67 | 1.43 | 1.197 | 1.096 | 0.960 | 0.859 | 0.756 | ..... |
| 3 1/8 | 2.54 | 2.30 | 2.05 | 1.74 | 1.49 | 1.247 | 1.143 | 1.001 | 0.895 | 0.788 | ..... |
| 3 1/4 | 2.64 | 2.39 | 2.14 | 1.81 | 1.55 | 1.298 | 1.189 | 1.041 | 0.931 | 0.820 | ..... |
| 3 3/8 | 2.74 | 2.48 | 2.22 | 1.88 | 1.62 | 1.348 | 1.235 | 1.082 | 0.967 | 0.851 | ..... |
| 3 1/2 | 2.85 | 2.58 | 2.30 | 1.95 | 1.68 | 1.399 | 1.281 | 1.122 | 1.003 | 0.883 | ..... |
| 3 5/8 | 2.95 | 2.67 | 2.39 | 2.02 | 1.74 | 1.449 | 1.327 | 1.162 | 1.039 | 0.915 | ..... |
| 3 3/4 | 3.06 | 2.76 | 2.47 | 2.09 | 1.80 | 1.50 | 1.373 | 1.203 | 1.075 | 0.946 | ..... |
| 3 7/8 | 3.16 | 2.86 | 2.56 | 2.16 | 1.86 | 1.55 | 1.42 | 1.243 | 1.111 | 0.978 | ..... |

To determine weight per foot of a tube of a given *inside diameter*, add to weights in above list the weights in pounds per foot given below under corresponding gage numbers.

| Gage No. | 15 | 16 | 17 | 18 | 19 | 20 | 21 | 22 | 23 | 24 | 25 |
|---|---|---|---|---|---|---|---|---|---|---|---|
| Weight Added | 0.120 | 0.097 | 0.078 | 0.055 | 0.041 | 0.028 | 0.024 | 0.018 | 0.014 | 0.011 | 0.009 |

Sizes and Weights in Pounds per Foot of Seamless Brass Tubes

| Outside Diam. of Tube, Inches | Thickness — Stub's or Birmingham Gage | | | | | | | | | |
|---|---|---|---|---|---|---|---|---|---|---|
| | 3 | 4 | 5 | 6 | 7 | 8 | 9 | 10 | 11 | 12 |
| | Decimal Equivalent of Gage Number, Inch | | | | | | | | | |
| | 0.259 | 0.238 | 0.220 | 0.203 | 0.180 | 0.165 | 0.148 | 0.134 | 0.120 | 0.109 |
| 4 | 11.19 | 10.33 | 9.60 | 8.90 | 7.94 | 7.31 | 6.58 | 5.98 | 5.37 | 4.89 |
| 4⅛ | 11.57 | 10.68 | 9.91 | 9.19 | 8.20 | 7.54 | 6.79 | 6.17 | 5.55 | 5.05 |
| 4¼ | 11.94 | 11.02 | 10.23 | 9.48 | 8.46 | 7.78 | 7.01 | 6.37 | 5.72 | 5.21 |
| 4⅜ | 12.32 | 11.36 | 10.55 | 9.77 | 8.72 | 8.02 | 7.22 | 6.56 | 5.89 | 5.37 |
| 4½ | 12.69 | 11.71 | 10.87 | 10.07 | 8.98 | 8.26 | 7.43 | 6.75 | 6.06 | 5.52 |
| 4⅝ | 13.06 | 12.05 | 11.18 | 10.36 | 9.24 | 8.50 | 7.65 | 6.94 | 6.24 | 5.68 |
| 4¾ | 13.44 | 12.39 | 11.50 | 10.65 | 9.50 | 8.73 | 7.86 | 7.14 | 6.41 | 5.84 |
| 4⅞ | 13.81 | 12.74 | 11.82 | 10.95 | 9.76 | 8.97 | 8.07 | 7.33 | 6.58 | 6.00 |
| 5 | 14.18 | 13.08 | 12.14 | 11.24 | 10.02 | 9.21 | 8.29 | 7.53 | 6.76 | 6.15 |
| 5⅛ | 14.56 | 13.42 | 12.45 | 11.53 | 10.28 | 9.45 | 8.50 | 7.72 | 6.93 | 6.31 |
| 5¼ | 14.93 | 13.77 | 12.77 | 11.82 | 10.53 | 9.69 | 8.71 | 7.91 | 7.10 | 6.47 |
| 5⅜ | 15.31 | 14.11 | 13.09 | 12.12 | 10.79 | 9.92 | 8.93 | 8.11 | 7.28 | 6.62 |
| 5½ | 15.68 | 14.45 | 13.41 | 12.41 | 11.05 | 10.16 | 9.14 | 8.30 | 7.45 | 6.78 |
| 5⅝ | 16.05 | 14.80 | 13.72 | 12.70 | 11.31 | 10.40 | 9.35 | 8.49 | 7.62 | 6.94 |
| 5¾ | 16.43 | 15.14 | 14.04 | 13.00 | 11.57 | 10.64 | 9.57 | 8.69 | 7.80 | 7.10 |
| 5⅞ | 16.80 | 15.48 | 14.36 | 13.29 | 11.83 | 10.88 | 9.78 | 8.88 | 7.97 | 7.25 |
| 6 | 17.17 | 15.83 | 14.67 | 13.58 | 12.09 | 11.12 | 9.99 | 9.07 | 8.14 | 7.41 |
| 6⅛ | 17.55 | 16.17 | 14.99 | 13.87 | 12.35 | 11.35 | 10.21 | 9.27 | 8.32 | 7.57 |
| 6¼ | 17.92 | 16.51 | 15.31 | 14.17 | 12.61 | 11.59 | 10.42 | 9.46 | 8.49 | 7.72 |
| 6⅜ | 18.30 | 16.86 | 15.63 | 14.46 | 12.87 | 11.83 | 10.64 | 9.65 | 8.66 | 7.88 |
| 6½ | 18.67 | 17.20 | 15.94 | 14.75 | 13.13 | 12.07 | 10.85 | 9.85 | 8.84 | 8.04 |
| 6⅝ | 19.04 | 17.54 | 16.26 | 15.05 | 13.39 | 12.31 | 11.06 | 10.04 | 9.01 | 8.20 |
| 6¾ | 19.42 | 17.89 | 16.58 | 15.34 | 13.65 | 12.54 | 11.28 | 10.23 | 9.18 | 8.35 |
| 6⅞ | 19.79 | 18.23 | 16.90 | 15.63 | 13.91 | 12.78 | 11.49 | 10.43 | 9.35 | 8.51 |
| 7 | 20.16 | 18.57 | 17.21 | 15.92 | 14.17 | 13.02 | 11.70 | 10.62 | 9.53 | 8.67 |
| 7⅛ | 20.54 | 18.92 | 17.53 | 16.22 | 14.43 | 13.26 | 11.92 | 10.81 | 9.70 | 8.83 |
| 7¼ | 20.91 | 19.26 | 17.85 | 16.51 | 14.69 | 13.50 | 12.13 | 11.01 | 9.87 | 8.98 |
| 7⅜ | 21.29 | 19.60 | 18.17 | 16.80 | 14.95 | 13.73 | 12.34 | 11.20 | 10.05 | 9.14 |
| 7½ | 21.66 | 19.95 | 18.48 | 17.10 | 15.21 | 13.97 | 12.56 | 11.39 | 10.22 | 9.30 |
| 7⅝ | 22.03 | 20.29 | 18.80 | 17.39 | 15.47 | 14.21 | 12.77 | 11.59 | 10.39 | 9.45 |
| 7¾ | 22.41 | 20.64 | 19.12 | 17.68 | 15.73 | 14.45 | 12.98 | 11.78 | 10.57 | 9.61 |
| 7⅞ | 22.78 | 20.98 | 19.44 | 17.98 | 15.99 | 14.69 | 13.20 | 11.97 | 10.74 | 9.77 |
| 8 | 23.15 | 21.32 | 19.75 | 18.27 | 16.25 | 14.93 | 13.41 | 12.17 | 10.91 | 9.93 |

To determine weight per foot of a tube of a given *inside diameter*, add to weights in above list the weights in pounds per foot given below under corresponding gage numbers.

| Gage No. | 3 | 4 | 5 | 6 | 7 | 8 | 9 | 10 | 11 | 12 |
|---|---|---|---|---|---|---|---|---|---|---|
| Weight Added | 1.549 | 1.308 | 1.117 | 0.951 | 0.748 | 0.628 | 0.506 | 0.414 | 0.332 | 0.274 |

### Sizes and Weights in Pounds per Foot of Seamless Brass Tubes

| Outside Diam. of Tube, Inches | Thickness — Stub's or Birmingham Gage | | | | | | | | | | | |
|---|---|---|---|---|---|---|---|---|---|---|---|---|
| | 13 | 14 | 15 | 16 | 17 | 18 | 19 | 20 | 21 | 22 | 23 | 24 |
| | Decimal Equivalent of Gage Number, Inch | | | | | | | | | | | |
| | 0.095 | 0.083 | 0.072 | 0.065 | 0.058 | 0.049 | 0.042 | 0.035 | 0.032 | 0.028 | 0.025 | 0.022 |
| 4 | 4.28 | 3.75 | 3.26 | 2.95 | 2.64 | 2.23 | 1.92 | 1.601 | 1.466 | 1.284 | 1.147 | 1.010 |
| 4⅛ | 4.42 | 3.87 | 3.37 | 3.05 | 2.72 | 2.30 | 1.98 | 1.651 | 1.512 | 1.324 | 1.183 | .... |
| 4¼ | 4.56 | 3.99 | 3.47 | 3.14 | 2.81 | 2.38 | 2.04 | 1.702 | 1.558 | 1.364 | 1.219 | .... |
| 4⅜ | 4.69 | 4.11 | 3.58 | 3.23 | 2.89 | 2.45 | 2.10 | 1.752 | 1.604 | 1.405 | 1.255 | .... |
| 4½ | 4.83 | 4.23 | 3.68 | 3.33 | 2.97 | 2.52 | 2.16 | 1.803 | 1.650 | 1.445 | 1.291 | .... |
| 4⅝ | 4.97 | 4.35 | 3.78 | 3.42 | 3.06 | 2.59 | 2.22 | 1.853 | 1.697 | 1.486 | ... | .... |
| 4¾ | 5.11 | 4.47 | 3.89 | 3.52 | 3.14 | 2.66 | 2.28 | 1.904 | 1.743 | 1.526 | ... | .... |
| 4⅞ | 5.24 | 4.59 | 3.99 | 3.61 | 3.22 | 2.73 | 2.34 | 1.954 | 1.789 | 1.566 | ... | .... |
| 5 | 5.38 | 4.71 | 4.09 | 3.70 | 3.31 | 2.80 | 2.40 | 2.005 | 1.835 | 1.607 | ... | .... |
| 5⅛ | 5.52 | 4.83 | 4.20 | 3.79 | 3.39 | 2.87 | 2.46 | 2.055 | 1.881 | .... | ... | .... |
| 5¼ | 5.65 | 4.95 | 4.30 | 3.89 | 3.48 | 2.94 | 2.52 | 2.106 | 1.928 | .... | ... | .... |
| 5⅜ | 5.79 | 5.07 | 4.41 | 3.98 | 3.56 | 3.01 | 2.58 | 2.156 | 1.974 | .... | ... | .... |
| 5½ | 5.93 | 5.19 | 4.51 | 4.08 | 3.64 | 3.08 | 2.65 | 2.207 | 2.02 | .... | ... | .... |
| 5⅝ | 6.07 | 5.31 | 4.61 | 4.17 | 3.73 | 3.15 | 2.71 | 2.257 | .... | .... | ... | .... |
| 5¾ | 6.20 | 5.43 | 4.72 | 4.26 | 3.81 | 3.22 | 2.77 | 2.308 | .... | .... | ... | .... |
| 5⅞ | 6.34 | 5.55 | 4.82 | 4.36 | 3.89 | 3.29 | 2.83 | 2.358 | .... | .... | ... | .... |
| 6 | 6.48 | 5.67 | 4.93 | 4.45 | 3.98 | 3.37 | 2.89 | 2.409 | .... | .... | ... | .... |
| 6⅛ | 6.61 | 5.79 | 5.03 | 4.54 | 4.06 | 3.44 | .... | .... | .... | .... | ... | .... |
| 6¼ | 6.75 | 5.91 | 5.13 | 4.64 | 4.15 | 3.51 | .... | .... | .... | .... | ... | .... |
| 6⅜ | 6.89 | 6.03 | 5.24 | 4.73 | 4.23 | 3.58 | .... | .... | .... | .... | ... | .... |
| 6½ | 7.03 | 6.15 | 5.34 | 4.83 | 4.31 | 3.65 | .... | .... | .... | .... | ... | .... |
| 6⅝ | 7.16 | 6.27 | 5.45 | 4.92 | 4.40 | 3.72 | .... | .... | .... | .... | ... | .... |
| 6¾ | 7.30 | 6.39 | 5.55 | 5.01 | 4.48 | 3.79 | .... | .... | .... | .... | ... | .... |
| 6⅞ | 7.44 | 6.51 | 5.65 | 5.11 | 4.56 | 3.86 | .... | .... | .... | .... | ... | .... |
| 7 | 7.57 | 6.63 | 5.76 | 5.20 | 4.65 | 3.93 | .... | .... | .... | .... | ... | .... |
| 7⅛ | 7.71 | 6.75 | 5.86 | 5.29 | .... | .... | .... | .... | .... | .... | ... | .... |
| 7¼ | 7.85 | 6.87 | 5.96 | 5.39 | .... | .... | .... | .... | .... | .... | ... | .... |
| 7⅜ | 7.99 | 6.99 | 6.07 | 5.48 | .... | .... | .... | .... | .... | .... | ... | .... |
| 7½ | 8.12 | 7.11 | 6.17 | 5.58 | .... | .... | .... | .... | .... | .... | ... | .... |
| 7⅝ | 8.26 | 7.23 | 6.28 | 5.67 | .... | .... | .... | .... | .... | .... | ... | .... |
| 7¾ | 8.40 | 7.35 | 6.38 | 5.76 | .... | .... | .... | .... | .... | .... | ... | .... |
| 7⅞ | 8.53 | 7.47 | 6.48 | 5.86 | .... | .... | .... | .... | .... | .... | ... | .... |
| 8 | 8.67 | 7.58 | 6.59 | 5.95 | .... | .... | .... | .... | .... | .... | ... | .... |

To determine weight per foot of a tube of a given *inside diameter*, add to weights in above list the weights in pounds per foot given below under corresponding gage numbers.

| Gage No. | 13 | 14 | 15 | 16 | 17 | 18 | 19 | 20 | 21 | 22 | 23 | 24 |
|---|---|---|---|---|---|---|---|---|---|---|---|---|
| Weight Added | 0.208 | 0.159 | 0.120 | 0.097 | 0.078 | 0.055 | 0.041 | 0.028 | 0.024 | 0.018 | 0.014 | 0.011 |

The number of chasers with which a die should be equipped depends upon the size of the die. The number recommended for different sizes is as follows:

| Size of Die | Number of Chasers | Size of Die | Number of Chasers |
|---|---|---|---|
| Up to 1¼ inch | 4 | 10 to 12 inches | 12 |
| 1¼ to 4 inches | 6 | 12 to 14 inches | 14 |
| 4 to 7 inches | 8 | 14 to 18 inches | 16 |
| 7 to 10 inches | 10 | 18 to 20 inches | 18 |

Pipe threading dies should be lubricated with a good quality of lard oil or crude cotton-seed oil, the lubricant being used in liberal quantities.

**Pipe and Tube Bending.** — In bending a pipe or tube, the outer part of the bend is stretched and the inner section compressed, and as the result of opposite and unequal stresses, the pipe or tube tends to flatten or collapse. To prevent such distortion, the common practice is to support the wall of the pipe or tube in some manner during the bending operation. This support may be in the form of a filling material, or, when a bending machine or fixture is used, an internal mandrel or ball-shaped member may support the inner wall when required. If a filling material is used, it is melted and poured into the pipe or tube. One filler material (a commercial alloy known as "Bendalloy") has a melting point of only 160 degrees F. and is composed of bismuth, lead, tin, and cadmium. With this material, tubes having very thin walls have been bent to small radii. The metal filler conforms to the inside of the tube so closely that the tube can be bent just as though it were a solid rod. The filler is removed readily by melting. This method has been applied to the bending of copper, brass, duralumin, plain steel, and stainless steel tubes with uniform success. Tubes plated with chromium or nickel can be bent without danger of the plate flaking off.

Other filling materials such as resin, tar, lead, and dry sand have also been used.

Pipes are often bent to avoid the use of fittings, thus eliminating joints, providing a smooth unobstructed passage for fluids, and resulting in certain other advantages.

*Minimum Radius:* The safe minimum radius for a given diameter, material, and method of bending depends upon the thickness of the pipe wall, it being possible, for example, to bend extra heavy pipe to a smaller radius than pipe of standard weight. As a general rule, wrought iron or steel pipe of standard weight may readily be bent to a radius equal to five or six times the nominal pipe diameter. The minimum radius for standard weight pipe should, as a rule, be three and one-half to four times the diameter. It will be understood, however, that the minimum radius may vary considerably, depending upon the method of bending. Extra heavy pipe may be bent to radii varying from two and one-half times the diameter for smaller sizes to three and one-half to four times the diameter for larger sizes.

*Rules for Finding Lengths of Bends:* In determining the required length of a pipe or tube before bending, the lengths of the straight sections are, of course, added to the lengths required for the curved sections in order to make the proper allowance for bends. The following rules are for finding the lengths of the curved sections.

*Rule for 90-Degree Bend:* To find the length of a 90-degree or right-angle bend, multiply the radius of the bend by 1.57 (the radius is measured to the center of the pipe or to a point midway between the inner and outer walls).

*Rule for 180-Degree Bend:* To find the length of a 180-degree or U bend, multiply the radius of the bend by 3.14.

*General Rule:* A general rule for finding the lengths of sections having degrees of curvature other than 90 and 180 is as follows: Multiply the radius of the bend by the included angle, and then multiply the product by the constant 0.01745. The result is the length of the curved section.

**Plastic Pipe.** — Shortly after World War II, plastic pipe became an acceptable substitute, under certain service conditions, for other piping materials. Now, however, plastic pipe is specified on the basis of its own special capabilities and limitations. The largest volume of application has been for water piping systems.

Besides being light in weight, plastic pipe performs well in resisting deterioration from corrosive or caustic fluids. Even if the fluid borne is harmless, the chemical resistance of plastic pipe offers protection against a harmful exterior environment, such as when buried in a corrosive soil.

Generally, plastic pipe is limited by its temperature and pressure capacities. The higher the operating pressure of the pipe system, the less will be its temperature capability. The reverse is true, also. Since it is formed from organic resins, plastic pipe will burn. For various piping compositions, ignition temperatures vary from 700° to 800°F (370° to 430°C).

The following are accepted methods for joining plastic pipe:

*Solvent Welding* is usually accomplished by brushing a solvent cement on the end of the length of pipe and into the socket end of a fitting or the flange of the next pipe section. A chemical weld then joins and seals the pipe after connection.

*Threading* is a procedure not recommended for thin-walled plastic pipe or for specific grades of plastic. During connection of thicker-walled pipe, strap wrenches are used to avoid damaging and weakening the plastic.

*Heat Fusion* involves the use of heated air and plastic filler rods to weld plastic pipe assemblies. A properly welded joint can have a tensile strength equal to 90 percent that of the pipe material.

*Elastomeric Sealing* is used with bell-end piping. It is a recommended procedure for large diameter piping and for underground installations. The joints are set quickly and have good pressure capabilities.

Table 1. Dimensions and Weights of Thermoplastic Pipe*

| Nominal Pipe Size | | Outside Diameter | | Schedule 40 | | | | Schedule 80 | | | |
|---|---|---|---|---|---|---|---|---|---|---|---|
| | | | | Nom. Wall Thickness | | Nominal Weight* | | Nom. Wall Thickness | | Nominal Weight* | |
| in. | cm | in. | cm | in. | cm | lb/100′ | kg/m | in. | cm | lb/100′ | kg/m |
| ⅛ | 0.3 | 0.405 | 1.03 | 0.072 | 0.18 | 3.27 | 0.05 | 0.101 | 0.256 | 4.18 | 0.06 |
| ¼ | 0.6 | 0.540 | 1.37 | 0.093 | 0.24 | 5.66 | 0.08 | 0.126 | 0.320 | 7.10 | 0.11 |
| ⅜ | 1.0 | 0.675 | 1.71 | 0.096 | 0.24 | 7.57 | 0.11 | 0.134 | 0.340 | 9.87 | 0.15 |
| ½ | 1.3 | 0.840 | 2.13 | 0.116 | 0.295 | 11.4 | 0.17 | 0.156 | 0.396 | 14.5 | 0.22 |
| ¾ | 2.0 | 1.050 | 2.67 | 0.120 | 0.305 | 15.2 | 0.23 | 0.163 | 0.414 | 19.7 | 0.29 |
| 1 | 2.5 | 1.315 | 3.34 | 0.141 | 0.358 | 22.5 | 0.33 | 0.190 | 0.483 | 29.1 | 0.43 |
| 1¼ | 3.2 | 1.660 | 4.22 | 0.148 | 0.376 | 30.5 | 0.45 | 0.202 | 0.513 | 40.1 | 0.60 |
| 1½ | 3.8 | 1.900 | 4.83 | 0.154 | 0.391 | 36.6 | 0.54 | 0.212 | 0.538 | 48.7 | 0.72 |
| 2 | 5.1 | 2.375 | 6.03 | 0.163 | 0.414 | 49.1 | 0.73 | 0.231 | 0.587 | 67.4 | 1.00 |
| 2½ | 6.4 | 2.875 | 7.30 | 0.215 | 0.546 | 77.9 | 1.16 | 0.293 | 0.744 | 103 | 1.5 |
| 3 | 7.6 | 3.500 | 8.89 | 0.229 | 0.582 | 102 | 1.5 | 0.318 | 0.808 | 138 | 2.1 |
| 3½ | 8.9 | 4.000 | 10.16 | 0.240 | 0.610 | 123 | 1.8 | 0.337 | 0.856 | 168 | 2.5 |
| 4 | 10.2 | 4.500 | 11.43 | 0.251 | 0.638 | 145 | 2.2 | 0.357 | 0.907 | 201 | 3.0 |
| 5 | 12.7 | 5.563 | 14.13 | 0.273 | 0.693 | 197 | 2.9 | 0.398 | 1.011 | 280 | 4.2 |
| 6 | 15.2 | 6.625 | 16.83 | 0.297 | 0.754 | 256 | 3.8 | 0.458 | 1.163 | 385 | 5.7 |
| 8 | 20.3 | 8.625 | 21.91 | 0.341 | 0.866 | 385 | 5.7 | 0.530 | 1.346 | 584 | 8.7 |
| 10 | 25.4 | 10.75 | 27.31 | 0.387 | 0.983 | 546 | 8.1 | 0.629 | 1.598 | 867 | 12.9 |
| 12 | 30.5 | 12.75 | 32.39 | 0.430 | 1.09 | 722 | 10.7 | 0.728 | 1.849 | 1192 | 17.7 |

* The nominal weights of plastic pipe given in this table are based on an empirically chosen material density of 1.00 g/cm³. The nominal unit weight for a specific plastic pipe formulation can be obtained by multiplying the weight values from the table by the density in g/cm³ or by the specific gravity of the particular plastic composition.

The following are ranges of density factors for various plastic pipe materials: PE, 0.93 to 0.96; PVC, 1.35 to 1.40; CPVC, 1.55; ABS, 1.04 to 1.08; SR, 1.05; PB, 0.91 to 0.92; and PP, 0.91. For meanings of abbreviations see Table 2.

Information supplied by the Plastics Pipe Institute.

*Insert Fitting* is particularly useful for PE and PB pipe. For joining pipe sections, insert fittings are pushed into the pipe and secured by stainless steel clamps.

*Transition Fitting* involves specially designed connectors to join plastic pipe with other materials, such as cast iron, steel, copper, clay, and concrete.

Plastic pipe can be specified by means of Schedules 40, 80, and 120, which conform dimensionally to metal pipe, or through a Standard Dimension Ratio (SDR). The SDR is a rounded value obtained by dividing the average outside diameter of the pipe by the wall thickness. Within an individual SDR series of pipe, pressure ratings are uniform, regardless of pipe diameter.

Table 1 provides the weights and dimensions for Schedule 40 and 80 thermoplastic pipe, while Table 3 gives ranges of water pressure ratings and pipe support centers for a variety of pipe grades under three generic descriptions: PE (polyethylene); PVC (polyvinyl chloride); and ABS (acrylonitrile-butadiene styrene).

For more detailed information concerning the properties of a particular plastic pipe formulation, consult the pipe manufacturer or The Plastics Pipe Institute, 250 Park Ave., New York, N. Y. 10017.

Table 2.   General Properties and Uses of Plastic Pipe*

| Plastic Pipe Material | Properties | Common Uses | Operating Temperature† | | Joining Methods |
|---|---|---|---|---|---|
| | | | With Pressure | Without Pressure | |
| ABS (Acrylonitrile-butadiene styrene) | Rigid; excellent impact strength at low temperatures; maintains rigidity at higher temperatures. | Water, Drain, Waste, Vent, Sewage. | 100°F (38°C) | 180°F (82°C) | Solvent cement, Threading, Transition fitting. |
| PE (Polyethylene) | Flexible; excellent impact strength; good performance at low temperatures. | Water, Gas, Chemical, Irrigation. | 100°F (38°C) | 180°F (82°C) | Heat fusion, Insert and Transition fitting. |
| PVC (Polyvinyl chloride) | Rigid; fire self-extinguishing; high impact and tensile strength. | Water, Gas, Sewage, Industrial process, Irrigation. | 100°F (38°C) | 180°F (82°C) | Solvent cement, Elastomeric seal, Mechanical coupling, Transition fitting. |
| CPVC (Chlorinated polyvinyl chloride) | Rigid; fire self-extinguishing; high impact and tensile strength. | Hot and cold water, Chemical. | 180°F (82°C) at 100 psig (690kPa) for SDR-11 | | Solvent cement, Threading, Mechanical coupling, Transition fitting. |
| PB (Polybutylene) | Flexible; good performance at elevated temperatures. | Water, Gas, Irrigation. | 180°F (82°C) | 200°F (93°C) | Insert fitting, Heat fusion, Transition fitting. |
| PP (Polypropylene) | Rigid; very light; high chemical resistance, particularly to sulfur-bearing compounds. | Chemical waste and processing. | 100°F (38°C) | 180°F (82°C) | Mechanical coupling, Heat fusion, Threading. |
| SR (Styrene rubber plastic) | Rigid; moderate chemical resistance; fair impact strength. | Drainage, Septic fields. | 150°F (66°C) | . . . . . | Solvent cement, Transition fitting, Elastomeric seal. |

* From information supplied by the Plastics Pipe Institute.
† The operating temperatures shown are general guide points. For specific operating temperature and pressure data for various grades of the types of plastic pipe given, please consult the pipe manufacturer or the Plastics Pipe Institute.

Table 3.  Thermoplastic Pipe Water Pressure Ratings at 73°F (23°C) and Horizontal Support-Center Distances — 1†

| Nominal Pipe Size | | Schedule 40 | | | | | | | | | | | |
|---|---|---|---|---|---|---|---|---|---|---|---|---|---|
| | | PE | | | | PVC | | | | ABS | | | |
| | | Pressure* Rating | | Support§ Center Dist. | | Pressure* Rating | | Support§ Center Dist. | | Pressure* Rating | | Support§ Center Dist. | |
| in. | cm | psi | kPa | ft | m | psi | kPa | ft | m | psi | kPa | ft | m |
| ⅛ | 0.3 | .... | .... | ... | ... | 400 to 810 | 2760 to 5580 | ... | ... | .... | .... | ... | ... |
| ¼ | 0.6 | .... | .... | 1.0 to 1.5 | 0.3 to 0.5 | 390 to 780 | 2690 to 5380 | ... | ... | .... | .... | ... | ... |
| ⅜ | 1.0 | .... | .... | 1.0 to 1.5 | 0.3 to 0.5 | 310 to 620 | 2140 to 4270 | ... | ... | .... | .... | ... | ... |
| ½ | 1.3 | 119 to 188 | 820 to 1300 | 1.2 to 2.0 | 0.4 to 0.6 | 300 to 600 | 2070 to 4140 | 3.5 | 1.1 | 298 to 476 | 2050 to 3280 | 4.0 | 1.2 |
| ¾ | 2.0 | 96 to 152 | 660 to 1050 | 1.2 to 2.0 | 0.4 to 0.6 | 240 to 480 | 1650 to 3310 | 3.5 | 1.1 | 241 to 385 | 1660 to 2650 | 4.0 | 1.2 |
| I | 2.5 | 90 to 142 | 620 to 980 | 1.2 to 2.0 | 0.4 to 0.6 | 220 to 450 | 1520 to 3100 | 4.0 | 1.2 | 225 to 360 | 1550 to 2480 | 4.5 | 1.4 |
| 1¼ | 3.2 | 74 to 116 | 510 to 800 | 1.5 to 2.2 | 0.5 to 0.7 | 180 to 370 | 1240 to 2550 | 4.5 | 1.4 | 184 to 294 | 1270 to 2030 | 4.5 | 1.4 |
| 1½ | 3.8 | 66 to 104 | 460 to 717 | 1.5 to 2.2 | 0.5 to 0.7 | 170 to 330 | 1170 to 2280 | 4.5 | 1.4 | 165 to 264 | 1140 to 1820 | 5.0 | 1.5 |
| 2 | 5.1 | 55 to 87 | 380 to 600 | 1.7 to 2.7 | 0.5 to 0.8 | 140 to 280 | 970 to 1930 | 4.5 | 1.4 | 139 to 222 | 960 to 1530 | 5.0 | 1.5 |
| 2½ | 6.4 | 61 to 96 | 420 to 660 | 1.7 to 2.7 | 0.5 to 0.8 | 150 to 300 | 1030 to 2070 | ... | ... | 152 to 243 | 1050 to 1680 | ... | ... |
| 3 | 7.6 | 53 to 83 | 370 to 570 | 2.0 to 3.0 | 0.6 to 0.9 | 130 to 260 | 900 to 1790 | 5.5 | 1.7 | 132 to 211 | 910 to 1450 | 6 | 1.8 |
| 3½ | 8.9 | 50 to 75 | 340 to 520 | ... | ... | 120 to 240 | 830 to 1650 | ... | ... | .... | .... | ... | ... |
| 4 | 10.2 | 55 to 70 | 380 to 480 | ... | ... | 110 to 220 | 760 to 1520 | 6.2 | 1.9 | 111 to 177 | 765 to 1220 | 6.2 | 1.9 |
| 5 | 12.7 | 50 to 61 | 340 to 420 | ... | ... | 100 to 190 | 690 to 1310 | ... | ... | .... | .... | ... | ... |
| 6 | 15.2 | 55 | 380 | ... | ... | 90 to 180 | 620 to 1240 | 6.7 | 2.0 | 88 to 141 | 610 to 972 | 6.7 | 2.0 |
| 8 | 20.3 | 50 | 340 | ... | ... | 80 to 160 | 550 to 1100 | 7.5 | 2.3 | .... | .... | ... | ... |
| 10 | 25.4 | .... | .... | ... | ... | 70 to 140 | 480 to 970 | 7.7 | 2.3 | .... | .... | ... | ... |
| 12 | 30.5 | .... | .... | ... | ... | 70 to 130 | 480 to 900 | 8.0 | 2.4 | .... | .... | ... | ... |

† From information provided by the Plastics Pipe Institute.
* The pressure ratings given apply only to unthreaded pipe.  Threading is not recommended for: (1) all PE; (2) Schedule 40 ABS; and (3) Schedule 40 PVC, 6 in. or less in nominal pipe diameter.    § Support-center values for PE are based on water being carried at 70°F (21°C).  From 70° to 100°F (21° to 38°C) closer support spacing is required. Above 100°F (38°C) support should be continuous.  All support centers given for PE pipe apply only to pipe grades having water pressure ratings of 80 psi (550 kPa) or more.

Table 3. Thermoplastic Pipe Water Pressure Ratings at 73°F (23°C) and Horizontal Support-Center Distances — 2†

| Nominal Pipe Size | | Schedule 80 | | | | | | | | | | | |
|---|---|---|---|---|---|---|---|---|---|---|---|---|---|
| | | PE | | | | PVC | | | | ABS | | | |
| | | Pressure* Rating | | Support§ Center Dist. | | Pressure* Rating | | Support§ Center Dist. | | Pressure* Rating | | Support§ Center Dist. | |
| in. | cm | psi | kPa | ft | m | psi | kPa | ft | m | psi | kPa | ft | m |
| 1/8 | 0.3 | .... | .... | ... | ... | 610 to 1230 | 4200 to 8480 | ... | ... | .... | .... | ... | ... |
| 1/4 | 0.6 | .... | .... | 1.0 to 1.5 | 0.3 to 0.5 | 570 to 1130 | 3900 to 7790 | ... | ... | .... | .... | ... | ... |
| 3/8 | 1.0 | .... | .... | 1.0 to 1.5 | 0.3 to 0.5 | 460 to 920 | 3200 to 6300 | ... | ... | .... | .... | ... | ... |
| 1/2 | 1.3 | 170 to 267 | 1170 to 1840 | 1.2 to 2.0 | 0.4 to 0.6 | 420 to 850 | 2900 to 5900 | 3.5 | 1.1 | 424 to 678 | 2920 to 4670 | 5.0 | 1.5 |
| 3/4 | 2.0 | 137 to 217 | 945 to 1500 | 1.2 to 2.0 | 0.4 to 0.6 | 340 to 690 | 2300 to 4800 | 4.0 | 1.2 | 344 to 550 | 2370 to 3790 | 5.0 | 1.5 |
| 1 | 2.5 | 126 to 199 | 869 to 1370 | 1.2 to 2.0 | 0.4 to 0.6 | 320 to 630 | 2200 to 4300 | 4.5 | 1.4 | 315 to 504 | 2170 to 3470 | 5.5 | 1.7 |
| 1¼ | 3.2 | 104 to 164 | 717 to 1130 | 1.5 to 2.2 | 0.5 to 0.7 | 260 to 520 | 1800 to 3600 | 5.0 | 1.5 | 260 to 416 | 1790 to 2870 | 5.5 | 1.7 |
| 1½ | 3.8 | 94 to 148 | 650 to 1020 | 1.5 to 2.2 | 0.5 to 0.7 | 240 to 470 | 1700 to 3200 | 5.0 | 1.5 | 235 to 376 | 1620 to 2590 | 6.0 | 1.8 |
| 2 | 5.1 | 81 to 127 | 560 to 876 | 1.7 to 2.7 | 0.5 to 0.8 | 200 to 400 | 1400 to 2800 | 5.0 | 1.5 | 202 to 323 | 1390 to 2230 | 6.0 | 1.8 |
| 2½ | 6.4 | 85 to 134 | 590 to 924 | 1.7 to 2.7 | 0.5 to 0.8 | 210 to 420 | 1500 to 2900 | ... | ... | 212 to 340 | 1460 to 2340 | ... | ... |
| 3 | 7.6 | 75 to 118 | 520 to 814 | 2.0 to 3.0 | 0.6 to 0.9 | 190 to 370 | 1300 to 2600 | 6.0 | 1.8 | 187 to 297 | 1290 to 2050 | 7 | 2.1 |
| 3½ | 8.9 | 69 to 109 | 480 to 752 | ... | ... | 170 to 350 | 1200 to 2400 | ... | ... | .... | .... | ... | ... |
| 4 | 10.2 | 65 to 102 | 450 to 703 | 2.2 to 3.5 | 0.7 to 1.1 | 160 to 320 | 1100 to 2200 | 7.5 | 2.3 | 162 to 259 | 1120 to 1790 | 7.5 | 2.3 |
| 5 | 12.7 | 58 to 91 | 400 to 630 | ... | ... | 140 to 290 | 970 to 2000 | ... | ... | .... | .... | ... | ... |
| 6 | 15.2 | 56 to 88 | 390 to 610 | 2.7 to 4.2 | 0.8 to 1.3 | 140 to 280 | 970 to 1900 | 8.5 | 2.6 | 139 to 222 | 958 to 1530 | 8.5 | 2.6 |
| 8 | 20.3 | .... | .... | ... | ... | 120 to 250 | 830 to 1720 | 9.0 | 2.7 | .... | .... | ... | ... |
| 10 | 25.4 | .... | .... | ... | ... | 120 to 230 | 830 to 1600 | 9.5 | 2.9 | .... | .... | ... | ... |
| 12 | 30.5 | .... | .... | ... | ... | 110 to 230 | 760 to 1600 | 10.0 | 3.0 | .... | .... | ... | ... |

† From information provided by the Plastics Pipe Institute.  * The pressure ratings given apply only to unthreaded pipe. Threading is not recommended for all PE. However, if threaded pipe is Schedule 80 PVC, one-half the pressure ratings shown in the table can be used.  § Support-center values for PE are based on water being carried at 70°F (21°C). From 70° to 100°F (21° to 38°C) closer support spacing is required. Above 100°F (38°C) support should be continuous. All support centers given for PE pipe apply only to pipe grades having water pressure ratings of 80 psi (550 kPa) or more.

## Definitions of Pipe Fittings

The following definitions for various pipe fittings are given by the National Tube Co.:

*Armstrong Joint.* — A two-bolt, flanged or lugged connection for high pressures. The ends of the pipes are peculiarly formed to properly hold a gutta-percha ring. It was originally made for cast-iron pipe. The two-bolt feature has much to commend it. There are various substitutes for this joint, many of which employ rubber in place of gutta-percha; others use more bolts in order to reduce the cost.

*Bell and Spigot Joint.* — (1) The usual term for the joint in cast-iron pipe. Each piece is made with an enlarged diameter or bell at one end into which the plain or spigot end of another piece is inserted when laying. The joint is then made tight by cement, oakum, lead, rubber or other suitable substance, which is driven in or calked into the bell and around the spigot. When a similar joint is made in wrought pipe by means of a cast bell (or hub), it is at times called hub and spigot joint (poor usage). Matheson joint is the name applied to a similar joint in wrought pipe which has the bell formed from the pipe. (2) Applied to fittings or valves, means that one end of the run is a "bell," and the other end is a "spigot," similar to those used on regular cast-iron pipe.

*Bonnet.* — (1) A cover used to guide and enclose the tail end of a valve spindle. (2) A cap over the end of a pipe (poor usage).

*Branch.* — The outlet or inlet of a fitting not in line with the run, but which may make any angle.

*Branch Ell.* — (1) Used to designate an elbow having a back outlet in line with one of the outlets of the "run." It is also called a heel outlet elbow. (2) Incorrectly used to designate side outlet or back outlet elbow.

*Branch Pipe.* — A very general term used to signify a pipe either cast or wrought, that is equipped with one or more branches. Such pipes are used so frequently that they have acquired common names such as tees, crosses, side or back outlet elbows, manifolds, double-branch elbows, etc. The term branch pipe is generally restricted to such as do not conform to usual dimensions.

*Branch Tee (Header).* — A tee having many side branches. (See Manifold.)

*Bull Head Tee.* — A tee the branch of which is larger than the run.

*Bushing.* — A pipe fitting for the purpose of connecting a pipe with a fitting of larger size, being a hollow plug with internal and external threads to suit the different diameters.

*Card Weight Pipe.* — A term used to designate standard or full weight pipe, which is the Briggs' standard thickness of pipe.

*Close Nipple.* — One the length of which is about twice the length of a standard pipe thread and is without any shoulder.

*Coupling.* — A threaded sleeve used to connect two pipes. Commercial couplings are threaded inside to suit the exterior thread of the pipe. The term coupling is occasionally used to mean any jointing device and may be applied to either straight or reducing sizes.

*Cross.* — A pipe fitting with four branches arranged in pairs, each pair on one axis, and the axes at right angles. When the outlets are otherwise arranged the fittings are branch pipes or specials.

*Cross-over.* — A small fitting with a double offset, or shaped like the letter *U* with the ends turned out. It is only made in small sizes and used to pass the flow of one pipe past another when the pipes are in the same plane.

*Cross-over Tee.* — A fitting made along lines similar to the cross-over, but having at one end two openings in a tee-head the plane of which is at right angles to the plane of the cross-over bend.

*Cross Valve.* — (1) A valve fitted on a transverse pipe so as to open communication at will between two parallel lines of piping. Much used in connection with oil and water pumping arrangements, especially on ship board. (2) Usually considered as an angle valve with a back outlet in the same plane as the other two openings.

*Crotch.* — A fitting that has the general shape of the letter *Y*. Caution should be exercised not to confuse the crotch and wye.

*Double-branch Elbow.* — A fitting that, in a manner, looks like a tee, or as if two elbows had been shaved and then placed together, forming a shape something like the letter *Y* or a crotch.

*Double Sweep Tee.* — A tee made with easy curves between body and branch, *i.e.*, the center of the curve between run and branch lies outside the body.

*Drop Elbow.* — A small sized ell that is frequently used where gas is put into a building. These fittings have wings cast on each side. The wings have small countersunk holes so that they may be fastened by wood screws to a ceiling or wall or framing timbers.

*Drop Tee.* — One having the same peculiar wings as the drop elbow.

*Dry Joint.* — One made without gasket or packing or smear of any kind, as a ground joint.

*Elbow (Ell).* — A fitting that makes an angle between adjacent pipes. The angle is always 90 degrees, unless another angle is stated. (See Branch, Service, and Union Ell.)

*Extra Heavy.* — When applied to pipe, means pipe thicker than standard pipe; when applied to valves and fittings, indicates goods suitable for a working pressure of 250 pounds per square inch.

*Header.* — A large pipe into which one set of boilers is connected by suitable nozzles or tees, or similar large pipes from which a number of smaller ones lead to consuming points. Headers are often used for other purposes — for heaters or in refrigeration work. Headers are essentially branch pipes with many outlets, which are usually parallel. Largely used for tubes of water-tube boilers.

*Hydrostatic Joint.* — Used in large water mains, in which sheet lead is forced tightly into the bell of a pipe by means of the hydrostatic pressure of a liquid.

*Kewanee Union.* — A patented pipe union having one pipe end of brass and the other of malleable iron, with a ring or nut of malleable iron, in which the arrangement and finish of the several parts is such as to provide a non-corrosive ball-and-socket joint at the junction of the pipe ends, and a non-corrosive connection between the ring and brass pipe end.

*Lead Joint.* — (1) Generally used to signify the connection between pipes which is made by pouring molten lead into the annular space between a bell and spigot, and then making the lead tight by calking. (2) Rarely used to mean the joint made by pressing the lead between adjacent pieces, as when a lead gasket is used between flanges.

*Lead Wool.* — A material used in place of molten lead for making pipe joints. It is lead fiber, about as coarse as fine excelsior, and when made in a strand, it can be calked into the joints, making them very solid.

*Line Pipe.* — Special brand of pipe that employs recessed and taper thread couplings, and usually greater length of thread than Briggs' standard. The pipe is also subjected to higher test.

*Lip Union.* — (1) A special form of union characterized by the lip that prevents the gasket from being squeezed into the pipe so as to obstruct the flow. (2) A ring union, unless flange is specified.

*Manifold.* — (1) A fitting with numerous branches used to convey fluids between a large pipe and several smaller pipes. (See Branch Tee.) (2) A header for a coil.

*Matheson Joint.* — A wrought pipe joint made by enlarging one end of the pipe to form a suitable lead recess, similar to the bell end of a cast-iron pipe, and which receives the male or spigot end of the next length. Practically the same style of a joint as used for cast-iron pipe.

*Medium Pressure.* — When applied to valves and fittings, means suitable for a working pressure of from 125 to 175 pounds per square inch.

*Needle Valve.* — A valve provided with a long tapering point in place of the ordinary valve disk. The tapering point permits fine graduation of the opening. At times called a needle point valve.

*Nipple.* — (1) A tubular pipe fitting usually threaded on both ends and under 12 inches in length. Pipe over 12 inches long is regarded as cut pipe. (See Close. Short, Shoulder and Space Nipple.)

*Reducer.* — (1) A fitting having a larger size at one end than at the other. Some have tried to establish the term "increaser" — thinking of direction of flow — but this has been due to a misunderstanding of the trade custom of always giving the largest size of run of a fitting first; hence, all fittings having more than one size are reducers. They are always threaded inside, unless specified flanged or for some special joint. (2) Threaded type, made with abrupt reduction. (3) Flanged pattern with taper body. (4) Flanged eccentric pattern with taper body, but flanges at 90 degrees to one side of body. (5) Misapplied at times, to a reducing coupling.

*Run.* — (1) A length of pipe that is made of more than one piece of pipe. (2) The portion of any fitting having its ends "in line" or nearly so, in contradistinction to the branch or side opening, as of a tee. The two main openings of an ell also indicate its run, and when there is a third opening on an ell, the fitting is a "side outlet" or "back outlet" elbow, except that when all three openings are in one plane and the back outlet is in line with one of the run openings, the fitting is a "heel outlet elbow" or a "single sweep tee" or sometimes a "branch tee."

*Rust Joint.* — Employed to secure rigid connection. The joint is made by packing an intervening space tightly with a stiff paste which oxidizes the iron, the whole rusting together and hardening into a solid mass. It generally cannot be separated except by destroying some of the pieces. One recipe is 80 pounds cast-iron borings or filings, 1 pound sal-ammoniac, 2 pounds flowers of sulphur, mixed to a paste with water.

*Service Ell.* — An elbow having an outside thread on one end. Also known as street ell.

*Service Pipe.* — A pipe connecting mains with a dwelling.

*Service Tee.* — A tee having inside thread on one end and on branch, but outside thread on other end of run. Also known as street tee.

*Short Nipple.* — One whose length is a little greater than that of two threaded lengths or somewhat longer than a close nipple. It always has some unthreaded portion between the two threads.

*Shoulder Nipple.* — A nipple of any length, which has a portion of pipe between two pipe threads. As generally used, however, it is a nipple about halfway between the length of a close nipple and a short nipple.

*Space Nipple.* — A nipple with a portion of pipe or shoulder between the two threads. It may be of any length long enough to allow a shoulder.

*Standard Pressure.* — A term applied to valves and fittings suitable for a working steam pressure of 125 pounds per square inch.

*Tee.* — A fitting, either cast or wrought, that has one side outlet at right angles to the run. A single outlet branch pipe. (See Branch, Bull Head, Cross-over, Double Sweep, Drop, Service and Union Tee.)

*Union.* — (1) The usual trade term for a device used to connect pipes. It commonly consists of three pieces which are, first, the thread end fitted with exterior and interior threads; second, the bottom end fitted with interior threads and a small exterior shoulder; and third, the ring which has an inside flange at one end while the other end has an inside thread like that on the exterior of the thread end. A gasket is placed between the thread and bottom ends, which are drawn together by the ring. Unions are very extensively used, because they permit of connections with little disturbance of the pipe positions.

*Union Ell.* — An ell with a male or female union at one end.

*Union Joint.* — A pipe coupling, usually threaded, which permits disconnection without disturbing other sections.

*Union Tee.* — A tee with male or female union at connection on one end of run.

*Wiped Joint.* — A lead joint in which the molten solder is poured upon the desired place, after scraping and fitting the parts together, and the joint is wiped up by hand with a moleskin or cloth pad while the metal is in a plastic condition.

*Wye (Y).* — A fitting either cast or wrought that has one side outlet at any angle other than 90 degrees. The angle is usually 45 degrees, unless another angle is specified. The fitting is usually indicated by the letter Y.

## Adhesives and Sealants

**Adhesives Bonding.** — By strict definition, an adhesive is any substance that fastens or bonds materials to be joined (adherends) by means of surface attachment. However, besides bonding a joint, an adhesive may serve as a seal against attack by or passage of foreign materials. When an adhesive performs both bonding and sealing functions, it is usually called an adhesive sealant.

Where the design of an assembly permits, bonding with adhesives can replace bolting, welding, and riveting. When considering other fastening methods for thin cross-sections, the joint loads might be of such an unacceptable concentration that adhesives bonding may provide the only viable alternative. Too, properly designed adhesive joints can minimize or eliminate irregularities and breaks in the contour of an assembly. Adhesives can also serve as dielectric insulation. An adhesive with dielectric properties can act as a barrier against galvanic corrosion when two dissimilar metals such as aluminum and magnesium are joined together. Conversely, adhesive products are available which also conduct electricity.

An adhesive can be classified as structural or non-structural. Agreement is not universal on the exact separation between both classifications. But, in a general way, an adhesive can be considered structural when it is capable of supporting heavy loads; non-structural when it cannot. Most adhesives are found in liquid, paste, or granular form, though film and fabric-backed tape varieties are available. Adhesive formulations are applied by brush, roller, trowel, or spatula. If application surfaces are particularly large or if high rates of production are required, power-fed flow guns, brushes, or sprays can be used.

The hot-melt adhesives are relatively new to the assembly field. In general, they permit fastening speeds that are much greater than water- or solvent-based adhesives. Supplied in solid form, the hot-melts liquefy when heated. After application, they cool quickly, solidifying and forming the adhesive bond. They have been used successfully for a wide variety of adherends, and can greatly reduce the need for clamping and lengths of time for curing storage.

If an adhesive bonding agent is to give the best results, time restrictions recommended by the manufacturer, such as shelf life and working life must be observed. The shelf life is considered as the period of time an adhesive can be stored after its manufacture. Working or "pot" life is the span of time between the mixing or making ready of an adhesive, on the job, and when it is no longer usable.

The actual performance of an adhesive-bonded joint depends on a wide range of factors, many of them quite complex. They include: the size and nature of the applied loads; environmental conditions such as moisture or contact with other fluids or vapors; the nature of prior surface treatment of adherends; temperatures, pressures and curing times in the bonding process.

A great number of adhesives, under various brand names, may be available for a particular bonding task. However, there can be substantial differences in the cost of purchase and difficulties in application. Therefore, it is always best to check with manufacturers' information before making a proper choice. Also, testing under conditions approximating those required of the assembly in service will help assure that joints meet expected performance.

Though not meant to be all-inclusive, the information which follows correlates classes of adherends and some successful adhesive compositions from the many that can be readily purchased.

*Bonding Metal:* Epoxy resin adhesives perform well in bonding metallic adherends. One type of epoxy formulation is a two-part adhesive which can be applied at room temperature. It takes, however, seven days at room temperature for full curing, achieving shear strengths as high as 2500 psi (17.2 MPa). Curing times for this adhesive can be greatly accelerated by elevating the bonding temperature. For example, curing takes only one hour at 160°F (71°C).

A structural adhesive-filler is available for metals which is composed of aluminum powder and epoxy resin. It is made ready by adding a catalyst to the base components, and can be used to repair structural defects. At a temperature of 140°F (60°C) it cures in approximately one hour. Depending on service temperatures and design of the joint, this adhesive-filler is capable of withstanding flexural stresses above 10,000 psi (69 MPa), tension above 5,000 psi (34 MPa), and compression over 30,000 psi (207 MPa).

Many non-structural adhesives for metal-to-metal bonding are also suitable for fastening combinations of types of materials. Polysulfide, neoprene, or rubber-based adhesives are used to bond metal foils. Ethylene cellulose cements, available in a selection of colors, are used to plug machined recesses in metal surfaces, such as with screw insets. They harden within 24 hours. Other, stronger adhesive fillers are available for the non-structural patching of defects in metallic parts. One variety, used for iron and steel castings, is a cement that combines powdered iron with water-activated binding agents. The consistency of the prepared mix is such that it can be applied with a trowel and sets within 24 hours at room temperature. The filler comes in types that can be applied to both dry and wet castings, and is able to resist the quick changes of temperature during quenching operations.

Polyester cement can replace lead and other fillers for dents and openings in sheet metal. One type, used successfully on truck and auto bodies, is a two-part cement consisting of a paste resin that can be combined with a paste or powder extender. It is brushed or trowelled on, and is ready for finishing operations in one hour.

Adhesives can be used for both structural and non-structural applications which combine metals with non-metals. Structural polyester-based adhesives can bond reinforced plastic laminates to metal surfaces. One type has produced joints, between glass reinforced epoxy and stainless steel, that have tensile strengths of over 3000 psi (21 MPa). Elevated temperature service is not recommended for this adhesive. However, it is easily brushed on and bonds under slight pressure at room temperature, requiring several days for curing. The curing process accelerates when heat is added in a controlled environment, but there results a moderate reduction in tensile strength.

Low-density epoxy adhesives are successful in structurally adhering light plastics, such as polyurethane foam, to various metals. Applied by brush or spatula, the bonds cure within 24 hours at room temperatures.

Metals can be bonded structurally to wood with a liquid adhesive made up of neoprene and synthetic resin. For the best surface coverage, the adhesive should be applied in a minimum of two coats. The joints formed are capable of reaching shear stresses of 125 psi, and can gain an additional 25 percent in shear strength with the passage of time. This adhesive also serves as a strong, general purpose bonding agent for other adherend combinations, including fabrics and ceramics.

For bonding strengths in shear over 500 psi (3.4 MPa) and at service temperatures slightly above 160°F (71°C), one- 'and two-part powder and jelly forms of metal-to-wood types are available.

Besides epoxy formulations, there are general purpose rubber, cellulose, and vinyl adhesives suitable for the non-structural bonding of metals to other adherends, which include glass and leather. These adhesives, however, are not limited only to applications in which one of the adherends is metal. The vinyl and cellulose types have similar bonding properties. But while the vinyls are less flammable, they are weaker in resistance to moisture than the comparable cellulosics. Rubber-based adhesives, in turn, have good resistance to moisture and lubricating oil. They can form non-structural bonds between metal and rubber.

One manufacturer has produced an acrylic-based adhesive that is highly suitable for rapidly bonding metal with other adherends at room temperature. For some applications it can be used as a structural adhesive, in the absence of moisture and high temperature. It cures within 24 hours and can be purchased in small bottles with dispenser tips.

A two-part epoxy adhesive is commercially available for non-structural bonding of joints or for patchwork in which one of the adherends is metal. Supplied in small tubes, it performs well even when temperatures vary between −50° to 200°F ( −46° to 93°C). However, it is not recommended for use on assemblies that may experience heavy vibrations.

*Bonding Plastic:* Depending on the type of resin compound used in its manufacture, a plastic material can be classified as one of two types: a thermoplastic or a thermoset.

Thermoplastic materials have the capability of being repeatedly softened by heat and hardened by cooling. Common thermoplastics are nylon, polyethylene, acetal, polycarbonate, polyvinyl chloride, cellulose nitrate and cellulose acetate. Also, solvents can easily dissolve a number of thermoplastic materials. Because of these physical and chemical characteristics of thermoplastics, heat or solvent welding may in many instances offer a better bonding alternative than adhesives.

Thermoplastics commonly require temperatures between 200° and 400°F (93° and 204°C) for successful heat welding. However, if the maximum temperature limit for a particular thermoplastic formulation is exceeded, the plastic material will experience permanent damage. Heat can be applied directly to thermoplastic adherends, as in hot-air welding. More sophisticated joining techniques employ processes in which the heat generated for fusing thermoplastics is activated by electrical, sonic, or frictional means.

In the solvent welding of thermoplastics, solvent is applied to the adherend surfaces with the bond forming as the solvent dries. Some common solvents for thermoplastics are: a solution of phenol and formic acid for nylon; methylene chloride for polycarbonate; and methyl alcohol for the cellulosics.

Many adhesive bonding agents for thermoplastics are "dope" cements. Dope or solvent cements combine solvent with a base material that is the same thermoplastic as the adherend. One type is used successfully on polyvinyl chloride water (PVC) pipe. This liquid adhesive, with a polyvinyl chloride base, is applied in at least two coats. The pipe joint, however, must be closed in less than a minute after the adhesive is applied. Resulting joint bonds can resist hydrostatic pressures over 400 psi (28 MPa), for limited periods, and also have good resistance to impact.

Previously mentioned general purpose adhesives, such as the cellulosics, vinyls, rubber cements, and epoxies are also used successfully on thermoplastics.

Thermoset plastics lack the fusibility and solubility of the thermoplastics and are usually joined by adhesive bonding. The phenolics, epoxies, and alkyds are common thermoset plastics. Epoxy-based adhesives can join most thermoset materials, as can neoprene, nitrile rubber, and polyester-based cements. Again, these adhesives are of a general purpose nature, and can bond both thermoplastics and thermosets to other materials which include ceramics, fabric, wood, and metal.

*Bonding Rubber:* Adhesives are available commercially which can bond natural, butyl, nitrile, neoprene, and silicone rubbers. Natural and synthetic rubber cements will provide flexible joints; some types resist lubricating and other oils. Certain general purpose adhesives, such as the acrylics or epoxies, can bond rubber to almost anything else, though joints will be rigid. Depending on the choice of adhesive as well as adherend types, the bonds can carry loadings that vary from weak non-structural to mild structural in description. One type of natural rubber with a benzene-naphtha solvent can resist shear stresses to 12.5 psi (83 kPa).

*Bonding Wood:* Animal glues, available in liquid and powder form, are familiar types of wood-to-wood adhesives, commonly used in building laminated assemblies. Both forms, however, require heavy bonding pressures for joints capable of resisting substantial loadings. Also, animal glues are very sensitive to variations in temperature and moisture.

Casein types of adhesive offer moderate resistance to moisture and high temperature, but also require heavy bonding pressures, as much as 200 psi, for strong joints. Urea resin adhesives also offer moderate weather resistance, but are good for bonding wood to laminated plastics as well as to other wooden adherends. For outdoor service, under severe weather conditions, phenol-resorcinol adhesives are recommended.

Vinyl-acetate emulsions are excellent for bonding wood to other materials that have especially non-porous surfaces, such as metal and certain plastic laminates. These adhesives, too, tend to be sensitive to temperature and moisture, but are recommended for wooden patternmaking.

Rubber, acrylic, and epoxy general-purpose adhesives also perform well with wood and other adherend combinations. Specific rubber-based formulations resist attack by oil.

*Fabric and Paper Bonding:* The general purpose adhesives, which include the rubber cements and epoxies previously mentioned, are capable of bonding fabrics together and fabrics with other adherend materials. A butadiene-acrylonitrile adhesive, suitable also for fastening metals, glass, plastic, and rubber, forms joints in fabric that are highly resistant to oil and which maintain bonding strength at temperatures up to 160°F (70°C). This adhesive, however, requires a long curing period, the first few hours of which are at an elevated temperature.

Commonly, when coated fabric materials must be joined, the base material forming the suitable adhesive is of the same type as that protecting the fabric. For example, a polyvinyl chloride-based adhesive is acceptable for vinyl-coated fabrics; and neoprene-based cements for neoprene-coated materials.

Rubber cements, gum mucilages, wheat pastes, and wood rosin adhesive can join paper as well as fabric assemblies. Solvent-based rosins can be used on glass and wood also. Rosin adhesives can also be treated as hot-melt adhesives for rapid curing. Generally, the rosins are water resistant, but usually weak against attack by organic solvents.

**Sealants.** — Normally, the primary role of a sealant composition is the prevention of leakage or access by dust, fluid, and other materials in assembly structures. Nevertheless, many products are currently being manufactured that are capable of performing additional functions. For example, though a sealant is normally not an adhesive, there exists a family of adhesive sealants which in varying degrees can bond structural joints as well. Besides resisting chemical attack, some sealant surface coatings can protect against physical wear. Sealants can also dampen noise and vibration, or restrict the flow of heat or electricity. Many sealant products are available in decorative tints that can help improve the appearance of an assembly.

Most sealants tend to be limited by the operating temperatures and pressures under which they are capable of sustained performance. Also, before a suitable choice of sealant formulation is made, other properties have to be examined; these include: strength of the sealant; its degree of rigidity; ease of repair; curing characteristics; and even shelf and working life.

Dozens of manufacturers supply hundreds of sealant compounds, a number of which may fill the requirements for a particular application. The following information, however, lists common uses for sealants, along with types of compositions that have been employed successfully within each category.

*Gasket Materials:* Silicone rubber gasket compositions are supplied in tubes in a semiliquid form ready for manual application. They can also be obtained in larger containers for power-fed applications. Suppliers offer a silicone rubber-based composition that can replace preformed paper, cork, and rubber gaskets for many manufacturing operations. This composition has performed successfully in sealing water pumps, engine filter housings, and oil pans. It can also seal gear housings and other joints that require a flexible gasket material that besides resisting shock can sustain large temperature changes. Silicone rubber compositions can withstand temperatures that vary from $-100°F$ to $450°F$ ($-73°C$ to $232°C$).

Gasket tapes, ropes, and strips can also be readily purchased to fit many assembly applications. One type of sealant tape combines a pressure-sensitive adhesive with a strip of silicone-rubber sponge. This tape has good cushioning properties for vibration damping and can stick to metal, plastic, ceramic, and glass combinations.

TFE-based gasketing strips are also available. This non-stick gasketing material can perform at pressures up to 200 psi (1.4 MPa) and temperatures to 250°F (120°C). Because of the TFE base, the strip does not adhere to or gum joint surfaces.

*Sealing Pipe Joints:* Phenolic-based sealants can seal threaded joints on high-pressure steam lines. One type, that is available in liquid or paste form, resists pressure up to 1200 psi (8.3 MPa) and temperatures to 950°F (510°C). This compound is brushed on and the joint closed and tightened to a torque of 135 in.-lb. (15.3 N · m). The connection is then subjected to a 24-hour cure with superheated steam.

The joining and sealing of plastic pipe is covered under the previous adhesives bonding section.

Sulfur-based compounds, though lacking the durability of caulking lead, can be used on bell and spigot sewer pipe. Available in a formulation that can resist temperatures up to 200°F (93°C), one sulfur-based sealant is applied as a hot-melt and allowed to flow into the bell and spigot connection. It quickly solidifies at room temperature, and can develop a joint tensile strength over 300 psi (2.1 MPa).

There are asphalt, coal-tar and plastic-based compositions that can be used on both cast-iron and ceramic bell and spigot pipe. Portland cement mortars also seal ceramic piping.

*Organic Coatings:* Organic coatings can be applied by brush, spray, and dipping. Polyurethane coatings are particularly resistant to salt sprays and therefore have applications as protective marine coatings. They are considered fire-retardant, but some polyurethane formulations are poorly resistant to alkalis and acids.

Cellulose nitrate coatings have very good adhesion to metals. Though they can resist acids, they are generally poor under attack by alkalis and organic solvents.

One typical application for cellulose nitrate coatings is as a fast-drying finish for machine parts.

The chlorinated polyether coating materials weather well and resist dilute acids, alkalis and organic solvents. They can withstand temperatures as high as 150°F (66°C), and perform well as coatings on pipes, tanks, pump housing, and impellers. The chlorinated fluorocarbon coating compounds offer tank and pipe protection up to 500°F (260°C). But while resistant to diluted acids and alkalis, they have poor chemical resistance to concentrated varieties of these solutions, as well as to organic solvents.

Silicone-based coatings, because of their high temperature resistance, are especially useful as protective coatings for furnace and hot-gas lines.

Among many other applications, the epoxy and alkyd coatings have excellent adhesion to metal surfaces, and often serve as protective machinery coatings. The epoxy coatings cured along with polyamide (nylon) agents offer good general purpose protection, but clear varieties tend to chalk under conditions of weather. While the phenolic-cured epoxies have the best chemical resistance among various organic coatings, and good abrasion properties, they must be bake-cured at temperatures between 300° to 400°F (150° to 200°C).

While neoprene synthetic-rubber coatings have been used as pipe and tank linings, they offer a good cross-section of coating properties that have enabled them to serve on equipment directly involved in manufacturing operations. The neoprenes can be applied by brush, or by trowel, and offer good adhesion to metal as well as resistance to abrasion. They can withstand attack by greases, oils, and other chemicals to continuous temperatures of 200°F (93°C) or less. The protection of electroplating racks is a typical use highlighted by these properties.

## STANDARDS FOR ELECTRIC MOTORS

**Classes of NEMA Standards.** — National Electrical Manufacturers Association Standards are of two classes: 1. *NEMA Standard*, which relates to a product commercially standardized and subject to repetitive manufacture, which standard has been approved by at least 90 per cent of the members of the Subdivision eligible to vote thereon; 2. *Suggested Standard for Future Design*, which may not have been regularly applied to a commercial product, but which suggests a sound engineering approach to future development, which standard has been approved by at least two-thirds of the members of the Subdivision eligible to vote thereon.

*Authorized Engineering Information* consists of explanatory data and other engineering information of an informative character not falling within the classification of NEMA Standard or Suggested Standard for Future Design.

**Mounting Dimensions and Frame Sizes for Electric Motors.** — The dimensions for foot-mounted electric motors as standardized in the United States by the National Electrical Manufacturers Association (NEMA) will be found in Tables 1 to 3, incl. These dimensions include the spacing of bolt holes in the feet of the motor, the distance from the bottom of the feet to the center-line of the motor shaft, the size of the conduit, the length and diameter of shaft, and other dimensions likely to be required by designers or manufacturers of motor-driven equipment. In Tables 4 and 5, incl., will be found NEMA standard dimensions for face-mounted and flange-mounted motors. In these tables the standard motor frame number is given in the first column and opposite this number the mounting and other essential dimensions.

Frame numbers for various sizes and two types of motors are given in Table 6. To find the mounting dimensions for a given size and type motor, first determine the frame number for the motor from Table 6, and then, using this frame number, find the mounting and other essential dimensions in Tables 1 to 5. If the motor is to be mounted upon a belt-tightening base or upon rails, the appropriate standard dimensions will be found in Table 3.

**Design Letters of Polyphase Integral-horsepower Motors.** — Designs A, B, C and D motors are squirrel-cage motors designed to withstand full voltage starting and developing locked-rotor torque and breakdown torque, drawing locked-rotor current and having a slip as specified below:

*Design A:* Locked-rotor torque as shown in Table 9, breakdown torque as shown in Table 10, locked-rotor current higher than the values shown in Table 8, and a slip at rated load of less than 5 per cent. Motors with 10 or more poles may have a slightly greater slip.

*Design B:* Locked-rotor torque as shown in Table 9, breakdown torque as shown in Table 10, locked-rotor current not exceeding that in Table 8, and a slip at rated load of less than 5 per cent. Motors with 10 or more poles may have a slightly greater slip.

*Design C:* Locked-rotor torque for special high-torque applications up to values shown in Table 9, breakdown torque up to values shown in Table 10, locked-rotor current not exceeding values shown in Table 8 and a slip at rated load of less than 5 per cent.

*Design D:* Locked-rotor torque as indicated in Table 9, locked-rotor current not greater than that shown in Table 8 and a slip at rated load of 5 per cent or more.

**Torque and Current Definitions.** — The definitions which follow have been adopted as standard by the National Electrical Manufacturers Association.

*Locked-Rotor or Static Torque:* The locked-rotor torque of a motor is the minimum torque which it will develop at rest for all angular positions of the rotor, with rated voltage applied at rated frequency.

*Breakdown Torque:* The breakdown torque of a motor is the maximum torque which the motor will develop, with rated voltage applied at rated frequency, without an abrupt drop in speed.

*Full-Load Torque:* The full-load torque of a motor is the torque necessary to produce its rated horsepower at full load speed. In pounds at 1-foot radius it is equal to the horsepower times 5252 divided by the full-load speed.

*Pull-out Torque:* The pull-out torque of a synchronous motor is the maximum sustained torque which the motor will develop at synchronous speed with rated voltage applied at rated frequency and with normal excitation.

*Pull-in Torque:* The pull-in torque of a synchronous motor is the maximum constant torque under which the motor will pull its connected inertia load into synchronism at rated voltage and frequency, when its field excitation is applied.

*Pull-Up Torque:* The pull-up torque of an alternating current motor is the minimum torque developed by the motor during the period of acceleration from rest to the speed at which breakdown torque occurs. For motors which do not have a definite breakdown torque, the pull-up torque is the minimum torque developed up to rated speed.

*Locked Rotor Current:* The locked rotor current of a motor is the steady-state current taken from the line with the rotor locked and with rated voltage (and rated frequency in the case of alternating-current motors) applied to the motor.

**Standard Direction of Motor Rotation.** — The standard direction of rotation for all non-reversing direct-current motors, all alternating-current single-phase motors, all synchronous motors, and all universal motors, is *counter clockwise* when facing that end of the motor opposite the drive.

This rule does not apply to two- and three-phase induction motors, as in most applications the phase sequence of the power lines is rarely known.

Table 1.    NEMA Standard Dimensions for Alternating-current Foot-mounted Motors with Single Straight-shaft Extension

| Frame No. | A Max | B Max | D* | E† | 2F† | BA | H† | U | N—W | V Min | Keyseat ES Min | Keyseat S |
|---|---|---|---|---|---|---|---|---|---|---|---|---|
| 42 | ... | ... | 2.62 | 1.75 | 1.69 | 2.06 | 0.28 | 0.3750 | 1.12 | ... | ... | flat |
| 48 | ... | ... | 3.00 | 2.12 | 2.75 | 2.50 | 0.34 | 0.5000 | 1.50 | ... | ... | flat |
| 48H | ... | ... | 3.00 | 2.12 | 4.75 | 2.50 | 0.34 | 0.5000 | 1.50 | ... | ... | flat |
| 56 | ... | ... | 3.50 | 2.44 | 3.00 | 2.75 | 0.34 | 0.6250 | 1.88 | ... | 1.41 | 0.188 |
| 56H | ... | ... | 3.50 | 2.44 | 5.00 | 2.75 | 0.34 | 0.6250 | 1.88 | ... | 1.41 | 0.188 |
| 143T | 7.0 | 6.0 | 3.50 | 2.75 | 4.00 | 2.25 | 0.34 | 0.8750 | 2.25 | 2.00 | 1.41 | 0.188 |
| 145T | 7.0 | 6.0 | 3.50 | 2.75 | 5.00 | 2.25 | 0.34 | 0.8750 | 2.25 | 2.00 | 1.41 | 0.188 |
| 182T | 9.0 | 6.5 | 4.50 | 3.75 | 4.50 | 2.75 | 0.41 | 1.1250 | 2.75 | 2.50 | 1.78 | 0.250 |
| 184T | 9.0 | 7.5 | 4.50 | 3.75 | 5.50 | 2.75 | 0.41 | 1.1250 | 2.75 | 2.50 | 1.78 | 0.250 |
| 213T | 10.5 | 7.5 | 5.25 | 4.25 | 5.50 | 3.50 | 0.41 | 1.3750 | 3.38 | 3.12 | 2.41 | 0.312 |
| 215T | 10.5 | 9.0 | 5.25 | 4.25 | 7.00 | 3.50 | 0.41 | 1.3750 | 3.38 | 3.12 | 2.41 | 0.312 |
| 254T | 12.5 | 10.8 | 6.25 | 5.00 | 8.25 | 4.25 | 0.53 | 1.625 | 4.00 | 3.75 | 2.91 | 0.375 |
| 256T | 12.5 | 12.5 | 6.25 | 5.00 | 10.00 | 4.25 | 0.53 | 1.625 | 4.00 | 3.75 | 2.91 | 0.375 |
| 284T | 14.0 | 12.5 | 7.00 | 5.50 | 9.50 | 4.75 | 0.53 | 1.875 | 4.62 | 4.38 | 3.28 | 0.500 |
| 284TS | 14.0 | 12.5 | 7.00 | 5.50 | 9.50 | 4.75 | 0.53 | 1.625 | 3.25 | 3.00 | 1.91 | 0.375 |
| 286T | 14.0 | 14.0 | 7.00 | 5.50 | 11.00 | 4.75 | 0.53 | 1.875 | 4.62 | 4.38 | 3.28 | 0.500 |
| 286TS | 14.0 | 14.0 | 7.00 | 5.50 | 11.00 | 4.75 | 0.53 | 1.625 | 3.25 | 3.00 | 1.91 | 0.375 |
| 324T | 16.0 | 14.0 | 8.00 | 6.25 | 10.50 | 5.25 | 0.66 | 2.125 | 5.25 | 5.00 | 3.91 | 0.500 |
| 324TS | 16.0 | 14.0 | 8.00 | 6.25 | 10.50 | 5.25 | 0.66 | 1.875 | 3.75 | 3.50 | 2.03 | 0.500 |
| 326T | 16.0 | 15.5 | 8.00 | 6.25 | 12.00 | 5.25 | 0.66 | 2.125 | 5.25 | 5.00 | 3.91 | 0.500 |
| 326TS | 16.0 | 15.5 | 8.00 | 6.25 | 12.00 | 5.25 | 0.66 | 1.875 | 3.75 | 3.50 | 2.03 | 0.500 |
| 364T | 18.0 | 15.2 | 9.00 | 7.00 | 11.25 | 5.88 | 0.66 | 2.375 | 5.88 | 5.62 | 4.28 | 0.625 |
| 364TS | 18.0 | 15.2 | 9.00 | 7.00 | 11.25 | 5.88 | 0.66 | 1.875 | 3.75 | 3.50 | 2.03 | 0.500 |
| 365T | 18.0 | 16.2 | 9.00 | 7.00 | 12.25 | 5.88 | 0.66 | 2.375 | 5.88 | 5.62 | 4.28 | 0.625 |
| 365TS | 18.0 | 16.2 | 9.00 | 7.00 | 12.25 | 5.88 | 0.66 | 1.875 | 3.75 | 3.50 | 2.03 | 0.500 |
| 404T | 20.0 | 16.2 | 10.00 | 8.00 | 12.25 | 6.62 | 0.81 | 2.875 | 7.25 | 7.00 | 5.65 | 0.750 |
| 404TS | 20.0 | 16.2 | 10.00 | 8.00 | 12.25 | 6.62 | 0.81 | 2.125 | 4.25 | 4.00 | 2.78 | 0.500 |
| 405T | 20.0 | 17.8 | 10.00 | 8.00 | 13.75 | 6.62 | 0.81 | 2.875 | 7.25 | 7.00 | 5.65 | 0.750 |
| 405TS | 20.0 | 17.8 | 10.00 | 8.00 | 13.75 | 6.62 | 0.81 | 2.125 | 4.25 | 4.00 | 2.78 | 0.500 |
| 444T | 22.0 | 18.5 | 11.00 | 9.00 | 14.50 | 7.50 | 0.81 | 3.375 | 8.50 | 8.25 | 6.91 | 0.875 |
| 444TS | 22.0 | 18.5 | 11.00 | 9.00 | 14.50 | 7.50 | 0.81 | 2.375 | 4.75 | 4.50 | 3.03 | 0.625 |
| 445T | 22.0 | 20.5 | 11.00 | 9.00 | 16.50 | 7.50 | 0.81 | 3.375 | 8.50 | 8.25 | 6.91 | 0.875 |
| 445TS | 22.0 | 20.5 | 11.00 | 9.00 | 16.50 | 7.50 | 0.81 | 2.375 | 4.75 | 4.50 | 3.03 | 0.625 |

All dimensions are in inches.    See Fig. 1 for diagram showing letter symbols.

*Dimension D will never be greater than the above values for rigid-base motors. However, it may be less, so that shims are usually required for coupled or geared motors. When the exact dimension is required, shims up to 0.03 inch may be necessary on frame sizes whose D dimension is 8.00 inches or less; on larger frames, shims up to 0.06 inch may be necessary. No tolerances have been established for the D dimension of resilient mounted motors.

†Frame Nos. 42, 48, 48H, 56 and 56H have a tolerance for the 2F dimension of ±0.03 inch and for the H dimension (width of slot) +0.02, −0 inch. For frame Nos. 143T through 445T, inclusive, the tolerance for the 2E and 2F dimensions is ±0.03 inch and for the H dimension (diameter of hole) is +0.05, −0 inch.

The minimum size of the threaded or clearance hole, AA, for external conduit entrance (expressed in conduit size) to the terminal housing is for frame Nos. 143T through 184T, ¾ inch; for frame Nos. 213T and 215T, 1 inch; for frame Nos. 254T and 256T, 1¼ inches; for frame Nos. 284T through 286TS, 1½ inches; for frame Nos. 324T through 326TS, 2 inches; and for frame Nos. 364T through 445TS, 3 inches.

For larger frame sizes see NEMA Standards.

## Motor Types According to Variability of Speed. —

Five types of motors classified according to variability of speed are:

*Constant-speed Motors:* In this type of motor the normal operating speed is constant or practically constant; for example, a synchronous motor, an induction motor with small slip, or a direct-current shunt-wound motor.

*Varying-speed Motor:* In this type of motor, the speed varies with the load, ordinarily decreasing when the load increases; such as a series-wound or repulsion motor.

*Adjustable-speed Motor:* In this type of motor, the speed can be varied gradually over a considerable range, but when once adjusted remains practically unaffected by

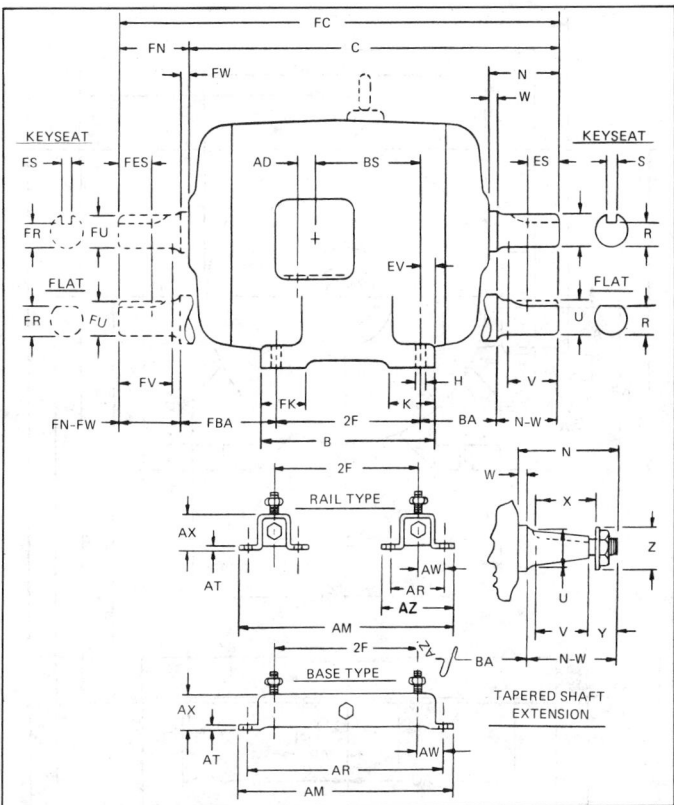

Fig. 1a.  NEMA Standard dimensional designations for alternating-
current and direct-current foot-mounted motors—side view.

the load; such as a direct-current shunt-wound motor with field resistance control
designed for a considerable range of speed adjustment.

The base speed of an adjustable-speed motor is the lowest rated speed obtained at rated
load and rated voltage at the temperature rise specified in the rating.

*Adjustable Varying-speed Motor:* This type of motor is one in which the speed can be
adjusted gradually, but when once adjusted for a given load will vary in considerable
degree with the change in load; such as a direct-current compound-wound motor
adjusted by field control or a wound-rotor induction motor with rheostatic speed
control.

*Multispeed Motor:* This type of motor is one which can be operated at any one of two or
more definite speeds, each being practically independent of the load; such as a direct-
current motor with two armature windings or an induction motor with windings
capable of various pole groupings. In the case of multispeed permanent-split capacitor
and shaded pole motors, the speeds are dependent upon the load.

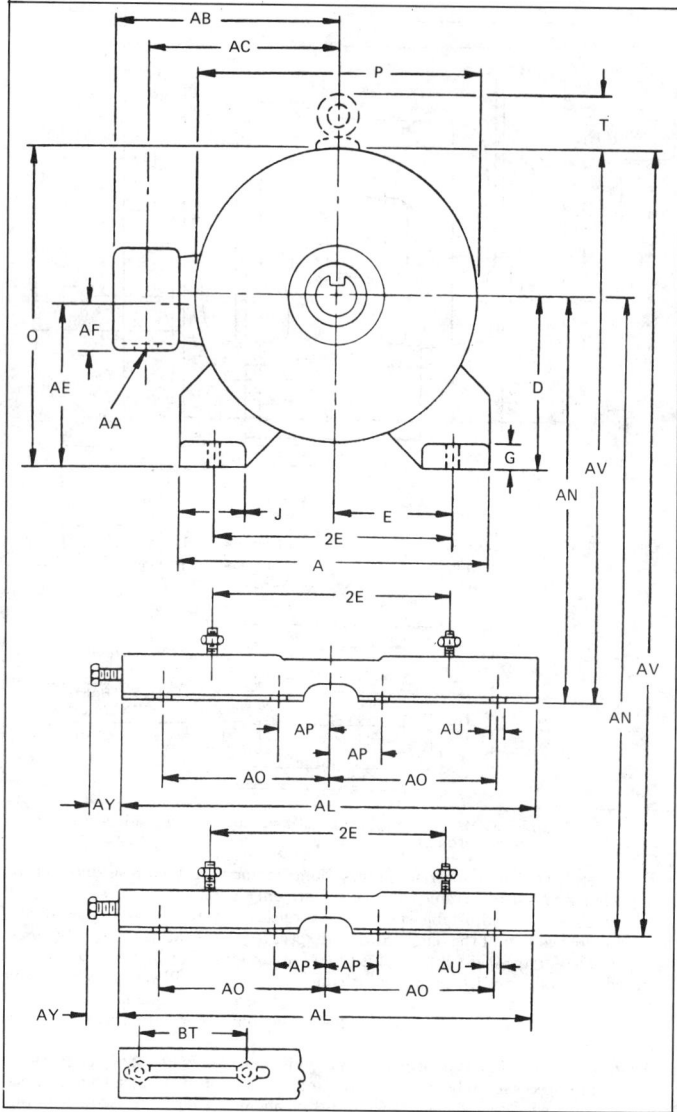

Fig. 1b.   NEMA Standard dimensional designations for alternating-current and direct-current foot-mounted motors—drive end view.

Table 2. NEMA Standard Shaft Extension and Keyseat Dimensions for
Alternating-current, Foot-mounted Motors with Single Tapered or
Double Straight/Tapered Shaft Extension

| Frame No. | Drive End—Tapered Shaft Extension† | | | | | | | | | | |
|---|---|---|---|---|---|---|---|---|---|---|---|
| | BA | U | N–W | V | X | Y | Z max | Shaft Threads | Keyseat | | Key Length‡ |
| | | | | | | | | | Width | Depth | |
| 143TR and 145TR | 2.25 | 0.8750 | 2.62 | 1.75 | 1.88 | 0.75 | 1.38 | ⅝–18 | 0.188 | 0.094 | 1.50 |
| 182TR and 184TR | 2.75 | 1.1250 | 3.38 | 2.25 | 2.38 | 0.88 | 1.50 | ¾–16 | 0.250 | 0.125 | 2.00 |
| 213TR and 215TR | 3.50 | 1.3750 | 4.12 | 2.62 | 2.75 | 1.25 | 2.00 | 1 –14 | 0.312 | 0.156 | 2.38 |
| 254TR and 256TR | 4.25 | 1.625 | 4.50 | 2.88 | 3.00 | 1.25 | 2.00 | 1 –14 | 0.375 | 0.188 | 2.62 |
| 284TR and 286TR | 4.75 | 1.875 | 4.75 | 3.12 | 3.25 | 1.25 | 2.38 | 1¼–12 | 0.500 | 0.250 | 2.88 |
| 324TR and 326TR | 5.25 | 2.125 | 5.25 | 3.50 | 3.62 | 1.38 | 2.75 | 1½–8 | 0.500 | 0.250 | 3.25 |
| 364TR and 365TR | 5.88 | 2.375 | 5.75 | 3.75 | 3.88 | 1.50 | 3.25 | 1¾–8 | 0.625 | 0.312 | 3.50 |
| 404TR and 405TR | 6.62 | 2.875 | 6.62 | 4.38 | 4.50 | 1.75 | 3.62 | 2 –8 | 0.750 | 0.375 | 4.12 |
| 444TR and 445TR | 7.50 | 3.375 | 7.50 | 5.00 | 5.12 | 2.00 | 4.12 | 2¼–8 | 0.875 | 0.438 | 4.75 |

Opposite Drive End—Straight Shaft Extension**

| Frame No. Series | FU | FN–FW | FV Min | Keyseat* | | |
|---|---|---|---|---|---|---|
| | | | | R | ES Min | S |
| 140 | 0.6250 | 1.62 | 1.38 | 0.517 | 0.91 | 0.188 |
| 180 | 0.8750 | 2.25 | 2.00 | 0.771 | 1.41 | 0.188 |
| 210 | 1.1250 | 2.75 | 2.50 | 0.986 | 1.78 | 0.250 |
| 250 | 1.3750 | 3.38 | 3.12 | 1.201 | 2.41 | 0.312 |
| 280 | 1.625 | 4.00 | 3.75 | 1.416 | 2.91 | 0.375 |
| 280 Short Shaft | 1.625 | 3.25 | 3.00 | 1.416 | 1.91 | 0.375 |
| 320 | 1.875 | 4.62 | 4.38 | 1.591 | 3.28 | 0.500 |
| 320 Short Shaft | 1.875 | 3.75 | 3.50 | 1.591 | 2.03 | 0.500 |
| 360 | 1.875 | 4.62 | 4.38 | 1.591 | 3.28 | 0.500 |
| 360 Short Shaft | 1.875 | 3.75 | 3.50 | 1.591 | 2.03 | 0.500 |
| 400 | 2.125 | 5.25 | 5.00 | 1.845 | 3.91 | 0.500 |
| 400 Short Shaft | 2.125 | 4.25 | 4.00 | 1.845 | 2.78 | 0.500 |
| 440 | 2.375 | 5.88 | 5.62 | 2.021 | 4.28 | 0.625 |
| 440 Short Shaft | 2.375 | 4.75 | 4.50 | 2.021 | 3.03 | 0.625 |

Opposite Drive End—Tapered Shaft Extension†**

| Frame No. Series | FU | FN–FW | FV | FX | FY | FZ Max | Shaft Threads | Keyseat | | Key Length‡ |
|---|---|---|---|---|---|---|---|---|---|---|
| | | | | | | | | Width | Depth | |
| 140 | 0.6250 | 2.00 | 1.38 | 1.50 | 0.50 | 1.12 | ⅜–24 | 0.188 | 0.094 | 1.12 |
| 180 | 0.8750 | 2.62 | 1.75 | 1.88 | 0.75 | 1.38 | ⅝–18 | 0.188 | 0.094 | 1.50 |
| 210 | 1.1250 | 3.38 | 2.25 | 2.38 | 0.88 | 1.50 | ¾–16 | 0.250 | 0.125 | 2.00 |
| 250 | 1.3750 | 4.12 | 2.62 | 2.75 | 1.25 | 2.00 | 1 –14 | 0.312 | 0.156 | 2.38 |
| 280 | 1.6250 | 4.50 | 2.88 | 3.00 | 1.25 | 2.00 | 1 –14 | 0.375 | 0.188 | 2.62 |
| 320 | 1.8750 | 4.75 | 3.12 | 3.25 | 1.25 | 2.38 | 1¼–12 | 0.500 | 0.250 | 2.88 |
| 360 | 1.8750 | 4.75 | 3.12 | 3.25 | 1.25 | 2.38 | 1¼–12 | 0.500 | 0.250 | 2.88 |
| 400 | 2.1250 | 5.25 | 3.50 | 3.62 | 1.38 | 2.75 | 1½–8 | 0.500 | 0.250 | 3.25 |
| 440 | 2.3750 | 5.75 | 3.75 | 3.88 | 1.50 | 3.25 | 1¾–8 | 0.625 | 0.312 | 3.50 |

All dimensions are in inches.
See Fig. 1 for diagram showing letter dimensions.
* Motors furnished with keyseats cut in the shaft extension for pulley, coupling, pinion, etc., are usually furnished with a key.
** For drive applications other than direct connected, the motor manufacturer should be consulted.
† The threaded end of the tapered shaft is furnished with a nut and suitable locking device. The taper of the shaft is at the rate of 1.25 inches in diameter per foot of length.
‡ The tolerance on the length of key is ±0.03 inch.

**Table 3. NEMA Standard Dimensions for Foot-mounted Industrial Direct-current Motors**

| Frame No. | A Max | B Max | D* | E† | 2F† | BA | H† Hole |
|---|---|---|---|---|---|---|---|
| 182AT | 9.00 | 6.50 | 4.50 | 3.75 | 4.50 | 2.75 | 0.41 |
| 183AT | 9.00 | 7.00 | 4.50 | 3.75 | 5.00 | 2.75 | 0.41 |
| 184AT | 9.00 | 7.50 | 4.50 | 3.75 | 5.50 | 2.75 | 0.41 |
| 185AT | 9.00 | 8.25 | 4.50 | 3.75 | 6.25 | 2.75 | 0.41 |
| 186AT | 9.00 | 9.00 | 4.50 | 3.75 | 7.00 | 2.75 | 0.41 |
| 187AT | 9.00 | 10.00 | 4.50 | 3.75 | 8.00 | 2.75 | 0.41 |
| 188AT | 9.00 | 11.00 | 4.50 | 3.75 | 9.00 | 2.75 | 0.41 |
| 189AT | 9.00 | 12.00 | 4.50 | 3.75 | 10.00 | 2.75 | 0.41 |
| 1810AT | 9.00 | 13.00 | 4.50 | 3.75 | 11.00 | 2.75 | 0.41 |
| 213AT | 10.50 | 7.50 | 5.25 | 4.25 | 5.50 | 3.50 | 0.41 |
| 214AT | 10.50 | 8.25 | 5.25 | 4.25 | 6.25 | 3.50 | 0.41 |
| 215AT | 10.50 | 9.00 | 5.25 | 4.25 | 7.00 | 3.50 | 0.41 |
| 216AT | 10.50 | 10.00 | 5.25 | 4.25 | 8.00 | 3.50 | 0.41 |
| 217AT | 10.50 | 11.00 | 5.25 | 4.25 | 9.00 | 3.50 | 0.41 |
| 218AT | 10.50 | 12.00 | 5.25 | 4.25 | 10.00 | 3.50 | 0.41 |
| 219AT | 10.50 | 13.00 | 5.25 | 4.25 | 11.00 | 3.50 | 0.41 |
| 2110AT | 10.50 | 14.50 | 5.25 | 4.25 | 12.50 | 3.50 | 0.41 |
| 253AT | 12.50 | 9.50 | 6.25 | 5.00 | 7.00 | 4.25 | 0.53 |
| 254AT | 12.50 | 10.75 | 6.25 | 5.00 | 8.25 | 4.25 | 0.53 |
| 255AT | 12.50 | 11.50 | 6.25 | 5.00 | 9.00 | 4.25 | 0.53 |
| 256AT | 12.50 | 12.50 | 6.25 | 5.00 | 10.00 | 4.25 | 0.53 |
| 257AT | 12.50 | 13.50 | 6.25 | 5.00 | 11.00 | 4.25 | 0.53 |
| 258AT | 12.50 | 15.00 | 6.25 | 5.00 | 12.50 | 4.25 | 0.53 |
| 259AT | 12.50 | 16.50 | 6.25 | 5.00 | 14.00 | 4.25 | 0.53 |
| 283AT | 14.00 | 11.00 | 7.00 | 5.50 | 8.00 | 4.75 | 0.53 |
| 284AT | 14.00 | 12.50 | 7.00 | 5.50 | 9.50 | 4.75 | 0.53 |
| 285AT | 14.00 | 13.00 | 7.00 | 5.50 | 10.00 | 4.75 | 0.53 |
| 286AT | 14.00 | 14.00 | 7.00 | 5.50 | 11.00 | 4.75 | 0.53 |
| 287AT | 14.00 | 15.50 | 7.00 | 5.50 | 12.50 | 4.75 | 0.53 |
| 288AT | 14.00 | 17.00 | 7.00 | 5.50 | 14.00 | 4.75 | 0.53 |
| 289AT | 14.00 | 19.00 | 7.00 | 5.50 | 16.00 | 4.75 | 0.53 |
| 323AT | 16.00 | 12.50 | 8.00 | 6.25 | 9.00 | 5.25 | 0.66 |
| 324AT | 16.00 | 14.00 | 8.00 | 6.25 | 10.50 | 5.25 | 0.66 |
| 325AT | 16.00 | 14.50 | 8.00 | 6.25 | 11.00 | 5.25 | 0.66 |
| 326AT | 16.00 | 15.50 | 8.00 | 6.25 | 12.00 | 5.25 | 0.66 |
| 327AT | 16.00 | 17.50 | 8.00 | 6.25 | 14.00 | 5.25 | 0.66 |
| 328AT | 16.00 | 19.50 | 8.00 | 6.25 | 16.00 | 5.25 | 0.66 |
| 329AT | 16.00 | 21.50 | 8.00 | 6.25 | 18.00 | 5.25 | 0.66 |
| 363AT | 18.00 | 14.00 | 9.00 | 7.00 | 10.00 | 5.88 | 0.81 |
| 364AT | 18.00 | 15.25 | 9.00 | 7.00 | 11.25 | 5.88 | 0.81 |
| 365AT | 18.00 | 16.25 | 9.00 | 7.00 | 12.25 | 5.88 | 0.81 |
| 366AT | 18.00 | 18.00 | 9.00 | 7.00 | 14.00 | 5.88 | 0.81 |
| 367AT | 18.00 | 20.00 | 9.00 | 7.00 | 16.00 | 5.88 | 0.81 |
| 368AT | 18.00 | 22.00 | 9.00 | 7.00 | 18.00 | 5.88 | 0.81 |
| 369AT | 18.00 | 24.00 | 9.00 | 7.00 | 20.00 | 5.88 | 0.81 |
| 403AT | 20.00 | 15.00 | 10.00 | 8.00 | 11.00 | 6.62 | 0.94 |
| 404AT | 20.00 | 16.25 | 10.00 | 8.00 | 12.25 | 6.62 | 0.94 |
| 405AT | 20.00 | 17.75 | 10.00 | 8.00 | 13.75 | 6.62 | 0.94 |
| 406AT | 20.00 | 20.00 | 10.00 | 8.00 | 16.00 | 6.62 | 0.94 |
| 407AT | 20.00 | 22.00 | 10.00 | 8.00 | 18.00 | 6.62 | 0.94 |
| 408AT | 20.00 | 24.00 | 10.00 | 8.00 | 20.00 | 6.62 | 0.94 |
| 409AT | 20.00 | 26.00 | 10.00 | 8.00 | 22.00 | 6.62 | 0.94 |
| 443AT | 22.00 | 16.50 | 11.00 | 9.00 | 12.50 | 7.50 | 1.06 |
| 444AT | 22.00 | 18.50 | 11.00 | 9.00 | 15.00 | 7.50 | 1.06 |
| 445AT | 22.00 | 20.50 | 11.00 | 9.00 | 16.50 | 7.50 | 1.06 |
| 446AT | 22.00 | 22.00 | 11.00 | 9.00 | 18.00 | 7.50 | 1.06 |

All dimensions are in inches. See Fig. 1 for diagram showing letter symbols.

* Dimension $D$ will never be greater than the values shown in this table, but it may be less so that shims are usually required for coupled or geared motors. When the exact dimension is required, shims up to 0.03 inch may be necessary on frame sizes whose dimension $D$ is 8 inches or less; on larger frame sizes, shims up to 0.06 inch may be necessary.

† The tolerance for the 2E and 2F dimensions is ±0.03 inch and for the H dimensions, +0.05, −0 inch.

For larger frame sizes see NEMA Standards.

**Table 3.** (*Continued*). **NEMA Standard Dimensions for Foot-mounted Industrial Direct-current Motors**

| Frame No. | AL | AM | AO | AR | AU | AX | AY Max Bases | BT |
|---|---|---|---|---|---|---|---|---|
| 182AT | 12.75 | 9.50 | 4.50 | 4.25 | 0.50 | 1.50 | 0.50 | 3.00 |
| 183AT | 12.75 | 10.00 | 4.50 | 4.50 | 0.50 | 1.50 | 0.50 | 3.00 |
| 184AT | 12.75 | 10.50 | 4.50 | 4.75 | 0.50 | 1.50 | 0.50 | 3.00 |
| 185AT | 12.75 | 11.25 | 4.50 | 5.12 | 0.50 | 1.50 | 0.50 | 3.00 |
| 186AT | 12.75 | 12.00 | 4.50 | 5.50 | 0.50 | 1.50 | 0.50 | 3.00 |
| 187AT | 12.75 | 13.00 | 4.50 | 6.00 | 0.50 | 1.50 | 0.50 | 3.00 |
| 188AT | 12.75 | 14.00 | 4.50 | 6.50 | 0.50 | 1.50 | 0.50 | 3.00 |
| 189AT | 12.75 | 15.00 | 4.50 | 7.00 | 0.50 | 1.50 | 0.50 | 3.00 |
| 1810AT | 12.75 | 16.00 | 4.50 | 7.50 | 0.50 | 1.50 | 0.50 | 3.00 |
| 213AT | 15.00 | 11.00 | 5.25 | 4.75 | 0.50 | 1.75 | 0.50 | 3.50 |
| 214AT | 15.00 | 11.75 | 5.25 | 5.12 | 0.50 | 1.75 | 0.50 | 3.50 |
| 215AT | 15.00 | 12.50 | 5.25 | 5.50 | 0.50 | 1.75 | 0.50 | 3.50 |
| 216AT | 15.00 | 13.50 | 5.25 | 6.00 | 0.50 | 1.75 | 0.50 | 3.50 |
| 217AT | 15.00 | 14.50 | 5.25 | 6.50 | 0.50 | 1.75 | 0.50 | 3.50 |
| 218AT | 15.00 | 15.50 | 5.25 | 7.00 | 0.50 | 1.75 | 0.50 | 3.50 |
| 219AT | 15.00 | 16.50 | 5.25 | 7.50 | 0.50 | 1.75 | 0.50 | 3.50 |
| 2110AT | 15.00 | 18.00 | 5.25 | 8.25 | 0.50 | 1.75 | 0.50 | 3.50 |
| 253AT | 17.75 | 13.88 | 6.25 | 6.00 | 0.62 | 2.00 | 0.62 | 4.00 |
| 254AT | 17.75 | 15.12 | 6.25 | 6.62 | 0.62 | 2.00 | 0.62 | 4.00 |
| 255AT | 17.75 | 15.88 | 6.25 | 7.00 | 0.62 | 2.00 | 0.62 | 4.00 |
| 256AT | 17.75 | 16.88 | 6.25 | 7.50 | 0.62 | 2.00 | 0.62 | 4.00 |
| 257AT | 17.75 | 17.88 | 6.25 | 8.00 | 0.62 | 2.00 | 0.62 | 4.00 |
| 258AT | 17.75 | 19.38 | 6.25 | 8.75 | 0.62 | 2.00 | 0.62 | 4.00 |
| 259AT | 17.75 | 20.88 | 6.25 | 9.50 | 0.62 | 2.00 | 0.62 | 4.00 |
| 283AT | 19.75 | 15.38 | 7.00 | 6.75 | 0.62 | 2.00 | 0.62 | 4.00 |
| 284AT | 19.75 | 16.88 | 7.00 | 7.50 | 0.62 | 2.00 | 0.62 | 4.00 |
| 285AT | 19.75 | 17.38 | 7.00 | 7.75 | 0.62 | 2.00 | 0.62 | 4.00 |
| 286AT | 19.75 | 18.38 | 7.00 | 8.25 | 0.62 | 2.00 | 0.62 | 4.00 |
| 287AT | 19.75 | 19.88 | 7.00 | 9.00 | 0.62 | 2.00 | 0.62 | 4.00 |
| 288AT | 19.75 | 21.38 | 7.00 | 9.75 | 0.62 | 2.00 | 0.62 | 4.00 |
| 289AT | 19.75 | 23.38 | 7.00 | 10.75 | 0.62 | 2.00 | 0.62 | 4.00 |
| 323AT | 22.75 | 17.75 | 8.00 | 7.75 | 0.75 | 2.50 | 0.75 | 5.25 |
| 324AT | 22.75 | 19.25 | 8.00 | 8.50 | 0.75 | 2.50 | 0.75 | 5.25 |
| 325AT | 22.75 | 19.75 | 8.00 | 8.75 | 0.75 | 2.50 | 0.75 | 5.25 |
| 326AT | 22.75 | 20.75 | 8.00 | 9.25 | 0.75 | 2.50 | 0.75 | 5.25 |
| 327AT | 22.75 | 22.75 | 8.00 | 10.25 | 0.75 | 2.50 | 0.75 | 5.25 |
| 328AT | 22.75 | 24.75 | 8.00 | 11.25 | 0.75 | 2.50 | 0.75 | 5.25 |
| 329AT | 22.75 | 26.75 | 8.00 | 12.25 | 0.75 | 2.50 | 0.75 | 5.25 |
| 363AT | 25.50 | 19.25 | 9.00 | 8.25 | 0.88 | 2.50 | 0.75 | 6.00 |
| 364AT | 25.50 | 20.50 | 9.00 | 9.12 | 0.88 | 2.50 | 0.75 | 6.00 |
| 365AT | 25.50 | 21.50 | 9.00 | 9.62 | 0.88 | 2.50 | 0.75 | 6.00 |
| 366AT | 25.50 | 23.25 | 9.00 | 10.50 | 0.88 | 2.50 | 0.75 | 6.00 |
| 367AT | 25.50 | 25.25 | 9.00 | 11.50 | 0.88 | 2.50 | 0.75 | 6.00 |
| 368AT | 25.50 | 27.25 | 9.00 | 12.50 | 0.88 | 2.50 | 0.75 | 6.00 |
| 369AT | 25.50 | 29.25 | 9.00 | 13.50 | 0.88 | 2.50 | 0.75 | 6.00 |
| 403AT | 28.75 | 21.12 | 10.00 | 9.25 | 1.00 | 3.00 | 0.88 | 7.00 |
| 404AT | 28.75 | 22.38 | 10.00 | 9.88 | 1.00 | 3.00 | 0.88 | 7.00 |
| 405AT | 28.75 | 23.88 | 10.00 | 10.62 | 1.00 | 3.00 | 0.88 | 7.00 |
| 406AT | 28.75 | 26.12 | 10.00 | 11.75 | 1.00 | 3.00 | 0.88 | 7.00 |
| 407AT | 28.75 | 28.12 | 10.00 | 12.75 | 1.00 | 3.00 | 0.88 | 7.00 |
| 408AT | 28.75 | 30.12 | 10.00 | 13.75 | 1.00 | 3.00 | 0.88 | 7.00 |
| 409AT | 28.75 | 32.12 | 10.00 | 14.75 | 1.00 | 3.00 | 0.88 | 7.00 |
| 443AT | 31.25 | 22.62 | 11.00 | 10.00 | 1.12 | 3.00 | 0.88 | 7.50 |
| 444AT | 31.25 | 24.62 | 11.00 | 11.00 | 1.12 | 3.00 | 0.88 | 7.50 |
| 445AT | 31.25 | 26.62 | 11.00 | 12.00 | 1.12 | 3.00 | 0.88 | 7.50 |
| 446AT | 31.25 | 29.12 | 11.00 | 12.75 | 1.12 | 3.00 | 0.88 | 7.50 |

All dimensions are in inches.
See Fig. 1 for diagram showing letter symbols.
For larger frame sizes, see NEMA Standards.

**Table 3.** (*Concluded*). **NEMA Standard Dimensions for Foot-mounted Industrial Direct-current Motors**

| Frame No.‡ | Drive End—For Belt Drive | | | | | End Opposite Drive—Straight | | | | |
|---|---|---|---|---|---|---|---|---|---|---|
| | U | N–W | V Min | Keyseat | | FU | FN–FW | FV Min | Keyseat | |
| | | | | ES Min | S | | | | FES Min | FS |
| 182AT-1810AT | 1.1250 | 2.25 | 2.00 | 1.41 | 0.250 | 0.8750 | 1.75 | 1.50 | 0.91 | 0.188 |
| 213AT-2110AT | 1.3750 | 2.75 | 2.50 | 1.78 | 0.312 | 1.1250 | 2.25 | 2.00 | 1.41 | 0.250 |
| 253AT-259AT | 1.625 | 3.25 | 3.00 | 2.28 | 0.375 | 1.3750 | 2.75 | 2.50 | 1.78 | 0.312 |
| 283AT-289AT | 1.875 | 3.75 | 3.50 | 2.53 | 0.500 | 1.625 | 3.25 | 3.00 | 2.28 | 0.375 |
| 323AT-329AT | 2.125 | 4.25 | 4.00 | 3.03 | 0.500 | 1.875 | 3.75 | 3.50 | 2.53 | 0.500 |
| 363AT-369AT | 2.375 | 4.75 | 4.50 | 3.53 | 0.625 | 2.125 | 4.25 | 4.00 | 3.03 | 0.500 |
| 403AT-409AT | 2.625 | 5.25 | 5.00 | 4.03 | 0.625 | 2.375 | 4.75 | 4.50 | 3.53 | 0.625 |
| 443AT-449AT | 2.875 | 5.75 | 5.50 | 4.53 | 0.750 | 2.625 | 5.25 | 5.00 | 4.03 | 0.625 |

All dimensions are in inches. See Fig. 1 for diagram showing letter symbols.

**Table 4.** **NEMA Standard Dimensions for Face-mounted and Flange-mounted Industrial Direct-current Motors**

| Frame No. | Type C Face Mounted | | | | | | | | | | |
|---|---|---|---|---|---|---|---|---|---|---|---|
| | AJ | AK | BA | BB Min | BC | BD Max | BF Tap Size* | U | AH | Keyseat | |
| | | | | | | | | | | ES Min | S |
| 182ATC-1810ATC | 7.250 | 8.500 | 2.75 | 0.25 | 0.12 | 9.00 | ½–13 | 1.1250 | 2.12 | 1.41 | 0.250 |
| 213ATC-2110ATC | 7.250 | 8.500 | 3.50 | 0.25 | 0.25 | 9.00 | ½–13 | 1.3750 | 2.50 | 1.78 | 0.312 |
| 253ATC-259ATC | 7.250 | 8.500 | 4.25 | 0.25 | 0.25 | 10.00 | ½–13 | 1.625 | 3.00 | 2.28 | 0.375 |
| 283ATC-289ATC | 9.000 | 10.500 | 4.75 | 0.25 | 0.25 | 11.25 | ½–13 | 1.875 | 3.50 | 2.53 | 0.500 |
| 323ATC-329ATC | 11.000 | 12.500 | 5.25 | 0.25 | 0.25 | 14.00 | ⅝–11 | 2.125 | 4.00 | 3.03 | 0.500 |

| Frame No. | Type D Flange Mounted | | | | | | | | | |
|---|---|---|---|---|---|---|---|---|---|---|
| | AJ | AK | BB Max | BC | BD Max | BE Nom | BF Size† | U | AH | Keyseat |
| | | | | | | | | | | S |
| 182ATD-1810ATD | 10.00 | 9.000 | 0.25 | o | 11.00 | 0.50 | 0.53 | 1.1250 | 2.25 | 0.250 |
| 213ATD-2110ATD | 12.50 | 11.000 | 0.25 | o | 14.00 | 0.75 | 0.81 | 1.3750 | 2.75 | 0.312 |
| 253ATD-259ATD | 16.00 | 14.000 | 0.25 | o | 18.00 | 0.75 | 0.81 | 1.625 | 3.25 | 0.375 |
| 283ATD-289ATD | 16.00 | 14.000 | 0.25 | o | 18.00 | 0.75 | 0.81 | 1.875 | 3.75 | 0.500 |
| 323ATD-329ATD | 16.00 | 14.000 | 0.25 | o | 18.00 | 0.75 | 0.81 | 2.125 | 4.25 | 0.500 |
| 363ATD-369ATD | 20.00 | 18.000 | 0.25 | o | 22.00 | 1.00 | 0.81 | 2.375 | 4.75 | 0.625 |
| 403ATD-409ATD | 22.00 | 18.000 | 0.19 | o | 24.00 | 1.00 | 0.81 | 2.625 | 5.25 | 0.625 |
| 443ATD-449ATD | 22.00 | 18.000 | 0.19 | o | 24.00 | 1.00 | 0.81 | 2.875 | 5.75 | 0.750 |

All dimensions are in inches.
See Figs. 2 and 3 for diagrams showing letter symbols.
* The number of holes is 4. The bolt penetration allowance is 0.75 inch for frame Nos. 182ATC through 289ATC and 0.94 inch for frame Nos. 323ATC through 329ATC.
† The number of holes is 4 for frame Nos. 182ATD through 329ATD and 8 for frame Nos. 363ATD through 449ASD. Recommended bolt lengths are 1.25 inches for frame Nos. 182ATD through 1810ATD; 2.00 inches for frame Nos. 213ATD through 329ATD; and 2.25 inches for frame Nos. 363ATD through 449ATD. For larger frame sizes, see NEMA Standards.

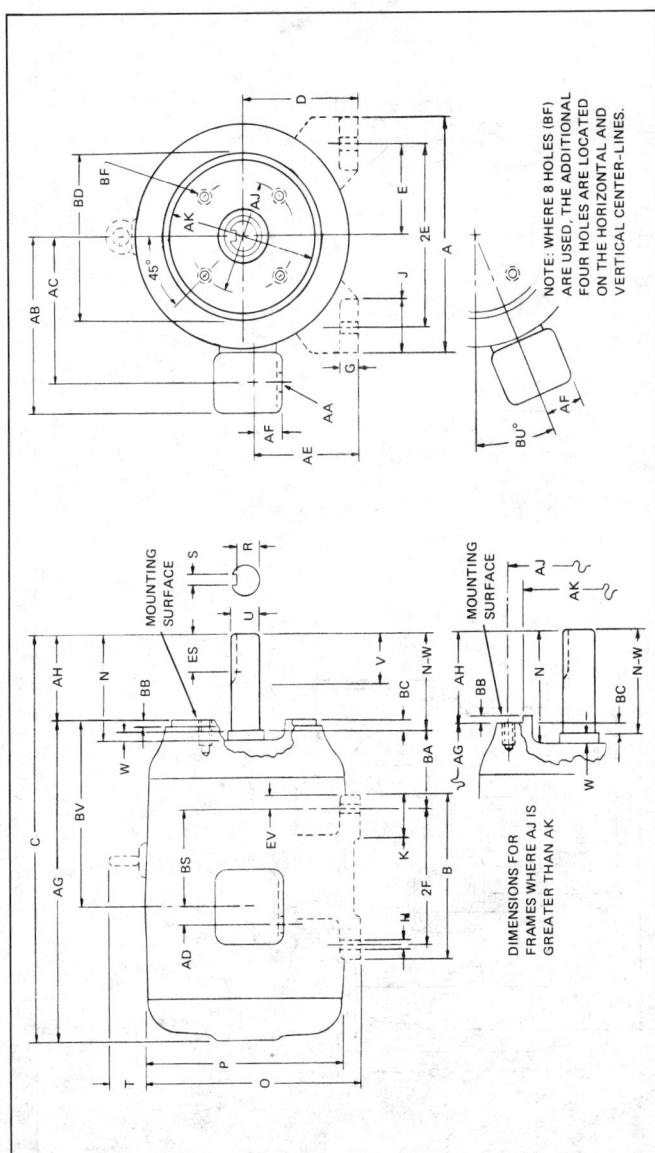

Fig. 2  NEMA Standard dimensions for Type C face-mounted foot or footless motors.

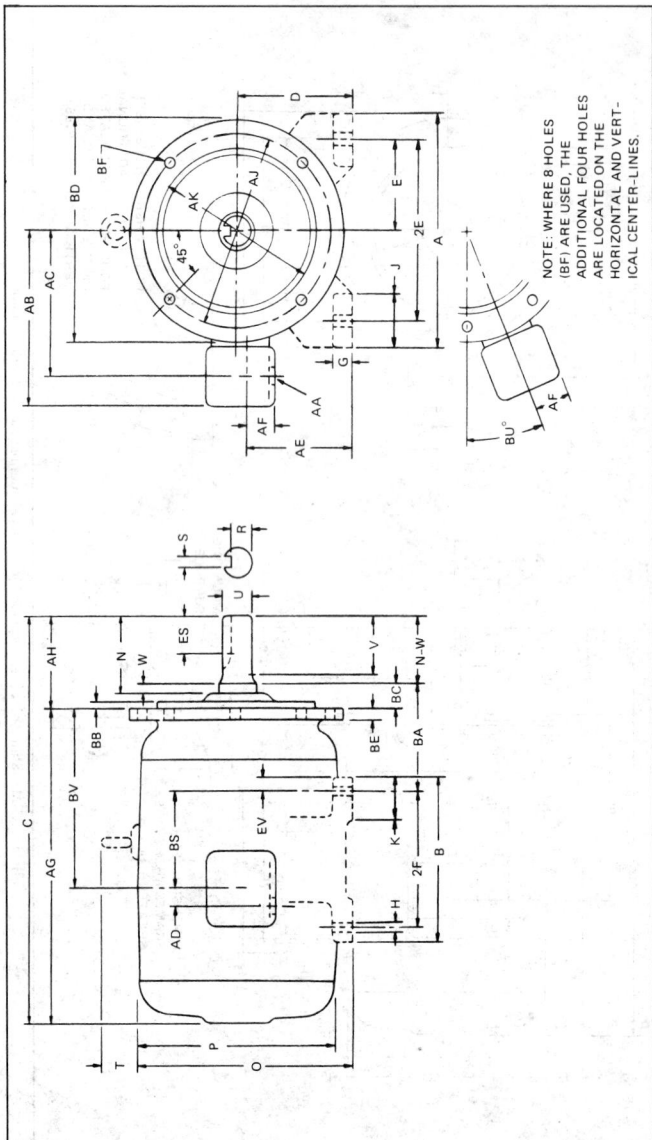

NOTE: WHERE 8 HOLES (BF) ARE USED, THE ADDITIONAL FOUR HOLES ARE LOCATED ON THE HORIZONTAL AND VERTICAL CENTER-LINES.

Fig. 3   NEMA Standard dimensions for Type D flange-mounted foot or footless motors.

**Table 5.  NEMA Standard Dimensions for Face-mounted and Flange-mounted Foot and Footless Alternating-current Motors**

| | | | | | | | | | | | Keyseat | |
|---|---|---|---|---|---|---|---|---|---|---|---|---|
| Frame No. | AJ** | AK | BA | BB Min | BC | BD Max | BF Tap Size§ | U | AH | | ES Min | S |
| **Type C Face Mounted** | | | | | | | | | | | | |
| 42C | 3.750 | 3.000 | 2.062 | 0.16† | −0.19 | 5.00‡ | ¼–20 | 0.3750 | 1.312* | | ... | flat |
| 48C | 3.750 | 3.000 | 2.50 | 0.16† | −0.19 | 5.625 | ¼–20 | 0.500 | 1.69* | | ... | flat |
| 56C | 5.875 | 4.500 | 2.75 | 0.16† | −0.19 | 6.50‡ | ⅜–16 | 0.6250 | 2.06* | | 1.41 | 0.188 |
| 143TC and 145TC | 5.875 | 4.500 | 2.75 | 0.16† | +0.12 | 6.50‡ | ⅜–16 | 0.8750 | 2.12 | | 1.41 | 0.188 |
| 182TC and 184TC | 7.250 | 8.500 | 3.50 | 0.25 | +0.12 | 9.00 | ½–13 | 1.1250 | 2.62 | | 1.78 | 0.250 |
| 182TCH and 184TCH | 5.875 | 4.500 | 3.50 | 0.16† | +0.12 | 6.50‡ | ⅜–16 | 1.1250 | 2.62 | | 1.78 | 0.250 |
| 213TC and 215TC | 7.250 | 8.500 | 4.25 | 0.25 | +0.25 | 9.00 | ½–13 | 1.3750 | 3.12 | | 2.41 | 0.312 |
| 254TC and 256TC | 7.250 | 8.500 | 4.75 | 0.25 | +0.25 | 10.00 | ½–13 | 1.625 | 3.75 | | 2.91 | 0.375 |
| 284TC and 286TC | 9.000 | 10.500 | 4.75 | 0.25 | +0.25 | 11.25 | ½–13 | 1.875 | 4.38 | | 3.28 | 0.500 |
| 284TSC and 286TSC | 9.000 | 10.500 | 4.75 | 0.25 | +0.25 | 11.25 | ½–13 | 1.625 | 3.00 | | 1.91 | 0.375 |
| 324TC and 326TC | 11.000 | 12.500 | 5.25 | 0.25 | +0.25 | 14.00 | ⅝–11 | 2.125 | 5.00 | | 3.91 | 0.500 |
| 324TSC and 326TSC | 11.000 | 12.500 | 5.25 | 0.25 | +0.25 | 14.00 | ⅝–11 | 1.875 | 3.50 | | 2.03 | 0.500 |
| 364TC and 365TC | 11.000 | 12.500 | 5.88 | 0.25 | +0.25 | 14.00 | ⅝–11 | 2.375 | 5.62 | | 4.28 | 0.625 |
| 364TSC and 365TSC | 11.000 | 12.500 | 5.88 | 0.25 | +0.25 | 14.00 | ⅝–11 | 1.875 | 3.50 | | 2.03 | 0.500 |
| 404TC and 405TC | 11.000 | 12.500 | 6.62 | 0.25 | +0.25 | 15.50 | ⅝–11 | 2.875 | 7.00 | | 5.65 | 0.750 |
| 404TSC and 405TSC | 11.000 | 12.500 | 6.62 | 0.25 | +0.25 | 15.50 | ⅝–11 | 2.125 | 4.00 | | 2.78 | 0.500 |
| 444TC and 445TC | 14.000 | 16.000 | 7.50 | 0.25 | +0.25 | 18.00 | ⅝–1: | 3.375 | 8.25 | | 6.91 | 0.875 |
| 444TSC and 445TSC | 14.000 | 16.000 | 7.50 | 0.25 | +0.25 | 18.00 | ⅝–11 | 2.375 | 4.50 | | 3.03 | 0.625 |

**Type D Flange Mounted**

| | | | | | | | | | | | Keyseat | |
|---|---|---|---|---|---|---|---|---|---|---|---|---|
| Frame No. | AJ | AK | BA | BB** | BC | BD Max | BE Nom | BF Size§ | U | AH | ES Min | S |
| 143TD and 145TD | 10.00 | 9.000 | 2.75 | 0.25 | 0.00 | 11.00 | 0.50 | 0.53 | 0.8750 | 2.25 | 1.41 | 0.188 |
| 182TD and 184TD | 10.00 | 9.000 | 3.50 | 0.25 | 0.00 | 11.00 | 0.50 | 0.53 | 1.1250 | 2.75 | 1.78 | 0.250 |
| 213TD and 215TD | 10.00 | 9.000 | 4.25 | 0.25 | 0.00 | 11.00 | 0.50 | 0.53 | 1.3750 | 3.38 | 2.41 | 0.312 |
| 254TD and 256TD | 12.50 | 11.000 | 4.75 | 0.25 | 0.00 | 14.00 | 0.75 | 0.81 | 1.625 | 4.00 | 2.91 | 0.375 |
| 284TD and 286TD | 12.50 | 11.000 | 4.75 | 0.25 | 0.00 | 14.00 | 0.75 | 0.81 | 1.875 | 4.62 | 3.28 | 0.500 |
| 284TSD and 286TSD | 12.50 | 11.000 | 4.75 | 0.25 | 0.00 | 14.00 | 0.75 | 0.81 | 1.625 | 3.25 | 1.91 | 0.375 |
| 324TD and 326TD | 16.00 | 14.000 | 5.25 | 0.25 | 0.00 | 18.00 | 0.75 | 0.81 | 2.125 | 5.25 | 3.91 | 0.500 |
| 324TSD and 326TSD | 16.00 | 14.000 | 5.25 | 0.25 | 0.00 | 18.00 | 0.75 | 0.81 | 1.875 | 3.75 | 2.03 | 0.500 |
| 364TD and 365TD | 16.00 | 14.000 | 5.88 | 0.25 | 0.00 | 18.00 | 0.75 | 0.81 | 2.375 | 5.88 | 4.28 | 0.625 |
| 364TSD and 365TSD | 16.00 | 14.000 | 5.88 | 0.25 | 0.00 | 18.00 | 0.75 | 0.81 | 1.875 | 3.75 | 2.03 | 0.500 |
| 404TD and 405TD | 20.00 | 18.000 | 6.62 | 0.25 | 0.00 | 22.00 | 1.00 | 0.81 | 2.875 | 7.25 | 5.65 | 0.750 |
| 404TSD and 405TSD | 20.00 | 18.000 | 6.62 | 0.25 | 0.00 | 22.00 | 1.00 | 0.81 | 2.125 | 4.25 | 2.78 | 0.500 |
| 444TD and 445TD | 20.00 | 18.000 | 7.50 | 0.25 | 0.00 | 22.00 | 1.00 | 0.81 | 3.375 | 8.50 | 6.91 | 0.875 |
| 444TSD and 445TSD | 20.00 | 18.000 | 7.50 | 0.25 | 0.00 | 22.00 | 1.00 | 0.81 | 2.375 | 4.75 | 3.03 | 0.625 |

All dimensions are in inches.
See Figs. 2 and 3 for diagrams showing letter dimensions.
*If the shaft extension length of the motor is not suitable for the application, it is recommended that deviations in length be in increments of 0.25 inch.
**For frames 182TC and 184TC, and 213TC through 500TC, the centerline of the bolt holes shall be within 0.025 inch of true location. True location is defined as angular and diametral location with reference to the centerline of the AK dimension.
†These BB dimensions are maximum dimensions.
‡These BD dimensions are nominal dimensions.
**The tolerance is +0.00, −0.06 inch.
§For Type C face-mounted motors, the number of holes is 4 for frame Nos. 42C through 326TSC and 8 for frame Nos. 364TC through 445TSC. The bolt penetration allowance is 0.56 inch for frame Nos. 143TC, 145TC, 182TCH, and 184TCH; 0.75 inch for frame Nos. 182TC, 184TC, and 213TC through 286TSC; and 0.94 inch for frame Nos. 324TC through 445TSC. For Type D flange-mounted motors, the number of holes is 4 for frame Nos. 143TD through 365TSD and 8 for frame Nos. 404TD through 445TSD. The recommended bolt length is 1.25 inches for frame Nos. 143TD through 215TD; 2.00 inches for frame Nos. 254TD through 365TSD; and 2.25 inches for frame Nos. 404TD through 445TSD.

**Table 6. NEMA Standard Frame Numbers for Polyphase, Squirrel-cage, Designs A and B, Horizontal and Vertical Motors, 60 Hertz, Class B Insulation System— 575 Volts and Less**

| Hp. | Totally Enclosed Fan-cooled Type* | | | | Open Type** | | | |
|---|---|---|---|---|---|---|---|---|
| | Speed, rpm | | | | Speed, rpm | | | |
| | 3600 | 1800 | 1200 | 900 | 3600 | 1800 | 1200 | 900 |
| ½ | ... | ... | ... | 143T | ... | ... | ... | 143T |
| ¾ | ... | ... | 143T | 145T | ... | ... | 143T | 145T |
| 1 | ... | 143T | 145T | 182T | ... | 143T | 145T | 182T |
| 1½ | 143T | 145T | 182T | 184T | 143T | 145T | 182T | 184T |
| 2 | 145T | 145T | 184T | 213T | 145T | 145T | 184T | 213T |
| 3 | 182T | 182T | 213T | 215T | 145T | 182T | 213T | 215T |
| 5 | 184T | 184T | 215T | 254T | 182T | 184T | 215T | 254T |
| 7½ | 213T | 213T | 254T | 256T | 184T | 213T | 254T | 256T |
| 10 | 215T | 215T | 256T | 284T | 213T | 215T | 256T | 284T |
| 15 | 254T | 254T | 284T | 286T | 215T | 254T | 284T | 286T |
| 20 | 256T | 256T | 286T | 324T | 254T | 256T | 286T | 324T |
| 25 | 284TS | 284T | 324T | 326T | 256T | 284T | 324T | 326T |
| 30 | 286TS | 286T | 326T | 364T | 284TS | 286T | 326T | 364T |
| 40 | 324TS | 324T | 364T | 365T | 286TS | 324T | 364T | 365T |
| 50 | 326TS | 326T | 365T | 404T | 324TS | 326T | 365T | 404T |
| 60 | 364TS | 364TS† | 404T | 405T | 326TS | 364TS† | 404T | 405T |
| 75 | 365TS | 365TS† | 405T | 444T | 364TS | 365TS† | 405T | 444T |
| 100 | 405TS | 405TS† | 444T | 445T | 365TS | 404TS† | 444T | 445T |
| 125 | 444TS | 444TS† | 445T | ... | 404TS | 405TS† | 445T | ... |
| 150 | 445TS | 445TS† | ... | ... | 405TS | 444TS† | ... | ... |
| 200 | ... | ... | ... | ... | 444TS | 445TS† | ... | ... |

The voltage rating of 115 volts applies only to motors rated 15 hp and smaller.
*1.00 Service Factor.    **1.15 Service Factor.
†When motors are to be used with V-belt or chain drives, the correct frame size is the frame size shown but with the suffix letter S omitted.

**Table 7. NEMA Standard Synchronous Speed Ratings of Polyphase Integral-horsepower Induction Motors**

| Hp. | 60 hertz | | | | | | | 50 hertz | | | |
|---|---|---|---|---|---|---|---|---|---|---|---|
| | Synchronous Speed, rpm | | | | | | | | | | |
| ½ | ... | ... | ... | 900 | 720 | 600 | 514 | ... | ... | ... | 750 |
| ¾ | ... | ... | 1200 | 900 | 720 | 600 | 514 | ... | ... | 1000 | 750 |
| 1 | ... | 1800 | 1200 | 900 | 720 | 600 | 514 | ... | 1500 | 1000 | 750 |
| 1½ | 3600* | 1800 | 1200 | 900 | 720 | 600 | 514 | 3000* | 1500 | 1000 | 750 |
| 2 | 3600* | 1800 | 1200 | 900 | 720 | 600 | 514 | 3000* | 1500 | 1000 | 750 |
| 3 | 3600* | 1800 | 1200 | 900 | 720 | 600 | 514 | 3000* | 1500 | 1000 | 750 |
| 5 | 3600* | 1800 | 1200 | 900 | 720 | 600 | 514 | 3000* | 1500 | 1000 | 750 |
| 7½ | 3600* | 1800 | 1200 | 900 | 720 | 600 | 514 | 3000* | 1500 | 1000 | 750 |
| 10 | 3600* | 1800 | 1200 | 900 | 720 | 600 | 514 | 3000* | 1500 | 1000 | 750 |
| 15 | 3600* | 1800 | 1200 | 900 | 720 | 600 | 514 | 3000* | 1500 | 1000 | 750 |
| 20 | 3600* | 1800 | 1200 | 900 | 720 | 600 | 514 | 3000* | 1500 | 1000 | 750 |
| 25 | 3600* | 1800 | 1200 | 900 | 720 | 600 | 514 | 3000* | 1500 | 1000 | 750 |
| 30 | 3600* | 1800 | 1200 | 900 | 720 | 600 | 514 | 3000* | 1500 | 1000 | 750 |
| 40 | 3600* | 1800 | 1200 | 900 | 720 | 600 | 514 | 3000* | 1500 | 1000 | 750 |
| 50 | 3600* | 1800 | 1200 | 900 | 720 | 600 | 514 | 3000* | 1500 | 1000 | 750 |
| 60 | 3600* | 1800 | 1200 | 900 | 720 | 600 | 514 | 3000* | 1500 | 1000 | 750 |
| 75 | 3600* | 1800 | 1200 | 900 | 720 | 600 | 514 | 3000* | 1500 | 1000 | 750 |
| 100 | 3600* | 1800 | 1200 | 900 | 720 | 600 | 514 | 3000* | 1500 | 1000 | 750 |
| 125 | 3600* | 1800 | 1200 | 900 | 720 | 600 | 514 | 3000* | 1500 | 1000 | 750 |
| 150 | 3600* | 1800 | 1200 | 900 | 720 | 600 | ... | 3000* | 1500 | 1000 | 750 |
| 200 | 3600* | 1800 | 1200 | 900 | 720 | ... | ... | 3000* | 1500 | 1000 | 750 |

*Applies to squirrel-cage motors only.
For motors above 200 horsepower, see NEMA Standard MG1-10.32.

Table 8. NEMA Standard Locked-rotor Current of 3-phase 60-hertz Integral-horsepower Squirrel-cage Induction Motors Rated at 230 Volts*

| Horse-power | Locked-rotor Current, Amps. | Design Letters | Horse-power | Locked-rotor Current, Amps. | Design Letters | Horse-power | Locked-rotor Current, Amps. | Design Letters |
|---|---|---|---|---|---|---|---|---|
| ½ | 20 | B, D | 7½ | 127 | B, C, D | 50 | 725 | B, C, D |
| ¾ | 25 | B, D | 10 | 162 | B, C, D | 60 | 870 | B, C, D |
| 1 | 30 | B, D | 15 | 232 | B, C, D | 75 | 1085 | B, C, D |
| 1½ | 40 | B, D | 20 | 290 | B, C, D | 100 | 1450 | B, C, D |
| 2 | 50 | B, D | 25 | 365 | B, C, D | 125 | 1815 | B, C, D |
| 3 | 64 | B, C, D | 30 | 435 | B, C, D | 150 | 2170 | B, C, D |
| 5 | 92 | B, C, D | 40 | 580 | B, C, D | 200 | 2900 | B, C |

*Note:* The locked-rotor current of a motor is the steady-state current taken from the line with the rotor locked and with rated voltage and frequency applied to the motor.

*For motors designed for voltages other than 230 volts, the locked-rotor current is inversely proportional to the voltages. For motors larger than 200 hp, see NEMA Standard MG 1-12.34.

Table 9. NEMA Standard Locked-rotor Torque of Single-speed Polyphase 60- and 50-hertz Squirrel-cage Integral-horsepower Motors with Continuous Ratings

| Hp | Designs A and B | | | | | | | Design C | | |
|---|---|---|---|---|---|---|---|---|---|---|
| | Synchronous Speed, rpm | | | | | | | | | |
| | 60 hertz 3600 | 1800 | 1200 | 900 | 720 | 600 | 514 | 1800 | 1200 | 900 |
| | 50 hertz 3000 | 1500 | 1000 | 750 | ... | ... | ... | 1500 | 1000 | 750 |
| | Percent of Full-load Torque* | | | | | | | | | |
| ½ | ... | ... | ... | 140 | 140 | 115 | 110 | ... | ... | ... |
| ¾ | ... | ... | 175 | 135 | 135 | 115 | 110 | ... | ... | ... |
| 1 | ... | 275 | 170 | 135 | 135 | 115 | 110 | ... | ... | ... |
| 1½ | 175 | 250 | 165 | 130 | 130 | 115 | 110 | ... | ... | ... |
| 2 | 170 | 235 | 160 | 130 | 125 | 115 | 110 | ... | ... | ... |
| 3 | 160 | 215 | 155 | 130 | 125 | 115 | 110 | ... | 250 | 225 |
| 5 | 150 | 185 | 150 | 130 | 125 | 115 | 110 | 250 | 250 | 225 |
| 7½ | 140 | 175 | 150 | 125 | 120 | 115 | 110 | 250 | 225 | 200 |
| 10 | 135 | 165 | 150 | 125 | 120 | 115 | 110 | 250 | 225 | 200 |
| 15 | 130 | 160 | 140 | 125 | 120 | 115 | 110 | 225 | 200 | 200 |
| 20 | 130 | 150 | 135 | 125 | 120 | 115 | 110 | | | |
| 25 | 130 | 150 | 135 | 125 | 120 | 115 | 110 | | | |
| 30 | 130 | 150 | 135 | 125 | 120 | 115 | 110 | 200 for all sizes above 15 hp. | | |
| 40 | 125 | 140 | 135 | 125 | 120 | 115 | 110 | | | |
| 50 | 120 | 140 | 135 | 125 | 120 | 115 | 110 | | | |
| 60 | 120 | 140 | 135 | 125 | 120 | 115 | 110 | | | |
| 75 | 105 | 140 | 135 | 125 | 120 | 115 | 110 | | | |
| 100 | 105 | 125 | 125 | 125 | 120 | 115 | 110 | | | |
| 125 | 100 | 110 | 125 | 120 | 115 | 115 | 110 | For Design D motors, see footnote. | | |
| 150 | 100 | 110 | 120 | 120 | 115 | 115 | ... | | | |
| 200 | 100 | 100 | 120 | 120 | 115 | ... | ... | | | |

*Note:* The locked-rotor torque of a motor is the minimum torque which it will develop at rest for all angular positions of the rotor, with rated voltage applied at rated frequency.

*These values represent the upper limit of application for these motors.

The locked-rotor torque of Design D, 60- and 50-hertz 4-, 6-, and 8-pole single-speed, polyphase squirrel-cage motors rated 150 hp and smaller, with rated voltage and frequency applied is 275 per cent of full-load torque, which represents the upper limit of application for these motors.

For motors larger than 200 hp, see NEMA Standard MG 1-12.37.

Table 10.    NEMA Standard Breakdown Torque of Single-speed Polyphase
Squirrel-cage, Integral-horsepower Motors with Continuous Ratings

| Horse-power | Synchronous Speed, rpm | | | | | | |
|---|---|---|---|---|---|---|---|
| | 60 hertz   3600 | 1800 | 1200 | 900 | 720 | 600 | 514 |
| | 50 hertz   3000 | 1500 | 1000 | 750 | ... | ... | ... |
| | Per Cent of Full Load Torque | | | | | | |
| | Designs A and B* | | | | | | |
| ½ | ... | ... | ... | 225 | 200 | 200 | 200 |
| ¾ | ... | ... | 275 | 220 | 200 | 200 | 200 |
| 1 | ... | 300 | 265 | 215 | 200 | 200 | 200 |
| 1½ | 250 | 280 | 250 | 210 | 200 | 200 | 200 |
| 2 | 240 | 270 | 240 | 210 | 200 | 200 | 200 |
| 3 | 230 | 250 | 230 | 205 | 200 | 200 | 200 |
| 5 | 215 | 225 | 215 | 205 | 200 | 200 | 200 |
| 7½ | 200 | 215 | 205 | 200 | 200 | 200 | 200 |
| 10–125, incl. | 200 | 200 | 200 | 200 | 200 | 200 | 200 |
| 150 | 200 | 200 | 200 | 200 | 200 | 200 | ... |
| 200 | 200 | 200 | 200 | 200 | 200 | ... | ... |
| | Design C | | | | | | |
| 3 | ... | ... | 225 | 200 | ... | ... | ... |
| 5 | ... | 200 | 200 | 200 | ... | ... | ... |
| 6½–200, incl. | ... | 190 | 190 | 190 | ... | ... | ... |

*Design A values are in excess of those shown.
These values represent the upper limit of the range of application for these motors. For above
200 hp, see NEMA Standard MG1-12.38.

Table 11.    NEMA Standard Breakdown Torque of Polyphase Wound-rotor
Motors with Continuous Ratings—60- and 50-hertz

| Horse-power | Speed, rpm | | | Horse-power | Speed, rpm | | |
|---|---|---|---|---|---|---|---|
| | 1800 | 1200 | 900 | | 1800 | 1200 | 900 |
| | Per cent of Full-load Torque | | | | Per cent of Full-load Torque | | |
| 1 | ... | ... | 250 | 7½ | 275 | 250 | 225 |
| 1½ | ... | ... | 250 | 10 | 275 | 250 | 225 |
| 2 | 275 | 275 | 250 | 15 | 250 | 225 | 225 |
| 3 | 275 | 275 | 250 | 20–200, incl. | 225 | 225 | 225 |
| 5 | 275 | 275 | 250 | ... | ... | ... | ... |

These values represent the upper limit of the range of application for these motors.

**Pull-up Torque.** — NEMA Standard pull-up torques for single-speed, polyphase, squirrel-cage integral-horsepower motors, Designs A and B, with continuous ratings and with rated voltage and frequency applied are as follows: When the locked-rotor torque given in Table 9 is 110 per cent or less, the pull-up torque is 90 per cent of the locked-rotor torque; when the locked-rotor torque is greater than 110 per cent but less than 145 per cent, the pull-up torque is 100 per cent of full-load torque; and when the locked-rotor torque is 145 per cent or more, the pull-up torque is 70 per cent of the locked-rotor torque. For Design C motors, with rated voltage and frequency applied, the pull-up torque is not less than 70 per cent of the locked-rotor torque as given in Table 9.

# TYPES AND CHARACTERISTICS
## OF ELECTRIC MOTORS

**Types of Direct-Current Motors.** — Direct-current motors may be grouped into three general classes: series-wound, shunt-wound and compound-wound. In the *series-wound motor* the field windings, which are fixed in the stator frame, and the armature windings, which are placed around the rotor, are connected in series so that all current passing through the armature also passes through the field. In the *shunt-wound motor*, both armature and field are connected across the main power supply so that the armature and field currents are separate. In the *compound-wound motor*, both series and shunt field windings are provided and these may be connected so that the currents in both are flowing in the same direction, called *cumulative compounding*, or so that the currents in each are flowing in opposite directions, called *differential compounding*.

**Characteristics of Series-wound Direct-Current Motors.** — In the series-wound motor, any increase in load results in more current passing through the armature and the field windings. As the field is strengthened by this increased current, the motor speed decreases. Conversely, as the load is decreased the field is weakened and the speed increases and at very light loads may become excessive. For this reason, series-wound direct-current motors are usually directly connected or geared to the load to prevent "runaway." (A series-wound motor designated as series-shunt wound, is sometimes provided with a light shunt field winding to prevent dangerously high speeds at light loads.) The increase in armature current with increasing load produces increased torque, so that the series-wound motor is particularly suited to heavy starting duty and where severe overloads may be expected. Its speed may be adjusted by means of a variable resistance placed in series with the motor, but due to variation with load, the speed cannot be held at any constant value. This variation of speed with load becomes greater as the speed is reduced. Series-wound motors are used where the load is practically constant and can easily be controlled by hand. They are usually limited to traction and lifting service.

**Shunt-wound Direct-Current Motors.** — In the shunt-wound motor, the strength of the field is not affected appreciably by change in the load, so that a fairly constant speed (about 10 to 12 per cent drop from no load to full load speed) is obtainable. This type of motor may be used for the operation of machines requiring an approximately constant speed and imposing low starting torque and light overload on the motor.

The shunt-wound motor becomes an adjustable-speed motor by means of field control or by armature control. If a variable resistance is placed in the field circuit, the amount of current in the field windings and hence the speed of rotation can be controlled. As the speed increases, the torque decreases proportionately, resulting in nearly constant horsepower. A speed range of 6 to 1 is possible using field control, but 4 to 1 is more common. Speed regulation is somewhat greater than in the constant-speed shunt-wound motors, ranging from about 15 to 22 per cent. If a variable resistance is placed in the armature circuit, the voltage applied to the armature can be reduced and hence the speed of rotation can be reduced over a range of about 2 to 1. With armature control, speed regulation becomes poorer as speed is decreased, and is about 100 per cent for a 2 to 1 speed range. Since the current through the field remains unchanged, the torque remains constant.

*Machine Tool Applications:* The adjustable-speed shunt-wound motors are useful on larger machines of the boring mill, lathe and planer type and are particularly adapted to spindle drives because constant horsepower characteristics permit heavy cuts at low speed and light or finishing cuts at high speed. They have long been used for planer drives because they can provide an adjustable low speed for the cutting stroke and a high speed for the return stroke. Their application has been limited, however, to plants in which direct-current power is available.

**Adjustable-voltage Shunt-wound Motor Drive.** — More extensive use of the shunt-wound motor has been made possible by a combination drive that includes a means of converting alternating-current to direct-current. This conversion may be effected by a self-contained unit consisting of a separately excited direct-current generator driven by a constant speed alternating-current motor connected to the regular alternating-current line, or by an electronic rectifier with suitable controls connected to the regular alternating-current supply lines. The latter has the advantage of causing no vibration when mounted directly on the machine tool, an important factor in certain types of grinders.

In this type of adjustable-speed, shunt-wound motor drive, speed control is effected by varying the voltage applied to the armature while supplying constant voltage to the field. In addition to providing for the adjustment of the voltage supplied by the conversion unit to the armature of the shunt-wound motor, the amount of current passing through the motor field may also be controlled. In fact, a single control may be provided to vary the motor speed from minimum to base speed (speed of the motor at full load with rated voltage on armature and field) by varying the voltage applied to the armature and from base speed to maximum speed by varying the current flowing through the field. When so controlled, the motor operates at constant torque up to base speed and at constant horsepower above base speed.

*Speed Range:* Speed ranges of at least 20 to 1 below base speed and 4 or 5 to 1 above base speed (a total range of 100 to 1, or more) are obtainable as compared with about 2 to 1 below normal speed and 3 or 4 to 1 above normal speed for the conventional type of control. Speed regulation may be as great as 25 per cent at high speeds. Special electronic controls, when used with this type shunt motor drive, make possible maintenance of motor speeds with as little variation as ½ to 1 per cent of full load speed from full load to no load over a line voltage variation of ±10 per cent and over any normal variation in motor temperature and ambient temperature.

*Applications:* These direct-current, adjustable-voltage drives, as they are sometimes called, have been applied successfully to such machine tools as planers, milling machines, boring mills and lathes, as well as to other industrial machines where wide, stepless speed control, uniform speed under all operating conditions, constant torque acceleration and adaptability to automatic operation are required.

**Compound-wound Motors.** — In the compound-wound motor, the speed variation due to load changes is much less than in the series-wound motor, but greater than in the shunt-wound motor (ranging up to 25 per cent from full load to no load). It has a greater starting torque than the shunt-wound motor, is able to withstand heavier overloads, but has a narrower adjustable speed range. Standard motors of this type have a cumulative-compound winding, the differential-compound winding being limited to special applications. They are used where the starting load is very heavy or where the load changes suddenly and violently as with reciprocating pumps, printing presses and punch presses.

**Types of Polyphase Alternating-Current Motors.** — The most widely used polyphase motors are of the induction type. The "*squirrel cage*" *induction motor* consists of a wound stator which is connected to an external source of alternating-current power and a laminated steel core rotor with a number of heavy aluminum or copper conductors set into the core around its periphery and parallel to its axis. These conductors are connected together at each end of the rotor by a heavy ring which provides closed paths for the currents induced in the rotor to circulate. This forms, in effect, a " squirrel-cage " from which the motor takes its name.

*Wound-rotor* type of *Induction motor:* This type has in addition to a squirrel cage, a series of coils set into the rotor which are connected through slip-rings to external variable resistors. By varying the resistance of the wound-rotor circuits, the amount of current flowing in these circuits and hence the speed of the motor can be controlled. Since the rotor of an induction motor is not connected to the power supply, the motor is said to operate by transfer action and is analogous to a transformer with a short-circuited secondary that is free to rotate. Induction motors are built with a wide range of speed and torque characteristics which are discussed under " Operating Characteristics of Polyphase Induction Motors."

*Synchronous Motor:* The other type of polyphase alternating-current motor used industrially is the *synchronous motor.* In contrast to the induction motor, the rotor of the synchronous motor is connected to a direct-current supply which provides a field that rotates in step with the alternating-current field in the stator. After having been brought up to synchronous speed, which is governed by the frequency of the power supply and the number of poles in the rotor, the synchronous motor operates at this constant speed throughout its entire load range.

**Operating Characteristics of Squirrel-cage Induction Motors.** — In general, squirrel-cage induction motors are simple in design and construction and offer rugged service. They are essentially constant-speed motors, their speed changing very little with load and not being subject to adjustment. They are used for a wide range of industrial applications calling for integral horsepower ratings. According to the NEMA (National Electrical Manufacturers Association) Standards, there are four classes of squirrel-cage induction motors designated respectively as *A, B, C* and *D.*

*Design A* motors are not commonly used since Design *B* has similar characteristics with the advantage of lower starting current.

*Design B* motors may be designated as a general purpose type suitable for the majority of polyphase alternating-current applications such as blowers, compressors, drill presses, grinders, hammer mills, lathes, planers, polishers, saws, screw machines, shakers, stokers, etc. The starting torque at 1800 R.P.M. is 250 to 275 per cent of full load torque for 3 H.P. and below; for 5 H.P. to 75 H.P. ratings the starting torque ranges from 185 to 150 per cent of full load torque. They have low starting current requirements, usually no more than 5 to 6 times full load current and can be started at full voltage. Their slip (difference between synchronous speed and actual speed at rated load) is relatively low.

*Design C* motors have high starting torque (up to 250 per cent of full load torque) but low starting current. They can be started at full voltage. Slip at rated load is relatively low. They are used for compressors requiring a loaded start, heavy conveyors, reciprocating pumps and other applications requiring high starting torque.

*Design D* motors have high slip at rated load, that is the motor speed drops off appreciably as the load increases, permitting use of the stored energy of a flywheel. They provide heavy starting torque, up to 275 per cent of full load torque, are quiet in operation and have relatively low starting current. Applications are for impact, shock and other high peak loads or flywheel drives such as trains, elevators, hoists, punch and drawing presses, shears, etc.

*Design F* motors are no longer standard. They had low starting torque, about 125 per cent of full-load torque, and low starting current. They were used to drive machines which required infrequent starting at no load or at very light load.

**Multiple Speed Induction Motors.** — This type has a number of windings in the stator so arranged and connected that the number of effective poles and hence the speed can be changed. These motors are for the same types of starting conditions as the conventional squirrel cage induction motors and are available in designs which provide constant horsepower at all rated speeds and in designs which provide constant torque at all rated speeds. Typical speed combinations obtainable in these motors are 600, 900, 1200 and 1800 R.P.M.; 450, 600, 900 and 1200 R.P.M.; and 600, 720, 900 and 1200 R.P.M. Where a gradual change in speed is called for, a wound rotor may be provided in addition to the multiple stator windings.

**Wound Rotor Induction Motors.** — These motors are designed for applications where extremely low starting current with high starting torque are called for, such as in blowers, conveyors, compressors, fans and pumps. They may be employed for adjustable-varying speed service where the speed range does not extend below 50 per cent of synchronous speed, as for steel plate-forming rolls, printing presses, cranes, blowers, stokers, lathes and milling machines of certain types. The speed regulation of a wound rotor induction motor ranges from 5 to 10 per cent at maximum speed and from 18 to 30 per cent at low speed. They are also employed for reversing service as in cranes, gates, hoists and elevators.

**High Frequency Induction Motors.** — This type is used in conjunction with frequency changers when very high speeds are desired, as on grinders, drills, routers, portable tools or woodworking machinery. These motors have an advantage over the series-wound or universal type of high speed motor in that they operate at a relatively constant speed over the entire load range. A motor-generator set, a two-unit frequency converter or a single unit inductor frequency converter may be used to supply three-phase power at the frequency required. The single unit frequency converter may be obtained for delivering any one of a number of frequencies ranging from 360 to 2160 cycles and it is self-driven and self-excited from the general polyphase power supply.

**Synchronous Motors.** — These are widely used in electric timing devices; to drive machines that must operate in synchronism; and also to operate compressors, rolling mills, crushers which are started without load, paper mill screens, shredders, vacuum pumps and motor-generator sets. They have an inherently high power factor and are often employed to make corrections for the low power factor of other types of motors on the same system.

**Types of Single Phase Alternating Current Motors.** — Most of the single-phase alternating-current motors are basically induction motors distinguished by different arrangements for starting. (A single-phase induction motor with only a squirrel cage rotor has no starting torque.) In the *capacitor-start* single-phase motor, an auxiliary winding in the stator is connected in series with a capacitor and a centrifugal switch. During the starting and accelerating period the motor operates as a two-phase induction motor. At about two-thirds full-load speed, the auxiliary circuit is disconnected by the switch and the motor then runs as a single phase induction motor. In the *capacitor-start, capacitor-run* motor, the auxiliary circuit is arranged to provide high effective capacity for high starting torque and to remain connected to the line but with reduced capacity during the running period. In the *single-value capacitor* or *capacitor split-phase* motor, a relatively small continuously-rated capacitor is permanently connected in one of the two stator windings and the motor both starts and runs like a two-phase motor.

In the *repulsion-start* single-phase motor, a drum-wound rotor circuit is connected to a commutator with a pair of short-circuited brushes set so that the magnetic axis of the rotor winding is inclined to the magnetic axis of the stator winding.

The current flowing in this rotor circuit reacts with the field to produce a starting and accelerating torque. At about two-thirds full load speed the brushes are lifted, the commutator is short circuited and the motor runs as a single-phase squirrel-cage motor. The *repulsion* motor employs a repulsion winding on the rotor for both starting and running. The *repulsion-induction* motor has an outer winding on the rotor acting as a repulsion winding and an inner squirrel cage winding. As the motor comes up to speed, the induced rotor current partially shifts from the repulsion winding to the squirrel cage winding and the motor runs partly as an induction motor.

In the *split-phase* motor, an auxiliary winding in the stator is used for starting with either a resistance connected in series with the auxiliary winding (*resistance-start*) or a reactor in series with the main winding (*reactor-start*).

The *series wound* single-phase motor has a rotor winding in series with the stator winding as in the series-wound direct-current motor. Since this motor may also be operated on direct-current, it is called a *universal* motor.

**Characteristics of Single-Phase Alternating Current Motors.** — Single-phase motors are used in sizes up to about 7½ horsepower for heavy starting duty chiefly in home and commercial appliances for which polyphase power is not available. The *capacitor-start* motor is available in normal starting torque designs for such applications as centrifugal pumps, fans, and blowers and in high-starting torque designs for reciprocating compressors, pumps, loaded conveyors, or belts. The *capacitor-start, capacitor-run* motor is exceptionally quiet in operation when loaded to at least 50 per cent of capacity. It is available in low-torque designs for fans and centrifugal pumps and in high-torque designs for applications similar to those of the capacitor-start motor.

The *capacitor split-phase* motor requires the least maintenance of all single-phase motors but has very low starting torque. Its high maximum torque makes it potentially useful in floor sanders or in grinders where momentary overloads due to excessive cutting pressure are experienced. It is also used for slow-speed direct connected fans.

The *repulsion-start, induction-run* motor has higher starting torque than the capacitor motors, although for the same current, the capacitor motors have equivalent pull-up and maximum torque. Electrical and mechanical noise and the extra maintenance sometimes required are disadvantages. These motors are used for compressors, conveyors and stokers starting under full load. The *repulsion-induction* motor has relatively high starting torque and low starting current. It also has a smooth speed-torque curve with no break and a greater ability to withstand long accelerating periods than capacitor type motors. It is particularly suitable for severe starting and accelerating duty and for high inertia loads such as laundry extractors. Brush noise is, however, continuous.

The *repulsion* motor has no limiting synchronous speed and the speed changes with the load. At certain loads, slight changes in load cause wide changes in speed. A brush shifting arrangement may be provided to adjust the speed which may have a range of 4 to 1 if full rated constant torque is applied but a decreasing range as the torque falls below this value. This type of motor may be reversed by shifting the brushes beyond the neutral point. These motors are suitable for machines requiring constant-torque and adjustable speed.

The *split-phase* and *universal* motors are limited to about ⅓ H.P. ratings and are used chiefly for small appliance and office machine applications.

**Motors with Built-in Speed Reducers.** — Electric motors having built-in speed-changing units are compact and the design of these motorized speed reducers tends to improve the appearance of the machines which they drive. There are several types of these speed reducers; they may be classified according to whether they are equipped with a worm-gear drive, a regular gear train with parallel shafts, or planetary gearing.

The claims made for the worm-gear type of reduction unit are that the drive is quiet in operation and well adapted for cases where the slow-speed shaft must be at right angles to the motor shaft and where a high speed ratio is essential.

For very low speeds, the double reduction worm-gear units are suitable. In these units two sets of worm-gearing form the gear train, and both the slow-speed shaft and the armature shaft are parallel. The intermediate worm-gear shaft can be built to extend from the housing, if required, so as to make two countershaft speeds available on the same unit.

In the parallel-shaft type of speed reducer, the slow-speed shaft is parallel with the armature shaft. The slow-speed shaft is rotated by a pinion on the armature shaft, this pinion meshing with a larger gear on the slow-speed shaft.

Geared motors having built-in speed-changing units are available with constant-mesh change-gears for varying the speed ratio.

Planetary gearing permits a large speed reduction with few parts; hence, it is well adapted for geared-head motor units where economy and compactness are essential. The slow-speed shaft is in line with the armature shaft.

## Factors Governing Motor Selection

**Speed, Horsepower, Torque and Inertia Requirements.** — Where more than one speed or a range of speeds are called for, one of the following types of motors may be selected, depending upon other requirements: For direct-current, the standard shunt-wound motor with field control has a 2 to 1 range in some cases; the adjustable speed motor may have a range of from 3 to 1 up to 6 to 1; the shunt motor with adjustable voltage supply has a range up to 20 to 1 or more below base speed and 4 or 5 to 1 above base speed, making a total range of up to 100 to 1 or more. For polyphase alternating current, multi-speed squirrel cage induction motors have 2, 3 or 4 fixed speeds; the wound-rotor motor has a 2 to 1 range. The two-speed wound rotor motor has a 4 to 1 range. The brush-shifting shunt motor has a 4 to 1 range. The brush-shifting series motor has a 3 to 1 range; and the squirrel-cage motor with a variable frequency supply has a very wide range. For single phase alternating current, the brush-shifting repulsion motor has a 2½ to 1 range; the capacitor motor with tapped winding has a 2 to 1 range and the multi-speed capacitor motor has 2 or 3 fixed speeds. Speed regulation (variation in speed from no load to full load) is greatest with motors having series field windings and entirely absent with synchronous motors.

*Horsepower:* Where the load to be carried by the motor is not constant but follows a definite cycle, a horsepower-time curve enables the peak horsepower to be determined as well as the root-mean-square average horsepower, which indicates the proper motor rating from a heating standpoint. Where the load is maintained at a constant value for a period of from 15 minutes to 2 hours depending on the size, the horsepower rating required will usually not be less than this constant value. When selecting the size of an induction motor, it should be kept in mind that this type of motor operates at maximum efficiency when it is loaded to full capacity. Where operation is to be at several speeds, the horsepower requirement for each should be determined.

*Torque:* Starting torque requirements may vary from 10 per cent of full load to 250 per cent of full load torque depending upon the type of machine being driven. Starting torque may vary for a given machine because of frequency of start, temperature, type and amount of lubricant, etc. and such variables should be taken into account. The motor torque supplied to the machine must be well above that required by the driven machine at all points up to full speed. The greater the excess torque, the more rapid the acceleration. The approximate time required for acceleration from rest to full speed is given by the formula:

$$\text{Time} = \frac{N \times WR^2}{T_a \times 308} \text{ seconds}$$

where    $N$ = Full load speed in R.P.M.

$T_a$ = Torque = average foot-pounds available for acceleration.

$WR^2$ = Inertia of rotating part in pounds feet squared ($W$ = weight and $R$ = radius of gyration of rotating part).

308 = Combined constant converting minutes into seconds, weight into mass and radius into circumference.

If the time required for acceleration is greater than 20 seconds, special motors or starters may be required to avoid overheating.

The running torque $T_r$ is found by the formula:

$$T_r = \frac{5250 \times \text{H.P.}}{N} \text{ foot pounds}$$

where H.P. = Horsepower being supplied to the driven machine

$N$ = Running speed in R.P.M.

5250 = Combined constant converting horsepower to foot-pounds per minute and work per revolution into torque.

The peak horsepower determines the maximum torque required by the driven machine and the motor must have a maximum running torque in excess of this.

*Inertia:* The inertia or flywheel effect of the rotating parts of a driven machine will, if large, appreciably affect the accelerating time and hence, the amount of heating in the motor. If synchronous motors are used, the inertia ($WR^2$) of both the motor rotor and the rotating parts of the machine must be known since the pull-in torque (torque required to bring the driven machine up to synchronous speed) varies approximately as the square root of the total inertia of motor and load.

**Space Limitations in Motor Selection.** — If the motor is to become an integral part of the machine which it drives and space is at a premium, a partial motor may be called for. A complete motor is one made up of a stator, a rotor, a shaft, and two end shields with bearings. A *partial motor* is without one or more of these elements. One common type is furnished without drive-end end shield and bearing and is directly connected to the end or side of the machine which it drives, such as the headstock of a lathe. A so-called *shaftless type of motor* is supplied without shaft, end shields or bearings and is intended for built-in application in such units as multiple drilling machines, precision grinders, deep well pumps, compressors and hoists where the rotor is actually made a part of the driven machine. Where partial motors are used, however, proper ventilation, mounting, alignment and bearings must be arranged for by the designer of the machine to which it is applied.

Sometimes it is possible to use a motor having a smaller frame size and wound with Class *B* insulation, permitting it to be subjected to a higher temperature rise than the larger frame Class *A* insulated motor having the same horsepower rating.

**Temperatures.** — The applicability of a given motor is limited not only by its load starting and carrying ability, but also by the temperature which it reaches under load. Motors are given temperature ratings which are based upon the type of insulation (Class A or Class B are the most common) used in their construction and their type of frame (open, semi-enclosed or enclosed).

*Insulating Materials:* Class A materials are: (1) Cotton, silk, paper and similar organic materials when either impregnated or immersed in a liquid dielectric; (2) molded and laminated materials with cellulose filler, phenolic resins and other resins of similar properties; (3) films and sheets of cellulose acetate and other cellulose derivatives of similar properties; (4) varnishes (enamel) as applied to conductors.

Class B insulating materials are: Materials or combinations of materials such as mica, glass fiber, asbestos, etc. with suitable bonding substances. Other materials shown capable of operation at Class B temperatures may be included.

*Ambient Temperature and Allowable Temperature Rise:* Normal ambient temperature is taken to be 40° C. (104° F.). For open general-purpose motors with Class A insulation, the normal temperature rise on which the performance guarantees are based is 40° C. (72° F.).

Motors with Class A insulation having protected, semi-protected, drip-proof, splash-proof, or drip-proof protected enclosures have a 50° C. (90° F.) rise rating.

Motors with Class A insulation and having totally enclosed, totally enclosed fan-cooled, explosion-proof, water-proof, dust-tight, submersible or dust-explosion-proof enclosures have a 55° C. (99° F.) rise rating.

Motors with Class B insulation are permissible for total temperatures up to 110 degrees C. (230° F.) for open motors and 115° C. (239° F.) for enclosed motors.

**Motors Exposed to Injurious Conditions.** — Where motors are to be used in locations imposing unusual operating conditions, the manufacturer should be consulted, especially where any of the following conditions apply: (1) exposed to chemical fumes; (2) operated in damp places; (3) operated at speeds in excess of specified overspeed; (4) expcsed to combustible or explosive dust; (5) exposed to gritty or conducting dust; (6) exposed to lint; (7) exposed to steam; (8) operated in poorly ventilated rooms; (9) operated in pits, or where entirely enclosed in boxes; (10) exposed to inflammable or explosive gases; (11) exposed to temperatures below 10° C. (50° F.); (12) exposed to oil vapor; (13) exposed to salt air; (14) exposed to abnormal shock or vibration from external sources; (15) where the departure from rated voltage is excessive; (16) where the alternating-current supply voltage is unbalanced.

Improved insulating materials and processes and greater mechanical protection against falling materials and liquids make it possible to use general-purpose motors in many locations where special purpose motors were previously considered necessary. *Splash-proof motors* having well protected ventilated openings and specially treated windings are used where they are to be subjected to falling and splashing water or are to be washed down as with a hose. Where climatic conditions are not severe, this type of motor is also successfully used in unprotected out-of-door installations.

If the surrounding atmosphere carries abnormal quantities of metallic, abrasive or non-explosive dust or acid or alkali fumes, a *totally enclosed fan-cooled motor* may be called for. In this type, the motor proper is completely enclosed but air is blown through an outer shell which completely or partially surrounds the inner case. Where the dust in the atmosphere is of a kind which tends to pack or solidify and close the air passages of open splash-proof or totally enclosed fan-cooled motors, *totally enclosed (non-ventilated) motors* are used. This type, which is limited to low horsepower ratings, is also used for outdoor service in mild or severe climates.

In addition to these special-purpose motors there are two types *of explosion-proof motors* designed for hazardous locations. One type is for operation in hazardous dust locations (Class II, Group *G* of the National Electrical Code) while the other is for atmospheres containing explosive vapors and fumes classified as Class I, Group *D* (gasoline, naptha, alcohols, acetone, lacquer-solvent vapors, natural gas).

**Table 1. Characteristics and Applications of D.C. Motors, 1-300H.P.**

| Type | Starting Duty | Maximum Momentary Running Torque | Speed Regulation | Speed Control† | Applications |
|---|---|---|---|---|---|
| Shunt-wound, constant-speed | Medium starting torque. Varies with voltage supplied to armature, and is limited by starting resistor to 125 to 200 per cent full-load torque | 125 to 200 per cent. Limited by commutation | 8 to 12 per cent | Basic speed to 200 per cent basic speed by field control | Drives where starting requirements are not severe. Use constant-speed or adjustable-speed, depending on speed required. Centrifugal pumps, fans, blowers, conveyors, elevators, wood- and metal-working machines |
| Shunt-wound, adjustable-speed | | | 10 to 20 per cent, increases with weak fields | Basic speed to 60 per cent basic speed (lower for some ratings) by field control | |
| Shunt-wound, adjustable voltage control | Heavy starting torque. Limited by starting resistor to 130 to 260 per cent of full-load torque | 130 to 260 per cent. Limited by commutation | Up to 25 per cent. Less than 5 per cent obtainable with special rotating regulator | Basic speed to 2 per cent basic speed and basic speed to 200 per cent basic speed | Drives where wide, stepless speed control, uniform speed, constant-torque acceleration and adaptability to automatic operation are required. Planers, milling machines, boring machines, lathes, etc. |
| Compound-wound, constant-speed | | | Standard compounding 25 per cent. Depends on amount of series winding | Basic speed to 125 per cent basic speed by field control | Drives requiring high starting torque and fairly constant speed. Pulsating loads. Shears, bending rolls, pumps, conveyors, crushers, etc. |
| Series-wound, varying-speed | Very heavy starting torque. Limited to 300 per cent to 350 per cent full-load torque | 300 to 350 per cent. Limited by commutation | Very high. Infinite no-load speed | From zero to maximum speed, depending on control and load | Drives where very high starting torque is required and speed can be regulated. Cranes, hoists, gates, bridges, car dumpers, etc. |

† Minimum speed below basic speed by armature control limited by heating.

Table 2. Characteristics and Applications of Polyphase A.C. Motors

| Polyphase Type | Ratings Hp | Speed Regulation | Speed Control | Starting Torque | Breakdown Torque | Applications |
|---|---|---|---|---|---|---|
| General-purpose squirrel cage, normal stg current, normal stg torque, Design B | 0.5 to 200 hp | Less than 5% | None, except multi-speed types, designed for 2 to 4 fixed speeds | 100 to 250% of full-load | 200 to 300% of full-load | Constant-speed service where starting torque is not excessive. Fans, blowers, rotary compressors, centrifugal pumps, wood-working machines, machine tools, line shafts |
| Full-voltage starting, high stg torque, normal stg current, squirrel-cage, Design C | 3 to 150 hp | Less than 5% | None, except multi-speed types, designed for 2 to 4 fixed speeds | 200 to 250% of full-load | 190 to 225% of full-load | Constant-speed service where fairly high starting torque is required at infrequent intervals with starting current of about 500% full load. Reciprocating pumps and compressors, conveyors, crushers, pulverizers, agitators, etc. |
| Full-voltage starting, high stg torque, high-slip squirrel cage, Design D | 0.5 to 150 hp | Drops about 7 to 12% from no load to full load | None, except multi-speed types, designed for 2 to 4 fixed speeds | 275% of full load, depending upon speed and rotor resistance | 275%. This motor will usually not stall until loaded to its maximum torque, which occurs at standstill | Constant-speed service and high starting torque if starting not too frequent, and for taking high-peak loads with or without flywheels. Punch presses, die stamping, shears, bulldozers, bailers, hoists, cranes, elevators, etc. |
| Wound-rotor, external-resistance starting | 0.5 to several thousand | With rotor rings short-circuited drops about 3% for large to 5% for small sizes | Speed can be reduced to 50% depending upon external resistance. Speed in rotor circuit varies inversely as the load | Up to 300% depending upon external resistance in rotor circuit and how distributed | 200% when rotor slip rings are short circuited | Where high-starting torque with low-starting current or where limited speed controls required. Fans, centrifugal and plunger pumps, compressors, conveyors, hoists, cranes, ball mills, gate hoists, etc. |
| Synchronous | 25 to several thousand | Constant | None, except special motors designed for 2 fixed speeds | 40% for slow speed to 160% for medium speed 80% p-f designs. Special high torque designs | Pull-out torque of unity-p-f motors 170%; 80%—p-f motors 225%; Special designs up to 300% | For constant-speed service, direct connection to slow-speed machines and where power-factor correction is required. |

General Electric Co.

# ELECTRIC MOTOR MAINTENANCE

**Electric Motor Inspection Schedule.** — Frequency and thoroughness of inspection depend upon such factors as; (1) importance of the motor in the production scheme; (2) percentage of days the motor operates; (3) nature of service; (4) winding conditions. The following schedules, recommended by the General Electric Company, and covering both A.C. and D.C. motors are based on average conditions in so far as duty and dirt are concerned.

**Weekly Inspection.** — (1) *Surroundings.* Check to see if the windings are exposed to any dripping water, acid or alcoholic fumes; also, check for any unusual amount of dust, chips or lint on or about the motor. See if any boards, covers, canvas, etc., have been left about that might interfere with the motor ventilation or jam moving parts.

(2) *Lubrication of sleeve bearing motors.* In sleeve-bearing motors check oil level, if a gage is used, and fill to the specified line. If the journal diameter is less than 2 inches, the motor should be stopped before checking the oil level. For special lubricating systems, such as wool-packed, forced lubrication, flood and disk lubrication, follow instruction book. Oil should be added to bearing housing only when motor is at rest. A check should be made to see if oil is creeping along the shaft toward windings where it may harm the insulation.

(3) *Mechanical condition.* — Note any unusual noise which may be caused by metal to metal contact or any odor as from scorching insulation varnish.

(4) *Ball or roller bearings.* Feel ball- or roller-bearing housings for evidence of vibration, and listen for any unusual noise. Inspect for creepage of grease on inside of motor.

(5) *Commutators and brushes.* Check brushes and commutator for sparking. If the motor is on cyclic duty it should be observed through several cycles. Note color and surface condition of the commutator. A stable copper oxide-carbon film (as distinguished from a pure copper surface) on the commutator is an essential requirement for good commutation. Such a film may vary in color all the way from copper to straw, chocolate to black. It should be clean and smooth and have a high polish. All brushes should be checked for wear and pigtail connections for looseness. The commutator surface may be cleaned by using a piece of dry canvas or other hard, nonlinting material which is wound around and securely fastened to a wooden stick, and held against the rotating commutator.

(6) *Rotors and armatures.* The air gap on sleeve bearing motors should be checked, especially if they have been recently overhauled. After installing new bearings, make sure that the average reading is within 10 per cent, provided reading should be less than 0.020 inch. Check air passages through punchings and make sure they are free of foreign matter.

(7) *Windings.* If necessary clean windings by suction or mild blowing. After making sure that the motor is dead, wipe off windings with dry cloth, note evidence of moisture and see if any water has accumulated in the bottom of frame. Check also and see if any oil or grease has worked its way up to the rotor or armature windings. Clean with carbon tetrachloride in a well-ventilated room.

(8) *General.* This is a good time to check the belt, gears, flexible couplings, chain and sprockets for excessive wear or improper location. The motor starting should be checked to make sure that it comes up to proper speed each time power is applied.

**Monthly or Bi-Monthly Inspection.** — (1) *Windings.* Check shunt, series and commutating field windings for tightness. Try to move field spools on the poles, as drying out may have caused some play. If this condition exists, a service shop should be consulted. The motor cable connections should be checked for tightness.

(2) *Brushes.* Check brushes in holders for fit and free play. Also check the brush-spring pressure. Tighten brush studs in holders to take up slack from drying out of washers, making sure that studs are not displaced, particularly on D.C. motors. Replace brushes that are worn down almost to the brush rivet, examine brush faces for chipped toes or heels, and for heat cracks. Damaged brushes should be replaced immediately.

(3) *Commutators.* Examine commutator surface for high bars and high mica, or evidence of scratches or roughness. See that the risers are clean and have not been damaged in any way.

(4) *Ball or roller bearings.* On hard-driven, 24-hour service ball- or roller-bearing motors, purge out old grease through drain hole and apply new grease. Check to make sure grease or oil is not leaking out of the bearing housing. If any leakage is present, correct the condition before continuing to operate.

(5) *Sleeve bearings.* Check sleeve bearings for wear, including end-play bearing surfaces. Clean out oil wells if there is evidence of dirt or sludge. Flush with lighter oil before refilling.

(6) *Enclosed gears.* For motors with enclosed gears, open drain plug and check oil flow for presence of metal scale, sand or water. If condition of oil is bad, drain, flush and refill as directed. Rock rotor to see if slack or backlash is increasing.

(7) *Loads.* Check loads for changed conditions, bad adjustment, poor handling or control.

(8) *Couplings and other drive details.* Note if belt-tightening adjustment is all used up. Shorten belt if this condition exists. See if belt runs steadily and close to inside (motor edge) of pulley. Chain should be checked for evidence of wear and stretch. Clean inside of chain housing. Check chain-lubricating system. Note incline of slanting base to make sure it does not cause oil rings to rub on housing.

**Annual or Bi-Annual Inspection.** — (1) *Windings.* Check insulation resistance by using either a megohmmeter or a volt meter having a resistance of about 100 ohms per volt. Check insulation surfaces for dry cracks and other evidence of need for coatings of insulating material. Clean surfaces and ventilating passages thoroughly if inspection shows accumulation of dust. Check for mold or water standing in frame to determine if windings need to be dried out, varnished and baked.

(2) *Air gap and bearings.* Check air gap to make sure that average reading is within 10 per cent, provided reading should be less than 0.020 inch. All bearings, ball, roller and sleeve should be thoroughly checked and defective ones replaced. Waste-packed and wick-oiled bearings should have waste or wicks renewed, if they have become glazed or filled with metal or dirt, making sure that new waste bears well against shaft.

(3) *Rotors (squirrel-cage).* Check squirrel-cage rotors for broken or loose bars and evidence of local heating. If fan blades are not cast in place, check for loose blades. Look for marks on rotor surface indicating foreign matter in air gap or a worn bearing.

(4) *Rotors (wound).* Clean wound rotors thoroughly around collector rings, washers and connections. Tighten connections if necessary. If rings are rough, spotted or eccentric, refer to service shop for refinishing. See that all top sticks or wedges are tight. If any are loose, refer to service shop.

(5) *Armatures.* Clean all armature air passages thoroughly if any are obstructed. Look for oil or grease creeping along shaft, checking back to bearing. Check commutator for surface condition, high bars, high mica or eccentricity. If necessary, turn down the commutator to secure a smooth fresh surface.

(6) *Loads.* Read load on motor with instruments at no load, full load or through an entire cycle, as a check on the mechanical condition of the driven machine.

## American National Standard Symbols for Mechanics (ANSI Y10.3·1968)

| | | | |
|---|---|---|---|
| Acceleration, angular...... | $\alpha$ (alpha) | Inertia, polar (area) | |
| Acceleration, due to gravity | $g$ | moment of............... | $J$ |
| Acceleration, linear......... | $a$ | Inertia, product (area) | |
| Amplitude................. | $A$ | moment of............... | $I_{xy}$ |
| Angle...................... | $\alpha$ (alpha) | Length..................... | $L$ |
| ......................... | $\beta$ (beta) | | $l$ |
| ......................... | $\gamma$ (gamma) | Load per unit distance..... | $q$ |
| ......................... | $\theta$ (theta) | | $w$ |
| ......................... | $\phi$ (phi) | Load, total................. | $P$ |
| ......................... | $\psi$ (psi) | | $W$ |
| Angular frequency......... | $\omega$ (omega) | Mass...................... | $m$ |
| Angular velocity........... | $\omega$ (omega) | Moment of area (statical or | |
| Arc length................. | $s$ | first moment)........... | $Q$ |
| Area...................... | $A$ | Moment of force, including | |
| Axes, through any point* | $X\text{-}X$ | bending moment....... | $M$ |
| ......................... | $Y\text{-}Y$ | Neutral axis, distance to | |
| ......................... | $Z\text{-}Z$ | extreme fiber from....... | $c$ |
| Breadth (width)........... | $b$ | Number................... | $N$ |
| Coefficient of expansion, | | ......................... | $n$ |
| linear.................... | $\alpha$ (alpha) | Period.................... | $T$ |
| Coefficient of friction....... | $f$ | Poisson's ratio............. | $\upsilon$ (upsilon) |
| ......................... | $\mu$ (mu) | Power..................... | $P$ |
| Concentrated load (same as | | Pressure, normal force per | |
| force)................... | $F$ | unit area............... | $p$ |
| ......................... | $P$ | Radius..................... | $r$ |
| ......................... | $Q$ | ......................... | $R$ |
| Deflection of beam, max... | $\delta$ (delta) | Revolutions per unit of time | $N$ |
| Density................... | $\rho$ (rho) | ......................... | $n$ |
| Depth..................... | $h$ | Section modulus.......... | $Z$ |
| Diameter................. | $D$ | Shear force in beam section | $V$ |
| ......................... | $d$ | Spring constant (load per | |
| Displacement............. | $u$ | unit deflection)......... | $k$ |
| ......................... | $v$ | Statical moment of any area | |
| ......................... | $w$ | about a given axis........ | $Q$ |
| Distance, linear........... | $s$ | Strain, normal............. | $\varepsilon$ (epsilon) |
| Eccentricity of application | | Strain, shear.............. | $\gamma$ (gamma) |
| of load.................. | $e$ | Stress, concentration | |
| Efficiency................. | $\eta$ (eta) | factor.................... | $K$ |
| Elasticity, modulus of....... | $E$ | Stress, normal............. | $s$ |
| Elasticity, modulus of, | | ......................... | $\sigma$ (sigma) |
| in shear................. | $G$ | Stress, shear.............. | $s_s$ |
| Elongation, total........... | $\delta$ (delta) | ......................... | $\tau$ (tau) |
| Energy, kinetic ............ | $E_k$ | Temperature, absolute..... | $T$ |
| ......................... | $T$ | Temperature............... | $t$ |
| Energy, potential .......... | $E_p$ | ......................... | $\vartheta$ (theta) |
| ......................... | $V$ | Thickness.................. | $h$ |
| Factor of safety............ | $N$ | Time...................... | $t$ |
| ......................... | $n$ | Torque.................... | $M$ |
| Force or load, concentrated | $F$ | ......................... | $T$ |
| ......................... | $P$ | Velocity, linear............. | $v$ |
| ......................... | $Q$ | Volume.................... | $V$ |
| Frequency................. | $f$ | Wave length............... | $\lambda$ (lambda) |
| Gyration, radius of......... | $k$ | Weight.................... | $W$ |
| Height.................... | $h$ | Weight per unit volume.... | $\gamma$ (gamma) |
| Inertia, area moment of.... | $I$ | Work...................... | $W$ |
| Inertia, mass moment of... | $I$ | ......................... | $W_k$ |

*Not specified in Standard

### American Standard Abbreviations for Scientific and Engineering Terms — I

Only the most commonly used terms have been included. These forms are recommended for those whose familiarity with the terms used makes possible a maximum of abbreviations. For others, less contracted combinations made up from this list may be used. For example, the list gives the abbreviation of the term "feet per second" as "fps." To some, however, ft per sec will be more easily understood.

| | |
|---|---|
| Absolute | abs |
| Acre | acre |
| Acre-foot | acre-ft |
| Air horsepower | air hp |
| Alternating-current (as adjective) | a-c |
| Ampere | amp |
| Ampere-hour | amp-hr |
| Angstrom unit | A |
| Antilogarithm | antilog |
| Atomic weight | at. wt |
| Arithmetical Average | AA |
| Atmosphere | atm |
| Avoirdupois | avdp |
| | |
| Barometer | bar. |
| Barrel | bbl |
| Baumé | Bé |
| Board feet (feet board measure) | fbm |
| Boiler pressure | bp |
| Boiling point | bp |
| Brake horsepower | bhp |
| Brake horsepower-hour | bhp-hr |
| Brinell hardness number | Bhn |
| British thermal unit | Btu or B |
| Bushel | bu |
| | |
| Calorie | cal |
| Candle | c |
| Candlepower | cp |
| Center to center | c to c |
| Centigram | cg |
| Centiliter | cl |
| Centimeter | cm |
| Centimeter-gram-second (system) | cgs |
| Cent | c or ¢ |
| Chemical | chem |
| Chemically pure | cp |
| Circular | cir |
| Circular mils | cir mils |
| Coefficient | coef |
| Cologarithm | colog |
| Concentrate | conc |
| Conductivity | cond |
| Constant | const |
| Continental horsepower | cont hp |
| Cord | cd |
| Cosecant | csc |
| Cosine | cos |
| Cost, insurance, and freight | cif |
| Cotangent | ctn |
| Coulomb | spell out |
| Counter electromotive force | counter emf |

| | |
|---|---|
| Cubic | cu |
| Cubic centimeter | cu cm, cm³, cc |
| (liquid, meaning milliliter, ml) | |
| Cubic foot | cu ft |
| Cubic feet per second | cfs |
| Cubic inch | cu in. |
| Cubic meter | cu m or m³ |
| Cubic micron | cu $\mu$ or cu mu or $\mu$³ |
| Cubic millimeter | cu mm or mm³ |
| Cubic yard | cu yd |
| Current density | spell out |
| Cylinder | cyl |
| | |
| Day | spell out |
| Decibel | db |
| Degree | deg or ° |
| Degree Centigrade | C |
| Degree Fahrenheit | F |
| Degree Kelvin | K |
| Degree Réaumur | R |
| Diameter | diam |
| Direct-current (as adjective) | d-c |
| Dozen | doz |
| Dram | dr |
| | |
| Efficiency | eff |
| Electric | elec |
| Electromotive force | emf |
| Elevation | el |
| Engine | eng |
| Engineer | engr |
| Engineering | engg |
| Equation | eq |
| External | ext |
| | |
| Farad | spell out |
| Feet board measure (board feet) | fbm |
| Feet per minute | fpm |
| Feet per second | fps |
| Fluid | fl |
| Foot | ft |
| Foot-candle | ft-c |
| Foot-Lambert | ft-L |
| Foot-pound | ft-lb |
| Foot-pound-second (system) | fps |
| Franc | fr |
| Free aboard ship | spell out |
| Free alongside ship | spell out |
| Free on board | f.o.b. |
| Freezing point | fp |
| Frequency | spell out |
| Furlong | fur. |
| Fusion point | fnp |

## American Standard Abbreviations for Scientific and Engineering Terms — 2

| | |
|---|---|
| Gallon | gal |
| Gallons per minute | gpm |
| Gallons per second | gps |
| Grain | spell out |
| Gram | g |
| Gram-calory | g-cal |
| Greatest common divisor | gcd |
| Hectare | ha |
| Henry | h |
| High-pressure (adjective) | h-p |
| Horsepower | hp |
| Horsepower-hour | hp-hr |
| Hour | hr |
| Hundred | C |
| Hundredweight (112 lb.) | cwt |
| Hyperbolic sine | sinh |
| Hyperbolic cosine | cosh |
| Hyperbolic tangent | tanh |
| Inch | in, |
| Inch-pound | in-lb |
| Inches per second | ips |
| Indicated horsepower | ihp |
| Indicated horsepower-hour | ihp-hr |
| Intermediate-pressure (adjective) | i-p |
| Internal | int |
| Joule | j |
| Kilocycle | kc |
| Kilogram | kg |
| Kilogram-meter | kg-m |
| Kilograms per cubic meter | kg per cu m or kg/m³ |
| Kilograms per second | kgps |
| Kiloliter | kl |
| Kilometer | km |
| Kilometers per second | kmps |
| Kilovar (reactive kilovolt-ampere) | kvar |
| Kilovarhour (reactive kilovolt-ampere-hour) | kvarh |
| Kilovolt | kv |
| Kilovolt-ampere | kva |
| Kilowatt | kw |
| Kilowatthour | kwhr |
| Lambert | L |
| Latitude | lat |
| Least common multiple | lcm |
| Linear foot | lin ft |
| Lira | spell out |
| Liter | l |
| Liquid | liq |
| Logarithm (common) | log |
| Logarithm (natural) | $\log_e$ or ln |
| Low-pressure (as adjective) | l-p |
| Lumen | l |
| Lumen-hour | l-hr |
| Lumens per watt | lpw |

| | |
|---|---|
| Magnetomotive force | mmf |
| Mark (German coinage) | M. |
| Mass | spell out |
| Mathematics (ical) | math |
| Maximum | max |
| Mean effective pressure | mep |
| Mean horizontal candlepower | mhcp |
| Megohm | spell out |
| Melting point | mp |
| Meter | m |
| Meter-kilogram | m-kg |
| Mho | spell out |
| Microampere | $\mu$a or mu a |
| Microfarad | $\mu$f or mu f |
| Micromicron | $\mu\mu$ or mu mu |
| Micron | $\mu$ or mu |
| Microwatt | $\mu$w or mu w |
| Mile | spell out |
| Miles per hour | mph |
| Miles per hour per second | mphps |
| Milliampere | ma |
| Millifarad | mf |
| Milligram | mg |
| Millihenry | mh |
| Millilambert | mL |
| Milliliter | ml |
| Millimeter | mm |
| Millimicron | m$\mu$ or m mu |
| Million | spell out |
| Million gallons per day | mgd |
| Millivolt | mv |
| Minimum | min |
| Minute | min |
| Minute (angular measure) | ' |
| Molecular weight | mol. wt |
| Mol | spell out |
| National Electric Code | NEC |
| Ohm | spell out |
| Ohm-centimeter | ohm-cm |
| Ounce | oz |
| Ounce-foot | oz-ft |
| Ounce-inch | oz-in. |
| Parts per million | ppm |
| Peck | pk |
| Pennyweight | dwt |
| Peso | spell out |
| Pint | pt |
| Potential | spell out |
| Potential difference | spell out |
| Pound | lb |
| Pound-foot | lb-ft |
| Pound-inch | lb-in. |
| Pounds per brake horsepower-hour | lb per bhp-hr |
| Pounds per square foot | lb per sq ft |
| Pounds per square inch | lb per sq in. |
| Pound sterling | £ |
| Power factor | spell out |

## American Standard Abbreviations for Scientific and Engineering Terms — 3

| | |
|---|---|
| Quart | qt |
| Radian | spell out |
| Reactive kilovolt-ampere | rkva |
| Reactive volt-ampere | rva |
| Revolutions per minute | rpm |
| Revolutions per second | rps |
| Rod | spell out |
| Root mean square | rms |
| Round | rd |
| Secant | sec |
| Second | sec |
| Second (angular measure) | ″ |
| Second-foot (see cubic feet per second) | |
| Shaft horsepower | shp |
| Shilling | s |
| Sine | sin |
| Specific gravity | sp gr |
| Specific heat | sp ht |
| Spherical candle power | scp |
| Square | sq |
| Square centimeter | sq cm or cm² |
| Square foot | sq ft |
| Square inch | sq in. |
| Square kilometer | sq km or km² |
| Square meter | sq m or m² |

| | |
|---|---|
| Square micron | sq $\mu$ or sq mu or $\mu^2$ |
| Square millimeter | sq mm or mm² |
| Square root of mean square | rms |
| Standard | std |
| Stere | s |
| Tangent | tan |
| Temperature | temp |
| Tensile strength | ts |
| Thousand | M |
| Ton | spell out |
| Ton-mile | spell out |
| Twaddell | Twad |
| Var (reactive volt-ampere) | var |
| Versed sine | vers |
| Volt | v |
| Volt-ampere | va |
| Volt-coulomb | spell out |
| Watt | w |
| Watthour | whr |
| Watts per candle | wpc |
| Week | spell out |
| Weight | wt |
| Yard | yd |

Alternative abbreviations conforming to the practice of the International Electrotechnical Commission.

| | | |
|---|---|---|
| Ampere | A | |
| Ampere-hour | Ah | |
| Coulomb | C | |
| Farad | F | |
| Henry | H | |
| Joule | J | |
| Kilovolt | kV | |
| Kilovolt-ampere | kVA | |
| Kilowatt | kW | |
| Kilowatthour | kWh | |
| Megawatt | MW | |
| Megohm | MΩ | |
| Microampere | $\mu$A | |
| Microfarad | $\mu$F | |
| Microwatt | $\mu$W | |
| Milliampere | mA | |
| Millifarad | mF | |
| Millihenry | mH | |
| Millivolt | mV | |
| Ohm | Ω | |
| Volt | V | |
| Volt-ampere | VA | |
| Volt-coulomb | VC | |
| Watt | W | |
| Watthour | Wh | |

Abbreviations should not be used where the meaning will not be clear. In case of doubt, spell out.

Abbreviations should be used sparingly in text and with regard to the context and to the training of the reader. Terms denoting units of measurement should be abbreviated in the text only when preceded by the amounts indicated in numerals; thus " several inches," " one inch," " 12 in." In tabular matter, specifications, maps, drawings, and texts for special purposes, the use of abbreviations should be governed only by the desirability of conserving space.

A sentence should not begin with a numeral followed by an abbreviation.

Short words such as ton, day, and mile should be spelled out.

The use of conventional signs for abbreviations in text is not recommended; thus " per," not /; " lb," not #; " in.," not ″. Such signs may be used sparingly in tables and similar places for conserving space.

The Committee endorses the movement which was begun by the International Committee on Weights and Measures in omitting the period in abbreviations of metric units and further endorses the growing tendency toward the omission in abbreviations of other origin. In the interests of economy and the reduction of waste the elimination of the period is recommended except where such an omission results in an English word. Exceptions to this practice will be found in a few mathematical and chemical terms, such as sin, tan, log, As, etc.

**American Standard Drafting-Room Practice.** — This standard (approved by American Standards Association) includes the lines for engineering drawings and the symbols for section lining as shown on the accompanying pages.

*Dimensioning:* Dimensions of parts that can be measured or that can be produced with sufficient accuracy by using an ordinary scale should be written in units and common fractions. Parts requiring greater accuracy should be dimensioned in decimal fractions. Dimensions up to and including 72 inches should preferably be expressed in inches, and those greater than this length, in feet and inches.

Where dimensions call for accurate machining with small tolerances it is recommended that the total dimension be given in inches and decimal fractions. In structural drawing all dimensions of 12 inches and over should be expressed in feet and inches. In automotive, locomotive, sheet metal and some other practices all dimensions are specified in inches.

The symbol (″) is used to indicate inches and common and decimal fractions of an inch. When all dimensions are given in inches the symbol is preferably omitted. A note may be placed on the drawing stating that all dimensions are given in inches. The symbol (′) is used to indicate feet and fractions of a foot. Dimensions in feet and inches should be hyphenated, thus 4′-3″; 4′-0½″; 4′-0″.

Fractions should be written with the division in line with the dimension line.

*Dimension Lines and Extension Lines:* Dimension lines should be fine full lines (broken where dimension is inserted) so as to contrast with the heavier outline of the drawing, and should be placed outside the figure or drawing outline wherever possible.

Extension lines indicate the distance measured when the dimension is placed outside the figure. They are made as light full lines starting ½₂ to ¹⁄₁₆ inch away from the outline and extending about ⅛ inch beyond the dimension line.

A center line should never be used as a dimension line. A line of the piece or part illustrated or an extension of such a line should never be used as a dimension line.

*Dimension Figures:* A dimension line must not pass through a dimension figure. If unbroken lines are used, as is common practice in structural drawing, the dimensions are placed above the line. When fractional dimensions of less than one inch are given, the numerator should be placed above the dimension line and the denominator below.

All dimension lines and their corresponding numbers should be placed so that they may be read from the bottom or right-hand edges of the drawing. All dimensions should be placed so as to read in the direction of the dimension lines.

When there are several parallel dimension lines the figures should be staggered to avoid confusion. Dimensions should be given from a base line, a center line or a finished surface that can be readily established. Over-all dimensions should be placed outside the intermediate dimensions. In dimensioning with tolerances, if an over-all dimension is used one intermediate distance should not be dimensioned.

In dimensioning angles an arc should be drawn and the dimension placed so as to read from the horizontal position. An exception is sometimes made in the dimensioning of large areas when the dimensions are placed along the arc.

*Dimensioning Circles:* A dimension indicating the diameter of a circle should be followed by the abbreviation "D" except when it is obvious from the drawing that the dimension is a diameter. The dimension of a radius should always be followed by the abbreviation "R." The center should be indicated by a cross or circle and the dimension line have one arrow-head.

*Dimensioning Holes:* Holes which are to be drilled, reamed, punched, swaged, cored, etc., should have the diameter, given preferably on a leader, followed by the word indicating the operation, and the number of holes to be so made. Holes which are to be machined after coring or casting should have finished marks and finished dimensions specified.

If needed by the shop on account of the method of laying out, as in the button method, the chordal distances between holes on a bolt circle or the center-to-center distances between holes located by coordinates, should be calculated and dimensioned in decimals.

*Dimensioning with Tolerances:* Accurate dimensions which are to be established with limit gage or micrometer should be expressed in decimals to at least three places and the drawing should give the maximum and minimum limits between which the actual measurements must come. For *external* dimensions the maximum limit is placed above the line and for *internal* dimensions the minimum limit is placed above the line. This method should be used for smaller parts and where gages are extensively employed.

A second method, used for larger parts and where few gages are employed, is to give the calculated size to the required number of decimal places, followed by the tolerances plus and minus, with the plus above the minus, as $8.625D \begin{smallmatrix} +.000 \\ -.002 \end{smallmatrix}$

*Changes in Dimensions:* On a drawing, if a dimension must be changed, the changed figures should be underlined or otherwise marked. It is customary to note changes in dimensions in a tabulation on the drawing and to refer to them by letters or symbols placed after the altered dimensions.

*Dimensioning Tapers:* At least three methods of dimensioning tapers are in general use.

*Standard Tapers:* Give one diameter or width, the length, and insert note on drawing designating the taper by number.

*Special Tapers:* In dimensioning a taper when the slope is specified, the length and only one diameter should be given or the diameters at both ends of the taper should be given and length omitted.

*Precision Work:* In certain cases where very precise measurements are necessary the taper surface, either external or internal, is specified by giving a diameter at a certain distance from a surface and the slope of the taper.

*Finish Marks:* A surface to be machined or "finished" from unfinished material such as a casting or a forging should be marked with a 60 degree "$\vee$," the bottom of the "$\vee$" touching the line representing the surface to be machined or finished. A code figure or letter should then be placed in the opening of the "$\vee$" to indicate the quality of the finish desired. For a more detailed explanation of the use of finish marks and code figures, see page 2384.

*Sizes of Drawing Paper and Cloth:* The recommended standard trimmed sheet sizes of drawing paper and cloth are shown by the table.

| Size, Inches | | | | Metric Size, mm. | | | |
|---|---|---|---|---|---|---|---|
| A | 8½ × 11 | D | 22 × 34 | A0 | 841 × 1189 | A4 | 210 × 297 |
| B | 11 × 17 | E | 34 × 44 | A1 | 594 × 841 | A5 | 148 × 210 |
| C | 17 × 22 | | | A2 | 420 × 594 | A6 | 105 × 148 |
| | | | | A3 | 297 × 420 | | |

The standard sizes shown by the left-hand section of the table, are based on the dimensions of the commercial letter head, 8½ × 11 inches, in general use in the United States. The use of the basic sheet size 8½ × 11 inches and its multiples permits filing of small tracings and folded blueprints in commercial standard letter files with or without correspondence. These sheet sizes also cut without unnecessary waste from the present 36 inch rolls of paper and cloth.

For drawings made in the metric system of units or for foreign correspondence it is recommended that the metric standard trimmed sheet sizes be used. (Right-hand section of table.) These sizes are based on the width to length ratio of 1 to $\sqrt{2}$.

American National Standard Lines for Engineering Drawings (ANSI Y14.2-1973)

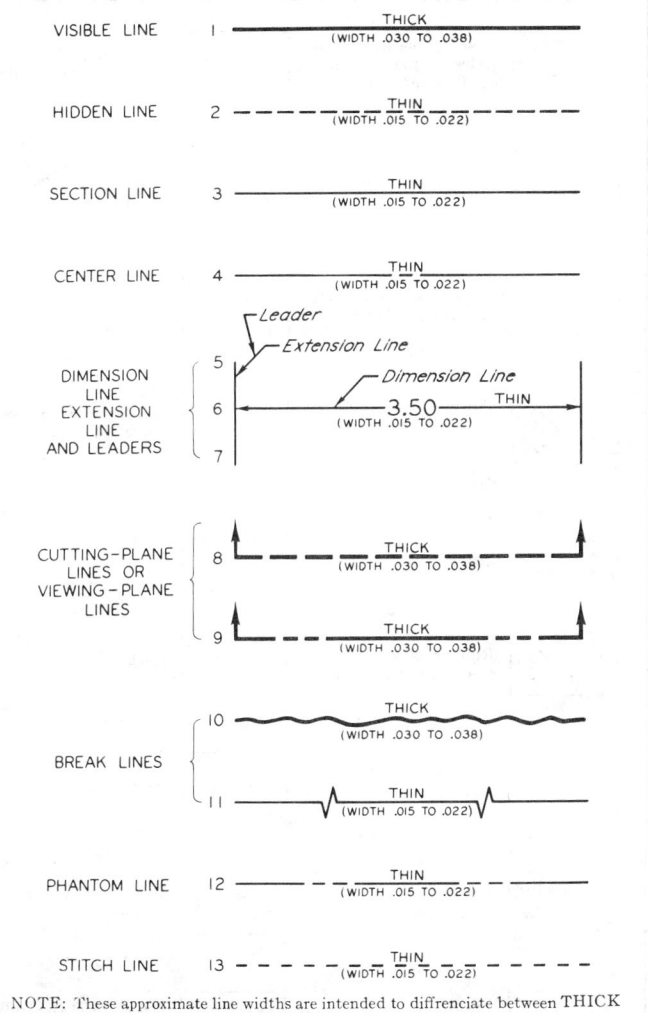

| | | |
|---|---|---|
| VISIBLE LINE | 1 | THICK (WIDTH .030 TO .038) |
| HIDDEN LINE | 2 | THIN (WIDTH .015 TO .022) |
| SECTION LINE | 3 | THIN (WIDTH .015 TO .022) |
| CENTER LINE | 4 | THIN (WIDTH .015 TO .022) |

Leader
Extension Line
Dimension Line

DIMENSION LINE
EXTENSION LINE
AND LEADERS
{ 5 6 7 }
3.50 THIN (WIDTH .015 TO .022)

CUTTING-PLANE LINES OR VIEWING-PLANE LINES
{ 8 THICK (WIDTH .030 TO .038)
9 THICK (WIDTH .030 TO .038) }

BREAK LINES
{ 10 THICK (WIDTH .030 TO .038)
11 THIN (WIDTH .015 TO .022) }

PHANTOM LINE   12   THIN (WIDTH .015 TO .022)

STITCH LINE   13   THIN (WIDTH .015 TO .022)

NOTE: These approximate line widths are intended to diffrenciate between THICK and THIN lines and are not values for control of acceptance or rejection of the drawings.

WIDTH AND TYPE OF LINES

American National Standard Symbols for Section Lining (ANSI Y14.2-1973)

| | | | |
|---|---|---|---|
| | Cast and malleable iron (Also for general use of all materials) | | Titanium and refractory material |
| | Steel | | Electric windings, electro-magnets, resistance, etc. |
| | Bronze, brass, copper, and compositions | | Concrete |
| | White metal, zinc, lead, babbitt, and alloys | | Marble, slate, glass, porcelain, etc. |
| | Magnesium, aluminum, and aluminum alloys | | Earth |
| | Rubber, plastic electrical insulation | | Rock |
| | Cork, felt, fabric, leather, fiber | | Sand |
| | Sound insulation | | Water and other liquids |
| | Thermal insulation | | Wood—across grain / Wood—with grain |

**American National Standard Geometric Characteristic Symbols for Engineering Drawings** (ANSI Y14.5-1973)

| | | CHARACTERISTIC | SYMBOL | NOTES |
|---|---|---|---|---|
| INDIVIDUAL FEATURES | FORM TOLERANCES | STRAIGHTNESS | — | 1 |
| | | FLATNESS | ▱ | 1 |
| | | ROUNDNESS (CIRCULARITY) | ○ | |
| | | CYLINDRICITY | ⌭ | |
| INDIVIDUAL OR RELATED FEATURES | | PROFILE OF A LINE | ⌒ | 2 |
| | | PROFILE OF A SURFACE | ⌓ | 2 |
| RELATED FEATURES | | ANGULARITY | ∠ | |
| | | PERPENDICULARITY (SQUARENESS) | ⊥ | |
| | | PARALLELISM | // | 3 |
| | LOCATION TOLERANCES | POSITION | ⌖ | |
| | | CONCENTRICITY | ◎ | 3,7 |
| | | SYMMETRY | ≡ | 5 |
| | RUNOUT TOLERANCES | CIRCULAR | ↗ | 4 |
| | | TOTAL | ↗ | 4,6 |

Note:- 1) The symbol ∼ formerly denoted flatness.

The symbol ⌒ or — formerly denoted flatness and straightness.

2) Considered "related" features where datums are specified.

3) The symbol ‖ and ◉ formerly denoted parallelism and concentricity, respectively.

4) The symbol ↗ without the qualifier "CIRCULAR" formerly denoted total runout.

5) Where symmetry applies, it is preferred that the position symbol be used.

6) "TOTAL" must be specified under the feature control symbol.

7) Consider the use of position or runout.

**American National Standard Miscellaneous Symbols for Engineering Drawings** (ANSI Y14.5-1973)

| TERM | ABBREVIATION | SYMBOL |
|---|---|---|
| Maximum Material Condition | MMC | Ⓜ |
| Regardless of Feature Size | RFS | Ⓢ |
| Diameter | DIA | ⌀ |
| Projected Tolerance Zone | TOL ZONE PROJ | Ⓟ |
| Reference | REF | (1.250) |
| Basic | BSC | 3.875 |

**Control and Production of Surface Texture.** — Surface characteristics should not be controlled on a drawing or specification unless such control is essential to functional performance or appearance of the product. Imposition of such restrictions when unnecessary may increase production costs and in any event will serve to lessen the emphasis on the control specified for important surfaces.

Smoothness and roughness are relative, i.e., surfaces may be either smooth or rough for the purpose intended; what is smooth for one purpose may be rough for another purpose. In the mechanical field comparatively few surfaces require any control of surface texture beyond that afforded by the processes required to obtain the necessary dimensional characteristics.

Working surfaces such as on bearings, pistons, and gears are typical of surfaces for which optimum performance may require control of the surface characteristics. Non-working surfaces such as on the walls of transmission cases, crankcases, or housings seldom require any surface control. Experimentation or experience with surfaces performing similar functions is the best criterion on which to base selection of optimum surface characteristics.

Determination of required characteristics for working surfaces may involve consideration of such conditions as the area of contact, the load, speed, direction of motion, type and amount of lubricant, temperature, material and physical characteristics of component parts, and that variations in any one of the conditions may require a change in the specified surface characteristics.

*Production:* Surface texture is a result of the processing method, the surface obtained from casting, forging, or burnishing is the result of plastic deformation; if machined or ground, lapped or honed, the surface obtained is the result of action of cutting tools, abrasives or other forces. It is important to understand that surfaces with like roughness average ratings may not have the same performance, due to tempering, sub-surface effects, different profile waveforms, etc.

**American National Standard Surface Texture (Surface Roughness, Waviness and Lay).** — American National Standard ANSI B46.1-1978 is concerned with the geometric irregularities of surfaces of solid materials, physical specimens for gaging roughness, and the characteristics of stylus instrumentation for measuring roughness. It defines surface texture and its constituents: roughness, waviness, lay and flaws. A set of symbols for drawings, specifications and reports is established. In order to assure a uniform basis for measurements, it also provides specifications for Precision Reference Specimens, and Roughness Comparison Specimens, and establishes requirements for stylus type instruments. The standard is not concerned with luster, appearance, color, corrosion resistance, wear resistance, hardness, subsurface micro-structure, surface integrity and many other characteristics which may be governing considerations in specific applications.

The standard does not define the degrees of surface roughness and waviness or type of lay suitable for specific purposes, nor does it specify the means by which any degree of such irregularities may be obtained or produced. However, criteria for selection of surface qualities and information on instrument techniques and methods of producing, controlling and inspecting surfaces are included in Appendixes attached to the standard. The Appendix sections are not considered a part of the standard; they are included for clarification or information purposes only.

Surfaces, in general, are very complex in character. The standard deals only with the height, width, and direction of surface irregularities since these are of practical importance in specific applications. Surface texture designations as delineated in this standard may not be a sufficient index to performance. Other part characteristics such as dimensional and geometrical relationships, material, metallurgy, and stress must also be controlled.

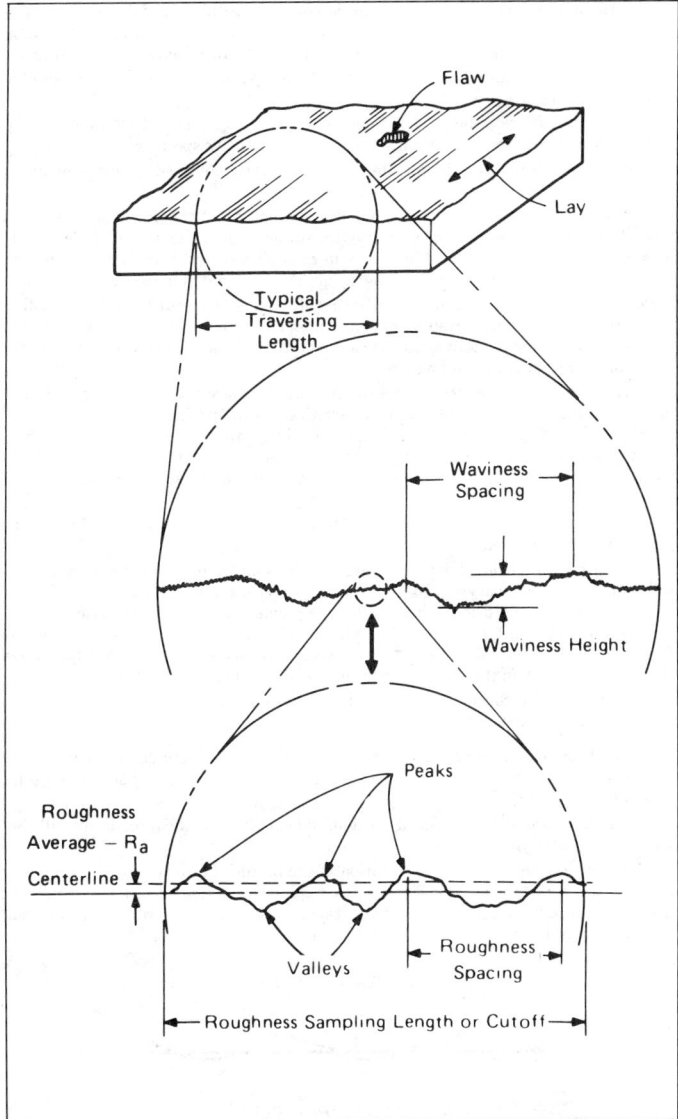

Fig. 1. Pictorial display of surface characteristics.

**Definitions of Terms Relating to the Surfaces of Solid Materials.** — The terms and ratings in the standard relate to surfaces produced by such means as abrading, casting, coating, cutting, etching, plastic deformation, sintering, wear, erosion, etc.

The *surface* of an object is the boundary which separates that object from another object, substance or space.

The *nominal surface* is the intended surface contour, the shape and extent of which is usually shown and dimensioned on a drawing or descriptive specification.

The *measured surface* is a representation of the surface obtained by instrumental or other means.

*Surface texture* is the repetitive or random deviations from the nominal surface which form the three-dimensional topography of the surface. Surface texture includes roughness, waviness, lay and flaws. Figure 1 is an example of a uni-directional lay surface. Roughness and waviness parallel to the lay are not represented in the expanded views.

*Roughness* consists of the finer irregularities of the surface texture, usually including those irregularities which result from the inherent action of the production process. These are considered to include traverse feed marks and other irregularities within the limits of the roughness sampling length.

*Waviness* is the more widely spaced component of surface texture. Unless otherwise noted, waviness is to include all irregularities whose spacing is greater than the roughness sampling length and less than the waviness sampling length. Waviness may result from such factors as machine or work deflections, vibration, chatter, heat-treatment or warping strains. Roughness may be considered superposed on a 'wavy' surface.

*Lay* is the direction of the predominant surface pattern, ordinarily determined by the production method used.

*Flaws* are unintentional irregularities which occur at one place or at relatively infrequent or widely varying intervals on the surface. Flaws include such defects as cracks, blow holes, inclusions, checks, ridges, scratches, etc. Unless otherwise specified, the effect of flaws shall not be included in the roughness average measurements. Where flaws are to be restricted or controlled, a special note as to the method of inspection should be included on the drawing or in the specifications.

The *error of form* is considered as being that deviation from the nominal surface which is not included in surface texture.

**Definitions of Terms Relating to the Measurement of Surface Texture.** — The *profile* is the contour of the surface in a plane perpendicular to the surface, unless some other angle is specified.

The *nominal profile* is a profile of the nominal surface; it is the intended profile. See Fig. 2.

The *measured profile* is a representation of the profile obtained by instrumental or other means. When the measured profile is a graphical representation, it will usually be distorted through the use of different vertical and horizontal magnifications but shall otherwise be as faithful to the profile as technically possible.

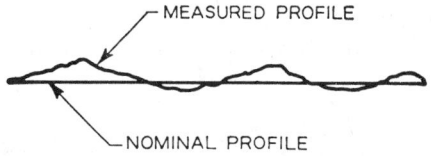

Fig. 2.   Nominal and measured profiles.

The *modified profile* is a measured profile where filter mechanisms (including the instrument datum) are used to minimize certain surface texture characteristics and emphasize others.

The *graphical centerline* is the line about which roughness is measured and is a line parallel to the general direction of the profile within the limits of the sampling length, such that the sums of the areas contained between it and those parts of the profile which lie on either side are equal. (See Fig. 3.)

The *electrical mean line* is the centerline established by the selected cutoff and its associated circuitry in an electric roughness average measuring instrument.

A *peak* is the point of maximum height on that portion of a profile which lies above the centerline and between the two intersections of the profile and the centerline.

A *valley* is the point of maximum depth on that portion of a profile which lies below the centerline and between two intersections of the profile and the centerline.

The *spacing* is the distance between specified points on the profile measured parallel to the nominal profile.

The *roughness spacing* is the average spacing between adjacent peaks of the measured profile within the roughness sampling length.

The *waviness spacing* is the average spacing between adjacent peaks of the measured profile within the waviness sampling length.

The *sampling length* is the nominal spacing within which a surface characteristic is determined.

The *roughness sampling length* is the sampling length within which the roughness average is determined. This length is chosen, or specified, to separate the profile irregularities which are designated as roughness from those irregularities designated as waviness.

The *cutoff* is the electrical response characteristic of the roughness average measuring instrument which is selected to limit the spacing of the surface irregularities to be included in the assessment of roughness average. The cutoff is rated in millimeters. In most electrical averaging instruments, the cutoff can be selected. It is a characteristic of the instrument rather than of the surface being measured. In specifying the cutoff, care must be taken to choose a value which will include all the surface irregularities which it is desired to assess.

The *waviness sampling length* is the sampling length within which the waviness height is determined.

The *traversing length* is the length of profile which is traversed by the stylus to establish a representative measurement. See Section 5.5 of the standard for values which must be used for different type measurements.

*Height* is considered to be those measurements of the profile in a direction normal to the nominal profile.

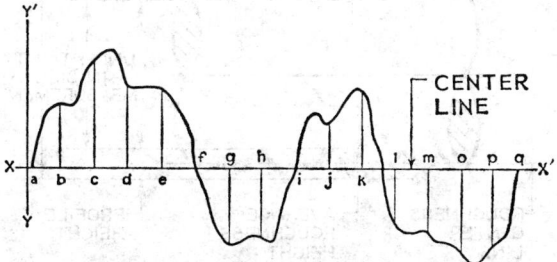

Fig. 3.   Short section of hypothetical profile divided into increments.

*Roughness average (R_a)*, also known as arithmetic average (AA) and centerline average (CLA), is the arithmetic average of the absolute values of the measured profile height deviations taken within the sampling length and measured from the graphical centerline. For graphical determinations of roughness average, the height deviations are measured normal to the chart centerline, as shown in Fig. 3. Roughness average is expressed in micrometers. A micrometer is one millionth of a meter, (0.000 001 meter). For written specifications or references as to surface requirements, micrometer can be abbreviated as $\mu$m. A microinch is one millionth of an inch, (0.000001 inch). For written specifications or reference to surface roughness requirements, microinch may be abbreviated as $\mu$in. One microinch equals 0.0254 micrometer, (1 $\mu$in. = 0.0254 $\mu$m).

*Roughness average value (R_a) from continuously averaging meter readings.* So that uniform interpretation may be made of readings from stylus-type instruments of the continuously averaging type, it should be understood that the reading which is considered significant is the mean reading around which the needle tends to dwell or fluctuate under small amplitude.

The *peak-to-valley height* is the maximum excursion below the centerline plus the maximum excursion below the centerline within the sampling length. This value is typically 3 or more times the roughness average.

The *waviness height* is the peak-to-valley height of the modified profile from which roughness and flaws have been removed by filtering, smoothing, or other means. The measurement is to be taken normal to the nominal profile within the limits of the waviness sampling length and expressed in millimeters.

**Relation of Surface Roughness to Tolerances.** — Since the measurement of surface roughness involves the determination of the average linear deviation of the actual surface from the nominal surface, there is a direct relationship between the dimensional tolerance on a part and the permissible surface roughness. It is evident that a requirement for the accurate measurement of a dimension is that the variations introduced by surface roughness should not exceed the tolerance placed on the dimension. If this is not the case, the measurement of the dimension will be subject to an uncertainty greater than the required tolerance as illustrated in Fig. 4. The standard

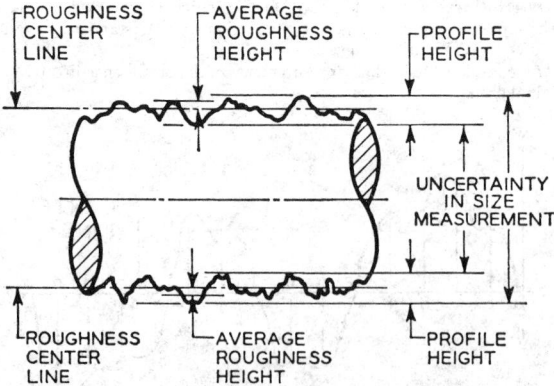

Fig. 4.   Uncertainty in dimensional measurement due to surface roughness.

method of measuring surface roughness involves the determination of the average deviation from the mean surface. On most surfaces the total profile height of the surface roughness (peak-to-valley height) will be approximately four times the measured average surface roughness in microinches. This factor will vary somewhat with the character of the surface under consideration, but the value of four may be used to establish approximate profile heights.

From these considerations it follows that if the arithmetical average value of surface roughness specified on a part exceeds one eighth of the dimensional tolerance, the whole tolerance will be taken up by the roughness height. In most cases, a smaller roughness specification than this will be found; but on parts where very small dimensional tolerances are given, it is necessary to specify a suitably small surface roughness in order that useful dimensional measurements can be made. The tables on pages 1539 and 1565 show the relation between machining processes and working tolerances. Values for surface roughness produced by common processing methods are shown in Fig. 5. The

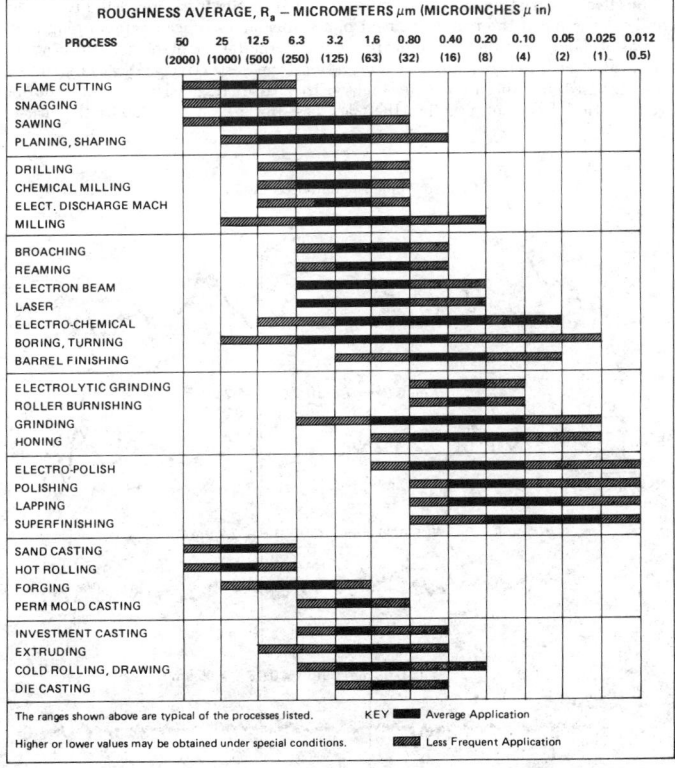

Fig. 5. Surface roughness produced by common production methods.

ability of a processing operation to produce a specific surface roughness depends on many factors. For example, in surface grinding, the final surface depends on the peripheral speed of the wheel, the speed of the traverse, the rate of feed, the grit size, bonding material and state of dress of the wheel, the amount and type of lubrication at the point of cutting, and the mechanical properties of the piece being ground. A small change in any of the above factors can have a marked effect on the surface produced.

**Selecting Cutoff for Roughness Measurements.** — In general, surfaces will contain irregularities with a large range of widths. Stylus-type instruments are designed to respond only to irregularity spacings less than a given value, called cutoff. In some cases, such as surfaces in which actual contact area with a mating surface is important, the largest convenient cutoff will be used. In other cases, such as surfaces subject to fatigue failure, only the irregularities of small width will be important, and more significant values will be obtained when a short cutoff is used. In still other cases, such as identifying chatter marks on machined surfaces, information is needed on only the widely spaced irregularities. For such measurements, a long cutoff instrument should be used.

The effect of variation in cutoff can be understood better by reference to Fig. 6. The profile at the top is the true movement of a stylus on a surface having a roughness spacing of about 1 mm and the profiles below are interpretations of the same surface with cutoff value settings of 0.8 mm, 0.25 mm and 0.08 mm, respectively. It can be seen that the trace based on 1 mm cutoff includes most of the coarse irregularities and all of the fine irregularities of the surface; that the trace based on 0.25 mm excludes the coarser

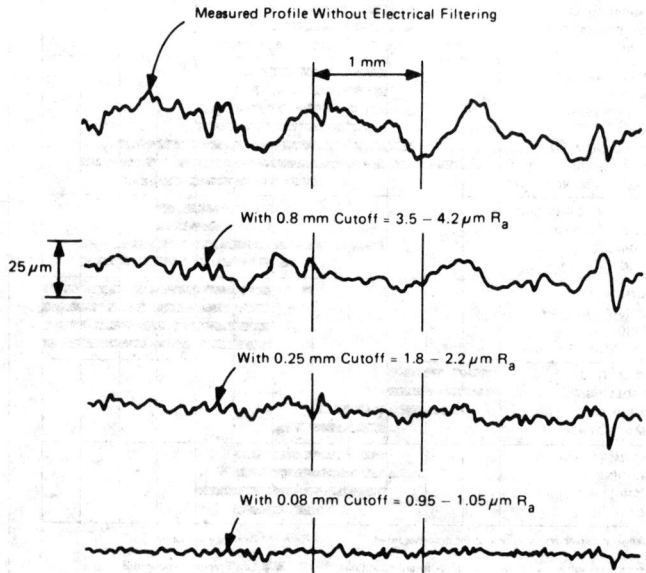

Fig. 6.    Effects of various cutoff values.

irregularities but includes the fine and medium fine; and that the trace based on 0.08 mm cutoff includes only the very fine irregularities. In this example the effect of reducing the cutoff has been to reduce the roughness height indication. However, had the surface been made up only of irregularities as fine as those of the bottom trace, the roughness height indications would have been the same for all three cutoff settings.

In other words, all irregularities having a spacing less than the value of the cutoff used are included in a measurement. Obviously, if the cutoff value is too small to include coarser irregularities of a surface, the measurements will not agree with those taken with a larger cutoff. For this reason, care must be taken to choose a cutoff value which will include all of the surface irregularities it is desired to assess.

To become proficient in the use of continuously averaging stylus-type instruments the inspector or machine operator must realize that for uniform interpretation, the reading which is considered significant is the mean reading around which the needle tends to dwell or fluctuate under small amplitude.

**Drawing Practices for Surface Texture Symbols.**—American National Standard ANSI Y14.36-1978 establishes the method to designate controls for surface texture of solid materials. It includes methods for controlling roughness, waviness, and lay, and provides a set of symbols for use on drawings, specifications, or other documents. The units, (metric or nonmetric) shall be consistent with the other units used on the drawing or documents. The numerical values expressed in the standard are stated in metric units and are to be regarded as standard. Approximate nonmetric equivalents are shown for reference.

**Surface Texture Symbol.**—The symbol used to designate control of surface irregularities is shown in Fig. 1(a). Where surface texture values other than roughness average are specified, the symbol must be drawn with the horizontal extension as shown in Fig. 1(e).

| | Symbol | Meaning |
|---|---|---|
| (a) |  | Basic Surface Texture Symbol. Surface may be produced by any method except when the bar or circle (Figure 1b or 1d) is specified. |
| (b) |  | Material Removal By Machining Is Required. The horizontal bar indicates that material removal by machining is required to produce the surface and that material must be provided for that purpose. |
| (c) | 3.5  | Material Removal Allowance. The number indicates the amount of stock to be removed by machining in millimeters (or inches). Tolerances may be added to the basic value shown or in a general note. |
| (d) |  | Material Removal Prohibited. The circle in the vee indicates that the surface must be produced by processes such as casting, forging, hot finishing, cold finishing, die casting, powder metallurgy or injection molding without subsequent removal of material. |
| (e) | 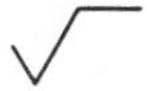 | Surface Texture Symbol. To be used when any surface characteristics are specified above the horizontal line or to the right of the symbol. Surface may be produced by any method except when the bar or circle (Figure 1b and 1d) is specified. |
| (f) | | |

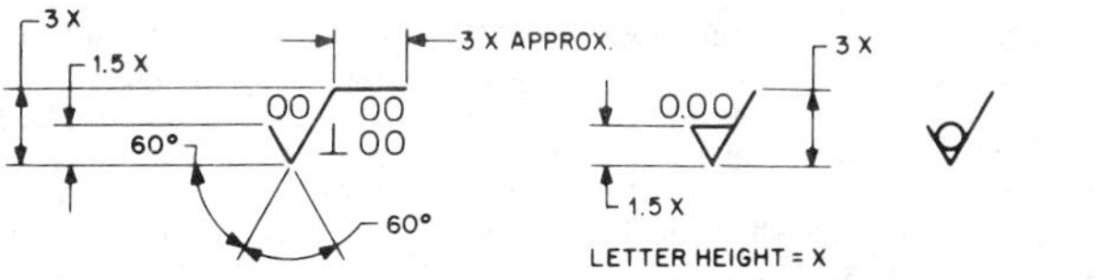


Fig. 1. Surface texture symbols and construction.

*Use of Surface Texture Symbols:* When required from a functional standpoint, the desired surface characteristics should be specified. Where no surface texture control is specified, the surface produced by normal manufacturing methods is satisfactory provided it is within the limits of size (and form) specified in accordance with ANSI Y14.5-1973, Dimensioning and Tolerancing. This is not viewed as good practice; there should always be some maximum value, either specifically or by default (for example, in the manner of the note shown in Fig. 2).

*Material Removal Required or Prohibited:* The surface texture symbol is modified when necessary to require or prohibit removal of material. When it is necessary to indicate that a surface must be produced by removal of material by machining, specify the symbol shown in Fig. 1(b). When required, the amount of material to be removed is specified as shown in Fig. 1(c), in millimeters for metric drawings and in inches for nonmetric drawings. Tolerance for material removal may be added to the basic value shown or specified in a general note. When it is necessary to indicate that a surface must be produced without material removal, specify the machining prohibited symbol as shown in Fig. 1(d).

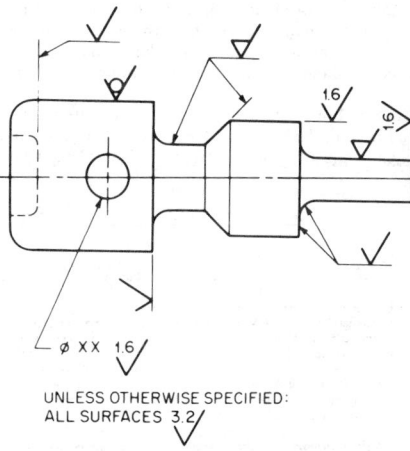

UNLESS OTHERWISE SPECIFIED:
ALL SURFACES 3.2

Fig. 2.    Application of surface texture symbols.

*Proportions of Surface Texture Symbols:* The recommended proportions for drawing the surface texture symbol are shown in Fig. 1(f). The letter height and line width should be the same as that for dimensions and dimension lines.

**Applying Surface Texture Symbols.**—The point of the symbol should be on a line representing the surface, an extension line of the surface, or a leader line directed to the surface, or to an extension line. The symbol may be specified following a diameter dimension. The long leg (and extension) shall be to the right as the drawing is read. For parts requiring extensive and uniform surface roughness control, a general note may be added to the drawing which applies to each surface texture symbol specified without values as shown in Fig. 2.

When the symbol is used with a dimension, it affects the entire surface defined by the dimension. Areas of transition, such as chamfers and fillets, shall conform with the roughest adjacent finished area unless otherwise indicated.

Table 1. Preferred Series Roughness Average
Values ($R_a$)

| $\mu$m | $\mu$in | $\mu$m | $\mu$in |
|---|---|---|---|
| 0.012 | 0.5 | 1.25 | 50 |
| 0.025* | 1* | 1.60* | 63* |
| 0.050* | 2* | 2.0 | 80 |
| 0.075* | 3 | 2.5 | 100 |
| 0.10* | 4* | 3.2* | 125* |
| 0.125 | 5 | 4.0 | 160 |
| 0.15 | 6 | 5.0 | 200 |
| 0.20* | 8* | 6.3* | 250* |
| 0.25 | 10 | 8.0 | 320 |
| 0.32 | 13 | 10.0 | 400 |
| 0.40* | 16* | 12.5* | 500* |
| 0.50 | 20 | 15 | 600 |
| 0.63 | 25 | 20 | 800 |
| 0.80* | 32* | 25* | 1000* |
| 1.00 | 40 | ... | ... |

*Recommended

Surface texture values, unless otherwise specified, apply to the complete surface. Drawings or specifications for plated or coated parts shall indicate whether the surface texture values apply before plating, after plating or both before and after plating.

Include in the symbol only those values required to specify and verify the required texture characteristics. Values should be in metric units for metric drawings and nonmetric units for nonmetric drawings.

Roughness and waviness measurements, unless otherwise specified, apply in a direction which gives the maximum reading; generally across the lay.

Table 2. Standard Roughness Sampling Length
(Cutoff) Values

| mm | in. | mm | in. |
|---|---|---|---|
| 0.08 | 0.003 | 2.5 | 0.1 |
| 0.25 | 0.010 | 8.0 | 0.3 |
| 0.80 | 0.030 | 25.0 | 1.0 |

*Roughness Average* ($R_a$): The preferred series of specified roughness average values is given in Table 1.

*Cutoff or Roughness Sampling Length:* Standard values are listed in Table 2. When no value is specified, the value 0.8 mm (0.030 in.) applies.

*Waviness Height:* The preferred series of maximum waviness height values is listed in Table 3. Waviness is not currently shown in ISO Standards. It is included here to follow present industry practice in the United States.

Table 3. Preferred Series Maximum Waviness
Height Values

| mm | in. | mm | in. | mm | in. |
|---|---|---|---|---|---|
| 0.0005 | 0.00002 | 0.008 | 0.0003 | 0.12 | 0.005 |
| 0.0008 | 0.00003 | 0.012 | 0.0005 | 0.20 | 0.008 |
| 0.0012 | 0.00005 | 0.020 | 0.0008 | 0.25 | 0.010 |
| 0.0020 | 0.00008 | 0.025 | 0.001 | 0.38 | 0.015 |
| 0.0025 | 0.0001 | 0.05 | 0.002 | 0.50 | 0.020 |
| 0.005 | 0.0002 | 0.08 | 0.003 | 0.80 | 0.030 |

| Lay Symbol | Meaning | Example Showing Direction of Tool Marks |
|:---:|---|:---:|
| — | Lay approximately parallel to the line representing the surface to which the symbol is applied. | 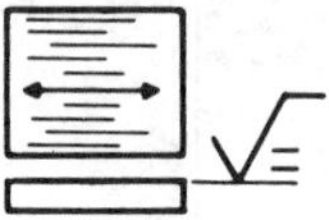 |
| ⊥ | Lay approximately perpendicular to the line representing the surface to which the symbol is applied. | 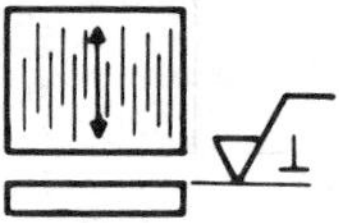 |
| X | Lay angular in both directions to line representing the surface to which the symbol is applied. | 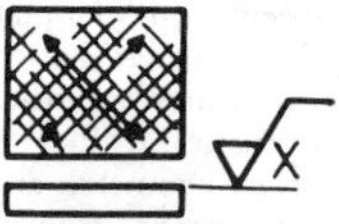 |
| M | Lay multidirectional. |  |
| C | Lay approximately circular relative to the center of the surface to which the symbol is applied. | 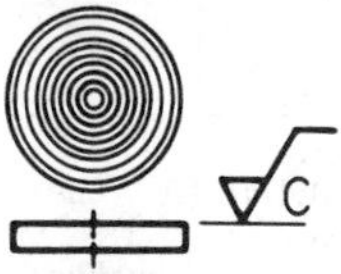 |
| R | Lay approximately radial relative to the center of the surface to which the symbol is applied. | 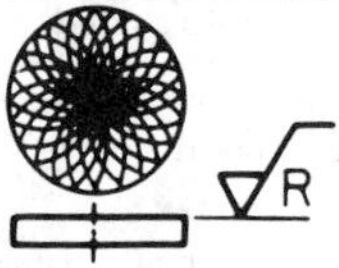 |
| P | Lay particulate, non-directional, or protuberant. | 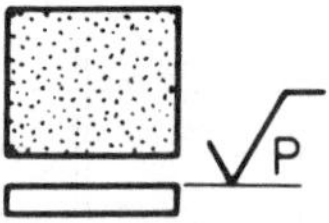 |

Fig. 3.   Lay symbols.

*Lay:* Symbols for designating the direction of lay are shown and interpreted in Fig. 3.

**Example Designations.** — Figure 4 illustrates examples of designations of roughness, waviness, and lay by insertion of values in appropriate positions relative to the symbol. Where surface roughness control of several operations is required within a given area, or on a given surface, surface qualities may be designated, as in Fig. 5(a). If a surface must be produced by one particular process or a series of processes, they should be

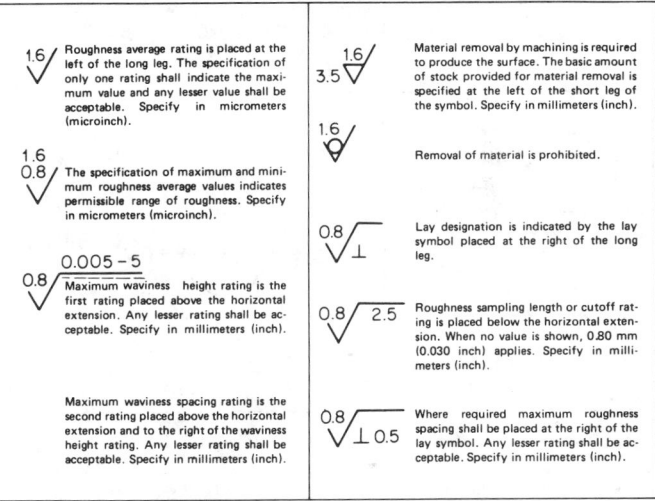

Fig. 4. Application of surface texture values to symbol.

specified as shown in Fig. 5(b). Where special requirements are needed on a designated surface, a note should be added at the symbol giving the requirements and the area involved. An example is illustrated in Fig. 5(c).

**Surface Texture of Castings.** — Surface characteristics should not be controlled on a drawing or specification unless such control is essential to functional performance or appearance of the product. Imposition of such restrictions when unnecessary may increase production costs and in any event will serve to lessen the emphasis on the control specified for important surfaces. Surface characteristics of castings should never be considered on the same basis as machined surfaces. Castings are characterized by random distribution of non-directional deviations from the nominal surface.

Surfaces of castings rarely need control beyond that provided by the production method necessary to meet dimensional requirements. Comparison specimens are frequently used for evaluating surfaces having specific functional requirements. Surface texture control should not be specified unless required for appearance or function of the surface. Specification of such requirements may increase cost to the user.

Engineers should recognize that different areas of the same castings may have different surface textures. It is recommended that specifications of the surface be

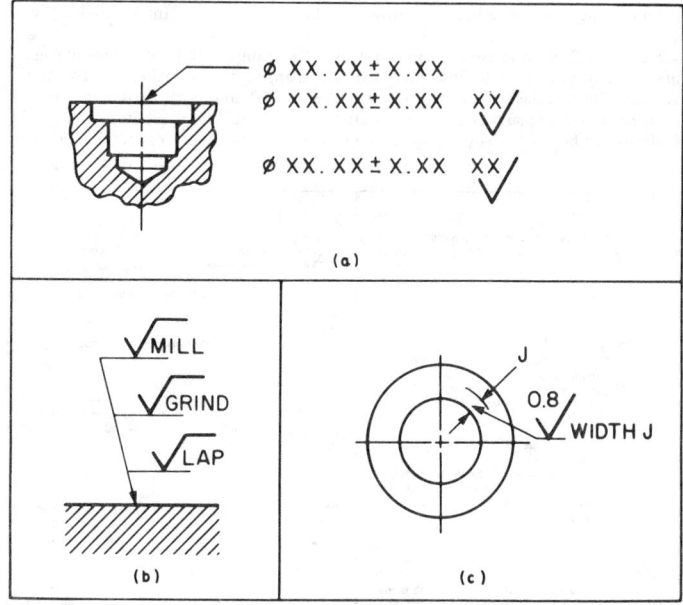

Fig. 5.    Examples of special designations.

limited to defined areas of the casting. Practicality of, and methods of determining that a casting's surface texture meets the specification shall be coordinated with the producer. The Society of Automotive Engineers standard J435 "Automotive Steel Castings" describes methods of evaluating steel casting surface texture used in the automotive and related industries.

**Checking Drawings.** — In order that the drawings may have a high standard of excellence, a set of instructions, as given in the following, has been issued to the checkers, and also to the draftsmen and tracers in the engineering department of a well-known machine-building company.

*Inspecting a New Design:* In case a new design is involved, first inspect the lay-outs carefully to see that the parts function correctly under all conditions, that they have the proper relative proportions, that the general design is correct in the matters of strength, rigidity, bearing areas, appearance, convenience of assembly, and direction of motion of the parts, and that there are no interferences. Consider the design as a whole to see if any improvements can be made. If the design appears to be unsatisfactory in any particular, or improvements appear to be possible, call the matter to the attention of the chief engineer.

*Checking for Strength:* Inspect the design of the part being checked for strength, rigidity, and appearance by comparing it with other parts for similar service whenever possible, giving preference to the later designs in such comparison, unless the later designs are known to be unsatisfactory. If there is any question regarding the matter, compute the stresses and deformations or find out whether the chief engineer has approved the stresses or deformations that will result from the forces applied to the part in service. In checking parts that are to go on a machine of increased size, be sure that standard parts used in similar machines and proposed for use on the larger machine, have ample strength and rigidity under the new and more severe service to which they will be put.

*Materials Specified.* — Consider the kind of material required for the part and the various possibilities of molding, forging, welding, or otherwise forming the rough part from this material. Then consider the machining operations to see whether changes in form or design will reduce the number of operations or the cost of machining.

See that parts are designed with reference to the economical use of material, and whenever possible, utilize standard sizes of stock and material readily obtainable from local dealers. In the case of alloy steel, special bronze, and similar materials, be sure that the material can be obtained in the size required.

*Method of Making Drawing.* — Inspect the drawing to see that the projections and sections are made in such a way as to show most clearly the form of the piece and the work to be done on it. Make sure that any workman looking at the drawing will understand what the shape of the piece is and how it is to be molded or machined. Make sure that the delineation is correct in every particular, and that the information conveyed by the drawing as to the form of the piece is complete.

*Checking Dimensions.* — Check all dimensions to see that they are correct. Scale all dimensions and see that the drawing is to scale. See that the dimensions on the drawing agree with the dimensions scaled from the lay-out. Wherever any dimension is out of scale, see that the dimension is so marked. Investigate any case where the dimension, the scale of the drawing, and the scale of the lay-out do not agree. All dimensions not to scale must be underlined on the tracing. In checking dimensions, note particularly the following points:

See that all figures are correctly formed and that they will print clearly, so that the workmen can easily read them correctly.

See that the over-all dimensions are given.

See that all witness lines go to the correct part of the drawing.

See that all arrow points go to the correct witness lines.

See that proper allowance is made for all fits.

See that the tolerances are correctly given where necessary.

See that all dimensions given agree with the corresponding dimensions of adjacent parts.

Be sure that the dimensions given on a drawing are those that the machinist will use, and that the workman will not be obliged to do addition or subtraction in order to obtain the necessary measurements for machining or checking his work.

Avoid strings of dimensions where errors can accumulate. It is generally better to give a number of dimensions from the same reference surface or center line.

When holes are to be located by boring on a horizontal spindle boring machine or other similar machine, give dimensions to centers of bored holes in rectangular coordinates and from the center lines of the first hole to be bored, so that the operator will not be obliged to add measurements or transfer gages.

*Checking Assembly.* — See that the part can readily be assembled with the adjacent parts. If necessary, provide tapped holes for eyebolts and cored holes for tongs, lugs, or other methods of handling.

Make sure that, in being assembled, the piece will not interfere with other pieces already in place and that the assembly can be taken apart without difficulty.

Check the sum of a number of tolerances; this sum must not be great enough to permit two pieces that should not be in contact to come together.

*Checking Castings.* — In the case of castings, study the form of the pattern, the methods of molding, the method of supporting and venting the cores, and the effect of draft and rough molding on clearances.

Avoid undue metal thickness, and especially avoid thick and thin sections in the same casting.

Indicate all metal thicknesses, so that the molder will know what chaplets to use for supporting the cores.

See that ample fillets are provided, and that they are properly dimensioned.

See that the cores can be assembled in the mold without crushing or interference.

See that swelling, shrinkage, or misalignment of cores will not make trouble in machining.

See that the amount of finish is indicated.

See that there is sufficient finish on large castings to permit them to be "cleaned up," even though they warp. In the case of such castings, make sure that the metal thickness will be sufficient after finishing, even though the castings do warp.

Make sure that sufficient sections are shown so that the patternmakers and molders will not be compelled to make assumptions about the form of any part of the casting. This is particularly important when a number of sections of the casting are similar in form, while others differ slightly.

*Checking Machined Parts.* — Study the sequences of operations in machining and see that all finish marks are indicated.

See that the finish marks are placed on the lines to which dimensions are given.

See that methods of machining are indicated where necessary.

Give all drill, reamer, tap, and rose bit sizes.

See that jig and gage numbers are indicated at the proper places.

See that all necessary bosses, lugs, and openings are provided for lifting, handling, clamping, and machining the piece.

See that adequate wrench room is provided for all nuts and bolt heads.

Avoid special tools, such as taps, drills, reamers, etc., unless such tools are specially authorized.

Where parts are right- and left-hand, be sure that the hand is correctly designated. When possible, make parts symmetrical, so as to avoid having them right- and left-hand, but do not sacrifice correct design or satisfactory operation on this account.

When heat-treatment is required, the heat-treatment should be specified.

Check the title, size of machine, the scale, and the drawing number on both the drawing and the drawing record card.

**Metric Dimensions on Drawings.** — The length units of the metric system that are most generally used in connection with any work relating to mechanical engineering are the meter (39.37 inches) and the millimeter (0.03937 inch). One meter equals 1000 millimeters. On mechanical drawings, all dimensions are generally given in millimeters, no matter how large the dimensions may be. In fact, dimensions of such machines as locomotives and large electrical apparatus are given exclusively in millimeters. This practice is adopted to avoid mistakes due to misplacing decimal points, or mis-reading dimensions as when other units are used as well. When dimensions are given in millimeters, many of them can be given without resorting to decimal points, as a millimeter is only a little more than $\frac{1}{32}$ inch. Only dimensions of precision need be given in decimals of a millimeter; such dimensions are generally given in hundredths of a millimeter — for example, 0.02 millimeter, which is equal to 0.0008 inch. As 0.01 millimeter is equal to 0.0004 inch, it is seldom that dimensions would be given with greater accuracy than to hundredths of a millimeter.

*Scales of Metric Drawings.* — Drawings made to the metric system are not made to scales of $\frac{1}{2}$, $\frac{1}{4}$, $\frac{1}{8}$, etc., as in the case of drawings made to the English system. If the object cannot be drawn full size, it may be drawn $\frac{1}{2.5}$, $\frac{1}{5}$, $\frac{1}{10}$, $\frac{1}{20}$, $\frac{1}{50}$, $\frac{1}{100}$, $\frac{1}{200}$, $\frac{1}{500}$ or $\frac{1}{1000}$ size. If the object is too small and has to be drawn larger it is drawn 2, 5, or 10 times its actual size.

# WEIGHTS AND MEASURES

## Measures of Length

1 mile = 1760 yards = 5280 feet.
1 yard = 3 feet = 36 inches.    1 foot = 12 inches.
1 mil = 0.001 inch.  1 fathom = 2 yards = 6 feet.
1 rod = 5.5 yards = 16.5 feet.   1 hand = 4 inches.   1 span = 9 inches.
1 micro-inch = one millionth inch or 0.000001 inch.   (1 micrometer or micron = one millionth meter = 0.00003937 inch.)

## Surveyor's Measure

1 mile = 8 furlongs = 80 chains.
1 furlong = 10 chains = 220 yards.
1 chain = 4 rods = 22 yards = 66 feet = 100 links.
1 link = 7.92 inches.

## Nautical Measure

1 league = 3 nautical miles.
1 nautical mile = 6076.11549 feet = 1.1508 statute miles.   (The *knot*, which is a nautical unit of speed, is equivalent to a speed of 1 nautical mile per hour.)
One degree at the equator = 60 nautical miles = 69.047 statute miles. 360 degrees = 21,600 nautical miles = 24,856.8 statute miles = circumference at equator.

## Square Measure

1 square mile = 640 acres = 6400 square chains.
1 acre = 10 square chains = 4840 square yards = 43,560 square feet.
1 square chain = 16 square rods = 484 square yards = 4356 square feet.
1 square rod = 30.25 square yards = 272.25 square feet = 625 square links.
1 square yard = 9 square feet.
1 square foot = 144 square inches.
An acre is equal to a square, the side of which is 208.7 feet.

## Measure used for Diameters and Areas of Electric Wires

1 circular inch = area of circle 1 inch in diameter = 0.7854 square inch.
1 circular inch = 1,000,000 circular mils.
1 square inch = 1.2732 circular inch = 1,273,239 circular mils.
A circular mil is the area of a circle 0.001 inch in diameter.

## Cubic Measure

1 cubic yard = 27 cubic feet.
1 cubic foot = 1728 cubic inches.
The following measures are also used for wood and masonry:
1 cord of wood = 4 × 4 × 8 feet = 128 cubic feet.
1 perch of masonry = $16\frac{1}{2} \times 1\frac{1}{2} \times 1$ foot = $24\frac{3}{4}$ cubic feet.

## Shipping Measure

For measuring entire internal capacity of a vessel:
1 register ton = 100 cubic feet.
For measurement of cargo:
Approximately 40 cubic feet of merchandise is considered a shipping ton, unless that bulk would weigh more than 2000 pounds, in which case the freight charge may be based upon weight.
40 cubic feet = 32.143 U. S. bushels = 31.16 Imperial bushels.

## Dry Measure

1 bushel (U. S. or Winchester struck bushel) = 1.2445 cubic foot = 2150.42 cubic inches.
1 bushel = 4 pecks = 32 quarts = 64 pints.
1 peck = 8 quarts = 16 pints.
1 quart = 2 pints.
1 heaped bushel = $1\frac{1}{4}$ struck bushel.
1 cubic foot = 0.8036 struck bushel.
1 British Imperial bushel = 8 Imperial gallons = 1.2837 cubic foot = 2218.19 cubic inches.

## Liquid Measure

1 U. S. gallon = 0.1337 cubic foot = 231 cubic inches = 4 quarts = 8 pints.
1 quart = 2 pints = 8 gills.
1 pint = 4 gills.
1 British Imperial gallon = 1.2009 U. S. gallon = 277.42 cubic inches.
1 cubic foot = 7.48 U. S. gallons.

## Old Liquid Measure

1 tun = 2 pipes = 3 puncheons.
1 pipe or butt = 2 hogsheads = 4 barrels = 126 gallons.
1 puncheon = 2 tierces = 84 gallons.
1 hogshead = 2 barrels = 63 gallons.
1 tierce = 42 gallons.
1 barrel = $31\frac{1}{2}$ gallons.

## Apothecaries' Fluid Measure

1 U. S. fluid ounce = 8 drachms = 1.805 cubic inch = $\frac{1}{128}$ U. S. gallon.
1 fluid drachm = 60 minims.
1 British fluid ounce = 1.732 cubic inch.

## Measures of Weight

### Avoirdupois or Commercial Weight

1 gross or long ton = 2240 pounds.
1 net or short ton = 2000 pounds.
1 pound = 16 ounces = 7000 grains.
1 ounce = 16 drachms = 437.5 grains.
The following measures for weight are now seldom used in the United States:
    1 hundred-weight = 4 quarters = 112 pounds (1 gross or long ton = 20 hundred-weights); 1 quarter = 28 pounds; 1 stone = 14 pounds; 1 quintal = 100 pounds.

### Troy Weight, used for Weighing Gold and Silver

1 pound = 12 ounces = 5760 grains.
1 ounce = 20 pennyweights = 480 grains.
1 pennyweight = 24 grains.
1 carat (used in weighing diamonds) = 3.086 grains.
1 grain Troy = 1 grain avoirdupois = 1 grain apothecaries' weight.

## Apothecaries' Weight

1 pound = 12 ounces = 5760 grains.
1 ounce = 8 drachms = 480 grains.
1 drachm = 3 scruples = 60 grains.
1 scruple = 20 grains.

## Measures of Pressure

1 pound per square inch = 144 pounds per square foot = 0.068 atmosphere = 2.042 inches of mercury at 62 degrees F. = 27.7 inches of water at 62 degrees F. = 2.31 feet of water at 62 degrees F.

1 atmosphere = 30 inches of mercury at 62 degrees F. = 14.7 pounds per square inch = 2116.3 pounds per square foot = 33.95 feet of water at 62 degrees F.

1 foot of water at 62 degrees F. = 62.355 pounds per square foot = 0.433 pound per square inch.

1 inch of mercury at 62 degrees F. = 1.132 foot of water = 13.58 inches of water = 0.491 pound per square inch.

## Miscellaneous

1 great gross = 12 gross = 144 dozen.
1 gross = 12 dozen = 144 units.
1 dozen = 12 units.
1 score = 20 units.

1 quire = 24 sheets.
1 ream = 20 quires = 480 sheets.
1 ream printing paper = 500 sheets.

### Decimal Equivalents of Fractions of an Inch

| Fraction | Decimal | Fraction | Decimal | Fraction | Decimal |
|---|---|---|---|---|---|
| 1/64 | 0.015 625 | 11/32 | 0.343 75 | 43/64 | 0.671 875 |
| 1/32 | 0.031 25 | 23/64 | 0.359 375 | 11/16 | 0.687 5 |
| 3/64 | 0.046 875 | 3/8 | 0.375 | 45/64 | 0.703 125 |
| 1/16 | 0.062 5 | 25/64 | 0.390 625 | 23/32 | 0.718 75 |
| 5/64 | 0.078 125 | 13/32 | 0.406 25 | 47/64 | 0.734 375 |
| 3/32 | 0.093 75 | 27/64 | 0.421 875 | 3/4 | 0.750 |
| 7/64 | 0.109 375 | 7/16 | 0.437 5 | 49/64 | 0.765 625 |
| 1/8 | 0.125 | 29/64 | 0.453 125 | 25/32 | 0.781 25 |
| 9/64 | 0.140 625 | 15/32 | 0.468 75 | 51/64 | 0.796 875 |
| 5/32 | 0.156 25 | 31/64 | 0.484 375 | 13/16 | 0.812 5 |
| 11/64 | 0.171 875 | 1/2 | 0.500 | 53/64 | 0.828 125 |
| 3/16 | 0.187 5 | 33/64 | 0.515 625 | 27/32 | 0.843 75 |
| 13/64 | 0.203 125 | 17/32 | 0.531 25 | 55/64 | 0.859 375 |
| 7/32 | 0.218 75 | 35/64 | 0.546 875 | 7/8 | 0.875 |
| 15/64 | 0.234 375 | 9/16 | 0.562 5 | 57/64 | 0.890 625 |
| 1/4 | 0.250 | 37/64 | 0.578 125 | 29/32 | 0.906 25 |
| 17/64 | 0.265 625 | 19/32 | 0.593 75 | 59/64 | 0.921 875 |
| 9/32 | 0.281 25 | 39/64 | 0.609 375 | 15/16 | 0.937 5 |
| 19/64 | 0.296 875 | 5/8 | 0.625 | 61/64 | 0.953 125 |
| 5/16 | 0.312 5 | 41/64 | 0.640 625 | 31/32 | 0.968 75 |
| 21/64 | 0.328 125 | 21/32 | 0.656 25 | 63/64 | 0.984 375 |

**Table of Decimal Equivalents of a Foot Corresponding to Inches and Fractions of Inches.** — Assume, for example, that it is required to find the equivalent of 6 7/32 inches in decimals of a foot. Locate 7/32 in the left-hand column and follow the horizontal line until the column headed "6" is reached. The figures 0.5182 read off in this column are the decimals of a foot corresponding to 6 7/32; in other words, 6 7/32 inches equals 0.5182 foot.

## Inches into Decimals of a Foot

| Inch | 0 | 1 | 2 | 3 | 4 | 5 | 6 | 7 | 8 | 9 | 10 | 11 |
|---|---|---|---|---|---|---|---|---|---|---|---|---|
| | | | | | | | Inches | | | | | |
| | | | | | | Decimals of a Foot | | | | | | |
| | | 0.0833 | 0.1667 | 0.2500 | 0.3333 | 0.4167 | 0.5000 | 0.5833 | 0.6667 | 0.7500 | 0.8333 | 0.9167 |
| 1/32 | 0.0026 | 0.0859 | 0.1693 | 0.2526 | 0.3359 | 0.4193 | 0.5026 | 0.5859 | 0.6693 | 0.7526 | 0.8359 | 0.9193 |
| 1/16 | 0.0052 | 0.0885 | 0.1719 | 0.2552 | 0.3385 | 0.4219 | 0.5052 | 0.5885 | 0.6719 | 0.7552 | 0.8385 | 0.9219 |
| 3/32 | 0.0078 | 0.0911 | 0.1745 | 0.2578 | 0.3411 | 0.4245 | 0.5078 | 0.5911 | 0.6745 | 0.7578 | 0.8411 | 0.9245 |
| 1/8 | 0.0104 | 0.0938 | 0.1771 | 0.2604 | 0.3438 | 0.4271 | 0.5104 | 0.5938 | 0.6771 | 0.7604 | 0.8438 | 0.9271 |
| 5/32 | 0.0130 | 0.0964 | 0.1797 | 0.2630 | 0.3464 | 0.4297 | 0.5130 | 0.5964 | 0.6797 | 0.7630 | 0.8464 | 0.9297 |
| 3/16 | 0.0156 | 0.0990 | 0.1823 | 0.2656 | 0.3490 | 0.4323 | 0.5156 | 0.5990 | 0.6823 | 0.7656 | 0.8490 | 0.9323 |
| 7/32 | 0.0182 | 0.1016 | 0.1849 | 0.2682 | 0.3516 | 0.4349 | 0.5182 | 0.6016 | 0.6849 | 0.7682 | 0.8516 | 0.9349 |
| 1/4 | 0.0208 | 0.1042 | 0.1875 | 0.2708 | 0.3542 | 0.4375 | 0.5208 | 0.6042 | 0.6875 | 0.7708 | 0.8542 | 0.9375 |
| 9/32 | 0.0234 | 0.1068 | 0.1901 | 0.2734 | 0.3568 | 0.4401 | 0.5234 | 0.6068 | 0.6901 | 0.7734 | 0.8568 | 0.9401 |
| 5/16 | 0.0260 | 0.1094 | 0.1927 | 0.2760 | 0.3594 | 0.4427 | 0.5260 | 0.6094 | 0.6927 | 0.7760 | 0.8594 | 0.9427 |
| 11/32 | 0.0286 | 0.1120 | 0.1953 | 0.2786 | 0.3620 | 0.4453 | 0.5286 | 0.6120 | 0.6953 | 0.7786 | 0.8620 | 0.9453 |
| 3/8 | 0.0313 | 0.1146 | 0.1979 | 0.2813 | 0.3646 | 0.4479 | 0.5313 | 0.6146 | 0.6979 | 0.7813 | 0.8646 | 0.9479 |
| 13/32 | 0.0339 | 0.1172 | 0.2005 | 0.2839 | 0.3672 | 0.4505 | 0.5339 | 0.6172 | 0.7005 | 0.7839 | 0.8672 | 0.9505 |
| 7/16 | 0.0365 | 0.1198 | 0.2031 | 0.2865 | 0.3698 | 0.4531 | 0.5365 | 0.6198 | 0.7031 | 0.7865 | 0.8698 | 0.9531 |
| 15/32 | 0.0391 | 0.1224 | 0.2057 | 0.2891 | 0.3724 | 0.4557 | 0.5391 | 0.6224 | 0.7057 | 0.7891 | 0.8724 | 0.9557 |
| 1/2 | 0.0417 | 0.1250 | 0.2083 | 0.2917 | 0.3750 | 0.4583 | 0.5417 | 0.6250 | 0.7083 | 0.7917 | 0.8750 | 0.9583 |
| 17/32 | 0.0443 | 0.1276 | 0.2109 | 0.2943 | 0.3776 | 0.4609 | 0.5443 | 0.6276 | 0.7109 | 0.7943 | 0.8776 | 0.9609 |
| 9/16 | 0.0469 | 0.1302 | 0.2135 | 0.2969 | 0.3802 | 0.4635 | 0.5469 | 0.6302 | 0.7135 | 0.7969 | 0.8802 | 0.9635 |
| 19/32 | 0.0495 | 0.1328 | 0.2161 | 0.2995 | 0.3828 | 0.4661 | 0.5495 | 0.6328 | 0.7161 | 0.7995 | 0.8828 | 0.9661 |
| 5/8 | 0.0521 | 0.1354 | 0.2188 | 0.3021 | 0.3854 | 0.4688 | 0.5521 | 0.6354 | 0.7188 | 0.8021 | 0.8854 | 0.9688 |
| 21/32 | 0.0547 | 0.1380 | 0.2214 | 0.3047 | 0.3880 | 0.4714 | 0.5547 | 0.6380 | 0.7214 | 0.8047 | 0.8880 | 0.9714 |
| 11/16 | 0.0573 | 0.1406 | 0.2240 | 0.3073 | 0.3906 | 0.4740 | 0.5573 | 0.6406 | 0.7240 | 0.8073 | 0.8906 | 0.9740 |
| 23/32 | 0.0599 | 0.1432 | 0.2266 | 0.3099 | 0.3932 | 0.4766 | 0.5599 | 0.6432 | 0.7266 | 0.8099 | 0.8932 | 0.9766 |
| 3/4 | 0.0625 | 0.1458 | 0.2292 | 0.3125 | 0.3958 | 0.4792 | 0.5625 | 0.6458 | 0.7292 | 0.8125 | 0.8958 | 0.9792 |
| 25/32 | 0.0651 | 0.1484 | 0.2318 | 0.3151 | 0.3984 | 0.4818 | 0.5651 | 0.6484 | 0.7318 | 0.8151 | 0.8984 | 0.9818 |
| 13/16 | 0.0677 | 0.1510 | 0.2344 | 0.3177 | 0.4010 | 0.4844 | 0.5677 | 0.6510 | 0.7344 | 0.8177 | 0.9010 | 0.9844 |
| 27/32 | 0.0703 | 0.1536 | 0.2370 | 0.3203 | 0.4036 | 0.4870 | 0.5703 | 0.6536 | 0.7370 | 0.8203 | 0.9036 | 0.9870 |
| 7/8 | 0.0729 | 0.1563 | 0.2396 | 0.3229 | 0.4063 | 0.4896 | 0.5729 | 0.6563 | 0.7396 | 0.8229 | 0.9063 | 0.9896 |
| 29/32 | 0.0755 | 0.1589 | 0.2422 | 0.3255 | 0.4089 | 0.4922 | 0.5755 | 0.6589 | 0.7422 | 0.8255 | 0.9089 | 0.9922 |
| 15/16 | 0.0781 | 0.1615 | 0.2448 | 0.3281 | 0.4115 | 0.4948 | 0.5781 | 0.6615 | 0.7448 | 0.8281 | 0.9115 | 0.9948 |
| 31/32 | 0.0807 | 0.1641 | 0.2474 | 0.3307 | 0.4141 | 0.4974 | 0.5807 | 0.6641 | 0.7474 | 0.8307 | 0.9141 | 0.9974 |

### Decimal Equivalents of 6ths, 12ths, and 24ths of an Inch

| | | | | | |
|---|---|---|---|---|---|
| 1/24 | 0.041 667 | 9/24 | 0.375 | 17/24 | 0.708 333 |
| 1/12... | 0.083 333 | 5/12.... | 0.416 667 | 9/12... | 0.75 |
| 3/24 | 0.125 | 11/24 | 0.458 333 | 19/24 | 0.791 667 |
| 1/6...... | 0.166 667 | 3/8...... | 0.5 | 5/6...... | 0.833 333 |
| 5/24 | 0.208 333 | 13/24 | 0.541 667 | 21/24 | 0.875 |
| 3/12... | 0.25 | 7/12... | 0.583 333 | 11/12... | 0.916 667 |
| 7/24 | 0.291 667 | 15/24 | 0.625 | 23/24 | 0.958 333 |
| 2/6...... | 0.333 333 | 4/6...... | 0.666 667 | ...... | ........ |

### Decimal Equivalents of 7ths, 14ths, and 28ths of an Inch

| | | | | | |
|---|---|---|---|---|---|
| 1/28 | 0.035 714 | 5/14.. | 0.357 143 | 19/28 | 0.678 571 |
| 1/14.. | 0.071 429 | 11/28 | 0.392 857 | 5/7.... | 0.714 286 |
| 3/28 | 0.107 143 | 3/7.... | 0.428 571 | 21/28 | 0.75 |
| 1/7.... | 0.142 857 | 13/28 | 0.464 286 | 11/14.. | 0.785 714 |
| 5/28 | 0.178 571 | 7/14.. | 0.5 | 23/28 | 0.821 429 |
| 3/14.. | 0.214 286 | 15/28 | 0.535 714 | 6/7.... | 0.857 143 |
| 7/28 | 0.25 | 4/7.... | 0.571 429 | 25/28 | 0.892 857 |
| 2/7.... | 0.285 714 | 17/28 | 0.607 143 | 13/14.. | 0.928 571 |
| 9/28 | 0.321 429 | 9/14.. | 0.642 857 | 27/28 | 0.964 286 |

### U. S. Gallons into Cubic Feet

| Gallons | Cubic Feet | Gallons | Cubic Feet | Gallons | Cubic Feet | Gallons | Cubic Feet |
|---|---|---|---|---|---|---|---|
| 1 | 0.134 | 20 | 2.674 | 300 | 40.10 | 4,000 | 534.72 |
| 2 | 0.267 | 30 | 4.010 | 400 | 53.47 | 5,000 | 668.40 |
| 3 | 0.401 | 40 | 5.347 | 500 | 66.84 | 6,000 | 802.08 |
| 4 | 0.535 | 50 | 6.684 | 600 | 80.21 | 7,000 | 935.76 |
| 5 | 0.668 | 60 | 8.021 | 700 | 93.58 | 8,000 | 1,069.44 |
| 6 | 0.802 | 70 | 9.358 | 800 | 106.94 | 9,000 | 1,203.12 |
| 7 | 0.936 | 80 | 10.694 | 900 | 120.31 | 10,000 | 1,336.81 |
| 8 | 1.069 | 90 | 12.031 | 1000 | 133.68 | 50,000 | 6,684.03 |
| 9 | 1.203 | 100 | 13.368 | 2000 | 267.36 | 100,000 | 13,368.06 |
| 10 | 1.337 | 200 | 26.736 | 3000 | 401.04 | 500,000 | 66,840.28 |

### Cubic Feet into Gallons

(1 cubic foot = 7.4805 U. S. gallons; 1 gallon = 231 cubic inches = 0.13368 cubic foot.)

| Cubic Feet | Gallons | Cubic Feet | Gallons | Cubic Feet | Gallons | Cubic Feet | Gallons |
|---|---|---|---|---|---|---|---|
| 0.1 | 0.75 | 2 | 14.96 | 30 | 224.4 | 400 | 2,992.2 |
| 0.2 | 1.50 | 3 | 22.44 | 40 | 299.2 | 500 | 3,740.3 |
| 0.3 | 2.24 | 4 | 29.92 | 50 | 374.0 | 600 | 4,488.3 |
| 0.4 | 2.99 | 5 | 37.40 | 60 | 448.8 | 700 | 5,236.4 |
| 0.5 | 3.74 | 6 | 44.88 | 70 | 523.6 | 800 | 5,984.4 |
| 0.6 | 4.49 | 7 | 52.36 | 80 | 598.4 | 900 | 6,732.5 |
| 0.7 | 5.24 | 8 | 59.84 | 90 | 673.2 | 1,000 | 7,480.5 |
| 0.8 | 5.98 | 9 | 67.32 | 100 | 748.1 | 5,000 | 37,402.6 |
| 0.9 | 6.73 | 10 | 74.81 | 200 | 1496.1 | 10,000 | 74,805.2 |
| 1.0 | 7.48 | 20 | 149.61 | 300 | 2244.2 | 50,000 | 374,025.9 |

## Contents in Cubic Feet and U. S. Gallons of Pipes and Cylinders
### One Foot in Length

| Diam. in Inches | For 1 Foot in Length | | Diam. in Inches | For 1 Foot in Length | | Diam. in Inches | For 1 Foot in Length | |
|---|---|---|---|---|---|---|---|---|
| | Cubic Feet | U. S. Gallons | | Cubic Feet | U. S. Gallons | | Cubic Feet | U. S. Gallons |
| ¼ | 0.0003 | 0.0025 | 6¾ | 0.2485 | 1.859 | 19 | 1.969 | 14.73 |
| 5⁄16 | 0.0005 | 0.0040 | 7 | 0.2673 | 1.999 | 19½ | 2.074 | 15.51 |
| 3⁄8 | 0.0008 | 0.0057 | 7¼ | 0.2867 | 2.145 | 20 | 2.182 | 16.32 |
| 7⁄16 | 0.0010 | 0.0078 | 7½ | 0.3068 | 2.295 | 20½ | 2.292 | 17.15 |
| ½ | 0.0014 | 0.0102 | 7¾ | 0.3276 | 2.450 | 21 | 2.405 | 17.99 |
| 9⁄16 | 0.0017 | 0.0129 | 8 | 0.3491 | 2.611 | 21½ | 2.521 | 18.86 |
| 5⁄8 | 0.0021 | 0.0159 | 8¼ | 0.3712 | 2.777 | 22 | 2.640 | 19.75 |
| 11⁄16 | 0.0026 | 0.0193 | 8½ | 0.3941 | 2.948 | 22½ | 2.761 | 20.66 |
| ¾ | 0.0031 | 0.0230 | 8¾ | 0.4176 | 3.125 | 23 | 2.885 | 21.58 |
| 13⁄16 | 0.0036 | 0.0269 | 9 | 0.4418 | 3.305 | 23½ | 3.012 | 22.53 |
| 7⁄8 | 0.0042 | 0.0312 | 9¼ | 0.4667 | 3.491 | 24 | 3.142 | 23.50 |
| 15⁄16 | 0.0048 | 0.0359 | 9½ | 0.4922 | 3.682 | 25 | 3.409 | 25.50 |
| 1 | 0.0055 | 0.0408 | 9¾ | 0.5185 | 3.879 | 26 | 3.687 | 27.58 |
| 1¼ | 0.0085 | 0.0638 | 10 | 0.5454 | 4.080 | 27 | 3.976 | 29.74 |
| 1½ | 0.0123 | 0.0918 | 10¼ | 0.5730 | 4.286 | 28 | 4.276 | 31.99 |
| 1¾ | 0.0167 | 0.1249 | 10½ | 0.6013 | 4.498 | 29 | 4.587 | 34.31 |
| 2 | 0.0218 | 0.1632 | 10¾ | 0.6303 | 4.715 | 30 | 4.909 | 36.72 |
| 2¼ | 0.0276 | 0.2066 | 11 | 0.6600 | 4.937 | 31 | 5.241 | 39.21 |
| 2½ | 0.0341 | 0.2550 | 11¼ | 0.6903 | 5.164 | 32 | 5.585 | 41.78 |
| 2¾ | 0.0412 | 0.3085 | 11½ | 0.7213 | 5.396 | 33 | 5.940 | 44.43 |
| 3 | 0.0491 | 0.3672 | 11¾ | 0.7530 | 5.633 | 34 | 6.305 | 47.16 |
| 3¼ | 0.0576 | 0.4309 | 12 | 0.7854 | 5.875 | 35 | 6.681 | 49.98 |
| 3½ | 0.0668 | 0.4998 | 12½ | 0.8522 | 6.375 | 36 | 7.069 | 52.88 |
| 3¾ | 0.0767 | 0.5738 | 13 | 0.9218 | 6.895 | 37 | 7.467 | 55.86 |
| 4 | 0.0873 | 0.6528 | 13½ | 0.9940 | 7.436 | 38 | 7.876 | 58.92 |
| 4¼ | 0.0985 | 0.7369 | 14 | 1.069 | 7.997 | 39 | 8.296 | 62.06 |
| 4½ | 0.1104 | 0.8263 | 14½ | 1.147 | 8.578 | 40 | 8.727 | 65.28 |
| 4¾ | 0.1231 | 0.9206 | 15 | 1.227 | 9.180 | 41 | 9.168 | 68.58 |
| 5 | 0.1364 | 1.020 | 15½ | 1.310 | 9.801 | 42 | 9.621 | 71.97 |
| 5¼ | 0.1503 | 1.125 | 16 | 1.396 | 10.44 | 43 | 10.085 | 75.44 |
| 5½ | 0.1650 | 1.234 | 16½ | 1.485 | 11.11 | 44 | 10.559 | 78.99 |
| 5¾ | 0.1803 | 1.349 | 17 | 1.576 | 11.79 | 45 | 11.045 | 82.62 |
| 6 | 0.1963 | 1.469 | 17½ | 1.670 | 12.49 | 46 | 11.541 | 86.33 |
| 6¼ | 0.2131 | 1.594 | 18 | 1.767 | 13.22 | 47 | 12.048 | 90.13 |
| 6½ | 0.2304 | 1.724 | 18½ | 1.867 | 13.96 | 48 | 12.566 | 94.00 |

One cubic foot of water at 39.1 degrees F. weighs 62.4245 pounds.

One cubic foot of air at 32 degrees F., atmospheric pressure, weighs 0.08073 pound.

One pound of water at 39.1 degrees F. has a volume of 0.01602 cubic foot.

One pound of air at 32 degrees F., atmospheric pressure, has a volume of 12.387 cubic feet.

One gallon of water at 62 degrees F. weighs 8.336 pounds.

One pound of water at 62 degrees F. has a volume of 0.1199 U. S. gallon.

## Contents of Cylindrical Tanks in U. S. Gallons

| Depth of Tank, Feet | Diameter of Tank, Feet | | | | | | | | |
|---|---|---|---|---|---|---|---|---|---|
| | 5 | 6 | 7 | 8 | 9 | 10 | 11 | 12 | 13 |
| | Contents of Tank, U. S. Gallons | | | | | | | | |
| 5 | 734 | 1058 | 1439 | 1880 | 2379 | 2,938 | 3,555 | 4,230 | 4,965 |
| 6 | 881 | 1269 | 1727 | 2256 | 2855 | 3,525 | 4,265 | 5,076 | 5,957 |
| 7 | 1028 | 1481 | 2015 | 2632 | 3331 | 4,113 | 4,976 | 5,922 | 6,950 |
| 8 | 1175 | 1692 | 2303 | 3008 | 3807 | 4,700 | 5,687 | 6,768 | 7,943 |
| 9 | 1322 | 1904 | 2591 | 3384 | 4283 | 5,288 | 6,398 | 7,614 | 8,936 |
| 10 | 1469 | 2115 | 2879 | 3760 | 4759 | 5,875 | 7,109 | 8,460 | 9,929 |
| 11 | 1616 | 2327 | 3167 | 4136 | 5235 | 6,463 | 7,820 | 9,306 | 10,922 |
| 12 | 1763 | 2538 | 3455 | 4512 | 5711 | 7,050 | 8,531 | 10,152 | 11,915 |
| 13 | 1909 | 2750 | 3742 | 4888 | 6187 | 7,638 | 9,242 | 10,998 | 12,808 |
| 14 | 2056 | 2961 | 4030 | 5264 | 6662 | 8,225 | 9,953 | 11,844 | 13,801 |
| 15 | 2203 | 3173 | 4318 | 5640 | 7138 | 8,813 | 10,664 | 12,690 | 14,894 |
| 16 | 2350 | 3384 | 4606 | 6016 | 7614 | 9,400 | 11,374 | 13,536 | 15,887 |
| 17 | 2497 | 3596 | 4894 | 6392 | 8090 | 9,988 | 12,085 | 14,383 | 16,879 |
| 18 | 2644 | 3807 | 5182 | 6768 | 8566 | 10,575 | 12,796 | 15,229 | 17,872 |
| 19 | 2791 | 4019 | 5480 | 7144 | 9042 | 11,163 | 13,507 | 16,075 | 18,865 |
| 20 | 2938 | 4230 | 5758 | 7520 | 9518 | 11,750 | 14,218 | 16,921 | 19,858 |

| Depth of Tank, Feet | Diameter of Tank, Feet | | | | | | | |
|---|---|---|---|---|---|---|---|---|
| | 14 | 15 | 16 | 18 | 20 | 22 | 24 | 25 |
| | Contents of Tank, U. S. Gallons | | | | | | | |
| 5 | 5,758 | 6,610 | 7,521 | 9,518 | 11,751 | 14,218 | 16,921 | 18,360 |
| 6 | 6,909 | 7,931 | 9,025 | 11,422 | 14,101 | 17,062 | 20,305 | 22,032 |
| 7 | 8,061 | 9,253 | 10,529 | 13,325 | 16,451 | 19,905 | 23,689 | 25,704 |
| 8 | 9,212 | 10,575 | 12,033 | 15,229 | 18,801 | 22,749 | 27,073 | 29,376 |
| 9 | 10,364 | 11,897 | 13,537 | 17,132 | 21,151 | 25,592 | 30,457 | 33,048 |
| 10 | 11,515 | 13,219 | 15,041 | 19,036 | 23,501 | 28,436 | 33,841 | 36,720 |
| 11 | 12,667 | 14,541 | 16,545 | 20,940 | 25,851 | 31,280 | 37,225 | 40,392 |
| 12 | 13,818 | 15,863 | 18,049 | 22,843 | 28,201 | 34,123 | 40,609 | 44,064 |
| 13 | 14,970 | 17,185 | 19,553 | 24,747 | 30,551 | 36,967 | 43,993 | 47,736 |
| 14 | 16,121 | 18,507 | 21,057 | 26,650 | 32,901 | 39,810 | 47,377 | 51,408 |
| 15 | 17,273 | 19,829 | 22,562 | 28,554 | 35,252 | 42,654 | 50,762 | 55,080 |
| 16 | 18,424 | 21,150 | 24,066 | 30,458 | 37,602 | 45,498 | 54,146 | 58,752 |
| 17 | 19,576 | 22,472 | 25,570 | 32,361 | 39,952 | 48,341 | 57,530 | 62,424 |
| 18 | 20,727 | 23,794 | 27,074 | 34,265 | 42,302 | 51,185 | 60,914 | 66,096 |
| 19 | 21,879 | 25,116 | 28,578 | 36,168 | 44,652 | 54,028 | 64,298 | 69,768 |
| 20 | 23,030 | 26,438 | 30,082 | 38,072 | 47,002 | 56,872 | 67,682 | 73,440 |

A cylinder 7 inches in diameter and 6 inches high contains one gallon within 0.1 of a cubic inch.

The volume, in U. S. gallons, of a cylinder, equals the square of the diameter in inches × height of cylinder in inches × 0.0034.

## Circular Mil Gage for Electrical Wires*

| A.W.G. or B. & S. Gage | Diam. Mils | Circu-lar Mils | A.W.G. or B. & S. Gage | Diam. Mils | Circu-lar Mils | A.W.G. or B. & S. Gage | Diam. Mils | Circu-lar Mils |
|---|---|---|---|---|---|---|---|---|
| 0000 | 460 | 212,000 | 12 | 81 | 6530 | 27 | 14.2 | 202. |
| 000 | 410 | 168,000 | 13 | 72 | 5180 | 28 | 12.6 | 160. |
| 00 | 365 | 133,000 | 14 | 64 | 4110 | 29 | 11.3 | 127. |
| 0 | 325 | 106,000 | 15 | 57 | 3260 | 30 | 10.0 | 101. |
| 1 | 289 | 83,700 | 16 | 51 | 2580 | 31 | 8.9 | 79.7 |
| 2 | 258 | 66,400 | 17 | 45 | 2050 | 32 | 8.0 | 63.2 |
| 3 | 229 | 52,600 | 18 | 40 | 1620 | 33 | 7.1 | 50.1 |
| 4 | 204 | 41,700 | 19 | 36 | 1290 | 34 | 6.3 | 39.8 |
| 5 | 182 | 33,100 | 20 | 32 | 1020 | 35 | 5.6 | 31.5 |
| 6 | 162 | 26,300 | 21 | 28.5 | 810 | 36 | 5.0 | 25.0 |
| 7 | 144 | 20,800 | 22 | 25.3 | 642 | 37 | 4.5 | 19.8 |
| 8 | 128 | 16,500 | 23 | 22.6 | 509 | 38 | 4.0 | 15.7 |
| 9 | 114 | 13,100 | 24 | 20.1 | 404 | 39 | 3.5 | 12.5 |
| 10 | 102 | 10,400 | 25 | 17.9 | 320 | 40 | 3.1 | 9.9 |
| 11 | 91 | 8,230 | 26 | 15.9 | 254 | .. | .... | .... |

* A circular mil is a unit of area that is applied to electrical wires and cables and is equal to the area of a circle one mil (.001 inch) in diameter. The area of any circle in circular mils is equal to the square of its diameter in mils.

# METRIC SYSTEMS OF MEASUREMENT

A metric system of measurement was first established in France in the years following the French Revolution, and various systems of metric units have been developed since that time. All metric unit systems are based, at least in part, on the International Metric Standards which are the meter and kilogram, or decimal multiples or sub-multiples of these standards.

In 1795, a metric system called the centimeter-gram-second (cgs) system was proposed, and was adopted in France in 1799. In 1873, the British Association for the Advancement of Science recommended the use of the cgs system, and since then it has been widely used in all branches of science throughout the world. From the base units in the cgs system are derived:

    Unit of velocity = 1 centimeter per second.
    Acceleration due to gravity (at Paris) = 981 centimeters per sec. per sec.
    Unit of force = 1 dyne = $\frac{1}{981}$ gram.
    Unit of work = 1 erg = 1 dyne-centimeter.
    Unit of power = 1 watt = 10,000,000 ergs per second.

Another metric system called the MKS (meter-kilogram-second) system of units was proposed by Professor G. Giorgi in 1902. In 1935, the International Electro-technical Commission (IEC) accepted his recommendation that this system of units of mechanics should be linked with the electro-magnetic units by the adoption of a fourth base unit. In 1950, the IEC adopted the ampere, the unit of electric current, as the fourth unit, and the MKSA system thus came into being.

A gravitational system of metric units, known as the technical system, is based on the meter, the kilogram as a force, and the second. It has been widely used in engineering. Because the standard of force is defined as the weight of the mass of the standard kilogram, the fundamental unit of force varies due to the difference in gravitational pull at different locations around the earth. By international agreement, a standard value for acceleration due to gravity was chosen (9.81 meters per second squared) which for all practical measurements is approximately the same as the local value at the point of measurement.

## The International System of Units (SI).

The Conference Generale des Poids et Mesures (CGPM) which is the body responsible for all international matters concerning the metric system, adopted in 1954, a rationalized and coherent system of units, based on the four MKSA units (see above), and including the *kelvin* as the unit of temperature and the *candela* as the unit of luminous intensity. In 1960, the CGPM formerly named this system the Systeme International d'Unites, for which the abbreviation is SI in all languages. In 1971, the 14th CGPM adopted a seventh base unit, the *mole* which is the unit of quantity ("amount of substance").

In the period since the first metric system was established in France towards the end of the 18th century, most of the countries of the world have adopted a metric system. At the present time most of the industrially advanced metric-using countries are changing from their traditional metric system to SI. Those countries which are currently changing or considering change from the English system of measurement to metric, have the advantage that they can convert directly to the modernized system. The United Kingdom, which can be said to have led the now worldwide move to change from the English system, went straight to SI.

The use of SI units instead of the traditional metric units has little effect on everyday life or trade. The units of linear measurement, mass, volume, and time remain the same, viz. meter, kilogram, liter, and second.

The SI, like the traditional metric system, is based on decimal arithmetic. For each physical quantity, units of different sizes are formed by multiplying or dividing a single base value by powers of 10. Thus, changes can be made very simply by adding zeros or shifting decimal points. For example, the meter is the basic unit of length; the kilometer is a multiple (1000 meters); and the millimeter is a sub-multiple (one-thousandth of a meter).

In the older metric systems, the simplicity of a series of units linked by powers of ten is an advantage for plain quantities such as length, but this simplicity is lost as soon as more complex units are encountered. For example, in different branches of science and engineering, energy may appear as the erg, the calorie, the kilogram meter, the liter atmosphere or the horsepower hour. In contrast, the SI provides only one basic unit for each physical quantity, and universality is thus achieved.

As mentioned above, there are seven base-units, which are for the basic quantities of length, mass, time, electric current, thermodynamic temperature, amount of substance and luminous intensity, expressed as the meter (m), the kilogram (kg), the second (s), the ampere (A), the kelvin (K), the mole (mol) and the candela (cd). The units are defined in the accompanying table.

The SI is a coherent system. A system is said to be coherent if the product or quotient of any two unit quantities in the system is the unit of the resultant quantity. For example, in a coherent system in which the foot is the unit of length, the square foot is the unit of area, whereas the acre is not.

Other physical quantities are derived from the base units. For example, the unit of velocity is the meter per second (m/s), which is a combination of the base units of length and time. The unit of acceleration is the meter per second squared (m/s$^2$). By applying Newton's second law of motion — force is proportional to mass multiplied by acceleration — the unit of force is obtained which is the kilogram meter per second squared (kgm/s$^2$). This unit is known as the newton or N. Work, or force times distance is the kilogram meter squared per second squared (kgm$^2$/s$^2$), which is the joule, (1 joule = 1 newton-meter), and energy is also expressed in these terms. The abbreviation for joule is J. Power or work per unit time is the kilogram meter squared per second cubed (kgm²/s³), which is the watt (1 watt = 1 joule per second = 1 newton-meter per second.) The abbreviation .or watt is W. The term horsepower is not used in the SI and is replaced by the watt which together with multiples and sub-multiples — kilowatt and milliwatt, for example — is the same unit as that used in electrical work.

The use of the newton as the unit of force is of particular interest to engineers. In practical work using the English or traditional metric systems of measurements, it is a common practice to apply weight units as force units. Thus, the unit of force in those systems is that force which when applied to unit mass produces an acceleration $g$, rather than unit acceleration. The value of gravitational acceleration $g$ varies around the earth, and thus the weight of a given mass also varies. In an effort to account for this minor error, the kilogram-force and pound-force were introduced, which are defined as the forces due to "standard gravity" acting on bodies of one kilogram or one pound mass respectively. The standard gravitational acceleration is taken as 9.80665 meters per second squared or 32.174 feet per second squared. The newton is defined as "that force which when applied to a body having a mass of one kilogram, gives it an acceleration of one meter per second squared." It is independent of $g$. As a result, the factor $g$ disappears from a wide range of formulas in dynamics. However, in some formulas in statics, where the weight of a body is important rather than its mass, $g$ does appear where it was formerly absent (the weight of a mass of $W$ kilograms is equal to a force of $Wg$ newtons, where $g$ = approximately 9.81 meter per second squared). Details concerning the use of SI units in mechanics calculations are given on page 285, and throughout the Mechanics section in this Handbook. The use of SI units in strength of materials calculations is covered in the section on that subject.

Decimal multiples and sub-multiples of the SI units are formed by means of the prefixes given in the following table, which represent the numerical factors shown.

**Factors and Prefixes for Forming Decimal Multiples and Sub-multiples of the SI Units**

| Factor by which the unit is multiplied | Prefix | Symbol | Factor by which the unit is multiplied | Prefix | Symbol |
|---|---|---|---|---|---|
| $10^{12}$ | tera | T | $10^{-2}$ | centi | c |
| $10^{9}$ | giga | G | $10^{-3}$ | milli | m |
| $10^{6}$ | mega | M | $10^{-6}$ | micro | $\mu$ |
| $10^{3}$ | kilo | k | $10^{-9}$ | nano | n |
| $10^{2}$ | hecto | h | $10^{-12}$ | pico | p |
| 10 | **deka** | da | $10^{-15}$ | femto | f |
| $10^{-1}$ | deci | d | $10^{-18}$ | atto | a |

**Standard of Length.** — In 1866 the United States, by act of Congress, passed a law making legal the meter, the only measure of length that has been legalized by the United States Government. The United States yard is defined by the relation: 1 yard = $\frac{3600}{3937}$ meter. The legal equivalent of the meter for commercial purposes was fixed as 39.37 inches, by law, in July, 1866, and experience having shown that this value was exact within the error of observation, the United States Office of Standard Weights and Measures was, in 1893, authorized to derive the yard from the meter by the use of this relation. The United States prototype meters Nos. 27 and 21 were received from the International Bureau of Weights and Measures in 1889. Meter No. 27, sealed in its metal case, is preserved in a fire-proof vault at the Bureau of Standards.

Comparisons made prior to 1893 indicated that the relation of the yard to the meter, fixed by the Act of 1866, was by chance the exact relation between the international meter and the British imperial yard, within the error of observation. A

Table 1. International System (SI) Units

| PHYSICAL QUANTITY | NAME OF UNIT | UNIT SYMBOL | DEFINITION |
|---|---|---|---|
| Basic SI Units | | | |
| Length | metre | m | 1,650,763.73 wavelengths in vacuo of the radiation corresponding to the transition between the energy levels 2p10 and 5d5 of the krypton — 86 atom. |
| Mass | kilogram | kg | Mass of the international prototype which is in the custody of the Bureau International des Poids et Mesures (BIPM) at Sèvres, near Paris. |
| Time | second | s | The duration of 9,192,631,770 periods of the radiation corresponding to the transition between the two hyperfine levels of the ground state of the cesium-133 atom. |
| Electric Current | ampere | A | The constant current which, if maintained in two parallel rectilinear conductors of infinite length, of negligible circular cross section, and placed at a distance of one metre apart in a vacuum, would produce between these conductors a force equal to $2 \times 10^{-7}$ N/m length. |
| Thermodynamic Temperature | degree kelvin | K | The fraction $1/273.16$ of the thermodynamic temperature of the triple point of water. |
| Amount of Substance | mole | mol | The amount of substance of a system which contains as many elementary entities as there are atoms in 0.012 kilogram of carbon 12. |
| Luminous Intensity | candela | cd | The luminous intensity, in the perpendicular direction, of a surface of $1/600,000$ square metre of a black body at the temperature of freezing platinum under a pressure of 101,325 newtons per square metre. |
| SI Units Having Special Names | | | |
| Force | newton | $N = kg \cdot m/s^2$ | That force which, when applied to a body having a mass of one kilogramme, gives it an acceleration of one metre per second squared. |
| Work, Energy, Quantity of Heat | joule | $J = N \cdot m$ | The work done when the point of application of a force of one newton is displaced through a distance of one metre in the direction of the force. |
| Power | watt | $W = J/s$ | One joule per second. |
| Electric Charge | coulomb | $C = A \cdot s$ | The quantity of electricity transported in one second by a current of one ampere. |
| Electric Potential | volt | $V = W/A$ | The difference of potential between two points of a conducting wire carrying a constant current of one ampere, when the power dissipated between these points is equal to one watt. |
| Electric Capacitance | farad | $F = C/V$ | The capacitance of a capacitor between the plates of which there appears a difference of potential of one volt when it is charged by a quantity of electricity equal to one coulomb. |

Table 1 *(Continued)*.  International System (SI) Units

| PHYSICAL QUANTITY | NAME OF UNIT | UNIT SYMBOL | DEFINITION |
|---|---|---|---|
| | | SI Units Having Special Names | |
| Electric Resistance | ohm | $\Omega = V/A$ | The resistance between two points of a conductor when a constant difference of potential of one volt, applied between these two points, produces in this conductor a current of one ampere, this conductor not being the source of any electromotive force. |
| Magnetic Flux | weber | $Wb = V \, s$ | The flux which, linking a circuit of one turn produces in it an electromotive force of one volt as it is reduced to zero at a uniform rate in one second. |
| Inductance | henry | $H = V \, s/A$ | The inductance of a closed circuit in which an electromotive force of one volt is produced when the electric current in the circuit varies uniformly at the rate of one ampere per second. |
| Luminous Flux | lumen | $lm = cd \, sr$ | The flux emitted within a unit solid angle of one steradian by a point source having a uniform intensity of one candela. |
| Illumination | lux | $lx = lm/m^2$ | An illumination of one lumen per square metre. |

Table 2.  International System (SI) Units with Complex Names

| PHYSICAL QUANTITY | SI UNIT | UNIT SYMBOL |
|---|---|---|
| | SI Units Having Complex Names | |
| Area | square metre | $m^2$ |
| Volume | cubic metre | $m^3$ |
| Frequency | hertz* | Hz |
| Density (Mass Density) | kilogram per cubic metre | $kg/m^3$ |
| Velocity | metre per second | $m/s$ |
| Angular Velocity | radian per second | $rad/s$ |
| Acceleration | metre per second squared | $m/s^2$ |
| Angular Acceleration | radian per second squared | $rad/s^2$ |
| Pressure | pascal‡ | Pa |
| Surface Tension | newton per metre | $N/m$ |
| Dynamic Viscosity | newton second per metre squared | $N \, s/m^2$ |
| Kinematic Viscosity } Diffusion Coefficient} | metre squared per second | $m^2/s$ |
| Thermal Conductivity | watt per metre degree Kelvin | $W/(m \, °K)$ |
| Electric Field Strength | volt per metre | $V/m$ |
| Magnetic Flux Density | tesla† | T |
| Magnetic Field Strength | ampere per metre | $A/m$ |
| Luminance | candela per square metre | $cd/m^2$ |

* Hz = cycle/second.
† T = weber/metre².
‡ Pa = newton/metre².

subsequent comparison made between the standards just mentioned indicates that
the legal relation adopted by Congress is in error 0.0001 inch; but, in view of the
fact that certain comparisons made by the English Standards Office between the
imperial yard and its authentic copies show variations as great if not greater than
this, it cannot be said with certainty that there is a difference between the imperial
yard of Great Britain and the United States yard derived from the meter.   The
bronze yard No. 11, which was an exact copy of the British imperial yard both in
form and material, had shown changes when compared with the imperial yard in
1876 and 1888, which could not reasonably be said to be entirely due to changes in
Bronze No. 11.   On the other hand, the new meters represented the most advanced
ideas of standards, and it therefore seemed that greater stability as well as higher
accuracy would be secured by accepting the international meter as a fundamental
standard of length.

### Traditional Metric Measures

#### Measures of Length

| | |
|---|---|
| 10 millimeters (mm) | = 1 centimeter (cm). |
| 10 centimeters | = 1 decimeter (dm). |
| 10 decimeters | = 1 meter (m). |
| 1000 meters | = 1 kilometer (km). |

#### Square Measure

| | |
|---|---|
| 100 square millimeters (mm²) | = 1 square centimeter (cm²). |
| 100 square centimeters | = 1 square decimeter (dm²). |
| 100 square decimeters | = 1 square meter (m²). |

#### Surveyor's Square Measure

| | |
|---|---|
| 100 square meters (m²) | = 1 are (a). |
| 100 ares | = 1 hectare (ha). |
| 100 hectares | = 1 square kilometer (km²). |

#### Cubic Measure

| | |
|---|---|
| 1000 cubic millimeters (mm³) | = 1 cubic centimeter (cm³). |
| 1000 cubic centimeters | = 1 cubic decimeter (dm³). |
| 1000 cubic decimeters | = 1 cubic meter (m³). |

#### Dry and Liquid Measure

| | |
|---|---|
| 10 milliliters (ml) | = 1 centiliter (cl). |
| 10 centiliters | = 1 deciliter (dl). |
| 10 deciliters | = 1 liter (l). |
| 100 liters | = 1 hectoliter (hl). |

1 liter = 1 cubic decimeter = the volume of 1 kilogram of pure water at a temperature of 39.2 degrees F.

#### Measures of Weight

| | |
|---|---|
| 10 milligrams (mg) | = 1 centigram (cg). |
| 10 centigrams | = 1 decigram (dg). |
| 10 decigrams | = 1 gram (g). |
| 10 grams | = 1 dekagram (dag). |
| 10 dekagrams | = 1 hectogram (hg). |
| 10 hectograms | = 1 kilogram (kg). |
| 1000 kilograms | = 1 (metric) ton (t). |

## Metric and English Conversion Table

### Linear Measure

1 kilometer = 0.6214 mile.

$$1 \text{ meter} = \begin{cases} 39.37 \text{ inches.} \\ 3.2808 \text{ feet.} \\ 1.0936 \text{ yards.} \end{cases}$$

1 centimeter = 0.3937 inch.

1 millimeter = 0.03937 inch.

1 mile = 1.609 kilometers.

1 yard = 0.9144 meter.

1 foot = 0.3048 meter.

1 foot = 304.8 millimeters.

1 inch = 2.54 centimeters.

1 inch = 25.4 millimeters.

### Square Measure

1 square kilometer = 0.3861 square mile = 247.1 acres.

1 hectare = 2.471 acres = 107,639 square feet.

1 are = 0.0247 acre = 1076.4 square feet.

1 square meter = 10.764 square feet = 1.196 square yards.

1 square centimeter = 0.155 square inch.

1 square millimeter = 0.00155 square inch.

1 square mile = 2.5899 square kilometers.

1 acre = 0.4047 hectare = 40.47 ares.

1 square yard = 0.836 square meter.

1 square foot = 0.0929 square meter = 929 square centimeters.

1 square inch = 6.452 square centimeters = 645.2 square millimeters.

### Cubic Measure

1 cubic meter = 35.315 cubic feet = 1.308 cubic yards.

1 cubic meter = 264.2 U. S. gallons.

1 cubic centimeter = 0.061 cubic inch.

1 liter (cubic decimeter) = 0.0353 cubic foot = 61.023 cubic inches.

1 liter = 0.2642 U. S. gallon = 1.0567 U. S. quarts.

1 cubic yard = 0.7646 cubic meter.

1 cubic foot = 0.02832 cubic meter = 28.317 liters.

1 cubic inch = 16.38706 cubic centimeters.

1 U. S. gallon = 3.785 liters.

1 U. S. quart = 0.946 liter.

### Weight

1 metric ton = 0.9842 ton (of 2240 pounds) = 2204.6 pounds.

1 kilogram = 2.2046 pounds = 35.274 ounces avoirdupois.

1 gram = 0.03215 ounce troy = 0.03527 ounce avoirdupois.

1 gram = 15.432 grains.

1 ton (of 2240 pounds) = 1.016 metric ton = 1016 kilograms.

1 pound = 0.4536 kilogram = 453.6 grams.

1 ounce avoirdupois = 28.35 grams.

1 ounce troy = 31.103 grams.

1 grain = 0.0648 gram.

1 kilogram per square millimeter = 1422.32 pounds per square inch

1 kilogram per square centimeter = 14.223 pounds per square inch.

1 kilogram-meter = 7.233 foot-pounds.

1 pound per square inch = 0.0703 kilogram per square centimeter.

1 calorie (kilogram calorie) = 3.968 B.T.U. (British thermal unit).

**Use of Conversion Tables.** — On this and following pages tables are given which permit conversion from English to metric units and vice versa over a wide range of values. Where the desired value cannot be obtained directly from these tables, a simple addition of two or more values taken directly from the table will suffice as shown in the following examples:

*Example* 1: Find the millimetre equivalent of 0.4476 inch.

$$
\begin{array}{rl}
.4 \quad \text{in.} & = 10.16000 \text{ mm} \\
.04 \quad \text{in.} & = 1.01600 \text{ mm} \\
.007 \quad \text{in.} & = .17780 \text{ mm} \\
.0006 \text{ in.} & = .01524 \text{ mm} \\
\hline
.4476 \text{ in.} & = 11.36904 \text{ mm}
\end{array}
$$

*Example* 2: Find the inch equivalent of 84.9 mm.

$$
\begin{array}{rl}
80. \quad \text{mm} & = 3.14961 \text{ in.} \\
4. \quad \text{mm} & = 0.15748 \text{ in.} \\
0.9 \text{ mm} & = 0.03543 \text{ in.} \\
\hline
84.9 \text{ mm} & = 3.34252 \text{ in.}
\end{array}
$$

### Inch—Millimetre and Inch—Centimetre Conversion Table*
(Based on 1 inch = 25.4 millimetres, exactly)

| INCHES TO MILLIMETRES | | | | | | | | | | | |
|---|---|---|---|---|---|---|---|---|---|---|---|
| in. | mm | in. | mm | in. | mm | in. | mm | in. | mm | in. | mm |
| 10 | 254.00000 | 1 | 25.40000 | .1 | 2.54000 | .01 | .25400 | .001 | .02540 | .0001 | .00254 |
| 20 | 508.00000 | 2 | 50.80000 | .2 | 5.08000 | .02 | .50800 | .002 | .05080 | .0002 | .00508 |
| 30 | 762.00000 | 3 | 76.20000 | .3 | 7.62000 | .03 | .76200 | .003 | .07620 | .0003 | .00762 |
| 40 | 1,016.00000 | 4 | 101.60000 | .4 | 10.16000 | .04 | 1.01600 | .004 | .10160 | .0004 | .01016 |
| 50 | 1,270.00000 | 5 | 127.00000 | .5 | 12.70000 | .05 | 1.27000 | .005 | .12700 | .0005 | .01270 |
| 60 | 1,524.00000 | 6 | 152.40000 | .6 | 15.24000 | .06 | 1.52400 | .006 | .15240 | .0006 | .01524 |
| 70 | 1,778.00000 | 7 | 177.80000 | .7 | 17.78000 | .07 | 1.77800 | .007 | .17780 | .0007 | .01778 |
| 80 | 2,032.00000 | 8 | 203.20000 | .8 | 20.32000 | .08 | 2.03200 | .008 | .20320 | .0008 | .02032 |
| 90 | 2,286.00000 | 9 | 228.60000 | .9 | 22.86000 | .09 | 2.28600 | .009 | .22860 | .0009 | .02286 |
| 100 | 2,540.00000 | 10 | 254.00000 | 1.0 | 25.40000 | .10 | 2.54000 | .010 | .25400 | .0010 | .02540 |

| MILLIMETRES TO INCHES | | | | | | | | | | | |
|---|---|---|---|---|---|---|---|---|---|---|---|
| mm | in. | mm | in. | mm. | in. | mm | in. | mm | in. | mm | in. |
| 100 | 3.93701 | 10 | .39370 | 1 | .03937 | .1 | .00394 | .01 | .00039 | .001 | .00004 |
| 200 | 7.87402 | 20 | .78740 | 2 | .07874 | .2 | .00787 | .02 | .00079 | .002 | .00008 |
| 300 | 11.81102 | 30 | 1.18110 | 3 | .11811 | .3 | .01181 | .03 | .00118 | .003 | .00012 |
| 400 | 15.74803 | 40 | 1.57480 | 4 | .15748 | .4 | .01575 | .04 | .00157 | .004 | .00016 |
| 500 | 19.68504 | 50 | 1.96850 | 5 | .19685 | .5 | .01969 | .05 | .00197 | .005 | .00020 |
| 600 | 23.62205 | 60 | 2.36220 | 6 | .23622 | .6 | .02362 | .06 | .00236 | .006 | .00024 |
| 700 | 27.55906 | 70 | 2.75591 | 7 | .27559 | .7 | .02756 | .07 | .00276 | .007 | .00028 |
| 800 | 31.49606 | 80 | 3.14961 | 8 | .31496 | .8 | .03150 | .08 | .00315 | .008 | .00031 |
| 900 | 35.43307 | 90 | 3.54331 | 9 | .35433 | .9 | .03543 | .09 | .00354 | .009 | .00035 |
| 1,000 | 39.37008 | 100 | 3.93701 | 10 | .39370 | 1.0 | .03937 | .10 | .00394 | .010 | .00039 |

\* For inches to centimetres, shift decimal point in mm column one place to left and read centimetres, thus:

$$40 \text{ in.} = 1016 \text{ mm} = 101.6 \text{ cm}$$

For centimetres to inches, shift decimal point of centimetre value one place to right and enter mm column, thus:

$$70 \text{ cm} = 700 \text{ mm} = 27.55906 \text{ inches}$$

### Decimals of an Inch to Millimeters
(Based on 1 inch = 25.4 millimeters, exactly)

| Inches | 0.000 | 0.001 | 0.002 | 0.003 | 0.004 | 0.005 | 0.006 | 0.007 | 0.008 | 0.009 |
|---|---|---|---|---|---|---|---|---|---|---|
| | | | | | Millimeters | | | | | |
| 0.000 | ... | 0.0254 | 0.0508 | 0.0762 | 0.1016 | 0.1270 | 0.1524 | 0.1778 | 0.2032 | 0.2286 |
| 0.010 | 0.2540 | 0.2794 | 0.3048 | 0.3302 | 0.3556 | 0.3810 | 0.4064 | 0.4318 | 0.4572 | 0.4826 |
| 0.020 | 0.5080 | 0.5334 | 0.5588 | 0.5842 | 0.6096 | 0.6350 | 0.6604 | 0.6858 | 0.7112 | 0.7366 |
| 0.030 | 0.7620 | 0.7874 | 0.8128 | 0.8382 | 0.8636 | 0.8890 | 0.9144 | 0.9398 | 0.9652 | 0.9906 |
| 0.040 | 1.0160 | 1.0414 | 1.0668 | 1.0922 | 1.1176 | 1.1430 | 1.1684 | 1.1938 | 1.2192 | 1.2446 |
| 0.050 | 1.2700 | 1.2954 | 1.3208 | 1.3462 | 1.3716 | 1.3970 | 1.4224 | 1.4478 | 1.4732 | 1.4986 |
| 0.060 | 1.5240 | 1.5494 | 1.5748 | 1.6002 | 1.6256 | 1.6510 | 1.6764 | 1.7018 | 1.7272 | 1.7526 |
| 0.070 | 1.7780 | 1.8034 | 1.8288 | 1.8542 | 1.8796 | 1.9050 | 1.9304 | 1.9558 | 1.9812 | 2.0066 |
| 0.080 | 2.0320 | 2.0574 | 2.0828 | 2.1082 | 2.1336 | 2.1590 | 2.1844 | 2.2098 | 2.2352 | 2.2606 |
| 0.090 | 2.2860 | 2.3114 | 2.3368 | 2.3622 | 2.3876 | 2.4130 | 2.4384 | 2.4638 | 2.4892 | 2.5146 |
| 0.100 | 2.5400 | 2.5654 | 2.5908 | 2.6162 | 2.6416 | 2.6670 | 2.6924 | 2.7178 | 2.7432 | 2.7686 |
| 0.110 | 2.7940 | 2.8194 | 2.8448 | 2.8702 | 2.8956 | 2.9210 | 2.9464 | 2.9718 | 2.9972 | 3.0226 |
| 0.120 | 3.0480 | 3.0734 | 3.0988 | 3.1242 | 3.1496 | 3.1750 | 3.2004 | 3.2258 | 3.2512 | 3.2766 |
| 0.130 | 3.3020 | 3.3274 | 3.3528 | 3.3782 | 3.4036 | 3.4290 | 3.4544 | 3.4798 | 3.5052 | 3.5306 |
| 0.140 | 3.5560 | 3.5814 | 3.6068 | 3.6322 | 3.6576 | 3.6830 | 3.7084 | 3.7338 | 3.7592 | 3.7846 |
| 0.150 | 3.8100 | 3.8354 | 3.8608 | 3.8862 | 3.9116 | 3.9370 | 3.9624 | 3.9878 | 4.0132 | 4.0386 |
| 0.160 | 4.0640 | 4.0894 | 4.1148 | 4.1402 | 4.1656 | 4.1910 | 4.2164 | 4.2418 | 4.2672 | 4.2926 |
| 0.170 | 4.3180 | 4.3434 | 4.3688 | 4.3942 | 4.4196 | 4.4450 | 4.4704 | 4.4958 | 4.5212 | 4.5466 |
| 0.180 | 4.5720 | 4.5974 | 4.6228 | 4.6482 | 4.6736 | 4.6990 | 4.7244 | 4.7498 | 4.7752 | 4.8006 |
| 0.190 | 4.8260 | 4.8514 | 4.8768 | 4.9022 | 4.9276 | 4.9530 | 4.9784 | 5.0038 | 5.0292 | 5.0546 |
| 0.200 | 5.0800 | 5.1054 | 5.1308 | 5.1562 | 5.1816 | 5.2070 | 5.2324 | 5.2578 | 5.2832 | 5.3086 |
| 0.210 | 5.3340 | 5.3594 | 5.3848 | 5.4102 | 5.4356 | 5.4610 | 5.4864 | 5.5118 | 5.5372 | 5.5626 |
| 0.220 | 5.5880 | 5.6134 | 5.6388 | 5.6642 | 5.6896 | 5.7150 | 5.7404 | 5.7658 | 5.7912 | 5.8166 |
| 0.230 | 5.8420 | 5.8674 | 5.8928 | 5.9182 | 5.9436 | 5.9690 | 5.9944 | 6.0198 | 6.0452 | 6.0706 |
| 0.240 | 6.0960 | 6.1214 | 6.1468 | 6.1722 | 6.1976 | 6.2230 | 6.2484 | 6.2738 | 6.2992 | 6.3246 |
| 0.250 | 6.3500 | 6.3754 | 6.4008 | 6.4262 | 6.4516 | 6.4770 | 6.5024 | 6.5278 | 6.5532 | 6.5786 |
| 0.260 | 6.6040 | 6.6294 | 6.6548 | 6.6802 | 6.7056 | 6.7310 | 6.7564 | 6.7818 | 6.8072 | 6.8326 |
| 0.270 | 6.8580 | 6.8834 | 6.9088 | 6.9342 | 6.9596 | 6.9850 | 7.0104 | 7.0358 | 7.0612 | 7.0866 |
| 0.280 | 7.1120 | 7.1374 | 7.1628 | 7.1882 | 7.2136 | 7.2390 | 7.2644 | 7.2898 | 7.3152 | 7.3406 |
| 0.290 | 7.3660 | 7.3914 | 7.4168 | 7.4422 | 7.4676 | 7.4930 | 7.5184 | 7.5438 | 7.5692 | 7.5946 |
| 0.300 | 7.6200 | 7.6454 | 7.6708 | 7.6962 | 7.7216 | 7.7470 | 7.7724 | 7.7978 | 7.8232 | 7.8486 |
| 0.310 | 7.8740 | 7.8994 | 7.9248 | 7.9502 | 7.9756 | 8.0010 | 8.0264 | 8.0518 | 8.0772 | 8.1026 |
| 0.320 | 8.1280 | 8.1534 | 8.1788 | 8.2042 | 8.2296 | 8.2550 | 8.2804 | 8.3058 | 8.3312 | 8.3566 |
| 0.330 | 8.3820 | 8.4074 | 8.4328 | 8.4582 | 8.4836 | 8.5090 | 8.5344 | 8.5598 | 8.5852 | 8.6106 |
| 0.340 | 8.6360 | 8.6614 | 8.6868 | 8.7122 | 8.7376 | 8.7630 | 8.7884 | 8.8138 | 8.8392 | 8.8646 |
| 0.350 | 8.8900 | 8.9154 | 8.9408 | 8.9662 | 8.9916 | 9.0170 | 9.0424 | 9.0678 | 9.0932 | 9.1186 |
| 0.360 | 9.1440 | 9.1694 | 9.1948 | 9.2202 | 9.2456 | 9.2710 | 9.2964 | 9.3218 | 9.3472 | 9.3726 |
| 0.370 | 9.3980 | 9.4234 | 9.4488 | 9.4742 | 9.4996 | 9.5250 | 9.5504 | 9.5758 | 9.6012 | 9.6266 |
| 0.380 | 9.6520 | 9.6774 | 9.7028 | 9.7282 | 9.7536 | 9.7790 | 9.8044 | 9.8298 | 9.8552 | 9.8806 |
| 0.390 | 9.9060 | 9.9314 | 9.9568 | 9.9822 | 10.0076 | 10.0330 | 10.0584 | 10.0838 | 10.1092 | 10.1346 |
| 0.400 | 10.1600 | 10.1854 | 10.2108 | 10.2362 | 10.2616 | 10.2870 | 10.3124 | 10.3378 | 10.3632 | 10.3886 |
| 0.410 | 10.4140 | 10.4394 | 10.4648 | 10.4902 | 10.5156 | 10.5410 | 10.5664 | 10.5918 | 10.6172 | 10.6426 |
| 0.420 | 10.6680 | 10.6934 | 10.7188 | 10.7442 | 10.7696 | 10.7950 | 10.8204 | 10.8458 | 10.8712 | 10.8966 |
| 0.430 | 10.9220 | 10.9474 | 10.9728 | 10.9982 | 11.0236 | 11.0490 | 11.0744 | 11.0998 | 11.1252 | 11.1506 |
| 0.440 | 11.1760 | 11.2014 | 11.2268 | 11.2522 | 11.2776 | 11.3030 | 11.3284 | 11.3538 | 11.3792 | 11.4046 |
| 0.450 | 11.4300 | 11.4554 | 11.4808 | 11.5062 | 11.5316 | 11.5570 | 11.5824 | 11.6078 | 11.6332 | 11.6586 |
| 0.460 | 11.6840 | 11.7094 | 11.7348 | 11.7602 | 11.7856 | 11.8110 | 11.8364 | 11.8618 | 11.8872 | 11.9126 |
| 0.470 | 11.9380 | 11.9634 | 11.9888 | 12.0142 | 12.0396 | 12.0650 | 12.0904 | 12.1158 | 12.1412 | 12.1666 |
| 0.480 | 12.1920 | 12.2174 | 12.2428 | 12.2682 | 12.2936 | 12.3190 | 12.3444 | 12.3698 | 12.3952 | 12.4206 |
| 0.490 | 12.4460 | 12.4714 | 12.4968 | 12.5222 | 12.5476 | 12.5730 | 12.5984 | 12.6238 | 12.6492 | 12.6746 |

Use previous table to obtain whole inch equivalents to add to decimal equivalents above. All values given in this table are exact; figures to the right of the last place figures are all zeros.

## Decimals of an Inch to Millimeters
(Based on 1 inch = 25.4 millimeters, exactly)

| Inches | 0.000 | 0.001 | 0.002 | 0.003 | 0.004 | 0.005 | 0.006 | 0.007 | 0.008 | 0.009 |
|---|---|---|---|---|---|---|---|---|---|---|
| | Millimeters | | | | | | | | | |
| 0.500 | 12.7000 | 12.7254 | 12.7508 | 12.7762 | 12.8016 | 12.8270 | 12.8524 | 12.8778 | 12.9032 | 12.9286 |
| 0.510 | 12.9540 | 12.9794 | 13.0048 | 13.0302 | 13.0556 | 13.0810 | 13.1064 | 13.1318 | 13.1572 | 13.1826 |
| 0.520 | 13.2080 | 13.2334 | 13.2588 | 13.2842 | 13.3096 | 13.3350 | 13.3604 | 13.3858 | 13.4112 | 13.4366 |
| 0.530 | 13.4620 | 13.4874 | 13.5128 | 13.5382 | 13.5636 | 13.5890 | 13.6144 | 13.6398 | 13.6652 | 13.6906 |
| 0.540 | 13.7160 | 13.7414 | 13.7668 | 13.7922 | 13.8176 | 13.8430 | 13.8684 | 13.8938 | 13.9192 | 13.9446 |
| 0.550 | 13.9700 | 13.9954 | 14.0208 | 14.0462 | 14.0716 | 14.0970 | 14.1224 | 14.1478 | 14.1732 | 14.1986 |
| 0.560 | 14.2240 | 14.2494 | 14.2748 | 14.3002 | 14.3256 | 14.3510 | 14.3764 | 14.4018 | 14.4272 | 14.4526 |
| 0.570 | 14.4780 | 14.5034 | 14.5288 | 14.5542 | 14.5796 | 14.6050 | 14.6304 | 14.6558 | 14.6812 | 14.7066 |
| 0.580 | 14.7320 | 14.7574 | 14.7828 | 14.8082 | 14.8336 | 14.8590 | 14.8844 | 14.9098 | 14.9352 | 14.9606 |
| 0.590 | 14.9860 | 15.0114 | 15.0368 | 15.0622 | 15.0876 | 15.1130 | 15.1384 | 15.1638 | 15.1892 | 15.2146 |
| 0.600 | 15.2400 | 15.2654 | 15.2908 | 15.3162 | 15.3416 | 15.3670 | 15.3924 | 15.4178 | 15.4432 | 15.4686 |
| 0.610 | 15.4940 | 15.5194 | 15.5448 | 15.5702 | 15.5956 | 15.6210 | 15.6464 | 15.6718 | 15.6972 | 15.7226 |
| 0.620 | 15.7480 | 15.7734 | 15.7988 | 15.8242 | 15.8496 | 15.8750 | 15.9004 | 15.9258 | 15.9512 | 15.9766 |
| 0.630 | 16.0020 | 16.0274 | 16.0528 | 16.0782 | 16.1036 | 16.1290 | 16.1544 | 16.1798 | 16.2052 | 16.2306 |
| 0.640 | 16.2560 | 16.2814 | 16.3068 | 16.3322 | 16.3576 | 16.3830 | 16.4084 | 16.4338 | 16.4592 | 16.4846 |
| 0.650 | 16.5100 | 16.5354 | 16.5608 | 16.5862 | 16.6116 | 16.6370 | 16.6624 | 16.6878 | 16.7132 | 16.7386 |
| 0.660 | 16.7640 | 16.7894 | 16.8148 | 16.8402 | 16.8656 | 16.8910 | 16.9164 | 16.9418 | 16.9672 | 16.9926 |
| 0.670 | 17.0180 | 17.0434 | 17.0688 | 17.0942 | 17.1196 | 17.1450 | 17.1704 | 17.1958 | 17.2212 | 17.2466 |
| 0.680 | 17.2720 | 17.2974 | 17.3228 | 17.3482 | 17.3736 | 17.3990 | 17.4244 | 17.4498 | 17.4752 | 17.5006 |
| 0.690 | 17.5260 | 17.5514 | 17.5768 | 17.6022 | 17.6276 | 17.6530 | 17.6784 | 17.7038 | 17.7292 | 17.7546 |
| 0.700 | 17.7800 | 17.8054 | 17.8308 | 17.8562 | 17.8816 | 17.9070 | 17.9324 | 17.9578 | 17.9832 | 18.0086 |
| 0.710 | 18.0340 | 18.0594 | 18.0848 | 18.1102 | 18.1356 | 18.1610 | 18.1864 | 18.2118 | 18.2372 | 18.2626 |
| 0.720 | 18.2880 | 18.3134 | 18.3388 | 18.3642 | 18.3896 | 18.4150 | 18.4404 | 18.4658 | 18.4912 | 18.5166 |
| 0.730 | 18.5420 | 18.5674 | 18.5928 | 18.6182 | 18.6436 | 18.6690 | 18.6944 | 18.7198 | 18.7452 | 18.7706 |
| 0.740 | 18.7960 | 18.8214 | 18.8468 | 18.8722 | 18.8976 | 18.9230 | 18.9484 | 18.9738 | 18.9992 | 19.0246 |
| 0.750 | 19.0500 | 19.0754 | 19.1008 | 19.1262 | 19.1516 | 19.1770 | 19.2024 | 19.2278 | 19.2532 | 19.2786 |
| 0.760 | 19.3040 | 19.3294 | 19.3548 | 19.3802 | 19.4056 | 19.4310 | 19.4564 | 19.4818 | 19.5072 | 19.5326 |
| 0.770 | 19.5580 | 19.5834 | 19.6088 | 19.6342 | 19.6596 | 19.6850 | 19.7104 | 19.7358 | 19.7612 | 19.7866 |
| 0.780 | 19.8120 | 19.8374 | 19.8628 | 19.8882 | 19.9136 | 19.9390 | 19.9644 | 19.9898 | 20.0152 | 20.0406 |
| 0.790 | 20.0660 | 20.0914 | 20.1168 | 20.1422 | 20.1676 | 20.1930 | 20.2184 | 20.2438 | 20.2692 | 20.2946 |
| 0.800 | 20.3200 | 20.3454 | 20.3708 | 20.3962 | 20.4216 | 20.4470 | 20.4724 | 20.4978 | 20.5232 | 20.5486 |
| 0.810 | 20.5740 | 20.5994 | 20.6248 | 20.6502 | 20.6756 | 20.7010 | 20.7264 | 20.7518 | 20.7772 | 20.8026 |
| 0.820 | 20.8280 | 20.8534 | 20.8788 | 20.9042 | 20.9296 | 20.9550 | 20.9804 | 21.0058 | 21.0312 | 21.0566 |
| 0.830 | 21.0820 | 21.1074 | 21.1328 | 21.1582 | 21.1836 | 21.2090 | 21.2344 | 21.2598 | 21.2852 | 21.3106 |
| 0.840 | 21.3360 | 21.3614 | 21.3868 | 21.4122 | 21.4376 | 21.4630 | 21.4884 | 21.5138 | 21.5392 | 21.5646 |
| 0.850 | 21.5900 | 21.6154 | 21.6408 | 21.6662 | 21.6916 | 21.7170 | 21.7424 | 21.7678 | 21.7932 | 21.8186 |
| 0.860 | 21.8440 | 21.8694 | 21.8948 | 21.9202 | 21.9456 | 21.9710 | 21.9964 | 22.0218 | 22.0472 | 22.0726 |
| 0.870 | 22.0980 | 22.1234 | 22.1488 | 22.1742 | 22.1996 | 22.2250 | 22.2504 | 22.2758 | 22.3012 | 22.3266 |
| 0.880 | 22.3520 | 22.3774 | 22.4028 | 22.4282 | 22.4536 | 22.4790 | 22.5044 | 22.5298 | 22.5552 | 22.5806 |
| 0.890 | 22.6060 | 22.6314 | 22.6568 | 22.6822 | 22.7076 | 22.7330 | 22.7584 | 22.7838 | 22.8092 | 22.8346 |
| 0.900 | 22.8600 | 22.8854 | 22.9108 | 22.9362 | 22.9616 | 22.9870 | 23.0124 | 23.0378 | 23.0632 | 23.0886 |
| 0.910 | 23.1140 | 23.1394 | 23.1648 | 23.1902 | 23.2156 | 23.2410 | 23.2664 | 23.2918 | 23.3172 | 23.3426 |
| 0.920 | 23.3680 | 23.3934 | 23.4188 | 23.4442 | 23.4696 | 23.4950 | 23.5204 | 23.5458 | 23.5712 | 23.5966 |
| 0.930 | 23.6220 | 23.6474 | 23.6728 | 23.6982 | 23.7236 | 23.7490 | 23.7744 | 23.7998 | 23.8252 | 23.8506 |
| 0.940 | 23.8760 | 23.9014 | 23.9268 | 23.9522 | 23.9776 | 24.0030 | 24.0284 | 24.0538 | 24.0792 | 24.1046 |
| 0.950 | 24.1300 | 24.1554 | 24.1808 | 24.2062 | 24.2316 | 24.2570 | 24.2824 | 24.3078 | 24.3332 | 24.3586 |
| 0.960 | 24.3840 | 24.4094 | 24.4348 | 24.4602 | 24.4856 | 24.5110 | 24.5364 | 24.5618 | 24.5872 | 24.6126 |
| 0.970 | 24.6380 | 24.6634 | 24.6888 | 24.7142 | 24.7396 | 24.7650 | 24.7904 | 24.8158 | 24.8412 | 24.8666 |
| 0.980 | 24.8920 | 24.9174 | 24.9428 | 24.9682 | 24.9936 | 25.0190 | 25.0444 | 25.0698 | 25.0952 | 25.1206 |
| 0.990 | 25.1460 | 25.1714 | 25.1968 | 25.2222 | 25.2476 | 25.2730 | 25.2984 | 25.3238 | 25.3492 | 25.3746 |
| 1.000 | 25.4000 | .. | .. | ... | ... | ... | ... | ... | ... | ... |

Use previous table to obtain whole inch equivalents to add to decimal equivalents above. All values given in this table are exact; figures to the right of the last place figures are all zeros.

### Millimeters to Inches
(Based on 1 inch = 25.4 millimeters, exactly)

| Milli-meters | 0 | 1 | 2 | 3 | 4 | 5 | 6 | 7 | 8 | 9 |
|---|---|---|---|---|---|---|---|---|---|---|
| | | | | | | Inches | | | | |
| 0 | ... | 0.03937 | 0.07874 | 0.11811 | 0.15748 | 0.19685 | 0.23622 | 0.27559 | 0.31496 | 0.35433 |
| 10 | 0.39370 | 0.43307 | 0.47244 | 0.51181 | 0.55118 | 0.59055 | 0.62992 | 0.66929 | 0.70866 | 0.74803 |
| 20 | 0.78740 | 0.82677 | 0.86614 | 0.90551 | 0.94488 | 0.98425 | 1.02362 | 1.06299 | 1.10236 | 1.14173 |
| 30 | 1.18110 | 1.22047 | 1.25984 | 1.29921 | 1.33858 | 1.37795 | 1.41732 | 1.45669 | 1.49606 | 1.53543 |
| 40 | 1.57480 | 1.61417 | 1.65354 | 1.69291 | 1.73228 | 1.77165 | 1.81102 | 1.85039 | 1.88976 | 1.92913 |
| 50 | 1.96850 | 2.00787 | 2.04724 | 2.08661 | 2.12598 | 2.16535 | 2.20472 | 2.24409 | 2.28346 | 2.32283 |
| 60 | 2.36220 | 2.40157 | 2.44094 | 2.48031 | 2.51969 | 2.55906 | 2.59843 | 2.63780 | 2.67717 | 2.71654 |
| 70 | 2.75591 | 2.79528 | 2.83465 | 2.87402 | 2.91339 | 2.95276 | 2.99213 | 3.03150 | 3.07087 | 3.11024 |
| 80 | 3.14961 | 3.18898 | 3.22835 | 3.26772 | 3.30709 | 3.34646 | 3.38583 | 3.42520 | 3.46457 | 3.50394 |
| 90 | 3.54331 | 3.58268 | 3.62205 | 3.66142 | 3.70079 | 3.74016 | 3.77953 | 3.81890 | 3.85827 | 3.89764 |
| 100 | 3.93701 | 3.97638 | 4.01575 | 4.05512 | 4.09449 | 4.13386 | 4.17323 | 4.21260 | 4.25197 | 4.29134 |
| 110 | 4.33071 | 4.37008 | 4.40945 | 4.44882 | 4.48819 | 4.52756 | 4.56693 | 4.60630 | 4.64567 | 4.68504 |
| 120 | 4.72441 | 4.76378 | 4.80315 | 4.84252 | 4.88189 | 4.92126 | 4.96063 | 5.00000 | 5.03937 | 5.07874 |
| 130 | 5.11811 | 5.15748 | 5.19685 | 5.23622 | 5.27559 | 5.31496 | 5.35433 | 5.39370 | 5.43307 | 5.47244 |
| 140 | 5.51181 | 5.55118 | 5.59055 | 5.62992 | 5.66929 | 5.70866 | 5.74803 | 5.78740 | 5.82677 | 5.86614 |
| 150 | 5.90551 | 5.94488 | 5.98425 | 6.02362 | 6.06299 | 6.10236 | 6.14173 | 6.18110 | 6.22047 | 6.25984 |
| 160 | 6.29921 | 6.33858 | 6.37795 | 6.41732 | 6.45669 | 6.49606 | 6.53543 | 6.57480 | 6.61417 | 6.65354 |
| 170 | 6.69291 | 6.73228 | 6.77165 | 6.81102 | 6.85039 | 6.88976 | 6.92913 | 6.96850 | 7.00787 | 7.04724 |
| 180 | 7.08661 | 7.12598 | 7.16535 | 7.20472 | 7.24409 | 7.28346 | 7.32283 | 7.36220 | 7.40157 | 7.44094 |
| 190 | 7.48031 | 7.51969 | 7.55906 | 7.59843 | 7.63780 | 7.67717 | 7.71654 | 7.75591 | 7.79528 | 7.83465 |
| 200 | 7.87402 | 7.91339 | 7.95276 | 7.99213 | 8.03150 | 8.07087 | 8.11024 | 8.14961 | 8.18898 | 8.22835 |
| 210 | 8.26772 | 8.30709 | 8.34646 | 8.38583 | 8.42520 | 8.46457 | 8.50394 | 8.54331 | 8.58268 | 8.62205 |
| 220 | 8.66142 | 8.70079 | 8.74016 | 8.77953 | 8.81890 | 8.85827 | 8.89764 | 8.93701 | 8.97638 | 9.01575 |
| 230 | 9.05512 | 9.09449 | 9.13386 | 9.17323 | 9.21260 | 9.25197 | 9.29134 | 9.33071 | 9.37008 | 9.40945 |
| 240 | 9.44882 | 9.48819 | 9.52756 | 9.56693 | 9.60630 | 9.64567 | 9.68504 | 9.72441 | 9.76378 | 9.80315 |
| 250 | 9.84252 | 9.88189 | 9.92126 | 9.96063 | 10.0000 | 10.0394 | 10.0787 | 10.1181 | 10.1575 | 10.1969 |
| 260 | 10.2362 | 10.2756 | 10.3150 | 10.3543 | 10.3937 | 10.4331 | 10.4724 | 10.5118 | 10.5512 | 10.5906 |
| 270 | 10.6299 | 10.6693 | 10.7087 | 10.7480 | 10.7874 | 10.8268 | 10.8661 | 10.9055 | 10.9449 | 10.9843 |
| 280 | 11.0236 | 11.0630 | 11.1024 | 11.1417 | 11.1811 | 11.2205 | 11.2598 | 11.2992 | 11.3386 | 11.3780 |
| 290 | 11.4173 | 11.4567 | 11.4961 | 11.5354 | 11.5748 | 11.6142 | 11.6535 | 11.6929 | 11.7323 | 11.7717 |
| 300 | 11.8110 | 11.8504 | 11.8898 | 11.9291 | 11.9685 | 12.0079 | 12.0472 | 12.0866 | 12.1260 | 12.1654 |
| 310 | 12.2047 | 12.2441 | 12.2835 | 12.3228 | 12.3622 | 12.4016 | 12.4409 | 12.4803 | 12.5197 | 12.5591 |
| 320 | 12.5984 | 12.6378 | 12.6772 | 12.7165 | 12.7559 | 12.7953 | 12.8346 | 12.8740 | 12.9134 | 12.9528 |
| 330 | 12.9921 | 13.0315 | 13.0709 | 13.1102 | 13.1496 | 13.1890 | 13.2283 | 13.2677 | 13.3071 | 13.3465 |
| 340 | 13.3858 | 13.4252 | 13.4646 | 13.5039 | 13.5433 | 13.5827 | 13.6220 | 13.6614 | 13.7008 | 13.7402 |
| 350 | 13.7795 | 13.8189 | 13.8583 | 13.8976 | 13.9370 | 13.9764 | 14.0157 | 14.0551 | 14.0945 | 14.1339 |
| 360 | 14.1732 | 14.2126 | 14.2520 | 14.2913 | 14.3307 | 14.3701 | 14.4094 | 14.4488 | 14.4882 | 14.5276 |
| 370 | 14.5669 | 14.6063 | 14.6457 | 14.6850 | 14.7244 | 14.7638 | 14.8031 | 14.8425 | 14.8819 | 14.9213 |
| 380 | 14.9606 | 15.0000 | 15.0394 | 15.0787 | 15.1181 | 15.1575 | 15.1969 | 15.2362 | 15.2756 | 15.3150 |
| 390 | 15.3543 | 15.3937 | 15.4331 | 15.4724 | 15.5118 | 15.5512 | 15.5906 | 15.6299 | 15.6693 | 15.7087 |
| 400 | 15.7480 | 15.7874 | 15.8268 | 15.8661 | 15.9055 | 15.9449 | 15.9843 | 16.0236 | 16.0630 | 16.1024 |
| 410 | 16.1417 | 16.1811 | 16.2205 | 16.2598 | 16.2992 | 16.3386 | 16.3780 | 16.4173 | 16.4567 | 16.4961 |
| 420 | 16.5354 | 16.5748 | 16.6142 | 16.6535 | 16.6929 | 16.7323 | 16.7717 | 16.8110 | 16.8504 | 16.8898 |
| 430 | 16.9291 | 16.9685 | 17.0079 | 17.0472 | 17.0866 | 17.1260 | 17.1654 | 17.2047 | 17.2441 | 17.2835 |
| 440 | 17.3228 | 17.3622 | 17.4016 | 17.4409 | 17.4803 | 17.5197 | 17.5591 | 17.5984 | 17.6378 | 17.6772 |
| 450 | 17.7165 | 17.7559 | 17.7953 | 17.8346 | 17.8740 | 17.9134 | 17.9528 | 17.9921 | 18.0315 | 18.0709 |
| 460 | 18.1102 | 18.1496 | 18.1890 | 18.2283 | 18.2677 | 18.3071 | 18.3465 | 18.3858 | 18.4252 | 18.4646 |
| 470 | 18.5039 | 18.5433 | 18.5827 | 18.6220 | 18.6614 | 18.7008 | 18.7402 | 18.7795 | 18.8189 | 18.8583 |
| 480 | 18.8976 | 18.9370 | 18.9764 | 19.0157 | 19.0551 | 19.0945 | 19.1339 | 19.1732 | 19.2126 | 19.2520 |
| 490 | 19.2913 | 19.3307 | 19.3701 | 19.4094 | 19.4488 | 19.4882 | 19.5276 | 19.5669 | 19.6063 | 19.6457 |

## Millimeters to Inches
(Based on 1 inch = 25.4 millimeters, exactly)

| Milli-meters | 0 | 1 | 2 | 3 | 4 | 5 | 6 | 7 | 8 | 9 |
|---|---|---|---|---|---|---|---|---|---|---|
| | Inches | | | | | | | | | |
| 500 | 19.6850 | 19.7244 | 19.7638 | 19.8031 | 19.8425 | 19.8819 | 19.9213 | 19.9606 | 20.0000 | 20.0394 |
| 510 | 20.0787 | 20.1181 | 20.1575 | 20.1969 | 20.2362 | 20.2756 | 20.3150 | 20.3543 | 20.3937 | 20.4331 |
| 520 | 20.4724 | 20.5118 | 20.5512 | 20.5906 | 20.6299 | 20.6693 | 20.7087 | 20.7480 | 20.7874 | 20.8268 |
| 530 | 20.8661 | 20.9055 | 20.9449 | 20.9843 | 21.0236 | 21.0630 | 21.1024 | 21.1417 | 21.1811 | 21.2205 |
| 540 | 21.2598 | 21.2992 | 21.3386 | 21.3780 | 21.4173 | 21.4567 | 21.4961 | 21.5354 | 21.5748 | 21.6142 |
| 550 | 21.6535 | 21.6929 | 21.7323 | 21.7717 | 21.8110 | 21.8504 | 21.8898 | 21.9291 | 21.9685 | 22.0079 |
| 560 | 22.0472 | 22.0866 | 22.1260 | 22.1654 | 22.2047 | 22.2441 | 22.2835 | 22.3228 | 22.3622 | 22.4016 |
| 570 | 22.4409 | 22.4803 | 22.5197 | 22.5591 | 22.5984 | 22.6378 | 22.6772 | 22.7165 | 22.7559 | 22.7953 |
| 580 | 22.8346 | 22.8740 | 22.9134 | 22.9528 | 22.9921 | 23.0315 | 23.0709 | 23.1102 | 23.1496 | 23.1890 |
| 590 | 23.2283 | 23.2677 | 23.3071 | 23.3465 | 23.3858 | 23.4252 | 23.4646 | 23.5039 | 23.5433 | 23.5827 |
| 600 | 23.6220 | 23.6614 | 23.7008 | 23.7402 | 23.7795 | 23.8189 | 23.8583 | 23.8976 | 23.9370 | 23.9764 |
| 610 | 24.0157 | 24.0551 | 24.0945 | 24.1339 | 24.1732 | 24.2126 | 24.2520 | 24.2913 | 24.3307 | 24.3701 |
| 620 | 24.4094 | 24.4488 | 24.4882 | 24.5276 | 24.5669 | 24.6063 | 24.6457 | 24.6850 | 24.7244 | 24.7638 |
| 630 | 24.8031 | 24.8425 | 24.8819 | 24.9213 | 24.9606 | 25.0000 | 25.0394 | 25.0787 | 25.1181 | 25.1575 |
| 640 | 25.1969 | 25.2362 | 25.2756 | 25.3150 | 25.3543 | 25.3937 | 25.4331 | 25.4724 | 25.5118 | 25.5512 |
| 650 | 25.5906 | 25.6299 | 25.6693 | 25.7087 | 25.7480 | 25.7874 | 25.8268 | 25.8661 | 25.9055 | 25.9449 |
| 660 | 25.9843 | 26.0236 | 26.0630 | 26.1024 | 26.1417 | 26.1811 | 26.2205 | 26.2598 | 26.2992 | 26.3386 |
| 670 | 26.3780 | 26.4173 | 26.4567 | 26.4961 | 26.5354 | 26.5748 | 26.6142 | 26.6535 | 26.6929 | 26.7323 |
| 680 | 26.7717 | 26.8110 | 26.8504 | 26.8898 | 26.9291 | 26.9685 | 27.0079 | 27.0472 | 27.0866 | 27.1260 |
| 690 | 27.1654 | 27.2047 | 27.2441 | 27.2835 | 27.3228 | 27.3622 | 27.4016 | 27.4409 | 27.4803 | 27.5197 |
| 700 | 27.5591 | 27.5984 | 27.6378 | 27.6772 | 27.7165 | 27.7559 | 27.7953 | 27.8346 | 27.8740 | 27.9134 |
| 710 | 27.9528 | 27.9921 | 28.0315 | 28.0709 | 28.1102 | 28.1496 | 28.1890 | 28.2283 | 28.2677 | 28.3071 |
| 720 | 28.3465 | 28.3858 | 28.4252 | 28.4646 | 28.5039 | 28.5433 | 28.5827 | 28.6220 | 28.6614 | 28.7008 |
| 730 | 28.7402 | 28.7795 | 28.8189 | 28.8583 | 28.8976 | 28.9370 | 28.9764 | 29.0157 | 29.0551 | 29.0945 |
| 740 | 29.1339 | 29.1732 | 29.2126 | 29.2520 | 29.2913 | 29.3307 | 29.3701 | 29.4094 | 29.4488 | 29.4882 |
| 750 | 29.5276 | 29.5669 | 29.6063 | 29.6457 | 29.6850 | 29.7244 | 29.7638 | 29.8031 | 29.8425 | 29.8819 |
| 760 | 29.9213 | 29.9606 | 30.0000 | 30.0394 | 30.0787 | 30.1181 | 30.1575 | 30.1969 | 30.2362 | 30.2756 |
| 770 | 30.3150 | 30.3543 | 30.3937 | 30.4331 | 30.4724 | 30.5118 | 30.5512 | 30.5906 | 30.6299 | 30.6693 |
| 780 | 30.7087 | 30.7480 | 30.7874 | 30.8268 | 30.8661 | 30.9055 | 30.9449 | 30.9843 | 31.0236 | 31.0630 |
| 790 | 31.1024 | 31.1417 | 31.1811 | 31.2205 | 31.2598 | 31.2992 | 31.3386 | 31.3780 | 31.4173 | 31.4567 |
| 800 | 31.4961 | 31.5354 | 31.5748 | 31.6142 | 31.6535 | 31.6929 | 31.7323 | 31.7717 | 31.8110 | 31.8504 |
| 810 | 31.8898 | 31.9291 | 31.9685 | 32.0079 | 32.0472 | 32.0866 | 32.1260 | 32.1654 | 32.2047 | 32.2441 |
| 820 | 32.2835 | 32.3228 | 32.3622 | 32.4016 | 32.4409 | 32.4803 | 32.5197 | 32.5591 | 32.5984 | 32.6378 |
| 830 | 32.6772 | 32.7165 | 32.7559 | 32.7953 | 32.8346 | 32.8740 | 32.9134 | 32.9528 | 32.9921 | 33.0315 |
| 840 | 33.0709 | 33.1102 | 33.1496 | 33.1890 | 33.2283 | 33.2677 | 33.3071 | 33.3465 | 33.3858 | 33.4252 |
| 850 | 33.4646 | 33.5039 | 33.5433 | 33.5827 | 33.6220 | 33.6614 | 33.7008 | 33.7402 | 33.7795 | 33.8189 |
| 860 | 33.8583 | 33.8976 | 33.9370 | 33.9764 | 34.0157 | 34.0551 | 34.0945 | 34.1339 | 34.1732 | 34.2126 |
| 870 | 34.2520 | 34.2913 | 34.3307 | 34.3701 | 34.4094 | 34.4488 | 34.4882 | 34.5276 | 34.5669 | 34.6063 |
| 880 | 34.6457 | 34.6850 | 34.7244 | 34.7638 | 34.8031 | 34.8425 | 34.8819 | 34.9213 | 34.9606 | 35.0000 |
| 890 | 35.0394 | 35.0787 | 35.1181 | 35.1575 | 35.1969 | 35.2362 | 35.2756 | 35.3150 | 35.3543 | 35.3937 |
| 900 | 35.4331 | 35.4724 | 35.5118 | 35.5512 | 35.5906 | 35.6299 | 35.6693 | 35.7087 | 35.7480 | 35.7874 |
| 910 | 35.8268 | 35.8661 | 35.9055 | 35.9449 | 35.9843 | 36.0236 | 36.0630 | 36.1024 | 36.1417 | 36.1811 |
| 920 | 36.2205 | 36.2598 | 36.2992 | 36.3386 | 36.3780 | 36.4173 | 36.4567 | 36.4961 | 36.5354 | 36.5748 |
| 930 | 36.6142 | 36.6535 | 36.6929 | 36.7323 | 36.7717 | 36.8110 | 36.8504 | 36.8898 | 36.9291 | 36.9685 |
| 940 | 37.0079 | 37.0472 | 37.0866 | 37.1260 | 37.1654 | 37.2047 | 37.2441 | 37.2835 | 37.3228 | 37.3622 |
| 950 | 37.4016 | 37.4409 | 37.4803 | 37.5197 | 37.5591 | 37.5984 | 37.6378 | 37.6772 | 37.7165 | 37.7559 |
| 960 | 37.7953 | 37.8346 | 37.8740 | 37.9134 | 37.9528 | 37.9921 | 38.0315 | 38.0709 | 38.1102 | 38.1496 |
| 970 | 38.1890 | 38.2283 | 38.2677 | 38.3071 | 38.3465 | 38.3858 | 38.4252 | 38.4646 | 38.5039 | 38.5433 |
| 980 | 38.5827 | 38.6220 | 38.6614 | 38.7008 | 38.7402 | 38.7795 | 38.8189 | 38.8583 | 38.8976 | 38.9370 |
| 990 | 38.9764 | 39.0157 | 39.0551 | 39.0945 | 39.1339 | 39.1732 | 39.2126 | 39.2520 | 39.2913 | 39.3307 |
| 1000 | 39.3701 | ... | ... | ... | ... | ... | ... | ... | ... | ... |

### Fractional Inch—Millimetre and Foot—Millimetre Conversion Tables
(Based on 1 inch = 25.4 millimetres, exactly)

#### FRACTIONAL INCH TO MILLIMETRES

| in. | mm | in. | mm | in. | mm | in. | mm |
|---|---|---|---|---|---|---|---|
| 1/64 | 0.397 | 17/64 | 6.747 | 33/64 | 13.097 | 49/64 | 19.447 |
| 1/32 | 0.794 | 9/32 | 7.144 | 17/32 | 13.494 | 25/32 | 19.844 |
| 3/64 | 1.191 | 19/64 | 7.541 | 35/64 | 13.891 | 51/64 | 20.241 |
| 1/16 | 1.588 | 5/16 | 7.938 | 9/16 | 14.288 | 13/16 | 20.638 |
| 5/64 | 1.984 | 21/64 | 8.334 | 37/64 | 14.684 | 53/64 | 21.034 |
| 3/32 | 2.381 | 11/32 | 8.731 | 19/32 | 15.081 | 27/32 | 21.431 |
| 7/64 | 2.778 | 23/64 | 9.128 | 39/64 | 15.478 | 55/64 | 21.828 |
| 1/8 | 3.175 | 3/8 | 9.525 | 5/8 | 15.875 | 7/8 | 22.225 |
| 9/64 | 3.572 | 25/64 | 9.922 | 41/64 | 16.272 | 57/64 | 22.622 |
| 5/32 | 3.969 | 13/32 | 10.319 | 21/32 | 16.669 | 29/32 | 23.019 |
| 11/64 | 4.366 | 27/64 | 10.716 | 43/64 | 17.066 | 59/64 | 23.416 |
| 3/16 | 4.762 | 7/16 | 11.112 | 11/16 | 17.462 | 15/16 | 23.812 |
| 13/64 | 5.159 | 29/64 | 11.509 | 45/64 | 17.859 | 61/64 | 24.209 |
| 7/32 | 5.556 | 15/32 | 11.906 | 23/32 | 18.256 | 31/32 | 24.606 |
| 15/64 | 5.953 | 31/64 | 12.303 | 47/64 | 18.653 | 63/64 | 25.003 |
| 1/4 | 6.350 | 1/2 | 12.700 | 3/4 | 19.050 | 1 | 25.400 |

#### INCHES TO MILLIMETRES

| in. | mm | in. | mm | in. | mm | in. | mm | in. | mm | in. | mm |
|---|---|---|---|---|---|---|---|---|---|---|---|
| 1 | 25.4 | 3 | 76.2 | 5 | 127.0 | 7 | 177.8 | 9 | 228.6 | 11 | 279.4 |
| 2 | 50.8 | 4 | 101.6 | 6 | 152.4 | 8 | 203.2 | 10 | 254.0 | 12 | 304.8 |

#### FEET TO MILLIMETRES

| ft | mm | ft | mm | ft | mm | ft | mm | ft | mm |
|---|---|---|---|---|---|---|---|---|---|
| 100 | 30,480 | 10 | 3,048 | 1 | 304.8 | 0.1 | 30.48 | 0.01 | 3.048 |
| 200 | 60,960 | 20 | 6,096 | 2 | 609.6 | 0.2 | 60.96 | 0.02 | 6.096 |
| 300 | 91,440 | 30 | 9,144 | 3 | 914.4 | 0.3 | 91.44 | 0.03 | 9.144 |
| 400 | 121,920 | 40 | 12,192 | 4 | 1,219.2 | 0.4 | 121.92 | 0.04 | 12.192 |
| 500 | 152,400 | 50 | 15,240 | 5 | 1,524.0 | 0.5 | 152.40 | 0.05 | 15.240 |
| 600 | 182,880 | 60 | 18,288 | 6 | 1,828.8 | 0.6 | 182.88 | 0.06 | 18.288 |
| 700 | 213,360 | 70 | 21,336 | 7 | 2,133.6 | 0.7 | 213.36 | 0.07 | 21.336 |
| 800 | 243,840 | 80 | 24,384 | 8 | 2,438.4 | 0.8 | 243.84 | 0.08 | 24.384 |
| 900 | 274,320 | 90 | 27,432 | 9 | 2,743.2 | 0.9 | 274.32 | 0.09 | 27.432 |
| 1,000 | 304,800 | 100 | 30,480 | 10 | 3,048.0 | 1.0 | 304.80 | 0.10 | 30.480 |

*Example* 1: Find millimetre equivalent of 293 feet, 5 47/64 inches.

$$
\begin{array}{rll}
200 \text{ ft} & = 60,960. & \text{mm} \\
90 \text{ ft} & = 27,432. & \text{mm} \\
3 \text{ ft} & = \phantom{0}914.4 & \text{mm} \\
5 \text{ in.} & = \phantom{0}127.0 & \text{mm} \\
47/64 \text{ in.} & = \phantom{00}18.653 & \text{mm} \\
\hline
293 \text{ ft, } 5\,47/64 \text{ in.} & = 89,452.053 & \text{mm}
\end{array}
$$

*Example* 2: Find millimetre equivalent of 71.86 feet.

$$
\begin{array}{rll}
70. \text{ ft} & = 21,336. & \text{mm} \\
1. \text{ ft} & = \phantom{00}304.8 & \text{mm} \\
.80 \text{ ft} & = \phantom{00}243.84 & \text{mm} \\
.06 \text{ ft} & = \phantom{000}18.288 & \text{mm} \\
\hline
71.86 \text{ ft} & = 21,902.928 & \text{mm}
\end{array}
$$

### Microinches to Micrometers (microns)
(Based on 1 microinch = 0.0254 micrometers, exactly)

| Micro-inches | 0 | 1 | 2 | 3 | 4 | 5 | 6 | 7 | 8 | 9 |
|---|---|---|---|---|---|---|---|---|---|---|
| | Micrometers (microns) | | | | | | | | | |
| 0 | ..... | 0.025 | 0.051 | 0.076 | 0.102 | 0.127 | 0.152 | 0.178 | 0.203 | 0.229 |
| 10 | 0.254 | 0.279 | 0.305 | 0.330 | 0.356 | 0.381 | 0.406 | 0.432 | 0.457 | 0.483 |
| 20 | 0.508 | 0.533 | 0.559 | 0.584 | 0.610 | 0.635 | 0.660 | 0.686 | 0.711 | 0.737 |
| 30 | 0.762 | 0.787 | 0.813 | 0.838 | 0.864 | 0.889 | 0.914 | 0.940 | 0.965 | 0.991 |
| 40 | 1.016 | 1.041 | 1.067 | 1.092 | 1.118 | 1.143 | 1.168 | 1.194 | 1.219 | 1.245 |
| 50 | 1.270 | 1.295 | 1.321 | 1.346 | 1.372 | 1.397 | 1.422 | 1.448 | 1.473 | 1.499 |
| 60 | 1.524 | 1.549 | 1.575 | 1.600 | 1.626 | 1.651 | 1.676 | 1.702 | 1.727 | 1.753 |
| 70 | 1.778 | 1.803 | 1.829 | 1.854 | 1.880 | 1.905 | 1.930 | 1.956 | 1.981 | 2.007 |
| 80 | 2.032 | 2.057 | 2.083 | 2.108 | 2.134 | 2.159 | 2.184 | 2.210 | 2.235 | 2.261 |
| 90 | 2.286 | 2.311 | 2.337 | 2.362 | 2.388 | 2.413 | 2.438 | 2.464 | 2.489 | 2.515 |
| 100 | 2.540 | 2.565 | 2.591 | 2.616 | 2.642 | 2.667 | 2.692 | 2.718 | 2.743 | 2.769 |
| 110 | 2.794 | 2.819 | 2.845 | 2.870 | 2.896 | 2.921 | 2.946 | 2.972 | 2.997 | 3.023 |
| 120 | 3.048 | 3.073 | 3.099 | 3.124 | 3.150 | 3.175 | 3.200 | 3.226 | 3.251 | 3.277 |
| 130 | 3.302 | 3.327 | 3.353 | 3.378 | 3.404 | 3.429 | 3.454 | 3.480 | 3.505 | 3.531 |
| 140 | 3.556 | 3.581 | 3.607 | 3.632 | 3.658 | 3.683 | 3.708 | 3.734 | 3.759 | 3.785 |
| 150 | 3.810 | 3.835 | 3.861 | 3.886 | 3.912 | 3.937 | 3.962 | 3.988 | 4.013 | 4.039 |
| 160 | 4.064 | 4.089 | 4.115 | 4.140 | 4.166 | 4.191 | 4.216 | 4.242 | 4.267 | 4.293 |
| 170 | 4.318 | 4.343 | 4.369 | 4.394 | 4.420 | 4.445 | 4.470 | 4.496 | 4.521 | 4.547 |
| 180 | 4.572 | 4.597 | 4.623 | 4.648 | 4.674 | 4.699 | 4.724 | 4.750 | 4.775 | 4.801 |
| 190 | 4.826 | 4.851 | 4.877 | 4.902 | 4.928 | 4.953 | 4.978 | 5.004 | 5.029 | 5.055 |
| 200 | 5.080 | 5.105 | 5.131 | 5.156 | 5.182 | 5.207 | 5.232 | 5.258 | 5.283 | 5.309 |
| 210 | 5.334 | 5.359 | 5.385 | 5.410 | 5.436 | 5.461 | 5.486 | 5.512 | 5.537 | 5.563 |
| 220 | 5.588 | 5.613 | 5.639 | 5.664 | 5.690 | 5.715 | 5.740 | 5.766 | 5.791 | 5.817 |
| 230 | 5.842 | 5.867 | 5.893 | 5.918 | 5.944 | 5.969 | 5.994 | 6.020 | 6.045 | 6.071 |
| 240 | 6.096 | 6.121 | 6.147 | 6.172 | 6.198 | 6.223 | 6.248 | 6.274 | 6.299 | 6.325 |
| 250 | 6.350 | 6.375 | 6.401 | 6.426 | 6.452 | 6.477 | 6.502 | 6.528 | 6.553 | 6.579 |
| 260 | 6.604 | 6.629 | 6.655 | 6.680 | 6.706 | 6.731 | 6.756 | 6.782 | 6.807 | 6.833 |
| 270 | 6.858 | 6.883 | 6.909 | 6.934 | 6.960 | 6.985 | 7.010 | 7.036 | 7.061 | 7.087 |
| 280 | 7.112 | 7.137 | 7.163 | 7.188 | 7.214 | 7.239 | 7.264 | 7.290 | 7.315 | 7.341 |
| 290 | 7.366 | 7.391 | 7.417 | 7.442 | 7.468 | 7.493 | 7.518 | 7.544 | 7.569 | 7.595 |

The following short table permits conversion of microinches to micrometers for ranges higher than in the main table given above. Appropriate quantities chosen from both tables are simply added to obtain the higher converted value:

| μin. | μm | μin. | μm | μin. | μm | μin. | μm | μin. | μm |
|---|---|---|---|---|---|---|---|---|---|
| 300 | 7.620 | 900 | 22.860 | 1500 | 38.100 | 2100 | 53.340 | 2700 | 68.580 |
| 600 | 15.240 | 1200 | 30.480 | 1800 | 45.720 | 2400 | 60.960 | 3000 | 76.200 |

*Example:* Convert 1375 μin. to μm:
From above table: 1200 μin. = 30.480 μm
From main table:    175 μin. =  4.445 μm
                   1375 μin. = 34.925 μm

### Micrometers (microns) to Microinches — 1
(Based on 1 microinch = 0.0254 micrometers, exactly)

| Micro-meters (microns) | 0 | 0.01 | 0.02 | 0.03 | 0.04 | 0.05 | 0.06 | 0.07 | 0.08 | 0.09 |
|---|---|---|---|---|---|---|---|---|---|---|
| | Microinches | | | | | | | | | |
| 0 | .... | 0.4 | 0.8 | 1.2 | 1.6 | 2.0 | 2.4 | 2.8 | 3.1 | 3.5 |
| 0.10 | 3.9 | 4.3 | 4.7 | 5.1 | 5.5 | 5.9 | 6.3 | 6.7 | 7.1 | 7.5 |
| 0.20 | 7.9 | 8.3 | 8.7 | 9.1 | 9.4 | 9.8 | 10.2 | 10.6 | 11.0 | 11.4 |
| 0.30 | 11.8 | 12.2 | 12.6 | 13.0 | 13.4 | 13.8 | 14.2 | 14.6 | 15.0 | 15.4 |
| 0.40 | 15.7 | 16.1 | 16.5 | 16.9 | 17.3 | 17.7 | 18.1 | 18.5 | 18.9 | 19.3 |

## Micrometers (microns) to Microinches — 2

| Micro-meters (microns) | 0 | 0.01 | 0.02 | 0.03 | 0.04 | 0.05 | 0.06 | 0.07 | 0.08 | 0.09 |
|---|---|---|---|---|---|---|---|---|---|---|
| | | | | | Microinches | | | | | |
| 0.50 | 19.7 | 20.1 | 20.5 | 20.9 | 21.3 | 21.7 | 22.0 | 22.4 | 22.8 | 23.2 |
| 0.60 | 23.6 | 24.0 | 24.4 | 24.8 | 25.2 | 25.6 | 26.0 | 26.4 | 26.8 | 27.2 |
| 0.70 | 27.6 | 28.0 | 28.3 | 28.7 | 29.1 | 29.5 | 29.9 | 30.3 | 30.7 | 31.1 |
| 0.80 | 31.5 | 31.9 | 32.3 | 32.7 | 33.1 | 33.5 | 33.9 | 34.3 | 34.6 | 35.0 |
| 0.90 | 35.4 | 35.8 | 36.2 | 36.6 | 37.0 | 37.4 | 37.8 | 38.2 | 38.6 | 39.0 |
| 1.00 | 39.4 | 39.8 | 40.2 | 40.6 | 40.9 | 41.3 | 41.7 | 42.1 | 42.5 | 42.9 |
| 1.10 | 43.3 | 43.7 | 44.1 | 44.5 | 44.9 | 45.3 | 45.7 | 46.1 | 46.5 | 46.9 |
| 1.20 | 47.2 | 47.6 | 48.0 | 48.4 | 48.8 | 49.2 | 49.6 | 50.0 | 50.4 | 50.8 |
| 1.30 | 51.2 | 51.6 | 52.0 | 52.4 | 52.8 | 53.1 | 53.5 | 53.9 | 54.3 | 54.7 |
| 1.40 | 55.1 | 55.5 | 55.9 | 56.3 | 56.7 | 57.1 | 57.5 | 57.9 | 58.3 | 58.7 |
| 1.50 | 59.1 | 59.4 | 59.8 | 60.2 | 60.6 | 61.0 | 61.4 | 61.8 | 62.2 | 62.6 |
| 1.60 | 63.0 | 63.4 | 63.8 | 64.2 | 64.6 | 65.0 | 65.4 | 65.7 | 66.1 | 66.5 |
| 1.70 | 66.9 | 67.3 | 67.7 | 68.1 | 68.5 | 68.9 | 69.3 | 69.7 | 70.1 | 70.5 |
| 1.80 | 70.9 | 71.3 | 71.7 | 72.0 | 72.4 | 72.8 | 73.2 | 73.6 | 74.0 | 74.4 |
| 1.90 | 74.8 | 75.2 | 75.6 | 76.0 | 76.4 | 76.8 | 77.2 | 77.6 | 78.0 | 78.3 |
| 2.00 | 78.7 | 79.1 | 79.5 | 79.9 | 80.3 | 80.7 | 81.1 | 81.5 | 81.9 | 82.3 |
| 2.10 | 82.7 | 83.1 | 83.5 | 83.9 | 84.3 | 84.6 | 85.0 | 85.4 | 85.8 | 86.2 |
| 2.20 | 86.6 | 87.0 | 87.4 | 87.8 | 88.2 | 88.6 | 89.0 | 89.4 | 89.8 | 90.2 |
| 2.30 | 90.6 | 90.9 | 91.3 | 91.7 | 92.1 | 92.5 | 92.9 | 93.3 | 93.7 | 94.1 |
| 2.40 | 94.5 | 94.9 | 95.3 | 95.7 | 96.1 | 96.5 | 96.9 | 97.2 | 97.6 | 98.0 |
| 2.50 | 98.4 | 98.8 | 99.2 | 99.6 | 100.0 | 100.4 | 100.8 | 101.2 | 101.6 | 102.0 |
| 2.60 | 102.4 | 102.8 | 103.1 | 103.5 | 103.9 | 104.3 | 104.7 | 105.1 | 105.5 | 105.9 |
| 2.70 | 106.3 | 106.7 | 107.1 | 107.5 | 107.9 | 108.3 | 108.7 | 109.1 | 109.4 | 109.8 |
| 2.80 | 110.2 | 110.6 | 111.0 | 111.4 | 111.8 | 112.2 | 112.6 | 113.0 | 113.4 | 113.8 |
| 2.90 | 114.2 | 114.6 | 115.0 | 115.4 | 115.7 | 116.1 | 116.5 | 116.9 | 117.3 | 117.7 |
| 3.00 | 118.1 | 118.5 | 118.9 | 119.3 | 119.7 | 120.1 | 120.5 | 120.9 | 121.3 | 121.7 |
| 3.10 | 122.0 | 122.4 | 122.8 | 123.2 | 123.6 | 124.0 | 124.4 | 124.8 | 125.2 | 125.6 |
| 3.20 | 126.0 | 126.4 | 126.8 | 127.2 | 127.6 | 128.0 | 128.3 | 128.7 | 129.1 | 129.5 |
| 3.30 | 129.9 | 130.3 | 130.7 | 131.1 | 131.5 | 131.9 | 132.3 | 132.7 | 133.1 | 133.5 |
| 3.40 | 133.9 | 134.3 | 134.6 | 135.0 | 135.4 | 135.8 | 136.2 | 136.6 | 137.0 | 137.4 |
| 3.50 | 137.8 | 138.2 | 138.6 | 139.0 | 139.4 | 139.8 | 140.2 | 140.6 | 140.9 | 141.3 |
| 3.60 | 141.7 | 142.1 | 142.5 | 142.9 | 143.3 | 143.7 | 144.1 | 144.5 | 144.9 | 145.3 |
| 3.70 | 145.7 | 146.1 | 146.5 | 146.9 | 147.2 | 147.6 | 148.0 | 148.4 | 148.8 | 149.2 |
| 3.80 | 149.6 | 150.0 | 150.4 | 150.8 | 151.2 | 151.6 | 152.0 | 152.4 | 152.8 | 153.1 |
| 3.90 | 153.5 | 153.9 | 154.3 | 154.7 | 155.1 | 155.5 | 155.9 | 156.3 | 156.7 | 157.1 |
| 4.00 | 157.5 | 157.9 | 158.3 | 158.7 | 159.1 | 159.4 | 159.8 | 160.2 | 160.6 | 161.0 |
| 4.10 | 161.4 | 161.8 | 162.2 | 162.6 | 163.0 | 163.4 | 163.8 | 164.2 | 164.6 | 165.0 |
| 4.20 | 165.4 | 165.7 | 166.1 | 166.5 | 166.9 | 167.3 | 167.7 | 168.1 | 168.5 | 168.9 |
| 4.30 | 169.3 | 169.7 | 170.1 | 170.5 | 170.9 | 171.3 | 171.7 | 172.0 | 172.4 | 172.8 |
| 4.40 | 173.2 | 173.6 | 174.0 | 174.4 | 174.8 | 175.2 | 175.6 | 176.0 | 176.4 | 176.8 |
| 4.50 | 177.2 | 177.6 | 178.0 | 178.3 | 178.7 | 179.1 | 179.5 | 179.9 | 180.3 | 180.7 |
| 4.60 | 181.1 | 181.5 | 181.9 | 182.3 | 182.7 | 183.1 | 183.5 | 183.9 | 184.3 | 184.6 |
| 4.70 | 185.0 | 185.4 | 185.8 | 186.2 | 186.6 | 187.0 | 187.4 | 187.8 | 188.2 | 188.6 |
| 4.80 | 189.0 | 189.4 | 189.8 | 190.2 | 190.6 | 190.9 | 191.3 | 191.7 | 192.1 | 192.5 |
| 4.90 | 192.9 | 193.3 | 193.7 | 194.1 | 194.5 | 194.9 | 195.3 | 195.7 | 196.1 | 196.5 |
| 5.00 | 196.9 | 197.2 | 197.6 | 198.0 | 198.4 | 198.8 | 199.2 | 199.6 | 200.0 | 200.4 |

The table given below can be used with the preceding main table to obtain higher converted values, simply by adding appropriate quantities chosen from each table:

| μm | μin. | μm | μin. | μm | μin. | μm | μin. | μm | μin. |
|---|---|---|---|---|---|---|---|---|---|
| 10 | 393.7 | 20 | 787.4 | 30 | 1,181.1 | 40 | 1,574.8 | 50 | 1,968.5 |
| 15 | 590.6 | 25 | 984.3 | 35 | 1,378.0 | 45 | 1,771.7 | 55 | 2,165.4 |

*Example:* Convert 23.55 μm to μin.:
From above table: 20.00 μm = 787.4 μin.
From main table: 3.55 μm = 139.8 μin.

23.55 μm = 927.2 μin.

### Foot—Metre and Mile—Kilometre Conversion Tables

(Based on 1 foot = 0.3048 metre, exactly)

#### FEET TO METRES
(1 ft = 0.3048 m, exactly)

| feet | metres | feet | metres | feet | metres | feet | metres | feet | metres |
|---|---|---|---|---|---|---|---|---|---|
| 100 | 30.480 | 10 | 3.048 | 1 | 0.305 | 0.1 | 0.030 | 0.01 | 0.003 |
| 200 | 60.960 | 20 | 6.096 | 2 | 0.610 | 0.2 | 0.061 | 0.02 | 0.006 |
| 300 | 91.440 | 30 | 9.144 | 3 | 0.914 | 0.3 | 0.091 | 0.03 | 0.009 |
| 400 | 121.920 | 40 | 12.192 | 4 | 1.219 | 0.4 | 0.122 | 0.04 | 0.012 |
| 500 | 152.400 | 50 | 15.240 | 5 | 1.524 | 0.5 | 0.152 | 0.05 | 0.015 |
| 600 | 182.880 | 60 | 18.288 | 6 | 1.829 | 0.6 | 0.183 | 0.06 | 0.018 |
| 700 | 213.360 | 70 | 21.336 | 7 | 2.134 | 0.7 | 0.213 | 0.07 | 0.021 |
| 800 | 243.840 | 80 | 24.384 | 8 | 2.438 | 0.8 | 0.244 | 0.08 | 0.024 |
| 900 | 274.320 | 90 | 27.432 | 9 | 2.743 | 0.9 | 0.274 | 0.09 | 0.027 |
| 1,000 | 304.800 | 100 | 30.480 | 10 | 3.048 | 1.0 | 0.305 | 0.10 | 0.030 |

#### METRES TO FEET
(1 m = 3.280840 ft)

| metres | feet | metres | feet | metres | feet | metres | feet | metres | feet |
|---|---|---|---|---|---|---|---|---|---|
| 100 | 328.084 | 10 | 32.808 | 1 | 3.281 | 0.1 | 0.328 | 0.01 | 0.033 |
| 200 | 656.168 | 20 | 65.617 | 2 | 6.562 | 0.2 | 0.656 | 0.02 | 0.066 |
| 300 | 984.252 | 30 | 98.425 | 3 | 9.843 | 0.3 | 0.984 | 0.03 | 0.098 |
| 400 | 1,312.336 | 40 | 131.234 | 4 | 13.123 | 0.4 | 1.312 | 0.04 | 0.131 |
| 500 | 1,640.420 | 50 | 164.042 | 5 | 16.404 | 0.5 | 1.640 | 0.05 | 0.164 |
| 600 | 1,968.504 | 60 | 196.850 | 6 | 19.685 | 0.6 | 1.969 | 0.06 | 0.197 |
| 700 | 2,296.588 | 70 | 229.659 | 7 | 22.966 | 0.7 | 2.297 | 0.07 | 0.230 |
| 800 | 2,624.672 | 80 | 262.467 | 8 | 26.247 | 0.8 | 2.625 | 0.08 | 0.262 |
| 900 | 2,952.756 | 90 | 295.276 | 9 | 29.528 | 0.9 | 2.953 | 0.09 | 0.295 |
| 1,000 | 3,280.840 | 100 | 328.084 | 10 | 32.808 | 1.0 | 3.281 | 0.10 | 0.328 |

#### MILES TO KILOMETRES
(1 mile = 1.609344 km, exactly)

| miles | km | miles | km | miles | km | miles | km | miles | km |
|---|---|---|---|---|---|---|---|---|---|
| 1,000 | 1,609.34 | 100 | 160.93 | 10 | 16.09 | 1 | 1.61 | 0.1 | 0.16 |
| 2,000 | 3,218.69 | 200 | 321.87 | 20 | 32.19 | 2 | 3.22 | 0.2 | 0.32 |
| 3,000 | 4,828.03 | 300 | 482.80 | 30 | 48.28 | 3 | 4.83 | 0.3 | 0.48 |
| 4,000 | 6,437.38 | 400 | 643.74 | 40 | 64.37 | 4 | 6.44 | 0.4 | 0.64 |
| 5,000 | 8,046.72 | 500 | 804.67 | 50 | 80.47 | 5 | 8.05 | 0.5 | 0.80 |
| 6,000 | 9,656.06 | 600 | 965.61 | 60 | 96.56 | 6 | 9.66 | 0.6 | 0.97 |
| 7,000 | 11,265.41 | 700 | 1,126.54 | 70 | 112.65 | 7 | 11.27 | 0.7 | 1.13 |
| 8,000 | 12,874.75 | 800 | 1,287.48 | 80 | 128.75 | 8 | 12.87 | 0.8 | 1.29 |
| 9,000 | 14,484.10 | 900 | 1,448.41 | 90 | 144.84 | 9 | 14.48 | 0.9 | 1.45 |
| 10,000 | 16,093.44 | 1,000 | 1,609.34 | 100 | 160.93 | 10 | 16.09 | 1.0 | 1.61 |

#### KILOMETRES TO MILES
(1 km = 0.6213712 mile)

| km | miles | km | miles | km | miles | km | miles | km | miles |
|---|---|---|---|---|---|---|---|---|---|
| 1,000 | 621.37 | 100 | 62.14 | 10 | 6.21 | 1 | 0.62 | 0.1 | 0.06 |
| 2,000 | 1,242.74 | 200 | 124.27 | 20 | 12.43 | 2 | 1.24 | 0.2 | 0.12 |
| 3,000 | 1,864.11 | 300 | 186.41 | 30 | 18.64 | 3 | 1.86 | 0.3 | 0.19 |
| 4,000 | 2,485.48 | 400 | 248.55 | 40 | 24.85 | 4 | 2.49 | 0.4 | 0.25 |
| 5,000 | 3,106.86 | 500 | 310.69 | 50 | 31.07 | 5 | 3.11 | 0.5 | 0.31 |
| 6,000 | 3,728.23 | 600 | 372.82 | 60 | 37.28 | 6 | 3.73 | 0.6 | 0.37 |
| 7,000 | 4,349.60 | 700 | 434.96 | 70 | 43.50 | 7 | 4.35 | 0.7 | 0.43 |
| 8,000 | 4,970.97 | 800 | 497.10 | 80 | 49.71 | 8 | 4.97 | 0.8 | 0.50 |
| 9,000 | 5,592.34 | 900 | 559.23 | 90 | 55.92 | 9 | 5.59 | 0.9 | 0.56 |
| 10,000 | 6,213.71 | 1,000 | 621.37 | 100 | 62.14 | 10 | 6.21 | 1.0 | 0.62 |

## Square Inch—Square Centimetre and Square Foot—Square Metre Conversion Tables
(Based on 1 inch = 2.54 centimetres, exactly)

### SQUARE INCHES TO SQUARE CENTIMETRES
(1 in.² = 6.4516 cm², exactly)

| in.² | cm² | in.² | cm² | in.² | cm² | in.² | cm² | in.² | cm² |
|---|---|---|---|---|---|---|---|---|---|
| 100 | 645.16 | 10 | 64.52 | 1 | 6.45 | 0.1 | 0.65 | 0.01 | 0.06 |
| 200 | 1,290.32 | 20 | 129.03 | 2 | 12.90 | 0.2 | 1.29 | 0.02 | 0.13 |
| 300 | 1,935.48 | 30 | 193.55 | 3 | 19.35 | 0.3 | 1.94 | 0.03 | 0.19 |
| 400 | 2,580.64 | 40 | 258.06 | 4 | 25.81 | 0.4 | 2.58 | 0.04 | 0.26 |
| 500 | 3,225.80 | 50 | 322.58 | 5 | 32.26 | 0.5 | 3.23 | 0.05 | 0.32 |
| 600 | 3,870.96 | 60 | 387.10 | 6 | 38.71 | 0.6 | 3.87 | 0.06 | 0.39 |
| 700 | 4,516.12 | 70 | 451.61 | 7 | 45.16 | 0.7 | 4.52 | 0.07 | 0.45 |
| 800 | 5,161.28 | 80 | 516.13 | 8 | 51.61 | 0.8 | 5.16 | 0.08 | 0.52 |
| 900 | 5,806.44 | 90 | 580.64 | 9 | 58.06 | 0.9 | 5.81 | 0.09 | 0.58 |
| 1,000 | 6,451.60 | 100 | 645.16 | 10 | 64.52 | 1.0 | 6.45 | 0.10 | 0.65 |

### SQUARE CENTIMETRES TO SQUARE INCHES
(1 cm² = 0.1550003 in.²)

| cm² | in.² | cm² | in.² | cm² | in.² | cm² | in.² | cm² | in.² |
|---|---|---|---|---|---|---|---|---|---|
| 100 | 15.500 | 10 | 1.550 | 1 | 0.155 | 0.1 | 0.016 | 0.01 | 0.002 |
| 200 | 31.000 | 20 | 3.100 | 2 | 0.310 | 0.2 | 0.031 | 0.02 | 0.003 |
| 300 | 46.500 | 30 | 4.650 | 3 | 0.465 | 0.3 | 0.047 | 0.03 | 0.005 |
| 400 | 62.000 | 40 | 6.200 | 4 | 0.620 | 0.4 | 0.062 | 0.04 | 0.006 |
| 500 | 77.500 | 50 | 7.750 | 5 | 0.775 | 0.5 | 0.078 | 0.05 | 0.008 |
| 600 | 93.000 | 60 | 9.300 | 6 | 0.930 | 0.6 | 0.093 | 0.06 | 0.009 |
| 700 | 108.500 | 70 | 10.850 | 7 | 1.085 | 0.7 | 0.109 | 0.07 | 0.011 |
| 800 | 124.000 | 80 | 12.400 | 8 | 1.240 | 0.8 | 0.124 | 0.08 | 0.012 |
| 900 | 139.500 | 90 | 13.950 | 9 | 1.395 | 0.9 | 0.140 | 0.09 | 0.014 |
| 1,000 | 155.000 | 100 | 15.500 | 10 | 1.550 | 1.0 | 0.155 | 0.10 | 0.016 |

### SQUARE FEET TO SQUARE METRES
(1 ft² = 0.09290304 m², exactly)

| ft² | m² | ft² | m² | ft² | m² | ft² | m² | ft² | m² |
|---|---|---|---|---|---|---|---|---|---|
| 1,000 | 92.903 | 100 | 9.290 | 10 | 0.929 | 1 | 0.093 | 0.1 | 0.009 |
| 2,000 | 185.806 | 200 | 18.581 | 20 | 1.858 | 2 | 0.186 | 0.2 | 0.019 |
| 3,000 | 278.709 | 300 | 27.871 | 30 | 2.787 | 3 | 0.279 | 0.3 | 0.028 |
| 4,000 | 371.612 | 400 | 37.161 | 40 | 3.716 | 4 | 0.372 | 0.4 | 0.037 |
| 5,000 | 464.515 | 500 | 46.452 | 50 | 4.645 | 5 | 0.465 | 0.5 | 0.046 |
| 6,000 | 557.418 | 600 | 55.742 | 60 | 5.574 | 6 | 0.557 | 0.6 | 0.056 |
| 7,000 | 650.321 | 700 | 65.032 | 70 | 6.503 | 7 | 0.650 | 0.7 | 0.065 |
| 8,000 | 743.224 | 800 | 74.322 | 80 | 7.432 | 8 | 0.743 | 0.8 | 0.074 |
| 9,000 | 836.127 | 900 | 83.613 | 90 | 8.361 | 9 | 0.836 | 0.9 | 0.084 |
| 10,000 | 929.030 | 1,000 | 92.903 | 100 | 9.290 | 10 | 0.929 | 1.0 | 0.093 |

### SQUARE METRES TO SQUARE FEET
(1 m² = 10.76391 ft²)

| m² | ft² | m² | ft² | m² | ft² | m² | ft² | m² | ft² |
|---|---|---|---|---|---|---|---|---|---|
| 100 | 1,076.39 | 10 | 107.64 | 1 | 10.76 | 0.1 | 1.08 | 0.01 | 0.11 |
| 200 | 2,152.78 | 20 | 215.28 | 2 | 21.53 | 0.2 | 2.15 | 0.02 | 0.22 |
| 300 | 3,229.17 | 30 | 322.92 | 3 | 32.29 | 0.3 | 3.23 | 0.03 | 0.32 |
| 400 | 4,305.56 | 40 | 430.56 | 4 | 43.06 | 0.4 | 4.31 | 0.04 | 0.43 |
| 500 | 5,381.96 | 50 | 538.20 | 5 | 53.82 | 0.5 | 5.38 | 0.05 | 0.54 |
| 600 | 6,458.35 | 60 | 645.83 | 6 | 64.58 | 0.6 | 6.46 | 0.06 | 0.65 |
| 700 | 7,534.74 | 70 | 753.47 | 7 | 75.35 | 0.7 | 7.53 | 0.07 | 0.75 |
| 800 | 8,611.13 | 80 | 861.11 | 8 | 86.11 | 0.8 | 8.61 | 0.08 | 0.86 |
| 900 | 9,687.52 | 90 | 968.75 | 9 | 96.88 | 0.9 | 9.69 | 0.09 | 0.97 |
| 1,000 | 10,763.91 | 100 | 1,076.39 | 10 | 107.64 | 1.0 | 10.76 | 0.10 | 1.08 |

## Acre — Hectare and U.K. Gallon — Litre Conversion Tables

### ACRES TO HECTARES
(1 acre = 0.4046856 hectare)

| acres | 0 | 10 | 20 | 30 | 40 | 50 | 60 | 70 | 80 | 90 |
|---|---|---|---|---|---|---|---|---|---|---|
| | | | | | hectares | | | | | |
| 0 | .... | 4.047 | 8.094 | 12.141 | 16.187 | 20.234 | 24.281 | 28.328 | 32.375 | 36.422 |
| 100 | 40.469 | 44.515 | 48.562 | 52.609 | 56.656 | 60.703 | 64.750 | 68.797 | 72.843 | 76.890 |
| 200 | 80.937 | 84.984 | 89.031 | 93.078 | 97.125 | 101.171 | 105.218 | 109.265 | 113.312 | 117.359 |
| 300 | 121.406 | 125.453 | 129.499 | 133.546 | 137.593 | 141.640 | 145.687 | 149.734 | 153.781 | 157.827 |
| 400 | 161.874 | 165.921 | 169.968 | 174.015 | 178.062 | 182.109 | 186.155 | 190.202 | 194.249 | 198.296 |
| 500 | 202.343 | 206.390 | 210.437 | 214.483 | 218.530 | 222.577 | 226.624 | 230.671 | 234.718 | 238.765 |
| 600 | 242.811 | 246.858 | 250.905 | 254.952 | 258.999 | 263.046 | 267.092 | 271.139 | 275.186 | 279.233 |
| 700 | 283.280 | 287.327 | 291.374 | 295.420 | 299.467 | 303.514 | 307.561 | 311.608 | 315.655 | 319.702 |
| 800 | 323.748 | 327.795 | 331.842 | 335.889 | 339.936 | 343.983 | 348.030 | 352.076 | 356.123 | 360.170 |
| 900 | 364.217 | 368.264 | 372.311 | 376.358 | 380.404 | 384.451 | 388.498 | 392.545 | 396.592 | 400.639 |
| 1000 | 404.686 | .... | | | | | | | | |

### HECTARES TO ACRES
(1 hectare = 2.471054 acres)

| hectares | 0 | 10 | 20 | 30 | 40 | 50 | 60 | 70 | 80 | 90 |
|---|---|---|---|---|---|---|---|---|---|---|
| | | | | | acres | | | | | |
| 0 | .... | 24.71 | 49.42 | 74.13 | 98.84 | 123.55 | 148.26 | 172.97 | 197.68 | 222.39 |
| 100 | 247.11 | 271.82 | 296.53 | 321.24 | 345.95 | 370.66 | 395.37 | 420.08 | 444.79 | 469.50 |
| 200 | 494.21 | 518.92 | 543.63 | 568.34 | 593.05 | 617.76 | 642.47 | 667.18 | 691.90 | 716.61 |
| 300 | 741.32 | 766.03 | 790.74 | 815.45 | 840.16 | 864.87 | 889.58 | 914.29 | 939.00 | 963.71 |
| 400 | 988.42 | 1013.13 | 1037.84 | 1062.55 | 1087.26 | 1111.97 | 1136.68 | 1161.40 | 1186.11 | 1210.82 |
| 500 | 1235.53 | 1260.24 | 1284.95 | 1309.66 | 1334.37 | 1359.08 | 1383.79 | 1408.50 | 1433.21 | 1457.92 |
| 600 | 1482.63 | 1507.34 | 1532.05 | 1556.76 | 1581.47 | 1606.19 | 1630.90 | 1655.61 | 1680.32 | 1705.03 |
| 700 | 1729.74 | 1754.45 | 1779.16 | 1803.87 | 1828.58 | 1853.29 | 1878.00 | 1902.71 | 1927.42 | 1952.13 |
| 800 | 1976.84 | 2001.55 | 2026.26 | 2050.97 | 2075.69 | 2100.40 | 2125.11 | 2149.82 | 2174.53 | 2199.24 |
| 900 | 2223.95 | 2248.66 | 2273.37 | 2298.08 | 2322.79 | 2347.50 | 2372.21 | 2396.92 | 2421.63 | 2446.34 |
| 1000 | 2471.05 | .... | | | | | | | | |

### U.K. GALLONS TO LITRES
(1 U.K. gallon = 4.546092 litres)

| Imp. gals | 0 | 1 | 2 | 3 | 4 | 5 | 6 | 7 | 8 | 9 |
|---|---|---|---|---|---|---|---|---|---|---|
| | | | | | | litres | | | | |
| 0 | .... | 4.546 | 9.092 | 13.638 | 18.184 | 22.730 | 27.277 | 31.823 | 36.369 | 40.915 |
| 10 | 45.461 | 50.007 | 54.553 | 59.099 | 63.645 | 68.191 | 72.737 | 77.284 | 81.830 | 86.376 |
| 20 | 90.922 | 95.468 | 100.014 | 104.560 | 109.106 | 113.652 | 118.198 | 122.744 | 127.291 | 131.837 |
| 30 | 136.383 | 140.929 | 145.475 | 150.021 | 154.567 | 159.113 | 163.659 | 168.205 | 172.751 | 177.298 |
| 40 | 181.844 | 186.390 | 190.936 | 195.482 | 200.028 | 204.574 | 209.120 | 213.666 | 218.212 | 222.759 |
| 50 | 227.305 | 231.851 | 236.397 | 240.943 | 245.489 | 250.035 | 254.581 | 259.127 | 263.673 | 268.219 |
| 60 | 272.766 | 277.312 | 281.858 | 286.404 | 290.950 | 295.496 | 300.042 | 304.588 | 309.134 | 313.680 |
| 70 | 318.226 | 322.773 | 327.319 | 331.865 | 336.411 | 340.957 | 345.503 | 350.049 | 354.595 | 359.141 |
| 80 | 363.687 | 368.233 | 372.780 | 377.326 | 381.872 | 386.418 | 390.964 | 395.510 | 400.056 | 404.602 |
| 90 | 409.148 | 413.694 | 418.240 | 422.787 | 427.333 | 431.879 | 436.425 | 440.971 | 445.517 | 450.063 |
| 100 | 454.609 | 459.155 | 463.701 | 468.247 | 472.794 | 477.340 | 481.886 | 486.432 | 490.978 | 495.524 |

### LITRES TO U.K. GALLONS
(1 litre = 0.2199692 U.K. gallons)

| litres | 0 | 1 | 2 | 3 | 4 | 5 | 6 | 7 | 8 | 9 |
|---|---|---|---|---|---|---|---|---|---|---|
| | | | | | Imperial gallons | | | | | |
| 0 | .... | 0.220 | 0.440 | 0.660 | 0.880 | 1.100 | 1.320 | 1.540 | 1.760 | 1.980 |
| 10 | 2.200 | 2.420 | 2.640 | 2.860 | 3.080 | 3.300 | 3.520 | 3.739 | 3.959 | 4.179 |
| 20 | 4.399 | 4.619 | 4.839 | 5.059 | 5.279 | 5.499 | 5.719 | 5.939 | 6.159 | 6.379 |
| 30 | 6.599 | 6.819 | 7.039 | 7.259 | 7.479 | 7.699 | 7.919 | 8.139 | 8.359 | 8.579 |
| 40 | 8.799 | 9.019 | 9.239 | 9.459 | 9.679 | 9.899 | 10.119 | 10.339 | 10.559 | 10.778 |
| 50 | 10.998 | 11.218 | 11.438 | 11.658 | 11.878 | 12.098 | 12.318 | 12.538 | 12.758 | 12.978 |
| 60 | 13.198 | 13.418 | 13.638 | 13.858 | 14.078 | 14.298 | 14.518 | 14.738 | 14.958 | 15.178 |
| 70 | 15.398 | 15.618 | 15.838 | 16.058 | 16.278 | 16.498 | 16.718 | 16.938 | 17.158 | 17.378 |
| 80 | 17.598 | 17.818 | 18.037 | 18.257 | 18.477 | 18.697 | 18.917 | 19.137 | 19.357 | 19.577 |
| 90 | 19.797 | 20.017 | 20.237 | 20.457 | 20.677 | 20.897 | 21.117 | 21.337 | 21.557 | 21.777 |
| 100 | 21.997 | 22.217 | 22.437 | 22.657 | 22.877 | 23.097 | 23.317 | 23.537 | 23.757 | 23.977 |

### Cubic Inch—Cubic Centimetre and Cubic Foot—Cubic Metre Conversion Tables

(Based on 1 inch = 2.54 centimetres, exactly)

#### CUBIC INCHES TO CUBIC CENTIMETRES
($1$ in.$^3$ = 16.38706 cm$^3$)

| in.$^3$ | cm$^3$ | in.$^3$ | cm$^3$ | in.$^3$ | cm$^3$ | in.$^3$ | cm$^3$ | in.$^3$ | cm$^3$ |
|---|---|---|---|---|---|---|---|---|---|
| 100 | 1,638.71 | 10 | 163.87 | 1 | 16.39 | 0.1 | 1.64 | 0.01 | 0.16 |
| 200 | 3,277.41 | 20 | 327.74 | 2 | 32.77 | 0.2 | 3.28 | 0.02 | 0.33 |
| 300 | 4,916.12 | 30 | 491.61 | 3 | 49.16 | 0.3 | 4.92 | 0.03 | 0.49 |
| 400 | 6,554.82 | 40 | 655.48 | 4 | 65.55 | 0.4 | 6.55 | 0.04 | 0.66 |
| 500 | 8,193.53 | 50 | 819.35 | 5 | 81.94 | 0.5 | 8.19 | 0.05 | 0.82 |
| 600 | 9,832.24 | 60 | 983.22 | 6 | 98.32 | 0.6 | 9.83 | 0.06 | 0.98 |
| 700 | 11,470.94 | 70 | 1,147.09 | 7 | 114.71 | 0.7 | 11.47 | 0.07 | 1.15 |
| 800 | 13,109.65 | 80 | 1,310.96 | 8 | 131.10 | 0.8 | 13.11 | 0.08 | 1.31 |
| 900 | 14,748.35 | 90 | 1,474.84 | 9 | 147.48 | 0.9 | 14.75 | 0.09 | 1.47 |
| 1,000 | 16,387.06 | 100 | 1,638.71 | 10 | 163.87 | 1.0 | 16.39 | 0.10 | 1.64 |

#### CUBIC CENTIMETRES TO CUBIC INCHES
($1$ cm$^3$ = 0.06102376 in.$^3$)

| cm$^3$ | in.$^3$ | cm$^3$ | in.$^3$ | cm$^3$ | in.$^3$ | cm$^3$ | in.$^3$ | cm$^3$ | in.$^3$ |
|---|---|---|---|---|---|---|---|---|---|
| 1,000 | 61.024 | 100 | 6.102 | 10 | 0.610 | 1 | 0.061 | 0.1 | 0.006 |
| 2,000 | 122.048 | 200 | 12.205 | 20 | 1.220 | 2 | 0.122 | 0.2 | 0.012 |
| 3,000 | 183.071 | 300 | 18.307 | 30 | 1.831 | 3 | 0.183 | 0.3 | 0.018 |
| 4,000 | 244.095 | 400 | 24.410 | 40 | 2.441 | 4 | 0.244 | 0.4 | 0.024 |
| 5,000 | 305.119 | 500 | 30.512 | 50 | 3.051 | 5 | 0.305 | 0.5 | 0.031 |
| 6,000 | 366.143 | 600 | 36.614 | 60 | 3.661 | 6 | 0.366 | 0.6 | 0.037 |
| 7,000 | 427.166 | 700 | 42.717 | 70 | 4.272 | 7 | 0.427 | 0.7 | 0.043 |
| 8,000 | 488.190 | 800 | 48.819 | 80 | 4.882 | 8 | 0.488 | 0.8 | 0.049 |
| 9,000 | 549.214 | 900 | 54.921 | 90 | 5.492 | 9 | 0.549 | 0.9 | 0.055 |
| 10,000 | 610.238 | 1,000 | 61.024 | 100 | 6.102 | 10 | 0.610 | 1.0 | 0.061 |

#### CUBIC FEET TO CUBIC METRES
($1$ ft$^3$ = 0.02831685 m$^3$)

| ft$^3$ | m$^3$ | ft$^3$ | m$^3$ | ft$^3$ | m$^3$ | ft$^3$ | m$^3$ | ft$^3$ | m$^3$ |
|---|---|---|---|---|---|---|---|---|---|
| 1,000 | 28.317 | 100 | 2.832 | 10 | 0.283 | 1 | 0.028 | 0.1 | 0.003 |
| 2,000 | 56.634 | 200 | 5.663 | 20 | 0.566 | 2 | 0.057 | 0.2 | 0.006 |
| 3,000 | 84.951 | 300 | 8.495 | 30 | 0.850 | 3 | 0.085 | 0.3 | 0.008 |
| 4,000 | 113.267 | 400 | 11.327 | 40 | 1.133 | 4 | 0.113 | 0.4 | 0.011 |
| 5,000 | 141.584 | 500 | 14.158 | 50 | 1.416 | 5 | 0.142 | 0.5 | 0.014 |
| 6,000 | 169.901 | 600 | 16.990 | 60 | 1.699 | 6 | 0.170 | 0.6 | 0.017 |
| 7,000 | 198.218 | 700 | 19.822 | 70 | 1.982 | 7 | 0.198 | 0.7 | 0.020 |
| 8,000 | 226.535 | 800 | 22.653 | 80 | 2.265 | 8 | 0.227 | 0.8 | 0.023 |
| 9,000 | 254.852 | 900 | 25.485 | 90 | 2.549 | 9 | 0.255 | 0.9 | 0.025 |
| 10,000 | 283.168 | 1,000 | 28.317 | 100 | 2.832 | 10 | 0.283 | 1.0 | 0.028 |

#### CUBIC METRES TO CUBIC FEET
($1$ m$^3$ = 35.31466 ft$^3$)

| m$^3$ | ft$^3$ | m$^3$ | ft$^3$ | m$^3$ | ft$^3$ | m$^3$ | ft$^3$ | m$^3$ | ft$^3$ |
|---|---|---|---|---|---|---|---|---|---|
| 100 | 3,531.47 | 10 | 353.15 | 1 | 35.31 | 0.1 | 3.53 | 0.01 | 0.35 |
| 200 | 7,062.93 | 20 | 706.29 | 2 | 70.63 | 0.2 | 7.06 | 0.02 | 0.71 |
| 300 | 10,594.40 | 30 | 1,059.44 | 3 | 105.94 | 0.3 | 10.59 | 0.03 | 1.06 |
| 400 | 14,125.86 | 40 | 1,412.59 | 4 | 141.26 | 0.4 | 14.13 | 0.04 | 1.41 |
| 500 | 17,657.33 | 50 | 1,765.73 | 5 | 176.57 | 0.5 | 17.66 | 0.05 | 1.77 |
| 600 | 21,188.80 | 60 | 2,118.88 | 6 | 211.89 | 0.6 | 21.19 | 0.06 | 2.12 |
| 700 | 24,720.26 | 70 | 2,472.03 | 7 | 247.20 | 0.7 | 24.72 | 0.07 | 2.47 |
| 800 | 28,251.73 | 80 | 2,825.17 | 8 | 282.52 | 0.8 | 28.25 | 0.08 | 2.83 |
| 900 | 31,783.19 | 90 | 3,178.32 | 9 | 317.83 | 0.9 | 31.78 | 0.09 | 3.18 |
| 1,000 | 35,314.66 | 100 | 3,531.47 | 10 | 353.15 | 1.0 | 35.31 | 0.10 | 3.53 |

## Cubic Foot—Litre and Gallon—Litre Conversion Tables

(Based on 1 litre = 1000 cubic centimetres)

### CUBIC FEET TO LITRES
(1 ft³ = 28.31685 litres)

| ft³ | litres | ft³ | litres | ft³ | litres | ft³ | litres | ft³ | litres |
|---|---|---|---|---|---|---|---|---|---|
| 100 | 2,831.68 | 10 | 283.17 | 1 | 28.32 | 0.1 | 2.83 | 0.01 | 0.28 |
| 200 | 5,663.37 | 20 | 566.34 | 2 | 56.63 | 0.2 | 5.66 | 0.02 | 0.57 |
| 300 | 8,495.06 | 30 | 849.51 | 3 | 84.95 | 0.3 | 8.50 | 0.03 | 0.85 |
| 400 | 11,326.74 | 40 | 1,132.67 | 4 | 113.27 | 0.4 | 11.33 | 0.04 | 1.13 |
| 500 | 14,158.42 | 50 | 1,415.84 | 5 | 141.58 | 0.5 | 14.16 | 0.05 | 1.42 |
| 600 | 16,990.11 | 60 | 1,699.01 | 6 | 169.90 | 0.6 | 16.99 | 0.06 | 1.70 |
| 700 | 19,821.80 | 70 | 1,982.18 | 7 | 198.22 | 0.7 | 19.82 | 0.07 | 1.98 |
| 800 | 22,653.48 | 80 | 2,263.35 | 8 | 226.53 | 0.8 | 22.65 | 0.08 | 2.27 |
| 900 | 25,485.16 | 90 | 2,548.52 | 9 | 254.85 | 0.9 | 25.49 | 0.09 | 2.55 |
| 1,000 | 28,316.85 | 100 | 2,831.68 | 10 | 283.17 | 1.0 | 28.32 | 0.10 | 2.83 |

### LITRES TO CUBIC FEET
(1 litre = 0.03531466 ft³)

| litres | ft³ | litres | ft³ | litres | ft³ | litres | ft³ | litres | ft³ |
|---|---|---|---|---|---|---|---|---|---|
| 1,000 | 35.315 | 100 | 3.531 | 10 | 0.353 | 1 | 0.035 | 0.1 | 0.004 |
| 2,000 | 70.629 | 200 | 7.063 | 20 | 0.706 | 2 | 0.071 | 0.2 | 0.007 |
| 3,000 | 105.944 | 300 | 10.594 | 30 | 1.059 | 3 | 0.106 | 0.3 | 0.011 |
| 4,000 | 141.259 | 400 | 14.126 | 40 | 1.413 | 4 | 0.141 | 0.4 | 0.014 |
| 5,000 | 176.573 | 500 | 17.657 | 50 | 1.766 | 5 | 0.177 | 0.5 | 0.018 |
| 6,000 | 211.888 | 600 | 21.189 | 60 | 2.119 | 6 | 0.212 | 0.6 | 0.021 |
| 7,000 | 247.203 | 700 | 24.720 | 70 | 2.472 | 7 | 0.247 | 0.7 | 0.025 |
| 8,000 | 282.517 | 800 | 28.252 | 80 | 2.825 | 8 | 0.283 | 0.8 | 0.028 |
| 9,000 | 317.832 | 900 | 31.783 | 90 | 3.178 | 9 | 0.318 | 0.9 | 0.032 |
| 10,000 | 353.147 | 1,000 | 35.315 | 100 | 3.531 | 10 | 0.353 | 1.0 | 0.035 |

### U. S. GALLONS TO LITRES
(1 U. S. gallon = 3.785412 litres)

| gals | litres | gals | litres | gals | litres | gals | litres | gals | litres |
|---|---|---|---|---|---|---|---|---|---|
| 1,000 | 3,785.41 | 100 | 378.54 | 10 | 37.85 | 1 | 3.79 | 0.1 | 0.38 |
| 2,000 | 7,570.82 | 200 | 757.08 | 20 | 75.71 | 2 | 7.57 | 0.2 | 0.76 |
| 3,000 | 11,356.24 | 300 | 1,135.62 | 30 | 113.56 | 3 | 11.36 | 0.3 | 1.14 |
| 4,000 | 15,141.65 | 400 | 1,514.16 | 40 | 151.42 | 4 | 15.14 | 0.4 | 1.51 |
| 5,000 | 18,927.06 | 500 | 1,892.71 | 50 | 189.27 | 5 | 18.93 | 0.5 | 1.89 |
| 6,000 | 22,712.47 | 600 | 2,271.25 | 60 | 227.12 | 6 | 22.71 | 0.6 | 2.27 |
| 7,000 | 26,497.88 | 700 | 2,649.79 | 70 | 264.98 | 7 | 26.50 | 0.7 | 2.65 |
| 8,000 | 30,283.30 | 800 | 3,028.33 | 80 | 302.83 | 8 | 30.28 | 0.8 | 3.03 |
| 9,000 | 34,068.71 | 900 | 3,406.87 | 90 | 340.69 | 9 | 34.07 | 0.9 | 3.41 |
| 10,000 | 37,854.12 | 1,000 | 3,785.41 | 100 | 378.54 | 10 | 37.85 | 1.0 | 3.79 |

### LITRES TO U. S. GALLONS
(1 litre = 0.2641720 U. S. gallon)

| litres | gals | litres | gals | litres | gals | litres | gals | litres | gals |
|---|---|---|---|---|---|---|---|---|---|
| 1,000 | 264.17 | 100 | 26.42 | 10 | 2.64 | 1 | 0.26 | 0.1 | 0.03 |
| 2,000 | 528.34 | 200 | 52.83 | 20 | 5.28 | 2 | 0.53 | 0.2 | 0.05 |
| 3,000 | 792.52 | 300 | 79.25 | 30 | 7.93 | 3 | 0.79 | 0.3 | 0.08 |
| 4,000 | 1,056.69 | 400 | 105.67 | 40 | 10.57 | 4 | 1.06 | 0.4 | 0.11 |
| 5,000 | 1,320.86 | 500 | 132.09 | 50 | 13.21 | 5 | 1.32 | 0.5 | 0.13 |
| 6,000 | 1,585.03 | 600 | 158.50 | 60 | 15.85 | 6 | 1.59 | 0.6 | 0.16 |
| 7,000 | 1,849.20 | 700 | 184.92 | 70 | 18.49 | 7 | 1.85 | 0.7 | 0.18 |
| 8,000 | 2,113.38 | 800 | 211.34 | 80 | 21.13 | 8 | 2.11 | 0.8 | 0.21 |
| 9,000 | 2,377.55 | 900 | 237.75 | 90 | 23.78 | 9 | 2.38 | 0.9 | 0.24 |
| 10,000 | 2,641.72 | 1,000 | 264.17 | 100 | 26.42 | 10 | 2.64 | 1.0 | 0.26 |

## Pound—Kilogram and Ounce—Gram Conversion Tables

### POUNDS TO KILOGRAMS (1 pound = 0.4535924 kilogram)

| lb | kg | lb | kg | lb | kg | lb | kg | lb | kg |
|---|---|---|---|---|---|---|---|---|---|
| 1,000 | 453.59 | 100 | 45.36 | 10 | 4.54 | 1 | 0.45 | 0.1 | 0.05 |
| 2,000 | 907.18 | 200 | 90.72 | 20 | 9.07 | 2 | 0.91 | 0.2 | 0.09 |
| 3,000 | 1,360.78 | 300 | 136.08 | 30 | 13.61 | 3 | 1.36 | 0.3 | 0.14 |
| 4,000 | 1,814.37 | 400 | 181.44 | 40 | 18.14 | 4 | 1.81 | 0.4 | 0.18 |
| 5,000 | 2,267.96 | 500 | 226.80 | 50 | 22.68 | 5 | 2.27 | 0.5 | 0.23 |
| 6,000 | 2,721.55 | 600 | 272.16 | 60 | 27.22 | 6 | 2.72 | 0.6 | 0.27 |
| 7,000 | 3,175.15 | 700 | 317.51 | 70 | 31.75 | 7 | 3.18 | 0.7 | 0.32 |
| 8,000 | 3,628.74 | 800 | 362.87 | 80 | 36.29 | 8 | 3.63 | 0.8 | 0.36 |
| 9,000 | 4,082.33 | 900 | 408.23 | 90 | 40.82 | 9 | 4.08 | 0.9 | 0.41 |
| 10,000 | 4,535.92 | 1,000 | 453.59 | 100 | 45.36 | 10 | 4.54 | 1.0 | 0.45 |

### KILOGRAMS TO POUNDS (1 kilogram = 2.204622 pounds)

| kg | lb | kg | lb | kg | lb | kg | lb | kg | lb |
|---|---|---|---|---|---|---|---|---|---|
| 1,000 | 2,204.62 | 100 | 220.46 | 10 | 22.05 | 1 | 2.20 | 0.1 | 0.22 |
| 2,000 | 4,409.24 | 200 | 440.92 | 20 | 44.09 | 2 | 4.41 | 0.2 | 0.44 |
| 3,000 | 6,613.87 | 300 | 661.39 | 30 | 66.14 | 3 | 6.61 | 0.3 | 0.66 |
| 4,000 | 8,818.49 | 400 | 881.85 | 40 | 88.18 | 4 | 8.82 | 0.4 | 0.88 |
| 5,000 | 11,023.11 | 500 | 1,102.31 | 50 | 110.23 | 5 | 11.02 | 0.5 | 1.10 |
| 6,000 | 13,227.73 | 600 | 1,322.77 | 60 | 132.28 | 6 | 13.23 | 0.6 | 1.32 |
| 7,000 | 15,432.35 | 700 | 1,543.24 | 70 | 154.32 | 7 | 15.43 | 0.7 | 1.54 |
| 8,000 | 17,636.98 | 800 | 1,763.70 | 80 | 176.37 | 8 | 17.64 | 0.8 | 1.76 |
| 9,000 | 19,841.60 | 900 | 1,984.16 | 90 | 198.42 | 9 | 19.84 | 0.9 | 1.98 |
| 10,000 | 22,046.22 | 1,000 | 2,204.62 | 100 | 220.46 | 10 | 22.05 | 1.0 | 2.20 |

### OUNCES TO GRAMS (1 ounce = 28.34952 grams)

| oz | g | oz | g | oz | g | oz | g | oz | g |
|---|---|---|---|---|---|---|---|---|---|
| 10 | 283.50 | 1 | 28.35 | 0.1 | 2.83 | 0.01 | 0.28 | 0.001 | 0.03 |
| 20 | 566.99 | 2 | 56.70 | 0.2 | 5.67 | 0.02 | 0.57 | 0.002 | 0.06 |
| 30 | 850.49 | 3 | 85.05 | 0.3 | 8.50 | 0.03 | 0.85 | 0.003 | 0.09 |
| 40 | 1,133.98 | 4 | 113.40 | 0.4 | 11.34 | 0.04 | 1.13 | 0.004 | 0.11 |
| 50 | 1,417.48 | 5 | 141.75 | 0.5 | 14.17 | 0.05 | 1.42 | 0.005 | 0.14 |
| 60 | 1,700.97 | 6 | 170.10 | 0.6 | 17.01 | 0.06 | 1.70 | 0.006 | 0.17 |
| 70 | 1,984.47 | 7 | 198.45 | 0.7 | 19.84 | 0.07 | 1.98 | 0.007 | 0.20 |
| 80 | 2,267.96 | 8 | 226.80 | 0.8 | 22.68 | 0.08 | 2.27 | 0.008 | 0.23 |
| 90 | 2,551.46 | 9 | 255.15 | 0.9 | 25.51 | 0.09 | 2.55 | 0.009 | 0.26 |
| 100 | 2,834.95 | 10 | 283.50 | 1.0 | 28.35 | 0.10 | 2.83 | 0.010 | 0.28 |

### GRAMS TO OUNCES (1 gram = 0.03527397 ounce)

| g | oz | g | oz | g | oz | g | oz | g | oz |
|---|---|---|---|---|---|---|---|---|---|
| 100 | 3.527 | 10 | 0.353 | 1 | 0.035 | 0.1 | 0.004 | 0.01 | 0.000 |
| 200 | 7.055 | 20 | 0.705 | 2 | 0.071 | 0.2 | 0.007 | 0.02 | 0.001 |
| 300 | 10.582 | 30 | 1.058 | 3 | 0.106 | 0.3 | 0.011 | 0.03 | 0.001 |
| 400 | 14.110 | 40 | 1.411 | 4 | 0.141 | 0.4 | 0.014 | 0.04 | 0.001 |
| 500 | 17.637 | 50 | 1.764 | 5 | 0.176 | 0.5 | 0.018 | 0.05 | 0.002 |
| 600 | 21.164 | 60 | 2.116 | 6 | 0.212 | 0.6 | 0.021 | 0.06 | 0.002 |
| 700 | 24.692 | 70 | 2.469 | 7 | 0.247 | 0.7 | 0.025 | 0.07 | 0.002 |
| 800 | 28.219 | 80 | 2.822 | 8 | 0.282 | 0.8 | 0.028 | 0.08 | 0.003 |
| 900 | 31.747 | 90 | 3.175 | 9 | 0.317 | 0.9 | 0.032 | 0.09 | 0.003 |
| 1,000 | 35.274 | 100 | 3.527 | 10 | 0.353 | 1.0 | 0.035 | 0.10 | 0.004 |

Pounds per Square Inch—Kilograms per Square Centimetre and
Pounds per Square Foot—Kilograms per Square Metre Conversion Tables

## POUNDS PER SQUARE INCH TO KILOGRAMS PER SQUARE CENTIMETRE
(1 lb/in.$^2$ = 0.07030697 kg/cm$^2$)

| lb/in.$^2$ | kg/cm$^2$ | lb/in.$^2$ | kg/cm$^2$ | lb/in.$^2$ | kg/cm$^2$ | lb/in.$^2$ | kg/cm$^2$ | lb/in.$^2$ | kg/cm$^2$ |
|---|---|---|---|---|---|---|---|---|---|
| 1,000 | 70.307 | 100 | 7.031 | 10 | 0.703 | 1 | 0.070 | 0.1 | 0.007 |
| 2,000 | 140.614 | 200 | 14.061 | 20 | 1.406 | 2 | 0.141 | 0.2 | 0.014 |
| 3,000 | 210.921 | 300 | 21.092 | 30 | 2.109 | 3 | 0.211 | 0.3 | 0.021 |
| 4,000 | 281.228 | 400 | 28.123 | 40 | 2.812 | 4 | 0.281 | 0.4 | 0.028 |
| 5,000 | 351.535 | 500 | 35.153 | 50 | 3.515 | 5 | 0.352 | 0.5 | 0.035 |
| 6,000 | 421.842 | 600 | 42.184 | 60 | 4.218 | 6 | 0.422 | 0.6 | 0.042 |
| 7,000 | 492.149 | 700 | 49.215 | 70 | 4.921 | 7 | 0.492 | 0.7 | 0.049 |
| 8,000 | 562.456 | 800 | 56.246 | 80 | 5.625 | 8 | 0.562 | 0.8 | 0.056 |
| 9,000 | 632.763 | 900 | 63.276 | 90 | 6.328 | 9 | 0.633 | 0.9 | 0.063 |
| 10,000 | 703.070 | 1,000 | 70.307 | 100 | 7.031 | 10 | 0.703 | 1.0 | 0.070 |

## KILOGRAMS PER SQUARE CENTIMETRE TO POUNDS PER SQUARE INCH
(1 kg/cm$^2$ = 14.22334 lb/in.$^2$)

| kg/cm$^2$ | lb/in.$^2$ | kg/cm$^2$ | lb/in.$^2$ | kg/cm$^2$ | lb/in.$^2$ | kg/cm$^2$ | lb/in.$^2$ | kg/cm$^2$ | lb/in.$^2$ |
|---|---|---|---|---|---|---|---|---|---|
| 100 | 1,422.33 | 10 | 142.23 | 1 | 14.22 | 0.1 | 1.42 | 0.01 | 0.14 |
| 200 | 2,844.67 | 20 | 284.47 | 2 | 28.45 | 0.2 | 2.84 | 0.02 | 0.28 |
| 300 | 4,267.00 | 30 | 426.70 | 3 | 42.67 | 0.3 | 4.27 | 0.03 | 0.43 |
| 400 | 5,689.34 | 40 | 568.93 | 4 | 56.89 | 0.4 | 5.69 | 0.04 | 0.57 |
| 500 | 7,111.67 | 50 | 711.17 | 5 | 71.12 | 0.5 | 7.11 | 0.05 | 0.71 |
| 600 | 8,534.00 | 60 | 853.40 | 6 | 85.34 | 0.6 | 8.53 | 0.06 | 0.85 |
| 700 | 9,956.34 | 70 | 995.63 | 7 | 99.56 | 0.7 | 9.96 | 0.07 | 1.00 |
| 800 | 11,378.67 | 80 | 1,137.87 | 8 | 113.79 | 0.8 | 11.38 | 0.08 | 1.14 |
| 900 | 12,801.01 | 90 | 1,280.10 | 9 | 128.01 | 0.9 | 12.80 | 0.09 | 1.28 |
| 1,000 | 14,223.34 | 100 | 1,422.33 | 10 | 142.23 | 1.0 | 14.22 | 0.10 | 1.42 |

## POUNDS PER SQUARE FOOT TO KILOGRAMS PER SQUARE METRE
(1 lb/ft$^2$ = 4.882429 kg/m$^2$)

| lb/ft$^2$ | kg/m$^2$ | lb/ft$^2$ | kg/m$^2$ | lb/ft$^2$ | kg/m$^2$ | lb/ft$^2$ | kg/m$^2$ | lb/ft$^2$ | kg/m$^2$ |
|---|---|---|---|---|---|---|---|---|---|
| 1,000 | 4,882.43 | 100 | 488.24 | 10 | 48.82 | 1 | 4.88 | 0.1 | 0.49 |
| 2,000 | 9,764.86 | 200 | 976.49 | 20 | 97.65 | 2 | 9.76 | 0.2 | 0.98 |
| 3,000 | 14,647.29 | 300 | 1,464.73 | 30 | 146.47 | 3 | 14.65 | 0.3 | 1.46 |
| 4,000 | 19,529.72 | 400 | 1,952.97 | 40 | 195.30 | 4 | 19.53 | 0.4 | 1.95 |
| 5,000 | 24,412.14 | 500 | 2,441.21 | 50 | 244.12 | 5 | 24.41 | 0.5 | 2.44 |
| 6,000 | 29,294.57 | 600 | 2,929.46 | 60 | 292.95 | 6 | 29.29 | 0.6 | 2.93 |
| 7,000 | 34,177.00 | 700 | 3,417.70 | 70 | 341.77 | 7 | 34.18 | 0.7 | 3.42 |
| 8,000 | 39,059.43 | 800 | 3,905.94 | 80 | 390.59 | 8 | 39.06 | 0.8 | 3.91 |
| 9,000 | 43,941.86 | 900 | 4,394.19 | 90 | 439.42 | 9 | 43.94 | 0.9 | 4.39 |
| 10,000 | 48,824.28 | 1,000 | 4,882.43 | 100 | 488.24 | 10 | 48.82 | 1.0 | 4.88 |

## KILOGRAMS PER SQUARE METRE TO POUNDS PER SQUARE FOOT
(1 kg/m$^2$ = 0.2048161 lb/ft$^2$)

| kg/m$^2$ | lb/ft$^2$ | kg/m$^2$ | lb/ft$^2$ | kg/m$^2$ | lb/ft$^2$ | kg/m$^2$ | lb/ft$^2$ | kg/m$^2$ | lb/ft$^2$ |
|---|---|---|---|---|---|---|---|---|---|
| 1,000 | 204.82 | 100 | 20.48 | 10 | 2.05 | 1 | 0.20 | 0.1 | 0.02 |
| 2,000 | 409.63 | 200 | 40.96 | 20 | 4.10 | 2 | 0.41 | 0.2 | 0.04 |
| 3,000 | 614.45 | 300 | 61.44 | 30 | 6.14 | 3 | 0.61 | 0.3 | 0.06 |
| 4,000 | 819.26 | 400 | 81.93 | 40 | 8.19 | 4 | 0.82 | 0.4 | 0.08 |
| 5,000 | 1,024.08 | 500 | 102.41 | 50 | 10.24 | 5 | 1.02 | 0.5 | 0.10 |
| 6,000 | 1,228.90 | 600 | 122.89 | 60 | 12.29 | 6 | 1.23 | 0.6 | 0.12 |
| 7,000 | 1,433.71 | 700 | 143.37 | 70 | 14.34 | 7 | 1.43 | 0.7 | 0.14 |
| 8,000 | 1,638.53 | 800 | 163.85 | 80 | 16.39 | 8 | 1.64 | 0.8 | 0.16 |
| 9,000 | 1,843.34 | 900 | 184.33 | 90 | 18.43 | 9 | 1.84 | 0.9 | 0.18 |
| 10,000 | 2,048.16 | 1,000 | 204.82 | 100 | 20.48 | 10 | 2.05 | 1.0 | 0.20 |

### Pounds per Square Inch — Kilopascals Conversion Table*

| lb/in.² | 0 | 1 | 2 | 3 | 4 | 5 | 6 | 7 | 8 | 9 |
|---|---|---|---|---|---|---|---|---|---|---|
| | kilopascals | | | | | | | | | |
| 0 | .... | 6.895 | 13.790 | 20.684 | 27.579 | 34.474 | 41.369 | 48.263 | 55.158 | 62.053 |
| 10 | 68.948 | 75.842 | 82.737 | 89.632 | 96.527 | 103.421 | 110.316 | 117.211 | 124.106 | 131.000 |
| 20 | 137.895 | 144.790 | 151.685 | 158.579 | 165.474 | 172.369 | 179.264 | 186.158 | 193.053 | 199.948 |
| 30 | 206.843 | 213.737 | 220.632 | 227.527 | 234.422 | 241.316 | 248.211 | 255.106 | 262.001 | 268.896 |
| 40 | 275.790 | 282.685 | 289.580 | 296.475 | 303.369 | 310.264 | 317.159 | 324.054 | 330.948 | 337.843 |
| 50 | 344.738 | 351.633 | 358.527 | 365.422 | 372.317 | 379.212 | 386.106 | 393.001 | 399.896 | 406.791 |
| 60 | 413.685 | 420.580 | 427.475 | 434.370 | 441.264 | 448.159 | 455.054 | 461.949 | 468.843 | 475.738 |
| 70 | 482.633 | 489.528 | 496.423 | 503.317 | 510.212 | 517.107 | 524.002 | 530.896 | 537.791 | 544.686 |
| 80 | 551.581 | 558.475 | 565.370 | 572.265 | 579.160 | 586.054 | 592.949 | 599.844 | 606.739 | 613.633 |
| 90 | 620.528 | 627.423 | 634.318 | 641.212 | 648.107 | 655.002 | 661.897 | 668.791 | 675.686 | 682.581 |
| 100 | 689.476 | 696.370 | 703.265 | 710.160 | 717.055 | 723.949 | 730.844 | 737.739 | 744.634 | 751.529 |

POUNDS PER SQUARE INCH TO KILOPASCALS
(1 lb/in.² = 6.894757 kPa)

| kPa | 0 | 1 | 2 | 3 | 4 | 5 | 6 | 7 | 8 | 9 |
|---|---|---|---|---|---|---|---|---|---|---|
| | lb/in.² | | | | | | | | | |
| 0 | .... | 0.145 | 0.290 | 0.435 | 0.580 | 0.725 | 0.870 | 1.015 | 1.160 | 1.305 |
| 10 | 1.450 | 1.595 | 1.740 | 1.885 | 2.031 | 2.176 | 2.321 | 2.466 | 2.611 | 2.756 |
| 20 | 2.901 | 3.046 | 3.191 | 3.336 | 3.481 | 3.626 | 3.771 | 3.916 | 4.061 | 4.206 |
| 30 | 4.351 | 4.496 | 4.641 | 4.786 | 4.931 | 5.076 | 5.221 | 5.366 | 5.511 | 5.656 |
| 40 | 5.802 | 5.947 | 6.092 | 6.237 | 6.382 | 6.527 | 6.672 | 6.817 | 6.962 | 7.107 |
| 50 | 7.252 | 7.397 | 7.542 | 7.687 | 7.832 | 7.977 | 8.122 | 8.267 | 8.412 | 8.557 |
| 60 | 8.702 | 8.847 | 8.992 | 9.137 | 9.282 | 9.427 | 9.572 | 9.718 | 9.863 | 10.008 |
| 70 | 10.153 | 10.298 | 10.443 | 10.588 | 10.733 | 10.878 | 11.023 | 11.168 | 11.313 | 11.458 |
| 80 | 11.603 | 11.748 | 11.893 | 12.038 | 12.183 | 12.328 | 12.473 | 12.618 | 12.763 | 12.908 |
| 90 | 13.053 | 13.198 | 13.343 | 13.489 | 13.634 | 13.779 | 13.924 | 14.069 | 14.214 | 14.359 |
| 100 | 14.504 | 14.649 | 14.794 | 14.939 | 15.084 | 15.229 | 15.374 | 15.519 | 15.664 | 15.809 |

KILOPASCALS TO POUNDS PER SQUARE INCH
(1 kPa = 0.1450377 lb/in.²)

* *Note:* 1 kilopascal = 1 kilonewton/metre².

## Power and Heat Equivalents

1 horsepower-hour = 0.746 kilowatt-hour = 1,980,000 foot-pounds = 2545 Btu (British thermal units) = 2.64 pounds of water evaporated at 212° F. = 17 pounds of water raised from 62° to 212° F.

1 kilowatt-hour = 1000 watt-hours = 1.34 horsepower-hour = 2,655,200 foot-pounds = 3,600,000 joules = 3415 Btu = 3.54 pounds of water evaporated at 212° F. = 22.8 pounds of water raised from 62° to 212° F.

1 horsepower = 746 watts = 0.746 kilowatt = 33,000 foot-pounds per minute = 550 foot-pounds per second = 2545 Btu per hour = 42.4 Btu per minute = 0.71 Btu per second = 2.64 lbs. of water evaporated per hour at 212° F.

1 kilowatt = 1000 watts = 1.34 horsepower = 2,655,200 foot-pounds per hour = 44,200 foot-pounds per minute = 737 foot-pounds per second = 3415 Btu per hour = 57 Btu per minute = 0.95 Btu per second = 3.54 pounds of water evaporated per hour at 212° F.

1 watt = 1 joule per second = 0.00134 horsepower = 0.001 kilowatt = 3.42 Btu per hour = 44.22 foot-pounds per minute = 0.74 foot-pounds per second = 0.0035 pound of water evaporated per hour at 212° F.

1 Btu (British thermal unit) = 1052 watt-seconds = 778 foot-pounds = 0.252 kilogram-calorie = 0.000292 kilowatt-hour = 0.000393 horsepower-hour = 0.00104 pound of water evaporated at 212° F.

1 foot-pound = 1.36 joule = 0.000000377 kilowatt-hour = 0.00129 Btu = 0.0000005 horsepower-hour.

1 joule = 1 watt-second = 0.000000278 kilowatt-hour = 0.00095 Btu = 0.74 foot-pound.

Pounds per Cubic Inch—Grams per Cubic Centimetre and Pounds per Cubic Foot—Kilograms per Cubic Metre Conversion Tables

## POUNDS PER CUBIC INCH TO GRAMS PER CUBIC CENTIMETRE
(1 lb/in.³ = 27.67990 g/cm³)

| lb/in.³ | g/cm³ | lb/in.³ | g/cm³ | lb/in.³ | g/cm³ | lb/in.³ | g/cm³ | lb/in.³ | g/cm³ |
|---|---|---|---|---|---|---|---|---|---|
| 100 | 2,767.99 | 10 | 276.80 | 1 | 27.68 | 0.1 | 2.77 | 0.01 | 0.28 |
| 200 | 5,535.98 | 20 | 553.60 | 2 | 55.36 | 0.2 | 5.54 | 0.02 | 0.55 |
| 300 | 8,303.97 | 30 | 830.40 | 3 | 83.04 | 0.3 | 8.30 | 0.03 | 0.83 |
| 400 | 11,071.96 | 40 | 1,107.20 | 4 | 110.72 | 0.4 | 11.07 | 0.04 | 1.11 |
| 500 | 13,839.95 | 50 | 1,384.00 | 5 | 138.40 | 0.5 | 13.84 | 0.05 | 1.38 |
| 600 | 16,607.94 | 60 | 1,660.79 | 6 | 166.08 | 0.6 | 16.61 | 0.06 | 1.66 |
| 700 | 19,375.93 | 70 | 1,937.59 | 7 | 193.76 | 0.7 | 19.38 | 0.07 | 1.94 |
| 800 | 22,143.92 | 80 | 2,214.39 | 8 | 221.44 | 0.8 | 22.14 | 0.08 | 2.21 |
| 900 | 24,911.91 | 90 | 2,491.19 | 9 | 249.12 | 0.9 | 24.91 | 0.09 | 2.49 |
| 1,000 | 27,679.90 | 100 | 2,767.99 | 10 | 276.80 | 1.0 | 27.68 | 0.10 | 2.77 |

## GRAMS PER CUBIC CENTIMETRE TO POUNDS PER CUBIC INCH
(1 g/cm³ = 0.03612730 lb/in.³)

| g/cm³ | lb/in.³ | g/cm³ | lb/in.³ | g/cm³ | lb/in.³ | g/cm³ | lb/in.³ | g/cm³ | lb/in.³ |
|---|---|---|---|---|---|---|---|---|---|
| 1,000 | 36.127 | 100 | 3.613 | 10 | 0.361 | 1 | 0.036 | 0.1 | 0.004 |
| 2,000 | 72.255 | 200 | 7.225 | 20 | 0.723 | 2 | 0.072 | 0.2 | 0.007 |
| 3,000 | 108.382 | 300 | 10.838 | 30 | 1.084 | 3 | 0.108 | 0.3 | 0.011 |
| 4,000 | 144.509 | 400 | 14.451 | 40 | 1.445 | 4 | 0.145 | 0.4 | 0.014 |
| 5,000 | 180.636 | 500 | 18.064 | 50 | 1.806 | 5 | 0.181 | 0.5 | 0.018 |
| 6,000 | 216.764 | 600 | 21.676 | 60 | 2.168 | 6 | 0.217 | 0.6 | 0.022 |
| 7,000 | 252.891 | 700 | 25.289 | 70 | 2.529 | 7 | 0.253 | 0.7 | 0.025 |
| 8,000 | 289.018 | 800 | 28.902 | 80 | 2.890 | 8 | 0.289 | 0.8 | 0.029 |
| 9,000 | 325.146 | 900 | 32.515 | 90 | 3.251 | 9 | 0.325 | 0.9 | 0.033 |
| 10,000 | 361.273 | 1,000 | 36.127 | 100 | 3.613 | 10 | 0.361 | 1.0 | 0.036 |

## POUNDS PER CUBIC FOOT TO KILOGRAMS PER CUBIC METRE
(1 lb/ft³ = 16.01846 kg/m³)

| lb/ft³ | kg/m³ | lb/ft³ | kg/m³ | lb/ft³ | kg/m³ | lb/ft³ | kg/m³ | lb/ft³ | kg/m³ |
|---|---|---|---|---|---|---|---|---|---|
| 100 | 1,601.85 | 10 | 160.18 | 1 | 16.02 | 0.1 | 1.60 | 0.01 | 0.16 |
| 200 | 3,203.69 | 20 | 320.37 | 2 | 32.04 | 0.2 | 3.20 | 0.02 | 0.32 |
| 300 | 4,805.54 | 30 | 480.55 | 3 | 48.06 | 0.3 | 4.81 | 0.03 | 0.48 |
| 400 | 6,407.38 | 40 | 640.74 | 4 | 64.07 | 0.4 | 6.41 | 0.04 | 0.64 |
| 500 | 8,009.23 | 50 | 800.92 | 5 | 80.09 | 0.5 | 8.01 | 0.05 | 0.80 |
| 600 | 9,611.08 | 60 | 961.11 | 6 | 96.11 | 0.6 | 9.61 | 0.06 | 0.96 |
| 700 | 11,212.92 | 70 | 1,121.29 | 7 | 112.13 | 0.7 | 11.21 | 0.07 | 1.12 |
| 800 | 12,814.77 | 80 | 1,281.48 | 8 | 128.15 | 0.8 | 12.81 | 0.08 | 1.28 |
| 900 | 14,416.61 | 90 | 1,441.66 | 9 | 144.17 | 0.9 | 14.42 | 0.09 | 1.44 |
| 1,000 | 16,018.46 | 100 | 1,601.85 | 10 | 160.18 | 1.0 | 16.02 | 0.10 | 1.60 |

## KILOGRAMS PER CUBIC METRE TO POUNDS PER CUBIC FOOT
(1 kg/m³ = 0.06242797 lb/ft³)

| kg/m³ | lb/ft³ | kg/m³ | lb/ft³ | kg/m³ | lb/ft³ | kg/m³ | lb/ft³ | kg/m³ | lb/ft³ |
|---|---|---|---|---|---|---|---|---|---|
| 1,000 | 62.428 | 100 | 6.243 | 10 | 0.624 | 1 | 0.062 | 0.1 | 0.006 |
| 2,000 | 124.856 | 200 | 12.486 | 20 | 1.249 | 2 | 0.125 | 0.2 | 0.012 |
| 3,000 | 187.284 | 300 | 18.728 | 30 | 1.873 | 3 | 0.187 | 0.3 | 0.019 |
| 4,000 | 249.712 | 400 | 24.971 | 40 | 2.497 | 4 | 0.250 | 0.4 | 0.025 |
| 5,000 | 312.140 | 500 | 31.214 | 50 | 3.121 | 5 | 0.312 | 0.5 | 0.031 |
| 6,000 | 374.568 | 600 | 37.457 | 60 | 3.746 | 6 | 0.375 | 0.6 | 0.037 |
| 7,000 | 436.996 | 700 | 43.700 | 70 | 4.370 | 7 | 0.437 | 0.7 | 0.044 |
| 8,000 | 499.424 | 800 | 49.942 | 80 | 4.994 | 8 | 0.499 | 0.8 | 0.050 |
| 9,000 | 561.852 | 900 | 56.185 | 90 | 5.619 | 9 | 0.562 | 0.9 | 0.056 |
| 10,000 | 624.280 | 1,000 | 62.428 | 100 | 6.243 | 10 | 0.624 | 1.0 | 0.062 |

## British Thermal Unit—Foot Pound and Horsepower—Kilowatt Conversion Tables

### BRITISH THERMAL UNITS TO FOOT-POUNDS
(1 Btu = 778.26 ft.-lb.)*

| Btu | Ft.-lb. | Btu | Ft.-lb. | Btu | Ft.-lb. | Btu | Ft.-lb. | Btu | Ft.-lb. |
|---|---|---|---|---|---|---|---|---|---|
| 100 | 77,826 | 10 | 7,783 | 1 | 778 | 0.1 | 78 | 0.01 | 8 |
| 200 | 155,652 | 20 | 15,565 | 2 | 1,557 | 0.2 | 156 | 0.02 | 16 |
| 300 | 233,478 | 30 | 23,348 | 3 | 2,335 | 0.3 | 233 | 0.03 | 23 |
| 400 | 311,304 | 40 | 31,130 | 4 | 3,113 | 0.4 | 311 | 0.04 | 31 |
| 500 | 389,130 | 50 | 38,913 | 5 | 3,891 | 0.5 | 389 | 0.05 | 39 |
| 600 | 466,956 | 60 | 46,696 | 6 | 4,670 | 0.6 | 467 | 0.06 | 47 |
| 700 | 544,782 | 70 | 54,478 | 7 | 5,448 | 0.7 | 545 | 0.07 | 54 |
| 800 | 622,608 | 80 | 62,261 | 8 | 6,226 | 0.8 | 623 | 0.08 | 62 |
| 900 | 700,434 | 90 | 70,043 | 9 | 7,004 | 0.9 | 700 | 0.09 | 70 |
| 1,000 | 778,260 | 100 | 77,826 | 10 | 7,783 | 1.0 | 778 | 0.10 | 78 |

### FOOT-POUNDS TO BRITISH THERMAL UNITS
(1 ft.-lb. = 0.00128492 Btu)*

| Ft.-lb. | Btu | Ft.-lb. | Btu | Ft.-lb. | Btu | Ft.-lb. | Btu | Ft.-lb. | Btu |
|---|---|---|---|---|---|---|---|---|---|
| 10,000 | 12.849 | 1,000 | 1.285 | 100 | 0.128 | 10 | 0.013 | 1 | 0.001 |
| 20,000 | 25.698 | 2,000 | 2.570 | 200 | 0.257 | 20 | 0.026 | 2 | 0.003 |
| 30,000 | 38.548 | 3,000 | 3.855 | 300 | 0.385 | 30 | 0.039 | 3 | 0.004 |
| 40,000 | 51.397 | 4,000 | 5.140 | 400 | 0.514 | 40 | 0.051 | 4 | 0.005 |
| 50,000 | 64.246 | 5,000 | 6.425 | 500 | 0.642 | 50 | 0.064 | 5 | 0.006 |
| 60,000 | 77.095 | 6,000 | 7.710 | 600 | 0.771 | 60 | 0.077 | 6 | 0.008 |
| 70,000 | 89.944 | 7,000 | 8.994 | 700 | 0.899 | 70 | 0.090 | 7 | 0.009 |
| 80,000 | 102.794 | 8,000 | 10.279 | 800 | 1.028 | 80 | 0.103 | 8 | 0.010 |
| 90,000 | 115.643 | 9,000 | 11.564 | 900 | 1.156 | 90 | 0.116 | 9 | 0.012 |
| 100,000 | 128.492 | 10,000 | 12.849 | 1,000 | 1.285 | 100 | 0.128 | 10 | 0.013 |

### HORSEPOWER TO KILOWATTS
(1 hp = 0.7456999 kW)†

| hp | kW | hp | kW | hp | kW | hp | kW | hp | kW |
|---|---|---|---|---|---|---|---|---|---|
| 1,000 | 745.7 | 100 | 74.6 | 10 | 7.5 | 1 | 0.7 | 0.1 | 0.1 |
| 2,000 | 1,491.4 | 200 | 149.1 | 20 | 14.9 | 2 | 1.5 | 0.2 | 0.1 |
| 3,000 | 2,237.1 | 300 | 223.7 | 30 | 22.4 | 3 | 2.2 | 0.3 | 0.2 |
| 4,000 | 2,982.8 | 400 | 298.3 | 40 | 29.8 | 4 | 3.0 | 0.4 | 0.3 |
| 5,000 | 3,728.5 | 500 | 372.8 | 50 | 37.3 | 5 | 3.7 | 0.5 | 0.4 |
| 6,000 | 4,474.2 | 600 | 447.4 | 60 | 44.7 | 6 | 4.5 | 0.6 | 0.4 |
| 7,000 | 5,219.9 | 700 | 522.0 | 70 | 52.2 | 7 | 5.2 | 0.7 | 0.5 |
| 8,000 | 5,965.6 | 800 | 596.6 | 80 | 59.7 | 8 | 6.0 | 0.8 | 0.6 |
| 9,000 | 6,711.3 | 900 | 671.1 | 90 | 67.1 | 9 | 6.7 | 0.9 | 0.7 |
| 10,000 | 7,457.0 | 1,000 | 745.7 | 100 | 74.6 | 10 | 7.5 | 1.0 | 0.7 |

### KILOWATTS TO HORSEPOWER
(1 kW = 1.341022 hp)†

| kW | hp | kW | hp | kW | hp | kW | hp | kW | hp |
|---|---|---|---|---|---|---|---|---|---|
| 1,000 | 1,341.0 | 100 | 134.1 | 10 | 13.4 | 1 | 1.3 | 0.1 | 0.1 |
| 2,000 | 2,682.0 | 200 | 268.2 | 20 | 26.8 | 2 | 2.7 | 0.2 | 0.3 |
| 3,000 | 4,023.1 | 300 | 402.3 | 30 | 40.2 | 3 | 4.0 | 0.3 | 0.4 |
| 4,000 | 5,364.1 | 400 | 536.4 | 40 | 53.6 | 4 | 5.4 | 0.4 | 0.5 |
| 5,000 | 6,705.1 | 500 | 670.5 | 50 | 67.1 | 5 | 6.7 | 0.5 | 0.7 |
| 6,000 | 8,046.1 | 600 | 804.6 | 60 | 80.5 | 6 | 8.0 | 0.6 | 0.8 |
| 7,000 | 9,387.2 | 700 | 938.7 | 70 | 93.9 | 7 | 9.4 | 0.7 | 0.9 |
| 8,000 | 10,728.2 | 800 | 1,072.8 | 80 | 107.3 | 8 | 10.7 | 0.8 | 1.1 |
| 9,000 | 12,069.2 | 900 | 1,206.9 | 90 | 120.7 | 9 | 12.1 | 0.9 | 1.2 |
| 10,000 | 13,410.2 | 1,000 | 1,341.0 | 100 | 134.1 | 10 | 13.4 | 1.0 | 1.3 |

* Conversion factor defined by International Steam Table Conference, 1929.
† Based on 1 horsepower = 550 foot-pounds per second.

British Thermal Unit — Kilojoule and Foot Pound — Joule Conversion Tables

## BRITISH THERMAL UNITS TO KILOJOULES
(1 Btu. = 1055.056 joules.)

| Btu. | 0 | 100 | 200 | 300 | 400 | 500 | 600 | 700 | 800 | 900 |
|---|---|---|---|---|---|---|---|---|---|---|
| | | | | | kilojoules | | | | | |
| 0 | .... | 105.51 | 211.01 | 316.52 | 422.02 | 527.53 | 633.03 | 738.54 | 844.04 | 949.55 |
| 1000 | 1055.06 | 1160.56 | 1266.07 | 1371.57 | 1477.08 | 1582.58 | 1688.09 | 1793.60 | 1899.10 | 2004.61 |
| 2000 | 2110.11 | 2215.62 | 2321.12 | 2426.63 | 2532.13 | 2637.64 | 2743.15 | 2848.65 | 2954.16 | 3059.66 |
| 3000 | 3165.17 | 3270.67 | 3376.18 | 3481.68 | 3587.19 | 3692.70 | 3798.20 | 3903.71 | 4009.21 | 4114.72 |
| 4000 | 4220.22 | 4325.73 | 4431.24 | 4536.74 | 4642.25 | 4747.75 | 4853.26 | 4958.76 | 5064.27 | 5169.77 |
| 5000 | 5275.28 | 5380.79 | 5486.29 | 5591.80 | 5697.30 | 5802.81 | 5908.31 | 6013.82 | 6119.32 | 6224.83 |
| 6000 | 6330.34 | 6435.84 | 6541.35 | 6646.85 | 6752.36 | 6857.86 | 6963.37 | 7068.88 | 7174.38 | 7279.89 |
| 7000 | 7385.39 | 7490.90 | 7596.40 | 7701.91 | 7807.41 | 7912.92 | 8018.43 | 8123.93 | 8229.44 | 8334.94 |
| 8000 | 8440.45 | 8545.95 | 8651.46 | 8756.96 | 8862.47 | 8967.99 | 9073.48 | 9178.99 | 9284.49 | 9390.00 |
| 9000 | 9495.50 | 9601.01 | 9706.52 | 9812.02 | 9917.53 | 10023.0 | 10128.5 | 10234.0 | 10339.5 | 10445.1 |
| 10000 | 10550.6 | .... | .... | .... | .... | .... | .... | .... | .... | .... |

## KILOJOULES TO BRITISH THERMAL UNITS
(1 joule = 0.0009478170 Btu.)

| kJ | 0 | 100 | 200 | 300 | 400 | 500 | 600 | 700 | 800 | 900 |
|---|---|---|---|---|---|---|---|---|---|---|
| | | | | | British Thermal Units | | | | | |
| 0 | .... | 94.78 | 189.56 | 284.35 | 379.13 | 473.91 | 568.69 | 663.47 | 758.25 | 853.04 |
| 1000 | 947.82 | 1042.60 | 1137.38 | 1232.16 | 1326.94 | 1421.73 | 1516.51 | 1611.29 | 1706.07 | 1800.85 |
| 2000 | 1895.63 | 1990.42 | 2085.20 | 2179.98 | 2274.76 | 2369.54 | 2464.32 | 2559.11 | 2653.89 | 2748.67 |
| 3000 | 2843.45 | 2938.23 | 3033.01 | 3127.80 | 3222.58 | 3317.36 | 3412.14 | 3506.92 | 3601.70 | 3696.49 |
| 4000 | 3791.27 | 3886.05 | 3980.83 | 4075.61 | 4170.39 | 4265.18 | 4359.96 | 4454.74 | 4549.52 | 4644.30 |
| 5000 | 4739.08 | 4833.87 | 4928.65 | 5023.43 | 5118.21 | 5212.99 | 5307.78 | 5402.56 | 5497.34 | 5592.12 |
| 6000 | 5686.90 | 5781.68 | 5876.47 | 5971.25 | 6066.03 | 6160.81 | 6255.59 | 6350.37 | 6445.16 | 6539.94 |
| 7000 | 6634.72 | 6729.50 | 6824.28 | 6919.06 | 7013.85 | 7108.63 | 7203.41 | 7298.19 | 7392.97 | 7487.75 |
| 8000 | 7582.54 | 7677.32 | 7772.10 | 7866.88 | 7961.66 | 8056.44 | 8151.23 | 8246.01 | 8340.79 | 8435.57 |
| 9000 | 8530.35 | 8625.13 | 8719.92 | 8814.70 | 8909.48 | 9004.26 | 9099.04 | 9193.82 | 9288.61 | 9383.39 |
| 10000 | 9478.17 | .... | .... | .... | .... | .... | .... | .... | .... | .... |

## FOOT POUNDS TO JOULES
(1 foot pound = 1.355818 joules)

| ft-lb | 0 | 1 | 2 | 3 | 4 | 5 | 6 | 7 | 8 | 9 |
|---|---|---|---|---|---|---|---|---|---|---|
| | | | | | joules | | | | | |
| 0 | .... | 1.356 | 2.712 | 4.067 | 5.423 | 6.779 | 8.135 | 9.491 | 10.847 | 12.202 |
| 10 | 13.558 | 14.914 | 16.270 | 17.626 | 18.981 | 20.337 | 21.693 | 23.049 | 24.405 | 25.761 |
| 20 | 27.116 | 28.472 | 29.828 | 31.184 | 32.540 | 33.895 | 35.251 | 36.607 | 37.963 | 39.319 |
| 30 | 40.675 | 42.030 | 43.386 | 44.742 | 46.098 | 47.454 | 48.809 | 50.165 | 51.521 | 52.877 |
| 40 | 54.233 | 55.589 | 56.944 | 58.300 | 59.656 | 61.012 | 62.368 | 63.723 | 65.079 | 66.435 |
| 50 | 67.791 | 69.147 | 70.503 | 71.858 | 73.214 | 74.570 | 75.926 | 77.282 | 78.637 | 79.993 |
| 60 | 81.349 | 82.705 | 84.061 | 85.417 | 86.772 | 88.128 | 89.484 | 90.840 | 92.196 | 93.551 |
| 70 | 94.907 | 96.263 | 97.619 | 98.975 | 100.331 | 101.686 | 103.042 | 104.398 | 105.754 | 107.110 |
| 80 | 108.465 | 109.821 | 111.177 | 112.533 | 113.889 | 115.245 | 116.600 | 117.956 | 119.312 | 120.668 |
| 90 | 122.024 | 123.379 | 124.735 | 126.091 | 127.447 | 128.803 | 130.159 | 131.514 | 132.870 | 134.226 |
| 100 | 135.582 | 136.938 | 138.293 | 139.649 | 141.005 | 142.361 | 143.717 | 145.073 | 146.428 | 147.784 |

## JOULES TO FOOT POUNDS
(1 joule = 0.7375621 foot pound)

| J | 0 | 1 | 2 | 3 | 4 | 5 | 6 | 7 | 8 | 9 |
|---|---|---|---|---|---|---|---|---|---|---|
| | | | | | foot-pounds | | | | | |
| 0 | .... | 0.7376 | 1.4751 | 2.2127 | 2.9502 | 3.6878 | 4.4254 | 5.1629 | 5.9005 | 6.6381 |
| 10 | 7.3756 | 8.1132 | 8.8507 | 9.5883 | 10.3259 | 11.0634 | 11.8010 | 12.5386 | 13.2761 | 14.0137 |
| 20 | 14.7512 | 15.4888 | 16.2264 | 16.9639 | 17.7015 | 18.4391 | 19.1766 | 19.9142 | 20.6517 | 21.3893 |
| 30 | 22.1269 | 22.8644 | 23.6020 | 24.3395 | 25.0771 | 25.8147 | 26.5522 | 27.2898 | 28.0274 | 28.7649 |
| 40 | 29.5025 | 30.2400 | 30.9776 | 31.7152 | 32.4527 | 33.1903 | 33.9279 | 34.6654 | 35.4030 | 36.1405 |
| 50 | 36.8781 | 37.6157 | 38.3532 | 39.0908 | 39.8284 | 40.5659 | 41.3035 | 42.0410 | 42.7786 | 43.5162 |
| 60 | 44.2537 | 44.9913 | 45.7289 | 46.4664 | 47.2040 | 47.9415 | 48.6791 | 49.4167 | 50.1542 | 50.8918 |
| 70 | 51.6293 | 52.3669 | 53.1045 | 53.8420 | 54.5796 | 55.3172 | 56.0547 | 56.7923 | 57.5298 | 58.2674 |
| 80 | 59.0050 | 59.7425 | 60.4801 | 61.2177 | 61.9552 | 62.6928 | 63.4303 | 64.1679 | 64.9055 | 65.6430 |
| 90 | 66.3806 | 67.1182 | 67.8557 | 68.5933 | 69.3308 | 70.0684 | 70.8060 | 71.5435 | 72.2811 | 73.0186 |
| 100 | 73.7562 | 74.4938 | 75.2313 | 75.9689 | 76.7065 | 77.4440 | 78.1816 | 78.9191 | 79.6567 | 80.3943 |

**Pounds-force — Newtons and Pound-Inches to Newton-Metres Conversion Tables**

## POUNDS-FORCE TO NEWTONS
(1 pound-force = 4.448222 newtons.)

| lbf | 0 | 1 | 2 | 3 | 4 | 5 | 6 | 7 | 8 | 9 |
|---|---|---|---|---|---|---|---|---|---|---|
| | | | | | newtons | | | | | |
| 0 | .... | 4.448 | 8.896 | 13.345 | 17.793 | 22.241 | 26.689 | 31.138 | 35.586 | 40.034 |
| 10 | 44.482 | 48.930 | 53.379 | 57.827 | 62.275 | 66.723 | 71.172 | 75.620 | 80.068 | 84.516 |
| 20 | 88.964 | 93.413 | 97.861 | 102.309 | 106.757 | 111.206 | 115.654 | 120.102 | 124.550 | 128.998 |
| 30 | 133.447 | 137.895 | 142.343 | 146.791 | 151.240 | 155.688 | 160.136 | 164.584 | 169.032 | 173.481 |
| 40 | 177.929 | 182.377 | 186.825 | 191.274 | 195.722 | 200.170 | 204.618 | 209.066 | 213.515 | 217.963 |
| 50 | 222.411 | 226.859 | 231.308 | 235.756 | 240.204 | 244.652 | 249.100 | 253.549 | 257.997 | 262.445 |
| 60 | 266.893 | 271.342 | 275.790 | 280.238 | 284.686 | 289.134 | 293.583 | 298.031 | 302.479 | 306.927 |
| 70 | 311.376 | 315.824 | 320.272 | 324.720 | 329.168 | 333.617 | 338.065 | 342.513 | 346.961 | 351.410 |
| 80 | 355.858 | 360.306 | 364.754 | 369.202 | 373.651 | 378.099 | 382.547 | 386.995 | 391.444 | 395.892 |
| 90 | 400.340 | 404.788 | 409.236 | 413.685 | 418.133 | 422.581 | 427.029 | 431.478 | 435.926 | 440.374 |
| 100 | 444.822 | 449.270 | 453.719 | 458.167 | 462.615 | 467.063 | 471.512 | 475.960 | 480.408 | 484.856 |

## NEWTONS TO POUNDS-FORCE
(1 newton = 0.2248089 pound-force.)

| N | 0 | 1 | 2 | 3 | 4 | 5 | 6 | 7 | 8 | 9 |
|---|---|---|---|---|---|---|---|---|---|---|
| | | | | | pounds-force | | | | | |
| 0 | .... | 0.22481 | 0.44962 | 0.67443 | 0.89924 | 1.12404 | 1.34885 | 1.57366 | 1.79847 | 2.02328 |
| 10 | 2.24809 | 2.47290 | 2.69771 | 2.92252 | 3.14732 | 3.37213 | 3.59694 | 3.82175 | 4.04656 | 4.27137 |
| 20 | 4.49618 | 4.72099 | 4.94580 | 5.17060 | 5.39541 | 5.62022 | 5.84503 | 6.06984 | 6.29465 | 6.51946 |
| 30 | 6.74427 | 6.96908 | 7.19388 | 7.41869 | 7.64350 | 7.86831 | 8.09312 | 8.31793 | 8.54274 | 8.76755 |
| 40 | 8.99236 | 9.21716 | 9.44197 | 9.66678 | 9.89159 | 10.1164 | 10.3412 | 10.5660 | 10.7908 | 11.0156 |
| 50 | 11.2404 | 11.4653 | 11.6901 | 11.9149 | 12.1397 | 12.3645 | 12.5893 | 12.8141 | 13.0389 | 13.2637 |
| 60 | 13.4885 | 13.7133 | 13.9382 | 14.1630 | 14.3878 | 14.6126 | 14.8374 | 15.0622 | 15.2870 | 15.5118 |
| 70 | 15.7366 | 15.9614 | 16.1862 | 16.4110 | 16.6359 | 16.8607 | 17.0855 | 17.3103 | 17.5351 | 17.7599 |
| 80 | 17.9847 | 18.2095 | 18.4343 | 18.6591 | 18.8839 | 19.1088 | 19.3336 | 19.5584 | 19.7832 | 20.0080 |
| 90 | 20.2328 | 20.4576 | 20.6824 | 20.9072 | 21.1320 | 21.3568 | 21.5817 | 21.8065 | 22.0313 | 22.2561 |
| 100 | 22.4809 | 22.7057 | 22.9305 | 23.1553 | 23.3801 | 23.6049 | 23.8297 | 24.0546 | 24.2794 | 24.5042 |

## POUND-INCHES TO NEWTON-METRES
(1 pound-force-inch = 0.1129848 newton-metre)

| lbf-in. | N·m | lbf-in. | N·m | lbf-in. | N·m | lbf-in. | N·m | lbf-in. | N·m |
|---|---|---|---|---|---|---|---|---|---|
| 100 | 11.298 | 10 | 1.130 | 1 | 0.113 | 0.1 | 0.011 | 0.01 | 0.001 |
| 200 | 22.597 | 20 | 2.260 | 2 | 0.226 | 0.2 | 0.023 | 0.02 | 0.002 |
| 300 | 33.895 | 30 | 3.390 | 3 | 0.339 | 0.3 | 0.034 | 0.03 | 0.003 |
| 400 | 45.194 | 40 | 4.519 | 4 | 0.452 | 0.4 | 0.045 | 0.04 | 0.005 |
| 500 | 56.492 | 50 | 5.649 | 5 | 0.565 | 0.5 | 0.056 | 0.05 | 0.006 |
| 600 | 67.791 | 60 | 6.779 | 6 | 0.678 | 0.6 | 0.068 | 0.06 | 0.007 |
| 700 | 79.089 | 70 | 7.909 | 7 | 0.791 | 0.7 | 0.079 | 0.07 | 0.008 |
| 800 | 90.388 | 80 | 9.039 | 8 | 0.904 | 0.8 | 0.090 | 0.08 | 0.009 |
| 900 | 101.686 | 90 | 10.169 | 9 | 1.017 | 0.9 | 0.102 | 0.09 | 0.010 |
| 1000 | 112.985 | 100 | 11.298 | 10 | 1.130 | 1.0 | 0.113 | 0.10 | 0.011 |

## NEWTON-METRES TO POUND-INCHES
(1 newton-metre = 8.850748 pound-force-inches)

| N·m | lbf-in. | N·m | lbf-in. | N·m | lbf-in. | N·m | lbf-in. | N·m | lbf-in. |
|---|---|---|---|---|---|---|---|---|---|
| 100 | 885.07 | 10 | 88.51 | 1 | 8.85 | 0.1 | 0.89 | 0.01 | 0.09 |
| 200 | 1770.15 | 20 | 177.01 | 2 | 17.70 | 0.2 | 1.77 | 0.02 | 0.18 |
| 300 | 2655.22 | 30 | 265.52 | 3 | 26.55 | 0.3 | 2.66 | 0.03 | 0.27 |
| 400 | 3540.30 | 40 | 354.03 | 4 | 35.40 | 0.4 | 3.54 | 0.04 | 0.35 |
| 500 | 4425.37 | 50 | 442.54 | 5 | 44.25 | 0.5 | 4.43 | 0.05 | 0.44 |
| 600 | 5310.45 | 60 | 531.04 | 6 | 53.10 | 0.6 | 5.31 | 0.06 | 0.53 |
| 700 | 6195.52 | 70 | 619.55 | 7 | 61.96 | 0.7 | 6.20 | 0.07 | 0.62 |
| 800 | 7080.60 | 80 | 708.06 | 8 | 70.81 | 0.8 | 7.08 | 0.08 | 0.71 |
| 900 | 7965.67 | 90 | 796.57 | 9 | 79.66 | 0.9 | 7.97 | 0.09 | 0.80 |
| 1000 | 8850.75 | 100 | 885.07 | 10 | 88.51 | 1.0 | 8.85 | 0.10 | 0.89 |

### Metric Conversion Factors

(Symbols of SI units, multiples and submultiples are
given in parentheses in the right-hand column)

| Multiply | By | To Obtain |
|---|---|---|
| LENGTH | | |
| centimetre | 0.03280840 | foot |
| centimetre | 0.3937008 | inch |
| fathom | 1.8288* | metre (m) |
| foot | 0.3048* | metre (m) |
| foot | 30.48* | centimetre (cm) |
| foot | 304.8* | millimetre (mm) |
| inch | 0.0254* | metre (m) |
| inch | 2.54* | centimetre (cm) |
| inch | 25.4* | millimetre (mm) |
| kilometre | 0.6213712 | mile [U. S. statute] |
| metre | 39.37008 | inch |
| metre | 0.5468066 | fathom |
| metre | 3.280840 | foot |
| metre | 0.1988388 | rod |
| metre | 1.093613 | yard |
| metre | 0.0006213712 | mile [U. S. statute] |
| microinch | 0.0254* | micrometre [micron] ($\mu$m) |
| micrometre [micron] | 39.37008 | microinch |
| mile [U. S. statute] | 1609.344* | metre (m) |
| mile [U. S. statute] | 1.609344* | kilometre (km) |
| millimetre | 0.003280840 | foot |
| millimetre | 0.03937008 | inch |
| rod | 5.0292* | metre (m) |
| yard | 0.9144* | metre (m) |
| AREA | | |
| acre | 4046.856 | metre$^2$ (m$^2$) |
| acre | 0.4046856 | hectare |
| centimetre$^2$ | 0.1550003 | inch$^2$ |
| centimetre$^2$ | 0.001076391 | foot$^2$ |
| foot$^2$ | 0.09290304* | metre$^2$ (m$^2$) |
| foot$^2$ | 929.0304* | centimetre$^2$ (cm$^2$) |
| foot$^2$ | 92,903.04* | millimetre$^2$ (mm$^2$) |
| hectare | 2.471054 | acre |
| inch$^2$ | 645.16* | millimetre$^2$ (mm$^2$) |
| inch$^2$ | 6.4516* | centimetre$^2$ (cm$^2$) |
| inch$^2$ | 0.00064516* | metre$^2$ (m$^2$) |
| metre$^2$ | 1550.003 | inch$^2$ |
| metre$^2$ | 10.763910 | foot$^2$ |
| metre$^2$ | 1.195990 | yard$^2$ |
| metre$^2$ | 0.0002471054 | acre |
| millimetre$^2$ | 0.00001076391 | foot$^2$ |
| millimetre$^2$ | 0.001550003 | inch$^2$ |
| yard$^2$ | 0.8361274 | metre$^2$ (m$^2$) |

* Where an asterisk is shown, the figure is exact.

**Metric Conversion Factors** (*Continued*)

| Multiply | By | To Obtain |
|---|---|---|
| VOLUME (including CAPACITY) | | |
| centimetre³ | 0.06102376 | inch³ |
| foot³ | 0.02831685 | metre³ (m³) |
| foot³ | 28.31685 | litre |
| gallon [U. K. liquid] | 0.004546092 | metre³ (m³) |
| gallon [U. K. liquid] | 4.546092 | litre |
| gallon [U. S. liquid] | 0.003785412 | metre³ (m³) |
| gallon [U. S. liquid] | 3.785412 | litre |
| inch³ | 16,387.06 | millimetre³ (mm³) |
| inch³ | 16.38706 | centimetre³ (cm³) |
| inch³ | 0.00001638706 | metre³ (m³) |
| litre | 0.001* | metre³ (m³) |
| litre | 0.2199692 | gallon [U. K. liquid] |
| litre | 0.2641720 | gallon [U. S. liquid] |
| litre | 0.03531466 | foot³ |
| metre³ | 219.9692 | gallon [U. K. liquid] |
| metre³ | 264.1720 | gallon [U. S. liquid] |
| metre³ | 35.31466 | foot³ |
| metre³ | 1.307951 | yard³ |
| metre³ | 1000.* | litre |
| metre³ | 61,023.76 | inch³ |
| millimetre³ | 0.00006102376 | inch³ |
| yard³ | 0.7645549 | metre³ (m³) |
| VELOCITY, ACCELERATION, and FLOW | | |
| centimetre/second | 1.968504 | foot/minute |
| centimetre/second | 0.03280840 | foot/second |
| centimetre/minute | 0.3937008 | inch/minute |
| foot/hour | 0.00008466667 | metre/second (m/s) |
| foot/hour | 0.00508* | metre/minute |
| foot/hour | 0.3048* | metre/hour |
| foot/minute | 0.508* | centimetre/second |
| foot/minute | 18.288* | metre/hour |
| foot/minute | 0.3048* | metre/minute |
| foot/minute | 0.00508* | metre/second (m/s) |
| foot/second | 30.48* | centimetre/second |
| foot/second | 18.288* | metre/minute |
| foot/second | 0.3048* | metre/second (m/s) |
| foot/second² | 0.3048* | metre/second² (m/s²) |
| foot³/minute | 28.31685 | litre/minute |
| foot³/minute | 0.0004719474 | metre³/second (m³/s) |
| gallon [U. S. liquid]/min. | 0.003785412 | metre³/minute |
| gallon [U. S. liquid]/min. | 0.00006309020 | metre³/second (m³/s) |
| gallon [U. S. liquid]/min. | 0.06309020 | litre/second |
| gallon [U. S. liquid]/min. | 3.785412 | litre/minute |
| gallon [U. K. liquid]/min. | 0.004546092 | metre³/minute |
| gallon [U. K. liquid]/min. | 0.00007576820 | metre³/second (m³/s) |
| inch/minute | 25.4* | millimetre/minute |
| inch/minute | 2.54* | centimetre/minute |
| inch/minute | 0.0254* | metre/minute |
| inch/second² | 0.0254* | metre/second² (m/s²) |

* Where an asterisk is shown, the figure is exact.

**Metric Conversion Factors** (*Continued*)

| Multiply | By | To Obtain |
|---|---|---|
| VELOCITY, ACCELERATION, and FLOW (*Continued*) | | |
| kilometre/hour | 0.6213712 | mile/hour [U. S. statute] |
| litre/minute | 0.03531466 | foot$^3$/minute |
| litre/minute | 0.2641720 | gallon [U. S. liquid]/minute |
| litre/second | 15.85032 | gallon [U. S. liquid]/minute |
| mile/hour | 1.609344* | kilometre/hour |
| millimetre/minute | 0.03937008 | inch/minute |
| metre/second | 11,811.02 | foot/hour |
| metre/second | 196.8504 | foot/minute |
| metre/second | 3.280840 | foot/second |
| metre/second$^2$ | 3.280840 | foot/second$^2$ |
| metre/second$^2$ | 39.37008 | inch/second$^2$ |
| metre/minute | 3.280840 | foot/minute |
| metre/minute | 0.05468067 | foot/second |
| metre/minute | 39.37008 | inch/minute |
| metre/hour | 3.280840 | foot/hour |
| metre/hour | 0.05468067 | foot/minute |
| metre$^3$/second | 2118.880 | foot$^3$/minute |
| metre$^3$/second | 13,198.15 | gallon [U. K. liquid]/minute |
| metre$^3$/second | 15,850.32 | gallon [U. S. liquid]/minute |
| metre$^3$/minute | 219.9692 | gallon [U. K. liquid]/minute |
| metre$^3$/minute | 264.1720 | gallon [U. S. liquid]/minute |
| MASS and DENSITY | | |
| grain [1/7000 lb avoirdupois] | 0.06479891 | gram (g) |
| gram | 15.43236 | grain |
| gram | 0.001* | kilogram (kg) |
| gram | 0.03527397 | ounce [avoirdupois] |
| gram | 0.03215074 | ounce [troy] |
| gram/centimetre$^3$ | 0.03612730 | pound/inch$^3$ |
| hundredweight [long] | 50.80235 | kilogram (kg) |
| hundredweight [short] | 45.35924 | kilogram (kg) |
| kilogram | 1000.* | gram (g) |
| kilogram | 35.27397 | ounce [avoirdupois] |
| kilogram | 32.15074 | ounce [troy] |
| kilogram | 2.204622 | pound [avoirdupois] |
| kilogram | 0.06852178 | slug |
| kilogram | 0.0009842064 | ton [long] |
| kilogram | 0.001102311 | ton [short] |
| kilogram | 0.001* | ton [metric] |
| kilogram | 0.001* | tonne |
| kilogram | 0.01968413 | hundredweight [long] |
| kilogram | 0.02204622 | hundredweight [short] |
| kilogram/metre$^3$ | 0.06242797 | pound/foot$^3$ |
| kilogram/metre$^3$ | 0.01002242 | pound/gallon [U. K. liquid] |
| kilogram/metre$^3$ | 0.008345406 | pound/gallon [U. S. liquid] |
| ounce [avoirdupois] | 28.34952 | gram (g) |
| ounce [avoirdupois] | 0.02834952 | kilogram (kg) |

* Where an asterisk is shown, the figure is exact.

**Metric Conversion Factors** (*Continued*)

| Multiply | By | To Obtain |
|---|---|---|
| MASS and DENSITY (*Continued*) | | |
| ounce [troy] | 31.10348 | gram (g) |
| ounce [troy] | 0.03110348 | kilogram (kg) |
| pound [avoirdupois] | 0.4535924 | kilogram (kg) |
| pound/foot³ | 16.01846 | kilogram/metre³ (kg/m³) |
| pound/inch³ | 27.67990 | gram/centimetre³ (g/cm³) |
| pound/gal [U. S. liquid] | 119.8264 | kilogram/metre³ (kg/m³) |
| pound/gal [U. K. liquid] | 99.77633 | kilogram/metre³ (kg/m³) |
| slug | 14.59390 | kilogram (kg) |
| ton [long 2240 lb] | 1016.047 | kilogram (kg) |
| ton [short 2000 lb] | 907.1847 | kilogram (kg) |
| ton [metric] | 1000.* | kilogram (kg) |
| tonne | 1000.* | kilogram (kg) |
| FORCE and FORCE/LENGTH | | |
| dyne | 0.00001* | newton (N) |
| kilogram-force | 9.806650* | newton (N) |
| kilopond | 9.806650* | newton (N) |
| newton | 0.1019716 | kilogram-force |
| newton | 0.1019716 | kilopond |
| newton | 0.2248089 | pound-force |
| newton | 100,000.* | dyne |
| newton | 7.23301 | poundal |
| newton | 3.596942 | ounce-force |
| newton/metre | 0.005710148 | pound/inch |
| newton/metre | 0.06852178 | pound/foot |
| ounce-force | 0.2780139 | newton (N) |
| pound-force | 4.448222 | newton (N) |
| poundal | 0.1382550 | newton (N) |
| pound/inch | 175.1268 | newton/metre (N/m) |
| pound/foot | 14.59390 | newton/metre (N/m) |
| BENDING MOMENT or TORQUE | | |
| dyne-centimetre | 0.0000001* | newton-metre (N · m) |
| kilogram-metre | 9.806650* | newton-metre (N · m) |
| ounce-inch | 7.061552 | newton-millimetre |
| ounce-inch | 0.007061552 | newton-metre (N · m) |
| newton-metre | 0.7375621 | pound-foot |
| newton-metre | 10,000,000.* | dyne-centimetre |
| newton-metre | 0.1019716 | kilogram-metre |
| newton-metre | 141.6119 | ounce-inch |
| newton-millimetre | 0.1416119 | ounce-inch |
| pound-foot | 1.355818 | newton-metre (N · m) |

* Where an asterisk is shown, the figure is exact.

**Metric Conversion Factors** (*Continued*)

| Multiply | By | To Obtain |
|---|---|---|
| MOMENT OF INERTIA and SECTION MODULUS | | |
| moment of inertia [kg · m²] | 23.73036 | pound-foot² |
| moment of inertia [kg · m²] | 3417.171 | pound-inch² |
| moment of inertia [lb · ft²] | 0.04214011 | kilogram-metre² (kg · m²) |
| moment of inertia [lb · inch²] | 0.0002926397 | kilogram-metre² (kg · m²) |
| moment of section [foot⁴] | 0.008630975 | metre⁴ (m⁴) |
| moment of section [inch⁴] | 41.62314 | centimetre⁴ |
| moment of section [metre⁴] | 115.8618 | foot⁴ |
| moment of section [centimetre⁴] | 0.02402510 | inch⁴ |
| section modulus [foot³] | 0.02831685 | metre³ (m³) |
| section modulus [inch³] | 0.00001638706 | metre³ (m³) |
| section modulus [metre³] | 35.31466 | foot³ |
| section modulus [metre³] | 61,023.76 | inch³ |
| MOMENTUM | | |
| kilogram-metre/second | 7.233011 | pound-foot/second |
| kilogram-metre/second | 86.79614 | pound-inch/second |
| pound-foot/second | 0.1382550 | kilogram-metre/second (kg · m/s) |
| pound-inch/second | 0.01152125 | kilogram-metre/second (kg · m/s) |
| PRESSURE and STRESS | | |
| atmosphere [14.6959 lb/inch²] | 101,325. | pascal (Pa) |
| bar | 100,000.* | pascal (Pa) |
| bar | 14.50377 | pound/inch² |
| bar | 100,000.* | newton/metre² (N/m²) |
| hectobar | 0.6474898 | ton [long]/inch² |
| kilogram/centimetre² | 14.22334 | pound/inch² |
| kilogram/metre² | 9.806650* | newton/metre² (N/m²) |
| kilogram/metre² | 9.806650* | pascal (Pa) |
| kilogram/metre² | 0.2048161 | pound/foot² |
| kilonewton/metre² | 0.1450377 | pound/inch² |
| newton/centimetre² | 1.450377 | pound/inch² |
| newton/metre² | 0.00001* | bar |
| newton/metre² | 1.0* | pascal (Pa) |
| newton/metre² | 0.0001450377 | pound/inch² |
| newton/metre² | 0.1019716 | kilogram/metre² |
| newton/millimetre² | 145.0377 | pound/inch² |
| pascal | 0.00000986923 | atmosphere |
| pascal | 0.00001* | bar |
| pascal | 0.1019716 | kilogram/metre² |
| pascal | 1.0* | newton/metre² (N/m²) |
| pascal | 0.02088543 | pound/foot² |
| pascal | 0.0001450377 | pound/inch² |

* Where an asterisk is shown, the figure is exact.

## Metric Conversion Factors (*Continued*)

| Multiply | By | To Obtain |
|---|---|---|
| PRESSURE and STRESS (*Continued*) | | |
| pound/foot² | 4.882429 | kilogram/metre² |
| pound/foot² | 47.88026 | pascal (Pa) |
| | | |
| pound/inch² | 0.06894757 | bar |
| pound/inch² | 0.07030697 | kilogram/centimetre² |
| pound/inch² | 0.6894757 | newton/centimetre² |
| pound/inch² | 6.894757 | kilonewton/metre² |
| pound/inch² | 6894.757 | newton/metre² (N/m²) |
| pound/inch² | 0.006894757 | newton/millimetre² (N/mm²) |
| pound/inch² | 6894.757 | pascal (Pa) |
| | | |
| ton [long]/inch² | 1.544426 | hectobar |
| ENERGY and WORK | | |
| Btu [International Table] | 1055.056 | joule (J) |
| Btu [mean] | 1055.87 | joule (J) |
| calorie [mean] | 4.19002 | joule (J) |
| foot-pound | 1.355818 | joule (J) |
| foot-poundal | 0.04214011 | joule (J) |
| joule | 0.0009478170 | Btu [International Table] |
| joule | 0.0009470863 | Btu [mean] |
| joule | 0.2386623 | calorie [mean] |
| joule | 0.7375621 | foot-pound |
| joule | 23.73036 | foot-poundal |
| joule | 0.9998180 | joule [International U. S.] |
| joule | 0.9999830 | joule [U. S. legal, 1948] |
| joule [International U. S.] | 1.000182 | joule (J) |
| joule [U. S. legal, 1948] | 1.000017 | joule (J) |
| joule | .0002777778 | watt-hour |
| watt-hour | 3600.* | joule (J) |
| POWER | | |
| Btu [International Table]/hour | 0.2930711 | watt (W) |
| foot-pound/hour | 0.0003766161 | watt (W) |
| foot-pound/minute | 0.02259697 | watt (W) |
| horsepower [550 ft-lb/s] | 0.7456999 | kilowatt (kW) |
| horsepower [550 ft-lb/s] | 745.6999 | watt (W) |
| horsepower [electric] | 746.* | watt (W) |
| horsepower [metric] | 735.499 | watt (W) |
| horsepower [U. K.] | 745.70 | watt (W) |
| kilowatt | 1.341022 | horsepower [550 ft-lb/s] |
| watt | 2655.224 | foot-pound/hour |
| watt | 44.25372 | foot-pound/minute |
| watt | 0.001341022 | horsepower [550 ft-lb/s] |
| watt | 0.001340483 | horsepower [electric] |
| watt | 0.001359621 | horsepower [metric] |
| watt | 0.001341022 | horsepower [U. K.] |
| watt | 3.412141 | Btu [International Table]/hour |

* Where an asterisk is shown, the figure is exact.

## Metric Conversion Factors (*Concluded*)

| Multiply | By | To Obtain |
|---|---|---|
| **VISCOSITY** | | |
| centipoise | 0.001* | pascal-second (Pa · s) |
| centistoke | 0.000001* | metre$^2$/second (m$^2$/s) |
| metre$^2$/second | 1,000,000.* | centistoke |
| metre$^2$/second | 10,000.* | stoke |
| pascal-second | 1000.* | centipoise |
| pascal-second | 10.* | poise |
| poise | 0.1* | pascal-second (Pa · s) |
| stoke | 0.0001* | metre$^2$/second (m$^2$/s) |

### TEMPERATURE

| To Convert From | To | Use Formula |
|---|---|---|
| temperature Celsius, $t_C$ | temperature Kelvin, $t_K$ | $t_K = t_C + 273.15$ |
| temperature Fahrenheit, $t_F$ | temperature Kelvin, $t_K$ | $t_K = (t_F + 459.67)/1.8$ |
| temperature Celsius, $t_C$ | temperature Fahrenheit, $t_F$ | $t_F = 1.8\, t_C + 32$ |
| temperature Fahrenheit, $t_F$ | temperature Celsius, $t_C$ | $t_C = (t_F - 32)/1.8$ |
| temperature Kelvin, $t_K$ | temperature Celsius, $t_C$ | $t_C = t_K - 273.15$ |
| temperature Kelvin, $t_K$ | temperature Fahrenheit, $t_F$ | $t_F = 1.8\, t_K - 459.67$ |
| temperature Kelvin, $t_K$ | temperature Rankine, $t_R$ | $t_R = 9/5\, t_K$ |
| temperature Rankine, $t_R$ | temperature Kelvin, $t_K$ | $t_K = 5/9\, t_R$ |

* Where an asterisk is shown, the figure is exact.

## Miscellaneous Conversion Factors
(English Units)

| Multiply | By | To Obtain |
|---|---|---|
| atmospheres | 29.92 | inches of mercury (32 deg. F.) |
| atmospheres | 14.70 | pounds/inch$^2$ |
| British thermal units/hour | 12.96 | foot-pounds/minute |
| circular mils | 0.7854 | square mils |
| feet of water (60 deg. F.) | 0.8843 | inches of mercury (60 deg. F.) |
| feet of water (60 deg. F.) | 0.4331 | pounds/inch$^2$ |
| feet/minute | 0.01136 | miles/hour |
| foot-pounds/second | 0.07716 | British thermal units/minute |
| gallons (U.S.) of water (60 deg. F.) | 8.337 | pounds of water (60 deg. F.) |
| gallons (U.S.)/second | 8.021 | feet$^3$/minute |
| inches of mercury (32 deg. F.) | 0.03342 | atmospheres |
| inches of mercury (60 deg. F.) | 1.131 | feet of water (60 deg. F.) |
| inches of mercury (60 deg. F.) | 0.4898 | pounds/inch$^2$ |
| inches of water (60 deg. F.) | 0.03609 | pounds/inch$^2$ |
| knots (International) | 1.151 | miles (statute)/hour |
| miles/hour | 88 | feet/minute |
| miles (statute)/hour | 0.8690 | knots (International) |
| ounces (avoirdupois) | 0.9115 | ounces (troy) |
| ounces (troy) | 1.097 | ounces (avoirdupois) |
| ounces (troy) | 0.06857 | pounds (avoirdupois) |
| pounds (avoirdupois) | 14.58 | ounces (troy) |
| pounds of water (60 deg. F.) | 0.01603 | feet$^3$ |
| pounds of water (60 deg. F.) | 0.1199 | gallons (U.S.) |
| pounds/inch$^2$ | 0.06805 | atmospheres |
| pounds/inch$^2$ | 2.309 | feet of water (60 deg. F.) |
| pounds/inch$^2$ | 2.042 | inches of mercury (60 deg. F.) |
| pounds/inch$^2$ | 27.71 | inches of water (60 deg. F.) |
| square mils | 1.273 | circular mils |

# INDEX

# SPECIAL METRIC INDEX

The following entries can also be found in the main index, but are given here separately for the convenience of readers who have a particular interest in metric information.